HAMMOND®
WORLD ATLAS

HAMMOND® AMBASSADOR WORLD

ATLAS

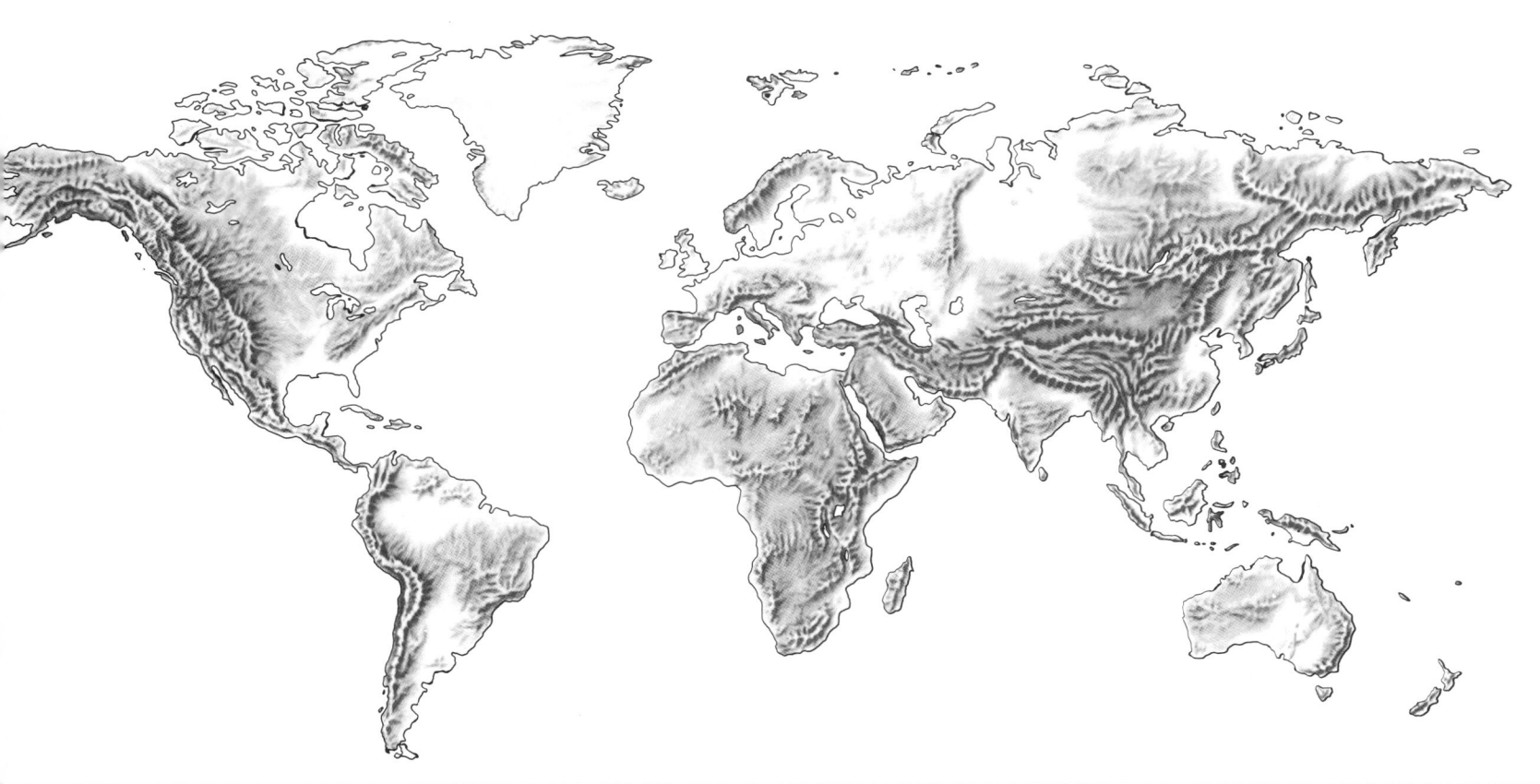

HAMMOND INCORPORATED   MAPLEWOOD, NEW JERSEY 07040

Library of Congress Cataloging-in-Publication Data

Hammond Incorporated.
   Hammond ambassador world atlas.
   Includes indexes.
   1. Atlases.   2. Zip code—United States.   I. Title:
Ambassador world atlas.
G1021.H265   1986   912   86-675338
ISBN 0-8437-1243-0

Contents

Gazetteer-Index of the World   IX

Introduction to the Maps and Indexes   XIV

Glossary of Abbreviations   XVI

World and Polar Regions

World   1
Arctic Ocean   4
Antarctica   5

Europe

Europe   6
United Kingdom and Ireland   10
Norway, Sweden, Finland, Denmark and Iceland   18
Germany   22
Netherlands, Belgium and Luxembourg   25
France   28
Spain and Portugal   31
Italy   34
Switzerland and Liechtenstein   37
Austria, Czechoslovakia and Hungary   40
Balkan States   43
Poland   46
Union of Soviet Socialist Republics   48

Asia

Asia   54
Near and Middle East   58
Turkey, Syria, Lebanon and Cyprus   61
Israel and Jordan   64
Iran and Iraq   66
Indian Subcontinent and Afghanistan   68
Burma, Thailand, Indochina and Malaya   72
China and Mongolia   75
Japan and Korea   79
Philippines   82
Southeast Asia   84

Pacific Ocean and Australia

Pacific Ocean   86
Australia   88
Western Australia   92
Northern Territory   93
South Australia   94
Queensland   95
New South Wales and Victoria   96
Tasmania   99
New Zealand   100

Africa

Africa   102
Western Africa   106
Northeastern Africa   110
Central Africa   113
Southern Africa   117

South America

South America   120
Venezuela   124
Colombia   126
Peru and Ecuador   128
Guianas   131
Brazil   132
Bolivia   136
Chile   138
Argentina   141
Paraguay and Uruguay   144

North America

North America   146
Mexico   150
Central America   153
West Indies   156

Canada

Canada   162
Newfoundland   166
Nova Scotia and Prince Edward Island   168
New Brunswick   170
Quebec   172
Ontario   175
Manitoba   178
Saskatchewan   180
Alberta   182
British Columbia   184
Yukon and Northwest Territories   186

United States

United States   188
Alabama   193
Alaska   196
Arizona   198
Arkansas   201
California   204
Colorado   207
Connecticut   210
Delaware   244
Florida   212
Georgia   215
Hawaii   218
Idaho   220
Illinois   222
Indiana   225
Iowa   228
Kansas   231
Kentucky   234
Louisiana   238
Maine   241
Maryland   244
Massachusetts   247
Michigan   250
Minnesota   253
Mississippi   256
Missouri   259
Montana   262
Nebraska   264
Nevada   266
New Hampshire   268
New Jersey   271
New Mexico   274
New York   276
North Carolina   279
North Dakota   282
Ohio   284
Oklahoma   287
Oregon   290
Pennsylvania   293
Rhode Island   247
South Carolina   296
South Dakota   298
Tennessee   234
Texas   301
Utah   304
Vermont   268
Virginia   306
Washington   309
West Virginia   312
Wisconsin   315
Wyoming   318
The Fifty States   320

World Index 325

GEOGRAPHICAL TERMS   474
MAP PROJECTIONS   475
WORLD STATISTICS   478
FOREIGN CITY WEATHER   480
U.S. CITY WEATHER   482
TABLES OF AIRLINE DISTANCES   484

Introduction to the World Atlas

As in previous editions, this Hammond World Atlas is organized to make the retrieval of information as simple and quick as possible. The guiding principle in organizing the atlas material has been to present separate subjects on *separate* maps. In this way, each individual map topic is shown with the greatest degree of clarity, unencumbered with extraneous information that is best revealed on separate maps. Of equal importance from the standpoint of good atlas design is the treatment of all current information on a given country or state as a single atlas unit. Thus, the basic reference map of an area is accompanied on adjacent pages by all supplementary information pertaining to that area. For example, the detailed index for a given map always appears on the same page as, or on the pages immediately following, the reference map. This same map index provides population data for the many cities, towns and villages shown on the map. Highlight information on the area, i.e., the total population and area, the capital, the highest point, is listed in the summary fact listings accompanying each unit. An adjacent locator map relates the subject area to the larger world beyond. A three-dimensional picture of the area is exhibited by means of the accompanying full-color topographic map. A separate economic map defines the vital agricultural, industrial and mineral resources of the area. In the case of the foreign maps, the flag of each independent nation appears on the appropriate page. Finally, certain country units contain special subject maps dealing with the history, climate, demography and vegetation of the area.

An outstanding feature of the atlas is the addition of ZIP codes to the index entries for each of the legion of communities shown on the state maps. With the exception of the U.S. Postal Service directories of limited availability, the ZIP code listings herein are the most extensive published. In addition to listing ZIP codes for the communities possessing post offices, ZIP codes of the nearest post offices are listed for communities without postal facilities. It may be said with a fair degree of certainty that this innovation in atlas content doubles the value of the work for home, office and school.

The back of the book contains a second type of index. This is a multi-paged "A-to-Z" index of all the world's places that appear on the maps. The use of this map index is essential when the name of a place is known but its country, state, or province is unknown. ZIP codes also are given here for each U.S. city, town, or community entry.

The numerous geographical changes of the decade are all recorded in the Hammond World Atlas. Over 8,000 changes, which occurred throughout the world since the last major revision, were entered on the maps. The state maps now reveal the many new towns and cities that have developed in the recent past. The majority of these are burgeoning suburbs on the fringes of our larger cities. On the other hand, large numbers of abandoned and defunct rural hamlets have been removed from the maps and map indexes. Hundreds of other changes are also recorded on the state maps: new national parks and monuments, new dams, new reservoirs, name changes, etc.

Of course, the maps of foreign areas have been thoroughly updated. These revisions echo the new nations, shifting boundaries and the fluid internal divisions of many countries. New communities generated by the opening up of resources in the developing nations are also noted.

In closing it may be said that the atlas has truly been designed for contemporary use. Just as the information presented on the following pages is as current and up to date as the editors and cartographers could issue it, so the design and organization has been as well planned as possible to create a work useful to present generations.

President
HAMMOND INCORPORATED

Gazetteer-Index of the World

This alphabetical list of continents, countries, states, colonial possessions and other major geographical areas provides a quick reference to their area in square miles and square kilometers, population, capital or chief town, map page number and index key thereon. The last name indicates the square on the respective page in which the name may be found. An indication of the population sources used is also included, and refers both to the total figures given in this Gazetteer-Index and to the populations appearing in greater detail with the maps throughout the atlas. The population figures used in each case are the latest reliable figures obtainable. A glance at the sources will show that the dates vary considerably throughout the world. In certain areas where no census has ever been taken, we must rely on official estimates. In other areas where censuses have been taken at infrequent intervals, we again rely on estimates. The key to the abbreviations used in the Gazetteer-Index follows:

aut = autonomous	est = estimates	reg = regions
boro = boroughs	excl = excluding	rep = republics
cap = capital	FC = final census	S.S.R. = Soviet Socialist Republic
CE = census (undetermined)	gov = governorates	terr = territories; territory
CIA = U.S. Central Intelligence Agency	incl = including	TP = total population
	isl = islands	U.K. = United Kingdom
cit = cities	met = metropolitan	UN = United Nations
co = counties	OE = official estimate	U.S.A. = United States of America
com = communes	oth = other populations	U.S.S.R. = Union of Soviet Socialist Republics
dept = departments	par = parishes	ws = with suburbs
dist = districts	PC = preliminary census	
div = divisions	prov = provinces; provincial	

Country	Square Miles	Area Square Kilometers	Population	Capital or Chief Town	Page and Index Ref.	Sources of Population Data
*Afghanistan	250,775	649,507	15,540,000	Kabul	68/A 2	79 PC
Africa	11,707,000	30,321,130	469,000,000	....	102/.....	80 UN est
Alabama, U.S.A.	51,705	133,916	3,893,888	Montgomery	195/.....	80 FC & OE
Alaska, U.S.A.	591,004	1,530,700	401,851	Juneau	196/.....	80 FC & OE
*Albania	11,100	28,749	2,590,600	Tiranë	45/E 5	TP—79 PC; cit over 6,000—70 OE; oth—63 OE
Alberta, Canada	255,285	661,185	2,237,724	Edmonton	182/.....	81 FC
*Algeria	919,591	2,381,740	17,422,000	Algiers	106/D 3	77 PC
American Samoa	77	199	32,297	Pago Pago	87/J 7; 86/.....	80 FC
Andorra	188	487	31,000	Andorra la Vella	33/G 1	TP—79 OE; cap—75 OE
*Angola	481,351	1,246,700	7,078,000	Luanda	114/C 6	TP—80 UN est; oth—70 FC
Anguilla	35	91	6,519	The Valley	156/F 3	74 FC
Antarctica	5,500,000	14,245,000	....		5/.....	....
*Antigua and Barbuda	171	443	75,000	St. John's	161/E11; 156/G 3	TP—80 OE; oth—70 FC
*Argentina	1,072,070	2,776,661	28,438,000	Buenos Aires	143/.....	TP—82 OE; oth—80 PC
Arizona, U.S.A.	114,000	295,260	2,718,425	Phoenix	198/.....	80 FC & OE
Arkansas, U.S.A.	53,187	137,754	2,286,435	Little Rock	202/.....	80 FC & OE
Armenian S.S.R., U.S.S.R.	11,506	29,800	3,031,000	Erivan	52/F 6	TP, cit over 50,000—79 PC; oth—70 FC
Aruba, Neth. Antilles	70	181	55,148	Oranjestad	161/E 9	TP—71 FC; cap—72 est
Ascension Island, St. Helena	34	88	719	Georgetown	102/A 5	76 FC
Ashmore & Cartier Islands, Australia	61	159	....	(Canberra, Austr.)	88/C 2	....
Asia	17,128,500	44,362,815	2,633,000,000	....	54/.....	80 est
*Australia	2,966,136	7,682,300	14,576,330	Canberra	88/.....	81 FC
Australian Capital Territory	927	2,400	221,609	Canberra	96/E 4	81 FC
*Austria	32,375	83,851	7,507,000	Vienna	40/B 3	TP—80 OE; cap, cit over 100,000—73 OE; oth—71 FC
Azerbaidzhan S.S.R., U.S.S.R.	33,436	86,600	6,028,000	Baku	52/G 6	TP, cit over 50,000—79 PC; oth—70 FC
Azores Islands, Portugal	902	2,335	264,400	Ponta Delgada; Angra do Heroismo; Horta	32/.....	TP—77 OE; oth—70 FC & PC
*Bahamas	5,382	13,939	209,505	Nassau	156/C 1	80 PC
*Bahrain	240	622	358,857	Manama	58/F 4	TP—81 PC; oth—71 FC
Baker Island, U.S.A.	1	2.6	....	....	87/J 5	....
Balearic Islands, Spain	1,936	5,014	558,287	Palma	33/H 3	70 FC
*Bangladesh	55,126	142,776	87,052,024	Dhaka	68/G 4	TP—81 PC; oth—74 FC
*Barbados	166	430	248,983	Bridgetown	161/B 8	80 PC
Belau (Palau)	188	487	12,116	Koror	86/D 5	80 FC
*Belgium	11,781	30,513	9,855,110	Brussels	27/E 7	TP—80 OE; oth—70 FC (com)
*Belize	8,867	22,966	144,857	Belmopan	154/C 2	TP, cap, cit over 1,000—80 PC; oth—70 PC
*Benin	43,483	112,620	3,338,240	Porto-Novo	106/E 6	TP—79 PC; cap, Cotonou—75 OE; oth—73 OE
Bermuda	21	54	67,761	Hamilton	156/H 3	80 PC
*Bhutan	18,147	47,000	1,298,000	Thimphu	68/G 3	TP—80 UN est; oth—70 OE
*Bolivia	424,163	1,098,582	5,600,000	La Paz; Sucre	136/.....	TP—80 OE; cap, dept, dept cap—76 FC; oth—50 FC
Bonaire, Neth. Antilles	112	291	8,087	Kralendijk	161/E 9	TP—71 FC; cap—72 est
Bophuthatswana (rep.), South Africa	15,570	40,326	1,200,000	Mmabatho	119/D 5	TP—78 est; oth—70 FC
*Botswana	224,764	582,139	819,000	Gaborone	119/C 4	TP—80 OE; cap, Francistown—74 OE; Selebi-Pikwe—75 FC; oth—71 FC
Bouvet Island	22	57	....	....	5/D 1	....
*Brazil	3,284,426	8,506,663	119,098,992	Brasília	132/.....	80 PC
British Columbia, Canada	366,253	948,596	2,744,467	Victoria	184/.....	81 FC
British Indian Ocean Terr.	29	75	2,000	(London, U.K.)	54/L10	78 est
British Virgin Islands	59	153	11,006	Road Town	157/H 1	TP—80 FC; oth—70 FC
Brunei	2,226	5,765	192,832	Bandar Seri Begawan	85/E 4	81 PC
*Bulgaria	42,823	110,912	8,862,000	Sofia	45/F 4	TP—80 OE; oth—75 PC
*Burkina Faso	105,869	274,200	6,908,000	Ouagadougou	106/D 6	TP—80 UN est; oth—75 FC, 73 OE
*Burma	261,789	678,034	32,913,000	Rangoon	72/B 2	TP—79 OE; states, div. cit over 100,000—73 PC; oth—53 FC
*Burundi	10,747	27,835	4,021,910	Bujumbura	114/E 4	79 PC
*Byelorussian S.S.R. (White Russian S.S.R.), U.S.S.R.	80,154	207,600	9,560,000	Minsk	52/C 4	TP, cit over 50,000—79 PC; oth—70 FC
California, U.S.A.	158,706	411,049	23,667,565	Sacramento	204/.....	80 FC & OE
*Cambodia (Kampuchea)	69,898	181,036	5,200,000	Phnom Penh	72/E 4	TP—79 CIA est; cap—80 est
*Cameroon	183,568	475,441	8,503,000	Yaoundé	114/B 2	TP—80 OE; cit over 21,000—76 FC; Ebolowa, oth—70 OE
*Canada	3,851,787	9,976,139	24,343,181	Ottawa	162/.....	81 FC
Canary Islands, Spain	2,808	7,273	1,170,224	Las Palmas; Santa Cruz	32/B 4	70 FC
Cape Province, South Africa	261,705	677,816	5,543,506	Cape Town	118/C 6	TP—80 PC; oth—70 FC
*Cape Verde	1,557	4,033	324,000	Praia	106/B 8	TP—80 UN est; oth—70 PC
Cayman Islands	100	259	18,000	Georgetown	156/B 3	TP—81 OE; oth—79 FC

*Member of the United Nations.

Country	Area Square Miles	Area Square Kilometers	Population	Capital or Chief Town	Page and Index Ref.	Sources of Population Data
Celebes, Indonesia	72,986	189,034	7,732,383	Ujung Pandang	85/G 6	71 PC
*Central African Republic	242,000	626,780	2,284,000	Bangui	114/C 2	TP—79 est; oth—75 FC
Central America	197,480	511,475	21,000,000	....	154/......	79 OE
Ceylon, see Sri Lanka						
*Chad	495,752	1,283,998	4,309,000	N'Djamena	111/C 4	TP—78 OE; oth—72 OE
Channel Islands	75	194	133,000	St. Helier; St. Peter Port	13/E 8	TP—81 OE; oth—71 FC
*Chile	292,257	756,946	11,275,440	Santiago	138/......	TP—82 PC; cit (part)—79 OE; oth—70 FC & PC
*China, People's Rep. of	3,691,000	9,559,690	958,090,000	Peking (Beijing)	77/......	TP, prov, Peking, Shanghai, Tianjin—78 OE; oth—70 est
China, Republic of (Taiwan)	13,971	36,185	16,609,961	Taipei	77/K 7	TP, cap, Penghu Isl., cit over 300,000—77 OE; oth—70 OE
Christmas Island, Australia	52	135	3,184	Flying Fish Cove	54/O11	80 OE
Ciskei (rep.), S. Africa	2,988	7,740	635,631	Bisho	119/D 6	80 PC
Clipperton Island	2	5.2	....	....	146/H 8	
Cocos (Keeling) Islands, Australia	5.4	14	555	West Island	54/N11	81 PC
*Colombia	439,513	1,138,339	27,520,000	Bogotá	126/......	TP—80 OE; oth—73 PC
Colorado, U.S.A.	104,091	269,596	2,889,735	Denver	208/......	80 FC & OE
*Comoros	719	1,862	290,000	Moroni	119/G 2	TP—78 est; cap—75 OE; oth—66 FC
*Congo, Republic of	132,046	342,000	1,537,000	Brazzaville	114/B 4	TP—80 UN est; cap—74 FC; oth—74 PC
Connecticut, U.S.A.	5,018	12,997	3,107,576	Hartford	210/......	80 FC & OE
Cook Islands	91	236	17,695	Avarua	87/K 7	81 PC
Coral Sea Islands, Australia	8.5	22	....	....	88/J 3	
Corsica, France	3,352	8,682	289,842	Ajaccio; Bastia	28/B 6	75 FC
*Costa Rica	19,575	50,700	2,245,000	San José	154/E 5	TP—80 OE; oth—73 FC
*Cuba	44,206	114,494	9,706,369	Havana	158/......	TP—81 PC; prov, cap—81 PC; oth—81 & 70 PC
Curaçao, Neth. Antilles	178	462	145,430	Willemstad	161/G 7	TP—71 FC; cap—75 OE
*Cyprus	3,473	8,995	629,000	Nicosia	62/E 5	TP—80 OE; oth—73 FC, 72 OE
*Czechoslovakia	49,373	127,876	15,276,799	Prague	41/C 2	TP—80 PC; cap, cit over 100,000—75 OE; rep, reg—74 OE; oth—75 OE, 70 OE
Delaware, U.S.A.	2,044	5,294	594,317	Dover	245/R 3	80 FC & OE
*Denmark	16,629	43,069	5,124,000	Copenhagen	21/......	TP—80 OE; oth—75 OE, 71 OE, 70 FC
District of Columbia, U.S.A.	69	179	638,432	Washington	244/F 5	80 FC
*Djibouti	8,880	23,000	386,000	Djibouti	111/H 5	TP—79 est; cap—73 OE
*Dominica	290	751	74,089	Roseau	161/E 7	TP—80 PC; oth—70 FC
*Dominican Republic	18,704	48,443	5,647,977	Santo Domingo	158/D 6	81 PC
*East Germany (German Democratic Republic)	41,768	108,179	16,737,000	Berlin (East)	22/......	TP—80 OE; oth—75 OE
*Ecuador	109,483	283,561	8,644,000	Quito	128/C 3	TP—81 OE; oth—74 FC
*Egypt	386,659	1,001,447	41,572,000	Cairo	110/E 2	TP—79 OE; oth—76 PC
*El Salvador	8,260	21,393	4,813,000	San Salvador	154/C 4	TP—80 OE; oth—71 FC
England, U.K.	50,516	130,836	46,220,955	London	13/......	TP—81 PC; co, cap (boro & ws)—76 OE; cit—76 & 73 OE; oth—71 FC
*Equatorial Guinea	10,831	28,052	244,000	Malabo	114/A 3	TP—79 est; terr—68 OE; oth—60 FC
Estonian S.S.R., U.S.S.R.	17,413	45,100	1,466,000	Tallinn	52/C 3; 53/......	TP, cit over 50,000—79 PC; oth—70 FC
*Ethiopia	471,776	1,221,900	31,065,000	Addis Ababa	110/G 5	TP—80 OE; cap, Asmara—78 OE; prov—72 OE; oth—72 & 71 OE
Europe	4,057,000	10,507,630	676,000,000	....	7/......	80 est
Faeroe Islands, Denmark	540	1,399	41,969	Tórshavn	21/B 2	77 FC
Falkland Islands & Dependencies	6,198	16,053	1,813	Stanley	120/E 8; 143/D 7	80 FC
*Fiji	7,055	18,272	588,068	Suva	87/H 8; 86/......	80 FC
*Finland	130,128	337,032	4,788,000	Helsinki	18/O 6	TP—80 OE; prov—75 OE; oth—75 OE, 70 FC
Florida, U.S.A.	58,664	151,940	9,746,342	Tallahassee	212/......	80 FC & OE
*France	210,038	543,998	53,788,000	Paris	28/......	TP—80 OE; oth—75 FC
French Guiana	35,135	91,000	73,022	Cayenne	131/E 3	82 FC
French Polynesia	1,544	4,000	137,382	Papeete	87/L 8	77 FC
*Gabon	103,346	267,666	551,000	Libreville	114/B 4	TP—80 UN est; oth—70 FC
*Gambia	4,127	10,689	601,000	Banjul	106/A 6	TP—80 OE; oth—73 FC
Gaza Strip	139	360	400,000	Gaza	65/A 4	TP—76 OE; oth—67 CE
Georgia, U.S.A.	58,910	152,577	5,463,105	Atlanta	217/......	80 FC & OE
Georgian S.S.R., U.S.S.R.	26,911	69,700	5,015,000	Tbilisi	52/F 6	TP, cit over 50,000—79 PC; oth—70 FC
*Germany, East (German Democratic Republic)	41,768	108,179	16,737,000	Berlin (East)	22/......	TP—80 OE; oth—75 OE
*Germany, West (Federal Republic)	95,985	248,601	61,658,000	Bonn	22/......	TP—80 OE; states, cap—76 OE; oth—76 OE, 70 FC
*Ghana	92,099	238,536	11,450,000	Accra	106/D 7	TP—80 OE; oth—70 OE
Gibraltar	2.28	5.91	29,760	Gibraltar	33/D 4	79 OE
*Great Britain & Northern Ireland (United Kingdom)	94,399	244,493	55,672,000	London	10/......	TP—81 OE (see England, Wales, Scotland, Northern Ireland)
*Greece	50,944	131,945	9,599,000	Athens	45/F 6	TP—80 OE; oth—71 FC
Greenland	840,000	2,175,600	49,773	Nuuk (Godthåb)	4/B12	TP—80 OE
*Grenada	133	344	103,103	St. George's	161/D 9; 156/G 4	TP, cap—81 OE; oth—70 FC
Guadeloupe & Dependencies	687	1,779	328,400	Basse-Terre	161/A 5; 156/F 4	82 FC
Guam	209	541	105,979	Agaña	87/E 4; 86/......	80 FC
*Guatemala	42,042	108,889	7,262,419	Guatemala	154/B 3	TP—80 OE; oth—73 FC
*Guinea	94,925	245,856	5,143,284	Conakry	106/B 6	TP, cap (ws), Kankan, Kindia, Labé—72 FC; oth—67 OE
*Guinea-Bissau	13,948	36,125	777,214	Bissau	106/A 6	79 PC
*Guyana	83,000	214,970	793,000	Georgetown	131/B 3	TP—80 OE; cap, cit over 10,000—70 FC; oth—60 FC
*Haiti	10,694	27,697	5,053,792	Port-au-Prince	158/C 5	82 PC
Hawaii, U.S.A.	6,471	16,760	964,691	Honolulu	218/......	80 FC & OE
Heard & McDonald Islands, Australia	113	293	....	....	2/N 8	
Holland, see Netherlands						
*Honduras	43,277	112,087	3,691,000	Tegucigalpa	154/D 3	TP—80 OE; oth—74 FC
Hong Kong	403	1,044	5,022,000	Victoria	77/H 7; 78/......	TP—81 PC; oth—76 FC
Howland Island, U.S.A.	1	2.6	....	....	87/J 5	
*Hungary	35,919	93,030	10,709,536	Budapest	41/D 3	TP, cap, co—80 PC; oth—80 PC, 70 FC
*Iceland	39,768	103,000	228,785	Reykjavík	21/B 1	TP—80 PC; oth—70 FC
Idaho, U.S.A.	83,564	216,431	944,038	Boise	220/......	80 FC & OE

Country	Area Square Miles	Area Square Kilometers	Population	Capital or Chief Town	Page and Index Ref.	Sources of Population Data
Illinois, U.S.A.	56,345	145,934	11,426,596	Springfield	222/......	80 FC & OE
*India	1,269,339	3,287,588	683,810,051	New Delhi	68/D 4	TP & states—81 PC; oth—71 FC
Indiana, U.S.A.	36,185	93,719	5,490,260	Indianapolis	227/......	80 FC & OE
*Indonesia	788,430	2,042,034	147,490,298	Jakarta	85/D 7	TP—80 PC; cit—80 PC & 71 PC; isls.—71 PC
Iowa, U.S.A.	56,275	145,752	2,913,808	Des Moines	229/......	80 FC & OE
*Iran	636,293	1,648,000	37,447,000	Tehran	66/F 4	TP—80 OE; div, cit over 50,000—76 PC; oth—66 FC & PC, 56 FC
*Iraq	172,476	446,713	12,767,000	Baghdad	66/C 4	TP—79 OE; oth—65 & 57 FC
*Ireland	27,136	70,282	3,440,427	Dublin	17/......	TP—81 PC; oth—71 FC
Ireland, Northern, U.K.	5,452	14,121	1,543,000	Belfast	17/F 2	TP—81 OE; dist—76 OE; cap, Londonderry—73 OE; oth—71 FC
Isle of Man	227	588	64,000	Douglas	13/C 3	TP—80 OE; oth—71 FC
*Israel	7,847	20,324	3,878,000	Jerusalem	65/B 4	TP—80 OE; cap, cit over 100,000—77 OE; dist, cit over 5,000—72 PC; oth—61 FC
*Italy	116,303	301,225	57,140,000	Rome	34/......	TP—80 OE; oth—71 FC
*Ivory Coast	124,504	322,465	7,920,000	Yamoussoukro	106/C 7	TP—79 OE; oth—75 PC
*Jamaica	4,411	11,424	2,184,000	Kingston	158/......	TP—80 OE; oth—70 & 60 FC
Jan Mayen	144	373	....	....	6/D 1	....
*Japan	145,730	377,441	117,057,485	Tokyo	81/......	TP—80 PC; oth—75 FC
Jarvis Island, U.S.A.	1	2.6	....	....	87/K 6	....
Java, Indonesia	48,842	126,500	73,712,411	Jakarta	85/J 2	71 PC
Johnston Atoll	.91	2.4	327	....	87/K 4	80 FC
*Jordan	35,000	90,650	2,152,273	Amman	65/D 3	TP—79 PC; cap, cit over 100,000—77 OE; gov, cit 9,000-100,000—73 OE; **oth—61 FC**
*Kampuchea (Cambodia)	69,898	181,036	5,200,000	Phnom Penh	72/E 4	TP—79 CIA est; cap—80 est
Kansas, U.S.A.	82,277	213,097	2,364,236	Topeka	232/......	80 FC & OE
Kazakh S.S.R., U.S.S.R.	1,048,300	2,715,100	14,684,000	Alma-Ata	48/G 5	TP, cit over 50,000—79 PC; oth—70 FC
Kentucky, U.S.A.	40,409	104,659	3,660,257	Frankfort	237/......	80 FC & OE
*Kenya	224,960	582,646	15,327,061	Nairobi	115/G 3	TP—79 PC; oth—69 FC
Kermadec Islands	13	33	5	....	87/J 9	81 FC
Kingman Reef	0.1	0.26	....	....	87/K 5	....
Kirgiz S.S.R., U.S.S.R.	76,641	198,500	3,529,000	Frunze	48/H 5	TP, cit over 50,000—79 PC; oth—70 FC
Kiribati	291	754	56,213	Bairiki	87/J 6	TP—78 FC; oth—73 FC
Korea, North	46,540	120,539	17,914,000	P'yŏngyang	80/D 3	TP—80 UN est; cap—76 OE; Hamhŭng—72 OE; oth—70 OE
Korea, South	38,175	98,873	37,448,836	Seoul	80/D 5	TP—80 PC; oth—75 FC & PC
*Kuwait	6,532	16,918	1,355,827	Al Kuwait	58/E 4	80 PC
*Laos	91,428	236,800	3,721,000	Vientiane	72/D 3	TP—80 UN est; cap—66 FC; oth—58 OE
Latvian S.S.R., U.S.S.R.	24,595	63,700	2,521,000	**Riga**	52/B 3; 53/......	TP, cit over 50,000—79 PC; oth—70 FC
*Lebanon	4,015	10,399	3,161,000	Beirut	62/F 6	TP—80 UN est; cap—70 FC; Tarabulus—64 OE; oth—61 OE
*Lesotho	11,720	30,355	1,339,000	Maseru	119/D 5	TP—80 OE; oth—80 est
*Liberia	43,000	111,370	1,873,000	Monrovia	106/C 7	TP—80 OE; oth—74 FC
*Libya	679,358	1,759,537	2,856,000	Tripoli	110/B 2	TP—79 OE; oth—73 FC & PC
Liechtenstein	61	158	25,220	Vaduz	39/J 2	80 PC
Lithuanian S.S.R., U.S.S.R.	25,174	65,200	3,398,000	**Vilna**	52/B 3; 53/......	TP, cit over 50,000—79 PC; oth—70 FC
Louisiana, U.S.A.	47,752	123,678	4,206,312	Baton Rouge	238/......	80 FC & OE
*Luxembourg	999	2,587	364,000	Luxembourg	27/J 9	TP—79 OE; cap—74 OE; oth—70 FC
Macau	6	16	271,000	Macau	77/H 7	TP—78 OE; cap—70 FC
*Madagascar	226,657	587,041	8,742,000	**Antananarivo**	**119/H 3**	TP—80 UN est; prov, cap, cit over 40,000—75 PC; oth—71 OE
Madeira Islands, Portugal	307	796	262,800	Funchal	32/A 2	TP—77 OE; oth—70 FC & PC
Maine, U.S.A.	33,265	86,156	1,125,027	Augusta	243/......	80 FC & OE
*Malawi	45,747	118,485	5,968,000	Lilongwe	114/F 6	TP—80 OE; oth—77 PC
Malaya, Malaysia	50,806	131,588	11,138,227	Kuala Lumpur	72/D 6	TP, states, Kuala Lumpur—80 PC; cit over 100,000—70 FC; oth—70 PC
*Malaysia	128,308	332,318	13,435,588	**Kuala Lumpur**	**72/D 6; 85/E 4**	TP, states, Kuala Lumpur—80 PC; Kuching, Kota Kinabalu, cit over 100,000—70 FC; oth—70 PC
*Maldives	115	298	143,046	Male	54/L 9	78 FC
*Mali	464,873	1,204,021	6,906,000	Bamako	106/C 6	TP—80 OE; oth—76 PC
*Malta	122	316	343,970	Valletta	34/E 7	TP, cit—79 OE; oth—73 OE
Man, Isle of	227	588	64,000	Douglas	13/C 3	TP—80 OE; oth—71 FC
Manitoba, Canada	250,999	650,087	1,026,241	Winnipeg	179/......	81 FC
Marquesas Islands, French Polynesia	492	1,274	5,419	Atuona	87/N 6	77 FC
Marshall Islands	70	181	30,873	Majuro	87/G 4	80 FC
Martinique	425	1,101	328,566	Fort-de-France	161/D 5	82 FC
Maryland, U.S.A.	10,460	27,091	4,216,975	Annapolis	245/......	80 FC & OE
Massachusetts, U.S.A.	8,284	21,456	5,737,037	Boston	249/......	80 FC & OE
*Mauritania	419,229	1,085,803	1,634,000	Nouakchott	106/B 5	TP—80 UN est; oth—76 PC
*Mauritius	790	2,046	959,000	Port Louis	119/G 5	TP—80 OE; cap—77 OE; Curepipe, Quatre Bornes—74 OE; oth—72 PC
Mayotte	144	373	47,300	Dzaoudzi	119/G 2	TP—78 CE; cap—66 FC
*Mexico	761,601	1,972,546	67,395,826	Mexico City	150/......	TP, states, cap—80 PC; cap (ws), Guadalajara (ws), Monterrey (ws)—78 OE; oth—70 FC
Michigan, U.S.A.	58,527	151,585	9,262,078	Lansing	250/......	80 FC & OE
Micronesia, Federated States of	....	....	73,160	Kolonia	87/E 5	TP—80 FC
Midway Islands	1.9	4.9	453	....	87/J 3	80 FC
Minnesota, U.S.A.	84,402	218,601	4,075,970	St. Paul	255/......	80 FC & OE
Mississippi, U.S.A.	47,689	123,515	2,520,638	Jackson	256/......	80 FC & OE
Missouri, U.S.A.	69,697	180,515	4,916,759	Jefferson City	261/......	80 FC & OE
Moldavian S.S.R., U.S.S.R.	13,012	33,700	3,947,000	Kishinev	52/C 5	TP, cit over 50,000—79 PC; oth—70 FC
Monaco	368 acres	149 hectares	25,029	Monaco	28/G 6	75 FC
*Mongolia	606,163	1,569,962	1,594,800	Ulaanbaatar	77/E 2	TP—79 PC; prov, cap, Darhan—77 OE; oth—69 FC
Montana, U.S.A.	147,046	380,849	786,690	Helena	262/......	80 FC & OE
Montserrat	40	104	12,073	Plymouth	157/G 3	80 PC
*Morocco	172,414	446,550	20,242,000	Rabat	106/C 2	TP—80 OE; oth—71 FC
*Mozambique	303,769	786,762	12,130,000	Maputo	119/E 4	TP, prov, cap—80 PC; oth—70 FC
Namibia (South-West Africa)	317,827	823,172	1,200,000	Windhoek	118/B 3	TP—74 est; oth—70 PC
Natal, South Africa	33,578	86,967	5,722,215	Pietermaritzburg	119/E 5	TP—80 PC; oth—70 PC
Nauru	7.7	20	7,254	Yaren (district)	87/G 6	77 PC
Navassa Island	2	5	....	....	156/C 3	....
Nebraska, U.S.A.	77,355	200,349	1,569,825	Lincoln	264/......	80 FC & OE
*Nepal	54,663	141,577	14,179,301	Kathmandu	68/E 3	TP—81 PC; oth—71 FC
*Netherlands	15,892	41,160	14,227,000	The Hague; Amsterdam	27/F 5	TP—81 OE; oth—76 OE (com)

Gazetteer-Index of the World

Country	Area Square Miles	Area Square Kilometers	Population	Capital or Chief Town	Page and Index Ref.	Sources of Population Data
Netherlands Antilles	390	1,010	246,000	Willemstad	156/E 4	TP—78 OE; Willemsted—75 OE; oth—72 est.
Nevada, U.S.A.	110,561	286,353	800,493	Carson City	266/......	80 FC & OE
New Brunswick, Canada	28,354	73,437	696,403	Fredericton	170/......	81 FC
New Caledonia & Dependencies	7,335	18,998	133,233	Nouméa	87/G 8	76 FC
Newfoundland, Canada	156,184	404,517	567,681	St. John's	166/......	81 FC
New Hampshire, U.S.A.	9,279	24,033	920,610	Concord	268/......	80 FC & OE
New Hebrides, see Vanuatu						
New Jersey, U.S.A.	7,787	20,168	7,364,823	Trenton	273/......	80 FC & OE
New Mexico, U.S.A.	121,593	314,926	1,302,981	Santa Fe	274/......	80 FC & OE
New South Wales, Australia	309,498	801,600	5.126,217	Sydney	96/B 2	81 FC
New York, U.S.A.	49,108	127,190	17,558,072	Albany	276/......	80 FC & OE
*New Zealand	103,736	268,676	3,175,737	Wellington	100/......	TP, inc. places, isls.—81 FC; oth—76 FC
*Nicaragua	45,698	118,358	2,703,000	Managua	154/D 4	TP—80 OE; oth—71 PC
*Niger	489,189	1,267,000	5,098,427	Niamey	106/F 5	TP, cap, Maradi, Tahoua, Zinder—77 PC; oth—72 OE
*Nigeria	357,000	924,630	82,643,000	Lagos	106/F 6	TP—79 OE; prov—63 FC; oth—75 & 71 OE
Niue	100	259	3,578	Alofi	87/K 7	79 OE
Norfolk Island, Australia	13.4	34.6	2,175	Kingston	88/L 5	81 FC
North America	9,363,000	24,250,170	370,000,000	....	146/......	80 UN est
North Carolina, U.S.A.	52,669	136,413	5,881,813	Raleigh	281/......	80 FC & OE
North Dakota, U.S.A.	70,702	183,118	652,717	Bismarck	282/......	80 FC & OE
Northern Ireland, U.K.	5,452	14,121	1,543,000	Belfast	17/F 2	TP—81 OE; dist—76 OE; cap, Londonderry—73 OE; oth—71 FC
Northern Marianas	184	477	16,780	Capitol Hill	87/E 4	80 FC
Northern Territory, Australia	519,768	1,346,200	123,324	Darwin	93/......	81 FC
North Korea	46,540	120,539	17,914,000	P'yŏngyang	80/D 3	TP—80 UN est; cap—76 OE; Hamhŭng—72 OE; oth—70 OE
Northwest Territories, Canada	1,304,896	3,379,683	45,741	Yellowknife	187/G 3	81 FC
*Norway	125,053	323,887	4,092,000	Oslo	18/F 7	TP—80 OE; co, Svalbard—76 OE; oth—76 OE, 70 OE
Nova Scotia, Canada	21,425	55,491	847,442	Halifax	168/......	81 FC
Oceania	3,292,000	8,526,280	23,000,000	....	87/......	80 UN est
Ohio, U.S.A.	41,330	107,045	10,797,624	Columbus	284/......	80 FC & OE
Oklahoma, U.S.A.	69,956	181,186	3,025,290	Oklahoma City	288/......	80 FC & OE
*Oman	120,000	310,800	891,000	Muscat	58/G 6	TP—80 UN est; cap, Matrah—66 OE; Salala—68 OE
Ontario, Canada	412,580	1,068,582	8,625,107	Toronto	175, 177/......	81 FC
Orange Free State, South Africa	49,866	129,153	1,833,216	Bloemfontein	119/D 5	TP—80 PC; oth—70 FC
Oregon, U.S.A.	97,073	251,419	2,633,149	Salem	291/......	80 FC & OE
Orkney Islands, Scotland	376	974	17,675	Kirkwall	15/E 1	TP—76 OE; oth—71 FC
Pacific Islands, Territory of the	533	1,380	132,929	Saipan	87/F 5	80 FC
*Pakistan	310,403	803,944	83,782,000	Islamabad	68/B 3	TP—81 PC; Abbottabad, Bannu, cit over 50,000—72 PC; oth—61 FC
Palau (Belau)	188	487	12,116	Koror	86/D 5	80 FC
Palmyra Atoll	3.85	1	....	....	87/K 5	
*Panama	29,761	77,082	1,830,175	Panamá	154/G 6	TP, cit over 1,600—80 PC; oth—70 FC
*Papua New Guinea	183,540	475,369	3,010,727	Port Moresby	85/B 7; 87/E 6	80 PC
Paracel Islands	....	....	....	....	85/E 2	....
*Paraguay	157,047	406,752	2,973,000	Asunción	144/......	TP—79 OE; oth—72 PC
Pennsylvania, U.S.A.	45,308	117,348	11,863,895	Harrisburg	294/......	80 FC & OE
Persia, see Iran						
*Peru	496,222	1,285,215	17,031,221	Lima	128/......	81 PC
*Philippines	115,707	299,681	48,098,460	Manila	82/......	80 FC
Pitcairn Islands	18	47	54	Adamstown	87/O 8	81 FC
*Poland	120,725	312,678	35,815,000	Warsaw	47/......	TP—81 OE; prov, cap, Cracow, Łódź—75 OE; oth—70
*Portugal	35,549	92,072	9,933,000	Lisbon	32/B 3	TP—80 OE; cap (ws)—76 OE; oth—70 FC & PC
Prince Edward Island, Canada	2,184	5,657	122,506	Charlottetown	168/E 2	81 FC
Puerto Rico	3,515	9,104	3,196,520	San Juan	161/......	80 FC
*Qatar	4,247	11,000	220,000	Doha	58/F 4	TP—80 UN est; cap—79 OE
Québec, Canada	594,857	1,540,680	6,438,403	Québec	172, 174/......	81 FC
Queensland, Australia	666,872	1,727,200	2,295,123	Brisbane	95/......	81 FC
Réunion	969	2,510	491,000	St-Denis	119/F 5	TP—80 OE; oth—74 FC
Rhode Island, U.S.A.	1,212	3,139	947,154	Providence	249/H 5	80 FC & OE
Rhodesia, see Zimbabwe						
*Romania	91,699	237,500	22,048,305	Bucharest	45/F 3	79 OE
Russian S.F.S.R., U.S.S.R.	6,592,812	17,075,400	137,551,000	Moscow	48/D 4	TP, cit over 50,000—79 PC; oth—70 FC
*Rwanda	10,169	26,337	4,819,317	Kigali	114/E 4	78 PC
Sabah, Malaysia	29,300	75,887	1,002,608	Kota Kinabalu	85/F 4	TP—80 PC; Kota Kinabalu—70 FC; oth—70 PC
*Saint Christopher and Nevis	104	269	44,404	Basseterre	156/F 3; 161/C11	TP, isl, cap—80 PC; oth—70 FC
Saint Helena & Dependencies	162	420	5,147	Jamestown	102/B 6	76 FC
*Saint Lucia	238	616	115,783	Castries	161/G 6	80 PC
Saint Pierre & Miquelon	93.5	242	6,034	Saint-Pierre	166/C 4	82 FC
*Saint Vincent & the Grenadines	150	388	124,000	Kingstown	161/A 8; 157/G 4	TP—80 OE; oth—70 FC
Sakhalin, U.S.S.R.	29,500	76,405	655,000	Yuzhno-Sakhalinsk	48/P 4	TP, cit over 50,000—79 PC; oth—70 FC
*Salvador, El	8,260	21,393	4,813,000	San Salvador	154/C 4	TP—80 OE; oth—71 FC
San Marino	23.4	60.6	19,149	San Marino	34/D 3	TP—76 FC; oth—77 OE
*São Tomé e Príncipe	372	963	85,000	São Tomé	106/F 8	TP—80 UN est; oth—70 PC
Sarawak, Malaysia	48,202	124,843	1,294,753	Kuching	85/E 5	TP—80 PC; Kuching—70 FC; oth—70 PC
Sardinia, Italy	9,301	24,090	1,450,483	Cagliari	34/B 4	71 FC
Saskatchewan, Canada	251,699	651,900	968,313	Regina	181/......	81 FC
*Saudi Arabia	829,995	2,149,687	8,367,000	Riyadh	58/D 4	TP—80 UN est; oth—74 PC
Scotland, U.K.	30,414	78,772	5,117,146	Edinburgh	15/......	TP—81 PC; reg—75 OE; cit—75 & 73 OE, 71 FC; oth—71 FC
*Senegal	75,954	196,720	5,508,000	Dakar	106/A 5	TP—79 OE; oth—76 PC
*Seychelles	145	375	63,000	Victoria	119/H 5	TP—79 OE; oth—77 FC
Shetland Islands, Scotland	552	1,430	18,494	Lerwick	15/G 2	TP—76 OE; oth—73 OE & 71 FC
Siam, see Thailand						
Sicily, Italy	9,926	25,708	4,628,918	Palermo	34/D 6	71 FC
*Sierra Leone	27,925	72,325	3,470,000	Freetown	106/B 7	TP—80 UN est; cap, Bo, Kenema, Makeni—74 PC; oth—63 FC
*Singapore	226	585	2,413,945	Singapore	72/F 6	80 FC
Society Islands, French Polynesia	677	1,753	117,703	Papeete	87/L 7	77 FC
*Solomon Islands	11,500	29,785	221,000	Honiara	87/G 6; 86/......	TP—79 OE; oth—76 FC
*Somalia	246,200	637,658	3,645,000	Mogadishu	115/H 3	TP—80 UN est; prov, cap—75 PC; oth—69, 68, 67, 63 & 62 OE

Gazetteer-Index of the World

Country	Area Square Miles	Area Square Kilometers	Population	Capital or Chief Town	Page and Index Ref.	Sources of Population Data
*South Africa	455,318	1,179,274	23,771,970	Cape Town; Pretoria	118/C 5	TP (excl Transkei, Bophuthatswana, Venda), prov—80 PC; Transkei, Bophuthatswana—78 est; Venda—79 est; oth—70 FC
South America	6,875,000	17,806,250	245,000,000	....	120/......	80 UN est
South Australia, Australia	379,922	984,000	1,285,033	Adelaide	94/......	81 FC
South Carolina, U.S.A.	31,113	80,583	3,121,833	Columbia	296/......	80 FC & OE
South Dakota, U.S.A.	77,116	199,730	690,768	Pierre	298/......	80 FC & OE
South Korea	38,175	98,873	37,448,836	Seoul	80/D 5	TP—80 PC; oth—75 FC & PC
South-West Africa (Namibia)	317,827	823,172	1,200,000	Windhoek	118/B 3	TP—74 est; oth—70 FC
*Spain	194,881	504,742	37,430,000	Madrid	33/......	TP—80 OE; met areas—75 OE; oth—70 FC
Spratly Island	....	....			85/E 4	....
*Sri Lanka	25,332	65,610	14,850,001	Colombo	68/E 7	TP—81 PC; cap, Jaffna—73 OE; oth—71 FC
*Sudan	967,494	2,505,809	18,691,000	Khartoum	110/E 4	TP—80 OE; cap, prov, prov cap—73 PC; oth—73 PC, 72 OE
Sumatra, Indonesia	164,000	424,760	19,360,400	Medan	84/B 5	71 PC
*Suriname	55,144	142,823	354,860	Paramaribo	131/C 3	TP, cap—80 PC; dist—71 PC; oth—64 FC
Svalbard, Norway	23,957	62,049	3,431	Longyearbyen	18/C 2	76 OE
*Swaziland	6,705	17,366	547,000	Mbabane	119/E 5	TP—80 OE; oth—76 FC
*Sweden	173,665	449,792	8,320,000	Stockholm	18/J 8	TP—81 OE; oth—75 FC
Switzerland	15,943	41,292	6,365,960	Bern	39/......	TP—80 FC; cantons—78 OE; cap, cit over 100,000 (& ws)—74 OE; cit (com) over 30,000 (& ws)—73 OE; oth—70 FC
*Syria	71,498	185,180	8,979,000	Damascus	62/G 5	TP—80 OE; oth—70 FC
Tadzhik S.S.R., U.S.S.R.	55,251	143,100	3,801,000	Dushanbe	48/G 6	TP, cit over 50,000—79 PC; oth—70 FC
Tahiti, French Polynesia	402	1,041	95,604	Papeete	87/L 7	77 FC
Taiwan	13,971	36,185	16,609,961	Taipei	77/K 7	TP, cap, Penghu Isl., cit over 300,000—77 OE; oth—70 OE
*Tanzania	363,708	942,003	17,527,560	Dar es Salaam	114/F 5	TP—78 PC; div, cap, cit over 17,000—78 PC; oth—67 FC
Tasmania, Australia	26,178	67,800	418,957	Hobart	99/......	81 FC
Tennessee, U.S.A.	42,144	109,153	4,591,120	Nashville	237/......	80 FC & OE
Texas, U.S.A.	266,807	691,030	14,229,288	Austin	303/......	80 FC & OE
*Thailand	198,455	513,998	46,455,000	Bangkok	72/D 3	TP—80 OE; oth—70 FC
Tibet, China	463,320	1,200,000	1,790,000	Lhasa	76/C 5	TP—78 OE; oth—70 est
*Togo	21,622	56,000	2,472,000	Lomé	106/E 7	TP—79 OE; oth—70 FC
Tokelau	3.9	10	1,575	Fakaofo	87/J 6	TP—76 FC; oth—72 FC
Tonga	270	699	90,128	Nuku'alofa	87/J 8	76 PC
Transkei (rep.), South Africa	16,910	43,797	2,000,000	Umtata	119/D 6	TP—80 est; oth—70 FC
Transvaal, South Africa	109,621	283,918	10,673,033	Pretoria	119/D 4	TP—80 PC; oth—70 FC
*Trinidad and Tobago	1,980	5,128	1,067,108	Port-of-Spain	157/G 5; 161/A10	80 PC
Tristan da Cunha, St. Helena	38	98	251	Edinburgh	2/J 7	79 OE
Tuamotu Archipelago, French Polynesia	341	883	9,052	Apataki	87/M 7	77 FC
*Tunisia	63,378	164,149	6,367,000	Tunis	106/F 1	TP—79 OE; oth—75 FC
*Turkey	300,946	779,450	45,217,556	Ankara	62/D 3	TP—80 PC; oth—75 FC
Turkmen S.S.R., U.S.S.R.	188,455	488,100	2,759,000	Ashkhabad	48/F 6	TP, cit over 50,000—79 PC; oth—70 FC
Turks and Caicos Islands	166	430	7,436	Cockburn Town, Grand Turk	156/D 2	80 PC
Tuvalu	9.78	25.33	7,349	Fongafale, Funafuti	87/H 6	79 FC
*Uganda	91,076	235,887	12,630,076	Kampala	114/F 3	TP, cap—80 PC; oth—69 FC
*Ukrainian S.S.R., U.S.S.R.	233,089	603,700	49,755,000	Kiev	52/D 5	TP, cit over 50,000—79 PC; oth—70 FC
*Union of Soviet Socialist Republics	8,649,490	22,402,179	262,436,227	Moscow	48/......	TP, S.S.R., cit over 50,000—79 PC; oth—70 FC
*United Arab Emirates	32,278	83,600	1,040,275	Abu Dhabi	58/F 5	TP—80 PC; oth—79 OE
*United Kingdom	94,399	244,493	55,672,000	London	10/......	TP—81 OE (see England, Wales, Scotland, Northern Ireland)
*United States of America	3,623,420	9,384,658	226,504,825	Washington	188/......	80 FC & OE
*Upper Volta (Burkina Faso)	105,869	274,200	6,908,000	Ouagadougou	106/D 6	TP—80 UN est; oth—75 FC, 73 OE
*Uruguay	72,172	186,925	2,899,000	Montevideo	145/......	TP—80 OE; oth—75 PC
Utah, U.S.A.	84,899	219,888	1,461,037	Salt Lake City	304/......	80 FC & OE
Uzbek S.S.R., U.S.S.R.	173,591	449,600	15,391,000	Tashkent	48/G 5	TP, cit over 50,000—79 PC; oth—70 FC
*Vanuatu	5,700	14,763	112,596	Vila	87/G 7	79 FC
Vatican City	108.7 acres	44 hectares	728	....	34/B 6	78 OE
Venda (rep.), South Africa	2,510	6,501	450,000	Thohoyandou	119/E 4	79 est
*Venezuela	352,143	912,050	14,313,000	Caracas	124/......	TP—81 OE; oth—71 FC
Vermont U.S.A.	9,614	24,900	511,456	Montpelier	268/......	80 FC & OE
Victoria, Australia	87,876	227,600	3,832,443	Melbourne	96/B 5	81 FC
*Vietnam	128,405	332,569	52,741,766	Hanoi	72/E 3	TP—79 FC; cap, Haiphong, Ho Chi Minh City—79 PC; oth cit over 100,000 (north)—70 est, (south)—73 & 71 OE; oth—69 OE, 60 FC
Virginia, U.S.A.	40,767	105,587	5,346,818	Richmond	307/......	80 FC & OE
Virgin Islands, British	59	153	11,006	Road Town	157/H 1	TP—80 FC; oth—70 FC
Virgin Islands, U.S.A.	132	342	96,569	Charlotte Amalie	161/A 4	80 FC
Wake Island	2.5	6.5	302	Wake Islet	87/G 4	80 FC
Wales, U.K.	8,017	20,764	2,790,462	Cardiff	13/D 5	TP—81 PC; co—76 OE; cit—76 & 73 OE; par—71 FC
Wallis and Futuna	106	275	9,192	Mata Utu	87/J 7	76 FC
Washington, U.S.A.	68,139	176,480	4,132,180	Olympia	310/......	80 FC & OE
West Bank	2,100	5,439	c. 800,000		65/C 3	TP—81 est; oth—67 CE & 61 FC
Western Australia, Australia	975,096	2,525,500	1,273,624	Perth	92/......	81 FC
Western Sahara	102,703	266,000	76,425	....	106/B 3	70 FC
*Western Samoa	1,133	2,934	158,130	Apia	87/J 7	81 PC
*West Germany (Federal Republic)	95,985	248,601	61,658,000	Bonn	22/......	TP—80 OE; states, cap—76 OE; oth—76 OE, 70 FC
West Virginia, U.S.A.	24,231	62,758	1,950,279	Charleston	312/......	80 FC & OE
*White Russian S.S.R. (Byelorussian S.S.R.), U.S.S.R.	80,154	207,600	9,560,000	Minsk	52/C 4	TP, cit over 50,000—79 PC; oth—70 FC
Wisconsin, U.S.A.	56,153	145,436	4,705,521	Madison	317/......	80 FC & OE
World	(land) 57,970,000	150,142,300	4,415,000,000	....	1, 2/......	80 UN est
Wyoming, U.S.A.	97,809	253,325	469,557	Cheyenne	319/......	80 FC & OE
*Yemen, People's Democratic Republic of	111,101	287,752	1,969,000	Aden	58/E 7	TP—81 PC; oth—75 FC
*Yemen Arab Republic	77,220	200,000	6,456,189	San a	58/D 6	TP—80 OE; Mukalla, Seiyun—76 OE; cap—73 OE; Saihut—60 OE
*Yugoslavia	98,766	255,804	22,471,000	Belgrade	45/C 3	TP—81 OE; oth—71 FC
Yukon Territory, Canada	207,075	536,324	23,153	Whitehorse	186/E 3	81 FC
*Zaire	905,063	2,344,113	28,291,000	Kinshasa	114/D 4	TP—80 OE; prov, cap—70 FC; oth—70 FC & PC
*Zambia	290,586	752,618	5,679,808	Lusaka	114/E 7	80 PC
*Zimbabwe	150,803	390,580	7,360,000	Harare	119/D 3	TP—80 OE; cap, cit over 12,000—77 OE; oth—69 FC

Introduction to the Maps and Indexes

The following notes have been added to aid the reader in making the best use of this atlas. Though he may be familiar with maps and map indexes, the publisher believes that a quick review of the material below will add to his enjoyment of this reference work.

Arrangement — The Plan of the Atlas. The atlas has been designed with maximum convenience for the user as its objective. All geographically related information pertaining to a country or region appears on adjacent pages, eliminating the task of searching throughout the entire volume for data on a given area. Thus, the reader will find, conveniently assembled, political, topographic, economic and special maps of a political area or region, accompanied by detailed map indexes, statistical data, and illustrations of the national flags of the area.

The sequence of country units in this American-designed atlas is international in arrangement. Units on the world as a whole are followed by a section on the polar regions which, in turn, is followed by pages devoted to Europe and its countries. Every continent map is accompanied by special population distribution, climatic and vegetation maps of that continent. Following the maps of the European continent and its countries, the geographic sequence plan proceeds as follows: Asia, the Pacific and Australia, Africa, South America, North America, and ends with detailed coverage on the United States.

Political Maps — The Primary Reference Tool. The most detailed maps in each country unit are the *political maps*. It is our feeling that the reader is likely to refer to these maps more often than to any other in the book when confronted by such questions as — Where? How big? What is it near? Answering these common queries is the function of the political maps. Each political map stresses *political* phenomena — countries, internal political divisions, boundaries, cities and towns. The major political unit or units, shown on the map, are banded in distinctive colors for easy identification and delineation. First-order political subdivisions (states, provinces, counties on the state maps) are shown, scale permitting.

The reader is advised to make use of the *legend* appearing under the title on each political map. Map *symbols,* the special "language" of maps, are explained in the legend. Each variety of dot, circle, star or interrupted line has a special meaning which should be clearly understood by the user so that he may interpret the map data correctly.

Each country has been portrayed at a *scale* commensurate with its political, areal, economic or tourist importance. In certain cases, a whole map unit may be devoted to a single nation if that nation is considered to be of prime interest to most atlas users. In other cases, several nations will be shown on a single map if, as separate entities, they are of lesser relative importance. Areas of dense settlement and important significance within a country have been enlarged and portrayed in inset maps inserted on the margins of the main map. The scale of each map is indicated as a fractional representation (1:1,000,000). The reader is advised to refer to the linear or "bar" scale appearing on each map or map inset in order to determine the distance between points.

The *projection* system used for each map is noted near the title of the map. Map projections are the special graphic systems used by cartographers to render the curved three-dimensional surface of the globe on a flat surface. Optimum map projections determined by the attributes of the area have been used by the publishers for each map in the atlas.

A word here as to the choice of place names on the maps. Throughout the atlas names appear, with a few exceptions, in their local official spellings. However, conventional Anglicized spellings are used for major geographical divisions and for towns and topographic features for which English forms exist; i.e., "Spain" instead of "España" or "Munich" instead of "München." Names of this type are normally followed by the local official spelling in parentheses. As an aid to the user the indexes are cross-referenced for all current and most former spellings of such names.

Names of cities and towns in the United States follow the forms listed in the *Post Office Directory* of the United States Postal Service. Domestic physical names follow the decisions of the Board on Geographic Names, U.S. Department of the Interior, and of various state geographic name boards.

It is the belief of the publishers that the boundaries shown in a general reference atlas should reflect current geographic and political realities. This policy has been followed consistently in the atlas. The presentation of *de facto* boundaries in cases of territorial dispute between various nations does not imply the political endorsement of such boundaries by the publisher, but simply the honest representation of boundaries as they exist at the time of the printing of the atlas maps.

Indexes — Pinpointing a Location. Each political map is accompanied by a comprehensive index of the place names appearing on the map. If you are unfamiliar with the location of a particular geographical place and wish to find its position within the confines of the subject area of the map, consult the map index as your first step. The name of the feature sought will be found in its proper alphabetical sequence with a key reference letter-number combination corresponding to its location on the map. After noting the key reference letter-number combination for the place name, turn to the map. The place name will be found within the square formed by the two lines of latitude and the two lines of longitude which enclose the co-ordinates — i.e., the marginal letters and numbers. The diagram below illustrates the system of indexing.

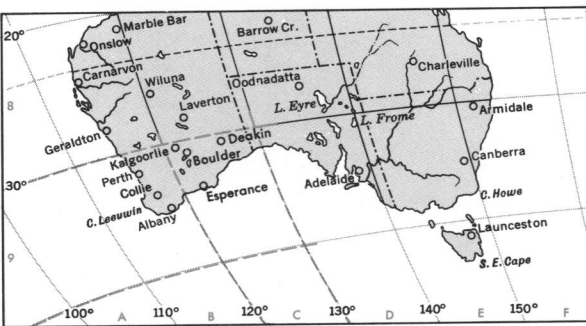

In the case of maps consisting entirely of insets, the place name is found near the intersection point of the imaginary lines connecting the co-ordinates at right angles. See below.

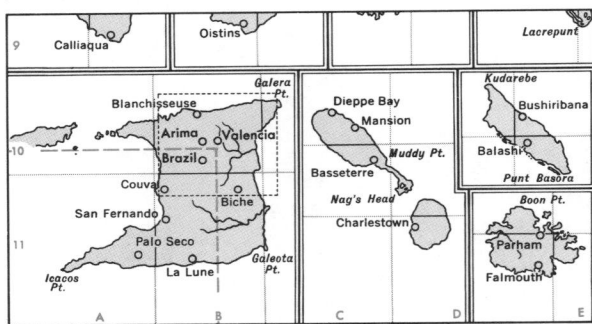

Where space on the map has not permitted giving the complete form of the place name, the complete form is shown in the index. Where a place is known by more than one name or by various spellings of the same name, the different forms have been included in the index. Physical features are listed under their proper names and not according to their generic terms; that is to say, Rio Negro will be found under Negro and not under Rio Negro. On the other hand, Rio Grande will be found under Rio Grande. Accompanying most index entries for cities and towns, and for other political units, are *population figures* for the particular entries. The large number of population figures in the atlas makes this work one of the most comprehensive statistical sources available to the public today. The population figures have been taken from the latest official censuses and estimates of the various nations. Dates and sources for the population figures are listed in the Gazetteer-Index of the World preceding this section.

Population and area figures for countries and major political units are listed in bold type *fact lists* on the margins of the indexes. In addition, the capital, largest city, highest point, monetary unit, principal languages and the prevailing religions of the country concerned are also listed. The Gazetteer-Index of the World on the preceding pages provides a quick reference index for countries and other important areas. Though population and area figures for each major unit area also found in the map section, the Gazetteer-Index provides a conveniently arranged statistical comparison contained in five pages. As mentioned, dates and sources of the population figures appearing in the country indexes are also listed in this section.

All index entries for cities and towns in the United States are preceded by a five-digit postal ZIP code number applying to the community. This useful feature permits the reader to address his mail so that it will be routed and delivered more efficiently and quickly by the U.S. Postal Service. A dagger (†) designates those places that do not possess a post office. The ZIP code number listed in such cases refers to that of the nearest post office. An asterisk (*) marks those larger cities which are divided into multiple ZIP code areas. Using the single ZIP code number listed in such cases will direct your letter to the proper city with dispatch. However, if the precise ZIP code number of the address within the city is needed, it is suggested that the reader refer to the latest National ZIP Code Directory at his local post office. This detailed guide lists every street in a multiple ZIP code city with the proper ZIP code for the street.

Relief Maps. Accompanying each political map is a relief map of the area. The purpose of the relief map is to illustrate the surface configuration (TOPOGRAPHY) of the region. A shading technique in color simulates the relative ruggedness of the terrain — plains, plateaus, valleys, hills and mountains. Graded colors, ranging from greens for lowlands, yellows for intermediate elevations to browns in the highlands, indicate the height above sea level of each part of the land. A vertical scale at the margin of the map shows the approximate height in meters and feet represented by each color.

Economic Maps — Agriculture, Industry and Resources. One of the most interesting features that will be found in each country unit is the economic map. From this map one can determine the basic activities of a nation as expressed through its economy. A perusal of the map yields a full understanding of the area's economic geography and natural resources.

The agricultural economy is manifested in two ways: color bands and commodity names. The color bands express broad categories of *dominant land use,* such as, cereal belts, forest lands, livestock range lands, nonagricultural wastes. The red commodity names, on the other hand, pinpoint the areas of production of *specific* crops; i.e., wheat, cotton, sugar beets, etc.

Major mineral occurrences are denoted by standard letter symbols appearing in blue. The relative size of the letter symbols signifies the relative importance of the deposit.

The manufacturing sector of the economy is presented by means of diagonal line patterns expressing the various *industrial areas* of consequence within a country.

The fishing industry is represented by names of commercial fish species appearing offshore in blue letters. Major waterpower sites are designated by blue symbols.

The publishers have tried to make this work the most comprehensive and useful atlas available, and it is hoped that it will prove a valuable reference work. Any constructive suggestions from the reader will be welcomed.

Sources and Acknowledgments

A multitude of sources goes into the making of a large-scale reference work such as this. To list them all would take many pages and would consume space better devoted to the maps and reference materials themselves. However, certain general sources were very useful in preparing this work and are listed below.

STATISTICAL OFFICE OF THE UNITED NATIONS.
Demographic Yearbook. New York. Issued annually.

STATISTICAL OFFICE OF THE UNITED NATIONS.
Statistical Yearbook. New York. Issued annually.

THE GEOGRAPHER, U.S. DEPARTMENT OF STATE.
International Boundary Study papers. Washington. Various dates.

THE GEOGRAPHER, U.S. DEPARTMENT OF STATE.
Geographic Notes. Washington. Various dates.

UNITED STATES BOARD ON GEOGRAPHIC NAMES.
Decisions on Geographic Names in the United States. Washington. Various dates.

UNITED STATES BOARD ON GEOGRAPHIC NAMES.
Official Standard Names Gazetteers. Washington. Various dates.

CANADIAN PERMANENT COMMITTEE ON GEOGRAPHICAL NAMES.
Gazetteer of Canada series. Ottawa. Various dates.

UNITED STATES POSTAL SERVICE.
National Five Digit ZIP Code and Post Office Directory. Washington. 1983.

UNITED STATES POSTAL SERVICE.
Postal Bulletin. Washington. Issued weekly.

UNITED STATES DEPARTMENT OF THE INTERIOR. BUREAU OF MINES.
Minerals Yearbook. 4 vols. Washington. Various dates.

UNITED STATES GEOLOGICAL SURVEY.
Elevations and distances in the United States. Reston, Va. 1980.

CARTACTUAL.
Cartactual — Topical Map Service. Budapest. Issued bi-monthly.

AMERICAN GEOGRAPHICAL SOCIETY.
Focus. New York. Issued ten times a year.

THE AMERICAN UNIVERSITY.
Foreign Area Studies. Washington. Various dates.

CENTRAL INTELLIGENCE AGENCY.
General reference maps. Washington. Various dates.

A sample list of sources used for specific countries follows:

Afghanistan
CENTRAL STATISTICS OFFICE.
Preliminary Results of the First Afghan Population Census 1979. Kabul.

Albania
DREJTORIA E STATISTIKES.
1979 Census. Tiranë.

Argentina
INSTITUTO NACIONAL DE ESTADISTICA Y CENSOS.
Censo Nacional de Población y Vivienda 1980. Buenos Aires.

Australia
AUSTRALIAN BUREAU OF STATISTICS.
Census of Population and Housing 1981. Canberra.

Brazil
FUNDAÇAO INSTITUTO BRASILEIRO DE GEOGRAFIA E ESTATISTICA.
IX Recenseamento Geral do Brasil 1980. Rio de Janeiro.

Canada
STATISTICS CANADA.
1981 Census of Canada. Ottawa.

Cuba
COMITE ESTATAL DE ESTADISTICAS.
Censo de Población y Viviendas 1981. Havana.

Hungary
HUNGARIAN CENTRAL STATISTICAL OFFICE.
1980 Census. Budapest.

Indonesia
BIRO PUSAT STATISTIK.
Sensus Penduduk 1980. Jakarta.

Kuwait
CENTRAL OFFICE OF STATISTICS.
1980 Census. Al Kuwait.

New Zealand
DEPARTMENT OF STATISTICS.
New Zealand Census of Population and Dwellings 1981. Wellington.

Panama
DIRECCIÓN DE ESTADISTICA Y CENSO.
Censos Nacionales de 1980. Panamá.

Papua New Guinea
BUREAU OF STATISTICS.
National Population Census 1980. Pòrt Moresby.

Philippines
NATIONAL CENSUS AND STATISTICS OFFICE.
1980 Census of Population. Manila.

Saint Lucia
CENSUS OFFICE.
1980 Population Census. Castries.

Singapore
DEPARTMENT OF STATISTICS.
Census of Population 1980. Singapore.

U.S.S.R.
CENTRAL STATISTICAL ADMINISTRATION.
1979 Census. Moscow.

United States
BUREAU OF THE CENSUS.
1980 Census of Population. Washington.

Vanuatu
CENSUS OFFICE.
1979 Population Census. Port Vila.

Zambia
CENTRAL STATISTICAL OFFICE.
1980 Census of Population and Housing. Lusaka.

Glossary of Abbreviations

A

A. A. F. — Army Air Field
Acad. — Academy
A. C. T. — Australian Capital Territory
adm. — administration; administrative
A. F. B. — Air Force Base
Afgh., Afghan. — Afghanistan
Afr. — Africa
Ala. — Alabama
Alb. — Albania
Alg. — Algeria
Alta. — Alberta
Amer. — American
Amer. Samoa — American Samoa
And. — Andorra
Ant., Antarc. — Antarctica
Ant. & Bar. — Antigua and Barbuda
Ar. — Arabia
arch. — archipelago
Arg. — Argentina
Ariz. — Arizona
Ark. — Arkansas
A. S. S. R. — Autonomous Soviet
    Socialist Republic
Aust. — Austria
Aust. Cap. Terr. — Australian Capital
    Territory
Austr., Austral. — Australian, Australia
aut. — autonomous
Aut. Obl. — Autonomous Oblast

B

B. — bay
Bah. — Bahamas
Barb. — Barbados
Battlef. — Battlefield
Bch. — Beach
Belg. — Belgium
Berm. — Bermuda
Bol. — Bolivia
Bots. — Botswana
Br. — Branch
Br. — British
Braz. — Brazil
Br. Col. — British Columbia
Br. Ind. Oc. Terr. — British Indian
    Ocean Territory
Bulg. — Bulgaria

C

C. — cape
Calif. — California
Can. — Canada
can. — canal
cap. — capital
Cent. Afr. Rep. — Central African
    Republic
Cent. Amer. — Central America
C. G. Sta. — Coast Guard Station
C. H. — Court House
chan. — channel
Chan. Is. — Channel Islands
Chem. Ctr. — Chemical Center
co. — county
C. of G. H. — Cape of Good Hope
Col. — Colombia
Colo. — Colorado
comm. — commissary
Conn. — Connecticut
cont. — continent
cord. — cordillera (mountain range)
C. Rica — Costa Rica
C. S. — County Seat
C. Verde — Cape Verde
Czech. — Czechoslovakia

D

D. C. — District of Columbia
Del. — Delaware
Dem. — Democratic
Den. — Denmark
depr. — depression
dept. — department
des. — desert
dist., dist's — district, districts
div. — division
Dom. Rep. — Dominican Republic

E

E. — East
Ec., Ecua. — Ecuador
E. Ger. — East Germany
elec. div. — electoral division
El Salv. — El Salvador
Eng. — England
Equat. Guinea, Eq. Guin — Equatorial
    Guinea

escarp. — escarpment
est. — estuary
Eth. — Ethiopia

F

Falk. Is. — Falkland Islands
Fin. — Finland
Fk., Fks. — Fork, Forks
Fla. — Florida
for. — forest
Fr. — France, French
Fr. Gui. — French Guiana
Fr. Poly. — French Polynesia
Ft. — Fort

G

G. — gulf
Ga. — Georgia
Game Res. — Game Reserve
Ger. — Germany
geys. — geyser
Gibr. — Gibraltar
glac. — glacier
gov. — governorate
Gr. — Group
Greenl. — Greenland
Gren. — Grenada
Gt. Brit. — Great Britain
Guad. — Guadeloupe
Guat. — Guatemala
Guinea-Biss. — Guinea-Bissau
Guy. — Guyana

H

har., harb., hbr. — harbor
hd. — head
highl. — highland, highlands
Hist. — Historic, Historical
Hond. — Honduras
Hts. — Heights
Hung. — Hungary

I

i., isl. — island, isle
I. C. — independent city
Ice., Icel. — Iceland
Ida. — Idaho
Ill. — Illinois
Ind. — Indiana
ind. city — independent city
Indon. — Indonesia
Ind. Res. — Indian Reservation
int. div. — internal division
inten. — intendency
Int'l — International
Ire. — Ireland
is., isls. — islands
Isr. — Israel
isth. — isthmus
Iv. Coast — Ivory Coast

J

Jam. — Jamaica
Jct. — Junction

K

Kans. — Kansas
Ky. — Kentucky

L

L. — Lake, Loch, Lough
La. — Louisiana
Lab. — Laboratory
lag. — lagoon
Ld. — Land
Leb. — Lebanon
Les. — Lesotho
Liecht. — Liechtenstein
Lux. — Luxembourg

M

Mad., Madag. — Madagascar
Man. — Manitoba
Mart. — Martinique
Mass. — Massachusetts
Maur. — Mauritania
Md. — Maryland
met. area — metropolitan area
Mex. — Mexico
Mich. — Michigan
Minn. — Minnesota
Miss. — Mississippi
Mo. — Missouri
Mon. — Monument
Mong. — Mongolia
Mont. — Montana
Mor. — Morocco

Moz., Mozamb. — Mozambique
mt. — mount
mtn. — mountain

N

N., No., North. — North, Northern
N. Amer. — North America
Nam., Namib. — Namibia
N. A. S. — Naval Air Station
Nat'l — National
Nat'l Cem. — National Cemetery
Nat'l Mem. Park — National Memorial
    Park
Nat'l Mil. Park — National Military
    Park
Nat'l Pkwy. — National Parkway
Nav. Base — Naval Base
Nav. Sta. — Naval Station
N. B., N. Br. — New Brunswick
N. C. — North Carolina
N. Dak. — North Dakota
Nebr. — Nebraska
Neth. — Netherlands
Neth. Ant. — Netherlands Antilles
Nev. — Nevada
New Bruns. — New Brunswick
New Cal., New Caled. — New Caledonia
Newf. — Newfoundland
New Hebr. — New Hebrides
N. H. — New Hampshire
Nic. — Nicaragua
N. Ire. — Northern Ireland
N. J. — New Jersey
N. Mex. — New Mexico
Nor. — Norway, Norwegian
North. — Northern
North. Terr., No. Terr. — Northern
    Territory
    (Australia)
N. S. — Nova Scotia
N. S. W., N.S. Wales — New South Wales
N. W. T., N. W. Terrs. — Northwest
    Territories
    (Canada)
N. Y. — New York
N. Z., N. Zealand — New Zealand

O

Obl. — Oblast
O. F. S. — Orange Free State
Okla. — Oklahoma
Okr. — Okrug
Ont. — Ontario
Ord. Depot — Ordnance Depot
Oreg. — Oregon

P

Pa. — Pennsylvania
Pac. Is. — Pacific Islands,
    Territory of the
Pak. — Pakistan
Pan. — Panama
Papua N. G. —Papua New Guinea
Par. — Paraguay
par. — parish
passg. — passage
P.D.R. Yemen — People's Democratic
    Republic of Yemen
P. E. I. — Prince Edward Island
pen. — peninsula
Phil., Phil. Is. — Philippines
Pk. — Park
pk. — peak
plat. — plateau
P. N. G. — Papua New Guinea
Pol. — Poland
Port. — Portugal, Portuguese
Pr. Edward I. — Prince Edward Island
pref. — prefecture
P. Rico — Puerto Rico
prom. — promontory
prov. — province, provincial
pt. — point

Q

Que. — Quebec
Queens. — Queensland

R

R. — River
ra. — range
Rec., Recr. — Recreation, Recreational
reg. — region
Rep. — Republic
res. — reservoir
Res. — Reservation, Reserve
R. i. — Rhode Island

riv. — river
Rom. — Romania

S

S. — South
Sa. — Sierra, Serra
S. Afr., S. Africa — South Africa
salt dep. — salt deposit
salt des. — salt desert
S. Amer. — South America
São T. & Pr. — São Tomé
    and Príncipe
Sask. — Saskatchewan
Saudi Ar. — Saudi Arabia
S. Aust., S. Austral. — South Australia
S. C. — South Carolina
Scot. — Scotland
Sd. — Sound
S. Dak. — South Dakota
Sen. — Senegal
sen. dist. — senatorial district
Seych. — Seychelles
S. F. S. R. — Soviet Federated Socialist
    Republic
Sing. — Singapore
S. Leone — Sierra Leone
S. Marino — San Marino
Sol. Is. — Solomon Islands
Sp. — Spanish
Spr., Sprs. — Spring, Springs
S. S. R. — Soviet Socialist Republic
St., Ste. — Saint, Sainte
Sta. — Station
St. Chris.-Nevis — Saint Christopher-
    Nevis
St. P. & M. — Saint Pierre and
    Miquelon
St. Vin. & Grens. — St. Vincent & The
    Grenadines
str., strs. — strait, straits
Sur. — Suriname
S. W. Afr. — South-West Africa
Swaz. — Swaziland
Switz. — Switzerland

T

Tanz. — Tanzania
Tas. — Tasmania
Tenn. — Tennessee
terr., terrs. — territory, territories
Tex. — Texas
Thai. — Thailand
trad. — traditional
Trin. & Tob. — Trinidad and Tobago
Tun. — Tunisia
twp. — township

U

U. A. E. — United
    Arab Emirates
U. K. — United Kingdom
Upp. Volta — Upper Volta
urb. area — urban area
Urug. — Uruguay
U. S. — United States
U. S. S. R. — Union of Soviet Socialist
    Republics

V

Va. — Virginia
Ven., Venez. — Venezuela
V. I. (Br.) — Virgin Islands (British)
V. I. (U. S.) — Virgin Islands (U. S.)
Vic. — Victoria
Viet. — Vietnam
Vill. — Village
vol. — volcano
Vt. — Vermont

W

W. — West, Western
Wash. — Washington
W. Aust., W. Austral. — Western
    Australia
W. Ger. — West Germany
W. Indies — West Indies
Wis. — Wisconsin
W. Samoa — Western Samoa
W. Va. — West Virginia
Wyo. — Wyoming

Y

Yugo. — Yugoslavia
Yukon — Yukon Territory

Z

Zim. — Zimbabwe

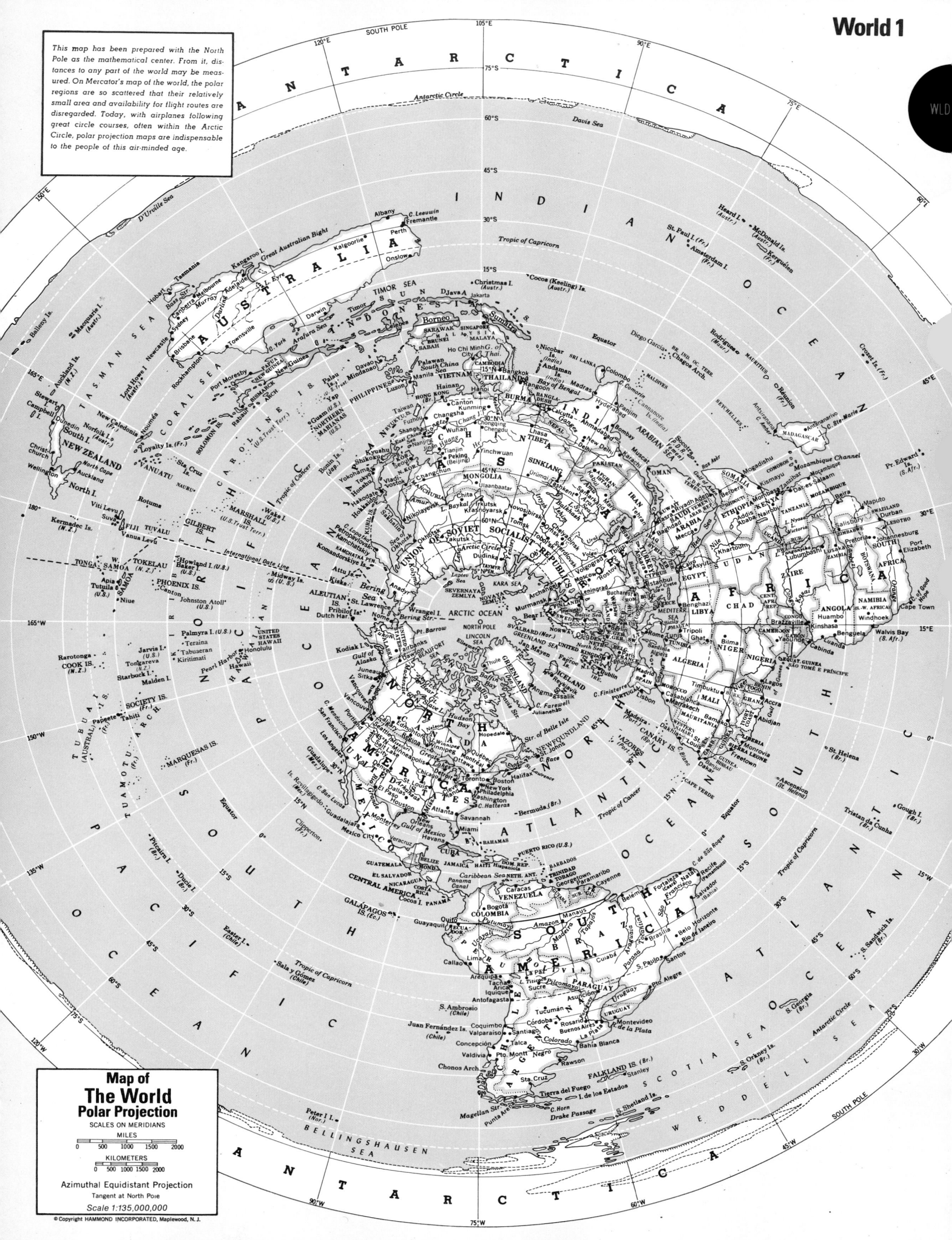

This map has been prepared with the North Pole as the mathematical center. From it, distances to any part of the world may be measured. On Mercator's map of the world, the polar regions are so scattered that their relatively small area and availability for flight routes are disregarded. Today, with airplanes following great circle courses, often within the Arctic Circle, polar projection maps are indispensable to the people of this air-minded age.

Map of
The World
Polar Projection

SCALES ON MERIDIANS

MILES

0   500   1000   1500   2000

KILOMETERS

0   500 1000 1500 2000

Azimuthal Equidistant Projection
Tangent at North Pole

Scale 1:135,000,000

© Copyright HAMMOND INCORPORATED, Maplewood, N.J.

The World

BRIESEMEISTER ELLIPTICAL
EQUAL-AREA PROJECTION

Capitals of Countries .... ⊗
Other Capitals .... ⊚
International Boundaries ..... ─ ─ ─

Scale 1:80,000,000

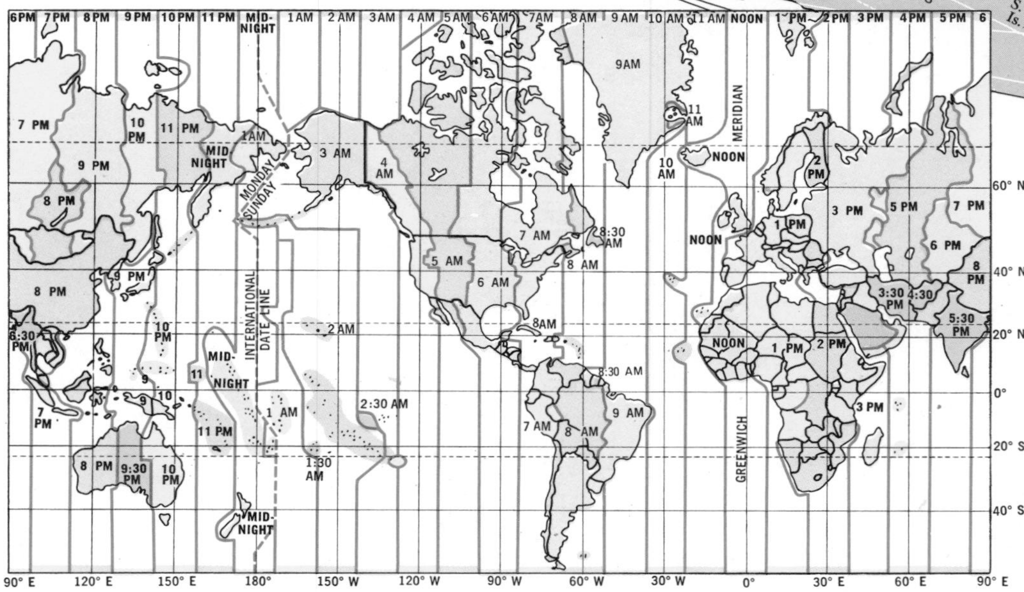

NORTH PACIFIC OCEAN

NORTH AMERICA

UNITED STATES
HAWAII
Honolulu

Palmyra I. (U.S.)

KIRIBATI
Line Is.
Tabuaeran

Anchorage
Whitehorse
Juneau
Fairbanks
ALASKA
Pt. Barrow
Yukon
Aleutian
UNITED STATES

Vancouver I.
Vancouver
Seattle
Portland
San Francisco
Los Angeles
Denver
Colo.
Lower California
Edmonton
Calgary
Yellowknife
Banks I.
Victoria
Sask.
CANADA
Queen Elizabeth Islands
Mackenzie

El Paso
Arkansas
Missouri
Winnipeg
Minneapolis
St. Louis
Chicago
Detroit
Great Lakes
Toronto
Ottawa
Montréal
Québec
Hudson Bay
Baffin
Baffin B.
Thule
GREENLAND (Den.)
Ellesmere I.

MEXICO
Guadalajara
Monterrey
Dallas
Houston
New Orleans
Atlanta
Ohio
Savannah
New York
Philadelphia
Washington
Boston
Halifax
Newfoundland
St. John's
C. Farewell
Julianehåb
Godthåb (Nûk)
ICELAND
Reykjavik
Arctic C.
60° N

Mexico City
Veracruz
CENTRAL AMERICA
GULF OF MEXICO
GUAT. HON.
BELIZE
EL SAL. NIC.
C.R. PAN.
Havana
CUBA
BAHAMAS
Miami
C. Canaveral
C. Hatteras
Bermuda
40° N
Azores (Port.)
UNITED KINGDOM
IRELAND
London
Paris
FRANCE
B. of Biscay
SPAIN
Madrid
Lisbon
PORT.
Gibraltar
Rabat
MOROCCO
Casablanca
ALGERIA

JAM.
HAITI
DOM. REP.
PUERTO RICO
CARIBBEAN SEA
West Indies
Tropic of Cancer
Canary Is. (Sp.)
WESTERN SAHARA
Madeira (Port.)

COLOMBIA
Bogotá
VENEZUELA
Orinoco
ST. VINCENT & GREN.
GRENADA
TRIN. & TOB.
DOMINICA
BARBADOS
ST. LUCIA
Georgetown
Paramaribo
SUR.
FR. GUI.
Cayenne
20° N
Dakar
SEN.
Nouakchott
MAURITANIA
MALI
GAMBIA
GUINEA-BISSAU
Bamako
SIERRA LEONE
GUINEA
IVORY COAST
LIBERIA
Monrovia
BURKINA FASO

ECUADOR
Quito
Guayaquil
Pta. Aguja
PERU
Lima
Callao
Ucayali
BRAZIL
Madeira
Manaus
Amazon
Negro
Equator
CAPE VERDE
C. Blanc
Accra
TOGO
GHANA
Abidjan
Gulf of Guinea
EQ. GUI.
Libreville
0°

Arequipa
BOLIVIA
La Paz
L. Titicaca
Sucre
Antofagasta
Juan Fernández Is. (Chile)
CHILE
Tapajós
Tocantins
São Francisco
Brasília
Natal
Fortaleza
C. de São Roque
Recife
40° W
Tropic of Capricorn
20° W
St. Helena (Br.)
Ascension (St. Hel.)

Valparaíso
Santiago
Valdivia
ARGENTINA
Córdoba
PARAGUAY
Asunción
Paraná
URUGUAY
Montevideo
R. de la Plata
São Paulo
Rio de Janeiro
Santos
Porto Alegre
Belo Horizonte
Salvador
Bahía Blanca
Colorado
20° S
Tristan da Cunha (St. Hel.)
Gough I. (St. Hel.)
40° S

Antarctic Circle
Str. of Magellan
Tierra del Fuego
Cape Horn
Falkland Is. (Br.)
Drake Passage
S. Georgia (Br.)
S. Orkney Is. (Br.)
S. Sandwich Is. (Br.)
SCOTIA SEA
60° S

MARIE BYRD LAND
ANTARCTICA
Ronne Ice Shelf
Berkner I.
Larsen Ice Shelf
Antarctic Pen.
COATS LAND
80° S

SOUTH PACIFIC OCEAN

Niue (N.Z.)
Cook Is. (N.Z.)
Australis. (Fr. Poly.)
Papeete
Tahiti (Fr. Poly.)
FRENCH POLYNESIA (Fr.)
Marquesas Is. (Fr. Poly.)
Tuamotu Arch. (Fr. Poly.)
Vostok I.
Equator
Tropic of Capricorn
Pitcairn I. (Br.)
Easter I. (Chile)
Sala y Gómez (Chile)
Antarctic

Is. Revillagigedo (Mex.)
C. San Lucas
Rio Grande
Clipperton (Fr.)
Galápagos Is. (Ec.)
Panama Can.
SOUTH AMERICA

NORTH ATLANTIC OCEAN
SOUTH ATLANTIC OCEAN

140° W
160° W
120° W
100° W
80° W
60° W

Time Zones

| STANDARD TIME ZONES | | Areas using half hour deviations. |
| Areas not using zone system. | | |

NOTE: Standard time zones in the U.S.S.R. are always advanced one hour.

6PM 7PM 8PM 9PM 10PM 11PM MIDNIGHT 1AM 2AM 3AM 4AM 5AM 6AM 7AM 8AM 9AM 10AM 11AM NOON 1PM 2PM 3PM 4PM 5PM 6

INTERNATIONAL DATE LINE
MONDAY SUNDAY
MERIDIAN
GREENWICH

90° E 120° E 150° E 180° 150° W 120° W 90° W 60° W 30° W 0° 30° E 60° E 90° E
60° N 40° N 20° N 20° S 40° S

LAND AREA 57,970,000 sq. mi.
(150,142,300 sq. km.)
WATER AREA 139,781,000 sq. km.
(362,032,790 sq. km.)
TOTAL SURFACE AREA 197,751,000 sq.mi.
(512,175,090 sq. km.)
POPULATION 4,415,000,000

Antarctica

AZIMUTHAL EQUIDISTANT PROJECTION

Scale 1:62,000,000

© Copyright HAMMOND INCORPORATED, Maplewood, N.J.

4 Arctic Ocean

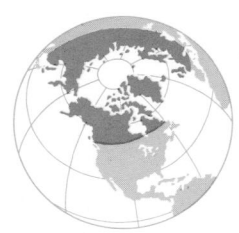

Alaska (gulf), U.S.	D17	
Alaska (pen.), U.S.	D18	
Alaska (range), U.S.	C17	
Alaska (reg.), U.S.	C17	
Aleutian (isls.), U.S.	D18	
Alexander (arch.), U.S.	D16	
Alexandra Land (isl.), U.S.S.R.	A8	
Amundsen (gulf), Canada	B16	
Anadyr', U.S.S.R.	C1	
Anadyr' (gulf), U.S.S.R.	C1	
Arctic Ocean	A15	
Atlantic Ocean	D11	
Attu (isl.), U.S.	D1	
Axel Heiberg (isl.), Canada	A14	
Baffin (bay)	B13	
Baffin (isl.), Canada	C13	
Banks (isl.), Canada	B16	
Barents (sea)	B8	
Barrow (pt.), U.S.	B18	
Bathurst (isl.), Canada	B14	
Bear (isl.), Norway	B9	
Bear (isl.), U.S.S.R.	B1	
Beaufort (sea)	B16	
Belyy (isl.), U.S.S.R.	B6	
Bering (str.)	C18	
Bilibino, U.S.S.R.	C1	
Bol'shevik (isl.), U.S.S.R.	A4	
Boothia (gulf), Canada	B14	
Boothia (pen.), Canada	B14	

Borden (isl.), Canada	B15
Bristol (bay), U.S.	D18
Brodeur (pen.), Canada	B14
Brooks (range), U.S.	C17
Canada	C13
Chelyuskin (cape), U.S.S.R.	B4
Cherskiy, U.S.S.R.	C1
Chukchi (pen.), U.S.S.R.	C18
Chukchi (sea)	C18
Columbia (cape), Canada	A13
Cook (inlet), U.S.	D17
Cumberland (sound), Canada	C13
Davis (str.)	C12
Denmark (strait)	C11
Devon (isl.), Canada	B14
Dezhnev (cape), U.S.S.R.	C18
Disko (isl.), Greenl.	C12
Dmitriya Lapteva (str.), U.S.S.R.	B2
Dvina, Northern (riv.), U.S.S.R.	C7
East (Dezhnev), (cape), U.S.S.R.	C18
East Siberian (sea), U.S.S.R.	B1
Edge (isl.), Norway	B8
Ellef Ringnes (isl.), Canada	B15
Ellesmere (isl.), Canada	B14
Faddeyevskiy (isl.), U.S.S.R.	B2

Farewell (cape), Greenl.	D12
Finland	C8
Foxe (basin), Canada	C13
Franz Josef Land (isls.), U.S.S.R.	A7
Frederikshåb, Greenl.	D12
Garry (lake), Canada	C14
George Land (isl.), U.S.S.R.	B7
Godhavn, Greenl.	C12
Godthåb (Nûk) (cap.)	C12
Graham Bell (isl.), U.S.S.R.	A6
Great Bear (lake), Canada	C16
Great Slave (lake), Canada	C15
Greenland	B12
Greenland (sea)	B10
Grise Fiord, Canada	B13
Grønnedal, Greenl.	D12
Gunnbjørn (mt.), Greenl.	C11
Gyda (pen.), U.S.S.R.	C6
Hammerfest, Norway	B9
Hayes (pen.), Greenl.	B13
Hekla (mt.), Ice.	C11
Holman Island, Canada	B15
Holsteinsborg, Greenl.	C12
Hope (isl.), Norway	B8
Iceland	C10
Igarka, U.S.S.R.	C5
Igloolik, Canada	B14
Indigirka (riv.), U.S.S.R.	C2

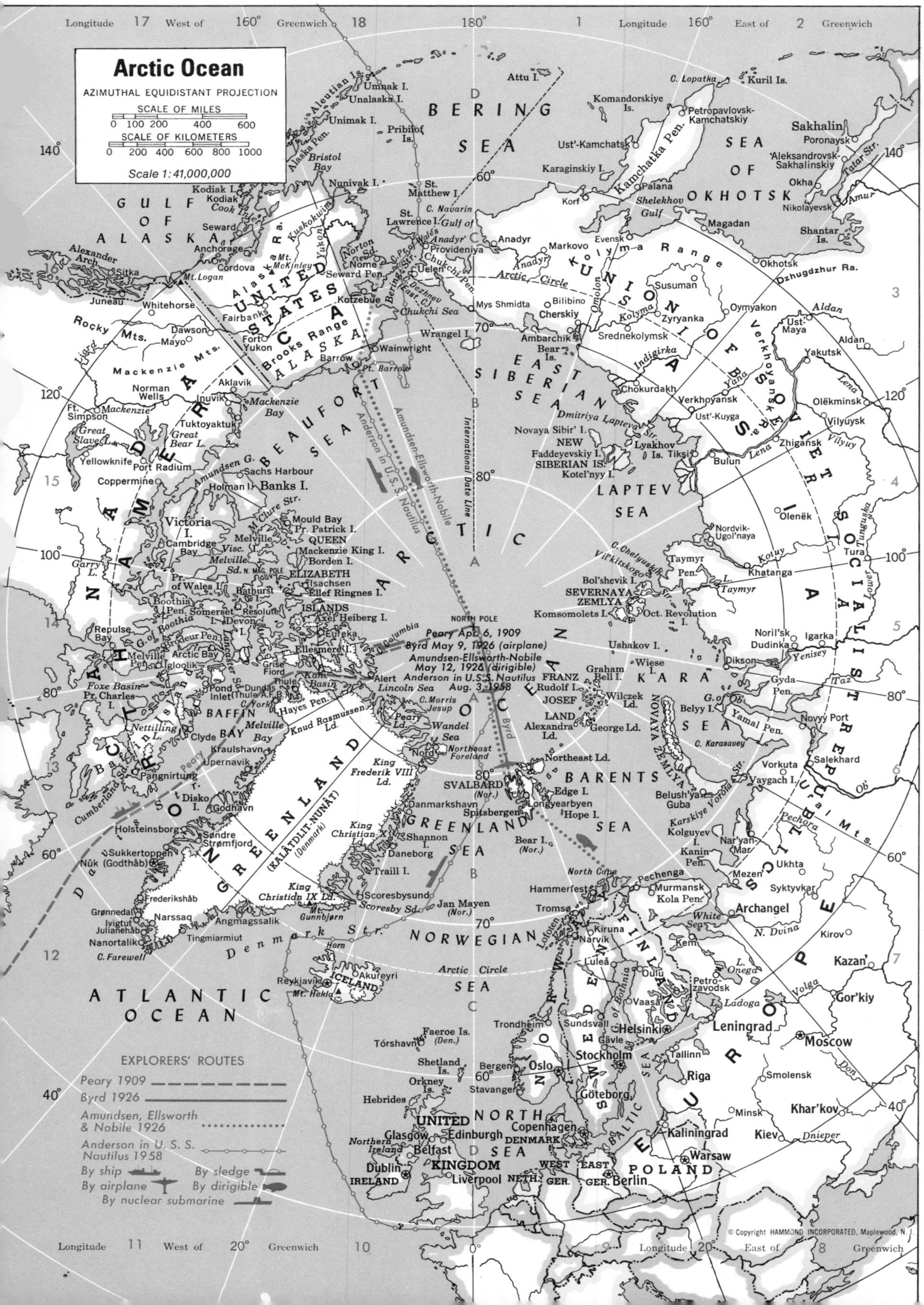

Arctic Ice

Approximate Limit of Pack Ice in September

Approximate Limit of Pack Ice in March

© C.S. Hammond & Co.

Inuvik, Canada	C16	October Revolution (isl.), U.S.S.R.	B5	
Ivigtut, Greenl.	D12	Olenek, U.S.S.R.	C4	
Isachsen, Canada	B15	Omolon (riv.), U.S.S.R.	C2	
Jan Mayen (isl.), Norway	B10	Oymyakon, U.S.S.R.	C2	
Julianehåb, Greenl.	D12	Pangnirtung, Canada	C13	
Juneau, U.S.	D16	Peary Land (reg.), Greenl.	A11	
Kalåtdlit-Nunåt (Greenland)	B12	Pechenga, U.S.S.R.	C8	
Kane (basin)	B13	Pechora (riv.), U.S.S.R.	C7	
Kanin (pen.), U.S.S.R.	C7	Pond Inlet, Canada	B13	
Kara (sea), U.S.S.R.	B6	Port Radium, Canada	C16	
Karasavey (cape), U.S.S.R.	B6	Pribilof (isls.), U.S.	D18	
Karskiye Vorota (str.), U.S.S.R.	B7	Prince Charles (isl.), Canada	C13	
Kem', U.S.S.R.	C8	Prince of Wales (cape), U.S.	C18	
Khatanga, U.S.S.R.	B4	Prince of Wales (isl.), Canada	B14	
King Christian IX Land (reg.), Greenl.	C11	Prince Patrick (isl.), Canada	B16	
King Christian X Land (reg.), Greenl.	B11	Queen Elizabeth (isls.), Canada	B15	
King Frederik VIII Land (reg.), Greenl.	B11	Repulse Bay, Canada	C14	
Kiruna, Sweden	C8	Resolute, Canada	B14	
Knud Rasmussen Land (reg.), Greenl.	B12	Reykjavik (cap.), Ice.	C11	
Kodiak, U.S.	D17	Rocky (mts.), Canada	D16	
Kodiak (isl.), U.S.	D17	Rudolf (isl.), U.S.S.R.	A7	
Kola (pen.), U.S.S.R.	C8	Sachs Harbour, Canada	B16	
Kolguyev (isl.), U.S.S.R.	B7	Saint Lawrence (isl.), U.S.	C18	
Kolyma (range), U.S.S.R.	C2	Saint Matthew (isl.), U.S.	C18	
Kolyma (riv.), U.S.S.R.	C1	Salekhard, U.S.S.R.	C6	
Komsomolets (isl.), U.S.S.R.	A5	Scoresby (sound), Greenl.	B10	
Korf, U.S.S.R.	C1	Scoresbysund, Greenl.	B10	
Kotel'nyy (isl.), U.S.S.R.	B2	Severnaya Zemlya (isls.), U.S.S.R.	A4	
Kotuy (riv.), U.S.S.R.	B4	Seward, U.S.	D17	
Kotzebue, U.S.	C18	Seward (pen.), U.S.	C18	
Kraulshavn, Greenl.	B13	Shannon (isl.), Greenl.	B10	
Kuskokwim (riv.), U.S.	C17	Siberia (reg.), U.S.S.R.	C2	
Lancaster (sound), Canada	B14	Sitka, U.S.	D16	
Laptev (sea), U.S.S.R.	B3	Somerset (isl.), Canada	B14	
Lena (riv.), U.S.S.R.	C3	Søndre Strømfjord, Greenl.	C12	
Lincoln (sea)	A4	Spitsbergen (isl.), Norway	B9	
Lofoten (isls.), Norway	C9	Srednekolymsk, U.S.S.R.	C2	
Logan (mt.), Canada	C17	Sukkertoppen, Greenl.	C12	
Longyearbyen, Norway	B8	Susuman, U.S.S.R.	C2	
Lyakhov (isl.), U.S.S.R.	B3	Svalbard (isls.), Norway	B9	
Mackenzie (bay), Canada	B16	Sweden	C9	
Mackenzie (isl.), Canada	C16	Taymyr (lake), U.S.S.R.	B5	
Mackenzie (riv.), Canada	C16	Taymyr (pen.), U.S.S.R.	B5	
Mackenzie King (isl.), Canada	B15	Taz (river), U.S.S.R.	C5	
Markovo, U.S.S.R.	C1	Thule, Greenl.	B13	
Mayo, Canada	C16	Thule A.F.B. (Dundas), Greenl.	B13	
McKinley (mt.), U.S.	C17	Tiksi, U.S.S.R.	B3	
Melville (bay), Greenl.	B13	Tingmiarmiut, Greenl.	C12	
Melville (isl.), Canada	B15	Traill (isl.), Greenl.	B10	
Melville (pen.), Canada	C14	Tromsø, Norway	B9	
Mezen', U.S.S.R.	C7	Tuktoyaktuk, Canada	C16	
Morris Jesup (cape), Greenl.	A11	Uelen, U.S.S.R.	C18	
Mould Bay, Canada	B15	Umnak (isl.), U.S.	D18	
Murmansk, U.S.S.R.	C8	Unalaska (isl.), U.S.	D18	
Mys Shmidta, U.S.S.R.	C1	Unimak (isl.), U.S.	D18	
Nanortalik, Greenl.	D12	Union of Soviet Socialist Republics	C2	
Narssaq, Greenl.	C12	United States	C17	
Narvik, Norway	C9	Upernavik, Greenl.	B12	
Nar'yan-Mar, U.S.S.R.	C7	Ural (mts.), U.S.S.R.	C6	
Navarin (cape), U.S.S.R.	C18	Ushakov (isl.), U.S.S.R.	A5	
Nettilling (lake), Canada	C13	Ust'-Kuyga, U.S.S.R.	C3	
New Siberian (isls.), U.S.S.R.	B2	Vaygach (isl.), U.S.S.R.	B7	
Nikolayevsk, U.S.S.R.	D2	Verkhoyansk, U.S.S.R.	C3	
Nome, U.S.	C18	Verkhoyansk (range), U.S.S.R.	C3	
Nord, Greenl.	A10	Victoria (isl.), Canada	B15	
Nordvik-Ugol'naya, U.S.S.R.	B4	Vil'kitskogo (str.), U.S.S.R.	B4	
Nor'il'sk, U.S.S.R.	B5	Viscount Melville (sound), Canada	B15	
Norman Wells, Canada	C16	Vorkuta, U.S.S.R.	C6	
North (cape), Ice.	C11	Wainwright, U.S.	B18	
North (cape), Norway	B8	Wandel (sea), Greenl.	A10	
Northeast Foreland (pen.), Greenl.	A10	White (sea), U.S.S.R.	C8	
Northeast Land (isl.), Norway	B8	Whitehorse, Canada	C16	
North Magnetic Pole, Canada	B15	Wiese (isl.), U.S.S.R.	B4	
North Pole	A1	Wilczek Land (isl.)	B6	
Norton (sound), U.S.	C18	Wrangel (isl.), U.S.S.R.	B18	
Norway	C9	Yamal (pen.), U.S.S.R.	B6	
Norwegian (sea)	C10	Yana (riv.), U.S.S.R.	C3	
Novaya Sibir' (isl.), U.S.S.R.	B2	Yellowknife, Canada	C15	
Novaya Zemlya (isls.), U.S.S.R.	B7	Yenisey (riv.), U.S.S.R.	C5	
Novyy Port, U.S.S.R.	C6	York (cape), Greenl.	B13	
Nûk (cap.), Greenl.	C12	Yukon (riv.)		
Nunivak (isl.), U.S.	D18	Zhigansk, U.S.S.R.	C3	
Ob' (gulf), U.S.S.R.	B6	Zyryanka, U.S.S.R.	C2	
Ob' (riv.), U.S.S.R.	C6			

Arctic Ocean

AZIMUTHAL EQUIDISTANT PROJECTION

SCALE OF MILES
0 100 200 400 600

SCALE OF KILOMETERS
0 200 400 600 800 1000

Scale 1:41,000,000

EXPLORERS' ROUTES

Peary 1909
Byrd 1926
Amundsen, Ellsworth & Nobile 1926
Anderson in U.S.S. Nautilus 1958

By ship
By airplane
By sledge
By dirigible
By nuclear submarine

Peary Apr. 6, 1909
Byrd May 9, 1926 (airplane)
Amundsen-Ellsworth-Nobile May 12, 1926 (dirigible)
Anderson in U.S.S. Nautilus Aug. 3, 1958

© Copyright HAMMOND INCORPORATED, Maplewood, N.J.

Antarctica
AZIMUTHAL EQUIDISTANT PROJECTION
SCALE OF MILES
0 200 400 600 800
KILOMETERS
0 200 400 600 800 1000
Scale 1:52,000,000
© Copyright HAMMOND INCORPORATED, Maplewood, N.J.

ATLANTIC OCEAN

INDIAN OCEAN

PACIFIC OCEAN

20° Longitude West 18 of Greenwich 0° Longitude East 1 of Greenwich 20°

160° Longitude West 10 of Greenwich 180° Longitude East 9 of Greenwich 160°

Map labels (selected): Bouvet I. (Bouvetøya) (Nor.), Prince Edward Is. (S. Afr.), South Sandwich Is. (Br.), Grytviken, South Georgia (Br.), Antarctic Circle, South Orkney Is. (Br.), Coronation I., Sanae, Lazarev, Riiser-Larsen Pen., Lützow-Holm Bay, Amundsen Bay, C. Batterbee, Enderby Land, KEMP COAST, Edward VIII Bay, Mawson, C. Daly, Mac-Robertson Land, C. Darnley, Mackenzie Bay, Amery Ice Shelf, Prydz Bay, Davis, American Highland, West Ice Shelf, WILHELM II COAST, Gaussberg, QUEEN MARY COAST, Mirny, Davis Sea, Farr Bay, Shackleton Ice Shelf, Mt. Barr Smith 4,108 ft. (1252 m.), KNOX COAST, Vincennes Bay, BUDD COAST, SABRINA COAST, C. Goodenough, BANZARE COAST, CLARIE COAST, C. Keltie, ADÉLIE COAST, GEORGE V COAST, Dumont d'Urville, SOUTH MAGNETIC POLE, OATES COAST, Mertz Glacier Tongue, Ninnis Glacier Tongue, Mt. Levick 9,101 ft. (2774 m.), Mt. Sabine 12,202 ft. (3719 m.), Adare, Balleny Is., Scott I., Macquarie I. (Australia), Campbell I. (N.Z.), Antipodes Is. (N.Z.), Bounty Is. (N.Z.), Auckland Is. (N.Z.), Stewart I., Dunedin, NEW ZEALAND, Tasman Sea, Hobart, Tasmania, King I., Furneaux Gr., Bass Str., Melbourne, AUSTRALIA, Stanley, Falkland Is. (Br.), SOUTH AMERICA, Drake Passage, C. Horn, Elephant I., King George I., Joinville I., Hope Bay, James Ross I., South Shetland, Palmer Arch., Larsen Ice Shelf, Hearst I., Palmer Land, Marguerite Bay, Charcot I., Alexander I., Ronne Ice Shelf, Berkner I., ENGLISH COAST, Vinson Massif 16,864 ft. (5140 m.), Ellsworth Land, EIGHTS COAST, Peter I. (Nor.), Thurston I., Amundsen Sea, Mt. Siple 10,171 ft. (3100 m.), C. Dart, Getz Ice Shelf, WALGREEN COAST, Hollick-Kenyon Plateau, Marie Byrd Land, Executive Comm. Ra., Mt. Sidley 13,717 ft. (4181 m.), Byrd Sta., Ford Ranges, HOBBS COAST, Roosevelt I., Little America, Ross Ice Shelf, Kainan Bay, Scott, Ross, McMurdo, Mt. Lister 13,205 ft. (4025 m.), Victoria Land, Weddell Sea, Filchner Ice Shelf, LUITPOLD COAST, CAIRD COAST, PRINCESS MARTHA COAST, PRINCESS ASTRID COAST, New Schwabenland, Queen Maud Land, PRINCESS RAGNHILD COAST, PRINCE OLAV COAST

Transantarctic Mts., SOUTH POLE, South Polar Plateau, AREA OF POLE OF INACCESSIBILITY, Amundsen-Scott Sta., Queen Maud Mts., Beardmore Glacier, Mt. Kirkpatrick 14,856 ft. (4528 m.), Mt. Markham 14,272 ft. (4350 m.)

Amundsen Dec. 14, 1911
Scott Jan. 18, 1912
Byrd Nov. 29, 1929 (airplane)
Fuchs Jan. 19, 1958

EXPLORERS' ROUTES
Palmer 1820 ————
Amundsen 1910-12 ••••••••
Scott 1910-13 ————
Byrd 1928-30 — — —
Fuchs 1957-58 ooooooo
By ship ⚓ By sledge By airplane ✈
By snow tractor

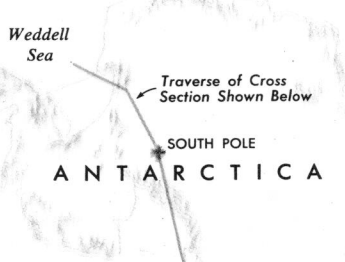

Weddell Sea
Traverse of Cross Section Shown Below
SOUTH POLE
ANTARCTICA
Ross Sea

Adare (cape) B9
Adelaide (isl.) C15
Adélie Coast (reg.) C7
Alexander (isl.) B15
American Highland B4
Amery Ice Shelf C4
Amundsen (bay) C3
Amundsen (sea) B13
Amundsen-Scott Station A14
Antarctic (pen.) C15
Balleny (isls.) C9
Banzare Coast (reg.) C7
Beardmore (glac.) A8
Bellingshausen (sea) C14
Berkner (isl.) B16
Biscoe (isls.) C15
Bouvet (isl.) D1
Bouvetøya (Bouvet) (isl.) D1
Bransfield (str.) C16
Budd Coast (reg.) C6
Byrd Station A12
Caird Coast (reg.) B17
Charcot (isl.) C14
Clarie Coast (reg.) C7
Coats Land (reg.) B17
Colbeck (cape) B10
Coronation (isl.) C16
Daly (cape) C4
Darnley (cape) C4
Dart (cape) B12
Davis (sea) C5
Davis Station C4
Drake (passage) C15
Dumont d'Urville Station C7
Edward VII (pen.) B11
Edward VIII (bay) C4
Eights Coast (reg.) B14
Elephant (isl.) D16
Ellsworth Land (reg.) B14
Enderby Land (reg.) B3
English Coast (reg.) B15
Executive Committee (range) B12
Farr (bay) C5
Filchner Ice Shelf B16
Ford Ranges (mts.) B11
Gaussberg (mt.) C5
George V Coast (reg.) C7
Getz Ice Shelf B12
Goodenough (cape) C7
Graham Land (reg.) C15
Grytviken D17
Hearst (isl.) B16
Hilton (inlet) B16
Hobbs Coast (reg.) B12
Hollick-Kenyon (plat.) B13
Hope (bay) C16
Indian Ocean C16
James Ross (isl.) C16
Joinville (isl.) C16
Kainan (bay) B10
Keltie (cape) C6
Kemp Coast (reg.) C3
King George (isl.) C16
Kirkpatrick (mt.) A8
Knox Coast (reg.) C6

Larsen Ice Shelf C16
Lazarev Station C1
Levick (mt.) B8
Lister (mt.) B8
Little America B10
Luitpold Coast (reg.) B17
Lützow-Holm (bay) C3
Mackenzie (bay) C4
Mac-Robertson Land (reg.) B4
Marguerite (bay) C15
Marie Byrd Land (reg.) B13
Markham (mt.) A8
Mawson C4
McMurdo (sound) B9
Mertz Glacier Tongue C8
Mirny C5
New Schwabenland (reg.) B1
Ninnis Glacier Tongue C8
Norvegia (cape) B18
Oates Coast (reg.) B8
Palmer (arch.) C15
Palmer Land (reg.) B15
Palmer Station C15
Peter I (isl.) B14
Prince Edward (isls.) E2
Prince Olav Coast (reg.) C3
Princess Astrid Coast (reg.) ... B1
Princess Martha Coast (reg.) ... B18
Princess Ragnhild Coast (reg.) . B2
Prydz C4
Queen Mary Coast (reg.) C5
Queen Maud (mts.) A12
Queen Maud Land (reg.) B1
Riiser-Larsen (pen.) C2
Ronne Entrance (inlet) B15
Ronne Ice Shelf B15
Roosevelt (isl.) A10
Ross (isl.) B9
Ross (sea) B10
Ross Ice Shelf A10
Sabine (mt.) B9
Sabrina Coast (reg.) C6
Sanae Station B18
Scotia (sea) D16
Scott (isl.) C10
Scott Station B9
Shackleton Ice Shelf C5
Sidley (mt.) B12
Siple (mt.) B12
South Georgia (isl.) D17
South Magnetic Pole C6
South Orkney (isls.) C16
South Polar (plat.) A1
South Pole A4
South Sandwich (isls.) D17
South Shetland (isls.) C15
Sulzberger (bay) B11
Thurston (isl.) C14
Transantarctic (mts.) B17
Victoria Land (reg.) B8
Vincennes (bay) C6
Vinson Massif (mt.) B14
Walgreen Coast (reg.) B13
Weddell (sea) C16
West Ice Shelf C5
Wilhelm II Coast (reg.) C5
Wilkes Land (reg.) B7

Antarctic Cross Section: Weddell Sea to Ross Sea

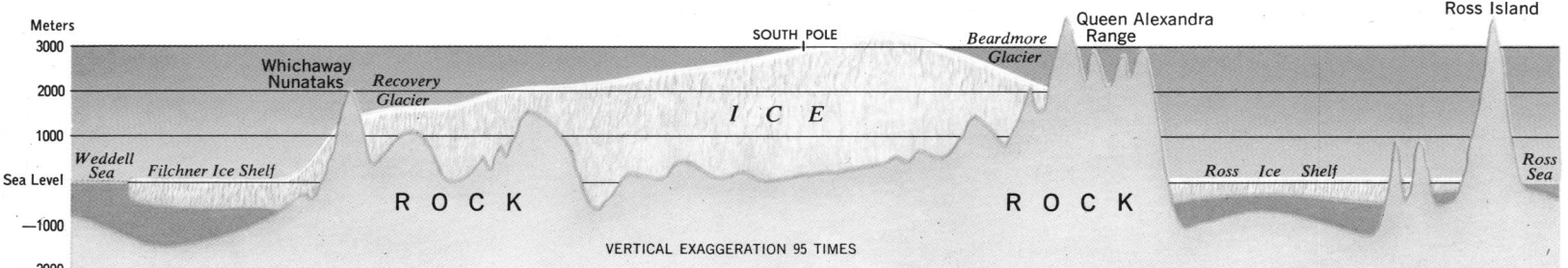

Meters 3000 2000 1000 Sea Level -1000 -2000

Weddell Sea, Filchner Ice Shelf, Whichaway Nunataks, Recovery Glacier, ICE, SOUTH POLE, Beardmore Glacier, Queen Alexandra Range, Ross Ice Shelf, Ross Island, Ross Sea, ROCK, ROCK

VERTICAL EXAGGERATION 95 TIMES

Information Based on American Geographical Society's "Antarctic Map Folio Series"

Aberdeen, Scotland D3
Adriatic (sea) F4
Aegean (sea), Greece G4
Albania G4
Ålborg, Denmark F3
Alps (mts.) F4
Amsterdam (cap.), Netherlands E3
Andorra E4
Antwerp, Belgium E3
Arad, Romania G4
Araks (riv.), U.S.S.R. J5
Archangel, U.S.S.R. J2
Armenian S.S.R., U.S.S.R. J4
Århus, Denmark E3
Astrakhan', U.S.S.R. J4
Athens (cap.), Greece G5
Atlantic Ocean A4
Austria F4
Azerbaidzhan S.S.R., U.S.S.R. ....
Azov (sea), U.S.S.R. H4
Baku, U.S.S.R. J4
Balaton (lake), Hungary F4
Balearic (isls.), Spain E5
Balkan (mts.) G4
Baltic (sea) F3
Barcelona, Spain E4
Barents (sea) J1
Bari, Italy F4
Basel, Switzerland E4
Belfast (cap.), N. Ireland D3
Belgium E3
Belgrade (cap.), Yugoslavia G4
Bergen, Norway E2
Berlin (cap.), E. Germany F3
Bern (cap.), Switzerland E4

Bilbao, Spain D4
Birmingham, England D3
Biscay (bay) D4
Black (sea) H4
Bologna, Italy F4
Bonn (cap.), W. Germany E3
Bordeaux, France D4
Bornholm (isl.), Denmark F3
Bosporus (str.), Turkey G4
Bothnia (gulf) G2
Braşov, Romania G4
Bratislava, Czech. F4
Bremen, W. Germany E3
Brest, France D4
Bristol, England D3
Brno, Czech. F4
Brussels (cap.), Belgium E3
Bucharest (cap.), Romania G4
Budapest (cap.), Hungary F4
Bug (riv.) G3
Bulgaria G4
Burgas, Bulgaria G4
Burgos, Spain D4
Cádiz, Spain C5
Calais, France E3
Cardiff, Wales D3
Carpathian (mts.) G4
Cartagena, Spain D5
Caspian (sea), U.S.S.R. J4
Caucasus (mts.), U.S.S.R. J4
Channel (isls.) D4
Cologne, W. Germany E3
Como, Italy E4
Constanţa, Romania G4
Copenhagen (cap.), Denmark F3
Córdoba, Spain D5
Cork, Ireland D3

Corsica (isl.), France E4
Cracow, Poland F3
Crete (isl.), Greece G5
Crimea (pen.), U.S.S.R. H4
Czechoslovakia F4
Danube (riv.) F4
Dardanelles (str.), Turkey G5
Debrecen, Hungary G4
Denmark E3
Denmark (str.) B2
Dnepropetrovsk, U.S.S.R. H4
Dnieper (riv.), U.S.S.R. H3
Dniester (riv.), U.S.S.R. G4
Don (riv.), U.S.S.R. J4
Donets (riv.), U.S.S.R. H4
Douro (riv.), Portugal D4
Dover, England E3
Drava (riv.) F4
Dresden, E. Germany F3
Dublin (cap.), Ireland D3
Dvina, Northern (riv.), U.S.S.R. J2
East Germany F3
Ebro (riv.), Spain D4
Edinburgh (cap.), Scotland D3
Edirne, Turkey G4
Elba (isl.), Italy F4
Elbe (riv.) F3
El'brus (mt.), U.S.S.R. J4
England (U.K.) D3
English (chan.) D4
Erfurt, E. Germany F3
Essen, W. Germany E3
Estonian S.S.R., U.S.S.R. G3
Etna (vol.), Italy F5
Faeroe (isls.), Denmark C4
Finisterre (cape), Spain C4

Finland G2
Finland (gulf) G3
Florence, Italy F4
France E4
Frankfurt, W. Germany E3
Frisian (isls.) E3
Gdansk, Poland F3
Garonne (riv.), France D4
Geneva, Switzerland E4
Geneva (lake) E4
Genoa, Italy E4
Georgian S.S.R., U.S.S.R. J4
Gibraltar D5
Gibraltar (str.) D5
Glasgow, Scotland D3
Göteborg, Sweden F3
Gor'kiy, U.S.S.R. J3
Granada, Spain D5
Graz, Austria F4
Greece G5
Guadalquivir (riv.), Spain D5
Guadiana (riv.) D5
Hague, The (cap.), Netherlands E3
Hamburg, W. Germany F3
Hammerfest, Norway G1
Helsinki (cap.), Finland G2
Hungary F4
Iceland B2
Ionian (sea) F5
Ireland D3
Irish (sea) D3
Istanbul, Turkey G4
Italy F4
Jan Mayen (isl.), Norway D1
Jönköping, Sweden F3
Kalinin, U.S.S.R. H3

Kaliningrad, U.S.S.R. G3
Kaluga, U.S.S.R. H3
Kama (riv.), U.S.S.R. K3
Karl-Marx-Stadt, E. Germany F3
Karlsruhe, W. Germany E4
Karlstad, Sweden F3
Kassel, W. Germany E3
Katowice, Poland F3
Kattegat (str.) E3
Kavála, Greece G4
Kazan', U.S.S.R. J3
Kecskemét, Hungary F4
Kharkov, U.S.S.R. H4
Kherson, U.S.S.R. H4
Kiel, W. Germany E3
Kielce, Poland F3
Kiev, U.S.S.R. H3
Kirov, U.S.S.R. J3
Kirovograd, U.S.S.R. H4
Kjølen (mts.) F2
Kola (pen.), U.S.S.R. H2
Krasnodar, U.S.S.R. H4
Kristiansand, Norway E3
Kristiansund, Norway E2
Krivoy Rog, U.S.S.R. H4
Kuopio, Finland G2
Kursk, U.S.S.R. H3
Kuybyshev, U.S.S.R. K3
Ladoga (lake), U.S.S.R. H2
La Coruña, Spain C4
Land's End (prom.), England D3
Lapland (reg.) G2
Latvian S.S.R., U.S.S.R. G3
Leeds, England D3
Leghorn, Italy E4

Le Havre, France E4
Leipzig, E. Germany F3
Leningrad, U.S.S.R. H3
León, Spain D4
Lésvos (isl.), Greece G4
Liechtenstein F4
Liège, Belgium E3
Lille, France E3
Limerick, Ireland D3
Limoges, France E4
Linköping, Sweden F3
Linz, Austria F4
Lions (gulf) E4
Lisbon (cap.), Portugal C5
Lithuanian S.S.R., U.S.S.R. G3
Liverpool, England D3
Ljubljana, Yugoslavia F4
Łódź, Poland F3
Lofoten (isls.), Norway F2
Loire (riv.), France E4
London (cap.), England D3
Luxembourg E4
Lyon, France E4
Madrid (cap.), Spain D4
Majorca (isl.), Spain E5
Málaga, Spain D5
Malta F5
Manchester, England D3
Marmara (sea), Turkey G4
Marseille, France E4
Mediterranean (sea) G5
Minsk, U.S.S.R. G3
Moldavian S.S.R., U.S.S.R. G4
Monaco E4
Moscow (cap.), U.S.S.R. H3
Munich, W. Germany E4
Murcia, Spain D5

Murmansk, U.S.S.R. H2
Nantes, France D4
Naples, Italy F4
Netherlands E3
Nice, France E4
North (Nordkapp) (cape), Norway G1
North (sea) D3
Northern Ireland, U.K. D3
Norway F2
Norwegian (sea) D2
Nuremberg, W. Germany F3
Odense, Denmark F3
Oder (riv.) F3
Odessa, U.S.S.R. H4
Onega (lake), U.S.S.R. H2
Orenburg, U.S.S.R. K3
Orkney (isls.), Scotland D3
Orléans, France E4
Oslo (cap.), Norway F2
Palermo, Italy F5
Paris (cap.), France E4
Perm', U.S.S.R. K3
Piraiévs, Greece G5
Ploieşti, Romania G4
Plovdiv, Bulgaria G4
Plymouth, England D3
Po (riv.), Italy F4
Poland F3
Portugal C5
Poznań, Poland F3
Prague (cap.), Czech. F3
Pyrenees (mts.) D4
Reykjavík (cap.), Iceland B2
Rhine (riv.) E3
Rhône (riv.), France E4

AREA 4,057,000 sq. mi.
(10,507,630 sq. km.)
POPULATION 676,000,000
LARGEST CITY Paris
HIGHEST POINT El'brus 18,510 ft.
(5,642 m.)
LOWEST POINT Caspian Sea -92 ft.
(-28 m.)

Population Distribution

DENSITY PER

SQ. KILOMETER	SQ. MILE
Over 100	Over 260
50-100	130-260
10-50	25-130
1-10	3-25
Under 1	Under 3

● Cities with over 2,000,000
inhabitants (including suburbs)

○ Cities with over 1,000,000
inhabitants (including suburbs)

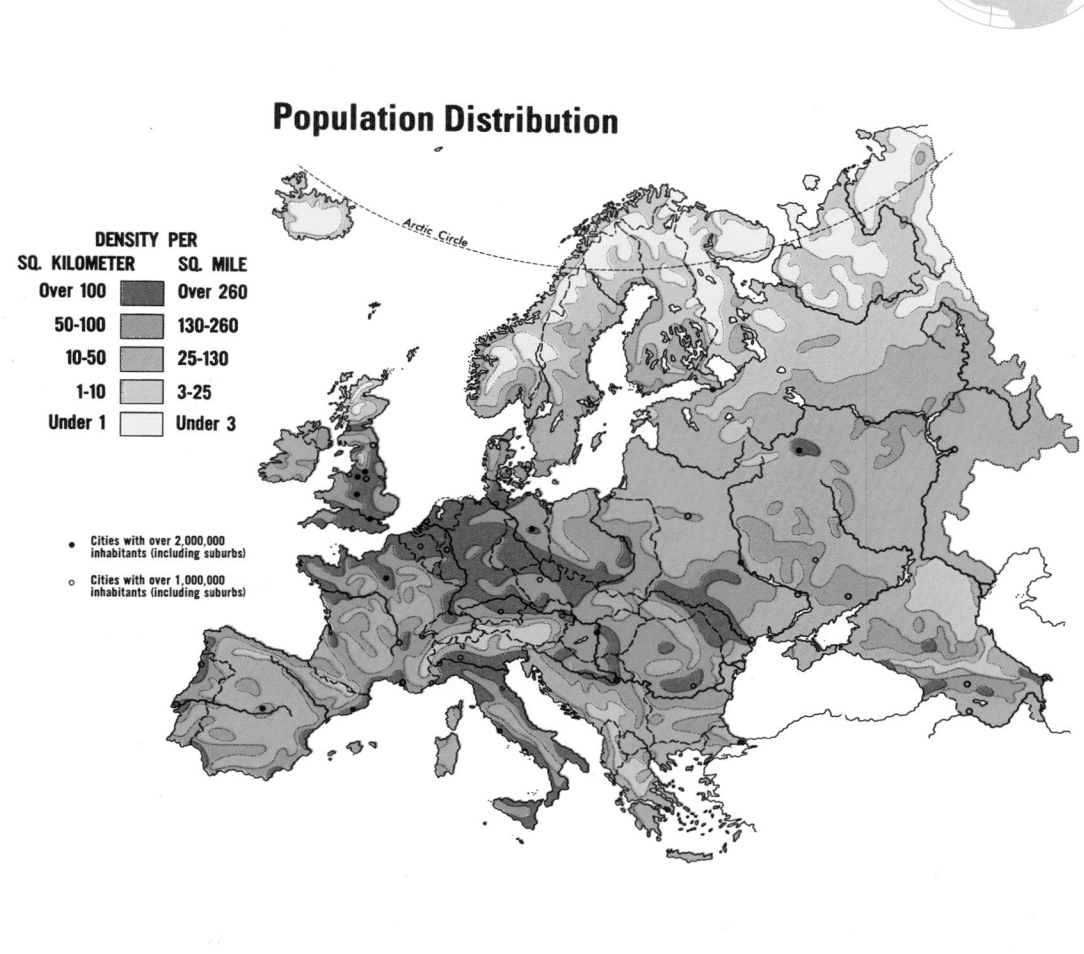

Vegetation

MID-LATITUDE FOREST

Coniferous Forest

Broadleaf Forest

Mixed Coniferous
and Broadleaf Forest

Woodland and Shrub
(Mediterranean)

MID-LATITUDE GRASSLAND

Short Grass (Steppe)

Wooded Steppe

HEATH AND MOOR

DESERT AND
DESERT SHRUB

TUNDRA AND ALPINE

PERMANENT ICE COVER

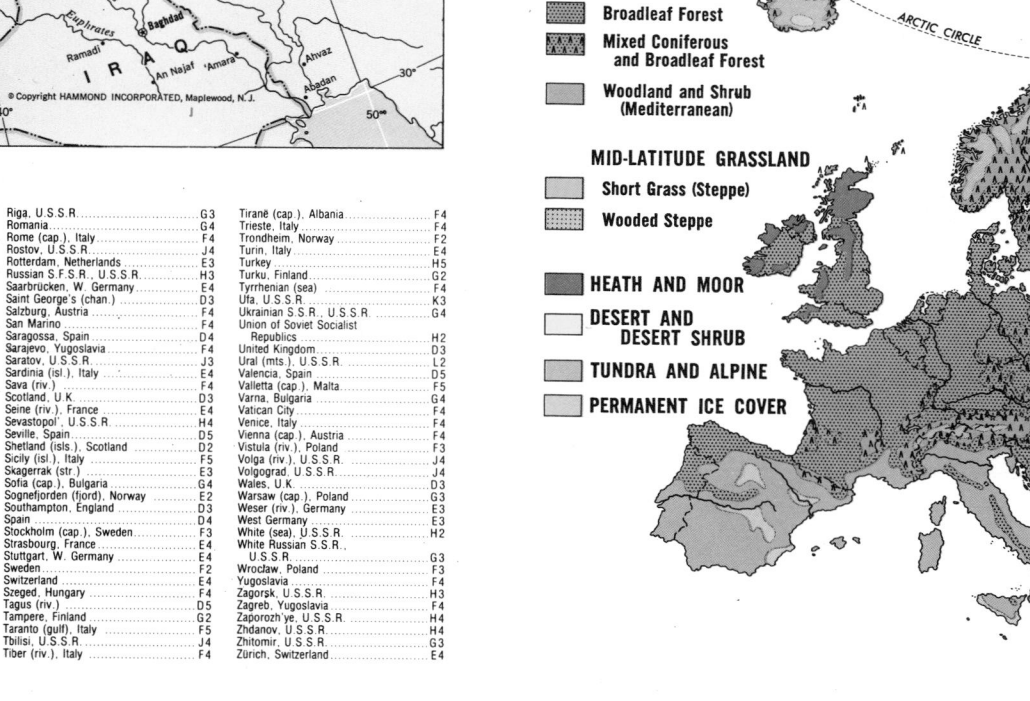

© Copyright HAMMOND INCORPORATED, Maplewood, N.J.

Riga, U.S.S.R.	G3	Tiranë (cap.), Albania	F4
Romania	G4	Trieste, Italy	F4
Rome (cap.), Italy	F4	Trondheim, Norway	F2
Rostov, U.S.S.R.	J4	Turin, Italy	E4
Rotterdam, Netherlands	E3	Turkey	H5
Russian S.F.S.R., U.S.S.R.	H3	Turku, Finland	G2
Saarbrücken, W. Germany	E4	Tyrrhenian (sea)	F4
Saint George's (chan.)	D3	Ufa, U.S.S.R.	K3
Salzburg, Austria	F4	Ukrainian S.S.R., U.S.S.R.	G4
San Marino	F4	Union of Soviet Socialist
Saragossa, Spain	D4	Republics	H2
Sarajevo, Yugoslavia	F4	United Kingdom	D3
Saratov, U.S.S.R.	J3	Ural (mts.) U.S.S.R.	L2
Sardinia (isl.), Italy	E4	Valencia, Spain	D5
Sava (riv.)	F4	Valletta (cap.), Malta	F5
Scotland, U.K.	D3	Varna, Bulgaria	G4
Seine (riv.), France	E4	Vatican City	F4
Sevastopol', U.S.S.R.	H4	Venice, Italy	F4
Seville, Spain	D5	Vienna (cap.), Austria	F4
Shetland (isls.), Scotland	D2	Vistula (riv.), Poland	F3
Sicily (isl.), Italy	F5	Volga (riv.), U.S.S.R.	J4
Skagerrak (str.)		Volgograd, U.S.S.R.	J4
Sofia (cap.), Bulgaria	G4	Wales, U.K.	D3
Sognefjorden (fjord), Norway	E2	Warsaw (cap.), Poland	F3
Southampton, England	D3	Weser (riv.), Germany	E3
Spain		West Germany	E3
Stockholm (cap.), Sweden	F3	White (sea), U.S.S.R.	H2
Strasbourg, France	E4	White Russian S.S.R.,
Stuttgart, W. Germany	E4	U.S.S.R.	G3
Sweden	F2	Wrocław, Poland	F3
Switzerland	E4	Yugoslavia	F4
Szeged, Hungary	F4	Zagorsk, U.S.S.R.	H3
Tagus (riv.)		Zagreb, Yugoslavia	F4
Tampere, Finland	G2	Zaporozh'ye, U.S.S.R.	H4
Taranto (gulf), Italy	F5	Zhdanov, U.S.S.R.	H4
Tbilisi, U.S.S.R.	J4	Zhitomir, U.S.S.R.	G3
Tiber (riv.), Italy	F4	Zürich, Switzerland	E4

8 Europe
(continued)

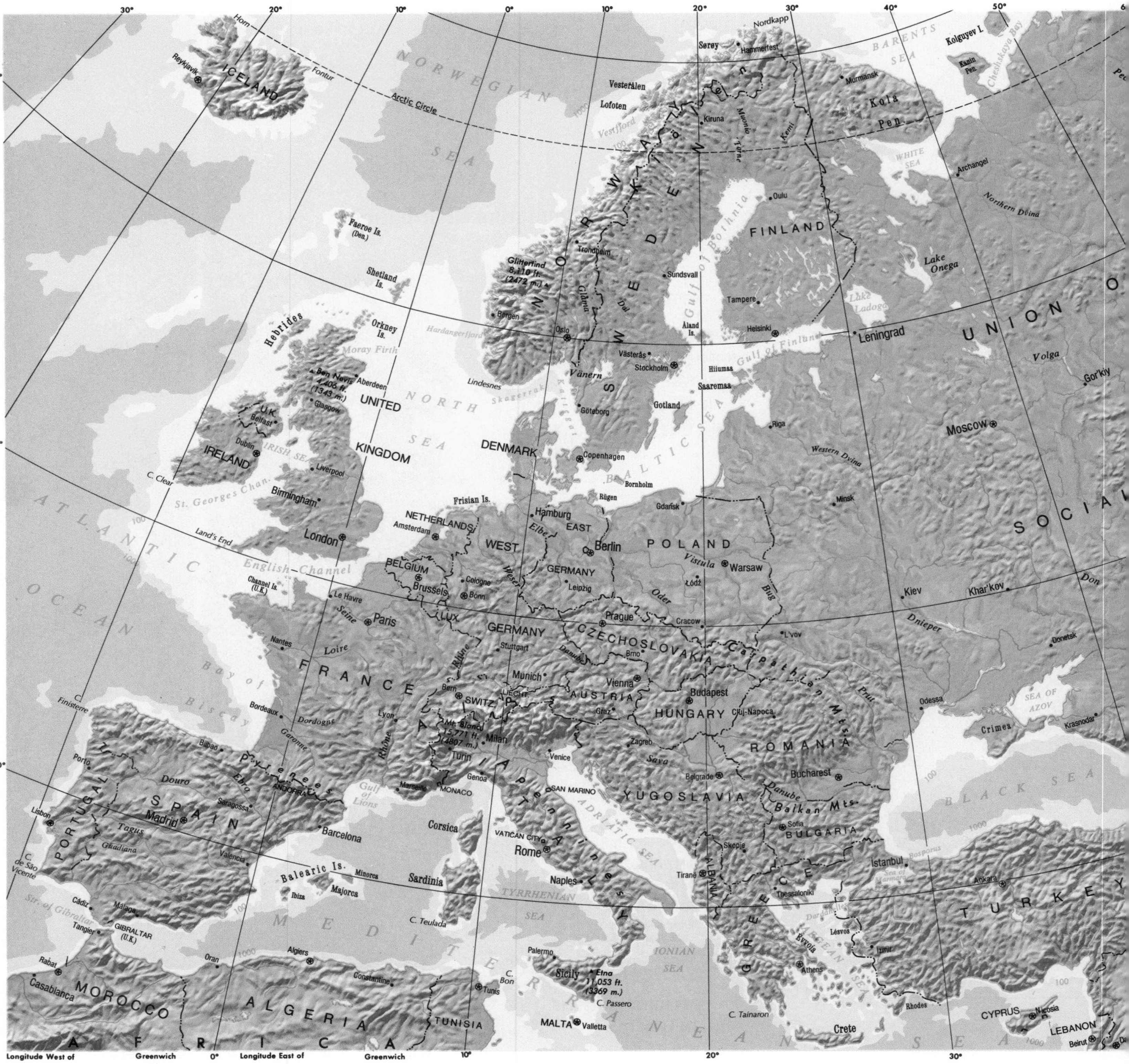

Vegetation/Relief

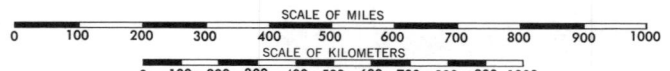

SCALE OF MILES

0 100 200 300 400 500 600 700 800 900 1000

SCALE OF KILOMETERS

0 100 200 300 400 500 600 700 800 900 1000

Capitals of Countries

International Boundaries

Canals

Depths in Fathoms

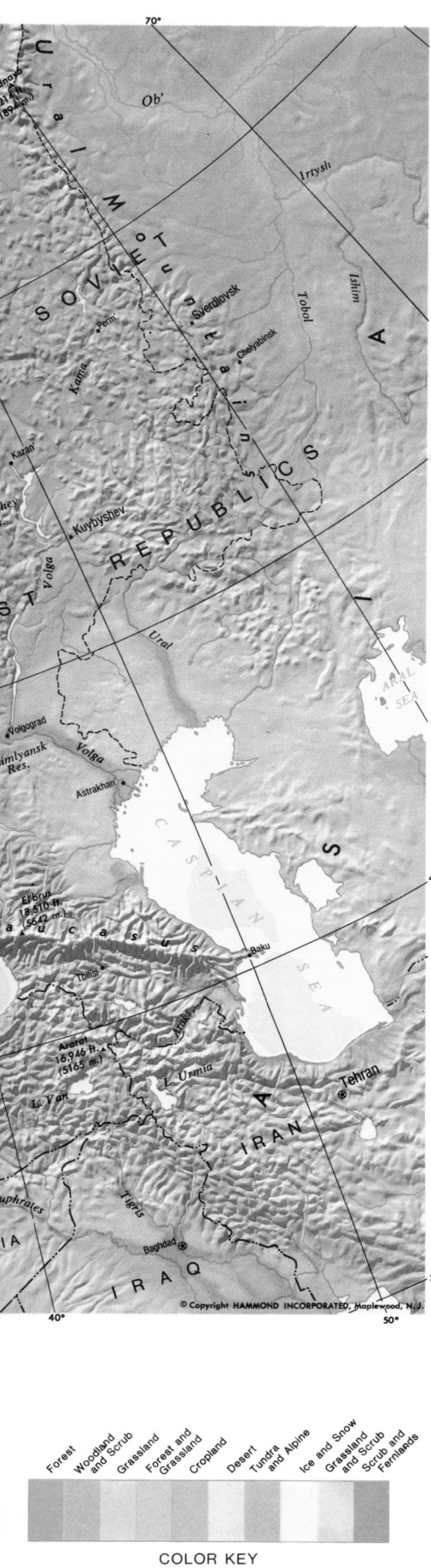

Rainfall

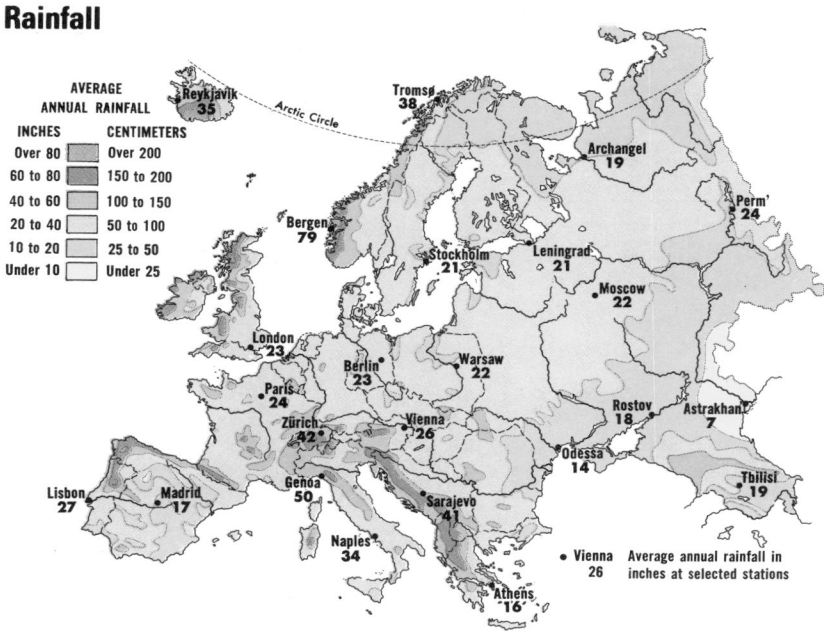

AVERAGE ANNUAL RAINFALL

INCHES	CENTIMETERS
Over 80	Over 200
60 to 80	150 to 200
40 to 60	100 to 150
20 to 40	50 to 100
10 to 20	25 to 50
Under 10	Under 25

Reykjavík 35
Tromsø 38
Archangel 19
Perm' 24
Bergen 79
Stockholm 21
Leningrad 21
Moscow 22
London 23
Berlin 23
Warsaw 22
Rostov 18
Astrakhan 7
Paris 24
Zürich 42
Vienna 26
Odessa 14
Tbilisi 19
Lisbon 27
Madrid 17
Genoa 50
Sarajevo 41
Naples 34
Athens 16

• Vienna 26 — Average annual rainfall in inches at selected stations

Average January Temperature

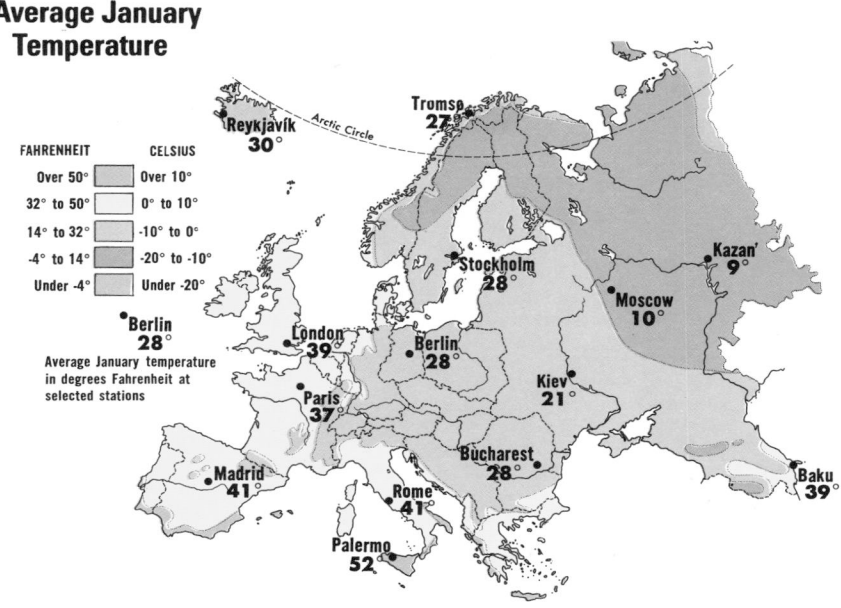

FAHRENHEIT	CELSIUS
Over 50°	Over 10°
32° to 50°	0° to 10°
14° to 32°	-10° to 0°
-4° to 14°	-20° to -10°
Under -4°	Under -20°

Reykjavík 30°
Tromsø 27°
Stockholm 28°
Kazan' 9°
Moscow 10°
• Berlin 28°
Average January temperature in degrees Fahrenheit at selected stations
London 39°
Berlin 28°
Kiev 21°
Paris 37°
Bucharest 28°
Baku 39°
Madrid 41°
Rome 41°
Palermo 52°

Average July Temperature

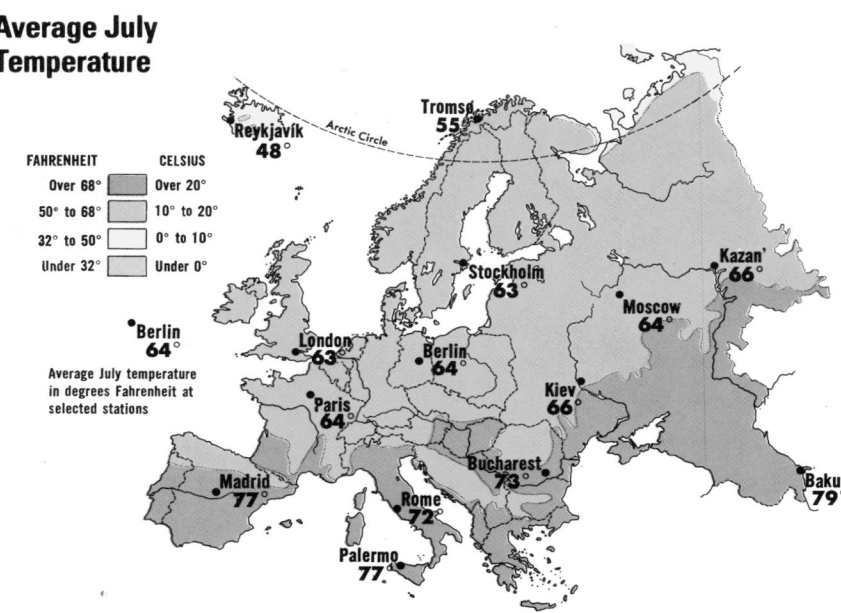

FAHRENHEIT	CELSIUS
Over 68°	Over 20°
50° to 68°	10° to 20°
32° to 50°	0° to 10°
Under 32°	Under 0°

Reykjavík 48°
Tromsø 55°
Stockholm 63°
Kazan' 66°
Moscow 64°
• Berlin 64°
Average July temperature in degrees Fahrenheit at selected stations
London 63°
Berlin 64°
Kiev 66°
Paris 64°
Bucharest 73°
Baku 79°
Madrid 77°
Rome 72°
Palermo 77°

COLOR KEY

Forest | Woodland and Scrub | Grassland | Forest and Grassland | Cropland | Desert | Tundra and Alpine | Ice and Snow | Grassland and Scrub and Fernlands

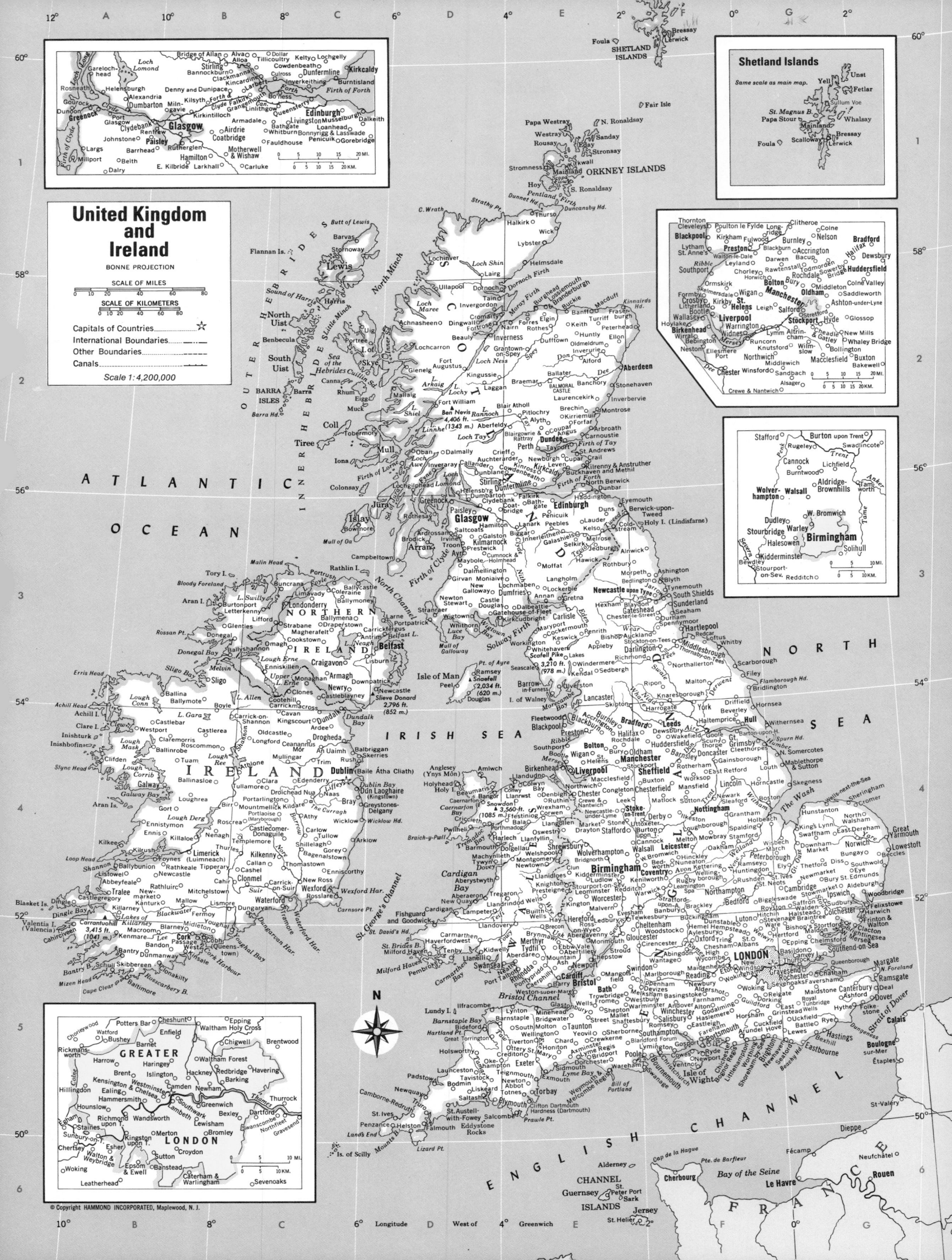

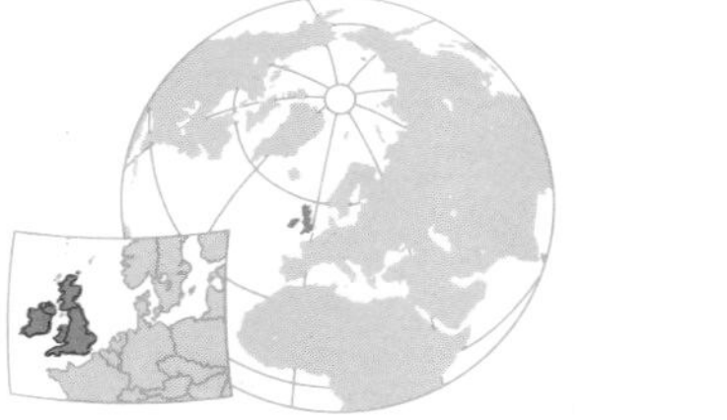

UNITED KINGDOM

AREA 94,399 sq. mi. (244,493 sq. km.)
POPULATION 55,672,000
CAPITAL London
LARGEST CITY London
HIGHEST POINT Ben Nevis 4,406 ft. (1,343 m.)
MONETARY UNIT pound sterling
MAJOR LANGUAGES English, Gaelic, Welsh
MAJOR RELIGIONS Protestantism, Roman Catholicism

IRELAND

AREA 27,136 sq. mi. (70,282 sq. km.)
POPULATION 3,440,427
CAPITAL Dublin
LARGEST CITY Dublin
HIGHEST POINT Carrantuohill 3,415 ft. (1,041 m.)
MONETARY UNIT Irish pound
MAJOR LANGUAGES English, Gaelic (Irish)
MAJOR RELIGION Roman Catholicism

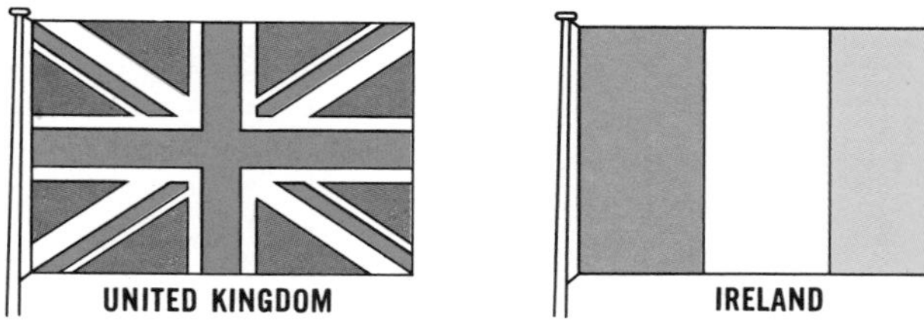

UNITED KINGDOM

IRELAND

ENGLAND

COUNTIES

Avon, 920,200 E 6
Bedfordshire, 491,700 G 5
Berkshire, 659,000 F 6
Buckinghamshire, 512,000 G 6
Cambridgeshire, 563,000 G 5
Cheshire, 916,400 E 4
Cleveland, 567,900 F 3
Cornwall, 405,200 C 7
Cumbria, 473,600 D 3
Derbyshire, 887,600 F 5
Devon, 942,100 D 7
Dorset, 575,800 E 7
Durham, 610,400 F 3
East Sussex, 655,600 H 7
Essex, 1,426,200 H 6
Gloucestershire, 491,500 E 6
Greater London, 7,028,200 H 8
Greater Manchester, 2,684,100 H 2
Hampshire, 1,456,100 F 6
Hereford and Worcester, 594,200 E 5
Hertfordshire, 937,300 G 6
Humberside, 848,600 G 4
Isle of Wight, 111,300 F 7
Isles of Scilly, 1,900 A 7
Kent, 1,448,100 H 6
Lancashire, 1,375,500 E 4
Leicestershire, 837,900 F 5
Lincolnshire, 524,500 G 4
London, Greater, 7,028,200 H 8
Manchester, Greater, 2,684,100 H 2
Merseyside, 1,578,000 G 2
Norfolk, 662,500 H 5
Northamptonshire, 505,900 G 5
Northumberland, 287,300 E 2
North Yorkshire, 653,000 F 3
Nottinghamshire, 977,500 F 4
Oxfordshire 541,800 F6
Shropshire (Salop) 359,000 E5
Somerset 404,400 E6
South Yorkshire 1,318,300 F4
Staffordshire 997,600 E5
Suffolk 577,600 H5
Surrey 1,002,900 G6
Sussex, East 655,600 H7
Sussex, West 623,400 G7
Tyne and Wear 1,182,900 H3
Warwickshire 471,000 F5
West Midlands 2,743,300 F5
West Sussex 623,400 G7
West Yorkshire 2,072,500 J1
Wiltshire 512,800 E6
Yorkshire, North 653,000 F3
Yorkshire, South 1,318,300 F4
Yorkshire, West 2,072,500 J1

CITIES and TOWNS

Abingdon, 20,130 F 6
Accrington, 36,470 H 1
Adwick le Street, 17,650 K 2
Aldeburgh, 2,750 J 5
Aldershot, 33,750 G 8
Aldridge Brownhills, 89,370 E 5
Alfreton, 21,560 F 4
Alnwick, 7,300 F 2
Altrincham, 40,800 H 2
Amersham, ⊙17,254 G 7
Andover, 27,620 F 6
Appleby, 2,240 E 3
Arnold, 35,090 F 4
Arundel, 2,390 G 7
Ashford, 36,380 H 6
Ashington, 24,720 F 2
Ashton-under-Lyne, 48,500 H 2
Axminster, ⊙4,515 D 7
Aycliffe, ⊙20,203 F 3
Aylesbury, 41,420 G 7
Bacup, 14,990 H 1
Bakewell, 4,100 J 2
Banbury, 31,060 F 5
Banstead, 44,100 H 8
Barking, 153,800 H 8
Barnet, 305,200 H 7
Barnsley, 74,730 J 2
Barnstaple, 17,820 D 6
Barrow-in-Furness, 73,400 D 3
Barton-upon-Humber, 7,750 G 4
Basildon, 135,720 J 8
Basingstoke, 60,910 F 6
Bath, 83,100 E 6
Batley, 41,630 J 1
Battle, ⊙4,987 H 7
Bebington, 62,500 G 2
Bedford, 74,390 G 5
Bedlington, 27,200 F 2
Bedworth, 41,600 F 5
Beeston and Stapleford, 65,360 F 5
Benfleet, 49,180 J 8
Bentley with Arksey, 22,320 F 4
Berkhamsted, 15,920 G 7
Beverley, 16,920 G 4
Bexhill, 34,680 H 7
Bexley, 213,500 H 8
Biddulph, 18,720 H 2
Birkenhead, 135,750 G 2
Birmingham, 1,058,800 F 5
Bishop Auckland, 32,940 E 3
Bishop's Stortford, 21,720 H 6
Blackburn, 101,670 H 1
Blackpool, 149,000 G 1
Blaydon, 31,940 H 3
Blyth, 35,390 F 2
Bodmin, 10,430 C 7
Bognor Regis, 34,620 G 7
Boldon, 24,430 J 3
Bolton, 154,480 H 2
Bootle 71,160 G2
Boston 26,700 G5
Bournemouth 144,100 F7
Bracknell† 34,067 G8
Bradford 458,900 J1
Braintree and Bocking 26,300 H6
Brent 256,500 H8
Brentwood 58,690 J8
Bridgwater 26,700 E6
Bridlington 26,920 G3
Bridport 6,660 E7
Brigg 4,870 G4
Brighouse 35,320 J1
Brightlingsea 7,170 J6
Brighton 156,500 G7
Bristol 416,300 E6
Broadstairs and Saint Peter's 21,670 J6
Bromley 299,100 H8
Bromsgrove 41,430 E5
Buckfastleigh 2,870 C7
Buckingham 5,290 G6
Bude-Stratton 5,750 C7
Bungay 4,120 J5
Burgess Hill 20,030 G7
Burnham-on-Crouch 4,920 H6
Burnley 74,300 H1
Burntwood† 23,088 F5
Burton upon Trent 49,480 F5
Bury 69,550 H2
Bury Saint Edmunds 26,800 H5
Bushey 24,500 H7
Buxton 20,050 J2
Caister-on-Sea† 6,287 J5
Camborne-Redruth 43,970 B7
Cambridge 106,400 G5
Camden 185,800 H8
Cannock 56,440 E5
Canterbury 115,600 H6
Canvey Island 29,550 J8
Carlisle, 99,600 D 3
Carlton, 46,690 F 5
Caterham and Warlingham, 35,840 H 8
Chatham, 59,550 J 8
Cheadle and Gatley, 62,460 H 2
Chelmsford, 58,320 J 7
Cheltenham, 75,910 E 6
Chertsey, 45,070 G 8
Chesham, 20,830 G 7
Cheshunt, 45,750 H 7
Chester, 117,200 G 2
Chesterfield, 69,480 J 2
Chester-le-Street, 20,720 J 3
Chichester, 20,940 G 7
Chigwell, 54,220 H 8
Chippenham, 18,550 E 6
Chorley, 31,800 G 2
Christchurch, 31,610 F 7
Cirencester, 14,500 E 6
Clacton, 39,380 J 6
Clay Cross, 9,630 J 2
Cleator Moor, ⊙7,686 D 3
Cleethorpes, 37,200 H 4
Clevedon, 15,140 D 6
Clun, ⊙1,261 D 6
Coalville, 28,740 F 5
Cockermouth, 6,480 D 3
Colchester, 79,600 H 6
Colne, 19,030 H 1
Colne Valley, 21,190 J 2
Congleton, 21,500 H 2
Consett, 35,080 H 3
Corby, 48,850 G 5
Coventry, 336,800 F 5
Cowes, 19,190 F 7
Crawley, 72,600 G 6
Crewe and Nantwich, 98,100 E 4
Cromer, 5,720 J 5
Crook and Willington, 21,120 E 3
Crosby, 56,750 G 2
Croydon, 330,600 H 8
Cuckfield, 26,500 G 6
Darlington, 85,120 F 3
Dartford, 44,130 J 8
Darton, 15,710 J 2
Darwen, 29,290 H 1
Deal, 26,840 J 6
Dearne, 24,780 K 2
Denton, 38,110 H 2
Derby, 213,700 F 5
Dewsbury, 50,560 J 1
Didcot, ⊙14,277 F 6
Doncaster, 81,530 F 4
Dorking, 22,410 G 8
Dover, 34,160 J 6
Downham Market, 4,120 H 5
Droitwich, 13,950 E 5
Dronfield, 20,000 J 2
Dudley, 187,110 E 5
Dunstable, 32,090 G 6
Durham, 88,800 J 3
Ealing, 293,800 H 8
Eastbourne, 73,200 H 7
East Grinstead, 19,420 G 6
Eastleigh, 46,340 F 7
East Retford, 18,260 G 4
Egham, 30,320 G 8
Egremont, ⊙7,253 D 3
Eling, ⊙20,006 F 7
Ellesmere, ⊙2,630 E 5
Ellesmere Port, 63,870 G 2
Enfield, 260,900 H 7
Epsom and Ewell, 70,700 G 8
Esher, 63,970 H 8
Eston, ⊙46,219 F 3
Eton, 4,950 G 8
Evesham, 14,090 F 5
Exeter, 93,300 D 7
Exminster, ⊙3,181 D 7
Exmouth, 26,840 D 7
Falmouth, 17,530 B 7
Fareham, 86,300 F 7
Farnborough, 43,520 G 8
Farnham, 33,140 G 8
Farnworth, 26,110 H 2
Faversham, 15,010 H 6
Felixstowe, 19,460 J 6
Felling, 38,990 J 3
Filey, 5,660 G 3
Fleet, 22,930 G 8
Fleetwood, 30,070 D 4
Folkestone, 45,610 J 6
Formby, 24,850 G 2
Framlingham, ⊙2,258 J 5
Frimley and Camberley, 47,390 G 8
Fulwood, 22,910 G 1
Gainsborough, 17,440 G 4
Gateshead, 91,230 J 3
Gillingham, Dorset, ⊙4,050 E 6
Gillingham, Kent, 93,900 J 8
Glastonbury, 6,580 E 6
Glossop, 24,820 J 2
Gloucester, 91,600 E 6
Godalming, 18,840 G 8
Golborne, 28,720 G 2
Goole, 17,920 G 4
Gosport, 82,300 F 7
Grange, 3,520 E 3
Grantham 27,830 G5
Gravesend 53,500 J8
Great Grimsby 93,800 G4
Great Torrington 3,430 C7
Great Yarmouth 49,410 J5
Greenwich 207,200 H8
Guildford 58,470 G8
Guisborough 14,860 F3
Hackney 192,500 H8
Hale 17,080 H2
Halesowen 54,120 E5
Halifax 88,580 J1
Haltemprice 54,850 G4
Haltwhistle† 3,511 E2
Hammersmith 170,000 H8
Haringey 228,200 H8
Harlow 79,160 H7
Harrogate 64,620 F4
Harrow 200,200 B5
Hartlepool 97,100 F3
Harwich 15,280 J6
Haslingden 15,140 H1
Hastings 74,600 H7
Hatfield† 25,359 H7
Havant and Waterloo 112,430 G7
Haverhill 14,550 H5
Havering 239,200 J8
Hayle† 5,378 B7
Hazel Grove and Bramhall 40,400 H2
Heanor 24,590 F4
Hebburn 23,150 J3
Hedon 3,010 G4
Hemel Hempstead 71,150 G7
Hereford 47,800 E5
Hertford 20,760 H7
Hetton 16,810 J3
Hexham 9,820 E3
Heywood 31,720 H2
High Wycombe 61,190 G8
Hillingdon 230,800 G8
Hinckley 49,310 F5
Hinderwell† 2,551 G3
Hitchin 29,190 G6
Hoddesdon 27,510 H7
Holmfirth 19,790 J2
Horley† 18,593 H8
Hornsea 7,280 G4
Horsham 26,770 G6
Horwich 16,670 G2
Houghton-le-Spring 33,150 J3
Hounslow, 199,100 G 8
Hove, 72,000 G 7
Hoylake, 32,000 G 2
Hoyland Nether, 15,500 J 2
Hucknall, 27,110 F 4
Huddersfield, 130,060 J 2
Hugh Town, ⊙1,958 A 8
Hull, 276,600 G 4
Hunstanton, 4,140 H 5
Huntingdon and Godmanchester, 17,200 G 5
Huyton-with-Roby, 65,950 G 2
Hyde, 37,040 H 2
Ilfracombe, 9,350 C 6
Ilkeston, 33,690 F 5
Immingham, ⊙10,259 G 4
Ipswich, 121,500 J 5
Islington, 171,600 H 8
Jarrow, 28,510 J 3
Kendal, 22,440 E 3
Kenilworth, 19,730 F 5
Kensington and Chelsea, 161,400 G 8
Keswick, 4,790 D 3
Kettering, 44,480 G 5
Keynsham, 18,970 E 6
Kidderminster, 49,960 E 5
Kidsgrove, 22,690 E 4
King's Lynn, 29,990 H 5
Kingston upon Thames, 135,600 H 8
Kingswood, 30,450 E 6
Kirkburton, 20,320 J 2
Kirkby, 59,100 G 2
Kirkby Lonsdale, ⊙1,506 E 3
Kirkby Stephen, ⊙1,539 E 3
Knutsford, 14,840 H 2
Lambeth, 290,300 H 8
Lancaster, 126,300 E 3
Leatherhead, 40,830 G 8
Leeds, 744,500 J 1
Leek, 19,460 H 2
Leicester, 289,400 F 5
Leigh, 46,390 H 2
Leighton-Linslade, 22,590 F 7
Letchworth, 31,520 G 6
Lewes, 14,170 H 7
Lewisham, 237,300 H 8
Leyland, 23,690 G 1
Lichfield, 23,690 F 5
Lincoln, 73,700 G 4
Liskeard, 5,360 C 7
Litherland, 23,530 G 2
Littlehampton, 20,320 G 7

(continued on following page)

ENGLAND

AREA 50,516 sq. mi. (130,836 sq. km.)
POPULATION 46,220,955
CAPITAL London
LARGEST CITY London
HIGHEST POINT Scafell Pike 3,210 ft. (978 m.)

WALES

AREA 8,017 sq. mi. (20,764 sq. km.)
POPULATION 2,790,462
CAPITAL Cardiff
LARGEST CITY Cardiff
HIGHEST POINT Snowdon 3,560 ft. (1,085 m.)

SCOTLAND

AREA 30,414 sq. mi. (78,772 sq. km.)
POPULATION 5,117,146
CAPITAL Edinburgh
LARGEST CITY Glasgow
HIGHEST POINT Ben Nevis 4,406 ft. (1,343 m.)

NORTHERN IRELAND

AREA 5,452 sq. mi. (14,121 sq. km.)
POPULATION 1,543,000
CAPITAL Belfast
LARGEST CITY Belfast
HIGHEST POINT Slieve Donard 2,796 ft. (852 m.)

Topography

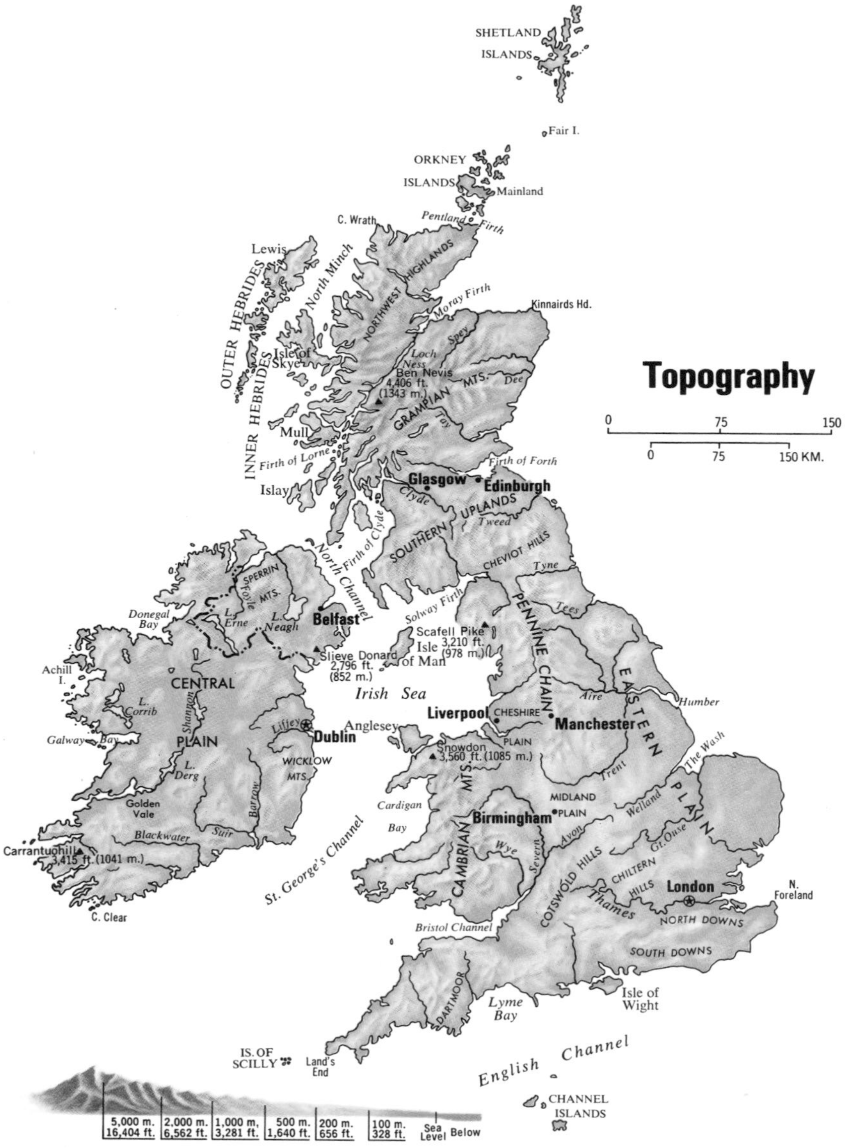
SHETLAND ISLANDS
Fair I.
ORKNEY ISLANDS
Mainland
C. Wrath
Pentland Firth
Lewis
OUTER HEBRIDES
North Minch
NORTHWEST HIGHLANDS
Moray Firth
Kinnairds Hd.
Isle of Skye
INNER HEBRIDES
Spey
Loch Ness
Ben Nevis 4,406 ft. (1343 m.)
GRAMPIAN MTS.
Dee
Tay
Mull
Firth of Lorne
Firth of Forth
Islay
Glasgow
Edinburgh
Clyde
SOUTHERN UPLANDS
Tweed
CHEVIOT HILLS
Tyne
Firth of Clyde
North Channel
SPERRIN MTS.
Donegal Bay
L. Erne
L. Neagh
Belfast
Solway Firth
Scafell Pike 3,210 ft. (978 m.)
Tees
PENNINE CHAIN
Slieve Donard 2,796 ft. (852 m.)
Isle of Man
Achill I.
CENTRAL
L. Corrib
Shannon
Irish Sea
Aire
Humber
EASTERN PLAIN
Liverpool
CHESHIRE PLAIN
Manchester
Galway Bay
PLAIN
Liffey
Dublin
Anglesey
WICKLOW MTS.
Snowdon 3,560 ft. (1085 m.)
Trent
The Wash
L. Derg
Barrow
Cardigan Bay
CAMBRIAN MTS.
MIDLAND PLAIN
Birmingham
Welland
Golden Vale
Suir
Blackwater
Wye
Severn
Avon
Gt. Ouse
Carrantuohill 3,415 ft. (1041 m.)
St. George's Channel
COTSWOLD HILLS
CHILTERN HILLS
London
N. Foreland
Thames
C. Clear
Bristol Channel
NORTH DOWNS
SOUTH DOWNS
DARTMOOR
Lyme Bay
Isle of Wight
IS. OF SCILLY
Land's End
English Channel
CHANNEL ISLANDS
0 75 150 MI.
0 75 150 KM.
5,000 m. 16,404 ft.
2,000 m. 6,562 ft.
1,000 m. 3,281 ft.
500 m. 1,640 ft.
200 m. 656 ft.
100 m. 328 ft.
Sea Level
Below

Liverpool, 539,700 G 2
Loftus, 7,850 G 3
London (cap.), 7,028,200 H 8
London, ★12,332,900 H 8
Long Eaton, 33,560 F 5
Longbenton, 50,120 J 3
Looe, 4,060 C 7
Loughborough, 49,010 F 5
Lowestoft, 53,260 J 5
Ludlow, ⊙7,466 E 5
Luton, 164,500 G 6
Lydd, 4,670 H 7
Lyme Regis, 3,460 E 7
Lymington, 36,780 F 7
Lynton, 1,770 D 6
Lytham Saint Anne's, 42,120 G 1
Mablethorpe and Sutton, 6,750 H 4
Macclesfield, 45,420 H 2
Maidenhead, 48,210 G 8
Maidstone, 72,110 J 8
Maldon, 14,350 H 6
Malmesbury, 2,550 E 6
Malton, 4,010 G 3
Malvern, 30,420 E 5
Manchester, 490,000 H 2
Mangotsfield, 23,000 E 6
Mansfield, 58,450 K 2
Mansfield Woodhouse, 25,400 F 4
March, 14,560 H 5
Margate, 50,290 J 6
Market Harborough, 15,230 G 5
Marlborough, 6,370 F 6
Matlock, 20,300 J 2
Melton Mowbray, 20,680 G 5
Merton, 169,400 H 8
Middlesbrough, 153,900 J 5
Middleton, 53,340 H 2
Middlewich, 7,600 H 2
Mildenhall, ⊙9,269 H 5
Millom, ⊙7,101 D 3
Milton Keynes, 89,900 F 5
Minehead, 8,230 D 6
Moretonhampstead, ⊙1,440 C 7
Morpeth, 14,450 F 2
Mundesley, ⊙1,536 J 5
Nelson, 31,220 H 1
Neston, 18,210 G 2
Newark, 24,760 G 4
Newbury, 24,850 F 6
Newcastle upon Tyne, 295,800 H 3
Newcastle-under-Lyme, 75,940 E 4
Newham, 228,900 H 8
Newhaven, 9,970 H 7
Newport, 22,430 F 7
New Romney, 3,830 J 7
Newton Abbot, 19,940 D 7
Newton-le-Willows, 21,780 H 2
New Windsor, 29,660 G 8
Northallerton F 3
Northam, 8,310 C 6
Northampton, 128,290 F 5
Northfleet, 27,150 J 8
North Sunderland, ⊙1,725 F 2
Northwich, 17,710 H 2
Norton, 5,580 G 3
Norton-Radstock, 15,900 E 6
Norwich, 119,200 J 5
Nottingham, 280,300 F 5
Nuneaton, 69,210 F 5
Oadby, 20,700 F 5
Oakham, 7,280 G 5
Okehampton, 4,000 D 7
Oldham, 103,690 H 2
Ormskirk, 28,860 G 2
Oswaldtwistle, 14,270 H 1
Oxford, 117,400 F 6
Padstow, ⊙2,802 B 7
Penryn, 5,660 B 7
Penzance, 19,360 B 7
Peterborough, 118,900 G 5
Peterlee, ⊙21,846 J 3
Plymouth, 259,100 C 7
Polperro, ⊙1,491 C 7
Poole, 110,600 E 7
Porlock, ⊙1,290 D 6
Portishead, 9,680 E 6
Portland, 14,860 E 7
Portslade-by-Sea, 18,040 G 7
Portsmouth, 198,500 F 7
Potters Bar, 24,670 H 7
Poulton-le-Fylde, 16,340 G 1
Preston, 94,760 G 1
Prestwich, 32,850 H 2
Queenborough, 31,550 H 6
Radcliffe, 29,630 H 2
Ramsbottom, 16,710 H 2
Ramsgate, 40,090 J 6
Rawtenstall, 20,950 H 1
Rayleigh, 26,740 J 8
Reading, 131,200 G 8
Redbridge, 231,600 H 8
Redcar, ⊙46,325 F 3
Redditch, 44,750 E 5
Reigate, 55,600 H 8
Richmond upon Thames, 166,800 H 8
Rickmansworth, 29,030 G 8
Ripley, 18,060 F 4
Rochdale, 93,780 H 2
Rochester, 56,030 J 8
Rothbury, ⊙1,818 E 2
Rotherham, 84,770 K 2
Royal Leamington Spa, 44,950 F 5
Royal Tunbridge Wells, 44,800 H 8
Rugby, 60,380 F 5
Rugeley, 24,440 E 5
Runcorn, 42,730 G 2
Rushden, 21,840 G 5
Ryde, 23,170 F 7
Rye, 4,530 H 7
Ryton, 15,170 H 3
Saddleworth, 21,340 J 2
Saint Agnes, ⊙4,747 B 7
Saint Albans, 123,800 H 7
Saint Austell-with-Fowey, 32,710 C 7
Saint Columb Major, ⊙3,953 B 7
Saint Helens, 104,890 G 2
Saint Ives, Cornwall, 9,760 B 7
Saint Neots, 17,940 G 5
Salcombe, 2,370 D 7
Sale, 59,060 H 2
Salford, 261,100 H 2
Salisbury, 35,460 F 6
Saltburn and Marske-by-the-Sea, 21,170 G 3
Sandbach, 14,280 H 2
Sandown-Shanklin, 14,800 F 7
Sandwich, 4,420 J 6
Saxmundham, 1,820 J 5
Scarborough, 43,300 G 3
Scunthorpe, 68,100 G 4
Seaford, 18,020 H 7
Seaham, 22,470 J 3
Seascale, ⊙2,106 D 3
Seaton, 4,500 D 7
Seaton Valley, 35,880 J 3
Sedbergh, ⊙2,741 E 3
Selsey, ⊙6,491 G 7
Sevenoaks, 18,160 J 8
Shaftesbury, 4,180 E 7
Sheffield, 558,000 J 2
Sherborne, 9,230 E 7
Sheringham, 4,940 J 5
Shildon, 15,360 F 3
Shoreham-by-Sea, 19,620 G 7
Shrewsbury, 56,120 E 5
Silloth, ⊙2,662 D 3
Sittingbourne and Milton, 32,830 H 6
Skelmersdale, 35,850 G 2
Skelton and Brotton, 15,930 G 3
Sleaford, 8,050 G 5
Slough, 89,060 G 8
Solihull, 108,230 F 5
Southampton, 213,700 F 7
Southend-on-Sea, 159,300 H 6
Southport, 86,030 G 1
South Shields, 96,900 J 3
Southwark, 224,900 H 8
Southwold, 1,960 J 5
Sowerby Bridge, 15,700 H 1
Spalding, 17,040 G 5
Spenborough, 41,460 J 1
Spennymoor, 19,050 F 3
Stafford, 54,860 E 5
Staines, 56,380 G 8
Stamford, 14,980 G 5
Stanley, 42,280 H 3
Staveley, 17,620 K 2
Stevenage, 72,600 G 6
Stockport, 138,350 H 2
Stockton-on-Tees, 165,400 F 3
Stoke-on-Trent, 256,200 E 4
Stourbridge, 56,530 E 5
Stourport-on-Severn, 19,430 E 5
Stowmarket, 9,020 J 5
Stratford-upon-Avon, 20,080 F 5
Stretford, 52,450 H 2
Stroud, 19,600 E 6
Sudbury, 8,860 H 5
Sunbury-on-Thames, 40,070 G 8
Sunderland, 214,820 J 3
Sutton, 166,700 H 8
Sutton Bridge, ⊙3,113 H 5
Sutton in Ashfield, 40,330 K 2
Swadlincote, 21,060 F 5
Swanage, 8,000 F 7
Swindon, 90,680 F 6
Tamworth, 46,960 F 5
Taunton, 37,570 D 6
Tavistock, ⊙7,620 C 7
Telford, ⊙79,451 E 5
Tenbury, ⊙2,151 E 5
Tewkesbury, 9,210 E 6
Thetford, 15,690 H 5
Thirsk, ⊙2,884 F 3
Thornaby-on-Tees, ⊙42,385 F 3
Thorne, ⊙16,694 F 4
Thornton Cleveleys, 27,090 G 1
Thurrock, 127,700 J 8
Tiverton, 16,190 D 7
Todmorden, 14,540 H 1
Tonbridge, 31,410 H 8
Torbay, 109,900 D 7
Torpoint, 6,840 C 7
Tower Hamlets, 146,100 H 8
Tow Law, 2,460 F 3
Trowbridge, 20,120 E 6
Truro, 15,690 B 7
Turton, 22,800 H 2
Tynemouth, 67,090 J 3
Upton upon Severn, ⊙2,048 E 5
Urmston, 44,130 H 2
Uttoxeter, 9,100 E 5
Ventnor, 6,980 F 7
Wainfleet All Saints, ⊙1,116 H 4
Wakefield, 306,500 J 2
Wallasey, 94,520 G 2
Wallsend, 45,490 J 3
Walsall, 182,430 E 5
Waltham Forest, 223,700 H 8
Waltham Holy Cross, 14,810 H 7
Walton and Weybridge, 51,270 G 8
Walton-le-Dale, 27,660 G 1
Wandsworth, 284,600 H 8
Wantage, 8,490 F 6
Ware, 14,900 H 7
Wareham, 4,630 E 7
Warley, 161,260 E 5
Warminster, 14,440 E 6
Warrington, 65,320 G 2
Warwick, 17,870 F 5
Washington, 27,720 J 3
Watchet, 2,980 D 6
Watford, 77,000 H 7
Wellingborough, 39,570 G 5
Wells, 8,960 E 6
Wells-next-the-Sea, 2,450 H 5
Welwyn, 39,900 H 7
Wem, ⊙3,411 E 5
West Bridgford, 28,340 F 5
West Bromwich, 162,740 F 5
West Mersea, 4,730 H 6
Westminster, 216,100 H 8
Weston-super-Mare, 51,960 D 6
Weymouth and Melcombe Regis, 41,080 E 7
Whickham, 29,710 J 3
Whitchurch, ⊙7,142 E 5
Whitehaven, 26,260 D 3
Whitley Bay, 37,010 J 3
Widnes, 58,330 G 2
Wigan, 80,920 G 2
Wigston, 31,650 F 5
Wilmslow, 31,250 H 2
Wilton, 4,090 F 6
Winchester, 88,900 F 6
Windermere, 7,860 E 3
Winsford, 26,920 G 2
Wirral, 27,510 G 2
Wisbech, 16,990 H 5
Witham, 19,730 H 6
Withernsea, 6,300 H 4
Wivenhoe, 5,630 J 6
Woking, 79,300 G 8
Wokingham, 22,390 G 8
Wolverhampton, 266,400 E 5
Wombwell, 17,850 K 2
Woodhall Spa, 2,420 G 4
Woodley and Sandford, ⊙24,581 G 8
Woodstock, 2,070 F 6
Wooler, ⊙1,833 F 2
Worcester, 73,900 E 5
Workington, 28,260 D 3
Worksop, 36,590 F 4
Worsbrough, 15,180 J 2
Worsley, 49,530 H 2
Worthing, 89,100 G 7
Wymondham, 9,390 J 5
Yateley, ⊙16,505 G 8
Yeovil, 26,180 E 7
York, 101,900 F 4

OTHER FEATURES

Aire (riv.) F 4
Atlantic Ocean A 7
Avon (riv.) F 5
Avon (riv.) F 7
Axe Edge (mt.) H 2
Barnstaple (bay) C 6
Beachy (head) H 7
Bigbury (bay) C 7
Blackwater (riv.) H 6
Bristol (chan.) C 6
Brown Willy (mt.) C 7
Cheviot (hills) E 2
Cheviot, The (mt.) E 2
Chiltern (hills) G 6
Cleveland (hills) F 3
Colne (riv.) G 8
Cornwall (cape) B 7
Cotswold (hills) E 6
Cross Fell (mt.) E 3
Cumbrian (mts.) D 3
Dart (riv.) D 7
Dartmoor National Park C 7
Dee (riv.) D 4
Derwent (riv.) G 3
Derwent (riv.) H 3
Don (riv.) F 4
Dorset Heights (hills) E 7
Dove (riv.) J 2
Dover (str.) J 7
Dungeness (prom.) J 7
Dunkery (hill) D 6
Eddystone (rocks) C 7
Eden (riv.) E 3
English (chan.) D 8
Esk (riv.) D 2
Exe (riv.) D 7
Exmoor National Park D 6
Fens, The (reg.) G 5
Formby (head) G 2
Foulness Island (pen.) J 6
Gibraltar (pt.) H 4
Great Ouse (riv.) H 5
Hartland (pt.) C 6
High Willhays (mt.) C 7
Hodder (riv.) H 1
Holderness (pen.), 43,900 G 4
Holy (isl.), 189 F 2
Humber (riv.) G 4
Irish (sea) B 4
Kennet (riv.) F 6
Lake District National Park E 3
Land's End (prom.) B 7
Lea (riv.) G 6
Lincoln Wolds (hills) G 4
Lindisfarne (Holy) (isl.), 189 F 2
Liverpool (bay) D 4
Lizard, The (pen.), 7,371 B 8
Lundy (isl.), 49 C 6
Lune (riv.) E 3
Lyme (bay) D 7
Manacle (pt.) C 7
Medway (riv.) H 6
Mendip (hills) E 6
Mersea (isl.), 4,423 J 6
Mersey (riv.) G 2
Morecambe (bay) D 3
Mounts (bay) B 7
Naze, The (prom.) J 6
Nene (riv.) H 5
New (for.) F 6
North (sea) J 4
North Downs (hills) G 6
North Foreland (prom.) J 6
Northumberland National Park E 2
North York Moors National Park G 3
Orford Ness (prom.) J 5
Ouse (riv.) G 4
Ouse (riv.) G 6
Parrett (riv.) E 6
Peak District National Park F 4
Peak, The (mt.) J 2
Peel Fell (mt.) E 2
Pennine Chain (range) E 3
Plymouth (sound) C 7
Portland, Bill of (pt.) E 7
Prawle (pt.) D 7
Purbeck, Isle of (pen.), 39,500 F 7
Ribble (riv.) E 4
Saint Alban's (head) F 7
Saint Bees (head) D 3
Saint Martin's (isl.), 106 A 8
Saint Mary's (isl.), 1,958 A 8
Scafell Pike (mt.) D 3
Scilly (isls.), 1,900 A 7
Selsey Bill (prom.) G 7
Severn (riv.) E 6
Sheppey (isl.), 31,550 J 6
Sherwood (for.) F 4
Skiddaw (mt.) D 3
Solent (chan.) F 7
Solway (firth) D 3
South Downs (hills) G 7
Spithead (chan.) F 7
Spurn (head) H 4
Stonehenge (ruins) F 6
Stour (riv.) E 7
Stour (riv.) H 6
Stour (riv.) J 6
Swale (riv.) F 3
Tamar (riv.) C 7
Taw (riv.) D 7
Tees (riv.) F 3
Test (riv.) F 6
Thames (riv.) H 6
Tintagel (head) C 7
Torridge (riv.) C 7
Trent (riv.) F 4
Tresco (isl.), 246 A 8
Tweed (riv.) E 2
Tyne (riv.) F 3
Ure (riv.) F 3
Ver (riv.) H 7
Walney, Isle of (isl.), 11,241 D 3
Wash, The (bay) H 5
Weald, The (reg.) H 6
Wear (riv.) F 3
Weaver (riv.) G 2
Welland (riv.) G 5
Wey (riv.) G 8
Wharfe (riv.) E 3
Wirral (pen.), 432,900 G 2
Witham (riv.) G 4
Wolds, The (hills) G 4
Wye (riv.) D 5
Wyre (riv.) G 1
Yare (riv.) J 5
Yorkshire Dales National Park E 3

CHANNEL ISLANDS

CITIES and TOWNS

Saint Anne E 8
Saint Helier (cap.), Jersey, ⊙28,135 E 8
Saint Peter Port (cap.), Guernsey, ⊙16,303 E 8
Saint Sampson's, ⊙6,534 E 8

OTHER FEATURES

Alderney (isl.), 1,686 E 8
Guernsey (isl.), 51,351 E 8
Herm (isl.), 96 E 8
Jersey (isl.), 72,629 E 8
Sark (isl.), 590 E 8

ISLE of MAN

CITIES and TOWNS

Castletown, 2,820 C 3
Douglas (cap.), 20,389 C 3
Laxey, 1,170 C 3
Michael, 408 C 3
Onchan, 4,807 C 3
Peel, 3,081 C 3
Port Erin, 1,714 C 3
Port Saint Mary, 1,508 C 3
Ramsey, 5,048 C 3

OTHER FEATURES

Ayre (pt.) C 3
Calf of Man (isl.) C 3
Langness (prom.) C 3
Snaefell (mt.) C 3
Spanish (head) C 3

WALES

COUNTIES

Clwyd, 376,000 D 4
Dyfed, 323,100 C 6
Gwent, 439,600 D 6
Gwynedd, 225,100 C 4
Mid Glamorgan, 540,400 D 6
Powys, 101,500 D 5
South Glamorgan, 389,200 A 7
West Glamorgan, 371,900 D 6

CITIES and TOWNS

Aberaeron, 1,340 C 5
Abercarn, 18,370 B 6
Aberdare, 38,030 A 6
Abertillery, 20,550 B 6
Amlwch, 3,630 C 4
Bala, 1,650 D 5
Bangor, 16,030 C 4
Barmouth, 2,070 C 5
Barry, 42,780 B 7
Beaumaris, 2,090 C 4
Bedwellty, 25,460 B 6
Bethesda, 4,180 D 4
Betws-y-Coed, 720 D 4
Brecknock (Brecon), 6,460 D 6
Brecon, 6,460 D 6
Bridgend, 14,690 A 7
Brynmawr, 5,990 B 6
Builth Wells, 1,480 D 5
Burry Port, 5,990 C 6
Caernarfon, 8,840 C 4
Caerphilly, 42,190 B 6
Cardiff, 281,500 B 7
Cardigan, 3,830 C 5
Chepstow, 8,260 E 6
Chirk, ⊙3,564 D 5
Colwyn Bay, 25,370 D 4
Criccieth, 1,590 C 5
Cwmamman, 3,950 D 6
Cwmbran, 32,980 B 6
Denbigh, 8,420 D 4
Dolgellau, 2,430 D 5
Ebbw Vale, 25,670 B 6
Ffestiniog, 5,510 D 5
Fishguard and Goodwick, 5,020 B 5
Flint, 15,070 G 2
Gelligaer, 33,820 A 6
Harlech, ⊙332 C 5
Haverfordwest, 8,930 B 6
Hawarden, ⊙20,389 G 2
Hay, 1,200 D 5
Holywell, 8,570 G 2
Kidwelly, 3,090 C 6
Knighton, 2,190 D 5
Llandeilo, 1,780 C 6
Llandovery, 2,040 D 5
Llandrindod Wells, 3,460 D 5
Llandudno, 17,700 D 4
Llanelli, 25,870 C 6
Llanfairfechan, 3,800 C 4
Llangefni, 4,070 C 4
Llangollen, 3,050 D 5
Llanguicke, ⊙15,029 D 6
Llanidloes, 2,390 D 5
Llantrisant, ⊙27,490 A 7
Llanwrtyd Wells, 460 D 5
Llwchwr, 27,530 D 6
Machynlleth, 1,830 D 5
Maesteg, 21,100 D 6
Menai Bridge, 2,730 C 4
Merthyr Tydfil, 61,500 A 6
Milford Haven, 13,960 B 6
Mold, 8,700 G 2
Montgomery, 1,000 D 5
Mountain Ash, 27,710 A 6
Mynyddislwyn, 15,590 B 6
Narberth, 970 C 6
Neath, 27,280 D 6
Nefyn, ⊙2,086 C 5
Newcastle Emlyn, 690 C 5
Newport, Dyfed, ⊙1,062 C 5
Newport, Gwent, 110,090 B 6
New Quay, 760 C 5
Newtown, 6,400 D 5
Neyland, 2,690 B 6
Ogmore and Garw, 19,680 A 6
Pembroke, 14,570 C 6
Penarth, 24,180 B 7
Penmaenmawr, 4,050 C 4
Pontypool, 36,710 B 6
Pontypridd, 34,180 A 6
Porthcawl, 14,980 D 6
Porthmadog, 3,900 C 5
Port Talbot, 58,200 D 6
Prestatyn, 15,480 D 4
Presteigne, 1,330 D 5
Pwllheli, 4,020 C 5
Rhondda, 85,400 A 6
Rhyl, 22,150 D 4
Risca, 15,780 B 6
Ruthin, 4,780 D 4
Saint David's, ⊙1,638 B 6
Swansea, 190,800 C 6
Tenby, 4,930 C 6
Tredegar, 17,450 B 6
Tywyn, 3,850 C 5
Welshpool, 7,370 D 5
Wrexham, 39,530 E 4

OTHER FEATURES

Anglesey (isl.), 64,500 C 4
Aran Fawddwy (mt.) C 5
Bardsey (isl.), 9 C 5
Berwyn (mts.) D 5
Black (mts.) D 6
Braich-y-Pwll (prom.) C 5
Brecon Beacons (mt.) D 6
Brecon Beacons National Park D 6
Caldy (isl.), 70 C 6
Cambrian (mts.) D 5
Cardigan (bay) C 5
Carmarthen (bay) C 6
Cemmaes (head) C 5
Dee (riv.) D 4
Dovey (riv.) D 5
Ely (riv.) B 7
Gower (pen.), 17,220 C 6
Great Ormes (head) C 4
Holy (isl.), 13,715 C 4
Lleyn (pen.), 25,800 C 5
Menai (str.) C 4
Milford Haven (inlet) B 6
Pembrokeshire Coast National Park C 6
Plynlimon (mt.) D 5
Preseli (mts.) C 6
Radnor (for.) D 5
Rhymney (riv.) B 6
Saint Brides (bay) B 6
Saint David's (head) B 6
Saint George's (chan.) B 5
Saint Gowans (head) C 6
Severn (riv.) E 5
Snowdon (mt.) D 4
Snowdonia National Park D 4
Taff (riv.) B 7
Teifi (riv.) C 5
Towy (riv.) D 6
Tremadoc (bay) C 5
Usk (riv.) B 6
Wye (riv.) D 5
Ynys Môn (Anglesey) (isl.), 64,500 C 4

★Population of met. area.
⊙Population of parish.

SCOTLAND
(map on page 15)

REGIONS

Borders, 99,409 E 5
Central, 269,281 D 4
Dumfries and Galloway, 143,667 E 5
Fife, 336,339 E 4
Grampian, 448,772 F 3
Highland, 182,044 D 3
Lothian, 754,008 E 5
Orkney (islands area), 17,675 E 1
Shetland (islands area), 18,494 F 2
Strathclyde, 2,504,909 C 4
Tayside, 401,987 E 4
Western Isles (islands area), 29,615 A 3

CITIES and TOWNS

Aberchirder, 877 F 3
Aberdeen, 210,362 F 3
Aberdour, 1,576 D 1
Aberfeldy, 1,552 E 4
Aberfoyle, 793 D 4
Aberlady, 737 F 4
Abernethy, 776 E 4
Aboyne, 1,040 F 3
Acharacle, ⊙764 C 4
Achiltibuie, ⊙1,564 C 3
Achnasheen, ⊙1,078 C 3
Ae, 239 E 5
Airdrie, 38,491 C 2
Alexandria, 9,758 A 1
Alford, 764 F 3
Alloa, 13,558 C 1
Alness, 2,560 D 3
Altnaharra, ⊙1,227 D 2
Alva, 4,593 C 1
Alyth, 1,738 E 4
Ancrum, 266 F 5
Annan, 6,250 E 6
Annat, ⊙550 C 3
Annbank Station, 2,530 D 5
Applecross, ⊙550 C 3
Arbroath, 22,706 F 4
Ardavasar, ⊙449 B 3
Ardersier, 942 E 3
Ardgay, 193 D 3
Ardrishaig, 946 C 4
Ardrossan, 11,072 D 5
Armadale, 7,200 C 2
Arrochar, 543 D 4
Ascog, 230 A 2
Auchenblae, 339 F 4
Auchencairn, 215 E 6
Auchinleck, 4,883 D 5
Auchterarder, 1,738 E 4
Auchtermuchty, 1,426 E 4
Auldearn, 405 E 3
Aviemore, 1,224 E 3
Avoch, 776 D 3
Ayr, 47,990 D 5
Ayton, 410 F 5
Bailivanish, 347 A 3
Baillieston, 7,671 B 2
Balallan, 283 B 2
Balerno, 3,576 D 2
Balfron, 1,149 B 1
Ballantrae, 262 C 5
Ballater, 981 F 3
Ballingry, 4,332 D 1
Ballinluig, 188 E 4
Balloch, Highland, 572 D 3
Balloch, Strathclyde, 1,484 B 1
Baltasound, 246 G 2
Banchory, 2,435 F 3
Banff, 3,832 F 3
Bankfoot, 868 E 4
Bankhead, 1,492 F 3
Bannockburn, 5,889 C 1
Barrhead, 18,736 B 2
Barrhill, 236 D 5
Barvas, 279 B 2
Bathgate, 14,038 C 2
Bayble, 543 B 2
Bearsden, 25,128 B 2
Beattock, 309 E 5
Beauly, 1,141 D 3
Beith, 5,859 D 5
Bellsbank, 3,066 D 5
Bellshill, 18,166 C 2
Berriedale, ⊙1,927 E 2
Bieldside, 1,137 F 3
Biggar, 1,718 E 5
Birnam, 659 E 4
Bishopbriggs, 21,570 B 2
Bishopton, 2,931 B 2
Blackburn, 7,636 C 2
Blackford, 529 E 4
Blair Atholl, 437 E 4
Blairgowrie and Rattray, 5,681 E 4
Blanefield, 835 B 1
Blantyre, 13,992 B 2
Blyth Bridge, ⊙441 E 5
Bo'ness, 12,959 C 1
Boat of Garten, 406 E 3
Boddam, 1,429 G 3
Bonar Bridge, 519 D 3
Bonhill, 4,385 B 1
Bonnybridge, 5,701 C 1
Bonnyrigg and Lasswade, 7,429 D 2
Bowmore, 947 B 5
Braemar, 394 E 3
Breasclete, 234 B 2
Brechin, 6,759 F 4
Bridge of Allan, 4,638 C 1
Bridge of Don, 4,086 F 3
Bridge of Weir, 4,724 A 2
Brightons, 3,106 C 1
Broadford, 310 B 3
Brodick, 630 C 5
Brora, 1,436 E 2
Broxburn, 7,776 D 1
Buchlyvie, 412 B 1
Buckhaven and Methil, 17,930 F 4
Buckie, 8,145 E 3
Bucksburn, 6,567 F 3
Bunessan, ⊙585 B 4
Burghead, 1,321 E 3
Burnmouth, 300 F 5
Burntisland, 5,626 D 1
Cairndow, ⊙874 D 4
Cairnryan, 199 D 6
Callander, 1,805 D 4
Cambuslang, 14,607 B 2
Campbeltown, 6,428 C 5
Cannich, 203 D 3
Canonbie, 234 F 5
Caol, 3,719 C 4
Carbost, ⊙772 B 3
Cardenden, 6,802 D 1
Carloway, 178 B 2
Carluke, 8,864 E 5
Carnoustie, 6,838 F 4
Carnwath, 1,246 E 5
Carradale, 262 C 5
Carrbridge, 416 E 3
Carron, 2,626 C 1
Carsphairn, 186 D 5
Castlebay, 284 A 4
Castle Douglas, 3,384 E 6
Castle Kennedy, 307 D 6
Castletown, 902 E 2
Catrine, 2,681 D 5
Cawdor, 111 E 3
Chirnside, 888 F 5
Chryston, 8,322 C 2
Clackmannan, 3,248 C 1
Clarkston, 8,404 B 2
Closeburn, 225 E 5
Clovulin, ⊙315 C 4
Clydebank, 47,538 B 2
Coalburn, 1,460 E 5
Coatbridge, 50,806 C 2
Cockburnspath, 233 F 5
Cockenzie and Port Seton, 3,539 D 1
Coldingham, 423 F 5
Coldstream, 1,393 F 5
Coll, 305 B 2
Colmonell, 218 D 5
Comrie, 1,119 E 4
Connel, 300 C 4
Cononbridge, 914 D 3
Corpach, 1,296 C 4
Coupar Angus, 2,010 E 4
Cove and Kilcreggan, 1,402 A 1
Cove Bay, 765 F 3
Cowdenbeath, 10,215 D 1
Cowie, 2,751 C 1
Craigellachie, 382 E 3
Craignure, ⊙544 C 4
Crail, 1,033 F 4
Crawford, 384 E 5
Creetown, 769 D 6
Crieff, 5,718 E 4
Crimond, 313 G 3
Crinan, ⊙462 C 4
Cromarty, 492 E 3
Crosshill, 535 D 5
Crossmichael, 317 D 6
Cruden Bay, 528 G 3
Cullen, 1,199 F 3
Culross, 504 C 1
Cults, 3,336 F 3
Cumbernauld, 41,200 C 1
Cumnock and Holmhead, 6,298 D 5
Cupar, 6,607 E 4
Currie, 6,764 D 2
Dailly, 1,258 D 5
Dalbeattie, 3,659 E 6
Dalburgh, 261 A 3
Dalkeith, 9,713 D 2
Dalmally, 283 D 4
Dalmellington, 1,949 D 5
Dalry, 5,833 D 5
Dalrymple, 1,336 D 5
Darvel, 3,177 D 5
Daviot, ⊙513 D 3
Denholm, 581 F 5
Denny and Dunipace, 10,424 C 1
Dervaig, ⊙1,081 B 4
Dingwall, 4,275 D 3
Dollar, 2,573 E 4
Dornoch, 880 D 3
Douglas, 1,843 E 5
Doune, 859 D 4
Drongan, 3,609 D 5
Drumbeg, ⊙833 C 2
Drummore, 336 D 6
Drumnadrochit, 359 D 3
Drymen, 659 B 1
Dufftown, 1,481 E 3
Dumbarton, 25,469 B 1
Dumfries, 29,259 E 5
Dunbar, 4,609 F 4
Dunbeath, 161 E 2
Dunbeg, 939 C 4
Dunblane, 5,222 E 4
Dundee, 194,732 F 4
Dundonald, 2,256 D 5
Dunfermline, 52,098 D 1
Dunkeld, 273 E 4
Dunning, 564 E 4
Dunoon, 8,759 A 2
Dunragit, 323 D 6
Duns, 1,812 F 5
Duntocher, 3,532 B 2
Dunure, 452 D 5
Dunvegan, 301 B 3
Dyce, 2,733 F 3
Eaglesfield, 581 E 5
Eaglesham, 2,788 D 5
Earlston, 1,415 F 5
East Calder, 2,690 D 2
East Kilbride, 71,200 B 2
East Linton, 882 F 5
Eastriggs, 1,455 E 5
Ecclefechan, 844 E 5
Edinburgh (cap.), 470,085 D 1
Edzell, 658 F 4
Elderslie, 5,204 B 2
Elgin, 17,042 E 3
Elie and Earlsferry, 807 F 4
Ellon, 2,855 F 3
Embo, 260 E 3
Errol, 762 E 4
Evanton, 562 D 3
Eyemouth, 2,704 F 5
Fairlie, 1,029 D 5
Falkirk, 36,901 C 1
Falkland, 998 E 4
Fallin, 3,159 C 1
Fauldhouse, 5,247 C 2
Ferness, ⊙287 E 3
Ferryden, 740 F 4
Findhorn, 664 E 3
Findochty, 1,229 E 3
Fintry, 296 B 1
Fochabers, 1,238 E 3
Forfar, 11,179 F 4
Forres, 5,317 E 3
Fort Augustus, 670 D 3
Forth, 2,929 C 2
Fortrose, 1,150 D 3
Fort William, 4,370 C 4
Foyers, 276 D 3
Fraserburgh, 10,930 G 3
Friockheim, 807 F 4
Furnace, 220 C 4
Fyvie, 405 F 3
Gairloch, 125 C 3
Galashiels, 12,808 F 5
Galston, 4,256 D 5
Gardenstown, 892 F 3
Garelochhead, 1,552 A 1
Gargunnock, 457 B 1
Garlieston, 385 D 6
Garmouth, 352 E 3
Garrabost, 307 B 2
Gartmore, 253 B 1
Gatehouse-of-Fleet, 835 D 6
Giffnock, 10,987 B 2
Gifford, 575 F 5
Girvan, 7,597 D 5
Glamis, 190 F 4
Glasgow, 880,617 B 2
Glasgow, ★1,674,789 B 2
Glenbarr, ⊙691 C 5
Glencaple, 275 E 5
Glencoe, 195 C 4
Glenelg, ⊙1,468 C 3
Glenluce, 725 D 6
Glenrothes, 31,400 E 4
Golspie, 1,374 E 3
Gordon, 320 F 5
Gorebridge, 3,426 D 2
Gourock, 11,192 A 1
Grangemouth, 24,430 C 1
Grantown-on-Spey, 1,578 E 3
Greenlaw, 574 F 5
Greenock, 67,275 A 2
Gretna, 1,907 E 5
Gullane, 1,701 F 4
Haddington, 6,767 F 5
Halkirk, 679 E 2
Hamilton, 45,495 C 2
Hamnavoe, 307 G 2
Harthill, 4,712 C 2
Hatton, 315 G 3
Hawick, 16,484 F 5
Heathhall, 1,365 E 5
Helensburgh, 13,327 A 1
Helmsdale, 727 E 2
Hill of Fearn, 233 D 3
Hillside, 692 F 4
Hillswick, ⊙696 G 2
Hopeman, 1,248 E 3
Huntly, 4,078 F 3
Hurlford, 4,294 D 5
Inchnadamph, ⊙833 D 2
Innellan, 922 A 2
Innerleithen, 2,293 E 5
Insch, 881 F 3
Inveraray, 473 C 4
Inverbervie, 853 F 4
Invercassley, ⊙1,067 D 3
Invergordon, 2,385 D 3
Invergowrie, 1,389 E 4
Inverie, ⊙1,468 C 3
Inverkeithing, 6,102 D 1
Inverness, 35,801 D 3
Inverurie, 5,534 F 3
Irvine, 48,500 D 5
Isle of Whithorn, 222 D 6
Jedburgh, 3,953 F 5
John O'Groats, 195 E 2
Johnshaven, 544 F 4
Johnstone, 23,251 B 2
Kames, 230 C 5
Keiss, 344 E 2
Keith, 4,192 F 3
Kelso, 4,934 F 5
Kelty, 6,573 D 1
Kemnay, 1,042 F 3
Kenmore, 211 E 4
Kilbarchan, 2,669 A 2
Kilbirnie, 8,259 A 2
Kilchoan, ⊙764 B 4
Kildonan, ⊙1,105 E 2
Killearn, 1,086 C 1
Killin, 600 D 4
Kilmacolm, 3,348 A 2
Kilmarnock, 50,175 D 5
Kilmaurs, 2,518 D 5
Kilninver, ⊙247 C 4
Kilrenny and Anstruther, 2,951 F 4
Kilsyth, 10,210 B 1
Kilwinning, 8,460 D 5
Kinbrace, ⊙1,105 E 2
Kincardine, 3,278 C 1
Kinghorn, 2,163 D 1
Kingussie, 1,036 D 3
Kinlochewe, ⊙1,794 C 3
Kinlochleven, 1,243 C 4
Kinloch Rannoch, 241 D 4
Kinloss, 2,378 E 3
Kinross, 2,829 E 4
Kintore, 970 F 3
Kippen, 529 B 1
Kirkcaldy, 50,207 D 1
Kirkcolm, 346 C 6
Kirkconnel, 3,318 E 5
Kirkcowan, 354 D 6
Kirkcudbright, 2,690 E 6
Kirkhill, 210 D 3
Kirkintilloch, 26,664 B 2
Kirkmuirhill, 2,575 E 5
Kirkton of Glenisla, ⊙331 E 4
Kirkwall, 4,777 E 1
Kirriemuir, 4,295 E 4
Kyleakin, 268 C 3
Kyle of Lochalsh, 687 C 3
Kylestrome, ⊙745 D 2
Ladybank, 1,216 E 4
Laggan, 393 D 3
Lairg, 572 D 2
Lamlash, 613 C 5
Lanark, 8,842 E 5
Langholm, 2,509 E 5
Larbert, 4,922 C 1
Largs, 9,461 A 2
Larkhall, 15,926 E 5
Lauder, 639 F 5
Laurencekirk, 1,416 F 4

(continued)

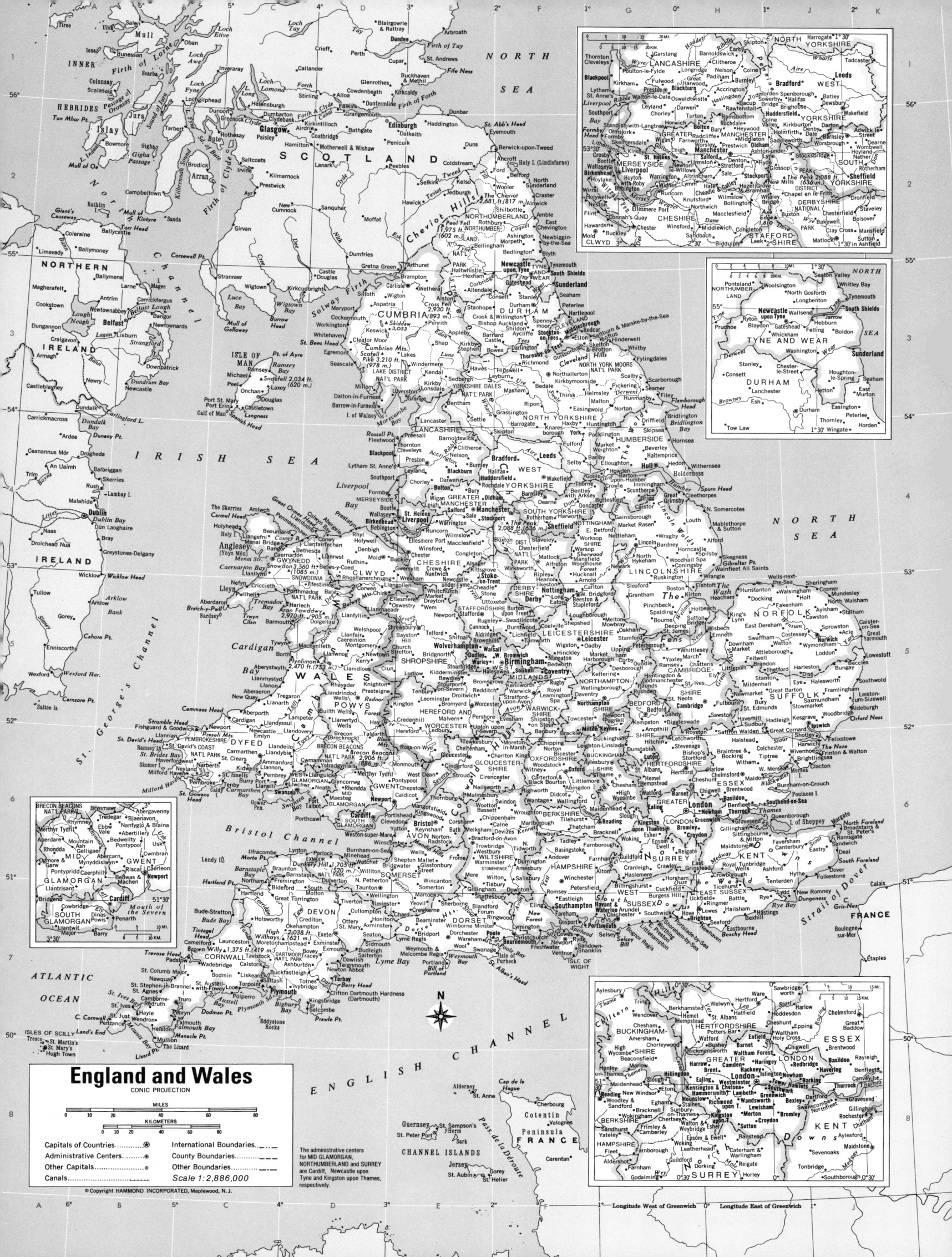

England and Wales

CONIC PROJECTION

MILES

KILOMETERS

Capitals of Countries............⊛
Administrative Centers............◉
Other Capitals............◉
Canals............

International Boundaries............
County Boundaries............
Other Boundaries............

The administrative centers for MID GLAMORGAN, NORTHUMBERLAND and SURREY are Cardiff, Newcastle upon Tyne and Kingston upon Thames, respectively.

Scale 1:2,886,000

© Copyright HAMMOND INCORPORATED, Maplewood, N.J.

United Kingdom and Ireland

(continued)

Lennoxtown, 3,070	B 1
Lerwick, 6,195	G 2
Leslie, 3,303	E 4
Lesmahagow, 3,906	E 5
Leswalt, 237	C 6
Letham, 804	F 4
Leuchars, 2,482	F 4
Leurbost, 461	B 2
Leven, 9,507	F 4
Leverburgh, 223	B 3
Lhanbryde, 1,184	E 3
Lilliesleaf, 212	F 5
Limekilns, 812	D 1
Linlithgow, 6,098	C 1
Linwood, 10,510	B 2
Lionel, 187	B 2
Livingston, 21,900	C 2
Loanhead, 5,971	D 2
Lochailort, ⊙673	C 4
Lochaline, 213	C 4
Lochans, 355	D 6
Locharbriggs, 2,561	E 5
Lochawe, 200	C 4
Lochboisdale, 382	A 3
Lochcarron, 204	C 3
Lochgelly, 7,754	D 1
Lochgilphead, 1,217	C 4
Lochgoilhead, 216	D 4
Lochinver, 283	C 2
Lochmaben, 1,304	E 5
Lochmaddy, 307	A 3
Lochore, 2,994	D 1
Lochwinnoch, 2,064	A 2
Lockerbie, 3,135	E 5
Lossiemouth and Branderburgh, 5,817	E 3
Lumsden, 248	F 3
Luncarty, 584	E 4
Lybster, 554	E 2
Lyness, ⊙454	E 2
Macduff, 3,682	F 3
Machrihanish, 212	C 5
Maidens, 536	D 5
Mallaig, 903	C 4
Markinch, 2,366	E 4
Mauchline, 3,612	D 5
Maud, 634	F 3
Maybole, 4,703	D 5
Mayfield, 8,232	D 2
Meigle, 357	E 4
Melrose, 2,197	F 5
Melvaig, ⊙1,794	C 3
Methlick, 315	F 3
Methven, 806	E 4
Mid Yell, 220	G 2
Millport, 1,161	A 2
Milnathort, 1,099	E 4
Milngavie, 10,846	B 1
Minnigaff, 658	D 6
Mintlaw, 657	F 3
Moffat, 2,041	E 5
Moniaive, 342	E 5
Monifieth, 7,100	F 4
Montrose, 4,704	F 4
Morar, 184	C 4
Motherwell and Wishaw, 72,991	C 2
Muirkirk, 2,607	E 5
Muir of Ord, 1,339	D 3
Musselburgh, 17,045	D 2
Muthill, 672	E 4
Nairn, 5,821	E 3
Neilston, 4,358	B 2
Nethy Bridge, 431	E 3
New Abbey, 339	E 6
Newarthill, 7,003	C 2
Newburgh, Fife, 2,124	E 4
Newburgh, Grampian, 447	G 3
Newcastleton, 903	F 5
New Cumnock, 5,077	D 5
New Deer, 601	F 3
New Galloway, 337	D 5
Newmains, 6,847	C 2
Newmarket, 613	B 2
Newmill, 449	E 3
Newmilns and Greenholm, 3,509	D 5
New Pitsligo, 1,125	F 3
Newport-on-Tay, 3,762	F 4
New Scone, 3,830	E 4
Newtongrange, 4,555	D 2
Newton Mearns, 6,901	B 2
Newtonmore, 894	D 3
Newton Stewart, 1,983	D 6
Newtown Saint Boswells, 1,101	F 5
Newtyle, 664	E 4
North Berwick, 4,317	F 4
North Tolsta, 527	B 2
Oakley, 3,499	C 1
Oban, 6,515	C 4
Old Kilpatrick, 3,256	B 2
Oldmeldrum, 1,103	F 3
Oykel Bridge, ⊙742	D 3
Paisley, 94,833	B 2
Palnackie, 225	E 6
Patna, 2,867	D 5
Peebles, 6,049	E 5
Penicuik, 10,476	D 2
Penpont, 364	E 5
Perth, 43,098	E 4
Peterculter, 3,226	F 3
Peterhead, 14,846	G 3
Pierowall, ⊙735	E 1
Pitlochry, 2,468	E 4
Pitmedden, 313	F 3
Pittenweem, 1,548	F 4
Plockton, 288	C 3
Poolewe, ⊙1,794	C 3
Port Appin, ⊙2,172	C 4
Port Askaig, ⊙1,795	B 5
Port Bannatyne, 730	A 2
Port Charlotte, 240	B 5
Port Ellen, 932	B 5
Port Glasgow, 22,189	A 2
Portgordon, 814	F 3
Portknockie, 1,217	F 3
Portmahomack, 226	E 3
Portpatrick, 643	C 6
Portree, 1,374	B 3
Portsoy, 1,717	F 3
Port William, 517	D 6
Prestonpans, 3,272	D 1
Prestwick, 13,218	D 5
Queensferry, 5,339	D 1
Reay, 283	E 2
Renfrew, 18,880	B 2
Renton, 3,443	A 1
Rhu, 1,540	A 1
Rhynie, 333	F 3
Rigside, 1,195	E 5
Rosehearty, 1,220	F 3
Rosneath, 946	A 1
Rothes, 1,240	E 3
Rothesay, 6,285	A 2
Rutherglen, 24,091	B 2
Saint Abbs, 203	F 5
Saint Andrews, 12,837	F 4
Saint Combs, 738	G 3
Saint Cyrus, 340	F 4
Saint Margaret's Hope, 210	F 2
Saint Monance, 1,205	F 4
Saline, 831	C 1
Saltcoats, 14,861	D 5
Sandbank, 850	A 1
Sandhead, 248	D 6
Sandwick, 603	B 2
Sanquhar, 2,030	D 5
Sauchie, 6,082	C 1
Scalasaig, ⊙137	B 4
Scalloway, 896	G 2
Scarinish, ⊙875	B 4
Scourie, ⊙745	C 2
Scrabster, 273	E 2
Selkirk, 5,635	F 5
Shader, 258	B 2
Shawbost, 458	B 2
Shieldaig, ⊙550	C 3
Shotts, 9,512	C 2
Skateraw, 674	F 3
Skelmorlie, 1,535	A 2
Skipness, ⊙765	C 5
Slamannan, 1,584	C 2
Spean Bridge, 235	D 4
Springholm, 340	E 5
Stanley, 1,385	E 4
Stenhousemuir, 8,203	C 1
Stevenston, 11,786	D 5
Stewarton, 5,165	D 5
Stirling, 29,799	C 1
Stonehaven, 4,837	F 4
Stonehouse, 7,900	D 5
Stornoway, 5,371	B 2
Stow, 485	F 5
Strachan, ⊙390	F 3
Strachur Bay, ⊙678	C 4
Stranraer, 10,174	C 6
Strathaven, 5,464	D 5
Strathpeffer, 874	D 3
Strichen, 962	F 3
Stromeferry, ⊙1,724	C 3
Stromness, 1,680	E 2
Strontian, ⊙764	C 4
Struan, ⊙772	B 3
Swinton, 235	F 5
Tain, 2,057	D 3
Tarbert, Strathclyde, 1,391	C 5
Tarbert, W. Isles, 479	B 3
Tarbolton, 2,224	D 5
Tarland, 452	F 3
Tayport, 2,848	F 4
Thornhill, Central, 443	D 4
Thornhill, Dumf. & Gall., 1,510	E 5
Thurso, 9,113	E 2
Tillicoultry, 4,320	C 1
Tobermory, 652	B 4
Tolob, ⊙2,033	G 2
Tomatin, 214	D 3
Tomintoul, 306	E 3
Torphins, 499	F 3
Tradespark, 425	E 3
Tranent, 7,212	F 5
Troon, 11,656	D 5
Tullibody, 6,082	C 1
Turriff, 3,051	F 3
Tweedsmuir, ⊙105	E 5
Twynholm, 274	D 6
Tyndrum, ⊙1,153	D 4
Uddingston, 5,278	B 2
Uig, Highland, 103	B 3
Uig, W. Isles, ⊙1,948	A 2
Ullapool, 807	C 3
Uphall, 3,035	C 1
Viewpark, 9,812	C 2
Walkerburn, 842	F 5
Watten, 347	E 2
Wemyss Bay, 323	A 2
West Barns, 659	F 5
West Calder, 2,005	C 2
West Kilbride, 3,883	D 5
West Linton, 705	D 2
Whitburn, 11,647	C 2
Whitehills, 875	F 3
Whithorn, 990	D 6
Whiting Bay, 352	C 5
Wick, 7,804	E 2
Wigtown, 1,118	D 6
Winchburgh, 2,409	D 1
Yetholm, 435	F 5

OTHER FEATURES

A'Chralaig (mt.)	C 3
Ailsa Craig (isl.), 3	C 5
Almond (riv.)	E 4
Annan (riv.)	E 5
Appin (dist.), 2,006	C 4
Ardgour (dist.), 315	C 4
Ardle (riv.)	E 4
Ardnamurchan (pen.), 764	B 4
Argyll (dist.), 4,940	C 4
Arkaig, Loch (lake)	C 4
Arran (isl.), 3,564	C 5
Askival (mt.)	B 4
Assynt (dist.), 833	C 2
Athol (dist.), 1,082	D 4
Atlantic Ocean	B 2
Avon (riv.)	C 1
Avon (riv.)	E 3
Awe, Loch (lake)	C 4
Ayr (riv.)	D 5
Ayr, Heads of (cape)	D 5
Badenoch (dist.), 2,717	D 4
Baleshare (isl.), 64	A 3
Balmoral Castle	E 3
Barra (sound)	A 3
Barra (isl.), 1,005	A 4
Barra (head)	A 4
Barra Isles (isls.), 1,092	A 4
Battock (mt.)	F 4
Beauly (riv.)	D 3
Beinn Dearg (mt.)	D 3
Beinn a Ghlo (mt.)	E 4
Bell Rock (isl.), 3	F 4
Ben Alder (mt.)	D 4
Ben Avon (mt.)	E 3
Benbecula (isl.), 1,355	A 3
Ben Cruachan (mt.)	C 4
Ben Lawers (mt.)	D 4
Ben Lui (mt.)	D 4
Ben Macdhui (mt.)	E 3
Ben Mhor (mt.)	A 3
Ben More (mt.)	B 4
Ben More (mt.)	D 4
Ben More Assynt (mt.)	D 2
Ben Nevis (mt.)	D 4
Bernera (isl.), 276	B 2
Berneray (isl.), 131	A 3
Berneray (isl.), 6	A 4
Bidean nam Bian (mt.)	D 4
Black Isle (pen.), 7,209	D 3
Blackwater (res.)	D 4
Boisdale, Loch (inlet)	A 3
Bracadale, Loch (inlet)	B 3
Braemar (dist.), 7,624	E 3
Breadalbane (dist.), 3,649	D 4
Bressay (isl.), 248	G 2
Broad (bay)	B 2
Broad Law (mt.)	E 5
Broom, Loch (inlet)	C 3
Brough Ness (prom.)	F 2
Buchan (dist.), 40,089	F 3
Buddon Ness (prom.)	F 4
Burray (isl.), 209	F 2
Burrow (head)	D 6
Bute (isl.), 8,423	C 5
Bute (sound)	C 5
Butt of Lewis (prom.)	B 2
Cairn Gorm (mt.)	E 3
Cairngorm (mts.)	E 3
Cairn Toul (mt.)	E 3
Caledonian (canal)	D 3
Canna (isl.), 22	B 3
Carn Ban (mt.)	D 3
Carn Eige (mt.)	C 3
Carrick (dist.), 21,425	D 5
Carron (riv.)	C 1
Carron (riv.)	C 3
Cheviot (hills)	F 5
Cheviot, The (mt.)	F 5
Clisham (mt.)	B 3
Clyde (riv.)	D 5
Clyde (firth)	D 5
Coll (isl.), 144	B 4
Colonsay (isl.), 137	B 4
Copinsay (isl.), 3	F 2
Cowal (dist.), 15,548	C 4
Creag Meagaidh (mt.)	D 4
Cromarty (firth)	D 3
Cuillin (hills)	B 3
Cuillin (sound)	B 3
Dee (riv.)	D 5
Dee (riv.)	F 3
Dennis (head)	F 1
Deveron (riv.)	F 3
Don (riv.)	F 3
Doon (riv.)	D 5
Dornoch (firth)	E 3
Duirinish (dist.), 1,085	B 3
Duncansby (head)	F 2
Dunnet (head)	E 2
Earn (riv.)	E 4
Earn, Loch (lake)	D 4
Eday (isl.), 179	F 1
Eddrachillis (bay)	C 2
Eden (riv.)	F 4
Egilsay (isl.), 39	F 1
Eigg (isl.), 69	B 4
Eil, Loch (lake)	C 4
Eishort, Loch (inlet)	B 3
Enard (bay)	C 2
Eriboll, Loch (inlet)	D 2
Ericht, Loch (lake)	D 4
Eriskay (isl.), 219	A 3
Erisort, Loch (inlet)	B 2
Esk (riv.)	F 5
Etive, Loch (inlet)	C 4
Ewe, Loch (inlet)	C 3
Eye (pen.), 850	B 2
Fair Isle (isl.), 65	F 3
Fetlar (isl.), 88	G 2
Fife Ness (prom.)	F 4
Findhorn (riv.)	E 3
Flannan (isls.), 3	A 2
Formartine (dist.), 10,768	F 3
Forth (riv.)	B 1
Forth (firth)	F 4
Forth and Clyde (canal)	B 2
Foula (isl.), 33	F 2
Fyne, Loch (inlet)	C 5
Galloway (dist.), 54,972	D 5
Galloway, Mull of (prom.)	D 6
Gare Loch (inlet)	A 1
Garioch (dist.), 6,863	F 3
Garry, Loch (lake)	D 3
Gigha (isl.), 174	C 5
Girdle Ness (prom.)	G 3
Glass (riv.)	D 3
Glen More (dist.), 55,035	D 3
Goat Fell (mt.)	C 5
Gometra (isl.), 10	B 4
Grampian (mts.)	D 4
Great Cumbrae (isl.), 1,296	A 2
Gruinard (bay)	C 3
Hallandale (riv.)	E 2
Harris (sound)	A 3
Harris (dist.), 2,175	B 3
Hebrides (sea)	B 3
Hebrides, Inner (isls.), 14,881	B 4
Hebrides, Outer (isls.), 29,615	A 3
Helmsdale (riv.)	E 2
Herma Ness (prom.)	G 2
Holy (isl.), 10	C 5
Holy Loch (inlet)	A 1
Hoy (isl.), 419	E 2
Inchcape (Bell Rock) (isl.), 3	F 4
Inchkeith (isl.), 3	D 1
Indaal, Loch (inlet)	B 5
Inner (sound)	C 3
Inner Hebrides (isls.), 14,881	B 4
Iona (isl.), 145	B 4
Isla (riv.)	E 4
Islay (isl.), 3,816	B 5
Jura (isl.), 210	C 5
Jura (sound)	C 5
Katrine, Loch (lake)	D 4
Kerrera (isl.), 27	C 4
Kilbrannan (sound)	C 5
Kinnairds (head)	G 3
Kintyre (pen.), 10,077	C 5
Kintyre, Mull of (prom.)	C 5
Knapdale (dist.), 4,082	C 5
Kyle of Tongue (inlet)	D 2
Laggan (bay)	B 5
Lammermuir (hills)	F 5
Lennox (hills)	B 1
Leven (lake)	E 4
Leven, Loch (inlet)	D 4
Lewis (dist.), 20,047	B 2
Liddel Water (riv.)	F 5
Linnhe, Loch (inlet)	C 4
Lismore (isl.), 166	C 4
Little Minch (sound)	B 3
Lochaber (dist.), 13,813	D 4
Lochnagar (mt.)	E 4
Lochy, Loch (lake)	D 3
Lomond, Loch (lake)	D 4
Long, Loch (inlet)	D 4
Lorne (dist.), 12,162	C 4
Lorne (firth)	C 4
Loyal, Loch (lake)	D 2
Luce (bay)	D 6
Luing (isl.), 151	C 4
Lyon (riv.)	D 4
Machers, The (pen.), 6,192	D 6
Mainland (isl.), 12,747	E 1
Mainland (isl.), 12,944	G 2
Mar (dist.), 23,931	F 3
Maree, Loch (lake)	C 3
May, Isle of (isl.), 10	F 4
Merrick (mt.)	D 5
Minginish (dist.), 772	B 3
Moidart (dist.), 155	C 4
Monach (sound)	A 3
Monadhliath (mts.)	D 3
Moorfoot (hills)	E 5
Moray (firth)	E 3
Moriston (riv.)	D 3
Morven (dist.), 398	C 4
Morven (mt.)	E 2
Muck (isl.), 24	B 4
Muckle Flugga (isl.), 3	G 2
Mull (isl.), 2,024	C 4
Mull (head)	F 1
Mull (sound)	C 4
Nairn (riv.)	D 3
na Keal, Loch (inlet)	B 4
Naver (riv.)	D 2
Ness, Loch (lake)	D 3
Nevis, Loch (inlet)	C 3
Nith (riv.)	E 5
North (chan.)	C 5
North (sound)	F 1
North (sound)	G 4
North Esk (riv.)	F 4
North Minch (sound)	B 3
North Ronaldsay (isl.), 134	F 1
North Uist (isl.), 1,469	A 3
Oa, Mull of (prom.)	B 5
Ochil (hills)	E 4
Oich (riv.)	D 3
Orchy (riv.)	D 4
Orkney (isls.), 17,675	F 1
Oronsay (isl.), 2	B 4
Outer Hebrides (isls.), 29,615	A 3
Oykel (riv.)	D 3
Pabbay (isl.), 4	A 3
Papa Stour (isl.), 24	F 2
Papa Westray (isl.), 106	F 1
Paps of Jura (mt.)	C 5
Park (dist.), 210	B 2
Peel Fell (mt.)	F 5
Pentland (hills)	D 2
Pentland (firth)	E 2
Pladda (isl.), 2	C 5
Quoich, Loch (lake)	C 3
Raasay (isl.), 163	C 3
Rannoch (dist.), 1,177	D 4
Rannoch, Loch (lake)	D 4
Rhinns, The (pen.), 8,295	C 6
Roag, Loch (inlet)	B 2
Rona (isl.), 3	B 3
Ross of Mull (pen.), 585	B 4
Rousay (isl.), 181	E 1
Rudha Hunish (cape)	B 3
Rudh Re (cape)	C 3
Rum (isl.), 40	B 3
Ryan, Loch (inlet)	C 5
Saint Kilda (isl.), 65	A 2
Saint Magnus (bay)	F 2
Sanda (isl.), 9	C 5
Sanday (isl.), 11	B 3
Sanday (isl.), 592	F 1
Scalpay (isl.), 483	B 3
Scalpay (isl.), 5	C 3
Scapa Flow (chan.)	E 2
Scarp (isl.), 12	A 2
Scridain, Loch (inlet)	B 4
Scurdie Ness (prom.)	F 4
Seaforth, Loch (inlet)	B 3
Seil (isl.), 326	C 4
Sgurr a Choire Ghlais (mt.)	D 3
Sgurr Alasdair (mt.)	B 3
Sgurr Mor (mt.)	C 3
Sgurr na Lapaich (mt.)	C 3
Shapinsay (isl.), 346	F 1
Shetland (isls.), 18,494	G 2
Shiant (sound)	B 3
Shiel, Loch (lake)	C 4
Shin (falls)	D 3
Shin, Loch (lake)	D 2
Shona (isl.), 17	C 4
Sidlaw (hills)	E 4
Sinclair's (bay)	E 2
Skye, Isle of (isl.), 7,183	B 3
Sleat (pt.)	B 4
Sleat (dist.), 449	C 3
Small Isles (isls.), 171	B 4
Snizort, Loch (inlet)	B 3
Soay (isl.), 5	B 3
Solway (firth)	E 6
South Esk (riv.)	F 4
South Ronaldsay (isl.), 776	F 2
South Uist (isl.), 2,281	A 3
Spean (riv.)	D 4
Spey (riv.)	E 3
Start (pt.)	F 1
Stinchar (riv.)	D 5
Strathbogie (dist.), 7,959	F 3
Strathmore (valley)	E 4
Strathspey (dist.), 6,668	E 3
Strathy (pt.)	E 2
Stroma (isl.), 8	E 2
Stronsay (isl.), 436	F 1
Sumburgh (head)	G 2
Sunart, Loch (inlet)	C 4
Swona (isl.), 3	E 2
Taransay (isl.), 5	A 3
Tarbat Ness (prom.)	E 3
Tarbert, East Loch (inlet)	B 3
Tarbert, Loch (inlet)	B 5
Tarbert, West Loch (inlet)	A 3
Tay (riv.)	E 4
Tay (firth)	F 4
Tay, Loch (lake)	D 4
Teith (riv.)	D 4
Teviot (riv.)	F 5
Thurso (riv.)	E 2
Tiree (isl.), 875	B 4
Tolsta (head)	B 2
Tor Ness (prom.)	E 2
Torridon, Loch (inlet)	C 3
Trossachs, The (valley)	D 4
Trotternish (dist.), 1,948	B 3
Tweed (riv.)	F 5
Tyne (riv.)	F 5
Ulva (isl.), 23	B 4
Unst (isl.), 1,124	G 2
Vaternish (dist.), 162	B 3
Vatersay (isl.), 77	A 4
West Burra (isl.), 501	G 2
Westray (firth)	E 1
Westray (isl.), 735	E 1
Whalsay (isl.), 870	G 2
White Coomb (mt.)	E 5
Wigtown (bay)	D 6
Wrath (cape)	C 2
Wyre (isl.), 36	F 1
Yarrow (riv.)	E 5
Yell (isl.), 1,143	G 2
Ythan (riv.)	F 3

★Population of met. area
⊙Population of parish.

Agriculture, Industry and Resources

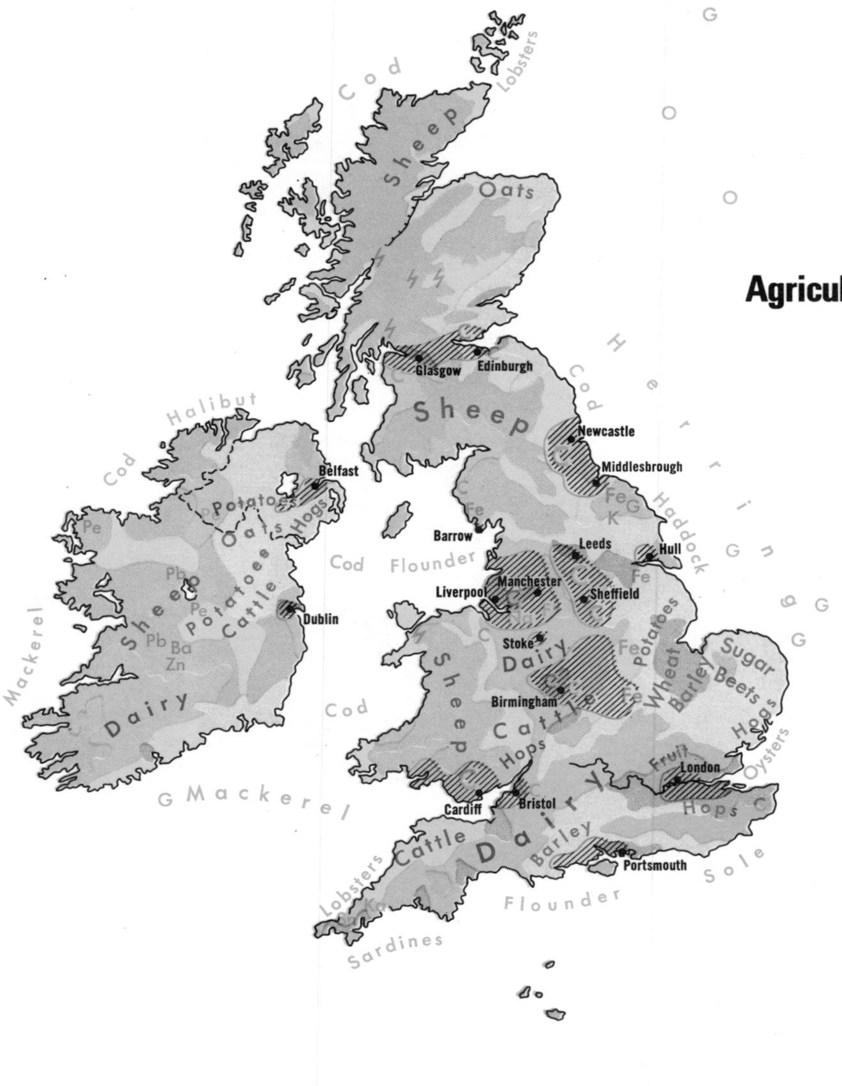

DOMINANT LAND USE

- Cereals (chiefly oats, barley)
- Truck Farming, Horticulture
- Dairy, Mixed Farming
- Livestock, Mixed Farming
- Pasture Livestock

MAJOR MINERAL OCCURRENCES

Ba	Barite	Na	Salt
C	Coal	O	Petroleum
F	Fluorspar	Pb	Lead
Fe	Iron Ore	Pe	Peat
G	Natural Gas	Sn	Tin
K	Potash	Zn	Zinc
Ka	Kaolin (china clay)		

Water Power

Major Industrial Areas

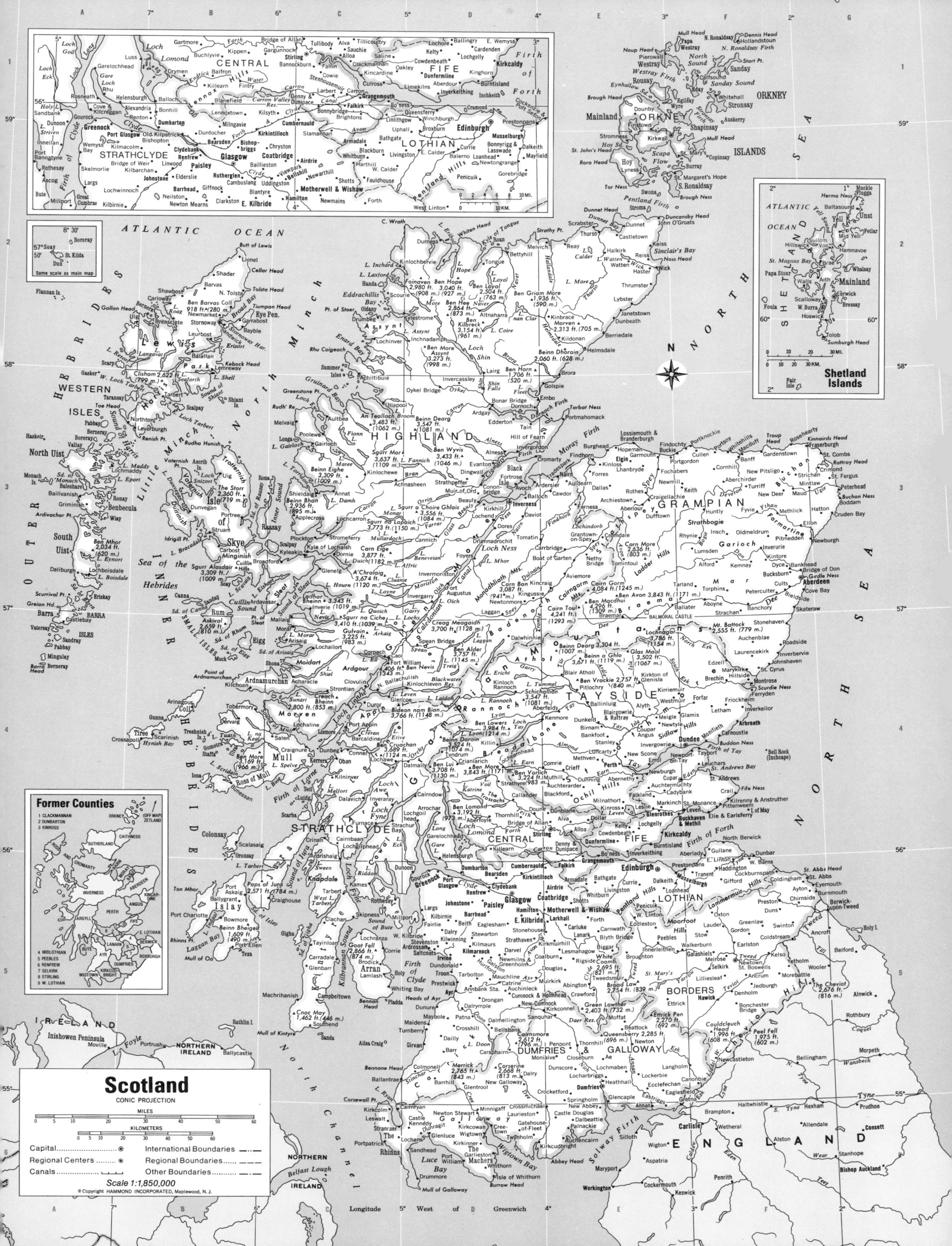

Scotland

CONIC PROJECTION

Capital.................⊛
Regional Centers.........◉
Canals...............

International Boundaries ———
Regional Boundaries ———
Other Boundaries ———

Scale 1:1,850,000

© Copyright HAMMOND INCORPORATED, Maplewood, N.J.

IRELAND

Carlow 34,237 H6
Cavan 52,618 G4
Clare 75,008 D6
Cork 352,883 D7
Donegal 108,344 K2
Dublin 852,219 J5
Galway 149,223 D5
Kerry 112,772 B7
Kildare 71,977 H5
Kilkenny 61,473 G6
Laois 45,259 G6
Leitrim 28,360 E3
Leix (Laois) 45,259 G6
Limerick 140,459 D7
Longford 28,250 F4
Louth 74,951 J4
Mayo 109,525 C4
Meath 71,729 H4
Monaghan 46,242 H3
Offaly 51,829 F5
Roscommon 53,519 E4
Sligo 50,275 D3
Tipperary 123,565 F6
Waterford 77,315 F7
Westmeath 53,570 G5
Wexford 86,351 H7
Wicklow 66,295 J5

CITIES and TOWNS

Abbeydorney, 188 B 7
Abbeyfeale, 1,337 C 7
Abbeylara, ‡290 F 4
Abbeyleix, 1,033 G 6
Achill Sound, ‡1,163 B 4
Aclare, ‡336 D 3
Adare, 545 D 6
Aghada-Farsid-Rostellan, 461 E 8
Aghadoe, ‡497 B 7
Aghagower, ‡693 C 4
Ahascragh, 221 E 5
Annagry, 201 E 1
Annascaul, 236 B 7
An Uaimh, 4,605 H 4
An Uaimh, *6,665 H 4
Ardagh, Limerick, 213 C 7
Ardagh, Longford, ‡974 F 4
Ardara, 683 E 2
Ardee, *3,183 H 4
Ardee, 3,096 H 4
Ardfert, 286 B 7
Ardfinnan, 510 F 7
Ardmore, 233 F 8
Ardrahan, ‡239 D 5
Arklow, 6,948 J 6
Arthurstown, 1,188 H 7
Arva, 370 F 4
Ashford, 341 J 5
Askeaton, 844 D 6
Athboy, 705 H 4
Athea, 328 C 7
Athenry, 1,240 D 5
Athleague, ‡955 E 4
Athlone, 9,825 F 5
Athlone, *11,611 F 5
Athy, 4,270 H 6
Athy, *4,654 H 6
Aughrim, 451 J 6
Avoca, ‡620 J 6
Bagenalstown (Muinebeag), 2,321 H 6
Baile Átha Cliath (Dublin) (cap.), 567,866 K 5
Bailieborough, 1,293 G 4
Balbriggan, 3,741 J 4
Balla, 293 C 4
Ballaghaderreen, 1,121 E 4
Ballina, Mayo, 6,063 C 3
Ballina, *6,369 C 3
Ballina, Tipperary, 336 E 6
Ballinagh, 459 G 4
Ballinakill, 300 G 6
Ballineen D 8
Ballinamore, 808 F 3
Ballinasloe, 5,969 E 5
Ballincollig-Carrigrohane, 2,110 D 8
Ballindine, 232 C 4
Ballingarry, Limerick, 422 D 7
Ballingarry, Tipperary, ‡574 F 6
Ballinlough, 242 D 4
Ballinrobe, 1,272 C 4
Ballintober, ‡867 E 4
Ballintra, 197 E 2
Ballisodare, 486 E 3
Ballivor, 287 H 4
Ballybay, 754 G 3
Ballybay, *1,159 G 3
Ballybofey-Stranorlar, 2,214 F 2
Ballybunion, 1,287 B 7
Ballycanew, ‡460 J 6
Ballycarney, ‡294 J 6
Ballycastle, ‡724 C 3
Ballyconnell, 421 F 3
Ballycotton, 389 F 8
Ballydehob, 253 C 8
Ballyduff, 406 B 7
Ballygar, 359 E 4
Ballygeary, 725 J 7
Ballyhaise, 274 G 3
Ballyhaunis, 1,093 D 4
Ballyheigue, 450 B 7
Ballyjamesduff, 673 G 4
Ballylanders, 266 E 7
Ballylongford, 504 B 6
Ballymahon, 707 F 4
Ballymakeery, 272 C 8
Ballymore, ‡447 F 5
Ballymore Eustace, 433 J 5
Ballymote, 952 D 3
Ballyporeen, ‡810 E 7
Ballyragget, 519 G 6
Ballyroan, ‡478 G 6
Ballyshannon, 2,325 E 3
Ballytore, ‡580 H 5
Baltimore, 200 C 9
Baltinglass, 909 H 6
Baltray, 236 J 4
Banagher, 1,052 F 5
Bandon, 2,257 D 8
Bandon, *4,071 D 8
Bannow, ‡798 H 7
Bansha, 184 E 7
Bantry, 2,579 C 8
Barna, ‡1,734 C 5
Belmullet, 744 B 3
Belturbet, 1,092 G 3
Bennettsbridge, 367 G 6
Birr, 3,319 F 5
Birr, *3,881 F 5
Blanchardstown, 3,279 H 5
Blarney, 1,128 D 8
Blessington, 637 J 5
Boherbue, 372 C 7
Borris, 430 H 6
Borris-in-Ossory, 276 F 6
Borrisokane, 769 E 6
Borrisoleigh, 471 E 6
Boyle, 1,727 E 4
Boyle, *1,939 E 4
Bray, 14,467 K 5
Bray, *15,841 K 5
Brí Chualann (Bray), 14,467 K 5
Broadford, 226 C 7
Brosna, 250 C 7
Bruff, 547 D 7
Bruree, 243 D 7
Bunbeg-Derrybeg, 878 E 1
Bunclody-Carrickduff, 929 H 6
Buncrana, 2,955 G 1
Buncrana, *3,334 G 1
Bundoran, 1,337 E 3
Burtonport, ‡1,288 E 2
Buttevant, 1,045 D 7
Cahir, 1,747 F 7
Cahirciveen, 1,547 A 8
Callan, 1,283 G 7
Camolin, 306 J 6
Campile, 231 H 7
Cappamore, 567 E 6
Cappawhite, 305 E 6
Cappoquin, 872 F 7
Carbury, ‡894 H 5
Carlingford, 559 J 3
Carlow, 9,588 H 6
Carlow, *10,399 H 6
Carndonagh, 1,146 G 1
Carnew, 570 H 6
Carrickmacross, 2,100 H 4
Carrickmacross, *2,475 H 4
Carrick-on-Shannon, 1,854 F 4
Carrick-on-Suir, 5,006 F 7
Carrigaholt, ‡493 B 6
Carrigaline, 951 E 8
Carrigallen, 230 F 4
Carrigart, ‡753 F 1
Carrigtwohill, 622 E 8
Carrowkeel, ‡326 G 1
Cashel, 2,692 F 7
Castlebar, 5,979 C 4
Castlebar, *6,476 C 4
Castlebellingham, 407 J 4
Castleblayney, 2,118 H 3
Castleblayney, *2,395 H 3
Castlecomer-Donaguile, 1,244 G 6
Castledermot, 583 H 6
Castlefin, 610 F 2
Castlegregory, 216 A 7
Castleisland, 1,929 B 7
Castlemartyr, 491 E 8
Castlepollard, 693 G 4
Castlerea, 1,752 D 4
Castletown, ‡504 F 6
Castletownbere, 812 B 8
Castletownroche, 399 D 7
Castletownshend, 170 C 9
Causeway, 215 B 7
Cavan, 3,273 G 3
Cavan, *4,312 G 3
Ceanannus Mór, 2,391 G 4
Ceanannus Mór, *2,653 G 4
Celbridge, 1,568 H 5
Charlestown-Bellahy, 677 D 4
Charleville (Rathluirc), 2,232 D 7
Clara, 2,156 F 5
Claregalway, ‡594 D 5
Claremorris, 1,718 C 4
Clashmore, ‡379 F 8
Clifden, 790 B 5
Cloghan, 404 F 5
Clogh-Chatsworth, 324 G 6
Clogheen, 530 F 7
Clogherhead, 649 J 4
Clonakilty, 2,430 D 8
Clonaslee, 285 F 5
Clondalkin, 7,009 J 5
Clonegal, 202 H 6
Clones, 2,164 G 3
Clonfert, ‡430 E 5
Clonmany, ‡936 G 1
Clonmel, 11,622 F 7
Clonmel, *12,291 F 7
Clonmellon, 328 H 4
Clonroche, 222 H 7
Clontuskert, 351 E 4
Cloone, ‡460 F 4
Cloughjordan, 480 E 6
Cloyne, 654 E 8
Coachford, 290 D 8
Cóbh, 6,076 E 8
Cóbh, *7,141 E 8
Coill Dubh, 920 H 5
Collon, 262 J 4
Collooney, 546 E 3
Cong, 233 C 4
Convoy, 654 F 2
Coolaney, ‡352 D 3
Coolgreany, ‡603 J 6
Cootehill, 1,415 G 3
Cootehill, *1,542 G 3
Cork, 128,645 E 8
Cork, *134,430 E 8
Corofin, 342 C 6
Courtmacsherry, 210 D 8
Courtown Harbour, 291 J 6
Creeslough, 269 F 1
Crookhaven, ‡400 B 9
Croom, 756 D 6
Crosshaven, 1,222 E 8
Crossmolina, 1,077 C 3
Crusheen, ‡405 D 6
Culdaff, ‡621 G 1
Daingean, 492 G 5
Delvin, 223 G 4
Dingle, 1,401 A 7
Doaghbeg, ‡701 F 1
Donabate, 426 J 5
Donegal, 1,725 F 2
Doneraile, 799 D 7
Dooagh-Keel, 649 A 4
Doon, 387 E 6
Douglas, ‡4,448 D 8
Drimoleague, 415 C 8
Drishane, ‡1,548 C 7
Drogheda, 19,762 J 4
Drogheda, *20,095 J 4
Droichead Nua, 5,053 H 5
Droichead Nua, *6,444 H 5
Dromahair, 177 E 3
Drumcar, ‡1,215 J 4
Drumconrath, ‡1,044 H 4
Drumkeerin, ‡467 E 3
Drumlish, 205 F 4
Drumshanbo, 576 E 3
Dublin (cap.), 567,866 K 5
Dublin, *679,748 K 5
Duleek, 658 J 4
Duncannon, 228 H 7
Dundalk, 21,672 H 3
Dundalk, *23,816 H 3
Dunfanaghy, 303 F 1
Dungarvan, 5,583 F 7
Dungloe, 940 E 2
Dunkineely, 288 E 2
Dún Laoghaire, 53,171 K 5
Dún Laoghaire, *98,379 K 5
Dunlavin, 423 H 5
Dunleer 855 J4
Dunmanway 1,392 C8
Dunmore 522 D4
Dunmore East 656 G7
Dunshaughlin⊙ 283 H5
Durrow, Laois 596 G6
Durrow, Offaly⊙ 441 F5
Easky 184 D3
Edenderry 2,953 G5
Edenderry* 3,116 G5
Elphin 489 E4
Emyvale 281 G3
Ennis 5,972 D6
Ennis* 10,840 D6
Enniscorthy 5,704 J7
Enniscorthy* 6,642 J7
Enniskerry 772 J5
Ennistymon 1,013 C6
Eyrecourt 314 E5
Fahan⊙ 1,023 G1
Falcarragh 506 E1
Feakle⊙ 398 D6
Fenit 360 B7
Ferbane 1,064 F5
Fermoy 3,237 E7
Fermoy* 4,033 E7
Ferns 712 J6
Fethard, Tipperary 1,064 F7
Fethard, Wexford⊙ 637 H7
Foxford 868 C4
Foynes 624 C6
Frankford (Kilcormac) 1,089 F5
Frenchpark⊙ 693 E4
Freshford 585 G6
Galbally 258 E7
Galway 27,726 C5
Galway* 29,375 C5
Geashill⊙ 751 G5
Glandore⊙ 695 C8
Glanmire-Riverstown 1,113 E8
Glanworth 335 E7
Glenamaddy 315 D4
Glenbeigh 266 B7
Glencolumbkille⊙ 787 D2
Glengarriff 244 C8
Glenties 734 E2
Glenville⊙ 264 D7
Glin 623 C6
Golden⊙ 640 F7
Gorey 2,946 J6
Gorey* 3,024 J6
Gormanston⊙ 1,384 J4
Gort 975 D5
Gowran 402 G6
Graiguenamanagh-Tinnahinch 1,303 H6
Granard 1,054 F4
Greencastle 322 H1
Greenore 882 J4
Greystones-Delgany 4,517 K5
Gurteen 165 D3
Hacketstown 574 H6
Headford 673 C5
Holycross⊙ 902 F6
Hospital 525 E7
Inchigeelagh⊙ 516 C8
Inishannon 190 D8
Inistioge 179 G7
Inniscrone 582 C3
Johnstown 303 G6
Kanturk 2,063 D7
Keel-Dooagh 649 A4
Kells⊙ 423 G6
Kells (Ceanannus Mór) 2,391 G4
Kenmare 903 B8
Kilbaha⊙ 471 B6
Kilbeggan 635 G5
Kilcar 273 D2
Kilcock 827 H5
Kilconnell⊙ 629 E5
Kilcoole 679 K5
Kilcormac 1,089 F5
Kilcullen 880 H5
Kildare 3,137 H5
Kildysart 239 C6
Kilfenora⊙ 441 C6
Kilfinane 561 D7
Kilgarvan 228 B8
Kilkee 1,287 B6
Kilkelly 225 D4
Kilkenny 9,838 G6
Kilkenny* 13,306 G6
Killala 368 C3
Killaloe 871 D6
Killarney 7,184 C7
Killarney* 7,541 C7
Killavullen 221 D7
Killenaule 592 F6
Killeshandra 432 F3
Killimor 221 E5
Killinaboy⊙ 297 C6
Killorglin 1,150 B7
Killucan-Rathwire 290 G4
Killybegs 1,094 E2
Kilmacrennan 274 F1
Kilmacthomas 396 G7
Kilmallock 1,170 D7
Kilmeadan⊙ 262 G7
Kilmihill 284 C6
Kilmoganny 181 G7
Kilmore Quay 273 H7
Kilmurry⊙ 387 C6
Kilnaleck 273 G4
Kilronan 243 B5
Kilrush 2,671 C6
Kilsheelan⊙ 665 F7
Kiltimagh 978 C4
Kilworth 360 E7
Kingscourt 1,016 H4
Kingstown (Dún Laoghaire) 53,171 K5
Kinlough 160 E3
Kinnegad 362 G5
Kinnitty⊙ 420 F5
Kinsale 1,622 D8
Kinsale* 1,989 D8
Kinvara 293 D5
Knightstown 236 A8
Knock⊙ 1,202 D4
Knocklong 248 D7
Knocknagashel 168 C7
Labasheeda⊙ 468 C6
Laghy⊙ 625 E2
Lahinch 455 C6
Lanesborough-Ballyleague 906 E4
Laracor⊙ 404 H4
Laytown-Bettystown-Mornington 1,882 J4
Leenane⊙ 271 B4
Leighlindridge 379 H6
Leitrim⊙ 544 F3
Leixlip 2,402 H5
Letterkenny 4,930 F2
Letterkenny* 5,207 F2
Lifford 1,121 F2
Limerick 57,161 D6
Limerick* 63,002 D6
Liscarroll 231 D7
Lisdoonvarna 459 C5
Lismore 884 F7
Lismore⊙ 1,041 F7
Listowel 3,021 C7
Littleton 322 F6
Longford 3,876 F4
Longford* 4,791 F4
Lorrha⊙ 685 E5
Loughrea 3,075 E5
Louisburgh 310 B4
Louth 208 J4
Lucan-Doddsborough 4,245 J5
Luimneach (Limerick) 57,161 D6
Lusk 553 J4
Macroom 2,256 C8
Malahide 3,834 J5
Malin⊙ 552 G1
Mallow 5,901 D7
Mallow* 6,506 D7
Manorhamilton 858 E3
Manulla⊙ 660 C4
Maryborough (Portlaoise) 3,902 G5
Maynooth 1,296 H5
Meathas Truim 546 G4
Midleton 3,075 E8
Midleton* 4,666 E8
Milford 763 F1
Millstreet 1,319 D7
Milltown 260 A7
Miltown-Malbay 677 C6
Minard⊙ 397 A7
Mitchelstown 2,783 E7
Moate 1,378 F5
Mohill 868 F4
Monaghan 5,256 G3
Monasterevan 1,619 H5
Moneygall 282 F6
Monivea⊙ 405 D5
Mooncoin 413 G7
Mount Bellew 275 D5
Mountcharles 445 E2
Mountmellick 2,595 G5
Mountmellick* 2,864 G5
Mountrath 1,098 F5
Moville 1,089 G1
Moycullen⊙ 498 C5
Moynalty⊙ 583 H4
Muff 240 G1
Muinebeag 2,321 H6
Mullagh 293 H4
Mullaghmore⊙ 629 D3
Mullinahone 262 F7
Mullinavat 343 G7
Mullingar 6,790 G4
Mullingar* 9,245 G4
Naas 5,078 H5
Navan (An Uaimh) 4,605 H4
Nenagh 5,085 E6
Nenagh* 5,174 E6
Newbliss⊙ 547 G3
Newbridge (Droichead Nua) 5,053 H5
Newcastle 2,549 D7
Newcastle* 2,680 D7
Newmarket 886 C7
Newmarket-on-Fergus 1,052 D6
New Pallas⊙ 1,271 E6
Newport, Mayo 420 C4
Newport, Tipperary 582 E6
New Ross 4,775 H7
New Ross* 5,153 H7
Newtown Forbes⊙ 495 F4
Newtownmountkennedy 882 J5
Newtownsandes 268 C6
O'Briensbridge-Montpelier 237 D6
Oldcastle 759 G4
Old Leighlin⊙ 309 G6
Oola 348 E6
Oranmore 440 D5
Oughterard 628 C5
Passage East 408 G7
Passage West 2,709 E8
Patrickswell 415 D6
Pettigo 332 F2
Piltown 456 G7
Portarlington 3,117 G5
Portlaoise 3,902 G5
Portlaoise* 6,470 G5
Portlaw 1,166 G7
Portmarnock 1,726 J5
Portumna 913 E5
Queenstown (Cóbh) 6,076 E8
Rahan⊙ 531 F5
Ramelton 807 F1
Raphoe 945 F2
Rathangan 868 G5
Rathcoole 1,740 J5
Rathcormac 191 E7
Rathdowney 892 F6
Rathdrum 1,141 J6
Rathgormuck⊙ 231 F7
Rathkeale 1,543 D7
Rathluirc 2,232 D7
Rathmore 437 J5
Rathmullen 486 F1
Rathnew-Merrymeeting 954 J6
Rathowen⊙ 294 F4
Rathvilly 230 H6
Ratoath 300 J5
Riverstown 236 E3
Rockcorry 233 H3
Rosapenna⊙ 822 F1
Roscommon 1,556 E4
Roscommon* 2,821 E4
Roscrea 3,855 F6
Rosscarbery 309 C8
Rosses Point 464 D3
Rosslare 588 J7
Rosslare Harbour (Ballygeary) 725 J7
Roundstone 204 A5
Roundwood 260 J5
Rush 2,633 J4
Saint Johnston 463 F2
Scarriff 619 E6
Schull 457 B8
Scotstown 264 H3
Shanagolden 231 C6
Shannon Airport 3,657 D6
Shannon Bridge 188 F5
Shercock 313 G4
Shillelagh 246 J6
Shinrone 365 F5
Shrule 288 C4
Sixmilebridge 567 D6
Skerries 3,044 J4
Skibbereen 2,104 C8
Slane 483 H4
Sligo 14,080 E3
Sligo* 14,456 E3
Sneem 285 B8
Spiddal⊙ 819 C5
Stepaside 748 J5
Stradbally, Laois 891 G5
Stradbally, Waterford 158 F7
Strokestown 563 E4
Swanlinbar 257 F3
Swinford 1,105 C4
Swords 4,133 J5
Taghmon 369 H7
Tallaght 6,174 J5
Tallow, 883 F 7
Tarbert, 485 C 6
Teltown, ‡739 H 4
Templemore, 2,174 F 6
Templetuohy, 197 F 6
Termonfeckin, 328 J 4
Thomastown, 1,270 G 7
Thurles, 6,840 F 6
Thurles, *7,087 F 6
Timoleague, 257 D 8
Tinahely, 450 H 6
Tipperary, 4,631 E 7
Tipperary, *4,717 E 7
Toomevara, 272 E 6
Tralee, 12,287 B 7
Tralee, *13,263 B 7
Tramore, 3,792 G 7
Trim, 1,700 H 4
Trim, *2,255 H 4
Tuam, 3,808 D 4
Tuam, *4,952 D 4
Tubbercurry, 959 D 3
Tulla, 415 D 6
Tullamore, 6,809 G 5
Tullamore, *7,474 G 5
Tullaroan, ‡301 G 6
Tullow, 1,838 H 6
Tullow, *1,945 H 6
Tynagh, ‡452 E 5
Tyrrellspass, 289 G 5
Urlingford, 652 F 6
Virginia, 583 G 4
Waterford, 31,968 G 7
Waterford, *33,676 G 7
Waterville, 547 A 8
Westport, 3,023 C 4
Wexford, 11,849 H 7
Wexford, *13,293 H 7
Whitegate, 370 E 8
Wicklow, 3,786 K 6
Wicklow, *3,915 K 6
Woodenbridge, ‡620 J 6
Woodford, 198 E 5
Youghal, 5,445 F 8
Youghal, *5,626 F 8

OTHER FEATURES

Achill (isl.), 3,129 A 4
Allen (lake) E 3
Allen, Bog of (marsh) H 5
Aran (isl.), 773 D 2
Aran (isls.), 1,499 B 5
Arklow (bank) K 6
Arrow (lake) E 3
Awbeg (riv.) D 7
Ballinskelligs (bay) A 8
Ballycotton (bay) F 8
Ballyheige (bay) B 7
Ballyhoura (hills) E 7
Ballyteige (bay) H 7
Bandon (riv.) D 8
Bann (riv.) J 6
Bantry (bay) B 8
Barrow (riv.) H 7
Baurtregaum (mt.) A 7
Bear (isl.), 288 B 8
Blacksod (bay) A 3
Blackstairs (mt.) H 6
Blackwater (riv.) D 7
Blackwater (riv.) H 4
Blasket (isls.) A 7
Bloody Foreland (prom.) E 1
Blue Stack (mts.) E 2
Boderg (lake) E 4
Boggeragh (mts.) D 7
Boyne (riv.) J 4
Brandon (head) A 7
Bride (riv.) E 7
Broad Haven (harb.) B 3
Brosna (riv.) F 5
Bull, The (isl.), 5 A 8
Caha (mts.) B 8
Carlingford (inlet) J 3
Carnsore (pt.) J 7
Carrantuohill (mt.) B 7
Clare (riv.) D 5
Clare (isls.), 168 A 4
Clear (cape) B 9
Clear (isl.), 192 C 9
Clew (bay) B 4
Comeragh (mts.) F 7
Conn (lake) C 3
Connacht (prov.), 390,902 C 4
Connemara (dist.), 7,599 B 5
Cork (harb.) E 8
Corrib (lake) C 5
Courtmacsherry (bay) D 8
Curragh, The H 5
Dee (riv.) H 4
Deel (riv.) D 7
Deele (riv.) F 2
Derg (lake) E 6
Derravaragh (lake) G 4
Derryveagh (mts.) E 2
Dingle (bay) A 7
Donegal (bay) D 3
Drum (hills) F 7
Dublin (bay) J 5
Dundalk (bay) J 4
Dunmanus (bay) B 8
Dursey (isl.), 38 A 8
Ennell (lake) G 5
Erne (riv.) E 3
Errigal (mt.) E 1
Erris (head) A 3
Fanad (head) F 1
Fastnet Rock (isl.), 3 B 9
Feale (riv.) C 7
Fergus (riv.) D 6
Finn (riv.) F 2
Finn (riv.) G 3
Flesk (riv.) C 7
Foyle (inlet) G 1
Foyle (riv.) G 2
Galley (head) D 9
Galtee (mts.) E 7
Galtymore (mt.) E 7
Galway (bay) C 5
Gara (lake) D 4
Garadice (lake) F 3
Gill (lake) E 3
Glyde (riv.) H 4
Golden Vale (plain) E 7
Gorumna (isl.), 1,108 B 5
Gowna (lake) G 4
Grand (canal) G 5
Greenore (pt.) J 7
Gweebarra (bay) D 2
Hags (head) B 6
Helvick (head) G 7
Hook (head) H 7
Horn (head) E 1
Iar Connacht (dist.), 10,774 C 5
Inishbofin (isl.), 236 A 4
Inishbofin (isl.), 103 E 1
Inisheer (isl.), 313 B 5
Inishmaan (isl.), 319 C 5
Inishmore (isl.), 864 B 5
Inishowen (head) H 1
Inishowen (pen.), 24,109 G 1
Inishtrahull (isl.), 3 G 1
Inishturk (isls.), 83 A 4
Inny (riv.) A 8
Inny (riv.) F 4
Inver (bay) E 2
Ireland's Eye (isl.) K 5
Irish (sea) K 4
Joyce's Country (dist.), 2,021 B 4
Kenmare (riv.) A 8
Kerry (head) A 7
Key (lake) E 3
Kilkieran (bay) B 5
Killala (bay) C 3
Killary (harb.) A 4
Kinsale (harb.) E 8
Kippure (mt.) J 5
Knockboy (mt.) B 8
Knockmealdown (mts.) F 7
Lady's Island Lake (inlet) J 7
Lambay (isl.), 24 K 4
Laune (riv.) B 7
Leane (lake) B 7
Leane (lake) G 4
Lee (riv.) D 8
Leinster (mt.) H 6
Leinster (prov.), 1,498,140 G 5
Lettermullan (isl.), 221 B 5
Liffey (riv.) H 5
Liscannor (bay) B 6
Long Island (bay) B 9
Loop (head) A 6
Lugnaquillia (mt.) J 5
Macgillicuddy's Reeks (mts.) B 7
Macnean (lake) F 3
Maigue (riv.) D 6
Maine (riv.) C 7
Malin (head) F 1
Mask (lake) C 4
Maumturk (mts.) B 5
Melvin (lake) E 3
Mizen (head) B 9
Moher (cliffs) B 6
Monavullagh (mts.) F 7
Moy (riv.) C 3
Mulkear (riv.) E 6
Mullaghareirk (mts.) C 7
Mulroy (bay) F 1
Munster (prov.), 882,002 D 7
Mweelrea (mt.) B 4
Mweenish (isl.), 198 B 5
Nagles (mts.) E 7
Nenagh (riv.) E 6
Nephin (mt.) C 3
Nore (riv.) G 7
North (sound) B 5
Omey (isl.), 34 A 5
Oughter (lake) G 3
Ovoca (riv.) J 6
Owenmore (riv.) D 3
Owey (isl.), 51 D 1
Paps, The (mt.) C 7
Partry (mts.) C 4
Pollaphuca (res.) J 5
Punchestown H 5
Rathlin O'Birne (isl.), 3 C 2
Ree (lake) F 5
Roaringwater (bay) B 9
Rosses (bay) D 1
Rosskeeragh (pt.) D 3
Royal (canal) G 4
Saint Finan's (bay) A 8
Saint George's (chan.) K 7
Saint John's (pt.) D 2
Saltee (isls.) H 7
Seven (heads) D 8
Seven Hogs, The (isls.) A 7
Shannon (riv.) E 6
Sheeffry (hills) B 4
Sheelin (lake) G 4
Sheep Haven (harb.) F 1
Sheeps (head) B 8
Sherkin (isl.), 82 C 9
Silvermine (mts.) E 6
Slaney (riv.) H 7
Slieve Aughty (mts.) D 5
Slieve Bloom (mts.) F 5
Slieve Gamph (mts.) D 3
Slievenaman (mt.) F 7
Sligo (bay) D 3
Slyne (head) A 5
South (sound) C 5
Stacks (mts.) B 7
Suck (riv.) E 5
Suir (riv.) G 7
Swilly (inlet) F 1
Tara (hill) H 4
Tory (isl.), 273 E 1
Tory (sound) E 1
Tralee (bay) B 7
Trawbreaga (bay) F 1
Ulster (part) (prov.), 207,204 G 2
Valencia (Valentia) (isl.), 770 A 8
Valentia (isl.), 770 A 8
Waterford (harb.) G 7
Wexford (bay) J 7
Wicklow (head) K 6
Wicklow (mts.) J 6
Youghal (bay) F 8

NORTHERN IRELAND

DISTRICTS

Antrim, 37,600 J 2
Ards, 52,100 K 2
Armagh, 47,500 H 3
Ballymena, 52,200 J 2
Ballymoney, 22,700 J 1
Bainbridge, 28,800 J 3
Belfast, 368,200 J 2
Carrickfergus, 27,500 K 2
Castlereagh, 63,600 K 2
Coleraine, 44,900 H 1
Cookstown, 27,500 H 2
Craigavon, 71,200 J 3
Down, 48,800 K 3
Dungannon, 43,000 H 3
Fermanagh, 50,900 F 3
Larne, 29,000 K 2
Limavady, 25,000 H 1
Lisburn, 80,800 J 2
Londonderry, 86,600 G 2
Magherafelt, 32,200 H 2
Mourne (Newry and Mourne), 75,300 J 3
Moyle, 13,400 J 1
Newtownabbey, 71,500 J 2
North Down, 59,600 K 2
Omagh, 41,800 G 2
Strabane, 35,500 G 2

CITIES and TOWNS

Ahoghill, ‡1,929 J 2
Annalong, 1,001 K 3
Antrim, 8,351 J 2
Ardglass, 1,052 K 3
Armagh, 13,606 H 3
Armoy, ‡1,051 J 1
Augher, ‡1,986 G 3
Aughnacloy, ‡1,885 H 3
Ballycastle, 2,899 J 1
Ballyclare, 5,155 J 2
Ballygawley, ‡2,165 G 3
Ballykelly, 1,116 G 1
Ballymena, 23,386 J 2
Ballymoney, 5,697 J 1
Ballynahinch, 3,485 J 3
Banbridge, 7,968 J 3
Bangor, 35,260 K 2
Belfast (cap.), 353,700 J 2
Belfast, *551,940 J 2
Bellaghy, ‡2,265 H 2
Belleek, ‡2,487 E 3
Beragh, ‡2,137 G 2
Bessbrook, 2,619 J 3
Brookeborough, ‡2,534 G 3
Broughshane, 1,288 J 2
Bushmills, 1,288 J 1
Caledon, ‡1,828 H 3
Carnlough, 1,416 J 2
Carrickfergus, 16,603 K 2
Carrowdore, 2,548 K 2
Castledawson, 1,162 H 2
Castlederg, 1,766 F 2
Castlewellan, 1,488 K 3
Claudy, ‡2,507 G 2
Clogher, ‡1,888 G 3
Coalisland, 3,614 H 2
Coleraine, 16,354 H 1
Comber, 5,575 K 2
Cookstown, 6,965 H 2
Craigavon, 12,740 J 3
Crossgar, 1,098 K 3
Crossmaglen, 1,085 H 3
Crumlin, 1,450 J 2
Cullybackey, 1,649 J 2
Derrygonnelly, ‡2,539 F 3
Dervock, ‡1,191 J 1
Donaghadee, 4,008 K 2
Downpatrick, 7,918 K 3
Draperstown, ‡2,247 H 2
Dromore, Bainbridge, 2,848 J 3
Dromore, Omagh, ‡2,224 G 3
Drumquin, ‡1,982 F 2
Dundrum, ‡2,245 K 3
Dungannon, 8,190 H 3
Dungiven, 1,536 H 2
Dunnamanagh, ‡2,242 G 2
Ederny and Kesh, ‡2,497 F 2
Enniskillen, 9,679 F 3
Feeny, ‡1,459 H 2
Fintona, 1,190 G 3
Fivemiletown, ‡1,649 G 3
Garvagh, ‡2,363 H 2
Gilford, 1,592 J 3
Glenarm, ‡1,728 J 2
Glenavy, ‡2,360 J 2
Glynn, ‡1,872 K 2
Gortin, ‡2,033 G 2
Greyabbey, ‡2,646 K 2
Hillsborough, 1,021 J 3
Holywood, 9,892 K 2
Irvinestown, 1,457 F 3
Keady, 2,145 H 3
Kells, ‡2,560 J 2
Kesh, ‡2,497 F 3
Kilkeel, 4,090 K 3
Killough, ‡3,295 K 3
Killyleagh, 2,359 K 3
Kilrea, 1,196 H 2
Kircubbin, 1,075 K 3
Larne, 18,482 K 2
Limavady, 6,004 H 1
Lisburn, 31,836 J 2
Lisnaskea, 1,443 G 3
Londonderry, 51,200 G 2
Loughbrickland, ‡2,056 J 3
Maghera, 2,085 H 2
Magherafelt, 4,704 H 2
Markethill, ‡2,352 H 3
Millisle, 1,172 K 2
Moneymore, 1,178 H 2
Moy, ‡2,349 H 3
Moygashel, 1,086 H 3
Newcastle, 4,647 J 3
Newry, 20,279 J 3
Newtownabbey, 58,114 K 2
Newtownards, 15,484 K 2
Newtownbutler, ‡2,663 G 3
Newtownhamilton, ‡1,936 H 3
Newtownstewart, 1,433 G 2
Omagh, 14,594 G 2
Pomeroy, ‡1,786 H 2
Portaferry, 1,730 K 3
Portavogie, 1,310 K 3
Portglenone, ‡2,061 H 2
Portrush, 5,376 H 1
Portstewart, 5,085 H 1
Randalstown, 2,799 J 2
Rathfriland, 1,886 J 3
Rostrevor, 1,617 J 3
Saintfield, ‡2,198 K 3
Sion Mills, 1,588 G 2
Sixmilecross, ‡1,980 G 2
Stewartstown, ‡1,759 H 2
Strabane, 9,413 G 2
Strangford, ‡1,987 K 3
Tandragee, 1,725 J 3
Tempo, ‡2,282 G 3
Trillick, ‡2,167 G 3
Warrenpoint, 4,291 J 3
Whitehead, 2,642 K 2

OTHER FEATURES

Bann (riv.) H 2
Belfast (inlet) K 2
Blackwater (riv.) H 3
Bush (riv.) H 1
Derg (riv.) F 2
Divis (mt.) J 2
Dundrum (bay) K 3
Erne (lake) F 3
Foyle (inlet) G 1
Foyle (riv.) G 2
Giant's Causeway H 1
Lagan (riv.) K 2
Larne (inlet) K 2
Magee, Island (pen.), 1,581 K 2
Magilligan (pt.) H 1
Main (riv.) J 2
Mourne (mts.) J 3
Mourne (riv.) G 2
Neagh (lake) J 2
North (chan.) K 1
Rathlin (isl.), 109 J 1
Red (bay) K 1
Roe (riv.) H 1
Saint John's (pt.) K 3
Slieve Donard (mt.) K 3
Sperrin (mts.) G 2
Strangford (inlet) K 3
Torr (head) K 1
Ulster (part) (prov.), 1,537,200 G 2
Upper Lough Erne (lake) F 3

*City and suburbs.
‡Population of district.

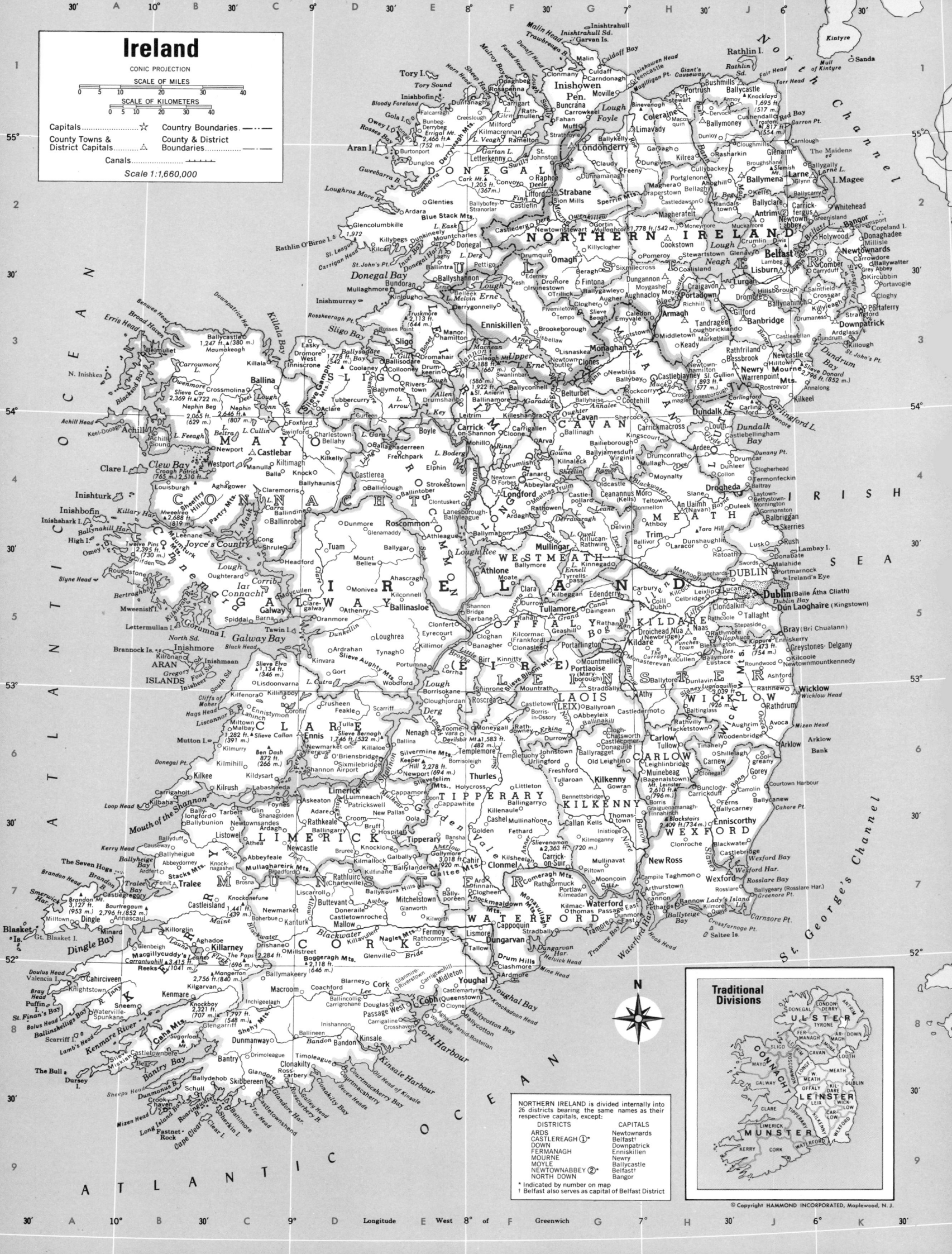

Ireland

CONIC PROJECTION

SCALE OF MILES

SCALE OF KILOMETERS

Capitals	☆	Country Boundaries	— ·· —
Capitals	Country Boundaries
County Towns & District Capitals	△	County & District Boundaries
Canals	

Scale 1:1,660,000

Traditional Divisions

ULSTER

CONNACHT

LEINSTER

MUNSTER

NORTHERN IRELAND is divided internally into 26 districts bearing the same names as their respective capitals, except B.

DISTRICTS	CAPITALS
ARDS	Newtownards
CASTLEREAGH ① *	Belfast †
DOWN	Downpatrick
FERMANAGH	Enniskillen
MOURNE	Newry
MOYLE	Ballycastle
NEWTOWNABBEY ② *	Belfast †
NORTH DOWN	Bangor

* Indicated by number on map
† Belfast also serves as capital of Belfast District

© Copyright HAMMOND INCORPORATED, Maplewood, N.J.

Norway, Sweden, Finland and Denmark

CONIC PROJECTION

SCALE OF MILES
0   50   100   150

SCALE OF KILOMETERS
0   50   100   150   200

Capitals of Countries .... ★
Administrative Centers .... △
International Boundaries .... —·—·—
Internal Boundaries .... —··—··—
Canals ....

SUBDIVISIONS
Indicated by Numbers

Counties in NORWAY
1 Akershus .... G 6
2 Vestfold .... G 7
3 Østfold .... G 7
4 Oslo .... G 7

Oslo is the administrative
center for Akershus and
Oslo County.

Counties in SWEDEN
5 Göteborg och
  Bohus .... G 7
6 Västmanland ...... H 7
7 Södermanland .... K 7
8 Östergötland ...... J 7
9 Malmöhus .... H 8
10 Kristianstad ...... J 8

© Copyright HAMMOND INCORPORATED, Maplewood, N.J.

Svalbard

NORWEGIAN SEA

Oslo

AREA 125,053 sq. mi.
(323,887 sq. km.)
POPULATION 4,092,000
CAPITAL Oslo
LARGEST CITY Oslo
HIGHEST POINT Glittertinden 8,110 ft. (2,472 m.)
MONETARY UNIT krone
MAJOR LANGUAGE Norwegian
MAJOR RELIGION Protestantism

AREA 173,665 sq. mi.
(449,792 sq. km.)
POPULATION 8,320,000
CAPITAL Stockholm
LARGEST CITY Stockholm
HIGHEST POINT Kebnekaise 6,946 ft. (2,117 m.)
MONETARY UNIT krona
MAJOR LANGUAGE Swedish
MAJOR RELIGION Protestantism

AREA 130,128 sq. mi.
(337,032 sq. km.)
POPULATION 4,788,000
CAPITAL Helsinki
LARGEST CITY Helsinki
HIGHEST POINT Haltiatunturi 4,343 ft. (1,324 m.)
MONETARY UNIT markka
MAJOR LANGUAGES Finnish, Swedish
MAJOR RELIGION Protestantism

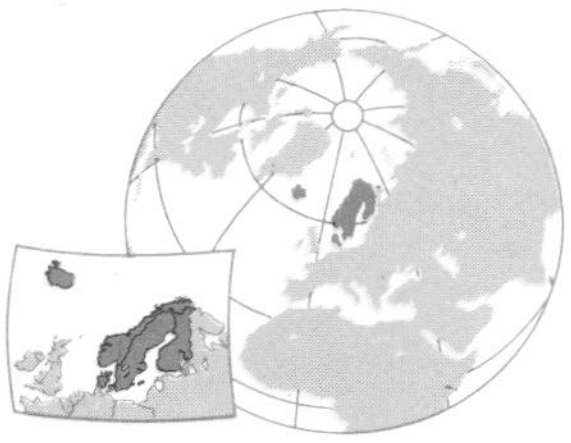

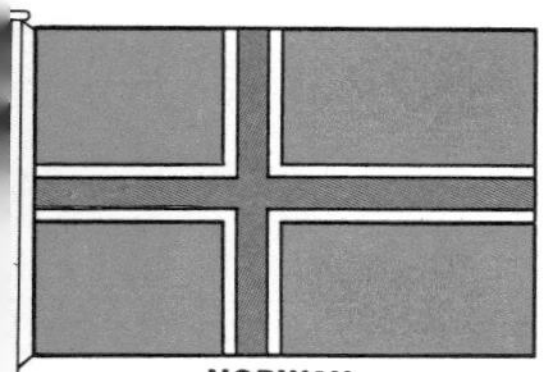

NORWAY

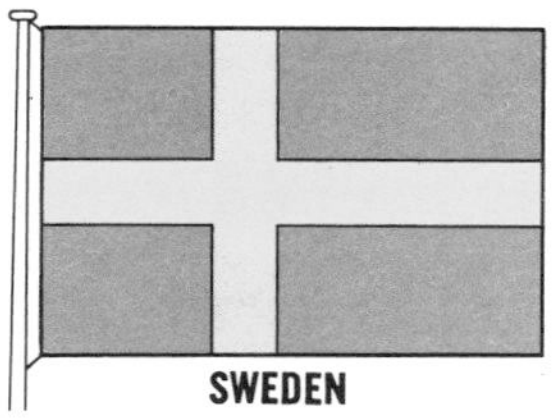

SWEDEN

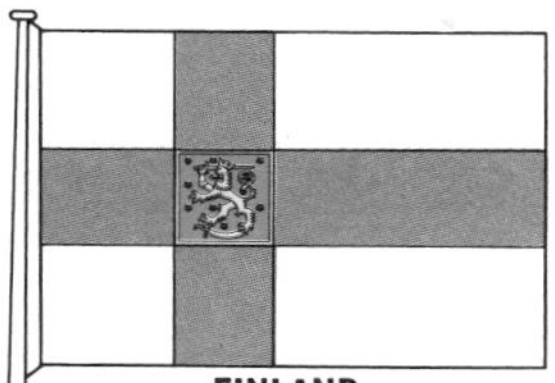

FINLAND

FINLAND

PROVINCES

venanmaa 22,380 ... L6
and (Ahvenanmaa) 22,380 ... L6
me 662,500 ... O6
ski-Suomi 241,770 ... O5
opio 252,023 ... P5
mi 346,478 ... Q6
ppi 196,792 ... P3
kkeli 211,453 ... P6
lu 406,309 ... P4
hjois-Karjala 179,065 ... Q5
rku ja Pori 697,988 ... N6
simaa 1,085,625 ... O6
asa 425,283 ... N5

CITIES and TOWNS

nekoski 10,725 ... O5
o (Turku) 164,857 ... N6
vus 10,285 ... N5
rgaå 18,740 ... O6
enäs 7,391 ... N6
Espoo 117,090 ... O6
Forssa 18,442 ... N6
Haapajärvi 7,791 ... O5
Hämeenlinna 40,761 ... O6
Hamina 11,055 ... P6
Hangö 10,374 ... N7
Hanko (Hangö) 10,374 ... N7
Harjavalta 8,445 ... M6
Heinola 15,350 ... P6
Helsinki (cap.) 502,961 ... O6
Helsinki* 794,746 ... O6
Huutokoski† 6,458 ... P5
Hyvinkää 35,865 ... O6
Iisalmi 21,159 ... P5
Ikaalinen 8,364 ... N6
Imatra 35,590 ... Q6
Ivalo 2,661 ... P2
Jakobstad 20,397 ... N5
Jämsä 12,526 ... O6
Järvenpää 16,259 ... O6
Joensuu 41,429 ... R5
Jyväskylä 61,209 ... O5
Jyväskylä* 84,185 ... O5
Kajaani 20,583 ... P4
Kalajoki 3,624 ... N4
Kankaanpää 12,564 ... M6
Karhula 21,834 ... P6
Karis 8,152 ... N6
Karjaa (Karis) 8,152 ... N6
Karkkila 8,678 ... N6
Kauniainien 6,219 ... O6
Kauttua 3,297 ... M6
Kelloselkä† 8,200 ... Q3
Kemi 27,893 ... O4
Kemijärvi 12,951 ... P3
Kerava 19,966 ... O6
Kokemäki 10,188 ... N6
Kokkola 22,096 ... N5
Kotka 34,026 ... P6
Kotka* 60,235 ... P6
Kouvola 29,383 ... P6
Kouvola* 59,507 ... P6
Kristiinankaupunki (Kristinestad) 9,331 ... N5
Kristinestad 9,331 ... N5
Kuhmo 4,150 ... Q4
Kuopio 71,684 ... Q5
Kurikka 11,177 ... M5
Kuusamo 4,449 ... Q4
Kuusankoski 22,342 ... P6
Lahti 94,864 ... O6
Lahti* 112,129 ... O6
Lappeenranta 52,682 ... P6
Lapua 15,189 ... N5
Lieksa 20,274 ... R5
Loimaa 6,575 ... N6
Lovisa 8,674 ... P6
Maarianhamina (Mariehamn) 9,574 ... M7
Mänttä 7,910 ... O6
Mariehamn 9,574 ... M7
Mikkeli 27,112 ... P6
Naantali 7,814 ... M6
Nokia 22,308 ... N6
Nurmes 11,721 ... Q5
Nykarleby 7,408 ... N5
Oulainen 7,322 ... O4
Oulu 93,707 ... O4
Oulu* 103,044 ... O4
Outokumpu 10,736 ... Q5
Parainen 10,170 ... M6
Parkano 8,518 ... N6
Pieksämäki 12,923 ... P5
Pietarsaari (Jakobstad) 20,397 ... N5
Pori 80,343 ... M6
Pori* 86,635 ... M6
Posio† 6,205 ... Q3
Pudasjärvi† 12,594 ... P4
Raahe 15,379 ... O4
Raisio 14,271 ... M6
Rauma 29,081 ... M6
Riihimäki 24,106 ... O6
Rovaniemi 28,411 ... O3
Saarijärvi 2,714 ... O5
Salo 19,176 ... N6
Savonlinna 28,336 ... Q6
Seinäjoki 22,123 ... N5
Sodankylä 3,304 ... P3
Sotkamo 2,316 ... Q4
Suolahti 5,936 ... O5
Suonenjoki 9,286 ... P5
Tammisaari (Ekenäs) 7,391 ... N6
Tampere 168,118 ... N6
Tampere* 220,920 ... N6
Toijala 8,080 ... N6
Tornio 19,971 ... O4
Turku 164,857 ... N6
Turku* 217,423 ... N6
Turtola† 5,852 ... O3
Ulvila† 8,040 ... N6
Uusikaarlepyy (Nykarleby) 7,408 ... N5
Uusikaupunki 11,915 ... M6
Vaasa 54,402 ... M5
Vaasa* 58,224 ... M5
Valkeakoski 22,588 ... N6
Vammala 16,363 ... N6
Varkaus 24,450 ... Q5
Vasa (Vaasa) 54,402 ... M5
Vuotso† 10,186 ... P2
Ylivieska 10,827 ... O4

OTHER FEATURES

Åland (isls.) ... L6
Baltic (sea) ... K9
Bothnia (gulf) ... M5
Finland (gulf) ... P7
Hailuoto (isl.) ... O4
Haltiatunturi (mt.) ... M2
Hangöudd (prom.) ... N7
Haukivesi (lake) ... Q5
Iijoki (riv.) ... O4
Inari (lake) ... P2
Ivalojoki (riv.) ... P2
Juojärvi (lake) ... Q5
Kalajoki (riv.) ... O4
Kallavesi (lake) ... P5
Karlö (Hailuoto) (isl.) ... O4
Keitele (lake) ... O5
Kemijärvi (lake) ... Q3
Kemijoki (riv.) ... O3
Kiantajärvi (lake) ... Q4
Kilpisjärvi (lake) ... M2
Kitinen (riv.) ... P3
Kivijärvi (lake) ... O5
Koitere (lake) ... R5
Kuusamojärvi (lake) ... Q4
Längelmävesi (lake) ... O6
Lapland (reg.) ... O2
Lappajärvi (lake) ... O5
Lapuanjoki (riv.) ... N5
Lestijärvi (lake) ... O5
Lokka (res.) ... Q3
Muojärvi (lake) ... R4
Muonio (riv.) ... M2
Näsijärvi (lake) ... O6
Onkivesi (lake) ... P5
Orihvesi (lake) ... Q5
Oulujärvi (lake) ... P4
Oulujoki (riv.) ... O4
Ounasjoki (riv.) ... O3
Päijänne (lake) ... O6
Pielinen (lake) ... Q5
Puruvesi (lake) ... Q6
Puulavesi (lake) ... P5
Pyhäjärvi (lake) ... O5
Pyhäjärvi (lake) ... M6
Saimaa (lake) ... Q6
Siikajoki (riv.) ... O4
Simojärvi (lake) ... P3
Simojoki (riv.) ... O4
Tana (riv.) ... P2
Tornio (riv.) ... O3
Vallgrund (isl.) ... M5
Ylikitka (lake) ... Q3

NORWAY

COUNTIES

Akershus 355,196 ... G6
Aust-Agder 86,216 ... E7
Buskerud 209,684 ... F6
Finnmark 79,373 ... O2
Hedmark 183,465 ... G6
Hordaland 386,492 ... E6
Møre og Romsdal 231,944 ... E5
Nordland 243,233 ... J3
Nord-Trøndelag 122,886 ... H4
Oppland 178,259 ... F6
Oslo (city) 462,732 ... D3
Østfold 228,546 ... G7
Rogaland 287,653 ... E7
Sogn og Fjordane 103,135 ... E6
Sør-Trøndelag 241,361 ... G5
Telemark 158,853 ... F7
Troms 144,111 ... L2
Vest-Agder 131,659 ... E7
Vestfold 182,433 ... G7

CITIES and TOWNS

Ålesund 40,868 ... D5
Ålgård 2,322 ... D7
Alta 5,582 ... N2
Åndalsnes 2,574 ... F5
Årdalstangen 2,360 ... F6
Arendal 11,701 ... F7
Arendal* 21,228 ... F7
Årnes 2,267 ... G6
Askim 8,413 ... E4
Bamble† 7,031 ... F7
Barentsburg ... C2
Bergen 213,434 ... D6
Bodø 31,077 ... J3
Borge† 3,294 ... H2
Brønnøysund 3,130 ... G4
Dombås 1,114 ... F5
Drammen 50,777 ... C4
Drammen* 56,521 ... C4
Drøbak 4,538 ... D4
Eidsvoll 2,906 ... G6
Eigersund 11,379 ... D7
Elverum 7,391 ... G6
Farsund 8,908 ... E7
Flekkefjord 8,750 ... E7
Flora 8,822 ... D6
Fredrikstad 29,024 ... D4
Fredrikstad* 51,141 ... D4
Gjøvik 25,963 ... G6
Grimstad 13,091 ... F7
Halden 27,087 ... G7
Hamar 16,418 ... G6
Hamar* 25,138 ... G6
Hammerfest 7,610 ... N1
Hammerfest* 8,005 ... N1
Harstad 21,125 ... K2
Haugesund 27,386 ... D7
Haugesund* 29,277 ... D7
Hermansverk 706 ... E6
Holmestrand 8,246 ... C4
Holmsbu 273 ... D4
Honningsvag 3,780 ... O1
Horten 13,746 ... D4
Horten* 17,246 ... D4
Kirkenes 4,466 ... Q2
Kongsberg 19,854 ... F7
Kongsvinger 16,146 ... H6
Kopervik 4,221 ... D7
Kornsjø† 6,079 ... G7
Kragerø 5,249 ... F7
Kristiansand 59,488 ... F8
Kristiansund 18,847 ... E5
Kvinnherad† 2,898 ... E6
Larvik 9,097 ... C4
Larvik* 19,202 ... C4
Lenvik† 11,098 ... L2
Levanger 5,066 ... G5
Lillehammer 21,248 ... F6
Lillesand 3,028 ... F7
Lillestrøm† 11,550 ... E3
Longyearbyen ... D2
Lysaker† 81,612 ... D3
Mandal 11,579 ... E7
Meråker† 2,907 ... G5
Mo 21,033 ... J3
Molde 20,334 ... E5
Mosjøen 9,341 ... H4
Moss 25,786 ... D4
Moss* 27,430 ... D4
Mysen 3,760 ... G7
Namsos 11,452 ... G4
Narvik 19,582 ... K2
Nesttun† 11,519 ... D6
Nittedal† 8,889 ... D3
Notodden 12,970 ... F7
Nøtterøy 11,944 ... D4
Ny-Ålesund ... C2
Odda 7,401 ... E6
Oppdal 2,173 ... F5
Orkanger 3,685 ... F5
Oslo (cap.) 462,732 ... D3
Oslo* 645,413 ... D3
Porsgrunn 31,709 ... G7
Rakkestad 2,392 ... G7
Ringerike 30,156 ... C3
Risør 6,560 ... F7
Rjukan 5,334 ... F7
Røros 3,041 ... G5
Sandefjord 33,350 ... C4
Sandnes 33,934 ... D7
Sandvika† 34,337 ... C3
Sarpsborg 12,889 ... D4
Sarpsborg* 36,449 ... D4
Selje† 3,386 ... D5
Ski 9,081 ... D4
Skien 47,105 ... F7
Stavanger 86,639 ... D7
Stavern 2,604 ... D4
Steinkjer 20,553 ... G4
Stor-Elvdal† 2,993 ... G6
Sunndalsøra 5,114 ... F5
Sveagruva ... D2
Svolvaer 3,942 ... J2
Tønsberg 9,964 ... D4
Tønsberg* 36,374 ... D4
Tromsø 43,830 ... L2
Trondheim 134,910 ... F5
Ullensvang† 2,326 ... E6
Vadsø 6,019 ... Q1
Vardø 3,875 ... R1
Vik 1,019 ... E6
Volda 3,511 ... E5
Voss 5,944 ... E6

OTHER FEATURES

Alsten (isl.) ... H4
Andøya (isl.) ... J2
Barduelv (riv.) ... L2
Bellsund ... C2
Bjørnafjorden (fjord) ... D6
Bjornøya (isl.) ... D3
Boknafjord (fjord) ... D7
Bremanger (isl.) ... D6
Dønna (isl.) ... H3
Dovrefjell (hills) ... F5
Edgeøya (isl.) ... E2
Femundsjø (lake) ... G5
Folda (fjord) ... G4
Folda (fjord) ... J3
Frohavet (bay) ... F5
Frøya (isl.) ... F5
Glittertinden (mt.) ... F6
Hardangervidda (plat.) ... E6
Hardangerfjord (fjord) ... D7
Hinlopenstreten (str.) ... C1
Hinnøya (isl.) ... K2
Hitra (isl.) ... F5
Hopen (isl.) ... E2
Isfjorden (fjord) ... C2
Jostedalsbreen (glac.) ... E6
Kjölen (mts.) ... K3
Kongsfjorden (fjord) ... B2
Kvaløya (isl.) ... O1
Lågen (riv.) ... G6
Laksefjorden (fjord) ... P1
Langøy (isl.) ... J2
Lapland (reg.) ... K2
Leka (isl.) ... G4
Lindesnes (cape) ... E8
Lista (pen.) ... E7
Lofoten (isls.) ... H2
Lopphavet (bay) ... M1
Magerøya (isl.) ... P1
Moskenesøya (isl.) ... H3
Namsen (riv.) ... H4
Nordaustlandet (isl.) ... D1
Nordfjord (fjord) ... E6
Nordkapp (pt.) ... C1
Nordkinn (headland) ... Q1
Nordkinn (pen.) ... P1
North Cape (Nordkapp) (pt.) ... P1
Norwegian (sea) ... F3
Ofotfjorden (fjord) ... K2
Oslofjord (fjord) ... D4
Otra (riv.) ... E7
Otterøya (isl.) ... E5
Pasvikelv (riv.) ... Q2
Platen, Kapp (pt.) ... D1
Porsangen (fjord) ... O1
Rana (fjord) ... H3
Rauma (riv.) ... F5
Ringvassøy (isl.) ... L2
Romsdalsfjorden (fjord) ... E5
Saltfjorden (fjord) ... J3
Seiland (isl.) ... N1
Senja (isl.) ... K2
Skagerrak (str.) ... F8
Smøla (isl.) ... E5
Sognafjorden (fjord) ... D6
Sørkapp (pt.) ... C2
Sorøya (isl.) ... N1
Spitsbergen (isl.) ... C2
Storfjorden (fjord) ... D2
Sulitjelma (mt.) ... J3
Svalbard (isls.) ... C3
Tana (riv.) ... P1
Tanafjord (fjord) ... Q1
Tokke (riv.) ... F7
Trondheimsfjorden (fjord) ... G5
Tyrifjord (lake) ... C3
Vaerøy (isl.) ... H3
Vagåvatn (lake) ... F6
Vannøy (isl.) ... L1
Varangerhalvøya (pen.) ... Q1
Varangerfjord (fjord) ... Q2
Vega (isl.) ... G4
Vesterålen (isls.) ... J2
Vestfjord (fjord) ... H3
Vestvågøya (isl.) ... H3
Vikna (isls.) ... G4

SWEDEN

COUNTIES

Älvsborg 418,150 ... H7
Blekinge 155,391 ... J8
Gävleborg 294,595 ... K6
Göteborg och Bohus 714,660 ... G7
Gotland 54,447 ... L8
Halland 219,767 ... H8
Jämtland 133,559 ... J5
Jönköping 301,905 ... H8
Kalmar 240,768 ... K8
Kopparberg 281,082 ... J6
Kristianstad 272,090 ... J8

(continued on following page)

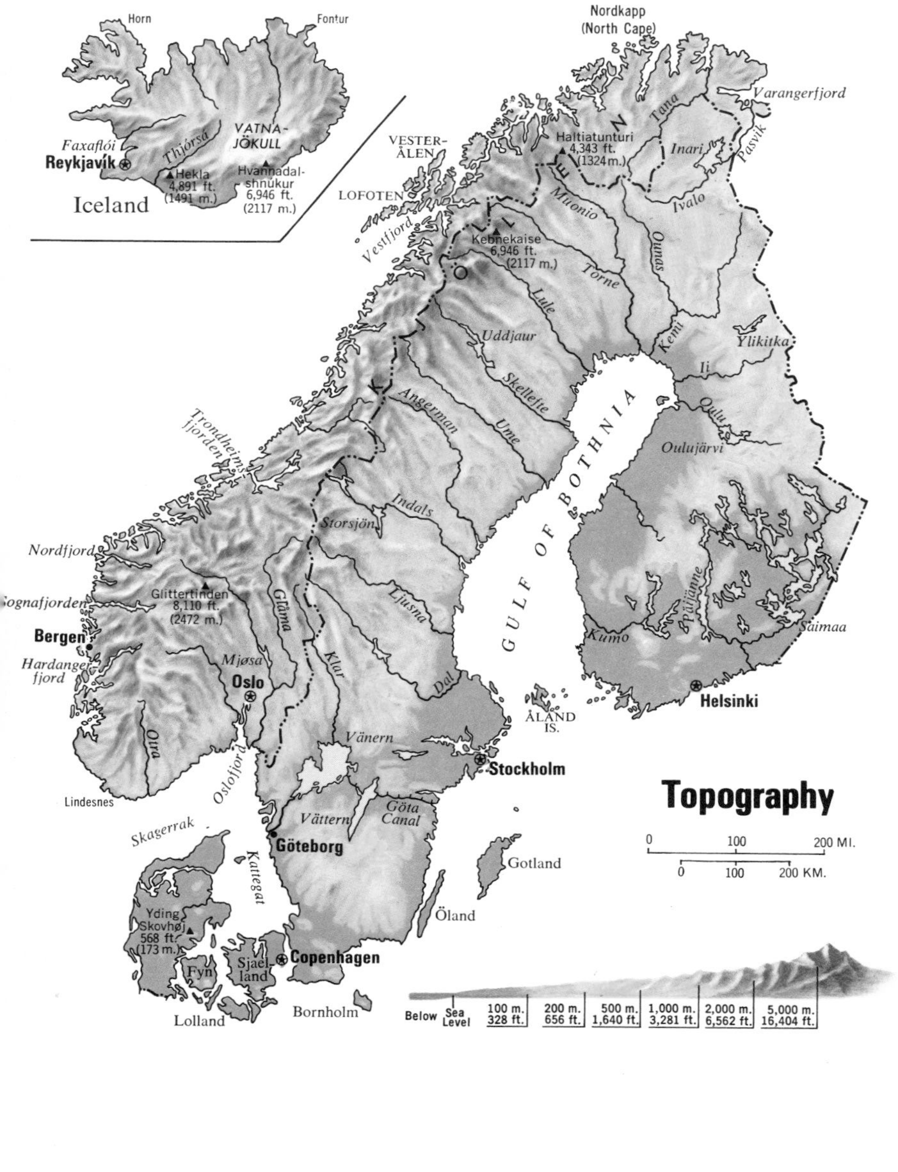

Horn
Fontur
Faxaflói
Reykjavík
Thjórsá
VATNA-JÖKULL
Hekla 4,891 ft. (1491 m.)
Hvannadal-shnúkur 6,946 ft. (2117 m.)
Iceland
VESTER-ÅLEN
LOFOTEN
Vestfjord
Nordkapp (North Cape)
Varangerfjord
Tana
Pasvik
Inari
Ivalo
Haltiatunturi 4,343 ft. (1324 m.)
Muonio
Ounas
Kebnekaise 6,946 ft. (2117 m.)
Torne
Lule
Uddjaur
Kemi
Ylikitka
Ii
Oulu
Oulujärvi
Skellefte
Ume
Ångerman
Trondheimsfjorden
GULF OF BOTHNIA
Indals
Storsjön
Ljusna
Nordfjord
Sognafjorden
Glittertinden 8,110 ft. (2472 m.)
Glåma
Klar
Bergen
Hardanger fjord
Mjøsa
Oslo
Dal
Päijänne
Kumo
Saimaa
Helsinki
ÅLAND IS.
Otra
Vänern
Stockholm
Oslofjord
Lindesnes
Vättern
Göta Canal
Skagerrak
Göteborg
Gotland
Kattegat
Öland
Yding Skovhøj 568 ft. (173 m.)
Fyn
Sjælland
Copenhagen
Lolland
Bornholm
Topography
0 100 200 MI.
0 100 200 KM.
Below
Sea Level
100 m. 328 ft.
200 m. 656 ft.
500 m. 1,640 ft.
1,000 m. 3,281 ft.
2,000 m. 6,562 ft.
5,000 m. 16,404 ft.

Kronoberg 169,454 .... J8
Malmöhus 740,137 .... H9
Norrbotten 264,215 .... L3
Örebro 273,994 .... J7
Östergötland 387,104 .... J7
Skaraborg 263,382 .... H7
Södermanland 252,030 .... K7
Stockholm 1,493,052 .... L7
Uppsala 229,879 .... K7
Värmland 284,442 .... H7
Västerbotten 236,367 .... K4
Västernorrland 268,202 .... K5
Västmanland 259,872 .... K7

CITIES and TOWNS

Åhus 6,125 .... J9
Alingsås 18,892 .... H7
Älmhult 7,390 .... H8
Alvesta 7,261 .... J8
Älvsbyn 4,707 .... M4
Åmål 9,556 .... H7
Ånge 3,760 .... J5
Ängelholm 16,016 .... H8
Arboga 11,819 .... J7
Arbrå 2,734 .... K6
Årjäng† 2,596 .... H7
Arvidsjaur 4,194 .... L4
Arvika 13,934 .... H7
Åseda 2,465 .... J8
Askim 17,609 .... G8
Åtvidaberg 8,436 .... K7
Avesta 19,095 .... J6
Bålsta 8,243 .... G1
Båstad 2,452 .... H8
Bengtsfors 3,535 .... H7
Boden 19,590 .... M4
Bollnäs 13,305 .... K6
Bollstabruk 3,548 .... L5
Borås 67,537 .... H8
Borås* 187,710 .... H8
Borgholm 2,789 .... K8
Borlänge 40,158 .... J6
Brunflo 3,460 .... J5
Dalby† 4,013 .... H6
Danderyd† 36,596 .... H1
Dannemora 291 .... K6
Edsbyn 4,388 .... J6
Eksjö 9,686 .... J8
Emmaboda 5,652 .... J8
Enköping 18,541 .... G1
Eskilstuna 66,409 .... K7
Eslöv 13,629 .... H9
Fagersta† 14,778 .... J6
Falkenberg 14,148 .... H8
Falköping 15,126 .... H7
Falun 30,073 .... J6
Färjestaden 2,995 .... K8
Filipstad 7,835 .... H7
Finspång 16,346 .... J7
Flen 6,770 .... K7
Forshaga 6,000 .... H7
Frösö 10,274 .... J5
Frövi 2,583 .... J7
Gällivare 8,669 .... M3
Gamleby 3,666 .... J8
Gävle 67,454 .... K6
Gimo 3,154 .... K6
Gislaved 8,564 .... H8
Gnesta 3,835 .... G2
Göteborg 444,540 .... G8
Göteborg* 690,767 .... G8
Hagfors 8,060 .... H6
Hällefors 7,862 .... J7
Hallsberg 6,799 .... J7
Hallstahammar 13,583 .... K7
Hallstavik 5,162 .... L6
Halmstad 49,558 .... H8
Haparanda 5,031 .... N4
Härnösand 18,971 .... L5
Hässleholm 16,813 .... H8
Hedemora 7,039 .... K6
Helsingborg 80,986 .... H8
Helsingborg* 215,894 .... H8
Hjo 4,615 .... J7
Hofors 11,459 .... K6
Höganäs 10,866 .... H8
Holmsund 5,467 .... M5
Hörnefors 2,441 .... L5
Huddinge 48,339 .... H1
Hudiksvall 15,004 .... K6
Hultsfred 5,763 .... K8
Husum 2,517 .... L5
Hyltebruk 3,469 .... H8
Iggesund 4,448 .... K6
Järna 6,237 .... G2
Jokkmokk 3,186 .... L3
Jönköping 78,650 .... H8
Jönköping* 131,499 .... H8
Kalix 7,668 .... N4
Kalmar 32,049 .... K8
Karlshamn 17,447 .... J8
Karlskoga† 35,425 .... J7
Karlskrona 33,414 .... J8
Karlstad 51,243 .... H7
Katrineholm 22,884 .... K7
Kinna 13,676 .... H8
Kiruna 25,410 .... L3
Kisa 4,323 .... J7
Köping 20,059 .... J7
Kopparberg 3,942 .... J7
Kramfors 7,719 .... L5
Kristianstad 30,780 .... J9
Kristinehamn 21,146 .... H7
Kumla 11,451 .... J7
Kungälv† 12,764 .... G8
Kungsbacka† 11,986 .... G8
Kvissleby 3,413 .... K5
Laholm 3,898 .... H8
Landskrona 29,486 .... H9
Långshyttan 2,744 .... K6
Laxå 5,166 .... J7
Leksand 4,410 .... J6
Lessebo 2,991 .... J8
Lidingö 30,098 .... H1
Lidköping 21,001 .... H7
Lindesberg 8,247 .... J7
Linköping 80,274 .... K7
Linköping* 132,839 .... K7
Ljungby 12,969 .... J8
Ljusdal 7,075 .... J6
Ljusne 3,578 .... K6
Ludvika 18,217 .... J6
Luleå 42,139 .... N4
Lund 55,047 .... H9
Lycksele 8,586 .... L4
Lysekil 7,815 .... G7
Malmberget 10,239 .... M3
Malmö 241,191 .... H9
Malmö* 453,339 .... H9
Malung 6,211 .... H6
Mariefred 2,553 .... F1
Mariestad 16,454 .... H7
Markaryd 4,266 .... H8
Märsta 17,066 .... K7
Marstrand 1,168 .... G8
Mellerud 3,579 .... H7
Mjölby 12,488 .... J7
Mölndal† 47,248 .... H8
Mönsterås 5,005 .... K8
Mora 8,772 .... J6
Motala 29,454 .... J7
Nacka 19,708 .... H1
Nässjö 18,634 .... J8
Nora 5,515 .... J7
Norberg 5,438 .... K6
Norrköping 85,244 .... K7
Norrköping* 163,206 .... K7
Norrtälje 12,784 .... L7
Nybro 13,010 .... J8
Nyköping 30,352 .... K7
Nynäshamn 11,070 .... L7
Ockelbo 2,810 .... K6
Olofström 10,096 .... J8
Örebro 117,877 .... J7
Örebro* 171,440 .... J7
Örnsköldsvik 29,514 .... L5
Orrefors 919 .... J8
Orsa 5,099 .... J6
Oskarshamn 19,021 .... K8
Östersund 40,056 .... J5
Östhammar 1,783 .... L6
Oxelösund 13,862 .... K7
Piteå 16,169 .... M4
Rättvik 4,087 .... J6
Rimbo 3,404 .... L7
Ronneby 12,086 .... J8
Säffle 11,428 .... H7
Sala 11,216 .... K7
Saltsjöbaden 8,113 .... J1
Sandviken 27,994 .... K6
Säter 4,297 .... J6
Sävsjö 4,913 .... J8
Sigtuna 4,780 .... H1
Simrishamn 5,834 .... J9
Skanör med Falsterbo 4,909 .... H9
Skara 10,138 .... H7
Skellefteå 29,353 .... M4
Skövde 29,945 .... H7
Skutskär 7,174 .... K6
Smedjebacken 8,418 .... J6
Söderhamn 14,673 .... K6
Söderköping 5,310 .... K7
Södertälje 58,408 .... G1
Sollefteå 8,923 .... K5
Sollentuna† 40,905 .... H1
Solna† 53,992 .... H1
Sölvesborg 7,292 .... J9
Stenungsund 8,361 .... G7
Stockholm (cap.) 665,550 .... G1
Stockholm* 1,357,183 .... G1
Storuman 2,587 .... K4
Storvik 2,748 .... K6
Strängnäs 10,255 .... F1
Strömstad 4,735 .... G7
Strömsund 4,119 .... K5
Sundbyberg† 27,058 .... G1
Sundsvall 52,268 .... K5
Sunne 4,273 .... H7
Surahammar 6,509 .... J7
Sveg 2,608 .... J5
Svenljunga 3,189 .... H8
Täby† 41,285 .... H1
Tibro 8,476 .... J7
Tidaholm 8,039 .... J7
Tierp 5,005 .... K6
Timrå 11,416 .... K5
Tomelilla 5,371 .... J9
Torsby 3,632 .... H6
Torshälla 8,231 .... K7
Tranås 14,854 .... J7
Trelleborg 22,559 .... H9
Trollhättan 42,499 .... H7
Trosa 3,128 .... K7
Uddevalla 32,700 .... G7
Ulricehamn 7,827 .... H8
Umeå 49,715 .... M5
Uppsala 101,850 .... K7
Uppsala* 157,202 .... L7
Vadstena 5,294 .... J7
Vaggeryd 3,974 .... J8
Valdemarsvik 3,558 .... K7
Vallentuna 10,477 .... H1
Vänersborg 20,510 .... G7
Vännäs 3,876 .... L5
Vansbro 2,708 .... H6
Vara 3,049 .... H7
Varberg 19,467 .... G8
Värnamo 15,726 .... J8
Västerås 98,858 .... K7
Västerås* 147,508 .... K7
Västerhaninge 14,125 .... H1
Västervik 21,239 .... K8
Vaxholm† 3,744 .... J1
Växjö 40,328 .... J8
Vetlanda 12,358 .... J8
Vilhelmina 4,060 .... K4
Vimmerby 7,405 .... J8
Virserum 2,495 .... J8
Visby 19,886 .... L8
Ystad 14,286 .... H9

OTHER FEATURES

Ångermanälven (riv.) K5
Åsnen (lake) J8
Baltic (sea) K9
Bolmen (lake) H8
Bothnia (gulf) N4
Dalälven (riv.) K6
Fårö (isl.) L8
Göta (canal) J7
Göta (riv.) H7
Gotland (isl.) L8
Gräsö (isl.) L6
Hanöbukten (bay) J9
Hjälmaren (lake) J7
Hoburgen (cliff) L8
Hornslandet (pen.) K6
Indalsälven (riv.) H5
Kalixälv (riv.) N3
Kalmarsund (sound) K8
Kattegat (str.) G8
Kebnekaise (mt.) L3
Kölen (mts.) K3
Klarälv (riv.) H6
Lapland (reg.) M2
Ljusnan (riv.) H5
Luleälv (riv.) M4
Mälaren (lake) G1
Muonioälv (riv.) M2
Öland (isl.) K8
Öresund (sound) H9
Ornö (isl.) J2
Österdalälven (riv.) H6
Piteälv (riv.) M4
Siljan (lake) J6
Skagerrak (str.) F8
Sommen (lake) J8
Stora Lulevatten (lake) L3
Storsjön (lake) J5
Sulitelma (mt.) K3
Torneälv (riv.) M3
Uddjaur (lake) L4
Umeälv (riv.) L4
Vänern (lake) H7
Västerdalälven (riv.) H6
Vättern (lake) J7

*City and suburbs.
†Population of commune.
†Population of parish.

DENMARK

COUNTIES

Århus 534,333 .... D5
Bornholm 47,241 .... F9
Copenhagen (commune) 622,612 .... F6
Faerøe Islands 41,969 .... B2
Frederiksberg (commune) 101,874 .... F6
Frederiksborg 260,825 .... E5
Fyn 433,765 .... D7
København (Copenhagen) (commune) 622,612 .... F6
Københaven 616,571 .... F6
Nordjylland 457,165 .... D4
Ribe 198,153 .... B7
Ringkøbing 242,006 .... B5
Roskilde 154,314 .... E6
Sønderjylland 238,502 .... C7
Storstrøm 252,780 .... E7
Vejle 306,809 .... C6
Vestsjaelland 259,484 .... E6
Viborg 221,002 .... C4

CITIES and TOWNS

Åbenrå 15,196 .... C7
Åbybro 2,897 .... C3
Åkirkeby 2,001 .... F9
Ålborg 154,582 .... D4
Ålestrup 1,926 .... C4
Århus 245,941 .... D5
Års 4,266 .... C4
Årup 1,675 .... D7
AErøskøbing 1,223 .... D8
Agerbaek 935 .... B6
Allingaåbro 1,385 .... D5
Allinge-Sandvig 1,991 .... F8
Ansager 1,157 .... B6
Arden 1,303 .... C4
Asaå 1,344 .... D3
Askov 904 .... C7
Asnaes 1,413 .... E6
Assens, Århus 1,341 .... D4
Assens, Fyn 5,139 .... D7
Augustenborg 2,628 .... D8
Auning 1,516 .... D5
Avlum 1,729 .... B5
Baelum 1,169 .... D4
Bagenkop 776 .... D8
Ballerup 50,673 .... F6
Bandholm 693 .... E8
Bedsted 965 .... B4
Birkerød 13,663 .... F6
Bjerringbro 4,761 .... C5
Bogense 2,861 .... D6
Bolderslev 774 .... C8
Børkop 1,410 .... C6
Borup 1,591 .... E7
Braedstrup 2,163 .... C6
Bramming 3,678 .... B7
Brande 4,784 .... B6
Bredebro 1,173 .... B7
Broager 2,143 .... C8
Brønderslev 10,247 .... C3
Brørup 2,584 .... C7
Brovst 4,200 .... C3
Bryrup 579 .... C5
Christiansfeld 1,994 .... C7
Copenhagen (cap.) 603,368 .... F6
Copenhagen* 1,327,940 .... F6
Dronninglund 4,661 .... D3
Dybvad 805 .... D3
Ebeltoft 3,017 .... D5
Egernsund 1,323 .... C8
Egtved 1,311 .... C6
Ejby 1,372 .... C7
Esbjerg 68,097 .... B7
Faåborg 6,495 .... D7
Fakse 2,720 .... F7
Fakse Ladeplads 1,799 .... F7
Farsø 2,821 .... C4
Farum 9,936 .... F6
Fjerritslev 2,134 .... C3
Fredensborg 4,709 .... F6
Fredericia 36,157 .... C6
Frederiksberg 101,874 .... F6
Frederikshavn 24,846 .... D3
Frederikssund 11,272 .... E6
Frederiksvaerk 8,903 .... E6
Fuglebjerg 1,094 .... E7
Gedser 1,200 .... F8
Gedsted 1,006 .... C4
Gelsted 1,307 .... C7
Gentofte 77,744 .... F6
Gilleleje 2,943 .... F5
Give 2,366 .... C6
Glamsbjerg 2,226 .... D7
Glostrup 28,326 .... F6
Glumsø 1,027 .... E7
Glyngøre 1,071 .... C4
Gørding 1,261 .... B7
Gørlev 1,542 .... E7
Graested 1,654 .... F5
Gram 2,061 .... C7
Graåsten 2,947 .... C8
Grenaå 12,569 .... D5
Grindsted 7,558 .... B6
Haårby 1,506 .... D7
Haderslev 20,042 .... C7
Hadsten 3,914 .... C5
Hadsund 3,652 .... D4
Hals 1,654 .... D3
Hammel 3,247 .... C5
Hammerum 3,227 .... C5
Hanstholm 1,716 .... B3
Harboør 1,369 .... B4
Haårlev 1,228 .... F7
Hasle 18
Haslev 6,925 .... E7
Havdrup 1,833 .... F6
Hedensted 2,659 .... C6
Hellebaek 2,911 .... F5
Helsinge 3,613 .... F6
Helsingør 42,425 .... F6
Herning 32,973 .... B5
Hillerød 23,963 .... F6
Hinnerup 2,061 .... D5
Hirtshals 6,861 .... C2
Hjallerup 1,573 .... C3
Hjerm 647 .... B5
Hjørring 19,692 .... C3
Hobro 8,737 .... C4
Højer 1,416 .... B8
Højslev 1,641 .... C4
Holbaek 19,485 .... E6
Holeby 1,434 .... E8
Holstebro 25,006 .... B5
Holsted 1,390 .... B6
Høng 2,488 .... E7
Hornslet 2,561 .... D5
Horsens 44,120 .... C6
Hørsholm 19,346 .... F6
Hørve 1,139 .... E6
Hov 635 .... D6
Humlum 546 .... B4
Hundested 5,443 .... E6
Hurup 2,287 .... B4
Hvidbjerg 994 .... B4
Hvide Sande 2,129 .... A6
Ikast 9,222 .... C5
Jelling 1,540 .... C6
Jerslev 798 .... D3
Juelsminde 1,991 .... D6
Jyderup 2,901 .... E6
Kalundborg 12,248 .... D6
Karise 1,184 .... F7
Karup 1,694 .... C5
Kastrup† 17,391 .... F6
Kerteminde 5,007 .... D7
Kibaek 1,279 .... B5
Kjellerup 3,245 .... C5
Klitmøller 542 .... B3
København (Copenhagen) (cap.) 603,368 .... F6
Køge 18,608 .... F7
Kolding 41,602 .... C7
Kolind 1,036 .... D5
Korsør 15,502 .... E7
Kvaerndrup 891 .... D7
Langaå 2,320 .... C5
Lem 1,026 .... B5
Lemvig 6,448 .... B4
Løgstør 3,633 .... C4
Løgumkloster 2,091 .... B7
Lohals 580 .... D7
Løjt Kirkeby 1,203 .... C7
Løkken 1,345 .... C3
Løsning 1,967 .... C6
Lundby 747 .... E7
Lunderskov 1,494 .... C7
Lyngby 61,516 .... F6
Malling 1,584 .... D5
Mariager 1,692 .... D4
Maribo 5,287 .... E8
Marstal 4,124 .... D8
Middelfart 13,315 .... C7

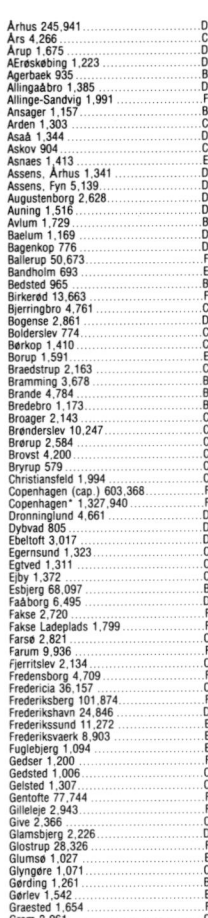

Agriculture, Industry and Resources

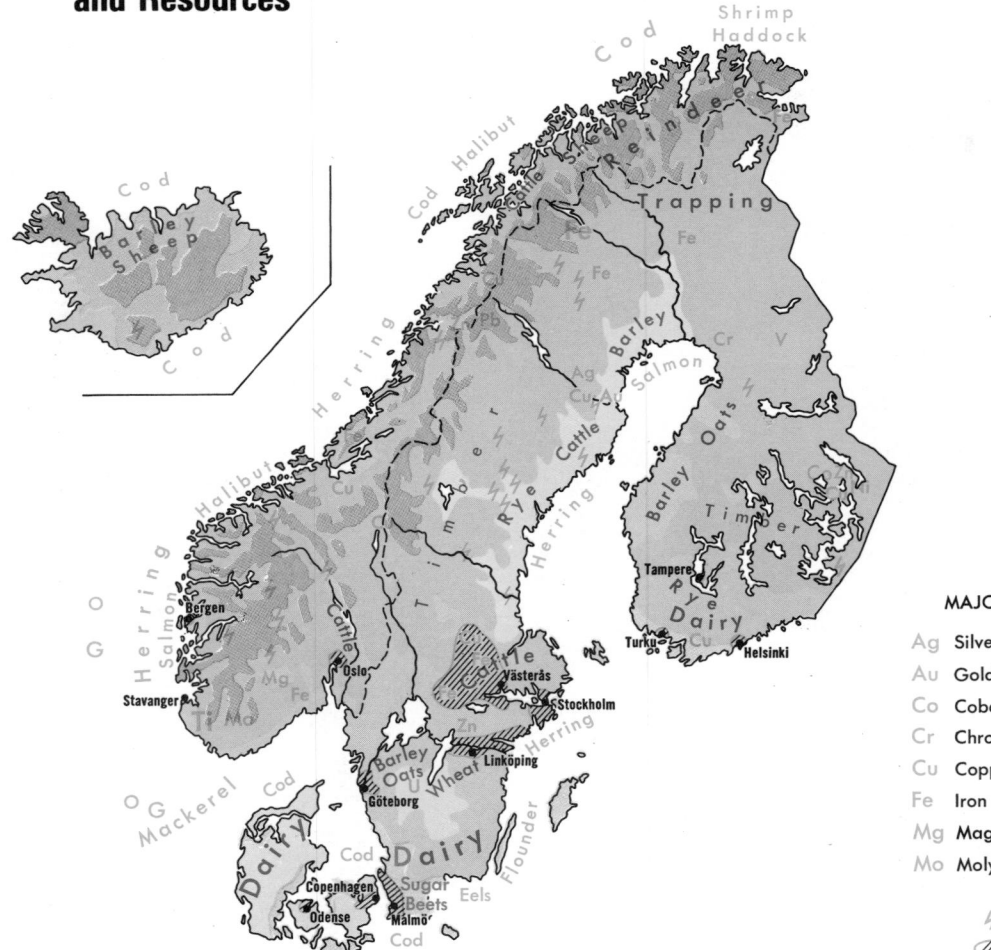

Shrimp
Haddock
Cod
Halibut
Reindeer
Trapping
Fe
Barley
Sheep
Herring
Pb
Salmon
Ag
Cu
Au
Cr
V
Cattle
Oats
Timber
Rye
Tampere
Turku
Helsinki
Dairy
Bergen
Oslo
Stavanger
Ti
Mo
Mg
Västerås
Stockholm
Zn
Linköping
Wheat
Göteborg
Mackerel
Flounder
Copenhagen
Odense
Sugar Beets
Malmö
Eels
DOMINANT LAND USE
Cash Cereals, Dairy
Dairy, Cattle, Hogs
Dairy, General Farming
General Farming (chiefly cereals)
Nomadic Sheep Herding
Forests, Limited Mixed Farming
Nonagricultural Land
MAJOR MINERAL OCCURRENCES
Ag Silver
Au Gold
Co Cobalt
Cr Chromium
Cu Copper
Fe Iron Ore
Mg Magnesium
Mo Molybdenum
Ni Nickel
O Petroleum
Pb Lead
Ti Titanium
U Uranium
V Vanadium
Zn Zinc
Water Power
Major Industrial Areas

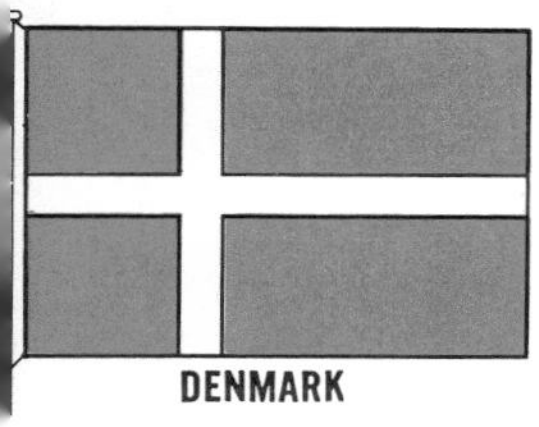

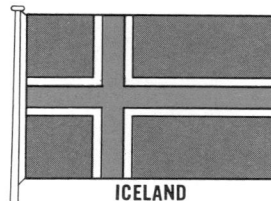

DENMARK

ICELAND

	DENMARK		ICELAND
AREA	16,629 sq. mi. (43,069 sq. km.)	**AREA**	39,768 sq. mi. (103,000 sq. km.)
POPULATION	5,124,000	**POPULATION**	228,785
CAPITAL	Copenhagen	**CAPITAL**	Reykjavík
LARGEST CITY	Copenhagen	**LARGEST CITY**	Reykjavík
HIGHEST POINT	Yding Skovhøj 568 ft. (173 m.)	**HIGHEST POINT**	Hvannadalshnúkur 6,952 ft. (2,119 m.)
MONETARY UNIT	krone	**MONETARY UNIT**	króna
MAJOR LANGUAGE	Danish	**MAJOR LANGUAGE**	Icelandic
MAJOR RELIGION	Protestantism	**MAJOR RELIGION**	Protestantism

øgeltønder 711 ... B8
estved 35,011 ... E7
kskov 16,393 ... E8
ksø 3,527 ... F9
e 2,796 ... C4
ordborg 4,132 ... C7
rdby, Ribe 2,084 ... B7
rre Aby 2,165 ... C8
rre Alslev 1,338 ... D7
rre Nebel 901 ... B6
rre Snede 1,461 ... C6
rre Vorupør 644 ... B4
borg 14,181 ... D7
købing, Storstrøm 20,059 ... F8
Vestsjælland 4,996 ... E6
købing, Viborg 9,066 ... E8
rsted 1,229 ... E8
lder 6,617 ... D6
lense 168,178 ... D7
god 2,258 ... B6
sted 1,093 ... D5
ter Vrå 906 ... D3
trup 2,673 ... B6
ndrup 1,525 ... C5
aesø 2,789 ... B4
mme 506 ... B4
nders 58,409 ... C5
num 1,472 ... C4
be 8,254 ... D7
nge 3,584 ... D7
købing 6,298 ... H6
ngsted 14,076 ... E8
adby 5,296 ... F8
adding 2,102 ... B7
adekro 2,246 ... C7
dkaersbro 1,098 ... C5
dvig 1,115 ... F7
mø 816 ... B7
nde 1,523 ... F9
skilde 44,248 ... E8
aslev 1,058 ... B4
adkøbing 4,080 ... C5
ds Vedby 1,071 ... E7
r 2,699 ... D5
romgård 1,000 ... D5
eby 5,430 ... E6
adding 2,102 ... F6
keborg 29,015 ... C5
ndal 2,406 ... D7
aelskør 4,585 ... E7
aerbaek 2,483 ... B7
ags 1,620 ... C4
als 960 ... C4
anderborg 11,344 ... D5
ärup 1,216 ... C7
ilby 1,549 ... F9
ive 17,015 ... C4
jern 6,056 ... B6
odborg 935 ... C7
ørping 1,675 ... C4
agelse 26,851 ... E7
angerup 3,036 ... E6
edsted 1,105 ... B4
llested 960 ... E8
nderborg 24,526 ... C8
nder Omme 1,393 ... B6
nderso 885 ... D7
re 8,683 ... F8
ege 3,869 ... F6
enslile 1,014 ... E6
anstrup 1,245 ... D7
oholm 1,224 ... C5
ore Heddinge 2,630 ... F7
øvring 2,366 ... C4
andby 1,017 ... C5
rue 10,848 ... B5
bbekøbing 2,031 ... F8
raneke 1,193 ... C7
rendborg 24,203 ... D7
rinninge 1,797 ... C5
rm 3,150 ... B6
arnby 45,661 ... F6
rhøj 30,608 ... C5
rem 511 ... B4
em 11,252 ... D5
nyborn 2,425 ... A4
yregod 1,001 ... C5
m 553 ... B5
nglev 1,524 ... D7
rfund 762 ... B6
rtlund 2,147 ... C4
illøse 1,982 ... E6
mmerup 1,439 ... D7
nder 7,469 ... C8
rring 1,537 ... C6
aneberg 657 ... C7
ense 771 ... D7
ustrup 794 ... D4
dum 885 ... C6
fborg 1,357 ... B5
ard 3,111 ... C7
arde 11,615 ... B7
be 6,213 ... D7
rd 43,976 ... C8
mb 989 ... B5
ester Skerninge 603 ... D7
esterv 747 ... B4
borg 27,441 ... C5
by 1,549 ... B5
borg 1,037 ... C4
idbjerg 1,500 ... B5
inderup 2,389 ... C5
jens 5,595 ... C7
orbasse 791 ... B6
ording 11,639 ... E7

Vraa 2.652 ... C3

OTHER FEATURES

AErø (isl.) ... D8
Als (isl.) ... C8
Amager (isl.) ... F6
Anholt (isl.) ... E4
Arø (isl.) ... C7
Baagø (isl.) ... C7
Baltic (sea) ... E9
Bornholm (isl.) ... F9
Endelave (isl.) ... D6
Falster (isl.) ... F8
Fanø (isl.) ... B7
Fehmarn (str.) ... E9
Fejø (isl.) ... E8
Femø (isl.) ... E8
Frisian, North (isls.) ... B7
Fyn (isl.) ... D7
Gelsaa (riv.) ... C7
Gudenaa (riv.) ... D5
Isefjord (fjord) ... E6
Jutland (pen.) ... C5
Jylland (Jutland) (pen.) ... C5
Kattegat (str.) ... E4
Laesø (isl.) ... D3
Langeland (isl.) ... D8
Lille Baelt (chan.) ... C7
Limfjorden (fjord) ... A4
Løgstør Bredning (fjord) ... C4
Lolland (isl.) ... E8
Møn (isl.) ... F8
Mors (isl.) ... B4
North (sea) ... B9
North Frisian (isls.) ... B7
 Omø (isl.) ... E7
Øresund (sound) ... F6
Rømø (isl.) ... B7
Samsø (isl.) ... D6
Sejerø (isl.) ... E6
Sjaelland (isl.) ... E6
Skagens Odde (cape) ... D2
Skagerrak (str.) ... C2
Skaw, The (Skagens Odde) (cape) ... D2
Storaa (riv.) ... B5
Store Baelt (chan.) ... D6
Susaa (riv.) ... E7
The Skaw (Skagens Odde) (cape) ... D2
Tranebjerg (mt.) ... C5
Yding Skovhøj (mt.) ... C6

FÆRØE ISLANDS

CITIES and TOWNS

Klaksvík 4,536 ... B2
Tórshavn (cap.), Faerøe Is. 11,618 ... A3

OTHER FEATURES

Faerøe (isls.) ... B2
Sandoy (isl.) ... B3
Streymoy (isl.) ... B3
Sudhuroy (isl.) ... B3

ICELAND

CITIES and TOWNS

Akranes 4,253 ... B1
Akureyri 10,755 ... C1
Hafnarfjordhur 9,696 ... B2
Húsavík 1,993 ... D1
Ísafjördhur 2,680 ... B1
Keflavík 5,663 ... B1
Kópavogur 11,165 ... B1
Nes (Neskaupstadhur) 1,552 ... D1
Neskaupstadhur 1,552 ... D1
Olafsfjördhur 1,086 ... C1
Reykjavík (cap.) 81,693 ... 70
Reykjavík* 98,521 ... 70
Saudharkrókur 1,600 ... B1
Seydhisfjördhur 884 ... D1
Siglufjördhur 2,161 ... C1
Vestmannaeyjar 5,186 ... B2

OTHER FEATURES

Bjartangar (pt.) ... A1
Breidhafjördhur (fjord) ... B1
Faxaflói (bay) ... B1
Fontur (pt.) ... D1
Gerpir (cape) ... D1
Grímsey (isl.) ... C1
Hekla (vol.) ... B1
Húnaflói (bay) ... B1
Hvannadalshnúkur (mt.) ... D1
North (Horn) (cape) ... B1
Reykjanesta (cape) ... A2
Surtsey (isl.) ... B2
Thjórsá (riv.) ... C1
Vatnajökull (glac.) ... C1

*City and suburbs.

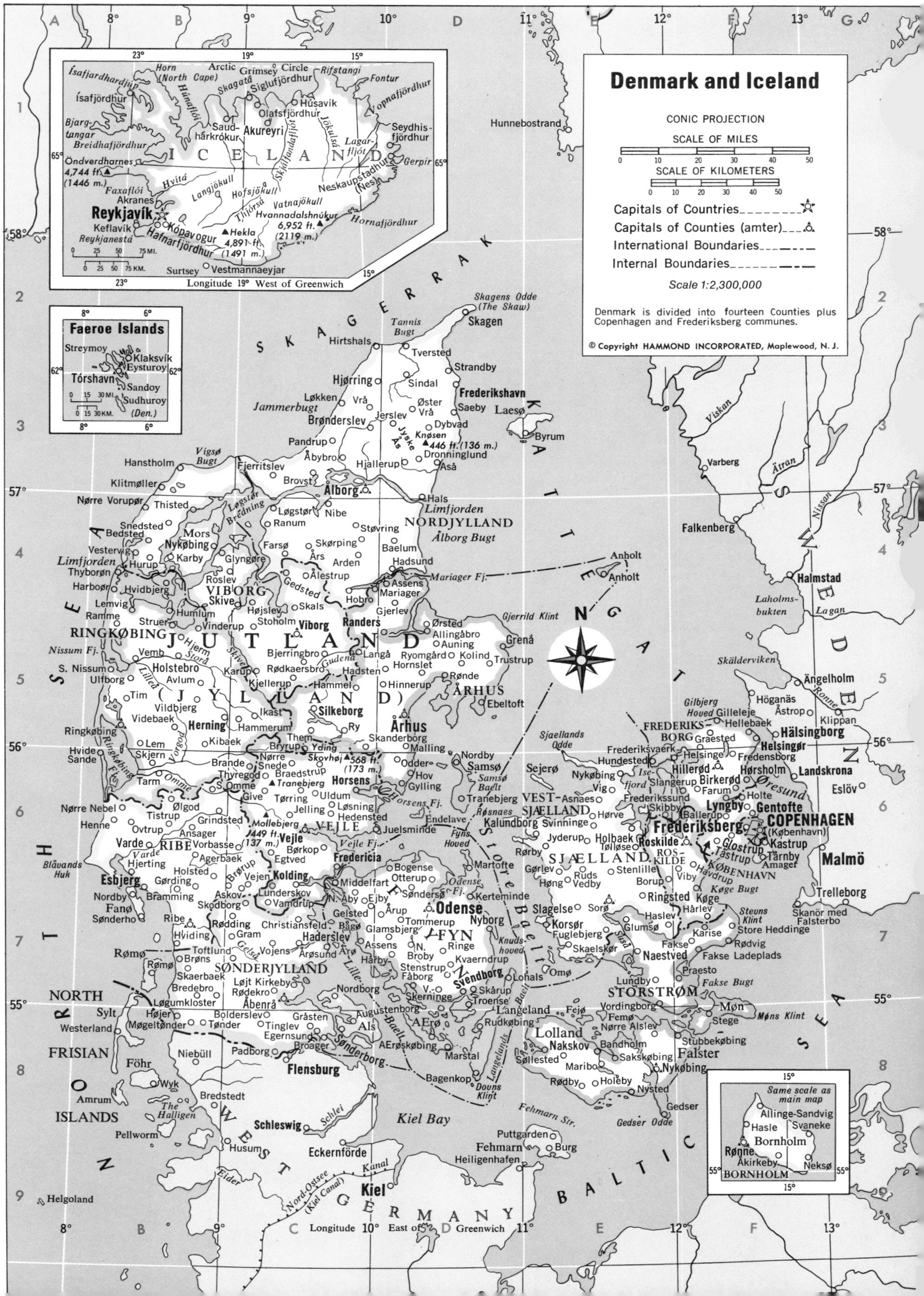

Denmark and Iceland

CONIC PROJECTION

SCALE OF MILES

SCALE OF KILOMETERS

Capitals of Countries ⭐
Capitals of Counties (amter) △
International Boundaries
Internal Boundaries

Scale 1:2,300,000

Denmark is divided into fourteen Counties plus Copenhagen and Frederiksberg communes.

© Copyright HAMMOND INCORPORATED, Maplewood, N.J.

ICELAND

Ísafjardhardjú
Ísafjördhur
Bjarg-tangar
Breidhafjördhur
Óndverdharnes
4,744 ft. (1446 m.)
Faxaflói
Akranes
Reykjavík
Keflavík
Kópavogur
Reykjanesta
Surtsey
Vestmannaeyjar
Horn (North Cape)
Arctic Circle
Grímsey
Skagafjördhur
Siglufjördhur
Saudhárkrókur
Akureyri
Jökulsá
Húsavík
Rifstangi
Fontur
Seydhis-fjördhur
Gerpir
Neskaupstadhur (Nes)
Hornafjördhur
Hvítá
Langjökull
Hofsjökull
Thjórsá
Hekla 4,891 ft. (1491 m.)
Vatnajökull
Hvannadalshnúkur 6,952 ft. (2119 m.)

Longitude 19° West of Greenwich

Faeroe Islands

Streymoy
Tórshavn
Klaksvík
Eysturoy
Sandoy
Sudhuroy
(Den.)

DENMARK MAP LABELS

SKAGERRAK
KATTEGAT
Skagen
Skagens Odde (The Skaw)
Hirtshals
Tversted
Strandby
Frederikshavn
Hjørring
Sindal
Saeby
Laesø
Løkken
Vrå
Øster Vrå
Jerslev
Brønderslev
Dybvad
Byrum
Pandrup
Åbybro
Hjallerup
Aså
Aalborg
Hals
Limfjorden
Nibe
Støvring
Støvring
Baelum
NORDJYLLAND
Ålborg Bugt
Anholt
Halmstad
Varberg
Falkenberg
Hanstholm
Fjerritslev
Klitmøller
Brovst
Nørre Vorupør
Thisted
Mors
Nykøbing
Fårsø
Ranum
Skørping
Års
Arden
Hadsund
Assens
Mariager
Snedsted
Bedsted
Vesterø
Karby
Glyngøre
Mariager Fj.
Anholt
Laholms-bukten
Limfjorden
Thyborøn
Hurup
Roslev
Gedsted
Ålestrup
Hobro
Ørsted
Gjerrild Klint
VIBORG
Skive
Højslev
Skals
Stoholm
Viborg
Randers
Allingåbro
Auning
Grenå
SWEDEN
Harboør
Hvidbjerg
Lemvig
Humlum
Struer
Vinderup
Rønde
Skälderviken
Ängelholm
RINGKØBING
JUTLAND (JYLLAND)
Holstebro
Bjerringbro
Rødkaersbro
Ulfborg
Vemb
Karup
Hammel
Kolind
Hornslet
Trustrup
Ebeltoft
Tim
Vildbjerg
Ikast
Silkeborg
Ry
Skanderborg
ÅRHUS
Ringkøbing
Videbaek
Herning
Hammerum
Them
Brande
Bryrup
Yding
Kibaek
Skjern
Hvide Sande
Tarm
Ølgod
Give
Grindsted
Ansager
Henne
Ovtrup
Varde
RIBE
Esbjerg
Nordby
Fanø
Sønderho
Rømø
SØNDERJYLLAND
NORTH
FRISIAN
ISLANDS
Sylt
Westerland
Föhr
Amrum
The Halligen
Pellworm
Wyk
Bredstedt
Niebüll
Husum
Schleswig
Eckernförde
Kiel Bay
Heiligenhafen
Puttgarden
Fehmarn
Burg
Helgoland
GERMANY
Kiel
Nord-Ostsee Kanal (Kiel Canal)
Eider
Flensburg
Tønder
Løgumkloster
Møgeltønder
Højer
Skaerbaek
Bredebro
Toftlund
Brøns
Gram
Rødding
Hviding
Ribe
Bramming
Skodborg
Vejen
Holsted
Agerbaek
Gørding
Bramming
Blåvands Huk
Kolding
Fredericia
Middelfart
Vejle
Børkop
Jelling
Løsning
Hedensted
Juelsminde
Horsens
Skanderborg
Malling
Odder
Hov
Samsø
Nordby
Gylling
VEST-SJAELLAND
VEJLE
Tørring
Uldum
Braedstrup
Thyregod
Skanderborg
Odense
FYN
Haderslev
Aabenraa
Gråsten
Sønderborg
Als
Augustenborg
AErøskøbing
AErø
Marstal
Langeland
Rudkøbing
Svendborg
Faaborg
Ringe
Kerteminde
Nyborg
Korsør
Slagelse
Sorø
Ringsted
Roskilde
Frederikssund
Hillerød
Helsingør
Hälsingborg
FREDERIKSBORG
Helsinge
Gilleleje
Hornbaek
Hellebaek
Frederiksvaerk
Landskrona
Eslöv
Malmö
Trelleborg
Copenhagen (København)
Frederiksberg
Lyngby
Gentofte
Glostrup
Tårnby
Kastrup
Amager
Køge
Køge Bugt
Haslev
Skaelskør
Naestved
Fakse
Fakse Ladeplads
Praestø
Store Heddinge
Fakse Bugt
STORSTRØM
Vordingborg
Møn
Møns Klint
Stege
Fejø
Femø
Nørre Alslev
Lolland
Nakskov
Maribo
Holeby
Nykøbing
Falster
Sakskøbing
Stubbekøbing
Gedser
Gedser Odde
Nysted
Rødby

BORNHOLM

Same scale as main map
Allinge-Sandvig
Hasle
Svaneke
Rønne
Akirkeby
Neksø
BORNHOLM

Germany

CONIC PROJECTION
SCALE OF MILES

SCALE OF KILOMETERS

Capitals of Countries .... ✪
State and District Capitals .... ◉
International Boundaries ....
State and District Boundaries ......
Canals ....

Scale 1:3,040,000

East Germany is divided into districts bearing the same name as their respective capitals.

Berlin

© Copyright HAMMOND INCORPORATED, Maplewood, N.J.

AREA 95,985 sq. mi. (248,601 sq. km.)
POPULATION 61,658,000
CAPITAL Bonn
LARGEST CITY Berlin (West)
HIGHEST POINT Zugspitze 9,718 ft. (2,962 m.)
MONETARY UNIT Deutsche mark
MAJOR LANGUAGE German
MAJOR RELIGIONS Protestantism, Roman Catholicism

AREA 41,768 sq. mi. (108,179 sq. km.)
POPULATION 16,737,000
CAPITAL Berlin (East)
LARGEST CITY Berlin (East)
HIGHEST POINT Fichtelberg 3,983 ft. (1,214 m.)
MONETARY UNIT East German mark
MAJOR LANGUAGE German
MAJOR RELIGIONS Protestantism, Roman Catholicism

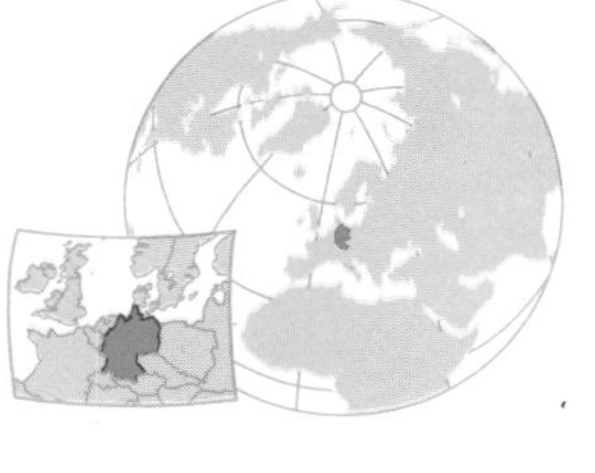

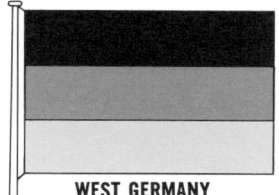

WEST GERMANY

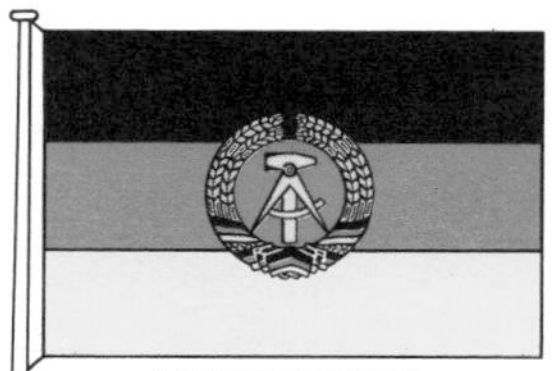
EAST GERMANY

Topography

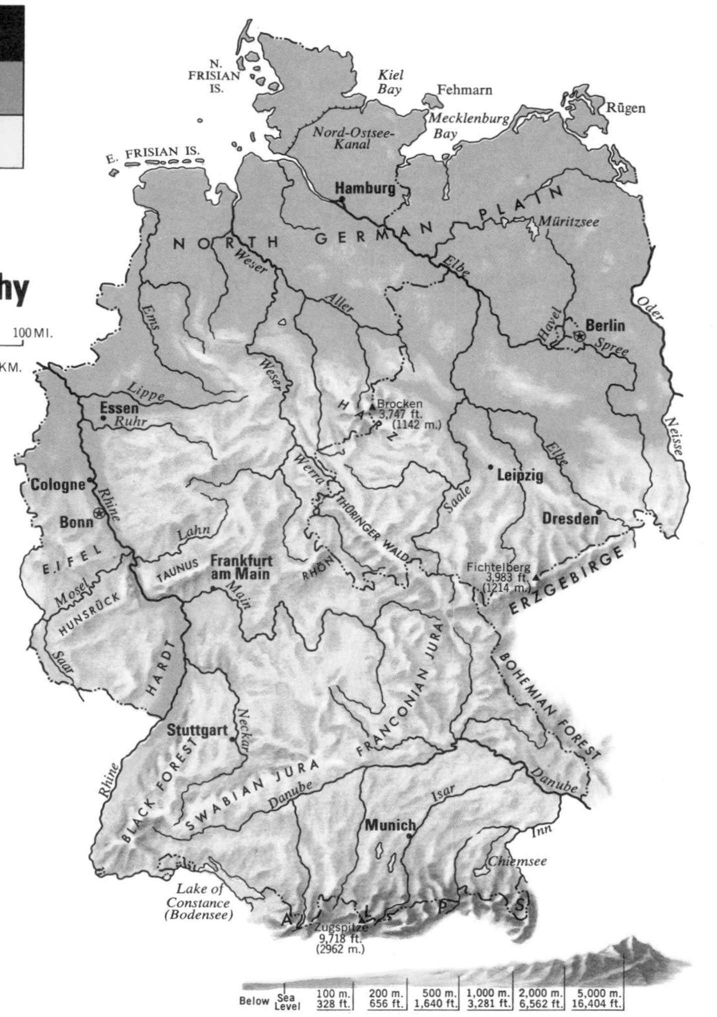

EAST GERMANY

DISTRICTS

Berlin 1,094,147 ..... F4
Cottbus 872,242 ..... F3
Dresden 1,845,459 ..... E3
Erfurt 1,247,213 ..... D3
Frankfurt 688,637 ..... F2
Gera 738,847 ..... D3
Halle 1,890,187 ..... D3
Karl-Marx-Stadt 1,994,115 ..... E3
Leipzig 1,457,817 ..... E3
Magdeburg 1,297,881 ..... D2
Neubrandenburg 628,686 ..... E2
Potsdam 1,124,892 ..... E2
Rostock 867,806 ..... E1
Schwerin 592,334 ..... D2
Suhl 550,497 ..... D3

CITIES and TOWNS

Aken 11,742 ..... D3
Altenburg 51,193 ..... E3
Angermünde 11,786 ..... E2
Anklam 19,099 ..... E2
Annaberg-Buchholz 26,561 ..... E3
Apolda 28,649 ..... D3
Arnstadt 29,462 ..... D3
Aschersleben 36,674 ..... D3
Aue 32,622 ..... E3
Auerbach 18,168 ..... E3
Bad Doberan 12,541 ..... D1
Bad Dürrenberg 15,192 ..... D3
Bad Langensalza 166,282 ..... D3
Bad Salzungen 17,277 ..... C3
Barth 12,069 ..... E1
Bautzen 45,851 ..... F3
Bergen 13,244 ..... E1
Berlin, East (cap.) 1,094,147 ..... F4
Bernau bei Berlin 15,749 ..... E2
Bernburg 44,428 ..... D3
Bischofswerda 11,540 ..... F3
Bitterfeld 27,062 ..... E3
Blankenburg am Harz 18,784 ..... D3
Boizenburg an der Elbe 12,428 ..... D2
Borna 21,807 ..... E3
Brandenburg 94,071 ..... E2
Burg bei Magdeburg 29,027 ..... D2
Calbe 15,976 ..... D3
Chemnitz (Karl-Marx-Stadt) 303,811 ..... E3
Coswig, Dresden 22,149 ..... E3
Coswig, Halle 12,473 ..... E3
Cottbus 94,293 ..... F3
Crimmitschau 28,845 ..... E3
Delitzsch 24,076 ..... E3
Demmin 17,270 ..... E2
Dessau 100,820 ..... E3
Döbeln 27,624 ..... E3
Dresden 507,692 ..... E3
Ebersbach 12,694 ..... F3
Eberswalde-Finow 47,141 ..... E2
Eilenburg 22,245 ..... E3
Eisenach 49,954 ..... D3
Eisenberg 13,450 ..... D3
Eisenhüttenstadt 46,455 ..... F2
Eisleben 29,297 ..... D3
Erfurt 202,979 ..... D3
Falkensee 25,295 ..... E3
Falkenstein 14,367 ..... E3
Finsterwalde 22,466 ..... E3
Forst 28,084 ..... F3
Frankfurt an der Oder 70,817 ..... F2
Freiberg 50,815 ..... E3
Freital 46,061 ..... E3
Friedland E2
Fürstenwalde 31,065 ..... F2
Gardelegen 12,987 ..... D2
Genthin 15,916 ..... E2
Gera 113,108 ..... E3
Glauchau 30,927 ..... E3
Görlitz 84,658 ..... F3
Gotha 59,243 ..... D3
Greifswald 53,940 ..... E1
Greiz 37,612 ..... E3
Grevesmühlen 12,005 ..... D2
Grimma 17,100 ..... E3
Grimmen 14,571 ..... E1
Grossenhain 18,712 ..... E3
Grossräschen 12,889 ..... E3
Guben (Wilhelm-Pieck-Stadt) 32,731 ..... F3
Güstrow 36,824 ..... E2
Halberstadt 46,669 ..... D3
Haldensleben 19,194 ..... D2
Halle 241,425 ..... D3
Halle-Neustadt 67,956 ..... D3
Havelberg D2
Heidenau 21,315 ..... E3
Heiligenstadt 13,931 ..... D3
Hennigsdorf bei Berlin 24,853 ..... E3
Hettstedt 20,291 ..... D3
Hildburghausen 11,372 ..... D3
Hoyerswerda 64,904 ..... F3
Ilmenau 22,021 ..... D3
Jena 99,431 ..... D3
Johanngeorgenstadt 10,328 ..... E3
Jüterbog 13,477 ..... E3
Kamenz 18,221 ..... F3
Karl-Marx-Stadt 303,811 ..... E3
Kleinmachnow 14,059 ..... E4
Klingenthal 13,614 ..... E3
Königs Wusterhausen 11,825 ..... E2
Köpenick 130,987 ..... F4
Köthen 35,451 ..... E3
Kühlungsborn D1
Lauchhammer 26,939 ..... E3
Leipzig 570,972 ..... E3
Lichtenberg 192,063 ..... F4
Limbach-Oberfrohna 25,706 ..... E3
Löbau 18,077 ..... F3
Lübben 14,224 ..... E3
Lübbenau 22,350 ..... F3
Luckenwalde 28,544 ..... E2
Ludwigslust 13,280 ..... D2
Magdeburg 276,089 ..... D2
Markkleeberg 22,380 ..... E3
Meerane 25,037 ..... E3
Meiningen 26,134 ..... D3
Meissen 43,561 ..... E3
Merseburg 54,269 ..... D3
Meuselwitz 13,585 ..... E3
Mittweida 19,259 ..... E3
Mühlhausen (Thomas-Müntzer-Stadt) 44,106 ..... D3
Nauen 11,940 ..... E2
Naumburg 36,358 ..... D3
Neubrandenburg 59,971 ..... E2
Neuenhagen bei Berlin 12,603 ..... F4
Neuruppin 24,888 ..... E2
Neustrelitz 27,074 ..... E2
Nordhausen 44,442 ..... D3
Oelsnitz 15,084 ..... E3
Oelsnitz im Erzgebirge 16,063 ..... E3
Olbernhau 13,479 ..... E3
Oranienburg 24,452 ..... E2
Oschatz 18,974 ..... E3
Oschersleben 17,377 ..... D2
Pankow 136,527 ..... F3
Parchim 22,927 ..... D2
Pasewalk 15,099 ..... F2
Peenemünde E1
Perleberg 15,029 ..... D2
Pirna 49,771 ..... E3
Plauen 80,353 ..... E3
Pössneck 18,648 ..... D3
Potsdam 117,236 ..... E2
Prenzlau 22,738 ..... E2
Pritzwalk 11,887 ..... E2
Quedlinburg 29,796 ..... D3
Radeberg 18,528 ..... E3
Radebeul 38,383 ..... E3
Rathenow 32,011 ..... E2
Reichenbach 27,440 ..... E3
Ribnitz-Damgarten 17,254 ..... E1
Riesa 49,989 ..... E3
Rosslau 16,520 ..... E3
Rostock 210,167 ..... E1
Rudolstadt 31,698 ..... D3
Saalfeld 33,648 ..... D3
Salzwedel 21,741 ..... D2
Sangerhausen 32,721 ..... D3
Sassnitz 13,857 ..... E1
Schkeuditz 15,585 ..... E3
Schmalkalden 15,017 ..... D3
Schmölln 13,406 ..... E3
Schneeberg 20,376 ..... E3
Schönebeck 45,197 ..... D2
Schwedt 45,729 ..... F2
Schwerin 104,984 ..... D2
Sebnitz 13,470 ..... F3
Senftenberg 29,953 ..... F3
Sömmerda 20,712 ..... D3
Sondershausen 23,383 ..... D3
Sonneberg 29,193 ..... D3
Spremberg 22,862 ..... F3
Stassfurt 26,225 ..... D3
Stendal 39,647 ..... D2
Stralsund 72,167 ..... E1
Strausberg 21,334 ..... F2
Suhl 36,642 ..... D3
Tangermünde 12,898 ..... D2
Teltow 16,171 ..... E4
Templin 11,718 ..... E2
Thale 17,248 ..... D3
Thomas-Müntzer-Stadt 44,106 ..... D3
Torgau 21,613 ..... E3
Torgelow 14,320 ..... F2
Treptow 127,448 ..... F4
Ueckermünde 11,423 ..... F2
Waldheim 11,925 ..... E3
Waltershausen 13,893 ..... D3
Waren 22,921 ..... E2
Weida 11,816 ..... D3
Weimar 63,144 ..... D3
Weissenfels 43,191 ..... D3
Weissensee 78,451 ..... F3
Weisswasser 25,910 ..... F3
Werdau 22,249 ..... E3
Wernigerode 34,658 ..... D3
Wilhelm-Pieck-Stadt 32,731 ..... F3
Wismar 56,765 ..... D2
Wittenberg 51,364 ..... E3
Wittenberge 32,907 ..... D2
Wolfen 27,570 ..... E3
Wolgast 16,384 ..... E1
Wurzen 20,501 ..... E3
Zehdenick 12,651 ..... E2
Zeitz 44,582 ..... E3
Zella-Mehlis 16,301 ..... D3
Zerbst 19,356 ..... E3
Zeulenroda 13,452 ..... D3
Zittau 42,298 ..... F3
Zwickau 123,069 ..... E3

OTHER FEATURES

Altmark (reg.) D2
Arkona (cape) E1
Baltic (sea) E1
Black Elster (riv.) E3
Brandenburg (reg.) E2
Elbe (riv.) D2
Elde (riv.) D2
Elster, Black (riv.) E3
Elster, White (riv.) E3
Erzgebirge (mts.) E3
Fichtelberg (mt.) E3
Harz (mts.) D3
Havel (riv.) E2
Lusatia (reg.) F3
Mecklenburg (bay) D1
Mecklenburg (reg.) E2
Mulde (riv.) E3
Neisse (riv.) F3
Oder (riv.) F2
Peene (riv.) E2
Pomerania (reg.) E2
Pomeranian (bay) F1
Rhön (mts.) D3
Rügen (isl.) E1
Saale (riv.) D3
Saxony (reg.) E3
Spree (riv.) F3
Spreewald (for.) F3
Thüringer Wald (for.) D3
Thuringia (reg.) D3
Ücker (riv.) E2
Unstrut (riv.) D3
Usedom (isl.) F1
Warnow (riv.) D2
Werra (riv.) D3
White Elster (riv.) E3

WEST GERMANY

STATES

Baden-Württemberg 9,152,700 ..... C4
Bavaria 10,810,400 ..... D4
Berlin (West) (free city) 1,984,800 ..... E4
Bremen 716,800 ..... C2
Hamburg 1,717,400 ..... D2
Hesse 5,549,800 ..... C3
Lower Saxony 7,238,500 ..... C2
North Rhine-Westphalia 17,129,600 ..... B3
Rhineland-Palatinate 3,665,800 ..... B4
Saarland 1,096,300 ..... B4
Schleswig-Holstein 2,582,400 ..... C1

CITIES and TOWNS

Aachen 242,453 ..... B3
Aalen 64,735 ..... D4
Ahaus 27,126 ..... B2
Ahlen 54,214 ..... B3
Ahrensburg 24,964 ..... D2
Alfeld 24,273 ..... C2
Alsdorf 47,473 ..... B3
Alsfeld 18,091 ..... C3
Altena 26,753 ..... B3
Altona C2
Alzey 15,190 ..... C4
Amberg 46,934 ..... D4
Andernach 27,132 ..... B3
Ansbach 39,117 ..... D4
Arnsberg 80,287 ..... C3
Arolsen 15,619 ..... C3
Aschaffenburg 55,398 ..... C4
Augsburg 249,943 ..... D4
Aurich 34,194 ..... B2
Backnang 29,614 ..... C4
Bad Berleburg 20,415 ..... C3
Bad Driburg 17,478 ..... C3
Bad Dürkheim 16,133 ..... C4
Bad Ems 10,487 ..... B3
Baden-Baden 49,718 ..... C4
Bad Gandersheim 11,614 ..... D3
Bad Harzburg 25,786 ..... D3
Bad Hersfeld 29,248 ..... C3
Bad Homburg vor der Höhe 51,196 ..... C3
Bad Honnef 20,903 ..... B3
Bad Kissingen 22,279 ..... D3
Bad Kreuznach 42,588 ..... B4
Bad Lauterberg im Harz 14,715 ..... D3
Bad Mergentheim 19,895 ..... C4
Bad Münstereifel 14,340 ..... B3
Bad Nauheim 25,916 ..... C3
Bad Neuenahr-Ahrweiler 26,371 ..... B3
Bad Oldesloe 19,640 ..... D2
Bad Pyrmont 21,896 ..... C3
Bad Reichenhall 13,048 ..... E5
Bad Salzuflen 50,924 ..... C2
Bad Schwartau 18,696 ..... D2
Bad Segeberg 13,320 ..... D2
Bad Tölz 12,458 ..... D5
Bad Vilbel 25,012 ..... C3
Bad Waldsee 14,296 ..... C5
Bad Wildungen 15,418 ..... C3
Bad Wimpfen 5,536 ..... C4
Baiersbronn 14,845 ..... C4
Balingen 29,310 ..... C4
Bamberg 74,236 ..... D4
Barsinghausen 32,873 ..... C2
Bassum 14,113 ..... C2
Bayreuth 67,035 ..... D4
Bayrischzell 1,639 ..... E5
Bebra 15,740 ..... C3
Bendorf 15,943 ..... B3
Bensheim 32,653 ..... C4
Bentheim 13,681 ..... B2
Berchtesgaden 8,558 ..... E5
Bergisch Gladbach 99,517 ..... B3
Berleburg (Bad Berleburg) 20,415 ..... C3
Berlin (West) 1,984,837 ..... E4
Biberach an der Riss 28,891 ..... C4
Bielefeld 316,058 ..... C2
Bietigheim-Bissingen 34,042 ..... C4
Bingen 24,541 ..... B4
Birkenfeld 5,883 ..... B4
Blaubeuren 11,652 ..... C4
Böblingen 40,547 ..... C4
Bocholt 65,460 ..... B3
Bochum 414,842 ..... B3
Bonn (cap.) 283,711 ..... B3
Boppard 16,888 ..... B3
Borghorst 17,238 ..... B2
Borken 30,212 ..... B3
Bornheim 32,847 ..... B3
Bottrop 101,495 ..... B3
Brake 18,089 ..... C2
Bramsche 24,119 ..... B2
Braunschweig (Brunswick) 268,519 ..... D2
Breisach am Rhein 9,230 ..... B4
Bremen 572,969 ..... C2
Bremerhaven 143,836 ..... C2
Bremervörde 17,565 ..... C2
Bretten 22,140 ..... C4
Brilon 24,595 ..... C3
Bruchsal 38,929 ..... C4
Brühl 44,305 ..... B3
Brunsbüttel 11,451 ..... C2
Brunswick 268,519 ..... D2
Buchholz in der Nordheide 25,713 ..... C2
Bückeburg 21,393 ..... C2
Büdingen 16,845 ..... C3
Bühl 21,596 ..... C4
Bünde 40,021 ..... C2
Büren 17,352 ..... C3
Burg auf Fehmarn 5,874 ..... D1
Burghausen 16,892 ..... E4
Burgsteinfurt 31,367 ..... B2
Butzbach 20,592 ..... C3
Buxtehude 30,249 ..... C2
Castrop-Rauxel 82,373 ..... B3
Celle 74,347 ..... D2
Cham 12,423 ..... E4
Charlottenburg 201,732 ..... E4
Clausthal-Zellerfeld 16,690 ..... D3
Cloppenburg 19,757 ..... B2
Coburg 46,244 ..... D3
Coesfeld 30,617 ..... B3
Cologne 1,013,771 ..... B3
Crailsheim 24,506 ..... D4
Cuxhaven 60,353 ..... C2
Dachau 33,207 ..... D4
Dahlem E4
Darmstadt 137,018 ..... C4
Deggendorf 25,188 ..... E4
Delmenhorst 71,488 ..... C2
Detmold 65,629 ..... C3
Diepholz 14,201 ..... C2
Dillenburg 14,068 ..... C3
Dillingen 21,369 ..... B4
Dillingen an der Donau 11,601 ..... D4
Dingolfing 13,325 ..... E4
Dinkelsbühl 10,034 ..... D4
Donaueschingen 17,578 ..... C5
Donauwörth 17,077 ..... D4
Dorsten 65,718 ..... B3
Dortmund 630,609 ..... B3
Duderstadt 23,255 ..... D3
Dudweiler 27,877 ..... B4
Duisburg 591,635 ..... B3
Dülmen 37,013 ..... B3
Düren 87,774 ..... B3
Düsseldorf 664,336 ..... B3
Eberbach 15,834 ..... C4
Ebingen 22,594 ..... C4
Eckernförde 22,938 ..... D1
Ehingen 21,600 ..... C4
Eichstätt 13,080 ..... D4
Einbeck 29,821 ..... C3
Eiserfeld 22,346 ..... C3
Ellwangen 21,994 ..... D4
Elmshorn 41,355 ..... C2
Emden 53,509 ..... B2
Emmendingen 24,722 ..... B4
Emmerich 29,113 ..... B3
Emsdetten 30,195 ..... B2
Erlangen 100,671 ..... D4
Eschwege 24,882 ..... C3
Eschweiler 53,603 ..... B3
Espelkamp 22,670 ..... C2
Essen 677,568 ..... B3
Esslingen am Neckar 95,298 ..... C4
Ettlingen 35,159 ..... C4
Euskirchen 43,558 ..... B3
Eutin 17,701 ..... D1
Fellbach 42,501 ..... C4
Flensburg 93,213 ..... C1
Forchheim 23,430 ..... D4
Frankenberg-Eder 15,337 ..... C3
Frankenthal 43,684 ..... C4
Frankfurt am Main 636,157 ..... C3
Frechen 41,453 ..... B3
Freiburg im Breisgau 175,371 ..... B5
Freising 31,524 ..... D4
Freudenstadt 19,454 ..... C4
Friedberg 24,762 ..... C3
Friedrichshafen 51,544 ..... C5
Fritzlar 15,079 ..... C3
Fulda 58,976 ..... C3
Fürstenfeldbruck 27,194 ..... D4
Fürth 101,639 ..... D4
Füssen 10,506 ..... D5
Gaggenau 28,846 ..... C4
Garbsen 56,337 ..... C2
Garmisch-Partenkirchen 26,831 ..... D5
Gatow E4
Geesthacht 24,745 ..... D2
Geislingen an der Steige 28,693 ..... C4
Geldern 24,082 ..... B3
Gelnhausen 17,889 ..... C3
Gelsenkirchen 322,584 ..... B3
Georgsmarienhütte 30,259 ..... B2
Geretsried 17,330 ..... D5
Germersheim 12,041 ..... C4
Gerolstein 6,857 ..... B3
Gifhorn 31,635 ..... D2
Glückstadt 12,159 ..... C2
Goch 28,213 ..... B3
Göggingen 15,980 ..... D4
Göppingen 54,365 ..... C4
Goslar 53,957 ..... D3
Göttingen 123,797 ..... D3
Greven 27,479 ..... B2
Grevenbroich 56,392 ..... B3
Griesheim 18,548 ..... C4
Gronau 40,527 ..... B2
Gummersbach 49,316 ..... B3
Günzburg 13,528 ..... D4
Gunzenhausen 13,565 ..... D4
Gütersloh 77,128 ..... C3
Haar 18,824 ..... D4
Hagen 229,224 ..... B3
Haltern 29,750 ..... B3
Hamburg 1,717,383 ..... D2
Hameln 61,066 ..... C2
Hamm 172,210 ..... B3
Hammelburg 12,350 ..... C3
Hanau 86,676 ..... C3
Hannover 552,955 ..... C2
Harburg-Wilhelmsburg C2
Hassloch 17,752 ..... C4
Haunstetten 21,810 ..... D4
Hechingen 15,926 ..... C4
Heide 21,918 ..... C1
Heidelberg 129,368 ..... C4
Heidenheim an der Brenz 49,943 ..... D4
Heilbronn 113,177 ..... C4
Helmstedt 28,095 ..... D2
Hennef 27,815 ..... B3
Herford 64,385 ..... C2
Herne 190,561 ..... B3
Hildesheim 105,290 ..... D2
Hockenheim 16,890 ..... C4
Hof 54,357 ..... D3
Hofgeismar 13,380 ..... C3
Holzminden 23,650 ..... C3
Homburg 41,861 ..... B4
Horn-Bad Meinberg 16,927 ..... C3
Höxter 32,759 ..... C3
Hückelhoven 34,865 ..... B3
Hünfeld 13,873 ..... C3
Hürth 51,692 ..... B3
Husum 24,984 ..... C1
Hüttental 39,561 ..... C3
Ibbenbüren 42,202 ..... B2
Idar-Oberstein 37,179 ..... B4
Immenstadt im Allgäu 13,720 ..... C5
Ingolstadt 88,500 ..... D4
Iserlohn 96,174 ..... B3
Isny im Allgäu 12,367 ..... D5
Itzehoe 35,077 ..... C2
Jever 12,096 ..... B2
Jülich 31,564 ..... B3
Kaiserslautern 100,886 ..... B4
Karlsruhe 280,448 ..... C4
Kassel 205,534 ..... C3
Kaufbeuren 42,224 ..... D5
Kehl 29,861 ..... B4
Kelheim 11,996 ..... D4
Kempten 56,944 ..... D5
Kevelaer 20,971 ..... B3
Kiel 262,164 ..... D1
Kirchheim unter Teck 31,666 ..... C4
Kitzingen 19,116 ..... C4
Kleve 44,043 ..... B3
Koblenz 118,394 ..... B3
Köln (Cologne) 1,013,771 ..... B3
Königswinter 34,586 ..... B3
Konstanz 70,152 ..... C5
Korbach 22,998 ..... C3
Kornwestheim 27,771 ..... C4
Krefeld 228,463 ..... B3
Kreutztal 30,473 ..... C3
Kronach 11,538 ..... D3
Kulmbach 25,711 ..... D3
Lage 31,724 ..... C3
Lahnstein 19,725 ..... B3
Lahr 35,570 ..... B4
Lampertheim 31,993 ..... C4
Landau in der Pfalz 37,661 ..... C4
Landsberg am Lech 15,862 ..... D4
Landshut 55,858 ..... E4
Langen 30,227 ..... C4
Langenhagen 47,092 ..... C2
Lauenburg an der Elbe 11,077 ..... D2
Lauf an der Pegnitz 19,443 ..... D4
Lauingen 8,778 ..... D4
Lauterbach 15,007 ..... C3
Leer 32,785 ..... B2
Lehrte 38,272 ..... D2
Lemgo 39,664 ..... C2
Lengerich 20,836 ..... B2
Leverkusen 165,947 ..... B3
Lichtenfels 13,719 ..... D3
Limburg an der Lahn 28,606 ..... C3
Lindau 23,930 ..... C5

(continued on following page)

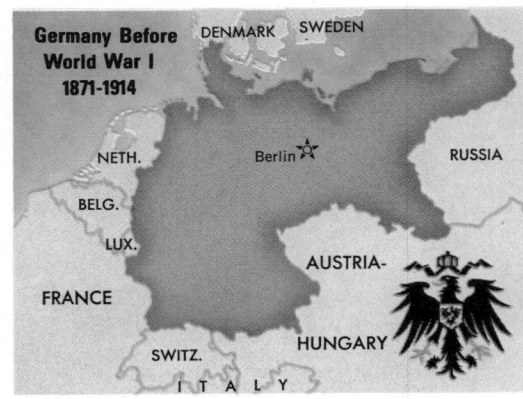

Germany Before World War I 1871-1914
DENMARK · SWEDEN · NETH. · Berlin ☆ · RUSSIA · BELG. · LUX. · FRANCE · AUSTRIA-HUNGARY · SWITZ. · ITALY

Germany Between Wars 1919-1937
DENMARK · SWEDEN · LITH. · DANZIG · NETH. · Berlin ☆ · POLAND · BELG. · LUX. · SAAR (To Germany 1935) · CZECHOSLOVAKIA · FRANCE · AUSTRIA · SWITZ. · ITALY · YUGO. · HUNG.

Occupied Germany 1945-1949
DENMARK · SWEDEN · U.S.S.R. · NETH. · BRITISH ZONE · BERLIN · RUSSIAN ZONE · POLAND · BELG. · LUX. · FRENCH ZONE · SAAR · AMERICAN ZONE · CZECHOSLOVAKIA · FRANCE · AUSTRIA · SWITZ. · ITALY · YUGO. · HUNG.

Lingen 43,785 ....B2	Oberpfalz 29,713 ....D4
Lippstadt 63,040 ....C3	Neumünster 84,777 ....C1
Löhne 17,859 ....C2	Neunkirchen 54,992 ....B4
Lohr am Main 16,435 ....C4	Neuss 148,198 ....B3
Lörrach 44,179 ....B5	Neustadt an der
Lübeck 232,270 ....C1	Weinstrasse 51,011 ....B4
Lüdenscheid 76,213 ....B3	Neustadt bei Coburg 12,665 ....D3
Ludwigsburg 83,622 ....C4	Neustadt in Holstein 15,333 ....D1
Ludwigshafen am Rhein 170,374 ....C4	Neu-Ulm 31,660 ....C4
Lüneburg 64,586 ....D2	Neuwied 62,029 ....B3
Lünen 85,685 ....B3	Nienburg 30,978 ....C2
Mainz 183,880 ....C4	Norden 24,207 ....B2
Mannheim 314,086 ....C4	Nordenham 31,457 ....C2
Marbach am Neckar 12,131 ....C4	Norderstedt 61,553 ....D2
Marburg an der Lahn 72,458 ....C3	Nordhorn 49,598 ....B2
Marktredwitz 16,404 ....E4	Nördlingen 16,480 ....D4
Marl 91,930 ....B3	Nordheim 32,665 ....D4
Mayen 21,018 ....B3	Nuremberg 499,060 ....D4
Meerbusch 52,872 ....B3	Nürnberg (Nuremberg) 499,060 ....D4
Meppen 27,308 ....B2	Nürtingen 34,333 ....C4
Merzig 30,197 ....B4	Oberammergau 4,704 ....D5
Meschede 32,472 ....C3	Oberhausen 237,147 ....B3
Metzingen 19,224 ....C4	Oberstdorf 11,687 ....D5
Michelstadt 13,591 ....C4	Obersulzel 39,802 ....C3
Minden 78,887 ....C2	Offenbach am Main 115,251 ....C3
Mittenwald 8,831 ....D5	Offenburg 51,553 ....B4
Mölln 15,780 ....D2	Oldenburg 134,706 ....C2
Mönchengladbach 261,367 ....B3	Oldenburg in Holstein 9,201 ....D1
Moosburg an der Isar 12,196 ....D4	Opladen 42,789 ....B3
Mosbach 23,663 ....C4	Osnabrück 161,671 ....C2
Mühldorf am Inn 12,638 ....E4	Osterholz-Scharmbeck 22,734 ....C2
Mülheim an der Ruhr 189,259 ....B3	Osterode am Harz 29,668 ....D3
Müllheim 12,183 ....B5	Paderborn 103,705 ....C3
München (Munich) 1,314,865 ....D4	Papenburg 27,039 ....B2
Münden 27,018 ....C3	Passau 50,920 ....E4
Munich 1,314,865 ....D4	Peine 49,450 ....D2
Münster 264,546 ....B3	Pfaffenhofen an der Ilm 13,684 ....D4
Nagold 19,047 ....C4	Pforzheim 108,635 ....C4
Neckarsulm 20,112 ....C4	Pfullingen 16,195 ....C4
Neheim-Hüsten 36,373 ....C3	Pinneberg 36,844 ....C2
Neuburg an der Donau 19,400 ....D4	Pirmasens 53,651 ....B4
Neu-Isenburg 35,631 ....C3	Piettenberg 29,273 ....C3
Neumarkt in der	Porz am Rhein 74,915 ....B3
	Preetz 15,305 ....D1
	Puttgarden D1
	Radolfzell 23,274 ....C5

Rastatt 38,030 ....C4	Schwetzingen 18,286 ....C4
Rastede 16,905 ....C2	Seesen 23,577 ....D3
Ratingen 86,028 ....B3	Selb 16,723 ....E3
Ratzeburg 12,189 ....D2	Sennestadt 20,187 ....C3
Ravensburg 42,725 ....C5	Siegburg 34,943 ....B3
Recklinghausen 122,437 ....B3	Siegen 116,552 ....C3
Regensburg 131,886 ....E4	Sigmaringen 15,437 ....C4
Remagen 14,627 ....B3	Singen 54,134 ....C4
Remscheid 133,145 ....B3	Singen 45,566 ....C5
Rendsburg 34,407 ....C1	Soest 40,308 ....C3
Reutlingen 95,289 ....C4	Solingen 171,810 ....B3
Rheda-Wiedenbrück 37,371 ....C3	Soltau 19,949 ....C2
Rheine 71,539 ....B2	Sonthofen 17,821 ....D5
Rheinfelden 27,500 ....B5	Spandau 197,687 ....E3
Rheydt 100,077 ....B3	Speyer 44,471 ....C4
Rietberg 22,421 ....C3	Springe 30,968 ....C2
Rinteln 25,595 ....C2	Stade 42,097 ....C2
Rosenheim 38,419 ....D5	Stadthagen 23,003 ....C2
Rotenburg 19,155 ....C2	Stolberg 57,379 ....B3
Rotenburg an der Fulda 14,438 ....C3	Straubing 43,774 ....E4
Roth bei Nürnberg 17,782 ....D4	Stuttgart 600,421 ....C4
Rothenburg ob der	Sulzbach 13,956 ....B4
Tauber 11,609 ....D4	Sulzbach-Rosenberg 18,596 ....D4
Rottenburg am Neckar 30,583 ....C4	Taifingen 17,278 ....C4
Rottweil 24,534 ....C4	Tegel E3
Rüsselsheim 62,067 ....C4	Tegle 15,165 ....B3
Saarbrücken 205,336 ....B4	Timmendorfer Strand 10,690 ....D1
Saarlouis 39,974 ....B4	Tönning 159,730 ....F4
Sackingen 13,956 ....C5	Traunstein 14,088 ....E5
Salzgitter 117,341 ....D2	Travemünde D1
Sankt Goar 3,511 ....B3	Treuchtlingen 11,939 ....D4
Sankt Ingbert 43,263 ....B4	Trier 100,338 ....B4
Sankt Wendel 27,558 ....B4	Troisdorf 56,402 ....B3
Saulgau 15,403 ....C5	Tuttlingen 32,342 ....C5
Schleswig 30,974 ....C1	Übach-Palenberg 22,403 ....B3
Schlüchtern 13,801 ....C3	Überlingen 17,735 ....C5
Schöneberg 169,835 ....E4	Uelzen 37,550 ....D2
Schöningen 16,348 ....D2	Uetersen 16,330 ....C2
Schramberg 19,677 ....C4	Ulm 98,237 ....C4
Schwabach 33,136 ....D4	Uslar 17,251 ....C3
Schwäbisch Gmünd 56,422 ....C4	Varel 24,435 ....C2
Schwäbisch Hall 32,129 ....C4	Vechta 21,786 ....C2
Schwalmstadt 17,800 ....C3	Verden 24,247 ....C2
Schwandorf in Bayern 22,547 ....E4	Viersen 84,220 ....B3
Schweinfurt 56,164 ....D3	Villingen-Schwenningen 80,646 ....C4
Schwelm 31,850 ....B3	

Völklingen 47,271 ....B4	Ammersee (lake) D4
Waldkirch 19,009 ....B4	Amrum (isl.) C1
Waldkraiburg 20,140 ....E4	Baltrum (isl.) B2
Waldshut-Tiengen 22,046 ....C4	Bavarian (for.) E4
Walsrode 23,423 ....C2	Bavarian Alps (range) D5
Wangen im Allgäu 23,127 ....C5	Black (for.) B4
Wanne-Eickel 99,156 ....B3	Bodensee (Constance) (lake) C5
Warburg 22,150 ....C3	Bohemian (for.) E4
Warendorf 32,273 ....B3	Borkum (isl.) B2
Wedel 30,045 ....C2	Breisgau (reg.) B5
Weiden in der Oberpfalz 42,697 ....D3	Chiemsee (lake) E5
Weilburg 12,652 ....C3	Constance (lake) C5
Weilheim im Oberbayern 15,347 ....D5	Danube (riv.) C4
Weingarten 21,143 ....C5	Donau (Danube) (riv.) C4
Weinheim 41,005 ....C4	East Friesland (reg.) B2
Weissenburg in Bayern 16,083 ....D4	East Frisian (isls.) B2
Wertheim 20,942 ....C4	Eder (res.) C3
Wesel 56,584 ....B3	Elbe (riv.) C2
Westerland 9,652 ....C1	Ems (riv.) B2
Westerstede 16,977 ....B2	Fehmarn (isl.) D1
Wiehl 19,004 ....B3	Feldberg (mt.) C5
Wiesbaden 250,592 ....C4	Fichtelgebirge (range) D3
Wildbad im Schwarzwald 11,611 ....C4	Föhr (isl.) C1
Wideshausen 12,055 ....C2	Franconian Jura (range) D4
Wilhelmshaven 103,417 ....B2	Frisian, East (isls.) B2
Witten 108,771 ....B3	Frisian, North (isls.) B1
Wittingen 12,189 ....D2	Grosser Arber (mt.) E4
Wittlich 15,321 ....B3	Halligen (isls.) C1
Witzenhausen 16,877 ....C3	Hardt (mts.) C4
Wolfenbüttel 51,386 ....D2	Harz (mts.) D3
Wolfsburg 126,298 ....D2	Hegau (reg.) C5
Worms 75,732 ....C4	Heligoland (isl.) C1
Wunstorf 36,795 ....C2	Heligoland (bay) B1
Wuppertal 405,369 ....B3	Helgoland (isl.) B1
Würzburg 112,584 ....C4	Hunsrück (mts.) B4
Xanten 15,688 ....B3	Hunte (riv.) C2
Zirndorf 13,661 ....D4	Iller (riv.) D4
Zülpich 16,171 ....B3	Inn (riv.) E4
Zweibrücken 35,978 ....B4	Isar (riv.) E4
Zwischenahn 22,581 ....B2	Juist (isl.) B2
	Kaiserstuhl (mt.) B4
OTHER FEATURES	Kiel (bay) D1
	Kiel (Nord-Ostsee) (canal) C1
Aller (riv.) C2	Königssee (lake) E5
Allgäu (reg.) D5	Lahn (riv.) C3
Altmühl (riv.) D4	Langeoog (isl.) B2

Lech (riv.) D4	
Leine (riv.) C2	
Lippe (riv.) B3	
Lüneburger Heide (dist.) C2	
Main (riv.) C4	
Mecklenburg (bay) D1	
Mosel (riv.) B4	
Naab (riv.) D4	
Neckar (riv.) C4	
Norderney (isl.) B2	
Nord-Ostsee (canal) C1	
Nordstrand (isl.) C1	
North (sea) C1	
North Friesland (reg.) C1	
North Frisian (isls.) C1	
Odenwald (for.) C4	
Oker (riv.) D2	
Pellworm (isl.) C1	
Regen (riv.) E4	
Regnitz (riv.) D4	
Rhine (riv.) B3	
Rhön (mts.) C3	
Ruhr (riv.) B3	
Saar (riv.) B4	
Sauer (riv.) B4	
Sauerland (reg.) C3	
Schneeberg (mt.) D3	
Schwarzwald (Black) (for.) C5	
Spessart (mts.) C4	
Spiekeroog (isl.) B2	
Starnbergersee (lake) D5	
Swabian Jura (range) C4	
Sylt (isl.) C1	
Tauber (riv.) C4	
Taunus (range) C3	
Tegernsee (lake) D5	
Teutoburger Wald (for.) C2	
Vogelsberg (mts.) C3	
Walchensee (lake) D5	
Wangerooge (isl.) B2	
Watzmann (mt.) E5	
Weser (riv.) C2	
Westerwald (for.) B3	
Würmsee (Starnbergersee) (lake) D5	
Zugspitze (mt.) D5	

Agriculture, Industry and Resources

DOMINANT LAND USE

- Wheat, Sugar Beets
- Cereals (chiefly rye, oats, barley)
- Potatoes, Rye
- Dairy, Livestock
- Mixed Cereals, Dairy
- Truck Farming
- Grapes, Fruit
- Forests

MAJOR MINERAL OCCURRENCES

Ag	Silver		K	Potash
Ba	Barite		Lg	Lignite
C	Coal		Na	Salt
Cu	Copper		O	Petroleum
Fe	Iron Ore		Pb	Lead
G	Natural Gas		U	Uranium
Gr	Graphite		Zn	Zinc

⚡ Water Power

▨ Major Industrial Areas

AREA 15,892 sq. mi. (41,160 sq. km.)
POPULATION 14,227,000
CAPITALS The Hague, Amsterdam
LARGEST CITY Amsterdam
HIGHEST POINT Vaalserberg 1,056 ft. (322 m.)
MONETARY UNIT guilder (florin)
MAJOR LANGUAGE Dutch
MAJOR RELIGIONS Protestantism, Roman Catholicism

AREA 11,781 sq. mi. (30,513 sq. km.)
POPULATION 9,855,110
CAPITAL Brussels
LARGEST CITY Brussels (greater)
HIGHEST POINT Botrange 2,277 ft. (694 m.)
MONETARY UNIT Belgian franc
MAJOR LANGUAGES French (Walloon), Flemish
MAJOR RELIGION Roman Catholicism

AREA 999 sq. mi. (2,587 sq. km.)
POPULATION 364,000
CAPITAL Luxembourg
LARGEST CITY Luxembourg
HIGHEST POINT Ardennes Plateau 1,825 ft. (556 m.)
MONETARY UNIT Luxembourg franc
MAJOR LANGUAGES Luxembourgeois (Letzeburgisch), French, German
MAJOR RELIGION Roman Catholicism

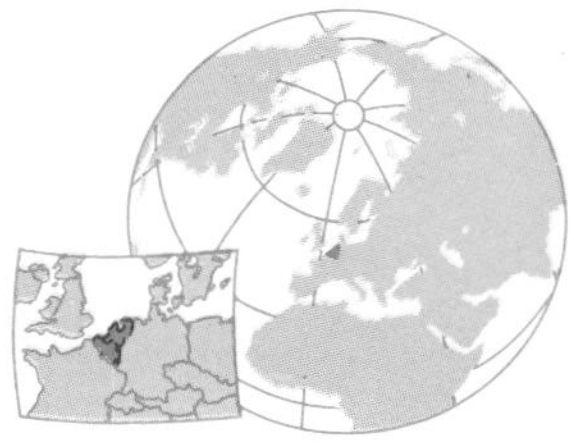

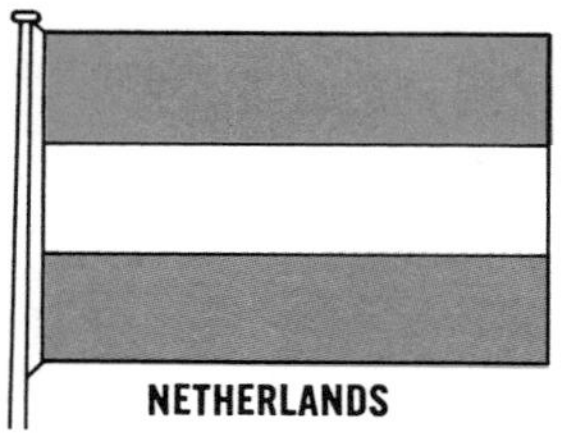

NETHERLANDS

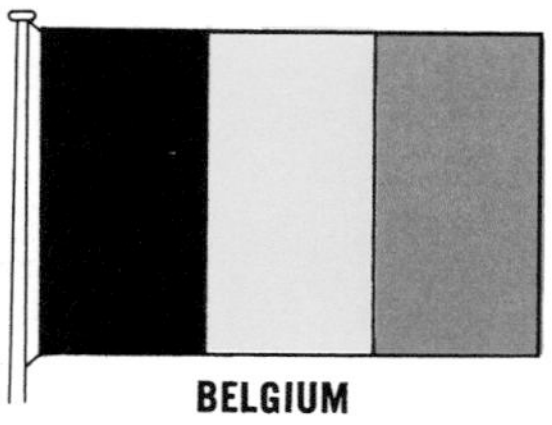

BELGIUM

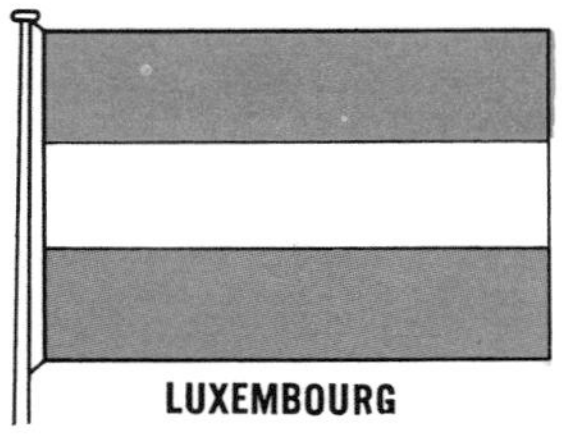

LUXEMBOURG

BELGIUM

PROVINCES

Antwerp 1,533,249 F6
Brabant 2,176,373 F7
East Flanders 1,310,117 D7
Hainaut 1,317,453 D7
Liège 1,008,905 H7
Limburg 652,547 G7
Luxembourg 217,310 G9
Namur 380,561 F8
West Flanders 1,054,429 B7

CITIES and TOWNS†

Aalst 46,659 D7
Aalter 9,173 C6
Aarlen (Arlon) 13,745 H9
Aarschot 12,474 F7
Aat (Ath) 11,842 D7
Alken 8,677 G7
Alost (Aalst) 46,659 D7
Amay 7,617 G7
Andenne 8,091 G8
Anderlecht 103,796 B9
Anderlues 12,176 E8
Ans H7
Antoing 3,426 C7
Antwerp 224,543 E6
Antwerp* 928,000 E6
Antwerpen (Antwerp) 224,543 E6
Ardooie 7,081 C7
Arendonk 9,919 G6
Arlon 13,745 H9
As 5,496 H6
Asse 6,583 E7
Ath 11,842 D7
Attert H9
Aubange 3,761 H9
Audenarde (Oudenaarde) 26,615 D7
Auderghem 34,546 C9
Auvelais 8,287 F8
Aywaille 3,850 H8
Baerle-Hertog F6
Balen 15,110 G6
Basse-Sambre F8
Bastenaken (Bastogne) 6,816 H9
Bastogne 6,816 H9
Beernem C6
Beloeil D7
Berchem 50,241 F6
Berchem-Sainte-Agathe 19,087 B9
Bergen (Mons) 59,362 E8
Beringen G6
Bertogne H8
Bertrix 4,562 G9
Beveren 15,913 E6
Bilzen 7,178 G7
Binche 10,098 E8
Blankenberge 13,969 C6
Bocholt 6,497 H6
Boom 16,584 E6
Borgerhout 49,002 E6
Borgloon 3,412 G7
Borgworm (Waremme) 10,956 G7
Bourg-Léopold (Leopoldsburg) 9,593 G6
Boussu 11,474 D8
Braine-l'Alleud 18,531 E7
Braine-le-Comte 11,957 D7
Brecht F6
Bredene 9,244 B6
Bree 10,389 H6
Bruges 117,220 C6
Brugge (Bruges) 117,220 C6
Brussels (cap.)* 1,054,970 C9
Bruxelles (Brussels) (cap.)* 1,054,970 C9
Cerfontaine E8
Charleroi 23,689 E8
Charleroi* 458,000 E8
Chastre F7
Châtelet 14,752 F8
Chièvres 3,283 D7
Chimay 3,288 E8
Chiny G9
Ciney 7,536 G8
Comblain-au-Pont 3,582 G8
Comines 8,192 B7
Courcelles 17,015 E8
Courtrai (Kortrijk) 44,961 C7
Couvin 4,234 F8
Damme C6
De Haan C6
Deinze 16,711 C7
Denderleeuw 9,925 E7
Dendermonde 22,119 E6
De Panne 6,985 B6
Dessel 7,505 G6
Destelbergen D6
Deurne 80,766 F6
Diest 10,799 F7
Diksmuide 6,669 B6
Dilbeek 15,108 B9
Dilsen H6
Dinant 9,747 G8
Dison 8,466 H7
Dixmude (Diksmuide) 6,669 B6
Doische F8
Doornik (Tournai) 32,794 C7
Dour 10,059 D8
Drogenbos 4,840 B10
Duffel 13,802 F6
Durbuy H8
Ecaussinnes 6,630 E7
Edingen (Enghien) 4,115 D7
Eeklo 19,144 D6
Éghezée F7
Eigenbrakel (Braine-l'Alleud) 18,531 E7
Ekeren 27,648 E6
Ellezelles 3,556 D7
Enghien 4,115 D7
Érezée G8
Erquelinnes 4,471 E8
Esneux 6,183 H7
Essen 10,795 F6
Estampuis C7
Etterbeek 51,030 B9
Eupen 14,879 J7
Evere 26,957 C9
Evergem 12,886 D6
Farciennes E8
Fernelmont F7
Ferrières H8
Flémalle 8,135 G7
Fleurus 8,523 E8
Florennes 4,107 F8
Forest 55,135 B9
Fosses-La-Ville 3,972 F8
Frameries 11,224 D8
Froidchapelle E8
Furnes (Veurne) 9,496 B6
Ganshoren 21,147 B9
Geel 29,346 F6
Geldenaken (Jodoigne) 4,132 F7
Gembloux-sur-Orneau 11,249 F7
Genk 57,913 H7
Gent (Ghent) 148,860 D6
Geraardsbergen 17,533 D7
Gerpinnes F8
Ghent 148,860 D6
Ghent* 477,000 D6
Gistel B6
Gooik E7
Gouvy H8
Grammont (Geraardsbergen) 17,533 D7
Grez-Doiceau F7
Grimbergen E7
Haacht 4,436 F7
Habay H9
Hal (Halle) 20,017 E7
Halen 5,322 G7
Halle 20,017 E7
Hamme 17,559 E6
Hamois G8
Hamont-Achel 6,893 H6
Hannuit (Hannut) 7,232 G7
Hannut 7,232 G7
Harelbeke 18,498 C7
Hasselt 39,663 G7
Hastière F8
Heist-Knokke 27,582 C6
Heist-op-den-Berg 13,472 F6
Hensies D8
Herentals 18,639 F6
Herne E7
Herselt 7,412 F6
Herstal 29,600 H7
Herve 4,118 H7
Heuvelland B7
Hoboken 33,693 E6
Hoei (Huy) 12,736 G8
Hoeselt 6,884 G7
Honnelles D8
Hoogstraten 4,381 F6
Hotton G8
Huy 12,736 G8
Ichtegem B6
Ieper 20,825 B7
Ingelmunster 10,245 C7
Ittre E7
Ixelles 86,450 C9
Izegem 22,928 C7
Jabbeke C6
Jemappes 18,632 D8
Jette 40,013 B9
Jodoigne 4,132 F7
Kalmthout 12,724 F6
Kapellen 13,352 E6
Kasterlee F6
Kinrooi H6
Knokke-Heist 27,582 C6
Koekelare 7,807 B6
Koekelberg 17,570 B9
Koksijde B6
Kontich 14,432 E6
Kortemark 5,904 C6
Kortrijk 44,961 C7
Kraainem 11,390 C9
La Louvière 23,310 E8
La Louvière* 113,259 E8
Lanaken 8,659 H7
Landen 5,740 G7
Langemark-Poelkapelle 5,457 B7
Lasne F7
Lede 10,316 D7
Léglise H9
Leopoldsburg 9,593 G6
Le Roeulx E8
Lessen (Lessines) 8,906 D7
Lessines 8,906 D7
Leuven 30,623 F7
Leuze-en-Hainaut 7,185 D7
Libin G9
Libramont-Chevigny 2,975 G9
Lichtervelde 7,459 C6
Liedekerke 10,482 D7
Liège 145,573 H7
Liège* 622,000 H7
Lier 28,416 F6
Lierre (Lier) 28,416 F6
Limbourg 3,762 J7
Limburg (Limbourg) 3,762 J7
Linkebeek 4,265 C10
Linter G7
Lochristi D6
Lokeren 26,740 D6
Lommel 21,984 G6
Lontzen H9
Looz (Borgloon) 3,412 G7
Lo-Reninge B7
Louvain (Leuven) 30,623 F7
Luik (Liège) 145,573 H7
Lummen G7
Maaseik 8,622 H6
Maasmechelen H7
Machelen 7,057 C9
Maldegem 14,474 C6
Malines (Mechelen) 65,466 F6
Malmédy 6,464 J8
Manage E7
Manhay H8
Marche-en-Famenne 4,567 G8
Marchin 4,206 G8
Mechelen 65,466 F6
Meerhout 8,567 G6
Meise E7
Menen 22,037 C7
Menin (Menen) 22,037 C7
Merchtem 8,998 E7
Merelbeke 13,837 D7
Merksem 39,768 E6
Merksplas 5,065 F6
Messancy 3,150 H9
Mettet 3,372 F8
Meulebeke 10,458 C7
Middelkerke B6
Moeskroen (Mouscron) 37,311 C7
Mol 28,823 G6
Molenbeek-Saint-Jean 68,411 B9
Momignies E8
Mons 59,362 E8
Montigny-le-Tilleul E8
Moorslede B7
Mortsel 28,012 E6
Mouscron 37,311 C7
Namen (Namur) 32,269 F8
Namur 32,269 F8
Nassogne G8
Nazareth D7
Neerpelt 8,771 G6
Neufchâteau 2,670 G9
Nevele D6
Nieuport (Nieuwpoort) 8,273 B6
Nieuwpoort 8,273 B6
Nijvel (Nivelles) 16,126 E7
Ninove 12,428 D7
Nivelles 16,126 E7
Ohey G8
Onhaye F8
Oostende (Ostend) 71,227 B6
Oostkamp 8,999 C6
Opwijk 9,699 E7
Ostend 71,227 B6
Oudenaarde 26,615 D7
Oudenburg B6
Oud-Turnhout 9,245 F6
Oupeye H7
Overijse 16,181 F7
Overpelt 10,470 G6
Paliseul G9
Peer 7,201 G6
Péruwelz 7,878 D8
Philippeville 2,076 E8
Plombières F7
Pont-à-Celles E8
Poperinge 12,671 B7
Profondeville F8
Putte 6,953 F6
Quaregnon 17,688 D8
Quévy D8
Quiévrain 5,510 D8
Raeren 3,655 J7
Ravels G6
Rebecq 3,744 E7
Renaix (Ronse) 25,056 D7
Rendeux H8
Retie 6,619 G6
Rochefort 4,357 G8
Roeselare 40,428 C7
Ronse 25,056 D7
Roulers (Roeselare) 40,428 C7
Rouvroy G9
Ruislede C6
Sainte-Ode H8
Saint-Georges-sur-Meuse 6,003 G7
Saint-Gilles 55,055 B9
Saint-Hubert 3,091 G8
Saint-Josse-ten-Noode 23,633 C9
Saint-Nicolas G7
Saint-Trond (Sint-Truiden) 21,473 G7
Saint-Vith (Sankt Vith) 3,001 J8
Sankt Vith 3,001 J8
Schaerbeek 118,950 C9
Schoten 29,914 F6
Seraing 40,545 G7
's-Gravenbrakel (Braine-le-Comte) 11,957 D7
Sint-Laureins D6
Sint-Niklaas 49,214 E6

(continued on following page)

Agriculture, Industry and Resources

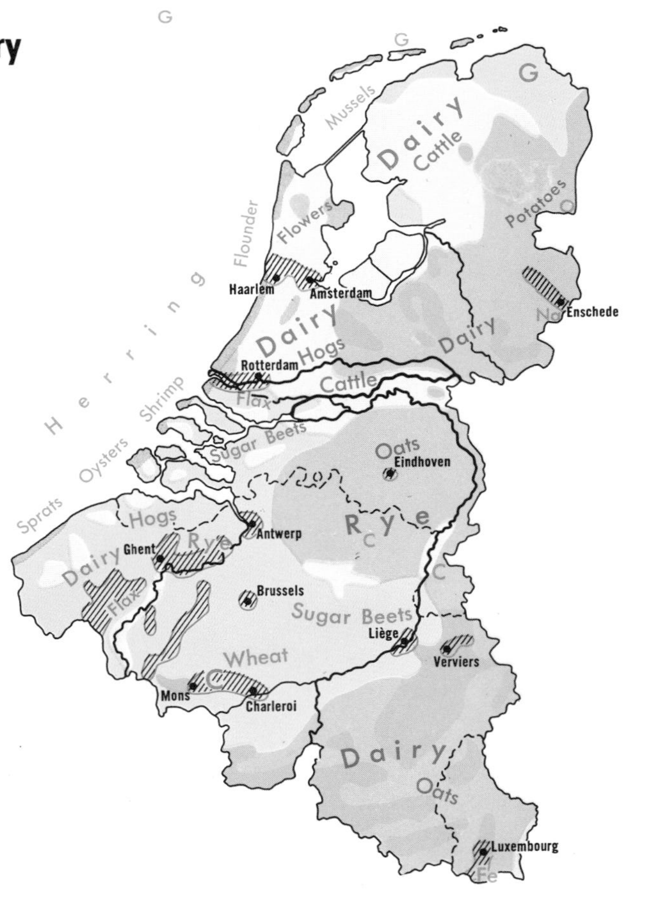

DOMINANT LAND USE

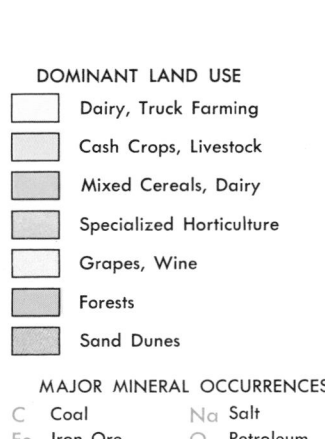

- Dairy, Truck Farming
- Cash Crops, Livestock
- Mixed Cereals, Dairy
- Specialized Horticulture
- Grapes, Wine
- Forests
- Sand Dunes

MAJOR MINERAL OCCURRENCES

C Coal
Fe Iron Ore
G Natural Gas
Na Salt
O Petroleum

Major Industrial Areas

Sint-Pieters-Leeuw 16,856 .... B9
Sint-Truiden 21,473 .... G7
Soignies 12,006 .... D7
Somme-Leuze H8
Spa 9,504 .... H8
Sprimont H8
Staden 5,499 .... B7
Stavelot 4,723 .... H8
Steenokkerzeel 4,037 .... C9
Stekene E6
Stoumont H8
Tamise (Temse) 14,950 .... E6
Tellin G8
Temse 14,950 .... E6
Tenneville H8
Termonde (Dendermonde) 22,119 .... E6
Tessenderlo 11,778 .... G6
Theux 5,316 .... H8
Thuin 5,777 .... E8
Tielt 14,077 .... C7
Tielt-Winge 3,743 .... F7
Tienen 24,134 .... F7
Tintigny G9
Tirlemont (Tienen) 24,134 .... F7
Tongeren 20,136 .... G7
Tongres (Tongeren) 20,136 .... G7
Torhout 15,156 .... C6
Tournai 32,794 .... C7
Trois-Ponts H8
Tubeke (Tubize) 11,507 .... E7
Tubize 11,507 .... E7
Turnhout 38,007 .... F6
Uccle 78,909 .... B9
Ukkel (Uccle) 78,909 .... B9
Vaux-eur-Sûre H8
Verviers 33,587 .... H7
Veurne 9,496 .... B6
Vielsalm 3,587 .... H8
Vilvoorde 34,633 .... C9
Vilvorde (Vilvoorde) 34,633 .... C9
Viroinval F8
Virton 3,558 .... G9
Visé 6,880 .... H7
Vleteren B6
Vorst (Forest) 55,135 .... B9
Vresse-sur-Semois F9
Waarschoot 7,905 .... D6
Wachtebeke D6
Waregem 17,725 .... C7
Waremme 10,956 .... G7
Waterloo 17,764 .... C9
Watermaal-Bosvoorde
  (Watermael-Boitsfort) C9
Watermael-Boitsfort 25,123 .... C9
Waver (Wavre) 11,767 .... F7
Wavre 11,767 .... F7
Wellin G8
Wemmel 12,631 .... B9
Wervik 12,672 .... B7
Westerlo 14,173 .... F6
Westmalle F6
Wetteren 20,816 .... D7
Wezembeek-Oppem 10,899 .... D9
Wezet (Visé) 6,880 .... H7
Willebroek 15,726 .... E6
Wilrijk 43,485 .... E6
Wingene 7,140 .... C6
Woluwe-Saint-Lambert 47,360 .... C9
Woluwe-Saint-Pierre 40,884 .... C9
Ypres (Ieper) 20,825 .... B7
Zaventem 10,625 .... C9
Zedelgem C6
Zeebrugge C6
Zele 18,585 .... E6
Zelzate 12,785 .... D6

Zemst E7
Zinnik (Soignies) 12,006 .... D7
Zonhoven 13,484 .... G6
Zottegem 21,461 .... D7
Zuienkerke C6

OTHER FEATURES

Albert, (canal) F6
Ardennes, (for.) F9
Botrange, (mt.) H8
Dender, (riv.) D7
Dedle, (riv.) F7
Dyle, (riv.) F7
Hohe Venn, (plat.) H8
Lesse, (riv.) F8
Lys, (riv.) B7
Mark, (riv.) F6
Meuse, (riv.) F8
Nethe, (riv.) F6
North, (sea) D4
Ourthe, (riv.) H8
Rupel, (riv.) E6
Sambre, (riv.) D8
Schelde (Scheldt), (riv.) C7
Scheldt, (riv.) C7
Schnee Eifel, (plat.) J8
Semois, (riv.) E7
Senne, (riv.) E7
Vaalserberg, (mt.) H7
Vesdre, (riv.) H7
Weisserstein, (mt.) J8
Yser, (riv.) B7
Zitterwald, (for.) J8

LUXEMBOURG

CITIES and TOWNS

Clervaux 916 .... J8
Diekirch† 5,059 .... J9
Differdange 9,287 .... H9
Dudelange† 14,615 .... J10
Echternach† 3,792 .... J9
Esch-sur-Alzette† 27,574 .... J9
Ettelbruck† 5,990 .... J9
Grevenmacher† 2,918 .... J9
Luxembourg (cap.) 78,272 .... H9
Mamer 3,123 .... H9
Mersch 1,869 .... J9
Pétange 6,234 .... H9
Remich† 12,138 .... J9
Vianden† 1,520 .... J9
Wiltz 1,601 .... H9

OTHER FEATURES

Alzette, (riv.) J9
Clerf, (riv.) J9
Eisling, (mts.) H9
Mosel, (riv.) J9
Our, (riv.) J7
Sauer, (riv.) J9

NETHERLANDS

PROVINCES

Drenthe 405,924 .... K3
Dronten 15,343 .... H4
Friesland 560,614 .... H2
Gelderland 1,639,997 .... H4
Groningen 540,062 .... K2
Lelystad H4
Limburg 1,051,620 .... H6

North Brabant 1,967,261 .... F5
North Holland 2,295,875 .... F3
Overijssel 985,569 .... J4
South Holland 3,048,648 .... G4
Utrecht 867,909 .... G4
Zeeland 332,286 .... D6
Zuidelijke
  IJsselmeerpolders 14,231 .... H4

CITIES and TOWNS†

Aalsmeer 20,779 .... F4
Aalten 17,486 .... K5
Aardenburg 3,869 .... C6
Akkrum 5,044 .... H2
Alkmaar 65,199 .... F3
Almelo 62,634 .... K4
Alphen aan de Rijn 46,065 .... F4
Amersfoort 87,784 .... G4
Amstelveen 71,803 .... B5
Amsterdam (cap.) 751,156 .... B4
Amsterdam* 987,205 .... B4
Andijk 5,301 .... G3
Apeldoorn 134,055 .... H4
Apeldoorn* 237,231 .... H4
Appingedam 13,295 .... K2
Arnhem 126,051 .... H4
Arnhem* 281,126 .... H4
Assen 43,783 .... K3
Asten 12,295 .... H6
Axel 12,072 .... D6
Baarle-Nassau 5,583 .... F6
Baarn 25,045 .... G4
Barneveld 34,189 .... H4
Bath D6
Beilen 12,948 .... K3
Bemmel 14,218 .... H5
Bergeijk 9,009 .... G6
Bergen 14,306 .... F3
Bergen op Zoom 40,770 .... E5
Bergum 28,047 .... H2
Berkel 9,367 .... F5
Berkhout 5,167 .... F3
Beverwijk 37,551 .... F4
Blerick J6
Bloemendaal 17,940 .... E4
Blokzijl H3
Bodegraven 15,848 .... F4
Bolsward 9,934 .... H2
Borculo 9,859 .... J4
Borger 12,077 .... K3
Borne 18,216 .... K4
Boskoop 12,985 .... F4
Boxmeer 12,662 .... H5
Boxtel 22,465 .... G5
Breda 118,086 .... F5
Breda* 151,182 .... F5
Breezand F3
Breskens C6
Brielle 15,620 .... E5
Brouwershaven 3,263 .... D5
Brummen 20,460 .... J4
Brunssum 26,116 .... J7
Buiksloot C4
Bussum 37,848 .... G4
Capelle 35,696 .... F5
Coevorden 13,089 .... K3
Colijnsplaat D5
Culemborg 17,682 .... G5
Cuyk 15,366 .... H5
Dalen 5,084 .... K3
De Bilt 32,588 .... G4
Dedemsvaart 12,975 .... J3
De Koog F2
Delft 86,103 .... E4

Delfzijl 23,316 .... K2
Den Burg 12,132 .... F2
Denekamp 11,533 .... L4
Den Helder 60,421 .... F3
Deurne 26,539 .... H6
Deventer 65,557 .... J4
Didam 14,263 .... J5
De Wijk 4,631 .... J3
Diemen 13,704 .... C5
Dieren J4
Diever 3,162 .... J3
Dinxperlo 7,296 .... K5
Dirksland 6,495 .... E5
Doesburg 9,759 .... J4
Doetinchem 34,915 .... J5
Dokkum 11,203 .... H2
Doml ·.org 3,874 .... C5
Dongen 19,219 .... F5
Doorn 11,966 .... G4
Dordrecht 101,840 .... F5
Dordrecht* 186,793 .... F5
Drachten 45,390 .... J2
Driebergen 17,022 .... G4
Dronten 16,544 .... H3
Druten 11,113 .... H5
Echt 17,035 .... H6
Edam-Volendam 21,507 .... G4
Ede 79,897 .... H4
Egmond aan Zee 5,734 .... E3
Eindhoven 192,562 .... G6
Eindhoven* 358,234 .... G6
Elburg 18,082 .... H4
Elst 16,686 .... H5
Emmeloord 34,467 .... H3
Emmen 86,090 .... L3
Enkhuizen 13,430 .... G3
Enschede 141,597 .... K4
Enschede* 239,015 .... K4
Epe 32,267 .... H4
Ermelo 23,835 .... H4
Etten-Leur 26,167 .... F5
Europoort E5
Flushing 43,806 .... C6
Franeker 11,415 .... H2
Geertruidenberg 6,185 .... F5
Geldermalsen 8,952 .... G5
Geldrop 25,879 .... H6
Geleen 35,910 .... H7
Gemert 15,267 .... H5
Gendringen 19,086 .... J5
Genemuiden 6,058 .... H3
Gennep 14,773 .... H5
Giessendam-Hardinxveld 15,523 .... F5
Giethoorn J3
Gilze 19,513 .... F5
Goes 28,505 .... D6
Goirle 13,447 .... G5
Goor 11,435 .... K4
Gorinchem 28,337 .... G5
Gorredijk J2
Gouda 56,403 .... F4
Graauw E6
Gramsbergen 5,866 .... K3
Grave 9,492 .... H5
Groenlo 8,693 .... K4
Groesbeek 18,094 .... H5
Groningen 163,357 .... K2
Groningen* 201,662 .... K2
Grouw 8,567 .... H2
Haamstede 4,575 .... D5
Haarlem 164,672 .... F4
Haarlem* 232,048 .... F4
Haarlemmermeer
  (Hoofddorp) 72,046 .... F4
Hague, The (cap.) 479,369 .... E4
Hague, The* 682,452 .... E4
Halfweg 4,456 .... B4
Hallum H2
Hardenberg 28,489 .... J3
Hardenberg 28,508 .... H4
Hardinxveld-Giessendam 15,523 .... G5
Harlingen 14,533 .... G2
Hasselt 5,817 .... J3
Hattem 11,074 .... H4
Heemskerk 31,728 .... F3
Heemstede 27,376 .... F4
Heer H7
Heerde 16,833 .... H4
Heerenveen 34,948 .... H3
Heerhugowaard 26,019 .... F3
Heerlen 71,500 .... J7
Heesch 8,659 .... G5
Heiloo 20,524 .... F3
Hellendoorn 32,068 .... J4
Hellevoetsluis 14,786 .... E5
Helmond 59,249 .... H6
Hengelo, Gelderland 8,015 .... J4
Hengelo, Overijssel 72,281 .... K4
Heusden 5,542 .... G5
Hillegom 17,489 .... E4
Hilvarenbeek 8,408 .... G6
Hilversum 94,041 .... G4
Hilversum* 110,498 .... G4
Hippolytushoef 7,847 .... G3
Hoek D6
Hoek van Holland (Hook of
  Holland) D4
Hoensbroek 22,441 .... H7
Holijsloot C4
Hollum H2
Holwerd H2
Hoofddorp
  (Haarlemmermeer) 72,046 .... F4
Hoogeveen 42,673 .... J3
Hoogezand-Sappemeer 33,860 .... K2
Hoogkarspel 5,112 .... G3
Hoorn 24,609 .... G3
Horst 16,242 .... H6
Huissen 11,049 .... H5
Huizen 25,603 .... G4
Hulst 17,283 .... E6
IJmuiden 6,633 .... F4
IJsselstein 15,450 .... F4
Joure 14,329 .... H3
Kampen 29,488 .... H3
Katwijk aan Zee 37,437 .... E4
Kerkdriel 7,584 .... G5
Kerkrade 46,609 .... J7
Kesteren 8,257 .... G5
Klazienaveen 9,520 .... L3
Kollum 11,887 .... J2
Krimpen aan den IJssel 26,396 .... F5
Landsmeer 8,082 .... C4
Laren 13,615 .... G4
Leek 15,713 .... J2
Leerdam 15,030 .... F5
Leeuwarden 85,074 .... H2
Leiden 99,891 .... E4
Leiden* 167,554 .... E4
Lelystad H3
Lemmer 10,013 .... H3
Lisse 19,182 .... E4
Lith 5,088 .... G5
Lochem 17,274 .... J4
Lonneker K4
Loon op Zand 18,000 .... G5
Losser 20,688 .... L4
Maarssen 18,346 .... F4
Maasbree 9,462 .... H6
Maassluis 28,170 .... E5
Slochteren 8,247 .... K3
Maastricht 111,044 .... H7
Maastricht* 145,862 .... H7
Margraten 3,318 .... H7
Medemblik 6,432 .... G3
Meerssen 8,414 .... H7
Meppel 21,057 .... J3
Middelburg 36,372 .... D6

Middelharnis 14,245 .... E5
Middenmeer F3
Millingen aan den Rijn 5,035 .... J5
Moerdijk F5
Monnickendam 8,127 .... C4
Montfoort 3,442 .... G4
Muiden 6,567 .... C4
Muntendam 4,147 .... K2
Naaldwijk 24,117 .... E4
Naarden 17,319 .... G4
Nagele H3
Needle 10,842 .... K4
Nes 3,012 .... H2
Nieuwegein 22,648 .... G4
Nieuwe-Pekela 5,086 .... L2
Nieuwkoop 8,923 .... F4
Nieuw-Schoonebeek 7,556 .... L3
Nijkerk 21,615 .... H4
Nijmegen 148,493 .... H5
Nijmegen* 213,981 .... H5
Noordwijk 22,386 .... E4
Norg 6,041 .... J2
Numansdorp 7,072 .... E5
Nunspeet 21,340 .... H4
Odoorn 11,973 .... K3
Oisterwijk 16,263 .... G5
Oldenzaal 26,624 .... K4
Olst 8,480 .... J4
Ommen 16,136 .... J3
Onstwedde K2
Oostburg 18,461 .... C6
Oosterhout 40,077 .... F5
Ooosterwolde 5,845 .... J2
Oostmarbern D4
Oostzaan 6,336 .... C4
Ootmarsum 3,901 .... K4
Oss 45,643 .... H5
Otterlo H4
Oud-Beijerland 14,251 .... E5
Oudenbosch 11,061 .... E5
Oude-Pekela 8,067 .... K2
Oudewater 6,870 .... F4
Purmerend 32,614 .... F4
Putten 18,243 .... H4
Raalte 23,988 .... J4
Renkum 34,547 .... H5
Reuse 6,901 .... J4
Rheden 49,755 .... J4
Rhenen 16,893 .... H5
Ridderkerk 45,069 .... F5
Rijnsburg 10,698 .... E4
Rijssen 20,008 .... J4
Rijswijk 54,123 .... E4
Roden 16,437 .... J2
Roermond 36,695 .... J6
Roosendaal 51,685 .... F5
Rotterdam 614,767 .... E5
Rotterdam* 1,016,505 .... E5
Rutten H3
Ruurlo 7,557 .... J4
Sappemeer-Hoogezand 33,860 .... K2
Schagen 13,929 .... F3
Scheveningen E4
Schiedam 78,068 .... E5
Schijndel 18,658 .... G5
Schiphol B5
Schoonhoven 10,753 .... F5
's Gravenhage 7,242 .... E4
's Gravenhage (The Hague)
  (cap.) 479,369 .... E4
's Gravenhage* 682,452 .... E4
's Gravenzande 15,833 .... E4
's Heerenberg 16,326 .... J5
's Hertogenbosch 86,184 .... H5
Simpelveld 6,783 .... H7
Sint Annaland E5
Sint Jacobiparochie H2
Sittard 34,278 .... H6
Sliedrecht 21,839 .... F5
Slochteren 13,447 .... K2
Sloten, North Holland B5
Slotdijk B4
Sluis 3,140 .... C6
Smilde 8,247 .... K3
Sneek 28,123 .... H2
Soest 40,165 .... G4

Soesterberg G4
Stadskanaal 13,946 .... L3
Staphorst 11,608 .... J3
Steenbergen 12,930 .... E5
Steenwijk 20,721 .... J3
Stiens 7,711 .... H2
Swifterbant H3
Tegelen 18,386 .... J6
Ter Apel L3
Termunten 4,803 .... K2
Terneuzen 33,731 .... D6
Tholen 17,213 .... E5
Tiel 24,974 .... G5
Tilburg 151,513 .... G5
Tilburg* 212,510 .... G5
Twello 22,542 .... J4
Uden 28,946 .... H5
Uithoorn 22,812 .... F4
Uithuizen 5,194 .... K2
Ulrum 3,665 .... J2
Urk 9,397 .... H3
Utrecht 250,887 .... G4
Utrecht* 464,357 .... G4
Vaals 11,057 .... H7
Vaassen 7,225 .... H4
Valkenswaard 27,121 .... H6
Veendam 26,168 .... K2
Veenendaal 35,845 .... H4
Veere 4,252 .... D5
Veghel 22,308 .... H5
Veldhoven 30,030 .... G6
Velp J5
Velsen 64,035 .... F4
Venlo 61,659 .... J6
Venraij 31,526 .... H6
Vianen 12,821 .... G5
Vlaardingen 78,311 .... E5
Vlagtwedde 16,719 .... L3
Vlijmen 13,515 .... G5
Vlissingen (Flushing) 43,806 .... C6
Volendam-Edam 21,507 .... G4
Voorburg 45,209 .... E4
Voorst 22,542 .... J4
Vorden 7,276 .... J4
Vriezenveen 16,025 .... K4
Vught 23,261 .... G5
Waalre 13,219 .... G6
Waalwijk 25,977 .... G5
Wageningen 28,659 .... H5
Wamel 8,979 .... H5
Warmenhuizen 3,818 .... F3
Weert 36,850 .... H6
Weesp 17,037 .... G4
West-Terschelling 4,542 .... H1
Wierden 20,618 .... K4
Wijhe 6,888 .... J4
Wijk bij Duurstede 7,927 .... G5
Wijk en Aalburg 9,266 .... G5
Winschoten 19,760 .... L2
Winsum 5,007 .... K2
Winterswijk 27,413 .... K5
Woensdrecht 9,101 .... E6
Woerden 22,064 .... F4
Wolvega 22,812 .... J3
Workum 4,135 .... G3
Zaandam (Zaanstad) 124,795 .... B4
Zaandam (Zaanstad)* 137,371 .... B4
Zaltbommel 8,010 .... G5
Zandvoort 16,289 .... E4
Zeist 58,630 .... G4
Zevenaar 26,560 .... J5
Zevenbergen 13,307 .... F5
Zierikzee 8,816 .... D5
Zundert 12,444 .... F6
Zutphen 29,188 .... J4
Zwartsluis 4,391 .... J3
Zwijndrecht 38,271 .... F5
Zwolle 77,826 .... J3

OTHER FEATURES

Alkmaardermeer (lake) F3
Ameland (isl.) H2
Bergumermeer (lake) J2
Beulaker Wijde (lake) H3

Borndiep (chan.) H2
De Fluessen (lake) H3
De Honte (bay) D6
De Peel (reg.) H6
De Twente (reg.) K4
De Zaan (riv.) B4
Dollard (bay) L2
Dommel (riv.) H6
Duiveland (isl.) D5
Eastern Scheldt (est.) D5
Eems (riv.) K2
Eijerlandsche Gat (str.) F2
Flevoland Polders 35,618 .... G4
Friesche Gat (chan.) J1
Frisian, West (isls.) H1
Galganberg (hill) H4
Goeree (isl.) D5
Grevelingen (str.) D5
Griend (isl.) G2
Groninger Wad (sound) J2
Groote IJ Polder B4
Haarlemmermeer Polder 72,046 .... F4
Haringvliet (riv.) E5
Het IJ (est.) C4
Hoek van Holland (cape) D4
Hondsrug (hills) K3
Houtrak Polder A4
Hunse (riv.) K2
IJmeer (bay) C4
IJssel (riv.) J4
IJsselmeer (lake) G3
Lauwers (chan.) J1
Lauwers Zee (bay) J1
Lek (riv.) F5
Lemelerberg (hill) J4
Lower Rhine (riv.) H5
Maas (riv.) G5
Marken (isl.) C4
Markerwaard Polder G4
Marsdiep (chan.) F3
North (sea) D3
North Beveland (isl.) D5
North East Polder 34,467 .... H3
North Holland (canal) F3
North Sea (canal) E4
Old Rhine (riv.) F4
Oostzaan Polder 6,336 .... C4
Orange (canal) J2
Overflakkee (isl.) E5
Pinke Gat (chan.) H2
Regge (riv.) K4
Rhine (riv.) J5
Roer (riv.) J6
Rottumeplaat (isl.) K1
Rottumeroog (isl.) K1
Schiermonnikoog (isl.) J1
Schouwen (isl.) D5
Slotermeer (lake) H3
Sneekermeer (lake) H2
South Beveland (isl.) D6
Terschelling (isl.) G1
Texel (isl.) F2
Tjeukemeer (lake) H3
Tjonger (riv.) J3
Vaalserberg (mt.) H7
Vecht (riv.) J3
Vechte (riv.) K4
Veersche Meer (lake) D6
Veluwe (reg.) H4
Vlieland (isl.) G1
Vliestroom (str.) G2
Voorne (isl.) E5
Waal (riv.) H5
Waddenzee (sound) G2
Walcheren (isl.) C6
Wester Eems (chan.) K1
Wilhelmina (canal) G5
Willems (canal) H6
West Frisian (isls.) H1
Westgat (chan.) J1
Wieringermeer Polder 11,870 .... G3
Wilhelmina (canal) G5
Willems (canal) H6

*City and suburbs.
†Population of cities in Belgium & Netherlands are communes.

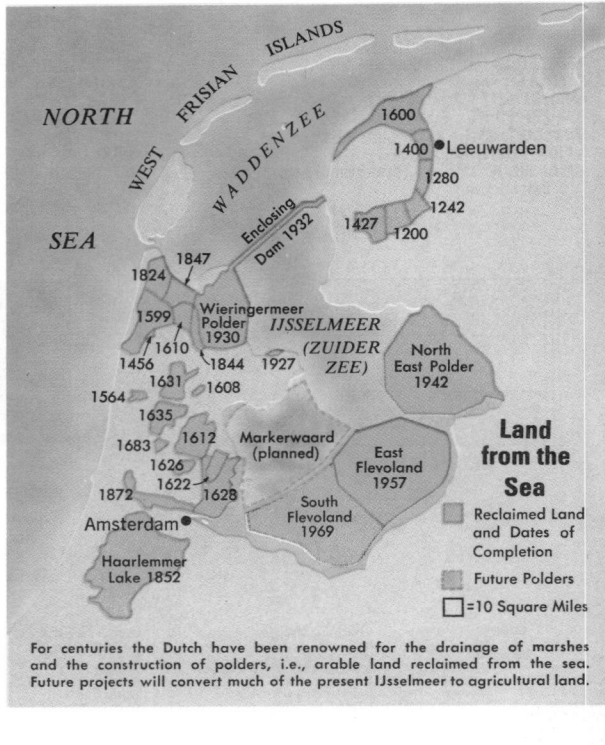

Land from the Sea

Reclaimed Land and Dates of Completion

Future Polders

=10 Square Miles

For centuries the Dutch have been renowned for the drainage of marshes and the construction of polders, i.e., arable land reclaimed from the sea. Future projects will convert much of the present IJsselmeer to agricultural land.

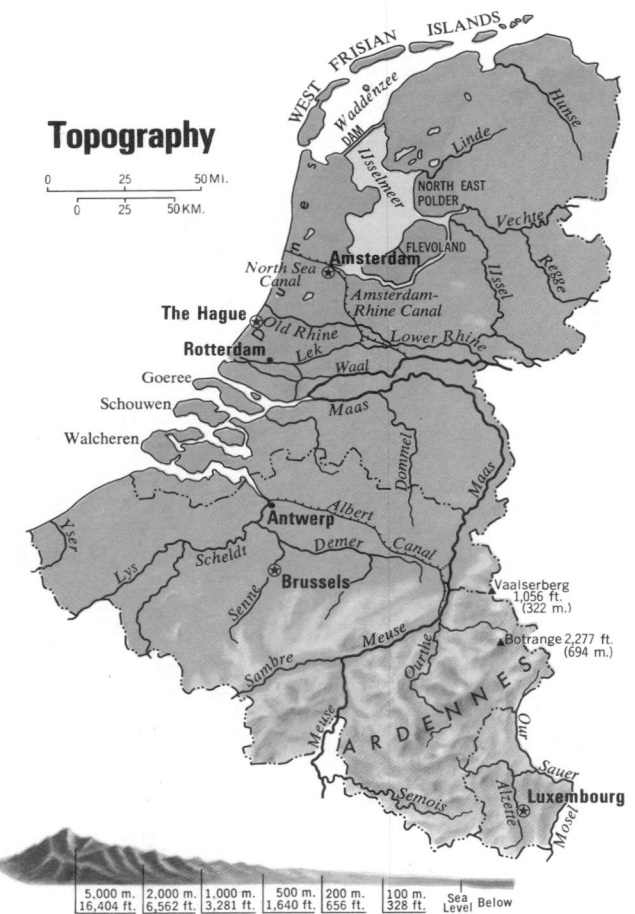

Topography

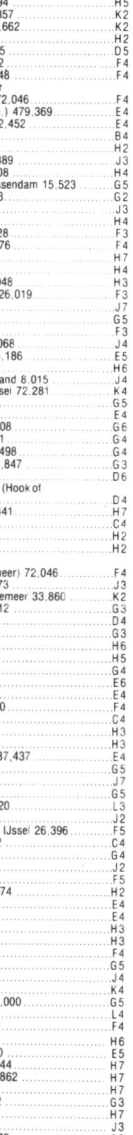

5,000 m. 16,404 ft. | 2,000 m. 6,562 ft. | 1,000 m. 3,281 ft. | 500 m. 1,640 ft. | 200 m. 656 ft. | 100 m. 328 ft. | Sea Level Below

Netherlands, Belgium and Luxembourg

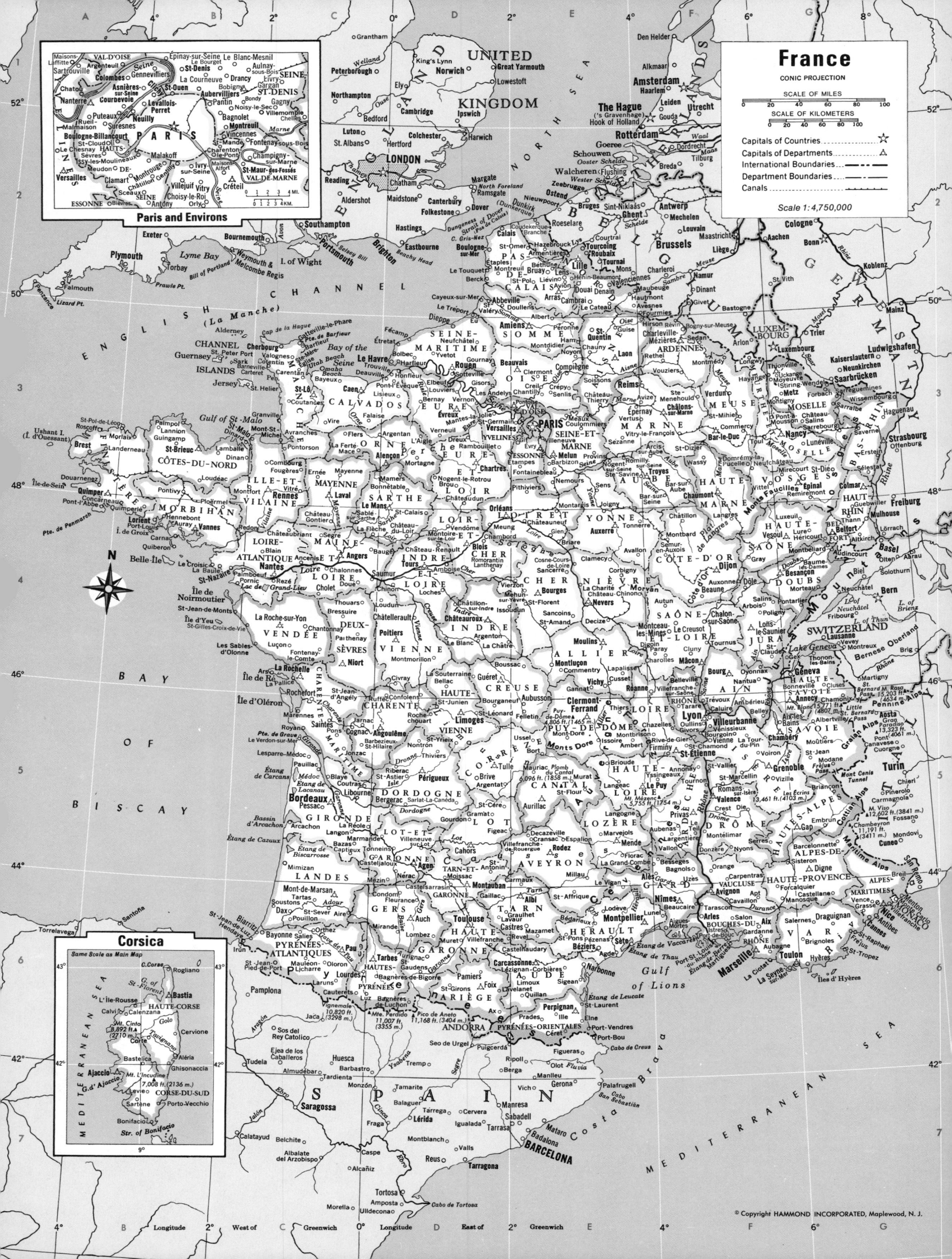

DEPARTMENTS

Ain 376,477 F4
Aisne 533,862 E3
Allier 378,406 E4
Alpes-de-Haute-Provence 112,178 G5
Alpes-Maritimes 816,681 G6
Ardèche 257,065 F5
Ardennes 309,306 F3
Ariège 137,857 D6
Aube 284,823 E3
Aude 272,366 E6
Aveyron 278,306 E5
Bas-Rhin 882,121 G3
Belfort (terr.) 128,125 G4
Bouches-du-Rhône 1,632,974 F6
Calvados 560,967 C3
Cantal 166,549 E5
Charente 337,064 D5
Charente-Maritime 497,859 C5
Cher 316,350 E4
Corrèze 240,363 D5
Corse du Sud 128,634 B6
Côte-d'Or 456,070 F4
Côtes-du-Nord 525,556 B3
Creuse 146,214 D4
Deux-Sèvres 335,829 C4
Dordogne 373,179 D5
Doubs 471,082 G4
Drôme 361,847 F5
Essonne 923,063 E3
Eure 422,952 D3
Eure-et-Loir 335,151 D3
Finistère 804,088 A3
Gard 494,575 F6
Gers 175,366 D6
Gironde 1,061,480 C5
Haute-Corse 161,208 B6
Haute-Garonne 777,431 D6
Haute-Loire 205,491 E5
Haute-Marne 212,304 F3
Hautes-Alpes 97,358 G5
Haute-Saône 222,254 G4
Haute-Savoie 447,795 G5
Hautes-Pyrénées 227,222 D6
Haute-Vienne 352,149 D5
Haut-Rhin 635,209 G4
Hauts-de-Seine 1,438,930 A2
Hérault 648,202 E6
Ille-et-Vilaine 702,199 C3
Indre 248,523 D4
Indre-et-Loire 478,601 D4
Isère 860,339 F5
Jura 238,856 F4
Landes 288,323 C5
Loire 742,396 F5
Loire-Atlantique 934,499 C4
Loiret 490,189 E4
Loir-et-Cher 283,686 D4
Lot 150,778 D5
Lot-et-Garonne 292,616 D5
Lozère 74,825 E5
Maine-et-Loire 629,849 C4
Manche 451,662 C3
Marne 530,399 F3
Mayenne 261,789 C3
Meurthe-et-Moselle 722,588 G3
Meuse 203,904 F3
Morbihan 563,588 B4
Moselle 1,006,373 G3
Nièvre 245,212 E4
Nord 2,510,738 E2
Oise 606,320 E3
Orne 293,523 C3
Paris (city) 2,299,830 B2
Pas-de-Calais 1,403,035 E2
Puy-de-Dôme 580,033 E5
Pyrénées-Atlantiques 534,748 C6
Pyrénées-Orientales 299,506 E6
Rhône 1,429,647 F5
Saône-et-Loire 569,810 F4
Sarthe 490,385 D3
Savoie 305,118 G5
Seine-et-Marne 755,762 E3
Seine-Saint-Denis 1,322,127 C1
Somme 538,462 E3
Tarn 338,024 E6
Tarn-et-Garonne 183,314 D5
Val-de-Marne 1,215,713 C1
Val-d'Oise 840,885 E3
Var 626,093 G6
Vaucluse 390,446 F6
Vendée 450,641 C4
Vienne 357,366 D4
Vosges 397,957 G3
Yonne 299,851 E4
Yvelines 1,082,255 D3

CITIES and TOWNS

Abbeville 25,252 D2
Agde 9,856 E6
Agen 33,763 D5
Aix-en-Provence 91,665 F6
Aix-les-Bains 21,884 G5
Ajaccio 47,065 B7
Albert 11,746 E2
Albertville 16,630 G5
Albi 43,942 E6
Alençon 32,917 D3
Alès 33,315 E5
Ambérieu-en-Bugey 9,294 F5
Amboise 10,498 D4
Amiens 129,453 D3
Ancenis 6,689 C4
Angers 136,603 C4
Angoulême 46,293 D5
Annecy 53,058 G5
Annonay 19,234 F5
Antibes 44,226 G6
Antony 57,450 B2
Apt 9,735 F6
Arcachon 13,856 C5
Argentan 16,063 D3
Argenteuil 101,542 A1
Arles 37,337 F6
Armentières 23,850 E2
Arras 45,804 E2
Asnières-sur-Seine 75,328 A1
Aubagne 26,145 F6
Aubenas 11,967 F5
Aubervilliers 72,859 B1
Auch 18,767 D6
Audincourt 18,570 G4
Aulnay-sous-Bois 77,982 B1
Auray 10,006 B4
Aurignac 744 D6
Aurillac 29,458 E5
Autun 19,441 F4
Auxerre 36,039 E4
Auxonne 6,414 F4
Avallon 8,518 E4
Avignon 73,482 F6
Avion 22,860 E2
Avranches 10,128 C3
Ax-les-Thermes 1,456 D6
Bagnères-de-Bigorre 9,080 D6
Bagnolet 35,858 B2
Bagnols-sur-Cèze 13,111 F5
Barbizon 1,189 E3
Barcelonnette 2,523 G5
Barfleur 701 C3
Bar-le-Duc 19,188 F3
Bar-sur-Aube 7,227 F3
Bastia 45,387 B6
Bayeux 13,381 C3
Bayonne 41,281 C6
Beaucaire 10,189 F6
Beaune 16,386 F4
Beauvais 53,493 E3
Belfort 54,469 G4
Belley 6,612 F5
Berck 14,104 D2
Bergerac 25,488 D5
Bernay 9,928 D3
Besançon 119,803 G4
Béthune 26,208 E2
Béziers 79,213 E6
Biarritz 27,453 C6
Blois 49,134 D4
Bobigny 43,041 B1
Bogny-sur-Meuse 6,845 F3
Bolbec 12,347 D3
Bondy 48,285 B1
Bonneville 6,717 G4
Bordeaux 220,830 C5
Boulogne-Billancourt 103,527 A2
Boulogne-sur-Mer 48,309 D2
Bourg-en-Bresse 40,052 F4
Bourges 75,200 E4
Bourgoin-Jallieu 18,504 F5
Bressuire 9,778 C4
Brest 163,940 A3
Briançon 8,523 G5

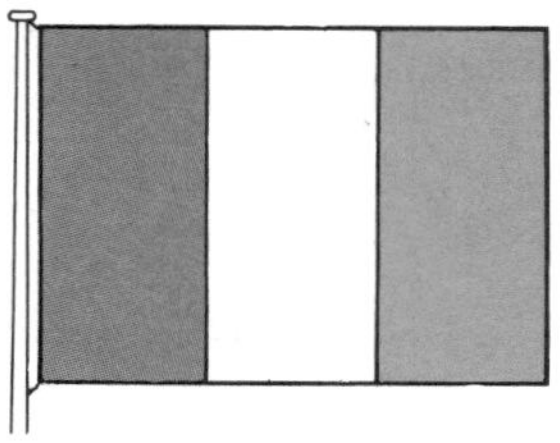

Brignoles 8,784 G6
Brioude 7,756 E5
Brive-la-Gaillarde 49,276 D5
Bruay-en-Artois 25,544 E2
Caen 116,987 C3
Cahors 19,288 D5
Calais 73,009 D2
Caluire-et-Cuire 43,024 F5
Cambrai 38,706 E2
Cannes 70,226 G6
Carcassonne 38,887 D6
Carmaux 11,970 E5
Carpentras 20,169 F5
Castelnaudary 8,947 E6
Castelsarrasin 6,562 D6
Castres 41,037 E6
Cavaillon 17,383 F6
Châlons-sur-Marne 50,870 F3
Chalon-sur-Saône 55,495 F4
Chambéry 52,286 F5
Chambord 166 D4
Chamonix-Mont-Blanc 6,246 G5
Champigny-sur-Marne 80,189 C2
Chantilly 10,517 E3
Charenton-le-Pont 20,383 B2
Charleville-Mézières 59,513 F3
Chartres 38,574 D3
Châteaubriant 12,417 C4
Château-du-Loir 5,598 D4
Châteaudun 14,634 D3
Château-Gontier 8,301 C4
Châteauroux 53,166 D4
Château-Thierry 13,379 E3
Châtellerault 33,811 D4
Châtillon 26,562 B2
Châtillon-sur-Seine 7,367 F4
Chatou 26,415 A1
Chaumont 26,568 F3
Chauny 14,324 E3
Chelles 24,192 C1
Cherbourg 31,333 C3
Chinon 5,378 D4
Choisy-le-Roi 38,629 B2
Cholet 49,887 C4
Clamart 52,881 A2
Clermont 7,834 E3
Clermont-Ferrand 153,379 E5
Clichy 47,731 B1
Cluny 4,335 F4
Cluses 12,713 G4
Cognac 21,567 C5
Colmar 58,585 G3
Colombes 83,241 A1
Commentry 8,074 E4
Commercy 6,918 F3
Compiègne 37,009 E3
Concarneau 15,096 A4
Cosne-Cours-sur-Loire 9,768 E4
Coudekerque-Branche 24,702 E2
Coulommiers 11,363 E3
Courbevoie 54,391 A1
Coutances 8,286 C3
Creil 31,893 E3
Crépy-en-Valois 10,661 E3
Créteil 58,665 B2
Cusset 13,672 E4
Dax 18,019 C6
Deauville 5,655 C3
Decazeville 9,318 E5
Decize 6,853 E4
Denain 26,096 E2
Dieppe 25,607 D3

AREA 210,038 sq. mi. (543,998 sq. km.)
POPULATION 53,788,000
CAPITAL Paris
LARGEST CITY Paris
HIGHEST POINT Mont Blanc 15,771 ft. (4,807 m.)
MONETARY UNIT franc
MAJOR LANGUAGE French
MAJOR RELIGION Roman Catholicism

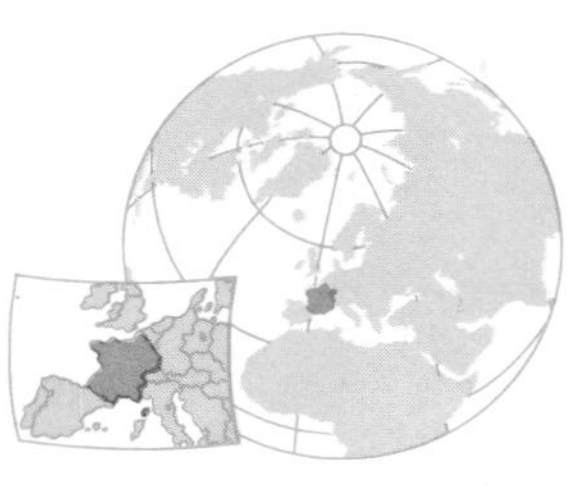

Topography

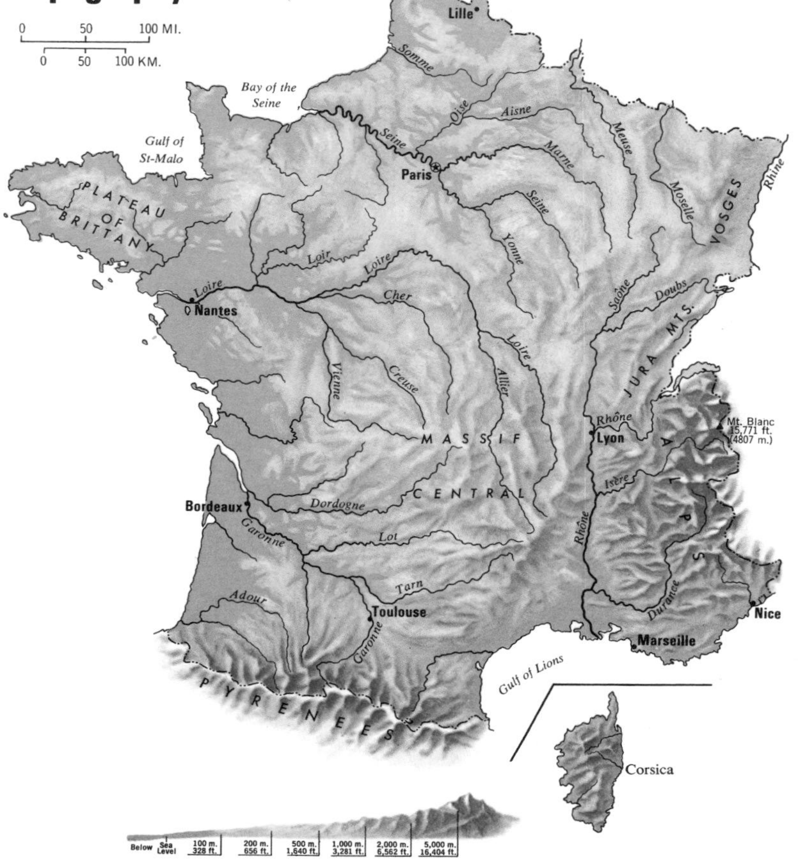

Historic Provinces

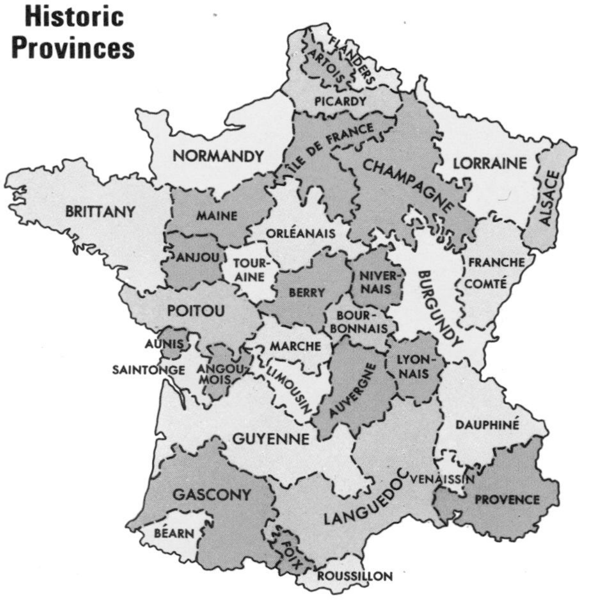

A resident of the city of Caen thinks of himself as a Norman rather than as a citizen of the modern department of Calvados. In spite of the passing of nearly two centuries, the historic provinces which existed before 1790 command the local patriotism of most Frenchmen.

Digne 13,140 G5
Digoin 10,449 F4
Dijon 149,899 F4
Dinan 13,303 B3
Dinard 9,211 B3
Dôle 28,109 F4
Domrémy-la-Pucelle 190 F3
Douai 43,954 E2
Douarnenez 17,851 A3
Doullens 6,806 E2
Draguignan 19,653 G6
Drancy 64,258 B1
Dreux 31,503 D3
Dunkirk (Dunkerque) 78,171 E2
Elbeuf 18,642 D3
Épernay 29,286 E3
Épinal 39,000 G3
Épinay-sur-Seine 46,458 B1
Erstein 6,494 G3
Étampes 18,810 E3
Étaples 10,423 D2
Eu 8,349 D3
Évreux 46,181 D3
Évry 15,300 E3
Falaise 8,133 C3
Fécamp 20,835 D3
Figeac 8,675 D5
Firminy 23,776 F5
Flers 18,590 C3
Foix 9,569 D6
Fontainebleau 16,436 E3
Fontenay-le-Comte 12,301 C4
Fontenay-sous-Bois 46,200 C2
Forbach 24,812 G3
Fougères 26,260 C3
Fourmies 15,318 F2
Fréjus 27,805 G6
Gagny 36,714 C1
Gaillac 7,653 D6
Gap 24,962 G5
Gardanne 8,175 F6
Gennevillers 50,154 B1
Gentilly 16,843 B2
Gex 3,959 G4
Gien 13,817 E4
Gif 10,866 E3
Gisors 7,591 D3
Givet 7,787 F2
Givors 19,356 F5
Granville 12,869 C3
Grasse 24,260 G6
Graulhet 11,099 E6
Gray 8,718 F4
Grenoble 165,431 F5
Guebwiller 10,477 G4
Guéret 14,418 D4
Guingamp 9,269 B3
Guise 6,642 E3
Haguenau 23,023 G3
Harfleur 9,857 D3
Hautmont 19,130 F2
Hayange 8,479 F3
Hazebrouck 18,867 E2
Hendaye 9,404 C6
Hénin-Beaumont 26,296 E2
Hennebont 8,978 B4
Héricourt 8,481 G4
Hirson 11,909 F3
Honfleur 8,995 D3
Hyères 29,366 G6
Issoire 13,560 E5
Issoudun 15,065 D4
Issy-les-Moulineaux 47,355 A2
Istres 10,127 F6
Ivry-sur-Seine 62,804 B2
Joigny 10,825 E3
La Baule-Escoublac 13,854 B4
La Ciotat 29,290 F6
La Courneuve 37,917 B1
La Flèche 12,743 C4
La Grand-Combe 9,406 E5
L'Aigle 9,198 D3
Landerneau 13,983 B3
Langres 10,745 F4
Lannion 13,692 B3
Laon 27,420 E3
La Pallice C4
La Rochelle 72,936 C4
La Roche-sur-Yon 40,789 C4
La Seyne-sur-Mer 50,059 F6
Laval 50,734 C3
Lavelanet 9,278 E6
Le Blanc 7,431 D4
Le Blanc-Mesnil 49,062 B1
Le Bourget 10,520 B1
Le Cateau 8,680 E2
Le Chesnay 24,590 A2
Le Creusot 31,643 F4
Le Havre 216,917 C3
Le Mans 150,289 C3
Lens 39,973 E2
Le Puy 24,793 F5
Les Andelys 7,524 D3
Les Sables-d'Olonne 17,157 B4
Le Teil 7,993 F5
Le Tréport 6,463 D2
Levallois-Perret 52,460 A1
Lézignan-Corbières 6,929 E6
Libourne 21,265 C5
Liévin 33,040 E2
Lille 171,010 E2
Limoges 136,059 D5
Limoux 9,595 E6
Lisieux 24,972 D3
Livry-Gargan 32,879 C1
Lodève 7,131 E6
Longwy 20,107 F3
Lons-le-Saunier 20,897 F4
Lorient 68,655 B4
Loudéac 7,173 B3
Loudun 7,060 D4
Lourdes 17,685 C6
Louviers 17,919 D3
Luçon 8,834 C4
Lunel 12,392 E6
Lunéville 22,438 G3
Lure 8,538 G4
Luxeuil-les-Bains 10,061 G4
Lyon 454,265 F5
Mâcon 39,130 F4
Maisons-Alfort 53,963 B2
Maisons-Laffitte 23,465 A1
Malakoff 34,100 A2
Manosque 17,256 G6
Mantes-la-Jolie 42,408 D3
Marmande 13,223 C5
Marseille 901,421 F6
Martigues 26,850 F6
Maubeuge 34,152 F2
Mayenne 11,278 C3
Mazamet 13,148 E6
Meaux 41,831 E3
Mehun-sur-Yèvre 6,533 E4
Melun 36,913 E3
Mende 10,040 E5
Menton 24,736 G6
Metz 110,939 G3
Meudon 31,294 A2
Millau 20,401 E5
Mimizan 6,826 C5
Mirecourt 7,160 G3
Moissac 7,403 D5
Montargis 18,021 E3
Montauban 35,344 D5
Montbard 7,477 F4
Montbéliard 29,968 G4
Montbrison 9,945 E5
Montceau-les-Mines 28,093 F4
Mont-de-Marsan 24,812 C6
Mont-Dore 2,074 E5
Montélimar 25,422 F5
Montfort 2,701 C3
Montigny-les-Metz 24,208 G3
Montluçon 56,337 E4
Montmédy 1,859 F3
Montpellier 178,136 E6
Montreuil,
Seine-Saint-Denis 96,441 B2
Montrouge 40,189 B2
Mont-Saint-Michel 88 C3
Morlaix 15,919 B3
Morteau 6,515 G4
Moulins 25,856 E4
Moyeuvre-Grande 12,448 G3
Mulhouse 116,494 G4
Muret 13,041 D6
Nancy 106,906 G3
Nanterre 94,441 A1
Nantes 252,537 C4
Narbonne 36,525 E6
Nemours 11,159 E3
Neufchâteau 8,582 F3
Neuilly-sur-Seine 65,941 A1
Nevers 45,122 E4
Nice 331,002 G6
Nîmes 123,914 F6
Niort 59,297 C4
Nogent-le-Rotrou 12,284 D3
Noisy-le-Sec 37,674 B1
Noyon 13,784 E3
Oloron-Sainte-Marie 11,616 C6
Orange 19,847 F5
Orléans 88,503 D3
Orly 26,090 B2
Orthez 9,639 C6
Oullins 27,731 F5
Oyonnax 22,548 F4
Pamiers 12,906 D6
Pantin 42,651 B1
Paray-le-Monial 11,523 F4
Paris (cap.) 2,291,554 B2
Parthenay 12,549 C4
Pau 81,560 C6
Périgueux 34,779 D5
Péronne 8,358 E3
Perpignan 101,198 E6
Pessac 50,333 C5
Pézenas 6,768 E6
Pithiviers 9,976 E3
Poitiers 78,739 D4
Pont-à-Mousson 14,461 G3
Pontarlier 17,778 G4
Pontivy 9,478 B3
Pont-l'Abbe 6,618 A4
Pontoise 26,702 E3
Port-de-Bouc 20,448 F6
Port-Saint-Louis-du-Rhône 9,649 F6
Port-Vendres 5,448 E6
Privas 9,385 F5
Provins 12,281 E3
Puteau 35,366 A2
Quimper 50,856 A4
Quimperlé 9,783 B4
Rambouillet 18,446 D3
Redon 9,528 C4
Reims 177,320 E3
Remiremont 10,250 G3
Rennes 194,094 C3

(continued on following page)

Rethel 8,189........................F3
Révin 11,459......................F3
Rezé 35,512.......................C4
Rive-de-Gier 17,369..............F5
Roanne 54,999....................E4
Rochechouart 2,953..............D5
Rochefort 27,264.................C4
Rodez 24,898......................E5
Romans-sur-Isère 30,974.........F5
Romilly-sur-Seine 17,276........F3
Romorantin-Lanthenay 15,727....D4
Roubaix 109,473...................E2
Rouen 113,536.....................D3
Royan 17,978......................C5
Rueil-Malmaison 62,504..........A2
Sablé-sur-Sarthe 9,913...........C4
Saint-Affrique 6,842..............E6
Saint-Amand-Mont-Rond 11,896...E4
Saint-Brieuc 51,838...............B3
Saint-Chamond 39,236............F5
Saint-Claude 12,651..............F4
Saint-Cloud 28,052...............A2
Saint-Denis 95,808...............B1
Saint-Dié 22,834..................G3
Saint-Dizier 36,377...............F3
Sainte-Mère-Église 1,041.........C3
Saintes 24,946.....................C5
Sainte-Savine 10,526.............E3
Saint-Étienne 218,289............F5
Saint-Florent-sur-Cher 6,385.....E4
Saint-Flour 6,900.................E5
Saint-Gaudens 12,103............D6
Saint-Germain-en-Laye 35,351....D3
Saint-Gilles-Croix-de-Vie 6,569...B4
Saint-Girons 7,259................D6
Saint-Jean-d'Angély 8,801........C4
Saint-Jean-de-Luz 10,921.........C6
Saint-Jean-de-Maurienne 9,525...G5
Saint-Jean-Pied-de-Port 1,725....C6
Saint-Junien 9,281................D5
Saint-Lô 21,670...................C3
Saint-Malo 43,277................B3
Saint-Mandé 20,714..............B2
Saint-Marcellin 6,768.............F5
Saint-Maur-des-Fossés 80,797....B2
Saint-Mihiel 5,544................F3
Saint-Nazaire 65,228.............B4
Saint-Omer 16,419...............E2
Saint-Ouen 43,569...............B1
Saint-Pol-de-Léon 6,571..........A3
Saint-Quentin 69,956.............E3
Saint-Raphaël 19,499.............G6
Saint-Tropez 4,484...............G6
Saint-Vallier 10,000..............F5
Salon-de-Provence 31,783.........F6
Sancerre 2,029....................E4
Sarlat-La-Canéda 8,191...........D5
Sarrebourg 12,442................G3
Sarreguemines 24,570............G3
Sartrouville 42,092...............A1
Saumur 30,984....................D4
Saverne 10,015...................G3
Sceaux 19,651.....................A2
Sedan 23,872......................F3
Sélestat 15,209...................G3
Senlis 13,481......................E3

Sens 25,621........................E3
Sèvres 21,100......................A2
Sisteron 6,434.....................G5
Soissons 29,694...................E3
Sotteville-les-Rouen 30,393.......D3
Stiring-Wendel 12,665.............G3
Strasbourg 251,520................H3
Suresnes 37,456...................A2
Tarare 11,931......................F5
Tarascon 8,522....................F6
Tarbes 54,286......................D6
Thann 8,508.......................G4
Thiers 14,534......................E5
Thionville 37,943..................G3
Thonon-les-Bains 24,673..........G4
Thouars 11,835....................C4
Tonneins 7,256....................D5
Toul 16,141.........................F3
Toulon 180,508...................F6
Toulouse 371,143.................D6
Tourcoing 102,092................E2
Tournon 8,568.....................F5
Tournus 7,284.....................F4
Tours 139,560......................D4
Troyes 71,600......................F3
Tulle 18,375........................D5
Uckange 11,552...................G3
Ussel 9,816.........................E5
Uzès 6,470..........................F5
Valence 67,101....................F5
Valenciennes 41,976..............E2
Vannes 36,722.....................B4
Vence 7,332........................G6
Vendôme 17,828...................D4
Vénissieux 74,264.................F5
Verdun-sur-Meuse 22,889.........F3
Vernon 21,184......................D3
Versailles 93,359..................A2
Vesoul 17,883......................F4
Vichy 32,107........................E4
Vienne 25,981......................F5
Vierzon 33,057.....................E4
Villefranche 6,600..................G6
Villefranche-de-Rouergue 10,848..E5
Villefranche-sur-Saône 29,996....F5
Villejuif 53,884.....................B2
Villemomble 28,684...............B2
Villeneuve-Saint-Georges 31,378..E3
Villeneuve-sur-Lot 17,818.........D5
Villeurbanne 115,913.............F5
Vincennes 44,256.................B2
Vire 12,832.........................C3
Vitré 10,989........................C3
Vitry-le-François 19,075...........F3
Vitry-sur-Seine 87,119............B2
Vittel 6,791.........................F3
Vizille 6,810........................F5
Voiron 17,879......................F5
Wissembourg 6,679..............G3
Yvetot 10,088......................D3

OTHER FEATURES

Adour (riv.)........................C6

Ain (riv.)............................F4
Aisne (riv.).........................E3
Ajaccio (gulf).......................B7
Allier (riv.).........................E5
Aube (riv.).........................F3
Auvergne (mts.)...................E5
Belle-Île (isl.)......................B4
Biscay (bay).......................B5
Blanc (mt.).........................G5
Bonifacio (str.)....................B7
Calais (Dover) (str.)...............D2
Causses (reg.).....................D5
Cévennes (mts.)...................E5
Charente (reg.)....................C5
Cher (riv.)..........................E4
Corse (cape)........................B6
Corsica (isl.)........................C4
Côte-d'Or (mts.)...................F4
Cotentin (pen.)....................C3
Cottian Alps (range)..............G5
Creuse (riv.).......................D4
Dordogne (riv.)....................D5
Dore Alps (mts.)...................E5
Doubs (riv.).........................G4
Drôme (riv.)........................F5
Dronne (riv.).......................D5
Durance (riv.)......................F5
English (chan.).....................B3
Eure (riv.)...........................D3
Faucilles (mts.).....................G3
Forez (mts.)........................E5
Fréjus (pass).......................G5
Gard (riv.)..........................F5
Garonne (riv.).....................C5
Gave de Pau (riv.).................C6
Geneva (lake).....................G4
Gers (riv.)..........................D6
Gironde (riv.)......................C5
Graian Alps (range)...............G5
Groix (isl.)..........................B4
Hague (cape).......................C3
Hérault (riv.).......................E6
Hyères (isl.)........................G6
Indre (riv.).........................D4
Isère (riv.)..........................F5
Isle (riv.)...........................D5
Langres (plat.).....................F4
Limousin (reg.)....................D5
Lions (gulf).........................F6
Little Saint Bernard (pass).......G5
Loir (riv.)...........................D4
Loire (riv.)..........................C4
Lot (riv.)............................D5
Manche, La (English) (chan.).....B3
Maritime Alps (range).............G5
Marne (riv.)........................C2
Mayenne (riv.).....................C4
Mediterranean (sea)..............E7
Médoc (reg.).......................C5
Meuse (riv.)........................F3
Morvan (plat.)......................F4
Moselle (riv.).......................G3
Noirmoutier (isl.)..................B4
North (sea)........................E1
Oise (riv.)...........................E3

Oléron (isl.)........................C5
Omaha (beach)....................C3
Orb (riv.)...........................E6
Orne (riv.)..........................C3
Ouessant (isl.).....................A3
Penmarch (pt.)....................A4
Perche (reg.).......................D3
Puy-de-Dôme (mt.)...............E5
Pyrenees (range)..................C6
Ré (isl.).............................C4
Rhine (riv.).........................G3
Rhône (riv.)........................F5
Risle (riv.)..........................D3
Riviera (reg.)......................G6
Saint-Florent (gulf)................B6
Saint-Malo (gulf)..................B3
Saône (riv.)........................F4
Sarthe (riv.)........................D4
Sein (isl.)...........................A3
Seine (bay)........................D3
Seine (riv.)..........................D3
Sologne (reg.).....................E4
Somme (reg.)......................D2
Tarn (riv.)...........................E6
Ushant (Ouessant) (isl.)..........A3
Utah (beach)......................C3
Vaccarès (lag.).....................F6
Vienne (riv.).......................D4
Vilaine (riv.)........................C4
Vosges (mts.)......................G3
Yonne (riv.)........................E3

MONACO

CITIES and TOWNS

Monte Carlo 11,599................G6

* City and suburbs

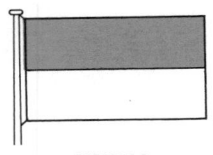

MONACO

AREA 368 acres
(149 hectares)
POPULATION 25,029

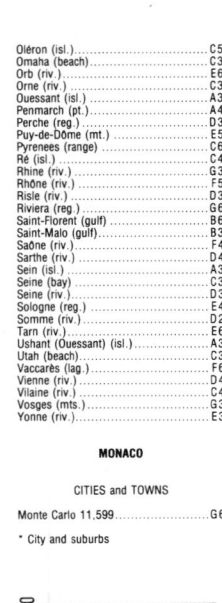

Wine Regions

Climate, soil and variety of grape planted determine the quality of wine. Long, hot and fairly dry summers with cool, humid nights constitute an ideal climate. The nature of the soil is such a determining influence that identical grapes planted in Bordeaux, Burgundy and Champagne, will yield wines of widely different types.

Agriculture, Industry and Resources

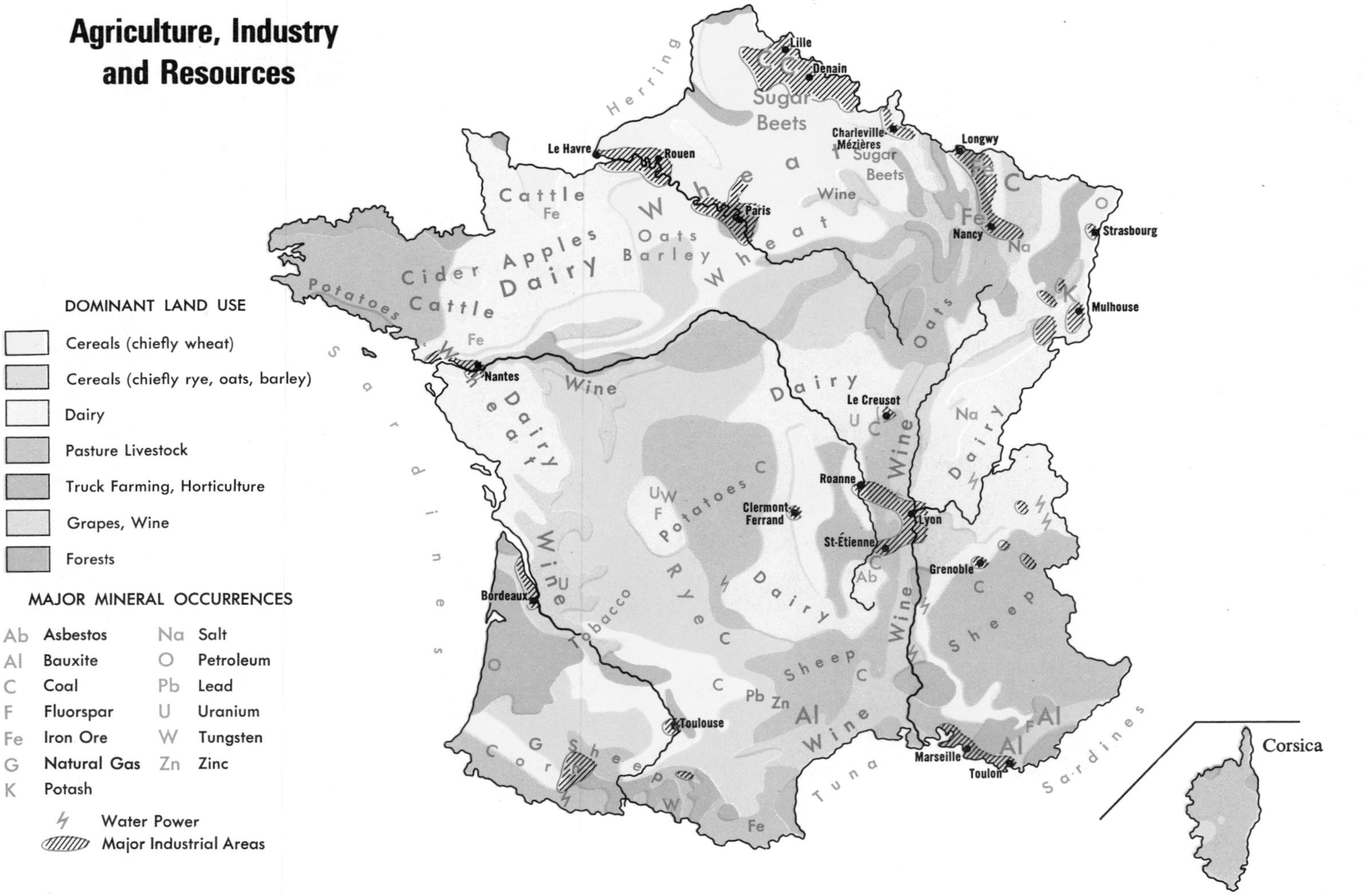

DOMINANT LAND USE

- Cereals (chiefly wheat)
- Cereals (chiefly rye, oats, barley)
- Dairy
- Pasture Livestock
- Truck Farming, Horticulture
- Grapes, Wine
- Forests

MAJOR MINERAL OCCURRENCES

Ab	Asbestos	Na	Salt
Al	Bauxite	O	Petroleum
C	Coal	Pb	Lead
F	Fluorspar	U	Uranium
Fe	Iron Ore	W	Tungsten
G	Natural Gas	Zn	Zinc
K	Potash		

⚡ Water Power
▨ Major Industrial Areas

SPAIN

PROVINCES

Álava 204,323 E1
Albacete 335,026 E3
Alicante 920,105 F3
Almería 375,004 E4
Ávila 203,798 D2
Badajoz 687,599 C3
Baleares 558,287 H3
Barcelona 3,929,194 G2
Burgos 358,075 E1
Cáceres 457,777 C3
Cádiz 885,433 D4
Castellón 385,823 G2
Ciudad Real 507,650 D3
Córdoba 724,116 D3
Cuenca 247,158 E2
Gerona 414,397 H1
Granada 733,375 E4
Guadalajara 147,732 E2
Guipúzcoa 631,003 E1
Huelva 397,683 C4
Huesca 222,238 F1
Jaén 661,146 E4
La Coruña 1,004,188 B1
Las Palmas 579,710 C4
León 548,721 C1
Lérida 347,015 G2
Logroño 235,713 E1
Lugo 415,052 C1
Madrid 3,792,561 E2
Málaga 867,330 D4
Murcia 832,313 F4
Navarra 464,867 F1
Orense 413,733 C1
Oviedo 1,045,635 C1
Palencia 198,763 D1
Pontevedra 750,701 B1
Salamanca 371,607 C2
Santa Cruz de Tenerife 590,514 B5
Santander 467,138 D1
Segovia 162,770 D2
Sevilla 1,327,190 D4
Soria 114,956 E2
Tarragona 431,961 G2
Teruel 170,284 F2
Toledo 468,925 D3
Valencia 1,767,327 F3
Valladolid 412,572 D2
Vizcaya 1,043,310 E1
Zamora 251,934 D2
Zaragoza 760,186 F2

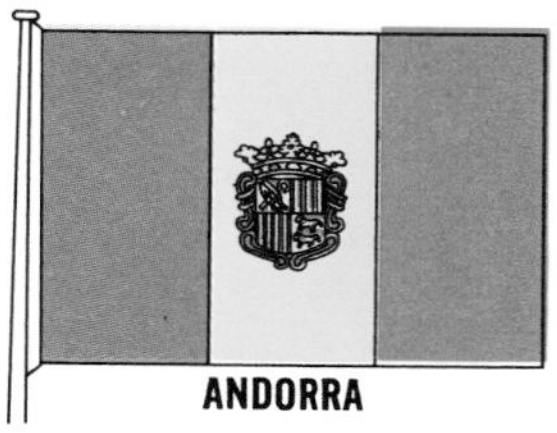

ANDORRA

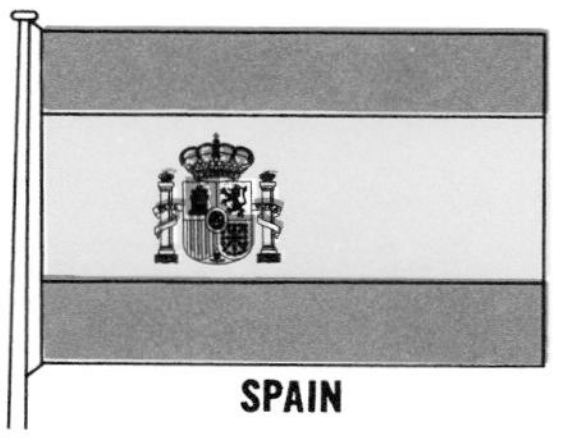

SPAIN

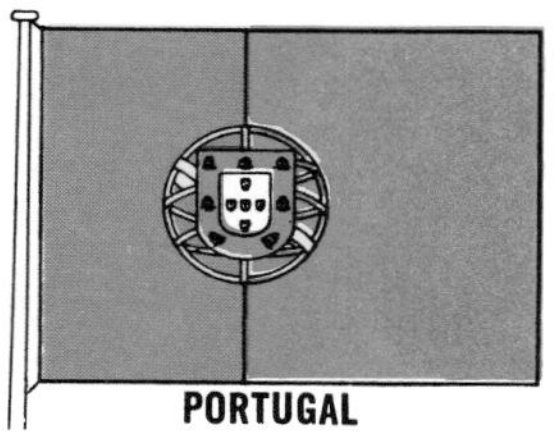

PORTUGAL

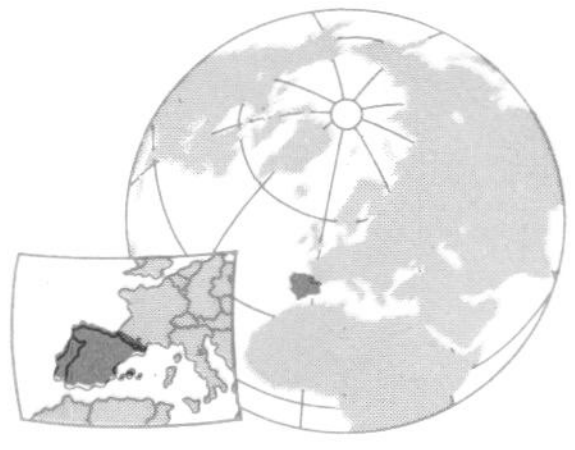

SPAIN

AREA 194,881 sq. mi. (504,742 sq. km.)
POPULATION 37,430,000
CAPITAL Madrid
LARGEST CITY Madrid
HIGHEST POINT Pico de Teide 12,172 ft. (3,710 m.) (Canary Is.); Mulhacén 11,411 ft. (3,478 m.) (mainland)
MONETARY UNIT peseta
MAJOR LANGUAGES Spanish, Catalan, Basque, Galician, Valencian
MAJOR RELIGION Roman Catholicism

PORTUGAL

AREA 35,549 sq. mi. (92,072 sq. km.)
POPULATION 9,933,000
CAPITAL Lisbon
LARGEST CITY Lisbon
HIGHEST POINT Malhão da Estrela 6,532 ft. (1,991 m.)
MONETARY UNIT escudo
MAJOR LANGUAGE Portuguese
MAJOR RELIGION Roman Catholicism

ANDORRA

AREA 188 sq. mi. (487 sq. km.)
POPULATION 31,000
CAPITAL Andorra la Vella
MONETARY UNITS French franc, Spanish peseta
MAJOR LANGUAGE Catalan
MAJOR RELIGION Roman Catholicism

GIBRALTAR

AREA 2.28 sq. mi. (5.91 sq. km.)
POPULATION 29,760
CAPITAL Gibraltar
MONETARY UNIT pound sterling
MAJOR LANGUAGES English, Spanish
MAJOR RELIGION Roman Catholicism

CITIES and TOWNS

Adra 10,851 E4
Aguilar 12,893 D4
Aguilas 15,525 F4
Alagón 5,114 F2
Alayor 5,124 J3
Albacete 82,607 F3
Albox 5,072 E4
Alburquerque 7,530 C3
Alcalá de Guadaira 28,781 D4
Alcalá de Henares 59,783 G4
Alcalá de los Gazules 5,262 D4
Alcalá la Real 9,849 E4
Alcanar 5,961 G2
Alcañiz 10,229 F2
Alcantarilla 19,895 F4
Alcaudete 8,557 E4
Alcázar de San Juan 24,620 E3
Alcira 30,493 F3
Alcora 6,711 F2
Alcoy 61,371 F3
Alfaro 8,766 F1
Algeciras 74,754 D4
Algemesí 21,158 F3
Alhama de Granada 6,148 E4
Alhama de Murcia 9,274 F4
Alicante 177,918 F3
Almadén 10,713 D3
Almagro 9,066 E3
Almansa 16,965 F3
Almendralejo 21,929 C3
Almería 104,008 E4
Almodóvar del Campo 7,310 D3
Almonte 9,960 C4
Almuñécar 7,812 E4
Álora 8,209 D4
Altea 7,262 G3
Amposta 11,767 G2
Andorra 6,485 F2
Andújar 25,962 D3
Antequera 28,039 D4
Aracena 5,390 C4
Aranda de Duero 18,183 E2
Aranjuez 28,559 E2
Archena 7,118 F3
Archidona 6,084 D4
Arcos de la Frontera 16,217 D4
Arenas de San Pedro 5,225 D2
Arenys de Mar 8,325 H2
Arévalo 5,807 D2
Argamasilla de Alba 6,192 E3
Arganda 11,876 G4
Arnedo 9,809 E1
Arrecife 21,310 C4
Arroyo de la Luz 8,130 C3
Artá 5,284 H3
Arucas 9,095 B5
Aspe 13,229 F3
Astorga 11,794 C1
Ávila de los Caballeros 30,958 D2
Avilés 67,186 C1
Ayamonte 9,897 C4
Ayora 5,249 F3
Azpeitia 7,835 E1
Azuaga 10,719 D3
Badajoz 80,793 C3
Badalona 162,888 H2
Baena 16,496 D4
Baeza 12,607 E4
Bailén 13,207 E3
Balaguer 11,676 G2
Bañolas 9,807 H1
Baracaldo 108,757 E1
Barbastro 13,243 F1
Barcarrota 5,012 C3
Barcelona 1,741,144 H2
Barcelona‡ 2,000,000 H2
Baza 14,290 E4
Beas de Segura 6,592 E3
Béjar 16,804 D2
Bélmez 5,161 D3
Benavente 11,779 D1
Benicarló 12,831 G2
Berga 11,163 G1
Berja 7,081 E4
Bermeo 16,714 E1
Betanzos 7,283 B1
Bilbao 393,179 E1
Bilbao‡ 450,000 E1
Binéfar 6,821 G2
Blanes 15,810 H2
Borjas Blancas 4,991 G2
Bujalance 8,236 D4
Bullas 8,131 F4
Burgos 118,366 E1
Burriana 21,298 G3
Cabeza del Buey 8,704 D3
Cabra 16,177 D4
Cáceres 53,108 C3
Cádiz 135,743 C4
Calahorra 16,315 E1
Calasparra 7,238 F3
Calatayud 16,524 F2
Calella 9,696 H2
Callosa de Ensarriá 5,701 G3
Calzada de Calatrava 5,751 E3
Campanario 7,722 D3
Campillos 7,014 D4
Campo de Criptana 12,604 E3
Candás 5,517 D1
Candeleda 5,153 D2
Cangas de Narcea 4,826 C1
Caniles 5,099 E4
Caravaca de le Cruz 10,411 E3
Carballo 5,542 B1
Carcagente 18,223 F3
Carmona 22,832 D4
Cartagena 52,312 F4
Caspe 8,766 G2
Cassá de la Selva 5,248 H2
Castellón de la Plana 79,773 G2
Castro del Río 10,087 D4
Castro-Urdiales 8,369 E1
Castuera 8,060 D3
Caudete 7,332 F3
Cazalla de la Sierra 5,382 D4
Cazorla 6,938 E4
Cehegín 9,661 F3
Cervera 5,693 G2
Ceuta 60,639 D5
Chiclana de la Frontera 22,986 C4
Chiva 5,394 F3
Ciempozuelos 9,185 F5
Cieza 22,929 F3
Ciudadela 13,701 H2
Ciudad Real 39,931 D3
Ciudad-Rodrigo 11,694 C2
Cocentaina 8,375 F3
Coín 14,190 D4
Colmenar de Oreja 4,930 G5
Colmenar Viejo 12,886 F4
Constantina 10,227 D4
Consuegra 10,026 E3
Córdoba 216,049 D4
Corella 5,850 F1
Coria 8,083 C3
Coria del Río 18,085 C4
Corral de Almaguer 8,006 E3
Crevillente 15,749 F3
Cuéllar 6,118 D2
Cuenca 33,980 E2
Cullera 15,128 F3
Daimiel 17,710 E3
Denia 14,514 G3
Dolores 5,420 F3
Don Benito 21,351 C3
Dos Hermanas 36,921 D4
Durango 20,403 E1
Écija 27,295 D4
Eibar 36,729 E1
Ejea de los Caballeros 9,766 F1
El Arahal 14,703 D4
Elche 101,271 F3
Elda 41,404 F3
Elizondo 2,516 F1
El Puerto de Santa María 36,451 C4
Espejo 5,925 D4
Estella 10,371 E1
Estepa 9,376 D4
Estepona 18,560 D4
Felanitx 9,100 H3
Ferrol del Caudillo 75,464 B1
Figueras 22,087 H1
Fraga 9,665 G2
Fregenal de la Sierra 6,826 C3
Fuengirola 20,597 D4
Fuente de Cantos 5,967 C3
Fuenterrabiá 2,350 E1
Fuentes de Andalucía 8,257 D4
Gandía 30,702 F3
Gerona 37,095 H2
Getafe 68,680 F4
Gijón 159,806 D1
Granada 185,799 E4
Granollers 30,066 H2
Guadalajara 30,924 E2
Guadix 15,311 E4
Guareña 7,706 C3
Guernica y Luno 12,046 E1
Haro 8,393 E1
Hellín 15,934 F3
Herencia 8,212 E3
Hinojosa del Duque 9,873 D3
Hortaleza G4
Hospitalet 241,978 H2
Huelma 5,260 E4
Huelva 96,689 C4
Huercal-Overa 5,158 F4
Huesca 33,076 F1
Huéscar 6,384 E4
Ibiza 16,943 G3
Igualada 27,941 G2
Inca 16,930 H3
Irún 38,014 F1
Iscar 5,192 D2
Isla Cristina 11,402 C4
Iznalloz 4,814 E4
Jaca 9,936 F1
Jaén 71,145 E4
Jaraíz de la Vera 6,379 D2
Játiva 20,934 F3
Jávea 6,228 G3
Jerez de la Frontera 112,411 C4
Jerez de los Caballeros 8,607 C3
Jijona 8,117 F3
Jódar 11,973 E4
Jumilla 16,407 F3
La Almunia de Doña Godina 4,835 F2
La Bañeza 8,480 C1
La Bisbal 6,374 H1
La Carolina 13,138 E3
La Coruña 184,372 B1
La Granja (San Ildefonso) 3,198 E2
La Guardia 4,967 B2
La Línea de la Concepción 51,021 D4
La Orotava 8,246 B4
La Palma del Condado 9,256 C4
La Puebla 9,923 H3
La Puebla de Montalbán 6,629 D3
La Rambla 6,525 D4
Laredo 9,114 E1
La Roda 11,460 E3
La Solana 13,894 E3
Las Palmas de Gran Canaria 260,368 B4
Las Pedroñeras 5,846 E3
La Unión 9,998 F4
Lebrija 15,081 D4
Leganés 57,537 F4
León 99,702 D1
Lérida 73,148 G2
Linares 45,330 E3
Liria 11,323 F3
Llerena 5,728 C3
Llivia 801 G1
Llodio 15,587 E1
Lluctmayor 9,630 H3
Logroño 83,117 E1
Loja 11,549 D4
Lora del Río 15,741 D4
Lorca 25,208 F4
Los Santos de Maimona 7,899 C3
Los Yébenes 5,477 E3
Lucena 21,527 D4
Lugo 53,504 C1
Madrid (cap.) 3,146,071 F4
Madrid‡ 3,500,000 F4
Madridejos 9,948 E3
Madroñera 5,397 D3
Mahón 17,802 J3
Málaga 334,988 D4
Málaga‡ 400,000 D4
Malagón 7,732 E3
Malpartida de Cáceres 5,054 C3
Manacor 20,266 H3
Mancha Real 7,547 E4
Manlleu 13,169 H1
Manresa 52,526 G2
Manzanares 15,024 E3
Marbella 19,648 D4
Marchena 16,227 D4
Marín 10,948 B1
Martos 16,395 E4
Mataró 73,129 H2
Medina del Campo 16,345 D2
Medina de Rioseco 4,874 D2
Medina-Sidonia 7,523 D4
Mérida 36,916 C3
Miajadas 8,042 D3
Mieres 22,790 D1
Minas de Ríotinto 3,939 C4
Miranda de Ebro 29,355 E1
Moguer 7,629 C4
Mollerusa 6,685 G2
Monesterio 5,923 C3
Monforte 14,002 C1
Monóvar 9,071 F3
Montehermoso 5,952 C2
Montellano 6,658 D4
Montijo 11,931 C3
Montilla 18,670 D4
Montoro 9,295 D3
Monzón 14,089 G2
Mora 10,523 E3
Moratalla 5,101 E3
Morón de la Frontera 25,662 D4
Mota del Cuervo 5,130 E3
Motril 25,121 E4
Mula 9,168 F3
Munera 5,003 E3
Murcia 102,242 F4
Navalcarnero 6,212 F4
Navalmoral de la Mata 9,650 D3
Nerja 7,413 E4
Nerva 10,830 C4
Novelda 16,867 F3
Nules 9,027 F3
Ocaña 5,603 E3
Oliva 16,717 F3
Oliva de la Frontera 8,560 C3
Olivenza 7,616 C3
Olot 18,062 H1
Olvera 9,825 D4
Onda 13,012 F3
Onteniente 23,685 F3
Orense 63,542 C1
Orihuela 17,610 F3
Osuna 17,384 D4
Oviedo 130,021 C1
Padul 6,377 E4
Palafrugell 10,421 H2
Palamós 7,679 H2
Palencia 58,327 D2
Palma 191,416 H3
Palma del Río 15,075 D4
Pamplona 142,686 F1
Pego 8,861 F3
Peñafiel 4,794 E2
Peñaranda de Bracamonte 6,094 D2
Peñarroya-Pueblonuevo 15,649 D3
Pinos-Puente 7,634 E4
Plasencia 26,897 C2
Pola de Lena 5,760 D1
Pollensa 7,625 H3
Ponferrada 22,838 C1
Pontevedra 27,118 B1
Porcuna 8,169 D4
Port-Bou 2,230 H1
Portugalete 45,589 E1
Posadas 7,245 D4
Pozoblanco 13,280 D3
Pozuelo de Alarcón 14,041 D2
Priego de Córdoba 12,676 D4
Puente-Genil 22,888 D4
Puertollano 50,609 D3
Puerto Real 13,993 D4
Puigcerdá 4,418 G1
Quesada 6,965 E4
Quintana de la Serena 5,171 D3
Quintanar de la Orden 7,764 E3
Reinosa 10,863 D1
Requena 9,836 F3
Reus 47,240 G2
Ripoll 9,283 H1
Ronda 22,094 D4
Roquetas 5,617 G2
Rosas 5,448 H1
Rota 20,021 C4
Rute 8,294 D4
Sabadell 148,223 H2
Sagunto 17,052 F3
Salamanca 125,132 D2
Sallent 7,118 H2
Salobreña 5,961 E4
Salt 5,572 H1
Sama 9,863 D1
San Carlos de la Rápita 8,946 G2
San Clemente 6,016 E3
San Feliu de Guíxols 12,006 H2
San Fernando 59,309 C4
San Ildefonso 3,198 E2

Agriculture, Industry and Resources

DOMINANT LAND USE

- Cereals (chiefly wheat)
- Livestock (chiefly sheep, goats)
- Mixed Cereals, Livestock
- Olives, Fruit
- Grapes, Fruit, Nuts, Mixed Cereals
- Forests
- Nonagricultural Land

MAJOR MINERAL OCCURRENCES

Ag Silver — Na Salt
C Coal — O Petroleum
Cu Copper — Pb Lead
Fe Iron Ore — Py Pyrites
G Natural Gas — Sb Antimony
Hg Mercury — Sn Tin
K Potash — U Uranium
Lg Lignite — W Tungsten
Mg Magnesium — Zn Zinc

Water Power
Major Industrial Areas

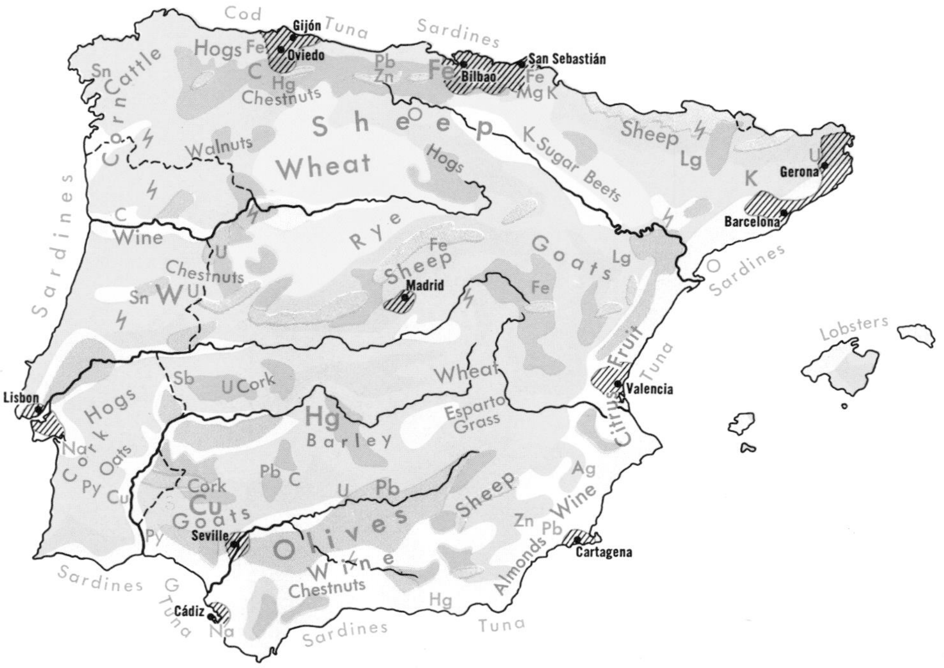

(continued on following page)

San Lorenzo de El		
Escorial 8,098.	E2	
Sanlúcar de Barrameda 29,483.	C4	
Sanlúcar la Mayor 6,121.	C4	
San Roque 8,224.	D4	
San Sebastián 159,557.	E1	
Santa Cruz de la Palma 10,393.	B4	
Santa Cruz de Mudela 6,354.	E3	
Santa Cruz de Tenerife 74,910.	B4	
Santa Eugenia 5,946.	B1	
Santa Fé 8,990.	E4	
Santander 130,019.	D1	
Santiago 51,620.	B1	
Santo Domingo de la		
Calzada 5,638.	E1	
Santoña 9,546.	E1	
San Vicente de		
Alcántara 7,006.	C3	
Saragossa 449,319.	F2	
Saragossa 500,000.	F2	
Segorbe 6,962.	F3	
Seo de Urgel 6,604.	G1	
Seville 511,447.	D4	
Seville 560,000.	D4	
Sitges 8,906.	G2	
Socuéllamos 12,610.	E3	
Sóller 6,470.	H3	
Solsona 5,346.	G2	
Sonseca 6,594.	D3	
Soria 24,744.	E2	
Sotrondio 5,914.	D1	
Sueca 20,019.	F3	
Tabernes de Valldigna 13,962.	G3	
Tafalla 8,858.	F1	
Talavera de la Reina 39,889.	D2	
Tarancón 6,238.	E3	
Tarazona 11,067.	E2	
Tarazona de la Mancha 5,952.	F3	
Tarifa 5,583.	D4	
Tarragona 53,548.	G2	
Tarrasa 134,481.	G2	
Tárrega 9,036.	G2	
Tauste 6,832.	F2	
Telde 13,257.	B5	
Teruel 20,614.	F2	

Tobarra 5,887.	F3	
Toledo 43,905.	D3	
Tolosa 15,164.	E1	
Tomelloso 26,041.	E3	
Tordesillas 5,815.	D2	
Toro 8,455.	D2	
Torredonjimeno 12,507.	D4	
Torrejón de Ardoz 21,081.	G4	
Torrelavega 19,933.	D1	
Torremolinos 20,484.	D4	
Torrente 38,397.	F3	
Torrevieja 9,431.	F4	
Torrijos 6,362.	D2	
Torrox 5,583.	E4	
Tortosa 20,030.	G2	
Totana 12,714.	F4	
Triguéros 6,280.	C4	
Trujillo 9,024.	D3	
Tudela 20,942.	F1	
Úbeda 28,306.	E3	
Ubrique 13,166.	D4	
Utrel 9,168.	F3	
Utrera 28,287.	D4	
Valdemoro 6,263.	F4	
Valdepeñas 24,018.	E3	
Valencia 626,675.	F3	
Valencia† 700,000.	F3	
Valencia de Alcántara 5,963.	C3	
Valladolid 227,511.	D2	
Vall de Uxó 23,976.	G4	
Vallecas.	G4	
Valls 14,189.	G2	
Valverde del Camino 10,566.	C4	
Vejer de la Frontera 6,184.	C4	
Vélez-Málaga 20,794.	E4	
Vendrell 7,951.	G2	
Vera 4,903.	F4	
Vergara 11,541.	E1	
Vich 23,449.	H2	
Vicálvaro.	G4	
Vigo 114,526.	B1	
Vilafranca del		
Penedés 16,875.	G2	
Villacañas 9,883.	E3	
Villacarrillo 9,452.	E3	
Villafranca de los		

Barros 12,610.	C3	
Villagarcía 6,601.	B1	
Villajoyosa 12,573.	F3	
Villanueva de Córdoba 11,270.	D3	
Villanueva del Arzobispo 8,076.	E3	
Villanueva de la Serena 16,687.	D3	
Villanueva de los		
Infantes† 8,154.	E3	
Villanueva y Geltrú 35,714.	G2	
Villarreal de los		
Infantes 29,482.	G3	
Villarrobledo 19,698.	E3	
Villarrubia de los Ojos 9,144.	E3	
Villaverde.	F3	
Villena 23,483.	F3	
Vinaroz 13,727.	G2	
Vitoria 124,791.	E1	
Yecla 19,352.	F3	
Zafra 11,583.	C3	
Zalamea de la Serena 6,017.	D3	
Zamora 48,791.	D2	
Zaragoza (Saragossa) 449,319.	F2	

OTHER FEATURES

Alborán (isl.).	E5	
Alcaraz, Sierra de (range).	E3	
Alcudia (bay).	H3	
Almanzor (mt.).	D2	
Almanzora (riv.).	F4	
Andalusia (reg.).	C4	
Aneto (peak).	G1	
Aragón (reg.).	F2	
Arosa, Ria de (est.).	B1	
Asturias (reg.).	C1	
Balaitous (mt.).	F1	
Balearic (Baleares)		
(isls.).	H3	
Barbate (riv.).	D4	
Biscay (bay).	E1	
Cabrera (isl.).	H3	
Cádiz (gulf).	C4	
Cala Burras (pt.).	D4	
Canary (isls.).	B4	
Cantabrian (range).	C1	
Catalonia (reg.).	G2	

Cinca (riv.).	G2	
Columbretes (isls.).	G3	
Costa Brava (reg.).	H2	
Costa de Sola (Costa del Sol)		
(reg.).	D4	
Creus (cape).	H1	
Cuenca, Sierra de (range).	F3	
Demanda, Sierra de la (range).	E1	
Duero, Sierra de la (range).	C2	
Duero (Douro) (riv.).	C2	
Ebro (riv.).	G2	
Eresma (riv.).	D2	
Esla (riv.).	D1	
Estats (peak).	G1	
Estremadura (reg.).	C3	
Finisterre (cape).	B1	
Formentera (isl.).	G3	
Formentor (cape).	H2	
Fuerteventura (isl.).	C4	
Galicia (reg.).	B1	
Gata (cape).	F4	
Gata (mts.).	C2	
Genil (riv.).	D4	
Gibraltar (str.).	D5	
Gomera (isl.).	B5	
Gran Canaria (isl.).	B5	
Gredos, Sierra de (range).	D2	
Guadalimar (riv.).	E3	
Guadalquivir (riv.).	C4	
Guadarrama, Sierra de (range).	E2	
Guadiana (riv.).	D3	
Gúdar, Sierra de (range).	F2	
Henares (riv.).	E2	
Hierro (isl.).	A5	
Ibiza (isl.).	G3	
Jalón (riv.).	E2	
Jarama (riv.).	E2	
Júcar (riv.).	F3	
Lanzarote (isl.).	C4	
La Palma (isl.).	A4	
León (riv.).	C1	
Llobregat (riv.).	G2	
Majorca (isl.).	H3	
Mallorca (Majorca)		
(isl.).	H3	

Mancha, La (reg.).	E3	
Manzanares (riv.).	F4	
Marismas, Las (marsh).	C4	
Mar Menor (lag.).	F4	
Mayor (cape).	D1	
Menorca (Minorca) (isl.).	J3	
Miño (riv.).	B1	
Minorca (isl.).	J3	
Moncayo, Sierra de (range).	F2	
Montserrat (mt.).	G2	
Morena, Sierra (range).	E3	
Mulhacén (mt.).	E4	
Murcia (reg.).	G3	
Navia (riv.).	C1	
Nevada, Sierra (mts.).	E4	
New Castile (reg.).	E3	
Odiel (riv.).	C4	
Old Castile (reg.).	D2	
Orbigo (riv.).	D1	
Palos (cape).	F4	
Peñalara (mt.).	D2	
Peñas (cape).	D1	
Penibética, Sistema (range).	E4	
Perdido (mt.).	G1	
Pyrenees (range).	F1	

Rosas (gulf).	H1	
San Jorge (gulf).	G2	
Segre (riv.).	G2	
Segura (riv.).	F3	
Sil (riv.).	C1	
Tagus (riv.).	D3	
Tajo (Tagus) (riv.).	E2	
Teide, Pico de (peak).	B5	
Tenerife (isl.).	B5	
Ter (riv.).	H1	
Tinto (riv.).	C4	
Toledo (mts.).	D3	
Tortosa (cape).	G2	
Trafalgar (cape).	C4	
Turia (riv.).	F3	
Ulla (riv.).	B1	
Urgel, Llanos de (plain).	G2	
Valencia (gulf).	G3	
Valencia (reg.).	F3	
Valencia, Albufera de (lag.).	G3	
Vascongadas (reg.).	E1	

PORTUGAL

DISTRICTS

Aveiro 545,230.	B2	

Beja 204,440.	C3	
Braga 609,415.	B2	
Bragança 180,395.	C2	
Castelo Branco 254,355.	C3	
Coimbra 399,380.	B2	
Évora 178,475.	C3	
Faro 268,040.	B4	
Guarda 210,720.	C2	
Leiria 376,940.	B3	
Lisbon 1,568,020.	A3	
Oporto (Porto)		
1,309,560.	A1	
Portalegre 145,545.	C3	
Porto 1,309,560.	A1	
Santarém 427,995.	B3	
Setúbal 469,555.	B3	
Viana do Castelo		
250,510.	A1	
Vila Real 265,605.	C2	
Viseu 410,795.	B2	

CITIES and TOWNS

Abrantes 11,775.	B3	
Águeda 9,343.	B2	
Albufeira 7,479.	B4	
Alcácer do Sal 13,187.	B3	
Alcântara 23,699.	A1	

Topography

0 ... 50 ... 100 MI.
0 ... 50 ... 100 KM.

Below Sea Level | 100 m. 328 ft. | 200 m. 656 ft. | 500 m. 1,640 ft. | 1,000 m. 3,281 ft. | 2,000 m. 6,562 ft. | 5,000 m. 16,404 ft.

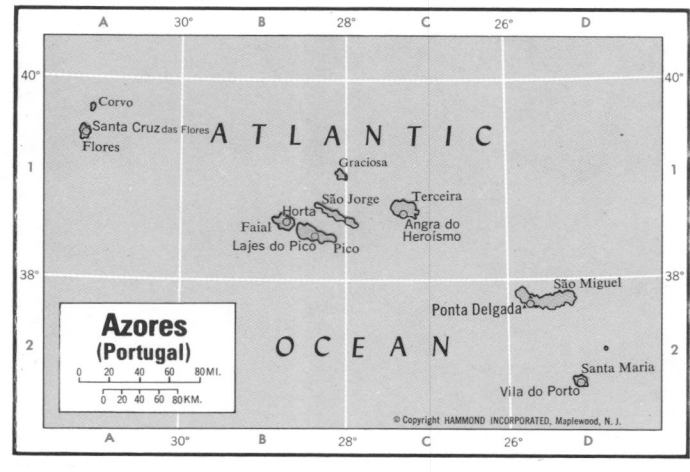

AZORES

INTERNAL DIVISIONS

Angra do Heroísmo		
(dist.) 83,500.	C1	
Horta (dist.) 38,700.	A1	
Ponta Delgada (dist.) 153,700.	D2	

CITIES and TOWNS

Angra do Heroísmo 13,795.	C1	
Horta 6,145.	B1	
Lajes do Pico 2,147.	B1	
Ponta Delgada 20,195.	C2	
Santa Cruz das Flores 1,880.	A1	
Vila do Porto 4,149.	D2	

OTHER FEATURES

Azores (isls.).	A2	
Corvo (isl.).	A1	
Faial (isl.).	B1	
Flores (isl.).	A1	
Graciosa (isl.).	C1	
Pico (isl.).	C1	
Santa Maria (isl.).	D2	
São Jorge (isl.).	B1	
São Miguel (isl.).	D2	
Terceira (isl.).	C1	

Azores (Portugal)
0 ... 20 ... 40 ... 60 ... 80 MI.
0 ... 20 ... 40 ... 60 ... 80 KM.

© Copyright HAMMOND INCORPORATED, Maplewood, N.J.

PORTUGAL is divided into 18 mainland districts bearing the same names as their respective capitals. The Azores and Madeira are offshore autonomous regions.

Alcobaça 4,799	B3	
Aldeia Nova de São Bento 5,228	C4	
Algés 18,010	A1	
Alhos Vedros 7,915	A3	
Aljustrel 7,473	B4	
Almada 38,990	A1	
Almeirim 8,780	B3	
Alpiarça 7,623	B3	
Alportel 7,632	C4	
Amadora 65,870	A1	
Amarante 6,067	B2	
Amora 10,330	A3	
Aveiro 19,905	B2	
Avis 1,686	C3	
Baixa da Banheira 18,550	B3	
Barreiro 53,690	B1	
Batalha 6,673	B3	
Beja 14,760	B4	
Belas 12,001	A1	
Belém 19,043	A1	
Benfica 39,459	A1	
Borba 4,879	C3	
Braga 48,735	B2	
Bragança 9,310	C2	
Caldas da Rainha 13,070	B3	
Câmara de Lobos 14,068	A2	

Campo Maior 7,405	C3	
Cantanhede 6,734	B2	
Caparica 13,315	A3	
Carnaxide 38,309	A1	
Cartaxo 6,628	B3	
Cascais 14,925	A1	
Castelo Branco 18,740	C3	
Cercal 5,021	B4	
Chaves 11,465	C2	
Coimbra 55,985	B2	
Coruche 17,461	B3	
Cova da Piedade 21,000	A3	
Covilhã 26,530	C2	
Elvas 10,305	C3	
Espinho 11,745	B2	
Estoril 15,740	A1	
Estremoz 9,565	C3	
Évora 23,665	C3	
Fafe 8,142	B2	
Faro 20,470	B4	
Fátima 6,433	B3	
Feira 5,222	B2	
Ferreira do Alentejo 6-153	B4	
Figueira da Foz 10,485	B3	
Funchal 38,340	A2	
Fundão 5,081	C2	
Gondomar 14,105	B2	

Grândola 9,698	B3	
Guarda 9,735	C2	
Guimarães 24,280	B2	
Ílhavo 11,083	B2	
Lagoa 5,694	B4	
Lagos 10,359	B4	
Lamego 10,350	C2	
Lavos 5,051	B3	
Leiria 7,540	B3	
Lisbon (Lisboa) (cap.) 769,410	A1	
Lisbon (Lisboa) 1,100,000	A1	
Loulé 12,777	B4	
Louriçal 6,087	B3	
Lourinhã 7,340	A3	
Lousã 7,341	B3	
Machico 10,905	A2	
Mafra 7,149	A3	
Manguable 4,839	C2	
Marinha Grande 18,548	B3	
Matosinhos 22,505	B2	
Mira 21,987	B2	
Mirandela 5,203	C2	
Monchique 4,835	B4	
Montargil 5,070	B3	
Montemor-o-Novo 9,284	B3	
Montijo 26,730	B3	
Moscavide 21,765	A1	

Moura 9,351	C3	
Nazaré 8,553	B3	
Odemira 6,793	B4	
Odivelas 26,020	A1	
Oeiras 14,880	A1	
Olhão 11,155	B4	
Olivais 55,138	A1	
Oporto (Porto) 300,925	B2	
Ovar 16,004	B2	
Paço de Arcos 11,791	A1	
Penafiel 6,463	C2	
Peniche 12,555	A3	
Peso da Régua 5,376	C2	
Pombal 12,508	B3	
Ponta do Sol 5,599	A2	
Ponte de Sor 9,951	B3	
Portalegre 10,970	C3	
Portimão 10,300	B4	
Porto 300,925	B2	
Póvoa de Varzim 17,415	B2	
Proença-a-Nova 4,792	C3	
Queluz 25,845	A1	
Redondo 6,858	C3	
Reguengos de Monsaraz 5,806	C3	
Ribeira Brava 7,416	A2	
Rio Maior 10,206	B3	
Sacavém 12,625	A1	

Salvaterra de Magos 6,265	B3	
Santa Cruz 6,348	A2	
Santarém 16,850	B3	
Santiago do Cacem 5,887	B4	
São Brás de Alportel 7,632	C4	
São João da Madeira 14,225	B2	
São Teotónio 6,146	B4	
São Vicente 5,147	A2	
Serpa 7,991	C4	
Serta 6,043	B3	
Sesimbra 16,614	B3	
Setúbal 49,670	B3	
Silves 9,493	B4	
Sines 6,996	B4	
Sintra 15,994	A3	
Soure 7,620	B3	
Tavira 10,263	C4	
Tomar 10,905	B3	
Torres Novas 13,806	B3	
Torres Vedras 14,833	A3	
Trafaria 6,145	A1	
Vagos 5,088	B2	
Vendas Novas 8,979	B3	
Viana do Castelo 12,510	B2	
Vila do Conde 16,485	B2	
Vila Franca de Xira 13,070	B3	

Vila Nova de Gaia 50,805	B2	
Vila Real 10,050	C2	
Vila Real de Santo Antonio 10,320	C4	
Viseu 16,140	C2	
OTHER FEATURES		
Atlantic Ocean	A3	
Carvoeiroeiro (cape)	A3	
Desertasrtas (isls.)	B2	
Douro (riv.)	B2	
Espichel (cape)	A3	
Estrela, Serra da (mts.)	C2	
Guadiana (riv.)	C4	
Lima (riv.)	B2	
Madeira (isl.)	A2	
Madeira (isls.)	A2	
Minho (riv.)	B4	
Monchique, Serra de (mts.)	B4	
Mondego (riv.)	B2	
Monsanto (riv.)		
Ossa, Serra da (mts.)	C3	
Palha, Mar da (bay)	A3	
Porto Santo (isl.)	B2	
Roca (cape)	B3	

Sadu (riv.)	B3	
São Vincent (cape)	A4	
Santa Marla (cape)	C4	
Setúbal (bay)	B3	
Tagus (riv.)	B3	
Tâmega (riv.)	B2	
Tejo (Tagus) (riv.)	B3	
Xarrama (riv.)	B3	

ANDORRA

CITIES and TOWNS

Andorra la Vella (cap.) 12,000 ... G1

GIBRALTAR

Gibraltar 29,760 ... D4

PHYSICAL FEATURES

Europa (pt.) ... D4

‡Population of metropolitan area

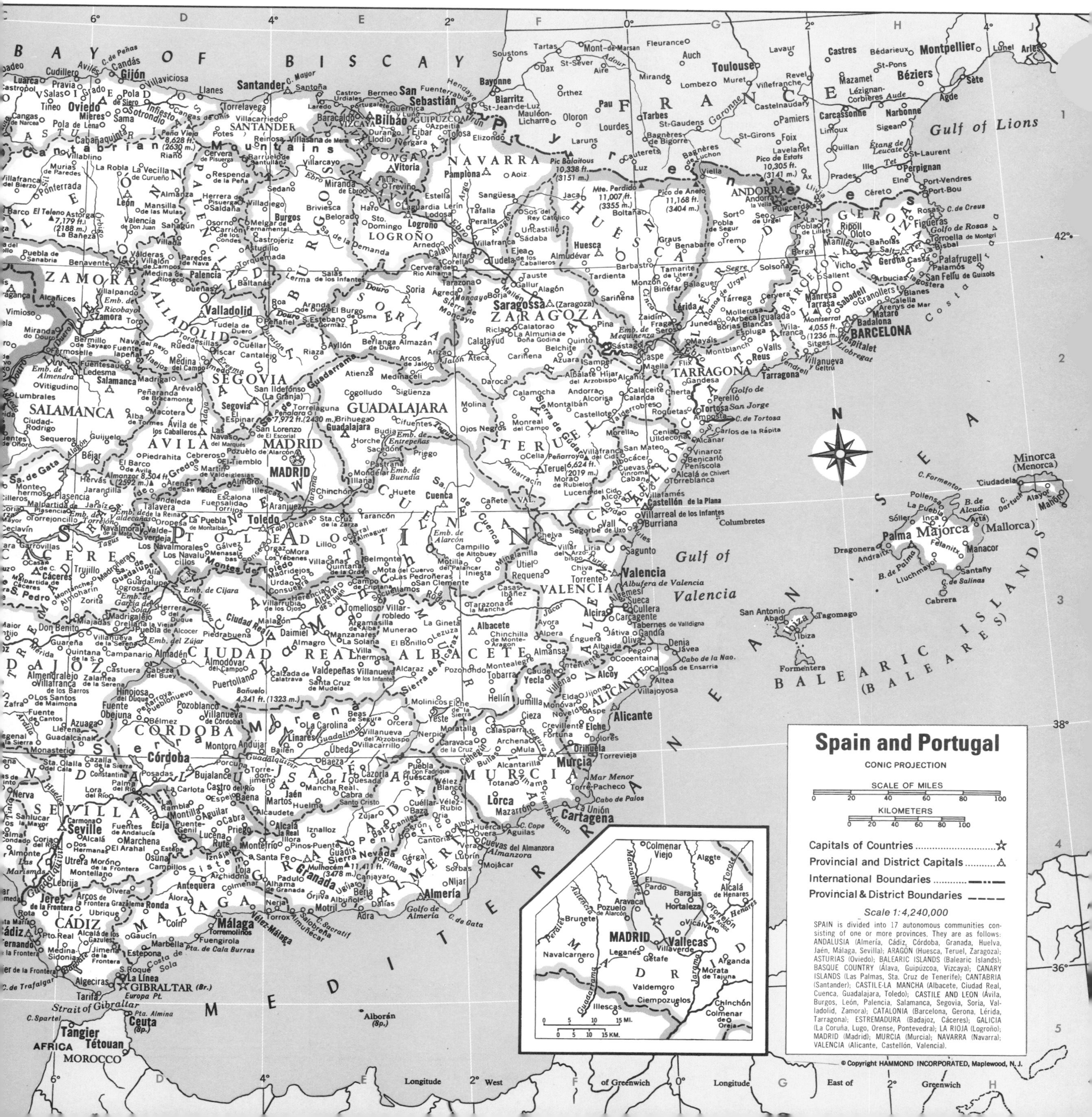

Spain and Portugal

CONIC PROJECTION

SCALE OF MILES

0 20 40 60 80 100

KILOMETERS

Capitals of Countries	☆
Provincial and District Capitals	△
International Boundaries	———
Provincial & District Boundaries	-----

Scale 1:4,240,000

SPAIN is divided into 17 autonomous communities consisting of one or more provinces. They are as follows: ANDALUSIA (Almería, Cádiz, Córdoba, Granada, Huelva, Jaén, Málaga, Sevilla); ARAGÓN (Huesca, Teruel, Zaragoza); ASTURIAS (Oviedo); BALEARIC ISLANDS (Balearic Islands); BASQUE COUNTRY (Álava, Guipúzcoa, Vizcaya); CANARY ISLANDS (Las Palmas, Sta. Cruz de Tenerife); CANTABRIA (Santander); CASTILE-LA MANCHA (Albacete, Ciudad Real, Cuenca, Guadalajara, Toledo); CASTILE AND LEON (Ávila, Burgos, León, Palencia, Salamanca, Segovia, Soria, Valladolid, Zamora); CATALONIA (Barcelona, Gerona, Lérida, Tarragona); ESTREMADURA (Badajoz, Cáceres); GALICIA (La Coruña, Lugo, Orense, Pontevedra); LA RIOJA (Logroño); MADRID (Madrid); MURCIA (Murcia); NAVARRA (Navarra); VALENCIA (Alicante, Castellón, Valencia).

© Copyright HAMMOND INCORPORATED, Maplewood, N.J.

Italy

CONIC PROJECTION

SCALE OF MILES

SCALE OF KILOMETERS

Capitals of Countries _____ ☆
Regional Capitals _____ ⊞
Provincial Capitals _____ △
International Boundaries _____ ▬ ▪ ▬ ▪
Regional Boundaries _____ ▬ ▪ ▪ ▬

Scale 1 : 4,710,000

The regions are subdivided into provinces bearing the same names as their respective capitals, except:

PROVINCE	CAPITAL
MASSA-CARRARA	Massa
PESARO-URBINO	Pesaro

Vatican City

SCALE

Rome and Environs

© Copyright HAMMOND INCORPORATED, Maplewood, N.J.

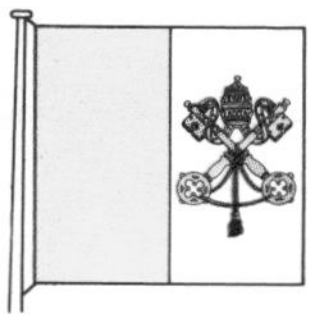

VATICAN CITY

AREA 108.7 acres (44 hectares)
POPULATION 728

SAN MARINO

AREA 23.4 sq. mi. (60.6 sq. km.)
POPULATION 19,149

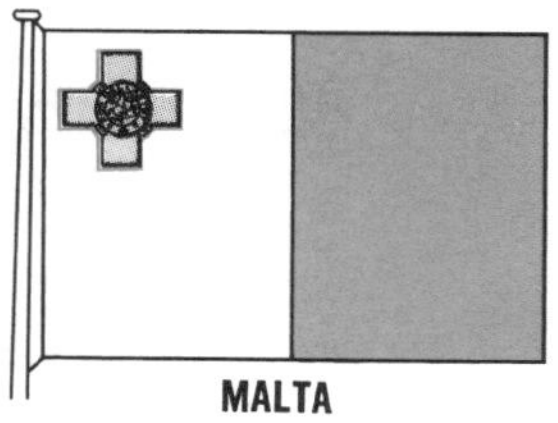

MALTA

AREA 122 sq. mi. (316 sq. km.)
POPULATION 343,970
CAPITAL Valletta
LARGEST CITY Sliema
HIGHEST POINT 787 ft. (240 m.)
MONETARY UNIT Maltese pound
MAJOR LANGUAGES Maltese, English
MAJOR RELIGION Roman Catholicism

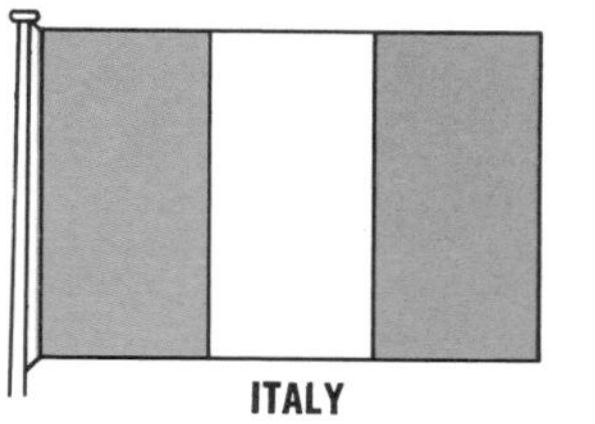

ITALY

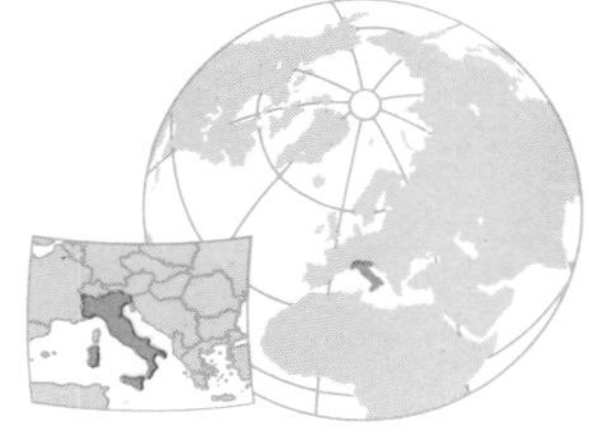

AREA 116,303 sq. mi. (301,225 sq. km.)
POPULATION 57,140,000
CAPITAL Rome
LARGEST CITY Rome
HIGHEST POINT Dufourspitze (Mte. Rosa) 15,203 ft. (4,634 m.)
MONETARY UNIT lira
MAJOR LANGUAGE Italian
MAJOR RELIGION Roman Catholicism

ITALY

REGIONS

Abruzzi 1,166,664	D3
Aosta 109,150	A2
Apulia (Puglia) 3,582,787	F4
Basilicata 603,064	F4
Calabria 1,988,051	F5
Campania 5,059,348	E4
Emilia-Romagna 3,846,755	C2
Friuli-Venezia Giulia 1,213,532	D1
Latium (Lazio) 4,689,482	D3
Liguria 1,853,578	B2
Lombardy 8,543,657	B2
Marche 1,359,907	D3
Molise 319,807	E4
Piedmont 4,432,313	A2
Sardinia 1,473,800	B4
Sicily 4,680,715	D6
Trentino-Alto Adige 841,886	C1
Tuscany 3,473,097	C3
Umbria 775,783	D3
Veneto 2,109,502	D3

PROVINCES

Agrigento 454,045	D6
Alessandria 483,183	B2
Ancona 416,611	D3
Aosta 109,150	A2
Arezzo 306,340	C3
Ascoli Piceno 340,758	D3
Asti 218,547	B2
Avellino 427,509	E4
Bari 1,351,288	F4
Belluno 221,155	D1
Benevento 286,499	E4
Bergamo 829,019	B2
Bologna 918,844	C2
Bolzano-Bozen 414,041	C1
Brescia 957,686	C2
Brindisi 366,027	G4
Cagliari 802,888	B5
Caltanissetta 282,069	D6
Campobasso 227,641	E4
Caserta 677,959	E4
Catania 938,273	E6
Catanzaro 718,069	F5
Chieti 351,567	E3
Como 720,463	B2
Cosenza 691,659	F5
Cremona 334,281	B2
Cuneo 540,504	A2
Enna 202,131	E6
Ferrara 383,639	C2
Florence 1,146,367	C3
Foggia 657,292	E4
Forlì 565,470	D2
Frosinone 422,630	D4
Genoa 1,087,973	B2
Gorizia 142,412	D2
Grosseto 216,315	C3
Imperia 225,127	B3
Isernia 92,166	E4
L'Aquila 293,066	D3
La Spezia 244,435	B2
Latina 376,238	D4
Lecce 696,503	G4
Leghorn 335,265	C3
Lucca 380,356	C3
Macerata 286,155	D3
Mantua 376,892	C2
Massa-Carrara 200,955	C2
Matera 194,629	F4
Messina 654,703	E5
Milan 3,903,685	B2
Modena 553,852	C2
Naples 2,709,929	E4
Novara 496,811	B2
Nuoro 273,021	B4
Padua 762,998	C2
Palermo 1,124,015	D5
Parma 395,497	C2
Pavia 526,389	B2
Perugia 552,936	D3
Pesaro e Urbino 316,383	D3
Pescara 264,981	E3
Piacenza 284,881	B2
Pisa 375,933	C3
Pistoia 254,335	C2
Pordenone 253,906	D2
Potenza 408,435	E4
Ragusa 255,047	E6
Ravenna 351,876	D2
Reggio di Calabria 578,323	E5
Reggio nell'Emilia 392,696	C2
Rieti 143,162	D3
Rome 3,490,377	F6
Rovigo 251,908	C2
Salerno 957,452	E4
Sassari 397,891	B4
Savona 296,043	B2
Siena 257,221	C3
Sondrio 169,149	B1
Syracuse 365,039	E6
Taranto 511,677	F4
Teramo 257,080	D3
Terni 222,847	D3
Trapani 405,393	D5
Trento 427,845	C1
Treviso 668,620	D2
Trieste 300,304	E2
Turin 2,287,016	A2
Udine 516,910	D1
Varese 725,823	B2
Venice 807,251	D2
Vercelli 406,252	B2
Verona 733,595	C2
Vicenza 677,884	C2
Viterbo 257,075	C3

CITIES and TOWNS

Acireale 34,081	E6
Acqui Terme 20,099	B2
Acri 8,150	F5
Adrano 31,988	E6
Adria 11,951	D2
Agira 11,262	E6
Agnone 3,965	E4
Agrigento 40,513	D6
Agropoli 9,413	E4
Alassio 13,512	A2
Alatri 5,710	D4
Alba 23,522	B2
Albano Laziale 15,561	F7
Albenga 13,397	B3
Albino 8,837	B2
Alcamo 41,448	D6
Alessandria 78,644	B2
Alghero 28,454	B4
Altamura 44,879	F4
Amalfi 4,205	E4
Amantea 6,132	E5
Amelia 4,331	D3
Ancona 88,427	D3
Andria 76,405	F4
Anguillara Sabazia 3,241	F6
Anzio 14,966	D4
Aosta 35,053	A2
Aprilia 18,412	D4
Aragona 11,213	D6
Arezzo 56,693	C3
Argenta 6,682	D2
Ariano Irpino 9,796	E4
Ariccia 7,287	F7
Artena 5,034	F7
Ascoli Piceno 43,041	D3
Assisi 4,630	D3
Asti 62,277	B2
Atessa 3,079	E3
Atri 4,686	D3
Augusta 32,501	E6
Avellino 44,750	E4
Aversa 46,536	E4
Avezzano 26,456	D3
Avigliano 5,400	E4
Avola 29,089	E6
Bagheria 32,465	D5
Barcellona Pozzo di Gotto 25,280	E5
Bari 339,110	F4
Barletta 75,116	F4
Bassano del Grappa 33,002	C2
Bellagio 3,258	B2
Belluno 22,180	D1
Benevento 48,523	E4
Bergamo 127,553	B2
Biancavilla 18,743	E6
Biella 46,453	B2
Bisceglie 45,014	F4
Bitonto 39,714	F4
Bitti 4,606	B4
Bologna 493,282	C2
Bolzano (Bolzen) 102,806	C1
Bondeno 7,451	C2
Bonorva 5,232	B4
Bordighera 8,994	A3
Borgo 4,013	C1
Borgomanero 16,655	B2
Bórgo San Lorenzo 7,699	C2
Bosa 8,045	B4
Boves 3,896	A2
Bra 18,399	A2
Bracciano 7,681	C3
Brescia 189,092	C2
Bressanone 12,261	C1
Brindisi 76,612	G4
Bronte 17,823	E6
Brunico 5,175	D1
Budrio 5,635	C2
Busto Arsizio 72,400	B2
Cagli 4,356	D3
Cagliari 211,015	B5
Caltagirone 34,444	E6
Caltanissetta 52,838	D6
Camaiore 8,578	C3
Camerino 4,644	D3
Campobasso 35,551	E4
Campo Tures 1,325	C1
Canicatti 28,761	E6
Canosa di Puglia 30,263	E4
Cantù 28,617	B2
Capua 13,938	E4
Caravaggio 11,298	B2
Carbonia 23,031	B5
Carini 14,255	D5
Carloforte 6,671	B5
Carmagnola 16,469	A2
Carpi 41,789	C2
Carrara 56,236	C2
Casale Monferrato 35,156	B2
Casalmaggiore 6,374	C2
Cascina-Navacchio 28,263	C3
Caserta 51,621	E4
Cassano allo Ionio 9,661	F5
Cassino 14,747	D4
Castelfranco Veneto 16,042	D2
Castel Gandolfo 2,965	F7
Castellammare del Golfo 13,144	D5
Castellammare di Stabia 64,341	E4
Castel San Pietro Terme 6,985	C2
Castelvetrano 29,167	D6
Castiglion Fiorentino 3,797	C3
Castrovillari 15,207	F5
Catania 403,390	E6
Catanzaro 52,054	F5
Caulonia 3,402	F5
Cava de'Tirreni 33,868	E4
Cavarzere 7,917	D2
Cecina 19,415	C3
Cefalù 11,043	E5
Ceglie Messapico 17,512	F4
Celano 9,531	D3
Cerignola 44,648	E4
Cernobbio 8,026	B2
Cerveteri 5,239	E6
Cesano 2,883	F6
Cesena 49,915	D2
Cesenatico 12,805	D2
Chiari 12,017	C2
Chiavari 29,950	B2
Chieri 27,548	A2
Chieti 31,895	E3
Chioggia 24,044	D2
Chivasso 21,369	A2
Ciampino 36,728	F7
Cittadella 9,321	C2
Città di Castello 18,880	C3
Cittanova 11,045	F5
Cividale del Friuli 8,345	D1
Civitavecchia 41,305	C3
Clusone-Fiorine 6,428	C2
Codroipo 6,117	D2
Colle di Val d'Elsa 8,657	C3
Comacchio 10,437	D2
Comiso 24,508	E6
Como 73,257	B2
Conegliano 28,635	D2
Conversano 16,805	F4
Corato 38,163	F4
Cori 6,829	F7
Corigliano Calabro 14,518	F5
Corleone 11,057	D6
Correggio 11,415	C2
Cortina d'Ampezzo 7,285	D1
Cortona 3,482	C3
Cosenza 94,565	F5
Courmayeur 1,401	A2
Crema 26,061	B2
Cremona 75,988	B2
Crotone 44,081	F5
Cuneo 41,633	A2
Cuorgnè 6,752	A2
Desenzano del Garda 14,624	C2
Diano Marina 6,001	B3
Domodossola 18,562	A1
Dorgali 6,714	B4
Eboli 19,787	E4
Edolo 3,707	C1
Empoli 30,526	C3
Enna 27,351	E6
Este 12,992	C2
Fabriano 18,355	D3
Faenza 36,241	D2
Fano 31,238	D3
Fasano 21,247	F4
Favara 27,940	D6
Feltre 11,806	C1
Fermo 17,521	D3
Ferrandina 8,372	F4
Ferrara 97,507	C2
Fidenza 18,064	B2
Fiesole 3,772	C3
Finale Emilia 7,474	C2
Finale Ligure 11,461	B2
Firenze (Florence) 441,654	C3
Fiumicino 13,180	F7
Florence 441,654	C3
Floridia 16,562	E6
Foggia 136,436	E4
Foligno 26,887	D3
Fondi 16,472	D4
Forlì 83,303	D2
Formia 18,978	D4
Fossano 15,857	A2
Fossombrone 5,882	D3
Francavilla Fontana 30,347	F4
Frascati 14,217	F7
Frosinone 34,066	D4
Gaeta 21,973	D4
Galatina 22,137	G4
Galatone 13,880	F4
Gallarate 43,773	B2
Gallipoli 16,878	F4
Garessio 3,359	A2
Gela 66,845	E6
Gemona 6,863	D1
Genoa 787,011	B2
Genova (Genoa) 787,011	B2
Genzano di Roma 14,147	F7
Giarre 18,233	E6
Gioia del Colle 23,299	F4
Gioiosa Ionica 3,811	F5
Giovinazzo 17,768	F4
Giulianova 17,926	E3
Gorizia 35,912	D2
Gravina in Puglia 32,006	F4
Grosseto 48,309	C3
Grottaferrata 10,639	F7
Grottaglie 23,556	F4
Guardiagrele 4,122	E3
Guastalla 7,639	C2
Gubbio 12,371	D3
Guidonia 8,413	F6
Iglesias 24,472	B5
Imola 42,111	C2
Imperia 37,585	B3
Isernia 12,290	E4
Ivrea 26,530	B2
Jesi 33,011	D3
Ladispoli 6,625	E6
Lagonegro 5,613	E4
La Maddalena 10,405	B4
Lanciano 19,652	E3
Lanusei 5,508	B5
Lanuvio 2,970	F7
L'Aquila 36,233	D3
Larino 5,166	E4
La Spezia 121,254	B2
Latina 53,003	D4
Lauria 4,927	E4
Lavello 11,486	E4
Lecce 80,114	G4
Lecco 53,165	B2
Leghorn 170,369	C3
Legnago 15,534	C2
Lendinara 7,079	C2
Lentini 31,429	E6
Leonforte 16,317	E6
Lerici 5,407	B2
Licata 40,997	D6
Lido di Ostia 61,492	F7
Lido di Venezia 18,794	D2
Lipari 3,886	E5
Livigno 2,135	C1
Livorno (Leghorn) 170,369	C3
Lodi 42,489	B2
Lonigo 6,368	C2
Lucca 54,280	C3
Lucera 29,355	E4
Lugo 19,497	D2
Macerata 33,470	D3
Macomer 9,433	B4
Maglie 13,326	G4
Manduria 25,194	F4
Manfredonia 44,463	F4
Mantua 59,529	C2
Marino 12,135	F7
Marsala 34,150	D6
Marsciano 5,372	D3
Martina Franca 31,811	F4
Massa 56,591	C2
Massafra 22,610	F4
Massa Marittima 6,438	C3
Matera 43,026	F4
Mazara del Vallo 37,441	D6
Mazzarino 14,981	E6
Melfi 13,355	E4
Menfi 12,386	D6
Merano 30,951	C1
Mesagne 26,955	G4
Messina 203,937	E5
Mestre 184,818	D2
Milan 1,724,557	B2
Milazzo 18,576	E5
Minturno 2,428	D4
Mirandola 11,551	C2
Mira Taglio 10,194	D2
Mistretta 6,631	E6
Modena 149,029	C2
Modica 31,074	E6
Mola di Bari 23,778	F4
Molfetta 63,250	F4
Moncalieri 49,953	A2
Mondovì Breo 12,524	A2
Monfalcone 29,589	D2
Monopoli 29,776	F4
Monreale 19,348	D5
Monselice 9,047	C2
Montalto Uffugo 3,173	E5
Montebelluna 9,573	D2
Montefiascone 6,885	D3
Montepulciano 4,069	C3
Monterotondo 15,869	F6
Monte Sant'Angelo 17,756	F4
Montevarchi 16,849	C3
Monza 110,735	B2
Mortara 13,929	B2
Naples 1,214,775	E4
Nardò 24,142	F4
Narni 6,213	D3
Naro 13,171	D6
Nettuno 20,927	D4
Nicastro 27,206	F5
Nicosia 13,982	E6
Niscemi 23,925	E6
Nizza Monferrato 7,532	B2
Nocera Inferiore 44,415	E4
Noto 21,606	E6
Novara 92,634	B2
Novi Ligure 29,944	B2
Nuoro 30,551	B4
Olbia 20,998	B4
Oliena 7,030	B4
Orbetello 6,884	C3
Oristano 20,966	B5
Ortona 11,966	E3
Orvieto 8,813	D3
Osimo 12,034	D3
Ostia Antica 2,583	F7
Ostuni 27,241	F4
Otranto 3,707	G4
Ozieri 9,149	B4
Pachino 20,427	E6
Padua 210,950	C2
Palazzolo Acreide 8,981	E6
Palermo 556,374	D5
Palestrina 9,239	F7
Palma di Montechiaro 22,381	D6
Palmi 14,405	E5
Palombara Sabina 5,292	F6
Pantelleria 3,116	C6
Paola 11,330	E5
Parma 151,967	C2
Partanna 10,303	D6
Partinico 25,447	D6
Paterno 41,504	E6
Patti 7,500	E5
Pavia 80,639	B2
Pavullo nel Frignano 5,026	C2
Penne 5,889	D3
Pergine Valsugana 6,248	C1
Pergola 3,866	D3
Perugia 65,975	D3
Pesaro 72,104	D3
Pescara 125,391	E3
Pescia 9,918	C3
Piacenza 100,001	B2
Piazza Armerina 21,754	E6
Pietrasanta 6,620	B3
Pinerolo 33,935	A2
Piombino 35,641	C3
Piove di Sacco 7,035	C2
Pisa 91,156	C3
Pisticci 11,239	F4
Pistoia 55,403	C2
Poggibonsi 21,271	C3
Pomezia 11,915	F7
Pont Canavese 4,075	A2
Pontecorvo 5,986	D4
Pontinia 3,166	D4
Pontremoli 5,222	B2
Popoli 5,372	E3
Pordenone 43,230	D2
Portocivitanova 25,773	D3
Porto Empedocle 15,986	D6
Portoferraio 7,579	C3
Portofino 720	B2
Portogruaro 12,258	D2
Portomaggiore 6,343	C2
Porto Recanati 5,389	D3
Porto Torres 15,422	B4
Potenza 46,869	E4
Pozzallo 12,199	E6
Pozzuoli 53,546	D4
Prato 108,385	C3
Prima Porta 11,393	F6
Priverno 9,950	D4
Putignano 19,290	F4
Quartu Sant'Elena 29,715	B5
Ragusa 55,751	E6
Rapallo 22,272	B2
Ravenna 75,153	D2
Recanati 10,176	D3
Reggio di Calabria 110,291	E5
Reggio nell'Emilia 102,337	C2
Rho 39,206	B2
Riesi 15,855	E6
Rieti 26,775	D3
Rimini 101,579	D2
Rionero in Vulture 11,230	E4
Riva del Garda 8,513	C2
Roccastrada 2,629	C3
Rome (cap.) 2,535,018	F6
Ronciglione 5,900	C3
Rossano 12,119	F5
Rovereto 26,827	C2
Rovigo 31,124	C2
Ruvo di Puglia 23,133	F4

(continued on following page)

Topography

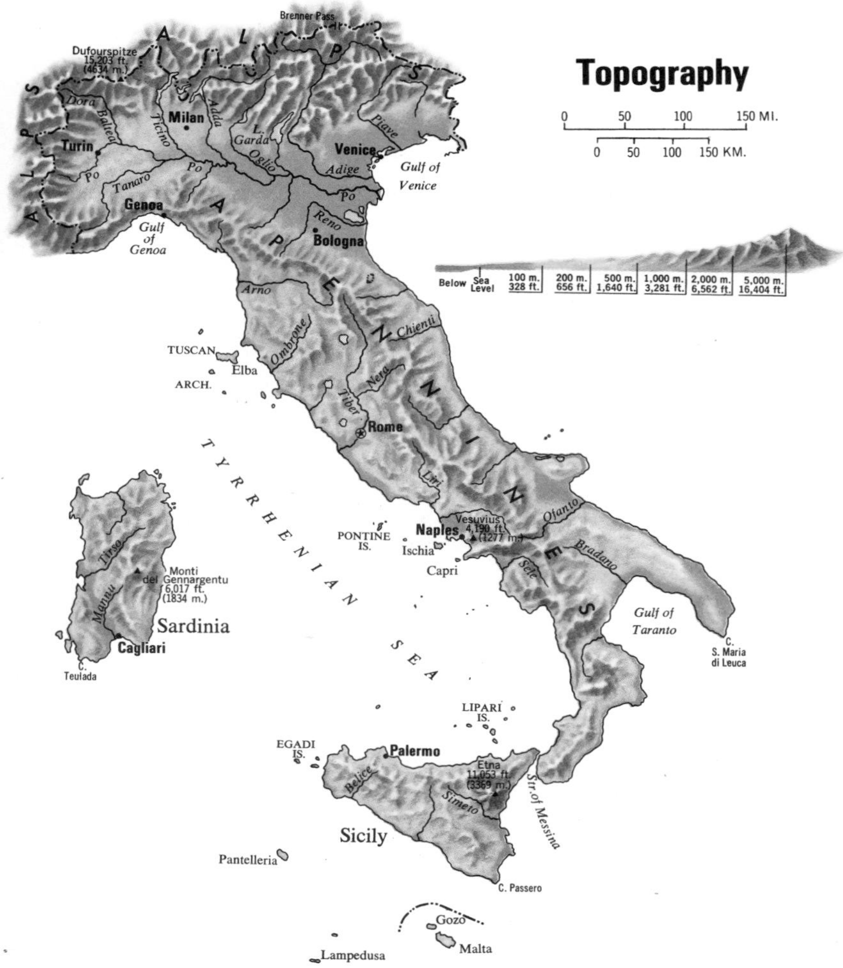

0 50 100 150 MI.
0 50 100 150 KM.
Below Sea Level
100 m. 328 ft.
200 m. 656 ft.
500 m. 1,640 ft.
1,000 m. 3,281 ft.
2,000 m. 6,562 ft.
5,000 m. 16,404 ft.
Brenner Pass
Dufourspitze 15,203 ft. (4634 m.)
A L P S
Dora Baltea
Ticino
Adda
Milan
Turin
Po
Tanaro
Garda
Oglio
Adige
Piave
Venice
Gulf of Venice
Genoa
Gulf of Genoa
Reno
Bologna
A P E N N I N E S
Arno
Ombrone
Chienti
Tiber
Nera
Rome
Liri
TUSCAN ARCH.
Elba
T Y R R H E N I A N S E A
PONTINE IS.
Naples
Vesuvius 4,190 ft. (1277 m.)
Ischia
Capri
Ofanto
Bradano
Sele
Gulf of Taranto
C. S. Maria di Leuca
Tirso
Mannu
Monti del Gennargentu 6,017 ft. (1834 m.)
Sardinia
Cagliari
C. Teulada
LIPARI IS.
EGADI IS.
Palermo
Etna 11,053 ft. (3369 m.)
Belice
Simeto
Str. of Messina
Sicily
C. Passero
Pantelleria
Gozo
Malta
Lampedusa

Sabaudia 4,501 .... D4
Saint Vincent 3,737 .... A2
Sala Consilina 8,177 .... E4
Salemi 10,180 .... D6
Salerno 146,534 .... E4
Salsomaggiore Terme 13,677 .... B2
Saluzzo 13,929 .... A2
Sambiase 10,567 .... F5
San Bartolomeo in Galdo 6,943 .... E4
San Benedetto del
  Tronto 40,108 .... E3
San Cataldo 19,609 .... D6
San Giovanni in Fiore 16,116 ..... F5
San Giovanni in
  Persiceto 12,151 .... C2
San Marco in Lamis 15,817 .... E4
San Miniato 3,245 .... C3
Sannicandro Garganico 17,939 ..... E4
San Remo 47,684 .... A3
Sansepolcro 11,443 .... C3
San Severino Marche 6,447 .... D3
San Severo 49,622 .... E4
Santa Maria Capua
  Vetere 31,077 .... E4
Sant'Elpidio a Mare 4,446 .... E3
Santeramo in Colle 19,758 .... F4
San Vito 3,901 .... B5
San Vito al Tagliamento 6,328 .... D2
San Vito dei Normanni 18,447 ..... F4
San Vito Romano 3,256 .... B2
Saronno 32,477 .... B2
Sarroch 3,560 .... B5
Sassari 94,312 .... B4
Sassuolo 33,451 .... C2
Savigliano 14,036 .... A2
Savona 76,274 .... B2
Schio 27,890 .... C2
Sciacca 29,803 .... D6
Scicli 18,405 .... E6
Segni 7,193 .... F7
Senigallia 25,413 .... D3
Sesto Fiorentino 41,636 .... C3
Sestri Levante 18,331 .... B2
Settebagni 5,022 .... D4
Sezze 7,043 .... D4
Siderno 8,023 .... F5
Siena 56,539 .... C3
Siniscola 6,149 .... B4
Sinnai 8,499 .... B5
Siracusa (Syracuse) 93,006 .... E6
Sondrio 19,724 .... B1
Sora 14,031 .... D4
Soresina 9,300 .... B2
Sorrento 13,078 .... E4
Sorso 10,741 .... B4
Spoleto 18,013 .... D3
Squinzano 14,053 .... G4
Stresa 3,758 .... B2
Sulmona 18,221 .... D3
Susa 5,773 .... A2
Suzzara 12,013 .... C2
Syracuse 93,006 .... E6
Taormina 6,696 .... E6
Taranto 205,158 .... F4
Tarquinia 10,300 .... C3
Taurianova 12,198 .... E5
Tempio Pausania 10,382 .... B4
Teramo 31,163 .... D3
Termini Imerese 24,085 .... D6
Termoli 13,986 .... E3
Terni 75,873 .... D3
Terracina 24,092 .... D4
Terralba 8,551 .... B5
Tirano 7,413 .... C1
Tivoli 28,393 .... F6
Todi 5,705 .... D3
Tolentino 11,642 .... D3
Torino (Turin) 1,181,698 .... A2
Torre Annunziata 71,068 .... E4
Torre del Greco 74,752 .... E4
Torremaggiore 16,171 .... E4
Tortona 24,165 .... B2
Trani 40,508 .... F4
Trapani 90,305 .... D5
Trento 64,272 .... C1
Treviglio 21,920 .... B2
Treviso 87,447 .... D2
Tricase 10,481 .... G5
Trieste 257,259 .... E2
Trino 8,722 .... A2
Turin 1,181,698 .... A2
Udine 97,544 .... D2
Umbertide 6,640 .... D3
Urbino 7,735 .... D3
Valdagno 20,342 .... C2
Valenza 20,533 .... B2
Valmontone 6,543 .... F7
Varallo Pombia 3,118 .... B2
Varazze 11,676 .... B2
Varese 65,978 .... B2
Vasto 17,295 .... E3
Velletri 22,020 .... F7
Venafro 5,156 .... E4
Venezia (Venice) 108,082 .... D2
Venice 108,082 .... D2
Venosa 10,993 .... F4
Ventimiglia 20,343 .... A3
Verbania 29,894 .... B2
Vercelli 54,934 .... B2
Veroli 2,793 .... D4
Verona 97,000 .... C2
Viadana 6,667 .... C2
Viareggio 49,965 .... C3
Vibo Valentia 18,005 .... F5
Vicenza 99,451 .... C2
Vicovaro 3,005 .... F6
Vigevano 62,855 .... B2
Villacidro 12,651 .... B5
Villafranca di Verona 11,762 ...... C2
Viterbo 39,291 .... C3
Vittoria 43,673 .... E6
Vittorio Veneto 25,476 .... C1
Vizzini 8,583 .... E6
Voghera 37,316 .... B2
Volterra 10,732 .... C3
Zagarolo 4,232 .... F7

OTHER FEATURES

Adda (riv.) B2
Adige (riv.) C1
Adriatic (sea) E3
Alicudi (isl.) E5
Apennines, Central (range) D3
Apennines, Northern (range) C2
Apennines, Southern (range) E4
Arno (riv.) C3
Asinara (isl.) B4
Bernina, Piz (peak) B1
Blanc (mt.) A2
Bolsena (lake) C3
Bonifacio (str.) B4
Bracciano (lake) D3
Brenner (pass) C1
Capraia (isl.) B3
Capri (isl.) E4
Carbonara (cape) B5
Carnic Alps (range) D1
Castellammare (gulf) D5
Circeo (cape) D4
Como (lake) B1
Cottian Alps (range) A2
Dolomite Alps (range) C1
Dora Baltea (riv.) A2
Dora Ripara (riv.) A2
Egadi (isls.) C6
Elba (isl.) C3
Etna (vol.) E6
Favignana (isl.) D6

Filicudi (isl.) E5
Gaeta (gulf) D4
Garda (lake) C2
Gennargentu, Monti del (mt.) B5
Genoa (gulf) B2
Giannutri (isl.) C3
Giglio (isl.) C3
Gorgona (isl.) B3
Graian Alps (range) A2
Gran Paradiso (mt.) A2
Great Saint Bernard (pass) A2
Ionian (sea) F6
Ischia (isl.) D4
Julian Alps (range) D1
Lampedusa (isl.) C3
Levanzo (isl.) D5
Lepontine Alps (range) B1
Levanto (isl.) C3
Ligurian (sea) B3
Linosa (isl.) D7
Lipari (isl.) E5
Lipari (isls.) E5
Maggiore (lake) B1
Manfredonia (gulf) F4
Marettimo (isl.) C6
Maritime Alps (range) A2
Marmolada (mt.) C1
Mediterranean (sea) B6
Messina (str.) E5
Metauro (riv.) D3
Mincio (riv.) C2
Montecristo (isl.) C3
Nera (riv.) D3
Oglio (riv.) C2
Ombrone (riv.) C3
Oristano (gulf) B5
Orosei (gulf) B4
Ortles (range) C1
Otranto (str.) G5
Öttzal Alps (range) C1
Panarea (isl.) E5
Panaro (riv.) C2
Pantelleria (isl.) D6
Pelagie (isls.) D7
Pennine Alps (range) A2
Pianosa (isl.) C3
Piave (riv.) D2
Po (riv.) C2
Pompeii (ruins) E4
Pontine (isls.) D4
Ponza (isl.) D4
Rosa (mt.) A1
Salina (isl.) E5
Salso (riv.) D6
San Pietro (isl.) B5
Santa Maria di Leuca (cape) G5
Sant'Antioco (pen.) B5
Sant'Eufemia (gulf) F5
Sardinia (isl.) B4
Sicily (isl.) E6
Sicily (str.) D6
Simplon (tunnel) A1
Spartivento (cape) B5
Spartivento (cape) F6
Squillace (gulf) F5
Stromboli (isl.) D1
Tagliamento (riv.) D1
Tanaro (riv.) B2
Taranto (gulf) F5
Testa del Gargano (cape) F4
Tiber (riv.) D3
Trasimeno (lake) D3
Tremiti (isls.) E3
Trieste (gulf) D2
Tuscan (arch.) C3
Tuscan (isl.) C3
Tyrrhenian (sea) C4
Ustica (isl.) D5
Vaticano (cape) E5
Venice (gulf) D2
Ventotene (isl.) D4

Vesuvius (vol.) E4
Viso (mt.) A2
Volturno (riv.) E4
Vulcano (isl.) E5

MALTA

CITIES and TOWNS

Sliema 20,095 .... E7
Valletta (cap.) 14,042 .... E7
Victoria 5,249 .... E6

SAN MARINO

CITIES and TOWNS

San Marino (cap.) 4,628 .... D3
San Marino* 5,410 .... D3

VATICAN CITY

Vatican City 728 .... B6

*City and suburbs.

Agriculture, Industry and Resources

DOMINANT LAND USE

- Wheat, Rice, Dairy
- Pasture Livestock
- Cereals, Livestock
- Fruit, Truck and Mixed Farming
- Grapes, Wine
- Forests
- Nonagricultural Land

MAJOR MINERAL OCCURRENCES

Ab	Asbestos	K	Potash	Pb	Lead
Al	Bauxite	Lg	Lignite	Py	Pyrites
C	Coal	Mr	Marble	S	Sulfur
Fe	Iron Ore	Na	Salt	Sb	Antimony
G	Natural Gas	O	Petroleum	Zn	Zinc
Hg	Mercury				

⚡ Water Power

▨ Major Industrial Areas

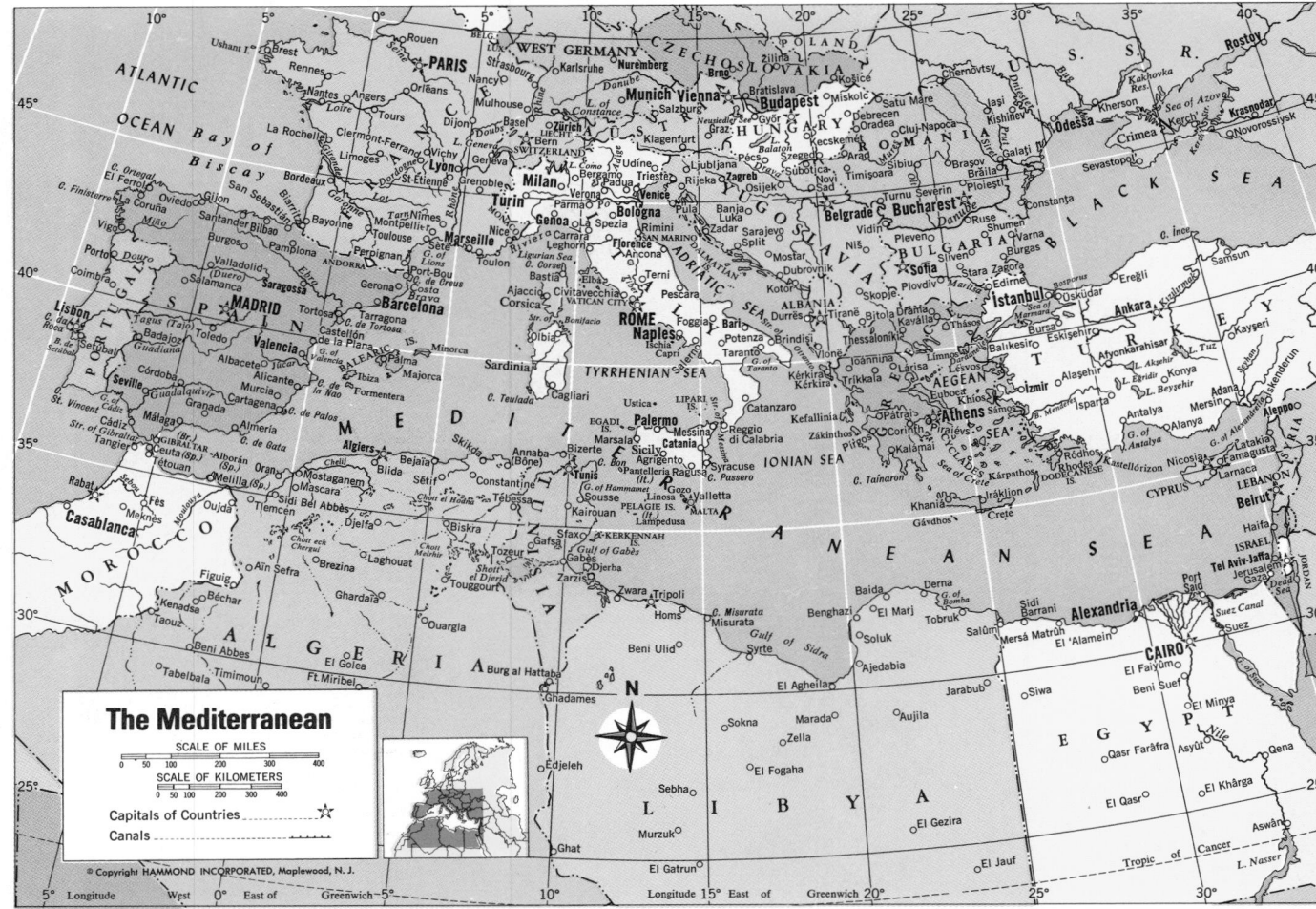

The Mediterranean

SCALE OF MILES
0 50 100 200 300 400

SCALE OF KILOMETERS
0 50 100 200 300 400

Capitals of Countries .... ☆

Canals

© Copyright HAMMOND INCORPORATED, Maplewood, N.J.

SWITZERLAND

AREA 15,943 sq. mi. (41,292 sq. km.)
POPULATION 6,365,960
CAPITAL Bern
LARGEST CITY Zürich
HIGHEST POINT Dufourspitze
(Mte. Rosa) 15,203 ft. (4,634 m.)
MONETARY UNIT Swiss franc
MAJOR LANGUAGES German, French,
Italian, Romansch
MAJOR RELIGIONS Protestantism,
Roman Catholicism

LIECHTENSTEIN

AREA 61 sq. mi. (158 sq. km.)
POPULATION 25,220
CAPITAL Vaduz
LARGEST CITY Vaduz
HIGHEST POINT Grauspitze 8,527 ft.
(2,599 m.)
MONETARY UNIT Swiss franc
MAJOR LANGUAGE German
MAJOR RELIGION Roman Catholicism

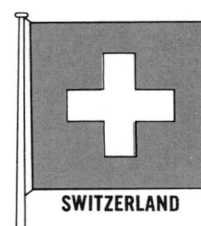

SWITZERLAND

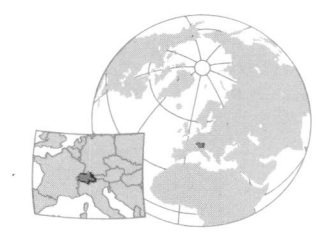

LIECHTENSTEIN

Languages

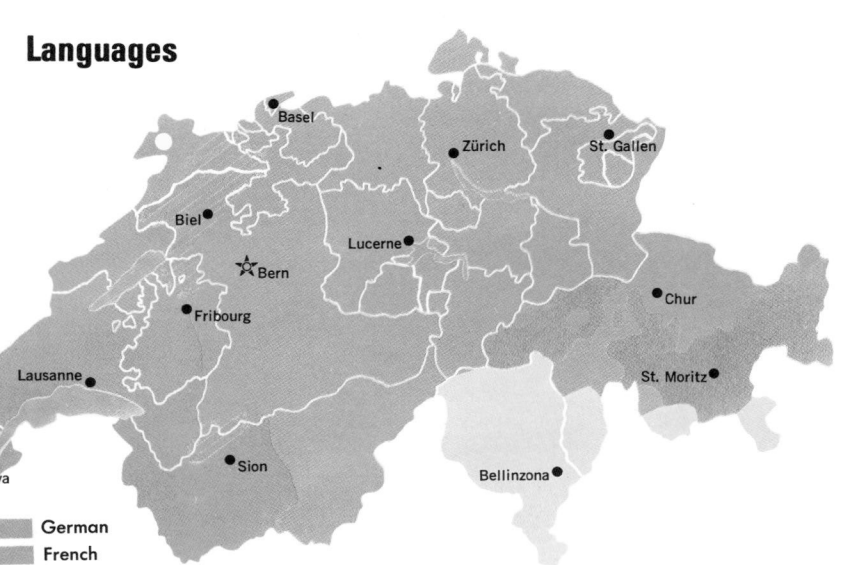

- German
- French
- Italian
- Romansch

Switzerland is a multilingual nation with four
official languages. 70% of the people speak
German, 19% French, 10% Italian and 1% Romansch.

Agriculture, Industry and Resources

DOMINANT LAND USE

- Cereals, Dairy
- Pasture Livestock
- General Farming, Livestock
- Fruit, Truck, Mixed Farming
- Forests
- Nonagricultural Land

⚡ Water Power
▨ Major Industrial Areas

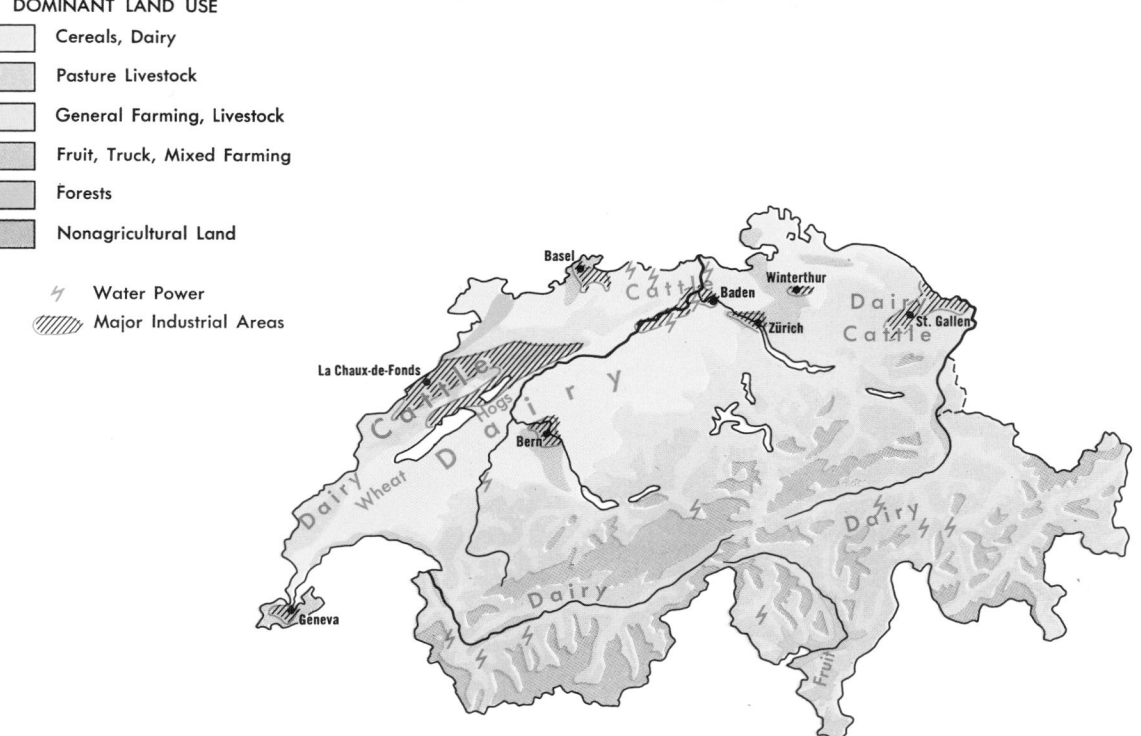

SWITZERLAND

CANTONS

Aargau 442,400 ....F2
Appenzell, Ausser
  Rhoden 46,700 ....H2
Appenzell, Inner Rhoden 13,500 ...H2
Baselland 219,500 ....E2
Baselstadt 209,700 ....E1
Bern 920,900 ....D2
Fribourg 181,600 ....D3
Geneva (Genève) 338,600 ....B4
Glarus 35,700 ....H3
Graubünden (Grisons) 164,300 ....H3
Grisons (Graubünden) 164,300 ....H3
Jura 67,200 ....D2
Lucerne (Luzern) 292,900 ....F2
Luzern 292,900 ....F2
Neuchâtel 162,200 ....C3
Nidwalden 26,900 ....F3
Obwalden 25,400 ....F3
Sankt Gallen 385,000 ....H2
Schaffhausen 69,300 ....G1
Schwyz 93,100 ....G2
Soleure (Solothurn) 221,800 ....E2
Solothurn 221,800 ....E2
Thurgau 183,500 ....H1
Ticino 264,400 ....G4
Uri 34,000 ....G3
Valais 214,000 ....D4
Vaud 523,500 ....B3
Zug 73,600 ....G2
Zürich 1,117,300 ....G2

CITIES and TOWNS

Aadorf 3,022 ....G2
Aarau 16,881 ....F2
Aarau* 51,800 ....F2

Aarberg 3,122 ....D2
Aarburg 5,943 ....E2
Adelboden 3,326 ....E3
Adliswil 15,920 ....G2
Aeschi bei Spiez 1,402 ....E3
Affoltern am Albis 7,363 ....G2
Affoltern im Emmental 1,223 ......E2
Aigle 6,532 ....C4
Airolo 2,140 ....G3
Alle 1,615 ....D1
Allschwil 17,638 ....D2
Alpnach 3,277 ....F3
Altdorf 8,647 ....G3
Altstätten 9,084 ....J2
Amriswil 7,601 ....H1
Andelfingen 1,453 ....G1
Andermatt 1,589 ....G3
Appenzell 5,217 ....H2
Arbedo-Castione 2,456 ....G4
Arbon 12,227 ....H1
Arbon* 15,400 ....H1
Ardon 1,498 ....D4
Arosa 2,717 ....J3
Arth 7,580 ....F2
Ascona 4,086 ....G4
Attalens 1,116 ....C3
Au 4,944 ....J2
Aubonne 1,983 ....B4
Avenches 2,235 ....D3
Baar 14,074 ....F2
Baden 14,115 ....F2
Baden* 66,800 ....F2
Bad Ragaz 3,713 ....H2
Balerna 3,885 ....G5
Balsthal 5,607 ....E2
Bäretswil 2,733 ....G2
Basel 199,600 ....E1
Basel* 379,700 ....E1
Bassecourt 2,985 ....D2
Bätterkinden 1,757 ....E2

Bauma 3,159 ....G2
Beatenberg 1,263 ....E3
Beinwil am See 2,520 ....F2
Belfaux 1,075 ....D3
Bellinzona 16,979 ....H4
Bellinzona* 31,000 ....H4
Belp 6,981 ....D3
Berg 1,039 ....H1
Bern (cap.) 154,700 ....D3
Bern* 285,300 ....D3
Beromünster 1,552 ....F2
Bettlach 4,046 ....D4
Bex 5,069 ....D4
Biasca 4,696 ....H4
Biberist 7,769 ....D2
Biel 63,400 ....D2
Biel**89,900 ....D2
Bière 1,252 ....B3
Binningen 15,344 ....D1
Bischofszell 4,233 ....H1
Blumenstein 1,049 ....E3
Bodio 1,425 ....G4
Bolligen 26,121 ....E3
Boltigen 1,519 ....D3
Bonaduz 1,289 ....H3
Boncourt 1,528 ....C2
Bönigen 1,738 ....F3
Boswil 1,904 ....F2
Boudry 4,372 ....C3
Bourg Saint-Pierre 236 ....D5
Breil-Brigels 1,215 ....H3
Breitenbach 2,455 ....E2
Bremgarten 4,873 ....F2
Brienz 2,796 ....F3
Brig 5,191 ....F4
Brissago 2,120 ....G4
Brittnau 2,888 ....E2
Broc 1,842 ....D3
Brugg 8,635 ....F2
Brusio 1,344 ....K4
Bubendorf 2,070 ....E2
Bubikon 3,244 ....G2
Buchs 8,454 ....H2
Bülach 11,043 ....G1
Bulle 7,556 ....D3
Buochs 3,232 ....F3
Büren an der Aare 3,085 ....D2
Burgdorf 15,888 ....E2
Burgdorf* 18,400 ....E2
Bürglen 1,920 ....H1
Bürglen, Thurgau 1,920 ....H1
Bürglen, Uri 3,401 ....G3
Bussigny-près-Lausanne 4,509 ..B3
Bütschwil 3,270 ....H2
Carouge 14,055 ....B4
Castagnola 4,430 ....G4
Cazis 1,687 ....H3
Cernier 1,717 ....C2
Chalais 1,651 ....E4
Cham 8,209 ....F2
Chamoson 2,049 ....D4
Charmey 1,155 ....D3
Château-d'Oex 3,203 ....D4
Châtel-Saint-Denis 2,842 ....C3
Chêne-Bougeries 8,670 ....B4
Chavornay 1,521 ....C3
Chexbres 1,607 ....C3
Chiasso 8,868 ....G5
Chippis 1,561 ....E4
Chur 32,400 ....J3
Churwalden 1,052 ....J3
Claro 1,143 ....G4
Collombey-Muraz 2,279 ....C4
Collonge-Bellerive 3,541 ....B4
Conthey 4,259 ....D4
Coppet 1,097 ....B4
Corcelles-près-Payerne 1,256 ...C3
Corgémont 1,645 ....D2
Cossonay 1,529 ....B3
Courgenay 1,954 ....D2
Courrendlin 2,656 ....D2
Courroux 1,788 ....D2
Courtelary 1,462 ....C2
Courtételle 1,864 ....D2
Couvet 3,481 ....C3
Cully 1,535 ....C4
Davos 10,238 ....J3
Degersheim 3,400 ....H2
Delémont 11,797 ....E2
Derendingen 4,917 ....E2
Dielsdorf 2,691 ....F1
Diemtigen 1,913 ....D3
Diepoldsau 3,311 ....J2
Diessenhofen 2,532 ....G1
Dietikon 22,705 ....F2
Disentis-Mustér 2,319 ....G3
Domat-Ems 5,701 ....H3
Dombresson 1,109 ....C2
Dornach 5,258 ....D2
Döttingen 3,380 ....F1
Dübendorf 19,639 ....G2
Düdingen 4,932 ....D3
Dürnten 4,820 ....G2
Dürrenroth 1,084 ....E2
Ebnat-Kappel 5,131 ....H2
Echallens 1,643 ....C3
Ecublens 6,379 ....B3
Egg 5,250 ....G2
Eggiwil 2,391 ....E3
Eglisau 2,160 ....G1
Egnach 3,466 ....H1

(continued on following page)

Topography

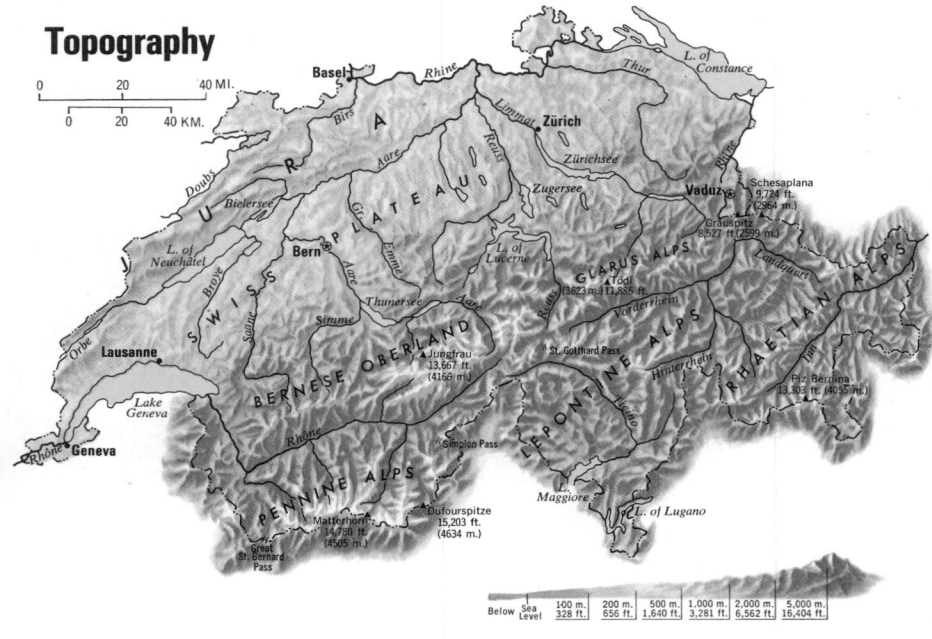

Below Sea Level | 100 m. 328 ft. | 200 m. 656 ft. | 500 m. 1,640 ft. | 1,000 m. 3,281 ft. | 2,000 m. 6,562 ft. | 5,000 m. 16,404 ft.

Einsiedeln 10,020 .... G2
Elgg 2,970 .... G2
Emmen 22,040 .... F2
Engelberg 2,841 .... F3
Ennenda 2,762 .... H2
Entlebuch 3,310 .... F3
Erlach 1,052 .... D2
Erlenbach im Simmental 1,436 .... E3
Ermatingen 1,787 .... H1
Erstfeld 4,516 .... G3
Escholzmatt 3,161 .... E3
Estavayer-le-Lac 3,439 .... C4
Evolène 1,403 .... D4
Faido 1,866 .... G4
Felsberg 1,321 .... H3
Feuerthalen 3,118 .... G1
Flawil 8,474 .... H2
Fleurier 4,124 .... C2
Flims 1,936 .... H3
Flüelen 1,731 .... G3
Flums 4,474 .... H2
Frauenfeld 17,576 .... G1
Freienbach 8,429 .... G2
Fribourg 41,600 .... D3
Fribourg* 53,500 .... D3
Frick 3,219 .... F1
Frutigen 5,796 .... E3
Fully 3,643 .... D4
Gais 2,344 .... H2
Gelterkinden 5,157 .... E2
Geneva (Genève) 163,100 .... B4
Geneva (Genève)* 320,200 .... B4
Gersau 1,753 .... G3
Gimel 1,205 .... B3
Giornico 1,389 .... G4
Giswil 2,760 .... F3
Giubiasco 5,796 .... H4
Gland 2,404 .... B3
Glarus 6,189 .... H2
Glattfelden 2,857 .... F1
Glis 3,389 .... E4
Gordola 2,586 .... G4
Gossau 12,793 .... H2
Grabs 4,245 .... H2
Grächen 1,063 .... E4
Grandson 2,135 .... C3
Grenchen 20,051 .... D2
Grenchen* 28,300 .... D2
Grindelwald 3,161 .... E3
Grosswangen 2,213 .... F2
Gruyères 1,234 .... D3
Gstaad 865 .... D4
Gsteig 865 .... D4
Guggisberg 1,739 .... D3
Gurtnellen 1,048 .... G3
Guttingen 1,060 .... H1
Hallau 1,836 .... F1
Heiden 3,716 .... H2
Heimberg 3,046 .... E3
Hérémence 1,484 .... D4
Hergiswil 4,364 .... F3
Herisau 14,597 .... H2
Herzogenbuchsee 5,140 .... E2
Hilterfingen 3,647 .... E3
Hinwil 8,547 .... G2
Hitzkirch 1,468 .... F2
Hochdorf 5,222 .... F2
Horgen 15,691 .... G2
Huttwil 4,800 .... E2
Igis 5,283 .... J3
Ilanz 2,435 .... H3
Illnau 13,693 .... G2
Ingenbohl 5,111 .... G2
Innertkirchen 1,064 .... F3
Ins 2,435 .... D2
Interlaken 4,735 .... E3
Jegenstorf 2,858 .... E2
Jenaz 1,124 .... J3
Jona 9,286 .... G2
Jungfraujoch E3
Kaltbrunn 2,751 .... H2
Kandersteg 957 .... E4
Kerns 3,807 .... F3
Kerzers 2,688 .... D2
Kilchberg, Bern 3,595 .... E2
Kirchberg, St. Gallen 6,309 .... H2
Kleinlützel 1,271 .... E2
Klingnau 2,545 .... F1
Klosters Dorf 3,534 .... J3
Kloten 16,388 .... G2
Koblenz 1,439 .... F1
Kölliken 3,219 .... F2
Köniz 33,800 .... D3
Konolfingen 4,137 .... E3
Kreuzlingen 15,760 .... H1
Kriens 20,409 .... F2
Krummenau 1,904 .... H2
Küsnacht 12,193 .... G2
Küssnacht am Rigi 7,956 .... F2

Küttigen 4,181 .... F2
L'Abbaye 1,319 .... B3
La Chaux-de-Fonds 42,500 .... C2
Lachen 4,914 .... G2
Lancy 20,523 .... B4
La Neuveville 3,917 .... D2
Langenthal 13,077 .... E2
Langenthal* 22,100 .... E2
Langnau am Albis 4,879 .... G2
Langnau im Emmental 8,950 .... E3
La Roche 1,069 .... D3
La Sarraz 1,190 .... C3
La Tour-de-Peilz 8,864 .... C4
Läufelfingen 1,243 .... E2
Laufen 4,723 .... D2
Laufenburg 2,128 .... F1
Laupen 2,139 .... D3
Lauperswil 2,542 .... E3
Lausanne 136,100 .... C3
Lausanne* 228,700 .... C3
Lauterbrunnen 3,431 .... E3
Le Brassus 5,465 .... B3
Le Châble 4,541 .... D4
Le Chenit (Le Brassus) 5,465 .... B3
Le Landeron 2,768 .... C2
Le Locle 14,452 .... C2
Le Mont-sur-Lausanne 2,692 .... C3
Lengau 4,736 .... D2
Lenk 1,876 .... D4
Le Noirmont 1,516 .... C2
Lens 2,052 .... D4
Lenzburg 7,594 .... F2
Les Bois 1,110 .... C2
Les Ponts-de-Martel 1,327 .... C2
Leuk 2,796 .... E4
Leukerbad 1,056 .... E4
Leysin 2,752 .... C4
Liechtensteig 3,842 .... H2
Liestal 12,500 .... E2
Liestal-Sissach* 40,800 .... E2
Linthal 1,458 .... H3
Littau 13,495 .... F2
Lotzwil 2,323 .... E2
Lucens 2,144 .... C3
Locarno 14,143 .... G4
Locarno* 39,200 .... G4
Lodrino 1,075 .... G4
Lottigna 2,323 .... E2
Lucens 2,144 .... C3
Lugano 22,280 .... G4
Lugano* 64,200 .... G4
Lungern 1,813 .... F3
Lupfig 1,706 .... F2
Lutry 4,994 .... C3
Lützelflüh 3,842 .... E3
Lyss 8,131 .... D2
Maienfeld 1,542 .... J3
Malans 1,294 .... J3
Malleray 1,969 .... D2
Malters 5,100 .... F2
Malvaglia 1,099 .... H4
Männedorf 7,419 .... G2
Marbach 1,265 .... E3
Martigny 10,478 .... C4
Meilen 9,881 .... G2
Meiringen 3,759 .... F3
Melide 1,315 .... G5
Mellingen 3,211 .... F2
Mels 5,969 .... H2
Mendrisio 6,223 .... G5
Menzingen 3,483 .... G2
Menznau 2,185 .... E2
Minusio 5,027 .... G4
Möhlin 6,003 .... E1
Mollis 2,628 .... H2
Montana 1,725 .... D4
Monthey 10,114 .... C4
Montreux 20,421 .... C4
Morges 17,200 .... B3
Morges 11,931 .... B3
Morges* 17,200 .... B3
Moudon 3,773 .... C3
Moutier 8,794 .... D2
Müllheim 1,468 .... G1
Mümliswil-Ramiswil 2,702 .... E2
Münsingen 8,350 .... E3
Muotathal 2,763 .... G3
Muri 4,853 .... F2
Muri bei Bern 3,057 .... E3
Murten 4,256 .... D3
Muttenz 15,518 .... E1
Näfels 3,739 .... H2
Naters 5,517 .... E4
Nebikon 1,378 .... F2
Nendaz 4,051 .... D4
Nesslau 1,934 .... H2

Netstal 2,771 .... H2
Neuchâtel 38,400 .... C3
Neuchâtel* 61,700 .... C3
Neunegg 3,452 .... D3
Neuhausen am Rheinfall 12,103 .... G1
Neunkirch 1,239 .... F1
Nidau 7,962 .... D2
Niederbipp 3,293 .... E2
Niederurnen 3,354 .... G2
Nunningen 1,450 .... E2
Nyon 11,424 .... B4
Oberägeri 2,992 .... G2
Oberburg 3,015 .... E2
Oberdiessbach 2,145 .... E3
Oberdorf 1,953 .... E2
Oberriet 6,123 .... J2
Obersiggenthal 6,623 .... F1
Oberwil 4,659 .... H2
Oensingen 3,387 .... E2
Oftringen 9,189 .... E2
Ollon 4,470 .... D4
Olten 21,209 .... E2
Olten* 49,000 .... E2
Opfikon 11,115 .... G2
Orbe 4,522 .... C3
Orsières 2,470 .... D4
Ouchy C3
Paradiso 3,101 .... G5
Payerne 6,899 .... C3
Penthalaz 1,701 .... C3
Péry 1,486 .... D2
Peseux 5,578 .... C2
Pfaffnau 2,584 .... E2
Pieterlen 3,485 .... D2
Pfäffikon 1,448 .... D3
Pontresina 1,646 .... J3
Porrentruy 7,827 .... C2
Port-Valais 1,363 .... C4
Poschiavo 3,563 .... J4
Prangins 1,466 .... B4
Pratteln 15,127 .... E1
Pully 15,917 .... C4
Quinto 1,490 .... G3
Rafz 2,215 .... G1
Ramsen 1,834 .... G1
Rapperswil 8,713 .... G2
Raron 1,257 .... E4
Regensdorf 8,566 .... F2
Reichenbach im Kandertal 2,900 .... E3
Reiden 3,275 .... F2
Reinach in Aargau 5,862 .... F2
Reinach in Baselland 13,419 .... E2
Renan 1,094 .... C2
Renens 17,391 .... C3
Rheinau 3,275 .... G1
Rheineck 3,275 .... J2
Rheinfelden 6,866 .... E1
Richterswil 7,380 .... G2
Riehen 21,026 .... E1
Riggisberg 2,193 .... E3
Riva San Vitale 1,607 .... G5
Rivera 1,146 .... G4
Roggwil 3,403 .... E2
Rolle 3,658 .... B4
Romanshorn 8,329 .... H1
Romont 3,276 .... C3
Rorschach 11,963 .... H2
Rorschach* 24,200 .... H2
Rosenlaui F3
Rothrist 5,883 .... E2
Roveredo 2,037 .... H4
Rüeggisberg 1,857 .... E3
Rumlang 5,677 .... F2
Rüschegg 1,346 .... D3
Ruswil 4,756 .... F2
Rüti 1,493 .... J2
Rüti, Zürich 9,546 .... G2
Saanen 3,481 .... D4
Sachseln 3,059 .... F3
Saignelégier 1,745 .... C2
Saint-Aubin-Sauges 2,058 .... C2
Saint-Blaise 2,586 .... D2
Sainte-Croix 6,240 .... C3
Saint-Imier 6,740 .... D2
Saint-Légier-La Chiésaz 2,222 .... C4
Saint-Martin 1,120 .... E4
Saint-Maurice 3,446 .... C4
Saint Moritz 5,699 .... J3
Saint-Prex 2,306 .... B4
Saint Stephan 1,213 .... D4
Saint-Ursanne 1,073 .... C2
Samedan 2,574 .... J3
Sankt Gallen 81,900 .... H2
Sankt Gallen* 90,400 .... H2
Sankt Margrethen 5,101 .... J2
Sargans 4,058 .... H2
Sarnen 6,952 .... F3
Satigny 1,877 .... A4

Savièse 3,585 .... D4
Saxon 2,409 .... D4
Schaffhausen 36,800 .... G1
Schaffhausen* 55,800 .... G1
Schänis 2,355 .... H2
Schattdorf 3,292 .... G3
Scherzingen 1,420 .... H1
Schiers 2,342 .... J3
Schinznach-Dorf 1,154 .... F2
Schleitheim 1,544 .... G1
Schlieren 11,869 .... F2
Schönenwerd 4,793 .... E2
Schübelbach 4,395 .... G2
Schüpfheim 3,773 .... F3
Schwanden 2,823 .... H2
Schwyz 12,194 .... G2
Scuol 1,686 .... K3
Sempach 1,619 .... F2
Seon 3,628 .... F2
Seuzach 3,258 .... G1
Sevelen 2,742 .... H2
Sierre 11,017 .... E4
Signau 2,642 .... E3
Sigriswil 3,540 .... E3
Silenen 2,338 .... G3
Sils im Domleschg 762 .... H3
Silvaplana 714 .... J4
Sins 2,435 .... F2
Sion 21,925 .... D4
Sirnach 3,706 .... G2
Sissach 4,938 .... E2
Solothurn (Soleure) 17,708 .... E2
Solothurn* 35,600 .... E2
Somvix 1,555 .... H3
Sonvico 1,129 .... G4
Spiez 9,911 .... E3
Stäfa 9,937 .... G2
Stalden 1,121 .... E4
Stans 5,180 .... F3
Steckborn 3,752 .... G1
Steffisburg 12,621 .... E3
Stein 1,763 .... E1
Stein am Rhein 2,751 .... G1
Suhr 7,223 .... F2
Sulgen 1,834 .... H1
Sumiswald 5,334 .... E2
Sursee 7,052 .... F2
Tafers 2,021 .... D3
Täuffelen 1,761 .... D2
Tavannes 3,869 .... D2
Tavetsch 1,273 .... H3
Teufen 5,300 .... H2
Thal 4,199 .... J2
Thalwil 13,591 .... G2
Thayngen 3,640 .... G1
Therwil 5,412 .... E1
Thun 37,000 .... E3
Thun* 63,600 .... E3
Thunstetten 2,483 .... E2
Thusis 2,381 .... H3
Trachselwald 1,199 .... E2
Tramelan 5,646 .... D2
Trimmis 1,109 .... J3
Troistorrents 2,208 .... C4
Trub 1,607 .... E3
Turbenthal 2,939 .... G2
Uetendorf 3,132 .... E3
Unterägeri 4,671 .... G2
Unteriberg 1,344 .... G2
Unterkulm 2,596 .... F2
Unterseen 4,192 .... E3
Untervaz 1,230 .... J3
Urnäsch 2,313 .... H2
Uster 21,819 .... G2
Utzenstorf 3,193 .... E2
Uznach 3,984 .... H2
Uzwil 9,133 .... H2
Vallorbe 4,028 .... B3
Vaz-Obervaz 2,003 .... H3
Vechigen 3,595 .... E3
Vernayaz 1,356 .... D4
Versoix 5,627 .... B4
Vevey 17,957 .... C4
Vevey-Montreux* 62,300 .... C4
Villeneuve 3,705 .... C4
Visp 5,252 .... E4
Vouvry 1,851 .... C4
Vuadens 1,278 .... D3
Wädenswil 15,695 .... G2
Wahlern 4,832 .... D3
Wald 8,185 .... G2
Waldenburg 1,449 .... E2
Waldkirch 2,669 .... H2
Walenstadt 3,446 .... H2
Wallisellen 10,415 .... G2
Walzenhausen 2,055 .... J2
Wangen an der Aare 2,013 .... E2
Wängi 2,730 .... G2
Wartau 3,604 .... H2

Wattwil 8,566 .... H2
Weesen 1,308 .... H2
Weggis 2,517 .... F2
Weinfelden 8,621 .... H1
Wettingen 19,900 .... F2
Wetzikon 13,469 .... G2
Wil 14,646 .... H2
Wil* 20,500 .... H2
Wilchingen 1,066 .... F1
Wilderswil 1,666 .... E3
Willisau 1,104 .... F2
Willisau 2,728 .... F2
Wimmis 1,833 .... E3
Windisch 7,444 .... F2
Winterthur 93,500 .... G1
Winterthur* 110,100 .... G1
Wohlen 12,024 .... F2
Wohlen 9,526 .... E3
Wohlen bei Bern 4,190 .... F3
Wolfenschiessen 1,470 .... F3
Wolhusen 3,556 .... F3
Worb 9,526 .... E3
Wünnewil 3,036 .... D3
Wynigen 1,986 .... E2
Yverdon 20,538 .... C3
Yvonand 1,321 .... C3

Zell, Luzern 1,590 .... E2
Zell, Zürich 4,008 .... G2
Zizers 1,913 .... J3
Zofingen 9,292 .... E2
Zollikofen 9,069 .... E3
Zollikon 12,117 .... G2
Zug 22,972 .... G2
Zug* 51,300 .... G2
Zürich 401,600 .... G2
Zürich* 718,100 .... G2
Zurzach 3,098 .... F1
Zweisimmen 2,823 .... D4

OTHER FEATURES

Aa (riv.) F3
Aare (riv.) F2
Ägerisee (lake) G2
Aiguille d'Argentière (mt.) C4
Aletschhorn (mt.) E4
Aroser Rothorn (mt.) H3
Ault (peak) D4
Balmhorn (mt.) E4
Bernese Oberland (reg.) E3

Bernina (peak) J4
Bernina (pass) J4
Bielersee (lake) D2
Bietschhorn (mt.) E4
Birs (riv.) D2
Blinnenhorn (mt.) F4
Blümlisalp (mt.) E3
Bodensee (Constance) (lake) H1
Borgne (riv.) D4
Breithorn (mt.) E5
Breithorn (mt.) E4
Brienzer Rothorn (mt.) F3
Brienzersee (lake) E3
Broye (riv.) D2
Buchegg (mts.) D2
Buin (peak) K3
Campo Tencia (peak) G4
Chasseron (mt.) C3
Churfirsten (mts.) H2
Clariden (mt.) G3
Constance (lake) H1
Cornettes de Bise (mts.) C4
Dammastock (mt.) F3
Davos (valley) J3
Dent Blanche (mt.) D4
Dent de Lys (mt.) D3

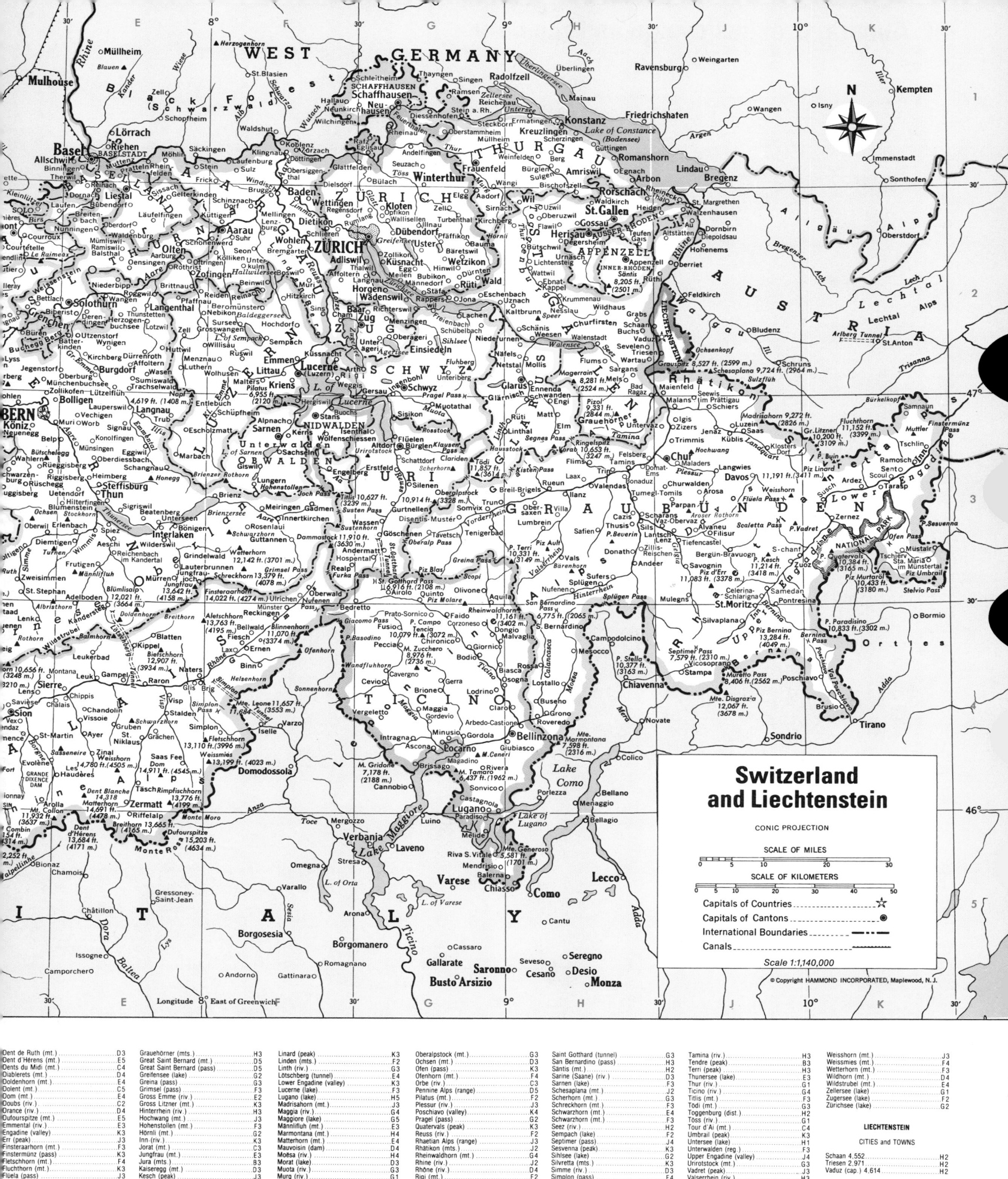

Switzerland and Liechtenstein

CONIC PROJECTION

SCALE OF MILES

SCALE OF KILOMETERS

Capitals of Countries .... ☆
Capitals of Cantons .... ◉
International Boundaries ....
Canals ....

Scale 1:1,140,000

© Copyright HAMMOND INCORPORATED, Maplewood, N.J.

Dent de Ruth (mt.)	D3	
Dent d'Hérens (mt.)	E5	
Dents du Midi (mt.)	C4	
Diablerets (mt.)	D4	
Doldenhorn (mt.)	E4	
Dolent (mt.)	D5	
Dom (mt.)	E4	
Doubs (riv.)	C2	
Drance (riv.)	D4	
Emmental (valley)	E3	
Engadine (valley)	K3	
Er (peak)	E3	
Finsteraarhorn (mt.)	E4	
Finstermünz (pass)	K3	
Fletschhorn (mt.)	E4	
Fluchthorn (mt.)	K3	
Flüela (pass)	J3	
Furka (pass)	F4	
Generoso (lake)	H5	
Geneva (lake)	C4	
Glarus Alps (mts.)	H3	
Grand Combin (mt.)	D5	
Grande Dixence (dam)	D5	
Grand Muveran (mt.)	D4	

Grauhörner (mts.)	H3
Great Saint Bernard (mt.)	D5
Great Saint Bernard (pass)	D5
Greifensee (lake)	G2
Greina (pass)	G3
Grimsel (pass)	E3
Gross Emme (riv.)	E2
Gross Litzner (mt.)	K3
Hinterrhein (riv.)	H3
Hochwang (mt.)	J3
Hohenstollen (mt.)	F3
Hörnli (mt.)	G2
Inn (riv.)	J3
Jorat (mt.)	C3
Jungfrau (mt.)	E3
Jura (mts.)	B3
Kaiseregg (mt.)	D3
Kesch (peak)	J3
Landquart (riv.)	J2
La Dôle (mt.)	B4
Le Chasseral (mt.)	C3
Le Gros Crêt (mt.)	B3
Léman (Geneva) (lake)	C4
Leone (mt.)	E4
Lepontine Alps (range)	F4
Limmat (riv.)	F2

Linard (peak)	K3
Linden (mts.)	F2
Linth (riv.)	G3
Lötschberg (tunnel)	E4
Lower Engadine (valley)	K3
Lucerne (lake)	F3
Lugano (lake)	H5
Maggia (riv.)	G4
Maggiore (lake)	G5
Männlifluh (mt.)	E3
Marmontana (mt.)	H4
Matterhorn (mt.)	E4
Mauvoisin (dam)	D4
Moésa (riv.)	H4
Morat (lake)	D3
Muota (riv.)	F3
Murg (riv.)	G1
Murtaröl (peak)	K3
Napf (mt.)	E3
National Park	K3
Neuchâtel (lake)	C3
Noirmont (mt.)	B4
Oberalp (pass)	G3

Oberalpstock (mt.)	G3
Ochsen (mt.)	D3
Ofen (pass)	K3
Ofenhorn (mt.)	F4
Orbe (riv.)	C3
Pennine Alps (range)	D5
Pilatus (mt.)	F3
Plessur (riv.)	J3
Poschiavo (valley)	J4
Pragel (pass)	G3
Quatervals (peak)	K3
Reuss (riv.)	F3
Rhaetian Alps (range)	J3
Rhätikon (mts.)	J3
Rhine (riv.)	J2
Rheinwaldhorn (mt.)	H4
Rigi (mt.)	F2
Rimpfischhorn (mt.)	E4
Ringelspitz (mt.)	H3
Risoux (mt.)	B3
Rosa (mt.)	E5
Rosstock (mt.)	F3
Rothorn (mt.)	E4
Saane (Sarine) (riv.)	D3
Saint Gotthard (pass)	G3

Saint Gotthard (tunnel)	G3
San Bernardino (pass)	H3
Santis (mt.)	H2
Sarine (Saane) (riv.)	D3
Sarnen (lake)	F3
Schesaplana (mt.)	J2
Schreckhorn (mt.)	E3
Schwarzhorn (mt.)	E4
Schwyz	G3
Seez (riv.)	H2
Sempach (lake)	F2
Septimer (pass)	J4
Sesvenna (reg.)	K3
Sihlsee (lake)	G2
Silvretta (mts.)	K3
Simme (riv.)	D4
Simplon (pass)	F4
Simplon (tunnel)	E4
Sonnenhorn (mt.)	F4
Stockhorn (mt.)	E3
Sulzfluh (mt.)	J3
Susten (pass)	F3
Sustenhorn (mt.)	F3
Tamaro (mt.)	G4

Tamina (riv.)	H3
Tendre (peak)	B3
Terri (peak)	H3
Thunersee (lake)	E3
Thur (riv.)	H2
Ticino (riv.)	G4
Titlis (mt.)	F3
Tödi (mt.)	G3
Toggenburg (dist.)	H2
Toss (riv.)	G2
Tour d'Ai (mt.)	C4
Umbrail (pass)	K3
Untersee (lake)	H1
Unterwalden (reg.)	F3
Upper Engadine (valley)	J4
Urirotstock (mt.)	F3
Vadret (mt.)	J3
Valserrhein (riv.)	H3
Vanil Noir (mt.)	D5
Vélan (mt.)	D5
Visp (riv.)	E4
Vorab (mt.)	H3
Vorderrhein (riv.)	G3
Wandfluhhorn (mt.)	F4
Weissenstein (mts.)	D2
Weisshorn (mt.)	E4

Weisshorn (mt.)	J3
Weissmies (mt.)	F4
Wetterhorn (mt.)	F3
Wildhorn (mt.)	D4
Wildstrubel (mt.)	D4
Zeller (lake)	G1
Zugersee (lake)	F2
Zürichsee (lake)	G2

LIECHTENSTEIN

CITIES and TOWNS

Schaan 4,552	H2
Triesen 2,971	H2
Vaduz (cap.) 4,614	H2

OTHER FEATURES

Grauspitz	J2
Ochsenkopf (mt.)	J2
Rhätikon (mts.)	J2
Rhine (riv.)	J2

*City and suburbs.

AUSTRIA

PROVINCES

Burgenland 272,119 ....D3
Carinthia 525,721 ....B3
Lower Austria 1,414,161 ....C2
Salzburg 401,766 ....B3
Styria 1,192,442 ....C3
Tirol 540,771 ....A3
Upper Austria 1,223,444 ....B2
Vienna (city) 1,614,841 ....D2
Vorarlberg 271,473 ....A3

CITIES and TOWNS†

Admont 3,126 ....C3
Allentsteig 2,783 ....C2
Altheim 4,766 ....B2
Althofen 3,886 ....C3
Amstetten 13,330 ....C2
Andau 3,058 ....D3
Arnoldstein 6,740 ....B3
Aspang Markt 2,316 ....D3
Attnang-Puchheim 7,837 ....B2
Bad Aussee 5,039 ....B3
Baden 22,631 ....D3
Badgastein 5,228 ....B3
Bad Goisern 6,360 ....B3
Bad Hofgastein 5,525 ....B3
Bad Ischl 12,740 ....B3
Bad Leonfelden 2,712 ....C2
Bad Sankt-Leonhard im
  Lavanttal 4,882 ....C3
Berndorf 8,371 ....C2
Bischofshofen 9,417 ....B3
Bludenz 12,050 ....A3
Bramberg am Wildkogel 3,129 ...B3
Braunau am Inn 16,432 ....B2
Bregenz 22,839 ....A3
Bruck an der Leitha 7,506 ....D2
Bruck an der Mur 16,359 ....C3
Deutsch Feistritz 3,820 ....C3
Deutschkreutz 3,673 ....D3
Deutsch Landsberg 6,614 ....C3
Deutsch Wagram 4,481 ....D2
Dornbirn 33,810 ....A3
Ebenfurth 2,272 ....D3
Ebensee 9,413 ....B3
Eferding 3,014 ....B2
Eggenburg 3,730 ....C2
Ehrwald 2,198 ....A3

Eisenerz 11,563 ....C3
Eisenkappel-Vellach 3,761 ....C3
Eisenstadt 10,059 ....D3
Enns 9,622 ....C2
Feldbach 3,887 ....C3
Feldkirch 21,214 ....A3
Feldkirchen in
  Kärnten 11,188 ....B3
Ferlach 7,621 ....C3
Fieberbrunn 3,651 ....B3
Fohnsdorf 11,169 ....C3
Frankenmarkt 2,960 ....B3
Frauenkirchen 2,749 ....D3
Freistadt 5,956 ....C2
Freidberg 2,504 ....C3
Friesach 7,257 ....C3
Frohnleiten 5,081 ....C3
Fulpmes 2,553 ....A3
Fürstenfeld 6,054 ....C3
Gaming 4,181 ....C2
Gänserndorf 4,211 ....D2
Gleisdorf 4,921 ....C3
Gloggnitz 7,078 ....D3
Gmünd, Carinthia 2,267 ....B3
Gmünd, Lower Austria 6,323 .....C2
Gmunden 12,270 ....B3
Golling an der Salzach 3,089 .....B3
Götzis 7,821 ....A3
Gratwein 2,747 ....C3
Graz 251,900 ....C3
Graz* 314,200 ....C3
Grein 2,767 ....C2
f21Grieskirchen 4,519 ....B2
Grosssiegharts 3,288 ....C2
Grünburg 3,775 ....C3
Güssing 3,675 ....C3
Haag 5,060 ....C2
Hainburg an der Donau 6,009 ...D2
Hainfeld 3,897 ....C2
Hallein 14,371 ....B3
Hallstadt 1,303 ....B3
Hartberg 5,702 ....C3
Haslach an der Mühl 2,636 ....C2
Heidenreichstein 4,340 ....C2
Heiligenblut 1,324 ....B3
Hermagor-Presseggersee 7,531 ...B3
Herzogenburg 7,299 ....C2
Hohenau an der March 3,591 ....D2
Hohenberg 2,016 ....C3
Hohenems 11,487 ....A3
Hollabrunn 6,563 ....D2
Hopfgarten in Nordtirol 4,784 ...B3

Horn 6,264 ....C2
Hüttenberg 3,251 ....C3
Imst 5,855 ....A3
Innsbruck 115,800 ....A3
Innsbruck* 167,200 ....A3
Jenbach 5,868 ....A3
Jennersdorf 4,210 ....C3
Judenburg 11,346 ....C3
Kapfenberg 26,001 ....C3
Kappl 2,156 ....A3
Kaprun 2,604 ....B3
Kindberg 6,128 ....C3
Kirchdorf an der Krems 3,471 ...C3
Kitzbühel 7,995 ....B3
Klagenfurt 74,326 ....C3
Klagenfurt* 112,600 ....C3
Klosterneuburg 21,912 ....D2
Knittelfeld 14,517 ....C3
Köflach 12,612 ....C3
Königswiesen 2,921 ....C2
Korneuburg 8,892 ....D2
Kössen 2,764 ....B3
Kötschach-Mauthen 3,740 ....B3
Krems an der Donau 21,733 .....C2
Kufstein 12,766 ....A3
Kundl 3,020 ....A3
Laa an der Thaya 5,455 ....D2
Laakirchen 7,664 ....C3
Lambach 3,301 ....C3
Landeck 7,388 ....A3
Längenfeld 2,838 ....A3
Langenlois 4,957 ....C2
Langenwang 4,071 ....C3
Lavamünd 4,120 ....C3
Leibnitz 6,646 ....C3
Lenzing 5,385 ....B3
Leoben 35,153 ....C3
Lienz 11,696 ....B3
Liezen 2,508 ....C3
Lilienfeld 3,126 ....C3
Linz 205,700 ....C2
Linz* 356,500 ....C2
Lustenau 15,239 ....A3
Mannersdorf am
  Leithagebirge 4,012 ....D3
Marchegg 2,678 ....D2
Mariazell 2,298 ....C3
Matrei in Osttirol 4,003 ....B3
Mattersburg 5,417 ....D3
Mattighofen 4,344 ....B2
Mauerkirchen 2,237 ....B2
Mautern in Steiermark 2,536 ...C3

Mauthausen 4,419 ....C2
Mauthen-Kötschach 3,750 ....B3
Mayrhofen 3,174 ....A3
Melk 5,108 ....C2
Mistelbach an der Zaya 6,306 ...D2
Mittersill 4,361 ....B3
Mödling 18,712 ....D2
Mondsee 2,141 ....B3
Murau 2,710 ....C3
Mürzzuschlag 11,564 ....C3
Neuberg an der Mürz 2,183 ......C3
Neumarkt am Wallersee 3,267 ...B3
Neunkirchen 10,922 ....D3
Neusiedl am See 3,999 ....D3
Neustift im Stubaital 2,789 ......A3
Ober Grafendorf 4,109 ....C2
Oberndorf bei Salzburg 3,293 ...B3
Obervellach 2,420 ....B3
Oberwart 5,661 ....C3
Paternion 5,805 ....B3
Perg 4,872 ....C2
Peuerbach 2,161 ....B2
Pfunds 2,043 ....A3
Pinkafeld 4,610 ....C3
Pöchlarn 3,199 ....C2
Pörtschach am
  Wörthersee 2,511 ....C3
Poysdorf 5,774 ....D2
Pregarten 3,244 ....C2
Raabs an der Thaya 4,194 ....C2
Radenthein 6,847 ....B3
Radkersburg 2,000 ....C3
Radstadt 3,585 ....B3
Rankweil 8,440 ....A3
Rechnitz 3,412 ....C3
Reichenau an der Rax 4,053 ...C3
Retz 4,780 ....C2
Ried im Innkreis 10,534 ....B2
Rottenmann 4,781 ....C3
Saalfelden am Steinernen
  Meer 10,172 ....B3
Salzburg 122,100 ....B3
Salzburg* 213,430 ....B3
Sankt Aegyd am Neuwalde 3,165 ...C3
Sankt Anton am Arlberg 2,086 ...A3
Sankt Johann in Tirol 5,942 .....B3
Sankt Michael in
  Obersteiermark 3,717 ....C3
Sankt Michael im Lungau 2,839 ...B3
Sankt Paul im Lavanttal 6,721 ...C3
Sankt Pölten 43,300 ....C2

Sankt Valentin 8,715 ....C2
Sankt Veit an der Glan 11,047 ...C3
Sankt Wolfgang im
  Salzkammergut 2,746 ....B3
Schärding 5,874 ....B2
Scheibbs 4,419 ....C2
Schladming 3,460 ....B3
Schrems 3,393 ....C2
Schruns 3,607 ....A3
Schwarzach im Pongau 3,616 ...B3
Schwaz 10,253 ....A3
Schwechat 14,997 ....D2
Schwertberg 3,881 ....C2
Sierning 8,162 ....C2
Sillian 1,988 ....B3
Solbad Hall in Tirol 12,335 ......A3
Spital am Pyhrn 2,315 ....C3
Spittal an der Drau 13,690 ......B3
Steinach 2,698 ....A3
Steyr 40,578 ....C2
Stockerau 12,634 ....D2
Strassburg 2,850 ....C3
Tamsweg 5,060 ....B3
Telfs 6,589 ....A3
Ternitz 10,287 ....C3
Traiskirchen 8,878 ....D2
Traun 20,843 ....C2
Trieben 4,639 ....C3
Trofaiach 8,731 ....C3
Tulln 7,705 ....D2
Velden am Wörthersee 7,306 ...C3
Vienna (cap.) 1,700,000 ....D2
Villach 50,979 ....B3
Vöcklabruck 10,627 ....B2
Voitsberg 11,094 ....C3
Völkermarkt 10,772 ....C3
Vordernberg 2,508 ....C3
Waidhofen an der Thaya 4,200 ...C2
Waidhofen an der Ybbs 5,218 ...C2
Weitensfeld-Flattnitz 5,206 ......B3
Weitra 3,250 ....C2
Weiz 8,241 ....C3
Wels 47,279 ....C2
Weyer Markt 2,518 ....C2
Wien (Vienna) (cap.) 1,700,000 ...D2
Wiener Neustadt 34,774 ....D3
Wildon 2,002 ....C3
Wilhelmsburg 6,307 ....C2
Wolfsberg 31,176 ....C3
Wörgl 7,811 ....A3
Ybbs an der Donau 6,422 ....C2

Zams 3,120 ....A3
Zell am See 7,456 ....B3
Zell am Ziller 1,882 ....A3
Zeltweg 8,431 ....C3
Zirl 4,157 ....A3
Zistersdorf 3,412 ....D2
Zwettl-Niederösterreich 11,624 ...C2

OTHER FEATURES

Allgäu Alps (mts.) A3
Bavarian Alps (mts.) A3
Bodensee (Constance) (lake) ...A3
Brenner (pass) A3
Carnic Alps (mts.) B3
Constance (lake) A3
Danube (riv.) C2
Drau (riv.) C3
Enns (riv.) C3
Grossglockner (mt.) B3
Hohe Tauern (range) B3
Inn (riv.) B2
Karawanken (range) D2
Mühlviertel (reg.) C2
Mur (riv.) C3
Neusiedler See (lake) D3
Niedere Tauern (range) B3
Ötztal Alps (mts.) A3
Raab (riv.) C3
Rhine (riv.) A3
Salzach (riv.) B3
Salzkammergut (reg.) B3
Semmering (pass) C3
Thaya (riv.) C2
Traun (riv.) C2
Wildspitze (mt.) A3
Zugspitze (mt.) A3

CZECHOSLOVAKIA

REPUBLICS

Czech Socialist Rep. 9,964,338 ...B1
Slovak Socialist Rep. 4,670,409 ...E2

REGIONS

Bratislava (city) 333,000 ....D2
Jihočeský 662,002 ....C2
Jihomoravský 1,966,850 ....D2
Praha (city) 1,161,200 ....C1

Severočeský 1,122,035 ....C1
Severomoravský 1,849,286 .....D1
Středočeský 1,193,041 ....C2
Středoslovenský 1,436,351 ......E2
Východočeský 1,214,581 ....C1
Východoslovenský 1,298,481 ...F2
Západočeský 865,094 ....C1
Západoslovenský 1,610,542 ....D2

CITIES and TOWNS

Aš 120,000 ....B1
Austerlitz (Slavkov) D2
Bánovce nad Bebravou 11,400 ...E2
Banská Bystrica 53,000 ....E2
Banská Štiavnica 7,486 ....E2
Bardejov 17,400 ....F2
Benešov 11,100 ....C2
Beroun 17,600 ....B1
Blina 17,800 ....B1
Blansko 13,800 ....D2
Boskovice 8,531 ....D2
Brandýs nad Labem-Stará
  Boleslav 333,000 ....C1
Bratislava 333,000 ....D2
Břeclav 21,100 ....D2
Brezno 14,800 ....E2
Brno 335,700 ....D2
Broumov 7,782 ....D1
Bruntál 12,300 ....D1
Frenštát pod
  Radhoštěm 8,516 ....D2
Fil'akovo 7,820 ....E2
Frýdlant v.
Bystřice nad
  Hostýnem 6,471 ....D2
Bystřice pod
  Hostýnem 6,681 ....D2
Bytča 6,922 ....E2

Čadca 16,800 ....E2
Čalovo 6,591 ....D3
Česká Lípa 18,600 ....C1
Česká Třebová 14,700 ....D2
České Budějovice 80,800 ....C2
Český Brod 6,640 ....C1
Český Krumlov 12,000 ....C2
Český Těšín 17,200 ....D2
Chеb 27,000 ....B1
Chocеň 8,198 ....D1
Chodov 14,400 ....B1
Chomutov 44,200 ....B1
Chotěboř 6,692 ....C2
Chrudim 18,800 ....C2
Cierny Balog 6,435 ....E2
Dečín 46,500 ....C1
Detva 13,100 ....E2
Dobříš 6,378 ....C2
Dobruška 5,779 ....D1
Dolný Kubín 9,900 ....E2
Domažlice 9,100 ....B2
Dubnica nad Váhom 11,300 ...E2
Duchcov 9,712 ....B1
Dunajská Streda 13,000 ....D3
Dvory nad Žitavou 5,847 ....D2
Dvě Králové nad
  Labem 16,800 ....C1
Falknov (Sokolov) 23,900 ......B1
Frýdek-Místek 43,800 ....E2
Frýdlant v.

Frýdlant nad
  Ostravicí 6,250 ....E2
Galanta 12,300 ....D2
Gottwaldov 84,300 ....D2
Handlová 16,200 ....E2
Havířov 85,000 ....E2
Havlíčkův Brod 19,200 ....C2
Hlinsko 8,890 ....C2
Hlohovec 15,200 ....D2
Hluřín 15,300 ....D1
Hnúšt'a-Likier E2
Hodonín 22,600 ....D2
Holešov 9,150 ....D2
Hollč 7,602 ....D2
Horažd'ovice C2
Hořice 6,151 ....C1
Hořice v
  Podkrkonoší 7,715 ....C1
Horná Štubňa E2
Horní Benešov D1
Horní Libina D1
Hořovice 5,665 ....C2
Horšovský Týn B2
Hostinné C1
Hradec Králové 85,600 ....C1
Hranice 13,300 ....D2
Hrinova 7,800 ....E2
Hronov 9,767 ....D1
Hrušovany D2
Humenné 22,200 ....F2
Humpolec 7,810 ....C2
Hurbanovo E3
Hustopeče D2
Ilava E2
Ivančice 7,314 ....D2

Topography

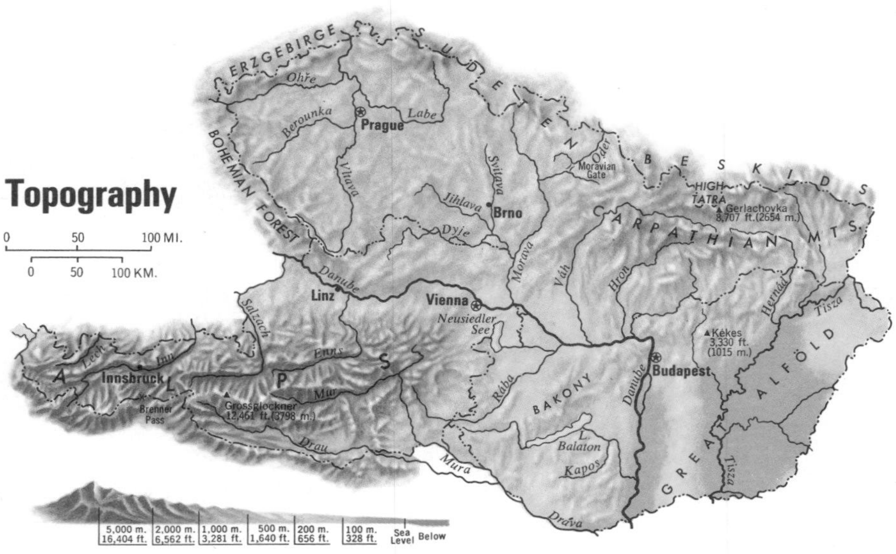

Topography

0   50   100 MI.
0   50   100 KM.

5,000 m. / 16,404 ft. 2,000 m. / 6,562 ft. 1,000 m. / 3,281 ft. 500 m. / 1,640 ft. 200 m. / 656 ft. 100 m. / 328 ft. Sea Level Below

AREA 32,375 sq. mi. (83,851 sq. km.)
POPULATION 7,507,000
CAPITAL Vienna
LARGEST CITY Vienna
HIGHEST POINT Grossglockner
  12,457 ft. (3,797 m.)
MONETARY UNIT schilling
MAJOR LANGUAGE German
MAJOR RELIGION Roman Catholicism

AREA 49,373 sq. mi. (127,876 sq. km.)
POPULATION 15,276,799
CAPITAL Prague
LARGEST CITY Prague
HIGHEST POINT Gerlachovka 8,707 ft.
  (2,654 m.)
MONETARY UNIT koruna
MAJOR LANGUAGES Czech, Slovak
MAJOR RELIGIONS Roman Catholicism,
  Protestantism

AREA 35,919 sq. mi. (93,030 sq. km.)
POPULATION 10,709,536
CAPITAL Budapest
LARGEST CITY Budapest
HIGHEST POINT Kékes 3,330 ft.
  (1,015 m.)
MONETARY UNIT forint
MAJOR LANGUAGE Hungarian
MAJOR RELIGIONS Roman Catholicism,
  Protestantism

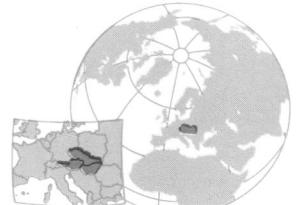

AUSTRIA

CZECHOSLOVAKIA

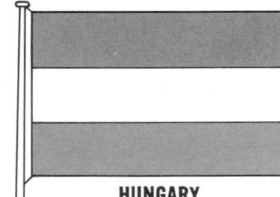

HUNGARY

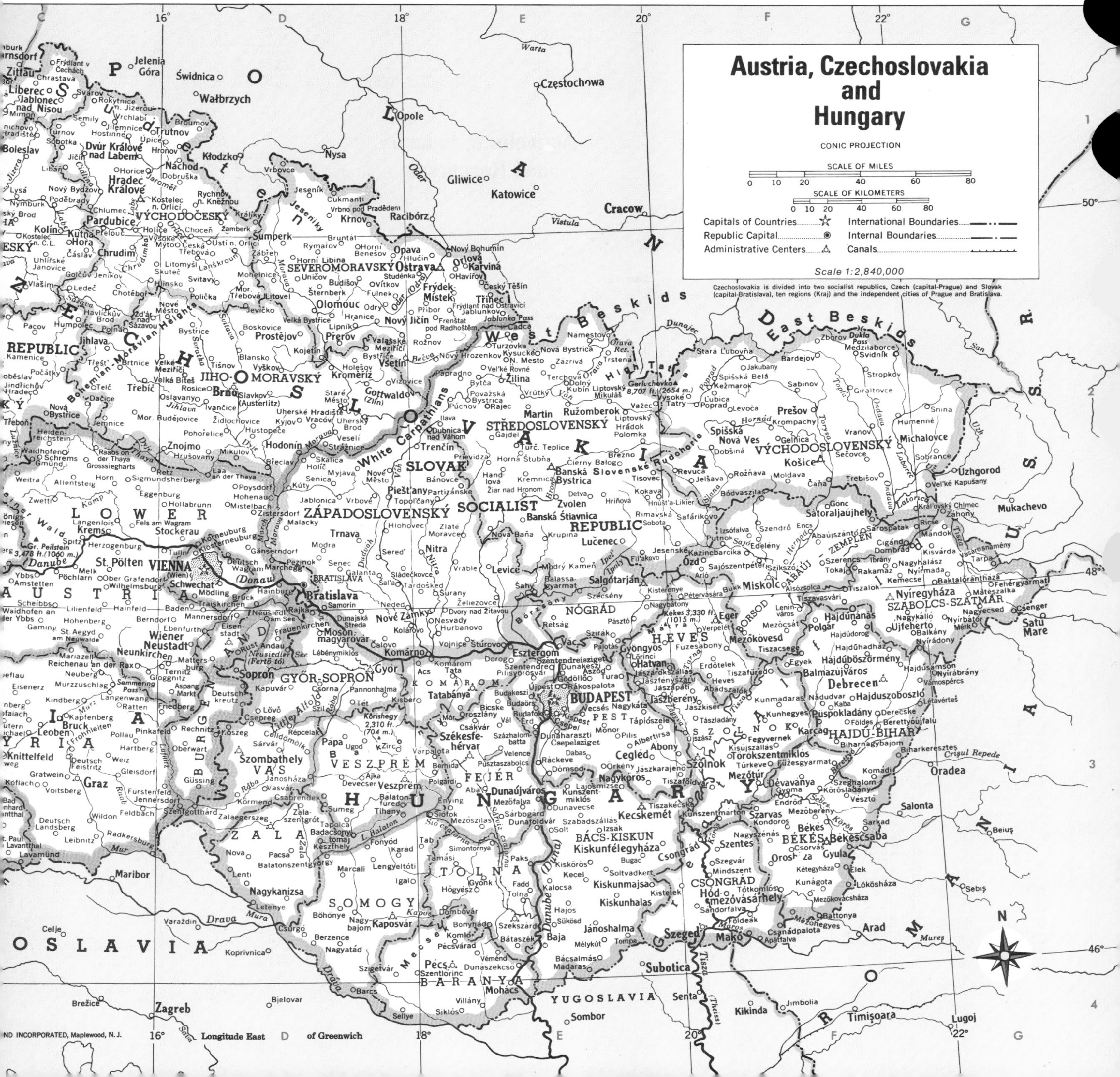

Austria, Czechoslovakia and Hungary

CONIC PROJECTION

SCALE OF MILES
0 10 20 40 60 80

SCALE OF KILOMETERS
0 10 20 40 60 80

Capitals of Countries...... ☆   International Boundaries......
Republic Capital........... ◉   Internal Boundaries.........
Administrative Centers..... △   Canals......................

Scale 1:2,840,000

Czechoslovakia is divided into two socialist republics, Czech (capital-Prague) and Slovak (capital-Bratislava), ten regions (Kraj) and the independent cities of Prague and Bratislava.

ND INCORPORATED, Maplewood, N.J.

(continued)

Jablonec nad Nisou 36,300 ..... C1
Jablonica D2
Jablunkov 9,405 ..... E2
Jáchymov B1
Jakubany F2
Jaroměř 11,600 ..... C1
Jelšava F2
Jemnice C2
Jeseník 10,900 ..... D1
Jesenské F2
Jevíčko D2
Jičín 13,200 ..... C1
Jihlava 44,500 ..... C2
Jilemnice C1
Jindřichův Hradec 15,700 ..... C2
Jiřkov 11,400 ..... B1
Kadaň 18,100 ..... B1
Kamenice C2
Kaplice C2
Karlovy Vary 43,300 ..... B1
Karviná 79,100 ..... E2
Kdyně B2
Kežmarok 11,000 ..... F2
Kladno 61,200 ..... B1
Klatovy 18,500 ..... B2
Kojetín 5,852 ..... D2
Kokava nad Rimavicou 5,391 ..... E2
Kolárovo 10,500 ..... D3
Kolín 29,100 ..... C1
Komárno 28,200 ..... D3
Košice 169,100 ..... F2
Kostelec nad Orlicí 5,575 ..... D1
Kráľovský Chlmec 5,329 ..... G2
Kralupy nad Vltavou 16,900 ..... C1
Kraslice 6,733 ..... B1
Kremnica 5,941 ..... E2
Krnov 25,000 ..... D1
Kroměříž 23,200 ..... D2
Krompachy 6,332 ..... F2
Krupina 6,627 ..... E2
Krupka 8,301 ..... B1
Kutná Hora 19,200 ..... C2
Kyjov 10,700 ..... D2
Kynšperk 5,524 ..... B1
Kysucké Nové Mesto 11,700 ..... E2
Lanškroun 8,683 ..... D2
Levice 19,000 ..... E2
Levoča 10,100 ..... F2
Libáň C1
Liberec 75,600 ..... C1
Lidice C1
Lipník nad Bečvou 7,358 ..... D2
Liptovský Mikuláš 19,400 ..... E2
Litoměřice 19,700 ..... C1
Litomyšl 8,112 ..... D2
Litovel 5,805 ..... D2
Litvínov 23,300 ..... B1
Lomnice C2
Louny 15,200 ..... B1
Lovosice 9,323 ..... C1
L'ubica F2
Lučenec 23,300 ..... E2
Lysá nad Labem 9,920 ..... C1
Malacky 13,200 ..... D2
Mariánské Lázně 14,600 ..... B2
Martin 47,800 ..... E2
Medzilaborce F2
Mělník 17,800 ..... C1
Michalovce 23,600 ..... G2
Mikulov 6,267 ..... D2
Milevsko 7,091 ..... C2
Mimoň 6,773 ..... C1
Mladá Boleslav 36,900 ..... C1
Mladá Vožice C2
Mnichovo Hradiště 5,239 ..... C1
Modra 7,219 ..... D2
Modrý Kameň 6,200 ..... E2
Mohelnice 6,050 ..... D2
Moldava nad Bodvou 5,397 ..... F2
Moravská Třebová 9,052 ..... D2
Moravské Budějovice 5,576 ..... D2
Most 59,400 ..... B1
Myjava 6,657 ..... D2
Náchod 19,300 ..... D1
Námestovo E2
Neded D2
Nejdek 8,187 ..... B1
Nepomuk B2
Nesvady 5,453 ..... E3
Netolice C2
Nitra 50,000 ..... E2
Nová Baňa 6,218 ..... E2
Nová Bystrica E2
Nová Bystřice C2
Nové Hrady C2
Nové Město na Moravě 6,581 ..... D2
Moravě 6,581 ..... D2
Nové Město nad
Váhom 15,900 ..... D2
Nové Strašecí B1
Nové Zámky 27,300 ..... D3
Nový Bohumín 16,700 ..... E2
Nový Bor 7,621 ..... C1
Nový Bydžov 6,824 ..... C1
Nový Hrozenkov E2
Nový Jičín 21,400 ..... E2
Nymburk 13,600 ..... C1
Nýřany 6,204 ..... B2
Nýrsko B2
Odry D2
Olomouc 82,800 ..... D2
Opava 53,800 ..... E2
Orlová 25,500 ..... E2
Ostrava 293,500 ..... E2
Ostrov 18,200 ..... B1
Pardubice 78,500 ..... C1
Partizánske 15,100 ..... E2
Pelhřimov 11,900 ..... C2
Pezinok 13,100 ..... D2
Piešťany 25,400 ..... D2
Písek 25,100 ..... C2
Plzeň 155,000 ..... B2
Počátky C2
Podbořany B1
Poděbrady 13,400 ..... C1
Pohořelice D2
Polička 6,529 ..... D2
Polná C2
Polomka E2
Poprad 25,800 ..... F2
Považská Bystrica 19,300 ..... E2
Prachatice 7,900 ..... B2
Prague (Praha) (cap.) 1,161,200 ..... C1
Přelouč 6,251 ..... C1
Přerov 43,500 ..... D2
Prešov 61,000 ..... F2
Přeštice B2
Příbor 7,726 ..... E2
Příbram 31,300 ..... B2
Prievidza 30,900 ..... E2
Prostějov 44,200 ..... D2
Protivín C2
Púchov 9,306 ..... E2
Radnice B2
Rajec E2
Rakovník 14,200 ..... B1
Revúca 5,901 ..... F2
Říčany u Prahy 8,407 ..... C2
Rimavská Sobota 5,800 ..... E2
Rokycany 12,800 ..... B2
Rokytnice nad Jizerou C1
Rosice D2
Roudnice nad Labem 11,800 ..... C1
Rožňava 12,400 ..... F2
Rožnov pod
Radhoštěm 11,600 ..... E2
Rumburk C1
Ružomberok 22,600 ..... E2
Rychnov nad Kněžnou 7,500 ..... D1
Rýmařov 7,522 ..... D2
Sabinov 5,473 ..... F2
Šafárikovo F2
Šahy 5,049 ..... E2
Šaľa 15,200 ..... D2
Samorín 8,287 ..... D2
Sečovce 5,744 ..... F2
Sedlčany C2
Semily 8,200 ..... C1
Senec 8,544 ..... D2
Senica 12,300 ..... D2
Sereď 12,500 ..... D2
Skalica 11,100 ..... D2
Skuteč D2
Sládečkovce 5,598 ..... D2
Slaný 13,200 ..... C1
Slavkov D2
Snina 10,900 ..... G2
Soběslav 6,140 ..... C2
Sobotka C1
Sobrance G2
Sokolov 23,900 ..... B1
Spišská Belá F2
Spišská Nová Ves 26,100 ..... F2
Stará L'ubovňa 5,800 ..... F2
Staré Město 6,293 ..... D2
Šternberk 13,700 ..... D2
Stod B2
Strakonice 19,000 ..... B2
Strážnice 5,482 ..... D2
Stříbro B2
Stropkov 5,645 ..... F2
Studénka 9,744 ..... D2
Štúrovo 8,287 ..... E3
Šumperk 25,900 ..... D1
Šurany 6,693 ..... E2
Sušice 10,300 ..... B2
Svárov C1
Svidník 4,600 ..... F2
Svitavy 15,000 ..... D2
Tábor 28,100 ..... C2
Tachov 11,400 ..... B2
Telč 5,285 ..... C2
Teplice 52,300 ..... B1
Tišnov 8,263 ..... D2
Topoľčany 17,500 ..... D2
Třebíč 23,900 ..... C2
Trebišov 13,700 ..... F2
Třeboň 6,068 ..... C2
Trenčín 38,800 ..... E2
Třešť 5,053 ..... C2
Třinec 32,000 ..... E2
Trnava 48,600 ..... D2
Trutnov 24,500 ..... D1
Turnov 13,600 ..... C1
Turzovka 6,107 ..... E2
Uherské Hradiště 32,100 ..... D2
Uherský Brod 12,800 ..... D2
Uničov 10,800 ..... D2
Úpice 6,323 ..... C1
Ústí nad Labem 74,900 ..... C1
Ústí nad Orlicí 13,700 ..... D2
Valašské
Meziříčí 19,400 ..... D2
Varnsdorf 14,700 ..... C1
Važec E2
Vejprty B1
Velká Bíteš D2
Velká Bystřice D2
Veľké Kapušany G2
Velké Meziříčí 7,590 ..... D2
Veľké Rovné E2
Veselí nad Lužnicí C2
Veselí nad Moravou 11,500 ..... D2
Vimperk 5,749 ..... B2
Vítkov 5,138 ..... D2
Vizovice D2
Vlašim 8,873 ..... C2
Vodňany 5,620 ..... C2
Vojnice E3
Volary B2
Volyně B2
Votice C2
Vráble E2
Vracov D2
Vranov nad Teplou 14,700 ..... F2
Vrbno pod Pradědem 5,594 ..... D1
Vrbovce D1
Vrbové D2
Vrchlabí 11,700 ..... C1
Vrútky 5,756 ..... E2
Vsetín 24,100 ..... D2
Vyškov 15,100 ..... D2
Vysoké Mýto 8,830 ..... D2
Vysoké Tatry F2
Vyšší Brod C2
Zábřeh 11,300 ..... D2
Žamberk 5,040 ..... D1
Žatec 17,400 ..... B1
Zázrivá E2
Zbiroh B2
Zborov F2
Žďár nad Sázavou 17,800 ..... C2
Želiezovce 5,478 ..... E2
Žiar nad Hronom 14,800 ..... E2
Židlochovice D2
Žilina 56,000 ..... E2
Zlaté Moravce 10,300 ..... E2
Zlín (Gottwaldov) 84,300 ..... D2
Žlutice B1
Znojmo 28,500 ..... D2
Zvolen 29,000 ..... E2

OTHER FEATURES

Berounka (riv.) C2
Beskids, East (mts.) F1
Beskids, West (mts.) E2
Bohemian (for.) B2
Bohemian-Moravian Heights
(hills) C2
Danube (riv.) C2
Dunajec (riv.) F2
Dyje (riv.) D2
Erzgebirge (mts.) B1
Gerlachovka (mt.) E2
Hornád (riv.) F2
Hron (riv.) E2
Ipeľ (riv.) E2
Jablunka (pass) E2
Jeseniky (mts.) D1
Jihlava (riv.) D2
Krušné Hory (Erzgebirge)
(mts.) B1
Labe (riv.) C1
Lipno (res.) C2
Lužnice (riv.) C2
Moldau (Vltava) (riv.) C2
Morava (riv.) D2
Nitra (riv.) E2
Oder (Odra) (riv.) D2
Ohře (riv.) B1
Ondava (riv.) F2
Orava (riv.) E2
Orlická (res.) C2
Sázava (riv.) C2
Slovenské Rudohorie (mts.) E2
Sudeten (mts.) C1
Svitava (riv.) D2
Svratka (riv.) D2
Tatra, High (mts.) E2
Torysa (riv.) F2
Úhlava (riv.) B2
Váh (riv.) D2
Vltava (riv.) C2
White Carpathians (mts.) E2

HUNGARY

COUNTIES

Bács-Kiskun 568,532 ..... E3
Baranya 434,030 ..... E4
Békés 436,987 ..... F3
Borsod-Abaúj-Zemplén 808,924 ..... F2
Budapest (city) 2,060,170 ..... E3
Csongrád 456,862 ..... F3
Fejér 421,568 ..... E3
Győr-Sopron 428,476 ..... D3
Hajdú-Bihar 552,417 ..... F3
Heves 350,874 ..... F3
Komárom 321,579 ..... E3
Nógrád 239,907 ..... E3
Pest 973,486 ..... E3
Somogy 360,308 ..... D3
Szabolcs-Szatmár 593,746 ..... G3
Szolnok 446,379 ..... F3
Tolna 266,414 ..... E3
Vas 285,527 ..... D3
Veszprém 386,740 ..... D3
Zala 316,610 ..... D3

CITIES and TOWNS

Aba 4,271 ..... E3
Abádszalók 6,386 ..... F3
Abaújszántó 4,209 ..... F2
Abony 15,624 ..... E3
Ács 8,423 ..... E3
Ajka 29,601 ..... D3
Albertirsa 11,252 ..... E3
Alsózsolca 5,045 ..... F2
Arló 4,203 ..... F2
Aszód 6,218 ..... E3
Bácsalmás 9,025 ..... E3
Badacsonytomaj 2,933 ..... D3
Baja 38,456 ..... E3
Baktalórántháza 3,736 ..... G2
Balassagyarmat 18,534 ..... E2
Balatonfüred 12,599 ..... D3
Balkány 7,667 ..... G3
Balmazújváros 17,371 ..... F3
Barcs 11,448 ..... D4
Bátaszék 7,274 ..... E3
Battonya 9,324 ..... F3
Békés 22,287 ..... F3
Békéscsaba 67,266 ..... F3
Berettyóújfalu 16,406 ..... F3
Berzence 3,406 ..... D3
Bicske 10,720 ..... E3
Biharkeresztes 4,788 ..... F3
Biharnagybajom 4,093 ..... F3
Böhönye 3,215 ..... D3
Bonyhád 14,841 ..... E3
Budafok 40,623 ..... E3
Budaörs 13,958 ..... E3
Budakeszi 10,429 ..... E3
Budapest (cap.) 2,060,170 ..... E3
Bugak 4,989 ..... E3
Cegléd 40,567 ..... E3
Celldömölk 12,533 ..... D3
Cigánd 4,767 ..... F2
Csabrendek 3,045 ..... D3
Csákvár 5,238 ..... E3
Csanádpalota 4,642 ..... F3
Csenger 4,792 ..... G3
Csepel 71,693 ..... E3
Csepreg 4,079 ..... D3
Csongrád 22,202 ..... E3
Csorna 12,131 ..... D3
Csorvás 6,826 ..... F3
Csurgó 5,463 ..... D3
Dabas 13,075 ..... E3
Debrecen 192,484 ..... F3
Derecske 9,579 ..... F3
Dévaványa 11,208 ..... F3
Devecser 5,482 ..... D3
Dombóvár 19,917 ..... E3
Dombrád 6,328 ..... F2
Dömsöd 6,545 ..... E3
Dorog 10,754 ..... E3
Dunaföldvár 10,318 ..... E3
Dunaharaszti 15,788 ..... E3
Dunakeszi 25,187 ..... E3
Dunaszekcső 2,999 ..... E3
Dunaújváros 60,694 ..... E3
Dunavecse 4,521 ..... E3
Edelény 9,559 ..... F2
Eger 61,283 ..... F3
Egyek 7,956 ..... F3
Elek 6,032 ..... F3
Enes 2,565 ..... F2
Endrőd 8,136 ..... F3
Enying 7,518 ..... E3
Érd 41,210 ..... E3
Erdőtelek 4,250 ..... F3
Esztergom 30,476 ..... E3
Fadd 4,805 ..... E3
Fegyvernek 8,421 ..... F3
Fehérgyarmat 6,729 ..... G3
Földeák 3,855 ..... F3
Földes 5,293 ..... F2
Fonyód 3,957 ..... D3
Füzesabony 6,965 ..... F3
Füzesgyarmat 7,097 ..... F3
Gödöllő 28,057 ..... E3
Gönc 2,875 ..... F2
Gyoma 10,392 ..... F3
Gyöngyös 36,927 ..... E3
Gyönk 2,507 ..... E3
Győr 123,618 ..... D3
Gyula 34,514 ..... F3
Hajdúböszörmény 32,145 ..... F3
Hajdúdorog 10,118 ..... F3
Hajdúhadház 13,626 ..... F3
Hajdúnánás 18,146 ..... F3
Hajdúsámson 7,492 ..... F3
Hajdúszoboszló 23,374 ..... F3
Hajós 5,113 ..... E3
Hatvan 24,790 ..... E3
Heves 10,943 ..... F3
Hódmezővásárhely 54,481 ..... F3
Hőgyész 3,534 ..... E3
Ibrány 7,037 ..... F2
Izsák 7,686 ..... E3
Izsófalva 6,816 ..... F2
Jánoshalma 12,534 ..... E3
Jánosháza 3,274 ..... D3
Jászapáti 10,424 ..... E3
Jászárokszállás 10,139 ..... E3
Jászberény 31,347 ..... E3
Jászfényszaru 6,869 ..... E3
Jászkarajenő 4,101 ..... E3
Jászkisér 6,816 ..... F3
Jászladány 7,823 ..... F3
Kaba 6,654 ..... F2
Kalocsa 18,613 ..... E3
Kaposvár 72,330 ..... D3
Kapuvár 11,243 ..... D3
Karád 2,754 ..... D3
Karcag 25,264 ..... F3
Kazincbarcika 37,481 ..... F2
Kecel 10,493 ..... E3
Kecskemét 91,929 ..... E3
Kemecse 4,583 ..... F2
Keszthely 21,671 ..... D3
Kétegyháza 4,728 ..... F3
Kisbér 4,562 ..... D3
Kiskőrös 15,499 ..... E3
Kiskunfélegyháza 35,339 ..... E3
Kiskunhalas 30,552 ..... E3
Kiskunmajsa 14,439 ..... E3
Kispest 65,106 ..... E3
Kistelek 8,544 ..... E3
Kisterenye 6,844 ..... E2
Kisújszállás 13,699 ..... F3
Kisvárda 17,828 ..... G2
Komádi 8,765 ..... F3
Komárom 19,955 ..... E3
Komló 30,301 ..... E3
Kondoros 7,319 ..... F3
Körmend 11,787 ..... D3
Körösladány 6,565 ..... F3
Kőszeg 12,705 ..... D3
Kunágota 4,622 ..... F3
Kunhegyes 10,116 ..... F3
Kunmadaras 7,343 ..... F3
Kunszentmárton 11,103 ..... F3
Kunszentmiklós 7,952 ..... E3
Lajosmizse 12,872 ..... E3
Lébénymiklós 6,190 ..... D3
Lengyeltóti 3,389 ..... D3
Leninváros 18,667 ..... F3
Lenti 8,106 ..... D3
Létavértes 9,106 ..... G3
Letenye 4,395 ..... D3
Lökösháza 2,514 ..... F3
Lőrinci 10,679 ..... E3
Madaras 4,519 ..... E3
Makó 29,943 ..... F3
Mándok 5,093 ..... G2
Marcali 12,485 ..... D3
Mátészalka 17,709 ..... G3
Mélykút 7,640 ..... E3
Mérk 3,211 ..... G3
Mezőberény 12,702 ..... F3
Mezőcsát 6,729 ..... F3
Mezőfalva 5,008 ..... E3
Mezőhegyes 8,631 ..... F3
Mezőkovácsháza 7,473 ..... F3
Mezőkövesd 18,435 ..... F3
Mezőszilas 2,792 ..... E3
Mezőtúr 22,018 ..... F3
Mindszent 8,730 ..... F3
Miskolc 206,727 ..... F2
Mohács 21,385 ..... E4
Monor 16,838 ..... E3
Mór 12,066 ..... E3
Mosonmagyaróvár 29,732 ..... D3
Nádudvar 9,447 ..... F3
Nagyatád 12,946 ..... D3
Nagybajom 4,402 ..... D3
Nagyecsed 8,225 ..... G3
Nagyhalász 6,437 ..... F2
Nagykálló 11,282 ..... F3
Nagykanizsa 48,494 ..... D3
Nagykáta 11,922 ..... E3
Nagykőrös 27,900 ..... E3
Nagyszénás 7,124 ..... F3
Nyírábrány 4,509 ..... G3
Nyíradony 7,146 ..... G3
Nyírbátor 13,388 ..... G3
Nyíregyháza 108,156 ..... F3
Nyírmada 4,744 ..... F2
Örkény 5,013 ..... E3
Orosháza 36,243 ..... F3
Oroszlány 20,604 ..... E3
Ózd 48,521 ..... F2
Pacsa 1,984 ..... D3
Paks 19,514 ..... E3
Pannonhalma 3,731 ..... D3
Pápa 32,202 ..... D3
Pásztó 7,962 ..... E3
Pécs 168,788 ..... E3
Pécsvárad 3,672 ..... E3
Pétervására 2,753 ..... F3
Pilis 9,055 ..... E3
Pilisvörösvár 10,217 ..... E3
Polgár 9,429 ..... F3
Polgárdi 5,767 ..... E3
Püspökladány 15,730 ..... F3
Pusztaszabolcs 5,794 ..... E3
Putnok 7,103 ..... F2
Ráckeve 7,534 ..... E3
Rajka 2,448 ..... D3
Rakamaz 5,407 ..... F2
Rákospalota 60,983 ..... E3
Répcelak 1,997 ..... D3
Ricse 2,992 ..... G2
Sajószentpéter 13,992 ..... F2
Salgótarján 49,320 ..... E2
Sándorfalva 5,949 ..... F3
Sárbogárd 11,178 ..... E3
Sarkad 11,937 ..... F3
Sárospatak 15,316 ..... F2
Sárvár 15,126 ..... D3
Sátoraljaújhely 19,252 ..... F2
Sellye 2,804 ..... D4
Siklós 10,567 ..... E4
Simontornya 4,892 ..... E3
Siófok 20,084 ..... E3
Solt 6,911 ..... E3
Soltvadkert 7,934 ..... E3
Sopron 53,930 ..... D3
Sükösd 4,430 ..... E3
Sümeg 6,229 ..... D3
Szabadszállás 8,223 ..... E3
Szarvas 20,598 ..... F3
Szécsény 5,690 ..... E2
Százhalombatta 13,963 ..... E3
Szeged 171,342 ..... E3
Szeghalom 9,736 ..... F3
Szegvár 6,395 ..... F3
Székesfehérvár 103,197 ..... E3
Szekszárd 34,592 ..... E3
Szendrő 4,098 ..... F2
Szentendre 16,844 ..... E3
Szentes 35,326 ..... F3
Szentgotthárd 5,837 ..... D3
Szentlőrinc 3,926 ..... E3
Szerencs 8,612 ..... F2
Szigetvár 12,114 ..... D3
Szikszó 6,419 ..... F2
Szil 2,073 ..... D3
Szolnok 75,203 ..... F3
Szombathely 82,830 ..... D3
Tab 3,922 ..... E3
Tamási 7,602 ..... E3
Tápiószele 5,575 ..... E3
Tapolca 17,161 ..... D3
Tarpa 3,436 ..... G2
Tata 24,114 ..... E3
Tatabánya 75,942 ..... E3
Tét 4,441 ..... D3
Tiszacsege 6,263 ..... F3
Tiszaföldvár 12,560 ..... F3
Tiszafüred 12,259 ..... F3
Tiszakécske 12,378 ..... E3
Tiszalök 6,230 ..... F2
Tiszavasvári 13,292 ..... F3
Tokaj 4,845 ..... F2
Tolna 8,997 ..... E3
Tompa 5,365 ..... E3
Törökszentmiklós 25,551 ..... F3
Tótkomlós 8,803 ..... F3
Tura 8,235 ..... E3
Túrkeve 11,393 ..... F3
Újfehértó 14,412 ..... F3
Újpest 80,384 ..... E3
Újszász 7,098 ..... F2
Vác 34,837 ..... E3
Vál 2,488 ..... E3
Vámospércs 5,213 ..... F3
Várpalota 28,293 ..... E3
Vásárosnamény 8,637 ..... G2
Vasvár 4,275 ..... D3
Vecsés 19,193 ..... E3
Velence 3,463 ..... E3
Véménd 2,293 ..... E3
Verpelét 4,622 ..... F2
Veszprém 54,898 ..... D3
Vésztő 9,815 ..... F3
Villány 2,764 ..... E4
Záhony 3,049 ..... G2
Zalaegerszeg 39,671 ..... D3
Zalaszentgrót 5,346 ..... D3
Zirc 5,980 ..... D3

OTHER FEATURES

Bakony (mts.) D3
Balaton (lake) D3
Berettyó (riv.) F3
Bükk (mts.) F2
Csepelsziget (isl.) E3
Danube (riv.) E3
Dráva (riv.) D3
Duna (Danube) (riv.) E3
Fertő tó (Neusiedler See)
(lake) D3
Great Alföld (plain) F3
Hernád (riv.) F2
Kapos (riv.) D3
Kékes (mt.) E3
Körös (riv.) F3
Maros (riv.) F3
Mátra (mts.) E3
Mecsek (mts.) D3
Mura (riv.) D3
Rába (riv.) D3
Sajó (riv.) F2
Sárviz csatorna (canal) E3
Sió csatorna (canal) E3
Szentendreisziget (isl.) E3
Tisza (riv.) F3
Zala (riv.) D3

*City and suburbs.
†Population of Austrian cities are communes.

Agriculture, Industry and Resources

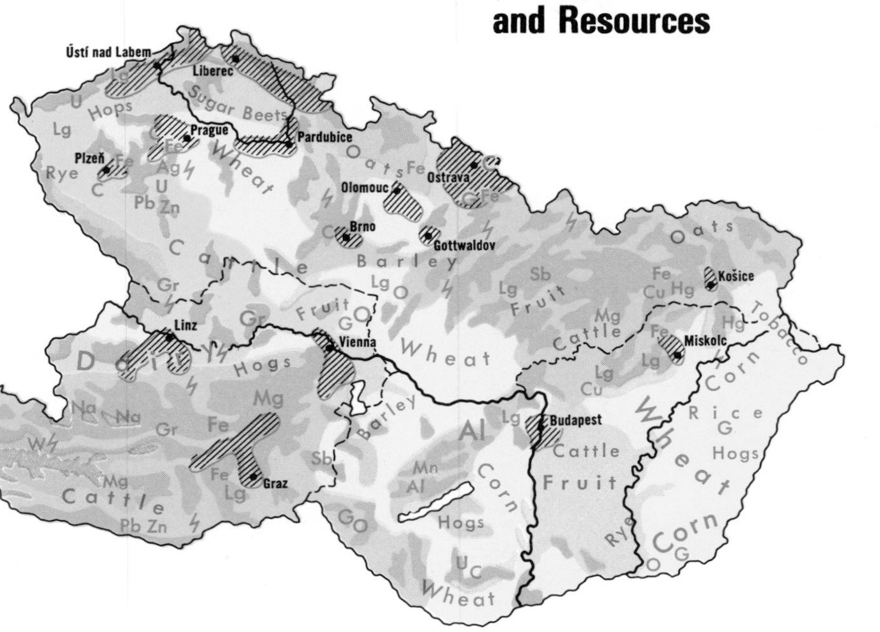

DOMINANT LAND USE

Cereals (chiefly wheat, corn)
Other Cereals, Livestock, Dairy
General Farming, Livestock
General Farming, Truck Farming
Pasture Livestock
Grapes, Wine
Forests
Nonagricultural Land

MAJOR MINERAL OCCURRENCES

Ag Silver
Al Bauxite
C Coal
Cu Copper
Fe Iron Ore
G Natural Gas
Gr Graphite
Hg Mercury
Lg Lignite
Mg Magnesium
Mn Manganese
Na Salt
O Petroleum
Pb Lead
Sb Antimony
U Uranium
W Tungsten
Zn Zinc

Water Power
Major Industrial Areas

YUGOSLAVIA

AREA 98,766 sq. mi. (255,804 sq. km.)
POPULATION 22,471,000
CAPITAL Belgrade
LARGEST CITY Belgrade
HIGHEST POINT Triglav 9,393 ft. (2,863 m.)
MONETARY UNIT Yugoslav dinar
MAJOR LANGUAGES Serbo-Croatian, Slovenian,
Macedonian, Montenegrin, Albanian
MAJOR RELIGIONS Eastern Orthodoxy,
Roman Catholicism, Islam

ALBANIA

AREA 11,100 sq. mi. (28,749 sq. km.)
POPULATION 2,590,600
CAPITAL Tiranë
LARGEST CITY Tiranë
HIGHEST POINT Korab 9,026 ft. (2,751 m.)
MONETARY UNIT lek
MAJOR LANGUAGE Albanian
MAJOR RELIGIONS Islam, Eastern Orthodoxy,
Roman Catholicism

ROMANIA

AREA 91,699 sq. mi. (237,500 sq. km.)
POPULATION 22,048,305
CAPITAL Bucharest
LARGEST CITY Bucharest
HIGHEST POINT Moldoveanul 8,343 ft.
(2,543 m.)
MONETARY UNIT leu
MAJOR LANGUAGES Romanian, Hungarian
MAJOR RELIGION Eastern Orthodoxy

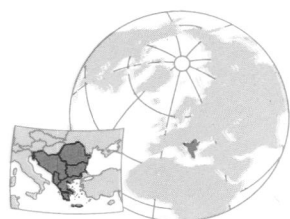

BULGARIA

AREA 42,823 sq. mi. (110,912 sq. km.)
POPULATION 8,862,000
CAPITAL Sofia
LARGEST CITY Sofia
HIGHEST POINT Musala 9,597 ft. (2,925 m.)
MONETARY UNIT lev
MAJOR LANGUAGE Bulgarian
MAJOR RELIGION Eastern Orthodoxy

GREECE

AREA 50,944 sq. mi. (131,945 sq. km.)
POPULATION 9,599,000
CAPITAL Athens
LARGEST CITY Athens
HIGHEST POINT Olympus 9,570 ft. (2,917 m.)
MONETARY UNIT drachma
MAJOR LANGUAGE Greek
MAJOR RELIGION Eastern (Greek) Orthodoxy

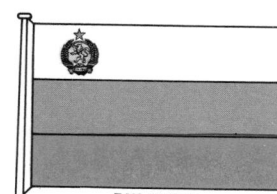
BULGARIA

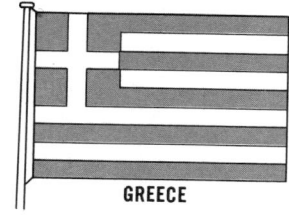
GREECE

YUGOSLAVIA

ALBANIA

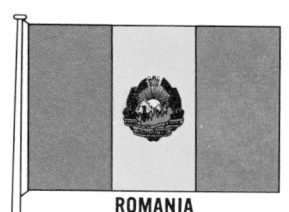
ROMANIA

Agriculture, Industry and Resources

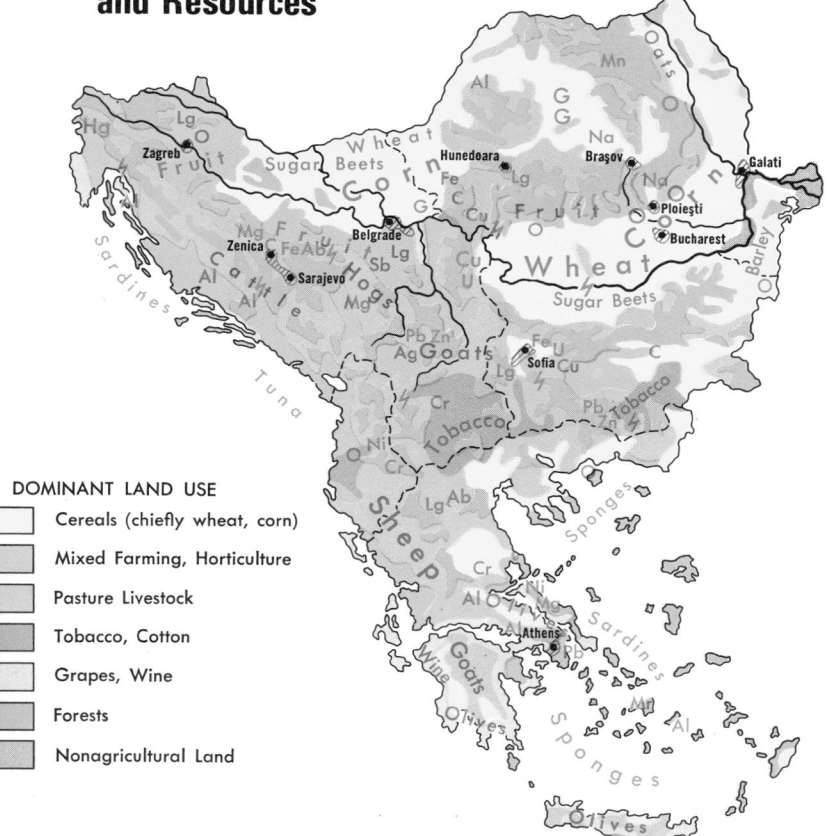

DOMINANT LAND USE

	Cereals (chiefly wheat, corn)
	Mixed Farming, Horticulture
	Pasture Livestock
	Tobacco, Cotton
	Grapes, Wine
	Forests
	Nonagricultural Land

MAJOR MINERAL OCCURRENCES

Ab	Asbestos	Mg	Magnesium
Ag	Silver	Mn	Manganese
Al	Bauxite	Mr	Marble
C	Coal	Na	Salt
Cr	Chromium	Ni	Nickel
Cu	Copper	O	Petroleum
Fe	Iron Ore	Pb	Lead
G	Natural Gas	Sb	Antimony
Hg	Mercury	U	Uranium
Lg	Lignite	Zn	Zinc

⚡ Water Power
▨ Major Industrial Areas

ALBANIA

CITIES and TOWNS

Berat 25,700	D5
Çorovodë	E5
Burrel	D5
Delvinë 6,000	D6
Durrës (Durazzo) 53,800	D5
Elbasan 41,700	E5
Erseke	E5
Fier 23,000	D5
Gjirokastër 17,100	D5
Kavajë 18,700	D5
Korçë 47,300	E5
Krujë 7,900	D5
Kuçovë (Stalin) 14,000	D5
Kukës 6,100	E4
Leskovik	E5
Lezhë	D5
Lushnjë 18,900	D5
Memaliaj	D5
Peqin	D5
Përmet	E5
Peshkopi 6,600	E5
Pogradec 10,100	E5
Pukë	E4
Sarandë 8,700	E6
Shëngjin	D5
Shijak 6,200	D5
Shkodër 55,300	D5
Stalin 14,000	D5
Tepelenë	D5
Tiranë (Tirana)	
(cap.) 171,300	E5
Vlorë 50,000	D5

OTHER FEATURES

Adriatic (sea)	B4
Drin (riv.)	E4
Korab (mt.)	E5
Ohrid (lake)	E4
Otranto (str.)	D5
Prespa (lake)	E5
Sazan (isl.)	D5
Scutari (lake)	D4
Vijosë (riv.)	D5

BULGARIA

CITIES and TOWNS

Akhtopol 938	H4
Aitatar 3,249	H4
Ardino 5,080	G5
Asenovgrad 43,049	G5
Aytos 20,967	H4
Balchik 11,070	H4
Bansko 10,011	F5
Belogradchik 6,892	F4
Berkovitsa 16,253	F4
Blagoevgrad 50,043	F5
Botevgrad 17,789	F4
Bregovo 5,567	F3
Breznik 4,699	F4
Burgas 144,449	H4
Byala 10,564	G4
Byala Slatina 15,788	F4
Chirpan 20,595	G4
Devin 7,120	G5
Dimitrovgrad 45,596	G4
Dobrich (Tolbukhin) 86,184	H4
Dryanovo 9,804	G4
Elena 7,008	G4
Elin Pelin 5,499	F4
Elkhovo 12,397	H4
Gabrovo 75,034	G4
General-Toshevo 8,928	H4
Godech 5,225	F4
Gorna Oryakhovitsa 34,157	G4
Gotse Delchev 17,015	F5
Grudovo 9,871	H4
Ikhtiman 11,482	F4
Isperikh 10,500	H3
Ivaylovgrad 3,900	H5
Karapelit	H4
Karlovo 25,472	G4
Karnobat 21,480	H4
Kavarna 10,872	J4
Kazanlŭk 53,607	G4
Kharmanli 19,240	H5
Khaskovo 75,031	G5
Kotel 8,229	H4
Krumovgrad 5,211	G5
Kula 5,667	F4
Kŭrdzhali 47,757	G5
Kyustendil 48,239	F4
Lom 30,538	F4
Lovech 43,858	G4

Lukovit 10,400	G4
Malko Tŭrnovo 4,233	H4
Maritsa 8,664	H4
Michurin 4,434	H4
Mikhaylovgrad 40,064	F4
Momchilgrad 8,185	G5
Nesebŭr 6,768	H4
Nikopol 5,563	G4
Nova Zagora 21,872	H4
Novi Pazar 15,751	H4
Omurtag 9,067	H4
Oryakhovo 14,012	F4
Panagyurishte 20,649	F4
Pazardzhik 65,577	F4
Pernik 87,432	F4
Peshtera 16,882	G4
Petrich 24,381	F5
Pirdop 8,248	G4
Pleven 107,567	G4
Plovdiv 300,242	G4
Pomorie 11,960	H4
Popina	H3
Popovo 19,428	H4
Provadiya 15,143	H4
Radomir 10,436	F4
Razgrad 42,486	H4
Razlog 13,690	F5
Rositsa	H4
Ruse 160,351	H4
Samokov 25,763	F4
Sandanski 19,003	F5
Sevlievo 24,421	G4
Shabla 4,471	J4
Shumen 83,525	H4
Silistra 58,270	H3
Simeonovgrad (Maritsa) 8,664	H4
Sliven 90,137	H4
Smolyan 29,032	G5
Smyadovo 5,020	H4
Sofia (cap.) 965,728	F4
Sozopol 3,877	H4
Stanke Dimitrov 42,034	F4
Stara Zagora 122,200	G4
Svilengrad 15,150	G5
Svishtov 29,412	G4
Teteven 12,555	G4
Tolbukhin 86,184	H4
Topolovgrad 7,230	H4
Troyan 23,692	G4
Trŭn 3,435	F4
Tŭrgovishte 38,796	H4
Tutrakan 11,447	H3
Varna 251,654	J4
Veliko Tŭrnovo 56,497	G4
Vidin 53,030	F4
Vratsa 61,265	F4
Yambol 75,861	H4
Zimnitsa	H4
Zlatograd 7,732	G5

OTHER FEATURES

Balkan (mts.)	G4
Black (sea)	J4
Danube (riv.)	H4
Dunav (Danube) (riv.)	H4
Emine (cape)	J4
Iskŭr (riv.)	G4
Kaliakra (cape)	J4
Maritsa (riv.)	F5
Mesta (riv.)	F5
Midzhur (mt.)	F4
Musala (mt.)	F4
Osŭm (riv.)	G4
Rhodope (mts.)	G5
Rujen (mt.)	F4
Struma (riv.)	F5
Timok (riv.)	F3
Tundzha (riv.)	G4
Vit (riv.)	G4

GREECE

REGIONS

Aegean Islands 417,813	G6
Athens, Greater 2,566,775	F7
Áyion Óros (aut. dist.) 1,732	G5
Central Greece and Euboea 966,543	F6
Crete 456,642	G8
Epirus 310,334	E6
Ionian Islands 184,443	D6
Macedonia 1,888,952	E5
Peloponnisos 986,912	F7
Thessaly 659,913	F6
Thrace 329,582	G5

CITIES and TOWNS

Agrínion 30,973	E6
Aíyina 5,704	F7

Alyión 18,829	F6
Alexandroúpolis 22,995	H5
AliVérion 4,414	G6
Almirós 5,680	F6
Amaliás 14,177	E7
Amfilokhía 4,668	E6
Ámfissa 6,605	F6
Andíssa 1,762	H6
Andravídha 3,046	E6
Ándros 1,827	G7
Áno Viánnos 1,431	G8
Ánoya 2,750	G8
Ardhéa 3,555	F5
Areópolis 674	F7
Argalastí 1,621	F6
Árgos 18,890	F7
Argostólion 7,060	E6
Arkhángelos 3,016	J7
Árta 2,424	E6
Ása 19,498	E6
Astipálaia 787	H7
Atalándi 4,581	F6
Athens (cap.) 867,023	F7
Athens* 2,566,775	F7
Ayiá 3,241	F6
Áyios Kírikos 1,083	H7
Áyios Matthaíos 1,596	D6
Áyios Nikólaos 5,002	G8
Candia (Iráklion) 77,506	G8
Canea (Khaniá) 40,564	F8
Corinth 20,773	F7
Delfí 1,185	F6
Delvinákion 1,067	E6
Dhidhimótikhon 8,388	H5
Dhíkaia 1,222	H5
Dhimitsána 996	F7
Dhomokós 1,991	F6
Dráma 29,692	G5
Édhessa 13,967	F5
Elassón 7,200	F6
Eleftheroúpolis 4,888	G5
Ermoúpolis 13,502	G7
Fársala 6,967	F6
Filiátes 2,579	E6
Filiatrá 5,919	E7
Filíppias 3,248	E6
Flórina 11,164	E5
Gargaliánoi 5,888	E7
Grevená 8,106	E5
Ídhra 2,381	F7
Ierápetra 7,055	G8
Igoumenítsa 4,109	E6
Ioánnina 40,130	E6
Íos 1,270	G7
Iráklion 77,506	G8
Istiaía 4,059	F6
Itháki 2,293	E6
Kalámai 39,133	F7
Kalampáka 5,453	E6
Kalávrita 1,948	F6
Kálimnos 6,492	H7
Kándanos 403	F8
Kardhítsa 25,685	F6
Kariá 1,342	E6
Karياí 301	G5
Káristos 3,550	G6
Kárpathos 1,363	H8
Karpenísion 4,414	F6
Kastéllion (Kíssamos) 2,996	F8
Kastéllion 1,152	G8
Kastoría 15,407	E5
Katákolon 690	E7
Kateríni 28,808	F5
Kaválla 46,234	G5
Kéa 693	G7
Kérkira 28,630	D6
Khalkís 36,300	F6
Khaniá 40,564	F8
Khíos 24,084	H6
Khóra Sfakíon 246	F8
Kiáton 7,392	F6
Kilkís 10,538	F5
Kími 2,772	G6
Kiparissía 3,882	E7
Kíssamos 2,996	F8
Kíthira 349	F7
Komotiní 28,896	G5
Kónitsa 3,150	E5
Korópi 9,367	G7
Kos 7,828	H7
Kozáni 23,240	E5
Kranídhion 3,657	F7
Lagkadá 1,350	E7
Lamía 37,872	F6
Langadhás 6,707	F5
Langádhia	F7
Lárisa 72,336	F6
Lávrion 8,283	G7
Leonídhion 3,181	F7
Levádhia 15,445	F6
Levkás 6,818	E6
Limenária 1,507	G5

(continued on following page)

44 Balkan States

(continued)

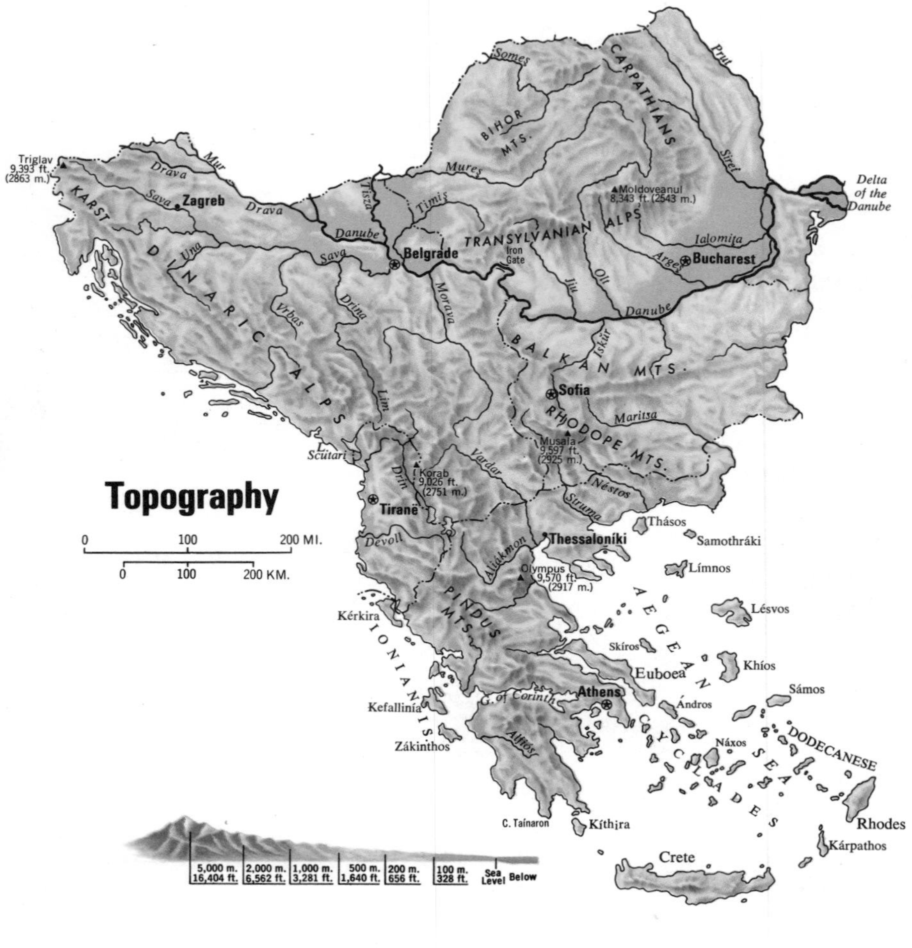

Topography
0 100 200 MI.
0 100 200 KM.
Triglav 9,393 ft. (2863 m.)
Mur
Drava
Sava
Zagreb
KARST
DINARIC ALPS
Una
Vrbas
Drina
Danube
Tisza
Belgrade
Morava
Timiş
Mureş
Someş
BIHOR MTS.
CARPATHIANS
Prut
Siret
Moldoveanul 8,343 ft. (2543 m.)
TRANSYLVANIAN ALPS
Iron Gate
Jiu
Olt
Argeş
Ialomiţa
Bucharest
Delta of the Danube
Iskŭr
BALKAN MTS.
Sofia
Maritsa
RHODOPE MTS.
Musala 9,597 ft. (2925 m.)
Néstos
Struma
Vardar
Lim
Scutari
Drin
Korab 9,026 ft. (2751 m.)
Tiranë
Devoll
Aliákmon
Thessaloníki
Olympus 9,570 ft. (2917 m.)
PINDUS MTS.
Kérkira
IONIAN SEA
Kefallinía
Zákinthos
G. of Corinth
Athens
Euboea
Skíros
Thásos
Samothráki
Límnos
Lésvos
Khíos
Sámos
Ándros
AEGEAN SEA
CYCLADES
Náxos
DODECANESE
Alfiós
C. Taínaron
Kíthira
Rhodes
Kárpathos
Crete
5,000 m. 16,404 ft.
2,000 m. 6,562 ft.
1,000 m. 3,281 ft.
500 m. 1,640 ft.
200 m. 656 ft.
100 m. 328 ft.
Sea Level
Below

Límni 2,394 F6
Líndos 700 J7
Litókhoron 5,561 F5
Lixoúrion 3,364 E6
Loutrá Aidhipsoú 2,195 F6
Marathón 1,976 G6
Megalópolis 3,357 E7
Mégara 17,294 F6
Meligalá 1,724 F7
Mesolóngion 11,614 E6
Messíni 6,625 E7
Métsovon 2,823 E6
Mikínai 390 F7
Mílos 850 G7
Mírina 3,982 G6
Míthimna 1,414 G6
Mitilíni 23,426 H6
Moîrai 2,948 G8
Moláoi 2,484 F7
Monólithos 247 H7
Moúdhros 1,024 G6
Náousa 17,375 F5
Návpaktos 8,170 F6
Návplion 9,281 F7
Náxos 2,892 G7
Neápolis 3,070 F7
Neméa 4,356 F7
Néon Karlóvasi 4,401 H7
Nestórion 1,143 E5
Nigríta 7,301 F5
Oinóî 188 F6
Orestías 10,727 H5
Paramithía 2,747 E6
Pátrai 111,607 E6
Pérdika 1,198 E6
Péta 2,116 E6
Pílos 2,258 E7
Piraiévs (Piraeus) 187,362 F7
Pírgos 20,599 E7
Piryí 1,455 G6
Píthion 1,047 H5
Plomárion 4,353 H6
Políkastron 5,279 F5
Políkhnitos 4,152 G6
Políyiros 3,707 F5
Póros 4,051 F7
Préveza 11,439 E6
Psakhná 4,650 F6
Psári 622 E7
Ptolemaís 16,588 E5
Réthimnon 14,969 G8
Rhodes (Ródhos) 32,092 J7
Salamís 18,256 F6
Salonika
(Thessaloníki) 345,799 F5
Sámi 957 E6
Sámos 5,146 H7
Samothráki 508 G5
Sápai 2,456 G5
Sérrai 39,897 F5
Sérvia 3,834 F5
Siátista 4,852 E5
Sidhirókastron 6,363 F5
Sími 2,344 H7
Sitía 6,167 H8
Skíathos 3,707 F6
Skíros 1,925 G6
Skópelos 2,545 F6
Souflíon 5,637 H5
Sparta 10,549 F7
Spétsai 3,427 F7
Spíli 789 G8
Stavrós 1,700 F5
Stilís 4,427 F6
Thásos 2,052 G5
Thessaloníki 345,799 F5
Thessaloníki* 482,361 F5
Thíra 1,322 G7
Thívai 15,971 F6
Timbákion 3,229 G8
Tínos 3,423 G7
Tírnavos 10,451 F6
Tríkkala 34,794 E6
Trípolis 20,209 F7
Vámos 652 F8
Vartholomión 3,015 E7
Vathí 2,491 H7
Velvendós 4,063 F5
Vérroia 29,528 F5
Vólos 51,290 F6
Vónitsa 3,324 E6
Vrondádhes 4,253 G6
Xánthi 24,867 G5
Yerolimín 73 F7
Yiannitsá 18,151 F5
Yíthion 4,915 F7
Zákinthos 9,339 E7
Zante (Zákinthos) 9,339 E7

OTHER FEATURES

Aegean (sea) G6
Akrítas (cape) E7
Aktí (pen.) G5
Amorgós (isl.) G7
Anáfi (isl.) G7
Andikíthira (isl.) F8
Ándros (isl.) G7
Arda (riv.) G5
Argolís (gulf) F7
Astipálaia (isl.) H7
Áthos (mt.) G5
Áyios Evstrátios
(isl.) G6
Áyios Yeóryios (cape) G5
Cephalonia (Kefallinía)
(isl.) E6
Corfu (Kérkira) (isl.) D6
Corinth (gulf) F6
Crete (isl.) G8
Crete (sea) G7
Cyclades (isls.) G7
Día (isl.) G8
Dodecanese (isls.) H8
Euboea (Évvoia)
(isl.) G6
Évros (riv.) H5
Évvoia (isl.) G6
Gávdhos (isl.) F8
Ídhi (mt.) G8
Ikaría (isl.) G7
Ionian (sea) D7
Íos (isl.) G7
Itháki (Ithaca) (isl.) E6
Kafirévs (cape) G6
Kálimnos (isl.) H7
Kárpathos (isl.) H8
Kásos (isl.) H8
Kassándra (pen.) F6
Kéa (isl.) G7
Kefallinía (isl.) E6
Kérkira (isl.) D6
Khálki (isl.) H7
Khaniá (gulf) G8
Khíos (isl.) G6
Kímolos (isl.) G7
Kiparissía (gulf) E7
Kíthira (isl.) F7
Kíthnos (isl.) G7
Kos (isl.) H7
Kriós (cape) F8
Kríti (Crete) (isl.) G8
Lakonía (gulf) F7
Léros (isl.) H7
Lésvos (isl.) G6
Levítha (isl.) H7
Levkás (isl.) E6
Límnos (isl.) G6
Maléa (cape) F7
Matapan (Taínaron) (cape) F7
Merabéllou (gulf) H8
Mesará (gulf) G8
Messíni (gulf) E7
Míkonos (isl.) G7
Mílos (isl.) G7
Mirtóön (sea) F7
Náxos (isl.) G7
Néstos (riv.) G5
Nísiros (isl.) H7
Northern Sporades (isls.) F6
Olympia (isls.) E7
Olympus (mt.) F5
Parnassus (mt.) F6
Páros (isl.) G7
Pátmos (isl.) H7
Paxoí (isl.) D6
Pindus (mts.) E6
Piniós (riv.) E6
Prespa (lake) E5
Psará (isl.) G6
Psevdhókavos (cape) G6
Rhodes (isl.) H7
Rhodope (mts.) G5
Salonika (Thermaic) (gulf) F6
Sámos (isl.) H7
Samothráki (isl.) G5
Saría (isl.) H8
Saronic (gulf) F7
Sérifos (isl.) G7
Sídheros (cape) H8
Sífnos (isl.) G7
Sími (isl.) H7
Síros (isl.) G7
Sithoniá (pen.) F5
Skíros (isl.) G6
Spátha (cape) F8
Strimón (gulf) G5
Strofádhes (isls.) E7
Taínaron (cape) F7
Thásos (isl.) G5
Thermaic (gulf) F6
Thíra (isl.) G7
Tílos (isl.) H7
Tínos (isl.) G7
Toronaic (gulf) F5
Vardar (riv.) E5
Voïvís (lake) F6
Vólvi (lake) F5
Voúxa (cape) F8
Zákinthos (Zante)
(isl.) E7

ROMANIA

CITIES and TOWNS

Aiud 25,173 F2
Alba Iulia 44,552 F2
Alexandria 38,296 G3
Anina 11,594 E3
Arad 161,568 E2
Babadag 8,423 J3
Bacău 131,413 H2
Baia de Aramă 5,065 F3
Baia Mare 112,893 F2
Băile Herculane 4,606 F3
Băileşti 21,246 F3
Balş 16,091 G3
Beiuş 9,992 F2
Bereşti Tîrg H2
Bicaz 9,490 G2
Bîrlad 59,059 H2
Bistriţa 47,562 G2
Bivolari H2
Blaj 21,678 F2
Borşa 25,287 G2
Botoşani 69,881 H2
Brad 18,391 F2
Brăila 203,983 H3
Braşov 259,108 G3
Bucharest (Bucureşti)
(cap.) 1,832,015 G3
Bucharest* 1,960,097 G3
Buhuşi 20,204 H2
Buzău 106,738 H3
Buziaş 8,310 E3
Calafat 16,421 F3
Călăraşi 58,960 H3
Caracal 31,159 G3
Caransebeş 27,429 F3
Carei 24,496 F2
Cernavodă 14,686 J3
Chişineu Criş 9,344 E2
Cîmpeni 7,722 F2
Cîmpia Turzii 23,745 F2
Cîmpina 33,259 H3
Cîmpulung 33,448 G3
Cîmpulung Moldovenesc 19,270 G2
Cisnădie 21,114 G3
Cluj-Napoca 274,095 F2
Cogealac J3
Comaneşti 18,177 H2
Constanţa 279,308 J3
Corabia 20,454 G4
Costeşti 10,446 G3
Craiova 220,893 F3
Cujmir F3
Curtea de Argeş 23,555 G3
Dăbuleni F4
Dăeni J3
Darabani 12,207 H1
Dej 35,396 F2
Deta 6,956 E3
Deva 68,290 F3
Dorohoi 23,121 H2
Drăgăneşti Olt 11,606 G3
Drăgăşani 16,290 F3
Drobeta-Turnu Severin 80,114 F3
Făgăraş 34,762 G3
Fălciu J2
Fălticeni 22,463 H2
Făurei 3,620 H3
Feteşti 28,730 H3
Focşani 62,275 H3
Folteşti H3
Găeşti 13,384 G3
Galaţi 252,884 H3
Gheorghe Gheorghiu-Dej 41,297 H2
Gheorghieni 20,592 G2
Gherla 19,303 G2
Giurgiu 53,241 G3
Haţeg 9,706 F3
Hîrlău 8,135 H2
Hîrşova 8,434 J3
Huedin 8,557 F2
Hunedoara 83,159 F3
Huşi 24,329 J2
Iaşi 262,493 H2
Ineu 10,414 E2
Isaccea 5,283 J3
Jibou F2
Jimbolia 15,325 E3
Lipova 12,427 E2
Luduş 15,771 G2
Lugoj 48,558 F3
Lupeni 28,251 F3
Mangalia 27,263 J4
Medgidia 43,691 J3
Mediaş 68,442 G2
Miercurea Ciuc 38,097 G2
Mizil 14,294 H3
Mociu G2
Moineşti 21,015 H2
Moldova Nouă 18,498 E3
Moreni 17,743 G3
Nădlac 8,407 E2
Năsăud 8,646 G2
Negreşti 7,435 H2
Ocna Mureş 16,381 G2
Odobeşti 8,440 H3
Odorheiu Secuiesc 33,392 G2
Olteniţa 25,536 H3
Oradea 175,400 E2
Orăştie 18,769 F3
Oraviţa 13,628 E3
Orşova 14,873 F3
Panciu 7,772 H3
Paşcani 26,937 H2
Pătulele F3
Pechea H3
Pecica E2
Periam E2
Petrila 25,087 F3
Petroşeni 42,316 F3
Piatra Neamţ 84,192 G2
Pincota 7,494 E2
Piteşti 125,029 G3
Pleniţa F3
Ploieşti 207,009 H3
Poenari Burchi G3
Poiana Mare F4
Pucioasa 14,056 G3
Rădăuţi 24,222 G2
Reghin 31,948 G2
Reşiţa 90,698 E3
Rîmnicu Sărat 29,815 H3
Rîmnicu Vîlcea 75,070 G3
Roman 56,466 H2
Roşiori de Vede 28,832 G3
Săcele 29,391 G3
Salonta 19,698 E2
Satu Mare 108,152 F2
Săveni 7,913 H1
Sebeş 27,448 F3
Sebiş 6,401 F2
Segarcea 8,783 F3
Sfîntu Gheorghe 51,210 G3
Sfîntu Gheorghe J3
Sibiu 156,854 G3
Sighetu Marmaţiei 38,879 F2
Sighişoara 32,296 G2
Şimleul Silvaniei 14,780 F2
Sinaia 14,215 G3
Sînnicolaul Mare 13,565 E2
Siret 6,677 G1
Slănic 8,017 G3
Slatina 54,954 G3
Slobozia 35,207 H3
Solca 4,835 G2
Sovata 10,745 G2
Ştefăneşti H2
Strehaia 11,431 F3
Suceava 66,857 G2
Sulina 5,240 J3
Tăşnad 10,441 F2
Techirghiol 11,228 J3
Tecuci 37,928 H3
Timişoara 281,320 E3
Tinca F2
Tîrgovişte 71,533 G3
Tîrgu Cărbuneşti 7,536 F3
Tîrgu Frumos 6,428 H2
Tîrgu Jiu 70,629 F3
Tîrgu Mureş 129,284 G2
Tîrgu Neamţ 15,756 G2
Tîrgu Ocna 12,960 H2
Tîrgu Secuiesc 18,265 H2
Tîrnăveni 27,799 F2
Topliţa 14,347 G2
Tulcea 67,091 J3
Turda 57,972 F2
Turnu Măgurele 30,003 G4
Urlaţa 10,900 H3
Urziceni 13,500 H3
Vasile Roaită J3
Vaslui 44,134 H2
Vatra Dornei 16,748 G2
Videle 11,323 G3
Vişeul de Sus 20,697 F2
Viziru H3
Zalău 36,158 F2
Zărneşti 23,378 G3
Zimnicea 15,111 G4

OTHER FEATURES

Argeş (riv.) G3
Bîrlad (riv.) H2
Black (sea) J4
Brăila (marshes) H3
Buzău (riv.) H3
Carpathian (mts.) G2
Crişul Alb (riv.) F2
Crişul Repede (riv.) F2
Danube (delta) J3
Danube (riv.) H4
Ialomiţa (marshes) J3
Ialomiţa (riv.) H3
Jijia (riv.) H2
Jiu (riv.) F3
Moldoveanul (mt.) G3
Mureş (riv.) E2
Olt (riv.) G3
Peleaga (mt.) F3
Pietrosul (mt.) G2
Prut (riv.) J2
Siret (riv.) H2
Someş (riv.) F2
Timiş (riv.) E3
Tîrnava Mare (riv.) G2
Transylvanian Alps (mts.) G3

YUGOSLAVIA

INTERNAL DIVISIONS

Bosnia and Hercegovina
(rep.) 3,710,965 C3
Croatia (rep.) 4,396,397 C3
Kosovo (aut. reg.) 1,240,919 E4
Macedonia (rep.) 1,623,598 E5
Montenegro (rep.) 527,207 D4
Serbia (rep.) 8,401,673 E3
Slovenia (rep.) 1,697,068 B2
Vojvodina (aut.
prov.) 1,953,980 D3

CITIES and TOWNS

Aleksinac 11,943 E4
Apatin 17,501 D3
Arendjelovac 15,659 E3
Bačka Topola 16,028 D3
Bakar B3
Banja Luka 85,786 C3
Bar 3,594 D4
Bečej 26,616 E3
Bela Crkva 11,137 E3
Belgrade (cap.) 727,945 E3
Beli Manastir 7,325 D3
Beograd (Belgrade)
(cap.) 727,945 E3
Berovo 5,053 F5
Bihać 24,155 B3
Bijeljina 24,888 D3
Bijelo Polje 9,298 D4
Bileća 4,083 D4
Biograd 3,595 B4
Bitola 64,467 E5
Bjelovar 21,019 C3
Blato 5,591 C4
Bled 4,710 A2
Bor 27,520 E3
Bosanska Dubica 9,191 C3
Bosanska Gradiška 9,742 C3
Bosanska Kostajnica 2,535 B3
Bosanska Krupa 8,947 C3
Bosanski Brod 10,113 D3
Bosanski Novi 9,861 C3
Bosanski Petrovac 4,113 C3
Bosanski Šamac 4,949 D3
Brčko 25,575 D3
Brežice 3,271 B3
Budva 2,483 D4
Bugojno 9,079 C3
Čačak 38,890 E4
Čakovec 11,766 C2
Čapljina 4,677 C4
Caribrod (Dimitrovgrad) 5,449 F4
Cazin 1,213 B3
Celje 30,827 B2
Cetinje 12,089 D4
Ćuprija 17,691 E4
Daruvar 8,478 C3
Debar 8,597 E5
Derventa 11,887 C3
Dimitrovgrad 5,449 F4
Djakovica 29,499 E4
Djakovo 15,833 D3
Doboj 18,073 C3
Donji Vakuf 4,928 C3
Drvar 6,237 C3
Dubrovnik 31,213 C4
Fiume (Rijeka) 128,883 B3
Foča 9,370 D4
Gacko 1,641 D4
Gevgelija 9,319 F5
Glamoč 2,627 C3
Gnjilane 21,359 E4
Gornji Milanovac 11,114 D3
Gornji Vakuf 2,429 C4
Gospić 8,238 B3
Gostivar 18,805 E5
Gračac 3,228 B3
Gračanica 9,302 D3
Gradačac 7,571 D3
Grubišno Polje 2,771 C3
Gusinje 2,616 D4
Herceg Novi 6,645 D4
Ivangrad 11,373 E4
Ivanjica 5,719 E4
Jajce 9,221 C3
Jesenice 16,163 A2
Kanjiža 11,348 D2
Karlovac 47,046 B3
Kavadarci 17,974 E5
Kičevo 14,189 E5
Kikinda 37,392 E3
Kladanj 3,255 D3
Ključ 3,466 C3
Knin 7,279 C3
Knjaževac 11,734 F4
Kočani 16,611 F5
Kočevje 7,277 B3
Kolašin 2,111 D4
Konjic 9,161 D4
Koper 16,683 A3
Koprivnica 16,398 C2
Kosovska Mitrovica 42,526 E4
Kostajnica 9,161 C3
Kotor 5,728 D4
Kragujevac 72,080 E3
Kraljevo 28,065 E4
Kranj 26,341 B2
Križevci 8,501 C2
Krk 1,500 B3
Krško 4,451 B3
Kruševac 29,902 E4
Kulen Vakuf 1,078 B3
Kumanovo 44,791 E4
Kutina 10,892 C3
Leskovac 46,050 E4
Livno 7,223 C4
Ljubinje 785 D4
Ljubljana 169,064 B3
Ljubuški 2,891 C4
Loznica 13,513 D3
Maglaj 5,869 D3
Makarska 6,589 C4
Maribor 94,976 B2
Modriča 7,406 D3
Mostar 47,821 D4
Murska Sobota 9,665 C2
Našice 5,836 C3
Negotin 11,325 F3
Nevesinje 3,077 D4
Nikšić 28,940 D4
Nin 1,782 B3
Niš 128,231 F4
Nova Gorizia A2
Nova Gradiška 11,765 C3
Novi 2,682 B3
Novi Pazar 28,696 E4
Novi Sad 143,591 D3
Novo Mesto 9,553 B3
Novska 5,168 C3
Ogulin 9,975 B3
Ohrid 26,352 E5
Omiš 3,515 C4
Opatija 9,238 A3
Osijek 94,989 D3
Pag 2,318 B3
Pančevo 53,979 E3
Paraćin 21,555 E4
Peć 41,783 E4
Petrinja 12,296 B3
Piran 5,485 A3
Pirot 29,658 F4
Plav 3,072 D4
Pljevlja 14,459 D4
Pločе 4,257 C4
Pola (Pula) 47,117 A3
Poreč 4,512 A3
Postojna 6,085 B3
Požarevac 33,336 E3
Preševo 7,634 E4
Priboj 12,556 D4
Prijedor 22,379 C3
Prijepolje 7,960 D4
Prilep 48,045 E5
Priština 71,264 E4
Prizren 41,875 E4
Prokuplje 20,617 E4
Prozor 1,420 C4
Ptuj 9,245 C2
Pula 47,117 A3
Rab 1,675 B3
Radoviš 9,373 F5
Ragusa (Dubrovnik) 31,213 C4
Raška 3,935 E4
Ravne na Koroškem 6,529 B2
Rijeka 128,883 B3
Rogatica 4,801 D4
Rovinj 8,998 A3
Rožaj E4
Ruma 24,180 D3
Šabac 43,539 D3
Samobor 7,821 B3
Sanski Most 8,718 C3
Sarajevo 245,058 D4
Senj 4,927 B3
Senta 24,694 D3
Šibenik 29,619 C4
Šid 11,867 D3
Sinj 4,705 C4
Sisak 37,215 C3
Sjenica 9,118 E4
Škofja Loka 4,971 A2
Skopje 308,117 E5
Skradin 893 B4
Slavonska Požega 18,160 C3
Slavonski Brod 38,829 D3
Smederevo 39,200 E3
Smederevska Palanka 18,837 E3
Sombor 44,210 D3
Split 150,739 C4
Srebrenica 3,101 D3
Sremska Mitrovica 32,569 D3
Štip 27,218 F5
Stolac 3,862 D4
Ston 407 C4
Struga 11,369 E5
Strumica 22,770 F5
Subotica 89,476 D2
Surdulica 7,048 F4
Svetozarevo 27,812 E4
Svilajnac 7,848 E3
Teslić 4,940 C3
Tetovo 35,293 E5
Titograd 54,639 D4
Titovo Užice 35,465 D4
Titov Veles 35,583 E5
Travnik 12,745 C3
Trbovlje 16,393 B2
Trebinje 3,553 D4
Trogir 6,162 C4
Trstenik 7,167 E4
Tržič 4,435 B2
Tuzla 53,836 D3
Ub 3,785 E3
Ulcinj 7,472 D5
Umag 3,228 A3
Uroševac E4
Valjevo 26,655 D3
Varaždin 34,662 B2
Vareš 7,632 D3
Velenje 11,225 B2
Velika Plana E3
Veliki Bečkerek
(Zrenjanin) 60,201 E3
Vinkovci 29,257 D3
Virovitica 16,389 C3
Višegrad 4,753 D4
Visoko 9,365 D4
Vlasenica 4,033 D3
Vranje 25,909 F4
Vrbas 22,502 D3
Vršac 33,573 E3
Vučitrn 11,701 E4
Vukovar 29,500 D3
Žabljak 1,023 D4
Zadar 43,588 B3
Zagreb 561,773 C3
Zaječar 27,724 E4
Zara (Zadar) 43,588 B3
Zenica 49,522 C3
Žepče 3,177 D3
Zrenjanin 60,201 E3
Zvornik 8,498 D3

OTHER FEATURES

Adriatic (sea) B4
Bobotov Kuk (mt.) D4
Bosna (riv.) D
Brač (isl.) C4
Cazma (riv.) C3
Cres (isl.) B3
Čvrsnica (mt.) C4
Dalmatia (reg.) C4
Danube (riv.) E
Dinaric Alps (mts.) B3
Drava (riv.) C3
Drina (riv.) D3
Dugi Otok (isl.) B3
Hvar (isl.) C4
Ibar (riv.) E
Istria (pen.) A
Kamenjak (cape) A3
Kladovo F3
Korab (mt.) E
Korčula (isl.) C
Kornat (isl.) B
Krk (isl.) B
Kupa (riv.) B3
Kvarner (gulf) B
Lastovo (Lagosta) (isl.) C4
Lim (riv.) D
Lošinj (isl.) B
Midzhur (mt.) F
Mljet (isl.) C4
Morava (riv.) E
Mur (riv.) B
Neretva (riv.) D
Ohrid (lake) E
Pag (isl.) B
Palagruža (Pelagosa) (isl.) C
Prespa (lake) E
Rab (isl.) B
Rujen (mt.) F
Sava (riv.) D
Scutari (lake) D
Slavonia (reg.) C
Šolta (isl.) C
Tara (riv.) D
Timok (riv.) F
Tisa (riv.) E
Triglav (mt.) A
Una (riv.) C
Vardar (riv.) E
Vis (isl.) C
Vrbas (riv.) C
Žirje (isl.) B

*City and suburbs.

The Balkan States

CONIC PROJECTION

SCALE OF MILES

0 25 50 75 100 125 150 175

SCALE OF KILOMETERS

0 25 50 75 100 125 150 175

Capitals of Countries	⭐
Administrative Centers	△
International Boundaries	—·—·—
Major Internal Boundaries	—·—·—
Minor Internal Boundaries	·········
Canals	

Scale 1:6,150,000

BULGARIA and GREECE are divided into counties and
departments, respectively. Because of the scale no
attempt has been made to delimit and name these subdivisions; their administrative centers have, however,
been designated.
    The larger divisions named in Greece are well-known
geographical regions, without administrative function.
    ROMANIA consists of thirty-nine counties and
three cities of regional status, Bucharest, Constanța
and Petroșeni. Scale does not permit delimiting
these counties.
    ALBANIA is divided into twenty-seven districts. Scale
does not permit the delimitation of these divisions.
    YUGOSLAVIA is a federation of six republics. The
Serbian republic includes an autonomous province
(Vojvodina), and an autonomous region (Kosovo).

© Copyright HAMMOND INCORPORATED, Maplewood, N.J.

Topography

0   50   100 MI.
0   50   100 KM.

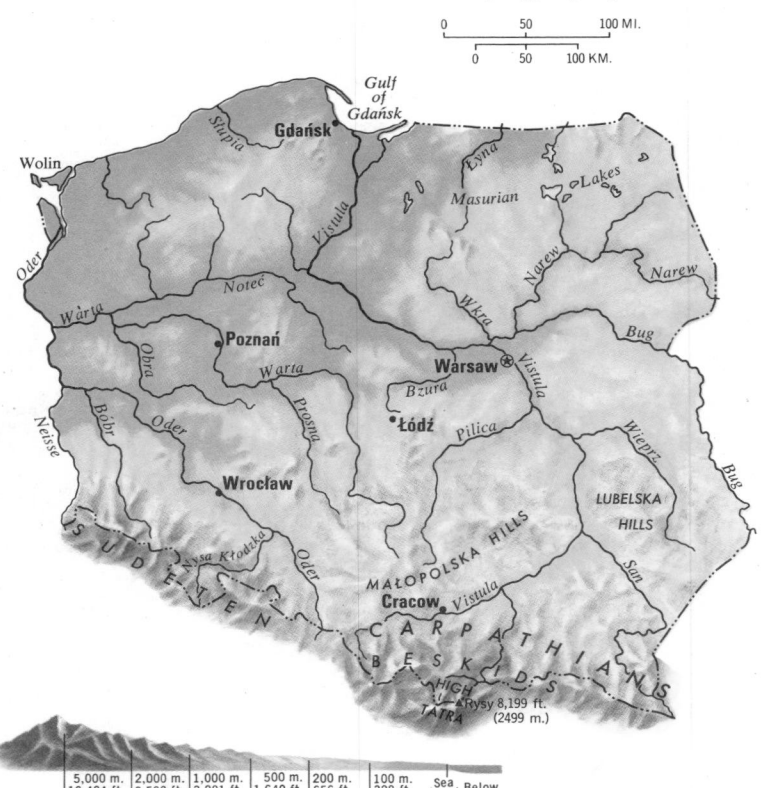

5,000 m. | 2,000 m. | 1,000 m. | 500 m. | 200 m. | 100 m. | Sea
16,404 ft. | 6,562 ft. | 3,281 ft. | 1,640 ft. | 656 ft. | 328 ft. | Level Below

Agriculture, Industry and Resources

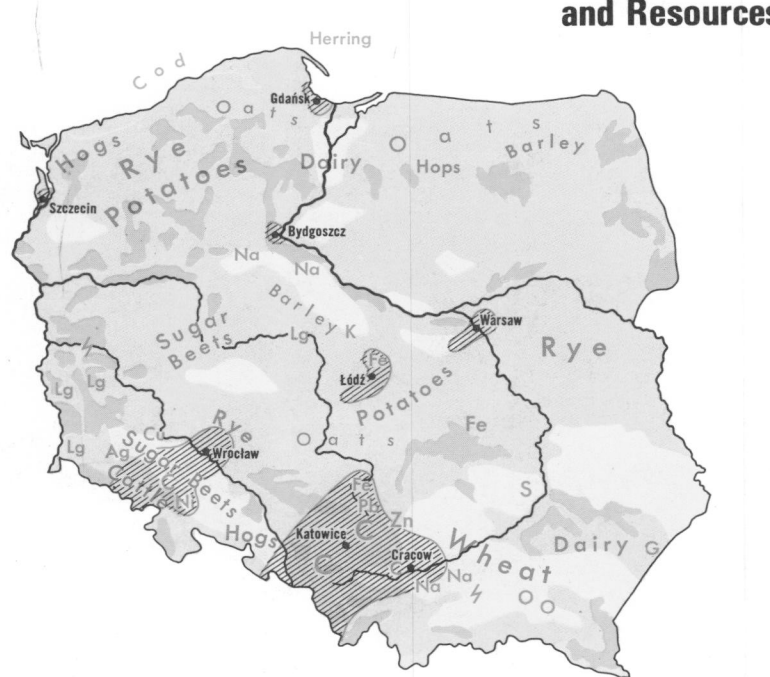

MAJOR MINERAL OCCURRENCES

Ag Silver   Na Salt
C Coal   Ni Nickel
Cu Copper   O Petroleum
Fe Iron Ore   Pb Lead
G Natural Gas S Sulfur
K Potash   Zn Zinc
Lg Lignite

⚡ Water Power
▨ Major Industrial Areas

DOMINANT LAND USE

☐ Cereals (chiefly wheat)

☐ Rye, Oats, Barley, Potatoes

☐ General Farming, Livestock

☐ Forests

PROVINCES

Biała Podlaska 283,200 ....F3
Białystok 613,800 ....F2
Bielsko 765,500 ....D4
Bydgoszcz 982,100 ....C2
Chełm 221,000 ....F3
Ciechanów 398,500 ....E2
Cracow 1,097,600 ....E4
Cracow (city) 651,300 ....E4
Częstochowa 723,200 ....D3
Elbląg 419,800 ....D1
Gdańsk 1,312,300 ....D1
Gorzów 428,700 ....C2
Jelenia Góra 483,400 ....B3
Kalisz 640,300 ....D3
Katowice 3,439,700 ....D3
Kielce 1,030,400 ....E3
Konin 423,700 ....D2
Koszalin 428,500 ....C1
Krosno 418,000 ....E4
Legnica 405,600 ....C4
Leszno 340,600 ....C3
Łódź 1,063,700 ....D3
Łódź (city) 777,800 ....D3
Łomża 320,600 ....F2
Lublin 875,300 ....F3

Nowy Sącz 600,300 ....E4
Olsztyn 654,400 ....E2
Opole 961,600 ....C3
Ostrołęka 360,700 ....E2
Piła 414,000 ....C2
Piotrków 581,900 ....D3
Płock 479,700 ....D2
Poznań 1,156,500 ....C2
Przemyśl 373,100 ....F4
Radom 674,400 ....E3
Rzeszów 602,200 ....F4
Siedlce 602,100 ....F2
Sieradz 388,000 ....D3
Skierniewice 388,300 ....E3
Słupsk 352,900 ....C1
Suwałki 412,700 ....F1
Szczecin 841,400 ....B2
Tarnobrzeg 532,200 ....E3
Tarnów 573,900 ....E4
Toruń 580,500 ....D2
Wałbrzych 709,600 ....C3
Warsaw 2,117,700 ....E2
Warsaw (city) 1,377,100 ....E2
Włocławek 402,000 ....D2
Wrocław 1,014,600 ....C3
Zamość 472,300 ....F3
Zielona Góra 575,000 ....B3

CITIES and TOWNS

Aleksandrów Kujawski 9,600 ....D2
Aleksandrów
  Łódzki 14,400 ....D3
Allenstein (Olsztyn) 94,119 .....E2
Andrespol 12,400 ....D4
Andrychów 14,300 ....D4
Augustów 19,784 ....F2
Auschwitz
  (Oświęcim) 39,600 ....D4
Bartoszyce 15,500 ....E1
Będzin 42,787 ....D3
Beuthen (Bytom) 186,993 ....A3
Biała Podlaska 26,100 ....F2
Białogard 20,500 ....C1
Białystok 166,619 ....F2
Bielawa 30,900 ....C3
Bielsk Podlaski 14,000 ....F2
Bielsko-Biała 105,601 ....D4
Biłgoraj 12,888 ....E4
Błonie 12,500 ....E2
Bochnia 14,500 ....E4
Bogatynia 11,800 ....B3
Boguszów-Gorce 11,900 ....B3
Bolesławiec 30,500 ....B3

Poland 1938

0   50   100
MILES

Poland 1945

0   50   100
MILES

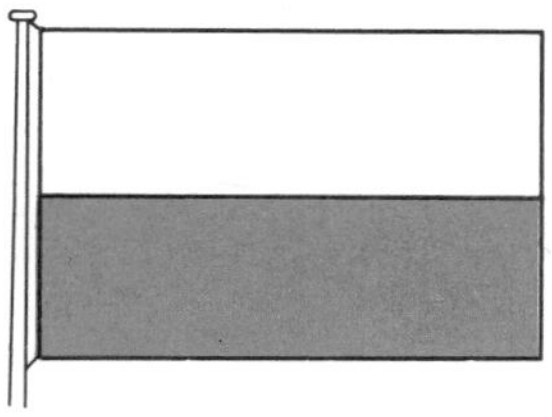

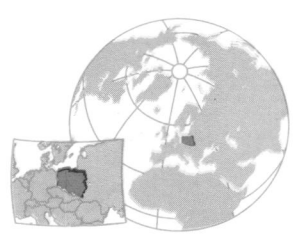

AREA 120,725 sq. mi. (312,678 sq. km.)
POPULATION 35,815,000
CAPITAL Warsaw
LARGEST CITY Warsaw
HIGHEST POINT Rysy 8,199 ft. (2,499 m.)
MONETARY UNIT zloty
MAJOR LANGUAGE Polish
MAJOR RELIGION Roman Catholicism

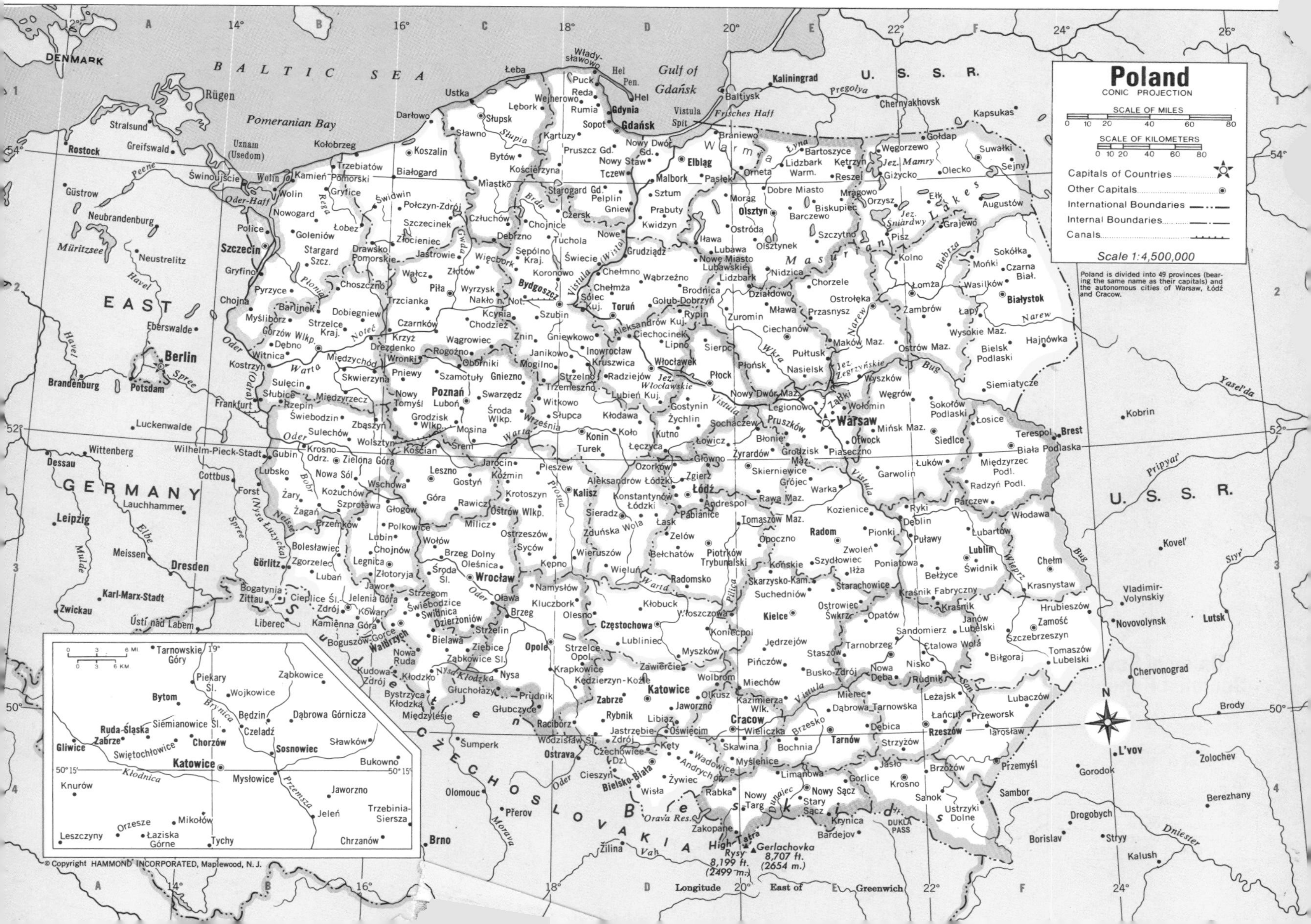

Braniewo 12,100	D1
Breslau (Wrocław) 461,900	C3
Brieg (Brzeg) 30,780	C3
Brodnica 17,300	D2
Brzeg 30,780	C3
Brzeg Dolny 10,800	C3
Brzesko 9,701	E3
Busko Zdrój 11,100	E3
Bydgoszcz 280,460	C2
Bytom 186,993	A3
Bytów 10,642	C1
Chełm 38,789	F3
Chełmno 17,906	D2
Chełmża 14,200	D2
Chodzież 14,100	C2
Chojnice 23,500	C2
Chojnów 11,000	B3
Chorzów 151,338	B4
Choszczno 9,800	B2
Chrzanów 29,300	D4
Ciechanów 28,500	E2
Cieplice Śląskie-Zdrój 15,400	B3
Cieszyn 25,234	E4
Cracow 651,300	E4
Czechowice-Dziedzice 25,400	D4
Czeladź 31,843	B3
Częstochowa 187,613	D3
Dąbrowa Górnicza 61,660	B4
Danzig (Gdańsk) 364,285	D1
Darłowo 11,200	C1
Dębica 22,900	E3
Dęblin 14,600	B2
Działdowo 10,100	E2
Dzierżoniów 32,800	C3
Ełk 27,188	F2
Elbląg 89,835	D1
Gdańsk 364,285	D1
Gdynia 190,125	D1
Gizycko 18,700	F1
Gleiwitz (Gliwice) 170,912	A4
Głogów (Głogau) 20,226	C3
Głowno 12,800	D2
Głubczyce 11,300	C4
Głuchołazy 13,200	C3
Gniezno 50,643	C2
Gorlice 15,200	E4
Góra Wielkopolski 74,267	C3
Gostyń 13,600	C3
Gostynin 12,000	D2
Grajewo 11,200	F2
Gryfice 13,200	B2
Grodzisk Mazowiecki 20,400	E2
Grójec 10,300	E2
Grudziądz 75,511	D2
Grünberg (Zielona Góra) 59,700	B3
Gryfów 13,200	B3
Hajnówka m4,345	F2
Hindenburg (Zabrze) 199,400	A4
Hirschberg (Jelenia Góra) 55,720	B3
Iława 16,400	D2
Inowrocław 54,817	D2

Jarocin 18,100	C3
Jarosław 29,000	F4
Jasło 17,025	E4
Jastrzębie Zdrój 34,400	D3
Jaworzno 63,271	B4
Jędrzejów 13,264	E3
Jelenia Góra 55,720	B3
Kalisz 81,227	D3
Kamienna Góra 21,000	C3
Kartuzy 10,558	C1
Katowice 303,264	B4
Kędzierzyn-Koźle 45,600	C3
Kępno 10,151	C3
Kętrzyn 19,300	E1
Kielce 125,952	E3
Kłobuck 12,600	D3
Kłodzko 26,000	C3
Kluczbork 15,600	D3
Knurów 28,400	A4
Kolberg (Kołobrzeg) 25,419	B1
Koło 13,100	D2
Kołobrzeg 25,419	B1
Konin 40,600	D2
Konstantynów Łódzki 12,800	D3
Kościan 18,700	C2
Kościerzyna 18,914	C1
Koslin (Koszalin) 64,414	C1
Kostrzyń 11,200	B2
Koszalin 64,414	C1
Kraków (Cracow) 651,300	E4
Krapkowice 13,800	D3
Kraśnik Fabryczny 14,600	F3
Krasnystaw 12,495	F3
Krosno 26,500	E4
Krotoszyn 21,900	C3
Krynica 10,200	E4
Küstrin 31,837	B2
Kutno 30,000	D2
Kwidzin 23,104	D2
Łańcut 12,049	F3
Landsberg (Gorzów Wielkopolski) 74,267	B2
Łaziska Górne 10,800	A4
Lębork 25,000	C1
Łęczyca 13,900	D2
Legionowo 20,800	E2
Legnica 75,843	C3
Leszno 33,128	C3
Libiąż 10,600	D3
Lidzbark Warmiński 12,900	E1
Liegnitz (Legnica) 75,843	C3
Lipno 10,900	D2
Lubań 17,200	B3
Lubartów 16,100	F3
Lubin 28,400	C3
Lublin 235,937	F3
Lubliniec 19,800	D3
Luboń 16,400	C2
Lubsko 14,999	B3
Łuków 15,500	F3

Międzyrzec Podlaski 13,500	F3
Międzyrzecz 14,900	B2
Mielec 26,800	E3
Mików 11,600	B4
Mińsk Mazowiecki 24,200	E2
Mława 20,007	E2
Mońki 9,560	F2
Morąg 9,681	D2
Mrągowo 13,400	E1
Myślenice 12,100	E4
Myszków 18,000	D3
Nakło nad Notecią 16,800	C2
Namysłów 11,076	C3
Neisse (Nysa) 31,837	C3
Nidzica 9,642	E2
Nisko 10,000	E3
Nowa Ruda 18,100	C3
Nowa Sól 33,300	B3
Nowy Dwór Mazowiecki 16,900	E2
Nowy Sącz 41,103	E4
Nowy Targ 21,900	E4
Nysa 31,837	C3
Oborniki 10,200	C2
Olawa 17,746	C3
Oleśnica 27,500	C3
Olkusz 16,900	D3
Olsztyn 94,119	E2
Opoczno 12,168	E3
Opole 86,510	C3
Oppeln 86,510	C3
Orzesze 9,600	A4
Ostróda 21,300	D2
Ostrołęka 21,981	E2
Ostrów Mazowiecka 15,000	E2
Ostrów Wielkopolski 49,530	C3
Ostrowiec Świętokrzyski 49,958	E3
Otwock 39,600	E2
Ozorków 18,200	D3
Pabianice 62,275	D3
Piekary Śląskie 36,300	B4
Piła 43,778	C2

Pionki 13,600	E3
Piotrków Trybunalski 59,683	D3
Pisz 11,100	F2
Pleszew 13,348	C3
Płock 71,727	D2
Płońsk 11,619	E2
Police 12,700	B2
Poznań 469,085	C2
Prudnik 20,300	C3
Pruszcz Gdański 13,000	D1
Pruszków 42,961	E2
Przasnysz 11,100	E2
Przemyśl 53,228	F4
Puck 9,500	D1
Puławy 34,800	F3
Pułtusk 12,800	E2
Rabka 10,700	D4
Raciborz 40,418	C3
Radom 158,640	E3
Radomsko 31,179	D3
Ratibor (Racibórz) 40,418	C3
Rawa Mazowiecka 9,800	E3
Rawicz 14,100	C3
Ruda Śląska 142,407	B4
Rumia 23,300	D1
Rybnik 43,415	D3
Rypin 10,029	D2
Rzeszów 82,192	F4
Sandomierz 16,800	E3
Sanok 21,600	F4
Schneidemühl (Piła) 36,600	C2
Schweidnitz (Świdnica) 47,542	C3
Siedlce 38,983	F2
Siemianowice 67,278	B4
Sieradz 38,500	D3
Sierpc 12,700	D2
Skarżysko-Kamienna 39,194	E3
Skawina 15,900	D4
Skierniewice 25,590	E2
Sławno 10,700	C1
Słubice 12,600	B2
Słupsk 68,311	C1

Sochaczew 20,500	E2
Sokółka 10,023	F2
Sokołów Podlaski 9,569	F2
Sopot 47,573	D1
Sosnowiec 144,652	B4
Śrem 15,600	C3
Środa Śląska 10,259	C3
Środa Wielkopolska 14,800	C2
Stalowa Wola 29,768	F3
Starachowice 42,807	E3
Stargard Szczeciński 44,400	B2
Stargard Gdański 33,400	D2
Stary Sącz 57,400	E4
Stettin (Szczecin) 337,294	B2
Stolp (Słupsk) 68,311	C1
Strzegom 14,000	C3
Strzelce Opolskie 14,700	D3
Strzelin 9,800	C3
Sulechów 10,200	B3
Suwałki 35,000	F1
Swarzędz 12,100	C2
Świdnica 47,542	C3
Świdnik 21,900	F3
Świdwin 12,500	B2
Świebodzice 18,500	C3
Świebodzin 14,900	B2
Świecie 17,900	D2
Świętochłowice 57,633	A4
Świnoujście (Swinemünde) 27,900	B1
Szamotuły 14,600	C2
Szczecin 337,204	B2
Szczecinek 28,600	C2
Szczytno 17,321	E2
Szprotawa 11,200	B3
Sztum 11,600	D2
Tarnów 85,514	E4
Tarnowskie Góry 34,200	A3
Tczew 40,794	D1
Tomaszów Lubelski 12,329	F3
Tomaszów Mazowiecki 54,911	E3
Toruń 129,152	D2
Trzcianka 10,900	C2
Trzebinia-Siersza	C4

Turek 18,500	D2
Tychy 71,384	B4
Ustka 9,900	C1
Wąbrzeźno 11,800	D2
Wadowice 11,700	D4
Wągrowiec 15,600	C2
Wałbrzych 125,048	C3
Wałcz 18,900	C2
Waldenburg (Wałbrzych) 125,048	C3
Warsaw (Warszawa) (cap.) 1,377,100	E2
Wejherowo 33,000	D1
Wieliczka 13,600	D4
Wieluń 14,800	D3
Wisła 9,800	D4
Włocławek 77,169	D2
Wodzisław Śląski 25,600	D3
Wolin 35,458	B2
Wołomin 24,000	E2
Wołów 10,500	C3
Wrocław 523,318	C3
Września 17,800	C2
Wschowa 10,000	C3
Ząbki 16,000	E2
Ząbkowice Śląskie 13,800	C3
Zabrze 197,214	A4
Żagań 21,400	B3
Zakopane 27,039	D4
Zambrów 14,082	F2
Zamość 34,734	F3
Żary 28,300	B3
Zawiercie 39,410	D3
Zduńska Wola 29,066	D3
Zgierz 42,838	D3
Zgorzelec 28,400	B3
Zielona Góra 73,156	B3
Złocieniec 10,100	C2
Złotoryja 12,200	C3
Złotów 11,600	C2
Żnin 9,600	C2
Żyrardów 33,196	E2
Żywiec 22,400	D4

OTHER FEATURES

Baltic (sea)	B1
Beskids (range)	D4
Brda (riv.)	C2
Brynica (riv.)	B4
Bug (riv.)	F2
Danzig (Gdańsk) (gulf)	D1
Dukla (pass)	E4
Dunajec (riv.)	E4
Gwda (riv.)	C2
Hel (pen.)	D1
High Tatra (range)	D4
Kłodnica (riv.)	A4
Łyna (riv.)	E1
Mamry, Jezioro (lake)	E1
Masurian (lkes)	E2
Narew (riv.)	E2
Neisse (riv.)	B3
Noteć (riv.)	B2
Nysa Kłodzka (riv.)	C3
Nysa Łużycka (Neisse) (riv.)	B3
Oder (riv.)	B2
Orava (riv.)	D4
Pilica (riv.)	E3
Pomeranian (bay)	B1
Prosna (riv.)	C3
Przemsza (riv.)	B4
Rysy (mt.)	D4
San (riv.)	F3
Słupia (riv.)	C1
Smardwy, Jezioro (lake)	D3
Sudeten (range)	B3
Uznam (Usedom) (isl.)	B1
Vistula (riv.)	D1
Warma (reg.)	E1
Warta (riv.)	C2
Wieprz (riv.)	F3
Wisła (Vistula) (riv.)	C2
Wkra (riv.)	E2
Wolin (Wollin) (isl.)	B1

UNION REPUBLICS

Armenian S.S.R. 3,031,000	E6
Azerbaidzhan S.S.R. 6,028,000	E5
Estonian S.S.R. 1,466,000	C4
Georgian S.S.R. 5,015,000	E5
Kazakh S.S.R. 14,684,000	G5
Kirgiz S.S.R. 3,529,000	H5
Latvian S.S.R. 2,521,000	C4
Lithuanian S.S.R. 3,398,000	C4
Moldavian S.S.R. 3,947,000	D4
Russian S.F.S.R. 137,551,000	D4
Tadzhik S.S.R. 3,801,000	H6
Turkmen S.S.R. 2,759,000	F6
Ukrainian S.S.R. 49,755,000	D5
Uzbek S.S.R. 15,391,000	G5
White Russian S.S.R. 9,560,000	C4

INTERNAL DIVISIONS

Abkhaz A.S.S.R. 505,000	E5
Adygey Aut. Obl. 405,000	D5
Adzhar A.S.S.R. 354,000	E5
Aginsk Buryat Aut. Okr. 69,000	M4
Bashkir A.S.S.R. 3,849,000	F4
Buryat A.S.S.R. 900,000	M4
Chechen-Ingush A.S.S.R. 1,154,000	E5
Chukchi Aut. Okr. 133,000	R3
Chuvash A.S.S.R. 1,292,000	E4
Dagestan A.S.S.R. 1,628,000	E5
Evenki Aut. Okr. 16,000	K3
Gorno-Altay Aut. Obl. 172,000	J4
Gorno-Badakhshan Aut. Obl. 127,000	H6
Jewish Aut. Obl. 190,000	O5
Kabardin-Balkar	

A.S.S.R. 674,000	E5
Kalmuck A.S.S.R. 294,000	E5
Karachay-Cherkess Aut. Obl. 368,000	E5
Karakalpak A.S.S.R. 904,000	G5
Karelian A.S.S.R. 736,000	D3
Khakass Aut. Obl. 500,000	J4
Khanty-Mansi Aut. Okr. 569,000	H3
Komi A.S.S.R. 1,119,000	F3
Komi-Permyak Aut. Okr. 173,000	F3
Koryak Aut. Okr. 34,000	R3
Mari A.S.S.R. 703,000	E4
Mordvinian A.S.S.R. 991,000	E4
Nagorno-Karabakh Aut. Obl. 161,000	E6
Nakhichevan' A.S.S.R. 239,000	E6
Nenets Aut. Okr. 47,000	F3
North Ossetian A.S.S.R. 597,000	E5
South Ossetian Aut. Obl. 98,000	E5
Tatar A.S.S.R. 3,436,000	F4
Taymyr Aut. Okr. 44,000	K2
Tuvinian A.S.S.R. 267,000	K4
Udmurt A.S.S.R. 1,494,000	F4
Ust'-Ordynskiy Buryat Aut. Okr. 133,000	L4
Yakut A.S.S.R. 839,000	N3
Yamal-Nenets Aut. Okr. 158,000	H3

CITIES and TOWNS

Abakan 128,000	K4
Abay 34,245	H5
Abaza 15,202	K4
Achinsk 117,000	K4

Agata	K3
Aginskoye 7,922	M4
Akmolinsk (Tselinograd) 234,000	H4
Aksay 10,010	F4
Aktas	G5
Aktash	K4
Aktyubinsk 191,000	F4
Aldan 17,689	N4
Aleksandrovsk-Sakhalinskiy 20,342	P5
Alekseyevka 18,041	E4
Aleysk 32,487	J4
Alga 12,000	F4
Aliskerovo	R3
Allakh-Yun'	N3
Alma-Ata 910,000	H5
Almaznyy	M3
Ambarchik	R3
Amderma	F3
Amursk 24,010	O4
Anadyr' 7,703	S3
Andizhan 230,000	H5
Andropov 239,000	E4
Angarsk 239,000	L4
Angren	H5
Anzhero-Sudzhensk	J4
Aral'sk 37,722	G5
Aral Sea 105,000	G5
Archangel (Arkhangel'sk) 385,000	E3
Arkalyk 15,108	G4
Armavir 162,000	E5
Arsen'yev 60,000	O5
Artem 69,000	O5
Artemovskiy	M4
Arys 26,414	G5
Arzamas 93,000	E4
Asbest 79,000	F4
Bira	O5

Ashkhabad 312,000	F6
Asino 29,395	J4
Astrakhan' 461,000	E5
Atbasar 37,228	H4
Atka	Q3
Ayaguz 35,827	H5
Ayan	O4
Aykhal	M3
Bagdarin	M4
Baku 1,022,000	F5
Baku* 1,550,000	E5
Balakovo 152,000	E4
Balashov 93,000	E4
Baley 27,215	M4
Balkhash 78,000	H5
Balykshi 22,397	F5
Bam	N4
Barabinsk 37,274	J4
Baranovichi 131,000	C4
Barnaul 533,000	J4
Batagay 10,000	O3
Batum 123,000	E5
Baykal	K4
Baykonur	G5
Bayram-Ali 31,987	G6
Belgorod 240,000	D4
Belogorsk 63,000	O4
Belomorsk 16,595	D3
Belovo 112,000	J4
Berdichev 80,000	C5
Berdsk 67,000	J4
Berezniki 185,000	F4
Berezovo 6,000	G3
Beringovskiy	T3
Bikin 17,473	O5

Birobidzhan 69,000	O5
Biruni	G5
Biysk 212,000	J4
Blagoveshchensk 172,000	N4
Bobruysk 192,000	C4
Bodaybo 19,000	M4
Borisoglebsk 68,000	E4
Borzya 27,815	M4
Bratsk 214,000	L4
Brest 177,000	C4
Brindakit	N4
Bryansk 394,000	D4
Bugul'ma 80,000	F4
Bukachacha 10,000	M4
Bukhara 185,000	G5
Bulun	N2
Buzuluk 76,000	F4
Chadan	K4
Chapayevsk 85,000	E4
Chara	M4
Chardzhou 140,000	G6
Chelsk 10,100	J4
Cheboksary 308,000	E4
Chegdomyn 16,499	O4
Chelkar 19,377	F5
Chelyabinsk 1,030,000	F4
Cheremkhovo 77,000	L4
Cherepovets 266,000	D4
Cherkessk 91,000	E5
Chernenko	K4
Chernigov 238,000	D4
Chernogorsk 71,000	K4
Chernovtsy 219,000	C5
Chernyshevsk 10,000	M4
Cherskiy	Q3
Chimbay 18,899	G5
Chimkent 322,000	H5
Chirchik 132,000	H5

Chita 303,000	M4
Chokurdakh	P2
Chumikan	O4
Dal'negorsk 33,506	O5
Dal'nerechensk 28,224	O5
Daugavpils 116,000	C4
Denau	G6
Dikson	J2
Dimitrovgrad 106,000	F4
Dnepropetrovsk 1,066,000	D5
Donetsk 1,021,000	D5
Drogobych 66,000	C5
Druzhba	G5
Druzhina	P3
Dudinka 19,701	J3
Dushanbe 494,000	G6
Dzerzhinsk 257,000	E4
Dzhalal-Abad 55,000	H5
Dzhalinda	N4
Dzhambul 264,000	H5
Dzhelinda	M2
Dzhetygara 32,169	G4
Dzhezkazgan 89,000	G5
Dzhusaly 20,658	G5
Egvekinot	O4
Ekibastuz 66,000	H4
Ekimchan	O4
El'dikan	N3
Elista 70,000	E5
Emba 17,820	F5
Engel's 161,000	E4
Erevan 1,019,000	E6
Evensk	Q3
Fergana 176,000	H5
Fort-Shevchenko 12,000	E5
Frolovo 33,398	E4
Frunze 533,000	H5

Gasan-Kuli	F6
Gomel' 383,000	D4
Gor'kiy 1,344,000	E4
Gorno-Altaysk 34,413	J4
Gornyak 16,643	J4
Grodno 195,000	C4
Groznyy 375,000	E5
Gubakha 33,243	F4
Gulistan 30,879	G5
Gur'yev 131,000	F5
Gusinoozersk 10,000	L4
Gyda	H2
Igarka 15,624	J3
Igrim	G3
Ilanskiy 22,852	K4
Indiga	F3
Inta 51,000	G3
Iolotan' 10,000	G6
Irkutsk 550,000	L4
Ishim 63,000	G4
Isil'kul' 25,958	H4
Iul'tin	T3
Ivano-Frankovsk 150,000	C5
Ivdel 15,308	G3
Izhevsk (Ustinov) 549,000	F4
Izmail 83,000	D5
Kachug	L4
Kagan 34,117	G6
Kalachinsk 20,809	H4
Kalakan	M4
Kalimykovo	F5
Kalinin 412,000	D4
Kaliningrad 355,000	B4
Kaluga 265,000	D4
Kamen'-na-Obi 35,604	J4

ADMINISTRATIVE DIVISIONS NOT NAMED ON MAP

Division	Ref.		Division	Ref.
1. Abkhaz A.S.S.R.	E5		13. Khakass Aut. Oblast	J4
2. Adygey Aut. Oblast	D5		14. Komi-Permyak Aut. Okrug	F4
3. Adzhar A.S.S.R.	E5		15. Mari A.S.S.R.	E4
4. Aginsk Buryat Autonomous Okrug	M4		16. Mordvinian A.S.S.R.	E4
5. Chechen-Ingush A.S.S.R.	E5		17. Nagorno-Karabakh Aut. Oblast	E6
6. Chuvash A.S.S.R.	E4		18. Nakhichevan' A.S.S.R.	E6
7. Gorno-Altay Aut. Oblast	J4		19. North Ossetian A.S.S.R.	E5
8. Gorno-Badakhshan Aut.Oblast	H6		20. South Ossetian Aut. Oblast	E5
9. Jewish Aut. Oblast	O5		21. Tatar A.S.S.R.	F4
10. Kabardin-Balkar A.S.S.R.	E5		22. Tuvinian A.S.S.R.	K4
11. Karachay-Cherkess Aut.Oblast	E5		23. Udmurt A.S.S.R.	F4
12. Karakalpak A.S.S.R.	G5		24. Ust-Ordynsk Buryat Autonomous Okrug	L4

Union of Soviet Socialist Republics

CONIC PROJECTION

SCALE OF MILES

SCALE OF KILOMETERS

Capitals — Boundaries
- National
- Union Republic
- A.S.S.R.
- Autonomous Oblast
- Autonomous Okrug

Scale 1:30,400,000

AREA 8,649,490 sq. mi. (22,402,179 sq. km.)
POPULATION 262,436,227
CAPITAL Moscow
LARGEST CITY Moscow
HIGHEST POINT Communism Peak 24,599 ft. (7,498 m.)
MONETARY UNIT ruble
MAJOR LANGUAGES Russian, Ukrainian, White Russian, Uzbek, Azerbaidzhani, Tatar, Georgian, Lithuanian, Armenian, Yiddish, Latvian, Mordvinian, Kirgiz, Tadzhik, Estonian, Kazakh, Moldavian (Romanian), German, Chuvash, Turkmenian, Bashkir
MAJOR RELIGIONS Eastern (Russian) Orthodoxy, Islam, Judaism, Protestantism (Baltic States)

Kamenskoye	R3	Kavalerovo 16.415	O5
Kamensk-Ural'skiy 187.000	G4	Kazan' 993.000	F4
Kamyshin 112.000	E4	Kem' 21.025	D3
Kandalaksha 42.656	C3	Kemerovo 471.000	J4
Kansk 101.000	K4	Kentau 52.000	G5
Kapchagay	H5	Kerki 10.000	G6
Kara	G3	Khabarovsk 528.000	O5
Karaganda 572.000	H5	Khandyga	O3
Karasuk 22.637	H4	Khanty-Mansiysk 24.754	H3
Karatau 26.962	H5	Khar'kov 1.444.000	D4
Karazhal 17.702	H5	Khatanga	L2
Kargasok	J4	Kherson 319.000	D5
Karpinsk	F4	Khilok 17.000	M4
Karshi 108.000	G6	Khiva 24.139	F5
Kartaly 42.801	G4	Khodzheyli 36.435	F5
Katangli	P4	Kholmsk 37.412	P5
Kattakurgan 53.000	G5	Khorog 12.295	H6
Kaunas 370.000	C4	Kiev 2.144.000	D4

UNION REPUBLICS

	AREA (sq. mi.)	AREA (sq. km.)	POPULATION	CAPITAL and LARGEST CITY
RUSSIAN S.F.S.R.	6,592,812	17,075,400	137,551,000	Moscow 7,831,000
KAZAKH S.S.R.	1,048,300	2,715,100	14,684,000	Alma-Ata 910,000
UKRAINIAN S.S.R.	233,089	603,700	49,755,000	Kiev 2,144,000
TURKMEN S.S.R.	188,455	488,100	2,759,000	Ashkhabad 312,000
UZBEK S.S.R.	173,591	449,600	15,391,000	Tashkent 1,780,000
WHITE RUSSIAN S.S.R.	80,154	207,600	9,560,000	Minsk 1,262,000
KIRGIZ S.S.R.	76,641	198,500	3,529,000	Frunze 533,000
TADZHIK S.S.R.	55,251	143,100	3,801,000	Dushanbe 494,000
AZERBAIDZHAN S.S.R.	33,436	86,600	6,028,000	Baku 1,022,000
GEORGIAN S.S.R.	26,911	69,700	5,015,000	Tbilisi 1,066,000
LITHUANIAN S.S.R.	25,174	65,200	3,398,000	Vilna 481,000
LATVIAN S.S.R.	24,595	63,700	2,521,000	Riga 835,000
ESTONIAN S.S.R.	17,413	45,100	1,466,000	Tallinn 430,000
MOLDAVIAN S.S.R.	13,012	33,700	3,947,000	Kishinev 503,000
ARMENIAN S.S.R.	11,506	29,800	3,031,000	Erivan 1,019,000

Kirensk 10.000	L4	Krasnokamsk 56.000	F4	Leninakan 207.000	E5	Miass 150.000	G4	Nazyvayevsk 15,792	H4
Kirov 390.000	E4	Krasnotur'insk 61.000	G3	Leningrad 4.073.000	D4	Michurinsk 101.000	E4	Nebit-Dag 71,000	F6
Kirovabad 232.000	E5	Krasnoural'sk 39.743	G4	Leningrad* 4.588.000	D4	Millerovo 34.627	E4	Nefteyugansk 52.000	H3
Kirovograd 237.000	D5	Krasnovodsk 53.000	F5	Leninogorsk 54.000	J5	Minsk 1.262.000	C4	Nel'kan	O4
Kirovskiy	H5	Krasnoyarsk 796.000	K4	Leninsk	G5	Minsk* 1.276.000	C4	Nepa	L4
Kiselevsk 122.000	J4	Kremenchug 210.000	D5	Leninsk-Kuznetskiy 132.000	J4	Minusinsk 56.000	K4	Neryungri	N4
Kishinev 503.000	C5	Krivoy Rog 650.000	D5	Leninskoye	O5	Mirnyy 23.826	M3	Nevel'sk 20,726	P5
Kizel 46.264	F4	Kudymkar 26.350	F4	Lenkoran' 35.505	E6	Mogilev 290.000	D4	Nikolayev 440,000	D5
Kizyl-Arvat 21.671	F6	Kul'sary 16.427	F5	Lensk 16.758	M3	Mogocha 17.884	N4	Nikolayevsk-na-Amure 30,082	P4
Klaipeda 176.000	B4	Kulunda 15.264	H4	Lesosibirsk	K4	Molodechno 73.000	C4	Nikol'skoye	R4
Kokand 153.000	H5	Kulyab 55.000	H6	Lesozavodsk 34.957	O5	Monchegorsk 51.000	C3	Nizhneudinsk 39,743	K4
Kokchetav 103.000	H4	Kum-Dag 10.000	F6	Liepaja 108.000	B4	Moscow (cap.) 7.831.000	D4	Nizhnevartovsk 109.000	H3
Kolomna 147.000	D4	Kungur 80.000	F4	Lipetsk 396.000	E4	Moscow* 8.011.000	D4	Nizhneyansk	O3
Kolpashevo 24.911	J4	Kupino 20.799	H4	Liski 31.905	E4	Motygino 10.000	K4	Nizhniy Tagil 398.000	G4
Komsomol'sk 15.385	G4	Kurgan 310.000	G4	Lutsk 137.000	C4	Mozyr' 73.000	C4	Nordvik-Ugol'naya	M2
Komsomol'sk-na-Amure 264.000	O4	Kurgan-Tyube 34.620	G6	L'vov 667.000	C4	Murgab	H6	Noril'sk 180.000	J3
Kondopoga 27.908	D3	Kursk 375.000	E4	Lys'va 75.000	F4	Murmansk 381.000	D3	Novaya Kazanka	F5
Kopeysk 146.000	G4	Kushka	G6	Magadan 121.000	P4	Muynak 12.000	F5	Novgorod 186.000	D4
Korf	R3	Kustanay 165.000	G4	Magdagachi 15.059	N4	Mys Shmidta	T3	Novoaltaysk 34.815	J4
Korsakov 38.210	P5	Kutaisi 194.000	E5	Magnitogorsk 406.000	G4	Nadym	H3	Novokazalinsk 34.815	G5
Koslan	E3	Kuybyshev 1.216.000	F4	Makhachkala 251.000	E5	Nagornyy	N4	Novokuznetsk 541.000	J4
Kostroma 255.000	E4	Kuybyshev 40.166	H4	Mak.sk 22.850	H4	Nakhichevan' 33.279	E6	Novomoskovsk 147.000	E4
Kotlas 61.000	E3	Kyakhta 15.316	L4	Mama	M4	Nakhodka 133.000	O5	Novorossiysk 159.000	D5
Kovel 33.351	C4	Kyusyur	N2	Markovo	S3	Nal'chik 207.000	E5	Novosibirsk 1.312.000	J4
Kovrov 143.000	E4	Kyzyl 66.000	K4	Mary (Merv) 74.000	G6	Namangan 227.000	H5	Novozybkov 34.433	D4
Kozhevnikovo	L2	Kyzyl-Orda 156.000	G5	Maykop 128.000	D5	Naminga	M4	Novyy Port	G3
Krasino	F2	Labytnangi	G3	Mednogorsk 38.024	F4	Nar'yan-Mar 16,864	F3	Novyy Uzen' 18,073	F5
Krasnodar 560.000	E5	Lebedinyy	N4	Medvezh'yegorsk 17.465	D3	Naryn 21.098	H5	Novyy Urengoy	H3
Krasnokamensk 51.000	M4	Leninabad 130.000	G5	Mezen'	E3	Navoi 84.000	G6	Nukus 109.000	G5

Topography

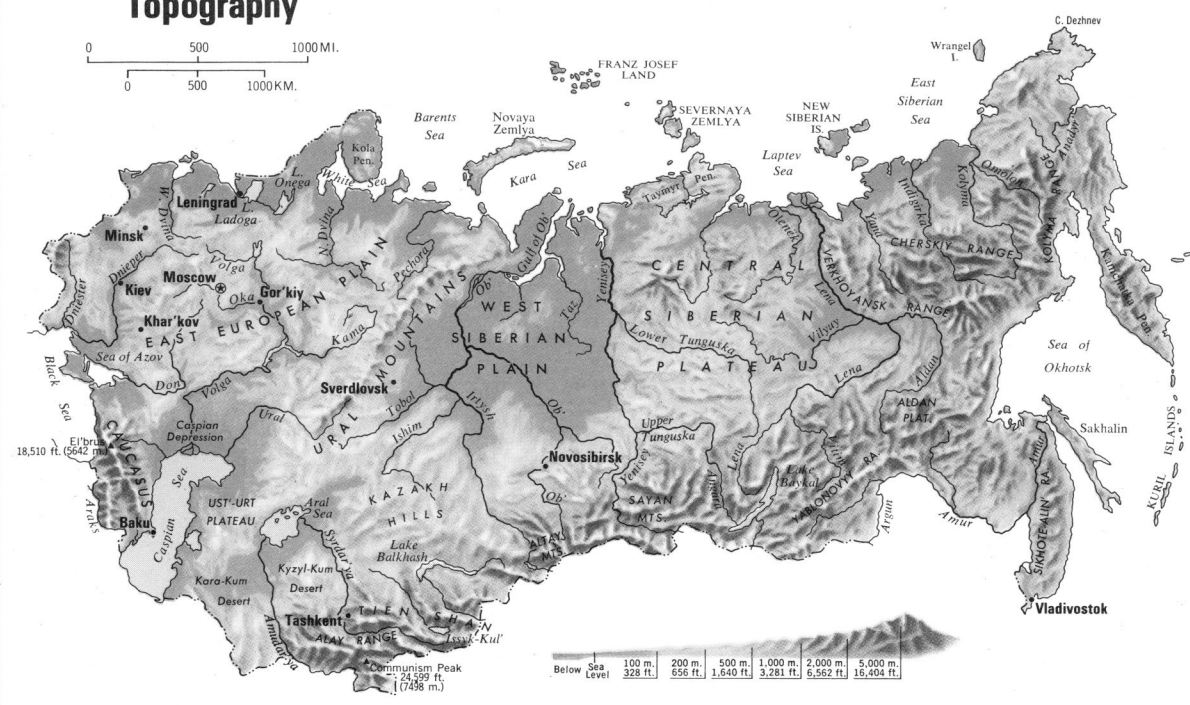

0	500	1000 MI.
0	500	1000 KM.

Below Sea Level	100 m. 328 ft.	200 m. 656 ft.	500 m. 1,640 ft.	1,000 m. 3,281 ft.	2,000 m. 6,562 ft.	5,000 m. 16,404 ft.

Communism Peak 24,599 ft. (7498 m.)

(continued)

Nyandoma 23,366 E3
Nyurba M3
Obluch'ye 17,000 N5
Odessa 1,046,000 D5
Okha 30,890 P4
Okhotsk P4
Olëkminsk N3
Olënek M3
Omsk 1,014,000 H4
Omsukchan Q3
Omutninsk 28,777 F4
Onega 25,047 D3
Ordzhonikidze 279,000 E5
Orel 305,000 D4
Orenburg 459,000 F4
Orotukan Q3
Orsk 247,000 F4
Osh 169,000 H5
Ostrogozhsk 29,921 D4
Oymyakon O3
Ozernovskiy Q4
Palana 2,735 R4
Panfilov 19,173 H5
Pärnu 51,000 C4
Partizansk 48,345 O5
Pavlodar 273,000 H4
Pechenga D2
Pechora 56,000 F3
Peleduy M4
Penza 483,000 E4
Perkatkin T2
Perm' 999,000 F4
Pervoural'sk 129,000 F4
Petropavlovsk 207,000 G4
Petropavlovsk-Kamchatskiy 215,000 R4
Petrovsk-Zabaykal'skiy 28,313 L4
Petrozavodsk 234,000 D3
Pevek S3
Pikol'skiy 32,862 G5
Pinsk 90,000 C4
Plastun O5
Podol'sk 202,000 D4
Pokrovsk N3
Poligus K3
Poltava 279,000 D5
Polyarnyy 15,321 D3
Ponoy E3
Poronaysk 23,610 P5
Prikumsk 35,768 E5
Progress 10,000 O5
Prokop'yevsk 266,000 J4
Provideniya T3
Przheval'sk 51,000 H5
Pskov 176,000 C4
Pushkin 90,000 C4
Raychikhinsk 25,157 N5
Riga 835,000 C4
Rostov-na-Donau 934,000 E5
Rovno 179,000 C4
Rubtsovsk 157,000 J4
Ruch'i E3
Rudnyy 110,000 G4
Ryazan' 453,000 E4
Rybach'ye H5
Rybinsk (Andropov) 239,000 D4
Rzhev 69,000 D4
Saksaul'sky F5
Salekhard 21,929 G3
Sal'sk 57,000 E5
Samagaltay K4
Samarkand 477,000 G6
Sangar N3
Saran' 55,000 H5
Saransk 263,000 E4
Sarapul 107,000 F4
Saratov 856,000 E4
Sarkand 18,296 J5
Segezha 28,810 D3
Semipalatinsk 283,000 H4
Serakhs G6
Serov 101,000 G4
Serpukhov 140,000 D4
Sevastopol' 301,000 D5
Severobaykal'sk M4
Severodvinsk 197,000 E3
Severo-Kuril'sk 8,000 Q4
Severoural'sk 29,880 G3
Severo-Yeniseysk K3
Shadrinsk 82,000 G4
Shakhtinsk 50,000 H5
Shakhty 209,000 E5
Shar'ya 25,788 E4
Shchuchinsk 40,432 H4
Shenkursk E3
Shevchenko 111,000 F5
Shilka 16,065 M4
Shimanovsk 16,880 N4
Shushenskoye 10,000 K4
Šiauliai 118,000 C4
Siktyakh N3
Simferopol' 302,000 D5
Skovorodino 10,000 N4
Slavgorod 32,908 H4
Slobodskoy 34,374 E4
Slyudyanka 20,639 L4
Smolensk 276,000 D4
Snezhnogorsk J3
Sochi 287,000 D5
Sokol 48,253 E4
Solikamsk 101,000 F3
Sortavala 22,188 C3
Sosnogorsk 24,688 F3
Sosnovo-Ozerskoye M4
Sovetskaya Gavan' 28,455 P5
Spassk-Dal'niy 53,000 O5
Srednekolymsk Q3
Sretensk 16,000 M4
Stalingrad (Volgograd) 929,000 E5
Stavropol' 258,000 E5
Stepanakert 30,293 E6
Sterlitamak 220,000 F4
Strezhevoy H3
Sukhana M3
Sukhumi 114,000 D5
Sumy 228,000 D4
Suntar M3
Surgut 107,000 H3
Susuman 12,000 P3
Sverdlovsk 1,211,000 F4
Svobodnyy 75,000 N4
Syktyvkar 171,000 F3
Syzran' 178,000 E4
Taganrog 276,000 D5
Takhiatash F5
Takhta-Bazar G6
Taksimo M4
Taldy-Kurgan 88,000 H5
Talgar 31,273 H5
Tallinn 430,000 C4
Tambey G2
Tambov 270,000 E4
Tara 22,358 H4
Tarko-Sale H3
Tartu 105,000 C4
Tashauz 84,000 F5
Tashkent 1,780,000 G5
Tatarsk 29,589 H4
Tavda G4
Tayshet 34,232 K4
Tazovskiy J3
Tbilisi 1,066,000 E5
Tedzhen 25,708 F6
Tekeli 29,846 H5
Temirtau 213,000 H4
Termez 57,000 G6
Ternopol' 144,000 C5
Tiksi N2
Tobol'sk 62,000 G4
Togliatti (Tol'yatti) 502,000 F4
Tokmak 59,000 H5
Tommot 8,000 N4
Tomsk 421,000 J4
Tot'ma E4
Troitsk 88,000 G4
Tselinograd 234,000 H4
Tskhinvali 30,311 E5
Tula 514,000 D4
Tulun 52,000 L4
Tura 3,528 L3
Turan K4
Turgay G5
Turkestan 67,000 G5
Tynda N4
Tyumen' 359,000 G4
Uelen T3
Ufa 969,000 F4
Uglegorsk 17,921 P5
Ukhta 87,000 F3
Ulan-Ude 300,000 L4
Ul'yanovsk 464,000 E4
Ural'sk 167,000 F4
Uray 17,385 G3
Urgench 100,000 G5
Ushtobe 24,484 H5
Usol'ye-Sibirskoye 103,000 L4
Ussuriysk 147,000 O5
Ust'-Ilimsk 69,000 L4
Ustinov 549,000 F4
Ust'-Kamchatsk 10,000 R4
Ust'-Kamenogorsk 274,000 J5
Ust'-Kut 50,000 L4
Ust'-Kuyga O3
Ust'-Maya O3
Ust'-Nera P3
Ust'-Omchug P3
Ust'-Olenëk M2
Ust'-Ordynskiy 10,693 L4
Ust'-Port J2
Vanavara L3
Vanino 15,401 P5
Velikiye Luki 102,000 D4
Velikiy Ustyug 36,737 E3
Vel'sk 21,899 E3
Ventspils 40,467 B4
Verkhnevilyuysk N3
Verkniy At-Uryakh Q3
Verkhoyansk 2,000 N3
Vilna (Vilnius) 481,000 C4
Vilyuysk N3
Vinnitsa 314,000 C5
Vitebsk 297,000 D4
Vitimskiy M4
Vladimir 296,000 D4
Vladivostok 550,000 O5
Volgograd 929,000 E5
Volochanka K2
Vologda 237,000 E4
Vorkuta 100,000 G3
Voronezh 783,000 E4
Voroshilovgrad 463,000 E5
Vostochnyy O5
Votkinsk 90,000 F4
Voy-Vozh 10,000 F3
Vyazemskiy 18,365 O5
Vyborg 76,000 C3
Vyshniy Volochek 72,000 D4
Yakutsk 152,000 N3
Yalutorsk 25,426 G4
Yamsk Q4
Yaroslavl' 597,000 D4
Yartsevo J4
Yelets 112,000 D4
Yelizovo 10,000 Q4
Yeniseysk 19,880 K4
Yermak 28,133 H4
Yermentau 15,276 H4
Yesil' 15,000 G4
Yessey L3
Yoshkar-Ola 201,000 E4
Yurga 78,000 J4
Yuzhno-Sakhalinsk 140,000 P5
Zabaykal'sk M5
Zakamensk 10,000 L4
Zaozernyy 27,216 K4
Zaporozh'ye 781,000 D5
Zarafshan G5
Zavitinsk 19,009 O4
Zaysan 10,000 J5
Zeya 16,684 N4
Zhatay O3
Zhdanov 503,000 D5
Zheleznogorsk-Ilimskiy 22,179 L4
Zhigalovo L4
Zhigansk N3
Zhitomir 244,000 C4
Zima 41,567 L4
Zlatoust 198,000 F4
Zyryanka Q3

OTHER FEATURES

Alakol' (lake) J5
Alazeya (riv.) Q3
Aldan (plat.) N4
Aldan (riv.) O3
Alexandra Land (isl.) E1
Altay (mts.) J5
Amga (riv.) O3
Amgun' (riv.) O4
Amudar'ya (riv.) G5
Amur (riv.) O4
Anabar (riv.) M2
Anadyr' (gulf) T3
Anadyr' (range) S3
Anadyr' (riv.) S3
Angara (riv.) K4
Aral (sea) F5
Arctic Ocean K1
Argun (riv.) M4
Arkticheskiy Institut (isls.) H2
Atrek (riv.) F6
Ayon (isl.) R2
Azov (sea) D5
Balkhash (lake) H5
Baltic (sea) B4
Barents (sea) D2
Baykal (lake) L4
Baykal (mts.) L4
Beloye (lake) D3
Belyy (isl.) G2
Bering (isl.) R4
Bering (sea) S4
Bering (strait) U3
Bet-Pak-Dala (des.) H5
Black (sea) D5
Bol'shevik (isl.) K2
Bol'shoy Lyakhovskiy (isl.) P2
Bolvanskiy Nos (cape) G2
Bratsk (res.) L4
Caspian (sea) F6
Caucasus (mts.) E5
Chelyuskin (cape) M2
Cherskiy (range) P3
Chu (riv.) H5
Chukchi (pen.) T3
Chukchi (sea) T2
Chulym (riv.) J4
Chuna (riv.) K4
Chunya (riv.) K3
Communism (peak) H6
Crimea (pen.) D5
Dezhnev (cape) T3
Dmitriya Lapteva (strait) O2
Dnieper (riv.) D5
Dniester (riv.) C5
Don (riv.) E5
Donets (riv.) D5
Dulgalakh (riv.) O3
Dvina, Northern (riv.) E3
Dvina, Western (riv.) C4
Dzhugdzhur (range) O4
East Siberian (sea) S2
Emba (riv.) F5
Etorofu (Iturup) (isl.) P5
Faddeyevskiy (isl.) P2
Finland (gulf) C4
Franz Josef Land (isls.) F1
George Land (isl.) E1
Gizhiga (bay) Q3
Govena (cape) R4
Graham Bell (isl.) G1
Gyda (pen.) H2
Gydan (Kolyma) (range) Q3
Hiiumaa (isl.) C4
Ili (riv.) H5
Imandra (lake) D3
Indigirka (riv.) P3
Irtysh (riv.) H4
Ishim (riv.) G4
Issyk-Kul' (lake) H5
Japan (sea) O6
Kakhovka (res.) D5
Kamchatka (pen.) Q4
Kanin (pen.) E3
Kanin Nos (cape) E3
Kara (sea) G2
Kara-Bogaz-Gol (gulf) F5
Karaginskiy (isl.) R4
Kara-Kum (canal) F6
Kara-Kum (des.) F5
Karskiye Vorota (strait) F2
Khanka (lake) O5
Kharasavey (cape) G2
Kheta (riv.) K2
Klyuchevskaya Sopka (vol.) Q4
Kola (pen.) D3
Kolguyev (isl.) E3
Kolyma (range) Q3
Kolyma (riv.) Q3
Komandorskiye (isls.) R4
Komsomolets (isl.) L1
Koni (pen.) Q4
Koryak (range) S3
Kotel'nyy (isl.) O2
Kotuy (riv.) L3
Kuma (riv.) E5
Kura (riv.) E6
Kuril (isls.) P5
Kuybyshev (res.) F4
Kyzyl-Kum (des.) G5
Ladoga (lake) D3
La Pérouse (str.) P5
Laptev (sea) N2
Lena (riv.) N3
Little Yenisey (riv.) K4
Long (str.) S2
Lopatka (cape) Q4
Lower Tunguska (riv.) K3
Lyatkhovskiye (isls.) O2
Mangyshlak (pen.) F5
Markha (riv.) M3
Matochkin Shar (str.) F2
Maya (riv.) O4
Mezen' (riv.) E3
Murgab (riv.) G6
Nadym (riv.) H3
Narodnaya (mt.) G3
Navarin (cape) T3
New Siberian (isls.) P2
Northern Dvina (riv.) E3
Novaya Sibir' (isl.) Q2
Novaya Zemlya (isls.) F2
Ob' (gulf) H3
Ob' (riv.) G3
October Revolution (isl.) L2
Oka (riv.) L4
Okhotsk (sea) P4
Olëkma (riv.) N4
Olënek (bay) N2
Olënek (riv.) M3
Oloy (range) R3
Olyutorskiy (cape) S4
Omolon (riv.) Q3
Omoloy (riv.) O3
Onega (lake) D3
Onega (riv.) D3
Ozernoy (cape) R4
Paramushir (isl.) Q4
Pechora (riv.) F3
Peipus (lake) C4
Penzhina (bay) R3
Pioner (isl.) J2
Pobeda (peak) J5
Pur (riv.) H3
Pyasina (riv.) J2
Riga (gulf) C4
Rybachiy (pen.) D2
Rybinsk (res.) D4
Saaremaa (isl.) B4
Sakhalin (gulf) P4
Sakhalin (isl.) P4
Sannikova (str.) O2
Sary Su (riv.) H5
Sayan (mts.) K4
Selemdzha (riv.) O4
Sergeya Kirova (isls.) J2
Severnaya Zemlya (isls.) L1
Shantar (isls.) O4
Shelagskiy (cape) R2
Shelekhov (gulf) Q4
Siberia (reg.) 38,524,000 M3
Sikhote-Alin' (range) O5
Stanovoy (range) N4
Stony Tunguska (riv.) K3
Syrdar'ya (riv.) G5
Tannu-Ola (range) K5
Tatar (str.) P4
Taymyr (lake) K2
Taymyr (pen.) L2
Taymyr (riv.) K2
Taz (riv.) J3
Tengiz (lake) G4
Terpeniye (cape) P5
Tobol (riv.) G4
Tsimlyansk (res.) E5
Tym (riv.) J3
Tyung (riv.) M3
Uda (riv.) Q4
Ulutau (mts.) G5
Ural (mts.) F4
Ural (riv.) F5
Urup (isl.) Q5
Ussuri (riv.) O5
Ust'-Urt (plat.) F5
Vakh (riv.) J3
Velikaya (riv.) S3
Verkhoyansk (range) N3
Vil'kitskogo (str.) L2
Vilyuy (range) M3
Vilyuy (res.) M3
Vilyuy (riv.) L3
Vitim (riv.) M4
Volga (riv.) E5
Western Dvina (riv.) C4
White (sea) D3
Wiese (isl.) H2
Wilczek Land (isl.) G1
Wrangel (isl.) T2
Yablonovyy (range) M4
Yamal (pen.) G2
Yana (riv.) O3
Yelizavety (cape) P4
Yenisey (riv.) J3
Zaysan (lake) J5
Zeya (riv.) N4
Zhelaniye (cape) H2

*City and suburbs.

Agriculture, Industry and Resources

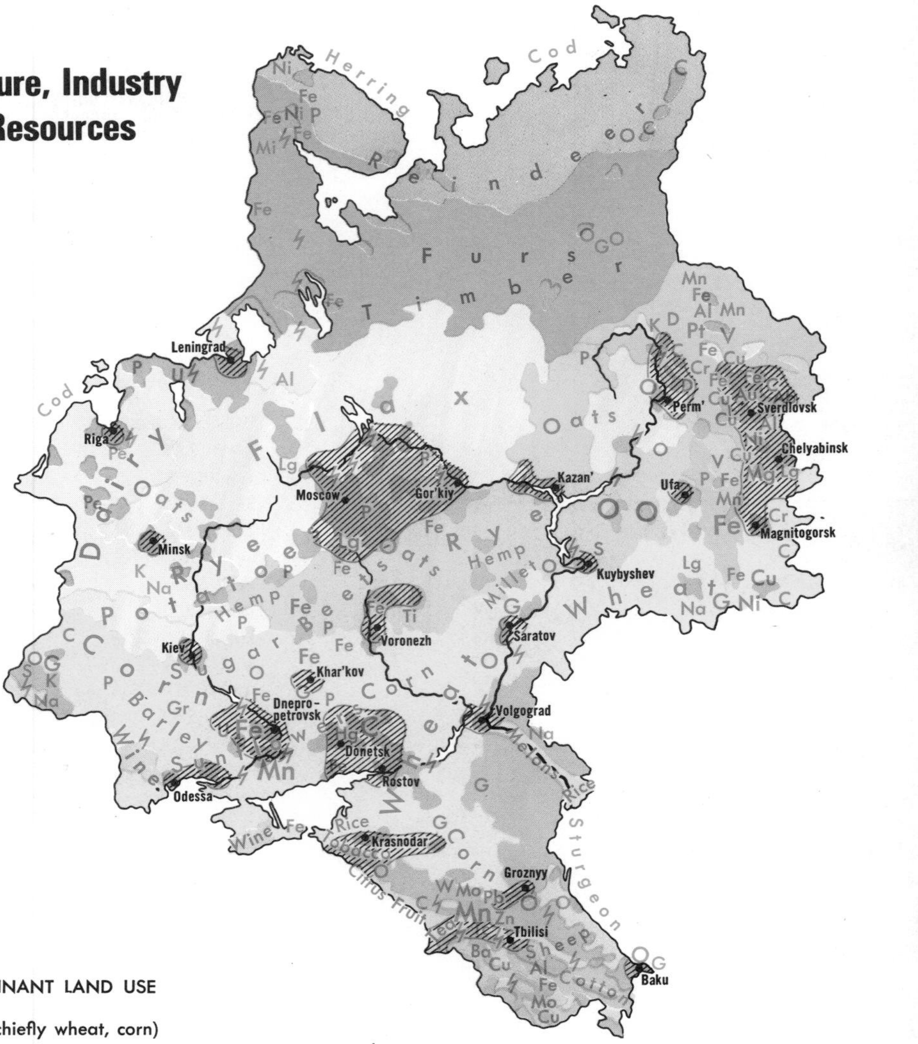

DOMINANT LAND USE

- Cereals (chiefly wheat, corn)
- Cereals (chiefly wheat, rye, oats)
- Dairy, Hogs, Livestock
- Livestock, Dairy
- Pasture Livestock
- Truck Farming, Potatoes, Vegetables, Dairy
- Flax, Dairy, Potatoes
- Cotton
- Vineyards, Orchards, Horticulture
- Sheep Herding, Limited Agriculture
- Forests
- Nonagricultural Land

MAJOR MINERAL OCCURRENCES

Ab	Asbestos	Hg	Mercury	Pb	Lead
Al	Bauxite	K	Potash	Pe	Peat
Au	Gold	Lg	Lignite	Pt	Platinum
Ba	Barite	Mg	Magnesium	S	Sulfur, Pyrites
C	Coal	Mi	Mica	Tc	Talc
Cr	Chromium	Mn	Manganese	Ti	Titanium
Cu	Copper	Mo	Molybdenum	U	Uranium
D	Diamonds	Na	Salt	V	Vanadium
Fe	Iron Ore	Ni	Nickel	W	Tungsten
G	Natural Gas	O	Petroleum	Zn	Zinc
Gr	Graphite	P	Phosphates		

Water Power

Major Industrial Areas

Agriculture, Industry and Resources

DOMINANT LAND USE

- Cereals (chiefly wheat, corn)
- Livestock, Dairy
- Truck Farming, Potatoes, Vegetables, Dairy
- Cotton
- Sheep Herding, Limited Agriculture
- Forests
- Nonagricultural Land

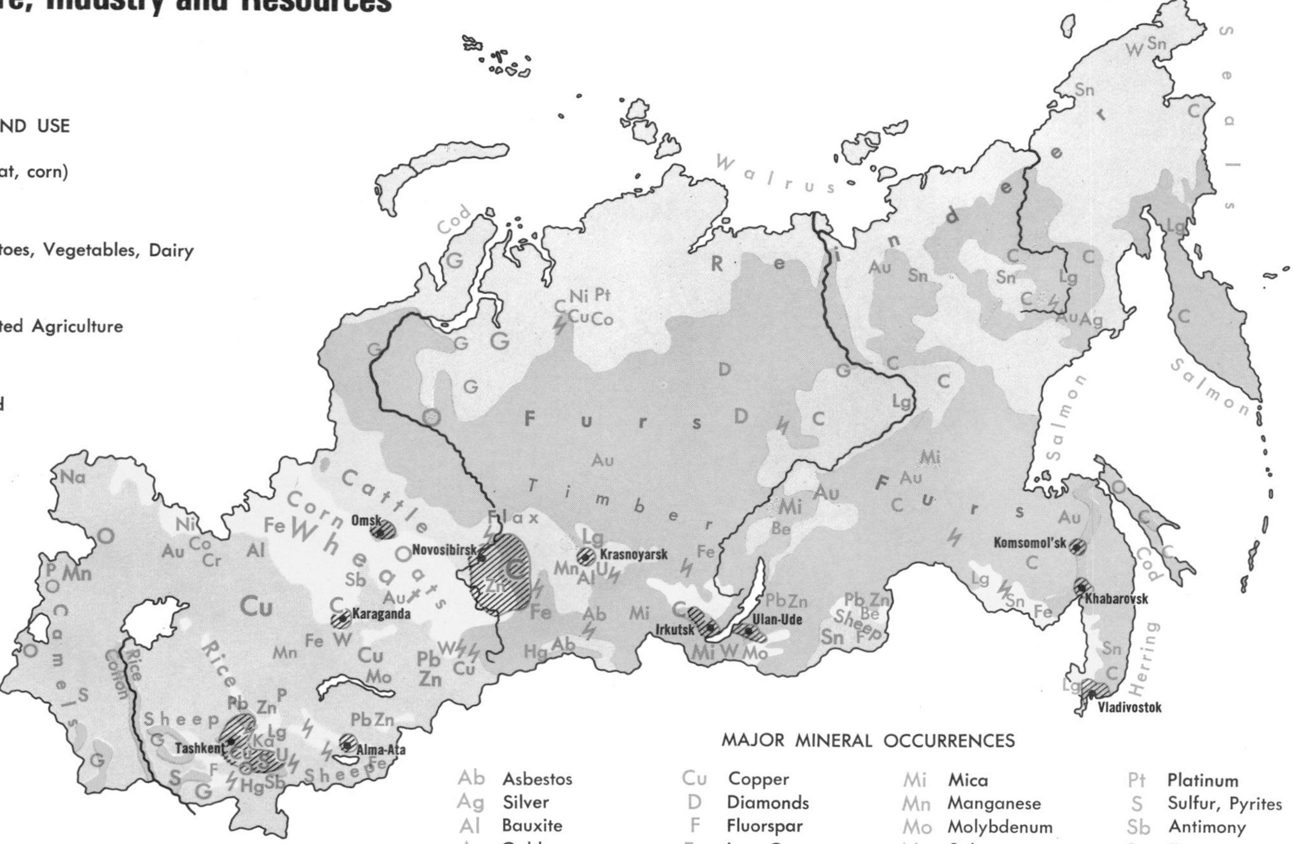

Walrus
Seals
Reindeer
Cod
Sn
W Sn
Sn
Sn
Au Ag
Lg
C
Salmon
Herring
G G G
Ni Pt
C CuCo
D
Lg
C
Furs
Timber
Au
Mi
Be
Mi
Au
Mi
Au
Au Ag
Be
F Curs
Salmon
Cod
Komsomol'sk
Khabarovsk
Sn Fe
Lg
Na
Cattle
Corn Wheat
Omsk
Novosibirsk
Flax
Lg
Krasnoyarsk
Mn
Al
U
Fe Ab
Mi
C
Pb Zn
Sheep
Sn F
Ni
Au Co
Fe W
Cr
Al
Oats
Au
C
Fe
Ab
Irkutsk
Ulan-Ude
Lg
Sn
Fe
C
P Mn
Cu
Karaganda
Fe W
Mn
Cu
Mo
C
Pb
Zn
Cu
Mi W Mo
Lg
C
Vladivostok
O
Ca mels
Rice
Pb Zn
P
Hg Ab
Ni
Cu
Rice Cotton
Sheep
Tashkent
Ka
Lg
Pb Zn
Alma-Ata
Sheep
G
S G
F
Hg Sb

MAJOR MINERAL OCCURRENCES

Ab	Asbestos	Cu	Copper	Mi	Mica	Pt	Platinum
Ag	Silver	D	Diamonds	Mn	Manganese	S	Sulfur, Pyrites
Al	Bauxite	F	Fluorspar	Mo	Molybdenum	Sb	Antimony
Au	Gold	Fe	Iron Ore	Na	Salt	Sn	Tin
Be	Beryl	G	Natural Gas	Ni	Nickel	U	Uranium
C	Coal	Hg	Mercury	O	Petroleum	W	Tungsten
Co	Cobalt	Ka	Kaolin	P	Phosphates	Zn	Zinc
Cr	Chromium	Lg	Lignite	Pb	Lead		

⚡ Water Power   ▨ Major Industrial Areas

U.S.S.R.—Railroads and Navigation

FRANCE
NORWAY
SWEDEN
SW.
W.GERMANY
DEN.
E.GER.
Berlin
Stockholm
AUST.
Vienna
CZ.
POLAND
Kaliningrad
Riga
FINLAND
Kandalaksha
Murmansk
ARCTIC OCEAN
Approximate
Ice
Limit of Permanent
Pevek
Anadyr'
Ambarchik
HUN.
Brest
Leningrad
Archangel
Nar'yan-Mar
YUGO.
L'vov
Minsk
Vologda
Vorkuta
Nordvik
Tiksi
Ust'-Kamchatsk
RUMANIA
Kiev
MOSCOW
Gor'kiy
Kirov
Ukhta
Salekhard
Dudinka
Noril'sk
BULG.
Odessa
Kazan'
Serginy
Magadan
Istanbul
Khar'kov
Kuybyshev
Surgut
Okhotsk
Petropavlovsk-Kamchatskiy
Black Sea
Rostov
Sverdlovsk
Ob'
TURKEY
Novorossiysk
Volgograd
Ural'sk
Tobol'sk
Ayan
Sea of Okhotsk
Batumi
Astrakhan'
Chelyabinsk
Irtysh
Chul'man
Tbilisi
Gur'yev
Orsk
Omsk
Trans-Siberian Railroad
Krasnoyarsk
Ust'-Kut
Baykal-Amur Mainline
Bam
Svobodnyy
Vanino
Korsakov
SYRIA
Shevchenko
Volga
Novosibirsk
Novokuznetsk
Bratsk
L. Baykal
Amur
Khabarovsk
IRAQ
Baku
Aral Sea
Aral'sk
Tselinograd
Semipalatinsk
Irkutsk
Chita
Zabaykal'sk
Tehran
Krasnovodsk
Dzhezkazgan
Karaganda
Harbin
Nakhodka
Mary
Tashkent
Alma-Ata
MONGOLIA
Ulaanbaatar
CHINA
Vladivostok
N. KOREA
SCALE OF MILES
0 500 1000
SCALE OF KILOMETERS
0 500 1000
Dushanbe
Osh
Kungrad
AFGHANISTAN
CHINA
Peking
Sea of Japan
JAPAN
S. KOREA
PACIFIC OCEAN

Principal Railroads ——
Navigable Rivers ......
Canals
Main Sea Routes ----
Major Russian Ports ⚓

(continued on following page)

Union of Soviet Socialist Republics
European Part

CONIC PROJECTION
SCALE OF MILES
0   50   100   300
SCALE OF KILOMETERS
0 50 100   200   300

National Capitals .... ☆
Capitals of Union Republics .... ⊠
Administrative Centers .... △
International boundaries ....
Union Republic boundaries ....
A.S.S.R., Oblast, Kray boundaries ...
Autonomous Oblast boundaries........
Autonomous Okrug boundaries........

Scale 1:13,250,000

The government of the United States has not recognized the incorporation of Estonia, Latvia and Lithuania into the Soviet Union, nor does it recognize as final the de facto western limit of Polish administration in Germany (the Oder-Neisse line).

Administrative Divisions bear same names as their respective Capitals or Centers, except:

Abkhaz A.S.S.R.	Sukhumi	F6
Adygey Aut. Oblast	Maykop	F6
Adzhar A.S.S.R.	Batumi	F6
Bashkir A.S.S.R.	Ufa	J4
Chechen-Ingush A.S.S.R.	Groznyy	G6
Chuvash A.S.S.R.	Cheboksary	G3
Crimean Oblast	Simferopol'	D6
Dagestan A.S.S.R.	Makhachkala	G6
Kabardin-Balkar A.S.S.R.	Nal'chik	F6
Kalmuck A.S.S.R.	Elista	F5
Karachay-Cherkess Aut. Obl.	Cherkessk	F6
Karelian A.S.S.R.	Petrozavodsk	D2
Komi A.S.S.R.	Syktyvkar	H2
Komi-Permyak Aut. Okrug	Kudymkar	H3
Mari A.S.S.R.	Yoshkar-Ola	G4
Mordvinian A.S.S.R.	Saransk	G4
Nagorno-Karabakh Aut. Obl.	Stepanakert	G7
Nenets Aut. Okrug	Nar'yan-Mar	H1
North Ossetian A.S.S.R.	Ordzhonikidze	F6
South Ossetian Aut. Obl.	Tskhinvali	F6
Tatar A.S.S.R.	Kazan'	G3
Trans-Carpathian Oblast	Uzhgorod	B5
Udmurt A.S.S.R.	Ustinov	H3
Volyn Oblast	Lutsk	C4

© Copyright HAMMOND INCORPORATED, Maplewood, N.J.

U.S.S.R. — EUROPEAN

UNION REPUBLICS

Armenian S.S.R. 3,031,000 F6
Azerbaidzhan S.S.R. 6,028,000 G6
Estonian S.S.R 1,466,000 C3
Georgian S.S.R. 5,015,000 F6
Latvian S.S.R 2,521,000 B3
Lithuanian S.S.R. 3,398,000 B3
Moldavian S.S.R. 3,947,000 C5
Russian S.F.S.R. 137,551,000 F3
Ukrainian S.S.R. 49,755,000 D5
White Russian S.S.R. 9,560,000 C4

INTERNAL DIVISIONS

Abkhaz A.S.S.R. 505,000 F6
Adygey Aut. Obl. 405,000 F6
Adzhar A.S.S.R. 354,000 F6
Bashkir A.S.S.R. 3,849,000 J4
Chechen-Ingush A.S.S.R. 1,154,000 G6
Chuvash A.S.S.R. 1,292,000 G3
Crimean Oblast 2,183,000 D6
Dagestan A.S.S.R. 1,628,000 G6
Kabardin-Balkar A.S.S.R. 674,000 F6
Kalmuck A.S.S.R. 294,000 F5
Karachay-Cherkess Aut. Obl. 368,000 F6
Karelian A.S.S.R. 736,000 D2
Komi A.S.S.R. 1,119,000 H2
Komi-Permyak Aut. Okr. 173,000 H3
Mari A.S.S.R. 703,000 G3
Mordvinian A.S.S.R. 991,000 G4
Nagorno-Karabakh Aut. Obl. 161,000 G7
Nakhichevan' A.S.S.R. 239,000 F7
Nenets Aut. Okr. 47,000 H1
North Ossetian A.S.S.R. 597,000 F6
South Ossetian Aut. Obl. 98,000 F6
Tatar A.S.S.R. 3,436,000 G3
Trans-Carpathian Oblast 1,155,000 B5
Udmurt A.S.S.R. 1,494,000 H3
Volyn Oblast 1,015,000 C4

CITIES and TOWNS

Abdulino 26,010 H4
Agdam 21,277 G6
Agryz 19,267 H3
Akhaltsikhe 18,972 F6
Akhtubinsk 43,466 G5
Akhty G6
Akhtyrka 41,354 E4
Akkerman (Belgorod-Dnestrovskiy) 32,928 D5
Alagir 18,161 F6
Alatyr' 43,499 G4
Alaverdi 21,311 F6
Aleksandriya 82,000 D5
Aleksandrovsk 18,286 J3
Alekseyevka 25,562 E4
Aleksin 67,000 E4
Ali-Bayramly 33,828 G7
Al'met'yevsk 110,000 H3
Alushta 22,016 D6
Amderma K1
Anapa 29,900 E6
Andropov 239,000 E3
Apatity 62,000 D1
Apsheronsk 32,867 F6
Archangel (Arkhangel'sk) 385,000 F2
Armavir 162,000 F5
Arzamas 93,000 F3
Astara G7
Astrakhan' 461,000 G5
Atkarsk 28,881 G4
Azov 75,000 E5
Bakhchisaray 15,912 D6
Baku 1,022,000 H6
Balakhna 36,542 F3
Balaklava D6
Balakovo 152,000 G4
Balashov 93,000 F4
Baltiysk 20,300 A4
Baranovichi 131,000 C4
Barysh 20,792 G4
Bataysk 90,000 E5
Batumi 123,000 F6
Belaya Tserkov' 151,000 C5
Belebey 32,460 H4
Belev 17,733 E4
Belgorod 240,000 E4
Belgorod-Dnestrovskiy 32,928 D5
Belomorsk 16,595 D2
Belorechensk 35,970 E6
Beloretsk 71,000 J4
Belozersk E3
Bel'tsy 125,000 C5
Belush'ya Guba H1
Bendery 101,000 C5
Berdichev 80,000 C5
Berdyansk 122,000 E5
Beregovo 27,308 B5
Berezniki 185,000 J3
Beslan 26,893 F6
Bezhetsk 30,030 E3
Birsk 29,607 J3
Bobrov 17,977 F4
Bobruysk 192,000 C4
Bologoye 33,949 D3
Bor 63,000 F3
Borislav 33,800 B5
Borisoglebsk 68,000 F4
Borisov 112,000 C4
Borovichi 60,000 D3
Brest 177,000 B4
Brezhnev 301,000 H3
Bryansk 394,000 D4
Bugul'ma 80,000 H4
Buguruslan 54,000 H4
Buturlinovka 21,643 F4
Buy 29,946 F3
Buynaksk 37,946 G6
Buzuluk 76,000 H4
Bykhov 17,371 C4
Cēsis 17,696 C3
Chadyr-Lunga 20,474 C5
Chapayevsk 85,000 G4
Chaykovskiy 48,034 H3
Cheboksary 308,000 G3
Cherepovets 266,000 E3
Cherkassy 228,000 D5
Cherkessk 91,000 F6
Chernigov 238,000 D4
Chernovtsy 219,000 C5
Chernushka 21,106 J3
Chervonograd 55,000 B4
Chiatura 25,474 F6
Chistopol' 64,000 H3
Chortkov 19,183 B5
Chudovo D3
Chusovoy 56,000 J3
Danilov 17,500 E3
Dankov 20,030 E4
Daugavpils 116,000 C3
Davlekanovo 20,123 H4
Derbent 70,000 G6
Dimitrovgrad 106,000 G4
Dneprodzerzhinsk 250,000 D5
Dnepropetrovsk 1,066,000 D5
Dobrush 16,809 D4
Dobryanka 18,349 J3
Donetsk 1,021,000 E5
Drogobych 66,000 B5
Dubna 55,000 E3
Dubna E4
Dubno 25,442 C4
Dvinsk (Daugavpils) 116,000 C3
Dyat'kovo 26,825 D4
Dzerzhinsk 257,000 F3
Dzhankoy 43,459 D5
Dzhul'fa G7
Echmiadzin 31,819 F6
Elektrostal' 139,000 E3
Elista 70,000 F5
El'ton G5
Engel's 161,000 G4
Erivan 1,019,000 F6
Fastov 51,000 C4
Feodosiya 76,000 D5
Frolovo 33,398 F5
Furmanov 40,155 F3
Gagra 23,025 E6
Galich 19,374 F3
Gandzha (Kirovabad) 232,000 G6
Gatchina 75,000 C3
Gay 28,250 J4
Gaysin 23,741 C5
Gdov C3
Gelendzhik 29,086 E6
Genichesk 20,031 E5
Georgiu-Dezh 52,000 E4
Glazov 81,000 H3
Glubokoye C3
Glukhov 27,096 D4
Gomel' 383,000 D4
Gori 56,000 F6
Gorki 22,117 D4
Gor'kiy 1,344,000 F3
Gorlovka 336,000 E5
Gorodets 34,229 F3
Gremikha E1
Gremyachinsk 29,975 J3
Grodno 195,000 B4
Groznyy 375,000 G6
Gryazi 41,292 F4
Gubakha 33,243 J3
Gubkin 65,000 E4
Gudauta F6
Gudermes 32,445 G6
Gukovo 68,000 F5
Gus'-Khrustal'nyy 72,000 F3
Imishli 17,839 G7
Inta 51,000 K1
Inza 19,060 G4
Ishimbay 57,000 J4
Ivano-Frankovsk 150,000 B5
Ivanovo 465,000 E3
Izberbash 17,299 G6
Izhevsk (Ustinov) 549,000 H3
Izmail 83,000 C5
Izyum 61,000 E5
Jēkabpils 22,440 C3
Jelgava 68,000 B3
Jurmala 61,000 B3
Kadiyevka (Stakhanov) 108,000 E5
Kafan 29,916 G7
Kagul 26,249 C5
Kakhovka 28,472 D5
Kalach 18,475 F4
Kalach-na-Donu 20,795 F5
Kalinin 412,000 E3
Kaliningrad, Kaliningrad 355,000 B4
Kaliningrad, Moscow Oblast 133,000 E3
Kalinkovichi 23,918 C4
Kaluga 265,000 E4
Kalush 60,000 B5
Kamenets-Podol'skiy 81,000 C5
Kamenka 30,067 F4
Kamensk-Shakhtinskiy 72,000 F5
Kamyshin 112,000 F4
Kanash 40,682 G3
Kandalaksha 42,656 D1
Kapsukas 28,763 B4
Karachayevsk F6
Karachev 15,972 E4
Kashin 17,678 E3
Kasimov 33,066 F4
Kaspiysk 38,990 G6
Kaunas 370,000 B4
Kazan' 993,000 G3
Kazatin 26,649 C5
Kem' 21,025 D2
Kerch' 157,000 E5
Keret' D1
Khachmas 22,313 G6
Khadyzhensk 17,856 E6
Khar'kov 1,444,000 E5
Khasavyurt 65,000 G6
Khashuri 24,469 F6
Kherson 319,000 D5
Khmel'nitskiy 172,000 C5
Khotin 10,319 C5
Khust 23,810 B5
Khvalynsk 16,249 G4
Kiev 2,144,000 D4
Kiliya 24,276 C5
Kimovsk 44,490 E4
Kimry 58,000 E3
Kinel' 39,373 H4
Kineshma 101,000 F3
Kirishi 27,252 D3
Kirov, Kaluga 29,355 D4
Kirov, Kirov 390,000 G3
Kirovabad 232,000 G6
Kirovakan 146,000 G6
Kirovo-Chepetsk 71,000 H3
Kirovograd 237,000 D5
Kirovsk 38,484 D1
Kirsanov 21,795 F4
Kishinev 503,000 C5
Kislovodsk 101,000 F6
Kizel 46,264 J3
Kizlyar 29,745 G6
Klaipeda 176,000 B3
Klintsy 67,000 D4
Kobrin 24,935 B4
Kobuleti 18,051 F6
Kohtla-Järve 73,000 C3
Kolomiya 52,000 B5
Kolomna 147,000 E4
Kolpino 114,000 D3
Kommunarsk 120,000 E5
Komrat 21,369 C5
Komsomol'skiy 17,078 K1
Kondopoga 27,908 D2
Königsberg (Kaliningrad) 355,000 B4
Konotop 82,000 D4
Konstantinovka 112,000 E5
Korenovsk 26,323 E5
Korosten' 65,000 C4
Korostyshev 21,153 C4
Koryazhma 33,230 G2
Kostopol' 17,548 C4
Kostroma 255,000 F3
Kotel'nich 29,196 G3
Kotel'nikovo 19,063 F5
Kotlas 61,000 G2
Kotovo 20,553 G4
Kotovsk, Odessa 36,463 C5
Kotovsk, Tambov 33,347 F4
Kovel' 33,351 C4
Kovrov 143,000 F3
Kovylkino 17,300 F4
Kramatorsk 178,000 E5
Krasnoarmeysk 60,000 G4
Krasnodar 560,000 E6
Krasnograd 18,386 E5
Krasnokamsk 56,000 H3
Krasnoslobodsk 17,749 G5
Krasnovishersk J2
Krasnyy Kut 17,087 G4

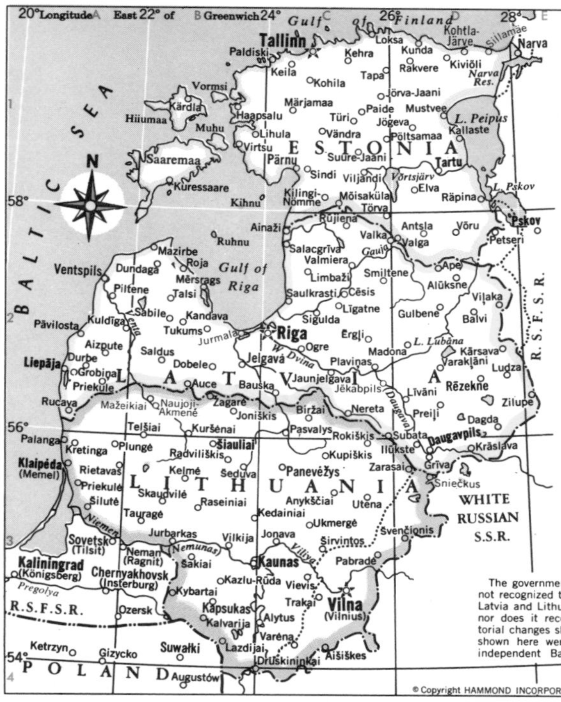

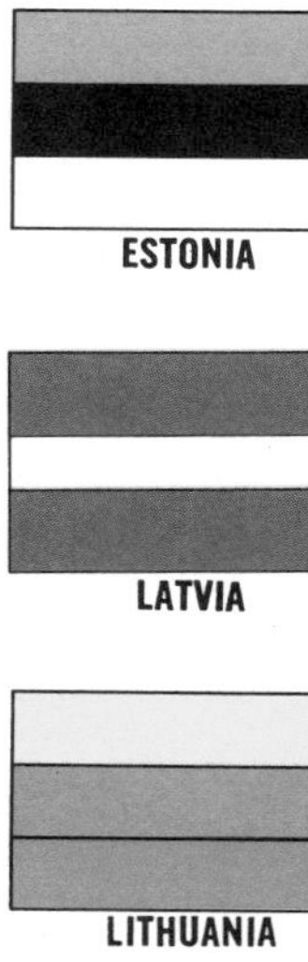

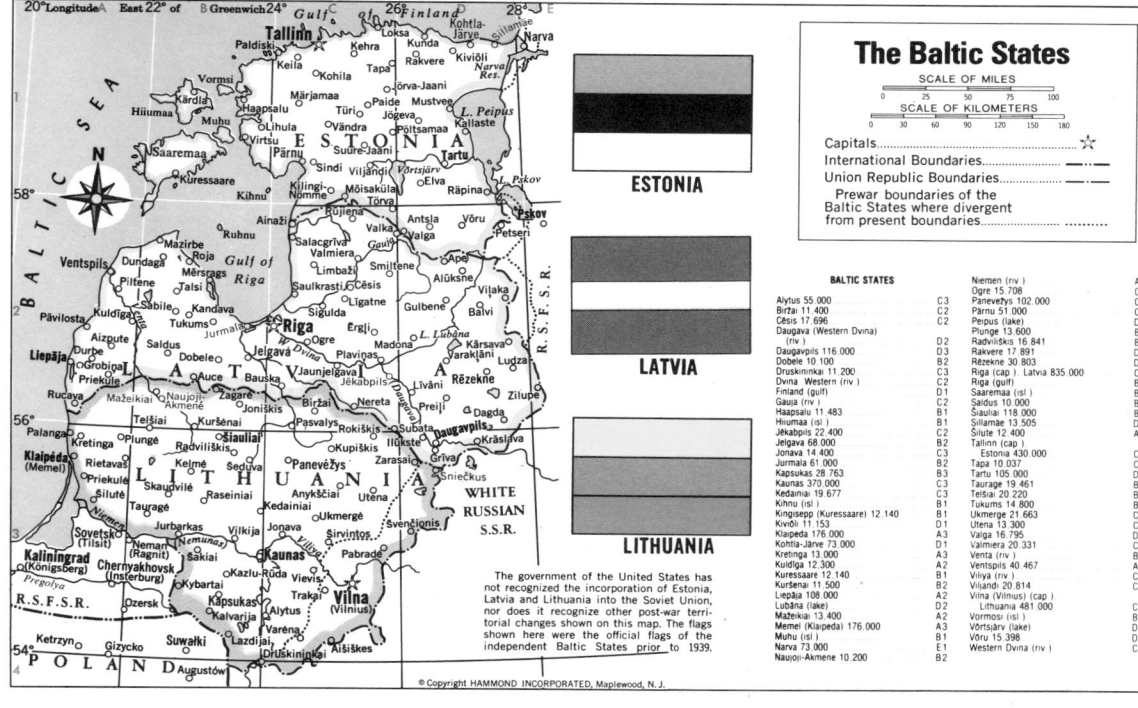

The Baltic States

SCALE OF MILES 0 25 50 75 100
SCALE OF KILOMETERS 0 30 60 90 120 150 180

Capitals ☆
International Boundaries
Union Republic Boundaries
Prewar boundaries of the Baltic States where divergent from present boundaries

BALTIC STATES

Alytus 55,000 C3
Biržai 11,400 C2
Cēsis 17,696 C2
Daugava (Western Dvina) (riv.) D2
Daugavpils 116,000 D3
Dobele 10,100 B2
Druskininkai 11,200 C3
Dvina, Western (riv.) C2
Finland (gulf) D1
Gauja (riv.) C2
Haapsalu 11,483 B1
Hiiumaa (isl.) B1
Jēkabpils 22,400 C2
Jelgava 68,000 B2
Jonava 14,400 C3
Jurmala 61,000 B2
Kapsukas 28,763 B3
Kaunas 370,000 C3
Kedainiai 19,677 C3
Kihnu (isl.) B1
Kingisepp (Kuressaare) 12,140 B1
Kiviõli 11,153 D1
Klaipeda 176,000 A3
Kohtla-Järve 73,000 D1
Kretinga 13,000 A3
Kuldīga 12,300 A2
Kuressaare 12,140 B1
Kuršenai 11,500 B2
Liepāja 108,000 A2
Lubāna (lake) D2
Mažeikiai 13,400 A2
Memel (Klaipeda) 176,000 A3
Muhu (isl.) B1
Narva 73,000 E1
Naujoji-Akmene 10,200 B2
Niemen (riv.) A3
Ogre 15,708 C2
Panevežys 102,000 C3
Pärnu 51,000 C1
Peipus (lake) D1
Plunge 13,600 B3
Radviliškis 16,841 B3
Rakvere 17,891 D1
Rēzekne 30,803 D2
Riga (cap.), Latvia 835,000 C2
Riga (gulf) B2
Saaremaa (isl.) B1
Saldus 10,000 B2
Šiauliai 118,000 B3
Sillamäe 13,505 D1
Šilute 12,400 A3
Tallinn (cap.), Estonia 430,000 C1
Tapa 10,037 C1
Tartu 105,000 D1
Taurage 19,461 B3
Telšiai 20,220 B2
Tukums 14,800 B2
Ukmerge 21,663 C3
Utena 13,300 C3
Valga 16,795 D2
Valmiera 20,331 C2
Venta (riv.) B2
Ventspils 40,467 A2
Viliya (riv.) C3
Viljandi 20,814 C1
Vilna (Vilnius) (cap.), Lithuania 481,000 C3
Vormosi (isl.) B1
Võrtsjärv (lake) D1
Võru 15,398 D2
Western Dvina (riv.) C2

The government of the United States has not recognized the incorporation of Estonia, Latvia and Lithuania into the Soviet Union, nor does it recognize other post-war territorial changes shown on this map. The flags shown here were the official flags of the independent Baltic States prior to 1939.

© Copyright HAMMOND INCORPORATED, Maplewood, N. J.

Krasnyy Luch 106,000 E5
Krasnyy Sulin 41,684 F5
Kremenchug 210,000 D5
Krichev 25,682 D4
Krivoy Rog 650,000 D5
Krolevets 18,307 D4
Kronshtadt 39,477 C3
Kropotkin 70,000 F5
Krymsk 41,430 E5
Kuba 18,871 G6
Kudymkar 26,350 H3
Kulebaki 46,252 F3
Kumertau 52,000 J4
Kunda C3
Kungur 80,000 J3
Kupyansk 30,055 E5
Kuressaare 12,140 B3
Kursk 375,000 E4
Kutaisi 194,000 F6
Kuvandyk 22,914 J4
Kuybyshev 1,216,000 H4
Kuznetsk 94,000 G4
Kuzomen' E1
Labinsk 54,000 F6
Lakhdenpokh'ya C2
Lebedin 29,240 D4
Leninakan 207,000 F6
Leningrad 4,073,000 C3
Leningrad* 4,588,000 C3
Leninogorsk 54,000 H4
Lenkoran' 35,505 G7
L'gov 25,110 E4
Lida 66,000 C4
Liepāja 108,000 B3
Likhoslavl' E3
Lipetsk 396,000 E4
Lisichansk 119,000 E5
Livny 37,290 E4
Lodeynoye Pole 19,632 D2
Lozovaya 53,000 E5
Lubny 54,000 D4
Luga 31,905 C3
Lutsk 137,000 B4
L'vov (Lwów) 667,000 B5
Lys'va 75,000 J3
Lyubertsy 160,000 E3
Lyubotin 33,324 E4
Lyudinovo 33,871 D4
Makeyevka 436,000 E5
Makhachkala 251,000 G6
Makharadze 21,679 F6
Malaya Vishera 15,381 D3
Malgobek 20,548 F6
Manturovo 21,510 F3
Marganets 50,000 E5
Mariupol' (Zhdanov) 503,000 E5
Marks 17,132 G4
Maykop 128,000 F6
Mednogorsk 38,024 J4
Medvezh'yegorsk 17,465 D2
Melenki 18,545 F3
Meleuz 24,851 J4
Melitopol' 161,000 D5
Memel (Klaipeda) 176,000 B3
Merefa 29,985 E5
Mezen' F1
Michurinsk 101,000 F4
Mikhaylovka 58,000 F4
Millerovo 34,627 F5
Mineral'nye Vody 67,000 F6
Mingechaur 60,000 G6
Minsk 1,262,000 C4
Minsk* 1,276,000 C4
Mirgorod 28,407 D5
Mogilev 290,000 C4
Mogilev-Podol'skiy 26,051 C5
Molodechno 73,000 C4
Molotov (Perm') 999,000 J3
Monchegorsk 51,000 D1
Morshansk 44,245 F4
Moscow (Moskva) (cap.) 7,831,000 E3
Moscow* 8,011,000 E3
Mozhaysk 20,321 E3
Mozhga 38,930 H3
Mozyr' 73,000 C4
Mtsensk 27,833 E4
Mukachevo 72,000 B5
Murmansk 381,000 D1
Murom 114,000 F3
Mytishchi 141,000 E3
Nakhichevan' 33,279 F7
Nal'chik 207,000 F6
Narva 73,000 C3
Nar'yan-Mar 16,864 H1
Neftekamsk 70,000 J3
Nelidovo 29,813 D3
Nerekhta 25,722 F3
Nevel' 17,804 D3
Nevinnomyssk 104,000 F6
Nezhin 70,000 D4
Nikel' 21,299 C1
Nikolayev 440,000 D5
Nikol'sk 20,740 G4
Nikopol' 146,000 D5
Nizhnekamsk 134,000 H3
Nizhniy Lomov 17,460 F4
Nizhniy Novgorod (Gor'kiy) 1,344,000 F3
Nosovka 19,430 D4
Novaya Kakhovka 52,000 D5
Novgorod 186,000 D3
Novgorod-Severskiy D4
Novoanninskiy 20,461 F4
Novocherkassk 183,000 F5
Novograd-Volynskiy 41,194 C4
Novogrudok 19,374 C4
Novokuybyshevsk 109,000 G4
Novomoskovsk 147,000 E4
Novopolotsk 67,000 C3
Novorossiysk 159,000 E6
Novoshakhtinsk 104,000 E5
Novotroitsk 95,000 J4
Novoukrainka 19,554 D5
Novouzensk G4
Novovolynsk 41,187 B4
Novovyatsk 26,408 G3
Novozybkov 34,433 D4
Nurlat 17,533 H4
Nyandoma 23,366 F2
Nytva 17,491 H3
Nyuvchim H2
Obninsk 73,000 E3
Ochamchira 18,718 F6
Odessa 1,046,000 D5
Oktyabr'sk 33,981 G4
Oktyabr'skiy 88,000 H4
Okulovka 19,194 D3
Olenegorsk 21,485 D1
Olonets D2
Omutninsk 28,777 H3
Onega 25,047 E2
Ordzhonikidze 279,000 F6
Orel 305,000 E4
Orenburg 459,000 J4
Orgeyev 25,798 C5
Orsha 112,000 C4
Orsk 247,000 J4
Osa 15,038 J3
Osipenko (Berdyansk) 122,000 E5
Osipovichi 19,705 C4
Ostashkov 23,419 D3
Ostrogozhsk 29,921 E4
Ostrov 22,369 C3
Otradnyy 44,426 H4
Panevežys 102,000 B3
Pärnu 51,000 C3
Pavlograd 107,000 E5
Pavlovo 68,000 F3
Pechenga D1
Pechora 56,000 J1
Penza 483,000 G4
Perm' 999,000 J3
Pervomaysk 72,000 D5
Petrokrepost' D3
Petrovsk 30,953 G4
Petrozavodsk 234,000 D2
Petsamo (Pechenga) D1
Pinsk 90,000 C4
Podol'sk 202,000 E3
Podporozh'ye 21,545 D2
Pokhvistnevo 26,125 H4
Polonnoye 22,484 C4
Polotsk 71,000 C3
Poltava 279,000 D5
Polyarnyy 15,321 D1
Ponoy F1
Poti 45,979 F6
Povenets E2
Povorino 20,591 F4
Prikumsk 35,768 F6
Priluki 65,000 D4
Primorsk C2
Primorsko-Akhtarsk 25,981 E5
Priozersk 16,652 D2
Privolzhskiy 23,041 G4
Priyutovo 21,051 H4
Prokhladnyy 40,074 F6
Pskov 176,000 C3
Pugachev 33,963 G4
Pushkin 90,000 C3
Pyatigorsk 110,000 F6
Rabocheostrovsk D1
Rakhov B5
Rakvere 17,891 C3
Rasskazovo 40,038 F4
Razdan 26,833 G6
Rechitsa 60,000 C4
Reni 19,625 C5
Revel (Tallinn) 430,000 B3
Rēzekne 30,803 C3
Riga 835,000 B3
Romny 53,000 D4
Roslavl' 56,000 D4
Rossosh' 36,438 E4
Rostov 30,815 E3
Rostov-na-Donu 934,000 F5
Rovno 179,000 C4
Rtishchevo 37,146 F4
Rubezhnoye 66,000 E5
Rustavi 129,000 G6
Ruzayevka 41,084 F4
Ryazan' 453,000 E4
Ryazhsk 25,425 F4
Rybinsk (Andropov) 239,000 E3
Rybnitsa 32,266 C5
Rzhev 69,000 D3
Safonovo 53,000 D3
Saki 24,208 D5
Salavat 137,000 H4
Sal'sk 57,000 F5
Sal'yany 24,228 G7
Samara (Kuybyshev) 1,216,000 H4
Sambor 29,253 B5
Saransk 263,000 G4
Sarapul 107,000 H3
Saratov 856,000 G4
Sasovo 27,228 F4
Segezha 28,810 D2
Semenov 23,633 F3
Semiluki 18,221 E4
Sengiley G4
Serdobol (Sortavala) 22,188 D2
Serdobsk 33,783 F4
Sergach 22,509 F3
Serpukhov 140,000 E4
Sevastopol' 301,000 D6
Severodonetsk 113,000 E5
Severodvinsk 197,000 E2
Severomorsk 50,000 D1
Shakhty 209,000 F5
Shakhun'ya 20,009 G3
Shar'ya 25,788 G3
Shchekino 70,000 E4
Shchigry 17,133 E4
Sheki 43,158 G6
Shemakha 17,986 G6
Shepetovka 38,707 C4
Shostka 82,000 D4
Shpola 19,806 D5
Shumerlya 33,816 G3
Shuya 72,000 F3
Šiauliai 118,000 B3
Sibay 37,656 J4
Simferopol' 302,000 D6
Skadovsk D5
Skopin 24,429 F4
Slantsy 41,146 C3
Slavuta 25,573 C4
Slavyansk 140,000 E5
Slavyansk-na-Kubani 54,000 E5
Slobodskoy 34,374 H3
Slonim 30,279 B4
Slutsk 35,609 C4
Smela 62,000 D5
Smolensk 276,000 D4
Sochi 287,000 E6
Sokol 48,243 F3
Soligorsk 65,000 C4
Solikamsk 101,000 J3
Sol'-Iletsk 22,227 J4
Sorochinsk 23,235 H4
Soroki 21,924 C5
Sortavala 22,188 D2
Sosnogorsk 24,688 H2
Sovetsk (Tilsit) 38,456 B4
Sovetsk 17,027 G3
Stakhanov 108,000 E5
Stalingrad (Volgograd) 929,000 F5
Staraya Russa 34,577 D3
Staryy Oskol 115,000 E4
Stavropol' 258,000 F5
Stepanakert 30,293 G7
Sterlitamak 220,000 J4
Stupino 70,000 E4
Sudak E6
Sukhumi 114,000 F6
Sumgait 190,000 G6
Sumy 228,000 E4
Svetlogorsk 55,000 C4
Svetlograd 40,265 F5
Syktyvkar 171,000 H2
Syzran' 178,000 G4
Taganrog 276,000 E5
Tallinn 430,000 B3
Tambov 270,000 F4
Tartu 105,000 C3
Taurage 19,461 B3
Tbilisi 1,066,000 F6
Telavi 21,179 G6
Telšiai 20,220 B3
Temryuk 23,172 E5
Ternopol' 144,000 C5
Teykovo 41,607 E3
Tiflis (Tbilisi) 1,066,000 F6
Tighina (Bendery) 101,000 C5
Tikhoretsk 64,000 F5
Tikhvin 59,000 D3
Tilsit (Sovetsk) 38,456 B4
Timashevsk 29,055 E5
Tiraspol' 139,000 D5
Togliatti (Tol'yatti) 502,000 G4
Tokmak 59,000 E5
Toropets 16,863 D3
Torzhok 45,443 D3
Troitsko-Pechorsk J2
Tskhinvali 30,311 F6
Tuapse 60,000 E6
Tula 514,000 E4
Tutayev 16,839 E3
Tuymazy 37,021 H4
Tver (Kalinin) 412,000 E3
Tyrnyauz 18,253 F6
Uchaly 21,808 J4
Ufa 969,000 J4
Uglich 35,463 E3
Ukmerge 21,663 C3
Ul'yanovsk 464,000 G4
Uman' 79,000 D5
Unecha 21,749 D4
Ungeny 17,228 C5
Uryupinsk 38,192 F4
Usinsk J1
Usman' 20,150 E4
Ustinov 549,000 H3
Uvarovo 24,946 F4
Uzhgorod 91,000 B5
Uzlovaya 65,000 E4
Valga 16,795 C3
Valmiera 20,331 C3
Valuyki 29,093 E4
Vasil'kov 26,741 D4
Velikiye Luki 102,000 D3
Velikiy Ustyug 36,737 F2
Vel'sk 21,899 F2
Ventspils 40,467 B3
Vereshchagino 23,585 H3
Vichuga 52,000 F3
Viipuri (Vyborg) 76,000 C2
Vileyka C4
Vilna (Vilnius) 481,000 C4
Vinnitsa 314,000 C5
Vinogradov 20,580 B5
Vitebsk 297,000 C3
Vladimir 296,000 F3
Vladimir-Volynskiy 28,412 B4
Volgodonsk 91,000 F5
Volgograd 929,000 F5
Volkhov 47,025 D3
Volkovysk 28,266 B4
Vologda 237,000 F3
Vol'sk 66,000 G4
Volzhsk 52,000 G3
Volzhskiy 209,000 G5
Vorkuta 100,000 K1
Voronezh 783,000 E4
Voroshilovgrad 463,000 E5
Voskresensk 76,000 E3
Votkinsk 90,000 H3
Voznesensk 36,457 D5
Vyatskiye Polyany 32,729 H3
Vyaz'ma 52,000 D3
Vyborg 76,000 C2
Vyksa 54,000 F3
Vyshniy Volochek 72,000 D3
Yalta 80,000 D6
Yanaul 20,115 J3
Yaroslavl' 597,000 E3
Yartsevo 36,662 D3
Yefremov 53,000 E4
Yelabuga 31,728 H3
Yelets 112,000 E4
Yenakiyevo 114,000 E5
Yershov 21,731 G4
Yessentuki 78,000 F6
Yevlakh 29,462 G6
Yevpatoria 93,000 D5
Yeysk 71,000 E5
Yoshkar-Ola 201,000 G3
Yur'yevets 20,144 F3
Zagorsk 107,000 E3
Zapolyarnyy 22,084 D1
Zaporozh'ye 781,000 E5
Zelenodol'sk 85,000 G3
Zelenokumsk 29,691 F6
Zernograd 20,324 F5
Zhdanov 503,000 E5
Zheleznodorozhnyy 76,000 H2
Zheleznogorsk 65,000 E4
Zhigulevsk 52,130 G4
Zhitomir 244,000 C4
Zhlobin 25,359 D4
Zhmerinka 36,195 C5
Zhodino 22,083 C4
Zhovtnevoye 31,102 D5
Znamenka 27,393 D5
Zolotonosha 27,639 D5
Zugdidi 39,896 F6
Zuyevka 17,001 H3

OTHER FEATURES

Apsheron (pen.) H6
Araks (riv.) G7
Azov (sea) E5
Baltic (sea) B3
Barents (sea) E1
Belaya (riv.) H3
Beloye (lake) E2
Black (sea) D6
Bug (riv.) B4
Bug (riv.) D5
Caspian (sea) G6
Caucasus (mts.) F6
Crimea (pen.) D5
Desna (riv.) D4
Dnieper (riv.) D5
Dniester (riv.) C5
Don (riv.) F5
Donets (riv.) E5
Dvina (bay) E2
Dvina, Northern (riv.) F2
Dvina, Western (riv.) C3
Dykh-Tau (mt.) F6
El'brus (mt.) F6
Finland (gulf) C3
Hiiumaa (isl.) B3
Il'men' (lake) D3
Imandra (lake) D1
Kakhovka (res.) D5
Kama (riv.) H2
Kandalaksha (gulf) D1
Kanin (pen.) G1
Kara (sea) K1
Karskiye Vorota (str.) J1
Kazbek (mt.) F6
Khoper (riv.) F4
Kola (pen.) E1
Kolguyev (isl.) G1
Kuban' (riv.) E5
Kura (riv.) G6
Kuybyshev (res.) G4
Ladoga (lake) D2
Lapland (reg.) D1
Mezen' (riv.) G1
Moksha (riv.) F4
Narodnaya (mt.) J1
Niemen (riv.) B4
Novaya Zemlya (isls.) H1
Oka (riv.) F4
Onega (bay) E2
Onega (lake) E2
Onega (riv.) E2
Pechora (riv.) H1
Peipus (lake) C3
Pripet (marshes) C4
Pripyat' (riv.) C4
Prut (riv.) C5
Riga (gulf) B3
Rybachiy (pen.) D1
Rybinsk (res.) E3
Saaremaa (isl.) B3
Samara (riv.) H4
Sevan (lake) G6
Seym (riv.) D4
Sura (riv.) G4
Svir' (riv.) D2
Timan (ridge) G1
Tsil'ma (riv.) H1
Tsimlyansk (res.) F5
Tuloma (riv.) D1
Ural (mts.) J2
Ural (riv.) J4
Usa (riv.) K1
Valday (hills) D3
Vaygach (isl.) K1
Velikaya (riv.) C3
Volga (riv.) G5
Volga-Don (canal) F5
Volgograd (res.) G5
Volkhov (riv.) D3
Vorskla (riv.) E4
Vyatka (riv.) H3
Vychegda (riv.) G2
Vyg (lake) E2
White (sea) E1
Yamantau (mt.) J4
Yugorskiy (pen.) K1

*City and suburbs.

Asia

LAMBERT AZIMUTHAL EQUAL-AREA PROJECTION

SCALE OF MILES

0 100 200   400   600   800   1000   1200

SCALE OF KILOMETERS

0   200 400   600   800 1000 1200

Capitals of Countries .... ⊛

Other Capitals .... ⊙

International Boundaries .... ▬·▬·▬

Other Boundaries......................... ▬··▬··▬

Canals ....

Scale 1:46,500,000

© Copyright HAMMOND INCORPORATED, Maplewood, N.J.

Population Distribution

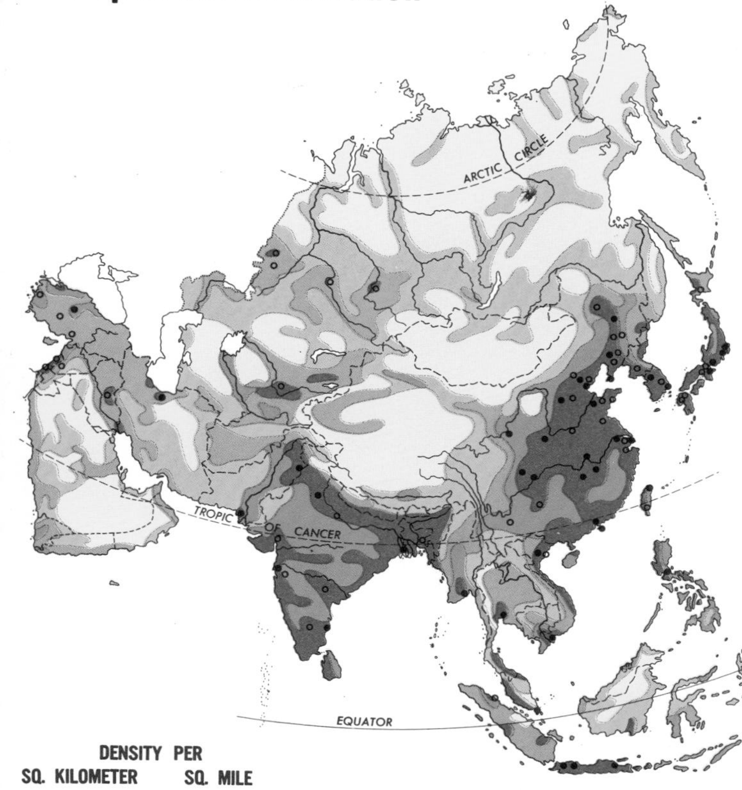

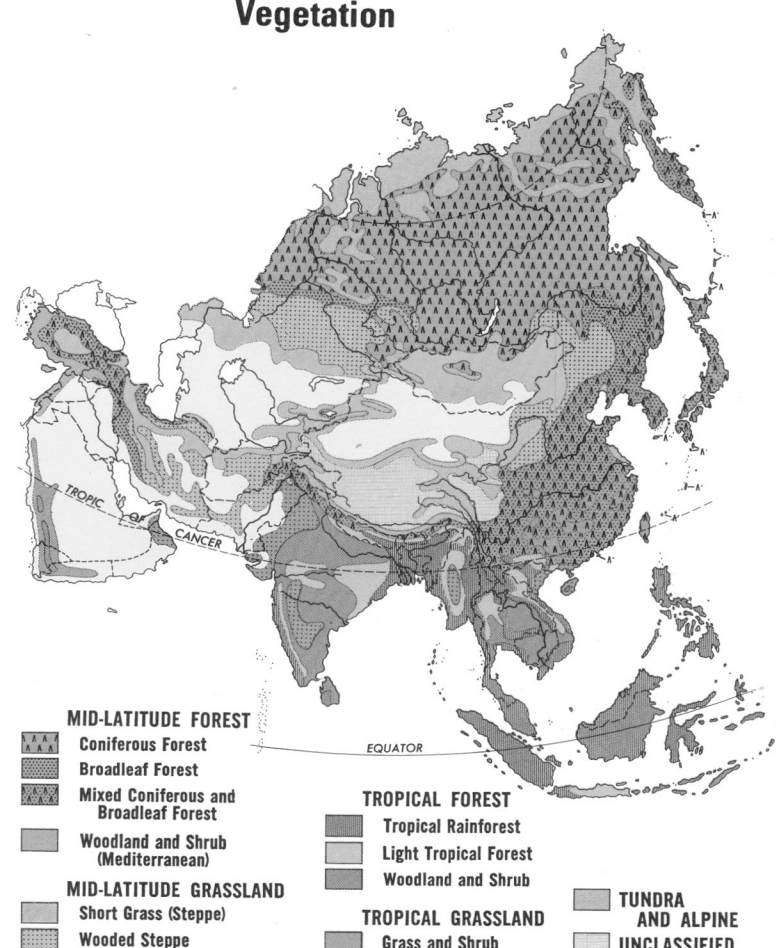

DENSITY PER

SQ. KILOMETER	SQ. MILE
Over 100	Over 260
50-100	130-260
10-50	25-130
1-10	3-25
Under 1	Under 3

• Cities with over 2,000,000 inhabitants (including suburbs)

∘ Cities with over 1,000,000 inhabitants (including suburbs)

AREA 17,128,500 sq. mi. (44,362,815 sq. km.)
POPULATION 2,633,000,000
LARGEST CITY Tokyo
HIGHEST POINT Mt. Everest 29,028 ft. (8,848 m.)
LOWEST POINT Dead Sea -1,296 ft. (-395 m.)

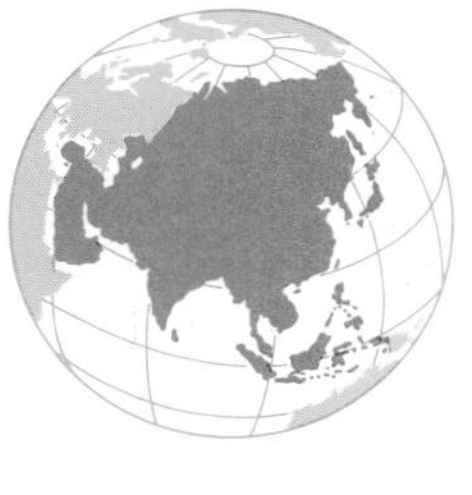

Vegetation

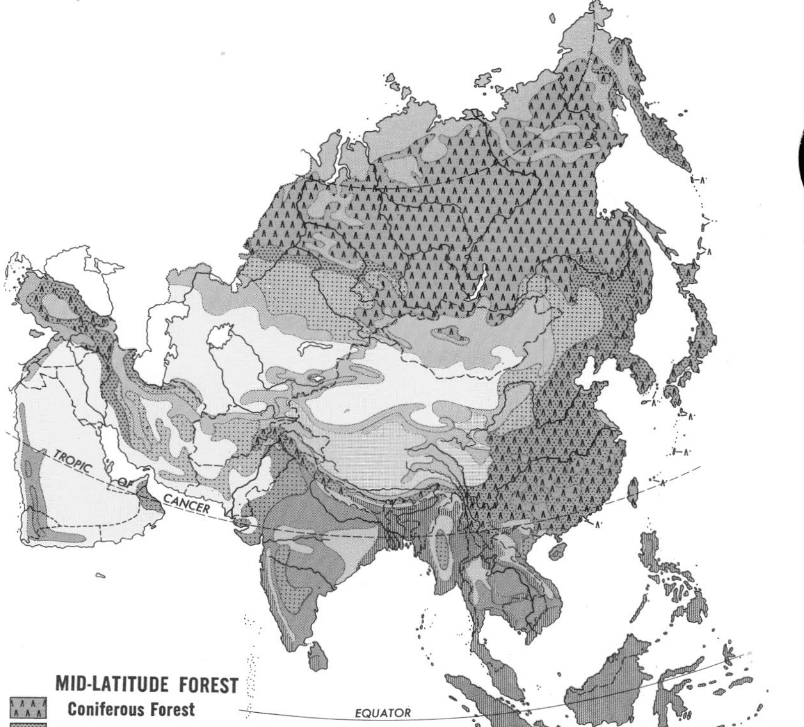

MID-LATITUDE FOREST
- Coniferous Forest
- Broadleaf Forest
- Mixed Coniferous and Broadleaf Forest
- Woodland and Shrub (Mediterranean)

MID-LATITUDE GRASSLAND
- Short Grass (Steppe)
- Wooded Steppe

DESERT AND DESERT SHRUB

TROPICAL FOREST
- Tropical Rainforest
- Light Tropical Forest
- Woodland and Shrub

TROPICAL GRASSLAND
- Grass and Shrub (Savanna)
- Wooded Savanna

TUNDRA AND ALPINE

UNCLASSIFIED HIGHLANDS

Abadan, Iran F6
Abu Dhabi (cap.), U.A.E. G7
Adana, Turkey E6
Aden (cap.), P.D.R. Yemen F8
Aden (gulf) F8
Afghanistan H6
Agra, India J7
Ahmadabad, India J7
Aleppo, Syria E6
Al Kuwait (cap.), Kuwait F7
Alma-Ata, U.S.S.R. J5
Altai (mts.) K5
Altun Shan (range), China K6
Amman (cap.), Jordan E6
Amudar'ya (riv.), U.S.S.R. H5
Amur (riv.) P5
Andaman (isls.), India L8
Ankara (cap.), Turkey E5
Aqaba (gulf) E7
Arabian (sea) H8
Araks (riv.) F6
Aral (sea), U.S.S.R. G5
Ararat (mt.), Turkey F6
Arctic (ocean) C1
Asahikawa, Japan P5
Ashkhabad, U.S.S.R. G6
Baghdad (cap.), Iraq F6
Bahawalpur, Pakistan J7
Bahrain G7
Bali (isl.), Indonesia N10
Balkhash (lake), U.S.S.R. J5
Bandar Seri Begawan (cap.), Brunei N9
Bandung, Indonesia M10
Bangalore, India J8
Bangkok (cap.), Thailand M8
Bangladesh L7
Basra, Iraq F6
Baykal (lake), U.S.S.R. N4
Beijing (Peking) (cap.), China N5
Beirut (cap.), Lebanon E6
Bengal (bay) K8
Bering (sea) V4
Bering (strait) W3
Bhutan L7
Black (sea) E5
Blagoveshchensk, U.S.S.R. O4
Bombay, India J8
Bonin (isls.), Japan E7
Borneo (isl.) N9
Brahmaputra (riv.) L7
British Indian Ocean Territory J10
Brunei N9
Bukhara, U.S.S.R. H5
Burma L7
Calcutta, India K7
Cambodia M8
Cannanore (isls.), India H8
Canton, China N7
Caspian (sea) G5
Celebes (isl.), Indonesia N10
Chang Jiang (Yangtze) (riv.), China N6
Chelyabinsk, U.S.S.R. H4
Chengdu, China M6
China L6
Chittagong, Bangladesh L7
Chongqing, China M7
Christmas (isl.), Australia M11
Chukchi (pen.), U.S.S.R. V3
Cocos (isls.), Australia L11
Colombo (cap.), Sri Lanka J9
Comorin (cape), India J9
Cyprus E6
Dacca (Dhaka) (cap.), Bangladesh L7
Dalian, China O6
Damascus (cap.), Syria E6
Da Nang, Vietnam M8
Delhi, India J7
Dhahran, Saudi Arabia F7
Dhaka (cap.), Bangladesh L7
Diego Garcia (isl.), Br. Ind. Ocean Terr. J10
Doha (cap.), Qatar G7
Dushanbe, U.S.S.R. H6
East China (sea) O7
Euphrates (riv.) F6
Everest (mt.) K7
Frunze, U.S.S.R. J5
Fukuoka, Japan O6
Ganges (riv.) K7
George Town, Malaysia M9
Gobi (des.) M5
Great Khingan (range), China O5
Great Wall (ruins), China N5
Guangzhou (Canton), China N7
Hadhramaut (reg.), P.D.R. Yemen F8
Hainan (isl.), China N8
Haiphong, Vietnam M7
Hakodate, Japan R5
Halmahera (isl.), Indonesia O9
Hangzhou, China N6
Hanoi (cap.), Vietnam M7
Harbin, China O5
Herat, Afghanistan H6
Himalaya (mts.) L7
Hindu Kush (mts.) J6
Hiroshima, Japan P6
Ho Chi Minh City, Vietnam M8
Hokkaido (isl.), Japan R5
Hong Kong N7
Honshu (isl.), Japan P6
Huang He (Hwang Ho) (riv.), China N6
Hue, Vietnam M8
Hyderabad, India J8
Hyderabad, Pakistan H7
Inch'ŏn, S. Korea O6
India J7
Indian (ocean) H10
Indonesia M10
Indus (riv.) H7
Inner Mongolia (reg.), China N5
Iran G7
Iraq F6
Irkutsk, U.S.S.R. M4
Irrawaddy (riv.), Burma L7
Irtysh (riv.), U.S.S.R. J4
Isfahan, Iran G6
Islamabad (cap.), Pakistan J6
Israel E6
Izmir, Turkey D6
Jaipur, India J7
Jakarta (cap.), Indonesia M10
Japan R6
Japan (sea) P6
Java (isl.), Indonesia M10
Jerusalem (cap.), Israel E6
Jidda, Saudi Arabia E7
Jordan E6
Kabul (cap.), Afghanistan H6
Kamchatka (pen.), U.S.S.R. S4
Kanpur, India K7
Kara (sea), U.S.S.R. H2
Karachi, Pakistan H7
Karakorum (ruins), Mongolia M5
Kathmandu (cap.), Nepal K7
Kazakh, S.S.R., U.S.S.R. H5
Kerman, Iran G6
Khabarovsk, U.S.S.R. P5
Khyber (pass) J6
Kirgiz S.S.R., U.S.S.R. J5
Kistna (riv.), India J8
Kitakyushu, Japan P6
Kobe, Japan P6
Kolyma (range), U.S.S.R. S3
Krasnoyarsk, U.S.S.R. L4
Kuala Lumpur (cap.), Malaysia M9
Kunlun (range), China K6
Kunming, China M7
Kuril (isls.), U.S.S.R. R5
Kuwait F7
Kyoto, Japan P6
Kyushu (isl.), Japan P6
Lahore, Pakistan J6
Lanzhou, China M6
Laos M8
Laptev (sea), U.S.S.R. O2
Latakia, Syria E6
Lebanon E6
Lena (riv.), U.S.S.R. O3
Leyte (isl.), Philippines O8
Lhasa, China L7
Lucknow, India K7
Luzon (isl.), Philippines O8
Macau N7
Madras, India K8
Magnitogorsk, U.S.S.R. H4
Makassar (str.), Indonesia N10
Malacca (str.) M9
Malaya (reg.), Malaysia M9
Malaysia M9
Maldives J9
Male (cap.), Maldives J9
Mandalay, Burma L7
Manila (cap.), Philippines N8
Mecca (cap.), Saudi Arabia F7
Medan, Indonesia L9
Medina, Saudi Arabia F7
Mekong (riv.) M8
Mindanao (isl.), Philippines O9
Molucca (isls.), Indonesia O10
Mongolia M5
Mosul, Iraq F6
Muscat (cap.), Oman G7
Mysore, India J8
Nagasaki, Japan O6
Nagoya, Japan P6
Naha, Japan O7
Nanjing, China N6
Nepal K7
New Delhi, India J7
New Guinea (isl.) P10
Nicosia (cap.), Cyprus E6
North Korea O5
Novosibirsk, U.S.S.R. J4
Ob' (riv.), U.S.S.R. H3
Okhotsk (sea), U.S.S.R. R4
Okinawa (isls.), Japan O7
Oman G8
Omsk, U.S.S.R. J4
Osaka, Japan P6
Pacific (ocean) T5
Pakistan H7
Pamir (plat.) J6
Peking (cap.), China N5
Persian (gulf) G7
Philippines O8
Phnom Penh (cap.), Cambodia M8
Poona, India J8
Pusan, S. Korea O6
P'yongyang (cap.), N. Korea O6
Qatar G7
Rangoon (cap.), Burma L8
Rawalpindi, Pakistan J6
Red (sea) E7
Riyadh (cap.), Saudi Arabia F7
Russian Soviet Federated Socialist Republic, U.S.S.R. L3
Ryukyu (isls.), Japan O7
Sabah (reg.), Malaysia N9
Saigon (Ho Chi Minh City), Vietnam M8
Sakhalin (isl.), U.S.S.R. R4
Salween (riv.) L8
Samarkand, U.S.S.R. H6
San'a (cap.), Yemen Arab Rep. F8
Sapporo, Japan P5
Sarawak (reg.), Malaysia N9
Saudi Arabia F7
Seoul (cap.), S. Korea O6
Severnaya Zemlya (isls.), U.S.S.R. M1
Shanghai, China O6
Shikoku, Japan P6
Siberia (reg.), U.S.S.R. M4
Singapore M9
Sinkiang (reg.), China K5
Socotra (isl.), P.D.R. Yemen G8
South China (sea) N8
South Korea O6
Sri Lanka K9
Sulu (sea), Philippines N9
Sumatra (isl.), Indonesia L9
Sunda (str.), Indonesia M10
Surabaya, Indonesia N10
Sverdlovsk, U.S.S.R. H4
Syrdar'ya (riv.), U.S.S.R. H5
Syria E6
Tabriz, Iran F6
Tadzhik S.S.R., U.S.S.R. H6
Taipei (cap.), Taiwan O7
Taiwan (isl.) N7
Ta'izz, Yemen Arab Rep. F8
Takla Makan (des.), China K6
Tashkent, U.S.S.R. H5
Taymyr (pen.), U.S.S.R. L2
Tehran (cap.), Iran G6
Tel Aviv-Jaffa, Israel E6
Thailand M8
Tianjin, China N6
Tibet (reg.), China K6
Tien Shan (range) K5
Tigris (riv.) F6
Timor (isl.), Indonesia O10
Tokyo (cap.), Japan R6
Tonkin (gulf) M8
Turkey E6
Turkmen S.S.R., U.S.S.R. G6
Ulaanbaatar (cap.), Mongolia M5
Union of Soviet Socialist Republics L3
United Arab Emirates G7
Ural (mts.), U.S.S.R. G4
Urmia (lake), Iran F6
Ürümqi, China K5
Uzbek S.S.R., U.S.S.R. H5
Varanasi, India K7
Verkhoyansk, U.S.S.R. P3
Vientiane (cap.), Laos M8
Vietnam M8
Vladivostok, U.S.S.R. P5
Wŏnsan, N. Korea O6
Wuhan, China N6
Xi'an, China M6
Yakutsk, U.S.S.R. O3
Yalu (riv.) O5
Yangtze (riv.), China N6
Yellow (sea) O6
Yemen Arab Republic F8
Yemen, People's Democratic Republic of F8
Yenisey (riv.), U.S.S.R. K3
Yogyakarta, Indonesia M10
Yokohama, Japan R6
Zhengzhou, China S3

Average January Temperature

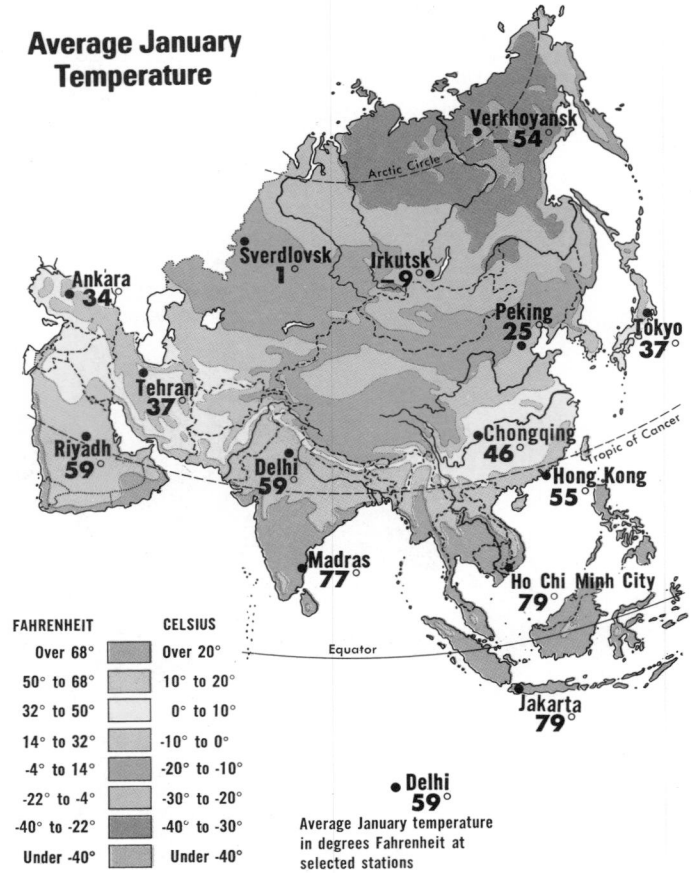

Verkhoyansk −54°
Sverdlovsk 1°
Irkutsk −9°
Ankara 34°
Peking 25°
Tokyo 37°
Tehran 37°
Chongqing 46°
Riyadh 59°
Delhi 59°
Hong Kong 55°
Madras 77°
Ho Chi Minh City 79°
Jakarta 79°

Arctic Circle
Tropic of Cancer
Equator

FAHRENHEIT	CELSIUS
Over 68°	Over 20°
50° to 68°	10° to 20°
32° to 50°	0° to 10°
14° to 32°	−10° to 0°
−4° to 14°	−20° to −10°
−22° to −4°	−30° to −20°
−40° to −22°	−40° to −30°
Under −40°	Under −40°

● Delhi 59°
Average January temperature in degrees Fahrenheit at selected stations

Average July Temperature

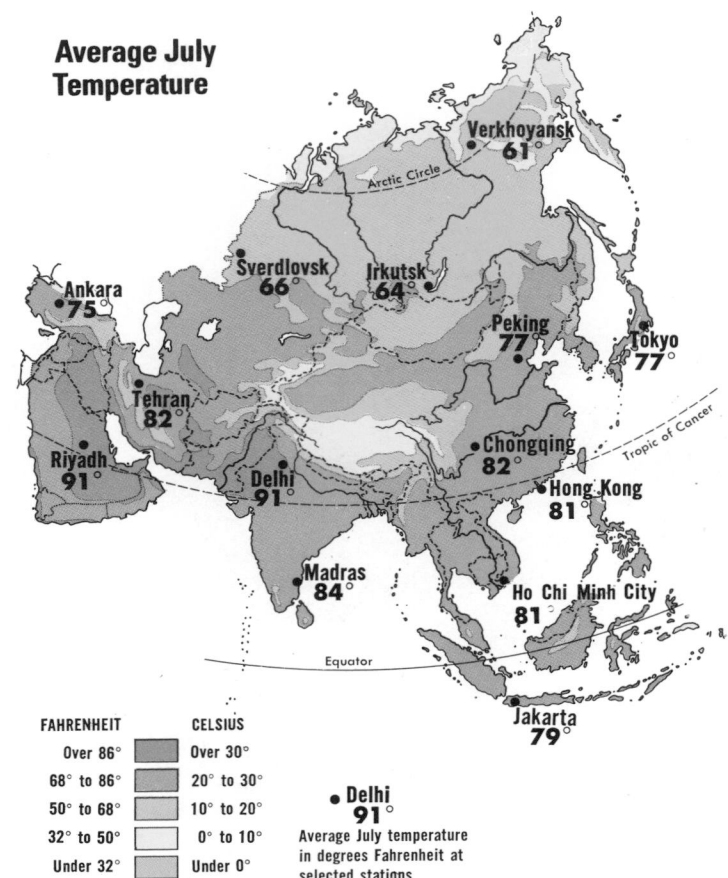

Verkhoyansk 61°
Sverdlovsk 66°
Irkutsk 64°
Ankara 75°
Peking 77°
Tokyo 77°
Tehran 82°
Chongqing 82°
Riyadh 91°
Delhi 91°
Hong Kong 81°
Madras 84°
Ho Chi Minh City 81°
Jakarta 79°

Arctic Circle
Tropic of Cancer
Equator

FAHRENHEIT	CELSIUS
Over 86°	Over 30°
68° to 86°	20° to 30°
50° to 68°	10° to 20°
32° to 50°	0° to 10°
Under 32°	Under 0°

● Delhi 91°
Average July temperature in degrees Fahrenheit at selected stations

Rainfall

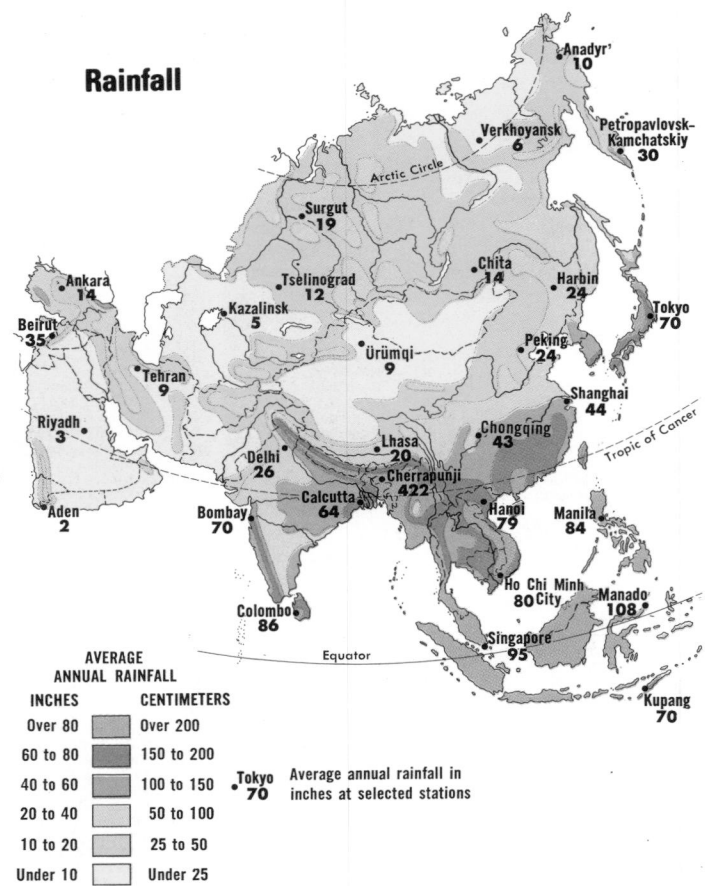

Anadyr' 10
Petropavlovsk-Kamchatskiy 30
Verkhoyansk 6
Surgut 19
Chita 14
Harbin 24
Tokyo 70
Tselinograd 12
Kazalinsk 5
Peking 24
Ankara 14
Beirut 35
Ürümqi 9
Shanghai 44
Tehran 9
Lhasa 20
Chongqing 43
Riyadh 3
Delhi 26
Cherrapunji 422
Calcutta 64
Hanoi 79
Manila 84
Aden 2
Bombay 70
Ho Chi Minh City 80
Manado 108
Colombo 86
Singapore 95
Kupang 70

Arctic Circle
Tropic of Cancer
Equator

AVERAGE ANNUAL RAINFALL

INCHES	CENTIMETERS
Over 80	Over 200
60 to 80	150 to 200
40 to 60	100 to 150
20 to 40	50 to 100
10 to 20	25 to 50
Under 10	Under 25

Tokyo ● 70
Average annual rainfall in inches at selected stations

Vegetation/Relief

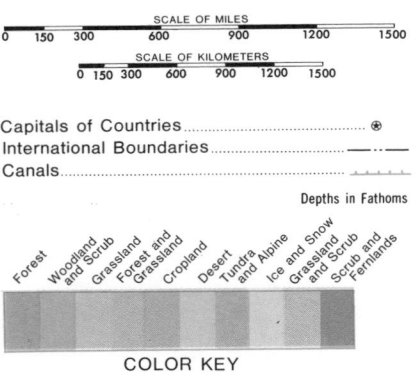

SCALE OF MILES
0 150 300 600 900 1200 1500

SCALE OF KILOMETERS
0 150 300 600 900 1200 1500

Capitals of Countries .... ⊛
International Boundaries .... — · —
Canals ....

Depths in Fathoms

Forest
Woodland and Scrub
Grassland
Forest and Grassland
Cropland
Desert
Tundra and Alpine
Ice and Snow
Grassland and Scrub
Scrub and Fernlands

COLOR KEY

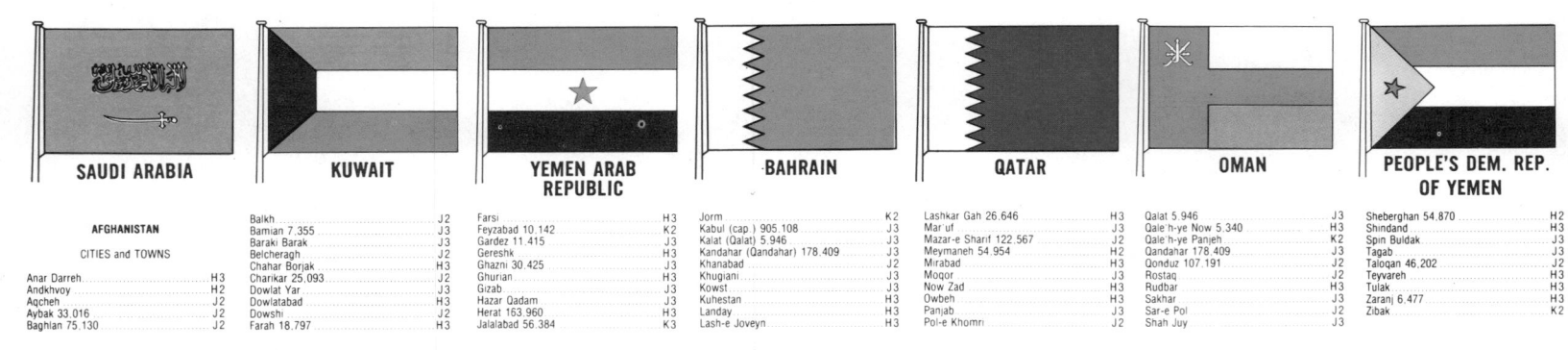

SAUDI ARABIA   KUWAIT   YEMEN ARAB REPUBLIC   BAHRAIN   QATAR   OMAN   PEOPLE'S DEM. REP. OF YEMEN

AFGHANISTAN

CITIES and TOWNS

Anar Darreh	H3	Farsi	H3
Andkhvoy	H2	Feyzabad 10.142	K2
Aqcheh	J2	Gardez 11.415	J3
Aybak 33.016	J2	Belcheragh	J2
Baghlan 75.130	J2	Gereshk	H3
Balkh 7.355	J2	Ghazni 30.425	J3
Bamian 7.355	J3	Ghurian	H3
Baraki Barak	J3	Gizab	J3
Belcheragh	J2	Hazar Qadam	J3
Chahar Borjak	H3	Herat 163.960	H3
Charikar 25.093	J2	Jalalabad 56.384	K3
Dowlat Yar	J3	Jorm	K2
Dowlatabad	H3	Kabul (cap.) 905.108	J3
Dowshi	J2	Kalat (Qalat) 5.946	J3
Farah 18.797	H3	Kandahar (Qandahar) 178.409	H3
		Khanabad	J2
		Khugiani	J3
		Kowst	J3
		Kuhestan	H3
		Landay	H3
		Lash-e Joveyn	H3
Lashkar Gah 26.646	H3	Qalat 5.946	J3
Mar uf	J3	Qale h-ye Now 5.340	H3
Mazar-e Sharif 122.567	J2	Qale h-ye Panjeh	K2
Meymaneh 54.954	H2	Qandahar 178.409	H3
Mirabad	H3	Qonduz 107.191	J2
Moqor	H3	Rostaq	J2
Now Zad	H3	Rudbar	H3
Owbeh	H3	Sakhar	J3
Panjab	J3	Sar-e Pol	J2
Pol-e Khomri	J2	Shah Juy	J3
Sheberghan 54.870	H2	
Shindand	H3	
Spin Buldak	J3	
Tagab	H3	
Taloqan 46.202	J2	
Teyvareh	H3	
Tulak	H3	
Zaranj 6.477	H3	
Zibak	K2	

UNITED ARAB EMIRATES

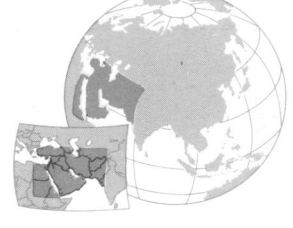

OTHER FEATURES

Farah Rud (riv.)	H3
Gowd-e Zerreh (depr.)	H4
Harirud (riv.)	H3
Helmand (riv.)	J3
Hindu Kush (mts.)	J2
Kabul (riv.)	K3
Konar (riv.)	K2
Lurah (riv.)	J3

Margow, Dasht-e (des.)	H3
Murghab (riv.)	H2
Namaksar (salt lake)	H3
Paropamisus (mts.)	H3
Rigestan (reg.)	H3

BAHRAIN

CITIES and TOWNS

Manama (cap.) 88,785	F4
Muharraq 37,732	F4

GAZA STRIP

CITIES and TOWNS

Gaza* 118,272	B3

IRAN

CITIES and TOWNS

Abadan 296,081	E3
Abadeh 16,000	F3
Abarqu 8,000	F3
Ahvaz 329,006	E3

Amol 68,782	F2
Anar 463	G3
Anarak 2,038	F3
Arak 114,507	E3
Ardabil 147,404	E2
Ardestan 5,868	F3
Asterabad (Gorgan) 88,348	F2
Babol 67,790	F2
Bafq 5,000	G3
Baft 6,000	G4

(continued on following page)

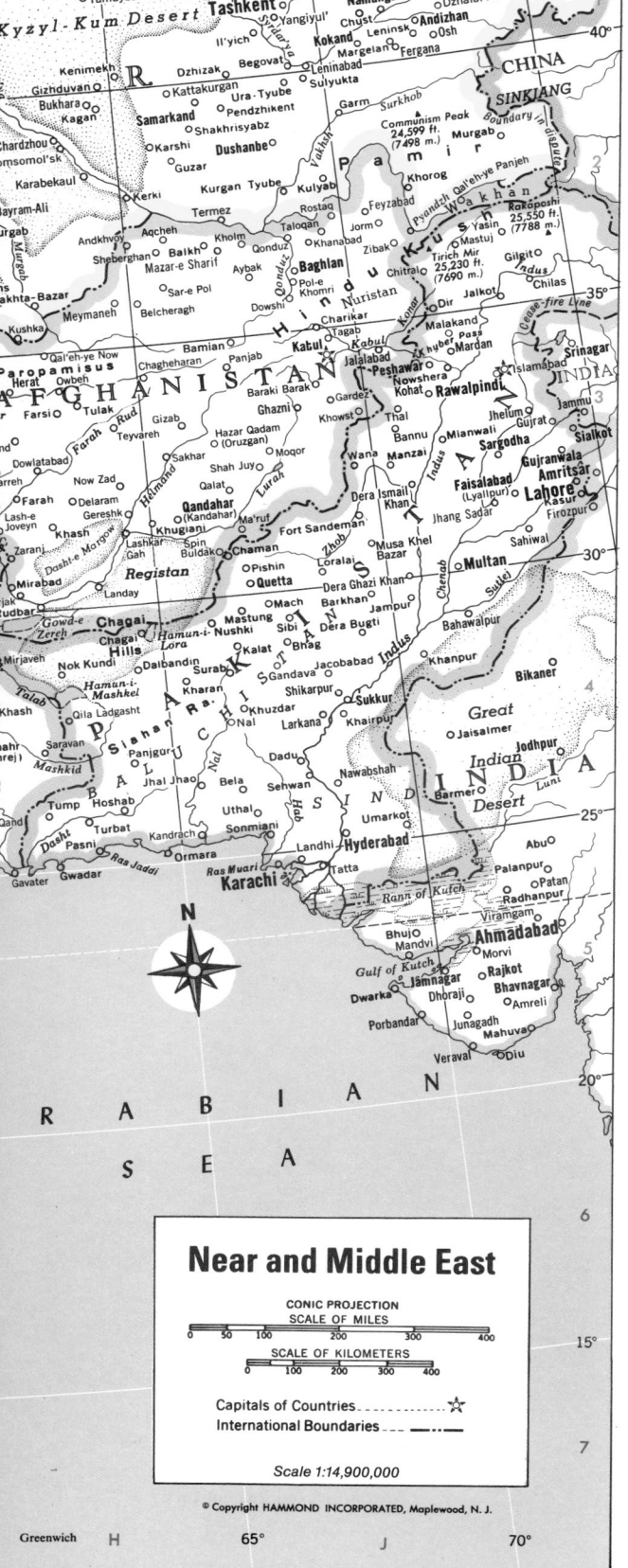

SAUDI ARABIA

AREA 829,995 sq. mi.
(2,149,687 sq. km.)
POPULATION 8,367,000
CAPITAL Riyadh
MONETARY UNIT Saudi riyal
MAJOR LANGUAGE Arabic
MAJOR RELIGION Islam

KUWAIT

AREA 6,532 sq mi. (16,918 sq. km.)
POPULATION 1,355,827
CAPITAL Al Kuwait
MONETARY UNIT Kuwaiti dinar
MAJOR LANGUAGE Arabic
MAJOR RELIGION Islam

YEMEN ARAB REPUBLIC

AREA 77,220 sq. mi. (200,000 sq. km.)
POPULATION 6,456,189
CAPITAL San'a
MONETARY UNIT Yemeni rial
MAJOR LANGUAGE Arabic
MAJOR RELIGION Islam

BAHRAIN

AREA 240 sq. mi. (622 sq. km.)
POPULATION 358,857
CAPITAL Manama
MONETARY UNIT Bahraini dinar
MAJOR LANGUAGE Arabic
MAJOR RELIGION Islam

QATAR

AREA 4,247 sq. mi. (11,000 sq. km.)
POPULATION 220,000
CAPITAL Doha
MONETARY UNIT Qatari riyal
MAJOR LANGUAGE Arabic
MAJOR RELIGION Islam

OMAN

AREA 120,000 sq. mi. (310,800 sq. km.)
POPULATION 891,000
CAPITAL Muscat
MONETARY UNIT Omani rial
MAJOR LANGUAGE Arabic
MAJOR RELIGION Islam

PEOPLE'S DEM. REP. OF YEMEN

AREA 111,101 sq. mi. (287,752 sq. km.)
POPULATION 1,969,000
CAPITAL Aden
MONETARY UNIT Yemeni dinar
MAJOR LANGUAGE Arabic
MAJOR RELIGION Islam

UNITED ARAB EMIRATES

AREA 32,278 sq. mi. (83,600 sq. km.)
POPULATION 1,040,275
CAPITAL Abu Dhabi
MONETARY UNIT dirham
MAJOR LANGUAGE Arabic
MAJOR RELIGION Islam

Topography

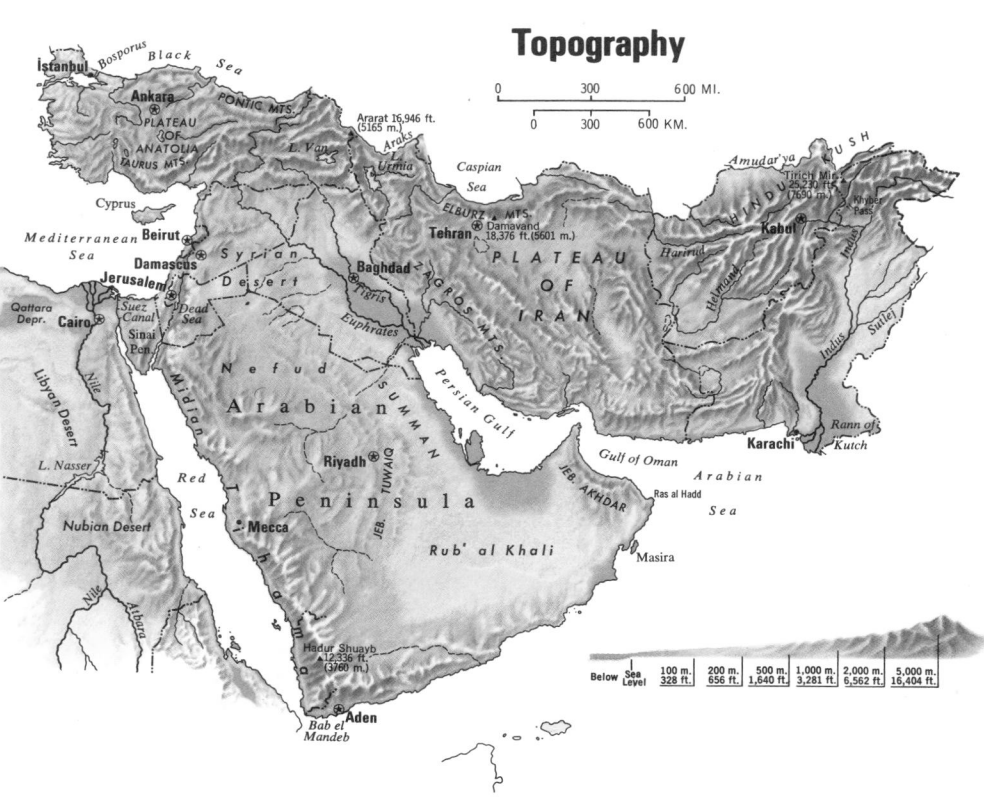

Near and Middle East

CONIC PROJECTION
SCALE OF MILES

SCALE OF KILOMETERS

Capitals of Countries .... ☆
International Boundaries ___ ___

Scale 1:14,900,000

© Copyright HAMMOND INCORPORATED, Maplewood, N.J.

Near and Middle East

(continued)

Bam 22,000 G4
Bampur 1,585 H4
Bandar A'bbas 89,103 G4
Bandar-e Anzali (Enzeli) 55,978 E2
Bandar-e Khomeyni 6,000 E3
Bandar-e Lengeh 4,920 F4
Bandar-e Rig 1,889 F4
Bandar-e Torkeman 13,000 F2
Bejestan 3,823 G3
Bir Bala 103 G4
Birjand 25,854 G3
Bojnurd 31,248 G2
Borazjan 20,000 F4
Borujerd 100,103 E3
Bushehr 57,681 F4
Chah Bahar 1,800 H4
Chalus 15,000 F2
Damghan 13,000 F2
Darab 13,000 G4
Dasht-e Azadegan 21,000 E3
Dashtiari H4
Dezful 110,287 E3
Dezh Shahpur 1,384 E2
Emamshahr 30,767 G2
Enzeli 55,978 E2
Estahbanat 18,187 F4
Fahrej (Iranshahr) 5,000 H4
Fasa 19,000 F4
Ferdows 11,000 G3
Gach Saran F3
Garmsar 4,723 F2
Golpayegan 20,515 F3
Gonabad 8,000 G3
Gorgan 88,348 F2
Hamadan 155,846 E3
Iranshahr 5,000 H4
Isfahan 617,825 F3
Jahrom 38,236 F4
Jask 1,078 G4
Kangan 2,682 F4
Kangavar 9,414 E3
Kashan 84,545 F3
Kashmar 17,000 G2
Kazerun 51,309 F4
Kerman 140,309 G3
Kermanshah 290,861 E3
Khash 7,439 H4
Khorramabad 104,928 E3
Khorramshahr 146,709 E3
Khvor 2,912 G3
Khvoy 70,040 E2
Lar 22,000 F4
Mahabad 28,610 E2
Maragheh 60,820 E2
Marand 24,000 E2
Meshed 670,180 H2
Mianeh 28,447 E2
Minab 4,228 G2
Mirjaveh 11,000 H4
Nahavand 24,000 E3
Na'in 5,925 F3
Najafabad 76,236 F3
Nasratabad (Zabol) 20,000 H3
Natanz 4,370 F3
Nehbandan 2,130 G3
Neyshabur 59,101 G2
Nikshahr H4
Pahlevi (Enzeli) 55,978 E2
Qasr-e Qand 1,879 H4
Qayen 6,000 G3
Qazvin 138,527 E2
Qom 246,831 F3
Quchan 29,133 G2
Qum (Qom) 246,831 F3
Rafsanjan 21,000 G3
Rasht 187,203 E2
Ravar 5,074 G3
Rey 102,825 F2
Reza'iyeh (Urmia) 163,991 D2
Sabzevar 69,174 G2
Sabzvaran 7,000 G4
Sai'dabad 20,000 G4
Sanandaj 95,834 E2
Saqqez 17,000 E2
Saravan H4
Sari 70,936 F2
Saveh 17,565 F2
Semnan 31,058 F2
Shahdad 2,777 G3
Shahreza 34,220 F3
Shiraz 416,408 F4
Shirvan 11,000 G2
Shustar 24,000 E3
Sirjan (Sai'dabad) 20,000 G4
Tabas 10,000 G3
Tabas-Masina (Tabas) 466 H3
Tabriz 598,576 E2
Tarom 394 G4
Tehran (cap.) 4,496,159 F2
Tonekabon 12,000 F2
Torbat-e Heydariyeh 30,106 G2
Torbat-e Jam 13,000 H2
Torud 721 F2
Turan G2
Turbat-i-Shaikh Jam 13,000 H2
Urmia 163,991 D2
Yazd 135,978 F3
Yazdan H3
Zabol 20,000 H3
Zahedan 92,628 H4
Zanjan 99,967 E2
Zarand 5,000 G3

OTHER FEATURES

Araks (riv.) E2
Atrek (riv.) G2
Bazman, Kuh-e (mt.) H4
Damavand (mt.) F2
Dez (riv.) E3
Elburz (mts.) F2
Gavkhuni (lake) F3
Gorgan (riv.) F2
Halil (riv.) G4
Jaz Murian, Hamun-e (marsh) G4
Karun (riv.) E3
Kavir, Dasht-e (salt des.) G3
Kavir-e Namak (salt des.) G3
Lut, Dasht-e (des.) G3
Maidani, Ras (cape) G4
Mand Rud (riv.) F4
Mashkid (riv.) H4
Mehran (riv.) F4
Namak, Daryacheh-ye (salt lake) F3
Namaksar (salt lake) H3
Namakzar-e Shahdad (salt lake) G3
Oman (gulf) G5
Persian (gulf) F4
Qeys (isl.) F4
Qezel Owzan (riv.) E2
Qeshm (isl.) G4
Safidar, Kuh-e (mt.) F4
Shaikh Shua'ib (isl.) F4
Shir Kuh (mt.) F3
Taftan, Kuh-e (mt.) H4
Talab (riv.) H4
Tashk (lake) F4
Urmia (lake) E2
Zagros (mts.) E3

IRAQ

CITIES and TOWNS

Al'Aziziya 7,450 E3
Al Falluja 38,072 D3
Al Fatha† 15,329 D2
Al Musaiyib 15,955 D3
Al Qurna 5,638 E3
'Amadiya 2,578 D2
'Amara 64,847 E3
'Ana 15,729 D3
An Najaf 128,096 D3
An Nasiriya 60,405 E3
Arbela (Erbil) 90,320 D2
Ar Rahhaliya 1,579 D3
As Salman 3,584 E3
Baghdad (cap.) 502,503 E3
Baghdad* 1,745,328 E3
Baq'uba 34,575 D3
Basra 313,327 E3
Erbil 90,320 D2
Habbaniya 14,405 D3
Haditha 6,870 D3
Hai 16,988 E3
Hilla 84,717 D3
Hit 9,131 D3
Karbal'a 83,301 D3
Khanaqin 23,522 D3
Kirkuk 167,413 D2
Kirkuk* 176,794 D2
Kut 42,116 E3
Maidan 354 E3
Mosul 315,157 D2
Qala' Sharqat 2,434 D2
Ramadi 28,723 D3
Rutba 5,091 D3
Samarra 24,746 D3
Samawa 33,473 D3
Shithatha 2,326 D3
Sulaimaniya 86,822 E2
Tikrit 9,921 D3

OTHER FEATURES

'Aneiza, Jebel (mt.) C3
'Ara'r, Wadi (dry riv.) D3
Batin, Wadi al (dry riv.) E4
Euphrates (riv.) E3
Hauran, Wadi (dry riv.) D3
Mesopotamia (reg.) D3
Syrian (El Hamad) (des.) D3
Tigris (riv.) E3

KUWAIT

CITIES and TOWNS

Al Kuwait (cap.) 181,774 E4
Mina al Ahmadi E4
Mina Saud E4

OTHER FEATURES

Bubiyan (isl.) E4
Persian (gulf) F4

OMAN

CITIES and TOWNS

Adam G5
Buraimi G5
Dhank G5
Ibra G5
I'bri G5
Juwara G6
Kamil G5
Khaluf G5
Khasab G4
Manah G5
Masqat (Muscat) (cap.) 7,500 G5
Matrah 15,000 G5
Mina al Fahal G5
Murbat G6
Muscat (cap.) 7,500 G5
Nizwa G5
Quryat G5
Raysut (Risut) F6
Salala 4,000 F6
Sarur G5
Shinas G5
Sohar G5
Sur G5
Suwaiq G5

OTHER FEATURES

Akhdar, Jebel (range) G5
Batina (reg.) G5
Dhofar (reg.) F6
Hadd, Ras al (cape) G5
Jibsh, Ras (cape) G5
Kuria Muria (isls.) G6
Madraka, Ras (cape) G6
Masira (gulf) G6
Masira (isl.) G5
Musandam, Ras (cape) G4
Nus, Ras (cape) G6
Oman (gulf) G5
Oman (reg.) G5
Ruus al Jibal (dist.) G4
Sauqira (bay) G6
Sauqira, Ras (cape) G6
Sham, Jebel (mt.) G5
Sharbatat, Ras (cape) G6

QATAR

CITIES and TOWNS

Doha (cap.) 150,000 F4
Dukhan F4
Umm Sai'd F5

OTHER FEATURES

Persian (gulf) F4
Rakan, Ras (cape) F4

SAUDI ARABIA

CITIES and TOWNS

Aba as Sau'd 47,501 D6
'Abaila F5
Abha 30,150 D6
Abqaiq E4
Abu 'Arish D6
Abu Hadriya E4
'Ain al Mubarrak C5
Al 'Ain C4
Al 'Ala C4
Al 'Auda E4
Al Birk D6
Al Hilla E5
Al Lidam D5
Al Lith C5
Al Muadhdam C4
'Anaiza D4
Artawiya E4
'Ashaira D5
Ayun D4
Badr C5
Buraida 69,940 D4
Dam D5
Dammam 127,844 F4
Dar al Hamra C4
Dhaba C4
Dhahran E4
Dharma E5
Dilam E5
Doqa D6
Duwadami D5
Er Ras D4
Faid D4
Gail E5
Haddar E5
Hadiya C4
Hafar al Batin E4
Hail 40,502 D4
Hamar E5
Hamda D6
Hanakiya D5
Haql C4
Harad E5
Haraja D6
Hariq E5
Hofuf 101,271 E4
Jabrin E5
Jauf C4
Jidda 561,104 C5
Jizan (Qizan) 32,812 D6
Jubail F4
Jubba D4
Junaina D5
Kaf C3
Khaibar, 'Asir D5
Khaibar, Hejaz C4
Khamis Mushait 49,581 D6
Khay D6
Khurma D5
Laila E5
Majmaa' D4
Maqna C4
Marib E6
Mastaba C5
Mastura C5
Mecca (cap.) 366,801 C5
Medain Salih C4
Medina 198,186 D5
Mendak D5
Mina Sau'd E4
Mubarraz 54,325 E4
Mudhnib D4
Muwailih C4
Najran (Aba as Sau'd) 47,501 D6
Nisab D4
O'qair E4
Qadhima C5
Qafar D4
Qasr al Haiyanya D4
Qatif E4
Qizan 32,812 D6
Qunfidha D6
Qusaiba D4
Rabigh C5
Ra's al Khafji E4
Ras Tanura F4
Riyadh (cap.) 666,840 E5
Rumah E4
Sabya D6
Sakaka D4
Salwa F5
Shaqra D4
Shuqaiq D6
Sufeina D5
Sulaiyil E5
Taif 204,857 D5
Taima C4
Tamra E5
Tathlith D6
Tebuk (Tabuk) 74,825 C4
Truba D4
Turaba D5
Umm Lajj C4
Wejh C4
Yamama E5
Yenbo C5
Zahran D6
Zalim D5
Zilfi E4

OTHER FEATURES

Abu-Mad, Ras (cape) C5
'Aneiza, Jebel (mt.) C3
'Aqaba (gulf) C4
Arafat, Jebel (mt.) D5
'Ara'r, Wadi (dry riv.) D3
Arma (plat.) E4
Aswad, Ras al (cape) C5
Bahr es Safi (des.) E6
Barida, Ras (cape) C5
Bisha, Wadi (dry riv.) D5
Dahana (des.) E4
Dawasir, Wadi (dry riv.) E5
Dawasir, Hadhb (range) D5
Farasan (isls.) D6
Hatiba, Ras (cape) C5
Jafura (des.) F5
Mashabi (isl.) C4
Midian (dist.) C4
Mishaa'b, Ras (cape) E4
Nefud (des.) D4
Nefud Dahi (des.) D5
Persian (gulf) F4
Ranya, Wadi (dry riv.) D5
Red (sea) C5
Rima, Wadi (dry riv.) D4
Rimal, Ar (des.) F5
Rub al Khali (des.) F5
Safaniya, Ras (cape) E4
Salma, Jebel (mts.) D4
Shaibara (isl.) C4
Shammar, Jebel (plat.) D4
Sirhan, Wadi (dry riv.) C3
Subh, Jebel (mt.) C5
Summan (plat.) E4
Tihama (reg.) C5
Tiran (isl.) C4
Tiran (str.) C4
Tuwaiq, Jebel (range) E5

UNITED ARAB EMIRATES

CITIES and TOWNS

Abu Dhabi (cap.) 347,000 F5
'AJman G4
'Aradah F5
Buraimi G5
Dubai F4
Fujairah G4
Jebel Dhanna F5
Ras al Khaimah G4
Ruwais F5
Sharjah F4
Umm al Qaiwain G4

OTHER FEATURES

Das (isl.) F4
Oman (gulf) G5
Yas (isl.) F5
Zirko (isl.) F5

WEST BANK

CITIES and TOWNS

Hebron 38,309 C3

OTHER FEATURES

Dead (sea) C3

YEMEN ARAB REP.

CITIES and TOWNS

'Amran D6
Bait al Faqih D7
Dhamar 19,467 D7
El Beida 5,975 E7
Hajja 5,814 D6
Harib E7
Hodeida 80,314 D7
Huth D6
Ibb 19,066 D7
Luhaiya D6
Marib 292 D6
Mocha D7
Saa'da 4,252 D6
Sana' (cap.) 134,588 D6
Sheikh Sai'd D7
Tai'zz 78,642 D7
Yarim D7
Zabid D7

OTHER FEATURES

Hanish (isls.) D7
Manar, Jebel (mt.) D7
Mandeb, Bab el (str.) D7
Red (sea) C5
Sabir, Jebel (mt.) D7
Tihama (reg.) C5
Zuqar (isl.) D7

YEMEN, PEOPLE'S DEM. REPUBLIC OF

CITIES and TOWNS

Aden (cap.) 240,370 E7
Ahwar E7
Balhaf E7
Bir 'Ali E7
Damqut F6
Ghaida F6
Habban E7
Hadibu F7
Hajarain E6
Haura E7
Hureidha E6
I'rqa E7
Lahej E7
Leijun E6
Lodar E7
Madinat ash Shab' E7
Meifa E7
Mukalla 45,000 E7
Nisab E7
Nuqub E6
Qishn F6
Riyan E7
Saihut F6
Seiyun 20,000 E6
Shabwa E6
Shibam E6
Shihr E7
Shuqra E7
Tarim E6
Yeshbum E7
Zinjibar E7

OTHER FEATURES

Fartak, Ras (cape) F6
Hadhramaut (dist.) E7
Hadhramaut, Wadi (dry riv.) F7
Kamaran (isl.) D6
Perim (isl.) D7
Socotra (isl.) F7

*City and suburbs.

Agriculture, Industry and Resources

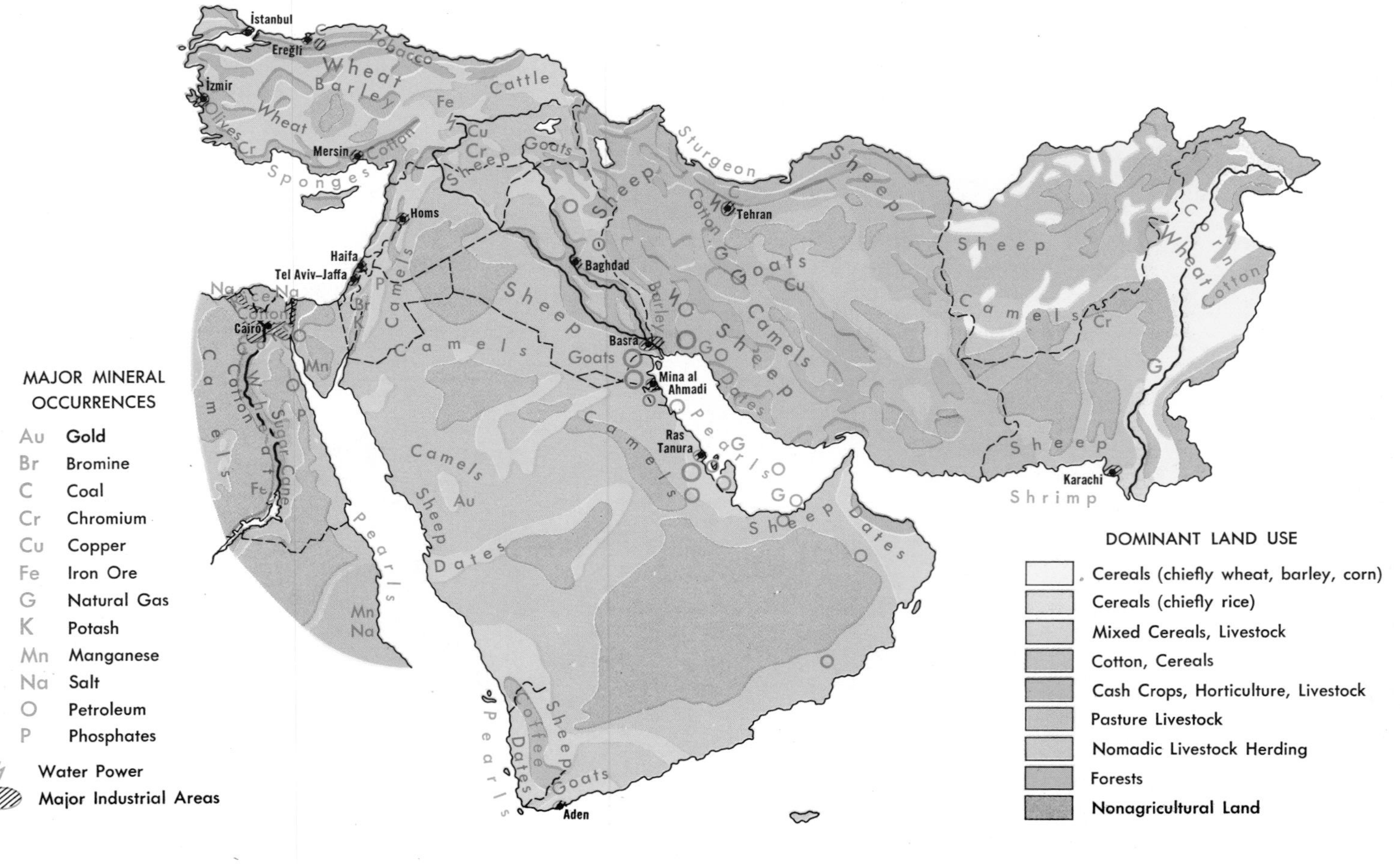

İstanbul
Ereğli
İzmir
Mersin
Homs
Haifa
Tel Aviv–Jaffa
Cairo
Baghdad
Tehran
Basra
Mina al Ahmadi
Ras Tanura
Karachi
Aden
Wheat
Barley
Cattle
Tobacco
Olives
Sponges
Sheep
Goats
Cotton
Sturgeon
Camels
Dates
Pearls
Corn
Shrimp
Coffee
MAJOR MINERAL OCCURRENCES
Au Gold
Br Bromine
C Coal
Cr Chromium
Cu Copper
Fe Iron Ore
G Natural Gas
K Potash
Mn Manganese
Na Salt
O Petroleum
P Phosphates
Water Power
Major Industrial Areas
DOMINANT LAND USE
Cereals (chiefly wheat, barley, corn)
Cereals (chiefly rice)
Mixed Cereals, Livestock
Cotton, Cereals
Cash Crops, Horticulture, Livestock
Pasture Livestock
Nomadic Livestock Herding
Forests
Nonagricultural Land

TURKEY

SYRIA

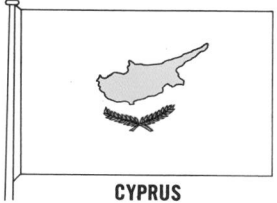

LEBANON

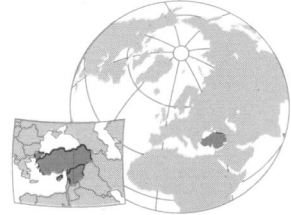

CYPRUS

AREA 300,946 sq. mi.
(779,450 sq. km.)
POPULATION 45,217,556
CAPITAL Ankara
LARGEST CITY Istanbul
HIGHEST POINT Ararat 16,946 ft.
(5,165 m.)
MONETARY UNIT Turkish lira
MAJOR LANGUAGE Turkish
MAJOR RELIGION Islam

AREA 71,498 sq. mi. (185,180 sq. km.)
POPULATION 8,979,000
CAPITAL Damascus
LARGEST CITY Damascus
HIGHEST POINT Hermon 9,232 ft.
(2,814 m.)
MONETARY UNIT Syrian pound
MAJOR LANGUAGES Arabic, French,
Kurdish, Armenian
MAJOR RELIGIONS Islam, Christianity

AREA 4,015 sq. mi. (10,399 sq. km.)
POPULATION 3,161,000
CAPITAL Beirut
LARGEST CITY Beirut
HIGHEST POINT Qurnet es Sauda
10,131 ft. (3,088 m.)
MONETARY UNIT Lebanese pound
MAJOR LANGUAGES Arabic, French
MAJOR RELIGIONS Christianity, Islam

AREA 3,473 sq. mi. (8,995 sq. km.)
POPULATION 629,000
CAPITAL Nicosia
LARGEST CITY Nicosia
HIGHEST POINT Troödos 6,406 ft. (1,953 m.)
MONETARY UNIT Cypriot pound
MAJOR LANGUAGES Greek, Turkish, English
MAJOR RELIGIONS Eastern (Greek) Orthodoxy,
Islam

CYPRUS

CITIES and TOWNS

...hali 2,970	E5
...ipiskopi 2,150	E5
...amagusta 38,960	E5
...tima	E5
...yrenia 3,892	E5
...ythrea 3,400	E5
...apithos 3,600	E5
...arnaca 19,608	E5
...efka 3,650	E5
...massol 79,641	E5
...Morphou 9,040	E5
...icosia (cap.) 115,718	E5
...aphos 8,984	E5
...reco (cape)	E5
...olis 2,200	E5
...izokarpasso 3,600	E5
...alousa 2,750	E5

OTHER FEATURES

...indreas (cape)	F5
...rnauti (cape)	E5
...ata (cape)	F5
...reco (cape)	F5
...ormakiti (cape)	E5
...roodos (mt.)	E5

LEBANON

CITIES and TOWNS

...leih 18,630	F6
...myun 7,926	F5
...aa'lbek 15,560	G5
...atrun 5,976	F5
...eirut (cap.) 474,870	F6
...eirut* 938,940	F6
...ermil 2,652	G5
...lerj U'yun 9,318	F6
...asheiya 6,731	F6
...ayak 1,480	F6
...aida 32,200	F6
...idon (Saida) 32,200	F6
...ur 16,483	F6
...ripoli (Tarabulus) 127,611	F5

Tyre (Sur) 16,483	F6
Zahle 53,121	F6
Zegharta 18,210	G5

OTHER FEATURES

Lebanon (mts.)	F6
Leontes (Litani) (riv.)	F6
Litani (riv.)	F6
Sauda, Qurnet es (mt.)	G5

SYRIA

PROVINCES

Aleppo 1,316,872	G4
Damascus 1,457,934	G6
Deir ez Zor 292,780	H5
Dera' 230,481	G6
El Quneitra 16,490	F6
Es Suweida 139,650	G6
Hama 514,748	G5
Haseke 468,506	J4
Homs 546,176	G5
Idlib 383,695	G5
Latakia 389,552	G5
Rashid 243,736	H5
Tartus 302,065	F5

CITIES and TOWNS

Abu Kemal 6,907	J5
A'in el A'rab 4,529	H4
Aleppo 639,428	G4
Azaz 13,923	G4
Baniyas 8,537	F5
Busra	G6
Damascus (cap.) 836,668	G6
Damascus* 923,253	G6
Deir ez Zor 66,164	H5
Dera' 27,651	G6
Dimashq (Damascus)
(cap.) 836,668	G6
Duma 30,050	G6
El Bab 27,366	G4
El Haseke 32,746	J4
El Ladhiqiya (Latakia) 125,716	F5
El Quryatein	G5
El Quneitra 17,752	F6
El Rashid 37,151	H5

En Nebk 16,334	G5
Es Suweide 29,524	G6
Et Tell el Abyad	H4
Haffe 4,656	G5
Haleb (Aleppo) 639,428	G4
Hama 137,421	G5
Harim 6,837	G4
Homs 215,423	G5
Idlib 34,515	G5
Izra 3,226	G6
Jeble 15,715	F5
Jerablus 8,610	G4
Jisr esh Shughur 13,131	G5
Khan Sheikhun	F5
Latakia 125,716	F5
Masyaf 7,058	G5
Membij 13,796	H4
Meskene	H5
Meyadin 12,515	J5
Qala'i es Salihiye	G6
Qamishliye 31,448	J4
Quteife 4,993	G6
Raqqa (El Rashid) 37,151	H5
Sabkha 3,375	H5
Safita 9,650	G5
Selemiya 21,677	G5
Tadmur 10,670	H5
Tartus 29,842	F5
Telkalakh 6,242	F5
Zebdani 10,010	G6

OTHER FEATURES

A'mrit (ruins)	F5
Arwad (Ruad) (isl.)	F5
Druz, Jebel ed (mts.)	G6
El Furat (riv.)	H4
Euphrates (El Furat) (riv.)	H4
Hermon (mt.)	J5
Khabur (riv.)	J5
Orontes (riv.)	G5
Palmyra (Tadmor) (ruins)	H5
Ruwaq, Jebel er (mts.)	G6

TURKEY

PROVINCES

Adana 1,240,475	F4

Adiyaman 346,892	H4
Afyonkarahisar 579,171	D3
Ağrı 330,201	K3
Amasya 322,806	F2
Ankara 2,585,293	E3
Antalya 669,357	D4
Artvin 228,026	J2
Aydın 609,869	B4
Balıkesir 789,255	B3
Bilecik 137,120	D2
Bingöl 210,804	H3
Bitlis 218,305	J3
Bolu 428,704	D2
Burdur 222,896	D4
Bursa 961,639	C2
Çanakkale 369,385	B2
Çankırı 265,468	E2
Çorum 547,580	F2
Denizli 560,916	C4
Diyarbakır 651,233	H4
Edirne 340,732	B2
Elazığ 417,924	H3
Erzincan 283,683	H3
Erzurum 746,666	J3
Eskişehir 495,097	D3
Gaziantep 715,939	G4
Giresun 463,587	H2
Gümüşhane 293,673	H2
Hakkâri 126,036	K4
Hatay 744,113	G4
İçel 714,817	F4
Isparta 322,685	D4
İstanbul 3,904,588	C2
İzmir 1,673,966	B3
Kahramanmaraş 641,480	G4
Kars 707,398	K2
Kastamonu 438,243	E2
Kayseri 676,809	F3
Kırklareli 268,399	B2
Kırşehir 232,853	F3
Kocaeli 477,736	C2
Konya 1,422,461	E4
Kütahya 470,423	C3
Malatya 574,558	H3
Manisa 872,375	B3
Mardin 519,687	J4
Muğla 400,796	C4
Muş 267,203	J3
Nevşehir 249,308	F3
Niğde 463,121	F4

CITIES and TOWNS

Acıgöl 3,934	F3
Acıpayam 5,046	C4
Adana (Antalya) 130,774	D4
Adalia 475,384	D4
Adapazarı 114,130	D2
Adilcevaz 9,022	K3
Adıyaman 43,782	H4
Afşin 18,231	G3
Afyonkarahisar 60,150	D3
Ağın 4,288	D4
Ağlasun 4,288	D4
Ağrı (Karaköse) 35,284	K3
Ahlat 7,995	J3
Akçaabat 10,756	H2
Akçadağ 7,366	G3
Akçakoca 9,066	D2
Akdağmadeni 7,909	F3
Akhisar 53,357	B3
Aksaray 45,564	F3
Akşehir 35,544	D3
Akseki 5,141	D4
Akviran 3,799	E4
Akyazı 12,438	D2
Alaca 12,552	F2
Alacahan 2,321	G3
Alaçam 10,013	F2
Alanya 18,520	D4
Alaşehir 23,243	C3
Alexandretta
(İskenderun) 107,437	G4
Aliağa 5,727	B3

Ordu 664,290	G2
Rize 336,278	J2
Sakarya 495,649	D2
Samsun 906,381	F2
Siirt 381,503	J4
Sinop 267,605	F2
Sivas 741,713	G3
Tekirdağ 319,987	B2
Tokat 599,166	G2
Trabzon 719,008	H2
Tunceli 164,591	H3
Urfa 597,277	H4
Uşak 229,679	C3
Van 386,314	K3
Yozgat 500,371	F3
Zonguldak 836,156	D2

CITIES and TOWNS

Alibeyköyü 33,387	D6
Almus 4,225	G2
Alpu 3,718	D3
Altındağ 512,392	E2
Altınova 6,980	B3
Altıntaş 3,386	C3
Altınözü 5,158	G4
Alucra 7,070	H2
Amasra 4,369	E2
Amasya 41,496	F2
Anamur 21,475	E4
Andırın 5,018	G4
Ankara (cap.) 1,701,004	E3
Antakya 77,518	G4
Antalya 130,774	D4
Antioch (Antakya) 77,518	G4
Araç 3,594	E2
Aralık 4,155	L3
Arapkir 8,436	H3
Ardahan 16,285	K2
Ardeşen 7,980	J2
Ardanuç 2,942	K2
Arguvan 2,461	H3
Arhavi 6,311	J2
Arpaçay 2,651	K2
Arsin 6,557	H2
Artova 2,813	G2
Artvin 13,390	J2
Aşkale 10,817	J3
Avanos 8,635	F3
Ayancık 7,202	F2
Ayaş 4,575	E2
Aybastı 13,180	G2
Aydın 59,579	B4
Aydıncık 6,739	E4
Ayrancı 2,664	E3
Ayvalık 18,041	B3
Ayvacık 3,120	B3
Babadağ 5,890	C4
Babaeski 17,090	B2
Bafra 34,288	F2
Bahçe 10,212	G4
Bakırköy 200,942	D6
Baklan 3,327	C4
Balâ 4,107	E3
Balıkesir 99,443	B3
Balya 2,362	B3
Bandırma 45,752	B2
Bartın 18,409	E2

Başkale 8,558	K3
Başmakçı 5,925	C4
Batman 64,384	J4
Bayat 4,671	F2
Bayburt 20,156	J2
Bayındır 14,078	B3
Baykan 2,690	J3
Bayramiç 6,385	B3
Bergama 29,749	B3
Beşiktaş 174,931	D6
Beşni 4,165	J4
Besni 16,313	G4
Beykoz 76,804	D6
Beyoğlu 230,532	D6
Beypazarı 14,963	D2
Beyşehir 15,060	D4
Beytüşşebap 2,766	K4
Biga 15,188	B2
Bigadiç 7,535	C3
Bilecik 11,269	C2
Bingöl (Çapakçur) 22,047	J3
Birecik 20,104	H4
Bismil 12,775	J4
Bitlis 25,054	J3
Bodrum 7,858	B4
Boğazlıyan 10,329	F3
Bolu 32,812	D2
Bolvadin 29,218	D3
Bor 16,560	F4
Borçka 4,636	J2
Bornova 45,096	B3
Boyabat 13,139	F2
Bozdoğan 7,218	C4
Bozkır 5,294	E4
Bozkurt 2,948	F2
Bozova 5,462	H4
Bozüyük 15,197	C3
Bucak 15,090	D4
Bulancak 14,153	H2
Bulanık 8,296	K3
Buldan 11,115	C3
Bünyan 12,277	G3
Burdur 36,633	D4
Burhaniye 12,800	B3
Bursa 346,103	C2
Büyükdere	D5
Çal 3,274	C3
Çala 2,450	K2
Çaldıran 3,366	K3

(continued on following page)

Agriculture, Industry and Resources

DOMINANT LAND USE

- Cereals (chiefly wheat, barley), Livestock
- Cash Crops, Horticulture, Livestock
- Pasture Livestock
- Nomadic Livestock Herding
- Forests
- Nonagricultural Land

MAJOR MINERAL OCCURRENCES

Ab	Asbestos		Na	Salt
Al	Bauxite		O	Petroleum
C	Coal		P	Phosphates
Cr	Chromium		Pb	Lead
Cu	Copper		Py	Pyrites
Fe	Iron Ore		Sb	Antimony
Hg	Mercury		Zn	Zinc
Mg	Magnesium			

⚡ Water Power

▨ Major Industrial Areas

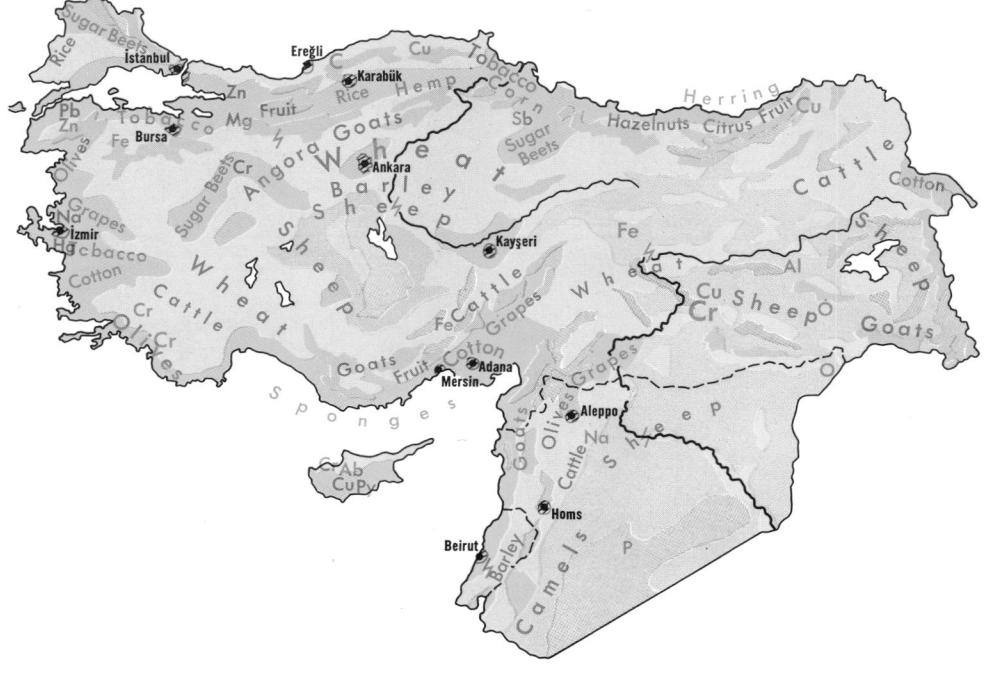

Çalköy 3,002 .... C3	Çeşme 5,284 .... B3	Demirkent 4,204 .... E4	Düzce 32,129 .... D2	Erzincan 60,351 .... H3	Gemerek 5,769 .... G3	Güdül 4,746 .... E2
Çamardi 2,419 .... F4	Çetinkaya 3,616 .... G3	Demirköy 4,257 .... B2	Eceabat 3,642 .... B6	Erzurum 162,973 .... J3	Gemlik 20,704 .... C2	Gülnar 6,344 .... E4
Çameli 2,502 .... C4	Ceyhan 62,909 .... F4	Denizli 106,902 .... C4	Edirne 63,001 .... B2	Eskimalatya 10,182 .... H3	Genç 7,671 .... J3	Gülşehir 6,188 .... F3
Çamlidere 4,386 .... E2	Ceylanpinar 20,171 .... H4	Dereli 4,188 .... H2	Edremit 26,110 .... B3	Eskipazar 2,865 .... E2	Genzein 4,925 .... F3	Gümüş 3,066 .... E2
Çan 11,797 .... B2	Çiçekdaği 3,203 .... F3	Derik 13,292 .... J4	Eğirdir 9,799 .... D4	Eskişehir 259,952 .... D3	Gerçüş 4,393 .... J4	Gümüşhacıköy 12,789 .... F2
Çanakkale 30,788 .... B6	Çide 3,520 .... E2	Derinkuyu 5,618 .... F3	Eğriani 1,793 .... H2	Esme 7,828 .... C3	Gerede 8,259 .... E2	Gümüşhane 11,166 .... H2
Çandir 6,986 .... F3	Çifteler 8,163 .... D3	Develi 17,323 .... F3	Elazığ 131,415 .... H3	Espiye 8,168 .... H2	Gerger 2,773 .... H3	Güney 7,154 .... C3
Çankaya 895,005 .... E3	Çihanbeyli 10,079 .... E3	Devrek 9,164 .... D2	Elbistan 26,048 .... G3	Eynesil 6,087 .... H2	Germencik 10,558 .... B4	Gürün 9,138 .... G3
Çankiri 28,512 .... E2	Cilvegözü E2	Devrekani 4,014 .... E2	Eldivan 3,392 .... E2	Eyüp 95,486 .... D6	Gerze 7,313 .... F2	Hacibektaş 5,032 .... F3
Çapakçur 22,047 .... J3	Çimin 5,341 .... H3	Dicle 5,247 .... J3	Eleşkirt 8,202 .... J3	Ezibider 3,631 .... H2	Gevaş 6,333 .... K3	Hacilar 15,622 .... F4
Çardak 4,232 .... C6	Çine 11,308 .... B4	Dikili 6,916 .... B3	Ezine 9,359 .... B3	Ezine 9,359 .... B3	Geyve 7,806 .... D2	Hadim 10,467 .... E4
Çarşamba 23,973 .... G2	Çivril 7,721 .... C3	Dinar 19,873 .... D3	Elmali 10,184 .... C4	Fakili 4,173 .... J3	Giresun 38,236 .... H2	Hafik 5,398 .... G3
Çatak 2,366 .... K3	Cizre 15,557 .... K4	Dirmil 3,476 .... C4	Emet 6,239 .... C3	Fatih 504,127 .... D6	Gökçe 4,470 .... B2	Hakkâri K4
Çatalca 7,693 .... C2	Çölemerik 11,735 .... K4	Divriği 12,302 .... H3	Fatsa 19,758 .... G2	Fatsa 19,758 .... G2	Gökçen 10,481 .... C3	(Çölemerik) 11,735 .... K4
Çatalzeytin 2,271 .... F1	Çorlu 40,134 .... C2	Diyadin 5,094 .... K3	Enez 2,486 .... B2	Feke 5,094 .... G3	Gölbaşı 15,103 .... H4	Halfeti 3,689 .... H3
Çay 12,200 .... D3	Çorum 64,852 .... F2	Diyarbakir 169,535 .... H4	Erbaa 20,315 .... G2	Fethiye 12,700 .... C4	Gölcük 33,279 .... C2	Hamur 2,267 .... K3
Çaycuma 8,118 .... D2	Doğanbey 3,077 .... D4	Doğanbey 3,077 .... D4	Erçis 22,351 .... K3	Fevzipaşa 5,576 .... G4	Göle 7,680 .... K2	Hanac 2,581 .... K4
Çayeli 13,480 .... J2	Dazkin 13,793 .... C3	Doğanhisar 9,487 .... D3	Erdek 8,685 .... B3	Findikli 5,008 .... J2	Gölhisar 7,095 .... C4	Hani 7,559 .... J3
Çayiralan 8,071 .... F3	Çukurca 3,019 .... K4	Doğanhisar 10,280 .... D3	Erdemli 19,936 .... F4	Finike 4,200 .... C4	Gölköy 10,022 .... G2	Harput 3,231 .... H3
Çayiri 4,580 .... J3	Çumra 19,225 .... E4	Döger 3,478 .... D3	Ereğli 45,992 .... D2	Foça 4,829 .... B3	Gölmarmara 11,982 .... C3	Haruniye 12,837 .... G4
Çekerek 3,796 .... F2	Çubuğlu 2,616 .... H3	Doğubeyazit 17,612 .... K3	Ereğli 50,354 .... F4	Gazipaşa 6,696 .... E4	Gölova 5,002 .... G2	Hassa 10,926 .... G4
Çelikhan 5,066 .... H3	Daday 2,528 .... E2	Domaniç 2,729 .... C3	Ergani 21,936 .... H3	Gaziantep 300,882 .... G4	Gönen 16,091 .... B3	Hatay (Antakya) 77,518 .... G4
Çemişkezek 3,048 .... H3	Darende 8,055 .... G3	Dortyol 19,390 .... F4	Erklet 3,924 .... F3	Gazipaşa 6,696 .... E4	Gördes 7,909 .... C3	Havran 7,552 .... B3
Çerkeş 3,780 .... E2	Çerkezköy 8,428 .... C2	Dumlu 3,259 .... J2	Ermenak 13,464 .... E4	Gebze 33,110 .... C2	Görele 8,079 .... H2	Havsa 4,298 .... B2
Çerkezköy 8,428 .... C2	Delice 3,462 .... F3	Dortyol 19,390 .... F4	Eruh 5,340 .... K4	Gediz 10,649 .... C3	Göynücek 2,600 .... F2	Havza 15,341 .... F2
Çermik 9,749 .... H3	Demirci 15,016 .... C3	Dursunbey 8,615 .... C3	Erzin 15,314 .... G4	Gelibolu (Gallipoli) 13,466 .... C5	Göynük 2,519 .... D2	Haymana 6,123 .... E3

Hayrabolu 12,331 B2
Hazro 4,896 J3
Hekimhan 11,818 G3
Hendek 15,291 D2
Hilvan 6,473 H4
Hinis 10,226 J3
Hisarönü 4,485 E2
Hizan 2,545 K3
Hopa 9,089 J2
Horasan 7,724 K2
Hozat 5,796 H3
çel (Mersin) 152,236 F4
dil 4,862 J4
ğdir 29,542 K3
lgaz 6,624 E2
lgin 11,830 D3
lica 8,947 J3
mranli 5,667 H2
ncesu 7,089 F3
nebolu 6,824 E2
negöl 37,805 C2
nönü 4,152 D3
psala 6,829 B2
psile 2,328 G2
skenderun 107,437 G4
skilip 16,588 F2
Islâhiye 20,683 G4
Isparta 62,870 D4
Ispir 3,929 J2
Istanbul 2,547,364 D6
Ivrindi 3,730 B3
Izmir 636,834 B3
Izmit 165,483 D2
Iznik 11,614 C2
Kadiköy 354,957 D6
Kadinhani 11,802 E3
Kadirli 34,779 F4
Kağithane 164,448 D6
Kağizman 11,517 K2
Kâhta 15,602 H4
Kalan 11,637 H3
Kale 3,399 C4
Kalecik 4,707 E2
Kaman 16,516 E3
Kandira 10,187 D2
Kangal 5,937 G3
Karabük 69,182 E2
Karacabey 21,648 C2
Karahalli 5,539 C3
Karaisali 2,316 F4
Karakoçan 5,604 H3
Karaköse (Ağri) 35,284 K3

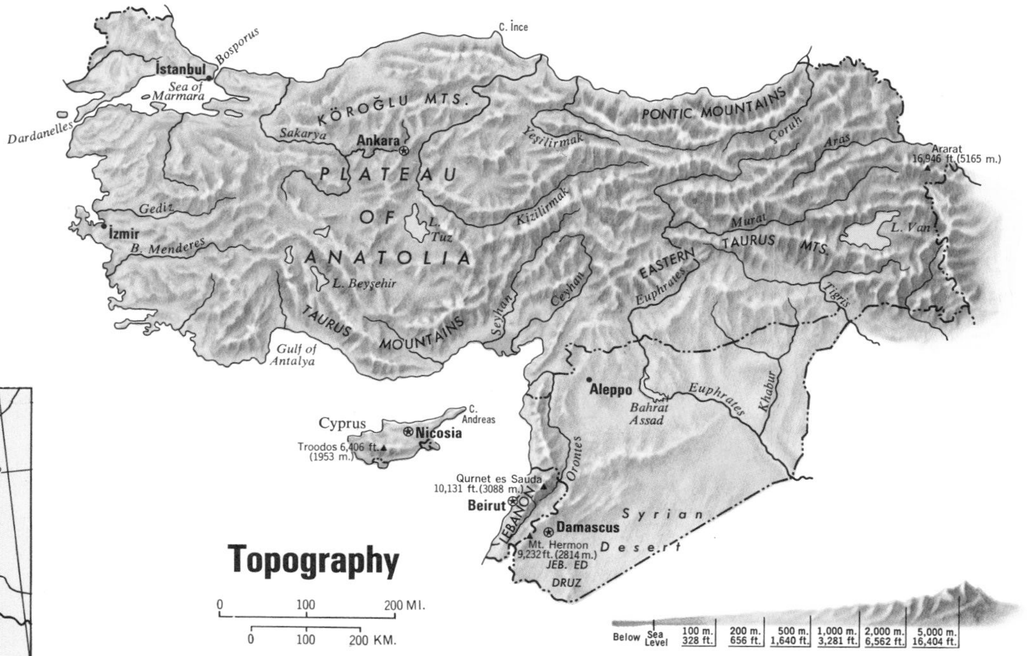

Topography

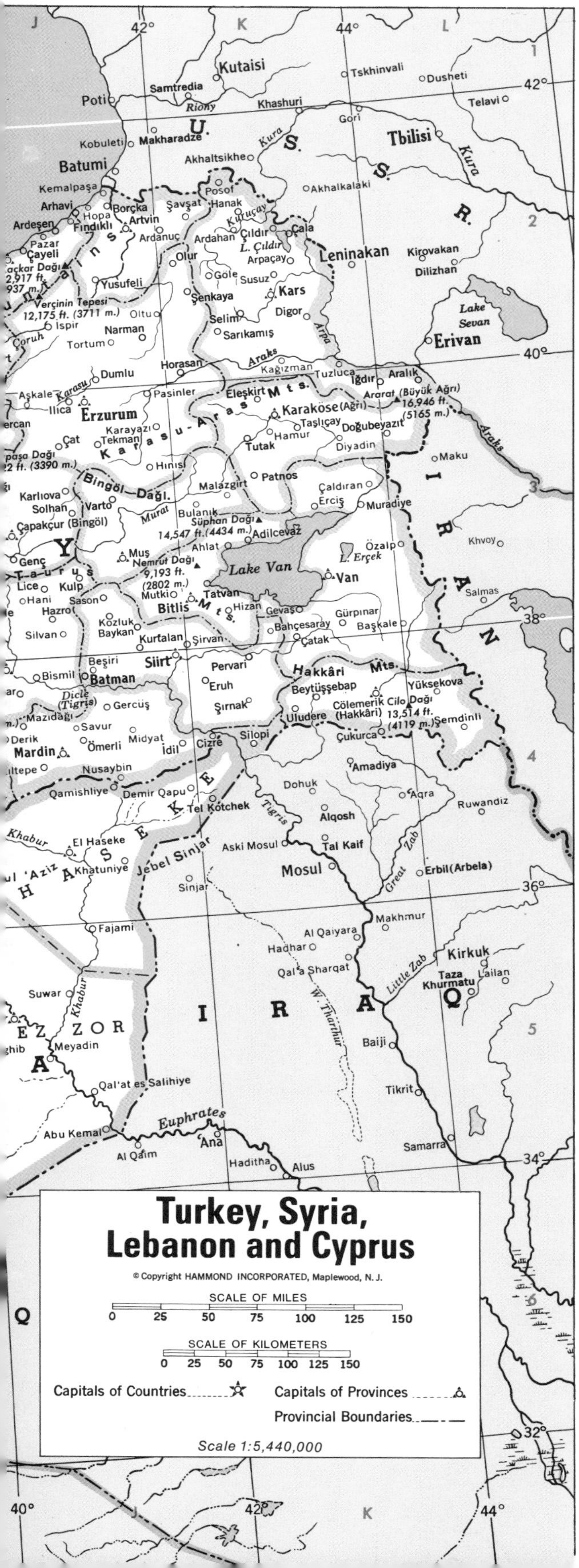

Karaman 43,759 E4
Karamanli 5,904 C4
Karapinar 19,589 E4
Karasu 11,600 D2
Karataş 5,598 F4
Karayaka 4,242 G2
Karayazi 3,595 J3
Kargi 5,021 F2
Karliova 3,631 J3
Kars 54,892 K2
Karşiyaka 171,600 B3
Kartal 53,073 D6
Kaş 2,493 C4
Kastamonu 29,993 F2
Kavak, Çanakkale 3,932 C5
Kavak, Samsun 3,964 F2
Kayseri 207,037 F3
Kazanli 4,461 F4
Kazimkarabekir 4,086 E4
Keban 5,800 H3
Keçiborlu 7,096 D4
Keles 2,423 C3
Kelkit 6,928 H2
Kemah 3,038 H3
Kemaliye 3,014 H3
Kemalpaşa 7,572 J2
Kemerburgaz 7,234 D5
Kemirhisar 6,205 F4
Kepsut 4,704 C3
Keşan 27,088 B2
Keşap 5,264 H2
Keskin 10,540 E3
Kiği 5,598 J3
Kilimli 26,649 D2
Kilis 54,055 G4
Kinik 11,785 B3
Kiraz 5,284 C3
Kirikhan 38,118 G4
Kirikkale 137,874 E3
Kirkağaç 15,078 B3
Kirklareli 33,265 B2
Kirşehir 41,415 F3
Kizilcahamam 7,050 E2
Kizilhisar 11,119 C4
Kiziltepe 21,531 J4
Kizilviran 3,260 E4
Kocaeli (Izmit) 165,483 D2
Koçarli 5,182 B4
Konya 246,727 E4
Korkuteli 10,334 D4
Köyceğiz 4,612 C4
Koyulhisar 3,861 G2
Kozakli 6,200 F3
Kozan 32,045 F4
Kozlu 27,322 D2
Kozluk 6,197 J3
Küçükköy 56,411 C6
Kula 10,807 C3
Kulp 4,474 J3
Kulu 11,707 E3
Kumkale 1,752 B6
Kumluca 7,704 D4
Küre 2,378 E2
Kurşunlu 6,562 E2
Kurtalan 7,001 J3
Kuşadasi 10,269 B4
Kütahya 82,442 C3
Kuyucak 6,039 C4
Lâdik 6,785 G2
Lâpseki 3,727 C6
Lice 8,625 J3
Lüleburgaz 32,401 B2
Maden 15,151 H3
Mağara 4,314 G3
Mahmudiye 5,240 D3
Malatya 154,505 H3
Malazgirt 13,094 K3
Malkara 14,399 B2
Maltepe 66,343 D6
Manavgat 10,804 D4
Manisa 78,114 B3
Manyas 4,410 B3
Maraş (Kahramanmaraş) 135,782 G4
Mardin 36,629 J4
Marmaris 5,596 C4
Mazgirt 3,141 H3
Mazidaği 4,842 J4
Mecitözü 6,066 F2
Menemen 18,464 B3
Mengen 2,459 D2
Meriç 3,922 B2
Mersin 152,236 F4
Merzifon 30,801 F2
Mesudiye 4,294 F4
Midyat 16,905 J4
Midye 2,003 C2
Mihaliççik 4,004 D3
Milâs 17,929 B4
Mucur 9,398 F3
Mudanya 8,399 C2
Mudurnu 3,905 D2
Muğla 24,178 C4
Muradiye 6,334 K3
Muş 27,761 J3
Mustafakemalpaşa 27,706 C3
Mut 11,466 E4
Mutki 2,815 J3
Muttalip 3,917 D3
Nallihan 7,883 D2
Narman 4,607 J2
Nazilli 52,176 C4
Nevşehir 30,203 F3
Niğde 31,844 F4
Niksar 19,156 G2
Nizip 36,190 G4
Nurhak 5,330 G4
Nusaybin 23,684 J4
Ödemiş 37,364 C3
Of 10,376 J2
Oğuzeli 7,194 G4
Oltu 10,093 J2
Ömerli 4,738 J4
Ordu 47,481 G2
Orhaneli 3,335 C3
Orhangazi 12,181 C2
Orta 3,596 E2
Ortaca 8,604 C4
Ortakaraviran 3,856 E4
Ortaköy, Çorum 2,657 F2
Ortaköy, Niğde 6,371 F3
Osmancik 11,921 F2
Osmaneli 4,789 D2
Osmaniye 61,581 G4
Ovacik, Tunceli 2,248 H3
Özalp 4,188 K3
Palu 5,489 H3
Pasinler 14,267 J3
Patnos 15,918 K3
Pazar, Rize 8,856 J2
Pazar, Tokat 4,337 G2
Pazarcik 15,943 G4
Pazaryeri 5,633 C3
Pera (Beyoğlu) 230,532 D6
Perşembe 6,701 G2
Pertek 4,176 H3
Pervari 4,126 K4
Pinarbaşi 9,503 G3
Pinarhisar 10,523 B2
Polatli 35,267 E3
Posof 2,209 K2
Pozanti 5,408 F4
Pülümür 3,442 H3
Pütürge 4,878 H3
Refahiye 6,570 H3
Reşadiye 9,022 G2
Reyhanli 25,749 G4
Rize 36,044 J2
Şabanözü 3,442 E2
Safranbolu 14,793 E2
Saimbeyli 3,622 G4
Sakarya (Adapazari) 114,130 D2
Salihli 45,514 C3
Samandaği 22,540 F4
Samsat 2,083 H4
Samsun 168,478 F2
Sandikli 13,181 D3
Sapanca 9,040 D2
Şaphane 3,919 C3
Sarayköy 10,513 C4
Sarayönü 8,946 E3
Sarigöl 6,979 C3
Sarikamiş 21,262 K2
Sarikaya 5,160 F3
Sariköy 4,695 B2
Sarioğlan 3,245 G3
Sariyer 79,329 D5
Sariz 3,591 G3
Şarkikaraağaç 4,772 D3
Şarkişla 12,763 G3
Şarköy 5,396 B2
Sason 3,211 J3
Savaştepe 7,179 B3
Şavşat 3,078 K2
Savur 4,983 J4
Seben 2,471 D2
Şebinkarahisar 10,214 H2
Şefaatli 6,769 F3
Seferihisar 6,484 B3
Selçuk 12,251 B3
Selendi 4,457 C3
Selim 3,569 K2
Selimiye 2,989 B4
Senirkent 8,247 D3
Şenkaya 3,190 K2
Şereflikoçhisar 20,523 E3
Serik 14,161 D4
Seydişehir 25,651 D4
Seyitgazi 2,819 D3
Siirt 35,654 J4
Şile 4,062 C2
Silifke 19,257 E4
Silivri 8,525 C2
Silopi 4,460 K4
Silvan 29,599 J3
Simav 11,601 C3
Sincanli 3,847 D3
Sindirgi 7,818 C3
Sinop 16,098 F2
Şiran 5,048 H2
Şirnak 10,587 K4
Şirvan 5,166 K3
Sivas 149,201 G3
Sivasli 4,394 C3
Siverek 40,990 H4
Sivrihisar 8,713 D3
Smyrna (Izmir) 636,834 B3
Söğüt 5,329 D3
Söke 35,407 B4
Solhan 7,014 J3
Soma 23,713 B3
Sorgun 14,081 F3
Şuhut 8,154 D3
Sulakyurt 4,311 E2
Sultandaği 4,017 D3
Sultanhani 5,112 E3
Suluova 21,278 F2
Sungurlu 21,641 F2
Sürmene 8,096 J2
Sürüç 20,395 H4
Suşehri 10,863 H2
Susurluk 14,000 C3
Susuz 5,006 K2
Sütçüler 2,721 D4
Tarsus 102,186 F4
Taşkent 7,098 E4
Taşköprü 8,146 F2
Taşliçay 3,684 K3
Taşova 6,516 F4
Tatvan 29,271 K3
Tavas 9,728 C4
Tavşanli 19,575 C3
Tefenni 4,280 C4
Tekirdağ 41,257 B2
Tercan 6,068 J3
Terme 15,660 G2
Tire 30,694 B3
Tirebolu 7,385 H2
Tokat 48,588 G2
Tomarza 6,548 F3
Tömük 7,660 F4
Tonya 10,544 H2
Torbali 17,237 B3
Tortum 4,110 J2
Torul 3,221 H2
Tosya 17,515 F2
Trabzon 97,210 H2
Trebizond (Trabzon) 97,210 H2
Tunceli (Kalan) 11,637 H3
Turgutlu 47,009 B3
Turhal 39,170 F2
Türkeli 2,194 F2
Türkoğlu 9,207 G4
Tutak 4,325 K3
Tuzluca 5,209 K3
Tuzlukçu 4,613 D3
Ula 5,117 C4
Ulaş 2,469 G3
Ulubey 4,214 C3
Uluborlu 10,016 D3
Uludere 4,050 K4
Ulukişla 6,336 F4
Umurbey 2,754 C6
Ünye 23,366 G2
Urfa 132,934 H4
Ürgüp 6,758 F3
Urla 13,903 B3
Uşak 58,578 C3
Üsküdar 202,957 D6
Üzümlü 4,365 D4
Uzunköprü 27,005 B2
Vakfikebir 12,556 H2
Van 63,663 K3
Varto 5,572 J3
Vezirköprü 11,705 F2
Viranşehir 26,244 H4
Vize 8,203 B2
Yahyali 13,738 F3
Yalova, Istanbul 27,289 C2
Yalvaç 18,305 D3
Yaprakli 3,020 E2
Yatağan 4,903 C4
Yayladaği 4,471 F5
Yenice, Çanakkale 4,004 B3
Yenice, Içel 4,106 F4
Yenice, Zonguldak 5,791 E2
Yeniceoba 5,740 E3
Yeniköy, Istanbul D6
Yenimahalle 198,643 E3
Yenişehir 15,188 C2
Yerkesik 2,381 C4
Yerköy 19,927 F3
Yeşilhisar 10,409 F3
Yeşilköy D6
Yeşilova, Burdur 3,685 C4
Yeşilova, Niğde 5,237 E3
Yeşilyurt 7,451 H3
Yildizeli 7,043 G3
Yozgat 32,501 F3
Yüksekova 7,329 L4
Yumurtalik 2,442 F4
Yunak 6,187 D3
Yusufeli 3,050 J2
Zara 10,376 G3
Zeytinburnu 123,548 D6
Zeytindağ 3,517 B3
Zile 32,157 G2
Zivarik 2,703 E3
Zonguldak 90,221 D2

OTHER FEATURES

Abydos (ruins) B6
Aci (lake) C4
Adalar (isl.) D6
Aegean (sea) A3
Ağri, Büyuk (Ararat) (mt.) L3
Akdağ (mt.) C4
Aladağ (mt.) F4
Alexandretta (gulf) F4
Amanos (mts.) G4
Anamur (cape) E5
Anatolia (reg.) D3
Ankara (riv.) D3
Antalya (gulf) D4
Anti-Taurus (mts.) G3
Araks (riv.) K2
Ararat (mt.) L3
Arpa (riv.) K2
Baba (cape) A3
Bati Firat (riv.) H3
Beyşehir (lake) D4
Black (sea) E1
Bosporus (str.) C2
Bozcaada (isl.) A3
Burgaz (isl.) D6
Büyük Ağri (Ararat) (mt.) L3
Çanakkale Boğazi (Dardanelles) (str.) B6
Çandarli (gulf) B3
Canik (mts.) G2
Ceyhan (riv.) F4
Cilo Daği (mt.) K4
Çoruh (riv.) J2
Dardanelles (str.) B6
Dicle (riv.) J4
Eastern Taurus (mts.) J3
Ephesus (ruins) B3
Erciyas Daği (mt.) F3
Ergene (riv.) B2
Euphrates (Firat) (riv.) G4
Firat (riv.) G4
Gediz (riv.) C3
Gelidonya (cape) D4
Gökçeada (isl.) A2
Göksu (riv.) E4
Helles (cape) B6
Heybeli (isl.) D6
Ilium (ruins) B6
Imroz (Gökçeada) (isl.) A2
Ince (cape) F1
Istranca (mts.) B2
Kaçkar Daği (mt.) J2
Karadeniz Boğazi (Bosporus) (str.) C2
Karasu-Aras (mts.) J3
Kelkit (riv.) G2
Kerme (gulf) B4
Keşiş Tepesi (mt.) H3
Kizilirmak (riv.) F2
Koca (riv.) C3
Köroğlu (mts.) E2
Küre (mts.) E2
Mandalya (gulf) B4
Marmara (isl.) B2
Marmara (sea) C2
Menderes, Büyük (riv.) C4
Meriç (riv.) B2
Murat (riv.) H3
Pontic (mts.) H2
Porsuk (riv.) D3
Prinkipo (Adalar) (isl.) D6
Sakarya (riv.) D2
Saros (gulf) B2
Seyhan (riv.) F4
Simav (riv.) C3
Sinop (cape) F1
Sultan (mts.) D3
Süphan Daği (mt.) K3
Taurus (mts.) D4
Tigris (Dicle) J4
Troy (Ilium) (ruins) B6
Tuz (lake) E3
Van (lake) K3
Yeşilirmak (riv.) G2

* City and suburbs

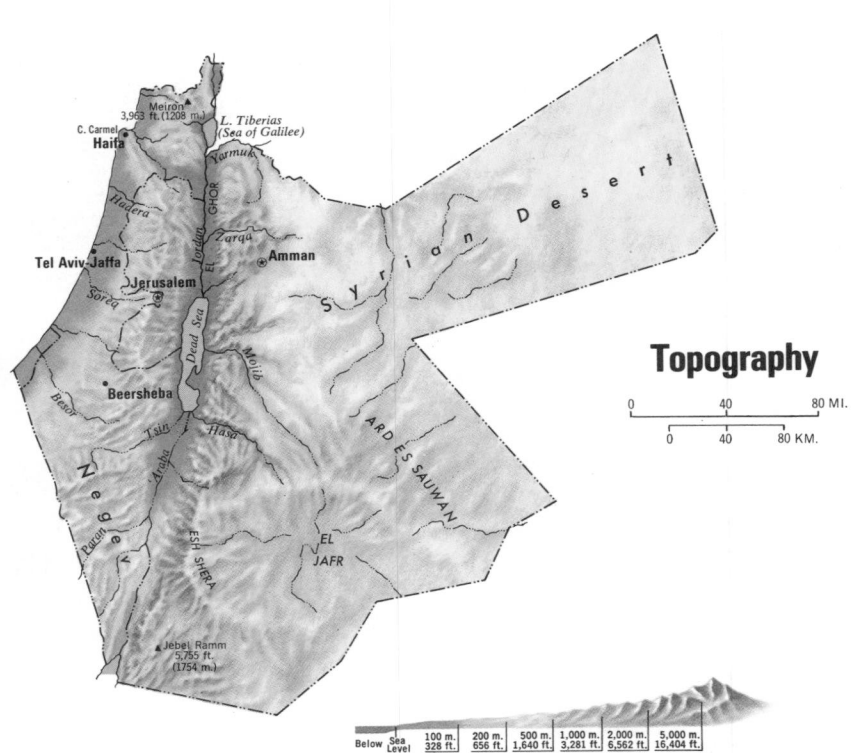

Topography

0   40   80 MI.

0   40   80 KM.

Below Sea Level	100 m. 328 ft.	200 m. 656 ft.	500 m. 1,640 ft.	1,000 m. 3,281 ft.	2,000 m. 6,562 ft.	5,000 m. 16,404 ft.

ISRAEL

DISTRICTS

Central 572,300 ....B3
Haifa 480,600 ....C2
Jerusalem 338,600 ....B4
Northern 473,700 ....C2
Southern 351,300 ....B5
Tel Aviv 905,100 ....B3

CITIES and TOWNS

Acre 34,400 ....C2
Afiqim 1,243 ....D2
'Afula 17,400 ....C2
Ahuzzam 407 ....B4
Akko (Acre) 34,400 ....C2
Arad 5,400 ....C5
'Arrabe 6,000 ....C2
Ashdod 40,500 ....B4
Ashdot Yaa'qov 1,197 ......D2
Ashqelon 43,100 ....A4
Atlit 1,516 ....B2
Avihayil 579 ....B3
Bat Shelomo 218 ....B2
Be'er Yam 124,100 ....B3
Be'er 390 ....A5
Be'er Menuha D5
Beersheba (Be'er
  Sheva) 101,000 ....B5
Be'er Tuveya 602 ....B4
Beit Guvrin B4
Bene Beraq 74,100 ....B3
Bet Qama 228 ....B5
Bet She'an 11,300 ....D3
Bet Shemesh 10,100 ....B4
Binyamina 2,701 ....B2
Carmiel C2
Dafna 577 ....D1
Dalyat al-Karmel 6,200 ....B2
Dan 498 ....D1
Dimona 23,700 ....D4
Dor 195 ....B2
E'in Gedi C5
E'in Harod 1,372 ....C2
Elat D6
Elath (Elat) 12,800 ....D6
El 'Auja D5
Elyakim 568 ....C2
Elyashiv 435 ....B3
Even Yehuda 3,464 ....B3
Gal'on 356 ....B4
Gat 430 ....B4
Gedera 5,400 ....B4
Gerofit D5
Gesher 360 ....B2
Gesher Haziv 238 ....C1
Gevara'm 283 ....B4
Gilat 561 ....B5
Ginnosar 473 ....D2
Giv'atayim 48,500 ....B3
Giva't Brenner 1,505 ....B4
Giv'at Hayyim 1,360 ....B3
Habonim 189 ....B2
Hadera 31,900 ....B3
Haifa 227,800 ....B2
Haifa* 367,400 ....B2
Hatseva D5
Hazerim 127 ....B5

Hazor Hagelilit D2
Helez 466 ....B4
Herzeliyya 41,200 ....B3
Hod Hasharon 13,500 ......C3
Hodiyya 400 ....B4
Holon 121,200 ....B3
Iksal 2,156 ....C2
Jerusalem (cap.) 376,000 ..C4
Jish 1,498 ....C1
Kafar Kanna 5,200 ....C2
Kafr Yasif 2,975 ....C2
Karkur-Pardes Hanna 13,600 ..C3
Kefar Blum 565 ....D1
Kefar Gila'di 701 ....C1
Kefar Ruppin 306 ....D3
Kefar Sava 26,500 ....B3
Kefar Vitkin 808 ....B3
Kefar Zekhariya 420 ....B4
Kinneret 909 ....D2
Lod (Lydda) 30,500 ....B4
Lydda 30,500 ....B4
Maa'lot-Tarshiha C1
Magen 149 ....A5
Maa'lot-Tarshiha C1
Malkiya D1
Mash 'Abbe Sade 238 .....B6
Mavqi'm 177 ....B4
Megiddo C2
Metula 261 ....D1
Migdal 688 ....C2
Migdal Ha E'meq C2
Mikhmoret 608 ....B3
Mishmar Hanegev 336 ....B5
Mishmar Hayarden D1
Mivtahim 398 ....A5
Mizpe Ramon 331 ....D5
Moza Illit 219 ....C4
Mughar 4,010 ....C2
Muqeible 459 ....C1
Nahariyya 24,000 ....C1
Nazareth 33,300 ....C2
Nazerat I'llit C2
Negba 453 ....B4
Nes Ziyyona 11,700 ....B4
Netanya 70,700 ....B3
Netivot B5
Nevatim 436 ....B5
Newe Yam 211 ....B2
Newe Zohar C5
Nir Yitzhaq 209 ....A5
Nizzanim 479 ....B4
Ofaqim B5
O'mer B5
Oron C6
Or Yehuda B4
Pardes Hanna-Karkur 13,600 ..B2
Peduyim 361 ....B5
Petah Tiqwa 112,000 ....B3
Qadima 2,937 ....B3
Qalansuwa B3
Qedma 157 ....B4
Qiryat Atta C2
Qiryat Bialik 18,000 ....C2
Qiryat Gat 19,200 ....B4
Qiryat Mal'akhi B4
Qiryat Motzkin 17,600 .....C2
Qiryat Shemona 15,200 ...C1
Qiryat Tivo'n 9,800 ....C2
Qiryat Yam 19,800 ....C2
Raa'nana 14,900 ....B3
Ramat Gan 120,900 ....B3

Ramat Hasharon 20,100 ...B3
Rame 2,986 ....C2
Ramla 34,100 ....B3
Rehovot 39,200 ....B4
Rei'm 155 ....B4
Revadim 175 ....B4
Revivim 258 ....B5
Rishon Le Ziyyon 51,900 ..B4
Rosh Ha 'Ayin B3
Rosh Pinna 700 ....C2
Ruhama 497 ....B4
Saa'd 418 ....B4
Safad (Zefat) 13,600 ....C1
Sakhnin 8,400 ....C2
Sede Boqer C5
Sederot B4
Sedot Yam 511 ....B2
Sedom C5
Shave Ziyyon 269 ....C1
Shefara'm 11,800 ....C2
Shefayim 614 ....B3
Shoval 393 ....B4
Tayibe 11,700 ....C3
Tel Aviv-Jaffa 343,300 .....B3
Tel Aviv-Jaffa* 1,219,900 ..B3
Tiberias 23,800 ....C2
Tirat Hakarmel 14,400 .....B2
Tirat Zevi 353 ....D3
Tur'an 2,304 ....C2
Umm el Fahm 13,300 ......C2
Urim 203 ....B5
Uzza 487 ....B4
Yad Mordekhai 416 ....A4
Yagur 1,266 ....C2
Yahav D5
Yavne 10,100 ....B4
Yavne'el 1,580 ....C2
Yehud 8,900 ....B3
Yeroham 5,800 ....C5
Yesodot 293 ....B4
Yesud Hamaa'la 428 ....C1
Yiftah C1
Yirka 2,715 ....C2
Yotvata D6
Zavdi'el 396 ....B4
Ze'elim 148 ....B5
Zefat 13,600 ....C1
Zikhron Yaa'qov 6,500 ....B2
Zippori 241 ....C2

OTHER FEATURES

Aqaba (gulf) C6
'Araba, Wadi (valley) C5
Beer Sheva' (dry riv.) B5
Besor (riv.) B5
Carmel (cape) B2
Carmel (mt.) C2
Dead (sea) C4
Galilee, Sea of (Tiberias)
  (lake) C2
Galilee (reg.) C1
Gerar (dry riv.) B4
Hadera (dry riv.) B3
Haniqra, Rosh (cape) C1
Jordan (riv.) C2
Judaea (reg.) C4
Lakhish (dry riv.) B4
Meiron (mt.) C1
Negev (reg.) C5

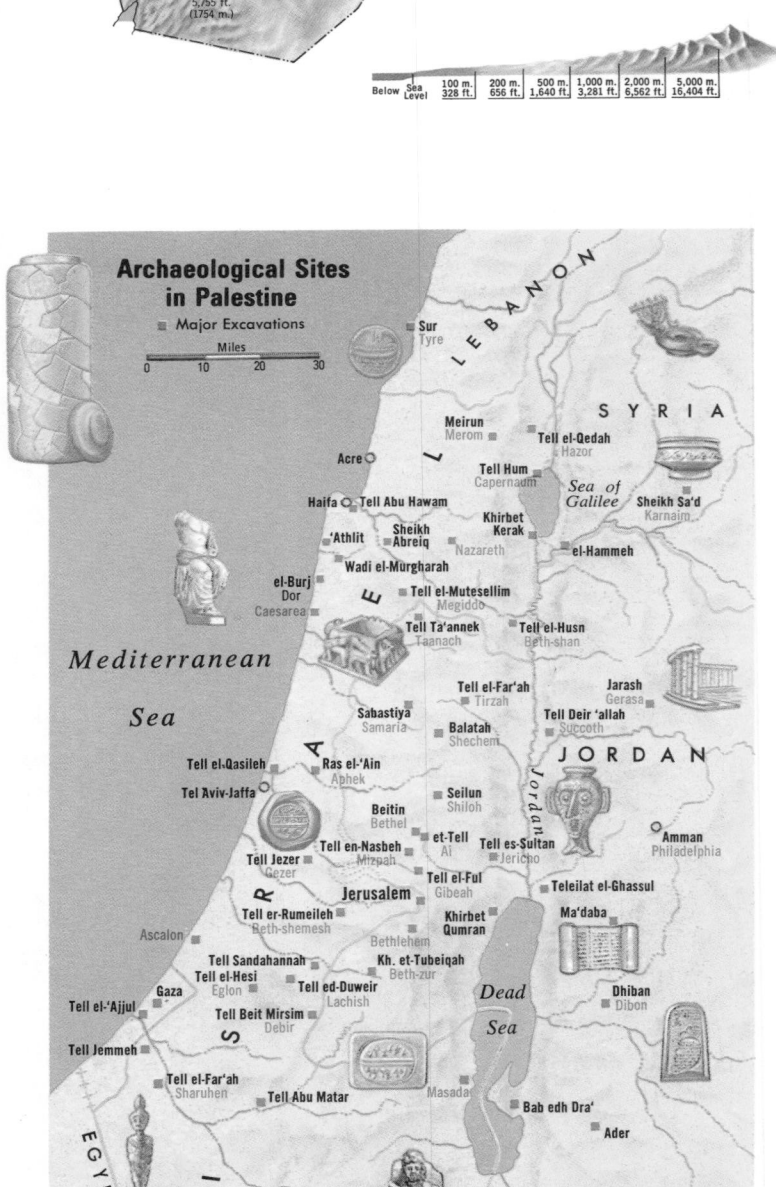

Archaeological Sites in Palestine

■ Major Excavations

Miles
0   10   20   30

LEBANON

SYRIA

Sur / Tyre
Meirun / Merom
Tell el-Qedah / Hazor
Acre
Tell Hum / Capernaum
Sea of Galilee
Sheikh Sa'd / Karnaim
Haifa ○ Tell Abu Hawam
Khirbet Kerak
'Athlit
Sheikh Abreiq
el-Hammeh
Nazareth
Wadi el-Murgharah
el-Burj / Dor / Caesarea
Tell el-Mutesellim / Megiddo
Tell el-Husn / Beth-shan
Tell Ta'annek / Taanach
Mediterranean Sea
Jarash / Gerasa
Tell el-Far'ah / Tirzah
Sabastiya / Samaria
Tell Deir 'alla / Succoth
Balatah / Shechem
JORDAN
Tell el-Qasileh
Ras el-'Ain / Aphek
Tel Aviv-Jaffa
Seilun / Shiloh
Beitin / Bethel
et-Tell / Ai
Tell en-Nasbeh / Mizpah
Tell es-Sultan / Jericho
Amman / Philadelphia
Tell Jezer / Gezer
Tell el-Ful / Gibeah
Jerusalem
Teleilat el-Ghassul
Tell er-Rumeileh / Beth-shemesh
Khirbet Qumran
Ma'daba
Ascalon
Tell Sandahannah
Kh. et-Tubeiqah / Beth-zur
Bethlehem
Dead Sea
Tell el-Hesi / Eglon
Tell ed-Duweir / Lachish
Dhiban / Dibon
Gaza
Tell Beit Mirsim / Debir
Tell el-'Ajjul
Tell Jemmeh
ISRAEL
Tell el-Far'ah / Sharuhen
Tell Abu Matar
Masada
Bab edh Dra'
Ader
EGYPT
Kurnub
Khirbet et-Tannur
Isbeita / Subaita

© Copyright HAMMOND INCORPORATED

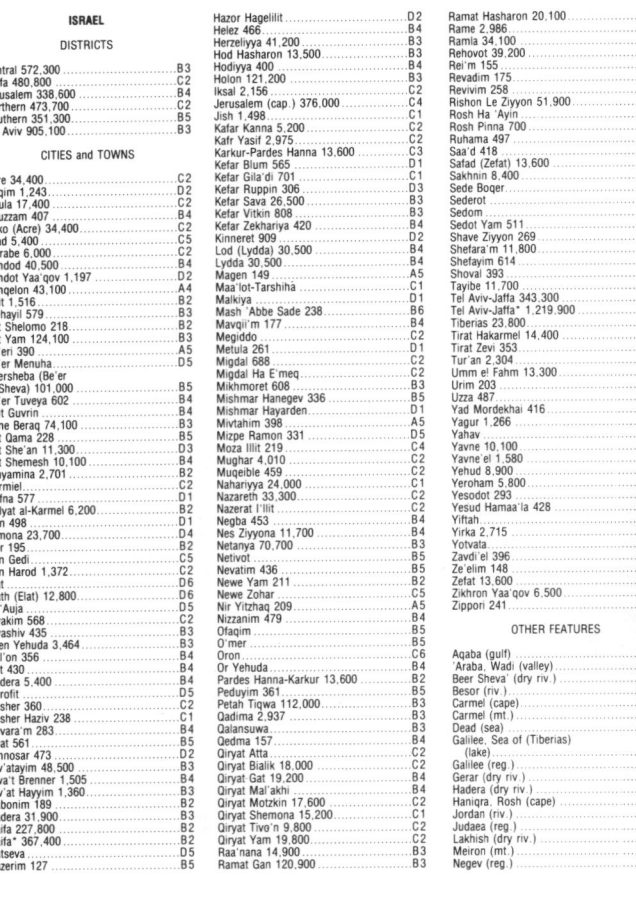

Agriculture, Industry and Resources

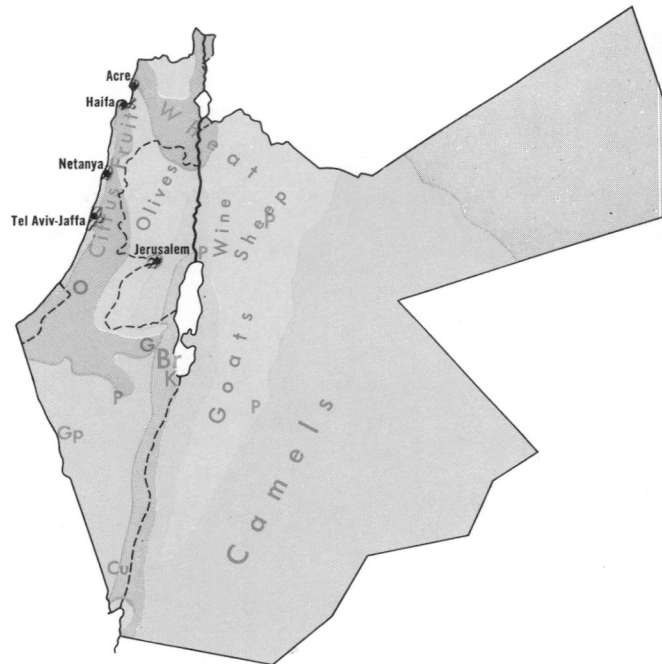

Acre
Haifa
Netanya
Tel Aviv-Jaffa
Jerusalem
Citrus Fruit
Wheat
Olives
Wine
Goats
Sheep
Camels

DOMINANT LAND USE

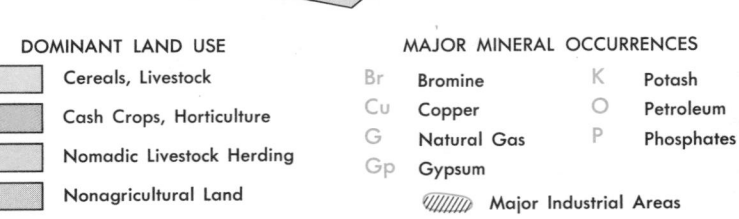

    Cereals, Livestock

    Cash Crops, Horticulture

    Nomadic Livestock Herding

    Nonagricultural Land

MAJOR MINERAL OCCURRENCES

Br Bromine   K Potash

Cu Copper   O Petroleum

G Natural Gas   P Phosphates

Gp Gypsum

////// Major Industrial Areas

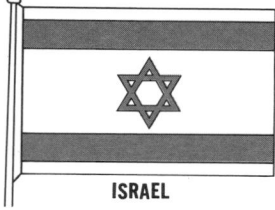

ISRAEL

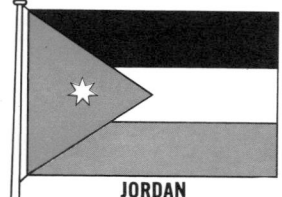

JORDAN

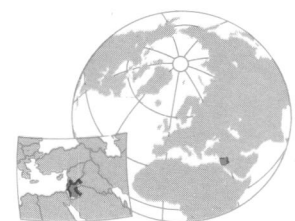

ISRAEL

AREA 7,847 sq. mi. (20,324 sq. km.)
POPULATION 3,878,000
CAPITAL Jerusalem
LARGEST CITY Tel Aviv-Jaffa
HIGHEST POINT Meiran 3,963 ft. (1,208 m.)
MONETARY UNIT shekel
MAJOR LANGUAGES Hebrew, Arabic
MAJOR RELIGIONS Judaism, Islam, Christianity

JORDAN

AREA 35,000 sq. mi. (90,650 sq. km.)
POPULATION 2,152,273
CAPITAL Amman
LARGEST CITY Amman
HIGHEST POINT Jeb. Ramm 5,755 ft. (1,754 m.)
MONETARY UNIT Jordanian dinar
MAJOR LANGUAGE Arabic
MAJOR RELIGION Islam

...ishon (riv.)	C2
...amon (mt.)	D5
...ubin (dry riv.)	B4
...abor (mt.)	C2
...iberas (lake)	D2
...armuk (riv.)	D2
...argon (riv.)	B3

GAZA STRIP

CITIES and TOWNS

Abasan 1,481	A5
...ani Suheila 7,561	A5
...eit Hanun 4,756	A4
...eir el Balah 10,854	A5
...eir el Balah* 18,118	A5
...aza 87,793	A5
...aza* 118,272	A5
...abaliya 10,508	A4
...abaliya* 43,604	A4
...han Yunis 29,522	A5
...han Yunis* 52,997	A5
...afah 10,812	A5
...afah* 49,812	A5

WEST BANK

CITIES and TOWNS

...kija 1,322	C3
...nabta 3,426	C3
...nza 807	C2
...in 914	C3
...riha (Jericho) 5,312	C4
...rraba 4,231	C3
...rura 849	C3
...ttil 3,808	C3
...eit Fajjar 2,474	C4
...eit Hanina 1,177	C4
...eit Jala 6,041	C4
...eit Lahm (Bethlehem) 14,439	C4
...eit Sahur 5,380	C4
...ethlehem 14,439	C4
...ddu 1,259	C4
...rqin 2,036	C3
...r Zeit 2,311	C4
...urqa 2,477	C3
...eir Ballut 1,058	C3
...eir Sharaf 973	C3
...ahiriya 4,875	C4
...uma 524	C3
...ra 4,954	C4
...ira 9,674	C4
...Bira* 13,037	C4
...Khalil (Hebron) 38,309	C4
...Rihiya 679	C5
...Zababida 1,474	C3
...lama 162	C3
...alul 6,041	C4
...rs 641	C3
...ebron 38,309	C4
...a 3,713	B4
...nwas 1,955	B4
...ana 784	C3
...bun 914	C3
...lud 221	C3
...nin 8,346	C3
...nin* 13,365	C3
...richo 5,312	C4
...richo* 6,931	C4
...na 655	C4
...karas 1,364	C3
...ablus (Nablus) 41,799	C3
...nulhin 1,109	C4
...l'in 1,227	C3
...abalan 1,970	C3
...alkiya 8,926	C3
...oya 926	C4
...idiya 1,123	C3
...mallah 12,134	C4
...mmun 1,198	C4
...antis 897	C3
...lfit 3,201	C3
...ma 3,784	C3
...af'at 14,000	C4
...uweika 2,332	C3
...at Dhahr 2,104	C3
...njil 1,823	C3
...is 1,285	C3
...mmun 2,952	C3
...rqumiya 2,412	C4
...bas 5,262	C3
...ikarm 10,255	C3
...lkarm* 15,275	C3
...r 12,200	C4
...b ad 4,857	C3
...brud 277	C4
...mun 4,384	C3
...na 7,281	C5
...ouba 633	C2

OTHER FEATURES

Golan Heights	D1
West Bank	C3

JORDAN

GOVERNORATES

El Asima 1,000,000	D4
El Balqa 113,000	D4
El Karak 93,000	E5
Irbid 506,000	D3
Ma'an 62,000	D5

CITIES and TOWNS

Ajlun⊙ 42,000	D3
Amman (cap.) 711,850	D4
'Anjara 3,163	D3
'Aqaba 15,000	D6
Bala'ma 769	E3
Baqura 3,042	D2
Damiya 483	D3
Dana 844	D5
Deir Abu Sai'd 1,927	D3
Dhira'	D5
El 'Al 492	D4
El Husn 3,728	D3
El Karak 10,000	E4
El Kitta 987	D3
El Madwar 164	D3
El Mafraq 15,500	E3
El Majdal 259	D3
El Quweira 268	E5
El Yaduda 251	D4
Er Rafid 787	D3
Er Ramtha 19,000	E2
Er Rumman 293	D3
Er Ruseifa 6,200	D4
Esh Shaubak 01	D5
Es Sahab 2,580	D4
Es Salt 24,000	D3
Es Sukhna 649	D3
Et Tafila* 17,000	E5
Et Taiyiba 2,606	D2
Ez Zarqa' 263,400	D4
Harima 635	D2
Hawara 2,342	D3
Hisban 718	D4
Ibbin 1,364	D3
Irbid 136,770	D2
Jabir 132	E2
Jarash⊙ 29,000	D3
Kitim 1,026	D3
Kufrinja 3,922	D3
Kuraiyima	D3
Maa'd 125	D2
Ma'in 9,500	E5
Ma'daba 22,600	D4
Ma'in 1,271	D4
Manja 353	C5
Mazra'	C5
Nau'r 2,382	D4
Nitil 348	D4
Qumeim 955	D2
Ra's en Naqb 225	E5
Safi	D5
Safut 4,210	D3
Samar 716	D3
Shunat Nimrin 109	D4
Subeihi 514	D3
Suf	D3
Suweilih 3,457	D3
Suweima 315	D4
Um Jauza 582	D3
Wadi es Sir 4,455	D4
Wadi Musa 654	E5
Waqqas 2,321	D3
Zuweiza 126	D4

OTHER FEATURES

'Ajlun (range)	D3
Aqaba (gulf)	D6
'Araba, Wadi (valley)	D5
Dead (sea)	C4
Ebal (mt.)	C3
El Ghor (reg.)	D3
El Lisan (pen.)	D4
Jordan (riv.)	C4
Judaea (reg.)	C4
Khirbet Qumran (site)	C4
Mashash, Wadi (dry riv.)	D4
Nebo (mt.)	D4
Petra (ruins)	D5
Samaria (reg.)	C3
Shallala, Wadi esh (dry riv.)	D2
Shu'eib, Wadi (dry riv.)	D3
Tell 'Asur (mt.)	C4
Yabis, Wadi (dry riv.)	D3
Zarqa' (riv.)	D3

*City and suburbs.
⊙Population of subdivision.

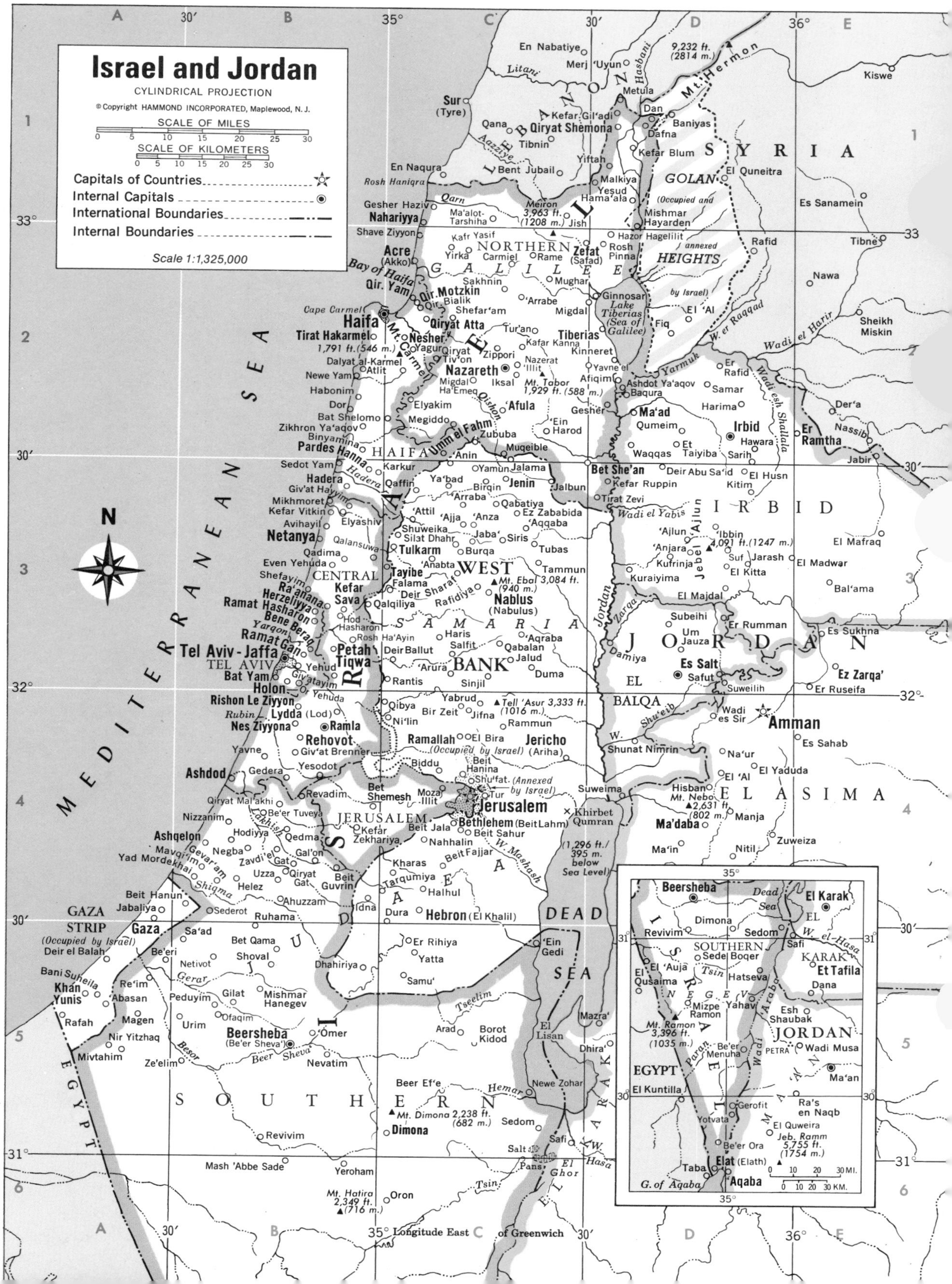

Israel and Jordan

CYLINDRICAL PROJECTION

® Copyright HAMMOND INCORPORATED, Maplewood, N.J.

SCALE OF MILES
0 5 10 15 20 25 30

SCALE OF KILOMETERS
0 5 10 15 20 25 30

Scale 1:1,325,000

Capitals of Countries ⋯⋯⋯⋯ ☆
Internal Capitals ⋯⋯⋯⋯⋯⋯ ⊙
International Boundaries ▬▬▬
Internal Boundaries ▬ ▬ ▬

IRAN

INTERNAL DIVISIONS

Azerbaijan, East (prov.) 3,194,543	E1
Azerbaijan, West (prov.) 1,404,875	D1
Bakhtiari (governorate) 394,300	F4
Boyer Ahmediyeh and Kohkiluyeh (governor 244,750	G5
Bushehr (governorate) 345,427	G6
Central (Markazi) (prov.) 6,921,283	G3
Esfahan (Isfahan) (prov.) 1,974,938	H4
Fars (prov.) 2,020,947	H6
Gilan (prov.) 1,577,800	F2
Hamadan (governorate) 1,086,512	F3
Hormozgan (governorate) 463,419	J7
Ilam (governorate) 244,222	E3
Isfahan (prov.) 1,974,938	H4
Kerman (prov.) 1,088,045	K6
Kermanshahan (prov.) 1,016,199	E3
Khorasan (prov.) 3,266,650	K3
Khuzestan (prov.) 2,176,612	F5
Kordestan (Kurdistan) (prov.) 781,889	E3
Lorestan (Luristan) (governorate) 924,848	F4
Mazandaran (prov.) 2,384,226	H2
Semnan (governorate) 485,875	J3
Sistan and Baluchestan (prov.) 659,297	M6
Yazd (governorate) 356,218	J5
Zanjan (governorate) 579,000	F2

CITIES and TOWNS

Abadan 296,081	F5
Abadeh 16,000	H5
Abarqu 8,000	H5
Abhar 24,000	F2
Ahar 24,000	E1
Agha Jari 24,195	F5
Ahvaz (Ahwaz) 329,006	F5
Amol 68,782	H2
Aradan 2,038	H4
Aradan 8,978	H3
Ardakan 16,000	J5
Ardabil 147,404	F1
Ardestan 5,868	H4
Asadabad 7,000	F3
Astarabad (Gorgan) 88,348	J2
Babol 67,790	H2
Babol Sar 7,237	H2
Bafq 5,000	J5
Bafq 5,000	K6
Baft 6,000	K6
Bajgiran 1,151	L2
Bam 22,000	L6
Bampur 1,585	M7
Bandar 'Abbas 89,103	J7
Bandar-e Khomeyni 6,000	F5
Bandar-e Lengeh 4,920	J7
Bandar-e Mas'hur 17,000	F5
Bandar-e Rig 1,889	G6
Bandar-e Torkeman 13,000	J2
Bandar Shahpur 6,000	F5
Bastak 2,473	J7
Bastam 3,296	J2
Behbehan 39,874	G5
Behshahr 26,032	H2
Bejestan 3,823	K3
Bijar 12,000	E3
Birjand 25,854	L4
Bojnurd 31,248	K2
Borazjan 20,000	G6
Borujerd 100,103	F4
Bostan 4,619	F5
Bowkan 9,000	D2
Bushehr (Bushire) 57,681	G6
Chah Bahar 1,800	M8
Chalus 15,000	G2
Damavand 5,319	H3
Damghan 13,000	J3
Darab 13,000	J6
Daran 4,609	G4
Darreh Gaz 12,000	L2
Dasht-e Azadegan 21,000	F5
Dehkhvaregan 6,000	D2
Delijan 6,000	G4
Dezful (Dezful) 110,287	F4
Duzdab (Zahedan) 92,628	M6
Emamshahr 30,767	J3
Enzeli 55,978	F2
Esfahan (Isfahan) 671,820	H4
Elamabad 12,000	E3
Estahabanat 18,187	H6
Evaz 6,064	J7
Ezna 5,000	F4
Fahrej (Iranshahr) 5,000	M7
Fariman 8,000	L3
Farrashband 3,532	G6
Fasa 19,000	H6
Ferdows 11,000	K3
Firuzabad 8,718	H6
Firuzkuh 4,684	H3
Fowman 9,000	F2
Ganaveh 9,000	G6
Garmsar 4,723	H3
Gavater	M8
Ghaemshahr 63,289	H2
Golpayegan 20,515	G4
Golshan (Tabas) 10,000	K4
Gomishan 6,000	J2
Gonabad 8,000	L3
Gonbad-e Kavus 59,868	J2
Gonbadli 531	M2
Gorgan (Gurgan) 88,348	J2
Haft Gel 10,000	F5
Hamadan 155,846	F3
Hashtpar 5,000	F2
Huzgan 4,722	F5
Ilam 15,000	E3
Iranshahr 5,000	M7
Isfahan 671,825	H4
Izeh 1,983	F5
Jahrom 38,236	H6
Jajarm 3,841	K2
Jask 1,078	K8
Kakhk 4,043	L3
Kangan 2,682	G7
Kangavar 9,414	F3
Karaj 138,774	G3
Kashan 84,545	G4
Kashmar 17,000	L3
Kazerun 51,309	G6
Kazvin (Qazvin) 138,527	F2
Kerman 140,309	K5
Kermanshah 290,861	E3
Khaf 5,000	L3
Khalkhal 5,422	F2
Khash 7,439	M7
Khiyav 9,000	E1
Khoman 3,054	G2
Khomeinishar 46,836	G4
Khorramabad 104,928	F4
Khorramshahr 146,709	F5
Khvaf 5,000	L3
Khvonsar 10,947	G4
Khvor 2,912	J4
Khvoy (Khoi) 70,040	D1
Kord Kuy 9,855	J2
Lahijan 25,725	G2
Lar 22,000	J7
Mahabad 28,610	D2
Mahallat 12,000	G4
Mahan 6,000	K5
Maku 7,000	D1
Malamir (Izeh) 1,983	F5
Malayer 28,431	F3
Marand 8,000	D1
Maragheh 60,820	E2
Marv Dasht 25,498	H6
Mashhad (Meshed) 670,180	L2
Masjed Soleyman 77,161	F5
Medishahr 9,000	H3
Mehran 664	E4
Meshed 670,180	L2
Meshed-i-Sar (Babol Sar) 12,000	H2
Meybod 15,000	J4
Miandoab 19,000	E2
Mianeh 28,447	E2
Minab 4,228	K7
Mirjaveh 11,000	M5
Naft-e Shah 3,043	F3
Nahavand 24,000	F3
Na'in 5,925	H4
Najafabad 76,236	G3
Naraz 2,725	G3
Nasratabad (Zabol) 20,000	M5
Natanz 4,370	H4
Neyriz 16,114	J6
Neyshabur 59,101	L2
Nishapur (Neyshabur) 59,101	L2
Nosratabad 20,000	L6
Now Shahr 8,000	G2
Orumiyeh (Urmia) 163,991	D2
Oshnoviyeh 5,000	D2
Pahlevi (Enzeli) 55,978	F2
Pazanan 81	E3
Qasr-e Shirin 15,094	E3
Qayen 6,000	L4
Qazvin 138,527	F2
Qom 246,831	G3
Qorveh 2,929	E3
Quchan 29,133	L2
Qum (Qom) 246,831	G3
Ramhormoz 9,000	F5
Rasht (Rasht) 187,203	F2
Ravar 5,074	K5
Resht (Rasht) 187,203	F2
Rey 102,825	G3
Reza Iyeh (Urmia) 163,991	D2
Rigan 8,255	L6
Rud Sar 7,460	G2
Sabzevar 69,174	K2
Sabzvaran 7,000	K6
Saeendey 4,195	E2
Sai (Sabzvar) 15,000	J6
Sakht-Sar 12,000	G2
Samlan 13,161	D1
Sanandaj 95,834	E3
Saqqez 17,000	E2
Sarab 16,000	E1
Sarakhs 8,000	M2
Saravan 4,012	N7
Sar Dasht 6,000	D2
Sari 70,936	H2
Savanat (Estahbanat) 18,187	J6
Saveh 17,565	G3
Semnan 31,058	H3
Shadegan 6,000	F5
Shahdad 2,777	K5
Shahistan (Saravan) 4,012	N7
Shahreza 34,220	H4
Shahr Kord 24,000	G4
Shahrud (Emamshahr) 30,767	J3
Sharafkhaneh 1,260	D1
Shiraz 416,408	H6
Shirvan 11,000	L2
Shush 1,433	F4
Shushtar 24,000	F5
Sinneh (Sanandaj) 95,834	E3
Sirjan (Sai'dabad) 20,000	K6
Sivand 1,811	H6
Songor 10,433	F3
Sufian 2,914	D1
Sultanabad (Kashmar) 17,000	L3
Tabas 10,000	K4
Tabriz 598,576	E2
Tajrish 157,486	G3
Takestan 13,485	F2
Tehran (cap.) 4,496,159	H3
Tonekabon 12,000	G2
Torbat-e Heydariyeh 30,106	L3
Torbat-e Jam 13,000	M3
Turbat-i-Shaikh Jam 13,000	M3
Tuysarkan 12,000	F3
Urmia 163,991	D2
Varamin 11,183	H3
Yazd (Yezd) 135,978	J5
Zabol 20,000	M5
Zahedan 92,628	M6
Zanjan 99,967	F2
Zarand 7,000	K5
Zarqam 7,000	H6
Zenjan (Zanjan) 99,967	F2

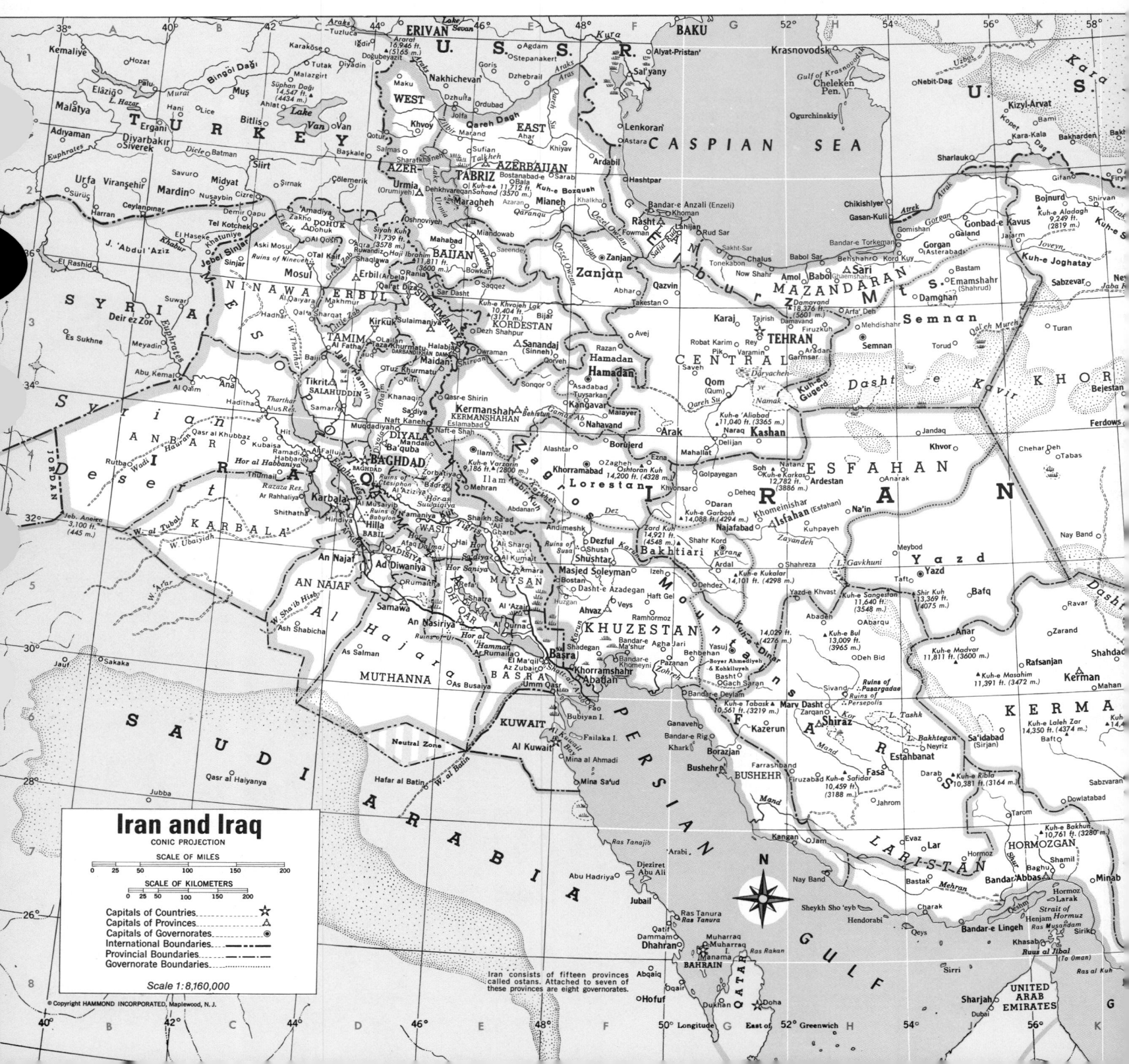

Iran and Iraq

CONIC PROJECTION

SCALE OF MILES
0 25 50 100 150 200

SCALE OF KILOMETERS
0 25 50 100 150 200

Capitals of Countries	★
Capitals of Provinces	△
Capitals of Governorates	◉
International Boundaries	— ·· —
Provincial Boundaries	— —
Governorate Boundaries	····

Scale 1:8,160,000

© Copyright HAMMOND INCORPORATED, Maplewood, N.J.

Iran consists of fifteen provinces called ostans. Attached to seven of these provinces are eight governorates.

OTHER FEATURES

Aji Chai (riv.)	E1
A'rabi (isl.)	G7
Araks (Aras) (riv.)	E1
Atrak (Atrek) (riv.)	J2
Bakhtegan (lake)	G6
Baluchistan (reg.)	M7
Bampur (riv.)	M7
Behistun (ruins)	E3
Caspian (sea)	G1
Damavand (Demavend) (mt.)	H3
Dez (riv.)	F4
Elburz (mts.)	G2
Farsi (isl.)	G7
Gorgan (riv.)	J2
Hari Rud (riv.)	M3
Karkheh (riv.)	E4
Karun (riv.)	F5
Kashaf Rud (riv.)	M2
Khark (Kharg) (isl.)	G6
Kuh (cape)	K8
Kurang (riv.)	G4
Laristan (reg.)	J7
Makran (reg.)	M8
Mand Rud (riv.)	G6
Mehran (riv.)	J7
Namaksar (lake)	M4
Nezwar (mt.)	H3
Oman (gulf)	M8
Pasargadae (ruins)	H5
Persepolis (ruins)	H6
Persian (gulf)	F6
Qareh (riv.)	E1
Qareh Su (riv.)	G3
Qeshm (isl.)	J7
Qezel Owzam (riv.)	F2
Safid Rud (riv.)	F2

Shaikh Shua'ib (isl.)	H7
Shelagh (riv.)	M5
Shirvan (riv.)	E3
Shur (riv.)	J7
Siah Kuh (mt.)	J3
Silup (riv.)	M8
Susa (ruins)	F4
Talab (riv.)	N6
Tashk (lake)	J6
Urmia (lake)	D2
Zagros (mts.)	E4
Zarineh (riv.)	E2
Zilbir (riv.)	D1
Zohreh (riv.)	F5

IRAQ

GOVERNORATES

Anbar	B4
An Najaf	C5
Babil	D4
Baghdad	D4
Basra	D5
Dhi Qar	E5
Diyala	D4
Dohuk	C2
Erbil	C3
Karbala	B4
Maysan	E5
Muthanna	D5
Ninawa	B3
Qadisiya	D4
Salahuddin	D3
Sulaimaniya	D3
Tamin	D3
Wasit	D4

CITIES and TOWNS

Ad Diwaniya 60,553	D5
A'faq 5,390	D4
Al A'ziziya 7,450	D4
Al Falluja 38,072	C4
Al Fathat 15,329	C3
A'li Gharbi 15,456	E4
A'li Sharqi 8,398	E4
Al Kufa 30,862	D4
Al Musaiyib 15,955	D4
Al Q'aim 3,372	B3
Al Qaiyara 3,060	C2
Al Qosh 3,863	C2
A'madiya 2,578	C2
A'mara 64,847	E5
A'na 15,729	B3
An Najaf 128,096	C4
An Nasiriya 60,405	D5
A'qra 8,659	D2
Arbela (Erbil) 90,320	C3
Aski Mosul 643	C2
As Salman 1,789	D5
Az Zubair 41,408	E5
Badra 3,564	D4
Baghdad (cap.) 502,503	D4
Baghdad* 1,745,328	D4
Baq'uba 34,575	D4
Baiji 6,785	C3
Basra 313,327	E5
Dohuk 16,998	C2
Erbil 90,320	C3
Fao 15,399	F6
Habbaniya 14,405	C4
Haditha 6,870	C3
Hai 16,988	E4
Halabja 11,206	D3
Hilla 84,717	D4
Hindiya 16,436	C4
Hit 9,131	C4
Karbal'a 83,301	C4
Khanaqin 23,522	D3
Kifri 8,500	D3
Kirkuk 167,413	C3
Kirkuk* 176,794	D3
Kubaisa 4,023	C4
Kut 42,116	D4
Makhmur 2,556	C3
Mandali 11,262	D4
Mosul 315,157	C2
Muqdadiyah 12,181	D3
Naft Kaneh	D3
Na'maniya 11,943	D4
Qal'at Diza 6,250	D2
Ramadi 28,723	C4
Rania 4,090	D2
Refai' 7,681	E5
Rumaitha 10,222	D5
Rutba 5,091	B4
Ruwandiz 5,801	D2
Sad'iya 5,285	D3
Samarra 24,746	C3
Samawa 33,473	D5
Shaikh Saa'd 2,958	E4
Shaqlawa 6,814	D2
Shatra 18,822	E5
Sinjar 7,942	B2
Sulaimaniya 86,822	D3
Tal Kaif 7,482	C2
Taza Khurmatu 2,681	C3
Tikrit 9,921	C3
Tuz Khurmatu 13,860	D3
Zakho 14,790	C2

OTHER FEATURES

Adhaim (riv.)	D3
Aneiza, Jebel (mt.)	A4
A'rab, Shatt-al- (riv.)	F5
A'ra'r, Wadi (dry riv.)	B5
Babylon (ruins)	D4
Batin, Wadi al (riv.)	E6
Ctesiphon (ruins)	D4
Darbandikhan (dam)	D3
Euphrates (riv.)	D4
Great Zab (riv.)	C2
Hauran, Wadi (dry riv.)	B4
Little Zab (riv.)	C3
Mesopotamia (reg.)	B3
Nineveh (ruins)	C2
Sad'iya, Hor (lake)	E4
Saniya, Hor (lake)	E4
Shai'b Hisb, Wadi (dry riv.)	C5
Sinjar, Jebel (mts.)	B2
Siyah Kuh (mt.)	D2
Syrian (des.)	B4
Tigris (riv.)	E4
Ubaiyidh, Wadi (dry riv.)	B5
Ur (ruins)	E5

*City and suburbs.
†Population of commune.

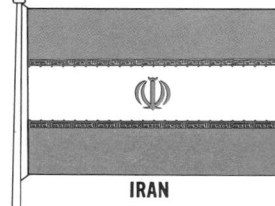

IRAN

IRAQ

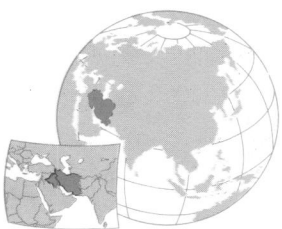

AREA 636,293 sq. mi. (1,648,000 sq. km.)
POPULATION 37,447,000
CAPITAL Tehran
LARGEST CITY Tehran
HIGHEST POINT Damavand 18,376 ft. (5,601 m.)
MONETARY UNIT Iranian rial
MAJOR LANGUAGES Persian, Azerbaijani, Kurdish
MAJOR RELIGION Islam

AREA 172,476 sq. mi. (446,713 sq. km.)
POPULATION 12,767,000
CAPITAL Baghdad
LARGEST CITY Baghdad
HIGHEST POINT Haji Ibrahim 11,811 ft. (3,600 m.)
MONETARY UNIT Iraqi dinar
MAJOR LANGUAGES Arabic, Kurdish
MAJOR RELIGION Islam

Topography

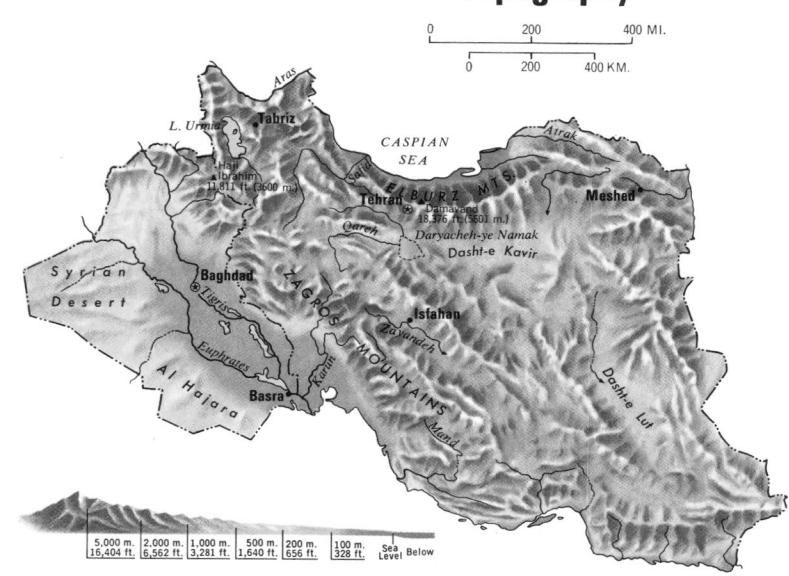

Agriculture, Industry and Resources

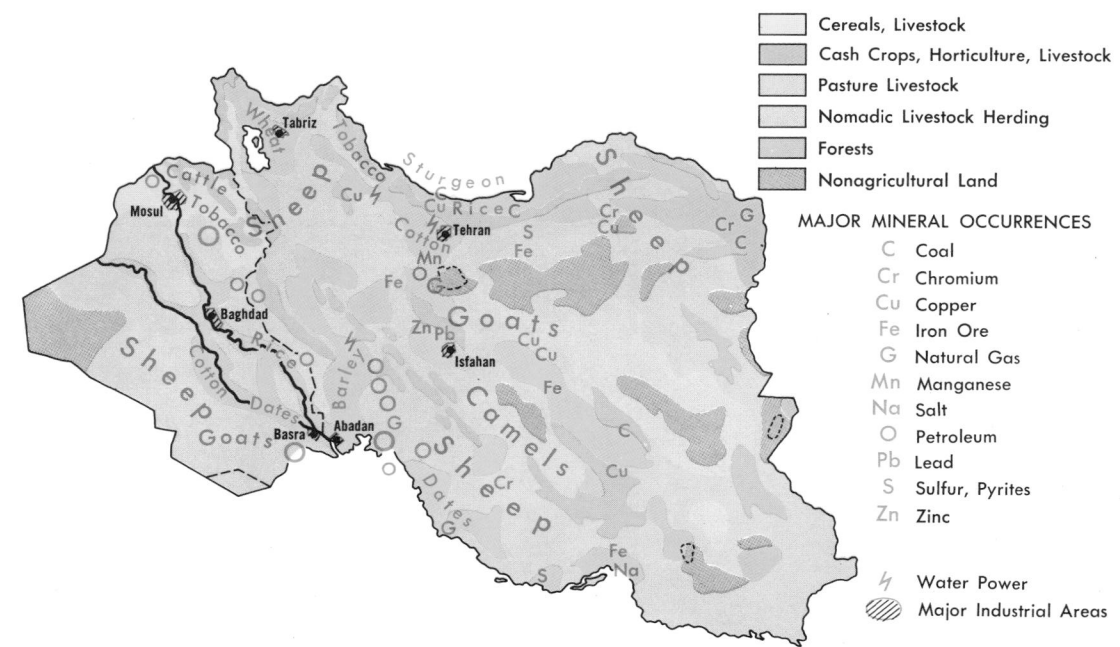

DOMINANT LAND USE

- Cereals, Livestock
- Cash Crops, Horticulture, Livestock
- Pasture Livestock
- Nomadic Livestock Herding
- Forests
- Nonagricultural Land

MAJOR MINERAL OCCURRENCES

C	Coal
Cr	Chromium
Cu	Copper
Fe	Iron Ore
G	Natural Gas
Mn	Manganese
Na	Salt
O	Petroleum
Pb	Lead
S	Sulfur, Pyrites
Zn	Zinc

Water Power

Major Industrial Areas

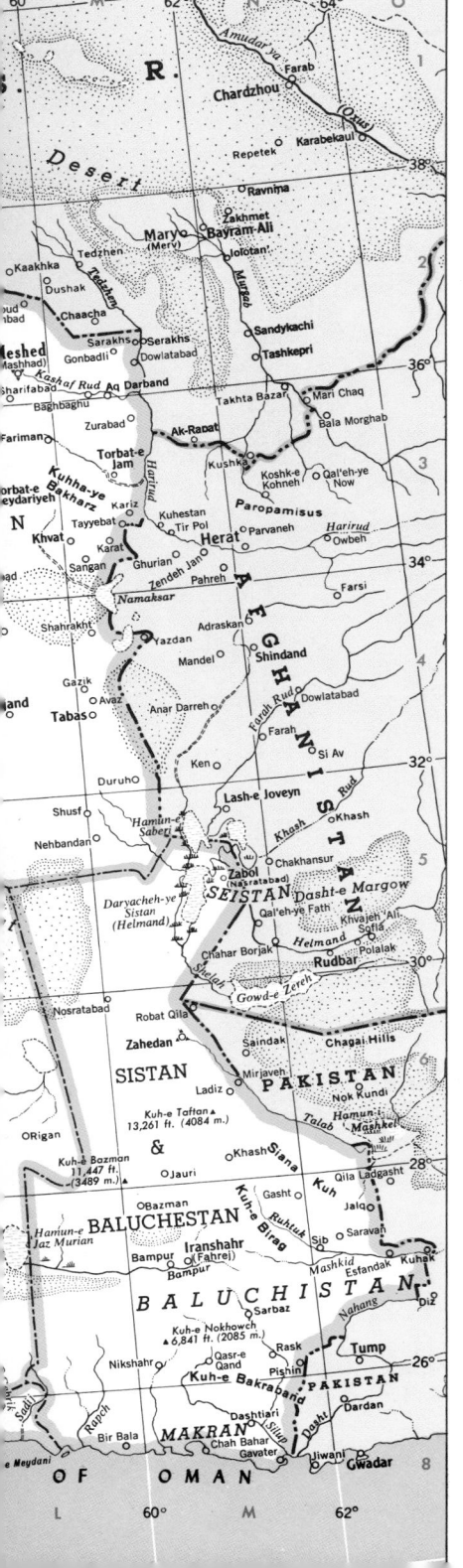

Indian Subcontinent and Afghanistan

CONIC PROJECTION

SCALE OF MILES

KILOMETERS

Capitals of Countries .... ☆
Provincial and State Capitals .... ◉
International Boundaries .... –––
Provincial and State Boundaries ... – · – ·
Canals ....

Scale 1:14,500,000

© Copyright HAMMOND INCORPORATED, Maplewood, N.J.

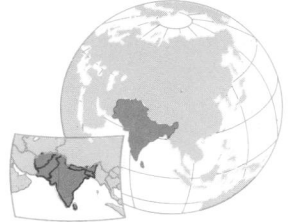

INDIA

AREA 1,269,339 sq. mi. (3,287,588 sq. km.)
POPULATION 683,810,051
CAPITAL New Delhi
LARGEST CITY Calcutta (greater)
HIGHEST POINT Nanda Devi 25,645 ft. (7,817 m.)
MONETARY UNIT Indian rupee
MAJOR LANGUAGES Hindi, English, Bihari, Telugu, Marathi, Bengali, Tamil, Gujarati, Rajasthani, Kanarese, Malayalam, Oriya, Punjabi, Assamese, Kashmiri, Urdu
MAJOR RELIGIONS Hinduism, Islam, Christianity, Sikhism, Buddhism, Jainism, Zoroastrianism, Animism

PAKISTAN

AREA 310,403 sq. mi. (803,944 sq. km.)
POPULATION 83,782,000
CAPITAL Islamabad
LARGEST CITY Karachi
HIGHEST POINT K2 (Godwin Austen) 28,250 ft. (8,611 m.)
MONETARY UNIT Pakistani rupee
MAJOR LANGUAGES Urdu, English, Punjabi, Pushtu, Sindhi, Baluchi, Brahui
MAJOR RELIGIONS Islam, Hinduism, Sikhism, Christianity, Buddhism

SRI LANKA (CEYLON)

AREA 25,332 sq. mi. (65,610 sq. km.)
POPULATION 14,850,001
CAPITAL Colombo
LARGEST CITY Colombo
HIGHEST POINT Pidurutalagala 8,281 ft. (2,524 m.)
MONETARY UNIT Sri Lanka rupee
MAJOR LANGUAGES Sinhala, Tamil, English
MAJOR RELIGIONS Buddhism, Hinduism, Christianity, Islam

AFGHANISTAN

AREA 250,775 sq. mi. (649,507 sq. km.)
POPULATION 15,540,000
CAPITAL Kabul
LARGEST CITY Kabul
HIGHEST POINT Nowshak 24,557 ft. (7,485 m.)
MONETARY UNIT afghani
MAJOR LANGUAGES Pushtu, Dari, Uzbek
MAJOR RELIGION Islam

NEPAL

AREA 54,663 sq. mi. (141,577 sq. km.)
POPULATION 14,179,301
CAPITAL Kathmandu
LARGEST CITY Kathmandu
HIGHEST POINT Mt. Everest 29,028 ft. (8,848 m.)
MONETARY UNIT Nepalese rupee
MAJOR LANGUAGES Nepali, Maithili, Tamang, Newari, Tharu
MAJOR RELIGIONS Hinduism, Buddhism

MALDIVES

AREA 115 sq. mi. (298 sq. km.)
POPULATION 143,046
CAPITAL Male
LARGEST CITY Male
HIGHEST POINT 20 ft. (6 m.)
MONETARY UNIT Maldivian rupee
MAJOR LANGUAGE Divehi
MAJOR RELIGION Islam

BHUTAN

AREA 18,147 sq. mi. (47,000 sq. km.)
POPULATION 1,298,000
CAPITAL Thimphu
LARGEST CITY Thimphu
HIGHEST POINT Kula Kangri 24,784 ft. (7,554 m.)
MONETARY UNIT ngultrum
MAJOR LANGUAGES Dzongka, Nepali
MAJOR RELIGIONS Buddhism, Hinduism

BANGLADESH

AREA 55,126 sq. mi. (142,776 sq. km.)
POPULATION 87,052,024
CAPITAL Dhaka
LARGEST CITY Dhaka
HIGHEST POINT Keokradong 4,034 ft. (1,230 m.)
MONETARY UNIT taka
MAJOR LANGUAGES Bengali, English
MAJOR RELIGIONS Islam, Hinduism Christianity

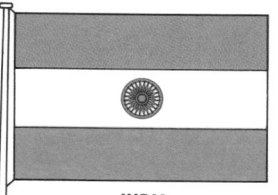

INDIA   **PAKISTAN**   **SRI LANKA (CEYLON)**   **BHUTAN**

AFGHANISTAN   **MALDIVES**   **BANGLADESH**   **NEPAL**

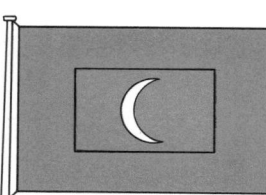
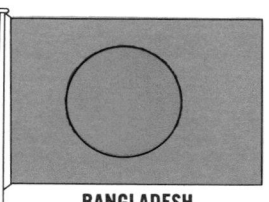

AFGHANISTAN

CITIES and TOWNS

Andkhvoy A1
Aqcheh B1
Aybak 33,016 ....B1
Baghlan 75,130 ....B1
Balkh B1
Bamian 7,355 ....B1
Belcheragh B1
Chaghcharan 2,974 ....B2
Chahar Borjak A2
Charikar 25,093 ....B1
Delaram A2
Dowlatabad A2
Dowlat Yar A2
Dowshi B1
Farah 18,797 ....A2
Farsi A2
Feyzabad 10,142 ....C1
Gardez 11,415 ....B2
Gereshk A2
Ghazni 30,425 ....B2
Ghurian A2
Gizab B2
Hazar Qadam B2
Herat 163,960 ....A2
Jalalabad 56,384 ....B2
Jorm C1
Kabul (cap.) 905,108 ....B2
Kalat (Qalat) 5,946 ....B2
Kandahar (Qandahar) 178,409 .B2
Ken B1
Khanabad B1
Khash A2
Kholm B1
Khowst B2
Khugiani A2
Koshke-e Kohneh A2
Kowt-e 'Ashrow B2
Kuhestan A2
Landay A2
Lash-e Joveyn A2
Lashkar Gah 26,646 ....A2
Mar'uf B2
Mazar-e Sharif 122,567 .....B1
Meymaneh 54,954 ....A1
Mirabad A2
Moqor B2
Now Zad A2
Oruzgan (Hazar Qadam) B2
Owbeh A2
Panjab B2
Pol-e Khomri B1
Qalat 5,946 ....B2
Qale'h-ye Now 5,340 ....A1
Qale'h-ye Panjeh C1
Qandahar 178,409 ....B2
Qonduz 107,191 ....B1
Rostaq B1
Rudbar A2
Sakhar B2
Sar-e Pol B1
Shay Juy B2
Sheberghan 54,870 ....B1
Shindand A2
Spin Buldak B2
Tagab B2
Taloqan 46,202 ....B1
Teyvareh A2
Towraghondi A1
Tulak A2
Zaranj 6,477 ....A2
Zibak C1

OTHER FEATURES

Farah Rud (riv.) A2

Harirud (riv.) A1
Helmand (riv.) B2
Hindu Kush (mts.) B1
Kabul (riv.) C2
Konar (riv.) C1
Lurah (riv.) B2
Margow, Dasht-e (des.) A2
Namaksar (salt lake) A2
Paropamisus (range) A2
Tarnak (riv.) B2

BANGLADESH

CITIES and TOWNS

Barisal 98,127 ....G4
Bogra 47,154 ....F4
Chalna Port 14,590 ....G4
Chittagong 889,760 ....G4
Comilla 86,446 ....G4
Cox's Bazar (Maheshkhali) 15,720 ..G4
Dhaka (Dacca) (cap.) 1,679,572 ..G4
Dinajpur 61,866 ....F3
Faridpur 46,232 ....F4
Habiganj 16,281 ....G4
Jamalpur 60,261 ....F4
Jessore 76,168 ....F4
Khulna 437,304 ....F4
Kishorganj 35,605 ....G4
Madaripur 32,488 ....G4
Maheshkhali 15,720 ....G4
Mymensingh (Nasirabad) 182,153 ..G4
Narayanganj 270,680 ....G4
Nasirabad 182,153 ....G4
Nawabganj 46,059 ....G4
Noakhali 32,490 ....G4
Pabna 62,254 ....F4
Rajshahi 132,909 ....F4
Rangamati 20,473 ....G4
Rangpur 72,829 ....F3
Sirajganj 74,457 ....F4
Sylhet 59,546 ....G4

OTHER FEATURES

Bengal, Bay of (sea) F5
Brahmaputra (riv.) G3
Ganges (riv.) F3
Sundarbans (reg.) F4

BHUTAN

CITIES and TOWNS

Bumthang 10,000 ....G3
Paro 35,000 ....F3
Punakha 12,000 ....G3
Taga Dzong 18,000 ....G3
Thimphu (cap.) 50,000 ......G3
Tongsa Dzong 2,500 ....G3

OTHER FEATURES

Chomo Lhari (mt.) F3
Himalaya (mts.) E2
Kula Kangri (mt.) G3

INDIA

INTERNAL DIVISIONS

Andaman and Nicobar Isls. (terr.) 188,254 ....G6
Andhra Pradesh (state) 53,403,619 ....D5
Arunachal Pradesh (terr.) 628,050 ....G3

(continued on following page)

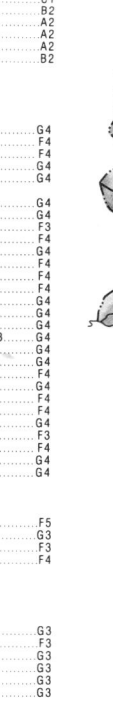

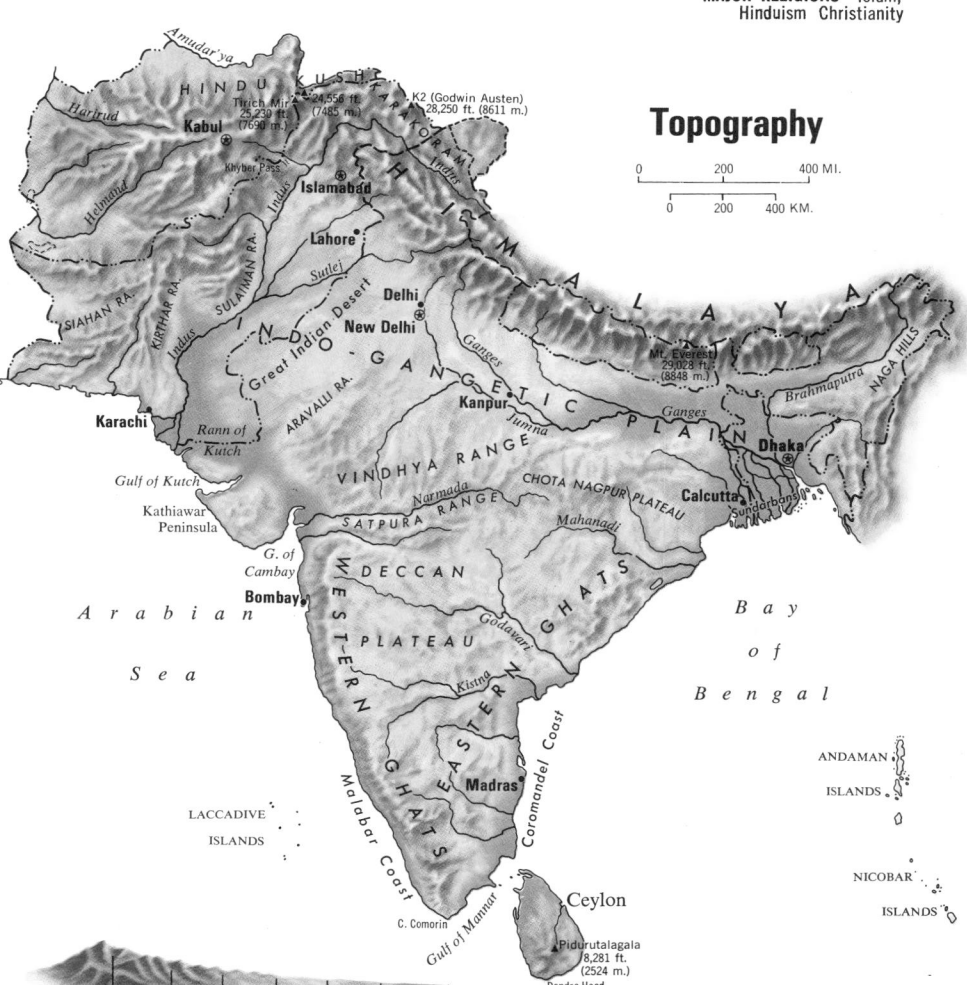

Topography

0   200   400 MI.
0   200   400 KM.

| 5,000 m. | 2,000 m. | 1,000 m. | 500 m. | 200 m. | 100 m. | Sea | Below |
| 16,404 ft. | 6,562 ft. | 3,281 ft. | 1,640 ft. | 656 ft. | 328 ft. | Level | |

Pidurutalagala 8,281 ft. (2,524 m.)
Dondra Head

(continued)

Assam (state) 19,902,826 G3
Bihar (state) 69,823,154 F4
Chandigarh (terr.) 450,061 D2
Dadra and Nagar Haveli (terr.) 103,677 C4
Delhi (terr.) 6,196,414 D3
Goa, Daman and Diu (terr.) 1,082,117 C4
Gujarat (state) 33,960,905 C4
Haryana (state) 12,850,902 D3
Himachal Pradesh (state) 4,237,569 D2
Jammu and Kashmir (state) 5,981,600 D2
Karnataka (state) 37,043,451 D6
Kerala (state) 25,403,217 D6
Lakshadweep (terr.) 40,237 C6
Madhya Pradesh (state) 52,131,717 D4
Maharashtra (state) 62,693,898 C5
Manipur (state) 1,433,691 G4
Meghalaya (state) 1,327,874 G3
Mizoram (terr.) 487,774 G4
Nagaland (state) 773,281 G3
Orissa (state) 26,272,054 E5
Pondicherry (terr.) 604,136 E6
Punjab (state) 16,669,755 D2
Rajasthan (state) 34,102,912 C3
Sikkim (state) 315,682 F3
Tamil Nadu (state) 48,297,456 D6
Tripura (state) 2,060,189 G4
Uttar Pradesh (state) 110,858,019 D3
West Bengal (state) 54,485,560 F4

CITIES and TOWNS

Abu 9,840 C4
Abu Road 25,331 C4
Achalpur 42,326 D4
Addanki 10,223 D5
Adilabad 30,368 D5
Adoni 85,311 D5
Agartala 59,625 G4
Agartala□ 100,264 G4
Agra 591,917 D3
Agra□ 634,622 D3
Ahmadabad 1,591,832 C4
Ahmadabad□ 1,741,522 C4
Ahmadnagar 118,236 C5
Ahmadnagar□ 148,405 C5
Aizwal 31,740 G4
Ajanta D4
Ajmer 262,851 C3
Akola 168,438 D4
Alibag 11,913 C5
Aligarh 252,314 D3
Alipore F2
Allahabad 490,622 E3
Allahabad□ 513,036 E3
Alleppey-Cochin 160,166 D7
Almora 19,671 D3
Along 3,524 G3
Alwar 100,378 D3
Amalner 55,544 C4
Ambala 83,633 D2
Ambala□ 186,168 D2
Ambikapur 23,087 E4
Amravati 193,800 D4
Amreli 39,520 C4
Amritsar 407,628 C2
Amritsar□ 458,029 C2
Anakapalle 57,273 E5
Anantapur 80,069 D6
Anantnag 27,643 D2
Andheri B7
Andul 3,602 F2
Arcot 30,230 D6
Arrah 92,919 E3
Aruppukkottai 62,223 D7
Arvi 26,494 D4
Asansol 155,968 F4
Asansol□ 241,792 F4
Aurangabad, Bihar 18,714 E4
Aurangabad, Maharashtra 150,483 D5
Aurangabad□ 165,253 D5
Azamgarh 40,963 E3
Badagara 53,938 D6
Bagalkot 51,746 D5
Bahraich 73,931 E3
Baidyabati 54,130 F1
Balaghat 27,872 E4
Balasore 46,239 F4
Ballia 47,101 E3
Bally 38,892 F1
Balotra 17,595 C3
Balrampur 36,191 E3
Balurghat 67,088 F3
Banda 50,575 D3
Bandar (Machilipatnam) 112,612 E5
Bandra B7
Bangalore 1,540,741 D6
Bangalore□ 1,653,779 D6
Bankura 79,129 F4
Bansberia 61,748 F1
Banswara 27,363 C4
Baramati 27,912 C5
Baramula 26,334 C2
Baranagar 136,842 F1
Barasat 42,642 F1
Barbil 24,342 F4
Bareilly 296,248 D3
Bareilly□ 326,106 D3
Baripada 28,725 F4
Barmer 38,630 C3
Baroda (Vadodara) 466,696 C4
Barpeta 26,479 G3
Barrackpore 96,889 F1
Barrackpore□ 198,255 F1
Barsi 62,374 D5
Baruipur 20,501 F2
Barwani 22,099 D4
Basim 32,496 D4
Basirhat 63,816 F4
Bassein 30,594 C5
Bastar E5
Batala 58,200 D2
Baudh 8,891 E4
Bauria 10,610 E2
Beawar 66,114 C3
Belgaum 192,427 C5
Belgaum□ 213,872 C5
Bellary 125,183 D5
Benares (Varanasi) 583,856 E3
Berhampore 72,605 F4
Berhampur 117,662 F5
Bettiah 51,018 E3
Betul 30,862 D4
Bhadrak 40,487 F4
Bhadravati 40,203 D6
Bhadravati□ 101,358 D6
Bhadreswar 45,586 F1
Bhagalpur 172,202 F4
Bhandara 39,423 E4
Bhandup B7
Bhanjanagar 12,353 E4
Bharatpur 68,036 D3
Bharuch 91,589 C4
Bhatapara 20,980 E4
Bhatinda 53,684 C2
Bhatkal 18,732 C6
Bhatpara 204,750 F1
Bhavnagar 225,358 C4
Bhavnagar□ 225,974 C4
Bhawanipatna 22,808 E5
Bhilai 157,173 E4
Bhilwara 82,155 C3
Bhimavaram 63,762 E5
Bhimunipatnam 14,291 E5
Bhind 42,371 D3
Bhinmal 14,050 C3
Bhir (Bir) 49,965 D5
Bhiwandi 79,576 C5
Bhiwani 73,086 D3
Bhopal 298,022 D4
Bhor 10,708 C5
Bhubaneswar 105,491 F4
Bhuj 52,177 B4
Bhusawal 96,800 D4
Bhusawal□ 104,708 D4
Bidar 50,670 D5
Bihar 100,046 F3
Bijapur, Karnataka 103,931 D5
Bijapur, Madhya Pradesh 5,289 E5
Bijnor 43,290 D3
Bikaner 188,518 C3
Bikaner□ 208,894 C3
Bilaspur 98,410 E4
Bina-Itawa 33,106 D4
Bir 49,965 D5
Birmitrapur 28,063 E4
Bobbili 30,649 E5
Bodhan 37,589 D5
Bodinayakkanur 54,176 D6
Bolangir 35,748 E4
Bombay (Greater)* 5,970,575 B7
Bomdila 2,264 G3
Broach (Bharuch) 91,589 C4
Budaun 72,204 D3
Budge-Budge 51,039 F2
Bundi 34,279 D3
Burdwan 143,318 F4
Burhanpur 105,246 D4
Calcutta 3,148,746 F2
Calcutta□ 7,031,382 F2
Calicut (Kozhikode) 333,979 D6
Cambay 62,097 C4
Cannanore 55,162 C6
Cawnpore (Kanpur) 1,154,388 E3
Chaibasa 35,386 F4
Chamba 11,814 D2
Champdani 58,596 F1
Chanderi 10,294 D4
Chandernagore 75,238 F1
Chandigarh 218,743 D2
Chandigarh□ 232,940 D2
Chandrapur 75,134 D5
Chapra 83,101 F3
Chatrapur 10,835 F5
Chembur B7
Cherrapunji⊙ 83,987 G3
Chhatarpur 32,271 D4
Chhindwara 53,492 D4
Chidambaram 48,811 E6
Chik Ballapur 29,227 D6
Chikmagalur 41,639 D6
Chingleput 38,419 E6
Chiplun 20,942 C5
Chirala 54,487 E5
Chitorgarh 25,917 C4
Chitradurga 50,254 D6
Chittoor 63,035 D6
Churachandpur 8,706 G4
Churu 52,502 C3
Chushul D2
Cocanada (Kakinada) 164,200 E5
Cochin-Alleppey 439,066 D6
Coimbatore 356,368 D6
Coimbatore□ 736,203 D6
Colachel 18,819 D7
Cooch Behar 53,684 F3
Coondapoor 23,831 C6
Cuddalore 101,335 E6
Cuddapah 66,195 D6
Cumbum 9,745 D5
Cuttack 194,068 F4
Cuttack□ 205,759 F4
Dabhoi 37,892 C4
Daltonganj 32,367 E4
Damoh 59,489 D4
Dapoli 6,296 C5
Darbhanga 132,059 F3
Darjeeling 42,873 F3
Datia 36,439 D3
Davangere 121,110 D6
Deesa 28,324 C4
Dehra Dun 166,073 D2
Dehra Dun□ 203,464 D2
Delhi 3,287,883 D3
Delhi□ 3,647,023 D3
Demchok D2
Deogarh, Orissa 8,906 E4
Deoghar, Bihar 40,356 F4
Deolali 55,436 C5
Deoria 38,161 E3
Dewas 51,545 D4
Dhamtari 34,546 E4
Dhanbad 79,838 F4
Dhanbad□ 434,031 F4
Dhar 36,172 C4
Dharmsala 10,939 D2
Dharwar-Hubli 379,166 C5
Dhenkanal 19,615 E4
Dholpur 31,865 D3
Dhond 16,583 C5
Dhoraji 59,773 C4
Dhubri 36,503 G3
Dhulia 137,129 C4
Dibrugarh 80,348 G3
Digboi 16,538 H3
Dindigul 128,429 D6
Diphu 10,200 G3
Dispur 1,725 G3
Diu 6,214 C4
Dohad 44,506 C4
Domjor 10,896 F1
Dudhi 5,084 E4
Dum Dum 31,363 F1
Dum Dum□ 273,812 F1
Dungarpur 19,773 C4
Durg 67,892 E4
Durgapur 206,638 F4
Dwarka 17,801 B4
Eluru 127,023 E5
English Bazar 61,335 F3
Erode 105,111 D6
Etawah 85,894 D3
Faizabad-cum-Ayodhya 102,835 E3
Faridabad 85,762 D3
Farrukhabad-cum-Fatehgarh 102,768 D3
Farrukhabad-cum-Fatehgarh□ 110,835 D3
Fatehpur, Rajasthan 34,929 C3
Fatehpur, Uttar Pradesh 54,665 E3
Firozabad 133,863 D3
Firozpur 49,545 C2
Gadag-Betgeri 95,426 D5
Gadwal 21,828 D5
Gandhinagar 24,055 C4
Ganganagar 90,042 C3
Gangapur 27,453 D3
Gangtok 12,000 F3
Garden Reach 154,913 F2
Garulia 44,271 F1
Gauhati 123,783 G3
Gauhati□ 200,377 G3
Gaya 179,884 F4
Ghat Kopar 34,256 B7
Ghaziabad 118,836 D3
Ghaziabad□ 127,700 D3
Ghazipur 45,635 E3
Goalpara 16,703 G3
Godhra 66,403 C4
Gonda 52,662 E3
Gondal 54,928 C4
Gondia 77,992 E4
Gorakhpur 230,911 E3
Goregaon B7
Gudur 33,778 D6
Gulbarga 145,588 D5
Guna 40,006 D4
Guntakal 66,320 D5
Guntur 269,991 D5
Gurais D2
Gwalior 384,772 D3
Gwalior□ 406,140 D3
Haflong 5,197 G3
Hanle D2
Hanumangarh 30,017 C3
Harda 28,504 D4
Hardoi 46,639 E3
Hardwar 77,864 D2
Hassan 51,325 D6
Hathras 74,349 D3
Hazaribagh 54,818 F4
Hindupur 42,959 D6
Hinganghat 44,349 D4
Hingoli 31,948 D5
Hissar 89,437 D3
Honavar 12,444 C6
Hooghly-Chinsura 105,241 F1
Hoshangabad 27,011 D4
Hospet 65,196 D5
Howrah 737,877 F2
Hubli-Dharwar 379,166 C5
Hyderabad 1,607,396 D5
Hyderabad□ 1,796,339 D5
Ichchapuram 15,850 F5
Ichhapur 11,975 F1
Imphal 100,366 G4
Indore 543,381 D4
Indore□ 560,936 D4
Itanagar⊙ 18,787 G3
Itarsi 44,191 D4
Jabalpur 426,224 D4
Jabalpur□ 534,845 D4
Jagdalpur 31,344 E5
Jagtial 30,900 D5
Jaipur 615,258 D3
Jaipur□ 636,768 D3
Jaisalmer 16,578 C3
Jajpur 16,707 F4
Jalgaon 106,711 D4
Jalna 91,099 D5
Jalor 15,478 C3
Jalpaiguri 55,159 F3
Jamalpur 61,731 F3
Jammu 155,338 D2
Jammu□ 164,207 D2
Jamnagar 214,816 B4
Jamnagar□ 227,640 B4
Jamshedpur 341,576 F4
Jamshedpur□ 456,146 F4
Jaora 37,235 D4
Jaunpur 80,737 E3
Jeypore 34,319 E5
Jhalawar 20,035 D4
Jhansi 173,292 D3
Jhansi□ 198,135 D3
Jharsuguda 24,727 E4
Jhunjhunu 32,024 D3
Jind 38,161 D3
Jodhpur 317,612 C3
Jorhat 30,247 G3
Jubbulpore (Jabalpur) 426,224 D4
Juhu B7
Jullundur 296,106 D2
Jullundur□ 329,830 D2
Junagadh 95,485 B4
Kadayanallur 50,295 D7
Kadiri 33,810 D6
Kakinada 164,200 E5
Kalyan 99,547 C5
Kamarhati 169,404 F1
Kamptee 53,412 D4
Kanchipuram 110,657 E6
Kanchrapara 78,768 F1
Kandla 17,995 C4
Kandukur 16,654 E5
Kanker 9,278 E4
Kannauj 28,187 D3
Kanpur 1,154,388 E3
Kanpur□ 1,275,242 E3
Karad 42,329 C5
Karaikudi 55,449 D7
Karanja 31,150 D4
Kargil 2,390 D2
Karikal 26,080 E6
Karkal 18,593 C6
Karnal 92,784 D3
Karwar 27,770 C6
Kasaragod 34,984 C6
Kasganj 46,467 D3
Katarnian Ghat E3
Katihar 67,014 F3
Katni (Murwara) 54,864 E4
Kavali 29,616 E6
Kavaratti 4,420 C6
Kawardha 11,226 E4
Kendrapara 20,079 F4
Keonjhar 19,340 F4
Khamgaon 53,692 D4
Khamman 56,919 D5
Khandwa 84,517 D4
Kharagpur 61,783 F4
Khardah 32,302 F1
Khurda 15,879 F4
Kirkee 65,497 C5
Kishangarh 37,405 D3
Kishtwar 5,276 D2
Kohima 21,545 G3
Kolar 43,418 D6
Kolar Gold Fields 76,112 D6
Kolhapur 259,050 C5
Konnagar 34,424 F1
Koppal 27,277 D5
Koraput 21,505 E5
Korba 30,963 E4
Kota 212,991 D3
Kottaguden 75,542 E5
Kottayam 59,714 D7
Kotturu 12,873 D6
Kovur 16,846 E6
Kozhikode 333,979 D6
Krishnanagar 85,923 F4
Kulu 8,958 D2
Kumbakonam 113,130 D6
Kumta 19,112 C6
Kurla B7
Kurnool 136,710 D5
Laful⊙ 8,161 G7
Lansdowne 6,670 D3
Latur 70,156 D5
Leh 5,519 D2
Lohardaga 17,087 E4
Lucknow 749,239 E3
Lucknow□ 813,982 E3
Ludhiana 397,850 D2
Ludhiana□ 401,176 D2
Lumding 29,253 G3
Lungleh 6,019 G4
Machilipatnam 112,612 E5
Madh B7
Madhubani 32,919 F3
Madras 2,469,449 E6
Madras□ 3,169,930 E6
Madugula 8,376 E5
Madurai 549,114 D7
Madurai□ 711,501 D7
Mahabaleshwar 7,318 C5
Mahbubnagar 51,756 D5
Mahe 8,972 D6
Mahim 11,344 C5
Mahoba 29,707 D3
Mahuva 39,497
Malad
Malakanagiri 7,494
Malegaon 191,847
Maler Kotla 48,536
Malkapur 35,476
Malvan 17,579
Mandi 16,849
Mandla 24,406
Mandsaur 52,347
Mandvi 27,849
Manendragarh 11,936
Mangalore 165,174
Mangrol 27,183
Manmad 29,571
Mannargudi 42,783
Manori
Margao 41,655
Marmagao 44,065
Mathura 132,028
Mau 64,058
Mayuram 60,195
Meerut 270,993
Mehsana 51,598
Mercara 19,357
Mhow 59,037
Midnapore 71,326
Miraj 77,606
Mirzapur-cum-Vindhyachal 105,939
Modasa 22,483
Mokokchung 17,423
Monghyr 102,474
Mora
Moradabad 258,590
Morena 44,901
Morvi 60,976
Mulund
Murud 11,210
Murwara 54,864
Muzaffarnagar 114,783
Muzaffarpur 126,379
Mysore 355,685
Nadiad 108,269
Nagapattinam 68,026
Nagaur 36,448
Nagercoil 141,288
Nagina 37,066
Nagpur 866,076
Nagpur□ 930,459
Nahan 16,017
Naihati 82,080
Naini Tal 23,986
Nainpur 14,683
Nalgonda 33,126
Nander 126,538
Nandurbar 54,070
Nandyal 63,193
Narayanpet 21,744
Narnaul 31,875
Narsimhapur 25,552
Narsinghgarh 13,814
Nasik 176,091
Nasirabad 25,732
Navsari 72,979
Nellore 133,590
New Delhi (cap.) 301,801
Nhava-Sheva
Nimach 47,113
Nipani 35,116
Nirmal 28,529
Nizamabad 115,640
North Lakhimpur 20,094
Nova Goa (Panaji) 34,953
Nowgong, Assam 56,537
Nowgong, Madhya Pradesh 10,248
Okha Port 10,687
Ongole 53,330
Ootacamund 63,310
Orai 42,513
Osmanabad 27,279
Pachmarhi 1,212
Palanpur 42,114
Palayankottai 70,070
Palghat 95,788
Pali 49,834
Palni 49,575
Panaji 34,953
Panchur 59,021
Pandharpur 53,638
Panihati 148,046
Panipat 87,981
Panna 22,316
Panruti 34,065
Paradip
Parbhani 61,570
Parlakhemundi 26,917
Partapgarh 17,402
Parvatipuram 30,025
Pasighat 5,116
Patan 64,519
Pathankot 76,355
Patiala 148,686
Patiala□
Patna 473,001
Patna□
Pauni 17,781
Phalodi 17,379
Phulbani 10,677
Pilibhit 68,273
Pokaran 7,769
Pondicherry 90,537
Ponnani 35,723
Poona (Pune)□
Porbandar 96,881
Porbandar□
Port Blair 26,218
Porto Novo 17,412
Proddatur 70,822
Puduchcheri (Pondicherry) 90,537
Pudukkottai 66,384
Pune 856,105
Puri 72,674
Purli 31,078
Purnea 56,484
Purulia 57,708
Puttur 17,483
Quilon 124,208
Radhanpur 18,360
Raichur 79,831
Raigarh 46,745
Raipur 174,518
Raipur□
Rajahmundry 165,912
Rajahmundry□
Rajapalaiyam 86,952
Rajapur 9,017
Rajgarh 11,475
Rajkot 300,612
Rajnandgaon 41,183
Rajpipla 25,769
Rajpur 34,393
Rajpura 14,840
Rameswarem 16,755
Rampur, Him. Pradesh 2,623
Rampur, Uttar Pradesh 161,417
Ranchi 175,934
Ratangarh 31,506
Ratlam 106,666
Ratnagiri 37,551
Raurkela 47,076
Raxaul 12,064
Rayagada 25,064
Renigunta 8,567
Rewa 69,182
Rishra 63,486
Robertsganj 7,093
Roha 8,631
Rohtak 124,783
Sadiya⊙ 64,252

British India
U.S.S.R.
GILGIT AGENCY
AFGHANISTAN
KASHMIR & JAMMU
N.W. FRONTIER PROV.
PUNJAB
PUNJAB STATES
BALUCHISTAN
TIBET
CHINA
IRAN
BAHAWALPUR (PUNJ. ST.)
DELHI
RAMPUR
UNITED PROVINCES
NEPAL
RAJPUTANA
PUNJ. ST.
SIND
AJMER-MERWARA
SIKKIM
BHUTAN
ASSAM
Brahmaputra
Ganges
E. ST.
KHASI HILLS
Gwadar (Oman)
GWALIOR
Indus
BENARES
BIHAR
MANIPUR
WESTERN INDIA
CENTRAL INDIA
GUJARAT ST.
EASTERN STATES
BENGAL
TRIPURA (E. ST.)
Arabian Sea
Chandernagore (Fr.)
BURMA
CENTRAL PROVINCES
BERAR
E. ST.
ORISSA
Diu (Port.)
Damão (Port.)
BOMBAY
EASTERN STATES
HYDERABAD
Bay of Bengal
DECCAN STATES
Yanaon (Fr.)
Gôa (Port.)
MADRAS
MYSORE
Bangalore (Br.)
COORG
Andaman Islands (Br.)
Mahé (Fr.)
Pondichéry (Fr.)
Karikal (Fr.)
M. ST.
Laccadive Islands (Madras)
Cochin (Br.)
MADRAS STATES
CEYLON
Nicobar Islands (Br.)
British India. The provinces of British India were directly administered by Britain. A few areas were leased from the Indian princes.
Indian States. The Indian States, sometimes referred to as the "Native" or "Princely States," were under the nominal control of maharajas or other hereditary princes.
Possessions of Other Countries in India
State or Provincial Boundaries
Other Internal Boundaries

Sagar 118,574 ... D4
Saharanpur 225,396 ... D3
Saharsa 23,217 ... F3
Salem 308,716 ... D6
Salem□ 416,440 ... D6
Samalkot 34,607 ... E5
Sambalpur 64,675 ... E4
Sambhal 86,323 ... D3
Sangamner 28,594 ... D5
Sangli 115,138 ... D6
Sankrail 11,300 ... F4
Santa Cruz ... B7
Santipur 61,166 ... F4
Sardarshahr 37,703 ... C3
Sarnath ... D3
Sasaram 48,282 ... E4
Satara 66,433 ... C5
Satna 57,531 ... D4
Savantvadi 16,873 ... C5
Savanur 18,302 ... D6
...awi□ 13,504 ... G7
Secunderabad 250,636 ... D5
Secunderabad□ 345,052 ... D5
Sehore 35,657 ... D4
Seoni 38,396 ... D4
Serampore 102,023 ... F1
Seringapatam 14,100 ... D6
Shahdol 28,490 ... E4
Shahjahanpur 135,604 ... D3
Shajapur 25,189 ... D4
Sheopur 16,418 ... D3
Shillong 87,659 ... G3
Shimoga 102,709 ... D6
Shivpuri 42,120 ... D3
Sholapur 398,361 ... D5
Shorapur 21,056 ... D5
...hyok ... D2
Sirsa 48,808 ... D3
Sirsi 28,576 ... D6
Sitapur 66,715 ... D3
Sonepur 8,084 ... E5
Srikakulam 45,179 ... E5
Srinagar 403,413 ... D2
Srinagar□ 423,253 ... D2
Srivardhan 12,342 ... C5
Sujangarh 17,244 ... E5
Surada 9,833 ... E5
Surat 471,656 ... C4
Surat□ 493,001 ... C4
Surendranagar 66,667 ... C4
...anda 41,611 ... E3
...heri 5,480 ... D2
...ellicherry 68,759 ... C6
...enali 102,937 ... E5
...ezpur 39,870 ... G3

Tezu 3,055 ... H3
Thana 170,675 ... B6
Thanjavur 140,547 ... D6
Tikamgarh 27,007 ... D4
Tinsukia 54,911 ... H3
Tiruchirappalli 307,400 ... D6
Tiruchirappalli□ 464,624 ... D6
Tiruchendur 18,126 ... D7
Tirunelveli 108,498 ... D7
Tirupati 65,843 ... D6
Tiruppattur 40,357 ... D6
Tiruppur 113,302 ... D6
Tiruvannamalai 61,370 ... D6
Titagarh 88,218 ... F1
Titlagarh 14,504 ... E4
Toibalawe ... G6
Tollygunge ... F2
Tonk 55,866 ... D3
Tranquebar 17,318 ... E6
Trichur 76,241 ... D6
Trivandrum 409,627 ... D7
Trombay ... B7
Tumkur 70,476 ... D6
Tura 28,344 ... E5
Tura 15,489 ... G3
Tuticorin 155,310 ... D7
Udaipur 161,278 ... C4
Udhampur 16,392 ... D2
Udipi 29,753 ... C6
Ujjain 203,278 ... D4
Umred 27,092 ... D4
Unnao 38,195 ... E3
Uran 12,616 ... B7
Uttarpara-Kotrung 67,568 ... F1
Vadodara 466,696 ... C4
Vadodara□ 467,487 ... C4
Valsad 43,254 ... C4
Vaniyambadi 51,810 ... D6
Varanasi 583,856 ... E3
Varanasi□ 606,721 ... E3
Vashi ... B7
Vedaranniyam 21,471 ... E6
Vellore 139,082 ... D6
Vellore□ 178,554 ... D6
Vengurla 11,805 ... C5
Venkatagiri 17,546 ... D6
Veraval 58,771 ... B4
Vesava ... B7
Vidisha 43,212 ... D4
Vijayawada 317,258 ... D5
Villupuram 60,242 ... D6
Vinukonda 16,259 ... D5
Virajpet 9,782 ... D6
Viramgam 43,790 ... C4
Visakhapatnam 352,504 ... E5
Visnagar 34,863 ... C4
Vizagapatam
(Visakhapatnam) 352,504 ... E5
Vizianagaram 86,608 ... E5
Warangal 207,520 ... D5
Wardha 69,037 ... D4
Wun 24,455 ... D5
Yadgir 32,756 ... D5
Yanam 8,291 ... E3
Yellamanchili 15,318 ... E5
Yeola 24,533 ... C5
Yeotmal 64,836 ... D5

OTHER FEATURES

Abor (hills) ... G3
Adam's Bridge (sound) ... D7
Agatti (isl.) ... C6
Amindivi (isls.) ... C6
Amini (Amindiri) (isl.) ... C6
Andaman (isls.) ... G6
Andaman (sea) ... G6
Androth (isl.) ... C6
Anjidiv (Angedeva) (isl.) ... C6
Arabian (sea) ... B5
Back (bay) ... B7
Bengal, Bay of (sea) ... F5
Berar (reg.) ... D4
Brahmaputra (riv.) ... G3
Butcher (isl.) ... B7
Cambay (gulf) ... C4
Cannanore (isls.) ... C6
Car Nicobar (isl.) ... G7
Chambal (riv.) ... D3
Chenab (riv.) ... C2
Chetlat (isl.) ... C6
Chilka (lake) ... E5
Coco (isl.) ... G6
Colaba (pt.) ... B7
Colair (lake) ... E5
Comorin (cape) ... D7
Coromandel Coast (reg.) ... E6
Daman (dist.) ... C4
Damodar (riv.) ... F4
Deccan (plat.) ... D5
Diu (dist.) ... B4
Eastern Ghats (mts.) ... D6
Elephanta (isl.) ... B7
Ganga (Ganges) (riv.) ... F3
Ganges, Mouths of the
(delta) ... F4
Ganges (riv.) ... F3
Ghaghra (riv.) ... E3
Goa (dist.) ... C5
Godavari (riv.) ... D5
Golconda (ruins) ... D5
Great (chan.) ... G7
Great Indian (des.) ... C3
Great Nicobar (isl.) ... G7
Himalaya (mts.) ... D2
Hindu Kush (mts.) ... C1
Hooghly (riv.) ... F2
Indus (riv.) ... B3
Jhelum (riv.) ... C2
Jumna (riv.) ... E3
Kadmat (isl.) ... C6
Kalpeni (isl.) ... C7
Kamet (mt.) ... D2
Kanchanjunga (mt.) ... F3
Karakoram (mts.) ... D1
Kaveri (riv.) ... D6
Khakchh (Kutch) (gulf) ... B4
Khasi (hills) ... G3
Kiltan (isl.) ... C6
Kistna (Krishna) (riv.) ... D5
Kunlun (range) ... D1
Kutch, Rann of
(salt marsh) ... B4
Laccadive (Cannanore)
(isls.) ... C6
Ladakh (reg.) ... D2
Little Andaman (isl.) ... G6
Little Nicobar (isl.) ... G7
Mahanadi (riv.) ... E4

Malabar (hill) ... B7
Malabar Coast (reg.) ... C6
Mannar (gulf) ... D7
Middle Andaman (isl.) ... G6
Minicoy (isl.) ... C7
Miri (hills) ... G3
Mishmi (hills) ... H3
Nancowry⊙ (isl.) ... G7
Nanda Devi (mt.) ... D2
Narmada (riv.) ... D4
North Andaman⊙ (isl.) ... G6
Palk (str.) ... D7
Penganga (riv.) ... D5
Periyar (lake) ... D7
Pitti (isl.) ... C6
Pulicat (lake) ... E6
Salsette (isl.) ... B7
Sambhar (lake) ... C3
Shipki (pass) ... D2
South Andaman (isl.) ... G6
Sundarbans (reg.) ... F4
Sutlej (riv.) ... C3
Ten Degree (chan.) ... G7
Towers of Silence ... B7
Travancore (reg.) ... D7
Tungabhadra (riv.) ... D5
Vindhya (range) ... D4
Western Ghats (mts.) ... C5
Zaskar (mts.) ... D2

MALDIVES

Maldives 143,046 ... C7

NEPAL

CITIES and TOWNS

Bhaktapur 40,112 ... F3
Bhaktapur⊙ 110,157 ... F3
Biratnagar 45,100 ... F3
Birganj 12,999 ... E3
Dhangarhi ... E3
Ilam 7,299 ... F3
Jaleswar ... F3
Janakpur 14,294 ... E3
Jumla⊙ 122,753 ... E3
Kathmandu (cap.) 150,402 ... E3
Kathmandu⊙ 353,752 ... E3
Lalitpur 59,049 ... E3
Lalitpur⊙ 154,998 ... E3
Mustang⊙ 26,944 ... E3
Nepalganj 23,523 ... E3
Pokhara 20,611 ... E3
Pyuthan⊙ 137,338 ... E3
Ridi ... E3
Sallyan⊙ 141,457 ... E3
Simikot ... E3

OTHER FEATURES

Annapurna (mt.) ... E3
Bheri (riv.) ... E3
Dhaulagiri (mt.) ... E3
Everest (mt.) ... F3
Himalaya (mts.) ... D2
Kanchenjunga (mt.) ... F3

PAKISTAN

PROVINCES

Azad Kashmir ... C2
Baluchistan 2,409,000 ... B3
Federal Administered Tribal Areas ... C2
Islamabad District 235,000 ... C2
Northern Areas ... D1
North-West Frontier 10,909,000 ... C2
Punjab 37,374,000 ... C2
Sind 13,965,000 ... B3

CITIES and TOWNS

Abbottabad 47,011 ... C2
Ahmadpur East 32,423 ... C3
Attock ... C2
Badin 6,387 ... B4
Bahawalnagar 36,290 ... C2
Bahawalpur 133,956 ... C1
Baltit ... C1
Bannu 43,795 ... C2
Barkhan 930 ... B3
Bhag 4,316 ... B3
Bhera 17,992 ... C2
Bostan ... B2
Bunji ... C1
Campbellpore 19,041 ... C2
Chachro ... C3
Chagai⊙ 41,263 ... A3
Chaman 12,208 ... B2
Chilas ... C1
Chiniot 69,124 ... C2
Dadu 19,142 ... B3
Dalbandin 1,724 ... A3
Dera Ghazi Khan 71,429 ... C3
Dera Ismail Khan 59,892 ... C2
Diplo ... B4
Dir ... C1
Duki 464 ... B2
Faisalabad 822,263 ... C2
Fort Sandeman 8,058 ... B2
Ghizar ... C1
Gilgit ... C1
Gujranwala 360,419 ... C2
Gujrat 100,581 ... C2
Gwadar 8,146 ... A4
Hindubagh 2,217 ... B2
Hoshab ... A3
Hyderabad 628,310 ... B3
Islamabad (cap.) 77,318 ... C2
Jacobabad 57,292 ... B3
Jhal Jhao ... B3
Jhang Sadar 135,722 ... C2
Jhelum 63,653 ... C2
Jhudo 6,950 ... B3
Kalam ... C1
Kalat 5,321 ... B3
Kandrach ... B3
Karachi 3,498,634 ... B4
Karachi* 3,650,000 ... B4
Kashmor ... C3
Kasur 102,531 ... C2
Khairpur 34,144 ... B3
Khanewal 67,617 ... C2
Khanpur 31,465 ... C3
Kharan Kalat 2,692 ... A3

Khushab 24,851 ... C2
Kohat 64,634 ... C2
Kotri 20,262 ... B3
Ladgasht (Qila Ladgasht) ... A3
Lahore 2,165,372 ... C2
Lahri ... B3
Larkana 71,943 ... B3
Leiah 19,608 ... C2
Loralai 5,519 ... B2
Lyallpur (Faisalabad) 822,263 ... C2
Mach 4,921 ... B3
Malakand ... C2
Mardan 115,218 ... C2
Mastung 5,962 ... B3
Mianwali 31,398 ... C2
Miram Shah ... C2
Mirpur ... C2
Mirpur Khas 81,617 ... B3
Misgar ... C1
Mithi ... C4
Multan 542,195 ... C2
Multan* 723,000 ... C2
Murree 13,486 ... C2
Musa Khel Bazar 429 ... B2
Muzaffarabad ... C2
Nagar ... C2
Nagar Parkar ... C4
Nal ... B3
Nawabshah 80,779 ... B3
Nok Kundi 861 ... A3
Nowshera 56,117 ... C2
Nushki 3,153 ... B3
Ormara ... A3
Panjgur 2,032 ... A3
Pasni 7,483 ... A4
Peshawar 268,366 ... C2
Pindi Gheb 12,416 ... C2
Pishin 2,906 ... B2
Qila Ladgasht ... A3
Quetta 156,000 ... B2
Rahimyar Khan 74,407 ... C3
Rawalpindi 615,392 ... C2
Ribat Qila ... A3
Risalpur Cantonment 11,291 ... C2
Rohri 19,072 ... B3
Rondu ... D1
Sahiwal 106,213 ... C2
Saidu 10,581 ... C2
Sargodha 201,407 ... C2
Sehwan 4,169 ... B3
Shahbandar ... B4
Shikarpur 70,301 ... B3
Sialkot 203,779 ... C2
Sibi 13,327 ... B3
Skardu ... D1
Sonmiani ... B3
Sorah ... B3
Sui 1,082 ... B3
Sukkur 158,876 ... B3
Surab ... B3
Tando Adam ... B3
Tando Muhammad Khan 31,246 ... B3
Tando Allahyar ... B3
Tatta 12,786 ... B4
Tump ... A3
Turbat 4,578 ... A3
Uch 5,483 ... B3
Uthal ... B3
Wah 107,671 ... C2
Wana ... C2
Yasin ... C1

OTHER FEATURES

Arabian (sea) ... B5
Bolan (pass) ... B3
Chagai (hills) ... A3
Chenab (riv.) ... C2
Hindu Kush (mts.) ... B1
Indus (riv.) ... B3
K2 (mt.) ... D1
Konar (riv.) ... C1
Kutch, Rann of (salt marsh) ... B4
Mohenjo Daro (ruins) ... B3
Muari, Ras (cape) ... B4
Ravi (riv.) ... C2
Siahan (range) ... A3
Sulaiman (range) ... B3
Sutlej (riv.) ... C3
Talab (riv.) ... A3
Taxila (ruins) ... C2
Tirich Mir (mt.) ... B2
Zhob (riv.) ... B2

SRI LANKA (CEYLON)

CITIES and TOWNS

Anuradhapura 34,836 ... E7
Badulla 34,658 ... E7
Batticaloa 36,761 ... E7
Colombo (cap.) 618,000 ... D7
Colombo* 852,098 ... D7
Dehiwala-Mt. Lavinia 154,785 ... D7
Galle 72,720 ... D7
Hambantota 6,908 ... D7
Jaffna 112,000 ... D7
Kalmunai 19,756 ... E7
Kalutara 28,748 ... D7
Kandy 93,602 ... D7
Kurunegala 25,189 ... D7
Mannar 11,157 ... D7
Matara 36,641 ... D7
Moratuwa 96,489 ... D7
Mullaittivu 4,930 ... E7
Negombo 57,115 ... D7
Nilaveli 4,556 ... E7
Nuwara Eliya 16,347 ... E7
Polonnaruwa 9,551 ... E7
Puttalam 17,982 ... D7
Ratnapura 29,116 ... D7
Sigiriya 1,446 ... E7
Tangalla 8,748 ... D7
Trincomalee 41,780 ... E7
Vavuniya 15,639 ... E7

OTHER FEATURES

Adam's (peak) ... D7
Adam's Bridge (shoals) ... D7
Dondra (head) ... E7
Kirigalpota (mt.) ... E7
Mannar (gulf) ... D7
Palk (str.) ... D7
Pedro (pt.) ... E6
Pidurutalagala (mt.) ... E7

*City and suburbs.
⊙Population of district.
□Population of urban areas.

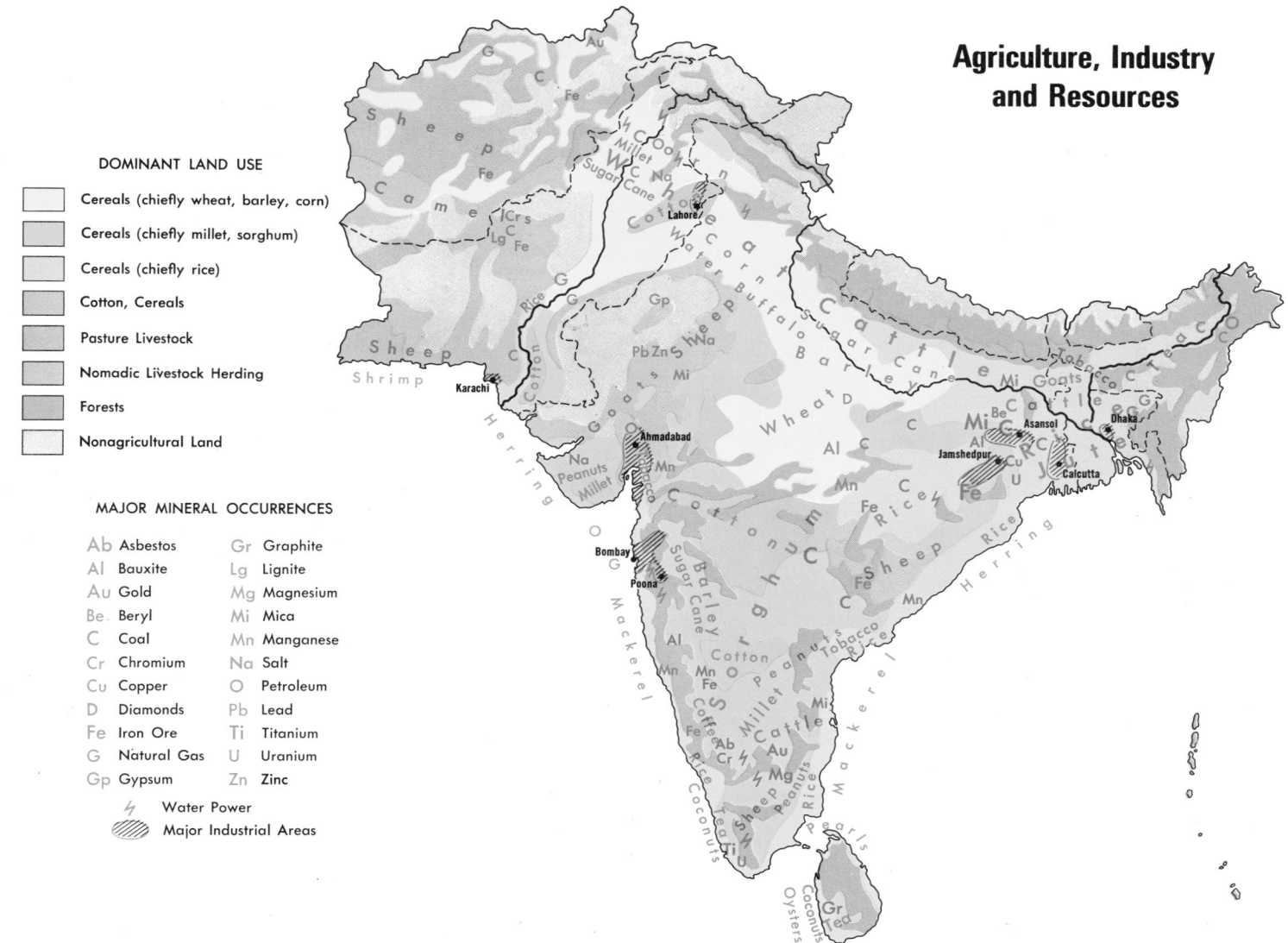

Agriculture, Industry and Resources

DOMINANT LAND USE

Cereals (chiefly wheat, barley, corn)
Cereals (chiefly millet, sorghum)
Cereals (chiefly rice)
Cotton, Cereals
Pasture Livestock
Nomadic Livestock Herding
Forests
Nonagricultural Land

MAJOR MINERAL OCCURRENCES

Ab Asbestos
Al Bauxite
Au Gold
Be Beryl
C Coal
Cr Chromium
Cu Copper
D Diamonds
Fe Iron Ore
G Natural Gas
Gp Gypsum

Gr Graphite
Lg Lignite
Mg Magnesium
Mi Mica
Mn Manganese
Na Salt
O Petroleum
Pb Lead
Ti Titanium
U Uranium
Zn Zinc

Water Power
Major Industrial Areas

Burma, Thailand, Indochina and Malaya

CONIC PROJECTION

SCALE OF MILES

SCALE OF KILOMETERS

International Boundaries ....
Division and State Boundaries ......
Capitals of Countries ....☆
Division and State Capitals ....◉

Scale 1:10,000,000

© Copyright HAMMOND INCORPORATED, Maplewood, N.J.

Longitude East 96° of Greenwich

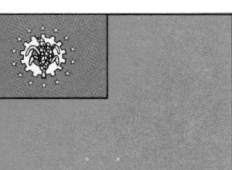

BURMA

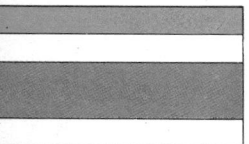

THAILAND

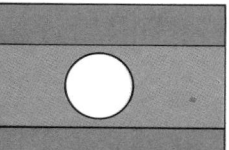

LAOS

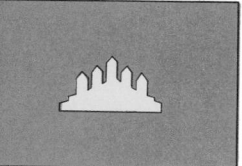

CAMBODIA

VIETNAM

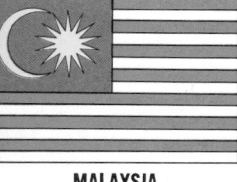

MALAYSIA

SINGAPORE

BURMA

AREA 261,789 sq. mi. (678,034 sq. km.)
POPULATION 32,913,000
CAPITAL Rangoon
LARGEST CITY Rangoon
HIGHEST POINT Hkakabo Razi 19,296 ft. (5,881 m.)
MONETARY UNIT kyat
MAJOR LANGUAGES Burmese, Karen, Shan, Kachin, Chin, Kayah, English
MAJOR RELIGIONS Buddhism, tribal religions

THAILAND

AREA 198,455 sq. mi. (513,998 sq. km.)
POPULATION 46,455,000
CAPITAL Bangkok
LARGEST CITY Bangkok
HIGHEST POINT Doi Inthanon 8,452 ft. (2,576 m.)
MONETARY UNIT baht
MAJOR LANGUAGES Thai, Lao, Chinese, Khmer, Malay
MAJOR RELIGIONS Buddhism, tribal religions

LAOS

AREA 91,428 sq. mi. (236,800 sq. km.)
POPULATION 3,721,000
CAPITAL Vientiane
LARGEST CITY Vientiane
HIGHEST POINT Phou Bia 9,252 ft. (2,820 m.)
MONETARY UNIT kip
MAJOR LANGUAGE Lao
MAJOR RELIGIONS Buddhism, tribal religions

CAMBODIA

AREA 69,898 sq. mi. (181,036 sq. km.)
POPULATION 5,200,000
CAPITAL Phnom Penh
LARGEST CITY Phnom Penh
HIGHEST POINT 5,948 ft. (1,813 m.)
MONETARY UNIT riel
MAJOR LANGUAGE Khmer (Cambodian)
MAJOR RELIGION Buddhism

VIETNAM

AREA 128,405 sq. mi. (332,569 sq. km.)
POPULATION 52,741,766
CAPITAL Hanoi
LARGEST CITY Ho Chi Minh City (Saigon)
HIGHEST POINT Fan Si Pan 10,308 ft. (3,142 m.)
MONETARY UNIT dong
MAJOR LANGUAGES Vietnamese, Thai, Muong, Meo, Yao, Khmer, French, Chinese, Cham
MAJOR RELIGIONS Buddhism, Taoism, Confucianism, Roman Catholicism, Cao-Dai

MALAYSIA

AREA 128,308 sq. mi. (332,318 sq. km.)
POPULATION 13,435,588
CAPITAL Kuala Lumpur
LARGEST CITY Kuala Lumpur
HIGHEST POINT Mt. Kinabalu 13,455 ft. (4,101 m.)
MONETARY UNIT ringgit
MAJOR LANGUAGES Malay, Chinese, English, Tamil, Dayak, Kadazan
MAJOR RELIGIONS Islam, Confucianism, Buddhism, tribal religions, Hinduism, Taoism, Christianity, Sikhism

SINGAPORE

AREA 226 sq. mi. (585 sq. km.)
POPULATION 2,413,945
CAPITAL Singapore
LARGEST CITY Singapore
HIGHEST POINT Bukit Timah 581 ft. (177 m.)
MONETARY UNIT Singapore dollar
MAJOR LANGUAGES Chinese, Malay, Tamil, English, Hindi
MAJOR RELIGIONS Confucianism, Buddhism, Taoism, Hinduism, Islam, Christianity

Topography

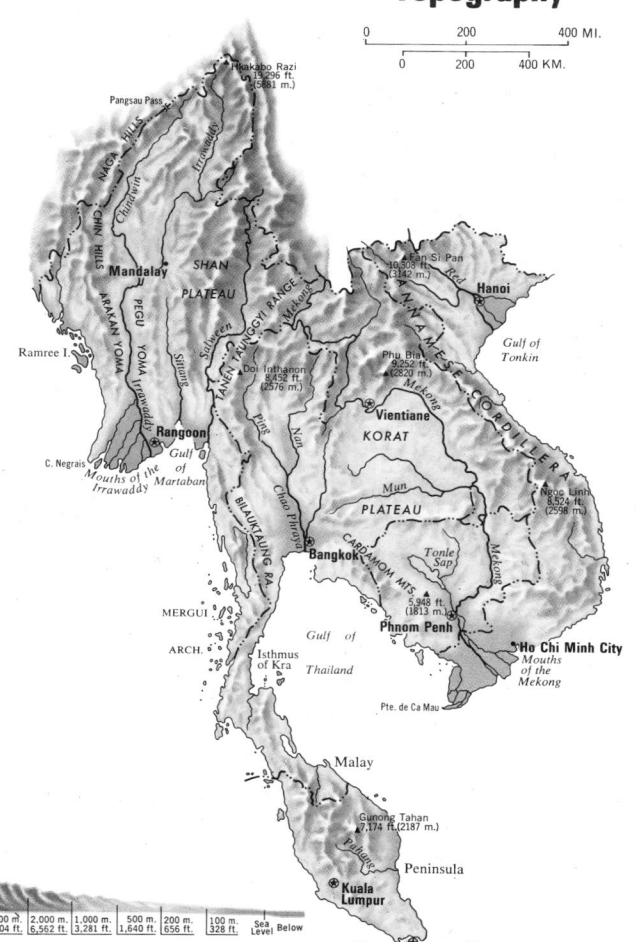

BURMA

INTERNAL DIVISIONS

Arakan (state) 1,710,913	B3
Chin (state) 323,094	B2
Irrawaddy (div.) 4,152,521	B3
Kachin (state) 735,144	C1
Karen (state) 865,218	C3
Kayah (state) 126,492	C3
Magwe (div.) 2,632,144	B2
Mandalay (div.) 3,662,312	B2
Mon (state) 1,313,111	C3
Pegu (div.) 3,174,109	C3
Rangoon (div.) 3,186,886	C3
Sagaing (div.) 3,115,502	B1
Shan (state) 3,178,214	C2
Tenasserim (div.) 717,607	C4

CITIES and TOWNS

Akyab (Sittwe) 42,329	B2
Allanmyo 15,580	B3
Amarapura 11,268	B2
Amherst 6,000	C3
An	B3
Anin	C4
Bassein 126,045	B3
Bhamo 9,821	C1
Chauk 24,466	B2
Danubyu	B3
Falam	B2
Fort Hertz (Putao)	C1
Gawai	C2
Gokteik	C2
Gwa	B3
Gyobingauk 9,922	C3
Haka	B2
Henzada 61,972	B3
Hmawbi 23,032	C3
Homalin	B1
Hsenwi	C2
Hsipaw	C2
Htawgaw	C1
Insein 143,625	C3
Kamaing	C1
Karathuri	C5
Katha 7,648	C1
Kawludo	C3
Kawthaung 1,520	C5
Keng Hkam	C2
Keng Tung	C2
Koma	C4
Kunlong	C2
Kyaikto 13,154	C3
Kya-in Seikkyi	C3
Kyangin 6,073	B3
Kyaukme	C2
Kyaukpadaung 5,480	B2
Kyaukpyu 7,335	B3
Kyaukse 8,659	C2
Labutta 12,982	B3
Lai-hka	C2
Lamu	B3
Lashio	C2
Lenya	C5
Letpadan 15,896	C3
Lewe	B3
Loi-kaw	C3
Lonton	B1
Magwe 13,270	B2
Maingkwan	C1
Maliwun	C5
Mandalay 418,008	C2
Man Hpang	C2
Martaban 5,661	C3
Ma-ubin 23,362	B3
Maungdaw 3,772	B2
Mawkmai	C2
Mawlaik 2,993	B2
Mawlu	C1
Maymyo 22,287	C2
Meiktila 19,474	B2
Mergui 33,697	C4
Minbu 9,096	B2

Minhla 6,470	B3
Mogaung 2,920	C1
Mogok 8,334	C2
Mohnyin	C1
Möng Hsat	C2
Möng Mau	C3
Möng Mit	C2
Möng Pan	C2
Möng Si	C2
Möng Tön	C2
Möng Tung	C2
Monywa 26,279	B2
Moulmein 171,977	C3
Mudon 20,136	C3
Myanaung 11,155	B3
Myaungmya 24,532	B3
Myingyan 36,439	B2
Myitkyina 12,382	C1
Myohaung 6,534	B2
Naba	B1
Namhkam	C2
Namlan	C2
Namtu	C2
Natmauk	B2
Okkan 14,443	B3
Okpo 12,155	C3
Pakokku 30,943	B2
Palaw 5,596	C4
Paletwa	B2
Pantha	B2
Papun	C3
Pasawng	C3
Paungde 17,286	B3
Pegu 47,378	C3
Prome (Pye) 36,997	B3
Putao	C1
Pyapon 19,174	B3
Pye 36,997	B3
Pyinmana 22,025	C3
Pyu 10,443	C3
Rangoon (cap.) 1,586,422	C3
Rangoon* 2,055,365	C3
Rathedaung 2,969	B2
Sadon	C1
Sagaing 15,382	B2
Samka	C2
Sandoway 5,172	B3
Shingbwiyang	B1
Shwebo 17,827	B2
Shwenyaung	C2
Singkaling Hkamti	B1
Singu 4,027	C1
Sinlumkaba	C1
Sittwe 42,329	B2
Sumprabum	C1
Syriam 15,296	C3
Taungdwingyi 16,233	C2
Taunggyi	C2
Tavoy 40,312	C4
Tharrawaddy 8,977	C3
Thaton 38,047	C3
Thaungdut	B1
Thayetmyo 11,649	B3
Thazi 7,531	C2
Thongwa 10,829	C3
Toungoo 31,589	C3
Wakema 20,716	B3
Yamethin 11,167	C2
Yandoon 15,245	B3
Ye 12,852	C4
Yenangyaung 24,416	B2
Yesagyo 7,880	B2
Ye-u 5,307	B2
Ywathit	C3
Zadi	C4
Zalun 899	B3

OTHER FEATURES

Amya (pass)	C4
Andaman (sea)	B4
Arakan Yoma (mts.)	B3
Ataran (riv.)	C4
Bengal, Bay of (sea)	B3
Bentinck (isl.)	C5

(continued on following page)

(continued)

Bilauktaung (range) C4
Chaukan (pass) C1
Cheduba (isl.) B3
Chin (hills) B2
Chindwin (riv.) B2
Coco (chan.) B4
Combermere (bay) B3
Daung Kyun (isl.) C4
Dawna (range) C3
Great Coco (isl.) B4
Great Tenasserim (riv.) C4
Heinze Chaung (bay) C4
Heywood (chan.) B3
Hka, Nam (riv.) C2
Hkakabo Razi (mt.) C1
Indawgyi (lake) C1
Inle (lake) C2
Irrawaddy (riv.) B3
Irrawaddy, Mouths of the (delta) B4
Kadan Kyun (isl.) C4
Kaladan (riv.) B2
Kalegauk (isl.) C4
Khao Luang (mt.) C5
Lanbi Kyun (isl.) C5
Launglon Bok (isls.) C4
Loi Leng (mt.) C2
Manipur (riv.) B2
Martaban (gulf) C4
Mekong (riv.) D2
Mergui (arch.) C5
Mon (riv.) B2
Mu (riv.) B2
Negrais (cape) B3
Pakchan (riv.) C5
Pangsau (pass) C1
Pawn, Nam (riv.) C2
Pegu Yoma (mts.) B3
Preparis (isl.) B4
Ramree (isl.) B3
Salween (riv.) C3
Shan (plat.) C2
Sittang (riv.) C3
Taungthonton (mt.) B1
Tavoy (pt.) C4
Tenasserim (isl.) C4
Teng, Nam (riv.) C2
Three Pagodas (pass) C4
Victoria (mt.) B2

CAMBODIA (KAMPUCHEA)

CITIES and TOWNS

Batdambang (Battambang) D4
Choam Khsant E4
Kampong Cham E4
Kampong Chhnang D4
Kampong Khleang E4
Kampong Saom D5
Kampong Spoe E5
Kampong Thum E4
Kampong Trabek E5
Kampot E5
Kaoh Nhek E4
Kracheh E4
Krong Kaoh Kong D4
Krong Keb E5
Kulen E4
Lumphat E4
Moung Roessei D4
Pailin D4
Paoy Pet D4
Phnom Penh (cap.) c. 300,000 E5
Phnum Tbeng Meanchey E4
Phsar Ream D5
Phumi Banam E5
Phumi Phsar E4
Phumi Prek Kak E4
Phumi Samraong D4
Pouthisat D4
Prek Pouthi E5
Prey Veng E5
Pursat (Pouthisat) D4
Rovieng Tbong E4
Sambor E4
Senmonoron E4
Siempang E4
Siemreab D4
Sisophon D4
Sre Ambel D5
Sre Khtum E4
Stoeng Treng E4
Suong E5
Svay Rieng E5
Takev E5
Virochey E4

OTHER FEATURES

Angkor Wat (ruins) E4
Dangrek (mts.) D4
Drang, Ia (riv.) E4
Joncs (plain) E5
Khong, Se (riv.) E4
Kong, Kaoh (isl.) D5
Mekong (riv.) E4
Rung, Kaoh (isl.) D5
San, Se (riv.) E4
Sen, Stoeng (riv.) E4
Srepok (riv.) E4
Tang, Kaoh (isl.) D5
Thailand (gulf) D5
Tonle Sap (lake) D4
Wai, Poulo (isls.) D5

LAOS

CITIES and TOWNS

Attapu 2,750 E4
Ban Khon E4
Ban Lahanam E3
Borikan D3
Champasak 3,500 E4
Dônghén E3
Khamkeut⊙ 31,206 E3
Louang Namtha 1,459 D2
Louangphrabang 7,596 D3
Muang Hinboun 1,750 E3
Muang Kènthao D3
Muang Khammoun 5,500 E3
Muang Không 1,750 E4
Muang Khôngxédôn 2,000 E4
Muang Khoua D2
Muang May E4
Muang Ou Tai D2
Muang Paktha D2
Muang Phin E3
Muang Tahoi E3
Muang Vapi E4
Muang Xaignabouri (Sayaboury) 2,500 D3
Mounlapamôk E4
Napè E3
Nong Het E3
Pakxé 8,000 E4
Phiafai⊙ 17,216 E4
Phôngsali 2,500 D2
San Nua (Sam Neua) 3,000 E2
Saravan 2,350 E4
Savannakhét 8,500 E3
Sayaboury (Muang Xaignabouri) 2,500 D3
Thakhek (Muang Khammouan) 5,500 E3
Tourakom D3
Viangchan (Vientiane) 132,253 D3
Vientiane (cap.) 132,253 D3
Xiangkhoang 3,500 D3

OTHER FEATURES

Bolovens (plat.) E4
Hou, Nam (riv.) D2
Jars (plain) D3
Mekong (riv.) D3
Ou, Nam (riv.) D2
Phou Bia (mt.) D3
Phou Cô Pi (mt.) E3
Phou Loi (mt.) D2
Rao Co (mt.) E3
Se Khong (riv.) E4
Tha, Nam (riv.) D2
Xianghoang (plat.) D3

MALAYA, MALAYSIA★

STATES

Federal Territory 937,875 D7
Johor (Johore) 1,601,504 D7
Kedah 1,102,200 D6
Kelantan 877,575 D6
Melaka 453,153 D7
Negeri Sembilan 563,955 D7
Pahang 770,644 D7
Perak 1,762,288 D6
Perlis 147,726 D6
Pinang (Penang) 911,586 D6
Selangor 1,467,441 D7
Terengganu 542,280 D6

CITIES and TOWNS

Alor Gajah 2,222 D7
Alor Setar 66,260 D6
Bandar Maharani (Muar) 61,218 D7
Bandar Penggaram (Batu Pahat) 53,291 D7
Batu Gajah 10,692 D6
Batu Pahat 53,291 D7
Bentong 22,683 D7
Butterworth 61,187 D6
Chukai 12,514 D6
Gemas 5,214 D7
George Town (Pinang) 269,603 C6
Ipoh 247,953 D6
Johor Baharu (Johore Bharu) 136,234 F5
Kampar 26,591 D6
Kangar 8,758 D6
Kelang 113,611 D7
Keluang 43,272 D7
Kota Baharu 55,124 D6
Kota Tinggi 8,725 F5
Kuala Dungun 17,560 D6
Kuala Lipis 9,270 D6
Kuala Lumpur (cap.) 451,977 D7
Kuala Lumpur* 937,875 D7
Kuala Pilah 12,508 D7
Kuala Rompin 1,384 D7
Kuala Selangor 3,132 D7
Kuala Terengganu 53,320 D6
Kuantan 43,358 D7
Kulai 11,841 F5
Lumut 3,255 D6
Malacca (Melaka) 87,160 D7
Mawai F5
Melaka 87,160 D7
Mersing 18,246 E7
Muar 61,218 D7
Pekan 4,682 D7
Pekan Nanas 9,003 E5
Pinang (George Town) 269,603 C6
Pontian Kechil 8,349 E5
Port Dickson 10,300 D7
Port Kelang D7
Port Weld 3,233 D6
Raub 18,433 D7
Segamat 17,796 D7
Seremban 80,921 D7
Sungai Petani 35,959 C6
Taiping 54,645 D6
Tanah Merah 7,012 D6
Telok Anson 44,524 D6
Tumpat 10,673 D6

OTHER FEATURES

Aur, Pulau (isl.) E7
Belumut, Gunong (mt.) D7
Gelang, Tanjong (pt.) D6
Johor, Sungai (riv.) F5
Johore (str.) E6
Kelantan, Sungai (riv.) D6
Langkawi, Pulau (isl.) C6
Ledang, Gunong (mt.) D7
Lima, Pulau (isl.) F6
Malacca (str.) D7
Malay (pen.) D6
Pahang, Sungai (riv.) D7
Pangkor, Pulau (isl.) D6
Perak, Gunong (mt.) D6
Perhentian, Kepulauan (isls.) D6
Pulai, Sungai (riv.) E5
Ramunia, Tanjong (pt.) F6
Redang, Pulau (isl.) D6
Sedili Kechil, Tanjong (pt.) F5
Tahan, Gunong (mt.) D6
Temiang, Bukit (mt.) D6
Tenggol, Pulau (isl.) D6
Tinggi, Pulau (isl.) E7

SINGAPORE

CITIES and TOWNS

Jurong 50,974 E6
Nee Soon 37,641 F6
Serangoon 89,558 F6
Singapore (cap.) 2,413,945 F6

OTHER FEATURES

Keppel (harb.) F6
Main (str.) F6
Singapore (str.) F6
Tekong Besar, Pulau (isl.) F6

THAILAND (SIAM)

CITIES and TOWNS

Ang Thong 7,267 C4
Ayutthaya (Phra Nakhon Si Ayutthaya) 37,213 D4
Ban Aranyaprathet 12,276 D4
Bangkok (cap.) 1,867,297 D4
Bangkok* 2,495,312 D4
Bang Lamung D4
Bang Saphan C5
Ban Kantang 9,247 C6
Ban Kapong C5
Ban Khlong Yai D5
Ban Kui Nua D4
Ban Ngon D3
Ban Pak Phanang 13,590 D5
Banphot Phisai C3
Ban Pua D3
Ban Sattahip D4
Ban Tha Uthen D3
Bua Chum D4
Buriram 16,431 D4
Chachoengsao 22,106 D4
Chai Badan D4
Chai Buri D3
Chainat 9,944 D4
Chaiya C5
Chaiyaphum 12,540 D4
Chang Khoeng C3
Chanthaburi 15,479 D4
Chiang Dao C3
Chiang Khan D3
Chiang Mai 83,729 C3
Chiang Rai 13,927 C3
Chiang Saen C2
Chon Buri 39,367 D4
Chumphon 11,643 C5
Den Chai C3
Hat Yai 47,953 C6
Hot C3
Hua Hin 21,426 D4
Kalasin 14,960 D3
Kamphaeng Phet 12,378 C3
Kanchanaburi 16,397 C4
Khanu C3
Khemmarat E4
Khon Kaen 29,431 D3
Khorat (Nakhon Ratchasima) 66,071 D4
Krabi 8,764 C5
Krung Thep (Bangkok) (cap.) 1,867,297 D4
Kumphawapi D3
Lae D3
Lampang 40,100 C3
Lamphun 11,309 C3
Lang Suan 4,020 C5
Loei 10,137 D3
Lom Sak 10,597 D3
Lop Buri 23,112 D4
Mae Hong Son 3,981 C3
Maha Sarakham 19,707 D3
Mukdahan E3
Nakhon Nayok 8,185 D4
Nakhon Pathom 34,300 C4
Nakhon Phanom 20,385 D3
Nakhon Ratchasima 66,071 D4
Nakhon Sawan 46,853 D4
Nakhon Si Thammarat 40,671 D5
Nan 17,738 D3
Nang Rong D4
Narathiwat 21,256 D6
Ngao D3
Nong Khai 21,150 D3
Pattani 21,938 D6
Phanat Nikhom 10,514 D4
Phangnga 5,738 C5
Phatthalung 13,336 D6
Phayao 20,346 C3
Phet Buri 27,755 C4
Phetchabun 6,240 D3
Phichai D3
Phichit 10,814 D3
Phitsanulok 33,883 D3
Phon Phisai D3
Phrae 17,555 D3
Phra Nakhon Si Ayutthaya 37,213 D4
Phuket 34,362 C6
Phutthaisong D4
Prachin Buri 14,167 D4
Prachuap Khiri Khan 9,075 D5
Pran Buri D4
Rahaeng (Tak) 16,317 C3
Ranong 10,301 C5
Rat Buri 32,271 C4
Rayong 14,846 D4
Roi Et 20,242 D4
Rong Kwang D3
Sakon Nakhon 18,943 E3
Samut Prakan 46,632 D4
Samut Sakhon 33,619 D4
Samut Songkhram 23,574 C4
Sara Buri 25,025 D4
Satun 7,315 C6
Sawankhalok 8,387 C3
Selaphum E3
Sing Buri 9,050 D4
Singora (Songkhla) 41,193 D6
Sisaket 13,662 E4
Songkhla 41,193 D6
Sukhothai 15,488 D3
Suphan Buri 18,768 C4
Surat Thani 24,923 C5
Surin 16,342 D4
Suwannaphum D4
Tak 16,317 C3
Takua Pa 7,825 C5
Thoen C3
Thon Buri 628,015 D4
To Mo D6
Trang 32,985 C6
Trat 7,917 D4
Ubon 40,650 E4
Udon Thani 56,218 D3
Uthai Thani 10,525 C4
Uttaradit 12,022 D3
Warin Chamrap 21,520 E4
Yala 30,051 D6
Yasothon 12,079 D4

OTHER FEATURES

Amya (pass) C4
Bilauktaung (range) C4
Chang, Ko (isl.) D4
Chao Phraya, Mae Nam (riv.) D4
Chi, Mae Nam (riv.) D3
Dangrek (Dong Rak) (mts.) D4
Doi Inthanon (mt.) C3
Doi Pha Hom Pok (mt.) C2
Doi Pia Fai (mt.) D4
Kao Prawa (mt.) C3
Khao Luang (mt.) C5
Khwae Noi, Mae Nam (riv.) C4
Kra (isth.) C5
Kut, Ko (isl.) D5
Laem Pho (cape) D6
Laem Talumphuk (cape) D5
Lanta, Ko (isl.) C6
Luang (mt.) C5
Mae Klong, Mae Nam (riv.) C4
Mekong (riv.) E3
Mun, Mae Nam (riv.) D4
Nan, Mae Nam (riv.) D3
Nong Lahan (lake) D3
Pakchan (riv.) C5
Pa Sak, Mae Nam (riv.) D4
Phangan, Ko (isl.) D5
Phuket, Ko (isl.) C5

Agriculture, Industry and Resources

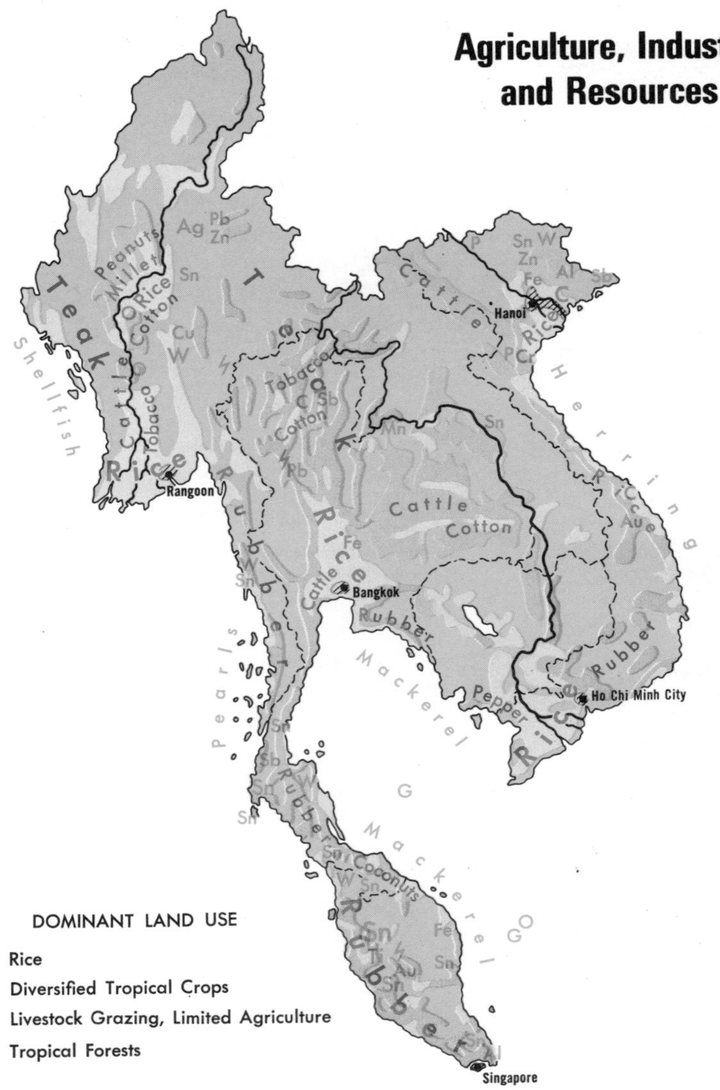

DOMINANT LAND USE

Rice
Diversified Tropical Crops
Livestock Grazing, Limited Agriculture
Tropical Forests

MAJOR MINERAL OCCURRENCES

Ag	Silver	Cu	Copper	O	Petroleum	Sn	Tin
Al	Bauxite	Fe	Iron Ore	P	Phosphates	Ti	Titanium
Au	Gold	G	Natural Gas	Pb	Lead	W	Tungsten
C	Coal	Mn	Manganese	Sb	Antimony	Zn	Zinc
Cr	Chromium						

Water Power

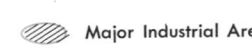

Major Industrial Areas

Ping, Mae Nam (riv.) C3
Samui (str.) D5
Samui, Ko (isl.) D5
Siam (Thailand) (gulf) D5
Tao, Ko (isl.) C5
Tapi, Mae Nam (riv.) C5
Terutao, Ko (isl.) C6
Tha Chin, Mae Nam (riv.) C4
Thale Luang (lag.) D6
Thalu, Ko (isls.) C5
Three Pagodas (pass) C4
Wang, Mae Nam (riv.) C3

VIETNAM

CITIES and TOWNS

An Loc (Binh Long) 15,276 E5
An Nhon F4
An Tuc (An Khe) F4
Ap Long Ha F5
Ap Vinh Hao F5
Bac Can E2
Bac Giang E2
Bac Lieu 53,841 E5
Bac Ninh 22,560 E2
Ba Don E3
Bai Thuong E3
Ban Me Thuot 68,771 F4
Bao Ha D2
Bao Lac E2
Bien Hoa 87,135 E5
Binh Long (An Loc) 15,276 E5
Binh Son F4
Bo Duc E4
Bong Son (Hoai Nhon) F4
Cam Ranh 118,111 F5
Can Tho 182,424 E5
Cao Bang E2
Cao Lanh 16,482 E5
Chau Phu 37,175 E5
Chu Lai F4
Con Cuong E3
Cua Rao E3
Da Lat 105,072 F5
Dam Doi E5
Da Nang 492,194 E3
Dien Bien Phu D2
Dong Hoi E3
Duong Dong D5
Gia Dinh E5
Go Cong 33,191 E5
Ha Giang E2
Haiphong* 1,279,067 E2
Hanoi (cap.)* 2,570,905 E2
Ha Tien E5
Ha Tinh E3
Hau Bon E4
Hoa Binh E2
Hoa Da F5
Hoai Nhon F4
Ho Chi Minh City (Saigon)* 3,419,678 E5
Hoi An 45,059 F4
Hoi Xuan E2
Hon Chong E5
Hon Gai 100,000 E2
Hue 209,043 E3
Huong Khe E3
Ke Bao E2
Khanh Hoa F4
Khanh Hung 59,015 E5
Khe Sanh E3
Kien Hung E5
Kontum 33,554 E4
Lac Giao (Ban Me Thuot) 68,771 E4
Lai Chau D2
Lang Son 15,071 E2
Lao Cai E2
Loc Ninh E5
Long Xuyen 72,658 E5
Mo Duc F4
Mong Cai E2
Muong Khuong E2
My Tho 119,892 E5
Nam Dinh E2
Nghia Lo D2
Nha Trang 216,227 F4
Ninh Binh E3
Phan Rang 33,377 F5
Phan Thiet 80,122 F5
Phu Cuong 28,267 E5
Phu Lang Thuong (Bac Giang) E2
Phuc Loi E3
Phu Dien E3
Phu Ly E2
Phu My F4
Phu Qui E3
Phu Rieng E5
Phu Tho 10,888 E2
Phu Vinh 48,485 E5
Pleiku 23,720 E4
Quang Nam E4
Quang Ngai 14,119 F4
Quang Tri 15,874 E3
Quang Yen E2
Quan Long 59,331 E5
Qui Nhon 213,757 F4
Rach Gia 104,161 E5
Ron E3
Sa Dec 51,867 E5
Saigon (Ho Chi Minh City)* 3,419,678 E5
Song Cau F4
Son Ha F4
Son La D2
Son Tay 19,213 E2
Tam Ky 38,532 F4
Tam Quan F4
Tan An 38,082 E5
Tay Ninh 22,957 E5
Thai Binh 14,739 E2
Thai Nguyen E2
Thanh Hoa 31,211 E3
Thanh Tri E5
That Khe E2
Tien Yen E2
Tra Vinh (Phu Vinh) 48,485 E5
Truc Giang 68,629 E5
Trung Khanh Phu E2
Tuyen Quang E2
Tuy Hoa 63,552 F4
Van Hoa E2
Van Ninh F4
Van Yen E2
Vinh 43,954 E3
Vinh Long 30,667 E5
Vinh Yen E2
Vu Liet E3
Vung Tau 108,436 E5
Xuan Loc E
Yen Bai E

OTHER FEATURES

Bach Long Vi, Dao (isl.) F
Ba Den, Nui (mt.) E
Bai Bung, Mui (Ca Mau) (pt.) E
Black (riv.) D
Ca Mau (Mui Bai Bung) (pt.) E
Cam Ranh, Vinh (bay) F
Cat Ba, Dao (isl.) E
Chon May, Vung (bay) F
Cu Lao, Hon (isls.) F
Deux Frères, Les (isls.) E
Dinh, Mui (cape) F
Fan Si Pan (mt.) D
Ia Drang (riv.) E
Joncs (plain) E
Kontum (plat.) E
Khoai, Hon (isl.) E
Lang Bian, Nui (mts.) E
Lay, Mui (cape) E
Mekong, Mouths of the (delta) E
Nam Tram, Mui (cape) F
Nightingale (Bach Long Vi) (isl.) F
Panjang, Hon (Hon Tho Chau) (isl.) D
Phu Quoc, Dao (isl.) D
Rao Co (mt.) E
Red (riv.) E
Se San (riv.) E
Sip Song Chau Thai (mts.) D
Song Ba (riv.) F
Song Ca (riv.) E
Song Cai (riv.) E
South China (sea) F
Tonkin (gulf) E
Varella, Mui (cape) F
Wai, Poulo (isls.) E
Yang Sin, Chu (mt.) F

★See Southeast Asia, p. 85 for other part of Malaysia.

*City and suburbs.
⊙Population of district.

CHINA (MAINLAND)

AREA 3,691,000 sq. mi. (9,559,690 sq. km.)
POPULATION 958,090,000
CAPITAL Peking (Beijing)
LARGEST CITY Shanghai
HIGHEST POINT Mt. Everest 29,028 ft. (8,848 m.)
MONETARY UNIT yuan
MAJOR LANGUAGES Chinese, Chuang, Uigur, Yi, Tibetan, Miao, Mongol, Kazakh
MAJOR RELIGIONS Confucianism, Buddhism, Taoism, Islam

CHINA (TAIWAN)

AREA 13,971 sq. mi. (36,185 sq. km.)
POPULATION 16,609,961
CAPITAL Taipei
LARGEST CITY Taipei
HIGHEST POINT Yü Shan 13,113 ft. (3,997 m.)
MONETARY UNIT new Taiwan yüan (dollar)
MAJOR LANGUAGES Chinese, Formosan
MAJOR RELIGIONS Confucianism, Buddhism, Taoism, Christianity, tribal religions

MONGOLIA

AREA 606,163 sq. mi. (1,569,962 sq. km.)
POPULATION 1,594,800
CAPITAL Ulaanbaatar
LARGEST CITY Ulaanbaatar
HIGHEST POINT Tabun Bogdo 14,288 ft. (4,355 m.)
MONETARY UNIT tughrik
MAJOR LANGUAGES Khalkha Mongolian, Kazakh (Turkic)
MAJOR RELIGION Buddhism

HONG KONG

AREA 403 sq. mi. (1,044 sq. km.)
POPULATION 5,022,000
CAPITAL Victoria
MONETARY UNIT Hong Kong dollar
MAJOR LANGUAGES Chinese, English
MAJOR RELIGIONS Confucianism, Buddhism, Christianity

MACAU

AREA 6 sq. mi. (16 sq. km.)
POPULATION 271,000
CAPITAL Macau
MONETARY UNIT pataca
MAJOR LANGUAGES Chinese, Portuguese
MAJOR RELIGIONS Confucianism, Buddhism, Taoism, Christianity

CHINA (MAINLAND)   CHINA (TAIWAN)   MONGOLIA

CHINA

PROVINCES

Anhui (Anhwei) 47,130,000	J5
Chekiang (Zhejiang) 37,510,000	K6
Fujian (Fukien) 24,500,000	J6
Gansu (Kansu) 18,730,000	E3
Guangdong (Kwangtung) 55,930,000	H7
Guangxi Zhuangzu (Kwangsi Chuang Aut. Reg. 34,020,000	G7
Guizhou (Kweichow) 26,860,000	G6
Heilongjiang (Heilungkiang) 33,760,000	K2
Hebei (Hopei) 50,570,000	J4
Henan (Honan) 70,660,000	H5
Hubei (Hupei) 45,750,000	H5
Hunan 51,660,000	H6
Inner Mongolian Aut. Reg. (Nei Monggol) 8,900,000	H3
Jiangxi (Kiangsi) 31,830,000	J6
Jiangsu (Kiangsu) 58,340,000	K5
Jilin (Kirin) 24,740,000	L3
Kansu (Gansu) 18,730,000	E3
Kiangsi (Jiangxi) 31,830,000	J6
Kiangsu (Jiangsu) 58,340,000	K5
Kirin (Jilin) 24,740,000	L3
Kwangsi Chuang Aut. Reg. (Guangxi Zhuang 34,020,000	G7
Kwangtung (Guangdong) 55,930,000	H7
Kweichow (Guizhou) 26,860,000	G6
Liaoning 37,430,000	K3
Nei Monggol (Inner Mongolian Aut. Reg.) 8,900,000	H3
Ningxia Huizu (Ningsia Hui Aut. Reg.) 3,660,000	F3
Qinghai (Tsinghai) 3,650,000	E4
Shaanxi (Shensi) 27,790,000	G5
Shanxi (Shansi) 24,340,000	H4
Shandong (Shantung) 71,600,000	J4
Sichuan (Szechwan) 97,070,000	F5
Sinkiang-Uigur Aut. Reg. (Xinjiang Uygur 12,330,000	B3
Taiwan 16,609,961	K7
Tibet Aut. Reg. (Xizang) 1,790,000	B5
Tsinghai (Qinghai) 3,650,000	E4
Xinjiang Uygur (Sinkiang-Uigur Aut. Reg. 12,330,000	B3
Xizang (Tibet Aut. Reg.) 1,790,000	B5
Yunnan 30,920,000	F7
Zhejiang (Chekiang) 37,510,000	K6

CITIES AND TOWNS†

Aba	F5
Abagnar (Silinhot)	J3
Aihui (Aigun) (Heihe)	L1
Aksu (Aqsu)	B3
Altay	C2
Alxa Youqi	F4
Alxa Zuoqi	F4
Amoy (Xiamen) 400,000	J7
Ankang	G5
Anqing (Anking) 160,000	J5
Anshan 1,500,000	K3
Anshun	G6
Antu	L3
Anxi	E3
Anyang 225,000	H4
Aqsu (Aksu)	B3
Aratürük (Yiwu)	D3
Ar Horqin	K3
Arixang (Wenquan)	B3
Artux (Atushi)	A4
Bachu (Maralwexi)	A4
Baicheng, Jilin	K2
Baicheng (Bay), Xinjiang Uygur	B3
Bairin Zuoqi	J3
Baoding (Paoting) 350,000	J4
Baoji (Paoki) 275,000	G5
Baoshan	E7
Baoting	G8
Baotou (Paotow) 800,000	G3
Bargrax (Bohu)	C3
Batang	E5
Bay (Baicheng)	B3
Bayan Obo	G3
Ba Xian	J4
Bei'an (Pehan) 130,000	L2
Beihai (Pakhoi) 175,000	G7
Beijing (Peking) (cap.) •8,000,000	J3
Bengbu (Pengpu) 400,000	J5
Benxi (Penki) 750,000	K3
Bohu (Bagrax)	C3
Bole	B3
Bortala (Bole)	B3
Boshan	J4
Bo Xian (Pohsien)	J5
Butha	K2
Cangzhou (Tsangchow)	J4
Canton (Guangzhou) 2,300,000	H7
Chamdo (Qamdo)	E5
Changchih (Changzhi)	H4
Changchow (Changzhou) 400,000	J5
Changchow (Zhangzhou)	J7
Changchun 1,500,000	K3
Changde (Changteh) 225,000	H6
Changhai 137,236	C3
Changji	C3
Changjiang	G8
Changsha 850,000	H6
Changteh (Changde) 225,000	H6
Changyeh (Zhangye)	F4
Changzhi (Changchih)	H4
Changzhou (Changchow) 400,000	J5
Chankiang (Zhanjiang) 220,000	H7
Chao'an (Chaochow)	J7
Chaotung (Zhaotong)	F6
Chaoyang, Liaoning	J3
Chaoyang, Guangdong	J7
Charkhlia (Ruoqiang)	C4
Chefoo (Yantai) 180,000	K4
Chengdu (Zhengzhou) 1,500,000	H5
Chengde (Chengteh) 200,000	J3
Chengdu (Chengtu) 2,000,000	F5
Chen Xian	H6
Cherchen (Qiemo)	C4
Chia 238,713	K7
Chifeng	J3
Chinchow (Jinzhou) 750,000	K3
Chindu	E5
Chinkiang (Zhenjiang) 250,000	J5
Chinsi (Jinxi)	K3
Chinwangtao (Qinhuangdao) 400,000	K4
Chishui	G6
Chongqing (Chungking) 3,500,000	G6
Chüanchow (Quanzhou) 130,000	J7
Chuchow (Zhuzhou) 350,000	H6
Chuguchak (Tacheng)	B2
Chumatien (Zhumadian)	H5
Chunghsing	K7
Chungking (Chongqing) 3,500,000	G6
Chungshan (Zhongshan) 135,000	H7
Da'an (Talai)	K2
Dali	E6
Dalian 1,480,240	K4
Dandong (Tantung) 450,000	K3
Dan Xian	G8
Da Qaidam	E4
Daqing 758,430	L2
Datong, Qinghai	F4
Datong (Tatung), Shanxi 300,000	H3
Da Xian	G5
Dazhai	H4
Dengkou	G3
Deyang	F5
Dezhou (Tehchow)	J4
Dingxing	H4
Dongchuan	F6
Dongfang	G8
Dongsheng	G3
Dongtai	K5
Dorbiljin (Emin)	B2
Dukou	F6
Dulan	E4
Dunhua (Tunhwa)	L3
Dunhuang	D3
Duolun	J3
Dushan	G6
Duyun (Tuyün)	G6
Ejin	F3
Emin (Dorbiljin)	B2
Erenhot	H3
Ergun Youqi	K1
Ergun Zuoqi	K1
Ertai	C2
Fatshan (Foshan)	H7

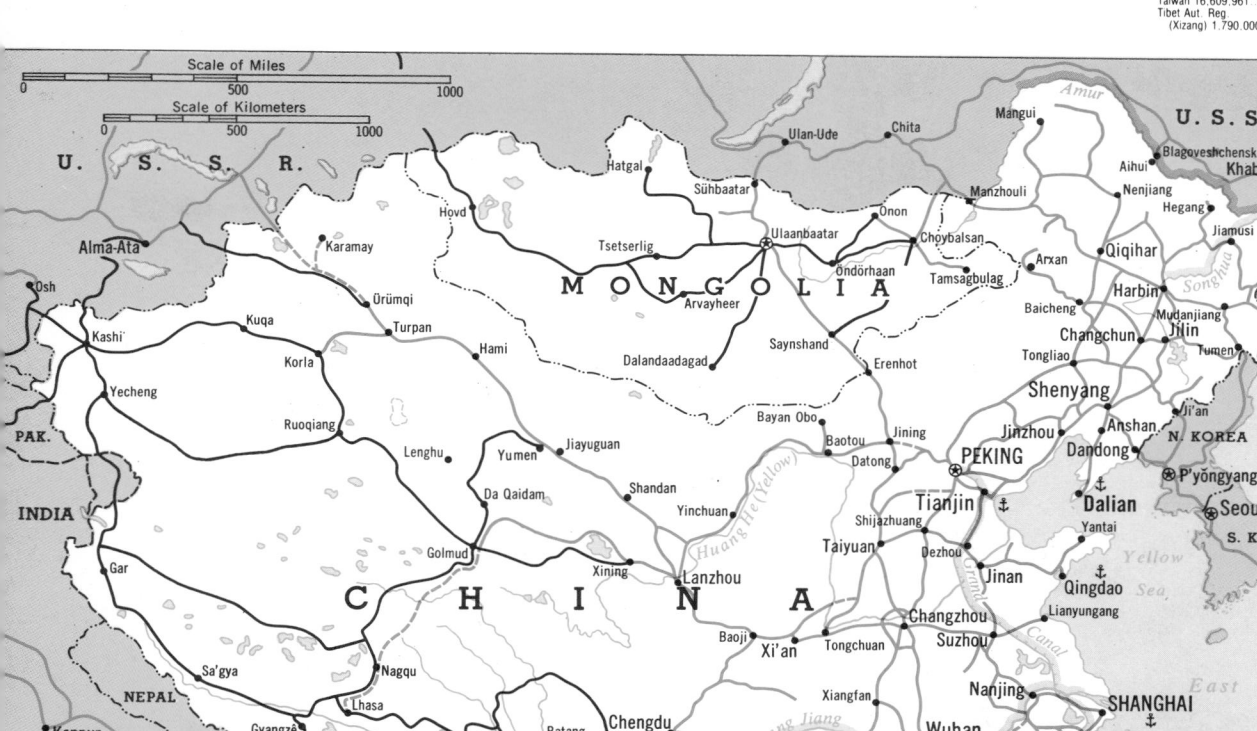

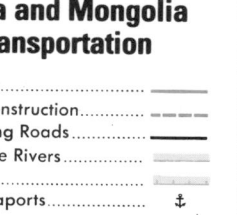
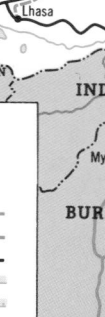

Scale of Miles
0   500   1000

Scale of Kilometers
0   500   1000

China and Mongolia Transportation

Railroads ....
Under Construction ....
Connecting Roads ....
Navigable Rivers ....
Canals ....
Major Seaports .... ⚓

© Copyright HAMMOND INCORPORATED, Maplewood, N.J.

(continued on following page)

Foochow (Fuzhou) 900,000.....J6
Foshan (Fatshan).....H7
Fowyang (Fuyang).....J5
Fushun 1,700,000.....K3
Fusingchen (Simao).....K4
Fu Xian, Liaoning.....K4
Fu Xian, Shaanxi.....K3
Fuxin 350,000.....K3
Fuyang (Fowyang).....J5
Fuyu, Heilongjiang.....K2
Fuyu, Jilin.....M2
Fuyuan, Heilongjiang.....F6
Fuyun.....C2
Fuzhou (Foochow),
 Fujian 900,000.....J6
Fuzhou, Jiangxi.....J6
Ganzhou (Kanchow) 135,000.....H6
Garyarsa (Gartok).....B5
Gejiu (Kokiu) 250,000.....F7
Golmud (Golmo).....E4
Gonghe.....G5
Guangyuan.....G5
Guan Xian.....F5
Guangzhou (Canton) 2,300,000.....H7
Guilin (Kweilin) 225,000.....H6
Guiyang (Kweiyang),
 Guizhou 1,500,000.....G6
Guiyang, Hunan.....H6
Gulja (Yining) 160,000.....B3
Guma (Pishan).....A4
Guyang.....G3
Guyuan.....G4
Gyaca.....D6
Gyangzê.....C6
Habahe.....C2
Haikou (Hoihow) 500,000.....H7
Hailar.....J2
Hami (Kumul).....D3
Hancheng.....H4
Hanchung (Hanzhong) 120,000.....G5
Handan (Hantan) 500,000.....H4
Hangzhou (Hangchow) 1,100,000.....J5
Hantan (Handan) 500,000.....H4
Hanzhong (Hanchung) 120,000.....G5
Harbin 2,750,000.....L2
Hebi.....H4
Hechuan (Hochwan).....G5
Hefei (Hofei) 400,000.....J5
Hegang (Hokang) 350,000.....L2
Heihe (Aihui) (Aigun).....L1
Hekou.....F7
Hengchun.....K7
Hengshan.....G4
Hengyang 310,000.....H6
Hepu (Hoppo).....G7
Hexigten.....J3
Hezuo.....F5
Hochwan (Hechuan).....G5
Hofei (Hefei) 400,000.....J5
Hoihow (Haikou) 500,000.....H7
Hokang (Hegang) 350,000.....L2
Hoppo (Hepu).....G7
Horqin Youyi Qianqi
 (Ulanhot) 100,000.....K2
Hotan.....B4
Houma.....H4
Hsüchang (Xuchang).....H5

Huadian.....L3
Huaibei.....J5
Huaide (Hwaiteh).....K3
Huainan 350,000.....J5
Hualien.....G4
Huangling.....G4
Huangshi 200,000.....J5
Huangzhong.....F4
Huhehot (Hohhot) 700,000.....H3
Hulin.....M2
Hunchun.....M3
Hunjiang.....L3
Hwainan (Huainan) 350,000.....J5
Hwaiteh (Huaide).....K3
Hwangshih (Huangshi) 200,000.....J5
Ichang (Yichang) 150,000.....H5
Ichun (Yichun) 200,000.....L2
Ilan.....K7
Ipin (Yibin) 275,000.....F6
Jeminay.....C2
Jiamusi (Kiamusze) 275,000.....M2
Ji'an (Kian) 100,000.....J6
Jiangmen (Kongmoon) 150,000.....H7
Jian'ou.....J6
Jiaozuo (Tsiaotso) 300,000.....H4
Jiaxing (Kashing).....K5
Jiayuguan.....E4
Jieyang.....J7
Jilin (Kirin) 1,200,000.....L3
Jinan (Tsinan) 1,500,000.....J4
Jingdezhen
 (Kingtehchen) 300,000.....J6
Jinghong.....F7
Jingle.....G7
Jing Xian, Anhui.....J5
Jing Xian, Hunan.....H6
Jingyuan.....F4
Jinhua (Kinhwa).....J6
Jining (Tsining), Nei
 Monggol 160,000.....H3
Jining (Tsining), Shandong.....J4
Jinshi (Tsingshih) 100,000.....H6
Jinxi (Chinsi).....K3
Jinzhou (Chinchow) 750,000.....K3
Jiujiang (Kiukiang) 120,000.....J6
Jiuquan (Kiuchüan).....E4
Jixi (Kisi) 350,000.....M2
Juichin (Ruijin).....J6
Jun Xian.....H5
Kaba (Habahe).....C2
Kaifeng 330,000.....H4
Kailu.....K3
Kaiyuan, Liaoning.....K3
Kaiyuan, Yunnan.....F7
Kalgan (Zhangjiakou) 1,000,000.....H3
Kanchow (Ganzhou) 135,000.....H6
Kanding.....F5
Kaohsiung 1,028,334.....J7
Karakax (Kara Kash) (Moyu).....A4
Karamay.....B2
Karghalik (Yecheng).....A4
Kashgar (Kashi) 175,000.....A4
Kashi (Kashgar) 175,000.....A4
Kashing (Jiaxing).....K5
Kaxgar (Kashi) 175,000.....A4
Keelung 342,604.....K6
Kenli.....J4
Keriya (Yutian).....B4
Khotan (Hotan).....B4

Kiamusze (Jiamusi)
 275,000.....M2
Kian (Ji'an) 100,000.....J6
Kienyang (Qianyang).....H6
Kingtehchen (Jingdezhen)
 300,000.....J6
Kinhwa (Jinhua).....J6
Kirin (Jilin) 1,200,000.....L3
Kisi (Jixi) 350,000.....M2
Kiuchüan (Jiuquan).....E4
Kiukiang (Jiujiang) 120,000.....J6
Kokiu (Gejiu) 250,000.....F7
Kongmoon (Jiangmen)
 150,000.....H7
Korla.....C3
Kuldja (Yining) 160,000.....B3
Kumul (Hami).....D3
Künes (Xinyuan).....B3
Kunming 1,700,000.....F6
Kuytun.....B3
Kwangchow (Canton).....H7
Kweilin (Guilin) 225,000.....H6
Kweisui (Hohhot) 700,000.....H3
Kweiyang (Guiyang)
 1,500,000.....G6
Lanzhou (Lanchow) 1,500,000.....F4
Lenghu.....D4
Lengshuijiang.....H6
Leshan (Loshan) 250,000.....F6
Lhasa 175,000.....D6
Lhazê (Lhatse).....C6
Lianyungang (Lienyünkang)
 300,000.....J5
Liaoyang 250,000.....K3
Liaoyuan 300,000.....K3
Lijiang.....F6
Linfen.....H4
Lingling.....H6
Linhe.....G3
Linqing (Lintsing).....J4
Linxi.....J3
Linxia (Linsia).....F4
Lizhou (Liuchow) 250,000.....G7
Loho (Luohe).....H5
Longjiang.....K2
Lopnur (Yuli).....D3
Loshan (Leshan) 250,000.....F6
Loyang (Luoyang) 750,000.....H5
Lu'an.....J5
Luchow (Luzhou) 225,000.....G6
Lüda (Dalian) 1,480,240.....K4
Luohe (Loho).....H5
Luoyang (Loyang) 750,000.....H5
Lüshun.....K4
Luxi.....F7
Luzhou (Luchow) 225,000.....G6
Ma'anshan.....J5
Manas.....C3
Manchouli (Manzhouli).....J2
Maoming (Mowming).....H7
Maralwexi (Bachu).....A4
Mengcheng.....J5
Mengzi.....F7
Mianyang, Hubei.....H5
Mianyang, Sichuan.....G5
Minfeng (Niya).....B4
Mingshui, Gansu.....E3
Mingshui, Heilongjiang.....L2
Minle.....F4
Mowming (Maoming).....H7
Moyu (Karakax).....A4

Mudanjiang (Mutankiang) 400,000.....M3
Mukden (Shenyang) 3,750,000.....K3
Muli.....D5
Nagqu.....D5
Nanchang 900,000.....J6
Nanchong (Nanchung) 275,000.....G5
Nanjing (Nanking) 2,000,000.....J5
Nanning 375,000.....G7
Nanping.....J6
Nantong 300,000.....K5
Nanyang.....H5
Napo.....G7
Neijiang (Neikiang) 240,000.....G6
Nenjiang.....L2
Ningbo (Ningpo) 350,000.....K6
Ningpo (Ningbo) 350,000.....K6
Ningxia (Yinchuan,
 Yinchwan) 175,000.....G4
Niya (Minfeng).....B4
Ongniud.....J3
Oroqen.....K1
Paicheng (Baicheng).....K2
Pakhoi (Beihai) 175,000.....G7
Paoki (Baoji) 275,000.....G5
Paoting (Baoding) 350,000.....J4
Paotow (Baotou) 800,000.....G3
Pehan (Bei'an) 130,000.....L2
Peking (Beijing)
 (cap.) ● 8,500,000.....J3
Pengpu (Bengbu) 400,000.....J5
Penki (Benxi) 750,000.....K3
Pingdingshan.....H5
Pingliang.....G4
Pingtung 165,360.....K7
Pingxiang, Guangxi Zhuangzu.....G7
Pingxiang, Jiangxi.....H6
Piqan (Shanshan).....D3
Pishan (Guma).....A4
Pohsien (Bo Xian).....J5
Qamdo.....E5
Qarkilik (Ruoqiang).....C4
Qarqan (Qiemo).....C4
Qiemo (Qarqan).....C4
Qingdao (Tsingtao) 1,900,000.....K4
Qingjiang, Jiangxi.....J6
Qingjiang 110,000.....J5
Qinhuangdao
 (Chinwangtao) 400,000.....K4
Qionghai.....H8
Qiqihar (Tsitsihar) 1,500,000.....K2
Qitai.....C3·
Qog.....G3
Qoqek (Tacheng).....B2
Quanzhou (Chüanchow) 130,000.....J7
Qu Xian, Sichuan.....G5
Qu Xian, Zhejiang.....J6
Quxü.....D6
Ruijin (Juichin).....J6
Ruoqiang (Qarkilik).....C4
Rutog.....A5
Sanmenxia.....H5
Sanming.....J6
Shache (Yarkand).....A4
Shandan.....F4
Shangdu.....H3
Shanghai 10,980,000.....K5
Shangqiu (Shangkiu) 250,000.....J5

Shangrao (Shangjao) 100,000.....J6
Shangshui 100,000.....J5
Shanshan (Piqan).....D3
Shantou (Swatow) 400,000.....J7
Shaoguan (Shiukwan) 125,000.....H7
Shaoxing (Shaohing) 225,000.....K5
Shaoyang 275,000.....H6
Shashi 125,000.....H5
Shenyang (Mukden) 3,750,000.....K3
Shigatse (Xigazê).....C6
Shijiazhuang
 (Shihkiachwang) 1,500,000.....J4
Shiquanhe.....A5
Shiukwan (Shaoguan) 125,000.....H7
Shiyan.....H5
Shizuishan (Shihsuishan).....G4
Shuangcheng.....L2
Shuangyashan 150,000.....M2
Shuo Xian.....H4
Siakwan (Xiaguan).....E6
Sian (Xi'an) 1,900,000.....G5
Siangtan (Xiangtan) 150,000.....H6
Siangtan (Xiangtan) 300,000.....H6
Sienyang (Xianyang) 125,000.....G5
Silinhot (Abnagar).....J3
Simao (Fusingchen).....K4
Sinchu 208,038.....K7
Sining (Xining) 250,000.....F4

Sinsiang (Xinxiang) 300,000.....H4
Sinyang (Xinyang) 125,000.....H5
Siping (Szeping) 180,000.....K3
Soche (Shache).....A4
Soochow (Suzhou) 1,300,000.....K5
Suao.....K7
Suchow (Xuzhou) 1,500,000.....J5
Suifenhe.....M3
Suihua.....L2
Suining.....G5
Suzhou (Soochow) 1,300,000.....K5
Swatow (Shantou) 400,000.....J7
Szeping (Siping) 180,000.....K3
Tacheng (Qoqek).....B2
Tai'an.....J4
Taibus.....J3
Taichow (Taizhou) 275,000.....K5
Taichung 565,255.....K7
Taigu.....H4
Tainan 581,390.....K7
Taipa 2,108,193.....K7
Taitung.....K7
Taiyuan 2,725,000.....H4
Taizhou (Taichow) 275,000.....K5
Talai (Da'an, Dalai).....K2
Tali (Dali).....E6
Tangshan 1,200,000.....K3
Tantung (Dandong) 450,000.....K3
Tao'an.....K2
Taoyuan 105,841.....K6

Tart.....D4
Tatung (Datong) 300,000.....H3
Taxkorgan.....A4
Tehchow (Dezhou).....J4
Tengchong.....E6
Tianjin (Tientsin) ● 7,210,000.....J3
Tianjun.....E4
Tianshui 100,000.....F5
Tieling.....K3
Tienshui (Tianshui) 100,000.....F5
Tientsin (Tianjin) ● 7,210,000.....J3
Tingri.....C6
Toksu (Xinhe).....C3
Toksun.....D3
Tongchuan (Tungchwan).....G4
Tonghua (Tunghwa) 275,000.....L3
Tongjiang (Tungkiang).....M2
Tongliao.....K3
Tongren.....G6
Tongxin.....G4
Tongyu.....K2
Tsaochow (Cangzhou).....H4
Tsiaotso (Jiaozuo) 300,000.....H4
Tsinan (Jinan) 1,500,000.....J4
Tsingkiang (Qingjiang) 110,000.....J5
Tsingshih (Jinshi) 100,000.....H6
Tsingtao (Qingdao) 1,900,000.....K4
Tsining (Jining), Nei
 Monggol 160,000.....H3

Topography

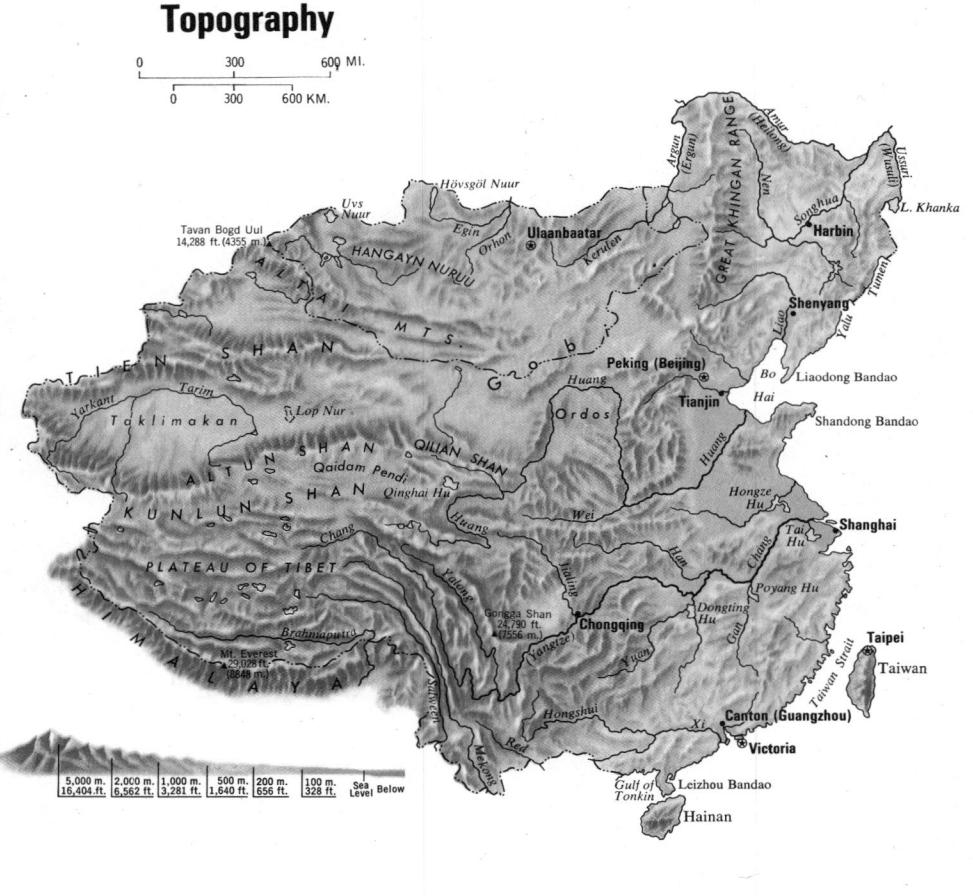

5,000 m. / 16,404 ft. — 2,000 m. / 6,562 ft. — 1,000 m. / 3,281 ft. — 500 m. / 1,640 ft. — 200 m. / 656 ft. — 100 m. / 328 ft. — Sea Level — Below

On this map Chinese place-names have been rendered according to the Pinyin spelling system within the area controlled by the People's Republic of China. Alphabetically listed below are selected Chinese place-names spelled in the traditional manner, followed by the equivalent Pinyin form.

Traditional	Pinyin	Traditional	Pinyin	Traditional	Pinyin
Amoy (Hsiamen)	Xiamen	Kirin	Jilin	Sian	Xi'an
Anhwei	Anhui	Kiukiang	Jiujiang	Siangtan	Xiangtan
Canton (Kwangchow)	Guangzhou	Kwangsi Chuang	Guangxi Zhuangzu	Sining	Xining
Chefoo (Yentai)	Yantai	Kwangtung	Guangdong	Sinkiang-Uighur	Xinjiang Uygur
Chekiang	Zhejiang	Kweichow	Guizhou	Soochow	Suzhou
Chengchow	Zhengzhou	Kweilin	Guilin	Süchow	Xuzhou
Chengtu	Chengdu	Kweiyang	Guiyang	Swatow	Shantou
Chinchow	Jinzhou	Lanchow	Lanzhou	Szechuan	Sichuan
Chungking	Chongqing	Liuchow	Liuzhou	Tachai	Dazhai
Foochow	Fuzhou	Loyang	Luoyang	Tatung	Datong
Fukien	Fujian	Lüta	Dalian	Tibet	Xizang
Hangchow	Hangzhou	Mutankiang	Mudanjiang	Tientsin	Tianjin
Heilungkiang	Heilongjiang	Nanking	Nanjing	Tsinan	Jinan
Hofei	Hefei	Ningpo	Ningbo	Tsinghai	Qinghai
Honan	Henan	Ningsia Hui	Ningxia Huizu	Tsingtao	Qingdao
Hopel	Hebei	Paoting	Baoding	Tsitsihar	Qiqihar
Huhehot	Hohhot	Paotow	Baotou	Tsunyi	Zunyi
Hupeh	Hubei	Peking	Beijing	Tungchwan	Tongchuan
Hwainan	Huainan	Pengpu	Bengbu	Tzepo	Zibo
Inner Mongolia	Nei Monggol	Penki	Benxi	Urumchi	Ürümqi
Kansu	Gansu	Shansi	Shanxi	Wusih	Wuxi
Kiangsi	Jiangxi	Shantung	Shandong	Yenan	Yan'an
Kiangsu	Jiangsu	Shensi	Shaanxi	Yinchwan	Yinchuan
Kingtehchen	Jingdezhen	Shihkiachwang	Shijiazhuang		

Tsining (Jining), Shandong	J4	
Tsitsihar (Qiqihar) 1,500,000	K2	
Tsunyi (Zunyi) 275,000	G6	
Tumen	M3	
Tungchwan (Tongchuan)	G5	
Tunghwa (Tonghua) 275,000	L3	
Tungkiang (Tongjiang)	M2	
Tungliao (Tongliao)	K3	
Tunhwa (Dunhua)	L3	
Tunxi (Tunki)	J6	
Turpan (Turfan)	C3	
Tuyün (Duyun)	G6	
Tzekung (Zigong) 350,000	F6	
Tzepo (Zibo) 1,750,000	J4	
Uch Turfan (Wushi)	A3	
Ulanhot (Horqin Youyi Qianqi) 100,000	K2	
Uluqhchat (Wuqia)	A4	
Ürümqi (Urumchi) 500,000	C3	
Usu	B3	
Wanning	G5	
Wanxian (Wanshien) 175,000	G5	
Weichang	J3	
Weifang 260,000	J4	
Weihai (Weihaiwei)	K4	
Weixi	E6	
Weixin	F6	
Wenchow (Wenzhou) 250,000	J6	
Wenquan, Qinghai	D5	
Wenquan, Xinjiang Uygur	B3	

Wenzhou 250,000	J6	
Wuchang	L3	
Wuchow (Wuzhou) 150,000	G6	
Wuchuan, Guizhou	G6	
Wuchuan, Nei Monggol	H3	
Wuchung (Wuzhong)	G4	
Wuda	G4	
Wuhan 4,250,000	H5	
Wuhing (Wuxing) 160,000	K5	
Wuhu 300,000	J5	
Wuqi	G4	
Wuqia	A4	
Wushi	A3	
Wusih (Wuxi) 900,000	K5	
Wutai	H4	
Wuwei	F4	
Wuxi (Wushi) 900,000	K5	
Wuxing (Wuhing) 160,000	K5	
Wuyuan	G3	
Wuzhong (Wuchung)	G4	
Wuzhou (Wuchow) 150,000	G6	
Xiaguan (Siakwan)	E6	
Xiamen (Amoy) 400,000	J7	
Xi'an (Sian) 1,900,000	G5	
Xiangfan (Siangfan) 150,000	H5	
Xiangtan (Siangtan) 300,000	H6	
Xianyang (Sienyang) 125,000	G5	
Xiapu (Siapu)	K6	
Xichang (Sichang)	F6	

Xigazê (Shigatse)	C6	
Ximiao	F3	
Xin Barag Zuoqi	J2	
Xinghai (Singai)	E4	
Xingtai (Singtai) 300,000	H4	
Xinhe (Toksu)	B3	
Xining (Sining) 250,000	F4	
Xinxiang (Sinsiang) 300,000	H4	
Xinyang (Sinyang) 125,000	H5	
Xinyuan (Künes)	B3	
Xuchang (Hsüchang)	H5	
Xuguit	K2	
Xuzhou (Süchow) 1,500,000	J5	
Ya'an 100,000	F6	
Yadong	C6	
Yan'an (Yenan)	G4	
Yangchow (Yangzhou) 210,000	J5	
Yangjiang	H7	
Yangquan (Yangchüan) 350,000	H4	
Yangzhou (Yangchow) 210,000	J5	
Yanji (Yenki) 130,000	L3	
Yanqi (Yenki)	C3	
Yantai (Chefoo) 180,000	J4	
Yarkant (Shache)	A4	
Ya Xian	G8	
Yecheng	A4	
Yeqing (Yan'an)	G4	
Yibin 275,000	F6	

Yichang (Ichang) 150,000	H5	
Yichun, Jiangxi	H6	
Yichun, Heilongjiang 200,000	L2	
Yidu, Hubei	H5	
Yidu, Shandong	J4	
Yinchuan (Yinchwan) 125,000	G4	
Yingkou 215,000	K3	
Yining 160,000	B3	
Yiwu (Aratürük)	D3	
Yiyang	H6	
Yiyuan	G6	
Yongji	B6	
Yuci (Yütze)	H4	
Yueyang	H6	
Yuli (Lopnur)	C3	
Yulin, Guangxi Zhuangzu	G7	
Yulin, Shanxi	G4	
Yumen 325,000	E4	
Yuncheng	H4	
Yungkia (Wenzhou) 250,000	J6	
Yushu, Jilin	L3	
Yushu, Qinghai	E5	
Yutian, Hebei	J4	
Yutian, Xinjiang Uygur	B4	
Yütze (Yuci)	H4	
Zadoi	E5	
Zanhuang	H4	
Zaozhuang	J5	
Zayü	D6	
Zêtang	D6	
Zhanghei	J3	
Zhangjiakou (Kalgan) 1,000,000	J3	
Zhangye (Changyeh)	F4	

Zhangzhou (Changchow)	J7	
Zhanjiang (Chankiang) 220,000	H7	
Zhaodong	K2	
Zhaojue	F6	
Zhaoqing	H7	
Zhaosu	B3	
Zhaotong (Chaotung)	F6	
Zhengzhou (Chengchow) 1,500,000	H5	
Zhenjiang (Chinkiang) 250,000	J5	
Zhenyuan	G6	
Zhenba	B6	
Zhongba	B6	
Zhongwei	G4	
Zhumadian (Chumatien)	H5	
Zhushan	H5	
Zhuzhou (Chuchow) 350,000	H6	
Zibo (Tzepo) 1,750,000	J4	
Zigong (Tzekung) 350,000	F6	
Zunhua	J3	
Zunyi (Tsunyi) 275,000	G6	

OTHER FEATURES

Altun Shan (range)	C4	
Aïxa Shamo (des.)	F4	
A nyêmaqên Shan (mts.)	E5	
Aqqikkol Hu (lake)	C4	
Argun (Ergun He) (riv.)	K1	

Bagrax (Bosten Hu) (lake)	C3	
Bangong Co (lake)	A5	
Bashi (chan.)	K7	
Bayan Har Shan (range)	E5	
Bo Hai (gulf)	J4	
Bosten (Bagrax) Hu (lake)	C3	
Chang Jiang (Yangtze) (riv.)	H5	
Da Hinggan Ling (range)	K2	
Dian Chi (lake)	F7	
Dongsha (isl.)	J7	
Dongting Hu (riv.)	H6	
East China (sea)	L6	
Ednur Hu (lake)	B2	
Ergun He (Argun) (riv.)	K1	
Er Hai (lake)	E6	
Everest (mt.)	C6	
Fen He (riv.)	H4	
Formosa (Taiwan) (str.)	J7	
Formosa (Taiwan) (isl.)	J7	
Gandisê Shan (range)	K7	
Gaoyou Hu (lake)	J5	
Genghis Khan Wall (ruin)	H2	
Gobi (des.)	G3	
Gongga Shan (mt.)	F6	
Grand (canal)	J4	
Great Wall (ruin)	G4,J	
Guria Mandhada (mt.)	B5	
Hailar He (riv.)	K2	
Hainan (isl.)	H8	

Hangzhou Wan (bay)	K5	
Han Shui (riv.)	H5	
Heilong Jiang (Amur) (riv.)	L2	
Himalaya (mts.)	C6	
Hongshui He (riv.)	G7	
Hongze Hu (lake)	J5	
Hotan He (riv.)	B4	
Huang He (Yellow) (riv.)	H4	
Hulun Nur (lake)	J2	
Hungtow (isl.)	K7	
Inner Mongolia (reg.)	H3	
Jinmen (Quemoy) (isl.)	J7	
Jinsha Jiang (Yangtze) (riv.)	E5	
Junggar Pendi (desert basin)	C3	
Kangrinboqê Feng (mt.)	K7	
Karakhoto (ruins)	F3	
Karamiran Shankou (pass)	B4	
Keriya Shankou (pass)	B4	
Khanka (lake)	M3	
Kongur Shan (mt.)	A4	
Kunlun Shan (range)	B3	
Kuruktag Shan (range)	C3	
Lancang Jiang (riv.)	F7	
Leizhou Bandao (pen.)	G7	
Liaodong Bandao (pen.)	K3	
Liao He (riv.)	K3	
Lop Nor (Lop Nur) (lake)	D3	
Manas He (riv.)	C3	
Manas Hu (lake)	C2	

(continued on following page)

China and Mongolia

SCALE OF MILES
0 100 200 300 400 500

SCALE OF KILOMETERS
0 100 200 300 400 500

Capitals of Countries ...... ⊛   International Boundaries ——·——
Provincial Capitals ...... ⊛   Provincial Boundaries ——·——
Canals ...... Walls ~~~~~~

Scale 1:19,100,000

© Copyright HAMMOND INCORPORATED, Maplewood, N.J.

Mazu (Matsu) (isl.)	K6	Tumen (riv.)	L3	Hovd 68,300	D2	Mönhhaan 400	H2
Mekong (Lancang Jiang) (riv.)	F7	Ulu Muztag (mt.)	C4	Hövsgöl 89,600	E1	Mörön (Muren) 10,700	F2
Min Jiang (riv.)	J6	Ulungur He (riv.)	C2	Ömnögovĭ 32,200	F3	Nalayh (Nalaikha) 14,000	G2
Mudan Jiang (riv.)	L3	Ulungur Hu (lake)	C2	Övörhangay 84,100	F2	Nomgon 500	G3
Muztag (mt.)	B4	Ussuri (Wusuli Jiang) (riv.)	M2	Selenge 53,700	H2	Noyon 300	F3
Muztagata (mt.)	A4	Wei He (riv.)	G5	Sühbaatar 44,100	H2	Ölgiy (Ulegei) 11,700	C2
Nam Co (lake)	D5	Wu Jiang (riv.)	G6	Töv 74,900	G2	Öndörhaan (Undur Khan) 7,900	G2
Namzha Parwa (mt.)	E6	Wusuli Jiang (Ussuri) (riv.)	M2	Uvs 76,300	D2	Onon 2,600	H2
Nan Ling (mts.)	H6	Wuyi Shan (range)	J6			Sayhan-Ovoo 400	F2
Nen Jiang (riv.)	K2	Xiang Jiang (riv.)	H6	CITIES AND TOWNS		Saynshand 10,000	H3
Ngangzê Co (lake)	C5	Xi Jiang (riv.)	H7			Selenge 1,300	F2
Ngoring Hu (lake)	E5	Yagradagzê Shan (mt.)	D4	Altay 10,000	E2	Sühbaatar (Sukhe Bator) 10,000	G1
Nu Jiang (riv.)	E6	Yalong Jiang (riv.)	F6	Arvayheer 9,100	F2	Sülanheer 300	G3
Nyainqêntanglha Shan (range)	D5	Yalu (riv.)	L3	Baatsagaan 800	F2	Tamsagbulag	J2
Olwampi (cape)	K7	Yangtze (Chang Jiang) (riv.)	K5	Baruun-Urt 8,200	H2	Tsagaannuur 2,000	C2
Ordos (reg.)	G4	Yarkant He (riv.)	A4	Bayanbaraat 400	F3	Tsagaan-Ovoo 900	H2
Penghu (Pescadores) (isls.) 113,397	J7	Yellow (Huang He) (riv.)	J4	Bayandalay 500	F3	Tsagaan-Uul 1,700	E2
Pingtan (isl.)	K6	Yellow (sea)	K4	Bayangovĭ 400	F2	Tsetseg 700	D2
Pobeda (peak)	A3	Yin Shan (mts.)	G3	Bayanhongor 11,300	F2	Tsetserleg 12,400	F2
Poyang Hu (lake)	J6	Yuhuan (isl.)	K6	Bayan-Öndör 300	E3		
Pratas (Dongsha) (isl.)	J7	Yu Shan (mt.)	K7	Bayan-Uul 1,200	E2	OTHER FEATURES	
Qaidam Pendi (basin) (swamp)	D4	Yushan (isls.)	K6	Beger 800	E2		
Qarqan He (riv.)	C4	Zhoushan (arch.)	K5	Bulgan, Bulgan 9,800	F2	Altai (mts.)	C2
Qilian Shan (range)	E4			Bulgan, Hovd 3,100	D2	Dörgön Nuur (lake)	D2
Qinghai Hu (lake)	E4	HONG KONG		Bulgan, Ömnögovĭ 700	F3	Dzavhan Gol (riv.)	E2
Qiongzhou Haixia (str.)	G7	CITIES and TOWNS		Bürentsogt 3,000	H2	Ghenghis Khan Wall (ruins)	H2
Quemoy (Jinmen) (isl.)	J7			Chandmanĭ 700	D2	Gobi (des.)	G3
Qumar He (riv.)	D4	Kowloon* 2,378,480	H7	Choybalsan 20,500	J2	Hangayn Nuruu (mts.)	E2
Salween (Nu Jiang) (riv.)	E6	Victoria (cap.)* 1,026,870	H7	Dalandzadgad 6,600	G3	Har Us Nuur (lake)	D2
Siling Co (lake)	C5			Darhan (Darkhan) 32,900	G2	Herlen Gol (Kerulen) (riv.)	H2
Songhua Hu (lake)	L3	MACAU (MACAO)		Dashbalbar 1,100	H2	Hovd Gol (riv.)	D2
Songhua Jiang (Sungari) (riv.)	M2	CITIES and TOWNS		Dashinchilen 600	F2	Hövsgöl Nuur (lake)	E1
South China (sea)	J7	Macau (Macao) (cap.) 226,880	H7	Deigertsogt 500	H2	Hyargas Nuur (lake)	D2
Tachen (Taizhou) (isls.)	K6			Dzamīn Üüd 1,500	H3	Ider Gol (riv.)	E2
Tai Hu (lake)	J5	MONGOLIA		Dzüünharaa 8,100	G2	Karakorum (ruins)	F2
Taiwan (Formosa) (isl.) 16,609,961	K7	PROVINCES		Dzuunmod 7,200	G2	Kerulen (riv.)	H2
Taiwan (Formosa) (str.)	J7			Erdentsagaan 1,500	J2	Munku-Sardyk (mt.)	F1
Taizhou (Tachen) (isls.)	K6	Arhangay 90,500	F2	Ereen 700	G3	Orhon Gol (riv.)	G2
Takla Makan (Taklimakan Shamo) (des.)	B	Bayamhongor 64,900	E2	Hanbogd 300	F1	Selenge Mörön (riv.)	G2
Tanggula Shan (range)	D5	Bayan-Ölgiy 73,300	C2	Hanh 500	E1	Tannu-Ola (mts.)	D1
Tangra Yumco (lake)	C5	Bulgan 45,700	F2	Hatgal 5,000	E1	Tavan Bogd Uul (mt.)	C2
Tarim He (riv.)	B3	Dornod 51,100	H2	Hongor	G2	Uvs Nuur (lake)	D1
Tarim Pendi (basin)	B4	Dornogovĭ 36,400	G3	Hovd (Kobdo, Jirgalanta) 12,400	D2		
Tian Shan (range)	C3	Dundgovĭ 37,800	G2	Hyargas 1,600	D2		
Tibet (reg.)	B5	Dzavhan 87,600	E2	Jargalant	D2		
Tongtian He (Zhi Qu) (riv.)	E5	Govĭ-Altay 59,200	E2	Jibhalanta (Uliastay) 13,000	E2		
Tonkin (gulf)	G7	Hentiy 49,900	H2	Jirgalanta (Hovd) 12,400	D2		
				Kobdo (Hovd) 12,400	D2		
				Mandah 200	G3		
				Mandalgovĭ 7,000	G2		
				Mandal-Ovoo 300	F3		

† Populations of mainland cities, excluding Peking (Beijing), Shanghai and Tianjin (Tientsin), courtesy of Kingsley Davis,
Office of Int'l Pop. and Research, Inst. of Int'l Studies Univ. of California.

• Population of municipality
*City and suburbs

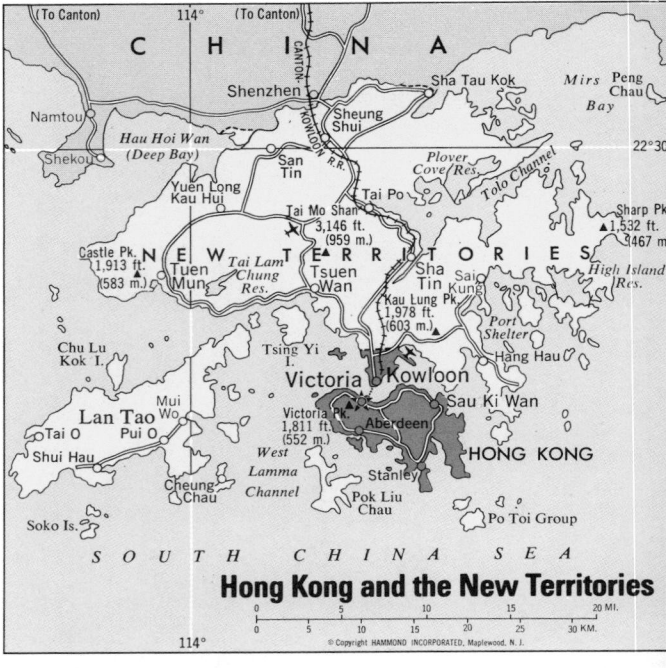

Hong Kong and the New Territories

Agriculture, Industry and Resources

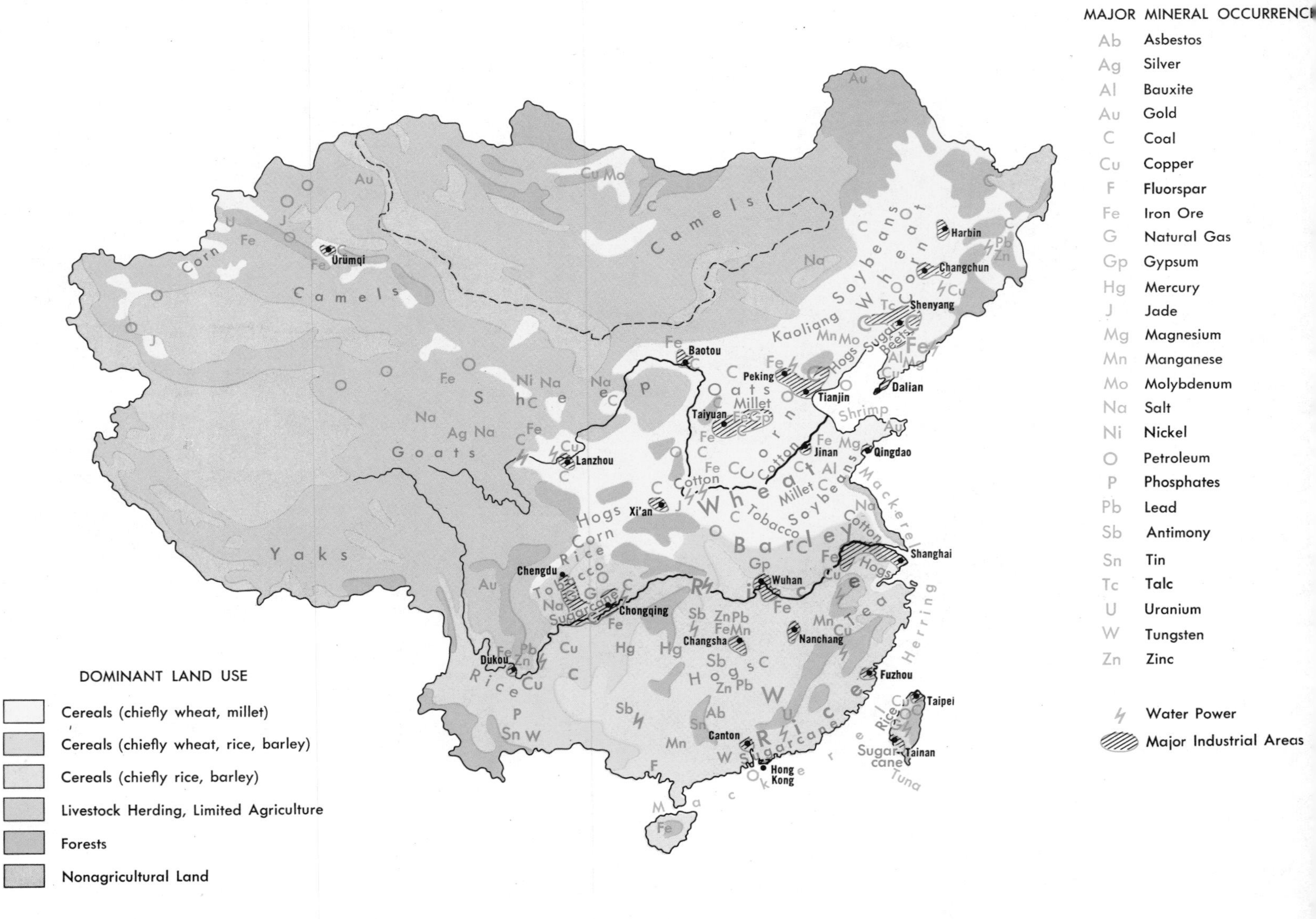

MAJOR MINERAL OCCURRENCE

Ab	Asbestos
Ag	Silver
Al	Bauxite
Au	Gold
C	Coal
Cu	Copper
F	Fluorspar
Fe	Iron Ore
G	Natural Gas
Gp	Gypsum
Hg	Mercury
J	Jade
Mg	Magnesium
Mn	Manganese
Mo	Molybdenum
Na	Salt
Ni	Nickel
O	Petroleum
P	Phosphates
Pb	Lead
Sb	Antimony
Sn	Tin
Tc	Talc
U	Uranium
W	Tungsten
Zn	Zinc

⚡ Water Power

▨ Major Industrial Areas

DOMINANT LAND USE

- Cereals (chiefly wheat, millet)
- Cereals (chiefly wheat, rice, barley)
- Cereals (chiefly rice, barley)
- Livestock Herding, Limited Agriculture
- Forests
- Nonagricultural Land

AREA 145,730 sq. mi. (377,441 sq. km.)
POPULATION 117,057,485
CAPITAL Tokyo
LARGEST CITY Tokyo
HIGHEST POINT Fuji 12,389 ft. (3,776 m.)
MONETARY UNIT yen
MAJOR LANGUAGE Japanese
MAJOR RELIGIONS Buddhism, Shintoism

AREA 46,540 sq. mi. (120,539 sq. km.)
POPULATION 17,914,000
CAPITAL P'yŏngyang
LARGEST CITY P'yŏngyang
HIGHEST POINT Paektu 9,003 ft. (2,744 m.)
MONETARY UNIT won
MAJOR LANGUAGE Korean
MAJOR RELIGIONS Confucianism, Buddhism, Ch'ondogyo

AREA 38,175 sq. mi. (98,873 sq. km.)
POPULATION 37,448,836
CAPITAL Seoul
LARGEST CITY Seoul
HIGHEST POINT Halla 6,398 ft. (1,950 m.)
MONETARY UNIT won
MAJOR LANGUAGE Korean
MAJOR RELIGIONS Confucianism, Buddhism, Ch'ondogyo, Christianity

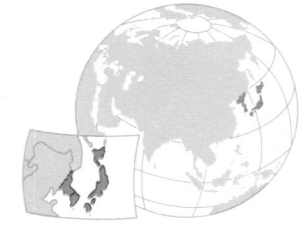

JAPAN

NORTH KOREA

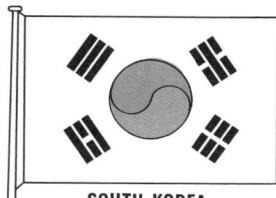

SOUTH KOREA

JAPAN

PREFECTURES

Aichi 5,923,569	H6
Akita 1,232,481	J4
Aomori 1,468,646	K3
Chiba 4,149,147	P2
Ehime 1,465,215	F7
Fukui 773,599	G5
Fukuoka 4,292,963	D7
Fukushima 1,970,616	K5
Gifu 1,867,978	H6
Gumma 1,756,480	J5
Hiroshima 2,646,324	E6
Hokkaido 5,338,206	K2
Hyogo 4,992,140	H7
Ibaraki 2,342,198	K5
Ishikawa 1,069,872	H5
Iwate 1,385,563	K4
Kagawa 961,292	G6
Kagoshima 1,723,902	E8
Kanagawa 6,397,748	O2
Kochi 808,397	F7
Kumamoto 1,715,273	E7
Kyoto 2,424,856	J7
Mie 1,626,002	H6
Miyagi 1,955,267	K4
Miyazaki 1,085,055	E8
Nagano 2,017,564	J5
Nagasaki 1,571,912	D7
Nara 1,077,491	J8
Niigata 2,391,938	J5
Oita 1,190,314	E7
Okayama 1,814,305	F6
Okinawa 1,042,572	N6
Osaka 8,278,925	J8
Saga 837,674	E7

Saitama 4,821,340	O2
Shiga 985,621	J7
Shimane 768,886	F6
Shizuoka 3,308,799	H6
Tochigi 1,698,003	K5
Tokushima 805,166	G7
Tokyo 11,673,554	O2
Tottori 581,311	G6
Toyama 1,070,791	H5
Wakayama 1,072,118	G6
Yamagata 1,220,302	K4
Yamaguchi 1,555,218	E6
Yamanashi 783,050	J6

CITIES and TOWNS

Abashiri 43,825	M1
Ageo 146,358	O2
Aikawa 13,546	H4
Aizuwakamatsu 108,650	J5
Aigasawa 18,086	J3
Akashi 234,905	H8
Aki 24,480	F7
Akita 261,246	J4
Akkeshi 16,778	M2
Akune 30,295	E7
Amagasaki 545,783	H8
Anan 42,725	G7
Anan 60,439	G7
Aomori 264,222	K3
Asahi 34,028	K6
Asahikawa 320,526	L2
Ashibetsu 36,520	L2
Ashikaga 162,359	J5
Ashiya 76,211	H8
Atami 51,437	J6
Atsugi 108,955	O2
Awaji 9,623	H8

Ayabe 43,490	G6
Beppu 133,894	E7
Bibai 38,416	L2
Biratori 9,331	L2
Chiba 659,356	P2
Chichibu 61,798	J5
Chigasaki 152,023	O3
Chitose 61,031	K2
Chofu 175,924	O2
Choshi 90,374	K6
Daito 110,829	J8
Ebetsu 77,624	K2
Eniwa 39,884	K2
Esashi, Hokkaido 10,172	L1
Esashi, Hokkaido 14,409	J3
Esashi, Iwate 36,336	K4
Fuchu, Hiroshima 50,217	F6
Fuchu, Tokyo 182,474	O2
Fuji 199,195	J6
Fujieda 90,358	J6
Fujisawa 265,975	O3
Fukagawa 36,000	L2
Fukuchiyama 60,003	G6
Fukue 32,018	D7
Fukui 231,364	G5
Fukuoka 1,002,201	D7
Fukushima 246,531	K5
Fukuyama 329,714	F6
Funabashi 423,101	P2
Furukawa 54,356	K4
Gifu 408,707	H6
Gobo 30,272	G7
Gose 37,554	J8
Gosen 39,376	J5
Goshogawara 49,040	K3
Gotsu 27,992	F6
Habikino 94,160	J8
Haboro 13,624	K1

Hachinohe 224,366	K3
Hachioji 322,580	O2
Iizuka 103,663	E7
Hagi 52,724	E6
Hakodate 307,453	L2
Hakui 28,726	H5
Hamada 50,316	E6
Hamamatsu 468,884	H6
Hanamaki 65,826	K4
Hanno 55,926	O2
Haramachi 43,483	K5
Hayama 24,026	O3
Higashiosaka 524,750	J8
Hikone 85,066	H6
Himeji 436,086	H6
Himi 61,789	H5
Hino 126,847	O2
Hirakata 297,618	J7
Hirara 29,301	L7
Hirata 30,942	F6
Hiratsuka 195,635	O3
Hiroo 11,399	L2
Hirosaki 164,911	K3
Hiroshima 852,611	E6
Hitachi 202,383	K5
Hitachiota 35,322	K5
Hitoyoshi 41,118	E6
Hofu 105,540	E6
Hondo 40,432	E7
Honjo 48,088	O2
Hyuga 53,448	E7
Ibaraki 210,286	J8
Ibusuki 32,339	E8
Ichihara 194,068	P3
Ichikawa 319,291	P2
Ichinohe 21,433	K3
Ichinomiya 238,463	H6
Ichinoseki 59,122	K4

Ide 9,112	J7
Iida 77,112	H6
Iizuka 75,417	E7
Ikeda, Hokkaido 12,306	L2
Ikeda, Osaka 100,268	H7
Ikoma 48,848	J8
Ikuno 6,658	G6
Imabari 119,726	F6
Imari 60,913	D7
Imazu 11,519	H6
Ina 54,468	H6
Isahaya 73,341	D7
Ise 104,967	H6
Ishigaki 34,657	L7
Ishige 19,220	P2
Ishinomaki 115,085	K4
Ishioka 43,679	K5
Itami 171,978	H7
Ito 68,072	J6
Itoigawa 36,646	H5
Itoman 39,363	N6
Iwaizumi 20,219	K4
Iwaki 330,213	K5
Iwakuni 111,069	E6
Iwami 16,063	G6
Iwamizawa 72,305	L2
Iwanai 25,823	K2
Iwasaki 4,437	J3
Iwata 67,665	H6
Iwatsuki 83,825	O2
Iyo 27,805	F7
Izuhara 18,460	D6
Izumi 118,237	J8
Izumiotsu 66,250	J8
Izumisano 86,139	J8
Izumo 71,568	F6
Joetsu 123,418	H5
Joyo 58,923	J7

Kadoma 143,238	J7
Kaga 61,599	H5
Kagoshima 456,827	E8
Kaizuka 79,506	H8
Kakegawa 169,293	G6
Kamaishi 68,981	L4
Kamakura 165,552	O3
Kameoka 58,184	J7
Kamiiso 27,229	K3
Kaminoyama 37,858	J4
Kamiyaku 8,668	E8
Kamo 8,953	J7
Kanazawa 395,263	H5
Kanonji 44,131	F6
Kanoya 67,951	E8
Kanuma 81,799	J5
Karatsu 75,224	D7
Kaseda 24,969	D8
Kashihara 95,701	J8
Kashiwa 203,065	P2
Kashiwara 63,586	J8
Kashiwazaki 80,351	J5
Kasugai 213,857	H6
Kasukabe 121,639	O2
Katsuta 79,996	K5
Katsuura 26,755	K6
Kawachinagano 66,936	J8
Kawagoe 225,465	O2
Kawaguchi 345,538	O2
Kawanishi 115,773	H7
Kawasaki 1,014,951	O2
Kesennuma 66,616	K4
Kikonai 10,034	K3
Kimitsu 76,016	O3
Kiryu 134,239	J5
Kisarazu 96,840	P3
Kishiwada 174,952	J8
Kitaibaraki 44,332	K5

Kitakami 48,759	K4
Kitakata 37,471	J5
Kitakyushu 1,058,058	E6
Kitami 91,519	L2
Kizu 11,890	J7
Kobayashi 38,325	E8
Kobe 1,360,605	H7
Kochi 280,962	F7
Kodaira 156,181	O2
Kofu 193,879	J6
Koga 55,973	J5
Koganei 102,714	O2
Kokubu 31,660	E8
Komagane 30,318	H6
Komatsu 100,273	H5
Koriyama 264,628	K5
Koshigaya 195,917	P2
Koyama 16,394	E8
Kudohama 17,817	F7
Kuji 38,122	K3
Kuki 45,797	O2
Kumagaya 131,485	J5
Kumamoto 488,166	E7
Kumano 27,026	G7
Kumiyama 11,540	J7
Kurashiki 392,755	F6
Kurayoshi 50,785	F6
Kure 242,655	E6
Kuroiso 42,349	K5
Kurume 204,474	E7
Kushima 30,438	E8
Kushikino 30,456	D8
Kushima 30,038	E8
Kushimoto 18,997	G7
Kushiro 206,840	M2
Kyonan 13,067	O3
Kyoto 1,461,059	J7
Machida 255,305	O2
Maebashi 250,241	J5
Maihara 12,845	G6
Maizuru 97,780	G6
Makubetsu 18,444	L2
Makurazaki 29,685	O3
Mashike 9,312	K2
Masuda 50,734	E6
Matsubara 132,662	H8
Matsue 127,440	F6
Matsumae 18,307	J3
Matsumoto 185,595	H5
Matsusaka 108,893	H6
Matsuto 36,170	H5
Matsuyama 367,323	F7
Mihara 83,679	F6
Miki 53,731	H7
Mikuni 21,602	G5
Minamata 36,782	E7
Minobu 10,345	J6
Minoo 79,621	J7
Misawa 37,437	K3
Mitaka 164,950	O2
Mito 197,953	K5
Mitsukaido 38,820	P2
Miura 47,888	O3
Miyako 61,912	L4
Miyakonojo 118,289	E8
Miyazaki 234,347	E8
Miyazu 30,194	G6
Miyoshi 37,193	F6
Mizusawa 52,266	K4
Mobara 64,942	K6
Mombetsu 32,825	L1
Monbetsu 15,029	L2
Mooka 47,345	K5
Mori 17,030	K2
Moriguchi 178,383	J7
Morioka 216,223	K4
Motobu 17,823	N6
Muko 45,886	J7
Murakami 32,939	J4
Muroran 158,715	K2
Muroto 26,660	G7
Musashino 139,508	O2
Mutsu 44,646	K3
Nachikatsuura 23,596	H7
Nagahama, Ehime 13,144	F7
Nagahama, Shiga 54,064	H6
Nagano 306,637	J5
Nagaoka, Kyoto 65,557	J7
Nagaoka, Niigata 171,742	J5
Nagaokakyo 65,557	J7
Nagasaki 450,194	D7
Nagato 27,327	E6
Nago 45,210	N6
Nagoya 2,079,740	H6
Naha 295,006	N6
Nakaminato 33,147	K5
Nakamura 34,437	F7
Nakasato 14,248	K3
Nakatsu 59,111	E7
Nanao 49,493	H5
Nankoku 42,832	F7
Nara 257,538	J8
Narashino 117,852	P2
Nayoro 35,145	L1
Naze 46,359	O5
Nemuro 45,817	M2
Neyagawa 254,311	J7
Nichinan 52,171	E8
Niigata 423,188	J5
Nihama 131,912	F6
Niimi 30,014	F6
Niitsu 58,970	J5
Nishinomiya 400,622	H8

(continued on following page)

Agriculture, Industry and Resources

DOMINANT LAND USE

- Cereals, Cash Crops
- Truck Farming, Horticulture
- Mixed Farming, Dairy
- Rice
- Forests, Scrub

MAJOR MINERAL OCCURRENCES

Ag	Silver	Mn	Manganese
Au	Gold	Mo	Molybdenum
C	Coal	O	Petroleum
Cu	Copper	Pb	Lead
Fe	Iron Ore	Py	Pyrites
G	Natural Gas	U	Uranium
Gr	Graphite	W	Tungsten
Mg	Magnesium	Zn	Zinc

⚡ Water Power

▨ Major Industrial Areas

(continued)

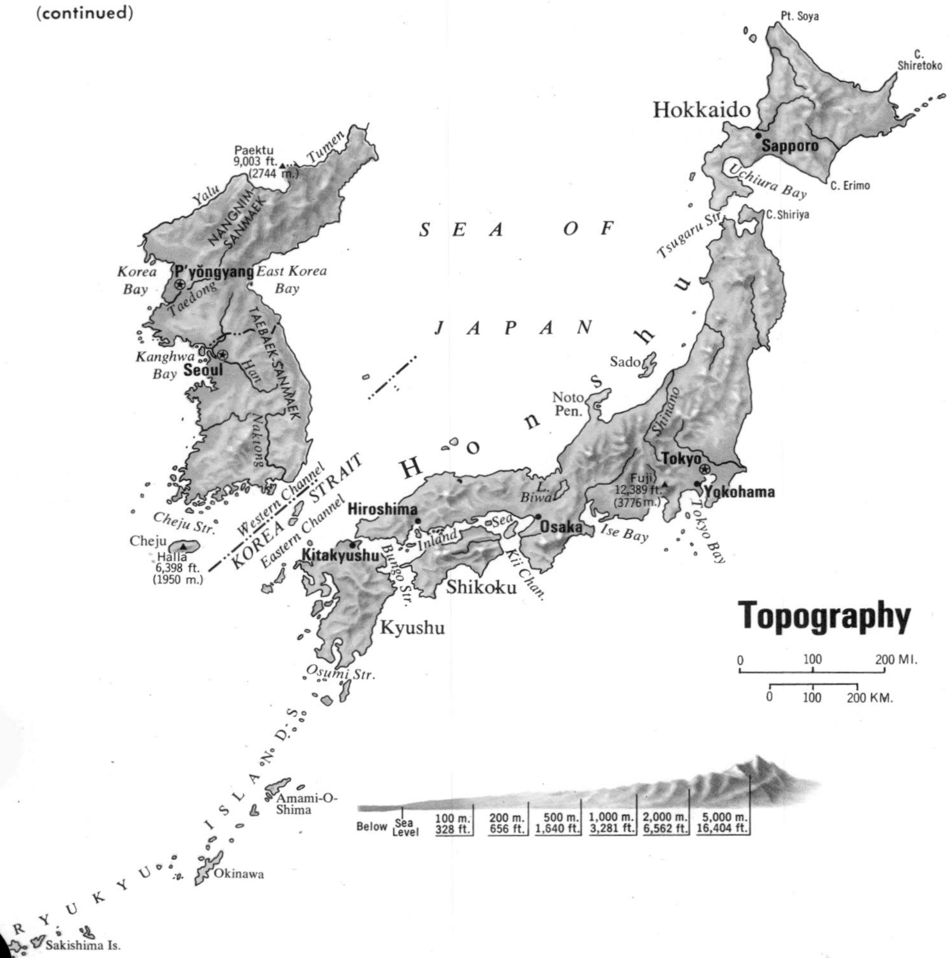

Topography
Pt. Soya
C. Shiretoko
Hokkaido
Sapporo
Uchiura Bay
C. Erimo
C. Shiriya
Tsugaru Str.
SEA OF JAPAN
Paektu 9,003 ft. (2744 m.)
Tumen
Yalu
NANGNIM SANMAEK
Korea Bay
P'yongyang
East Korea Bay
Taedong
TAEBAEK SANMAEK
Kanghwa Bay
Seoul
Han
Naktong
Sado
Noto Pen.
Honshu
Shinano
Tokyo
Yokohama
Fuji 12,389 ft. (3776 m.)
L. Biwa
Osaka
Ise Bay
Tokyo Bay
Kii Chan.
Inland Sea
Hiroshima
Kitakyushu
Bungo Str.
Shikoku
Kyushu
Osumi Str.
Cheju Str.
Western Channel
KOREA STRAIT
Eastern Channel
Cheju
Halla 6,398 ft. (1950 m.)
RYUKYU ISLANDS
Amami-O-Shima
Okinawa
Sakishima Is.
0 100 200 MI.
0 100 200 KM.
Below Sea Level
100 m. 328 ft.
200 m. 656 ft.
500 m. 1,640 ft.
1,000 m. 3,281 ft.
2,000 m. 6,562 ft.
5,000 m. 16,404 ft.

Nishinoomote 24,266 E8
Nobeoka 134,521 E7
Noboribetsu 50,885 K2
Noda 78,193 P2
Nogata 58,551 E7
Nose 9,749 J7
Noshiro 59,215 J3
Noto 15,815 H5
Numata 45,255 J5
Numazu 199,325 J6
Obama 33,890 G6
Obihiro 141,774 L2
Oda 37,449 F6
Odate 71,828 K3
Odawara 173,519 J6
Ofunato 39,632 K4
Oga 39,619 J4
Ogaki 140,424 H6
Ogi 4,717 J5
Ohata 12,632 K3
Oita 320,237 E7
Ojiya 44,375 J5
Okawa 50,395 E7
Okaya 61,776 H5
Okayama 513,471 F6
Okazaki 234,510 H6
Omagari 40,581 K4
Omiya 327,698 O2
Omu 7,407 L1
Omura, Bonin Is. 1,507 M3
Omura, Nagasaki 60,919 E7
Omuta 165,969 E7
Onagawa 16,945 K4
Ono 41,918 H6
Onoda 43,804 E6
Onomichi 102,951 F6
Osaka 2,778,987 J8
Ota 110,723 J5
Otaru 184,406 K2
Otawara 42,332 K5
Otofuke 26,933 L2
Otsu 191,481 J7
Owase 31,797 H6
Oyabe 35,791 H5
Oyama 120,264 J5
Ozu 37,294 F7
Rausu 8,249 M1
Rikuzentakata 29,439 K4
Rumoi 36,882 K2
Ryotsu 22,110 J4
Ryugasaki 40,565 P2
Sabae 57,252 H5
Saga 152,258 E7
Sagamihara 377,398 O2
Saigo 14,409 F5
Saiki 52,863 E7
Saito 37,054 E7
Sakado 51,232 O2
Sakai, Ibaraki 24,347 P1
Sakai, Osaka 750,688 J8
Sakaide 67,624 G6
Sakaiminato 35,821 F6
Sakata 97,723 J4
Saku 56,143 J5
Sakurai 54,314 J8
Sanda 35,261 H7
Sanjo 81,806 J5
Sapporo 1,240,613 K2
Sarufutsu 3,552 L1
Sasebo 250,729 D7
Satte 43,083 O1
Sawara 48,670 K6
Sayama 98,548 O2
Sendai, Kagoshima 61,788 E8
Sendai, Miyagi 615,473 K4
Setouchi 15,290 O5
Settsu 76,704 J8
Shari 15,996 M2
Shibata 74,025 J5
Shibetsu 30,028 M2
Shimabara 45,179 E7
Shimamoto 22,404 J7
Shimizu 243,049 J6
Shimoda 31,700 J6
Shimonoseki 266,593 E6
Shingu 39,023 H7
Shinjo 42,227 K4
Shiogama 59,235 K4
Shirakawa 42,685 K5
Shiranuka 14,897 M2
Shiroishi 40,862 K4
Shizunai 24,833 L2
Shizuoka 446,952 H6
Shobara 23,867 F6
Soka 167,177 O2
Soma 37,551 K5
Sonobe 14,827 J7
Suita 300,956 J7
Sukagawa 54,922 K5
Sukumo 25,340 F7
Sumoto 44,137 G6
Sunagawa 26,023 K2
Susaki 31,019 F7
Suttsu 6,511 J2
Suwa 49,594 H6
Suzu 28,238 H5
Suzuka 141,829 H6
Tachikawa 138,129 O2
Tagawa 61,464 E7
Tajimi 68,901 H6
Takaishi 66,824 H8
Takamatsu 298,999 F6
Takaoka 169,621 H5
Takarazuka 162,624 H7
Takasaki 211,348 J5
Takatsuki 330,570 J7
Takayama 60,504 H5
Takefu 65,012 G6
Takikawa 50,090 K2
Tanabe, Kyoto 30,022 J7
Tanabe, Wakayama 66,999 G7
Tateyama 56,139 K6
Tendo 48,082 K4
Tenri 62,909 J8
Teshio 6,509 K1
Toba 29,346 H6
Tobetsu 17,351 K2
Togane 33,406 K6
Toi 6,983 J6
Tojo 13,796 F6
Tokamachi 50,211 J5
Tokorozawa 196,870 O2
Tokunoshima 35,391 O5
Tokushima 239,281 G7
Tokuyama 106,967 F6
Tokyo (cap.) 8,646,520 O2
Tokyo* 11,673,554 O2
Tomakomai 132,477 K2
Tomiyama 7,389 O3
Tondabayashi 91,393 J8
Tosa 30,679 F7
Tosashimizu 24,856 F7
Tosu 50,733 E7
Tottori 122,312 G6
Towada 54,365 K3
Toyama 290,143 H5
Toyohashi 284,585 H6
Toyonaka 398,384 J7
Toyooka 46,210 G6
Toyota 248,774 H6
Tsu 139,538 H6
Tsubame 43,265 J5
Tsuchiura 104,028 J5
Tsuruga 60,205 G6
Tsuruoka 95,932 J4
Tsuyama 79,907 F6
Ube 161,969 E6
Ueda 105,151 J5
Ugo 21,956 K4
Uji 133,405 J7
Uozu 48,419 H5
Urakawa 20,213 L2
Urawa 331,145 O2
Ushibuka 24,250 D7
Usuki 39,163 F7
Utsunomiya 344,420 K5
Uwajima 70,428 F7
Wajima 33,234 H5
Wakasa 6,989 G6
Wakayama 389,717 G6
Wakkanai 55,464 K1
Warabi 76,311 O2
Yaizu 94,102 J6
Yakumo 19,260 J2
Yamagata 219,773 K4
Yamaguchi 106,099 E6
Yamato 145,881 O2
Yamatokoriyama 71,001 J8
Yamatotakada 58,637 J8
Yao 261,639 J8
Yatabe 22,225 P2
Yatsushiro 103,691 E7
Yawata 50,132 J7
Yawatahama 45,259 F7
Yoichi 25,816 K2
Yokawa 8,015 H7
Yokkaichi 247,001 H6
Yokohama 2,621,771 O3
Yokosuka 389,557 O3
Yokote 43,030 K4
Yonago 118,332 F6
Yonezawa 91,974 K5
Yono 71,044 O2
Yubari 50,131 L2
Yubetsu 6,693 L1
Yukuhashi 53,750 E7
Yuzawa 38,005 K4
Zushi 56,298 O3

OTHER FEATURES

Abashiri (riv.) M1
Agano (riv.) J4
Akan National Park M2
Amakusa (isls.) D7
Amami (isls.) N5
Amami-O-Shima (isl.) N5
Ara (riv.) O2
Asahi (mt.) J4
Asama (mt.) J5
Ashizuri (cape) F7
Aso (mt.) E7
Aso National Park E7
Atsumi (bay) H6
Awa (isl.) J4
Awaji (isl.) H8
Bandai (mt.) K5
Bandai-Asahi National Park J4
Biwa (lake) H6
Bonin (isls.) M3
Boso (pen.) K6
Bungo (strait) F7
Chichi (isl.) M3
Chichibu-Tama National Park J6
Chokai (mt.) J4
Chubu-Sangaku National Park H5
Dai (mt.) F6
Daimanji (mt.) F5
Daio (cape) H6
Daisen-Oki National Park F6
Daisetsu (mt.) L2
Daisetsu-Zan National Park L2
Dogo (isl.) F5
Dozen (isls.) F5
East China (sea) C8
Edo (riv.) P2
Erimo (cape) L3
Esan (pt.) K3
Etorofu (isl.) N1
Fuji (mt.) J6
Fuji (riv.) J6
Fuji-Hakone-Izu National Park H6
Gassan (mt.) J4
Goto (isls.) D7
Habomai (isls.) N2
Hachiro (lag.) J3
Haha (isl.) M3
Hakken (mt.) H6
Haku (mt.) H5
Hakusan National Park H5
Harima (sea) G6
Hida (riv.) H6
Hodaka (mt.) H5
Hokkaido (isl.) L2
Honshu (isl.) J5
Ie (isl.) N6
Iheya (isl.) N6
Iki (isl.) D7
Ina (riv.) H7
Inawashiro (lake) K5
Inubo (cape) K6
Iriomote (isl.) K7
Iro (cape) J6
Ise (bay) H6
Ise-Shima National Park H6
Ishigaki (isl.) L7
Ishikari (bay) K2
Ishikari (riv.) L2
Ishizuchi (mt.) F7
Iwaki (mt.) K3
Iwate (mt.) K4
Iwo (isl.) M4
Iyo (sea) E7
Izu (isls.) J6
Izu (pen.) J6
Japan (sea) G4
Joshinetsu-Kogen National Park J5
Kagoshima (bay) E8
Kamui (cape) K2
Kariba (mt.) K2
Kasumiga (lag.) K5
Kazan-retto (Volcano) (isls.) M4
Kerama (isls.) M6
Kii (chan.) G7
Kikai (isl.) O5
Kino (riv.) G6
Kirishima-Yaku National Park E7
Kita Iwo (isl.) M4
Kitakami (riv.) K4
Komaga (mt.) K2
Koshiki (isls.) D8
Kuchino (isl.) O4
Kuju (mt.) E7
Kume (isl.) M6
Kunashiri (isl.) N1
Kutcharo (lake) M2
Kyushu (isl.) E7
Meakan (mt.) L2
Minami Iwo (isl.) M5
Miura (pen.) O3
Miyako (isl.) L7
Miyako (isls.) L7
Mogami (riv.) K4
Motsuta (cape) J2
Muko (isl.) M3
Muko (riv.) H7
Muroto (pt.) G7
Mutsu (bay) K3
Naka (riv.) K5
Nampo-Shoto (isls.) M3
Nansei Shoto (Ryukyu) (isls.) M6
Nantai (mt.) J5
Nasu (mt.) J5
Nemuro (strait) M1
Nii (isl.) J6
Nikko National Park J5
Nishino (isl.) M3
Nojima (cape) K6
Nosappu (pt.) N2
Noto (pen.) H5
Nyudo (cape) J4
Oani (riv.) K3
Obitsu (riv.) P3
Oga (pen.) J4
Ogasawara-gunto (Bonin) (isls.) M3
Okhotsk (sea) M1
Oki (isls.) F5
Okinawa (isl.) N6
Okinawa (isls.) N6
Okinoerabu (isl.) N5
Okushiri (isl.) J2
Oma (cape) K3
Omono (riv.) J4
Ono (riv.) E7
Ontake (mt.) H6
Osaka (bay) H8
O-Shima (isl.) J6
Osumi (isls.) E8
Osumi (pen.) E8
Osumi (str.) E8
Otakine (mt.) K5
Rikuchu-Kaigan National Park L4
Rishiri (isl.) K1
Ryukyu (isls.) L7
Sado (isl.) J4
Sagami (bay) O3
Sagami (riv.) O2
Sagami (sea) J6
Saikai National Park D7
Sakishima (isls.) K7
San'in Kaigan National Park G6
Sata (cape) E8
Setonaikai National Park H7
Shikoku (isl.) F7
Shikotan (isl.) N2
Shikotsu (lake) K2
Shikotsu-Toya National Park K2
Shimane (pen.) F6
Shimokita (pen.) K3
Shinano (riv.) J5
Shiono (cape) H7
Shiragami (cape) J3
Shirane (mt.) J5
Shirane (mt.) H6
Shiretoko (cape) M1
Shiriya (cape) K3
Soya (pt.) L1
Suo (sea) E7
Suruga (bay) J6
Suwanose (isl.) O4
Suzu (pt.) H5
Takeshima (isls.) F5
Tama (riv.) O2
Tanega (isl.) E8
Tappi (cape) K3

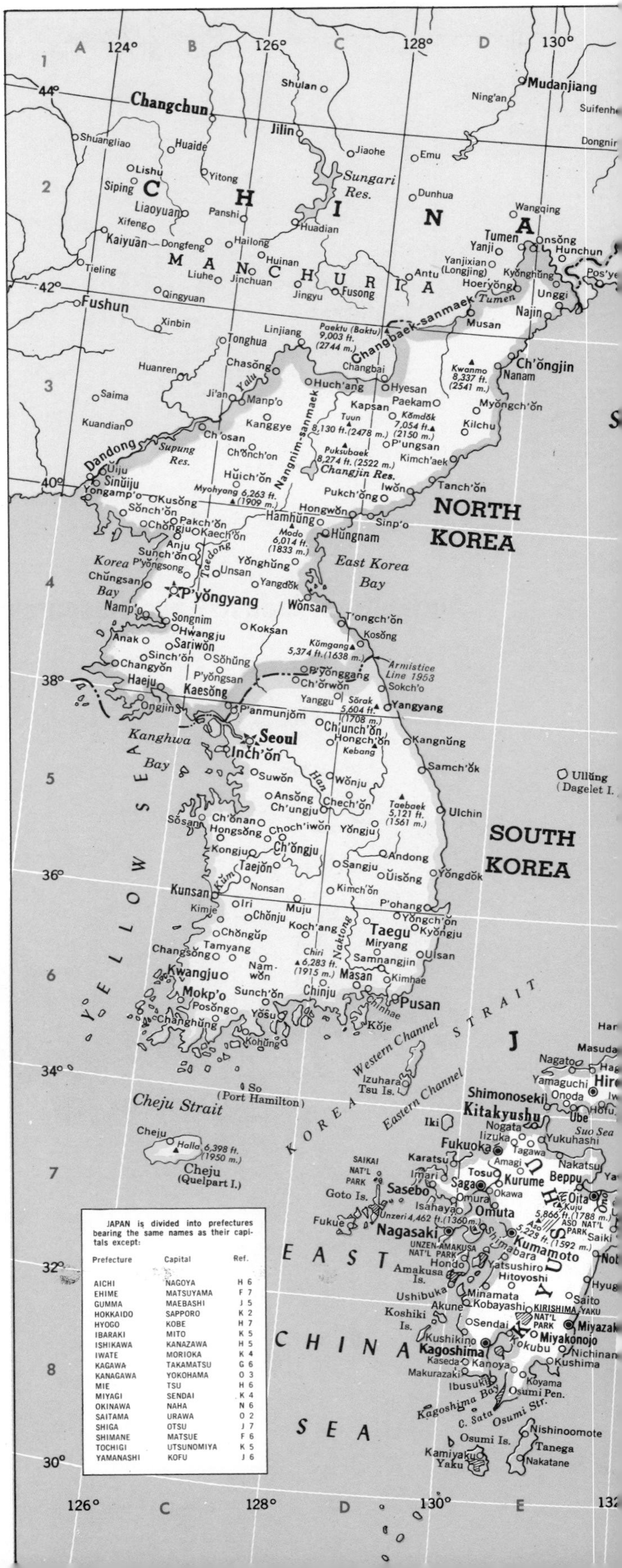

124° 126° 128° 130°
44° 42° 40° 38° 36° 34° 32° 30°
Mudanjiang
Changchun
Jilin
Shulan
Sungari Res.
CHINA
MANCHURIA
Siping
Liaoyuan
Kaiyuan
Tieling
Fushun
Yanji
Tumen
Dandong
Sinuiju
Paektu (Baektu) 9,003 ft. (2744 m.)
Changbaek-sanmaek
Nangnim-sanmaek
Ch'ongjin
Kapsan
Hyesan
Kimch'aek
Hamhung
Hungnam
NORTH KOREA
Korea Bay
P'yongyang
East Korea Bay
Wonsan
Sariwon
Haeju
Kaesong
Armistice Line 1953
P'anmunjom
Seoul
Inch'on
Kanghwa Bay
Ch'unch'on
Kangnung
Samch'ok
Ulchin
Ullung (Dagelet I.)
SOUTH KOREA
YELLOW SEA
Ch'ongju
Taejon
Kunsan
Chonju
Taegu
Kwangju
Mokp'o
Masan
Pusan
Ulsan
Kyongju
KOREA STRAIT
Western Channel
Eastern Channel
Cheju Strait
Cheju
Halla 6,398 ft. (1950 m.)
Cheju (Quelpart I.)
Tsu Is.
Izuhara
Shimonoseki
Kitakyushu
Fukuoka
Nagasaki
Sasebo
Kumamoto
Kagoshima
Miyazaki
EAST CHINA SEA

JAPAN is divided into prefectures bearing the same names as their capitals except:

Prefecture	Capital	Ref.
AICHI	NAGOYA	H 6
EHIME	MATSUYAMA	F 7
GUMMA	MAEBASHI	J 5
HOKKAIDO	SAPPORO	K 2
HYOGO	KOBE	H 7
IBARAKI	MITO	K 5
ISHIKAWA	KANAZAWA	H 5
IWATE	MORIOKA	K 4
KAGAWA	TAKAMATSU	G 6
KANAGAWA	YOKOHAMA	O 3
MIE	TSU	H 6
MIYAGI	SENDAI	K 4
OKINAWA	NAHA	N 6
SAITAMA	URAWA	O 2
SHIGA	OTSU	J 7
SHIMANE	MATSUE	F 6
TOCHIGI	UTSUNOMIYA	K 5
YAMANASHI	KOFU	J 6

Tarama (isl.) L7
Tazawa (lake) K4
Teshio (mt.) L1
Teshio (riv.) L1
Tobi (isl.) J4
Tokachi (mt.) L2
Tokachi (riv.) L2
Tokara (isls.) O5
Tokuno (isl.) O5
Tokyo O2
Tokyo (bay) O2
Tone (riv.) K6
Tosa (bay) F7
Towada (lake) K3
Towada-Hachimantai National
　Park K3
Toya (lake) K2
Toyama H5
Toyama (bay) H5
Tsu (isls.) D6
Tsugaru (str.) K3
Tsurugi (mt.) G7
Uchiura (bay) K2
Unzen (mt.) D7
Unzen-Amakusa National Park D7
Volcano (isls.) M4
Wakasa (bay) G6

Yaeyama (isls.) K7
Yaku (isl.) E8
Yodo (riv.) J7
Yonaguni (isl.) K7
Yoron (isl.) N6
Yoshino (riv.) G6
Yoshino-Kumano National Park H7
Zao (mt.) K5

KOREA (NORTH)

CITIES and TOWNS

Ch'ŏngjin 306.000 E3
Chŏngju B4
Haeju 140.000 B4
Hamhŭng 484.000 C3
Heijo (P'yŏngyang) C4
Hongwŏn C3
Huch'ang C3
Hŭich'ŏn C3
Hyesan C3
Iwŏn D3
Kaech'ŏn B4
Kaesŏng 175.000 C4
Kanggye C3
Kapsan C3
Kilchu D3
Kimch'aek 100.000 D3
Najin E2
Namp'o 140.000 B4
Onsŏng D2
P'anmunjom C5
P'yŏngyang (cap.) 1.250.000 C4
Sariwŏn B4
Sindŭiju 300.000 B3
Songnim C4
Wŏnsan 275.000 C4

OTHER FEATURES

Baktu (Paektu) (mt.) C3
Changjin (res.) C3
East Korea (bay) D4
Japan (sea) D4
Kanghwa (bay) B5
Kŏmdŏk (mt.) D3
Kŭmgang (mt.) D4

KOREA (SOUTH)

CITIES and TOWNS

Andong 95.364 D5
Ansŏng 27.723 C5
Changhŭng 22.227 C6
Changsŏng 26.266 C6
Chech'ŏn 74.239 D5
Chinju 135.081 C7
Chinhae 103.640 D7

Chinju 154.646 D6
Choch'iwŏn 29.198 C5
Ch'ŏngju 214.864 C5
Ch'ŏngju 311.393 C5
Ch'ŏrwŏn 8.180 C4
Ch'unch'ŏn 140.530 C4
Ch'ungju 105.274 C5
Hongch'ŏn 29.499 C4
Hongsŏng 26.995 C5
Inch'ŏn 800.007 C5
Iri 117.155 C6
Kangnŭng 84.981 D4
Kimch'ŏn 67.078 C5
Kimje 221.414 C6
Koch'ang 23.721 C6
Kohŭng 217.446 C6
Kongju 39.756 C5
Kunsan 154.780 C6
Kwangju 607.011 C6
Kyŏngju 108.431 D6
Masan 371.917 D6
Miryang 42.951 D6

Mokp'o 192.958 C6
Muju 18.130 D5
Namwŏn 50.857 C6
Nonsan 226.429 C5
P'anmunjom C5
P'ohang 134.418 D5
Posŏng 20.256 C6
Pusan 2.453.173 D6
Samch'ŏk 42.526 D5
Samnangjin 19.374 D6
Sangju 52.839 D5
Seoul (cap.) 6.889.502 C5
Sŏkch'o 71.387 D4
Sŏsan 38.081 C5
Sunch'ŏn 108.063 C6
Suwŏn 224.145 C5
Taegu 1.310.768 D6
Taejŏn 506.708 C5
Tamyang 15.494 C6
Ŭisŏng 26.480 D5
Ulchin 27.607 D5
Ulsan 252.570 D6
Wŏnju 120.276 D5
Yanggu 277.986 C4
Yangyang 10.819 D4

Yŏngch'ŏn 50.765 D6
Yŏngdŏk 18.671 D5
Yŏngju 70.793 D5
Yŏsu 130.623 C6

OTHER FEATURES

Cheju C7
Cheju (str.) C7
Chiri (mt.) C6
Dagelet (Ullŭng) (isl.) E5
East China (sea) C8
Halla (mt.) C7
Han (riv.) C5
Japan (sea) G4
Kanghwa (bay) B5
Kebang (mt.) D5
Kŏje (isl.) D6
Korea (str.) D6
Naktong (riv.) D6
Port Hamilton (So) (isl.) C7
Quelpart (Cheju) (isl.) C7
So (isl.) C6

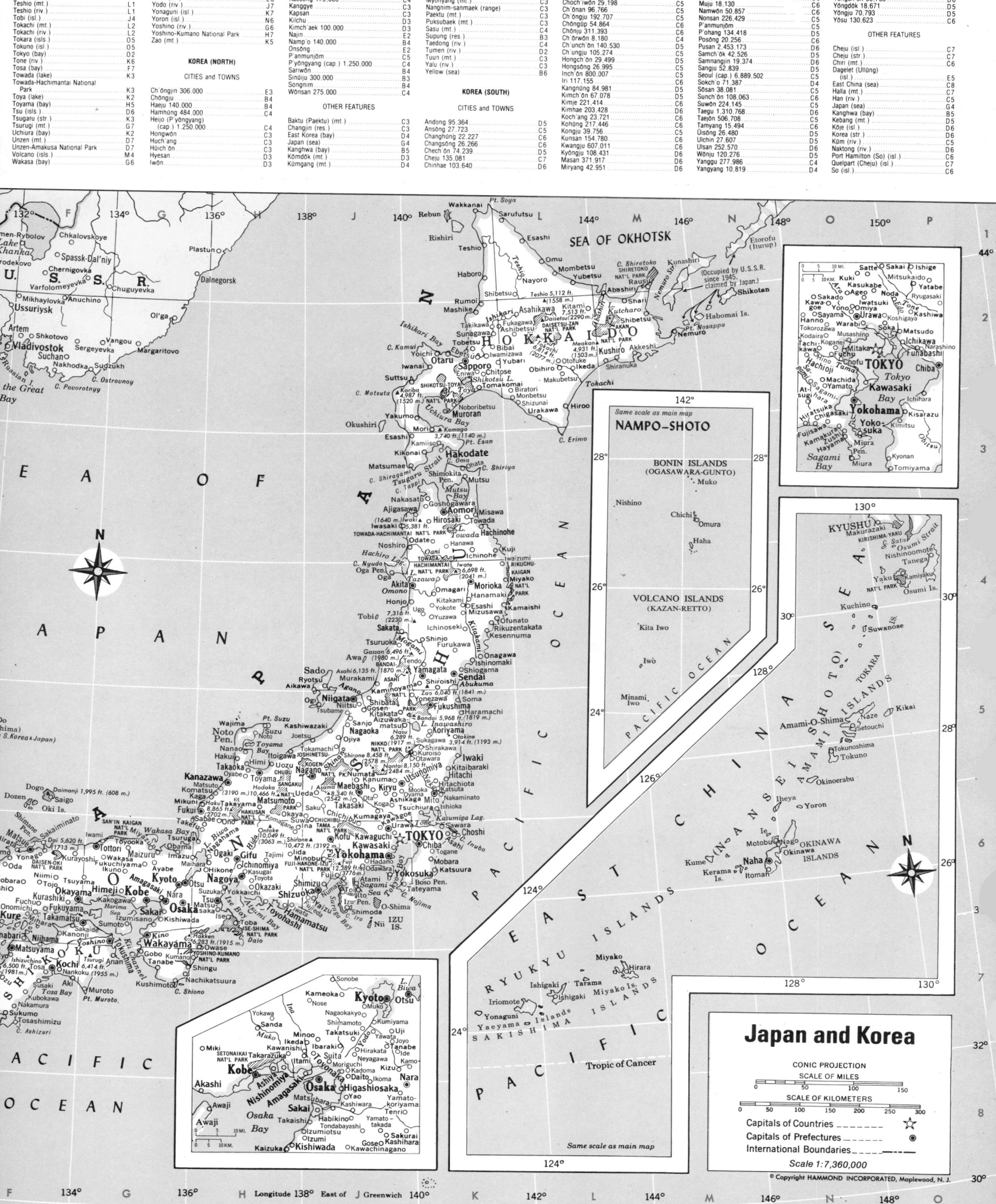

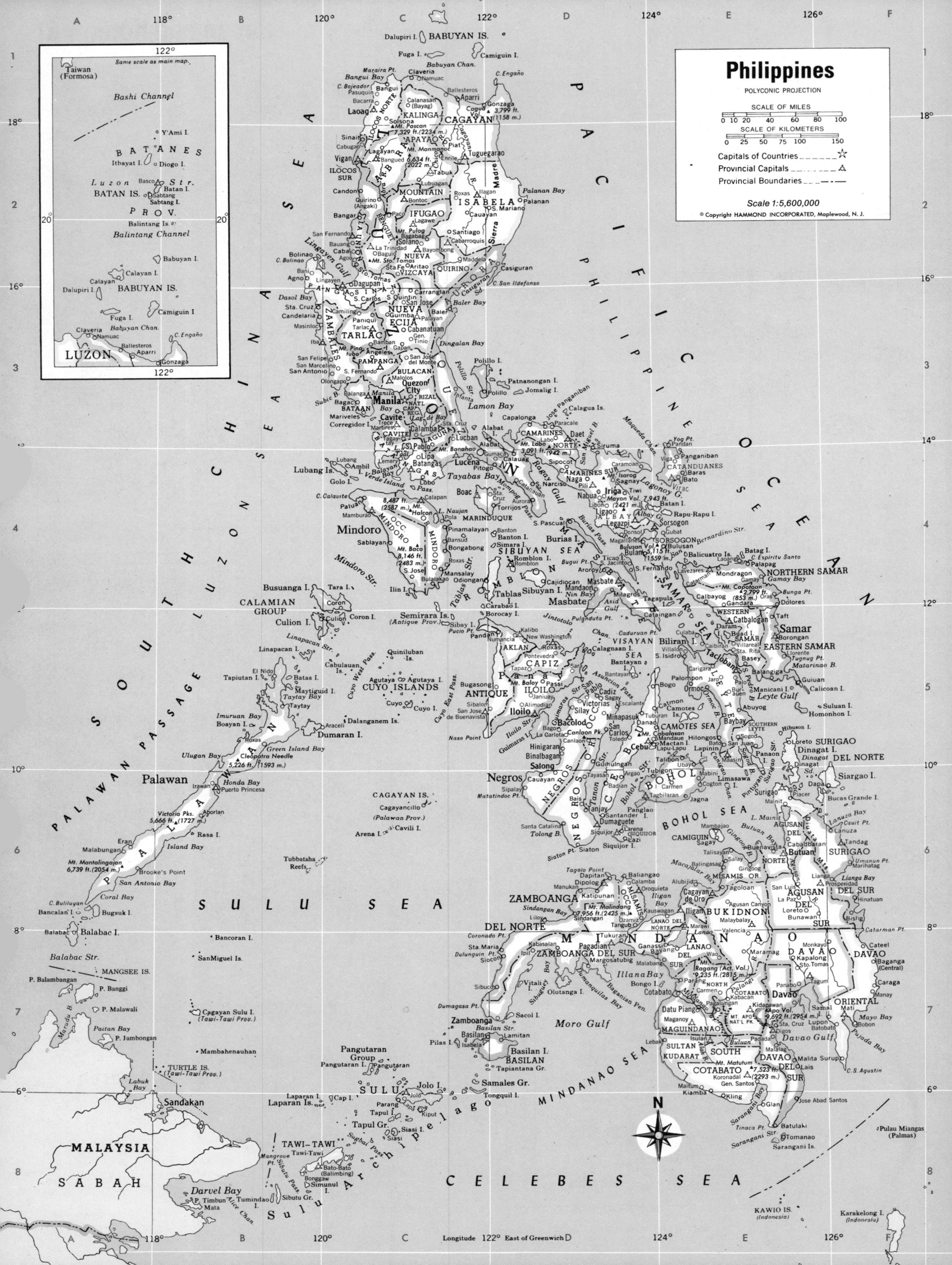

Philippines

POLYCONIC PROJECTION

SCALE OF MILES

0 10 20 40 60 80 100

SCALE OF KILOMETERS

0 25 50 75 100 150

Capitals of Countries ⎯⎯⎯ ☆
Provincial Capitals ⎯ · ⎯ △
Provincial Boundaries ⎯ · · ⎯

Scale 1:5,600,000

© Copyright HAMMOND INCORPORATED, Maplewood, N. J.

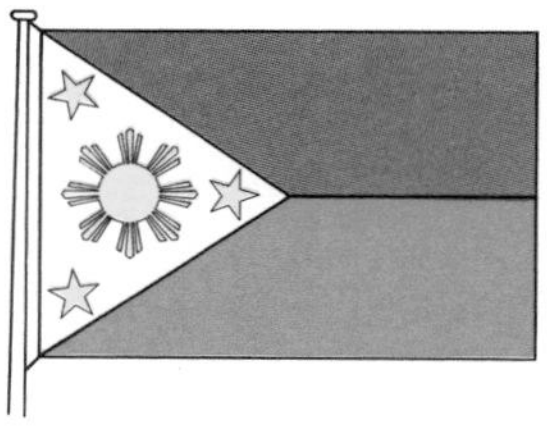

AREA 115,707 sq. mi. (299,681 sq. km.)
POPULATION 48,098,460
CAPITAL Manila
LARGEST CITY Manila
HIGHEST POINT Apo 9,692 ft. (2,954 m.)
MONETARY UNIT piso
MAJOR LANGUAGES Pilipino (Tagalog), English, Spanish, Bisayan, Ilocano, Bikol
MAJOR RELIGIONS Roman Catholicism, Islam, Protestantism, tribal religions

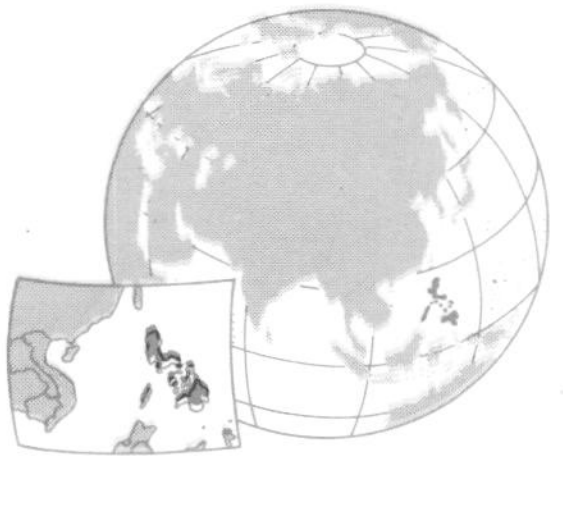

PROVINCES

Abra 160,198 C2
Agusan del Norte 365,421 E6
Agusan del Sur 631,634 E6
Aklan 324,563 D5
Albay 809,177 D4
Antique 344,879 D5
Aurora 107,145 C3
Basilan 201,407 D7
Bataan 323,254 C3
Batanes 12,091 A2
Batangas 1,174,201 C4
Benguet 354,751 C2
Bohol 806,031 E6
Bukidnon 631,634 E6
Bulacan 1,098,046 C3
Cagayan 711,476 C1
Camarines Norte 368,007 D3
Camarines Sur 1,099,346 D4
Camiguin 57,126 E6
Capiz 492,231 D5
Catanduanes 175,247 E4
Cavite 771,320 C3
Cebu 2,091,602 D5
Davao 725,153 E7
Davao del Sur 1,133,599 E7
Davao Oriental 339,931 F7
Eastern Samar 320,637 E5
Ifugao 111,368 C2
Ilocos Norte 390,666 C1
Ilocos Sur 443,591 C2
Iloilo 1,433,641 D5
Isabela 870,604 C2
Kalinga-Apayao 185,063 C1
Laguna 973,104 C3
Lanao del Norte 461,049 E6
Lanao del Sur 404,971 E7
La Union 452,578 C2
Leyte 1,302,648 E5
Maguindanao 536,546 E7
Manila 5,925,884 C3
Marinduque 173,715 C4
Masbate 584,526 D4
Misamis Occidental 386,328 D6
Misamis Oriental 690,032 E6
Mountain 103,052 C2
National Capital Region (Manila) 5,925,884 C3
Negros Occidental 1,930,301 D6
Negros Oriental 819,399 D6
North Cotabato 564,599 E7
Northern Samar 378,516 E4
Nueva Ecija 1,069,409 C3
Nueva Vizcaya 241,690 C2
Occidental Mindoro 222,431 C4
Oriental Mindoro 448,938 C4
Palawan 371,782 B6
Pampanga 1,181,590 C3
Pangasinan 1,636,057 C3
Quezon 1,129,277 C3
Quirino 83,230 C2
Rizal 555,533 C3
Romblon 193,174 D4
Siquijor 70,300 D6
Sorsogon 500,685 E4
South Cotabato 770,473 E7
Southern Leyte 298,294 E5
Sultan Kudarat 303,784 E7
Sulu 360,588 C7
Surigao del Norte 363,414 F5
Surigao del Sur 377,647 F6
Tarlac 638,457 C3
Tawi-Tawi 194,651 B8
Western Samar 501,439 E5
Zambales 444,037 C3
Zamboanga del Norte 588,015 D6
Zamboanga del Sur 1,183,845 D7

CITIES and TOWNS

Angeles 188,834 C3
Aparri 45,070 C1
Bacolod 262,415 D5
Bagac 13,109 C3
Bago 99,631 D5
Baguio 119,009 C2
Balanga 39,132 C3
Baler 18,349 C3
Balimbing (Bato-Bato) 22,189 C8
Bamban 26,072 C3
Basco 4,341 A2
Batangas 143,570 C4
Bato-Bato 22,189 C8
Baybay 74,640 E5
Bislig 81,615 F6
Boac 37,005 C4
Bontoc 17,091 C2
Burauen 48,058 E5
Butuan 172,489 E6
Cabanatuan 138,298 C3
Cabarroquis 17,450 C2
Cadiz 129,632 D5
Cagayan de Oro 227,312 E6
Calamba 121,175 C3
Calbayog 106,719 E4
Carigara 34,377 E5
Cauayan 70,017 D6
Cavite 87,666 C3
Cebu 490,281 D5
Cotabato 83,871 D7
Dagupan 98,344 C2
Davao 610,375 E7
Digos 70,065 E7
Escalante 71,293 D5
General Santos 149,396 E7
Gingoog 79,937 E6
Guihulngan 84,156 D5
Iba 22,791 B3
Ilagan 79,336 C2
Iligan 167,358 E6
Iloilo 244,827 D5
Infanta 27,914 C3
Jaro 29,739 E5
Jolo 52,429 C8
Koronadal 80,566 E7
Lagawe 15,075 C2
Lapu-Lapu 98,723 E5
Legazpi 99,766 D4
Ligao 69,860 D4
Lingayen 65,187 C2
Lipa 121,166 C4
Lucena 107,880 C4
Maganoy 45,845 E7
Mainit 18,078 E6
Malabang 18,955 D7
Malolos 95,699 C3
Mandaue 110,590 E5
Manila (cap.) 1,630,485 C3
Mariveles 48,594 C3
Mati 78,178 F7
Naga 90,712 D4
Olongapo 156,430 C3
Ormoc 104,978 E5
Ozamiz 77,832 D6
Pagadian 80,861 D7
Palo 31,124 E5
Palompon 40,242 E5
Panabo 71,098 E7
Prosperidad 33,824 F6
Puerto Princesa 60,234 B6
Quezon City 1,165,865 C3
Romblon 24,251 D4
Roxas 81,183 D5
Sagay 99,118 D5
San Antonio 42,969 B3
San Carlos, Negros Occ. 91,627 D5
San Carlos, Pangasinan 101,243 C3
San Fernando, La Union 68,410 C2
San Fernando, Pampanga 110,891 C3
San Jose 64,254 C3
San Jose del Monte 90,732 C3
San Pablo 131,655 C3
Santa Fe 6,338 C2
Santiago 69,877 C2
Silay 111,131 D5
Siquijor 17,533 D6
Surigao 79,745 E6
Tacloban 102,523 E5
Tagaytay 16,322 C3
Tagum 86,201 E7
Tarlac 175,691 C3
Toledo 91,668 D5
Tuguegarao 73,507 C2
Zamboanga 343,722 C7

OTHER FEATURES

Agusan (riv.) E6
Alabat (isl.) D3
Apo (vol.) E7
Babuyan (isl.) B2
Balabac (isl.) A7
Balayan (bay) C4
Balintang (chan.) A2
Baloy (mt.) D5
Bantayan (isl.) D5
Banton (isl.) D4
Bashi (chan.) A1
Basilan (isl.) D7
Batan, Albay (isl.) E4
Batan, Batanes (isl.) B2
Batan (isls.) A2
Bay, Laguna de (lake) C3
Biliran (isl.) E5
Bohol (isl.) E6
Bojeador (cape) C1
Borocay (isl.) D5
Bucas Grande (isl.) F6
Bugsuk (isl.) A6
Bulíluyan (cape) A6
Bunga (pt.) E4
Burias (isl.) D4
Busuanga (isl.) B4
Cabalasan (mt.) E5
Cabulauan (isls.) C5
Cagayan (isls.) C6
Cagayan (riv.) C2
Cagayan Sulu (isl.) B7
Cagua (vol.) D1
Calagua (isls.) D3
Calamian Group (isls.) B4
Calayan (isl.) A2
Calicoan (isl.) E5
Camiguin, Cagayan (isl.) B3
Camiguin, Camiguin (isl.) E6
Camotes (isls.) E5
Camotes (sea) E5
Canigao (chan.) E5
Canlaon (peak) D5
Capotoan (mt.) E4
Carabao (isl.) D4
Catanduanes (isl.) E4
Cebu (isl.) D5
Celebes (sea) D8
Cleopatra Needle (mt.) B5
Coron (isl.) C5
Corregidor (isl.) C3
Culion (isl.) B5
Cuyo (isl.) C5
Cuyo (isls.) C5
Daram (isl.) E5
Davao (gulf) E7
Dinagat (isl.) E5
Diuata (mts.) E6
Dumanquilas (bay) D7
Dumaran (isl.) C5
Engaño (cape) D1
Espiritu Santo (cape) E4
Fuga (isl.) A3
Guimaras (isl.) D5
Halcon (mt.) C4
Hibuson (isl.) E5
Homonhon (isl.) E5
Honda (bay) B6
Iligan (bay) E6
Ilin (isl.) C4
Illana (bay) D7
Imuruan (bay) B5
Island (bay) B6
Itbayat (isl.) A2
Jintotolo (chan.) D5
Jolo (isl.) C7
Jomalig (isl.) D3
Lagonoy (gulf) E4
Lamon (bay) C3
Lanao (lake) E7
Laparan (isls.) B8
Lapinin (isl.) E5
Leyte (gulf) E5
Leyte (isl.) E5
Limasawa (isl.) E6
Linapacan (isl.) B5
Lingayen (gulf) C2
Lubang (isls.) B4
Luzon (isl.) C3
Luzon (str.) A2
Macajalar (bay) E6
Malindang (mt.) D6
Mangsee (isls.) A7
Manila (bay) C3
Mantalingajan (mt.) A6
Maqueda (chan.) D3
Maraira (pt.) C1
Marinduque (isl.) C4
Masbate (isl.) D4
Mayon (vol.) D4
Maytiguid (isl.) B5
Mindanao (isl.) D7
Mindanao (riv.) E7
Mindoro (isl.) C4
Mindoro (str.) C4
Mompog (passg.) D4
Moro (gulf) D7
Mount Apo National Park E7
Naso (pt.) C5
Negros (isl.) D6
Olutanga (isl.) D7
Pacsan (mt.) C2
Palawan (isl.) B6
Palawan (passg.) A6
Panaon (isl.) E5
Panay (isl.) D5
Panglao (isl.) D6
Pangutaran (isl.) C7
Pangutaran Group (isls.) C7
Patnanongan (isl.) D3
Philippine (sea) D3
Pilas (isl.) C7
Pinatubo (mt.) C3
Polillo (isl.) C3
Pujada (bay) F7
Pulangi (riv.) E7
Ragang (vol.) E7
Ragay (gulf) D4
Rapu-Rapu (isl.) E4
Romblon (isl.) D4
Sabtang (isl.) B2
Sacol (isl.) D7
Samal (isl.) E7
Samales Group (isls.) D7
Samar (isl.) E5
Samar (sea) E4
San Agustin (cape) F7
San Bernardino (str.) E4
San Miguel (bay) D3
San Pedro (bay) E5
Santo Tomas (mt.) C2
Semirara (isls.) C5
Siargao (isl.) F6
Sibay (isl.) C5
Sibuguey (bay) D7
Sibutu Group (isls.) B8
Sibuyan (isl.) D4
Sibuyan (sea) D4
Sierra Madre (mt.) D2
Simunul (isl.) B8
Siquijor (isl.) D6
South China (sea) B3
Subic (bay) C3
Sulu (arch.) B8
Sulu (sea) B6
Suluan (isl.) F5
Surigao (str.) E6
Taal (lake) C4
Tablas (isl.) D4
Tablas (str.) C4
Tagapula (isl.) E4
Tagolo (pt.) D6
Tanon (str.) D6
Tapul (isl.) C8
Tapul Group (isls.) C8
Tara (isl.) C4
Tawi-Tawi (isl.) B8
Tayabas (bay) C4
Ticao (isl.) D4
Tinaca (pt.) E8
Tongquil (isl.) D8
Tumindao (isl.) B8
Turtle (isls.) B7
Verde Island (passg.) C4
Victoria (peaks) B6
Visayan (sea) D5

Topography

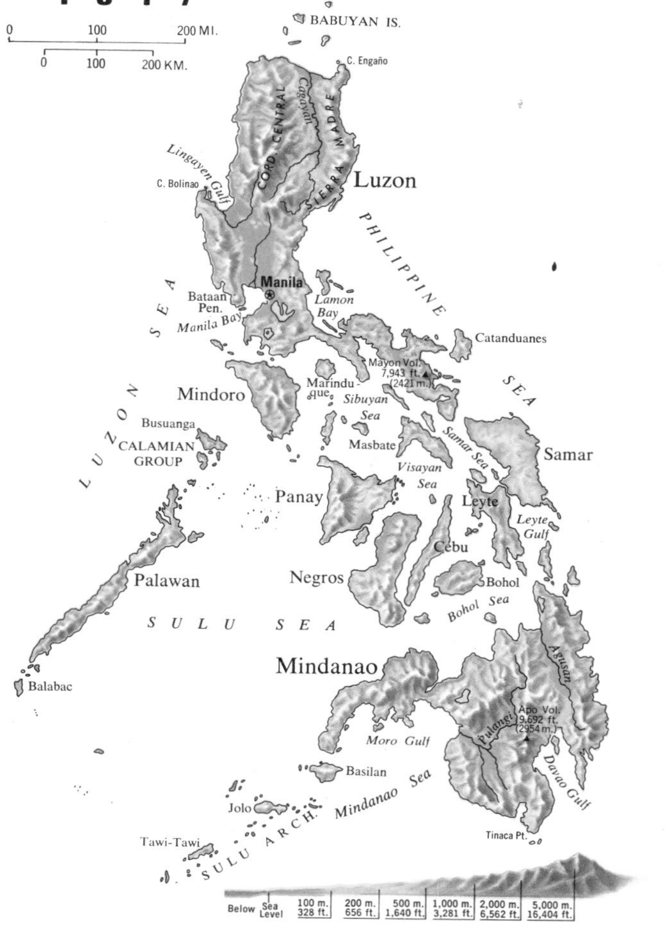

0 100 200 MI.
0 100 200 KM.
BABUYAN IS.
C. Engaño
Luzon
C. Bolinao
Lingayen Gulf
PHILIPPINE SEA
Manila
Bataan Pen.
Manila Bay
Lamon Bay
Catanduanes
Mayon Vol. 7,943 ft. (2421 m.)
Mindoro
Marinduque
Sibuyan Sea
Busuanga
CALAMIAN GROUP
LUZON SEA
Samar Sea
Masbate
Samar
Visayan Sea
Panay
Leyte
Leyte Gulf
Cebu
Palawan
Negros
Bohol
Bohol Sea
SULU SEA
Mindanao
Agusan
Balabac
Apo Vol. 9,692 ft. (2954 m.)
Pulangi
Moro Gulf
Davao Gulf
Basilan
Jolo
Mindanao Sea
SULU ARCH.
Tawi-Tawi
Tinaca Pt.
Below Sea Level
100 m. 328 ft.
200 m. 656 ft.
500 m. 1,640 ft.
1,000 m. 3,281 ft.
2,000 m. 6,562 ft.
5,000 m. 16,404 ft.

Agriculture, Industry and Resources

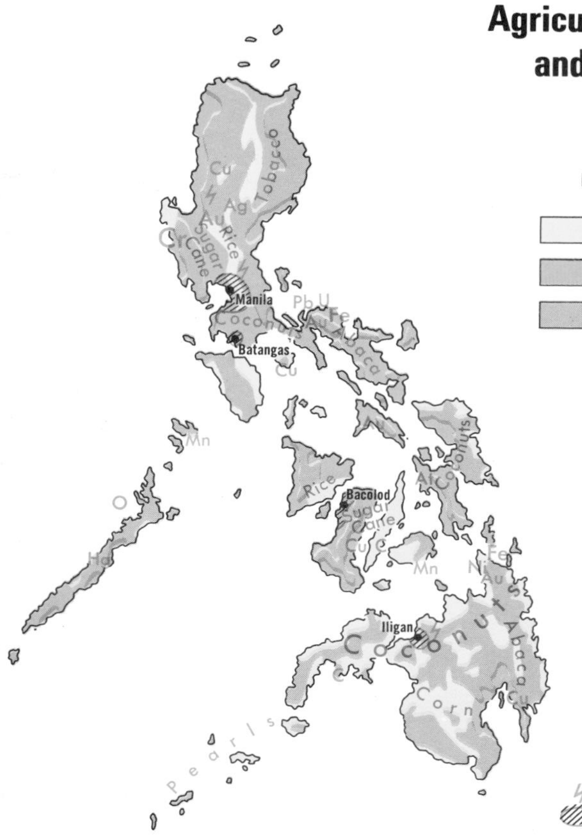

Tobacco
Rice
Sugar Cane
Manila
Batangas
Coconuts
Bacolod
Iligan
Abaca
Corn
Pearls

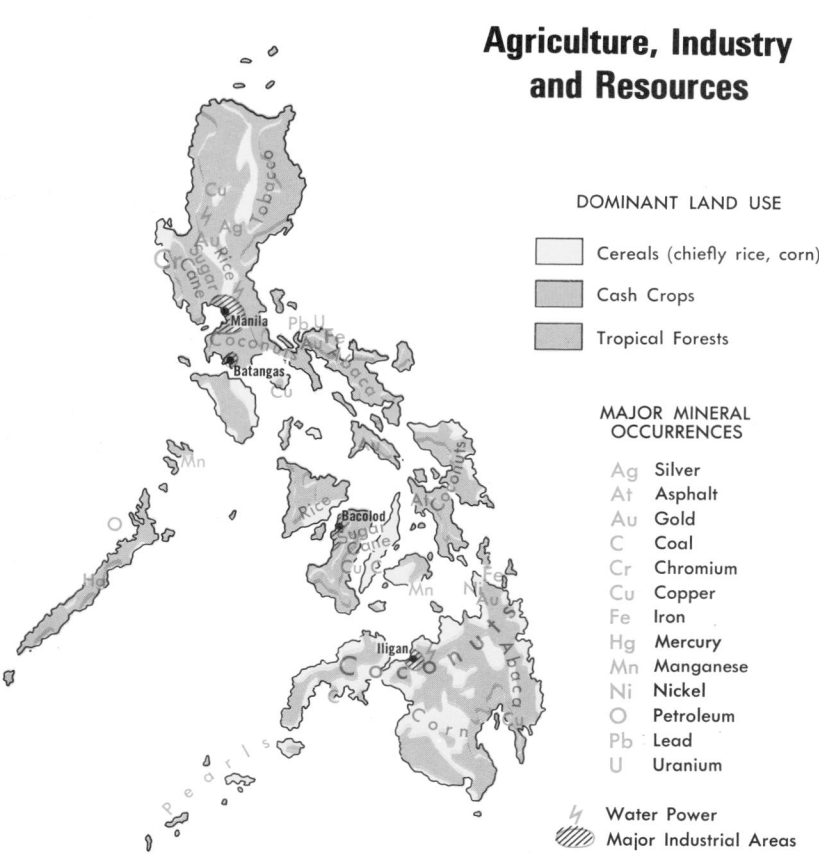

DOMINANT LAND USE

- Cereals (chiefly rice, corn)
- Cash Crops
- Tropical Forests

MAJOR MINERAL OCCURRENCES

- Ag Silver
- At Asphalt
- Au Gold
- C Coal
- Cr Chromium
- Cu Copper
- Fe Iron
- Hg Mercury
- Mn Manganese
- Ni Nickel
- O Petroleum
- Pb Lead
- U Uranium

- Water Power
- Major Industrial Areas

BRUNEI

CITIES and TOWNS

Bandar Seri Begawan 63,868 E4
Seria 23,511 . . . . E5

INDONESIA

CITIES and TOWNS

Adaut J7
Ayats K7
Ambon (Amboina) 208,898 . . . H6
Amuntai F6
Amurang G5
Atambua G7
Aubá H7
Baa G8
Bagansiapiapi C5
Balikpapan 280,675 . . . . F6
Banda Aceh 72,090 . . . . A4
Bandanaira H6
Bandung 1,462,637 . . . . H2
Banggai G6
Banjarmasin 381,286 . . . . E6
Banyumas J2
Batang J2
Batavia (Jakarta) (cap.) 6,503,449 . . . . H1
Baukau H7
Bekasi H2
Belawan B5
Bengkulu 64,783 . . . . C6
Beo H5
Biak K6
Binjai 76,464 . . . . B5
Bintuhan C6
Blitar 78,503 . . . . K2
Bogor 247,409 . . . . H2
Bojonegoro J2
Bukittinggi 70,771 . . . . B6
Bula J6
Bulukumba G7
Buntok F6
Cianjur H2
Cimahi H2
Cirebon 223,776 . . . . H2
Demta L6
Denpasar E7
Dili H7
Djambi (Jambi) 230,373 . . . . . C6
Djokjakarta (Yogyakarta) 398,727 . . . . J2
Dobo J7
Donggala F6
Enaratoli K6
Ende G7
Fakfak J6
Garut H2
Gorontalo 97,628 . . . . G5
Hollandia (Jayapura) K6
Indramayu H2
Jailolo H5
Jakarta (cap.) 6,503,449 . . . . . H1
Jambi 230,373 . . . . C6
Jayapura (Hollandia) K6
Jogjakarta (Yogyakarta) 398,727 . . . . J2
Jombang K2
Kaimana J6
Kampung Baru (Tolitoli) G5
Kediri 221,820 . . . . K2
Kendari G6
Kepi K7
Ketapang E6
Kokonau K6
Kolonodale G6
Kotabaharu E6
Kotabaru F6
Kotawaringin E6
Kragen K2
Kupang G8
Kutaraja (Banda Aceh) 72,090 . . . . A4
Labuha H6
Labuhan G2
Laiwui H6
Larantuka G7
Lekitobi G6
Longiram F5
Madiun 150,562 . . . . K2
Magelang 123,484 . . . . J2
Majalengka H2
Makassar (Ujung Pandang) 709,038 . . . . F7
Malang 511,780 . . . . K2
Malili G6
Manado 217,159 . . . . G5
Manokwari J6
Maumere G7
Medan 1,378,955 . . . . B5
Menggala D6
Merauke K7
Mindiptana L7
Mojokerto 68,849 . . . . K2
Muarasiberut B6
Nangatayap E6
Pacitan J2
Padang 480,922 . . . . B6
Padangpanjang 34,517 . . . . . . B6
Padangsidempuan B5
Pakanbaru 186,262 . . . . C5
Palangkaraya 60,447 . . . . E6
Palembang 787,187 . . . . D6
Pangkalanbuun E6
Pangkalpinang 90,096 . . . . D6
Parepare 86,450 . . . . F6
Pasangkayu F6
Pasuruan 95,864 . . . . K2
Payakumbuh 78,836 . . . . C6
Pekalongan 132,558 . . . . J2
Pemalang J2
Pematangsiantar 150,376 . . . . B5
Pinrang F6
Plaju D6
Pontianak 304,778 . . . . D6
Probolinggo 100,296 . . . . K2
Purbolinggo J2
Raha G6
Rantauprapat C5
Rembang K2
Sabang, Celebes F5
Sabang, Weh 23,821 . . . . B4
Salatiga 85,849 . . . . J2
Samarinda 264,718 . . . . F6
Sampit E6
Sarmi K6
Sawahlunto 13,561 . . . . C6
Seba G8
Semarang 1,026,671 . . . . J2
Semitau E5
Serui K6
Sibolga 59,897 . . . . B5
Sigli B4
Sinabang B5
Singaraja F7
Solo (Surakarta) 469,888 . . . . J2
Solok 31,724 . . . . C6
Sorong J6
Sragen J2
Subang H2
Sukabumi 109,994 . . . . H2
Sumbawa Besar F7
Sumedang H2
Surabaya 2,027,913 . . . . K2
Surakarta 469,888 . . . . J2
Tanahmerah K7
Tanjungbalai 41,894 . . . . C5
Tanjungkarang 284,275 . . . . . . D7
Tanjungpinang C5
Tanjungselor F5
Tarakan F5
Tebingtinggi 92,087 . . . . B5
Tegal 131,728 . . . . J2
Telukbayur C6
Tepa H7
Terempa D5
Tjilatjap (Cilicap) H2
Tjirebon (Cirebon) 223,776 . . . H2
Tolitoli G5
Tuban K2
Ujung Pandang 709,038 . . . . . F7
Vikeke H7
Wahai H6
Waigama H6
Wajabula H5
Waren K6
Weda H5
Wonreli H7
Yogyakarta 398,727 . . . . J2

OTHER FEATURES

Anambas (isls.) 29,572 . . . . . D5
Arafura (sea) J8
Aru (isls.) 34,195 . . . . K7
Babar (isl.) H7
Bali (isl.) 2,074,438 . . . . F7
Banda (sea) H7
Banggai (arch.) 169,025 . . . G6
Bangka (isl.) 298,017 . . . . . . D6
Banyak (isls.) 1,980 . . . . B5
Barisan (mts.) C6
Barito (riv.) E6
Batu (isls.) 16,390 . . . . B6
Bawean (isl.) 64,551 . . . . K1
Belitung (Billiton) (isl.) 128,694 . . . . D6
Berau (bay) J6
Biak (isl.) K6
Billiton (isl.) 128,694 . . . . . . D6
Binongko (isl.) 11,549 . . . . . G7
Bone (gulf) G7
Borneo (isl.) E5
Bosch, van den (cape) J6
Bunguran (Great Natuna) (isl.) D5
Buru (isl.) 23,034 . . . . H6
Butung (isl.) 188,173 . . . . . . G6
Celebes (Sulawesi) (isl.) 7,732,383 . . . . G5
Celebes (sea) G5
Cenderawasih (bay) K6
Dampier (str.) J6
Digul (riv.) K7
Doberai (pen.) J6
Enggano (isl.) 1,082 . . . . C7
Ewab (Kai) (isls.) 108,328 . . J7
Flores (isl.) 860,328 . . . . G7
Flores (sea) F7
Frederik Hendrik (Kolepom) (isl.) K7
Geelvink (Cenderawasih) (bay) K6
Great Kai (isl.) 38,748 . . . . . . J7
Halmahera (isl.) 122,521 . . . H5
Irian Jaya (reg.) 923,440 . . . J6
Jambuair (cape) B4
Jamursba (cape) J5
Java (head) C7
Java (isl.) 73,712,411 . . . . . . J2
Java (sea) D6
Jaya, Puncak (mt.) K6
Jayawijaya (range) K6
Jemaja (isl.) 5,628 . . . . D5
Kabaena (isl.) G7
Kai (isls.) 108,328 . . . . J7
Kalao (isl.) G7
Kalaotoa (isl.) G5
Kalimantan (reg.) 4,956,865 . E5
Kangean (isl.) F7
Kapuas (riv.) D6
Karakelong (isl.) H5
Karimata (arch.) 9,398 . . . . . D6
Karimunjawa (isls.) 5,025 . . . J1
Kerinci (mt.) C6
Kisar (isl.) H7
Komodo (isl.) 30,407 . . . . F7
Krakatau (Rakata) (isl.) C7
Laut (isl.) 55,711 . . . . F6
Leuser (mt.) B5
Lingga (arch.) 46,658 . . . . . . D5
Lingga (isl.) 18,027 . . . . D6
Lombok (isl.) 1,581,193 . . . . F7
Madura (isl.) 1,509,774 . . . . K2
Mahakam (riv.) F6
Makassar (str.) F6
Malacca (str.) C5
Mamberamo (riv.) K6
Maoke (mts.) K6
Mapia (isls.) J5
Mentawai (isls.) 30,107 . . . . B6
Misool (isl.) J6
Molucca (sea) H6
Moluccas (isls.) 944,240 . . . H6
Morotai (isl.) 27,333 . . . . H5
Muli (str.) K7
Müller (mts.) E5
Muna (isl.) 156,186 . . . . G7
Musi (riv.) C6
Natuna (isls.) 23,893 . . . . D5
Ngunju (cape) F8
Nias (isl.) 356,093 . . . . B5
Numfoor (isl.) J6
Obi (isls.) 12,437 . . . . H6
Ombai (str.) H7
Pantar (isl.) 28,259 . . . . G7
Perkam (cape) K6
Puting, Borneo (cape) E6
Puting, Sumatra (cape) C7
Raja Ampat Group (isls.) . . . H6
Rakata (isl.) C7
Rantekombola (mt.) F6
Raya (mt.) E6
Riau (arch.) 483,230 . . . . C5
Rokan (riv.) C5
Roti (isl.) 76,270 . . . . G8
Salawati (isl.) J6
Sangihe (isl.) H5
Sangihe (isls.) 183,000 . . . . . G5
Sawu (isls.) 51,002 . . . . G8
Sawu (sea) G7
Schouten (isls.) 110,148 . . . K6
Schwaner (mts.) E6
Sebuku (bay) F5
Selatan (cape) E6
Selayar (isl.) 92,342 . . . . G7
Semeru (mt.) K2
Siau (isl.) 46,801 . . . . H5
Siberut (str.) B6
Simeulue (isl.) 29,147 . . . . . A5
Singkep (isl.) 28,631 . . . . . . D6
Sipura (isl.) 6,051 . . . . B6
Slamet (mt.) J2
Sorikmerapi (mt.) B5
South Natuna (isls.) D5
Sudirman (range) K6
Sula (isls.) 36,922 . . . . H6
Sulawesi (isl.) 7,732,383 . . . G6
Sumatra (isl.) 19,360,400 . . B5
Sumba (isl.) 291,190 . . . . . . F7
Sumba (str.) F7
Sumbawa (isl.) 621,140 . . . . F7
Sunda (str.) C7
Tahulandang (isl.) 21,493 . . H5
Talaud (isls.) 46,395 . . . . . . H5
Taliabu (isl.) 18,303 . . . . . . G6
Tambelan (isls.) 4,032 . . . . . D5
Tanimbar (isls.) 55,405 . . . . . J7
Tariku (riv.) K6
Tidore (isl.) 28,655 . . . . H5
Timor (reg.) 1,435,527 . . . . H7
Timor (sea) H7
Toba (lake) B5
Tolo (gulf) G6
Tomini (gulf) G6
Tukangbesi (isls.) 73,106 . . . G7
Vals (cape) K7
Vogelkop (Doberai) (pen.) . . . J6
Waigeo (isl.) J5
Wakde (isl.) K6
Wangiwangi (isl.) 28,469 . . . G7
We (isl.) B4
Wetar (isl.) H7
Yapen (isl.) 50,888 . . . . K6

MALAYSIA

STATES

North Borneo (Sabah) 1,002,608 . . . . F3
Sarawak 1,294,753 . . . . E5

CITIES and TOWNS

Beaufort 2,709 . . . . F4
Bintulu 4,424 . . . . E5
Kabong E5
Kampong Sibuti E5
Kapit 1,929 . . . . E5
Keningau 2,037 . . . . F4
Kota Kinabalu 40,939 . . . . . . F4
Kuching 63,535 . . . . E5
Kudat 5,089 . . . . F4
Labuan 7,216 . . . . E4
Lahad Datu 5,169 . . . . F5
Lamag F4
Marudi 4,700 . . . . E5
Miri 35,702 . . . . E5
Mukah 1,717 . . . . E5

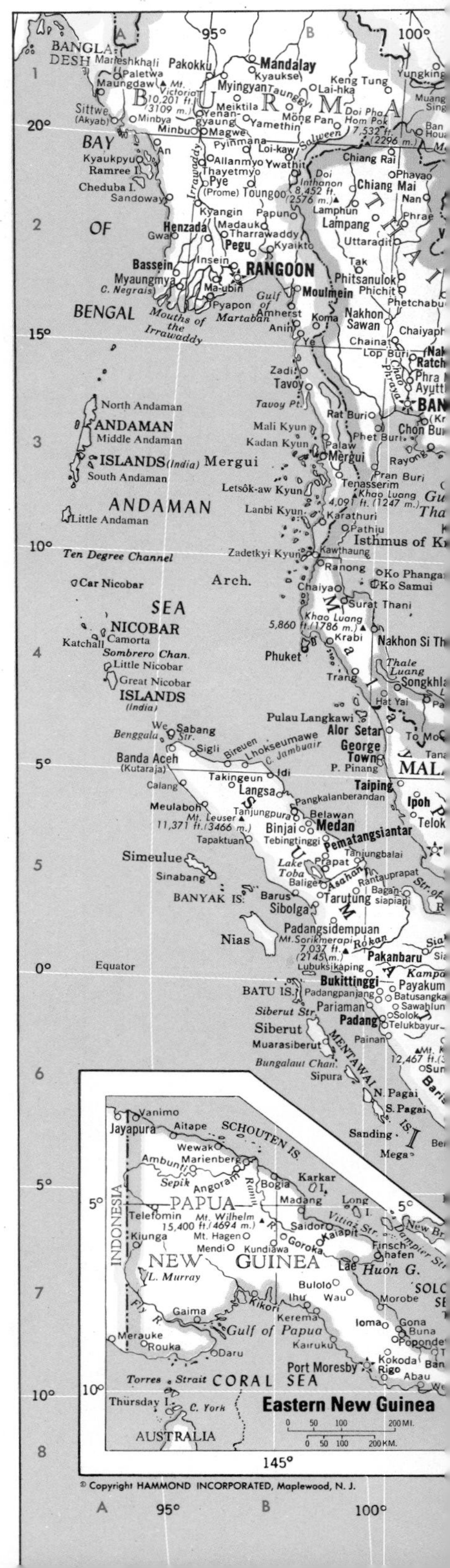
BANGLADESH
Mandalay
BAY OF BENGAL
RANGOON
ANDAMAN ISLANDS (India)
NICOBAR ISLANDS (India)
Medan
Padang
Eastern New Guinea

Topography

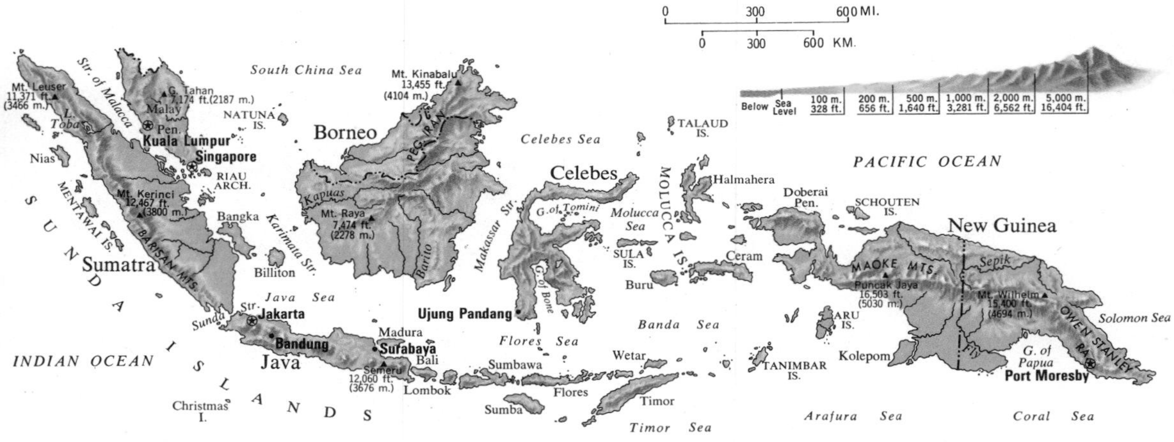
0 300 600 MI.
0 300 600 KM.
Below Sea Level
100 m. 328 ft.
200 m. 656 ft.
500 m. 1,640 ft.
1,000 m. 3,281 ft.
2,000 m. 6,562 ft.
5,000 m. 16,404 ft.
South China Sea
Borneo
Celebes Sea
PACIFIC OCEAN
Celebes
New Guinea
Sumatra
Java Sea
Java
INDIAN OCEAN
SUNDA ISLANDS
Kuala Lumpur
Singapore
Jakarta
Bandung
Surabaya
Ujung Pandang
Banda Sea
Flores Sea
Timor Sea
Arafura Sea
Coral Sea
Solomon Sea
Port Moresby

Agriculture, Industry and Resources

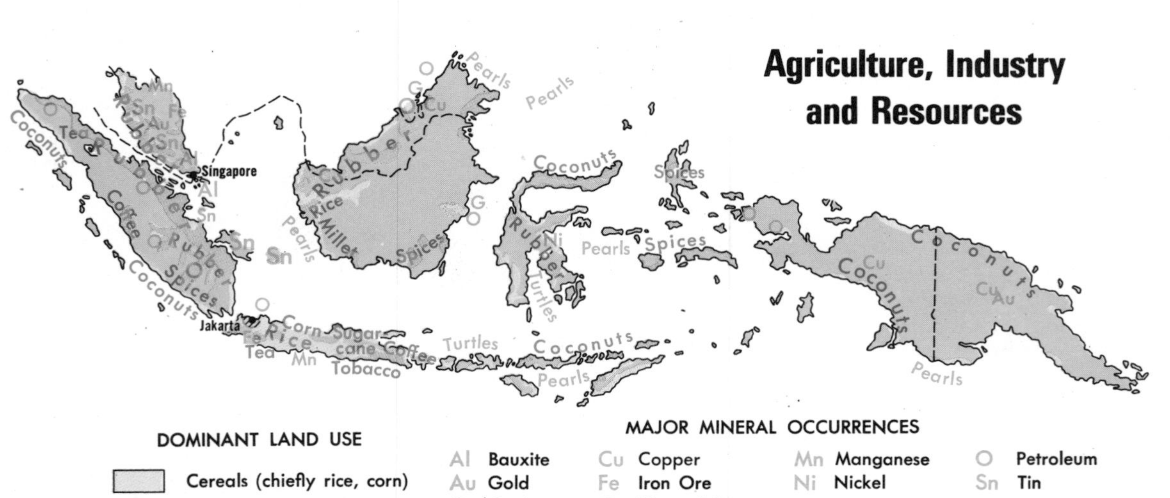
Singapore
Jakarta

DOMINANT LAND USE

- Cereals (chiefly rice, corn)
- Diversified Tropical Crops
- Forests

MAJOR MINERAL OCCURRENCES

Al	Bauxite	Cu	Copper	Mn	Manganese	O	Petroleum
Au	Gold	Fe	Iron Ore	Ni	Nickel	Sn	Tin
C	Coal	G	Natural Gas				

Major Industrial Areas

Papar 1,855 ....F4
Ranau 2,024 ....F4
Sandakan 42,413 ....F4
Sematan D5
Semporna 3,371 ....F5
Serian 2,209 ....E5
Sibu 50,635 ....E5
Simanggang 8,445 ......E5
Suai E5
Tawau 24,247 ....F4
Weston F4

OTHER FEATURES

Balambangan (isl.) F4
Banggi (isl.) F4
Iran (mts.) E5
Kinabalu (mt.) F4
Labuan (isl.) 17,189 ...F4
Labuk (bay) F4
Rajang (riv.) E5
Sirik (cape) E5

PAPUA NEW GUINEA

CITIES and TOWNS

Abau 3,368 ....C7
Aitape 3,368 ....B6
Ambunti 1,035 ....B6
Angoram 1,846 ....B6

Baniara C7
Bogia 755 ....B6
Bulolo 6,730 ....B7
Buna C7
Daru 7,127 ....C7
Finschhaffen 756. .....B7
Gaima B7
Gehua C8
Gona B7
Goroka 18,511 ....B7
Ihu 541 ....C7
Ioma C7
Kaiapit 515 ....B7
Kairuku C7
Kerema 3,389 ....B7
Kikori 763 ....C7
Kiunga 1,407 ....B7
Kokoda C7
Kundiawa 4,299 ....B7
Lae 61,617 ....B7
Madang 21,335 ....B7
Marienberg B6
Mendi 4,130 ....C7
Morobe C7
Mount Hagen 13,441 ....B7
Popondetta 6,429 ......C7
Port Moresby
  (cap.) 123,624 ......B7
Rouka B7
Saidor 500 ....B7
Samarai 864 ....C8

Telefomin B7
Vanimo 3,071 ....B6
Wau 2,349 ....B7
Wedau C7
Wewak 19,890 ....B6

OTHER FEATURES

Dampier (str.) C7
D'Entrecasteaux (isls.) C7
Fly (riv.) A7
Huon (gulf) C7
Karkar (isl.) B6
Kiriwina (isl.) C7
Long (isl.) B7
Louisiade (arch.) D8
Milne (bay) C8
Misima (isl.) C8
New Britain (isl.) 148,773 C7
Ramu (riv.) B7
Rossel (isl.) D8
Schouten (isls.) B6
Sepik (riv.) B6
Solomon (sea) C7
Tagula (isl.) C8
Torres (str.) A7
Trobriand (isls.) C7
Vitiaz (str.) B7
Woodlark (isl.) C7

★See page 74 for other
  Malaysian entries.

INDONESIA

AREA 788,430 sq. mi. (2,042,034 sq. km.)
POPULATION 147,490,298
CAPITAL Jakarta
LARGEST CITY Jakarta
HIGHEST POINT Puncak Jaya 16,503 ft.
  (5,030 m.)
MONETARY UNIT rupiah
MAJOR LANGUAGES Bahasa Indonesia,
  Indonesian and Papuan languages,
  English
MAJOR RELIGIONS Islam, tribal religions,
  Christianity, Hinduism

PAPUA NEW GUINEA

AREA 183,540 sq. mi. (475,369 sq. km.)
POPULATION 3,010,727
CAPITAL Port Moresby
LARGEST CITY Port Moresby
HIGHEST POINT Mt. Wilhelm 15,400 ft.
  (4,694 m.)
MONETARY UNIT kina
MAJOR LANGUAGES pidgin English,
  Hiri Motu, English
MAJOR RELIGIONS Tribal religions,
  Christianity

BRUNEI

AREA 2,226 sq. mi. (5,765 sq. km.)
POPULATION 192,832
CAPITAL Bandar Seri Begawan
LARGEST CITY Bandar Seri Begawan
HIGHEST POINT Pagon 6,070 ft.
  (1,850 m.)
MONETARY UNIT Brunei Dollar
MAJOR LANGUAGES Malay, English,
  Chinese
MAJOR RELIGIONS Islam, Buddhism,
  Christianity, tribal religions

INDONESIA   PAPUA NEW GUINEA   BRUNEI

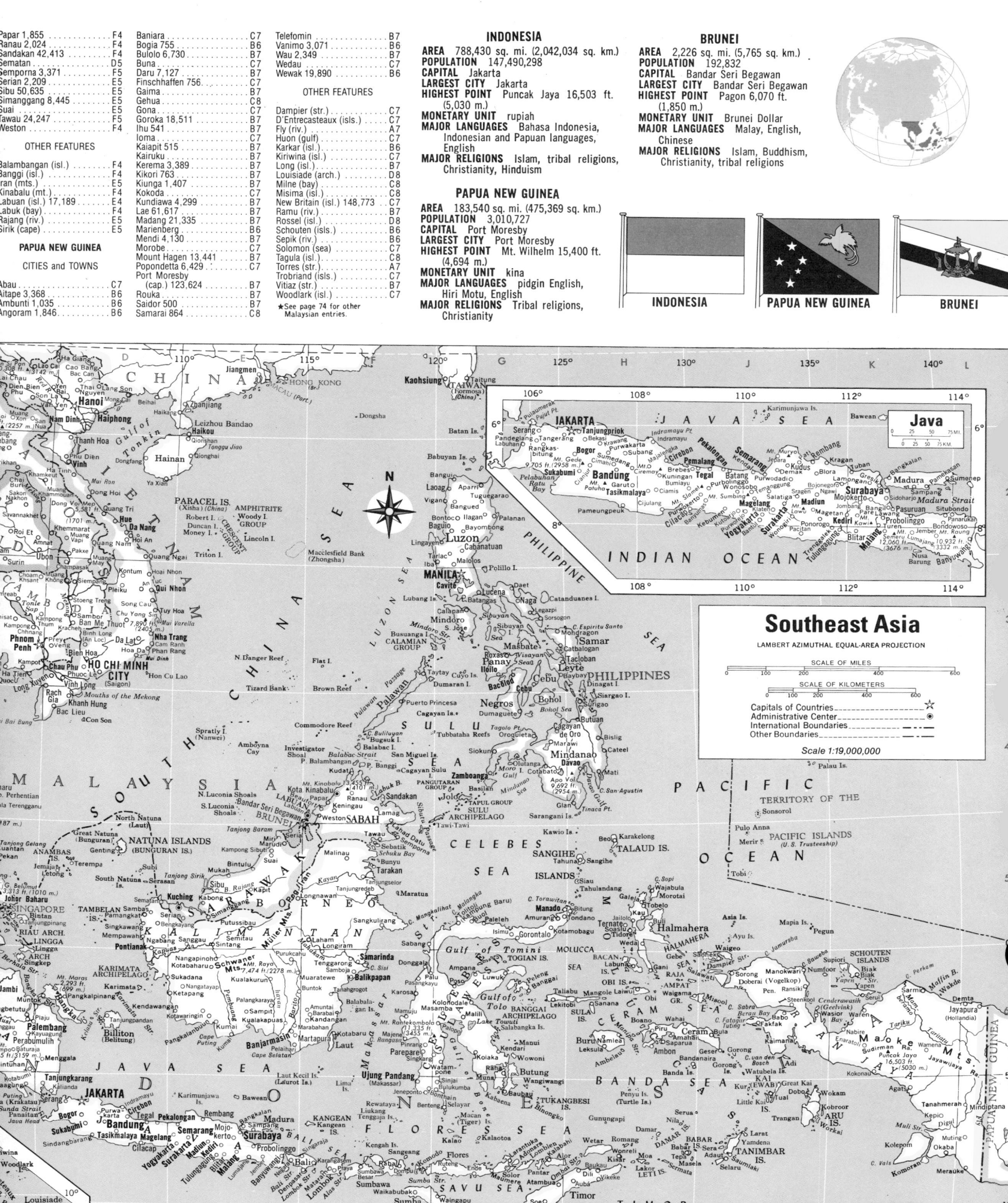

FIJI

AREA 7,055 sq. mi. (18,272 sq. km.)
POPULATION 588,068
CAPITAL Suva
LARGEST CITY Suva
HIGHEST POINT Tomaniivi 4,341 ft.
   (1,323 m.)
MONETARY UNIT Fijian dollar
MAJOR LANGUAGES Fijian, Hindi, English
MAJOR RELIGIONS Protestantism, Hinduism

KIRIBATI

AREA 291 sq. mi. (754 sq. km.)
POPULATION 56,213
CAPITAL Bairiki (Tarawa)
HIGHEST POINT (on Banaba I.) 285 ft. (87 m.)
MONETARY UNIT Australian dollar
MAJOR LANGUAGES I-Kiribati, English
MAJOR RELIGIONS Protestantism, Roman
  Catholicism

NAURU

AREA 7.7 sq. mi. (20 sq. km.)
POPULATION 7,254
CAPITAL Yaren (district)
MONETARY UNIT Australian dollar
MAJOR LANGUAGES Nauruan, English
MAJOR RELIGION Protestantism

SOLOMON ISLANDS

AREA 11,500 sq. mi. (29,785 sq. km.)
POPULATION 221,000
CAPITAL Honiara
HIGHEST POINT Mount Popomanatseu
  7,647 ft. (2,331 m.)
MONETARY UNIT Solomon Islands dollar
MAJOR LANGUAGES English, pidgin English,
  Melanesian dialects
MAJOR RELIGIONS Tribal religions,
  Protestantism, Roman Catholicism

TONGA

AREA 270 sq. mi. (699 sq. km.)
POPULATION 90,128
CAPITAL Nuku'alofa
LARGEST CITY Nuku'alofa
HIGHEST POINT 3,389 ft. (1,033 m.)
MONETARY UNIT pa'anga
MAJOR LANGUAGES Tongan, English
MAJOR RELIGION Protestantism

TUVALU

AREA 9.78 sq. mi. (25.33 sq. km.)
POPULATION 7,349
CAPITAL Fongafale (Funafuti)
HIGHEST POINT 15 ft. (4.6 m.)
MONETARY UNIT Australian dollar
MAJOR LANGUAGES English, Tuvaluan
MAJOR RELIGION Protestantism

Abaiang (atoll) 3,296 . . . . H 5
Abemama (atoll) 2,300 . . . . . . H 5
Adamstown (cap.), Pitcairn Is.
  54 . . . . N 8
Admiralty (isls.) . . . . E 6
Agaña (cap.), Guam 896 . . . . E 4
Agrihan (isl.) . . . . E 4
Ailinglapalap (atoll) 1,385 . . . . H 4
Ailuk (atoll) 413 . . . . H 4
Aitutaki (atoll) 2,348 . . . . K 7
Alofi (cap.), Niue 960 . . . . . . K 7
Alotau 4,310 . . . . E 7
Ambrym (isl.) 6,324 . . . . G 7
American Samoa 32,297 . . . . . J 7
Anaa (atoll) 444 . . . . M 7
Angaur 243 . . . . D 5
Apataki (atoll) . . . . M 7
Apia (cap.), W. Samoa 33,100 . J 7
Arno (atoll) 1,487 . . . . H 5
Arorae (atoll) 1,626 . . . . H 6
Atafu (atoll) 577 . . . . J 6
Atiu (isl.) -1,225 . . . . L 8
Austral (isls.) 5,208 . . . . L 8
Avarua (cap.), Cook Is. . . . . . L 8
Babelthuap (isl.) 10,391 . . . . . D 5
Bairiki (cap.), Kiribati 1,777 . . H 5
Baker (isl.) . . . . J 5
Banaba (isl.) 2,314 . . . . G 6
Banks (isls.) 3,158 . . . . G 7
Belau (Palau) 12,116 . . . . D 5
Belep (isls.) 624 . . . . G 7
Bellona (reefs) . . . . G 8
Beru (atoll) 2,318 . . . . H 6
Bikini (atoll) . . . . G 4
Bismarck (arch.) 218,339 . . . . E 6
Bonin (isls.) 1,879 . . . . E 3
Bora-Bora (isl.) 2,572 . . . . . . L 7
Bougainville (isl.) 71,761 . . . . F 6
Bounty (isls.) . . . . H 10
Bourail 3,149 . . . . G 8
Butaritari (atoll) 2,971 . . . . . . H 5
Canton (isl.) . . . . J 6
Capitol Hill (cap.), No.
  Marianas 592 . . . . E 4
Caroline (isl.) . . . . M 7
Caroline (isls.) . . . . E 5
Chichi (isl.) 1,879 . . . . E 3
Choiseul (isl.) 10,349 . . . . F 6
Christmas (Kiritimati) (isl.) 674 . L 5
Cook (isls.) 17,695 . . . . K 7
Coral (sea) . . . . F 7
Danger (Pukapuka) (atoll)
  797 . . . . K 7
Daru 7,127 . . . . E 6
Disappointment (isls.) 373 . . . M 7
Ducie (isl.) . . . . O 8
Easter (isl.) 1,598 . . . . Q 8
Ebon (atoll) 887 . . . . G 5
Efate (isl.) 18,038 . . . . G 7
Enderbury (isl.) . . . . J 6
Enewetak (Eniwetok) (atoll)
  542 . . . . G 4
Erromanga (isl.) 945 . . . . H 7
Espiritu Santo (isl) 16,220 . . . G 7
Fais (isl.) 207 . . . . E 5
Fakaofo (atoll) 654 . . . . J 6
Fanning (Tabuaeran) (isl.) 340 . L 5
Faraulep (atoll) 132 . . . . E 5
Fatuhiva (isl.) 386 . . . . N 7
Fiji 588,068 . . . . H 8
Flint (isl.) . . . . L 7
Fly (riv.) . . . . E 6
Fongafale (cap.), Tuvalu . . . . . H 6
French Polynesia 137,382 . . . . L 8
Funafuti (atoll) 2,120 . . . . H 6
Futuna (Hoorn) (isls.) 3,173 . . J 7
Gambier (isls.) 556 . . . . N 8
Gardner (isl.) . . . . J 6
Gilbert (isls.) 47,711 . . . . H 5
Greenwich (Kapingamarangi)
  (atoll) 508 . . . . F 5
Guadalcanal (isl.) 46,619 . . . . F 7
Guam (isl.) 105,979 . . . . E 4
Hall (isls.) 647 . . . . F 5
Hawaiian (isls.) 964,691 . . . . J 3
Henderson (isl.) . . . . O 8
Hivaoa (isl.) 1,159 . . . . N 6
Honiara (cap.), Solomon Is.
  14,942 . . . . F 6
Hoorn (isls.) 3,173 . . . . J 7
Howland (isl.) . . . . J 5
Huahine (isl.) 3,140 . . . . L 7
Hull (isl.) . . . . J 6
Huon (gulf) . . . . E 6
Ifalik (atoll) 389 . . . . E 5
Iwo (isl.) . . . . E 3
Jaluit (atoll) 1,450 . . . . G 5
Jarvis (isl.) . . . . K 6
Johnston (atoll) 327 . . . . K 4
Kadavu (Kandavu) (isl.) 8,699 . H 7
Kapingamarangi (atoll) 508 . . . F 5
Kavieng 4,633 . . . . E 6
Kermadec (isls.) 5 . . . . J 9
Kieta 3,491 . . . . F 6
Kimbe 4,662 . . . . F 6
Kingman (reef) . . . . K 5
Kiribati 57,500 . . . . J 6
Kirimati (isl.) 674 . . . . L 5
Kolonia (cap.), Micronesia
  5,549 . . . . F 5
Koror (cap.), Belau 6,222 . . . . D 5
Kosrae (isl.) 5,491 . . . . G 5
Kwajalein (atoll) 6,624 . . . . . . G 5
Lae 61,617 . . . . E 6
Lau Group (isls.) 14,452 . . . . . J 7
Lavongai (isl.) . . . . F 6
Lifu (isl.) 7,585 . . . . G 8
Line (isls.) . . . . K 5
Little Makin (atoll) 1,445 . . . . . H 5
Lord Howe (Ontong Java) (isl.)
  1,082 . . . . G 6
Lord Howe (isl.) 287 . . . . G 9
Lorengau 3,986 . . . . E 6
Louisiade (arch.) . . . . F 7
Loyalty (isls.) 14,518 . . . . G 8
Luganville 4,935 . . . . G 7
Madang 21,335 . . . . E 6

Majuro (atoll) (cap.), Marshall
  Is. 8,583 . . . . H 5
Makin (Butaritari) (atoll) 2,971 . H 5
Malaita (isl.) 50,912 . . . . G 6
Malden (isl.) . . . . L 6
Malekula (isl.) 15,931 . . . . . . G 7
Maloelap (atoll) 763 . . . . H 5
Mangaia (isl.) 1,364 . . . . L 8
Mangareva (isl.) 556 . . . . N 8
Manihiki (atoll) 405 . . . . K 7
Manua (isls.) 1,459 . . . . K 7
Manus (isl.) 25,844 . . . . E 6
Marcus (isl.) . . . . F 3
Maré (isl.) 4,156 . . . . G 8
Marianas, Northern 16,780 . . . E 4
Mariana Trench . . . . E 4
Marquesas (isls.) 5,419 . . . . . N 6
Marshall Islands 30,873 . . . . . G 4
Marutea (atoll) . . . . N 8
Mata Utu (cap.), Wallis and
  Futuna 558 . . . . J 7
Mauke (isl.) 684 . . . . L 8
Melanesia (reg.) . . . . E 5
Micronesia (reg.) . . . . E 4
Micronesia, Federated States
  of 73,160 . . . . F 5
Midway (isls.) 453 . . . . J 3
Mili (atoll) 763 . . . . H 5
Moen (isl.) 10,351 . . . . F 5
Moorea (isl.) 5,788 . . . . L 7

Mururoa (isl.) . . . . M 8
Nadi 6,938 . . . . H 7
Namonuito (atoll) 783 . . . . . . E 5
Namorik (atoll) 617 . . . . G 5
Nanumea (atoll) 844 . . . . H 6
Nauru 7,254 . . . . G 6
Ndeni (isl.) 4,854 . . . . G 7
New Britain (isl.) 148,773 . . . . F 6
New Caledonia 133,233 . . . . . G 8
New Caledonia (isl.) 118,715 . . G 8
New Georgia (isl.) 16,472 . . . . F 6
New Guinea (isl.) . . . . D 6
New Ireland (isl.) 65,657 . . . . F 6
Ngatik (atoll) 560 . . . . F 5
Ngulu (atoll) 21 . . . . D 5
Niuatoputapu (isl.) 1,650 . . . . J 7
Niue (isl.) 3,578 . . . . K 7
Niutao (atoll) 866 . . . . H 6
Nomoi (isls.) 1,879 . . . . F 5
Nonouti (atoll) 2,223 . . . . H 6
Norfolk Island (terr.) 2,175 . . . G 8
Northern Marianas 16,780 . . . E 4
Nouméa (cap.), New Caled.
  56,078 . . . . G 8
Nouméa *74,335 . . . . G 8
Nui (atoll) 603 . . . . H 6
Nuku'alofa (cap.), Tonga
  18,356 . . . . J 8
Nukuhiva (isl.) 1,484 . . . . M 6
Ocean (Banaba) (isl.) 2,314 . . G 6

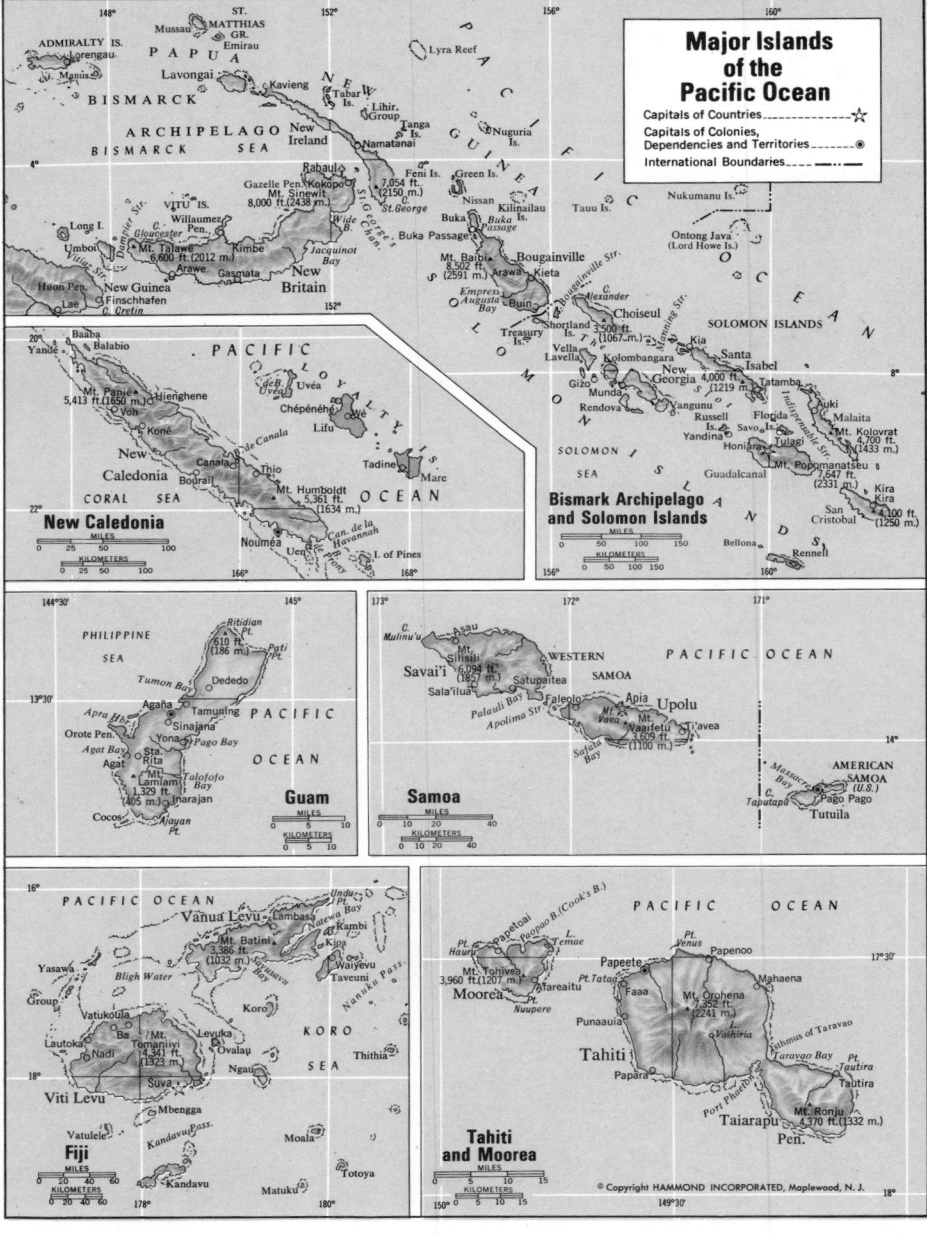

Major Islands of the Pacific Ocean
Capitals of Countries .... ☆
Capitals of Colonies,
Dependencies and Territories ...... ⊛
International Boundaries ......_____

New Caledonia

Bismark Archipelago and Solomon Islands

Guam

Samoa

Fiji

Tahiti and Moorea

© Copyright HAMMOND INCORPORATED, Maplewood, N.J.

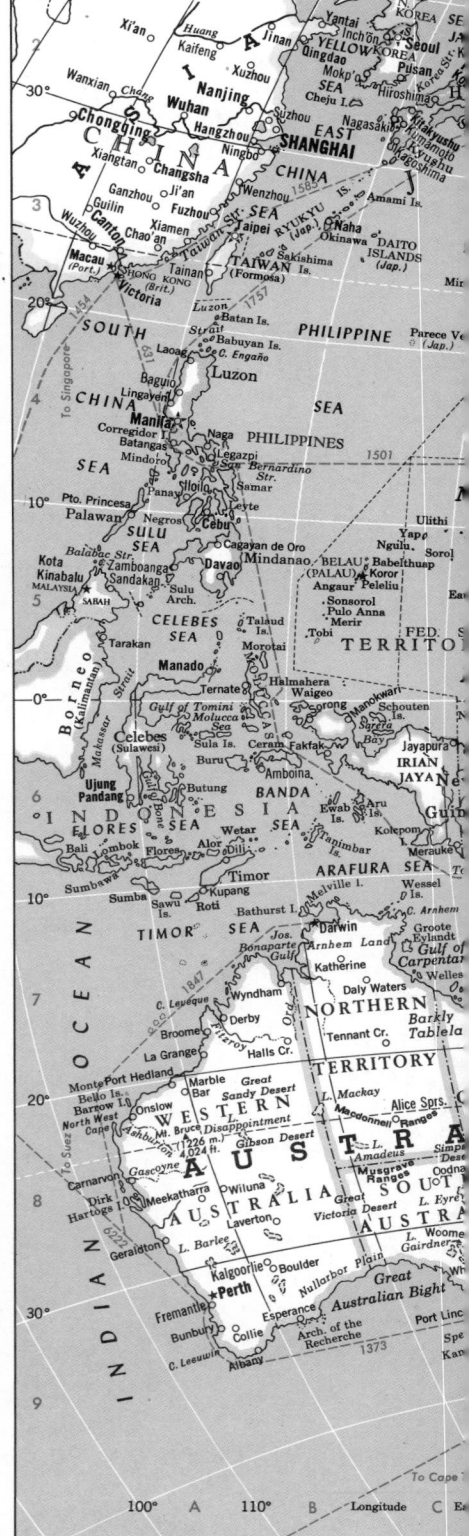

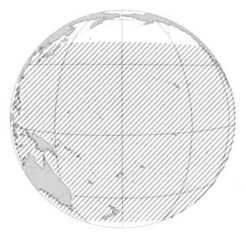

Oeno (isl.) O8
Onotoa (atoll) 1,997 .... H6
Ontong Java (isl.) 1,082....... G6
Pacific Islands, Terr. of the 132,929.................. F5
Pagan (isl.).................. E4
Pago Pago (cap.), Amer. Samoa 3,075............. J7
Palau (Belau) 12,116 .... D5
Palmyra (atoll) K5
Papeete (cap.), Fr. Poly. 22,967.................. M7
Papeete •51,987 .... M7
Papua (gulf).................. E6
Papua New Guinea 3,010,727 E6
Peleliu (isl.) 609............. D5
Penrhyn (Tongareva) (atoll) 608 .... L6
Phoenix (isls.) J6
Pines (isl.) 1,095............. G8
Pitcairn (isl.) 54............. O8
Pohnpei (isl.) 19,935 .... F5
Polynesia (reg.).............. K7
Popondetta 6,429 .... E6
Port Moresby (cap.), Papua N.G. 123,624.............. E6
Pukapuka (atoll) 797 .... K7
Pulap (atoll) 427 .... E5
Puluwat (atoll) 441 .... E5
Rabaul 14,954............... F6
Raiatea (isl.) 2,517.......... L 7
Raivavae (isl.) 1,023......... M 8
Rakahanga (atoll) 269 .... K 7
Ralik Chain (isls.) .... G 5
Rangiroa (atoll) .... M 7
Rapa (isl.) 398............... M 8
Rarotonga (isl.) 9,477 .... K 8
Ratak Chain (isls.) .... G 5
Reao (atoll) 424.............. N 7
Rennell (isl.) 1,132 .... F 7
Rikitea...................... N 8
Rimatara (isl.) 813 .... L 8
Rongelap (atoll) 235.......... G 4
Rota (isl.) 1,261............. E 4
Rotuma (isl.) 2,805........... H 7
Rurutu (isl.) 1,555........... L 8
Saipan (isl.) 14,549 .... E 4
Sala y Gómez (isl.)........... P 8
Samarai 869 .... E 7
Samoa (isls.) .... J 7
San Cristobal (isl.) 11,212.... G 7
Santa Cruz (isls.) 5,421...... G 6
Santa Isabel (isl.) 10,420..... G 6
Savai'i (isl.) 43,150 .... J 7
Senyavin (isls.) 20,035....... F 5
Society (isls.) 117,703 .... L 7
Solomon (isls.) .... F 6
Solomon (sea)............... F 6
Solomon Islands 221,000 .... G 6
Starbuck (isl.) .... L 6
Suva (cap.), Fiji 63,628........ H 7
Suva *117,827............... H 7
Swains (isl.) 27 .... K 7
Sydney (isl.)................ K 6
Tabiteuea (atoll) 3,942........ H 6
Tabuaeron (isl.) 340 .... L 5
Tahaa (isl.) 3,513 .... L 7
Tahiti (isl.) 95,604 .... L 7
Takaroa (atoll) 337 .... M 7
Tanna (isl.) 15,715 .... H 7
Tarawa (atoll) 17,129 .... H 5
Tasman (sea)................ G 9
Teraina (isl.) 458 .... L 5
Tinian (isl.) 866 .... E 4
Tokelau (isls.) 1,575 .... J 6
Tonga 90,128................ J 8
Tongareva (atoll) 608 .... L 6
Tongatapu (isls.) 57,130 ..... J 8
Torres (isls.) 325 .... G 7
Torres (strait)............... E 7
Trobriand (isls.) .... F 6
Truk (isls.) 37,488........... F 5
Tuamotu (arch.) 9,052 .... M 7
Tubuai (Austral) (isls.) 5,208.. M 8
Tubuai (isl.) 1,419........... M 8
Tutuila (isl.) 30,538.......... J 7
Tuvalu 7,349 .... H 6
Uapou (isl.) 1,563 .... M 6
Ujelang (atoll) 309 .... F 5
Ulithi (atoll) 710............. D 4

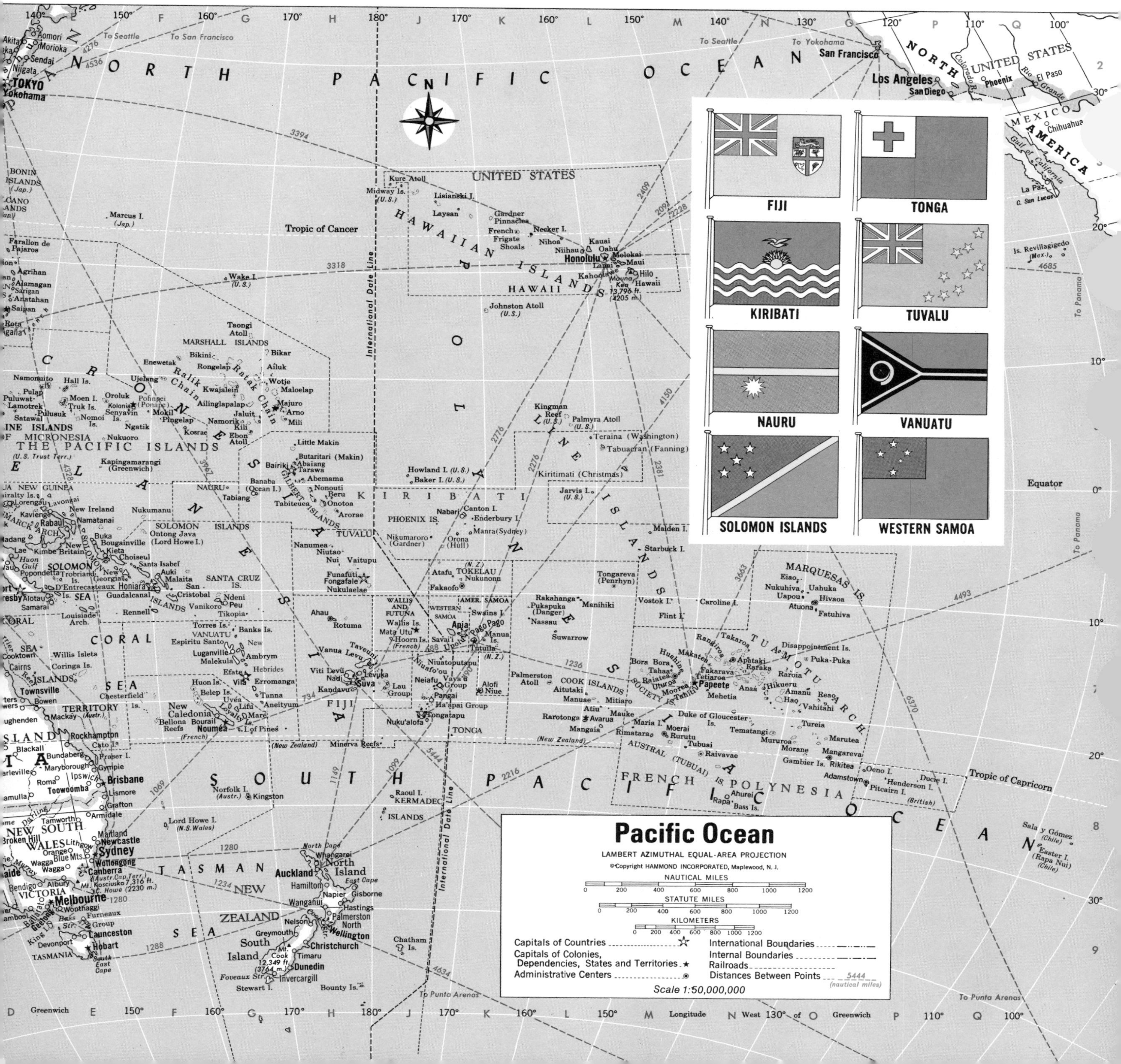

VANUATU

AREA 5,700 sq. mi. (14,763 sq. km.)
POPULATION 112,596
CAPITAL Vila
HIGHEST POINT Mt. Tabwemasana 6,165 ft. (1,879 m.)
MONETARY UNIT vatu
MAJOR LANGUAGES Bislama, English, French
MAJOR RELIGIONS Christian, animist

WESTERN SAMOA

AREA 1,133 sq. mi. (2,934 sq. km.)
POPULATION 158,130
CAPITAL Apia
LARGEST CITY Apia
HIGHEST POINT Mt. Silisili 6,094 ft. (1,857 m.)
MONETARY UNIT tala
MAJOR LANGUAGES Samoan, English
MAJOR RELIGIONS Protestantism, Roman Catholicism

Upolu (isl.) 114,620 .... J 7
Uturoa 2,517 .... L 7
Uvéa (isl.) 2,777 .... G 7
Vaitupu (atoll) 1,273......... H 6
Vanikoro (isl.) 267........... G 7
Vanimo 3,071................ E 6
Vanua Levu (isl.) 103,122 .... H 7
Vanuatu 112,596............. G 7
Vila (cap.), Vanuatu 4,729 .... G 7
Vila *14,797................. G 7
Viti Levu (isl.) 445,422....... H 7
Volcano (isls.) .... E 3
Vostok (isl.) .... L 7
Wake (isl.) 302.............. G 4
Wallis (isls.) 6,019........... J 7
Wallis and Futuna 9,192 ...... J 7
Washington (Teraina) (isl.) 458 .... L 5
Wau 2,349 .... E 6
Western Samoa 158,130..... J 7
Wewak 19,890............... E 6
Woleai (isl.) 638 .... E 5
Wotje (atoll) 535 .... H 5
Yap (isl.) 6,670 .... D 5

*City and suburbs.
•Population of urban area.

NORTH PACIFIC OCEAN
SOUTH PACIFIC OCEAN
FIJI
TONGA
KIRIBATI
TUVALU
NAURU
VANUATU
SOLOMON ISLANDS
WESTERN SAMOA
Pacific Ocean
LAMBERT AZIMUTHAL EQUAL-AREA PROJECTION
©Copyright HAMMOND INCORPORATED, Maplewood, N. J.
NAUTICAL MILES
STATUTE MILES
KILOMETERS
Capitals of Countries
Capitals of Colonies, Dependencies, States and Territories
Administrative Centers
International Boundaries
Internal Boundaries
Railroads
Distances Between Points 5444 (nautical miles)
Scale 1:50,000,000
Tropic of Cancer
Tropic of Capricorn
Equator
International Date Line
HAWAIIAN ISLANDS
MICRONESIA
POLYNESIA
MELANESIA
MARSHALL ISLANDS
TRUST TERRITORY OF THE PACIFIC ISLANDS
FRENCH POLYNESIA
TUAMOTU ARCH.
MARQUESAS IS.
COOK ISLANDS
SOCIETY IS.
LINE ISLANDS
PHOENIX IS.
GILBERT ISLANDS
CORAL SEA
TASMAN SEA
NEW ZEALAND
North Island
South Island
NEW SOUTH WALES
VICTORIA
TASMANIA
UNITED STATES
MEXICO
NORTH AMERICA
Longitude West of Greenwich

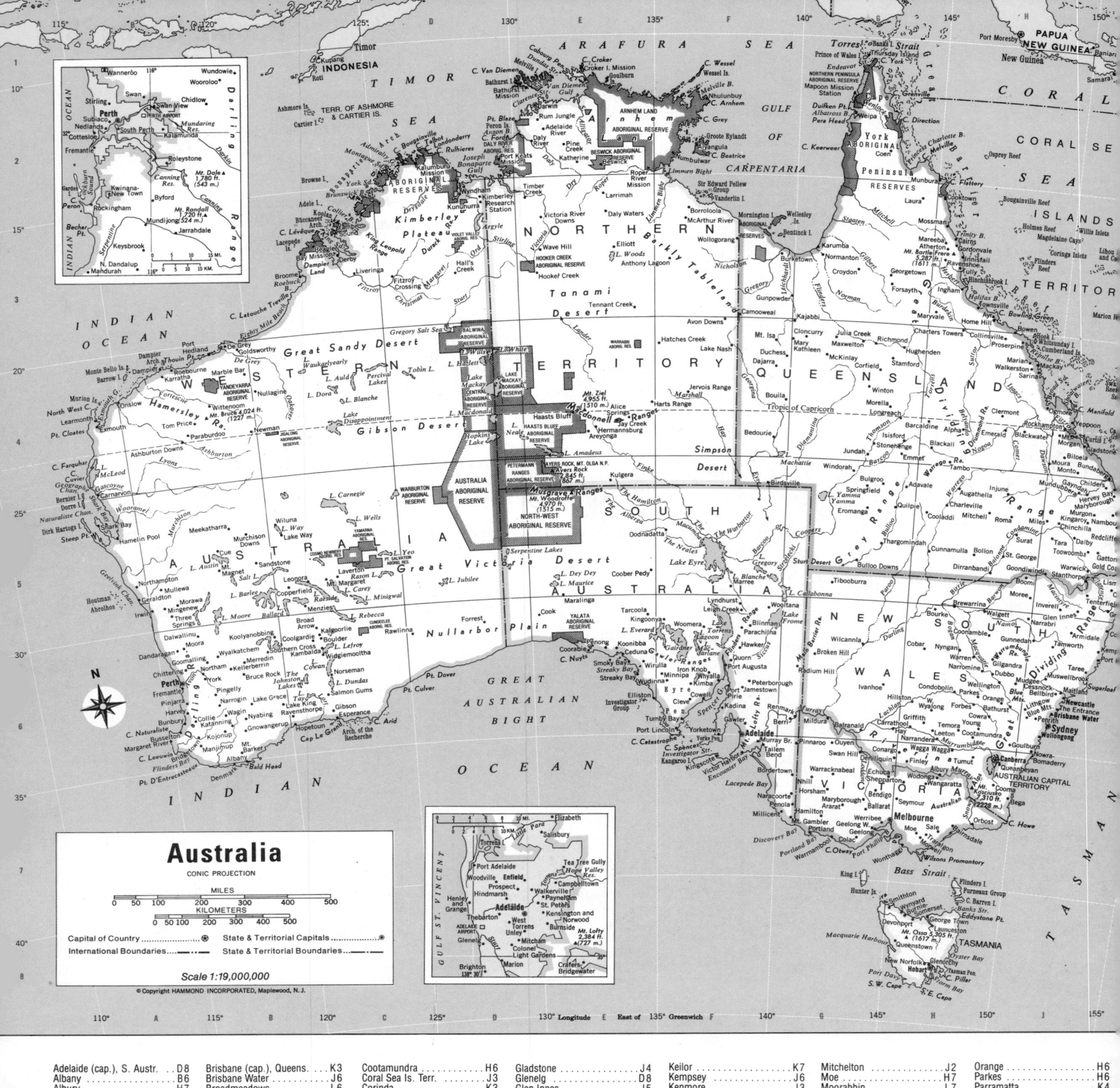

Australia

CONIC PROJECTION

MILES
0 50 100 200 300 400 500

KILOMETERS
0 50 100 200 300 400 500

Capital of Country .... ⊛
State & Territorial Capitals .... ⊛
International Boundaries ....
State & Territorial Boundaries ....

Scale 1:19,000,000

© Copyright HAMMOND INCORPORATED, Maplewood, N.J.

Adelaide (cap.), S. Austr. . .D8	Brisbane (cap.), Queens. . .K3	Cootamundra H6	Gladstone J4	Keilor K7	Mitchelton J2	Orange H6
Albany B6	Brisbane Water J6	Coral Sea Is. Terr. . . .J3	Glenelg D8	Kempsey J6	Moe H7	Parkes H6
Albury H7	Broadmeadows L6	Corinda K3	Glen Innes J5	Kenmore J3	Moorabbin L7	Parramatta K4
Alice Springs E4	Brunswick K7	Cowra H6	Glenorchy H8	Kensington and Norwood .E8	Moorooka K3	Payneham E8
Altona K7	Bunbury B6	Crafters-Bridgewater . . .E8	Gold Coast J5	Kew L7	Mordialloc L7	Penrith J5
Ararat G7	Bundaberg J4	Cranbourne M8	Goulburn J6	Knox M7	Moree H5	Perth (cap.), W. Austr. . .B2
Armidale J6	Burnie H8	Croydon M7	Graceville K3	Kogarah K4	Morningside K3	Port Adelaide D7
Ashfield K4	Burnside E8	Dalby J5	Grafton J5	Ku-ring-gai K4	Morwell H7	Port Augusta F6
Ashgrove K2	Burwood K4	Dandenong L7	Greenslopes K3	Kwinana-New Town B2	Mosman L4	Port Hedland B3
Ashmore and Cartier Is.,	Busselton A6	Darwin (cap.), North. Terr. .E2	Griffith H6	Lane Cove L4	Mount Gambier G7	Portland G7
Terr. of C2	Cairns H3	Deniliquin G7	Gunnedah H6	Launceston H8	Mount Isa G4	Port Lincoln E6
Aspley K2	Caloundra J5	Devonport H8	Gympie J5	Leeton H6	Mudgee J6	Port Macquarie J6
Auburn K4	Camberwell L7	Doncaster and Templestowe .L7	Hamilton G7	Leichhardt L4	Murray Bridge F7	Port Melbourne K7
Australian Capital Territory .H7	Campbelltown E8	Drummoyne K4	Hawthorn L7	Lismore J5	Murwillumbah J5	Port Pirie F6
Ayr H3	Camp Hill K3	Dubbo H6	Heidelberg L7	Lithgow J6	Muswellbrook J6	Prahran L7
Bairnsdale G7	Canberra (cap.), Australia .H7	Echuca G7	Hervey Bay J5	Liverpool K4	Narrabri H6	Preston L7
Ballarat G7	Canterbury K4	Elizabeth E7	Hindmarsh D8	Mackay H4	Narrandera H6	Prospect D8
Bankstown K4	Carina K3	Eltham L6	Hobart (cap.), Tasmania .H8	Maitland J6	Nedlands B2	Queanbeyan H7
Banyo K2	Casino J5	Enfield D7	Holyroyd K4	Malvern L7	Newcastle J6	Queensland G4
Bathurst H6	Caulfield L7	Enoggera K2	Hornsby K3	Mandurah B6	Newman C4	Randwick L4
Baulkham Hills J5	Cessnock-Bellbird J6	Esperance C6	Horsham G7	Manly L4	New Norfolk H8	Red Cliffe J5
Beenleigh J5	Charters Towers H4	Essendon K7	Hunters Hill K4	Mareeba G3	New South Wales H5	Richmond M7
Bendigo G7	Chelsea L8	Fairfield K4	Hurstville K4	Marion D8	Norfolk Island 2,180 . . . .L5	Ringwood M7
Blacktown J6	Chermside K2	Fitzroy L7	Inala K3	Maroochydore-Mooloolaba .J5	Northam B2	Rockdale K4
Blue Mountains J6	Coburg K7	Footscray K7	Indooroopilly K3	Marrickville L4	Northcote L7	Rockhampton H4
Botany L4	Coffs Harbour J6	Forbes H6	Innisfail H3	Maryborough J5	Northern Territory E4	Rockingham B2
Boulder-Kalgoorlie C6	Colac G7	Fremantle B6	Inverell J5	Maryborough G7	North Sydney L4	Roma H5
Bowen H3	Collingwood L7	Gawler F6	Ipswich J3	Melbourne (cap.),	Nowra-Bomaderry J6	Ryde K4
Box Hill L7	Concord K4	Geelong G7	Kalgoorlie C6	Victoria H7	Nunawading L7	Saint Kilda L7
Brighton D8	Cooma H7	Geelong West G7	Kalgoorlie-Boulder C6	Mildura G6	Nundah K2	Saint Lucia K3
Brighton L7	Coorparoo K3	Geraldton A5	Karratha B4	Mitcham D8	Oakleigh L7	Sale H7

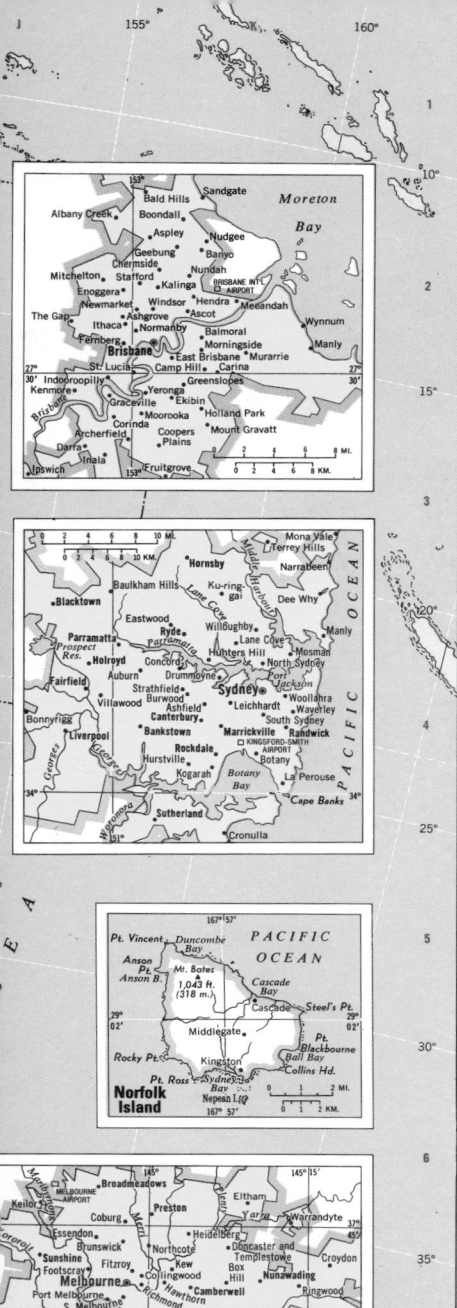

AREA 2,966,136 sq. mi. (7,682,300 sq. km.)
POPULATION 14,576,330
CAPITAL Canberra
LARGEST CITY Sydney
HIGHEST POINT Mt. Kosciusko 7,310 ft.
(2,228 m.)
LOWEST POINT Lake Eyre -39 ft. (-12 m.)
MONETARY UNIT Australian dollar
MAJOR LANGUAGE English
MAJOR RELIGIONS Protestantism,
Roman Catholicism

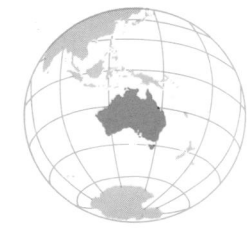

Population Distribution

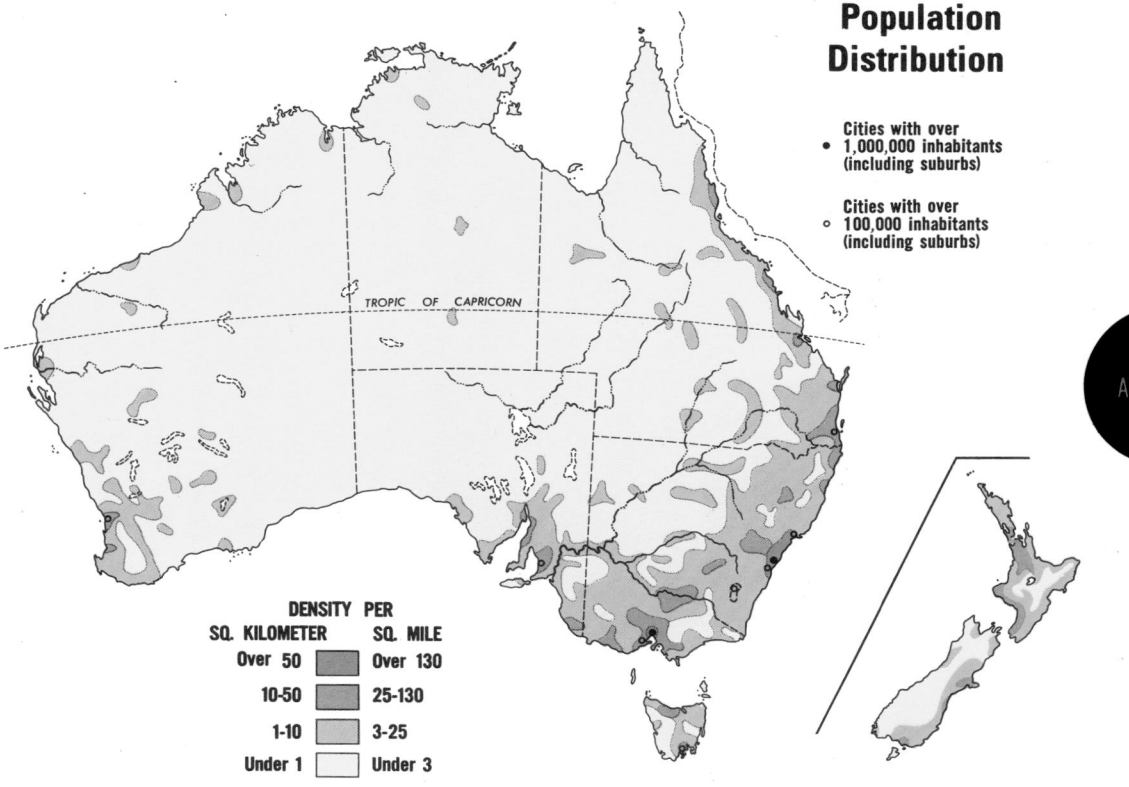

• Cities with over
1,000,000 inhabitants
(including suburbs)

○ Cities with over
100,000 inhabitants
(including suburbs)

DENSITY PER

SQ. KILOMETER	SQ. MILE
Over 50	Over 130
10-50	25-130
1-10	3-25
Under 1	Under 3

Vegetation

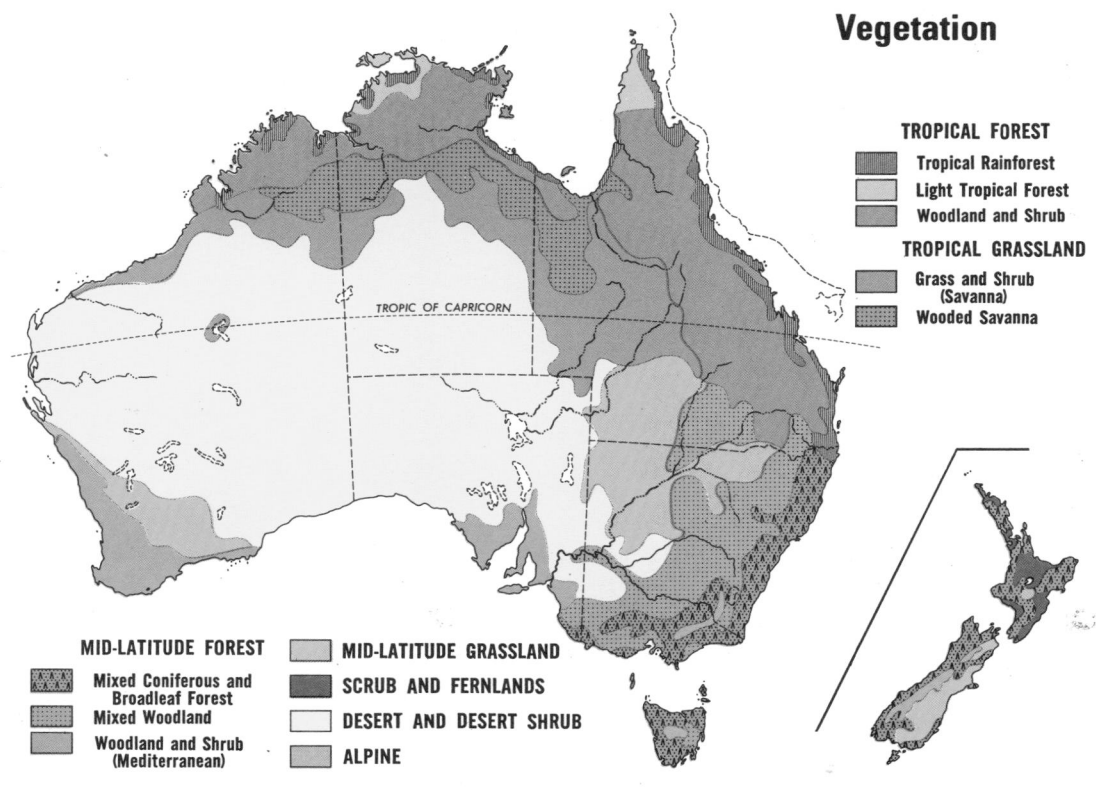

TROPICAL FOREST
Tropical Rainforest
Light Tropical Forest
Woodland and Shrub

TROPICAL GRASSLAND
Grass and Shrub
(Savanna)
Wooded Savanna

MID-LATITUDE FOREST
Mixed Coniferous and
Broadleaf Forest
Mixed Woodland
Woodland and Shrub
(Mediterranean)

MID-LATITUDE GRASSLAND
SCRUB AND FERNLANDS
DESERT AND DESERT SHRUB
ALPINE

Salisbury E7
Sandgate K2
Sandringham L7
Seymour H7
Shepparton (Shepparton-
Mooroopna) G7
South Australia F6
South Melbourne K7
South Perth B2
South Sydney L4
Springvale L7
Stafford (Stafford-Stafford
Heights) K2
Stirling B2
Strathfield K4
Subiaco B2
Sunshine K7
Sutherland K5
Swan Hill G7
Sydney (cap.), N. S. Wales . L4
Tamworth J6
Taree J6
Tasmania H8
Tea Tree Gully E7
Temora H6
Thebarton D8
The Entrance (The Entrance-
Terrigal) J6
The Gap J2
Toowoomba J5
Townsville H3

Traralgon H7
Tumut H7
Unley D8
Victoria G7
Wagga Wagga H7
Wangaratta H7
Wanneroo B2
Warrnambool G7
Warwick J5
Waverley L4
Waverley L7
Werribee G7
Western Australia B5
West Torrens D8
Whyalla F6
Williamstown K7
Willoughby K3
Windsor K2
Wodonga H7
Wollongong J6
Woodville D7
Woollahra L4
Wynnum L2
Yeppoon J4
Young H6

*City and suburbs.
†Population of met. area.
‡Population of urban area.

Average January Temperature

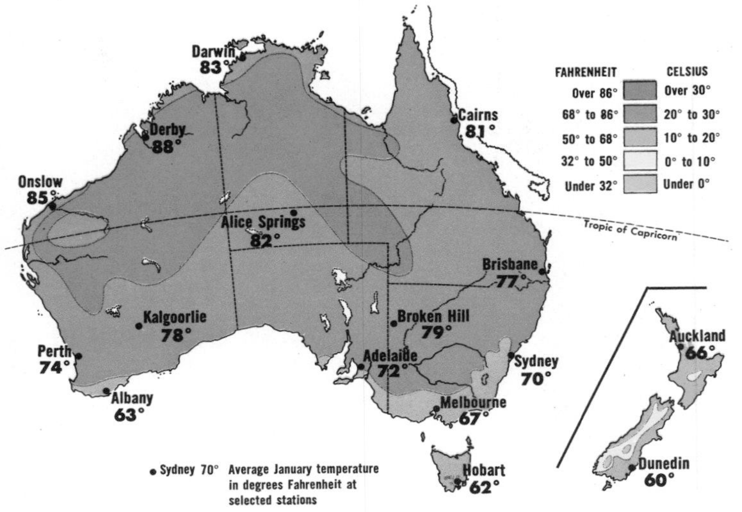
FAHRENHEIT
CELSIUS
Over 86°
Over 30°
68° to 86°
20° to 30°
50° to 68°
10° to 20°
32° to 50°
0° to 10°
Under 32°
Under 0°
Darwin 83°
Derby 88°
Cairns 81°
Onslow 85°
Alice Springs 82°
Tropic of Capricorn
Brisbane 77°
Kalgoorlie 78°
Broken Hill 79°
Perth 74°
Adelaide 72°
Sydney 70°
Auckland 66°
Albany 63°
Melbourne 67°
Hobart 62°
Dunedin 60°
• Sydney 70° Average January temperature in degrees Fahrenheit at selected stations

Average July Temperature

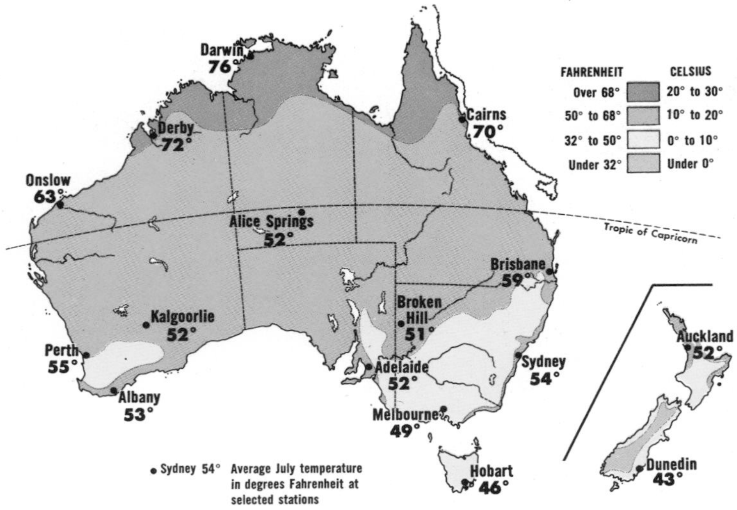
FAHRENHEIT
CELSIUS
Over 68°
20° to 30°
50° to 68°
10° to 20°
32° to 50°
0° to 10°
Under 32°
Under 0°
Darwin 76°
Derby 72°
Cairns 70°
Onslow 63°
Alice Springs 52°
Tropic of Capricorn
Brisbane 59°
Kalgoorlie 52°
Broken Hill 51°
Perth 55°
Adelaide 52°
Sydney 54°
Auckland 52°
Albany 53°
Melbourne 49°
Hobart 46°
Dunedin 43°
• Sydney 54° Average July temperature in degrees Fahrenheit at selected stations

Rainfall

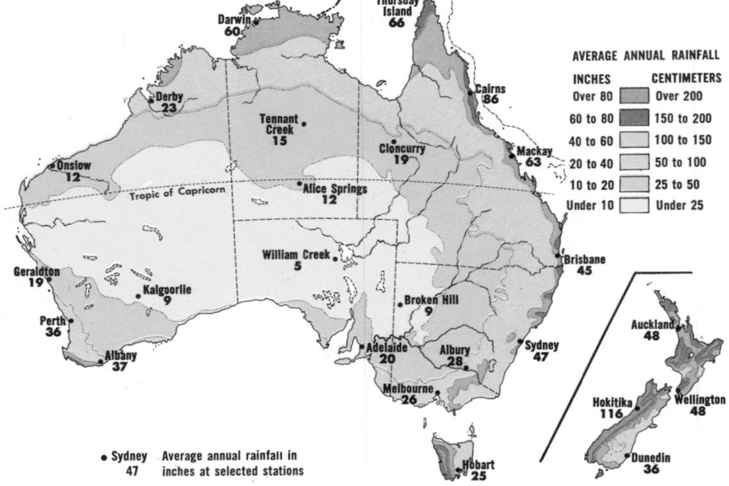
AVERAGE ANNUAL RAINFALL
INCHES
CENTIMETERS
Over 80
Over 200
60 to 80
150 to 200
40 to 60
100 to 150
20 to 40
50 to 100
10 to 20
25 to 50
Under 10
Under 25
Thursday Island 66
Darwin 60
Cairns 86
Derby 23
Tennant Creek 15
Cloncurry 19
Mackay 63
Onslow 12
Tropic of Capricorn
Alice Springs 12
William Creek 5
Brisbane 45
Geraldton 19
Kalgoorlie 9
Broken Hill 9
Auckland 48
Perth 36
Adelaide 20
Albury 28
Sydney 47
Albany 37
Melbourne 26
Hokitika 116
Wellington 48
Hobart 25
Dunedin 36
• Sydney 47 Average annual rainfall in inches at selected stations

DOMINANT LAND USE

- Cereals (chiefly wheat), Livestock
- Dairy, Truck Farming
- Cash Crops, Horticulture, Fruit
- Pasture Livestock
- Range Livestock
- Forests
- Nonagricultural Land

MAJOR MINERAL OCCURRENCES

Ab	Asbestos	Na	Salt
Ag	Silver	Ni	Nickel
Al	Bauxite	O	Petroleum
Au	Gold	Op	Opals
C	Coal	P	Phosphates
Cu	Copper	Pb	Lead
D	Diamonds	S	Sulfur, Pyrites
Fe	Iron Ore	Sb	Antimony
G	Natural Gas	Sn	Tin
Gp	Gypsum	Ti	Titanium
Lg	Lignite	U	Uranium
Ls	Limestone	W	Tungsten
Mg	Magnesium	Zn	Zinc
Mi	Mica	Zr	Zirconium
Mn	Manganese		

Water Power

Major Industrial Areas

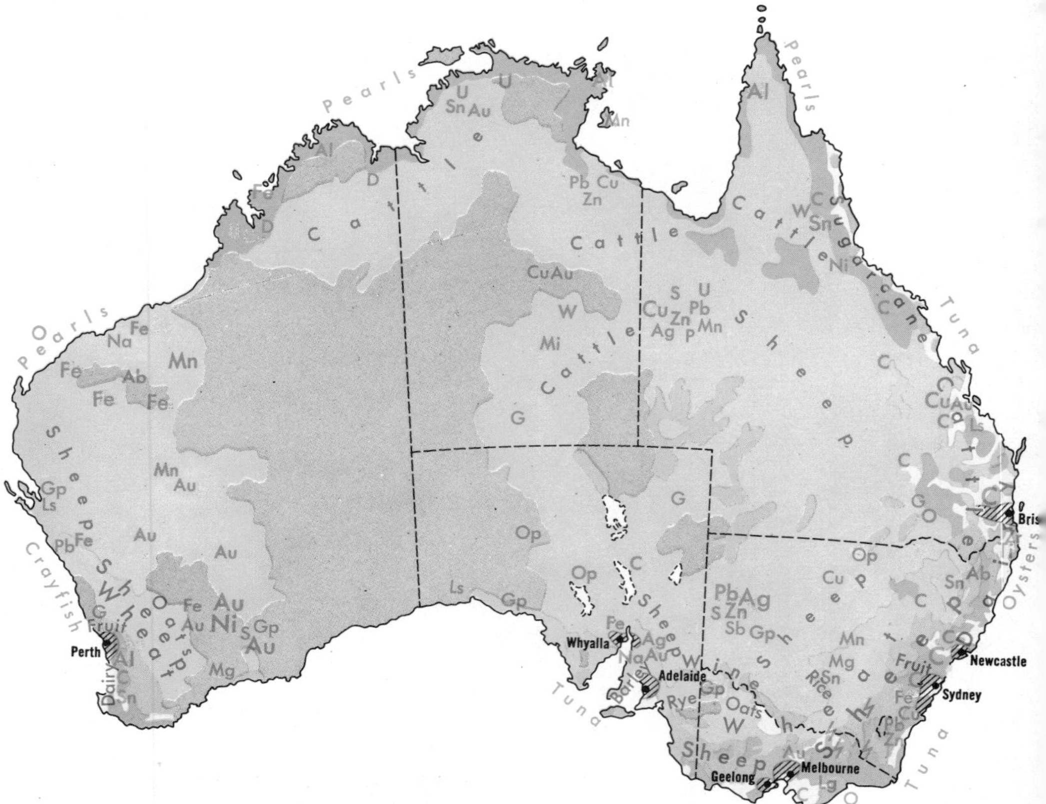
Pearls
Cattle
Sheep
Crayfish
Tuna
Oysters
Salmon
Sugarcane
Fruit
Dairy
Rice
Oats
Barley
Rye
Wine
Wheat
Perth
Whyalla
Adelaide
Geelong
Melbourne
Sydney
Newcastle
Bris

Agriculture, Industry and Resources

120° 125° 130° 135° 140° 145° 150°

INDONESIA

Sumba Timor ARAFURA SEA

New Guinea PAPUA NEW GUINEA

Port Moresby

10°

TIMOR SEA

Ashmore Is. TERR. OF ASHMORE & CARTIER IS.
Cartier I.

Melville I. Cobourg Pen. C. Wessel

Darwin Arnhem Land

Torres Strait C. York

Gulf of Carpentaria

Cape York Peninsula

INDIAN OCEAN

Kimberley Plateau

Derby Ord

Fitzroy Victoria Daly Groote Eylandt

Great Barrier Reef

Mitchell

Mt. Bartle Frere 5,287 ft. (1611 m.) Cairns

15°

NORTHERN

Tanami Desert

Barkly Tableland

Townsville

CORAL SEA

Port Hedland

Fortescue

North West

Hamersley Ra. Mt. Bruce 4,024 ft. (1227 m.)

WESTERN

Great Sandy Desert

Lake Disappointment Tropic of Capricorn

Lake Mackay

TERRITORY

Flinders

Mt. Isa

QUEENSLAND

Georgina

Mackay

20°

Gibson Desert

Macdonnell Ranges Alice Springs

Simpson

Rockhampton

Lake Carnegie

AUSTRALIA

Finke

Ayers Rock 2,845 ft. (867 m.)

Desert

Diamantina

Barcoo

Bundaberg

25°

Geraldton

Lake Barlee

Musgrave Ranges

SOUTH

Lake Eyre

Barcoo

Grey Range Warrego Range

Toowoomba Brisbane
Gold Coast

Great Victoria Desert

AUSTRALIA

Lake Torrens

Sturt Desert

Range

Lake Frome

Darling

Tamworth

Great Dividing Range

Kalgoorlie-Boulder

Nullarbor Plain

Lake Gairdner

Flinders Ra.

Broken Hill

NEW SOUTH

30°

Perth Darling Ra.
Fremantle

Whyalla Eyre Pen.

Lachlan

Newcastle

Bunbury

Great Australian Bight

Spencer Gulf

Adelaide

WALES

Sydney
Wollongong

35°

C. Leeuwin

Albany

Kangaroo I.

Mt. Lofty Ra.

Murray

Wagga Wagga

Albury

Canberra
AUSTRALIAN CAPITAL TERRITORY

INDIAN OCEAN

Mt. Gambier

Bendigo
Ballarat

VICTORIA

Geelong Melbourne

Mt. Kosciusko 7,316 ft. (2230 m.)

C. Howe

OCEAN

King I.

Bass Strait

TASMAN SEA

40°

Furneaux Group

Launceston

TASMANIA

Hobart

© Copyright HAMMOND INCORPORATED, Maplewood, N. J.

South Cape

110° 115° 120° 125° 130° Longitude 140°East of Greenwich 145° 150° 155°

Vegetation/Relief

SCALE OF MILES
0 100 200 300 400 500 600

SCALE OF KILOMETERS
0 100 200 300 400 500 600

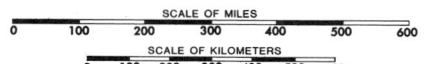

Capital of Country.................................⊛
State and Territorial Capitals........................◉
International Boundaries................................
State and Territorial Boundaries..............

Depths in Fathoms

Forest Woodland and Scrub Grassland Forest and Grassland Cropland Desert Tundra and Alpine Ice and Snow Grassland and Scrub Scrub and Fernlands

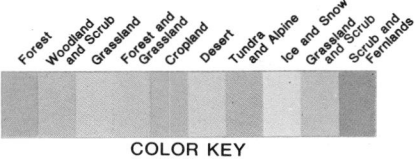

COLOR KEY

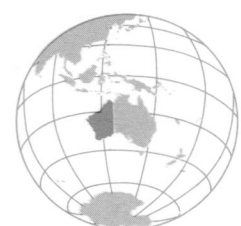

AREA 975,096 sq. mi.
(2,525,500 sq. km.)
POPULATION 1,273,624
CAPITAL Perth
LARGEST CITY Perth
HIGHEST POINT Mt. Bruce 4,024 ft.
(1,227 m.)

Topography

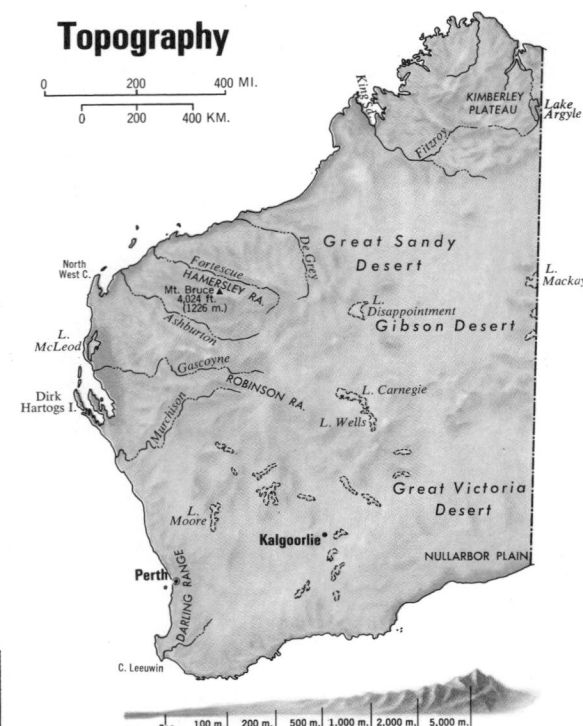

Below Sea Level	100 m. 328 ft.	200 m. 656 ft.	500 m. 1,640 ft.	1,000 m. 3,281 ft.	2,000 m. 6,562 ft.	5,000 m. 16,404 ft.

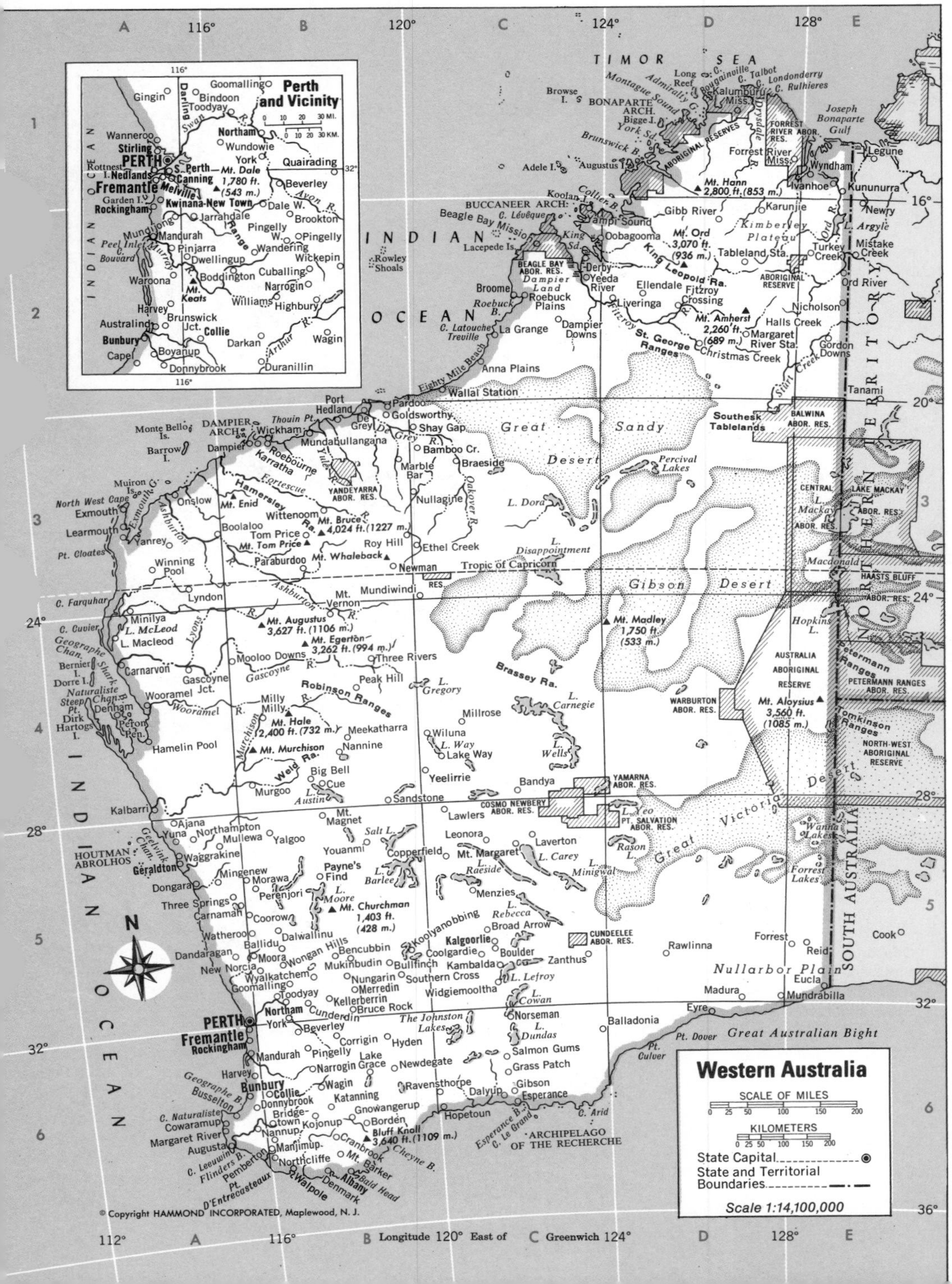

CITIES and TOWNS

Albany 15,222 . . . . B6
Augusta 588 . . . . A6
Australind 1,681 . . . . . . A2
Balladonia D6
Beverley 756 . . . . B1
Boddington 367 . . . . . . B2
Boulder-Kalgoorlie 19,848 . C5
Boyanup 365 . . . . A2
Bridgetown 1,521 . . . . . B6
Brookton 595 . . . . B2
Broome 3,666 . . . . C2
Bruce Rock 565 . . . . . . B5
Brunswick Junction 889 . . A2
Bunbury 21,749 . . . . . . A2
Busselton 6,463 . . . . . . A6
Canning 52,816 . . . . . . A1
Capel 680 . . . . A2
Carnamah 422 . . . . A5
Carnarvon 5,053 . . . . . . A4
Collie 7,667 . . . . B2
Coolgardie 891 . . . . . . C5
Coorow 226 . . . . B5
Corrigin 841 . . . . B6
Cranbrook 316 . . . . B6
Cue 320 . . . . B4
Cuballing ○647 . . . . . . B2
Cunderdin 731 . . . . B5
Dalwallinu 639 . . . . . . B5
Dampier 2,471 . . . . B3
Dandaragan ○1,748 . . . . A5
Darkan 242 . . . . B2
Denham 402 . . . . A4
Denmark 985 . . . . B6
Derby 2,933 . . . . C2
Dongara-Port Denison 1,155 A5
Donnybrook 1,197 . . . . . A2
Dwellingup 453 . . . . . . B2
Esperance 6,375 . . . . . . C6
Eucla E5
Exmouth 2,583 . . . . . . A3
Fitzroy Crossing D2
Fremantle 22,484 . . . . . A1
Geraldton 20,895 . . . . . A5
Gingin 382 . . . . A1
Gnowangerup 872 . . . . . B6
Goldsworthy 923 . . . . . . B3
Goomalling 600 . . . . . . B1
Halls Creek 966 . . . . . . D2
Harvey 2,479 . . . . A2
Hopetoun C6
Hyden B6
Jarrahdale 315 . . . . B2
Kalbarri 820 . . . . A4
Kalgoorlie 9,145 . . . . . . C5
Kalgoorlie-Boulder 19,848 . C5
Kambalda 4,463 . . . . . . C5
Karratha 8,341 . . . . . . B3
Katanning 4,413 . . . . . . B6
Kellerberrin 1,091 . . . . . B5
Kojonup 544 . . . . B6
Koolyanobbing 277 . . . . B5
Kununurra 2,081 . . . . . E2
Kwinana New Town 12,355 . A1
Lake Grace 575 . . . . . . B6
Laverton 872 . . . . C5
Learmonth A3
Leonora 524 . . . . C5
Madura D5
Mandurah 10,978 . . . . . A2
Manjimup 4,150 . . . . . . B6
Marble Bar 357 . . . . . . C3
Margaret River 798 . . . . A6
Meekatharra 989 . . . . . B4
Melville 61,211 . . . . . . A1
Menzies 232 . . . . C5
Merredin 3,520 . . . . . . B5
Mingenew 368 . . . . A5
Moora 1,677 . . . . B5
Morawa 694 . . . . B5
Mount Barker 1,519 . . . . B6
Mount Magnet 618 . . . . B4
Mukinbudin 370 . . . . . . B5
Mullewa 918 . . . . A5
Mundijong 356 . . . . A2
Nannup 552 . . . . B6
Narrogin 4,969 . . . . . . B2
Nedlands 20,257 . . . . . A1
Newman 5,466 . . . . . . B3
New Norcia A5
Norseman 1,895 . . . . . . C6
Northam 6,791 . . . . . . B1
Northampton 750 . . . . . A5
Northcliffe B6
Nungarin ○332 . . . . . . B5
Onslow 594 . . . . A3
Pannawonica 1,170 . . . . B3
Paraburdoo 2,357 . . . . . B3
Pardoo C3
Pemberton 871 . . . . . . A6
Perenjori 257 . . . . B5
Perth (cap.) 809,035 . . . A1
Perth *898,918 . . . . . . A1
Pingelly 937 . . . . B2
Pinjarra 1,336 . . . . A2
Port Denison-Dongara 1,155 A5
Port Hedland 12,948 . . . B3
Quairading 741 . . . . . . B1
Ravensthorpe 327 . . . . . B6
Rockingham 24,932 . . . . A2
Roebourne 1,688 . . . . . B3
Sandstone ○133 . . . . . . B4
Shay Gap 853 . . . . C3
Southern Cross 798 . . . . B5
South Perth 31,524 . . . . A1
Stirling 161,858 . . . . . . A1
Three Springs 638 . . . . . A5
Tom Price 3,540 . . . . . . B3
Toodyay 560 . . . . B1
Turkey Creek 212 . . . . . E2
Wagin 1,488 . . . . B2
Walpole 291 . . . . B6
Wandering ○470 . . . . . . B2
Wanneroo 6,745 . . . . . . A1
Waroona 1,462 . . . . A2
Wickepin 267 . . . . B5
Wickham 2,387 . . . . . . B3
Williams 453 . . . . B2
Wiluna 221 . . . . C4
Wittenoom 247 . . . . . . B3
Wongan Hills 947 . . . . . B5
Wundowie 720 . . . . B1
Wyalkatchem 453 . . . . . B5
Wyndham 1,509 . . . . . . E1
Yalgoo ○315 . . . . B5
Yampi Sound C2
York 1,136 . . . . B1

OTHER FEATURES

Adele (isl.) C1
Admiralty (gulf) D1
Aloysius (mt.) E4
Argyle (lake) E2
Arid (cape) C6
Ashburton (riv.) A3
Augustus (mt.) B4
Austin (lake) B4
Australia Aboriginal Res. . . E4
Balwina Aboriginal Res. . . E3
Bald (head) B6
Barlee (lake) B5
Barrow (isl.) A3
Beaglebay Aboriginal Res. . C2
Bluff Knoll (mt.) B6
Bonaparte (arch.) D1
Bougainville (cape) D1
Brassey (range) C4
Bruce (mt.) B3
Brunswick (bay) D1
Buccaneer (arch.) C2
Carey (lake) C5
Carnegie (lake) C4
Central Aboriginal Res. . . E3
Churchman (mt.) C1
Collier (bay) C1
Cosmo Newbery Aboriginal
Res. C5
Cowan (lake) C5
Cundeelee Aboriginal Res. . C5
Dale (lake) B1
Dampier (arch.) B3
Dampier Land (reg.) C2
Darling (range) A1
De Grey (riv.) B3
D'Entrecasteaux (pt.) . . . A6
Dirk Hartogs (isl.) A4
Disappointment (lake) . . . C3
Drysdale (riv.) D1
Dundas (lake) C6
Egerton (mt.) B4
Eighty Mile (beach) C2
Enid (mt.) B3
Esperance (bay) C6
Exmouth (gulf) A3
Fitzroy (riv.) D2
Flinders (bay) A6
Forrest River Aboriginal Res. D1
Fortescue (riv.) B3
Garden (isl) A1
Gascoyne (riv.) A4
Geelvink (chan.) A5
Geographe (bay) A6
Geographe (chan.) A4
Gibson (des.) D3
Great Australian (bight) . . E6
Great Sandy (des.) C3
Great Victoria (des.) D5
Hamersley (range) B3
Hann (mt.) D1
Hopkins (lake) E4
Houtman Abrolhos (isls.) . . A5
Indian Ocean A3
Johnston, The (lakes) . . . C6
Joseph Bonaparte (gulf) . . E1
Kimberley (plat.) C2
King (sound) C2
King Leopold (range) C2
Koolan (isl.) C1
Leeuwin (cape) A6
Le Grand (cape) C6
Lévêque (cape) C2
Londonderry (cape) D1
Lyons (riv.) A4
Macdonald (lake) E3
Mackay (lake) E3
McLeod (lake) A4
Minigwal (lake) C5
Monte Bello (isls.) A3
Moore (lake) B5
Murchison (riv.) A4
Murray (riv.) A2
Naturaliste (cape) A6
Naturaliste (chan.) A4
North West (cape) A3
North-West Aboriginal Res. . E4
Nullarbor (plain) C6
Oakover (riv.) C3
Ord (mt.) D1
Ord (riv.) E2
Percival (lakes) D3
Peron (pen.) A4
Petermann (ranges) E4
Rason (lake) D5
Rebecca (lake) C5
Recherche (arch.) C6
Robinson (ranges) B4
Roebuck (bay) C2
Rottnest (isl.) A1
Saint George (ranges) . . . D2
Shark (bay) A4
Southesk Tablelands D2
Sturt (creek) D2
Swan (riv.) A1
Timor (sea) D1
Tomkinson (ranges) E4
Wanna (lakes) D5
Warburton Aboriginal Res. . D4
Way (lake) C4
Weld (range) B4
Wells (lake) C4
Whaleback (mt.) B3
Wooramel (riv.) A4
York (sound) D1

○ Population of district.
*Population of met. area.

Western Australia

SCALE OF MILES
KILOMETERS

State Capital . . . . ⊙
State and Territorial
Boundaries . . . . —

Scale 1:14,100,000

© Copyright HAMMOND INCORPORATED, Maplewood, N.J.

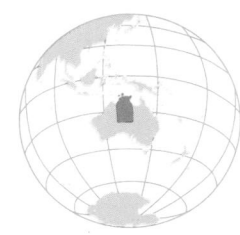

CITIES and TOWNS

Adelaide River	B2
Aileron	C7
Alice Springs 18,395	D7
Alyangula 1,181	E2
Angurugu 597	E3
Anthony Lagoon	D4
Areyonga	C8
Arltunga	D7
Avon Downs	E5
Bamyili-Beswick 685	C3
Banka Banka	C4
Barrow Creek	D6
Batchelor	B2
Bathurst Island 1,032	B1
Birdum	C3
Birrimbah	C3
Borroloola 420	E4
Bundooma	D8
Burramurra	D5
Charlotte Waters	D8
Claravale	B3
Coniston	C7
Coolibah	B3
Creswell Downs	E4
Croker Island Mission	C1
Daly River	B2
Daly Waters	C3
Darwin (cap.) 56,482	B2
Docker River 217	A8
Elliott	C3
Epenarra	D6
Erldunda	C8
Eva Downs	D5
Ewaninga	D7
Goulburn Island 277	C1
Gove (Nhulunbuy) 3,879	E2
Harts Range	D7
Hatches Creek	D6
Helen Springs	C5
Henbury	C8
Hermannsburg 541	C7
Hooker Creek 671	B5
Humpty Doo	B2
Katherine 3,737	B3
Kildurk	A4
Koolpinyah	B2
Kulgera	C8
Kurundi	D6
Lake Nash	E6
Larrimah	C3
Legune	A3
Limbunya	B4
Lucy Creek	E7
Mainoru	C3
Maningrida 702	C2
Mataranka	C3
Milingimbi 564	D2
Mistake Creek	A4
Montejinnie	C4
Mount Cavenagh	C8
Mount Doreen	B7
Murray Downs	D6
Napperby	C7
Newcastle Waters	C4
Nhulunbuy 3,879	E2
Numbulwar 422	D3
Oenpelli 452	C2
O. T. Downs	D4
Papunya 635	B7
Pine Creek 214	C2
Plenty River Mine	D7
Port Keats 819	A3
Powell Creek	C5
Rankine Store	E5
Robinson River	E4
Rockhampton Downs	D5
Rodinga	D8
Rum Jungle	B2
Santa Teresa 479	D8
Soudan	E6
Stirling Station	C6
Tanami	A5
Tarlton Downs	E7
Tea Tree Well	C7
Tempe Downs	C8
Tennant Creek 3,118	C5
The Granites	B6
Top Springs	C4
Ucharonidge	D4
Umbakumba 247	E3
Umbeara	C8
Urapunga	D3
Utopia	D7
Victoria River Downs	B4
Warrabri 459	D6
Warrego 991	C5
Wave Hill	B4
White Quartz Hill	D7
Willeroo	B3
Willowra	C6
Wollogorang	F4
Yambah	C7
Yirrkala 543	E2
Yuendumu 687	B7

OTHER FEATURES

Amadeus (lake)	B8
Arafura (sea)	D1
Arnhem (cape)	E2
Arnhem Land (reg.)	D2
Arnhem Land Aboriginal Res.	C2
Arnold (riv.)	D3
Ayers Rock Nat'l Park	B8
Barkly Tableland	D4
Bathurst (isl.)	A1
Beagle (gulf)	A2
Beatrice (cape)	E3
Bennett (lake)	B7
Beswick Aboriginal Res.	C3
Bickerton (isl.)	E2
Blaze (pt.)	A2
Carpentaria (gulf)	E3
Central Wedge (mt.)	C7
Clarence (str.)	B2
Cobourg (pen.)	C1
Conner (mt.)	B8
Croker (cape)	C1
Daly (riv.)	B2
Daly River Aboriginal Res.	A2
Davenport (mt.)	B7
Dundas (str.)	B1
East Alligator (riv.)	C2
Ehrenberg (range)	B7
Elcho (isl.)	D1
Finke (riv.)	C8
Fitzmaurice (riv.)	B3
Ford (cape)	A2
Georgina (riv.)	E6
Goulburn (isls.)	C1
Goyder (riv.)	D2
Groote Eylandt (isl.) 2,230	E3
Haasts Bluff Aboriginal Res.	B7
Hale (riv.)	D8
Hanson (riv.)	C6
Hay (dry riv.)	E7
Hogarth (mt.)	E6
Hopkins (lake)	A8
Joseph Bonaparte (gulf)	A3
Katherine (riv.)	C3
Lake MacKay Aboriginal Res.	A6
Lander (riv.)	C6
Leisler (mt.)	A7
Limmen (bight)	D3
Limmen Bight (riv.)	D4
Macdonald (lake)	B7
Macdonnell (ranges)	B7
MacKay (lake)	A7
Mann (riv.)	D2
Marshall (riv.)	D7
Melville (bay)	E2
Melville (isl.)	B1
Mount Olga Nat'l Park	B8
Murchison (range)	D6
Napier (mt.)	A4
Neale (lake)	A8
Newcastle (creek)	C4
Nicholson (riv.)	E5
Olga (mt.)	B8
Peron (isls.)	A2
Petermann (ranges)	A8
Petermann Ranges Aboriginal Res.	A8
Port Darwin (inlet)	B2
Ranken (riv.)	E6
Robinson (riv.)	E4
Roper (riv.)	C3
Sandover (riv.)	D6
Simpson (des.)	E8
Singleton (mt.)	B6
Sir Edward Pellew Group (isls.)	E3
South Alligator (riv.)	C2
Stanley (mt.)	B7
Stewart (cape)	D1
Stirling (creek)	A4
Sturt (plain)	C4
Tanami (des.)	C5
Timor (sea)	A1
Todd (riv.)	D8
Vanderlin (isl.)	E3
Van Diemen (cape)	A1
Van Diemen (gulf)	B1
Victoria (riv.)	B3
Wagait Aboriginal Res.	B2
Warwick (chan.)	E3
Wessel (cape)	E1
Wessel (isls.)	E1
West Baines (riv.)	A4
White (lake)	A6
Woods (lake)	C4
Young (mt.)	D3
Ziel (mt.)	C7

AREA 519,768 sq. mi.
(1,346,200 sq. km.)
POPULATION 123,324
CAPITAL Darwin
LARGEST CITY Darwin
HIGHEST POINT Mt. Ziel 4,955 ft.
(1,510 m.)

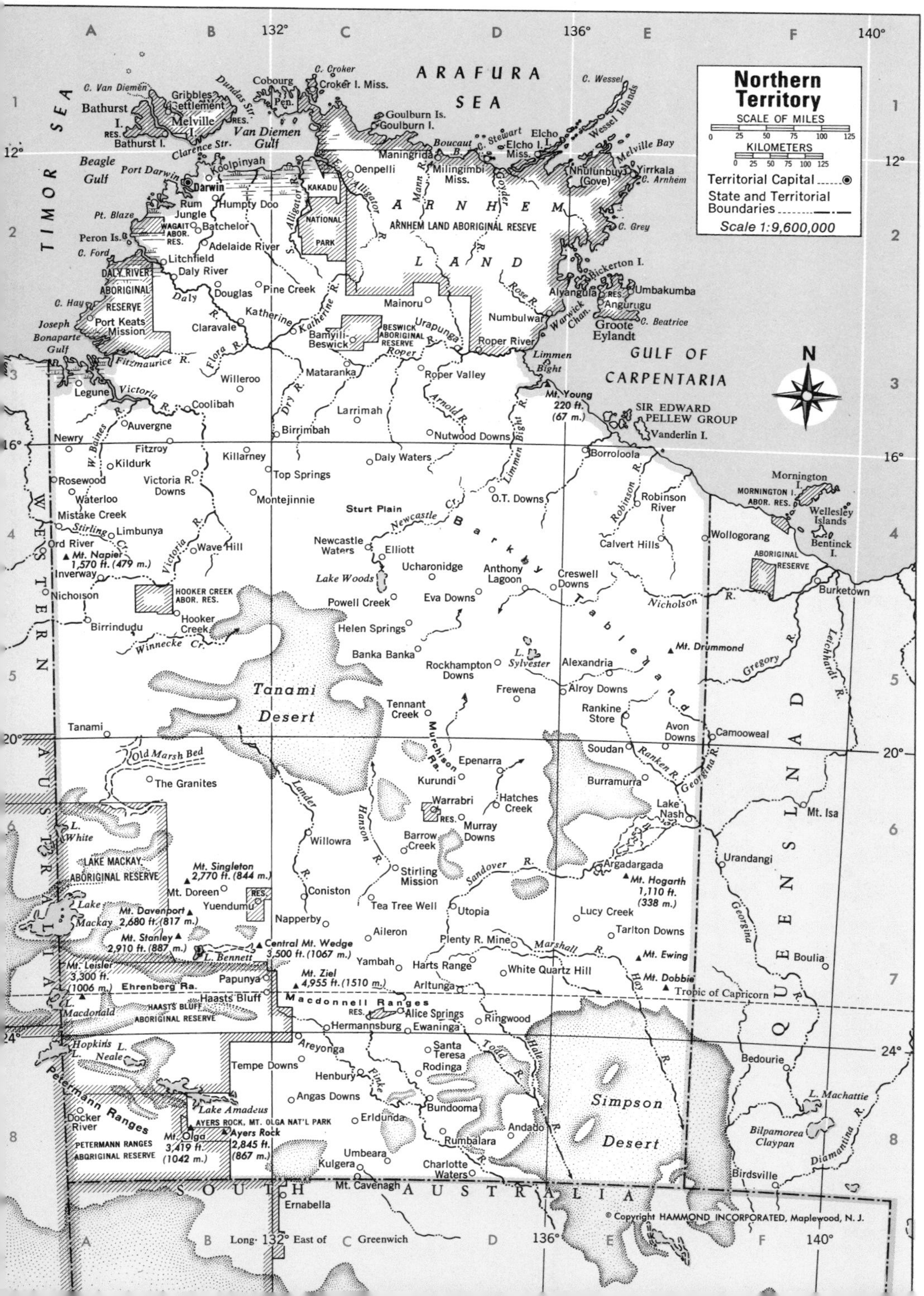

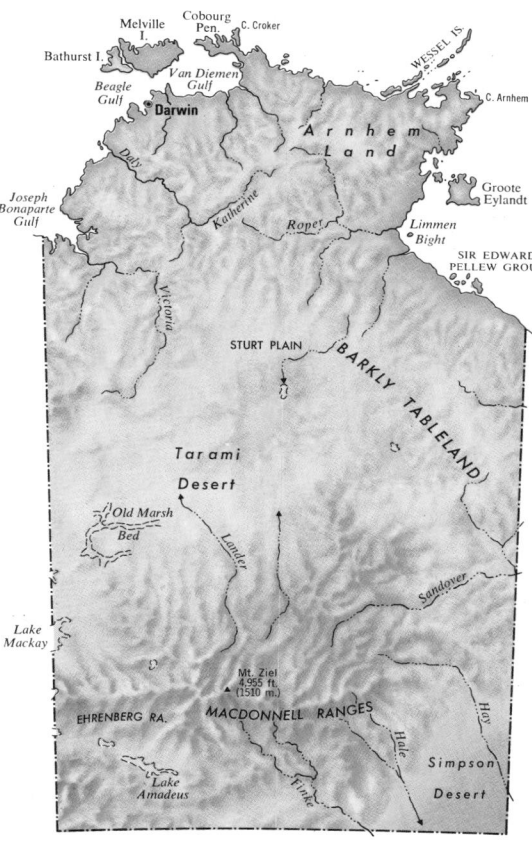

Topography

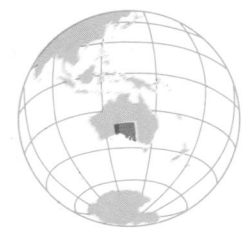

AREA 379,922 sq. mi. (984,000 sq. km.)
POPULATION 1,285,033
CAPITAL Adelaide
LARGEST CITY Adelaide
HIGHEST POINT Mt. Woodroffe 4,970 ft.
(1,515 m.)

CITIES and TOWNS

Adelaide (cap.) 882,520 .... B6
Adelaide *931,886 .... B6
Andamooka 402 .... E4
Angaston 1,753 .... F6
Balaklava 1,306 .... F6
Barmera 2,014 .... G6
Beachport 357 .... F7
Berri 3,419 .... G6
Birdwood 397 .... C7
Blinman F4
Bordertown 2,138 .... G7
Brighton 19,441 .... A8
Burnside 37,593 .... B8
Burra 1,222 .... F5
Campbelltown 43,084 .... B7
Ceduna 2,794 .... D5
Clare 2,381 .... F5
Cleve 827 .... E5
Coober Pedy 2,078 ..... D3
Cowell 626 .... E5
Crafters-Bridgewater 9,764 .. B8
Crystal Brook 1,240 ... E5
Cummins 767 .... D6
Edithburgh 359 .... E6
Elizabeth 32,608 ...... B7
Elliston ○1,345 .... D5
Enfield 66,797 .... B7
Gawler 9,433 .... B6
Gladstone 680 .... F5
Glenelg 13,306 .... A8
Gumeracha 387 .... C7
Hahndorf 1,274 ...... C8
Hawker 351 .... F4
Hindmarsh 7,593 ..... A7
Iron Knob 398 .... E5
Jamestown 1,384 .... F5
Kadina 2,943 .... E5
Kapunda 1,340 ...... F6
Keith 1,147 .... G7
Kensington and Norwood
8,950 .... B8
Kimba 862 .... E5
Kingscote 1,236 ... E6
Kingston 1,325 .... G7
Lameroo 599 .... G6
Laura 504 .... F5
Leigh Creek 1,635 ...... F4
Lobethal 1,522 .... C7
Lock 213 .... D5
Loxton 3,100 .... G6
Lyndoch 539 .... C6
Maitland 1,085 .... E6
Mannum 1,984 .... F6
Marion 66,580 .... A8
Marree E3
Meadows 388 .... B8
Meningie 807 .... F6
Millicent 5,255 .... F7
Minlaton 865 ...... E6
Moonta 1,751 ...... E5
Mount Barker 4,190 . C8
Mount Gambier 18,193 . G7
Murray Bridge 8,664 . F6
Nairne 706 .... C8
Nangwarry 758 ...... G7

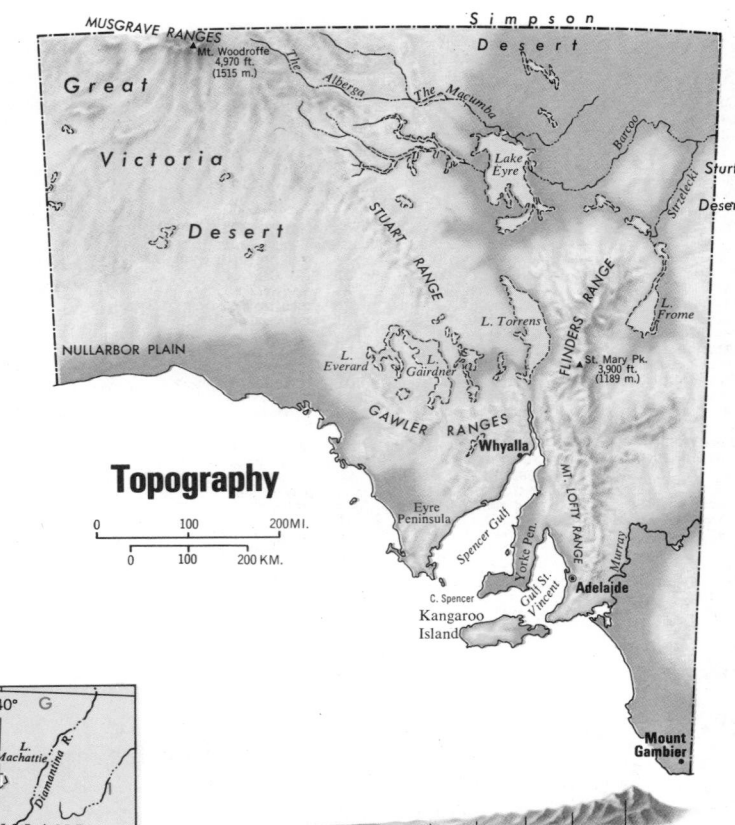

Topography

Naracoorte 4,758 .... G7
Noarlunga 60,928 .... A8
Nuriootpa 2,851 .... F6
Oodnadatta D2
Orroroo 604 .... F5
Paynehan 16,502 ...... B7
Penola 1,205 .... G7
Peterborough 2,575 ... F5
Pinnaroo 731 .... G6
Port Adelaide 35,407 . A7
Port Augusta 15,566 .. E5
Port Broughton 587 ... F5
Port Lincoln 9,846 ... E6
Port Pirie 14,695 .... E5
Prospect 18,591 ...... B7
Quorn 1,049 .... F5
Renmark 3,475 .... G5
Robe 590 .... F7
Salisbury 86,451 .... B7
Snowtown 492 .... F5
Strathalbyn 1,756 ... F6
Streaky Bay 985 .... D5
Tailem Bend 1,677 .. F6
Tanunda 2,621 ..... C6
Tea Tree Gully 67,237 . B7
Thebarton 9,208 .... A7
Tumby Bay 933 ..... E6
Unley 35,844 ...... B8
Uraidla 303 .... B8
Victor Harbor 4,522 . F6
Virginia 353 ...... B7
Waikerie 1,629 .... F6
Wallaroo 2,043 .... E5
West Torrens 45,099 . A8
Whyalla 30,518 .... E5
Williamstown 495 .. C7
Willunga 667 ...... F6
Wilmington 227 .... F6
Woodside 724 ...... C8
Woodville 77,634 .. A7
Woomera 1,658 .... E4
Wudinna 572 ...... D5
Yorketown 713 .... E6

OTHER FEATURES

Acraman (lake) D5
Alberga, The (riv.) D2
Alexandrina (lake) F6
Anxious (bay) D5
Arckaringa (creek) D2
Barcoo (creek) F3
Birksgate (range) A2
Blanche (lake) F3
Brady (mt.) D3
Cadibarrawirracanna (lake) . D3
Callabonna (lake) ... F3
Catastrophe (cape) .. D6
Coffin (bay) D6
Coffin Bay (pen.) ... D6
Coopers (Barcoo) (creek) . F3
Coorong, The (lag.) . F6
Dey Dey (lake) B3
Encounter (bay) F6
Everard (lake) D4
Everard (ranges) ... C2
Eyre (pen.) D5
Eyre North (lake) .. E3
Eyre South (lake) .. E3
Finke (riv.) C1

Flinders (range) F4
Frome (lake) G4
Gairdner (lake) D4
Gawler (ranges) E5
Gawler (riv.) B6
Gilles (lake) E5
Goyders (lag.) F3
Great Australian (bight) .. A5
Great Victoria (des.) B3
Gregory (lake) F3
Hack (mt.) F4
Hamilton, The (riv.) D2
Harris (lake) D4
Head of Bight (bay) B4
Indian Ocean E7
Investigator (str.) E6
Investigator Group (isls.) . D5
Island (lag.) E4
Jaffa (cape) F7
Kangaroo (isl.) 3,515 .... E6
Lacepede (bay) F7
Lofty (mt.) F6
Macfarlane (lake) E5
Macumba, The (riv.) D2
Maurice (lake) B3
Meramangye (lake) C3
Morris (mt.) C1
Murray (res.) G6
Musgrave (ranges) B2
Neales, The (riv.) D2
Northumberland (cape) .. F8
Nukey Bluff (mt.) D5
Nullarbor (plain) C4
Nuyts (arch.) D6
Nuyts (cape) D6
Peera Peera Poolanna (lake) . F2
Saint Mary (peak) F4
Saint Vincent (gulf) ... E6
Serpentine (lakes) A2
Simpson (des.) E1
Sir Joseph Banks Group
(isls.) E6
Spencer (cape) E6
Spencer (gulf) E6
Stevenson, The (riv.) .. D2
Streaky (bay) D5
Strzelecki (creek) G3
Stuart (range) D3
Sturt (des.) G2
The Alberga (riv.) ... D2
The Coorong (lag.) .. F6
The Hamilton (riv.) . D2
The Macumba (riv.) . D2
The Neales (riv.) ... D2
The Stevenson (riv.) . D2
The Warburton (riv.) . F2
Thistle (isl.) E6
Torrens (lake) E4
Torrens (riv.) B7
Warburton, The (riv.) . F2
Wilkinson (lakes) ... C3
Woodroffe (mt.) B2
Yalata Aboriginal Res. . C4
Yarle (lakes) C3
Yorke (pen.) E6

○ Population of district.
*Population of met. area.

Adelaide and Vicinity

South Australia

SCALE OF MILES
0 25 50 75 150

KILOMETERS
0 25 50 75 150

State Capital _____ ◉
State and Territorial
Boundaries _____ —.—.—
Scale 1:9,790,000

© Copyright HAMMOND INCORPORATED, Maplewood, N.J.

Longitude D East 136° of E Greenwich

CITIES and TOWNS

Aramac 428 .... C4
Archerfield 785 .... D3
Ascot 4,298 .... E2
Atherton 4,196 .... C3
Ayr 8,787 .... C3
Balmoral 2,915 .... E2
Barcaldine 1,432 .... C5
Beaudesert 3,780 .... E6
Biloela 4,643 .... D5
Birdsville A5
Blackall 1,609 .... C5
Blackwater 5,434 .... D4
Boulia 292 .... A4
Bowen 7,663 .... D3
Brisbane (cap.) 689,378 .. D2
Brisbane *1,028,527 ...... D2
Bucasia 1,356 .... D4
Bundaberg 32,560 .... D5
Burketown 210 .... A3
Cairns 48,557 .... C3
Caloundra 16,758 .... E5
Camooweal 251 .... A3
Camp Hill 8,999 .... E3
Capella 660 .... D4
Cardwell 1,249 .... C3
Charleville 3,523 .... C5
Charters Towers 6,823 ... C4
Cherbourg 963 .... D5
Chermside 6,892 .... D2
Clermont 1,659 .... C4
Cloncurry 1,961 .... B4
Collinsville 2,756 .... C4
Cooktown 913 .... C2
Coopers Plains 4,492 .... D3
Corinda 4,894 .... D2
Croydon ○255 .... B3
Cunnamulla 1,627 .... C6
Dalby 8,784 .... D5
Dirranbandi 480 .... D6
East Brisbane 4,853 ..... E3
Eidsvold 613 .... D5
Emerald 4,628 .... C4
Esk 676 .... E5
Gatton 4,190 .... E5
Gayndah 1,708 .... D5
Geebung 4,850 .... E2
Georgetown 319 .... B3
Gladstone 22,083 .... D4
Gold Coast 135,437 ...... E6
Goondiwindi 3,576 .... D6
Gordonvale 2,375 .... C3
Greenslopes 7,219 .... E3
Gympie 10,768 .... E5

Hervey Bay 13,569 .... E5
Holland Park 7,363 ...... E3
Home Hill 3,138 .... C3
Hughenden 1,657 .... B4
Inala 17,383 .... D3
Indooroopilly 7,959 ..... D2
Ingham 5,598 .... C3
Injune 407 .... D5
Innisfail 7,933 .... C3
Ipswich 68,297 .... D2
Isisford ○605 .... C5
Jandowae 781 .... D5
Jericho ○1,177 .... C4
Julia Creek 602 .... B4
Karumba 670 .... B3
Kilcoy 1,257 .... E5
Kingaroy 5,134 .... D5
Longreach 2,971 .... B4
Mackay 35,361 .... D4
Mareeba 6,309 .... C3
Marian 796 .... D4
Maroochydore-Mooloolaba
    17,460 .... E5
Maryborough 20,111 ..... E5
Mary Kathleen 830 ...... A4
McKinlay ○1,477 .... B4
Millmerran 1,107 .... D5
Mitchell 1,171 .... C5
Mitchelton 5,810 .... D2
Monto 1,397 .... D5
Moorooka 8,740 .... D3
Moranbah 4,362 .... C9
Mossman 1,614 .... C3
Mount Isa 23,679 .... A4
Moura 2,871 .... D5
Murgon 2,327 .... D5
Nambour 7,965 .... E5
Newmarket 3,520 .... D2
Normanton 926 .... B3
Nundah 7,358 .... E2
Proserpine 3,058 .... D4
Quilpie 694 .... C5
Ravenshoe 915 .... C3
Redcliffe 42,223 .... E5
Richmond 784 .... B4
Rockhampton 50,146 ... D4
Roma 5,706 .... D5
Saint George 2,204 .... D5
Saint Lucia 6,075 ..... D3
Sandgate 6,776 .... E2
Sarina 2,815 .... D4
Springsure 774 .... D5
Stafford (Stafford Heights)
    13,731 .... D2
Stanthorpe 3,966 .... D6
Tara 864 .... D5

Taroom 688 .... D5
Tewantin-Noosa 9,965 . E5
Theodore 643 .... D5
Thursday Island 2,283 . B1
Toowoomba 63,401 .... D5
Townsville 86,112 .... C3
Tully 2,728 .... C3
Walkerston 1,277 .... D4
Warwick 8,853 .... D6
Weipa 2,433 .... B2
Windsor 6,119 .... D2
Winton 1,259 .... B4
Wynnum 10,794 ...... E5

Yeppoon 6,447 .... D4
Yeronga 4,579 .... D3

OTHER FEATURES

Albatross (bay) B2
Archer (riv.) B2
Balonne (riv.) D6
Banks (isl.) B1
Barcoo (creek) B5
Barkly Tableland A4
Bartle Frere (mt.) ... C3
Beal (range) B5

AREA 666,872 sq. mi. (1,727,200 sq. km.)
POPULATION 2,295,123
CAPITAL Brisbane
LARGEST CITY Brisbane
HIGHEST POINT Mt. Bartle Frere 5,287 ft.
    (1,611 m.)

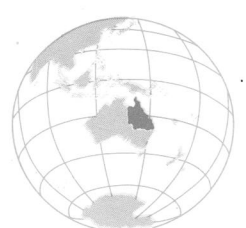

Belyando (riv.) C4
Broad (sound) D4
Bulloo (lake) B6
Bulloo (riv.) B6
Bunker Group (isls.) ... E4
Burdekin (riv.) C3
Cape York (pen.) B2
Capricorn (chan.) D4
Capricorn Group (isls.) . E4
Carnarvon (range) D5
Carpentaria (gulf) A2
Cloncurry (riv.) B4
Coopers (Barcoo) (creek) . B5
Coral (sea) C1
Culgoa (riv.) C6
Cumberland (isls.) D4
Curtis (isl.) D4
Darling Downs D5
Dawson (riv.) D5
Diamantina (riv.) B4
Drummond (range) C5
Duifken (pt.) B2
Endeavour (str.) B1

Fitzroy (riv.) D4
Flinders (riv.) B3
Fraser (isl.) E5
Georgina (riv.) A4
Gilbert (riv.) B3
Great Dividing (range) . C4
Gregory (range) B3
Gregory (riv.) A3
Grey (range) B5
Hamilton (riv.) B4
Hervey (bay) E5
Hinchinbrook (isl.) .. C3
Hook (isl.) D4
Leichhardt (riv.) A3
Machattie (lake) B5
Macintyre (riv.) D6
Maranoa (riv.) C5
Mary (riv.) E5
Melville (cape) C2
Mitchell (riv.) B2
Moreton (bay) E5
Moreton (isl.) E5
Mornington (isl.) A3

Norman (riv.) B3
Northern Peninsula
    Aboriginal Res. .. B1
Prince of Wales (isl.) . B1
Princess Charlotte (bay) . C2
Sandy (cape) E5
Selwyn (range) B4
Simpson (des.) A5
Sturt (des.) B3
Suttor (riv.) C4
Swain (reefs) E4
Thompson (riv.) B5
Torres (str.) B1
Warrego (range) ... C5
Warrego (riv.) C5
Wellesley (isls.) .. A3
Whitsunday (isl.) .. D4
Willies (range) C6
Yamma Yamma (lake) . B5
York (cape) B1

○ Population of district.
*Population of met. area.

Topography

5,000 m. | 2,000 m. | 1,000 m. | 500 m. | 200 m. | 100 m. | Sea
16,404 ft. | 6,562 ft. | 3,281 ft. | 1,640 ft. | 656 ft. | 328 ft. | Level
| | | | | | Below

New South Wales and Victoria

AUSTRALIAN CAPITAL TERRITORY

CITIES and TOWNS

Canberra (cap.), Australia 220,423 . . . . . E4
Canberra †220,822 . . . . . E4
Jervis Bay 787 . . . . . F4

NEW SOUTH WALES

CITIES and TOWNS

Aberdeen 1,410 . . . . . F3
Adaminaby 378 . . . . . E5
Adelong 806 . . . . . D4
Albury 35,072 . . . . . D5
Alstonville 2,936 . . . . . G1
Ardlethan 602 . . . . . D4
Ariah Park 301 . . . . . D4
Armidale 18,922 . . . . . F2
Ashfield 41,253 . . . . . J3
Ashford 594 . . . . . F1
Attunga 282 . . . . . F2
Auburn 46,622 . . . . . J3
Avondale 2,265 . . . . . F3
Ballina 9,738 . . . . . G1
Balranald 1,442 . . . . . B4
Bangalow 614 . . . . . G1
Bankstown 152,636 . . . . . J3
Baradine 723 . . . . . E2
Barellan 380 . . . . . D4
Bargo 868 . . . . . F4
Barham-Koondrook 1,039 . . . . . C4
Barmedman 223 . . . . . D4
Barooga 556 . . . . . C4
Barraba 1,679 . . . . . F2
Barringun C1
Batemans Bay 4,924 . . . . . F4
Bathurst 19,640 . . . . . E3
Batlow 1,354 . . . . . E4
Baulkham Hills 93,084 . . . . . H3
Beechwood 258 . . . . . G2
Bega 4,388 . . . . . E5
Bellbird-Cessnock 16,916 . . . . . F3
Bellbird-Cessnock *38,724 . . . . . F3
Bellingen 1,593 . . . . . G2
Bendemeer 231 . . . . . F2
Bermagui 827 . . . . . F5
Berridale 695 . . . . . E5
Berrigan 960 . . . . . C4
Berry 1,220 . . . . . F4
Binalong 212 . . . . . E4
Bingara 1,257 . . . . . F1
Binnaway 583 . . . . . E2
Blacktown 181,139 . . . . . H3
Blayney 2,694 . . . . . E3
Blue Mountains 55,877 . . . . . F3
Bodalla 224 . . . . . F5
Bogan Gate D3
Boggabilla 528 . . . . . F1
Boggabri 1,023 . . . . . F2
Bomaderry-Nowra 17,887 . . . . . F4
Bombala 1,504 . . . . . E5
Bonalbo 466 . . . . . G1
Booligal C3
Boorooban C4
Boorowa 1,183 . . . . . E4
Botany 34,703 . . . . . J4
Bourke 3,326 . . . . . D2
Bowral 6,862 . . . . . F4
Bowraville 848 . . . . . G2
Braidwood 944 . . . . . E4
Branxton-Greta 2,849 . . . . . F3
Bredbo E4
Brewarrina 1,236 . . . . . D1
Bribbaree D4
Brisbane Water 71,984 . . . . . F3
Broken Hill 26 913 . . . . . A3
Browning E4
Brunswick Heads 1,877 . . . . . G1
Budgewoi Lake 25,474 . . . . . F3
Bugaldie E2
Bulahdelah 972 . . . . . G3
Bundanoon 1,018 . . . . . F4
Bundarra 377 . . . . . F2
Bungendore 683 . . . . . E4
Burcher D3
Burns A3
Burraboi C4
Burwood 28,896 . . . . . J3
Byrock D2
Byron Bay 3,187 . . . . . G1
Camden 9,000 . . . . . F4
Camden Haven 3,161 . . . . . G2
Campbelltown 91,525 . . . . . F4
Candelo 255 . . . . . E5
Canowindra 1,720 . . . . . E3
Canterbury 126,741 . . . . . J3
Captains Flat 309 . . . . . E4
Caragabal D3
Carcoar 277 . . . . . E3
Carinda D2
Carrathool ○1,210 . . . . . C4
Carroll F2
Casino 9,743 . . . . . G1
Cassilis E3
Cessnock-Bellbird 16,916 . . . . . F3
Cessnock-Bellbird *38,724 . . . . . F3
Cobar 3,583 . . . . . C2
Cobargo 285 . . . . . E5

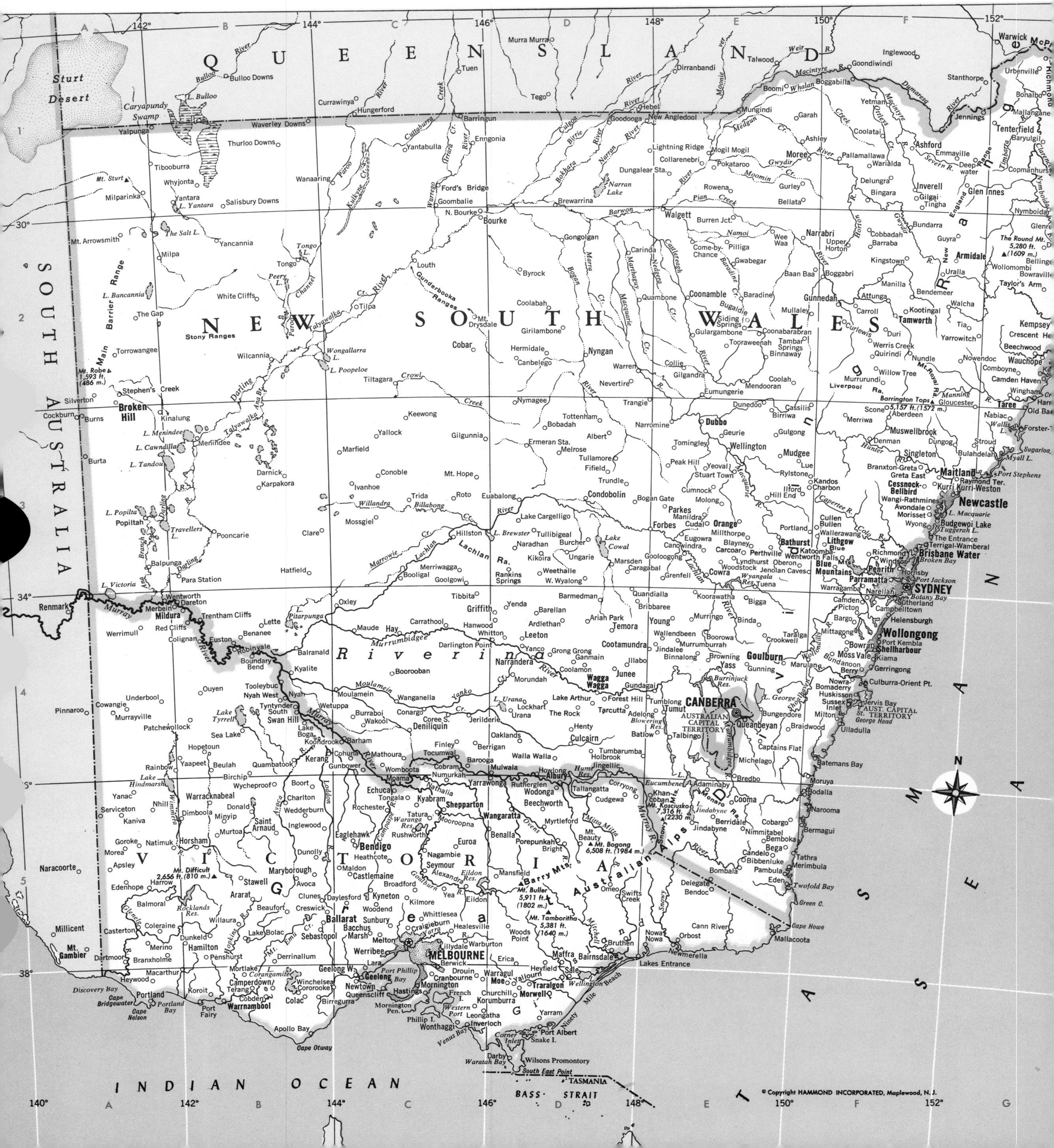

Cobbadah	F2	Coraki 895	G1	Dee Why	K3
Coffs Harbour 16,020	G2	Coramba 202	G2	Delegate 347	E5
Collarenebri 602	E1	Corowa-Wahgunyah 3,390	D4	Delungra 332	F1
Collie	E2	Cowra 7,900	E3	Deniliquin 7,354	C4
Comboyne	G2	Crescent Head 944	G2	Denman 1,122	F3
Concord 23,926	J3	Cronulla	J4	Dorrigo 1,192	G2
Condobolin 3,355	D3	Crookwell 2,063	E4	Drummoyne 30,961	J3
Conoble	C3	Cudal 400	E3	Dubbo 23,986	E3
Coogee	K3	Cullburra-Orient Point 2,068	F4	Dudedoo 836	E3
Coolah 878	E2	Culcairn 1,027	D4	Dungalear Station	D1
Coolamon 1,088	D4	Cullen Bullen 231	E3	Dungog 2,126	F3
Cooma 7,978	E5	Cumnock 252	E3	Eden 3,107	E5
Coonabarabran 3,001	E2	Curlewis 487	E2	Emmaville 503	F1
Coonamble 3,090	E2	Dareton 612	B4	Ermeran Station	D3
Cootamundra 6,540	D4	Darlington Point 599	C4	Eugowra 577	E3
Copmanhurst ○2,857	G1	Deepwater 260	F1	Evans Head 1,802	G1

NEW SOUTH WALES

AREA 309,498 sq. mi.
(801,600 sq. km.)
POPULATION 5,126,217
CAPITAL Sydney
LARGEST CITY Sydney
HIGHEST POINT Mt. Kosciusko
7,310 ft. (2,228 m.)

VICTORIA

AREA 87,876 sq. mi.
(227,600 sq. km.)
POPULATION 3,832,443
CAPITAL Melbourne
LARGEST CITY Melbourne
HIGHEST POINT Mt. Bogong
6,508 ft. (1,984 m.)

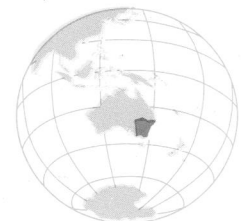

Topography

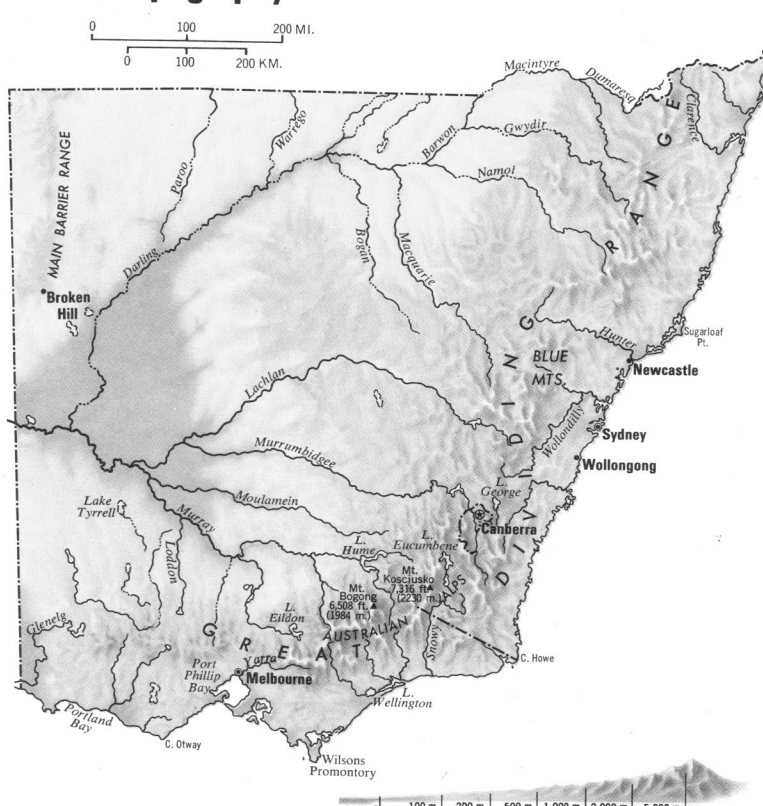

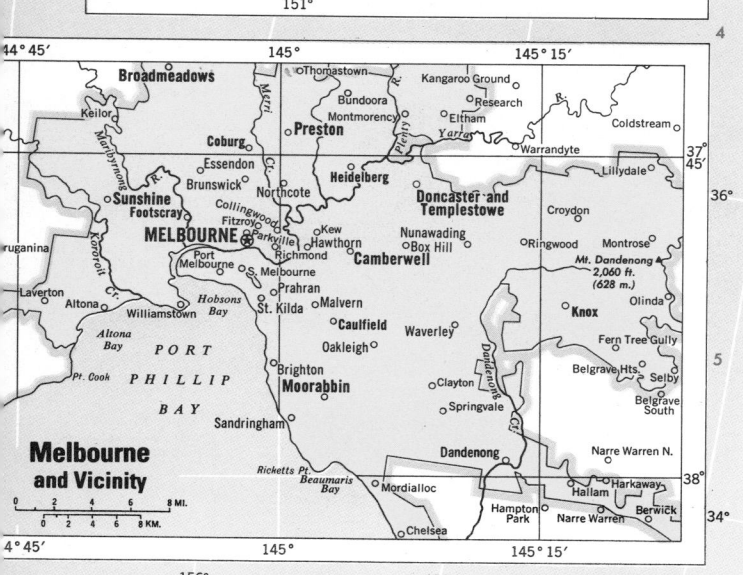

Fairfield 129,557	H3	Keewong	C3	Moss Vale 4,415	F4
Finley 2,193	C4	Kempsey 9,037	G2	Mossgiel	C4
Forbes 8,029	E3	Kendall 522	G2	Moulamein 396	C4
Forest Hill 1,977	D4	Khancoban 515	E5	Mount Arrowsmith	A2
Forster-Tuncurry 9,261	G3	Kiama 7,717	F4	Mount Drysdale	C2
Frederickton 616	G2	Kinalung	B3	Mount Hope	C3
Ganmain 650	D4	Kogarah 46,322	J4	Mudgee 6,015	E3
Gerringong 1,775	F4	Koorawatha 262	E4	Mullumbimby 2,234	G1
Geurie 290	E3	Kootingal 731	F2	Mungindi 707	E1
Gilgai 257	F1	Ku-ring-gai 101,501	J3	Murrumburrah 2,070	E4
Gilgandra 2,700	E2	Kurri Kurri-Weston 12,795	F3	Murrurundi 861	F2
Gilgunnia	C3	Kyogle 3,070	G1	Murwillumbah 7,807	G1
Glen Innes 6,052	F2	Lake Cargellico 1,240	D3	Muswellbrook 8,548	F3
Glenreagh 233	G2	Lane Cove 29,113	J3	Nabiac 363	G3
Gloucester 2,488	F2	Leeton 6,498	D4	Nambucca Heads 4,053	G2
Goodooga 248	D1	Leichhardt 57,332	J3	Narellan 2,104	F3
Goolgowi 245	C3	Lette	B4	Narooma 2,758	F5
Gooloogong 221	E3	Lidcombe	J3	Narrabeen	K3
Goombalie	C1	Lightning Ridge 1,112	E1	Narrabri 7,296	E2
Goulburn 21,755	E4	Lismore 24,033	G1	Narrandera 5,013	D4
Grafton 17,005	G1	Lithgow 12,793	F3	Narromine 2,994	E3
Grenfell 2,070	E3	Liverpool 92,715	H4	New Angledool	E1
Greta-Branxton 2,849	F3	Lockhart 923	D4	Newcastle 135,207	F3
Greta East 398	F3	Macksville 2,352	G2	Newcastle †389,237	F3
Griffith 13,187	C4	Maclean 2,593	G1	Nimmitabel 257	E5
Gulargambone 457	E2	Maitland 38,865	F3	North Sydney 48,500	J3
Gulgong 1,740	E3	Manildra 520	E3	Nowra-Bomaderry 17,887	F4
Gundagai 2,308	D4	Manilla 1,884	F2	Nundle 235	F2
Gunnedah 8,909	F2	Manly 37,080	K3	Nymboida ○2,044	G1
Gunning 438	E4	Maroubra	K3	Nyngan 2,485	D2
Guyra 1,840	F2	Marrickville 83,448	J3	Oaklands 283	D4
Hanwood 306	C4	Marsden	D3	Oberon 1,937	E3
Harrington 1,183	G2	Marsden Park 518	H3	Old Bar 970	G3
Hay 2,958	C4	Marulan 330	E4	Orange 27,626	E3
Henty 883	D4	Mathoura 582	C4	Pallamallawa 340	E1
Hillston 999	C3	Melrose	D3	Pambula 604	E5
Holbrook 1,276	D4	Mendooran 325	E2	Parkes 9,047	E3
Holroyd 80,116	H3	Menindee 455	B3	Parramatta 130,943	H3
Hornsby 111,081	J3	Merimbula 2,899	F5	Peak Hill 1,037	E3
Howlong 1,072	D4	Merriwa 943	F3	Penrith 108,720	J3
Hunters Hill 12,537	J3	Milpa	D1	Perthville 639	E3
Hurstville 64,910	J4	Milperra	H4	Picton 1,817	F3
Huskisson 2,296	F4	Milton 740	F4	Piliga 206	E2
Iluka 1,362	G1	Mittagong 4,266	F4	Popiltah	A3
Inverell 9,734	F1	Moama	C5	Portland 1,980	E3
Ivanhoe 517	C3	Mogil Mogil	E1	Port Kembla	F4
Jerilderie 1,075	C4	Molong 1,374	E3	Port Macquarie 19,581	G2
Jindabyne 1,602	E5	Mona Vale	K2	Queanbeyan 19,383	E4
Junee 3,993	D4	Moree 10,459	E1	Quirindi 2,851	F2
Kandos 1,626	E3	Morisset 1,593	F3	Randwick 116,202	J3
Karpakora	B3	Moruya 2,003	F4	Raymond Terrace 7,548	F3
Katoomba-Wentworth Falls		Mosman 26,200	J3	Richmond-Windsor 15,491	F3
13,942	F3			Rockdale 83,719	J4

(continued on following page)

Ryde 88,948 . . . J3
Rylstone 651 . . . E3
Salisbury Downs . . . B1
Sawtell 5,970 . . . G2
Scone 3,949 . . . F3
Shellharbour 41,790 . . . F4
Singleton 9,572 . . . F3
Smithtown-Gladstone 953 . . . G2
South Sydney 30,776 . . . J3
South West Rocks 1,314 . . . G2
Stephen's Creek . . . A2
Strathfield 25,882 . . . J3
Stroud 522 . . . G3
Sussex Inlet 1,293 . . . F4
Sutherland 165,336 . . . J4
Sydney (cap.) 2,876,508 . . . J3
Sydney †3,204,696 . . . J3
Talbingo 481 . . . E4
Tamworth 29,657 . . . F2
Taralga 272 . . . E4
Tarcutta 263 . . . D4
Taree 14,697 . . . G2
Tathra 1,077 . . . F5
Temora 4,350 . . . D4
Tenterfield 3,402 . . . G1
Terrigal-The Entrance 37,891 F3
The Rock 693 . . . D4
Thurloo Downs . . . B1
Tibbita . . . C4
Tibooburra . . . B1
Tiltagara . . . C2
Tingha 886 . . . F1
Tocumwal 1,174 . . . C4
Tongo . . . B2
Torrowangee . . . A2
Tottenham 366 . . . D3
Trangie 977 . . . D3
Trundle 515 . . . D3
Tullamore 324 . . . D3
Tumbarumba 1,536 . . . D4
Tumut 5,816 . . . E4
Tweed Heads . . . G1
Ulladulla 6,018 . . . F4
Ulmarra 395 . . . G1
Ungarie 428 . . . D3
Uralla 2,090 . . . F2
Urana 419 . . . D4
Urbenville 282 . . . G1
Urunga 2,045 . . . G2
Villawood . . . H3
Wagga Wagga 36,837 . . . D4
Wakool 278 . . . C4
Walcha 1,674 . . . F2
Walgett 2,157 . . . E2
Walla Walla 593 . . . D4

Wallerawang 1,855 . . . F3
Wangi-Rathmines 5,106 . . . F3
Warialda 1,340 . . . F1
Warragamba 1,406 . . . F3
Warren 2,153 . . . D2
Warringah ○172,653 . . . K3
Wauchope 3,645 . . . G2
Waverley 61,575 . . . K3
Waverley Downs . . . B1
Wee Waa 1,904 . . . E2
Wellington 5,280 . . . E3
Wentworth 1,180 . . . B4
Werris Creek 1,924 . . . F2
West Wyalong 3,778 . . . D3
Wetuppa . . . B4
White Cliffs . . . B2
Whitton 344 . . . D4
Whyjonta . . . B1
Wilcannia 982 . . . B2
Willoughby 52,120 . . . J3
Willow Tree 258 . . . F2
Wingham 3,937 . . . G2
Wollongong 169,381 . . . F4
Wollongong †222,539 . . . F4
Woodburn 647 . . . G1
Woodenbong 409 . . . G1
Woodstock 266 . . . E3
Woolgoolga 2,081 . . . G2
Wooli 457 . . . G1
Woollahra 51,659 . . . K3
Wyong 3,902 . . . F3
Yallock . . . C3
Yalpunga . . . A1
Yamba 2,528 . . . G1
Yancannia . . . B2
Yanco 415 . . . D4
Yantara . . . B1
Yass 4,283 . . . E4
Yenda 697 . . . D4
Yeoval 288 . . . E3
Young 6,906 . . . E4

OTHER FEATURES

Ana Branch, Darling (riv.) . . . A3
Australian Alps (mts.) . . . D5
Barrington Tops (mt.) . . . F2
Barwon (riv.) . . . D2
Blue (mts.) . . . F3
Bogan (riv.) . . . D2
Bondi (beach) . . . K3
Botany (bay) . . . J4
Broken (bay) . . . F3
Burrinjuck (res.) . . . E4
Byron (cape) . . . G1

Caryapundy (swamp) . . . B1
Castlereagh (riv.) . . . E2
Cawndilla (lake) . . . A3
Clarence (riv.) . . . G1
Colo (riv.) . . . F3
Cowal (lake) . . . D3
Culgoa (riv.) . . . D1
Cuttaburra (creek) . . . C1
Darling (riv.) . . . B3
Dumaresq (riv.) . . . F1
Eucumbene (lake) . . . E5
George (lake) . . . E4
Georges (riv.) . . . H4
Gower (mt.) . . . J2
Great Dividing (range) . . . E3
Green (cape) . . . F5
Gunderbooka (ranges) . . . C2
Gwydir (riv.) . . . E1
Howe (cape) . . . F5
Hume (res.) . . . D4
Hunter (riv.) . . . F3
Kosciusko (mt.) . . . E5
Kurnell (pen.) . . . J4
Lachlan (range) . . . C3
Lachlan (riv.) . . . C3
Liverpool (range) . . . F2
Lord Howe (isl.) 287 . . . J2
Macintyre (riv.) . . . E1
Macquarie (lake) . . . F3
Macquarie (riv.) . . . D2
Main Barrier (range) . . . A2
Manning (riv.) . . . F2
Marthaguy (creek) . . . D2
McPherson (range) . . . G1
Menindee (lake) . . . B3
Monaro (range) . . . E5
Moonie (riv.) . . . E1
Moulamein (creek) . . . C4
Mount Royal (range) . . . F2
Murray (riv.) . . . A4
Murrumbidgee (riv.) . . . C4
Myall (lake) . . . G3
Namoi (riv.) . . . E2
Narran (lake) . . . D1
New England (range) . . . F1
Paroo (riv.) . . . C1
Parramatta (riv.) . . . J3
Poopeloe (lake) . . . C2
Port Jackson (inlet) . . . J3
Port Stephens (inlet) . . . G3
Richmond (range) . . . G1
Richmond (riv.) . . . G1
Riverina (reg.) . . . C4
Robe (mt.) . . . A2
Round, The (mt.) . . . G2

Salt, The (lake) . . . B2
Shoalhaven (riv.) . . . E4
Smoky (cape) . . . G2
Snowy (mts.) . . . E5
Snowy (riv.) . . . E5
Stony (ranges) . . . B2
Sturt (mt.) . . . A1
Sugarloaf (pt.) . . . G3
Talyawalka (creek) . . . B2
Tandou (lake) . . . A3
Tasman (sea) . . . F5
The Round (mts.) . . . C4
The Salt (lake) . . . B2
Timbarra (riv.) . . . G1
Tuggerah (lake) . . . F3
Victoria (lake) . . . A3
Warrego (riv.) . . . C1
Willandra Billabong (creek) . . . C3
Wollondilly (riv.) . . . F4

VICTORIA
CITIES and TOWNS

Alexandra 1,756 . . . C5
Altona 30,909 . . . H5
Apollo Bay 921 . . . B6
Ararat 8,336 . . . B5
Avoca 1,032 . . . B5
Bacchus Marsh 6,224 . . . C5
Bairnsdale 9,459 . . . D5
Ballarat 35,681 . . . C5
Ballarat †71,930 . . . C5
Balmoral 257 . . . A5
Beaufort 1,214 . . . B5
Beechworth 3,154 . . . D5
Belgrave Heights . . . J5
Belgrave South . . . K5
Benalla 8,151 . . . C5
Bendigo 31,841 . . . C5
Bendigo †58,818 . . . C5
Berwick 36,181 . . . K6
Beulah 290 . . . B4
Birchip 895 . . . B4
Birregurra 416 . . . B6
Boort 863 . . . B5
Box Hill 47,579 . . . J5
Bright 1,545 . . . D5
Brighton 33,697 . . . J5
Broadford 1,580 . . . C5
Broadmeadows 103,540 . . . H4
Brunswick 44,464 . . . J5
Bruthen 449 . . . D5
Bundoora . . . J4
Camberwell 85,883 . . . J5

Camperdown 3,545 . . . C6
Cann River 345 . . . E5
Casterton 1,945 . . . A5
Castlemaine 7,583 . . . C5
Caulfield 69,922 . . . J5
Charlton 1,377 . . . B5
Chelsea 26,034 . . . J6
Churchill 4,796 . . . D6
Clunes 761 . . . B5
Cobden 1,453 . . . B6
Cobram 3,817 . . . C4
Coburg 55,035 . . . H5
Cohuna 2,178 . . . C4
Colac 10,587 . . . B6
Coldstream 1,395 . . . K4
Coleraine 1,232 . . . A5
Collingwood 15,089 . . . J5
Corryong 1,320 . . . D4
Craigieburn 4,296 . . . C5
Cranbourne 9,400 . . . C6
Creswick 2,036 . . . B5
Croydon 36,210 . . . K5
Dandenong 54,962 . . . K5
Darby . . . D6
Dartmoor 349 . . . A5
Daylesford 2,883 . . . C5
Derrinallum 287 . . . B5
Dimboola 1,675 . . . B5
Donald 1,609 . . . B5
Doncaster and Templestowe
90,660 . . . J5
Drouin 3,492 . . . D5
Dunkeld 402 . . . B5
Dunolly 621 . . . B5
Eaglehawk 7,355 . . . C5
Echuca 7,943 . . . C5
Edenhope 827 . . . A5
Eildon 737 . . . C5
Eltham 34,648 . . . J4
Erica 236 . . . D5
Essendon 56,380 . . . H5
Euroa 2,640 . . . C5
Fitzroy 19,112 . . . H5
Footscray 49,756 . . . H5
Geelong 14,471 . . . C6
Geelong †137,173 . . . C6
Geelong West 14,823 . . . C6
Goroke 370 . . . A5
Gunbower 259 . . . C4
Hamilton 9,751 . . . A5
Hawthorn 30,689 . . . J5
Healesville 4,526 . . . C5
Heathcote 1,213 . . . C5
Heidelberg 64,757 . . . J5
Heyfield 1,635 . . . D6

Heywood 1,266 . . . A6
Hopetoun 1,832 . . . B4
Horsham 12,034 . . . B5
Inglewood 674 . . . B5
Inverloch 1,523 . . . C6
Kaniva 956 . . . A5
Keilor 81,762 . . . H5
Kerang 4,049 . . . B4
Kew 28,870 . . . J5
Kilmore 1,728 . . . C5
Knox 88,902 . . . K5
Koroit 1,988 . . . B6
Korumburra 2,798 . . . D6
Kyabram 5,414 . . . C5
Kyneton 3,185 . . . C5
Lake Boga 502 . . . B4
Lake Bolac 211 . . . B5
Lakes Entrance 3,414 . . . E5
Lara 4,231 . . . C6
Leongatha 3,736 . . . D6
Lillydale 62,077 . . . J4
Macarthur 322 . . . A6
Maffra 3,822 . . . D5
Maldon 1,009 . . . C5
Mallacoota 726 . . . E5
Malvern 43,211 . . . J5
Mansfield 1,920 . . . D5
Maryborough 7,858 . . . B5
Melbourne (cap.)
2,578,759 . . . H5
Melbourne †2,722,817 . . . H5
Melton 20,599 . . . C5
Merbein 1,735 . . . A4
Merino 298 . . . A5
Mildura 15,763 . . . A4
Minyip 567 . . . B5
Moe 16,649 . . . D6
Montmorency . . . J4
Montrose . . . K5
Moorabbin 97,810 . . . J5
Mooroopna . . . C5
Mordialloc 27,869 . . . J6
Morea . . . A5
Mornington 23,512 . . . C6
Mortlake 1,056 . . . B6
Morwell 16,491 . . . D5
Mount Beauty 1,509 . . . D5
Murrayville 313 . . . A4
Murtoa 946 . . . B5
Myrtleford 2,815 . . . D5
Nagambie 1,102 . . . C5
Narre Warren North 761 . . . K5
Nathalia 1,222 . . . C5
Natimuk 482 . . . A5
Newtown 10,210 . . . C6

Nhill 1,567 . . . A5
Northcote 51,235 . . . J5
Numurkah 2,713 . . . C5
Nunawading 97,052 . . . J5
Nyah 351 . . . B4
Nyah West 535 . . . B4
Oakleigh 55,612 . . . J5
Omeo 272 . . . D5
Orbost 2,586 . . . E5
Ouyen 1,527 . . . B4
Penshurst 558 . . . B5
Porepunkah 268 . . . D6
Port Albert 267 . . . D6
Port Fairy 2,276 . . . A6
Portland 9,353 . . . A6
Port Melbourne 8,585 . . . H5
Prahran 45,018 . . . J5
Preston 84,519 . . . J5
Quambatook 359 . . . B4
Queenscliff 3,420 . . . C6
Rainbow 700 . . . A4
Red Cliffs 2,409 . . . A4
Richmond 24,506 . . . J5
Ringwood 38,665 . . . K5
Robinvale 1,751 . . . B4
Rochester 2,399 . . . C5
Rushworth 994 . . . C5
Rutherglen 1,454 . . . C4
Saint Arnaud 2,721 . . . B5
Saint Kilda 49,366 . . . J5
Sale 12,968 . . . D6
Sandringham 31,175 . . . J5
Sea Lake 943 . . . B4
Sebastopol 6,462 . . . B5
Seymour 6,494 . . . C5
Shepparton-Mooroopna
‡28,373 . . . C5
South Barwon 35,307 . . . C6
South Melbourne 19,955 . . . J5
Springvale 80,186 . . . J5
Stawell 6,160 . . . B5
Sunbury 11,085 . . . C5
Sunshine 94,419 . . . H5
Swan Hill 8,398 . . . B4
Swifts Creek 288 . . . D5
Tallangatta 950 . . . D5
Tatura 2,697 . . . C5
Templestowe and Doncaster
90,660 . . . J5
Terang 2,111 . . . B6
Tongala 994 . . . C5
Traralgon 18,057 . . . D6
Underbool 274 . . . A4
Wangaratta 16,202 . . . D5
Warburton 2,009 . . . D5
Warracknabeal 2,735 . . . B5
Warragul 7,712 . . . D6
Warrnambool 21,414 . . . A6
Waverley 122,471 . . . J5
Wedderburn 868 . . . B5
Werrimull . . . A4
Whittlesea 65,657 . . . C5
Willaura 377 . . . B5
Williamstown 25,554 . . . H5
Winchelsea 825 . . . B6
Wodonga 19,208 . . . D5
Wonthaggi 4,797 . . . C6
Woodend 1,785 . . . C5
Wycheproof 938 . . . B5
Yallourn 26 . . . D6
Yarram 2,085 . . . D6
Yarrawonga 3,442 . . . C5
Yea 996 . . . C5

OTHER FEATURES

Australian Alps (mts.) . . . D5
Avoca (riv.) . . . B5
Barry (mts.) . . . D5
Bogong (mt.) . . . D5
Bridgewater (cape) . . . A6
Buller (mt.) . . . D5
Campaspe (riv.) . . . C5
Corangamite (lake) . . . B6
Corner (inlet) . . . D6
Dandenong (mt.) . . . K5
Difficult (mt.) . . . B5
Discovery (bay) . . . A6
Eildon (lake) . . . C5
French (isl.) 123 . . . C6
Gippsland (reg.) . . . D5
Glenelg (riv.) . . . A5
Goulburn (riv.) . . . C5
Hindmarsh (lake) . . . A5
Hobsons (bay) . . . H5
Hopkins (riv.) . . . B5
Hume (lake) . . . D4
Indian Ocean . . . B6
Loddon (riv.) . . . B5
Mitchell (riv.) . . . D5
Mitta Mitta (riv.) . . . D5
Mornington (pen.) . . . C6
Mount Emu (creek) . . . B5
Murray (riv.) . . . A4
Nelson (cape) . . . A6
Ninety Mile (beach) . . . D6
Otway (cape) . . . B6
Ovens (riv.) . . . D5
Phillip (isl.) 2,832 . . . C6
Portland (bay) . . . A6
Port Phillip (bay) . . . C6
Rocklands (res.) . . . B5
Snowy (riv.) . . . E5
South East (pt.) . . . D6
Tasman (sea) . . . F5
Tyrrell (lake) . . . B4
Waratah (bay) . . . D6
Wellington (lake) . . . D6
Western Port (inlet) . . . C6
Wilsons (prom.) . . . D6
Wimmera (riv.) . . . A5
Yarra (riv.) . . . C5

*City and suburbs.
○ Population of district.
†Population of met. area.
‡Population of urban area.

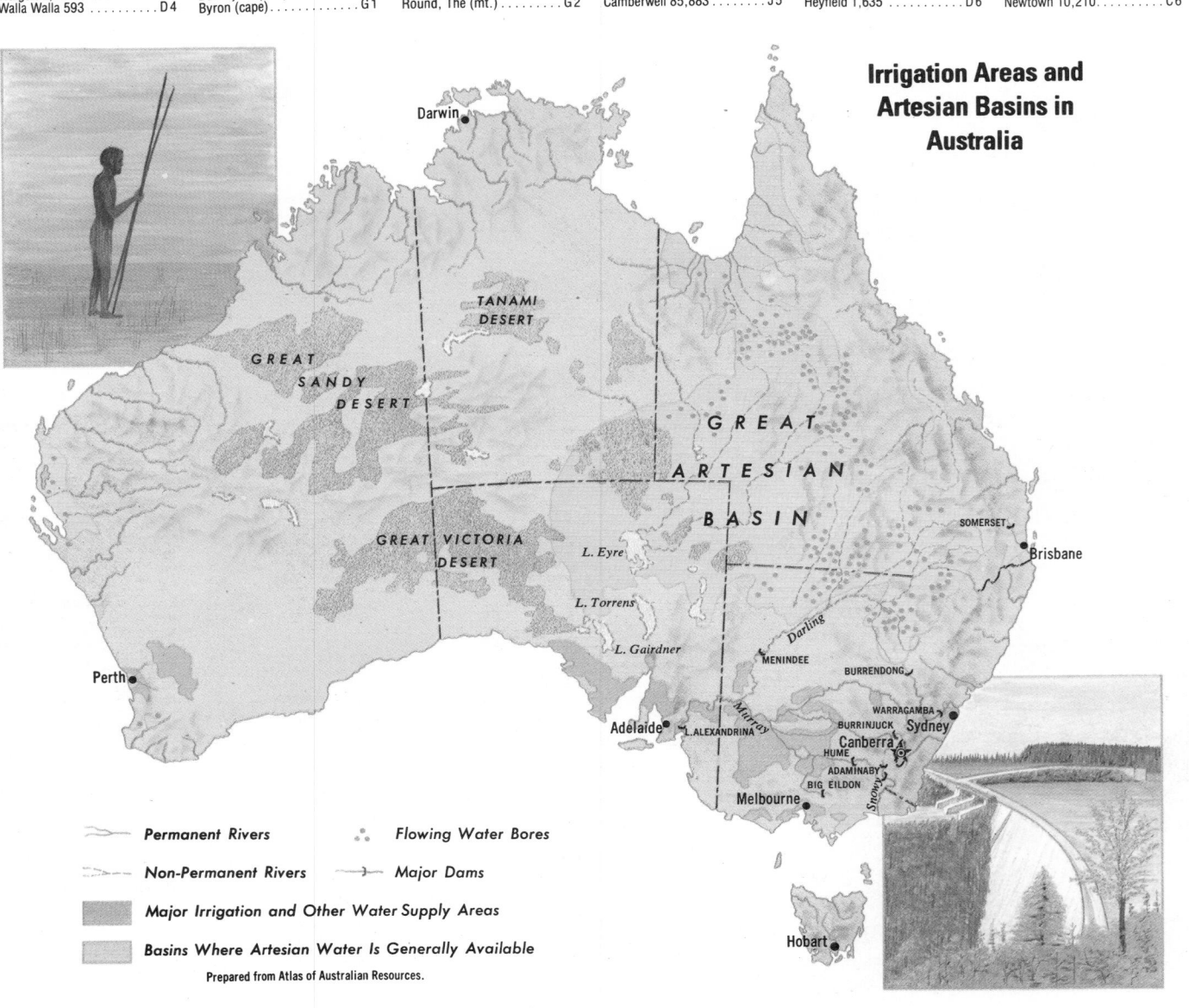

Irrigation Areas and Artesian Basins in Australia

Permanent Rivers

Non-Permanent Rivers

Major Irrigation and Other Water Supply Areas

Basins Where Artesian Water Is Generally Available

Flowing Water Bores

Major Dams

Prepared from Atlas of Australian Resources.

Topography

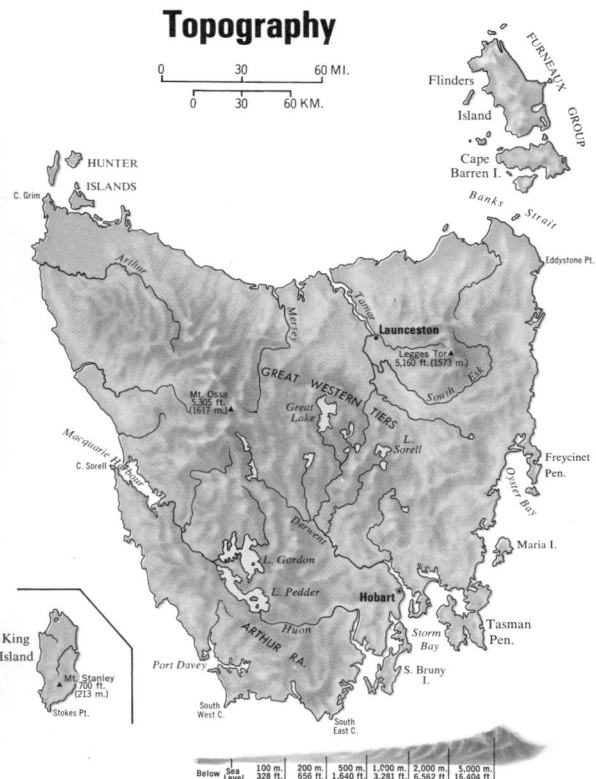

0   30   60 MI.
0   30   60 KM.

FURNEUX GROUP
Flinders Island
Cape Barren I.
Banks Strait
HUNTER ISLANDS
C. Grim
Eddystone Pt.
Launceston
Legges Tor 5,160 ft. (1573 m.)
GREAT WESTERN TIERS
Mt. Ossa 5,305 ft. (1617 m.)
Great Lake
C. Sorell
L. Sorell
Freycinet Pen.
Macquarie Harbour
Oyster Bay
L. Gordon
L. Pedder
Derwent
Hobart
Maria I.
King Island
Mt. Stanley 700 ft. (213 m.)
Stokes Pt.
ARTHUR RA.
Huon
Port Davey
Storm Bay
Tasman Pen.
S. Bruny I.
South West C.
South East C.

Below Sea Level | 100 m. 328 ft. | 200 m. 656 ft. | 500 m. 1,640 ft. | 1,000 m. 3,281 ft. | 2,000 m. 6,562 ft. | 5,000 m. 16,404 ft.

TASMANIA

AREA 26,178 sq. mi. (67,800 sq. km.)
POPULATION 418,957
CAPITAL Hobart
LARGEST CITY Hobart
HIGHEST POINT Mt. Ossa 5,305 ft. (1,617 m.)

Forth (riv.)	C3
Frankland (cape)	D1
Frankland (range)	B4
Franklin (riv.)	B4
Frenchmans Cap (mt.)	B4
Freycinet (pen.)	E4
Furneaux Group (isls.) 1,039	E1
Gordon (lake)	C4
Gordon (riv.)	B4
Great (lake)	C3
Great Western Tiers (mts.)	C3
Grim (cape)	A2
Hartz (mt.)	C5
Hibbs (pt.)	B4
Hogan Group (isl.)	D1
Hummock (isl.)	D2
Hunter (isl.)	A2
Hunter (isls.)	B2
Huon (riv.)	C5
Indian Ocean	A4
Kent Group (isls.)	D1
King (isl.) 2,592	A1

King (riv.)	B4
King William (lake)	B4
Lake (riv.)	D3
Legges Tor (mt.)	D3
Leven (riv.)	B3
Lofty (range)	E3
Low Rocky (pt.)	B4
Lyell (mt.)	B4
Maatsuyker (isls.)	C5
Macquarie (harb.)	B4
Macquarie (riv.)	D3
Maria (isl.)	E4
Marion (bay)	E4
Mersey (riv.)	C3
Munro (mt.)	E2
Naturaliste (cape)	E2
Nive (riv.)	C4
Norfolk (bay)	D4
North (pt.)	E1
North Bruny (isl.)	D5
North Esk (riv.)	D3

Ouse (riv.)	C4
Oyster (bay)	E4
Pedder (riv.)	B4
Phoques (bay)	A1
Picton (mt.)	C5
Pieman (riv.)	B3
Pillar (cape)	E5
Port Davey (inlet)	B5
Portland (cape)	D2
Ramsey (mt.)	B3
Raoul (cape)	D5
Reid (rapid)	B1
Ringarooma (bay)	D2
Robbins (isl.)	B2
Saint Clair (lake)	C4
Saint Helens (pt.)	E3
Saint Vincent (cape)	B5
Savage (riv.)	B3
Schouten (isl.)	E4
Sorell (cape)	B4
Sorell (lake)	D4
South (cape)	C5

South Bruny (isl.)	D5
South East (cape)	C5
South Esk (riv.)	D3
South West (cape)	B5
Stanley (mt.)	A1
Stokes (pt.)	A1
Storm (bay)	D5
Strzelecki (mt.)	D2
Tamar (riv.)	D3
Tasman (head)	D5
Tasman (pen.)	E5
Tasman (sea)	E4
Three Hummock (isl.)	B2
Vansittart (isl.)	E2
West (pt.)	A2
West Sister (isl.)	D1
Wickham (cape)	A1

○ Population of district.
*Population of met. area.

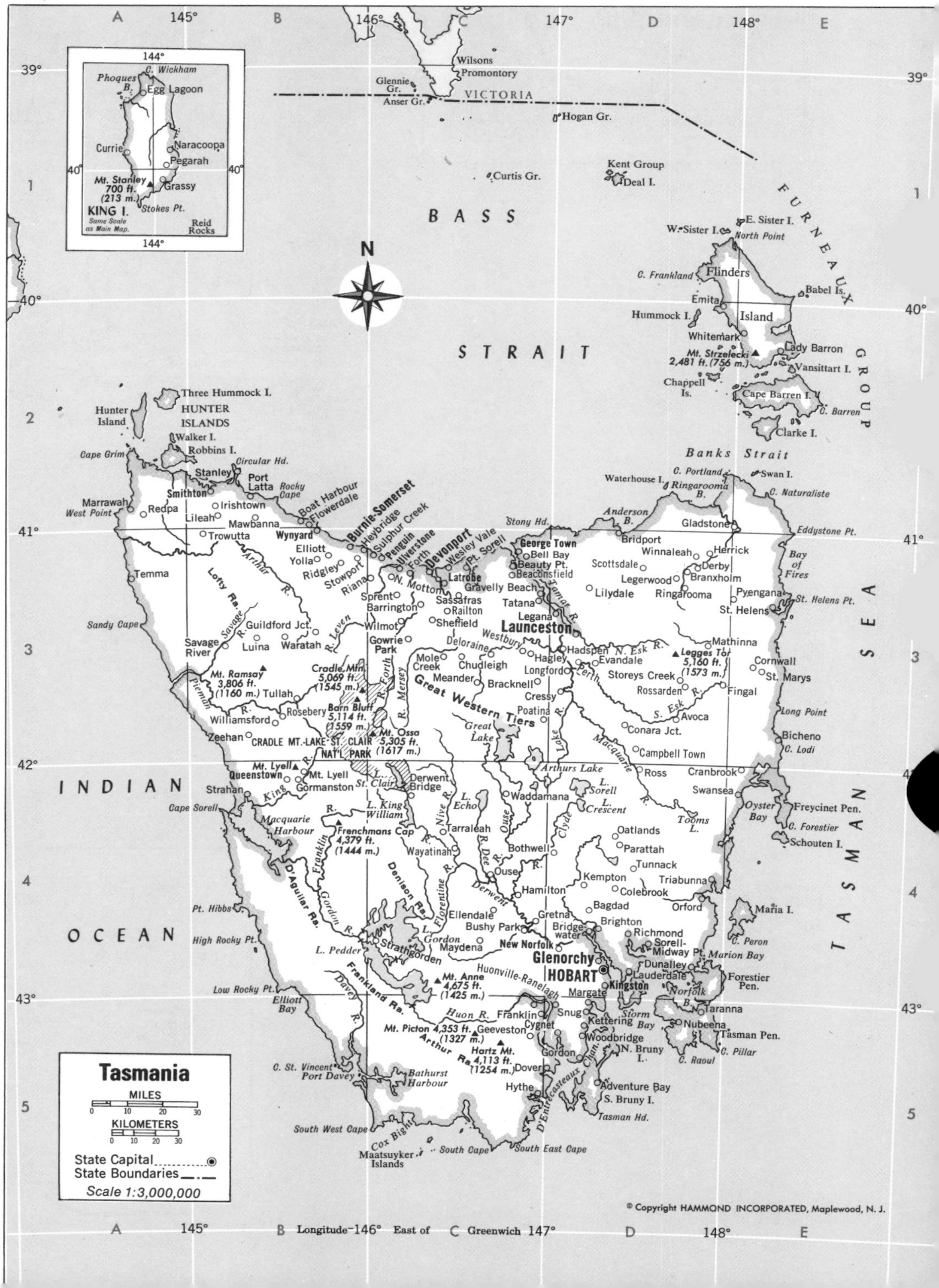

CITIES and TOWNS

Adventure Bay	D5
Avoca	D3
Bagdad	D4
Beaconsfield 898	C3
Beauty Point 998	C3
Bell Bay	C3
Bicheno 674	E3
Boat Harbour	B2
Bothwell 356	C4
Bracknell 347	D3
Branxholm 273	D3
Bridgewater 6,880	D4
Bridport 885	D3
Brighton 9,441	D4
Burnie 19,994	B3
Campbell Town 879	D3
Chudleigh	C3
Colebrook	D4
Cressy 640	D3
Currie 859	A1
Cygnet 715	C5
Deloraine 1,923	C3
Derwent Bridge	C4
Devonport 21,424	C3
Dover 570	C5
Dunalley 203	D4
Evandale 614	D3
Exeter 353	C3
Fingal 424	E3
Forth 273	C3
Franklin 479	C5
Geeveston 860	C5
George Town 5,592	C3
Glenorchy 41,019	C4
Gormanston 126	B4
Gowrie Park	C3
Grassy 780	B1
Gravelly Beach 535	C3
Hadspen 908	D3
Hagley 232	C3
Hamilton 2,488	C4
Heybridge 395	C3
Hobart (cap.) 128,603	D4
Hobart *168,359	D4
Huonville-Ranelagh 1,347	C5
Kettering 288	D5
Kingston 8,556	C4
Latrobe 2,401	C3
Lauderdale 2,117	D4
Launceston 31,273	D3
Launceston *64,555	C3
Legana 964	C3
Lilydale 308	D3
Longford 2,027	C3
Luina 522	B3
Margate 476	C4
Maydena 461	C4
Meander	C3
Mole Creek 303	C3
New Norfolk 6,243	C4
Oatlands 545	D4
Orford 378	D4
Penguin 2,616	C3
Perth 1,229	D3
Poatina	D3
Port Sorell 859	C3
Queenstown 3,714	B4
Railton 857	C3
Richmond 587	D4
Ridgley 452	B3

Ringarooma 223	D3
Rosebery 2,675	B3
Ross 289	D4
Rossarden 365	D3
Saint Helens 1,005	E3
Saint Marys 653	E3
Sassafras	C3
Savage River 1,141	B3
Scottsdale 2,002	D3
Sheffield 945	C3
Smithton 3,378	A2
Snug 684	D5
Sorell-Midway Point 2,544	D4
Stanley 603	B2
Storeys Creek	D3
Strahan 402	B4
Strathgordon	C4
Sulphur Creek 367	C3
Swansea 428	D4
Tarraleah 498	C4
Temma	A3
Triabunna 924	D4
Tullah 1,894	B3
Ulverstone 9,413	C3
Waratah 342	B3
Wesley Vale	C3
Westbury 1,161	C3
Whitemark	D2
Woodbridge 259	D5
Wynyard 4,582	B3
Zeehan 1,750	B3

OTHER FEATURES

Anderson (bay)	D2
Anne (mt.)	C4
Anser Group (isls.)	C1
Arthur (lake)	D4
Arthur (range)	C5
Arthur (riv.)	A2
Babel (isl.)	E1
Banks (str.)	D2
Barn Bluff (mt.)	B3
Barren (cape)	E2
Bass (str.)	C1
Bathurst (gulf)	C5
Cape Barren (isl.)	E2
Chappell (isls.)	D2
Circular (gulf)	B2
Clarke (isl.)	E2
Clyde (riv.)	D4
Cox (bight)	C5
Cradle (mt.)	B3
Cradle Mt. Lake St. Clair Nat'l Park	B3
Crescent (lake)	D4
Curtis Group (isls.)	C1
D'Aguilar (range)	B4
Davey (riv.)	B4
Deal (isl.)	D1
Dee (riv.)	C4
Denison (range)	C4
D'Entrecasteaux (chan.)	D5
Derwent (riv.)	C4
East Sister (isl.)	E1
Echo (lake)	C4
Eddystone (pt.)	E2
Elliott (bay)	B5
Fires (bay)	E3
Flinders (isl.) 2,150	D1
Florence (riv.)	C4
Forestier (chan.)	E4
Forestier (pen.)	E4

Tasmania

MILES
0 10 20 30
KILOMETERS
0 10 20 30

State Capital ⊙
State Boundaries – ⋅ –
Scale 1:3,000,000

© Copyright HAMMOND INCORPORATED, Maplewood, N.J.

Main Map Labels

KING I. (inset)
C. Wickham
Phoques B.
Egg Lagoon
Currie
Naracoopa
Pegarah
Grassy
Mt. Stanley 700 ft. (213 m.)
Stokes Pt.
Reid Rocks
Same Scale as Main Map

Wilsons Promontory
VICTORIA
Glennie Gr.
Anser Gr.
Hogan Gr.
Curtis Gr.
Kent Group
Deal I.

BASS STRAIT

N

FURNEAUX GROUP
W. Sister I.
E. Sister I.
North Point
C. Frankland
Flinders
Emita
Island
Babel Is.
Hummock I.
Whitemark
Mt. Strzelecki 2,481 ft. (756 m.)
Lady Barron
Chappell Is.
Vansittart I.
Cape Barren I.
C. Barren
Clarke I.

Hunter Island
HUNTER ISLANDS
Walker I.
Robbins I.
Cape Grim
Circular Hd.
Port Latta
Rocky Cape
Stanley
Banks Strait
Waterhouse I.
C. Portland
Ringarooma B.
Swan I.
C. Naturaliste

West Point
Marrawah
Redpa
Lileah
Irishtown
Smithton
Trowutta
Boat Harbour
Wynyard
Elliott
Yolla
Ridgley
Heybridge
Penguin
Ulverstone
Somerset
Burnie
Devonport
Stony Hd.
Anderson
Gladstone
Eddystone Pt.

Temma
Lorty Ra.
Guildford Jct.
Luina
Waratah
Savage River
Sandy Cape
Arthur R.
Savage R.
Williamsford
Zeehan
Rosebery
Tullah
Mt. Ramsay 3,806 ft. (1160 m.)
Barn Bluff 5,114 ft. (1559 m.)
Cradle Mtn. 5,069 ft. (1545 m.)
CRADLE MT.-LAKE ST. CLAIR NAT'L PARK
Mt. Lyell 5,305 ft. (1617 m.)
Mt. Ossa

Yolla
Sulphur Creek
Natone
Wilmot
Sprent
Barrington
Sheffield
Railton
Latrobe
Port Sorell
Pt. Sorell
Bell Bay
Beauty Pt.
Beaconsfield
Gravelly Beach
Legana
George Town
Scottsdale
Winnaleah
Herrick
Derby
Branxholm
Lilydale
Ringarooma
Pyengana
St. Helens
St. Helens Pt.

Mole Creek
Chudleigh
Deloraine
Meander
Westbury
Hagley
Hadspen
Longford
Bracknell
Cressy
Poatina
Launceston
Evandale
Legges Tor 5,160 ft. (1573 m.)
Mathinna
Cornwall
St. Marys
Great Western Tiers
Great Lake
Storeys Creek
Rossarden
Conara Jct.
Avoca
Fingal
Long Point
Bicheno
C. Lodi

Mt. Lyell
Queenstown
Gormanston
Strahan
Cape Sorell
Macquarie Harbour
King R.
Franklin R.
Gordon R.
Mt. Lyell
St. Clair
L. St. Clair
Derwent Bridge
Waddamana
Arthurs Lake
L. Sorell
L. Crescent
L. Echo
Ross
Cranbrook
Swansea
Oyster Bay
Freycinet Pen.
C. Forestier
Schouten I.

Campbell Town
Oatlands
Parattah
Tunnack
Kempton
Colebrook
Orford
Maria I.
C. Peron
Marion Bay

Frenchmans Cap 4,379 ft. (1444 m.)
Jarralton
Bothwell
Hamilton
Ellendale
Bushy Park
Bridge water
Bagdad
Brighton
Richmond
Sorell-Midway Pt.
Dunalley
Forestier Pen.

INDIAN OCEAN

Pt. Hibbs
High Rocky Pt.
L. Pedder
L. Gordon
Strathgorden
Maydena
New Norfolk
Huonville-Ranelagh
Mt. Anne 4,675 ft. (1425 m.)
Glenorchy
HOBART
Kingston
Lauderdale
Norfolk
Tasman Pen.

Low Rocky Pt.
Elliott Bay
Davey R.
Frankland Ra.
Denison Ra.
Gordon R.
Arthur Ra. 4,113 ft. (1254 m.)
Mt. Picton 4,353 ft. (1327 m.)
Hartz Mt.
Huon R.
Franklin
Geeveston
Cygnet
Snug
Woodbridge
Kettering
Storm Bay
S. Bruny I.
C. Raoul
C. Pillar

C. St. Vincent
Port Davey
Bathurst Harbour
Hythe
Dover
Nubeena
Adventure Bay
Tasman Hd.

South West Cape
Maatsuyker Islands
Cox Bight
South Cape
South East Cape

D'Entrecasteaux Chan.

TASMAN SEA

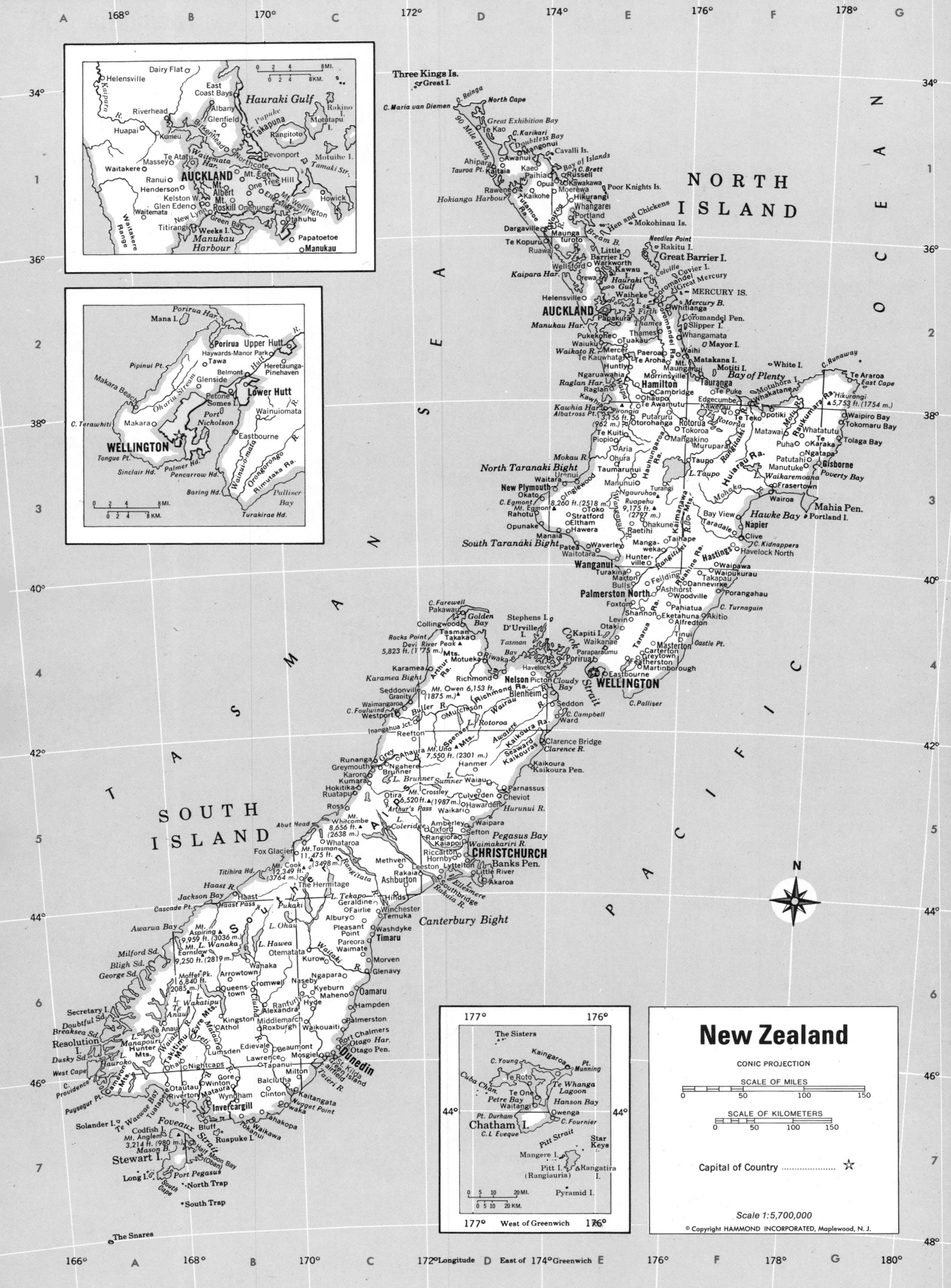

New Zealand

CONIC PROJECTION

SCALE OF MILES

0   50   100   150

SCALE OF KILOMETERS

0   50   100   150

Capital of Country .... ☆

Scale 1:5,700,000

© Copyright HAMMOND INCORPORATED, Maplewood, N.J.

Topography

Three Kings Is.
C. Maria van Diemen — North Cape
Bay of Islands
Great Barrier I.
Kaipara Har.
Coromandel Pen.
Auckland
Bay of Plenty — East Cape

North Island

0 75 150 MI.
0 75 150 KM.

L. Taupo
Ruapehu 9,175 ft. (2796 m.)
C. Egmont
Mt. Egmont 8,260 ft. (2518 m.)
Mahia Pen.
Hawke Bay

C. Farewell
Tasman Bay
Cook Strait
Wellington
C. Foulwind
Wairau
C. Palliser

South Island

SOUTHERN ALPS
Pegasus Bay
Christchurch
Banks Pen.
Mt. Cook 12,349 ft. (3764 m.)
Cascade Pt.
CANTERBURY PLAINS
Canterbury Bight

Below Sea Level | 100 m. 328 ft. | 200 m. 656 ft. | 500 m. 1,640 ft. | 1,000 m. 3,281 ft. | 2,000 m. 6,562 ft. | 5,000 m. 16,404 ft.

Awau
Clutha
Otago Pen.
Dunedin
West Cape
Foveaux Str.
Stewart I.

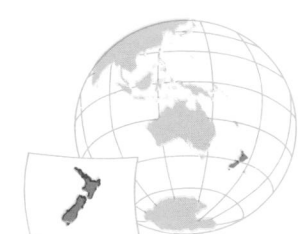

AREA 103,736 sq. mi. (268,676 sq. km.)
POPULATION 3,175,737
CAPITAL Wellington
LARGEST CITY Auckland
HIGHEST POINT Mt. Cook 12,349 ft. (3,764 m.)
MONETARY UNIT New Zealand dollar
MAJOR LANGUAGES English, Maori
MAJOR RELIGIONS Protestantism, Roman Catholicism

Wellington †321,004	A3
Wellsford 1,621	E2
Westport 4,686	C4
Whakatane 12,286	F2
Whangamata 1,566	F2
Whangarei 36,550	E1
Whangarei †40,212	E1
Whitianga 1,960	E2
Winton 2,035	B7
Woodville 1,647	F4

OTHER FEATURES

Arthur's (pass)	C5
Aspiring (mt.)	B6
Banks (pen.)	D5
Bream (bay)	E1
Brett (cape)	E1
Buller (riv.)	D4
Campbell (cape)	E4
Canterbury (bight)	D6
Cascade (pt.)	B6
Chatham (isls.) 751	D7
Cloudy (bay)	E4
Clutha (riv.)	B6
Coleridge (lake)	C5
Colville (cape)	E2
Cook (mt.)	C5
Cook (str.)	E4
Coromandel (pen.)	F2
Devil River (peak)	D4
D'Urville (isl.)	D4
Dusky (sound)	A6
East (cape)	G2
Egmont (cape)	D3
Egmont (mt.)	D3
Ellesmere (lake)	D5
Farewell (cape)	D4
Foulwind (cape)	C4
Fournier (cape)	E7
Foveaaux (str.)	A7
Golden (bay)	D4
Great Barrier (isl.) 572	E2
Hauraki (gulf)	C1
Hawke (bay)	F3
Hikurangi (mt.)	G2
Hokianga (harb.)	D1
Huiarau (range)	F3
Hutt (riv.)	C2
Islands (bay)	E1
Jackson (bay)	B5
Kaikoura (range)	D5
Kaimanawa (range)	E3
Kaipara (harb.)	D2
Karamea (bight)	C4
Kawhia (harb.)	E3
Kidnappers (cape)	F3
Mahia (pen.)	G3
Manapouri (lake)	A6
Manukau (harb.)	B1
Maria van Diemen (cape)	D1
Mataura (riv.)	B6
Mercury (isls.)	F2
Milford (sound)	A6
Needles (pt.)	E2
Nicholson, Port (inlet)	B3
Ninety Mile (beach)	D1
North (cape)	D1
North (isl.) 2,322,989	F1
North Taranaki (bight)	D3
Otago (pen.)	C6
Owen (mt.)	D4
Palliser (cape)	E4
Pegasus (bay)	D5
Pitt (isl.)	E7
Plenty (bay)	F2
Port Nicholson (inlet)	B3
Port Pegasus (inlet)	B7
Pukaki (lake)	B6
Puysegur (pt.)	A7
Rakaia (riv.)	C5
Rangitata (riv.)	C5
Rangitikei (riv.)	E3
Raukumara (range)	F3
Reinga (cape)	D1
Resolution (isl.)	A6
Richmond (range)	D4
Rocks (pt.)	C4
Rotorua (lake)	F3
Ruahine (range)	F4
Ruapehu (mt.)	E3
Ruapuke (isl.)	B7
South (cape)	A7
South (isl.) 852,748	B5
Southern Alps (range)	C5
South Taranaki (bight)	D3
Spenser (mts.)	D5
Stewart (isl.) 600	A7
Tararua (range)	E4
Tasman (bay)	D4
Tasman (mt.)	C5
Tasman (mts.)	D4
Tasman (sea)	B4
Taupo (lake)	F3
Tauroa (pt.)	D1

Te Anau (lake)	A6
Tekapo (lake)	C5
Terawhiti (cape)	A3
Thames (firth)	E2
Three Kings (isls.)	D1
Turakirae (head)	B3
Una (mt.)	D5
Waiheke (isl.) 3,223	E2
Waikato (riv.)	E2
Waimakariri (riv.)	D5
Waipa (riv.)	E2
Wairau (riv.)	D4
Waitaki (riv.)	C6
Waitemata (harb.)	B1
Wakatipu (lake)	B6
Wanaka (lake)	B6
Wanganui (riv.)	E3
West (cape)	A6
Whitcombe (mt.)	C5

†Population of urban area.

Agriculture, Industry and Resources

Snapper
Fruit
Auckland
Snapper
Sheep
Dairy
Wellington
Christchurch
Sheep
Wheat
Crayfish
Soles
Sheep
Oysters
Dunedin
Crayfish

CITIES and TOWNS

Albany 2,001	B1
Alexandra 4,348	B6
Ashburton 14,151	C5
Ashhurst 1,906	E4
Auckland 144,963	B1
Auckland †769,558	B1
Balclutha 4,495	B7
Belmont 2,402	B2
Birkenhead 21,324	B1
Blenheim 17,849	D4
Bluff 2,720	B7
Bulls 1,839	E4
Cambridge 8,514	E2
Carterton 3,971	E4
Christchurch 164,680	D5
Christchurch †289,959	D5
Cromwell 2,364	B6
Dannevirke 5,663	F4
Dargaville 4,747	D1
Devonport 10,410	C1
Dunedin 77,176	C6
Dunedin †107,445	C6
Eastbourne 4,561	B3
East Coast Bays 28,866	B1
Edgecumbe 1,929	F2
Ellerslie 5,404	C1
Eltham 2,411	E3
Fairfield 1,849	C6
Featherston 2,458	E4
Feilding 11,522	E4
Foxton 2,719	E4
Geraldine 2,128	C6
Gisborne 29,986	G3
Gisborne †32,062	G3
Glen Eden 9,406	B1
Glenfield 3,691	B1
Gore 9,185	B7
Green Bay 3,035	B1
Green Island 6,899	C7
Greymouth 8,103	C5
Greytown 1,797	E4
Half Moon Bay (Oban) 2,448	B7
Hamilton 91,109	E2
Hamilton †97,907	E2
Hastings 36,083	F3
Hastings †52,563	F3
Havelock North 8,507	F3
Hawera 8,400	E3
Helensville 1,360	B1
Henderson 6,645	B1
Heretaunga-Pinehaven 6,171	C2
Hokitika 3,414	C5
Hornby 8,215	D5
Howick 13,866	C1
Huntly 6,534	E2
Hutt (Upper and Lower) †131,257	B2
Inglewood 2,839	E3

Invercargill 49,446	B7
Invercargill †53,868	B7
Kaiapoi 4,894	D5
Kaikohe 3,663	D1
Kaikoura 2,180	D5
Kaitaia 4,737	D1
Kawerau 8,593	F3
Kumeu 3,414	B1
Levin 14,652	E4
Lower Hutt 63,245	B2
Lyttelton 3,184	D5
Manukau 159,362	C1
Marton 4,858	E4
Masterton 18,785	E4
Mataura 2,345	B7
Milton 2,193	B7
Morrinsville 5,080	E2
Mosgiel 9,264	C6
Motueka 4,693	D4
Mount Albert 26,462	B1
Mount Eden 18,305	B1
Mount Maunganui 11,391	E2
Mount Roskill 33,577	B1
Mount Wellington 19,528	C1
Murupara 2,964	F3
Napier 48,314	F3
Napier †51,330	F3
Nelson 33,304	D4
Nelson †43,121	D4
New Lynn 10,445	B1
New Plymouth 36,048	D3
New Plymouth †44,095	D3
Ngaruawahia 4,435	E2
Northcote 10,061	B1
Oamaru 13,043	C6
Oban (Half Moon Bay) 2,448	B7
Onehunga 15,386	B1
One Tree Hill 11,078	B1
Opotiki 3,388	F3
Orewa 5,552	E2
Otahuhu 10,298	C1
Otaki 4,301	E4
Otorohanga 2,574	E3
Paeroa 3,702	E2
Pahiatua 2,599	F4
Paihia 1,740	D1
Palmerston North 60,105	E4
Palmerston North †66,691	E4
Papakura 22,473	E2
Papatoetoe 21,700	C1
Patea 1,938	E3
Petone 8,113	B2
Picton 3,220	D4
Pinehaven (Heretaunga-Pinehaven) 6,171	C2
Porirua 41,104	B2
Port Chalmers 2,917	C6
Pukekohe 9,070	E2
Putaruru 4,222	E3
Queenstown 3,367	B6

Raetihi 1,247	E3
Raglan 1,414	E2
Rangiora 6,385	D5
Reefton 1,200	C5
Riccarton 6,709	D5
Richmond 6,847	D4
Riverton 1,479	B7
Rotorua 38,157	F3
Rotorua †48,314	F3
Runanga 1,264	C5
Russell 932	E1
Saint Kilda 6,147	C7
Shannon 1,465	E4
Stratford 5,518	E3
Taihape 2,586	E3
Takapuna 64,844	B1
Tapanui 1,042	B6
Taradale 4,681	F3
Taumarunui 6,541	E3
Taupo 13,651	F3
Tauranga 37,099	F2
Tauranga †53,097	F2
Tawa 12,216	B2
Te Anau 2,610	A6
Te Aroha 3,331	E2
Te Atatu 14,713	B1
Te Awamutu 7,922	E3
Te Kauwhata 842	E2
Te Kuiti 4,795	E3
Temuka 3,771	C6
Te Puke 4,577	F2
Thames 6,456	E2
The Hermitage	C5
Timaru 28,412	C6
Timaru †29,225	C6
Titirangi 8,426	B1
Tokoroa 18,713	F3
Tuakau 1,982	E2
Tuatapere 884	A7
Turangi 5,517	E3
Upper Hutt 31,405	B2
Waihi 3,538	E2
Waikanae 4,818	E4
Waikouaiti 858	C6
Waimate 3,393	C6
Wainuiomata 19,192	B3
Waipawa 1,732	F4
Waipukurau 3,648	F4
Wairoa 5,439	F3
Waitangi	D7
Waitara 6,012	E3
Waitemata 87,452	B1
Waiuku 3,654	E2
Waanaka 1,155	B6
Wanganui 37,012	E3
Wanganui †39,595	E3
Warkworth 1,734	E2
Washdyke 949	C6
Waverley 1,239	E3
Wellington (cap.) 135,688	A3

DOMINANT LAND USE

- Mixed Farming, Livestock
- Dairy
- Truck Farming, Horticulture
- Pasture Livestock (chiefly sheep)
- Livestock Herding
- Forests
- Nonagricultural Land

MAJOR MINERAL OCCURRENCES

- C Coal
- G Natural Gas
- J Jade
- Ka Kaolin
- Lg Lignite
- O Petroleum
- U Uranium

⚡ Water Power
▨ Major Industrial Areas

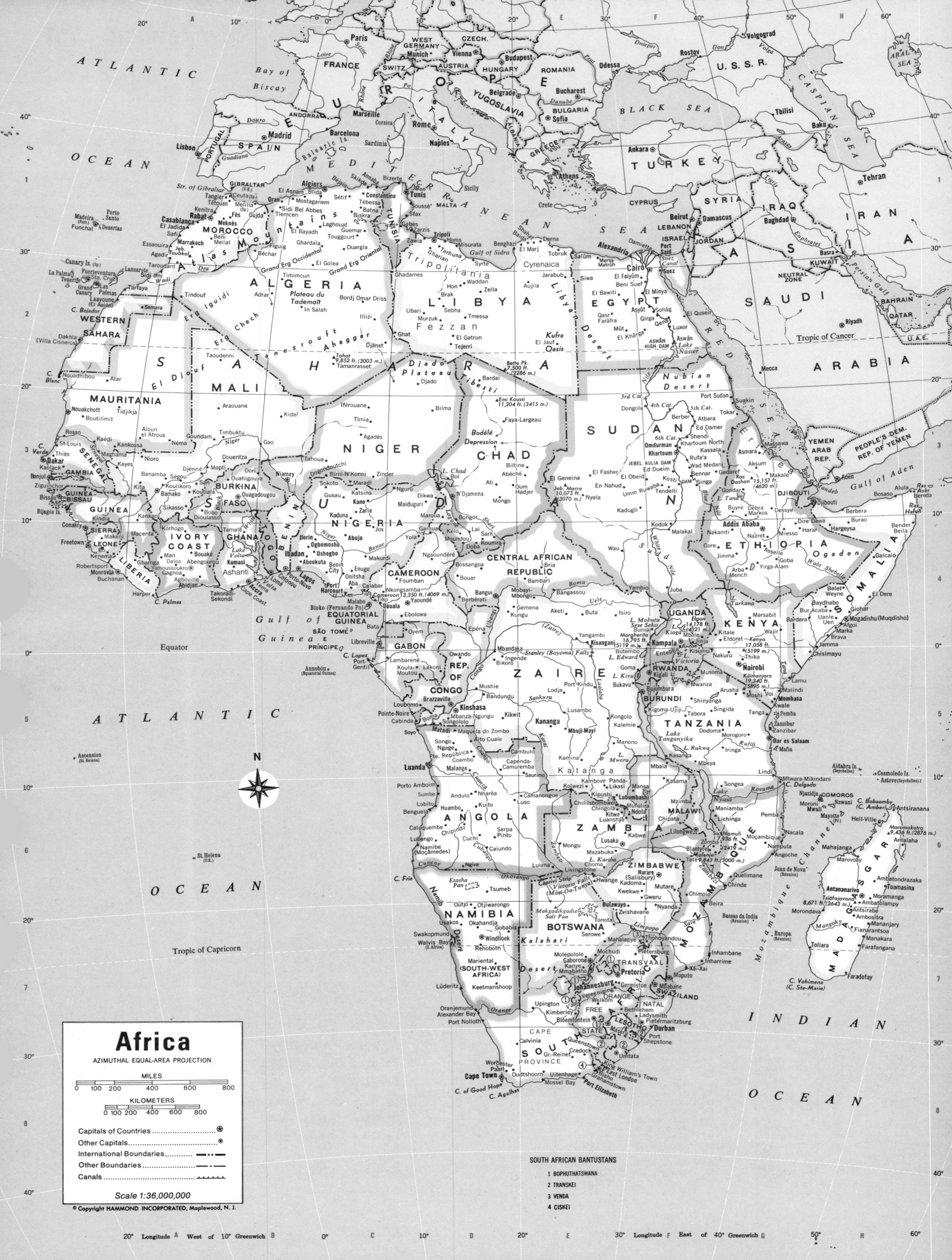

Africa

AZIMUTHAL EQUAL-AREA PROJECTION

MILES
0 100 200 400 600 800

KILOMETERS
0 100 200 400 600 800

Capitals of Countries .... ⊛
Other Capitals .... ⊛
International Boundaries .... ▬ ▪ ▬ ▪ ▬
Other Boundaries .... ▬ ▪ ▪ ▬ ▪ ▪
Canals .... ⊢⊢⊢⊢⊢

Scale 1:36,000,000

© Copyright HAMMOND INCORPORATED, Maplewood, N.J.

SOUTH AFRICAN BANTUSTANS

1 BOPHUTHATSWANA
2 TRANSKEI
3 VENDA
4 CISKEI

AREA 11,707,000 sq. mi. (30,321,130 sq. km.)
POPULATION 469,000,000
LARGEST CITY Cairo
HIGHEST POINT Kilimanjaro 19,340 ft. (5,895 m.)
LOWEST POINT Lake Assal, Djibouti -512 ft. (-156 m.)

Population Distribution

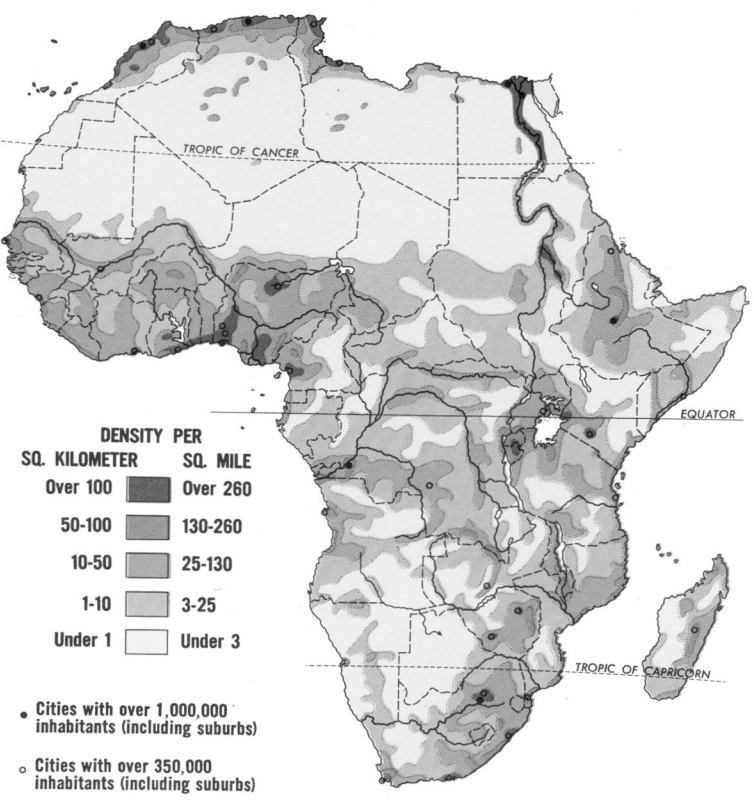

DENSITY PER

SQ. KILOMETER	SQ. MILE
Over 100	Over 260
50-100	130-260
10-50	25-130
1-10	3-25
Under 1	Under 3

● Cities with over 1,000,000 inhabitants (including suburbs)

○ Cities with over 350,000 inhabitants (including suburbs)

Vegetation

TROPICAL FOREST
- Tropical Rainforest
- Light Tropical Forest
- Woodland and Shrub

TROPICAL GRASSLAND
- Grass and Shrub (Savanna)
- Wooded Savanna

MID-LATITUDE FOREST
- Mixed Coniferous and Broadleaf Forest
- Woodland and Shrub (Mediterranean)

MID-LATITUDE GRASSLAND
- Short Grass (Steppe)

RIVER VALLEY AND OASIS

DESERT AND DESERT SHRUB

UNCLASSIFIED HIGHLANDS

TROPIC OF CANCER
EQUATOR
TROPIC OF CAPRICORN

Abéché, Chad ... D3
Abidjan, Ivory Coast ... B4
Accra (cap.), Ghana ... B4
Addis Ababa (cap.), Ethiopia ... F4
Aden (gulf) ... G3
Agulhas (cape), S. Africa ... D8
Ahaggar (range), Algeria ... C2
Alexandria, Egypt ... E1
Algeria ... C2
Algiers (cap.), Algeria ... C1
Amber(Bobaomby) (cape), Madagascar ... G6
Angola ... D6
Annaba, Algeria ... C1
Annobon (isl.), Equat. Guinea ... C5
Antananarivo (cap.), Madagascar ... G6
Arusha, Tanzania ... F5
Ascension (isl.), St. Helena ... A5
Asmara, Ethiopia ... F3
Aswân, Egypt ... F2
Asyût, Egypt ... F2
Atbara, Sudan ... F3
Atlas (mts.) ... B1
Bab el Mandeb (strait) ... G3
Bamako (cap.), Mali ... B3
Bangui (cap.), Cent. Afr. Rep. ... D4
Banjul (cap.), Gambia ... A3
Béchar, Algeria ... B1
Beira, Mozambique ... F7
Benghazi, Libya ... D1
Benguela, Angola ... D6
Benin ... C4
Bioko (isl.), Equat. Guinea ... C4
Biskra, Algeria ... C1
Bissau (cap.), Guinea-Biss. ... A3
Bizerte, Tunisia ... C1
Blanc (cape), Mauritania ... A2
Blantyre, Malawi ... F6
Bloemfontein, S. Africa ... E7
Blue Nile (riv.) ... F4
Bobaomby (cape), Madagascar ... G6
Bobo Dioulasso, Burkina Faso ... B3
Bon (cape), Tunisia ... D1
Bophuthatswana, S. Africa ... E7
Botswana ... E7
Bouaké, Ivory Coast ... B4
Boyoma (Stanley) (falls), Zaire ... E5
Brazzaville (cap.), Congo ... D5
Buchanan, Liberia ... A4
Bujumbura (cap.), Burundi ... F5
Bukavu, Zaire ... E5
Bulawayo, Zimbabwe ... E7
Burkina Faso ... B3

Burundi ... F5
Cabinda, Angola ... D5
Cairo (cap.), Egypt ... F2
Cameroon ... D4
Canary (isls.), Spain ... A2
Cape (prov.), S. Africa ... E8
Cape Town (cap.), S. Africa ... D8
Casablanca, Morocco ... B1
Central African Republic ... D4
Ceuta, Spain ... B1
Chad ... D3
Chad (lake) ... D3
Comoros ... G6
Conakry (cap.), Guinea ... A4
Congo, Rep. of ... D5
Congo (riv.) ... D5
Constantine, Algeria ... C1
Cotonou, Benin ... C4
Cyrenaica (reg.), Libya ... E1
Dakar (cap.), Senegal ... A3
Damietta, Egypt ... F1
Dar es Salaam (cap.), Tanzania ... F5
Dire Dawa, Ethiopia ... G4
Djibouti ... G3
Djibouti (cap.), Djibouti ... G3
Dodoma, Tanzania ... F5
Douala, Cameroon ... D4
Durban, S. Africa ... F7
East London, S. Africa ... E8
Edward (lake) ... E5
Egypt ... E2
El Bayadh, Algeria ... C1
El Faiyûm, Egypt ... F2
El Fasher, Sudan ... E3
Elgon (mt.) ... F4
El Minya, Egypt ... F2
El Obeid, Sudan ... E3
Emi Koussi (mt.), Chad ... D3
Entebbe, Uganda ... F5
Equatorial Guinea ... C4
Eritrea (reg.), Ethiopia ... F3
Ethiopia ... F4
Etosha Salt Pan, Namibia ... D6
Fernando Po (Bioko) (isl.), Equat. Guinea ... C4
Fès, Morocco ... B1
Fezzan (reg.), Libya ... D2
Fianarantsoa, Madagascar ... G7
Freetown (cap.), S. Leone ... A4
Funchal (cap.), Madeira, Portugal ... A1
Gabon ... D4
Gaborone (cap.), Botswana ... E7
Gambia ... A3
Garoua, Cameroon ... D4
Germiston, S. Africa ... E7
Ghana ... B4
Gharian, Libya ... D1

Gibraltar (str.) ... B1
Good Hope (cape), S. Africa ... D8
Grand Erg Occidental (des.), Algeria ... C1
Grand Erg Oriental (des.), Algeria ... C1
Guinea ... A3
Guinea (gulf) ... C4
Guinea-Bissau ... A3
Gweru, Zimbabwe ... F6
Harar, Ethiopia ... G4
Harare (cap.), Zimbabwe ... E6
Hargeysa, Somalia ... G4
Huambo, Angola ... D6
Ibadan, Nigeria ... C4
Iguidi, Erg (des.) ... B2
Ilorin, Nigeria ... C4
Ivory Coast ... B4
Jinja, Uganda ... F4
Johannesburg, S. Africa ... E7
Kalahari (des.) ... E7
Kampala (cap.), Uganda ... F4
Kananga, Zaire ... E5
Kankan, Guinea ... B3
Kano, Nigeria ... C3
Kaolack, Senegal ... A3
Kariba (lake) ... E6
Kasai (riv.) ... E5
Kassala, Sudan ... F3
Katanga (reg.), Zaire ... E5
Kayes, Mali ... A3
Kenitra, Morocco ... B1
Kenya ... F4
Kenya (mt.), Kenya ... F4
Khartoum (cap.), Sudan ... F3
Kigali (cap.), Rwanda ... F5
Kilimanjaro (mt.), Tanzania ... F5
Kimberley, S. Africa ... E7
Kinshasa (cap.), Zaire ... D5
Kisangani, Zaire ... E4
Kisumu, Kenya ... F5
Kitwe, Zambia ... E6

Lubumbashi, Zaire ... E6
Lusaka (cap.), Zambia ... E6
Luxor, Egypt ... F2
Madagascar ... G7
Madeira (isls.), Portugal ... A1
Malabo (cap.), Equat. Guinea ... C4
Malawi ... F6
Mali ... B2
Maputo (cap.), Mozambique ... F7
Marrakech, Morocco ... B1
Maseru (cap.), Lesotho ... E8
Massawa, Ethiopia ... F3
Matadi, Zaire ... D5
Mauritania ... A3
Mayotte (isl.), France ... G6
Mbabane (cap.), Swaziland ... F7
Mbuji-Mayi, Zaire ... E5
Meknès, Morocco ... B1
Melilla, Spain ... B1
Misurata, Libya ... D1
Mmabatho (cap.), Bophuthatswana, S. Africa ... E7
Mobutu Sese Seko (lake) ... F4
Moçambique, Mozambique ... F6
Mogadishu (cap.), Somalia ... G4
Mombasa, Kenya ... G5
Monrovia (cap.), Liberia ... A4
Morocco ... B1
Moroni (cap.), Comoros ... G6
Mosi-Oa-Tunya (Victoria) (falls) ... E6
Mostaganem, Algeria ... C1
Mozambique ... F6
Mozambique (chan.) ... G6
Muqdisho (Mogadishu) (cap.), Somalia ... G4
Nairobi (cap.), Kenya ... F5
Nakuru, Kenya ... F5
Namibia ... D7
Nasser (lake), Egypt ... F2
Natal (prov.), S. Africa ... F7

N'Djamena (cap.), Chad ... D3
Ndola, Zambia ... E6
Niamey (cap.), Niger ... C3
Niger ... C3
Niger (riv.) ... C4
Nigeria ... C4
Nile (riv.) ... F2
Nouadhibou, Mauritania ... A2
Nouakchott (cap.), Mauritania ... A3
Nubian (des.), Sudan ... F2
Nyasa (lake) ... F6
Ogaden (reg.), Ethiopia ... G4
Ogbomosho, Nigeria ... C4
Okovango (riv.) ... D6
Omdurman, Sudan ... F3
Oran, Algeria ... B1
Orange (riv.) ... D7
Orange Free State (prov.), S. Africa ... E7
Oshogbo, Nigeria ... C4
Ouagadougou (cap.), Burkina Faso ... B3
Oujda, Morocco ... B1
Palmas (cape) ... B4
Pemba (isl.), Tanzania ... G5
Pietermaritzburg (cap.), Natal, S. Africa ... F7
Pointe-Noire, Congo ... D5
Port Elizabeth, S. Africa ... E8
Port Harcourt, Nigeria ... C4
Porto-Novo (cap.), Benin ... C4
Port Said, Egypt ... F1
Port Sudan, Sudan ... F3
Pretoria (cap.), S. Africa ... E7
Rabat (cap.), Morocco ... B1
Ras Asèr (cape), Somalia ... H3
Red (sea) ... F2
Rufiji (riv.), Tanzania ... F5
Rwanda ... F5
Sahara (des.) ... C2
Saint Helena (isl.), U.K. ... B6
Saint-Louis, Senegal ... A3

Sainte-Marie (Vohimena) (cape), Madagascar ... G7
Salisbury (Harare) (cap.), Zimbabwe ... E6
Santa Cruz (cap.), Canary Is., Spain ... A2
São Tomé e Príncipe ... C4
Ségou, Mali ... B3
Senegal ... A3
Sétif, Algeria ... C1
Sfax, Tunisia ... D1
Sidi Bel Abbes, Algeria ... C1
Sidi Ifni, Morocco ... A2
Sierra Leone ... A4
Skikda, Algeria ... C1
Sohâg, Egypt ... F2
Somalia ... G4
Sousse, Tunisia ... D1
South Africa ... E7
South-West Africa (Namibia) ... D7
Stanley (falls), Zaire ... E5
Sudan ... E3
Sudan (reg.) ... D3
Suez, Egypt ... F2
Suez (canal), Egypt ... F1
Swaziland ... F7
Takoradi-Sekondi, Ghana ... B4
Tamale, Ghana ... B4
Tana (lake), Ethiopia ... F3
Tanezrouft (des.), Algeria ... C2
Tanga, Tanzania ... F5
Tanganyika (lake) ... F5
Tangier, Morocco ... B1
Tanzania ... F5
Tenerife (isl.), Spain ... A2
Thiès, Senegal ... A3
Thohoyandou (cap.), Venda, S. Africa ... F7
Tibesti (mts.) ... D2
Tidjikja, Mauritania ... A3
Timbuktu, Mali ... B3
Tlemcen, Algeria ... B1
Toamasina, Madagascar ... G6

Tobruk, Libya ... E1
Togo ... C4
Toubkal, Jebel (mt.), Morocco ... B1
Touggourt, Algeria ... C1
Transkei, S. Africa ... E8
Transvaal (prov.), S. Africa ... E7
Tripoli (cap.), Libya ... D1
Tripolitania (reg.), Libya ... D1
Tunis (cap.), Tunisia ... D1
Tunisia ... C1
Turkana (lake), Kenya ... F4
Ubangi (riv.) ... D4
Uganda ... F4
Uitenhage, S. Africa ... E8
Umtata (cap.), Transkei, S. Africa ... E8
Upper Volta (Burkina Faso) ... B3
Vaal (riv.), S. Africa ... E7
Venda, S. Africa ... F7
Verde (cape), Senegal ... A3
Victoria (lake) ... F5
Victoria (falls) ... E6
Volta (lake), Ghana ... B4
Volta (riv.), Ghana ... C4
Wabi Shebelle (riv.) ... G4
Wad Medani, Sudan ... F3
Walvis Bay, S. Africa ... D7
Western Sahara ... A2
White Nile (riv.) ... F3
Windhoek (cap.), Namibia ... D7
Yaoundé (cap.), Cameroon ... D4
Yamoussoukro (cap.), Ivory Coast ... B4
Zaire ... E5
Zaire (Congo) (riv.) ... E4
Zambezi (riv.) ... F6
Zambia ... E6
Zannzibar, Tanzania ... F5
Zanzibar (isl.), Tanzania ... F5
Zimbabwe ... E6
Zomba, Malawi ... F6

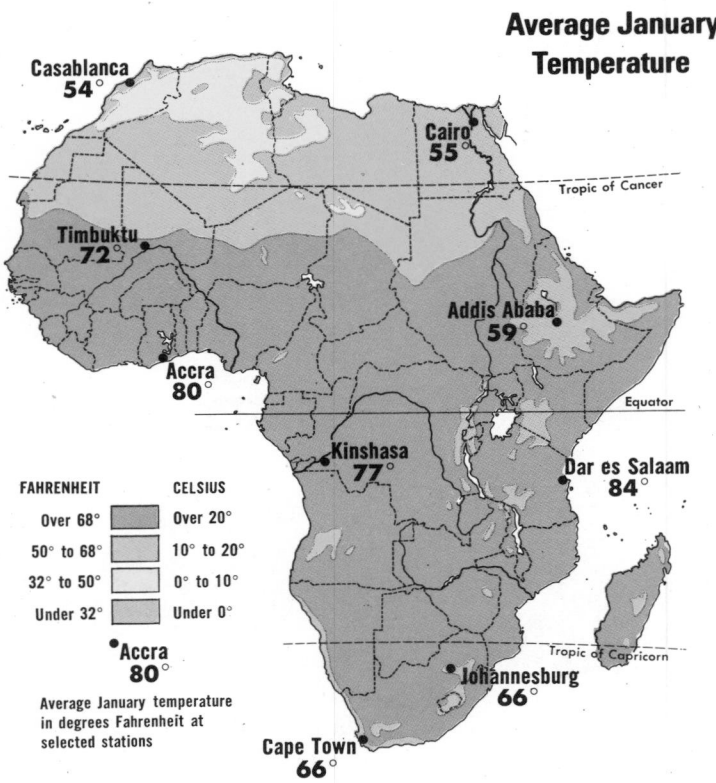

Average January Temperature

Casablanca 54°
Cairo 55°
Timbuktu 72°
Addis Ababa 59°
Accra 80°
Kinshasa 77°
Dar es Salaam 84°
Johannesburg 66°
Cape Town 66°

Tropic of Cancer
Equator
Tropic of Capricorn

FAHRENHEIT	CELSIUS
Over 68°	Over 20°
50° to 68°	10° to 20°
32° to 50°	0° to 10°
Under 32°	Under 0°

• Accra 80°
Average January temperature in degrees Fahrenheit at selected stations

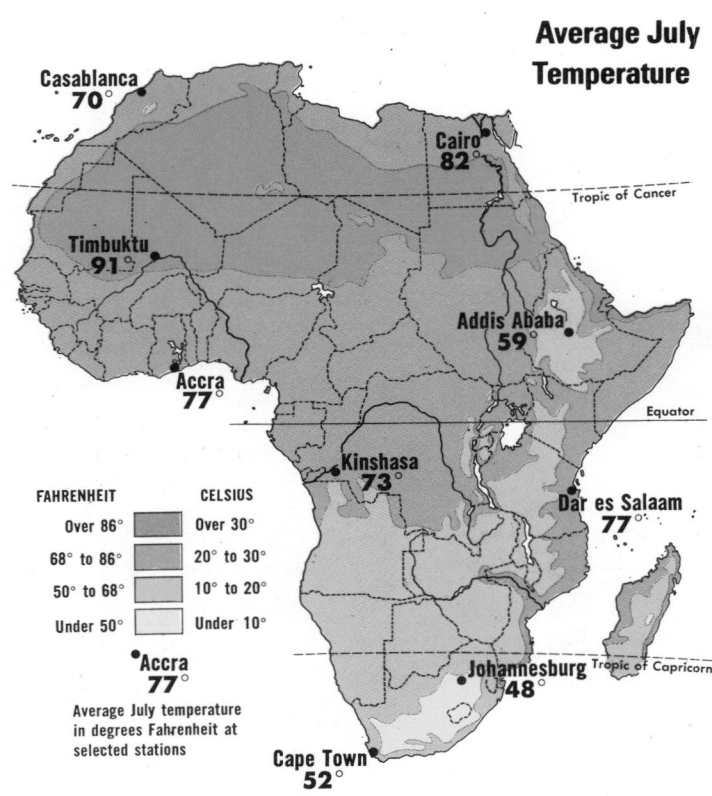

Average July Temperature

Casablanca 70°
Cairo 82°
Timbuktu 91°
Addis Ababa 59°
Accra 77°
Kinshasa 73°
Dar es Salaam 77°
Johannesburg 48°
Cape Town 52°

Tropic of Cancer
Equator
Tropic of Capricorn

FAHRENHEIT	CELSIUS
Over 86°	Over 30°
68° to 86°	20° to 30°
50° to 68°	10° to 20°
Under 50°	Under 10°

• Accra 77°
Average July temperature in degrees Fahrenheit at selected stations

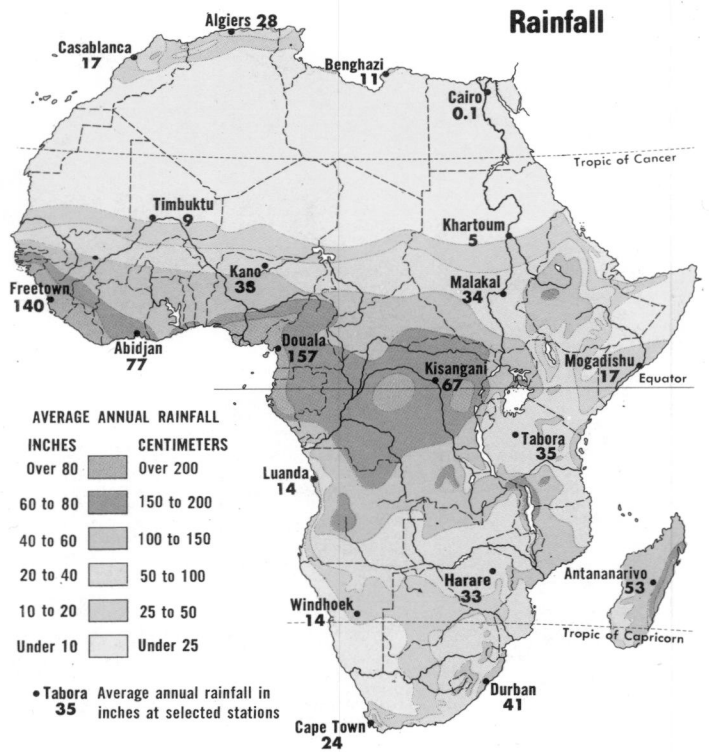

Rainfall

Algiers 28
Casablanca 17
Benghazi 11
Cairo 0.1
Timbuktu 9
Khartoum 5
Kano 35
Malakal 34
Freetown 140
Abidjan 77
Douala 157
Kisangani 67
Mogadishu 17
Luanda 14
Tabora 35
Harare 33
Antananarivo 53
Windhoek 14
Durban 41
Cape Town 24

Tropic of Cancer
Equator
Tropic of Capricorn

AVERAGE ANNUAL RAINFALL

INCHES	CENTIMETERS
Over 80	Over 200
60 to 80	150 to 200
40 to 60	100 to 150
20 to 40	50 to 100
10 to 20	25 to 50
Under 10	Under 25

• Tabora 35 Average annual rainfall in inches at selected stations

Vegetation/Relief

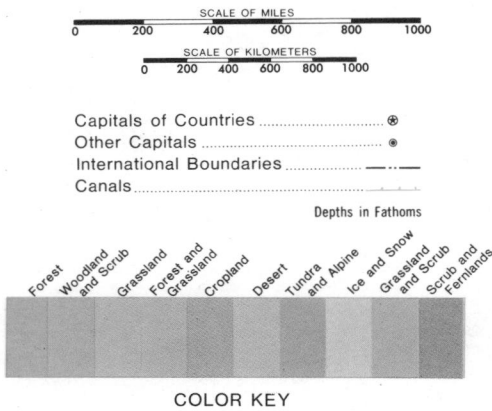

SCALE OF MILES
0   200   400   600   800   1000

SCALE OF KILOMETERS
0   200   400   600   800   1000

Capitals of Countries .... ⊗
Other Capitals .... ⊛
International Boundaries .... ——
Canals ....

Depths in Fathoms

Forest
Woodland and Scrub
Grassland
Forest and Grassland
Cropland
Desert
Tundra and Alpine
Ice and Snow
Grassland and Scrub
Scrub and Fernlands

COLOR KEY

Longitude 10° West of Greenwich   Longitude 10° East of Greenwich

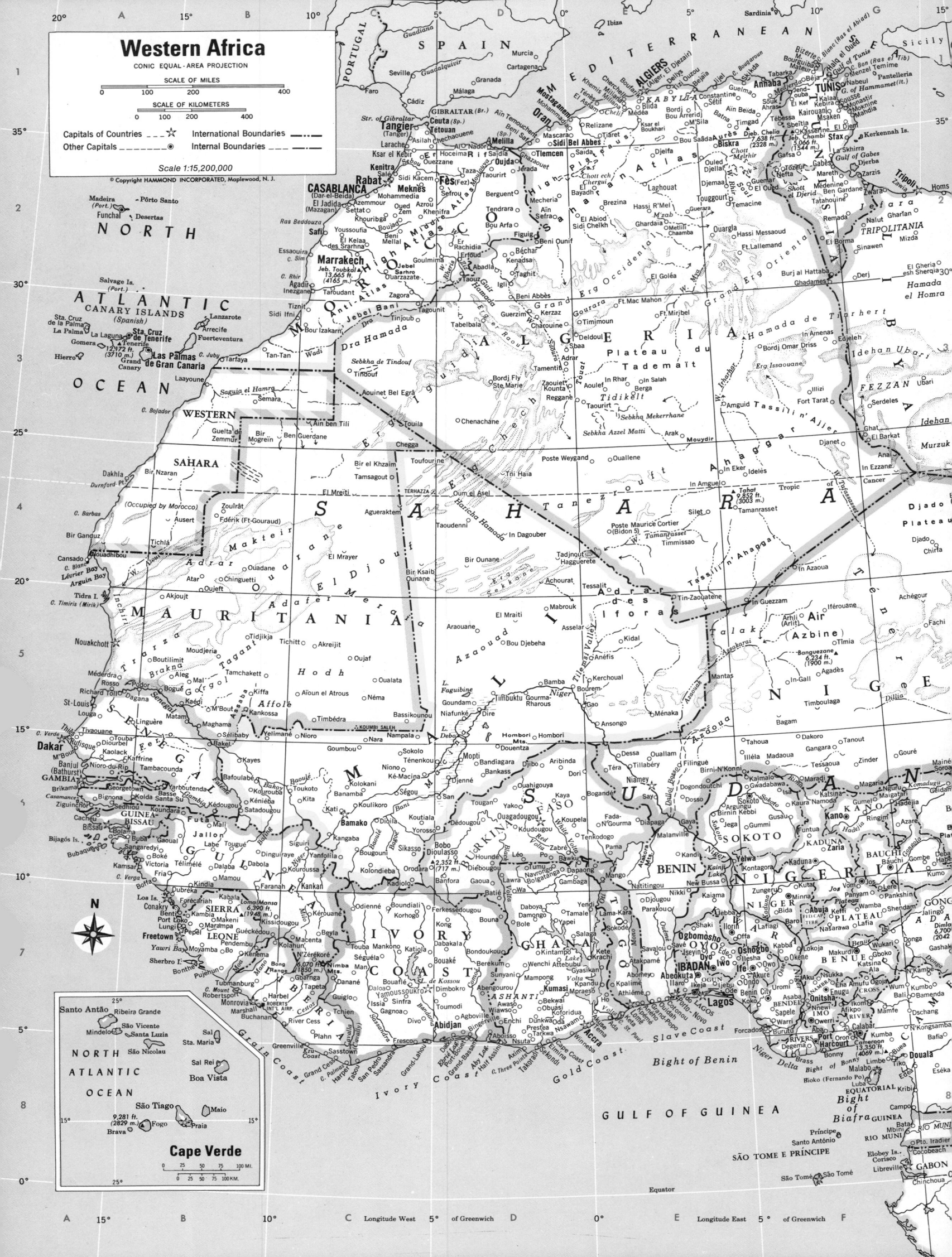

ALGERIA

AREA 919,591 sq. mi. (2,381,740 sq. km.)
POPULATION 17,422,000
CAPITAL Algiers
LARGEST CITY Algiers
HIGHEST POINT Tahat 9,852 ft. (3,003 m.)
MONETARY UNIT Algerian dinar
MAJOR LANGUAGES Arabic, Berber, French
MAJOR RELIGION Islam

BENIN

AREA 43,483 sq. mi. (112,620 sq. km.)
POPULATION 3,338,240
CAPITAL Porto-Novo
LARGEST CITY Cotonou
HIGHEST POINT Atakora Mts. 2,083 ft. (635 m.)
MONETARY UNIT CFA franc
MAJOR LANGUAGES Fon, Somba, Yoruba, Bariba, French, Mina, Dendi
MAJOR RELIGIONS Tribal religions, Islam, Roman Catholicism

CAPE VERDE

AREA 1,557 sq. mi. (4,033 sq. km.)
POPULATION 324,000
CAPITAL Praia
LARGEST CITY Praia
HIGHEST POINT 9,281 ft. (2,829 m.)
MONETARY UNIT Cape Verde escudo
MAJOR LANGUAGE Portuguese
MAJOR RELIGION Roman Catholicism

GAMBIA

AREA 4,127 sq. mi. (10,689 sq. km.)
POPULATION 601,000
CAPITAL Banjul
LARGEST CITY Banjul
HIGHEST POINT 100 ft. (30 m.)
MONETARY UNIT dalasi
MAJOR LANGUAGES Mandingo, Fulani, Wolof, English, Malinke
MAJOR RELIGIONS Islam, tribal religions, Christianity

GHANA

AREA 92,099 sq. mi. (238,536 sq. km.)
POPULATION 11,450,000
CAPITAL Accra
LARGEST CITY Accra
HIGHEST POINT Togo Hills 2,900 ft. (884 m.)
MONETARY UNIT cedi
MAJOR LANGUAGES Twi, Fante, Dagbani, Ewe, Ga, English, Hausa, Akan
MAJOR RELIGIONS Tribal religions, Christianity, Islam

GUINEA

AREA 94,925 sq. mi. (245,856 sq. km.)
POPULATION 5,143,284
CAPITAL Conakry
LARGEST CITY Conakry
HIGHEST POINT Nimba Mts. 6,070 ft. (1,850 m.)
MONETARY UNIT syli
MAJOR LANGUAGES Fulani, Mandingo, Susu, French
MAJOR RELIGIONS Islam, tribal religions

GUINEA-BISSAU

AREA 13,948 sq. mi. (36,125 sq. km.)
POPULATION 777,214
CAPITAL Bissau
LARGEST CITY Bissau
HIGHEST POINT 689 ft. (210 m.)
MONETARY UNIT Guinea-Bissau escudo
MAJOR LANGUAGES Balante, Fulani, Crioulo, Mandingo, Portuguese
MAJOR RELIGIONS Islam, tribal religions, Roman Catholicism

IVORY COAST

AREA 124,504 sq. mi. (322,465 sq. km.)
POPULATION 7,920,000
CAPITAL Yamoussoukro
LARGEST CITY Abidjan
HIGHEST POINT 5,745 ft. (1,751 m.)
MONETARY UNIT CFA franc
MAJOR LANGUAGES Bale, Bete, Senufu, French, Dioula
MAJOR RELIGIONS Tribal religions, Islam

LIBERIA

AREA 43,000 sq. mi. (111,370 sq. km.)
POPULATION 1,873,000
CAPITAL Monrovia
LARGEST CITY Monrovia
HIGHEST POINT Wutivi 5,584 ft. (1,702 m.)
MONETARY UNIT Liberian dollar
MAJOR LANGUAGES Kru, Kpelle, Bassa, Vai, English
MAJOR RELIGIONS Christianity, tribal religions, Islam

MALI

AREA 464,873 sq. mi. (1,204,021 sq. km.)
POPULATION 6,906,000
CAPITAL Bamako
LARGEST CITY Bamako
HIGHEST POINT Hombori Mts. 3,789 ft. (1,155 m.)
MONETARY UNIT Mali franc
MAJOR LANGUAGES Bambara, Senufu, Fulani, Soninke, French
MAJOR RELIGIONS Islam, tribal religions

MAURITANIA

AREA 419,229 sq. mi. (1,085,803 sq. km.)
POPULATION 1,634,000
CAPITAL Nouakchott
LARGEST CITY Nouakchott
HIGHEST POINT 2,972 ft. (906 m.)
MONETARY UNIT ouguiya
MAJOR LANGUAGES Arabic, Wolof, Tukolor, French
MAJOR RELIGION Islam

MOROCCO

AREA 172,414 sq. mi. (446,550 sq. km.)
POPULATION 20,242,000
CAPITAL Rabat
LARGEST CITY Casablanca
HIGHEST POINT Jeb. Toubkal 13,665 ft. (4,165 m.)
MONETARY UNIT dirham
MAJOR LANGUAGES Arabic, Berber, French
MAJOR RELIGIONS Islam, Judaism, Christianity

NIGER

AREA 489,189 sq. mi. (1,267,000 sq. km.)
POPULATION 5,098,427
CAPITAL Niamey
LARGEST CITY Niamey
HIGHEST POINT Banguezane 6,234 ft. (1,900 m.)
MONETARY UNIT CFA franc
MAJOR LANGUAGES Hausa, Songhai, Fulani, French, Tamashek, Djerma
MAJOR RELIGIONS Islam, tribal religions

NIGERIA

AREA 357,000 sq. mi. (924,630 sq. km.)
POPULATION 82,643,000
CAPITAL Lagos
LARGEST CITY Lagos
HIGHEST POINT Dimlang 6,700 ft. (2,042 m.)
MONETARY UNIT naira
MAJOR LANGUAGES Hausa, Yoruba, Ibo, Ijaw, Fulani, Tiv, Kanuri, Ibibio, English, Edo
MAJOR RELIGIONS Islam, Christianity, tribal religions

SÃO TOMÉ E PRÍNCIPE

AREA 372 sq. mi. (963 sq. km.)
POPULATION 85,000
CAPITAL São Tomé
LARGEST CITY São Tomé
HIGHEST POINT Pico 6,640 ft. (2,024 m.)
MONETARY UNIT dobra
MAJOR LANGUAGES Bantu languages, Portuguese
MAJOR RELIGIONS Tribal religions, Roman Catholicism

SENEGAL

AREA 75,954 sq. mi. (196,720 sq. km.)
POPULATION 5,508,000
CAPITAL Dakar
LARGEST CITY Dakar
HIGHEST POINT Futa Jallon 1,640 ft. (500 m.)
MONETARY UNIT CFA franc
MAJOR LANGUAGES Wolof, Peul (Fulani), French, Mende, Mandingo, Dida
MAJOR RELIGIONS Islam, tribal religions, Roman Catholicism

SIERRA LEONE

AREA 27,925 sq. mi. (72,325 sq. km.)
POPULATION 3,470,000
CAPITAL Freetown
LARGEST CITY Freetown
HIGHEST POINT Loma Mts. 6,390 ft. (1,947 m.)
MONETARY UNIT leone
MAJOR LANGUAGES Mende, Temne, Vai, English, Krio (pidgin)
MAJOR RELIGIONS Tribal religions, Islam, Christianity

TOGO

AREA 21,622 sq. mi. (56,000 sq. km.)
POPULATION 2,472,000
CAPITAL Lomé
LARGEST CITY Lomé
HIGHEST POINT Agou 3,445 ft. (1,050 m.)
MONETARY UNIT CFA franc
MAJOR LANGUAGES Ewe, French, Twi, Hausa
MAJOR RELIGIONS Tribal religions, Roman Catholicism, Islam

TUNISIA

AREA 63,378 sq. mi. (164,149 sq. km.)
POPULATION 6,367,000
CAPITAL Tunis
LARGEST CITY Tunis
HIGHEST POINT Jeb. Chambi 5,066 ft. (1,544 m.)
MONETARY UNIT Tunisian dinar
MAJOR LANGUAGES Arabic, French
MAJOR RELIGION Islam

BURKINA FASO (UPPER VOLTA)

AREA 105,869 sq. mi. (274,200 sq. km.)
POPULATION 6,908,000
CAPITAL Ouagadougou
LARGEST CITY Ouagadougou
HIGHEST POINT 2,352 ft. (717 m.)
MONETARY UNIT CFA franc
MAJOR LANGUAGES Mossi, Lobi, French, Samo, Gourounsi
MAJOR RELIGIONS Islam, tribal religions, Roman Catholicism

WESTERN SAHARA

AREA 102,703 sq. mi. (266,000 sq. km.)
POPULATION 76,425
HIGHEST POINT 2,700 ft. (823 m.)
MAJOR LANGUAGE Arabic
MAJOR RELIGION Islam

Topography

0   200   400   600 MI.
0   200   400   600 KM.

5,000 m. | 2,000 m. | 1,000 m. | 500 m. | 200 m. | 100 m. | Sea Level | Below
16,404 ft. | 6,562 ft. | 3,281 ft. | 1,640 ft. | 656 ft. | 328 ft. | |

(continued)

ALGERIA

CITIES and TOWNS

Abadla 12,200 D2
Adrar 22,800 D3
Aîn Beïda 26,976 F1
Aîn Sefra 22,400 D2
Aîn Temouchent 42,000 D1
Algiers (cap.) 1,365,400 E1
Amguid F3
Annaba 255,900 F1
Aoulef 17,200 E3
Arak E3
Batna 112,100 F1
Béchar 72,800 D2
Bejaïa 89,500 F1
Beni Abbès 5,000 D2
Beni Ounif 7,500 D2
Beni Saf 30,700 D1
Berga E3
Bidon 5 (Poste Maurice Cordier) E4
Biskra 90,500 F2
Blida 160,900 E1
Bône (Annaba) 255,900 F1
Bordj Bou Arreridj 65,000 E1
Bordj Fly Sainte Marie D3
Bordj Omar Driss 1,900 F3
Boufarik 50,000 E1
Bougie (Béjaïa) 89,500 F1
Bou Saâda 50,000 E1
Brezina 10,000 E2
Charouine D3
Chenachane D3
Cherchell 36,800 E1
Constantine 335,100 F1
Deldoul E3
Dellys 29,700 E1
Djanet 5,300 F4
Djelfa 51,000 E2
Djemaa 34,600 F2
Edjeleh F3
El Abiod Sidi Cheikh 15,300 E2
El Asnam 106,100 E1
El Bayadh 38,500 E2
El Djezair (Algiers) (cap.) 1,365,400 E1
El Goléa 24,400 E2
El Oued 72,100 F2
Fort Lallemand F2
Fort MacMahon E3
Fort Miribel E3
Fort Tarat F3
Ghardaïa 70,500 E2
Ghazaouet 25,900 D2
Guelma 60,100 F1
Guemar F2
Guerara 22,300 E2
Guerzim D3
Hassi Messaoud F2
Hassi R'Mel E2
Idelès F4
Igli 3,400 D2
Illizi 4,600 F3
In Amenas 4,200 F3
In Amguel E4
In Eker F4
In Guezzam F5
In Rhar E3
In Salah 18,800 E3
Jijel 49,800 F1
Kenadsa 7,600 D2
Kerzaz 2,900 D3
Khemis Miliana 57,800 E1
Ksar el Boukhari 41,200 E1
Laghouat 59,200 E2
Mascara 62,300 D1
Mecheria 22,600 D2
Médéa 72,300 E1
Metlili Chaamba 21,300 E2
Miliana 36,400 E1
Mohammadia 53,700 D1
Mostaganem 101,600 D1
M'Sila 49,100 E1
Oran 491,900 D1
Orléansville (El Asnam) 106,100 E1
Ouallene E4
Ouargla 77,400 F2
Ouled Djellal 22,700 F2
Philippeville (Skikda) 107,700 F1
Poste Maurice Cortier E4
Poste Weygand D4
Reggane 11,300 D3
Relizane 60,000 E1
Saïda 62,100 E2
Sbaa D3
Sétif 144,200 F1
Sidi Bel-Abbes 116,000 D1
Silet E4
Skikda 107,700 F1
Souk Ahras 60,200 F1
Tabelbala 3,100 D3
Taghit 3,500 D2
Tamanrasset 23,200 F4
Tamentit D3
Taourirt E3
Tébessa 67,200 F1
Temacine F2
Ténès 30,100 E1
Tiaret 62,900 E1
Tiguentourine F3
Timgad 9,800 F1
Timimoun 20,500 E3
Tindouf 6,500 C3
Tinjoub C3
Tin-Zaouatene E5
Tizi Ouzou 73,100 E1
Tlemcen 109,400 D2
Touggourt 75,600 F2
Zaouiet Kounta 13,800 D3

OTHER FEATURES

Adrar des Iforas (plat.) E5
Ahaggar (range) F4
Anaî (well) G4
Aouinet Bel Egrâ (well) C3
Atlas (mts.) E2
Aurès (lag.) F1
Azzel Mati, Sebkha (lake) E3
Bougaroun (cape) F1
Chech, Erg (des.) D3
Chelia (mt.) F1
Chelif (riv.) E1
Chergui, Chott Ech (salt lake) E2
Gourara (oasis) E3
Grand Erg Occidental (des.) E2
Grand Erg Oriental (des.) F2
Guir Hamada (des.) D2
High Plateaus (ranges) D2
Iguidi, Erg (des.) C3
In Ezzane (well) G4
Irharhar, Wadi (dry riv.) F3
Issaouane Erg (des.) F3
Kabylia (reg.) E1
Mediterranean (sea) E1
Medjerda (riv.) F1
Melrhir, Chott (salt lake) F2
Mouydir (mts.) E3
Mya, Wadi (dry riv.) E2
M'zab (oasis) E2
Raoui, Erg er (des.) D3
Rhir, Wadi (dry riv.) F2
Sahara (des.) E4
Saharan Atlas (ranges) E2
Saoura, Wadi (dry riv.) D3
Souf (oasis) F2
Tademaït, Plateau du (plat.) E3
Tafassasset, Wadi (dry riv.) F4
Tahat (mt.) F4
Tamanrasset, Wadi (dry riv.) E4
Tanezrouft (des.) E4
Tassili N'Ahagger (plat.) E4
Tassili N'Ajjer (plat.) F3
Tidikelt (oasis) E3
Timmissao (well) E4
Tindouf, Sebkha de (salt lake) C3
Tinrhert, Hamada de (des.) F3
Tni Haïa (well) D4
Touat (oasis) E3
Touila (well) C3

BENIN

CITIES and TOWNS

Abomey 38,000 E7
Cotonou 178,000 E7
Djougou E7
Grand-Popo E7
Kandi E6
Lokossa 6,000 E7
Malanville E6
Natitingou 49,000 E6
Nikki E7
Ouidah E7
Parakou 21,000 E7
Porto-Novo (cap.) 104,000 E7
Savalou E7
Savé E7

OTHER FEATURES

Atakora (mts.) E6
Benin (bight) E8
Guinea (gulf) E8
Mono (riv.) E7
Niger (riv.) E6
Ouémé (riv.) E7
Slave Coast (reg.) E7
Sudan (reg.) E6

CAPE VERDE

CITIES and TOWNS

Mindelo 28,797 A7
Praia (cap.) 21,494 B8
Ribeira Grande 1,892 B7
Sal Rei 1,296 B8
Santa Maria 956 B8

OTHER FEATURES

Boa Vista (isl.) B8
Brava (isl.) B8
Fogo (isl.) B8
Maio (isl.) B8
Sal (isl.) B7
Santa Luzia (isl.) B8
Santo Antão (isl.) A7
São Nicolau (isl.) B8
São Tiago (isl.) B8
São Vicente (isl.) B7

GAMBIA

CITIES and TOWNS

Banjul (cap.) 39,476 A6
Basse Santa Su 2,899 B6
Brikama 9,483 A6
Georgetown 2,510 A6

GHANA

CITIES and TOWNS

Accra (cap.) 564,194 D7
Accra* 738,498 D7
Ada 4,285 E7
Akuse 3,791 E7
Attebubu 6,630 D7
Awaso 5,449 D7
Axim 8,107 D8
Bawku 20,567 D6
Bekwai 11,287 D7
Berekum 14,296 D7
Bole 4,772 D7
Bolgatanga 18,896 D6
Cape Coast 51,653 D7
Daboya 1,872 D7
Damongo 7,760 D7
Dunkwa 15,437 D7
Elmina 11,401 D8
Enchi 4,382 D7
Gambaga 3,730 D6
Gyasikan 6,403 D7
Half Assini 5,429 D8
Ho 24,199 E7
Keta 14,446 E7
Kete Krachi 5,097 E7
Kintampo 7,149 D7
Koforidua 46,235 D7
Kpandu 12,842 D7
Kumasi 260,286 D7
Kumasi* 345,117 D7
Lawra 2,709 D6
Mampong 13,895 D7
Mpraeso 5,908 D7
Navrongo D6
Nsawam 25,518 D7
Nsuta 3,854 D7
Obuasi 31,005 D7
Oda 20,957 D7
Prestea 15,143 D7
Salaga 6,413 D7
Sekondi 33,713 D8
Sekondi-Takoradi* 160,868 D8
Sunyani 23,780 D7
Takoradi 58,161 D8
Tamale 83,653 D7
Tarkwa 14,702 D7
Tema 60,767 E7
Tumu 4,366 D6
Wa 21,374 D6
Wenchi 13,836 D7
Wiawso 5,558 D7
Winneba 30,778 D7
Yapei 1,203 D7
Yendi 22,072 D7

OTHER FEATURES

Ashanti (reg.) D7
Benin (bight) E8
Black Volta (riv.) D6
Gold Coast (reg.) D8
Guinea (gulf) E8
Oti (riv.) E7
Red Volta (riv.) D6
Saint Paul (cape) E7
Three Points (cape) D8
Volta (lake) D7
Volta (riv.) E7
White Volta (riv.) D6

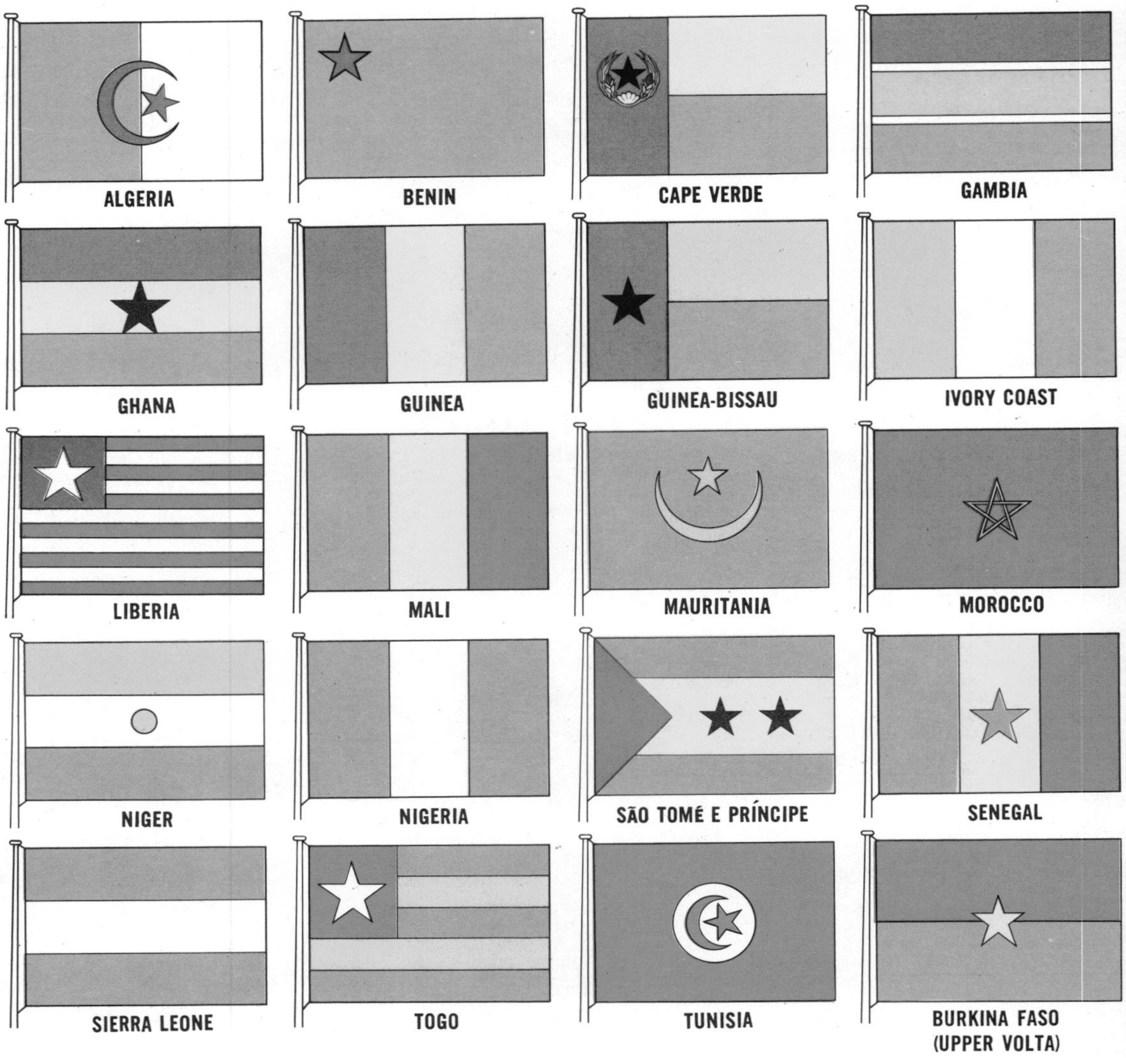

ALGERIA
BENIN
CAPE VERDE
GAMBIA
GHANA
GUINEA
GUINEA-BISSAU
IVORY COAST
LIBERIA
MALI
MAURITANIA
MOROCCO
NIGER
NIGERIA
SÃO TOMÉ E PRÍNCIPE
SENEGAL
SIERRA LEONE
TOGO
TUNISIA
BURKINA FASO (UPPER VOLTA)

GUINEA

CITIES and TOWNS

Beyla C7
Boffa B6
Boké B6
Conakry (cap.)* 525,671 B7
Dabola B6
Dalaba B6
Dinguiraye B6
Dubréka B7
Faranah B6
Forécariah B7
Fria B6
Gaoual B6
Guéckédou B7
Kamsar B6
Kankan 85,310 C6
Kérouané C7
Kindia 79,861 B6
Kissidougou B7
Koundara 6,000 B6
Kouroussa C6
Labé 79,670 B6
Macenta C7
Mali B6
Mamou B6
N'Zérékoré 23,000 C7
Sangaredyi B6
Siguiri C6
Télimélé 12,000 B6
Tougué B6
Victoria B6

OTHER FEATURES

Bafing (riv.) B6
Bakoy (riv.) B6
Futa Jallon (lag.) B6
Los (isls.) B7
Milo (riv.) C7
Moa (riv.) B7
Niger (riv.) C6
Nimba (lag.) C7
Verga (cape) B6

GUINEA-BISSAU

CITIES and TOWNS

Bissau (cap.) 109,486 A6
Bolama○ 9,133 A6
Buba○ 6,706 B6
Bubaque○ 8,441 A6
Cacheu○ 15,194 A6

OTHER FEATURES

Bijagós (isls.) A6

IVORY COAST

CITIES and TOWNS

Abengourou 31,239 D7
Abidjan 685,828 D7
Aboisso 14,272 D7
Agboville 27,192 D7
Bingerville 18,218 D7
Bondoukou 19,111 D7
Bouaflé 15,917 C7
Bouaké 173,248 D7
Bouna 5,787 D7
Boundiali 9,869 C7
Dabakala 3,272 D7
Dabou 23,870 D7
Daloa 60,958 C7
Danané 19,872 C7
Dimbokro 30,986 D7
Divo 37,896 C7
Ferkessédougou 25,307 D7
Fresco 1,865 C7
Gagnoa 42,362 C7
Grand-Bassam 25,808 D7
Grand-Lahou 4,070 C8
Guiglo 10,441 C7
Issia 11,143 C7
Katiola 21,559 C7
Kong 2,551 D7
Korhogo 47,657 C7
Man 50,315 C7
Mankono 6,570 C7
Odienné 13,864 C7
Port-Bouet 72,616 D7
San Pedro 27,616 C8
Sassandra 9,404 C8
Séguéla 12,587 C7
Sinfra 16,399 C7
Tabou 7,255 C8
Touba 5,256 C7
Toumodi 12,983 D7
Yamoussoukro (cap.) 50,000 C7

OTHER FEATURES

Aby (lag.) D8
Bagoé (riv.) C6
Bandama (riv.) C7
Baoulé (riv.) C6
Black Volta (riv.) D6
Cavally (riv.) C7
Comoé (riv.) D7
Ebrié (lag.) D8
Guinea (gulf) E8
Ivory Coast (reg.) C8
Kossou, Lac de (lake) C7
Nimba (lag.) C7
Sassandra (riv.) C7

LIBERIA

CITIES and TOWNS

Buchanan 23,999 B7
Gbarnga 6,896 C7
Grand Cess C8
Greenville 8,462 C8
Harbel 11,445 B7
Harper 10,627 C8
Kolahun C7
Marshall B7
Monrovia (cap.) 166,507 B7
Plahn C7
River Cess 2,041 C7
Robertsport 2,562 B7
Sasstown C8
Tapeta 3,927 C7
Tchien 6,094 C7
Tubmanburg 14,089 B7

OTHER FEATURES

Bong (range) B7
Cavalla (riv.) C7
Cestos (riv.) C7
Grain Coast (reg.) B8
Kru Coast (reg.) C8
Mano (riv.) B7
Mount (cape) B7
Nimba (lag.) C7
Palmas (cape) C8
Roberts Field Int'l Airport C7

MALI

CITIES and TOWNS

Anéfis E5
Ansongo 3,485 E5
Araouane D5
Bafoulabé 2,163 B6
Bamako (cap.) 404,022 C6
Bamba D5
Banamba 6,776 C6
Bandiagara 8,920 D6
Bankass 3,229 D6
Bou Djebeha D5
Bougouni 17,246 C6
Bourem 4,538 E5
Dioila 4,953 C6
Dire 8,941 D5
Djenné 10,251 D6
Douentza 6,746 D6
Gao 30,714 E5
Goundam 10,262 D5
Gourma-Rharous 4,671 D5
Hombori D5
Kadiolo 3,991 C6
Kangaba 3,184 C6
Kati 24,991 C6
Kayes 44,736 B6
Ké-Macina 5,426 C6
Kéniéba 4,510 B6
Kerchoual E5
Kidal 3,308 E5
Kita 17,538 C6
Kolokani 8,923 C6
Kolondiéba 5,882 C6
Koulikoro 16,376 C6
Kourouba B6
Koutiala 27,497 C6
Mabrouk D5
Ménaka 3,693 E5
Mopti 53,885 D6
Nampala C5
Nara 6,091 C5
Niafunké 6,399 D5
Niono 12,290 C6
Nioro 11,617 C5
San 22,962 D6
Satadougou B6
Ségou 64,890 C6
Sikasso 47,030 C6
Sokolo C6
Taoudenni D4
Ténenkou 4,708 C6
Tessalit E4
Timbuktu (Tombouctou) 20,483 D5
Toukoto C6
Yanfolila 3,809 C6
Yelimané 1,481 B5
Yorosso 2,390 C6

OTHER FEATURES

Achourat (well) D4
Adrar des Iforas (plat.) E5
Asselar (well) D5
Azaouad (reg.) D5
Azaouak (dry riv.) E5
Bafing (riv.) B6
Bagoé (riv.) C6
Bakoy (riv.) B6
Bani (riv.) C6
Baoulé (dry riv.) C6
Baoulé (riv.) C6
Bir Ounane (well) D4
Chech, Erg (des.) D4
Debo (lake) D5
El Mraiti (well) D5
Faguibine (lake) D5
Falémé (riv.) B6
Haricha Hamada (des.) D4
Hombori (mts.) D5
In Dagouber (well) D4
Macina (depr.) D6
Niger (riv.) D5
Oum el Asel (well) D4
Sahara (des.) D4
Sekkane, Erg (des.) D4
Senegal (riv.) B5
Sudan (reg.) D6
Tadjnout Hagguerete (well) D4
Terhazza (ruins) C4
Tilemsi (valley) E5
Toufourine (well) C4

MAURITANIA

CITIES and TOWNS

Aîoun el Atrous C5
Akjoujt 8,044 B5
Akreïjit C5
Aleg 6,415 B5
Atar 16,326 B4
Bassikounou C5
Bir Mogreîn B3
Boutilimit 7,261 B5
Bogué 8,056 B5
Chinguetti B4
Fderik (Fort-Gouraud) 2,160 B4
Kaédi 20,848 B5
Kankossa B5
Kiffa 10,629 B5
Maghama B5
M'Bout B5
Méderdra A5
Néma 8,232 C5
Nouakchott (cap.) 134,986 A5
Nouadhibou 21,961 A4
Ouadane B4
Oualata C5
Oujaf C5
Oujeft B4
Rosso 16,466 A5
Sélibaby 5,994 B5
Tamchakett B5
Tamsagout C4
Tazadit B4
Tichitt C5
Tidjikja 7,870 B5
Timbédra 5,317 C5
Zouîrât 17,474 B4

OTHER FEATURES

Adafer (reg.) B5
Adrar (reg.) B4
Affolé (reg.) B5
Agueraktem (well) C4
Aîn ben Tili (well) C3
Arguin (bay) A4
Assaba (reg.) B5
Atoui, Wadi (dry riv.) B4
Ben Guerdane (well) B3
Bir el Khzaim (well) C4
Blanc (cape) A4
Brakna (reg.) B5
Chegga (well) C3
Djouf, El (des.) C4
El Mrayer (well) C4
El Mreîti (well) C4
Gorgol (reg.) B5
Hodh (reg.) C5
Iguidi, Erg (des.) C3
Inchiri (reg.) A5
Koumbi Saleh (ruins) C5
Lévrier (bay) A4
Makteîr (des.) B4
Meraia (reg.) C5
Mirik (Timiris) (cape) A5
Ouarane (reg.) B4
Sahara (des.) C4
Senegal (riv.) B5
Tagant (reg.) B5
Tidra (isl.) A5
Timiris (cape) A5
Touila (well) C3
Trarza (reg.) A5

MOROCCO

CITIES and TOWNS

Agadir 61,192 C2
Al Hoceima 18,686 D1
Asilah 14,074 C1
Azemmour 17,182 C2
Azrou 20,756 C2
Beni Mellal 53,826 C2
Berguent 3,356 D2
Bou Arfa D2
Bou Izakarn 2,342 C3
Boujad 18,838 C2
Casablanca 1,506,373 C2
Chechaouene 15,362 D1
Dar-el-Beida (Casablanca) 1,506,373 C2
El Jadida 55,501 C2
El Kelaa des Srarhna 17,163 C2
Erfoud 5,400 D2
Er Rachidia 16,775 D2
Essaouira 30,061 B2
Fédala (Mohammedia) 70,392 C2
Fès (Fez) 325,327 D2
Figuig 13,660 D2
Goulmima 4,056 C2
Inezgane 11,495 C2

Jerada 30,633 D2
Kenitra 139,206 C2
Khenifra 25,526 C2
Khouribga 73,667 C2
Ksar el Kebir 48,262 C2
Larache 45,710 C1
Marrakech 332,741 C2
Mazagan (El Jadida) 55,501 C2
Meknès 248,369 C2
Mogador (Essaouira) 30,061 B2
Mohammedia 70,392 C2
Nador 32,490 D1
Ouarzazate 11,142 C2
Oued Zem 33,323 C2
Ouezzane 33,267 C2
Oujda 175,532 D2
Petitjean (Sidi Kacem) 26,831 C2
Port-Lyautey (Kénitra) 139,206 C2
Rabat (cap.) 367,620 C2
Safi 129,113 C2
Saïdia D2
Salé 155,557 C2
Sefrou 28,607 D2
Settat 42,325 C2
Sidi Ifni 13,650 B3
Sidi Kacem 26,831 C2
Tagounite C3
Tangier (Tanger) 187,894 C1
Tan-Tan 10,772 B3
Taourirt 15,580 D2
Taouz D2
Tarfaya 1,104 B3
Taroudant 22,272 C2
Taza 55,157 D2
Tendrara D2
Tétouan 139,105 C1
Tiznit 11,391 B3
Youssoufia 22,435 C2
Zagora 5,306 C2

OTHER FEATURES

Anti-Atlas (ranges) C3
Atlas (mts.) C2
Bani, Jebel (mts.) C3
Beddouza, Ras (cape) C2
Dra, Wadi (dry riv.) C3
Er Rif (range) D2
Gibraltar (str.) C1
High Atlas (ranges) C2
Juby (cape) B3
Mediterranean (sea) D1
Middle Atlas (ranges) C2
Moulouya (riv.) D2
Rhéris, Wadi (dry riv.) D2
Rhir (cape) B2
Rif, Er (range) D2
Sarhro, Jebel (mts.) C2
Sebou (riv.) C2
Sim (cape) B2
Toubkal, Jebel (mt.) C2
Ziz, Wadi (dry riv.) D2

NIGER

CITIES and TOWNS

Agadès 11,000 F5
Arhli (Arlit) F4
Bilma G5
Birni-N'Konni 10,000 E6
Bosso G6
Chirfa G4
Dakoro F6
Dessa E6
Diffa G6
Djado G4
Dogondoutchi 9,000 E6
Dosso E6
Fachi G5
Filingué 10,000 E6
Gangara F6
Gaya 5,000 E6
Gouré G6
Iférouane F5
Illela 9,000 F6
In-Gall F5
Madama G4
Madaoua F6
Magaria F6
Mainé-Soroa G6
Maradi 45,852 F6
N'Guigmi G6
Niamey (cap.) 225,314 E6
Ouallam E6
Say E6
Tahoua 31,265 F6
Tanout F6
Téra 8,000 E6
Tessaoua 5,000 F6
Tillabéry E6
Timia F4
Zinder 58,436 F6

OTHER FEATURES

ichégour (well) G5
igadem (well) G5
r (mts.) F5
naye (well) G5
ssakarai (dry riv.) F5
zaoua (reg.) E5
zbine (Air) (mts.) F5
agam (well) F5
anguezane (mt.) F5
edouaram (well) G5
had (lake) G6
allol Bosso (dry riv.) E6
llia (dry riv.) G5
jado (plat.) G4
War (well) G4
Azaoua (well) F4
omadugu Yobe (riv.) G6
lantas (well) E5
iger (riv.) E6
ahara (des.) F4
udan (reg.) F6
afassasset, Wadi (dry riv.) F4
alak (reg.) E5
énéré (des.) G5
mboulaga (well) F5
ummo (El War) (well) G4
o Baba (well) G5

NIGERIA

STATES

nambra 2,300,000 F7
auchi 2,496,329 F6
endel 2,336,000 F7
enue 2,641,496 F7
orno 2,853,553 G6
ross River 3,633,582 F7
ongola 1,585,200 G7
no 5,000,000 F7
aduna 4,098,303 F6
ano 5,775,000 F6
wara 1,600,600 E7
agos 1,100,000 E7
ger 2,900,000 E7
gun 1,448,966 E7
ndo 2,727,676 E7
yo 5,208,884 E7
ateau 1,367,450 F7
vers 1,544,314 F8
okoto 1,367,450 F6

CITIES and TOWNS

Aba 177,000 F7
Abeokuta 253,000 E7
Abuja F7
Ado 213,000 E7
Afikpo F7
Aku F7
Akure F7
Argungu E6
Asaba F7
Azare G6
Baga G6
Bama G6
Baro F7
Bauchi F6
Benin City 136,000 F7
Bida F7
Birnin Kebbi E6
Biu G6
Bonny F8
Brass F8
Burutu F7
Calabar 103,000 F7
Deba Habe G6
Degema F8
Dikwa G6
Donga G7
Ede 182,000 E7
Eha Amufu F7
Enugu 187,000 F7
Forcados E7
Funtua F6
Gashaka G7
Gbogo F7
Geidam G6
Gombe G6
Gumel F6
Gummi E6
Gusau F6
Gwadabawa F6
Hadejia G6
Ibadan 847,000 E7
Ibi F7
Ife 176,000 E7
Ijebu-Ode E7
Ikeja E7
Ikom F7
Ilesha 224,000 E7
Ilorin 282,000 E7
Isa F6
Iseyin 115,083 E7
Iwo 214,000 E7
Jalingo G7
Jebba E7
Jega E6
Jos F7
Kabba F7
Kaduna 202,000 F6
Kaiama E7
Kalmalo F6
Kano 399,000 F6
Katsina 109,424 F6
Katsina Ala F7
Kaura Namoda F6
Keffi F7
Koko E7
Kontagora F6
Kukawa G6
Kumo G7
Kuta F7
Lafia F7
Lafiagi E7
Lagos (cap.) 1,060,848 E7
Laro G7
Lere F7
Lokoja F7
Maiduguri 189,000 G6
Maigatari F6
Makurdi F7
Minna F7
Mubi G6
Nasarawa F7
New Bussa E6
Nguru G6
Nnewi F7
Nsukka F7
Offa E7
Ogbomosho 432,000 E7
Ogoja F7
Okene F7
Ondo E7
Onitsha 220,000 F7
Oron F8
Oshogbo 282,000 E7
Owerri F7
Owo F7
Oyo 152,000 E7
Pankshin F7
Panyam F7
Port Harcourt 242,000 F8
Ringim F6
Sapele F7
Shaki E7
Shendam F7
Sokoto F6
Toungo G7
Uromi F7
Vom F7
Wamba F7
Warri F7
Wukari F7
Yan G7
Yelwa F6
Yola G7
Zaria 224,000 F6
Zungeru F7

OTHER FEATURES

Adamawa (reg.) G7
Benin (bight) E8
Benue (riv.) F7
Biafra (bight) F8
Biu (plat.) G6
Bonny (bight) F8
Chad (lake) G6
Cross (riv.) F7
Dimlang (mt.) G7
Donga (riv.) G7
Foge (isl.) E6
Gongola (riv.) G6
Guinea (gulf) E8
Hadejia (riv.) F6
Jos (plat.) F7
Kaduna (riv.) F7
Kainji (res.) E6
Kebbi (riv.) E6
Komadugu Yobe (riv.) G6
Niger (delta) F8
Niger (riv.) F7
Osse (riv.) F7
Slave Coast (reg.) E7
Sokoto (riv.) F6
Sudan (reg.) F6

PORTUGAL-Madeira

CITIES and TOWNS

Funchal (cap.) 38,340 A2

OTHER FEATURES

Desertas (isls.) A2
Madeira (isl.) A2
Pôrto Santo (isl.) A2
Salvage (isls.) A2

SÃO TOMÉ E PRINCIPE

CITIES and TOWNS

Santo Antônio 1,618 F8
São Tomé (cap.) 7,681 F8

OTHER FEATURES

Guinea (gulf) E8
Príncipe (isl.) F8
São Tomé (isl.) F8

SENEGAL

CITIES and TOWNS

Bakel 6,339 B6
Bignona 14,537 A6
Dagana 10,506 A5
Dakar (cap.) 798,792 A6
Diourbel 50,618 A6
Kaolack 106,899 A6
Kédougou 7,575 B6
Kaffrine 11,211 A6
Kolda 19,302 B6
Linguère 7,890 B5
Louga 35,063 A5
Matam 10,002 B5
M'Bour 37,663 A6
Nioro-du-Rip 7,824 A6
Podor 6,914 B5
Richard Toll A5
Rufisque A6
Saint-Louis 88,404 A5
Sedhiou 9,421 A6
Tambacounda 25,147 B6
Thiès 117,333 A6
Tivaouane 17,351 A5
Touba A6
Yarboutenda B6
Ziguinchor 72,726 A6

OTHER FEATURES

Casamance (riv.) A6
Falémé (riv.) B6
Ferlo (reg.) B6
Gambia (riv.) B6
Senegal (riv.) B5
Verde (cape) A6

SIERRA LEONE

CITIES and TOWNS

Bo 42,216 B7
Bonthe 6,230 B7
Freetown (cap.) 274,000 B7
Kabala 4,610 B7
Kambia 3,700 B7
Kenema 33,880 B7
Lungi 2,170 B7
Marampa B7
Makeni 28,684 B7
Moyamba 4,564 B7
Pendembu 2,696 B7
Pepel 3,793 B7
Port Loko 5,809 B7
Pujehun 1 B7

OTHER FEATURES

Loma, Mansa (lag.) B7
Mano (riv.) B7
Moa (riv.) B7
Sherbro (isl.) B7
Yawri (bay) B7

SPAIN-Canary Islands, Ceuta and Melilla

CITIES and TOWNS

Arrecife 21,310 B3
Ceuta 60,639 C1
La Laguna A3
Las Palmas de Gran Canaria 260,368 B3
Melilla 64,942 D1
Santa Cruz de la Palma 10,393 A3
Santa Cruz de Tenerife 74,910 A3

OTHER FEATURES

Canary (isls.) A3
Fuerteventura (isl.) B3
Gomera (isl.) A3
Grand Canary (isl.) A3
Hierro (isl.) A3
Lanzarote (isl.) B3
La Palma (isl.) A3
Tenerife (isl.) A3

TOGO

CITIES and TOWNS

Aného (Anécho) 10,889 E7
Atakpamé 17,440 E7
Dapaong 10,100 E6
Kpalimé 19,801 E7
Kpémé 3,600 E7
Lama-Kara 9,400 E7
Lomé (cap.) 148,443 E7
Mango 9,600 E6
Sokodé 29,623 E7

OTHER FEATURES

Benin (bight) E8
Guinea (gulf) E8
Mono (riv.) E7
Oti (riv.) E7
Slave Coast (reg.) E7

TUNISIA

CITIES and TOWNS

Béja 39,226 F1
Ben Gardane 6,593 G2
Bizerte 62,856 F1
Burj al Hattaba F2
El Borma F2
El Djem 10,666 G1
El Kef 27,939 F1
Gabès 40,585 F2
Gafsa 42,225 F2
Halq el Oued 41,912 G1
Jendouba 18,127 F1
Kairouan 54,546 F1
Kalaa-Kebira 23,508 F1
Kasserine 22,594 F1
La Goulette (Halq el Oued) 41,912 G1
La Skhirra 4,565 G2
Le Kef (El Kef) 27,939 F1
Mahdia 25,711 G1
Mareth 2,185 F2
Mateur 19,645 F1
Médenine 15,826 F2
Menzel Bourguiba 42,111 F1
Menzel Temime 18,857 G1
Moknine 26,035 G1
Monastir 26,759 G1
Msaken 33,559 G1
Nabeul 30,476 G1
Nefta 12,476 F2
Remada 6,100 F2
Sbeitla 8,039 F1
Sfax 171,297 G2
Sousse 69,530 G1
Tabarka 3,140 F1
Tatahouine 10,399 G2
Tozeur 16,772 F2
Tunis (cap.) 550,404 G1
Tunis* 873,515 G1
Zarzis 14,420 G2

OTHER FEATURES

Abiad, Ras el (Blanc) (cape) G1
Blanc (cape) G1
Bon (cape) G1
Chambi, Jebel (mt.) F2
Djerba (isl.) G2
Djerid, Shott el (salt lake) F2
Gabès (gulf) G2
Grand Erg Oriental (des.) F2
Hammamet (gulf) G1
Jefara (reg.) G2
Kerkennah (isls.) G2
Mediterranean (sea) F1
Medjerda (riv.) F1
Tib, Ras el (Bon) (cape) G1
Tunis (gulf) G1

BURKINA FASO (UPPER VOLTA)

CITIES and TOWNS

Aribinda D6
Banfora 12,358 D6
Batié D7
Bobo Dioulasso 115,063 D6
Bogandé E6
Dédougou D6
Diapaga E6
Diébougou D6
Djibo D6
Dori D6
Fada-N'Gourma 12,000 E6
Gaoua D6
Houndé D6
Kaya 18,000 D6
Koudougou 36,838 D6
Koupela D6
Léo D6
Ouagadougou (cap.) 172,661 D6
Ouahigouya 25,690 D6
Pama E6
Po D6
Tenkodogo E6
Tougan D6
Yako D6
Zabré D6

OTHER FEATURES

Black Volta (riv.) D6
Comoé (riv.) D7
Oti (riv.) E7
Red Volta (riv.) D6
Sudan (reg.) D6
White Volta (riv.) D6

WESTERN SAHARA

CITIES and TOWNS

Dakhla 6,554 A4
El Aaiún (Laayoune) 24,519 B3
Semara 2,655 B3
Villa Cisneros (Dakhla) 6,554 A4

OTHER FEATURES

Atoui, Wadi (dry riv.) B4
Ausert (well) B4
Barbas (cape) A4
Bir Ganduz (well) A4
Bir Nzaran (well) B4
Blanc (cape) A4
Bojador (cape) B3
Durnford (pt.) A4
Guelta de Zemmur (well) B3
Saguia el Hamra (dry riv.) B3
Tichlá (well) B4

*City and suburbs.
○Population of sub-district or division.

Agriculture, Industry and Resources

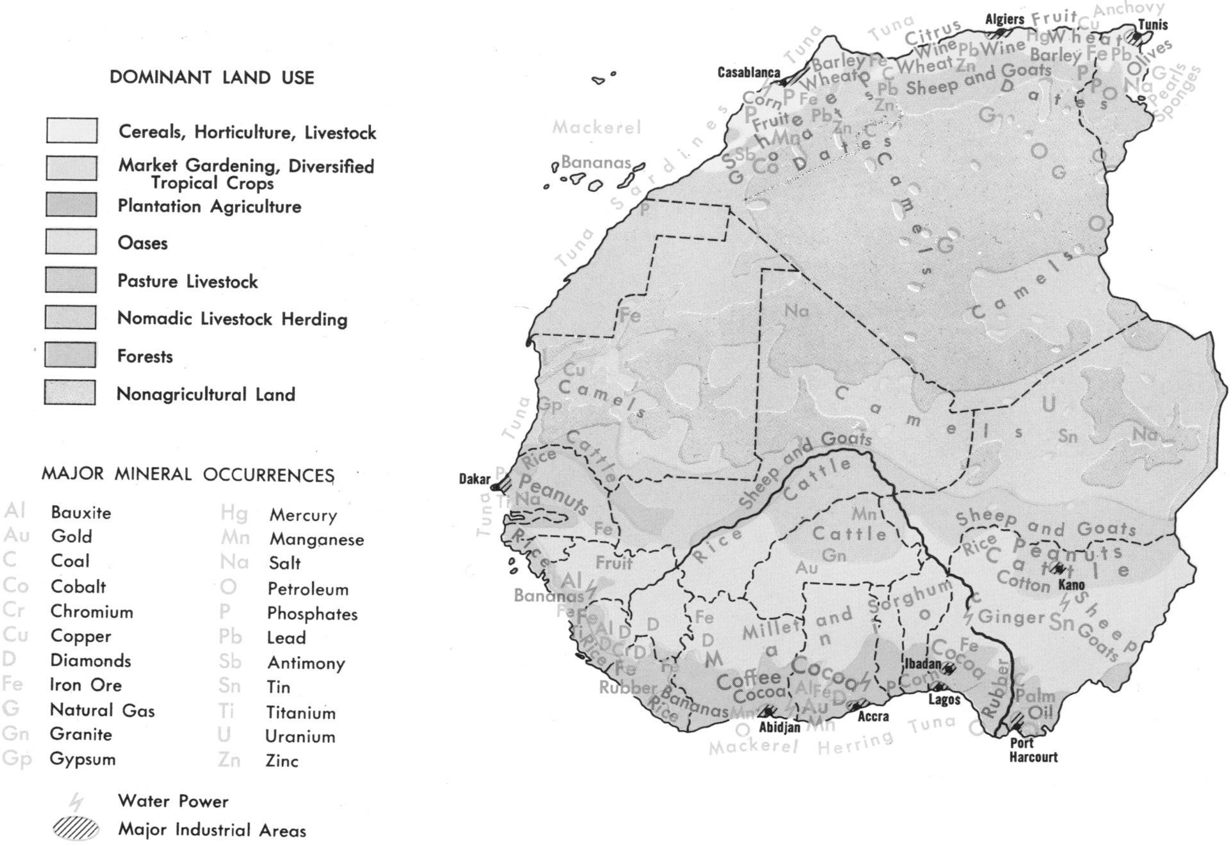
DOMINANT LAND USE
Cereals, Horticulture, Livestock
Market Gardening, Diversified Tropical Crops
Plantation Agriculture
Oases
Pasture Livestock
Nomadic Livestock Herding
Forests
Nonagricultural Land
MAJOR MINERAL OCCURRENCES
Al Bauxite
Au Gold
C Coal
Co Cobalt
Cr Chromium
Cu Copper
D Diamonds
Fe Iron Ore
G Natural Gas
Gn Granite
Gp Gypsum
Hg Mercury
Mn Manganese
Na Salt
O Petroleum
P Phosphates
Pb Lead
Sb Antimony
Sn Tin
Ti Titanium
U Uranium
Zn Zinc
Water Power
Major Industrial Areas
Algiers
Tunis
Casablanca
Dakar
Kano
Ibadan
Lagos
Abidjan
Accra
Port Harcourt
Mackerel
Bananas
Sardines
Tuna
Anchovy
Citrus
Wine
Wheat
Barley
Fruit
Olives
Pearls
Sponges
Sheep and Goats
Dates
Camels
Corn
Peanuts
Rice
Cattle
Millet and Sorghum
Cotton
Ginger
Coffee
Cocoa
Rubber
Palm Oil
Herring

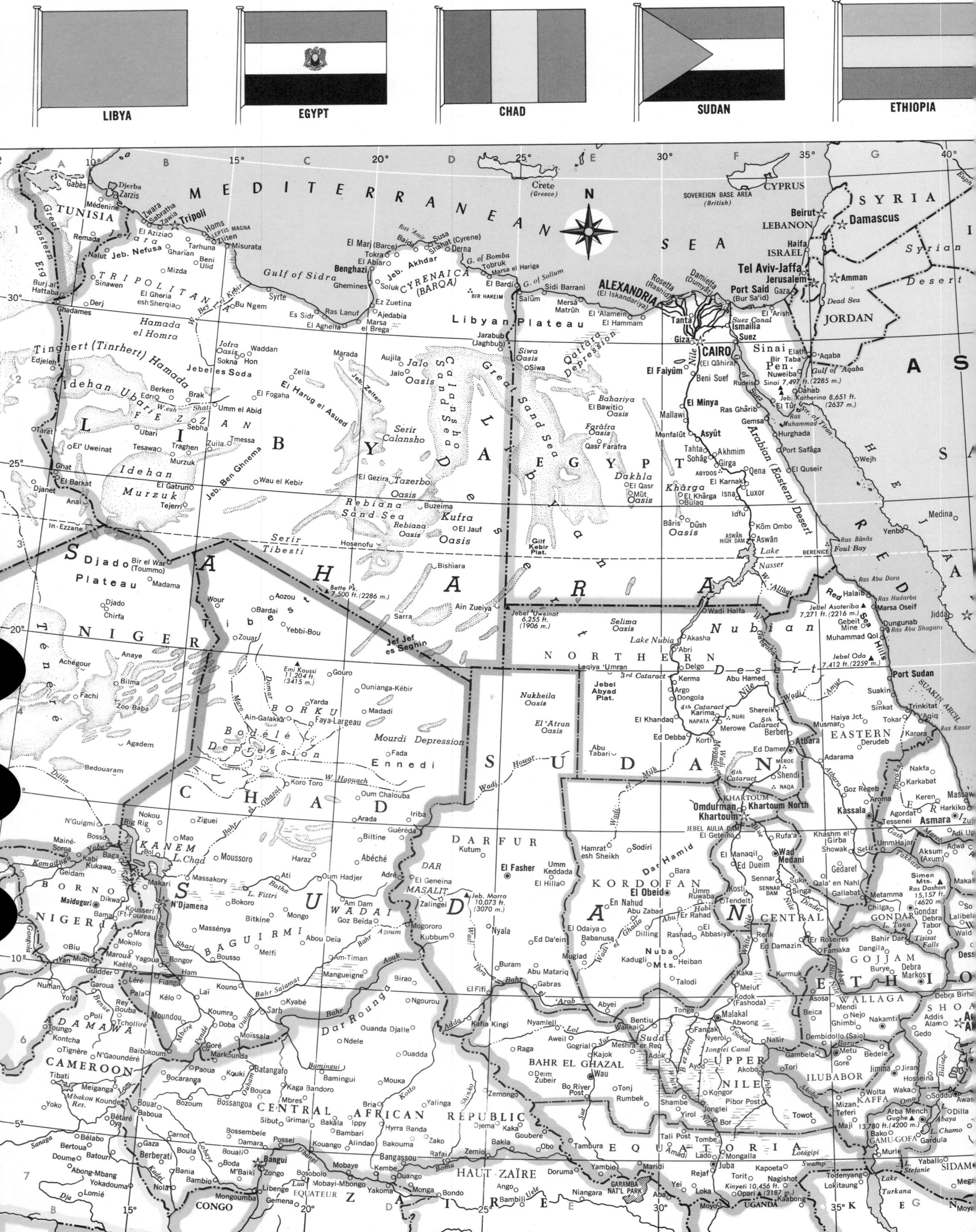

LIBYA **EGYPT** **CHAD** **SUDAN** **ETHIOPIA**

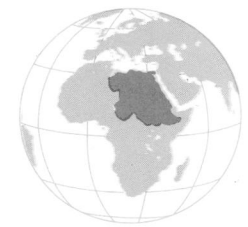

LIBYA

AREA 679,358 sq. mi. (1,759,537 sq. km.)
POPULATION 2,856,000
CAPITAL Tripoli
LARGEST CITY Tripoli
HIGHEST POINT Bette Pk. 7,500 ft. (2,286 m.)
MONETARY UNIT Libyan dinar
MAJOR LANGUAGES Arabic, Berber
MAJOR RELIGION Islam

EGYPT

AREA 386,659 sq. mi. (1,001,447 sq. km.)
POPULATION 41,572,000
CAPITAL Cairo
LARGEST CITY Cairo
HIGHEST POINT Jeb. Katherina 8,651 ft. (2,637 m.)
MONETARY UNIT Egyptian pound
MAJOR LANGUAGE Arabic
MAJOR RELIGIONS Islam, Coptic Christianity

CHAD

AREA 495,752 sq. mi. (1,283,998 sq. km.)
POPULATION 4,309,000
CAPITAL N'Djamena
LARGEST CITY N'Djamena
HIGHEST POINT Emi Koussi 11,204 ft. (3,415 m.)
MONETARY UNIT CFA franc
MAJOR LANGUAGES Arabic, Bagirmi, French, Sara, Massa, Moudang
MAJOR RELIGIONS Islam, tribal religions

SUDAN

AREA 967,494 sq. mi. (2,505,809 sq. km.)
POPULATION 18,691,000
CAPITAL Khartoum
LARGEST CITY Khartoum
HIGHEST POINT Jeb. Marra 10,073 ft. (3,070 m.)
MONETARY UNIT Sudanese pound
MAJOR LANGUAGES Arabic, Dinka, Nubian, Beja, Nuer
MAJOR RELIGIONS Islam, tribal religions

ETHIOPIA

AREA 471,776 sq. mi. (1,221,900 sq. km.)
POPULATION 31,065,000
CAPITAL Addis Ababa
LARGEST CITY Addis Ababa
HIGHEST POINT Ras Dashan 15,157 ft. (4,620 m.)
MONETARY UNIT birr
MAJOR LANGUAGES Amharic, Gallinya, Tigrinya, Somali, Sidamo, Arabic, Ge'ez
MAJOR RELIGIONS Coptic Christianity, Islam

DJIBOUTI

AREA 8,880 sq. mi. (23,000 sq. km.)
POPULATION 386,000
CAPITAL Djibouti
LARGEST CITY Djibouti
HIGHEST POINT Moussa Ali 6,768 ft. (2,063 m.)
MONETARY UNIT Djibouti franc
MAJOR LANGUAGES Arabic, Somali, Afar, French
MAJOR RELIGIONS Islam, Roman Catholicism

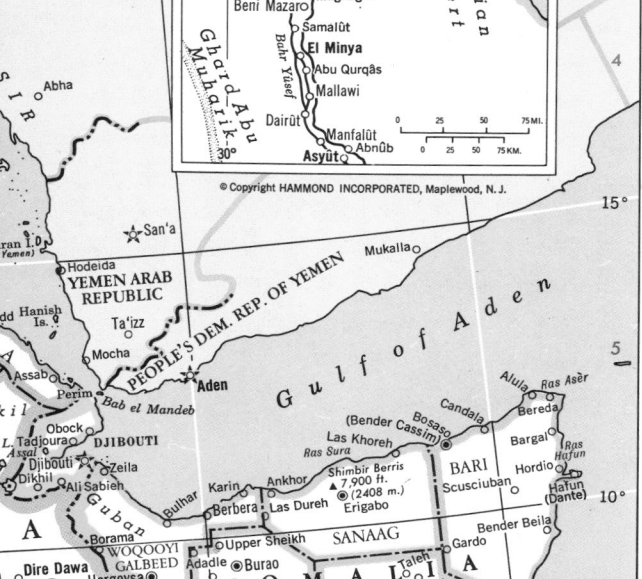

Northeastern Africa

CONIC EQUAL-AREA PROJECTION

SCALE OF MILES
0 50 100 200 300

SCALE OF KILOMETERS
0 50 100 200 300

Capitals of Countries .... ☆
Other Capitals .... ◉
International Boundaries ....
Internal Boundaries ....

Scale 1:14,300,000

© Copyright HAMMOND INCORPORATED, Maplewood, N.J.

CHAD

CITIES and TOWNS

Abéché 28,100	D5
Abou Dela	C5
Adré	D5
Ain-Galakka	C4
Am-Dam	D5
Am-Timan 4,200	D5
Arada	D4
Ati 7,500	C5
Baïbokoum 5,500	C6
Bardaï	C3
Biltine 3,900	D5
Bitkine 5,000	C5
Bokoro 6,500	C5
Bol 2,500	B5
Bongor 14,300	C5
Bousso 4,500	C5
Doba 13,300	C6
Fada	D4
Faya-Largeau 6,800	C4
Fianga 10,000	C6
Goré	C6
Gouro	C4
Goz Belda	D5
Guéréda	D5
Ham	D5

Haraz	C5
Iriba	D4
Kélo 16,800	C6
Koro Toro	C4
Koumra 17,000	C6
Kouno	C5
Kyabé 5,000	C6
Lal 10,400	C6
Léré	B6
Mangalmé	D5
Mangueigne	D5
Mao 4,900	C5
Massakory	C5
Massénya	C5
Melfi	C5
Mogororo	D5
Moïssala 5,500	C6
Mongo 8,300	C5
Moundou 39,600	C6
Moussoro 7,700	C5
N'Djamena (cap.) 179,000	C5
Nokou	B5
Oum Chalouba	D4
Oum Hadjer 5,600	D5
Pala 13,200	C6
Rig Rig	B5
Sarh 43,700	C6
Wour	C3
Yarda	C4

Yebbi-Bou	C3
Ziguei	C5
Zouar	C3

OTHER FEATURES

Azoum, Bahr	D5
Baguirmi (reg.)	C5
Bahr el Ghazal (dry riv.)	C5
Batha (riv.)	C5
Bodélé (depr.)	C4
Borku 72	C4
Chad (lake)	B5
Domar (dry riv.)	C4
Emi Koussi (mt.)	C3
Ennedi (plat.)	D4
Fittri (lake)	C5
Haouach, Wadi (dry riv.)	C4
Jef Jef es Seghin (plat.)	D3
Kanem (reg.)	C5
Logone (riv.)	C5
Maro (dry riv.)	C6
Mbéré (riv.)	C6
Mourdi (depr.)	D4
Ouham (riv.)	C6
Pendé (riv.)	C6
Sahara (des.)	C3
Salamat, Bahr (riv.)	C6
Sara (riv.)	C5
Shari (riv.)	C5

Sudan (reg.)	C5
Tibesti (mts.)	C3
Wadai (reg.)	D5

DJIBOUTI

CITIES and TOWNS

Ali Sabieh	H5
Dikhil	H5
Djibouti (cap.) 96,000	H5
Obock	H5
Tadjoura	H5

OTHER FEATURES

Abbe (lake)	H5
Aden (gulf)	J5
Bab el Mandeb (str.)	H5

EGYPT

CITIES and TOWNS

Abnûb 39,343	J4
Abu Qurqas	J4
Akhmim 53,234	F2
Alexandria 2,318,655	J2

(continued on following page)

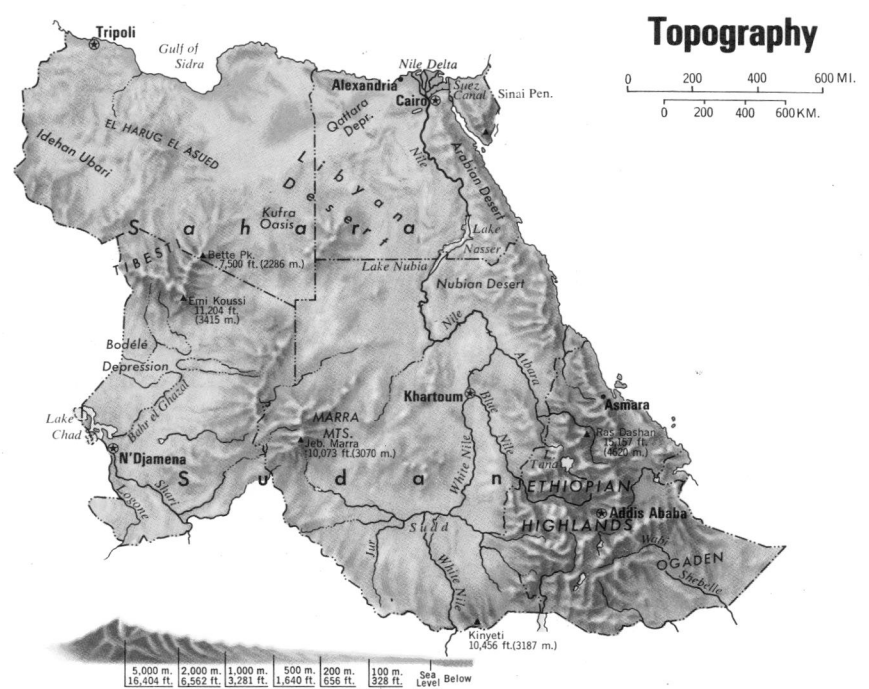

Topography

0 200 400 600 MI.
0 200 400 600 KM.

5,000 m.	2,000 m.	1,000 m.	500 m.	200 m.	100 m.	Sea
16,404 ft.	6,562 ft.	3,281 ft.	1,640 ft.	656 ft.	328 ft.	Level Below

(continued on following page)

112 Northeastern Africa

(continued)

Aswan 144,377 F3
Asyût 213,983 J4
Bâris F3
Benha 88,992 J3
Beni Mazar 39,373 J4
Beni Suef 118,148 J3
Biba 33,074 J4
Bûlaq F2
Bur Sai'd (Port Said) 262,620 K2
Cairo (cap.) 5,084,463 J3
Dahab F2
Dairût 31,624 J4
Damanhur 188,927 J3
Damietta 93,546 J3
Disûq 58,650 J3
Dumyât (Damietta) 93,546 J3
Dûsh F3
El A'lamein E1
El A'rish F1
El Bawiti E2
El Faiyûm 167,081 J3
El Fashn 33,506 J4
El Hammam 6,588 E1
El Iskandariya (Alexandria) 2,318,655 J2
El Karnak F2
El Khârga 26,375 F2
El Mahalla el Kubra 292,853 J3
El Mansûra 257,866 K3
El Minya 146,423 J4
El Qâhira (Cairo) (cap.) 5,084,463 J3
El Qantara 919 K3
El Qasr E2
El Quseir 12,297 F2
El Tûr F2
El Wasta 17,659 J3
Gemsa F2
Girga 51,110 F2
Giza 1,246,713 J3
Heliopolis J3
Helwân J3
Hurghada F2
Idfu 34,858 F3
Imbâba J3
Ismailia 145,978 K3
Isna 34,186 F2
Karnak (El Karnak) F2
Kôm Ombo 44,531 F3
Luxor 92,748 F2
Maghâgha 40,802 J4
Mallawi 74,256 J4
Manfalût 41,126 J4
Mersá Matrûh 27,857 E1
Minûf 55,131 J3
Mût 8,032 E2
Nuweiba F2
Port Fuad K3
Port Safâga F2
Port Said 262,620 K2
Port Taufiq K3
Qalyub 62,739 J3
Qasr Farâfra E2
Qena 94,013 F2
Ras Ghârib F2
Rashid (Rosetta) 42,962 J2
Rudeis F2
Salûm 4,161 E1
Samalût 48,146 J4
Shibin el Kom 102,844 J3
Sidi Barrani 1,574 E1
Sinnûris 42,022 J3
Siwa 4,999 E2
Sohâg 101,758 F2
Suez 194,001 K3
Tahta 45,242 F2
Tanta 284,636 J3
Zagazig 202,637 K3
Zifta 50,410 J3

OTHER FEATURES

Abu Qir (bay) J2
Abydos (ruins) F2
A'llaqi, Wadi (dry riv.) F3
A'qaba (gulf) G2
Arabian (des.) F2
Aswân (dam) F3
Aswân High (dam) F3
Bahariya (oasis) E2
Bahr Yusef (stream) J4
Bânâs, Ras (cape) G3
Berenice (ruins) F3
Birket Qârûn (lake) J3
Bir Taba (well) F2
Bitter (lkes) K3
Dakhla (oasis) E2
Eastern (Arabian) (des.) F2
El Sollum (gulf) E1
Farâfra (oasis) E2
Foul (bay) G3
Ghard Abu Muharik (des.) J4
Gilf Kebir (plat.) E3
Great Sand Sea (des.) D2
Katherina, Jebel (mt.) F2
Khârga (oasis) F2
Libyan (des.) E2
Libyan (plat.) E1
Mediterranean (sea) E1
Memphis (ruins) J3
Muhammad, Ras (cape) F2
Nasser (lake) F3
Nile (riv.) F2
Pyramids (ruins) J3
Qattara (depr.) E2
Red (sea) G3
Sahara (des.) E3
Sinai (mt.) F2
Sinai (pen.) F2
Siwa (oasis) E2
Suez (canal) K3
Suez (gulf) F2
Tiran (str.) F2
U'weinat, Jebel (mt.) E3

ETHIOPIA

PROVINCES

Arusi 852,900 G6
Bale 707,800 H6
Eritrea 1,947,600 G4
Gamu-Gofa 698,800 G6
Gojjam 1,750,100 G5
Gondar 1,355,800 G5
Harar 3,359,200 H6
Ilubabor 688,800 F6
Kaffa 1,693,000 G6
Shoa 5,369,500 G6
Sidamo 2,479,800 G7
Tigre 1,828,900 H5
Wallaga 1,269,100 G6
Wallo 2,459,900 H5

CITIES and TOWNS

Addis Ababa (cap.) 1,196,300 G6
Addis Alam 5,500 G6
Adigrat 9,400 G5
Adi Ugri 12,800 G5
Adwa 16,400 G5
Afdem H6
Agordat G4
Aksum 12,800 G5
Ankober H6
Arba Mench 7,660 G6
Asmara 393,800 G4
Asosa F5
Assab 16,000 H5
Asselle 19,390 G6
Awareh H6
Awasa 16,790 G6
Awash H6
Axum (Aksum) 12,800 G5
Bahir Dar 25,100 G5
Burye G5
Callafo H6
Chilga G5
Dagabur H6
Dallol G5
Dangila G5
Debra Birhan 16,700 G6
Debra Markos 30,260 G5
Debra Tabor 8,700 G5
Dembidollo 7,600 F6
Dessye 49,750 G5
Dilla 13,800 G6
Dire Dawa 63,700 H6
Dolo H7
Domo J6
Edd H5
El Carre H6
El Der H6
Filtu H6
Gabredarre H6
Galadi J6
Gambela F6
Gardula 5,800 G6
Gedo G6
Gerlogubi H6
Ghimbi 8,300 G6
Ginir H6
Goba 13,500 H6
Gondar 38,600 G5
Gore 8,500 G6
Gorrahei H6
Harar 48,440 H6
Harkiko G4
Hosseina 8,500 G6
Imi H6
Jijiga 8,000 H6
Jimma 47,360 G6
Jiran G6
Karkabat G4
Keren G4
Kibre Mengist 8,300 G6
Lalibela G5
Magdala G5
Maji G6
Makale 30,780 G5
Massawa 19,800 G4
Mega G7
Mendi G6
Mersa Fatma H5
Metamma G5
Metu G6
Miesso H6
Mizan Teferi G6
Moyale G7
Murle G6
Mustahil H6
Nakamti 18,310 G6
Nakfa G4
Nazret 42,900 G6
Negelli 8,800 G6
Nejo G6
Saio (Dembidollo) 7,600 F6
Soddu 11,900 G6
Sokota G5
Tessenei G4
Thio H5
Tori F6
Umm Hajar G5
Waka G6
Waldia 9,600 G5
Wardere J6
Wolta G6
Yaballo G6
Zula G4

OTHER FEATURES

Abay (riv.) G5
Abaya (lake) G6
Akobo (riv.) F6
Assale (lake) H5
Atbara (riv.) G4
Awash (riv.) H5
Bale (mt.) G6
Baraka (riv.) G4
Baro (riv.) G6
Billate (riv.) G6
Blue Nile (Abay) (riv.) G5
Buri (pen.) H4
Chamo (lake) G6
Dahlak (arch.) H4
Dahlak (isl.) H4
Danakil (reg.) H5
Dawa (riv.) G7
Fafan (riv.) H6
Ganale Dorya (riv.) H6
Gash Mareb (riv.) G5
Gughe (mt.) G6
Haud (reg.) J6
Kasar, Ras (cape) G4
Ogaden (reg.) H6
Omo (riv.) G6
Ras Dashan (mt.) G5
Red (sea) H4
Rudolf (Turkana) (lake) G7
Simen (mts.) G5
Stefanie (lake) G7
Takkaze (riv.) G5
Tana (lake) G5
Tisisat (fall) G5
Turkana (lake) G7
Wabi (riv.) H6
Wabi Shebelle (riv.) H6
Zwai (lake) G6

LIBYA

CITIES and TOWNS

Ajedabia○ 53,170 D1
Aujila○ 6,695 D2
Baida○ 59,765 D1
Barce (El Marj)○ 55,444 D1
Benghazi (cap.)○ 286,943 C1
Beni Ulid○ 19,113 B1
Berken B2
Brak○ 12,507 B2
Bu Ngem C1
Cyrene (Shahat)○ 17,157 D1
Derj○ 2,152 B1
Derna○ 44,145 D1
Edri B2
El Abiar○ 17,685 D1
El Agheila C1
El Azizia○ 34,077 B1
El Bardi○ 4,330 D1
El Barkat○ 2,139 B3
El Fogaha C2
El Gatrun B3
El Gezira D2
El Jauf○ 6,481 D3
El Marj○ 55,444 D1
El' Uweinat B2
Es Sidr○ 706 C1
Ez Zuetina○ 7,256 D1
Ghadames○ 6,172 A2
Gharfan○ 65,224 B1
Ghat○ 6,924 B3
Ghemines○ 4,313 C1
Homs○ 66,890 B1
Hon○ 2,766 C2
Jaghbub (Jarabub)○ 1,436 D2
Jalo D2
Jarabub○ 1,436 D2
Marada○ 3,201 C2
Marsa el Brega○ 2,618 D1
Marsa el Hariga○ 5,043 D1
Mekili D1
Misurata○ 102,439 C1
Mizda○ 11,472 B1
Murzuk○ 22,185 B2
Nalut○ 23,535 B1
Ras Lanuf○ 1,990 C1
Sabrathaa○ 30,836 B1
Sebha○ 35,879 B2
Shahat○ 17,157 D1
Sinawen○ 1,549 B1
Sokna○ 3,757 C2
Soluk○ 6,501 D1
Susa D1
Syrte○ 22,797 C1
Tarhuna○ 52,657 B1
Tejerri B3
Tesawa B2
Tmessa C2
Tobruk○ 58,384 D1
Tokra○ 10,714 D1
Traghen B2
Tripoli (cap.)○ 550,438 B1
Ubari○ 19,132 B2
Umm el Abid C2
Waddan○ 5,347 C2
Wau el Kebir C2
Zawia○ 72,092 B1
Zella○ 4,835 C2
Zliten○ 58,981 C1
Zuila C2
Zwara○ 15,078 B1

OTHER FEATURES

Ain Zueiya (well) D3
Akhdar, Jebel (mts.) D1
A'mir, Ras (cape) D1
Barqa (Cyrenaica) (reg.) D1
Ben Ghnema, Jebel (mts.) C2
Bette (peak) C3
Bey el Kebir, Wadi (dry riv.) B1
Bir Hakeim (ruins) D1
Bishiara (well) D3
Bomba (gulf) D1
Buzeima (well) D3
Calansho Sand Sea (des.) D2
Calansho, Serir (des.) D2
Cyrenaica (reg.) D1
Fezzan (reg.) B2
Great Sand Sea (des.) D2
Harug el Asued, El (mts.) C2
Homra, Hamada el (des.) B2
Hosenofu (well) D3
Idehan Ubari (des.) B2
Idehan Murzuk (des.) B2
Jalo (oasis) D2
Jefara (reg.) B1
Jef Jef es Seghin (plat.) D3
Jofra (oasis) C2
Kufra (oasis) D3
Leptis Magna (ruins) B1
Libyan (des.) D2
Libyan (plat.) D1
Mediterranean (sea) C1
Nefusa, Jebel (mts.) B1
Rebiana○ (oasis) D3
Rebiana Sand Sea (des.) D3
Sahara (des.) C3
Sarra (well) D3
Shati, Wadi esh (dry riv.) B2
Sidra (gulf) C1
Soda, Jebel es (mts.) C2
Tazerbo○ (oasis) D2
Tibesti, Serir (des.) C3
Tinghert Hamada (Tinrhert) (des.) B2
Tripolitania (reg.) B1
U'weinat, Jebel (mt.) E3
Zelten, Jebel (mts.) D2

SUDAN

PROVINCES

Bahr el Ghazal E6
Central F5
Darfur D5
Eastern G4
Equatoria E6
Khartoum F4
Kordofan E5
Northern E3
Upper Nile F6

CITIES and TOWNS

A'bri F3
Abu Hamed F4
Abu Matariq E5
Abu Zabad E5
Abwong F6
Abyei E6
Adarama G4
Adok F6
Akasha F3
Akobo F6
Amadi F6
A'qiq G4
Argo F4
Aroma G4
Atbara 66,000 F4
Aweil E6
Ayod F6
Babanusa E5
Bara F5
Bentiu E6
Berber F4
Bor F6
Bo River Post E6
Buram E5
Damazin (Ed Damazin) 12,000 F5
Deim Zubeir E6
Delgo F3
Derudeb G4
Dilling E5
Dongola 6,000 F4
Dungunab G3
Ed Dae'in E5
Ed Damer 17,000 F4
Ed Damazin 12,000 F5
Ed Debba F4
Ed Dueim 27,000 F5
El Abbasiya F5
El Fasher 52,000 E5
El Fifi D5
El Geneina 33,000 D5
El Geteina F5
El Hilla E5
El Khandaq E4
El Manaqil F5
El Obeid 90,000 E5
El Odaiya E5
En Nahud 23,000 E5
Er Rahad F5
Er Roseires F5
Famaka F5
Fangak F6
Fashoda (Kodok) F6
Gabras E5
Gallabat G5
Gebeit Mine G3
Gedaref 92,000 G5
Gogrial E6
Goz Regeb G4
Haiya Junction G4
Halaib G3
Heiban F5
Jonglei F
Juba 57,000 F
Kadugli 18,000 E
Kafia Kingi D
Kajok E
Kaka F
Kapoeta F
Karima F
Karora G
Kassala 99,000 G
Kerma F
Khartoum (cap.) 334,000 F
Khartoum North 151,000 F
Khashm el Girba G
Kodok F
Kongor F
Korti F
Kosti 57,000 F
Kubbum D
Kurmuk F
Kutum D
Lado F
Loka F
Malakal 35,000 F
Maridi E
Marsa Oseif G
Melut F
Merowe F
Meshra er Req E
Mongalla F
Muglad E
Muhammad Qol G
Musmar G
Nagishot F
Nasir F
Nimule F
Nyala 60,000 E
Nyamlell E
Nyerol F
Omdurman 299,000 F
Opari F
Pibor Post F
Port Sudan 133,000 G
Qalae'n Nahl F
Raga E
Rashad F
Rejaf F
Renk F
Rufaa' F
Rumbek 17,000 E
Sennar F
Shambe F
Shendi F
Shereik F
Showak G
Singa F
Sinkat G
Sodiri E
Suakin G
Suki F
Tali Post F
Talodi F
Tambura E
Tendelti F
Tokar G
Tombe F
Tonga F
Tonj E
Torit F
Towot F
Trinkitat G
Umm Keddada E
Umm Ruwaba F
Wadi Halfa F
Wad Medani 107,000 F
Wankai E
Wau 53,000 E
Yambio 7,000 E
Yei F
Yirol F
Zalingei D

OTHER FEATURES

Abu Dara, Ras (cape) G
Abu Habl, Wadi (dry riv.) F
Abu Shagara, Ras (cape) G
Abu Tabari (well) E
Adda (riv.) D
Akobo (riv.) F
A'mur, Wadi (dry riv.) G
Asoteriba, Jebel (mt.) G
Atbara (riv.) G
Bahr Azoum (riv.) D
Bahr el A'rab (riv.) E
Bahr ez Zeraf (riv.) F
Baraka (riv.) G
Blue Nile (riv.) F
Dar Hamid (reg.) F
Dar Masalit (reg.) D
Dinder (riv.) F
El A'trun (oasis) E
Fifth Cataract F
Fourth Cataract F
Gabgaba, Wadi (dry riv.) F
Gezira, El (reg.) F
Ghalla, Wadi el (dry riv.) E
Hadarba, Ras (cape) G
Howar, Wadi (dry riv.) E
Ibra, Wadi (dry riv.) D
Jebel Abyad (plat.) E
Jebel Aulia (dam) F
Jur (riv.) E
Kasar, Ras (cape) G
Kinyeti (mt.) F
Laqiya U'mran (well) E
Libyan (des.) E
Lol (dry riv.) E
Lotagipi Swamp (plain) F
Marra, Jebel (mt.) D
Meroe (ruins) F
Milk, Wadi el (dry riv.) E
Muqaddam, Wadi (dry riv.) F
Napata (ruins) F
Naqa (ruins) F
Nile (riv.) F
Nuba (mts.) E
Nubia (lake) F
Nubian (des.) F
Nukheila (oasis) E
Nuri (ruins) F
Oda, Jebel (mt.) G
Pibor (riv.) F
Red (sea) G
Sahara (des.) E
Second Cataract F
Selima (oasis) E
Sennar (dam) F
Setit (riv.) G
Sixth Cataract F
Sobat (riv.) F
Suakin (arch.) G
Sudan (reg.) E
Sudd (swamp) F
Sue (riv.) E
Third Cataract E
U'weinat, Jebel (mt.) E
White Nile (riv.) F

○Population of sub-district or division.

Agriculture, Industry and Resources

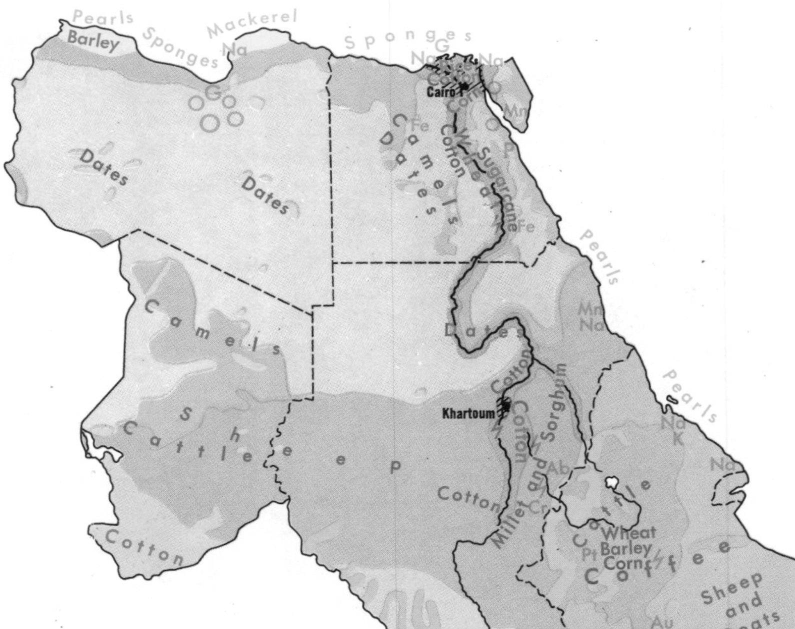

DOMINANT LAND USE

Cereals, Horticulture, Livestock
Cash Crops, Mixed Cereals
Cotton, Cereals
Market Gardening, Diversified Tropical Crops
Plantation Agriculture
Oases
Pasture Livestock
Nomadic Livestock Herding
Forests
Nonagricultural Land

MAJOR MINERAL OCCURRENCES

Ab Asbestos
Au Gold
Cr Chromium
Fe Iron Ore
G Natural Gas
K Potash
Mn Manganese
Na Salt
O Petroleum
P Phosphates
Pt Platinum

Water Power
Major Industrial Areas

ANGOLA
AREA 481,351 sq. mi. (1,246,700 sq. km.)
POPULATION 7,078,000
CAPITAL Luanda
LARGEST CITY Luanda
HIGHEST POINT Mt. Moco 8,593 ft. (2,620 m.)
MONETARY UNIT kwanza
MAJOR LANGUAGES Mbundu, Kongo, Lunda, Portuguese
MAJOR RELIGIONS Tribal religions, Roman Catholicism

BURUNDI
AREA 10,747 sq. mi. (27,835 sq. km.)
POPULATION 4,021,910
CAPITAL Bujumbura
LARGEST CITY Bujumbura
HIGHEST POINT 8,858 ft. (2,700 m.)
MONETARY UNIT Burundi franc
MAJOR LANGUAGES Kirundi, French, Swahili
MAJOR RELIGIONS Tribal religions, Roman Catholicism, Islam

CAMEROON
AREA 183,568 sq. mi. (475,441 sq. km.)
POPULATION 8,503,000
CAPITAL Yaoundé
LARGEST CITY Douala
HIGHEST POINT Cameroon 13,350 ft. (4,069 m.)
MONETARY UNIT CFA franc
MAJOR LANGUAGES Fang, Bamileke, Fulani, Duala, French, English
MAJOR RELIGIONS Tribal religions, Christianity, Islam

CENTRAL AFRICAN REP.
AREA 242,000 sq. mi. (626,780 sq. km.)
POPULATION 2,284,000
CAPITAL Bangui
LARGEST CITY Bangui
HIGHEST POINT Gao 4,659 ft. (1,420 m.)
MONETARY UNIT CFA franc
MAJOR LANGUAGES Banda, Gbaya, Sangho, French
MAJOR RELIGIONS Tribal religions, Christianity, Islam

CONGO
AREA 132,046 sq. mi. (342,000 sq. km.)
POPULATION 1,537,000
CAPITAL Brazzaville
LARGEST CITY Brazzaville
HIGHEST POINT Leketi Mts. 3,412 ft. (1,040 m.)
MONETARY UNIT CFA franc
MAJOR LANGUAGES Kikongo, Bateke, Lingala, French
MAJOR RELIGIONS Christianity, tribal religions, Islam

EQUATORIAL GUINEA
AREA 10,831 sq. mi. (28,052 sq. km.)
POPULATION 244,000
CAPITAL Malabo
LARGEST CITY Malabo
HIGHEST POINT 9,868 ft. (3,008 m.)
MONETARY UNIT ekuele
MAJOR LANGUAGES Fang, Bubi, Spanish
MAJOR RELIGIONS Tribal religions, Christianity

GABON
AREA 103,346 sq. mi. (267,666 sq. km.)
POPULATION 551,000
CAPITAL Libreville
LARGEST CITY Libreville
HIGHEST POINT Ibounzi 5,165 ft. (1,574 m.)
MONETARY UNIT CFA franc
MAJOR LANGUAGES Fang and other Bantu languages, French
MAJOR RELIGIONS Tribal religions, Christianity, Islam

KENYA
AREA 224,960 sq. mi. (582,646 sq. km.)
POPULATION 15,327,061
CAPITAL Nairobi
LARGEST CITY Nairobi
HIGHEST POINT Kenya 17,058 ft. (5,199 m.)
MONETARY UNIT Kenya shilling
MAJOR LANGUAGES Kikuyu, Luo, Kavirondo, Kamba, Swahili, English
MAJOR RELIGIONS Tribal religions, Christianity, Hinduism, Islam

MALAWI
AREA 45,747 sq. mi. (118,485 sq. km.)
POPULATION 5,968,000
CAPITAL Lilongwe
LARGEST CITY Blantyre
HIGHEST POINT Mulanje 9,843 ft. (3,000 m.)
MONETARY UNIT Malawi kwacha
MAJOR LANGUAGES Chichewa, Yao, English, Nyanja, Tumbuka, Tonga, Ngoni
MAJOR RELIGIONS Tribal religions, Islam, Christianity

RWANDA
AREA 10,169 sq. mi. (26,337 sq. km.)
POPULATION 4,819,317
CAPITAL Kigali
LARGEST CITY Kigali
HIGHEST POINT Karisimbi 14,780 ft. (4,505 m.)
MONETARY UNIT Rwanda franc
MAJOR LANGUAGES Kinyarwanda, French, Swahili
MAJOR RELIGIONS Tribal religions, Roman Catholicism, Islam

SOMALIA
AREA 246,200 sq. mi. (637,658 sq. km.)
POPULATION 3,645,000
CAPITAL Mogadishu
LARGEST CITY Mogadishu
HIGHEST POINT Surud Ad 7,900 ft. (2,408 m.)
MONETARY UNIT Somali shilling
MAJOR LANGUAGES Somali, Arabic, Italian, English
MAJOR RELIGION Islam

TANZANIA
AREA 363,708 sq. mi. (942,003 sq. km.)
POPULATION 17,527,560
CAPITAL Dar es Salaam
LARGEST CITY Dar es Salaam
HIGHEST POINT Kilimanjaro 19,340 ft. (5,895 m.)
MONETARY UNIT Tanzanian shilling
MAJOR LANGUAGES Nyamwezi-Sukuma, Swahili, English
MAJOR RELIGIONS Tribal religions, Christianity, Islam

UGANDA
AREA 91,076 sq. mi. (235,887 sq. km.)
POPULATION 12,630,076
CAPITAL Kampala
LARGEST CITY Kampala
HIGHEST POINT Margherita 16,795 ft. (5,119 m.)
MONETARY UNIT Ugandan shilling
MAJOR LANGUAGES Luganda, Acholi, Teso, Nyoro, Soga, Nkole, English, Swahili
MAJOR RELIGIONS Tribal religions, Christianity, Islam

ZAIRE
AREA 905,063 sq. mi. (2,344,113 sq. km.)
POPULATION 28,291,000
CAPITAL Kinshasa
LARGEST CITY Kinshasa
HIGHEST POINT Margherita 16,795 ft. (5,119 m.)
MONETARY UNIT zaire
MAJOR LANGUAGES Tshiluba, Mongo, Kikongo, Kingwana, Zande, Lingala, Swahili, French
MAJOR RELIGIONS Tribal religions, Christianity

ZAMBIA
AREA 290,586 sq. mi. (752,618 sq. km.)
POPULATION 5,679,808
CAPITAL Lusaka
LARGEST CITY Lusaka
HIGHEST POINT Sunzu 6,782 ft. (2,067 m.)
MONETARY UNIT Zambian kwacha
MAJOR LANGUAGES Bemba, Tonga, Lozi, Luvale, Nyanja, English
MAJOR RELIGIONS Tribal religions

ANGOLA | BURUNDI | CAMEROON | CENTRAL AFRICAN REP. | CONGO

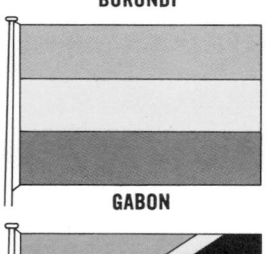

EQUATORIAL GUINEA | GABON | KENYA | MALAWI | RWANDA

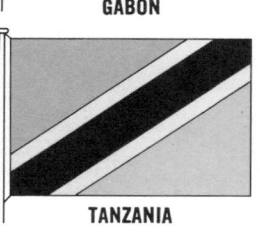 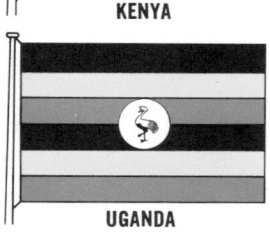 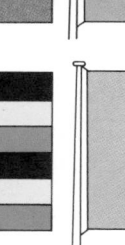 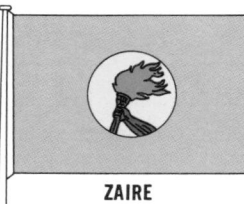  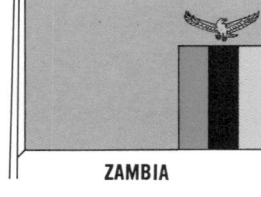

SOMALIA | TANZANIA | UGANDA | ZAIRE | ZAMBIA

ANGOLA
DISTRICTS
go 68,885 .... B5
guela 474,897 .... B6
550,337 .... C6
inda 80,857 .... B7
ndo Cubango 112,073 .... C7
anza-Norte 298,062 .... B5
nza-Sul 458,592 .... C6
ene 147,394 .... C7
mbo 837,627 .... C6
a 497,470 .... B7
ada 491,704 .... B5
da Norte 210,000 .... C5
da Sul 98,000 .... D6
ange 558,630 .... C6
nico 213,119 .... D6
nibe 53,058 .... B7
e 386,037 .... B5
e 41,766 .... B5

CITIES and TOWNS
Chicapa C6
Cuale B5
oriz B5
ulo B7
dos Tigres B7
Farta B6
ombo B7
Vista C6
guela 40,996 .... B6
a 8,894 .... B7
inda 21,124 .... B5
onda B7
uso C7
ando B6
ulo D6
uquembe B6
nacupa 5,740 .... D6
anongue D6
abulo B7
gamba D6
elongo C6
enda-Camulemba D5
sai D6
samba B5
ete B5
ngula C6
lto B5
ombo D6
2,784 .... C6
ange B7
guar C6
pindo B7
ado C5
mbo D6
ngo C5
hi C6
 e C5
o-Cuanavale C7
na B6
ba B5
co D7
mbe Grande B6
ado C5
gares C6
e República C5

Foz do Cunene B7
Gabela 6,930 .... B6
Gambos B6
Golungo Alto B5
Huambo 61,885 .... C6
Iona B7
Kalandula C5
Kassinga C7
Kuito 18,941 .... C6
Lobito 59,528 .... B6
Lóvua D5
Longa C6
Luacano D6
Luachimo D5
Luanda (cap.) 475,328 .... B5
Lubango 31,674 .... B6
Lucira B6
Luiana D7
Lukapa D5
Macondo D6

Malange 31,599 .... C5
Maquela do Zombo C5
Massango (Forte República) C5
Mavinga D7
Mbanza Congo 4,002 .... B5
Menongue 3,023 .... C6
Moçâmedes (Namibe) 12,076 .... B7
Muconda D6
Mucope B7
Mucusso D7
Munhango C6
Muxima B5
Namibe 12,076 .... B7
Nana Candundo D6
Ndalatando 7,342 .... B5
N'gage 2,548 .... C5
Ngiva C7
Ngunza (Sumbe) 7,911 .... B6
Nharêa C6
Nóqui B5

Nova Gaia C5
Nzeto B5
Oncócua B7
Porto Alexandre 8,235 .... B7
Porto Amboim B6
Quela C6
Quibala C6
Quibaxe B5
Quinzau B5
Sanza Pombo C5
São Nicolau B6
Saurimo 12,901 .... D5
Songo B5
Soyo B5
Sumbe 7,911 .... B6
Uíge 11,972 .... C5
Vila Guilherme Capelo C7
Xangongo C7

OTHER FEATURES
Bero (riv.) B7
Chicapa (riv.) D5
Chiumbe (riv.) D5
Coango (riv.) C4
Coporolo (riv.) B6
Cuando (riv.) C7
Cuango (riv.) C5
Cuanza (riv.) C5
Cubango (riv.) C7
Cuito (riv.) C7
Cunene (riv.) B7
Cunene (dam) C7
Cuvo (riv.) B6
Kasai (riv.) D5
Kwilu (riv.) C5
Loange (riv.) D5
Loge (riv.) B5
Lungwebungu (riv.) D6
Matala (dam) C7

M'Bridge (riv.) B5
Moco (mt.) C6
Negro (cape) B7
Palmeirinhas (pt.) B5
Ruacana Falls (dam) B7
Santa Maria (cape) B6
Zambezi (riv.) C7

BURUNDI
CITIES and TOWNS
Bujumbura (cap.) 141,040 .... E4
Bururi 7,800 .... F4
Gitega 19,500 .... F4

OTHER FEATURES
Ruzizi (riv.) E4

Tanganyika (lake) E5

CAMEROON
CITIES and TOWNS
Abong-Mbang 6,000 .... B3
Ambam 4,000 .... B3
Bafia 12,000 .... B3
Bafoussam 62,239 .... B2
Bali A2
Bamenda 48,111 .... B2
Banyo B2
Batouri 7,000 .... B3
Bélabo B3
Bengbis B3
Bertoua 10,000 .... B3
Bétaré-Oya B2
Bonabéri A3

(continued on following page)

Buea 24,584	A3	Kontcha	B2	Mora 6,000	B1	Yaoundé (cap.) 313,706	B3	Logone (riv.)	C2	Bambari 31,285	D2	Bozoum 13,573	C2
Campo	B3	Kousseri 11,627	B1	Nanga-Eboko	B3	Yokadouma 6,000	B3	Lom (riv.)	B2	Bangassou 21,773	D3	Bria 14,786	D2
Djoum	B3	Kribi 10,512	A3	Ngaoundéré 38,992	B2	Yoko	B2	Mbéré (riv.)	B2	Bangui (cap.) 279,792	C3	Carnot 17,863	C3
Douala 458,426	B3	Kumba 44,175	A3	Nkambe 5,000	B2			Mbakou (res.)	B2	Bania	C3	Damara 2,556	C2
Doumé	B3	Kumbo 11,699	B2	Nkongsamba 71,298	A3	OTHER FEATURES		Sanaga (riv.)	B3	Batangafo 12,543	C2	Dekoa 7,663	C2
Dschang 20,000	A2	Limbe 27,016	A3	Poli	B2			Sanga (riv.)	C3	Berberati 27,285	C3	Gaza	C2
Ebolowa 24,000	B3	Lomié	B3	Rey Bouba	B2	Adamawa (reg.)	B2			Birao 3,317	D1	Goubere	E2
Edéa 25,493	B3	Makari	B1	Sangmelima 13,776	B3	Benué (riv.)	A2			Bocaranga 6,202	C2	Grimari 7,308	C3
Esséka 10,000	B3	Mamfé 10,000	A2	Tchollíré	B2	Biafra (bight)	A3	CENTRAL AFRICAN REPUBLIC		Boda 8,771	C2	Hyrra Banda	D2
Essé	B3	Maroua 67,187	B1	Tibati	B2	Cameroon (mt.)	B3			Bossangoa 25,150	C2	Ippy 10,816	D2
Fort-Foureau (Kousseri) 7,000	B1	Mbalmayo 22,106	B3	Tignéré	B2	Cross (riv.)	A2	CITIES and TOWNS		Bossembele 5,091	C2	Kaga Bandoro 11,876	C2
Foumban 33,944	B2	Meiganga 15,906	B2	Tiko 13,824	A3	Dja (riv.)	B3			Bouali	C2	Kaka	E2
Garoua 63,900	B2	Mokolo 7,000	B1	Wum 15,149	A2	Ivindo (riv.)	B3	Alindao 12,295	D2	Bouar 29,528	C2	Kembe 5,051	D2
Guidder 10,000	B2	Moloundou	B3	Yabassi 13,000	A3	Kadei (riv.)	C3	Babaoua 3,999	C2	Bouca 8,874	C2	Kouango 2,602	D2
Kaélé 17,000	B1	Monatélé	B3	Yagoua 13,541	B1			Bakia	E2	Boula	T3	Kouki	C2

Kounde	B2
Mbalki 12,346	C3
Mbres 2,622	D3
Mobaye 4,220	D3
Mouka	C2
Ndele 5,858	D2
Ngourou	D2
Nola 6,703	B3
Obo 3,978	E2
Ouadda 3,009	D2
Paoua 7,052	C2
Possel	C2
Sibut 13,341	D2
Zako	D2
Zemio 3,259	D2

Zemongo	E2

OTHER FEATURES

Bamingui (riv.)	C2
Bomu (riv.)	D3
Dar Rounga (reg.)	D2
Gao (mt.)	B3
Kadei (riv.)	C3
Kotto (riv.)	D2
Lobaye (riv.)	C3
Mbéré (riv.)	B2
Ouham (riv.)	C2
Pendé (riv.)	C2
Sanga (riv.)	C3

Sara (riv.)	C2
Shari (riv.)	C2
Shinko (riv.)	D2
Ubangi (riv.)	C2

CONGO

CITIES and TOWNS

Abala	C4
Boko	B4
Brazzaville (cap.) 298,967	C4
Boundji	C4
Djambala	B4

Dongou	C3
Enyellé	C3
Epéna	C3
Etoumbi	B3
Ewo	C4
Gamboma	C4
Ikelemba	B3
Impfondo	C3
Kellé	B4
Kibangou	B4
Kindama	B4
Kinkala	C4
Komono	B4
Loubomo 29,600	B4
Loudima	B4

Madingo-Kayes	B4
Madingo	B4
Makoua	C3
Mbinda	B4
Mindouli	C4
Mossaka	C3
Mossendjo	B4
M'Pouya	C4
M'Vouti	B4
Nkayi 30,600	B4
Okoyo	C4
Ouesso	C3
Owando	C4
Oyo	C4
Pangala	C4
Pointe-Noire 141,700	B4
Sembé	B3
Sibiti	B4
Souanké	B3
Zanaga	B4

Tchibanga 14,001	B4

OTHER FEATURES

Crystal (mts.)	B4
Ibounzi (mt.)	B4
Ivindo (riv.)	B3
Lopez (cape)	A4
N'Dogo (lag.)	B4
N'Gounié (riv.)	B4
N'Komi (lag.)	A4
Ogooué (riv.)	A4
Onangué (lake)	A4
Pongara (pt.)	A3

KENYA

PROVINCES

Central 1,675,647	G4
Coast 944,082	G4
Eastern 1,907,301	G4
Nairobi 509,286	G4
North-Eastern 245,757	G4
Nyanza 2,122,045	F4
Rift Valley 2,210,289	G3
Western 1,328,298	G3

CITIES and TOWNS

Buna	G3
Bunyala	F4
Bura	H3
Eldoret 18,196	G3
El Wak	H3
Embu 3,928	G4
Fort Hall 4,750	G4
Galole 3,609	G4
Garba Tula	G3
Garissa	G4
Garsen	G4
Gilgil 4,178	G4
Isiolo 8,201	G4
Kakamega 6,244	F3
Kaningo	G4
Kericho 10,144	F4
Kiambu 2,776	G4
Kilifi 2,662	G4
Kipini	G4
Kisii 6,080	F4
Kisumu 32,431	F3
Kitale 11,573	G3
Kitui 3,071	G4
Kolbio	H4
Konza	G4
Laisamis	G3
Lamu 7,403	H4
Lodwar	G3
Lokitaung 4,090	G3
Lolgorien	G4
Machakos 6,312	G4
Magadi	G4
Malindi 10,757	H4
Mambrui	H4
Maralal 3,878	G3
Marsabit 6,635	G3
Meru 4,475	G4
Mombasa 247,073	G4
Moyale	G3
Nairobi (cap.) 509,286	G4
Naivasha 6,920	G4
Nakuru 47,151	G4
Namanga	G4
Nanyuki 11,624	G4
Narok 2,608	G4
North Horr	G3
South Horr	G3
Taveta	G4
Thika 18,387	G4
Thomson's Falls 7,602	G4
Todenyang	G3
Tsavo	G4
Vanga	G4
Voi 5,313	G4
Wajir	H3
Wamba 2,650	G3

OTHER FEATURES

Daua (riv.)	H3
Elgon (mt.)	F3
Formosa (bay)	H4
Galana (riv.)	G4
Gedi (ruins)	G4
Kavirondo (gulf)	F4
Kenya (mt.)	G4
Lak Dera (dry riv.)	H3
Lorian (swamp)	G3
Natron (lake)	G4
Nyiru (mt.)	G3
Patta (isl.)	H4

Rudolf (Turkana) (lake)	G3
Tana (riv.)	G4
Tsavo Nat'l Park	G4
Turkana (lake)	G3
Victoria (lake)	F4
Winam (bay)	F4

MALAWI

CITIES and TOWNS

Bandawe	F6
Blantyre 222,153	F7
Chilumba	F6
Chipoka	F7
Chiromo	F7
Chitipa 3,079	F5
Dedza 5,448	F6
Karonga 11,873	F5
Kasungu	F6
Lilongwe (cap.) 102,924	F6
Livingstonia	F6
Mangochi 3,341	G6
Mzimba 4,962	F6
Nkhata Bay 4,024	F6
Nkhotakota 10,312	F6
Nsanje 6,091	G7
Rumphi 3,998	F6
Salima 4,646	F6
Thyolo 4,186	F7
Zomba 21,000	G7

OTHER FEATURES

Chilwa (lake)	G7
Malawi (Nyasa) (lake)	F6
Mulanje (mts.)	G7
Nyasa (lake)	F6
Shire (riv.)	G7

RWANDA

CITIES and TOWNS

Butare 21,691	E4
Cyangugu 7,042	E4
Gisenyi 12,436	E4
Kigali (cap.) 117,749	F4
Nyabisindu 8,587	F4

OTHER FEATURES

Kagera Nat'l Park	F4
Karisimbi (mt.)	E4
Kivu (lake)	E4
Ruzizi (riv.)	E4
Virunga (range)	E4

EQUATORIAL GUINEA

TERRITORIES

Bioko 78,000	A3
Rio Muni 203,000	B3

CITIES and TOWNS

Bata 27,024	B3
Luba 19,933	A3
Malabo (cap.) 37,237	A3
Mbini 14,503	A3

OTHER FEATURES

Biafra (bight)	A3
Bioko (isl.)	A3
Corisco (isl.)	A3
Elobey (isls.)	A3
Fernando Po (Bioko) (isl.)	A3

GABON

CITIES and TOWNS

Banda	B4
Bitam 5,936	B3
Booué	B3
Chinchoua	B3
Cocobeach	A4
Fougamou	B4
Franceville 9,345	B4
Iguéla	A4
Kango	B3
Kemboma	B3
Koula-Moutou 8,032	B4
Lalara	B3
Lambaréné 17,770	B4
Lastoursville	B4
Lekoni	B4
Libreville (cap.) 105,080	A3
Makokou 5,005	B3
Mayumba	B4
M'Bigou	B4
Médouneu	B3
Mekambo	B3
Mimongo	B4
Minvoul	B3
Mitzic	B3
Moanda 10,709	B4
Mouila 15,016	B4
Mounana 4,000	B4
N'Dendé	B4
N'Djolé	B4
Nyanga	B4
Okondja	B4
Omboué	A4
Owendo	A4
Oyem 12,455	B3
Port-Gentil 48,190	A4
Setté-Cama	A4

SOMALIA

PROVINCES

Bakool 100,000	H3
Bari 155,000	J1
Bay 302,000	H3
Galguduud 182,000	J2
Gedo 212,000	H3
Hiiraan 147,000	J3
Jubbada Hoose 246,000	H4
Mogadiscio 371,000	J3
Mudug 215,000	J2
Nugaal 85,000	J2
Sanaag 146,000	J1
Shabeellaha Dhexe 237,000	J3
Shabeellaha Hoose 398,000	H3
Togdheer 258,000	J2
Woqooyi Galbeed 440,000	H1

CITIES and TOWNS

Adadle	H2
Afgoi	J3
Afmadu 2,580	H3
Alula	K1
Ankhor	J1
Audegle	J3
Baduen	J2
Barawa (Brava)	H3
Bardera	H3
Bargal	K1
Baydhabo 14,962	H3
Belet Weyne 11,426	J3
Bender Beila	K2
Bender Cassim (Bosaso)	J1
Berbera 12,219	J1
Bereda	K1
Bircao	H4
Bohodleh	J2
Borama 3,244	H1

(continued on following page)

Central Africa

CYLINDRICAL EQUAL-AREA PROJECTION

SCALE OF MILES

0 50 100   200   300

SCALE OF KILOMETERS

0 50 100   200   300

Capitals of Countries — — — — ☆

Other Capitals — — — — ◉

International Boundaries ——————

Internal Boundaries — — — — —

Scale 1:13,800,000

© Copyright HAMMOND INCORPORATED, Maplewood, N.J.

Topography

0	200	400	600 MI.
0	200	400	600 KM.

Below Sea Level	100 m. 328 ft.	200 m. 656 ft.	500 m. 1,640 ft.	1,000 m. 3,281 ft.	2,000 m. 6,562 ft.	5,000 m. 16,404 ft.

Bosaso J1
Brava 6,167 H3
Bulhar H1
Bulo Burti 5,247 J3
Bur Acaba H3
Burao 12,617 J2
Callis J2
Candala J1
Chisimayu 17,872 H4
Chiambone H4
Coriole 4,341 H3
Dante (Hafun) K1
Dif H3
Dinsor H3
Dusa Marreb J2
Eil J2
El Athale (Itala) J3
El Bur J3
El Dere J3
El Hamurre J2
Erigabo 4,279 J1
Ferfer J2
Galcaio J2
Garad J2
Garbahaarrey H3
Gardo J2
Garoe J2
Giohar 13,156 J3
Gobwen H4
Hafun K1
Halin J2
Harardera J3
Hargeysa 40,254 H2
Hordio K1
Iddan J2
Iet H3
Itala J3
Jamama 5,408 H3
Jilib 3,232 H3
Karin J1
Kismayu (Chisimayu) 17,872 H4
Las Dureh J1
Luuq H3
Margherita (Jamama) H3
Marka (Merka) 17,708 H3
Mogadishu (cap.) 371,000 J3
Muqdisho (Mogadishu) (cap.) 371,000 J3
Obbia J2
Oddur H3
Taleh J2
Uanle Uen H3
Upper Sheikh J2
Villabruzzi (Johar) J3
Zeila 1,226 H1

OTHER FEATURES

Aden (gulf) J1
Asèr, Ras (cape) K1
Giuba (riv.) H3
Guban (reg.) H1
Hafun, Ras (cape) K1
Haud (plat.) J2
Lak Dera (dry riv.) H3
Negro (bay) J2
Nogal (reg.) J2
Shimbir Berris (mt.) J1
Sura, Ras (cape) J1
Surud Ad (mt.) J1
Webi Shabelle (riv.) H3

TANZANIA

REGIONS

Arusha 928,478 G4
Dar es Salaam 851,222 G5
Dodoma 971,921 G5
Iringa 922,801 G5
Kagera 1,009,379 F4
Kigoma 648,950 F4
Kilimanjaro 902,394 G4
Lindi 527,902 G5
Mara 723,295 F4
Mbeya 1,080,241 F5
Morogoro 939,190 G5
Mtwara 771,726 G5
Mwanza 1,443,418 F4
Pemba 205,870 H5
Pwani (Coast) 516,949 G5
Rukwa 451,897 F5
Ruvuma 564,113 G6
Shinyanga 1,323,482 F4
Singida 614,030 F5
Tabora 818,049 F5
Tanga 1,088,592 G5
Zanzibar Mjini 143,616 G5
Zanzibar Shambani North 77,424 G5
Zanzibar Shambani South 52,325 G5

CITIES and TOWNS

Arusha 55,281 G4
Babati G4
Bagamoyo 5,112 G5
Bukoba 20,430 F4
Chake Chake 4,862 H5
Dar es Salaam (cap.) 757,346 G5
Dodoma 45,703 G5
Geita 3,066 F4
Handeni G5
Ifakara G5
Iringa 57,182 G5
Itigi F5
Kahama 3,211 F4
Kaliua F5
Kanga G5
Karema F5
Kasanga F5
Kasulu F4
Kibara F4
Kibaya G5
Kibondo F4
Kigoma-Ujiji 50,044 E4
Kilosa 4,458 G5
Kilwa Kivinje 2,790 G5
Kilwa Masoko G5
Kinyangiri G4
Kipili F5
Kisiju G5
Kitunda F5
Kizimkazi G5
Kondoa 4,514 G4
Kongwa G5
Korogwe 6,675 G5
Lindi 27,308 G5
Liuli F6
Liwale G5
Longido G4
Mahenge G5
Makumbako G5
Manda F6
Manyoni G5
Masasi G6
Mbamba Bay G6
Mbeya 76,606 F5
Mbulu G4
Mchinga H5
Mohoro G5
Mombo G4
Morogoro 61,890 G5
Moshi 52,223 G4
Mpanda F5
Mtakuja F5
Mtwara-Mikindani 48,510 H6
Murongo F4
Musoma 32,658 F4
Muwale F5
Mwadui 7,383 F4
Mwanza 110,611 F4
Mwaya F5
Mwesi F5
Nachingwea 3,751 G6
Newala G6
Ngara F4
Njombe F5
Pangani 2,955 G5
Rungwa F5
Sadani G5
Same G4
Sekenke F4
Shinyanga 21,703 F4
Singida 29,252 F4
Songea 17,954 G6
Sumbawanga 28,586 F5
Tabora 67,392 F5
Tanga 103,409 G4
Tukuyu 4,089 F5
Tunduru G6
Urambo F4
Utete G5
Uvinza F5
Wete 8,469 G4
Zanzibar 110,669 G5

OTHER FEATURES

Eyasi (lake) F4
Great Ruaha (riv.) G5
Juani (isl.) G5
Kalambo (falls) F5
Kanzi (cape) G5
Kilimanjaro (mt.) G4
Kilombero (riv.) G5
Mafia (isl.) H5
Manyara (lake) G4
Masai (steppe) G4
Mbarangandu (riv.) G5
Mbemkuru (riv.) G5
Meru (mt.) G4
Mikumi Nat'l Park G5
Natron (lake) G4
Ngorongoro (crater) F4
Njombe (riv.) F5
Nyasa (lake) F6
Olduvai Gorge (canyon) G4
Pangani (riv.) G4
Pemba (isl.) H5
Rovuma (riv.) G6
Rufiji (riv.) G5
Ruaha Nat'l Park F5
Rukwa (lake) F5
Rungwa (riv.) F5
Rungwe (mt.) F5
Serengeti Nat'l Park F4
Tanganyika (lake) E5
Tarangire Nat'l Park G4
Victoria (lake) F4
Wami (riv.) G5
Wembere (riv.) F4
Zanzibar (isl.) G5

UGANDA

CITIES and TOWNS

Arua 10,837 F3
Atura F3
Butiaba 261 F3
Entebbe 21,096 F4
Fort Portal 7,947 F3
Gulu 18,170 F3
Hoima 2,339 F3
Jinja 52,509 F3
Kabale 8,234 E4
Kampala (cap.) 478,895 F3
Kasese 7,213 F3
Kilembe F3
Kitgum 3,242 F3
Lira 7,340 F3
Masaka 12,987 F4
Masindi 2,100 F3
Mbale 23,544 F3
Mbarara 16,078 F4
Moroto 5,488 F3
Moyo 2,656 F3
Mubende 6,004 F3
Rhino Camp 198 F3
Soroti 8,130 F3
Tororo 15,977 F3

OTHER FEATURES

Albert (Mobuto Sese Seko) (lake) F3
Edward (lake) E4
Elgon (mt.) F3
George (lake) F4
Kabalega (falls) F3
Kagalega Nat'l Park F3
Kidepo Nat'l Park F3
Kioga (lake) F3
Margherita (mt.) E3
Mobuto Sese Seko (lake) F3
Owen Falls (dam) F3
Ruwenzori (range) E3
Sese (isls.) F4
Victoria (lake) F4
Virunga (range) E4
Virunga Nat'l Park F4

ZAIRE

PRUVINCES

Bandundu 2,600,556 C4
Bas-Zaïre 1,504,361 B4
Equateur 2,431,812 D3
Haut-Zaïre 3,356,419 E3
Kasai-Occidental 2,433,861 D4
Kasai-Oriental 1,872,231 D5
Kinshasa 1,323,039 C4
Kivu 3,361,883 E4
Shaba 2,753,714 E5

CITIES and TOWNS

Aba 7,600 F3
Abumombazi D3
Aketi 17,200 D3
Andoma E3
Ango E3
Ankoro E5
Bagata C4
Balangala D3
Bambesa E3
Bambili E3
Banalia E3
Banana B5
Bandundu 74,467 C4
Baraka E4
Basankusu C3
Basoko 9,100 D3
Basongo D4
Befale D3
Bena-Dibele D4
Beni 22,800 E3
Bikoro C4
Boende 12,800 D4
Bokote D4
Bokungu D4
Bolobo 10,300 C4
Bolomba 7,200 C3
Boma 61,100 B5
Bomboma C3
Bomongo C3
Bondo 10,000 D3
Bongandanga 12,900 D3
Bosobolo 11,100 D3
Budjala C3
Bukama E5
Bukavu 134,861 E4
Bulungu 16,300 C4
Bumba 34,700 D3
Bunia 28,800 E3
Bunkeya 5,100 E6
Businga 11,000 D3
Busu-Djanoa D3
Buta 19,800 D3
Butembo 27,800 E3
Dekese D4
Demba 22,000 D5
Dibaya 11,400 D5
Dibaya-Lubue 7,900 C4
Dilolo 14,000 D6
Dimbelenge D5
Djolu D3
Djugu F3
Dongo C3
Doruma E3
Dungu 9,100 E3
Etoile E6
Faradje 10,400 E3
Feshi C5
Fizi E4
Gandajika 60,100 D5
Gemena 37,300 D3
Goma 48,600 E4
Gungu C5
Idiofa C4
Ikela D4
Ilebo 32,200 D4
Imese C3
Ingende C4
Inongo 14,800 C4
Irumu 9,300 E3
Isangi D3
Isiro 49,300 E3
Kabalo 22,600 E5
Kabambare E4
Kabare 12,600 E4
Kabinda 60,500 D5
Kabongo 6,500 E5
Kahemba C5
Kalehe E4
Kalemie 62,300 E5
Kalima 27,500 E4
Kama 17,700 E4
Kambove 18,900 E6
Kamina 56,300 D5
Kampene 14,600 E4
Kananga 428,960 D5
Kanda-Kanda D5
Kaniama D5
Kapanga D5
Kasaji D6
Kasangulu 11,900 C4
Kasenga E6
Kasenyi E3
Kasese E4
Kasongo 37,800 E4
Kasongo-Lunda C5
Katako-Kombe D4
Katenga E5
Kazumba D5
Kenge 17,500 C4
Kiambi E5
Kibombo E4
Kikwit 111,960 C5
Kilembe C5
Kilwa E5
Kilo E3
Kinda D5
Kiniama E6
Kinshasa (cap.) 1,323,039 C4
Kipushi 32,900 E6
Kiri C4
Kirundu E4
Kisangani 229,596 E3
Kole, Kasai-Oriental D4
Kole, Haut-Zaïre E3
Kolwezi 81,600 E6
Komba D3
Kongolo 14,800 E5
Kungu C3
Kutu 10,000 C4
Kwamouth C4
Libenge 12,500 C3
Likasi, Panda- 146,394 E6
Likati D3
Lisala D3
Lodja 20,300 D4
Lokolama D4
Lomela D4
Loto D4
Luashi D6
Lubefu D4
Lubero E4
Lubudi 6,000 E5
Lubumbashi 318,000 E6
Lubutu E4
Luebo 21,800 D5
Luishia E6
Luiza D5
Lukolela, Equateur C4
Lukolela, Kasai-Oriental D5
Lukula 9,400 B5
Luozi 7,000 B5
Lusambo 13,100 D4
Makanza C3
Malemba-Nkulu E5
Mambasa 7,400 E3
Mangai 15,200 C4
Manono 44,500 E5
Masi-Manimba 6,300 C4
Masisi E4
Matadi 110,436 B5
Mbandaka 107,910 C3
Mbanza-Ngungu 55,800 C5
Mbuji-Mayi 256,154 D5
Mitwaba E5
Moanda 6,400 B5
Mobayi-Mbongo D3
Moliro E5
Monga D3
Monkoto D4
Mulongo E5
Mungbere E3
Mushie 13,700 C4
Mutshatsha D6
Muyumba E5
Mwadingusha E6
Mwanza E5
Mweka 24,900 D4
Mwene-Ditu 71,200 D5
Mwenga E4
Niangara 9,200 E3
Niemba E5
Nyunzu 11,300 E5
Opala D4
Oshwe C4
Panda-Likasi 146,394 E6
Pangi E4
Penge D5
Poko E3
Popokabaka C5
Port Kindu 42,800 E4
Punia E4
Pweto E5
Rutshuru E4
Sakania E6
Sampwe E5
Sandoa D5
Seke-Banza B5
Sentery 24,300 E5
Shabunda 6,900 E4
Songololo 4,600 B5
Tenke E6
Titule E3
Tshela 10,700 B4
Tshikapa 38,900 D5
Tshofa D5
Ubundu 6,300 E4
Uvira 15,900 E4
Virunga 21,900
Waka
Walikale
Wamba 11,500
Watsa 21,300
Yahuma
Yakoma
Yangambi 22,600
Zongo

OTHER FEATURES

Albert (Mobuto Sese Seko) (lake)
Aruwimi (riv.)
Bomu (riv.)
Boyoma (Stanley) (falls)
Chicapa (riv.)
Congo (riv.)
Edward (lake)
Elila (riv.)
Fimi (riv.)
Garamba Nat'l Park
Giri (riv.)
Itimbiri (riv.)
Ituri (for.)
Karisimbi (mt.)
Kasai (riv.)
Kivu (lake)
Kwa (riv.)
Kwango (riv.)
Kwilu (riv.)
Lindi (riv.)
Livingstone (falls)
Loange (riv.)
Lokoro (riv.)
Lomami (riv.)
Lomela (riv.)
Lowa (riv.)
Lua (riv.)
Lualaba (riv.)
Luapula (riv.)
Lubilash (riv.)
Lufira (riv.)
Luilaka (riv.)
Lukenie (riv.)
Lukuga (riv.)
Lulua (riv.)
Luvua (riv.)
Mai-Ndombe (lake)
Malebo (Stanley Pool) (lake)
Margherita (mt.)
Marungu (mts.)
Mobuto Sese Seko (lake)
Mweru (lake)
Ruwenzori (range)
Ruzizi (riv.)
Salonga Nat'l Park
Sankuru (riv.)
Stanley (falls)
Stanley Pool (lake)
Tanganyika (lake)
Tshuapa (riv.)
Tumba (lake)
Ubangi (riv.)
Uele (riv.)
Ulindi (riv.)
Upemba (lake)
Upemba Nat'l Park
Virunga (range)
Virunga Nat'l Park
Zaire (Congo) (riv.)

ZAMBIA

CITIES and TOWNS

Abercorn (Mbala) 11,179
Bancroft (Chililabombwe) 61,928
Broken Hill (Kabwe) 143,635
Chibwe
Chilanga 12,503
Chililabombwe 61,928
Chingola 145,869
Chinsali 4,211
Chipata 32,291
Choma 17,943
Fort Rosebery (Mansa) 34,801
Isoka 6,832
Kabompo 5,357
Kabwe 143,635
Kafue 29,794
Kalabo 7,398
Kalomo 5,878
Kaoma 6,731
Kapiri Mposhi 13,677
Kasama 38,093
Kasempa 3,063
Kataba
Kawambwa 7,235
Kitwe 314,794
Lealui
Livingstone 71,987
Luanshya 132,164
Lundazi 4,083
Lusaka (cap.) 538,469
Luwingu 3,763
Mansa 34,801
Mazabuka 29,602
Mbala 11,179
Mkushi 4,104
Mongu 24,919
Monze 13,141
Mpika 25,880
Mporokoso 6,008
Mpulungu 6,354
Mufulira 149,778
Mulobezi 2,589
Mumbwa 7,570
Mwinilunga 3,169
Nakonde 4,599
Namwala 3,008
Ndola 282,439
Petauke 7,531
Senanga 7,204
Serenje 6,008
Sesheke 3,500
Solwezi 15,032
Zambezi 8,166

OTHER FEATURES

Bangweulu (lake)
Barotseland (reg.)
Chambeshi (riv.)
Cuando (riv.)
Dongwe (riv.)
Kabompo (riv.)
Kafue (riv.)
Kafue Nat'l Park
Kalambo (falls)
Kariba (dam)
Kariba (lake)
Luangwa (riv.)
Luapula (riv.)
Lungwebungu (riv.)
Mosi-Oa-Tunya (Victoria) (falls)
Mulungushi (dam)
Mweru (lake)
Sunzu (mt.)
Tanganyika (lake)
Victoria (falls)
Zambezi (riv.)

Agriculture, Industry and Resources

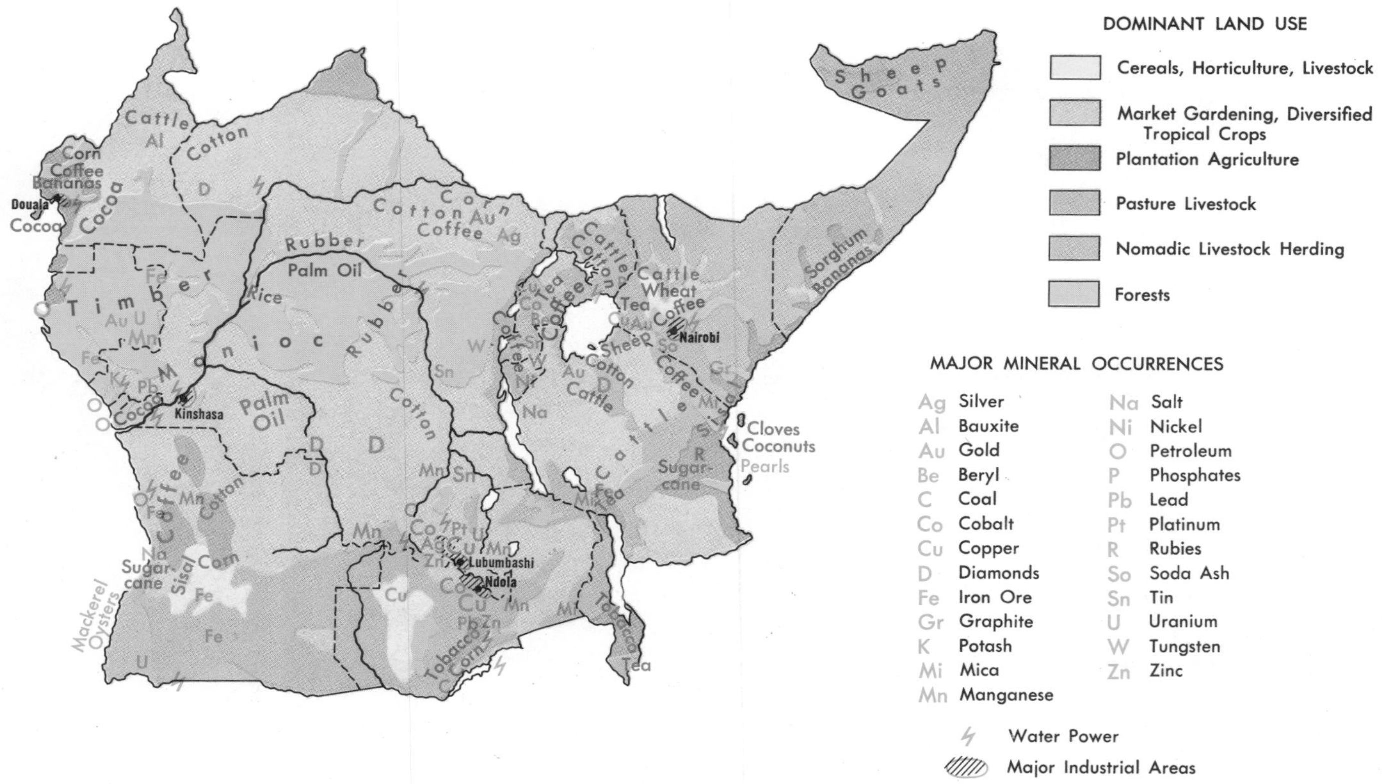

Sheep
Goats
Cattle
Al
Cotton
Corn
Coffee
Bananas
Douala
Cocoa
Corn
Cotton Au
Coffee Ag
Rubber
Palm Oil
Timber
Rice
Manioc
Rubber
Cattle
Cotton
Cattle
Wheat
Tea
Coffee
Nairobi
Sorghum
Bananas
Kinshasa
Palm Oil
Cotton
Cloves
Coconuts
Pearls
Sugar-cane
Lubumbashi
Ndola
Sugar-cane
Sisal
Corn
Mackerel
Oysters
Tobacco
Corn
Tea
DOMINANT LAND USE
Cereals, Horticulture, Livestock
Market Gardening, Diversified Tropical Crops
Plantation Agriculture
Pasture Livestock
Nomadic Livestock Herding
Forests
MAJOR MINERAL OCCURRENCES
Ag Silver
Al Bauxite
Au Gold
Be Beryl
C Coal
Co Cobalt
Cu Copper
D Diamonds
Fe Iron Ore
Gr Graphite
K Potash
Mi Mica
Mn Manganese
Na Salt
Ni Nickel
O Petroleum
P Phosphates
Pb Lead
Pt Platinum
R Rubies
So Soda Ash
Sn Tin
U Uranium
W Tungsten
Zn Zinc
Water Power
Major Industrial Areas

NAMIBIA (SOUTH-WEST AFRICA)

AREA 317,827 sq. mi. (823,172 sq. km.)
POPULATION 1,200,000
CAPITAL Windhoek
LARGEST CITY Windhoek
HIGHEST POINT Brandberg 8,550 ft.
(2,606 m.)
MONETARY UNIT rand
MAJOR LANGUAGES Ovambo, Hottentot,
Herero, Afrikaans, English
MAJOR RELIGIONS Tribal religions,
Protestantism

SOUTH AFRICA

AREA 455,318 sq. mi. (1,179,274 sq. km.)
POPULATION 23,771,970
CAPITALS Cape Town, Pretoria
LARGEST CITY Johannesburg
HIGHEST POINT Injasuti 11,182 ft. (3,408 m.)
MONETARY UNIT rand
MAJOR LANGUAGES Afrikaans, English,
Xhosa, Zulu, Sesotho
MAJOR RELIGIONS Protestantism,
Roman Catholicism, Islam, Hinduism,
tribal religions

LESOTHO

AREA 11,720 sq. mi. (30,355 sq. km.)
POPULATION 1,339,000
CAPITAL Maseru
LARGEST CITY Maseru
HIGHEST POINT 11,425 ft. (3,482 m.)
MONETARY UNIT loti
MAJOR LANGUAGES Sesotho, English
MAJOR RELIGIONS Tribal religions,
Christianity

BOTSWANA

AREA 224,764 sq. mi. (582,139 sq. km.)
POPULATION 819,000
CAPITAL Gaborone
LARGEST CITY Francistown
HIGHEST POINT Tsodilo Hill 5,922 ft.
(1,805 m.)
MONETARY UNIT pula
MAJOR LANGUAGES Setswana, Shona,
Bushman, English, Afrikaans
MAJOR RELIGIONS Tribal religions,
Protestantism

MOZAMBIQUE

AREA 303,769 sq. mi. (786,762 sq. km.)
POPULATION 12,130,000
CAPITAL Maputo
LARGEST CITY Maputo
HIGHEST POINT Mt. Binga 7,992 ft.
(2,436 m.)
MONETARY UNIT metical
MAJOR LANGUAGES Makua, Thonga,
Shona, Portuguese
MAJOR RELIGIONS Tribal religions,
Roman Catholicism, Islam

SWAZILAND

AREA 6,705 sq. mi. (17,366 sq. km.)
POPULATION 547,000
CAPITAL Mbabane
LARGEST CITY Manzini
HIGHEST POINT Emlembe 6,109 ft.
(1,862 m.)
MONETARY UNIT lilangeni
MAJOR LANGUAGES siSwati, English
MAJOR RELIGIONS Tribal religions,
Christianity

ZIMBABWE

AREA 150,803 sq. mi. (390,580 sq. km.)
POPULATION 7,360,000
CAPITAL Harare
LARGEST CITY Harare
HIGHEST POINT Mt. Inyangani 8,517 ft.
(2,596 m.)
MONETARY UNIT Zimbabwe dollar
MAJOR LANGUAGES English, Shona,
Ndebele
MAJOR RELIGIONS Tribal religions,
Protestantism

MADAGASCAR

AREA 226,657 sq. mi. (587,041 sq. km.)
POPULATION 8,742,000
CAPITAL Antananarivo
LARGEST CITY Antananarivo
HIGHEST POINT Maromokotro 9,436 ft.
(2,876 m.)
MONETARY UNIT Madagascar franc
MAJOR LANGUAGES Malagasy, French
MAJOR RELIGIONS Tribal religions,
Roman Catholicism, Protestantism

COMOROS

AREA 719 sq. mi. (1,862 sq. km.)
POPULATION 290,000
CAPITAL Moroni
LARGEST CITY Moroni
HIGHEST POINT Karthala 7,746 ft.
(2,361 m.)
MONETARY UNIT CFA franc
MAJOR LANGUAGES Arabic, French,
Swahili
MAJOR RELIGION Islam

MAURITIUS

AREA 790 sq. mi. (2,046 sq. km.)
POPULATION 959,000
CAPITAL Port Louis
LARGEST CITY Port Louis
HIGHEST POINT 2,711 ft. (826 m.)
MONETARY UNIT Mauritian rupee
MAJOR LANGUAGES English, French,
French Creole, Hindi, Urdu
MAJOR RELIGIONS Hinduism, Christianity,
Islam

SEYCHELLES

AREA 145 sq. mi. (375 sq. km.)
POPULATION 63,000
CAPITAL Victoria
LARGEST CITY Victoria
HIGHEST POINT Morne Seychellois
2,993 ft. (912 m.)
MONETARY UNIT Seychellois rupee
MAJOR LANGUAGES English, French,
Creole
MAJOR RELIGION Roman Catholicism

RÉUNION

AREA 969 sq. mi. (2,510 sq. km.)
POPULATION 491,000
CAPITAL St-Denis

MAYOTTE

AREA 144 sq. mi. (373 sq. km.)
POPULATION 47,300
CAPITAL Dzaoudzi

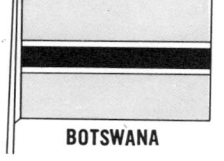
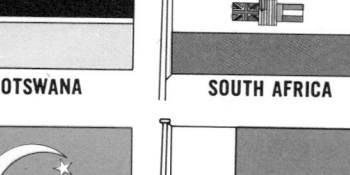
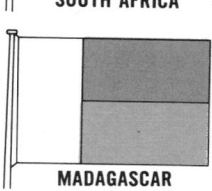
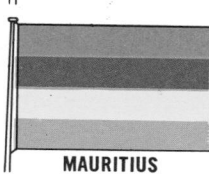
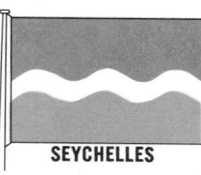

ZIMBABWE · BOTSWANA · SOUTH AFRICA · LESOTHO · SWAZILAND
MOZAMBIQUE · COMOROS · MADAGASCAR · MAURITIUS · SEYCHELLES

Agriculture, Industry and Resources

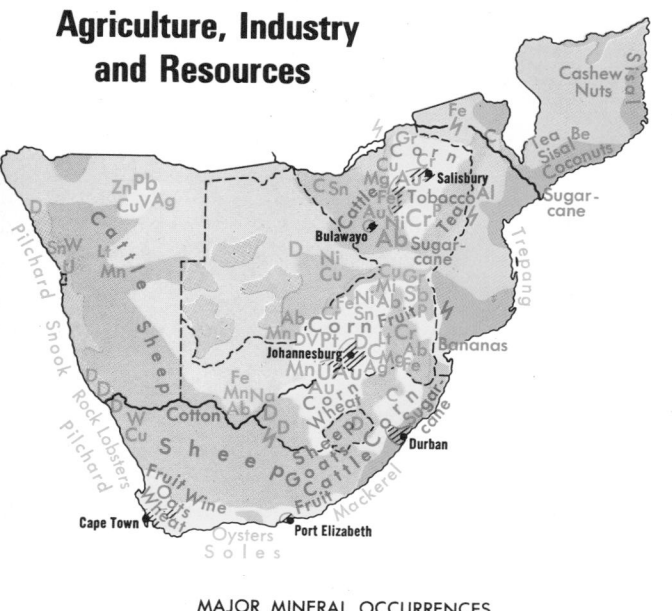
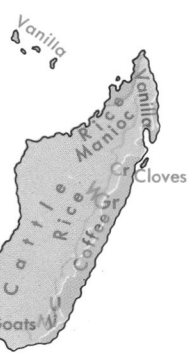
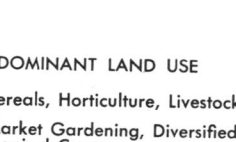
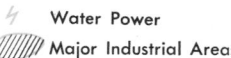

DOMINANT LAND USE

Cereals, Horticulture, Livestock

Market Gardening, Diversified Tropical Crops

Plantation Agriculture

Pasture Livestock

Nomadic Livestock Herding

Forests

Nonagricultural Land

↯ Water Power
▨ Major Industrial Areas

MAJOR MINERAL OCCURRENCES

Ab	Asbestos	Cu	Copper	Pb	Manganese
Ag	Silver	D	Diamonds	Pt	Salt
Al	Bauxite	Fe	Iron Ore	Mn	Nickel
Au	Gold	Gr	Graphite	Na	Phosphates
Be	Beryl	Lt	Lithium	Ni	Lead
C	Coal	Mg	Magnesium	P	Platinum
Cr	Chromium	Mi	Mica	Zn	Zinc
				Sb	Antimony
				Sn	Tin
				U	Uranium
				V	Vanadium
				W	Tungsten

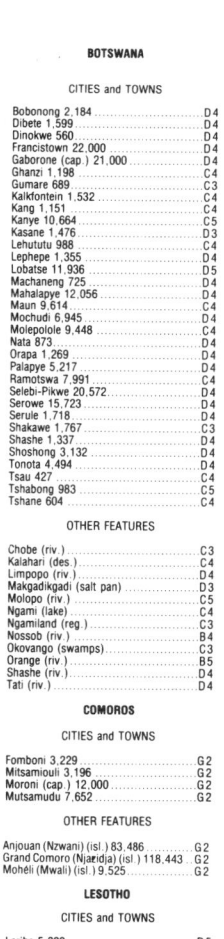
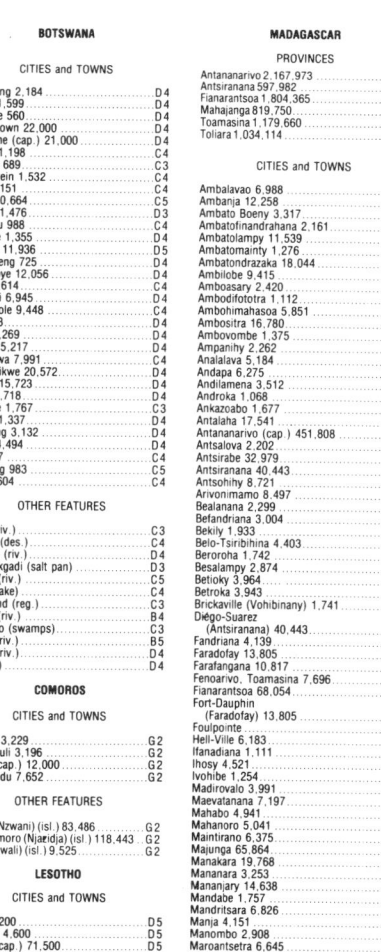

BOTSWANA

CITIES and TOWNS

Bobonong 2,184	D4
Dibete 1,599	D4
Dinokwe 560	D4
Francistown 22,000	D4
Gaborone (cap.) 21,000	D4
Ghanzi 1,198	C4
Gumare 689	C3
Kalkfontein 1,532	C4
Kang 1,151	C4
Kanye 10,664	C5
Kasane 1,476	D3
Lehututu 988	C4
Lephepe 1,355	D4
Lobatse 11,936	D5
Machaneng 725	D4
Mahalapye 12,056	D4
Maun 9,614	C4
Mochudi 6,945	D4
Molepolole 9,448	C4
Nata 873	D4
Orapa 1,269	D4
Palapye 5,217	D4
Ramotswa 7,991	C4
Selebi-Pikwe 20,572	D4
Serowe 15,723	D4
Serule 1,718	D4
Shakawe 1,767	C3
Shashe 1,337	D4
Shoshong 3,132	D4
Tonota 4,494	D4
Tsau 427	C4
Tshabong 983	C5
Tshane 604	C4

OTHER FEATURES

Chobe (riv.)	C3
Kalahari (des.)	C4
Limpopo (riv.)	D4
Makgadikgadi (salt pan)	D3
Molopo (riv.)	C5
Ngami (lake)	C4
Ngamiland (reg.)	C3
Nossob (riv.)	B4
Okovango (swamps)	C3
Orange (riv.)	B5
Shashe (riv.)	D4
Tati (riv.)	D4

COMOROS

CITIES and TOWNS

Fomboni 3,229	G2
Mitsamiouli 3,196	G2
Moroni (cap.) 12,000	G2
Mutsamudu 7,652	G2

OTHER FEATURES

Anjouan (Nzwani) (isl.) 83,486	G2
Grand Comoro (Njaridja) (isl.) 118,443	G2
Mohéli (Mwali) (isl.) 9,525	G2

LESOTHO

CITIES and TOWNS

Leribe 5,200	D5
Mafeteng 4,600	D5
Maseru (cap.) 71,500	D5
Mohaleshoek 3,600	D6

MADAGASCAR

PROVINCES

Antananarivo 2,167,973	H3
Antsiranana 597,982	H2
Fianarantsoa 1,804,365	H4
Mahajanga 819,750	H3
Toamasina 1,179,660	H3
Toliara 1,034,114	G4

CITIES and TOWNS

Ambalavao 6,988	H4
Ambania 12,258	H2
Ambato Boeny 3,317	H3
Ambatofinandrahana 2,161	H3
Ambatolampy 11,539	H3
Ambatomainty 1,276	H3
Ambatondrazaka 18,044	H3
Ambilobe 9,415	H2
Amboasary 2,420	H4
Ambodifototra 1,112	J3
Ambohimahasoa 5,851	H4
Ambositra 16,780	H4
Ambovombe 1,375	H5
Ampanihy 2,262	G4
Analalava 5,184	H2
Andapa 6,275	H2
Andilamena 3,512	H3
Androka 1,068	G5
Ankazoabo 1,677	G4
Antalaha 17,541	J2
Antananarivo (cap.) 451,808	H3
Antsalova 2,202	G3
Antsirabe 32,979	H3
Antsiranana 40,443	H2
Antsohihy 8,721	H2
Arivonimamo 8,497	H3
Bealanana 2,299	H2
Bekily 1,933	G4
Belo-Tsiribihina 4,403	G4
Beroroha 1,742	H4
Besalampy 2,874	G3
Betioky 3,964	G4
Betroka 3,943	H4
Brickaville (Vohibinany) 1,741	H3
Diego-Suarez	
(Antsiranana) 40,443	H2
Fandriana 4,139	H4
Faradofay 13,805	H5
Farafangana 10,817	H4
Fenoarivo, Toamasina 7,696	H3
Fianarantsoa 68,054	H4
Fort-Dauphin	
(Faradofay) 13,805	H5
Foulpointe	H3
Hell-Ville 6,183	H2
Ifanadiana 1,111	H4
Ihosy 4,521	H4
Ivohibe 1,254	H4
Madirovalo 3,991	H3
Maevatanana 7,197	H3
Mahabo 4,941	G4
Mahanoro 5,041	H3
Maintirano 6,375	G3
Majunga 65,864	H3
Manakara 19,768	H4
Mananara 3,253	J3
Mananjary 14,638	H4
Mandabe 1,757	G4
Mandritsara 6,826	H3
Manja 4,151	G4
Manombo 2,908	G4
Maroantsetra 6,645	J3
Marovoay 20,253	H3

(continued on following page)

Topography

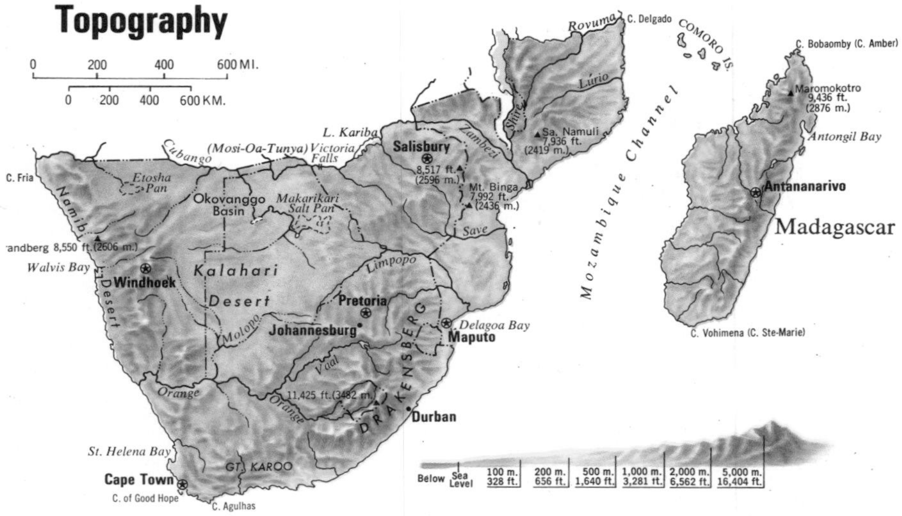
0 200 400 600 MI.
0 200 400 600 KM.
C. Delgado
COMORO IS.
C. Bobaomby (C. Amber)
Rovuma
Lurio
Maromokotro 9,436 ft. (2876 m.)
Antongil Bay
L. Kariba
Salisbury
(Mosi-Oa-Tunya) Victoria Falls
Cubango
Sa. Namuli 7,936 ft. (2419 m.)
C. Fria
Etosha Pan
8,517 ft. (2596 m.)
Mt. Binga 7,992 ft. (2436 m.)
Antananarivo
Okovanggo Basin
Makarikari Salt Pan
Save
Madagascar
Mozambique Channel
Walvis Bay
Windhoek
Kalahari Desert
Limpopo
Pretoria
Johannesburg
Delagoa Bay
Maputo
C. Vohimena (C. Ste-Marie)
Orange
Vaal
11,425 ft. (3482 m.)
DRAKENSBERG
Durban
St. Helena Bay
GT. KAROO
Cape Town
C. of Good Hope
C. Agulhas
Below Sea Level
100 m. 328 ft.
200 m. 656 ft.
500 m. 1,640 ft.
1,000 m. 3,281 ft.
2,000 m. 6,562 ft.
5,000 m. 16,404 ft.

Miandrivazo 2,371 .... G3
Midongy Atsimo 1,068 .... H4
Mitsinjo 3,118 .... H3
Moramanga 10,806 .... H3
Morombe 6,967 .... G4
Morondava 19,061 .... G3
Nosy-Varika 1,252 .... H4
Port-Bergé 4,734 .... H3
Sambava 6,215 .... J2
Soanierana-Ivongo 2,876 .... H3
Sosumav 10,946 .... H2
Tamatave (Toamasina) 77,395 .... H3
Tambohorano 1,383 .... G3
Tananarive (Antananarivo) (cap.) 451,808 .... H3
Tangainony 6,952 .... H4
Toamasina 77,395 .... H3
Toliara (Tuléar) 45,676 .... G4
Tsihombe 1,008 .... H5
Tsiroanomandidy 11,444 .... H3
Tsivory 1,036 .... H4
Vangaindrano 3,249 .... H4
Vatomandry 4,202 .... H3
Vohibinany 1,741 .... H3
Vohimarina (Vohémar) 4,289 .... J2
Vohipeno 2,736 .... H4

OTHER FEATURES

Alaotra (lake) H3
Amber (Bobaomby) (cape) H2
Antongil (bay) J3
Betsiboka (riv.) H3
Bobaomby (Amber) (cape) H2
Mangoky (riv.) G4
Mangoro (riv.) H3
Maromokotro (mt.) H2
Masoala (pen.) J3
Mozambique (chan.) G3
Nosy Be (isl.) H2
Nosy Boraha (isl.) J3
Onilahy (riv.) G4
Saint-André (cape) G3
Sainte-Marie (Vohimena) (cape) G5
Sainte-Marie (Nosy Boraha) (isl.) J3
Tsiafajavona (mt.) H3
Tsiribihina (riv.) G3
Vohimena (cape) G5

MAURITIUS

CITIES and TOWNS

Curepipe 52,709 .... G5
Mahébourg 15,463 .... G5
Port Louis (cap.) 141,022 .... G5
Poudre d'Or 1,799 .... G5
Quatre Bornes 51,638 .... G5
Souillac 3,361 .... G5

OTHER FEATURES

Mascarene (isls.) F5

MAYOTTE

CITIES and TOWNS

Dzaoudzi (cap.) 196 .... H2

MOZAMBIQUE

PROVINCES

Cabo Delgado 940,000 .... F2
Gaza 999,900 .... E4
Inhambane 977,000 .... E4
Manica 541,200 .... E4
Maputo 491,800 .... E5
Maputo (city) 755,300 .... E5
Nampula 2,402,700 .... F2
Niassa 514,100 .... F2
Sofala 1,055,200 .... E3
Tete 831,000 .... E3
Zambézia 2,500,000 .... F3

CITIES and TOWNS

Alto Molócuè 415 .... F3
Angoche 1,714 .... G3
Bartolomeu Dias○ 6,102 .... F4
Beira 46,293 .... F3
Beira* 130,398 .... F3
Bela Vista 851 .... E5
Benga 1,398 .... E3
Caia 1,363 .... E3
Catandica 663 .... E3
Chemba 588 .... E3
Chibuto 23,763 .... E4
Chicualacuala 2,050 .... E4
Chimoio 4,507 .... E3
Chinde 742 .... F3
Cóbuè○ 770 .... F2
Cuamba 1,416 .... F2
Dona Ana (Mutarara) 686 .... F3
Dondo 2,112 .... F3
Errego 418 .... F3
Espungabera 405 .... E4
Fíngoè 1,137 .... E2
Funhalouro○ 42,366 .... E4
Gorongoza 435 .... E3
Guija 530 .... E4
Homoíne 1,122 .... E4
Ibo 1,015 .... G2
Inhambane 4,975 .... F4
Inhaminga 1,607 .... F3
Inharrime 856 .... F4
Lichinga 3,011 .... F2
Lumbo○ 11,080 .... G3
Lúrio○ 13,417 .... G2
Mabalane○ 13,158 .... E4
Mabote○ 28,970 .... E4
Machanga○ 15,754 .... F4
Machaze○ 42,255 .... E4
Macia 1,203 .... E4
Macomia 730 .... F2
Magude 1,502 .... E5
Malema 430 .... F2
Mandié○ 24,382 .... E3
Mandimba○ 7,634 .... F2
Manhiça 1,680 .... E5
Maniamba○ 2,045 .... F2
Manica 1,529 .... E3
Manjacaze 641 .... E5
Maputo (cap.) 755,300 .... E5
Marracuene 1,342 .... E5
Marromeu 1,330 .... F3
Marrupa 824 .... F2
Massangena○ 3,301 .... E4
Massinga 517 .... F4
Maxixe 902 .... F4
Meconta 1,051 .... F3
Memba 379 .... G2
Metangula 1,502 .... F2
Milanje 1,048 .... F3
Moamba 643 .... E5
Moçambique 1,730 .... G3
Mocímboa da Praia 935 .... G2
Mocuba 2,293 .... F3
Moma 433 .... F3
Monapo 902 .... G2
Montepuez 2,837 .... F2
Morrumbala 415 .... F3
Morrumbene 1,121 .... F4
Mualama○ 34,992 .... F3
Mucojo○ 15,867 .... G2
Mueda 1,583 .... F2
Murrupula 444 .... F3
Mutarara (Dona Ana) 686 .... F3
Nacala 4,601 .... G2
Namacurra 399 .... F3
Namapa 440 .... F2
Nametil 453 .... F3
Nampula 23,072 .... F3
Negomane○ 656 .... F2
Nova Lusitânia 1,363 .... E3
Nova Mambone 883 .... F4
Nova Sofala 274 .... F4
Pafúri○ 2,599 .... E4
Pemba 3,629 .... G2
Quelimane 10,522 .... F3
Quionga○ 3,181 .... G2
Quissico 2,615 .... F4
Ribáuè 437 .... F3
Songo 1,350 .... E3
Tete 4,549 .... E3
Ulongue 451 .... E2
Vila de Sena○ 21,074 .... E3
Vilanculos 887 .... E4
Xai-Xai 5,234 .... E5

OTHER FEATURES

Angoche (isl.) G3
Bazaruto, Ilha do (isl.) F4
Binga (mt.) E3
Changane (riv.) E4
Chilwa (lake) F3
Delagoa (bay) E5
Delgado (cape) G2
Ligonha (riv.) F3
Limpopo (riv.) E4
Lugenda (riv.) F2
Lúrio (riv.) F2
Mazoe (riv.) E3
Mozambique (chan.) G3
Namuli, Serra (mt.) F3
Nyasa (lake) E2
Olifants (riv.) D4
Rovuma (riv.) F2
São Sebastião (pt.) F4
Save (riv.) E4
Shire (riv.) E3
Zambezi (riv.) E3

NAMIBIA (SOUTH-WEST AFRICA)

CITIES and TOWNS

Aroab 783 .... B5
Aus 767 .... B5
Berseba B5
Bethanie 1,207 .... B5
Gibeon B5
Gobabis 4,428 .... B4
Grootfontein 4,627 .... B3
Kalkfeld 587 .... B4
Kamanjab 713 .... A3
Karasburg 2,693 .... B5
Karibib 1,653 .... B4
Katima Mulilo C3
Keetmanshoop 10,297 .... B5
Khorixas 1,299 .... A4
Koes 514 .... B5
Lüderitz 6,642 .... A5
Maltahöhe 1,313 .... B4
Mariental 4,629 .... B4
Ohopoho A3
Okahandja 1,688 .... B4
Omaruru 2,783 .... B4
Ondangua B3
Ongwediva B3
Oranjemund 2,594 .... B5
Otavi 1,814 .... B3
Otjiwarongo 8,018 .... B4
Outjo 2,545 .... B4
Rehoboth 5,363 .... B4
Runtu 521 .... B3
Stampriet 271 .... B4
Swakopmund 5,681 .... A4
Tsumeb 12,338 .... B3
Usakos 2,334 .... B4
Warmbad 810 .... B5
Windhoek (cap.) 61,369 .... B4
Witvlei 303 .... B4

OTHER FEATURES

Brandberg (mt.) A4
Caprivi Strip (reg.) C3
Chobe (riv.) C3
Cubango (riv.) B3
Damaraland (reg.) B4
Diamond Coast (reg.) A5
Elephant (riv.) B5
Etosha Pan (salt pan) B3
Fish (riv.) B5
Great Namaland (reg.) B4
Hottentot (bay) A5
Kalahari (des.) C4
Kaokoveld (reg.) A3
Kaukauveld (mts.) C3
Namib (des.) A3
Nossob (riv.) B4
Okovango (riv.) C3
Ovamboland (reg.) B3
Skeleton Coast (reg.) A3
Swakop (riv.) B4
Zambezi (riv.) C3

REUNION

CITIES and TOWNS

Le Port 21,564 .... F5
Saint-André 6,584 .... G5
Saint-Benoît 7,778 .... G5
Saint-Denis (cap.) 80,075 .... F5
Saint-Denis* 104,603 .... F5
Saint-Joseph 8,928 .... G6
Saint-Louis 10,252 .... F5
Saint-Pierre 21,817 .... F6

OTHER FEATURES

Bassas da India (isl.) F4
Europa (isl.) G4
Glorioso (isls.) H2
Juan de Nova (isl.) G3
Piton des Neiges (mt.) G5

SEYCHELLES

CITIES and TOWNS

Anse Boileau† 3,420 .... H5
Anse Royale† 3,182 .... H5
Cascade† 2,600 .... H5
Victoria (cap.) 15,559 .... H5
Victoria* 23,012 .... H5

OTHER FEATURES

Aldabra (isls.) H1
Assumption (isl.) H1
Astove (isl.) H2
Cosmoledo (isls.) H1
Frigate (isl.) J5
La Digue (isl.) J5
Mahé (isl.) H5
North (isl.) H5
Praslin (isl.) H5
Silhouette (isl.) H5

SOUTH AFRICA

PROVINCES

Cape Province 5,543,506 .... C6
Natal 5,722,215 .... E5
Orange Free State 1,833,216 .... D5
Transvaal 10,673,033 .... D4

AUTONOMOUS REPUBLICS

Bophuthatswana 1,200,000 .... D5
Ciskei 345,191 .... D6
Transkei 2,000,000 .... D6
Venda 450,000 .... E4

CITIES and TOWNS

Aberdeen 4,968 .... C6
Adelaide 7,227 .... D6
Alberton 23,988 .... H6
Alexandra 57,040 .... H6
Alexander Bay 2,675 .... B5
Aliwal North 12,311 .... D6
Barberton 12,382 .... E5
Barkly East 4,023 .... D6
Beaufort West 17,862 .... C6
Bellville 49,026 .... F6
Benoni 151,294 .... J6
Benoni□ 164,543 .... J6
Bethlehem 29,918 .... D5
Bethulie 4,918 .... D6
Bloemfontein 149,836 .... C5
Bloemfontein□ 182,329 .... C5
Bloubergstrand 378 .... E6
Boksburg 106,126 .... J6
Botrivier 743 .... F7
Brakpan 73,210 .... J6
Brandvlei 1,337 .... B6
Bredasdorp 5,264 .... B6
Brentwood Park 5,296 .... J6
Brits 12,182 .... D5
Britstown 3,039 .... C6
Burgersdorp 8,340 .... D6
Butterworth (Gcuwa) 2,769 .... D6
Caledon 5,406 .... G7
Calvinia 6,386 .... B6
Cape Town (cap.) 697,514 .... E6
Cape Town□ 833,731 .... E6
Carltonville 40,641 .... G7
Carnarvon 5,199 .... C6
Ceres 9,230 .... B6
Christiana 6,882 .... D5
Clanwilliam 2,724 .... B6
Clayville 3,994 .... H6
Colesberg 7,088 .... D6
Constantia 7,220 .... E6
Cradock 20,822 .... D6
De Aar 18,057 .... C6
Delmas 6,424 .... J6
Dibeng 945 .... C5
Douglas 4,335 .... C5
Dundee 17,162 .... E5
Dunnottar 3,089 .... J6
Durban 736,852 .... E5
Durban□ 975,494 .... E5
Durbanville 7,438 .... F6
East London 119,727 .... D6
East London□ 126,671 .... D6
Edenburg 3,710 .... D5
Edendale 41,194 .... D5
Edenvale 25,126 .... H6
Eersterivier 1,459 .... F6
Elliot 3,739 .... D6
Eloff 1,134 .... J6
Elsburg 3,501 .... H6
Elsiesrivier 63,706 .... F6
Empangeni 7,532 .... E5
Ermelo 19,036 .... E5
Eshowe 4,552 .... E5
Estcourt 10,922 .... D5
Ficksburg 9,504 .... D5
Firgrove 2,551 .... F6
Fort Beaufort 11,640 .... D6
Franschhoek 1,216 .... F6
Garies 1,339 .... B6
Gcuwa 2,769 .... D6
George 24,625 .... C6
Germiston 221,972 .... H6
Germiston□ 293,257 .... H6
Glencoe 10,513 .... E5
Goodwood 31,592 .... F6
Gordon's Bay 1,112 .... F7
Graaff-Reinet 22,392 .... C6
Grabouw 4,286 .... F7
Grahamstown 41,302 .... D6
Grassy Park 32,709 .... E6
Greytown 9,028 .... E5
Griquatown 2,996 .... C5
Halfway House 3,639 .... H6
Harrismith 16,082 .... D5
Hawston 2,501 .... G7
Heidelberg 12,521 .... J7
Heilbron 8,258 .... D5
Hermanus 4,956 .... G7
Hopetown 3,273 .... C5
Houtbaai 5,691 .... E6
Howick 12,429 .... E5
Humansdorp 4,215 .... C6
Ingwavuma 718 .... E5
Jagersfontein 4,142 .... D5
Jameson Park 2,280 .... J7
Johannesburg 654,232 .... H6
Johannesburg□ 1,417,818 .... H6
Keimoes 4,534 .... C5
Kempton Park 37,205 .... J6
Kenhardt 3,230 .... C5
Kimberley 105,258 .... C5
Kimberley□ 108,609 .... C5
King William's Town 15,798 .... D6
Kirkwood 5,151 .... D6
Kleinmond 1,115 .... F7
Klerksdorp 63,558 .... D5
Knysna 13,479 .... C6
Koffiefontein 3,672 .... D5
Kokstad 10,227 .... D6
Kraaifontein 10,286 .... F6
Kroonstad 51,988 .... D5
Krugersdorp 92,725 .... H6
Kuilsrivier 8,132 .... F6
Kuruman 5,758 .... C5
Ladybrand 8,757 .... D5
Ladysmith 28,920 .... D5
Lambert's Bay 3,247 .... B6
Lombardy 3,395 .... H6
Louis Trichardt 8,906 .... E4
Lydenburg 7,427 .... E4
Macassar 882 .... F6
Maclear 3,279 .... D6
Mafikeng (Mafeking) 6,515 .... C5
Malmesbury 9,314 .... B6
Margate 4,410 .... E6
Matatiele 3,853 .... D6
Melkbosstrand 453 .... E6
Messina 12,121 .... D4
Meyerton 8,654 .... H7
Middelburg, C. of Good Hope 11,121 .... D6
Middelburg, Transvaal 26,942 .... D5
Milnerton 10,893 .... F6
Modderfontein 8,538 .... H6
Molteno 5,825 .... D6
Montagu 5,504 .... C6
Moorreesburg 4,945 .... B6
Mossel Bay 17,574 .... C6
Nababeep 8,293 .... B5
Nelspruit 25,092 .... E5
Newcastle 14,407 .... E5
Nigel 41,179 .... J7
Noupoort 7,403 .... C6
Nyanga 15,655 .... F6
Nylstroom 6,906 .... D4
Odendaalsrus 15,603 .... D5
Okiep 4,983 .... B5
Oudtshoorn 26,907 .... C6
Paarl 49,244 .... F6
Parow 60,768 .... F6
Parys 17,447 .... D5
Phalaborwa 7,543 .... E4
Pietermaritzburg 114,822 .... E5
Pietermaritzburg□ 174,179 .... E5
Pietersburg 27,174 .... D4
Piet Retief 10,056 .... D5
Piketberg 3,638 .... B6
Pinelands 11,769 .... F6
Pinetown 22,721 .... E6
Pniel 1,596 .... F6
Port Alfred 8,640 .... D6
Port Elizabeth 392,231 .... D6
Port Elizabeth□ 413,961 .... D6
Port Nolloth 2,893 .... B5
Port Saint Johns (Umzimbuvu) 1,817 .... D6
Port Shepstone 5,581 .... E6
Postmasburg 9,020 .... C5

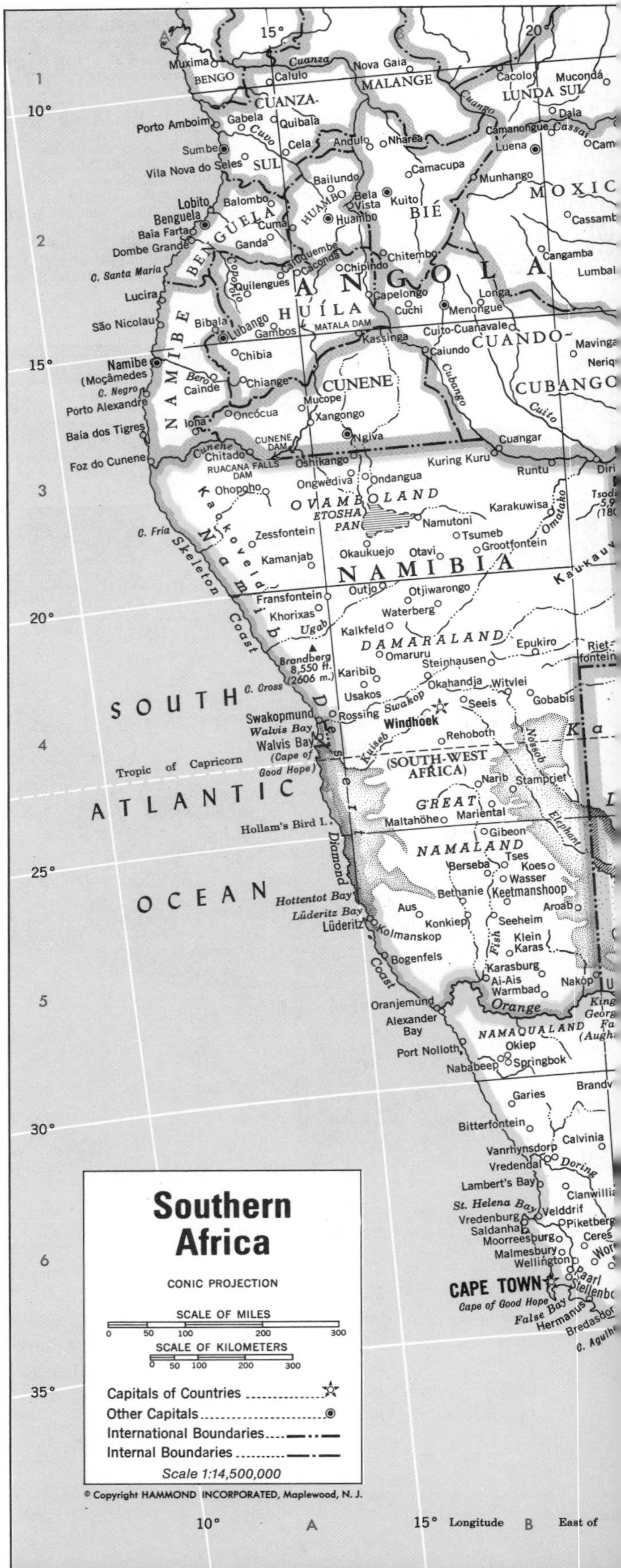
Southern Africa
CONIC PROJECTION
SCALE OF MILES
0 50 100 200 300
SCALE OF KILOMETERS
0 50 100 200 300
Capitals of Countries
Other Capitals
International Boundaries
Internal Boundaries
Scale 1:14,500,000
© Copyright HAMMOND INCORPORATED, Maplewood, N. J.
SOUTH ATLANTIC OCEAN
ANGOLA
NAMIBIA (SOUTH-WEST AFRICA)
Windhoek
Walvis Bay
Swakopmund
Lüderitz
Tropic of Capricorn
CAPE TOWN
Cape of Good Hope
10° A 15° Longitude B East of

Potchefstroom 57,443 ... D5
Potgietersrus 6,667 ... D4
Pretoria (cap.) 545,450 ... D5
Pretoria★ 573,283 ... D5
Prieska 8,521 ... C5
Prince Albert 3,346 ... C5
Queenstown 39,304 ... D6
Randburg 43,257 ... H6
Randfontein 50,481 ... G6
Reitz 5,650 ... D5
Rensburg 2,042 ... H6
Richards Bay 598 ... J7
Richmond 3,185 ... E6
Riversdale 6,165 ... C6
Robertson 10,237 ... C6
Roodepoort 115,366 ... H6
Saldanha 4,994 ... B6
Senekal 9,124 ... D5
Sesfontein 2,731 ... D5
Simonstown 12,137 ... E7
Sishen 2,692 ... C5
Somerset East 10,383 ... D6
Somerset West 11,828 ... F6
Soweto 602,043 ... H5
Springbok 4,922 ... B5
Springs 142,812 ... J6
□ 146,831 ... J6

Standerton 21,038 ... D5
Stanger 11,064 ... E5
Stellenbosch 29,955 ... D5
Strand 24,503 ... F7
Stutterheim 12,077 ... D6
Sundra 2,088 ... J6
Swellendam 6,423 ... C6
Taung 1,316 ... C5
Tembisa 81,821 ... H6
Thabazimbi 6,711 ... D4
Thohoyandou ... E4
Tzaneen 4,331 ... E4
Ubombo 3,697 ... E5
Uitenhage 70,517 ... D6
Ulundi ... E5
Umtata 25,216 ... D6
Umzimkulu 1,817 ... D6
Umzinto 5,272 ... E6
Vaalwater 3,740 ... D4
Vanderbijl Park 78,754 ... C5
Vanderbijlpark 2,279 ... H6
Veldrif 3,361 ... B6
Ventersdorp ... D5
Vereeniging 172,549 ... D5
□ 200,078 ... D5
Victoria West 3,949 ... C6
Villiersdorp 2,349 ... G6

Vishoek 6,721 ... E7
Volksrust 10,238 ... D5
Vrede 6,309 ... D5
Vredenburg 6,094 ... B6
Vredendal 5,377 ... B5
Vryburg 16,916 ... C5
Vryheid 16,992 ... E5
Walvis Bay 21,725 ... A4
Warmbad 8,343 ... D4
Warrenton 9,614 ... C5
Waterval-Bo 6,951 ... E5
Welkom 67,472 ... D5
Wellington 17,092 ... B6
Westonaria 36,253 ... H7
Willowmore 3,740 ... C6
Winburg 6,761 ... D5
Witbank 37,456 ... D5
Wolmaransstad 7,219 ... D5
Worcester 41,198 ... B6
Zastron 4,483 ... D6
Zeerust 6,972 ... D5
Zwelitsha 22,131 ... D6

OTHER FEATURES
Addo Nat'l Park ... D6
Agulhas (cape) ... B6
Bot (riv.) ... G7

Bredasdorp Nat'l Park ... C6
Cape (riv.) ... E7
Crocodile (riv.) ... H6
Drakensberg (range) ... F6
False (bay) ... E7
Good Hope (cape) ... E7
Great Fish (riv.) ... D6
Great Karoo (reg.) ... C6
Great Kei (riv.) ... D6
Griqualand West (reg.) ... C5
Groote (riv.) ... C6
Hartbees (riv.) ... C5
Kalahari Gemsbok Nat'l Park ... B5
King George's (falls) ... B5
Klip (riv.) ... D4
Kruger Nat'l Park ... E4
Limpopo (riv.) ... D4
Molopo (riv.) ... C5
Mountain Zebra Nat'l Park ... D6
Olifants (riv.) ... B5
Orange (riv.) ... B5
Palmiet (riv.) ... D6
Plettenberg (bay) ... C6
Pondoland (reg.) ... D6
Robben (isl.) ... E6
Royal Natal Nat'l Park ... E5
Saint Helena (bay) ... B6
Saint Lucia (lake) ... E5

Sak (riv.) ... C6
Sand (riv.) ... D4
Slangkop (pt.) ... E7
Sneeuwkop (mt.) ... F6
Table (bay) ... E6
Table (mt.) ... E6
Vaal (riv.) ... D5
Walvis (bay) ... A4
Witwatersberg (range) ... G6
Witwatersrand (reg.) ... H7
Zonderend (riv.) ... G6
Zululand (reg.) ... E5

SWAZILAND
CITIES and TOWNS
Manzini 28,837 ... E5
Mbabane (cap.) 23,109 ... E5
Siteki 1,362 ... E5

ZIMBABWE
CITIES and TOWNS
Beitbridge 1,986 ... E4
Bindura 17,000 ... E3

Bulawayo 359,000 ... D3
Chegutu 12,000 ... E3
Chimanimani 667 ... E3
Chinhoyi 25,000 ... E3
Chipinge 2,350 ... E4
Chivhu 1,669 ... E3
Dete 2,473 ... D3
Gwaai 2,710 ... D3
Gwanda 2,049 ... D3
Gweru 68,000 ... E3
Harare (Salisbury) (cap.) 601,000 ... E3
Hwange 33,000 ... D3
Inyanga 733 ... E3
Kadoma 32,000 ... D3
Kariba 3,943 ... D3
Kwekwe 54,000 ... E3
Marondera 23,000 ... E3
Masvingo 22,000 ... E3
Matopos 11,330 ... D3
Mount Darwin 904 ... E3
Mutare 61,000 ... E3
Mvuma 1,525 ... E3
Mwenezi 7,830 ... D3
Plumtree 2,041 ... D3
Rusape 5,286 ... E3
Salisbury (Harare) (cap.) 601,000 ... E3
Shamva 785 ... E3
Shurugwi 8,387 ... E3

Tuli 340 ... D4
West Nicholson 1,929 ... D4
Zvishavane 20,000 ... E4

OTHER FEATURES
Inyanga Nat'l Park ... E3
Kariba (lake) ... D3
Lundi (riv.) ... E4
Mashonaland (reg.) 1,875,700 ... E3
Matabeleland (reg.) 969,220 ... D3
Mazoe (riv.) ... E3
Sabi (riv.) ... E4
Shangani (riv.) ... D3
Shashe (riv.) ... D4
Umvukwe (range) ... E3
Victoria (falls) ... D3
Zambezi (riv.) ... D3
Zimbabwe Nat'l Park ... E4

★City and suburbs.
†Population of parish.
○Population of subdivision.
□Population of urban area.

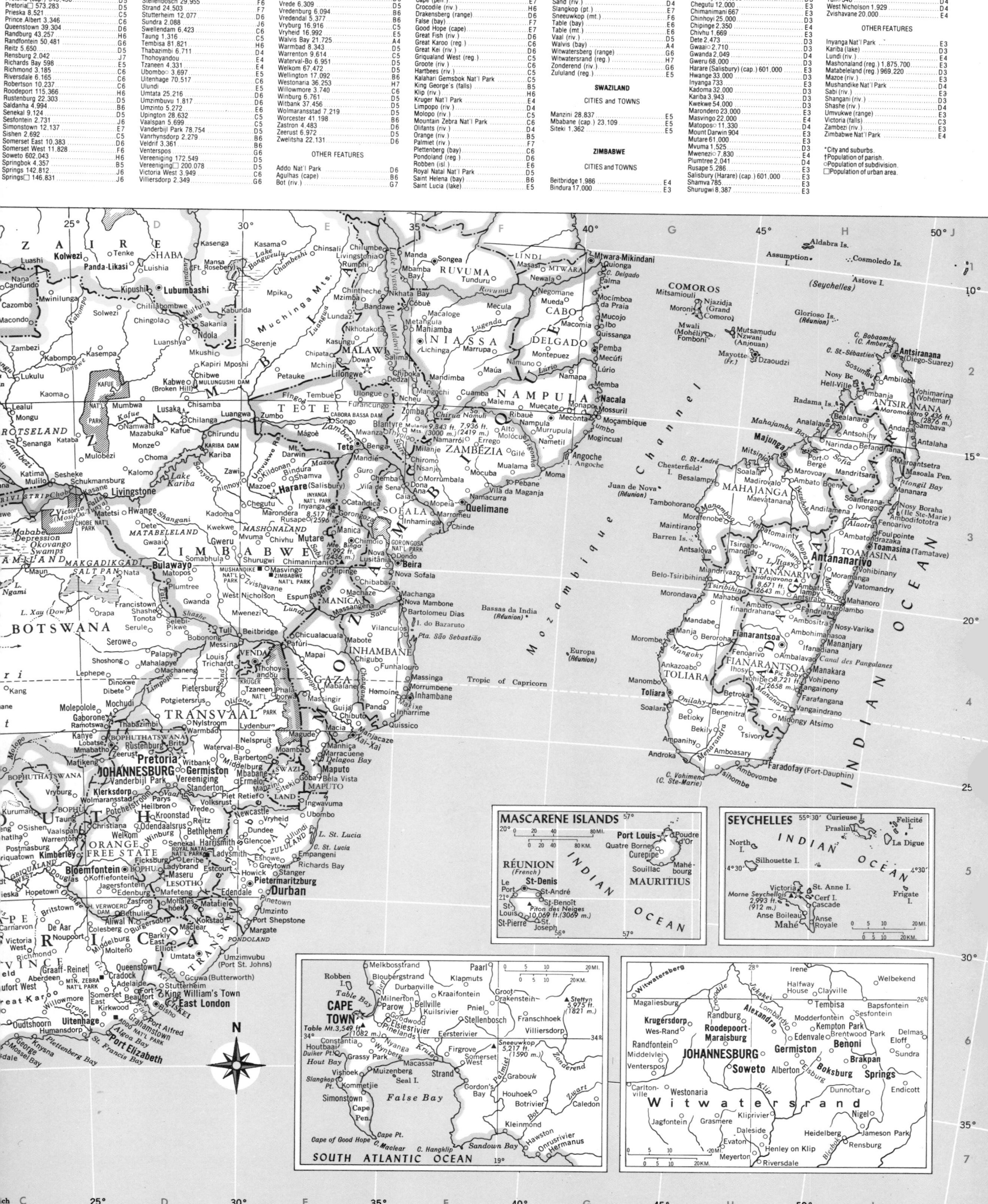

Population Distribution

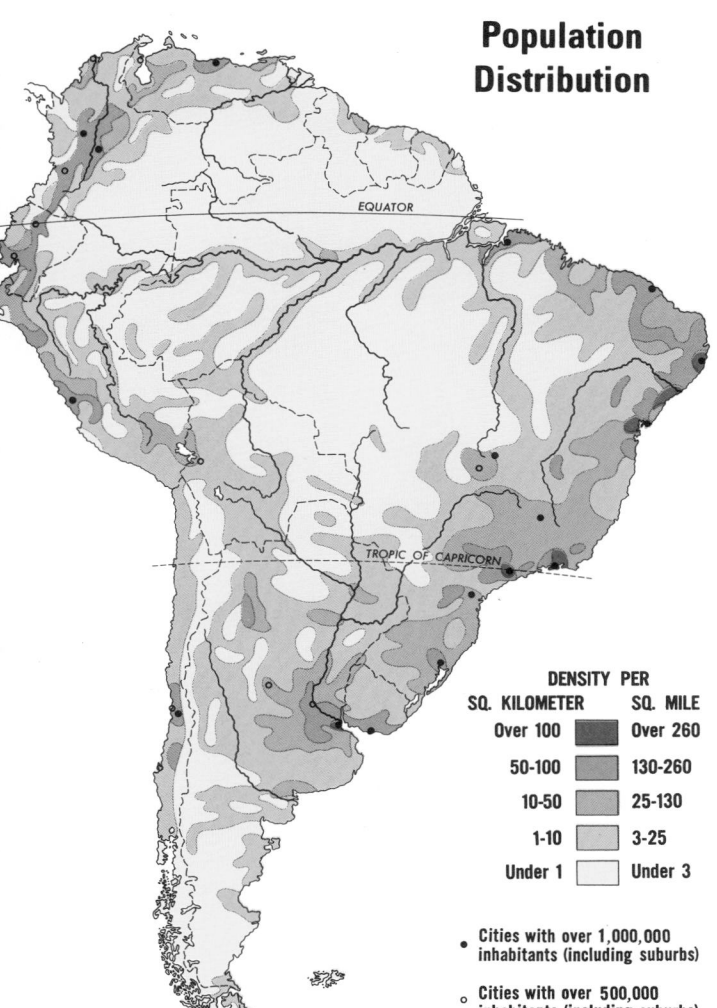

AREA 6,875,000 sq. mi. (17,806,250 sq. km.)
POPULATION 245,000,000
LARGEST CITY São Paulo
HIGHEST POINT Cerro Aconcagua 22,831 ft. (6,959 m.)
LOWEST POINT Salina Grande -131 ft. (-40 m.)

Vegetation

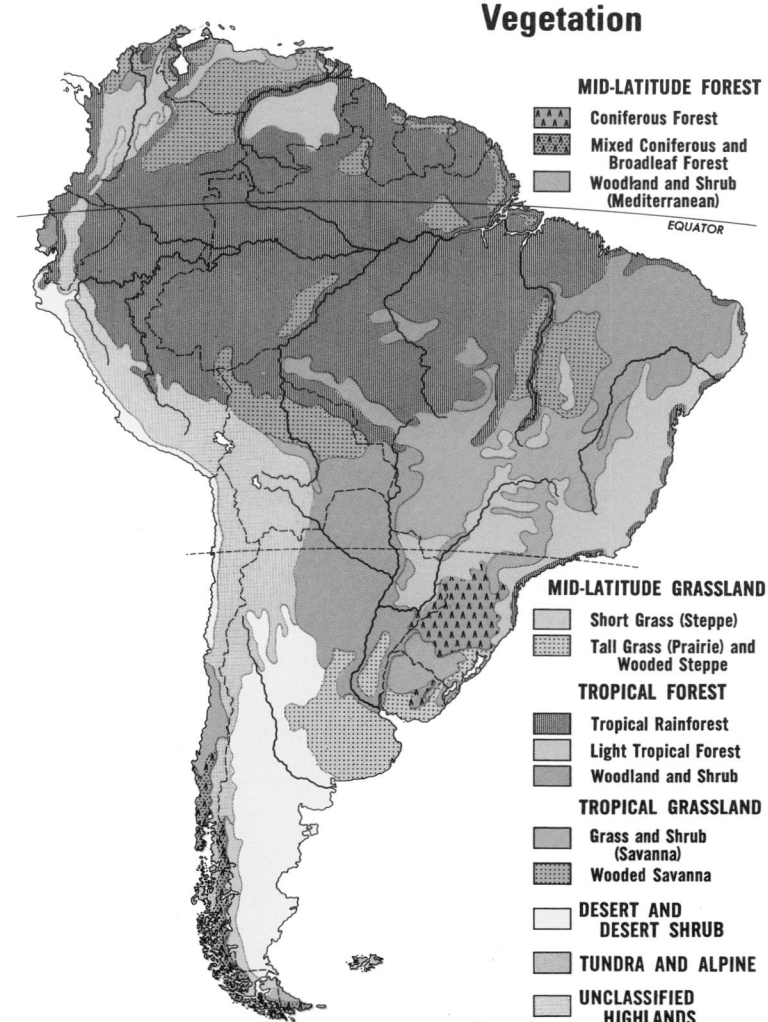

DENSITY PER	
SQ. KILOMETER	SQ. MILE
Over 100	Over 260
50-100	130-260
10-50	25-130
1-10	3-25
Under 1	Under 3

● Cities with over 1,000,000 inhabitants (including suburbs)

○ Cities with over 500,000 inhabitants (including suburbs)

MID-LATITUDE FOREST
- Coniferous Forest
- Mixed Coniferous and Broadleaf Forest
- Woodland and Shrub (Mediterranean)

MID-LATITUDE GRASSLAND
- Short Grass (Steppe)
- Tall Grass (Prairie) and Wooded Steppe

TROPICAL FOREST
- Tropical Rainforest
- Light Tropical Forest
- Woodland and Shrub

TROPICAL GRASSLAND
- Grass and Shrub (Savanna)
- Wooded Savanna

DESERT AND DESERT SHRUB

TUNDRA AND ALPINE

UNCLASSIFIED HIGHLANDS

...cagua (mt.) ...C6
...a (pt.), Peru ...A3
...inhas, Brazil ...F4
...ete, Brazil ...D5
...ndro Selkirk (isl.) ...A6
...ile ...C4
...ano (plat.) ...
...itacuva (mt.) ...
...ombia ...B2
...cun (riv.), Brazil ...D3
...olis, Brazil ...
...huma (mt.), Bolivia ...C4
...s (range) ...B-26
...l (falls), Venezuela ...C2
...stura (falls), Colombia ...B2
...agasta, Chile ...B5
...lmac (riv.), Peru ...B4
...m, Brazil ...F4
...guaia (riv.), Brazil ...E3
...juana, Brazil ...E4
...ca (riv.) ...C2
...ca, Colombia ...B2
...uipa, Peru ...B4
...ntina ...C6
...na (riv.), Brazil ...B4
...nã (riv.), Brazil ...D3
...(cap.), Paraguay ...D5
...des), Chile ...C5
...Peru ...B4
...D6
...ca, Argentina ...C6
...), Brazil ...D4
...dera (isl.), Brazil ...E5
...acena, Brazil ...E3
...elona, Venezuela ...C2
...elos, Brazil ...C3
...uisimeto, Venezuela ...C2
...anquilla, Colombia ...B1
...m, Brazil ...E3
...Horizonte, Brazil ...E4
...ca (bay), Argentina ...C6
...Vista, Brazil ...C2
...atá (cap.), Colombia ...B2
...via ...C4
...ança, Brazil ...D3
...lia (cap.), Brazil ...E4
...lia (cap.), Brazil ...D4
...nado, Brazil ...E4
...aramanga, Colombia ...B2
...naventura, Colombia ...B2
...nos Aires (cap.) ...B7
...rgentina ...C6
...ingas (for.), Brazil ...E3
...marca, Peru ...B3
...ma, Chile ...C5
...Colombia ...B2
...o, Peru ...B4
...pana (isl.), Chile ...B7
...pinas, Brazil ...E5
...pinas, Brazil ...
...po Grande, Brazil ...D5
...pos, Brazil ...E5
...etá (riv.), Colombia ...B2
...cas (cap.), Venezuela ...C2
...n (riv.), Venezuela ...C1
...gena, Colombia ...B1
...pano, Venezuela ...C1
...marca, Argentina ...C5
...ra (riv.), Colombia ...B2

Caviana (isl.), Brazil ...E2
Caxias, Brazil ...E3
Caxias do Sul, Brazil ...D5
Cayenne (cap.), Fr. Guiana ...D2
Cerro de Pasco, Peru ...B4
Chiclayo, Peru ...B3
Chile ...B5
Chillán, Chile ...B6
Chiloé (isl.), Chile ...B7
Chimbote, Peru ...B3
Chiquita (lake), Argentina ...C6
Chivilcoy, Argentina ...C6
Chonos (arch.), Chile ...B7
Chubut (riv.), Argentina ...C7
Ciénaga, Colombia ...B1
Ciudad Bolívar, Venezuela ...C2
Cochabamba, Bolivia ...C4
Colombia ...B2
Colorado (riv.), Argentina ...C6
Comodoro Rivadavia, Argentina ...C7
Concepción, Chile ...B6
Concepción, Paraguay ...D5
Concordia, Argentina ...D6
Copiapó, Chile ...B5
Coquimbo, Chile ...B6
Corcovado (gulf), Chile ...B7
Córdoba, Argentina ...C6
Coro, Venezuela ...C1
Corrientes, Argentina ...D5
Corrientes (cape), Colombia ...B2
Corumbá, Brazil ...D4
Courantyne (riv.) ...D2
Cruz Alta, Brazil ...D5
Cúcuta, Colombia ...B2
Cuenca, Ecuador ...B3
Cuiabá, Brazil ...D4
Cumaná, Venezuela ...C1
Curitiba, Brazil ...F5
Cuyuni (riv.) ...C2
Deseado (riv.), Argentina ...C7
Devils (isl.), Fr. Guiana ...D2
Ecuador ...B3
Encarnación, Paraguay ...D5
Esmeraldas, Ecuador ...A2
Essequibo (riv.), Guyana ...D2
Estados (isl.), Argentina ...C8
Falkland Islands ...D8
Florianópolis, Brazil ...E5
Formosa, Argentina ...D5
Fortaleza, Brazil ...F3
French Guiana ...D2
Frio (cape), Brazil ...E5
Georgetown (cap.), Guyana ...D2
Goiânia, Brazil ...D4
Gran Chaco (reg.) ...C4
Grande (bay), Argentina ...C8
Grande (riv.), Brazil ...E5
Guajira (pen.) ...B1
Guaporé (riv.) ...C4
Guaviare (riv.), Colombia ...C3
Guayaquil, Ecuador ...A3
Guri (res.), Venezuela ...C2
Guyana ...D2
Hanover (isl.), Chile ...B8
Horn (cape), Chile ...C8
Hoste (isl.), Chile ...B8
Huacho, Peru ...B4
Huaráz, Peru ...B3
Huascarán (mt.), Peru ...B3

Huila (mt.), Colombia ...B2
Humaitá, Brazil ...C3
Ibagué, Colombia ...B2
Ibarra, Ecuador ...B2
Ica, Peru ...B4
Iguaçu (riv.) ...D5
Ilhéus, Brazil ...F4
Iquique, Chile ...B5
Iquitos, Peru ...B3
Japurá (riv.), Brazil ...C3
Jequié, Brazil ...E4
João Pessoa, Brazil ...F3
Joinvile, Brazil ...E5
Juan Fernández (isls.), Chile ...B6
Juàzeiro, Brazil ...E3
Juiz de Fora, Brazil ...E5
Jujuy, Argentina ...C5
Juliaca, Peru ...B4
Juruá (riv.), Brazil ...C3
Juruena (riv.), Brazil ...D4
La Guaira, Venezuela ...C1
La Oroya, Peru ...B4
La Paz (cap.), Bolivia ...C4
La Plata, Argentina ...D6
La Rioja, Argentina ...C5
La Serena, Chile ...B5
Leticia, Colombia ...B3
Lima (cap.), Peru ...B4
Llanos (plain) ...B2
Llullaillaco (mt.) ...C5
Loja, Ecuador ...B3
Londrina, Brazil ...D5
Macapá, Brazil ...D2
Maceió, Brazil ...F3
Madeira (riv.), Brazil ...C3
Madre de Dios (isl.), Chile ...B8
Magdalena (riv.), Colombia ...B2
Magellan (str.) ...C8
Maipelo (isl.), Colombia ...A2
Mamoré (riv.), Bolivia ...C4
Manaus, Brazil ...D3
Manicoré, Brazil ...C3
Manizales, Colombia ...B2
Manta, Ecuador ...A3
Mar (mts.), Brazil ...E5
Maracaibo, Venezuela ...B2
Maracaibo (lake), Venezuela ...B2
Marajó (isl.), Brazil ...E3
Marañón (riv.), Peru ...B3
Mar del Plata, Argentina ...D6
Margarita (isl.), Venezuela ...C1
Mariscal Estigarribia, Paraguay ...C5
Maroni (riv.) ...D2
Mato Grosso, Brazil ...D4
Mato Grosso (plat.), Brazil ...D4
Medellín, Colombia ...B2
Melo, Uruguay ...D6
Mendoza, Argentina ...C6
Mercedes, Argentina ...C6
Mercedes, Uruguay ...C6
Mérida, Venezuela ...B2
Meta (riv.) ...C2
Mirim (lake) ...D6
Misti, El (mt.), Peru ...B4
Mollendo, Peru ...B4
Montes Claros, Brazil ...E4
Montevideo (cap.), Uruguay ...D6
Moquegua, Peru ...B4
Morawhanna, Guyana ...D2

Nahuel Huapi (lake), Argentina ...B7
Nassau (bay), Chile ...C8
Natal, Brazil ...F3
Negro (riv.), Argentina ...C6
Negro (riv.), Brazil ...C3
Neiva, Colombia ...B2
Netherlands Antilles ...C1
New Amsterdam, Guyana ...D2
Nieuw-Nickerie, Suriname ...D2
Niterói, Brazil ...E5
Ojos del Salado (mt.) ...C5
Orinoco (riv.) ...C2
Oruro, Bolivia ...C4
Ovalle, Chile ...B6
Oyapock (riv.) ...D2
Pampas (plain), Argentina ...C6
Pará (est.), Brazil ...E3
Paraguay (riv.) ...D5
Paraguay ...D5
Paramaribo (cap.), Suriname ...D2
Paraná (riv.) ...D5
Paraná, Argentina ...C6
Parecis (mts.), Brazil ...D4
Parnaíba, Brazil ...E3
Parnaíba (riv.), Brazil ...E3
Pasto, Colombia ...B2
Patagonia (reg.), Argentina ...C7
Paysandú, Uruguay ...D6
Pelotas, Brazil ...D6
Penas (gulf), Chile ...B7
Pergamino, Argentina ...C6
Peru ...B4
Pilcomayo (riv.) ...C5
Pisco, Peru ...B4
Piura, Peru ...A3
Poopó (lake), Bolivia ...C4
Popayán, Colombia ...B2
Porto Alegre, Brazil ...D6
Porto Nacional, Brazil ...E4
Posadas, Argentina ...D5
Potosí, Bolivia ...C4
Puerto Cabello, Venezuela ...C1
Puerto Carreño, Colombia ...C2
Puerto Aisén, Chile ...B7
Puerto Madryn, Argentina ...C7
Puerto Montt, Chile ...B7
Puno, Peru ...B4
Punta Arenas, Chile ...B8

Purus (riv.), Brazil ...C3
Putumayo (riv.) ...B3
Quito (cap.), Ecuador ...B2
Rancagua, Chile ...B6
Rawson, Argentina ...C7
Recife, Brazil ...F3
Reina Adelaida (arch.), Chile ...B8
Resistencia, Argentina ...D5
Ribeirão Preto, Brazil ...E5
Riobamba, Ecuador ...B3
Rio de Janeiro, Brazil ...E5
Rio Grande, Brazil ...D6
Rivera, Uruguay ...D6
Robinson Crusoe (isl.), Chile ...B6
Roraima (mt.) ...C2
Rosario, Argentina ...C6
Salado (riv.), Argentina ...C6
Salta, Argentina ...C5
Salto, Uruguay ...D6
Salto Grande (falls) ...
Salvador, Brazil ...F4
San Ambrosio (isl.), Chile ...B5
San Antonio (cape), Argentina ...C6
San Carlos de Bariloche, Argentina ...B7
San Félix (isl.), Chile ...A5
San Fernando, Venezuela ...C2
San Jorge (gulf), Argentina ...C7
San Juan, Argentina ...C6
San Julián, Argentina ...C7
San Luis, Argentina ...C6
San Martín (lake) ...B7
San Rafael, Argentina ...C6
Santa Catarina (isl.), Brazil ...E5
Santa Cruz, Argentina ...C8
Santa Cruz, Bolivia ...C4
Santa Fe, Argentina ...C6
Santa Marta, Colombia ...B1
Santarém, Brazil ...D3
Santiago (cap.), Chile ...B6
Santiago del Estero, Argentina ...C5
Santos, Brazil ...E5
São Francisco (riv.) ...

Brazil ...E4
São Luís, Brazil ...E3
São Paulo, Brazil ...E5
São Roque (cape), Brazil ...F3
Sarmiento, Argentina ...C7
Selvas (for.), Brazil ...C3
Senhor do Bonfim, Brazil ...F4
Serrinha, Brazil ...
Sete Lagoas, Brazil ...E4
Sete Quedas (falls) ...D5
Sobral, Brazil ...E3
Sorocaba, Brazil ...E5
Sousa, Brazil ...F3
Stanley (cap.), Falk. Is. ...D8
Sucre (cap.), Bolivia ...C4
Sullana, Peru ...A3
Suriname ...D2
Tacna, Peru ...B4
Tafí Viejo, Argentina ...C5
Taguatinga, Brazil ...E4
Taitao (pen.), Chile ...B7
Talara, Peru ...A3
Talca, Chile ...B6
Talcahuano, Chile ...B6
Taltal, Chile ...B5
Tandil, Argentina ...C6
Tapajós (riv.), Brazil ...D3
Tarapoto, Peru ...B3
Tarauacá, Brazil ...C3
Tarija, Bolivia ...C5
Tartagal, Argentina ...C5
Tefé, Brazil ...C3
Temuco, Chile ...B6
Teófilo Otoni, Brazil ...E4
Teresina, Brazil ...E3
Tierra del Fuego (isl.) ...C8
Titicaca (lake) ...C4
Tocantinópolis, Brazil ...E3
Tocopilla, Chile ...B5
Tolima (mt.), Colombia ...B2
Tres Arroyos, Argentina ...C6
Três Lagoas, Brazil ...D5
Tres Montes (cape), Chile ...B7
Tres Puntas (cape), Argentina ...C7
Trinidad, Bolivia ...C4
Trujillo, Peru ...B3
Trujillo, Venezuela ...C2
Tubarão, Brazil ...E5

Tumaco, Colombia ...B2
Tumbes, Peru ...A3
Tunja, Colombia ...B2
Tupiza, Bolivia ...C5
Uberaba, Brazil ...E4
Uberlândia, Brazil ...E4
Ucayali (riv.), Peru ...B3
Uruaçu, Brazil ...E4
Urubupungá (dam), Brazil ...D5
Uruguaiana, Brazil ...D6
Uruguay ...D6
Uruguay (riv.) ...C8
Ushuaia, Argentina ...C8
Uyuni, Bolivia ...C5
Valdés (pen.), Argentina ...C7
Valdivia, Chile ...B6
Valença, Brazil ...F4
Valencia, Venezuela ...C2
Valera, Venezuela ...B1
Valledupar, Colombia ...B1
Vallegrande, Bolivia ...C4
Valparaíso, Chile ...B6
Van Blommenstein (lake), Suriname ...D2
Vaupés (riv.), Colombia ...B2
Venezuela ...C2
Venezuela (gulf), Venezuela ...B1
Viacha, Bolivia ...C4
Viedma, Argentina ...C7
Viedma (lake), Argentina ...B7
Villa María, Argentina ...C6
Villa Montes, Bolivia ...C5
Villarrica, Paraguay ...D5
Villavicencio, Colombia ...B2
Villazón, Bolivia ...C5
Viña del Mar, Chile ...B6
Vitória, Brazil ...F5
Vitória da Conquista, Brazil ...E4
Vitória de Sto. Antao, Brazil ...F3
Volta Redonda, Brazil ...E5
Xingu (riv.), Brazil ...D3
Yacimientos de Río Turbio, Argentina ...B8
Yacuíba, Bolivia ...C5
Zanderij, Suriname ...D2
Zapala, Argentina ...B6

Average January Temperature

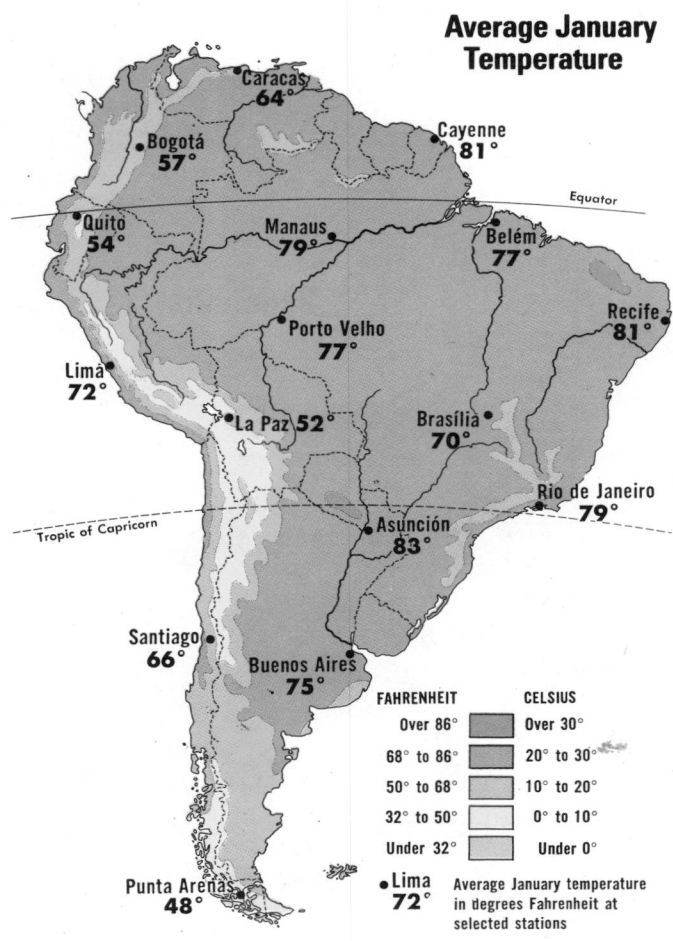

Caracas **64°**
Bogotá **57°**
Cayenne **81°**
Equator
Quito **54°**
Manaus **79°**
Belém **77°**
Recife **81°**
Porto Velho **77°**
Lima **72°**
La Paz **52°**
Brasília **70°**
Rio de Janeiro **79°**
Tropic of Capricorn
Asunción **83°**
Santiago **66°**
Buenos Aires **75°**
Punta Arenas **48°**

FAHRENHEIT	CELSIUS
Over 86°	Over 30°
68° to 86°	20° to 30°
50° to 68°	10° to 20°
32° to 50°	0° to 10°
Under 32°	Under 0°

•Lima **72°** Average January temperature in degrees Fahrenheit at selected stations

Average July Temperature

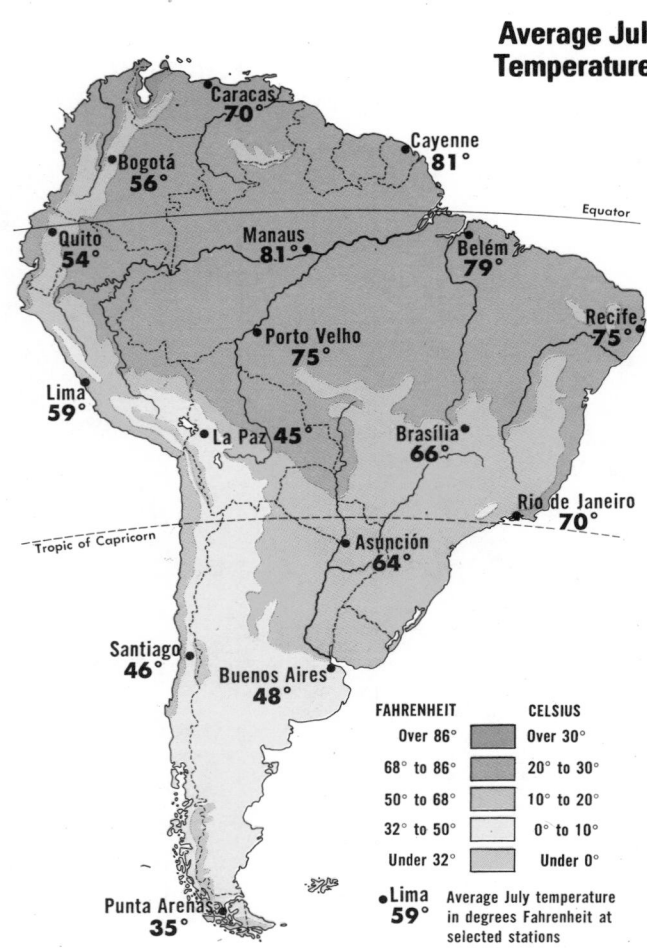

Caracas **70°**
Bogotá **56°**
Cayenne **81°**
Equator
Quito **54°**
Manaus **81°**
Belém **79°**
Recife **75°**
Porto Velho **75°**
Lima **59°**
La Paz **45°**
Brasília **66°**
Rio de Janeiro **70°**
Tropic of Capricorn
Asunción **64°**
Santiago **46°**
Buenos Aires **48°**
Punta Arenas **35°**

FAHRENHEIT	CELSIUS
Over 86°	Over 30°
68° to 86°	20° to 30°
50° to 68°	10° to 20°
32° to 50°	0° to 10°
Under 32°	Under 0°

•Lima **59°** Average July temperature in degrees Fahrenheit at selected stations

Rainfall

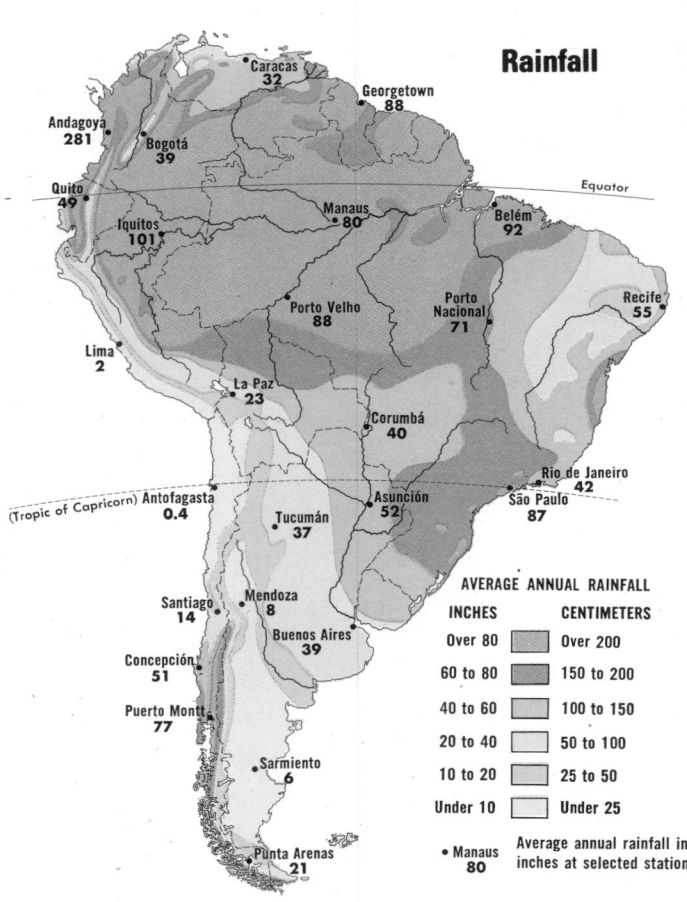

Caracas **32**
Georgetown **88**
Andagoyá **281**
Bogotá **39**
Quito **49**
Iquitos **101**
Manaus **80**
Belém **92**
Equator
Porto Velho **88**
Porto Nacional **71**
Recife **55**
Lima **2**
La Paz **23**
Corumbá **40**
Rio de Janeiro **42**
(Tropic of Capricorn) Antofagasta **0.4**
Tucumán **37**
Asunción **52**
São Paulo **87**
Santiago **14**
Mendoza **8**
Buenos Aires **39**
Concepción **51**
Puerto Montt **77**
Sarmiento **6**
Punta Arenas **21**

AVERAGE ANNUAL RAINFALL

INCHES	CENTIMETERS
Over 80	Over 200
60 to 80	150 to 200
40 to 60	100 to 150
20 to 40	50 to 100
10 to 20	25 to 50
Under 10	Under 25

•Manaus **80** Average annual rainfall in inches at selected stations

Vegetation/Relief

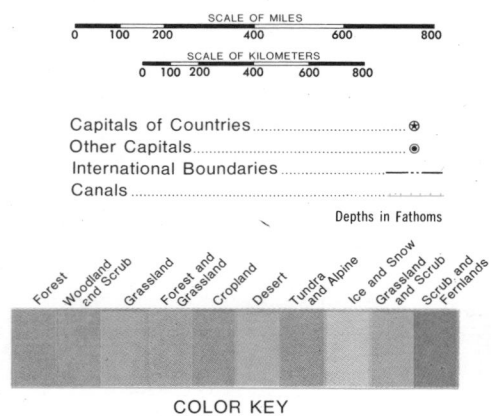

SCALE OF MILES
0 100 200 400 600 800

SCALE OF KILOMETERS
0 100 200 400 600 800

Capitals of Countries .... ⊛
Other Capitals .... ⊚
International Boundaries .... —··—
Canals ....

Depths in Fathoms

Forest | Woodland and Scrub | Grassland | Forest and Grassland | Cropland | Desert | Tundra and Alpine | Ice and Snow | Grassland and Scrub | Scrub and Fernlands

COLOR KEY

30° Longitude West of Greenwich 20°

STATES

Amazonas (terr.) 21,696 ....E5
Anzoátegui 506,297 ....F3
Apure 164,705 ....D4
Aragua 543,170 ....E3
Barinas 231,046 ....D4
Bolívar 391,665 ....F7
Carabobo 659,339 ....D2
Cojedes 94,351 ....D3
Delta Amacuro (terr.) 48,139 ....H3
Dependencias Federales (terr.) 463 ....E2
Distrito Federal 1,860,637 ....E2
Falcón 407,957 ....D2
Guárico 318,905 ....E3
Lara 671,410 ....C3
Mérida 347,095 ....C3
Miranda 856,272 ....E2
Monagas 298,239 ....G3
Nueva Esparta 118,830 ....G2
Portuguesa 297,047 ....D3
Sucre 469,004 ....G2
Táchira 511,346 ....C4
Trujillo 381,334 ....C3
Yaracuy 223,545 ....D2
Zulia 1,299,030 ....B2

CITIES and TOWNS

Acarigua 56,743 ....D3
Achaguas 4,633 ....D4
Adícora 707 ....D2
Aguada Grande 2,901 ....D2
Agua Fría D2
Agua Linda E5
Aguasay 1,752 ....G3
Altagracia 11,116 ....C2
Altagracia de Orituco 18,717 ....E3
Amuay C2
Anaco 29,487 ....G3
Aparurén G5
Apurito 740 ....D4
Arabopó H5
Aragua de Barcelona 9,107 ....F3
Aragua de Maturín 4,051 ....G3
Aragua 22,466 ....D2
Aricagua 231 ....C3
Arichuna 1,204 ....E3
Aripao 296 ....F4
Arismendi 1,257 ....E3
Aroa 5,418 ....D2
Atapirire 337 ....F3
Bachaquero C2
Baragua 659 ....D2
Barbacoas 2,513 ....E3
Barcelona 78,201 ....F2
Barinas 56,329 ....D3
Barinitas 9,644 ....D3
Barquisimeto 330,815 ....D2
Barrancas, Barinas 4,489 ....D3
Barrancas, Monagas 5,738 ....G3
Betijoque 5,851 ....C3
Biruaca 2,266 ....E4
Biscucuy 6,114 ....D3
Bobare 1,204 ....D2
Bobures 2,468 ....C3
Boca de Aroa 2,756 ....D2
Boca del Mangle E2
Boca del Pao 403 ....F3
Bocono 15,915 ....C3
Borbón C2
Borojó 423 ....C2
Bruzual 941 ....D4
Buena Vista, Anzoátegui G3
Buena Vista, Apure D4
Buena Vista, Falcón 944 ....D2
Cabimas 118,037 ....C2
Cabruta 1,927 ....E4
Cabudare 14,593 ....D2
Cabure 1,673 ....D2
Cachipo G3
Cacuri F5
Cagua 29,601 ....E2
Caicara 6,092 ....E3
Caicara de Orinoco 6,867 ....E4
Calabozo 37,282 ....E3
Calderas 1,195 ....D3
Camaguán 4,143 ....E3
Camatagua 3,335 ....E2
Campo Claro 1,832 ....F4
Candelaria E3
Cantaura 15,839 ....F3
Capatárida 1,375 ....D2
Capibara E6
Carabobo, Bolívar H4

Carabobo, Carabobo D3
Caracas 3,966 ....E2
Caracas (cap.) 1,035,499 ....E2
Caracas* 2,183,935 ....E2
Carapa 119 ....E3
Cariaco 6,549 ....F2
Caribén D3
Caripe 4,729 ....G2
Caripito 19,053 ....G2
Carirubana 15,701 ....C2
Carmelo 2,556 ....C3
Carora 36,115 ....C3
Carrasquero 2,193 ....C2
Carúpano 50,935 ....G2
Casanay 4,985 ....G2
Casigua, Falcón 460 ....C2
Casigua, Zulia 3,665 ....B3
Caucagua 6,218 ....E2
Cazorla 700 ....E3
Chaguaramas 2,748 ....E3
Chichiriviche 3,236 ....D2
Chivacoa 19,210 ....D2
Choroní 534 ....D2
Churuguara 6,636 ....D2
Ciudad Bolívar 103,728 ....F3
Ciudad Bolivia 4,864 ....C3
Ciudad de Nutrias 769 ....D3
Ciudad Guayana 143,540 ....G3
Ciudad Ojeda 83,083 ....C2
Ciudad Piar 3,965 ....G4
Clarines 2,099 ....F3
Cojoro C2

Colón E6
Comunidad E6
Coporito H3
Coro 68,701 ....D2
Corozo Pando E3
Cúa 9,953 ....E2
Cubiro 1,988 ....D3
Cuchivero F3
Cumaná 119,751 ....F2
Cumanacoa 9,179 ....G2
Cunaviche 795 ....E4
Curiapo H3
Dabajuro 4,516 ....C2
Delicias 1,616 ....B4
Democracia E6
Dolores 1,454 ....D3
Duaca 7,519 ....D2
Ejido 11,170 ....C3
El Almacén G4
El Amparo de Apure 2,015 ....C4
El Baúl 1,715 ....D3
El Callao 4,270 ....G4
El Calvario 384 ....E3
El Chaparro 3,768 ....F3
El Cristo G4
El Dorado 1,888 ....H4
El Empedrado 1,788 ....C3
El Guapo 1,231 ....F2
El Manteco 1,962 ....G4
El Miamo 335 ....H4
Elorza 3,184 ....D4
El Oso H5

El Palmar 2,758 ....G4
El Pao, Anzoátegui 761 ....F3
El Pao, Bolívar 1,259 ....G3
El Pao, Cojedes 1,715 ....D3
El Perú H4
El Pilar 3,278 ....G2
El Rastro 903 ....E3
El Samán de Apure 1,399 ....D4
El Socorro E3
El Sombrero 8,373 ....E3
El Tigre 49,801 ....F3
El Tocuyo 19,351 ....C3
El Toro H3
El Vigía 20,970 ....C3
El Vínculo D1
El Yagual 699 ....D4
Encontrados 5,607 ....B3
Espino 559 ....F3
Garcitas D2
Guacara 35,111 ....D2
Guachara 577 ....D4
Guadarrama 334 ....E3
Guaina G5
Guana G5
Guanare 34,148 ....D3
Guanarito 3,150 ....D3
Guardatinajas 1,206 ....E3
Guarero B2

Guárico 3,259 ....D3
Guariquén 619 ....G2
Guasdualito 7,793 ....C4
Guasimal 582 ....D3
Guasipati 4,807 ....G4
Guayabal, Amazonas E5
Guayabal, Guárico 1,403 ....E3
Güiria 13,905 ....G2
Guri G4
Guzmán Blanco C3
Higuerote 5,008 ....E2
Icabarú G5
Independencia 4,897 ....C3
Irapa 4,470 ....G2
Juangriego 6,062 ....G2
Judibana D1
Jusepín G2
Kavanayén G4
La Aduana E3
La Asunción 6,381 ....G2
La Canoa F3
La Ceiba, Apure D4
La Ceiba, Trujillo 212 ....C3
La Concepción 13,885 ....B2
La Esmeralda F5
La Esperanza F2
La Fría 8,134 ....B3
La Grita 9,954 ....C3
La Guaira 20,344 ....E2
Lagunetas C3
Lagunillas C3

Venezuela

MERCATOR PROJECTION

SCALE OF MILES
0 25 50 75 100 125

KILOMETERS
0 25 50 75 100 200

Capitals of Countries .... ☆
State Capitals .... ◉
International Boundaries ....
State Boundaries ....
Canals ....

Scale 1:6,120,000

La Horqueta....G3
La Inglesa....G3
La Leona....G3
La Luz 672....D3
La Margarita....H3
La Paragua 1,676....G4
Las Bonitas 343....F4
Las Lajitas....F4
Las Mercedes 6,739....E3
Las Piedras, Falcón....C2
Las Piedras, Zulia 4,583....B2
Las Trincheras....F4
Las Vegas 3,212....D3
La Tigra....H4
La Trinidad 129....D3
La Trinidad de Arauca....D4
La Trinidad de Orichuna 665....D4
La Unión 713....E3
La Urbana 661....E4
La Vela de Coro 7,172....D2
La Victoria, Apure 689....D4
La Victoria, Apure....C4
La Victoria, Aragua 40,731....E2
Libertad, Barinas 2,072....D3
Libertad, Cojedes 1,919....D3
Los Castillos....G3
Los Taques 1,160....C2
Los Teques 63,106....E2
Macareo Santo Niño....H3
Machiques 18,898....B3
Macuro 1,122....H2
Macuto 11,704....E2

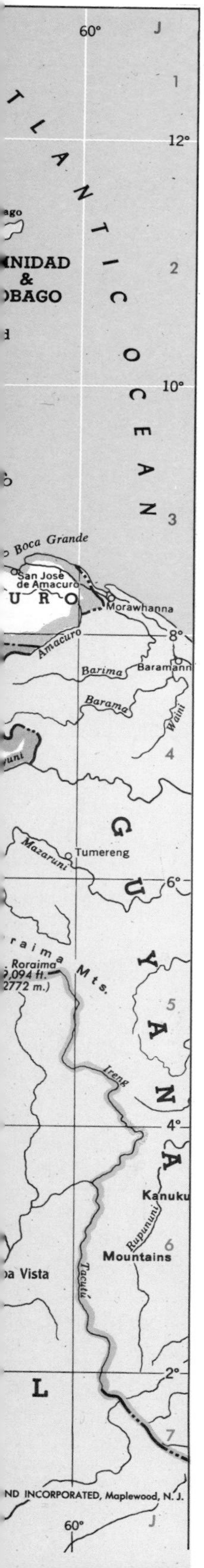
60°
J
1
12°
2
10°
Boca Grande
San José de Amacuro
Morawhanna
8°
Amacuro
Barima
Waini
4
Tumereng
6°
5
Ireng
4°
Rupununi
6
Mountains
Takutu
2°
L
7
ND INCORPORATED, Maplewood, N.J.
60°
J

Maiquetía 59,238....E2
Mantecal, Apure 1,136....D4
Mantecal, Bolívar....F4
Maparari 1,376....D2
Mapire 1,195....F4
Maporal 249....C4
Maracaibo 651,574....C2
Maracay 255,134....E2
Marigüitar 5,645....G2
Maripa 913....F4
Maroa 408....E6
Matu....F4
Maturín 98,188....G3
Mene de Mauroa 4,336....C2
Mene Grande 11,498....C3
Mérida 74,214....C3
Mesa Bolívar 956....C3
Mirimire 3,424....D2
Moitaco 458....F4
Morganito....E5
Morón 19,451....D2
Mucuchachí 472....C3
Mucuchíes 1,625....C3
Naricual 1,047....F2
Nirgua 11,918....D2
Nuevo Mamo....G3
Obispos 1,140....D3
Ocumare de la Costa 2,840....E2
Ocumare del Tuy 24,229....E2
Onoto 1,991....F3
Ortiz 1,793....E3
Ospino 3,544....D3
Palmarejo....C2
Palmarito, Apure 926....D4
Palmarito, Guárico....F3
Palmarito, Mérida 988....C3
Papelón 774....D3
Paraguaipoa 3,850....C2
Paraíso de
Chabasquén 2,094....D3
Pariaguán 8,173....F3
Parmana....F4
Pedernales....G3
Pedregal 1,317....C2
Peraitepuí....H5
Piacoa....H3
Pimichín....E6
Píritu, Anzoátegui 2,479....F2
Píritu, Falcón 1,186....D2
Píritu, Portuguesa 8,128....D3
Platanal....F6
Porlamar 31,985....G2
Pozuelos 45,391....F2
Pregonero 3,598....C3
Pueblo Hondo....B3
Pueblo Nuevo 3,426....D1
Puerto Ayacucho 10,417....E5
Puerto Cabello 72,103....E2
Puerto Cumarebo 10,064....D2
Puerto de Nutrias 675....D3
Puerto Hierro....H2
Puerto La Cruz 63,276....F2
Puerto Miranda....E4
Puerto Páez 954....E4
Puerto Píritu 3,495....F2
Punta Cardón 18,182....C2
Punta de Mata 7,777....G3
Punta de Piedras 2,826....F2
Punto Fijo 5,548....D2
Puruey....F4
Puruname....E6
Quibor 12,216....D3
Quiriquire 7,304....G3
Quisiro 1,383....C2
Río Caribe 8,963....G2
Río Chico 4,491....F2
Río Claro 2,460....D3
Río Tocuyo 916....C2
Rosario....B2
Rubio 19,156....B4
Sabaneta, Barinas 4,680....D3
Sabaneta, Falcón 650....D2
Samariapo....E5
San Antonio, Amazonas....E6
San Antonio, Monagas 4,235....G2
San Antonio, Zulia....C3
San Antonio de Caparo 289....C4
San Antonio del
Táchira 20,342....B4
San Antonio de Tabasca....G3
Sanare 6,717....D3
San Carlos, Cojedes 21,029....D3
San Carlos, Zulia 749....C2

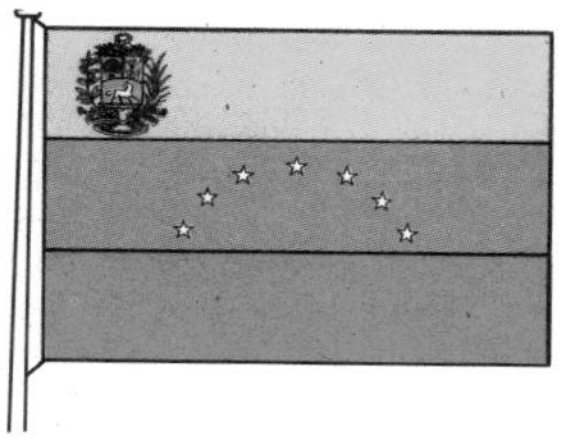

San Carlos del Zulia 26,762....C3
San Carlos de Río Negro 515....E7
San Casimiro 4,843....E3
San Cristóbal 151,717....B4
San Diego de Cabrutica 432....F3
San Felipe, Yaracuy 43,801....D2
San Felipe, Zulia....B3
San Félix 379....C2
San Fernando de Apure 38,960....E4
San Fernando de Atabapo 1,537....E5
San Francisco, Lara 861....C2
San Ignacio....B2
San José, Amazonas....E5
San José, Zulia 4,498....B3
San José de Amacuro....H3
San José de Areocuar 985....G2
San José de Guanipa 22,530....G3
San José de la Costa....D2
San José de Río
Chico 3,600....F2
San José de Tiznados 666....E3
San Juan de Colón....B3
San Juan de las Galdonas 1,196....G2
San Juan de los Cayos 1,692....D2
San Juan de los Morros 38,265....E3
San Juan de Manapiare....E5
San Juan de Payara 1,018....E4
San Lorenzo, Falcón 716....D2
San Lorenzo, Zulia....C3
San Luis 1,405....D2
San Mateo 2,424....F3
San Mauricio....E3
San Pedro de las Bocas....G4
San Rafael 10,910....C2
San Rafael de Atamaica 635....E4
San Rafael de Orituco 1,378....E3
San Sebastián 5,582....E2
San Simón del Cocuy....E3
Santa Ana, Anzoátegui 3,558....F3
Santa Ana, Táchira 5,116....B4
Santa Bárbara, Amazonas....E6
Santa Bárbara, Barinas 6,155....C4
Santa Bárbara, Monagas 2,034....G3
Santa Bárbara, Zulia....C3
Santa Catalina, Barinas 1,077....D4
Santa Catalina, Delta
Amacuro....H3
Santa Cruz....C3
Santa Cruz de Bucaral 2,904....D2
Santa Cruz del Zulia 4,221....B3
Santa Cruz de Mara 5,773....C2
Santa Cruz de Orinoco 513....F3
Santa Elena 608....H5
Santa Inés,
Anzoátegui 1,049....F3
Santa Inés, Barinas 391....C3
Santa Isabel....F7
Santa Lucía 619....D3
Santa María, Bolívar....G3
Santa María de Erebató....F5
Santa María de Ipire 3,307....F3
Santa María del Orinoco....E4
Santa Rita, Guárico....E3
Santa Rita, Zulia 15,668....C2
Santa Rosa, Anzoátegui 954....F3

AREA 352,143 sq. mi. (912,050 sq. km.)
POPULATION 14,313,000
CAPITAL Caracas
LARGEST CITY Caracas
HIGHEST POINT Pico Bolívar 16,427 ft. (5,007 m.)
MONETARY UNIT Bolívar
MAJOR LANGUAGE Spanish
MAJOR RELIGION Roman Catholicism

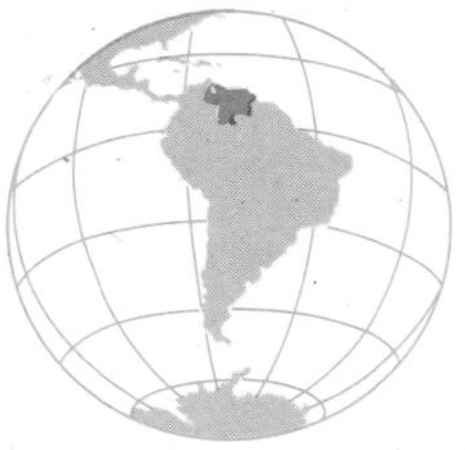

Topography

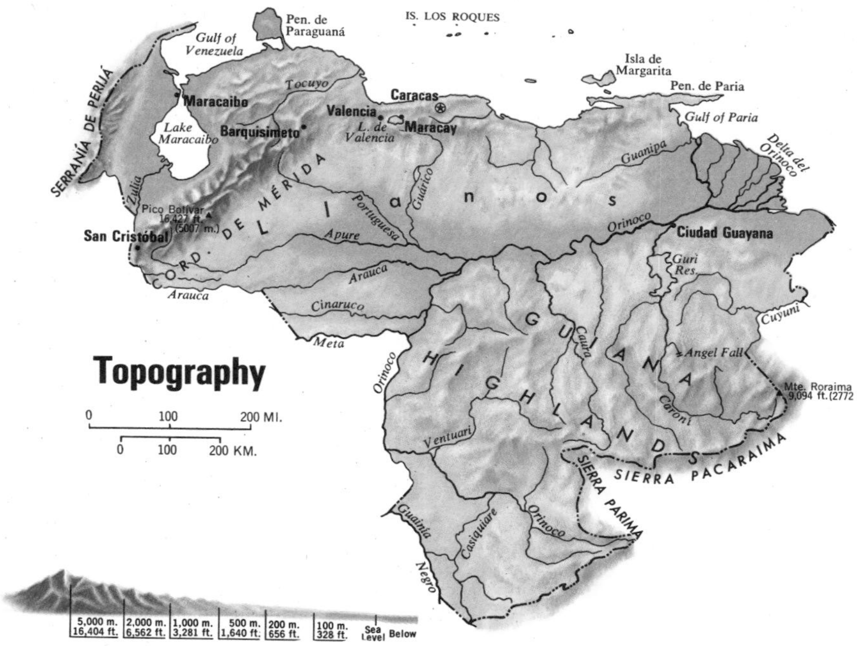
Pen. de Paraguaná
IS. LOS ROQUES
Gulf of Venezuela
Isla de Margarita
Pen. de Paria
Gulf of Paria
Maracaibo
Tocuyo
Caracas
Valencia
Maracay
L. de Valencia
Lake Maracaibo
Barquisimeto
SERRANÍA DE PERIJÁ
Delta del Orinoco
Guanipa
Zulia
Pico Bolívar 16,427 ft. (5,007 m.)
CORD. DE MÉRIDA
L l a n o s
Portuguesa
Guárico
Orinoco
Ciudad Guayana
San Cristóbal
Apure
Arauca
Guri Res.
Arauca
Cinaruco
Cuyuni
Meta
Angel Fall
GUIANA HIGHLANDS
Caura
Caroní
Mte. Roraima 9,094 ft. (2772 m.)
Orinoco
Ventuari
SIERRA PACARAIMA
SIERRA PARIMA
Casiquiare
Guainía
Negro
0 100 200 MI.
0 100 200 KM.
5,000 m. 16,404 ft.
2,000 m. 6,562 ft.
1,000 m. 3,281 ft.
500 m. 1,640 ft.
200 m. 656 ft.
100 m. 328 ft.
Sea Level
Below

Santa Rosa, Apure....D4
Santa Rosa, Barinas 1,514....D3
Santa Rosa de Amanadona....E7
Santa Rosalía 513....F4
Santa Teresa 10,220....E2
San Timoteo 3,635....C3
San Tomé....F3
San Vicente, Amazonas....E5
San Vicente, Apure 365....D4
Sarare 4,236....D3
Seboruco 2,616....B3
Simaraña....G5
Sinamaica....B2
Siquisique 3,821....D2
Solano....E6

Soledad 7,108....G3
Sucre 608....D3
Suripa....D4
Tamatama....F6
Táriba 15,683....B4
Temblador 5,380....G3
Tia Juana....C2
Timotes 3,229....C3
Tinaco 7,263....D3
Tinaquillo 12,015....D3
Tocópero 1,033....D2
Tocuyo de la Costa 4,023....D2
Torunos 739....C3
Tovar 12,814....C3
Trujillo 25,921....C3

Tucacas 4,780....D2
Tucupido 9,522....F3
Tucupita 21,417....H3
Tumeremo 5,036....H4
Tupí 88....D2
Turén....D3
Turiamo....E2
Turmero 43,832....E2
Upata 22,793....G3
Urachiche 4,759....D2
Uracoa 1,165....G3
Urica 1,881....F3
Urimán....G5
Urumaco 829....C2
Uruyén....G5
Uverito 468....F3
Valencia 367,171....E2
Valera 76,740....C3
Valle de Guanape 3,468....F3
Valle de la Pascua 36,809....F3
Vara de María....C4
Villa Bruzual 14,003....D3
Villa de Cura 27,832....E2
Villa Frontado 1,600....G2
Yaguaraparo 3,931....G2
Yaritagua 21,363....D2
Yavita....E6
Yerichaña....F5
Yoco 2,196....G2
Zanja de Lira....E3
Zaraza 15,480....F3
Zuata 914....F3

OTHER FEATURES

Amacuro (riv.)....H4
Angel (fall)....G5
Aponguao (riv.)....H5
Apure (riv.)....E4
Arauca (riv.)....E4
Arichuna (riv.)....D4
Aro (riv.)....F4
Atabapo (riv.)....E6
Auyantepui (mt.)....G5
Baria (riv.)....E7
Bolívar, Cerro (mt.)....G4
Bolívar, Pico (peak)....C3
Canagua (riv.)....C3
Caño Capure (riv.)....H3
Caño Macareo (riv.)....H3
Caño Mánamo (riv.)....G3
Capanaparo (riv.)....E4
Caparo (riv.)....C4
Caroní (riv.)....G4
Carrao (riv.)....G5
Caruai (riv.)....H5
Casiquiare, Brazo (riv.)....E6
Catatumbo (riv.)....B3
Caura (riv.)....F5
Chicanán (riv.)....H4
Chimantá-tepuí (mt.)....G5
Chivapure (riv.)....E4
Cinaruco (riv.)....D4
Coche (isl.)....F2
Codera (cape)....F2
Cojedes (riv.)....D3
Cuao (riv.)....E5
Cubagua (isl.)....F2
Cuchivero (riv.)....F4
Cuquenán (riv.)....H5
Curutú (riv.)....G5
Cuyuni (riv.)....H4
Delgado Chalbaud, Cerro (mt.)....G6
Dragons Mouth (str.)....H2
Duida, Cerro (mt.)....F6
Erebato (riv.)....F5
Gran Sabana, La (plain)....G5
Guainía (riv.)....E6
Guampí, Sierra de (mts.)....F4

Guanare (riv.)....D3
Guanare Viejo (riv.)....D3
Guanipa (riv.)....G3
Guárico (res.)....E3
Guárico (riv.)....E3
Guayapo, Serranía (mts.)....E5
Güere (riv.)....F3
Guri (res.)....G4
Icabarú (riv.)....G5
Imataca, Serranía (mts.)....H4
Imeri, Sierra (mts.)....F7
La Blanquilla (isl.)....F2
La Gran Sabana (plain)....G5
La Orchila (isl.)....F2
Las Aves (isls.)....E2
La Tortuga (isl.)....F2
Los Hermanos (isls.)....F2
Los Monjes (isls.)....C1
Los Roques (isls.)....E2
Los Testigos (isls.)....G2
Macanao (pen.)....F2
Maigualida, Sierra (range)....F4
Manapire (riv.)....E3
Maracaibo (lake)....C3
Margarita (isl.)....F2
Mavaca (riv.)....F6
Médanos (isth.)....D2
Merevari (riv.)....F5
Mérida, Cordillera de
(range)....C3
Meta (riv.)....E4
Morichal Largo (riv.)....G3
Neblina (Phelps) (peak)....E7
Negro (riv.)....E7
Nuria, Sierra de (mts.)....H4
Ocamo (riv.)....F6
Orinoco (delta)....H3
Orinoco (riv.)....G3
Orituco (riv.)....E3
Pacaraima, Sierra (mts.)....G5
Pao (riv.)....D3
Pao (riv.)....F3
Paragua (riv.)....G4
Paraguaná (pen.)....C1
Paria (gulf)....H2
Paria (pen.)....G2
Parima, Sierra (mts.)....F6
Perijá, Sierra de (mts.)....B2
Phelps (peak)....E7
Portuguesa (riv.)....D3
Roraima (mt.)....H5
Salto Angel (fall)....G5
Sarare (riv.)....C4
Serpents Mouth (passage)....H3
Siapa (riv.)....E7
Sipapo (riv.)....E5
Suapure (riv.)....E4
Suripá (riv.)....C4
Tapirapecó, Sierra (mts.)....F7
Tigre (riv.)....G3
Tocuco (riv.)....B3
Tocuyo (riv.)....D2
Tramán-tepuí (mt.)....G5
Triste (gulf)....D2
Turagua, Serranía (mts.)....F4
Tuy (riv.)....E2
Unare (riv.)....F3
Valencia (lake)....E2
Venamo, Cerro (mt.)....H4
Venamo (riv.)....H4
Venezuela (gulf)....C2
Ventuari (riv.)....E5
Votamo (riv.)....F6
Yatua (riv.)....E7
Yuruari (riv.)....H4
Zuata (riv.)....F3
Zulia (riv.)....B3

*City and suburbs

Agriculture, Industry and Resources

MAJOR MINERAL OCCURRENCES

- Al Bauxite
- Au Gold
- C Coal
- D Diamonds
- Fe Iron Ore
- G Natural Gas
- Mn Manganese
- Na Salt
- O Petroleum

- Water Power
- Major Industrial Areas

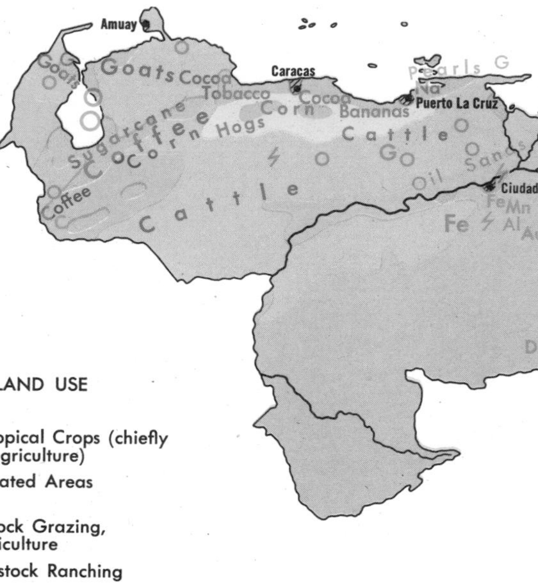
Amuay
Caracas
Puerto La Cruz
Ciudad Guayana
Goats
Cocoa
Tobacco
Cocoa
Pearls
Corn
Bananas
Sugarcane
Coffee
Corn
Hogs
Cattle
Coffee
Oil Sands
Cattle
Fe Al Au
D

DOMINANT LAND USE

- Diversified Tropical Crops (chiefly plantation agriculture)
- Upland Cultivated Areas
- Upland Livestock Grazing, Limited Agriculture
- Extensive Livestock Ranching
- Forests

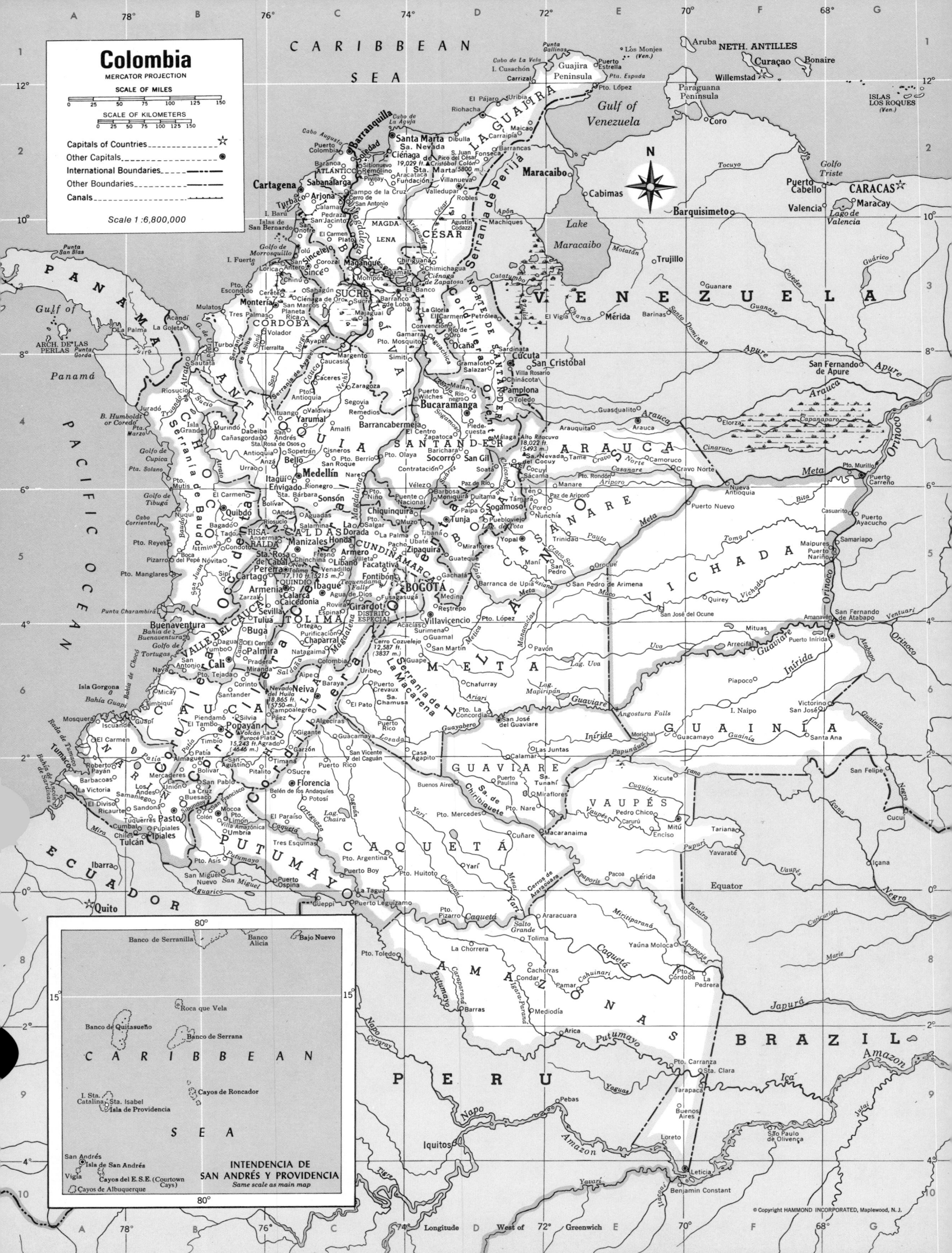

Colombia

MERCATOR PROJECTION

SCALE OF MILES

0 25 50 75 100 125 150

SCALE OF KILOMETERS

0 25 50 75 100 125 150

Capitals of Countries _____ ☆
Other Capitals _____ ◉
International Boundaries _____
Other Boundaries _____
Canals _____

Scale 1:6,800,000

INTENDENCIA DE
SAN ANDRÉS Y PROVIDENCIA
Same scale as main map

© Copyright HAMMOND INCORPORATED, Maplewood, N.J.

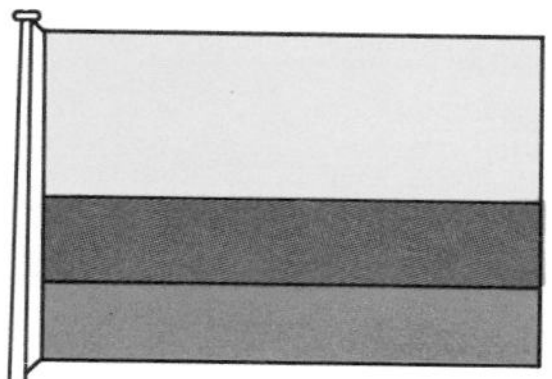

AREA 439,513 sq. mi. (1,138,339 sq. km.)
POPULATION 27,520,000
CAPITAL Bogotá
LARGEST CITY Bogotá
HIGHEST POINT Pico Cristóbal Colón 19,029 ft. (5,800 m.)
MONETARY UNIT Colombian peso
MAJOR LANGUAGE Spanish
MAJOR RELIGION Roman Catholicism

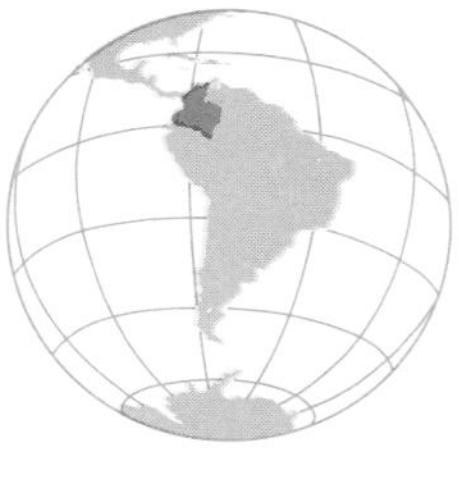

INTERNAL DIVISIONS

Amazonas (comm.) 6,825	D8
Antioquia (dept.) 2,976,153	B4
Arauca (inten.) 19,884	E4
Atlántico (dept.) 958,560	C2
Bolívar (dept.) 802,407	C3
Boyacá (dept.) 1,084,766	D5
Caldas (dept.) 700,954	C5
Caquetá (inten.) 57,103	C7
Casanare (inten.)	B3
Cauca (dept.) 603,894	B6
César (dept.) 339,843	D3
Chocó (dept.) 201,915	B4
Córdoba (dept.) 645,478	C3
Cundinamarca (dept.) 1,106,626	C5
Distrito Especial 2,855,065	C5
Guainía (comm.) 1,792	F6
Guajira, La (dept.) 180,520	D2
Guaviare (comm.)	D7
Huila (dept.) 469,834	C6
La Guajira (dept.) 180,520	D2
Magdalena (dept.) 536,122	C3
Meta (dept.) 245,176	D6
Nariño (dept.) 807,112	B7
Norte de Santander (dept.) 693,298	D3
Putumayo (inten.) 22,916	C7
Quindío (dept.) 321,677	C5
Risaralda (dept.) 452,626	B5
San Andrés y Providencia (inten.) 22,719	B10
Santander (dept.) 1,130,977	D4
Sucre (dept.) 354,412	C3
Tolima (dept.) 903,520	C5
Valle del Cauca (dept.) 2,204,722	B6
Vaupés (comm.) 6,923	E7
Vichada (comm.) 2,172	F5

CITIES and TOWNS

Acacías 9,238	D6
Acandí 2,358	B3
Agrado 2,771	C6
Aguachica 16,771	D3
Aguadas 9,995	C5
Agua de Dios 9,689	C5
Agustín Codazzi 21,932	D3
Aipe 3,794	C6
Algeciras 5,022	C6
Almaguer 1,518	B7
Amalfi 6,494	C4
Andes 14,957	C5
Anserma 15,559	B5
Antioquia 6,841	B4
Anzá 647	C4
Aracataca 7,511	D2
Arauca 7,613	E4
Arauquita 1,096	E4
Arjona 20,571	C2
Armenia 135,615	B5
Armero 19,567	C5
Ayapel 7,475	C3
Bagadó 1,575	B5
Baranoa 18,397	C2
Baraya 2,581	C6
Barbacoas 4,653	A7
Barbosa 7,960	D5
Barichara 2,548	D4
Barrancabermeja 87,191	C4
Barrancas 2,979	D2
Barranco de Loba 2,215	C3
Barranquilla 661,009	C2
Belén de los Andaquíes 2,190	C7
Bello 115,119	C4
Bogotá (cap.) 2,696,270	D5
Bogotá* 2,855,065	D5
Bolívar, Antioquia 13,259	C5
Bucaramanga 291,661	D4
Buenaventura 115,770	B6
Buesaco 2,763	B7
Buga 71,016	B6
Cáceres 7,154	C4
Caicedonia 23,567	C5
Calamar, Bolívar 5,867	C2
Calarcá 29,349	C5
Cali 898,253	B6
Campoalegre 11,799	C6
Campo de la Cruz 13,137	C2
Cañasgordas 3,900	B4
Cartagena 292,512	C2
Cartago 69,154	B5
Caucasia 19,348	C4
Cereté 18,788	C3
Cerro de San Antonio 3,394	C2
Chaparral 14,546	C6
Chimichagua 6,382	D3
Chinácota 4,478	D4
Chinchiná 24,891	C5
Chinú 10,023	C3
Chiquinquirá 21,727	C5
Chiriguaná 6,611	D3
Ciénaga 42,546	C2
Ciénaga de Oro 10,607	C3
Cisneros 7,226	C4
Colombia 2,903	C6
Colón 1,306	B7
Condoto 4,798	B5
Contratación 3,057	D4
Convención 7,545	D3
Corinto 6,933	B6
Corozal 17,419	C3
Cravo Norte 771	F4
Cúcuta 219,772	D4
Cumbal 2,891	B7
Dabeiba 7,600	B4
Dagua 5,392	B6
Duitama 36,551	D5
El Banco 20,756	D3
El Carmen, Chocó 1,879	B5
El Carmen, Norte de Santander 2,362	D3
El Carmen de Bolívar 23,392	C3
El Cerrito 17,357	B6
El Cocuy 2,740	D4
El Tambo 2,179	B6
Envigado 63,584	C4
Espinal 32,475	C5
Facatativá 27,892	C5
Florencia 31,817	C7
Fonseca 9,988	D2
Fresno 8,141	C5
Fundación 17,497	C2
Fusagasugá 25,456	C5
Gachalá 1,364	D5
Gamarra 5,071	D3
Garzón 13,783	C6
Gigante 4,880	C6
Girardot 59,165	C5
Gramalote 2,880	D4
Guamal, Magdalena 4,986	C3
Guamal, Meta 2,854	D6
Guapí 5,005	B6
Guateque 6,032	D5
Honda 21,506	C5
Ibagué 176,223	C5
Inírida 1,792	F6
Ipiales 30,871	B7
Iscuandé 561	A6
Istmina 5,575	B5
Itagüí 96,972	C4
Ituango 5,561	C4
Juradó 935	B4
La Cruz 4,353	B7
La Dorada 30,962	C5
La Gloria 2,632	D3
La Palma 5,430	C5
La Plata 8,047	C6
La Unión 5,392	B7
Leticia 6,285	F10
Líbano 19,132	C5
Lorica 18,251	C3
Los Andes 1,414	B7
Magangué 34,396	C3
Maicao 21,645	D2
Majagual 2,329	C3
Málaga 10,645	D4
Maní 951	D5
Manizales 199,904	C5
Matanza 1,211	D4
Medellín 1,070,924	C4
Medina 1,436	D5
Mercaderes 3,877	B7
Miraflores, Boyacá 3,584	D5
Miraflores, Vaupés 536	D7
Miranda 6,439	B6
Mitú 1,637	E7
Mocoa 6,221	B7
Mompós 14,076	C3
Moniquirá 5,711	D5
Montería 89,583	B3
Morichal	E6
Mosquera 594	A6
Murindó 485	B4
Muzo 1,823	D5
Natagaima 7,772	C6
Neiva 105,476	C6
Nóvita 802	B5
Nunchía 437	D5
Nuquí 1,115	B5
Ocaña 38,352	D3
Orocué 1,011	E5
Ortega 5,150	C6
Pacho 6,786	C5
Páez 2,098	C6
Paipa 4,260	D5
Palmira 140,481	B6
Pamplona 31,817	D4
Pasto 119,339	B7
Patía 5,306	B6
Paz de Ariporo 2,584	E5
Paz de Río 3,464	D4
Pedraza 1,872	C2
Pereira 174,128	C5
Piedecuesta 17,308	D4
Piendamó 5,046	B6
Pitalito 15,049	B7
Pivijay 10,172	C2
Planeta Rica 12,932	C3
Plato 18,589	C3
Popayán 77,669	B6
Pore 389	D5
Pradera 15,732	B6
Puente Nacional 4,317	D5
Puerto Asís 6,364	B7
Puerto Berrío 19,579	C4
Puerto Carreño 2,172	G4
Puerto Colombia 9,255	C2
Puerto Escondido 1,368	B3
Puerto Leguízamo 3,179	C8
Puerto López, Meta 4,948	D5
Puerto Murillo	G4
Puerto Mutis	B4
Puerto Nare	D7
Puerto Paulina	D7
Puerto Rico, Caquetá 4,853	C7
Puerto Rondón 1,010	E4
Puerto Salgar 6,396	C5
Puerto Tejada 18,315	B6
Puerto Wilches 5,282	D4
Pupiales 2,723	B7
Purificación 8,164	C6
Quibdó 28,040	B5
Remedios 4,681	C4
Remolino 3,408	C2
Restrepo 2,704	D5
Ricaurte 1,205	A7
Río de Oro 2,985	D3
Riohacha 19,604	D2
Rionegro, Antioquia 22,654	C4
Rionegro, Santander 3,491	D4
Riosucio, Caldas 11,619	C5
Riosucio, Chocó 2,184	B4
Roberto Payán 445	A7
Robles 5,422	D2
Rovira 5,105	C5
Sabanalarga 26,542	C2
Sácama 69	D4
Sahagún 18,717	C3
Salamina 12,136	C5
Salazar 2,791	D4
Samaniego 4,790	B7
San Agustín 4,532	B7
San Andrés, Antioquia 2,003	C4
San Andrés, San Andrés y Providencia 14,428	A9
San Antero 7,129	C3
Sandoná 7,222	B7
San Francisco 1,654	B7
San Gil 21,679	D4
San Jacinto 13,459	C3
San José del Guaviare 4,138	D6
San Juan del César 9,468	D2
San Marcos 10,415	C3
San Martín 8,281	D6
San Onofre 7,899	C3
San Pablo 3,662	B7
San Roque 4,972	C4
Santa Bárbara 11,848	C5
Santa Marta 102,484	D2
Santander 13,625	B6
Santa Rosa de Cabal 28,368	C5
Santa Rosa de Osos 8,593	C4
San Vicente del Caguán 3,182	C6
Sardinata 3,726	D3
Segovia 10,000	C4
Sevilla 31,143	C5
Sibundoy 2,853	B7
Silvia 3,045	B6
Simití 3,062	C4
Sincé 11,909	C3
Sincelejo 68,797	C3
Sipí 153	B5
Sitionuevo 5,919	C2
Soatá 4,294	D4
Socorro 15,596	D4
Sogamoso 48,891	D5
Soledad 64,469	C2
Sonsón 15,990	C5
Sopetrán 5,223	C4
Tadó 3,102	B5
Támara 947	D5
Tame 4,811	E4
Tibaná 1,100	D5
Tierralta 7,950	C3
Timaná 4,262	C7
Timbío 4,755	B6
Timbiquí 1,048	B6
Toledo 2,942	D4
Tolú 9,118	C3
Trinidad 729	E5
Tuluá 86,736	B5
Tumaco 38,742	A7
Tunja 51,620	D5
Túquerres 12,058	B7
Turbaco 19,360	C2
Turbo 16,070	B3
Ubaté 7,716	D5
Uribia 2,193	D2
Urrao 8,577	B4
Valdivia 4,318	C4
Valledupar 87,425	D2
Vélez 8,241	D4
Venadillo 8,383	C5
Villanueva 9,836	D2
Villa Rosario 8,668	D4
Villavicencio 82,869	D5
Villeta 6,507	C5
Yarumal 21,333	C4
Yopal 5,851	D5
Yumbo 28,011	B6
Zapatoca 6,258	D4
Zaragoza 9,660	C4
Zarzal 21,370	B5
Zipaquirá 25,413	D5

OTHER FEATURES

Abibe, Serranía de, (mts.)	B3
Aguarico, (riv.)	B7
Aguja, La, (cape)	C2
Albuquerque, (cays)	A10
Alicia, (bank)	B8
Alto Ritacuva, (mt.)	D4
Amazon, (riv.)	E9
Ancón de Sardinas, (bay)	A7
Angostura, (falls)	E6
Apaporis, (riv.)	F8
Araracuara, Cerros de, (mts.)	E7
Arauca, (riv.)	E4
Ariari, (riv.)	D6
Ariguaní, (riv.)	D3
Ariporo, (riv.)	E4
Atabapo, (riv.)	G6
Atrato, (riv.)	B4
Augusta, (cape)	C2
Ayapel, Serranía de, (mts.)	C4
Bajo Nuevo, (shoal)	C8
Barú, (isl.)	C2
Baudó, Serranía de, (mts.)	B5
Baudó, (riv.)	B5
Bita, (riv.)	F5
Buenaventura, (bay)	B6
Caguán, (riv.)	C7
Cahuinarí, (riv.)	E8
Caquetá, (riv.)	E8
Caraparaná, (riv.)	D8
Casanare, (riv.)	E4
Catatumbo, (riv.)	D3
Cauca, (riv.)	C4
Cazueleja, Cerro, (mt.)	C6
Central, Cordillera, (range)	C5
César, (riv.)	D2
Chaira, Laguna, (lake)	C7
Chamusa, Sierra, (mts.)	C6
Charambirá, (pt.)	B5
Chicamocha, (riv.)	D4
Chiribiquete, Sierra de, (mts.)	D7
Cinaruco, (riv.)	F4
Chocó, (bay)	B6
Cocuy, Sierra Nevada del, (mts.)	D4
Coredó (Humboldt), (bay)	B4
Corrientes, (cape)	B5
Courtown (Este Sudeste), (cays)	A10
Cravo Norte, (riv.)	E4
Cravo Sur, (riv.)	E5
Cristóbal Colón, Pico, (peak)	D2
Cuemaní, (riv.)	D7
Cupica, (gulf)	B4
Cuquiari, (riv.)	E7
Cusachón, (isl.)	D1
Cusiana, (riv.)	D5
Espada, (pt.)	E1
Este Sudeste, (cays)	A10
Fuerte, (isl.)	B3
Gallinas, (pt.)	E1
Gorgona, (isl.)	A6
Grande, (isl.)	B4
Grande, Salto, (falls)	D8
Guainía, (riv.)	F6
Guajira, (pen.)	E1
Guapi, (bay)	A6
Guaviare, (riv.)	F6
Guayabero, (riv.)	D6
Huila, Nevado del, (mt.)	C6
Humboldt, (bay)	B4
Igara-Paraná, (riv.)	D8
Inírida, (riv.)	F6
Isana, (riv.)	F7
La Aguja, (cape)	C2
La Macarena, Serranía de, (mts.)	D6
La Vela, (cape)	D1
Lebrija, (riv.)	D4
Llanos, (plains)	D5
Losada, (riv.)	C6
Macarena, Serranía de La, (mts.)	D6
Magdalena, (riv.)	C3
Manacacías, (riv.)	D6
Mapiripán, Laguna, (lake)	E6
Marzo, (pt.)	B4
Mesai, (riv.)	D7
Meta, (riv.)	E5
Metica, (riv.)	D6
Mira, (riv.)	A7
Miritiparaná, (riv.)	E8
Morrosquillo, (gulf)	C3
Muco, (riv.)	E5
Naipo, (isl.)	F6
Nechí, (riv.)	C4
Negro, (riv.)	G7
Occidental, Cordillera, (range)	B5
Oriental, Cordillera, (range)	D5
Orinoco, (riv.)	G5
Orteguaza, (riv.)	C7
Papunáua, (riv.)	E6
Papurí, (riv.)	F7
Patía, (riv.)	B6
Pauto, (riv.)	E5
Perijá, Serranía de, (mts.)	D2
Providencia, (isl.)	B9
Puracé, (vol.)	B6
Putumayo, (riv.)	E9
Quitasueño, (bank)	A8
Roca que Vela, (cay)	B8
Roncador, (cays)	B9
Saldaña, (riv.)	C6
Salto Grande, (falls)	D8
San Andrés, (isl.)	A10
San Bernardo, (isls.)	C3
San Jorge, (riv.)	C3
San Juan, (riv.)	B5
San Miguel, (riv.)	B7
Santa Catalina, (isl.)	A9
Santa Marta, Sierra Nevada de, (range)	D2
Serrana, (bank)	B9
Serranilla, (bank)	B8
Sinú, (riv.)	B3
Sogamoso, (riv.)	D4
Solano, (pt.)	B4
Suárez, (riv.)	D4
Sucio, (riv.)	B4
Taraíra, (riv.)	F8
Tequendama, (falls)	C5
Tibugá, (gulf)	B5
Tolima, Nevada del, (mt.)	C5
Tomo, (riv.)	F5
Tortugas, (gulf)	B6
Tota, Laguna de, (lake)	D5
Truandó, (riv.)	B4
Tumaco, Rada de, (bay)	A6
Tunahí, Sierra, (mts.)	E7
Upía, (riv.)	D5
Urabá, (gulf)	B3
Uva, Laguna, (lake)	E6
Uva, (riv.)	E6
Vaupés, (riv.)	E7
Vela, La, (cape)	D1
Vela, Roca que, (cay)	B8
Vichada, (riv.)	F5
Vigía, (cay)	A10
Yarí, (riv.)	D8
Zapatosa, Ciénaga de, (swamp)	D3

*City and suburbs.

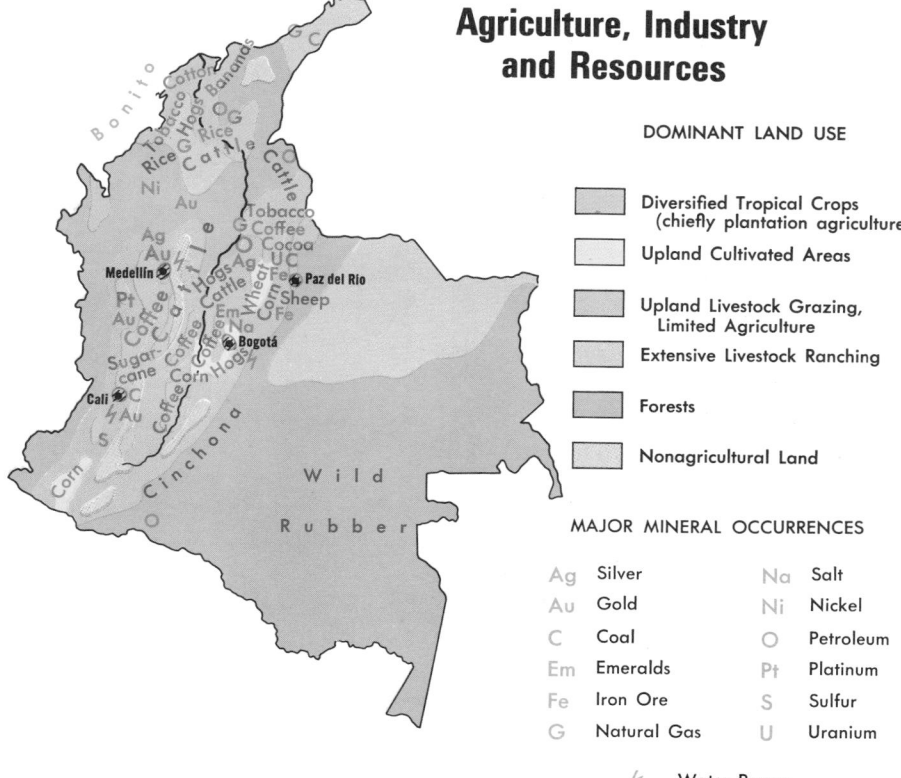

Agriculture, Industry and Resources

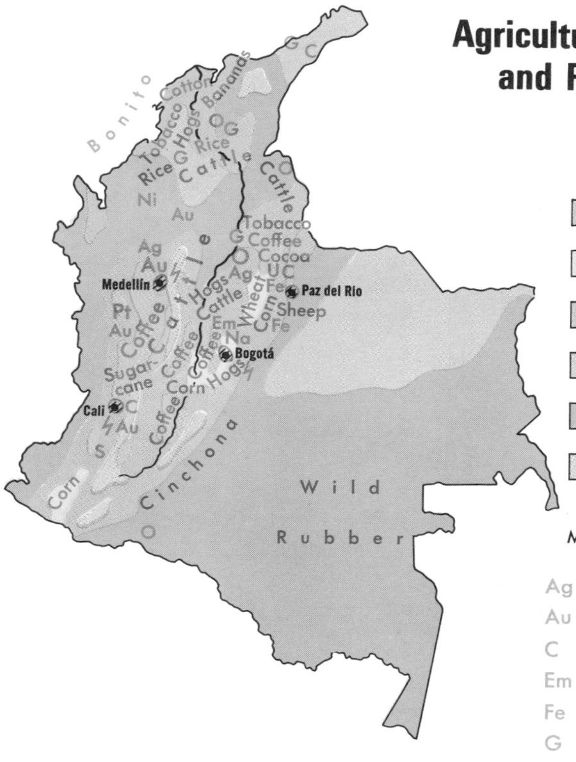

Bonito
Tobacco
Cotton
Hogs
Bananas
Rice
Cattle
Ni
Au
Ag
Coffee
Cocoa
Medellín
Paz del Río
Sheep
Wheat
Bogotá
Sugar-cane
Corn
Cali
Cinchona
Wild
Rubber

DOMINANT LAND USE

- Diversified Tropical Crops (chiefly plantation agriculture)
- Upland Cultivated Areas
- Upland Livestock Grazing, Limited Agriculture
- Extensive Livestock Ranching
- Forests
- Nonagricultural Land

MAJOR MINERAL OCCURRENCES

Ag	Silver	Na	Salt
Au	Gold	Ni	Nickel
C	Coal	O	Petroleum
Em	Emeralds	Pt	Platinum
Fe	Iron Ore	S	Sulfur
G	Natural Gas	U	Uranium

- Water Power
- Major Industrial Areas

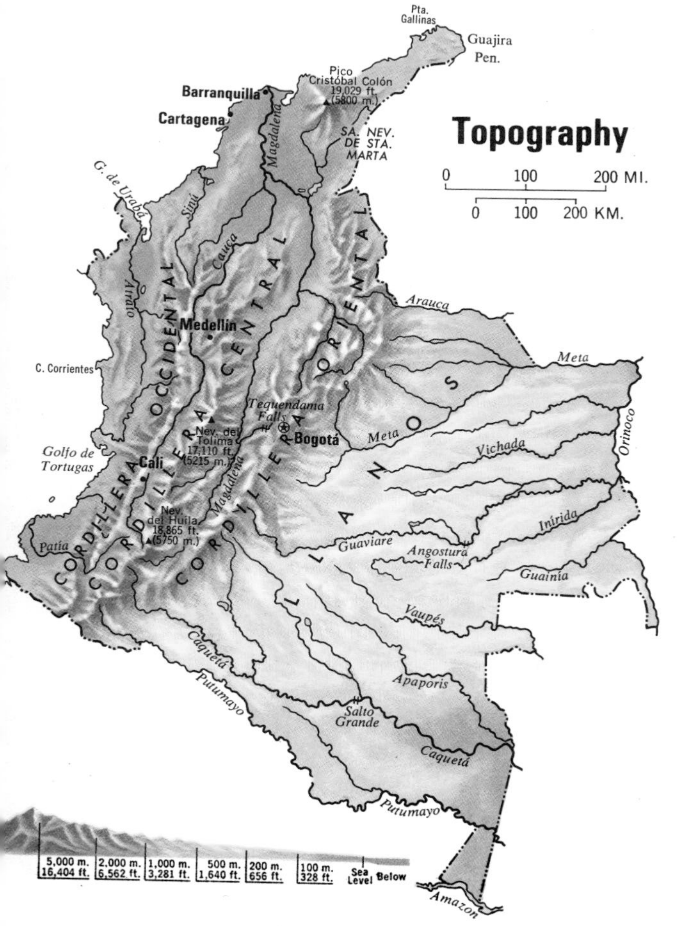

Topography
0 100 200 MI.
0 100 200 KM.
Pta. Gallinas
Guajira Pen.
Pico Cristóbal Colón 19,029 ft. (5800 m.)
Barranquilla
Cartagena
SA. NEV. DE STA. MARTA
Magdalena
G. de Urabá
Sinú
Atrato
Cauca
Medellín
Arauca
Meta
C. Corrientes
CORDILLERA OCCIDENTAL
CORDILLERA CENTRAL
CORDILLERA ORIENTAL
Tequendama Falls
Bogotá
Nev. del Tolima 17,110 ft. (5215 m.)
LLANOS
Vichada
Orinoco
Golfo de Tortugas
Cali
Nev. del Huila 18,865 ft. (5750 m.)
Patía
Guaviare
Angostura Falls
Inírida
Guainía
Vaupés
Caquetá
Apaporis
Putumayo
Salto Grande
Amazon
5,000 m. 16,404 ft.
2,000 m. 6,562 ft.
1,000 m. 3,281 ft.
500 m. 1,640 ft.
200 m. 656 ft.
100 m. 328 ft.
Sea Level
Below

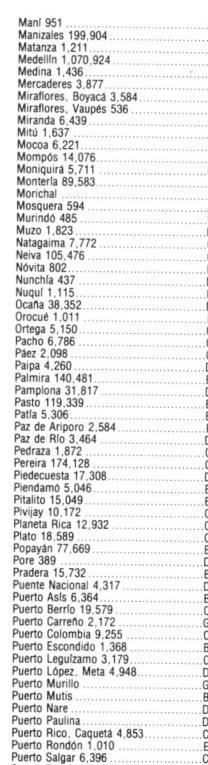

PERU

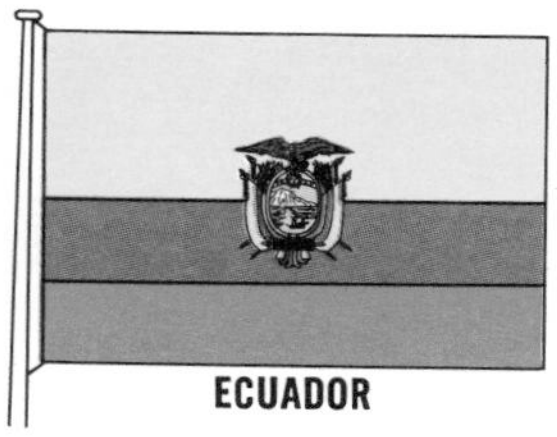

ECUADOR

PERU

AREA 496,222 sq. mi. (1,285,215 sq. km.)
POPULATION 17,031,221
CAPITAL Lima
LARGEST CITY Lima
HIGHEST POINT Huascarán 22,205 ft. (6,768 m.)
MONETARY UNIT sol
MAJOR LANGUAGES Spanish, Quechua, Aymara
MAJOR RELIGION Roman Catholicism

ECUADOR

AREA 109,483 sq. mi. (283,561 sq. km.)
POPULATION 8,644,000
CAPITAL Quito
LARGEST CITY Guayaquil
HIGHEST POINT Chimborazo 20,561 ft. (6,267 m.)
MONETARY UNIT sucre
MAJOR LANGUAGES Spanish, Quechua
MAJOR RELIGION Roman Catholicism

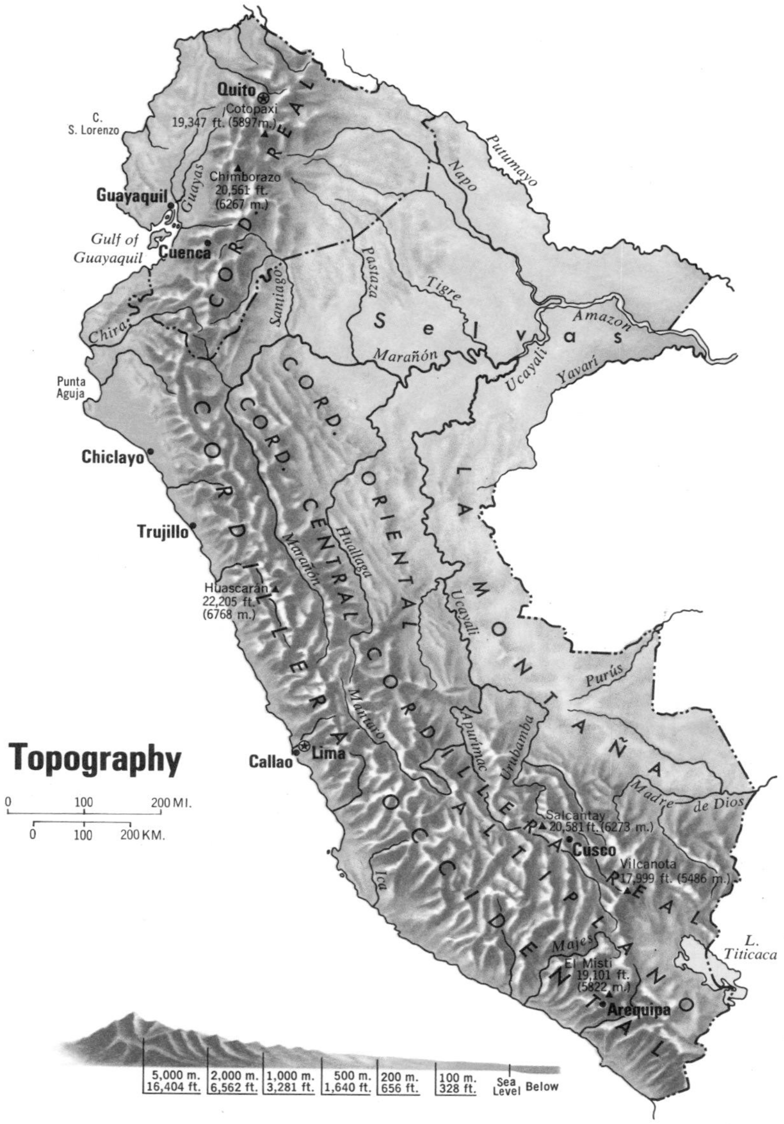
Topography
0 100 200 MI.
0 100 200 KM.
Quito
Cotopaxi 19,347 ft. (5897 m.)
C. S. Lorenzo
Chimborazo 20,561 ft. (6267 m.)
Guayaquil
Gulf of Guayaquil
Cuenca
Guayas
Napo
Putumayo
Pastaza
Tigre
Santiago
Chira
Marañón
Amazon
Ucayali
Yavarí
S e l v a s
Punta Aguja
Chiclayo
Trujillo
Huascarán 22,205 ft. (6768 m.)
Huallaga
Marañón
Mantaro
Apurímac
Urubamba
Purús
Madre de Dios
Salcantay 20,581 ft. (6273 m.)
Cusco
Vilcanota 17,999 ft. (5486 m.)
L. Titicaca
Majes
El Misti 19,101 ft. (5822 m.)
Arequipa
Ica
Callao
Lima
CORD. OCCIDENTAL
CORD. CENTRAL
CORD. ORIENTAL
CORDILLERA OCCIDENTAL
LA MONTAÑA
ALTIPLANO
5,000 m. 16,404 ft.
2,000 m. 6,562 ft.
1,000 m. 3,281 ft.
500 m. 1,640 ft.
200 m. 656 ft.
100 m. 328 ft.
Sea Level
Below

PERU

DEPARTMENTS

Amazonas 256,460 ... C5
Ancash 815,646 ... D7
Apurímac 321,936 ... F10
Arequipa 702,308 ... F10
Ayacucho 500,732 ... E9
Cajamarca 1,044,689 ... C6
Callao (prov.) 446,730 ... D9
Cusco 829,294 ... F9
Huancavelica 346,460 ... E9
Huánuco 481,924 ... D7
Ica 431,442 ... E10
Junín 848,993 ... E8
La Libertad 960,537 ... C6
Lambayeque 683,425 ... B6
Lima 4,738,266 ... D8
Loreto 446,316 ... E5
Madre de Dios 36,555 ... G8
Moquegua 99,287 ... G11
Pasco 221,219 ... E8
Piura 1,168,442 ... B5
Puno 893,586 ... G10
San Martín 319,670 ... D6
Tacna 133,240 ... G11
Tumbes 103,979 ... B4
Ucayali 200,085 ... E6

CITIES and TOWNS

Abancay 19,807 ... F9
Acarí 4,907 ... E10
Acobamba 2,156 ... E9
Acolla 5,717 ... E8
Acomayo, Cusco 1,419 ... G9
Acomayo, Huánuco 2,883 ... E7
Acora 1,910 ... H11
Acuracay 1,282 ... F5
Aija 1,843 ... D7
Alca 755 ... F10
Ambo 3,060 ... D8
Ananea 668 ... H10
Ancón 8,610 ... D8
Andahuaylas 7,654 ... F9
Andamarca 470 ... E8
Anta 3,703 ... F9
Antabamba 2,223 ... F10
Aplao 1,941 ... F11
Aquia 970 ... D8
Arequipa 107,858 ... G11
Arequipa* 447,431 ... G11
Ascope 12,070 ... C6
Astillero ... H9
Atalaya 2,229 ... E8
Atico 2,316 ... F11
Ayabaca 4,543 ... C5
Ayacucho 68,535 ... F9
Ayaviri 11,067 ... G10
Azángaro 7,658 ... H10
Bagua 9,735 ... C5
Balsapuerto 164 ... D5
Bambamarca 6,867 ... C6
Barranca, Lima 31,312 ... C8
Barranca, Loreto 1,351 ... D5
Bartra Antiguo ... E4
Bartra Nuevo ... E4
Bayóvar ... B5
Bellavista 4,906 ... C5
Bolívar 1,106 ... D6
Bolognesi ... F6
Bolognesi 661 ... F8
Borja 215 ... D5
Bretaña 1,035 ... E5
Buldibuyo 582 ... D7
Cabana 1,804 ... C7
Cabo Blanco ... B5
Cahuapanas 304 ... D5
Cailloma 1,187 ... G10
Cajabamba 7,282 ... C6
Cajacay 668 ... D8
Cajamarca 60,280 ... C6
Cajatambo 1,721 ... D8
Calca 6,112 ... G9
Callalli 819 ... G10
Callao 260,581 ... D9
Callao* 441,374 ... D9
Camaná 11,386 ... F11
Candarave 1,207 ... G11
Cangallo 1,584 ... E9
Canta 3,431 ... D8
Capachica 307 ... H10
Caraz 6,376 ... D7
Caravelí 1,827 ... F10
Carhuás 3,147 ... D7
Carumás 1,031 ... G11
Cascas 2,638 ... C6
Casma 12,725 ... C7
Castrovirreyna 1,749 ... E9
Catacaos 30,927 ... B5
Celendín 8,538 ... D6
Cerro Azul 2,314 ... D9
Cerro de Pasco 71,558 ... D8
Chachapoyas 11,919 ... D6
Chala 1,646 ... E10
Chalhuanca 3,071 ... F10
Chancay 18,993 ... D8
Chao ... C7
Chepén 29,919 ... C6
Chicama 11,160 ... C6
Chiclayo 280,244 ... C6
Chilca (Pucusana) 3,329 ... D9
Chilete 2,537 ... C6
Chimbote 216,406 ... C7
Chincha Alta 237,475 ... D9
Chiquián 3,521 ... D8
Chirinos 1,061 ... C5
Chivay 3,296 ... G10
Chosica ... D8
Chota 8,299 ... C6
Chulucanas 34,977 ... B5
Chupaca 5,422 ... E9
Chuquibamba 2,630 ... F10
Chuquibambilla 2,147 ... F9
Churín 1,801 ... D8
Cocachacra 5,985 ... G11
Cocama ... G8
Cojata 888 ... H10
Colasay 721 ... C5
Colcamar 1,216 ... D6
Conaica 1,154 ... E9
Concepción 7,129 ... E8
Concordia 1,372 ... E5
Contamana 5,718 ... E6
Contumazá 2,491 ... C6
Coracora 4,598 ... F10
Córdova 453 ... E10
Corongo 1,762 ... D7
Cotahuasi 1,301 ... F10
Culebras ... C7
Cumaría ... F7
Cusco (Cuzco) 85,044 ... F9
Cusco* 181,604 ... F9
Cutervo 6,890 ... C6
Cuyocuyo 1,101 ... H10
Desaguadero 2,682 ... H11
Deustua 544 ... G10
Dos de Mayo 574 ... E6
Echarate 1,071 ... F9
El Portugués ... C7
Esperanza 375 ... G7
Espinar 6,381 ... G10
Ferreñafe 22,200 ... C6
Francisco de Orellana 445 ... F4
Guadalupe 7,613 ... E9
Güeppi ... E3
Huacho 43,402 ... D8
Huacrachuco 1,210 ... D7
Hualgayoc 1,691 ... C6
Hualla 4,042 ... F9
Huallanca, Ancash 930 ... D7
Huallanca, Huánuco 4,806 ... D7
Huamachuco 8,273 ... D6
Huancabamba 4,393 ... C5
Huancané 5,227 ... H10
Huancapi 2,539 ... E9
Huancavelica 20,889 ... E9
Huancayo 165,132 ... E9
Huanchaco 6,005 ... C7
Huanta 11,213 ... E9
Huánuco 52,628 ... E7
Huaral 34,235 ... D8
Huaráz 45,116 ... D7
Huari 2,344 ... D7
Huariaca 2,671 ... E8
Huarmey 11,094 ... C8
Huarochirí 1,828 ... D9
Huarocondo 2,498 ... F9
Huaura 9,338 ... D8
Huaylas 1,344 ... C7
Iberia 2,307 ... F5
Ica 111,087 ... E10
Ichuña 277 ... G11
Ilave 9,891 ... H11
Ilo 31,549 ... G11
Imperial 20,894 ... D9
Iñapari 188 ... H8
Intutu 746 ... E4
Iparia 278 ... E7
Iquitos 173,629 ... F4
Jaén 24,356 ... C5
Jauja 14,630 ... E8
Jayanca 6,401 ... B6
Jeberos 1,493 ... D5
Juanjuí 9,324 ... D6
Juli 5,575 ... H11
Juliaca 77,976 ... G10
Jumbilla 1,035 ... C5
Junín 8,988 ... E8
Lagunas 4,601 ... E5
La Huaca 5,161 ... B5
La Jalca 1,769 ... D6
La Joya 5,000 ... G11
Lamas 8,937 ... D6
Lambayeque 23,746 ... B6
Lampa 4,319 ... G10
Lamud 2,405 ... C6
Lanlacuni Bajo 405 ... G9
La Oroya 33,305 ... D8
Las Piedras ... H9
Las Yaras 759 ... G11
La Unión 2,828 ... D7
Leimebamba 1,957 ... D6
Lima (cap.) 375,957 ... D8
Lima* 3,968,972 ... D8
Limbani 728 ... H10
Lircay 5,213 ... E9
Llata 2,922 ... D7
Lobitos 2,975 ... B5
Locumba 369 ... G11
Lomas 287 ... E10
Lucerna ... H9
Lurín 14,405 ... D9
Machupicchu 544 ... F9
Macusani 3,389 ... G10
Madre de Dios 660 ... G9
Máncora 5,358 ... B5
Manú 234 ... G9
Marcapata 369 ... G9
Marcona 25,962 ... E10
Margos 1,622 ... D8
Masisea 1,586 ... E7
Matarani ... F11
Matucana 4,196 ... D8
Mavila ... H8
Mazán 281 ... F4
Mazocruz 1,580 ... H11
Mendoza 1,902 ... D6
Mishagua ... F8
Moho 2,560 ... H10
Mollendo 21,206 ... F11
Monsefú 17,186 ... C6
Moquegua 21,488 ... G11
Morales 4,370 ... D6
Morococha 11,234 ... D8
Morropón 7,611 ... C5
Motupe 3,411 ... C6
Moyobamba 14,319 ... D6
Nauta 4,083 ... F5
Nazca 22,756 ... E10
Negritos 12,476 ... B5
Nuñoa 3,613 ... G10
Ocoña 1,062 ... F11
Ocros 1,037 ... D8
Ollachea 1,308 ... G9
Ollantaytambo 1,500 ... F9
Olmos 7,946 ... C5
Omaguas ... F5
Omas 249 ... D9
Omate 1,131 ... G11
Orcotuna 3,359 ... E8
Orellana 2,886 ... E6
Otuzco 5,765 ... C6
Oxapampa 5,233 ... E8
Oyón 6,279 ... D8
Pacasmayo 17,588 ... C6
Pachiza 889 ... D6
Paiján 12,699 ... C6
Paita 18,749 ... B5
Palpa 3,393 ... E10
Pampachiri 428 ... F10
Pampacolca 2,010 ... F10
Pampas 3,850 ... E9
Panao 1,363 ... E7
Pantoja 457 ... E3
Parinari 375 ... E5
Paruro 1,727 ... F9
Pataz 759 ... D6
Paucarbamba 534 ... E9
Paucartambo, Cusco 1,620 ... G9
Paucartambo, Pasco 3,497 ... E8
Pevas 1,325 ... G4
Picota 2,288 ... D6
Pimentel 9,129 ... B6
Pinquén ... G9
Pisac 1,566 ... G9
Pisco 53,414 ... D9
Piura 186,354 ... B5
Pizacoma 400 ... H11
Pomabamba 2,489 ... D7
Porvenir ... E5
Pozuzo 326 ... E8
Puca Barranca ... E4
Pucallpa 91,953 ... E7
Pucará 2,268 ... G10
Pucaurco 628 ... G4
Pucusana 3,329 ... D9
Puerto Alianza ... D5
Puerto América 240 ... D5
Puerto Arturo ... F3
Puerto Bermúdez 1,133 ... E8
Puerto Caballas ... E10
Puerto Chicama 3,136 ... C6
Puerto Eten 2,575 ... B6
Puerto Inca 1,286 ... E7
Puerto José Pardo ... D4
Puerto Leguía, Loreto ... D4
Puerto Leguía, Puno ... G9
Puerto Maldonado 12,609 ... H9
Puerto Morín ... C7
Puerto Ocopa 1,088 ... E8
Puerto Pardo ... F7
Puerto Pizarro ... B4
Puerto Portillo 86 ... F7
Puerto Prado 328 ... E8
Puerto Samanco 1,435 ... C7
Puerto Tahuantinsuyo ... G9
Puerto Victoria ... E7
Puno 66,477 ... G10
Punta de Bombón 4,647 ... F11
Punta Moreno ... C6
Puquina 1,026 ... G11
Puquio 8,099 ... F10
Putina 5,414 ... H10
Querecotillo 10,637 ... B5
Quicacha 255 ... F10
Quilca 235 ... F11
Quillabamba 16,837 ... F9
Quince Mil ... G9
Ramón Castilla 1,811 ... G5
Recuay 2,764 ... D7
Requena 8,270 ... F5
Reventazón ... B6
Rioja 9,876 ... D6
Salaverry 5,539 ... C7
Saña 40,144 ... C6
Sandia 1,682 ... H10
San José 4,070 ... B6
San José de Sisa 3,782 ... D6
San Juan ... E10
San Lorenzo 124 ... H8
San Martín ... E3
San Miguel, Ayacucho 1,440 ... F9
San Miguel, Cajamarca 1,798 ... C6
San Pedro de Lloc 11,463 ... C6
San Ramón 7,145 ... E8
Santa 20,490 ... C7
Santa Clotilde 1,068 ... E4
Santa Cruz, Cajamarca 2,739 ... C6
Santa Cruz, Loreto 449 ... E5
Santa Elena 368 ... F5
Santa María de Nanay 294 ... F4
Santiago 5,092 ... E10
Santiago de Cao 22,119 ... C6
Santiago de Chocorvos 525 ... E9
Santiago de Chuco 5,189 ... C7
Santo Tomás, Amazonas 1,093 ... C6
Santo Tomás, Cusco 2,755 ... G10
Santo Tomás de Andoas 272 ... D4
San Vicente de Cañete 15,277 ... D9
Saposoa 4,541 ... D6
Saquena 2,755 ... F5
Satipo 9,208 ... E8
Sauce 2,263 ... D6
Sayán 5,129 ... D8
Sechura 11,724 ... B5
Sicuani 21,176 ... G10
Sihuas 2,178 ... D7
Sullana 80,947 ... B5
Sumbay ... G10
Sumbilca 1,155 ... D8
Supe 10,061 ... D8
Tacna 92,640 ... G11
Tahuamanu 2,619 ... H8
Talara 55,122 ... B5
Tambo de Mora 2,790 ... D9
Tambo Grande 10,087 ... B5
Tamshiyacu 2,040 ... F5
Tarapoto 33,429 ... D6
Tarata 2,624 ... H11
Tarma 34,369 ... E8
Tarqui ... E3
Tayabamba 1,649 ... D7
Ticaco 781 ... H11
Tingo María 25,030 ... D7
Tiruntán 723 ... E6
Tocache 5,940 ... D7
Tonegrama ... D4
Topará ... D9
Toquepala ... G11
Torata 6,320 ... G11
Tournavista ... E7
Trujillo 354,557 ... C7
Tumbes 48,187 ... B4
Ubinas 422 ... G11
Uchiza 2,471 ... D7
Unini ... F8
Urcos 4,155 ... G9
Urubamba 4,686 ... F9
Vinchos 735 ... E9
Virú 6,587 ... C7
Vitor 416 ... G11
Yambrasbamba 277 ... D5
Yanahuanca 5,109 ... D8
Yanaoca 1,152 ... G10
Yauca 1,805 ... E10
Yauli 1,020 ... D8
Yauyos 1,296 ... E9
Yunguyo 7,253 ... H11
Yurimaguas 22,858 ... E5
Zarumilla 9,713 ... B4
Zorritos 4,497 ... B4

OTHER FEATURES

Acarí (riv.) ... E10
Aguaytía (riv.) ... E7
Aguja (pt.) ... B5
Amazon (riv.) ... F4
Andes, Cordillera de los (mts.) ... F10
Apurímac (riv.) ... F9
Azángaro (riv.) ... G10
Azul, Cordillera (mts.) ... E7
Blanca, Cordillera (mts.) ... D7
Blanco (cape) ... B5
Blanco (riv.) ... F6
Boquerón, El (pass) ... E7
Cañete (riv.) ... D9
Casma (riv.) ... C7
Chimbote (bay) ... C7
Chincha (isls.) ... D9
Coles (pt.) ... G11
Cóndor, Cordillera del (range) ... C5
Coropuna, Nudo (mt.) ... F10
Corrientes (riv.) ... E4
El Boquerón (pass) ... E7
El Misti (mt.) ... G11
Ene (riv.) ... E8
Ferrol (pen.) ... C7
Grande (riv.) ... E10
Guañape (isls.) ... C7
Heath (riv.) ... H9
Huallaga (riv.) ... D5
Huasaga (riv.) ... D4
Huascarán (mt.) ... D7
Huayabamba (riv.) ... D6
Ica (riv.) ... E10
Inambari (riv.) ... H9
Independencia (bay) ... D10
Independencia (isl.) ... D10
Junín (lake) ... E8
Juruá (riv.) ... F7
Lachay (pt.) ... D8
Lobos de Afuera (isls.) ... B6
Lobos de Tierra (isl.) ... B6
Locumba (riv.) ... G11
Madre de Dios (riv.) ... G9
Majes (riv.) ... F11
Mantaro (riv.) ... E8
Manú (riv.) ... G8
Marañón (riv.) ... E5
Mayo (riv.) ... D6
Misti, El (mt.) ... G11
Montaña, La (reg.) ... F8
Morona (riv.) ... D5
Nanay (riv.) ... E4
Napo (riv.) ... F4
Negra, Cordillera (mts.) ... D7
Negra (pt.) ... B6
Ñermete (pt.) ... B5
Occidental, Cordillera (range) ... F10
Ocoña (riv.) ... F11
Oriental, Cordillera (range) ... H10
Pachitea (riv.) ... E7
Paita (bay) ... B5
Pampas (riv.) ... E9
Paracas (pen.) ... D9
Parinacochas (lake) ... F10
Pariñas (pt.) ... B5
Pastaza (riv.) ... D5
Pativilca (riv.) ... D8
Perené (riv.) ... E8
Pichis (riv.) ... E8
Piedras, Las (riv.) ... G8
Pisco (bay) ... D9
Pisco (riv.) ... D9
Piura (riv.) ... B5
Puinagua, Canal de (riv.) ... E5
Purús (riv.) ... G8
Putumayo (riv.) ... G4
Rímac (riv.) ... D9
Salcantay (mt.) ... F9
Sama (riv.) ... G11
San Gallán (isl.) ... D9
San Lorenzo (isl.) ... D9
San Nicolás (bay) ... E10
Santa (riv.) ... C7
Santiago (riv.) ... D4
Sechura (bay) ... B5
Tahuamanu (riv.) ... H8
Tambo (riv.) ... G11
Tambopata (riv.) ... H9
Tapiche (riv.) ... E6
Tigre (riv.) ... E4
Titicaca (lake) ... H10
Tumbes (riv.) ... B4
Ucayali (riv.) ... F5

(continued on following page)

Urituyacu (riv.)	D5
Urubamba (riv.)	F8
Vilcabamba, Cordillera (mts.)	F9
Vilcanota (mt.)	G10
Vitor (riv.)	F11
Yaguas (riv.)	G4
Yavarí (riv.)	G5
Yavero (riv.)	F9
Yuruá (riv.)	F7

ECUADOR

PROVINCES

Azuay 367,324	C4
Bolívar 144,593	C3
Cañar 146,570	C3
Carchi 120,857	C2
Chimborazo 304,316	C3
Colón, Archipiélago de (terr.) 4,037	C8
Cotopaxi 236,313	C3
El Oro 262,564	C4
Esmeraldas 203,151	C2
Guayas 1,512,333	B4
Imbabura 216,027	C2
Loja 342,339	C4
Los Ríos 383,432	C3
Manabí 817,966	B3
Morona-Santiago 53,325	C4
Napo 62,186	D3
Pastaza 23,465	D3
Pichincha 988,306	C3
Tungurahua 279,920	C3
Zamora-Chinchipe 34,493	C5

CITIES and TOWNS

Alausí 7,137	C4
Ambato 77,955	C3
Andoas Nuevo	D4
Arapicos	C3
Archidona	C3
Arenillas 5,862	B4
Atuntaqui 9,907	C2
Azogues 10,953	C4
Baba 953	C3
Babahoyo 28,914	C3
Baeza 253	D3
Bahía de Caráquez 11,258	C3
Balao	C4
Balzar 10,924	C3
Baquerizo Moreno 1,311	C9
Bolívar 410	C2
Cajabamba 2,318	C3
Calceta 7,152	C3
Cañar 6,727	C3
Cariamanga 6,682	C5
Catacocha	C5
Catamayo	C4
Catarama 2,868	C3
Cayambe 11,199	D3
Celica 3,081	C4
Chone 23,627	C3
Chunchi 2,802	C4
Coca 1,211	D3
Cojimíes	B2
Cuenca 104,470	C4
Cuyabeno	E3
Daule 13,170	C4
Edén	E3
El Ángel 3,660	C2
El Corazón 1,073	C3
El Progreso	C9
El Pun	D2
Esmeraldas 60,364	B2
Farfán	D2
Floreana (Sta María)	B10
Girón 2,361	C4
Gualaceo 4,575	C4
Gualaquiza 1,519	C4
Guale	B3
Guamote 2,438	C4
Guano 5,389	C3
Guaranda 11,364	B4
Guayaquil 823,219	B4
Ibarra 41,335	C2
Jama	B3
Jipijapa 19,996	B3
La Bonita 184	D2
La Libertad 26,078	B4
La Tola	C2
Latacunga 21,921	C3
La Tola	C2
Loja 47,697	C4
Loreto	D3
Macará 8,063	C5
Macas 1,934	D4
Machachi 4,745	C3
Machala 69,170	B4
Machalilla	B3
Mangiaralto	B3
Manta 64,519	B3
Méndez 1,043	C4
Mera 631	C3
Miazal	D4
Milagro 53,106	C4
Montecristi 6,386	B3
Morona	D4
Mulaló	C3
Nuevo Rocafuerte 198	E3
Otavalo 13,605	C2
Paján 2,610	B3
Palanda	C5
Papallacta	D3
Pasaje 20,790	C4
Paute 1,998	C4
Pedernales	B2
Pelileo 3,754	C3
Píllaro 4,052	C3
Piñas 5,770	C4
Playas	B4
Portoviejo 59,550	B3
Posorja	B4
Puerto Ayora 900	B9
Puerto de Cayo	B3
Puerto El Carmen 308	E3
Puerto Napo	D3
Pujilí 2,510	C3
Putumayo	E3
Puyo 4,730	D3
Quevedo 43,101	C3
Quito (cap.) 599,828	C3
Riobamba 58,087	C3
Río Tigre	D4
Rocafuerte 5,519	B3
Rosa Zárate 4,847	C3
Salinas 12,409	B4
San Gabriel 10,036	D2
Sangolquí 10,554	C3
San Lorenzo	C2
San Miguel 2,743	C4
San Miguel de Salcedo 4,159	C3
Santa Ana 5,004	B3
Santa Elena 7,687	B4
Santa Isabel 2,068	C4
Santa Rosa 19,696	C4
Santo Domingo de los Colorados 30,523	C3
Saraguro 1,739	C4
Sarayacu	D3
Sigsig 2,021	C4
Sigüe	D2
Sucre 2,929	B3
Sucúa 9,694	C4
Tabacundo 1,942	C2
Tachina	C2
Tena	D3
Tulcán 24,398	D2
Valdez 3,837	C2
Veinticinco de Mayo 266	C4
Viche	C2
Villamil	B9
Vinces 10,126	C3
Yaguachi 3,816	C4
Yaupi	D4
Zamora 2,667	C5
Zapotillo	B5
Zaruma	C4
Zumba 905	C5

OTHER FEATURES

Aguarico (riv.)	D3
Albemarle (pt.)	B9
Ancón de Sardinas (bay)	C2
Antisana (mt.)	D3
Baltra (isl.)	B9
Banks (bay)	B9
Bobonaza (riv.)	D3
Cayambe (mt.)	D2
Chaves (Santa Cruz) (isl.)	C9
Chimborazo (mt.)	C3
Chira (riv.)	B5
Cóndor, Cordillera del (range)	C5
Cotopaxi (mt.)	C3
Cristóbal (pt.)	B9
Culpepper (isl.)	B8
Curaray (riv.)	D3
Darwin (Culpepper) (isl.)	B8
Esmeraldas (riv.)	C2
Española (isl.)	C10
Fernandina (isl.)	B9
Floreana (Santa María) (isl.)	B10
Galápagos (isls.)	C8
Galera (pt.)	B2
Genovesa (isl.)	C9
Guayas (riv.)	B4
Isabel (bay)	B9
Isabela (isl.)	B9
La Puntilla (cape)	B4
Manta (bay)	B3
Marchena (isl.)	C9
Mira (riv.)	C2
Napo (riv.)	D3
Naranjal (riv.)	C4
Pasado (cape)	B3
Pastaza (riv.)	D4
Pindo (riv.)	D3
Pinta (isl.)	B9
Pinzón (isl.)	B9
Puná (isl.)	B4
Putumayo (riv.)	E2
Rosa (cape)	B10
San Cristóbal (isl.)	C9
San Francisco (cape)	B2
Sangay (mt.)	D4
San Lorenzo (cape)	B3
San Miguel (riv.)	D2
San Salvador (isl.)	B9
Santa Cruz (isl.)	C9
Santa Elena (bay)	B3
Santa Fe (isl.)	C9
Santa María (isl.)	B10
Santiago (San Salvador) (isl.)	B9
Tumbes (riv.)	B4
Wenman (isl.)	B8
Wolf (Wenman) (isl.)	B8
Zamora	B4

FRENCH GUIANA

DISTRICTS

Cayenne 45,987	E3
Saint-Laurent du Maroni 9,270	E4

CITIES and TOWNS

Aouara	E3
Bienvenue	E4
Camopi 228	E4
Cayenne (cap.) 37,097	E4
Clément	E4
Counamama	E3
Délices	E3
Edmondt	E3
Grand Santi 305	D4
Guisanbourg	E4
Inini	E4
Iracoubo 483	E3
Kaw	E4
Kourou 6,465	E3
Macouria 94	E3
Malmanoury	E3
Mana 623	E3
Maripa	E4
Maripasoula 556	D4
Matoury 586	E3
Montsinéry 94	E3
Organabo	E3
Oscar	E4
Ouanary 61	F3
Ouaqui	D4
Paul Isnard	D3
Régina 258	E3
Rémire 5,421	E3
Roura 160	E3
Saint-Élie 57	E3
Saint-Georges 921	F4
Saint-Jean	D3
Saint-Laurent du Maroni 5,042	E3
Saint-Nazaire	E3
Sadó 60	E4
Saut-Tigre	E3
Sinnamary 1,669	E3
Sophie	E4
Tonate	E3

OTHER FEATURES

Approuague (riv.)	E4
Araoua (mts.)	E4
Béhague (pt.)	F4
Camopi (riv.)	E4
Comté (riv.)	E4
Connétable (isls.)	F3
Devil's (isl.)	D4
Granitique, Chaîne (range)	E4
Inini (riv.)	D4
Lawa (riv.)	D4
Litani (riv.)	D4
Mana (riv.)	E3
Maroni (riv.)	D3
Marouini (riv.)	D4
Oyapock (riv.)	E4
Rémire (isls.)	F3
Saint-Marcel (mt.)	E4
Salut (isls.)	E3
Sinnamary (riv.)	E3
Tampoc (riv.)	D4
Tumuc-Humac (mts.)	D3

GUYANA

DISTRICTS

East Berbice-Corantyne	C3
East Demerara-West Coast Berbice	
Mazaruni-Potaro	A2
North West	A1
Rupununi	B3
West Demerara-Essequibo Coast	B2

CITIES and TOWNS

Adventure 645	B2
Annai 569	B3
Anna Regina 1,124	B2
Apoteri 74	B3
Aurora 210	B2
Baramanni 231	A1
Baramita	A1
Bartica 4,087	A2
Biloku 290	B3
Charity 1,175	B2
Corriverton 10,502	C2
Dadanawa	B3
Danielstown 861	B2
Enmore	B2
Enterprise	B2
Epira 230	C2
Five Stars	C2
Fort Wellington	B2
Georgetown (cap.) 63,184	B2
Georgetown 164,039	B2
Imbaimadai 270	A2
Isherton	B3
Issano 207	B2
Issineru 124	A2
Ituni	B2
Kamakusa 211	B2
Kamarang 308	A2
Kangaruma	B2
Kumaka	B1
Kurupukari 284	B2
Kwakwani	B2
Lethem 645	B3
Linden 23,956	B2
Mabaruma 391	A1
Mahaica 6,967	B2
Mahaicony 4,665	B2
Mahaicony Village 4,665	B2
Mahdia 147	B2
Marao 203	B2
Matthews Ridge	A1
Morawhanna 292	A1
Mount Everard 369	A2
New Amsterdam 17,782	C2
Orealla 674	C2
Paradise	B2
Parika 1,101	B2
Pickersgill 508	A1
Port Kaituma	A1
Queenstown 1,211	B2
Rockstone 728	B2
Rosignol 2,001	C2
Suddie 705	B2
Takama	C2
Towakaima	A1
Tumatumari 353	B2
Tumerengo 238	B2
Vreed-en-Hoop 3,054	B2
Wichabai 216	B3

OTHER FEATURES

Acarai (mts.)	B3
Amakura (riv.)	A1
Amuku (mts.)	B3
Atkinson Field	B2
Barama (riv.)	A1
Barima (riv.)	A1
Berbice (riv.)	B2
Burro-Burro (riv.)	B2
Caburai (riv.)	B3
Canje (riv.)	C2

Agriculture, Industry and Resources

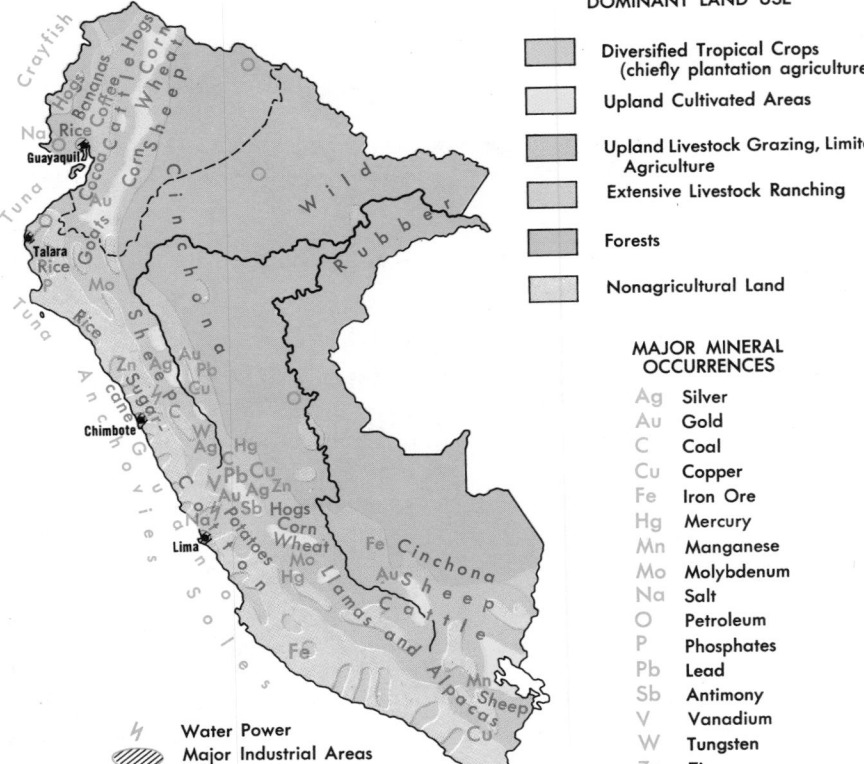

DOMINANT LAND USE

- Diversified Tropical Crops (chiefly plantation agriculture)
- Upland Cultivated Areas
- Upland Livestock Grazing, Limited Agriculture
- Extensive Livestock Ranching
- Forests
- Nonagricultural Land

MAJOR MINERAL OCCURRENCES

Ag	Silver
Au	Gold
C	Coal
Cu	Copper
Fe	Iron Ore
Hg	Mercury
Mn	Manganese
Mo	Molybdenum
Na	Salt
O	Petroleum
P	Phosphates
Pb	Lead
Sb	Antimony
V	Vanadium
W	Tungsten
Zn	Zinc

⚡ Water Power

▨ Major Industrial Areas

Agriculture, Industry and Resources

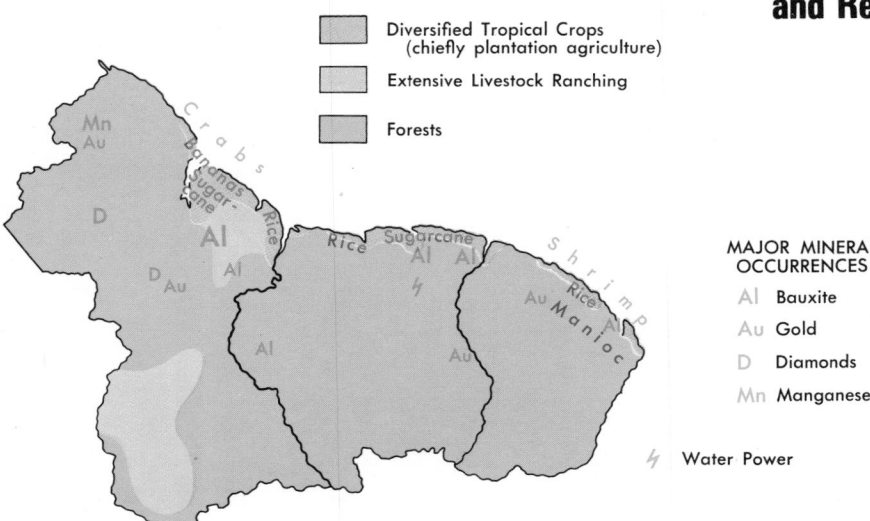

DOMINANT LAND USE

- Diversified Tropical Crops (chiefly plantation agriculture)
- Extensive Livestock Ranching
- Forests

MAJOR MINERAL OCCURRENCES

Al	Bauxite
Au	Gold
D	Diamonds
Mn	Manganese

⚡ Water Power

* City and suburbs
○ Population of district

GUYANA
AREA 83,000 sq. mi. (214,970 sq. km.)
POPULATION 793,000
CAPITAL Georgetown
LARGEST CITY Georgetown
HIGHEST POINT Mt. Roraima 9,094 ft. (2,772 m.)
MONETARY UNIT Guyana dollar
MAJOR LANGUAGES English, Hindi
MAJOR RELIGIONS Christianity, Hinduism, Islam

SURINAME
AREA 55,144 sq. mi. (142,823 sq. km.)
POPULATION 354,860
CAPITAL Paramaribo
LARGEST CITY Paramaribo
HIGHEST POINT Julianatop 4,200 ft. (1,280 m.)
MONETARY UNIT Suriname guilder
MAJOR LANGUAGES Dutch, Hindi, Indonesian
MAJOR RELIGIONS Christianity, Islam, Hinduism

FRENCH GUIANA
AREA 35,135 sq. mi. (91,000 sq. km.)
POPULATION 73,022
CAPITAL Cayenne
LARGEST CITY Cayenne
HIGHEST POINT 2,723 ft. (830 m.)
MONETARY UNIT French franc
MAJOR LANGUAGE French
MAJOR RELIGIONS Roman Catholicism, Protestantism

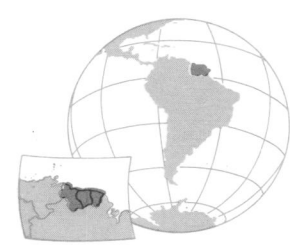

Courantyne (riv.) C3
Cuyuni (riv.) B2
Demerara (riv.) B3
Enwarak (mt.) B3
Essequibo (riv.) B3
Great (fall) B3
Ireng (riv.) B3
Kaieteur (fall) B3
Kamaria (falls) B2
Kuyuwini (riv.) B4
Kwitaro (riv.) B4
Leguan (isl.) B2
Marudi (mts.) B5
Mazaruni (riv.) A2
Moruka (riv.) B2
New (riv.) C4
Pakaraima (mts.) A3
Playa (pt.) B1
Pomeroon (riv.) B2
Potaro (riv.) B3
Puruni (riv.) B2
Roraima (mt.) A3
Rupununi (riv.) B4
Sororieng (mt.) B2
Surwakwima (fall) A2
Takutu (riv.) B4
Venamo (mt.) A3
Waini (riv.) B2
Wenamu (riv.) A2

SURINAME
DISTRICTS
Brokopondo 17,763 ....D4
Commewijne 18,740 ....D3
Coronie 3,251 ....C3
Marowijne 25,911 ....D4
Nickerie 35,178 ....C3
Para 16,635 ....D3
Paramaribo 102,297 ......D2
Saramacca 13,554 ....C3
Suriname 151,585 ....D3

CITIES and TOWNS
Ajoewa C4
Alalapadu C4

Albina 1,000 ....D3
Asidonhoppo D4
Berg en Dal D3
Bitagron C3
Brokopondo D3
Burnside C2
Calcutta 1,100 ....D3
Cottica D4
Domburg 1,200 ....D3
Groningen 600 ....D2
Huwelijkszorg D3
Kwakoegron D3
Lelydorp 300 ....D3
Majoli D4
Mariënburg 3,500 ....D2
Moengo 2,100 ....D3
Nieuw-Amsterdam 1,400 ..D2
Nieuw-Nickerie 7,400 ....C2
Paramaribo (cap.) ⊙ 167,905 .D2
Paranam D3
Totness 1,300 ....C3
Uitkijk D3
Wageningen 800 ....C3
Zanderij D3

OTHER FEATURES
Bakhuys (mts.) C3
Coeroeni (riv.) C4
Commewijne (riv.) D3
Coppename (riv.) C3
Corantijn (riv.) C4
Cottica (riv.) D3
Eilerts de Haan (mts.) ...C4
Frederik Willem IV (falls) .C4
Julianatop (mt.) C4
Kayser (mts.) C4
Lely (mts.) D3
Litani (riv.) D3
Marowijne (riv.) D3
Nickerie (riv.) C3
Orange (mts.) .j.D4
Saramacca (riv.) D3
Sipwliwini (riv.) C4
Suriname (riv.) D3
Tapanahoni (riv.) C3
Toekomstig (res.) C3
Van Blommestein (lake) ..D3
Wilhelmina (mts.) C4

Topography

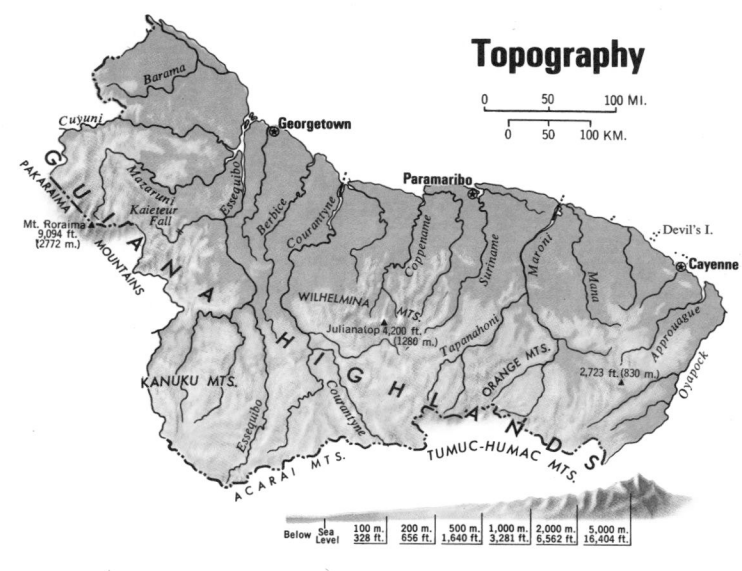

Below Sea Level | 100 m. 328 ft. | 200 m. 656 ft. | 500 m. 1,640 ft. | 1,000 m. 3,281 ft. | 2,000 m. 6,562 ft. | 5,000 m. 16,404 ft.

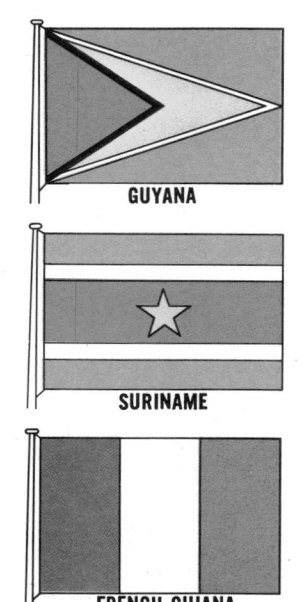

GUYANA
SURINAME
FRENCH GUIANA

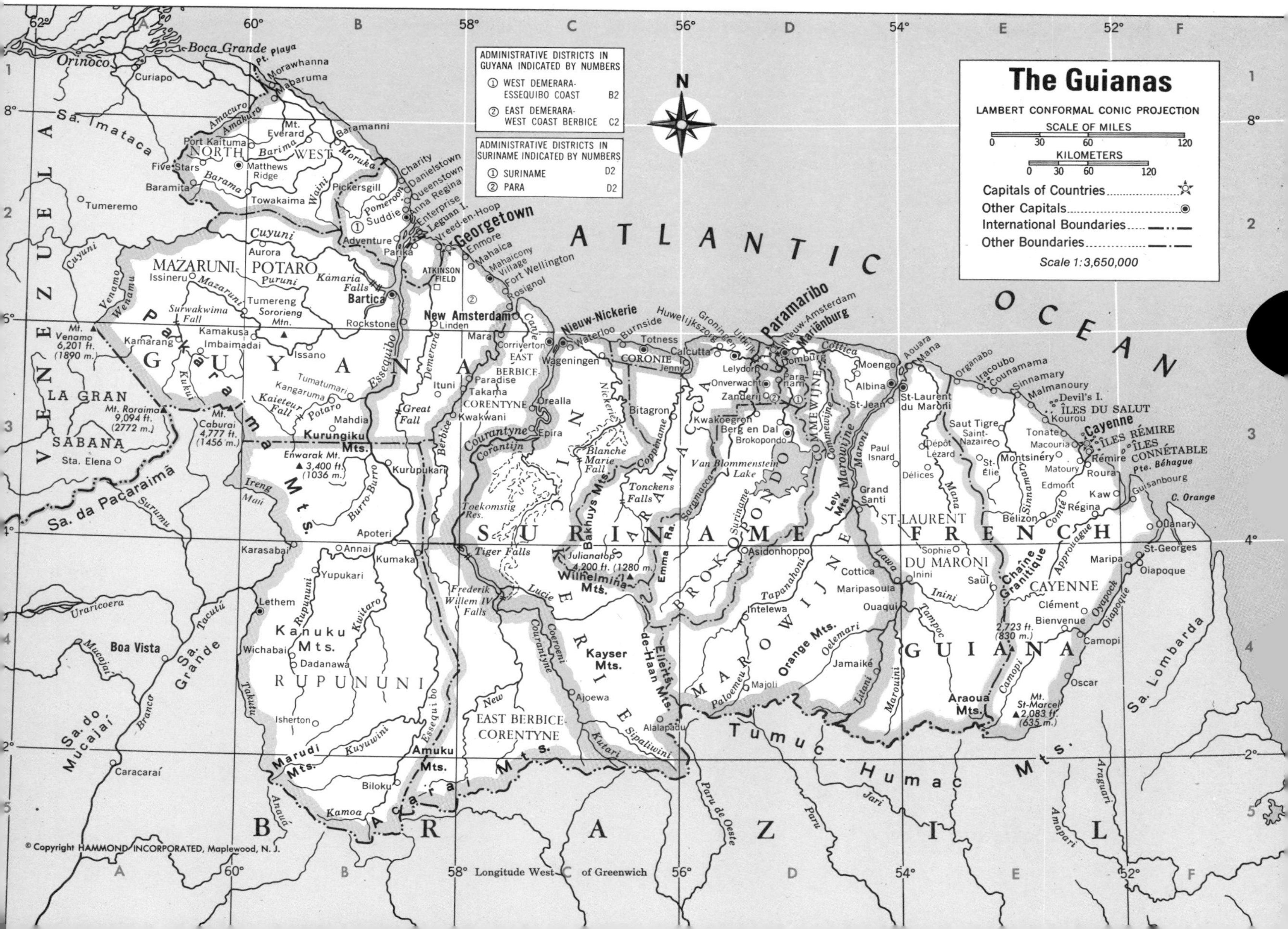

© Copyright HAMMOND INCORPORATED, Maplewood, N. J.

Brazil

BIPOLAR OBLIQUE CONIC CONFORMAL PROJECTION

SCALE OF MILES

KILOMETERS

Capitals of Countries ⊛
State Capitals ⊚
International Boundaries
State Boundaries

Scale 1:14,700,000

© Copyright HAMMOND INCORPORATED, Maplewood, N.J.

BRAZIL
WESTERN PART

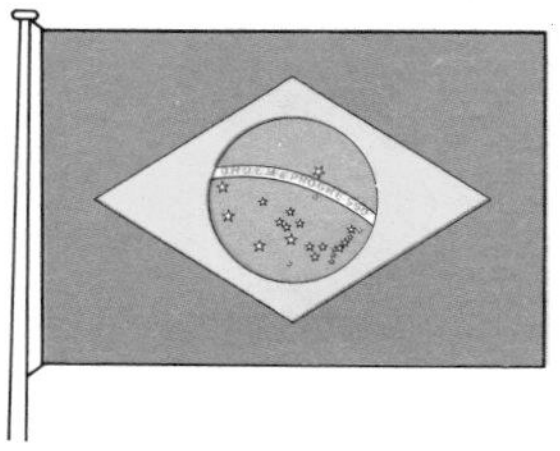

AREA 3,284,426 sq. mi. (8,506,663 sq. km.)
POPULATION 119,098,992
CAPITAL Brasília
LARGEST CITY São Paulo (greater)
HIGHEST POINT Pico da Neblina 9,889 ft. (3,014 m.)
MONETARY UNIT cruzeiro
MAJOR LANGUAGE Portuguese
MAJOR RELIGION Roman Catholicism

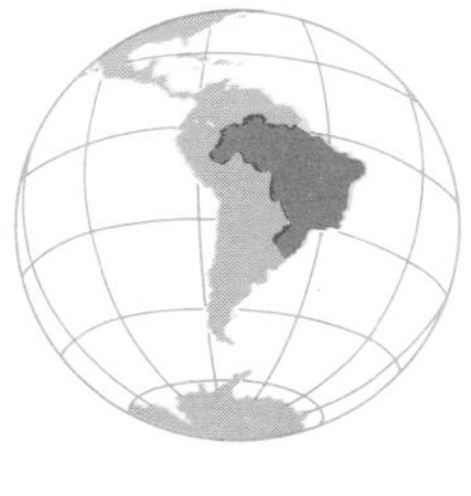

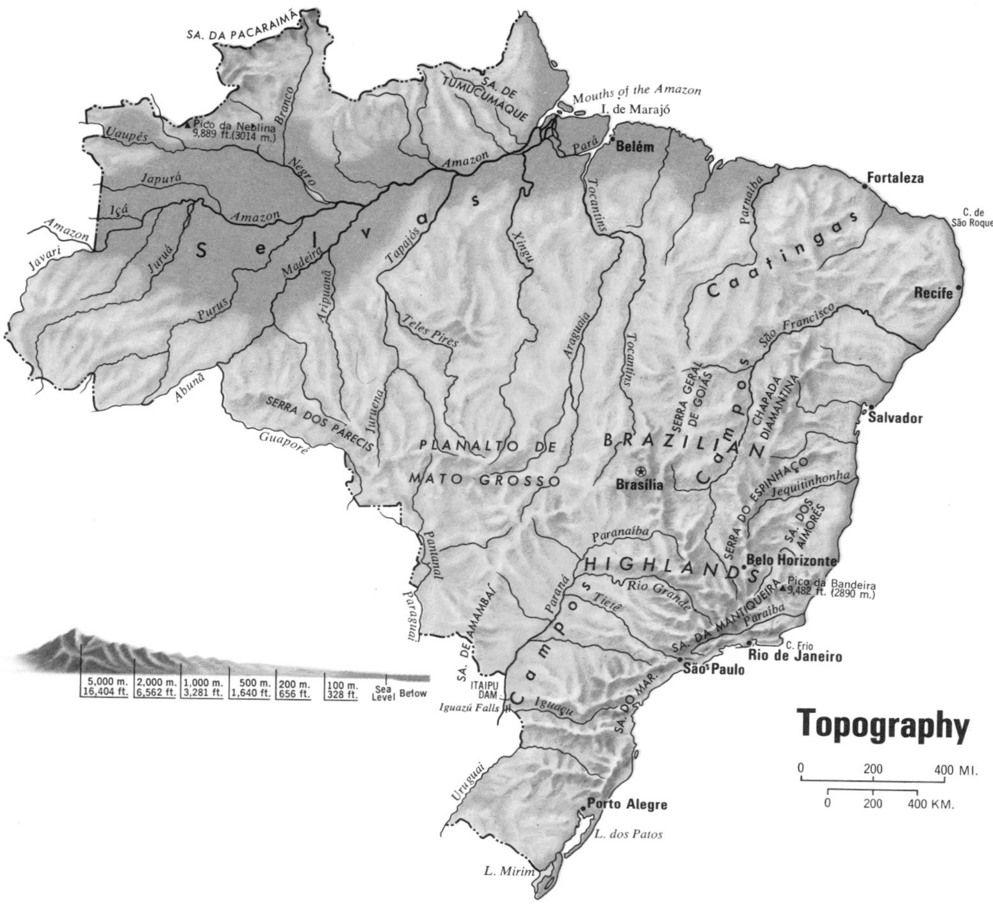
SA. DA PACARAIMÃ
SA. DE TUMUCUMAQUE
Mouths of the Amazon
I. de Marajó
Pico da Neblina 9,889 ft. (3014 m.)
Uaupés
Branco
Negro
Amazon
Japurá
Içá
Amazon
Javari
Juruá
Purus
Madeira
Belém
Pará
Tocantins
Xingu
Tapajós
Selvas
Parnaíba
Fortaleza
C. de São Roque
Caatingas
Recife
São Francisco
Teles Pires
Aripuanã
Araguaia
Tocantins
Abunã
Juruena
SERRA DOS PARECIS
Guaporé
PLANALTO DE MATO GROSSO
SERRA GERAL DE GOIÁS
CHAPADA DIAMANTINA
Salvador
BRAZILIAN
Campos
Brasília
SERRA DO ESPINHAÇO
Jequitinhonha
SA. DOS AIMORÉS
Paranaíba
HIGHLANDS
Belo Horizonte
Pico da Bandeira 9,482 ft. (2890 m.)
Rio Grande
Tietê
SA. DA MANTIQUEIRA
Paraíba
C. Frio
Rio de Janeiro
São Paulo
SA. DE AMAMBAÍ
Paraguai
Paraná
Campos
SA. DO MAR
ITAIPU DAM
Iguaçú Falls
Iguaçu
Uruguai
Porto Alegre
L. dos Patos
L. Mirim
5,000 m. 16,404 ft.
2,000 m. 6,562 ft.
1,000 m. 3,281 ft.
500 m. 1,640 ft.
200 m. 656 ft.
100 m. 328 ft.
Sea Level
Below

Topography

0 200 400 MI.
0 200 400 KM.

STATES and TERRITORIES

Acre 301,605 ... G10
Alagoas 1,987,581 ... G5
Amapá (terr.) 175,634 ... D2
Amazonas 1,432,066 ... G9
Bahia 9,474,263 ... F6
Ceará 5,294,876 ... G4
Espírito Santo 2,023,821 ... F7
Federal District 1,177,393 ... E6
Goiás 3,865,482 ... D6
Maranhão 4,002,599 ... E4
Mato Grosso 1,141,661 ... B6
Mato Grosso do Sul 1,370,333 ... C7
Minas Gerais 13,390,805 ... E7
Pará 3,411,868 ... C4
Paraíba 2,772,600 ... G4
Paraná 7,630,466 ... D9
Pernambuco 6,147,102 ... G5
Piauí 2,140,066 ... F4
Rio de Janeiro 11,297,327 ... F8
Rio Grande do Norte 1,899,720 ... G4
Rio Grande do Sul 7,777,212 ... C10
Rondônia (terr.) 492,810 ... H10
Roraima (terr.) 79,153 ... H8
Santa Catarina 3,628,751 ... D9
São Paulo 25,040,698 ... D8
Sergipe 1,141,834 ... G5

CITIES and TOWNS

Abaeté 12,861 ... E7
Abaetetuba 33,031 ... D3
Acaraú 7,144 ... F3
Acopiara 10,747 ... G4
Açu 20,544 ... G4
Agudos 18,790 ... *B3
Alagoa Grande 14,204 ... H4
Alagoinhas 76,377 ... G6
Alcobaça 3,430 ... G7
Alegre 9,441 ... *F2
Alegrete 54,786 ... B10
Além Paraíba 23,028 ... *E2
Alenquer 16,477 ... C3
Alfenas 31,815 ... *D2
Altamira 24,846 ... C3
Altos 13,621 ... F4
Amambaí 12,507 ... C8
Amapá 2,676 ... D2
Amarante 6,848 ... F4
Amargosa 11,118 ... F6
Americana 121,794 ... *C3
Amparo 26,970 ... *C3
Anápolis 160,520 ... D7
Anchieta 5,741 ... F8
Andaraí 2,476 ... F6
Andradina 42,036 ... D8
Andrelândia 8,737 ... *D2
Angra dos Reis 24,894 ... *D3
Antonina 11,950 ... *B4
Aparecida 27,265 ... *D3
Apiaí 7,809 ... *B4
Aquidauana 21,514 ... C8
Aracaju 288,106 ... G5
Aracati 20,282 ... G4
Araçatuba 113,486 ... *A2
Araçuaí 12,292 ... F7
Araguari 73,302 ... D7
Araranguá 22,468 ... D10
Araraquara 77,202 ... *B2
Araras 54,323 ... *C3
Araxá 51,339 ... E7
Arcoverde 40,646 ... G5
Areia Branca 12,979 ... G4
Assis 57,217 ... *A3
Avaré 40,716 ... *B3
Bacabal 43,229 ... E4
Bagé 66,743 ... C10
Bahia (Salvador) 1,496,276 ... G6
Baixo Guandu 13,714 ... F7
Balsas 13,566 ... E4
Bambuí 14,172 ... *C2
Barão de Cocais 11,950 ... *E1
Barbacena 69,675 ... *E2
Barcelos 1,846 ... H9
Bariri 15,372 ... *B3
Barra 10,809 ... F5
Barra do Corda 19,280 ... E4
Barra do Piraí 51,214 ... *E3
Barra Mansa 123,421 ... *D3
Barras 8,904 ... F4
Barreiras 30,355 ... E6
Barreiros 19,419 ... H5
Barretos 65,294 ... *B2
Batatais 30,478 ... *C2
Baturité 12,388 ... G4
Bauru 178,861 ... *B3
Bebedouro 39,070 ... *B2
Bela Vista 11,936 ... C8
Belém 758,117 ... E3
Belém †1,000,349 ... E3
Belo Horizonte 1,442,483 ... *D1
Belo Horizonte †2,541,788 ... *D1
Benjamin Constant 6,563 ... G9
Bento Gonçalves 40,323 ... C10
Betim 71,599 ... *D2
Bicas 8,611 ... *E2
Birigui 45,348 ... *A2
Blumenau 144,819 ... D9
Boa Esperança 17,394 ... *D2
Boa Vista 43,131 ... H8
Bocaiúva 16,616 ... E7
Bom Conselho 13,196 ... G5
Bom Despacho 22,941 ... *D1
Bom Jesus da Lapa 19,978 ... F6
Bom Sucesso 10,331 ... *D2
Borba 5,366 ... H9
Bragança Paulista 61,021 ... *C3
Brasiléia 4,835 ... G10
Brasília (cap.) 411,305 ... E6
Brasília de Minas 10,171 ... F7
Brejo 5,859 ... F3
Breves 31,452 ... D3
Brumado 24,663 ... F6
Brusque 37,898 ... D9
Cabedelo 18,581 ... H4
Cabo Frio 40,668 ... *F3
Caçador 25,287 ... D9
Caçapava 45,258 ... *D3
Caçapava do Sul 15,180 ... C10
Cáceres 33,472 ... B7
Cachoeira 11,520 ... G6
Cachoeira do Sul 59,967 ... C10
Cachoeiro de Itapemirim 84,994 ... G8
Caeté 23,331 ... *E1
Caetité 8,823 ... F6
Caiapônia 9,358 ... C7
Caicó 30,777 ... G4
Cajazeiras 30,834 ... G4
Cajuru 9,670 ... *C2
Camaquã 28,078 ... C10
Cambará 13,218 ... *A3
Cambuí 8,552 ... *C3
Cametá 15,539 ... D3
Camocim 19,921 ... F3
Campina Grande 222,229 ... G4
Campinas 566,517 ... *C3
Campo Belo 30,392 ... *D2
Campo Formoso 10,324 ... F5
Campo Grande 282,844 ... C8
Campo Largo 34,506 ... *B4
Campo Maior 24,009 ... F4
Campos 174,218 ... *F2
Cananéia 5,581 ... *C4
Canavieiras 14,076 ... G6
Canindé 18,573 ... G4
Canoas 214,115 ... D10
Canoinhas 25,880 ... D9
Capanema 28,272 ... E3
Capão Bonito 24,081 ... *B4
Caraguatatuba 22,932 ... *D3
Carangola 15,621 ... *E2
Caratinga 39,621 ... *E1
Caravelas 3,704 ... G7
Carazinho 41,913 ... C10
Carolina 10,136 ... E4
Caruaru 137,636 ... G5
Casa Branca 13,739 ... *C2
Cascavel 16,238 ... G4
Cássia 10,701 ... *C2
Castanhal 51,797 ... E3
Castelo 9,162 ... F8
Castro 21,079 ... *B4
Castro Alves 11,286 ... G6
Cataguases 40,659 ... *E2
Catalão 30,516 ... E7
Catanduva 64,813 ... *B2
Catolé do Rocha 12,165 ... G4
Caxambu 16,221 ... *D2
Caxias 56,755 ... F4
Caxias do Sul 198,824 ... D10
Ceará (Fortaleza) 648,815 ... G3
Ceará-Mirim 17,097 ... H4
Ceres 13,671 ... D6
Chapecó 53,198 ... C9
Coari 14,841 ... H9
Codajás 4,923 ... H9
Codó 11,593 ... E4
Colatina 61,057 ... F7
Conceição do Araguaia 18,143 ... D5
Concórdia 17,973 ... D9
Conselheiro Lafaiete 66,262 ... *E2
Corinto 17,056 ... E7
Cornélio Procópio 31,201 ... D8
Coroatá 16,070 ... F3
Coromandel 11,604 ... E7
Corumbá 66,014 ... B7
Coxim 14,876 ... C7
Crateús 29,905 ... F4
Crato 49,244 ... G4
Criciúma 74,003 ... D10
Cristalina 10,521 ... E7
Cruz Alta 53,315 ... C10
Cruzeiro 55,175 ... *D3
Cruzeiro do Sul 11,189 ... G10
Cubatão 78,327 ... *C3
Cuiabá 167,894 ... C6
Curitiba 843,733 ... *B4
Curitiba †1,441,743 ... *B4
Currais Novos 25,663 ... G4
Cururupu 10,358 ... E3
Curvelo 37,734 ... E7
Diamantina 20,197 ... F7
Divinópolis 108,344 ... *D2
Dois Córregos 11,811 ... *B3
Dom Pedrito 25,773 ... C10
Dores do Indaiá 13,058 ... E7
Dourados 76,838 ... C8
Duque de Caxias 306,057 ... *E3
Erexim 46,927 ... C9
Esperança 12,964 ... G4
Esplanada 9,822 ... G5
Estância 28,250 ... G5
Feira de Santana 225,003 ... G5
Fernandópolis 39,737 ... *A2
Floriano 35,761 ... F4
Florianópolis 153,547 ... E9
Fonte Boa 3,278 ... G9
Formiga 36,681 ... *D2
Formosa 29,304 ... E6
Fortaleza 648,815 ... G3
Fortaleza †1,581,588 ... G3
Foz do Iguaçu 93,619 ... C9
Franca 143,630 ... *C2
Frutal 22,955 ... *B2
Garanhuns 64,854 ... G5
Garça 26,527 ... *B3
Goiana 30,108 ... H4
Goiânia 703,263 ... D7
Goiás 15,768 ... D6
Governador Valadares 173,699 ... F7
Grajaú 11,147 ... E4
Guacuí 12,715 ... *F2
Guajará-Mirim 19,992 ... H10
Guarapuava 17,189 ... C9
Guaratinguetá 68,370 ... *D3
Guarujá 67,730 ... *C4
Guarulhos 395,117 ... *C3
Guaxupé 23,637 ... *C2
Guiratinga 8,981 ... C7
Gurupi 27,319 ... D5
Humaitá 10,004 ... H10
Ibaiti 11,352 ... *A3
Ibiá 11,161 ... E7
Ibicaraí 18,202 ... G6
Ibitinga 23,359 ... *B2
Icó 13,007 ... G4
Igarapava 15,342 ... *C2
Igarapé-Miri 12,172 ... D3
Iguape 16,827 ... *C4
Iguatu 39,611 ... G4
Ijuí 51,925 ... C10
Ilhéus 71,240 ... G6
Imbituba 9,998 ... D10
Imperatriz 111,818 ... E4
Inhumas 23,455 ... D7
Ipameri 14,163 ... E7
Ipu 12,787 ... F4
Irati 21,956 ... *A4
Itabaiana, Paraíba 17,843 ... H4
Itabaiana, Sergipe 26,055 ... G5
Itaberaba 27,590 ... F6
Itabira 57,691 ... F7
Itabirito 22,978 ... *E2
Itabuna 129,938 ... G6
Itacoatiara 26,737 ... B3
Itaituba 19,644 ... C4
Itajaí 78,867 ... D9
Itajubá 53,506 ... *D3
Itanhaem 26,181 ... *C4
Itapecerica 10,234 ... *D2
Itapecuru-Mirim 12,216 ... F3
Itapemirim 16,829 ... F8
Itaperuna 34,644 ... *F2
Itapetinga 36,897 ... G6
Itapetininga 61,344 ... *B3
Itapeva 36,551 ... *B3
Itapipoca 19,463 ... G3
Itapira 36,308 ... *C3
Itápolis 13,750 ... *B2
Itaporanga 8,988 ... G4
Itaqui 23,136 ... B10
Itararé 24,368 ... *B4
Itatiba 35,537 ... *C3
Itaúna 49,372 ... *D2
Itu 62,211 ... *C3
Ituaçu 1,749 ... F6
Ituiutaba 65,178 ... D7
Itumbiara 56,602 ... D7
Iturama 12,363 ... *A1
Ituverava 21,323 ... *C2
Jaboatão 67,120 ... H5
Jaboticabal 40,276 ... *B2
Jacareí 103,652 ... *D3
Jacarezinho 23,684 ... *A3
Jacobina 26,723 ... F5
Jacupiranga 7,044 ... *B4
Jaguaquara 11,336 ... F6
Jaguarão 18,165 ... C11
Jaguariaíva 8,566 ... *B4
Januária 20,484 ... E6
Jataí 40,957 ... D7
Jaú 59,522 ... *B3
Jequié 84,792 ... F6
Jequitinhonha 10,900 ... F7
Ji-Paraná 31,724 ... H10
Joaçaba 16,195 ... D9
João Pessoa 290,424 ... H4
João Pinheiro 17,013 ... E7
Joinvile 217,074 ... D9
Juazeiro 60,940 ... G5
Juazeiro do Norte 125,248 ... F4
Juiz de Fora 299,728 ... *E2
Jundiaí 210,015 ... *C3
Lages 108,768 ... D9
Laguna 27,743 ... D10
Lambari 9,722 ... *D2
Lapa 13,314 ... D9
Laranjeiras do Sul 19,329 ... C9
Lavras 35,345 ... *D2
Leme 40,155 ... *C3
Leopoldina 28,554 ... *E2
Limeira 137,812 ... *C3
Limoeiro 36,088 ... H4
Limoeiro do Norte 13,112 ... G4
Linhares 51,575 ... F7
Lins 44,633 ... *B2
Londrina 258,054 ... D8
Lorena 51,276 ... *D3
Luís Correia 3,576 ... F3
Luz 10,068 ... *D1
Luziânia 67,284 ... E7
Macaé 39,644 ... *F3
Macaíba 17,036 ... H4
Macapá 89,081 ... D2
Macau 17,543 ... G4
Maceió 376,479 ... H5
Machado 16,164 ... *C2
Mafra 26,226 ... D9
Magé 37,597 ... *E3
Mamanguape 16,321 ... H4
Manacapuru 17,016 ... H9
Manaus 613,068 ... H9
Manhuaçu 22,678 ... *E2
Manhumirim 11,085 ... *E2
Manicoré 9,532 ... H9
Marabá 41,564 ... D4
Maracaju 9,699 ... C8
Maragogipe 13,512 ... G6
Maranguape 20,098 ... G3
Marechal Deodoro 9,400 ... H5
Mariana 11,785 ... *E2
Marília 103,904 ... *A3
Maringá 158,047 ... D8
Mata de São João 23,741 ... G6
Mato Grosso (Vila Bela da Santissima Trindade) 1,401 ... B6
Maués 10,846 ... B3
Mazagão 1,824 ... D3
Mineiros 16,844 ... C7
Miracema 15,545 ... *E2
Mirassol 25,173 ... *B2
Mococa 33,682 ... *C2
Mogi das Cruzes 122,265 ... *C3
Mogi-Mirim 41,827 ... *C3
Monte Alegre 10,646 ... C3
Monte Aprazível 9,767 ... *A2
Monteiro 11,051 ... G4
Montenegro 27,246 ... D10
Montes Claros 151,881 ... E7
Morrinhos 20,154 ... D7
Mossoró 118,007 ... G4
Muriaé 50,040 ... *E2
Muzambinho 8,803 ... *C2
Nanuque 34,445 ... F7
Natal 376,552 ... H4
Nazaré 18,068 ... G6
Niquelândia 8,828 ... D6
Niterói 386,185 ... *E3
Nova Cruz 12,824 ... H4
Nova Era 11,126 ... *E1
Nova Friburgo 88,943 ... *E3
Nova Iguaçu 491,802 ... *E3
Nova Lima 35,035 ... *E2
Nova Russas 10,021 ... F4
Novo Hamburgo 132,066 ... D10
Novo Horizonte 18,439 ... *B2
Óbidos 17,143 ... C3
Oeiras 12,406 ... F4
Olímpia 24,376 ... *B2
Olinda 266,392 ... H4
Oliveira 22,642 ... *D2
Oriximiná 12,078 ... C3
Orlândia 22,924 ... *C2
Osasco 376,689 ... *C3
Ourinhos 52,698 ... *B3
Ouro Preto 27,821 ... *E2
Palmares 40,624 ... H5
Palmas 15,823 ... C9
Palmeira 11,521 ... *B4
Palmeira das Missões 23,943 ... C9
Pará (Belém) 758,117 ... E3
Paracatu 29,911 ... E7
Pará de Minas 37,127 ... *D1
Paraguaçu Paulista 17,399 ... D8
Paraíba do Sul 13,510 ... *E3
Paranaíba 21,305 ... D7
Paranaguá 68,366 ... *B4
Parati 8,684 ... *D3
Parintins 29,369 ... B3
Parnaíba 78,718 ... F3
Passo Fundo 103,121 ... D10
Passos 56,998 ... *C2
Patos 58,735 ... G4
Patos de Minas 59,896 ... E7
Patrocínio 29,520 ... E7
Pau dos Ferros 12,985 ... G4
Paulo Afonso 62,066 ... G5
Pederneiras 18,864 ... *B3
Pedra Azul 13,615 ... F6
Pedreiras 30,843 ... E4
Pedro Segundo 9,693 ... F4
Pelotas 197,092 ... C10
Penápolis 32,168 ... *A2
Penedo 27,064 ... G5
Pernambuco (Recife) 1,184,215 ... H5
Petrolina 73,436 ... G5
Petrópolis 149,427 ... *E3
Picos 33,098 ... F4
Piedade 13,054 ... *C3
Pilar 14,778 ... H5
Pindamonhangaba 51,174 ... *D3

(continued on following page)

Pinhal (Espírito Santo do
Pinhal) 23,235 . . . . *C3
Pinheiro 19,556 . . . . E3
Piquete 10,316 . . . . D3
Piracanjuba 11,151 . . . . . . D7
Piracicaba 179,395 . . . . . *C3
Piracuruca 9,419 . . . . F3
Piraí do Sul 13,709 . . . . . . B4
Piraju 16,288 . . . . *B3
Pirapora 31,533 . . . . E7
Pirassununga 32,510 . . . . *C2
Pires do Rio 16,659 . . . . . . D7
Piripiri 29,497 . . . . F4
Pitangui 12,116 . . . . *D1
Piúí 17,327 . . . . D2
Poções 16,036 . . . . F6
Poconé 12,960 . . . . B7
Poços de Caldas 81,448 . . *C2
Pombal 14,831 . . . . G4
Pompéia 11,282 . . . . *A3
Ponta Grossa 171,111 . . . . B4
Ponta Porã 25,807 . . . . C8
Ponte Nova 34,807 . . . . . . *E2
Porangatu 21,192 . . . . D6
Porto Alegre 1,108,883 . . D10
Porto Alegre †2,232,370 . . D10
Porto Feliz 19,680 . . . . . . *C3
Porto Nacional 19,052 . . . . E5
Porto Seguro 5,007 . . . . . . G7
Porto União 19,426 . . . . . . D9
Porto Velho 101,644 . . . . H10
Pouso Alegre 50,517 . . . . *D3
Presidente Dutra 14,506 . . E4
Presidente Prudente
127,623 . . . . D8
Presidente Venceslau 26,720 D8
Propriá 19,034 . . . . G5
Promissão 15,333 . . . . . . *B2
Prudentópolis 8,645 . . . . . D9
Quaral 15,091 . . . . C10
Quixadá 25,149 . . . . G4
Quixeramobim 14,387 . . . . F4
Raposos 11,078 . . . . *E2
Raul Soares 10,055 . . . . . *E2
Recife 1,184,215 . . . . H5
Recife †2,348,362 . . . . . . H5
Registro 28,702 . . . . *C4
Remanso 13,067 . . . . F5
Resende 36,633 . . . . *D3
Ribamar (São José de
Ribamar) 17,560 . . . . . . F3
Ribeirão Preto 300,704 . . . *C2
Rio Bonito 20,561 . . . . . . *E3
Rio Branco 87,462 . . . . . G10
Rio Claro 103,174 . . . . . . *C3
Rio de Janeiro 5,093,237 . *E3
Rio de Janeiro †9,018,637 . *E3
Rio do Sul 33,408 . . . . . . D9
Rio Grande 124,706 . . . . D11
Rio Negro 15,851 . . . . D9
Rio Pardo 18,370 . . . . . . C10
Rio Pomba 9,319 . . . . *E2
Rio Tinto 12,511 . . . . H4
Rio Verde 47,639 . . . . D7
Rio Verde de Mato Grosso
10,001 . . . . C7
Rosário 11,669 . . . . F3
Rosário do Sul 30,753 . . . C10
Russas 16,259 . . . . G4
Sabará 22,883 . . . . *E1
Sacramento 10,524 . . . . . *C1
Salgueiro 25,915 . . . . G5
Salinas 12,613 . . . . F7
Salinópolis 10,395 . . . . . . E3
Salto 42,351 . . . . *C3
Salvador 1,496,276 . . . . . G6
Salvador †1,772,018 . . . . . G6
Santa Cruz 13,172 . . . . . . G4
Santa Cruz do Rio Pardo
20,507 . . . . *B3
Santa Cruz do Sul 52,050 . C10
Santa Helena de Goiás
20,067 . . . . D7
Santa Leopoldina 1,217 . . G7
Santa Maria 151,202 . . . . C10
Santa Maria da Vitória
16,294 . . . . F6
Santana do Ipanema 15,311 . G5
Santana do Livramento
58,165 . . . . C10
Santarém 101,534 . . . . . . C3
Santa Rita do Sapucaí
15,005 . . . . *D3
Santa Vitória do Palmar
14,758 . . . . C11
Santiago 30,406 . . . . C10
Santo Amaro 29,627 . . . . G6
Santo Ângelo 50,161 . . . . C10
Santo André 549,278 . . . *C3
Santo Antônio da Platina
21,284 . . . . *A3
Santos 411,023 . . . . *C3
Santos Dumont 31,053 . . *E2
São Bento 9,607 . . . . E3
São Bernardo do Campo
381,261 . . . . *C3
São Borja 41,598 . . . . . . C10
São Carlos 109,231 . . . . *C3
São Cristóvão 11,720 . . . G5
São Fidélis 11,713 . . . . . *F2
São Francisco 12,011 . . . E6
São Francisco do Sul
13,914 . . . . E9
São Gabriel 40,497 . . . . . C10
São Gonçalo 221,278 . . . F8
São João da Boa Vista
45,712 . . . . *C2
São João del Rei 53,401 . D2
São João dos Patos 12,848 . F4
São João Nepomuceno
12,752 . . . . *E2
São Joaquim da Barra
26,273 . . . . *C2

São José 37,562 . . . . . . D9
São José do Rio Pardo
21,914 . . . . *C2
São José do Rio Preto
171,982 . . . . *B2
São José dos Campos
268,073 . . . . *D3
São José dos Pinhais
53,422 . . . . D9
São Leopoldo 94,864 . . D10
São Lourenço 23,047 . . . *D3
São Lourenço do Sul
13,251 . . . . C10
São Luís 182,466 . . . . . . F3
São Luís Gonzaga 29,188 . C10
São Manuel 17,028 . . . . *B3
São Mateus 22,522 . . . . G7
São Miguel de Guamá 9,929 E3
São Miguel dos Campos
18,495 . . . . G5
São Paulo 7,033,529 . . . *C3
São Paulo †12,588,439 . . *C3
São Paulo de Olivença 3,102 G9
São Raimundo Nonato 8,574 F5
São Roque 26,118 . . . . . *C3
São Sebastião 11,065 . . . *D3
São Sebastião do Paraíso
28,482 . . . . *C2
São Vicente 192,770 . . . *C4
Senador Pompeu 10,109 . . G4
Sena Madureira 6,668 . . . H9
Senhor do Bonfim 33,811 . . F5
Serra do Navio 415 . . . . . C2
Serra Talhada 28,912 . . . G4
Serrinha 23,920 . . . . G5
Sertânia 11,410 . . . . G5
Sete Lagoas 94,502 . . . . E7
Sobral 69,072 . . . . G3
Socorro 12,111 . . . . *C3
Sorocaba 254,718 . . . . . *C3
Soure 11,306 . . . . D3
Taguatinga 480,109 . . . . D6
Taquaritinga 28,018 . . . . *B2
Tarauacá 6,889 . . . . . . G10
Tatuí 44,816 . . . . *C3
Taubaté 155,371 . . . . . . E8
Tefé 14,670 . . . . G9
Teófilo Otoni 83,108 . . . F7
Teresina 339,264 . . . . . F4
Teresópolis 78,782 . . . . *E3
Tijucas 8,979 . . . . D9
Timon 55,318 . . . . F4
Tocantinópolis 8,427 . . . D4
Touros H4

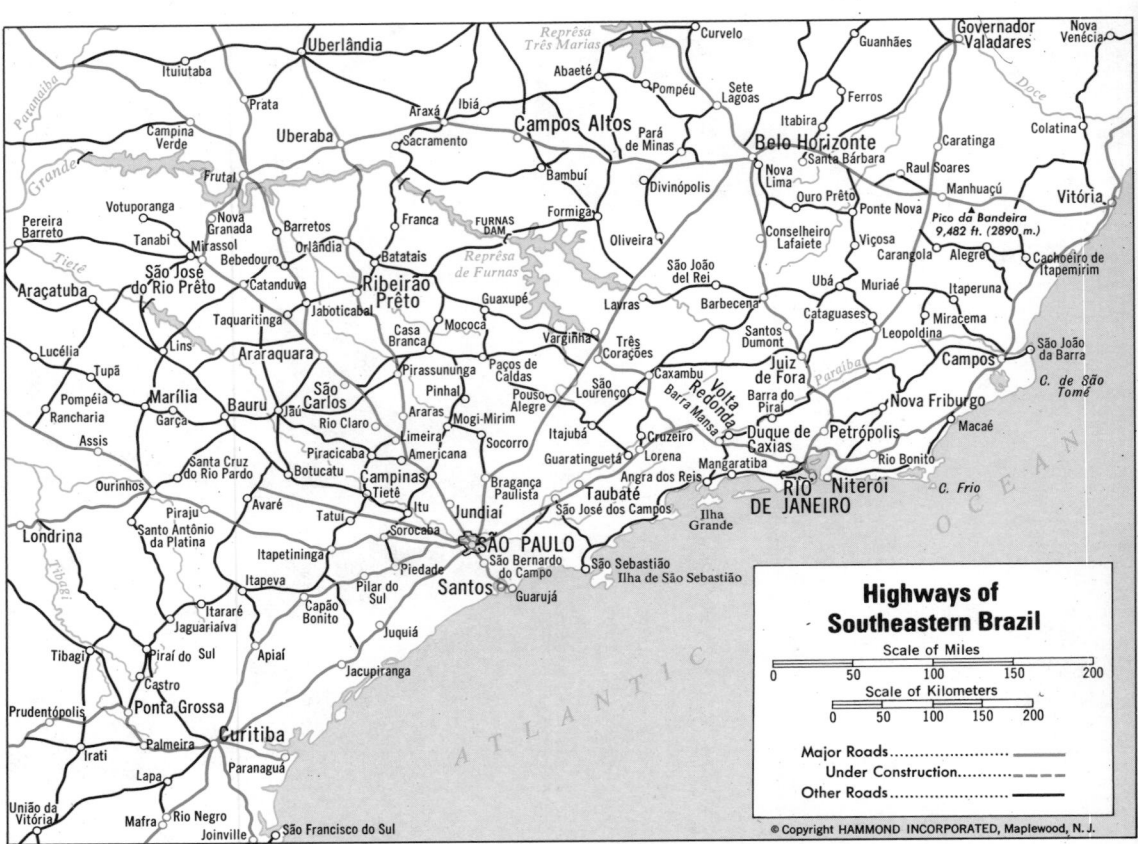

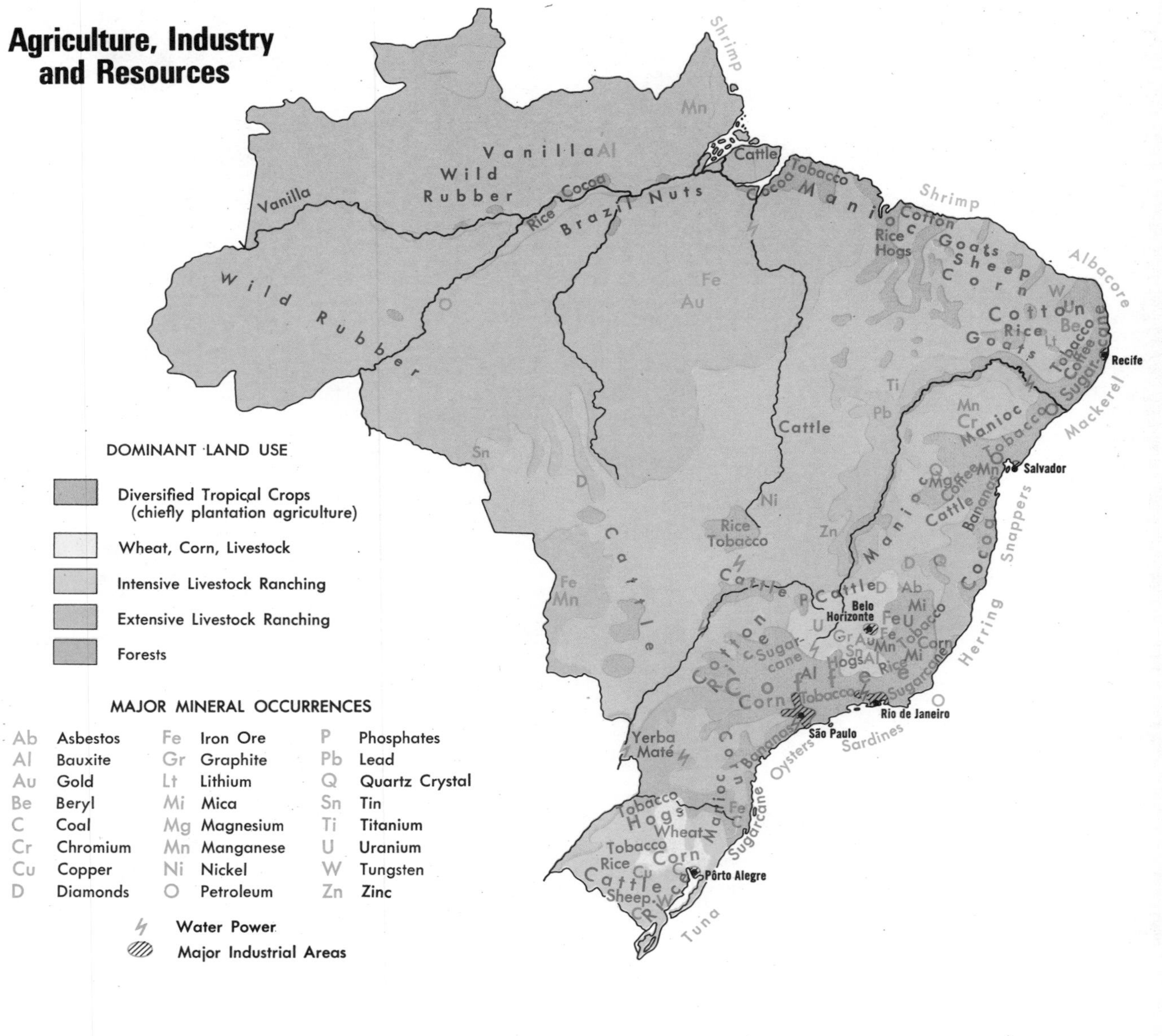

Highways of Southeastern Brazil

Scale of Miles
0   50   100   150   200

Scale of Kilometers
0   50   100   150   200

Major Roads . . . .
Under Construction . . . .
Other Roads . . . .

© Copyright HAMMOND INCORPORATED, Maplewood, N.J.

Agriculture, Industry and Resources

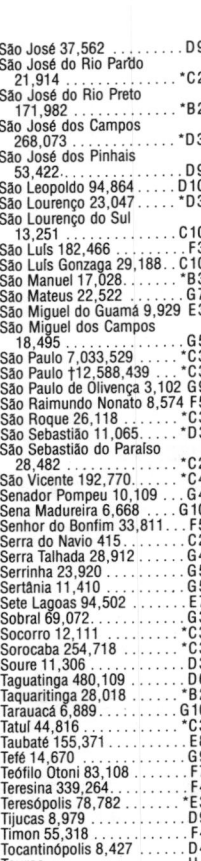

DOMINANT LAND USE

Diversified Tropical Crops
(chiefly plantation agriculture)

Wheat, Corn, Livestock

Intensive Livestock Ranching

Extensive Livestock Ranching

Forests

MAJOR MINERAL OCCURRENCES

Ab	Asbestos	Fe	Iron Ore	P	Phosphates
Al	Bauxite	Gr	Graphite	Pb	Lead
Au	Gold	Lt	Lithium	Q	Quartz Crystal
Be	Beryl	Mi	Mica	Sn	Tin
C	Coal	Mg	Magnesium	Ti	Titanium
Cr	Chromium	Mn	Manganese	U	Uranium
Cu	Copper	Ni	Nickel	W	Tungsten
D	Diamonds	O	Petroleum	Zn	Zinc

⚡ Water Power

▨ Major Industrial Areas

Três Corações 36,179 . . . . . *D2
Três Lagoas 45,171 . . . . C8
Três Pontas 24,225 . . . . . . *D2
Três Rios 47,497 . . . . *E3
Trindade 22,321 . . . . D7
Tubarão 64,585 . . . . D10
Tucuruí 27,209 . . . . D3
Tupã 44,450 . . . . *A2
Tupanciretã 13,103 . . . . . . C10
Tutóia 4,766 . . . . F3
Ubá 43,080 . . . . *E2
Ubaitaba 9,413 . . . . G6
Ubatuba 23,078 . . . . *D3
Uberaba 180,296 . . . . *C1
Uberlândia 230,400 . . . . . . E7
Unaí 28,148 . . . . E7
União 9,396 . . . . F4
União da Vitória 22,682 . . . . D9
União dos Palmares 20,876 . . H5
Uruaçu 19,607 . . . . D6
Uruçuí 6,047 . . . . E4
Uruguaiana 79,059 . . . . . . B10
Vacaria 37,370 . . . . D10
Valença 34,231 . . . . *E3
Varginha 57,448 . . . . *D2
Viana 9,753 . . . . E3
Viçosa 9,843 . . . . G5
Viçosa 29,198 . . . . *E2
Vigia 14,749 . . . . E3
Vila Velha Argolas 74,166 . . F8
Vilhena 12,565 . . . . H10
Viscondé dos Rio Branco
  17,295 . . . . *E2
Vitória 144,143 . . . . G8
Vitória da Conquista 125,717 H6
Vitória de Santo Antão
  62,890 . . . . G4
Volta Redonda 177,772 . . . *D3
Votuporanga 44,169 . . . . . . *A2
Xapuri 3,122 . . . . G10
Xique-Xique 17,625 . . . . . . F5

OTHER FEATURES

Abacaxis (riv.) B4
Abunã (riv.) G10
Acaraí, Serra do (range) . . . B2
Acre (riv.) G10
Aiama (lake) H9
Amambaí, Serra de (range) . C7
Amapari (riv.) C2
Amazon (riv.) C3

Anauá (riv.) B2
Aporé (riv.) D7
Araguaia (riv.) D4
Araguari (riv.) D2
Arinos (riv.) B5
Aripuanã (riv.) A4
Armando Laydner (res.) . . . *B3
Bailique (isl.) D2
Balsas (riv.) E5
Bananal (isl.) D5
Bandeira, Pico da (mt.) . *E2, F8
Braço Maior do Araguaia
  (riv.) D5
Braço Menor do Araguaia
  (riv.) D6
Branco (riv.) H8
Buzios (cape) *F3
Canumã (riv.) B4
Carajás, Serra dos (range) . . D4
Cardoso (isl.) *C4
Cassiporé (cape) D2
Caviana (isl.) D2
Chavantes, Serra dos
  (range) D5
Claro (riv.) D7
Comprida (isl.) *C4
Cuiabá (riv.) B7
Culuene (riv.) C6
Curuá (riv.) C4
Doce (riv.) *E2, F7
Dois Irmãos, Serra (range) . . F5
Espigão Mestre (Geral
  de Goiás) (range) E6
Espinhaço, Serra da (range) . F7
Estrondo, Serra do (range) . D4
Feia (lake) *F3
Feio (riv.) *B2
Formosa, Serra (range) C5
Frio (cape) *F3
Furnas (dam) *C2
Geral de Goiás, Serra
  (range) E6
Gi-Paraná (riv.) H10
Gradaús, Serra dos (range) . D4
Grajaú (riv.) E4
Grande (isl.) *D3
Grande (riv.) *B2, E8
Guanabara (bay) *E3
Guaporé (riv.) H10

Gurguéia (riv.) E5
Gurupi, Serra do (range) . . . E4
Gurupi (riv.) E3
Ibicuí (riv.) C10
Içá (riv.) G9
Iguaçu (riv.) C9
Iguazú (falls) C9
Ilha Grande (bay) *D3
Iriri (riv.) C4
Itaipu (dam) C9
Itaipu (res.) C9
Itapecuru (riv.) F3
Itapi (riv.) F3
Itapicuru (riv.) G5
Itararé (riv.) *B3
Ival (riv.) C8
Jaculpe (riv.) F5
Jaguaribe (riv.) G4
Jamanxim (riv.) C4
Japurá (riv.) G9
Jari (riv.) C3
Jauari, Serra (mts.) C3
Javari (riv.) F9
Jequitinhonha (riv.) F7
Juruá (riv.) G10
Juruena (riv.) B5
Jutaí (riv.) G9
Lombarda, Serra (mts.) C3
Madeira (riv.) A4
Mangueira (lag.) D11
Manso (riv.) C6
Mantiqueira (range) *D3
Mapuera (riv.) B3
Mar, Serra do (range) . . *C4, E9
Maracá (isl.) D2
Marajó (bay) E2
Marajó (isl.) 147,895 . . . . D3
Mato Grosso, Planalto de
  (plat.) B6
Maués-Açu (riv.) B4
Mearim (riv.) E4
Mexiana (isl.) D2
Miranda (riv.) B8
Mirim (lag.) C11
Mogi Guaçu (riv.) *C2
Mortes (Manso) (riv.) D6
Neblina, Pico da (peak) . . . G8
Negro (riv.) H9
Nhamundá (riv.) B3
Norte, Serra do (range) . . . B5
Oiapoque (Oyapock) (riv.) . . C2

Orange (cape) D1
Órgãos (range) *E3
Oyapock (riv.) C2
Pacajá Grande (riv.) D4
Pacaraimã, Serra da (mts.) . H8
Papagaio (riv.) B6
Pará (riv.) D3
Paracatu (riv.) E7
Paraguaçu (riv.) F6
Paraguai (riv.) B8
Paraíba (riv.) *E2
Paraná (riv.) C8
Paraná (riv.) C7
Paranapanema (riv.) . . . *B3, C8
Paranapiacaba (range) *B4
Paranatinga (range) C6
Pardo (riv.) *B2, D8
Pardo (riv.) C8
Pardo (riv.) F6
Parecis, Serra dos (range) . . B6
Parnaíba (riv.) F3
Paru (riv.) C3
Patos (lag.) D10
Penitente, Serra do (range) . E5
Piauí, Serra do (range) F5
Piauí (riv.) F5
Purus (riv.) H9
Ribeira (riv.) *B4
Roncador, Serra do (range) . D5
Ronuro (riv.) C5
Roosevelt (riv.) A4
Santa Catarina (isl.) 138,556 E9
São Lourenço (riv.) C7
São Marcos (bay) E3
São Roque (cape) F4
São Francisco (riv.) . . *D2, G5
São Sebastião (isl.) 5,724 . *D3,
  . . . . E8
São Tomé (cape) F8
Sapucaí (riv.) *D2
Sepetiba (bay) *D3
Sete Quedas (falls) C9
Sete Quedas (Grande) (isl.) . C8
Sobradino (res.) F5
Sono (riv.) E5
Sul (chan.) D2
Tacutu (riv.) B2
Tapajós (riv.) B4
Taquari (riv.) C7
Tefé (riv.) G9
Teles Pires (riv.) B5

Tibagi (riv.) *A4
Tietê (riv.) *B2, D8
Tiracambu, Serra (range) . . E3
Tocantins (riv.) D3
Tombador, Serra do (range) . B6
Trombetas (riv.) B3
Tucuruí (res.) D4
Tumucumaque, Serra de
  (range) C2

Turvo (riv.) *B2
Uaupés (riv.) G9
Uraricoera (riv.) H8
Urubu (riv.) A3
Urubupungá (dam) C8
Urucún, Morro do (mt.) . . . B7
Uruguai (riv.) D4
Vasa Barris (riv.) G5
Velhas (riv.) E7

Verde (riv.) C7
Verdinho (riv.) D7
Xavantes (res.) *B3
Xingu (riv.) C3

†Population of met. area.
*preceding reference indicates
  that the name will be found on
  S.E. Brazil map, page 135.

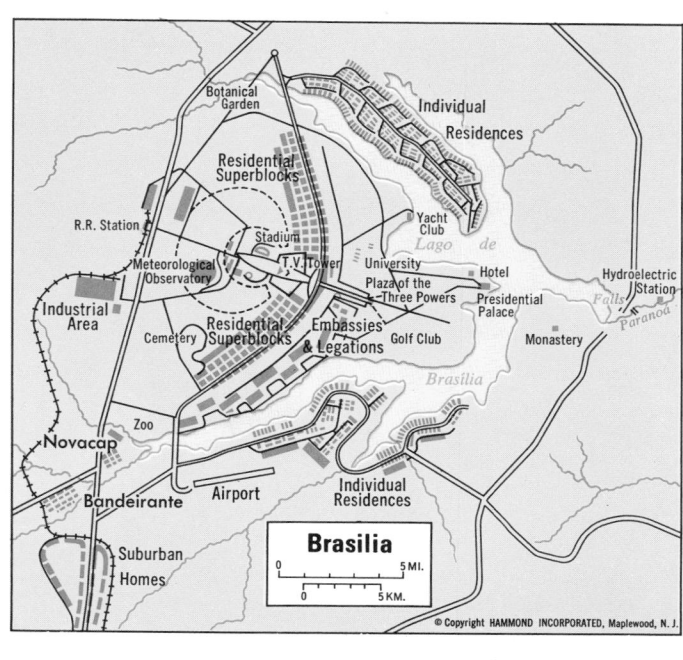

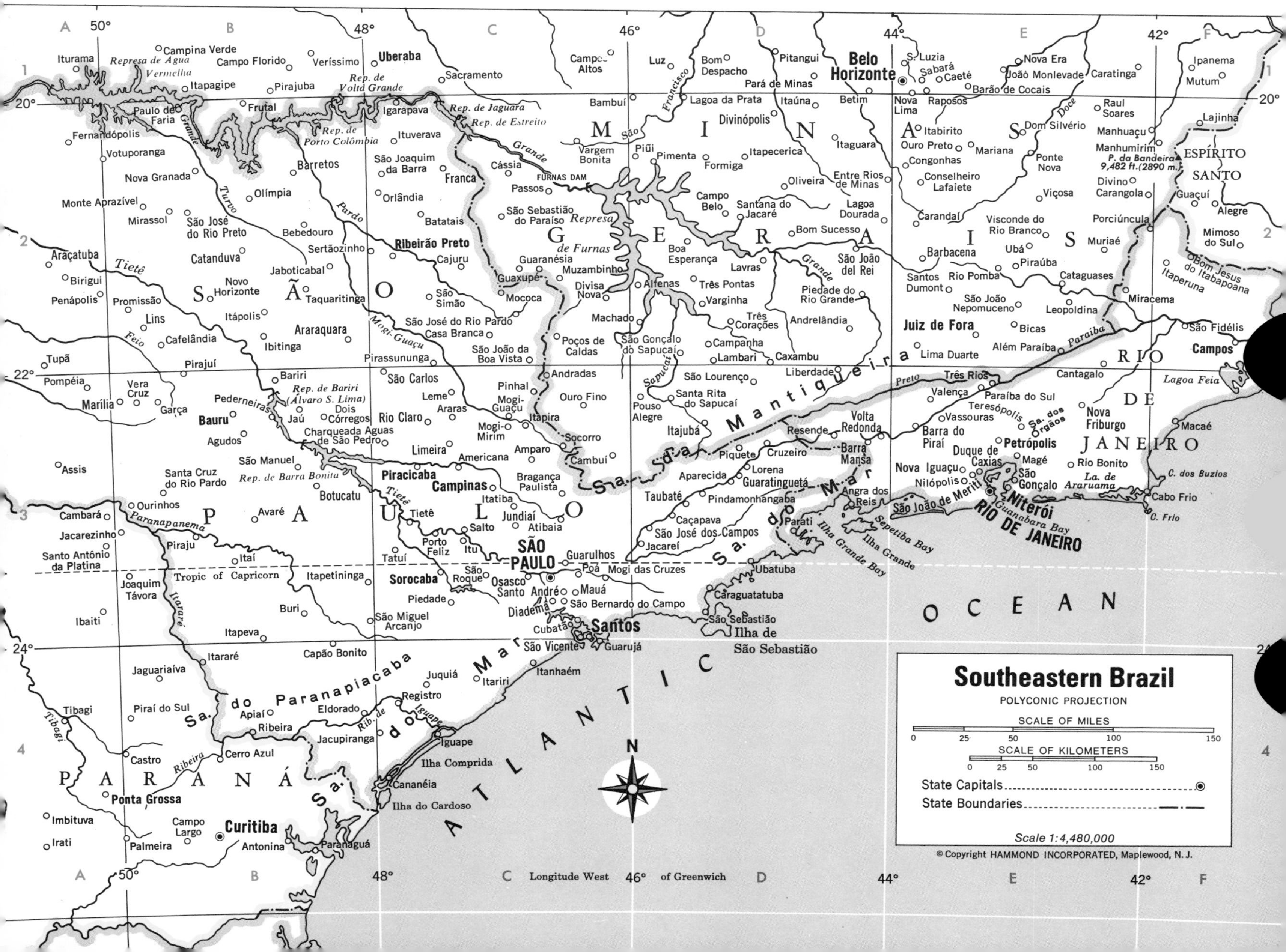

Brasilia

© Copyright HAMMOND INCORPORATED, Maplewood, N.J.

Southeastern Brazil

POLYCONIC PROJECTION

SCALE OF MILES

SCALE OF KILOMETERS

State Capitals

State Boundaries

Scale 1:4,480,000

© Copyright HAMMOND INCORPORATED, Maplewood, N.J.

DEPARTMENTS

Beni, El 168,367 ....C3
Chuquisaca 358,516 ....C6
Cochabamba 720,952 ....C5
El Beni 168,367 ....C3
La Paz 1,465,078 ....A4
Oruro 310,409 ....A5
Pando 34,493 ....B2
Potosí 657,743 ....B7
Santa Cruz 710,724 ....E5
Tarija 187,204 ....D7

CITIES and TOWNS

Abapó 466 ....D6
Acchilla 208 ....C7
Achacachi 3,621 ....A5
Aiquile 3,465 ....C6
Alcalá 236 ....C6
Alejandría‡ 198 ....C3
Alto Secoz 3,414 ....D4
Amarete 902 ....A4
Ananea 302 ....A4
Ancoraimes 769 ....A4
Añimbo 443 ....C7
Anzaldo 1,056 ....C5

Apolo 1,043 ....A4
Aracat 3,537 ....B5
Arampampa 829 ....C5
Arani 2,200 ....C5
Arcopongo‡ 2,223 ....B5
Aroma‡ 873 ....B6
Arque 1,254 ....B5
Arroyo Grande B5
Ascención (Añez) D4
Asunción ....
Asunta 45 ....A5
Atén 199 ....A4
Atocha‡ 3,964 ....B7
Ayacucho 729 ....D5
Ayata 479 ....A4

Azurduy 1,234 ....C6
Barrera B3
Baures 592 ....D3
Bella Flor B2
Bella Vista E3
Berenguela‡ 2,412 ....A5
Betanzos 1,097 ....B6
Bolívar B5
Bolpebra A2
Boyuibe 537 ....D7
Buena Vista, Santa Cruz ...D5
Cabezas 298 ....D6
Cachuela Esperanza 1,073 ..C2
Caiza 838 ....C7
Cajuata 447 ....B5

Calacoto 415 ....A5
Calamarca 802 ....A5
Callapa 636 ....A5
Camacho‡ 875 ....A5
Camargo 1,609 ....C7
Camatindi‡ 297 ....D7
Camiri 4,969 ....D7
Candelaria 468 ....F5
Canquella 148 ....A7
Capinota 1,734 ....B5
Capirenda D7
Caquiaviri 760 ....A5
Carabuco 626 ....A5
Caracollo 909 ....B5
Caranavi‡ 525 ....B4

Carandaiti 1,403 ....D7
Caraparí 351 ....D7
Carmen‡ 845 ....B2
Cavari 249 ....B5
Cataricahua 3,240 ....B6
Cavinas‡ 1,011 ....A3
Chachacomani 159 ....A6
Chaguaya 643 ....C7
Challacollo 284 ....A5
Challana‡ 1,206 ....A4
Challapata 2,529 ....B6
Chapacura‡ 152 ....C5
Chaqui 291 ....C6
Charagua 1,185 ....D6

Charaña 794 ....A5
Chayanta 1,272 ....B6
Chiguana 154 ....A7
Chijlijo 27 ....A4
Chivet 336 ....A3
Chocaya 444 ....B7
Choquecotat 1,976 ....A4
Chulumani 2,362 ....B5
Chuma 931 ....A4
Chuquichambi 1,094 ....A5
Chuquichuquit 1,892 ....B6
Cliza 3,121 ....C5
Cobija 3,650 ....B2
Cocaní 658 ....A4
Cocapata 2,855 ....B5

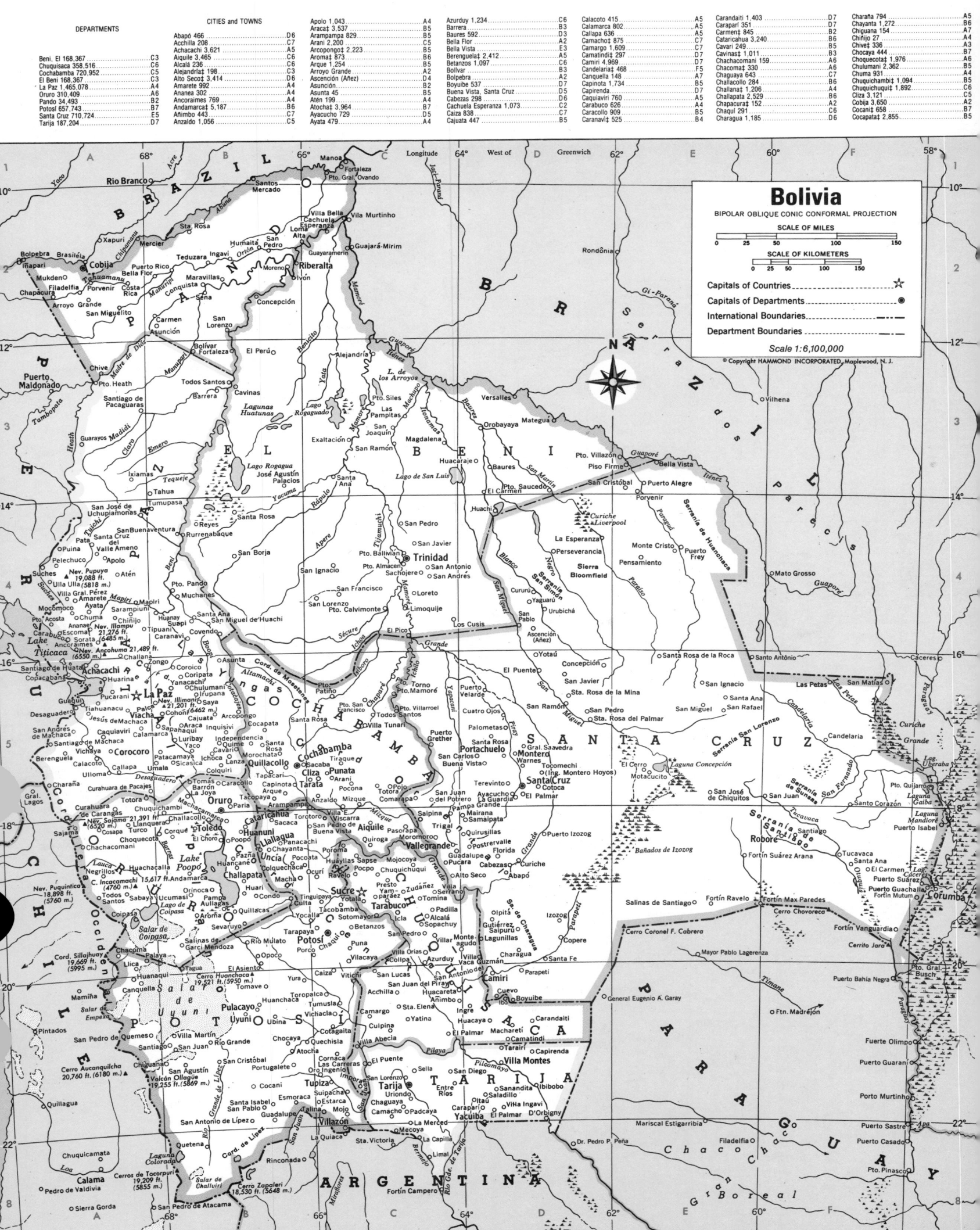

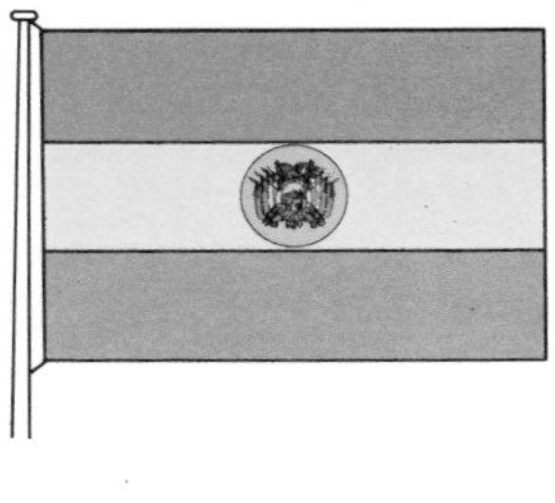

AREA 424,163 sq. mi. (1,098,582 sq. km.)
POPULATION 5,600,000
CAPITALS La Paz, Sucre
LARGEST CITY La Paz
HIGHEST POINT Nevada Ancohuma 21,489 ft. (6,550 m.)
MONETARY UNIT Bolivian peso
MAJOR LANGUAGES Spanish, Quechua, Aymara
MAJOR RELIGION Roman Catholicism

ochabamba 204,684 C5
ohoni 890 B5
oipasa‡ 202 A6
ollpa 481 C6
olquechaca 1,070 B6
olquiri 806 B5
omarapa 1,096 C5
oncepción, El Beni‡ 61 B2
oncepción, Santa Cruz 1,056 D5
ondo‡ 5,525 B6
onquista‡ 1,162 B2
opacabana 1,981 A5
opere D6
oripata 1,647 B5
ornaca 264 C7
orocoro 4,431 A5
oroico 2,235 B5
orque 423 B6
osapa 297 A6
osta Rica‡ 43 A2
otagaita 1,353 C7
otoca 915 D5
ovendo 71 B4
uatro Ojos‡ 465 D5
uevo 902 D7
ulpina 981 C7
ulta‡ 4,412 B6
urahuara de Carangas 235 A5
urahuara de Pacajes 510 A5
uriche 257 D6
urirú D4
esaguadero 201 A5
'Orbigny‡ 214 D7
Asiento B6
Carmen, El Beni 232 D3
Carmen, Santa Cruz F6
Cerro 117 E5
Choro 224 B6
Palmar, Chuquisaca‡ 772 D7
Palmar, Santa Cruz 437 D5
Palmar, Tarija 832 D7
Perú B3
Pico C4
Puente, Santa Cruz‡ 1,185 D5
Puente, Tarija‡ 1,310 C7
ntre Ríos 1,011 C7
scoma 220 A4
smoraca‡ 1,137 B7
starca‡ 2,331 C7
xaltación, El Beni 405 C3
ladelfia‡ 942 A2
orida, Santa Cruz 128 D6
ortaleza‡ 765 B3
ortaleza C1
ortín Campero‡ 87 C8
ortín Max Paredes F6
ortín Mutum F6
ortín Ravelo E6
ortín Suárez Arana F6
ortín Vanguardia F6
eneral Saavedra 1,006 D5
uadalupe, Potosí 71 B7
uadalupe, Santa Cruz 2,355 C6
uaqui 2,266 A5
uayaramerín 1,470 C2
uacaraje 673 D3
uacareta 239 C7
uacaya 229 D7
uachacalla 801 A6
uachi D4
uanaqui 359 A7
uanay 574 B4
uancané 148 B6
uanchaca B7
uanuni 5,696 B6
uari 1,070 B6
uarina 1,151 A5
uayllas 206 C6
umaitá‡ 429 B2
ibobo D7
o D7
hoca 591 B5
la 196 C6
npora 274 C7
dependencia 1,742 B5
gavi‡ 111 B2
geniero Montero Hoyos (Tocomechi) 575 D5
gre 162 D7
quisivi 530 B5
upana 1,937 B5
aú 102 D7
ón‡ 772 C2
iamas 292 A3
ozog‡ 2,759 D6
esús de Machaca 529 A5
osé Agustín Palacios‡ 2,273 B3
a Capilla‡ 1,870 C8
a Esmeralda D8
a Esperanza D4
a Guardia 470 D5
agunillas 840 D6
a Joya 401 B5

La Merced‡ 688 C8
Lanza 526 B5
La Paz (cap.) 635,283 B5
Las Carreras 155 C7
Las Pampitas‡ 71 C3
Las Petas‡ 383 F5
Limal‡ 524 C8
Limoquije C4
Llallagua 6,719 B6
Llanquera 613 A6
Llica 560 A6
Loma Alta B2
Loreto 589 C4
Los Cusis D4
Luribay 392 B5
Macha 1,050 B6
Machacamarca 1,746 B5
Macharetí‡ 1,164 D7
Magdalena 1,724 C3
Mairana 508 D6
Manoa C1
Mapiri 289 B4
Maravillas B2
Mateguá 38 D3
Mecoya‡ 585 C8
Mercier‡ 272 B2
Mizque 870 C5
Mocomoco 977 A4
Mojo 469 C7
Mojocoya 498 C6
Monteagudo 971 D6
Monte Cristo E4
Montero 2,713 D5
Moreno B2
Morochata 461 B5
Moromoro 556 C6
Motacucito‡ 585 E5
Muchanes B4
Mukden‡ 84 A2
Negrillos 85 A6
Ocurí 1,531 C6
Opoco B6
Orinoca‡ 2,380 B6
Orobayaya‡ 1,132 D3
Oro Ingenio‡ 945 C7
Oruro 124,213 B5
Padcaya 324 C7
Padilla 2,462 C6
Palaya 300 A6
Palca 887 A5
Palometas‡ 3,453 D5
Pampa Aullagas‡ 1,834 B6
Pampa Grande 727 D5
Panacachi 952 B6
Paria 335 B5
Pasorapa 1,016 C6
Pata 122 A4
Patacamaya 1,278 B5
Pazña 671 B6
Pelechuco 873 A4
Pensamiento E4
Perseverancia D4
Piso Firme D3
Pocoata 859 B6
Pocona 518 C5
Pocpo‡ 2,791 C6
Pojo 1,047 C5
Poopó 736 B6
Porco 817 B6
Poroma 171 C6
Portachuelo 2,456 D5
Portugalete‡ 1,590 B7
Porvenir, Pando‡ 846 A2
Porvenir, Santa Cruz E4
Postrervalle 750 D6
Potosí 77,397 C6
Presto 725 C6
Pucara 762 C6
Pucarani 1,041 A5
Puerto Acosta 1,302 A4
Puerto Alegre E3
Puerto Almacen 358 C4
Puerto Ballivián C4
Puerto Calvimonte C4
Puerto Frey E4
Puerto General Ovando 50
Puerto Grether C5
Puerto Guachalla F6
Puerto Heath‡ 570 A3
Puerto Isabel F6
Puerto Izozog D6
Puerto Mamoré C5
Puerto Pando B4
Puerto Patiño C5
Puerto Quijarro G5
Puerto Rico‡ 539 B2
Puerto San Francisco C5
Puerto Saucedo D3
Puerto Siles 357 C3
Puerto Suárez 1,159 F6
Puerto Torno C5
Puerto Velarde D5
Puerto Villarroel C5
Puerto Villazón D3

Puina A4
Pulacayo 7,984 B7
Puna 852 C6
Punata 5,014 C5
Quechisla 171 C7
Queteña 183 B8
Quillacas 1,170 B6
Quillacollo 9,123 B5
Quime 1,256 B5
Quiroga‡ 3,467 C6
Quirusillas 433 D6
Ravelo 907 C6
Reyes 1,404 B4
Riberalta 6,549 C2
Río Grande 281 B7
Río Mulato 381 B6
Roboré 3,715 F6
Rurrenabaque 1,225 B4
Sabaya 649 A6
Sacaba 2,752 C5
Sacaca 1,778 B6
Sachojere 401 C4
Saipina 573 C6
Sajama 231 A6
Saladillo‡ 1,315 D7
Salinas de Garci Mendoza 335 B6
Salinas de Santiago E6
Samaipata 1,656 D6
San Agustín‡ 810 B7
Sanandita 379 D7
San Andrés 399 C4
San Andrés de Machaca 101 A5
San Antonio, El Beni 436 C4
San Antonio de Lípez‡ 177 B7
San Antonio del Parapetí 497 D7
San Borja 708 B4
San Buenaventura 307 A4
San Carlos 570 D5
San Cristóbal, Potosí‡ 1,200 B7
San Cristóbal, Santa Cruz E3
San Diego‡ 773 D7
San Francisco, El Beni 185 C4
San Ignacio, El Beni 1,757 C4
San Ignacio, Santa Cruz 1,819 E5
San Javier, El Beni 233 C4
San Javier, Santa Cruz 564 D5
San Joaquín 1,959 C3
San José de Chiquitos 1,933 E5
San José de Uchupiamonas 277 A4
San Juan, Potosí 131 B7
San Juan, Santa Cruz‡ 1,482 F5
San Juan del Piray 541 C7
San Juan del Potrero 263 C5
San Lorenzo, El Beni 496 C4
San Lorenzo, Pando‡ 317 B2
San Lorenzo, Tarija 785 C7
San Lucas 925 C7
San Matías 887 F5
San Miguel 502 E5
San Miguel de Huachi 25 B4
San Miguelito A2
San Pablo, Potosí 11 B7
San Pablo, Santa Cruz D4
San Pedro, Chuquisaca 182 C6
San Pedro, El Beni 262 C4
San Pedro, Pando‡ 312 B2
San Pedro, Santa Cruz 80 D5
San Pedro de Buena Vista 1,094 C6
San Pedro de Quemes‡ 290 A7
San Rafael‡ 1,282 E5
San Ramón, El Beni 1,161 C3
San Ramón, Santa Cruz 379 D5
Santa Ana, El Beni 2,225 C3
Santa Ana, La Paz 171 B4

Santa Ana, Santa Cruz 275 E5
Santa Ana, Santa Cruz 663 F6
Santa Cruz, Santa Cruz 254,682 D5
Santa Cruz del Valle Ameno 442 A4
Santa Elena‡ 4,474 C7
Santa Fe D6
Santa Isabel‡ 323 B7
Santa Rosa, Cochabamba‡ 942 B5
Santa Rosa, Cochabamba‡ 276 C5
Santa Rosa, El Beni 765 B4
Santa Rosa, Pando‡ 105 B2
Santa Rosa, Santa Cruz 995 D5
Santa Rosa de la Mina 99 D5
Santa Rosa de la Roca 101 E5
Santa Rosa del Palmar 441 E5
Santiago, Potosí 172 A7
Santiago, Santa Cruz 765 F6
Santiago de Huata 948 A5
Santiago de Machaca 218 A5
Santiago de Pacaguaras A3
Santo Corazón‡ 963 F5
Santos Mercado B1
Sapahaqui 55 B5
Sapse‡ 89 C6
Sarampiuni 138 A4
Saya 339 B5
Sella C7
Sena‡ 660 B2
Sevaruyo 475 B6
Sicasica 1,486 B5
Sopachuy 713 C6
Sorata 2,087 A4
Sotomayor 510 C6
Suapi‡ 1,750 B4
Suches‡ 231 A4
Sucre (cap.) 63,625 C6
Suipacha‡ 2,701 C7
Tacobamba‡ 6,933 C6
Tacopaya 795 B5
Tagua B6
Tahua B3
Talina 122 B7
Tapacarí 980 B5
Tarabuco 2,833 C6
Tarairí‡ 394 D7
Tarapaya 357 B6
Tarata 3,016 C5
Tarija 38,916 C7
Teduzara‡ 271 B2
Terevinto‡ 3,790 D5
Tiahuanacu 1,227 A5
Tinguipaya 766 C6
Tipuani‡ 1,216 B4
Tiraque 1,390 C5
Tocomechi 575 D5
Todos Santos, Cochabamba 408 C5
Todos Santos, La Paz B3
Todos Santos, Oruro 68 A6
Toledo 3,273 B6
Tomás Barrón 1,852 B5
Tomave 201 B7
Tomina 708 C6
Toropalca‡ 199 B7
Torotoro 1,233 C6
Totora, Cochabamba C5
Totora, Oruro A5
Trigal 749 C6
Trinidad, El Beni 27,487 C4
Trinidad, Pando‡ 332 B2
Tucavaca F6
Tumupasa 349 B4
Tumusla‡ 526 C7
Tupiza 8,248 C7
Turco 131 A6
Ubina‡ 462 B7
Ucumasi‡ 1,040 B6

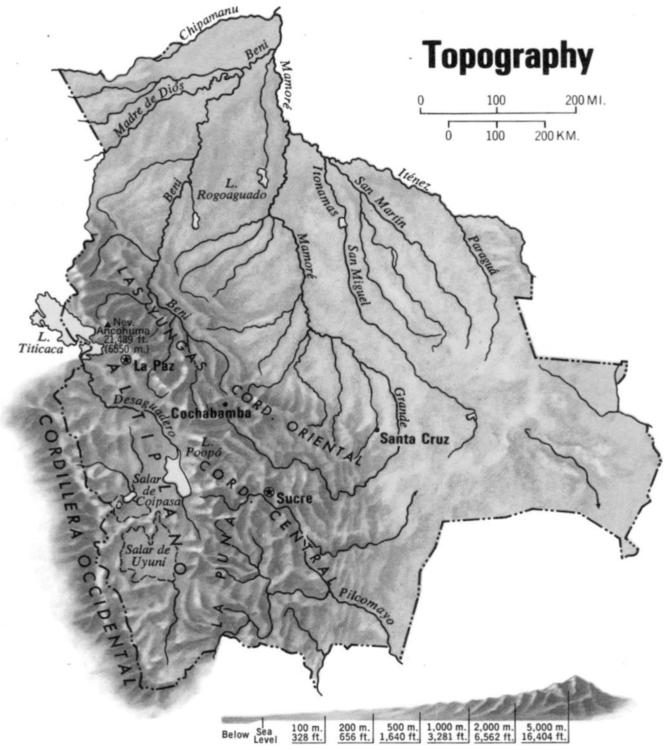

Ulla Ulla 52 A4
Ulloma 116 A5
Umala 481 B5
Uncía 4,507 B6
Uriondo 860 C7
Urubichá 1,369 D4
Uyuni 6,968 B7
Vallegrande 5,094 C6
Versalles 83 D3
Viacha 6,607 A5
Vichacla 317 C7
Vichaya 422 A5
Vilacaya 200 C6
Villa Abecia 539 C7
Villa Bella 88 C2
Villa E. Viscarra 658 C6
Villa General Pérez 802 A4
Villa Ingavi 122 D7
Villa Martín 543 B7
Villa Montes 3,105 D7

Villa Orías 404 C6
Villar 322 C6
Villa Serrano 1,570 C6
Villa Tunari 510 C5
Villa Vaca Guzmán 699 D6
Villazón 6,261 C7
Vitichi 1,515 C7
Warnes 1,571 D5
Yaco 835 B5
Yacuiba 5,027 D7
Yaguarú D4
Yamparaéz 725 C6
Yanacachi‡ 1,964 B5
Yatina‡ 1,850 C7
Yocalla‡ 1,814 B6
Yotala 1,554 C6
Yotaú D5
Yura 136 B7
Zongo 141 B5
Zudáñez 1,868 C6

OTHER FEATURES

Abuná (riv.) B2
Altamachi (riv.) B5
Ancohuma, Nevada (mt.) A4
Apere (riv.) C4
Arroyas, Los (lake) C3
Barras (riv.) B6
Baures (riv.) D3
Beni (riv.) B2
Benicito (riv.) C3
Bermejo (riv.) C8
Blanco (riv.) D4
Bloomfield, Sierra (mts.) D4
Boopi (riv.) B4
Cáceres (lag.) G6
Candelaria (riv.) F5
Capitán Ustarés, Cerro (mt.) E6
Central, Cordillera (range) C6
Challviri (salt dep.) B8
Chaparé (riv.) C5
Charagua, Sierra de (mts.) D6
Chipamanu (riv.) A2
Chovoreca, Cerro (mt.) F6
Claro (riv.) A3
Coipasa (lake) B6
Coipasa (salt dep.) A6
Colorada (lag.) A8
Concepción (lag.) E5
Coronel F. Gabrera E6
Cotacajes (riv.) B5
Desaguadero (riv.) B5
Emero (riv.) B3
Empexa (salt dep.) A7
Gaiba (lag.) F5
Grande (marsh) F5
Grande (riv.) C4
Grande (riv.) C6
Grande de Lípez (riv.) B7
Guaporé (riv.) C3
Heath (riv.) A3
Huanchaca, Cerro (mt.) B7
Huanchaca, Serranía de (mts.) E4
Huatunas (lag.) B3
Ichilo (riv.) C5
Ichoa (riv.) C4
Illampu, Nevada (mt.) A4
Illimani, Nevada (mt.) B5
Incacamachi, Cerro (mt.) A6

Isiboro (riv.) C5
Iténez (Guaporé) (riv.) C3
Itonamas (riv.) C3
Izozog (swamp) E6
Jara, Cerrito (mt.) F6
Las Yungas (reg.) B5
Lauca (riv.) A6
Lípez, Cordillera de (range) B8
Liverpool (swamp) D4
Machupo (riv.) C3
Madidi (riv.) A3
Madre de Diós (riv.) A3
Mamoré (riv.) C2
Mandioré (lag.) F6
Manuripi (riv.) B2
Mizque (riv.) C6
Mosetenes, Cordillera de (range) B5
Negro (riv.) D4
Occidental, Cordillera (range) A6
Ollagüe (vol.) B7
Oriental, Cordillera (range) C5
Ortón (riv.) B2
Otuquis (riv.) F6
Paraguá (riv.) E4
Paraguay (riv.) F7
Parapetí (riv.) D6
Petas, Las (riv.) F5
Pilaya (riv.) C7
Pilcomayo (riv.) D7
Piray (riv.) D5
Poopó (lake) B6
Pupuya, Nevada (mt.) A4
Puquintica, Nevado (mt.) A6
Rápulo (riv.) C4
Real, Cordillera (range) A5
Rogagua (lake) B3
Rogaguado (lake) C3
Sajama, Nevada (mt.) A6
San Fernando (riv.) F5
San Juan (riv.) C7
San Lorenzo, Serranía (mts.) E5
San Luis (lake) C3
San Martín (riv.) D3
San Miguel (riv.) D4
San Simón, Serranía (mts.) D4
Santiago, Serranía de (mts.) F6
Sécure (riv.) C4
Sillajhuay, Cordillera (mt.) A6
Suches (riv.) A4
Sunsas, Serranía de (mts.) F5
Tahuamanu (riv.) A2
Tarija, Rio Grande de (riv.) C8
Tequeje (riv.) B3
Tijamuchi (riv.) C4
Titicaca (lake) A4
Tocorpuri, Cerros de (mt.) A8
Tucavaca (riv.) F6
Tuichi (riv.) A4
Uberaba (lag.) G5
Uyuni (salt dep.) B7
Yacuma (riv.) B3
Yapacani (riv.) C5
Yata (riv.) C3
Yungas, Las (reg.) B5
Zapaleri, Cerro (mt.) B8

‡Population of canton.

Agriculture, Industry and Resources

DOMINANT LAND USE

- Diversified Tropical Crops (chiefly plantation agriculture)
- Upland Cultivated Areas
- Upland Livestock Grazing, Limited Agriculture
- Extensive Livestock Ranching
- Forests
- Nonagricultural Land

MAJOR MINERAL OCCURRENCES

Ag	Silver	G	Natural Gas	Sb	Antimony
Au	Gold	O	Petroleum	Sn	Tin
Cu	Copper	Pb	Lead	W	Tungsten
Fe	Iron Ore	S	Sulfur	Zn	Zinc

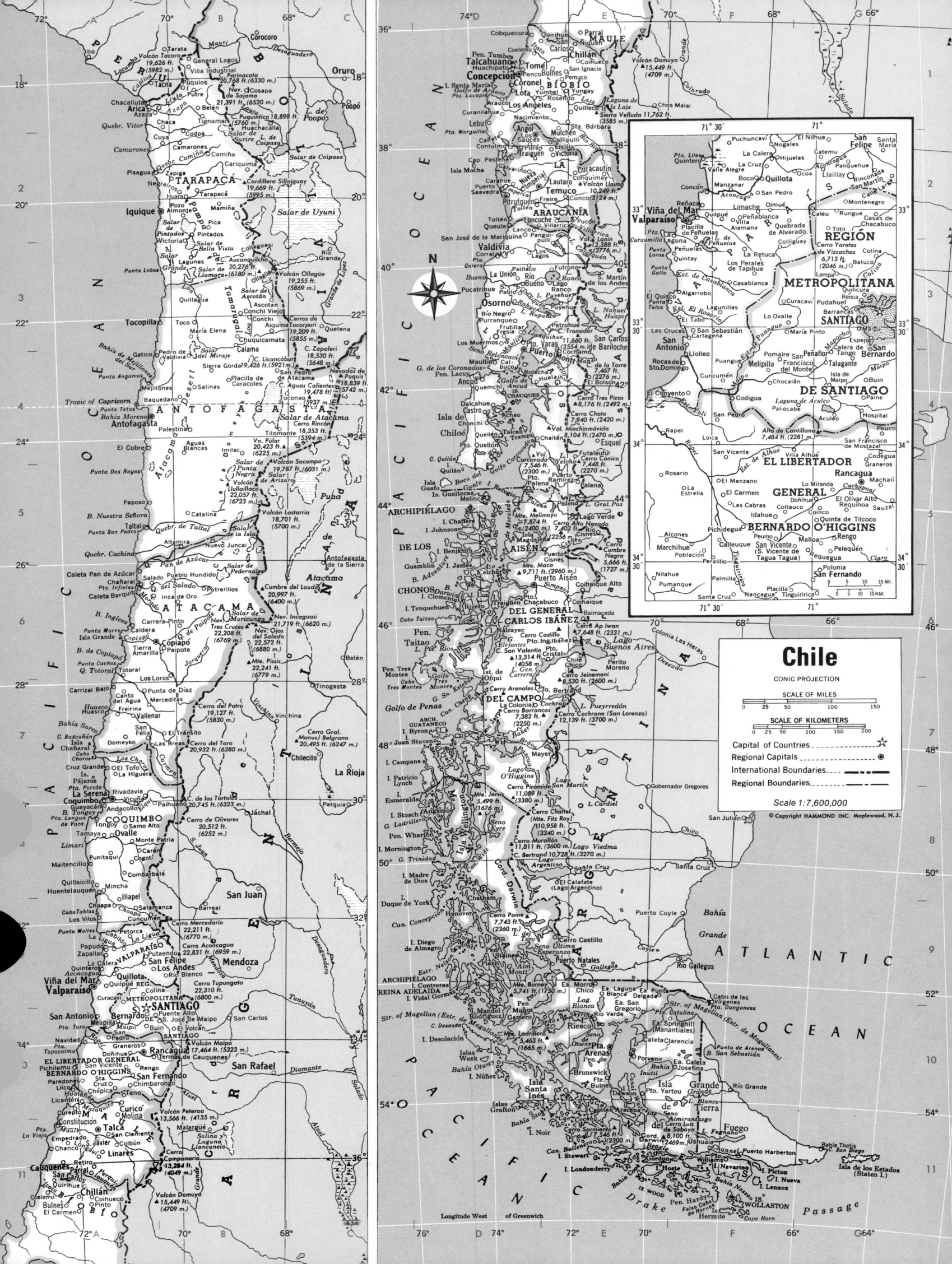

Chile

CONIC PROJECTION

SCALE OF MILES

SCALE OF KILOMETERS

Capital of Countries .... ☆
Regional Capitals .... ◉
International Boundaries .... ─ ─ ─
Regional Boundaries .... ─ · ─ · ─

Scale 1:7,600,000

© Copyright HAMMOND INC. Maplewood, N.J.

Topography

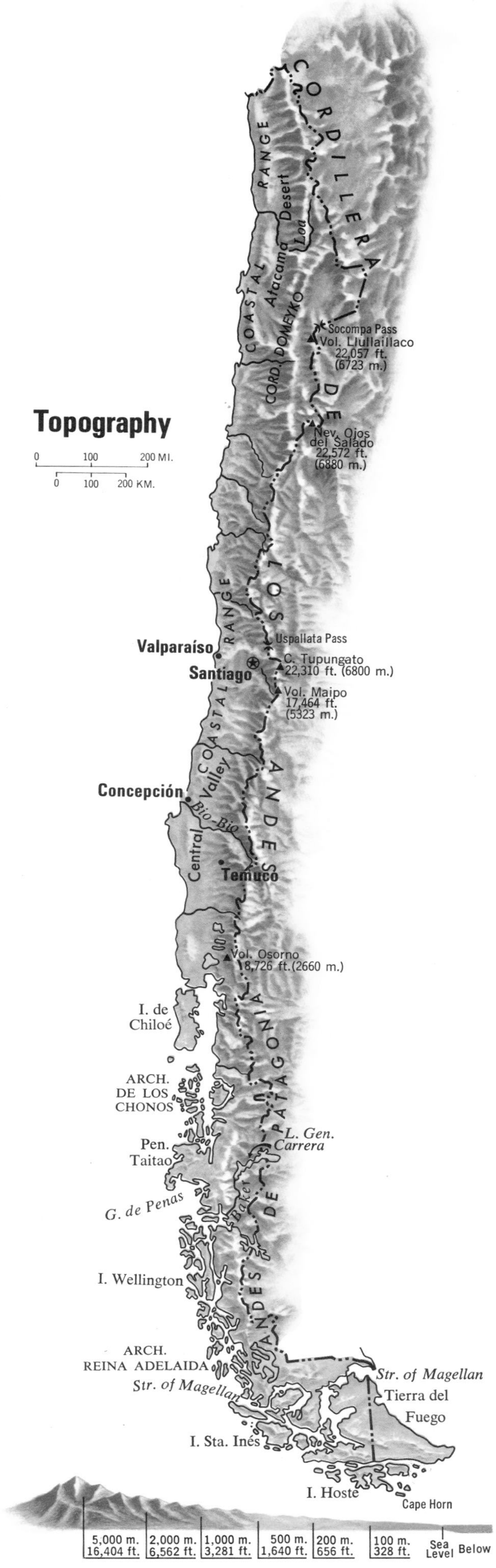
0 100 200 MI.
0 100 200 KM.
CORDILLERA DE LOS ANDES
COASTAL RANGE
Atacama Desert
Loa
CORD. DOMEYKO
Socompa Pass
Vol. Llullaillaco 22,057 ft. (6723 m.)
Nev. Ojos del Salado 22,572 ft. (6880 m.)
Valparaíso
Santiago
Uspallata Pass
C. Tupungato 22,310 ft. (6800 m.)
Vol. Maipo 17,464 ft. (5323 m.)
Central Valley
Concepción
Bío-Bío
Temuco
Vol. Osorno 8,726 ft. (2660 m.)
I. de Chiloé
ARCH. DE LOS CHONOS
Pen. Taitao
L. Gen. Carrera
Baker
G. de Penas
ANDES DE PATAGONIA
I. Wellington
ARCH. REINA ADELAIDA
Str. of Magellan
Str. of Magellan
Tierra del Fuego
I. Sta. Inés
I. Hoste
Cape Horn
5,000 m. 16,404 ft.
2,000 m. 6,562 ft.
1,000 m. 3,281 ft.
500 m. 1,640 ft.
200 m. 656 ft.
100 m. 328 ft.
Sea Level
Below

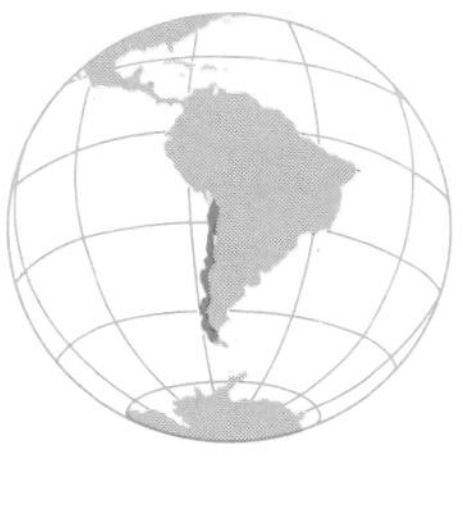

AREA 292,257 sq. mi. (756,946 sq. km.)
POPULATION 11,275,440
CAPITAL Santiago
LARGEST CITY Santiago
HIGHEST POINT Ojos del Salado 22,572 ft. (6,880 m.)
MONETARY UNIT Chilean escudo
MAJOR LANGUAGE Spanish
MAJOR RELIGION Roman Catholicism

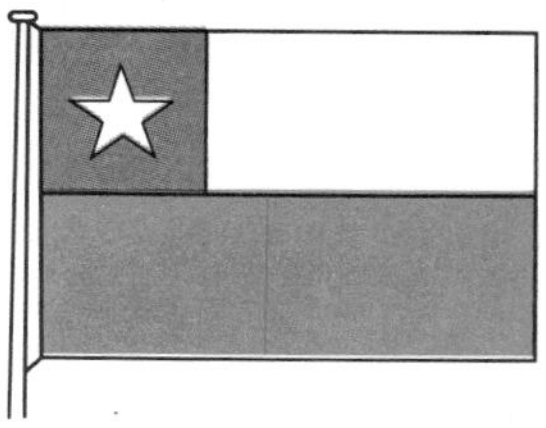

REGIONS

Aisén del General Carlos Ibáñez del Campo 65,478 . . . . . E6
Antofagasta 341,203 . . . . . B4
Atacama 183,071 . . . . . B6
Bíobío 1,516,552 . . . . . E1
Coquimbo 419,178 . . . . . A8
El Libertador General Bernardo O'Higgins 584,989 . . . . . A10
La Araucanía 692,924 . . . . . E2
Los Lagos 843,430 . . . . . D3
Magallanes 132,333 . . . . . E10
Maule 723,224 . . . . . A11
Santiago, Región Metropolitana de (Santiago Metropolitan Region) 4,294,938 . . . . . A9
Tarapacá 273,427 . . . . . B2
Valparaíso 1,204,693 . . . . . A9

CITIES and TOWNS

Achao ○11,501 . . . . . D4
Aguas Blancas ○203 . . . . . B4
Algarrobo ○3,941 . . . . . F3
Ancud 11,900 . . . . . D4
Andacollo 6,000 . . . . . A8
Angol 42,670 . . . . . D1
Antofagasta 125,100 . . . . . A4
Arauco 5,400 . . . . . D1
Arica 87,700 . . . . . A1
Ascotán B3
Barrancas ○184,241 . . . . . G3
Belén ○925 . . . . . B1
Buin 11,800 . . . . . G4
Bulnes 6,900 . . . . . E1
Cabildo 5,800 . . . . . A9
Calama 45,900 . . . . . B3
Calbuco ○21,673 . . . . . D4
Caldera ○3,268 . . . . . A6
Calera de Tango ○6,198 . . . G4
Calle Larga ○7,172 . . . . . G2
Cañete 7,900 . . . . . D2
Carahue ○12,733 . . . . . D2
Cartagena ○7,124 . . . . . F3
Casablanca 5,500 . . . . . F3
Casas de Chacabuco G2
Castro 11,200 . . . . . D4
Catalina ○1,637 . . . . . B5
Catemu ○8,728 . . . . . G2
Cauquenes 20,200 . . . . . A11
Cerro Castillo ○537 . . . . . E9
Cerro Manantiales F10
Chaitén ○4,067 . . . . . E4
Chañaral ○36,949 . . . . . A6
Chanco ○12,433 . . . . . A11
Chépica ○11,199 . . . . . A10
Chillán 128,515 . . . . . A11
Chimbarongo 5,300 . . . . . A10
Chonchi ○8,911 . . . . . D4
Chuquicamata 22,100 . . . . . B3
Cobquecura ○6,298 . . . . . D1
Cochamó ○5,042 . . . . . E3
Codegua ○6,757 . . . . . G4
Codpa ○950 . . . . . B1
Coelemu 5,400 . . . . . D1
Coihaique 32,129 . . . . . E6
Coihueco ○17,276 . . . . . A11
Coinco ○4,942 . . . . . G5
Colbún ○12,924 . . . . . A11
Colina 7,400 . . . . . G3
Collipulli 7,200 . . . . . E2
Coltauco ○11,857 . . . . . F5
Combarbalá ○17,332 . . . . . A8
Concepción 206,226 . . . . . D1
Constitución 11,500 . . . . . A11
Contulmo ○13,987 . . . . . D2
Copiapó 45,200 . . . . . B6
Coquimbo 73,953 . . . . . A8
Coronel 37,300 . . . . . D1
Corral ○5,533 . . . . . D3
Cunco ○18,836 . . . . . E2
Curacautín 9,800 . . . . . E2
Curacaví 5,800 . . . . . G3
Curanilahue 13,200 . . . . . D1
Curepto ○13,020 . . . . . A10
Curicó 41,300 . . . . . A10
Dalcahue ○7,084 . . . . . D4
Domeiko A7
Doñihue ○8,837 . . . . . G5
El Carmen ○13,226 . . . . . A11
El Monte 7,000 . . . . . G4
El Quisco ○2,152 . . . . . E3
El Tabo ○2,180 . . . . . F3
El Tofo A7
Empedrado ○7,887 . . . . . A11
Ercilla ○8,061 . . . . . E2
Estancia Caleta Josefina ○1,042 . . . . . F10
Estancia Morro Chico ○785 . E9
Estancia San Gregorio ○1,156 . . . . . E9
Estancia Springhill (Cerro Manantiales) F10
Freire ○23,313 . . . . . E2
Freirina ○5,523 . . . . . A7
Fresia ○15,359 . . . . . D3
Frutillar ○12,721 . . . . . D3
Futaleufú ○2,366 . . . . . E4
Futrono ○7,109 . . . . . E3
Galvarino ○9,495 . . . . . D2
General Lagos ○810 . . . . . B1
Graneros 8,900 . . . . . G5
Guayacán A8
Hijuelas ○7,128 . . . . . F2
Hualañé ○6,912 . . . . . A10
Huara ○1,934 . . . . . B2
Huasco ○4,971 . . . . . A7
Illapel 12,200 . . . . . A8
Inca de Oro 1,406 . . . . . B6
Iquique 64,500 . . . . . A2
Isla de Maipo ○12,903 . . . . G4
La Calera 24,600 . . . . . F2
La Cruz ○8,907 . . . . . F2
La Estrella ○3,707 . . . . . F5
Lago Ranco ○12,767 . . . . . E3
Lagunas ○5,653 . . . . . B3
La Higuera ○6,991 . . . . . A7
La Ligua 7,500 . . . . . A9
Lampa ○10,220 . . . . . G3
Lanco 5,200 . . . . . D2
Las Cabras ○12,119 . . . . . F5
La Serena 99,908 . . . . . A8
La Unión 15,200 . . . . . D3
Lautaro 11,900 . . . . . E2
Lebu 12,500 . . . . . D1
Licantén ○6,354 . . . . . A10
Limache 15,200 . . . . . F2
Linares 37,900 . . . . . A11
Llay-Llay 9,700 . . . . . G2
Loica F4
Loncoche ○17,539 . . . . . D2
Longaví ○15,909 . . . . . A11
Lonquimay ○9,524 . . . . . E2
Los Andes 23,500 . . . . . B9
Los Ángeles 49,500 . . . . . D1
Los Lagos ○14,934 . . . . . D3
Los Muermos ○9,296 . . . . . D3
Los Sauces ○7,613 . . . . . D2
Los Vilos ○10,453 . . . . . A9
Lota 48,100 . . . . . D1
Machalí 5,800 . . . . . G5
Maipú ○117,872 . . . . . G3
Malloa ○9,742 . . . . . G5
Marchigüe ○4,451 . . . . . F5
María Elena 5,900 . . . . . B3
María Pinto ○5,980 . . . . . G3
Maullín ○14,544 . . . . . D4
Mejillones ○3,333 . . . . . A4
Melipilla 23,900 . . . . . F4
Mincha ○11,329 . . . . . A8
Molina 9,400 . . . . . A10
Monte Patria ○18,927 . . . . . A8
Mulchén 13,700 . . . . . E1
Nacimiento ○17,651 . . . . . D1
Nancagua ○11,076 . . . . . F6
Navidad ○6,618 . . . . . A10
Negreiros ○1,144 . . . . . B2
Ñiquén ○13,640 . . . . . E1
Nogales ○18,529 . . . . . F2
Nueva Imperial 8,000 . . . . . D2
Olivar Alto ○5,414 . . . . . G5
Ollagüe B3
Olmué ○8,804 . . . . . F2
Osorno 68,800 . . . . . D3
Ovalle 31,700 . . . . . A8
Paihuano ○6,048 . . . . . B8
Paillaco 5,200 . . . . . D3
Paine ○21,876 . . . . . G4
Palena ○2,508 . . . . . E5
Palmilla ○7,965 . . . . . F6
Panguipulli 5,700 . . . . . E2
Panquehue ○4,230 . . . . . G2
Papudo ○2,594 . . . . . A9
Paredones ○7,404 . . . . . A10
Parral 17,000 . . . . . A11
Pedro de Valdivia 6,200 . . . . B4
Pemuco ○7,577 . . . . . E1
Peñaflor 15,500 . . . . . G4
Penco ○33,962 . . . . . D1
Peñuelas F3
Petorca ○8,343 . . . . . A9
Petrohué E3
Peumo ○11,308 . . . . . F5
Pica ○1,487 . . . . . B2
Pichidegua ○13,550 . . . . . F5
Pichilemu ○8,042 . . . . . A10
Pinto ○8,687 . . . . . A11
Pisagua ○1,880 . . . . . A2
Pitrufquén 7,800 . . . . . D2
Placilla ○6,441 . . . . . F6
Porvenir ○4,000 . . . . . E10
Potrerillos 5,800 . . . . . B6
Pozo Almonte ○1,798 . . . . . B2
Puchuncaví ○7,542 . . . . . F2
Pucón 18,000 . . . . . E2
Pudahuel G3
Pueblo Hundido 6,200 . . . . . B6
Puente Alto 65,100 . . . . . B10
Puerto Aisén 17,848 . . . . . E6
Puerto Cisnes ○2,800 . . . . . E5
Puerto Ingeniero Ibáñez ○1,900 . . . . . E6
Puerto Montt 119,059 . . . . . E4
Puerto Natales 17,280 . . . . . E9
Puerto Quellón ○7,734 . . . . D4
Puerto Varas 10,900 . . . . . E3
Puerto Williams ○949 . . . . F11
Pumanque ○3,137 . . . . . F6
Punitaqui ○16,167 . . . . . A8
Punta Arenas 2,140 . . . . . E10
Purén ○11,604 . . . . . D2
Purranque 5,900 . . . . . D3
Putaendo ○12,806 . . . . . A9
Putre ○855 . . . . . B1
Puyehue E3
Queilén ○6,055 . . . . . D4
Quemchi ○6,707 . . . . . D4
Quilicura 8,100 . . . . . G3
Quillagua B3
Quilleco ○16,043 . . . . . E1
Quillota 36,500 . . . . . F2
Quilpué 40,600 . . . . . F2
Quinta de Tilcoco ○6,513 . . G5
Quintero 9,900 . . . . . F2
Quirihue ○11,178 . . . . . E1
Rancagua 140,589 . . . . . G5
Renca ○67,168 . . . . . G3
Rengo 12,400 . . . . . G5
Requínoa ○10,730 . . . . . G5
Retiro ○15,146 . . . . . A11
Rinconada San Martín ○4,118 . . . . . G2
Río Blanco B9
Río Bueno 9,600 . . . . . D3
Río Negro 5,100 . . . . . D3
Río Verde ○554 . . . . . E10
Rocas de Santo Domingo ○4,114 . . . . . F4
Rosario ○3,383 . . . . . F5
Salamanca ○18,741 . . . . . A9
Samo Alto ○5,689 . . . . . A8
San Antonio 46,700 . . . . . F3
San Bernardo ○117,766 . . . G4
San Carlos 17,000 . . . . . E1
San Clemente ○23,273 . . . A11
San Felipe 26,100 . . . . . G2
San Fernando 23,600 . . . . . G6
San Francisco de Mostazal ○11,439 . . . . . G4
San Ignacio ○13,523 . . . . . E1
San Javier 10,800 . . . . . A11
San José de Maipo ○9,601 . . . . . B10
San Pablo ○7,978 . . . . . D3
San Pedro ○8,255 . . . . . F4
San Pedro de Atacama C4
San Rosendo ○14,337 . . . . . E1
Santa Bárbara ○14,345 . . . . E1
Santa Cruz 8,600 . . . . . F6
Santa María ○8,162 . . . . . G2
Santiago (cap.) 3,614,947 . . G3
Santiago *3,672,374 . . . . . G3
San Vicente F4
San Vicente (San Vicente de Tagua Tagua) ○28,333 F5
Sierra Gorda ○8,805 . . . . . B4
Talagante 16,500 . . . . . G4
Talca 133,160 . . . . . A11
Talcahuano 148,300 . . . . . D1
Taltal 6,400 . . . . . A5
Tamaya A8
Tarapacá B2
Temuco 197,232 . . . . . E2
Teno ○17,675 . . . . . A10
Termas de Cauquenes B10
Tierra Amarilla ○7,899 . . . . A6
Tiltil ○9,198 . . . . . G2
Toco ○8,734 . . . . . B3
Toconao C4
Tocopilla 22,000 . . . . . A3
Toltén ○16,265 . . . . . D2
Tomé 29,600 . . . . . D1
Traiguén 11,400 . . . . . D2
Valdivia 115,536 . . . . . D3
Vallenar 26,800 . . . . . A7
Valparaíso 271,580 . . . . . E2
Victoria 16,500 . . . . . D2
Vicuña 5,100 . . . . . A8
Villa Alemana 29,600 . . . . . F2
Villa Alhué ○5,078 . . . . . G4
Villarrica 25,091 . . . . . E2
Viña del Mar 281,361 . . . . . F2
Yumbel ○21,858 . . . . . E1
Yungay ○10,725 . . . . . E1
Zapallar ○2,894 . . . . . A9
Zapiga B2

OTHER FEATURES

Aconcagua (riv.) F2
Aculeo (lag.) G4
Adventure (bay) D5
Aguas Calientes, Cerro (mt.) C4
Almirantazgo (bay) F11
Almirante Montt (gulf) D9
Ancud (gulf) D4
Angamos (isl.) D8
Angamos (pt.) A4
Ap Iwan, Cerro (mt.) E6
Arauco (gulf) D1
Arenales, Cerro (mt.) D7
Atacama (des.) B4
Atacama, Salar de (salt dep.) C4
Aucanquilcha, Cerro (mt.) . . B3
Azapa, Quebrada (riv.) B1
Baker (riv.) D7
Ballenero (chan.) E11
Bascuñán (cape) A7
Beagle (chan.) E11
Bella Vista, Salar de (salt dep.) B3
Benjamín (isl.) D5
Bío-Bío (riv.) E2
Blanca (lag.) E10
Blanco (lake) F10
Bravo (riv.) D7
Brunswick (pen.) E10
Bueno (riv.) D3
Buenos Aires (lake) E6
Byron (isl.) D7
Cachapoal (riv.) G5
Cachina, Quebrada (riv.) A5
Cachos (pt.) A6
Calafquén (lake) E3
Camarones (riv.) A2
Camiña, Quebrada (riv.) B2
Campana (isl.) D7
Campanario, Cerro (mt.) . . A10
Capitán Aracena (isl.) E10
Carmen (riv.) B7
Castillo, Cerro (mt.) E6
Catalina (pt.) F10
Chaffers (isl.) D5
Chaltel, Cerro (mt.) E8
Chañaral (isl.) A7
Chatham (isl.) D9
Chauques (isls.) D4
Cheap (chan.) D7
Chiloé (isl.) 119,286 . . . . . D4
Choapa (riv.) A9
Chonos (arch.) D6
Choros (cape) A7
Cisnes (riv.) E5
Clarence (isl.) E10
Clemente (isl.) D6
Cochrane (lake) E7
Cochrane, Cerro (mt.) E7
Cockburn (chan.) E11
Concepción (chan.) D9
Cónico, Cerro (mt.) E4
Contreras (isl.) D9
Cook (bay) E11
Copiapó (bay) A6
Copiapó (riv.) A6
Corcovado (gulf) D4
Corcovado (vol.) D5
Coronados (gulf) D4
Curaumilla (pt.) E2
Darwin (bay) D6
Darwin, Cordillera (mts.) . . . D8
Darwin, Cordillera (mts.) . . E11

(continued on following page)

Dawson (isl.) E10	Guafo (gulf) D5	Laja (riv.) E1
Deseado (cape) D10	Guafo (isl.) D5	La Laja (lag.) E1
Desolación (isl.) D10	Guaitecas (isls.) D5	La Ligua (riv.) A9
Diego de Almagro (isl.) . D9	Guamblin (isl.) D5	Lanín (vol.) E2
Domeyko, Cordillera (mts.) . B4	Guayaneco (arch.) D7	Lastarria (vol.) B5
Dos Reyes (pt.) A5	Hanover (isl.) D9	Lauca (riv.) B1
Drake (passg.) E11	Hardy (pen.) F11	Lavapié (pt.) D1
Dungeness (pt.) F10	Hermite (isls.) F11	La Vieja (isl.) A11
Duque de York (isl.) ... C9	Horn (cape) F11	Lengua de Vaca (pt.) . A8
Elefantes (gulf) D6	Hornos, Falso (cape) . F11	Lennox (isl.) F11
Elqui (riv.) A8	Hoste (isl.) F11	Liles (pt.) F2
Esmeralda (isl.) C8	Huasco (riv.) A7	Limari (riv.) A8
Eyre (bay) D8	Imperial (riv.) D2	Llaima (vol.) E2
Fagnano (lake) F11	Incaguasi, Nevada (mt.) . C6	Llamara, Salar de (salt dep.) . B3
Fitz Roy (Chaltel) (mt.) . E8	Infieles (pt.) A6	Llanquihue (lake) E3
Galera (pt.) D3	Inglesa (bay) A6	Llullaillaco (vol.) B5
General Paz (lake) E5	Inútil (bay) E10	Lluta (riv.) B1
Gordon (isl.) E11	James (isl.) D5	Loa (riv.) B3
Grafton (isls.) D10	Johnson (isl.) D5	Lobos (pt.) A3
Grande (isl.) A6	Jorge Montt (isl.) D9	Londonderry (isl.) E11
Grande (riv.) F10	Juan Stuven (isl.) D7	Loros (pt.) E3
Grande, Salar (salt dep.) . B3	Lacuy (pen.) D4	Luz (isl.) D6
Grande de Tierra	Ladrillero (gulf) C8	Macá (mt.) D5
del Fuego (isl.) E11	Ladrillero (mt.) E10	Madre de Dios (isl.) . D8

Agriculture, Industry and Resources

DOMINANT LAND USE

- Cereals, Livestock
- Mediterranean Agriculture (cereals, fruit, livestock)
- Pasture Livestock
- Extensive Livestock Ranching
- Limited Seasonal Grazing
- Forests
- Nonagricultural Land

MAJOR MINERAL OCCURRENCES

Ag	Silver	Hg	Mercury
Au	Gold	Id	Iodine
C	Coal	Mn	Manganese
Cu	Copper	Mo	Molybdenum
Fe	Iron Ore	N	Nitrates
G	Natural Gas	Na	Salt
Gp	Gypsum	O	Petroleum
		S	Sulfur

⚡ Water Power   ▨ Major Industrial Areas

Highways of Central Chile

SCALE OF MILES

0   25   50   75

SCALE OF KILOMETERS

0   50   100   150

Major Roads ....
Other Roads ....
Trails ....

© Copyright HAMMOND INCORPORATED, Maplewood, N.J.

Magallanes (Magellan)	Riesco (isl.) E10
(str.) D10	Rincón, Cerro (mt.) C4
Magdalena (isl.) D5	Rivero (isl.) D6
Magellan (str.) D10	Rupanco (lake) D3
Maipo (riv.) F4	San Esteban (gulf) D7
Maipú (vol.) B10	San Lorenzo, Cerro
Manso (riv.) E4	(Cochrane) (mt.) .. E7
Manuel Rodríguez (isl.) . D10	San Martín (lake) E7
Mapocho (riv.) G3	San Pedro (pt.) A5
Mataquito (riv.) A10	Santa Inés (isl.) D10
Maule (riv.) A11	Santa María (isl.) D1
Maullín (riv.) D3	San Valentín, Cerro (mt.) . D6
Mejillones del Sur (bay) . A4	Sarco (bay) A7
Melchor (isl.) D6	Sarmiento, Cerro (mt.) . E11
Melimoyu (mt.) D5	Sillajguay, Cordillera (range) . B2
Merino Jarpa (isl.) D7	Simpson (riv.) E6
Minchinmávida (vol.) .. E4	Skyring (bay) E10
Miraje, Salar del (salt dep.) . B3	Socompa (vol.) B4
Mocha (isl.) A9	Staines (pen.) D9
Molles (pt.) A9	Stewart (isl.) E11
Morado, Quebrada (riv.) . A6	Stokes (bay) D10
Moraleda (chan.) D5	Stosch (isl.) C8
Moreno (bay) A4	Surire, Salar de (salt dep.) . B2
Mornington (isl.) D8	Tablas (cape) A9
Morro (pt.) A6	Tacora (vol.) B1
Muñoz Gamero (pen.) . D10	Taitao (cape) D6
Murallón, Cerro (mt.) .. D8	Taitao (pen.) D6
Nalcayec (isl.) D6	Talca (pt.) E3
Nassau (bay) F11	Taltal, Quebrada de (riv.) . B5
Navarino (isl.) F11	Tamarugal, Pampa del
Nelson (str.) D9	(plain) B3
Nuestra Señora (bay) .. A5	Tenquehuen (isl.) D6
Nueva (isl.) F11	Tetas (pt.) A4
Núñez (isl.) D10	Tierra del Fuego,
O'Higgins (lake) D7	Grande de (isl.) ... E11
Ojos del Salado,	Tinguiririca (riv.) F5
Nevado (mt.) B6	Toltén (riv.) D2
Otway (bay) D10	Tongoy (bay) A8
Otway (sound) E10	Topocalma (pt.) A10
Oyahue (vol.) C3	Toro (lake) D9
Paine, Cerro (mt.) D9	Toro, Cerro del (mt.) .. B7
Pájaros (isls.) A7	Toro (pt.) A10
Palena (lake) E5	Tórtolas, Cerro de las (mt.) . B8
Palena (riv.) D5	Totoral, Quebrada (riv.) . A6
Pan de Azúcar,	Traiguén (isl.) D6
Quebrado (riv.) B5	Tranqui (isl.) D4
Parinacota, Cerro (mt.) . B1	Tres Cruces, Nevada (mt.) . B6
Pascua (riv.) D7	Tres Montes (cape) C7
Patricio Lynch (isl.) ... D7	Tres Montes (gulf) D6
Penas (gulf) D7	Tres Montes (pen.) C6
Perquilauquén (riv.) ... A11	Trinidad (gulf) D8
Piazzi (isl.) D9	Tronador, Cerro (mt.) .. E3
Picton (isl.) F11	Tumbes (pen.) D1
Pilmaiquén (riv.) D3	Tupungato, Cerro (mt.) . B9
Pintados, Salar de	Última Esperanza (sound) . E9
(salt dep.) B2	Vidal Gormaz (riv.) ... D9
Poroto (pt.) A7	Villarrica (lake) E2
Potro, Cerro del (mt.) .. B7	Vitor, Quebrada (riv.) .. A1
Prat (pt.) D7	Week (isls.) D10
Presidente Ríos (lake) .. D6	Wellington (isl.) D8
Puangue, Estero de (riv.) . F3	Wharton (pen.) D8
Puelo (riv.) E4	Whiteside (chan.) E10
Púlar, Cerro (mt.) B4	Wollaston (isls.) F11
Puquintica, Cerro (mt.) . B1	Wood (isls.) E11
Puyehue (lake) E3	Yelcho (lake) E4
Quilán (cape) E3	Zapaleri, Cerro (mt.) .. C4
Ranco (lake) E3	
Rapel (riv.) F4	
Refugio (isl.) D5	*City and suburbs.
Reina Adelaida (arch.) . D9	○ Population of commune.
Reloncaví (bay) D4	

PROVINCES

Buenos Aires 10,796,036 . . . D4
Catamarca 206,204 . . . . C2
Chaco 692,410 . . . . D2
Chubut 262,196 . . . . C5
Córdoba 2,407,135 . . . . D3
Corrientes 657,716 . . . . E2
Distrito Federal 2,908,001 . . H7
Entre Ríos 902,241 . . . . E3
Formosa 292,479 . . . . D1
Jujuy 408,514 . . . . C1
La Pampa 207,132 . . . . C4
La Rioja 163,342 . . . . C2
Mendoza 1,187,305 . . . . C4
Misiones 579,579 . . . . F2
Neuquén 241,904 . . . . C4
Río Negro 383,896 . . . . C5
Salta 662,369 . . . . D1
San Juan 469,973 . . . . C3
San Luis 212,837 . . . . C3
Santa Cruz 114,479 . . . . C6
Santa Fe 2,457,188 . . . . D3
Santiago del Estero 652,318 D2
Tierra del Fuego, Antártida, e Islas del Atlántico Sur 29,451 . . . . C7
Tucumán 968,066 . . . . C2

CITIES and TOWNS

Abra Pampa 2,929 . . . . C1
Adolfo Alsina 7,707 . . . . . . D4
Aguaray 4,802 . . . . D1
Aguilares 20,286 . . . . C2
Aimogasta 4,640 . . . . C2
Alberti 6,440 . . . . G7
Alcorta 5,818 . . . . F6
Algarrobo del Águila C4
Allen 14,041 . . . . C4
Alpachiri 1,657 . . . . D4
Alta Gracia 30,628 . . . . D3
Aluminé 1,560 . . . . B4
Alvear 5,419 . . . . E2
Ameghino 2,775 . . . . D3
Añatuya 15,025 . . . . D2
Andalgalá 6,853 . . . . C2
Antofagasta de la Sierra C2
Apóstoles 11,252 . . . . E2
Arrecifes 17,719 . . . . F7
Arroyo Seco 12,886 . . . . F6
Ascensión 3,031 . . . . F7
Avellaneda 330,654 . . . . G7
Ayacucho 12,363 . . . . E4
Azul 43,582 . . . . E4
Bahía Blanca 220,765 . . . . . . D4
Bahía Bustamante C6
Bahía Thetis C7
Balcarce 28,985 . . . . E4
Balnearia 4,531 . . . . D3
Baradero 20,103 . . . . G6
Barrancas 3,602 . . . . F6
Barranqueras E2
Barreal 2,739 . . . . C3
Basavilbaso 7,657 . . . . G6
Belén 7,411 . . . . C2
Bella Vista, Corrientes 14,229 . . . . E2
Bella Vista, Tucuman 9,177 . D2
Bell Ville 26,559 . . . . D3
Bolívar 16,382 . . . . D4
Bovril 4,735 . . . . G5
Bragado 27,101 . . . . F7
Buenos Aires (cap.) 2,908,001 . . . . H7
Buenos Aires *9,927,404 . . . H7
Cafayate 5,048 . . . . C2
Calafate B7
Calchaquí 5,958 . . . . F5
Caleta Olivia 20,141 . . . . C6
Camarones C5
Campana 51,498 . . . . G6
Cañada de Gómez 24,706 . . . F6
Canals 6,627 . . . . D3
Cañuelas 14,831 . . . . G7
Carcarañá 11,121 . . . . F6
Carlos Casares 13,286 . . . . . F7
Carlos Tejedor 4,421 . . . . . . D4
Carmen de Areco 7,882 . . . . F7
Carmen de Patagones 13,981 . . . . D5
Casilda 23,492 . . . . F6
Castelli 4,507 . . . . H7
Catamarca 88,432 . . . . C2
Caucete 14,512 . . . . C3
Ceres 10,743 . . . . D2
Chabás 5,156 . . . . F6
Chacabuco 26,492 . . . . F7
Chajarí 15,242 . . . . G5
Chamical 6,333 . . . . C3
Charadai 1,078 . . . . D2
Charata 13,070 . . . . D2
Chascomús 21,864 . . . . H7
Chepes 4,775 . . . . C3
Chicoana 1,844 . . . . C2
Chilecito 14,010 . . . . C2
Chivilcoy 43,779 . . . . F7
Choele-Choel 6,191 . . . . C4
Chos-Malal 4,823 . . . . C4
Cinco Saltos 15,094 . . . . C4
Cipolletti 40,123 . . . . C4
Clorinda 21,008 . . . . E2
Colón, Buenos Aires 16,070 . F6
Colón, Entre Ríos 11,648 . . . G6
Colonia Las Heras 3,176 . . . . C6
Comandante Fontana 4,468 . D2
Comandante Luis Piedrabuena 2,492 . . . . C6
Comodoro Rivadavia 96,865 . C6
Concepción 29,359 . . . . C2
Concepción de la Sierra 2,778 . . . . E2
Concepción del Uruguay 46,065 . . . . G6
Concordia 93,618 . . . . G5
Constanza 1,313 . . . . G6
Córdoba 982,018 . . . . D3
Coronda 11,554 . . . . F6
Coronel Brandsen 10,484 . . . H7
Coronel Dorrego 10,661 . . . . D4
Coronel Pringles 16,592 . . . . D4
Coronel Suárez 16,359 . . . . . D4

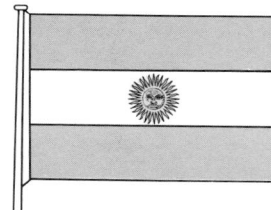

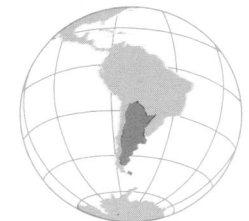

AREA 1,072,070 sq. mi. (2,776,661 sq. km.)
POPULATION 28,438,000
CAPITAL Buenos Aires
LARGEST CITY Buenos Aires
HIGHEST POINT Cerro Aconcagua 22,831 ft. (6,959 m.)
MONETARY UNIT austral
MAJOR LANGUAGE Spanish
MAJOR RELIGION Roman Catholicism

Agriculture, Industry and Resources

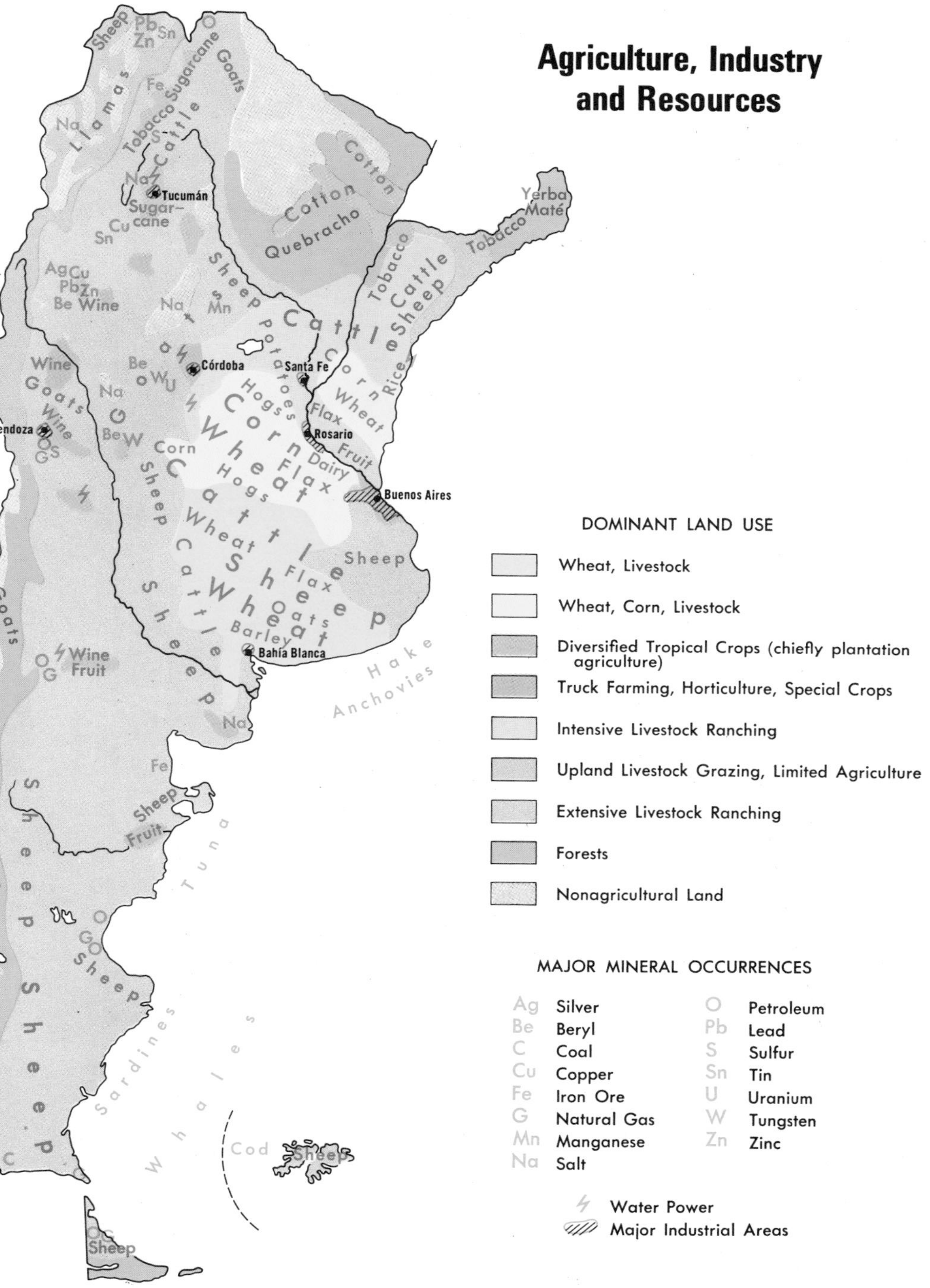

Tucumán
Córdoba
Santa Fe
Mendoza
Rosario
Buenos Aires
Bahía Blanca
Sheep
Llamas
Tobacco
Sugarcane
Goats
Cattle
Cotton
Quebracho
Sugar-cane
Yerba Maté
Wine
Corn
Wheat
Flax
Hogs
Potatoes
Dairy
Fruit
Rice
Oats
Barley
Hake
Anchovies
Tuna
Sardines
Whales
Cod
DOMINANT LAND USE
Wheat, Livestock
Wheat, Corn, Livestock
Diversified Tropical Crops (chiefly plantation agriculture)
Truck Farming, Horticulture, Special Crops
Intensive Livestock Ranching
Upland Livestock Grazing, Limited Agriculture
Extensive Livestock Ranching
Forests
Nonagricultural Land
MAJOR MINERAL OCCURRENCES
Ag Silver
Be Beryl
C Coal
Cu Copper
Fe Iron Ore
G Natural Gas
Mn Manganese
Na Salt
O Petroleum
Pb Lead
S Sulfur
Sn Tin
U Uranium
W Tungsten
Zn Zinc
Water Power
Major Industrial Areas

Coronel Vidal 4,774 . . . . E4
Corral de Bustos 8,613 . . . . . D3
Corrientes 179,590 . . . . E2
Cosquín 13,929 . . . . D3
Crespo 10,668 . . . . F6
Cruz del Eje 23,473 . . . . C3
Curuzú Cuatiá 24,955 . . . . . . G5
Cutral-Có 25,870 . . . . C4
Daireaux 8,150 . . . . D4
Deán Funes 16,306 . . . . D3
Diamante 13,464 . . . . F6
Dolavon 1,778 . . . . C5
Dolores 19,307 . . . . E4
Eduardo Castex 5,397 . . . . . D4
El Bolsón 5,001 . . . . B5
Eldorado 22,821 . . . . F2
El Maitén 2,350 . . . . B5
Elortondo 4,939 . . . . F6
El Quebrachal 2,202 . . . . D2
Embarcación 9,016 . . . . D1
Empedrado 4,732 . . . . E2
Escobar 70,829 . . . . G7
Esperanza 22,838 . . . . F5
Esquel 17,228 . . . . B5
Esquina 10,380 . . . . G5
Famatina 1,237 . . . . C2
Federación 7,259 . . . . G5
Felipe Yofré 1,140 . . . . G4
Fernández 6,062 . . . . D2
Fiambalá 1,201 . . . . C2
Firmat 13,588 . . . . F6
Formosa 95,067 . . . . E2
Fortín Olmos 1,101 . . . . F4
Frías 20,901 . . . . D2
Gaiman 2,651 . . . . C5
Gálvez 14,711 . . . . F6
General Acha 7,647 . . . . C4
General Alvear, Buenos Aires 5,481 . . . . F7
General Alvear, Mendoza 21,250 . . . . C3
General Arenales 3,332 . . . . . F7
General Belgrano 10,909 . . . G7
General Conesa 3,566 . . . . . . C5
General Galarza 3,057 . . . . . C6
General Güemes 15,534 . . . . D1
General José de San Martín 16,296 . . . . . . E2
General Juan Madariaga 13,409 . . . . E4
General La Madrid 5,154 . . . D4
General Las Heras 6,005 . . . G7
General Paz 5,127 . . . . H7
General Pico 30,180 . . . . D4
General Ramírez 5,393 . . . . . F6
General Roca 38,296 . . . . . . C4
General San Martín, Buenos Aires 384,306 . . . . G7
General San Martín, La Pampa 2,168 . . . . D4
General Viamonte 10,112 . . . F7
General Villegas 11,307 . . . . D4
Gobernador Crespo 2,972 . . . F5
Godoy Cruz 141,553 . . . . C3
Goya 47,357 . . . . G4
Gualeguay 24,883 . . . . G6
Gualeguaychú 51,057 . . . . . . G6
Guandacol 1,351 . . . . C2
Hasenkamp 2,804 . . . . F5
Helvecia 3,927 . . . . F5
Hernandarias 3,002 . . . . F5
Hernando 8,619 . . . . D3
Huinca Renancó 7,187 . . . . . D3
Humahuaca 3,963 . . . . C1
Humberto (Humberto Primo) 4,163 . . . . F5
Ibarreta 5,262 . . . . D2
Ibicuy 3,082 . . . . G6
Ingeniero Huergo 3,385 . . . . C4
Ingeniero Jacobacci 4,045 . . C5
Ingeniero Luiggi 3,002 . . . . . D4
Intendente Alvear 3,640 . . . . D4
Itatí 3,269 . . . . E2
Ituzaingó 8,687 . . . . E2
Jáchal 8,832 . . . . C3
Jesús María 17,594 . . . . D3
Joaquín V. González 6,054 . . D2
Juárez 11,798 . . . . E4
Jujuy 124,487 . . . . C1
Junín 62,080 . . . . F7
Junín de los Andes 5,638 . . . B4
La Banda 46,994 . . . . D2
Laboulaye 16,883 . . . . D3
La Carlota 8,614 . . . . D3
La Cruz 4,132 . . . . E2
La Cumbre 6,110 . . . . C3
La Falda 12,502 . . . . D3
Laguna Paiva 11,129 . . . . F5
Lanús 465,891 . . . . H7
La Paz, Entre Ríos 14,920 . . G5
La Paz, Mendoza 4,604 . . . . C3
La Plata 560,341 . . . . H7
Laprida 6,495 . . . . D4
La Quiaca 8,289 . . . . C1
La Rioja 66,826 . . . . C2
Larroque 3,147 . . . . F5
Las Flores 18,287 . . . . E4
Las Lomitas 4,047 . . . . D1
Las Palmas 5,061 . . . . E2
Las Parejas 7,430 . . . . F6
Las Rosas 9,725 . . . . F6
Las Varillas 10,605 . . . . D3
La Toma 4,325 . . . . C3
Lincoln 19,009 . . . . F7
Lobería 8,898 . . . . E4
Lobos 20,798 . . . . G7
Lomas de Zamora 508,620 . G7
Lucas González 3,015 . . . . . . G6
Luján 38,919 . . . . G7
Lules 11,391 . . . . C2
Maciel 4,066 . . . . F6
Magdalena 7,135 . . . . H7
Maipú 7,289 . . . . E4
Malabrigo 3,294 . . . . F4
Malargüe 9,496 . . . . C4
Maquinchao 1,299 . . . . C5
Marcos Juárez 19,827 . . . . . D3
Mar del Plata 407,024 . . . . . . E4
Máximo Paz 3,216 . . . . F6
Mburucuya 3,044 . . . . E2
Médanos 4,511 . . . . D4
Mendoza 596,796 . . . . C3
Mercedes, Buenos Aires 46,581 . . . . G7
Mercedes, Corrientes 20,603 . . . . G4
Mercedes, San Luis 50,856 . C3
Merlo 293,059 . . . . G7
Metán 18,928 . . . . D2
Miramar 15,473 . . . . E4
Monte Caseros 18,247 . . . . . G5
Monte Quemado 4,707 . . . . . D2
Monteros 15,832 . . . . C2
Morón 596,769 . . . . G7
Morteros 11,456 . . . . D3
Navarro 7,176 . . . . G7
Necochea 50,939 . . . . E4
Neuquén 90,037 . . . . C4
Nogoyá 15,862 . . . . F6
Ñorquincó B5
Nueve de Julio 26,608 . . . . . F7
Oberá 27,311 . . . . F2
Olavarría 63,686 . . . . D4
Oliva 9,231 . . . . D3
Palo Santo 3,088 . . . . E2
Paraná 159,581 . . . . F5
Paso de Los Libres 24,112 . . E2
Pedro Luro 3,142 . . . . D4
Pehuajó 25,613 . . . . D4
Pellegrini 3,940 . . . . D4
Pergamino 68,989 . . . . F6
Pico Truncado 9,626 . . . . C6
Pigüé 10,793 . . . . D4
Pilar 3,805 . . . . F5
Piranė 9,039 . . . . E2
Plaza Huincul 7,988 . . . . B4

(continued on following page)

Posadas 139,941	E2
Presidencia de la Plaza 4,904	D2
Presidencia Roque Sáenz Peña 49,261	D2
Puán 4,148	D4
Puerto Deseado 4,017	D6
Puerto Harberton	C7
Puerto Iguazú 10,250	F2
Puerto Madryn 20,995	C5
Puerto Rico 8,195	D1
Punta Alta 54,375	E4
Quequén 11,737	E4
Quimili 8,972	D2
Quines 3,352	C3
Quitilipi 9,937	D2
Rafaela 53,132	F5
Ramallo 8,248	F6
Rauch 8,348	E4
Rawson 12,981	D5
Reconquista 32,442	F4
Recreo 3,502	C2
Resistencia 218,438	E2
Rinconada	C1
Río Colorado 7,361	D4
Río Cuarto 110,148	D3
Río Gallegos 43,479	C7
Río Grande 13,271	C7
Río Segundo 12,839	D3
Río Tercero 34,735	D3
Rivadavia 10,953	C3
Rojas 14,247	F7
Romang 4,017	F4
Roque Pérez 5,434	G7
Rosario 954,606	F6
Rosario de la Frontera 13,531	D2
Rosario de Lerma 9,540	C1
Rosario del Tala 9,552	G6
Rufino 15,306	D3
Saladas 7,345	E2
Saladillo 14,806	G7
Salliqueló 5,479	D4
Salta 260,323	C1
Salto 18,462	F7
San Antonio de Areco 12,932	G7
San Antonio de los Cobres 2,357	C1
San Antonio Oeste 8,690	C5
San Carlos 7,613	F6
San Carlos de Bariloche 48,222	B5
San Cayetano 5,960	E4

San Cristóbal 13,345	F5
San Fernando 128,939	G7
San Francisco, Córdoba 58,616	D3
San Francisco, San Luis 2,448	C3
San Genaro 2,977	F6
San Ignacio 3,437	E2
San Jaime de la Frontera 2,811	G5
San Javier 7,557	F5
San José de Feliciano 4,986	G5
San Juan 290,479	C3
San Julián 4,278	C6
San Justo 14,135	F5
San Luis 70,632	C3
San Martín 29,746	C3
San Martín de los Andes 9,507	C5
San Miguel de Tucumán 496,914	D2
San Nicolás 96,313	F6
San Pedro, Buenos Aires 27,058	F6
San Pedro, Jujuy 36,907	D1
San Rafael 70,477	C3
San Ramón de la Nva. Orán 32,955	D1
San Salvador 4,342	G5
San Sebastián	C7
Santa Cruz 2,353	C7
Santa Elena 14,655	F5
Santa Fe 287,240	F5
Santa Lucía 4,452	E2
Santa María 5,380	C2
Santa Rosa, Córdoba 4,306	D3
Santa Rosa, La Pampa 51,689	C4
Santa Rosa, San Luis 2,878	C3
Santa Victoria	D1
Santiago del Estero 148,357	D2
Santo Tomé, Corrientes 14,352	E2
Santo Tomé, Santa Fe 35,363	F5
Sarmiento 6,313	B6
Sauce 4,677	G5
Sierra Grande 9,585	C5
Suipacha 4,505	G7
Sunchales 12,493	F5
Suncho Corral 3,837	D2
Tafí Viejo 26,625	C2
Tandil 78,821	E4

Tapalqué 5,356	E4
Tartagal 31,367	D1
Tigre 199,366	G7
Tinogasta 7,829	C2
Toay 3,617	D4
Tornquist 4,696	D4
Tostado 10,492	D2
Trelew 52,073	C5
Trenque Lauquen 22,504	D4
Tres Arroyos 42,118	D4
Trevelín 2,935	B5
Tunuyán 14,665	C3
Urdinarrain 5,472	G6
Ushuaia 10,988	C7
Valcheta 2,994	C5
Vedia 6,273	F7
Veinticinco de Mayo 18,936	F7
Venado Tuerto 46,775	D3
Vera 13,555	F5
Verónica 5,657	H7
Viale 5,635	F5
Vicente López 289,815	G7
Victoria 18,883	F6
Victorica 3,895	C4
Viedma 24,338	D5
Villa Ángela 25,586	D2
Villa Atuel 2,774	C3
Villa Cañas 7,303	F6
Villa Constitución 36,157	F6
Villa del Rosario 10,133	D3
Villa Dolores 21,508	C3
Villa Elisa 4,106	G6
Villa Federal 9,222	G5
Villaguay 18,699	G5
Villa Guillermina 2,971	D2
Villa Huidobro 4,154	D3
Villa María 67,490	D3
Villa María Grande 4,517	F5
Villa Nueva 4,604	C3
Villa Ocampo 9,162	D2
Villa Regina 14,017	C4
Villa San José 6,800	G6
Villa San Martín 6,237	C2
Vinchina 1,070	C2
Zapala 18,293	B4
Zárate 65,504	G6
Zavalla 3,800	F6

OTHER FEATURES

Aconcagua, Cerro (mt.)	C3
Andes, Cordillera de los (mts.)	C2

Argentino (lake)	B7
Arizaro, Salar de (salt dep.)	C2
Arrecifes (riv.)	G6
Atacama, Puna de (reg.)	C2
Atuel (riv.)	C4
Bermejo (riv.)	E2
Blanca (bay)	D4
Brazo Sur, Pilcomayo (riv.)	E1
Buenos Aires (lake)	B6
Campanario, Cerro (mt.)	C4
Chaco Austral (reg.)	D2
Chaco Central (reg.)	D1
Chico (riv.)	C6
Chico (riv.)	C6
Chubut (riv.)	C5
Colhué Huapi (lake)	C6
Colorado (riv.)	C4
Cónico, Cerro (mt.)	B5
Corrientes (riv.)	E2
Coyle (riv.)	B7
Delgada (pt.)	D5
Desaguadero (riv.)	C3
Deseado (riv.)	C6
Diamante (riv.)	C3
Domuyo (vol.)	B4
Dos Bahías (cape)	D5
Dulce (riv.)	D2
Dungeness (pt.)	C7
El Chocón (res.)	C4
Estados, Los (isl.)	D7
Fagnano (lake)	C7
Famatina, Sierra de (mts.)	C2
Feliciano (riv.)	G5
Gallegos (riv.)	B7
General Manuel Belgrano, Cerro (mt.)	C2
Gran Chaco (reg.)	D1
Grande (bay)	C7
Grande (falls)	E3
Grande de Tierra del Fuego (isl.)	C7
Gualeguay (riv.)	G5
Guayaquilaró (riv.)	G5
Iguazú (falls)	F2
Iguazú Nat'l Park	E2
Lanín (vol.)	B4
Lanín Nat'l Park	B4
Lechiguanas (isls.)	G6
Lennox (isl.)	C8
Limay (riv.)	C4
Llancanelo, Salina y Laguna (salt lake)	C4
Llullaillaco (vol.)	C1
Magallanes (Magellan) (str.)	C7

Topography

0	150	300 MI.
0	150	300 KM.

5,000 m. 16,404 ft.	2,000 m. 6,562 ft.	1,000 m. 3,281 ft.	500 m. 1,640 ft.	200 m. 656 ft.	100 m. 328 ft.	Sea Level	Below

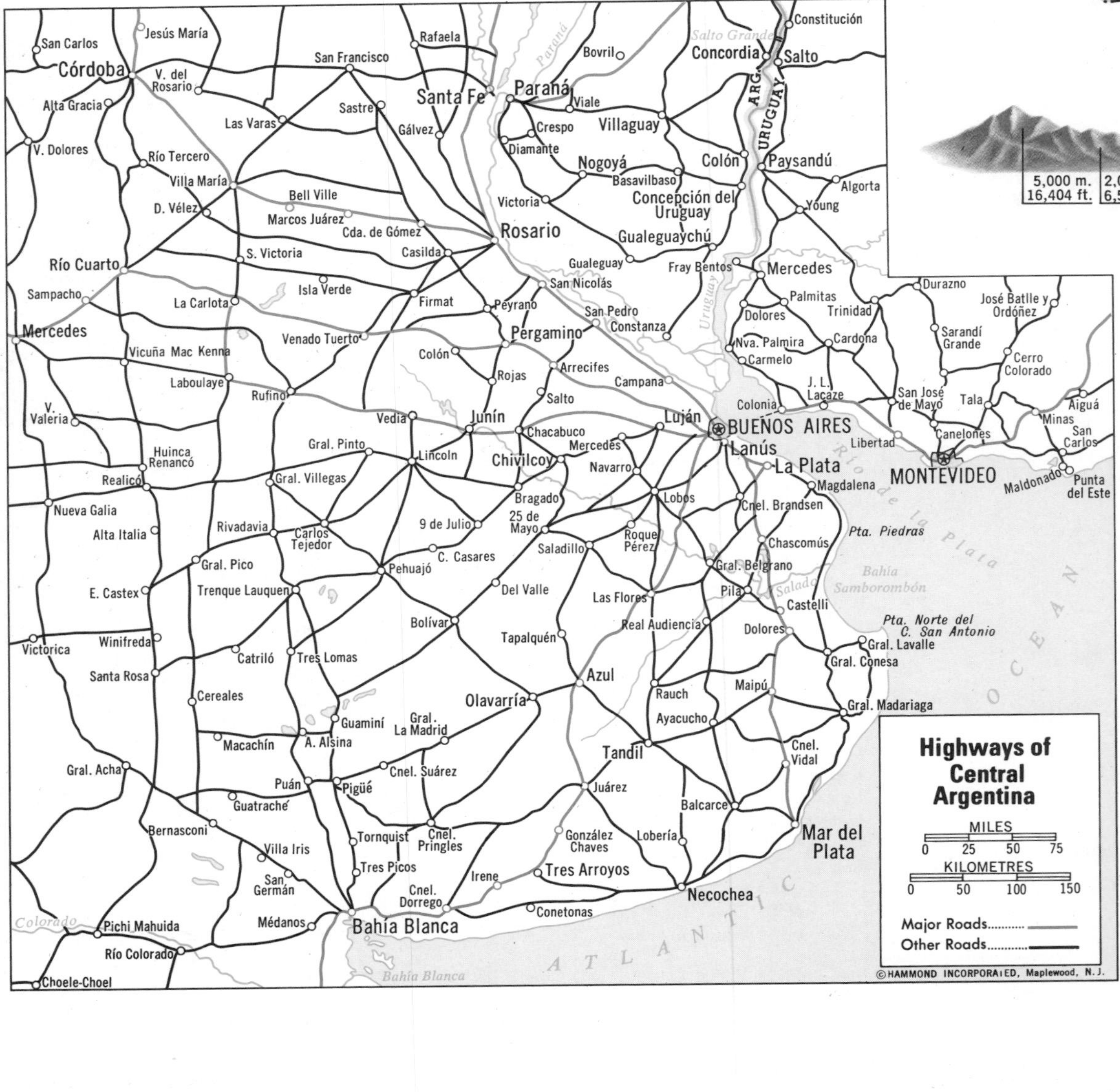

Highways of Central Argentina

MILES

| 0 | 25 | 50 | 75 |

KILOMETRES

| 0 | 50 | 100 | 150 |

Major Roads ———
Other Roads ———

© HAMMOND INCORPORATED, Maplewood, N.J.

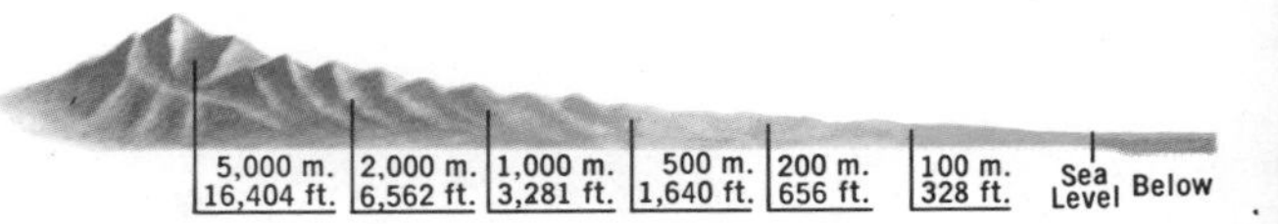

Maipo (vol.)	C3
Mar Chiquita (lake)	D3
Mendoza (riv.)	C3
Mercedario, Cerro (mt.)	B3
Mogotes (pt.)	E4
Montemayor (plat.)	C5
Nahuel Huapi (lake)	B5
Nahuel Huapi Nat'l Park	B5
Negro (riv.)	C5
Neuquén (riv.)	C4
Ninfas (pt.)	D5
Norte (pt.)	D5
Ojos del Salado, Cerro (mt.)	C2
Pampa de las Tres Hermanas (plain)	C6
Pampas (plain)	D4
Paraná (riv.)	E2
Patagonia (reg.)	C5
Peteroa (vol.)	B4
Pilcomayo (riv.)	E1
Pissis (mt.)	C2
Plata, Río de la (est.)	E4
Pueyrredón (lake)	B6
Puna de Atacama (reg.)	C2
Quinto (riv.)	D3
Rincón, Cerro (mt.)	C1
Saladillo (riv.)	D4
Salado (riv.)	C4
Salado (riv.)	H7
Salado del Norte (riv.)	D2
Salí (riv.)	C2
Salto (riv.)	F7
Samborombón (bay)	E4
San Antonio (cape)	E4
San Diego (cape)	D7
San Jorge (gulf)	C6
San Juan (riv.)	C3
San Lorenzo, Cerro (mt.)	B6
San Martín (lake)	B6
San Matías (gulf)	D5
Santa Cruz (riv.)	B7

Senguerr (riv.)	B6
Staten (Los Estados) (isl.)	D7
Tarija (riv.)	D1
Tercero (riv.)	D3
Teuco (riv.)	D1
Tierra del Fuego, Grande de (isl.)	C7
Toro, Cerro del (mt.)	B2
Tres Puntas (cape)	D4
Trinidad (isl.)	D4
Tronador (mt.)	B5
Tunuyán (riv.)	C3
Tupungato, Cerro (mt.)	C3
Uruguay (riv.)	E3
Valdés (pen.)	D5
Viedma (lake)	B6
Zapaleri, Cerro (mt.)	C1

FALKLAND ISLANDS

CITIES and TOWNS

Stanley (cap.) 1,050 | E7

OTHER FEATURES

Adventure (sound)	E7
Choiseul (sound)	E7
East Falkland (isl.) 1,491	H7
Falkland (isls.)	D7
Falkland (sound)	D7
George (isl.)	D7
Jason (isls.)	D7
Lively (isl.)	E7
Malvinas (Falkland) (isls.)	D7
Pebble (isl.)	D7
Saunders (isl.)	D7
Weddel (isl.)	D7
West Falkland (isl.) 322	D7

*City and suburbs.

Argentina
CONIC PROJECTION

SCALE OF MILES

0   50   100   200   300

SCALE OF KILOMETERS

0   50   100   200   300

Capitals of Countries .... ☆
Capitals of Provinces .... ◉
International Boundaries .... —·—·—
Boundaries of Provinces .... ------

Scale 1:13,000,000

© Copyright HAMMOND INCORPORATED, Maplewood, N.J.

Paraguay

CONIC PROJECTION

SCALE OF MILES
0 20 40 60 80 100 120 140

SCALE OF KILOMETERS
0 20 40 60 80 100 140

Capitals of Countries .... ★
Capitals of Departments .... ◉
International Boundaries .... — — —
Department Boundaries .... — ·· — ··

Scale 1:6,740,000

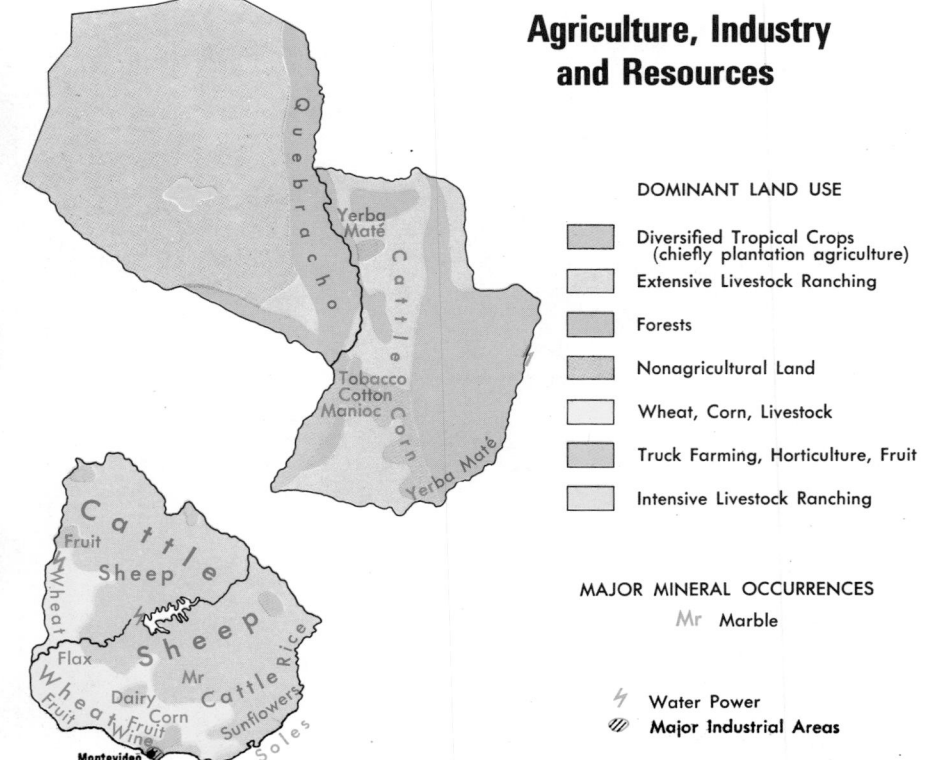

Agriculture, Industry and Resources

DOMINANT LAND USE

- Diversified Tropical Crops (chiefly plantation agriculture)
- Extensive Livestock Ranching
- Forests
- Nonagricultural Land
- Wheat, Corn, Livestock
- Truck Farming, Horticulture, Fruit
- Intensive Livestock Ranching

MAJOR MINERAL OCCURRENCES

Mr Marble

⚡ Water Power

▨ Major Industrial Areas

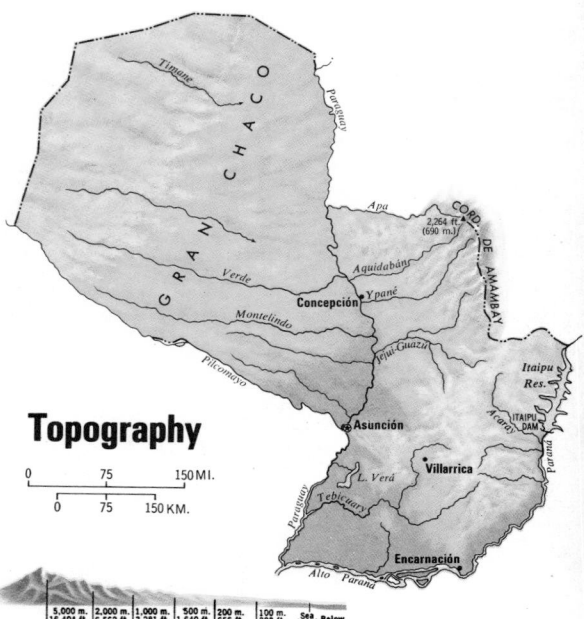

Topography

0 75 150 MI.

0 75 150 KM.

| 5,000 m. 16,404 ft. | 2,000 m. 6,562 ft. | 1,000 m. 3,281 ft. | 500 m. 1,640 ft. | 200 m. 656 ft. | 100 m. 328 ft. | Sea Level | Below |

PARAGUAY

DEPARTMENTS

Alto Paraguay C2
Alto Paraná E4
Amambay D3
Asunción B3
Boquerón D-E4
Caaguazú D-E5
Caazapá D5
Canendiyu E4
Central D4
Chaco B-C2
Concepción D3
Cordillera D4
Guairá D4
Itapúa E5
Misiones D5
Ñeembucú C-D5
Nueva Asunción B2
Paraguarí D4-5
Presidente Hayes D4-5
San Pedro D4

CITIES and TOWNS

Abal 1,507 .... E4
Acahay 1,937 .... B5
Alberdi 2,346 .... B4
Altos 1,441 .... B4
Antequera 1,281 .... D4
Areguá 3,941 .... B4
Arroyos y Esteros 1,253 .... B4
Asunción (cap.) 387,676 .... A4
Atyrá 1,427 .... B4
Ayolas 309 .... D5
Belén 1,219 .... D3
Bella Vista 3,101 .... D3
Bella Vista 1,421 .... E5
Benjamín Aceval 2,877 .... C4
Buena Vista 1,353 .... B5
Caacupé 7,278 .... D4
Caaguazú 7,950 .... D4
Caapucú 1,400 .... D5
Caazapá 3,132 .... D5
Caballero 1,225 .... B4
Capiatá 2,827 .... B4
Capitán Bado 915 .... E3
Capitán Meza 375 .... E5
Caraguatay 1,439 .... D4
Carapeguá 3,416 .... C4
Carayaó 1,190 .... C4
Carmen del Paraná 1,980 .... D5
Cerrito 958 .... D5
Ciudad Presidente
  Stroessner 7,085 .... E4
Concepción 19,392 .... D3
Coronel Bogado 3,973 .... D4
Coronel Martínez 1,598 .... B5
Coronel Oviedo 13,786 .... E4
Curuguaty 1,112 .... E4
Desmochados 551 .... C5
Doctor Cecilio Báez 1,300 .... D4
Doctor Juan L. Mallorquín 1,913 .... E4
Doctor Juan Manuel
  Frutos 1,494 .... E4
Doctor M. Irala 468 .... E4
Emboscada 1,222 .... B4
Encarnación 23,343 .... E5
Escobar 548 .... B5
Eusebio Ayala 4,328 .... B4
Fernando de la Mora 36,834 .... A4
Filadelfia 1,438 .... B3
Fram 1,090 .... E5
Fuerte Olimpo 3,063 .... C2
General Artigas 3,542 .... D5
General Elizardo Aquino 1,304 .... D4
General Eugenio A. Garay 740 .... A2
Guarambaré 3,640 .... B5
Hernandarias 3,898 .... E4
Hohenau 1,121 .... E5
Horqueta 4,328 .... D3
Hugo Stroessner 536 .... C4
Humaitá 938 .... C5
Isla Pucú 1,766 .... B4
Isla Umbú 236 .... B5
Itá 7,041 .... B5
Itacurubí 1,997 .... B5
Itacurubí del Rosario 2,467 .... D4
Itapé 1,376 .... D4
Itaquyry E4
Itauguá 3,767 .... B5
Iturbe 3,413 .... D5
Jesús 1,495 .... E5
Juan de Mena 1,027 .... B5
La Colmena 1,804 .... B5
Lambaré 31,656 .... A4
Laureles 435 .... D5
Lima 1,098 .... D3
Limpio 2,219 .... B4
Loreto 1,258 .... D3
Luque 13,921 .... B4
Maciel 376 .... D5
Mariano Roque Alonso 1,492 .... A4
Mariscal Estigarribia 3,150 .... B3
Mayor Martínez 324 .... C5

Mayor Pablo Lagerenza B1
Mbocayaty 925 .... C5
Mbuyapey 1,560 .... D5
Ñacunday 380 .... E5
Natalicio Talavera 1,228 .... D4
Nueva Germania 572 .... D3
Nueva Italia 1,517 .... B5
Numí 941 .... D4
Paraguarí 5,036 .... D4
Paso de Patria 698 .... C5
Pedro Juan Caballero 21,033 .... E3
Pilar 12,506 .... C5
Pirayú 2,698 .... B4
Piribebuy 4,497 .... B4
Primero de Marzo 696 .... B4
Puerto Casado 4,078 .... C3
Puerto Guaraní 302 .... C3
Puerto Pinasco 5,477 .... C3
Puerto Presidente Franco 4,152 .... E4
Puerto Sastre 160 .... C3
Quiindy 2,664 .... B5
Quyquyó 928 .... D5
Roque González de Santa
  Cruz 1,375 .... B5
Rosario 4,165 .... D4
Salto del Guairá E4
San Antonio 4,906 .... A5
San Bernardino 949 .... B4
San Cosme y Damián 602 .... D5
San Estanislao 4,753 .... D4
San Ignacio 6,116 .... D5
San Joaquín 536 .... E4
San José 3,102 .... B5
San Juan Bautista 6,457 .... D5
San Juan Bautista de
  Neembucú 688 .... D5
San Juan Nepomuceno 2,974 .... D5
San Lázaro 1,767 .... D3
San Lorenzo 11,616 .... A4
San Miguel 1,030 .... D5
San Patricio 1,130 .... D5
San Pedro 3,186 .... D4
San Pedro del Paraná 2,723 .... D5
San Salvador 1,393 .... C5
Santa Elena 1,439 .... B5
Santa María 793 .... D5
Santa Rosa 3,736 .... D5
Santiago 1,265 .... D5
Sapucaí 1,864 .... B5
Tacuaras 193 .... C5
Tacuatí 836 .... D3
Tavaí 472 .... E5
Tebicuary MI 183 .... D4
Tobatí 4,983 .... B4
Trinidad 837 .... E5
Unión 1,286 .... D4
Valenzuela 1,108 .... B5
Valle MI 1,318 .... B5
25 de Diciembre 1,261 .... D4
Villa Florida 1,261 .... D5
Villa Franca 359 .... C5
Villa Hayes 4,749 .... A4
Villa Oliva 564 .... C5
Villarrica 17,687 .... D4
Villeta 3,156 .... A4
Yabebyry 797 .... D5
Yaguarón 3,368 .... B5
Ybycuí 1,736 .... B5
Ybytymí 816 .... B4
Yegros 1,051 .... D5
Ygatimí 396 .... E4
Yhú 964 .... E4
Ypacaraí 5,195 .... B4
Ypané 1,474 .... B5
Ypé Jhú 645 .... E3
Yuty 2,392 .... D5

OTHER FEATURES

Acaray (riv.) E4
Alto Paraná (riv.) D5
Amambay, Cordillera de (mts.) D-E3
Apa (riv.) D3
Aquidabán (riv.) D3
Chaco Boreal (reg.) B2-3
Chovoreca (mt.) C4
Coronel F. Cabrera (mt.) B1
González, Riacho (riv.) C3
Gran Chaco (reg.) B2-3
Iguazú (falls) E4
Itaipú (res.) E4
Jara (hill) C3
Mbaracayú, Cordillera de (mts.) E3
Monday (riv.) E4
Montelindo (riv.) C3
Mosquito, Riacho (riv.) C3
Negro (riv.) C4
Paraguay (riv.) C4
Pilcomayo (riv.) C4
Tebicuary (riv.) C5
Timane (riv.) B2
Vera (lag.) D5
Verde (riv.) C3

© Copyright HAMMOND INCORPORATED, Maplewood, N.J.

URUGUAY

DEPARTMENTS

Artigas 52,843 .... B1
Canelones 258,195 .... D5
Cerro Largo 71,023 .... E3
Colonia 105,350 .... B5
Durazno 53,635 .... C3
Flores 23,530 .... C4
Florida 63,987 .... D4
Lavalleja 65,823 .... D5
Maldonado 61,259 .... E5
Montevideo 1,202,757 .... B7
Paysandú 88,029 .... B3
Río Negro 46,861 .... B3
Rivera 77,086 .... D2
Rocha 55,097 .... E4
Salto 92,183 .... B2
San José 79,563 .... C5
Soriano 77,906 .... B4
Tacuarembó 76,964 .... D3
Treinta y Tres 43,419 .... E4

CITIES and TOWNS

Aceguá 930 .... E2
Achar 606 .... C3
Agraciada 638 .... A4
Aguas Corrientes 992 .... A6
Aiguá 2,470 .... E5
Algorta 1,372 .... B3
Artigas 29,256 .... C1
Atlántida 2,268 .... B6
Balneario El Tesoro E5
Balneario La Barra E5
Balneario Solís 288 .... D5
Baltasar Brum 1,753 .... B1
Belén 2,129 .... B1
Bella Unión 7,778 .... B1
Bernabé Rivera 540 .... B1
Blanquillo 1,053 .... D3
Cañada Nieto 503 .... B4
Canelones 15,938 .... B6
Cardal 847 .... C5
Cardona 4,126 .... B4
Cardozo 143 .... C3
Carlos Reyles 961 .... C4
Carmelo 13,631 .... A4
Carmen 2,318 .... D4
Carrasco B7
Castillos 6,446 .... F5
Casupá 2,265 .... D5
Cebollatí 1,233 .... F4
Cerrillos 1,690 .... A6
Cerro Chato, Treinta y Tres 1,850 .... D4
Chamizo 486 .... D5
Chuy 4,472 .... F4
Colón, Lavalleja 367 .... E4
Colonia 16,895 .... B5
Colonia Lavalleja C2
Colonia Rossel y Rius 130 .... D4
Colonia Valdense 2,113 .... B5
Conchillas 748 .... B5
Constitución 3,217 .... A2
Costa Azul 453 .... E5
Cufré 430 .... B5
Cuñapirú D2
Curtina 723 .... C3
Diez y Nueve (19) de Abril 308 .... E5
Diez y Ocho (18) de Julio 742 .... F4
Dolores 12,771 .... A4
Durazno 25,811 .... C4
Egaña 667 .... B4
Empalme Olmos 2,084 .... B6
Estación Atlántida 1,845 .... B6
Estación Migues 241 .... C6
Florida 25,030 .... C5
Fortaleza de Santa Teresa F5
Fraile Muerto 2,468 .... E3
Fray Bentos 19,569 .... A4
Fray Marcos 1,573 .... D5
Garzón 258 .... E5
General Enrique Martínez 973 .... F4
Goñi 278 .... C4
Grecco 447 .... B3
Guichón 4,720 .... B3
Ituzaingó 717 .... A6
Javier de Viana 286 .... C1
Joanico 692 .... B6
Joaquín Suárez, Canelones 3,517 .... B6
José Batlle y Ordóñez 2,044 .... D4
José Enrique Rodó 1,334 .... B4
José Pedro Varela 3,541 .... E4
Juan L. Lacaze 11,133 .... B5
Julio María Sanz E4
La Bolsa C1
La Coronilla 571 .... F4
La Cruz 633 .... C4
La Cuchilla F3
La Floresta C7
La Lata E2
La Paloma 1,558 .... F5
La Paz, Canelones 14,402 .... B6
La Paz, Colonia B5
La Pedrera 116 .... F5
Lascano 6,043 .... E4
Las Flores 403 .... D5
La Sierra D5
Las Piedras 53,983 .... B6
Las Toscas 893 .... E3
Libertad 6,071 .... C5
Lorenzo Geyres 474 .... B3
Mal Abrigo 209 .... C5
Maldonado 22,159 .... D6
Mariscala 1,393 .... E5
Mazangano E3
Melo 38,260 .... E3
Mercedes 34,667 .... B4
Merinos 403 .... C3
Miguelete 533 .... B5
Migues 2,183 .... C6
Minas 35,433 .... D5
Minas de Corrales 2,518 .... D2
Montes 2,217 .... D5
Montevideo (cap.) 1,173,254 .... B7
Nico Pérez D4
Nueva Helvecia 8,598 .... B5
Nueva Palmira 6,934 .... A4
Nuevo Berlín 1,970 .... B3
Ombúes de Lavalle 1,689 .... B4
Ombúes de Oribe C4
Palmitas 1,332 .... B4
Pan de Azúcar 4,862 .... D5
Pando 16,184 .... B6
Paso de la Laguna, Salto B2
Paso de la Laguna, Tacuarembó D3
Paso de León B1
Paso del Borracho D2
Paso del Cerro 317 .... C2
Paso de los Toros 13,178 .... C3
Paso Potrero C2
Paysandú 62,412 .... A3
Peralta C3
Piedra Sola 233 .... C3
Piedras Coloradas 487 .... B3
Piñera 261 .... C3
Pintado, Artigas C1
Piraraja 774 .... E4
Piriápolis 5,221 .... D5
Porvenir 705 .... B3
Progreso 8,257 .... B6
Pueblo del Sauce E4
Pueblo Nuevo B2
Punta del Este 6,914 .... E6
Quebracho 1,514 .... B2
Reboledo 373 .... D4
Rio Branco 5,697 .... F3
Rivera 49,013 .... D1
Rocha 21,612 .... E5
Rodríguez 1,575 .... C5
Rosario 8,302 .... B5
Salto 72,94 .... B2
San Antonio, Canelones 1,122 .... B6
San Bautista 1,472 .... B6
San Carlos 16,883 .... E5
San Gregorio, San José C4
San Gregorio, Tacuarembó 2,892 .... D3
San Jacinto 2,292 .... C6
San Javier 1,583 .... A3
San José de Mayo 28,427 .... C5
San Ramón 6,570 .... D5
San Servando F3
Santa Catalina 885 .... B4
Santa Clara de Olimar 2,867 .... D3
Santa Lucía 14,101 .... B6
Santa Rosa 2,736 .... B6
Santiago Vázquez 1,323 .... A7
Sarandí del Yi 6,326 .... D4
Sarandí de Navarro 259 .... C3
Sarandí Grande 5,598 .... C4
Sauce, Canelones 3,942 .... B6
Sauce del Yi D4
Saucedo B2
Sequeira C1
Solís 356 .... D5
Solís de Mataojo 1,763 .... D5
Soriano 1,125 .... A4
Tacuarembó 34,152 .... D2
Tala 3,611 .... D5
Tambores 1,534 .... C2
Toledo 3,127 .... B6
Tomás Gomensoro 2,105 .... B1
Totoral C3
Tranqueras 3,922 .... D2
Treinta y Tres 25,757 .... E4
Tres Bocas B2
Tres Islas E3
Trinidad 17,598 .... B4
Tupambaé 1,039 .... E3
Unión B7
Valentines 153 .... E4
Veinticinco (25) de Agosto 1,891 .... A6
Veinticinco (25) de Mayo 1,744 .... C5
Velázquez 1,042 .... E5
Vergara 2,822 .... E3
Vichadero 1,989 .... E2
Villa Darwin 507 .... B4
Villa del Cerro A7
Young 11,080 .... B3
Zapicán 764 .... E4
Zapucay D2

OTHER FEATURES

Aiguá (riv.) E4
Alférez (riv.) E5
Arapey Chico (riv.) B1
Arapey Grande (riv.) B2
Belén (range) C1
Bonete (dam) C3
Brava (pt.) B7
Cañas (range) C2
Caraguatá (riv.) D3
Castillos (lag.) F5
Cebollatí (riv.) F4
Cordobés (riv.) D3
Cuareim (riv.) B1
Cuñapirú, Arroyo (riv.) D2
Daymán (range) B2
Daymán (riv.) B2
Durazno, Grande del (range) D4
Espinillo (pt.) A7
Este (pt.) D6
Flores (isl.) D5
Garzón (lag.) E5
Grande (range) D4
Grande, Arroyo (riv.) B4
Grande Inferior (range) C4
Haedo (range) C2
India Muerta (riv.) E4
José Ignacio (lag.) E5
Lobos (isl.) E6
Maciel, Arroyo (riv.) C4
Merín (lag.) F4
Mirador Nacional (mt.) D5
Negra (lag.) F5
Negra (range) D2
Negro (riv.) B4
Negro, Arroyo (riv.) B3
Olimar Grande (riv.) E4
Pando (riv.) B6
Parao (riv.) E3
Plata, La (riv.) B5
Polonio (cape) F5
Quequay Chico (riv.) B3
Quequay Grande (riv.) B3
Río Negro (res.) D3
Rocha (lag.) E5
Salto Grande (falls) A2
San José (riv.) C4
San Miguel (swamp) F4
San Salvador (riv.) B4
Santa Ana (range) D2
Santa Lucía (riv.) D5
Santa Lucía Chico (riv.) D4
Santa María (cape) F5
Sauce (lag.) D5
Sopas, Arroyo (riv.) C2
Tacuarembó (riv.) D2
Tacuarí (riv.) E3
Tigre (isl.) A7
Uruguay (riv.) A3
Yaguarón (riv.) F3
Yi (riv.) B4

PARAGUAY

AREA 157,047 sq. mi. (406,752 sq. km.)
POPULATION 2,973,000
CAPITAL Asunción
LARGEST CITY Asunción
HIGHEST POINT Amambay Range 2,264 ft. (690 m.)
MONETARY UNIT guaraní
MAJOR LANGUAGES Spanish, Guaraní
MAJOR RELIGION Roman Catholicism

URUGUAY

AREA 72,172 sq. mi. (186,925 sq. km.)
POPULATION 2,899,000
CAPITAL Montevideo
LARGEST CITY Montevideo
HIGHEST POINT Mirador Nacional 1,644 ft. (501 m.)
MONETARY UNIT Uruguayan peso
MAJOR LANGUAGE Spanish
MAJOR RELIGION Roman Catholicism

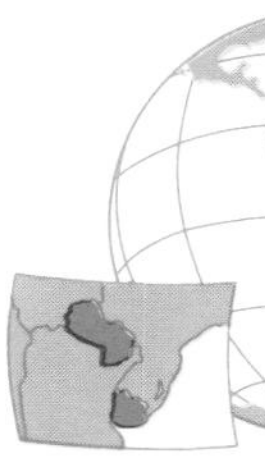

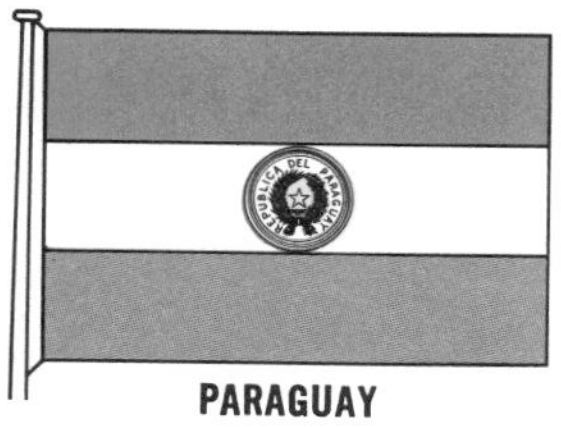

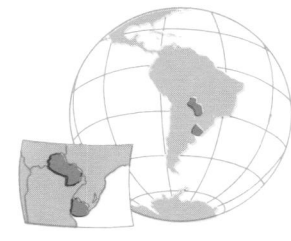

PARAGUAY

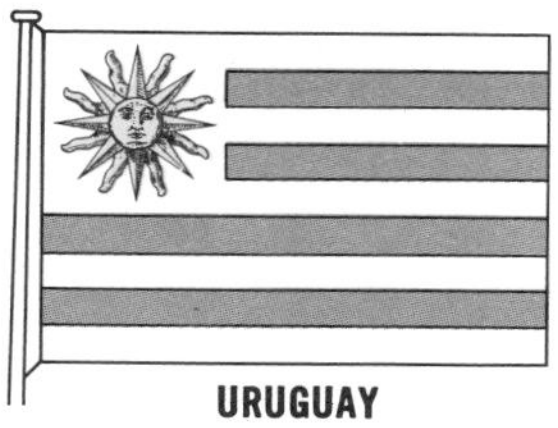

URUGUAY

Topography

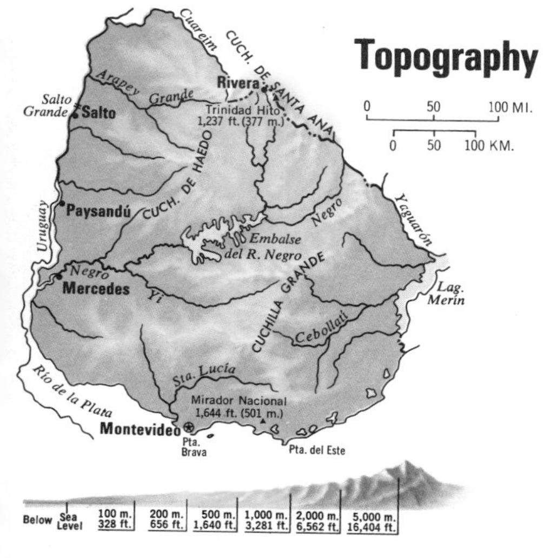

Rivera
Salto
Paysandú
Mercedes
Montevideo
Mirador Nacional 1,644 ft. (501 m.)
0 50 100 MI.
0 50 100 KM.
Below Sea Level
100 m. 328 ft.
200 m. 656 ft.
500 m. 1,640 ft.
1,000 m. 3,281 ft.
2,000 m. 6,562 ft.
5,000 m. 16,404 ft.

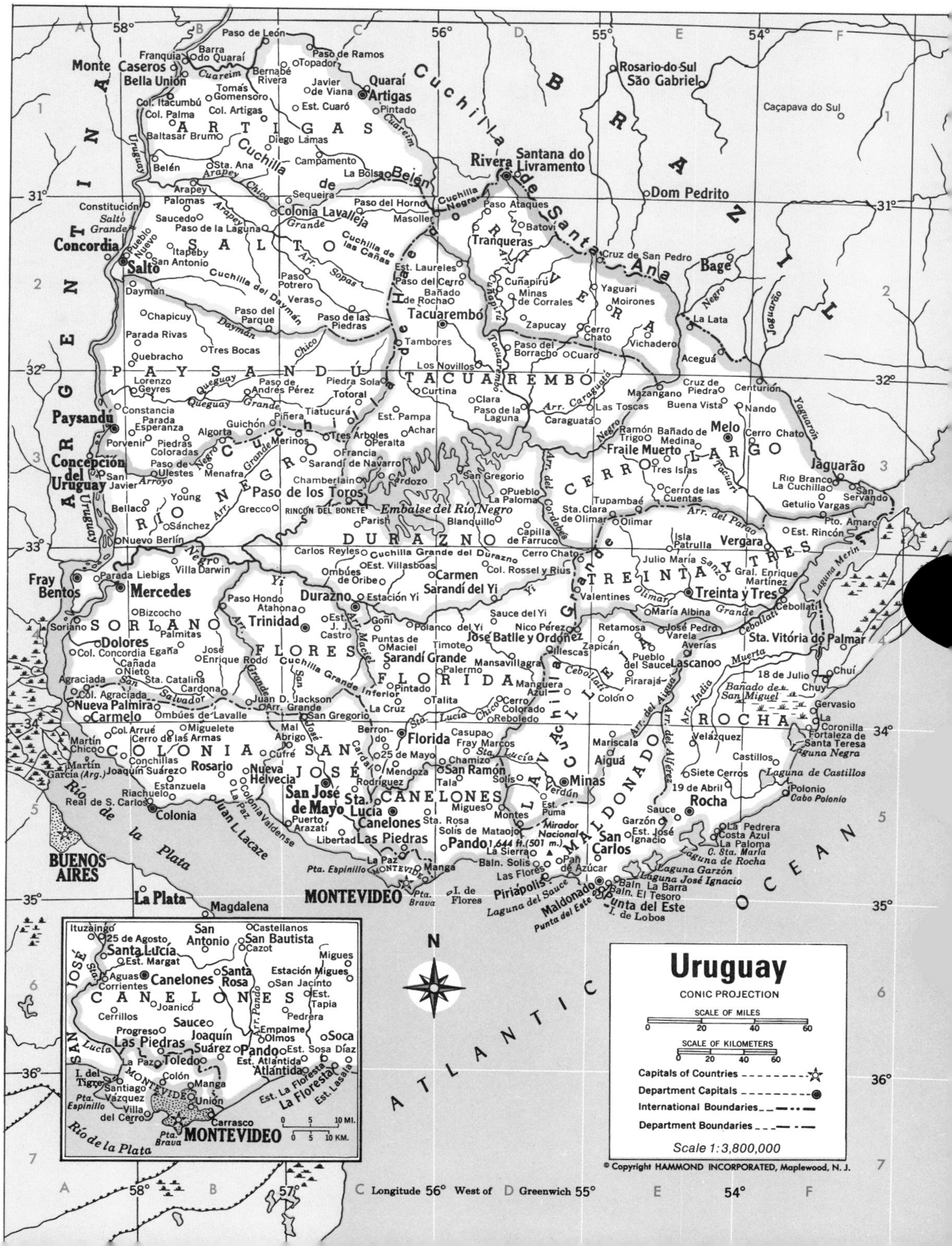

Uruguay
CONIC PROJECTION
SCALE OF MILES
SCALE OF KILOMETERS
Capitals of Countries
Department Capitals
International Boundaries
Department Boundaries
Scale 1:3,800,000
MONTEVIDEO
ATLANTIC OCEAN
Longitude 56° West of Greenwich 55°

© Copyright HAMMOND INCORPORATED, Maplewood, N. J.

North America

LAMBERT AZIMUTHAL EQUAL-AREA PROJECTION

MILES
0 100 200 400 600 800

KILOMETERS
0 100 200 400 600 800

Capitals of Countries .... ⊛
Other Capitals .... ⊛
International Boundaries .... — · — · —
Other Boundaries .... — · · — · · —

Scale 1:36,600,000

© Copyright HAMMOND INCORPORATED, Maplewood, N.J.

Population Distribution

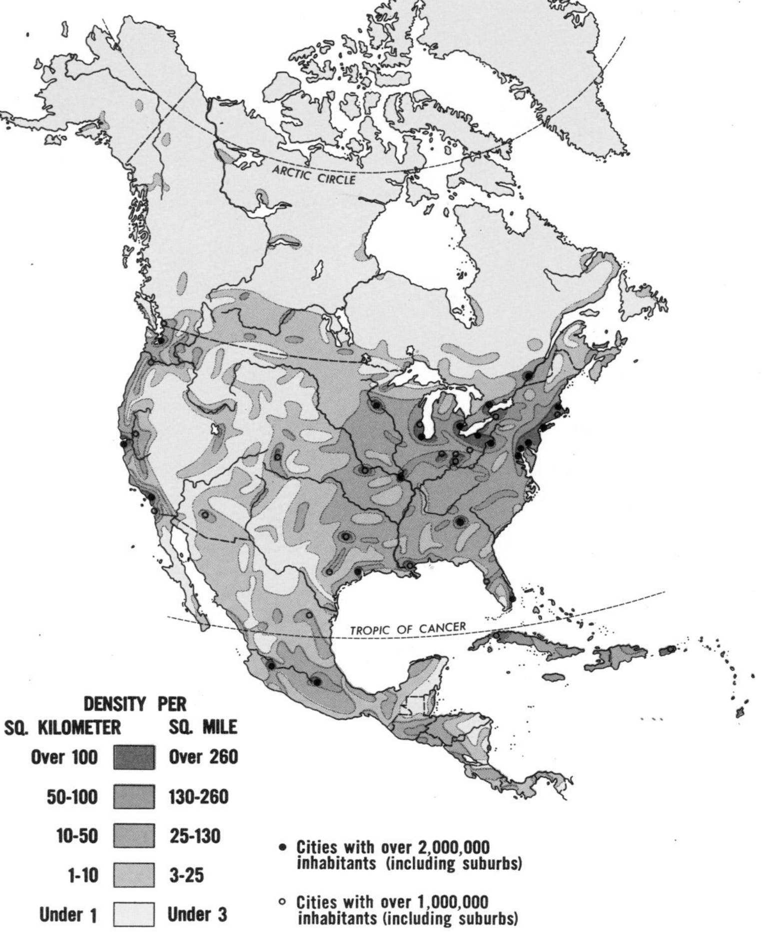
ARCTIC CIRCLE
TROPIC OF CANCER
DENSITY PER
SQ. KILOMETER
SQ. MILE
Over 100
Over 260
50-100
130-260
10-50
25-130
1-10
3-25
Under 1
Under 3
• Cities with over 2,000,000 inhabitants (including suburbs)
○ Cities with over 1,000,000 inhabitants (including suburbs)

AREA 9,363,000 sq. mi. (24,250,170 sq. km.)
POPULATION 370,000,000
LARGEST CITY New York
HIGHEST POINT Mt. McKinley 20,320 ft. (6,194 m.)
LOWEST POINT Death Valley -282 ft. (-86 m.)

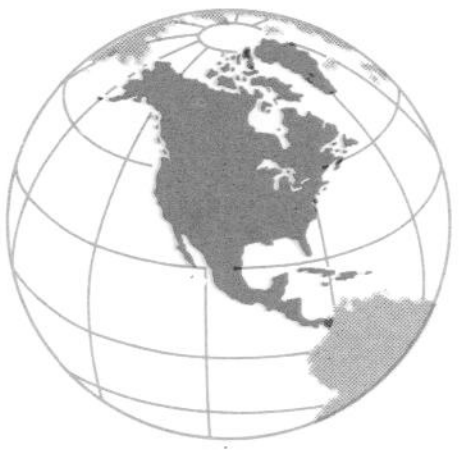

Vegetation

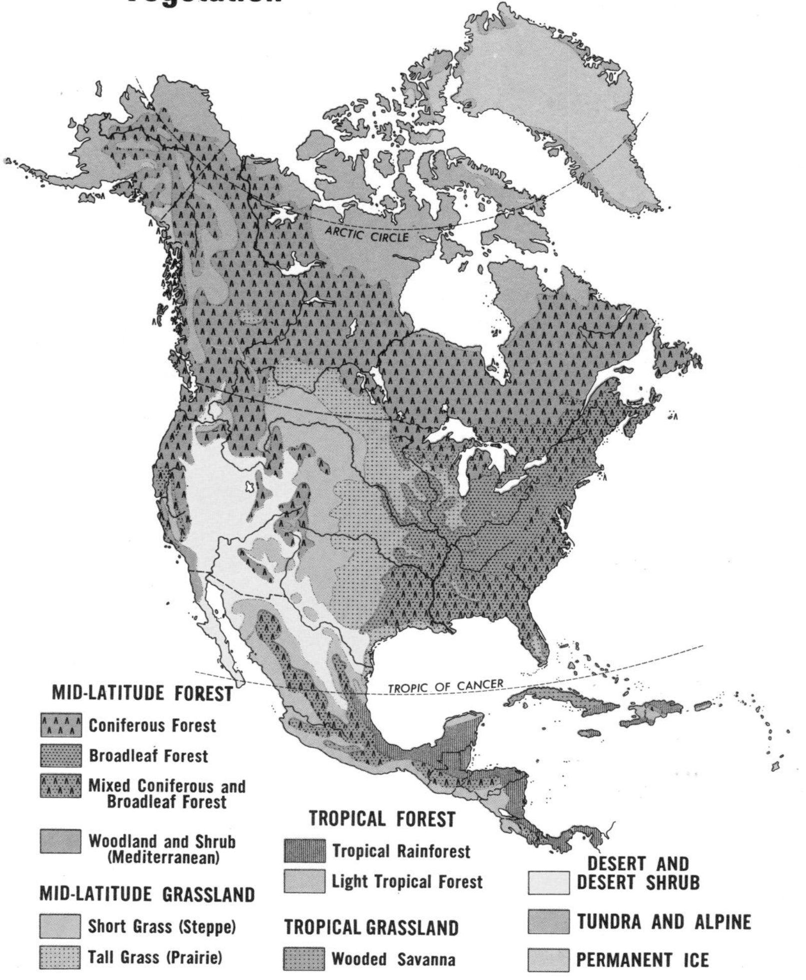
ARCTIC CIRCLE
TROPIC OF CANCER
MID-LATITUDE FOREST
Coniferous Forest
Broadleaf Forest
Mixed Coniferous and Broadleaf Forest
Woodland and Shrub (Mediterranean)
MID-LATITUDE GRASSLAND
Short Grass (Steppe)
Tall Grass (Prairie)
TROPICAL FOREST
Tropical Rainforest
Light Tropical Forest
TROPICAL GRASSLAND
Wooded Savanna
DESERT AND DESERT SHRUB
TUNDRA AND ALPINE
PERMANENT ICE

Acapulco de Juárez, Mexico H8
Akron, Ohio K5
Alabama (state), U.S. K6
Alaska (gulf) D4
Alaska (pen.), Alaska C4
Alaska (state), U.S. C3
Albany (cap.), N.Y. L5
Alberta (prov.) Canada G4
Albuquerque, N. Mex. H6
Amarillo, Texas H6
Anchorage, Alaska D3
Angmagssalik, Greenland Q3
Anguilla (isl.) M8
Antigua (isl.) M8
Arctic (ocean) B2
Arizona (state), U.S. G6
Arkansas (riv.), U.S. J6
Arkansas (state), U.S. J6
Athabasca (lake), Canada H4
Atlanta (cap.), Ga. K6
Augusta, Ga. K6
Augusta (cap.), Maine M5
Austin (cap.), Texas J6
Baffin (bay) M2
Baffin (isl.), N.W.T. L2
Bahamas L7
Bakersfield, Calif. G6
Baltimore, Md. L6
Barbados N8
Barrow (pt.), Alaska C2
Baton Rouge (cap.), Louisiana J6
Beaufort (sea) D2
Belize K8
Belle Isle (str.), Canada N5
Belmopan (cap.), Belize K8
Bering (sea) A3
Bering (str.) B3
Bermuda (isls.) M6
Birmingham, Ala. K6
Bismarck (cap.), N. Dak. H5
Boise (cap.), Idaho G5
Boston (cap.), Mass. L5
British Columbia (prov.), Canada F4
Buffalo, N.Y. L5
Butte, Mont. G5
Calgary, Alta. G4
California (gulf), Mexico G7
California (state), U.S. G6
Canada G4
Canadian (riv.), U.S. H6
Canaveral (cape), Fla. L7
Caribbean (sea) K8
Carson City (cap.), Nev. G6
Cascade (range), U.S. F5
Charleston, S.C. L6
Charleston (cap.), W. Va. K6
Charlotte, N.C. K6
Charlottetown (cap.), P.E.I. M5
Chattanooga, Tenn. K6
Chesapeake (bay), U.S. L6
Cheyenne (cap.), Wyo. H5
Chicago, Ill. K5
Chihuahua, Mexico H7
Cincinnati, Ohio K6
Ciudad Juárez, Mexico H6
Cleveland, Ohio K5
Clipperton (isl.) H8
Coast (mts.), Br. Col. E4
Coast (ranges), U.S. F5
Cocos (isl.), C. Rica K9
Cod (cape), Mass. M5
Colón, Pan. K9
Colorado (riv.), Texas H6
Colorado (riv.), U.S. G6
Colorado (state), U.S. H6
Colorado Springs, Colo. H6
Columbia (cap), S.C. K6
Columbia (riv.), U.S. F5
Columbus, Ga. K6
Columbus (cap.), Ohio K6
Concord (cap.), N.H. L5
Connecticut (state), U.S. L5
Costa Rica K8
Cuba L7
Dallas, Texas J6
Dayton, Ohio K6
Delaware (state), U.S. L6
Denmark (str.) S3
Denver (cap.), Colo. H6
Des Moines (cap.), Iowa J5
Detroit, Mich. K5
District of Columbia L6
Dominica M8
Dominican Republic L8
Dover (cap.), Del. L6
Duluth, Minn. J5
Edmonton (cap.), Alta. G4
Ellesmere (isl.), N.W.T. K1
El Paso, Texas H6
El Salvador J8
Erie (lake) K5
Fairbanks, Alaska D3
Farewell (cape), Greenland P4
Flin Flon, Man. H4
Florida (state), U.S. K7
Florida (strs.) K7
Fort McPherson, N.W.T. E3
Fort Smith, Ark. J6
Fort Worth, Texas J6
Foxe (basin), N.W.T. L3
Frankfort (cap.), Ky. K6
Fraser (riv.), Br. Col. F4
Fredericton (cap.), N. Br. M5
Fresno, Calif. G6
Georgia (state), U.S. K6
Godhavn, Greenland N3
Godthab (Nûk) (cap.), Greenland N3
Goose Bay, Newf. M4
Gracias a Dios (cape), Nic. K8
Grand Bahama (isl.), Bahamas L7
Grand Rapids, Mich. K5
Great Bear (lake), N.W.T. F3
Greater Antilles (isls.) K7
Great Falls, Mont. G5
Great Salt (lake), Utah G6
Great Slave (lake), N.W.T. H3
Green Bay, Wis. J5
Greenland P2
Greenland (sea) T2
Grenada M8
Guadalajara, Mexico H7
Guadeloupe (isl.) M8
Guantánamo, Cuba L7
Guatemala J8
Guatemala (cap.), Guatemala J8
Haiti L8
Halifax (cap.), N.S. M5
Hamilton, Ont. K5
Harrisburg (cap.), Pa. L5
Hartford (cap.), Conn. L5
Hatteras (cape), N.C. L6
Havana (cap.), Cuba K7
Helena (cap.), Mont. G5
Hispaniola (isl.) L7
Honduras K8
Honduras (gulf) K8
Houston, Texas J7
Hudson (bay), Canada K3
Hudson (str.), Canada L3
Huron (lake) K5
Idaho (state), U.S. G5
Illinois (state), U.S. K6
Indiana (state), U.S. K6
Indianapolis (cap.), Ind. K5
Inuvik, N.W.T. E3
Iowa (state), U.S. J5
Jackson (cap.), Miss. K6
Jacksonville, Fla. K6
Jamaica L8
James (bay), Canada K4
Jefferson City (cap.), Mo. J6
Juan de Fuca (str.) F5
Juneau, Alaska E4
Kansas (state), U.S. J6
Kansas City, Mo. J6
Kentucky (state), U.S. K6
Ketchikan, Alaska E4
Key West, Fla. K7
Kingston (cap.), Jamaica L8
Kingston, Ont. L5
Knoxville, Tenn. K6
Kodiak (isl.), Alaska C4
Labrador (reg.), Newf. M4
Lansing (cap.), Mich. K5
Lesser Antilles (isls.) M8
Lexington, Ky. K6
Liard (riv.), Canada F3
Limón, C. Rica K8
Lincoln (cap.), Nebr. J5
Lincoln (sea) M1
Little Rock (cap.), Ark. J6
London, Ont. K5
Los Angeles, Calif. G6
Louisiana (state), U.S. J6
Louisville, Ky. K6
Lower California (pen.), Mexico G7
Mackenzie (riv.), N.W.T. F3
Macon, Ga. K6
Madison (cap.), Wis. K5
Maine (state), U.S. M5
Managua (cap.), Nic. K8
Manitoba (prov.), Canada J4
Manitoba (lake), Man. H4
Marquette, Mich. K5
Martinique (isl.) M8
Maryland (state), U.S. L6
Massachusetts (state), U.S. L5
Matanzas, Cuba K7
Mazatlán, Mexico H7
McKinley (mt.), Alaska C3
Mead (lake), U.S. G6
Medicine Hat, Alta. H4
Memphis, Tenn. K6
Mendocino (cape), Calif. F5
Mérida, Mexico J7
Mexico H7
Mexico (gulf) K7
Mexico City (cap.), Mexico J7
Miami, Fla. K7
Michigan (lake), U.S. K5
Michigan (state), U.S. K5
Milwaukee, Wis. K5
Minneapolis, Minn. J5
Minnesota (state), U.S. J5
Mississippi (riv.), U.S. J6
Mississippi (state), U.S. K6
Missouri (riv.), U.S. J5
Missouri (state), U.S. J6
Mobile, Ala. K6
Montana (state), U.S. H5
Monterrey, Mexico J7
Montgomery (cap.), Ala. K6
Montpelier (cap.), Vt. L5
Montréal, Que. L5
Moose Jaw, Sask. H4
Nares (str.) L2
Nashville (cap.), Tenn. K6
Nassau (cap.), Bahamas L7
Nebraska (state), U.S. J5
Nelson (riv.), Man. J4
Netherlands Antilles M8
Nevada (state), U.S. G6
New Brunswick (prov.), Canada M5
Newfoundland (prov.), Canada M4
New Hampshire (state), U.S. L5
New Jersey (state), U.S. L5
New Mexico (state), U.S. H6
New Orleans, La. K7
New York, N.Y. L5
New York (state), U.S. L5
Nicaragua K8
Nicaragua (lake), Nic. K8
Nome, Alaska B3
Norfolk, Va. L6
North Bay, Ont. L5
North Carolina (state), U.S. K6
North Dakota (state), U.S. H5
North Saskatchewan (riv.), Canada G4
Northwest Territories (prov.), Canada G3
Nova Scotia (prov.), Canada M5
Nûk (cap.), Greenland N3
Oakland, Calif. F6
Oaxaca, Mexico J8
Ohio (riv.), U.S. K6
Ohio (state), U.S. K5
Oklahoma (state), U.S. J6
Oklahoma City (cap.), Okla. J6
Olympia (cap.), Wash. F5
Omaha, Nebr. J5
Ontario (lake) L5
Ontario (prov.), Canada K4
Oregon (state), U.S. F5
Ottawa (cap.), Canada L5
Ottawa (riv.), Canada L5
Panama K9
Panama (canal), Pan. L8
Panamá (cap.), Pan. L9
Panamá (gulf), Pan. L9
Parry (chan.), N.W.T. G2
Peace (riv.), Canada G4
Pennsylvania (state), U.S. L5
Pensacola, Fla. K6
Philadelphia, Pa. L6
Phoenix (cap.), Ariz. G6
Pierre (cap.), S. Dak. J5
Pittsburgh, Pa. K5
Port-au-Prince (cap.), Haiti L8
Portland, Maine M5
Portland, Oreg. F5
Prince Edward Island (prov.), Canada M5
Providence (cap.), R.I. L5
Prudhoe (bay), Alaska D2
Puebla, Mexico J8
Puerto Rico M8
Québec (cap.), Que. L5
Québec (prov.), Que. L4
Queen Charlotte (isls.), Br. Col. E4
Queen Elizabeth (isls.), N.W.T. G2
Raleigh (cap.), N.C. L6
Red (riv.) J5
Red (riv.), U.S. J6
Regina (cap.), Sask. H4
Reno, Nev. G6
Rhode Island (state), U.S. M5
Richmond (cap.), Va. L6
Rio Grande (riv.) H7
Rochester, N.Y. L5
Rocky (mts.) F4
Sable (cape), N.S. M5
Sacramento (cap.), Calif. F6
Saint Augustine, Fla. K7
Saint Christopher & Nevis M8
Saint John, N. Br. M5
St. John's (cap.), Newf. N5
Saint Lawrence (gulf) M5
Saint Lawrence (riv.) L5
Saint Louis, Mo. K6
Saint Lucia M8
Saint Paul (cap.), Minn. J5
Saint Pierre and Miquelon (isls.) N5
Saint Vincent and the Grenadines M8
Salem (cap.), Oreg. F5
Salt Lake City (cap.), Utah G6
Saltillo, Mexico H7
San Antonio, Texas J7
San Diego, Calif. G6
San Francisco, Calif. F6
San José (cap.), C. Rica K9
San Juan (cap.), P. Rico M8
San Salvador (cap.), El Salvador J8
Santa Fe (cap.), N. Mex. H6
Santiago de Cuba, Cuba L8
Santo Domingo (cap.), Dom. Rep. L8
Saskatchewan (prov.), Canada H4
Saskatoon, Sask. H4
Sault Ste. Marie, Ont. K5
Savannah, Ga. K6
Seattle, Wash. F5
Shreveport, La. J6
Sierra Madre Occidental (mts.), Mexico H7
Sierra Nevada (range), California F6
Sioux City, Iowa J5
Sioux Falls, S. Dak. J5
Sitka, Alaska E4
Slave (riv.), Canada G3
Snake (riv.), U.S. G5
South Carolina (state), U.S. K6
South Dakota (state), U.S. H5
South Platte (riv.), U.S. H6
South Saskatchewan (riv.), Canada G4
Springfield (cap.), Ill. J6
Sudbury, Ont. K5
Superior (lake) K5
Sverdrup (isls.), N.W.T. J2
Tacoma, Wash. F5
Tallahassee (cap.), Fla. K6
Tampa, Fla. K7
Tampico, Mexico J7
Tegucigalpa (cap.), Hond. K8
Tennessee (state), U.S. K6
Texas (state), U.S. J6
Thule, Greenland M2
Thunder Bay, Ont. K5
Toledo, Ohio K5
Topeka (cap.), Kansas J6
Toronto (cap.), Ont. K5
Trenton (cap.), N.J. L5
Trinidad and Tobago N8
Tucson, Ariz. G6
Tulsa, Okla. J6
Ungava (bay), Canada M4
United States H5
Utah (state), U.S. G6
Vancouver, Br. Col. F4
Vancouver (isl.), Br. Col. F5
Veracruz, Mexico J8
Vermont (state), U.S. L5
Victoria (cap.), Br. Col. F5
Victoria (isl.), N.W.T. G2
Virgin Islands (isls.) M8
Virginia (state), U.S. L6
Washington (state), U.S. F5
Washington, D.C. (cap.), U.S. L6
West Indies (isls.) M7
West Virginia (state), U.S. K6
Whitehorse (cap.), Yukon E3
Wichita, Kansas J6
Windward (passg.) L8
Winnipeg (cap.), Man. J5
Winnipegosis (lake), Man. H4
Wisconsin (state), U.S. K5
Woods (lake) J5
Wyoming (state), U.S. H5
Yellowknife, N.W.T G3
Yucatán (pen.), Mexico K7
Yukon (riv.) C3
Yukon Territory (terr.), Canada E3

Average January Temperature

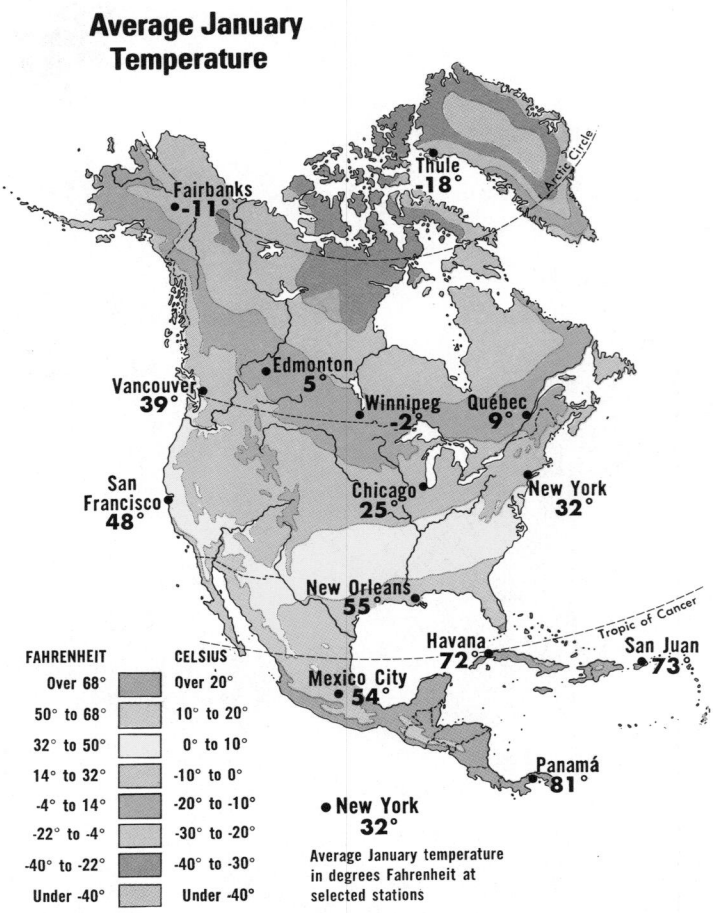

Fairbanks -11°
Thule -18°
Vancouver 39°
Edmonton 5°
Winnipeg -2°
Québec 9°
San Francisco 48°
Chicago 25°
New York 32°
New Orleans 55°
Havana 72°
San Juan 73°
Mexico City 54°
Panamá 81°

FAHRENHEIT

FAHRENHEIT	CELSIUS
Over 68°	Over 20°
50° to 68°	10° to 20°
32° to 50°	0° to 10°
14° to 32°	-10° to 0°
-4° to 14°	-20° to -10°
-22° to -4°	-30° to -20°
-40° to -22°	-40° to -30°
Under -40°	Under -40°

● New York
   32°

Average January temperature
in degrees Fahrenheit at
selected stations

Average July Temperature

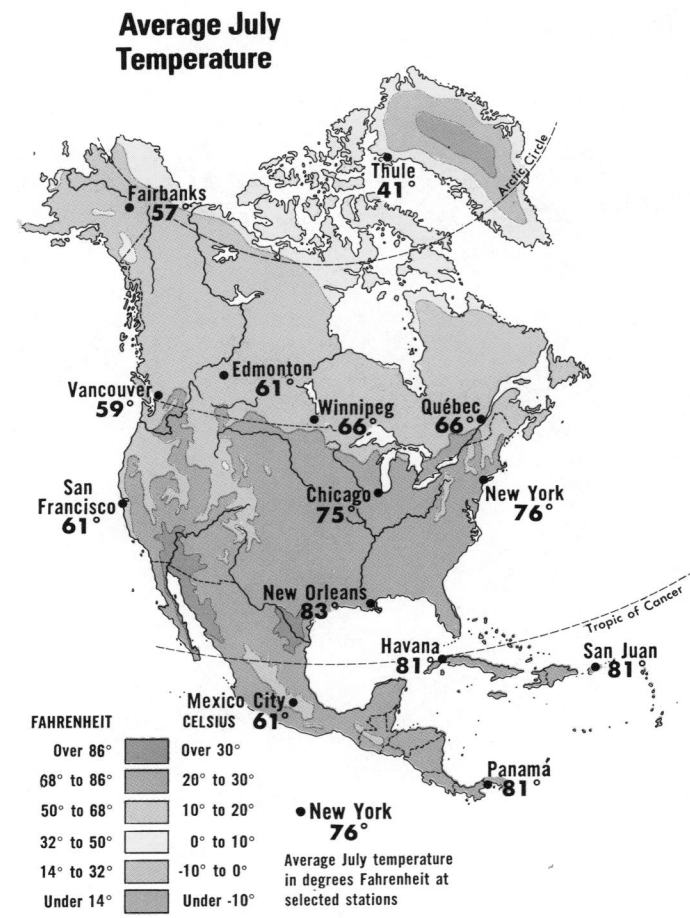

Fairbanks 57°
Thule 41°
Vancouver 59°
Edmonton 61°
Winnipeg 66°
Québec 66°
San Francisco 61°
Chicago 75°
New York 76°
New Orleans 83°
Havana 81°
San Juan 81°
Mexico City 61°
Panamá 81°

FAHRENHEIT	CELSIUS
Over 86°	Over 30°
68° to 86°	20° to 30°
50° to 68°	10° to 20°
32° to 50°	0° to 10°
14° to 32°	-10° to 0°
Under 14°	Under -10°

● New York
   76°

Average July temperature
in degrees Fahrenheit at
selected stations

Rainfall

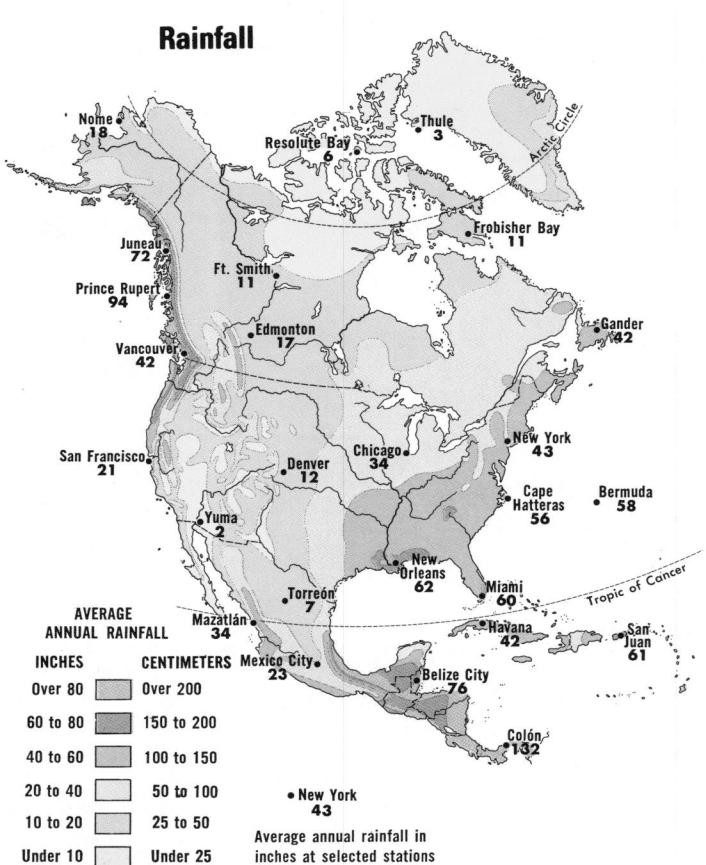

Nome 18
Resolute Bay 6
Thule 3
Frobisher Bay 11
Juneau 72
Ft. Smith 11
Prince Rupert 94
Gander 42
Vancouver 42
Edmonton 17
San Francisco 21
Chicago 34
Denver 12
New York 43
Yuma 2
Cape Hatteras 56
Bermuda 58
New Orleans 62
Miami 60
Torreón 7
Mazatlán 34
Havana 42
San Juan 61
Mexico City 23
Belize City 76
Colón 132

AVERAGE ANNUAL RAINFALL

INCHES	CENTIMETERS
Over 80	Over 200
60 to 80	150 to 200
40 to 60	100 to 150
20 to 40	50 to 100
10 to 20	25 to 50
Under 10	Under 25

● New York
   43

Average annual rainfall in
inches at selected stations

Vegetation/Relief

SCALE OF MILES
0   200   400   600   800   1000

SCALE OF KILOMETERS
0   200   400   600   800   1000

Capitals of Countries...........................⊛
Other Capitals ....⊙
International Boundaries ....—--—
Canals...

Depths in Fathoms

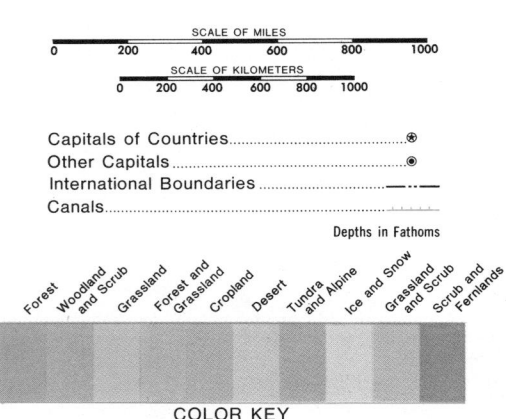

Forest | Woodland and Scrub | Grassland | Forest and Grassland | Cropland | Desert | Tundra and Alpine | Ice and Snow | Grassland and Scrub | Scrub and Fernlands

COLOR KEY

Longitude 90° West of Greenwich

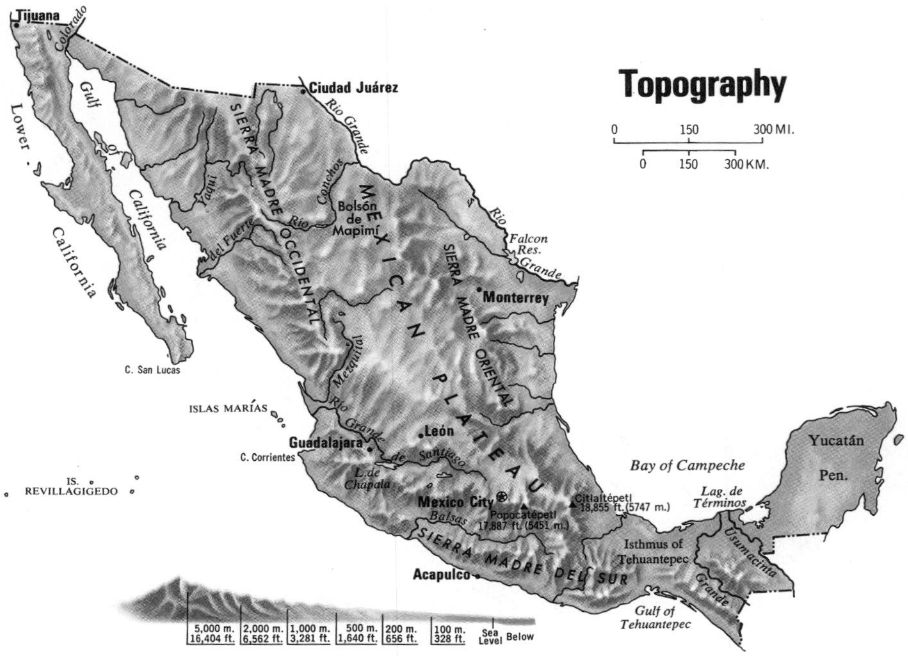
Topography
0 150 300 MI.
0 150 300 KM.
Tijuana
Colorado
Gulf of California
Lower California
C. San Lucas
Ciudad Juárez
Río Grande
SIERRA MADRE OCCIDENTAL
Yaqui
Conchos
Bolsón de Mapimí
MEXICAN PLATEAU
Falcon Res.
Monterrey
SIERRA MADRE ORIENTAL
ISLAS MARÍAS
IS. REVILLAGIGEDO
C. Corrientes
Guadalajara
León
L. de Chapala
Mexico City
Citlaltépetl 18,855 ft. (5747 m.)
Popocatépetl 17,887 ft. (5451 m.)
Balsas
SIERRA MADRE DEL SUR
Acapulco
Bay of Campeche
Lag. de Términos
Isthmus of Tehuantepec
Gulf of Tehuantepec
Yucatán Pen.
5,000 m. 16,404 ft.
2,000 m. 6,562 ft.
1,000 m. 3,281 ft.
500 m. 1,640 ft.
200 m. 656 ft.
100 m. 328 ft.
Sea Level
Below

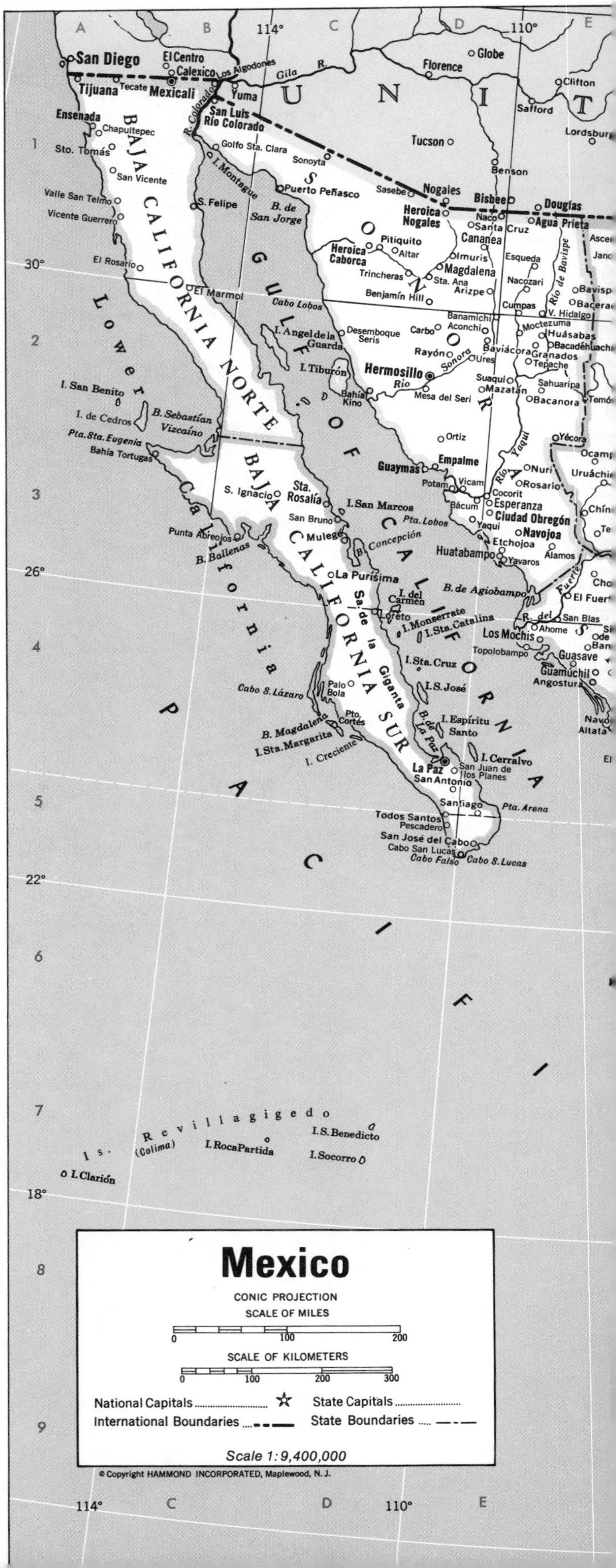
San Diego
El Centro
Calexico
Tijuana
Tecate
Mexicali
Yuma
San Luis Río Colorado
Ensenada
Golfo Sta. Clara
Sonoyta
Puerto Peñasco
Nogales
Heroica Nogales
Bisbee
Douglas
Naco
Agua Prieta
Cananea
Tucson
Benson
Safford
Globe
Florence
Heroica Caborca
Magdalena
Hermosillo
Guaymas
Empalme
Ciudad Obregón
Navojoa
Huatabampo
Los Mochis
Guasave
Guamúchil
BAJA CALIFORNIA NORTE
BAJA CALIFORNIA SUR
GULF OF CALIFORNIA
Lower California
PACIFIC
Sta. Rosalía
Mulegé
La Purísima
Loreto
La Paz
Santiago
Todos Santos
San José del Cabo
Cabo San Lucas
I. Tiburón
I. Ángel de la Guarda
I. de Cedros
B. Sebastián Vizcaíno
I. San Benito
B. Magdalena
I. Sta. Margarita
I. Cerralvo
I. Espíritu Santo
I. San José
I. Sta. Cruz
Is. Revillagigedo
I. S. Benedicto
I. Roca Partida
I. Socorro
I. Clarión
114°
110°
30°
26°
22°
18°
Mexico
CONIC PROJECTION
SCALE OF MILES
0 100 200
SCALE OF KILOMETERS
0 100 200 300
National Capitals ☆
State Capitals
International Boundaries
State Boundaries
Scale 1:9,400,000

© Copyright HAMMOND INCORPORATED, Maplewood, N. J.

STATES

Aguascalientes 504,300 H6
Baja California 1,227,400 B1
Baja California Sur 221,000 C3
Campeche 371,800 O7
Chiapas 2,097,500 N8
Chihuahua 1,935,100 F2
Coahuila 1,561,000 H3
Colima 339,400 G7
Distrito Federal 9,377,300 L1
Durango 1,160,300 G4
Guanajuato 3,045,600 J6
Guerrero 2,174,200 J8
Hidalgo 1,518,200 K6
Jalisco 4,296,500 H6
México 7,542,300 K7
Michoacán 3,049,400 H7
Morelos 931,400 K7
Nayarit 729,500 G6
Nuevo León 2,463,500 K4
Oaxaca 2,517,500 L8
Puebla 3,285,300 L7
Querétaro 730,900 J6
Quintana Roo 209,900 P7
San Luis Potosí 1,669,900 J5
Sinaloa 1,882,200 F4
Sonora 1,498,100 D2
Tabasco 1,150,000 N7
Tamaulipas 1,924,900 K4
Tlaxcala 548,500 N1
Veracruz 5,263,800 L7
Yucatán 1,034,300 P6
Zacatecas 1,144,700 H5

CITIES and TOWNS

Acala 11,483 N8
Acámbaro 32,257 J7
Acaponeta 11,844 G5
Acapulco de Juárez 309,254 K8
Acatlán de Osorio 7,624 K7
Acatzingo de Hidalgo 6,905 N2
Acayucan 21,173 M8
Aconchi 1,596 D2
Actopan, Hidalgo 11,037 K6
Actopan, Veracruz 2,265 Q1
Agua Dulce 21,060 M7
Agualeguas 2,502 J3
Agua Prieta 20,754 E1
Aguascalientes 181,277 H6
Aguililla 5,715 H7
Ahome 4,182 E4
Ahuacatitlán 6,436 L1
Ahuacatlán 5,350 G6
Ahumada 6,466 F1
Ajalpan 8,238 L7
Alamo 9,954 L6
Álamos 4,269 E3
Aldama, Chihuahua 6,047 G2
Aldama, Tamaulipas 3,033 L5
Aljojuca 3,204 O1
Allende, Coahuila 11,076 J2
Allende, Nuevo León 9,914 J4
Almoloya del Río 3,714 K1
Altamira 6,053 L5
Altar 2,519 D1
Altepexi 6,661 L7
Alto Lucero 3,698 P1
Altotonga 6,754 P1
Alvarado 15,592 M7
Amatlán de los Reyes 3,664 P2
Amealco 2,960 K6
Ameca 21,018 H6
Amecameca de Juárez 16,276 L1
Amozoc de Mota 9,203 N2
Anáhuac, Chihuahua 10,886 F2
Anáhuac, Nuevo León 8,168 J3
Angostura 2,663 E4
Antiguo Morelos 1,569 K5
Apan 13,705 M1
Apatzingán de la Constitución 44,849 H7
Apizaco 21,189 N1
Aquiles Serdán 2,565 G2
Aramberri 1,786 J5
Arandas 18,934 H6
Arcelia 10,024 J7
Ario de Rosales 8,774 J7
Arizpe 1,736 D1
Armería 10,616 G7
Arriaga 13,193 N8
Arteaga 5,324 H7
Ascensión 4,104 E1
Asunción Nochixtlán 3,235 L8
Atlixco 41,967 M2
Atotonilco el Alto 16,271 H6
Atoyac de Álvarez 8,874 J8
Autlán de Navarro 20,398 G7
Axochiapan 8,283 M2
Ayutla de los Libres 3,618 K8
Azcapotzalco 534,554 L1
Azoyú 3,446 K8
Bacadéhuachi 1,514 E2
Bacalar 2,121 P7
Bachíniva 1,809 F2
Bácum 2,668 D3
Bahía Tortugas 1,457 B3
Balancán de Domínguez 3,669 O8
Bamoa 5,866 E4
Banderilla 3,488 P1
Baviácora 2,049 E2
Benjamín Hill 5,366 D1
Bernardino de Sahagún 12,327 M1
Boca del Río 2,354 Q2
Bolonchén de Rejón 2,342 O7
Buenaventura 3,924 F2
Burgos 673 K4
Cabo San Lucas 1,534 E5
Cacahoatán 5,079 N9
Cadereyta Jiménez 13,586 K4
Calkiní 6,870 O6
Calnali 3,318 K6
Calpulálpan 8,659 M1
Calvillo 6,453 H6
Campeche 69,506 O7
Cananea 17,518 D1
Canatlán 5,983 G4
Cancún 326 Q6
Candela 1,689 J3
Candelaria 1,982 O7
Cañitas de Felipe Pescador 4,885 H5
Capulhuac de Mirafuentes 8,289 K1
Carbo 2,804 D2
Cárdenas, San Luis Potosí 12,020 K6
Cárdenas, Tabasco 15,643 N8
Carichic 1,520 F2
Castaños 8,996 J3
Catemaco 11,786 M7
Ceballos 2,937 H3
Cedral 4,057 J5
Celaya 79,977 J6
Celestún 1,490 O6
Cerritos 10,421 J5
Cerro Azul 20,259 L6
Chahuites 5,218 M8
Chalchihuites 1,894 G5
Chalco de Díaz Covarrubias 12,172 M1
Champotón 6,606 O7
Charcas 10,491 J5
Chetumal 23,685 Q7
Chiapa de Corzo 8,571 N8
Chiautempan 12,327 N1
Chietla 4,602 M2
Chignahuapan 3,805 N1
Chihuahua 327,313 F2
Chilapa de Álvarez 9,204 K8
Chilpancingo de los Bravos 36,193 K8
China, Nuevo León 4,958 K4
Chocomán 5,114 P2
Choix 2,503 E3
Cholula de Rivadavia 15,399 M1
Cihuatlán 9,451 G7
Cintalapa de Figueroa 12,036 N8
Ciudad Acuña (Villa Acuña) 30,276 J2
Ciudad Altamirano 8,694 J7
Ciudad Camargo, Chihuahua 24,030 G3
Ciudad Camargo, Tamaulipas 5,953 K3
Ciudad del Carmen 34,656 N7
Ciudad Delicias 52,446 G2
Ciudad del Maíz 5,241 K5
Ciudad de Río Grande 11,651 H5
Ciudad Guerrero 3,110 F2
Ciudad Guzmán 48,166 H7
Ciudad Hidalgo, Chiapas 4,105 N9
Ciudad Hidalgo, Michoacán 24,692 J7
Ciudad Juárez 424,135 F1
Ciudad Lerdo 19,803 H4
Ciudad Madero 115,302 L5
Ciudad Mante 51,247 K5
Ciudad Mendoza 18,696 O2
Ciudad Miguel Alemán 11,259 K3
Ciudad Obregón 144,795 E3
Ciudad Río Bravo 39,018 K4
Ciudad Satélite 35,083 L1
Ciudad Serdán 9,581 O2
Ciudad Valles 47,587 K5
Ciudad Victoria 83,897 K5
Coalcomán de Matamoros 4,875 H7
Coatepec 21,542 P1
Coatetelco 5,268 L2
Coatzacoalcos 69,753 M7
Coatzingo 3,038 N2
Cocorit 4,478 E3
Colima 58,450 H7
Colón 3,346 K6
Colotlán 6,135 H5
Comala 5,592 H7
Comalcalco 14,963 N7
Comitán de Domínguez 21,249 O8
Compostela 9,801 G6
Concepción del Oro 8,144 J4
Concordia 3,947 G5
Contla 7,517 N1
Copala 3,783 K8
Coquimatlán 6,212 G7
Córdoba 78,495 P2
Cosalá 2,279 F4
Cosamaloapan de Carpio 19,766 M7
Cosautlán de Carvajal 2,039 P1
Coscomatepec de Bravo 6,023 P2
Cosío 2,680 H5
Costa Rica 11,795 F4
Cotija de la Paz 9,178 H7
Coyoacán 339,446 L1
Coyotepec 8,888 L1
Coyuca de Benítez 6,328 J8
Coyuca de Catalán 2,926 J7
Coyutla 3,726 L6
Cozumel 5,858 Q6
Creel 2,449 E3
Cuatrociénagas de Carranza 5,523 H3
Cuauhtémoc 26,598 F2
Cuautepec de Hinojosa 5,501 K6
Cuautitlán de Romero Rubio 11,439 L1
Cuautla Morelos 13,946 L2
Cuencamé de Ceniceros 3,774 H4
Cuernavaca 239,813 L2
Cuicatlán 2,733 L8
Cuitlahuac 4,813 P2
Culiacán 228,001 F4
Cumpas 2,395 E1
Cunduacán 4,397 N7
Dimas 2,194 F5
Doctor Arroyo 4,290 K5
Dolores Hidalgo de la Independencia Naci 16,849 J6
Durango 182,633 G1
Dzibalchén 1,917 P7
Dzidzantún 7,064 P6
Dzitbalché 4,393 P6
Ebano 17,489 K5
Ecatepec de Morelos 11,899 L1
Ejutla de Crespo 5,263 L8
Eldorado 8,115 E4
El Fuerte 7,179 E3
El Porvenir 3,030 G1
El Potosí 2,032 J4
El Salto 7,818 G5
El Zacatón 2,686 J5
Empalme 24,927 D2
Encarnación de Díaz 10,474 H6
Ensenada 77,687 A1
Escalón 2,998 G3
Escarcega 7,248 O7
Escuinapa de Hidalgo 16,442 G5
Escuintla 4,111 N9
Esperanza, Puebla 4,258 O2
Esperanza, Sonora 11,762 E3
Espita 5,394 Q6
Esqueda 1,458 E1
Etchojoa 4,398 E3
Ezequiel Montes 3,139 K6
Fortín de las Flores 9,358 P2
Francisco I. Madero 12,613 H4
Fresnillo de González Echererría 44,475 H5
Frontera 10,066 N7
Galeana, Nuevo León 3,429 J4
General Bravo 2,894 K4
General Cepeda 3,486 J4
General Terán 5,354 K4
Gómez Farías 3,030 F2
Gómez Palacio 79,650 G4
González 6,440 K5
Guadalajara 1,478,383 H6
Guadalajara* 2,343,034 H6
Guadalupe, Nuevo León 51,899 K4
Guadalupe, Zacatecas 13,246 H5
Guadalupe Bravo 3,333 F1
Guadalupe Victoria, Durango 7,931 H4
Guadalupe Victoria, Puebla 3,946 O1
Guamúchil 17,151 E4
Guanajuato 36,809 J6
Guasave 26,080 E4
Guaymas 57,492 D3
Gustavo Díaz Ordaz 10,154 K3
Gutiérrez Zamora 9,099 L6
Halachó 4,804 O6
Hecelchakán 4,279 O6
Hermosillo 232,691 D2
Heroica Caborca 20,771 C1
Heroica Nogales 52,108 D1
Hidalgo, Tamaulipas 2,450 K4
Hidalgo del Parral (Parral) 57,619 G3
Hopelchén 3,699 P7
Huajuapan de León 13,822 L8
Huamantla 15,565 N1
Huaquechula 2,294 M2
Huatabampo 18,506 D3
Huatusco de Chicuellar 9,501 P2
Huauchinango 16,826 L6
Huautla de Jiménez 6,132 L7
Huehuetlán el Chico 2,667 M2
Huejotzingo 8,552 M1
Huejutla 6,854 K6
Huetamo 9,333 J7
Hueyotlipan de Hidalgo 2,353 M1
Huimanguillo 7,075 N8
Huitzilán 3,573 O1
Huitzuco de los Figueroa 9,406 K7
Huixcolotla 4,039 N2
Huixtepec 5,927 L8
Huixtla 15,737 N9
Hunucmá 8,020 O6
Ignacio de la Llave 3,962 Q2
Iguala de la Independencia 45,355 K7
Imuris 1,958 D1
Irapuato 135,596 J6
Isla Mujeres 2,663 Q6
Isla, Veracruz 8,075 M7
Ixmiquilpan 6,048 K6
Ixtapa J8
Ixtapalapa 522,095 L1
Ixtenco 5,035 N1
Ixtepec 14,025 M8
Ixtlán del Río 10,986 G6
Izamal 9,749 P6
Izúcar de Matamoros 21,164 M2
Jala 4,535 G6
Jalacingo 3,427 P1
Jalapa Enríquez 161,352 P1
Jalpa 9,904 H6
Jalpa de Méndez 4,785 N7
Jalpan 1,878 K6
Jáltipan de Morelos 15,170 M8
Jantetelco 2,015 L2
Jaumave 3,072 K5
Jerez de García Salinas 20,325 H5
Jico 7,269 P1
Jilotepec de Abasolo 4,252 K7
Jiménez, Chihuahua 18,095 G3
Joachín 3,918 Q2
Jojutla de Juárez 14,438 L2
Jonacatepec 3,868 M2
Jonuta 2,746 N7
José Cardel 5,396 Q1
Juan Aldama 9,667 H4
Juchipila 6,328 H6
Juchitán de Zaragoza 30,218 M8
Kantunilkin 1,970 Q6
La Barca 18,055 H6
La Barra de Navidad 1,829 G7
La Concordia 3,559 N9
La Cruz, Sinaloa 4,218 F5
Lagos de Moreno 33,782 J6
La Huerta 4,328 G7
La Paz, Baja California Sur 46,011 D5
La Paz, San Luis Potosí 3,735 J5
La Piedad Cavadas 34,963 H6
Las Choapas 20,166 M7
Las Hadas G7
Las Nieves 2,262 G3
Las Rosas 7,658 N8
León 468,887 J6
Lerdo de Tejada 11,628 M8
Lerma 4,158 O7
Libres 4,830 O1
Linares 24,456 K4
Llera de Canales 3,564 K5
Loma Bonita 15,804 M7
Loreto, Baja California 2,570 D4
Loreto, Zacatecas 7,132 J5
Los Mochis 67,953 E4
Los Reyes de Salgado 19,452 H7
Macuspana 12,293 N8
Madera 9,759 F2
Magdalena de Kino 10,281 D1
Maltrata 5,457 O2
Manzanillo 20,777 G7
Mapastepec 5,907 N9
Mapimí 2,737 G4
Martínez de la Torre 17,203 L6
Mascota 5,674 G6
Matamoros, Coahuila 15,125 H4
Matamoros, Tamaulipas 165,124 L4
Matehuala 28,799 J5
Matías Romero 13,200 M8
Maxcanú 6,505 O6
Mazatlán 147,010 F5
Melchor Múzquiz 18,868 H3
Melchor Ocampo del Balsas 4,766 H8
Meoquí 12,308 G2
Mérida 233,912 P6
Metepec 4,625 M2
Metlatonoc 1,870 K8
Mexicali 317,228 B1
Mexico City (cap.) 9,377,300 L1
Mexico City* 13,993,866 L1
Miacatlán 3,980 K2
Mier 5,636 K3
Miguel Auza 9,303 H4
Minatitlán 68,397 M8
Mineral del Monte 8,887 K6
Miquihuana 1,971 J5
Misantla 8,799 P1
Miahuatlán de Porfirio Díaz 5,714 L8
Mocorito 3,993 F4
Moctezuma, San Luis Potosí 1,734 J5
Moctezuma, Sonora 2,700 E2
Monclova 78,134 J3
Montemorelos 18,642 K4
Monterrey 1,006,221 J4
Monterrey* 1,923,402 J4
Morelia 199,099 J7
Morelos 4,241 J2
Morelos Cañada 2,288 O2
Moroleón 25,620 J6
Motozintla de Mendoza 4,682 N9
Motul de Felipe Carillo Puerto 12,949 P6
Muna 5,491 P6
Naco 3,580 D1
Nacozari 2,976 E1
Nadadores 2,461 H3
Naica 7,190 G2
Namiquipa 4,875 F2
Nanacamilpa 6,356 M1
Naolinco de Victoria 4,365 P1
Naranjos 14,732 L6
Naucalpan de Juárez 9,425 L1
Nautla 1,935 L6
Nava 4,097 J2
Navojoa 43,817 E3
Navolato 12,799 E4
Nazas 2,881 G4
Netzahualcóyotl 580,436 L1
Nieves 3,966 H5
Nochistlán 8,780 H6
Nogales 14,254 P2
Nombre de Dios 3,188 G5
Nopalucan de la Granja 3,002 O1
Nueva Casas Grandes 20,023 F1
Nueva Ciudad Guerrero 3,300 K3
Nueva Italia de Ruiz 14,718 J7
Nueva Rosita 34,706 J2
Nuevo Ideal 5,252 G4
Nuevo Laredo 184,622 J3
Oaxaca de Juárez 114,948 L8
Ocampo, Coahuila 1,613 H3
Ocampo, Tamaulipas 4,801 K5
Ocosingo 2,946 O8
Ocotlán 35,361 H6
Ocotlán de Morelos 5,882 L8
Ojinaga 12,757 G2
Ojocaliente 7,582 H5
Ometepec 7,342 K8
Oriental 6,009 O1
Orizaba 105,150 P2
Otumba de Gómez Farías 3,198 M1
Oxkutzcab 8,182 P6
Ozuluama 2,851 L6
Ozumba de Alzate 6,876 M1
Pachuca de Soto 83,892 K6
Padilla 4,581 K5
Palenque 2,595 O8
Palizada 2,332 O7
Palomas 2,129 F1

Panabá 3,056	P6	Profesor Rafael	
Pánuco 14,277	K6	Ramírez 5,338	O1
Papanoa 3,033	J8	Progreso 17,518	P6
Papantla de Olarte 26,773	L6	Puebla de Zaragoza 465,985	N2
Paraíso 7,561	N7	Puente de Ixtla 10,435	K2
Parral 57,619	G3	Puerto Ángel 1,489	L9
Parras de la Fuente 18,207	H4	Puerto Escondido 3,845	L9
Paso de Ovejas 4,371	Q2	Puerto Juárez 100	Q6
Pátzcuaro 17,299	J7	Puerto Madero 1,908	N9
Pedro Montoya 4,563	K6	Puerto Peñasco 8,452	C1
Pénjamo 9,245	J6	Puerto Vallarta 24,155	G6
Peñón Blanco 2,726	H4	Purificación 3,311	G7
Pericos 4,445	H4	Puruándiro 9,956	J7
Perote 12,742	F4	Putla de Guerrero 3,572	L8
Petatlán 9,419	J8	Quecholac 3,374	O2
Peto 8,362	P6	Querétaro 142,448	J6
Pichucalco 4,615	N8	Ramos Arizpe 6,205	J4
Piedras Negras,		Rayón, San Luis	
Coahuila 41,033	J2	Potosí 4,451	K6
Piedras Negras, Veracruz 4,099	Q2	Rayón, Sonora	D2
Pijijiapan 5,053	N9	Reynosa 181,646	K3
Pitiquito 2,268	D1	Rincón de Romos 8,348	H5
Potam 2,825	D3	Ríoverde 16,804	J6
Poza Rica de Hidalgo 152,276	L6	Rodeo 2,584	G4
Praxedis G. Guerrero 2,399	G1	Rosamorada 2,635	G5

(continued on following page)

AREA 761,601 sq. mi. (1,972,546 sq. km.)
POPULATION 67,395,826
CAPITAL Mexico City
LARGEST CITY Mexico City
HIGHEST POINT Citlaltépetl 18,855 ft. (5,747 m.)
MONETARY UNIT Mexican peso
MAJOR LANGUAGE Spanish
MAJOR RELIGION Roman Catholicism

States Indicated by Numbers

1 Tlaxcala	6 Querétaro
2 Morelos	7 Guanajuato
3 Distrito Federal	8 Aguascalientes
4 México	9 Nayarit
5 Hidalgo	10 Colima

Rosario, Sinaloa 10,276 ....G5
Rosario, Sonora 1,887 ....E3
Ruiz 8,954 ....G6
Sabancuy 1,819 ....O7
Sabinas 20,538 ....J3
Sabinas Hidalgo 17,439 ....J3
Sahuaripa 4,710 ....E2
Sahuayo de Díaz 28,727 ....H7
Saín Alto 3,628 ....H5
Salamanca 61,039 ....J6
Salina Cruz 22,004 ....M9
Salinas 7,471 ....J5
Saltillo 200,712 ....J4
Salvatierra 18,975 ....J6
San Andrés Tuxtla 24,267 ....M7
San Blas, Nayarit 3,443 ....G6
San Blas, Sinaloa 6,222 ....E3
San Buenaventura 9,188 ....J3
San Carlos, Coahuila 1,960 ....J2
San Cristóbal de las
  Casas 25,700 ....N8
San Felipe, Baja
  California 160 ....B1
San Felipe, Guanajuato 10,129 ....J6
San Fernando,
  Tamaulipas 27,656 ....L4
San Francisco del Oro 12,116 .....F3
San Francisco del
  Rincón 27,079 ....H6
San Gabriel Chilac 6,707 ....K7
San Ignacio, Sinaloa 1,804 ....F5
San Jerónimo de
  Juárez 5,204 ....J8
San José del Cabo 2,571 ....D5
San Juan 15,422 ....K6
San Juan de los Lagos 19,570 .....H6
San Juan Xiutetelco 3,306 ....O1
San Luis de la Paz 12,654 ....J6
San Luis del Cordero 2,203 ....H4
San Luis Potosí 271,123 ....J5
San Luis Río Colorado 49,990 .....B1
San Marcos 5,861 ....K8
San Martín de las
  Pirámides 4,575 ....M1
San Martín Texmelucan 23,355 .....M1
San Miguel de Allende 24,286 .....J6
San Nicolás de los
  Garza 28,803 ....J3
San Pedro de las
  Colonias 26,882 ....H4
San Pedro Pochutla 4,395 ....L9
San Rafael 8,974 ....M1
San Salvador del Seco 7,729 ......O1
Santa Ana 7,020 ....D1
Santa Ana Chiautempan
  (Chiautempan) 12,327 ....N1
Santa Bárbara 16,978 ....F3
Santa Clara 3,449 ....H4
Santa María del Oro 4,231 ....G3
Santa María del Río 4,972 ....J6
Santa María del Tule 1,674 ....L8
Santander Jiménez 3,586 ....K4
Santa Rosalía 7,356 ....C3
Santiago Ixcuintla 17,321 ....G6
Santiago Jamiltepec 5,280 ....K8
Santiago Juxtlahuaca 2,923 ....K8
Santiago Miahuatlán 4,917 ....O2
Santiago Papasquiaro 6,636 ....F4
Santiago Pinotepa
  Nacional 9,382 ....K8
Santiago Tuxtla 9,426 ....M7
Saucillo 8,467 ....G2
Sayula 14,339 ....H7
Sayula de Alemán 4,896 ....N8
Seybaplaya 4,439 ....O7
Silao 31,825 ....J6
Simojovel de Allende 3,779 ....N8
Sinaloa de Leyva 1,998 ....E4
Soledad de Doblado 6,612 ....Q2
Soledad Díez

Gutiérrez 9,622 ....J5
Sombrerete 11,077 ....H5
Sonoyta 2,463 ....C1
Sotuta 3,772 ....P6
Tabasco 2,019 ....H6
Tacámbaro de Codallos 9,695 ......J7
Tacotalpa 2,019 ....N8
Tala 15,744 ....H6
Talpa de Allende 4,264 ....G6
Tamahua 6,264 ....L6
Tamazulapan del Progreso 2,870 ...L8
Tamazunchale 12,302 ....K6
Tamiahua 6,264 ....L6
Tampico 212,188 ....L6
Tamuín 7,251 ....K6
Tantoyuca 11,902 ....L6
Tapachula 60,620 ....N9
Taxco de Alarcón 27,089 ....K7
Tayoltita 2,697 ....G4
Teapa 6,534 ....N8
Tecamachalco 3,319 ....O2
Tecate 14,738 ....A1
Tecomán 31,625 ....H7
Tecpan de Galeana 8,095 ....J8
Tecuala 12,461 ....G5
Tehuacán 47,497 ....L7
Tehuantepec 16,179 ....M8
Teíxa de Alaro
  Obregón 10,326 ....P6
Teloloapan 10,335 ....J7
Temax 4,915 ....P6
Temósachic 1,738 ....E2
Tenabo 3,278 ....P6
Tenancingo de Degollado 12,807 ...K7
Tenango de Río Blanco 12,302 .....O2
Tenosique de Pino
  Suárez 11,393 ....O8
Teocaltiche 13,745 ....H6
Teocelo 4,572 ....P1
Teotihuacán de Arista 2,238 ......L1
Teotitlán del Camino 3,106 ....L8
Tepache 1,591 ....E2
Tepalcingo 5,968 ....M2
Tepeaca 7,466 ....N2
Tepeapulco 7,027 ....M1
Tepehuanes 2,531 ....G4
Tepeji del Río 10,365 ....L1
Tepexi de Rodríguez 2,618 ....N2
Tepic 108,924 ....G6
Tepoztlán 6,851 ....L1
Tequixquitla 4,825 ....O1
Terán 5,215 ....N8
Terrenate 1,515 ....N1
Texcoco de Mora 18,044 ....M1
Teziutlán 23,948 ....O1
Tezonapa 3,506 ....P2
Tezontepec 2,762 ....M1
Ticul 14,341 ....P6
Tierra Blanca 22,727 ....L7
Tila 2,633 ....N8
Tixtla de Guerrero 10,334 ....A1
Tizayuca 6,262 ....L1
Tizimín 18,343 ....Q6
Tlachichuca 3,721 ....O1
Tlacolula de Matamoros 8,300 .....L8
Tlacotepec de Mejía 1,595 ....P1
Tlahualilo de Zaragoza 8,951 .....H3
Tlalancalanca 5,090 ....M1
Tlalixcoyan 3,211 ....Q2
Tlalmanalco de
  Velásquez 5,744 ....L1
Tlalnepantla de
  Comonfort 45,575 ....L1
Tlalpan 130,092 ....L1
Tlapa de Comonfort 6,676 ....K8

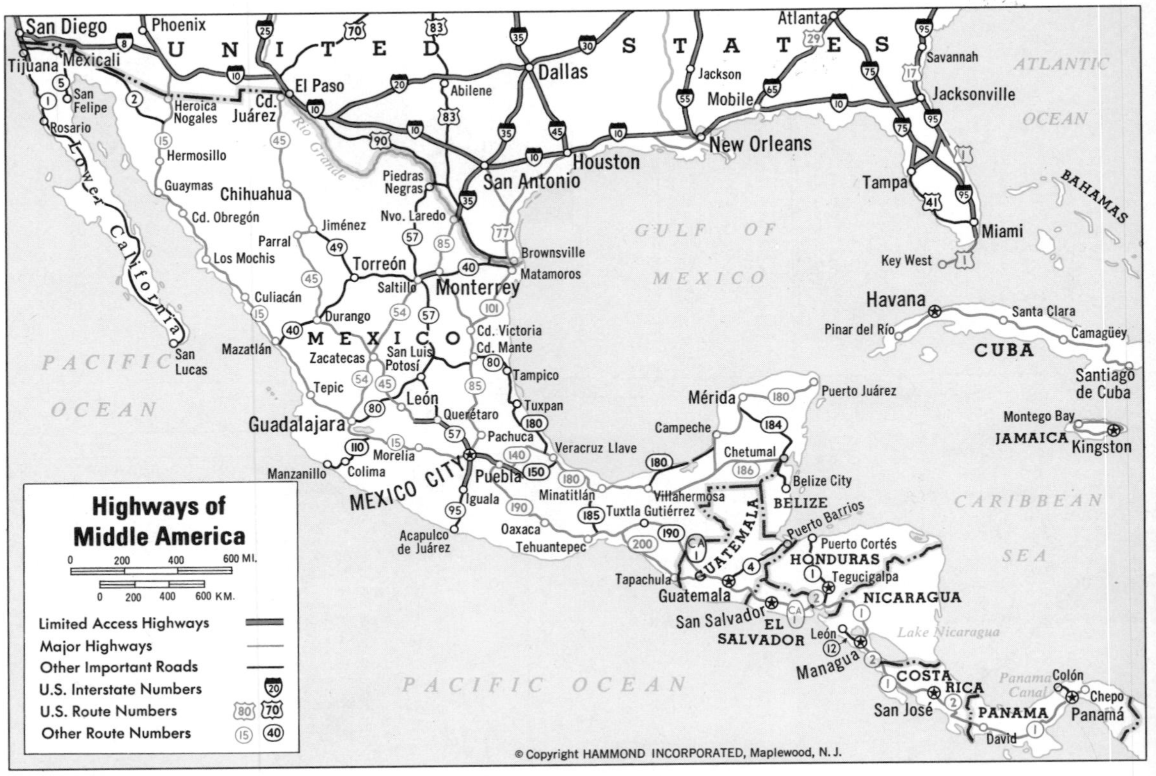

Tlaquepaque 59,760 ....G6
Tlatlauquitepec 4,272 ....O1
Tlaquiltenango 8,625 ....L2
Tlaxcala de Xicotencatl 9,972 ....M1
Tlaxco 4,969 ....N1
Tlaxiaco 4,477 ....L8
Tlayacapan 3,538 ....L1
Tochimilco 3,190 ....M2
Todos Santos 2,400 ....D5
Tojolapan 2,695 ....G6
Tonalá 15,611 ....N8
Topolobampo 4,685 ....E4
Torreón 244,309 ....H4
Tula, Tamaulipas 5,407 ....K5

Tula de Allende 10,720 ....K6
Tulancingo 35,799 ....K7
Tulcingo del Valle 2,983 ....M2
Tultepec 8,321 ....L1
Tuxpan, Jalisco 14,693 ....H7
Tuxpan, Nayarit 20,322 ....G6
Tuxpan de Rodríguez L6
Tuxtepec 17,701 ....L7
Tuxtla Gutiérrez 66,851 ....N8
Tzucabab 4,876 ....P7
Umán 8,371 ....P6
Unión de Tula 6,399 ....G7
Unión Hidalgo 8,658 ....M8
Ures 3,681 ....D2

Úrsulo Galván 2,637 ....Q1
Uruapan del Progreso 108,124 .....H7
Valladolid 14,663 ....P6
Valle de Allende 4,973 ....G3
Valle de Bravo 7,628 ....J7
Valle Hermoso 19,278 ....L4
Vanegas 2,042 ....J5
Venado 2,790 ....J5
Venustiano Carranza 23,624 ....N8
Veracruz 255,646 ....Q1
Vicam 4,104 ....D3
Vicente Guerrero,
  Durango 8,451 ....G5
Víctor Rosales 7,629 ....H5
Viesca 2,923 ....H4

Villa Acuña 30,276 ....J2
Villa Cuauhtémoc 6,611 ....L5
Villa de Cos 1,850 ....H5
Villa de Guadalupe
  Hidalgo 88,583 ....L1
Villa Frontera 25,761 ....J3
Villa García 2,765 ....J5
Villahermosa 133,181 ....N8
Villa Hidalgo 2,126 ....E1
Villaldama 2,350 ....J3
Villa Matamoros 1,998 ....G3
Villanueva 5,895 ....H5
Villa Unión, Coahuila 4,058 ......J2
Villa Unión, Durango 4,042 ....H5
Villa Unión, Sinaloa 6,789 ....F5
Villa Vicente Guerrero 18,280 ....N1
Xaltocan 2,524 ....N1
Xicoténcatl 6,374 ....K5
Xicotepec de Juárez 12,656 ....L6
Xochihuehuetlán 3,268 ....K8
Xochimilco 116,493 ....L1
Xochitlán 3,312 ....N2
Yajalón 4,506 ....N8
Yanga 3,843 ....P2
Yaqui 8,061 ....D3
Yautepec 13,952 ....L2
Yavaros 1,959 ....E3
Yécora 2,816 ....E2
Yecuatla 2,816 ....P1
Yehualtepec 2,558 ....O2
Zaachila 7,270 ....L8
Zacapoaxtla 4,527 ....O1
Zacapu 31,989 ....J7
Zacatepec 16,839 ....L2
Zacatecas 50,251 ....H5
Zacatelco 14,117 ....N1
Zacatlán 7,909 ....N1
Zacoalco de Torres 11,343 ....H6
Zamora de Hidalgo 5,775 ....H7
Zaragoza, Coahuila 6,797 ....J2
Zaragoza, Chihuahua 3,984 ....F1
Zaragoza, Puebla 4,754 ....O1
Zempoala 5,064 ....Q1
Zihuatenejo 4,879 ....J8
Zimatlán de Álvarez 5,746 ....L8
Zitácuaro 36,911 ....J7
Zongolica 2,378 ....P2
Zumpango de Ocampo 12,923 ....L1
Zumpango del Río 8,162 ....J8

OTHER FEATURES

Agiobampo (bay) E3
Aguanaval (riv.) H4
Amistad (res.) J2
Ángel de la Guarda (isl.) C2
Antigua (riv.) Q1
Arena (pt.) E5
Arenas (cay) O5
Atoyac (riv.) N2
Atoyac (riv.) Q2
Babia (riv.) J2
Bacalar (lake) P7
Ballenas (bay) C2
Balsas (riv.) J7
Banderas (bay) G6
Bavispe, Río de (riv.) E1
Blanco (riv.) Q2
Bravo (Grande) (riv.) J2
Burro (mts.) J2
California (gulf) D3
Campeche (bank) O6
Campeche (bay) N7
Candelaria (riv.) O8
Carmen (isl.) D3
Casas Grandes (riv.) F1
Catoche (cape) Q6
Cedros (isl.) B3
Cerralvo (isl.) E4
Chamela (bay) G7
Chapala (lake) H6
Chetumal (bay) P8
Chichén-Itzá (ruin) P6
Clarión (isl.) B7
Colorado (riv.) B1
Conchos (riv.) G2
Corrientes (cape) F6
Coyuca (riv.) O1
Creciente (isl.) D4
Cuitzeo (lake) J7
Delgada (pt.) N2
Dzibanché (ruin) P6
El Azúcar (res.) K3

Falcón (res.) K3
Falso (cape) E5
Fuerte (riv.) E3
Giganta, Sierra de la (mts.) D4
Grande (riv.) D4
Grande (riv.) N8
Grande de Santiago (riv.) N7
Grijalva (riv.) N8
Guzmán (lake) E1
Herrero (pt.) Q7
Hondo (riv.) P8
Jesús María (reef) D4
La Boquilla (res.) F3
La Paz (bay) D4
Lobos (cape) D2
Lobos (isl.) L6
Lobos (pt.) L1
Lower California (pen.) B2
Madre (lag.) L4
Madre del Sur, Sierra (mts.) K8
Madre Occidental, Sierra
  (mts.) F3
Madre Oriental, Sierra (mts.) J4
Magdalena (bay) C4
Maldonado (pt.) K8
Mapimí (depr.) G3
María Cleofas (isl.) F6
María Madre (isl.) F6
María Magdalena (isl.) F6
Mexico (gulf) N4
Mezquital (riv.) G5
Mita (pt.) G6
Mitla (ruin) M8
Moctezuma (riv.) K6
Monserrate (isl.) D4
Montague (isl.) C1
Muerto, Mar (lag.) M8
Nauhcampatépetl (mt.) O2
Nayarit, Sierra (mts.) G4
Nazas (riv.) G4
Nuevo, Bajo (reef) G4
Orizaba (Citlaltépetl) O2
Palenque (ruin) O8
Palmito de la Vírgen
  (isl.) F5
Palmito del Verde (isl.) F5
Pánuco (riv.) L6
Paricutín (vol.) H7
Pátzcuaro (lake) J7
Pérez (isl.) O5
Petacalco (bay) H8
Popocatépetl (mt.) M2
Ramos (riv.) G4
Revillagigedo (isls.) C7
Roca Partida (isl.) C7
Sabinas (riv.) J3
San Antonio (reef) O5
San Benedicto (isl.) D7
San Benito (isl.) C1
San Jorge (bay) C1
San José (isl.) C4
San Lázaro (cape) C4
San Lucas (cape) D5
San Marcos (isl.) D3
San Rafael (isl.) D4
Santa Ana (reef) N4
Santa Catalina (isl.) D4
Santa Cruz (isl.) D4
Santa Eugenia (pt.) B3
Santa Margarita (isl.) C4
Santa María (isl.) F6
Santiaguillo (lake) G4
Sebastián Vizcaíno (bay) B3
Socorro (isl.) C7
Sonora (riv.) D2
Superior (lag.) M9
Teacapán (inlet) F5
Tehuantepec (gulf) M9
Tehuantepec (isth.) M8
Teotihuacán (ruin) M1
Términos (lag.) N7
Tiburón (isl.) C2
Triángulo Este (isl.) N6
Triángulo Oeste (isl.) N6
Tula (riv.) L1
Urique (riv.) F3
Usumacinta (riv.) O8
Uxmal (ruins) P6
Valsequillo (res.) N2
Verde (riv.) N2
Verde (riv.) H6
Yaqui (riv.) D2

*City and suburbs.

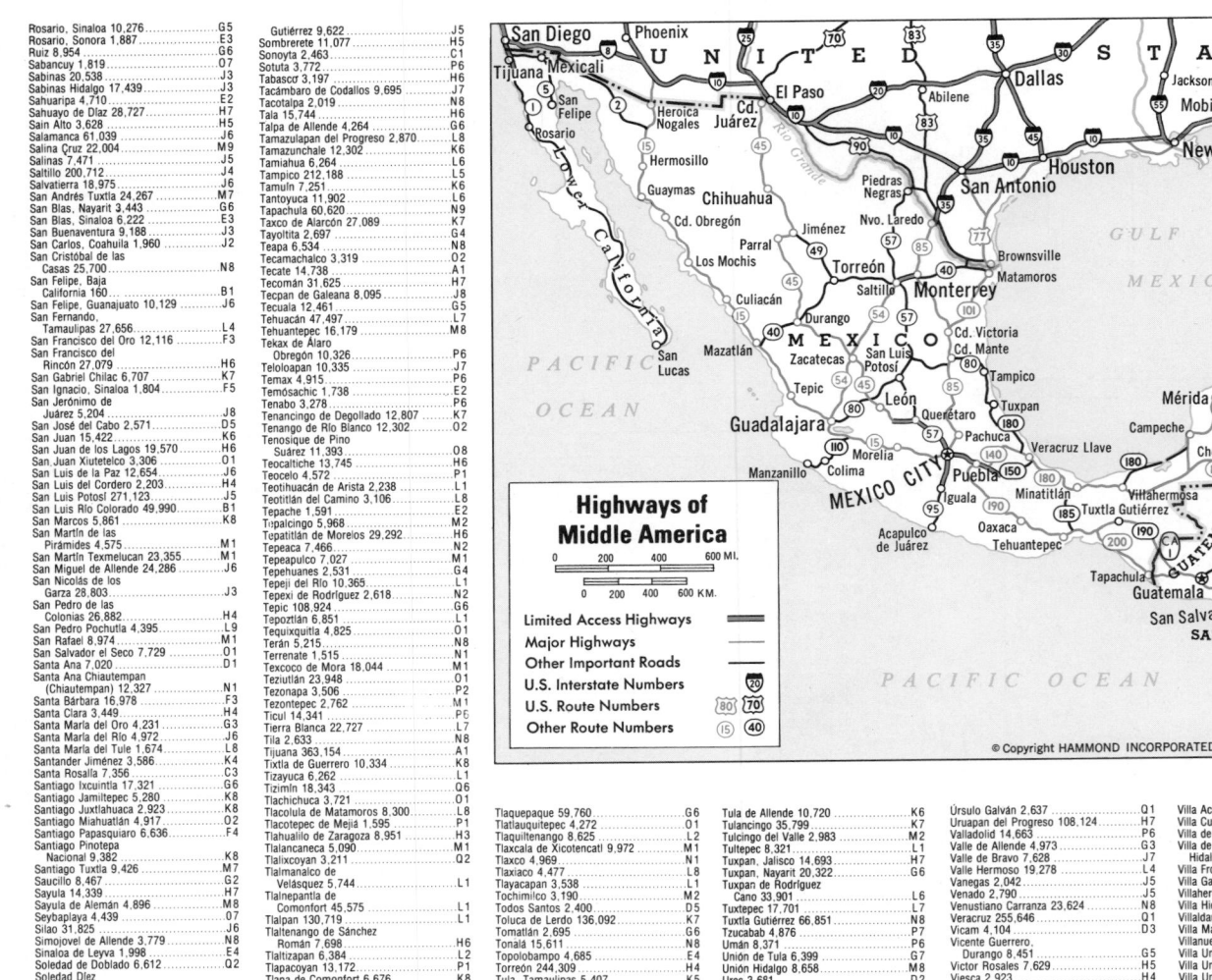

Highways of Middle America

```
0   200   400      600 MI.
0   200   400   600 KM.
```

Limited Access Highways ▬▬▬
Major Highways ────
Other Important Roads ────
U.S. Interstate Numbers [20]
U.S. Route Numbers [80] [70]
Other Route Numbers [15] [40]

© Copyright HAMMOND INCORPORATED, Maplewood, N.J.

Agriculture, Industry and Resources

DOMINANT LAND USE

Wheat, Livestock

Cereals (chiefly corn), Livestock

Diversified Tropical Cash Crops

Cotton, Mixed Cereals

Livestock, Limited Agriculture

Range Livestock

Forests

Nonagricultural Land

MAJOR MINERAL OCCURRENCES

Ag	Silver	G	Natural Gas	O	Petroleum
Au	Gold	Gr	Graphite	Pb	Lead
C	Coal	Hg	Mercury	S	Sulfur
Cu	Copper	Mn	Manganese	Sb	Antimony
F	Fluorspar	Mo	Molybdenum	Sn	Tin
Fe	Iron Ore	Na	Salt	W	Tungsten
				Zn	Zinc

⚡ Water Power

▨ Major Industrial Areas

GUATEMALA

AREA 42,042 sq. mi. (108,889 sq. km.)
POPULATION 7,262,419
CAPITAL Guatemala
LARGEST CITY Guatemala
HIGHEST POINT Tajumulco 13,845 ft.
(4,220 m.)
MONETARY UNIT quetzal
MAJOR LANGUAGES Spanish, Quiché
MAJOR RELIGION Roman Catholicism

BELIZE

AREA 8,867 sq. mi. (22,966 sq. km.)
POPULATION 144,857
CAPITAL Belmopan
LARGEST CITY Belize City
HIGHEST POINT Victoria Peak 3,681 ft. (1,122 m.)
MONETARY UNIT Belize dollar
MAJOR LANGUAGES English, Spanish, Mayan
MAJOR RELIGIONS Roman Catholicism, Protestantism

EL SALVADOR

AREA 8,260 sq. mi. (21,393 sq. km.)
POPULATION 4,813,000
CAPITAL San Salvador
LARGEST CITY San Salvador
HIGHEST POINT Santa Ana 7,825 ft.
(2,385 m.)
MONETARY UNIT colón
MAJOR LANGUAGE Spanish
MAJOR RELIGION Roman Catholicism

HONDURAS

AREA 43,277 sq. mi. (112,087 sq. km.)
POPULATION 3,691,000
CAPITAL Tegucigalpa
LARGEST CITY Tegucigalpa
HIGHEST POINT Las Minas 9,347 ft.
(2,849 m.)
MONETARY UNIT lempira
MAJOR LANGUAGE Spanish
MAJOR RELIGION Roman Catholicism

NICARAGUA

AREA 45,698 sq. mi. (118,358 sq. km.)
POPULATION 2,703,000
CAPITAL Managua
LARGEST CITY Managua
HIGHEST POINT Cerro Mocotón 6,913 ft.
(2,107 m.)
MONETARY UNIT córdoba
MAJOR LANGUAGE Spanish
MAJOR RELIGION Roman Catholicism

COSTA RICA

AREA 19,575 sq. mi. (50,700 sq. km.)
POPULATION 2,245,000
CAPITAL San José
LARGEST CITY San José
HIGHEST POINT Chirripó Grande
12,530 ft. (3,819 m.)
MONETARY UNIT colón
MAJOR LANGUAGE Spanish
MAJOR RELIGION Roman Catholicism

PANAMA

AREA 29,761 sq. mi. (77,082 sq. km.)
POPULATION 1,830,175
CAPITAL Panamá
LARGEST CITY Panamá
HIGHEST POINT Vol. Baru 11,401 ft.
(3,475 m.)
MONETARY UNIT balboa
MAJOR LANGUAGE Spanish
MAJOR RELIGION Roman Catholicism

Agriculture, Industry and Resources

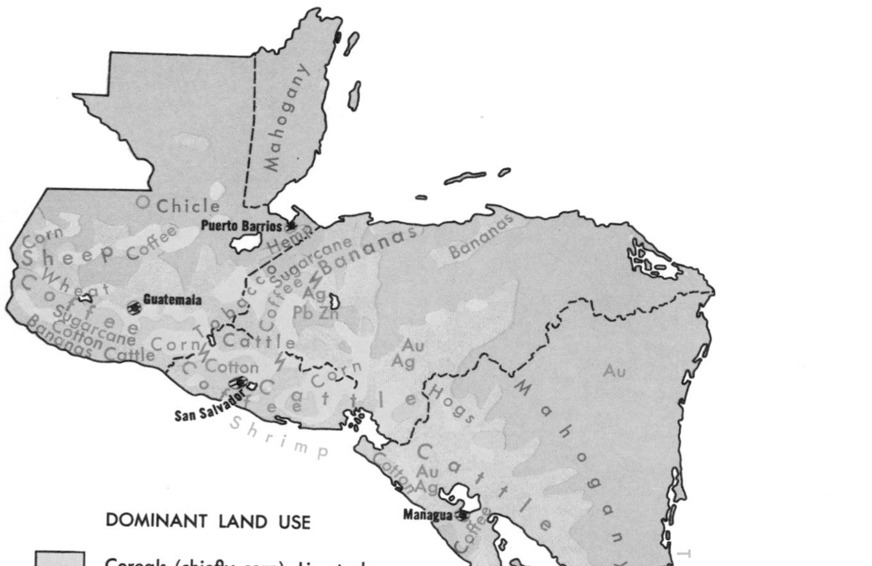

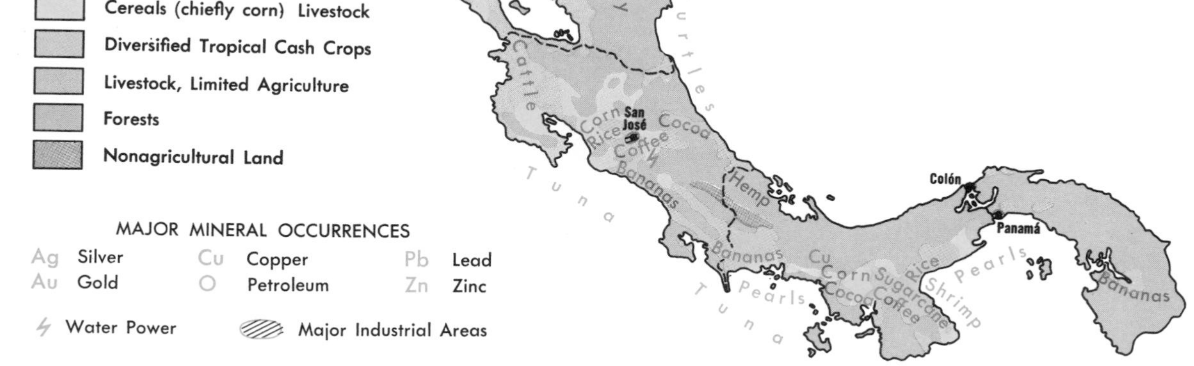

DOMINANT LAND USE

- Cereals (chiefly corn) Livestock
- Diversified Tropical Cash Crops
- Livestock, Limited Agriculture
- Forests
- Nonagricultural Land

MAJOR MINERAL OCCURRENCES

Ag Silver
Au Gold
Cu Copper
O Petroleum
Pb Lead
Zn Zinc

⚡ Water Power   ▨ Major Industrial Areas

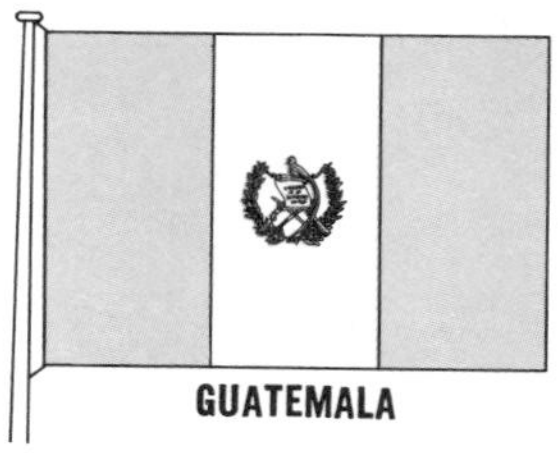

GUATEMALA
HONDURAS 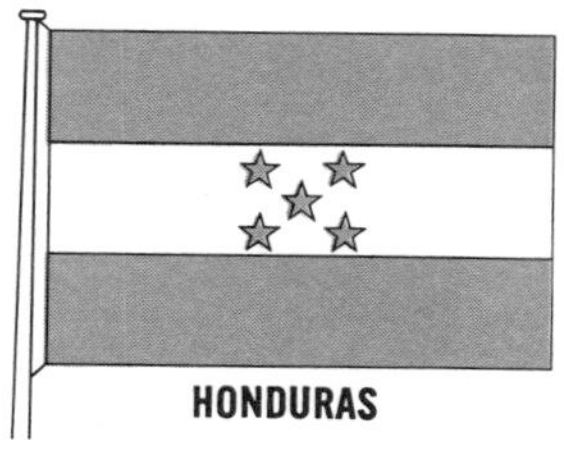
BELIZE 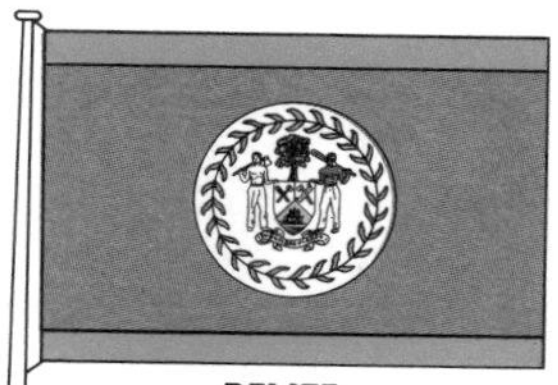
NICARAGUA 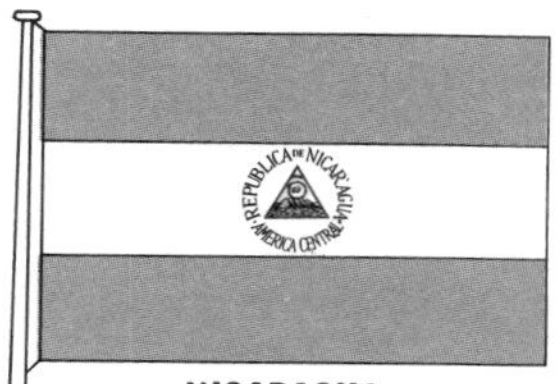
EL SALVADOR 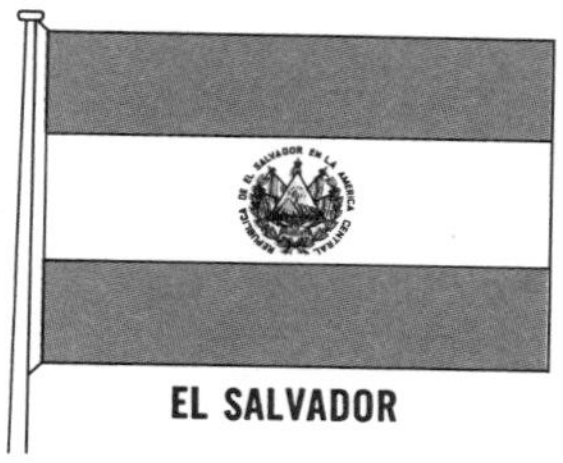
COSTA RICA 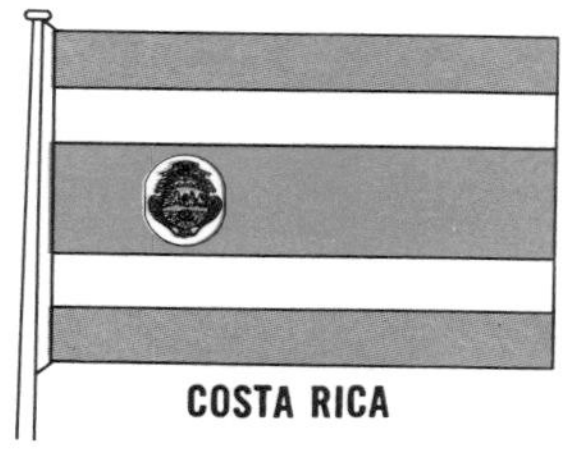
PANAMA 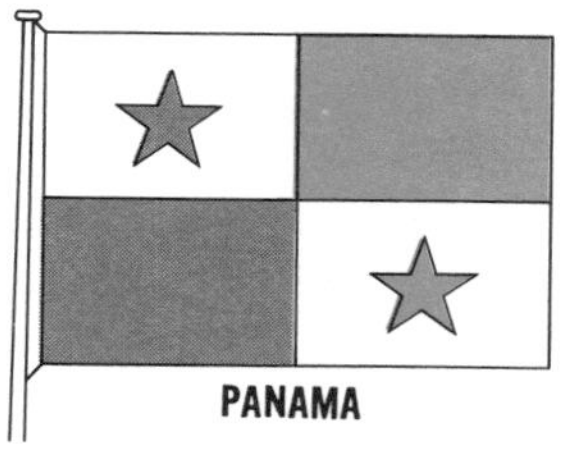

BELIZE

CITIES and TOWNS

Belize City 39,887 ....C2
Belize City* 50,925 ....C2
Belmopan (cap.) 2,932 ....C2
Corozal Town 6,862 ....C1
Hattieville 904 ....C2
Libertad 856 ....C1
Orange Walk Town 8,441 ....C1
Punta Gorda 2,219 ....C2
San Ignacio 5,606 ....C2
Stann Creek Town 6,627 ....C2

OTHER FEATURES

Ambergris (cay) D1
Belize (riv.) C2

Bokel (cay) D2
Glover (reef) D2
Half Moon (cay) D2
Hondo (riv.) C1
Honduras (gulf) D2
Mauger (cay) D2
New (riv.) C2
Saint Georges (cay) D2
Sarstún (riv.) C3
Turneffe (isls.) D2

COSTA RICA

CITIES and TOWNS

Alajuela 33,122 ....E6
Atenas 1,728 ....E6
Bagaces 2,129 ....E5
Boruca⊙ 1,892 ....F6
Buenos Aires⊙ 302 ....F6
Cañas 6,053 ....E5
Cartago 21,753 ....F6
Ciudad Quesada 9,754 ....E5
Esparta 4,699 ....E5
Filadelfia 2,958 ....E5
Golfito 6,962 ....F6
Grecia 8,355 ....E5
Guácimo 1,168 ....F5
Guápiles 3,524 ....F5
Heredia 22,700 ....E5
Las Juntas 1,129 ....E5
Liberia 10,802 ....E5
Limón 29,621 ....F6
Miramar 1,673 ....E5
Nicoya 7,474 ....E5
Orotina 3,170 ....E5
Palmares 3,083 ....E5
Paraíso 8,446 ....F6
Puerto Cortés 2,070 ....F6
Puntarenas 26,331 ....E6
Quepos 2,155 ....E6
San José (cap.) 215,441 ....F5
San José* 391,107 ....F5
San Marcos 917 ....E6
San Ramón 9,245 ....E5
Santa Cruz 5,777 ....E5
Santo Domingo 5,148 ....F6
Siquirres 4,361 ....F6
Turrialba 12,151 ....F6

(continued on following page)

OTHER FEATURES

Blanca (pt.)	F5
Blanco (cape)	E6
Blanco (peak)	F6
Burica (pt.)	F6
Cahuita (pt.)	F6
Caño (isl.)	F6
Carreta (pt.)	F6
Chirripó Grande (mt.)	F6
Coronada (bay)	F6
Cuilapa Miravalles (vol.)	F5
Dulce (gulf)	F6
Góngora (mt.)	E6
Guionos (pt.)	E6
Irazú (mt.)	F6
Judas (pt.)	E6
Llerena (pt.)	F6
Matapalo (cape)	E6
Nicoya (gulf)	E6
Nicoya (pen.)	E6
Papagayo (gulf)	D5
Salinas (bay)	D5
San Juan (riv.)	E5
Santa Elena (cape)	D5

Talamanca (range)	F6
Velas (cape)	D5

EL SALVADOR

CITIES and TOWNS

Acajutla 8,598	B4
Ahuachapán 17,242	B4
Atiquizaya 7,035	C3
Chalatenango 7,633	C3
Chinameca 6,303	C4
Cojutepeque 20,615	C4
Estanzuelas 2,548	C4
Ilobasco 6,572	C4
Intipucá 3,469	D4
Jucuarán 1,443	C4
La Libertad	C4
La Palma 1,998	C3
La Unión 17,207	D4
Metapán 7,704	C3
Nueva San Salvador 35,106	C4
Puerto de la Concordia	C4
San Francisco Gotera 4,725	C4

San Miguel 59,304	D4
San Salvador (cap.) 337,171	C4
Santa Ana 96,306	C4
Santa Rosa de Lima 5,707	C4
San Vicente 18,872	C4
Sensuntepeque 7,226	C4
Sonsonate 33,562	C4
Suchitoto 5,540	C4
Texistepeque 1,722	C3
Usulután 19,616	C4
Zacatecoluca 15,718	C4

OTHER FEATURES

Fonseca (gulf)	D4
Güija (lake)	C3
Lempa (riv.)	C4
Remedios (pt.)	B4
Santa Ana (mt.)	C4

GUATEMALA

CITIES and TOWNS

Amatitlán 15,251	B3

Antigua 17,994	B3
Asunción Mita 7,477	C3
Cahabón 1,344	C3
Chajul 4,329	B3
Champerico 5,722	A3
Chichicastenango 2,635	B3
Chimaltenango 12,860	B3
Chiquimula 16,126	C3
Coatepeque 15,979	A3
Cobán 11,418	B3
Comalapa 10,980	B3
Cubulco 2,021	B3
Cuilapa 4,287	B3
Cuilco 862	A3
Dolores 973	C2
El Estor 2,324	B3
El Progreso 4,009	B3
Escuintla 33,205	B3
Flores 1,477	C2
Gualán 5,169	C3
Guatemala (cap.) 700,538	B3
Huehuetenango 12,570	A3
Ipala 3,386	C3
Iztapa 1,237	B4

Jacaltenango 4,517	A3
Jalapa 15,788	B3
Jutiapa 8,210	B3
La Gomera 2,394	B3
La Libertad 908	B2
Livingston 2,898	C3
Los Amates 1,383	C3
Masagua 1,178	B3
Mazatenango 23,285	B3
Momostenango 5,210	B3
Morales 2,113	C3
Ocós 741	A3
Panzós 1,643	C3
Puerto Barrios 22,598	C3
Quezaltenango 53,021	B3
Quezaltepeque 2,222	C3
Rabinal 4,625	B3
Retalhuleu 19,060	B3
Río Hondo 1,416	C3
Sacapulas 1,439	B3
Salamá 5,529	B3
San Andrés 1,066	B2
San Felipe 3,210	B3
San José 9,402	B4
San Luis 1,136	C2

San Luis Jilotepeque 6,055	C3
San Marcos 5,700	B3
San Martín Jilotepeque 3,770	B3
San Mateo Ixtatán 1,834	B3
San Pedro Carchá 4,465	B3
Santa Cruz del Quiché 7,651	B3
Santa Rosa de Lima 1,161	B3
Sololá 3,960	B3
Tacaná 1,280	A3
Tejutla 1,205	B3
Tikal	B3
Totonicapán 8,568	B3
Zacapa 2,936	C3

OTHER FEATURES

Atitlán (lake)	B3
Atitlán (vol.)	B3
Azul (riv.)	B2
Chixoy (riv.)	B2
Güija (lake)	D2
Honduras (gulf)	C3
Izabal (lake)	C3
Minas (mts.)	C3
Motagua (riv.)	C3

Pasión (riv.)	B2
Petén-Itzá (lake)	B2
San Pedro (riv.)	B2
Sarstún (riv.)	C3
Tacaná (vol.)	A3
Tajumulco (vol.)	A3
Tres Puntas (cape)	C3
Usumacinta (riv.)	B2

HONDURAS

CITIES and TOWNS

Amapala 2,274	D4
Brus Laguna 933	E3
Catacamas 9,134	D3
Cedros 917	D3
Choloma 961	D3
Choluteca 26,152	D4
Comayagua 15,941	D3
Corquín 2,629	D3
Danlí 10,825	D3
El Dulce Nombre 1,297	E3

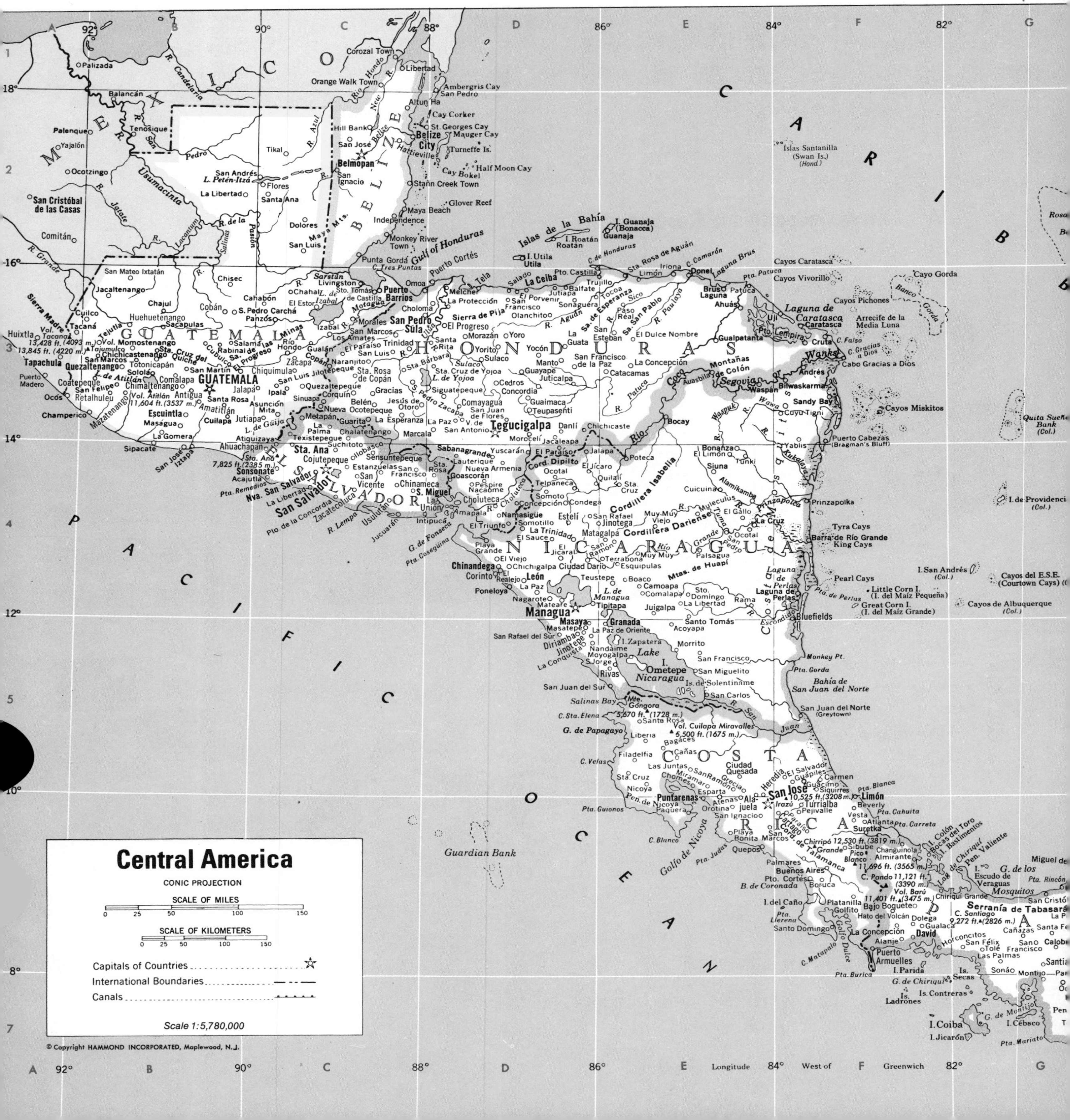

Central America

CONIC PROJECTION

SCALE OF MILES
0 25 50 100 150

SCALE OF KILOMETERS
0 25 50 100 150

Capitals of Countries .... ☆
International Boundaries ....
Canals ....

Scale 1:5,780,000

El Paraíso, Copán 2,164 .... C3
El Paraíso, El Paraíso 6,709 .... D4
El Porvenir 1,076 .... D3
El Progreso 28,105 .... D3
El Triunfo 2,925 .... D4
Goascorán 996 .... D4
Gracias 2,299 .... C3
Guaimaca 3,953 .... D3
Guanaja 1,947 .... E2
Guayape 804 .... D3
Iriona 26 .... E2
Jacaleapa 1,609 .... D3
Jesús de Otoro 2,976 .... C3
Jutiapa 1,126 .... D3
Juticalpa 10,075 .... D3
La Ceiba 38,788 .... D3
La Esperanza 2,146 .... C3
La Paz 6,811 .... D3
Limón 1,704 .... E3
Manto 689 .... D3
Marcala 3,183 .... C3
Morazán 4,367 .... D3
Morocelí 1,442 .... D3
Nacaome 6,159 .... D4
Namasigüe 816 .... D4
Naranjito 2,770 .... C3
Nueva Armenia 670 .... D4
Nueva Ocotepeque 4,724 .... C3
Olanchito 7,411 .... D3
Omoa 9,161 .... C3
Pespire 1,895 .... D4
Puerto Castilla D3
Puerto Cortés 25,817 .... D2
Puerto Lempira 727 .... F3
Roatán 1,943 .... D2
Sabanagrande 1,446 .... D4
Salado D3
San Francisco 1,557 .... D3
San Francisco de la Paz 2,291 .... D3
San Juan de Flores 1,184 .... D3
San Luis 2,237 .... C3
San Marcos 2,499 .... C3
San Pedro Sula 150,991 .... C3
Santa Bárbara 5,883 .... C3
Santa Cruz de Yojoa 1,848 .... D3
Santa Rita 5,298 .... D3
Santa Rosa de Aguán 1,622 .... E2
Santa Rosa de Copán 12,413 .... C3
Siguatepeque 12,456 .... D3

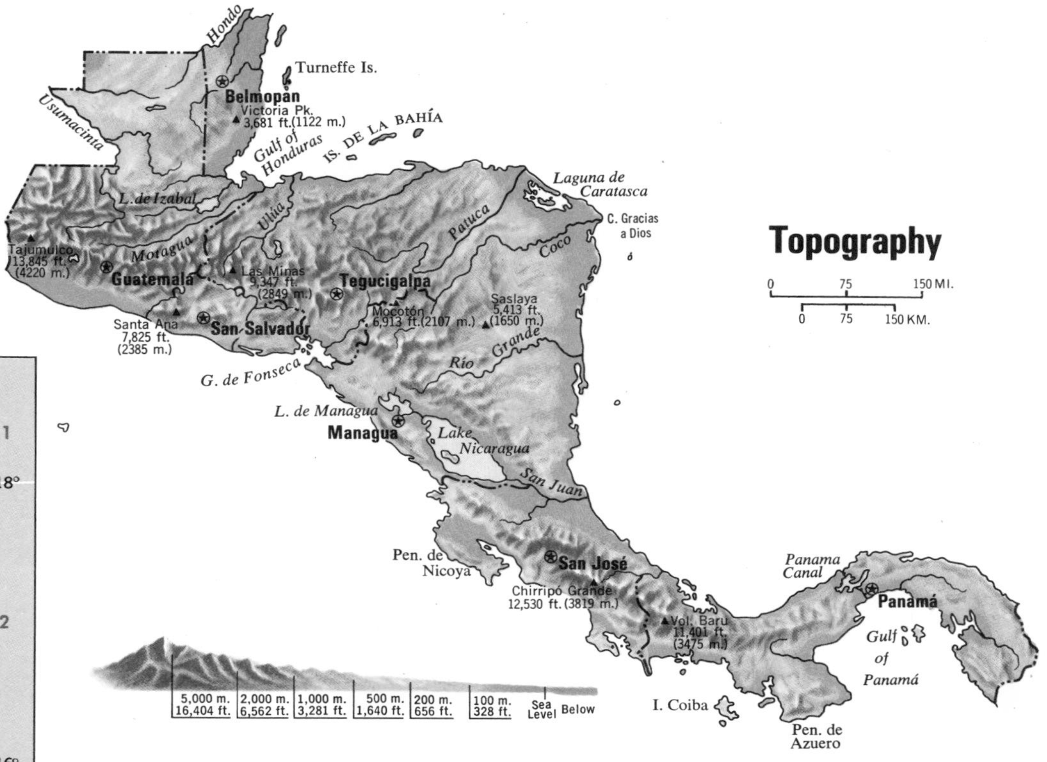

Topography
0 75 150 MI.
0 75 150 KM.
Hondo
Belmopan
Turneffe Is.
Usumacinta
Victoria Pk. 3,681 ft.(1122 m.)
IS. DE LA BAHÍA
Gulf of Honduras
L. de Izabal
Laguna de Caratasca
C. Gracias a Dios
Tajumulco 13,845 ft.(4220 m.)
Motagua
Ulúa
Las Minas 9,347 ft.(2849 m.)
Patuca
Coco
Guatemala
Tegucigalpa
Santa Ana 7,825 ft.(2385 m.)
San Salvador
Mogotón 6,913 ft.(2107 m.)
Saslaya 5,413 ft.(1650 m.)
Río Grande
G. de Fonseca
L. de Managua
Managua
Lake Nicaragua
San Juan
Pen. de Nicoya
San José
Chirripó Grande 12,530 ft.(3819 m.)
Vol. Baru 11,401 ft.(3475 m.)
Panama Canal
Panamá
Gulf of Panamá
I. Coiba
Pen. de Azuero
5,000 m. 16,404 ft.
2,000 m. 6,562 ft.
1,000 m. 3,281 ft.
500 m. 1,640 ft.
200 m. 656 ft.
100 m. 328 ft.
Sea Level
Below

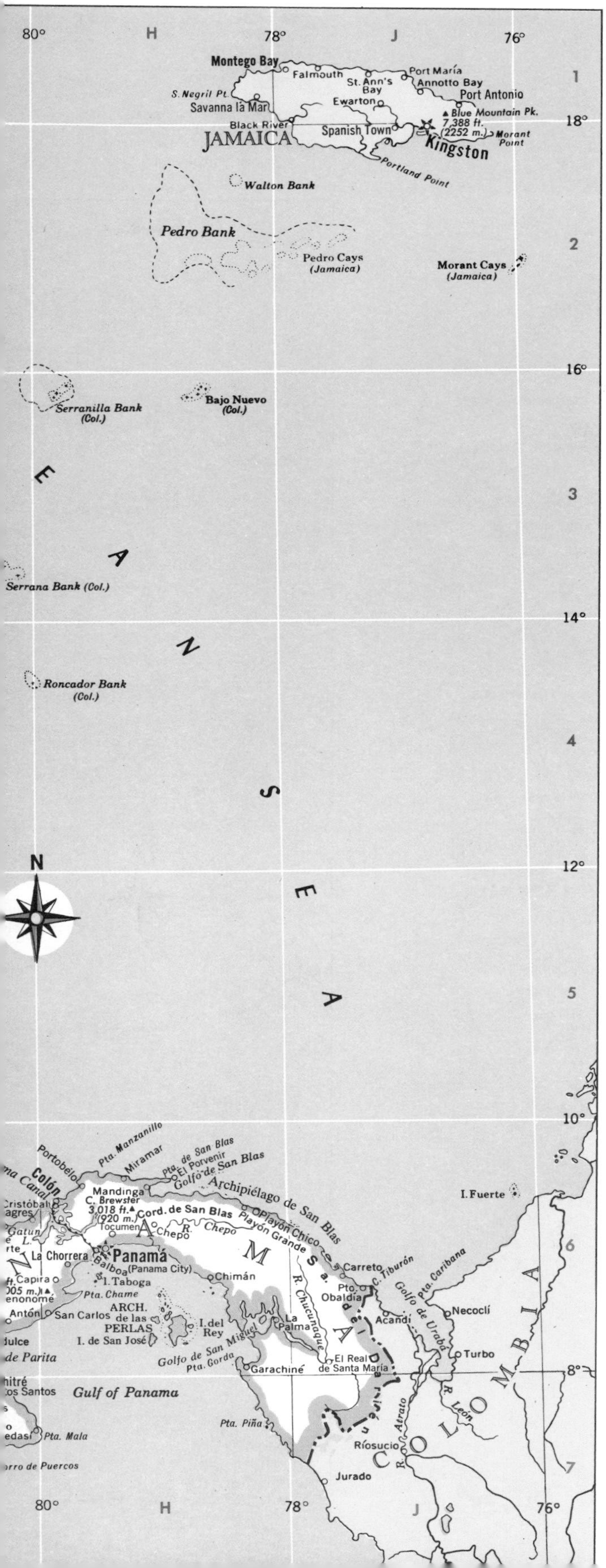

80° H 78° J 76°
Montego Bay
Falmouth
Port María
Annotto Bay
St. Ann's Bay
Port Antonio
S. Negril Pt.
Savanna la Mar
Ewarton
Black River
Blue Mountain Pk. 7,388 ft.(2252 m.)
JAMAICA
Spanish Town
Kingston
Morant Point
Portland Point
Walton Bank
Pedro Bank
Pedro Cays (Jamaica)
Morant Cays (Jamaica)
Serranilla Bank (Col.)
Bajo Nuevo (Col.)
Serrana Bank (Col.)
Roncador Bank (Col.)
CARIBBEAN SEA
N
18° 16° 14° 12° 10° 8°
1 2 3 4 5 6 7
Pta. Manzanillo
Portobelo
Miramar
Colón
Pta. de San Blas
El Porvenir
Golfo de San Blas
Archipiélago de San Blas
Mandinga
C. Brewster 3,018 ft.(920 m.)
Cord. de San Blas
Playón Chico
Playón Grande
Balboa
Tocumen
Chepo
La Chorrera
Panamá (Panama City)
Capira
Taboga
Chimán
Antón
San Carlos
ARCH. de las PERLAS
I. del Rey
I. de San José
La Palma
Pto. Obaldía
Carreto
Pta. Tiburón
Acandí
Necoclí
Golfo de Urabá
Turbo
Golfo de San Miguel
Pta. Gorda
Garachiné
El Real de Santa María
R. Chucunaque
Gulf of Panama
Pta. Piña
Pta. Mala
Los Santos
Río Atrato
R. León
Ríosucio
I. Fuerte
COLOMBIA
Jurado

Sinuapa 831 .... C3
Sonaguera 2,264 .... D3
Sulaco 1,121 .... D3
Tegucigalpa (cap.) 273,894 .... D3
Tegucigalpa* 305,387 .... D3
Tela 19,055 .... D3
Teupasenti 2,003 .... D3
Tocoa 2,803 .... E3
Trinidad 1,598 .... C3
Trujillo 3,961 .... E3
Utila 1,177 .... D2
Villa de San Antonio 2,359 .... D3
Yorito 770 .... D3
Yoro 4,449 .... D3
Yuscarán 1,835 .... D4

OTHER FEATURES

Aguán (riv.) D3
Bahía (isls.) D2
Bonacca (Guanaja) (isl.) E2
Brus (lag.) E2
Camarón (cape) E2
Caratasca (cays) F2
Caratasca (lag.) F3
Choluteca (riv.) D4
Coco (riv.) E3
Colón (mts.) E3
Esperanza (mts.) E3
Falso (cape) F3
Fonseca (gulf) D4
Gorda (bank) F3
Guanaja (isl.) E2
Honduras (cape) E2
Honduras (gulf) D2
Patuca (pt.) E3
Patuca (riv.) E3
Paulaya (riv.) E3
Pija, Sierra de (mts.) D3
Roatán (isl.) D2
San Pablo, Sierra (mts.) E3
Santanilla (isls.) F2
Segovia (Coco) (riv.) E3
Sico (riv.) E3
Sulaco (riv.) D3
Swan (Santanilla) (isls.) F2
Ulúa (riv.) D3
Utila (isl.) D2
Wanks (Coco) (riv.) E3
Yojoa (lake) D3

NICARAGUA

CITIES and TOWNS

Acoyapa 2,588 .... E5
Barra de Río Grande F4
Bluefields 14,252 .... F4
Boaco 6,372 .... E4
Bonanza E4
Bragman's Bluff (Puerto Cabezas) 5,457 .... F3
Cabo Gracias a Dios○ 3,846 .... F3
Camoapa 4,385 .... E4
Chichigalpa 14,498 .... D4
Chinandega 30,441 .... D4
Ciudad Darío 5,304 .... D4
Condega 3,414 .... D4
Corinto 13,404 .... D4
Diriamba 10,085 .... D5
El Jícaro 1,669 .... E4
El Limón E4
El Realejo 2,229 .... D4
El Sauce 3,202 .... D4
El Viejo 8,507 .... D4
Esquipulas 2,232 .... E4
Estelí 20,222 .... D4
Granada 34,976 .... E5
Greytown (San Juan del Norte) 294 .... F5
Jalapa 3,633 .... E4
Jinotega 9,506 .... E4
Jinotepe 12,473 .... D5
Juigalpa 8,497 .... E4
La Cruz 150 .... E4
La Libertad 1,286 .... E4
La Paz Central 6,175 .... D4
La Paz de Oriente 957 .... E5
La Trinidad 3,548 .... D4
León 55,625 .... D4
Managua (cap.) 398,514 .... D4
Managua* 404,634 .... D4
Masatepe 6,307 .... D5
Masaya 30,753 .... D5
Matagalpa 21,385 .... E4
Mateare 1,405 .... D4
Morrito 368 .... E5
Moyogalpa 1,551 .... E5
Muy Muy 1,373 .... E4
Nagarote 7,185 .... D4
Nandaime 5,631 .... E5
Ocotal, Segovia 8,215 .... D4
Poneloya D4
Prinzapolka 8,979 .... F4
Puerto Cabezas 5,457 .... F3
Quilalí 1,245 .... E4
Rama 1,341 .... E4
Rivas 10,125 .... E5
San Carlos 2,022 .... E5
San Jorge 2,874 .... E5
San Juan del Norte 294 .... F5
San Juan del Sur 2,393 .... D5
San Miguelito 1,312 .... E5
San Rafael del Norte 1,938 .... E4
San Rafael del Sur 2,914 .... D5
San Ramón 477 .... E4
Santo Domingo 1,949 .... E4
Santo Tomás 2,309 .... E5
Siuna E4
Somotillo 1,864 .... D4
Somoto 5,847 .... D4
Telpaneca 991 .... D4
Terrabona 904 .... E4
Teustepe 1,060 .... E4
Tipitapa 5,758 .... E4
Waspán 1,246 .... E3

OTHER FEATURES

Coco (riv.) E3
Cosegüina (pt.) D4
Dariense, Cordillera (range) E4
Dipilto, Cordillera (range) D4
Escondido (riv.) F4
Fonseca (gulf) D4
Gorda (pt.) F5
Gracias a Dios (cape) F3
Grande (riv.) E4
Great Corn (isl.) F4
Huapí (mts.) E4
Isabelia, Cordillera (range) E4
King (cays) F4
Kukalaya (riv.) F4
Little Corn (isl.) F4
Maíz Grande (Great Corn) (isl.) F4
Maíz Pequeña (Little Corn) (isl.) F4
Managua (lake) E4
Miskitos (cays) F3
Monkey (pt.) F5
Mosquitos, Costa de (reg.) E4
Nicaragua (lake) E5
Ometepe (isl.) E5
Pearl (cays) F4
Perlas (pt.) F4
Prinzapolca (riv.) F4
Salinas (bay) D5
San Juan (riv.) E5
San Juan del Norte (bay) F5
Solentiname (isls.) E5
Tuma (riv.) E4
Tyra (cays) F4
Waspuk (riv.) E3
Wawa (riv.) E3
Zapatera (isl.) E5

PANAMA

CITIES and TOWNS

Aguadulce 10,659 .... G6
Alanje 866 .... F6
Almirante 4,664 .... F6
Antón 4,259 .... G6
Bajo Boquete 2,831 .... F6
Balboa 1,952 .... H6
Bocas del Toro 2,515 .... F6
Cañazas 1,526 .... G6
Capira 1,749 .... G6
Changuinola 9,528 .... F6
Chepo 4,529 .... H6
Chiriquí Grande F6
Chitré 17,156 .... G7
Coclé del Norte G6
Colón 59,832 .... H6
Cristóbal⊙ 7,959 .... G6
David 50,621 .... F6
Dolega 1,019 .... F6
El Real de Santa María 912 .... J6
Garachiné 1,116 .... H6
Gualaca 1,510 .... F6
Horconcitos 1,090 .... F6
La Chorrera 36,971 .... H6
La Concepción 10,460 .... F6
La Palma 1,634 .... H6
La Pintada 1,100 .... G6
Las Palmas 738 .... G6
Las Tablas 5,230 .... G7
Los Santos 4,644 .... G7
Mandinga 81 .... H6
Montijo 1,152 .... G6
Natá 5,603 .... G6
Nuevo Chagres 306 .... G6
Ocú 2,353 .... G7
Panamá (cap.) 388,638 .... H6
Panamá* 498,624 .... H6
Parita 1,616 .... G6
Pedasí 934 .... G7
Penonomé 7,389 .... G6
Playón Chico 1,395 .... H6
Portobelo 551 .... H6
Puerto Armuelles 12,488 .... F6
Puerto Obaldía 491 .... J6
San Francisco 990 .... G6
Santa Fe 490 .... G6
Santiago 21,809 .... G6
Soná 4,471 .... G6
Tocumen⊙ 21,745 .... H6
Tolé 1,052 .... G6
Tonosí 891 .... G7

OTHER FEATURES

Azuero (pen.) G7
Bastimentos (isl.) G6
Brewster, Cerro (mt.) H6
Burica, Punta (cape) F6
Cébaco (isl.) G7
Chepo (riv.) H6
Chiriquí (gulf) F7
Chiriquí (lag.) G6
Chucunaque (riv.) J6
Coiba, Isla de (isl.) F7
Colón, Isla de (isl.) G6
Contreras (isls.) F7
Darién (mts.) J6
Escudo de Veraguas (isl.) G6
Gatun (lake) G6
Gorda (pt.) H6
Jicarón (isl.) F7
Ladrones (isls.) F7
Manzanillo (pt.) H6
Montijo (gulf) G7
Mosquitos, Golfo de los (gulf) G6
Panama (gulf) H7
Pando, Cerro (mt.) F6
Parida (isl.) F6
Parita (bay) G6
Perlas (arch.) H6
Puercos, Morro de (head) H7
Rey (isl.) H6
Rincón (pt.) G6
San Blas, Golfo de (bay) H6
San Blas, Pta. de (pt.) H6
San Blas, Cordillera de (mts.) H6
San José (isl.) H6
San Miguel, Golfo de (bay) H6
Santiago, Cerro (mt.) G6
Secas (isls.) G7
Tabasará (mts.) G6
Taboga (isl.) H6
Tiburón (pt.) J6
Valiente (pen.) G6

*City and suburbs.
○Population of sub-district or division.
⊙Population of district.

CUBA

HAITI

DOMINICAN REPUBLIC

JAMAICA

TRINIDAD AND TOBAGO

BARBADOS

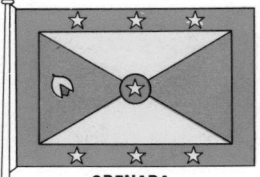
GRENADA

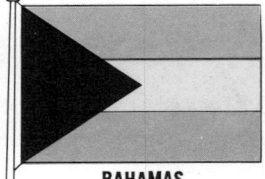
BAHAMAS

DOMINICA

ST. LUCIA

ST. VINC. & GRENS.

ANTIGUA AND BARBUDA

CUBA

AREA 44,206 sq. mi. (114,494 sq. km.)
POPULATION 9,706,369
CAPITAL Havana
LARGEST CITY Havana
HIGHEST POINT Pico Turquino
6,561 ft. (2,000 m.)
MONETARY UNIT Cuban peso
MAJOR LANGUAGE Spanish
MAJOR RELIGION Roman Catholicism

HAITI

AREA 10,694 sq. mi. (27,697 sq. km.)
POPULATION 5,053,792
CAPITAL Port-au-Prince
LARGEST CITY Port-au-Prince
HIGHEST POINT Pic La Selle 8,793 ft. (2,680 m.)
MONETARY UNIT gourde
MAJOR LANGUAGES Creole French, French
MAJOR RELIGION Roman Catholicism

DOMINICAN REPUBLIC

AREA 18,704 sq. mi. (48,443 sq. km.)
POPULATION 5,647,977
CAPITAL Santo Domingo
LARGEST CITY Santo Domingo
HIGHEST POINT Pico Duarte
10,417 ft. (3,175 m.)
MONETARY UNIT Dominican peso
MAJOR LANGUAGE Spanish
MAJOR RELIGION Roman Catholicism

JAMAICA

AREA 4,411 sq. mi. (11,424 sq. km.)
POPULATION 2,184,000
CAPITAL Kingston
LARGEST CITY Kingston
HIGHEST POINT Blue Mountain Peak
7,402 ft. (2,256 m.)
MONETARY UNIT Jamaican dollar
MAJOR LANGUAGE English
MAJOR RELIGIONS Protestantism,
Roman Catholicism

PUERTO RICO

AREA 3,515 sq. mi. (9,104 sq. km.)
POPULATION 3,196,520
CAPITAL San Juan
MONETARY UNIT U.S. dollar
MAJOR LANGUAGES Spanish, English
MAJOR RELIGION Roman Catholicism

NETHERLANDS ANTILLES

AREA 390 sq. mi. (1,010 sq. km.)
POPULATION 246,000
CAPITAL Willemstad
MONETARY UNIT Antilles guilder
MAJOR LANGUAGES Dutch, Papiamento, English
MAJOR RELIGIONS Roman Catholicism,
Protestantism

BERMUDA

AREA 21 sq. mi. (54 sq. km.)
POPULATION 67,761
CAPITAL Hamilton
MONETARY UNIT Bermuda dollar
MAJOR LANGUAGE English
MAJOR RELIGION Protestantism

ANGUILLA

Anguilla (isl.) 6,519 . . . . F3

ANTIGUA and BARBUDA

Antigua (isl.) 76,213 . . . . G3
Barbuda (isl.) 1,071 . . . . G3
Caribbean (sea) B4
Codrington 1,071 . . . . G3
Falmouth 1,134 . . . . F3
Redonda (isl.) F3
Saint John's (cap.) 21,814 . . . . . . G3

BAHAMAS

Acklins (isl.) 616 . . . . C2
Andros (isl.) 8,397 . . . . B1
Atwood (Samana) (cay) C2
Berry (isls.) 509 . . . . B1
Biminis, The (isls.) 1,432 . . . . B1
Caicos (passg.) D2
Cat (isl.) 2,143 . . . . C1
Cay Sal (bank) B1
Crooked (isl.) 517 . . . . D2
Crooked Island (passg.) C2
Eleuthera (isl.) 8,326. . . . . C1
Exuma (cays) C1
Exuma (sound) C1
Flamingo (cay) C2
Freeport 22,301 . . . . B1
Grand Bahama (isl.) 33,102 . . . . . B1
Great Abaco (isl.) 7,324 . . . . B1
Great Bahama (bank) B1
Great Exuma (isl.) C2
Great Inagua (isl.) 939 . . . . D2
Great Isaac (isl.) B1
Gun (cay) B1
Harbour (isl.) C1
Little Inagua (isl.) D2

CAYMAN ISLANDS

Bartlett Deep B3
Cayman Brac (isl.) 1,603 . . . . B3
George Town (cap.) 7,617. . . . . . . B3
Grand Cayman (isl.) 15,000 . . . . . B3
Little Cayman (isl.) 74 . . . . A3
Misteriosa (bank). A3

CUBA

Bayamo 109,201 . . . . C2
Camagüey 245,235 . . . . B2
Cienfuegos 107,396 . . . . B2
Florida (str.) B1
Guanabacoa 89,741. . . . . B2
Guantánamo 178,129 . . . . C2
Havana (cap.) 1,924,886. . . . . A2
Holguín 190,155 . . . . C2
Juventud (Pines) (isl.) 57,879 . . . . A2
Manzanillo 95,420 . . . . C2
Marianao ○127,563 . . . . A2
Matanzas 103,302. . . . . B2
Pinar del Río 104,598 . . . . A2
San Felipe (isl.) A2
Santa Clara 175,113 . . . . B2
Santiago de Cuba 362,432 . . . . . . C3
Windward (passg.) C3

BARBADOS

Bridgetown (cap.) 7,552 . . . . G4
Speightstown G4

BERMUDA

Bermuda (isl.) H3
Castle (harb.) H2
Great (sound) G3
Hamilton (cap.) 1,617 . . . . G3
Harrington (sound) H3
Ireland (isl.) G3
North (rapid) H2
Saint Davids (isl.) H2
Saint George 1,647 . . . . H2
Saint George's (isl.) H2
Somerset (isl.) G3

DOMINICA

Portsmouth 2,329 . . . . G4
Roseau (cap.) 9,968 . . . . G4

DOMINICAN REPUBLIC

La Romana 91,571 . . . . E3
San Francisco de Macorís 64,906 . . E3
San Pedro de Macorís 78,562 . . . . E3
Santiago 278,638 . . . . D3
Santo Domingo (cap.) 1,313,172 . . E3

GRENADA

Carriacou (isl.) 6,052 . . . . G4
Gouyave 2,498 . . . . F4
Grenadines (isls.) G4
Saint George's (cap.) 6,463. . . . . . F5

GUADELOUPE

Basse-Terre (cap.) 13,397 . . . . . . F4
Saint-Barthélemy (isl.) 3,059 . . . . F3
Saint Martin (isl.) 8,072 . . . . F3

HAITI

Cap-Haïtien 64,406 . . . . D3
Gonaïves 34,209 . . . . D3
Port-au-Prince (cap.) 449,831 . . . . D3
Gonâve (isl.) D3
Jamaica (chan.) C3
Tortuga (isl.) D2

JAMAICA

Blue Mountain (peak) C3
Jamaica (isl.) 106,791 . . . . C3
Kingston (cap.) 106,791 . . . . C3
Montego Bay 43,521 . . . . B3
Pedro (cays) C3
Savanna-la-Mar 11,759 . . . . B3

MARTINIQUE

Fort-de-France 96,649 . . . . G4
Saint-Pierre 4,923 . . . . G4
Pelée (vol.) G4

MONTSERRAT

Plymouth (cap.) 1,623. . . . . F3

NETHERLANDS ANTILLES

Aruba (isl.) E4
Bonaire (isl.) E4
Curaçao (isl.) E4
Oranjestad 10,100 . . . . D4
Saba (isl.) F3
Saint Eustatius (isl.) F3
Saint Martin (Sint Maarten) (isl.) . . F3
Willemstad (cap.) 95,000 . . . . E4

PUERTO RICO

Bayamón 185,087 . . . . G1
Caguas 87,214 . . . . G1
Culebra (isl.) 1,265 . . . . G1
Mayagüez 82,968 . . . . F1
Mona (passg.) E3
Ponce 161,739 . . . . F1

San Juan (cap.) 424,600 . . . . G1
Vieques (isl.) 7,662. . . . . G1

SAINT CHRISTOPHER and NEVIS

Basseterre (cap.) 14,725 . . . . F3
Nevis (isl.) 9,300 . . . . F3
Saint Christopher (isl.) 35,104 . . . . F3

SAINT LUCIA

Castries (cap.) ●42,770 . . . . G4
Vieux Fort ●10,675 . . . . G4

**SAINT VINCENT and
THE GRENADINES**

Bequia (isl.) G4
Georgetown 1,100 . . . . G4
Grenadines (isls.) 8,371 . . . . G4
Kingstown 17,117. . . . . G4

TRINIDAD and TOBAGO

Port-of-Spain (cap.) 67,978 . . . . . . G5
Scarborough 6,057 . . . . G5
Tobago (isl.) 39,695 . . . . G5
Trinidad (isl.) 1,020,130. . . . . G5

TURKS and CAICOS ISLANDS

Caicos (isls.) 4,008. . . . . D2
Cockburn Harbour D2
Grand Caicos (isl.) 371 . . . . D2
Grand Turk (isl.) 3,146 . . . . D2
Providenciales (isl.) 979 . . . . D2
Turks (isls.) 3,348. . . . . D2

VIRGIN ISLANDS (British)

Anegada (isl.) 89 . . . . H1
Jost Van Dyke (isl.) 135 . . . . G1
Road Town (cap.) 2,200 . . . . H1
Tortola (isl.) 9,257 . . . . H1
Virgin Gorda (isl.) 1,443 . . . . H1

VIRGIN ISLANDS (U.S.)

Charlotte Amalie (cap.) 11,842 . . . . H1
Christiansted 2,914 . . . . H2
Fredriksted 1,046 . . . . H2
Saint Croix (isl.) 49,725 . . . . H2
Saint John (isl.) 2,472 . . . . H1
Saint Thomas (isl.) 44,372 . . . . . . G1

WEST INDIES

Antilles, Greater (isls.) B2
Antilles, Lesser (isls.) E4
Aves (Bird) (isl.) F4
Hispaniola (isl.) D2
Leeward (isls.) F3
Navassa (isl.) C3
Windward (isls.) G4

● Population of district.
○ Population of municipality.

Topography

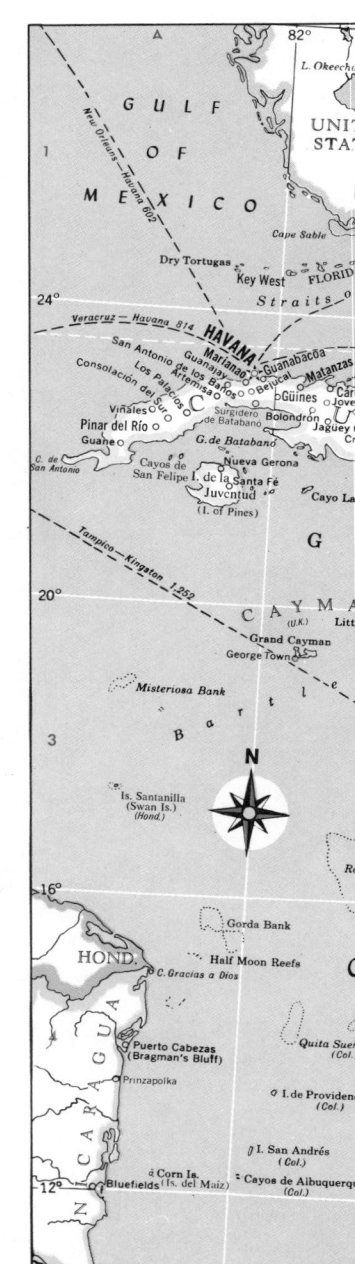

TRINIDAD AND TOBAGO

AREA 1,980 sq. mi. (5,128 sq. km.)
POPULATION 1,067,108
CAPITAL Port of Spain
LARGEST CITY Port of Spain
HIGHEST POINT Mt. Aripo 3,084 ft. (940 m.)
MONETARY UNIT Trinidad and Tobago dollar
MAJOR LANGUAGES English, Hindi
MAJOR RELIGIONS Roman Catholicism,
Protestantism, Hinduism, Islam

BARBADOS

AREA 166 sq. mi. (430 sq. km.)
POPULATION 248,983
CAPITAL Bridgetown
LARGEST CITY Bridgetown
HIGHEST POINT Mt. Hillaby 1,104 ft. (336 m.)
MONETARY UNIT Barbadian dollar
MAJOR LANGUAGE English
MAJOR RELIGION Protestantism

GRENADA

AREA 133 sq. mi. (344 sq. km.)
POPULATION 103,103
CAPITAL St. George's
LARGEST CITY St. George's
HIGHEST POINT Mt. St. Catherine 2,757 ft. (840 m.)
MONETARY UNIT East Caribbean dollar
MAJOR LANGUAGES English, French patois
MAJOR RELIGIONS Roman Catholicism, Protestantism

BAHAMAS

AREA 5,382 sq. mi. (13,939 sq. km.)
POPULATION 209,505
CAPITAL Nassau
LARGEST CITY Nassau
HIGHEST POINT Mt. Alvernia 206 ft. (63 m.)
MONETARY UNIT Bahamian dollar
MAJOR LANGUAGE English
MAJOR RELIGIONS Roman Catholicism, Protestantism

DOMINICA

AREA 290 sq. mi. (751 sq. km.)
POPULATION 74,089
CAPITAL Roseau
HIGHEST POINT Morne Diablotin 4,747 ft. (1,447 m.)
MONETARY UNIT Dominican dollar
MAJOR LANGUAGES English, French patois
MAJOR RELIGIONS Roman Catholicism, Protestantism

SAINT LUCIA

AREA 238 sq. mi. (616 sq. km.)
POPULATION 115,783
CAPITAL Castries
HIGHEST POINT Mt. Gimie 3,117 ft. (950 m.)
MONETARY UNIT East Caribbean dollar
MAJOR LANGUAGES English, French patois
MAJOR RELIGIONS Roman Catholicism, Protestantism

SAINT VINCENT AND THE GRENADINES

AREA 150 sq. mi. (388 sq. km.)
POPULATION 124,000
CAPITAL Kingstown
HIGHEST POINT Soufrière 4,000 ft. (1,219 m.)
MONETARY UNIT East Caribbean dollar
MAJOR LANGUAGE English
MAJOR RELIGIONS Protestantism, Roman Catholicism

ANTIGUA AND BARBUDA

AREA 171 sq. mi. (443 sq. km.)
POPULATION 75,000
CAPITAL St. John's
HIGHEST POINT Boggy Peak 1,319 ft. (402 m.)
MONETARY UNIT East Caribbean dollar
MAJOR LANGUAGE English
MAJOR RELIGION Protestantism

SAINT CHRISTOPHER & NEVIS

AREA 104 sq. mi. (269 sq. km.)
POPULATION 44,404
CAPITAL Basseterre
HIGHEST POINT Mt. Misery 4,314 ft. (1,315 m.)
MONETARY UNIT East Caribbean dollar
MAJOR LANGUAGE English
MAJOR RELIGIONS Protestantism, Roman Catholicism

ST. CHRISTOPHER & NEVIS

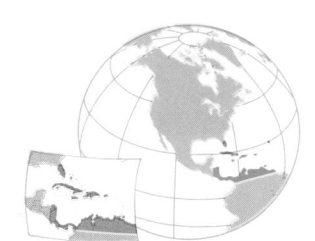

The West Indies

CONIC PROJECTION

SCALE OF MILES
0   50   100   200

SCALE OF KILOMETERS
0   50   100   200   300

Capitals ⎯⎯⎯⎯⎯⎯⎯ ☆

Scale 1:11,200,000
Distances are given in Nautical Miles

Puerto Rico

Bermuda Islands

© Copyright HAMMOND INCORPORATED, Maplewood, N.J.

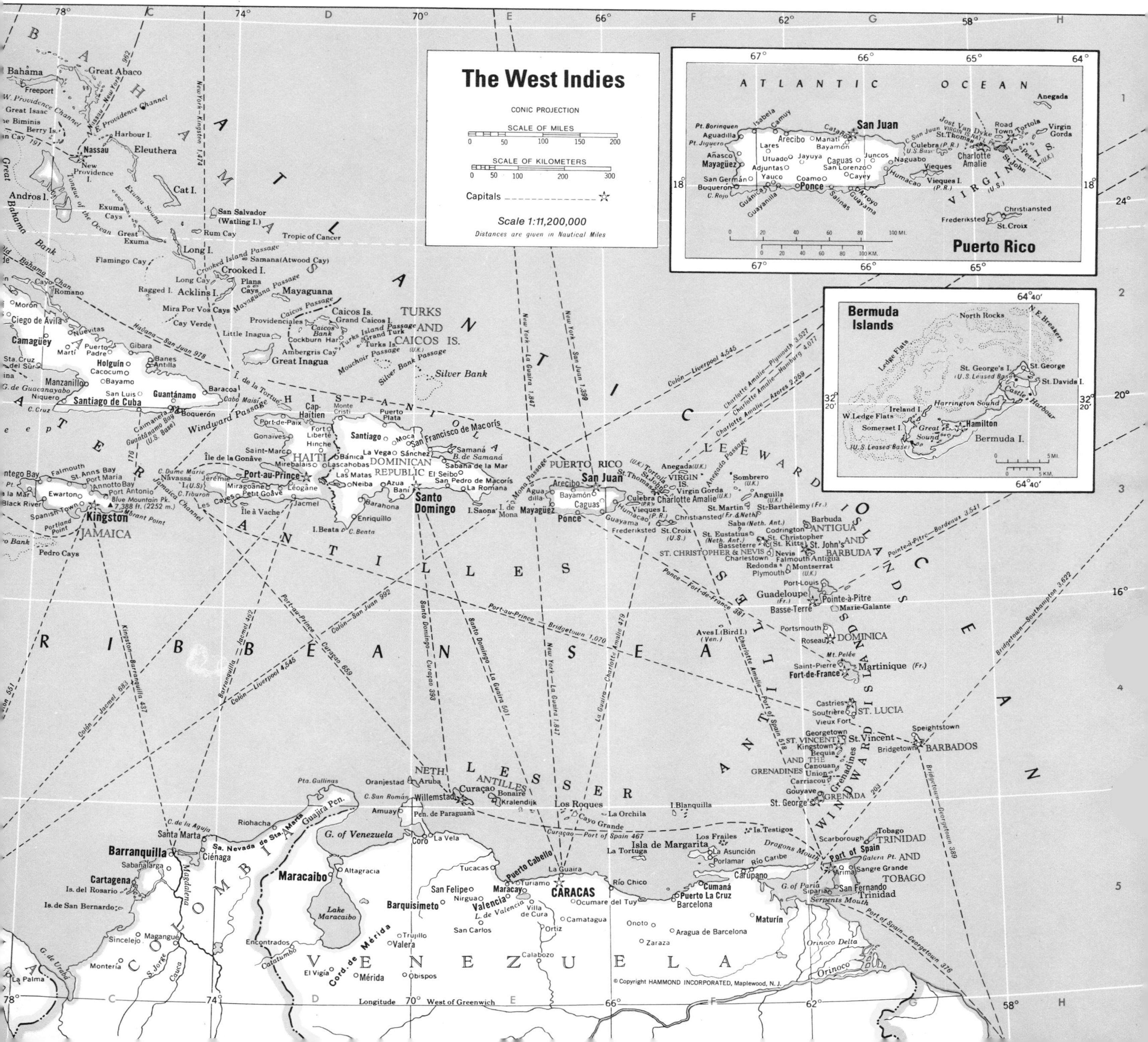

CUBA

PROVINCES

Camagüey 664,566 .... G2
Ciego de Ávila 320,961 .... E2
Cienfuegos 326,412 .... E2
Granma 739,335 .... H4
Guantánamo 466,609 ...... K4
Habana 1,924,886 .... C1
Habana, La (Havana)
  586,029 .... C1
Holguín 911,034 .... J3
Juventud (municipio
  especial) 57,879 ...... C2
Las Tunas 436,341 .... H3
Matanzas 557,628 .... D1
Pinar del Río 640,740 ..... A2
Sancti Spíritus 399,700 ... F2
Santiago de Cuba 909,506 . H4
Villa Clara 764,743 .... E1

CITIES and TOWNS

Abreus 14,267 .... D2
Agramonte 4,603 .... D1
Aguada de Pasajeros 20,219 D2
Alacranes 4,959 .... D1
Alonso Rojas 1,427 .... A2
Alquízar 12,691 .... C1
Altagracia 1,722 .... G3
Alto Songo-La Maya 25,188 J4

Amarillas 2,767 .... D2
Amazonas 1,066 .... F2
Antilla 10,052 .... J3
Arroyo Blanco 1,431 ...... E2
Artemisa 45,689 .... B1
Báez 4,178 .... E2
Báguanos 12,678 .... J3
Bahía Honda 16,901 ...... B1
Banao 803 .... F2
Banes 38,905 .... J3
Baracoa 36,702 .... K4
Baraguá 12,633 .... F2
Bauta 26,826 .... C1
Bayamo 109,201 .... H3
Bejucal 15,649 .... C1
Bolondrón 5,840 .... D1
Buenaventura 4,711 ...... H3
Buey Arriba 8,017 .... H4
Cabaiguán 36,544 .... F2
Cabañas 4,897 .... B1
Cabezas 5,262 .... D1
Cacocum 14,145 .... H3
Caibarién 32,094 .... E1
Caimanera 6,664 .... J4
Calabazar de Sagua 9,023 E1
Calimete 19,925 .... D1
Camagüey 245,235 ...... G3
Camajuaní 26,653 .... E1
Campechuela 20,743 ..... G4
Canasí 1,637 .... C1

Candelaria 10,810 .... B1
Cárdenas 65,585 .... D1
Cartagena 2,166 .... D2
Cascajal 3,530 .... E1
Cauto Embarcadero 949 .. H4
Cauto el Cristo 1,626 .... J3
Central Amancio Rodríguez
  22,506 .... G3
Central Bolivia 6,301 .... G2
Central Brasil 4,904 ..... G2
Central Cándido González
  3,414 .... G3
Central Colombia 16,799 . G3
Central Frank País 9,066 . K3
Central Guatemala 5,584 . J4
Central Haití 3,609 ...... G3
Central Los Reynaldos 3,997 J4
Central Loynaz Echevarría
  3,245 .... J3
Central Manuel Tames 7,864 K4
Céspedes 6,634 .... F2
Chambas 19,877 .... F2
Chaparra 8,428 .... H3
Cidra 3,567 .... D1
Ciego de Ávila 80,010 ... F2
Cienfuegos 107,396 ...... E2
Colón 47,010 .... D1
Condado 33,115 .... D1
Consolación del Norte 4,681 B1
Consolación del Sur 34,334 B2
Contramaestre 44,991 ... G3
Corralillo 15,822 .... D1

Cruces 20,324 .... E2
Cueto 23,183 .... J3
Cumanayagua 25,338 .... E2
Daiquirí J4
Delicias 10,562 .... H3
Dos Caminos 3,772 ..... J4
Dos Ríos 1,786 .... J4
El Caney 3,921 .... J4
El Cobre 3,952 .... J4
El Santo 2,473 .... E1
Encrucijada 23,029 ..... E1
Esmeralda 17,205 ...... G1
Esperanza 9,241 .... E1
Florencia 6,979 .... F2
Florida 43,881 .... G2
Fomento 17,310 .... F2
Gaspar 2,682 .... F2
Gibara 23,137 .... J3
Guáimaro 29,712 ..... G3
Guanabacoa 89,741 .... C1
Guanajay 21,042 ...... B1
Guane 14,126 .... A2
Guantánamo 178,129 .. K4
Guaro 3,086 .... J3
Guasimal 3,057 ...... F2
Guayabal 3,703 ...... G3
Guayos 6,753 .... F2
Güines 51,691 .... C1
Güira de Melena 19,851 . C1
Guisa 15,182 .... H4
Herradura 3,762 ..... B1

Holguín 190,155 .... J3
Ignacio Agramonte 1,487 . G3
Imías 4,491 .... K4
Isabela de Sagua 3,721 .. E1
Jagüey Grande 30,205 ... D2
Jamaica 5,128 .... J4
Jaruco 16,844 .... C1
Jatibonico 17,047 .... F2
Jíbaro 1,263 .... F2
Jiguaní 25,069 .... H3
Jobabo 14,899 .... H3
Jovellanos 35,043 .... D1
La Coloma 3,462 ..... B2
La Maya-Alto Songo 25,188 J4
Las Martinas 4,511 ... A2
Limonar 5,095 .... D1
Limonar 9,629 .... F2
Los Arabos 10,664 ... E1
Los Palacios 21,884 .. B1
Lugareño 4,396 ...... G2
Mabay 6,176 .... H4
Maceo 2,652 .... J4
Majagua 9,110 ...... F2
Manacas 5,914 ...... E1
Manatí 11,054 ...... H3
Manguito 2,739 ..... D1
Manicaragua 33,900 . E2
Mantua 9,165 .... A2
Mapos (Amazonas) 1,066 . F2
Manzanillo 95,420 .. H4
Marianao ○127,563 .. C1
Mariel 24,115 ...... B1
Martí 11,474 .... D1

Matanzas 103,302 ..... C1
Máximo Gómez, Ciego
  de Ávila 5,116 ...... F2
Máximo Gómez, Matanzas
  4,970 .... D1
Mayajigua 4,425 ...... F2
Mayarí 54,699 .... J3
Mayarí Arriba 2,302 .. J4
Media Luna 13,794 ... G4
Mendoza 2,914 ...... A2
Meneses 4,768 ...... F2
Minas 17,675 .... G2
Minas de Matahambre
  14,976 .... A1
Moa 28,696 .... K3
Morón 40,396 ...... F2
Nicaro 9,506 ...... J3
Niquero 15,544 .... G4
Nueva Gerona 17,175 . B2
Nuevitas 35,103 ... G2
Orozco 4,256 ...... B1
Palma Soriano 66,222 . J4
Pedro Betancourt 22,915 . D1
Perico 20,633 ..... D1
Pilón 10,194 ...... H4
Pinar del Río 104,598 . B2
Placetas 46,038 ... E1
Primero Enero 14,807 . F2
Puerto Esperanza 3,499 . A1
Puerto Padre 46,806 . H3
Quemado de Güines 11,208 E1

Rancho Veloz 3,966 .... D1
Ranchuelo 34,255 ..... E2
Regla 38,491 .... C1
Remedios 27,722 ..... E2
República Dominicana
  2,540 .... F2
Río Cauto 19,550 .... H4
Rodas 16,350 .... E2
Sagua de Tánamo 15,327 . K3
Sagua la Grande 52,315 . E1
San Andrés 2,127 .... J4
San Antonio de los Baños
  28,137 .... C1
San Cristóbal 30,769 . C1
Sancti Spíritus 79,542 . F2
San Diego de los Baños
  1,430 .... B1
San Germán 12,362 ... J3
San José de las Lajas
  37,149 .... C1
San José de los Ramos
  1,726 .... D1
San Juan y Martínez 13,227 B2
San Luis, Pinar del Río
  5,677 .... B2
San Luis, Santiago de Cuba
  32,826 .... J4
San Nicolás 12,368 .. C1
San Ramón 2,676 ... H4
Santa Clara 175,113 . E2
Santa Cruz del Norte
  15,239 .... C1

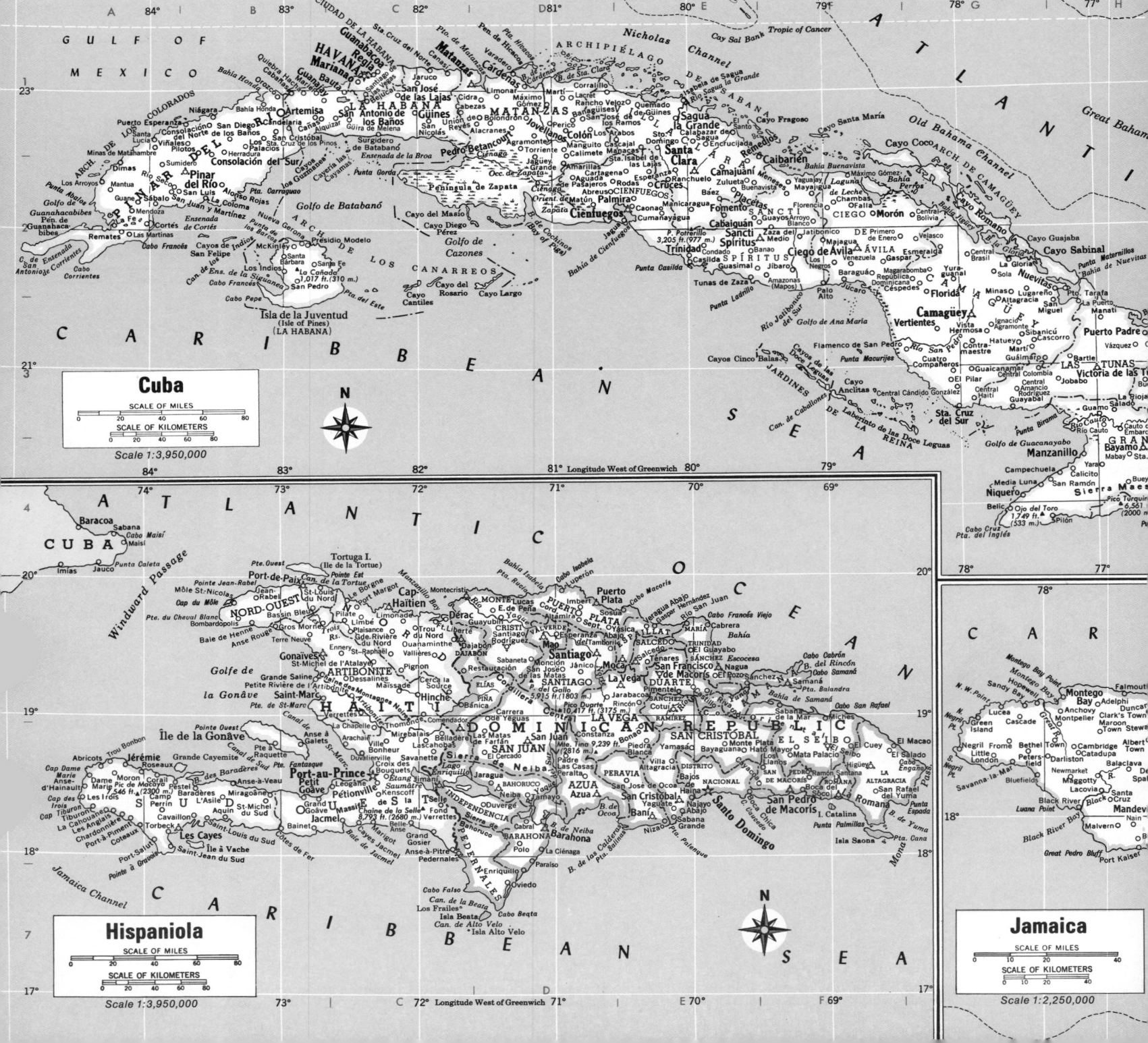

Santa Cruz de los Pinos 3,545 B1
Santa Cruz del Sur 27,142 G3
Santa Fe 3,925 B2
Santa Isabel de las Lajas 7,279 E2
Santa Lucía 3,734 J3
Santa Rita 6,358 H4
Santiago de Cuba 362,432 J4
Santiago de las Vegas 29,325 C1
Santo Domingo 32,950 E1
Sibanicú 14,252 G3
Sola 2,436 G2
Sumidero 980 A2
Surgidero de Batabanó 11,533 C1
Tacajó 4,469 J3
Torriente 1,759 D1
Trinidad 42,080 E2
Unión de Reyes 28,422 C1
Varadero 14,737 D1
Vázquez 3,851 H3
Velasco 5,618 H3
Venezuela 13,744 F2
Vertientes 25,178 G3
Victoria de las Tunas 87,522 H3
Viñales 2,049 A1
Yaguajay 30,720 F2
Yara 238,879 H4
Zaza del Medio 7,495 F2
Zulueta 5,425 E2

OTHER FEATURES

Abalos (pt.) A2
Ana María (gulf) F3
Anclitas (cay) F3
Batabanó (gulf) C2
Birama (pt.) G4
Broa (inlet) C1
Buenavista (bay) F2
Caballones (chan.) F3
Camagüey (arch.) G2
Cantiles (cay) C3
Cárdenas (bay) D1
Carraguao (pt.) B2
Casilda (pt.) E2
Cauto (riv.) H3
Cayamas (cays) C2
Cazones (gulf) C2
Cienfuegos (bay) D2
Cinco Balas (cays) E3
Cochinos (bay) D2
Coco (cay) G1
Corrientes (cape) A2
Corrientes (inlet) A2
Cortés (inlet) B2
Cristal, Sierra del (mts.) J3
Cruz (cape) G4
Diego Pérez (cay) C2
Doce Leguas (cays) F3
Este (pt.) C3
Fragoso (cay) F1
Francés (cape) A2
Gorda (pt.) C2
Gran Piedra (mt.) J4
Guacanayabo (gulf) G4
Guajaba (cay) G2
Guanahacabibes (gulf) A2
Guanahacabibes (pen.) A2
Guantánamo (bay) J4
Guantánamo Bay U.S. Nav. Reserve K4
Guarico (pt.) K3
Guzmanes (cays) B2
Hicacos (pen.) D1
Hicacos (pt.) D1
Honda (bay) B1
Indios (chan.) B2
Inglés (pt.) G4
Jardines de la Reina (arch.) F3
Jatibonico del Sur (riv.) F3
Jigüey (bay) G2
Juventud, Isla de la (Pines) (isl.) 57,879 B3
Laberinto de las Doce Leguas (cays) F3
Ladrillo (pt.) E3
Largo (cay) D2
Leche (lag.) F2
Los Barcos (pt.) B2
Los Canarreos (arch.) C2
Los Colorados (arch.) A1
Lucrecia (cape) J3
Macurijes (pt.) F3
Maestra, Sierra (mts.) H4
Maisí (cape) K4
Mangle (pt.) J3
Masío (cay) C2
Matanzas (bay) D1
Nicholas (chan.) E1
Nipe (bay) J3
Nuevitas (bay) H2
Ojo del Toro (mt.) G4
Old Bahama (chan.) G1
Pepe (cape) B3
Perros (bay) G2
Pigs (Cochinos) (bay) D2
Pines (Isla de la Juventud) (isl.) 7,879 B3
Potrerillo (peak) E2
Quemado (pt.) K4
Romano (cay) G2
Rosario (cay) C2
Sabana (arch.) E1
Sabinal (cay) H2
Sagua la Grande (riv.) E1
San Antonio (cape) A2
San Felipe (cays) B2
San Pedro (riv.) G3
Santa Clara (bay) D1
Santa María (cay) F1
Siguanea (bay) B2
Tabacal (pt.) H4
Toa, Cuchillas de (mts.) K4
Tortuguilla (pt.) K4
Turquino (peak) H4
Zapata (pen.) C2
Zapata Occidental (swamp) D2
Zapata Oriental (swamp) D2

DOMINICAN REPUBLIC

PROVINCES

Azua 142,770 D6
Bahoruco 78,636 D6
Barahona 137,160 D6
Dajabón 57,709 D5
Distrito Nacional 1,550,739 E6
Duarte 235,544 E5
Elías Piña 65,384 C5
El Seibo 157,866 F6
Espaillat 164,017 E5
Independencia 38,768 D6
La Altagracia 100,112 F6
La Romana 109,769 F6
La Vega 385,043 D6
María Trinidad Sánchez 112,629 E5
Monte Cristi 83,407 D5
Pedernales 17,006 D7
Peravia 168,123 E6
Puerto Plata 206,757 D5
Salcedo 99,191 E5
Samaná 65,699 E5
Sánchez Ramírez 126,567 E5
San Cristóbal 446,132 E6
San Juan 239,957 D6
San Pedro de Macorís 152,890 F6
Santiago 550,372 D5
Santiago Rodríguez 55,411 D5
Valverde 100,319 D5

CITIES and TOWNS

Altamira 2,759 D5
Azua 31,481 D6
Bajos de Haina 33,135 E6
Baní 36,705 E6
Barahona 49,334 D6
Bonao 44,486 E6
Cabrera 2,542 E5
Comendador 5,962 C6
Constanza 15,141 D6
Cotuí 16,688 E5
Dajabón 8,808 D5
El Seibo 13,511 F6
Hato Mayor 17,859 F6
Higüey 33,501 F6
Imbert 5,315 D5
Jarabacoa 13,416 E5
Jimaní 3,327 C6
La Romana 91,571 F6
La Vega 52,432 E5
Luperón 2,500 D5
Mao 33,527 D5
Moca 31,176 D5
Monción 3,344 D5
Nagua 20,912 E5
Puerto Plata 45,348 D5
Sabana de la Mar 9,983 F5
Sabaneta 9,170 D5
Samaná 5,023 F5
Sánchez 7,919 E5
San Cristóbal 58,520 E6
San Francisco de Macorís 64,906 E5
San Juan 49,764 D6
San Pedro de Macorís 78,562 F6
Santiago 278,638 D5
Santo Domingo (cap.) 1,313,172 E6
Tenares 4,065 E5
Villa Altagracia 20,890 E6

OTHER FEATURES

Alto Velo (chan.) C7
Alto Velo (isl.) D7
Balandra (pt.) F5
Beata (cape) D7
Beata (chan.) C7
Beata (isl.) C7
Cabrón (cape) F5
Calderas (bay) D6
Cana (pt.) F6
Catalina (isl.) F6
Caucedo (capee) E6
Central, Cordillera (range) D5
Duarte (peak) D5
Engaño (cape) F6
Enriquillo (lake) C6
Escocesa (bay) E5
Espada (pt.) F6
Falso (cape) C7
Francés Viejo (cape) E5
Gallo (mt.) D5
Isabela (bay) D5
Isabela (cape) D5
Los Frailes (isl.) C7
Macorís (cape) E5
Manzanillo (bay) C5
Mona (passg.) F6
Neiba (bay) D6
Neiba, Sierra de (mts.) D6
Ocoa (bay) D6
Oriental, Cordillera (range) F6
Palenque (pt.) E6
Palmillas (pt.) F6
Rincón (bay) F5
Rucia (pt.) D5
Salinas (pt.) E6
Samaná (bay) F5
Samaná (cape) F5
San Rafael (cape) F5
Saona (isl.) F6
Septentrional, Cordillera (range) D5
Tina (mt.) D6
Yaque del Norte (riv.) D5
Yaque del Sur (riv.) D6
Yuma (bay) F6
Yuna (riv.) E5

HAITI

DEPARTMENTS

Artibonite C5
Nord C5
Nord-Ouest B5
Ouest C6
Sud A6

CITIES and TOWNS

Anse-à-Galets 3,623 B6
Anse-d'Hainault 5,220 A6
Aquin 3,820 B6
Cap-Haïtien 64,406 C5
Croix des Bouquets 4,365 C6
Dame Marie 4,320 A6
Dérac 1,300 C5
Dessalines 7,984 C5
Fort Liberté 5,012 C5
Gonaïves 34,209 B5
Grande Rivière du Nord 6,007 C5
Gros Morne 4,739 B5
Hinche 10,070 C5
Jacmel 13,730 C6
Jérémie 18,493 A6
Kenscoff 2,605 C6
Lascahobas 3,805 C6
Léogâne 5,782 C6
Les Cayes 34,090 B6
Limbé 10,476 C5
Miragoâne 4,327 B6
Mirebalais 6,069 C6
Ouanaminthe 7,276 C5
Pétionville 35,333 C6
Petite Rivière de l'Artibonite 10,099 B5
Petit Goâve 7,310 B6
Pignon 4,576 C5
Port-au-Prince (cap.) 449,831 C6
Port-de-Paix 15,540 B5
Saint-Louis du Nord 7,203 B5
Saint-Marc 24,165 B5
Saint-Michel de l'Atalaye 7,559 C5
Saint-Raaphaël 3,889 C5
Trou du Nord 7,637 C5
Verrettes 3,670 C5

OTHER FEATURES

Artibonite (riv.) C5
Baradères (bay) B6
Cheval Blanc (pt.) B5
Dame Marie (cape) A6
Est (pt.) C4
Fantasque (pt.) B6
Gonâve (gulf) B5
Gonâve (isl.) B6
Grande Cayemite (isl.) B6
Gravois (pt.) A7
Irois (cape) A6
Jean-Rabel (pt.) B5
Macaya (mt.) A6
Manzanillo (bay) C5
Môle (cape) B5
Noires (mts.) C5
Ouest (pt.) B4
Ouest (pt.) B6
Saint-Marc (chan.) B6
Saint-Marc (pt.) B5
Saumâtre (lake) C6
Selle (peak) C6
Sud (chan.) B6
Tortue (chan.) C5
Tortue (Tortuga) (isl.) C4
Tortuga (isl.) C4
Trois-Rivières (riv.) B5
Vache (isl.) B6
Windward (passg.) A5

JAMAICA

CITIES and TOWNS

Alley J7
Alligator Pond H6
Anchovy 2,558 H5
Annotto Bay K6
Bamboo 2,971 J6
Bath K6
Black River 2,701 H6
Bog Walk J6
Bowden K6
Browns Town 5,479 J6
Bull Savanna-Junction 5,110 H6
Cambridge 2,449 H6
Catadupa H6
Christiana H6
Discovery Bay 1,814 J5
Falmouth 3,937 H5
Green Island G6
Hope Bay K6
Kingston (cap.) 106,791 K6
Kingston *516,865 J7
Linstead J6
Lucea 3,635 G5
Mandeville 14,421 H6
Maroon Town 2,717 H6
May Pen 26,074 J6
Montego Bay 43,521 H5
Montpelier H6
Morant Bay 7,465 K7
Negril G6
Ocho Rios 5,851 J6
Oracabessa J5
Port Antonio 10,538 K6
Port Kaiser H7
Port Maria 5,259 J6
Port Morant K6
Saint Ann's Bay 7,101 J5
Saint Margaret's Bay K6
Savanna-la-Mar 11,759 G6
Spanish Town 40,731 J6
Williamsfield H6

OTHER FEATURES

Black (riv.) H6
Black River (bay) G6
Blue (mts.) J6
Blue Mountain (peak) K6
Galina (pt.) J6
Grande (riv.) K6
Great (riv.) H6
Great Pedro Bluff (prom.) H6
Long (bay) H7
Luana (pt.) G6
Minho (riv.) J6
Montego (bay) G5
Montego Bay ((pt.) G5
North East (pt.) K6
North Negril (pt.) G6
North West (pt.) G5
Old Harbour (bay) J6
Portland (pt.) J7
Sir John's (peak) K6
South East (pt.) K6
South Negril (pt.) G6

*City and Suburbs.
○ Population of municipality.

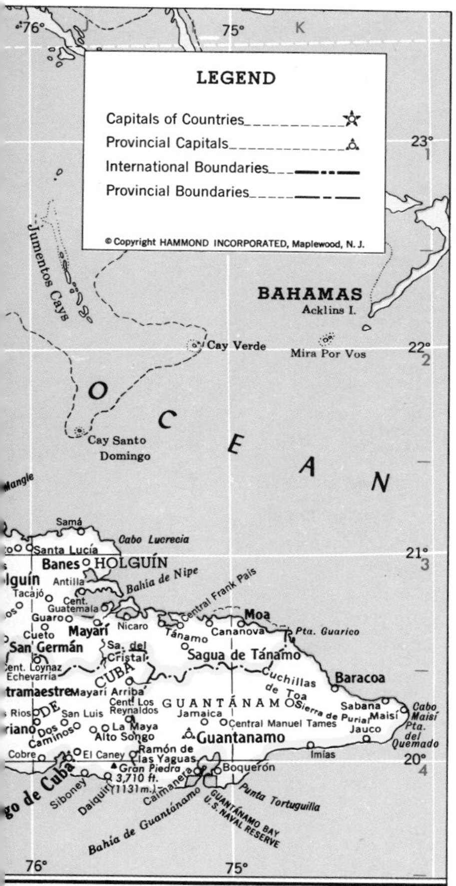
LEGEND
Capitals of Countries
Provincial Capitals
International Boundaries
Provincial Boundaries
© Copyright HAMMOND INCORPORATED, Maplewood, N. J.
BAHAMAS
Acklins I.
Cay Verde
Mira Por Vos
Cay Santo Domingo
O C E A N
Cabo Lucrecia
Banes
HOLGUÍN
Bahía de Nipe
Mayarí
Moa
Sagua de Tánamo
Baracoa
Guantanamo
Boquerón
Punta Tortuguilla
Bahía de Guantánamo
U.S. NAVAL RESERVE

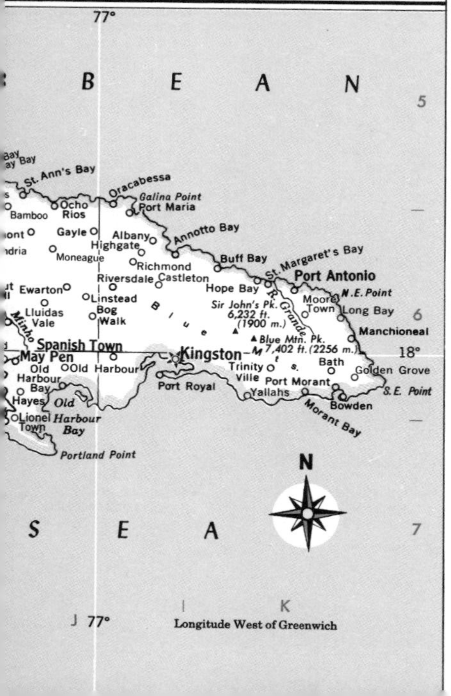
B E A N
St. Ann's Bay
Oracabessa
Galina Point
Port Maria
Annotto Bay
Buff Bay
St. Margaret's Bay
Port Antonio
Spanish Town
Kingston
Port Royal
May Pen
Old Harbour
Morant Bay
Portland Point
S E A
Longitude West of Greenwich

Agriculture, Industry and Resources

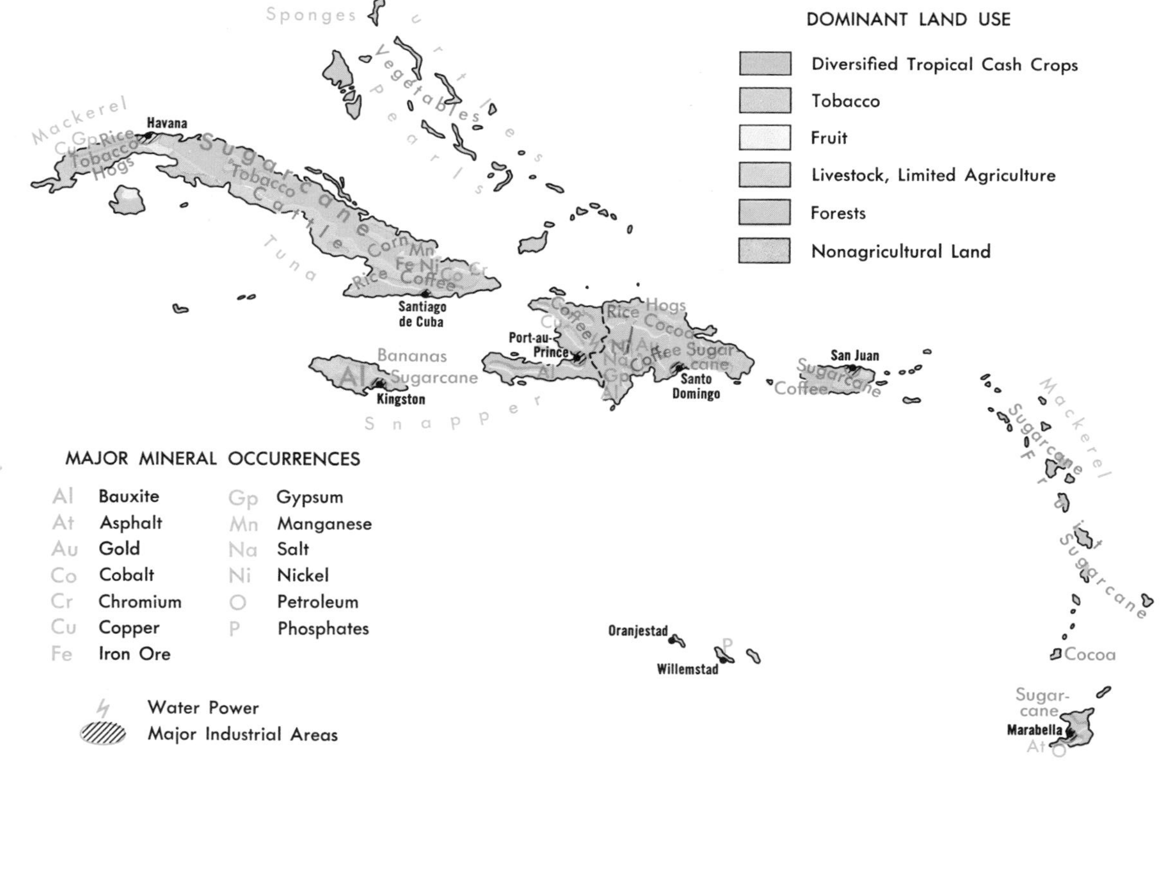
DOMINANT LAND USE
Diversified Tropical Cash Crops
Tobacco
Fruit
Livestock, Limited Agriculture
Forests
Nonagricultural Land
MAJOR MINERAL OCCURRENCES
Al Bauxite
At Asphalt
Au Gold
Co Cobalt
Cr Chromium
Cu Copper
Fe Iron Ore
Gp Gypsum
Mn Manganese
Na Salt
Ni Nickel
O Petroleum
P Phosphates
Water Power
Major Industrial Areas
Havana
Santiago de Cuba
Port-au-Prince
Santo Domingo
San Juan
Kingston
Oranjestad
Willemstad
Marabella
Sponges
Mackerel
Tuna
Snapper
Sugarcane
Tobacco
Cattle
Coffee
Bananas
Cocoa

(continued)

PUERTO RICO

DISTRICTS

Aguadilla A1
Arecibo C1
Bayamón D1
Guayama D2
Humacao E2
Mayagüez B2
Ponce C2
San Juan D1

CITIES and TOWNS

Adjuntas 5,239 B2
Aguada 5,025 A1
Aguadilla 22,039 A1
Aguas Buenas 3,766 E2
Aibonito 9,331 D2
Añasco 5,646 A1
Ángeles ○2,817 B2
Arecibo 48,779 B1
Arroyo 8,435 E3
Bahomamey A1
Bajadero 3,678 C1
Barceloneta 4,502 C1
Barranquitas 3,618 D2
Bayamón 185,087 D1
Boquerón ○3,675 A3
Cabo Rojo 10,292 A2
Caguas 87,214 E2
Caguas †156,819 E2
Camuy 3,834 B1
Carolina 147,835 E1
Cataño 26,243 D1
Cayey 23,305 D2
Ceiba 4,973 F2
Central Aguirre 1,049 D3
Ciales 3,582 C1
Cidra 6,069 D2
Coamo 12,851 D2
Comerío 5,736 D2
Coquí 3,018 D3
Corozal 5,889 D1
Coto Laurel ○5,192 C2
Culebra (Dewey) 938 G1
Dorado 10,203 D1
Ensenada B3
Esperanza 1,130 G2
Fajardo 26,928 F1
Florida 3,641 C1
Guánica 9,628 B3
Guayama 21,097 E3
Guayanilla 6,163 B3
Guaynabo 65,075 D1
Gurabo 7,645 E2
Hatillo 5,019 B1
Hato Rey E1
Hormigueros 12,031 A2
Humacao 19,147 E2
Isabela 12,087 A1
Isabel Segunda 2,330 G2
Jayuya 3,588 C2
Jobos 4,194 D3
Juana Díaz 10,469 C2
Juncos 7,851 E2
Lajas 4,275 A2
Lares 5,224 B2
Las Piedras 4,857 E2
Levittown 31,613 D1
Loíza 3,932 E1
Loíza Aldea E1
Luquillo 4,531 F1
Manatí 17,347 C1
Maricao 1,390 B2
Mayagüez 82,968 A2
Mayagüez †98,155 A2
Moca 3,960 A1
Naguabo 4,135 F2
Naranjito 2,849 D1
Palmer 1,566 F1
Palo Seco 1,172 D1
Parguera A3
Patillas 3,172 E2
Peñuelas 4,235 B2
Playa de Fajardo F1
Playa de Humacao ○5,573 F2
Ponce 161,739 C3
Ponce †168,272 C3
Puerto Nuevo D1
Puerto Real 2,390 A2
Puerto Real (Playa de Fajardo) F1
Punta Santiago (Playa de Humacao) ○5,573 F2
Quebradillas 3,770 B1
Río Blanco 1,433 F2
Río Grande 12,047 E1
Río Piedras E1
Rosario A2
Sabana Grande 7,435 B2
Sabana Seca 11,431 D1
Salinas 6,220 D3
San Antonio 2,681 A1
San Germán 13,054 A2
San Juan (cap.) 424,600 E1
San Juan †1,081,193 E1
San Lorenzo 8,880 E2
San Sebastián 10,619 B1
Santa Isabel 6,948 C3
Santurce E1
Tallaboa 1,059 B3
Toa Alta 4,427 D1
Toa Baja 1,992 D1
Trujillo Alto 41,141 E1
Utuado 11,113 B2
Vega Alta 10,582 D1
Vega Baja 18,233 D1
Vieques (Isabel Segunda) 2,330 G2
Villalba 3,469 C2
Yabucoa 6,797 E2
Yauco 14,594 B2

OTHER FEATURES

Aguadilla (bay) A1
Algarrobo (pt.) A2
Añasco (bay) A1
Arenas (pt.) F2
Bauta (riv.) C2
Bayamón (riv.) D1
Boquerón (bay) A3
Borinquen (pt.) A1
Cabullones (pt.) C3
Caja de Muertos (isl.) C3
Camuy (riv.) B1
Canovanas (riv.) E1
Caonillas (lake) C2
Carite (lake) E2
Carraízo (lake) E1
Cayey, Sierra de (mts.) D2
Central, Cordillera (range) C2
Cerro Gordo (pt.) D1
Coamo (res.) D3
Coamo (riv.) D3
Culebra (isl.) 1,265 G1
Culebrinas (riv.) A1
Culebrita (isl.) G2
El Toro (mt.) F2
El Yunque (mt.) F1
Este (pt.) G2
Fajardo (riv.) F1
Figuras (pt.) E3
Fosforescente (bay) A3
Grande de Añasco (riv.) A2
Grande de Arecibo (riv.) C1
Grande de Loíza (riv.) E1
Grande de Manatí (riv.) C1
Guajataca (lake) B1
Guanajibo (pt.) A2
Guanajibo (riv.) A2
Guánica (lake) B3
Guaniquilla (pt.) A2
Guayabal (lake) C2
Guayanés (pt.) F2
Guayanés (riv.) E2
Guayanilla (bay) B3
Guayo (lake) B2
Guilarte (mt.) B2
Honda (bay) F2
Jacaguas (riv.) C2
Jaicoa, Cordillera (mts.) B1
Jiguero (pt.) A1
Jobos (bay) D3
Lima (pt.) F2
Luquillo, Sierra de (mts.) E2
Manglillo (pt.) B3
Mayagüez (bay) A2
Miquillo (pt.) F1
Molinos (pt.) G1
Mona (passg.) A2
Negra (pt.) G2
Nigua (riv.) D2
Ola Grande (pt.) D3
Palmas Altas (pt.) C1
Patillas (lake) E2
Petrona (pt.) D3
Pirata (mt.) F2
Plata (riv.) D2
Puerca (pt.) F2
Puerto Medio Mundo (bay) F2
Punta, Cerro de (mt.) C2
Ramey A.F.B. A1
Rincón (bay) D3
Rojo (cape) A3
Roosevelt Road Naval Res. F2
Salinas (pt.) D1
San José (lag.) E1
San Juan, Cabezas de (prom.) F1
San Juan Nat'l Hist. Site D1
Soldado (pt.) G2
Sucia (bay) A3
Tanamá (riv.) B1
Toro, El (mt.) F2
Torrecilla (lag.) E1
Tortuguero (lag.) D1
Tuna (pt.) E3
Vacía Talega (pt.) E1
Vieques (isl.) 7,662 G2
Vieques (passg.) F2
Vieques (sound) G2
Yagüez (riv.) A2
Yauco (lake) B2
Yeguas ((pt.) F3

ANTIGUA

CITIES and TOWNS

All Saints 1,796 E11
Cedar Grove 1,460 E11
Falmouth 1,134 E11
Freetown 1,250 E11
Jennings 1,370 D11
Liberta 2,394 E11
Old Road 1,244 D11
Parham 1,570 E11
Saint John's (cap.) 21,814 E11
Willikies 1,843 E11

OTHER FEATURES

Antigua (isl.) 76,213 E11
Boggy (peak) D11
Boon (pt.) E11
Green (isl.) E11
Guiana (isl.) E11
Long (isl.) E11
Saint John's (harb.) C11
Standfast (pt.) E11
Willoughby (bay) E11

BARBADOS

CITIES and TOWNS

Bathsheba B8
Belleplaine B8
Bridgetown (cap.) 7,552 B9
Carlton B8
Cave Hill B9
Checker Hall B8
Codrington B8
Crab Hill B8
Crane C9
Drax Hall B8
Ellerton B9
Greenland B8
Holetown B8
Kendal B8
Lodge Hill B8
Marchfield B9
Mount Standfast B8
Oistins B9
Rose Hill B8
Rouen B9
Saint Lawrence B9
Saint Martins C9
Scarboro B9
Seawell B9
Six Mens B8
Speightstown B8
Spring Hall B8
Welchman Hall B8

OTHER FEATURES

Carlisle (bay) B9
Hillaby (mt.) B8
Long (bay) B9
North (pt.) B8
Oistins (bay) B9
Pelican (isl.) B9
Ragged (pt.) C8
Sam Lord's Castle C9
South (pt.) B9

DOMINICA

CITIES and TOWNS

Barroui 1,480 E6
Castle Bruce 1,975 F6
Coulihaut 1,735 E6
Delice F7
Grand Bay 3,152 F7
Hampstead E5
La Plaine F6
Mahout 2,095 E6
Marigot 3,183 F6
Petit Soufrière F6
Portsmouth 2,329 E5
Rosalie F6
Roseau (cap.) 9,968 E7
Roseau *16,035 E7
Saint Joseph 2,643 E6
Salybia F6
Soufrière E7
Vieille Case E5
Wesley 2,002 F5

OTHER FEATURES

Capuchin (cape) E5
Carib Reserve F6
Clyde (riv.) F6
Crumpton (pt.) F5
Diablotin, Morne (mt.) E6
Dominica (passg.) E5
Douglas (bay) E5
Grand (bay) F7
Jaquet (pt.) E5
Layou (riv.) E6
Martinique (passg.) E7
Micotrin (mt.) F6
Pagoua (bay) F6
Prince Rupert (bay) E5
Scotts (head) E7
Soufrière (bay) E7
Trois Pitons, Morne (mt.) E6

GRENADA

CITIES and TOWNS

Gouyave 2,498 C8
Grand Roy C8
Grenville 1,723 D8
Hermitage D8
La Taste D8
Marquis D8
Mount Tivoli D8
Saint George's (cap.) 6,463 C9
Saint George's *34,624 C9
Sauteurs 605 D8
Victoria 1,673 D8
Woodford C8

OTHER FEATURES

Bedford (pt.) D8
David (pt.) D8
Great Bacolet (pt.) D8
Green (isl.) D8
Grenville (bay) D8
Gros (pt.) C8
Halifax (harb.) C8
Irvin's (bay) D8
Les Tantes (isls.) D7
Molinière (pt.) C8
Prickly (pt.) C9
Ronde (isl.) D7
Saint Catherine (mt.) D8
Saline (pt.) C9
Sinai (mt.) D8
Telescope (pt.) D8

GUADELOUPE

Total Population 329,017

CITIES and TOWNS

Anse-Bertrand 1,921 A5
Baie-Mahault 5,874 A6
Baillif 3,844 A7
Bananier A7
Basse-Terre (cap.) 13,397 A7
Bouillante 1,821 A6
Bourg-des-Saintes 907 A7
Capesterre 7,541 A7
Ferry A6
Gosier 13,741 B6
Gourbeyre 5,637 A7
Goyave 1,709 A6
Grand-Bourg 3,249 B7
Lamentin 2,319 A6
Les Abymes 51,837 B6
Morne-à-l'Eau 9,457 A6
Moule 9,800 B6
Petit-Bourg 5,097 A6
Petit-Canal 1,581 A6
Pigeon A6
Pointe-à-Pitre 25,151 B6
Pointe-Noire 2,180 A6
Port-Louis 4,517 B5
Saint-Claude 6,755 A7
Sainte-Anne 11,527 B6
Sainte-Marguerite B6
Sainte-Marie A6
Sainte-Rose 4,805 A6
Saint-François 3,141 B6
Trois-Rivières 7,881 A7
Vieux-Fort 1,073 B7
Vieux-Habitants 4,065 A7

OTHER FEATURES

Allègre (pt.) A6
Antigues (pt.) A5
Basse-Terre (isl.) 138,777 A6
Châteaux (pt.) B6
Constant, Morne (hill) B7
Désirade, La (isl.) 1,602 B6
Fajou (isl.) A6
Grand Cul-de-Sac Marin (bay) A6
Grande-Terre (isl.) B6
Grande Vigie (pt.) B5
Grand-Îlet (isl.) A7
Guadeloupe (isl.) 167,896 B6
Guadeloupe (passg.) A5
Guadeloupe Nat'l Park A6
Kahouanne (isl.) A6
Marie-Galante (isl.) 13,757 B7
Nord (pt.) B7
Nord-Est (bay) B6
Petit Cul-de-Sac Marin (bay) A6
Petite-Terre (isls.) B6
Saintes (chan.) A7
Saintes (isls.) 2,901 A7
Salée (riv.) A6
Sans Toucher (mt.) A6
Soufrière (mt.) A7
Terre-de-Bas (isl.) 1,427 A7
Terre-de-Haut (isl.) 1,453 A7
Vieux-Fort (pt.) A7

MARTINIQUE

Total Population 330,220

CITIES and TOWNS

Ajoupa-Bouillon 1,569 C5
Basse-Pointe 2,163 C5
Bellefontaine 818 C6
Case-Pilote 1,776 C6
Ducos 4,429 D6
Fond-Saint-Denis 962 C6
Fort-de-France (cap.) 96,649 C6
Grand' Rivière 1,053 C5
Gros-Morne 1,976 D6
La Trinité 3,380 D6
Le Carbet 2,321 C6
Le François 2,940 D6
Le Lamentin 6,872 D6
Le Lorrain 2,024 D6
Le Marin 2,651 D7
Le Morne-Rouge 2,650 C5
Le Prêcheur 1,350 C5
Le Robert 3,610 D6
Le Saint-Esprit 3,947 D6
Les Trois-Îlets 1,484 D6
Le Vauclin 3,054 D6
Macouba 1,142 C5
Marigot 1,765 D5
Rivière-Pilote 1,587 D7
Rivière-Salée 1,859 D7
Sainte-Luce 1,502 D7
Sainte-Marie 3,966 D5
Saint-Joseph 2,052 D6
Saint-Pierre 4,923 C6
Schoelcher 16,412 C6

OTHER FEATURES

Cabet, Pitons du (mt.) C6
Cabrits (isl.) D7
Caravelle (pen.) D6
Cul-de-Sac du Marin (bay) D7
Diable (pt.) D5
Ferré (cape) E7
Fort-de-France (bay) C6
Galion (bay) D6
Lézarde (riv.) D6
Long (isl.) D6
Lorrain (riv.) D5
Martinique (passg.) C5
Pelée (vol.) C5
Pilote (riv.) D7
Ramiers (isl.) C6
Ramville (isl.) D6
Robert (harb.) D6
Rose (pt.) D6
Saint-Martin (cape) C5
Saint-Pierre (bay) C6
Salines (pt.) D7
Salomon (pt.) C7
Vauclin (mt.) D6

NETHERLANDS ANTILLES

CITIES and TOWNS

Aresji D9
Ascension F8
Bacuna E8
Balashi E10
Boven Bolivia E8
Bubali D10
Bushiribana E10
Dokterstuin F8
Druif D1
Emmastad F9
Entrejo E8
Fontein E8
Fuik G9
Groot Sint Joris G9
Hato G8
Kralendijk (cap.), Bonaire 2,500 E8
Lago E10
Lagoen F8
Montanja di Reij G9
New Port G9
Noord di Salinja E8
Onima E8
Oranjestad (cap.), Aruba 10,100 D10
Otrabanda F9
Patrick F8
Rincon E8
Rooi E8
Santa Barbara G9
Santa Catharina G9
Savaneta E10
Savonet F8
Sint Anna D1
Sint Jan D8
Sint Kruis F8
Sint Martha F8
Sint Michiel F9
Sint Nicolaas E10
Sint Willebrordus F8
Terra Corra E8
Westpunt, Aruba D10
Westpunt, Curaçao F8
Willemstad (cap.) 95,000 F9
Willemstad *130,000 F9

OTHER FEATURES

Aruba (isl.) 55,148 E9
Basora (pt.) E10
Bonaire (isl.) 8,087 E9
Bullen (bay) F8
Caracas (bay) G9
Curaçao (isl.) 145,430 G7
Goto (lake) D8
Jamanota (mt.) E10
Kanon (pt.) G9
Klein Bonaire (isl.) E8
Kudarebe (pt.) D9
Lac (bay) D9
Lacre (pt.) E9
Malmok (mt.) E8
Noord (pt.) D8
Noord (pt.) F8
Paarden (bay) D10
Palm (beach) D10
Pekelmeer (lake) E9
Piscadera (bay) F9
Schottegat (bay) G9
Sint Anna (bay) F9
Sint Christoffel (mt.) F8
Sint Joris (bay) G9
Slag (bay) D8
Vierkant (pt.) E8

SAINT CHRISTOPHER and NEVIS

CITIES and TOWNS

Basseterre (cap.) 14,725 C10
Cayon C10
Charlestown 1,326 C11
Cotton Ground 471 C11
Dieppe Bay C10
Frigate Bay C10
Gingerland D11
Golden Rock C10
Newcastle D11
Old Road Town C10
Sadlers Village C10
Sandy Point 862 C10
Tabernacle C10
Zion Hill D11

OTHER FEATURES

Brimstone (hill) C10
Dogwood (pt.) D11
Fort (pt.) C11
Great Salt (pond) D10
Heldens (pt.) C10
Horse Shoe (pt.) C11
Misery (mt.) C10
Monkey (hill) C10
Narrows, The (str.) D11
Nevis (isl.) 9,300 D11
Nevis (peak) D11
North Friars (bay) D10
Pinney's (beach) D11
Saint Christopher (Saint Kitts) (isl.) 35,104 D10
South Friars (bay) C10

SAINT LUCIA

CITIES and TOWNS

Anse la Raye •5,007 F6
Canaries •2,075 G6
Castries (cap.) •42,770 G6
Choc G5
Choiseul •6,382 F7
Dauphin G5
Dennery •9,654 G6
Gros Islet •10,329 G5
Laborie •6,944 G7
Marigot G6
Marquis G6
Micoud •12,264 G6
Praslin G6
Soufrière •7,456 F6
Vieux Fort •10,675 G7

OTHER FEATURES

Beaumont (pt.) F6
Canaries, Piton (mt.) G6
Cannelles (pt.) G7
Cannelles (riv.) G6
Cap (pt.) G5
Choc (bay) G5
Fond d'Or (bay) G6
Gimie (mt.) G6
Grand Caille (pt.) F6
Grand Cul de Sac (riv.) G6
Gros Islet (bay) G5
Gros Piton (mt.) G6
La Sorcière (mt.) G6
Maria (isls.) G7
Ministre (pt.) G7
Moule-à-Chique (cape) G7
Petit Piton (mt.) G6
Pigeon (isl.) G5
Port Castries (harb.) G6
Port Praslin (bay) G6
Roseau (riv.) G6
Saint Lucia (chan.) G5
Saint Vincent (chan.) G7
Savannes (bay) G7
Sorcière, La (mt.) G6
Soufrière (bay) F6
Vierge (pt.) G6

SAINT VINCENT and THE GRENADINES

CITIES and TOWNS

Barrouallie 1,298 A9
Calliaqua 627 A9
Camden Park A9
Colonarie A9
Georgetown 1,100 A8
Kingstown (cap.) 17,117 A9
Kingstown *23,330 A9
Layou 1,147 A9
Wallibu A8

OTHER FEATURES

Colonarie (pt.) A9
Cumberland (bay) A8
Dark (head) A8
De Volet (pt.) A8
Espagnol (pt.) A8
Greathead (bay) A9
Kingstown (bay) A9
Owia (bay) A8
Porter (pt.) A8
Richmond (peak) A8
Saint Andrew (mt.) A9
Saint Vincent (passg.) A8
Soufrière (mt.) A8
Yambou (head) A9

TRINIDAD and TOBAGO

CITIES and TOWNS

Arima 11,390 B10
Arouca B10
Basse Terre B11
Biche B10
Blanchisseuse B10
California A11
Carapichaima B10
Caroni B10
Cedros A11
Chaguanas 6,122 B10
Chaguaramas A10
Couva 3,635 B10
Cunapo B10
Ecclesville B11
Flanagin Town B10
Fullarton A11
Fyzabad 1,564 A11
Grande Rivière B10
Guaico B10
Guayaguayare B11
La Brea 1,487 A11
Marabella 18,158 A11
Matelot B10
Matura B10
Mayaro 2,638 B11
Moruga B11
Mucurapo A10
Palo Seco A11
Peñal 3,606 B11
Point Fortin 6,538 A11
Port-of-Spain (cap.) 67,978 A10
Princes Town 8,288 B11
Redhead B10
Rio Claro 2,423 B11
Saint Joseph 4,132 B10
Saint Joseph B11
San Fernando 33,490 A11
San Francique A11
Sangre Grande 8,948 B1
San Juan A10
Sans Souci B10
Siparia 5,773 B11
Tabaquite 2,309 B11
Tacarigua B10
Talparo B10
Toco 1,287 B10
Tunapuna 10,251 B10
Upper Manzanilla B1
Valencia B10
Waterloo A10

OTHER FEATURES

Aripo, El Cerro del (mt.) B10
Boca Grande (passg.) A10
Chacachacare (isl.) A10
Chupara (pt.) B1
Cocos (bay) B1
Dragons Mouth (str.) A1
El Tucuche (mt.) B1
Erin (bay) A1
Galeota (pt.) B1
Galera (pt.) C1
Guapo (bay) A1
Guatuaro (pt.) B1
Icacos (pt.) A1
Maracas (bay) C1
Pitch (lake) A1

VIRGIN ISLANDS (Br.)

CITIES and TOWNS

Road Town (cap.) 2,200 D
West End C

OTHER FEATURES

Flanagan (passg.) D
Frenchman (cay) C
Great Thatch (isl.) C
Great Tobago (isl.) B
Jost Van Dyke (isl.) 135 C
Little Tobago (isl.) B
Narrows, The (str.) C
Norman (isl.) D
Peter (isl.) D
Road (bay) D
Sage (mt.) D
Sir Francis Drake (chan.) D
Tortola (isl.) 9,257 D

VIRGIN ISLANDS (U.S.)

CITIES and TOWNS

Bethlehem E
Canebay E
Charlotte Amalie (cap.) 11,842 B
Christiansted 2,914 F
Cruz Bay 1,928 C
Diamond F
Eastend D
Emmaus C
Fredensdal F
Frederiksted 1,046 E
Grove Place 3,599 E
Kingshill F
Longford F
Negro Bay E

OTHER FEATURES

Altona (lag.) F
Annaly (bay) E
Baron Bluff (prom.) E
Bordeaux (mt.) C
Brass (isls.) A
Buck (isl.) G
Buck Island (chan.) F
Buck Island Reef Nat'l Mon. G
Butler (bay) E
Caneel (bay) B
Capella (isls.) B
Christiansted Nat'l Hist. Site F
Coral (bay) C
Crown (mt.) A
Dutch Cap (cay) A
Eagle (mt.) E
East (pt.) G
Flanagan (passg.) D
Flat (cays) A
Grass (pt.) G
Great (pond)
Great Pond (bay)
Green (cay)
Hams Bluff (prom.) E
Hans Lollik (isls.) E
Hassel (isl.) E
Jersey (bay) E
Krause Lagoon (chan.)
Leeward (passg.) E
Long (pt.) E
Long (pt.) E
Lovango (cay) C
Magens (bay) E
Maho (bay) C
Narrows, The (str.) C
Nulliberg (mt.)
Perseverance (bay) A
Picara (pt.)
Pillsbury (sound)
Privateer (pt.)
Pull (pt.)
Ram (head) C
Red (pt.)
Reef (bay) C
Saba (isl.) A
Saint Croix (isl.) 49,725
Saint James (isls.)
Saint John (isl.) 2,472
Saint Thomas (harb.)
Saint Thomas (isl.) 44,372 A
Salt (cay) A
Salt (riv.)
Salt River (bay)
Sandy (pt.)
Savana (isl.)
Southwest (cape)
Tague (bay) C
Thatch (cay)
Turner Hole (bay)
U.S. Nav. Air Sta.
Virgin (str.)
Virgin Isls. Nat'l Park
Water (isl.)
Westend Saltpond (lag.)

*City and suburbs.
• Population of district.
†Population of met. area.
○ Population of municipality.

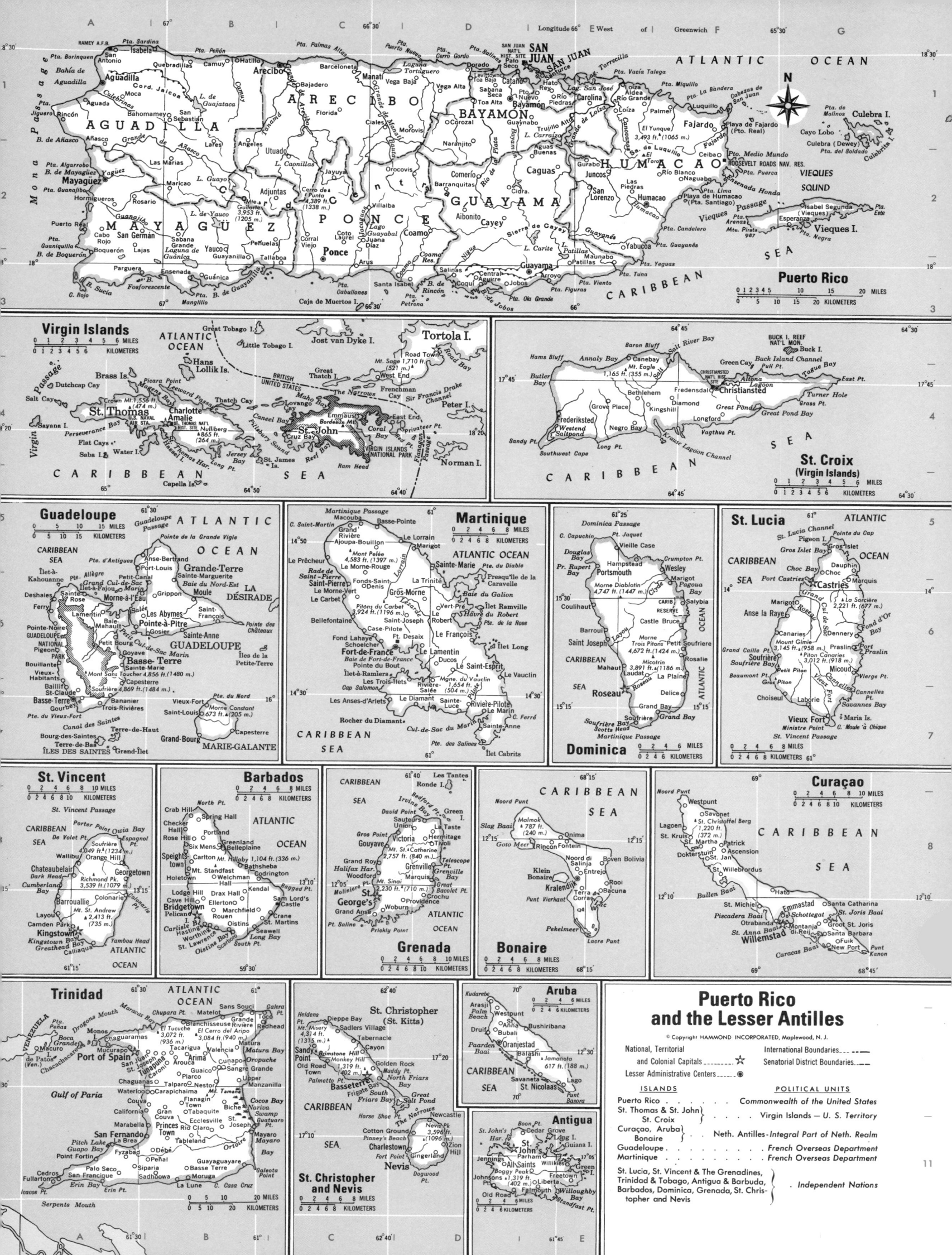

Puerto Rico and the Lesser Antilles

© Copyright HAMMOND INCORPORATED, Maplewood, N.J.

National, Territorial
and Colonial Capitals .... ☆ International Boundaries ......

Lesser Administrative Centers ..... ⊙ Senatorial District Boundaries ...

ISLANDS   **POLITICAL UNITS**

Puerto Rico . . . . . . Commonwealth of the United States

St. Thomas & St. John } . . . Virgin Islands – U. S. Territory
St. Croix

Curaçao, Aruba } . . Neth. Antilles-Integral Part of Neth. Realm
Bonaire

Guadeloupe . . . . French Overseas Department

Martinique . . . . French Overseas Department

St. Lucia, St. Vincent & The Grenadines,
Trinidad & Tobago, Antigua & Barbuda, } . . Independent Nations
Barbados, Dominica, Grenada, St. Chris-
topher and Nevis

Canada

CONIC PROJECTION

SCALE OF MILES
0   50   100   200   300

SCALE OF KILOMETERS
0 50 100 200   300   400   500

Capitals of Countries .... ☆
Provincial & Territorial Capitals ...... △
Administrative Centers .... ◉
International Boundaries ....
Provincial Boundaries ....
Regional Boundaries ....

Scale 1:19,600,000

© Copyright HAMMOND INCORPORATED, Maplewood, N.J.

Abitibi (lake), Ont. .... H 6
Aklavik, N.W.T. 721 .... C 2
Albany (riv.), Ont. .... H 5
Alberta (prov.) 2,237,724 ..... E 5
Amherst, N.S. 9,684 .... K 6
Amos, Que. 9,421 .... J 6
Anticosti (isl.), Que. .... K 6
Athabasca (lake) .... F 4
Athabasca (riv.), Alta. .... E 4
Axel Heiburg (isl.), N.W.T. .... N 3
Baffin (reg.), N.W.T. 8,300 ... G 1
Baffin (bay), N.W.T. .... J 1
Baffin (isl.), N.W.T. .... J 1
Baker Lake, N.W.T. 954 ...... G 3
Banff Nat'l Park, Alta. .... E 5
Banks (isl.), N.W.T. .... D 1
Bathurst, N. Br. 15,705 .... K 6
Belle Isle (str.), Newf. .... L 5
Bonavista, Newf. 4,460 .... L 6
Boothia (pen.), N.W.T. .... G 1
Brandon, Man. 36,242 .... F 6
British Columbia (prov.)
    2,744,467 .... D 4
Cabot (str.) .... K 6
Calgary, Alta. 592,743 .... E 5
Cambridge Bay, N.W.T. 815 ... F 2
Campbellton, N. Br. 9,818 .... K 6
Camrose, Alta. 12,570 .... E 5
Cape Breton (isl.), N.S. .... L 6
Cartwright, Newf. 658 .... L 5
Channel-Port aux Basques,
    Newf. 5,988 .... L 6
Charlottetown (cap.), P.E.I.
    15,282 .... K 6
Chatham, N. Br. 6,779 .... K 6

Chesterfield Inlet, N.W.T. 249 . G 3
Chibougamau, Que. 10,732 ... J 6
Chicoutimi, Que. 60,064 ...... J 6
Chidley (cape), Newf. .... K 3
Chilliwack, Br. Col. ◉40,642 .. D 6
Churchill, Man. 1,186 .... G 4
Coast (mts.) .... C 4
Coppermine, N.W.T. 809 ..... E 2
Corner Brook, Newf. 24,339 .. K 6
Cornwall, Ont. 46,144 .... J 7
Cranbrook, Br. Col. 15,915 ... E 6
Cree (lake), Sask. .... F 4
Dartmouth, N.S. 62,277 ...... K 7
Davis (str.), N.W.T. .... K 1
Dawson, Yukon 697 .... C 3
Devon (isl.), N.W.T. .... M 3
Drumheller, Alta. 6,508 ...... E 5
Edmonton (cap.), Alta.
    532,246 .... E 5
Edmundston, N. Br. 12,044 ... K 6
Ellesmere (isl.), N.W.T. .... M 3
Eskimo Point, N.W.T. 1,022 .. G 3
Estevan, Sask. 9,174 .... F 6
Finlay (riv.), Br. Col. .... D 4
Flin Flon, Man.-Sask. 8,261 .. F 4
Fogo (isl.), Newf. .... L 6
Fort-Chimo, Que. .... K 4
Fort Frances, Ont. 8,906 ..... G 6
Fort Franklin, N.W.T. 521 .... D 3
Fort-George, Que. 2,222 ...... J 5
Fort McMurray, Alta. 31,000 . E 4
Fort McPherson, N.W.T. 632 .. C 2
Fort Nelson, Br. Col. 3,724 ... D 4
Fort Providence, N.W.T. 605 .. E 3

Fort Saskatchewan, Alta.
    12,169 .... E 5
Fort Simpson, N.W.T. 980 .... D 3
Fort Smith (reg.), N.W.T.
    22,384 .... E 3
Fort Smith, N.W.T. 2,298 .... E 3
Foxe (basin), N.W.T. .... H 2
Franklin (dist.), N.W.T. .... H 1
Fraser (riv.), Br. Col. .... D 5
Fredericton, N. Br. 43,723 ... K 6
Frobisher Bay, N.W.T. 2,333 . K 3
Fundy (bay) .... K 6
Gander, Newf. 10,404 .... L 6
Gaspé, Que. 17,261 .... K 6
Georgian (bay), Ont. .... H 6
Geraldton, Ont. 2,956 .... H 6
Glace Bay, N.S. 21,466 ...... L 6
Goose Bay, Newf. 7,103 ...... K 5
Gouin (res.), Que. .... J 6
Grand Falls, Newf. 8,765 ..... L 6
Grande Prairie, Alta. 24,263 .. E 4
Great Bear (lake), N.W.T. .... D 2
Great Slave (lake), N.W.T. ... E 3
Guelph, Ont. 71,207 .... H 7
Halifax (cap.), N.S. 114,594 .. K 7
Hay River, N.W.T. 2,863 ..... E 3
Hearst, Ont. 5,533 .... H 6
Hecate (str.), Br. Col. .... C 5
Hull, Que. 56,225 .... J 6
Humboldt, Sask. 4,705 .... F 5
Iroquois Falls, Ont. 6,339 .... H 6

Jasper Nat'l Park, Alta. .... E 5
Jonquière, Que. 60,354 ...... J 6
Juan de Fuca (str.), Br. Col. .. D 6
Kamloops, Br. Col. 64,048 .... D 5
Kane (basin), N.W.T. .... N 3
Kapuskasing, Ont. 12,014 .... H 6
Keewatin (reg.), N.W.T. 4,327 . G 2
Kelowna, Br. Col. 59,196 ..... D 6
Kenora, Ont. 9,817 .... G 5
Kingston, Ont. 52,616 .... J 7
Kirkland Lake, Ont. 12,219 ... H 6
Kitikmeot (reg.), N.W.T. 3,245 . F 1
Kitimat, Br. Col. 12,462 ...... D 5
Kluane (lake), Yukon .... C 3
Kootenay (lake), Br. Col. ..... E 6
Labrador (reg.), Newf. .... K 4
Lacombe, Alta. 5,591 .... E 5
Lake Harbour, N.W.T. 252 ... J 3
Lake Louise, Alta. 355 .... E 5
Lancaster (sound), N.W.T. .... H 1
Leduc, Alta. 12,471 .... E 5
Lesser Slave (lake), Alta. ..... E 4
Lethbridge, Alta. 54,072 ..... E 6
Liard (riv.) .... D 3
Lloydminster, Alta.-Sask.
    15,031 .... E 5
Logan (mt.), Yukon .... B 3
London, Ont. 254,280 .... H 7
Lunenburg, N.S. 3,014 .... K 7
Mackenzie (dist.), N.W.T. ..... C 2
Mackenzie (riv.), N.W.T. ..... C 2
Magdalen (isls.), Que. .... L 6
Manicouagan (riv.), Que. ..... K 5
Manitoba (lake), Man. .... G 5
Manitoba (prov.) 1,026,241 .. G 5

Manitoulin (isl.), Ont. .... H 6
Maple Creek, Sask. 2,470 .... F 6
Marathon, Ont. 2,271 .... H 6
Mayo, Yukon 398 .... C 3
M'Clintock (chan.), N.W.T. ... F 1
Medicine Hat, Alta. 40,380 ... E 5
Melville, Sask. 5,092 .... F 5
Melville (isl.), N.W.T. .... E 1
Merritt, Br. Col. 6,110 .... D 5
Minto (lake), Que. .... J 4
Mistassibi (riv.), Que. .... J 5
Mistassini (lake), Que. .... J 5
Moncton, N. Br. 54,743 ...... K 6
Mont-Joli, Que. 6,359 .... K 6
Mont-Laurier, Que. 8,405 .... J 6
Montréal, Que. 980,354 ...... J 7
Moose Jaw, Sask. 33,941 .... F 6
Moosomin, Sask. 2,579 .... F 5
Moosonee, Ont. 1,433 .... H 5
Morden, Man. 4,579 .... G 6
Nain, Newf. 938 .... K 4
Nanaimo, Br. Col. 47,069 .... D 6
Nares (str.), N.W.T. .... M 3
Nelson, Br. Col. 9,143 .... E 6
Nelson (riv.), Man. .... G 4
New Brunswick (prov.)
    696,423 .... K 6
Newfoundland (isl.) .... L 6
Newfoundland (prov.) 567,681 L 5
New Westminster, Br. Col.
    38,550 .... D 6
Niagara Falls, Ont. 70,960 ... J 7
Norman Wells, N.W.T. 420 ... D 2
North Battleford, Sask.

North Bay, Ont. 51,268 .... J 6
North Magnetic Pole .... F 1
North Saskatchewan (riv.) .... E 5
N. Vancouver, Br. Col. 33,952 . D 6
Northwest Territories 45,741 .. E 2
Nova Scotia (prov.) 847,422 .. K 7
Okanagan (lake), Br. Col. ..... D 6
Ontario (prov.) 8,625,107 .... H 5
Ottawa (cap.), Canada
    295,163 .... J 6
Ottawa (riv.) .... J 6
Owen Sound, Ont. 19,883 .... H 7
Pangnirtung, N.W.T. 839 ..... K 2
Parry (chan.), N.W.T. .... E-H 1
Parry Sound, Ont. 6,124 ..... J 6
Peace (riv.), Alta. .... E 4
Peace River, Alta. 5,907 ..... E 4
Peel (riv.) .... C 2
Pelly (riv.), Yukon .... C 3
Pembroke, Ont. 14,026 .... J 6
Péribonca (riv.), Que. .... J 5
Peterborough, Ont. 60,620 ... J 7
Pond Inlet, N.W.T. 705 ...... J 1
Portage la Prairie, Man.
    13,086 .... G 5
Port Radium, N.W.T. 56 ...... E 2
Poste-de-la-Baleine, Que.
    435 .... J 4
Povungnituk, Que. .... J 3
Prince Albert, Sask. 31,380 .. F 5
Prince Albert Nat'l Park, Sask. F 5
Prince Edward Island (prov.)
    122,506 .... K 6
Prince George, Br. Col.
    67,559 .... D 5

Prince Patrick (isl.), N.W.T. ... M 3
Prince Rupert, Br. Col. 16,197 . C 5
Québec (prov.) 6,438,403 .... J 5
Québec (cap.), Que. 166,474 . J 6
Queen Charlotte (isls.), Br.
    Col. .... C 5
Queen Elizabeth (isls.),
    N.W.T. .... M 3
Quesnel, Br. Col. 8,240 ...... D 5
Race (cape), Newf. .... L 6
Rainy (lake), Ont. .... G 6
Rainy River, Ont. 1,061 ...... G 6
Rankin Inlet, N.W.T. 1,109 ... G 3
Ray (cape), Newf. .... K 6
Red Deer, Alta. 46,393 .... E 5
Regina (cap.), Sask. 162,613 . F 5
Reindeer (lake) .... F 4
Revelstoke, Br. Col. 5,544 ... E 5
Riding Mountain Nat'l Park,
    Man. .... F 5
Rimouski, Que. 29,120 .... K 6
Rivière-du-Loup, Que. 13,459 . K 6
Roberval, Que. 11,429 .... J 6
Robson (mt.), Br. Col. .... D 5
Rocky (mts.) .... D 4
Rocky Mountain House, Alta.
    4,698 .... E 5
Rouyn, Que. 17,224 .... J 6
Sable (cape), N.S. .... K 7
Sable (isl.), N.S. .... L 7
Saint Elias (mt.), Yukon ...... B 3
Saint John, N. Br. 80,521 .... K 6
St. John's (cap.), Newf.
    83,770 .... L 6
Saint Lawrence (riv.) .... L 6

AREA 3,851,787 sq. mi. (9,976,139 sq. km.)
POPULATION 24,343,181
CAPITAL Ottawa
LARGEST CITY Montréal
HIGHEST POINT Mt. Logan 19,524 ft. (5,951 m.)
MONETARY UNIT Canadian dollar
MAJOR LANGUAGES English, French
MAJOR RELIGIONS Protestantism, Roman Catholicism

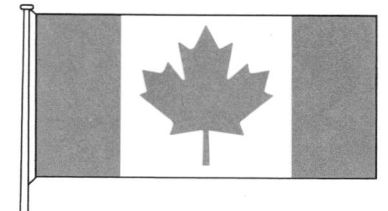

Population Distribution

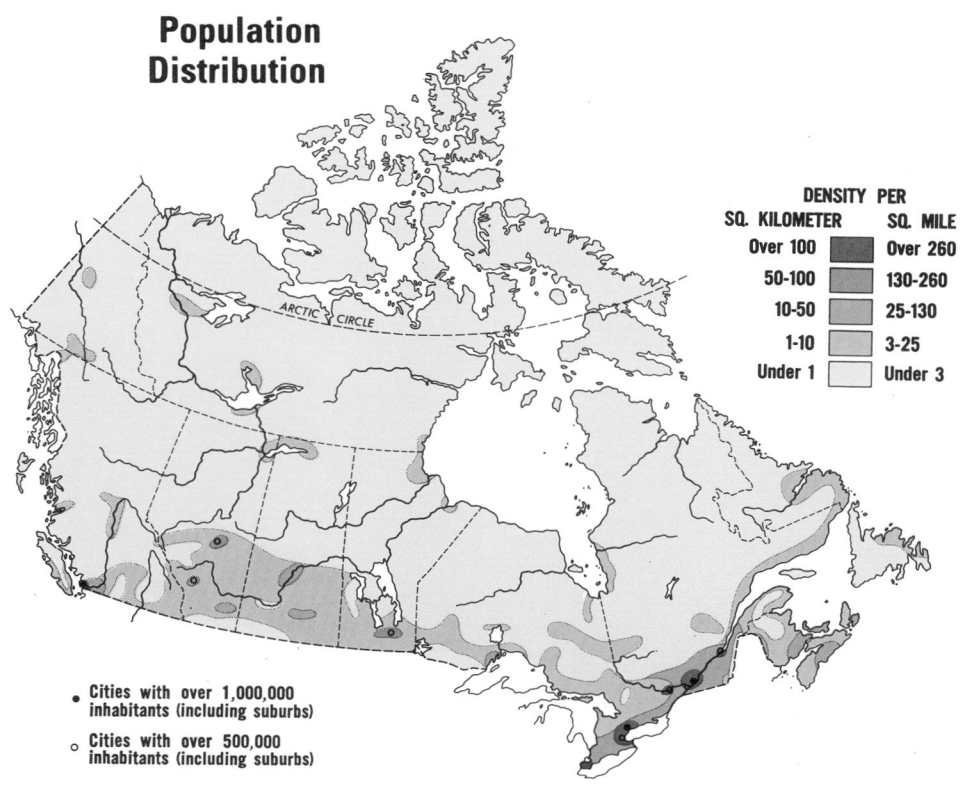

DENSITY PER

SQ. KILOMETER	SQ. MILE
Over 100	Over 260
50-100	130-260
10-50	25-130
1-10	3-25
Under 1	Under 3

• Cities with over 1,000,000 inhabitants (including suburbs)

○ Cities with over 500,000 inhabitants (including suburbs)

Vegetation

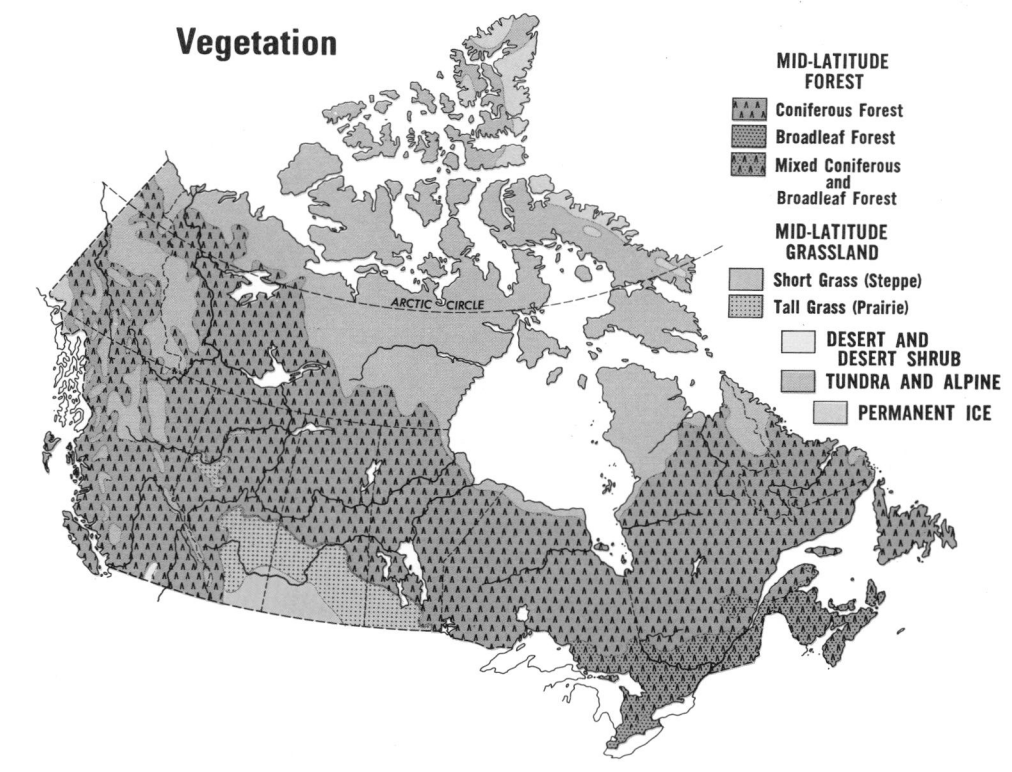

MID-LATITUDE FOREST
- Coniferous Forest
- Broadleaf Forest
- Mixed Coniferous and Broadleaf Forest

MID-LATITUDE GRASSLAND
- Short Grass (Steppe)
- Tall Grass (Prairie)

- DESERT AND DESERT SHRUB
- TUNDRA AND ALPINE
- PERMANENT ICE

CAN

Saint Pierre & Miquelon (isls.) 6,041 .... L 6
Sarnia, Ont. 50,892 .... H 7
Saskatchewan (prov.) 968,313 F 5
Saskatchewan (riv.). .... F 5
Saskatoon, Sask. 154,210 ... F 5
Sault Sainte Marie, Ont. 82,697 .... H 6
Schefferville, Que. 1,997 ... K 5
Selkirk, Man. 10,037 .... G 5
Sept-Îles (Seven Is.), Que. 29,262 .... K 5
Shawinigan, Que. 23,011 ... J 6
Sherbrooke, Que. 74,075 ... J 7
Sioux Lookout, Ont. 3,074... G 5
Skeena (riv.), Br. Col. .... D 5
Slave (riv.) .... E 3
Smallwood (res.), Newf. .... K 5
Southampton (isl.), N.W.T. .. H 2
Stettler, Alta. 5,136 .... E 5
Stewart (riv.), Yukon .... C 3
Stikine (riv.), Br. Col. .... C 4
Sudbury, Ont. 91,829 .... H 6
Swift Current, Sask. 14,747 . F 5
Sydney, N.S. 29,444. .... K 6
Terrace, Br. Col. ○10,914. ... D 5
The Pas, Man. 6,390 .... F 5
Thompson, Man. 14,288 .... G 4
Thunder Bay, Ont. 112,486 . H 6
Timmins, Ont. 46,114. .... H 6
Toronto (cap.), Ont. 599,217 . H 7
Trail, Br. Col. 9,599 .... E 6
Trois-Rivières, Que. 50,466.. J 6
Truro, N.S. 12,552. .... K 6
Tuktoyaktuk, N.W.T. 772. .... C 2

Val-d'Or, Que. 21,371 .... J 6
Vancouver, Br. Col. 414,281 .. D 6
Vancouver (isl.), Br. Col. .... D 6
Vanderhoof, Br. Col. 2,323. ... D 5
Végreville, Alta. 5,251 .... E 5
Vernon, Br. Col. 19,987 ...... E 5
Victoria (cap.), Br. Col. 64,379 D 6
Victoria (isl.), N.W.T. .... E 1
Wabush, Newf. 3,155 .... K 5
Waterton-Glacier International Peace Park, Alta. .... E 6
Wetaskiwin, Alta. 9,597 ..... E 5
Weyburn, Sask. 9,523 .... F 6
Whitehorse (cap.), Yukon 14,814 .... C 3
Williams Lake, Br. Col. 8,362 . D 5
Williston (lake), Br. Col. ..... D 4
Windsor, N.S. 3,646 .... K 7
Windsor, Ont. 192,083 .... H 7
Winnipeg (cap.), Man. 564,473. .... G 6
Winnipeg (lake), Man. .... G 5
Winnipegosis (lake), Man. ... F 5
Wood Buffalo Nat'l Park, Alta.. E 4
Woods (lake) .... G 6
Wrigley, N.W.T. 137. .... D 3
Yarmouth, N.S. 7,475. .... K 7
Yellowknife (cap.), N.W.T. 9,483 .... E 3
Yoho Nat'l Park, Br. Col. .... E 5
Yorkton, Sask. 15,339 .... F 5
Yukon Territory 23,153. ...... C 3

○Population of municipality.

Average January Temperature

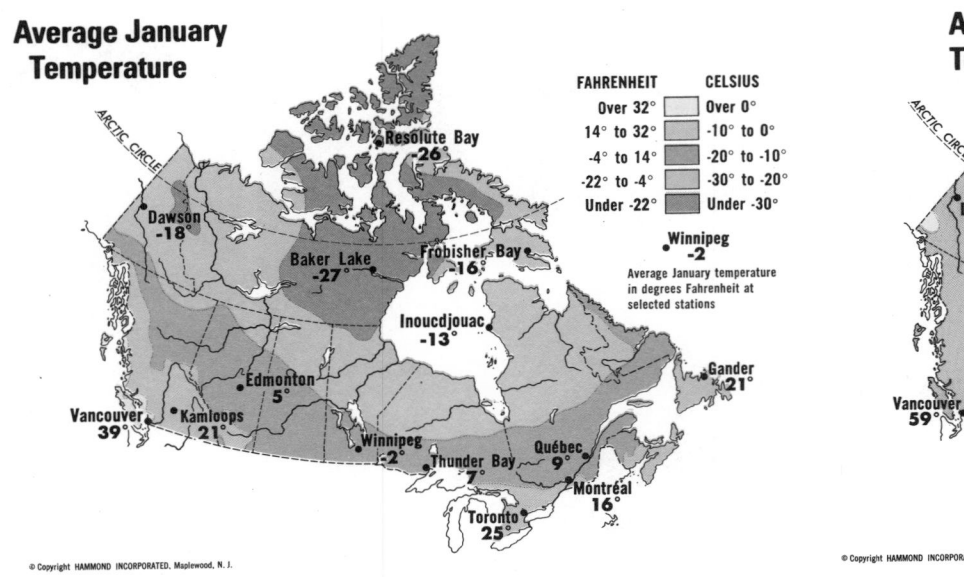

FAHRENHEIT	CELSIUS
Over 32°	Over 0°
14° to 32°	-10° to 0°
-4° to 14°	-20° to -10°
-22° to -4°	-30° to -20°
Under -22°	Under -30°

Winnipeg
-2

Average January temperature in degrees Fahrenheit at selected stations

Resolute Bay -26

Dawson -18

Baker Lake -27°

Frobisher Bay -16

Inoucdjouac -13°

Edmonton 5°

Gander 21°

Vancouver 39°

Kamloops 21°

Winnipeg -2°

Thunder Bay 7°

Québec 9°

Montréal 16°

Toronto 25°

© Copyright HAMMOND INCORPORATED, Maplewood, N. J.

Average July Temperature

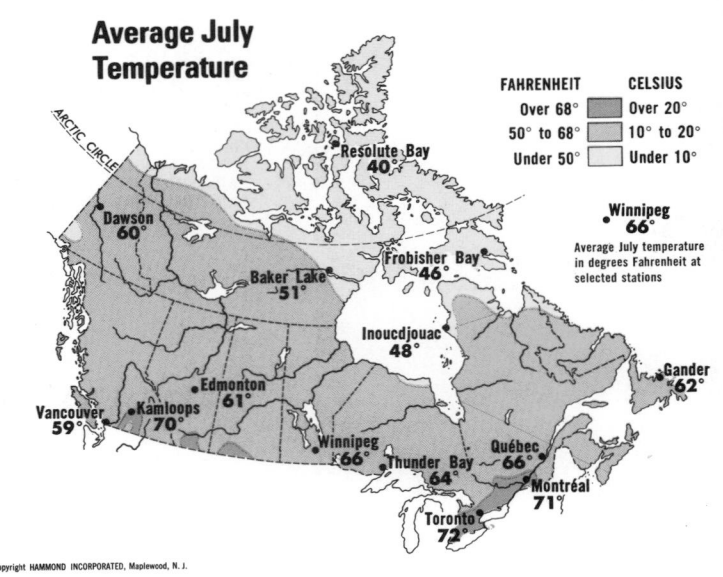

FAHRENHEIT	CELSIUS
Over 68°	Over 20°
50° to 68°	10° to 20°
Under 50°	Under 10°

Winnipeg
66

Average July temperature in degrees Fahrenheit at selected stations

Resolute Bay 40

Dawson 60°

Baker Lake 51°

Frobisher Bay 46°

Inoucdjouac 48°

Edmonton 61°

Gander 62°

Vancouver 59°

Kamloops 70°

Winnipeg 66°

Thunder Bay 64°

Québec 66°

Montréal 71°

Toronto 72°

© Copyright HAMMOND INCORPORATED, Maplewood, N. J.

Agriculture, Industry and Resources

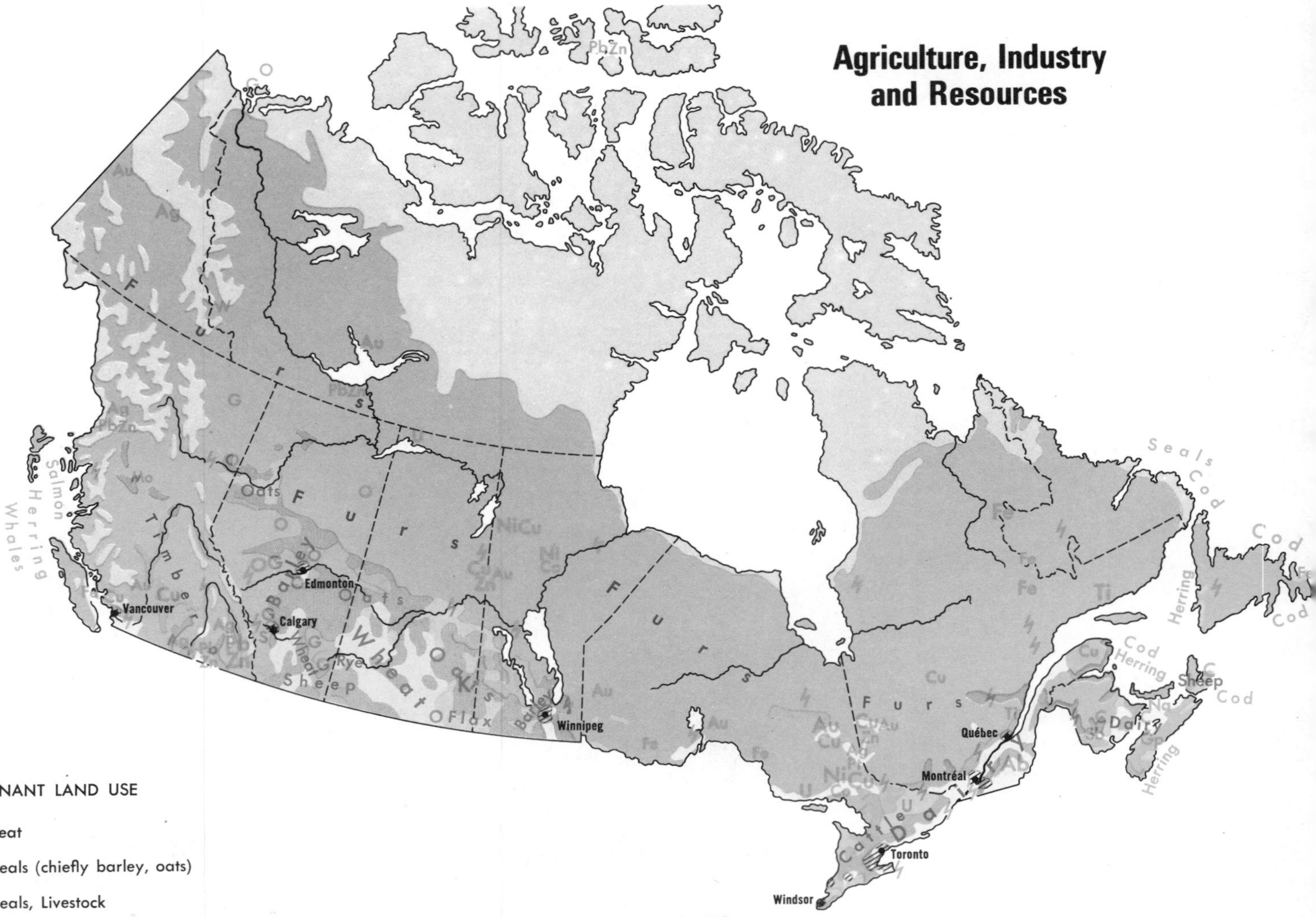

Vancouver

Calgary

Edmonton

Winnipeg

Québec

Montréal

Toronto

Windsor

DOMINANT LAND USE

- Wheat
- Cereals (chiefly barley, oats)
- Cereals, Livestock
- General Farming, Livestock
- Dairy
- Fruit, Vegetables
- Pasture Livestock
- Range Livestock
- Forests
- Nonagricultural Land

MAJOR MINERAL OCCURRENCES

Ab	Asbestos	Fe	Iron Ore	Ni	Nickel	Sb	Antimony
Ag	Silver	G	Natural Gas	O	Petroleum	Ti	Titanium
Au	Gold	Gp	Gypsum	Pb	Lead	U	Uranium
C	Coal	K	Potash	Pt	Platinum	W	Tungsten
Co	Cobalt	Mo	Molybdenum	S	Sulfur	Zn	Zinc
Cu	Copper	Na	Salt				

⚡ Water Power

▨ Major Industrial Areas

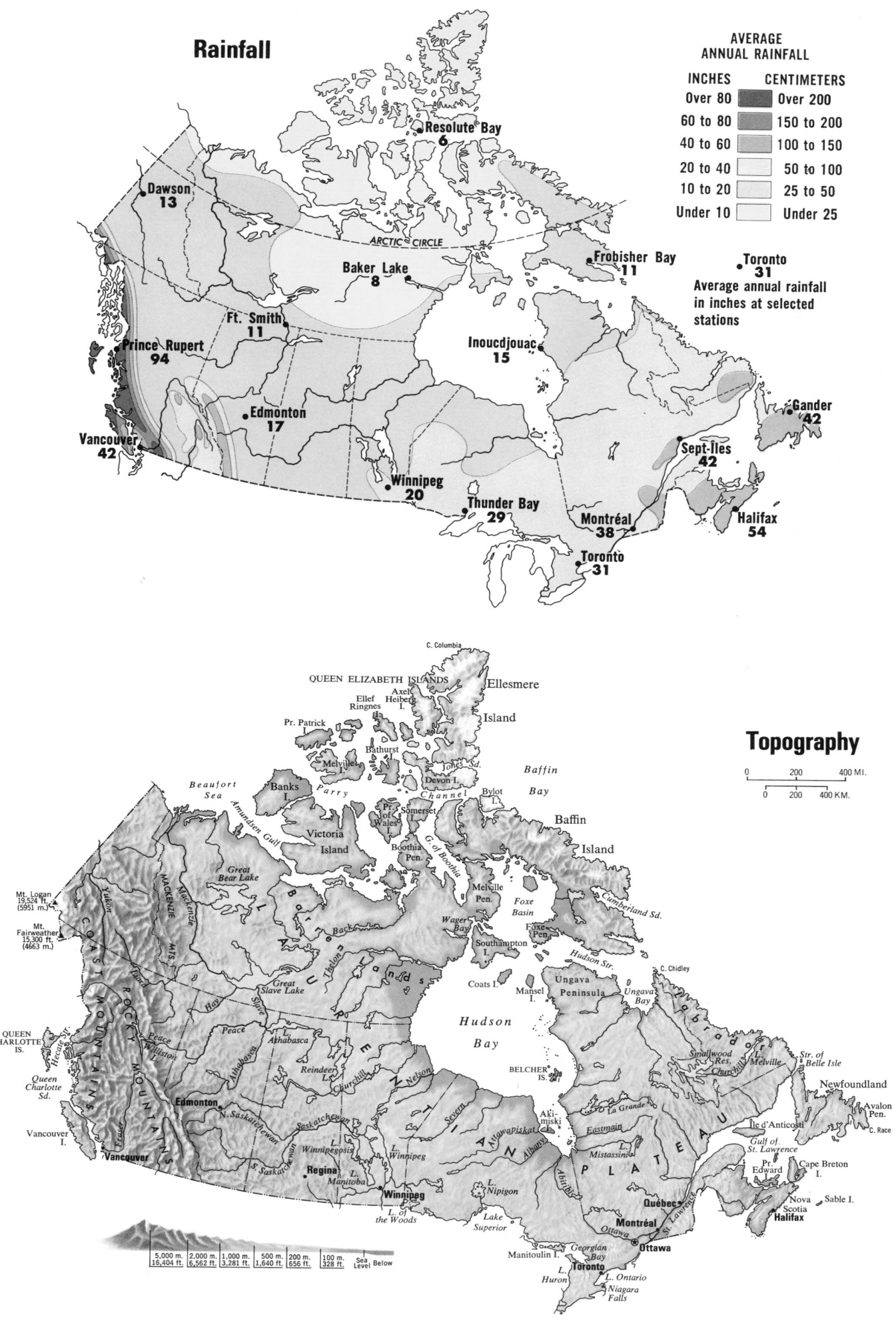

Rainfall

AVERAGE ANNUAL RAINFALL

INCHES	CENTIMETERS
Over 80	Over 200
60 to 80	150 to 200
40 to 60	100 to 150
20 to 40	50 to 100
10 to 20	25 to 50
Under 10	Under 25

ARCTIC CIRCLE

Resolute Bay
6

Dawson
13

Baker Lake
8

Frobisher Bay
11

Toronto
31

Average annual rainfall in inches at selected stations

Ft. Smith
11

Prince Rupert
94

Inoucdjouac
15

Gander
42

Edmonton
17

Vancouver
42

Sept-Îles
42

Winnipeg
20

Thunder Bay
29

Halifax
54

Montréal
38

Toronto
31

Topography

C. Columbia

QUEEN ELIZABETH ISLANDS   Ellesmere

Ellef   Axel
Ringnes   Heiberg   Island
Pr. Patrick   I.

Bathurst

Melville   Jones Sd.   Baffin
Devon I.   Bay

Beaufort   Banks   Parry   Channel   Bylot
Sea   I.   I.

Amundsen Gulf   Pr.   Somerset   Baffin
of   I.
Wales

Victoria   Boothia   Island
Island   Pen.

Great Bear Lake   G. of Boothia

Mt. Logan   Melville   Cumberland Sd.
19,524 ft.   MACKENZIE   Pen.
(5951 m.)   Back

Mt. Fairweather   Foxe   Foxe
15,300 ft.   Wager   Basin   Pen.
(4663 m.)   Bay

Thelon   Southampton   Hudson Str.
I.   C. Chidley

Great   Coats I.   Ungava
Slave Lake   Mansel   Peninsula   Ungava
I.   Bay

QUEEN   Peace   Hudson
CHARLOTTE   Williston   Bay
IS.   Peace

Athabasca   BELCHER   Smallwood   L.
IS.   Res.   Melville   Str. of
Queen   Reindeer   Churchill   Belle Isle
Charlotte   Akimiski   Newfoundland
Sd.   Churchill   La Grande
Nelson   Avalon
Pen.

Vancouver   Edmonton   Eastmain   Île d'Anticosti   C. Race
I.   N. Saskatchewan   Saskatchewan   Albany   L.
Vancouver   Athabasca   Seven   Mistassini   Gulf of
St. Lawrence
Winnipegosis   L.   Abitibi   PLATEAU   Pr.
Winnipeg   Attawapiskat   Edward   Cape Breton
Regina   L.   I.
Manitoba   L.   Québec   Nova   Sable I.
Nipigon   Scotia
Winnipeg   Montréal   Halifax
L. of   Lake   St. Lawrence
the Woods   Superior   Ottawa
Manitoulin I.   Georgian   Ottawa
Bay
Toronto
L. Ontario
L.   Niagara
Huron   Falls

Hecate

ROCKY MOUNTAINS   COAST MOUNTAINS   MTS.   Yukon   Liard   Mackenzie   Great Slave   Hay   Slave   Fraser   Peace

LAURENTIAN

| 5,000 m. | 2,000 m. | 1,000 m. | 500 m. | 200 m. | 100 m. | Sea | Below |
| 16,404 ft. | 6,562 ft. | 3,281 ft. | 1,640 ft. | 656 ft. | 328 ft. | Level | |

| 0 | 200 | 400 MI. |
| 0 | 200 | 400 KM. |

Newfoundland including Labrador

SCALE

0   25   50   100   150 MI.

0   25   50   100   150 KM.

Capitals of Provinces .... ⊕

Provincial Boundaries .... —··—··—

Provincial Boundary according to
Imperial Privy Council decision, 1927 .... — — —

Scale 1:5,200,000

© Copyright HAMMOND INCORPORATED, Maplewood, N.J.

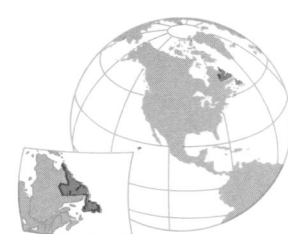

NEWFOUNDLAND

CITIES and TOWNS

Admiral's Beach 362 . . . . D2
Admiral's Cove 99 . . . . D2
Anchor Point 368 . . . . C3
Aquaforte 200 . . . . D2
Argentia 93 . . . . C2
Arnold's Cove 1,124 . . . . C2
Avondale 890 . . . . D2
Badger 1,090 . . . . C4
Badger's Quay-Valleyfield-
Pool's Island 1,566 . . . . . . D4
Baie Verte 2,491 . . . . C4
Battle Harbour C3
Bauline 423 . . . . D2
Bay Bulls 1,081 . . . . D2
Bay de Verde 786 . . . . D2
Bay L'Argent 483 . . . . D4
Bay Roberts 4,512 . . . . D2
Bellburns 147 . . . . C3
Belleoram 565 . . . . C4
Bellevue 286 . . . . D2
Bide Arm 339 . . . . C3
Big Pond 167 . . . . D2
Birchy Bay 707 . . . . D4
Bird Cove 400 . . . . C3
Bishop's Falls 4,395 . . . . C4
Black Tickle 194 . . . . C3
Blackhead Road 1,855 . . . . . D2
Blaketown 617 . . . . D2
Bloomfield 715 . . . . D2
Bonavista 4,460 . . . . D2
Botwood 4,074 . . . . C4
Branch 462 . . . . D2
Brigus 898 . . . . D2
Broad Cove 198 . . . . D2
Brooklyn 197 . . . . D2
Brownsdale 199 . . . . D2
Buchans 1,655 . . . . C4
Bunyan's Cove 590 . . . . C2
Burgeo 2,504 . . . . C4
Burin 2,904 . . . . C4
Burnt Islands 991 . . . . C4
Burnt Point 260 . . . . D2
Calvert 482 . . . . D2
Campbellton 703 . . . . D4
Cape Broyle 698 . . . . D2
Cape Ray 484 . . . . C4
Caplin Cove 150 . . . . D2
Carbonear 5,335 . . . . D2
Carmanville 966 . . . . D4
Cartwright 658 . . . . C3
Catalina 1,162 . . . . D2
Cavendish 343 . . . . D2
Champney's West 141 . . . . . D2
Chance Cove 498 . . . . D2
Change Islands 580 . . . . D4
Channel-Port aux
Basques 5,988 . . . . C4
Chapel Arm 689 . . . . D2
Charlottetown 330 . . . . D2
Charlottetown 250 . . . . C3
Churchill Falls 936 . . . . B3
Clarenville 2,878 . . . . C2
Clarke's Beach 1,009 . . . . . . D2
Codroy 346 . . . . C4
Colinet 318 . . . . D2
Colliers 819 . . . . D2
Come By Chance 337 . . . . . . C2
Conception Harbour 917 . . . . D2
Conche 464 . . . . C3
Cook's Harbour 388 . . . . C3
Corner Brook 24,339 . . . . . . C4
Cow Head 695 . . . . C4
Cox's Cove 980 . . . . C4
Cupids 706 . . . . D2
Daniell's Harbour 614 . . . . . . C3
Dark Cove 1,344 . . . . D4
Davis Inlet 240 . . . . B2
Deep Bight 243 . . . . C2
Deer Lake 4,348 . . . . C4
Dildo 877 . . . . D2
Dunville 1,817 . . . . D2
Durrell 1,145 . . . . D4
Eastport 597 . . . . D1
Elliston 527 . . . . D2
Embree 846 . . . . C4
Englee 998 . . . . C3
English Harbour 118 . . . . D2
English Harbour West 327 . . C4
Fermeuse 584 . . . . D2
Ferryland 795 . . . . D2
Flat Bay 322 . . . . C4
Flat Rock 808 . . . . D2
Fleur de Lys 616 . . . . C3
Flowers Cove 459 . . . . C3
Fogo 1,105 . . . . D4
Forteau 520 . . . . C3
Fortune 2,473 . . . . C4
Fox Harbour 280 . . . . C3
Fox Harbour 538 . . . . D2
François 219 . . . . C4
Freshwater 1,276 . . . . C2
Freshwater 209 . . . . D2
Gambo 2,932 . . . . D4
Gander 10,404 . . . . D4
Garnish 761 . . . . C4
Gaskiers-Point la Haye 505 . . D2
Gaultois 558 . . . . C4
Georges Brook 356 . . . . D2
Glenwood 1,129 . . . . D4
Glovertown 2,165 . . . . C1
Goobies 185 . . . . D2
Goose Bay-Happy
Valley 7,103 . . . . B3
Gooseberry Cove 195 . . . . . . C2
Goose Cove 134 . . . . C2
Goose Cove 368 . . . . C3
Goulds 4,242 . . . . D2
Grand Bank 3,901 . . . . C4
Grand Falls 8,765 . . . . C4
Grates Cove 275 . . . . D2
Green Island Cove 222 . . . . . C3
Green's Harbour 785 . . . . . . D2
Greenspond 423 . . . . D4
Grey River 234 . . . . C4
Gull Island 362 . . . . D2
Hampden 838 . . . . C4
Hant's Harbour 542 . . . . D2
Happy Adventure 352 . . . . . . D2
Happy Valley-
Goose Bay 7,103 . . . . B3
Harbour Breton 2,464 . . . . . . C4
Harbour Deep 278 . . . . C3
Harbour Grace 2,988 . . . . . . D2
Harbour Main-Chapel
Cove-Lakeview 1,303 . . . . D2
Hare Bay 1,520 . . . . D4
Hawke's Bay 555 . . . . C3
Head of Bay d'Espoir 586 . . . C4
Heart's Content 625 . . . . D2
Heart's Delight-Islington 899 D2
Heart's Desire 416 . . . . D2
Heatherton 328 . . . . C4
Hermitage 863 . . . . C4
Hickman's Harbour 479 . . . . . D2
Hillview 295 . . . . D2
Hodge's Cove 438 . . . . D2
Holyrood 1,789 . . . . D2
Hopedale 425 . . . . B2
Howley 456 . . . . C4
Isle aux Morts 1,238 . . . . C4
Jackson's Arm 623 . . . . C4
Jeffrey's 276 . . . . C4
Jerseyside 641 . . . . B3
Job's Cove 201 . . . . D2
Joe Batt's Arm-
Barr'd Islands 1,155 . . . . . D4
Keels 129 . . . . D1
Kelligrews (Foxtrap-
Greeleytown-Peachtown-
Kelligrews) 2,292 . . . . D2
Kilbride 5,014 . . . . D2
King's Cove 253 . . . . D1
King's Point 825 . . . . C4
Kippens 1,219 . . . . C4
Labrador City 11,538 . . . . . . A3
Lamaline 548 . . . . C4
L'Anse-au-Clair 267 . . . . C3
L'Anse-au-Loup 589 . . . . C3
L'Anse au Meadow 66 . . . . . . C3
La Poile 186 . . . . C4
Lark Harbour 783 . . . . C4
La Scie 1,422 . . . . C4
Lawn 999 . . . . C4
Lethbridge 686 . . . . D2
Lewisporte 3,963 . . . . C4
Little Bay Islands 407 . . . . . . C4
Little Catalina 750 . . . . D2
Little Heart's Ease 467 . . . . . D2
Lodge Bay 124 . . . . C3
Long Harbour-Mount Arlington
Heights 660 . . . . D2
Lourdes 932 . . . . C4
Lower Island Cove 415 . . . . . D2
Lumsden 645 . . . . D4
Main Brook 514 . . . . C3
Makkovik 347 . . . . C2
Markland 344 . . . . D2
Mary's Harbour 408 . . . . C3
Marystown 6,299 . . . . D4
McCallum 243 . . . . C4
Melrose 416 . . . . D2
Middle Arm, Green Bay 575 . C4
Millertown 228 . . . . C4
Milltown-Head of Bay
d'Espoir 1,376 . . . . C4
Milton 258 . . . . C2
Mobile 171 . . . . D2
Mount Carmel-Mitchell's Brook-
St. Catherine's 699 . . . . . . D2
Mount Pearl 11,543 . . . . D2
Musgrave Harbour 1,554 . . . D4
Musgravetown 635 . . . . C2
Nain 938 . . . . B2
New Bonaventure 106 . . . . . . D2
New Chelsea 144 . . . . D2
New Harbour 777 . . . . D2
Newmans Cove 231 . . . . D2
New Perlican 350 . . . . D2
Newtown 511 . . . . D4
Nippers Harbour 259 . . . . . . C4
Norman's Cove-
Long Cove 1,152 . . . . D2
Norris Arm 1,216 . . . . C4
Norris Point 1,033 . . . . C4
North Harbour 151 . . . . D2
North River 245 . . . . D2
North West Brook 279 . . . . . C2
North West River 515 . . . . . B3
O'Donnells 280 . . . . D2
Old Bonaventure 111 . . . . . . D2
Old Perlican 709 . . . . D2
Paradise 2,861 . . . . D2
Parkers Cove 424 . . . . D4
Parson's Pond 605 . . . . C3
Pasadena 2,685 . . . . C4
Patrick's Cove 155 . . . . C2
Perry's Cove 141 . . . . D2
Peterview 1,119 . . . . C4
Petites 108 . . . . C4
Petley 147 . . . . D2
Petty Harbour-Maddox
Cove 853 . . . . D2
Picadilly 524 . . . . C4
Pinware River 201 . . . . C3
Placentia 2,204 . . . . C2
Plate Cove 474 . . . . D2
Point La Haye 195 . . . . D2
Point Lance 141 . . . . C2
Point Leamington 848 . . . . . . C4
Point Verde 296 . . . . C2
Pollards Point 502 . . . . C4
Port au Bras 366 . . . . D4
Port au Choix 1,311 . . . . C3
Port au Port 603 . . . . C4
Port Blandford 702 . . . . C2
Port Hope Simpson 581 . . . . C3
Port Kirwan 164 . . . . D2
Port Rexton 489 . . . . D2
Port Saunders 769 . . . . C3
Portugal Cove 2,361 . . . . D2
Portugal Cove South 371 . . . D2
Port Union 671 . . . . D2
Postville 223 . . . . B3
Pouch Cove 1,522 . . . . D2
Princeton 204 . . . . D2
Raleigh 373 . . . . C3
Ramea 1,386 . . . . C4
Red Bay 316 . . . . C3
Red Head Cove 225 . . . . D2
Rencontre East 230 . . . . C4
Renews-Cappahayden 578 . . D2
Rigolet 271 . . . . C3
Riverhead 431 . . . . D2
River of Ponds 304 . . . . C3
Robert's Arm 1,005 . . . . C4
Rocky Harbour 1,273 . . . . . . C4
Roddickton 1,142 . . . . C3
Rose Blanche-Harbour
le Cou 975 . . . . C4
Rushoon 520 . . . . D4
Saint Alban's 1,968 . . . . C4
Saint Andrew's 262 . . . . C4
Saint Anthony 3,107 . . . . C3
Saint Brendan's 468 . . . . D4
Saint Bride's 599 . . . . C2
Saint George's 1,756 . . . . . . C4
St. John's (cap.) 83,770 . . . . D2
Saint Joseph's 262 . . . . D2
Saint Lawrence 2,012 . . . . . . C4
Saint Lunaire-Griquet 1,010 . . C3
Saint Mary's 701 . . . . D2
Saint Paul's 454 . . . . C4
Saint Phillips 1,365 . . . . D2
Saint Shotts 239 . . . . D2
Saint Vincent's-Saint
Stephens-Peter's
River 796 . . . . D2
Sally's Cove 100 . . . . C4
Salmon Cove 786 . . . . D2
Seal Cove 751 . . . . C3
Seal Cove-White Bay 498 . . . C4
Seldom-Little Seldom 560 . . D4
Ship Harbour 265 . . . . D2
Shoal Cove 223 . . . . C3
Shoal Harbour 1,000 . . . . C2
South Branch 264 . . . . C4
South Brook, Hall's
Bay Dist. 786 . . . . C4
South Brook, Humber
Dist. 477 . . . . C4
Southern Harbour 772 . . . . . C2
South River 645 . . . . D2
Spaniard's Bay 2,125 . . . . D2
Springdale 3,501 . . . . C4
Stephenville 8,876 . . . . C4
Stephenville Crossing 2,172 . C4
Summerford 1,198 . . . . C4
Summerville 346 . . . . D2
Sunnyside 703 . . . . D2
Sweet Bay 204 . . . . D2
Swift Current 329 . . . . C2
Terrenceville 796 . . . . D4
Tilting 427 . . . . D4
Torbay 3,394 . . . . D2
Tors Cove 355 . . . . D2
Traytown 383 . . . . D1
Trepassey 1,473 . . . . D2
Trinity 522 . . . . D2
Trinity 375 . . . . D4
Trout River 759 . . . . C4
Twillingate 1,506 . . . . C4
Upper Island Cove 2,025 . . . D2
Victoria 1,870 . . . . D2
Wabana 4,254 . . . . D2
Wabush 3,155 . . . . A3
Wesleyville 1,125 . . . . D4
Western Bay 463 . . . . D2
West Saint Modeste 273 . . . . C3
Whitbourne 1,233 . . . . D2
Wild Cove 152 . . . . C3
Windsor 5,747 . . . . C4
Winterton 753 . . . . D2
Witless Bay 907 . . . . D2

OTHER FEATURES

Alexis (riv.) C3
Anguille (cape) C4
Annieopsquotch (mts.) C4
Ashuanipi (lake) A3
Ashuanipi (riv.) A3
Atikonak (lake) B3
Attikamagen (lake) A3
Avalon (pen.) D2
Barachois Pond Prov. Park . . C4
Bauld (cape) C3
Bell (isl.) C3
Bell (isl.) D2
Belle Isle (isl.) C3
Belle Isle (str.) C3
Blackhead (bay) D2
Bonavista (bay) D1
Bonavista (cape) D1
Bonne (bay) C4
Branch (riv.) C2
Broyle (cape) D2
Bull Arm (inlet) D2
Burin (pen.) C4
Butter Pot Prov. Park D2
Cabot (str.) B4
Canada (bay) C3
Chidley (cape) B1
Churchill (falls) B3
Churchill (riv.) B3
Cirque (mt.) B2
Clode (sound) D2
Conception (bay) D2
Deep (inlet) B2
Double Mer (lake) C3
Dyke (lake) A3
Eagle (riv.) C3
Espoir (bay) C4
Exploits (riv.) C4
Fogo (isl.) D4
Fortune (bay) C4
Freels (cape) D3
Gander (lake) D4
Gander (riv.) D4
Glover (isl.) C4
Goose (riv.) B3
Grand (lake) B3
Grand (lake) C4
Grates (pt.) D2
Great Colinet (isl.) D2
Grey (isls.) C3
Groais (isl.) C3
Gros Morne (mt.) C4
Gros Morne Nat'l Park C4
Groswater (bay) C3
Hamilton (inlet) C3
Hamilton (sound) D4
Hare (bay) C3
Hawke (hills) D2
Hebron (fjord) B2
Hermitage (bay) C4
Holyrood (bay) D2
Horse (isls.) C3
Horse Chops (head) D2
Humber (riv.) C4
Ingornachoix (bay) C3
Ireland's Eye (isl.) D2
Islands (bay) C4
Kaipokok (bay) B2
Kanairiktok (riv.) B3
Kaumajet (mts.) B2
Kingurutik (mesa) B2
Labrador (reg.) B2
Labrador (sea) C2
La Manche Valley Prov. Park D2
La Poile (bay) C4
Little Mecatina (riv.) B3
Long (isl.) C2
Long (lake) A3
Long (pt.) C4
Long Range (mts.) C4
Main Topsail (mt.) C4
Makkovik (cape) C2
McLelan (str.) B1
Mealy (lake) C3
Meelpaeg (lake) C4
Melville (lake) C3
Menihek (lakes) A3
Merasheen (isl.) C2
Mistaken (pt.) D2
Mistastin (lake) B2
Nachvak (fjord) B2
Naskaupi (riv.) B3
Newfoundland (isl.) C4
Newman (sound) D2
New World (isl.) C4
Norman (cape) C3
North Aulatsivik (isl.) B2
Notre Dame (bay) C4
Okak (bay) B2
Ossokmanuan (res.) B3
Petitsikapau (lake) A3
Pine (cape) D2
Pinware (riv.) C3
Pistolet (bay) C3
Placentia (bay) C2
Ponds (isl.) C3
Port au Port (bay) C4
Port au Port (pen.) C4
Port Manvers (harb.) B2
Race (cape) D2
Ramah (bay) B2
Ramea (isls.) C4
Random (isl.) D2
Random (sound) D2
Ray (cape) C4
Red (isl.) C2
Red Indian (lake) C4
Red Wine (riv.) B3
Rocky (riv.) D2
Round (pond) C4
Saglek (bay) B2
Saint Francis (cape) D2
Saint George (cape) C4
Saint George's (bay) C4
Saint John (bay) C3
Saint John (cape) C3
Saint Lawrence (gulf) B4
Saint Lewis (cape) C3
Saint Mary's (bay) C2
Saint Mary's (cape) C2
Saint Michaels (bay) C3
Salmonier (riv.) D2
Sandwich (bay) C3
Shabogamo (lake) A3
Shoal (bay) D2
Smallwood (res.) B3
Smith (sound) D2
South Aulatsivik (isl.) B2
Spear (cape) D2
Squires Mem. Park C4
Swale (isl.) D1
Terra Nova (riv.) C2
Terra Nova Nat'l Park D2
Territok (cape) B2
Thoresby (mt.) B2
Torbay (pt.) D2
Torngat (mts.) B2
Trespassey (bay) D2
Trinity (bay) D2
Tunungayualok (isl.) B2
Ukasiksalik (isl.) B2
Victoria (lake) C4
White (bay) C3
White Bear (lake) C4
White Handkerchief (cape) . . B2

SAINT PIERRE and MIQUELON

CITIES and TOWNS

Saint-Pierre (cap.) 5,415 . . . C4

OTHER FEATURES

Miquelon (isl.) 626 . . . . C4
Saint Pierre (isl.) 5,415 . . . . C4

AREA 156,184 sq. mi. (404,517 sq. km.)
POPULATION 567,681
CAPITAL St. John's
LARGEST CITY St. John's
HIGHEST POINT in Torngat Mountains 5,420 ft. (1,652 m.)
SETTLED IN 1610
ADMITTED TO CONFEDERATION 1949
PROVINCIAL FLOWER Pitcher Plant

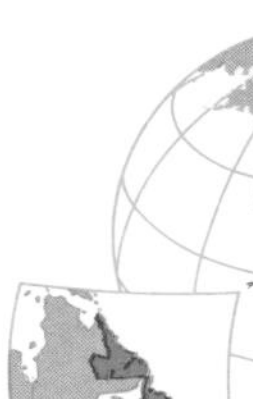

Agriculture, Industry and Resources

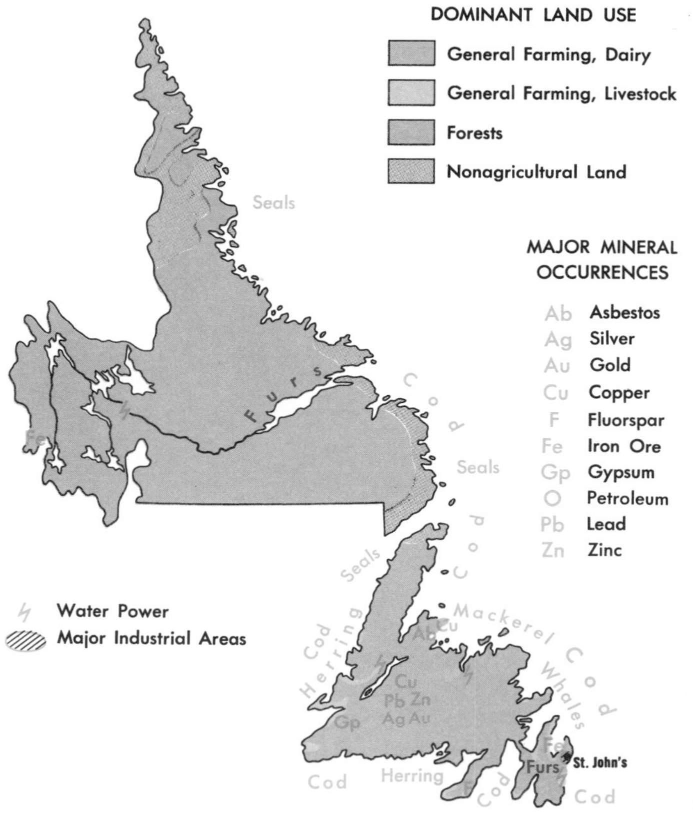

DOMINANT LAND USE
General Farming, Dairy
General Farming, Livestock
Forests
Nonagricultural Land
MAJOR MINERAL OCCURRENCES
Ab Asbestos
Ag Silver
Au Gold
Cu Copper
F Fluorspar
Fe Iron Ore
Gp Gypsum
O Petroleum
Pb Lead
Zn Zinc
Water Power
Major Industrial Areas
Seals
Furs
Cod
Herring
Mackerel
Whales
St. John's

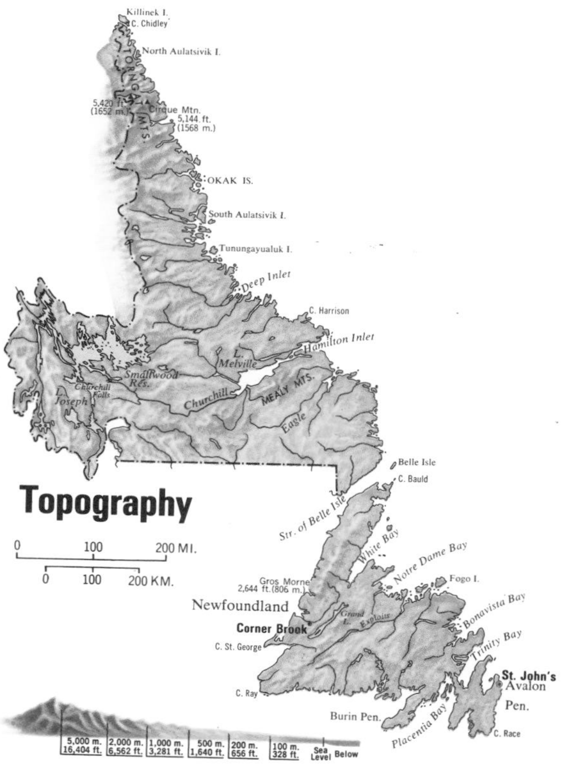

Killinek I.
C. Chidley
North Aulatsivik I.
Cirque Mtn.
5,144 ft. (1568 m.)
5,420 ft. (1652 m.)
OKAK IS.
South Aulatsivik I.
Tunungayualuk I.
Deep Inlet
C. Harrison
Hamilton Inlet
Melville
Smallwood Res.
Churchill
MEALY MTS.
Eagle
Belle Isle
C. Bauld
Str. of Belle Isle
White Bay
Notre Dame Bay
Fogo I.
Gros Morne 2,644 ft. (806 m.)
Newfoundland
Corner Brook
C. St. George
Bonavista Bay
Trinity Bay
St. John's
Avalon Pen.
C. Ray
Burin Pen.
Placentia Bay
C. Race
5,000 m. 16,404 ft.
2,000 m. 6,562 ft.
1,000 m. 3,281 ft.
500 m. 1,640 ft.
200 m. 656 ft.
100 m. 328 ft.
Sea Level
Below
0 100 200 MI.
0 100 200 KM.

Topography

NOVA SCOTIA

COUNTIES

Annapolis 22,522 .... C 4
Antigonish 18,110 .... F 3
Cape Breton 127,035 .... H 3
Colchester 43,224 .... E 3
Cumberland 35,231 .... D 3
Digby 21,689 .... C 4
Guysborough 12,752 .... F 3
Halifax 288,126 .... E 4
Hants 33,121 .... D 4
Inverness 22,337 .... G 2
Kings 49,739 .... D 4
Lunenburg 45,746 .... D 4
Pictou 50,350 .... F 3
Queens 13,126 .... C 4
Richmond 12,284 .... H 3
Shelburne 17,328 .... C 5
Victoria 8,432 .... H 2
Yarmouth 26,290 .... C 5

CITIES and TOWNS

Alder Point 651 .... H 2
Aldershot .... D 3
Amherst◎ 9,684 .... D 3
Annapolis Royal◎ 631 .... C 4
Antigonish◎ 5,205 .... F 3
Arichat 824 .... H 3
Aylesford 744 .... D 3
Baddeck◎ 972 .... H 2
Barrington Passage 722 .... C 5
Bear River-Sissiboo 854 .... C 4
Beaverbank 1,322 .... E 4
Berwick 1,699 .... D 4
Bridgetown 1,047 .... C 4
Bridgewater 6,669 .... D 4
Brookfield 619 .... E 3
Brooklyn 1,269 .... D 4
Cambridge Station 799 .... D 3
Canning 763 .... D 3
Canso 1,255 .... H 3
Centreville 765 .... D 3
Chéticamp 1,022 .... G 2
Chester 1,131 .... D 4
Chester Basin 639 .... D 4
Church Point 318 .... B 4
Clark's Harbour 1,059 .... C 5
Coldbrook Station 617 .... D 3
Cow Bay 670 .... E 4
Dartmouth 62,277 .... E 4
Debert 618 .... E 3
Digby◎ 2,558 .... C 4
Dominion 2,856 .... J 2
Donkin 873 .... J 2
Ellershouse-Hartville 662 .... D 4
Elmsdale 1,172 .... E 4
Enfield 1,510 .... E 4
Fall River 1,897 .... E 4
Falmouth 1,110 .... D 3
Glace Bay 21,466 .... J 2
Guysborough◎ 496 .... G 3
Halifax (cap.)◎ 114,594 .... E 4
Halifax *277,727 .... E 4
Hantsport 1,395 .... D 3
Herring Cove 1,323 .... E 4
Hilden 1,262 .... E 3
Ingonish 471 .... H 2
Inverness 2,013 .... G 2
Judique 925 .... G 3
Kentville◎ 4,974 .... D 3
Kingston 1,612 .... D 4
Lakeside 936 .... E 4
Lantz 1,172 .... E 4
Liverpool◎ 3,304 .... D 4
Lockeport 929 .... C 5
Louisbourg 1,410 .... J 3
Louisdale 979 .... G 3
Lower West Pubnico 790 .... C 5
Lunenburg◎ 3,014 .... D 4
Mahone Bay 1,228 .... D 4
Meteghan 890 .... B 4
Middleton 1,834 .... C 4
Milford Station 748 .... E 3
Milton 1,678 .... D 4
Mount Uniacke 1,145 .... E 4
Mulgrave 1,099 .... G 3
Musquodoboit Harbour 936 .... E 4
New Glasgow 10,464 .... F 3
New Victoria 1,374 .... H 2
New Waterford 8,808 .... J 2
North Sydney 7,820 .... H 2
Oxford 1,470 .... E 3
Parrsboro 1,799 .... D 3
Pictou◎ 4,628 .... F 3
Porters Lake 893 .... E 4
Port Hastings 312 .... G 3
Port Hawkesbury 3,850 .... G 3
Port Hood◎ 701 .... G 2
Port Morien 717 .... J 2
Port Williams 1,227 .... D 3
Prospect 693 .... E 4
Pugwash 648 .... E 3
Reserve Mines 2,472 .... H 2
River Hébert 835 .... D 3
Saint Peters 669 .... H 3
Sandy Point 691 .... C 5
Scotchtown 2,037 .... H 2
Sheet Harbour 819 .... F 4
Shelburne◎ 2,303 .... C 5
Shubenacadie 984 .... E 3
Springhill 4,896 .... E 3
Stellarton 5,435 .... F 3
Stewiacke 1,174 .... E 3
Sydney◎ 29,444 .... H 2
Sydney Mines 8,501 .... H 2
Terence Bay 960 .... E 4
Thorburn 1,014 .... F 3
Three Mile Plains 1,355 .... D 4
Timberlea 1,159 .... E 4
Trenton 3,154 .... F 3
Truro◎ 12,552 .... E 3
Waterville 687 .... D 3
Waverley 1,699 .... E 4
Wedgeport 827 .... C 5
Western Shore 1,712 .... D 4
Westmount 3,097 .... H 2
Westville 4,522 .... F 3
Wileville 746 .... D 4
Windsor◎ 3,646 .... D 3
Wolfville 3,235 .... D 3
Yarmouth◎ 7,475 .... B 5

OTHER FEATURES

Advocate (bay) .... D 3
Ainslie (lake) .... G 2
Amet (sound) .... E 3
Andrew (isl.) .... H 3
Annapolis (basin) .... C 4
Annapolis (riv.) .... C 4
Antigonish (harb.) .... G 3
Argos (cape) .... G 3
Aspy (bay) .... H 2
Avon (riv.) .... D 4
Baccaro (pt.) .... C 5
Baddeck (riv.) .... H 2
Barachois (pt.) .... G 4
Barren (isl.) .... G 4
Barrington (bay) .... C 5
Bedford (basin) .... E 4
Berry (head) .... G 3
Boularderie (isl.) .... H 2
Bras d'Or (lake) .... H 3
Breton (cape) .... J 3
Brier (isl.) .... B 4
Canso (cape) .... H 3
Canso (str.) .... G 3
Cap d'Or (cape) .... D 3

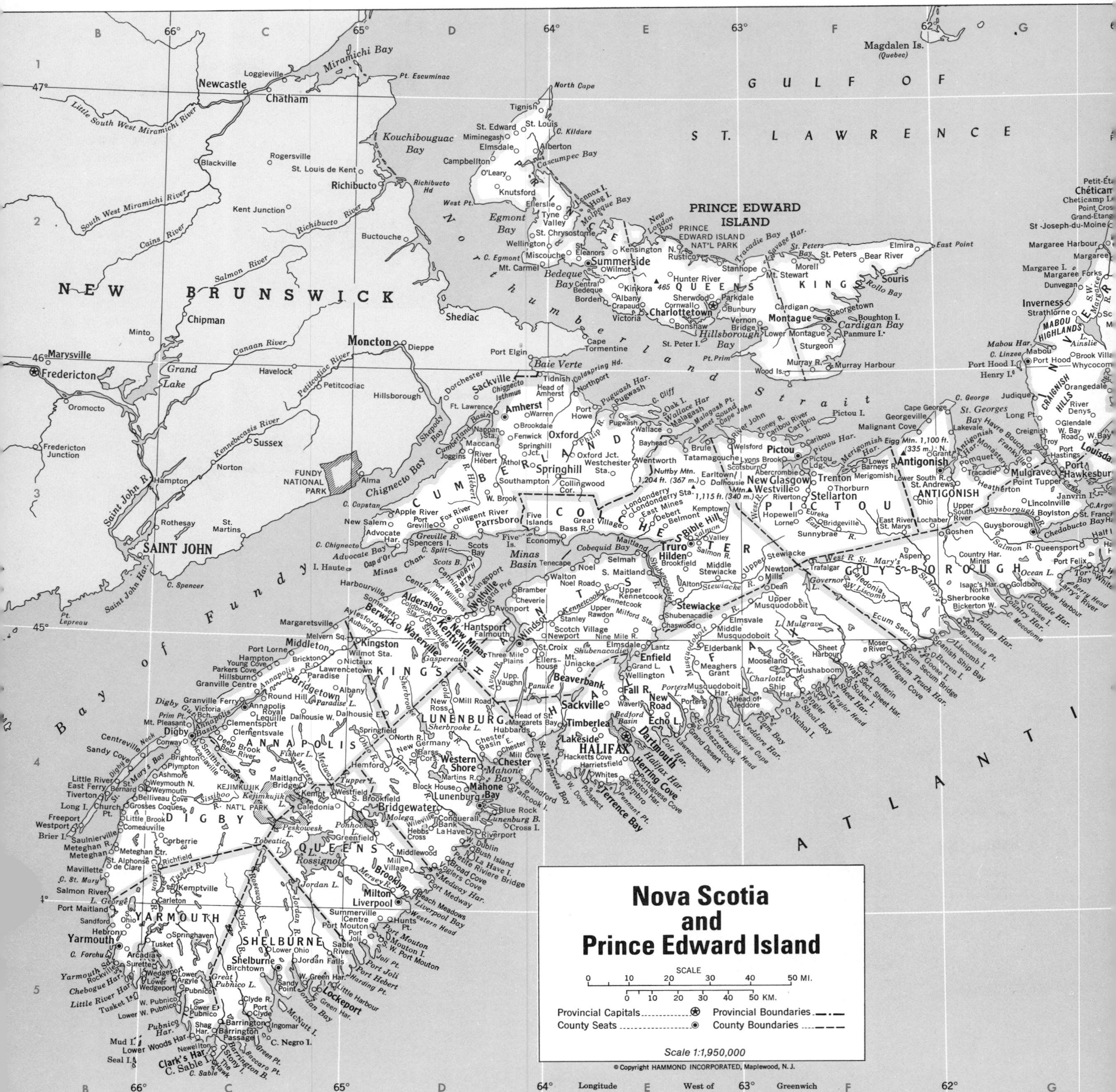

Cape Breton (isl.) .... J 2
Cape Breton Highlands Nat'l
  Park .... H 2
Cape Negro (isl.) .... C 5
Cape Sable (isl.) .... C 5
Capstan (cape) .... D 3
Caribou (isl.) .... F 3
Carleton (riv.) .... C 4
Charlotte (lake) .... F 4
Chebogue (harb.) .... B 5
Chedabucto (bay) .... G 3
Chéticamp (isl.) .... G 2
Chignecto (bay) .... D 3
Chignecto (cape) .... C 3
Chignecto (isth.) .... D 3
Clam (bay) .... F 4
Cliff (cape) .... E 3
Clyde (riv.) .... C 5
Cobequid (bay) .... E 3
Coddle (harb.) .... E 3
Coldspring (head) .... E 3
Cole (harb.) .... E 4
Country (harb.) .... G 3

Craignish (hills) .... G 3
Cross (isl.) .... D 4
Cumberland (basin) .... D 3
Dalhousie (mt.) .... E 3
Dauphin (cape) .... H 2
Digby Gut (chan.) .... C 4
Digby Neck (pen.) .... B 4
East (bay) .... H 3
East (isl.) .... F 3
East Bay (hills) .... H 3
Egmont (cape) .... H 2
Eigg (mt.) .... F 3
Fisher (lake) .... C 4
Five (isls.) .... D 3
Forchu (bay) .... H 3
Forchu (harb.) .... B 5
Framboise Cove (bay) ..... H 3
Fundy (bay) .... C 3
Gabarus (bay) .... H 3
Gabarus (cape) .... J 3
Gaspereau (lake) .... D 4
George (cape) .... G 3
George (lake) .... B 5

Gold (riv.) .... D 4
Goose (isl.) .... F 4
Goose (isl.) .... G 3
Governor (lake) .... D 4
Great Bras d'Or (chan.) .. H 2
Great Pubnico (lake) ...... C 5
Green (pt.) .... C 5
Greville (harb.) .... D 3
Guysborough (riv.) .... G 3
Halifax (harb.) .... E 4
Harding (pt.) .... D 5
Haute (isl.) .... C 3
Hébert (riv.) .... D 3
Henry (isl.) .... G 3
Indian (harb.) .... G 3
Ingonish North (bay) ...... H 2
Janvrin (isl.) .... G 3
Jeddore (harb.) .... F 4
John (cape) .... E 3
Joli (pt.) .... D 5
Jordan (bay) .... C 5
Jordan (lake) .... C 4
Jordan (riv.) .... C 5
Kejimkujik (lake) .... C 4
Kejimkujik Nat'l Park ...... C 4
Kennetcook (riv.) .... E 3
La Have (isl.) .... D 4
La Have (riv.) .... D 4
Linzee (cape) .... G 2
Liscomb (isl.) .... G 4
Little River (harb.) .... B 5
Liverpool (bay) .... D 5
Lomond, Loch (lake) ...... H 3
Long (isl.) .... B 4
Louisbourg Nat'l Hist. Park J 3
Lunenburg (bay) .... D 4
Mabou (harb.) .... G 2
Mabou Highlands (hills) .. G 2
Madame (isl.) .... H 3
Mahone (bay) .... D 4
Malagash (pt.) .... E 3
Margaree (riv.) .... F 4
McNutt (isl.) .... C 5
Medway (harb.) .... D 4
Medway (riv.) .... D 4
Merigomish (harb.) .... F 3
Mersey (riv.) .... C 4
Michaud (pt.) .... H 3
Minas (basin) .... D 3
Minas (chan.) .... D 3
Mira (bay) .... J 2
Mira (riv.) .... H 3
Mocodome (cape) .... G 3
Molega (lake) .... D 4
Morien (cape) .... J 2
Mouton (isl.) .... D 5
Mud (isl.) .... B 5
Mulgrave (pt.) .... F 3
Musquodoboit (riv.) .... E 4
Necum Teuch (harb.) ...... F 4
Nichol (isl.) .... F 4
North (cape) .... H 1
North (mt.) .... D 3
North Aspy (riv.) .... H 2
North Bay Ingonish (bay) . H 2
North East Margaree (riv.) H 2
Northumberland (str.) ..... E 3
Nuttby (mt.) .... E 3
Oak (isl.) .... E 3
Ocean (lake) .... G 3
Ohio (riv.) .... D 4
Panuke (lake) .... D 4
Paradise (lake) .... C 4
Pennant (pt.) .... E 4
Percé (cape) .... J 2
Peskowesk (lake) .... C 4
Petit-de-Grat (isl.) .... H 3
Petpeswick (head) .... E 4
Philip (riv.) .... E 3
Pictou (harb.) .... F 3
Pictou (isl.) .... F 3
Pleasant (bay) .... H 2
Ponhook (lake) .... D 4
Porters (lake) .... E 4
Port Hebert (harb.) .... D 5
Port Hood (isl.) .... G 2
Port Joli (harb.) .... D 5
Port Mouton (harb.) .... D 5
Poulet Cove (bay) .... H 2
Prim (pt.) .... C 4
Pubnico (lake) .... C 5
Pugwash (harb.) .... E 3
Roseway (riv.) .... C 4
Rossignol (lake) .... C 4
Sable (cape) .... C 5
Sable (isl.) .... J 5
Saint Andrews (chan.) ..... H 2
Saint Anns (bay) .... H 2
Saint Georges (bay) .... H 2
Saint Lawrence (bay) ...... H 1
Saint Lawrence (cape) .... H 1
Saint Margarets (bay) ..... E 4
Saint Mary (cape) .... B 4
Saint Marys (bay) .... B 4
Saint Mary's (riv.) .... F 3
Saint Patrick (chan.) ...... H 2
Saint Paul (isl.) .... H 1
Saint Peters (bay) .... H 3

Salmon (riv.) .... E 3
Salmon (riv.) .... G 3
Scatarie (isl.) .... J 2
Scots (bay) .... D 3
Seall (isl.) .... B 5
Sheet (harb.) .... F 4
Sherbrooke (lake) .... D 4
Sherbrooke (riv.) .... D 4
Shoal (bay) .... F 4
Shubenacadie (lake) .... E 4
Shubenacadie (riv.) .... E 3
Sissiboo (riv.) .... C 4
Smoky (cape) .... H 2
Sober (isl.) .... F 4
South West Margaree (riv.) G 2
Split (cape) .... D 3
Spry (harb.) .... F 4
Stewiacke (riv.) .... E 3
Sydney (harb.) .... H 2
Tangier (riv.) .... F 4
Taylor (head) .... F 4
Tobeatic (lake) .... C 4
Tor (bay) .... G 3
Tupper (lake) .... D 4
Tusket (isl.) .... B 5

Tusket (riv.) .... C 4
Verte (bay) .... D 2
Wallace (harb.) .... E 3
West (bay) .... G 3
West (pt.) .... H 5
West (riv.) .... F 3
Western (head) .... D 5
West Liscomb (riv.) .... F 3
West Saint Mary's (riv.) .. F 3
Whitehaven (harb.) .... G 3
Yarmouth (sound) .... B 5

PRINCE EDWARD ISLAND

COUNTIES

Kings 19,215 .... F 2
Prince 42,821 .... D 2
Queens 60,470 .... E 2

CITIES and TOWNS

Alberton 1,020 .... E 2
Bunbury 1,024 .... E 2
Charlottetown (cap.)⊙ 15,282 . E 2

Cornwall 1,838 .... E 2
Georgetown⊙ 737 .... F 2
Kensington 1,143 .... E 2
Miscouche 752 .... D 2
Montague 1,957 .... F 2
Murray Harbour 443 ...... F 2
North Rustico 688 .... E 2
O'Leary 736 .... D 2
Parkdale 2,018 .... E 2
Saint Edward 650 .... D 2
Saint Eleanors 2,716 ..... E 2
Sherwood 5,681 .... E 2
Souris 1,413 .... F 2
Summerside⊙ 7,828 ..... E 2
Tignish 982 .... D 2
Wilmot 1,563 .... E 2

OTHER FEATURES

Bedeque (bay) .... E 2
Boughton (riv.) .... F 2
Cardigan (bay) .... F 2
Cascumpeque (bay) .... E 2
East (pt.) .... G 2
Egmont (bay) .... D 2

Egmont (cape) .... D 2
Hillsborough (bay) .... E 2
Hog (isl.) .... E 2
Kildare (cape) .... E 2
Lennox (isl.) .... E 2
Malpeque (bay) .... E 2
New London (bay) .... E 2
North (pt.) .... E 1
Northumberland (str.) ..... D 2
Panmure (isl.) .... F 2
Prim (pt.) .... E 2
Prince Edward Island Nat'l
  Park .... E 2
Rollo (bay) .... F 2
Saint Lawrence (gulf) ..... F 2
Saint Peters (bay) .... F 2
Saint Peters (isl.) .... E 2
Savage (harb.) .... F 2
Tracadie (bay) .... F 2
West (pt.) .... D 2
Wood (isls.) .... F 3

⊙County seat.
*Population of metropolitan area.

PRINCE EDWARD ISLAND

AREA   2,184 sq. mi. (5,657 sq. km.)
POPULATION   122,506
CAPITAL   Charlottetown
LARGEST CITY   Charlottetown
HIGHEST POINT   465 ft. (142 m.)
SETTLED IN   1720
ADMITTED TO CONFEDERATION   1873
PROVINCIAL FLOWER   Lady's Slipper

NOVA SCOTIA

AREA   21,425 sq. mi. (55,491 sq. km.)
POPULATION   847,442
CAPITAL   Halifax
LARGEST CITY   Halifax
HIGHEST POINT   Cape Breton Highlands
  1,747 ft. (532 m.)
SETTLED IN   1605
ADMITTED TO CONFEDERATION   1867
PROVINCIAL FLOWER   Trailing Arbutus or
  Mayflower

Topography

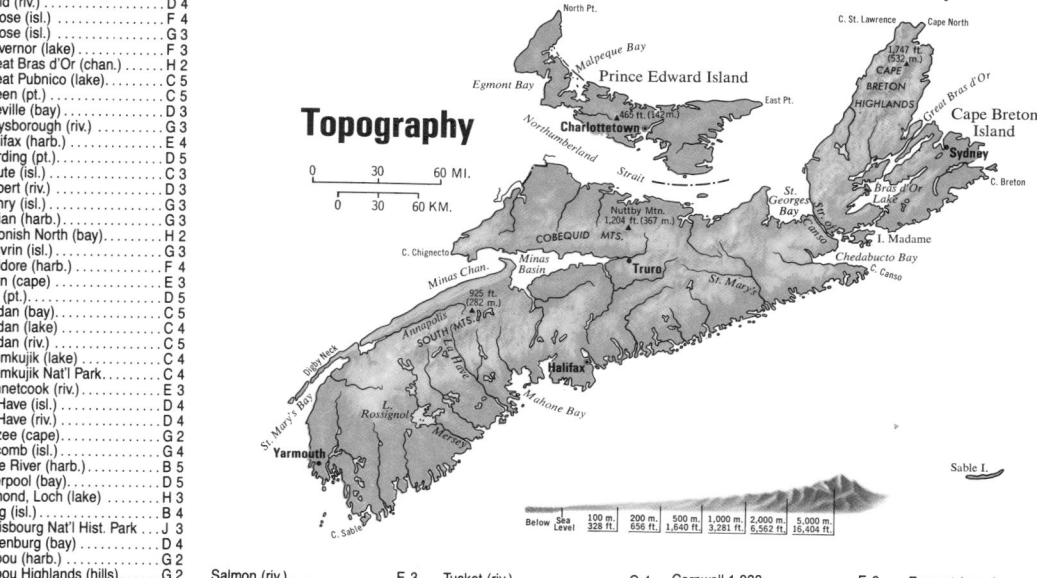

Agriculture, Industry and Resources

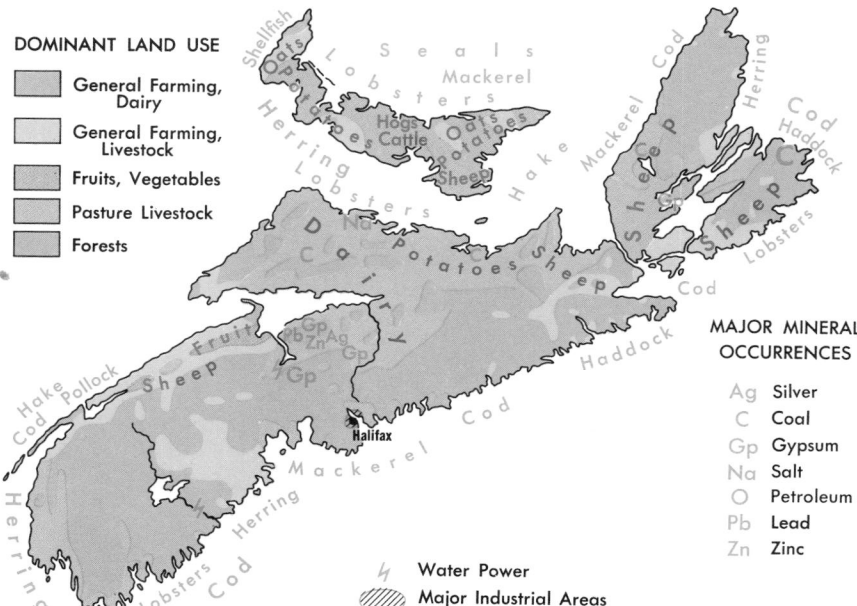

DOMINANT LAND USE

General Farming, Dairy
General Farming, Livestock
Fruits, Vegetables
Pasture Livestock
Forests

MAJOR MINERAL OCCURRENCES

Ag   Silver
C   Coal
Gp   Gypsum
Na   Salt
O   Petroleum
Pb   Lead
Zn   Zinc

Water Power
Major Industrial Areas

COUNTIES

Albert 23,632 .... F 3
Carleton 24,659 .... C 2
Charlotte 26,571 .... C 3
Gloucester 86,156 .... E 1
Kent 30,799 .... E 2
King's 51,114 .... D 3
Madawaska 34,892 .... B 1
Northumberland 54,134 .... D 2
Queen's 12,485 .... D 3
Restigouche 40,593 .... C 1
Saint John 86,148 .... D 3
Sunbury 21,012 .... D 3
Victoria 20,815 .... C 2
Westmorland 107,640 ..... F 2
York 74,213 .... C 3

CITIES and TOWNS

Acadie Siding 64 .... E 2
Acadieville 176 .... E 2
Adamsville 94 .... E 2
Albert Mines 120 .... F 3
Alcida 174 .... E 1
Aldouane 64 .... E 2
Allardville 478 .... E 1
Alma 329 .... F 3
Anagance 114 .... E 2
Anse-Bleue 562 .... E 1

Apohaqui 341 .... E 3
Argyle 63 .... C 2
Armstrong Brook 191 ..... E 1
Aroostook 403 .... C 2
Arthurette 178 .... C 2
Astle 201 .... D 1
Atholville 1,694 .... D 1
Aulac 113 .... F 3
Back Bay 455 .... D 3
Baie-Sainte-Anne 709 .... F 1
Baie-Verte 175 .... C 2
Bairdsville 81 .... C 2
Baker Brook 527 .... B 1
Balmoral 1,823 .... D 1
Barachois 686 .... F 2
Barnaby River 38 .... E 2
Barnettville 117 .... E 1
Bartibog Bridge 122 ...... E 1
Bas-Caraquet 1,859 ...... F 1
Bass River 112 .... E 2
Bath 794 .... C 2
Bathurst⊛ 15,705 .... E 1
Bayfield 81 .... G 2
Bayside .... C 3
Beaubois 211 .... F 2
Beaver Brook Station 95 . E 2
Beaver Harbour 316 ...... D 3
Beechwood 111 .... C 2
Beersville 52 .... E 2
Belledune 690 .... E 1

Bellefleur 83 .... C 1
Bellefond 243 .... E 1
Belleisle Creek 145 ...... E 3
Benjamin River 171 ...... D 1
Ben Lomond .... E 3
Benton 101 .... C 3
Beresford 3,652 .... E 1
Berry Mills 238 .... E 2
Bertrand 1,268 .... E 1
Berwick 147 .... E 3
Black Point 131 .... D 1
Black River 150 .... E 3
Blacks Harbour 1,356 .... D 3
Blackville 892 .... E 2
Blissfield 119 .... D 2
Bloomfield Ridge 153 .... D 2
Bloomfield Station 62 .... E 3
Bocabec 34 .... C 3
Boiestown 299 .... D 2
Bonny River 153 .... D 2
Bosse 193 .... B 1
Bourgeois 215 .... F 2
Brantville 1,066 .... E 1
Breau-Village 293 .... F 2
Brest 94 .... E 2
Brewers Mills 199 .... C 2
Briggs Corner 89 .... E 2
Bristol 824 .... C 2
Brockway (Lower Brockway-
  Brockway) 97 .... C 3

Browns Flat 295 .... D 3
Buctouche 2,476 .... F 2
Burnsville 156 .... E 1
Burton⊛ 291 .... D 3
Burtts Corner 484 ...... D 2
Cambridge-Narrows 433 .. E 3
Campbellton 9,818 ...... D 1
Canaan 115 .... E 2
Canaan Forks 78 .... E 2
Canaan Road 86 .... E 2
Canterbury 474 .... C 3
Cap-Bateau 417 .... F 1
Cape Tormentine 229 ... G 2
Cap-Lumière 262 .... F 2
Cap-Pelé 2,199 .... F 2
Caraquet 4,315 .... E 1
Carlingford 229 .... C 2
Carlisle 75 .... C 2
Caron Brook 171 ...... B 1
Carrolls Crossing 119 .. D 2
Castalia 115 .... D 4
Central Blissville 155 .. D 3
Centre-Saint-Simon (St.
  Simon) 991 .... E 1
Centreville 577 .... C 2
Chance Harbour 63 .... D 3
Charlo 1,603 .... D 1
Chatham 6,779 .... E 1
Chatham Head .... E 2
Chipman 1,829 .... E 2

Clair 915 .... B 1
Clarendon 80 .... D 3
Cliffordvale (Limestone-
  Cliffordvale) 69 ...... C 2
Clifton 194 .... E 1
Coal Branch 90 .... E 2
Coal Creek 61 .... E 2
Cocagne Cape 278 .... F 2
Cocagne-Cocagne Sud 600 F 2
Codys 125 .... E 3
Coldstream 217 .... C 2
Coles Island 150 ...... E 3
College Bridge 536 .... F 3
Collette 198 .... E 2
Connell 58 .... C 2
Connors 96 .... B 1
Cork 54 .... E 3
Cornhill 111 .... E 3
Coughlan 181 .... D 2
Cross Creek 192 ..... D 2
Cumberland Bay 231 .. E 2
Dalhousie⊛ 4,958 .... D 1
Dalhousie Junction 105 . D 1
Darlington 749 .... D 1
Daulnay 398 .... F 1
Dawsonville 278 ..... C 1
Debec 200 .... C 2
Dieppe 8,511 .... F 2
Dipper Harbour 166 .. D 3
Doaktown 1,009 ..... D 2

Dorchester⊛ 1,101 ... F 3
Dorchester Crossing 605 . F 2
Douglastown 1,091 .... E 1
Drummond 849 .... C 1
Duguayville 337 ...... E 1
Dupuis Corner 303 .... F 2
Durham Bridge 255 ... D 2
East Riverside-Kingshurst
  989 .... F 3
Edmundston⊛ 12,044 .. B 1
Eel River Bridge 377 ... D 1
Eel River Crossing 1,431 . D 1
Elgin 301 .... E 3
Enniskillen 63 .... E 3
Escuminac 194 .... F 1
Evandale 58 .... E 3
Evangeline 356 ...... F 1
Everett 48 .... C 1
Fairfield 250 .... E 3
Fairhaven 142 .... C 4
Fairisle 415 .... E 1
Fairvale 3,960 ...... E 3
Ferry Road 325 ..... E 1
Fielding 197 .... C 2
Five Fingers 189 .... C 1
Flatlands 249 .... D 1
Florenceville 709 .... C 2
Forest City 25 ...... C 3
Fosterville 58 ...... C 3

Four Falls 69 .... C 2
Fredericton (cap.)⊛ 43,723 . D 3
Fredericton Junction 711 . D 3
Gagetown⊛ 618 ...... D 3
Gardner Creek 56 ..... D 3
Geary 654 .... D 3
Germantown 62 ...... F 3
Gillespie 96 .... C 2
Glassville 147 ...... C 2
Glencoe 147 .... E 3
Glenlivet 284 .... D 1
Gondola Point 3,076 .. E 3
Grafton 385 .... C 2
Grand Bay 3,173 .... E 3
Grande-Anse 817 .... E 1
Grand Falls 6,203 ... C 2
Grand Falls Hill 152 .. C 2
Grand Harbour 614 ... D 4
Gray Rapids 266 .... E 2
Hammondvale 72 ..... F 3
Hampstead 87 .... D 3
Hampton⊛ 3,141 ..... E 3
Harcourt 127 .... E 2
Hardwicke 114 ...... F 1
Hardwood Ridge 191 . E 2
Hartland 846 .... C 2
Harvey, Albert 58 ... F 3
Harvey, York 356 ... D 3
Hatfield Point 176 ... E 3

New Brunswick

SCALE

0 5 10 20 30 40 MI.

0 5 10 20 30 40 KM.

Provincial Capitals .... ⊛
County Seats .... ◉
International Boundaries .... _._._._
Provincial Boundaries .... _ _ _ _
County Boundaries .... _ . _ . _

Scale 1:1,900,000

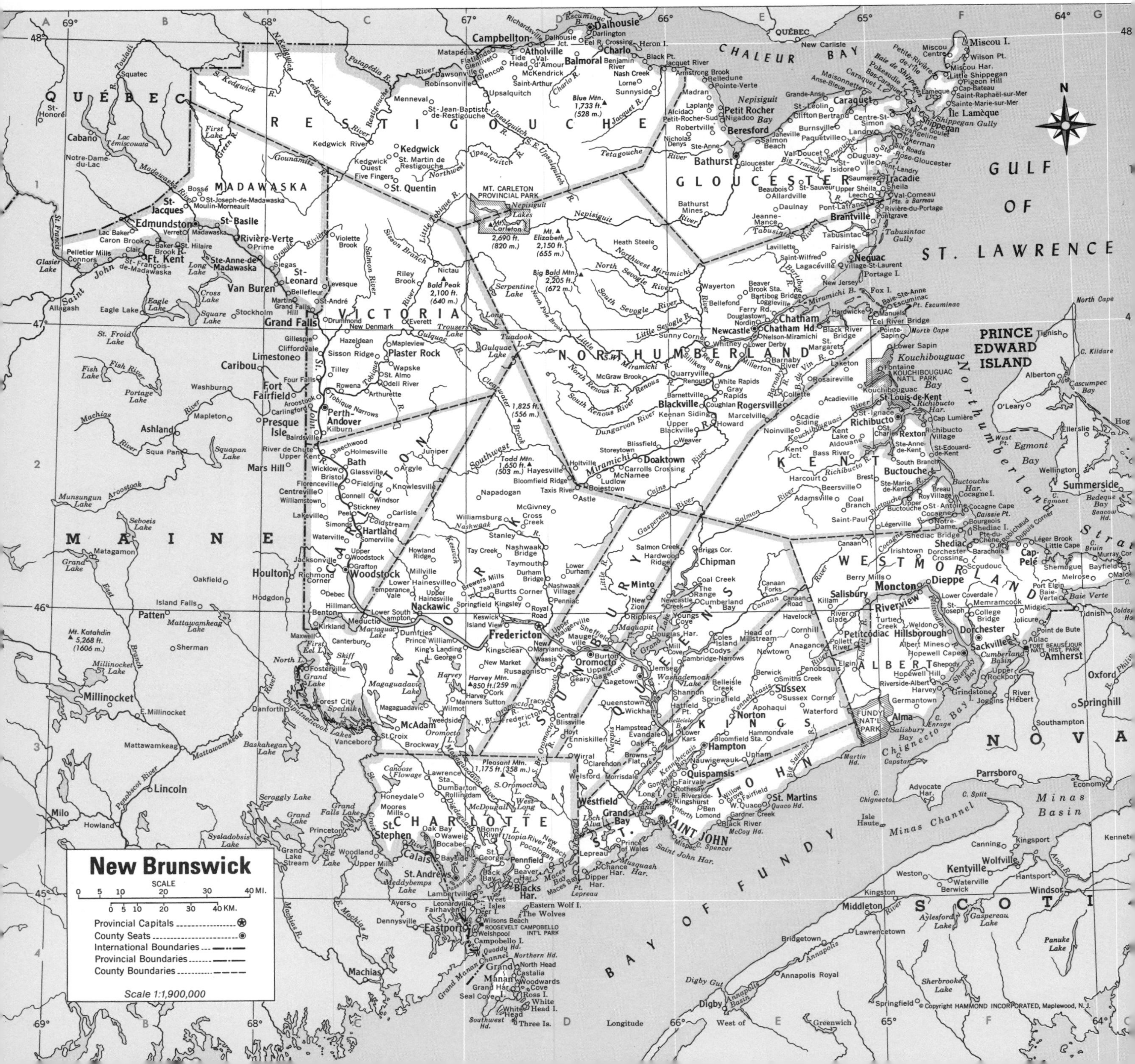

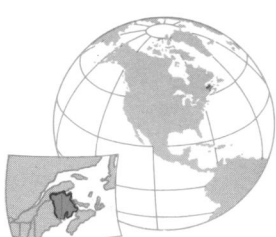

Havelock 439 . . . . . E 3
Hayesville 107 . . . . . D 2
Hazeldean 108 . . . . . C 2
Head of Millstream 61 . . . . . E 3
Hillman 69 . . . . . C 2
Hillsborough 1,239 . . . . . F 3
Holmesville 146 . . . . . C 2
Holtville 222 . . . . . D 2
Honeydale 77 . . . . . C 3
Hopewell Cape⊙ 144 . . . . . F 3
Hopewell Hill 172 . . . . . F 3
Howard 77 . . . . . E 2
Howland Ridge 55 . . . . . C 2
Hoyt 114 . . . . . D 3
Inkerman 396 . . . . . F 1
Irishtown 605 . . . . . F 2
Island View 240 . . . . . D 3
Jacksonville 363 . . . . . C 2
Jacquet River 778 . . . . . E 1
Janeville 204 . . . . . E 1
Jeanne Mance 89 . . . . . E 1
Jemseg 228 . . . . . D 3
Jolicure 96 . . . . . F 3
Juniper 525 . . . . . C 2
Kedgwick 1,222 . . . . . C 1
Keenan Siding 86 . . . . . E 2
Kent Junction 112 . . . . . E 2
Kent Lake 57 . . . . . E 2
Keswick 260 . . . . . D 3
Kilburn 134 . . . . . C 2
Killam 60 . . . . . E 2
Kingsclear 250 . . . . . D 3
Kingsley 145 . . . . . D 2
Kirkland 69 . . . . . C 3
Knowlesville 82 . . . . . C 2
Kouchibouguac 213 . . . . . F 2
Lac Baker 292 . . . . . B 1
Lagacéville 227 . . . . . E 1
Lake George 170 . . . . . C 3
Laketon 81 . . . . . E 2
Lakeville 201 . . . . . C 2
Lambertville 109 . . . . . C 3
Lamèque 1,571 . . . . . F 1
Landry 281 . . . . . E 1
Laplante 197 . . . . . E 1
Lavillette 576 . . . . . E 1
Lawrence Station 229 . . . . . C 3
Leech 584 . . . . . E 1
Léger Brook . . . . . F 2
Légerville 184 . . . . . F 2
Le Goulet 1,173 . . . . . F 1
Leonardville 158 . . . . . C 4
Lepreau 208 . . . . . D 3
Levesque 77 . . . . . C 1
Little Cape 513 . . . . . F 2
Little Shippegan 131 . . . . . F 1
Loggieville 781 . . . . . E 1
Lorne 937 . . . . . D 1

Lower Coverdale 616 . . . . . F 2
Lower Derby 206 . . . . . E 2
Lower Durham 52 . . . . . D 2
Lower Hainesville 66 . . . . . C 2
Lower Kars 30 . . . . . E 3
Lower Millstream 184 . . . . . E 3
Lower Sapin . . . . . F 2
Lower Southampton . . . . . C 3
Ludlow 100 . . . . . D 2
Maces Bay 182 . . . . . D 3
Madran 247 . . . . . E 1
Magaguadavic 126 . . . . . C 3
Maisonnette 757 . . . . . E 1
Malden 93 . . . . . G 2
Manners Sutton 159 . . . . . D 3
Manuels 332 . . . . . F 1
Mapleview 65 . . . . . C 2
Marcelville 61 . . . . . E 2
Martin 104 . . . . . C 1
Maugerville 249 . . . . . D 3
Maxwell 64 . . . . . C 3
McAdam 1,837 . . . . . C 3
McGivney 156 . . . . . D 2
McKendrick 608 . . . . . D 1
McNamee 147 . . . . . D 2
Meductic 234 . . . . . C 3
Melrose 121 . . . . . F 2
Memramcook 276 . . . . . F 2
Menneval 110 . . . . . C 1
Midgic Station 208 . . . . . F 3
Mill Cove 253 . . . . . D 3
Millerton 130 . . . . . E 2
Millville 309 . . . . . C 2
Minto 3,399 . . . . . D 2
Miscou Centre 554 . . . . . F 1
Miscou Harbour 106 . . . . . F 1
Mispec 180 . . . . . E 3
Moncton 54,743 . . . . . F 2
Moores Mills 117 . . . . . C 3
Morrisdale 202 . . . . . D 3
Moulin-Morneault 459 . . . . . B 1
Murray Corner 233 . . . . . G 2
Nackawic 1,357 . . . . . C 2
Napadogan 103 . . . . . D 2
Nash Creek 235 . . . . . D 1
Nashwaak Bridge 142 . . . . . D 2
Nashwaak Village 258 . . . . . D 2
Nauwigewauk 139 . . . . . E 3
Neguac 1,755 . . . . . E 1
Nelson-Miramichi 1,452 . . . . . E 2
Newcastle⊙ 6,284 . . . . . E 2
Newcastle Creek 210 . . . . . D 2
New Denmark 112 . . . . . C 1
New Jersey 65 . . . . . E 1
New Market 143 . . . . . D 3
New Maryland 485 . . . . . D 3
New River Beach 33 . . . . . D 3
Newtown 154 . . . . . E 3

New Zion 171 . . . . . D 2
Nicholas Denys 170 . . . . . D 1
Nictau 30 . . . . . C 1
Nigadoo 1,075 . . . . . E 1
Noinville 50 . . . . . E 2
Nordin 393 . . . . . E 1
North Head 661 . . . . . D 4
Norton 1,372 . . . . . E 3
Notre-Dame 344 . . . . . F 2
Oak Bay 183 . . . . . C 3
Oak Point 83 . . . . . D 3
Oromocto 9,064 . . . . . D 3
Paquetville 626 . . . . . E 1
Peel 117 . . . . . C 2
Pelletier Mills 88 . . . . . B 1
Pennfield . . . . . D 3
Penniac 179 . . . . . D 2
Penobsquis 259 . . . . . E 3
Perth-Andover⊙ 1,872 . . . . . C 2
Petitcodiac 1,401 . . . . . E 3
Petite-Rivière-de-l'Île 549 . . . . . F 1
Petit Rocher 1,860 . . . . . E 1
Petit Rocher Sud . . . . . E 1
Pigeon Hill 595 . . . . . F 1
Plaster Rock 1,222 . . . . . C 2
Pocologan 150 . . . . . D 3
Point de Bute 155 . . . . . F 3
Pointe-du-Chêne 482 . . . . . F 2
Pointe-Sapin 331 . . . . . F 2
Pointe-Verte 1,335 . . . . . E 1
Pollett River 73 . . . . . E 3
Pontgrave 229 . . . . . F 1
Pont-Lafrance 875 . . . . . E 1
Pont-Landry 444 . . . . . F 1
Port Elgin 504 . . . . . F 2
Prime 89 . . . . . B 1
Prince of Wales 138 . . . . . D 3
Prince William 225 . . . . . C 3
Quarryville 205 . . . . . E 2
Queenstown 112 . . . . . D 3
Quispamsis 6,022 . . . . . E 3
Red Bank 141 . . . . . E 2
Renforth 1,490 . . . . . E 3
Renous 192 . . . . . E 2
Rexton 928 . . . . . F 2
Richardsville . . . . . D 1
Richibucto⊙ 1,722 . . . . . F 2
Richibucto Village 442 . . . . . F 2
Richmond Corner 84 . . . . . C 2
Riley Brook 126 . . . . . C 1
Ripples 233 . . . . . D 2
River de Chute 22 . . . . . C 2
River Glade 268 . . . . . E 3
Riverside-Albert 478 . . . . . F 3
Riverview 14,907 . . . . . F 2
Rivière-du-Portage 661 . . . . . F 1
Rivière Verte 1,054 . . . . . B 1
Robertville 733 . . . . . E 1

Robichaud 485 . . . . . F 2
Robinsonville 206 . . . . . C 1
Rogersville 1,237 . . . . . E 2
Rollingdam 65 . . . . . C 3
Rosaireville 86 . . . . . E 2
Rothesay 1,764 . . . . . E 3
Rowena 73 . . . . . C 2
Roy 173 . . . . . F 2
Royal Road 41 . . . . . D 2
Rusagonis 231 . . . . . D 3
Sackville 5,654 . . . . . F 3
Saint Almo 17 . . . . . C 2
Saint-André 385 . . . . . C 1
Saint Andrews⊙ 1,760 . . . . . C 3
Saint-Antoine 1,217 . . . . . F 2
Saint Arthur 369 . . . . . D 1
Saint-Basile 3,214 . . . . . B 1
Saint-Charles 355 . . . . . F 2
Saint Croix 86 . . . . . C 3
Sainte-Anne 329 . . . . . E 1
Sainte-Anne-de-Kent 337 . . . . . F 2
Sainte-Anne-de-Madawaska
  1,332 . . . . . B 1
Saint-Édouard-de-Kent 157 . . . . . F 2
Sainte-Marie-de-Kent 283 . . . . . F 2
Sainte-Marie-sur-Mer 539 . . . . . F 1
Sainte-Rose-Gloucester 410 . . . . . F 1
Saint-François-de-Madawaska
  753 . . . . . B 1
Saint George 1,163 . . . . . D 3
Saint Hilaire 244 . . . . . B 1
Saint-Ignace 96 . . . . . F 2
Saint-Isidore 794 . . . . . E 1
Saint-Jacques 2,297 . . . . . B 1
Saint-Jean-Baptiste-de-
  Restigouche 228 . . . . . C 1
Saint John⊙ 80,521 . . . . . E 3
Saint-Joseph 630 . . . . . F 3
Saint-Joseph-de-Madawaska
  173 . . . . . B 1
Saint-Léolin 799 . . . . . E 1
Saint Leonard 1,566 . . . . . C 1
Saint-Louis-de-Kent 1,166 . . . . . F 2
Saint Margarets 63 . . . . . E 2
Saint Martin de Restigouche
  124 . . . . . C 1
Saint Martins 530 . . . . . E 3
Saint-Paul 365 . . . . . E 2
Saint Quentin 2,334 . . . . . C 1
Saint-Raphaël-sur-Mer 562 . . . . . F 1
Saint Sauveur 252 . . . . . E 1
Saint Stephen 5,120 . . . . . C 3
Saint Wilfred . . . . . E 1
Salisbury 1,672 . . . . . E 2
Salmon Beach 277 . . . . . E 1
Salmon Creek 38 . . . . . E 2
Saumarez 690 . . . . . E 1
Scoudouc 207 . . . . . F 2
Seal Cove 548 . . . . . D 4
Shannon 39 . . . . . E 3
Shediac 4,285 . . . . . F 2
Shediac Bridge 441 . . . . . F 2
Sheffield 112 . . . . . D 3
Sheila 1,172 . . . . . F 1
Shemogue 199 . . . . . F 2
Shepody 86 . . . . . F 3
Shippegan 2,471 . . . . . F 1
Siegas 227 . . . . . C 1
Sillikers 292 . . . . . E 2
Simonds 221 . . . . . C 2
Sisson Ridge 170 . . . . . C 2
Six Roads 239 . . . . . F 1
Smiths Creek 163 . . . . . E 3
Somerville 326 . . . . . C 2
South Branch 86 . . . . . F 2
Springfield, King's 116 . . . . . E 3
Springfield, York 130 . . . . . C 2
Stanley 432 . . . . . D 2
Stickney 232 . . . . . C 2
Storeytown 140 . . . . . D 2
Sunny Corner 405 . . . . . E 2
Sunnyside 87 . . . . . D 1
Sussex 3,972 . . . . . E 3
Sussex Corner 1,023 . . . . . E 3
Tabusintac 231 . . . . . E 1

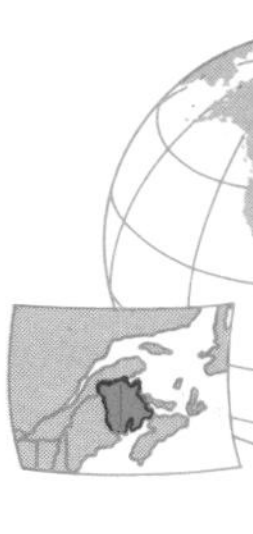

AREA 28,354 sq. mi. (73,437 sq. km.)
POPULATION 696,403
CAPITAL Fredericton
LARGEST CITY Saint John
HIGHEST POINT Mt. Carleton 2,690 ft. (820 m.)
SETTLED IN 1611
ADMITTED TO CONFEDERATION 1867
PROVINCIAL FLOWER Purple Violet

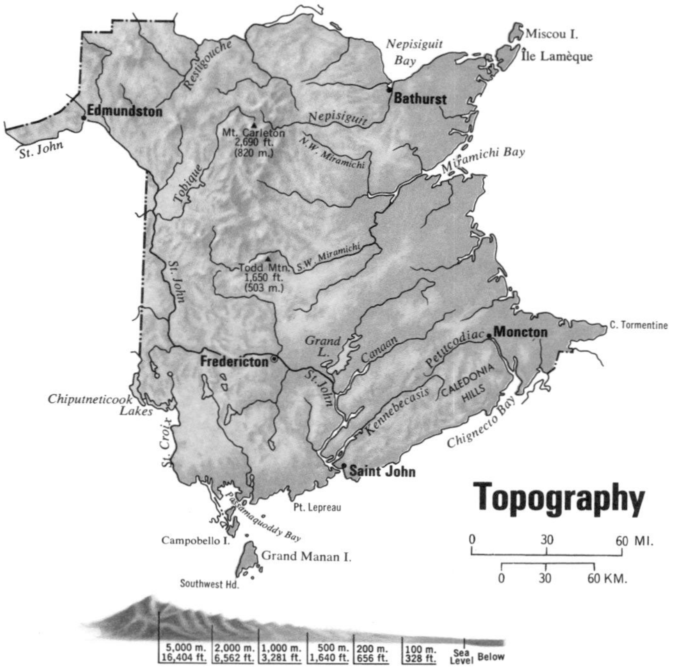

Topography
Miscou I.
Île Lamèque
Nepisiguit Bay
Bathurst
Edmundston
Restigouche
Nepisiguit
Mt. Carleton 2,690 ft. (820 m.)
N.W. Miramichi
Miramichi Bay
St. John
Tobique
Todd Mtn. 1,650 ft. (503 m.)
S.W. Miramichi
St. John
Grand L.
Canaan
Petitcodiac
Moncton
C. Tormentine
Fredericton
CALEDONIA HILLS
Chiputneticook Lakes
St. Croix
Kennebecasis
Chignecto Bay
Saint John
Pt. Lepreau
Passamaquoddy Bay
Campobello I.
Grand Manan I.
Southwest Hd.
0 30 60 MI.
0 30 60 KM.
5,000 m. 16,404 ft.
2,000 m. 6,562 ft.
1,000 m. 3,281 ft.
500 m. 1,640 ft.
200 m. 656 ft.
100 m. 328 ft.
Sea Level
Below

Agriculture, Industry and Resources

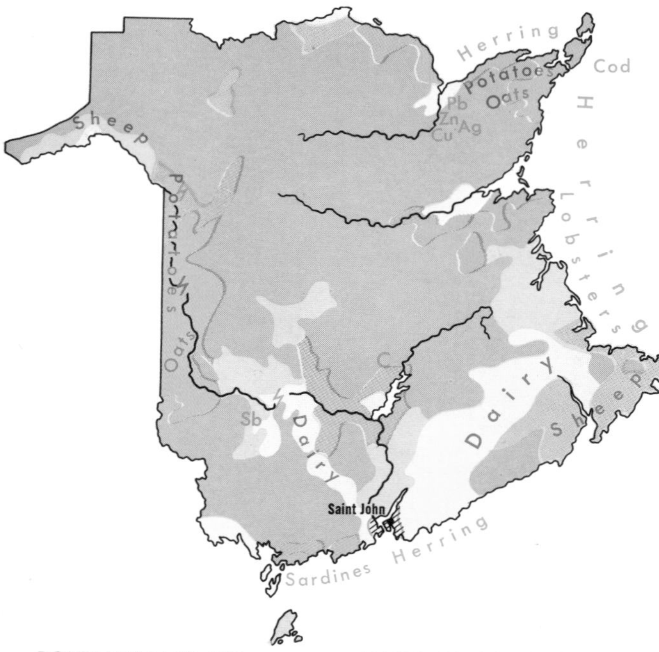

Herring
Potatoes
Cod
Pb
Oats
Zn
Cu
Ag
Sheep
Potatoes
Herring
Lobsters
Oats
C
Sb
Dairy
Dairy
Sheep
Saint John
Herring
Sardines

DOMINANT LAND USE

- Cereals, Livestock
- Dairy
- Potatoes
- General Farming, Livestock
- Pasture Livestock
- Forests

MAJOR MINERAL OCCURRENCES

Ag Silver
C Coal
Cu Copper
Pb Lead
Sb Antimony
Zn Zinc

Water Power
Major Industrial Areas

Taxis River 118 . . . . . D 2
Tay Creek 161 . . . . . D 2
Taymouth 301 . . . . . D 2
Temperance Vale 357 . . . . . C 2
The Range 58 . . . . . E 2
Thibault 306 . . . . . C 1
Tide Head 952 . . . . . D 1
Tilley 95 . . . . . C 2
Tobique Narrows 140 . . . . . C 2
Tracadie 2,452 . . . . . F 1
Tracy 636 . . . . . D 3
Turtle Creek 81 . . . . . F 3
Tweedside 87 . . . . . C 3
Upham 107 . . . . . E 3
Upper Blackville 60 . . . . . E 2
Upper Buctouche 158 . . . . . F 2
Upper Gagetown 236 . . . . . D 3
Upper Hainesville 189 . . . . . C 2
Upper Kent 203 . . . . . C 2
Upper Maugerville 543 . . . . . D 3
Upper Mills 153 . . . . . C 3
Upper Rockport 18 . . . . . F 3
Upper Sheila 706 . . . . . E 1
Upper Woodstock 257 . . . . . C 2
Upsalquitch 112 . . . . . D 1
Val-Comeau 534 . . . . . F 1
Val d'Amour 462 . . . . . D 1
Val Doucet 505 . . . . . E 1
Verret 637 . . . . . B 1
Village-Saint-Laurent 187 . . . . . E 1
Waasis 264 . . . . . D 3
Wapske 195 . . . . . C 2
Waterford 120 . . . . . E 3
Waterville 181 . . . . . C 2
Waweig . . . . . C 3
Wayerton 188 . . . . . E 1
Weaver 86 . . . . . E 2
Weldon 227 . . . . . F 3
Welsford 230 . . . . . D 3
Welshpool 260 . . . . . D 4
Westfield 1,100 . . . . . D 3
West Quaco 48 . . . . . E 3
White Head 185 . . . . . D 4
White Rapids 238 . . . . . E 2
Whitney 216 . . . . . E 2
Wickham 72 . . . . . D 3
Wicklow 143 . . . . . C 2
Williamsburg 258 . . . . . D 2
Williamstown 156 . . . . . C 2
Willow Grove 509 . . . . . E 3
Wilmot 57 . . . . . C 3
Wilsons Point 45 . . . . . F 1
Wilsons Beach 844 . . . . . D 4
Windsor 43 . . . . . C 2
Wirral 110 . . . . . D 3

Woodstock⊙ 4,649 . . . . . C 2
Woodwards Cove 146 . . . . . D 4
Youngs Cove 65 . . . . . E 3
Zealand 458 . . . . . D 2

OTHER FEATURES

Bald (mt.) . . . . . C 1
Bartibog (riv.) . . . . . E 1
Bay du Vin (riv.) . . . . . E 2
Big Tracadie (riv.) . . . . . E 1
Buctouche (harb.) . . . . . F 2
Buctouche (riv.) . . . . . F 2
Campobello (isl.) . . . . . D 4
Canaan (riv.) . . . . . E 2
Carleton (mt.) . . . . . D 1
Chaleur (bay) . . . . . E 1
Chignecto (bay) . . . . . F 3
Chiputneticook (lakes) . . . . . C 3
Cocagne (isl.) . . . . . F 2
Cumberland (basin) . . . . . F 3
Deer (isl.) . . . . . D 4
Digdeguash (riv.) . . . . . C 3
Escuminac (bay) . . . . . D 1
Escuminac (pt.) . . . . . F 1
Fundy (bay) . . . . . E 3
Fundy Nat'l Park . . . . . E 3
Gaspereau (riv.) . . . . . D 2
Grand (bay) . . . . . D 3
Grand (lake) . . . . . C 3
Grand (lake) . . . . . D 3
Grand Manan (chan.) . . . . . C 4
Grand Manan (isl.) . . . . . D 4
Grande (riv.) . . . . . C 1
Green (riv.) . . . . . B 1
Hammond (riv.) . . . . . E 3
Harvey (lake) . . . . . C 3
Heron (isl.) . . . . . D 1
Kedgwick (riv.) . . . . . C 1
Kennebecasis (riv.) . . . . . E 3
Keswick (riv.) . . . . . C 2
Kouchibouguac (bay) . . . . . F 2
Kouchibouguacis (riv.) . . . . . E 2
Kouchibouguac Nat'l Park . . . . . F 2
Lamèque (isl.) . . . . . F 1
Lepreau (riv.) . . . . . D 3
Little (riv.) . . . . . D 2
Long (isl.) . . . . . D 3
Long Reach (inlet) . . . . . D 3
Maces (bay) . . . . . D 3
Mactaquac (lake) . . . . . C 3
Madawaska (riv.) . . . . . B 1
Magaguadavic (lake) . . . . . C 3
Magaguadavic (riv.) . . . . . C 3
Miramichi (bay) . . . . . E 1

Miscou (isl.) . . . . . F 1
Miscou (pt.) . . . . . F 1
Mount Carleton Prov. Park . . . . . D 1
Musquash (harb.) . . . . . D 3
Nashwaak (riv.) . . . . . D 2
Nepisiguit (bay) . . . . . E 1
Nepisiguit (riv.) . . . . . D 1
Nerepis (riv.) . . . . . D 3
Northern (head) . . . . . D 4
North Sevogle (riv.) . . . . . D 1
Northumberland (str.) . . . . . F 2
Northwest Miramichi (riv.) . . . . . D 1
Oromocto (lake) . . . . . C 3
Oromocto (riv.) . . . . . D 3
Passamaquoddy (bay) . . . . . C 3
Patapédia (riv.) . . . . . C 1
Petitcodiac (riv.) . . . . . F 3
Pokemouche (riv.) . . . . . E 1
Pokesudie (isl.) . . . . . F 1
Pollett (riv.) . . . . . E 3
Quaco (head) . . . . . E 3
Renous (riv.) . . . . . D 2
Restigouche (riv.) . . . . . C 1
Richibucto (harb.) . . . . . F 2
Richibucto (riv.) . . . . . E 2
Roosevelt Campobello Int'l
  Park . . . . . D 4
Saint Croix (riv.) . . . . . C 3
Saint Francis (riv.) . . . . . A 1
Saint John (harb.) . . . . . E 3
Saint John (riv.) . . . . . C 2
Saint Lawrence (gulf) . . . . . F 1
Salisbury (bay) . . . . . F 3
Salmon (riv.) . . . . . C 1
Salmon (riv.) . . . . . E 2
Shediac (isl.) . . . . . F 2
Shepody (bay) . . . . . F 3
Shippegan (bay) . . . . . E 1
Shippegan Gully (str.) . . . . . F 1
South Sevogle (riv.) . . . . . D 1
Southwest (head) . . . . . D 4
Southwest Miramichi (riv.) . . . . . D 2
Spear (cape) . . . . . G 2
Spednik (lake) . . . . . C 3
Spencer (cape) . . . . . E 3
Tabusintac (riv.) . . . . . E 1
Tabusintac Gully (str.) . . . . . F 1
Tetagouche (riv.) . . . . . D 1
Tobique (riv.) . . . . . C 2
Upsalquitch (riv.) . . . . . D 1
Utopia (lake) . . . . . D 3
Verte (bay) . . . . . G 2
Washademoak (lake) . . . . . E 3
West (isls.) . . . . . D 4
White Head (isl.) . . . . . D 4

⊙County seat.

Topography

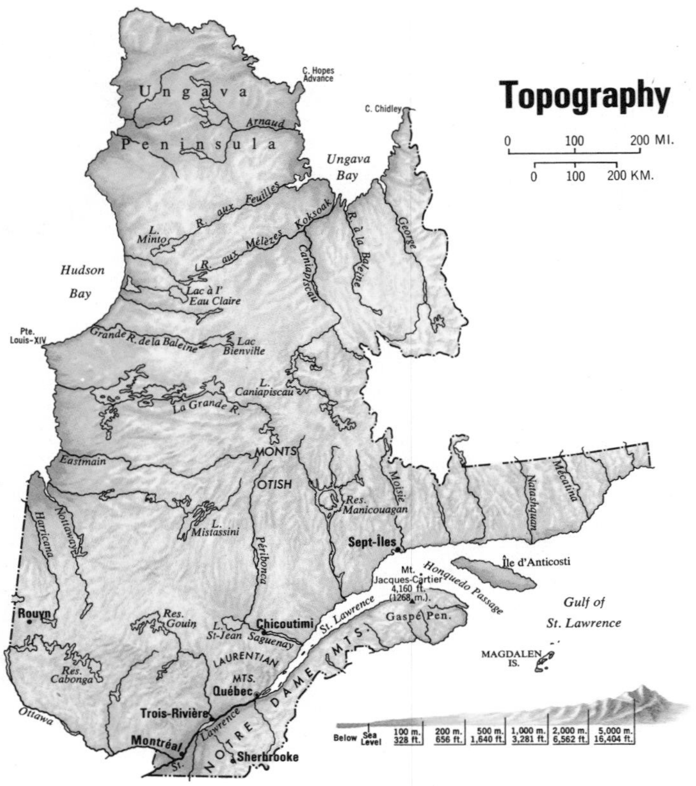
Topography
0 100 200 MI.
0 100 200 KM.
Ungava Peninsula
C. Hopes Advance
C. Chidley
Ungava Bay
Hudson Bay
Pte. Louis-XIV
L. Minto
R. aux Feuilles
Arnaud
Koksoak
R. aux Mélèzes
Lac à l'Eau Claire
Grande R. de la Baleine
Lac Bienville
L. Caniapiscau
La Grande R.
Eastmain
MONTS OTISH
Res. Manicouagan
L. Mistassini
Sept-Îles
Île d'Anticosti
Mt. Jacques-Cartier 4,160 ft. (1268 m.)
Gulf of St. Lawrence
Gaspé Pen.
Rouyn
Res. Gouin
Chicoutimi
Saguenay
LAURENTIAN MTS.
Québec
Res. Cabonga
Trois-Rivières
Montréal
Sherbrooke
NOTRE DAME MTS.
MAGDALEN IS.
Below Sea Level 100 m. 328 ft. 200 m. 656 ft. 500 m. 1,640 ft. 1,000 m. 3,281 ft. 2,000 m. 6,562 ft. 5,000 m. 16,404 ft.

COUNTIES

Argenteuil 32,454 C 4
Arthabaska 59,277 E 4
Bagot 26,840 E 4
Beauce 73,427 G 3
Beauharnois 54,034 C 4
Bellechasse 23,559 G 3
Berthier 31,096 C 3
Bonaventure 40,487 C 2
Brome 17,436 E 4
Chambly 307,090 J 4
Champlain 119,595 E 2
Charlevoix-Est 17,448 G 2
Charlevoix-Ouest 14,172 G 2
Châteauguay 59,968 D 4
Chicoutimi 174,441 G 1
Compton 20,536 F 4
Deux-Montagnes 71,252 C 4
Dorchester 33,949 C 3
Drummond 69,770 E 4
Frontenac 26,814 G 4
Gaspé-Est 41,173 D 1
Gaspé-Ouest 18,943 C 1
Gatineau 54,229 B 3
Hull 131,213 B 4
Huntingdon 16,953 C 4
Iberville 23,180 D 4
Île-de-Montréal 1,760,122 H 4
Île-Jésus 268,335 H 4
Joliette 60,384 C 3
Kamouraska 28,642 H 2
Labelle 34,395 B 3
Lac-Saint-Jean-Est 47,891 F 1
Lac-Saint-Jean-Ouest 62,952 E 1
Laprairie 105,962 H 4
L'Assomption 109,705 D 4
Lévis 94,104 J 3
L'Islet 22,062 G 2
Lotbinière 29,653 F 3
Maskinongé 20,763 D 3
Matane 29,955 B 1
Matapédia 23,715 B 2
Mégantic 57,892 F 3
Missisquoi 36,161 D 4
Montcalm 27,557 C 3
Montmagny 25,622 G 3
Montmorency No. 1 23,048 F 2
Montmorency No. 2 6,436 G 3
Napierville 13,562 D 4
Nicolet 33,513 E 3
Papineau 37,975 B 4
Pontiac 20,283 A 3
Portneuf 58,843 E 3
Québec 458,980 F 3
Richelieu 53,058 D 4
Richmond 40,871 E 4
Rimouski 69,099 J 1
Rivière-du-Loup 41,250 H 2
Rouville 42,391 D 4
Saguenay 115,881 H 1
Saint-Hyacinthe 55,888 D 4
Saint-Jean 55,576 D 4
Saint-Maurice 107,703 D 3
Shefford 70,733 E 4
Sherbrooke 115,983 E 4
Soulanges 15,429 C 4
Stanstead 38,186 F 4
Témiscouata 52,570 J 2
Terrebonne 193,865 H 4
Vaudreuil 50,043 C 4
Verchères 63,353 J 4
Wolfe 15,635 F 4
Yamaska 14,797 E 3

CITIES and TOWNS

Acton Vale 4,371 E 4
Albanel 992 E 1
Alma⊙ 26,322 F 1
Amqui⊙ 4,048 B 2
Ancienne-Lorette 12,935 H 3
Angers B 4
Anjou 37,346 H 4
Annaville 712 E 3
Armagh 878 G 3
Arthabaska⊙ 6,827 F 3
Arvida F 1
Asbestos 7,967 F 4
Ascot Corner 847 F 4
Audet 760 G 4
Ayer's Cliff⊙ 810 E 4
Aylmer 26,695 B 4
Baie-Comeau 12,866 A 1
Baie-d'Urfé 3,674 G 4
Baie-Saint-Paul⊙ 3,961 G 2
Baie-Trinité 749 B 1
Beaconsfield 19,613 H 4
Beauceville 4,302 G 3
Beauharnois⊙ 7,025 D 4
Beaumont 791 F 3
Beauport 60,447 J 3
Beaupré 2,740 G 2
Bécancour⊙ 10,247 E 3
Bedford⊙ 2,832 E 4
Beebe Plain 1,072 E 4
Bélair (Val-Bélair) 12,695 H 3
Beloeil 17,540 D 4
Bernierville 2,120 F 3
Berthier-en-Bas 562 G 3
Berthierville⊙ 4,049 D 3
Bic 2,994 J 1
Biencourt 824 J 2
Black Lake 5,148 F 3
Blainville 14,682 H 4
Boischatel 3,345 J 3
Bois-des-Filion 4,943 H 4
Bolduc 1,565 G 4
Bonaventure 1,371 C 2
Boucherville 29,704 J 4
Bromont 2,731 E 4
Bromptonville 3,035 F 4
Brossard 52,232 H 4
Brownsburg 2,875 C 4
Buckingham 7,992 B 4
Cabano 3,291 J 2
Cacouna 1,160 H 2
Calumet 729 C 4
Candiac 8,502 J 4
Cap-à-l'Aigle 819 G 2
Cap-Chat 3,464 B 1
Cap-de-la-Madeleine 32,626 E 3
Caplan-Rivière Caplan 1,139 C 2
Cap-Saint-Ignace 1,485 G 2
Cap-Santé⊙ 671 F 3
Carignan 4,544 J 4
Carleton 2,710 C 2
Causapscal 2,501 B 2
Chambly 12,190 J 4
Chambord 961 E 1
Chandler 3,946 D 2
Charlemagne 4,827 H 4
Charlesbourg 68,326 J 3
Charny 8,240 J 3
Châteauguay 36,928 H 4
Château-Richer⊙ 3,628 F 3
Chénéville 633 B 4
Chicoutimi⊙ 60,064 G 1
Chicoutimi-Jonquière *135,172 G 1
Chute-aux-Outardes 2,280 A 1
Clermont 3,621 G 2
Coaticook 6,271 F 4
Coleraine 1,660 F 4
Compton 728 F 4
Contrecoeur 5,449 D 4
Cookshire⊙ 1,480 F 4
Coteau-du-Lac 1,247 C 4
Coteau-Landing⊙ 1,386 C 4
Côte-Saint-Luc 27,531 H 4
Courcelles 608 G 4
Courville J 3
Cowansville 12,240 E 4
Crabtree 1,950 D 4
Danville 2,200 E 4
Daveluyville 1,257 E 3
Deauville 942 E 4
Dégelis 3,477 J 2
Delisle 4,011 F 1
Delson 4,935 H 4
Desbiens 1,541 E 1
Deschaillons-sur-Saint-Laurent 950 E 3
Deschambault 977 E 3
Deschênes B 4
Deux-Montagnes 9,944 H 4
Didyme 667 E 1
Disraëli 3,181 F 4
Dolbeau 8,766 E 1
Dollard-des-Ormeaux 39,940 H 4
Donnacona 5,731 F 3
Dorion 5,749 C 4
Dorval 17,727 H 4
Dosquet 703 F 3
Douville D 4
Drummondville⊙ 27,347 E 4
Drummondville-Sud 9,220 E 4
Dunham 2,887 E 4
Durham-Sud 1,045 E 4
East Angus 4,016 F 4
East Broughton 1,397 F 3
East Broughton Station 1,302 F 3
Eastman 612 E 4
Entrelacs 1,735 C 3
Farnham 6,498 E 4
Ferme-Neuve 2,266 B 3
Forestville 4,271 H 1
Frampton 684 G 3
Francoeur 1,422 F 3
Gaspé 17,261 D 1
Gatineau 74,988 B 4
Giffard J 3
Girardville 1,128 E 1
Gracefield 869 A 3
Granby 38,069 E 4
Grand'Mère 15,442 E 3
Grande-Rivière 4,420 D 2
Grandes-Bergeronnes 748 H 1
Grande-Vallée 700 D 1
Greenfield Park 18,527 J 4
Grenville 1,417 C 4
Gros-Morne 672 C 1
Hampstead 7,598 H 4
Ham-Sud⊙ 62 F 4
Hauterive 13,995 A 1
Hébertville 2,515 F 1
Hébertville-Station 1,442 F 1
Hemmingford 737 D 4
Henryville 595 D 4
Howick 639 D 4
Hudson 4,414 C 4
Hull⊙ 56,225 B 4
Huntingdon⊙ 3,018 C 4
Île-Perrot 5,945 G 4
Iberville⊙ 8,587 D 4
Inverness⊙ 329 F 3
Joliette⊙ 16,987 D 3
Jonquière 60,354 F 1
Jonquière-Chicoutimi *135,172 F 1
Kingsey Falls 818 E 4
Kirkland 10,476 H 4
Knowlton (Lac-Brome)⊙ 4,316 E 4
La Baie 20,935 G 1
Labelle 1,534 C 3
Lac-à-la-Croix 1,017 F 1
Lac-Alouette-Lac-Brière 1,356 D 4
Lac-au-Saumon 1,332 B 2
Lac-aux-Sables 838 E 3
Lac-Beaufort F 3
Lac-Bouchette 1,703 E 1
Lac-Carré 717 C 3
Lac-des-Écorces 766 B 3
Lac-Drolet 1,120 G 4
Lac-Etchemin 2,729 G 3
Lachenaie 8,631 D 4
Lachine 37,521 H 4
Lachute⊙ 11,729 C 4
Lac-Mégantic⊙ 6,119 G 4
Lacolle 1,319 D 4
Lac-Saint-Charles 5,837 H 3
Lafontaine 4,877 C 4
La Guadeloupe 1,692 F 4
La Malbaie⊙ 4,030 G 2
Lambton 1,559 F 4
L'Annonciation 2,384 C 3
Lanoraie (Lanoraie-d'Autry) 1,613 D 4
La Pêche 4,977 B 4
La Pérade 1,039 E 3
La Pocatière 4,560 H 2
La Prairie⊙ 10,627 J 4
La Providence E 4
Larouche 662 F 1
La Salle 76,299 H 4
L'Ascension 1,287 F 1
L'Assomption⊙ 4,844 D 4
La Station-du-Coteau 892 C 4
Laterrière 788 F 1
La Tuque 11,556 E 2
Laurentides 1,947 D 4
Laurier-Station 1,123 F 3
Laurierville 939 F 3
Lauzon 13,362 J 3
Laval 268,335 H 4
Lavaltrie 2,053 D 4
L'Avenir 1,116 E 4
Lawrenceville 562 E 4
Le Moyne 6,137 J 4
L'Épiphanie 2,971 D 4
Léry 2,239 H 4
Lévis 17,895 J 3
Lennoxville 3,922 F 4
Les Méchins 803 B 1
Linière 1,168 G 3
L'Islet 1,070 G 2
L'Islet-sur-Mer 774 G 2
L'Isle-Verte 1,142 G 1
Longueuil⊙ 124,320 J 4
Loretteville⊙ 15,060 H 3
Lorraine 6,881 H 4
Louiseville⊙ 3,735 E 3
Luceville 1,524 J 1
Lyster 830 F 3
Magog 13,604 E 4
Maniwaki⊙ 5,424 B 3
Manseau 626 E 3
Maple Grove 2,009 H 4
Maria 1,178 C 2
Marieville⊙ 4,877 D 4
Mascouche 20,345 H 4
Maskinongé 1,005 E 3
Masson 4,264 B 4
Massueville 671 E 4
Matane⊙ 13,612 B 1
Matapédia 586 B 2
Melocheville 1,892 C 4
Mercier 6,352 H 4
Metabetchouan 3,406 F 1
Mirabel⊙ 14,080 H 4
Mistassini 6,682 E 1
Montauban 557 E 3
Mont-Carmel 807 H 2
Montcerf 570 A 3
Montebello 1,229 B 4
Mont-Joli 6,359 J 1
Mont-Laurier⊙ 8,405 B 3
Mont-Louis 756 C 1
Montmagny⊙ 12,405 G 3
Montréal⊙ 980,354 H 4
Montréal *2,828,349 H 4
Montréal-Est 3,778 J 4
Montréal-Nord 94,914 H 4
Mont-Rolland 1,517 C 4
Mont-Royal 19,247 H 4
Mont-Saint-Hilaire 10,066 D 4
Morin Heights 592 C 4
Murdochville 3,396 C 1
Nantes 1,167 F 4

Agriculture, Industry and Resources

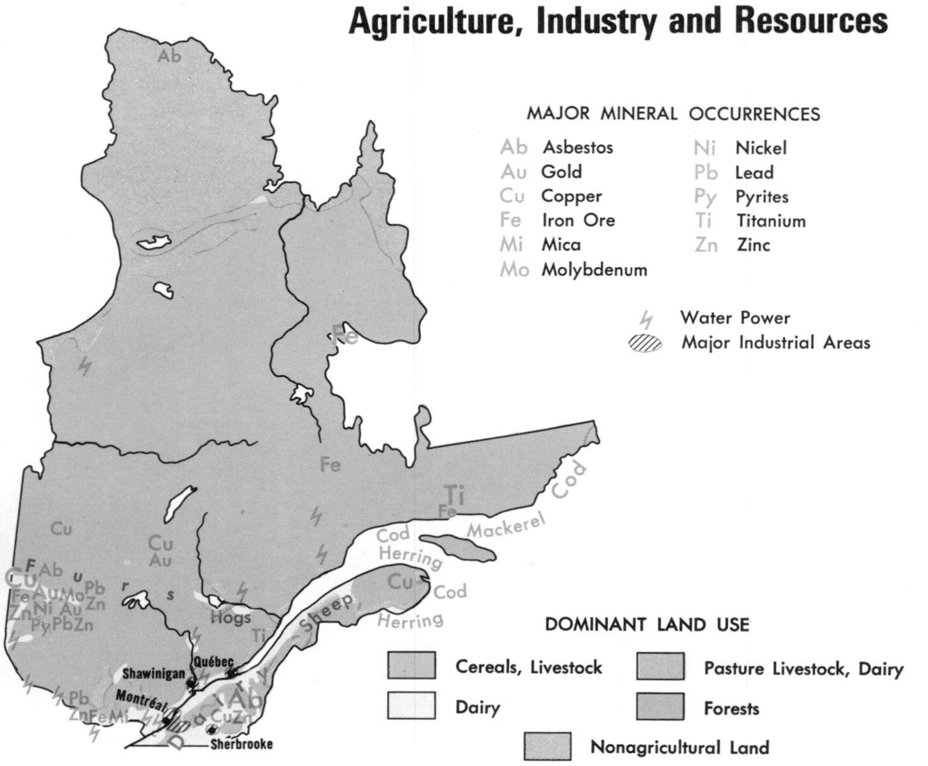
MAJOR MINERAL OCCURRENCES
Ab Asbestos
Au Gold
Cu Copper
Fe Iron Ore
Mi Mica
Mo Molybdenum
Ni Nickel
Pb Lead
Py Pyrites
Ti Titanium
Zn Zinc
Water Power
Major Industrial Areas
DOMINANT LAND USE
Cereals, Livestock
Pasture Livestock, Dairy
Dairy
Forests
Nonagricultural Land
Cod
Mackerel
Herring
Hogs
Sheep
Shawinigan
Québec
Montréal
Sherbrooke

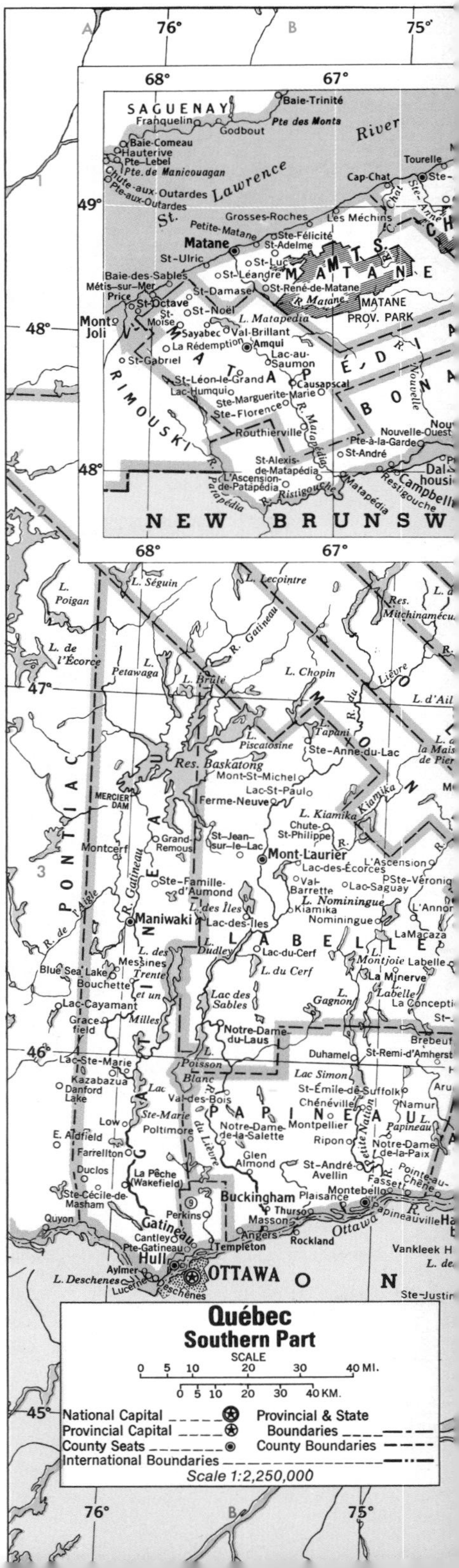
Québec Southern Part
SCALE
0 5 10 20 30 40 MI.
0 5 10 20 30 40 KM.
National Capital
Provincial Capital
County Seats
International Boundaries
Provincial & State Boundaries
County Boundaries
Scale 1:2,250,000
SAGUENAY
Baie-Trinité
Franquelin
Godbout
Pte des Monts
Baie-Comeau
Hauterive
St. Lawrence River
Matane
MATANE PROV. PARK
RIMOUSKI
NEW BRUNSWICK
Amqui
Causapscal
Mont-Laurier
Maniwaki
Buckingham
Gatineau
Hull
OTTAWA
Aylmer
Montebello
Ottawa

Napierville© 2,343 . . . . D 4
Neuville 996 . . . . F 3
New Carlisle© 1,292 . . . . D 2
New Richmond 4,257 . . . C 2
Nicolet 4,880 . . . . B 3
Nominingue 881 . . . . . . B 3
Normandin 4,041 . . . . . E 1
North Hatley 689 . . . . . F 4
Notre-Dame-de-la-Doré 1,064 E 1
Notre-Dame-des-Laurentides H 3
Notre-Dame-des-Prairies
6,150 . . . . D 3
Notre-Dame-du-Bon-Conseil
1,089 . . . . E 4
Notre-Dame-du-Lac 2,258 . J 2
Nouvelle 669 . . . . C 2
Oka 1,538 . . . . C 4
Omerville 1,398 . . . . . . E 4
Ormstown 1,659 . . . . . . D 4
Orsainville . . . . H 3
Otis 673 . . . . G 1
Otterburn Park 4,268 . . . D 4
Outremont 24,338 . . . . . H 4
Pabos 1,295 . . . . D 2
Pabos-Mills 1,565 . . . . . D 2
Papineauville 1,481 . . . . C 4
Paspébiac 1,914 . . . . . . D 2
Percé© 4,839 . . . . D 1
Petit-Cap 1,023 . . . . . . D 1
Petite-Matane 1,065 . . . . B 1
Petit-Saguenay (Saint-
François-d'Assise) 804 . . G 1
Pierrefonds 38,390 . . . . H 4
Pierreville 1,212 . . . . . . E 3

Pincourt 8,750 . . . . D 4
Pintendre 1,849 . . . . . . J 3
Plaisance 748 . . . . B 4
Plessisville 7,249 . . . . . F 3
Pohénégamooke 3,702 . . H 2
Pointe-à-la-Croix 1,481 . . C 2
Pointe-au-Père 796 . . . . J 1
Pointe-au-Pic 1,054 . . . . G 2
Pointe-aux-Outardes 1,056 . A 1
Pointe-aux-Trembles 36,270 J 4
Pointe-Calumet 2,935 . . . G 4
Pointe-Claire 24,571 . . . H 4
Pointe-du-Lac 5,359 . . . E 3
Pointe-Gatineau . . . . . . H 4
Pointe-Lebel 1,573 . . . . A 1
Port-Alfred 8,621 . . . . . G 1
Portneuf 1,333 . . . . F 3
Portneuf-sur-Mer (Rivière-
Portneuf-sur-Mer) 1,255 . H 1
Price 2,273 . . . . A 1
Princeville 4,023 . . . . . . F 3
Proulxville 588 . . . . E 3
Québec (cap.) 166,474 . . H 3
Québec ©576,075 . . . . . H 3
Quyon 744 . . . . A 4
Rawdon 2,958 . . . . D 3
Repentigny 34,419 . . . . J 4
Richelieu 1,832 . . . . . . D 4
Richmond© 3,568 . . . . . E 4
Rigaud 2,268 . . . . C 4
Rimouski© 29,120 . . . . . J 1
Rimouski-Est 2,506 . . . . J 1
Ripon 620 . . . . B 4

Rivière-à-Pierre 615 . . . . E 3
Rivière-au-Renard 2,211 . D 1
Rivière-Bleue 1,690 . . . . J 2
Rivière-Bois-Clair 604 . . . F 3
Rivière-du-Loup 13,459 . . H 2
Rivière-du-Moulin . . . . . G 1
Rivière-Éternité 659 . . . . G 1
Rivière-Portneuf-Portneuf-sur-
Mer 1,255 . . . . H 1
Robertsonville 1,987 . . . F 3
Roberval© 11,429 . . . . . E 1
Rock Island 1,179 . . . . . E 4
Rosemère 7,778 . . . . . . H 4
Rougemont 972 . . . . . . D 4
Roxboro 6,292 . . . . H 4
Roxton Falls 1,245 . . . . . E 4
Sacré-Coeur-de-Saguenay
1,678 . . . . H 1
Saint-Adelme 618 . . . . . B 1
Saint-Adelphe 1,159 . . . . E 3
Saint-Adolphe-d'Howard
1,686 . . . . C 4
Saint-Adrien 597 . . . . . . E 3
Saint-Agapitville 2,954 . . F 3
Saint-Aimé-des-Lacs 861 . G 2
Saint-Alban 673 . . . . . . E 3
Saint-Alexandre-de-
Kamouraska 1,048 . . . . H 2
Saint-Alexis-des-Monts 1,984 D 3
Saint-Amable 2,424 . . . . J 4
Saint-Ambroise 3,606 . . . F 1
Saint-Anaclet 1,377 . . . . J 1
Saint-André-Avellin 1,312 . B 4
Saint-André-Est 1,293 . . C 4

Saint-Anselme 1,808 . . . F 3
Saint-Antoine 7,012 . . . . H 4
Saint-Antonin 941 . . . . . H 2
Saint-Aubert 884 . . . . . G 2
Saint-Augustin-de-Québec
2,475 . . . . E 3
Saint-Basile-Sud 1,719 . . E 3
Saint-Basile-le-Grand 7,658 J 4
Saint-Benjamin 1,027 . . . G 3
Saint-Bernard 585 . . . . . F 3
Saint-Bernard-sur-Mer 711 . G 2
Saint-Boniface-de-Shawinigan
3,164 . . . . D 3
Saint-Bruno 2,580 . . . . . F 1
Saint-Bruno-de-Montarville
22,880 . . . . J 4
Saint-Camille-de-Bellechasse
1,744 . . . . G 3
Saint-Casimir 1,133 . . . . E 3
Saint-Césaire 2,935 . . . . D 4
Saint-Charles 1,019 . . . . F 3
Saint-Charles-de-Mandeville
1,392 . . . . D 3
Saint-Chrysostome 1,018 . D 4
Sainte-Côme 660 . . . . . D 3
Saint-Constant 9,938 . . . H 4
Saint-Cyprien 860 . . . . . J 2
Saint-Cyrille 1,041 . . . . . E 4
Saint-Damien-de-Buckland
1,522 . . . . G 3
Saint-David 5,380 . . . . . E 3
Saint-David-de-Falardeau
1,876 . . . . F 1
Saint-Denis 861 . . . . . . D 4

Saint-Dominique 2,068 . . . E 4
Saint-Donat-de-Montcalm
1,521 . . . . C 3
Sainte-Adèle 4,675 . . . . C 4
Sainte-Agathe 709 . . . . . F 3
Sainte-Agathe-des-Monts
5,641 . . . . C 3
Sainte-Anne-de-Beaupré
3,292 . . . . F 2
Sainte-Anne-de-Bellevue
3,981 . . . . H 4
Sainte-Anne-des-Monts©
6,062 . . . . C 1
Sainte-Anne-des-Plaines
4,258 . . . . H 4
Sainte-Anne-du-Lac 686 . . B 3
Sainte-Aurélie 1,045 . . . . G 3
Sainte-Blandine 849 . . . . J 1

Sainte-Catherine 1,474 . . F 3
Sainte-Claire 1,566 . . . . G 3
Sainte-Croix 1,814 . . . . . F 3
Sainte-Félicité 711 . . . . . B 1
Sainte-Foy 68,883 . . . . . H 3
Sainte-Geneviève 2,573 . . H 4
Sainte-Geneviève-de-
Batiscan© 356 . . . . . . E 3
Sainte-Hélène-de-Bagot
1,328 . . . . E 4
Sainte-Hénédine© 639 . . . F 3
Sainte-Julie-de-Verchères
14,243 . . . . J 4
Sainte-Julienne© 750 . . . G 3
Sainte-Justine 1,080 . . . . G 3
Saint-Élie 639 . . . . E 3
Saint-Elzéar 743 . . . . . . F 3
Sainte-Marie 8,937 . . . . G 3

Sainte-Martine© 2,196 . . . D 4
Saint-Émile 5,216 . . . . . H 3
Sainte-Monique 705 . . . . F 1
Sainte-Pétronille 982 . . . J 3
Sainte-Perpétue-de-L'Islet
1,232 . . . . H 2
Saint-Éphrem-de-Tring 973 . G 3
Saint-Épiphane 647 . . . . H 2
Saint-Pudentienne 866 . . . E 4
Sainte-Rosalie 2,862 . . . . E 4
Saint-Esprit 1,068 . . . . . D 4
Sainte-Thérèse 18,750 . . H 4
Sainte-Thérèse-Ouest
(Boisbriand) 13,471 . . . . H 4
Sainte-Thècle 1,703 . . . . E 3
Saint-Étienne-de-Grès 845 . E 3
Saint-Étienne-de-Lauzon
1,218 . . . . J 3

AREA 594,857 sq. mi. (1,540,680 sq. km.)
POPULATION 6,438,403
CAPITAL Québec
LARGEST CITY Montréal
HIGHEST POINT Mont D'Iberville 5,420 ft. (1,652 m.)
SETTLED IN 1608
ADMITTED TO CONFEDERATION 1867
PROVINCIAL FLOWER White Garden Lily

Gaspé Peninsula

COUNTIES
indicated by numbers:
1 Iberville . . . . D 4
2 Napierville . . . . . . D 4
3 Rouville . . . . E 4
4 St-Hyacinthe . . . . . D 4
5 Île-de-Montréal . . . H 4
6 Deux-Montagnes . . . C 4
7 Soulanges . . . . C 4
8 Beauharnois . . . . . C 4
9 Hull . . . . B 4
10 Jésus . . . . H 4
11 Richelieu . . . . D 4
12 Vaudreuil . . . . C 4

Internal divisions represent Municipal Counties

© Copyright HAMMOND INCORPORATED, Maplewood, N.J.

Saint-Eustache 29,716......H 4
Saint-Fabien 1,361.........J 1
Saint-Félicien 9,058.......E 1
Saint-Félix-de-Valois 1,462...D 3
Saint-Ferréol-les-Neiges
1,758.................G 2
Saint-Flavien 734..........E 3
Saint-François-de-Sales 831...E 1
Saint-François-du-Lac® 942...E 3
Saint-Fulgence 950.........G 1
Saint-Gabriel 3,161........E 3
Saint-Gabriel-de-Rimouski
779...................J 1
Saint-Gédéon, Frontenac
1,569.................G 4
Saint-Gédéon, Lac-St-Jean-E.®
1,000.................F 1
Saint-Georges, Beauce
10,342................G 3
Saint-Georges, Champlain
3,344.................E 3
Saint-Georges-Ouest 6,378...G 3
Saint-Germain-de-Grantham
1,373.................E 4
Saint-Gervais 973..........F 3
Saint-Gilles 912...........F 3
Saint-Grégoire (Mont-St-
Grégoire)® 740.........D 4
Saint-Henri 1,970..........J 3
Saint-Honoré, Beauce 1,116...G 4
Saint-Honoré, Chicoutimi
1,790.................F 1
Saint-Hubert 60,573........J 4
Saint-Hubert-de-Témiscouata
871...................J 2
Saint-Hyacinthe® 38,246....D 4
Saint-Isidore 811..........J 3
Saint-Isidore-de-Laprairie 769 D 4
Saint-Jacques 2,152........D 3
Saint-Jacques-le-Mineur
1,203.................H 4
Saint-Jean-Chrysostome
6,930.................J 3
Saint-Jean-de-Dieu 1,377...J 1
Saint-Jean-de-Matha 931....D 3
Saint-Jean-Port-Joli® 1,813...G 2
Saint-Jean-sur-Richelieu®
35,640................D 4
Saint-Jérôme 25,123........H 4
Saint-Joachim 1,139........G 2
Saint-Joseph-de-Beauce
3,216.................G 3
Saint-Joseph-de-Sorel 2,545...D 3
Saint-Jovite 3,841.........C 3
Saint-Lambert 20,557.......J 4
Saint-Laurent 65,900.......H 4

Saint-Lazare 731...........G 3
Saint-Léonard 79,429.......H 4
Saint-Léonard-d'Aston 992...E 3
Saint-Léon-de-Chicoutimi 749 F 1
Saint-Léon-de-Standon 816...G 3
Saint-Léon-le-Grand 722....B 2
Saint-Liboire® 746.........E 4
Saint-Louis-de-Gonzague
615...................D 4
Saint-Louis-de-Terrebonne
14,172................H 4
Saint-Louis-du-Ha! Ha! 809...H 2
Saint-Luc 8,815............D 4
Saint-Luc-de-Matane 598....B 1
Saint-Marc-des-Carrières
2,822.................E 3
Saint-Méthode-de-Frontenac
925...................F 3
Saint-Michel-de-Bellechasse
963...................G 3
Saint-Michel-des-Saints
1,584.................D 3
Saint-Nazaire-de-Chicoutimi
962...................F 1
Saint-Nérée 970...........G 3
Saint-Nicolas 5,074........H 3
Saint-Noël 666............B 1
Grande-Odilon 580.........G 3
Saint-Omer 718............C 2
Saint-Ours 625............D 4
Saint-Pacôme 1,996........G 2
Saint-Pamphile 3,428.......H 3
Saint-Pascal® 2,763........H 2
Saint-Paul-de-Montminy 602...G 3
Saint-Paulin 663...........E 3
Saint-Paul-l'Ermite (Le
Gardeur) 8,312.........J 4
Saint-Philippe-de-Néri 715...H 2
Saint-Pie 1,725............E 4
Saint-Pierre 5,305.........H 4
Saint-Pierre-d'Orléans 880...G 4
Saint-Polycarpe 602........C 4
Saint-Prime 2,522.........E 1
Saint-Prosper-de-Dorchester
2,150.................G 3
Saint-Raphaël 1,346.......G 3
Saint-Raymond 3,605.......F 3
Saint-Rédempteur 4,463....J 3
Saint-Régis 1,370.........C 4
Saint-Rémi 5,146..........D 4
Saint-Roch-de-l'Achigan
1,160.................H 4
Saint-Roch-de-Richelieu®
1,650.................D 3
Saint-Romuald-d'Etchemin®
9,849.................J 3

Saint-Sauveur-des-Monts
2,348.................C 4
Saint-Siméon 1,152........H 1
Saint-Simon 602...........E 3
Saint-Stanislas 1,443......E 3
Saint-Sylvère 1,006.......E 3
Saint-Timothée 2,113......J 4
Saint-Tite 3,031..........E 3
Saint-Tite-des-Caps 626....G 2
Saint-Ubald 1,605.........E 3
Saint-Ulric 792...........B 1
Saint-Urbain-de-Charlevoix
1,079.................G 2
Saint-Victor 1,104........G 3
Saint-Zacharie 1,284......G 3
Saint-Zotique 1,774.......C 4
Sault-au-Mouton 828.......H 1
Sawyerville 939...........F 4
Sayabec 1,721............B 2
Scotstown 762............F 4
Senneville 1,221..........G 4
Shannon 3,488............F 3
Shawbridge 942...........C 4
Shawinigan 23,011.........E 3
Shawinigan-Sud 11,325....E 3
Shawville 1,608...........A 4
Sherbrooke® 74,075.......E 4
Sherrington 614...........D 4
Sillery 12,825............J 3
Sorel® 20,347............D 4
Squatec 1,000............J 2
Stanstead Plain 1,093.....F 4
Sutton 1,599.............E 4
Tadoussac® 900..........H 1
Templeton................H 4
Terrebonne 11,769........H 4
Thetford Mines 19,965....F 3
Thurso 2,780............B 4
Tourelle (Tourelle-Grand-
Tourelle) 942..........C 1
Tourville 659............H 2
Tracy 12,843............D 3
Tring-Jonction 1,315......F 3
Trois-Pistoles 4,445......H 1
Trois-Rivières 50,466.....E 3
Trois-Rivières *111,453....E 3
Trois-Rivières-Ouest 13,107...E 3
Upton 926...............E 4
Val-Barrette 609.........B 3
Val-Brillant 687.........B 1
Valcourt 2,601..........E 4
Val-David 2,336.........C 3
Vallée-Jonction 1,200.....G 3
Valleyfield (Salaberry-de-
Valleyfield) 29,574.....C 4
Vanier 10,725...........J 3

Saint-Sauveur-des-Monts

Varennes 8,764...........J 4
Vaudreuil® 7,608.........J 4
Verchères® 4,473.........J 4
Verdun 61,287...........H 4
Victoriaville 21,838.......F 3
Villeneuve..............J 3
Warwick 2,847...........F 3
Waterloo® 4,664.........E 4
Waterville 1,397.........F 4
Weedon-Centre 1,263.....F 4
Westmount 20,480........H 4
Wickham 2,043...........E 4
Windsor 5,233...........F 4
Wottonville 673..........F 4
Yamachiche® 1,258.......E 3

OTHER FEATURES

Alma (isl.)...............F 1
Aylmer (lake)...........F 4
Baskatong (res.)........B 3
Batiscan (riv.)..........E 3
Bécancour (riv.)........F 3
Bonaventure (isl.).......C 1
Bonaventure (riv.).......C 1
Brome (lake)...........E 4
Brompton (lake)........E 4
Cascapédia (riv.).......C 1
Chaleur (bay)..........C 2
Champlain (lake).......D 4
Chaudière (riv.)........G 4
Chic-Chocs (mts.).......C 1
Chicoutimi (riv.)........F 2
Coudres (isl.)..........G 2
Deschênes (lake).......A 4
Deux Montagnes (lake)...C 4
Ditton (riv.)............F 4
Forillon Nat'l Park......D 1
Fort Chambly Nat'l Hist. Park. J 4
Gaspé (bay)............D 2
Gaspé (cape)..........D 1
Gaspé (pen.)...........D 2
Gaspésie Prov. Park.....C 1
Gatineau (riv.).........B 3
Îles (pen.)............B 3
Jacques-Cartier (mt.)....C 1
Jacques-Cartier (riv.)...F 1
Kénogami (lake)........F 1
Kiamika (lake).........B 3
La Maurice Nat'l Park....E 3
Laurentides Prov. Park...F 2
Lièvre (riv.)...........B 3
Lièvres (isl.)..........H 2
Maskinongé (riv.).......E 3
Matane (riv.)..........B 1
Matane Prov. Park.......B 1

Matapédia (riv.).........B 2
Mégantic (lake).........G 4
Memphremagog (lake)....E 4
Mercier (dam)..........A 3
Métabetchouane (riv.)...F 1
Mille Îles (riv.)........H 4
Montmorency (riv.)......F 2
Mont-Tremblant Prov. Park. C 3
Nicolet (riv.)...........E 3
Nominingue (lake)......B 3
Nord (riv.)............H 4
Orléans (isl.)..........F 3
Ottawa (riv.)..........B 4
Ouareau (riv.).........D 3
Ouelle (riv.)..........H 2
Patapédia (riv.).......B 2
Péribonca (riv.).......E 1
Petite Nation (riv.)....B 4
Prairies (riv.).........H 4
Rimouski (riv.).........J 1
Ristigouche (riv.)......B 2
Saguenay (riv.).........G 1
Sainte-Anne (riv.)......F 3
Sainte-Anne (riv.)......G 2
Saint-François (lake)...F 4
Saint-François (riv.)...E 4
Saint-Jean (lake).......E 1
Saint Lawrence (gulf)...D 2
Saint Lawrence (riv.)...H 1
Saint-Louis (lake)......H 4
Saint-Maurice (riv.)....E 2
Saint-Pierre (lake).....E 3
Shawinigan (riv.).......E 3
Shipshaw (riv.).........F 1
Soeurs (isl.)..........H 4
Témiscouata (lake)......H 2
Tremblant (lake).......C 3
Trente et un Milles (lake). B 3
Verte (isl.)............J 1
Yamaska (riv.).........E 4
York (riv.)............D 1

QUÉBEC, NORTHERN
INTERNAL DIVISIONS

Abitibi (county) 93,529.....B 2
Abitibi (terr.)..........B 3
Berthier (county) 31,096...B 3
Bonaventure (county) 40,487. C 3
Champlain (county) 119,595.. C 3
Charlevoix-Est (co.) 17,448.. C 3

Charlevoix-Ouest (county)
14,172................C 3
Chicoutimi (county) 174,441.. C 2
Gaspé-Est (county) 41,173...C 2
Gaspé-Ouest (county) 18,943 D 3
Gatineau (county) 54,229....B 3
Joliette (county) 60,384....C 3
Lac-Saint-Jean-Est (county)
47,891................C 3
Lac-Saint-Jean-Ouest
(county) 62,952........C 3
Maskinongé (county) 20,763.. C 3
Matane (county) 29,955....D 3
Matapédia (county) 23,715.. C 3
Mistassini (terr.).......B 3
Montcalm (county) 27,557.. B 3
Montmorency No. 1 (county)
23,048................C 3
Nouveau-Québec (terr.)...E 1
Pontiac (county) 20,283....C 3
Portneuf (county) 58,843...C 3
Québec (county) 458,980...C 3
Rimouski (county) 69,099...D 3
Saguenay (county) 115,881.. D 2
Saint-Maurice (co.) 107,703.. C 3
Témiscamingue (co.) 52,570.. B 3

CITIES and TOWNS

Alma® 26,322...........C 3
Amos® 9,421............B 3
Baie-Comeau 12,866......D 3
Baie-du-Poste 1,690......C 2
Chicoutimi® 60,064.....C 3
Gaspé 17,261...........E 3
Hauterive 13,995........D 3
Jonquière 60,354........C 3
Lévis 17,895...........C 3
La Tuque 11,556........C 3
Manicouagan..........D 2
Maniwaki 5,424.........B 3
Matane® 13,612........D 2
Mistassini (Baie-du-Poste)
1,690.................C 2
Mont-Laurier® 8,405.....B 3
Montmagny 12,405......C 3
New Carlisle® 781......E 3
Nouveau-Comptoir......C 2
Percé® 4,839..........E 3
Port-Cartier-Ouest......D 3
Port-Menier 275........D 3
Povungnituk 745........E 1
Québec (cap.)® 166,474..C 3
Rimouski® 29,120.......D 3
Rivière-au-Tonnerre 480...D 3
Rivière-du-Loup 13,459...D 3

Rouyn 17,224...........B 3
Sept-Îles 29,262.........D 2
Seven Islands (Sept-Îles)
29,262................D 2
Shawinigan 23,011......C 3
Tadoussac 900.........C 3
Val d'Or 21,371.........C 3
Ville-Marie 2,651.......B 3

OTHER FEATURES

Allard (lake)..........E 2
Anticosti (isl.).........D 3
Baleine, Grand Rivière de la
(riv.)................B 1
Bell (riv.)............B 3
Betsiamites (riv.)......C 2
Bienville (lake)........C 2
Broadback (riv.).......B 3
Cabonga (res.).........B 3
Caniapiscau (riv.)......D 1
Eastmain (riv.)........C 2
Eau Claire (lake).......C 1
Feuilles (riv.)........C 1
Gaspésie Prov. Park.....D 3
George (riv.)..........C 1
Gouin (res.)...........C 3
Grande Rivière, La (riv.)..E 3
Honguedo (passage)....E 3
Hudson (bay)..........A 1
Hudson (str.)..........F 1
Jacques-Cartier (passage). D 3
James (bay)...........A 2
Koksoak (riv.).........D 1
Laurentides Prov. Park...C 3
Louis-XIV (pt.)........B 2
Manicouagan (res.).....D 2
Minto (lake)..........E 1
Mistassini (riv.).......C 2
Mistassini (lake)......C 2
Moisie (riv.)..........D 2
Natashquan (riv.)......D 2
Nottaway (riv.)........B 3
Nouveau-Québec (crater).. F 1
Otish (mts.)..........C 2
Ottawa (riv.)..........B 3
Péribonca (riv.).......C 2
Plétipi (lake).........C 2
Saguenay (riv.)........C 3
Saint-Jean (lake).......C 3
Saint Lawrence (gulf)...E 3
Saint Lawrence (riv.)...D 3
Ungava (pen.)........E 1

® County seat.   ® County seat.
*Population of metropolitan area. *Population of metropolitan area.

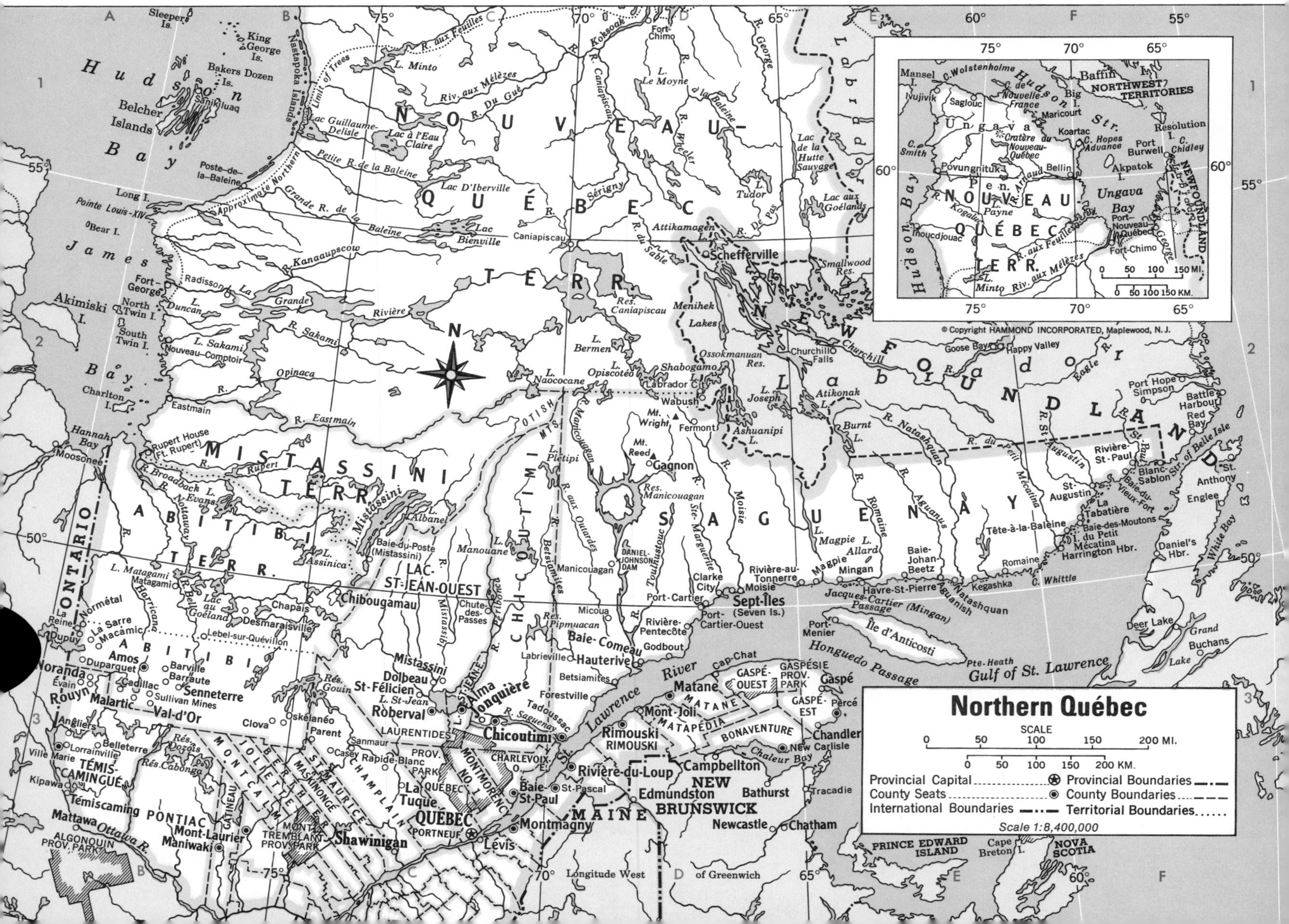

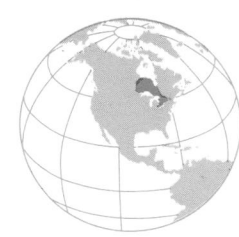

ONTARIO, NORTHERN

INTERNAL DIVISIONS

Algoma (terr. dist.) 133,553...D 3
Cochrane (terr. dist.) 96,875...D 2
Kenora (terr. dist.) 59,421...C 2
Manitoulin (terr. dist.) 11,001...D 3
Nipissing (terr. dist.) 80,268...E 3
Parry Sound (terr. dist.)
33,528...E 3
Rainy River (terr. dist.) 22,798 B 3
Renfrew (county) 87,484...E 3
Sudbury (reg. munic.)
159,779...D 3
Sudbury (terr. dist.) 27,068...D 3
Thunder Bay (terr. dist.)
153,997...C 3
Timiskaming (terr. dist.)
41,288...D 3

CITIES and TOWNS

Chalk River 1,010...E 3
Elliot Lake 16,723...D 3
Fort Albany 482...D 2
Fort Frances® 8,906...C 2
Kapuskasing 12,014...D 3
Kenora® 9,817...C 3
Kirkland Lake 12,219...D 3
Moose Factory 1,452...D 2
Moosonee 1,433...D 2
Nickel Centre 12,318...D 3
North Bay® 51,268...E 3
Pembroke 14,026...E 3
Sault Sainte Marie® 82,697...D 3
Sudbury 91,829...D 3
Thunder Bay® 112,486...C 3
Timmins 46,114...D 3
Valley East 20,433...D 3

OTHER FEATURES

Abitibi (lake)...E 3
Abitibi (riv.)...D 2
Albany (riv.)...C 2
Algonquin Prov. Park...E 3
Asheweig (riv.)...C 2
Attawapiskat (lake)...C 2
Attawapiskat (riv.)...C 2
Basswood (lake)...B 3
Berens (riv.)...A 2
Big Trout (lake)...B 2
Black Duck (riv.)...C 1
Bloodvein (riv.)...A 2
Caribou (isl.)...C 3

Cobham (riv.)...A 2
Eabamet (lake)...C 2
Ekwan (riv.)...C 2
English (riv.)...B 2
Fawn (riv.)...C 2
Finger (lake)...B 2
Georgian (bay)...D 3
Hannah (bay)...D 2
Henrietta Maria (cape)...D 1
Hudson (bay)...D 1
Huron (lake)...D 3
James (bay)...D 2
Kapiskau (riv.)...D 2
Kapuskasing (riv.)...D 2
Kenogami (riv.)...C 2
Kesagami (riv.)...E 2
Lake of the Woods (lake)...B 3
Lake Superior Prov. Park...D 3
Little Current (riv.)...C 2
Long (lake)...C 3
Manitoulin (isl.)...D 3
Mattagami (riv.)...D 3
Michipicoten (isl.)...C 3
Mille Lacs (lake)...B 3
Missinaibi (lake)...D 3
Missinaibi (riv.)...D 2
Missisa (lake)...D 2
Nipigon (lake)...C 3
Nipissing (lake)...E 3
North (chan.)...D 3
North Caribou (lake)...B 2
Nungesser (lake)...B 2
Ogidaki (mt.)...D 3
Ogoki (riv.)...D 3
Opazatika (riv.)...D 3
Opinnagau (riv.)...D 2
Otoskwin (riv.)...C 2
Ottawa (riv.)...E 3
Pipestone (riv.)...C 2
Polar Bear Prov. Park...D 2
Pukaskwa Prov. Park...C 3
Quetico Prov. Park...B 3
Rainy (lake)...B 3
Red (lake)...B 2
Sachigo (riv.)...B 2
Saganaga (lake)...B 3
Saint Ignace (isl.)...C 3
Saint Joseph (lake)...B 2
Sandy (lake)...B 2
Savant (lake)...B 2
Seine (riv.)...B 3
Seul (lake)...B 2
Severn (lake)...B 2
Severn (riv.)...B 2
Shamattawa (riv.)...C 1
Shibogama (lake)...C 2

Sibley Prov. Park...C 3
Slate (isls.)...C 3
Stout (lake)...B 2
Superior (lake)...C 3
Sutton (lake)...D 2
Sutton (riv.)...D 2
Timagami (lake)...D 3
Timiskaming (lake)...E 3
Trout (lake)...B 2
Wabuk (pt.)...D 1
Winisk (lake)...C 2
Winisk (riv.)...C 2
Winnipeg (riv.)...A 2
Woods (lake)...B 3

ONTARIO

INTERNAL DIVISIONS

Algoma (terr. dist.) 133,553...J 5
Brant (county) 104,427...D 4
Bruce (county) 60,020...C 3
Cochrane (terr. dist.) 96,875...J 4
Dufferin (county) 31,145...D 3
Dundas (county) 18,946...J 2
Durham (reg. munic.) 283,639 F 3
Elgin (county) 69,707...C 5
Essex (county) 312,467...B 5
Frontenac (county) 108,133...H 3
Glengarry (county) 20,254...K 2
Grenville (county) 27,176...J 3
Grey (county) 73,824...D 3
Haldimand-Norfolk (reg.
munic.) 89,456...E 5
Haliburton (county) 11,361...F 2
Halton (reg. munic.) 253,883 E 4
Hamilton-Wentworth (reg.
munic.) 411,445...D 4
Hastings (county) 106,883...G 3
Huron (county) 56,127...C 4
Kenora (terr. dist.) 59,421...G 5
Kent (county) 107,022...B 5
Lambton (county) 123,445...B 5
Lanark (county) 45,676...H 3
Leeds (county) 53,765...H 3
Lennox and Addington
(county) 33,040...H 3
Manitoulin (terr. dist.) 11,001..B 2
Middlesex (county) 318,184...C 4
Muskoka (dist. munic.)
38,370...E 3
Niagara (reg. munic.) 368,288 E 4
Nipissing (terr. dist.) 80,268..F 2
Northumberland (county)
64,966...G 3

Ottawa-Carleton (reg. munic.)
546,849...J 2
Oxford (county) 85,920...D 4
Parry Sound (terr. dist.)
33,528...E 2
Peel (reg. munic.) 490,731...E 4
Perth (county) 66,096...C 4
Peterborough (county)
102,452...F 3
Prescott (county) 30,365...K 2
Prince Edward (county)
22,336...G 3
Rainy River (terr. dist.) 22,798 G 5
Renfrew (terr. dist.) 87,484...J 2
Russell (county) 22,412...J 2
Simcoe (county) 225,071...E 3
Stormont (county) 61,927...K 2
Sudbury (reg. munic.)
159,779...K 6
Sudbury (terr. dist.) 27,068...J 5
Thunder Bay (terr. dist.)
153,997...H 5
Timiskaming (terr. dist.)
41,288...H 5
Toronto (metro. munic.)
2,137,395...K 4
Victoria (county) 47,854...F 3
Waterloo (reg. munic.)
305,496...D 4
Wellington (county) 129,432...D 4
York (reg. munic.) 252,053...E 4

CITIES and TOWNS

Ailsa Craig 765...C 4
Ajax 25,475...E 4
Alban 342...D 1
Alexandria 3,271...K 2
Alfred 1,057...K 2
Alliston 4,712...E 3
Almonte 3,855...H 2
Alvinston 736...B 5
Amherstburg 5,685...A 5
Amherst View 6,110...H 3
Ancaster 14,428...D 4
Angus 3,085...E 3
Apsley 264...F 3
Arkona 473...C 4
Armstrong 378...H 4
Arnprior 5,828...H 2
Aroland 291...H 4
Arthur 1,700...D 4
Astorville 340...E 1
Athens 948...J 3
Atherley 366...E 3
Atikokan 4,452...G 5

Atwood 723...D 4
Aurora 16,267...J 3
Avonmore 273...K 2
Aylmer 5,254...C 5
Ayr 1,295...D 4
Ayton 424...D 3
Baden 945...D 4
Bala 577...E 2
Bancroft 2,329...G 2
Barrie® 38,423...E 3
Barry's Bay 1,216...G 2
Batawa 430...G 3
Bath 1,071...H 3
Bayfield 649...C 4
Beachburg 682...H 2
Beachville 917...D 4
Beardmore 583...H 5
Beaverton 1,952...E 3
Beeton 1,989...E 3
Belle River 3,568...B 5
Belleville® 34,881...G 3
Belmont 831...C 5
Bethany 365...F 3
Bewdley 508...F 3
Binbrook 306...E 4
Blackstock 720...F 3
Blenheim 4,044...C 5
Blind River 3,444...J 5
Bloomfield 718...G 4
Blyth 926...C 4
Bobcaygeon 1,625...F 3
Bonfield 540...E 1
Bothwell 915...C 5
Bourget 1,057...J 2
Bracebridge® 9,063...E 2
Bradford 7,370...E 3
Braeside 492...H 2
Brampton® 149,030...J 4
Brantford® 74,315...D 4
Bridgenorth 1,633...F 3

Brigden 635...B 5
Brighton 3,147...G 3
Britt 419...D 2
Brockville® 19,896...J 3
Bruce Mines 635...J 5
Brussels 962...C 4
Burford 1,461...D 4
Burgessville 302...D 4
Burk's Falls 922...E 2
Burlington 114,853...E 4
Cache Bay 665...D 1
Caesarea 551...F 3
Calabogie 256...H 2
Caledon 26,645...E 4
Caledonia 3,409...E 4
Callander 1,158...E 1
Cambridge 77,183...D 4
Campbellford 3,409...G 3
Cannington 1,623...E 3
Capreol 3,845...K 5
Caramat 265...H 5
Cardinal 1,753...J 3
Carleton Place 5,626...H 2
Carlisle 781...D 4
Carlsbad Springs 616...J 2
Carp 707...H 2
Cartier 590...J 5
Casselman 1,675...J 2
Castleton 346...F 3
Chalk River 1,010...G 1
Chapleau 3,243...J 5
Charing Cross 443...B 5
Chatham® 40,952...B 5
Chatsworth 383...D 3
Cherry Valley 289...G 4
Chesley 1,840...C 3
Chesterville 1,430...J 2
Chute-à-Blondeau 365...K 2
City View...J 2
Clarence Creek 796...J 2
Clarksburg 508...D 3

Clifford 645...D 4
Clinton 3,081...C 4
Cobalt 1,759...K 5
Cobden 997...H 2
Coboconk 426...F 3
Cochrane® 4,848...K 5
Colborne 1,796...G 4
Colchester 711...B 5
Coldwater 964...E 3
Collingwood 12,064...D 3
Comber 667...B 5
Consecon 295...G 3
Cookstown 918...E 3
Cornwall® 46,144...K 2
Cottam 404...B 5
Courtland 647...D 5
Courtright 1,024...B 5
Crediton 370...C 4
Creemore 1,182...D 3
Crysler 540...J 2
Cumberland 518...J 2
Cumberland Beach-Bramshot-
Buena Vista 679...E 3
Dashwood 426...C 4
Deep River 5,095...G 1
Delaware 481...C 5
Delhi 4,043...D 5
Delta 360...H 3
Deseronto 1,740...G 3
Douglas 303...H 2
Drayton 809...D 4
Dresden 2,550...B 5
Drumbo 476...D 4
Dryden 6,640...G 4
Dublin 295...C 4
Dubreuilville △988...J 5
Dundalk 1,250...D 3
Dundas 19,586...D 4
Dungannon 284...C 4
Dunnville 11,353...E 5
Durham 2,458...D 3
Dutton 1,115...C 5
Earlton 1,028...K 5
East York 101,974...J 4
Echo Bay 786...J 5
Eden Mills 318...D 4
Eganville 1,245...G 2
Egmondville 465...C 4
Elgin 327...H 3
Elk Lake 526...K 5
Elliot Lake 16,723...B 1
Elmira 7,063...D 4
Elmvale 1,183...E 3
Elmwood 364...C 3
Elora 2,666...D 4
Embro 727...C 4
Embrun 1,883...J 2
Emeryville-Puce 1,611...B 5
Emo 762...F 5
Englehart 1,689...K 5
Enterprise 357...H 3
Erieau 430...C 5
Erin 2,313...D 4
Espanola 5,836...J 5
Essex 6,295...B 5
Etobicoke 298,713...J 4
Everett 570...E 3
Exeter 3,732...C 4
Fauquier 561...J 5
Fenelon Falls 1,701...F 3
Fergus 6,064...D 4
Field 462...E 1
Finch 353...J 2
Fingal 380...C 5
Fitzroy Harbour 446...H 2
Flesherton 565...D 3
Foleyet 484...J 5
Fordwich 365...C 4
Forest 2,671...C 4
Formosa 393...C 3
Fort Erie 24,096...E 5
Fort Frances® 8,906...F 5
Foxboro 597...G 3
Frankford 1,919...G 3
Fraserdale 303...J 5
Freelton 307...D 4
Gananoque 4,863...H 3
Garden Village 270...E 1
Geraldton 2,956...H 5
Glencoe 1,694...C 5
Glen Miller 639...G 3
Glen Robertson 378...K 2
Glen Walter 710...K 2
Goderich® 7,322...C 4
Gogama 652...J 5
Goodwood 335...E 3
Gore Bay® 777...B 2
Gorrie 468...C 4
Grafton 409...G 4
Grand Bend 680...C 4
Grand Valley 1,226...D 4
Granton 315...C 4
Gravenhurst 8,532...E 3
Greely 567...J 2
Green Valley 459...K 2
Grimsby 15,797...E 4
Guelph® 71,207...D 4

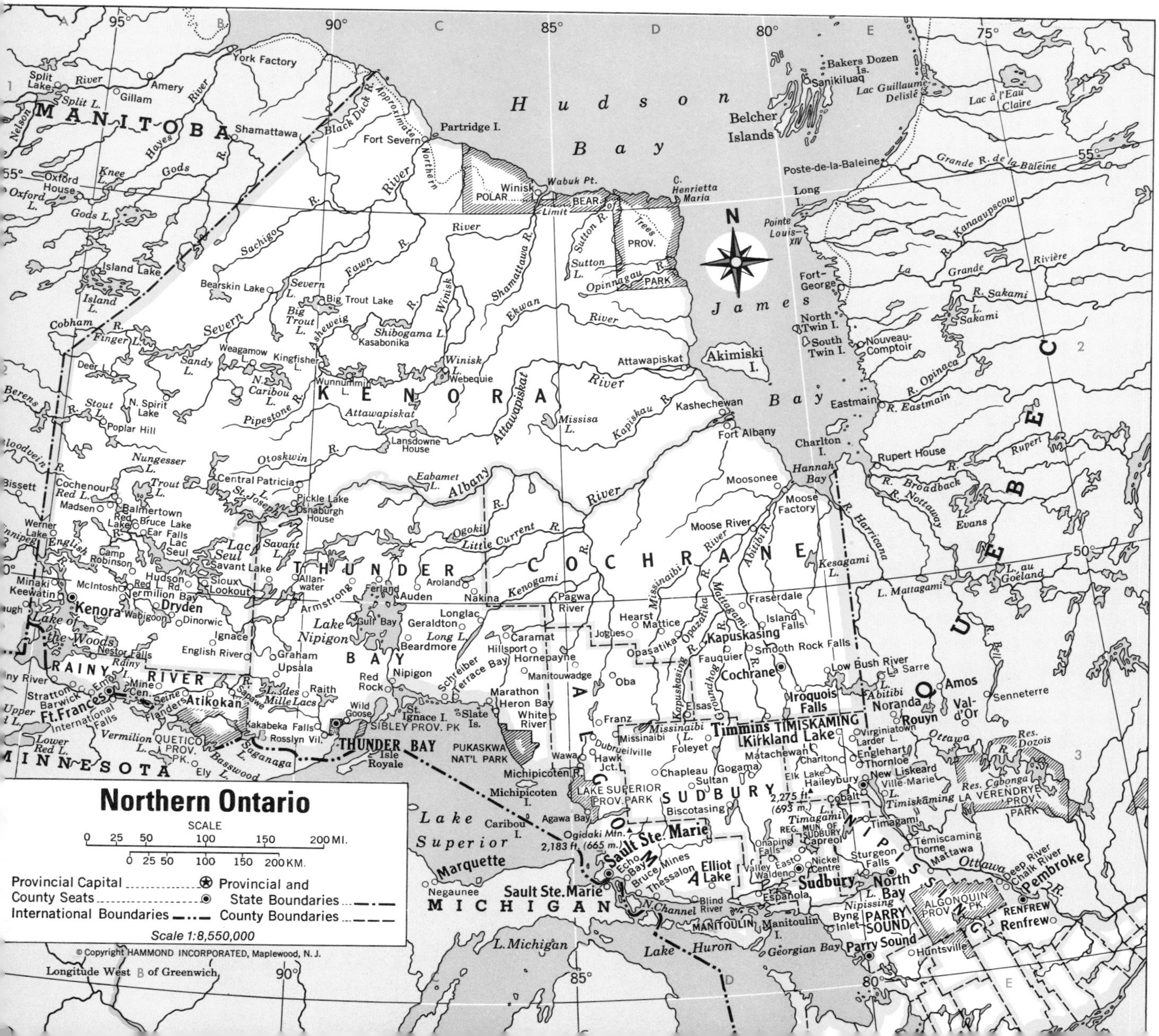

(continued on following page)

Northern Ontario

SCALE
0   25   50   100   150   200 MI.
0 25 50   100   150   200 KM.

Provincial Capital............⊛ Provincial and
County Seats.................● State Boundaries ___
International Boundaries ___ County Boundaries ___

Scale 1:8,550,000

© Copyright HAMMOND INCORPORATED, Maplewood, N.J.

Longitude West B of Greenwich

AREA 412,580 sq. mi. (1,068,582 sq. km.)
POPULATION 8,625,107
CAPITAL Toronto
LARGEST CITY Toronto
HIGHEST POINT in Timiskaming Dist.
2,275 ft. (693 m.)
SETTLED IN 1749
ADMITTED TO CONFEDERATION 1867
PROVINCIAL FLOWER White Trillium

Haileybury 4,925 ....K 5	Iroquois 1,211 ....J 3	Lisle 265 ....E 3
Haldimand 16,866 ....E 5	Iroquois Falls 6,339 .....J 5	Listowel 5,026 ....D 4
Haliburton 1,443 ....F 2	Johnstown 789 ....J 3	Little Britain 265 ....F 3
Halton Hills 35,190 .....E 4	Kakabeka Falls 300 ......G 5	Little Current 1,507 .....B 2
Hamilton 306,434 ....E 4	Kanata 19,728 ....J 2	London® 254,280 ....C 5
Hamilton *542,095 ....E 4	Kapuskasing 12,014 ......J 5	London *283,668 ....C 5
Hanover 6,316 ....C 3	Kars 449 ....J 2	Longlac 2,431 ....H 5
Harriston 1,954 ....D 4	Kearney 538 ....E 2	Long Sault 1,227 ....K 2
Harrow 2,274 ....B 5	Keene 353 ....F 3	L'Orignal 1,819 ....K 2
Harrowsmith 599 ....H 3	Keewatin 1,863 ....F 5	Lucan 1,616 ....D 4
Hastings 975 ....F 3	Kemptville 2,362 ....J 2	Lucknow 1,088 ....C 3
Havelock 1,385 ....F 3	Kenora® 9,817 ....F 5	Lyn 518 ....J 3
Hawkesbury 9,877 ....K 2	Killaloe Station 634 ....G 2	Lynden 451 ....D 4
Hawkestone 275 ....E 3	Killarney 433 ....C 2	Lynhurst 685 ....C 5
Hawk Junction 349 ....J 5	Kincardine 5,778 ....C 3	MacGregor's Bay 861 .....G 2
Hearst 5,533 ....J 5	Kingston® 52,616 ....H 3	MacTier 647 ....E 2
Hensall 973 ....C 4	Kingsville 5,134 ....B 6	Madawaska 264 ....F 2
Hepworth 393 ....C 3	Kinmount 262 ....F 3	Madoc 1,249 ....G 3
Hickson 263 ....C 5	Kirkland Lake 12,219 ....K 5	Maitland 667 ....J 3
Highgate 435 ....C 5	Kitchener® 139,734 ......D 4	Mallorytown 368 ....J 3
Hillsburgh 1,065 ....D 4	Kitchener *287,801 ......D 4	Manitouwadge 3,155 ......H 5
Hillsdale 370 ....E 3	Komoka 1,152 ....C 5	Manitowaning 518 ....C 2
Holland Landing 2,771 ...E 3	Lakefield 2,374 ....F 3	Manotick-Hillside Gardens
Honey Harbour 505 ....E 2	Lanark 753 ....H 2	2,694 ....J 2
Hornepayne 1,848 ....J 5	Lancaster 637 ....K 2	Marathon 2,271 ....H 5
Hudson 515 ....G 4	Langton 348 ....D 5	Markdale 1,289 ....D 3
Huntsville 11,467 ....E 2	Lansdowne 540 ....J 3	Markham 77,037 ....E 4
Huron Park 1,104 ....C 4	Larder Lake 1,084 ....K 5	Markstay 444 ....D 1
Ignace 2,499 ....G 5	Latchford 397 ....K 5	Marmora 1,304 ....G 3
Ilderton 301 ....C 4	Leamington 12,528 ....B 5	Martintown 388 ....K 2
Ingersoll 8,494 ....C 4	Limoges 930 ....J 2	Massey 1,274 ....C 1
Ingleside 1,400 ....J 2	Lincoln 14,196 ....E 4	Matachewan 444 ....D 1
Innerkip 715 ....D 4	Linden Beach 579 ....B 6	Matheson 966 ....K 5
Inverhuron 438 ....C 3	Lindsay® 13,596 ....F 3	Mattawa 2,652 ....F 1
Iron Bridge 821 ....A 1	Linwood 450 ....D 4	Mattice 803 ....J 5
	Lion's Head 467 ....C 2	Maxville 836 ....K 2

Maynooth 277 ....G 2	Napanee 4,803 ....G 3	Ottawa® (cap.), Canada	Port Rowan 811 ....D 5
McGregor 1,145 ....B 5	Navan 419 ....J 2	295,163 ....J 2	Port Stanley 1,891 .....C 5
McKerrow 260 ....C 1	Neustadt 511 ....D 3	Ottawa-Hull *717,978 ...J 2	Pottageville 286 ....J 3
Meaford 4,367 ....D 3	Newboro 260 ....H 3	Otterville 776 ....D 5	Powassan 1,169 ....E 1
Melbourne 346 ....C 5	Newburgh 617 ....H 3	Owen Sound® 19,883 .....D 3	Prescott® 4,670 ....J 3
Merlin 745 ....B 5	Newbury 441 ....C 5	Paincourt 414 ....B 5	Princeton 462 ....D 4
Merrickville 984 ....J 3	Newcastle 32,229 ....F 4	Paisley 1,039 ....C 3	Puce-Emeryville 1,611 ..B 5
Metcalfe 687 ....J 2	New Hamburg 3,923 ......D 4	Pakenham 367 ....H 2	Rainy River 1,061 ......F 5
Midhurst 1,457 ....E 3	New Liskeard 5,551 .....K 5	Palmerston 1,989 ....D 4	Ramore 382 ....K 5
Midland 12,132 ....E 3	Newmarket 29,753 ....E 3	Paris 7,485 ....D 4	Rayside-Balfour 15,017 .D 1
Mildmay 928 ....C 3	Niagara Falls 70,960 ...E 4	Parkhill 1,358 ....C 4	Red Rock 1,260 ....H 5
Milford Bay 401 ....E 2	Niagara-on-the-Lake 12,186..E 4	Parry Sound® 6,124 .....E 2	Renfrew 8,283 ....H 2
Millbank 327 ....D 4	Nickel Centre 12,318 ...D 1	Pefferlaw 857 ....E 3	Richards Landing 405 ...B 1
Millbrook 927 ....F 3	Nipigon 2,377 ....H 5	Pelham 11,104 ....E 4	Richmond 2,880 ....J 2
Milton® 28,067 ....E 4	Nobel 386 ....D 2	Pembroke® 14,026 ....G 2	Richmond Hill 37,778 ...E 4
Milverton 1,463 ....D 4	Nobleton 1,861 ....J 3	Penetanguishene 5,315 ..D 3	Ridgetown 3,062 ....C 5
Minaki 319 ....F 4	Noelville 702 ....D 1	Perth® 5,655 ....H 3	Ripley 591 ....C 3
Mindemoya 376 ....B 2	North Bay® 51,268 ......E 1	Petawawa 5,520 ....G 2	River Valley 275 ....D 1
Minden 838 ....F 3	North Gower 818 ....J 2	Peterborough® 60,620 ...F 3	Rockcliffe Park 1,869 ..J 2
Mississauga 315,056 ....J 4	North York 559,521 .....J 4	Petrolia 4,234 ....B 5	Rockland 3,961 ....J 2
Mitchell 2,777 ....C 4	Norwich 2,117 ....D 5	Pickering 37,754 ....K 4	Rockwood 1,068 ....D 4
Monkton 325 ....C 4	Norwood 1,278 ....F 3	Pioton® 4,361 ....G 3	Rodney 1,007 ....C 5
Moonbeam 838 ....J 5	Nottawa 360 ....D 3	Plantagenet 870 ....K 2	Rosslyn Village 362 ....G 5
Moorefield 308 ....D 4	Oakville 75,773 ....E 4	Plattsville 495 ....D 4	Round Lake Centre 255 ..G 2
Mooretown 344 ....B 5	Oakwood 404 ....F 3	Point Edward 2,383 .....B 4	Russell 1,099 ....J 2
Moose Creek 393 ....K 2	Odessa 849 ....H 3	Pontypool 759 ....F 3	Ruthven 649 ....B 6
Morewood 264 ....J 2	Oil City 266 ....B 5	Port Burwell 655 ....D 5	Saint Albert 254 ....J 2
Morpeth 284 ....C 5	Oil Springs 627 ....B 5	Port Carling 629 ....E 2	Saint Catharines® 124,018..E 4
Morrisburg 2,308 ....J 3	Omemee 819 ....F 3	Port Colborne 19,225 ...E 5	Saint Catharines-Niagara
Mount Albert 1,165 .....E 3	Onaping Falls 6,198 ....J 5	Port Elgin 6,131 ....C 3	*304,353 ....E 4
Mount Brydges 1,557 ....C 5	Opasatika 413 ....J 5	Port Franks 547 ....C 4	Saint Charles 382 ......D 1
Mount Forest 3,474 .....D 4	Orangeville® 13,740 ....D 4	Port Hope 9,992 ....F 4	Saint Clair Beach 2,845 .B 5
Mount Hope 557 ....E 4	Orillia 23,955 ....E 3	Port Lambton 921 ....B 5	Saint Clements 890 .....D 4
Munster 1,531 ....J 2	Osgoode 1,138 ....J 2	Portland 271 ....H 3	Saint-Eugène 470 ....K 2
Nakina 936 ....H 4	Oshawa 117,519 ....F 4	Port McNicoll 1,883 ....E 3	Saint George 865 ....D 4
Nanticoke® 19,816 ......E 5	Oshawa *154,217 ....F 4	Port Perry 4,712 ....F 3	Saint Isidore de Prescott 746 .K 2

Saint Jacobs 1,189	D 4	Spencerville 438	J 3
Saint Mary's 4,883	C 4	Springfield 555	C 5
Saint Thomas⊛ 28,165	C 5	Springford 309	D 5
Saint Williams 442	D 5	Stayner 2,530	E 3
Salem 825	D 4	Stirling 1,638	G 3
Sarnia⊛ 50,892	B 5	Stittsville 2,652	J 2
Sauble Beach 729	C 3	Stoney Creek 36,762	E 4
Sault Sainte Marie 82,697	J 5	Stoney Point 1,090	B 5
Scarborough 443,353	K 4	Straffordville 752	D 5
Schomberg 923	J 3	Stratford⊛ 26,262	C 4
Schreiber 1,968	H 5	Strathroy 8,748	C 5
Scotland 600	D 4	Sturgeon Falls 6,045	E 1
Seaforth 2,114	C 4	Sudbury⊛ 91,829	K 5
Searchmont 384	J 5	Sudbury *149,923	K 5
Sebringville 579	C 4	Sunderland 703	E 3
Seeleys Bay 503	H 3	Sundridge 734	E 2
Shakespeare 602	D 4	Sydenham 595	H 3
Shallow Lake 418	C 3	Tamworth 402	H 3
Shannonville 314	G 3	Tara 687	C 3
Shanty Bay 358	E 3	Tavistock 1,885	D 4
Sharbot Lake 495	H 3	Tecumseh 6,364	B 5
Shedden 292	C 5	Teeswater 1,026	C 3
Shelburne 2,862	D 3	Terrace Bay 2,639	H 5
Simcoe⊛ 14,326	D 5	Thamesford 1,920	C 4
Sioux Lookout 3,074	G 4	Thamesville 961	C 4
Sioux Narrows 394	F 5	Thedford 694	C 4
Smithfield 349	G 3	Thessalon 1,620	J 5
Smiths Falls 8,831	H 3	Thornbury 1,435	D 3
Smithville 1,936	E 5	Thorndale 581	C 4
Smooth Rock Falls 2,352	J 5	Thornton 414	E 3
Sombra 420	B 5	Thorold 15,412	E 4
Southampton 2,830	C 3	Thunder Bay⊛ 112,486	H 5
South Mountain 285	J 3	Thunder Bay *121,379	H 5
South River 1,109	E 2	Tilbury 4,298	B 5
Spanish 1,063	J 5	Tillsonburg 10,487	D 5
Sparta 283	C 5	Timmins 46,114	J 5

Tiverton 806	C 3		
Tobermory 282	C 2		
Toronto (cap.)⊛ 599,217	K 4		
Toronto *2,998,947	K 4		
Tottenham 3,022	E 3		
Trenton 15,085	G 3		
Trout Creek 652	E 2		
Turkey Point 407	D 5		
Tweed 1,574	G 3		
Udora 375	E 3		
Union 485	C 5		
Uxbridge 4,209	E 3		
Valley East 20,433	J 5		
Vanier 18,792	J 2		
Vankleek Hill 1,774	K 2		
Vars 527	J 2		
Vaughan 29,674	J 4		
Vermilion Bay 505	G 4		
Verner 1,076	D 1		
Vernon 303	J 2		
Verona 754	H 3		
Victoria Harbour 1,125	E 3		
Vienna 369	D 5		
Virginiatown 1,010	K 5		
Vittoria 420	D 5		
Wabigoon 268	G 5		
Walden 10,139	J 5		
Walkerton⊛ 4,682	C 3		
Wallaceburg 11,506	B 5		
Wardsville 450	C 5		
Warkworth 618	G 3		
Warren 579	D 1		
Warsaw 314	F 3		
Wasaga Beach 4,705	D 3		
Washago 569	E 3		
Waterloo 49,428	D 4		
Watford 1,402	C 5		
Waubaushene 878	E 3		
Wawa 4,206	J 5		
Webbwood 519	C 1		
Welcome 293	F 4		
Welland 454,448	E 5		
Wellesley 997	D 4		
Wellington 1,082	G 4		
Wendover 326	J 2		
West Lorne 1,258	C 5		
Westmeath 262	H 2		
Westport 621	H 3		
Wheatley 1,638	B 5		
Whitby⊛ 36,698	F 4		
Whitchurch-Stouffville 13,557	J 3		
White River △1,006	J 5		
Whitney 766	F 2		
Wiarton 2,074	C 3		
Wikwemikong 1,030	C 2		
Williamsburg 407	J 3		
Williamsford 256	D 3		
Williamstown 328	K 2		
Winchester 2,001	J 2		
Windsor⊛ 192,083	B 5		
Windsor *246,110	B 5		
Wingham 2,897	C 4		
Wolfe Island 271	H 3		
Woodstock⊛ 26,603	D 4		
Woodville 575	F 3		
Wroxeter 350	C 4		
Wyoming 1,682	B 5		
Yarker 319	H 3		
York 134,617	J 4		
Zephyr 330	E 3		
Zurich 795	C 4		

Topography

0   100   200 MI.

0   100   200 KM.

C. Henrietta Maria

Severn · Winisk · Attawapiskat · Sandy L. · Albany · Missinaibi · Abitibi · English · L. St. Joseph · Lac Seul · Lake Nipigon · Kenogami · L. Abitibi · Lake of the Woods · Rainy Lake · Thunder Bay · 2,275 ft. (693 m.) · Ogidaki Mtn. 2,183 ft. (665 m.) · Sault Ste. Marie · Sudbury · Ottawa · L. Nipissing · Manitoulin I. · C. Hurd · Georgian Bay · L. Simcoe · London · Toronto · Niagara Falls · Windsor · Thames · Long Pt. · Pt. Pelee · St. Lawrence

Below Sea Level	100 m. 328 ft.	200 m. 656 ft.	500 m. 1,640 ft.	1,000 m. 3,281 ft.	2,000 m. 6,562 ft.	5,000 m. 16,404 ft.

OTHER FEATURES

Abitibi (riv.)	J 5	Don (riv.)	J 4
Algonquin Prov. Park	F 2	Doré (lake)	G 2
Amherst (isl.)	H 3	Douglas (pt.)	C 3
Balsam (lake)	F 3	Erie (lake)	E 5
Barrie (isl.)	B 1	Flowerpot (isl.)	C 2
Bays (lake)	F 2	French (riv.)	D 1
Big Rideau (lake)	H 3	Georgian (bay)	D 2
Black (riv.)	E 3	Georgian Bay Is.	
Bruce (pen.)	C 2	Nat'l Park	C 2, D 3
Buckhorn (lake)	F 3	Georgina (isl.)	E 3
Cabot (head)	C 2	Grand (riv.)	D 4
Charleston (lake)	J 3	Humber (riv.)	J 3
Christian (isl.)	D 3	Hurd (cape)	C 2
Clear (lake)	F 3	Ipperwash Prov. Park	C 4
Cockburn (isl.)	A 2	Joseph (lake)	E 2
Couchiching (lake)	E 3	Killarney Prov. Park	C 1
Croker (cape)	D 3	Killbear Point Prov. Park	D 2
		Lake of the Woods (lake)	F 5

Lake Superior Prov. Park	J 5	Rainy (lake)	G 5
Lonely (isl.)	C 2	Rice (lake)	F 3
Long (pt.)	D 5	Rideau (lake)	H 3
Long Point (bay)	D 5	Rondeau Prov. Park	C 5
Madawaska (riv.)	G 2	Rosseau (lake)	E 2
Magnetawan (riv.)	D 2	Saint Clair (lake)	B 5
Main (chan.)	C 2	Saint Clair (riv.)	B 5
Manitou (lake)	C 2	Saint Lawrence (lake)	K 3
Manitoulin (isl.)	B 2	Saint Lawrence (riv.)	J 3
Mattagami (riv.)	J 5	Saint Lawrence Is. Nat'l Park	J 3
Michipicoten (isl.)	H 5	Saugeen (riv.)	C 3
Missinaibi (riv.)	J 5	Scugog (lake)	F 3
Mississagi (riv.)	A 1	Seul (lake)	G 4
Mississippi (lake)	H 2	Severn (riv.)	E 3
Muskoka (lake)	E 2	Sibley Prov. Park	H 5
Niagara (riv.)	E 4	Simcoe (lake)	E 3
Nipigon (lake)	H 5	South (bay)	C 2
Nipissing (lake)	E 1	Spanish (riv.)	C 1
North (chan.)	A 1	Stony (lake)	G 3
Nottawasaga (bay)	D 3	Superior (lake)	H 5
Ogidaki (mt.)	J 5	Sydenham (riv.)	B 5
Ontario (lake)	G 4	Thames (riv.)	B 5
Opeongo (lake)	F 2	Theano (pt.)	J 5
Ottawa (riv.)	H 2	Thousand (isls.)	H 3
Owen (sound)	D 3	Timagami (lake)	K 5
Panache (lake)	C 1	Trout (lake)	E 1
Parry (isl.)	D 2	Vernon (lake)	E 2
Parry (sound)	D 2	Walpole (isl.)	B 5
Pelee (pt.)	B 6	Welland (canal)	E 5
Petre (pt.)	G 4	Woods (lake)	F 5
Point Pelee Nat'l Park	B 5		
Presqu'ile Prov. Park	G 4		
Pukaskwa Prov. Park	H 5		
Quetico Prov. Park	G 5		

⊛County seat.
*Population of metropolitan area.
△Population of town or township.

Ontario
Southern Part

SCALE

0   10   20   30   40   50 MI.

0   10   20   30   40   50 KM.

National Capital	⊛
Provincial Capital	⊛
County Seats	⊛
International Boundaries	
Provincial & State Boundaries	
County Boundaries	- - - -
Canals	

Scale 1:2,620,000

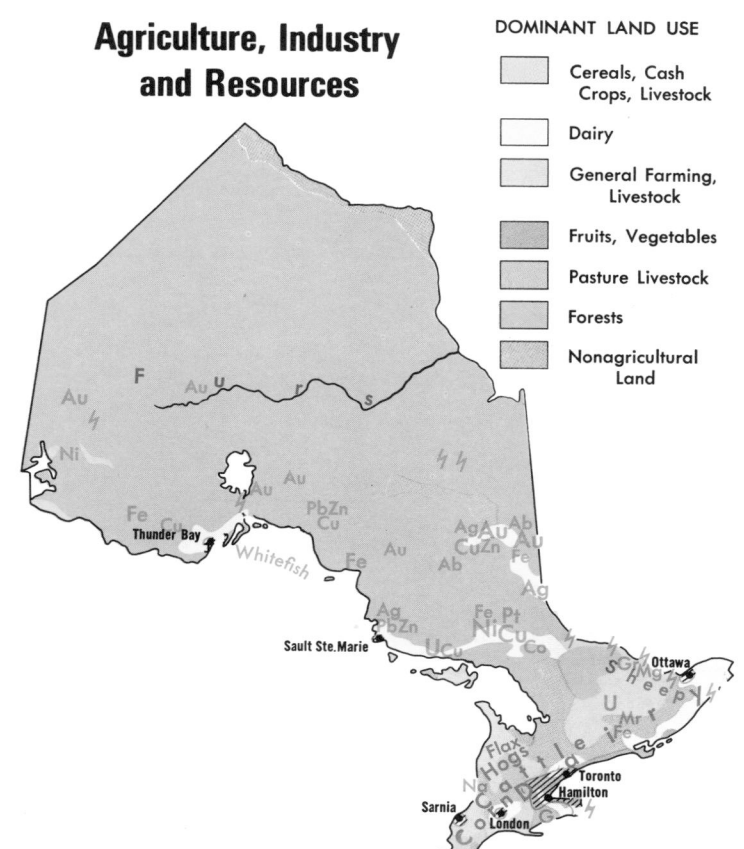

Agriculture, Industry and Resources

DOMINANT LAND USE

- Cereals, Cash Crops, Livestock
- Dairy
- General Farming, Livestock
- Fruits, Vegetables
- Pasture Livestock
- Forests
- Nonagricultural Land

MAJOR MINERAL OCCURRENCES

Ab	Asbestos	Mg	Magnesium
Ag	Silver	Mr	Marble
Au	Gold	Na	Salt
Co	Cobalt	Ni	Nickel
Cu	Copper	Pb	Lead
Fe	Iron Ore	Pt	Platinum
G	Natural Gas	U	Uranium
Gr	Graphite	Zn	Zinc

⚡ Water Power

▨ Major Industrial Areas

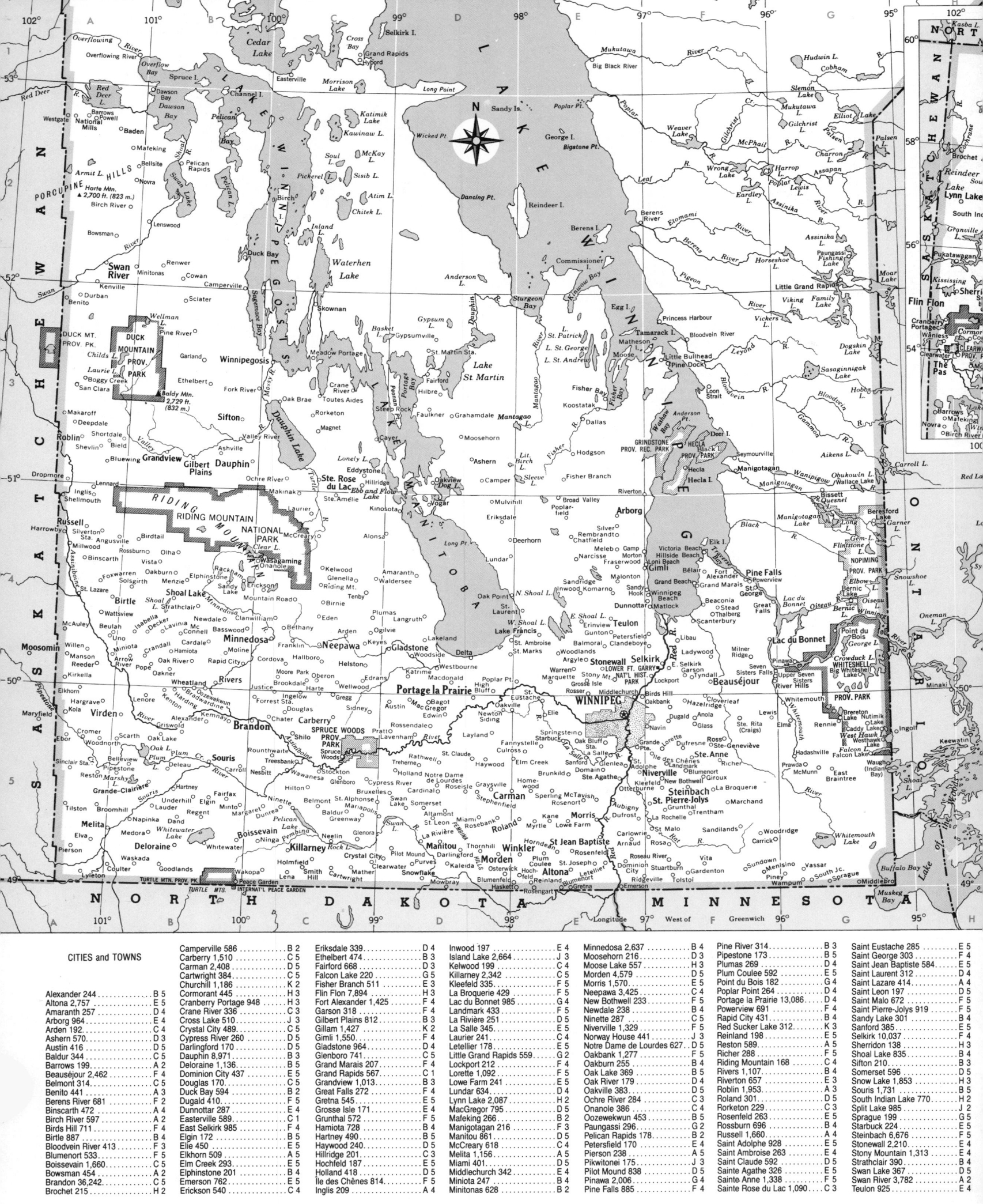

CITIES and TOWNS

Alexander 244 .... B 5
Altona 2,757 .... E 5
Amaranth 257 .... D 4
Arborg 964 .... E 4
Arden 192 .... C 4
Ashern 570 .... D 3
Austin 416 .... D 5
Baldur 344 .... C 5
Barrows 199 .... A 2
Beauséjour 2,462 .... F 4
Belmont 314 .... C 5
Benito 441 .... A 2
Berens River 681 .... F 2
Binscarth 472 .... A 4
Birch River 597 .... A 2
Birds Hill 711 .... F 4
Birtle 887 .... B 4
Bloodvein River 413 .... F 3
Blumenort 533 .... E 5
Boissevain 1,660 .... C 5
Bowsman 454 .... A 2
Brandon 36,242 .... C 5
Brochet 215 .... H 2
Camperville 586 .... B 2
Carberry 1,510 .... C 5
Carman 2,408 .... D 5
Cartwright 384 .... C 5
Churchill 1,186 .... K 2
Cormorant 445 .... H 3
Cranberry Portage 948 .... H 3
Crane River 336 .... C 3
Cross Lake 510 .... J 3
Crystal City 489 .... C 5
Cypress River 260 .... D 5
Darlingford 170 .... D 5
Dauphin 8,971 .... B 3
Deloraine 1,136 .... B 5
Dominion City 437 .... E 5
Douglas 170 .... C 5
Duck Bay 594 .... B 2
Dugald 410 .... F 5
Dunnottar 287 .... E 4
Easterville 589 .... C 1
East Selkirk 985 .... F 4
Elgin 172 .... C 5
Elie 450 .... E 5
Elkhorn 509 .... A 5
Elm Creek 293 .... E 5
Elphinstone 201 .... C 4
Emerson 762 .... E 5
Erickson 540 .... C 4
Eriksdale 339 .... D 4
Ethelbert 474 .... B 3
Fairford 668 .... D 3
Falcon Lake 220 .... G 5
Fisher Branch 511 .... D 4
Flin Flon 7,894 .... H 3
Fort Alexander 1,425 .... F 4
Garson 318 .... F 4
Gilbert Plains 812 .... B 3
Gillam 1,427 .... K 2
Gimli 1,550 .... E 4
Gladstone 964 .... D 4
Glenboro 741 .... C 5
Grand Marais 207 .... F 4
Grand Rapids 567 .... C 1
Grandview 1,013 .... B 3
Great Falls 272 .... F 4
Gretna 545 .... E 5
Grosse Isle 171 .... E 4
Grunthal 572 .... F 5
Hamiota 728 .... B 4
Hartney 490 .... B 5
Haywood 240 .... D 5
Hillridge 201 .... C 3
Hochfeld 187 .... E 5
Holland 418 .... D 5
Île des Chênes 814 .... F 5
Inglis 209 .... A 4
Inwood 197 .... E 4
Island Lake 2,664 .... J 3
Kelwood 199 .... C 4
Killarney 2,342 .... C 5
Kleefeld 335 .... F 5
La Broquerie 429 .... F 5
Lac du Bonnet 985 .... G 4
Landmark 433 .... F 5
La Rivière 251 .... D 5
La Salle 345 .... E 5
Laurier 241 .... C 4
Letellier 178 .... E 5
Little Grand Rapids 559 .... G 2
Lockport 212 .... F 4
Lorette 1,092 .... F 5
Lowe Farm 241 .... E 5
Lundar 634 .... D 4
Lynn Lake 2,087 .... H 2
MacGregor 795 .... D 5
Mafeking 266 .... A 2
Manigotagan 216 .... F 3
Manitou 861 .... D 5
McCreary 618 .... C 4
Melita 1,156 .... A 5
Miami 401 .... D 5
Middlechurch 342 .... F 4
Miniota 247 .... B 4
Minitonas 628 .... B 2
Minnedosa 2,637 .... B 4
Moosehorn 216 .... D 3
Moose Lake 557 .... H 3
Morden 4,579 .... D 5
Morris 1,570 .... E 5
Neepawa 3,425 .... C 4
New Bothwell 233 .... F 5
Newdale 238 .... B 4
Ninette 287 .... C 5
Niverville 1,329 .... F 5
Norway House 441 .... J 3
Notre Dame de Lourdes 627 .... D 5
Oakbank 1,277 .... F 5
Oakburn 255 .... B 4
Oak Lake 369 .... B 5
Oak River 179 .... B 4
Oakville 383 .... D 5
Ochre River 284 .... C 3
Onanole 386 .... C 4
Oozewekwun 453 .... B 5
Paungassi 296 .... G 2
Pelican Rapids 178 .... B 2
Petersfield 170 .... E 4
Pikwitonei 175 .... J 3
Pilot Mound 838 .... D 5
Pinawa 2,006 .... F 4
Pine Falls 885 .... F 4
Pine River 314 .... B 3
Pipestone 173 .... B 5
Plumas 269 .... D 4
Plum Coulee 592 .... E 5
Point du Bois 182 .... G 4
Poplar Point 264 .... D 4
Portage la Prairie 13,086 .... D 4
Powerview 691 .... F 4
Rapid City 431 .... B 4
Red Sucker Lake 312 .... K 3
Reinland 198 .... E 5
Reston 589 .... A 5
Richer 288 .... F 5
Riding Mountain 168 .... C 4
Rivers 1,107 .... B 4
Riverton 657 .... E 3
Roblin 1,953 .... A 3
Roland 301 .... D 5
Rorketon 229 .... C 3
Rosenfeld 263 .... E 5
Rossburn 696 .... B 4
Russell 1,660 .... A 4
Saint Adolphe 928 .... E 5
Saint Ambroise 263 .... E 4
Saint Claude 592 .... D 5
Sainte Agathe 326 .... E 5
Sainte Anne 1,338 .... F 5
Sainte Rose du Lac 1,090 .... C 3
Saint Eustache 285 .... E 5
Saint George 303 .... F 4
Saint Jean Baptiste 584 .... E 5
Saint Laurent 312 .... D 4
Saint Lazare 414 .... A 4
Saint Leon 197 .... D 5
Saint Malo 672 .... E 5
Saint Pierre-Jolys 919 .... F 5
Sandy Lake 301 .... B 4
Sanford 385 .... E 5
Selkirk 10,037 .... F 4
Sherridon 138 .... H 3
Shoal Lake 835 .... B 4
Sifton 210 .... B 3
Snow Lake 1,853 .... H 3
Souris 1,731 .... B 5
South Indian Lake 770 .... H 2
Split Lake 985 .... J 2
Sprague 199 .... G 5
Starbuck 224 .... E 5
Steinbach 6,676 .... F 5
Stonewall 2,210 .... E 4
Stony Mountain 1,313 .... E 4
Strathclair 390 .... B 4
Swan Lake 367 .... D 5
Swan River 3,782 .... A 2
Teulon 925 .... E 4

Manitoba
Northern Part

0 40 80 120 MI.
0 40 80 120 KM.

HUDSON BAY

ONTARIO

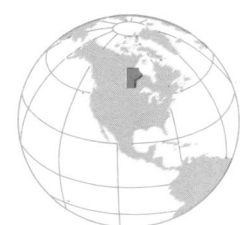

Manitoba
Southern Part

SCALE

0 5 10 20 40 60 MI.
0 5 10 20 40 60 KM.

Provincial Capital ⊛
International Boundaries —··—··—
Provincial Boundaries —·—·—

Scale 1:2,340,000

© Copyright HAMMOND INCORPORATED, Maplewood, N.J.

The Pas 6,390	H 3
Thicket Portage 195	J 3
Thompson 14,288	J 2
Treherne 743	D 5
Tyndall 421	F 4
Virden 2,940	A 5
Vita 364	F 5
Wabowden 655	J 3
Wallace Lake ●2,044	G 3
Wanless 193	H 3
Warren 459	E 4
Waskada 239	B 5
Wawanesa 492	C 5
Whitemouth 320	G 5
Whitewater ●856	B 5
Winkler 5,046	E 5
Winnipeg (cap.) 564,473	E 5
Winnipeg *584,842	E 5
Winnipeg Beach 565	F 4
Winnipegosis 855	B 3
Woodlands 185	E 4
Wooodridge 170	G 5
York Landing 229	J 2

AREA etc.

AREA 250,999 sq. mi. (650,087 sq. km.)
POPULATION 1,026,241
CAPITAL Winnipeg
LARGEST CITY Winnipeg
HIGHEST POINT Baldy Mtn. 2,729 ft.
(832 m.)
SETTLED IN 1812
ADMITTED TO CONFEDERATION 1870
PROVINCIAL FLOWER Prairie Crocus

OTHER FEATURES

Aikens (lake)	G 3
Anderson (lake)	D 2
Anderson (pt.)	F 3
Armit (lake)	A 2
Assapan (riv.)	G 2
Assiniboine (riv.)	C 5
Assinika (lake)	G 2
Assinika (riv.)	G 2
Atim (lake)	C 2
Baldy (mt.)	B 3
Basket (lake)	C 3
Beaverhill (lake)	J 3
Berens (isl.)	E 2
Berens (riv.)	F 2
Bernic (lake)	G 4
Big Sand (lake)	H 2
Bigstone (lake)	J 3
Bigstone (pt.)	E 2
Bigstone (riv.)	J 3
Birch (isl.)	C 2
Black (isl.)	F 3
Black (riv.)	F 4
Bloodvein (riv.)	F 3
Bonnet (lake)	G 4
Buffalo (bay)	G 5
Burntwood (riv.)	J 2
Caribou (riv.)	J 1
Carroll (lake)	G 3
Cedar (lake)	B 1
Channel (isl.)	B 2
Charron (lake)	G 2
Childs (lake)	A 3
Chitek (lake)	C 3
Churchill (cape)	K 2
Churchill (riv.)	J 2
Clear (lake)	C 4
Clearwater Lake Prov. Park	H 3
Cobham (riv.)	G 1
Cochrane (riv.)	H 2
Commissioner (isl.)	E 2
Cormorant (lake)	H 3
Cross (bay)	C 1
Cross (lake)	J 3
Crowduck (lake)	G 4
Dancing (pt.)	D 2
Dauphin (lake)	C 3
Dauphin (riv.)	D 3
Dawson (bay)	B 2
Dog (lake)	D 3
Dogskin (lake)	G 3
Duck Mountain Prov. Park	B 3
Eardley (lake)	F 2

East Shoal (lake)	E 4
Ebb and Flow (lake)	C 3
Egg (isl.)	E 3
Elbow (lake)	G 4
Elk (isl.)	F 4
Elliot (lake)	G 2
Etawney (lake)	J 2
Etomami (riv.)	F 2
Falcon (lake)	G 5
Family (lake)	G 3
Fisher (bay)	E 3
Fisher (riv.)	E 3
Fishing (lake)	C 3
Flintstone (lake)	G 4
Fox (riv.)	K 2
Gammon (riv.)	G 3
Garner (lake)	G 4
Gem (lake)	G 4
George (isl.)	E 2
George (lake)	G 4
Gilchrist (creek)	F 2
Gilchrist (lake)	G 2
Gods (lake)	K 3
Gods (riv.)	K 3
Granville (lake)	H 2
Grass (riv.)	J 3
Grass River Prov. Park	H 3
Grindstone Prov. Rec. Park	F 3
Gunisao (lake)	J 3
Gypsum (lake)	D 3
Harrop (lake)	G 2
Harte (mt.)	A 2
Hayes (riv.)	K 3
Hecla (isl.)	F 3
Hecla Prov. Park	F 3
Hobbs (lake)	G 3
Horseshoe (lake)	G 2
Hubbart (pt.)	K 2
Hudson (bay)	K 2
Hudwin (lake)	G 1
Inland (lake)	C 2
International Peace Garden	B 5
Island (lake)	K 3
Katikim (lake)	C 2
Kawinaw (lake)	C 2
Kinwow (bay)	E 2
Kississing (lake)	H 2
Knee (lake)	J 3
Lake of the Woods (lake)	H 5
La Salle (riv.)	E 5
Laurie (lake)	A 3
Leaf (riv.)	F 2
Lewis (lake)	G 2
Leyond (riv.)	F 3
Little Birch (lake)	E 3
Lonely (lake)	C 3
Long (lake)	G 4
Long (pt.)	D 1
Long (riv.)	D 4
Manigotagan (lake)	G 4

Manigotagan (riv.)	G 3
Manitoba (lake)	D 4
Mantagao (riv.)	E 3
Marshy (lake)	B 5
McKay (lake)	C 2
McPhail (riv.)	F 2
Minnedosa (riv.)	B 4
Moar (lake)	G 2
Molson (lake)	J 3
Moose (isl.)	E 3
Morrison (lake)	C 1
Mossy (riv.)	C 3
Mukutawa (lake)	G 2
Mukutawa (riv.)	E 1
Muskeg (bay)	G 6
Nejanilini (lake)	J 1
Nelson (riv.)	J 2
Nopiming Prov. Park	G 4
Northern Indian (lake)	J 2
North Knife (lake)	J 2
North Seal (riv.)	H 2
North Shoal (lake)	E 4
Nueltin (lake)	H 1
Oak (lake)	B 5
Obukowin (lake)	G 3
Oiseau (lake)	G 4
Oiseau (riv.)	G 4
Overflow (bay)	A 1
Overflowing (riv.)	A 1
Owl (riv.)	K 2
Oxford (lake)	J 3
Paint (lake)	J 2
Palsen (riv.)	G 2
Pelican (bay)	B 2
Pelican (lake)	B 2
Pelican (lake)	C 5
Pembina (hills)	D 5
Pembina (riv.)	C 5
Peonan (pt.)	D 3
Pickerel (lake)	C 2
Pigeon (riv.)	F 2
Pipestone (creek)	A 5
Plum (creek)	B 5
Plum (lake)	B 5
Poplar (riv.)	E 2
Porcupine (hills)	A 2
Portage (bay)	D 3
Punk (isl.)	F 3
Quesnel (lake)	G 4
Rat (riv.)	F 5
Red (riv.)	F 4
Red Deer (lake)	A 2
Red Deer (riv.)	A 2
Reindeer (isl.)	E 2
Reindeer (lake)	H 2
Riding (mt.)	B 4
Riding Mountain Nat'l Park	B 4
Rock (lake)	C 5
Ross (isl.)	J 3
Sagemace (bay)	B 3

Saint Andrew (lake)	E 3
Saint George (lake)	E 3
Saint Martin (lake)	D 3
Saint Patrick (lake)	E 3
Sale (riv.)	E 5
Sandy (isls.)	D 2
Sasaginnigak (lake)	G 3
Seal (riv.)	J 2
Selkirk (isl.)	C 1
Setting (lake)	H 3
Shoal (lake)	G 5
Shoal (riv.)	B 2
Sipiwesk (lake)	J 3
Sisib (lake)	C 2
Sleeve (lake)	E 3
Slemon (lake)	G 1
Snowshoe (lake)	G 4
Soul (lake)	C 2
Souris (riv.)	B 5
Southern Indian (lake)	H 2
South Knife (riv.)	J 2
South Seal (riv.)	J 2
Split (lake)	J 2
Spruce (isl.)	B 1
Spruce Woods Prov. Park	C 5
Stevenson (lake)	J 3
Sturgeon (bay)	E 3
Swan (lake)	B 2
Swan (lake)	D 5
Swan (riv.)	A 3
Tadoule (lake)	J 2
Tamarack (isl.)	F 3
Tatnam (cape)	K 2
Traverse (bay)	F 4
Turtle (mts.)	B 5
Turtle (riv.)	C 3
Turtle Mountain Prov. Park	B 5
Valley (riv.)	B 3
Vickers (lake)	F 3
Viking (lake)	G 3
Wanipigow (riv.)	G 3
Washow (bay)	F 3
Waterhen (lake)	C 2
Weaver (lake)	F 2
Wellman (lake)	B 3
West Hawk (lake)	G 5
West Shoal (lake)	E 4
Whitemouth (lake)	G 5
Whitemouth (riv.)	G 5
Whiteshell Prov. Park	G 4
Whitewater (lake)	B 5
Wicked (pt.)	D 2
Winnipeg (lake)	E 2
Winnipeg (riv.)	G 4
Winnipegosis (lake)	C 2
Woods (lake)	H 5
Wrong (lake)	F 2

*Population of metropolitan area.
●Population of rural municipality.

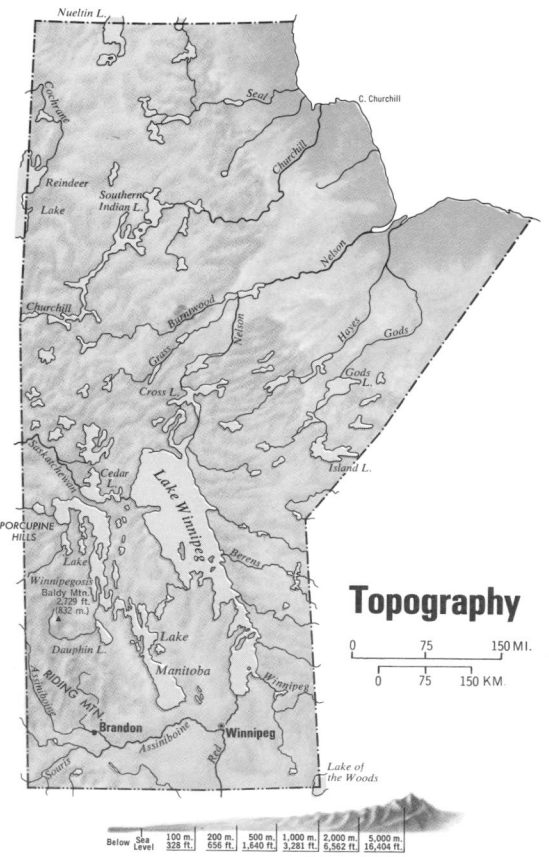

Topography

0 75 150 MI.
0 75 150 KM.

Below Sea Level | 100 m. 328 ft. | 200 m. 656 ft. | 500 m. 1,640 ft. | 1,000 m. 3,281 ft. | 2,000 m. 6,562 ft. | 5,000 m. 16,404 ft.

Agriculture, Industry and Resources

DOMINANT LAND USE

Cereals (chiefly barley, oats)
Cereals, Livestock
Dairy
Livestock
Forests
Nonagricultural Land

MAJOR MINERAL OCCURRENCES

Au Gold
Co Cobalt
Cu Copper
Na Salt
Ni Nickel
O Petroleum
Pb Lead
Pt Platinum
Zn Zinc

⚡ Water Power
▨ Major Industrial Areas

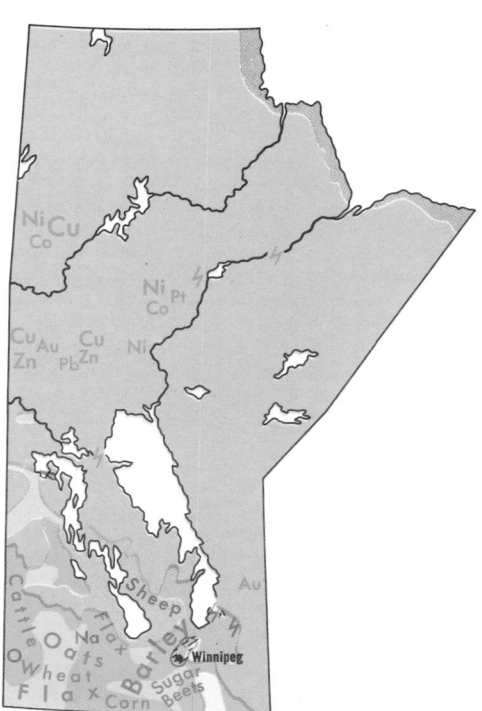

Topography

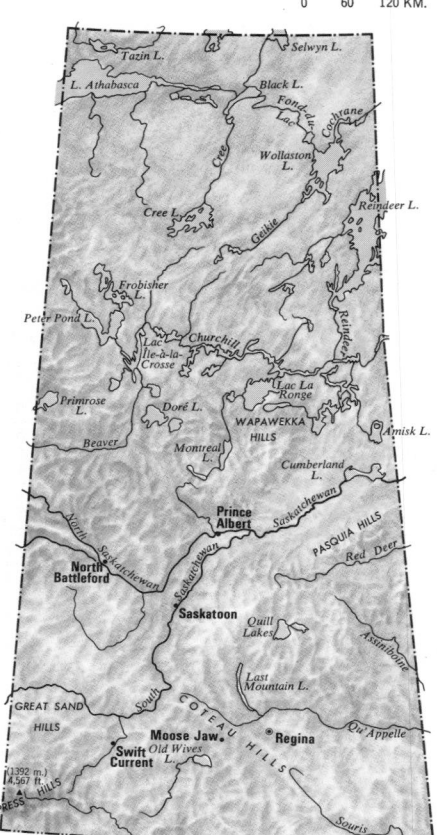

0 60 120 MI.
0 60 120 KM.

5,000 m. 2,000 m. 1,000 m. 500 m. 200 m. 100 m. Sea Level Below
16,404 ft. 6,562 ft. 3,281 ft. 1,640 ft. 656 ft. 328 ft.

CITIES and TOWNS

Abbey 218	C 5
Aberdeen 496	E 3
Abernethy 300	H 5
Air Ronge 557	M 3
Alameda 318	J 6
Alida 169	K 6
Allan 871	E 4
Alsask 652	B 4
Annaheim 209	G 3
Antelope ●231	C 5
Arborfield 439	H 2
Archerwill 286	H 3
Arcola 493	J 6
Arlington Beach ●432	F 4
Asquith 507	D 3
Assiniboia 2,924	E 6
Avonlea 442	G 5
Baildon ●799	F 5
Balcarres 739	H 5
Balgonie 777	G 5
Batoche	E 3
Battleford 3,565	C 3
Beauval 606	L 3
Beechy 275	D 5
Bengough 536	F 6
Bethune 369	F 5
Bienfait 835	J 6
Big River 819	D 2
Biggar 2,561	C 3
Birch Hills 957	F 3
Bjorkdale 269	H 3
Blaine Lake 653	D 3
Borden 197	D 3
Brabant Lake 245	M 3
Bradwell 168	E 4
Bredenbury 467	K 5
Briercrest 151	F 5
Broadview 840	J 5
Brock 184	C 4
Browning ●687	J 6
Bruno 772	F 3
Buchanan 392	J 4
Buffalo Gap ●598	F 6
Buffalo Narrows 1,088	L 3
Burstall 550	B 5
Cabri 632	C 5
Cadillac 173	D 6
Calder 164	K 4
Cana ●1,238	J 5
Candle Lake 219	F 2
Cando 163	C 3
Canoe Lake 182	L 3
Canora 2,667	J 4
Canwood 340	E 2
Carievale 246	K 6
Carlyle 1,074	J 6
Carnduff 1,043	K 6
Carrot River 1,169	H 2

Central Butte 548	E 5
Ceylon 184	G 6
Chaplin 389	E 5
Chitek Lake 170	D 2
Choiceland 543	G 2
Christopher Lake 227	F 2
Churchbridge 972	J 5
Clavet 234	E 4
Climax 293	C 6
Cochin 221	C 2
Codette 236	H 2
Coleville 383	B 4
Colonsay 594	F 4
Connaught Heights ●982	G 3
Conquest 256	D 4
Consul 153	B 6
Coronach 1,032	F 6
Craik 565	F 4
Craven 206	G 5
Creelman 184	H 6
Creighton 1,636	N 4
Cudworth 947	F 3
Cumberland House 831	J 2
Cupar 669	G 5
Cut Knife 624	B 3
Dalmeny 1,064	E 3
Davidson 1,166	E 4
Debden 403	E 2
Delisle 980	D 4
Denare Beach 592	M 4
Denzil 199	B 3
Deschambault Lake 386	M 3
Dinsmore 398	D 4
Dodsland 272	C 4
Domremy 209	F 3
Drake 211	G 4
Duck Lake 699	E 3
Dundurn 531	E 4
Dysart 275	H 5
Earl Grey 303	G 4
Eastend 723	C 6
Eatonia 528	B 4
Ebenezer 164	J 4
Edam 384	C 2
Edenwold 143	G 5
Elbow 313	E 4
Eldorado 229	L 2
Elfros 199	H 4
Elrose 624	D 4
Elstow 143	E 4
Endeavour 199	J 3
Englefeld 271	G 3
Erwood 149	J 3
Esterhazy 3,065	K 5
Eston 1,413	C 4
Eyebrow 168	E 5
Fillmore 396	H 5
Fleming 141	K 5
Flin Flon 367	N 4

Foam Lake 1,452	H 4
Fond du Lac 494	L 2
Fort Qu'Appelle 1,827	H 5
Fox Valley 380	B 5
Francis 182	H 5
Frobisher 166	J 6
Frontier 619	C 6
Gainsborough 308	K 6
Gerald 197	K 5
Glaslyn 430	C 2
Glenavon 284	J 5
Glen Ewen 168	K 6
Goodsoil 263	L 3
Govan 394	G 4
Grand Coulee 208	G 5
Gravelbourg 1,338	E 6
Grayson 264	J 5
Green Acres 139	F 2
Green Lake 634	L 4
Grenfell 1,307	J 5
Guernsey 198	F 4
Gull Lake 1,095	C 5
Hafford 557	D 3
Hague 625	E 3
Hanley 484	E 4
Harris 259	D 4
Hawarden 137	E 4
Hearts Hill ●552	B 3
Hepburn 411	E 3
Herbert 1,019	D 5
Hodgeville 329	E 5
Holdfast 297	F 5
Hudson Bay 2,361	J 3
Humboldt 4,705	F 3
Hyas 165	J 4
Ile-à-la-Crosse 1,035	L 3
Imperial 501	F 4
Indian Head 1,889	H 5
Invermay 353	J 4
Ituna 870	H 4
Jansen 223	G 4
Jasmin ●14	H 4
Kamsack 2,688	K 4
Kelliher 397	H 4
Kelvington 1,054	H 3
Kenaston 345	E 4
Kennedy 215	J 5
Kerrobert 1,141	C 4
Kincaid 256	D 6
Kindersley 3,969	B 4
Kinistino 783	F 3
Kipling 1,016	J 5
Kisbey 429	J 6
Kronau 154	G 5
Kyle 516	D 4
Lac Pelletier ●586	C 6
Lafleche 583	E 6
Laird 233	E 3
Lake Lenore 361	G 3
La Loche 1,632	L 3
Lampman 651	J 6
Lancer 156	C 5
Landis 277	C 3
Lang 219	G 6
Langenburg 1,324	K 5
Langham 1,151	D 3
Lanigan 1,732	F 4
La Ronge 2,579	L 3
Lashburn 813	B 2
Leader 1,108	B 5
Leask 478	E 2
Lebret 274	H 5
Lemberg 414	H 5
Leoville 393	D 2
Leroy 504	G 4
Lestock 402	H 4
Limerick 164	E 6
Lintlaw 234	H 3

Lipton 364	H 5
Lloydminster 6,034	A 2
Loon Lake 369	B 1
Loreburn 201	E 4
Lucky Lake 333	D 5
Lumsden 1,303	G 5
Luseland 704	B 3
Macdowall 171	E 3
Macklin 976	A 3
Macoun 190	H 6
Maidstone 1,001	B 2
Mankota 375	D 6
Manor 368	K 6
Maple Creek 2,470	B 6
Marcelin 238	E 3
Margo 153	H 4
Marriott ●627	D 4
Marsden 229	B 3
Marshall 453	B 2
Martensville 1,966	E 3
Maryfield 431	K 6
Maymont 212	D 3
McLean 189	G 5
Meacham 178	F 3
Meadow Lake 3,857	C 1
Meath Park 262	F 2
Medstead 163	C 2
Melfort 6,010	G 3
Melville 5,092	J 5
Meota 235	C 2
Mervin 155	C 2
Midale 564	H 6
Middle Lake 275	F 3
Milden 251	D 4
Milestone 602	G 5
Montmartre 544	H 5
Montreal Lake 448	F 1
Moose Jaw 33,941	F 5
Moose Range ●679	H 2
Moosomin 2,579	K 5
Morse 416	D 5
Mortlach 293	E 5
Mossbank 464	E 6
Muenster 385	F 4
Naicam 886	G 3
Neilburg 354	B 3
Neuanlage 144	E 3
Neudorf 425	J 5
Neuhorst 146	E 3
Nipawin 4,376	H 2
Nokomis 524	F 4
Norquay 552	J 4
North Battleford 14,030	C 3
North Portal 164	J 6
Odessa 232	H 5
Ogema 441	G 6
Osler 527	E 3
Outlook 1,976	D 4
Oxbow 1,191	J 6
Paddockwood 211	F 2
Pangman 227	G 6
Paradise Hill 421	B 2
Patuanak 173	L 3
Paynton 203	B 2
Pelican Narrows 331	N 3
Pelly 391	K 4
Pennant 202	C 5
Pense 472	G 5
Perdue 407	D 3
Pierceland 425	K 4
Pilger 150	F 3
Pilot Butte 1,255	G 5
Pine House 612	M 3
Plenty 175	C 4
Plunkett 150	F 4
Ponteix 769	D 6
Porcupine Plain 937	H 3
Preeceville 1,243	J 4

Prelate 317	B 5
Prince Albert 31,380	F 2
Prud'homme 222	F 3
Punnichy 394	G 4
Qu'Appelle 653	G 5
Quill Lake 514	G 3
Quinton 169	G 4
Rabbit Lake 159	D 2
Radisson 439	D 3
Radville 1,012	G 6
Rama 133	H 4
Raymore 635	G 4
Redvers 859	K 6
Regina (cap.) 162,613	G 5
Regina *164,313	G 5
Regina Beach 603	G 5
Rhein 271	J 4
Richmound 188	B 5
Riverhurst 193	E 5
Rocanville 934	K 5
Roche Percé 142	J 6
Rockglen 511	F 6
Rosetown 2,664	D 4
Rose Valley 538	H 3
Rosthern 1,609	E 3
Rouleau 443	G 5
Saint Benedict 157	F 3
Saint Brieux 401	G 3
Saint Louis 448	F 3
Saint Philips ●538	K 4
Saint Walburg 802	B 2
Saltcoats 549	J 4
Sandy Bay 756	N 3
Saskatoon 154,210	E 3
Saskatoon *154,210	E 3
Sceptre 169	C 5
Scott 203	C 3
Sedley 373	H 5
Semans 344	G 4
Shaunavon 2,112	C 6
Sheho 285	H 4
Shell Lake 220	D 2
Shellbrook 1,228	E 2
Simpson 231	F 4
Sintaluta 215	H 5
Smeaton 246	G 2
Southey 697	G 5
Spalding 337	G 3
Spiritwood 926	D 2
Springside 533	J 4
Spy Hill 354	K 5
Star City 527	G 3
Stenen 143	J 4
Stockholm 391	J 5
Stonehenge ●701	F 6
Storthoaks 142	K 6
Stoughton 716	J 6
Strasbourg 842	G 4
Sturgis 789	J 4
Swift Current 14,747	D 5
Tantallon 196	K 5
Theodore 473	J 4
Timber Bay 152	F 1
Tisdale 3,107	H 3
Togo 181	K 4
Tompkins 275	C 5
Torch River ●2,440	G 2
Torquay 311	H 6
Tramping Lake 178	B 3
Tugaske 175	E 5
Turnor Lake 166	L 3
Turtleford 505	B 2
Unity 2,408	C 3
Uranium City 2,507	L 2
Val Marie 236	D 6
Vanguard 292	D 6
Vanscoy 298	D 4
Vibank 369	H 5

Viscount 386	F 4
Vonda 313	F 3
Wadena 1,495	H 4
Wakaw 1,030	F 3
Waldeck 292	D 5
Waldheim 758	E 3
Walpole ●711	K 6
Wapella 487	K 5
Warman 2,076	E 3
Waseca 169	C 2
Waskesiu Lake 176	E 2
Watrous 1,830	F 4
Watson 901	G 3
Wawota 622	K 6
Weldon 279	F 3
Welwyn 170	K 5
Weyburn 9,523	H 6
White City 602	G 5
White Fox 394	H 2
Whitewood 1,003	J 5
Wilcox 202	G 5
Wilkie 1,501	C 3
Willow Bunch 494	F 6
Willow Creek ●1,218	B 5
Windthorst 254	J 5
Wiseton 195	D 4
Wishart 212	H 4
Wollaston Lake 248	N 2
Wolseley 904	H 5
Wymark 162	D 5
Wynyard 2,147	G 4
Yarbo 158	K 5

Yellow Grass 477	H 6
Yorkton 15,339	J 4
Young 456	F 4
Zenon Park 273	H 2

OTHER FEATURES

Allan (hills)	E 4
Amisk (lake)	M 4
Antelope (lake)	C 5
Antler (riv.)	K 6
Arm (riv.)	F 3
Assiniboine (riv.)	J 5
Athabasca (lake)	L 2
Bad (lake)	C 4
Bad (riv.)	F 3
Basin (lake)	F 3
Batoche Nat'l Hist. Site	E 3
Battle (creek)	B 6
Battle (riv.)	B 3
Bear (hills)	C 4
Beaver (hills)	H 4
Beaver (lake)	F 1
Beaver (riv.)	L 4
Beaverlodge (lake)	L 2
Big Muddy (lake)	G 6
Bigstick (lake)	B 5
Birch (lake)	C 3
Bitter (lake)	B 5
Black (lake)	M 2
Boundary (plat.)	B 6
Brightsand (lake)	B 2
Bronson (lake)	B 2

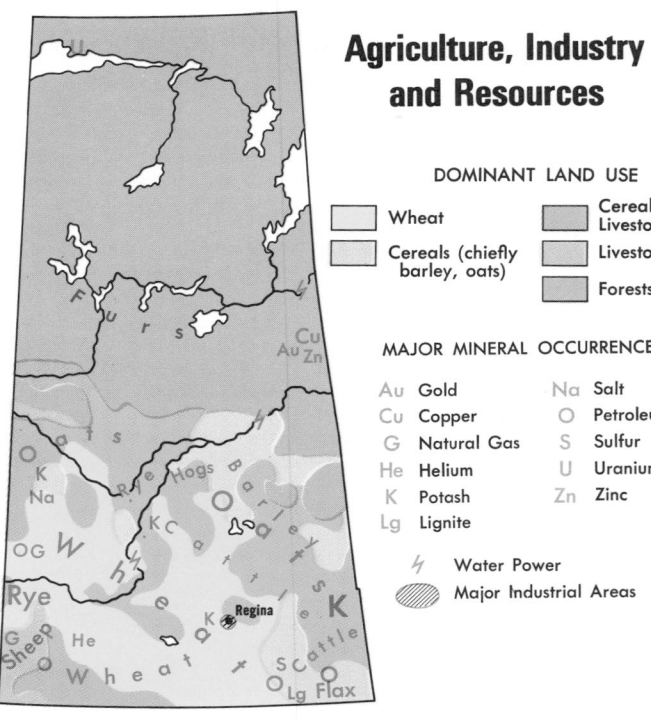

Agriculture, Industry and Resources

DOMINANT LAND USE

▢ Wheat	▨ Cereals, Livestock
▢ Cereals (chiefly barley, oats)	▨ Livestock
	▨ Forests

MAJOR MINERAL OCCURRENCES

Au	Gold	Na	Salt
Cu	Copper	O	Petroleum
G	Natural Gas	S	Sulfur
He	Helium	U	Uranium
K	Potash	Zn	Zinc
Lg	Lignite		

⚡ Water Power

▨ Major Industrial Areas

Buffalo Pound Prov. Park .... F 5
Cabri (lake) .... B 4
Cactus (hills) .... F 5
Candle (lake) .... L 3
Cannington Manon Hist. Park J 6
Canoe (lake) .... L 3
Carrot (riv.) .... J 2
Chaplin (lake) .... E 5
Chipman (riv.) .... M 2
Chitek (lake) .... D 2
Churchill (riv.) .... M 3
Clearwater (riv.) .... L 3
Cochrane (riv.) .... N 2
Coteau (hills) .... D 4
Cowan (lake) .... D 2
Crane (lake) .... B 5
Crean (lake) .... E 1
Cree (lake) .... L 3
Cree (riv.) .... M 2
Cumberland (lake) .... J 1
Cypress (hills) .... B 6
Cypress (lake) .... B 6
Cypress Hills Prov. Park ... B 6
Danielson Prov. Park .... E 4
Delaronde (lake) .... E 1
Diefenbaker (lake) .... E 4
Doré (lake) .... L 3
Douglas Prov. Park .... E 4
Duck Lake Hist. Park .... E 3
Duck Mountain Prov. Park ... K 4
Eagle (hills) .... C 3
Eaglehill (creek) .... D 4

Ear (lake) .... B 3
Echo Valley Prov. Park ..... G 5
Etomami (riv.) .... J 3
Eyebrow (lake) .... E 5
Eyehill (creek) .... B 3
Fife (lake) .... E 6
File (hills) .... H 5
Fir (riv.) .... J 2
Fond du Lac (riv.) .... M 2
Forrest (lake) .... L 3
Fort Battleford Nat'l Hist. Park C 3
Fort Carlton Hist. Park .... E 3
Fort Pitt Hist. Park. ...... C 2
Fort Walsh Nat'l Hist. Park. A 6
Foster (riv.) .... M 3
Frenchman (riv.) .... C 6
Frobisher (lake) .... L 3
Gap (creek) .... B 6
Gardiner (dam) .... D 4
Geikie (riv.) .... M 3
Good Spirit (lake) .... J 4
Goodspirit Lake Prov. Park . J 4
Great Sand (hills) .... B 5
Green (lake) .... D 1
Greenwater Lake Prov. Park . H 3
Haultain (riv.) .... L 3
Ile-à-la-Crosse (lake) ..... L 3
Ironspring (creek) .... G 3
Jackfish (lake) .... C 2
Katepwa Prov. Park .... H 5
Kingsmere (lake) .... E 1
Kiyiu (lake) .... C 4

Lac La Ronge Prov. Park .... M 3
Lanigan (creek) .... F 4
Last Mountain (lake) .... F 4
Leaf (lake) .... J 2
Leech (lake) .... J 4
Lenore (lake) .... G 3
Little Manitou (lake) ...... F 4
Lodge (creek) .... A 5
Long (creek) .... H 6
Loon (creek) .... G 4
Makwa (lake) .... B 1
Makwa (riv.) .... B 1
Manito (lake) .... B 3
Maple (creek) .... B 5
McFarlane (riv.) .... L 2
Meadow (lake) .... C 1
Meadow Lake Prov. Park ..... K 4
Meeting (lake) .... D 2
Midnight (lake) .... C 2
Ministikwan (lake) .... B 1
Missouri Coteau (hills) .... F 5
Montreal (lake) .... F 1
Moose (mt.) .... J 6
Moose Jaw (lake) .... G 5
Moose Mountain (creek) ..... J 6
Moose Mountain Prov. Park .. J 6
Mossy (riv.) .... H 1
Muddy (lake) .... B 3
Mudjatik (riv.) .... L 3
Nipawin (lake) .... G 1
North Saskatchewan (riv.) .. D 3
Notukeu (creek) .... D 6

Oldman (riv.) .... L 2
Old Wives (lake) .... E 5
Opuntia (lake) .... C 4
Overflowing (riv.) .... K 2
Pasquia (hills) .... J 2
Pasquia (riv.) .... K 2
Pelican (lake) .... E 2
Peter Pond (lake) .... L 3
Pheasant (hills) .... H 5
Pine Lake Prov. Park .... E 4
Pinto (lake) .... D 6
Pipestone (creek) .... K 5
Pipestone (riv.) .... L 3
Ponass (lakes) .... H 3
Poplar (riv.) .... F 6
Porcupine (hills) .... K 3
Primrose (lake) .... L 3
Primrose Lake Air Weapons
  Range .... L 2
Prince Albert Nat'l Park ... E 1
Qu'Appelle (riv.) .... G 4
Quill (lake) .... G 4
Red Deer (lake) .... A 5
Red Deer (riv.) .... G 5
Reindeer (lake) .... N 3
Reindeer (riv.) .... M 3
Riou (lake) .... M 2
Rivers (lake) .... J 6
Ronge, La (lake) .... M 3
Rowans Ravine Prov. Park ... F 4
St. Victor Petroglyphs Hist.
  Park .... E 6

Saskatchewan (riv.) .... H 2
Saskatchewan Landing Prov.
  Park .... C 5
Saskeram (riv.) .... K 2
Scott (lake) .... M 2
Selwyn (lake) .... M 2
Souris (riv.) .... J 6
South Saskatchewan (riv.) .. C 5
Steele Narrows Hist. Park .. B 2
Stripe (lake) .... C 4
Sturgeon (riv.) .... E 2
Swan (riv.) .... J 3
Swift Current (creek) ...... D 5
Tazin (lake) .... L 2
The Battlefords Prov. Park . C 2

Thickwood (hills) .... D 2
Thunder (hills) .... L 4
Tobin (lake) .... H 2
Torch (riv.) .... H 2
Touchwood (hills) .... G 4
Tramping (lake) .... C 3
Trout (lake) .... L 2
Turtle (lake) .... C 2
Twelvemile (lake) .... E 6
Vermilion (hills) .... E 5
Wapawekka (hills) .... M 4
Waskana (creek) .... G 5
Waskesiu (lake) .... E 1
Watham (riv.) .... M 3
Weed (hills) .... J 5

White Fox (riv.) .... G 2
White Gull (creek) .... G 2
Whiteshore (lake) .... C 3
Whiteswan (lakes) .... F 1
William (riv.) .... L 2
Willow Bunch (lake) .... F 6
Witchekan (lake) .... D 2
Wollaston (lake) .... N 2
Wood (mt.) .... E 6
Wood (riv.) .... E 6
Wood Mountain Hist. Park ... E 6

*Population of metropolitan area.
•Population of rural municipality.

AREA 251,699 sq. mi. (651,900 sq. km.)
POPULATION 968,313
CAPITAL Regina
LARGEST CITY Regina
HIGHEST POINT Cypress Hills 4,567 ft. (1,392 m.)
SETTLED IN 1774
ADMITTED TO CONFEDERATION 1905
PROVINCIAL FLOWER Prairie Lily

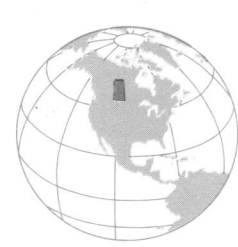

© Copyright HAMMOND INCORPORATED, Maplewood, N.J.

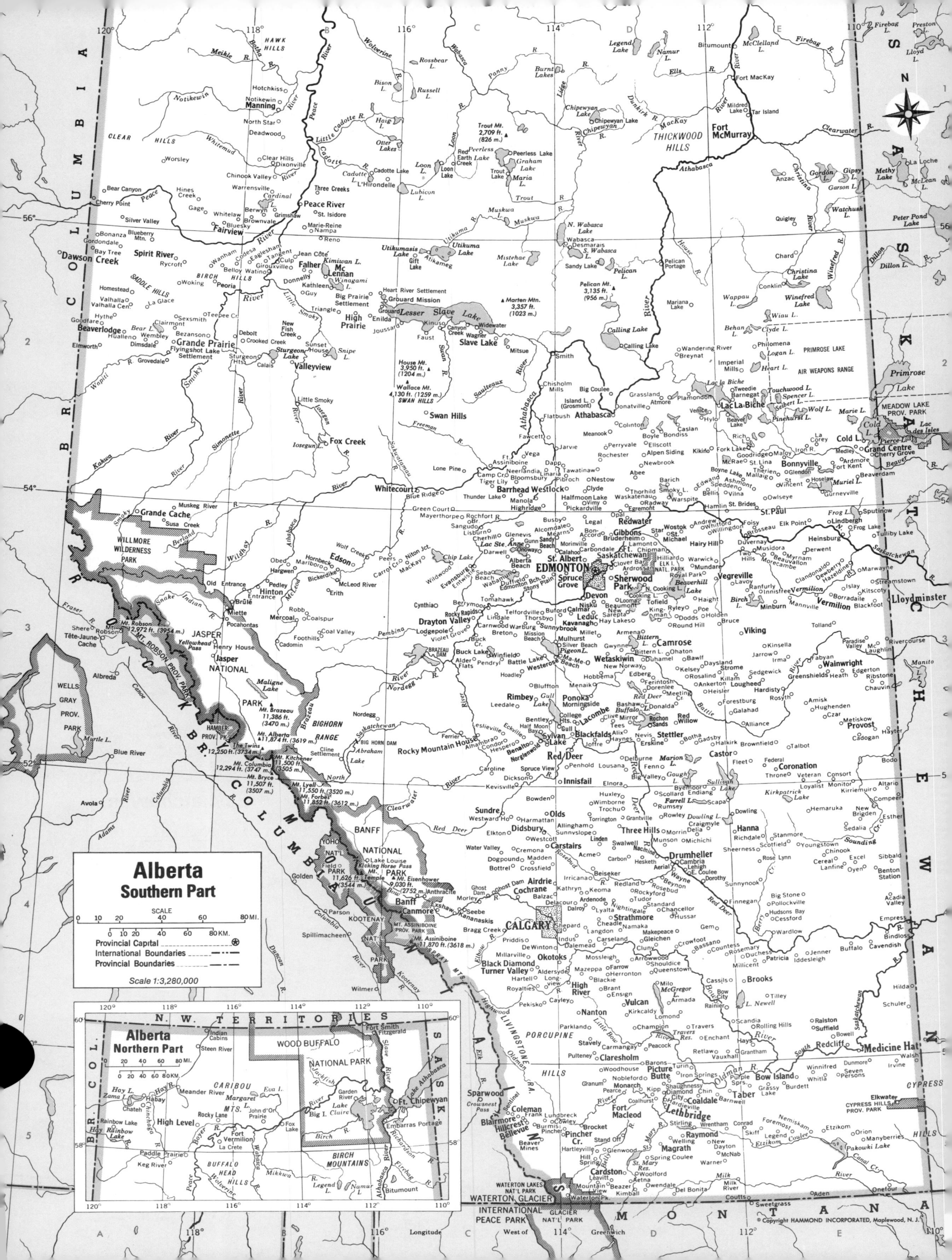

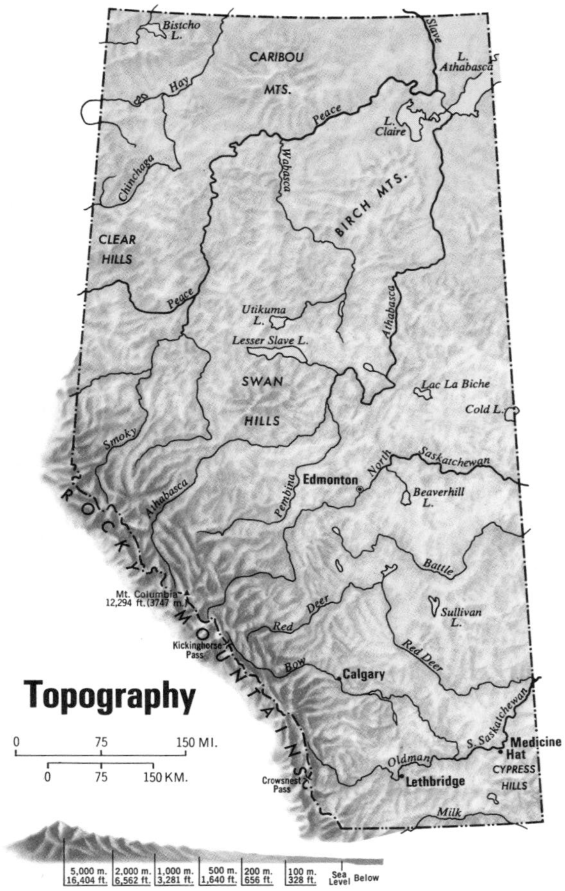

Bistcho L.
CARIBOU MTS.
Hay
Slave
L. Athabasca
Peace
L. Claire
Wabasca
BIRCH MTS.
Chinchaga
CLEAR HILLS
Utikuma L.
Lesser Slave L.
Athabasca
SWAN HILLS
Lac La Biche
Cold L.
Smoky
North Saskatchewan
Edmonton
Pembina
Beaverhill L.
ROCKY MOUNTAINS
Battle
Mt. Columbia 12,294 ft. (3747 m.)
Kickinghorse Pass
Red Deer
Sullivan L.
Bow
Calgary
Red Deer
South Saskatchewan
Medicine Hat
Oldman
Lethbridge
CYPRESS HILLS
Crowsnest Pass
Milk
0 75 150 MI.
0 75 150 KM.
5,000 m. 16,404 ft.
2,000 m. 6,562 ft.
1,000 m. 3,281 ft.
500 m. 1,640 ft.
200 m. 656 ft.
100 m. 328 ft.
Sea Level
Below

Topography

AREA 255,285 sq. mi. (661,185 sq. km.)
POPULATION 2,237,724
CAPITAL Edmonton
LARGEST CITY Edmonton
HIGHEST POINT Mt. Columbia 12,294 ft. (3,747 m.)
SETTLED IN 1861
ADMITTED TO CONFEDERATION 1905
PROVINCIAL FLOWER Wild Rose

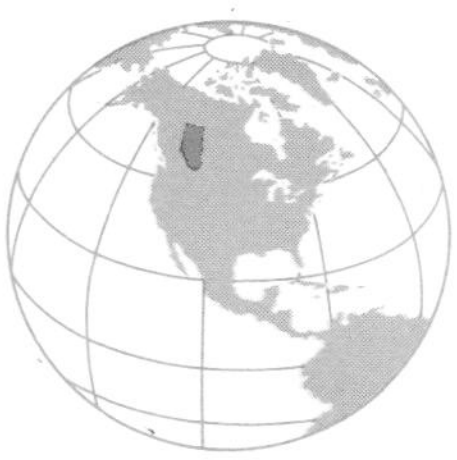

CITIES and TOWNS

cme 457 D 4
irdrie 8,414 C 4
lberta Beach 485 C 3
lix 837 D 3
ndrew 548 D 3
ntler Lake 334 D 3
rdmore 224 E 2
rrowwood 156 D 4
thabasca 1,731 D 2
anff 4,208 C 4
arnwell 359 D 5
arons 315 D 4
arrhead 3,736 C 2
ashaw 875 D 3
assano 1,200 D 4
awlf 350 D 3
eaumont 2,638 D 3
eaverlodge 1,937 A 2
eiseker 580 D 4
entley 823 C 3
erwyn 557 B 1
ig Valley 360 D 3
lack Diamond 1,444 C 4
lackfalds 1,488 D 3
lackfoot 220 E 3
lackie 298 D 4
on Accord 1,376 D 3
onnyville 4,454 E 2
owden 989 C 4
ow Island 1,491 E 5
oyle 638 D 2
ragg Creek 505 C 4
reton 552 C 3
rooks 9,421 E 4
ruce 88 E 3
ruderheim 1,136 D 3
urdett 220 E 5
algary 592,743 C 4
algary *592,743 C 4
almar 1,003 D 3
amrose 12,570 D 3
anmore 3,484 C 4
arbon 434 D 4
ardston 3,267 D 5
armangay 266 D 4
aroline 436 C 3
arseland 484 D 4
arstairs 1,587 D 4
astor 1,123 D 3
ereal 249 E 4
hampion 339 D 4
hauvin 298 E 3
hipman 266 D 3
lairmont 469 A 2
laresholm 3,493 D 4
live 364 D 3
lyde 364 D 2
oaldale 4,579 D 5
oalhurst 882 D 5
ochrane 3,544 C 4
old Lake 2,110 E 2
ollege Heights 267 D 3
onsort 632 E 3
ooking Lake 218 D 3
Coronation 1,309 E 3
Coutts 400 D 5
Cowley 304 D 5
Cremona 382 C 4
Crossfield 1,217 C 4
Daysland 679 D 3
Delburne 574 D 3
Desmarais 260 D 2
Devon 3,885 D 3
Didsbury 3,095 C 4
Donalda 280 D 3
Donnelly 336 B 2
Drayton Valley 5,042 C 3
Drumheller 6,508 D 4
Duchess 429 E 4
East Coulee 218 D 4
Eckville 870 C 3
Edgerton 387 E 3
Edmonton (cap.) 532,246 D 3
Edmonton *657,057 D 3
Edmonton Beach 280 C 3
Edson 5,835 B 3
Elk Point 1,022 E 3
Elnora 249 D 3
Entwistle 462 C 3
Erskine 259 D 3
Evansburg 779 C 3
Exshaw 353 C 4
Fairview 2,869 A 1
Falher 1,102 B 2
Faust 399 C 2
Foremost 568 E 5
Forestburg 924 E 3
Fort Assiniboine 207 C 2
Fort Chipewyan 944 C 5
Fort Macleod 3,139 D 5
Fort McKay 267 E 1
Fort McMurray 31,000 E 1
Fort Saskatchewan 12,169 D 3
Fort Vermilion 752 B 5
Fox Creek 1,978 B 2
Fox Lake 634 B 5
Gibbons 2,276 D 3
Gift Lake 428 C 2
Girouxville 325 B 2
Gleichen 381 D 4
Glendon 430 E 2
Glenwood 259 D 5
Grand Centre 3,146 E 2
Grande Cache 4,523 A 3
Grande Prairie 24,263 A 2
Granum 399 D 5
Grimshaw 2,316 B 1
Grouard Mission 221 C 2
Hanna 2,806 E 4
Hardisty 641 E 3
Hay Lakes 302 D 3
Heisler 212 D 3
High Level 2,194 A 5
High Prairie 2,506 B 2
High River 4,792 D 4
Hines Creek 575 A 1
Hinton 8,342 B 3
Holden 430 D 3
Hughenden 267 E 3
Hythe 639 A 2
Innisfail 5,247 D 3
Innisfree 255 E 3
Irma 474 E 3
Irricana 558 D 4
Irvine 360 E 5
Jasper 3,269 B 3
John d'Or Prairie 437 B 5
Joussard 330 B 2
Killam 1,005 E 3
Kinuso 285 C 2
Kitscoty 497 E 3
Lac La Biche 2,007 E 2
Lacombe 5,591 D 3
La Crete 479 B 5
Lake Louise 355 B 4
Lamont 1,563 D 3
Leduc 12,471 D 3
Legal 1,022 D 3
Lethbridge 54,072 D 5
Linden 407 D 4
Little Buffalo Lake 253 B 1
Lloydminster 8,997 E 3
Longview 301 C 4
Lougheed 226 E 3
Lundbreck 244 C 5
Magrath 1,576 D 5
Manning 1,173 B 1
Mannville 788 E 3
Marlboro 211 B 3
Marwayne 500 E 3
Mayerthorpe 1,475 C 3
McLennan 1,125 B 2
Medicine Hat 40,380 E 4
Milk River 894 D 5
Millet 1,120 D 3
Mirror 507 D 3
Monarch 212 D 5
Morinville 4,657 D 3
Morrin 244 D 4
Mundare 604 D 3
Myrnam 397 E 3
Nacmine 369 D 4
Nampa 334 B 1
Nanton 1,641 D 4
New Norway 291 D 3
New Sarepta 417 D 3
Nobleford 534 D 5
North Calling Lake 234 D 2
Okotoks 3,847 C 4
Olds 4,813 D 4
Onoway 621 C 3
Oyen 975 E 4
Peace River 5,907 B 1
Penhold 1,531 D 3
Picture Butte 1,404 D 5
Pincher Creek 3,757 D 5
Plamondon 259 D 2
Pollockville 19 E 4
Ponoka 5,221 D 3
Provost 1,645 E 3
Rainbow Lake 504 A 5
Ralston 357 E 4
Raymond 2,837 D 5
Redcliff 3,876 E 4
Red Deer 46,393 D 3
Redwater 1,932 D 3
Rimbey 1,685 C 3
Robb 230 B 3
Rockyford 329 D 4
Rocky Mountain House 4,698 C 3
Rosemary 328 E 4
Rycroft 649 A 2
Ryley 483 D 3
Saint Albert 31,996 D 3
Saint Paul 4,884 E 3
Sangudo 398 C 3
Sedgewick 879 E 3
Sexsmith 1,180 A 2
Shaughnessy 270 D 5
Sherwood Park 29,285 D 3
Slave Lake 4,506 C 2
Smith 216 D 2
Smoky Lake 1,074 D 2
Spirit River 1,104 A 2
Spruce Grove 10,326 D 3
Standard 379 D 4
Stavely 504 D 4
Stettler 5,136 D 3
Stirling 688 D 5
Stony Plain 4,839 C 3
Strathmore 2,986 D 4
Strome 281 E 3
Sundre 1,742 C 4
Swan Hills 2,497 C 2
Sylvan Lake 3,779 C 3
Taber 5,988 E 5
Thorhild 576 D 2
Thorsby 737 C 3
Three Hills 1,787 D 4
Tilley 345 E 4
Tofield 1,504 D 3
Trochu 880 D 4
Turner Valley 1,311 C 4
Two Hills 1,193 E 3
Valleyview 2,061 B 2
Vauxhall 1,049 D 4
Vegreville 5,251 E 3
Vermilion 3,766 E 3
Veteran 314 E 3
Viking 1,232 E 3
Vilna 345 E 2
Vulcan 1,489 D 4
Wabamun 662 C 3
Wabasca 701 D 2
Wainwright 4,266 E 3
Warburg 501 C 3
Warner 477 D 5
Waskatenau 290 D 2
Wembley 1,169 A 2
Westlock 4,424 C 2
Wetaskiwin 9,597 D 3
Whitecourt 5,585 C 2
Wildwood 441 C 3
Willingdon 366 E 3
Youngstown 297 E 4

OTHER FEATURES

Abraham (lake) B 3
Alberta (mt.) B 3
Assiniboine (mt.) C 4
Athabasca (lake) C 5
Athabasca (riv.) D 1
Banff Nat'l Park B 4
Battle (riv.) D 3
Bear (lake) A 2
Beaver (riv.) E 2
Beaverhill (lake) D 3
Behan (lake) E 2
Belly (riv.) D 5
Berland (riv.) A 3
Berry (creek) E 4
Biche (lake) E 2
Big (isl.) B 5
Big Horn (dam) B 3
Bighorn (range) B 3
Birch (hills) A 2
Birch (lake) E 3
Birch (mts.) B 5
Birch (riv.) B 5
Bison (lake) B 1
Bittern (lake) D 3
Botha (riv.) B 1
Bow (riv.) D 4
Boyer (riv.) A 5
Brazeau (mt.) B 3
Brazeau (riv.) B 3
Buffalo (lake) D 3
Buffalo Head (hills) B 5
Burnt (lakes) C 1
Cadotte (lake) B 1
Cadotte (riv.) B 1
Calling (lake) D 2
Canal (creek) E 5
Cardinal (lake) B 1
Caribou (mts.) B 5
Chinchaga (riv.) A 5
Chip (lake) C 3
Chipewyan (lake) D 1
Chipewyan (riv.) D 1
Christina (lake) E 2
Christina (riv.) E 1
Claire (lake) B 5
Clear (hills) A 1
Clearwater (riv.) C 4
Clearwater (riv.) E 1
Clyde (lake) E 2
Cold (lake) E 2
Columbia (mt.) B 3
Crowsnest (pass) C 5
Cypress (hills) E 5
Cypress Hills Prov. Park E 5
Dillon (riv.) E 2
Dowling (lake) D 4
Dunkirk (riv.) D 1
Eisenhower (mt.) C 4
Elbow (riv.) C 4
Elk Island Nat'l Park D 3
Ells (riv.) D 1
Etzikom Coulee (riv.) E 5
Eva (lake) B 5
Farrell (lake) D 4
Firebag (riv.) E 1
Forbes (mt.) B 4
Freeman (riv.) C 2
Frog (lake) E 3
Garson (lake) E 1
Gipsy (lake) E 1
Gordon (lake) E 1
Gough (lake) D 3
Graham (lake) C 1
Gull (lake) C 3
Haig (lake) B 1
Hawk (hills) B 1
Hay (lake) A 5
Hay (riv.) A 5
Heart (lake) E 2
Highwood (riv.) C 4
House (mt.) C 2
House (riv.) D 2
Iosegun (lake) B 2
Iosegun (riv.) B 2
Jackfish (riv.) B 5
Jasper Nat'l Park A 3
Kakwa (riv.) A 2
Kickinghorse (pass) B 4
Kimiwan (lake) B 2
Kirkpatrick (lake) E 4
Kitchener (mt.) B 3
Legend (lake) D 1
Lesser Slave (lake) C 2
Liége (riv.) D 1
Little Bow (riv.) D 4
Little Cadotte (riv.) B 1
Little Smoky (riv.) B 2
Livingstone (range) C 4
Logan (lake) E 2
Loon (lake) C 1
Loon (riv.) C 1
Lubicon (lake) C 1
Lyell (mt.) B 4
MacKay (riv.) D 1
Maligne (lake) B 3
Margaret (lake) B 5
Marie (lake) E 2
Marion (lake) D 3
Marten (mt.) C 2
McClelland (lake) E 1
McGregor (lake) D 4
McLeod (riv.) B 3
Meikle (riv.) A 1
Mikkwa (riv.) B 5
Milk (riv.) D 5
Mistehae (lake) C 2
Muriel (lake) E 2
Muskwa (lake) C 1
Muskwa (riv.) C 1
Namur (lake) D 1
Newell (lake) E 4
Nordegg (riv.) C 3
North Saskatchewan (riv.) E 3
North Wabasca (lake) D 1
Notikewin (riv.) A 1
Oldman (riv.) D 5
Otter (lakes) B 1
Pakowki (lake) E 5
Panny (riv.) C 1
Peace (riv.) B 1
Peerless (lake) C 1
Pelican (lake) D 2
Pelican (mts.) D 2
Pembina (riv.) C 3
Pigeon (lake) D 3
Pinehurst (lake) E 2
Porcupine (hills) C 4
Primrose (lake) E 2
Rainbow (lake) A 5
Red Deer (lake) D 3
Red Deer (riv.) D 4
Richardson (riv.) C 5
Rocky (mts.) B-C 4
Rosebud (riv.) D 4
Russell (lake) C 1
Saddle (hills) A 2
Sainte Anne (lake) C 3
Saint Mary (res.) D 5
Saint Mary (riv.) D 5
Saulteaux (riv.) C 2
Seibert (lake) E 2
Simonette (riv.) A 2
Slave (riv.) C 5
Smoky (riv.) A 2
Snake Indian (riv.) A 3
Snipe (lake) B 2
Sounding (creek) E 4
South Saskatchewan (riv.) E 4
South Wabasca (lake) D 2
Spencer (lake) E 2
Spray (mts.) C 4
Sturgeon (lake) B 2
Sullivan (lake) D 3
Swan (hills) C 2
Swan (riv.) C 2
Temple (mt.) B 4
The Twins (mt.) B 3
Thickwood (hills) D 1
Touchwood (lake) E 2
Travers (res.) D 4
Trout (mt.) C 1
Trout (riv.) C 1
Utikuma (lake) C 2
Utikuma (riv.) C 1
Utikumasis (lake) C 2
Vermilion (riv.) E 3
Wabasca (riv.) C 1
Wallace (mt.) C 2
Wapiti (riv.) A 2
Wappau (lake) E 2
Watchusk (lake) E 1
Waterton-Glacier Int'l Peace Park C 5
Waterton Lakes Nat'l Park C 5
Whitemud (riv.) A 1
Wildhay (riv.) B 3
Willmore Wilderness Prov. Park A 3
Winagami (lake) B 2
Winefred (lake) E 2
Winefred (riv.) E 2
Wolf (lake) E 2
Wolverine (riv.) B 1
Wood Buffalo Nat'l Park B 5
Yellowhead (pass) A 3
Zama (lake) A 5

*Population of metropolitan area.

Agriculture, Industry and Resources

DOMINANT LAND USE

Wheat
Cereals (chiefly barley, oats)
Cereals, Livestock
Dairy
Pasture Livestock
Range Livestock
Forests
Nonagricultural Land

MAJOR MINERAL OCCURRENCES

C Coal
O Petroleum
G Natural Gas
S Sulfur
Na Salt

Water Power
Major Industrial Areas

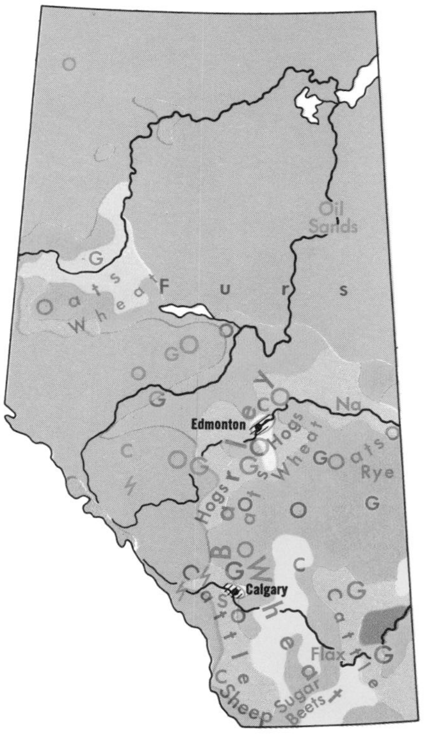

Oil Sands
Oats
Wheat
Furs
Edmonton
Na
Hogs
Oats
Rye
Calgary
Flax
Sugar Beets
Sheep
Cattle

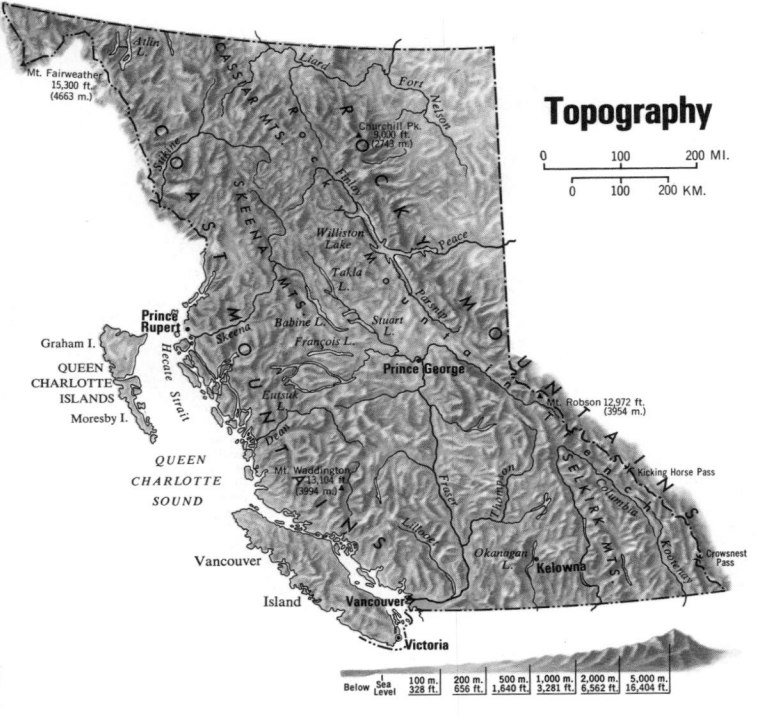

Topography

0 — 100 — 200 MI.
0 — 100 — 200 KM.

Below Sea Level	100 m. 328 ft.	200 m. 656 ft.	500 m. 1,640 ft.	1,000 m. 3,281 ft.	2,000 m. 6,562 ft.	5,000 m. 16,404 ft.

CITIES and TOWNS

Abbotsford 12,745 .... L 3
Alert Bay 626 .... D 5
Armstrong 2,683 .... H 5
Ashcroft 2,156 .... G 5
Ashton Creek 452 .... H 5
Balfour 472 .... J 5
Barlow 441 .... F 3
Barrière 1,370 .... H 4
Blueberry Creek 635 .... J 5
Blue River 384 .... H 4
Boston Bar 498 .... G 5
Bowen Island 1,125 .... K 3
Brackendale 1,719 .... F 5
Burnaby ○136,494 .... K 3
Burns Lake 1,777 .... D 3
Cache Creek 1,308 .... G 5
Campbell River 15,370 .... E 5
Canal Flats 919 .... J 5
Canyon 698 .... J 5
Cassiar 1,045 .... K 2
Castlegar 6,902 .... J 5

Cawston 785 .... H 5
Central Saanich ○9,890 .... K 3
Chase 1,777 .... H 5
Chemainus 2,069 .... J 3
Cherry Creek 450 .... G 5
Chetwynd 2,553 .... G 2
Chilliwack ○40,642 .... M 3
Clearwater 1,461 .... G 4
Clinton 804 .... G 4
Coldstream ○6,450 .... H 5
Comox 6,607 .... H 2
Coquitlam ○61,077 .... K 3
Courtenay 8,992 .... E 5
Cranbrook 15,915 .... K 5
Creston 4,190 .... J 5
Crofton 1,303 .... J 3
Cultus Lake 481 .... M 3
Cumberland 1,947 .... E 5
Dawson Creek 11,373 .... G 2
Delta ○74,692 .... K 3
Duncan 4,228 .... J 3
Elkford 3,126 .... K 5
Enderby 1,816 .... H 5
Erickson 972 .... J 5

Errington 609 .... J 3
Esquimalt ○15,870 .... K 4
Falkland 478 .... H 5
Fernie 5,444 .... K 5
Forest Grove 444 .... G 4
Fort Fraser 574 .... E 3
Fort Langley 2,284 .... L 3
Fort Nelson 3,724 .... M 2
Fort Saint James 2,284 .... E 3
Fort Saint John 13,891 .... G 2
Fraser Lake 1,543 .... E 3
Fruitvale 1,904 .... J 5
Gabriola 1,627 .... J 3
Galiano 669 .... K 3
Ganges 1,118 .... K 3
Gibsons 2,594 .... K 3
Gold River 2,225 .... D 5
Golden 3,476 .... J 4
Grand Forks 3,486 .... H 6
Granisle 1,430 .... D 3
Greenwood 856 .... H 5
Hagensborg 350 .... D 4
Harrison Hot Springs 569 .... M 3
Hatzic 1,055 .... L 3

Hazelton 393 .... D 2
Hedley 426 .... G 5
Holberg 444 .... C 5
Honeymoon Bay 474 .... J 3
Hope 3,205 .... M 3
Hornby Island 474 .... H 2
Horsefly 430 .... G 4
Houston 1,714 .... D 3
Hudson Hope 984 .... F 2
Invermere 1,969 .... J 5
Kaleden 998 .... H 5
Kamloops 64,048 .... G 5
Kaslo 854 .... J 5
Kelowna 59,196 .... H 5
Kent ○3,394 .... M 3
Keremeos 830 .... G 5
Kimberley 7,375 .... K 5
Kitimat 12,462 .... C 3
Kitsault 554 .... C 2
Kitwanga 369 .... D 2
Lac La Hache 647 .... G 4
Ladysmith 4,558 .... J 3
Lake Cowichan 2,391 .... J 3
Langley 15,124 .... L 3
Lantzville 969 .... J 3
Likely 425 .... G 4
Lillooet 1,725 .... G 5
Lion's Bay 1,078 .... K 3
Logan Lake 2,637 .... G 5
Lumby 1,266 .... H 5
Lytton 428 .... G 5
Mackenzie 5,797 .... F 2
Mackenzie ○5,890 .... F 2
Malakwa 392 .... H 5
Maple Bay 393 .... K 3
Maple Ridge ○32,232 .... L 3
Masset 1,569 .... B 3
Matsqui ○42,001 .... L 3
Mayne 546 .... K 3
McBride 641 .... G 3
Merritt 6,110 .... G 5
Midway 633 .... H 6
Mill Bay 583 .... K 3
Mission ○20,056 .... L 3
Mission City 9,948 .... L 3
Montrose 1,229 .... J 5
Nakusp 1,495 .... J 5
Nanaimo 47,069 .... J 3
Naramata 876 .... H 5
Nelson 9,143 .... J 5
New Denver 642 .... J 5
New Hazelton 792 .... D 2
New Westminster 38,550 .... K 3
Nicomen Island 360 .... L 3
Nootka .... D 5
North Cowichan ○18,210 .... J 3
North Pender Island 906 .... K 3
North Saanich ○6,117 .... K 3
North Vancouver 33,952 .... K 3
North Vancouver ○65,367 .... K 3
Oak Bay ○16,990 .... K 4
Okanagan Falls 1,030 .... H 5
Okanagan Landing 834 .... H 5
Okanagan Mission .... H 5
Old Barkerville 11 .... G 3
Oliver 1,893 .... H 5
One Hundred Mile House 1,925 .... G 4
Osoyoos 2,738 .... H 5
Oyama 430 .... H 5
Parksville 5,216 .... J 3
Peachland ○2,865 .... G 5

Penticton 23,181 .... H 5
Pitt Meadows ○6,209 .... L 3
Port Alberni 19,892 .... H 3
Port Alice 1,668 .... D 5
Port Clements 380 .... B 3
Port Coquitlam 27,535 .... L 3
Port Edward 989 .... B 3
Port Hardy ○3,778 .... D 5
Port McNeill 2,474 .... D 5
Port Moody 14,917 .... L 3
Pouce-Coupé 821 .... G 2
Powell River ○13,423 .... E 5
Prince George 67,559 .... F 3
Prince Rupert 16,197 .... B 3
Princeton 3,051 .... G 5
Qualicum Beach 2,844 .... J 3
Queen Charlotte 1,070 .... A 3
Quesnel 8,240 .... F 4
Radium Hot Springs 419 .... J 5
Revelstoke 5,544 .... J 5
Richmond ○96,154 .... K 3
Roberts Creek 926 .... J 3
Robson 1,008 .... J 5
Rossland 3,967 .... H 6
Royston 754 .... H 2
Saanich ○78,710 .... K 3
Salmo 1,169 .... J 5
Salmon Arm 1,946 .... H 5
Salmon Arm ○10,780 .... H 5
Saltair 1,356 .... J 3
Sandspit 794 .... B 3
Sayward 482 .... D 5
Sechelt 1,096 .... J 2
Shawnigan Lake 419 .... J 3
Shoreacres 555 .... J 5
Sicamous 1,057 .... H 5
Sidney 7,946 .... K 3
Slocan 351 .... J 5
Slocan Park 414 .... J 5
Smithers 4,570 .... D 3
Sointula 567 .... D 5
Sooke 852 .... J 4
Sorrento 659 .... H 5
South Hazelton 500 .... D 2
South Wellington 620 .... J 3
Spallumcheen 4,213 .... H 5
Sparwood 3,267 .... K 5
Sproat Lake 440 .... H 3
Squamish 1,590 .... F 5
Stewart 1,456 .... C 2
Summerland ○7,473 .... G 5
Surrey ○147,138 .... K 3
Tahsis 1,739 .... D 5
Taylor 966 .... G 2
Telkwa 840 .... D 3
Terrace 8,893 .... C 3
Terrace ○10,914 .... C 3
Thornhill 4,281 .... C 3
Thrums 360 .... J 5
Tofino 705 .... E 5
Trail 9,599 .... J 5
Ucluelet 1,593 .... E 6
Union Bay 601 .... H 2
Valemount 1,130 .... H 4
Vancouver 414,281 .... K 3
Vancouver (Greater) *1,169,831 .... K 3
Vanderhoof 2,323 .... E 3
Vavenby 479 .... H 4
Vernon 19,987 .... H 5
Victoria (cap.) 64,379 .... K 4
Victoria *233,481 .... K 3
Warfield 1,969 .... J 5
Wasa 345 .... K 5
Wells 417 .... G 3
Westbank 1,271 .... H 5
West Vancouver ○35,728 .... K 3
Westwold 409 .... G 5
Whistler ○1,365 .... F 5
White Rock 13,550 .... K 3
Williams Lake 8,362 .... F 4
Wilson Creek 611 .... J 2
Windermere 611 .... K 5
Winlaw 435 .... J 5
Woss Lake 395 .... D 5
Wynndel 566 .... J 5
Yarrow 1,201 .... M 3
Youbou 965 .... J 3

OTHER FEATURES

Adams (lake) .... H 4
Adams (riv.) .... H 4
Alberni (inlet) .... H 3
Alsek (riv.) .... H 1
Aristazabal (isl.) .... C 4
Assiniboine (mt.) .... K 5
Atlin (lake) .... J 1
Azure (lake) .... G 4
Babine (lake) .... E 3
Babine (riv.) .... D 2
Banks (isl.) .... B 3
Barkley (sound) .... E 6
Beale (riv.) .... E 6
Beatton (riv.) .... G 1
Bella Coola (riv.) .... D 4
Bennett, W.A.C. (dam) .... F 2
Birkenhead Lake Prov. Park .... F 5
Bowron Lake Prov. Park .... G 3
Bowser (lake) .... C 2
Brooks (pen.) .... D 5
Browning Entrance (str.) .... B 3
Bryce (mt.) .... J 4
Bugaboo Glacier Prov. Park .... J 5
Bulkley (riv.) .... D 2
Burke (chan.) .... D 4
Burnaby (isl.) .... B 4
Bute (inlet) .... E 5
Caamaño (sound) .... C 3
Calvert (isl.) .... C 4
Canim (lake) .... G 4
Canoe (riv.) .... H 4
Cariboo (mts.) .... G 3
Carpenter (lake) .... F 5
Carp Lake Prov. Park .... F 3
Cassiar (mts.) .... K 2
Castle (mt.) .... A 2

Charlotte (lake) .... E 4
Chatham (sound) .... B 3
Chehalis (lake) .... L 3
Chilcotin (riv.) .... E 4
Chilko (lake) .... F 4
Chilko (riv.) .... E 4
Chilkoot (pass) .... J 1
Chuchi (lake) .... E 2
Churchill (peak) .... L 2
Clayoquot (sound) .... D 5
Clearwater (lake) .... G 4
Clearwater (riv.) .... G 4
Coast (mts.) .... D 3
Columbia (lake) .... K 5
Columbia (mt.) .... J 4
Columbia (riv.) .... J 5
Cook (cape) .... C 5
Cowichan (lake) .... J 3
Crowsnest (pass) .... K 5
Cypress Prov. Park .... K 3
Dean (chan.) .... D 4
Dean (riv.) .... D 4
Dease (lake) .... K 2
Dease (riv.) .... K 2
Devils Thumb (mt.) .... A 1
Dixon Entrance (chan.) .... A 3
Douglas (chan.) .... C 3
Duncan (riv.) .... J 5
Dundas (isl.) .... B 3
Elk (riv.) .... K 5
Elk Lakes Prov. Park .... K 5
Eutsuk (lake) .... D 3
Fairweather (mt.) .... H 1

Finlay (riv.) .... E 1
Fitzhugh (sound) .... D 4
Flathead (riv.) .... K 6
Flores (isl.) .... D 5
Fontas (riv.) .... M 2
Forbes (mt.) .... J 4
Fort Nelson (riv.) .... M 2
François (lake) .... D 3
Fraser (lake) .... E 3
Fraser (riv.) .... F 4
Fraser Reach (chan.) .... C 3
Galiano (isl.) .... K 3
Gardner (canal) .... D 3
Garibaldi Prov. Park .... F 5
Georgia (str.) .... J 3
Germansen (lake) .... E 2
Gil (isl.) .... C 3
Glacier Nat'l Park .... J 4
Golden Ears Prov. Park .... L 2
Gordon (riv.) .... H 3
Graham (isl.) .... A 3
Graham Reach (chan.) .... C 3
Grenville (chan.) .... C 3
Halfway (riv.) .... F 2
Hamber Prov. Park .... H 4
Harrison (lake) .... M 2
Hawkesbury (isl.) .... C 3
Hazelton (mts.) .... C 2
Hecate (str.) .... B 3
Hobson (lake) .... H 4
Homathko (riv.) .... E 4
Horsefly (lake) .... G 4

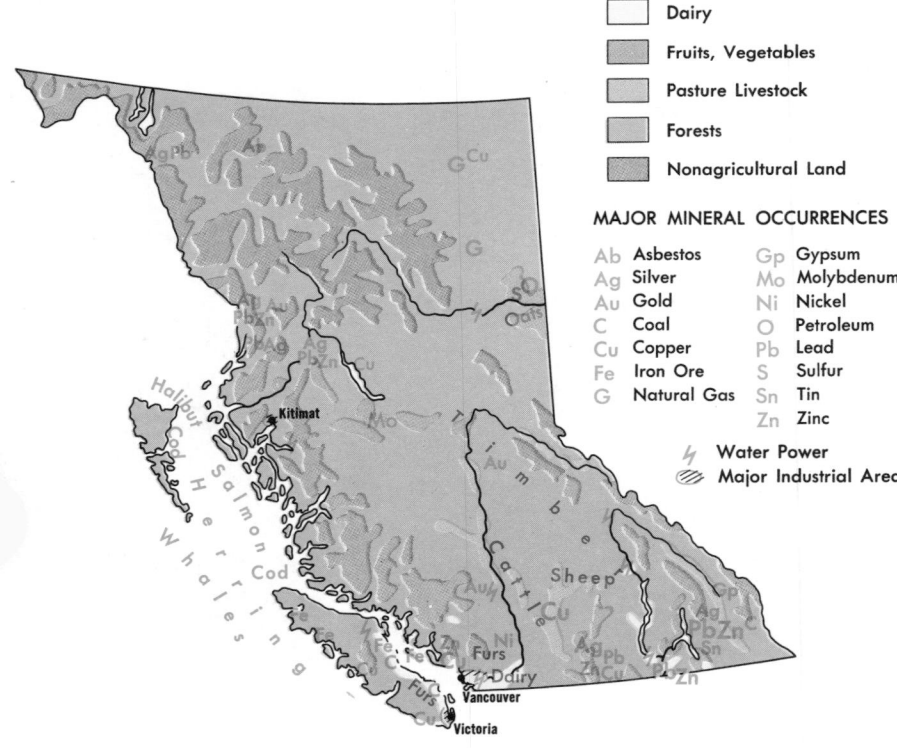

Agriculture, Industry and Resources

DOMINANT LAND USE

- Cereals, Livestock
- Dairy
- Fruits, Vegetables
- Pasture Livestock
- Forests
- Nonagricultural Land

MAJOR MINERAL OCCURRENCES

Ab	Asbestos	Gp	Gypsum
Ag	Silver	Mo	Molybdenum
Au	Gold	Ni	Nickel
C	Coal	O	Petroleum
Cu	Copper	Pb	Lead
Fe	Iron Ore	S	Sulfur
G	Natural Gas	Sn	Tin
		Zn	Zinc

⚡ Water Power
Major Industrial Areas

British Columbia

SCALE
0 15 30 60 90 120 MI.
0 15 30 60 90 120 KM.

Provincial Capital .... ⊛
State Capital .... ⊛
International Boundaries .... ▬ ▬
Provincial Boundaries .... ▬ ▬

Scale 1:5,200,000

® Copyright HAMMOND INCORPORATED, Maplewood, N.J.

Howe (sound)	K 2	Louise (isl.)	B 4

Howe (sound) .... K 2
Hunter (isl.) .... C 4
Inklin (riv.) .... J 2
Inzana (lake) .... E 3
Isaac (lake) .... G 3
Iskut (riv.) .... B 2
Jervis (inlet) .... E 5
John Jay (mt.) .... B 2
Johnstone (str.) .... D 5
Juan de Fuca (str.) .... J 4
Kates Needle (mt.) .... A 1
Kechika (riv.) .... L 2
Kenney (dam) .... E 3
Kettle (riv.) .... H 5
Kicking Horse (pass) .... J 4
King (isl.) .... D 4
Klinaklini (riv.) .... E 4
Kloch (lake) .... E 2
Knight (inlet) .... E 5
Knox (cape) .... A 3
Kokanee Glacier Prov. Park .... J 5
Koocanusa (lake) .... K 6
Kootenay (lake) .... J 5
Kootenay (riv.) .... K 5
Kootenay Nat'l Park .... J 4
Kotcho (lake) .... M 2
Kotcho (riv.) .... M 2
Kunghit (isl.) .... B 4
Kyuquot (sound) .... D 5
Langara (isl.) .... A 3
Laredo (sound) .... C 4
Liard (riv.) .... L 2
Lillooet (riv.) .... F 5

Louise (isl.) .... B 4
Lower Arrow (lake) .... H 5
Lyell (isl.) .... B 4
Lyell (mt.) .... J 4
Mabel (lake) .... H 5
Mahood (lake) .... G 4
Malaspina (str.) .... J 2
Manning Prov. Park .... H 6
Masset (inlet) .... A 3
McCauley (isl.) .... B 3
McGregor (riv.) .... G 3
Meziadin (lake) .... C 2
Milbanke (sound) .... C 4
Moberly (lake) .... F 2
Monashee (mts.) .... H 4
Moresby (isl.) .... B 4
Morice (lake) .... D 3
Morice (riv.) .... D 3
Mount Assiniboine Prov. Park .... K 5
Mount Edziza Prov. Park and
  Rec. Area. .... B 1
Mount Revelstoke Nat'l Park .... H 4
Mount Robson Prov. Park .... H 3
Muncho Lake Prov. Park .... L 2
Murray (riv.) .... G 3
Murtle (lake) .... H 4
Muskwa (riv.) .... M 2
Nanika (dam) .... D 3
Nass (riv.) .... C 2
Nation (riv.) .... F 2
Nechako (riv.) .... E 3
Nitinat (lake) .... H 3
Nootka (isl.) .... D 5

Nootka (sound) .... D 5
North Thompson (riv.) .... G 4
Observatory (inlet) .... C 2
Okanagan (lake) .... H 5
Okanagan Mtn. Prov. Park .... G 5
Okanagan (riv.) .... H 6
Omineca (mts.) .... E 2
Omineca (riv.) .... E 2
Ootsa (lake) .... D 3
Owikeno (lake) .... D 4
Pacific Rim Nat'l Park .... D 5
Parsnip (riv.) .... F 3
Peace (riv.) .... F 2
Pend Oreille (riv.) .... J 6
Petitot (riv.) .... L 2
Pinchi (lake) .... E 3
Pine (riv.) .... G 2
Pitt (isl.) .... C 3
Pitt (lake) .... L 2
Porcher (isl.) .... B 3
Portland (canal) .... B 2
Portland (inlet) .... B 2
Price (isl.) .... C 4
Princess Royal (isl.) .... C 3
Principe (chan.) .... C 3
Prophet (riv.) .... M 2
Purcell (mts.) .... J 5
Quatsino (sound) .... D 5
Queen Charlotte (isls.) .... B 3
Queen Charlotte (sound) .... C 4
Queen Charlotte (str.) .... D 5
Queens (sound) .... C 4
Quesnel (lake) .... G 4

Quesnel (riv.) .... F 4
Rivers (inlet) .... D 4
Robson (mt.) .... H 3
Rocky (mts.) .... F 2
Roderick (isl.) .... C 4
Rose (pt.) .... B 3
Saint James (cape) .... B 4
Salmon (riv.) .... F 3
Salmon Arm (inlet) .... J 2
Schoen Lake Prov. Park .... E 5
Scott (cape) .... C 5
Scott (isls.) .... C 5
Seechelt (inlet) .... J 2
Seechelt (pen.) .... J 2
Selkirk (mts.) .... J 4
Seymour (inlet) .... D 4
Sheslay (riv.) .... J 2
Shuswap (lake) .... H 4
Sikanni Chief (riv.) .... F 1
Silver Star Prov. Park .... H 4
Sir Sandford (mt.) .... H 4
Skagit (riv.) .... G 6
Skeena (mts.) .... C 2
Skeena (riv.) .... C 2
Skidegate (inlet) .... B 3
Slocan (lake) .... J 5
Smith (sound) .... C 4
South Bentinck Arm (inlet) .... D 4
Stave (lake) .... L 2
Stephens (isl.) .... B 3
Stikine (riv.) .... B 1
Stone Mountain Prov. Park .... L 2

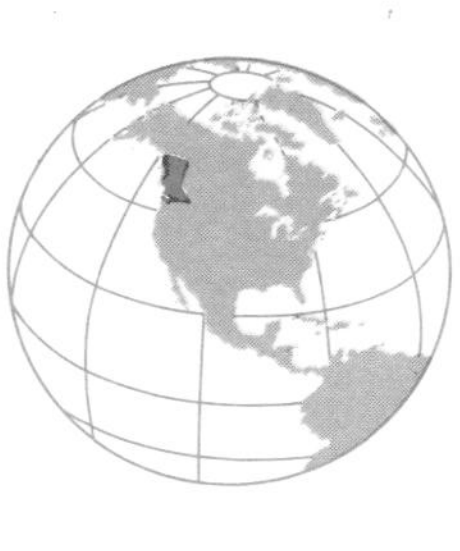

Strathcona Prov. Park .... E 5
Stuart (lake) .... E 3
Sustut (riv.) .... D 2
Tagish (lake) .... J 1
Tahtsa (lake) .... D 3
Takla (lake) .... D 2
Taku (riv.) .... J 2
Tatlatui (lake) .... D 2
Tatlayoko (lake) .... E 4
Tchentlo (lake) .... E 2
Teslin (lake) .... K 1
Tetachuck (lake) .... E 3
Tezzeron (lake) .... E 3
Thompson (riv.) .... G 5

Three Guardsmen (mt.) .... H 1
Thutade (lake) .... D 2
Tiedemann (mt.) .... E 4
Toad (riv.) .... L 2
Toba (inlet) .... E 5
Tochcha (lake) .... E 3
Top Of The World Prov. Park .... K 5
Trembleur (lake) .... E 3
Troitsa (lake) .... D 3
Tumeka (lake) .... C 1
Turnagain (riv.) .... K 2
Tuya (riv.) .... K 2
Tweedsmuir Prov. Park .... D 3
Upper Arrow (lake) .... H 5
Valdes (isl.) .... K 2

Vancouver (isl.) .... D 5
Virago (sound) .... A 3
Waddington (mt.) .... E 4
Wapiti (riv.) .... H 3
Wells Gray Prov. Park .... H 4
West Road (riv.) .... E 3
Whitesail (lake) .... D 3
Williston (lake) .... F 2
Work (chan.) .... C 3
Yellowhead (pass) .... H 4
Yoho Nat'l Park .... J 4

AREA 366,253 sq. mi. (948,596 sq. km.)
POPULATION 2,744,467
CAPITAL Victoria
LARGEST CITY Vancouver
HIGHEST POINT Mt. Fairweather 15,300 ft.
  (4,663 m.)
SETTLED IN 1806
ADMITTED TO CONFEDERATION 1871
PROVINCIAL FLOWER Dogwood

*Population of metropolitan area.
○Population of municipality.

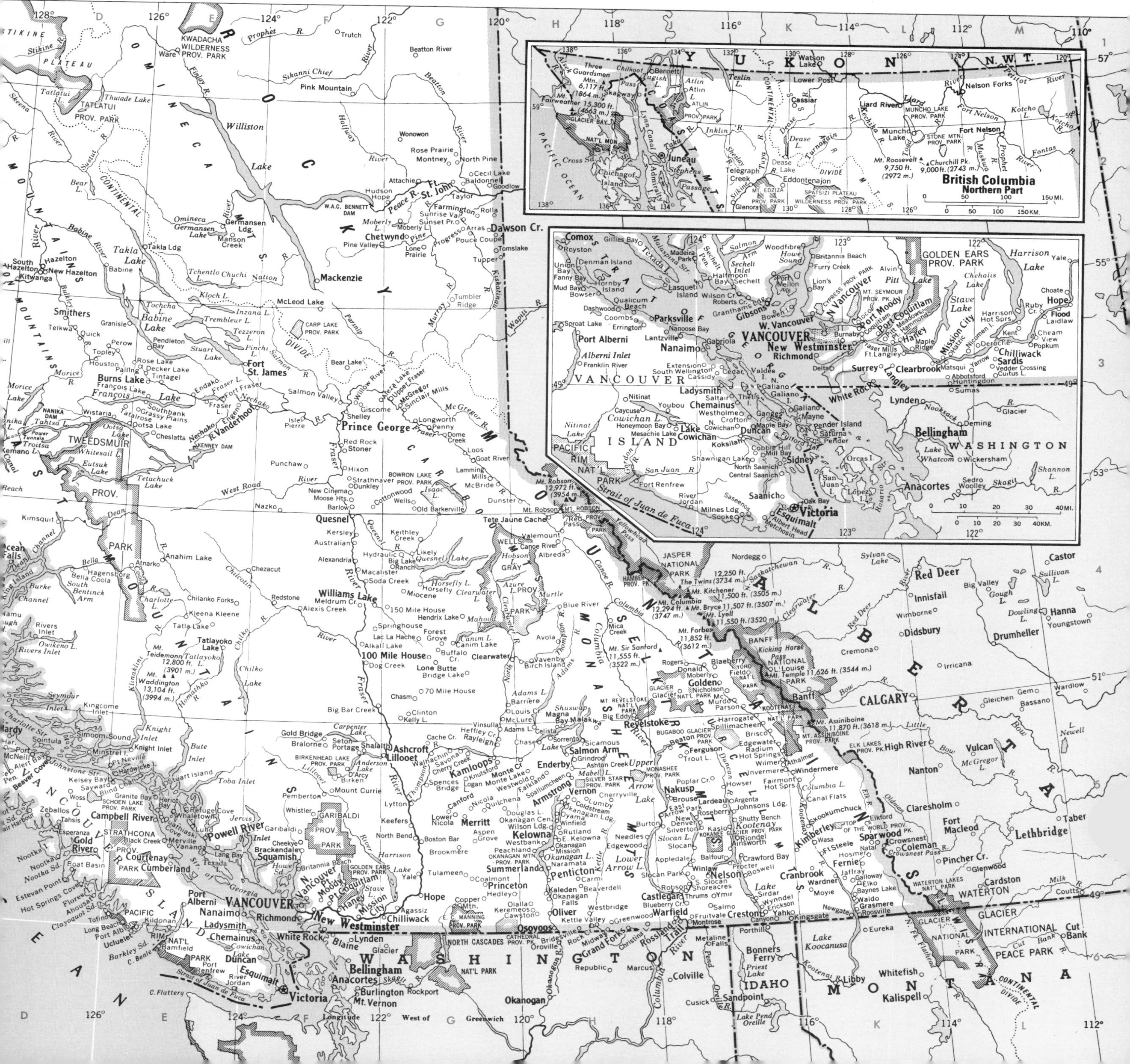

NORTHWEST TERRITORIES

DISTRICTS

Baffin 8,300 . . . . J2
Fort Smith 22,384 . . . . . . G3
Inuvik 7,485 . . . . F3
Keewatin 4,327 . . . . J3
Kitikmeot 3,245 . . . . G2

CITIES and TOWNS

Aklavik 721 . . . . E3
Alert M1
Amadjuak L3
Arctic Bay 375 . . . . K2
Arctic Red River 120 . . . . . E3
Baker Lake 954 . . . . J3
Bathurst Inlet 20 . . . . H3
Bay Chimo 60 . . . . H3
Bell Rock G3
Broughton Island 378 . . . . M3
Buffalo River Junction G3
Cambridge Bay 815 . . . . . H3
Cape Dorset 784 . . . . L3
Cape Dyer M3
Cape Smith L3
Chesterfield Inlet 249 . . . . K3
Clyde (Clyde River) 443 . . M2
Colville Lake 57 . . . . F3
Coppermine 809 . . . . G3
Coral Harbour 429 . . . . . . K3
Detah 143 . . . . G3
Dory Point G3
Enterprise 46 . . . . G3
Eskimo Point 1,022. . . . . . J3
Eureka K2
Fort Franklin 521 . . . . F3
Fort Good Hope 463 . . . . . F3
Fort Liard 405 . . . . F3
Fort McPherson 632 . . . . . E3
Fort Norman 286 . . . . F3
Fort Providence 605 . . . . . G3
Fort Resolution 480 . . . . . G3
Fort Simpson 980 . . . . . . F3
Fort Smith 2,298 . . . . G4
Frobisher Bay 2,333. . . . . M3
Gjoa Haven 523 . . . . J3
Grise Fiord 106 . . . . K2
Hall Beach 349 . . . . K3
Hay River 2,863 . . . . G3
Holman Island 300 . . . . . . G2
Igloolik 746 . . . . K3
Inuvik 3,147 . . . . E3
Isachsen H2
Jean-Marie River 69 . . . . . F3
Kakisa 36 . . . . G3
Kipisa 43 . . . . M3
Lac la Martre 268 . . . . . . G3
Lake Harbour 252 . . . . . . L3
Mould Bay F2
Nahanni Butte 85. . . . . . . F3
Nanisivik 261 . . . . K2
Norman Wells 420. . . . . . . F3
Pangnirtung 839 . . . . M3
Paulatuk 174 . . . . F3
Pelly Bay 257 . . . . K3
Pine Point 1,861 . . . . G3
Pond Inlet 705 . . . . L2
Port Burwell M3
Port Radium 56 . . . . G3
Rae-Edzo 1,378 . . . . G3
Rae Lakes 200 . . . . G3
Rankin Inlet 1,109 . . . . . . G3
Reliance 15. . . . . H3
Repulse Bay 352. . . . . . . K3
Resolute Bay 168 . . . . . . J2

Resolution Island M3
Rocher River. G3
Sachs Harbour 161 . . . . . F2
Salt River G3
Sawmill Bay G3
Snare Lake 69. . . . . G3
Snowdrift 253. . . . . G3
Spence Bay 431 . . . . J3
Trout Lake 59 . . . . F3
Tuktoyaktuk 772 . . . . E3
Tungsten 320 . . . . F3
Whale Cove 188 . . . . J3
Wrigley 137 . . . . F3
Yellowknife (cap.) 9,483 . . G3

OTHER FEATURES

Adelaide (pen.) J3
Admiralty (inlet) K2
Air Force (isl.). L3
Akpatok (isl.) M3
Amadjuak (lake) L3
Amund Ringnes (isl.) J2
Amundsen (gulf) F2
Anderson (riv.) F3
Arctic Red (riv.). E3
Artillery (lake) H3
Auyuittug Nat'l Park M3
Axel Heiberg (isl.). J2
Aylmer (lake) H3
Back (riv.) J3
Baffin (bay) M2
Baffin (isl.) L2
Baker (lake) J3
Banks (isl.) F2
Barbeau (peak) L1
Barrow (str.) J2
Bathurst (cape). F2
Bathurst (inlet) H3
Bathurst (isl.) H2
Beaufort (sea) D2
Bellot (str.) J2
Boothia (gulf) K3
Boothia (pen.) J2
Borden (isl.) G2
Borden (pen.) K2
Brodeur (pen.) K2
Bruce (mts.) L2
Buchan (gulf) L2
Burnside (riv.) G3
Byam Martin (chan.). H2
Byam Martin (isl.) H2
Bylot (isl.) L2
Camsell (riv.) G3
Challenger (mts.) L1
Chantrey (inlet) J3
Chesterfield (inlet) J3
Chidley (cape) M3
Clinton-Colden (lake) H3
Clyde (inlet) M2
Coats (isl.) K3
Coburg (isl.) L2
Columbia (cape) M1
Colville (lake). F3
Committee (bay) K3
Contwoyto (lake). H3
Coppermine (riv.) G3
Cornwall (isl.) J2
Cornwallis (isl.) J2
Coronation (gulf) G3
Croker (isl.) K2
Crown Prince Frederik (isl.). K3
Cumberland (pen.) M3
Cumberland (sound) M3
Dalhousie (cape) E2
Davis (str.) M3
Dease (str.) H3

Denmark (bay) H2
Devon (isl.) K2
Dolphin and Union (str.). . . G3
Dubawnt (lake) H3
Dubawnt (riv.) H3
Dundas (pen.) G2
Dyer (cape) M3
Eclipse (sound) L2
Eglinton (isl.) F2
Ellef Ringnes (isl.) H2
Ellesmere (isl.) K2
Ennadai (lake) H3
Eskimo (lakes) E3
Eureka (sound) K2
Evans (str.) K3
Exeter (sound) M3
Fisher (str.) K3
Fosheim (pen.) K1
Foxe (basin). L3
Foxe (chan.) K3
Foxe (pen.) L3
Franklin (bay) F2
Franklin (mts.) F3
Franklin (str.) J2
Frobisher (bay) M3
Frozen (str.) K3
Fury and Hecla (str.) K3
Gabriel (str.) M3
Garry (lake) H3
Gods Mercy (bay) K3
Great Bear (lake) F3
Great Bear (riv.) F3
Great Slave (lake) G3
Greely (fjord) K1
Grinnell (pen.) J2
Hadley (bay) H2
Hall (basin) M1
Hall (pen.) M3
Hayes (riv.) J3
Hazen (lake) L1
Hazen (str.) G2
Henik (lakes) J3
Henry Kater (cape) M3
Home (bay) M3
Hood (riv.) G3
Horn (mts.) G3
Hornaday (riv.) F3
Horton (riv.) F3
Hottah (lake) G3
Hudson (bay) K3
Hudson (str.) L3
Isachsen (cape) H2
James Ross (str.) J3
Jenny Lind (isl.) H3
Jens Munk (isl.) K3
Jones (sound) K2
Kaminuriak (lake) J3
Kane (basin) L2
Kasba (lake) H3
Kazan (riv.) H3
Keele (riv.) F3
Keith Arm (inlet) F3
Kellett (cape) F2
Kellett (str.) G2
Kennedy (chan.) M1
Kent (pen.) H3
King Christian (isl.) H2
King William (isl.) J3
Lady Ann (str.) L2
La Martre (lake) G3
Lancaster (sound) K2
Lands End (cape) F2
Larsen (sound) J2
Liard (riv.) F4
Lincoln (sea) M1
Liverpool (bay) E2
Lockhart (riv.) H3

Lougheed (isl.) H2
Lyon (inlet) K2
MacKay (lake) G3
Mackenzie (bay) E3
Mackenzie (mts.) E3
Mackenzie (riv.) F3
Mackenzie King (isl.) G2
Macmillan (pass) F3
Maguse (bay) J3
Makinson (inlet) L2
Mansel (isl.) K3
Marian (lake) G3
Markham (inlet) L1
McLeod (bay) G3
M'Clintock (chan.) H2
M'Clure (str.) G2
McTavish Arm (inlet) G3
Meighen (isl.) H1
Melville (isl.) G2
Melville (pen.) K3
Mercy (cape) M3
Mills (lake) G3
Minto (inlet) G2
Mistake (bay) J3
Nahanni Nat'l Park F3
Nansen (sound) J1
Nares (str.) L2
Navy Board (inlet) K2
Nelson Head (prom.) F2
Nettilling (lake) L3

Nonacho (lake) H3
North Arm (inlet) G3
North Magnetic Pole H2
Norwegian (bay) J2
Nottingham (isl.) K3
Nueltin (lake) H3
Ommanney (bay) H2
Padloping (isl.) M3
Parry (bay) K3
Parry (chan.) G2
Parry (isle.) G2
Parry (pen.) F2
Peary (chan.) H2
Peel (sound) J2
Pelly (bay) J3
Penny (str.) J2
Point (lake) G3
Pond (inlet) L2
Prince Albert (pen.) G2
Prince Albert (sound) G2
Prince Charles (isl.) L3
Prince Gustav Adolf (sea) . . H2
Prince of Wales (isl.) J2
Prince of Wales (str.) G2
Prince Patrick (isl.) F2
Prince Regent (inlet) K2
Queen Elizabeth (isls.) . . . H1
Queen Maud (gulf) H3
Queens (chan.) J2
Raanes (pen.) K2

Topography

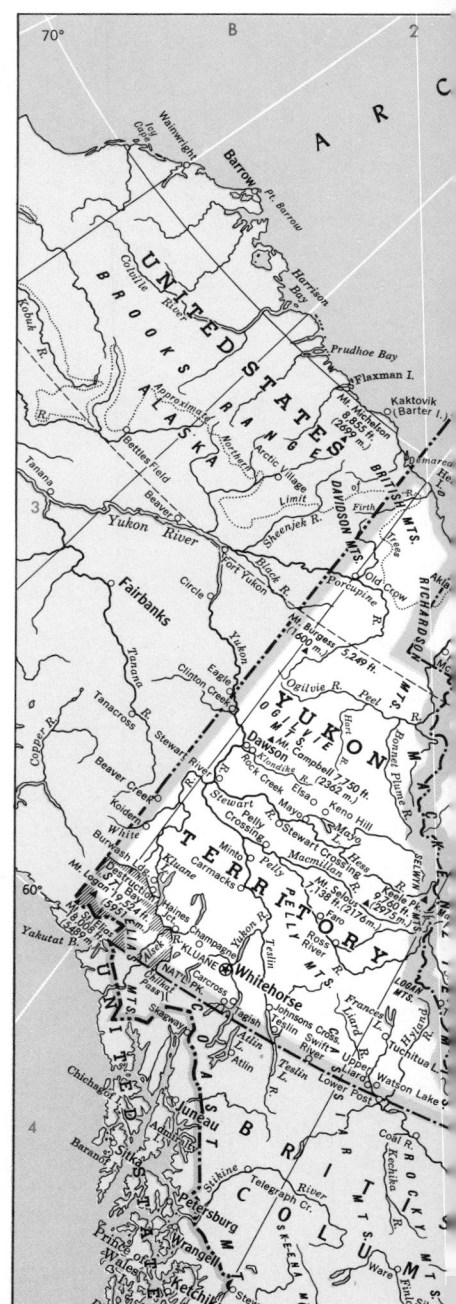

Agriculture, Industry and Resources

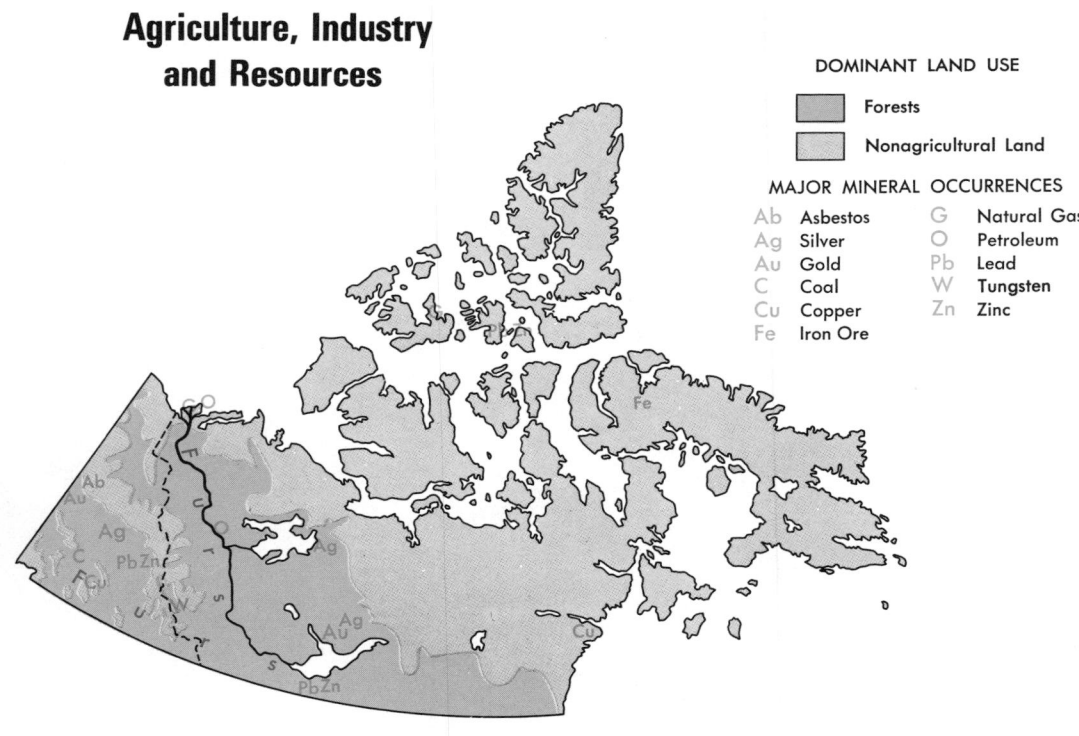

DOMINANT LAND USE

Forests

Nonagricultural Land

MAJOR MINERAL OCCURRENCES

Ab Asbestos   G Natural Gas
Ag Silver   O Petroleum
Au Gold   Pb Lead
C Coal   W Tungsten
Cu Copper   Zn Zinc
Fe Iron Ore

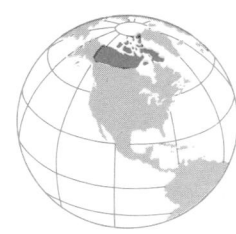

Rae (isth.) K3
Rae (riv.) G3
Rae (str.) J3
Ramparts (riv.) E3
Resolution (isl.) M3
Richard Collinson (inlet) G2
Richards (isl.) E3
Richardson (mts.) E3
Robeson (chan.) M1
Roes Welcome (sound) K3
Rowley (isl.) K3
Royal Geographical Society (isls.) J3
Russell (isl.) J2
Sabine (pen.) H2
Salisbury (isl.) L3
Seahorse (pt.) L3
Selwyn (lake) H4
Sherman (inlet) J3
Simpson (pen.) K3
Sir James MacBrien (mt.) F3
Slave (riv.) G3
Smith (bay) L2
Smith (cape) K3
Smith (sound) L2
Snare (riv.) G3
Snowbird (lake) H3
Somerset (isl.) J2
South (bay) K3
South Nahanni (riv.) F3
Southampton (isl.) K3
Stallworthy (cape) J1
Steensby (inlet) L2
Stefansson (isl.) H2

Sverdrup (chan.) J1
Sverdrup (isls.) J2
Talbot (inlet) L2
Taltson (riv.) G3
Tathlina (lake) G3
Tha-anne (riv.) J3
Thelon (riv.) H3
Thlewiaza (riv.) J3
Trout (lake) G3
Ungava (bay) M4
Vansittart (isl.) K3
Victoria (isl.) G2
Victoria (str.) H3
Viscount Melville (sound) G2
Wager (bay) K3
Wales (isl.) K3
Walsingham (cape) M3
Wellington (chan.) J2
Wholdaia (lake) H3
Winter (harb.) H2
Wollaston (pen.) G3
Wood Buffalo Nat'l Park G3
Wynniatt (bay) G2
Yathkyed (lake) J3
Yellowknife (riv.) G3

YUKON TERRITORY

AREA 207,075 sq. mi. (536,324 sq. km.)
POPULATION 23,153
CAPITAL Whitehorse
LARGEST CITY Whitehorse
HIGHEST POINT Mt. Logan 19,524 ft. (5,951 m.)
SETTLED IN 1897
ADMITTED TO CONFEDERATION 1898
PROVINCIAL FLOWER Fireweed

NORTHWEST TERRITORIES

AREA 1,304,896 sq. mi. (3,379,683 sq. km.)
POPULATION 45,741
CAPITAL Yellowknife
LARGEST CITY Yellowknife
HIGHEST POINT Mt. Sir James MacBrien 9,062 ft. (2,762 m.)
SETTLED IN 1800
ADMITTED TO CONFEDERATION 1870
PROVINCIAL FLOWER Mountain Avens

YUKON TERRITORY

CITIES and TOWNS

Beaver Creek 90 .... D3
Burwash Landing 73 .... D3
Carcross 216 .... E3
Carmacks •256 .... E3
Champagne E3
Clinton Creek D3
Cowley E3
Dawson 697 .... E3
Destruction Bay 45 .... E3
Elsa 336 .... E3
Faro 1,652 .... E3
Haines Junction •366 .... E3
Johnson's Crossing 13 .... E3
Keno Hill 88 .... E3
Koidern D3
Mayo 398 .... E3
Minto E3

Old Crow 243 .... E3
Pelly Crossing 182 .... E3
Rock Creek 59 .... E3
Ross River 294 .... E3
Stewart Crossing 20 .... E3
Stewart River D3
Swift River 24 .... E3
Tagish 89 .... E3
Teslin •310 .... E3
Tuchitua Lake E3
Upper Liard 130 .... F3
Watson Lake •748 .... E3
Whitehorse (cap.) 14,814 .... E3

OTHER FEATURES

Alsek (riv.) E3
Bonnet Plume (riv.) E3
British (mts.) D3
Campbell (mt.) E3
Cassiar (mts.) E3
Frances (lake) F3
Herschel (isl.) E3
Hess (riv.) E3
Hyland (riv.) F3
Keele (peak) E3
Keele (riv.) E3
Klondike (riv.) E3
Kluane (lake) E3

Kluane Nat'l Park E3
Liard (riv.) E3
Logan (mt.) D3
Logan (mt.) F3
Mackenzie (bay) E3
Mackenzie (mts.) E3
Macmillan (riv.) E3
Mayo (lake) E3
Ogilvie (mts.) E3
Ogilvie (riv.) E3
Peel (riv.) E3
Pelly (mts.) E3
Pelly (riv.) E3

Porcupine (riv.) E3
Richardson (mts.) E3
Rocky (mts.) F4
Saint Elias (mt.) D3
Saint Elias (mts.) E3
Selous (mt.) E3
Selwyn (riv.) E3
Stewart (riv.) E3
Teslin (lake) E4
Teslin (riv.) E3
White (riv.) D3
Yukon (riv.) E3

• Population of district.

Yukon and Northwest Territories

SCALE

0   50   100   200   300 MI.
0   50   100   200   300 KM.

Territorial Capitals ....
Regional Capitals ....
International Boundaries ....
Provincial & Territorial Boundaries ....
Regional Boundaries ....

Scale 1:14,000,000

All islands in Hudson and James Bay lie within the Northwest Territories

© Copyright HAMMOND INCORPORATED, Maplewood, N.J.

United States

POLYCONIC PROJECTION

SCALE OF MILES

SCALE OF KILOMETERS

Capitals of Countries .... ☆
State Capitals .... △
International Boundaries .... —

Scale 1:17,400,000

© Copyright HAMMOND INCORPORATED, Maplewood, N.J.

Akron, Ohio‡ 660,328 .... K2
Alabama (state) 3,890,061 ...... J4
Alaska (state) 400,481 .... C5
Alaska (gulf), Alaska D6
Alaska (range), Alaska C6
Albany (cap.), N.Y.‡ 795,019 ... M2
Albuquerque, N. Mex.‡ 454,499 .. E3
Aleutian (isls.), Alaska D6
Allentown, Pa.‡ 636,714 .... L2
Anchorage, Alaska‡ 173,017 ..... D6
Annapolis (cap.), Md. 31,740 ... L3
Ann Arbor, Mich.‡ 264,748 ...... K2
Appalachian (mts.) K3
Appleton, Wis.‡ 291,325 .... J2
Arizona (state) 2,717,866 ...... D4
Arkansas (state) 2,285,513 ..... H3
Arkansas (riv.) H3
Atlanta (cap.), Ga.‡ 2,029,618 . K4
Atlantic City, N.J.‡ 194,119 ... M3
Attu (isl.), Alaska D6
Augusta, Ga.‡ 327,372 .... K4
Augusta (cap.), Maine 21,819 ... N2
Austin (cap.), Texas‡ 536,450 .. G4
Bakersfield, Calif.‡ 403,089 ... C3
Baltimore, Md.‡ 2,174,023 ...... L3
Baton Rouge (cap.),
 La.‡ 493,973 .... H4
Beaumont, Texas‡ 375,497 .... H4
Bering (sea), Alaska C6
Bering (str.), Alaska C5
Bighorn (riv.) E2
Binghamton, N.Y.‡ 301,336 ...... L2
Birmingham, Ala.‡ 847,360 ...... J4
Bismarck (cap.), N.
 Dak.‡ 79,988 .... G1
Bitterroot (range) D1
Black Hills (mts.) F2
Boise (cap.), Idaho 173,076 .... D2
Borah (peak), Idaho D2

Boston (cap.), Mass.‡ 2,763,357 . M2
Bridgeport, Conn.‡ 395,455 ..... M2
Brazos (riv.), Texas G4
Brooks (range), Alaska C5
Buffalo, N.Y.‡ 1,242,573 .... L2
California (state) 23,668,562 .. B3
Canadian (riv.) F3
Canaveral (Kennedy) (cape),
 Fla. L5
Canton, Ohio‡ 404,421 .... K2
Cape Fear (riv.), N.C. L4
Carson City (cap.),
 Nev. 32,022 .... C3
Cascade (range) B1
Cedar Rapids, Iowa‡ 169,775 .... H2
Champlain (lake) M2
Charleston, S.C.‡ 430,301 ...... L4
Charleston (cap.), W.
 Va.‡ 269,595 .... K3
Charlotte, N.C.‡ 637,218 .... K3
Chattahoochee (riv.) K4
Chattanooga, Tenn.‡ 426,540 .... J3
Chesapeake (bay) L3
Cheyenne (cap.), Wyo. 47,283 ... F2
Chicago, Ill.‡ 7,102,328 .... J2
Cimarron (riv.) F3
Cincinnati, Ohio‡ 1,401,403 .... K3
Cleveland, Ohio‡ 1,898,720 ..... K2
Columbia (cap.), S.C.‡ 408,176 . K4
Columbia (riv.) B1
Columbus, Ga.‡ 239,196 .... K4

Columbus (cap.), Ohio‡ 1,093,293 . K3
Concord (cap.), N.H. 30,400 .... M2
Connecticut (state) 3,107,576 .. M2
Connecticut (riv.) M2
Corpus Christi, Texas‡ 326,228 . G5
Cumberland (riv.) J3
Dallas, Texas‡ 2,974,878 .... G4
Davenport, Iowa‡ 383,958 .... H2
Dayton, Ohio‡ 830,070 .... K3
Death Valley (depr.), Calif. ... C3
Delaware (state) 595,225 .... L3
Delaware (bay) M3
Denver (cap.), Colo.‡ 1,619,921 . F3
Des Moines (cap.),
 Iowa‡ 338,048 .... H2
Detroit, Mich.‡ 4,352,762 ...... K2
District of Columbia 637,651 ... L3
Dover (cap.), Del. 23,512 ...... L3
Duluth, Minn.‡ 266,650 .... H1
Durham, N.C.‡ 530,673 .... L3
Elbert (mt.) Colo. E3
El Paso, Texas‡ 479,899 .... E4
Erie, Pa.‡ 279,780 .... K2
Erie (lake) K2
Eugene, Ore.‡ 275,226 .... B2
Evansville, Ind.‡ 309,408 ...... J3
Everglades, The (swamp),
 Fla. K5
Fayetteville, N.C.‡ 247,160 .... L3
Flint, Mich.‡ 521,589 .... K2
Florida (state) 9,739,992 ...... K5
Florida (keys), Fla. K6
Fort Smith, Ark.‡ 203,269 ...... H3
Fort Wayne, Ind.‡ 382,961 ...... J2
Fort Worth, Texas 385,141 ...... G4
Frankfort (cap.), Ky. 25,973 ... K3
Fresno, Calif.‡ 515,013 .... C3
Galveston, Texas‡ 195,940 ...... H5
Gary, Ind.‡ 642,781 .... J2

Georgia (state) 5,464,265 ...... K4
Gila (riv.) D4
Glacier Nat'l Park, Mont. D1
Golden Gate (chan.), Calif. B3
Grand Canyon Nat'l Park,
 Ariz. D3
Grand Rapids, Mich.‡ 601,680 ... J2
Great Salt (lake), Utah D2
Greensboro, N.C.‡ 827,385 ...... K3
Greenville, S.C.‡ 568,758 ...... K3
Hamilton, Ohio‡ 258,787 .... K3
Harrisburg (cap.), Pa.‡ 446,072 . L2
Hartford (cap.), Conn.‡ 726,114 . M2
Hatteras (cape), N.C. M3
Havasu (lake) D3
Hawaii (state) 965,000 .... C4
Hawaii (isl.), Hawaii F6
Helena (cap.), Mont. 23,938 .... D1
Honolulu (cap.),
 Hawaii‡ 762,874 .... F5
Houston, Texas‡ 2,905,350 ...... G5
Huntington, W. Va.‡ 311,350 .... K3
Huntsville, Ala.‡ 308,593 ...... J3
Huron (lake), Mich. K2
Idaho (state) 943,935 .... D2
Illinois (state) 11,418,461 .... J3
Indiana (state) 5,490,179 ...... J3
Aleutian (isls.), Alaska D6
Iowa (state) 2,913,387 .... H2
Jackson (cap.), Miss.‡ 320,425 . J4
Jacksonville, Fla.‡ 737,519 .... K4
Jefferson City (cap.),
 Mo. 33,619 .... H3
Jersey City, N.J.‡ 556,972 ..... M2
Johnstown, Pa.‡ 664,506 .... L2
Juneau (cap.), Alaska 19,528 ... E6
Kalamazoo, Mich.‡ 279,192 ...... J2
Kansas (state) 2,363,208 .... G3
Kansas City.

Kans.-Mo.‡ 1,327,020 .... G3
Kauai (isl.), Hawaii E5
Kentucky (state) 3,661,433 ..... J3
Kentucky (lake) J3
Knoxville, Tenn.‡ 476,517 ...... K3
Lancaster, Pa.‡ 362,346 .... L2
Lansing (cap.), Mich.‡ 476,517 . K2
Las Vegas, Nev.‡ 461,816 .... C3
Lawrence, Mass.‡ 281,981 .... M2
Lexington, Ky.‡ 318,136 .... K3
Lima, Ohio‡ 218,244 .... K2
Lincoln (cap.), Nebr.‡ 192,884 . G2
Little Rock (cap.),
 Ark.‡ 393,494 .... H4
Long (isl.), N.Y M2
Long Beach, Calif. 361,334 ..... C4
Los Angeles, Calif.‡ 7,477,657 . C4
Louisiana (state) 4,203,972 .... H4
Louisville, Ky.‡ 906,240 .... J3
Lowell, Mass.‡ 233,410 .... M2
Lubbock, Texas‡ 211,651 .... F4
Macon, Ga.‡ 254,623 .... K4
Madison (cap.), Wis.‡ 323,545 .. J2
Maine (state) 1,124,660 .... N1
Maryland (state) 4,216,446 ..... L3
Massachusetts (state) 5,737,037 . M2
Maui (isl.), Hawaii F5
Mauna Kea (mt.), Hawaii G6
Mauna Loa (mt.), Hawaii F6
May (cape), N.J. M3
McKinley (mt.), Alaska D5
Memphis, Tenn.‡ 912,887 .... J3
Mendocino (cape), Calif. A2
Mexico (gulf) J5
Miami, Fla.‡ 1,625,979 .... K5
Michigan (state) 9,258,344 ..... J2
Michigan (lake) J2
Milwaukee, Wis.‡ 1,397,143 ..... J2
Minneapolis, Minn.‡ 2,114,256 .. H1

Minnesota (state) 4,077,148 .... H1
Mississippi (state) 2,520,638 .. J4
Mississippi (riv.) H4
Missouri (state) 4,917,444 ..... H3
Missouri (riv.) G2
Mitchell (mt.), N.C. K3
Mobile, Ala.‡ 442,819 .... J4
Montana (state) 786,690 .... E1
Montgomery (cap.),
 Ala.‡ 272,687 .... J4
Nantucket (isl.), Mass. N3
Nashville (cap.),
 Tenn.‡ 850,505 .... J3
Nebraska (state) 1,570,006 ..... F2
Nevada (state) 799,184 .... C3
Newark, N.J.‡ 1,965,304 .... M2
New Hampshire (state) 920,610 .. M2
New Haven, Conn.‡ 417,592 ...... M2
New Jersey (state) 7,364,158 ... M3
New Mexico (state) 1,299,968 ... E4
New Orleans, La.‡ 1,186,725 .... H5
Newport News, Va.‡ 364,449 ..... L3
New York (state) 17,557,288 .... L2
New York, N.Y.‡ 9,119,737 ...... M2
Norfolk, Va.‡ 806,691 .... L3
North Carolina
 (state) 5,874,429 .... L3
North Dakota (state) 652,695 ... F1
Oahu (isl.), Hawaii E5
Oakland, Calif.‡ 3,252,721 ..... B3
Ohio (state) 10,797,419 .... J3
Ohio (riv.) J3
Oklahoma (state) 3,025,266 ..... G3
Oklahoma City (cap.),
 Okla.‡ 834,088 .... G3
Olympia (cap.), Wash.‡ 124,264 . B1
Olympic Nat'l Park, Wash. A1
Omaha, Nebr.‡ 570,399 .... G2
Ontario (lake), N.Y. L2

Oregon (state) 2,632,663 .... B2
Orlando, Fla.‡ 700,699 .... K5
Ozark (mts.) H3
Paterson, N.J.‡ 447,585 .... M2
Pennsylvania (state) L2
Pensacola, Fla.‡ 289,782 .... J4
Peoria, Ill.‡ 365,864 .... J2
Philadelphia, Pa.‡ 4,716,818 ... M2
Phoenix (cap.), Ariz.‡ 1,508,030 . D4
Pierre (cap.), S. Dak. 11,793 .. F2
Pikes (peak), Colo. F3
Pittsburgh, Pa.‡ 2,263,894 ..... L2
Platte (riv.), Nebr. G2
Pontchartrain (lake), La. J5
Portland, Maine‡ 183,625 .... N2
Portland, Oreg.‡ 1,242,187 ..... B1
Potomac (riv.) L3
Providence (cap.),
 R.I.‡ 919,216 .... M2
Racine, Wis.‡ 173,132 .... J2
Raleigh (cap.), N.C.‡ 530,673 .. L3
Rainier (mt.), Wash. B1
Reading, Pa.‡ 312,509 .... L2
Red (riv.) H4
Red River of the North
 (riv.) G1
Rhode Island (state) 947,154 ... M2
Richmond (cap.), Va.‡ 632,015 .. L3
Rio Grande (riv.) E3
Roanoke, Va.‡ 218,244 .... K3
Rochester, N.Y.‡ 971,079 .... L2
Rockford, Ill.‡ 279,514 .... J2
Rocky (mts.) E3
Sacramento (cap.),
 Calif.‡ 1,014,002 .... B3
Saginaw, Mich.‡ 224,548 .... K2
Saint Clair (lake), Mich. K2
Saint Lawrence (riv.), N.Y. N1
Saint Louis, Mo.‡ 2,355,276 .... H3

AREA 3,623,420 sq. mi.
    (9,384,658 sq. km.)
POPULATION 226,504,825
CAPITAL Washington
LARGEST CITY New York
HIGHEST POINT Mt. McKinley 20,320 ft.
    (6,194 m.)
MONETARY UNIT U.S. dollar
MAJOR LANGUAGE English
MAJOR RELIGIONS Protestantism,
    Roman Catholicism, Judaism

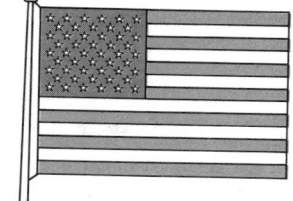

Population Distribution

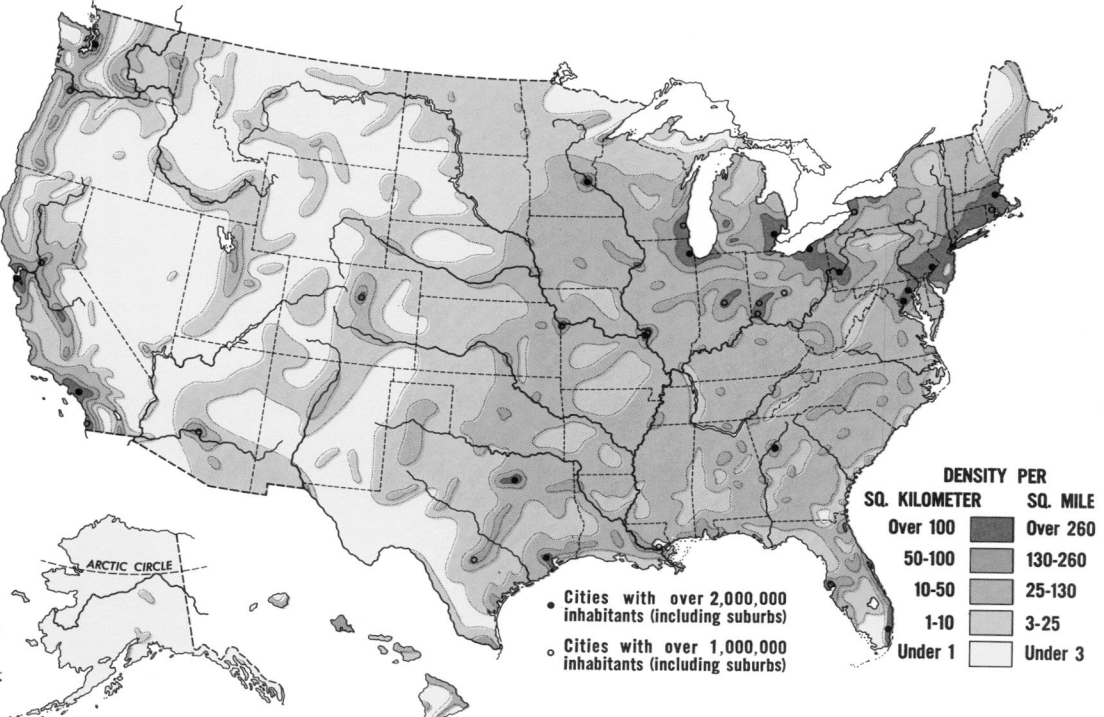

DENSITY PER

SQ. KILOMETER	SQ. MILE
Over 100	Over 260
50-100	130-260
10-50	25-130
1-10	3-25
Under 1	Under 3

• Cities with over 2,000,000 inhabitants (including suburbs)

○ Cities with over 1,000,000 inhabitants (including suburbs)

Vegetation

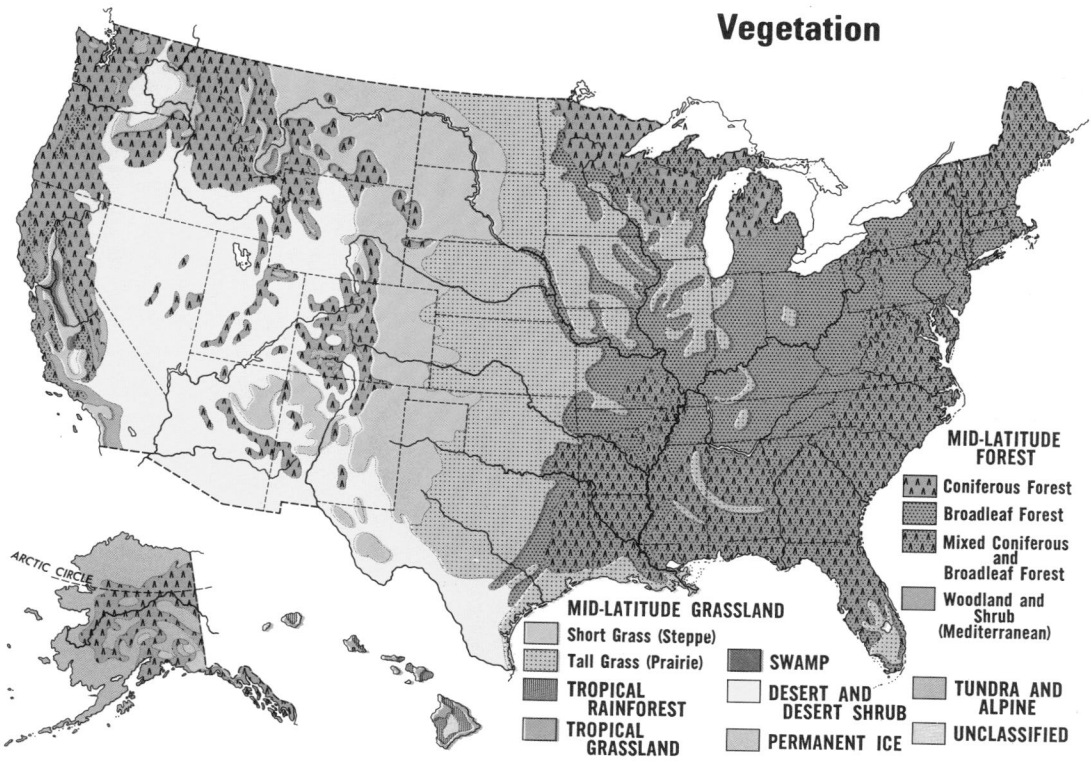

MID-LATITUDE FOREST

Coniferous Forest

Broadleaf Forest

Mixed Coniferous and Broadleaf Forest

Woodland and Shrub (Mediterranean)

MID-LATITUDE GRASSLAND

Short Grass (Steppe)

Tall Grass (Prairie)

TROPICAL RAINFOREST

TROPICAL GRASSLAND

SWAMP

DESERT AND DESERT SHRUB

PERMANENT ICE

TUNDRA AND ALPINE

UNCLASSIFIED

Saint Paul (cap.),
  Minn. 270,230 ....H1
Sakakawea (lake), N. Dak. F1
Salem (cap.), Oreg.‡ 249,895 ....B1
Salinas, Calif.‡ 290,444 ....B3
Salt Lake City (cap.),
  Utah‡ 936,255 ....D2
Salton Sea (lake), Calif. D4
San Antonio, Texas‡ 1,071,954 ..G5
San Bernardino,
  Calif.‡ 1,557,080 ....C4
San Diego, Calif.‡ 1,861,846 .....C4
San Francisco, Calif.‡ 325,721 ...B3
San Joaquin (riv.), Calif. C3
San Jose, Calif.‡ 1,295,071 ......B3
Santa Ana, Calif.‡ 1,931,570 .....C4
Santa Barbara, Calif.‡ 298,660 ..C4
Santa Fe (cap.), N.
  Mex. 48,899 ....E3
Savannah, Ga.‡ 228,178 ....K4
Scranton, Pa. 88,117 ....L2
Seattle, Wash.‡ 1,606,765 ....B1
Shasta (mt.), Calif. B2
Shreveport, La.‡ 376,646 ....H4
Sierra Nevada (mts.) B3
Snake (riv.) C1
South Bend, Ind.‡ 280,772 ....J2
Spokane, Wash.‡ 341,835 ....C1
South Carolina
  (state) 3,119,208 ....K4
South Dakota (state) 690,178 ....F2
Springfield (cap.),
  Ill.‡ 187,338 ....H3
Springfield, Mass.‡ 530,668 ......M2
Springfield, Mo.‡ 207,704 ....H3
Springfield, Ohio‡ 183,885 ....K3
Stockton, Calif.‡ 347,342 ....B3
Superior (lake) J1
Syracuse, N.Y.‡ 642,375 ....L2

Tacoma, Wash.‡ 485,643 ....B1
Tahoe (lake) C3
Tallahassee (cap.),
  Fla.‡ 159,542 ....K4
Tampa, Fla.‡ 1,569,492 ....K5
Tennessee (state) 4,590,750 .....J3
Terre Haute, Ind‡ 176,583 ....J3
Texas (state) 14,228,383 ....G4
Toledo, Ohio‡ 791,599 ....K2
Topeka (cap.), Kans.‡ 185,442 ..G3
Trenton (cap.), N.J.‡ 307,863 ....M2
Tucson, Ariz.‡ 531,263 ....D4
Tulsa, Okla.‡ 689,628 ....G3
Utah (state) 1,461,037 ....D3
Utica, N.Y.‡ 320,180 ....M2
Vermont (state) 511,456 ....M2
Virginia (state) 5,346,279 ....L3
Wabash (riv.) J3
Washington (state) 4,130,163 ....B1
Washington, D.C. (cap.),
  U.S.‡ 3,060,240 ....L3
Waterbury, Conn.‡ 228,059 ......M2
Waterloo, Iowa 75,985 ....H2
West Palm Beach, Fla.‡ 573,125 .K5
West Virginia (state) 1,949,644 ..K3
Wheeling, W. Va.‡ 185,566 ....K2
Whitney (mt.), Calif. C3
Wichita, Kans.‡ 411,313 ....G3
Wilkes-Barre, Pa. 51,551 ....L2
Wilmington, Del.‡ 524,108 ....M3
Wisconsin (state) 4,705,335 .....J2
Woods (lake), Minn. G1
Worcester, Mass.‡ 372,940 ......M2
Wyoming (state) 470,816 ....E2
Yellowstone Nat'l Park, Wyo. E2
York, Pa.‡ 381,255 ....L3
Yosemite Nat'l Park, Calif. C3
Youngstown, Ohio‡ 531,350 .....K2
Yukon (riv.), Alaska C5

‡ Population of metropolitan area.

USA

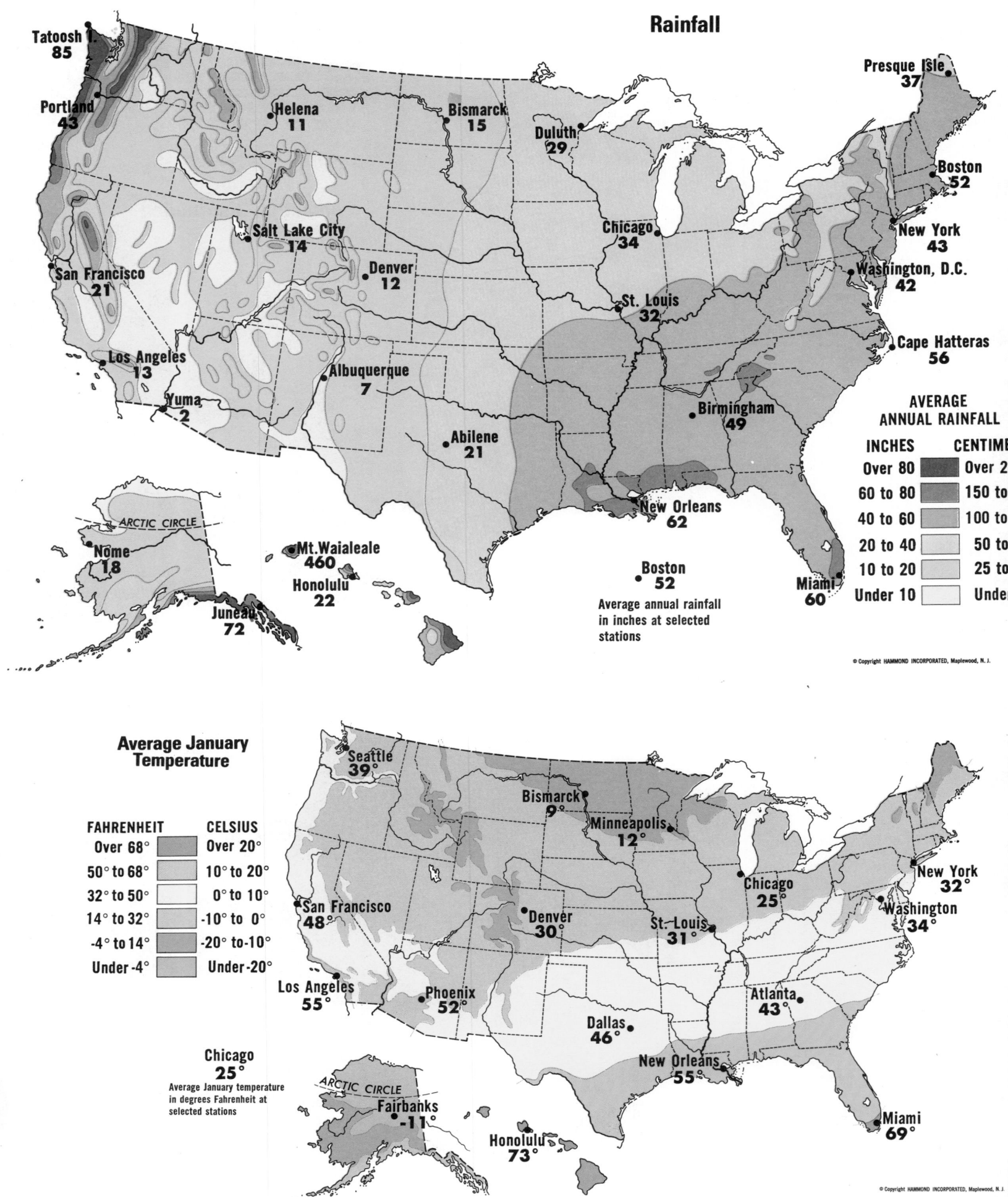

Rainfall

Tatoosh I. 85

Portland 43

Helena 11

Bismarck 15

Duluth 29

Presque Isle 37

Boston 52

Salt Lake City 14

Denver 12

Chicago 34

St. Louis 32

New York 43

Washington, D.C. 42

San Francisco 21

Los Angeles 13

Albuquerque 7

Yuma 2

Abilene 21

Birmingham 49

Cape Hatteras 56

New Orleans 62

ARCTIC CIRCLE

Nome 18

Mt. Waialeale 460

Honolulu 22

Juneau 72

Boston 52

Average annual rainfall in inches at selected stations

Miami 60

AVERAGE ANNUAL RAINFALL

INCHES	CENTIME?
Over 80	Over 20
60 to 80	150 to
40 to 60	100 to
20 to 40	50 to
10 to 20	25 to
Under 10	Under

© Copyright HAMMOND INCORPORATED, Maplewood, N.J.

Average January Temperature

FAHRENHEIT	CELSIUS
Over 68°	Over 20°
50° to 68°	10° to 20°
32° to 50°	0° to 10°
14° to 32°	-10° to 0°
-4° to 14°	-20° to -10°
Under -4°	Under -20°

Seattle 39

Bismarck 9°

Minneapolis 12°

New York 32°

Chicago 25°

Washington 34°

San Francisco 48°

Denver 30°

St. Louis 31°

Los Angeles 55°

Phoenix 52°

Dallas 46°

Atlanta 43°

New Orleans 55°

Chicago 25°

Average January temperature in degrees Fahrenheit at selected stations

ARCTIC CIRCLE

Fairbanks -11°

Honolulu 73°

Miami 69°

© Copyright HAMMOND INCORPORATED, Maplewood, N.J.

Topography

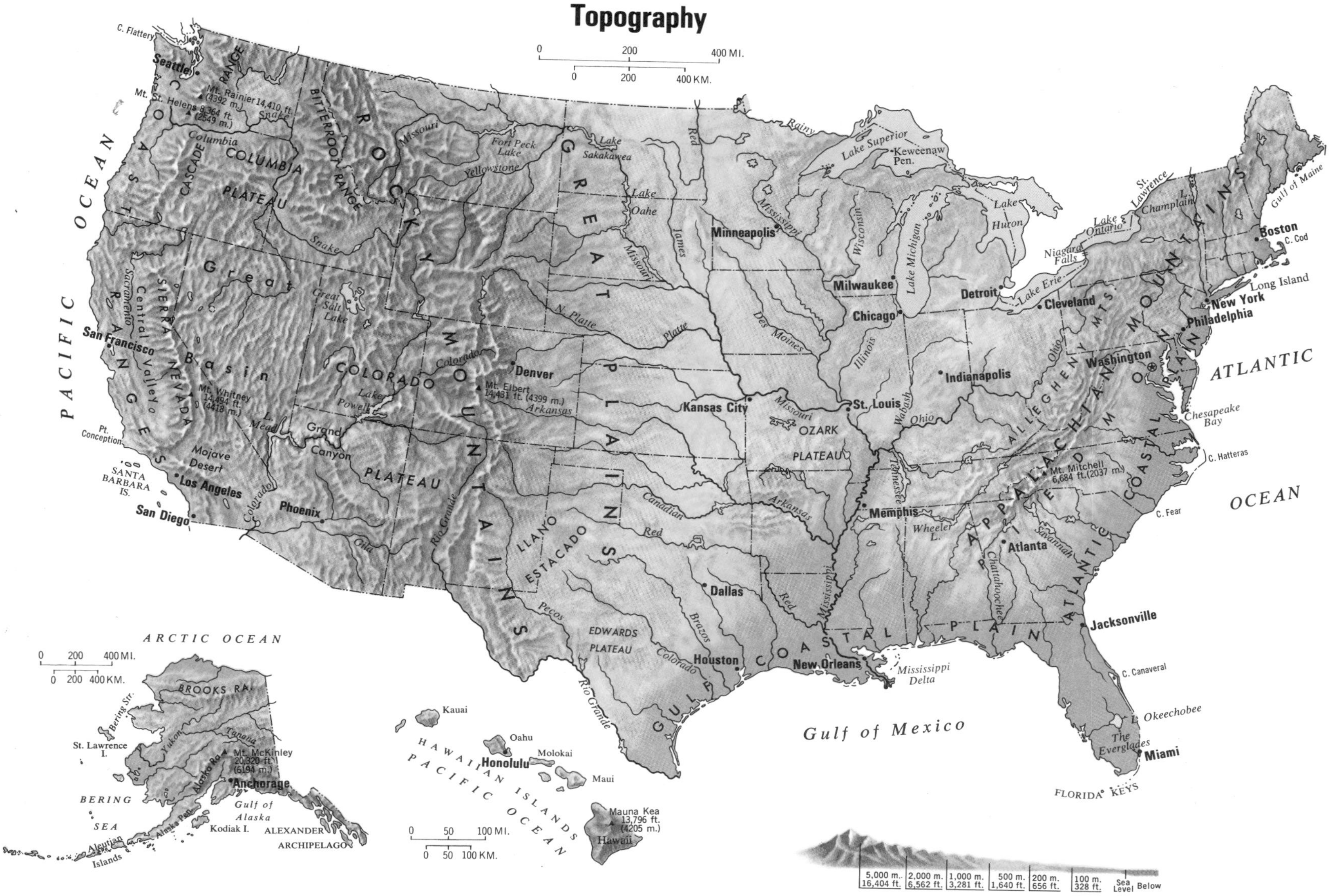

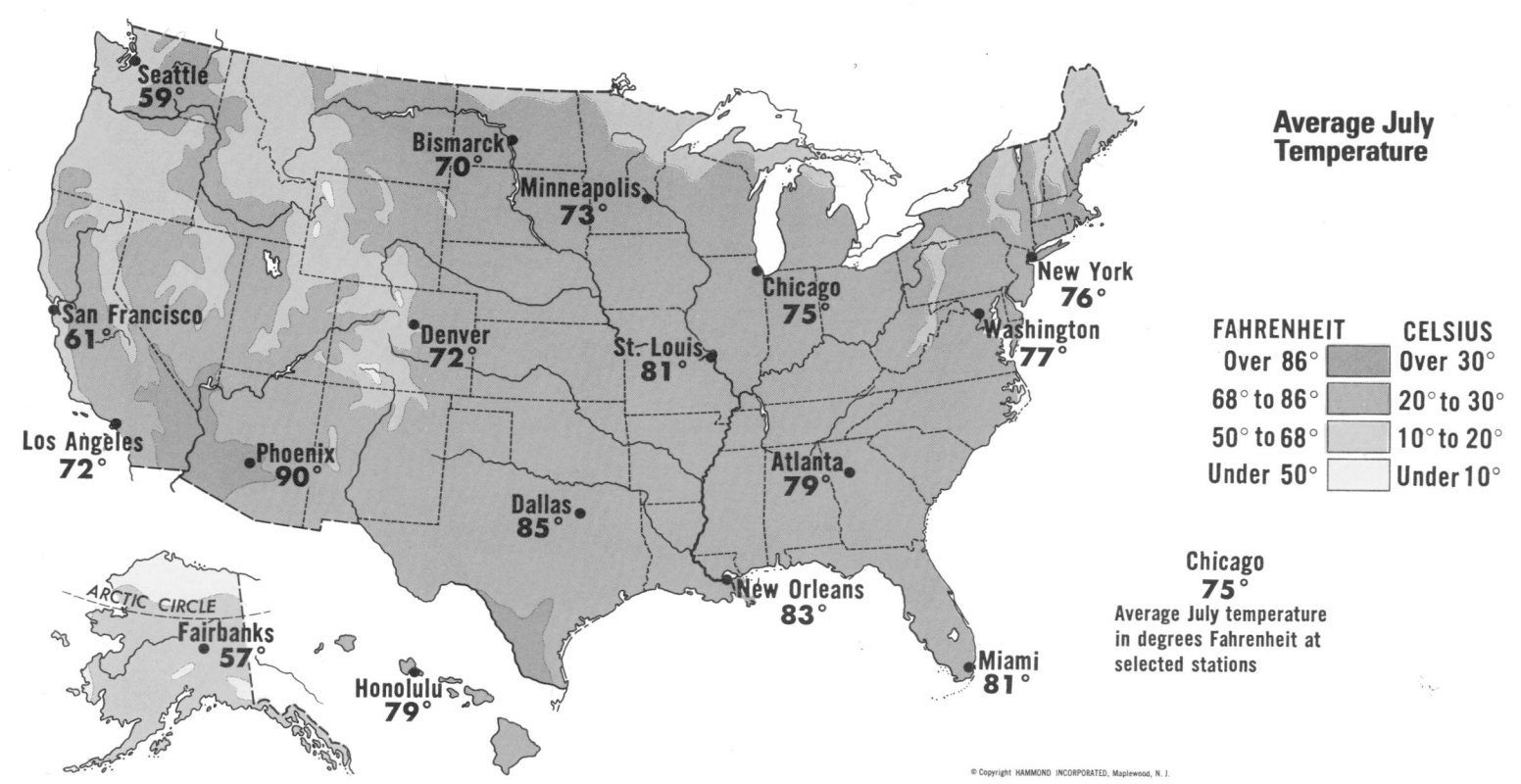

0   200   400 MI.
0   200   400 KM.

C. Flattery
Seattle
Mt. St. Helens 9,764 ft. (2549 m.)
Mt. Rainier 14,410 ft. (4392 m.)

PACIFIC OCEAN

COAST RANGE
CASCADE RANGE
BITTERROOT RANGE
COLUMBIA PLATEAU
ROCKY MOUNTAINS
GREAT PLAINS

Columbia
Snake
Yellowstone
Fort Peck Lake
Missouri
Lake Sakakawea
Red
Rainy
Lake Superior
Keweenaw Pen.

Minneapolis
Milwaukee
Chicago
Detroit
Cleveland
Lake Huron
Lake Erie
Lake Michigan
Lake Ontario

St. Lawrence
L. Champlain
Boston
C. Cod
New York
Long Island
Philadelphia

San Francisco
SIERRA NEVADA
Great Basin
Great Salt Lake
COLORADO PLATEAU
Mt. Whitney 14,494 ft. (4418 m.)
Central Valley
Sacramento
Lake Powell
Lake Mead

Denver
Mt. Elbert 14,431 ft. (4399 m.)
N. Platte
Platte
Colorado
Arkansas
Kansas City
Missouri
St. Louis
Ohio
Illinois
Des Moines
James
Wisconsin

Indianapolis
Washington
ALLEGHENY MTS.
ATLANTIC

Pt. Conception
SANTA BARBARA IS.
Mojave Desert
Los Angeles
San Diego
Phoenix
Gila
Colorado
Rio Grande
Grand Canyon
LLANO ESTACADO
PLATEAU
Pecos
Dallas
Red
Canadian

Memphis
Wabash
Tennessee
OZARK PLATEAU
Wheeler
Chattahoochee
Savannah
Atlanta
Mt. Mitchell 6,684 ft. (2037 m.)
APPALACHIAN MOUNTAINS
ATLANTIC COASTAL PLAIN

Chesapeake Bay
C. Hatteras
OCEAN
C. Fear

EDWARDS PLATEAU
Brazos
Colorado
Houston
New Orleans
Mississippi Delta
GULF COASTAL PLAIN
Jacksonville
C. Canaveral

Gulf of Mexico
L. Okeechobee
The Everglades
Miami
FLORIDA KEYS

ARCTIC OCEAN
0   200   400 MI.
0   200   400 KM.
BROOKS RA.
BERING STR.
St. Lawrence I.
Yukon
Tanana
Mt. McKinley 20,320 ft. (6194 m.)
Anchorage
Kuskokwim
BERING SEA
Gulf of Alaska
Alaska Pen.
Kodiak I.
ALEXANDER ARCHIPELAGO
Aleutian Islands

Kauai
Oahu
Honolulu
Molokai
Maui
HAWAIIAN ISLANDS
PACIFIC OCEAN
Mauna Kea 13,796 ft. (4205 m.)
Hawaii
0   50   100 MI.
0   50   100 KM.

5,000 m. 16,404 ft. | 2,000 m. 6,562 ft. | 1,000 m. 3,281 ft. | 500 m. 1,640 ft. | 200 m. 656 ft. | 100 m. 328 ft. | Sea Level | Below

Average July Temperature

Seattle **59°**
Bismarck **70°**
Minneapolis **73°**
Chicago **75°**
New York **76°**
Washington **77°**
San Francisco **61°**
Denver **72°**
St. Louis **81°**
Los Angeles **72°**
Phoenix **90°**
Dallas **85°**
Atlanta **79°**
New Orleans **83°**
Miami **81°**
ARCTIC CIRCLE
Fairbanks **57°**
Honolulu **79°**

FAHRENHEIT
Over 86°
68° to 86°
50° to 68°
Under 50°

CELSIUS
Over 30°
20° to 30°
10° to 20°
Under 10°

Chicago
75°
Average July temperature in degrees Fahrenheit at selected stations

United States Standard Time Zones

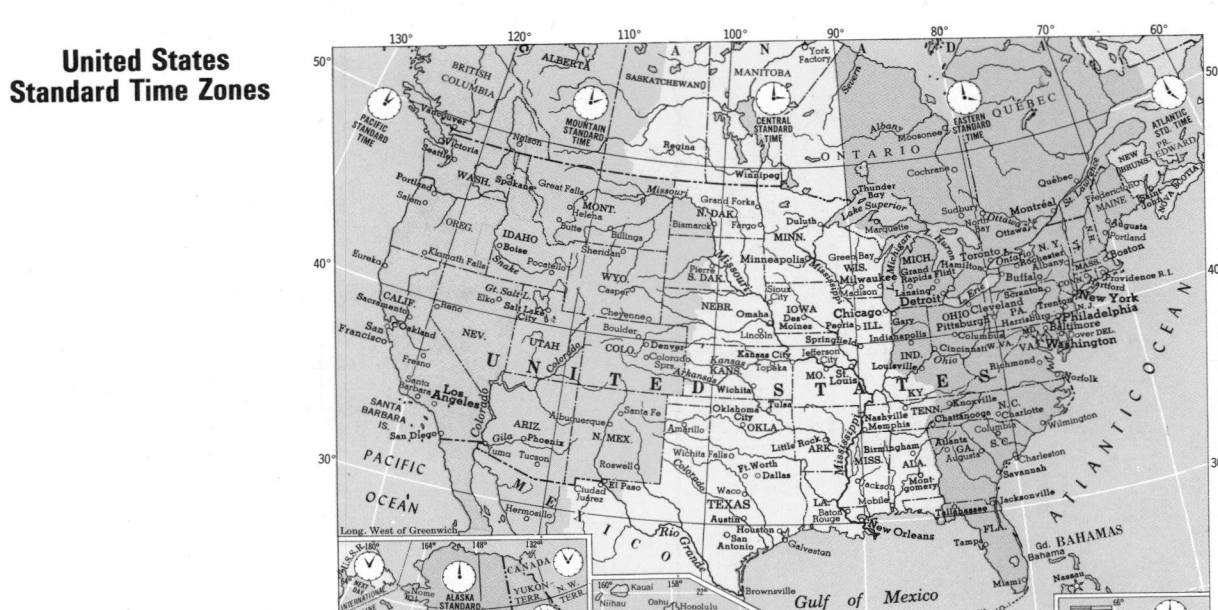

U. S. STANDARD TIME ZONES
Established by the Uniform Time Act
SCALE OF MILES
PACIFIC STANDARD TIME
MOUNTAIN STANDARD TIME
CENTRAL STANDARD TIME
EASTERN STANDARD TIME
ATLANTIC STD. TIME
ALASKA STANDARD TIME
HAWAII-ALEUTIAN STD. TIME
HAWAII-ALEUTIAN STANDARD TIME
INTERNATIONAL DATE LINE
PACIFIC OCEAN
ATLANTIC OCEAN
Gulf of Mexico
UNITED STATES
CANADA
MEXICO
ONTARIO
QUEBEC
MANITOBA
SASKATCHEWAN
ALBERTA
BRITISH COLUMBIA
ALASKA
HAWAII
PUERTO RICO
Gd. BAHAMAS
Long. West of Greenwich

Agriculture, Industry and Resources

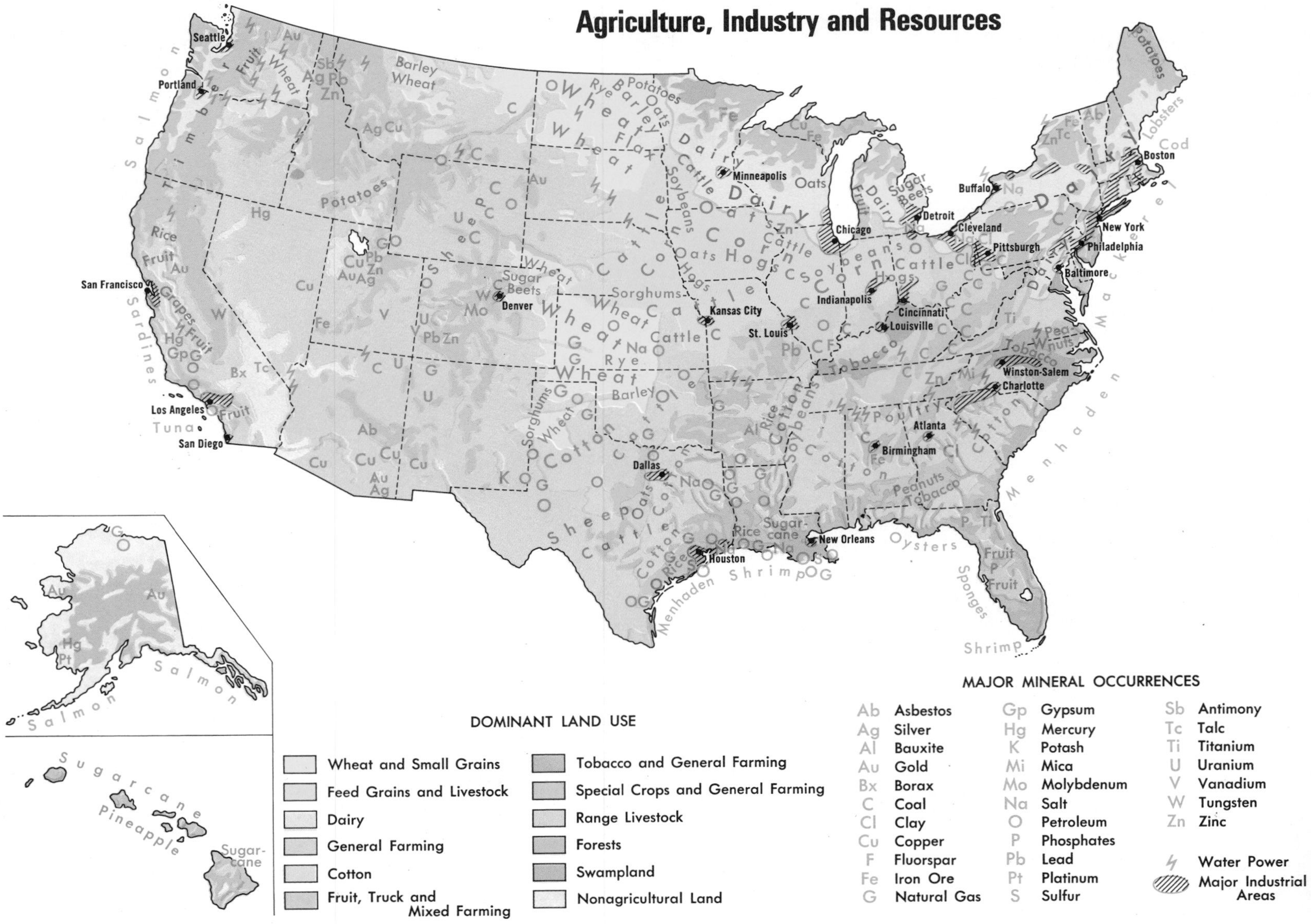

Seattle
Portland
San Francisco
Los Angeles
San Diego
Denver
Minneapolis
Kansas City
St. Louis
Chicago
Detroit
Buffalo
Cleveland
Pittsburgh
Indianapolis
Cincinnati
Louisville
Boston
New York
Philadelphia
Baltimore
Winston-Salem
Charlotte
Atlanta
Birmingham
Dallas
Houston
New Orleans
DOMINANT LAND USE
Wheat and Small Grains
Feed Grains and Livestock
Dairy
General Farming
Cotton
Fruit, Truck and Mixed Farming
Tobacco and General Farming
Special Crops and General Farming
Range Livestock
Forests
Swampland
Nonagricultural Land
MAJOR MINERAL OCCURRENCES
Ab Asbestos
Ag Silver
Al Bauxite
Au Gold
Bx Borax
C Coal
Cl Clay
Cu Copper
F Fluorspar
Fe Iron Ore
G Natural Gas
Gp Gypsum
Hg Mercury
K Potash
Mi Mica
Mo Molybdenum
Na Salt
O Petroleum
P Phosphates
Pb Lead
Pt Platinum
S Sulfur
Sb Antimony
Tc Talc
Ti Titanium
U Uranium
V Vanadium
W Tungsten
Zn Zinc
Water Power
Major Industrial Areas

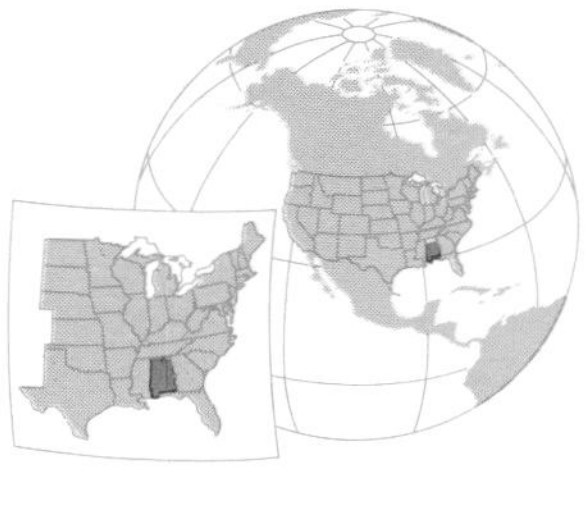

AREA 51,705 sq. mi. (133,916 sq. km.)
POPULATION 3,893,888
CAPITAL Montgomery
LARGEST CITY Birmingham
HIGHEST POINT Cheaha Mtn. 2,407 ft. (734 m.)
SETTLED IN 1702
ADMITTED TO UNION December 14, 1819
POPULAR NAME Heart of Dixie; Cotton State; Yellowhammer State
STATE FLOWER Camellia
STATE BIRD Yellowhammer

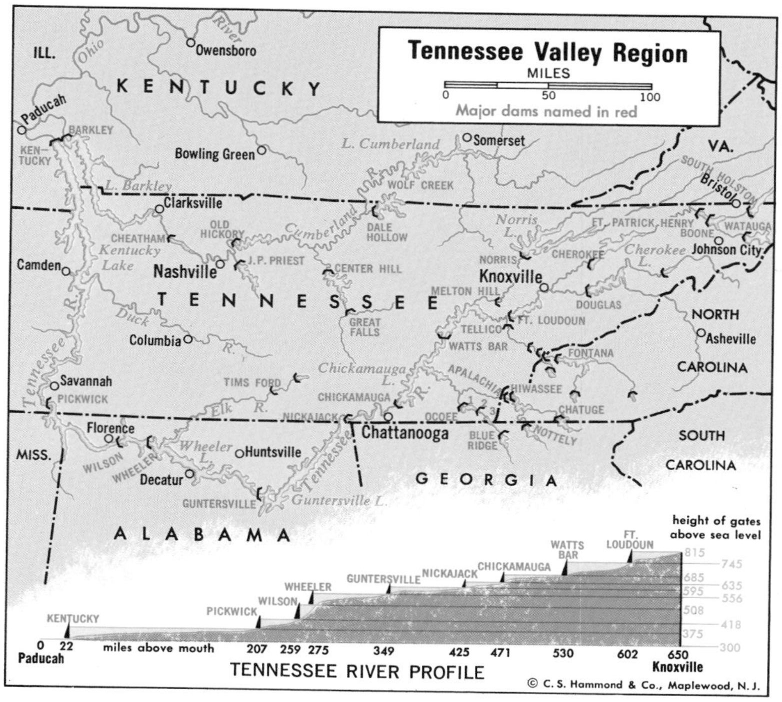

COUNTIES

Autauga 32,259 .... E5
Baldwin 78,556 .... C9
Barbour 24,756 .... H7
Bibb 15,723 .... D5
Blount 36,459 .... E2
Bullock 10,596 .... G6
Butler 21,680 .... E7
Calhoun 119,761 .... G3
Chambers 39,191 .... H5
Cherokee 18,760 .... G2
Chilton 30,612 .... E5
Choctaw 16,839 .... B6
Clarke 27,702 .... C7
Clay 13,703 .... G4
Cleburne 12,595 .... G3
Coffee 38,533 .... G8
Colbert 54,519 .... C1
Conecuh 15,884 .... E8
Coosa 11,377 .... F5
Covington 36,850 .... F8
Crenshaw 14,110 .... F7
Cullman 61,642 .... E2
Dale 47,821 .... G8
Dallas 53,981 .... D6
De Kalb 53,658 .... G2
Elmore 43,390 .... F5
Escambia 38,440 .... D8
Etowah 103,057 .... F2
Fayette 18,809 .... C3
Franklin 28,350 .... C2
Geneva 24,253 .... G8
Greene 11,021 .... C5
Hale 15,604 .... C5
Henry 15,302 .... H7
Houston 74,632 .... H8
Jackson 51,407 .... F1
Jefferson 671,324 .... E3
Lamar 16,453 .... B3
Lauderdale 80,546 .... C1
Lawrence 30,170 .... D1
Lee 76,283 .... H5
Limestone 46,005 .... E1
Lowndes 13,253 .... E6
Macon 26,829 .... G6
Madison 196,966 .... E1
Marengo 25,047 .... C6
Marion 30,041 .... C2
Marshall 65,622 .... F2
Mobile 364,980 .... B9
Monroe 22,651 .... D7
Montgomery 197,038 .... F6
Morgan 90,231 .... E2
Perry 15,012 .... D5
Pickens 21,481 .... B4
Pike 28,050 .... G7
Randolph 20,075 .... H4
Russell 47,356 .... H6
Saint Clair 41,205 .... F3
Shelby 66,298 .... E4
Sumter 16,908 .... B5
Talladega 73,826 .... F4
Tallapoosa 38,676 .... G5
Tuscaloosa 137,541 .... C4
Walker 68,660 .... D3
Washington 16,821 .... B8
Wilcox 14,755 .... D7
Winston 21,953 .... D2

CITIES and TOWNS

Zip Name/Pop. Key

36310 Abbeville⊙ 3,155 .... H7
35440 Abernant 405 .... D4
35005 Adamsville 2,498 .... D3
35540 Addison 746 .... D2
35006 Adger 400 .... D4
35441 Akron 604 .... C5
35007 Alabaster 7,079 .... E4
35950 Albertville 12,039 .... F2
35115 Aldrich 500 .... E4
35010 Alexander City 13,807 .... G5
36250 Alexandria 600 .... G3
35442 Aliceville 3,207 .... B4
35013 Allgood 387 .... F3
36501 Alma 500 .... C8
35952 Altoona 928 .... F2
36420 Andalusia⊙ 10,415 .... E8
35610 Anderson 405 .... D1
36201 Anniston⊙ 29,523 .... G3
Anniston‡ 116,936 .... G3
35016 Arab 5,967 .... E2
35805 Ardmore 1,096 .... E1
35173 Argo 600 .... E3
36311 Ariton 844 .... G7
35033 Arkadelphia 150 .... E3
35541 Arley 276 .... D2
35035 Ashby 500 .... E4
36312 Ashford 2,165 .... H8
36251 Ashland⊙ 2,052 .... G4
35953 Ashville⊙ 1,489 .... F3
35611 Athens⊙ 14,558 .... E1
36503 Atmore 8,789 .... C8
35954 Attalla 7,737 .... F2
36830 Auburn 28,471 .... H5
36003 Autaugaville 843 .... E6
†36312 Avon 433 .... H8
36505 Axis 500 .... B9
†36420 Babbie 553 .... F8
35019 Baileyton 396 .... E2
36005 Banks 160 .... G7
†36532 Barnwell 700 .... C10
36507 Bay Minette⊙ 7,455 .... C9
36509 Bayou La Batre 2,005 .... B10
35543 Bear Creek 353 .... C2
36425 Beatrice 558 .... D7
35544 Beaverton 360 .... B3
†35653 Belgreen 500 .... C2
35545 Belk 308 .... C3
36901 Bellamy 700 .... B6
35615 Belle Mina 675 .... E1
36313 Bellwood 400 .... G8
36785 Benton 74 .... E6
35546 Berry 916 .... C3
35020 Bessemer 31,729 .... D4
†36872 Beulah 500 .... H5
36006 Billingsley 106 .... E5
*35201 Birmingham⊙ 284,413 .... D3
Birmingham‡ 847,360 .... D3
36314 Black 156 .... G8
35031 Blountsville 1,509 .... E2
36201 Blue Mountain 284 .... G3
†36017 Blue Springs 112 .... G7
35957 Boaz 7,151 .... F2
35443 Boligee 164 .... C5
35032 Bon Air 118 .... F4
36511 Bon Secour 850 .... C10
†35120 Branchville 365 .... F3
36009 Brantley 1,151 .... F7
35034 Brent 2,862 .... D5
36426 Brewton⊙ 6,680 .... D8
35740 Bridgeport 2,974 .... G1
35020 Brighton 5,308 .... D4
35548 Brilliant 871 .... C2
35036 Brookside 1,409 .... E3
35444 Brookwood 492 .... D4
36010 Brundidge 3,213 .... G7
36725 Burkville 250 .... E6
36431 Burnt Corn 60 .... D7
36904 Butler⊙ 1,882 .... B6
†36767 Cahaba 75 .... D6
35040 Calera 2,035 .... E4
†36047 Calhoun 950 .... F6
36513 Calvert 600 .... B8
36726 Camden⊙ 2,406 .... D7
36850 Camp Hill 1,628 .... G5
†36502 Canoe 560 .... D8
†36726 Canton Bend 300 .... D6
35549 Carbon Hill 2,452 .... D3
35041 Cardiff 140 .... E3
†36420 Carolina 203 .... E8
35447 Carrollton⊙ 1,104 .... B4
†36023 Carrville 820 .... G5
†36548 Carson 400 .... C8
36432 Castleberry 847 .... D8
35959 Cedar Bluff 1,129 .... G2
35960 Centre⊙ 2,351 .... G2
35042 Centreville⊙ 2,504 .... D5
36518 Chatom⊙ 1,122 .... B8
35043 Chelsea 600 .... E4
35616 Cherokee 1,589 .... C1
36611 Chickasaw 7,402 .... B9
35044 Childersburg 5,084 .... F4
36254 Choccolocco 500 .... G3
36905 Choctaw 600 .... B6
†36550 Chrysler 400 .... C8
36521 Chunchula 700 .... B9
36522 Citronelle 2,841 .... B8
35045 Clanton⊙ 5,832 .... E5
†36322 Clayhatchee 560 .... G8
36015 Clayton⊙ 1,589 .... G7
35049 Cleveland 487 .... E3
36017 Clio 1,224 .... G7
35449 Coaling 400 .... D4
36523 Coden 600 .... B10
36318 Coffee Springs 339 .... G8
36524 Coffeeville 448 .... B7
35452 Coker 800 .... C4
35961 Collinsville 1,383 .... G2
36319 Columbia 881 .... H8
35051 Columbiana⊙ 2,655 .... E4
36020 Coosada 980 .... F5
35550 Cordova 3,123 .... D3
35453 Cottondale 500 .... D4
36320 Cottonwood 1,352 .... H8
†35172 County Line 199 .... E3
†36467 County Line 124 .... F8
35618 Courtland 456 .... D1
36321 Cowarts 418 .... H8
36435 Coy 950 .... D7
36525 Creola 1,652 .... B9
36906 Cromwell 650 .... B6
35962 Crossville 1,222 .... G2
36907 Cuba 486 .... B6
35055 Cullman⊙ 13,084 .... E2
36852 Cusseta 650 .... H5
36853 Dadeville⊙ 3,263 .... G5
36322 Daleville 4,250 .... G8
36526 Daphne 3,406 .... C9
36528 Dauphin Island 950 .... B10
36256 Daviston 334 .... G4
36731 Dayton 113 .... C6
*35601 Decatur⊙ 42,002 .... D1
36257 De Armanville 350 .... G3
36732 Demopolis 7,678 .... C6
35552 Detroit 326 .... B2
35062 Dora 2,327 .... D3
*36303 Dothan⊙ 48,750 .... H8
35553 Double Springs⊙ 1,057 .... D2
35964 Douglas 116 .... F2
36028 Dozier 494 .... F7
35744 Dutton 276 .... G1
36426 East Brewton 3,012 .... E8
36024 Eclectic 1,124 .... F5
36261 Edwardsville 207 .... H3
36323 Elba⊙ 4,355 .... F8
36530 Elberta 491 .... C10
35554 Eldridge 230 .... C3
35620 Elkmont 429 .... E1
36025 Elmore 600 .... F5
35458 Elrod 746 .... C4
35063 Empire 600 .... D3
36330 Enterprise 18,033 .... G8
35460 Epes 399 .... B5
35461 Ethelsville 95 .... B4
36027 Eufaula 12,097 .... H7
†36340 Eunola 169 .... G8
35462 Eutaw⊙ 2,444 .... C5
35621 Eva 185 .... E2
36401 Evergreen⊙ 4,171 .... E8
36439 Excel 385 .... D8
35746 Fackler 250 .... G1
36854 Fairfax 3,776 .... H5
35064 Fairfield 13,242 .... E4
36532 Fairhope 7,286 .... C10
35208 Fairview 450 .... E2
35622 Falkville 1,310 .... E2
36738 Faunsdale 174 .... C6
35555 Fayette⊙ 5,287 .... C3
36855 Five Points 197 .... H4
35966 Flat Rock 750 .... G1
†35601 Flint City 673 .... D1
36441 Flomaton 1,882 .... D8
36442 Florala 2,165 .... F8
*35630 Florence⊙ 37,029 .... C1
Florence‡ 135,023 .... C1
36535 Foley 4,003 .... C10
35214 Forestdale 10,814 .... E3
36740 Forkland 429 .... C5
36031 Fort Davis 500 .... G6
36032 Fort Deposit 1,519 .... E7
36856 Fort Mitchell 900 .... H6
35967 Fort Payne⊙ 11,485 .... G2
35463 Fosters 400 .... C4
36444 Franklin 133 .... G6
36445 Frisco City 1,424 .... D8
36539 Fruitdale 500 .... B8
36262 Fruithurst 239 .... G3
36446 Fulton 606 .... C7
35068 Fultondale 6,217 .... E3
35971 Fyffe 1,305 .... G2
*35901 Gadsden⊙ 47,565 .... G2
Gadsden‡ 103,057 .... G2
35464 Gainesville 207 .... B5
35972 Gallant 475 .... F2
36038 Gantt 314 .... E8
35070 Garden City 655 .... E2
35071 Gardendale 7,928 .... E3
35973 Gaylesville 192 .... G2
†35459 Geiger 200 .... B5
36340 Geneva⊙ 4,866 .... G8
36033 Georgiana 1,993 .... E7
35974 Geraldine 911 .... G2
36908 Gilbertown 218 .... B7
35559 Glen Allen 312 .... C3
35905 Glencoe 4,648 .... G3
36034 Glenwood 341 .... F7
†35010 Goldville 89 .... G4
†36024 Good Hope 1,442 .... E2
35072 Goodwater 1,895 .... F4
35466 Gordo 2,112 .... C4
36343 Gordon 362 .... H8
†35580 Gorgas 500 .... D3
36035 Goshen 365 .... F7
†36482 Gosport 500 .... C7
36541 Grand Bay 3,185 .... B10
35747 Grant 632 .... F1
35073 Graysville 2,642 .... D3
35074 Green Pond 750 .... D4
36744 Greensboro⊙ 3,248 .... C5
36037 Greenville⊙ 7,807 .... E7
†36350 Grimes 298 .... H8
36451 Grove Hill⊙ 1,912 .... C7
35563 Guin 2,418 .... C3
36542 Gulf Shores 1,349 .... C10
35976 Guntersville⊙ 7,041 .... F2
35748 Gurley 735 .... F1
†35563 Gu-Win 266 .... C3
35564 Hackleburg 883 .... C2
†36319 Haleburg 106 .... H8
35565 Haleyville 5,306 .... C2
35570 Hamilton⊙ 5,093 .... C2
†35989 Hammondville 369 .... G1
35077 Hanceville 2,220 .... E2
36039 Hardaway 600 .... G6
35078 Harpersville 934 .... F4
36344 Hartford 2,647 .... G8
35640 Hartselle 8,858 .... E2
36858 Hatchechubbee 840 .... H6
†35672 Hatton 950 .... D1
35079 Hayden 268 .... E3
36040 Hayneville⊙ 592 .... E6
35750 Hazel Green 1,503 .... E1
36345 Headland 3,327 .... H8
†36558 Healing Springs 100 .... B7
†36420 Heath 354 .... F8
36264 Heflin⊙ 3,014 .... G3
35080 Helena 2,130 .... E4
35978 Henagar 1,188 .... G1
35979 Higdon 925 .... G1
†35013 Highland Lake 210 .... F3
35643 Hillsboro 278 .... D1
†36201 Hobson City 1,268 .... G3
35571 Hodges 250 .... C2
35903 Hokes Bluff 3,216 .... G3
35082 Hollins 500 .... F4
35083 Holly Pond 493 .... E2
35752 Hollywood 1,110 .... G1
35209 Homewood 21,412 .... E4
†35216 Hoover 19,792 .... E4
36043 Hope Hull 975 .... F6
†36467 Horn Hill 186 .... F8
35020 Hueytown 13,478 .... D4
*35801 Huntsville⊙ 142,513 .... E1
Huntsville‡ 308,593 .... E1
36860 Hurtsboro 752 .... H6
35981 Ider 698 .... G1
35210 Irondale 6,510 .... E3
36545 Jackson 6,073 .... C8
36861 Jacksons Gap 800 .... G5
36265 Jacksonville 9,735 .... G3
35501 Jasper⊙ 11,894 .... D3
35085 Jemison 1,828 .... E5
35573 Kansas 267 .... C3
35574 Kennedy 604 .... B3
35645 Killen 747 .... D1
35091 Kimberly 1,043 .... E3
†36301 Kinsey 1,239 .... H8
36453 Kinston 604 .... F8
36862 Lafayette⊙ 3,647 .... H5
†35986 Lakeview 441 .... G2
36863 Lanett 6,897 .... H5
36864 Langdale 2,034 .... H5
†35768 Larkinsville 425 .... F1
36911 Lavaca 500 .... B6
35094 Leeds 8,638 .... E3
35983 Leesburg 116 .... G2
35646 Leighton 1,218 .... D1
36548 Leroy 699 .... B8
35647 Lester 117 .... D1
†36322 Level Plains 867 .... G8
35648 Lexington 884 .... D1
†36420 Libertyville 141 .... F8
35096 Lincoln 2,081 .... F3
36748 Linden⊙ 2,773 .... C6
36266 Lineville 2,257 .... G4
35020 Lipscomb 3,741 .... E4
36912 Lisman 638 .... B6
†36876 Little Shawmut 2,793 .... H5
†35653 Littleville 1,262 .... C1
35470 Livingston⊙ 3,187 .... B5
36865 Loachapoka 335 .... G5
36455 Lockhart 547 .... F8
35097 Locust Fork 488 .... E3
†35137 Longview 475 .... E4
36048 Louisville 791 .... G7
36751 Lower Peach Tree 926 .... C7
36752 Lowndesboro 207 .... E6
36551 Loxley 804 .... C9
36049 Luverne⊙ 2,639 .... F7
35575 Lynn 554 .... C2
35758 Madison 4,057 .... E1
36348 Madrid 172 .... H8
36555 Magnolia Springs 800 .... C10
36349 Malvern 558 .... G8
36750 Maplesville 754 .... E5
35112 Margaret 757 .... F3
36756 Marion⊙ 4,467 .... D5
35114 Maylene 500 .... E4
35111 McCalla 657 .... E4
36552 McCullough 500 .... D8
36553 McIntosh 319 .... B8
36456 McKenzie 605 .... E7
†35442 Memphis 95 .... B4
35984 Mentone 476 .... G1
35759 Meridianville 1,403 .... F1
35228 Midfield 6,203 .... E4
36350 Midland City 1,903 .... H8
36053 Midway 593 .... H6
†35150 Mignon 2,054 .... F4
36054 Millbrook 3,101 .... F6
35576 Millport 1,287 .... B3
36558 Millry 956 .... B7
36761 Minter 450 .... D6
*36601 Mobile⊙ 200,452 .... B9
Mobile‡ 442,819 .... B9
36460 Monroeville⊙ 5,674 .... D7
†35804 Monrovia 500 .... E1
35115 Montevallo 3,965 .... E4
*36101 Montgomery (cap.)⊙ 178,857 .... F6
Montgomery‡ 272,687 .... F3
36559 Montrose 750 .... C9
†35125 Moody 1,840 .... F3
35649 Mooresville 58 .... E1
35116 Morris 623 .... E3
35650 Moulton⊙ 3,197 .... D2
35474 Moundville 1,310 .... C5
†35957 Mountainboro 266 .... F2
35223 Mountain Brook 19,718 .... E4
36560 Mount Vernon 1,038 .... B8
36268 Munford 700 .... F3
35660 Muscle Shoals 8,911 .... C1
36763 Myrtlewood 252 .... C6
36764 Nanafalia 500 .... B6
36303 Napier Field 493 .... H8
35578 Nauvoo 259 .... D3
†35049 Nectar 367 .... E3
36765 Newbern 307 .... C5
36351 New Brockton 1,392 .... G8
35760 New Hope 1,546 .... F1
35761 New Market 680 .... F1
†35010 New Site 340 .... G4
36352 Newton 1,540 .... G8
36353 Newville 814 .... H8
35086 North Johns 243 .... D4
35476 Northport 14,291 .... C4
36866 Notasulga 876 .... G5
35006 Oak Grove 638 .... F4
36766 Oak Hill 63 .... D7
35579 Oakman 770 .... D3
35120 Odenville 724 .... F3
36271 Ohatchee 860 .... G3
35121 Oneonta⊙ 4,824 .... E3
†36467 Onycha 147 .... F8
36801 Opelika⊙ 21,896 .... H5
36467 Opp 7,204 .... F8
36561 Orange Beach 600 .... C10
36767 Orrville 349 .... D6
35763 Owens Cross Roads 804 .... E1
36203 Oxford 8,939 .... G3
36360 Ozark⊙ 13,188 .... G8
35764 Paint Rock 221 .... F1
35580 Parrish 1,583 .... D3
35124 Pelham 6,759 .... E4
35125 Pell City⊙ 6,616 .... F3
36916 Pennington 355 .... B6
36562 Perdido 500 .... C8
36471 Peterman 600 .... D7
36062 Petrey 93 .... F7
36867 Phenix City⊙ 26,928 .... H6
35581 Phil Campbell 1,549 .... C2
†35447 Pickensville 132 .... B4
36272 Piedmont 5,544 .... G3
36371 Pinckard 771 .... G8
36768 Pine Apple 298 .... E7
36769 Pine Hill 510 .... C7
35765 Pisgah 699 .... G1
36758 Plantersville 650 .... E5
35127 Pleasant Grove 7,102 .... D4
36564 Point Clear 1,812 .... C10
†36441 Pollard 144 .... D8

(continued on following page)

Agriculture, Industry and Resources

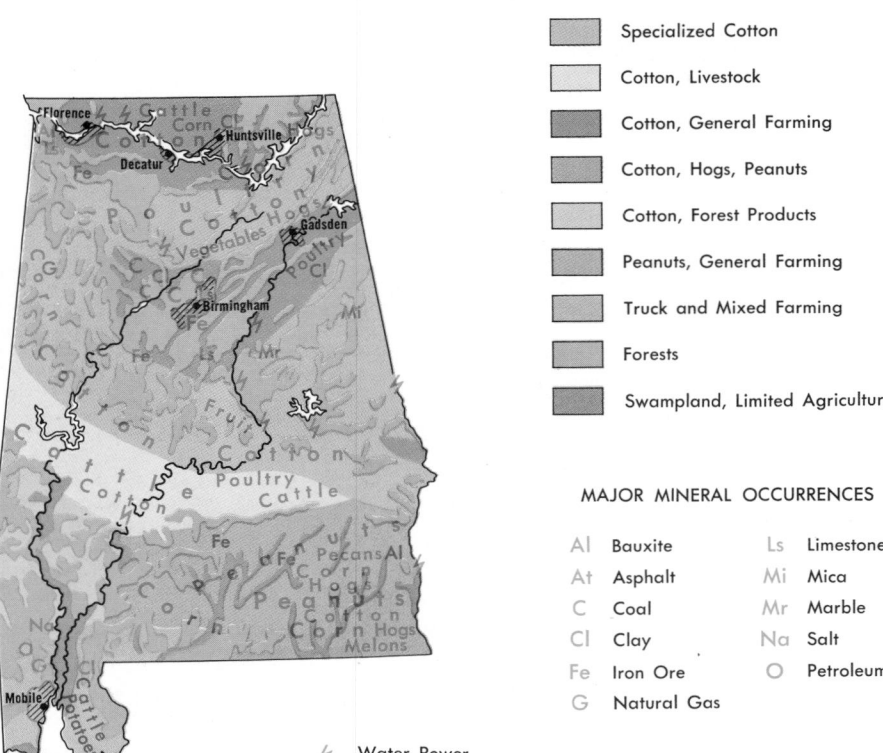

DOMINANT LAND USE

- Specialized Cotton
- Cotton, Livestock
- Cotton, General Farming
- Cotton, Hogs, Peanuts
- Cotton, Forest Products
- Peanuts, General Farming
- Truck and Mixed Farming
- Forests
- Swampland, Limited Agriculture

MAJOR MINERAL OCCURRENCES

Al	Bauxite	Ls	Limestone
At	Asphalt	Mi	Mica
C	Coal	Mr	Marble
Cl	Clay	Na	Salt
Fe	Iron Ore	O	Petroleum
G	Natural Gas		

Water Power

Major Industrial Areas

Topography

0   30   60 MI.

0   30   60 KM.

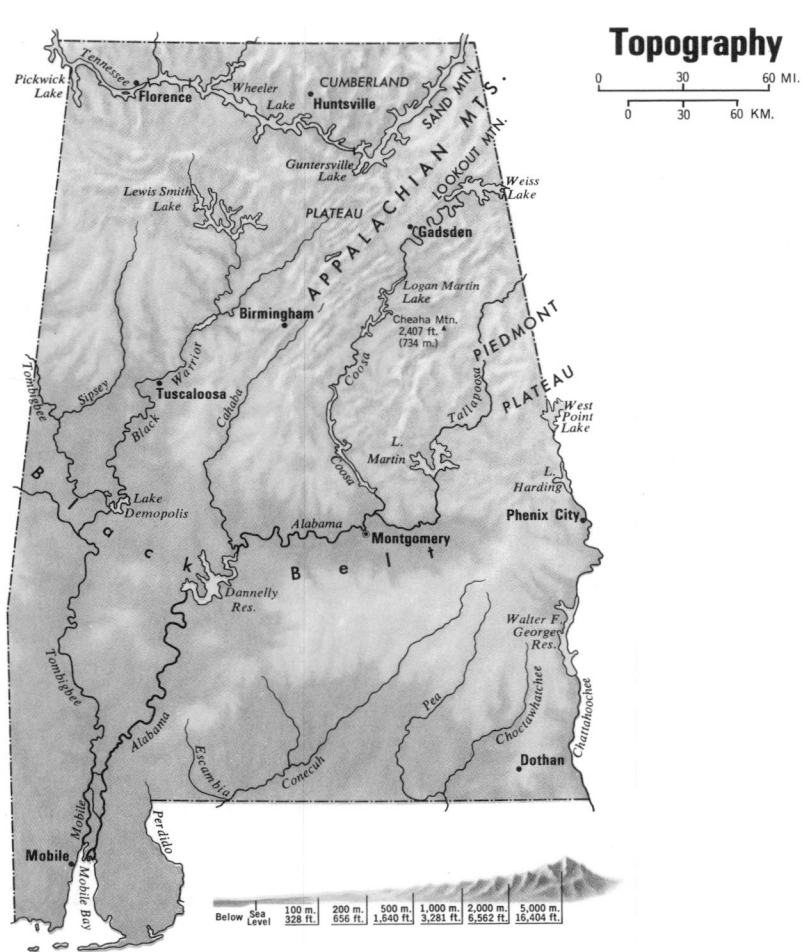

Below Sea Level | 100 m. 328 ft. | 200 m. 656 ft. | 500 m. 1,640 ft. | 1,000 m. 3,281 ft. | 2,000 m. 6,562 ft. | 5,000 m. 16,404 ft.

†35986 Powell's Crossroads 636 ...G1
36067 Prattville⊙ 18,647 ....E6
†35601 Priceville 966 ....E1
36610 Prichard 39,541 ....B9
†36748 Providence 363 ....C6
35131 Ragland 1,860 ....F3
35901 Rainbow City 6,299 ....F3
35986 Rainsville 3,907 ....G2
36069 Ramer 680 ....F6
36273 Ranburne 417 ....H3
35582 Red Bay 3,232 ....B2
36474 Red Level 504 ....E8
†35954 Reece City 718 ....G2
35481 Reform 2,245 ....C4
35133 Remlap 800 ....E3
36475 Repton 313 ....D8
†35203 Republic 500 ....E3
36476 River Falls 669 ....E8
35135 Riverside 849 ....F3
†36426 Riverview 132 ....D8
36274 Roanoke 5,896 ....H4
36567 Robertsdale 2,306 ....C9
35136 Rockford⊙ 494 ....F5
36274 Rock Mills 600 ....H4
35652 Rogersville 1,224 ....D1
35020 Roosevelt City 3,352 ......E4
†35049 Rosa 204 ....E3
35653 Russellville⊙ 8,195 ....C2
35228 Rutledge 496 ....F7
35137 Saginaw 475 ....E4
36568 Saint Elmo 700 ....B10
†35630 Saint Florian 305 ....C1
36569 Saint Stephens 700 ....B7
36570 Salitpa 550 ....C7
36477 Samson 2,402 ....F8
†36420 Sanford 250 ....F8
36571 Saraland 9,833 ....B9
†35957 Sardis 883 ....F2
36572 Satsuma 3,822 ....B9
35139 Sayre 700 ....E3
35768 Scottsboro⊙ 14,758 ....F1
35771 Section 821 ....G1
36701 Selma⊙ 26,684 ....E6
36701 Selmont-West
       Selmont 5,255 ....E6
36876 Shawmut 2,284 ....H5
35660 Sheffield 11,903 ....C1
35143 Shelby 500 ....E4
†35979 Shiloh 297 ....G2
36576 Silverhill 624 ....C9
36375 Slocomb 2,153 ....G8
36877 Smiths 950 ....H5
35952 Snead 667 ....F2
35670 Somerville 140 ....E2
†35901 Southside 5,141 ....F3
36527 Spanish Fort 3,415 ....C9
†35674 Spring Valley 600 ....C1
35146 Springville 1,476 ....E3
36578 Stapleton 975 ....C9
35987 Steele 795 ....F3
35772 Stevenson 2,568 ....G1
35484 Stewart 450 ....C5
36579 Stockton 500 ....C9
35586 Sulligent 2,130 ....B3
35148 Sumiton 2,815 ....D3
36580 Summerdale 546 ....C10
36581 Sunflower 100 ....B8
36782 Sweet Water 253 ....C6
35149 Sycamore 800 ....F4
35150 Sylacauga 12,708 ....F4
35988 Sylvania 1,156 ....G1
35160 Talladega⊙ 19,128 ....F4
†35150 Talladega Springs 196 ...F4
36078 Tallassee 4,763 ....G5
35671 Tanner 600 ....E1
35217 Tarrant City 8,148 ....E3
†36301 Taylor 1,003 ....H8
36582 Theodore 6,392 ....B9
36783 Thomaston 679 ....C6
36784 Thomasville 4,387 ....C7
35171 Thorsby 1,422 ....E5
36583 Tibbie 675 ....B8
35672 Town Creek 1,201 ....D1
35587 Townley 500 ....D3
36921 Toxey 265 ....B7
35172 Trafford 673 ....E3
†35758 Triana 285 ....E1
35673 Trinity 1,328 ....D1
36081 Troy⊙ 12,945 ....G7
35173 Trussville 3,507 ....E3
*35401 Tuscaloosa⊙ 75,211 ......C4
       Tuscaloosa‡ 137,473 ......C4
35674 Tuscumbia⊙ 9,137 ....C1
36083 Tuskegee⊙ 13,327 ....G6
36088 Tuskegee Institute G6
†35462 Union 358 ....C5
35175 Union Grove 127 ....E2
36089 Union Springs⊙ 4,431 ......G6
36786 Uniontown 2,112 ....D6
36480 Uriah 450 ....D8
35775 Valhermoso Springs 500 ..E2
35989 Valley Head 609 ....G1
35490 Vance 254 ....D4
35176 Vandiver 700 ....F4
36091 Verbena 500 ....E5
35592 Vernon⊙ 2,609 ....B3
35216 Vestavia Hills 15,722 ......E4
35593 Vina 346 ....B2
35178 Vincent 1,652 ....F4
35179 Vinemont 500 ....E2
36481 Vredenburgh 433 ....D7
36276 Wadley 532 ....G4
†36022 Wadsworth 500 ....E5
36585 Wagarville 550 ....B8
†35150 Waldo 231 ....F4
36586 Walker Springs 500 ....C7
35990 Walnut Grove 510 ....F2
35180 Warrior 3,260 ....E3
35677 Waterloo 260 ....B1
35182 Wattsville 550 ....F3
36879 Waverly 228 ....G5
36277 Weaver 2,765 ....G3
36376 Webb 448 ....H8
36278 Wedowee⊙ 908 ....H4
35183 Weogufka 500 ....F4

35184 West Blocton 1,147 ....D4
†36201 West End-Cobb
       Town 5,189 ....G3
†35005 West Jefferson 357 ....D4
†35570 Weston 350 ....B2
35185 Westover 500 ....E4
†35179 West Point 248 ....D2
36092 Wetumpka⊙ 4,341 ....E5
36482 Whatley 800 ....C7
†36040 White Hall 195 ....E6
†35094 Whites Chapel 336 ....F3
36587 Wilmer 581 ....B9
35186 Wilsonville 914 ....E4
35187 Wilton 642 ....E4
35594 Winfield 3,781 ....C3
36280 Woodland 192 ....H4
35776 Woodville 609 ....F1
36924 Yantley 500 ....B6
36925 York 3,392 ....B6

OTHER FEATURES

Alabama (riv.) C8
Aliceville (dam) B4
Anniston Army Depot G3
Bankhead (lake) D2
Bartletts Ferry (dam) H5
Big Canoe (creek) F2
Big Creek (lake) B5
Black Warrior (riv.) C5
Bon Secour (bay) C10
Brookley Air Force Base B9
Buttahatchee (riv.) B3
Cahaba (riv.) D6
Cedar (pt.) B10
Chattahoochee (riv.) H6
Chattooga (riv.) H3
Cheaha (mt.) G4
Choctawhatchee (riv.) F8
Coffeeville (dam) B7
Conecuh (riv.) D8
Coosa (riv.) F3
Cowikee, North Fork (creek) H6
Cumberland (plat.) F2
Dannelly (res.) D6
Demopolis (dam) C5
Elk (riv.) D1
Escambia (creek) D8
Escambia (riv.) D8
Escatawpa (riv.) B8
Eufaula (Walter F. George Res.)
       (lake) H7
Fort Gaines C10
Fort McClellan Mil. Res. 7,605 ...G3
Fort Morgan C10
Fort Rucker 8,932 ....F8
Gainesville (dam) B5
Goat Rock (dam) H5
Goat Rock (lake) H5
Grants Pass (chan.) B10
Gunter Air Force Base E5
Guntersville (dam) F2
Guntersville (lake) F1
Harding (lake) F5
Herbes (isl.) B10
Holt (dam) D3
Horseshoe Bend Nat'l Mil. Park ...G4
Inland (lake) E3
Jordan (dam) F5
Jordan (lake) F5
Lay (lake) E4
Lewis Smith (dam) D2
Lewis Smith (lake) D2
Little (riv.) G1
Little (riv.) G6
Locust Fork (riv.) E3
Logan Martin (lake) F3
Lookout (mt.) G1
Martin (dam) F5
Martin (lake) F5
Maxwell Air Force Base E5
Mexico (gulf) E10
Mississippi (sound) B10
Mitchell (lake) E5
Mobile (bay) B10
Mobile (pt.) B10
Mobile (riv.) C9
Mulberry (creek) D6
Mulberry Fork (riv.) D3
Neely Henry (lake) F3
Oakmulgee (creek) D5
Oliver (dam) H5
Paint Rock (riv.) F1
Patsaliga (creek) E7
Pea (riv.) F7
Perdido (bay) D10
Perdido (riv.) D9
Pickwick (lake) B1
Pigeon (creek) E7
Redstone Arsenal 5,728 ....E1
Russell Cave Nat'l Mon. G1
Sand (mt.) G1
Sandy (creek) E5
Sepulga (riv.) D8
Sipsey (riv.) C5
Sipsey Fork (riv.) D2
Tallapoosa (riv.) G5
Tennessee (riv.) D1
Tennessee-Tombigbee Waterway ...B5
Tensaw (riv.) C9
Thurlow (dam) F5
Tombigbee (riv.) B7
Town (creek) D2
Tuscaloosa (lake) C4
Tuskegee Institute Nat'l Hist. Park ...G6
Walter F. George (dam) H7
Walter F. George (res.) H7
Warrior (riv.) C5
Weiss (lake) G2
West Point (lake) H5
Wheeler (dam) D1
Wheeler (lake) E1
Wilson (dam) C1
Yates (dam) F5

⊙County seat.
‡Population of metropolitan area.
† Zip of nearest p.o.   * Multiple zip

Alabama

SCALE

0 5 10   20   30   40 MI.

0 5 10   20   30   40 KM.

State Capitals .... ⊛
County Seats .... ●
Major Limited Access Hwys. ━━━

Scale 1:1,930,000

© Copyright Hammond Incorporated, Maplewood, N.J.

CITIES and TOWNS

Zip   Name/Pop.   Key

†99609 Akolmiut (Kasigluk) 641 . . F2
99554 Alakanuk 522 . . . . E2
*99501 Anchorage⊙ 174,431 . . B1
     Anchorage‡ 174,431 . . . . B1
†99760 Anderson 517 . . . . . . H2
99723 Barrow 2,207 . . . . G1
99559 Bethel 3,576 . . . . F2
99704 Clear 504 . . . . J2
99701 College 4,043 . . . . . . J1
99574 Cordova 1,879 . . . . . . D1
99921 Craig 527 . . . . M2
99737 Delta Junction 945 . . . . J2
99576 Dillingham 1,563 . . . . . F3
†99685 Dutch Harbor 250 . . . . E4
99581 Emanguk (Emmonak) 567 E2
99701 Fairbanks⊙ 22,645 . . . . J2
99740 Fort Yukon 619 . . . . . . J1
99741 Galena 765 . . . . G2
99588 Glennallen 511 . . . . . . D1
99827 Haines 993 . . . . M1
99603 Homer 2,209 . . . . B2
99829 Hoonah 680 . . . . M1

99604 Hooper Bay 627 . . . . E2
99801 Juneau (cap.)⊙ 19,528 . . N1
99830 Kake 555 . . . . M1
99609 Kasigluk 641 . . . . F2
99611 Kenai 4,324 . . . . B1
99901 Ketchikan 7,198 . . . . N2
99615 Kodiak 4,756 . . . . H3
99752 Kotzebue 2,054 . . . . F1
99926 Metlakatla 1,056 . . . . . . N2
†99901 Mountain Point 396 . . . . N2
99632 Mountain Village 583 . . . E2
99762 Nome⊙ 2,301 . . . . E2
99763 Noorvik 492 . . . . F1
99645 Palmer 2,141 . . . . C1
99833 Petersburg 2,821 . . . . . . N2
99660 Saint Paul Island 551 . . . D3
99661 Sand Point 625 . . . . G3
99664 Seward 1,843 . . . . C1
99835 Sitka 7,803 . . . . M1
99840 Skagway 768 . . . . M1
99669 Soldotna 2,320 . . . . B1
99503 Spenard C1
99672 Sterling 919 . . . . B1
99780 Tok 589 . . . . K2
99684 Unalakleet 623 . . . . G2
99685 Unalaska 1,322 . . . . E4
99686 Valdez 3,079 . . . . D1
99929 Wrangell 2,184 . . . . N2
99689 Yakutat 449 . . . . L3

OTHER FEATURES

Adak (isl.) L4
Admiralty (isl.) M1

Afognak (isl.) H3
Agattu (isl.) J3
Akutan (isl.) E4
Alaska (gulf) K3
Alaska (range) H2
Aleutian (isls.) J4
Aleutian (range) L1
Alexander (arch.) L1
Amchitka (isl.) K4
Amlia (passage) L4
Amukta (isl.) D4
Andreanof (isls.) L4
Atka (isl.) L4
Attu (isl.) J3
Baird (mts.) F1
Baranof (isl.) M1
Barrow (pt.) G1
Bear (mt.) K2
Beaufort (sea) K1
Becharof (lake) G3
Bering (glac.) K2
Bering (sea) D2
Bering (str.) E1
Blackburn (mt.) K2
Bona (mt.) K2
Bristol (bay) F3
British (mts.) K1
Brooks (range) G1
Chandalar (riv.) J1
Chatham (str.) M1
Chichagof (isl.) M1
Chignik (bay) G3
Chilkoot (pass) M1
Chirikof (isl.) G3

Chitina (riv.) K2
Christian (sound) M2
Chugash (mts.) C1
Chukchi (sea) E1
Clarence (str.) N2
Clark (lake) H2
Coast (mts.) N1
Columbia (glac.) C1
Colville (riv.) G1
Constantine (cape) G3
Cook (inlet) B1
Cook (mt.) K2
Copper (riv.) J2
Cordova (bay) M2
Coronation (isl.) M2
Cross (sound) L1
Dease (inlet) H1
Decision (cape) M2
Denali Nat'l Park H2
Devils Paw (mt.) N1
Dixon Entrance (chan.) M2
Douglas (mt.) H3
Dry (bay) L3
Eielson A.F.B. 5,232 . . . . . . J2
Elmendorf A.F.B. B1
Endicott (mts.) H1
Etolin (isl.) N2
Fairweather (cape) L1
Fairweather (mt.) L1
Firth (riv.) K1
Foraker (mt.) H2
Fort Davis E2
Fort Greely 1,635 . . . . J2
Fort Richardson C1

Fort Wainwright J1
Four Mountains (isls.) E4
Fox (isls.) E4
Frederick (sound) N1
Gates of the Arctic Nat'l
     Park H1
Glacier (bay) M1
Glacier Bay Nat'l Park M1
Goodhope (bay) F1
Great Sitkin (isl.) L4
Guyot (glac.) K2
Hagemeister (isl.) F3
Halkett (cape) H1
Hall (mt.) D2
Harding Icefield C2
Harrison (bay) H1
Hayes (mt.) J2
Hazen (bay) E3
Hinchinbrook (isl.) D1
Hoonah (sound) M1
Hope (pt.) E1
Howard (pass) G1
Icy (bay) L3
Icy (cape) F1
Icy (pt.) L1
Icy (str.) M1
Iliamna (lake) G3
Iliamna (vol.) H2
Innoko (riv.) G2
Kachemak (bay) B2
Kanaga (isl.) K4
Kates Needle (mt.) N1
Katmai (vol.) H3
Katmai Nat'l Park H3

Kayak (isl.) . . . . K
Kenai (lake) C
Kenai (mt.) C
Kenai (pen.) C
Kenai Fjords Nat'l Park C
Kennedy Entrance (str.) C
King (isl.) . . . . E
Kiska (isl.) . . . . K
Kiska (vol.) . . . . K
Klondike Gold Rush Nat'l Hist.
     Park . . . . M
Knight (isl.) C
Knik Arm (inlet) C
Kobuk (riv.) . . . . F
Kobuk Valley Nat'l Park . . . . F
Kodiak (isl.) . . . . H
Kotzebue (sound) . . . . F
Koyukuk (riv.) . . . . G
Krusenstern (cape) . . . . F
Kuiu (isl.) . . . . M
Kuskokwim (bay) . . . . F
Kuskokwim (mts.) . . . . G
Kuskokwim (riv.) . . . . G
Kvichak (bay) . . . . G
Lake Clark Nat'l Park . . . . . H
Lisburne (cape) . . . . F
Little Diomede (isl.) . . . . . . E
Little Sitkin (isl.) . . . . J
Lynn Canal (inlet) . . . . M
Makushin (vol.) . . . . E
Malaspina (glac.) . . . . K
Marcus Baker (mt.) C
Marmot (isl.) . . . . H
Matanuska (riv.) C

Agriculture, Industry and Resources

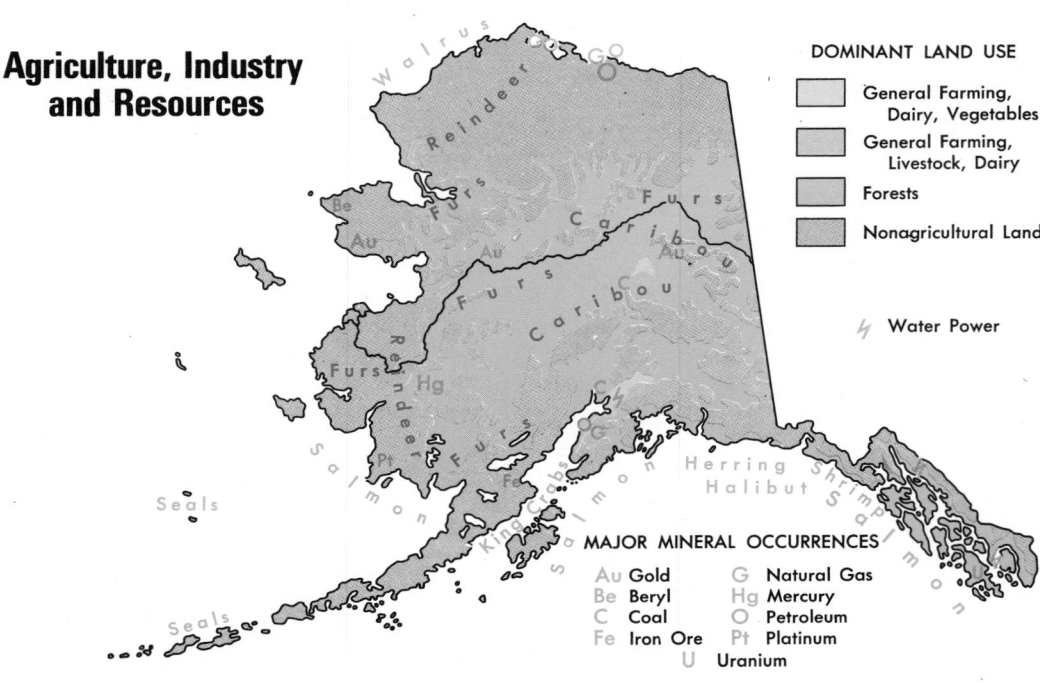

DOMINANT LAND USE

- General Farming, Dairy, Vegetables
- General Farming, Livestock, Dairy
- Forests
- Nonagricultural Land

⚡ Water Power

MAJOR MINERAL OCCURRENCES

Au	Gold	G	Natural Gas
Be	Beryl	Hg	Mercury
C	Coal	O	Petroleum
Fe	Iron Ore	Pt	Platinum
		U	Uranium

Topography

Scale
0   200   400 MI.
0   200   400 KM.

Below Sea Level   100 m. 328 ft.   200 m. 656 ft.   500 m. 1,640 ft.   1,000 m. 3,281 ft.   2,000 m. 6,562 ft.   5,000 m. 16,404 ft.

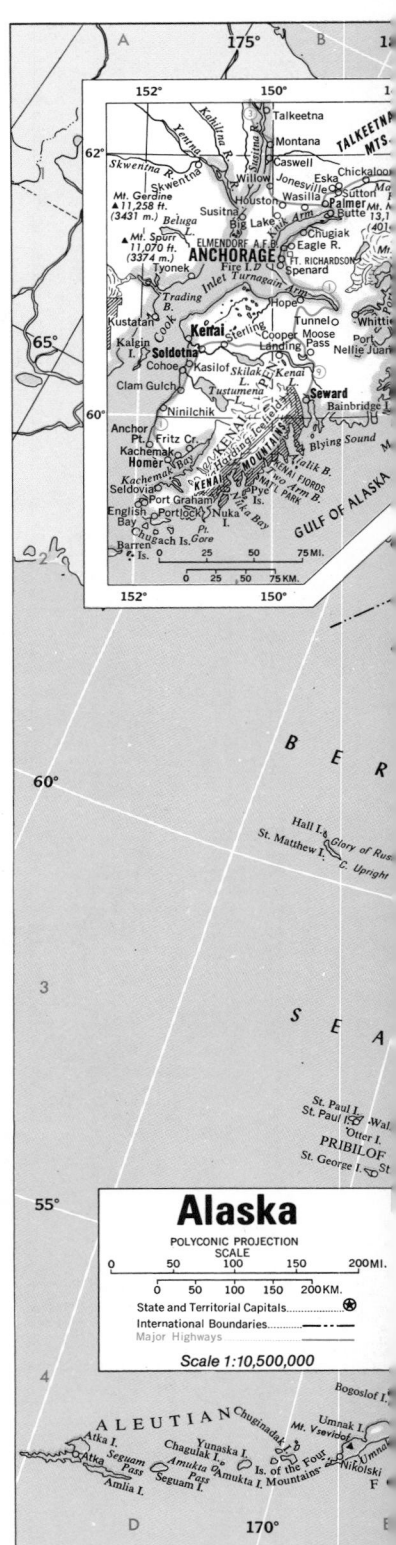

Alaska

POLYCONIC PROJECTION

SCALE
0   50   100   150   200 MI.
0   50   100   150   200 KM.

State and Territorial Capitals . . . . ⊛
International Boundaries . . . .
Major Highways

Scale 1:10,500,000

McKinley (mt.) H2
Meade (riv.) G1
Mendenhall (cape) E3
Mentasta (pass) K2
Merrill (pass) H2
Michelson (mt.) K1
Middleton (isl.) J3
Misty Fjords Nat'l Mon. N2
Mitkof (isl.) D1
Montague (isl.) D1
Muir (glac.) M1
Mulchatna (riv.) G2
Muzon (cape) M2
Naknek (lake) H3
Near (isls.) H3
Nelson (isl.) E2
Newenham (cape) F3
Noatak (riv.) F1
Norton (bay) F2
Norton (sound) E2
Nowitna (riv.) H2
Nuka (bay) C2
Nunivak (isl.) E3
Nushagak (riv.) G2
Nuyakuk (lake) G2
Nunimaney (cape) M2
Otter (isl.) D3
Pastol (bay) F2
Pavlof (riv.) F3
Pavlof (vol.) F3
Philip Smith (mts.) J1
Porcupine (riv.) K1
Port Clarence (inlet) E1
Port Heiden (inlet) G3

Portland Canal (inlet) N2
Port Moller (inlet) F3
Port Wells (inlet) C1
Pribilof (isls.) D3
Prince of Wales (cape) E1
Prince of Wales (isl.) N2
Prince William (sound) D1
Prudhoe (bay) J1
Rat (isls.) K4
Redoubt (vol.) H2
Revillagigedo (chan.) N2
Revillagigedo (isl.) N2
Romanzof (cape) E2
Sagavanirktok (riv.) J1
Saint Elias (cape) K3
Saint Elias (mts.) L2
Saint George (isl.) D3
Saint Lawrence (isl.) D2
Saint Matthew (isl.) D2
Saint Paul (isl.) D3
Salisbury (sound) M1
Sanak (isl.) F4
Sanford (mt.) K2
Schwatka (mts.) G1
Seguam (isl.) D4
Selawik (lake) F1
Semichi (isls.) J3
Semidi (isls.) G3
Semisopochnoi (isl.) K4
Seward (pen.) E1
Seymour (canal) N1
Sheenjek (riv.) K1
Shelikof (str.) H3
Shemya (isl.) J3

Shishaldin (vol.) E4
Shumagin (isls.) G4
Shuyak (isl.) H3
Sitka (sound) M1
Sitka Nat'l Hist. Park M1
Sitkin (str.) H3
Skilak (lake) C1
Skwentna (riv.) A1
Smith (bay) H1
Spencer (cape) L1
Stephens (passage) N1
Stevenson Entrance (str.) H3
Stikine (riv.) N2
Stikine (str.) N2
Stony (riv.) G2
Stuart (isl.) F2
Suemez (isl.) M2
Sumner (str.) M2
Susitna (riv.) B1
Sutwik (isl.) G3
Taku (glac.) N1
Taku (riv.) N1
Talkeetna (mts.) J2
Tanaga (isl.) K4
Tanaga (vol.) K4
Tanana (riv.) J2
Taylor (mts.) G2
Tazlina (lake) D1
Tazlina (riv.) D1
Teshekpuk (lake) H1
Tigalda (isl.) F4
Tikchik (lkes) G2
Togiak (bay) F3
Tugidak (isl.) G3

Turnagain Arm (inlet) B1
Tustumena (lake) C1
Two Arm (bay) C2
Ugashik (lkes) G3
Umnak (isl.) E4
Umnak (passage) E4
Unalaska (isl.) E4
Unga (isl.) F3
Unimak (isl.) E4
Unimak (passage) F4
Utukok (riv.) F1
Valley of Ten Thousand Smokes .G3
Vancouver (cape) L2
Veniaminof (crater) F3
Vsevidof (isl.) E4
Walrus (isl.) E3
Walrus (isls.) E3
Waring (mts.) G1
West Point (mt.) K2
White (pass) N1
White (riv.) N1
White Mountains Nat'l Rec. Area .J1
Witherspoon (mt.) C1
Wrangell (cape) H3
Wrangell (isl.) N2
Wrangell (mts.) K2
Wrangell-St. Elias Nat'l Park ...K2
Yakobi (isl.) M1
Yakutat (bay) K3
Yentna (riv.) A1
Yukon (riv.) F2

○ Court House
‡Population of metropolitan area.
† Zip of nearest p.o.
* Multiple zips.

AREA 591,004 sq. mi. (1,530,700 sq. km.)
POPULATION 401,851
CAPITAL Juneau
LARGEST CITY Anchorage
HIGHEST POINT Mt. McKinley 20,320 ft. (6194 m.)
SETTLED IN 1801
ADMITTED TO UNION January 3, 1959
POPULAR NAME Great Land; Last Frontier
STATE FLOWER Forget-me-not
STATE BIRD Willow Ptarmigan

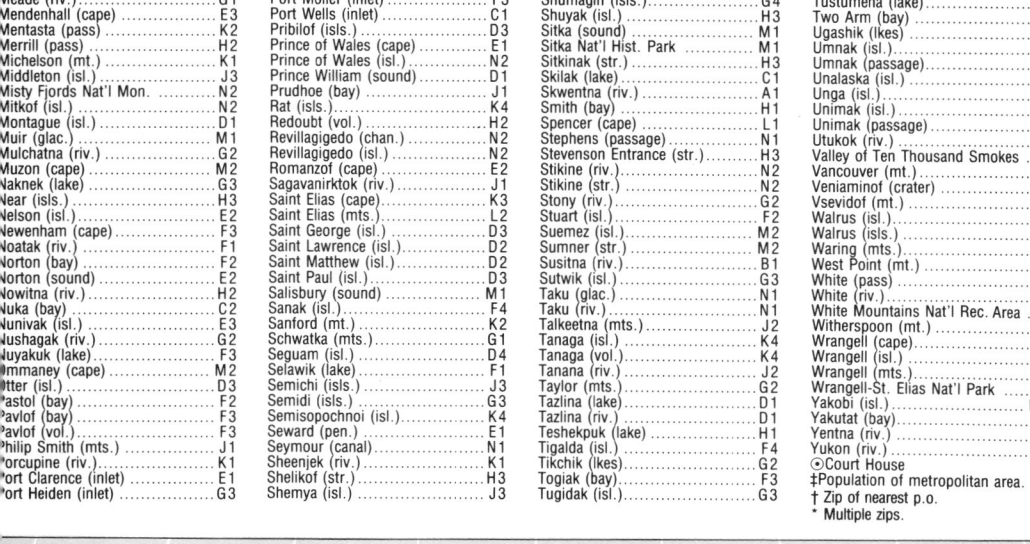

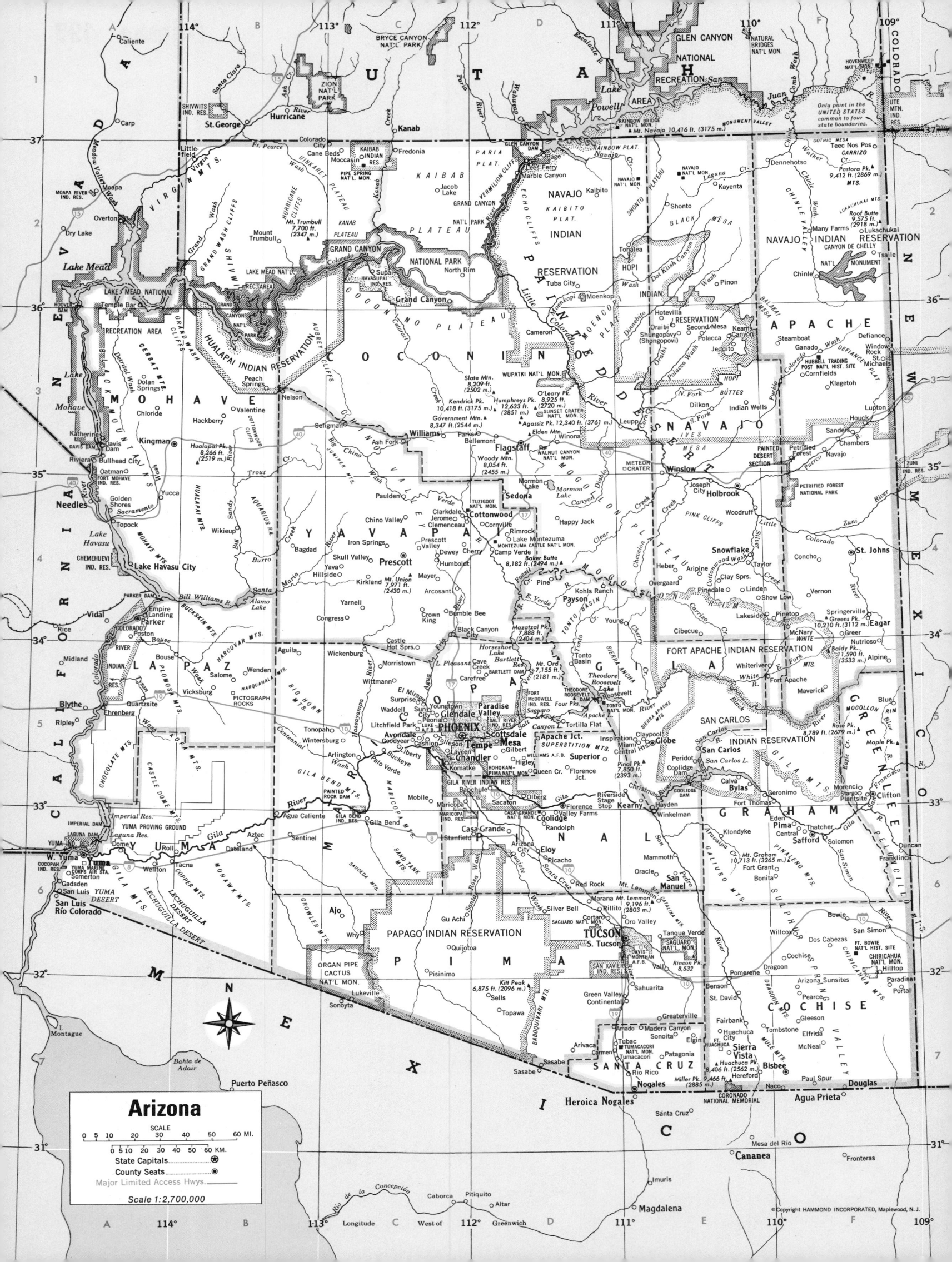

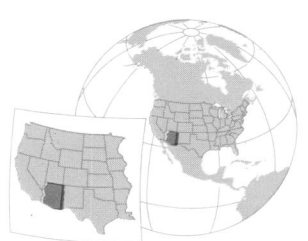

AREA 114,000 sq. mi. (295,260 sq. km.)
POPULATION 2,718,425
CAPITAL Phoenix
LARGEST CITY Phoenix
HIGHEST POINT Humphreys Pk. 12,633 ft.
(3851 m.)
SETTLED IN 1752
ADMITTED TO UNION February 14, 1912
POPULAR NAME Grand Canyon State
STATE FLOWER Saguaro Cactus Blossom
STATE BIRD Cactus Wren

Agriculture, Industry and Resources

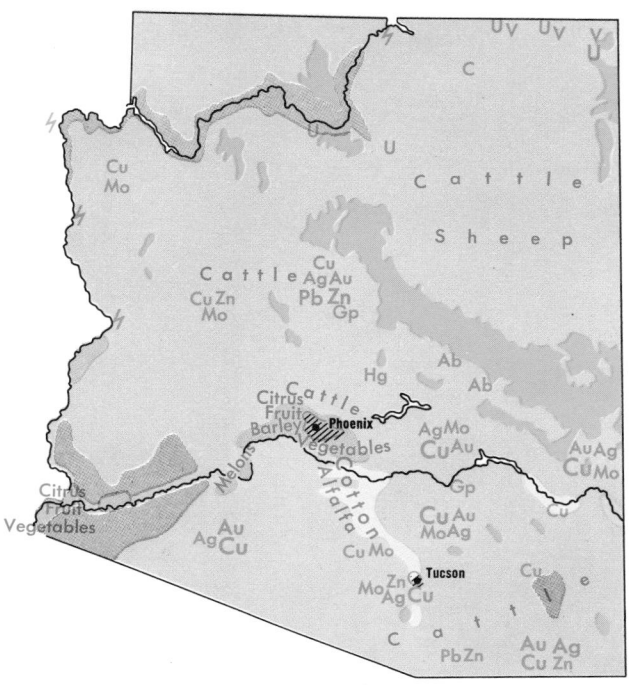

MAJOR MINERAL OCCURRENCES

Ab	Asbestos	Cu	Copper	Pb	Lead
Ag	Silver	Gp	Gypsum	U	Uranium
Au	Gold	Hg	Mercury	V	Vanadium
C	Coal	Mo	Molybdenum	Zn	Zinc

DOMINANT LAND USE

- Fruit, Truck and Mixed Farming
- Cotton and Alfalfa
- General Farming, Livestock, Special Crops
- Range Livestock
- Forests
- Nonagricultural Land

⚡ Water Power
▨ Major Industrial Areas

COUNTIES

Apache 52,108 .... F3
Cochise 85,686 .... F7
Coconino 75,008 .... C3
Gila 37,080 .... E5
Graham 22,862 .... E6
Greenlee 11,406 .... F5
La Paz▪ 13,100 .... A5
Maricopa 1,509,052 .... C5
Mohave 55,865 .... A3
Navajo 67,629 .... E3
Pima 531,443 .... D6
Pinal 90,918 .... D6
Santa Cruz 20,459 .... E7
Yavapai 68,145 .... C4
Yuma▪ 81,800 .... A6

▪1982 official estimate.

CITIES and TOWNS

Zip	Name/Pop.	Key
†85333	Agua Caliente 60	B6
85320	Aguila 900	B5
85321	Ajo 5,189	C6
85920	Alpine 450	F5
85640	Amado 75	D7
85220	Apache Junction 9,935	D5
†85901	Aripine 25	E4
85601	Arivaca 400	D7
85223	Arizona City 825	D6
85625	Arizona Sunsites 825	F7
85322	Arlington 950	C5
86320	Ash Fork 800	C3

85323	Avondale 8,168	C5
†85333	Aztec 20	B6
86321	Bagdad 2,331	B4
85221	Bapchule 400	D5
86015	Bellemont 210	D3
85602	Benson 4,190	E7
85603	Bisbee⊙ 7,154	F7
85324	Black Canyon City 600	C4
85922	Blue 50	F5
†85643	Bonita 20	E6
85325	Bouse 500	A5
85605	Bowie 600	F6
85326	Buckeye 3,434	C5
86430	Bullhead	
	City-Riviera 10,364	A3
†86301	Bumble Bee 15	C4
85530	Bylas 1,175	E5
†85530	Calva 10	E5
86020	Cameron 600	D3
86322	Camp Verde 1,125	D4
†86022	Cane Beds 30	B2
85331	Carefree 986	C5
†85640	Carmen 200	D7
85222	Casa Grande 14,971	D6
85329	Cashion 3,014	C5
†85342	Castle Hot Springs 50	C5
85331	Cave Creek 1,589	D5
†85501	Central Heights-Midland	
	City 2,791	E5
86502	Chambers 500	F3
85224	Chandler 29,673	D5
†86327	Cherry 20	C4
86503	Chinle 2,815	F2

86323	Chino Valley 2,858	C4
86431	Chloride 225	A3
†85292	Christmas 201	E5
85911	Cibecue 100	E4
86324	Clarkdale 1,512	C4
85532	Claypool 2,362	E5
†85934	Clay Springs 500	E4
†86326	Clemenceau 300	C4
85533	Clifton⊙ 4,245	F5
85606	Cochise 150	F6
86021	Colorado City 350	B2
85924	Concho 100	F4
85332	Congress 800	C4
†85640	Continental 250	D7
85228	Coolidge 6,851	D6
†85542	Coolidge Dam 42	E5
†86505	Cornfields 200	F3
86325	Cornville 425	D4
85230	Cortaro 375	D6
86326	Cottonwood 4,550	D4
86333	Crown King 100	C4
85333	Dateland 100	B6
†86430	Davis Dam 125	A3
86327	Dewey 100	C4
†86047	Dilkon 90	E3
86441	Dolan Springs 870	A3
†85364	Dome 48	A6
†85643	Dos Cabezas 30	F6
85607	Douglas 13,058	F7
85609	Dragoon 150	F6
85534	Duncan 603	F6
85925	Eagar 2,791	F4
85535	Eden 89	F6
85334	Ehrenburg 93	A5

(continued on following page)

Topography

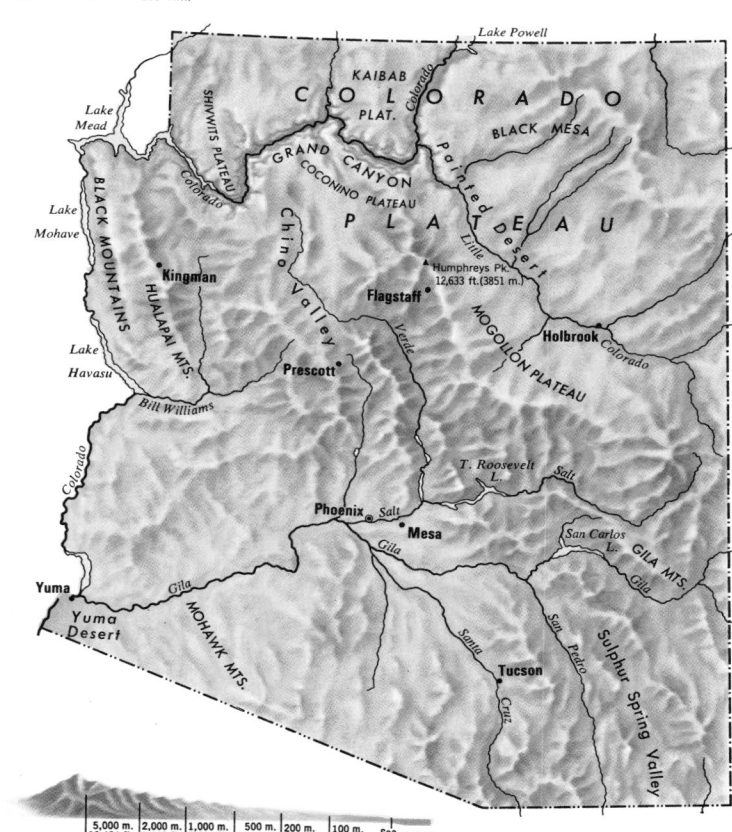
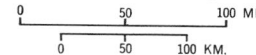

†85617 Elfrida 700 .... F7
†85637 Elgin 525 .... E7
85335 El Mirage 4,307 .... C5
85231 Eloy 6,240 .... D6
85612 Fairbank 100 .... E7
86001 Flagstaff⊙ 34,743 .... D3
85232 Florence⊙ 3,391 .... D5
†85220 Florence Junction 35 .... D5
85926 Fort Apache 500 .... F5
86504 Fort Defiance 3,431 .... F3
85643 Fort Grant 240 .... E6
85536 Fort Thomas 450 .... E5
85534 Franklin 300 .... F6
86022 Fredonia 1,040 .... C2
85336 Gadsden 250 .... A6
86505 Ganado 816 .... F3
†85536 Geronimo 25 .... F5
85337 Gila Bend 1,585 .... C6
85234 Gilbert 5,717 .... D5
†85617 Gleeson 15 .... F7
*85301 Glendale 97,172 .... C5
85501 Globe⊙ 6,886 .... E5
85323 Goodyear 2,747 .... C5
86023 Grand Canyon 1,348 .... C2
†85637 Greaterville 15 .... E7
85614 Green Valley 7,999 .... D7
85927 Greer 385 .... F4
†85634 Gu Achi 339 .... C6
86411 Hackberry 250 .... B3
86024 Happy Jack 50 .... D4
85235 Hayden 1,205 .... E5
85928 Heber 750 .... E4
85615 Hereford 10 .... E7
85236 Higley 500 .... D5
†86301 Hillside 100 .... B4
†85632 Hilltop 9 .... F6
86025 Holbrook⊙ 5,785 .... E4
86030 Hotevilla 3,009 .... E3
86506 Houck 900 .... F3
85616 Huachuca City 1,661 .... E7
86329 Humboldt 787 .... C4
86031 Indian Wells 150 .... E3
85537 Inspiration 500 .... D5
86330 Iron Springs 175 .... C4
86051 Jacob Lake 16 .... C2
†86025 Jeddito 20 .... E3
86331 Jerome 420 .... C4
86032 Joseph City 650 .... E4
86053 Kaibito 275 .... D2
†86430 Katherine 102 .... A3
86033 Kayenta 3,343 .... E2
86042 Keams Canyon 400 .... E3
85237 Kearny 2,646 .... E5
86401 Kingman⊙ 9,257 .... A3
86332 Kirkland 100 .... C4
†86505 Klagetoh 200 .... F3
85643 Klondyke 86 .... E6
85538 Kohls Ranch 100 .... D4
†85339 Komatke 300 .... C5
86403 Lake Havasu City 15,909 .... A4
86342 Lake Montezuma 900 .... D4
85929 Lakeside 1,333 .... E4
85339 Laveen 800 .... C5
†86036 Lees Ferry 10 .... D2
86035 Leupp 150 .... E3
†85326 Liberty 150 .... C5
†85901 Linden 50 .... E4
85340 Litchfield Park 3,657 .... C5
86432 Littlefield 40 .... B2
86507 Lukachukai 1,049 .... F2
85341 Lukeville 50 .... C7
86508 Lupton 250 .... F3
†85637 Madera Canyon 75 .... E7
85618 Mammoth 1,906 .... E6
86538 Many Farms 1,364 .... F2
85238 Marana 1,674 .... D6
86036 Marble Canyon 6 .... D2
85239 Maricopa 750 .... C5

†85920 Maverick 50 .... F5
86333 Mayer 810 .... C4
85930 McNary 1,320 .... F4
85617 McNeal 100 .... F7
*85201 Mesa 152,453 .... D5
85539 Miami 2,716 .... E5
†85239 Mobile 100 .... C5
†86022 Moccasin 150 .... C2
†86045 Moenkopi D2
85540 Morenci 2,736 .... F5
86038 Mormon Lake 20 .... D4
85342 Morristown 400 .... C5
85619 Mount Lemmon 400 .... E6
†84770 Mount Trumbull 14 .... B2
85620 Naco 750 .... E7
86509 Navajo 100 .... F3
†86434 Nelson 39 .... B3
85621 Nogales⊙ 15,683 .... E7
86052 North Rim 50 .... C2
85932 Nutrioso 500 .... F5
86433 Oatman 175 .... A3
†85247 Olberg 65 .... D5
85623 Oracle 2,484 .... E6
86039 Oraibi 600 .... E3
†85704 Oro Valley 1,489 .... E6
85933 Overgaard 750 .... E4
86040 Page 4,907 .... D2
85343 Palo Verde 500 .... C5
†85632 Paradise 15 .... F7
85253 Paradise Valley 11,085 .... D5
85344 Parker⊙ 2,542 .... A4
86018 Parks 175 .... C3
85624 Patagonia 980 .... E7
86334 Paulden 350 .... C4
†85607 Paul Spur 34 .... F7
85541 Payson 5,068 .... D4
86434 Peach Springs 900 .... B3
85625 Pearce 700 .... F7
85345 Peoria 12,307 .... C5
85542 Peridot 950 .... E5
86028 Petrified Forest 80 .... F3
*85001 Phoenix (cap.)⊙ 789,704 .... C5
Phoenix‡ 1,508,030 .... C5
85241 Picacho 850 .... D6
85543 Pima 1,599 .... F6
85544 Pine 800 .... D4
85934 Pinedale 400 .... E4
85935 Pinetop 1,527 .... F4
86510 Pinon 100 .... E2
85634 Pisinimo 187 .... C6
†85540 Plantsite F5
86042 Polacca 500 .... E3
85627 Pomerene 365 .... E6
85632 Portal 72 .... F7
85371 Poston 500 .... A4
86301 Prescott⊙ 20,055 .... C4
†86301 Prescott Valley 2,284 .... C4
85346 Quartzsite 255 .... A5
85242 Queen Creek 600 .... D5
†85634 Quijotoa 200 .... C6
85243 Randolph 350 .... D6
85245 Red Rock 250 .... D6
85246 Rillito 400 .... D6
86335 Rimrock 217 .... D4
85237 Riverside Stage Stop 418 .... D5
86440 Riviera-Bullhead
City 10,364 .... A3
85347 Roll 700 .... A6
85545 Roosevelt 125 .... D5
85247 Sacaton 1,951 .... D5
85546 Safford⊙ 7,010 .... F6
85629 Sahuarita 200 .... E7
85630 Saint David 800 .... E7
85936 Saint Johns⊙ 3,368 .... F4
86511 Saint Michaels 250 .... F3
85348 Salome 800 .... B5
85550 San Carlos 2,668 .... E5
86512 Sanders 900 .... F3

85349 San Luis 1,946 .... A6
85631 San Manuel 5,443 .... E6
85632 San Simon 400 .... F6
85633 Sasabe 50 .... D7
*85251 Scottsdale 88,622 .... D5
86043 Second Mesa 450 .... E3
86336 Sedona 5,368 .... D4
86337 Seligman 510 .... B3
85634 Sells 1,864 .... D7
†85333 Sentinel 40 .... B6
86054 Shonto 700 .... E2
85901 Show Low 4,298 .... F4
†86043 Shungopavy
(Shongopovi) 570 .... E3
85635 Sierra Vista 24,937 .... E7
85270 Silver Bell 900 .... D6
86338 Skull Valley 250 .... C4
85937 Snowflake 3,510 .... E4
85551 Solomon 700 .... F6
85350 Somerton 5,761 .... A6
85637 Sonoita 220 .... E7
85713 South Tucson 6,554 .... D6
85938 Springerville 1,452 .... F4
85272 Stanfield 150 .... C6
†85540 Stargo 1,038 .... F5
†86505 Steamboat 100 .... F3
*85351 Sun City 40,505 .... C5
86435 Supai 350 .... C2
85273 Superior 4,600 .... D5
85345 Surprise 3,723 .... C5
85352 Tacna 950 .... B6
†85701 Tanque Verde 850 .... E6
85939 Taylor 1,915 .... E4
86514 Teec Nos Pos 550 .... F2
*85282 Tempe 106,743 .... D5
86443 Temple Bar 84 .... A2
85552 Thatcher 3,374 .... F6
85353 Tolleson 4,433 .... C5
85638 Tombstone 1,632 .... F7
86044 Tonalea 125 .... E2
85354 Tonopah 54 .... B5
85553 Tonto Basin 250 .... D5
85639 Topawa 500 .... D7
86436 Topock 325 .... A4
85290 Tortilla Flat 37 .... D5
85640 Tubac 140 .... E7
86045 Tuba City 5,045 .... D2
*85701 Tucson⊙ 330,537 .... D6
Tucson‡ 531,263 .... D6
85640 Tumacacori 100 .... D7
85641 Vail 175 .... E6
86437 Valentine 120 .... B3
85291 Valley Farms 240 .... D6
85940 Vernon 75 .... F4
†85348 Vicksburg 16 .... B5
85355 Waddell 100 .... C5
85356 Wellton 911 .... A6
85357 Wenden 400 .... B5
85941 Whiteriver 2,256 .... E5
85321 Why 65 .... C6
85358 Wickenburg 3,535 .... C5
85360 Wikieup 150 .... B4
85643 Willcox 3,243 .... F6
86046 Williams 2,266 .... C3
86515 Window Rock 2,230 .... F3
85292 Winkelman 1,060 .... E6
†86001 Winona 25 .... D3
86047 Winslow 7,921 .... E3
†85322 Wintersburg 400 .... B5
85361 Wittmann 600 .... C5
85942 Woodruff 280 .... E4
85362 Yarnell 800 .... C4
†86301 Yava 40 .... C4
85554 Young 500 .... D4
85363 Youngtown 2,254 .... C5
86438 Yucca 250 .... A4
85364 Yuma⊙ 42,481 .... A6

OTHER FEATURES

Agassiz (peak) D3
Agua Fria (riv.) C5
Alamo (lake) B4
Apache (lake) D5
Aquarius (range) B4
Aravaipa (creek) E6
Aubrey (cliffs) B3
Baboquivari (mts.) D7
Baker Butte (mt.) D4
Balakai (mesa) F3
Baldy (peak) F5
Bartlett (dam) D5
Bartlett (res.) D5
Big Chino Wash (dry riv.) C3
Big Horn (mts.) B5
Big Sandy (riv.) B4
Bill Williams (riv.) B4
Black (mesa) E2
Black (mts.) A3
Black (riv.) E5
Blue (riv.) F5
Bouse Wash (dry riv.) A4
Buckskin (mts.) B4
Burro (creek) B4
Canyon (lake) D5
Canyon de Chelly Nat'l Mon. F2
Carrizo (creek) E4
Carrizo (mts.) G2
Casa Grande Ruins Nat'l Mon. D6
Castle Dome (mts.) A5
Cataract (creek) C3
Centennial Wash (dry riv.) B5
Cerbat (mts.) A3
Cherry (creek) E4
Chevelon (creek) E4
Chinle (creek) F2
Chinle (valley) F2
Chinle Wash (dry riv.) F2
Chino (valley) C3
Chiricahua (mts.) F6
Chiricahua Nat'l Mon. F6
Chocolate (mts.) A5
Clear (creek) E4
Coconino (plat.) C3
Cocopah Ind. Res. 355 .... A6
Colorado (riv.) A5
Colorado River Ind. Res. 6,640 .... A5
Coolidge (dam) E5
Copper (mts.) B6
Corn (creek) E3
Coronado Nat'l Memorial E7
Cottonwood (cliffs) B3
Cottonwood Wash (dry riv.) E4
Davis (dam) A3
Davis-Monthan A.F.B. 6,279 .... E6
Defiance (plat.) F3
Detrital Wash (dry riv.) A3
Diablo (canyon) D3
Dinnebito Wash (dry riv.) E3
Dot Klish (canyon) E2
Dragoon (mts.) F7
Eagle (creek) F5
East Verde (riv.) D4
Echo (cliffs) D2
Elden (mt.) D3
Fort Apache Ind. Res. 7,774 .... E5
Fort Bowie Nat'l Hist. Site F6
Fort Huachuca E7
Fort McDowell Ind. Res. 349 .... D5
Fort Mohave Ind. Res. 183 .... A4
Fort Pearce Wash (dry riv.) B2
Fossil (creek) D4
Four Peaks (mt.) D5
Galiuro (mts.) E6
Gila (mts.) A6
Gila (mts.) F5

Gila (riv.) B6
Gila Bend (mts.) B5
Gila Bend Ind. Res. 353 .... C6
Gila River Ind. Res. 7,445 .... C5
Glen Canyon (dam) D2
Glen Canyon Nat'l Rec. Area D1
Gothic (mesa) F2
Government (mt.) C3
Graham (mt.) F6
Grand Canyon Nat'l Park C2
Grand Wash (butte) B2
Grand Wash (riv.) B2
Greens (peak) F4
Growler (mts.) B6
Harcuvar (mts.) B5
Harquahala (mts.) B5
Hassayampa (riv.) C5
Havasu (lake) A4
Havasupai Ind. Res. 282 .... C2
Hohokam Pima Nat'l Mon. D5
Hoover (dam) A2
Hopi (buttes) E3
Hopi Ind. Res. 6,896 .... E2
Horseshoe (lake) D5
Huachuca (peak) E7
Hualapai (mts.) B4
Hualapai (peak) B3
Hualapai Ind. Res. 849 .... B3
Hubbell Trading Post Nat'l Hist.
Site F3
Humphreys (peak) D3
Hurricane (cliffs) B2
Imperial (res.) A6
Ives (mesa) E3
Juniper (mts.) C3
Kaibab (plat.) C2
Kaibab Ind. Res. 173 .... C2
Kaibito (plat.) D2
Kanab (creek) C2
Kanab (plat.) C2
Kellogg (mt.) E6
Kendrick (peak) D3
Kitt (peak) D7
Kofa (mts.) B5
Laguna (creek) E2
Laguna (res.) A6
Lake Mead Nat'l Rec. Area A2
Lechuguilla (des.) A6
Lemmon (mt.) E6
Little Colorado (riv.) D3
Lukachukai (mts.) F2
Luke A.F.B. 3,515 .... C5
Maple (peak) F5
Marble Canyon Nat'l Mon. D2
Maricopa (mts.) C5
Maricopa Ind. Res. 397 .... C6
Mazatzal (peak) D4
Mead (lake) A2
Meteor (crater) E3
Miller (peak) E7
Moencopi (plat.) D3
Moenkopi Wash (dry riv.) D2
Mogollon (plat.) D4
Mogollon Rim (cliffs) D4
Mohave (lake) A3
Mohave (mts.) A4
Mohawk (mts.) B6
Montezuma Castle Nat'l Mon. D4
Mormon (lake) D4
Mule (mts.) E7
Navajo (creek) D2
Navajo Ind. Res. 76,173 .... D2
Navajo Nat'l Mon. E2
Navajo Ord. Depot D3
O'Leary (peak) D3
Oraibi Wash (dry riv.) E3
Ord (mt.) D5
Organ Pipe Cactus Nat'l Mon. C6

Painted (des.) D2
Painted Desert Section (Petrified
Forest F3
Painted Rock (dam) C5
Papago Ind. Res. 7,171 .... C6
Paria (plat.) D2
Paria (riv.) D1
Parker (dam) A4
Pastora (peak) F2
Peloncillo (mts.) F6
Petrified Forest Nat'l Park F4
Pictograph (rocks) B5
Pinal (peak) E5
Pinaleno (mts.) F6
Pink (cliffs) E4
Pipe Spring Nat'l Mon. C2
Pleasant (lake) C5
Plomosa (mts.) A5
Polacca Wash (dry riv.) E3
Powell (lake) E1
Pueblo Colorado Wash (dry riv.) F3
Puerco (riv.) F3
Quajote Wash (dry riv.) D6
Rainbow (plat.) D2
Rincon (peak) E6
Roof Butte (mt.) F2
Rose (peak) F5
Sacramento Wash (dry riv.) A4
Saguaro (lake) D5
Saguaro Nat'l Mon. E6
Salt (riv.) D5
Salt River Ind. Res. 4,089 .... D5
San Carlos (lake) E5
San Carlos (riv.) E5
San Carlos Ind. Res. 6,104 .... E5
Sand Tank (mt.) C6
San Francisco (riv.) F5
San Pedro (riv.) E6
San Simon (riv.) F6
Santa Catalina (mts.) E6
Santa Cruz (riv.) D6
Santa Maria (riv.) B4
Santa Rosa Wash (dry riv.) D6
San Xavier Ind. Res. 875 .... D6
Sauceda (mts.) C6
Shivwits (plat.) B2
Shonto (plat.) E2
Sierra Ancha (mts.) D5
Sierra Apache (mts.) E5
Silver (creek) E4
Slate (mt.) D3
Sulphur Spring (valley) F6
Sunset Crater Nat'l Mon. D3
Superstition (mts.) D5
Theodore Roosevelt (lake) D5
Tonto (creek) D4
Tonto Nat'l Mon. D5
Trout (creek) B3
Trumbull (mt.) B2
Tumacacori Nat'l Mon. E7
Tuzigoot Nat'l Mon. D4
Tyson Wash (dry riv.) A5
Uinkaret (plat.) B2
Union (mt.) C4
Verde (riv.) D5
Vermilion (cliffs) D2
Virgin (mts.) B2
Walker (creek) F2
Walnut Canyon Nat'l Mon. D3
White (riv.) E5
Williams A.F.B. 3,435 .... D5
Woody (mt.) D3
Wupatki Nat'l Mon. D3
Yuma (des.) A6
Yuma Proving Ground 1,098 .... A6
Zuni (riv.) F4

⊙County seat.
‡Population of metropolitan area.
† Zip of nearest p.o. * Multiple zips.

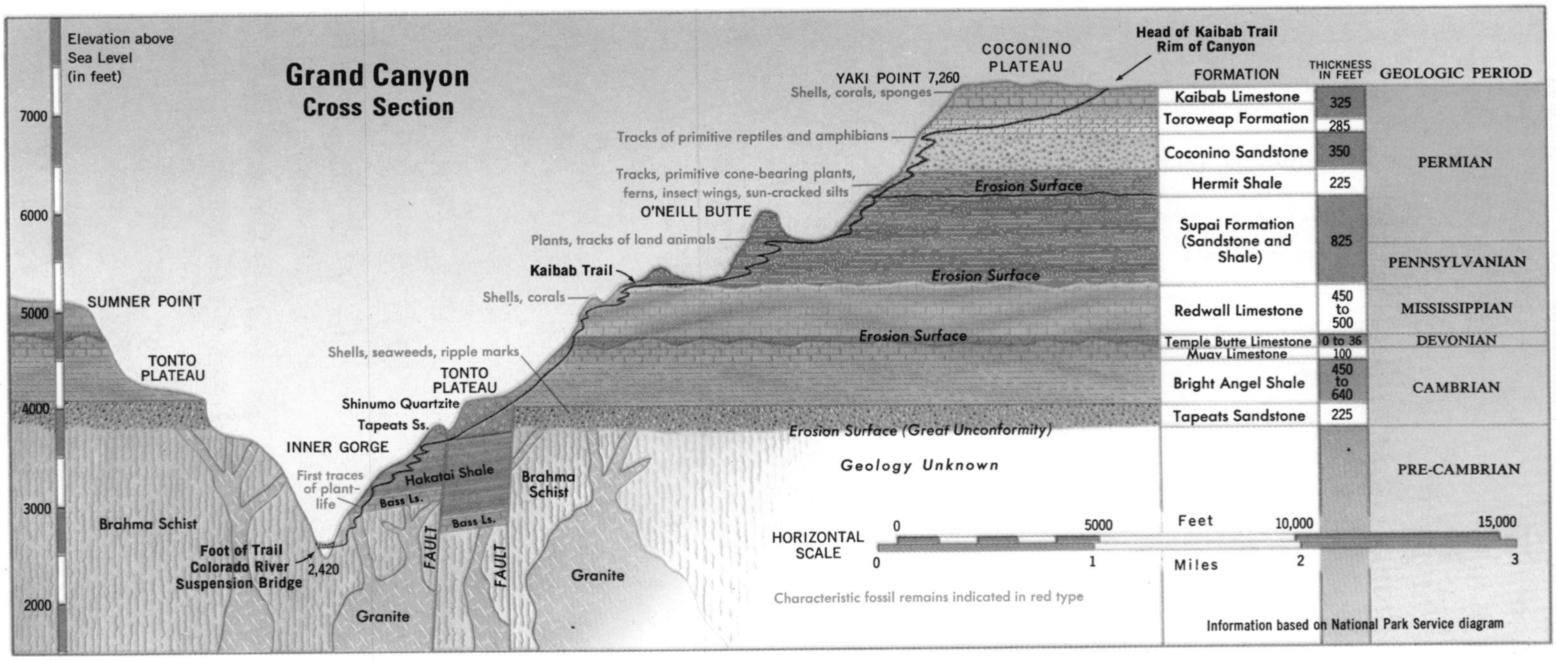
Grand Canyon
Cross Section
Elevation above
Sea Level
(in feet)
7000
6000
5000
4000
3000
2000
COCONINO
PLATEAU
Head of Kaibab Trail
Rim of Canyon
YAKI POINT 7,260
Shells, corals, sponges
Tracks of primitive reptiles and amphibians
Tracks, primitive cone-bearing plants,
ferns, insect wings, sun-cracked silts
O'NEILL BUTTE
Plants, tracks of land animals
Kaibab Trail
Shells, corals
Shells, seaweeds, ripple marks
SUMNER POINT
TONTO
PLATEAU
TONTO
PLATEAU
Shinumo Quartzite
Tapeats Ss.
INNER GORGE
First traces
of plant-
life
Hakatai Shale
Bass Ls.
Bass Ls.
Brahma
Schist
Brahma Schist
FAULT
FAULT
Foot of Trail
Colorado River
Suspension Bridge
2,420
Granite
Granite
Erosion Surface
Erosion Surface
Erosion Surface
Erosion Surface (Great Unconformity)
Geology Unknown
FORMATION
THICKNESS IN FEET
GEOLOGIC PERIOD
Kaibab Limestone 325
Toroweap Formation 285
Coconino Sandstone 350
Hermit Shale 225
PERMIAN
Supai Formation (Sandstone and Shale) 825
PENNSYLVANIAN
Redwall Limestone 450 to 500
MISSISSIPPIAN
Temple Butte Limestone 0 to 36
DEVONIAN
Muav Limestone 100
Bright Angel Shale 450 to 640
CAMBRIAN
Tapeats Sandstone 225
PRE-CAMBRIAN
HORIZONTAL
SCALE
Feet
0
5000
10,000
15,000
Miles
0
1
2
3
Characteristic fossil remains indicated in red type
Information based on National Park Service diagram

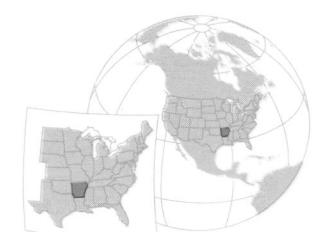

AREA 53,187 sq. mi. (137,754 sq. km.)
POPULATION 2,286,435
CAPITAL Little Rock
LARGEST CITY Little Rock
HIGHEST POINT Magazine Mtn. 2,753 ft. (839 m.)
SETTLED IN 1685
ADMITTED TO UNION June 15, 1836
POPULAR NAME Land of Opportunity
STATE FLOWER Apple Blossom
STATE BIRD Mockingbird

COUNTIES

Arkansas 24,175 ....H5
Ashley 26,538 ....G7
Baxter 27,409 ....F1
Benton 78,115 ....B1
Boone 26,067 ....D1
Bradley 13,803 ....F7
Calhoun 6,079 ....E6
Carroll 16,203 ....C1
Chicot 17,793 ....H7
Clark 23,326 ....D5
Clay 20,616 ....K1
Cleburne 16,909 ....F2
Cleveland 7,868 ....F6
Columbia 26,644 ....D7
Conway 19,505 ....E3
Craighead 63,239 ....J2
Crawford 36,892 ....B2
Crittenden 49,499 ....K3
Cross 20,434 ....J3
Dallas 10,515 ....E6
Desha 19,760 ....H6
Drew 17,910 ....G6
Faulkner 46,192 ....F3
Franklin 14,705 ....C2
Fulton 9,975 ....G1
Garland 70,531 ....D4
Grant 13,008 ....F5
Greene 30,744 ....J1
Hempstead 23,635 ....C6
Hot Spring 26,819 ....E5
Howard 13,459 ....C5
Independence 30,147 ....G2
Izard 10,768 ....G1
Jackson 21,646 ....H2
Jefferson 90,718 ....G5
Johnson 17,423 ....C2
Lafayette 10,213 ....C7
Lawrence 18,447 ....H1
Lee 15,539 ....J4
Lincoln 13,369 ....G6
Little River 13,952 ....B6
Logan 20,144 ....C3
Lonoke 34,518 ....G4
Madison 11,373 ....C1
Marion 11,334 ....E1
Miller 37,766 ....C7
Mississippi 59,517 ....K2
Monroe 14,052 ....H4
Montgomery 7,771 ....C4
Nevada 11,097 ....D6
Newton 7,756 ....D2
Ouachita 30,541 ....E6
Perry 7,266 ....E4
Phillips 34,772 ....J5
Pike 10,373 ....C5
Poinsett 27,032 ....J2
Polk 17,007 ....B5
Pope 39,021 ....D3
Prairie 10,140 ....G4
Pulaski 340,613 ....F4
Randolph 16,834 ....H1
Saint Francis 30,858 ....J3
Saline 53,161 ....E4
Scott 9,685 ....B4
Searcy 8,847 ....E2
Sebastian 95,172 ....B3
Sevier 14,060 ....B6
Sharp 14,607 ....G1
Stone 9,022 ....F2
Union 48,573 ....E7
Van Buren 13,357 ....E2
Washington 100,494 ....B2
White 50,835 ....G3
Woodruff 11,222 ....H3
Yell 17,026 ....D3

CITIES and TOWNS

Zip Name/Pop.   Key

72001 Adona 230 ....E3
72002 Alexander 223 ....F4
72410 Alicia 246 ....H2
72820 Alix 225 ....C3
†72004 Allport 295 ....G4
72921 Alma 2,755 ....B3
72003 Almyra 294 ....H5
72611 Alpena 344 ....D1
72004 Altheimer 1,231 ....G5
72821 Altus 441 ....C3
72005 Amagon 126 ....H2
71921 Amity 859 ....D5
71922 Antoine 194 ....D5
71923 Arkadelphia⊙ 10,005 ....D5
71630 Arkansas City⊙ 668 ......H6
72310 Armorel 500 ....L2
71822 Ashdown⊙ 4,218 ....B6
72513 Ash Flat⊙ 524 ....G1
72823 Atkins 3,002 ....E3
72311 Aubrey 267 ....J4
72006 Augusta⊙ 3,496 ....H3
72007 Austin 269 ....G4
72711 Avoca 256 ....B1
72010 Bald Knob 2,756 ....G3
71631 Banks 216 ....F6
72922 Barber 35 ....B3
72923 Barling 3,761 ....B3
72313 Bassett 243 ....K2
72924 Bates B4
72501 Batesville⊙ 8,263 ....G2
72411 Bay 1,605 ....J2
71720 Bearden 1,191 ....E6
72613 Beaver C1
72012 Beebe 3,599 ....G3
72014 Beedeville 183 ....H3
†72712 Bella Vista 2,589 ......B1
†72601 Bellefonte 393 ....D1
72824 Belleville 571 ....D3
71823 Ben Lomond 155 ....B6
72015 Benton⊙ 17,717 ....E4
72712 Bentonville⊙ 8,756 .....B1
72615 Bergman 320 ....E1
72616 Berryville⊙ 2,966 ......C1
†72764 Bethel Heights 296 ....B1
72016 Bigelow 373 ....E3
72617 Big Flat 150 ....F1
72413 Biggers 363 ....H1
72017 Biscoe 486 ....H4
72414 Black Oak 309 ....K2
72415 Black Rock 848 ....H1
†71960 Black Springs 92 ....C5
71825 Blevins 314 ....C6
65611 Blue Eye 43 ....D1
72826 Blue Mountain 112 ......C3
71722 Bluff City 292 ....D6
72315 Blytheville⊙ 23,844 ....L2
†71858 Bodcaw 197 ....D6
†72901 Bonanza 553 ....B3
72416 Bono 967 ....J2
72927 Booneville⊙ 3,718 ......C3
72020 Bradford 950 ....G3
71826 Bradley 790 ....C7
72928 Branch 353 ....C3
72021 Brinkley 4,909 ....H4
72417 Brookland 840 ....J2
72022 Bryant 2,682 ....F4
71827 Buckner 436 ....D7
72619 Bull Shoals 1,312 ......E1
72321 Burdette 328 ....L2
72023 Cabot 4,806 ....F4
72322 Caldwell 283 ....J3
71828 Cale 110 ....D6
72519 Calico Rock 1,046 ......F1
71724 Calion 638 ....E7
71701 Camden⊙ 15,356 ....E6
†72201 Cammack Village 920 ....E4
†72473 Campbell Station 297 ...H2
72419 Caraway 1,165 ....K2
72024 Carlisle 2,567 ....G4
71725 Carthage 568 ....E5
72025 Casa 179 ....D3
72421 Cash 285 ....J2
72026 Casscoe 297 ....H4
72521 Cave City 1,634 ....G2
72718 Cave Springs 429 ....B1
72932 Cedarville 375 ....B2
72719 Centerton 425 ....B1
72829 Centerville 300 ....D3
72933 Charleston⊙ 1,748 ......B3
†72525 Cherokee Village-Hidden
       Valley 4,058 ....G1
72324 Cherry Valley 729 ......J3
72934 Chester 139 ....B2
71726 Chidester 342 ....D6
72029 Clarendon⊙ 2,361 ....H4
72325 Clarkedale 300 ....K3
72830 Clarksville⊙ 5,237 .....D3
72031 Clinton⊙ 1,284 ....F2
72832 Coal Hill 859 ....C3
72476 College City 432 ....J1
72326 Colt 378 ....J3
71831 Columbus 265 ....C6
72523 Concord 234 ....G2
72032 Conway⊙ 20,375 ....F3
72524 Cord 250 ....H2
72422 Corning⊙ 3,650 ....J1
72626 Cotter 920 ....E1
72036 Cotton Plant 1,323 .....H3
71937 Cove 391 ....B5
72037 Coy 183 ....G4
72327 Crawfordsville 685 .....K3
71635 Crossett 6,706 ....G7
71728 Curtis 300 ....D6
72526 Cushman 556 ....G2
†71950 Daisy 177 ....C5
72039 Damascus 307 ....F3
72833 Danville⊙ 1,698 ....D3
72834 Dardanelle⊙ 3,621 ......D3
72424 Datto 112 ....J1
72722 Decatur 1,013 ....A1
72425 Delaplaine 161 ....J1
71940 Delight 431 ....C5
72426 Dell 310 ....K2
†72821 Denning 238 ....C3
71832 De Queen⊙ 4,594 ....B5
71638 Dermott 4,731 ....H7
72040 Des Arc⊙ 2,001 ....G4
72041 De Valls Bluff⊙ 738 ....H4
72042 De Witt⊙ 3,928 ....H5
72644 Diamond City 650 ....E1
72043 Diaz 1,192 ....H2
71833 Dierks 1,249 ....B5
71941 Donaldson 300 ....E5
72837 Dover 948 ....D3
71639 Dumas 6,091 ....H6
72935 Dyer 608 ....B3
72330 Dyess 446 ....K2
72331 Earle 3,517 ....K3
71701 East Camden 632 ....E6
72332 Edmondson 344 ....K3
72333 Elaine 991 ....J5
71730 El Dorado⊙ 25,270 .....E7
72727 Elkins 579 ....C1
72728 Elm Springs 781 ....B1
71740 Emerson 444 ....D7
71835 Emmet 475 ....D6
72046 England 3,081 ....G4
72047 Enola 186 ....F3
71640 Eudora 3,840 ....H7
72632 Eureka Springs⊙ 1,989 .C1
72532 Evening Shade 397 .....G1
72633 Everton 134 ....E1
72730 Farmington 1,283 ......B1
72701 Fayetteville⊙ 36,608 ..B1
      Fayetteville-Springdale
      07 ....B1
†71747 Felsenthal 220 ....F7
72429 Fisher 302 ....J2
71742 Flippin 1,072 ....E1
71742 Fordyce⊙ 5,175 ....F6
71836 Foreman 1,377 ....B6
72335 Forrest City⊙ 13,803 ..J3
*72901 Fort Smith⊙ 71,626 ....B3
      Fort Smith‡ 203,269 ...B3
71837 Fouke 614 ....C7
71642 Fountain Hill 352 .....G7
†72016 Fourche 51 ....E4
72536 Franklin 253 ....G1
72017 Fredonia (Biscoe) 486 .H4
71942 Friendship 163 ....E5
71838 Fulton 326 ....C6
72732 Garfield 187 ....C1
71839 Garland 660 ....C7
72052 Garner 216 ....G3
72635 Gassville 859 ....F1
72733 Gateway 75 ....B1
71840 Genoa 350 ....C7
72734 Gentry 1,468 ....A1
72636 Gilbert 43 ....E2
72055 Gillett 927 ....H5
71841 Gillham 252 ....B5
72339 Gilmore 503 ....K3
71943 Glenwood 1,402 ....C5
72340 Goodwin 225 ....J4
†72315 Gosnell 3,215 ....K2
71643 Gould 1,671 ....G6
71644 Grady 488 ....G5
72838 Gravelly 300 ....C4
72736 Gravette 1,218 ....B1
72058 Greenbrier 1,423 ......F3
72638 Green Forest 1,609 ....D1
72737 Greenland 622 ....B1
72430 Greenway 317 ....K1
72936 Greenwood⊙ 3,317 ......B3
†72067 Greers Ferry 558 ......F2
72060 Griffithville 254 .....G3
72431 Grubbs 546 ....H2
72540 Guion 177 ....G2
†71923 Gum Springs 255 ....D5
71743 Gurdon 2,707 ....D6
72061 Guy 209 ....F3
72937 Hackett 505 ....B3
†71638 Halley H6
71646 Hamburg⊙ 3,394 ....G7
71744 Hampton⊙ 1,627 ....F6
72542 Hardy 643 ....H1
71745 Harrell 302 ....F7
72432 Harrisburg⊙ 1,921 .....J2
72601 Harrison⊙ 9,567 ....D1
72938 Hartford 613 ....B3
72840 Hartman 517 ....C3
†72015 Haskell 1,074 ....E4
71945 Hatfield 410 ....B5
72842 Havana 352 ....D3
72341 Haynes 359 ....J4
72064 Hazen 1,636 ....G4
72543 Heber Springs⊙ 4,589 ..G2
72843 Hector 449 ....E3
72342 Helena⊙ 9,598 ....J4
72065 Hensley 500 ....F4
71647 Hermitage 378 ....F7
72347 Hickory Ridge 478 .....J3
72068 Higden 45 ....F2
72068 Higginson 333 ....G3
†72734 Highfill 92 ....B1
72738 Hindsville C1
72069 Holly Grove 754 ....H4
†72958 Hon 250 ....B4
71801 Hope⊙ 10,290 ....C6
71842 Horatio 989 ....B3
72512 Horseshoe Bend 1,909 ..G1
71901 Hot Springs National
      Park⊙ 35,781 ....D4
72070 Houston 183 ....E3

(continued on following page)

Agriculture, Industry and Resources

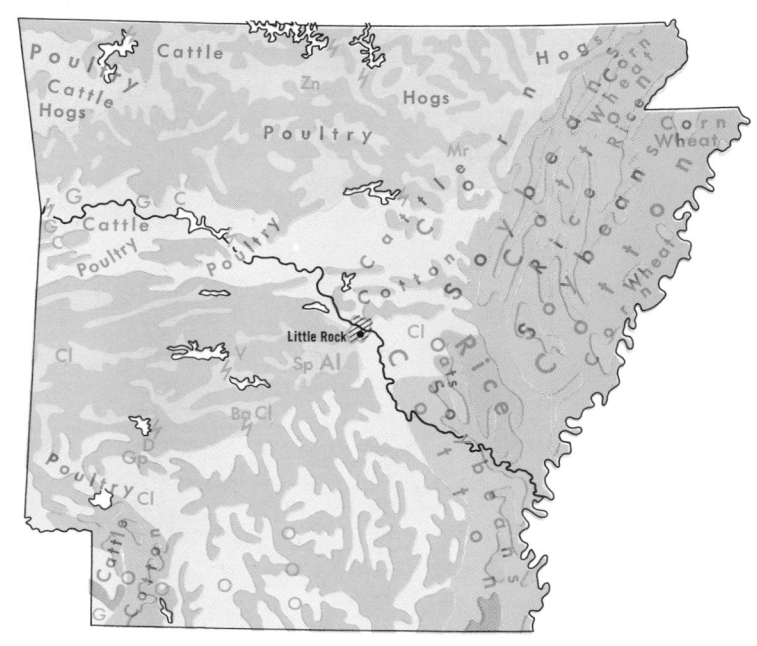

Little Rock

DOMINANT LAND USE

Fruit and Mixed Farming

Specialized Cotton

Cotton, General Farming

Rice, General Farming

General Farming, Livestock, Truck Farming, Cotton

Forests

Swampland, Limited Agriculture

MAJOR MINERAL OCCURRENCES

Al	Bauxite	Gp	Gypsum
Ba	Barite	Mr	Marble
C	Coal	O	Petroleum
Cl	Clay	Sp	Soapstone
D	Diamonds	V	Vanadium
G	Natural Gas	Zn	Zinc
	Water Power	▨	Major Industrial Areas

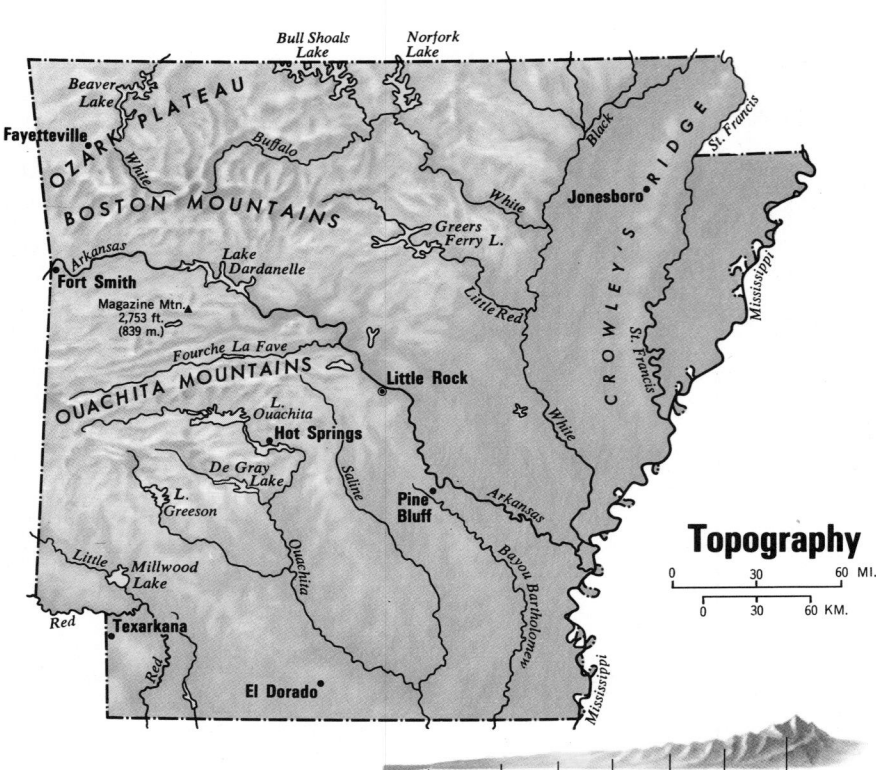

Topography

0 30 60 MI.

0 30 60 KM.

Below Sea Level | 100 m. 328 ft. | 200 m. 656 ft. | 500 m. 1,640 ft. | 1,000 m. 3,281 ft. | 2,000 m. 6,562 ft. | 5,000 m. 16,404 ft.

71764 Stephens 1,366 ....E7
72159 Steprock 600 ....G3
72469 Strawberry 280 ....H2
71765 Strong 785 ....F7
72160 Stuttgart⊙ 10,941 ......H4
72865 Subiaco 744 ....C3
72470 Success 223 ....J1
72579 Sulphur Rock 316 ......H2
72768 Sulphur Springs 496 ..B1
72677 Summit 506 ....E1
72471 Swifton 859 ....H2
72351 Taylor 657 ....D7
71861 Taylor 657 ....D7
75502 Texarkana⊙ 21,459 ....C7
Texarkana‡ 127,019 .....C7
71766 Thornton 711 ....F6
72166 Tichnor 350 ....H5
71670 Tillar 280 ....H6
71767 Tinsman 112 ....F6
71851 Tollette 407 ....C6

72770 Tontitown 615 ....B1
72167 Traskwood 459 ....E5
72472 Trumann 6,405 ....J2
72168 Tucker 375 ....G5
72473 Tuckerman 2,078 ......H2
†72015 Tull 281 ....E5
72169 Tupelo 248 ....H3
72384 Turrell 1,041 ....K3
72386 Tyronza 777 ....K3
72170 Ulm 201 ....H4
72955 Uniontown 600 ....B3
71768 Urbana 500 ....E7
72682 Valley Springs 190 ....C1
72956 Van Buren⊙ 12,020 ...B3
71972 Vandervoort 98 ....B5
72370 Victoria 175 ....K2
72173 Vilonia 736 ....F3
†72002 Vimy Ridge 600 ......F4
72583 Viola 362 ....G1

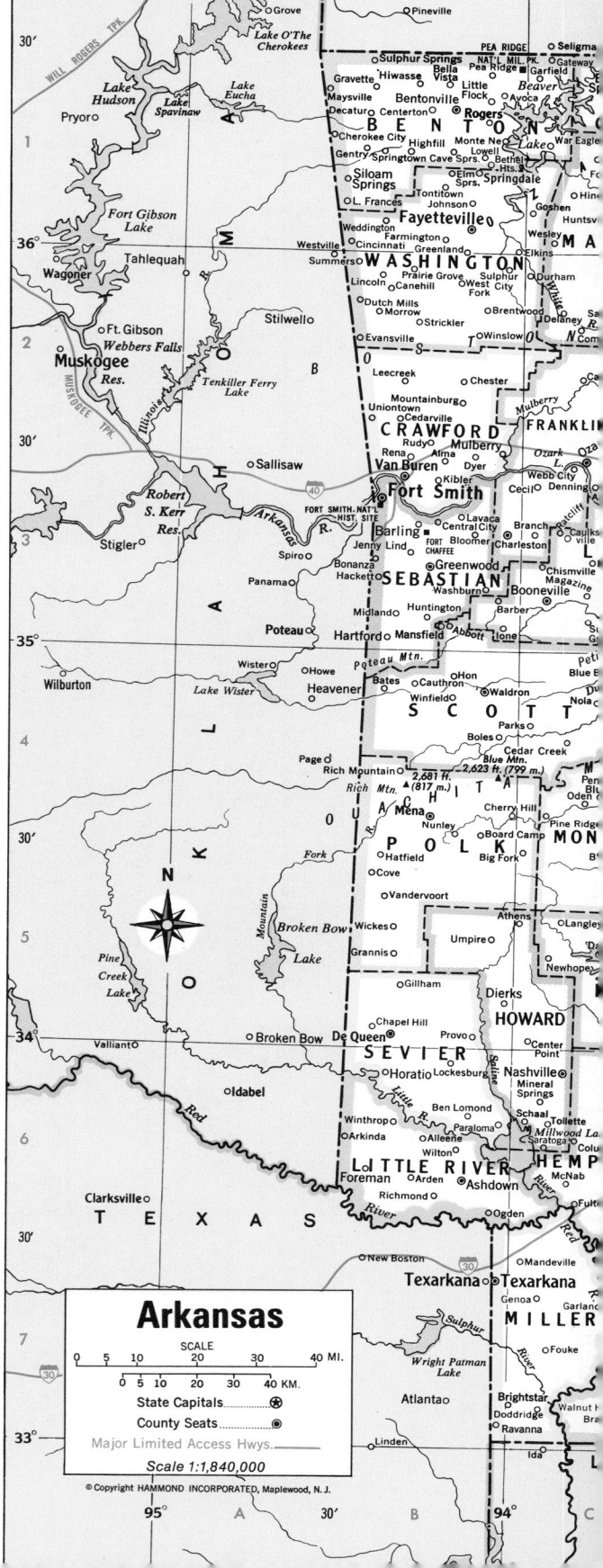

Arkansas

SCALE
0 5 10 20 30 40 MI.

0 5 10 20 30 40 KM.

State Capitals ....⊛
County Seats ....⊙
Major Limited Access Hwys.
Scale 1:1,840,000

© Copyright HAMMOND INCORPORATED, Maplewood, N.J.

72433 Hoxie 2,961 ....H1
72348 Hughes 1,919 ....J4
72072 Humnoke 442 ....G4
72073 Humphrey 872 ....G5
72074 Hunter 170 ....H3
72940 Huntington 662 ....B3
72740 Huntsville⊙ 1,394 .....C1
71747 Huttig 976 ....F7
72434 Imboden 661 ....H1
72075 Jacksonport 288 ....H2
72076 Jacksonville 27,589 ...F4
†72501 Jamestown G2
72641 Jasper⊙ 519 ....D1
72079 Jefferson 250 ....F5
71650 Jerome 54 ....G7
72080 Jerusalem 300 ....E3
71949 Jessieville 350 ....D4
72741 Johnson 519 ....B1
72350 Joiner 725 ....K3
72401 Jonesboro⊙ 31,530 ...J2
72081 Judsonia 2,025 ....G3
71749 Junction City 813 .....E7
72351 Keiser 962 ....K2
72082 Kensett 1,751 ....G3
72083 Keo 208 ....G4
†72956 Kibler 798 ....B3
71652 Kingsland 320 ....F6
71950 Kirby 800 ....C5
72435 Knobel 503 ....J1
72845 Knoxville 264 ....D3
72436 Lafe 215 ....J1
72437 Lake City⊙ 1,842 .....K2
72642 Lakeview 512 ....E1
†72389 Lake View 609 ....J5
71653 Lake Village⊙ 3,088 .H7
72846 Lamar 708 ....D3
72941 Lavaca 1,092 ....B3
71750 Lawson 250 ....F7
72438 Leachville 1,882 ......K2
72644 Lead Hill 247 ....D1
72084 Leola 481 ....E5
72354 Lepanto 1,964 ....K2
72645 Leslie 501 ....E2
72085 Letona 231 ....G3
71845 Lewisville⊙ 1,476 ....C7
72355 Lexa 500 ....J4
72744 Lincoln 1,422 ....B2
†72712 Little Flock 663 ......B1
*72201 Little Rock
(cap.)⊙ 158,461 ....F4
Little Rock-North Little
Rock‡ 393,494 ....F4
71846 Lockesburg 616 ....B6
72847 London 859 ....D3
72086 Lonoke⊙ 4,128 ....G4
72087 Lonsdale 117 ....E4
71751 Louann 282 ....E7
72745 Lowell 1,078 ....B1
†72856 Lurton 38 ....D2
72358 Luxora 1,739 ....K2
72440 Lynn 345 ....J4
72359 Madison 1,238 ....J4
72943 Magazine 799 ....C3
72553 Magness 196 ....H2
71753 Magnolia⊙ 11,909 ....D7
72104 Malvern⊙ 10,163 .....E5
72554 Mammoth Spring 1,158 .G1
72442 Manila 2,553 ....K2
72360 Marianna⊙ 6,220 .....J4
†72395 Marie 287 ....K2
72364 Marion⊙ 2,996 ....K3

72365 Marked Tree 3,201 ...K2
72443 Marmaduke 1,168 ...K1
72650 Marshall⊙ 1,595 .....E2
72366 Marvell 1,724 ....J4
72106 Mayflower 1,381 ......F4
72444 Maynard 381 ....J1
71847 McCaskill 87 ....C6
72101 McCrory 1,942 ....H3
72441 McDougal 239 ....K1
71654 McGehee 5,671 ....H6
71752 McNeil 725 ....D7
72102 McRae 641 ....G3
72556 Melbourne⊙ 1,619 ...G1
72367 Mellwood 250 ....H5
71953 Mena⊙ 5,154 ....B4
72107 Menifee 368 ....E3
72945 Midland 286 ....B3
71851 Mineral Springs 936 ..C6
72445 Minturn 169 ....H2
†71639 Mitchellville 618 .....H6
72447 Monette 1,165 ....K1
72108 Monroe 250 ....H4
71655 Monticello⊙ 8,259 ...G6
71658 Montrose 641 ....H7
†72501 Moorefield 129 ....G2
72368 Moro 327 ....H4
72110 Morrilton⊙ 7,355 .....E3
71659 Moscow 325 ....G5
72946 Mountainburg 595 ...B2
72653 Mountain Home⊙ 8,066 .F1
71956 Mountain Pine 1,068 .D4
72560 Mountain View⊙ 2,147 .F2
71758 Mount Holly 250 ......E7
72561 Mount Ida⊙ 1,023 ....C4
72111 Mount Pleasant 438 ..G2
72947 Mulberry 1,444 ....B2
71958 Murfreesboro 1,883 ..C5
71852 Nashville⊙ 4,554 .....C5
72562 Newark 1,128 ....H2
72851 New Blaine 200 ....D3
71959 Newhope 300 ....C5
72112 Newport⊙ 8,339 ......H2
72461 Nimmons 112 ....K1
†71601 Noble Lake 250 ......G5
72658 Norfork 399 ....F1
71960 Norman 539 ....C4
71759 Norphlet 756 ....E7
†72801 Norristown 625 ......D3
71635 North Crossett 3,513 .G7
*72114 North Little Rock 64,288 .F4
72660 Oak Grove 265 ....C1
†71801 Oakhaven 72 ....C6
71961 Oden 186 ....C4
71853 Ogden 334 ....B6
72564 Oil Trough 280 ....G2
72449 O'Kean 291 ....J1
71962 Okolona 200 ....D5
72853 Ola 1,121 ....D3
72662 Omaha 191 ....D1
†72110 Oppelo 486 ....E3
72370 Osceola⊙ 8,881 ......K2
72565 Oxford 520 ....G1
71855 Ozan 111 ....C6
72949 Ozark⊙ 3,597 ....C3
72372 Paiestine 976 ....J4
72121 Paragon 673 ....G3
72450 Paragould⊙ 15,248 ..J1
72855 Paris⊙ 3,991 ....C3
71661 Parkdale 471 ....H7
72373 Parkin 2,035 ....J3
72950 Parks 600 ....B4

†71801 Patmos 88 ....C7
72123 Patterson 567 ....H3
72453 Peach Orchard 243 ...J1
71964 Pearcy 400 ....D5
72751 Pea Ridge 1,488 ......B1
†72104 Perla 149 ....E5
72125 Perry 254 ....E3
71801 Perrytown 282 ....C6
72126 Perryville⊙ 1,058 .....E3
72454 Piggott⊙ 3,762 ....K1
*71601 Pine Bluff⊙ 56,636 ..F5
Pine Bluff‡ 90,718 .....F5
†72847 Piney 2,283 ....D3
72857 Plainview 752 ....D4
72568 Pleasant Plains 267 ..G2
72127 Plumerville 785 ....E3
72455 Pocahontas⊙ 5,995 ..H1
72456 Pollard 298 ....K1
72374 Poplar Grove 300 .....J4
72457 Portia 480 ....H1
71663 Portland 701 ....H7
72258 Pottsville 564 ....D3
72458 Powhatan 49 ....H1
72128 Poyen 329 ....E5
72753 Prairie Grove 1,708 ..B2
72129 Prattsville 317 ....F5
71857 Prescott⊙ 4,103 ......D6
72672 Pyatt 217 ....E1
72131 Quitman 556 ....F3
72951 Ratcliff 197 ....C3
72333 Ratio 250 ....J5
72459 Ravenden 338 ....H1
72460 Ravenden Springs 230 .H1
71726 Reader 127 ....D6
72132 Redfield 745 ....F5
71670 Reed 395 ....H6
72462 Reyno 521 ....J1
71665 Rison⊙ 1,325 ....F6
†72104 Rockport 231 ....E5
72134 Roe 136 ....H4
72756 Rogers 17,429 ....B1
†72355 Rondo 330 ....J4
72137 Rose Bud 202 ....F3
71858 Rosston 274 ....D6
72952 Rudy 79 ....B2
72139 Russell 232 ....G3
72801 Russellville⊙ 14,031 .D3
72140 Saint Charles 199 ....H5
72464 Saint Francis 266 ....K1
72760 Saint Paul 198 ....C2
72576 Salem⊙ 1,424 ....G1
†72658 Salesville 406 ....F1
72863 Scranton 244 ....C3
72143 Searcy⊙ 13,612 ......G3
72465 Sedgwick 205 ....J2
72103 Shannon Hills 1,656 .F4
72150 Sheridan⊙ 3,042 .....F5
72152 Sherrill 161 ....G5
72116 Sherwood 10,406 ....F4
72153 Shirley 354 ....F2
72577 Sidney 270 ....G1
72761 Siloam Springs 7,940 .B1
71762 Smackover 2,453 .....E7
72466 Smithville 113 ....H1
†71658 Snyder 700 ....G7
71763 Sparkman 622 ....E6
72764 Springdale 23,458 ....B1
Springdale-Fayetteville‡
177,850 ....B1
71860 Stamps 2,859 ....D7
71667 Star City⊙ 2,066 .....G6

72389 Wabash 300 ....J5
72175 Wabbaseka 428 ....G5
72475 Waidenburg 124 ....J2
71770 Waldo 1,685 ....D7
72958 Waldron⊙ 2,642 ....B4
72476 Walnut Ridge⊙ 4,152 ..J1
72176 Ward 981 ....F3
71671 Warren⊙ 7,646 ....F6
71862 Washington 265 ....C6
71674 Watson 433 ....H6
72479 Weiner 750 ....J2
72177 Weldon 161 ....H3
†71635 West Crossett 1,466 ..F7
72685 Western Grove 378 .....D1
72774 West Fork 1,526 ....B2
72390 West Helena 11,367 ...J4
72301 West Memphis 28,138 .K3
72178 West Point 226 ....G3
72391 West Ridge 300 ....K2

72392 Wheatley 523 ....H4
72482 Wiseman 327 ....G1
71602 White Hall 2,214 ....F5
71973 Wickes 464 ....B5
72394 Widener 316 ....J3
72482 Williford 169 ....H1
71675 Wilmar 747 ....G6
71676 Wilmot 1,227 ....G7
72395 Wilson 1,115 ....K2
71865 Wilton 495 ....B6
71677 Winchester 279 ....G6
72959 Winslow 247 ....B2
71866 Winthrop 238 ....B6
72587 Wiseman 327 ....G1
72180 Woodson 450 ....F4
72181 Wooster 398 ....F3
71983 Wrightsville-Tafton 1,434 ..K3
72396 Wynne⊙ 7,805 ....J3
72687 Yellville⊙ 1,044 ....E1

†72601 Zinc 113 ....E1

OTHER FEATURES

Arkansas (riv.) G5
Arkansas Post Nat'l Mem. ..H5
Bartholomew (bayou) G6
Bayou Bodcau (res.) C7
Bayou Des Arc (riv.) G3
Beaver (lake) C1
Black (riv.) H2
Blue Mountain (lake) C3
Blytheville A.F.B. K2
Boston (mts.) B2
Buffalo (riv.) E2
Bull Shoals (lake) E1
Cache (riv.) H3
Caddo (riv.) D5
Catherine (lake) E5

Chinkapin Knob (mt.) E2
Conway (lake) F3
Current (riv.) J1
Cypress (bayou) F3
Dardanelle (lake) D3
De Gray (lake) D5
Des Arc (bayou) G3
De View (bayou) J3
Erling (lake) C7
Fort Smith Nat'l Hist. Site ...B3
Fourche LaFave (riv.) D4
Greers Ferry (lake) G2
Greeson (lake) C5
Hamilton (lake) D5
Hot Springs Nat'l Park D4
Illinois (bayou) D3
L'Anguille (riv.) J3
La Grue (bayou) H5
Little (riv.) B6

Little Missouri (riv.) D6
Little Red (riv.) G3
Little Rock A.F.B. F4
Magazine (mt.) C3
Meto (bayou) H5
Millwood (lake) C6
Mississippi (riv.) H7
Mountain Fork (riv.) A5
Mulberry (riv.) C2
Nebo (mt.) D3
Nimrod (lake) D4
Norfork (lake) F1
Ouachita (lake) C4
Ouachita (mts.) B4
Ouachita (riv.) E7
Ozark (plat.) C1
Pea Ridge Nat'l Mil. Park B1
Peckerwood (lake) G4
Petit Jean (mt.) C3

Petit Jean (riv.) D3
Pine Bluff Arsenal F5
Poteau (mt.) B4
Red (riv.) C6
Reeves Knob (mt.) E2
Saint Francis (riv.) J4
Saline (riv.) B5
Saline (riv.) E5
Seven Devils (res.) G6
Spring (riv.) H1
Sulphur (riv.) B7
Table Rock (riv.) D1
Tyronza (riv.) K2
Wattensaw (bayou) G4
White (riv.) H5
White Oak (lake) D6
Winona (lake) E4

⊙County seat.
‡Population of metropolitan area.
† Zip of nearest p.o.   * Multiple zips.

COUNTIES

Alameda 1,105,379 ... D6
Alpine 1,097 ... F5
Amador 19,314 ... E5
Butte 143,851 ... D4
Calaveras 20,710 ... E5
Colusa 12,791 ... C4
Contra Costa 656,380 ... D6
Del Norte 18,217 ... B2
El Dorado 85,812 ... E5
Fresno 514,229 ... E7
Glenn 21,350 ... C4
Humboldt 108,514 ... B3
Imperial 92,110 ... K10
Inyo 17,895 ... H7
Kern 403,089 ... G8
Kings 73,738 ... G8
Lake 36,366 ... C4
Lassen 21,661 ... E3
Los Angeles 7,477,503 ... G9
Madera 63,116 ... F6
Marin 222,592 ... C5
Mariposa 11,108 ... E6
Mendocino 66,738 ... B4
Merced 134,558 ... E6
Modoc 8,610 ... E2
Mono 8,577 ... F5
Monterey 290,444 ... D7
Napa 99,199 ... C5
Nevada 51,645 ... E4
Orange 1,932,709 ... H10
Placer 117,247 ... E4
Plumas 17,340 ... E4
Riverside 663,199 ... J10
Sacramento 783,381 ... D5
San Benito 25,005 ... D7
San Bernardino 895,016 ... J9
San Diego 1,861,846 ... J10
San Francisco (city county) 678,974 ... J2
San Joaquin 347,342 ... D6
San Luis Obispo 155,435 ... E8
San Mateo 587,329 ... J3
Santa Barbara 298,694 ... E9
Santa Clara 1,295,071 ... D6
Santa Cruz 188,141 ... C6
Shasta 115,715 ... C3
Sierra 3,073 ... E4
Siskiyou 39,732 ... C2
Solano 235,203 ... D5
Sonoma 299,681 ... C5
Stanislaus 265,900 ... D6
Sutter 52,246 ... D4
Tehama 38,888 ... C3
Trinity 11,858 ... B3
Tulare 245,738 ... G7
Tuolumne 33,928 ... F5
Ventura 529,174 ... F9
Yolo 113,374 ... D5
Yuba 49,733 ... D4

CITIES and TOWNS

Zip *Name/Pop.* *Key*

94501 Alameda 63,852 ... J2
94507 Alamo 8,505 ... K2
94706 Albany 15,130 ... J2
*91801 Alhambra 64,615 ... C10
92001 Alpine 5,368 ... J11
91001 Altadena 40,983 ... C10
96101 Alturas⊙ 3,025 ... E2
†95116 Alum Rock 16,890 ... L3
*92801 Anaheim 219,494 ... D11
Anaheim-Santa Ana-Garden Grove‡ 1,931,570 ... D11
96007 Anderson 7,381 ... C3
95222 Angels Camp 2,302 ... E5
94508 Angwin 3,526 ... C5
94509 Antioch 42,683 ... L1
92307 Apple Valley 14,305 ... H9
95003 Aptos 7,039 ... K4
91006 Arcadia 45,994 ... C10
95521 Arcata 12,850 ... A3
95825 Arden-Arcade 87,570 ... B8
93420 Arroyo Grande 11,290 ... E8
90701 Artesia 14,301 ... C11
93203 Arvin 6,863 ... G8
†94577 Ashland 13,983 ... K2
95413 Asti 75 ... C5
93422 Atascadero 16,232 ... E8
94025 Atherton 7,797 ... K3
95301 Atwater 17,530 ... E6
95603 Auburn⊙ 7,540 ... C8
90704 Avalon 2,022 ... G10
93204 Avenal 4,137 ... E8
91702 Azusa 29,380 ... D10
*93301 Bakersfield⊙ 105,735 ... G8
Bakersfield‡ 403,089 ... G8
91706 Baldwin Park 50,554 ... D10
92220 Banning 14,020 ... J10
92311 Barstow 17,690 ... H9
†93402 Baywood Park-Los Osos 10,933 ... E8
92223 Beaumont 6,818 ... J10
90201 Bell 25,450 ... C11
90706 Bellflower 53,441 ... C11
90201 Bell Gardens 34,117 ... C11
94002 Belmont 24,505 ... J3
94510 Benicia 15,376 ... K1
95005 Ben Lomond 7,238 ... K4
*94701 Berkeley 103,328 ... J2
*90210 Beverly Hills 32,367 ... B10
92315 Big Bear Lake ... J9
93920 Big Sur 500 ... D7
93514 Bishop 3,333 ... G6
92316 Bloomington 18,888 ... E10
92225 Blythe 6,805 ... L10
94923 Bodega Bay 800 ... B5
93516 Boron 2,040 ... H8
92004 Borrego Springs 1,405 ... J10
95006 Boulder Creek 5,662 ... J4
92227 Brawley 14,946 ... K11
92621 Brea 27,913 ... D11
94513 Brentwood 4,434 ... L2
93517 Bridgeport⊙ 525 ... F5
94005 Brisbane 2,969 ... J2
95605 Broderick-Bryte 10,194 ... B8
*90622 Buena Park 64,165 ... D11
*91501 Burbank 84,625 ... C10
94010 Burlingame 26,173 ... J2
96013 Burney 3,187 ... D3
92231 Calexico 14,412 ... K11
93505 California City 2,743 ... H8
94515 Calistoga 3,879 ... C5
93745 Calwa 6,640 ... F7
93010 Camarillo 37,797 ... F9
95008 Campbell 26,910 ... K3
*91303 Canoga Park ... B10
92624 Capistrano Beach 6,168 ... H10
95010 Capitola 9,095 ... K4
92007 Cardiff-by-the-Sea 10,054 ... H10
92008 Carlsbad 35,490 ... H10
93923 Carmel 4,707 ... D7
93924 Carmel Valley 4,013 ... D7
95608 Carmichael 43,108 ... C8
93013 Carpinteria 10,835 ... F9
90745 Carson 81,221 ... C11
94546 Castro Valley 44,011 ... K2
95012 Castroville 4,396 ... D7
92234 Cathedral City 4,130 ... J10
96019 Central Valley 3,424 ... C3
95307 Ceres 13,281 ... D6
†90701 Cerritos 53,020 ... C11
†94541 Cherryland 9,425 ... K2
95926 Chico 26,603 ... D4
Chico‡ 143,851 ... D4
†93555 China Lake 4,275 ... H8
95309 Chinese Camp 150 ... E6
91710 Chino 40,165 ... D10
93610 Chowchilla 5,122 ... E6
*92010 Chula Vista 83,927 ... J11
95610 Citrus Heights 85,911 ... C8
91711 Claremont 30,950 ... D10
95425 Cloverdale 3,989 ... B5
93612 Clovis 33,021 ... F7
92236 Coachella 9,129 ... J10
93210 Coalinga 6,593 ... E7
95713 Colfax 981 ... E4
92324 Colton 15,201 ... E10
95932 Colusa⊙ 4,075 ... C4
90040 Commerce 10,509 ... C10
*90220 Compton 81,286 ... C11
*94520 Concord 103,255 ... K1
93212 Corcoran 6,454 ... F7
96021 Corning 4,745 ... C4
91720 Corona 37,791 ... E11
92118 Coronado 18,790 ... H11
94925 Corte Madera 8,074 ... J2
*92626 Costa Mesa 82,562 ... D11
94928 Cotati 3,346 ... C5
*91722 Covina 33,751 ... D10
95531 Crescent City⊙ 3,075 ... A2
92325 Crestline 6,715 ... H9
90201 Cudahy 17,984 ... C11
90230 Culver City 38,139 ... B10
95014 Cupertino 34,265 ... K3
93615 Cutler 3,149 ... F7
90630 Cypress 40,391 ... D11
*94014 Daly City 78,519 ... H2
92629 Dana Point 10,602 ... H10
94526 Danville 26,446 ... K2
95616 Davis 36,640 ... B8
93215 Delano 16,491 ... F8
95315 Delhi 2,832 ... E6
92014 Del Mar 5,017 ... H11
92240 Desert Hot Springs 5,941 ... J9
93618 Dinuba 9,907 ... F7
95620 Dixon 7,541 ... B9
93620 Dos Palos 3,121 ... E6
*90240 Downey 82,602 ... C11
95936 Downieville⊙ 500 ... E4

AREA 158,706 sq. mi. (411,049 sq. km.)
POPULATION 23,667,565
CAPITAL Sacramento
LARGEST CITY Los Angeles
HIGHEST POINT Mt. Whitney 14,494 ft. (4418 m.)
SETTLED IN 1769
ADMITTED TO UNION September 9, 1850
POPULAR NAME Golden State
STATE FLOWER Golden Poppy
STATE BIRD California Valley Quail

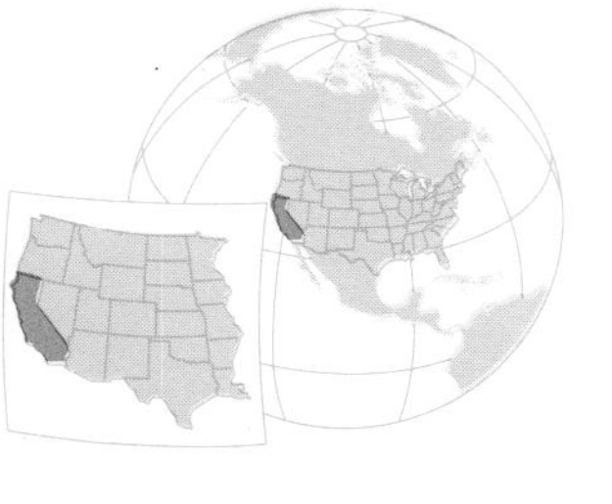

91010 Duarte 16,766 ... D10
94566 Dublin 13,496 ... K2
93219 Earlimart 4,578 ... F8
90022 East Los Angeles 100,017 ... C10
*92020 El Cajon 73,892 ... J11
92243 El Centro⊙ 23,996 ... K11
94530 El Cerrito 22,731 ... J2
95630 El Dorado Hills 3,453 ... C8
94018 El Granada 3,582 ... H3
95624 Elk Grove 10,959 ... B9
*91731 El Monte 79,494 ... D10
†93030 El Rio 5,674 ... F9
90245 El Segundo 13,752 ... B11
92630 El Toro 38,153 ... E11
94608 Emeryville 3,714 ... J2
92024 Encinitas 10,796 ... H10
91316 Encino ... B10
95320 Escalon 3,127 ... E6
*92025 Escondido 64,355 ... J10
95501 Eureka⊙ 24,153 ... A3
93221 Exeter 5,606 ... F7
94930 Fairfax 7,391 ... H1
94533 Fairfield⊙ 58,099 ... K1
95628 Fair Oaks 22,602 ... C8
92028 Fallbrook 14,041 ... H10
93223 Farmersville 5,544 ... F7
95018 Felton 4,564 ... K4
93015 Fillmore 9,602 ... G9
93622 Firebaugh 3,740 ... E7
95828 Florin 16,523 ... B8
95630 Folsom 11,003 ... C8
92335 Fontana 37,107 ... E10
†93268 Ford City 3,392 ... F8
95437 Fort Bragg 5,019 ... B4
†95421 Fort Ross 30 ... B5
95540 Fortuna 7,591 ... A3
94404 Foster City 23,287 ... J3
92708 Fountain Valley 55,080 ... D11
95019 Freedom 6,416 ... L4
*94536 Fremont 131,945 ... K3
*93706 Fresno⊙ 217,289 ... F7
Fresno‡ 515,013 ... F7
*92631 Fullerton 102,034 ... D11
95632 Galt 5,514 ... C9
*90747 Gardena 45,165 ... C11
*92640 Garden Grove 123,307 ... D11
95020 Gilroy 21,641 ... D6
92509 Glen Avon Heights 8,444 ... E10
*91201 Glendale 139,060 ... C10
91740 Glendora 38,500 ... D10
93926 Gonzales 2,891 ... D7
91344 Granada Hills ... B10
92324 Grand Terrace 8,498 ... E10
95945 Grass Valley 6,697 ... D4
93308 Greenacres 5,381 ... F8
93927 Greenfield 4,181 ... D7
95948 Gridley 3,982 ... D4
93433 Grover City 8,827 ... E8
93434 Guadalupe 3,629 ... E9
95322 Gustine 3,142 ... D6
94019 Half Moon Bay 7,282 ... H3
93230 Hanford⊙ 20,958 ... F7
90250 Hawthorne 56,447 ... C11
*94541 Hayward 94,342 ... K2
95448 Healdsburg 7,217 ... B5
92343 Hemet 22,454 ... H10
94547 Hercules 5,963 ... J1
90254 Hermosa Beach 18,070 ... B11
92345 Hesperia 13,540 ... H9
92346 Highland 10,908 ... H9
94010 Hillsborough 10,372 ... J2
95023 Hollister⊙ 11,488 ... D7
90028 Hollywood ... C10
92250 Holtville 4,399 ... K11
†91720 Home Gardens 5,783 ... E11
95326 Hughson 2,943 ... E6
*92646 Huntington Beach 170,505 ... C11
90255 Huntington Park 46,223 ... C11
92251 Imperial 3,451 ... K11
92032 Imperial Beach 22,689 ... H11
93526 Independence⊙ 748 ... H7
92201 Indio 21,611 ... J10
*90301 Inglewood 94,245 ... B11
92713 Irvine 62,134 ... D11
95642 Jackson⊙ 2,331 ... C9
†94701 Kensington 5,342 ... J2
93600 Kerman 4,002 ... E7
93930 King City 5,495 ... D7
93631 Kingsburg 5,115 ... F7
91011 La Canada 20,153 ... C10
91214 La Crescenta-Montrose 16,531 ... C10
94549 Lafayette 20,879 ... K2
*92651 Laguna Beach 17,901 ... G10
92653 Laguna Hills 33,600 ... D11
92677 Laguna Niguel 12,237 ... H10
90631 La Habra 45,232 ... D11
92037 La Jolla ... H11
92352 Lake Arrowhead 6,272 ... H9
92330 Lake Elsinore 5,982 ... F11
93240 Lake Isabella 3,428 ... G8
95453 Lakeport⊙ 3,675 ... C4
*90712 Lakewood 74,654 ... C11
92041 La Mesa 50,308 ... H11
90638 La Mirada 40,986 ... D11
93241 Lamont 9,616 ... G8
93534 Lancaster 48,027 ... G9
*91744 La Puente 30,882 ... D10
94939 Larkspur 11,064 ... H1
95330 Lathrop 3,717 ... D6
91750 La Verne 23,508 ... D10
90260 Lawndale 23,460 ... B11
92045 Lemon Grove 20,780 ... J11
93245 Lemoore 8,832 ... F7
†92311 Lenwood 2,974 ... H9
92024 Leucadia 9,478 ... H10
95648 Lincoln 4,132 ... B8
†95901 Linda 10,225 ... D4
93247 Lindsay 6,924 ... F7
95953 Live Oak 3,103 ... D4
†95073 Live Oak 11,482 ... K4
94550 Livermore 48,349 ... L2
95334 Livingston 5,326 ... E6
95240 Lodi 35,221 ... C9
92354 Loma Linda 10,694 ... F10
90717 Lomita 18,807 ... C11
93436 Lompoc 26,267 ... E9
*90801 Long Beach 361,334 ... C11
90720 Los Alamitos 11,529 ... D11
94022 Los Altos 25,769 ... K3
94022 Los Altos Hills 7,421 ... J3
*90001 Los Angeles⊙ 2,966,850 ... C10
Los Angeles-Long Beach‡ 7,477,657 ... C10
93635 Los Banos 10,341 ... E6
95030 Los Gatos 26,906 ... K4
†93402 Los Osos-Baywood Park 10,933 ... E8
90262 Lynwood 48,548 ... C11
93637 Madera⊙ 21,732 ... E7
90265 Malibu ... B10
93546 Mammoth Lakes 3,929 ... G6
90266 Manhattan Beach 31,542 ... B11
95336 Manteca 24,925 ... D6
93933 Marina 20,647 ... D7
95338 Mariposa⊙ 1,150 ... F6
94553 Martinez⊙ 22,582 ... K1
95901 Marysville⊙ 9,898 ... D4
90201 Maywood 21,810 ... C10
93250 McFarland 5,151 ... F8
93023 Meiners Oaks-Mira Monte 9,512 ... F9
93640 Mendota 5,038 ... E7
94025 Menlo Park 26,369 ... J3
95340 Merced⊙ 36,499 ... E6
94030 Millbrae 20,058 ... J2
94941 Mill Valley 12,967 ... H2
95035 Milpitas 37,820 ... L3
91752 Mira Loma 8,707 ... E10
92691 Mission Viejo 50,666 ... D11
*95350 Modesto⊙ 106,602 ... D6
Modesto‡ 265,902 ... D6
93501 Mojave 2,886 ... G8
91016 Monrovia 30,531 ... D10
91763 Montclair 22,628 ... D10
90640 Montebello 52,929 ... C10
93940 Monterey 27,558 ... D7
91754 Monterey Park 54,338 ... C10
95030 Monte Sereno 3,434 ... K4
91214 Montrose-La Crescenta 16,531 ... C10
93021 Moorpark 4,030 ... G9
94556 Moraga 15,014 ... K2
95037 Morgan Hill 17,060 ... L4
93442 Morro Bay 9,064 ... D8
*94042 Mountain View 58,655 ... K3
96067 Mount Shasta 2,837 ... C5
92405 Muscoy 6,188 ... E10
94558 Napa⊙ 50,879 ... C5
92050 National City 48,772 ... J11
92363 Needles 4,120 ... L9
95959 Nevada City⊙ 2,431 ... D4
94560 Newark 32,126 ... K3
91321 Newhall 12,029 ... G9
95360 Newman 2,785 ... D6
*92660 Newport Beach 62,556 ... D11
93444 Nipomo 5,247 ... E8
91760 Norco 21,126 ... E11
95660 North Highlands 37,825 ... B8
*91601 North Hollywood ... B10
90650 Norwalk 85,286 ... C11
94947 Novato 43,916 ... H1
95361 Oakdale 8,474 ... E6
*94601 Oakland⊙ 339,337 ... J2
93022 Oak View 4,671 ... F9
93445 Oceano 4,478 ... E8
92054 Oceanside 76,698 ... H10
93308 Oildale 23,382 ... F8
93023 Ojai 6,816 ... F9
*91761 Ontario 88,820 ... D10
†95060 Opal Cliffs 5,041 ... K4
*92666 Orange 91,450 ... D11
93646 Orange Cove 4,026 ... F7
94563 Orinda 16,825 ... J2
95963 Orland 4,031 ... C4
93647 Orosi 4,076 ... F7
95965 Oroville⊙ 8,683 ... D4
93030 Oxnard 108,195 ... F9
Oxnard-Simi Valley-Ventura‡ 529,899 ... F9
94553 Pacheco-Vine Hill 6,129 ... K1
94044 Pacifica 36,866 ... H2
93950 Pacific Grove 15,755 ... C7
93550 Palmdale 12,277 ... G9
92260 Palm Desert 11,801 ... J10
92262 Palm Springs 32,366 ... J10
*94301 Palo Alto 55,225 ... K3
90274 Palos Verdes Estates 14,376 ... B11
95969 Paradise 22,571 ... D4
90723 Paramount 36,407 ... C11
93648 Parlier 2,902 ... F7
*91101 Pasadena 118,072 ... C10
93446 Paso Robles 9,163 ... E8
95363 Patterson 3,908 ... D6
93953 Pebble Beach ... C7
92370 Perris 6,827 ... F11
94952 Petaluma 33,834 ... H1
90660 Pico Rivera 53,387 ... C10
94611 Piedmont 10,498 ... J2
94564 Pinole 14,253 ... J1
93449 Pismo Beach 5,364 ... E8
94565 Pittsburg 33,034 ... L1
92670 Placentia 35,041 ... D11
95667 Placerville⊙ 6,739 ... C8
94523 Pleasant Hill 25,124 ... K2
94566 Pleasanton 35,160 ... L2
*91766 Pomona 92,742 ... D10
93257 Porterville 19,707 ... G7
93041 Port Hueneme 17,803 ... F9
94025 Portola Valley 3,939 ... J3
92064 Poway 32,263 ... J11
93534 Quartz Hill 7,421 ... G9
95971 Quincy⊙ 4,451 ... E4
92065 Ramona 8,173 ... J10
95670 Rancho Cordova 42,881 ... C8
91730 Rancho Cucamonga 55,250 ... E10
92270 Rancho Mirage 6,281 ... J10
90274 Rancho Palos Verdes 36,577 ... B11
92067 Rancho Santa Fe 4,014 ... H10
96080 Red Bluff⊙ 9,490 ... C3
96001 Redding⊙ 41,995 ... C3
Redding○ ... 80
92373 Redlands 43,619 ... H9
*90277 Redondo Beach 57,102 ... B11
*94061 Redwood City⊙ 54,951 ... J3
93654 Reedley 11,071 ... F7
92376 Rialto 37,474 ... E10
*94801 Richmond 74,676 ... J1
93555 Ridgecrest 15,929 ... H8
95562 Rio Dell 2,687 ... A3
95673 Rio Linda 7,359 ... B8
94571 Rio Vista 3,142 ... L1
95366 Ripon 3,509 ... D6
95367 Riverbank 5,695 ... E6
*92501 Riverside⊙ 170,591 ... E11
Riverside-San Bernardino-Ontario‡ 1,557,080 ... E11
95677 Rocklin 7,344 ... B8
94572 Rodeo 8,286 ... J1
94928 Rohnert Park 22,965 ... C5
90274 Rolling Hills 2,049 ... B11
90274 Rolling Hills Estates 7,701 ... B11
91770 Rosemead 42,604 ... C10
95678 Roseville 24,347 ... B8
94957 Ross 2,801 ... H1
92509 Rubidoux 17,048 ... E10

(continued on following page)

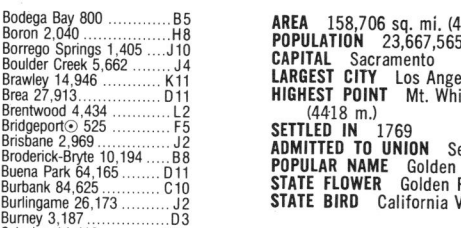

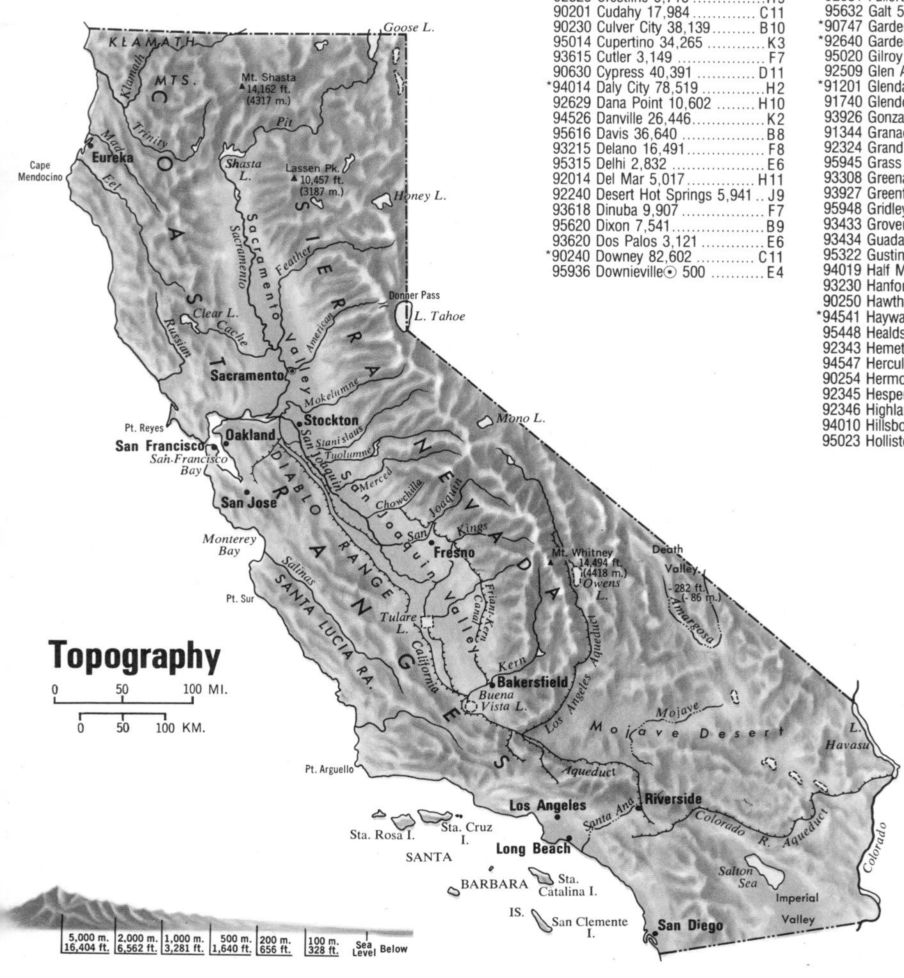
Topography
0 50 100 MI.
0 50 100 KM.
KLAMATH
MTS.
Goose L.
Mt. Shasta 14,162 ft. (4317 m.)
Lassen Pk. 10,457 ft. (3187 m.)
Honey L.
Cape Mendocino
Eureka
Shasta L.
Pit
Trinity
Eel
Sacramento
Feather
Donner Pass
L. Tahoe
Clear L.
Cache
American
Russian
Sacramento
Pt. Reyes
San Francisco
San Francisco Bay
Oakland
Stockton
Mokelumne
Stanislaus
Tuolumne
Merced
Chowchilla
San Joaquin
Mono L.
San Jose
DIABLO RANGE
Monterey Bay
Kings
Fresno
Mt. Whitney 14,494 ft. (4418 m.)
Death Valley
-282 ft. (-86 m.)
Owens L.
Amargosa
Pt. Sur
Salinas
SANTA LUCIA RA.
Tulare L.
Kern
Bakersfield
Buena Vista L.
Los Angeles Aqueduct
California Aqueduct
Mojave
Mojave Desert
Pt. Arguello
Aqueduct
Los Angeles
Riverside
Santa Ana
Colorado R. Aqueduct
Havasu L.
Colorado
Sta. Rosa I.
Sta. Cruz I.
Long Beach
SANTA BARBARA IS.
Sta. Catalina I.
San Clemente I.
Salton Sea
Imperial Valley
San Diego
5,000 m. 16,404 ft.
2,000 m. 6,562 ft.
1,000 m. 3,281 ft.
500 m. 1,640 ft.
200 m. 656 ft.
100 m. 328 ft.
Sea Level
Below

*95801 Sacramento (cap.)⊙ 275,741 .... B8
Sacramento‡ 1,014,002 .... B8
94574 Saint Helena 4,898 .... C5
93901 Salinas⊙ 80,479 .... D7
Salinas-Seaside-Monterey‡ 290,444 .... D7
95249 San Andreas⊙ 1,912 .... E5
94960 San Anselmo 12,067 .... H1
*92401 San Bernardino⊙ 118,794 E10
94066 San Bruno 35,417 .... J2
94070 San Carlos 24,710 .... J3
92672 San Clemente 27,325 ..... H10
*92101 San Diego⊙ 875,538 ..... H11
San Diego‡ 1,861,846 ... H11
91773 San Dimas 24,014 .... D10
*91340 San Fernando 17,731 ..... C10
*94101 San Francisco⊙ 678,974 .. H2
San Francisco-Oakland‡ 3,252,721 .... H2
*91775 San Gabriel 30,072 .... C10
93657 Sanger 12,542 .... F7
92383 San Jacinto 7,098 .... H10
*95101 San Jose ⊙ 629,546 .... L3
San Jose‡ 1,295,071 .... L3
†92691 San Juan Capistrano 18,959 .... H10
*94577 San Leandro 63,952 .... J2
94580 San Lorenzo 20,545 .... K2
93401 San Luis Obispo⊙ 34,252 . E8
92069 San Marcos 17,479 .... H10
91108 San Marino 13,307 .... D10
*94401 San Mateo 77,640 .... J3
94806 San Pablo 19,750 .... J1
94964 San Quentin 450 .... H1
*94901 San Rafael⊙ 44,700 .... J1
94583 San Ramon 22,356 .... K2
93452 San Simeon 350 .... D8
*92701 Santa Ana⊙ 204,023 ..... D11
*93101 Santa Barbara⊙ 74,414 .... F9
Santa Barbara-Santa Maria-Lompoc‡ 298,660 F9
*95050 Santa Clara 87,700 .... K3
*95060 Santa Cruz⊙ 41,483 .... K4
Santa Cruz‡ 188,141 .... K4
90670 Santa Fe Springs 14,520 . C11
93454 Santa Maria 39,685 .... E9
*90401 Santa Monica 88,314 ..... B10
93060 Santa Paula 20,552 .... F9
*95401 Santa Rosa⊙ 83,320 .... C5
Santa Rosa‡ 299,827 .... C5
92071 Santee 47,080 .... J11
95070 Saratoga 29,261 .... K4
94965 Sausalito 7,338 .... H2
95060 Scotts Valley 6,891 .... K4
90740 Seal Beach 25,975 .... C11
93955 Seaside 36,567 .... D7
95472 Sebastopol 5,595 .... C5
93662 Selma 10,942 .... F7
93263 Shafter 7,010 .... F8
96125 Sierra City 500 .... E4
91024 Sierra Madre 10,837 ...... D10
†90806 Signal Hill 5,734 .... C11
*93065 Simi Valley 77,500 .... G9
92075 Solana Beach 13,047 ..... H11
93960 Soledad 5,928 .... D7
93463 Solvang 3,091 .... E9
95476 Sonoma 6,054 .... C5
95370 Sonora⊙ 3,247 .... E6
95073 Soquel 6,212 .... K4
91733 South El Monte 16,623 ... C10
90280 South Gate 66,784 .... C11
95705 South Lake Tahoe 20,681 .. F5
†95965 South Oroville 7,246 .... D4
91030 South Pasadena 22,681 .. C10
94080 South San Francisco 49,393 .... J2
94305 Stanford 11,045 .... J3
90680 Stanton 23,723 .... D11
*95201 Stockton⊙ 149,779 .... D6
Stocton‡ 347,342 .... D6
94585 Suisun City 11,087 .... K1
92381 Sun City 8,460 .... F11
92388 Sunnymead 11,554 .... F11
*94086 Sunnyvale 106,618 .... K3
96130 Susanville⊙ 6,520 .... E3
95685 Sutter Creek 1,705 .... C9
93268 Taft 5,316 .... F8
95730 Tahoe City E4
93561 Tehachapi 4,126 .... G8
91780 Temple City 28,972 .... D10
†95965 Thermalito 4,961 .... D4
*91360 Thousand Oaks 77,072 ..... G9
92276 Thousand Palms 1,718 .... J10
94920 Tiburon 6,685 .... J2
90290 Topanga B10
*90501 Torrance 129,881 .... C11
95376 Tracy 18,428 .... D6
93274 Tulare 22,526 .... F7
95380 Turlock 26,287 .... E6
92680 Tustin 32,317 .... D11
92277 Twentynine Palms 7,465 ... K9
†95060 Twin Lakes 4,502 .... K4
95482 Ukiah⊙ 12,035 .... B4
94587 Union City 39,406 .... K2
91786 Upland 47,647 .... E10
95688 Vacaville 43,367 .... D5
91355 Valencia 12,163 .... G9
94590 Vallejo 80,303 .... J1
Vallejo-Fairfield-Napa‡ 334,402 .... J1
*91401 Van Nuys B10
90291 Venice B11
*93001 Ventura⊙ 74,393 .... F9
92392 Victorville 14,220 .... H9
92667 Villa Park 7,137 .... D11
93277 Visalia⊙ 49,729 .... F7
Visalia-Tulare-Porterville‡ 245,738 .... F7
92083 Vista 35,834 .... H10
91789 Walnut 12,478 .... D10
*94595 Walnut Creek 53,643 .... K2
93280 Wasco 9,613 .... F8
95386 Waterford 2,683 .... E6
95076 Watsonville 23,663 .... D7
96093 Weaverville⊙ 2,787 .... B3
96094 Weed 2,879 .... C2
*91790 West Covina 80,291 .... D10
†90069 West Hollywood 35,703 .. B10
90025 West Los Angeles B10
92683 Westminster 71,133 .... D11
†90047 Westmont 27,916 .... C11
†94565 West Pittsburg 8,773 .... K1
95691 West Sacramento 10,875 .. B8
*90601 Whittier 69,717 .... D11
95490 Willits 4,008 .... B4
95988 Willows⊙ 4,777 .... C4
*90744 Wilmington C11
95388 Winton 4,995 .... E6
93286 Woodlake 4,343 .... G7
95695 Woodland⊙ 30,235 .... B8
*91364 Woodland Hills B10
94062 Woodside 5,291 .... J3
95697 Yolo 600 .... B8
92686 Yorba Linda 28,254 .... D11
94599 Yountville 2,893 .... C5
96097 Yreka⊙ 5,916 .... C2
95991 Yuba City⊙ 18,736 .... D4
Yuba City‡ 101,979 .... D4
92399 Yucaipa 23,345 .... J9
92284 Yucca Valley 8,294 .... J9

OTHER FEATURES

Agua Caliente Ind. Res. J10
Alameda (creek) K3
Alamo (riv.) K10
Alcatraz (isl.) J2
Alkali (lkes) E2
All American (canal) K11
Almanor (lake) D3
Amargosa (range) J7
Amargosa (riv.) J7
American (riv.) C8
Anacapa (isl.) F10
Angel (isl.) J2
Ano Nuevo (pt.) J4
Arena (pt.) B5
Arguello (pt.) E9
Argus (range) H7
Arroyo del Valle (dry riv.) L3
Arroyo Hondo (dry riv.) L3
Arroyo Mocho (dry riv.) L2
Arroyo Seco (dry riv.) K10
Beale A.F.B. D4
Berryessa (lake) D5
Bethany (res.) L2
Big Sage (res.) E2
Black Butte (lake) C4
Bodega (bay) B5
Bonita (pt.) H2
Bristol (lake) K9
Buchon (pt.) D8
Buena Vista (lake) F8
Cabrillo Nat'l Mon. H11
Cachuma (lake) F9
Cadiz (lake) K9
Cahuilla Ind. Res. J10
Calaveras (res.) L3
California Aqueduct E7
Camanche (res.) C9
Camp Pendleton 10,017 .... H10
Campo Ind. Res. J11
Capitan Grande Ind. Res. J11
Cascade (range) D1
Castle A.F.B. E6
Channel Islands Nat'l Park E11
China Lake Naval Weapons Center . H8
Chemehuevi Valley Ind. Res. L9
Chocolate (mts.) K10
Clair Engle (lake) C3
Clear (lake) C4
Clear Lake (res.) D2
Coachella (canal) K10
Coast (ranges) D7
Colorado (riv.) L8
Colorado River Aqueduct K10
Colorado River Ind. Res. L10
Conception (pt.) E9
Cooper (pt.) D7
Copco (lake) C2
Cosumnes (riv.) C9
Cottonwood (creek) C3
Coyote (res.) L4
Crowley (lake) G6
Crystal Springs (res.) J3
Cuyama (riv.) E8
Cuyapaipe Ind. Res. J11
Danby (lake) K9
Death (valley) H7
Death Valley Nat'l Mon. H7
Delgada (pt.) A3
Del Valle (lake) L3
Devils Postpile Nat'l Mon. F6
Donner (pass) E4
Dume (pt.) G10
Duxbury (pt.) H2
Eagle (lake) E3
Eagle (peak) E2
Eagle Crags (mt.) J8
Edison (lake) F6
Edwards A.F.B. 8,554 .... H9
Eel (riv.) B4
Elsinore (lake) E11
El Toro Marine Air Sta. 7,632 ..... D11
Estero (bay) D8
Estero (pt.) D8
Estrella (riv.) E8
Eugene O'Neill Nat'l Hist. Site K2
Farallon (isls.) B6
Farallons, The (gulf) H2
Feather (riv.) D4
Florence (lake) G6
Folsom (lake) C8
Fort Bidwell Ind. Res. E2
Fort Hunter Liggett D8
Fort Independence Ind. Res. G7
Fort MacArthur C11
Fort Mohave Ind. Res. L9
Fort Ord D7
Fort Point Nat'l Hist. Site J2
Freel (peak) F5

Agriculture, Industry and Resources

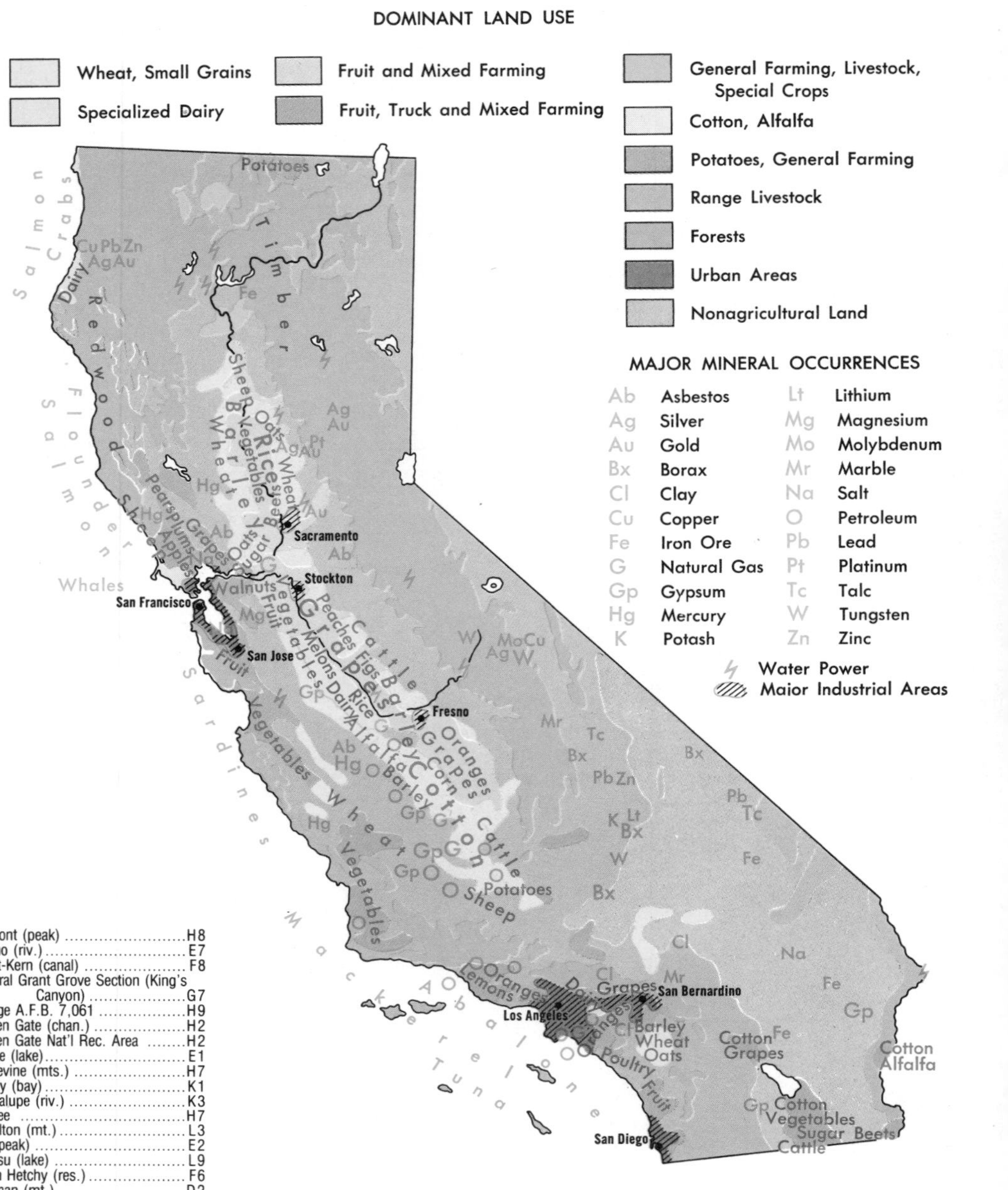

DOMINANT LAND USE
Wheat, Small Grains
Specialized Dairy
Fruit and Mixed Farming
Fruit, Truck and Mixed Farming
General Farming, Livestock, Special Crops
Cotton, Alfalfa
Potatoes, General Farming
Range Livestock
Forests
Urban Areas
Nonagricultural Land
MAJOR MINERAL OCCURRENCES
Ab Asbestos
Ag Silver
Au Gold
Bx Borax
Cl Clay
Cu Copper
Fe Iron Ore
G Natural Gas
Gp Gypsum
Hg Mercury
K Potash
Lt Lithium
Mg Magnesium
Mo Molybdenum
Mr Marble
Na Salt
O Petroleum
Pb Lead
Pt Platinum
Tc Talc
W Tungsten
Zn Zinc
Water Power
Major Industrial Areas
Sacramento
Stockton
San Francisco
San Jose
Fresno
Los Angeles
San Bernardino
San Diego
Whales
Salmon
Crabs
Sardines
Mackerel
Tuna
Timber
Potatoes
Redwood
Dairy
Sheep
Vegetables
Barley
Rice
Grapes
Fruit
Walnuts
Cattle
Oranges
Lemons
Cotton
Alfalfa
Poultry
Wheat
Oats
Sugar Beets

Fremont (peak) H8
Fresno (riv.) E7
Friant-Kern (canal) F8
General Grant Grove Section (King's Canyon) G7
George A.F.B. 7,061 .... H9
Golden Gate (chan.) H2
Golden Gate Nat'l Rec. Area H2
Goose (lake) E1
Grapevine (mts.) H7
Grizzly (bay) K1
Guadalupe (riv.) K3
Haiwee H7
Hamilton (mt.) L3
Hat (peak) E2
Havasu (lake) L9
Hetch Hetchy (res.) F6
Hoffman (mt.) D2
Honey (lake) E3
Hoopa Valley Ind. Res. A2
Humboldt (bay) A3
Imperial (res.) L10
Imperial (valley) K10
Ingalls (mt.) E3
Inyo (mts.) G6
Iron Gate (res.) C2
Isabella (lake) G8
John Muir Nat'l Hist. Site K1
Joshua Tree Nat'l Mon. J10
Kern (riv.) G8
Kings (riv.) F7
Kings Canyon Nat'l Park G7
Klamath (riv.) B2
Laguna (res.) L11
La Jolla Ind. Res. J10
Lassen (peak) D3
Lassen Volcanic Nat'l Park D3
Lava Beds Nat'l Mon. D2
Lemoore N.A.S. 5,888 .... F7
Leroy Anderson (res.) L4
Lopez (pt.) D7
Los Angeles Aqueduct G8
Los Coyotes Ind. Res. J10
Lost (riv.) D1
Lower Alkali (lake) E2
Lower Klamath (lake) D2
Mad (riv.) B3
Manzanita Ind. Res. J11
March A.F.B. 3,607 .... E11
Mare Island Navy Yard J1
Mather A.F.B. 5,245 .... C8
Mathews (lake) E11
McClellan A.F.B. B8
McClure (lake) E6
Mendocino (cape) A3
Merced (riv.) E6
Middle Alkali (lake) E2
Mill (creek) D3
Millerton (lake) F6
Moffett Nav. Air Sta. K3
Mojave (des.) H9
Mojave (riv.) J9
Mokelumne (riv.) C9
Mono (lake) G5
Monterey (bay) K4
Moon (lake) E2
Morongo Ind. Res. J10
Mountain Meadows (res.) E3
Muir Woods Nat'l Mon. H2
Nacimiento (riv.) D8
Navarro (riv.) B4
Nevada, Sierra (mts.) E4
New (riv.) K11
Norton A.F.B. F10
Noyo (riv.) B4
Oakland Army Base J2
Old (riv.) L1
Oroville (lake) D4
Owens (lake) H7
Owens (peak) H8
Owens (riv.) G6
Oxnard A.F.B. F9
Paiute Ind. Res. G6
Pala Ind. Res. H10
Palomar (mt.) J10
Panamint (range) H7
Panamint (valley) H7
Pescadero (pt.) J3
Piedras Blancas (pt.) D8
Pillar (pt.) H3
Pillsbury (lake) C4
Pine (creek) D3
Pine Flat (lake) F7
Pinnacles Nat'l Mon. D7
Pit (riv.) D2
Point Mugu Pacific Missile Test Center F9
Point Reyes Nat'l Seashore H1
Presidio J2
Providence (mts.) K8
Punta Gorda (pt.) A3
Quartz (peak) L11
Railroad Canyon (res.) E11
Redwood Nat'l Park A2
Reyes (pt.) B6
Rogers (lake) H9
Rosamond (lake) G9
Round Valley Ind. Res. B4
Russian (riv.) B4
Sacramento (riv.) D5
Sacramento Army Depot B8
Saint George (pt.) A2
Salinas (riv.) D7
Salmon (riv.) B2
Salton Sea (lake) K10
San Andreas (lake) H2
San Antonio (lake) E8
San Benito (riv.) D7
San Bernardino (mts.) J10
San Clemente (isl.) G11
San Diego (bay) H11
San Francisco (bay) J2
San Gabriel (res.) D10
San Joaquin (riv.) E6
San Joaquin (valley) D6
San Lorenzo (riv.) K4
San Luis (res.) E7
San Martin (cape) D8
San Miguel (isl.) E9
San Nicolas (isl.) F10
San Pablo (bay) J1
San Pedro (bay) C11
Santa Ana (riv.) E11
Santa Barbara (chan.) E9
Santa Barbara (isl.) G10
Santa Barbara (isls.) F10
Santa Catalina (gulf) G11
Santa Catalina (isl.) G10
Santa Cruz (chan.) F10
Santa Cruz (isl.) F10
Santa Maria (riv.) E9
Santa Monica (bay) B11
Santa Rosa (isl.) E10
Santa Rosa Ind. Res. J10
Santa Ynez (riv.) E9
Santa Ysabel Ind. Res. J10
Searles (lake) H8
Sequoia Nat'l Park G7
Sharpe Army Depot D6
Shasta (lake) C3
Shasta (mt.) C2
Shasta (riv.) C
Sierra Army Depot .... E
Sierra Nevada (mts.) .... E
Siskiyou (mts.) C
Smith (riv.) .... A
Soda (lake) .... K
South Bay Aqueduct L
South Cow (creek) C
Stony Gorge (res.) C
Suisun (bay) .... K
Sur (pt.) .... D
Tahoe (lake) .... F
Tamalpais (mt.) .... H
Tehachapi (mts.) .... G
Telescope (peak) .... H
Tomales (pt.) .... B
Torres Martinez Ind. Res. J1
Travis A.F.B. L
Trinidad (head) .... A
Trinity (riv.) .... B
Truckee (riv.) .... F
Tulare (lake) .... F
Tule (lake) .... D
Tule River Ind. Res. .... G
Twentynine Palms Marine Base 7,079 .... J
Twitchell (res.) .... E
Upper Alkali (lake) .... E
Vandenberg A.F.B. 8,136 .... E
Vizcaino (cape) .... B
Walnut (creek) .... K
Wheeler (peak) .... F
Whipple (mts.) L
Whiskeytown-Shasta-Trinity Nat'l Rec Area C
Whitney (mt.) .... G
Willow (creek) .... E
Wilson (mt.) D1
Yosemite Nat'l Park .... F
Yuba (riv.) .... D
Yuma Ind. Res. L1

⊙County seat.
‡Population of metropolitan area.
† Zip of nearest p.o. * Multiple zip

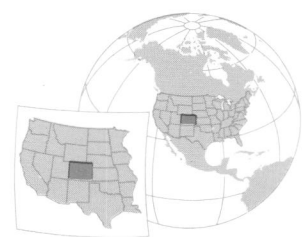

AREA 104,091 sq. mi. (269,596 sq. km.)
POPULATION 2,889,735
CAPITAL Denver
LARGEST CITY Denver
HIGHEST POINT Mt. Elbert 14,433 ft. (4399 m.)
SETTLED IN 1858
ADMITTED TO UNION August 1, 1876
POPULAR NAME Centennial State
STATE FLOWER Rocky Mountain Columbine
STATE BIRD Lark Bunting

COUNTIES

Adams 245,944 L3
Alamosa 11,799 H7
Arapahoe 293,621 L3
Archuleta 3,664 E8
Baca 5,419 O8
Bent 5,945 N7
Boulder 189,625 J2
Chaffee 13,227 G5
Cheyenne 2,153 O5
Clear Creek 7,308 H3
Conejos 7,794 G8
Costilla 3,071 J8
Crowley 2,988 M6
Custer 1,528 J6
Delta 21,225 D5
Denver 492,365 K3
Dolores 1,658 C7
Douglas 25,153 K4
Eagle 13,320 F3
Elbert 6,850 L4
El Paso 309,424 K5
Fremont 28,676 J5
Garfield 22,514 C3
Gilpin 2,441 H3
Grand 7,475 G2
Gunnison 10,689 E5
Hinsdale 408 E7
Huerfano 6,440 K7
Jackson 1,863 G1
Jefferson 371,741 J3
Kiowa 1,936 O6
Kit Carson 7,599 O4
Lake 8,830 G4
La Plata 27,195 D8
Larimer 149,184 H1
Las Animas 14,897 L8
Lincoln 4,663 M5
Logan 19,800 N1
Mesa 81,530 B5
Mineral 804 F7
Moffat 13,133 C1
Montezuma 16,510 B8
Montrose 24,352 C6
Morgan 22,513 M2
Otero 22,567 M7
Ouray 1,925 D6
Park 5,333 H4
Phillips 4,542 P1
Pitkin 10,338 F4
Prowers 13,070 P7
Pueblo 125,972 K6
Rio Blanco 6,255 C3
Rio Grande 10,511 G7
Routt 13,404 E1
Saguache 3,935 G6
San Juan 833 D7
San Miguel 3,192 C6
Sedgwick 3,266 P1
Summit 8,848 G3
Teller 8,034 J5
Washington 5,304 N3
Weld 123,438 L1
Washington 5,304 N3
Weld 123,438 L1
Yuma 9,682 P2

CITIES and TOWNS

Zip *Name/Pop.* *Key*

80101 Agate 90 M4
81020 Aguilar 624 K8
80720 Akron⊙ 1,716 N2
81101 Alamosa⊙ 6,830 H8
80510 Allenspark 200 J2
80420 Alma 132 G4
81210 Almont 135 F5
80721 Amherst 85 P1
80801 Anton 55 N3
81120 Antonito 1,103 H8
80802 Arapahoe 300 P5
81021 Arlington 37 N6
80804 Arriba 236 N4
†81323 Arriola 56 B8
*80001 Arvada 84,576 J3
81611 Aspen⊙ 3,678 F4
80722 Atwood 100 N1
80610 Ault 1,056 K1
*80010 Aurora 158,588 K3
81410 Austin D5
81620 Avon 640 F3
81022 Avondale 750 L6
80421 Bailey 150 H4
†80624 Barnesville 20 L2
81621 Basalt 529 E4
81122 Bayfield 724 D8
81411 Bedrock 45 B6
†80758 Beecher Island 5 P3
80512 Bellvue 250 J1
80102 Bennett 942 L3
80513 Berthoud 2,362 J2
†80438 Berthoud Pass 40 H3
80805 Bethune 149 P4
81023 Beulah 650 K6
80908 Black Forest 3,372 K4
80422 Black Hawk 232 J3
81123 Blanca 252 H8
†80424 Blue River 230 G4
†81155 Bonanza 8 G6
81024 Boncarbo 200 K8
80423 Bond 65 F3
81025 Boone 431 L6
*80301 Boulder⊙ 76,685 J2
†81428 Bowie 18 D5
80821 Boyero 12 N5
81026 Brandon 30 P6
81027 Branson 73 M8
80424 Breckenridge⊙ 818 G4
80611 Briggsdale 85 L1
80601 Brighton⊙ 12,773 K3
81028 Bristol 200 P6
†81212 Brookside 178 J6
80020 Broomfield 20,730 J3
80723 Brush 4,082 M2
†80742 Buckingham 5 L1
81211 Buena Vista 2,075 G5
80425 Buffalo Creek 150 J4
80807 Burlington⊙ 3,107 P4
80426 Burns 100 F3
80103 Byers 490 L3
81320 Cahone 200 B7
80808 Calhan 541 L4
81029 Campo 185 O8
81212 Canon City⊙ 13,037 J6
81124 Capulin 600 G8
81623 Carbondale 2,084 E4
80612 Carr 49 K1
80909 Cascade 950 K5
80104 Castle Rock⊙ 3,921 K4
81413 Cedaredge 1,184 D5
81125 Center 1,630 G7
80427 Central City⊙ 329 J3
81126 Chama 239 J8
81030 Cheraw 233 N6
80810 Cheyenne Wells⊙ 950 P5
81127 Chimney Rock 76 E8
81031 Chivington 20 O6
81128 Chromo 115 F8
81220 Cimarron 50 D6
80428 Clark 20 F1
81520 Clifton 5,223 C4
80429 Climax 975 G4
81221 Coal Creek 190 J6
81222 Coaldale 153 H6
80430 Coalmont 50 F1
81032 Cokedale 90 K8
81624 Collbran 344 C4
†81401 Colona 54 D6
81019 Colorado City 411 K6
*80901 Colorado Springs⊙ 214,821 K5
Colorado Springs‡ 317,458 K5
†80428 Columbine 12 E1
80022 Commerce City 16,234 K3
80432 Como 30 H4
81129 Conejos⊙ 200 G8
80812 Cope 110 O3
†80611 Cornish 15 L2
81321 Cortez⊙ 7,095 B8
81223 Cotopaxi 250 H6
80434 Cowdrey 80 G1
81625 Craig⊙ 8,133 D2
81415 Crawford 268 D5
81130 Creede⊙ 610 E7
81224 Crested Butte 959 E5
81131 Crestone 54 H7
80813 Cripple Creek⊙ 655 J5
80726 Crook 177 O1
81033 Crowley 192 M6
81055 Cuchara 43 J8
80514 Dacono 2,321 K2
†80728 Dailey 20 O1
81630 De Beque 279 C4
†80135 Deckers 4 J4
80105 Deer Trail 463 M3
†81059 Delhi 10 M7
81132 Del Norte⊙ 1,709 G7
81416 Delta⊙ 3,931 D5
*80201 Denver (cap.)⊙ 492,365 K3
Denver‡ 1,619,921 K3
†81054 Deora 2 O7
80435 Dillon 337 H3
81610 Dinosaur 313 B2
80814 Divide 700 J5
81323 Dolores 802 C8
81324 Dove Creek⊙ 826 A7
†81239 Doyleville 75 F6
80515 Drake 300 J2
81301 Durango⊙ 11,649 D8
81036 Eads⊙ 878 O6
81631 Eagle⊙ 950 F3
80615 Eaton 1,932 K1
80727 Eckley 262 P2
80214 Edgewater 4,766 J3
81632 Edwards 250 F3
81325 Egnar 50 B7
80106 Elbert 200 L4
†80466 Eldora 100 H3
80107 Elizabeth 789 K4
81633 Elk Springs 18 C2
80438 Empire 423 H3
†80110 Englewood 30,021 K3
80516 Erie 1,254 K2
80517 Estes Park 2,703 J2
†81433 Eureka 25 D7
80620 Evans 5,063 K2
80439 Evergreen 6,376 J3
80440 Fairplay⊙ 421 H4
81037 Farisita 116 J7
†80221 Federal Heights 7,846 J3
80520 Firestone 1,204 K2
†80810 Firstview 6 O5
80815 Flagler 550 N4
80728 Fleming 388 O1
81226 Florence 2,987 J6
80816 Florissant 130 J5
80521 Fort Collins⊙ 65,092 J1
Fort Collins‡ 149,184 J1
81133 Fort Garland 700 J8
80621 Fort Lupton 4,251 K2
81038 Fort Lyon 500 N6
80701 Fort Morgan⊙ 8,768 M2
80817 Fountain 8,324 K5
81039 Fowler 1,227 L6
80441 Foxton 12 J4
80116 Franktown 200 K4
80442 Fraser 470 H3
80530 Frederick 855 K2
80820 Freshwater (Guffey) 24 H5
80443 Frisco 1,221 G3
81521 Fruita 2,810 B4
80622 Galeton 200 K1
81134 Garcia 75 J8
81040 Gardner 100 J7
81227 Garfield 30 G5
81522 Gateway 350 B5
80818 Genoa 165 N4
80444 Georgetown⊙ 830 H3
80623 Gilcrest 1,025 K2
80624 Gill 250 L2
81634 Gilman 160 G3
81523 Glade Park 100 B5
†80485 Glendevey 50 H1
80532 Glen Haven 110 H2
81601 Glenwood Springs⊙ 4,637 E4
80401 Golden⊙ 12,237 J3
†80653 Goodrich 85 M2
†80480 Gould 12 G2
81041 Granada 557 P6
80446 Granby 963 H2
81501 Grand Junction⊙ 27,956 B4
80447 Grand Lake 382 H2
81228 Granite 47 G4
80448 Grant 50 H4
80631 Greeley⊙ 53,006 K2
Greeley‡ 123,438 K2
†80118 Greenland 21 K4
80819 Green Mountain Falls 607 K5
†81640 Greystone 2 B1
80729 Grover 158 L1
80820 Guffey 24 H5
81042 Gulnare 6 K8
81230 Gunnison⊙ 5,785 E5
81637 Gypsum 743 F3
80730 Hale 4 P3
81638 Hamilton 100 D2
81043 Hartman 122 P6
80449 Hartsel 69 H4
81044 Hasty 150 O6
81045 Haswell 126 N6
80731 Haxtun 1,014 O1
81639 Hayden 1,720 E2
80732 Hereford 50 L1
81326 Hesperus 250 C8
80733 Hillrose 213 N2
81232 Hillside 79 H6
81046 Hoehne 400 L8
81047 Holly 969 P6
80734 Holyoke⊙ 2,092 P1
81136 Hooper 71 H7
81419 Hotchkiss 849 D5
80451 Hot Sulphur Springs⊙ 405 H2
81233 Howard 200 H6
80641 Hoyt 60 L2
80642 Hudson 698 K2
80821 Hugo⊙ 776 N4
80533 Hygiene 450 J2
80452 Idaho Springs 2,077 H3
80735 Idalia 125 P3
81137 Ignacio 667 D8
80736 Iliff 218 N1
80455 Jamestown 223 J2
†81082 Jansen 267 K8
81138 Jaroso 50 H8
80456 Jefferson 50 H4
80822 Joes 100 O3
80534 Johnstown 1,535 K2
80737 Julesburg⊙ 1,528 P1
80823 Karval 51 N5
80643 Keenesburg 541 L2
†80729 Keota 4 L1
80644 Kersey 913 L2
81049 Kim 100 N8
80117 Kiowa⊙ 206 L4
80824 Kirk 30 P3
80825 Kit Carson 278 O5
80459 Kremmling 1,296 G2
†80832 Kutch 2 M5
80026 Lafayette 8,935 K3
†81132 La Garita 10 G7
80739 Laird 105 P2
81140 La Jara 858 H8
81050 La Junta⊙ 8,388 M7
81235 Lake City⊙ 206 E6
80827 Lake George 500 J5
80215 Lakewood 113,808 J3
81052 Lamar⊙ 7,713 O6
80535 Laporte 950 J1
80118 Larkspur 141 K4
80645 La Salle 1,929 K2
81054 Las Animas⊙ 2,818 N6
†81151 Lasauces 150 H8
†81153 Lavalley 237 J8
81055 La Veta 611 J8
†80452 Lawson 108 H3
†81625 Lay 40 D2
81420 Lazear 60 D5
80461 Leadville⊙ 3,879 G4
†81323 Lebanon 50 B8
81327 Lewis 150 B8
80828 Limon 1,805 M4
†81212 Lincoln Park 2,984 J6
80740 Lindon 60 N3
*80120 Littleton⊙ 28,631 K3
80536 Livermore 150 J1
†80601 Lochbuie 895 K2
†80701 Log Lane Village 709 M2
81524 Loma 265 B4
80501 Longmont 42,942 J2
†80135 Longview 10 J4
80027 Louisville 5,593 J3
80131 Louviers 300 K4
80537 Loveland 30,244 J2
80646 Lucerne 135 K2
†81054 Lycan 4 P7
80540 Lyons 1,137 J2
81525 Mack 380 B4
81421 Maher 75 D5
†80461 Malta 200 G4
81141 Manassa 945 H8
81328 Mancos 870 C8
80829 Manitou Springs 4,475 J5
81058 Manzanola 459 M6
†81623 Marble 30 E4
81329 Marvel 176 C8
80541 Masonville 200 J2
†80649 Masters 50 L2
80830 Matheson 120 M4
81640 Maybell 130 C2
81057 McClave 125 O6
80463 McCoy 62 F3
80542 Mead 356 K2
81641 Meeker⊙ 2,356 D2
81642 Meredith 47 F4
80741 Merino 255 N2
81005 Mesa 120 C4
81330 Mesa Verde National Park 45 C8
81142 Mesita 70 H8
80543 Milliken 1,506 K2
80477 Milner 196 F2
81645 Minturn 1,060 G3

(continued on following page)

Agriculture, Industry and Resources

DOMINANT LAND USE

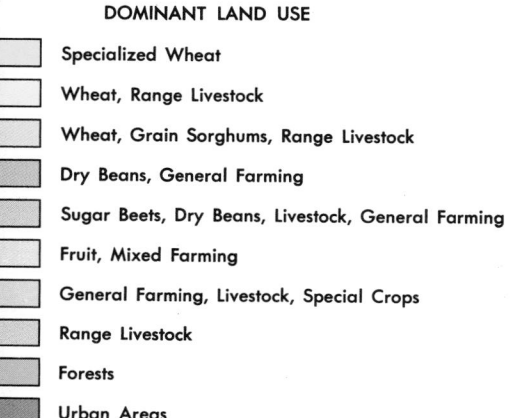

- Specialized Wheat
- Wheat, Range Livestock
- Wheat, Grain Sorghums, Range Livestock
- Dry Beans, General Farming
- Sugar Beets, Dry Beans, Livestock, General Farming
- Fruit, Mixed Farming
- General Farming, Livestock, Special Crops
- Range Livestock
- Forests
- Urban Areas
- Nonagricultural Land

MAJOR MINERAL OCCURRENCES

Ag	Silver	Mi	Mica
Au	Gold	Mo	Molybdenum
Be	Beryl	Mr	Marble
C	Coal	O	Petroleum
Cl	Clay	Pb	Lead
Cu	Copper	U	Uranium
F	Fluorspar	V	Vanadium
Fe	Iron Ore	W	Tungsten
G	Natural Gas	Zn	Zinc

Water Power
Major Industrial Areas

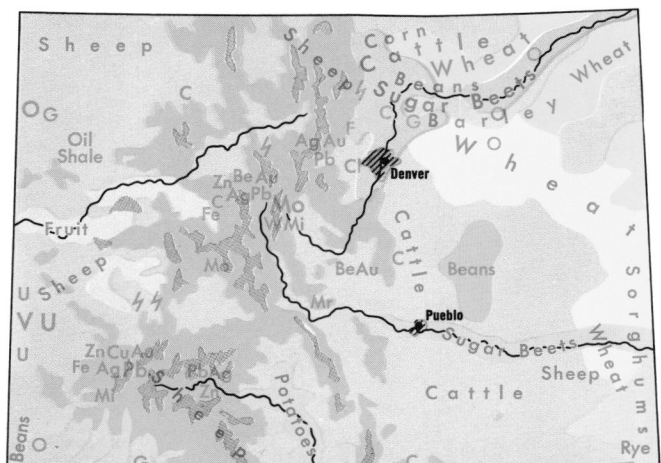

Topography

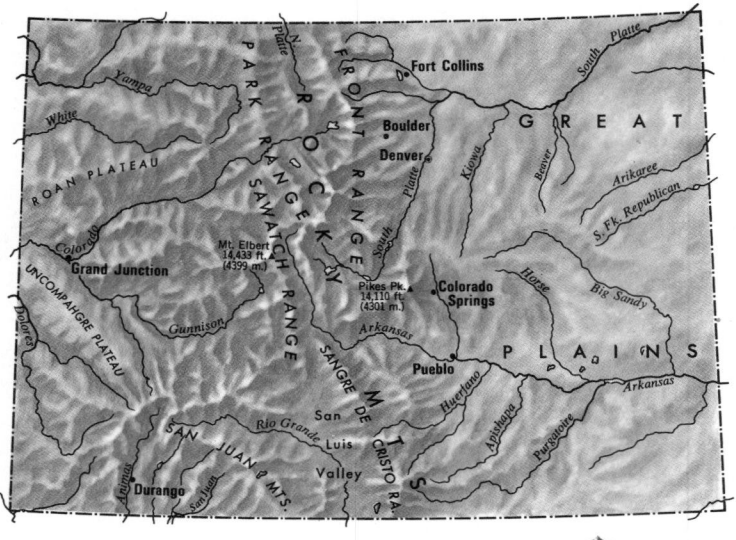

Below Sea Level	100 m. 328 ft.	200 m. 656 ft.	500 m. 1,640 ft.	1,000 m. 3,281 ft.	2,000 m. 6,562 ft.	5,000 m. 16,404 ft.

81646 Molina 200 ....D4
81144 Monte Vista 3,902 ....G7
†80435 Montezuma 6 ....H3
81401 Montrose⊙ 8,722 ....D6
80132 Monument 690 ....K4
80465 Morrison 478 ....J3
81146 Mosca 100 ....H7
81236 Nathrop 150 ....H5
81422 Naturita 819 ....B6
80466 Nederland 1,212 ....H3
81647 New Castle 563 ....E3
80742 New Raymer 80 ....M1
†81054 Ninaview 2 ....N7
80544 Niwot 500 ....J2
†81022 North Avondale 110 ....L6
80233 Northglenn 29,847 ....K3
†81050 North La Junta 1,076 ....N7
81423 Norwood 478 ....C6
81424 Nucla 1,027 ....B6
80648 Nunn 295 ....K1
80467 Oak Creek 929 ....F2
81237 Ohio 100 ....F5
81425 Olathe 1,262 ....D5
81062 Olney Springs 253 ....M6
81426 Ophir 38 ....D7
80649 Orchard 79 ....L2
†81501 Orchard Mesa 4,876 ....C4
81063 Ordway⊙ 1,135 ....M6
†81120 Ortiz 163 ....H8
80743 Otis 534 ....O2
81427 Ouray⊙ 684 ....D6
80744 Ovid 439 ....P1
80745 Padroni 100 ....N1
†81147 Pagosa Junction 15 ....E8
81147 Pagosa Springs⊙ 1,331 ....E8
81526 Palisade 1,551 ....C4
80133 Palmer Lake 1,130 ....J4
80746 Paoli 81 ....P1
81428 Paonia 1,425 ....D5
81635 Parachute 338 ....C4
81429 Paradox 250 ....B6
†81212 Parkdale 21 ....H6
80134 Parker 200 ....K4
81239 Parlin 494 ....F6
80468 Parshall 80 ....G2
80747 Peetz 220 ....N1
81240 Penrose 500 ....K6
80831 Peyton 500 ....K4
80469 Phippsburg 300 ....F2
80650 Pierce 878 ....K1
80470 Pine 100 ....J4
80471 Pinecliffe 375 ....J3
†81001 Pinon 50 ....K6
81241 Pitkin 59 ....F5
81430 Placerville 50 ....D6
†81624 Plateau City 35 ....D4
†80743 Platner 30 ....N2
80651 Platteville 1,662 ....K2
81331 Pleasant View 300 ....B7
81242 Poncha Springs 321 ....G6
†81226 Portland 17 ....H6
†81427 Portland D6
81243 Powderhorn 100 ....E6
81064 Pritchett 183 ....O8
†80736 Proctor 25 ....N1
81065 Pryor 50 ....K8
*81001 Pueblo⊙ 101,686 ....K6
Pueblo‡ 125,972 ....K6
80472 Radium 22 ....G3
80832 Ramah 119 ....L4
80473 Rand 50 ....G2
81648 Rangely 2,113 ....B2

80473 Rand 50 ....G2
81648 Rangely 2,113 ....B2
80742 Raymer (New Raymer) 80 .M1
81649 Red Cliff 409 ....G4
80545 Red Feather Lakes 150 ....H1
†81326 Red Mesa 100 ....C8
†81623 Redstone 115 ....E4
81431 Redvale 300 ....B6
81066 Red Wing 200 ....J7
81332 Rico 76 ....C7
81432 Ridgway 369 ....D6
81650 Rifle 3,215 ....D3
†81650 Rio Blanco 4 ....C3
81244 Rockvale 338 ....J6
81067 Rocky Ford 4,804 ....M6
80652 Roggen 100 ....L2
81148 Romeo 308 ....G8
80833 Rush 40 ....L5
81069 Rye 232 ....K7
81149 Saguache⊙ 656 ....G6
†81236 Saint Elmo 75 ....G5
81201 Salida⊙ 44,870 ....H6
81150 San Acacio 50 ....H8
81151 Sanford 687 ....H8
†81069 San Isabel 8 ....K7
81152 San Luis⊙ 842 ....J8
81153 San Pablo 150 ....J8
81248 Sargents 31 ....F6
†81430 Sawpit 41 ....D7
80911 Security-Widefield 18,768 ....K5
80135 Sedalia 200 ....K4
80749 Sedgwick 258 ....O1
81070 Segundo 200 ....K8
80834 Seibert 180 ....O4
80546 Severance 100 ....K1
80475 Shawnee 100 ....H4
†80110 Sheridan 5,377 ....J3
81071 Sheridan Lake 87 ....P6
81652 Silt 923 ....D4
81249 Silver Cliff 280 ....J6
80476 Silver Plume 140 ....H3
80498 Silverthorne 989 ....G3
81433 Silverton⊙ 794 ....D7
80835 Simla 494 ....M4
81653 Slater 10 ....E1
81654 Snowmass 999 ....F4
80750 Snyder 200 ....M2
81434 Somerset 200 ....E5
81154 South Fork 500 ....F7
81073 Springfield⊙ 1,657 ....O8
81074 Starkville 127 ....L8
80477 Steamboat Springs⊙ 5,098 F2
80751 Sterling⊙ 11,385 ....N1
80754 Stoneham 35 ....M1
81075 Stonington 27 ....P8
81076 Strasburg 1,005 ....L4
80836 Stratton 705 ....O4
81076 Sugar City 306 ....M6
†81640 Sunbeam 19 ....E1
†80027 Superior 208 ....J3
81077 Swink 668 ....M7
80478 Tabernash 250 ....H3
81435 Telluride⊙ 1,047 ....D7
81250 Texas Creek 80 ....H6
†81082 Thatcher 50 ....L7
80229 Thornton 40,343 ....K3
†81137 Tiffany 24 ....D8
†81034 Timnath 185 ....J2
†81034 Timpas 25 ....M7
†81210 Tincup 8 ....F5
80479 Toponas 55 ....F2

81334 Towaoc 300 ....B8
81080 Towner 61 ....P6
81081 Trinchera 30 ....M8
81082 Trinidad⊙ 9,663 ....L8
†80864 Truckton 10 ....L5
81251 Twin Lakes 40 ....G4
81084 Two Buttes 84 ....P7
81059 Tyrone 9 ....L8
81436 Uravan 500 ....B6
†81064 Utleyville 2 ....O8
81657 Vail 2,261 ....G3
†81082 Valdez 12 ....K8
80755 Vernon 50 ....P3
80860 Victor 265 ....J5
81087 Vilas 118 ....P8
81155 Villa Grove 37 ....G6
81088 Villegreen 6 ....M8
†81001 Vineland 100 ....K6
80548 Virginia Dale 2 ....J1
80861 Vona 94 ....O4
†81130 Wagon Wheel Gap 20 ....F7
80480 Walden⊙ 947 ....G1
81089 Walsenburg⊙ 3,945 ....K7
81090 Walsh 884 ....P8
80481 Ward 129 ....H2
80653 Weldona 200 ....M2
80549 Wellington 1,215 ....K1
81252 Westcliffe⊙ 324 ....H6
†80135 Westcreek 2 ....J4
80030 Westminster 50,211 ....J3
81091 Weston 150 ....K8
81253 Wetmore 150 ....J6
80033 Wheat Ridge 30,293 ....J3
81527 Whitewater 300 ....C5
80654 Wiggins 531 ....L2
80862 Wild Horse 13 ....N5
81092 Wiley 425 ....O6
†81226 Williamsburg 72 ....H6
80550 Windsor 4,277 ....J2
80482 Winter Park 480 ....H3
81655 Wolcott 30 ....F3
80863 Woodland Park 2,634 ....J4
80757 Woodrow 24 ....M3
81656 Woody Creek 400 ....F4
80758 Wray⊙ 2,131 ....P2
80483 Yampa 472 ....F2
81335 Yellow Jacket 115 ....B7
80864 Yoder 25 ....L5
80759 Yuma 2,824 ....O2

OTHER FEATURES

Adams (mt.) H6
Adobe Creek (res.) N6
Air Force Academy 8,655 ....K5
Alamosa (creek) G8
Alva B. Adams (tunnel) H2
Animas (riv.) D8
Antero (mt.) B2
Antero (peak) H5
Antero (res.) H5
Antora (peak) G6
Apishapa (riv.) L8
Arapaho Nat'l Rec. Area G2
Arapahoe (peak) H2
Arikaree (riv.) O3
Arkansas (riv.) P6
Arkansas Divide (mts.) L4
Baker (mt.) H2
Bald (mt.) H4
Bear (riv.) P8
Beaver (creek) M3
Bennett (peak) G7

Bent's Old Fort Nat'l Hist.
Site M6
Big Grizzly (creek) G1
Big Sandy (creek) N4
Big Thompson (riv.) H2
Bijou (creek) L3
Black Canyon of the Gunnison Nat'l Mon. D5
Black Squirrel (creek) L5
Blanca (peak) H7
Blue (mt.) B2
Blue (riv.) G2
Blue Mesa (res.) E5
Bonny (res.) P3
Box Elder (creek) K4
Cache la Poudre (riv.) H1
Cameron (peak) H1
Camp Hale G4
Carbon (peak) E5
Castle (peak) F5
Cebolla (creek) E6
Chacuaco (creek) M8
Cheesman (lake) J4
Clay (creek) O7

Cochetopa (creek) F6
Colorado (riv.) A5
Colorado Nat'l Mon. B4
Conejos (peak) G8
Conejos (riv.) G8
Crestone (peak) H7
Crow (creek) L1
Culebra (creek) H8
Culebra (peak) J8
Curecanti Nat'l Rec. Area F6
Del Norte (peak) F7
De Weese (plat.) J6
Dinosaur Nat'l Mon. B2
Disappointment (creek) B7
Dolores (riv.) B5
Douglas (creek) B3
Eagle (riv.) E3
Elbert (mt.) G4
El Diente (peak) C7
Eleven Mile Canyon (res.) H5
Elk (riv.) F1
Empire (res.) L2
Ent A.F.B. K5
Ethel (mt.) F1

Evans (mt.) H3
Florissant Fossil Beds Nat'l Mon. J5
Fort Carson 19,399 ....K5
Fountain (creek) K5
Frenchman (creek) O1
Frenchman, North Fork (creek) O1
Frenchman, South Fork (creek) O1
Front (range) H1
Gore (range) G3
Graham (peak) H7
Granby (lake) G2
Great Sand Dunes Nat'l Mon. H7
Green (riv.) A2
Green Mountain (res.) G3
Gunnison (riv.) C5
Gunnison (tunnel) D6
Gunnison, North Fork (riv.) D5
Hale, Camp G4
Handies (peak) E7
Harvard (mt.) H5
Hermosa (peak) D7
Hesperus (mt.) C8
Holy Cross (mt.) F4

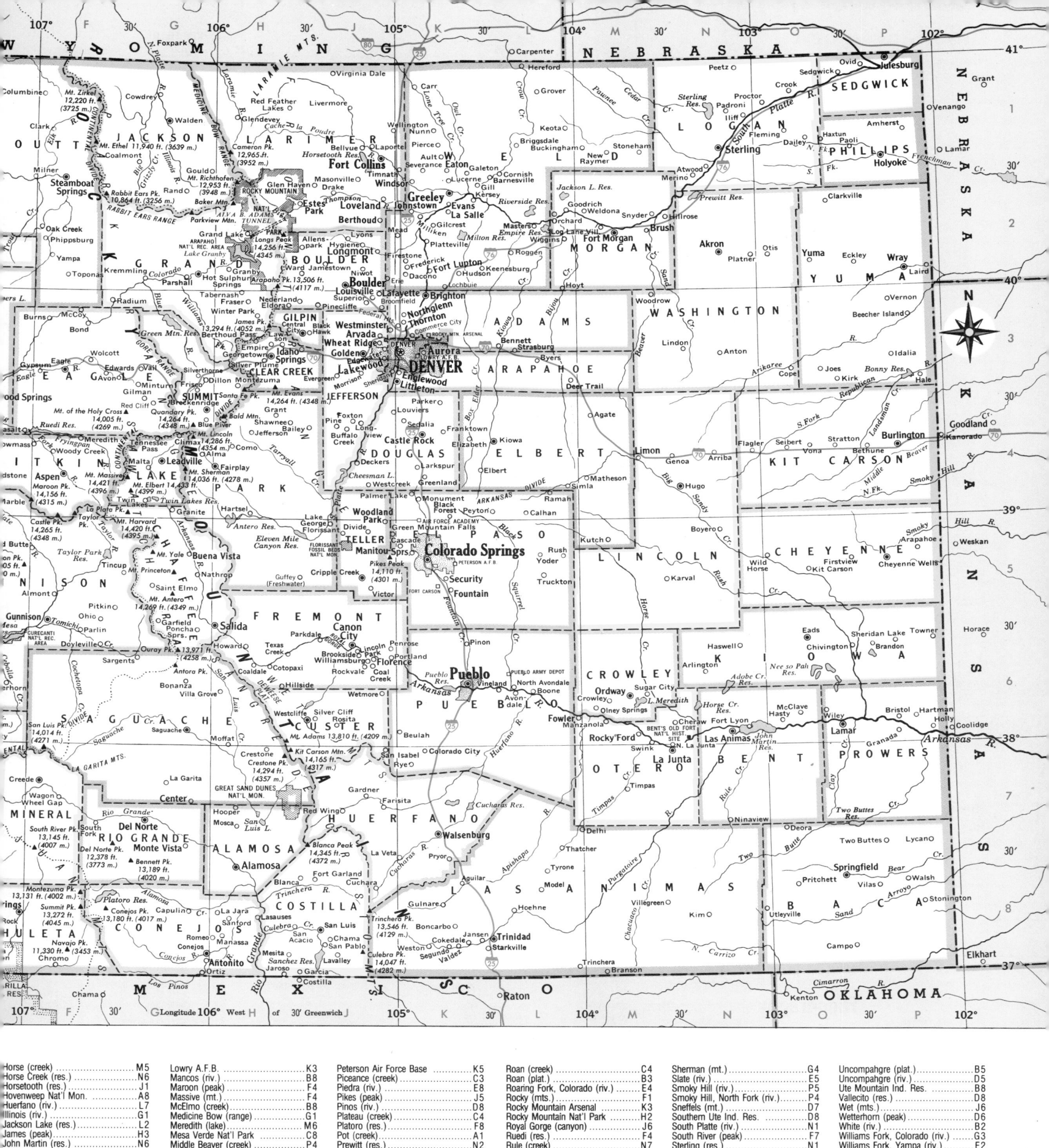

Horse (creek) M5
Horse Creek (res.) N6
Horsetooth (res.) J1
Hovenweep Nat'l Mon. A8
Huerfano (riv.) L7
Illinois (riv.) G1
Jackson Lake (res.) L2
James (peak) H3
John Martin (res.) N6
Juniper (mt.) C1
Kiowa (creek) L3
Kit Carson (mt.) H7
La Garita (mts.) F7
Lake Fork, Gunnison (riv.) E6
Landsman (creek) P4
Laramie (mts.) H1
Laramie (riv.) H1
Lincoln (mt.) G4
Lone Cone (mt.) C7
Lone Tree (creek) K1
Los Pinos (riv.) G8

Lowry A.F.B. K3
Mancos (riv.) B8
Maroon (peak) F4
Massive (mt.) F4
McElmo (creek) B8
Medicine Bow (range) G1
Meredith (lake) M6
Mesa Verde Nat'l Park C8
Middle Beaver (creek) P4
Milton (res.) K2
Montezuma (peak) F8
Morrow Point (res.) E6
Muddy (creek) D3
Navajo (peak) F8
Navajo (riv.) E8
Nee so Pah (res.) D6
North Carrizo (creek) N8
North Platte (riv.) G1
Ouray (peak) G6
Owl (creek) K1

Peterson Air Force Base K5
Piceance (creek) C3
Piedra (riv.) E8
Pikes (peak) J5
Pinos (riv.) D8
Plateau (creek) C4
Platoro (res.) F8
Pot (creek) A1
Prewitt (res.) N2
Princeton (mt.) G5
Pueblo (res.) K6
Pueblo Army Depot L6
Purgatoire (riv.) M8
Quandary (peak) G4
Quaternary (riv.) L2
Rabbit Ears (peak) G2
Rabbit Ears (range) F2
Redcloud (peak) E6
Republican (riv.) P3
Richthofen (mt.) G2
Rifle (creek) D3
Rio Grande (res.) E7
Rio Grande (riv.) H8
Rio Grande Pyramid (mt.) E7
Riverside (res.) L2

Roan (creek) C4
Roan (plat.) B3
Roaring Fork, Colorado (riv.) . E4
Rocky (mts.) F1
Rocky Mountain Arsenal K3
Rocky Mountain Nat'l Park ... H2
Royal Gorge (canyon) J6
Ruedi (res.) F4
Rule (creek) N7
Rush (creek) N5
Saguache (creek) F7
San Juan (mts.) F7
San Juan (riv.) E8
San Luis (creek) H6
San Luis (lake) H7
San Luis (peak) F6
San Miguel (mts.) C7
San Miguel (riv.) B6
Santa Fe (peak) G4
Sawatch (range) G4
Sheep (mt.) E6

Sherman (mt.) G4
Slate (riv.) E5
Smoky Hill (riv.) P5
Smoky Hill, North Fork (riv.) . P4
Sneffels (mt.) D7
Southern Ute Ind. Res. D8
South Platte (riv.) N1
South River (peak) F7
Sterling (res.) N1
Summit (peak) F8
Tarryall (creek) H4
Taylor (peak) F5
Taylor Park (res.) F5
Timpas (creek) M7
Tomichi (creek) F5
Trappers (lake) E3
Trinchera (creek) J8
Trinchera (riv.) H8
Trout (creek) E2
Twin Lakes (res.) G4
Two Butte (creek) N7
Two Buttes (res.) O7
Uncompahgre (peak) E6

Uncompahgre (plat.) B5
Uncompahgre (riv.) D5
Ute Mountain Ind. Res. D8
Vallecito (res.) D8
Wet (mts.) J6
Wetterhorn (peak) D6
White (riv.) B2
Williams Fork, Colorado (riv.) . G3
Williams Fork, Yampa (riv.) ... E2
Wilson (mt.) C7
Windom (peak) E7
Yale (mt.) G5
Yampa (riv.) C3
Yellow (creek) C3
Yucca House Nat'l Mon. B8
Zenobia (peak) B1
Zirkel (mt.) F1

⊙County seat.
‡Population of metropolitan area.
† Zip of nearest p.o. * Multiple zips.

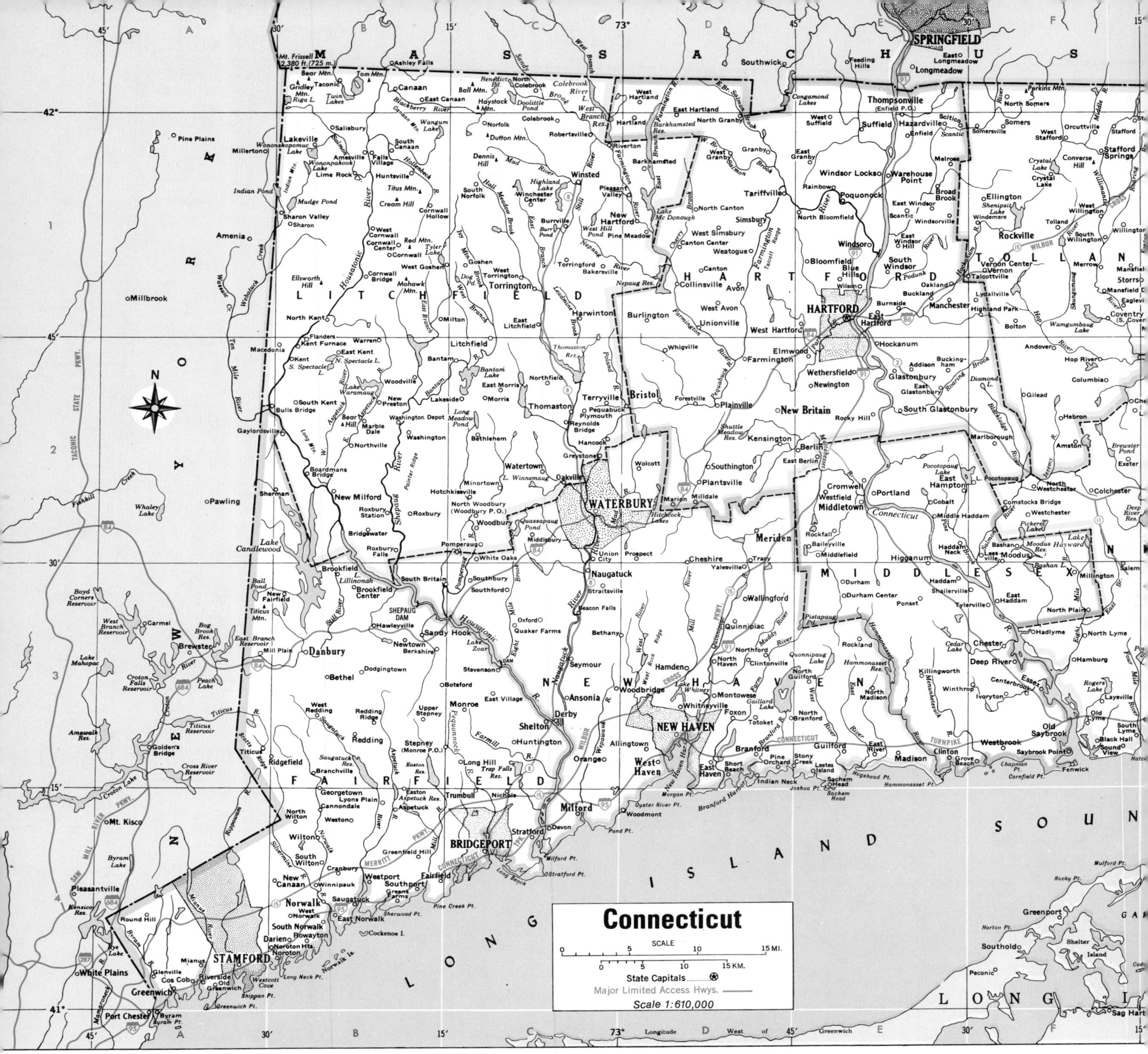

Connecticut

SCALE

0 ... 5 ... 10 ... 15 MI.

0 ... 5 ... 10 ... 15 KM.

State Capitals .... ⊛

Major Limited Access Hwys. ————

Scale 1:610,000

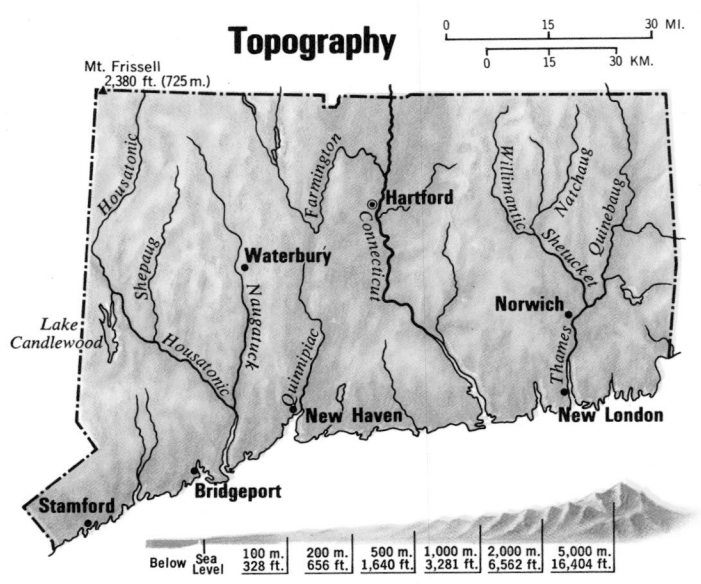

Topography

Mt. Frissell
2,380 ft. (725 m.)

0 ... 15 ... 30 MI.

0 ... 15 ... 30 KM.

Housatonic • Shepaug • Naugatuck • Quinnipiac • Farmington • Connecticut • Willimantic • Natchaug • Shetucket • Quinebaug • Thames

Hartford • Waterbury • Norwich • New Haven • New London • Bridgeport • Stamford

Lake Candlewood

Below Sea Level	100 m. 328 ft.	200 m. 656 ft.	500 m. 1,640 ft.	1,000 m. 3,281 ft.	2,000 m. 6,562 ft.	5,000 m. 16,404 ft.

COUNTIES

County	Pop.	Key
Fairfield 807,143		B3
Hartford 807,766		D1
Litchfield 156,769		B1
Middlesex 129,017		E3
New Haven 761,337		D3
New London 238,409		G2
Tolland 114,823		F1
Windham 92,312		H1

CITIES and TOWNS

Zip	Name/Pop.	Key
06230	Abington 600	G1
06231	Amston 900	F2
06232	Andover 2,144	F2
06401	Ansonia 19,039	C3
06278	Ashford 3,221	G1
06278	Ashford P.O. (Warrenville) 500	G1
†06241	Attawaugan 400	H1
06001	Avon 11,201	D1
06001	Avon 1,434	D1
06233	Ballouville 800	H1
06330	Baltic	G2
06750	Bantam 860	B2
†06423	Bashan 90	F2
06403	Beacon Falls 3,995	C3
06037	Berlin 15,121	E2
†06501	Bethany‡ 4,330	C3
06801	Bethel○ 16,004	B3
06801	Bethel 8,755	B3
06751	Bethlehem○ 2,573	C2
06751	Bethlehem 1,762	C2
06002	Bloomfield○ 18,608	E1
06112	Blue Hills	E1
06040	Bolton 3,951	F1
06404	Botsford 400	C3
06016	Broad Brook	E1
06804	Brookfield 12,872	B3
06234	Brooklyn 5,691	H1
06013	Burlington 5,660	D1
06830	Byram	A4
06018	Canaan○ 1,002	B1
06018	Canaan 1,160	B1
†06897	Cannondale 400	B4
06331	Canterbury○ 3,426	H2
06019	Canton○ 7,635	D1
06019	Canton 1,680	D1
06409	Centerbrook 800	F3
06332	Central Village 950	H2
06235	Chaplin 1,793	G2
06410	Cheshire○ 21,788	D2
06410	Cheshire 5,722	D2
06412	Chester 3,068	F3
06412	Chester 1,388	F3
06413	Clinton○ 11,195	E3
06413	Clinton 3,168	E3
06414	Cobalt 700	E2
06415	Colchester○ 7,761	F2
06415	Colchester 3,190	F2
06021	Colebrook○ 1,221	C1
06022	Collinsville 2,555	D1
06237	Columbia○ 3,386	F2
06753	Cornwall○ 1,288	B1
06807	Cos Cob	A4
06238	Coventry 8,895	F1
06416	Cromwell○ 10,265	E2
06810	Danbury 60,470	B3
	Danbury‡ 146,405	B3
06239	Danielson 4,553	H1
06820	Darien○ 18,892	B4
06241	Dayville	H1
06417	Deep River○ 3,994	F3
06417	Deep River 2,495	F3
06418	Derby 12,346	C3
06422	Durham○ 5,143	E3
06422	Durham 2,641	E3
06023	East Berlin 950	E2
†06239	East Brooklyn 1,251	H1
06024	East Canaan 800	B1
06242	Eastford 1,028	G1
06025	East Glastonbury 300	E2
06026	East Granby 4,102	E1
06423	East Haddam 5,621	F3
06424	East Hampton‡ 8,572	E2

* City and town figures cross-referenced.
○ County seat.
† Other symbol.
‡ Metropolitan area.

AREA 5,018 sq. mi. (12,997 sq. km.)
POPULATION 3,107,576
CAPITAL Hartford
LARGEST CITY Bridgeport
HIGHEST POINT Mt. Frissell (S. Slope) 2,380 ft. (725 m.)
SETTLED IN 1635
ADMITTED TO UNION January 9, 1788
POPULAR NAME Constitution State; Nutmeg State
STATE FLOWER Mountain Laurel
STATE BIRD Robin

© Copyright HAMMOND INCORPORATED, Maplewood, N.J.

06351 Lisbon○ 3,279	G2
06759 Litchfield 7,605	C2
06759 Litchfield 1,489	C2
†06378 Lords Point 500	H3
06443 Madison○ 14,031	E3
06443 Madison 2,069	E3
06040 Manchester○ 49,761	E1
06040 Manchester 31,058	E1
†06250 Mansfield 20,634	F1
06250 Mansfield Center 1,043	G1
06777 Marble Dale 300	B2
06444 Marion 900	D2
06447 Marlborough○ 4,746	F2
06447 Marlborough 1,039	F2
†06382 Massapeag 350	G3
06252 Mechanicsville 425	H1
06450 Meriden 57,118	D2
Meriden‡ 57,118	D2
06762 Middlebury○ 5,995	C2
06455 Middlefield○ 3,796	E2
06456 Middle Haddam 325	E2
06457 Middletown 39,040	E2
06460 Milford 49,101	C4
06467 Milldale 975	D2
†06759 Milton 600	C1
06468 Monroe○ 14,010	C3
06468 Monroe P.O. (Stepney)	B3
06353 Montville○ 16,455	G3
06353 Montville 1,711	G3
06469 Moodus 1,179	F2
06354 Moosup 3,308	H2
06763 Morris○ 1,899	C2
06355 Mystic 2,333	H3
06770 Naugatuck 26,456	C3
*06050 New Britain 73,840	E2
New Britain‡ 142,241	E2
06840 New Canaan○ 17,931	B4
06810 New Fairfield○ 11,260	B3
06057 New Hartford 4,884	C1
06057 New Hartford 1,310	C1
*06501 New Haven 126,109	D3
New Haven-West Haven‡ 417,592	D3
06111 Newington 28,841	E2
06320 New London 28,842	G3
New London-Norwich‡ 248,554	G3
06776 New Milford 19,420	B2
06776 New Milford 5,186	B2
06777 New Preston 1,209	B2
06470 Newtown○ 19,107	B3
06470 Newtown 2,022	B3
06357 Niantic 3,151	G3
06340 Noank 1,406	G3
06058 Norfolk○ 2,156	C1
06471 North Branford 11,554	E3
06778 Northfield 600	C2
06254 North Franklin 500	G2
06060 North Granby 450	D1
06255 North Grosvenor Dale 1,856	H1
†06437 North Guilford 500	E3
06473 North Haven○ 22,080	D3
06359 North Stonington○ 4,219	H3
06256 North Windham 200	G1
*06850 Norwalk 77,767	B4
06360 Norwich 38,074	G2
06370 Oakdale 600	G3
06779 Oakville 8,737	C2
06371 Old Lyme○ 6,159	F3

06372 Old Mystic 600	H3
06475 Old Saybrook○ 9,287	F3
06475 Old Saybrook 1,857	F3
06373 Oneco 550	H2
06477 Orange○ 13,237	C3
06483 Oxford○ 6,634	C3
06379 Pawcatuck 5,216	H3
06781 Pequabuck 642	C2
06061 Pine Meadow 400	D1
†06405 Pine Orchard 300	D3
06374 Plainfield○ 12,774	H1
06374 Plainfield 2,799	H1
06062 Plainville○ 16,401	D2
06063 Pleasant Valley 300	C1
†06385 Pleasure Beach 1,356	G3
06782 Plymouth○ 10,732	C2
06258 Pomfret○ 2,775	H1
†06340 Poquonock Bridge 2,549	G3
06480 Portland○ 8,383	E2
06480 Portland 5,914	E2
06712 Prospect○ 6,807	D2
06260 Putnam 8,580	H1
06260 Putnam 6,855	H1
06375 Quaker Hill 2,052	G3
06262 Quinebaug 1,088	H1
06875 Redding 7,272	B3
06876 Redding Ridge 550	B3
06877 Ridgefield○ 20,120	B3
06877 Ridgefield 6,066	B3
06065 Riverton 250	D1
06481 Rockfall 900	E2
†06066 Rockville	F1
06067 Rocky Hill○ 14,559	E2
06263 Rogers 650	H1
06783 Roxbury○ 1,468	B2
†06415 Salem○ 2,335	F3
06068 Salisbury○ 3,896	B1
06264 Scotland○ 1,072	G1
06483 Seymour○ 13,434	C3
06069 Sharon○ 2,623	A1
06484 Shelton 31,314	C3
06784 Sherman○ 2,281	A2
06070 Simsbury○ 21,161	D1
06070 Simsbury 5,488	D1
06071 Somers○ 8,473	F1
06071 Somers 1,643	F1
06072 Somersville 750	F1
06487 South Britain 390	B3
06488 Southbury○ 14,156	C3
†06238 South Coventry (Coventry) 3,769	F1
06073 South Glastonbury	E2
06489 Southington○ 36,879	D2
06785 South Kent 450	B2
06265 South Willington 450	F1
06266 South Windham 1,399	G2
06074 South Windsor○ 17,198	E1
06267 South Woodstock 1,319	G1
06075 Stafford○ 9,268	F1
06076 Stafford Springs 3,392	F1
06077 Staffordville 500	G1
*06901 Stamford 102,453	A4
Stamford‡ 198,854	A4
†06468 Stepney	B3
06377 Sterling○ 1,791	H2
06491 Stevenson 300	C3
06378 Stonington 16,220	H3
06378 Stonington 1,228	H3
06268 Storrs 11,394	F1
06497 Stratford○ 50,541	C4

06078 Suffield○ 9,294	E1
06078 Suffield 1,122	E1
06079 Taconic 400	B1
06380 Taftville	G2
06081 Tariffville 1,324	D1
06786 Terryville 5,634	C2
06787 Thomaston○ 6,276	C2
06277 Thompson○ 8,141	H1
†06082 Thompsonville	E1
06084 Tolland○ 9,694	F1
06790 Torrington 30,987	C1
06611 Trumbull○ 32,989	C4
06382 Uncasville 1,597	G3
†06076 Union○ 546	G1
06066 Vernon○ 27,974	F1
06383 Versailles 540	G2
06384 Voluntown○ 1,637	H2
06492 Wallingford○ 37,274	D3
06492 Wallingford 17,821	D3
06754 Warren○ 1,027	B2
†06278 Warrenville 500	G1
06793 Washington○ 3,657	B2
06794 Washington Depot 900	B2
*06701 Waterbury 103,266	C2
Waterbury‡ 228,178	C2
06385 Waterford○ 17,843	G3
06385 Waterford 2,736	G3
06795 Watertown○ 19,489	C2
06089 Weatogue 2,140	D1
06498 Westbrook○ 5,216	F3
06498 Westbrook 2,035	F3
06796 West Cornwall 425	B1
06090 West Granby 567	D1
06107 West Hartford○ 61,301	D1
06516 West Haven 53,184	D3
06388 West Mystic 3,364	H3
06883 Weston○ 8,284	B4
06880 Westport○ 25,290	B4
06896 West Redding 500	B3
06092 West Simsbury 2,140	D1
06109 Wethersfield○ 26,013	E2
06517 Whitneyville	D3
06226 Willimantic 14,652	G2
†06279 Willington○ 4,694	F1
06897 Wilton○ 15,351	B4
06094 Winchester○ 10,841	C1
06094 Winchester Center 350	C1
06280 Windham○ 21,062	G2
06095 Windsor○ 25,204	E1
06095 Windsor 17,517	E1
06096 Windsor Locks○ 12,190	E1
06097 Windsorville 450	E1
06098 Winsted 8,092	C1
†06417 Winthrop 750	E3
06716 Wolcott○ 13,008	D2
†06515 Woodbridge○ 7,761	D3
06798 Woodbury○ 6,942	C2
06798 Woodbury 1,290	C2
06460 Woodmont 1,797	D4
06281 Woodstock○ 5,117	H1

OTHER FEATURES

Aspetuck (res.)	B4
Bantam (lake)	C2
Barkhamsted (res.)	D1
Bear (mt.)	B1
Byram (riv.)	A4
Candlewood (lake)	A2
Coast Guard Academy	G3

Colebrook River (lake)	C1
Congamond (lkes)	E1
Connecticut (riv.)	E2
Dennis (hill)	C1
Easton (res.)	B3
Eight Mile (riv.)	F3
Farmington (riv.)	D1
French (riv.)	H1
Frissell (mt.)	B1
Gaillard (lake)	D3
Gardner (lake)	G2
Hammonasset (pt.)	E3
Hammonasset (riv.)	E3
Haystack (mt.)	C1
Highland (lake)	C1
Hockanum (riv.)	E1
Hop (riv.)	F1
Housatonic (riv.)	C3
Lillinonah (lake)	B3
Little (riv.)	G2
Long Island (sound)	C4
Mad (riv.)	C1
Mashapaug (lake)	G1
Mason (isl.)	H3
Mattabesset (riv.)	E2
Mianus (riv.)	A4
Mohawk (mt.)	B1
Moosup (riv.)	H2
Mount Hope (riv.)	F1
Mudge (pond)	B1
Mystic (riv.)	H3
Natchaug (riv.)	G1
Naugatuck (riv.)	C3
Nepaug (riv.)	D1
Niantic (riv.)	G3
Norwalk (riv.)	B4
Pachaug (pond)	H2
Pawcatuck (riv.)	H3
Pequabuck (riv.)	D2
Pequonnock (riv.)	C3
Pocotopaug (lake)	E2
Quaddick (res.)	H1
Quinebaug (riv.)	H2
Quinnipiac (riv.)	D3
Rippowam (riv.)	A4
Sachem (head)	E3
Salmon (brook)	D1
Salmon (riv.)	F2
Saugatuck (res.)	B3
Scantic (riv.)	E1
Shenipsit (lake)	F1
Shepaug (riv.)	B2
Shetucket (riv.)	G2
Silvermine (riv.)	B4
Spectacle (lkes)	B2
Still (riv.)	B3
Still (riv.)	C1
Talcott (range)	D1
Thames (riv.)	G3
Thomaston (res.)	C2
Titicus (riv.)	A3
Trap Falls (res.)	C3
Twin (lkes)	B1
Wamgumbaug (lake)	F1
Waramaug (lake)	B2
West Rock Ridge (hills)	D3
Willimantic (riv.)	F1
Wononskopomuc (lake)	B1
Yantic (riv.)	G2

‡Population of metropolitan area.
○Population of town or township.
† Zip of nearest p.o.   * Multiple zips.

06424 East Hampton 2,152	E2
06108 East Hartford 52,563	E1
06027 East Hartland 900	D1
06512 East Haven 25,028	D3
06243 East Killingly 900	H1
06333 East Lyme○ 13,870	G3
†06763 East Morris 800	C2
06612 Easton○ 5,962	B4
†06088 East Windsor 8,925	E1
06028 East Windsor Hill 500	E1
06244 East Woodstock 400	H1
06029 Ellington○ 9,711	F1
06082 Enfield 42,695	E1
06082 Enfield 8,151	E1
06426 Essex○ 5,078	F3
06426 Essex 2,501	F3
06245 Fabyan 600	H1
06430 Fairfield○ 54,849	B4
06031 Falls Village 600	B1
06032 Farmington○ 16,407	D2
06334 Fitchville 400	G2
06035 Granby 7,956	D1
06035 Granby 1,912	D1
06830 Greenwich○ 59,578	A4
06246 Grosvenor Dale 700	H1
06340 Groton○ 41,062	G3
06340 Groton 10,086	G3
06437 Guilford○ 17,375	E3
06437 Guilford 2,555	E3
06438 Haddam○ 6,383	E3
06439 Hadlyme 450	F3
06514 Hamden○ 51,071	D3
06247 Hampton○ 1,322	G1
06350 Hanover 500	G2
*06101 Hartford (cap.) 136,392	E1
Hartford‡ 726,114	E1
†06091 Hartland○ 1,416	D1
06791 Harwinton○ 4,889	C1
06791 Harwinton 3,293	C1
06440 Hawleyville 600	B3
06082 Hazardville 5,436	E1
06248 Hebron○ 5,453	F2
06441 Higganum 1,660	E2
†06040 Highland Park 500	E1
06351 Jewett City 3,294	H2
06037 Kensington 7,502	D2
06757 Kent○ 2,505	B2
†06241 Killingly○ 14,519	H1
06413 Killingworth○ 3,976	E3
†06424 Lake Pocotopaug 2,137	F2
06758 Lakeside 350	B2
06249 Lebanon○ 4,762	G2
06339 Ledyard○ 13,735	G3
†06437 Leetes Island 500	E3
†06039 Lime Rock 350	B1
06335 Gales Ferry 1,191	G3
06755 Gaylordsville 960	A2
06829 Georgetown 1,834	B4
06336 Gilman 350	G2
06336 Glasgo 450	H2
06033 Glastonbury 24,327	E2
06033 Glastonbury 7,049	E2
06756 Goshen○ 1,706	C1

Agriculture, Industry and Resources

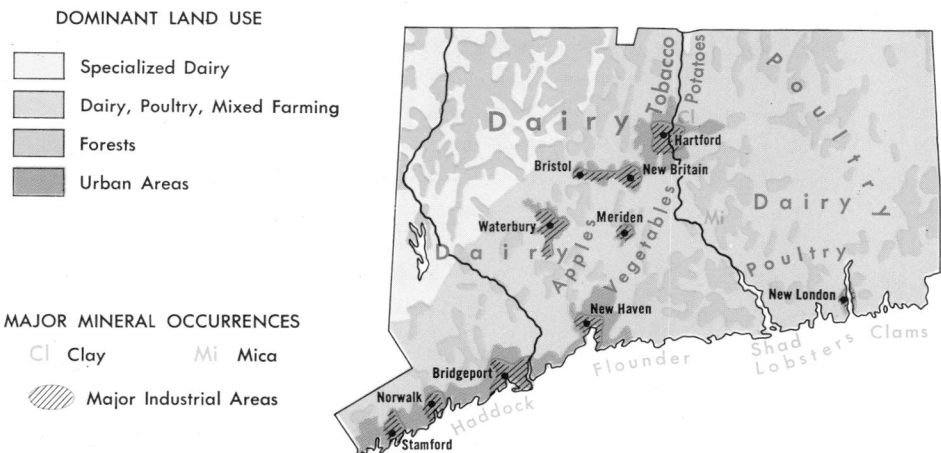

DOMINANT LAND USE

- Specialized Dairy
- Dairy, Poultry, Mixed Farming
- Forests
- Urban Areas

MAJOR MINERAL OCCURRENCES

Cl Clay   Mi Mica

Major Industrial Areas

AREA 58,664 sq. mi. (151,940 sq. km.)
POPULATION 9,746,342
CAPITAL Tallahassee
LARGEST CITY Jacksonville
HIGHEST POINT (Walton County) 345 ft. (105 m.)
SETTLED IN 1565
ADMITTED TO UNION March 3, 1845
POPULAR NAME Sunshine State; Peninsula State
STATE FLOWER Orange Blossom
STATE BIRD Mockingbird

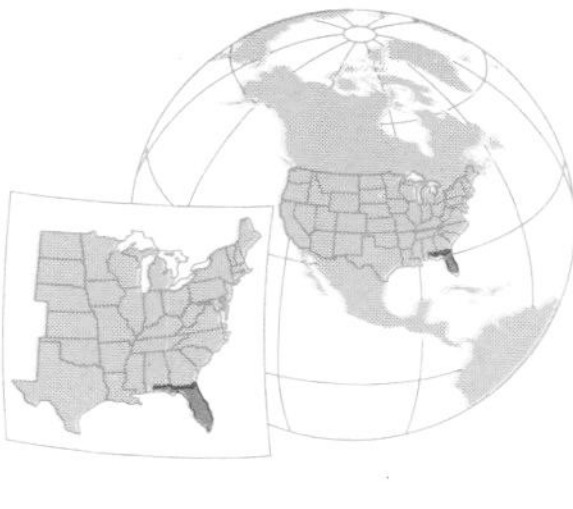

Topography

0 50 100 MI.
0 50 100 KM.

COUNTIES

Alachua 151,348 .... D2
Baker 15,289 .... D1
Bay 97,740 .... C6
Bradford 20,023 .... D2
Brevard 272,959 .... F3
Broward 101,820 .... F5
Calhoun 9,294 .... D6
Charlotte 58,460 .... E5
Citrus 54,703 .... D3
Clay 67,052 .... E2
Collier 85,791 .... E5
Columbia 35,399 .... D1
Dade 1,625,781 .... F6
De Soto 19,039 .... E4
Dixie 7,751 .... C2
Duval 571,003 .... E1
Escambia 233,794 .... B6
Flagler 10,913 .... E2
Franklin 7,661 .... B2
Gadsden 41,565 .... B1
Gilchrist 5,767 .... D2
Glades 5,992 .... E5
Gulf 10,658 .... D7
Hamilton 8,761 .... D1
Hardee 19,379 .... E4
Hendry 18,599 .... E5
Hernando 44,4693
Highlands 47,526 .... E4
Hillsborough 646,960 .... D4
Holmes 14,723 .... C5
Indian River 59,896 .... F4
Jackson 39,154 .... D5
Jefferson 10,703 .... C1
Lafayette 4,035 .... C2
Lake 104,870 .... E3
Lee 205,266 .... E5
Leon 148,655 .... B1
Levy 19,870 .... D2
Liberty 4,260 .... B1
Madison 14,894 .... C1
Manatee 148,442 .... D4
Marion 122,488 .... D2
Martin 64,014 .... F4
Monroe 63,188 .... E7
Nassau 32,894 .... E1
Okaloosa 109,920 .... C6
Okeechobee 20,264 .... F4
Orange 471,016 .... E3
Osceola 49,287 .... E3
Palm Beach 576,863 .... F5
Pasco 193,643 .... D3
Pinellas 728,531 .... D4
Polk 321,652 .... E4
Putnam 50,549 .... E2
Saint Johns 51,303 .... E2
Saint Lucie 87,182 .... F4
Santa Rosa 55,988 .... B6
Sarasota 202,251 .... D4
Seminole 179,752 .... E3
Sumter 24,272 .... D3
Suwannee 22,287 .... C1
Taylor 16,532 .... C1
Union 10,166 .... D1
Volusia 258,762 .... E2
Wakulla 10,887 .... B1
Walton 21,300 .... C6
Washington 14,509 .... C6

CITIES and TOWNS

Zip *Name/Pop.* *Key*

32615 Alachua 3,561 .... D2
32420 Alford 548 .... D6
32701 Altamonte Springs 22,028 . E3
32421 Altha 478 .... A1
33820 Alturas 900 .... E4
33501 Anna Maria 1,537 .... D4
32320 Apalachicola⊙ 2,565 .... A2
33570 Apollo Beach 4,014 .... C3
32703 Apopka 6,019 .... E3
33821 Arcadia⊙ 6,002 .... E4
32618 Archer 1,230 .... D2
33502 Aripeka 450 .... D3
32705 Astatula 755 .... E3
32233 Atlantic Beach 7,847 .... E1
33823 Auburndale 6,501 .... E3
33825 Avon Park 8,026 .... E4
32807 Azalea Park 8,301 .... E3
32530 Bagdad 1,479 .... B6
32234 Baldwin 1,526 .... E1
†33101 Bal Harbour 2,973 .... C4
33830 Bartow⊙ 14,780 .... E4
32423 Bascom 134 .... A1
†33101 Bay Harbor Islands 4,869 .. B4
†32786 Bay Lake 74 .... E3
33504 Bay Pines 5,757 .... B3
33507 Bayshore Gardens 14,945 .. D4
†33578 Bee Ridge 3,313 .... D4
32619 Bell 227 .... D2
33540 Belleair 3,673 .... B2
†33540 Belleair Beach 1,643 .... B2
33540 Belleair Bluffs 2,522 .... B3
†33540 Belleair Shores 80 .... B3
33430 Belle Glade 16,535 .... F5
†33430 Belle Glade Camp 1,645 .... F5
†32801 Belle Isle 2,848 .... E3
32620 Belleview 1,913 .... D2
†32036 Beverly Beach 217 .... E2
33152 Biscayne Park 3,088 .... B4
†32801 Bithlo 3,143 .... E3
32424 Blountstown⊙ 2,632 .... A1
33921 Boca Grande 900 .... D5
*33432 Boca Raton 49,505 .... F5
32425 Bonifay⊙ 2,534 .... C5
33923 Bonita Springs 5,435 .... E5
33834 Bowling Green 2,310 .... E4
*33435 Boynton Beach 35,624 .... F5
*33506 Bradenton⊙ 30,170 .... D4
Bradenton‡ 148,442 .... D4
33510 Bradenton Beach 1,595 .... D4
33835 Bradley 1,108 .... D4
33511 Brandon 41,826 .... D4
32008 Branford 622 .... D2
†33435 Briny Breezes 387 .... G5
32321 Bristol⊙ 1,044 .... B1
†33314 Broadview Park 6,022 .... B4
32621 Bronson⊙ 853 .... D2
32622 Brooker 429 .... D2
33512 Brooksville⊙ 5,582 .... D3
†33311 Browardale 7,409 .... B4
32010 Bunnell⊙ 1,816 .... E2
33513 Bushnell⊙ 983 .... D3
32011 Callahan 869 .... E1
32401 Calloway 7,154 .... D6
32426 Campbellton 336 .... D5
32624 Candler 275 .... E2
32920 Cape Canaveral 5,733 .... F3
33904 Cape Coral 32,103 .... E5
33055 Carol City 47,349 .... B4
Carrabelle 1,304 .... B2
32427 Caryville 633 .... C6
32707 Casselberry 15,247 .... E3
†32401 Cedar Grove 1,104 .... D6
32625 Cedar Key 700 .... C2
33514 Center Hill 751 .... D3
32535 Century 495 .... B5
†33950 Charlotte Harbor 2,084 .... E5
32324 Chattahoochee 5,332 .... B1
32626 Chiefland 1,986 .... D2
32428 Chipley⊙ 3,330 .... D6
†32548 Cinco Bayou 202 .... B6
*33515 Clearwater⊙ 85,528 .... B2
32711 Clermont 5,461 .... E3
†33950 Cleveland 2,417 .... E5
33440 Clewiston 5,219 .... E5
32922 Cocoa 16,096 .... F3
32931 Cocoa Beach 10,926 .... F3
†33060 Coconut Creek 6,288 .... F5
33521 Coleman 1,022 .... D3
33328 Cooper City 10,140 .... B4
†33559 Coral Cove 2,042 .... D4
33134 Coral Gables 43,241 .... B5
33060 Coral Springs 37,349 .... F5
33522 Cortez 3,821 .... D4
32431 Cottondale 1,056 .... D6
32327 Crawfordville⊙ 1,110 .... B1
32012 Crescent City 1,722 .... E2
32536 Crestview⊙ 7,617 .... C6
32628 Cross City⊙ 2,154 .... C2
32629 Crystal River 2,778 .... D3
33157 Cutler Ridge 20,886 .... F6
33880 Cypress Gardens 8,043 .... E3
†33472 Cypress Quarters 1,479 .... F4
33525 Dade City⊙ 4,923 .... D3
33004 Dania 11,811.B4
33837 Davenport 1,509 .... E3
33314 Davie 20,877 .... B4
*32014 Daytona Beach 54,176 .... F2
Daytona Beach‡ 258,762 .. F2
32016 Daytona Beach Shores 1,324 .... F2
32713 De Bary 4,980 .... E3
33441 Deerfield Beach 39,193 .... F5
32433 De Funiak Springs⊙ 5,563 C6
32720 De Land⊙ 15,354 .... E2
32028 De Leon Springs 1,669 .... E2
*33444 Delray Beach 34,325 .... F5
32725 Deltona 15,710 .... E3
32541 Destin 3,672 .... C6
33527 Dover 2,354 .... D4
33838 Dundee 2,227 .... E3
33528 Dunedin 30,203 .... B2
32630 Dunnellon 1,427 .... D2
33839 Eagle Lake 1,678 .... E4
†33601 East Lake-Orient Park 5,612 .... C2
†33940 East Naples 12,127 .... E5
32031 East Palatka 1,613 .... E2
32328 Eastpoint 1,246 .... B2
32751 Eatonville 2,185 .... E3
32437 Ebro 233 .... C6
32032 Edgewater 6,726 .... F3
†32801 Edgewood 1,034 .... E3
†33614 Egypt Lake 11,932 .... C2
33531 Elfers 11,396 .... D3
†33101 El Portal 1,819 .... B4
33533 Englewood 9,633 .... D5
32504 Ensley 14,422 .... B6
32425 Esto 304 .... C5
32726 Eustis 9,453 .... E3
33929 Everglades City 524 .... E6
32634 Fairfield 450 .... D2
†32693 Fanning Springs (Suwannee Riv.) 314 .... D2
32948 Fellsmere 1,161 .... F4
32034 Fernandina Beach⊙ 7,224 . E1
32922 Five Points 1,691 .... D1
32036 Flagler Beach 2,208 .... E2
32636 Floral City 1,181 .... D3
33034 Florida City 6,174 .... F6
†32960 Florida Ridge 4,988 .... F4
†33472 Fort Drum 70 .... F4
*33301 Fort Lauderdale⊙ 153,279. C4
Fort Lauderdale-Hollywood‡ 1,014,043 .. C4
33841 Fort Meade 5,546 .... E4
*33901 Fort Myers⊙ 36,638 .... E5
Fort Myers-Cape Coral‡ 205,266 .... E5
33931 Fort Myers Beach 5,753 .... E5
33842 Fort Ogden 900 .... E4
*33450 Fort Pierce⊙ 33,802 .... F4
32548 Fort Walton Beach 20,829 . C6
Fort Walton Beach‡ 109,920 .... C6
32038 Fort White 386 .... D2
32438 Fountain 900 .... D6
32439 Freeport 669 .... C6
33843 Frostproof 2,995 .... E4
32731 Fruitland Park 2,259 .... D3
33578 Fruitville 3,070 .... D4
*32601 Gainesville⊙ 81,371 .... D2
Gainesville‡ 151,348 .... D2
32732 Geneva 1,120 .... E3
33534 Gibsonton 7,219 .... C3
32960 Gifford 6,240 .... F4
32040 Glen Saint Mary 462 .... D1
†33160 Golden Beach 612 .... C4
33999 Golden Gate 4,327 .... E5
†33444 Golf 110 .... F5
32560 Gonzalez 6,084 .... B6
33933 Goodland 600 .... E6
†32502 Goulding 5,352 .... B6
33170 Goulds 7,078 .... F6
32440 Graceville 2,918 .... D5
32442 Grand Ridge 591 .... A1
33463 Greenacres City 8,843 .... F5
32043 Green Cove Springs⊙ 4,154 .... E2
32330 Greensboro 562 .... B1
32331 Greenville 1,096 .... C1
32443 Greenwood 577 .... A1
32332 Gretna 1,448 .... B1
33533 Grove City 1,932 .... D5
32736 Groveland 1,992 .... E3
32561 Gulf Breeze 5,478 .... B6
33737 Gulfport 11,180 .... B3
†33444 Gulf Stream 475 .... F5
†33301 Hacienda Village 126 .... B4
33844 Haines City 10,799 .... E3
33009 Hallandale 36,517 .... B4
32044 Hampton 466 .... D2
33440 Harlem 2,669 .... F5
32045 Hastings 636 .... E2
32333 Havana 2,782 .... B1
32640 Hawthorne 1,303 .... D2
32642 Hernando 1,653 .... D3
*33010 Hialeah 145,254 .... B4
†33010 Hialeah Gardens 2,700 .... B4
33431 Highland Beach 2,030 .... F5
33846 Highland City 1,555 .... E4
32401 Highland Park 184 .... E4
32643 High Springs 2,491 .... D2
32405 Hiland Park 4,763 .... C6
†33827 Hillcrest Heights 177 .... E4
32046 Hilliard 1,869 .... E1
†33060 Hillsboro Beach 1,554 .... F5
33455 Hobe Sound 6,822 .... F4
32047 Hollister 980 .... E2
32017 Holly Hill 9,953 .... E2
*33020 Hollywood 121,323 .... B4
33509 Holmes Beach 4,023 .... D4
*33030 Homestead 20,668 .... F6
32646 Homosassa 1,426 .... D3
32648 Horseshoe Beach 304 .... C2
32334 Hosford 750 .... B1
32737 Howey In The Hills 626 .... E3
33568 Hudson 5,799 .... D3
†33460 Hypoluxo 573 .... F5
33934 Immokalee 11,038 .... E5
32903 Indialantic 2,883 .... F3
†33139 Indian Creek 103 .... B4
†32901 Indian Harbour Beach 5,967 .... F3
32960 Indian River Shores 1,254 . F4
33535 Indian Rocks Beach 3,717 . B3
†33535 Indian Shores 984 .... B3
33456 Indiantown 3,383 .... F4
32649 Inglis 1,173 .... D2
32048 Interlachen 848 .... E2
32650 Inverness⊙ 4,095 .... D3
33036 Islamorada 1,441 .... F7
†33101 Islandia 12 .... F6
*32201 Jacksonville⊙ 540,920 .... E1
Jacksonville‡ 737,519 .... E1
32250 Jacksonville Beach 15,462 . E1
†33568 Jasmine Estates 11,995 .... D3
32052 Jasper⊙ 2,093 .... D1
32565 Jay 633 .... B5
32053 Jennings 749 .... C1
33457 Jensen Beach 6,639 .... F4
†32901 June Park 4,051 .... F3
†33404 Juno Beach 1,142 .... F5
33458 Jupiter 9,868 .... F5
†33455 Jupiter Island 364 .... F4
33849 Kathleen 1,866 .... D3
33156 Kendall 73,758 .... B5
33709 Kenneth City 4,344 .... B3
33149 Key Biscayne 6,313 .... B5
33051 Key Colony Beach 977 .... F7
33037 Key Largo 7,447 .... F6
32656 Keystone Heights 1,056 .... E2
33040 Key West⊙ 24,382 .... E7
32741 Kissimmee⊙ 15,487 .... E3
33935 La Belle⊙ 2,287 .... E5
33537 Lacoochee 1,720 .... D3
32658 La Crosse 170 .... D2
32659 Lady Lake 1,193 .... E3
33850 Lake Alfred 3,134 .... E3
†32830 Lake Buena Vista 98 .... E3
32054 Lake Butler⊙ 1,830 .... D1
†33601 Lake Carroll 13,012 .... C2
32055 Lake City⊙ 9,257 .... D1
32744 Lake Helen 2,047 .... E3
*33801 Lakeland 47,406 .... D3
Lakeland-Winter Haven‡ 321,652 .... D3
†33612 Lake Magdalene 13,331 .... D3
32746 Lake Mary 2,853 .... E3
33403 Lake Park 6,909 .... F5
33852 Lake Placid 963 .... E4
33853 Lake Wales 8,466 .... E4
*33460 Lake Worth 27,048 .... G5
33539 Land O'Lakes 4,515 .... D3
33462 Lantana 8,048 .... F5
*33540 Largo 58,977 .... B3
33308 Lauderdale-by-the-Sea 2,639 .... C3
†33313 Lauderdale Lakes 25,426 ... B3
33313 Lauderhill 37,271 .... B3
33545 Laurel 6,368 .... D4
32567 Laurel Hill 610 .... C5
32058 Lawtey 692 .... D1
†33050 Layton 88 .... F7
†33301 Lazy Lake 31 .... B3
32059 Lee 297 .... C1
32748 Leesburg 13,191 .... E3
33936 Lehigh Acres 9,604 .... E5
33033 Leisure City 17,905 .... F6
†33614 Leto 9,003 .... C2
33064 Lighthouse Point 11,488 ... F5
32060 Live Oak⊙ 6,732 .... D1
32662 Lochloosa 450 .... E2
33548 Longboat Key 4,843 .... D4
32750 Longwood 10,029 .... E3
33549 Lutz 5,555 .... D3
32444 Lynn Haven 6,239 .... C6
32063 Macclenny⊙ 3,851 .... D1

(continued on following page)

33738 Madeira Beach 4,520 .... B3
32340 Madison⊙ 3,487 .... C1
32751 Maitland 8,763 .... E3
32950 Malabar 1,118 .... F3
32445 Malone 897 .... A1
33550 Mango 6,493 .... D4
33050 Marathon 7,568 .... E7
33937 Marco (Marco Island) 4,679 .... E6
33063 Margate 35,900 .... F5
32446 Marianna⊙ 7,006 .... A1
†32084 Marineland 31 .... E2
32569 Mary Esther 3,530 .... B6
32753 Mascotte 1,112 .... E3
32066 Mayo⊙ 891 .... C1
32664 McIntosh 404 .... D2
†33101 Medley 537 .... B4
*32901 Melbourne 46,536 .... F3
Melbourne-Titusville-Cocoa‡ 272,959 .... F3
32951 Melbourne Beach 2,713 .... F3
†33301 Melrose Park 5,672 .... B4
†33561 Memphis 5,501 .... D4
32952 Merritt Island 30,708 .... F3
32410 Mexico Beach 632 .... D6
*33101 Miami⊙ 346,931 .... B5
Miami‡ 1,625,979 .... B5
33139 Miami Beach 96,298 .... C5
†33101 Miami Lakes 9,809 .... B4
33153 Miami Shores 9,244 .... B4
33166 Miami Springs 12,350 .... B5
32667 Micanopy 737 .... D2
†32960 Micco 3,585 .... F4
32343 Midway 950 .... B1
32570 Milton⊙ 7,206 .... B6
32754 Mims 7,583 .... F3
32755 Minneola 851 .... E3
33023 Miramar 32,813 .... B4
32577 Molino 1,456 .... B6
32344 Monticello⊙ 2,994 .... C1
32756 Montverde 397 .... E3
33471 Moore Haven⊙ 1,250 .... E5
32757 Mount Dora 5,883 .... E3
33860 Mulberry 2,932 .... E4
33938 Murdock 272 .... D4
32506 Myrtle Grove 14,238 .... B6
*33940 Naples⊙ 17,581 .... E5
†33940 Naples Park 5,438 .... E5
33032 Naranja 10,381 .... F6
32233 Neptune Beach 5,248 .... E1
32669 Newberry 1,826 .... D2
*33552 New Port Richey 11,196 .... D3
32069 New Smyrna Beach 13,557 F2
32578 Niceville 8,543 .... C6
33555 Nokomis 3,108 .... D4
32452 Noma 113 .... C5
†33169 Norland 19,471 .... B4
33141 North Bay Village 4,920 .... B4
33903 North Fort Myers 22,808 .... E5
†33010 North Lauderdale 18,653 .... B3
33161 North Miami 42,566 .... B4
33161 North Miami Beach 36,481 C4
33940 North Naples 7,950 .... E5
33403 North Palm Beach 11,344 .... F5
33595 North Port 6,205 .... D4
†33708 North Redington Beach 1,156 .... B3
32759 Oak Hill 938 .... F3
32760 Oakland 658 .... E3
33334 Oakland Park 23,035 .... B3
*32670 Ocala⊙ 37,170 .... D2
Ocala‡ 122,488 .... D2
†33457 Ocean Breeze Park 469 .... F4
33444 Ocean Ridge 1,355 .... F5
32761 Ocoee 7,803 .... E3
33163 Ojus 17,344 .... B4
32762 Okahumpka 900 .... D3
33472 Okeechobee⊙ 4,225 .... F4
33557 Oldsmar 2,608 .... B2
33558 Oneco 6,417 .... D4
33054 Opa Locka 14,460 .... B4
32763 Orange City 2,795 .... E3
32073 Orange Park 8,766 .... E1
†32970 Orchid 42 .... F4
*32801 Orlando⊙ 128,291 .... E3
Orlando‡ 700,699 .... E3
32074 Ormond Beach 21,378 .... E2
32074 Ormond-by-the-Sea 7,665 . E2
33559 Osprey 1,660 .... D4
32683 Otter Creek 167 .... D2
32765 Oviedo 3,074 .... E3
32570 Pace 5,006 .... B6
33476 Pahokee 6,346 .... F5
†32036 Painters Hill 40 .... E2
32077 Palatka⊙ 10,175 .... E2
32905 Palm Bay 18,560 .... F3
33480 Palm Beach 9,729 .... G4
†33403 Palm Beach Gardens 14,407 .... F5
†33404 Palm Beach Shores 1,232 . G5
33490 Palm City 2,177 .... F4
32037 Palm Coast 2,837 .... E2
33561 Palmetto 8,637 .... D4
33563 Palm Harbor 5,215 .... D3
33619 Palm River-Clair Mel 14,447 .... C3
†32901 Palm Shores 77 .... F3
33460 Palm Springs 8,166 .... F5
*32401 Panama City⊙ 33,346 .... C6
Panama City‡ 97,740 .... C6
32407 Panama City Beach 2,148 .... C6
32401 Parker 4,298 .... C6
†33441 Parkland 545 .... F5
32538 Paxton 659 .... C5
†33023 Pembroke Park 4,783 .... B4
33024 Pembroke Pines 35,776 .... B4
32079 Penney Farms 630 .... E2
†33010 Pennsuco 15 .... B4
*32501 Pensacola⊙ 57,619 .... B6
Pensacola‡ 289,782 .... B6
33157 Perrine 16,129 .... F6
32347 Perry⊙ 8,254 .... C1
32080 Pierson 1,085 .... E2
32808 Pine Hills 35,771 .... E3
33565 Pinellas Park 32,811 .... B3
33317 Plantation 48,653 .... B4
33566 Plant City 17,064 .... D3
33868 Polk City 576 .... E3
32081 Pomona Park 791 .... E2
*33060 Pompano Beach 52,618 .... F5
32455 Ponce de Leon 454 .... C6
†32019 Ponce Inlet 1,003 .... F2
33952 Port Charlotte 25,770 .... D5
32019 Port Orange 18,756 .... F2
33568 Port Richey 2,165 .... D3
32456 Port Saint Joe 4,027 .... D6
33452 Port Saint Lucie 14,690 .... F4
33492 Port Salerno 4,511 .... F4
33032 Princeton 10,381 .... F6
*33950 Punta Gorda⊙ 6,797 .... E5
32351 Quincy⊙ 8,591 .... B1
32083 Raiford 259 .... D1
32686 Reddick 657 .... D2
33708 Redington Beach 1,708 .... B3
†33708 Redington Shores 2,142 .... B3
†33158 Richmond Heights 8,577 .... F6
†33301 Riverland 5,919 .... B4
33404 Riviera Beach 26,489 .... G5
32955 Rockledge 11,877 .... F3
32957 Roseland 1,607 .... F4
33570 Ruskin 5,117 .... C3
33572 Safety Harbor 6,461 .... B2
32084 Saint Augustine⊙ 11,985 .... E2
32084 Saint Augustine Beach 1,289 .... E2
32769 Saint Cloud 7,840 .... E3
33956 Saint James City 1,298 .... D5
33574 Saint Leo 917 .... D3
33452 Saint Lucie 593 .... F4
32355 Saint Marks 286 .... B1
*33701 Saint Petersburg 238,647 .... B3
33736 Saint Petersburg Beach 9,354 .... B3
†33508 Samoset 5,747 .... D4
†32069 Samsula 1,971 .... E2
33576 San Antonio 529 .... D3
32771 Sanford⊙ 23,176 .... E3
33957 Sanibel 3,363 .... D5
*33577 Sarasota⊙ 48,868 .... D4
Sarasota‡ 202,251 .... D4
†33577 Sarasota Springs 13,860 .... D4
32935 Satellite Beach 9,163 .... F3
32775 Scottsmoor 900 .... F3
†33301 Sea Ranch Lakes 584 .... C3
32958 Sebastian 2,831 .... F4
33870 Sebring⊙ 8,736 .... E4
33584 Seffner 6,493 .... D4
33542 Seminole 4,586 .... B3
†33457 Sewalls Point 1,187 .... F4
32579 Shalimar 390 .... C6
32959 Sharpes 4,149 .... F3
32688 Silver Springs 1,082 .... D2
32460 Sneads 1,690 .... B1
32358 Sopchoppy 444 .... B1
33493 South Bay 3,886 .... F5
32021 South Daytona 11,252 .... F2
33143 South Miami 10,944 .... B5
†33157 South Miami Heights 23,559 .... F6
33707 South Pasadena 4,188 .... B3
†32901 South Patrick Shores 9,816 .... F3
†32401 Southport 1,992 .... C6
33452 South Port Saint Lucie (Port Saint Lucie 14,690 .... F4
33595 South Venice 8,075 .... D4
32690 Sparr 902 .... D2
32401 Springfield 7,220 .... D6
32091 Starke⊙ 5,306 .... D2
33494 Stuart⊙ 9,467 .... F4
33586 Sun City D4
†33570 Sun City Center 5,605 .... C3
†33450 Sunland Gardens F4
33160 Sunny Isles 12,564 .... C4
33313 Sunrise 39,681 .... B4
33154 Surfside 3,763 .... B4
32692 Suwannee (Fanning Sprs.) 314 .... C2
†33144 Sweetwater 8,251 .... B5
†32043 Switzerland 3,906 .... E1
32809 Taft 900 .... E3
*32301 Tallahassee (cap.)⊙ 81,548 .... B1
Tallahassee‡ 159,542 .... B1
†33321 Tamarac 29,376 .... B3
*33601 Tampa⊙ 271,523 .... C2
Tampa-Saint Petersburg‡ 1,569,492 .... C2
*33589 Tarpon Springs 13,251 .... D3
32778 Tavares⊙ 4,103 .... E3
33070 Tavernier 1,834 .... F6
33617 Temple Terrace 11,097 .... C2
33458 Tequesta 3,685 .... F5
33905 Tice 6,645 .... E5
32780 Titusville⊙ 31,910 .... F3
33740 Treasure Island 6,316 .... B3
32693 Trenton⊙ 1,131 .... D2
32784 Umatilla 1,872 .... E3
33620 University 24,514 .... C2
32580 Valparaiso 6,142 .... C6
*33595 Venice 12,153 .... D4
32462 Vernon 885 .... C6
32960 Vero Beach⊙ 16,176 .... F4
†33166 Virginia Gardens 2,098 .... B5
32970 Wabasso 2,157 .... F4
†32327 Wakulla 225 .... B1
32694 Waldo 993 .... D2
†32456 Ward Ridge 104 .... D6
32507 Warrington 15,792 .... B6
†32055 Watertown 3,804 .... D1
33873 Wauchula⊙ 2,986 .... E4
32463 Wausau 347 .... D6
33877 Waverly 1,208 .... E4
33597 Webster 856 .... D3
†33512 Weeki Wachee 8 .... D3
32093 Welaka 492 .... E2
32935 West Eau Gallie 2,591 .... F3
†32901 West Melbourne 5,078 .... F3
†33101 West Miami 6,076 .... B5
*33401 West Palm Beach⊙ 63,305 F5
West Palm Beach-Boca Raton‡ 573,125 .... F5
†32502 West Pensacola 24,371 .... B6
32464 Westville 343 .... C6
†33165 Westwood Lakes 11,478 .... B5
32465 Wewahitchka⊙ 1,742 .... D6
†32465 White City 4,110 .... F4
32096 White Springs 781 .... D1
32785 Wildwood 2,665 .... D3
32696 Williston 2,240 .... D2
33334 Wilton Manors 12,742 .... B3
33598 Wimauma 1,477 .... D4
32786 Windermere 1,302 .... E3
33880 Winter Haven 21,119 .... E3
*32789 Winter Park 22,339 .... E3
†32801 Winter Springs 10,475 .... E3
32362 Woodville 1,768 .... B1
32697 Worthington Springs 220 .... D2
32698 Yankeetown 600 .... D2
32097 Yulee 3,168 .... E1
32798 Zellwood 1,760 .... E3
33599 Zephyrhills 5,742 .... D3
33890 Zolfo Springs 1,495 .... E4

OTHER FEATURES

Alapaha (riv.) C1
Alligator (lake) E3
Amelia (isl.) E1
Anastasia (isl.) E2
Anclote (keys) D3
Apalachee (bay) B2
Apalachicola (bay) B2
Apalachicola (riv.) A1
Apopka (lake) E3
Arbuckle (lake) E4
Aucilla (riv.) C1
Banana (riv.) F3
Beresford (lake) E3
Big Cypress (swamp) E5
Big Cypress Nat'l Preserve E5
Biscayne (bay) F6
Biscayne (key) B5
Biscayne Nat'l Park F6
Blackwater (riv.) B6
Blue Cypress (lake) F4
Boca Chica (key) E7
Boca Ciega (bay) B3
Boca Grande (key) D7
Bryant (lake) E2
Caloosahatchee (riv.) E5
Captiva (isl.) D5
Casey (key) D4
Castillo de San Marcos Nat'l Mon. E2
Cecil Field Naval Air Sta. E1
Charlotte (harb.) D5
Chattahoochee (riv.) B1
Chipola (riv.) D6
Choctawhatchee (riv.) C6
Crescent (lake) E2
Cumberland Island Nat'l Seashore E1
Cypress (lake) E3
De Soto Nat'l Mem. D4
Dead (lake) D6
Dexter (lake) E2
Dog (isl.) B2
Dorr (lake) E2
Dry Tortugas (keys) D7
Dumfoundling (bay) C4
East (pt.) E6
Eglin A.F.B. 7,574 .... C6
Egmont (key) D4
Elliott (key) F6
Escambia (riv.) B6
Estero (isl.) E5
Eureka (res.) E2
Everglades, The (swamp) F6
Everglades Nat'l Park F6
Fenholloway (riv.) C1
Florida (bay) F6
Florida (cape) F6
Florida (keys) E7
Florida (strs) F7
Fort Caroline Nat'l Mem. E1
Fort Jefferson Nat'l Mon. C7
Fort Matanzas Nat'l Mon. E2
Gasparilla (isl.) D5
George (lake) E2
Grassy (key) F7
Gulf Island Nat'l Seashore B6
Harney (lake) F3
Hart (lake) E3
Hillsborough (bay) C3
Hillsborough (canal) F5
Hillsborough (riv.) C2
Homosassa (isls.) D3
Homestead A.F.B. 7,594 .... F6
Iamonia (lake) B1
Indian (riv.) F3
Iron (mt.) E4
Istokpoga (lake) E4
Jackson (lake) B1
Jackson (lake) E4
Jacksonville Naval Air Sta. E1
John F. Kennedy Space Center F3
June in Winter (lake) E4
Kennedy (Canaveral) (cape) F3
Kerr (lake) ....
Key Largo (key) ....
Key Vaca (key) ....
Key West Naval Air Sta. ....
Kissimmee (lake) ....
Kissimmee (riv.) ....
Largo (key) ....
Levy (lake) ....
Lochloosa (lake) ....
Long (key) ....
Long (key) ....
Longboat (key) ....
Lower Matecumbe (key) ....
Lowery (lake) ....
MacDill A.F.B. ....
Manatee (riv.) ....
Marco (isl.) ....
Marian (lake) ....
Marquesas (keys) ....
Matanzas (inlet) ....
Mayport Naval Air Sta. ....
McCoy A.F.B. ....
Merritt (isl.) ....
Mexico (gulf) ....
Miami (canal) ....
Miami (riv.) ....
Miccosukee (lake) ....
Monroe (lake) ....
Mosquito (lag.) ....
Mullet (key) ....
Myakka (riv.) ....
Nassau (riv.) ....
Nassau (sound) ....
New (riv.) ....
New (riv.) ....
Newnans (lake) ....
North Merritt (isl.) ....
North New River (canal) ....
Ochlockonee (riv.) ....
Okaloacoochee Slough (swamp) ....
Okeechobee (lake) ....
Okefenokee (swamp) ....
Oklawaha (riv.) ....
Old Rhodes (key) ....
Old Tampa (bay) ....
Olustee (riv.) ....
Orange (lake) ....
Patrick A.F.B. 2,843 ....
Peace (riv.) ....
Pensacola (bay) ....
Pensacola Naval Air Sta. ....
Perdido (riv.) ....
Pine (isl.) ....
Pine Island (sound) ....
Pine Log (creek) ....
Pinellas (pt.) ....
Piney (isl.) ....
Piney (pt.) ....
Placid (lake) ....
Plantation (key) ....
Poinsett (lake) ....
Ponce de Leon (bay) ....
Port Everglades (harb.) ....
Port Tampa (harb.) ....
Reedy (lake) ....
Romano (cape) ....
Sable (cape) ....
Saint Andrew (pt.) ....
Saint George (cape) ....
Saint George (isl.) ....
Saint George (sound) ....
Saint Johns (riv.) ....
Saint Joseph (bay) ....
Saint Joseph (pt.) ....
Saint Lucie (canal) ....
Saint Lucie (inlet) ....
Saint Marys (riv.) ....
Saint Marys Entrance (inlet) ....
Saint Vincent (isl.) ....
San Blas (cape) ....
Sand (key) ....
Sands (key) ....
Sanibel (isl.) ....
Santa Fe (lake) ....
Santa Fe (riv.) ....
Santa Rosa (isl.) ....
Santa Rosa (sound) ....
Sarasota (pt.) ....
Seminole (lake) ....
Seminole Ind. Res. ....
Seminole Ind. Res. ....
Shark (pt.) ....
Shoal (riv.) ....
Snake Creek (canal) ....
South New River (canal) ....
Stafford (lake) ....
Sugarloaf (key) ....
Suwannee (riv.) ....
Suwannee (sound) ....
Talbot (isl.) ....
Talquin (lake) ....
Tamiami (canal) ....
Tampa (bay) ....
Ten Thousand (isls.) ....
Torch (key) ....
Treasure (isl.) ....
Tsala Apopka (lake) ....
Tyndall A.F.B. 4,542 ....
Upper Matecumbe (key) ....
Vaca (key) ....
Virginia (key) ....
Waccasassa (bay) ....
Waccasassa (riv.) ....
Washington (lake) ....
Weir (lake) ....
Weohyakapka (lake) ....
West Palm Beach (canal) ....
Whitewater (bay) ....
Whiting Field Naval Air Sta. ....
Wimico (lake) ....
Winder (lake) ....
Withlacoochee (riv.) ....
Withlacoochee (riv.) ....
Yale (lake) ....
Yellow (riv.) ....

⊙County seat.
‡Population of metropolitan area.
† Zip of nearest p.o. * Multiple zip

Agriculture, Industry and Resources

Cotton
Peanuts
Corn
Hogs
Pensacola
Oysters
Red Snapper
Mullet
Tobacco
Peanuts
Jacksonville
Menhaden
Potatoes
Hogs
Melons
Vegetables
Oranges
Groupers
Crabs
Mackerel
Orlando
Grapefruit
Sponges
Tampa
St. Petersburg
Cattle
Oranges
Citrus Fruit
Kingfish
W. Palm Beach
Sugar-cane
Vegetables
Shrimp
Clams
Miami
Lobsters
Shrimp
Sponges
DOMINANT LAND USE
Fruit, Truck & Mixed Farming
Truck & Mixed Farming
Truck Farming
Cotton, Tobacco, Hogs, Peanuts
Peanuts, General Farming
General Farming, Forest Products, Truck Farming, Cotton
Livestock Grazing
Forests
Swampland, Limited Agriculture
Urban Areas
Nonagricultural Land
MAJOR MINERAL OCCURRENCES
Cl Clay
Ls Limestone
O Petroleum
P Phosphates
Pe Peat
Ti Titanium
Zr Zirconium
Water Power
Major Industrial Areas

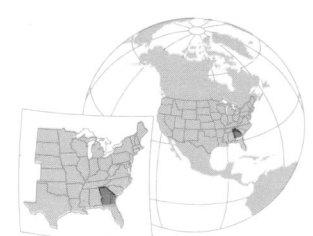

AREA 58,910 sq. mi. (152,577 sq. km.)
POPULATION 5,463,105
CAPITAL Atlanta
LARGEST CITY Atlanta
HIGHEST POINT Brasstown Bald 4,784 ft.
(1458 m.)
SETTLED IN 1733
ADMITTED TO UNION January 2, 1788
POPULAR NAME Empire State of the South;
Peach State
STATE FLOWER Cherokee Rose
STATE BIRD Brown Thrasher

COUNTIES

ppling 15,565	H7	
tkinson 6,141	G8	
acon 9,379	G7	
aker 3,808	D8	
aldwin 34,686	F4	
anks 8,702	E2	
arrow 21,293	E2	
artow 40,760	C2	
en Hill 16,000	F7	
errien 13,525	F8	
ibb 151,085	E5	
leckley 10,767	F6	
rantley 8,701	J8	
rooks 15,255	E9	
ryan 10,175	K6	
ulloch 35,785	J6	
urke 19,349	J4	
utts 13,665	E4	
alhoun 5,717	C7	
amden 13,371	J9	
andler 7,518	H6	
arroll 56,346	B3	
atoosa 36,991	B1	
harlton 7,343	H9	
hatham 202,226	K6	
hattooga 21,856	B1	
hattahoochee 21,732	C6	
herokee 51,699	D2	
larke 74,498	F3	
lay 3,553	B7	
layton 150,357	D3	
inch 6,660	G9	

Cobb 297,694	C3	
Coffee 26,894	G8	
Colquitt 35,376	E8	
Columbia 40,118	H3	
Cook 13,490	F8	
Coweta 39,268	C4	
Crawford 7,684	E5	
Crisp 19,489	E7	
Dade 12,318	A1	
Dawson 4,774	D2	
Decatur 25,495	C9	
De Kalb 483,024	D3	
Dodge 16,955	F6	
Dooly 10,826	E6	
Dougherty 100,978	D7	
Douglas 54,573	C3	
Early 13,158	C8	
Echols 2,297	G9	
Effingham 18,327	K6	
Elbert 18,758	G2	
Emanuel 20,795	H5	
Evans 8,428	J6	
Fannin 14,748	D1	
Fayette 29,043	C4	
Floyd 79,800	B2	
Forsyth 27,958	D2	
Franklin 15,185	F2	
Fulton 589,904	D3	
Gilmer 11,110	D1	
Glascock 2,382	G4	
Glynn 54,981	J8	
Gordon 30,070	C2	
Grady 19,845	D9	
Greene 11,391	F3	

Gwinnett 166,903	D2	
Habersham 25,020	E1	
Hall 75,649	E2	
Hancock 9,466	G4	
Haralson 18,422	B3	
Harris 15,464	C5	
Hart 18,585	G2	
Heard 6,520	B4	
Henry 36,309	D4	
Houston 77,605	E6	
Irwin 8,988	F7	
Jackson 25,343	E2	
Jasper 7,553	E4	
Jeff Davis 11,473	G7	
Jefferson 18,403	H4	
Jenkins 8,841	J5	
Johnson 8,660	G5	
Jones 16,579	E5	
Lamar 12,215	D4	
Lanier 5,654	F8	
Laurens 36,990	G6	
Lee 11,684	D7	
Liberty 37,583	J7	
Lincoln 6,949	H3	
Long 4,524	J7	
Lowndes 67,972	F9	
Lumpkin 10,762	D1	
Macon 14,003	D6	
Madison 17,747	F2	
Marion 5,297	C6	
McDuffie 18,546	H4	
McIntosh 8,046	K7	
Meriwether 21,229	C4	
Miller 7,038	C8	

Mitchell 21,114	D8	
Monroe 14,610	E4	
Montgomery 7,011	G6	
Morgan 11,572	F3	
Murray 19,685	C1	
Muscogee 170,108	C6	
Newton 34,489	E3	
Oconee 12,427	F3	
Oglethorpe 8,929	F3	
Paulding 26,042	C3	
Peach 19,151	E5	
Pickens 11,652	D2	
Pierce 11,897	H8	
Pike 8,937	D4	
Polk 32,386	B3	
Pulaski 8,950	E6	
Putnam 10,295	F4	
Quitman 2,357	B7	
Rabun 10,466	F1	
Randolph 9,599	C7	
Richmond 181,629	H4	
Rockdale 36,747	D3	
Schley 3,433	D6	
Screven 14,043	J5	
Seminole 9,057	C9	
Spalding 47,899	D4	
Stephens 21,763	F1	
Stewart 5,896	C6	
Sumter 29,360	D6	
Talbot 6,536	C5	
Taliaferro 2,032	G3	
Tattnall 18,134	J6	
Taylor 7,902	D5	
Telfair 11,445	G7	

Terrell 12,017	D7	
Thomas 38,098	E9	
Tift 32,862	E7	
Toombs 22,592	H6	
Towns 5,638	E1	
Treutlen 6,087	G6	
Troup 50,003	B4	
Turner 9,510	E7	
Twiggs 9,354	F5	
Union 9,390	E1	
Upson 25,998	D5	
Walker 56,470	B1	
Walton 31,211	E3	
Ware 37,180	H8	
Warren 6,583	G4	
Washington 18,842	G4	
Wayne 20,750	J7	
Webster 2,341	C6	
Wheeler 5,155	G6	
White 10,120	E1	
Whitfield 65,780	B1	
Wilcox 7,682	F7	
Wilkes 10,951	G3	
Wilkinson 10,368	F5	
Worth 18,064	E8	

CITIES and TOWNS

Zip	Name/Pop.	Key
31001	Abbeville⊙ 985	F7
30101	Acworth 3,648	C2
30103	Adairsville 1,739	C2
31620	Adel⊙ 5,592	F8
31002	Adrian 756	G5
30410	Ailey 579	G6
30411	Alamo⊙ 993	G6
31622	Alapaha 771	F8
*31701	Albany⊙ 74,550	D7
	Albany‡ 112,456	D7
†30204	Aldora 139	D4
31301	Allenhurst 606	J7
31003	Allentown 321	F5
31510	Alma⊙ 3,819	G7
30201	Alpharetta 3,128	D2
30412	Alston 111	H6
30510	Alto 618	E2
†30161	Alto Park	B2
31512	Ambrose 360	G7
31709	Americus⊙ 16,120	D6
31711	Andersonville 267	D6
30802	Appling⊙ 150	H3
31712	Arabi 376	E7
30104	Aragon 855	B2
†30549	Arcade 223	E2
†31520	Arco	J8
31623	Argyle 206	G8
31713	Arlington 1,572	C8
30619	Arnoldsville 187	F3
31714	Ashburn⊙ 4,766	E7
*30601	Athens⊙ 42,549	F3
	Athens‡ 130,015	F3
*30301	Atlanta (cap.)⊙ 425,022	K1
	Atlanta‡ 2,029,618	K1
31715	Attapulgus 623	D9
30203	Auburn 692	E2
*30901	Augusta⊙ 47,532	J4
	Augusta‡ 327,372	J4
30001	Austell 3,939	J1
†30557	Avalon 200	F1
30803	Avera 248	G4
30002	Avondale Estates 1,313	L1
31716	Baconton 763	D8
31717	Bainbridge⊙ 10,553	C9
30511	Baldwin 1,080	E2
30107	Ball Ground 640	D2
30204	Barnesville⊙ 4,887	D4
31625	Barney 146	E8
30413	Bartow 357	G5
31720	Barwick 413	E9
31513	Baxley⊙ 3,586	H7
†31554	Beach	G8
30414	Bellville 173	H6
31721	Benevolence 138	C7
†30136	Berkeley Lake 503	D3
31722	Berlin 538	E8
30620	Bethlehem 281	E3
†31901	Bibb City 667	B5
30621	Bishop 172	F3
31516	Blackshear⊙ 3,222	H8
30512	Blairsville⊙ 530	E1
31723	Blakely⊙ 5,880	C8
31302	Bloomingdale 1,855	K6
30513	Blue Ridge⊙ 1,376	D1
31724	Bluffton 132	C7
30805	Blythe 367	H4
30622	Bogart 819	E3
31626	Boston 1,424	E9
30623	Bostwick 357	E3
30108	Bowdon 1,743	B3
30516	Bowersville 318	G2
30624	Bowman 890	G2
30517	Braselton 308	E2
†30153	Braswell 282	C3
30110	Bremen 3,966	B3
31725	Brinson 274	C9
31726	Bronwood 524	D7

Zip	Name/Pop.	Key
30415	Brooklet 1,035	J6
30205	Brooks 199	D4
31519	Broxton 1,117	G7
31520	Brunswick⊙ 17,605	K8
30113	Buchanan⊙ 1,019	B3
30625	Buckhead 219	F3
31803	Buena Vista⊙ 1,544	C6
30518	Buford 6,578	D2
31006	Butler⊙ 1,959	D5
31007	Byromville 567	E6
31008	Byron 1,661	E5
31009	Cadwell 353	G6
31728	Cairo⊙ 8,777	D9
30701	Calhoun⊙ 5,335	C1
30807	Camak 283	G4
31730	Camilla⊙ 5,414	D8
30520	Canon 704	F2
30114	Canton⊙ 3,601	C2
30203	Carl 239	E3
30627	Carlton 291	F2
30521	Carnesville⊙ 465	F2
30117	Carrollton⊙ 14,078	C3
30120	Cartersville⊙ 9,247	C2
30124	Cave Spring 883	B2
31627	Cecil 280	F8
30125	Cedartown⊙ 8,619	B2
†30601	Center 330	F2
31028	Centerville 2,622	E5
†30217	Centralhatchee 240	B4
†31816	Chalybeate Springs 265	C5
30341	Chamblee 7,137	K1
30705	Chatsworth⊙ 2,493	C1
31011	Chauncey 350	F6
31012	Chester 409	F6
30707	Chickamauga 2,232	B1
30523	Clarkesville⊙ 1,348	F1
30021	Clarkston 4,539	L1
30417	Claxton⊙ 2,694	J6
30525	Clayton⊙ 1,838	F1
30527	Clermont 300	E2
30528	Cleveland⊙ 1,578	E1
31735	Cobb	E7
30420	Cobbtown 494	H6
31014	Cochran⊙ 5,121	F6
30710	Cohutta 407	C1
30628	Colbert 498	F2
31736	Coleman 164	C7
30337	College Park 24,632	K2
30421	Collins 639	H6
31737	Colquitt⊙ 2,065	C8
*31901	Columbus⊙ 169,441	C6
	Columbus‡ 239,196	C6
30629	Comer 930	F2
30529	Commerce 4,092	E2
30206	Concord 317	D4
*30207	Conyers⊙ 6,567	D3
31738	Coolidge 736	E8
31015	Cordele⊙ 11,184	E7
30531	Cornelia 3,203	E1
31739	Cotton 122	D8
30209	Covington⊙ 10,586	E3
30711	Crandall	C1
30630	Crawford 498	F3
30631	Crawfordville⊙ 594	G3
†31771	Crosland	E8
31016	Culloden 281	D5
30130	Cumming⊙ 2,094	D2
31805	Cusseta⊙ 1,218	C6
31740	Cuthbert⊙ 4,340	C7
30211	Dacula 1,577	E3
30533	Dahlonega⊙ 2,844	D1
30423	Daisy 174	J6
30132	Dallas⊙ 2,440	C3
30720	Dalton⊙ 20,743	C1
31741	Damascus 403	C8
30633	Danielsville⊙ 354	F2
31017	Danville 529	F5
31305	Darien⊙ 1,731	K8
31601	Dasher 659	F9
31018	Davisboro 433	G5
31742	Dawson⊙ 5,699	D7
30534	Dawsonville⊙ 342	D2
30808	Dearing 539	H4
*30030	Decatur⊙ 18,404	K1
†31501	Deenwood	H8
31082	Deepstep 120	G4
30535	Demorest 1,130	F1
31532	Denton 286	G7
31743	De Soto 248	D7
31019	Dexter 527	G6
30537	Dillard 238	F1
31629	Dixie 259	E9
†31520	Dock Junction (Arco)	J8
31744	Doerun 1,062	E8
31745	Donalsonville⊙ 3,320	C8
30340	Doraville 7,414	K1
31533	Douglas⊙ 10,980	G7
*30133	Douglasville⊙ 7,641	C3
31021	Dublin⊙ 16,083	G5
31022	Dudley 425	F5
30136	Duluth 2,956	D2
31630	Du Pont 267	G9
†31830	Durand 206	C5
31021	East Dublin 2,916	G5
30539	East Ellijay 469	C1

(continued on following page)

Agriculture, Industry and Resources

DOMINANT LAND USE

- Specialized Cotton
- Cotton, General Farming
- Cotton, Tobacco, Hogs, Peanuts
- Peanuts, General Farming
- General Farming, Livestock, Fruit, Tobacco
- General Farming, Forest Products, Cotton, Truck Farming
- Forests
- Swampland, Limited Agriculture
- Urban Areas

MAJOR MINERAL OCCURRENCES

Al	Bauxite
Ba	Barite
C	Coal
Cl	Clay
Fe	Iron Ore
Gn	Granite
Mi	Mica
Mn	Manganese
Mr	Marble
Sl	Slate
Tc	Talc
Ti	Titanium

⚡ Water Power   ▨ Major Industrial Areas

(continued)

†31046 East Juliette E4
31023 Eastman⊙ 5,330 F6
†30263 East Newnan C4
30344 East Point 37,486 K2
†30677 Eastville E3
31024 Eatonton⊙ 4,833 F4
31307 Eden 990 K6
31746 Edison 1,128 C7
30635 Elberton⊙ 5,686 G2
31806 Ellaville⊙ 1,684 D6
31747 Ellenton 277 E8
31807 Ellerslie 700 C5
30540 Ellijay⊙ 1,507 C1
30137 Emerson 1,110 C2
31749 Enigma 574 F8
†30217 Ephesus 184 B4
30724 Eton 301 C1
†30120 Euharlee 477 C2
30809 Evans H3
30212 Experiment D4
30213 Fairburn 3,466 J2
30139 Fairmount 842 C2
30214 Fayetteville⊙ 2,715 C4
†31071 Finleyson 101 F6
31750 Fitzgerald⊙ 10,187 F7
†31313 Flemington 440 K7
30216 Flovilla 458 E4
30542 Flowery Branch 755 E2
31537 Folkston⊙ 2,243 H9
30050 Forest Park 18,782 K2
31029 Forsyth⊙ 4,624 E4
31751 Fort Gaines⊙ 1,260 C7
30742 Fort Oglethorpe 5,443 B1
31030 Fort Valley⊙ 9,000 E5
30217 Franklin⊙ 711 B4
30639 Franklin Springs 797 F2
31753 Funston 337 E8
30501 Gainesville⊙ 15,280 E2
31408 Garden City 6,895 K6
30425 Garfield 222 H5
30218 Gay 175 C4
31810 Geneva 232 C5
31754 Georgetown⊙ 935 B7
30810 Gibson⊙ 730 G4
30426 Girard 225 J4
30427 Glennville 4,144 J7
30428 Glenwood 824 L1
30641 Good Hope 200 E3
31031 Gordon 2,768 F5
30220 Grantville 1,110 C4
31032 Gray⊙ 2,145 F4
30221 Grayson 464 E3
30726 Graysville 193 B1
30642 Greensboro⊙ 2,985 F3
30222 Greenville⊙ 1,213 C4
30223 Griffin⊙ 20,728 D4
30813 Grovetown 3,491 H4
31312 Guyton 749 K6
31033 Haddock 800 F4
30429 Hagan 880 J6

31632 Hahira 1,534 F9
31811 Hamilton⊙ 506 C5
30228 Hampton 2,059 D4
30354 Hapeville 6,166 K2
30229 Haralson 123 C4
31034 Hardwick F4
30814 Harlem 1,485 H4
31035 Harrison 456 G5
30643 Hartwell⊙ 4,855 G2
31036 Hawkinsville⊙ 4,372 E6
31539 Hazlehurst⊙ 4,249 G7
30545 Helen 265 E1
31037 Helena 1,390 G6
30815 Hephzibah 1,452 H4
30546 Hiawassee⊙ 491 E1
†30410 Higgston 152 G6
30467 Hilltonia 515 J5
31313 Hinesville⊙ 11,309 J7
30141 Hiram 711 C3
31542 Hoboken 514 H8
30230 Hogansville 3,362 C4
30142 Holly Springs 687 D2
†31537 Homeland 683 H9
30547 Homer⊙ 734 F2
31634 Homerville⊙ 3,112 G8
30548 Hoschton 490 E2
30646 Hull 188 F2
31041 Ideal 619 D6
30647 Ila 287 F2
†30705 Industrial City 1,054 C1
31759 Iron City 367 C8
31042 Irwinton⊙ 841 F5
†31031 Ivey 455 F5
30233 Jackson⊙ 4,133 E4
31544 Jacksonville 206 G7
31761 Jakin 194 C8
30143 Jasper⊙ 1,556 D2
30549 Jefferson⊙ 1,820 F2
31044 Jeffersonville⊙ 1,473 F5
30234 Jenkinsburg 360 E4
30235 Jersey 201 E3
31545 Jesup⊙ 9,418 J7
30236 Jonesboro⊙ 4,132 D4
31812 Junction City 254 C5
30144 Kennesaw 5,095 C2
31548 Kingsland 2,008 J9
30145 Kingston 733 C2
31049 Kite 328 G5
31050 Knoxville⊙ 75 E5
30728 La Fayette⊙ 6,517 B1
30240 La Grange⊙ 24,204 B4
30252 Lake 2,963 D3
31635 Lakeland⊙ 2,647 F8
31636 Lake Park 448 F9
30553 Lavonia 2,024 F2
30245 Lawrenceville⊙ 8,928 D3
31762 Leary 783 C8
30146 Lebanon 800 D2
31763 Leesburg⊙ 1,301 D7
31637 Lenox 965 F8

31764 Leslie 470 D7
30648 Lexington⊙ 278 F3
30247 Lilburn 3,765 D3
31051 Lilly 202 E6
†30286 Lincoln Park D5
30817 Lincolnton⊙ 1,406 G3
30147 Lindale B2
†30728 Linwood 417 B1
30058 Lithonia 2,637 D3
30248 Locust Grove 1,479 D4
30249 Loganville 1,841 E3
30433 Lollie G6
†30230 Lone Oak 119 C4
†30741 Lookout Mountain 1,505 B1
30434 Louisville⊙ 2,823 H4
30250 Lovejoy 205 D4
31316 Ludowici⊙ 1,286 J7
30554 Lula 857 E2
31549 Lumber City 1,426 G7
31815 Lumpkin⊙ 1,335 C6
30251 Luthersville 597 C4
30730 Lyerly 482 B2
30436 Lyons⊙ 4,203 H6
30059 Mableton J1
*31201 Macon⊙ 116,860 E5
Macon‡ 254,623 E5
30650 Madison⊙ 2,954 F3
30438 Manassas 116 H6
31816 Manchester 4,796 C5
30255 Mansfield 435 E4
31057 Marshallville 1,540 D6
*30060 Marietta⊙ 30,805 J1
30557 Martin 305 F2
30671 Maxeys 205 F3
30558 Maysville 619 E2
30555 McCaysville 1,219 D1
30253 McDonough⊙ 2,778 D4
31054 McIntyre 386 F5
31055 McRae⊙ 3,409 G6
30256 Meansville 303 D4
30040 Mechanicsville L1
31765 Meigs 1,231 D8
30731 Menlo 611 B2
†31792 Metcalf E9
30439 Metter⊙ 3,531 H6
30441 Midville 670 H5
31320 Midway 457 K7
31060 Milan 1,115 G6
31061 Milledgeville⊙ 12,176 F4
30442 Millen⊙ 3,988 J5
30257 Milner 320 D4
30207 Milstead D3
30559 Mineral Bluff 130 D1
30820 Mitchell 214 G4
30258 Molena 379 D4
30655 Monroe⊙ 8,854 E3
31063 Montezuma 4,830 E6
31064 Monticello⊙ 2,382 E4
31065 Montrose 170 F5
30259 Moreland 358 C4

31766 Morgan⊙ 364 C7
30560 Morganton 263 D1
30260 Morrow 3,791 K2
31638 Morven 471 E9
31768 Moultrie⊙ 15,708 E8
30562 Mountain City 701 F1
†30075 Mountain Park 378 D2
30563 Mount Airy 670 F1
30149 Mount Berry B2
30445 Mount Vernon⊙ 1,737 G6
30261 Mountville 168 C4
30150 Mount Zion 445 B3
31553 Nahunta⊙ 951 H8
31639 Nashville⊙ 4,831 F8
31641 Naylor 228 F9
30151 Nelson 562 D2
30262 Newborn 391 E3
30446 Newington 402 J5
30263 Newnan⊙ 11,449 C4
31770 Newton⊙ 711 D8
31554 Nicholls 1,114 G7
30565 Nicholson 491 F2
*30071 Norcross 3,317 D3
31771 Norman Park 757 E8
†30645 North High Shoals 256 E3
30821 Norwood 306 G4
30448 Nunez 168 H5
31772 Oakfield 113 E7
30732 Oakman 150 C1
31903 Oak Park 256 H6
30566 Oakwood 723 E2
31773 Ochlocknee 627 E9
31774 Ocilla⊙ 3,436 F7
31067 Oconee 306 G5
†30222 Odessadale 142 C5
31555 Odum 401 H7
31406 Oglethorpe⊙ 1,305 D6
30449 Oliver 239 J5
31821 Omaha 169 C6
31775 Omega 996 E8
30266 Orchard Hill 162 D4
30267 Oxford 1,750 E3
30268 Palmetto 2,086 C3
31777 Parrott 222 D7
31557 Patterson 763 H8
31778 Pavo 830 E9
†31201 Payne 196 E5
30269 Peachtree City 6,429 C4
31642 Pearson⊙ 1,827 G8
31779 Pelham 4,306 D8
31321 Pembroke⊙ 1,400 J6
30567 Pendergrass 302 E2
31069 Perry⊙ 9,453 E6
†31794 Phillipsburg E8
31070 Pinehurst 431 E6
30072 Pine Lake 901 D3
31822 Pine Mountain 984 C5
†31312 Pineora 387 K6
†31728 Pine Park D9
31071 Pineview 564 F6

31072 Pitts 384 E7
31073 Plainfield 128 F6
31780 Plains 651 D6
30733 Plainville 281 C2
31322 Pooler 2,543 K6
30450 Portal 694 J5
30270 Porterdale 1,451 E3
31407 Port Wentworth 3,947 K6
31781 Poulan 818 E8
30073 Powder Springs 3,381 C3
31824 Preston⊙ 429 C6
30451 Pulaski 257 J6
31643 Quitman⊙ 5,188 E9
30734 Ranger 171 C2
31645 Ray City 658 F8
30660 Rayle 177 G3
31783 Rebecca 272 E7
30453 Reidsville⊙ 2,296 H6
31601 Remerton 443 F9
31075 Rentz 337 G6
†30518 Rest Haven 231 E2
31076 Reynolds 1,298 D5
31077 Rhine 590 F7
31323 Riceboro 216 K7
31825 Richland 1,802 C6
31324 Richmond Hill 1,177 K7
†31018 Riddleville 154 G5
31326 Rincon 1,988 K6
30736 Ringgold⊙ 1,821 B1
*30274 Riverdale 7,121 K2
†31768 Riverside 99 E8
†30759 Riverside B2
31078 Roberta 859 D5
31079 Rochelle 1,626 F7
30153 Rockmart 3,645 B2
30455 Rocky Ford 223 J5
30161 Rome⊙ 29,654 B2
30170 Roopville 229 B4
30741 Rossville 3,851 B1
*30075 Roswell 23,337 D2
30662 Royston 2,404 F2
†30680 Russell 378 E3
30663 Rutledge 694 E3
31558 Saint Marys 3,596 J9
31522 Saint Simons Island K8
31784 Sale City 336 D8
31082 Sandersville⊙ 6,137 G5
†20436 Santa Claus 167 H6
30456 Sardis 1,180 J5
30275 Sargent 800 C4
31785 Sasser 407 D7
*31401 Savannah⊙ 141,634 L6
Savannah‡ 230,728 L6
31083 Scotland 222 G6
31095 Scott 139 G5
31560 Screven 872 H7
30276 Senoia 900 C4
31084 Seville 209 E7
31085 Shady Dale 155 E4
30172 Shannon B2
30664 Sharon 140 G3
30277 Sharpsburg 194 C4
31786 Shellman 1,254 C7
31826 Shiloh 392 C5
30665 Siloam 446 F3
31787 Smithville 867 D7
30080 Smyrna 20,312 K1
30278 Snellville 8,514 D3
30279 Social Circle 2,591 E3
30457 Soperton⊙ 2,981 G6
31647 Sparks 1,353 F8
31087 Sparta⊙ 1,754 F4
31329 Springfield⊙ 1,075 K6
†30705 Spring Place 246 C1
30823 Stapleton 388 H4
31648 Statenville⊙ 700 G9
30458 Statesboro⊙ 14,866 J6
30666 Statham 1,101 E3
30464 Stillmore 527 H6
30281 Stockbridge 2,103 D3
*30083 Stone Mountain 4,867 D3
†30518 Sugar Hill 2,473 E2
30746 Sugar Valley C1
30466 Summertown 215 H5
30747 Summerville⊙ 4,878 B2
31789 Sumner 213 E7
30284 Sunny Side 338 D4
31563 Surrency 368 H7
30174 Suwanee 1,026 D2
30401 Swainsboro⊙ 7,602 H5
31790 Sycamore 474 E7
30467 Sylvania⊙ 3,352 J5
31791 Sylvester⊙ 5,860 E7
31827 Talbotton⊙ 1,140 C5
30176 Tallapoosa 2,647 B3
30573 Tallulah Falls 162 F1
30575 Talmo E2
30470 Tarrytown 145 H6
30178 Taylorsville 266 C2
30179 Temple 1,520 B3
31089 Tennille 1,709 G5
30285 The Rock 78 D5
30286 Thomaston⊙ 9,682 D5
31792 Thomasville⊙ 18,463 E9
30824 Thomson⊙ 7,001 H4
†31404 Thunderbolt 2,165 K6
31794 Tifton⊙ 13,749 F8
30576 Tiger 299 F1
30668 Tignall 733 G3
30577 Toccoa⊙ 9,104 F1
31090 Toomsboro 673 F5
30752 Trenton⊙ 1,636 A1
30753 Trion 1,732 B1
30755 Tunnel Hill 867 C1
30289 Turin 260 C4
30471 Twin City 1,402 H5
31328 Tybee Island 2,240 L6
30290 Tyrone 1,038 C4
31795 Ty Ty 618 E8
31091 Unadilla 1,566 E6
30291 Union City 4,780 J2
30669 Union Point 1,750 F3
†31794 Unionville F8
30473 Uvalda 646 H6
31601 Valdosta⊙ 37,596 F9
30672 Vanna F2
†30153 Van Wert 303 B3

30756 Varnell 288 C
†31401 Vernonburg 178 K
30474 Vidalia 10,393 H
†30830 Vidette H
31092 Vienna⊙ 2,886 E
30180 Villa Rica 3,420 C
30182 Waco 471 E
30477 Wadley 2,438 H
30183 Waleska 450 D
†30209 Walnut Grove 387 E
31333 Walthourville 905
31830 Warm Springs 425 C
31093 Warner Robins 39,893 E
30828 Warrenton⊙ 2,172 G
31796 Warwick 488 E
30673 Washington⊙ 4,662 G
30677 Watkinsville⊙ 1,240 E
31831 Waverly Hall 913 C
31501 Waycross⊙ 19,371 H
30830 Waynesboro⊙ 5,760 J
31832 Weston 109 C
31833 West Point 4,294 B
31797 Whigham 507 D
30184 White 501 C
31568 White Oak 450 J
30678 White Plains 231 F
30185 Whitesburg 775 B
31650 Willacoochee 1,166 G
30292 Williamson 250 D
31410 Wilmington Island L
30680 Winder⊙ 6,705 E
31406 Windsor Forest K
30683 Winterville 621 F
31569 Woodbine⊙ 910 J
30293 Woodbury 1,738 C
31836 Woodland 664 D
30188 Woodstock 2,699 D
30670 Woodville 455 F
30833 Wrens 2,415 H
31096 Wrightsville⊙ 2,526 G
31097 Yatesville 390 D
30582 Young Harris 687 E
30295 Zebulon⊙ 995 D

OTHER FEATURES

Alapaha (riv.) F
Allatoona (lake) C
Altamaha (riv.) H
Andersonville Nat'l Hist. Site D
Atlanta Nav. Air Sta. J
Banks (lake) F
Bartletts Ferry (dam) B
Blackshear (lake) E
Blue Ridge (mts.) D
Brasstown Bald (mt.) E
Burton (lake) E
Carters (lake) C
Chattahoochee (riv.) B
Chattahoochee River Nat'l Rec. Area K
Chattooga (riv.) A
Chattooga (riv.) F
Chatuge (lake) E
Chickamauga and Chattanooga Nat'l Mil. Park B
Clark Hill (lake) H
Coosa (riv.) A
Coosawattee (riv.) C
Cumberland (isl.) K
Cumberland Island Nat'l Seashore K
Dobbins A.F.B. J
Doboy (sound) K
Etowah (riv.) C
Eufaula (Walter F. George Res.) (lake) B
Flint (riv.) D
Fort Benning B
Fort Frederica Nat'l Mon. K
Fort Gordon H
Fort McPherson K
Fort Pulaski Nat'l Mon. L
Fort Stewart J
Goat Rock (lake) B
Harding (lake) B
Hartwell (lake) G
Jekyll (isl.) K
Kennesaw Mtn. Nat'l Battlefield Park J
Lawson A.A.F. B
Martin Luther King, Jr., Nat'l Hist. Site K
Moody A.F.B. F
Nottely (lake) D
Ochlockonee (riv.) C1
Ocmulgee (riv.) E
Ocmulgee Nat'l Mon. F
Oconee (riv.) F
Ogeechee (riv.) J
Okefenokee (swamp) H
Oliver (lake) B
Oostanaula (riv.) B
Ossabaw (sound) K
Rabun (lake) E
Robins A.F.B. F
Saint Andrew (sound) K
Saint Catherines (isl.) K
Saint Marys (riv.) J
Saint Simons K
Sapelo (isl.) K
Satilla (riv.) G
Savannah (riv.) K
Sea (isls.) K
Seminole (lake) B
Sidney Lanier (lake) D
Sinclair (lake) F
Skidaway (isl.) L
Springer (mt.) D
Suwannee (riv.) G1
Tugaloo (riv.) F
Walter F. George (res.) B
Wassaw (sound) L
Weiss (lake) A
West Point (lake) B

⊙County seat.
‡Population of metropolitan area.
† Zip of nearest p.o. * Multiple zip

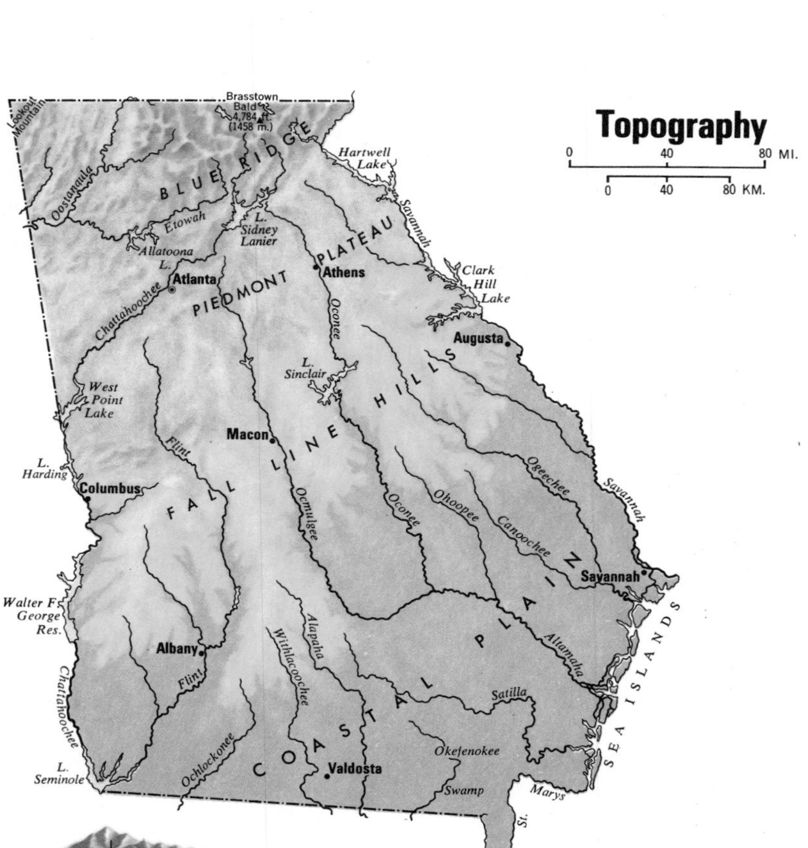
Topography
0 40 80 MI.
0 40 80 KM.
Brasstown Bald 4,784 ft. (1458 m.)
Lookout Mountain
Hartwell Lake
BLUE RIDGE
Oostanaula
Etowah
Allatoona L.
L. Sidney Lanier
Savannah
PLATEAU
Atlanta
Athens
Clark Hill Lake
PIEDMONT
Chattahoochee
Oconee
Augusta
L. Sinclair
West Point Lake
Macon
FALL LINE HILLS
Flint
Ogeechee
L. Harding
Columbus
Ocmulgee
Ohoopee
Canoochee
Savannah
Walter F. George Res.
Albany
Alapaha
Withlacoochee
COASTAL PLAIN
Altamaha
Satilla
SEA ISLANDS
L. Seminole
Ochlockonee
Valdosta
Okefenokee Swamp
St. Marys
5,000 m. 16,404 ft.
2,000 m. 6,562 ft.
1,000 m. 3,281 ft.
500 m. 1,640 ft.
200 m. 656 ft.
100 m. 328 ft.
Sea Level
Below

Georgia

SCALE
0 5 10   20   30   40 MI.
0 5 10   20   30   40 KM.

State Capitals........................⊛
County Seats.........................◉
Major Limited Access Hwys.

Scale 1:2,210,000

℗ Copyright HAMMOND INCORPORATED, Maplewood, N.J.

COUNTIES

Hawaii 92,053 ....K7
Honolulu 762,565 ....D3
Kalawao 144 ....G1
Kauai 39,082 ....A1
Maui 70,847 ....J1

CITIES and TOWNS

Zip   Name/Pop.   Key

96701 Aiea 32,879 ....B3
96821 Aina Haina F2
  Ala Moana 96,820...........C4
96703 Anahola 915 ....C1
†96706 Barbers Point 1,373.....E2
96704 Captain Cook 2,008 .....G5
96705 Eleele 580 ....C2
96706 Ewa 2,637 ....A4
96706 Ewa Beach 14,369.......A4
96701 Foster Village B3
†96714 Haena 200 ....C1
96708 Haiku 619 ....J2
96710 Hakalau J4
†96711 Halawa, Hawaii 50 .....G3
†96748 Halawa, Molokai 15 ...H1
†96701 Halawa Heights B3
96712 Haleiwa 2,412 ....E1
†96718 Halfway House 150.....H6
96787 Haliimaile 741 ....J2
†96713 Hamoa 35 ....K2
96713 Hana 643 ....K2
96714 Hanalei 483 ....C1
96715 Hanamaulu 3,227 ....C1
96716 Hanapepe 1,417 ....C2
96717 Hauula 2,997 ....E1
96825 Hawaii Kai F2
96718 Hawaii Nat'l Park 250 ...J6
96719 Hawi 795 ....G3
96824 Hickam Housing 4,425 ..B4
96720 Hilo⊙ 35,269 ....J5
96725 Holualoa 1,243 ....G5
96726 Honaunau 950 ....G6
96727 Honokaa 1,936 ....H4
†96761 Honokahua 309 ....H1
96728 Honomu 559 ....J4
96729 Hoolehua G1
†96706 Iroquois Point 3,915 ...A4
96730 Kaaawa 295 ....F1
†96761 Kaanapali 541 ....H2
†96793 Kahakuloa 75 ....J1
96801 Kahala D5
96744 Kahaluu 2,925 ....E2
96731 Kahuku 935 ....E1
96732 Kahului 12,978 ....J2
96740 Kailua (Kailua Kona),
  Hawaii 4,751 ....F5
96734 Kailua, Oahu 35,812 ....F2
96740 Kailua Kona 4,751 ....F5
96750 Kainaliu 512 ....G5
96741 Kalaheo 2,500 ....C2
96742 Kalaupapa⊙ 170 ....G1
†96754 Kalihiwai 40 ....C1
†96748 Kaluaaha 20 ....H1
†96748 Kamalo 60 ....H1
96743 Kamuela 1,179 ....G3
96744 Kaneohe 29,919 ....F2
96746 Kapaa 4,467 ....D1
96755 Kapaau 612 ....G3
96817 Kapalama C4
96747 Kaumakani 888 ....C2
96748 Kaunakakai 2,231 ....G1
96708 Kaupakalua 600 ....K2
†96713 Kaupo 65 ....K2
96743 Kawaihae 50 ....G4
96712 Kawailoa 200 ....E1
96749 Keaau 775 ....J5
96750 Kealakekua 1,033 ....G5
96751 Kealia, Kauai 300 ....D1
96708 Keanae 280 ....K2
96752 Kekaha 3,260 ....C2
96753 Kihei 5,644 ....J2
96754 Kilauea 895 ....C1
†96713 Kipahulu 75 ....K2
†96713 Koali 60 ....K2
96708 Kokomo 500 ....K2
96756 Koloa 1,457 ....C2
†96756 Koloa Landing C2
†96757 Kualapuu 502 ....G1
†96775 Kukaiau 75 ....H4
96727 Kukuihaele 332 ....H3
96790 Kula 800 ....J2
96759 Kunia 550 ....E2
96760 Kurtistown 900 ....J5
96761 Lahaina 6,095 ....H2
96763 Laie 4,643 ....E1
96763 Lanai City 2,092 ....H2
96764 Laupahoehoe 500 ......J4
96765 Lawai 950 ....C2
96766 Lihue⊙ 4,000 ....C2
96779 Lower Paia 1,500 ....J1
96719 Mahukona 2 ....G3
96708 Makakilo 600 ....K2
96768 Makawao 2,900 ....K2
96769 Makaweli 500 ....B2
96790 Makena 100 ....J2
96822 Makiki C4
96770 Maunaloa 633 ....G1
96744 Maunawili 5,239 ....F2
96789 Mililani Town 21,365 ...E2
96828 Moiliili C4
96734 Mokapu 11,615 ....F2
96771 Mountainview 540 ....J5
96772 Naalehu 1,168 ....H7
†96713 Nahiku 50 ....K2
96792 Nanakuli 8,185 ....D2
†96761 Napili-Honokowai 2,446 ...H1
96773 Ninole 75 ....J4
†96781 Onomea 10 ....J4
96774 Ookala 401 ....J4
96775 Paauhau 350 ....H4
96776 Paauilo 755 ....H4
96777 Pahala 1,619 ....H6
96778 Pahoa 923 ....J5
96779 Paia J2
96780 Papaaloa J4
96781 Papaikou 1,567 ....J5
96781 Paukaa 544 ....J5
96708 Pauwela 468 ....K2
96708 Peahi 308 ....K2
96782 Pearl City 42,575 ....B3
96783 Pepeekeo J4
96756 Poipu 685 ....C2
96714 Princeville 500 ....C1
96766 Puhi 991 ....C2
96788 Pukalani 3,950 ....J2
96748 Pukoo 50 ....H1
†96713 Puuiki 75 ....K2
96784 Puunene 572 ....J2
†96801 Puunui C4
96786 Schofield Barracks 18,851 ...E2
96779 Spreckelsville 350 ......J1
†96801 Ulumalu 201 ....K2
96776 Umikoa 25 ....H4
96785 Volcano 400 ....J6
96786 Wahiawa 16,911 ....
†96788 Waiakoa ....
96816 Waialae ....
†96781 Waialua 10 ....J4
†96748 Waialua, Molokai 30 ...H1
96791 Waialua, Oahu 4,051 ...
96792 Waianae 7,941 ....
†96793 Waihee 413 ....
96815 Waikiki ....
†96748 Wailau 20 ....H1
†96710 Wailea, Hawaii 150 .....H1
96790 Wailea, Maui 1,124 ....
96746 Wailua 1,587 ....
96793 Wailuku⊙ 10,260 ....
†96743 Waimea (Kamuela),
  Hawaii 1,179 ....C1
96796 Waimea, Kauai 1,569 ...
96720 Wainaku 1,045 ....
†96714 Wainiha 175 ....
96797 Waipahu 29,139 ....
†96786 Waipio Acres 4,091 .....
†96786 Whitmore Village 2,318....

OTHER FEATURES

Alalakeiki (chan.) ....
Alenuihaha (chan.) ....

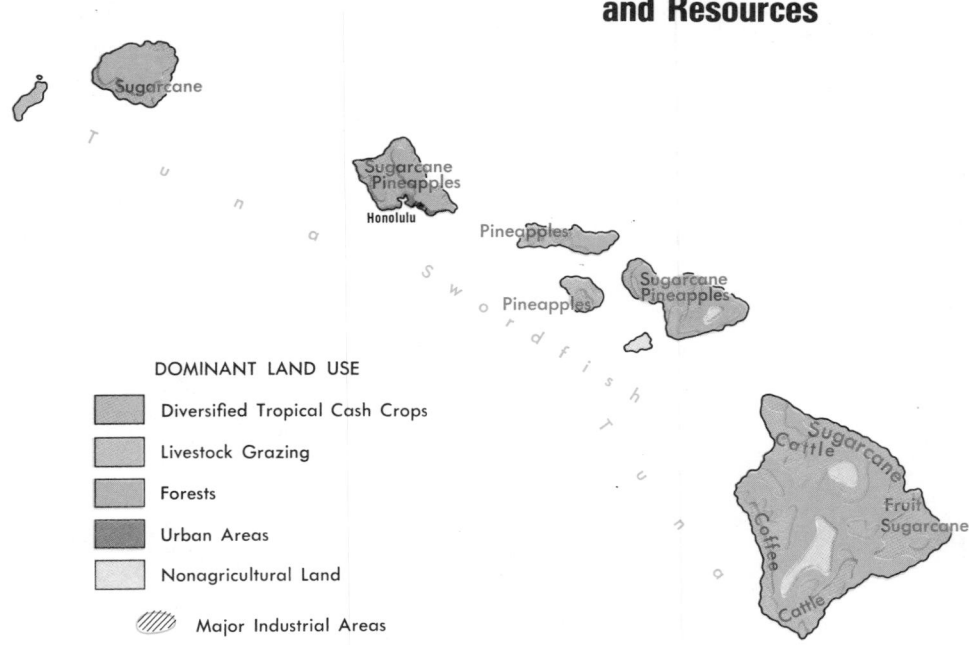

Topography

0   40   80 MI.

0   40   80 KM.

Lehua Kaulakahi Channel Kauai Lihue
Niihau C. Kawaihoa Kauai Channel
Kaena Pt. Kahuku Pt. Oahu Honolulu
Pearl Harbor Diamond Head Kaiwi Channel Molokai
Lanai Maui Wailuku Kauiki Head
Kahoolawe Alenuihaha Channel Upolu Pt.
PACIFIC OCEAN
Keahole Pt. Mauna Kea 13,796 ft. (4205 m.) Hawaii Hilo C. Kumukahi
Mauna Loa 13,677 ft. (4169 m.)
Ka Lae (South Cape)

5,000 m. 16,404 ft. | 2,000 m. 6,562 ft. | 1,000 m. 3,281 ft. | 500 m. 1,640 ft. | 200 m. 656 ft. | 100 m. 328 ft. | Sea Level | Below

Agriculture, Industry and Resources

Sugarcane

Sugarcane Pineapples Honolulu

Pineapples

Pineapples Sugarcane Pineapples

Sugarcane Cattle Fruit Sugarcane Coffee Cattle

Tuna Swordfish Tuna

DOMINANT LAND USE

Diversified Tropical Cash Crops
Livestock Grazing
Forests
Urban Areas
Nonagricultural Land
Major Industrial Areas

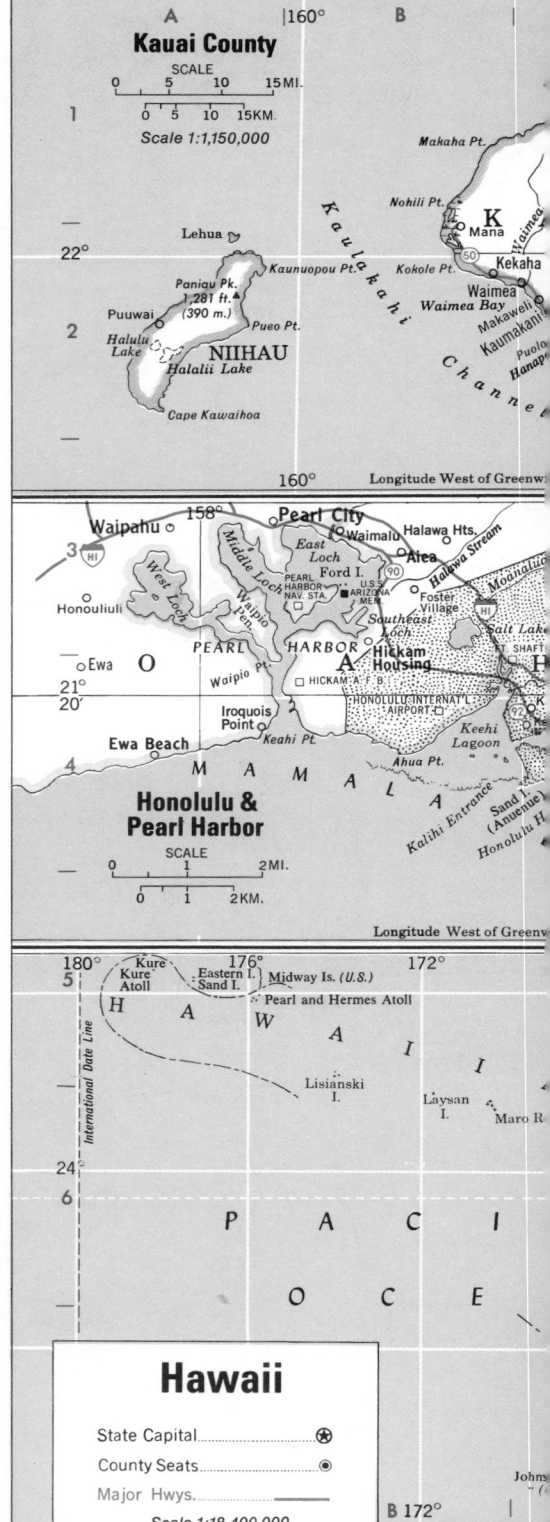

Kauai County

SCALE
0 5 10 15 MI.
0 5 10 15 KM.
Scale 1:1,150,000

A |160° B
Makaha Pt.
Nohili Pt. K Mana
Lehua Kaunuopou Pt. Kokole Pt. Kekaha
Puuwai Paniau Pk. 1,281 ft. (390 m.) Waimea Waimea Bay
Halulu Lake NIIHAU Pueo Pt. Kaumakani
Halalii Lake Kaulakahi Channel
Cape Kawaihoa
160° Longitude West of Greenwich

Honolulu & Pearl Harbor

Waipahu 158° Pearl City Halawa Hts.
East Loch Waimalu Aiea
West Loch Middle Loch Ford I. Halawa Stream
Honouliuli PEARL HARBOR NAV. STA. U.S.S. ARIZONA MEM. Foster Village
Ewa O PEARL HARBOR Southeast Loch Salt Lake SHAFT
21° 20' Waipio Hickam Housing HICKAM A.F.B. HONOLULU INTERNAT'L AIRPORT Keehi Lagoon
Iroquois Point Keahi Pt.
Ewa Beach MAMALA Ahua Pt. Kalihi Entrance Sand I.

SCALE
0 1 2 MI.
0 1 2 KM.

Longitude West of Greenwich

180° 176° 172°
Kure Atoll Eastern I. Midway Is. (U.S.)
Kure Sand I. Pearl and Hermes Atoll
H A W A I I
International Date Line Lisianski I. Laysan I. Maro R
24° P A C I
O C E

Hawaii

State Capital ....⊛
County Seats ....⊙
Major Hwys. ....—

Scale 1:18,400,000

© Copyright HAMMOND INC.

B 172°

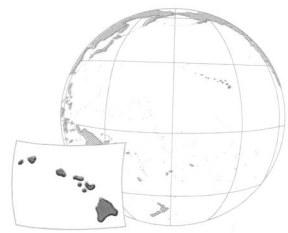

Anuenue (Sand) (isl.)	C4	Kanapou (bay)	J3	Lisianski (isl.)	B5	Puolo (pt.)	C2
Auau (chan.)	H2	Kaneohe Bay U.S.M.C. Air		Lua Makika (mt.)	J3	Puuhonua O Honaunau Nat'l Hist.	
Barbers Point Nav. Air Sta.	E2	Station	F2	Maalaea (bay)	J2	Park	F6
Diamond (head)	C5	Kau (des.)	J6	Makaha (pt.)	B1	Puu Keahiakahoe (mt.)	D3
East Loch (inlet)	B3	Kauai (chan.)	E6	Makahuena (pt.)	C2	Puukohola Heiau Nat'l Hist.	
Ford (isl.)	B3	Kauai (isl.)	C1	Makapuu (pt.)	F2	Site	G4
Fort Shafter	C3	Kauiki (head)	K2	Mamala (bay)	B4	Puu Kukui (mt.)	J2
French Frigate (shoals)	C6	Kaula (isl.)	D6	Mana (isl.)	F2	Red Hill (mt.)	K2
Gardner Pinnacles (isls.)	C6	Kaulakahi (chan.)	B2	Manana (isl.)	F2	Roundtop (mt.)	C4
Halalii (lake)	A2	Kawaihae (bay)	G4	Maro (reef)	C6	Salt (lake)	B3
Halawa (bay)	H1	Kawaihoa (cape)	A2	Maui (isl.)	J2	Sand (isl.)	B4
Haleakala (crater)	K2	Kawaihoa (cape)	A2	Mauna Kea (mt.)	H4	South (Ka Lae) (cape)	G7
Haleakala Nat'l Park	K2	Kawaikini (peak)	C1	Mauna Loa (mt.)	G6	Southeast Loch (inlet)	B3
Hawaii (isl.)	H5	Keahi (pt.)	A4	Moanalua (stream)	B3	Sugarloaf (hill)	C4
Hawaii Volcanoes Nat'l Park	H6	Kealaikahiki (chan.)	H3	Mokapu (pen.)	F2	Tantalus (mt.)	D4
Hickam A.F.B.	B4	Kealakekua (bay)	F6	Mokuaweoweo (crater)	H6	Upolu (pt.)	G3
Hilo (bay)	J5	Keanapapa (pt.)	G2	Molokai (isl.)	G1	U.S.S. Arizona Memorial	B3
Honolulu Int'l Airport	B4	Keehi (lag.)	B4	Molokini (isl.)	J2	Waialeale (mt.)	C1
Honolulu (harb.)	C4	Kiholo (bay)	F4	Nawiliwili (bay)	D2	Waikiki (beach)	C4
Ilio (pt.)	G1	Kilauea (crater)	H6	Necker (isl.)	D6	Wailuku (riv.)	J5
Kaala (mt.)	D1	Koko (head)	F2	Nihoa (isl.)	D6	Waimea (bay)	B2
Kahala (pt.)	D1	Konahuanui (peaks)	C3	Niihau (isl.)	A2	Waimea (bay)	B2
Kahana (bay)	F1	Koolau (range)	E2	Oahu (isl.)	E2	Wainiha (riv.)	C1
Kaiwi (chan.)	E6	Kumukahi (cape)	K5			Waipio (bay)	H3
Ka Lae (cape)	G7	Kure (atoll)	A5	Pailolo (chan.)	H1	Waipio (pen.)	A3
Kalaupapa Nat'l Hist. Park	H1	Kure (isl.)	A5	Palolo (stream)	D4	Waipio (pt.)	A4
Kalohi (chan.)	G1	Laau (pt.)	G1	Paniau (peak)	A2	West Loch (inlet)	A3
Kaloko-Honokohau Nat'l Hist.		Lanai (isl.)	H2	Pearl (harb.)	A3	Wheeler A.F.B.	E1
Park	F6	Lanaihale (mt.)	H2	Pearl and Hermes (atoll)	B5		
Kamakou (peak)	H1	Laysan (isl.)	B5	Pearl Harbor Naval Sta.	B3	⊙County seat.	
				Punchbowl (hill)	C4	‡Population of metropolitan area.	
						† Zip of nearest p.o	
						* Multiple zips.	

AREA 6,471 sq. mi. (16,760 sq. km.)
POPULATION 964,691
CAPITAL Honolulu
LARGEST CITY Honolulu
HIGHEST POINT Mauna Kea 13,796 ft. (4205 m.)
SETTLED IN —
ADMITTED TO UNION August 21, 1959
POPULAR NAME Aloha State
STATE FLOWER Hibiscus
STATE BIRD Nene (Hawaiian Goose)

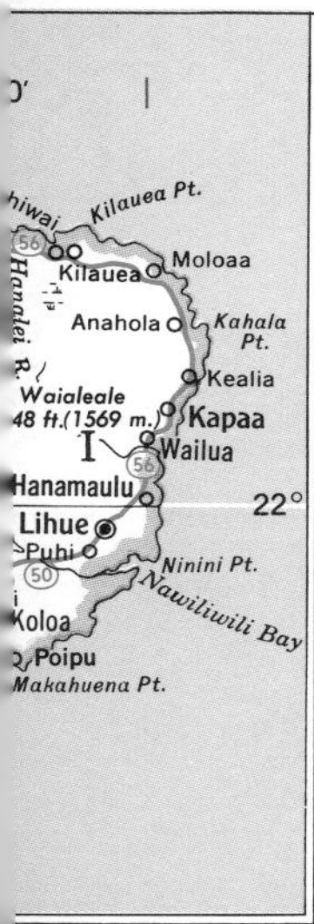

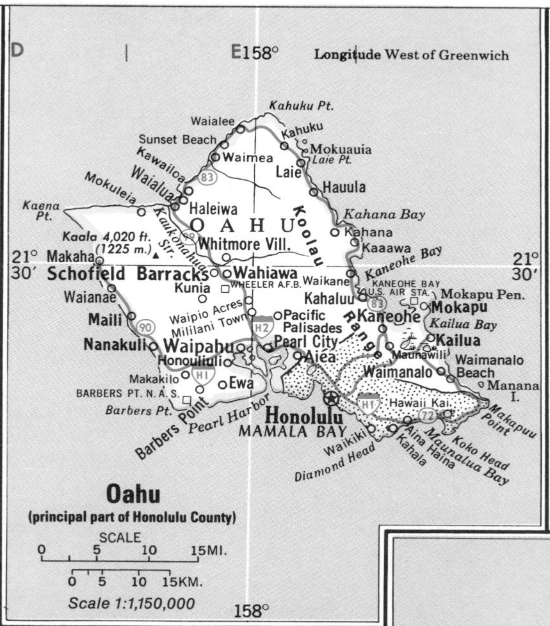

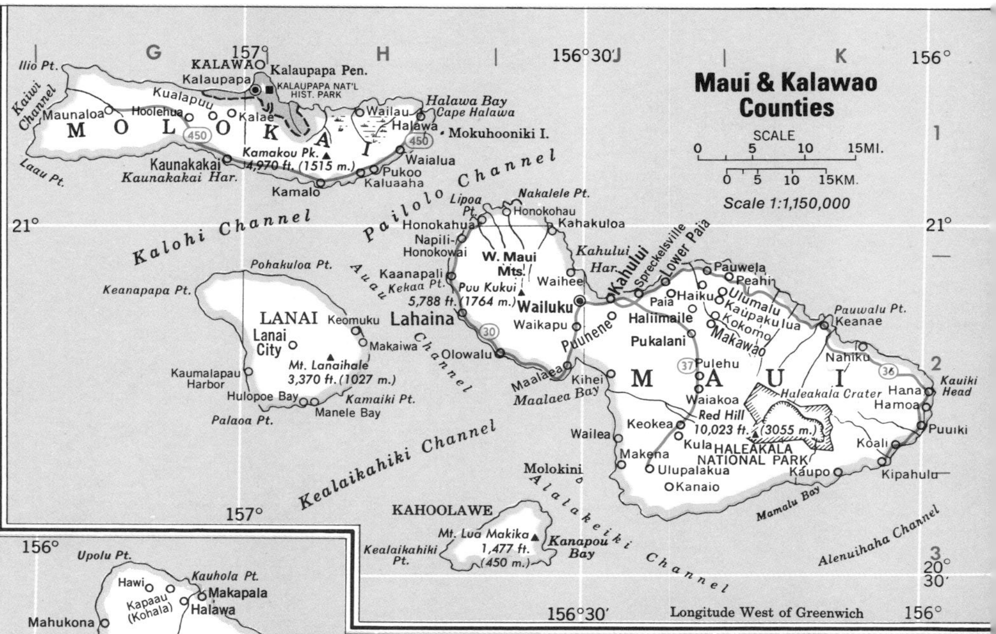

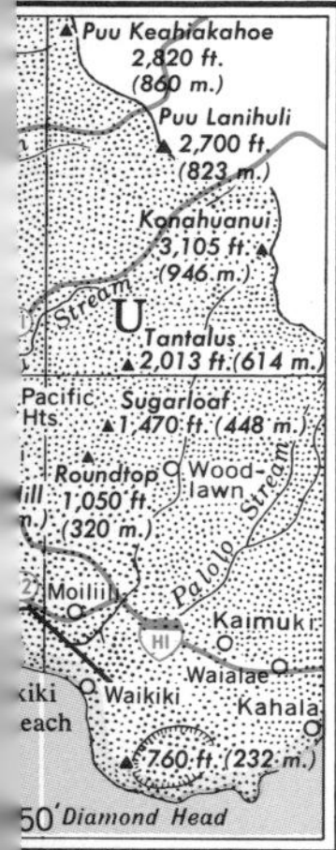

Oahu
(principal part of Honolulu County)

SCALE
0   5   10   15MI.
0   5   10   15KM.
Scale 1:1,150,000

Maui & Kalawao Counties
SCALE
0   5   10   15MI.
0   5   10   15KM.
Scale 1:1,150,000

Map below shows relative position of the islands comprising the State of Hawaii. The other maps show the more important island counties in detail.

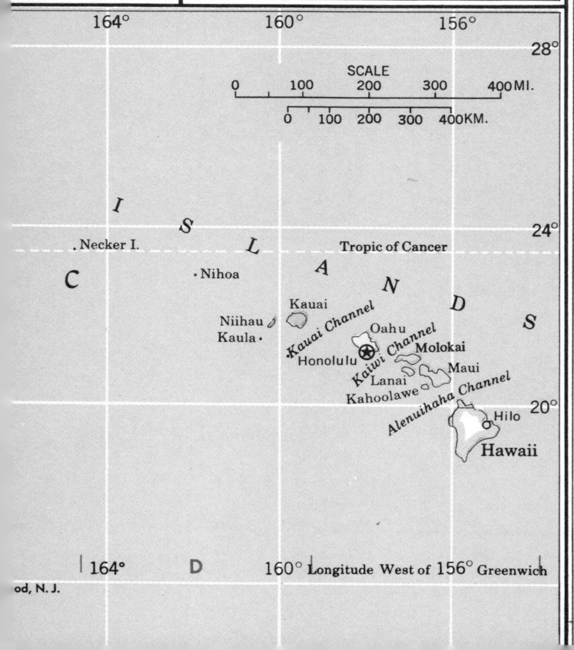

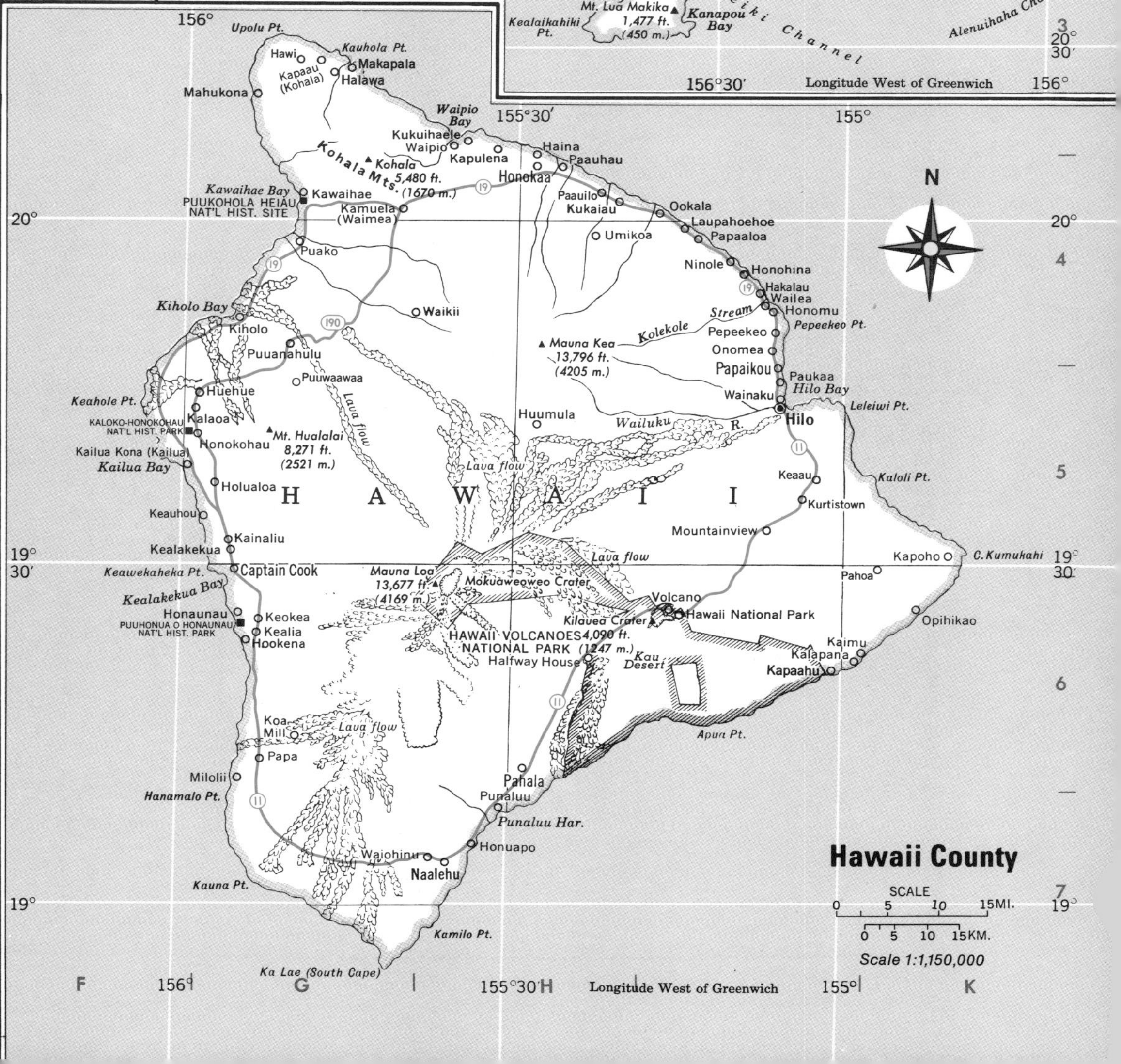

Hawaii County
SCALE
0   5   10   15MI.
0   5   10   15KM.
Scale 1:1,150,000

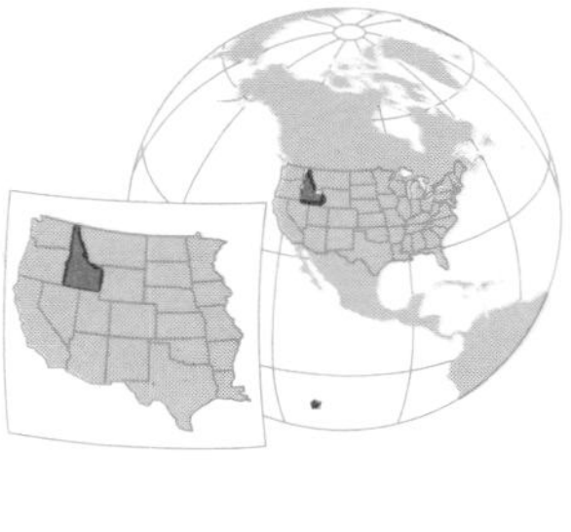

AREA 83,564 sq. mi. (216,431 sq. km.)
POPULATION 944,038
CAPITAL Boise
LARGEST CITY Boise
HIGHEST POINT Borah Pk. 12,662 ft. (3859 m.)
SETTLED IN 1842
ADMITTED TO UNION July 3, 1890
POPULAR NAME Gem State
STATE FLOWER Syringa
STATE BIRD Mountain Bluebird

COUNTIES

da 173,036 B6
dams 3,347 B5
annock 65,421 F7
ear Lake 6,931 G7
enewah 8,292 B2
ingham 36,489 F6
laine 9,841 D6
oise 2,999 C6
onner 24,163 B1
onneville 65,980 G6
oundary 7,289 B1
utte 3,342 E6
amas 818 D6
anyon 83,756 B6
aribou 8,695 G7
assia 19,427 E7
ark 798 F5
earwater 10,390 C3
uster 3,385 D5
more 21,565 C6
anklin 8,895 G7
emont 10,813 G5
em 11,972 B6
ooding 11,874 D6
aho 14,769 C4
fferson 15,304 F6
rome 14,840 D7
ootenai 59,770 B2
atah 28,749 B3
emhi 7,460 D4
ewis 4,118 B3
ncoln 3,436 D6
adison 19,480 G6
inidoka 19,718 E7
ez Perce 33,220 B3
neida 3,258 F7
wyhee 8,272 B7
ayette 15,825 B5
ower 6,844 F7
hoshone 19,226 B2
eton 2,897 G6
win Falls 52,927 D7
alley 5,604 C5
ashington 8,803 B5

CITIES and TOWNS

p Name/Pop. Key

3210 Aberdeen 1,528 F7
3350 Acequia 100 E7
3311 Albion 286 E7
3211 American Falls⊙ 3,626 E7
3401 Ammon 4,669 G6
3213 Arco⊙ 1,241 E6
3214 Arimo 338 F7
3420 Ashton 1,219 G5
3801 Athol 312 B2
3217 Bancroft 505 G7
3218 Basalt 414 F6
3313 Bellevue 1,016 D6
3221 Blackfoot⊙ 10,065 F6
3314 Bliss 208 D7
3223 Bloomington 212 G7
3701 Boise (cap.)⊙ 102,160 B6
Boise‡ 173,036 B6
3805 Bonners Ferry⊙ 1,906 B1
3806 Bovill 289 B3
3316 Buhl 3,629 D7
3318 Burley⊙ 8,761 E7
3213 Butte City 93 E6
3605 Caldwell⊙ 17,699 B6
3610 Cambridge 428 B5
3611 Cascade⊙ 945 C5
3321 Castleford 191 C7
3226 Challis⊙ 758 D5
3851 Chatcolet 181 B2
3202 Chubbuck 7,052 F7
3811 Clark Fork 449 B1
3227 Clayton 43 D5
3228 Clifton 208 F7
3814 Coeur d'Alene⊙ 20,054 B2
3522 Cottonwood 941 B3
3612 Council⊙ 917 B5
3523 Craigmont 617 B3
3622 Crouch 69 B5
3524 Culdesac 261 B3
3814 Dalton Gardens 1,795 B2
3232 Dayton 368 F7
3823 Deary 539 B3
3323 Declo 276 E7
3324 Dietrich 101 D7
3615 Donnelly 139 B5
3234 Downey 645 F7
3422 Driggs⊙ 727 G6
3423 Dubois⊙ 413 F5
3616 Eagle 2,620 B6
3836 East Hope 258 B1
3325 Eden 355 D7
3827 Elk River 265 B3
3617 Emmett⊙ 4,605 B6
3327 Fairfield⊙ 404 D6
3526 Ferdinand 144 B3
†83814 Fernan Lake 178 B2
83328 Filer 1,645 D7
83236 Firth 460 F6
83203 Fort Hall 750 F6
83237 Franklin 423 G7
83619 Fruitland 2,456 B6
†83704 Garden City 4,571 B6
83832 Genesee 791 B3
83239 Georgetown 544 G7
83623 Glenns Ferry 1,374 C7
83330 Gooding⊙ 2,949 D7
83241 Grace 1,216 G7
83624 Grand View 366 B7
83530 Grangeville⊙ 3,666 B4
83626 Greenleaf 663 B6
83332 Hagerman 602 D7
83333 Hailey⊙ 2,109 D6
83425 Hamer 93 F6
83334 Hansen 1,078 D7
83833 Harrison 260 B2
†83854 Hauser 305 A2
†83835 Hayden 2,586 B2
83835 Hayden Lake 273 B2
83335 Hazelton 496 E7
83336 Heyburn 2,889 E7
†83301 Hollister 167 D7
83628 Homedale 2,078 A6
83836 Hope 106 B1
83629 Horseshoe Bend 700 B6
†83854 Huetter 65 B2
83631 Idaho City⊙ 300 C6
*83401 Idaho Falls⊙ 39,590 F6
83245 Inkom 830 F7
83427 Iona 1,072 G6
83428 Irwin 113 G6
83429 Island Park 154 G5
83338 Jerome⊙ 6,891 D7
83535 Juliaetta 522 B3
83536 Kamiah 1,478 B3
83837 Kellogg 3,417 B2
83537 Kendrick 395 B3
83340 Ketchum 2,200 D6
83341 Kimberly 2,307 D7
83539 Kooskia 784 C3
83840 Kootenai 280 B1
83634 Kuna 1,767 B6
83540 Lapwai 1,043 B3
83246 Lava Hot Springs 467 F7
83464 Leadore 114 E5
83501 Lewiston⊙ 27,986 A3
83431 Lewisville 502 F6
83251 Mackay 541 E6
83252 Malad City⊙ 1,915 F7
83342 Malta 196 E7
83639 Marsing 786 B6
83638 McCall 2,188 C5
83250 McCammon 770 F7
83641 Melba 276 B6
83434 Menan 605 F6
83642 Meridian 6,658 B6
83644 Middleton 1,901 B6
83645 Midvale 205 B5
83343 Minidoka 101 E7
83254 Montpelier 3,107 G7
83255 Moore 210 E6
83843 Moscow⊙ 16,513 B3
83647 Mountain Home⊙ 7,540 C6
83845 Moyie Springs 386 B1
†83450 Mud Lake 243 F6
83846 Mullan 1,269 C2
83650 Murphy⊙ 200 B6
83344 Murtaugh 114 D7
83651 Nampa 25,112 B6
83436 Newdale 329 G6
83654 New Meadows 576 B4
83655 New Plymouth 1,186 B6
83543 Nezperce⊙ 517 B3
83656 Notus 437 B6
83346 Oakley 663 D7
†99156 Oldtown 257 A1
†83855 Onaway 254 B3
83544 Orofino⊙ 3,711 B3
83849 Osburn 2,220 B2
†83263 Oxford 66 F7
83261 Paris⊙ 707 G7
83438 Parker 262 G6
83660 Parma 1,820 B6
83347 Paul 940 E7
83661 Payette⊙ 5,448 B5
83545 Peck 209 B3
83546 Pierce 1,060 C3
83850 Pinehurst 2,183 B2
83851 Plummer 634 B2
*83201 Pocatello⊙ 46,340 F7
83852 Ponderay 399 B1
83854 Post Falls 5,736 A2
83855 Potlatch 819 A3
83263 Preston⊙ 3,759 G7
83856 Priest River 1,639 A1
83858 Rathdrum 1,369 A2
83548 Reubens 87 B3
83440 Rexburg⊙ 11,559 G6
83349 Richfield 357 D6
83442 Rigby⊙ 2,624 F6
83549 Riggins 527 B4
83443 Ririe 555 G6
83444 Roberts 466 F6
83271 Rockland 283 F7
83350 Rupert⊙ 5,476 E7
83445 Saint Anthony⊙ 3,212 G6
83272 Saint Charles 211 G7
83861 Saint Maries⊙ 2,794 B2
83467 Salmon⊙ 3,308 D4
83864 Sandpoint⊙ 4,460 B1
83274 Shelley 3,300 F6
83352 Shoshone⊙ 1,242 D7
†83650 Silver City 1 B6
83868 Smelterville 776 B2
83276 Soda Springs⊙ 4,051 G7
83869 Spirit Lake 834 A2
83278 Stanley 99 D5
83552 Stites 253 C3
83448 Sugar City 1,022 G6
83353 Sun Valley 545 D6
83449 Swan Valley 135 G6
83870 Tensed 113 B2
83451 Teton 559 G6
83452 Tetonia 191 G6
83871 Troy 820 B3
83301 Twin Falls⊙ 26,209 D7
83454 Ucon 833 F6
83455 Victor 323 G6
83873 Wallace⊙ 1,736 C2
†83837 Wardner 423 B2
83553 Weippe 828 C3
83672 Weiser⊙ 4,771 B5
83355 Wendell 1,974 D7
83286 Weston 310 F7
83554 White Bird 154 B4
83676 Wilder 1,260 A6
83555 Winchester 343 B3
83876 Worley 206 B2

OTHER FEATURES

Albeni Falls (dam) B1
Albion (mts.) E7
Allan (mt.) D4
American Falls (res.) F6
Anderson Ranch (res.) C6
Antelope (creek) E6
Arrowrock (res.) C6
Auger (falls) D7
Badger (peak) E7
Bald (mt.) D5
Bannock (creek) F7
Bannock (peak) F7
Bannock (range) F7
Bargamin (creek) C4
Battle (creek) B7
Bear (lake) G7
Bear (riv.) G7
Beaver (creek) F5
Beaverhead (mts.) E4
Big (creek) C4
Big Boulder (creek) B7
Big Elk (peak) G6
Big Hole (mts.) G6
Big Lost (riv.) E6
Big Southern (butte) E6
Big Wood (riv.) D6
Birch (creek) F5
Birch Creek (valley) E5
Bitterroot (range) D3
Blackfoot (res.) G7
Black Pine (mts.) E7
Blue Nose (mt.) D4
Boise (mts.) B6
Boise (riv.) B6
Borah (peak) E5
Boulder (mts.) D6
Brownlee (dam) B5
Bruneau (riv.) C7
Camas (creek) D5
Camas (creek) D6
Camas (creek) F5
Canyon (creek) C6
Cape Horn (mt.) C5
Caribou (mt.) G6
Caribou (range) G6
Cascade (res.) C5
Castle (creek) B7
Castle (peak) D5
Cedar Creek (peak) E7
Cedar Creek (res.) D7
Centennial (mts.) F5
Clearwater (mts.) C3
Clearwater (riv.) B3
Coeur d'Alene (lake) B2
Coeur d'Alene (mts.) C2
Coeur d'Alene (riv.) B2
Cottonwood (butte) C4
Craig (mts.) B4
Crane Creek (res.) B5
Craters of the Moon Nat'l Mon. E6
Deadwood (res.) C5
Deep (creek) B7
Deep (creek) F7
Deep Creek (mts.) F7
Diamond (peak) E5
Dworshak (res.) C3
East Sister (peak) C2
Eighteen Mile (peak) E5
Fish Creek (res.) E6
Fort Hall Ind. Res. F6
Goldstone (mt.) E4
Goose (creek) E7
Goose Creek (mts.) E7
Grand Canyon of the Snake River (canyon) B4
Grays (lake) G6
Grays Lake Outlet (creek) G6
Greylock (mt.) C6
Hayden (lake) B2
Hells (canyon) B4
Hells Canyon Nat'l Rec. Area B4
Henrys (lake) G5
Henrys Fork, Snake (riv.) G5
Hunter (peak) D3
Hyndman (peak) D6
Indian (creek) C5
Island Park (res.) G5
Jarbidge (riv.) C7
Johnson (creek) C5
Jordan (creek) A7
Kootenai (riv.) C1
Lemhi (pass) E5
Lemhi (range) E5
Lemhi (riv.) E5
Little Lost (riv.) E5
Little Owyhee (riv.) B7
Little Salmon (riv.) B4
Little Weiser (riv.) B5
Little Wood (riv.) D6
Lochsa (riv.) C3
Lolo (creek) C3
Lolo (pass) D3
Lone Pine (peak) D5
Lookout (mt.) D5
Lookout (mt.) F5
Lost River (range) E5
Lost Trail (pass) E4
Lowell (lake) B6
Lower Goose Creek (res.) D7
Lower Granite (lake) A3
Lucky Peak (lake) B6
Mackay (res.) E6
Magic (res.) D6
Malad (riv.) F7
Marsh (creek) F7
McGuire (mt.) D4
Meade (peak) G7
Meadow (creek) C4
Medicine Lodge (creek) F5
Middle Fork (peak) D5
Monument (peak) B4
Moose (creek) D3
Mores (creek) C6
Mormon (mt.) D4
Mountain Home (res.) C6
Mountain Home A.F.B. 6,403 C6
Moyie (riv.) B1
Mud (lake) F6
National Reactor Testing Sta. F6
Nez Perce Nat'l Hist. Park B-C3
North Fork (riv.) B7
Norton (peak) D6
Orofino (creek) C3
Owyhee (mts.) B6
Owyhee, East Fork (riv.) B7
Oxbow (dam) B5
Pack (riv.) B1
Pahsimeroi (riv.) E5
Palisades (res.) G6
Palouse (riv.) B3
Panther (creek) D4
Payette (lake) C4
Payette (mts.) B5
Payette (riv.) B6
Peale (mts.) G7
Pend Oreille (lake) B1
Pend Oreille (mt.) B1
Pend Oreille (riv.) A1
Pilot (peak) C4
Pilot (peak) C6
Pilot Knob (mt.) C4
Pinyon (peak) C5
Pioneer (mts.) D6
Portneuf (res.) F7
Pot (mt.) C3
Potlatch (riv.) B3
Priest (lake) B1
Priest (riv.) B1
Purcell (mts.) B1
Pyramid (peak) E4
Raft (riv.) E7
Rainbow (mt.) C4
Ranger (peak) D3
Rays (lake) F6
Red (riv.) C4
Redfish (lake) D5
Reynolds (creek) B6
Rhodes (peak) D3
Rocky (mts.) D1
Rocky Ridge (mt.) C3
Ryan (peak) D6
Saddle (mt.) D3
Saddle (mt.) F6
Sailor (creek) C7
Saint Joe (riv.) B2
Saint Maries (riv.) B2
Salmon (falls) C7
Salmon (riv.) B4
Salmon Falls (creek) D7
Salmon Falls Creek (res.) D7
Salmon River (mts.) C5
Sawtooth (range) C6
Sawtooth Nat'l Rec. Area D5
Secesh (riv.) C4
Selkirk (mts.) B1
Selway (riv.) C3
Seven Devils (mts.) B4
Shoshone (falls) D7
Sleeping Deer (mt.) D5
Smith (creek) B1
Smoky (mts.) D6
Snake (riv.) A3
Snake River (plain) D7
Snake River (range) G6
Spirit (lake) B2
Squaw (creek) B5
Squaw (peak) D4
Steamboat (mt.) C4
Steel (mt.) C6
Strike, C.J. (res.) C7
Sublett (mts.) E7
Sunset (peak) E6
Taylor (mt.) D5
Teton (riv.) G6
Thompson (peak) C5
Trinity (mt.) C6
Trout (creek) B1
Twin (falls) D7
Twin Peaks (mt.) D5
Walcott (lake) E7
Wasatch (range) G7
Waugh (mt.) D4
Weiser (riv.) B5
Western Shoshone Ind. Res. B7
White Knob (mts.) E6
Wickahoney (creek) C7
Willow (creek) G6
Wilson Lake (res.) D7
Yankee Fork, Salmon (riv.) D5
Yellowstone Nat'l Park G5

⊙County seat.
‡Population of metropolitan area.
† Zip of nearest p.o.
* Multiple zips.

Agriculture, Industry and Resources

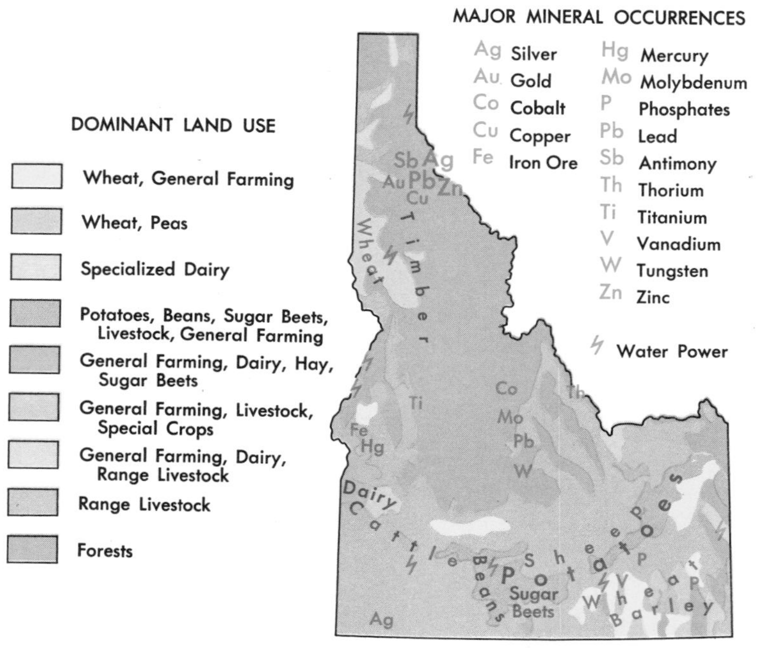

MAJOR MINERAL OCCURRENCES
Ag Silver
Au Gold
Co Cobalt
Cu Copper
Fe Iron Ore
Hg Mercury
Mo Molybdenum
P Phosphates
Pb Lead
Sb Antimony
Th Thorium
Ti Titanium
V Vanadium
W Tungsten
Zn Zinc
Water Power
DOMINANT LAND USE
Wheat, General Farming
Wheat, Peas
Specialized Dairy
Potatoes, Beans, Sugar Beets, Livestock, General Farming
General Farming, Dairy, Hay, Sugar Beets
General Farming, Livestock, Special Crops
General Farming, Dairy, Range Livestock
Range Livestock
Forests
Wheat
Timber
Dairy
Cattle
Sheep
Potatoes
Beans
Sugar Beets
Wheat
Barley

AREA 56,345 sq. mi. (145,934 sq. km.)
POPULATION 11,426,596
CAPITAL Springfield
LARGEST CITY Chicago
HIGHEST POINT Charles Mound 1,235 ft. (376 m.)
SETTLED IN 1720
ADMITTED TO UNION December 3, 1818
POPULAR NAME Prairie State; Land of Lincoln
STATE FLOWER Native Violet
STATE BIRD Cardinal

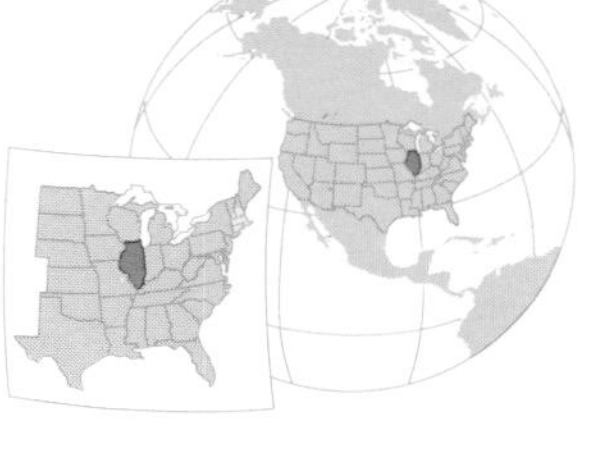

COUNTIES

ams 71,622 ....B4
xander 12,264 ....D6
nd 16,224 ....D5
one 28,630 ....E1
wn 5,411 ....C4
reau 39,114 ....D2
houn 5,867 ....C4
roll 18,779 ....D1
ss 15,084 ....C4
ampaign 168,392 ....E3
istian 36,446 ....D4
rk 16,913 ....F4
y 15,283 ....E5
nton 32,617 ....D5
es 52,260 ....E4
ok 5,253,655 ....F2
wford 20,818 ....F4
mberland 11,062 ....E4
Kalb 74,624 ....E2
Witt 18,108 ....E3
uglas 19,774 ....E4
Page 658,835 ....E2
ar 21,725 ....F4
wards 7,961 ....E5
ngham 30,944 ....E4
ette 22,167 ....D4
d 15,265 ....E3
nklin 43,201 ....E5
ton 43,687 ....C3
latin 7,590 ....E6
ene 16,661 ....C4
ndy 30,582 ....E2
milton 9,172 ....E5
cock 23,877 ....B3
din 5,383 ....E6
nderson 9,114 ....C3
ry 57,968 ....C2
uois 32,976 ....F3
Jackson 61,522 ....D6
Jasper 11,318 ....E4
Jefferson 36,354 ....E5
Jersey 20,538 ....C4
Jo Daviess 23,520 ....C1
Johnson 9,624 ....E6
Kane 278,405 ....E2
Kankakee 102,926 ....F2
Kendall 37,202 ....E2
Knox 61,607 ....C3
Lake 440,372 ....E1
La Salle 112,033 ....E2
Lawrence 17,807 ....F5
Lee 36,328 ....D2
Livingston 41,381 ....E3
Logan 31,802 ....D3
Macon 131,375 ....E4
Macoupin 49,384 ....D4
Madison 247,691 ....D5
Marion 43,523 ....E5
Marshall 14,479 ....D2
Mason 19,492 ....D3
Massac 14,990 ....E6
McDonough 37,467 ....C3
McHenry 147,897 ....E1
McLean 119,149 ....E3
Menard 11,700 ....D3
Mercer 19,286 ....C2
Monroe 20,117 ....C5
Montgomery 31,686 ....D4
Morgan 37,502 ....C4
Moultrie 14,546 ....E4
Ogle 46,338 ....D1
Peoria 200,466 ....D3
Perry 21,714 ....D5
Piatt 16,581 ....E4
Pike 18,896 ....C4
Pope 4,404 ....E6
Pulaski 8,840 ....D6
Putnam 6,085 ....D2
Randolph 35,652 ....D5
Richland 17,587 ....E5
Rock Island 165,968 ....C2
Saint Clair 267,531 ....D5
Saline 28,448 ....E6
Sangamon 176,089 ....D4
Schuyler 8,365 ....C3
Scott 6,142 ....C4
Shelby 23,923 ....E4
Stark 7,389 ....D2
Stephenson 49,536 ....D1
Tazewell 132,078 ....D3
Union 17,765 ....D6
Vermilion 95,222 ....F3
Wabash 13,713 ....F5
Warren 21,943 ....C3
Washington 15,472 ....D5
Wayne 18,059 ....E5
White 17,864 ....E5
Whiteside 65,970 ....D2
Will 324,460 ....F2
Williamson 56,538 ....E6
Winnebago 250,884 ....D1
Woodford 33,320 ....D3

CITIES and TOWNS

Zip *Name/Pop.* *Key*

61410 Abingdon 4,210 ....C3
60101 Addison 29,826 ....B5
61230 Albany 1,014 ....C2
62806 Albion⊙ 2,285 ....E5
61231 Aledo⊙ 3,881 ....C2
61412 Alexis 1,076 ....C2
60102 Algonquin 5,834 ....E1
62207 Alorton 2,237 ....B2
61413 Alpha 815 ....C2
†60658 Alsip 17,134 ....B6
62411 Altamont 2,389 ....E4
62002 Alton 34,171 ....A2
61310 Amboy 2,377 ....D2
61232 Andalusia 1,238 ....C2
62906 Anna 5,408 ....D6
61234 Annawan 908 ....C2
60002 Antioch 4,419 ....E1
61910 Arcola 2,714 ....E4
62501 Argenta 994 ....E4
*60004 Arlington Heights 66,116 ....B5
61911 Arthur 2,122 ....E4
60911 Ashkum 735 ....E3
62612 Ashland 1,351 ....C4
62808 Ashley 658 ....D5
61912 Ashmore 883 ....F4
61006 Ashton 1,140 ....D2
62510 Assumption 1,283 ....E4
61501 Astoria 1,370 ....C3
62613 Athens 1,371 ....D4
61235 Atkinson 1,138 ....C2
61723 Atlanta 1,807 ....D3
61913 Atwood 1,464 ....E4
62615 Auburn 3,616 ....D4
62311 Augusta 764 ....C3
*60504 Aurora 81,293 ....E2
62907 Ava 811 ....D6
62216 Aviston 846 ....D5
61415 Avon 1,019 ....C3
†60015 Bannockburn 1,316 ....B5
60010 Barrington 9,029 ....A5
†60010 Barrington Hills 3,631 ....A5
62312 Barry 1,487 ....B4
60103 Bartlett 13,254 ....A5
61607 Bartonville 6,137 ....D3
60510 Batavia 12,574 ....E2
62618 Beardstown 6,338 ....C3
62219 Beckemeyer 1,119 ....D5
60401 Beecher 2,024 ....F2
*62220 Belleville⊙ 41,580 ....B3
60104 Bellwood 19,811 ....B5
61008 Belvidere⊙ 15,176 ....E1
61813 Bement 1,770 ....E4
62009 Benld 1,638 ....D4
60106 Bensenville 16,124 ....B5
62812 Benton⊙ 7,778 ....E6
60162 Berkeley 5,467 ....B5
60402 Berwyn 46,849 ....B6
62010 Bethalto 8,630 ....B2
61914 Bethany 1,550 ....E4
61420 Blandinsville 886 ....C3
60108 Bloomingdale 12,659 ....A5
61701 Bloomington⊙ 44,189 ....D3
Bloomington-Normal‡ 119,149 ....D3
60406 Blue Island 21,855 ....B6
62513 Blue Mound 1,338 ....D4
62621 Bluffs 821 ....C4
60439 Bolingbrook 37,261 ....A6
60914 Bourbonnais 13,280 ....F2
60407 Braceville 721 ....E2
61421 Bradford 924 ....D2
60915 Bradley 11,008 ....F2
60408 Braidwood 3,429 ....E2
62230 Breese 3,516 ....D5
62417 Bridgeport 2,281 ....F5
60455 Bridgeview 14,155 ....B6
62012 Brighton 2,364 ....C4
61517 Brimfield 890 ....D3
60153 Broadview 8,618 ....B6
60513 Brookfield 19,395 ....B6
†62059 Brooklyn (Lovejoy) 1,233 ....A2
62910 Brookport 1,128 ....E6
61314 Buda 668 ....D2
†60090 Buffalo Grove 22,230 ....B5
62014 Bunker Hill 1,700 ....D4
60459 Burbank 28,462 ....B6
†60601 Burnham 4,030 ....C6
†60558 Burr Ridge 3,833 ....B6
61422 Bushnell 3,811 ....C3
61010 Byron 2,035 ....D1
62206 Cahokia 18,904 ....A3
62914 Cairo⊙ 5,931 ....D6
60409 Calumet City 39,697 ....C6
†60643 Calumet Park 8,788 ....C6
62915 Cambria 1,090 ....D6
61238 Cambridge⊙ 2,217 ....C2
62320 Camp Point 1,285 ....B3
61520 Canton 14,626 ....C3
61239 Carbon Cliff 1,578 ....C2
62901 Carbondale 26,414 ....D6
62626 Carlinville⊙ 5,439 ....D4
62231 Carlyle⊙ 3,388 ....D5
62821 Carmi⊙ 6,264 ....E5
†60187 Carol Stream 15,472 ....A5
60110 Carpentersville 23,272 ....E1
62917 Carrier Mills 2,268 ....E6
62016 Carrollton⊙ 2,816 ....C4
62918 Carterville 3,445 ....D6
62321 Carthage⊙ 2,978 ....B3
60013 Cary 6,640 ....E1
62420 Casey 3,026 ....F4
62232 Caseyville 4,308 ....B2
61817 Catlin 2,226 ....F3
61013 Cedarville 766 ....D1
†62801 Central City 1,505 ....D5
62801 Centralia 15,126 ....D5
62206 Centreville 9,747 ....B3
61818 Cerro Gordo 1,553 ....E4
61820 Champaign 58,133 ....E3
Champaign-Urbana-Rantoul‡ 168,392 ....E3
62627 Chandlerville 842 ....C3
60410 Channahon 3,734 ....E2
61920 Charleston⊙ 19,355 ....E4
62629 Chatham 5,597 ....D4
60921 Chatsworth 1,187 ....E3
60922 Chebanse 1,191 ....F3
61726 Chenoa 1,847 ....E3
61016 Cherry Valley 946 ....D1
62233 Chester⊙ 8,401 ....D6
*60601 Chicago⊙ 3,005,072 ....C5
Chicago‡ 7,102,328 ....C5
60411 Chicago Heights 37,026 ....C6
60415 Chicago Ridge 13,473 ....B6
61523 Chillicothe 6,176 ....D3
61924 Chrisman 1,413 ....F4
62822 Christopher 3,086 ....D6
60650 Cicero 61,232 ....B5
60924 Cissna Park 825 ....F3
60514 Clarendon Hills 6,870 ....B6
62824 Clay City 1,038 ....E5
62324 Clayton 889 ....B3
60927 Clifton 1,390 ....F3
61727 Clinton⊙ 8,014 ....E3
60416 Coal City 3,028 ....E2
61240 Coal Valley 3,800 ....C2
62920 Cobden 1,210 ....D6
62017 Coffeen 842 ....D4
62326 Colchester 1,729 ....C3
61728 Colfax 920 ....E3
62234 Collinsville 19,613 ....B2
61241 Colona 2,172 ....C2
62236 Columbia 4,269 ....C5
60112 Cortland 1,019 ....E2
62018 Cottage Hills B2
62237 Coulterville 1,118 ....D5
†60525 Countryside 6,538 ....B6
62922 Creal Springs 845 ....E6
60431 Crest Hill 9,252 ....E2
†60445 Crestwood 10,852 ....B6
60417 Crete 5,417 ....F2
61611 Creve Coeur 6,851 ....D3
62827 Crossville 944 ....F5
60014 Crystal Lake 18,590 ....E1
61427 Cuba 1,648 ....C3
62330 Dallas City 1,408 ....B3
61320 Dalzell 824 ....D2
61732 Danvers 921 ....D3
61832 Danville⊙ 38,985 ....F3
†60559 Darien 14,536 ....B6
*62521 Decatur⊙ 94,081 ....E4
Decatur‡ 131,375 ....E4
60015 Deerfield 17,430 ....B5
†60010 Deer Park 1,368 ....A5
60115 De Kalb 33,099 ....E2
61734 Delavan 1,973 ....D3
61322 Depue 1,873 ....D2
62924 De Soto 1,589 ....D6
*60016 Des Plaines 53,568 ....B5
62530 Divernon 1,081 ....D4
†60469 Dixmoor 4,175 ....C6
61021 Dixon⊙ 15,701 ....D2
60419 Dolton 24,766 ....C6
62926 Dongola 886 ....D6
60515 Downers Grove 42,572 ....A6
60118 Dundee (East and West Dundee) 6,169 ....E1
61525 Dunlap 824 ....D3
62239 Dupo 3,039 ....A3
62832 Du Quoin 6,594 ....D5
61024 Durand 1,073 ....D1
60420 Dwight 4,146 ....E2
60518 Earlville 1,382 ....E2
62024 East Alton 7,096 ....A2
†60411 East Chicago Heights 5,347 ....C6
61025 East Dubuque 2,194 ....C1
†60118 East Dundee (Dundee) 2,618 ....E1
61430 East Galesburg 928 ....C3
†60429 East Hazelcrest 1,362 ....C6
61244 East Moline 20,907 ....C2
61611 East Peoria 22,385 ....D3
*62201 East Saint Louis 55,200 ....A2
62531 Edinburg 1,231 ....D4
62025 Edwardsville⊙ 12,480 ....B2
62401 Effingham⊙ 11,270 ....E4
60119 Elburn 1,224 ....E2
62930 Eldorado 5,198 ....E6
60120 Elgin 63,981 ....E1
61028 Elizabeth 772 ....C1
62931 Elizabethtown⊙ 478 ....E6
60007 Elk Grove Village 28,907 ....B5
62932 Elkville 973 ....D6
60126 Elmhurst 44,276 ....B5
61529 Elmwood 2,117 ....D3
60635 Elmwood Park 24,016 ....B5
61738 El Paso 2,676 ....D3
62028 Elsah 990 ....C5
60421 Elwood 814 ....E2
62933 Energy 1,138 ....E6
62835 Enfield 890 ....E5
62934 Equality 831 ....E6
61250 Erie 1,725 ....C2
61530 Eureka⊙ 4,306 ....D3
*60201 Evanston 73,706 ....B5
62242 Evansville 863 ....D5
60642 Evergreen Park 22,260 ....B6
61739 Fairbury 3,544 ....E3
62837 Fairfield⊙ 5,954 ....E5
†62201 Fairmont City 2,313 ....B2
61841 Fairmount 851 ....F3
62208 Fairview Heights 12,414 ....B3
61842 Farmer City 2,252 ....E3
61531 Farmington 3,118 ....C3
62534 Findlay 868 ....E4
61843 Fisher 1,572 ....E3
61740 Flanagan 978 ....E3
62839 Flora 5,379 ....E5
60422 Flossmoor 8,423 ....B6
60130 Forest Park 15,177 ....B5
†60402 Forest View 764 ....B6
61741 Forrest 1,246 ....E3
61030 Forreston 1,384 ....D1
60020 Fox Lake 6,831 ....A4
60021 Fox River Grove 2,515 ....A5
60423 Frankfort 4,357 ....B6
61031 Franklin Grove 965 ....D2
60131 Franklin Park 17,507 ....B5
62243 Freeburg 2,989 ....D5
61032 Freeport⊙ 26,266 ....D1
61252 Fulton 3,936 ....C2
62935 Galatia 1,042 ....E6
61036 Galena⊙ 3,876 ....C1
61401 Galesburg⊙ 35,305 ....C3
61434 Galva 3,185 ....D2
60424 Gardner 1,322 ....E2
61254 Geneseo 6,373 ....C2
60134 Geneva⊙ 9,881 ....E2
60135 Genoa 3,276 ....E1
61846 Georgetown 4,220 ....F4
62245 Germantown 1,191 ....D5
60936 Gibson City 3,498 ....E3
61847 Gifford 848 ....E3
62033 Gillespie 3,740 ....D4
60938 Gilman 1,913 ....E3
62640 Girard 2,246 ....D4
61533 Glasford 1,201 ....D3
62034 Glen Carbon 5,197 ....B2
60022 Glencoe 9,200 ....B5
†60108 Glendale Heights 23,163 ....A5
60137 Glen Ellyn 23,717 ....A5
60025 Glenview 32,060 ....B5
60425 Glenwood 10,538 ....C6
62035 Godfrey A2
62938 Golconda⊙ 960 ....E6
62939 Goreville 978 ....E6
62037 Grafton 1,024 ....C5
62942 Grand Tower 748 ....D6
†62701 Grandview 1,794 ....D4
62040 Granite City 36,815 ....A2
60940 Grant Park 1,038 ....F2
61326 Granville 1,537 ....D2
60030 Grayslake 5,260 ....B4
62844 Grayville 2,313 ....B4
62044 Greenfield 1,090 ....C4
†60048 Green Oaks 1,415 ....B4
†61241 Green Rock 3,324 ....C2
62428 Greenup 1,655 ....E4
61534 Green Valley 768 ....D3
62642 Greenview 830 ....D3
62246 Greenville⊙ 5,271 ....D5
61744 Gridley 1,246 ....E3
62340 Griggsville 1,301 ....C4
60031 Gurnee 7,179 ....B4
62341 Hamilton 3,509 ....B3
60140 Hampshire 1,735 ....E1
61256 Hampton 1,873 ....C2
61536 Hanna City 1,361 ....D3
61041 Hanover 1,069 ....C1
60103 Hanover Park 28,719 ....A5
62047 Hardin⊙ 1,107 ....C4
62946 Harrisburg⊙ 10,410 ....E6
62537 Harristown 1,456 ....D4
62048 Hartford 1,887 ....A2
60033 Harvard 5,126 ....E1
60426 Harvey 35,810 ....B6
60656 Harwood Heights 8,228 ....B5
62644 Havana⊙ 4,277 ....D3
†60047 Hawthorn Woods 1,658 ....B5
60429 Hazel Crest 13,973 ....B6
60034 Hebron 786 ....E1
†61832 Hegeler 1,853 ....F3
61327 Hennepin⊙ 716 ....D2
61537 Henry 2,740 ....D2
62948 Herrin 10,708 ....E6
60941 Herscher 1,214 ....E2
61745 Heyworth 1,598 ....E3
60457 Hickory Hills 13,778 ....B6
62249 Highland 7,122 ....D5
60035 Highland Park 30,611 ....B5
60040 Highwood 5,452 ....B5
62049 Hillsboro⊙ 4,408 ....D4
60162 Hillside 8,279 ....B5
60520 Hinckley 1,447 ....E2
60521 Hinsdale 16,726 ....B6
60525 Hodgkins 2,005 ....B6
60195 Hoffman Estates 37,272 ....A5
61849 Homer 1,279 ....F3
60456 Hometown 5,324 ....B6
60430 Homewood 19,724 ....B6
60942 Hoopeston 6,411 ....F3
61747 Hopedale 913 ....D3
61748 Hudson 929 ....E3

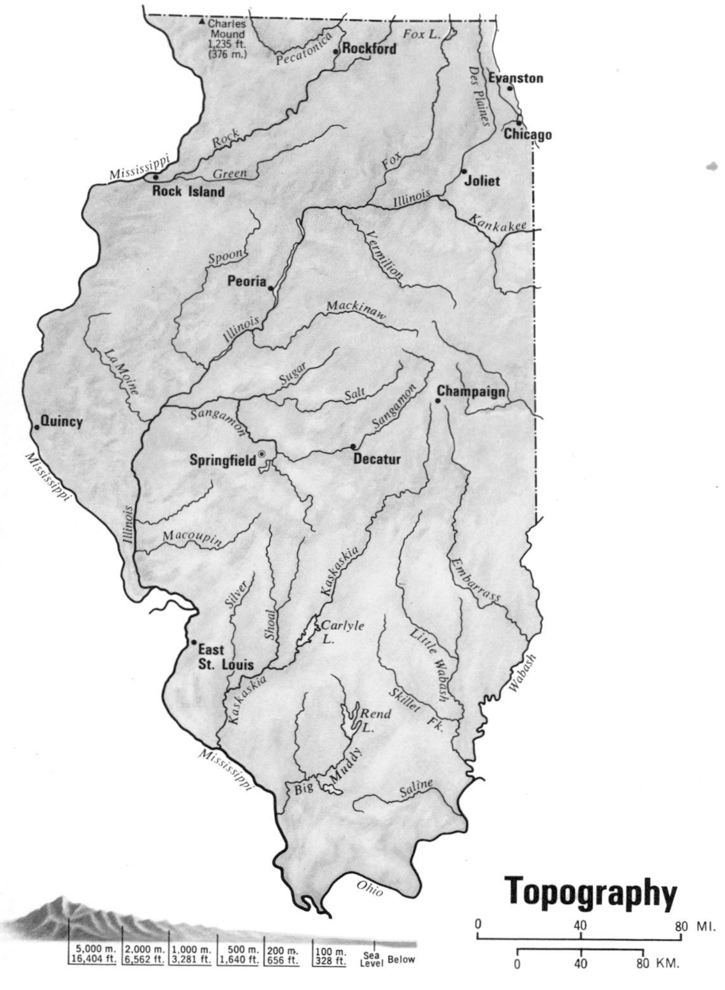
Charles Mound 1,235 ft. (376 m.)
Rockford
Pecatonica
Fox L.
Des Plaines
Evanston
Chicago
Rock
Mississippi
Green
Rock Island
Fox
Joliet
Illinois
Kankakee
Spoon
Vermilion
Peoria
Mackinaw
Illinois
La Moine
Sugar
Salt
Champaign
Sangamon
Sangamon
Quincy
Springfield
Decatur
Mississippi
Illinois
Macoupin
Embarrass
Kaskaskia
Silver
Shoal
Carlyle L.
Little Wabash
East St. Louis
Wabash
Kaskaskia
Rend L.
Skillet Fk.
Mississippi
Big Muddy
Saline
Ohio
5,000 m. 16,404 ft.
2,000 m. 6,562 ft.
1,000 m. 3,281 ft.
500 m. 1,640 ft.
200 m. 656 ft.
100 m. 328 ft.
Sea Level
Below
0 40 80 MI.
0 40 80 KM.

Topography

(continued on following page)

Agriculture, Industry and Resources

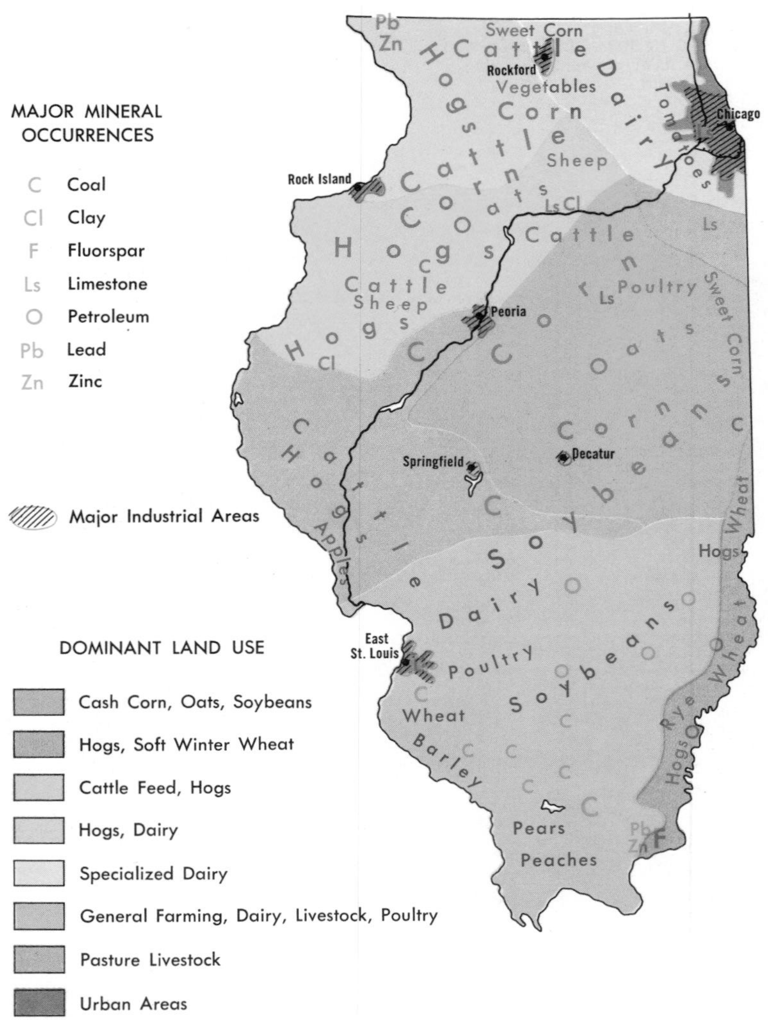

MAJOR MINERAL OCCURRENCES
C Coal
Cl Clay
F Fluorspar
Ls Limestone
O Petroleum
Pb Lead
Zn Zinc
Major Industrial Areas
DOMINANT LAND USE
Cash Corn, Oats, Soybeans
Hogs, Soft Winter Wheat
Cattle Feed, Hogs
Hogs, Dairy
Specialized Dairy
General Farming, Dairy, Livestock, Poultry
Pasture Livestock
Urban Areas
Rockford
Chicago
Rock Island
Peoria
Springfield
Decatur
East St. Louis
Sweet Corn
Cattle
Hogs
Vegetables
Corn
Dairy
Tomatoes
Sheep
Oats
Poultry
Soybeans
Wheat
Apples
Rye
Barley
Pears
Peaches

60142 Huntley 1,646 ... E1
62949 Hurst 938 ... D6
62539 Illiopolis 1,118 ... D4
†60067 Inverness 4,046 ... A5
62848 Irvington 789 ... D5
60042 Island Lake 2,293 ... A4
60143 Itasca 7,129 ... B5
62650 Jacksonville⊙ 20,284 ... C4
†62701 Jerome 1,374 ... D4
62052 Jerseyville⊙ 7,506 ... C4
62436 Jewett 230 ... E4
62951 Johnston City 3,873 ... E6
*60431 Joliet⊙ 77,956 ... E2
62952 Jonesboro⊙ 1,842 ... D6
†60458 Justice 10,552 ... B6
60901 Kankakee⊙ 30,141 ... F2
Kankakee‡ 102,926 ... F2
61933 Kansas 791 ... F4
†63673 Kaskaskia 33 ... C6
61442 Keithsburg 936 ... B2
60043 Kenilworth 2,708 ... B5
61443 Kewanee 14,508 ... C2
†60069 Kildeer 1,609 ... A5
62540 Kincaid 1,591 ... D4
62854 Kinmundy 945 ... E5
60146 Kirkland 1,155 ... E1
61447 Kirkwood 1,008 ... C3
61448 Knoxville 3,432 ... C3
61540 Lacon⊙ 2,135 ... D2
61329 Ladd 1,337 ... D2
60525 La Grange 15,445 ... B6
60525 La Grange Park 13,359 ... B5
61450 La Harpe 1,471 ... C3
†60010 Lake Barrington 2,320 ... A5
60044 Lake Bluff 4,434 ... B4
†60002 Lake Catherine 1,335 ... E1
60045 Lake Forest 15,245 ... B4
†60102 Lake in the Hills 5,651 ... E1
60046 Lake Villa 1,462 ... A4
62438 Lakewood 1,254 ... E4
60047 Lake Zurich 8,225 ... A5
61330 La Moille 734 ... D2
61046 Lanark 1,483 ... D1
60438 Lansing 29,039 ... C6
61301 La Salle 10,347 ... E2
62439 Lawrenceville⊙ 5,652 ... F5
62254 Lebanon 3,245 ... D5
60531 Leland 775 ... E2

60439 Lemont 5,640 ... B6
61048 Lena 2,295 ... D1
61752 Le Roy 2,870 ... E3
61542 Lewistown⊙ 2,758 ... C3
61753 Lexington 1,806 ... E3
60048 Libertyville 16,520 ... B4
62656 Lincoln⊙ 16,327 ... D3
†60015 Lincolnshire 4,151 ... B5
†60645 Lincolnwood 11,921 ... B5
†60046 Lindenhurst 6,220 ... B4
60532 Lisle 13,625 ... A6
62056 Litchfield 7,204 ... D4
62058 Livingston 949 ... D5
62661 Loami 770 ... D4
60441 Lockport 9,170 ... B6
60148 Lombard 36,897 ... B5
60047 Long Grove 2,013 ... B5
62858 Louisville⊙ 1,166 ... E5
62059 Lovejoy 1,233 ... A2
61111 Loves Park 13,192 ... E1
61937 Lovington 1,313 ... E4
61261 Lyndon 777 ... D2
†60411 Lynwood 4,195 ... C6
60534 Lyons 9,925 ... B6
61755 Mackinaw 1,354 ... D3
61455 Macomb⊙ 19,863 ... C3
62544 Macon 1,300 ... E4
62060 Madison 5,915 ... A2
61853 Mahomet 1,986 ... E3
60150 Malta 995 ... E2
60442 Manhattan 1,944 ... F2
61546 Manito 1,869 ... D3
61854 Mansfield 921 ... E3
60950 Manteno 3,155 ... F2
60152 Marengo 4,361 ... E1
62061 Marine 957 ... D5
62959 Marion⊙ 14,031 ... E6
62257 Marissa 2,568 ... D5
60426 Markham 15,172 ... B6
61756 Maroa 1,760 ... E3
†61554 Marquette Heights 3,386 ... D3
61341 Marseilles 4,766 ... E2
62441 Marshall⊙ 3,655 ... F4
62442 Martinsville 1,298 ... F4
62062 Maryville 1,937 ... B2
62258 Mascoutah 4,962 ... D5
62664 Mason City 2,719 ... D3
61263 Matherville 793 ... C2

60443 Matteson 10,223 ... B6
61938 Mattoon 19,055 ... E4
60153 Maywood 27,998 ... B5
60444 Mazon 828 ... E2
60050 McHenry 10,908 ... E1
†60050 McHenry Shores 1,041 ... E1
61754 McLean 836 ... D3
62859 McLeansboro⊙ 2,960 ... E5
†62010 Meadowbrook 1,082 ... B2
*60160 Melrose Park 20,735 ... B5
62351 Mendon 979 ... B3
61342 Mendota 7,134 ... D2
62665 Meredosia 1,272 ... C4
†60601 Merrionette Park 2,054 ... B6
61548 Metamora 2,482 ... D3
62960 Metropolis⊙ 7,171 ... E6
60445 Midlothian 14,274 ... B6
61264 Milan 6,264 ... C2
60953 Milford 1,716 ... F3
61051 Milledgeville 1,209 ... D1
62260 Millstadt 2,736 ... B3
61759 Minier 1,261 ... D3
61760 Minonk 2,039 ... D3
60447 Minooka 1,565 ... E2
60448 Mokena 4,578 ... B6
61265 Moline 46,278 ... C2
60954 Momence 3,297 ... F2
60449 Monee 993 ... F2
61462 Monmouth⊙ 10,706 ... C3
60538 Montgomery 3,369 ... E2
61856 Monticello⊙ 4,753 ... E3
60450 Morris⊙ 8,833 ... E2
61270 Morrison⊙ 4,605 ... C2
62546 Morrisonville 1,208 ... D4
61550 Morton 14,178 ... D3
60053 Morton Grove 23,747 ... B5
62963 Mound City⊙ 1,102 ... D6
62964 Mounds 1,669 ... D6
62863 Mount Carmel⊙ 8,908 ... F5
61053 Mount Carroll⊙ 1,936 ... D1
61054 Mount Morris 2,989 ... D1
62069 Mount Olive 2,357 ... D4
60056 Mount Prospect 52,634 ... B5
62548 Mount Pulaski 1,783 ... D3
62353 Mount Sterling⊙ 2,186 ... C4
62864 Mount Vernon⊙ 17,193 ... E5
62549 Mount Zion 4,563 ... E4
62550 Moweaqua 1,922 ... E4

60060 Mundelein 17,053 ... A4
62966 Murphysboro⊙ 9,866 ... D6
60540 Naperville 42,601 ... A6
62263 Nashville⊙ 3,186 ... D5
62354 Nauvoo 1,133 ... B3
62447 Neoga 1,736 ... E4
60541 Newark 798 ... E2
62264 New Athens 1,937 ... D5
62265 New Baden 2,476 ... D5
62670 New Berlin 834 ... D4
61272 New Boston 731 ... B2
60451 New Lenox 5,792 ... B6
61942 Newman 1,079 ... B6
62448 Newton⊙ 3,186 ... E5
61465 New Windsor 863 ... C2
62551 Niantic 761 ... D4
60648 Niles 30,363 ... B5
62868 Noble 832 ... E5
62075 Nokomis 2,656 ... D4
61761 Normal 35,672 ... E3
†60656 Norridge 16,483 ... B5
62869 Norris City 1,515 ... E6
60542 North Aurora 5,205 ... E2
†60010 North Barrington 1,475 ... A5
60062 Northbrook 30,778 ... B5
60064 North Chicago 38,774 ... B4
60093 Northfield 5,807 ... B5
60164 Northlake 12,166 ... B5
†61111 North Park 15,806 ... D1
†61554 North Pekin 1,824 ... D3
60546 North Riverside 6,764 ... B5
†61373 North Utica (Utica) 1,067 ... E2
60521 Oak Brook 6,641 ... B6
†60181 Oakbrook Terrace 2,285 ... B5
60452 Oak Forest 26,096 ... B6
61943 Oakland 1,035 ... F4
*60453 Oak Lawn 60,590 ... B6
*60303 Oak Park 54,887 ... B5
61858 Oakwood 1,627 ... F3
62449 Oblong 1,840 ... F5
60460 Odell 1,083 ... E2
62870 Odin 1,285 ... D5
62269 O'Fallon 12,241 ... B2
61859 Ogden 818 ... F3
61348 Oglesby 3,979 ... D2
62271 Okawville 1,337 ... D5
62450 Olney⊙ 9,026 ... E5
60461 Olympia Fields 4,146 ... B6
60955 Onarga 1,269 ... F3
61467 Oneida 765 ... C2
61469 Oquawka⊙ 1,533 ... C3
62554 Oreana 999 ... E4
61061 Oregon⊙ 3,559 ... D1
61273 Orion 2,013 ... C2
60462 Orland Park 23,045 ... B6
60543 Oswego 3,021 ... E2
61350 Ottawa⊙ 18,166 ... E2
60067 Palatine 32,166 ... B5
62451 Palestine 1,718 ... F4
62674 Palmyra 864 ... C4
60463 Palos Heights 11,096 ... B6
60465 Palos Hills 16,654 ... B6
60464 Palos Park 3,150 ... B6
62557 Pana 6,040 ... D4
61944 Paris⊙ 9,885 ... F4
†60085 Park City 3,673 ... B4
60466 Park Forest 26,222 ... B6
60466 Park Forest South 6,245 ... F2
60068 Park Ridge 38,704 ... B5
62558 Pawnee 2,577 ... D4
61353 Pawpaw 839 ... E2
60957 Paxton⊙ 4,258 ... E3
62360 Payson 1,065 ... B4
61063 Pecatonica 1,732 ... D1
61554 Pekin⊙ 33,967 ... D3
*61601 Peoria⊙ 124,160 ... D3
Peoria‡ 365,864 ... D3
61614 Peoria Heights 7,453 ... D3
60468 Peotone 2,832 ... F2
62272 Percy 1,053 ... D5
61354 Peru 10,886 ... D2
62675 Petersburg⊙ 2,419 ... D4
61864 Philo 973 ... E3
†60426 Phoenix 2,850 ... C6
62274 Pinckneyville⊙ 3,319 ... D5
60959 Piper City 905 ... E3
62363 Pittsfield⊙ 4,170 ... C4
60544 Plainfield 3,767 ... A6
60545 Plano 4,875 ... E2
62366 Pleasant Hill 1,112 ... C4
62275 Pocahontas 866 ... D5
61074 Polo 2,643 ... D1
61764 Pontiac⊙ 11,227 ... E3
†62040 Pontoon Beach 3,336 ... A2
61065 Poplar Grove 818 ... E1
61275 Port Byron 1,289 ... C2
60469 Posen 4,642 ... B6
61865 Potomac 874 ... F3
61470 Prairie City 580 ... C3
61356 Princeton⊙ 7,342 ... D2
61559 Princeville 1,712 ... D3
61277 Prophetstown 2,141 ... D2
60070 Prospect Heights 11,808 ... B5
62301 Quincy⊙ 42,554 ... B4
62080 Ramsey 1,058 ... D4
60960 Rankin 727 ... F3
61866 Rantoul 20,161 ... E3
61278 Rapids City 1,058 ... C2
62560 Raymond 957 ... D4
62278 Red Bud 2,850 ... D5
60071 Richmond 1,068 ... E1
60471 Richton Park 9,403 ... B6
61870 Ridge Farm 1,096 ... F4
62979 Ridgway 1,245 ... E6
60627 Riverdale 13,233 ... C6
60305 River Forest 12,392 ... B5
60171 River Grove 10,368 ... B5
60546 Riverside 9,236 ... B6
62561 Riverton 2,783 ... D4
†60015 Riverwoods 2,804 ... B5
61561 Roanoke 2,001 ... D3
60472 Robbins 8,853 ... B6
62454 Robinson⊙ 7,285 ... F5
61068 Rochelle 8,982 ... D2
62563 Rochester 2,488 ... D4
60436 Rockdale 1,913 ... E2
61071 Rock Falls 10,633 ... D2

*61101 Rockford⊙ 139,712 ... D1
Rockford‡ 279,514 ... D1
61201 Rock Island⊙ 46,928 ... C2
Rock Island-Moline-
Davenport‡ 383,958 ... C2
61072 Rockton 2,313 ... E1
60008 Rolling Meadows 20,167 ... A5
61562 Rome 2,744 ... D3
60441 Romeoville 15,519 ... B6
62082 Roodhouse 2,364 ... C4
61073 Roscoe 1,388 ... D1
60172 Roselle 16,948 ... A5
60018 Rosemont 4,137 ... B5
61473 Roseville 1,254 ... C3
†62024 Rosewood Heights 5,085 ... B2
62982 Rosiclare 1,441 ... E6
60963 Rossville 1,363 ... F3
60673 Round Lake 2,644 ... A4
†60673 Round Lake Beach 12,921 ... A4
†60673 Round Lake Heights 1,192 ... E1
†60673 Round Lake Park 4,032 ... A4
62084 Roxana 1,587 ... B2
62983 Royalton 1,320 ... D6
62681 Rushville⊙ 3,348 ... C3
60964 Saint Anne 1,421 ... F2
60174 Saint Charles 17,492 ... E2
61563 Saint David 786 ... C3
62458 Saint Elmo 1,611 ... E4
62460 Saint Francisville 1,040 ... F5
62281 Saint Jacob 792 ... D5
61873 Saint Joseph 1,900 ... E3
62881 Salem⊙ 7,813 ... E5
62882 Sandoval 1,734 ... D5
60548 Sandwich 5,244 ... E2
62682 San Jose 784 ... D3
60411 Sauk Village 10,906 ... C6
61074 Savanna 4,529 ... C1
61874 Savoy 2,126 ... E3
61770 Saybrook 882 ... E3
60194 Schaumburg 53,305 ... A5
60176 Schiller Park 11,458 ... B5
61360 Seneca 2,098 ... E2
62884 Sesser 2,238 ... D5
60550 Shabbona 851 ... E2
61078 Shannon 938 ... D1
62984 Shawneetown⊙ 1,841 ... E6
61361 Sheffield 1,130 ... D2
62565 Shelbyville⊙ 5,259 ... E4
60966 Sheldon 1,215 ... F3
62684 Sherman 1,501 ... D4
61281 Sherrard 811 ... C2
†62220 Shiloh 1,045 ... B3
60435 Shorewood 4,714 ... E2
61877 Sidney 886 ... E3
61282 Silvis 7,130 ... C2
*60076 Skokie 60,278 ... B5
†60118 Sleepy Hollow 2,000 ... E1
62285 Smithton 1,447 ... C5
60552 Somonauk 1,344 ... E2
†60010 South Barrington 1,168 ... A5
61080 South Beloit 4,088 ... E1
60411 South Chicago
Heights 3,932 ... C6
60177 South Elgin 5,970 ... E2
60473 South Holland 24,977 ... C6
†62650 South Jacksonville 3,382 ... C4
61564 South Pekin 1,243 ... D3
62087 South Roxana 2,286 ... B2
60474 South Wilmington 747 ... E2
62286 Sparta 4,957 ... D5
*62701 Springfield (cap.)⊙
100,054 ... D4
Springfield‡ 187,789 ... D4
61362 Spring Valley 5,822 ... D2
61774 Stanford 720 ... D3
62088 Staunton 4,744 ... D5
62288 Steeleville 2,240 ... D6
60475 Steger 9,269 ... F2
61081 Sterling 16,281 ... D2
62463 Stewardson 745 ... E4
60402 Stickney 5,893 ... B6
61084 Stillman Valley 961 ... D1
61085 Stockton 1,872 ... C1
†60160 Stone Park 4,273 ... B5
62567 Stonington 1,184 ... D4
60103 Streamwood 23,456 ... A5
61364 Streator 14,795 ... E2
61480 Stronghurst 865 ... C3
60554 Sugar Grove 1,366 ... E2
61951 Sullivan⊙ 4,526 ... E4
60501 Summit-Argo 10,110 ... B6
62466 Sumner 1,238 ... F5
†60050 Sunnyside 1,432 ... A4
62221 Swansea 5,347 ... B3
60178 Sycamore⊙ 9,219 ... E2
62888 Tamaroa 885 ... D5
62988 Tamms 826 ... D6
61283 Tampico 966 ... D2
62568 Taylorville⊙ 11,386 ... D4
62467 Teutopolis 1,414 ... E4
62689 Thayer 759 ... D4
61878 Thomasboro 1,242 ... E3
61285 Thomson 911 ... C2
60476 Thornton 3,024 ... C6
62292 Tilden 1,025 ... D5
†61832 Tilton 2,405 ... F3
60477 Tinley Park 26,171 ... B6
61368 Tiskilwa 990 ... D2
62468 Toledo⊙ 1,284 ... E4
61880 Tolono 2,434 ... E3
61369 Toluca 1,471 ... D2
61483 Toulon⊙ 1,390 ... D2
†60010 Tower Lakes 1,177 ... A4
61568 Tremont 2,096 ... D3
62293 Trenton 2,504 ... D5
62294 Troy 3,772 ... B2
61953 Tuscola⊙ 3,839 ... E4
61801 Urbana⊙ 35,978 ... E3
61373 Utica 1,067 ... E2
62891 Valier 729 ... D5
†60120 Valley View 2,112 ... E2
62295 Valmeyer 898 ... C5
62471 Vandalia⊙ 5,338 ... D5
62090 Venice 3,480 ... A2
61484 Vermont 885 ... C3
60061 Vernon Hills 9,827 ... B4
62995 Vienna⊙ 1,420 ... E6

61956 Villa Grove 2,707 ... E
60181 Villa Park 23,185 ... B
61486 Viola 1,144 ... C
62690 Virden 3,899 ... D
62691 Virginia⊙ 1,825 ... C
60083 Wadsworth 1,104 ... B
61376 Walnut 1,513 ... D
†62801 Wamac 1,665 ... D
61777 Wapella 768 ... E
61087 Warren 1,595 ... C
62573 Warrensburg 1,372 ... D
60555 Warrenville 7,519 ... A
62379 Warsaw 1,842 ... B
61570 Washburn 1,206 ... D
61571 Washington 10,364 ... D
62204 Washington Park 8,223 ... E
61488 Wataga 996 ... C
62298 Waterloo⊙ 4,646 ... C
60556 Waterman 943 ... E
60970 Watseka⊙ 5,543 ... F
60084 Wauconda 5,688 ... A
60085 Waukegan⊙ 67,653 ... E
62692 Waverly 1,537 ... D
60184 Wayne 940 ...
62895 Wayne City 1,132 ...
61377 Wenona 1,025 ...
60153 Westchester 17,730 ...
60185 West Chicago 12,550 ... A
†60118 West Dundee
(Dundee) 3,551 ...
60558 Western Springs 12,876 ...
62474 Westfield 733 ...
62896 West Frankfort 9,437 ...
†60462 Westhaven 2,784 ...
60559 Westmont 16,718 ...
62476 West Salem 1,145 ...
61883 Westville 3,573 ...
60187 Wheaton⊙ 43,043 ...
60090 Wheeling 23,266 ...
62092 White Hall 2,935 ...
62693 Williamsville 996 ...
†60521 Willowbrook 4,953 ...
60480 Willow Springs 4,147 ...
60091 Wilmette 28,229 ...
60481 Wilmington 4,424 ...
62694 Winchester⊙ 1,716 ...
61957 Windsor 1,228 ...
†61465 Windsor (New
Windsor) 863 ...
60190 Winfield 4,422 ...
61088 Winnebago 1,644 ...
60093 Winnetka 12,772 ...
60096 Winthrop Harbor 5,431 ...
62094 Witt 1,205 ...
60191 Wood Dale 11,251 ...
61490 Woodhull 901 ...
†60517 Woodridge 22,561 ...
62095 Wood River 12,446 ...
60098 Woodstock⊙ 11,725 ...
62097 Worden 953 ...
60482 Worth 11,592 ...
61379 Wyanet 1,069 ...
61491 Wyoming 1,614 ...
61572 Yates City 860 ...
60560 Yorkville⊙ 3,422 ...
62999 Zeigler 1,858 ...
60099 Zion 17,861 ...

OTHER FEATURES

Apple (creek) ...
Apple (riv.) ...
Argonne Nat'l Laboratory ...
Big Bureau (riv.) ...
Big Muddy (riv.) ...
Bonpas (creek) ...
Cache (riv.) ...
Calumet (lake) ...
Carlyle (lake) ...
Chanute A.F.B. ...
Charles Mound (hill) ...
Chicago Portage Nat'l Hist. Site ...
Crab Orchard (lake) ...
Des Plaines (riv.) ...
Du Page (riv.) ...
Edwards (riv.) ...
Embarras (riv.) ...
Fort Sheridan ...
Fox (lake) ...
Fox (riv.) ...
Fox (riv.) ...
Glenview Nav. Air. Sta. ...
Granite City Army Depot ...
Great Lakes Nav. Trng. Ctr. ...
Green (riv.) ...
Henderson (riv.) ...
Illinois (riv.) ...
Illinois - Mississippi (canal) ...
Iroquois (riv.) ...
Kankakee (riv.) ...
Kaskaskia (riv.) ...
La Moine (riv.) ...
Little Wabash (riv.) ...
Mackinaw (riv.) ...
Macoupin (riv.) ...
Michigan (lake) ...
Mississippi (riv.) ...
O'Hare Field-Chicago International
Airport ...
Ohio (riv.) ...
Plum (riv.) ...
Pope (creek) ...
Rend (lake) ...
Rock (creek) ...
Rock (riv.) ...
Rock Island Arsenal ...
Saline (riv.) ...
Salt (creek) ...
Sangamon (riv.) ...
Savanna Army Depot ...
Scott A.F.B. 8,648 ...
Shelbyville (lake) ...
Spoon (riv.) ...
Wabash (riv.) ...

⊙County seat.
‡Population of metropolitan area.
† Zip of nearest p.o. * Multiple

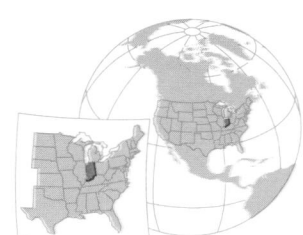

AREA 36,185 sq. mi. (93,719 sq. km.)
POPULATION 5,490,260
CAPITAL Indianapolis
LARGEST CITY Indianapolis
HIGHEST POINT 1,257 ft. (383 m.) (Wayne County)
SETTLED IN 1730
ADMITTED TO UNION December 11, 1816
POPULAR NAME Hoosier State
STATE FLOWER Peony
STATE BIRD Cardinal

COUNTIES

ams 29,619	H3
en 294,335	G2
rtholomew 65,088	F6
nton 10,218	C3
ckford 15,570	G4
one 36,446	E4
own 12,377	E6
rroll 19,722	D3
ss 40,936	E3
rk 88,838	F8
nton 31,545	E4
awford 9,820	E8
viess 27,836	C7
arborn 34,291	H6
catur 23,841	G6
Kalb 33,606	H2
bois 34,238	D8
hart 137,330	F1
yette 28,272	G5
yd 61,169	F8
untain 19,033	C4
nklin 19,612	G6
ton 19,335	E2
son 33,156	B8
ant 80,934	F3
eene 30,416	D6
milton 82,027	E4
ncock 43,939	F5
rrison 27,276	E8
ndricks 69,804	D5
nry 53,336	G5
ward 86,896	E4
ntington 35,596	G3
ckson 36,523	E7
sper 26,138	C2
23,239	G4
fferson 30,419	G7
nnings 22,854	F7
hnson 77,240	E6
ox 41,838	C7
sciusko 59,555	F2
grange 25,550	G1
ke 522,965	C2
Porte 108,632	D1
wrence 42,472	E7
dison 139,336	F4
rion 765,233	E5
rshall 39,155	E2
rtin 11,001	D7
ami 39,820	E3
nroe 98,785	D6
ntgomery 35,501	D4
rgan 51,999	E6
wton 14,844	C3
ble 35,443	G2
o 5,114	H7
ange 18,677	E7
en 15,841	D6
ke 16,372	C5
rry 19,346	D8
e 13,465	C8
ter 119,816	C2
sey 26,414	B8
aski 13,258	D2
nam 29,163	D5
ndolph 29,997	G4
ley 24,398	G6
sh 19,604	G5
nt Joseph 241,617	E1
ott 20,422	F7
elby 39,887	F5
encer 19,361	C9
rke 21,997	D2
uben 24,694	G1
livan 21,107	C6
itzerland 7,153	G7
pecanoe 121,702	D4
on 16,819	E4
on 6,860	H5
derburgh 167,515	B8
million 18,229	C5
o 112,385	C6
bash 36,640	F3
rren 8,976	C4
rrick 41,474	C8
shington 21,932	E7
yne 76,058	G5
lls 25,401	G3
ite 23,867	D3
tley 26,215	F2

CITIES and TOWNS

Name/Pop.	Key
7240 Adams 250	F6
9947 Adamsboro 325	E3
5102 Advance 559	D5
5910 Akron 1,045	E2
3320 Albany 2,625	G4
5701 Albion⊙ 1,637	G2
7283 Alert 102	F6
6001 Alexandria 6,028	F4
6738 Altona 263	G2
47917 Ambia 274	C4
46911 Amboy 450	F3
†46131 Amity 200	E6
46103 Amo 444	D5
*46011 Anderson⊙ 64,695	F4
Anderson‡ 139,336	F4
†47024 Andersonville 225	G5
46702 Andrews 1,243	F3
46703 Angola⊙ 5,486	G1
46030 Arcadia 1,801	E4
46704 Arcola 300	G2
†46624 Ardmore 800	E1
46501 Argos 1,547	E2
46104 Arlington 500	F5
46705 Ashley 841	G1
46031 Atlanta 657	E4
46502 Atwood 300	F2
46706 Auburn⊙ 8,122	G2
47001 Aurora 3,816	H6
47102 Austin 4,857	F7
46710 Avilla 1,272	G2
47420 Avoca 400	D7
46105 Bainbridge 644	D5
46106 Bargersville 1,647	E5
47006 Batesville 4,152	G6
47920 Battle Ground 812	D3
46107 Beech Grove 13,196	E5
47421 Bedford⊙ 14,410	E7
†46526 Benton 220	F2
46711 Berne 3,300	H3
†46111 Bethany 127	E5
46301 Beverly Shores 864	C1
47512 Bicknell 4,713	C7
46713 Bippus 300	F3
47513 Birdseye 533	D8
†46406 Black Oak	C1
47831 Blanford 500	B5
47138 Blocher 400	F7
47424 Bloomfield⊙ 2,705	D6
47832 Bloomingdale 409	C5
47401 Bloomington⊙ 52,044	D6
Bloomington‡ 98,387	D6
†47360 Blountsville 213	G4
†46176 Blue Ridge 219	F5
46714 Bluffton⊙ 8,705	G3
46110 Boggstown 200	F5
46302 Boone Grove 220	C2
47601 Boonville⊙ 6,300	C8
47106 Borden 384	F8
47324 Boston 189	H5
47921 Boswell 810	C3
46504 Bourbon 1,522	E2
47833 Bowling Green 200	D6
47107 Bradford 350	E8
47834 Brazil⊙ 7,852	C5
46506 Bremen 3,565	E2
47836 Bridgeton 250	C5
†45030 Bright 450	H6
46720 Brimfield 292	G2
46913 Bringhurst 275	E3
46507 Bristol 1,203	F1
47922 Brook 926	C3
46111 Brooklyn 889	E5
†47250 Brooksburg 132	G7
47923 Brookston 1,701	D3
47012 Brookville⊙ 2,874	G6
46112 Brownsburg 6,242	E5
47220 Brownstown⊙ 2,704	F7
47325 Brownsville 250	H5
47516 Bruceville 646	C7
47326 Bryant 277	G3
47924 Buck Creek 225	D4
47647 Buckskin 200	C8
47925 Buffalo 500	D3
46914 Bunker Hill 984	E3
46508 Burket 260	F2
46915 Burlington 680	E4
47926 Burnettsville 496	D3
47222 Burney 300	F6
†46401 Burns Harbor 920	C1
46916 Burrows 250	E3
46721 Butler 2,509	H2
47223 Butlerville 300	F6
†46371 Byron 200	C5
†47362 Cadiz 180	G5
47327 Cambridge City 2,407	G5
46917 Camden 618	D3
47108 Campbellsburg 695	E7
47224 Canaan 90	G7
47519 Cannelburg 152	C7
47520 Cannelton⊙ 2,373	D9
47837 Carbon 307	C5
47838 Carlisle 717	C7
46032 Carmel 18,272	E5
47114 Cartersburg 300	E5
46115 Carthage 886	F5
47927 Cates 125	C4
47928 Cayuga 1,258	C5
47016 Cedar Grove 217	H6
46303 Cedar Lake 8,754	C2
47521 Celestine 150	D8
†47842 Centenary 150	B5
†46901 Center 310	E4
47840 Centerpoint 242	C6
46116 Centerton 250	E5
47330 Centerville 2,284	H5
47929 Chalmers 554	D3
47610 Chandler 3,043	C8
47111 Charlestown 5,596	F8
46117 Charlottesville 300	F5
†47138 Chelsea 200	F7
46017 Chesterfield 2,701	F4
46304 Chesterton 8,531	D1
47611 Chrisney 537	C8
46723 Churubusco 1,638	G2
47225 Clarksburg 300	G6
46930 Clarks Hill 653	D4
47130 Clarksville 15,164	F8
47841 Clay City 883	C6
46510 Claypool 464	F2
46118 Clayton 703	D5
47226 Clifford 310	F6
47842 Clinton 5,267	C5
46120 Cloverdale 1,357	D5
†47834 Cloverland 175	C6
47427 Coal City 225	D6
47845 Coalmont 450	C6
46121 Coatesville 474	D5
47931 Colburn 300	D3
46035 Colfax 823	D4
47978 Collegeville 1,059	C3
46725 Columbia City⊙ 5,091	G2
47201 Columbus⊙ 30,614	E6
47331 Connersville⊙ 17,023	G5
46919 Converse 1,279	F3
47228 Cortland 175	F7
47112 Corydon⊙ 2,724	E8
47932 Covington⊙ 2,883	C4
†47302 Cowan 428	G4
47114 Crandall 176	E8
47522 Crane	D7
47933 Crawfordsville⊙ 13,325	D4
46732 Cromwell 458	F2
47229 Crothersville 1,747	F7
46307 Crown Point⊙ 16,455	C2
46511 Culver 1,601	E2
47612 Cynthiana 874	B8
47523 Dale 1,693	C8
47334 Daleville	F4
47847 Dana 803	C5
46122 Danville⊙ 4,220	D5
47940 Darlington 811	D4
47618 Darmstadt 1,280	B8
47941 Dayton 781	D4
46733 Decatur⊙ 8,649	H3
47524 Decker 256	B7
†46917 Deer Creek 250	E3
46923 Delphi⊙ 3,042	D3
46310 Demotte 2,559	C2
46926 Denver 589	E3
47230 Deputy 200	F7
47302 Desoto 385	G4
47018 Dillsboro 1,038	G7
46513 Donaldson 320	E2
†47118 Doolittle Mills 200	D8
47335 Dublin 979	G5
47525 Dubois 550	D8
47848 Dugger 1,118	C6
†46304 Dune Acres 291	C1
47336 Dunkirk 3,180	G4
†46514 Dunlap 5,397	F1
47337 Dunreith 184	F5
47231 Dupont 392	G7
46311 Dyer 9,555	C1
†46704 Eagletown 306	E4
47942 Earl Park 469	C3
46312 East Chicago 39,786	C1
47019 East Enterprise 250	H7
†47370 East Germantown (Pershing) 438	G5
46732 Eaton 1,804	G4
47116 Eckerty 108	D8
47339 Economy 237	G5
†46011 Edgewood 2,215	F4
46124 Edinburgh 4,856	E6
47528 Edwardsport 459	C7
†47150 Edwardsville 700	F8
47613 Elberfeld 640	C8
47117 Elizabeth 178	F8
47232 Elizabethtown 603	F6
46514 Elkhart 41,305	F1
Elkhart‡ 137,330	F1
47429 Ellettsville 3,328	D6
47529 Elnora 756	C7
†47018 Elrod 200	G6
†47901 Elston 500	D4
46036 Elwood 10,867	F4
46125 Eminence 200	D5
47118 English⊙ 633	E8
46524 Etna Green 522	E2
†47928 Eugene 400	B5
*47701 Evansville⊙ 130,496	C9
Evansville‡ 309,408	C9
†47331 Everton 500	G5
46126 Fairland 950	F5
46928 Fairmount 3,286	F4
†47842 Fairview Park 1,545	C5
47850 Farmersburg 1,240	C6
47340 Farmland 1,560	G4
†47421 Fayetteville 180	D7
47532 Ferdinand 2,192	D8
46128 Fillmore 550	D5
46129 Finly 400	F5
46038 Fishers 2,008	E5
47234 Flat Rock 323	F6
46929 Flora 2,303	E3
47119 Floyds Knobs 500	F8
47851 Fontanet 325	C5
46039 Forest 400	E4
47648 Fort Branch 2,504	B8
46040 Fortville 2,787	F5
*46801 Fort Wayne⊙ 172,028	G2
Fort Wayne‡ 382,961	G2
47341 Fountain City 839	H5
46130 Fountaintown 225	F5
47944 Fowler⊙ 2,319	C3
46930 Fowlerton 300	F4
47946 Francesville 944	D3
47649 Francisco 612	B8
46041 Frankfort⊙ 15,168	E4
46131 Franklin⊙ 11,563	E6
46044 Frankton 2,080	F4
47120 Fredericksburg 233	E8
47431 Freedom 100	D6
47535 Freelandville 600	C7
47235 Freetown 600	E7
46737 Fremont 1,180	H1
47432 French Lick 2,265	D7
46931 Fulton 393	E3
†47119 Galena 1,186	F8
46932 Galveston 1,822	E3
46738 Garrett 4,751	G2
*46401 Gary 151,953	C1
Gary-Hammond-East Chicago‡ 642,781	C1
46933 Gas City 6,370	F4
47342 Gaston 1,150	G4
46740 Geneva 1,430	H3
47537 Gentryville 299	C8
47122 Georgetown 1,494	F8
46133 Glenwood 370	G5
†47567 Glezen 300	C8
46045 Goldsmith 235	E4

(continued on following page)

Agriculture, Industry and Resources

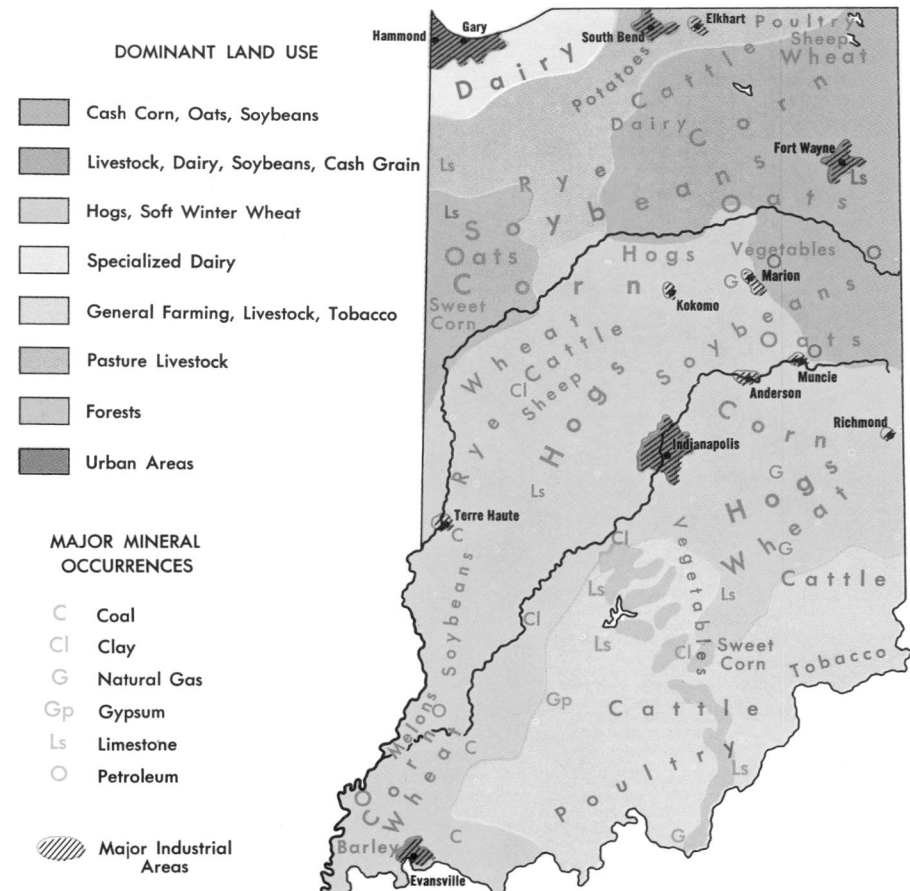

DOMINANT LAND USE
- Cash Corn, Oats, Soybeans
- Livestock, Dairy, Soybeans, Cash Grain
- Hogs, Soft Winter Wheat
- Specialized Dairy
- General Farming, Livestock, Tobacco
- Pasture Livestock
- Forests
- Urban Areas

MAJOR MINERAL OCCURRENCES
- C Coal
- Cl Clay
- G Natural Gas
- Gp Gypsum
- Ls Limestone
- ○ Petroleum

Major Industrial Areas

47948 Goodland 1,200 ......C3
46526 Goshen⊙ 19,665 ......F1
47433 Gosport 729 ......D6
46741 Grabill 658 ......H2
47615 Grandview 670 ......C9
46530 Granger 350 ......E1
46135 Greencastle⊙ 8,403 ......D5
†47025 Greendale 3,795 ......H6
46140 Greenfield⊙ 11,299 ......F5
47344 Greensboro 175 ......G5
47240 Greensburg⊙ 9,254 ......G6
47345 Greens Fork 426 ......H5
46936 Greentown 2,265 ......E4
47124 Greenville 537 ......F8
46142 Greenwood 19,327 ......E5
47616 Griffin 192 ......B8
46319 Griffith 17,026 ......C1
46144 Gwynneville 250 ......F5
47346 Hagerstown 1,950 ......G5
46742 Hamilton 587 ......H1
46532 Hamlet 738 ......D2
*46320 Hammond 93,714 ......B1
46340 Hanna 550 ......D2
47243 Hanover 4,054 ......F7
47125 Hardinsburg 298 ......E8
46743 Harlan 840 ......H2
47853 Harmony 613 ......C5
47434 Harrodsburg 400 ......D6
47348 Hartford City⊙ 7,622 ......G4
47244 Hartsville 379 ......F6
47617 Hatfield 800 ......C9
47639 Haubstadt 1,389 ......B8
†47546 Haysville 600 ......D8
47640 Hazleton 368 ......B8
46341 Hebron 2,696 ......C2
47436 Heltonville 400 ......E7
46937 Hemlock 300 ......F4
47126 Henryville 1,132 ......F7
46322 Highland 25,935 ......B1
47949 Hillsboro 561 ......C4
47854 Hillsdale 500 ......C5
46745 Hoagland 600 ......H3
46342 Hobart 22,987 ......C1
46047 Hobbs 200 ......F4
47541 Holland 683 ......C8
47023 Holton 487 ......G6
46146 Homer 235 ......F5
47246 Hope 2,185 ......F6
†46069 Hortonville 240 ......E4
46746 Howe 800 ......G1
46747 Hudson 447 ......G1
46552 Hudson Lake 1,347 ......D1
46748 Huntertown 1,265 ......G2
47542 Huntingburg 5,376 ......D8
46750 Huntington⊙ 16,202 ......G3
†46064 Huntsville 120 ......G4
47437 Huron 250 ......D7
47855 Hymera 1,054 ......C6
47950 Idaville 655 ......D3
*46201 Indianapolis (cap.)⊙ 700,807 ......E5
Indianapolis‡ 1,166,929 ......E5
†46601 Indian Village 151 ......E1
46048 Ingalls 909 ......F5
47545 Ireland 600 ......C8
46147 Jamestown 924 ......D5
47438 Jasonville 2,497 ......C6
47546 Jasper⊙ 9,097 ......D8
47130 Jeffersonville⊙ 21,220 ......F8
†47565 Johnson 100 ......B8
†46074 Jolietville 300 ......E4
46938 Jonesboro 2,279 ......F4
47247 Jonesville 213 ......F6
46049 Kempton 410 ......E4
46755 Kendallville 7,299 ......G2
47351 Kennard 441 ......G5
47951 Kentland⊙ 1,936 ......C3
46939 Kewanna 711 ......E2
46759 Keystone 204 ......G3
46760 Kimmell 250 ......F2
47952 Kingman 566 ......C5
46345 Kingsbury 329 ......D1
46346 Kingsford Heights 1,618 ......D2
46050 Kirklin 662 ......E4
46148 Knightstown 2,325 ......F5
47857 Knightsville 763 ......C5
46534 Knox⊙ 3,674 ......D2
46901 Kokomo⊙ 47,808 ......E4
Kokomo‡ 103,715 ......E4
†46574 Koontz Lake 1,436 ......D2
46347 Kouts 1,619 ......C2
46348 La Crosse 713 ......D2
47954 Ladoga 1,151 ......D5
*47901 Lafayette⊙ 43,011 ......D4
Lafayette-West Lafayette‡ 121,702 ......D4
46940 La Fontaine 946 ......F3
46761 Lagrange⊙ 2,164 ......F1
46941 Lagro 549 ......F3
†46157 Lake Hart 231 ......E5
†46703 Lake James 400 ......H1
46943 Laketon 500 ......F3
46349 Lake Village 900 ......C2
46536 Lakeville 629 ......E1
46944 Landess 150 ......F3
47136 Lanesville 570 ......E8
46763 Laotto 361 ......G2
46537 Lapaz 651 ......E2
46051 Lapel 1,881 ......F4
46350 LaPorte⊙ 21,796 ......D1
46764 Larwill 286 ......F2
47024 Laurel 819 ......G6
46226 Lawrence 25,591 ......E5
47025 Lawrenceburg⊙ 4,403 ......H6
47137 Leavenworth 356 ......E8
46052 Lebanon⊙ 11,456 ......D4
46538 Leesburg 629 ......F2
46945 Leiters Ford 280 ......E2
46765 Leo 500 ......G2
47551 Leopold 175 ......D8
46355 Leroy 400 ......C2
†47240 Letts 247 ......F6
47352 Lewisville 577 ......G5
47138 Lexington F7
47353 Liberty⊙ 1,844 ......H5
46766 Liberty Center 275 ......G3
46946 Liberty Mills 200 ......F2
46767 Ligonier 3,134 ......F2
47955 Linden 700 ......D4
46769 Linn Grove 175 ......H3
47441 Linton 6,315 ......C6
†46755 Lisbon 200 ......G2
47139 Little York 150 ......F7
46149 Lizton 456 ......D5
46947 Logansport⊙ 17,731 ......E3
†46360 Long Beach 2,262 ......D1
47553 Loogootee 3,100 ......D7
47354 Losantville 306 ......G4
46356 Lowell 5,827 ......C2
46950 Lucerne 135 ......E3
†46601 Lydick E1
†47874 Lyford 400 ......C5
47355 Lynn 1,250 ......H4
47619 Lynnville 566 ......C8
47443 Lyons 782 ......C7
46951 Macy 282 ......E3
47250 Madison⊙ 12,472 ......G7
47555 Magnet 75 ......D8
†47001 Manchester 250 ......H6
46150 Manilla 350 ......F5
†47872 Mansfield 200 ......C5
†47443 Marco 150 ......C7
47140 Marengo 892 ......E8
47556 Mariah Hill 300 ......D8
†46176 Marietta 234 ......F6
46952 Marion⊙ 35,874 ......F3
46770 Markle 975 ......G3
46056 Markleville 427 ......F5
47859 Marshall 413 ......C5
46151 Martinsville⊙ 11,311 ......D6
46957 Matthews 745 ......F4
46154 Maxwell 300 ......F5
46055 McCordsville 600 ......F5
47860 Mecca 482 ......C5
47957 Medaryville 731 ......D2
47260 Medora 853 ......E7
47958 Mellott 294 ......C4
47143 Memphis 300 ......F8
46539 Mentone 973 ......E2
47861 Merom 360 ......B6
46410 Merrillville 27,677 ......C2
47030 Metamora 350 ......G6
†46703 Metz 200 ......H1
46958 Mexico 850 ......E3
46959 Miami 350 ......E3
†49117 Michiana Shores 464 ......D1
46360 Michigan City 36,850 ......C1
46057 Michigantown 453 ......E4
46540 Middlebury 1,665 ......F1
47356 Middletown 2,978 ......F4
47445 Midland 250 ......C6
47031 Milan 1,566 ......G6
46542 Milford 1,153 ......F2
†47240 Milford 177 ......F6
46543 Millersburg 809 ......F1
47261 Millhousen 214 ......G6
47145 Milltown 1,006 ......E8
†47362 Millville 275 ......G5
46156 Milroy 750 ......G6
47357 Milton 729 ......G5
46544 Mishawaka 40,201 ......E1
47446 Mitchell 4,641 ......E7
47358 Modoc 243 ......G4
46771 Mongo 225 ......G1
47959 Monon 1,540 ......D3
46772 Monroe 739 ......H3
47557 Monroe City 569 ......C7
46773 Monroeville 1,372 ......H3
46157 Monrovia 800 ......E5
46960 Monterey 236 ......D2
47862 Montezuma 1,352 ......C5
47558 Montgomery 390 ......C7
47960 Monticello⊙ 5,162 ......D3
47962 Montmorenci 300 ......D4
47359 Montpelier 1,995 ......G3
47360 Mooreland 479 ......G5
47032 Moores Hill 566 ......G6
46158 Mooresville 5,349 ......E5
46160 Morgantown 897 ......E6
47963 Morocco 1,348 ......C3
47033 Morris 350 ......G6
46161 Morristown 989 ......F5
†47327 Mount Auburn 192 ......G5
47964 Mount Ayr 207 ......C3
47361 Mount Summit 357 ......G4
47620 Mount Vernon⊙ 7,656 ......B9
46058 Mulberry 1,225 ......D4
*47302 Muncie⊙ 77,216 ......G4
Muncie‡ G4
46321 Munster 20,671 ......B1
47147 Nabb 150 ......F7
47034 Napoleon 246 ......G6
46550 Nappanee 4,694 ......F2
47448 Nashville⊙ 705 ......E6
†47421 Needmore 200 ......E7
47150 New Albany⊙ 37,103 ......F8
47449 Newberry 246 ......C7
47630 Newburgh 2,906 ......C9
46552 New Carlisle 1,439 ......E1
47362 New Castle⊙ 20,056 ......G5
†46342 New Chicago 3,284 ......C1
47863 New Goshen 500 ......B5
47631 New Harmony 945 ......B8
46774 New Haven 6,714 ......H2
47366 New Lisbon 300 ......G5
†46979 New London 200 ......E4
47965 New Market 608 ......D5
46163 New Palestine 749 ......F5
46553 New Paris 1,062 ......F2
†47165 New Pekin 1,125 ......F7
47263 New Point 296 ......G6
47966 Newport⊙ 704 ......C5
†47106 New Providence (Borden) 384 ......F8
47967 New Richmond 403 ......D4
47968 New Ross 306 ......D5
†46173 New Salem 200 ......G5
47161 New Salisbury 350 ......E8
47632 Newtonville 136 ......D8
47969 Newtown 277 ......C4
47035 New Trenton 200 ......H6
47162 New Washington 800 ......F7
46961 New Waverly 162 ......E3
46184 New Whiteland 4,502 ......E5
†46122 New Winchester 180 ......D5
46060 Noblesville⊙ 12,056 ......F4
46366 North Judson 1,653 ......D2
46554 North Liberty 1,211 ......E1
46962 North Manchester 5,998 ......F3
46165 North Salem 581 ......D5
47805 North Terre Haute C5
47265 North Vernon 5,768 ......F6
46555 North Webster 709 ......F2
†47960 Norway 300 ......D3
46556 Notre Dame E1
†47331 Nulltown 235 ......G5
46965 Oakford 325 ......E4
47660 Oakland City 3,301 ......C8
47561 Oaktown 776 ......C7
47367 Oakville 220 ......G4
47562 Odon 1,463 ......C7
†46401 Ogden Dunes 1,489 ......C1
47036 Oldenburg 770 ......G6
47451 Oolitic 1,495 ......E7
†47343 Orange 200 ......G5
46063 Orestes 539 ......F4
46776 Orland 424 ......G1
47452 Orleans 2,161 ......D7
46561 Osceola 1,990 ......E1
47037 Osgood 1,554 ......G6
46777 Ossian 1,945 ......G3
46367 Otis 250 ......D1
47163 Otisco 425 ......F7
47970 Otterbein 1,118 ......C4
47564 Otwell 600 ......C8
47453 Owensburg 785 ......D7
47665 Owensville 1,261 ......B8
47971 Oxford 1,327 ......C3
†46508 Palestine 800 ......F2
47164 Palmyra 692 ......E8
47454 Paoli⊙ 3,637 ......E7
46166 Paragon 538 ......D6
47368 Parker City 1,414 ......G4
47666 Patoka 832 ......B8
47455 Patricksburg 250 ......D6
47038 Patriot 265 ......H7
47865 Paxton 200 ......C6
47165 Pekin 950 ......E7
46064 Pendleton 2,130 ......F5
47369 Pennville 805 ......G4
†46011 Perkinsville 175 ......F4
47974 Perrysville 532 ......C4
47370 Pershing 438 ......G5
†46975 Pershing 425 ......E2
46970 Peru⊙ 13,764 ......E3
47567 Petersburg⊙ 2,987 ......C7
46778 Petroleum 212 ......G3
46562 Pierceton 1,086 ......F2
47866 Pimento 150 ......C6
†46350 Pine Lake 1,676 ......D1
47975 Pine Village 257 ......C4
46167 Pittsboro 891 ......D5
†46923 Pittsburg 175 ......D3
46168 Plainfield 9,191 ......E5
47568 Plainville 556 ......C7
46779 Pleasant Lake 800 ......H1
46563 Plymouth⊙ 7,693 ......E2
47868 Poland 230 ......C6
46781 Poneto 250 ......G3
46368 Portage 27,409 ......C1
46304 Porter 2,988 ......C1
47371 Portland⊙ 7,074 ......H4
47633 Poseyville 1,247 ......B8
†46360 Pottawattamie Park 284 ......C1
47869 Prairie Creek 275 ......C6
47870 Prairieton 200 ......B6
46782 Preble 150 ......H3
†46164 Princes Lakes 937 ......E6
47670 Princeton⊙ 8,976 ......B8
46170 Putnamville 250 ......D5
47456 Quincy 250 ......D6
47573 Ragsdale 135 ......C7
46737 Ray 200 ......H1
†47274 Reddington 400 ......F6
46171 Reelsville 210 ......D5
47977 Remington 1,268 ......C3
47978 Rensselaer⊙ 4,944 ......C3
47980 Reynolds 632 ......D3
47634 Richland 500 ......C9
47374 Richmond⊙ 41,349 ......H5
47380 Ridgeville 933 ......G4
47871 Riley 269 ......C6
47040 Rising Sun⊙ 2,478 ......H7
46172 Roachdale 958 ......D5
46974 Roann 548 ......F3
46783 Roanoke 891 ......G3
46975 Rochester⊙ 5,050 ......E2
46977 Rockfield 300 ......D3
47635 Rockport⊙ 2,590 ......C9
47872 Rockville⊙ 2,785 ......C5
46371 Rolling Prairie 550 ......D1
47574 Rome 50 ......D9
46784 Rome City 1,319 ......G1
47981 Romney 250 ......D4
47874 Rosedale 744 ......C5
†46601 Roseland 832 ......E1
†46310 Roselawn 200 ......C2
46065 Rossville 1,148 ......D4
46978 Royal Center 908 ......E3
†47302 Royerton 300 ......G4
46173 Rushville⊙ 6,113 ......G5
46175 Russellville 376 ......D5
46975 Russiaville 973 ......E4
47575 Saint Anthony 470 ......D8
47875 Saint Bernice 500 ......C5
46785 Saint Joe 546 ......H2
46383 Saint John 3,974 ......C2
46373 Saint Leon 515 ......H6
47876 Saint Mary-of-the-Woods 920 ......B6
†46556 Saint Marys E1
47577 Saint Meinrad 910 ......D8
47272 Saint Paul 976 ......F6
†47012 Saint Peter 175 ......H6
†47620 Saint Philip 400 ......B9
†47638 Saint Wendel 250 ......B8
47167 Salem⊙ 5,290 ......E7
47578 Sandborn 576 ......C7
†47401 Sanders 65 ......E6
46374 San Pierre 325 ......D2
47579 Santa Claus 514 ......D8
47382 Saratoga 338 ......H4
†47283 Sardinia 133 ......F6
46375 Schererville 13,209 ......C2
46376 Schneider 364 ......C2
47580 Schnellville 250 ......D8
47273 Scipio 200 ......F6
46066 Scircleville 125 ......E4
47170 Scottsburg⊙ 5,068 ......F7
47878 Seelyville 1,374 ......C6
47172 Sellersburg 3,211 ......F8
47383 Selma 1,056 ......G4
47274 Seymour 15,050 ......F7
46068 Sharpsville 617 ......E4
47879 Shelburn 1,259 ......C6
46377 Shelby 700 ......C2
46176 Shelbyville⊙ 14,989 ......F6
47880 Shepardsville 325 ......B5
46069 Sheridan 2,200 ......E4
†47338 Shideler 275 ......G4
46565 Shipshewana 466 ......F1
47384 Shirley 919 ......F5
†46797 Shirley City (Woodburn) 1,002 ......H2
47581 Shoals⊙ 967 ......D7
46566 Sidney 194 ......F2
46982 Silver Lake 576 ......F2
46983 Sims 250 ......F3
†46142 Smith Valley E5
47458 Smithville 500 ......D6
46984 Somerset 350 ......F3
47683 Somerville 340 ......C8
*46601 South Bend⊙ 109,727 ......E1
South Bend‡ 280,772 ......E1
46786 South Milford 270 ......G1
†46201 Southport 2,266 ......E5
46787 South Whitley 1,575 ......F2
†47355 Spartanburg 201 ......H4
47172 Speed 800 ......F8
46224 Speedway 12,641 ......E5
†47808 Spelterville 200 ......C5
47460 Spencer⊙ 2,732 ......D6
46788 Spencerville 400 ......G2
47385 Spiceland 940 ......F5
†47374 Spring Grove 469 ......H5
†46140 Spring Lake 236 ......F5
47386 Springport 221 ......G4
47462 Springville 279 ......D7
47584 Spurgeon 250 ......C8
47463 Stanford 200 ......D6
46985 Star City 351 ......D3
47982 State Line 233 ......C4
47881 Staunton 607 ......C6
47585 Stendal 175 ......C8
47636 Stewartsville 225 ......B8
46180 Stilesville 350 ......D5
46351 Stillwell 225 ......D1
47464 Stinesville 227 ......D6
47983 Stockwell 310 ......D4
47387 Straughn 331 ......G5
46789 Stroh 350 ......G1
47882 Sullivan⊙ 4,774 ......C6
47388 Sulphur Springs 345 ......G4
46379 Sumava Resorts 300 ......C2
46070 Summitville 1,085 ......F4
47041 Sunman 924 ......G6
46987 Sweetser 944 ......F3
47465 Switz City 300 ......C6
46567 Syracuse 2,579 ......F2
47280 Taylorsville 1,247 ......F6
47586 Tell City 8,704 ......D9
47637 Tennyson 331 ......C8
*47801 Terre Haute⊙ 61,125 ......C6
Terre Haute‡ 176,583 ......C6
46381 Thayer 350 ......C2
46071 Thorntown 1,468 ......D4
†46975 Tiosa 100 ......E2
46570 Tippecanoe 320 ......E2
46072 Tipton⊙ 5,004 ......E4
46571 Topeka 876 ......F1
†46360 Town of Pines 962 ......D1
46181 Trafalgar 466 ......E6
†46360 Trail Creek 2,581 ......D1
†46725 Tri Lakes 1,356 ......G2
47588 Troy 550 ......D9
46988 Twelve Mile 240 ......E3
46572 Tyner 245 ......E2
47177 Underwood 550 ......F7
47390 Union City 3,908 ......H4
46791 Uniondale 303 ......G3
46382 Union Mills 650 ......D2
47468 Unionville 225 ......E6
47884 Universal 428 ......C5
46989 Upland 3,335 ......F4
46990 Urbana 400 ......F3
†47130 Utica 501 ......F8
47281 Vallonia 550 ......F7
46383 Valparaiso⊙ 22,247 ......C2
46991 Van Buren 935 ......F3
47987 Veedersburg 2,261 ......C4
47590 Velpen 375 ......C8
47282 Vernon⊙ 329 ......F7
47042 Versailles⊙ 1,560 ......G6
47043 Vevay⊙ 1,343 ......G7
†47441 Vicksburg 175 ......C6
†47170 Vienna 175 ......F7
47591 Vincennes⊙ 20,857 ......C7
46992 Wabash⊙ 12,985 ......F3
47638 Wadesville 450 ......B8
46573 Wakarusa 1,281 ......F1
46182 Waldron 850 ......F6
†47201 Walesboro 214 ......F6
46574 Walkerton 2,051 ......E2
†46802 Wallen 945 ......G2
46994 Walton 1,202 ......E3
46390 Wanatah 879 ......D2
46792 Warren 1,254 ......G3
46580 Warsaw⊙ 10,647 ......F2
47501 Washington⊙ 11,325 ......C7
46793 Waterloo 1,951 ......G2
†47130 Watson 200 ......F8
47989 Waveland 559 ......D5
46794 Wawaka 320 ......F2
47990 Waynetown 915 ......C4
47392 Webster 350 ......H5
47469 West Baden Springs 796 ......D7
†47353 West College Corner 614 ......H5
46074 Westfield 2,783 ......E4
†45030 West Harrison 328 ......H6
47906 West Lafayette 21,247 ......D4
47991 West Lebanon 946 ......C4
46995 West Middleton 327 ......E4
47596 Westphalia 300 ......C7
47992 Westpoint 375 ......C4
47283 Westport 1,450 ......F6
47885 West Terre Haute 2,806 ......B6
46391 Westville 2,887 ......D1
46392 Wheatfield 755 ......C2
47597 Wheatland 532 ......C7
46393 Wheeler 540 ......C1
†47342 Wheeling 180 ......G4
46184 Whiteland 1,956 ......E5
46075 Whitestown 497 ......E5
46394 Whiting 5,630 ......C1
46186 Wilkinson 493 ......F5
47470 Williams 350 ......D7
47993 Williamsport⊙ 1,747 ......C4
46996 Winamac⊙ 2,370 ......D2
47394 Winchester⊙ 5,659 ......G4
46076 Windfall 911 ......F4
47994 Wingate 373 ......C4
46590 Winona Lake 2,827 ......F2
47598 Winslow 1,017 ......C8
47995 Wolcott 923 ......C3
46795 Wolcottville 890 ......G1
46796 Wolflake 230 ......F2
46797 Woodburn 1,002 ......H2
†46624 Woodland 400 ......E1
47471 Worthington 1,574 ......C6
46595 Wyatt 250 ......E1
†47630 Yankeetown 250 ......C9
46798 Yoder 250 ......G3
47396 Yorktown 3,945 ......G4
46998 Young America 259 ......E3
†47808 Youngstown 350 ......C6
46799 Zanesville 575 ......G3
46077 Zionsville 3,948 ......E5

OTHER FEATURES

Anderson (riv.) D8
Bass (lake) D2
Beanblossom (creek) D6
Big (creek) B8
Big Blue (riv.) F5
Big Pine (creek) C3
Big Raccoon (creek) C5
Big Walnut (creek) D5
Blue (riv.) E8
Brookville (lake) G6
Buck (creek) E8
Busseron (creek) C7
Camp (creek) E6
Cedar (creek) G2
Clifty (creek) F6
Coal (creek) C4
Crooked (creek) D2
Cypress (pond) B8
Deer (creek) E3
Deer (creek) D5
Eagle (creek) E4
Eel (riv.) C6
Eel (riv.) F3
Elkhart (riv.) F1
Fawn (riv.) ......
Flatrock (creek) ......
Fort Benjamin Harrison ......
Freeman (lake) ......
Geist (res.) ......
George Rogers Clark Nat'l Hist. Park ......
Graham (creek) ......
Grissom A.F.B. 4,676 ......
Huntington (lake) ......
Indian (creek) ......
Indian (creek) ......
Indiana Dunes Nat'l Lakeshore ......
Iroquois (riv.) ......
Jefferson Proving Ground ......
Kankakee (riv.) ......
Lemon (lake) ......
Lincoln Boyhood Nat'l Mem. ......
Little (riv.) ......
Little Elkhart (riv.) ......
Little Pigeon (creek) ......
Little Vermilion (riv.) ......
Lost (riv.) ......
Maria (creek) ......
Maumee (riv.) ......
Maxinkuckee (lake) ......
Michigan (lake) ......
Mill (creek) ......
Mississinewa (lake) ......
Mississinewa (riv.) ......
Monroe (lake) ......
Morse (res.) ......
Muscatatuck (riv.) ......
Ohio (riv.) ......
Patoka (riv.) ......
Pigeon (creek) ......
Pigeon (riv.) ......
Pipe (creek) ......
Prairie (creek) ......
Richland (creek) ......
Saint Joseph (riv.) ......
Saint Joseph (riv.) ......
Saint Marys (lake) ......
Saint Marys (riv.) ......
Salamonie (lake) ......
Salamonie (riv.) ......
Salt (creek) ......
Sand (creek) ......
Shafer (lake) ......
Silver (creek) ......
Sugar (creek) ......
Sugar (creek) ......
Sugar (creek) ......
Tippecanoe (riv.) ......
Vermilion (riv.) ......
Vernon Fork (creek) ......
Wabash (riv.) ......
Wawasee (lake) ......
White (riv.) ......
White, East Fork (riv.) ......
White, West Fork (riv.) ......
Whitewater (riv.) ......
Wildcat (creek) ......

⊙County seat.
‡Population of metropolitan area.
† Zip of nearest p.o. * Multiple z

Topography

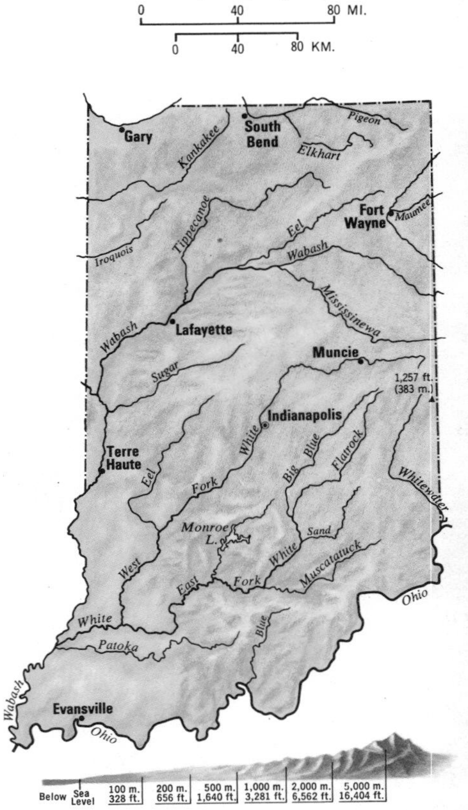

0 40 80 MI.
0 40 80 KM.
Gary
South Bend
Pigeon
Kankakee
Elkhart
Fort Wayne
Maumee
Tippecanoe
Eel
Iroquois
Wabash
Mississinewa
Wabash
Lafayette
Muncie
1,257 ft. (383 m.)
Sugar
Indianapolis
Terre Haute
Eel
White
Big Blue
Flatrock
Whitewater
Fork
Monroe L.
Sand
White
Muscatatuck
West
East
Fork
Ohio
White
Patoka
Blue
Wabash
Evansville
Ohio
Below
Sea Level
100 m. 328 ft.
200 m. 656 ft.
500 m. 1,640 ft.
1,000 m. 3,281 ft.
2,000 m. 6,562 ft.
5,000 m. 16,404 ft.

Indiana

SCALE

0 5 10 20 30 40 MI.

0 5 10 20 30 40 KM.

State Capitals .... ✳

County Seats .... ◉

Major Limited Access Hwys. _____

Scale 1:1,570,000

© Copyright HAMMOND INCORPORATED, Maplewood, N.J.

COUNTIES

Adair 9,509	..... E6	
Adams 5,731	..... D6	
Allamakee 15,108	..... L2	
Appanoose 15,511	..... H7	
Audubon 8,559	..... D5	
Benton 23,649	..... J4	
Black Hawk 137,961	..... J4	
Boone 26,184	..... F5	
Bremer 24,820	..... J3	
Buchanan 22,900	..... K4	
Buena Vista 20,774	..... C3	
Butler 17,668	..... H3	
Calhoun 13,542	..... D4	
Carroll 22,951	..... D4	
Cass 16,932	..... D6	
Cedar 18,635	..... L5	
Cerro Gordo 48,458	..... G2	
Cherokee 16,238	..... B3	
Chickasaw 15,437	..... J2	
Clarke 8,612	..... F6	
Clay 19,576	..... C2	
Clayton 21,098	..... L3	
Clinton 57,122	..... M5
Crawford 18,935	..... C4
Dallas 29,513	..... E5
Davis 9,104	..... J7
Decatur 9,794	..... F7
Delaware 18,933	..... L4
Des Moines 46,203	..... L7
Dickinson 15,629	..... C2
Dubuque 93,745	..... M4
Emmet 13,336	..... D2
Fayette 25,488	..... K3
Floyd 19,597	..... H2
Franklin 13,036	..... G3
Fremont 9,401	..... B7
Greene 12,119	..... E5
Grundy 14,366	..... H4
Guthrie 11,983	..... D5
Hamilton 17,862	..... F4
Hancock 13,833	..... F2
Hardin 21,776	..... G4
Harrison 16,348	..... B5
Henry 18,890	..... K6
Howard 11,114	..... J2
Humboldt 12,246	..... E3
Ida 8,908	..... C4
Iowa 15,429	..... J5
Jackson 22,503	..... M4
Jasper 36,425	..... G5
Jefferson 16,316	..... K6
Johnson 81,717	..... K5
Jones 20,401	..... L4
Keokuk 12,921	..... J6
Kossuth 21,891	..... E2
Lee 43,106	..... L7
Linn 169,775	..... K4
Louisa 12,055	..... L6
Lucas 10,313	..... G6
Lyon 12,896	..... A2
Madison 12,597	..... E6
Mahaska 22,867	..... H6
Marion 29,669	..... G6
Marshall 41,652	..... G4
Mills 13,406	..... B6
Mitchell 12,329	..... H2
Monona 11,692	..... B4
Monroe 9,209	..... H7
Montgomery 13,413	..... C6
Muscatine 40,436	..... L5
O'Brien 16,972	..... B2
Osceola 8,371	..... B2
Page 19,063	..... C7
Palo Alto 12,721	..... D2
Plymouth 24,743	..... A3
Pocahontas 11,369	..... D3
Polk 303,170	..... F5
Pottawattamie 86,561	..... B6
Poweshiek 19,306	..... H5
Ringgold 6,112	..... E7
Sac 14,118	..... C4
Scott 160,022	..... M5
Shelby 15,043	..... C5
Sioux 30,813	..... A2
Story 72,326	..... G4
Tama 19,533	..... H4
Taylor 8,353	..... C7
Union 13,858	..... E7
Van Buren 8,626	..... K7
Wapello 40,241	..... J6
Warren 34,878	..... F6
Washington 20,141	..... K6
Wayne 8,199	..... G7
Webster 45,953	..... E4
Winnebago 13,010	..... F2
Winneshiek 21,876	..... K2
Woodbury 100,884	..... B4
Worth 9,075	..... G2
Wright 16,319	..... F3

CITIES and TOWNS

Zip	Name/Pop.	Key
50601	Ackley 1,900	G3
50002	Adair 883	D6
50003	Adel⊙ 2,846	E5
50830	Afton 985	E6
52530	Agency 657	J7
52201	Ainsworth 547	K6
51001	Akron 1,517	A3
50510	Albert City 818	C3
52531	Albia⊙ 4,184	H6
50005	Albion 739	H4
50006	Alburnett 411	K4
52202	Alburnett 411	K4
50511	Algona⊙ 6,289	E2
50007	Alleman 307	F5
50008	Allerton 670	G7
50602	Allison⊙ 1,132	H3
51002	Alta 1,720	C3
50603	Alta Vista 314	J2
51003	Alton 986	A3
50009	Altoona 5,764	G5
51230	Alvord 246	A2
52203	Amana 300	K5
50010	Ames 45,775	F4
52205	Anamosa⊙ 4,958	L4
52030	Andrew 349	M4
50020	Anita 1,153	D6
50021	Ankeny 15,429	F5
51004	Anthon 687	B4
50604	Aplington 1,027	H3
51430	Arcadia 454	C4
50606	Arlington 498	K3
51003	Armstrong 1,153	D2
51331	Arnolds Park 1,051	C2
51431	Arthur 288	C4
†52001	Asbury 2,017	M4
51232	Ashton 441	B2
52720	Atalissa 360	L5
52206	Atkins 678	K4
50022	Atlantic⊙ 7,789	D6

51433 Auburn 320	D4	
50025 Audubon⊙ 2,841	D5	
51005 Aurelia 1,143	C3	
50607 Aurora 248	K3	
51521 Avoca 1,650	C6	
50515 Ayrshire 243	D2	
50516 Badger 653	E3	
50026 Bagley 370	E5	
50517 Bancroft 1,082	E2	
50027 Barnes City 266	H6	
52533 Batavia 525	J7	
51006 Battle Creek 919	B4	
50028 Baxter 951	G5	
50029 Bayard 637	D5	
52534 Beacon 530	H6	
50833 Bedford⊙ 1,692	D7	
52208 Belle Plaine 2,903	J5	
52031 Bellevue 2,450	M4	
50421 Belmond 2,505	F3	
52721 Bennett 458	L5	
50032 Berwick 600	G5	
52722 Bettendorf 27,381	N5	
52535 Birmingham 410	K7	
50034 Blairsburg 288	F4	

52209 Blairstown 695	J5	
52536 Blakesburg 404	H7	
51523 Blencoe 247	A5	
50836 Blockton 280	D7	
52537 Bloomfield⊙ 2,849	J7	
52726 Blue Grass 1,377	M5	
50519 Bode 406	E3	
52620 Bonaparte 489	K7	
50035 Bondurant 1,283	G5	
50036 Boone⊙ 12,602	F4	
50040 Boxholm 267	E4	
51234 Boyden 708	B2	
52210 Brandon 337	K4	
51436 Breda 502	C4	
50837 Bridgewater 233	D6	
52540 Brighton 804	K6	
50611 Bristow 252	H3	
50423 Britt 2,185	F2	
51007 Bronson 289	A4	
52211 Brooklyn 1,509	J5	
52728 Buffalo 1,569	M6	
50424 Buffalo Center 1,233	F2	
52601 Burlington⊙ 29,529	L7	
50522 Burt 689	E2	

AREA 56,275 sq. mi. (145,752 sq. km.)
POPULATION 2,913,808
CAPITAL Des Moines
LARGEST CITY Des Moines
HIGHEST POINT (Osceola Co.) 1670 ft. (509 m.)
SETTLED IN 1788
ADMITTED TO UNION December 28, 1846
POPULAR NAME Hawkeye State
STATE FLOWER Wild Rose
STATE BIRD Eastern Goldfinch

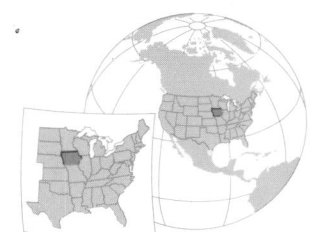

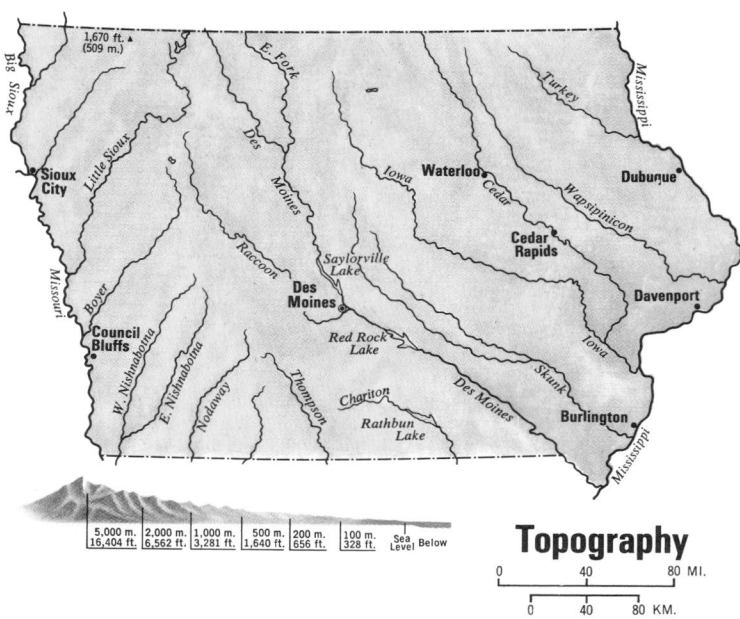

Topography

50044 Bussey 579	H6	
52729 Calamus 452	M5	
50523 Callender 446	E4	
52132 Calmar 1,053	K2	
52730 Camanche 4,725	N5	
50046 Cambridge 732	G5	
50047 Carlisle 3,073	G6	
51525 Carson 716	C6	
52213 Center Point 1,591	K4	
52544 Centerville⊙ 6,558	H7	
52214 Central City 1,067	K4	
50049 Chariton⊙ 4,987	G6	
50616 Charles City⊙ 8,778	H2	
52731 Charlotte 442	M5	
51439 Charter Oak 615	C4	
52215 Chelsea 376	J5	
51012 Cherokee⊙ 7,004	B3	
52216 Clarence 1,001	M5	
51632 Clarinda⊙ 5,458	C7	
50525 Clarion⊙ 3,060	F3	
50619 Clarksville 1,424	H3	
50840 Clearfield 433	D7	
50428 Clear Lake 7,458	G2	
51014 Cleghorn 275	B3	
52732 Clinton⊙ 32,828	N5	
50318 Clive 6,064	F5	
52217 Clutier 249	J4	
52218 Coggon 639	L4	
51636 Coin 316	C7	
52035 Colesburg 463	L3	
50054 Colfax 2,234	G5	
51637 College Springs 307	C7	
50055 Collins 451	G5	
50056 Colo 808	G4	
52737 Columbus City 367	L6	
52738 Columbus Junction 1,429	L6	
52739 Conesville 301	L6	
50631 Conrad 1,133	H4	
52220 Conroy 250	J5	
50058 Coon Rapids 1,448	D5	
52241 Coralville 7,687	K5	
50841 Corning⊙ 1,939	D7	
51016 Correctionville 935	B4	
50430 Corwith 480	F3	
50060 Corydon⊙ 1,818	G7	
50431 Coulter 264	G3	
51501 Council Bluffs⊙ 56,449	B6	
52621 Crawfordsville 290	K6	
51526 Crescent 547	B6	
52136 Cresco⊙ 3,860	J2	
50801 Creston⊙ 8,429	E6	
50432 Crystal Lake 314	F2	
50843 Cumberland 351	D6	

51018 Cushing 270	B4	
50529 Dakota City⊙ 1,072	E3	
50062 Dallas 451	G6	
50063 Dallas Center 1,360	E5	
51019 Danbury 492	B4	
52623 Danville 994	L7	
*52801 Davenport⊙ 103,264	M5	
Davenport-Rock Island-Moline‡ 383,958	M5	
50065 Davis City 327	F7	
50530 Dayton 941	E4	
52101 Decorah⊙ 7,991	K2	
51440 Dedham 321	D5	
52222 Deep River 323	J5	
51527 Defiance 383	C5	
52223 Delhi 511	L4	
52037 Delmar 633	M4	
52550 Delta 482	J6	
51442 Denison⊙ 6,675	C4	
52624 Denmark 480	L7	
50622 Denver 1,647	J3	
*50301 Des Moines (cap.)⊙ 191,003	G5	
Des Moines‡ 338,048	G5	
50069 De Soto 1,035	E5	
50623 Dewar 230	J3	
52742 De Witt 4,512	N5	
50070 Dexter 678	E5	
50845 Diagonal 362	E7	
51333 Dickens 289	C2	
50624 Dike 987	H4	
52745 Dixon 312	M5	
52746 Donahue 289	M5	
52625 Donnellson 972	K7	
51235 Doon 537	A2	
52551 Douds 425	J7	
51528 Dow City 616	B5	
50071 Dows 771	F3	
52001 Dubuque⊙ 62,321	M3	
Dubuque‡ 93,745	M3	
50625 Dumont 815	H3	
50532 Duncombe 504	E4	
50626 Dunkerton 718	J3	
51529 Dunlap 1,374	B5	
52747 Durant 1,583	M5	
52040 Dyersville 3,825	L3	
52224 Dysart 1,355	J4	
50533 Eagle Grove 4,324	F3	
50072 Earlham 1,140	E6	
52041 Earlville 844	L4	
50535 Early 670	C4	
52553 Eddyville 1,116	H6	
52042 Edgewood 900	K3	
52554 Eldon 1,255	J7	
50627 Eldora⊙ 3,063	G4	
52748 Eldridge 3,279	M5	
52141 Elgin 702	K3	
52043 Elkader⊙ 1,688	L3	
50073 Elkhart 256	F5	
51531 Elk Horn 746	C5	
†50700 Elk Run Heights 1,186	J4	
51532 Elliott 493	C6	

50075 Ellsworth 480	F4	
50628 Elma 714	J2	
52227 Ely 425	K5	
51533 Emerson 502	C6	
50536 Emmetsburg⊙ 4,621	D2	
52045 Epworth 1,380	M4	
51638 Essex 1,001	C7	
51334 Estherville⊙ 7,518	D2	
50707 Evansdale 4,798	J4	
51338 Everly 796	C2	
50076 Exira 978	D5	
50629 Fairbank 980	K3	
52228 Fairfax 683	K5	
52556 Fairfield⊙ 9,428	J6	
52046 Farley 1,287	L4	
52047 Farmersburg 276	L3	
52626 Farmington 869	K7	
50538 Farnhamville 461	D4	
51639 Farragut 603	C7	
52142 Fayette 1,515	K3	
50539 Fenton 394	E2	
50434 Fertile 372	G2	
50435 Floyd 408	H2	
50540 Fonda 816	D3	
50846 Fontanelle 805	E6	
50436 Forest City⊙ 4,270	F2	
52144 Fort Atkinson 374	J2	
50501 Fort Dodge⊙ 29,423	E3	
52627 Fort Madison⊙ 13,520	L7	
51340 Fostoria 261	C2	
50630 Fredericksburg 1,075	J3	
50631 Frederika 223	J3	
52561 Fremont 730	H6	
52749 Fruitland 461	L6	
51020 Galva 420	C3	
50103 Garden Grove 297	F7	
52049 Garnavillo 723	L3	
50438 Garner⊙ 2,908	F2	
52229 Garrison 411	J4	
50632 Garwin 626	H4	
51237 George 1,241	B2	
50105 Gilbert 805	F4	
50634 Gilbertville 740	J4	
50106 Gilman 642	H5	
50541 Gilmore City 626	D3	
50635 Gladbrook 970	H4	
51534 Glenwood⊙ 5,280	B6	
51443 Glidden 1,076	D4	
50542 Goldfield 789	F3	
52750 Goose Lake 274	N5	
50543 Gowrie 1,089	E4	
51342 Graettinger 923	D2	
50440 Grafton 255	G2	
50107 Grand Junction 970	E4	
52751 Grand Mound 674	M5	
52752 Grandview 473	L6	
50109 Granger 619	F5	
51022 Granville 336	B3	
50848 Gravity 245	D7	
52050 Greeley 313	L3	
50636 Greene 1,332	H3	
50849 Greenfield⊙ 2,243	D6	
50111 Grimes 1,973	F5	
50112 Grinnell 8,868	H5	

(continued on following page)

Agriculture, Industry and Resources

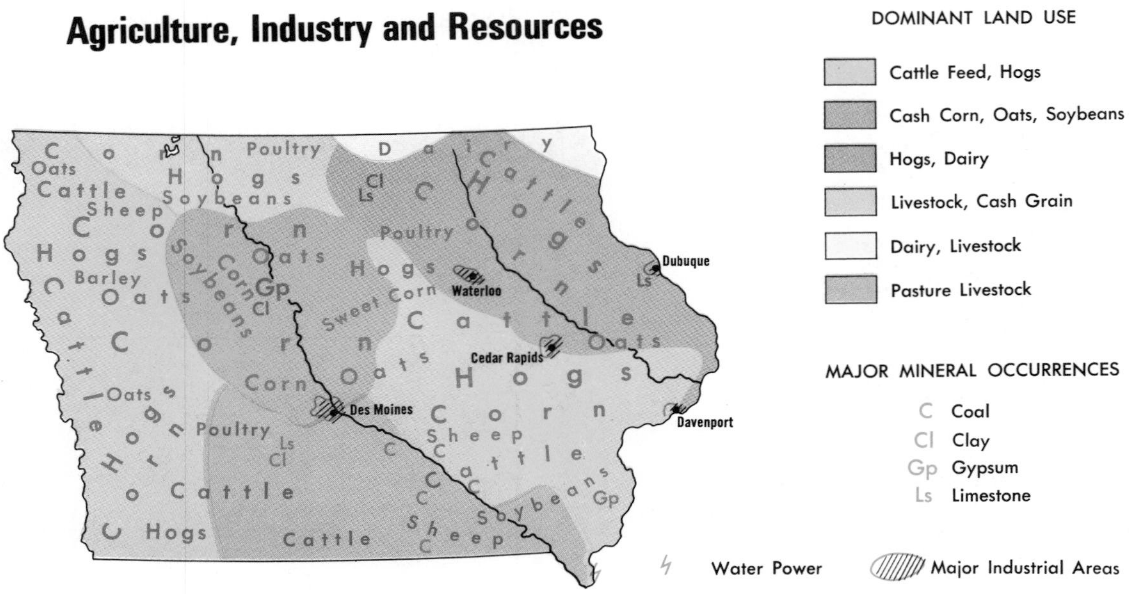
Dubuque
Waterloo
Cedar Rapids
Des Moines
Davenport
Poultry
Dairy
Oats
Cattle
Sheep
Hogs
Soybeans
Barley
Corn
Sweet Corn
Gp
Cl
Ls
C
DOMINANT LAND USE
Cattle Feed, Hogs
Cash Corn, Oats, Soybeans
Hogs, Dairy
Livestock, Cash Grain
Dairy, Livestock
Pasture Livestock
MAJOR MINERAL OCCURRENCES
C Coal
Cl Clay
Gp Gypsum
Ls Limestone
Water Power
Major Industrial Areas

51535 Griswold 1,176 ... C6
50638 Grundy Center⊙ 2,880 ... H4
50115 Guthrie Center⊙ 1,713 ... D5
52052 Guttenberg 2,428 ... L3
51640 Hamburg 1,597 ... B7
50441 Hampton⊙ 4,630 ... G3
51536 Hancock 254 ... C6
50544 Harcourt 347 ... E4
51537 Harlan⊙ 5,357 ... C5
52146 Harpers Ferry 258 ... L2
50118 Hartford 761 ... G6
51346 Hartley 1,700 ... C2
50119 Harvey 275 ... H6
50546 Havelock 279 ... D3
51023 Hawarden 2,722 ... A2
52147 Hawkeye 512 ... J3
50641 Hazleton 877 ... K3
52563 Hedrick 847 ... J6
51541 Henderson 236 ... B6
52233 Hiawatha 4,825 ... K4
52235 Hills 547 ... K5
52630 Hillsboro 208 ... K7
51024 Hinton 659 ... A3
50642 Holland 278 ... H4
51025 Holstein 1,477 ... B4
52053 Holy Cross 310 ... L3
52237 Hopkinton 774 ... L4
51026 Hornick 239 ... A4
51238 Hospers 655 ... B2
50122 Hubbard 852 ... G4
50643 Hudson 2,267 ... H4
51239 Hull 1,714 ... A2
50548 Humboldt 4,794 ... E3
50123 Humeston 671 ... G7
50124 Huxley 1,884 ... F5
51445 Ida Grove⊙ 2,285 ... B4
50644 Independence⊙ 6,392 ... K4
50125 Indianola⊙ 10,843 ... F6
51240 Inwood 755 ... A2
50645 Ionia 350 ... J2
52240 Iowa City⊙ 50,508 ... L5
Iowa City○‡ 81,717 ... L5
50126 Iowa Falls 6,174 ... G3
51027 Ireton 588 ... A3
51446 Irwin 427 ... C5
50128 Jamaica 275 ... E5
50647 Janesville 840 ... J3
50129 Jefferson⊙ 4,854 ... E4
50648 Jesup 2,343 ... J4
50130 Jewell 1,145 ... F4
50131 Johnston 2,617 ... F5
52247 Kalona 1,862 ... K6
50447 Kanawha 756 ... F3
50133 Kellerton 278 ... E7
50134 Kelley 237 ... F5
50135 Kellogg 654 ... H5
50448 Kensett 360 ... G2
52632 Keokuk⊙ 13,536 ... L8
52565 Keosauqua⊙ 1,003 ... J7
52248 Keota 1,034 ... K6
50136 Keswick 300 ... J6
52249 Keystone 618 ... J5
51543 Kimballton 362 ... D5
51028 Kingsley 1,209 ... A3
51448 Kiron 317 ... C4
50449 Klemme 620 ... F3
50138 Knoxville⊙ 8,143 ... G6
50139 Lacona 376 ... G6
52251 Ladora 289 ... J5
51449 Lake City 2,006 ... D4
50450 Lake Mills 2,281 ... F2
51347 Lake Park 1,123 ... C2
50588 Lakeside 589 ... C3
51450 Lake View 1,291 ... C4
50451 Lakota 330 ... E2
50140 Lamoni 2,705 ... E7
50650 Lamont 554 ... K3
52054 La Motte 322 ... M4
52151 Lansing 1,181 ... L2
50651 La Porte City 2,324 ... J4
51241 Larchwood 701 ... A2
50452 Latimer 441 ... G3
50141 Laurel 278 ... H5
50554 Laurens 1,606 ... D3
52154 Lawler 534 ... J2
51030 Lawton 447 ... A4
52753 Le Claire 2,899 ... N5
50142 Le Grand 921 ... H5
50557 Lehigh 654 ... E4
50453 Leland 274 ... F2
51031 Le Mars⊙ 8,276 ... A3
50851 Lenox 1,338 ... D7
50144 Leon⊙ 2,094 ... F7
51242 Lester 274 ... A2
52754 Letts 473 ... L6
51544 Lewis 497 ... C6
52567 Libertyville 281 ... K7
52155 Lime Springs 476 ... J2
50146 Linden 264 ... E5
50147 Lineville 319 ... G7
52253 Lisbon 1,458 ... L5
50148 Liscomb 296 ... H4
51243 Little Rock 490 ... B2
51545 Little Sioux 251 ... B5
50558 Livermore 490 ... E3
52635 Lockridge 271 ... K7
51546 Logan⊙ 1,540 ... B5
51453 Lohrville 521 ... D4
52755 Lone Tree 1,014 ... L6
52756 Long Grove 596 ... M5
50149 Lorimor 405 ... E6
52254 Lost Nation 524 ... M5
50150 Lovilia 637 ... H6
52255 Lowden 717 ... L5
52757 Low Moor 346 ... N5
52156 Luana 246 ... K2
50151 Lucas 292 ... G6
50560 Lu Verne 418 ... E3
52056 Luxemburg 271 ... L3
50153 Lynnville 406 ... H5
50561 Lytton 377 ... D4
51549 Macedonia 279 ... C6
50156 Madrid 2,281 ... F5
50157 Malcom 418 ... H5
50562 Mallard 407 ... D3
51551 Malvern 1,244 ... B7
52057 Manchester⊙ 4,942 ... L3
51454 Manilla 1,020 ... C5
50456 Manly 1,496 ... G2
51455 Manning 1,609 ... C5
50563 Manson 1,924 ... D3
51034 Mapleton 1,495 ... B4
52060 Maquoketa⊙ 6,313 ... M4
50565 Marathon 442 ... C3
50653 Marble Rock 419 ... H3
51035 Marcus 1,206 ... B3
52301 Marengo⊙ 2,308 ... J5
52302 Marion 19,474 ... K4
52158 Marquette 528 ... L2
50158 Marshalltown⊙ 26,938 ... G4
52305 Martelle 316 ... L4
50160 Martensdale 438 ... F6
50401 Mason City⊙ 30,144 ... G2
50853 Massena 518 ... D6
51036 Maurice 288 ... A3
50161 Maxwell 783 ... G5
50655 Maynard 561 ... K3
50154 McCallsburg 304 ... G4
52758 McCausland 381 ... M5
52157 McGregor 945 ... L2
52306 Mechanicsville 1,166 ... L5
52637 Mediapolis 1,685 ... L6
50162 Melbourne 732 ... G5
50163 Melcher 953 ... G6
51350 Melvin 277 ... B2
50164 Menlo 410 ... E5
51037 Meriden 233 ... B3
51038 Merrill 737 ... A3
50457 Meservey 324 ... G3
52307 Middle 335 ... K5
52638 Middletown 487 ... L7
52064 Miles 398 ... N4
51351 Milford 2,076 ... C2
50166 Milo 778 ... G6
52570 Milton 567 ... J7
50167 Minburn 390 ... E5
51553 Minden 419 ... C6
50168 Mingo 303 ... G5
51555 Missouri Valley 3,107 ... B5
50169 Mitchellville 1,530 ... G5
51556 Modale 373 ... B5
51557 Mondamin 423 ... B5
52159 Monona 1,530 ... L2
50170 Monroe 1,875 ... G5
50171 Montezuma⊙ 1,485 ... H5
52310 Monticello 3,641 ... L4
50173 Montour 387 ... H5
52759 Montpelier 250 ... M6
52639 Montrose 1,038 ... L7
51558 Moorhead 264 ... B5
50566 Moorland 257 ... E4
52571 Moravia 706 ... H7
52640 Morning Sun 959 ... L6
52760 Moscow 350 ... L5
52572 Moulton 762 ... H7
50854 Mount Ayr⊙ 1,938 ... E7
52641 Mount Pleasant⊙ 7,322 ... L7
52314 Mount Vernon 3,325 ... K5
51039 Moville 1,273 ... A4
50174 Murray 703 ... F6
52761 Muscatine⊙ 23,467 ... L6
52574 Mystic 665 ... H7
50658 Nashua 1,846 ... J3
51559 Neola 839 ... B6
50201 Nevada⊙ 5,912 ... G5
52160 New Albin 609 ... L2
50568 Newell 913 ... D3
52315 Newhall 899 ... K5
50660 New Hartford 764 ... H3
52645 New London 2,043 ... L7
51646 New Market 554 ... D7
50206 New Providence 249 ... G4
50207 New Sharon 1,225 ... H6
50208 Newton⊙ 15,292 ... H5
52065 New Vienna 430 ... L3
50210 New Virginia 512 ... F6
52766 Nichols 375 ... L6
50458 Nora Springs 1,572 ... H2
52316 North English 990 ... J5
52317 North Liberty 2,046 ... K5
50459 Northwood⊙ 2,193 ... G2
50211 Norwalk 2,676 ... F6
52318 Norway 633 ... K5
52319 Oakdale 300 ... K5
51560 Oakland 1,552 ... C6
52646 Oakville 470 ... L6
51354 Ocheyedan 599 ... B2
51458 Odebolt 1,299 ... C4
50662 Oelwein 7,564 ... K3
50212 Ogden 1,953 ... E4
51355 Okoboji 559 ... C2
52320 Olin 735 ... L5
52576 Ollie 232 ... J6
51040 Onawa⊙ 3,283 ... A4
51041 Orange City⊙ 4,588 ... A2
50858 Orient 416 ... E6
†51360 Orleans 546 ... C2
50461 Osage⊙ 3,718 ... H2
50213 Osceola⊙ 3,750 ... F6
52577 Oskaloosa⊙ 10,984 ... H6
52161 Ossian 829 ... K2
50569 Otho 692 ... E4
52501 Ottumwa⊙ 27,381 ... J6
52322 Oxford 676 ... K5
52323 Oxford Junction 600 ... M4
51561 Pacific Junction 511 ... B6
50571 Palmer 288 ... D3
52324 Palo 529 ... K4
51562 Panama 229 ... B5
50216 Panora 1,211 ... E5
50665 Parkersburg 1,968 ... H3
52325 Parnell 234 ... J5
50217 Paton 291 ... E4
51046 Paullina 1,224 ... B3
50219 Pella 8,349 ... H6
50220 Perry 7,053 ... E5
50221 Pershing 325 ... G6
51563 Persia 355 ... B5
51047 Peterson 470 ... C3
51048 Pierson 408 ... B3
51564 Pisgah 307 ... B5
50666 Plainfield 469 ... J3
50225 Pleasantville 1,531 ... G6
50464 Plymouth 463 ... G2
50574 Pocahontas⊙ 2,352 ... D3
50226 Polk City 1,658 ... F5
50575 Pomeroy 895 ... D3
51565 Portsmouth 240 ... C5
52162 Postville 1,475 ... K2
50228 Prairie City 1,278 ... G5
50859 Prescott 349 ... D6
52069 Preston 1,120 ... N4
Primghar⊙ 1,050 ... B2
52768 Princeton 965 ... N5
52163 Protivin 368 ... J2
52584 Pulaski 267 ... J7
52326 Quasqueton 599 ... K4
51049 Quimby 424 ... B3
50230 Radcliffe 593 ... G4
50465 Rake 283 ... F2
50667 Raymond 655 ... J4
50668 Readlyn 858 ... J3
50232 Reasnor 277 ... G5
50233 Redfield 959 ... E5
51566 Red Oak⊙ 6,810 ... C6
50669 Reinbeck 1,808 ... H4
50576 Rembrandt 291 ... C3
51050 Remsen 1,592 ... B3
50577 Renwick 410 ... E3
50234 Rhodes 367 ... G5
50466 Riceville 919 ... H2
52585 Richland 600 ... K6
52165 Ridgeway 308 ... K2
50578 Ringsted 557 ... D2
50235 Rippey 304 ... E5
†52722 Riverdale 462 ... N5
52327 Riverside 826 ... K6
51650 Riverton 342 ... B7
52328 Robins 726 ... K4
50468 Rockford 1,012 ... H2
51246 Rock Rapids⊙ 2,693 ... A2
51247 Rock Valley 2,706 ... A2
50469 Rockwell 1,039 ... G3
50579 Rockwell City⊙ 2,276 ... D4
50236 Roland 1,005 ... F4
50581 Rolfe 796 ... D3
50470 Rowan 259 ... F3
52329 Rowley 275 ... K4
51357 Royal 522 ... C2
50471 Rudd 460 ... H2
50237 Runnells 377 ... G5
50238 Russell 593 ... G7
51358 Ruthven 769 ... D2
52330 Ryan 390 ... K4
52070 Sabula 824 ... N4
50583 Sac City⊙ 3,000 ... C4
†52001 Sageville 291 ... M3
50472 Saint Ansgar 1,100 ... H2
50240 Saint Charles 507 ... F6
52649 Salem 463 ... K7
51052 Salix 429 ... A4
51248 Sanborn 1,398 ... B2
51053 Schaller 832 ... C4
51461 Schleswig 868 ... B4
51462 Scranton 748 ... D4
51054 Sergeant Bluff 2,416 ... A4
52590 Seymour 1,036 ... G7
50475 Sheffield 1,224 ... G3
51570 Shelby 665 ... C5
50243 Sheldahl 315 ... F5
51201 Sheldon 5,003 ... B2
50670 Shell Rock 1,478 ... H3
52332 Shellsburg 771 ... K4
51601 Shenandoah 6,274 ... C7
†52401 Shueyville 287 ... K5
51249 Sibley⊙ 3,051 ... B2
51652 Sidney⊙ 1,308 ... B7
52591 Sigourney⊙ 2,330 ... J6
51571 Silver City 291 ... B6
51250 Sioux Center 4,588 ... A2
*51101 Sioux City⊙ 82,003 ... A3
Sioux City‡ 117,457 ... A3
50585 Sioux Rapids 897 ... C3
50244 Slater 1,312 ... F5
51055 Sloan 978 ... A4
51056 Smithland 282 ... B4
51572 Soldier 257 ... B5
52333 Solon 969 ... L5
51301 Spencer⊙ 11,726 ... C2
52168 Spillville 415 ... J2
51360 Spirit Lake⊙ 3,976 ... C2
52336 Springville 1,165 ... L4
50476 Stacyville 538 ... H2
50246 Stanhope 492 ... F4
51573 Stanton 747 ... C7
52337 Stanwood 705 ... L5
50247 State Center 1,292 ... G5
50672 Steamboat Rock 387 ... G4
52651 Stockport 272 ... K7
52769 Stockton 240 ... M5
50588 Storm Lake⊙ 8,814 ... C3
50248 Story City 2,762 ... F4
50249 Stratford 806 ... F4
52076 Strawberry Point 1,463 ... K3
50250 Stuart 1,650 ... E6
50251 Sully 828 ... H5
50674 Sumner 2,335 ... J3
51058 Sutherland 897 ... B3
50590 Swea City 813 ... E2
52338 Swisher 654 ... K5
51653 Tabor 1,088 ... B7
52339 Tama 2,968 ... H5
51463 Templeton 319 ... D5
51364 Terril 420 ... C2
50478 Thompson 668 ... F2
50479 Thornton 442 ... G3
52340 Tiffin 413 ... K5
52772 Tipton⊙ 3,055 ... L5
50480 Titonka 607 ... E2
52342 Toledo⊙ 2,445 ... H4
50675 Traer 1,703 ... J4
51575 Treynor 981 ... B6
50676 Tripoli 1,280 ... J3
50257 Truro 407 ... F6
51576 Underwood 448 ... B6
50258 Union 515 ... G4
†52240 University Heights 1,069 ... K5
52595 University Park 645 ... H6
52345 Urbana 574 ... K4
50322 Urbandale 17,869 ... F5
51060 Ute 479 ... B4
51465 Vail 490 ... C4
52346 Van Horne 682 ... J4
50261 Van Meter 747 ... E5
50262 Van Wert 245 ... F7
50482 Ventura 614 ... F2
52347 Victor 1,046 ... J5
50864 Villisca 1,434 ... C7
52349 Vinton⊙ 5,040 ... J4
52077 Volga 310 ... L3
52169 Wadena 230 ... K3
†51360 Wahpeton 372 ... C2
52773 Walcott 1,425 ... M5
52351 Walford 285 ... K5
52352 Walker 733 ... K4
51365 Wallingford 256 ... D2
51466 Wall Lake 892 ... C4
51577 Walnut 897 ... C6
52653 Wapello⊙ 2,011 ... L6
52353 Washington⊙ 6,584 ... K6
51061 Washta 320 ... B3
*50701 Waterloo⊙ 75,985 ... J4
Waterloo-Cedar Falls‡ 137,961 ... J4
52171 Waucoma 308 ... J2
50263 Waukee 2,227 ... F5
52172 Waukon⊙ 3,983 ... L2
50677 Waverly⊙ 8,444 ... J3
52654 Wayland 720 ... K6
52356 Wellman 1,125 ... K6
50680 Wellsburg 761 ... H4
50483 Wesley 598 ... E2
50597 West Bend 941 ... D3
52358 West Branch 1,867 ... L5
52655 West Burlington 3,371 ... L7
50318 West Des Moines 21,894 ... F5
50681 Westgate 263 ... K3
52776 West Liberty 2,723 ... L5
†51351 West Okoboji 435 ... C2
52656 West Point 1,133 ... K7
51467 Westside 387 ... C4
52175 West Union⊙ 2,783 ... K3
50268 What Cheer 803 ... J6
52777 Wheatland 840 ... M5
51063 Whiting 734 ... A4
50598 Whittemore 647 ... E2
50271 Williams 410 ... F3
52361 Williamsburg 2,033 ... J5
52778 Wilton 2,502 ... M5
50311 Windsor Heights 5,474 ... F5
52659 Winfield 1,042 ... L6
50273 Winterset⊙ 4,021 ... E6
50682 Winthrop 767 ... K4
50484 Woden 287 ... F2
51579 Woodbine 1,463 ... B5
50276 Woodward 1,212 ... E5
50599 Woolstock 235 ... F3
52078 Worthington 432 ... L4
52362 Wyoming 702 ... L4
50277 Yale 299 ... E5
50278 Zearing 630 ... G4

OTHER FEATURES

Big Sioux (riv.) ... A3
Boyer (riv.) ... B5
Cedar (riv.) ... K4
Chariton (riv.) ... G7
Clear (lake) ... G2
Eagle (lake) ... F2
East Nishnabotna (riv.) ... C6
Effigy Mounds Nat'l Mon. ... L2
Five Island (lake) ... D2
Floyd (riv.) ... A3
Herbert Hoover Nat'l Hist. Site ... L5
Iowa (riv.) ... H4
Little Sioux (riv.) ... B3
Lost Island (lake) ... D2
Mississippi (riv.) ... L7
Missouri (riv.) ... A4
Nodaway (riv.) ... D7
Palo Alto (lake) ... D2
Platte (riv.) ... D8
Raccoon (riv.) ... D4
Rathbun (lake) ... G7
Red Rock (lake) ... G6
Rock (riv.) ... A2
Sac and Fox Ind. Res. ... H5
Saylorville (lake) ... F5
Skunk (riv.) ... K6
Spirit (lake) ... C2
Storm (lake) ... C3
Thompson (riv.) ... E7
Trumbull (lake) ... C2
Turkey (riv.) ... K2
Upper Iowa (riv.) ... K2
Wapsipinicon (riv.) ... J3
West Nishnabotna (riv.) ... C6

⊙County seat.
‡Population of metropolitan area.
† Zip of nearest p.o. * Multiple zips.

AREA 82,277 sq. mi. (213,097 sq. km.)
POPULATION 2,364,236
CAPITAL Topeka
LARGEST CITY Wichita
HIGHEST POINT Mt. Sunflower 4,039 ft. (1231 m.)
SETTLED IN 1831
ADMITTED TO UNION January 29, 1861
POPULAR NAME Sunflower State
STATE FLOWER Sunflower
STATE BIRD Western Meadowlark

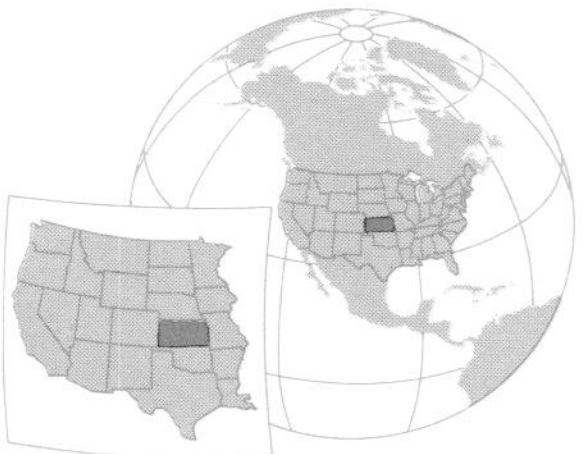

Agriculture, Industry and Resources

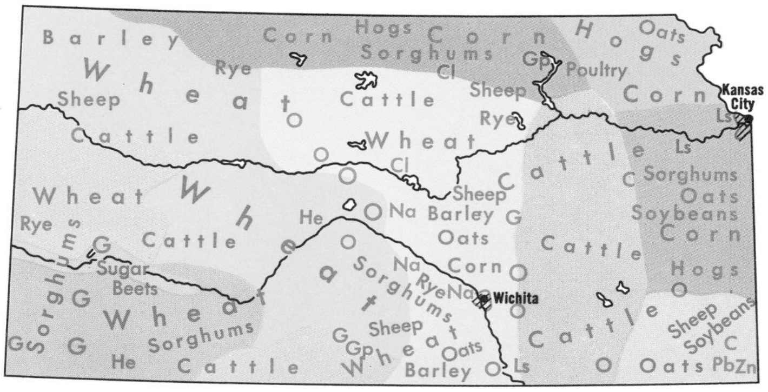

DOMINANT LAND USE

- Specialized Wheat
- Wheat, General Farming
- Wheat, Range Livestock
- Wheat, Grain Sorghums, Range Livestock
- Cattle Feed, Hogs
- Livestock, Cash Grain
- Livestock, Cash Grain, Dairy
- General Farming, Livestock, Cash Grain
- General Farming, Livestock, Special Crops
- Range Livestock

MAJOR MINERAL OCCURRENCES

C	Coal	Ls	Limestone
Cl	Clay	Na	Salt
G	Natural Gas	O	Petroleum
Gp	Gypsum	Pb	Lead
He	Helium	Zn	Zinc

Major Industrial Areas

COUNTIES

Allen 15,654 ..... G4
Anderson 8,749 ..... G3
Atchison 18,397 ..... G2
Barber 6,548 ..... D4
Barton 31,343 ..... D3
Bourbon 15,969 ..... H4
Brown 11,955 ..... G2
Butler 44,782 ..... F4
Chase 3,309 ..... F3
Chautauqua 5,016 ..... F4
Cherokee 22,304 ..... H4
Cheyenne 3,678 ..... A2
Clark 2,599 ..... C4
Clay 9,802 ..... E2
Cloud 12,494 ..... E2
Coffey 9,370 ..... G3
Comanche 2,554 ..... C4
Cowley 36,824 ..... F4
Crawford 37,916 ..... H4
Decatur 4,509 ..... B2
Dickinson 20,175 ..... E3
Doniphan 9,268 ..... G2
Douglas 67,640 ..... G3
Edwards 4,271 ..... C4
Elk 3,918 ..... F4
Ellis 26,098 ..... C3
Ellsworth 6,640 ..... D3
Finney 23,825 ..... B3
Ford 24,315 ..... C4
Franklin 22,062 ..... G3
Geary 29,852 ..... F3
Gove 3,726 ..... B3
Graham 3,995 ..... C2
Grant 6,977 ..... A4
Gray 5,138 ..... B4
Greeley 1,845 ..... A3
Greenwood 8,764 ..... F4
Hamilton 2,514 ..... A3
Harper 7,778 ..... D4
Harvey 30,531 ..... E3
Haskell 3,814 ..... B4
Hodgeman 2,269 ..... C3
Jackson 11,644 ..... G2
Jefferson 15,207 ..... G2
Jewell 5,241 ..... D2
Johnson 270,269 ..... H3
Kearny 3,435 ..... A3
Kingman 8,960 ..... D4
Kiowa 4,046 ..... C4
Labette 25,682 ..... G4
Lane 2,472 ..... B3
Leavenworth 54,809 ..... G2
Lincoln 4,145 ..... D2
Linn 8,234 ..... H3
Logan 3,478 ..... A3
Lyon 35,108 ..... F3
Marion 13,522 ..... E3
Marshall 12,787 ..... F2
McPherson 26,855 ..... E3
Meade 4,788 ..... B4
Miami 21,618 ..... H3
Mitchell 8,117 ..... D2
Montgomery 42,281 ..... G4
Morris 6,419 ..... F3
Morton 3,454 ..... A4
Nemaha 11,211 ..... F2
Neosho 18,967 ..... G4
Ness 4,498 ..... C3
Norton 6,689 ..... C2
Osage 15,319 ..... G3
Osborne 5,959 ..... D2
Ottawa 5,971 ..... E2
Pawnee 8,065 ..... C3
Phillips 7,406 ..... C2
Pottawatomie 14,782 ..... F2
Pratt 10,275 ..... D4
Rawlins 4,105 ..... A2
Reno 64,983 ..... D4
Republic 7,569 ..... E2
Rice 11,900 ..... D3
Riley 63,505 ..... F2
Rooks 7,006 ..... C2
Rush 4,516 ..... C3
Russell 8,868 ..... D3
Saline 48,905 ..... E3
Scott 5,782 ..... B3
Sedgwick 367,088 ..... E4
Seward 17,071 ..... B4
Shawnee 154,916 ..... G2
Sheridan 3,544 ..... B2
Sherman 7,759 ..... A2
Smith 5,947 ..... D2
Stafford 5,694 ..... D3
Stanton 2,339 ..... A4
Stevens 4,736 ..... A4
Sumner 24,928 ..... E4
Thomas 8,451 ..... A2
Trego 4,165 ..... C3
Wabaunsee 6,867 ..... F3
Wallace 2,045 ..... A3
Washington 8,543 ..... E2
Wichita 3,041 ..... A3
Wilson 12,128 ..... G4
Woodson 4,600 ..... G4
Wyandotte 172,335 ..... H2

CITIES and TOWNS

Zip Name/Pop. Key

67510 Abbyville 123 ..... D4
67410 Abilene⊙ 6,572 ..... E3
66830 Admire 158 ..... F3
66930 Agenda 106 ..... E2
67621 Agra 321 ..... C2
67511 Albert 236 ..... C3
67512 Alden 214 ..... D3
67513 Alexander 116 ..... C3
66833 Allen 205 ..... F3
66401 Alma⊙ 925 ..... F2
67622 Almena 517 ..... C2
67330 Altamont 1,054 ..... G4
66834 Alta Vista 430 ..... F3
67623 Alton 135 ..... D2
66710 Altoona 564 ..... G4
66835 Americus 915 ..... F3
67001 Andale 538 ..... E4
67002 Andover 2,801 ..... E4
67003 Anthony⊙ 2,661 ..... D4
66711 Arcadia 460 ..... H4
67004 Argonia 587 ..... E4
67005 Arkansas City 13,201 ..... E4
67514 Arlington 631 ..... D4
66712 Arma 1,676 ..... H4
67831 Ashland⊙ 1,096 ..... C4
67416 Assaria 414 ..... E3
66002 Atchison⊙ 11,407 ..... G2
66932 Athol 90 ..... D2
67008 Atlanta 256 ..... F4
67009 Attica 730 ..... D4
67730 Atwood⊙ 1,665 ..... B2
66402 Auburn 890 ..... G3
67010 Augusta 6,968 ..... F4
67417 Aurora 130 ..... E2
66403 Axtell 470 ..... F2
66404 Baileyville 130 ..... F2
66006 Baldwin City 2,829 ..... G3
67418 Barnard 163 ..... D2
66933 Barnes 257 ..... F2
67332 Bartlett 163 ..... G4
66007 Basehor 1,483 ..... G2
†66749 Bassett 31 ..... G4
66713 Baxter Springs 4,730 ..... H4
67516 Bazine 385 ..... C3
66406 Beattie 316 ..... F2
67013 Belle Plaine 1,706 ..... E4
66935 Belleville⊙ 2,805 ..... E2
67420 Beloit⊙ 4,367 ..... D2
67519 Belpre 154 ..... C4
66407 Belvue 212 ..... F2
66714 Benedict 111 ..... G4
67422 Bennington 579 ..... E2
67016 Bentley 311 ..... E4
67017 Benton 609 ..... E4
66408 Bern 220 ..... F2
67423 Beverly 171 ..... E2
67731 Bird City 546 ..... A2
67520 Bison 279 ..... C3
66010 Blue Mound 319 ..... H3
66411 Blue Rapids 1,280 ..... F2
67018 Bluff City 95 ..... E4
67625 Bogue 197 ..... C2
66012 Bonner Springs 6,266 ..... H2
67732 Brewster 327 ..... A2
66716 Bronson 414 ..... H4
67425 Brookville 259 ..... E3
67521 Brownell 92 ..... C3
67834 Bucklin 786 ..... C4
66717 Buffalo 386 ..... G4
67522 Buhler 1,188 ..... E3
67626 Bunker Hill 124 ..... D3
67019 Burden 518 ..... F4
67523 Burdett 275 ..... C3
66413 Burlingame 1,239 ..... G3
66839 Burlington⊙ 2,901 ..... G3
66840 Burns 224 ..... F3
66936 Burr Oak 366 ..... D2
67020 Burrton 976 ..... E3
66841 Bushong 62 ..... F3
67427 Bushton 388 ..... D3
67021 Byers 47 ..... D4
67022 Caldwell 1,401 ..... E4
67023 Cambridge 113 ..... F4
67333 Caney 2,284 ..... G4
67428 Canton 926 ..... E3
66414 Carbondale 1,518 ..... G3
67429 Carlton 49 ..... E3
66842 Cassoday 122 ..... F3
67430 Cawker City 640 ..... D2
67628 Cedar 53 ..... D2
66843 Cedar Point 66 ..... F3
67024 Cedar Vale 848 ..... F4
66415 Centralia 486 ..... F2
66720 Chanute 10,506 ..... G4
67431 Chapman 1,255 ..... E3
67524 Chase 753 ..... D3
67334 Chautauqua 156 ..... F4
67025 Cheney 1,404 ..... E4
66724 Cherokee 775 ..... H4
67335 Cherryvale 2,769 ..... G4
67336 Chetopa 1,751 ..... G4
67835 Cimarron⊙ 1,491 ..... B4
66416 Circleville 164 ..... G2
67525 Claflin 764 ..... D3
67432 Clay Center⊙ 4,948 ..... E2
67629 Clayton 102 ..... B2
67026 Clearwater 1,684 ..... E4
66937 Clifton 695 ..... E2
67027 Climax 81 ..... F4
66938 Clyde 909 ..... E2
67028 Coats 153 ..... D4
67337 Coffeyville 15,185 ..... G4
67701 Colby⊙ 5,544 ..... A2
67029 Coldwater⊙ 989 ..... C4
67631 Collyer 151 ..... B2
66015 Colony 474 ..... G3
66725 Columbus⊙ 3,426 ..... H4
67030 Colwich 935 ..... E4
66901 Concordia⊙ 6,847 ..... E2
67031 Conway Springs 1,313 ..... E4
67836 Coolidge 82 ..... A3
67837 Copeland 323 ..... B4
66417 Corning 158 ..... F2
66845 Cottonwood Falls⊙ 954 ..... F3
66846 Council Grove⊙ 2,381 ..... F3
66939 Courtland 377 ..... E2
66727 Coyville 98 ..... G4
66940 Cuba 286 ..... E2
†67124 Cullison 154 ..... D4
67435 Culver 167 ..... E3
67035 Cunningham 540 ..... D4
67632 Damar 204 ..... C2
67036 Danville 71 ..... E4
67340 Dearing 475 ..... G4
67838 Deerfield 538 ..... A4
66418 Delia 181 ..... G2
67436 Delphos 570 ..... E2
66419 Denison 231 ..... G2
66017 Denton 156 ..... G2
67037 Derby 9,786 ..... E4
66018 De Soto 2,061 ..... H3
67038 Dexter 366 ..... F4
67839 Dighton⊙ 1,390 ..... B3
67801 Dodge City⊙ 18,001 ..... B4
67634 Dorrance 220 ..... D3
67039 Douglass 1,450 ..... F4
67437 Downs 1,324 ..... D2
67635 Dresden 84 ..... B2
66848 Dunlap 82 ..... F3
67438 Durham 130 ..... E3
66849 Dwight 320 ..... F3
†66720 Earlton 79 ..... G4
†67201 Eastborough 854 ..... E4
66020 Easton 460 ..... G2
66021 Edgerton 1,214 ..... H3
67636 Edmond 56 ..... C2
67342 Edna 537 ..... G4
66113 Edwardsville 3,364 ..... H2
66023 Effingham 634 ..... G2
67041 Elbing 175 ..... E3
67042 El Dorado⊙ 10,510 ..... F4
†67361 Elgin 139 ..... F4
67344 Elk City 404 ..... G4
67345 Elk Falls 151 ..... F4
67950 Elkhart⊙ 2,243 ..... A4
67526 Ellinwood 2,508 ..... D3
67637 Ellis 2,062 ..... C3
67439 Ellsworth⊙ 2,465 ..... D3
66850 Elmdale 109 ..... F3
66732 Elmore 104 ..... G4
66024 Elwood 1,275 ..... H2
66422 Emmett 223 ..... F2
66801 Emporia⊙ 25,287 ..... F3
67840 Englewood 111 ..... C4
67841 Ensign 209 ..... B4
67441 Enterprise 839 ..... E3
66733 Erie⊙ 1,415 ..... G4
66941 Esbon 234 ..... D2
66423 Eskridge 603 ..... F3
66025 Eudora 2,934 ..... G3
67045 Eureka⊙ 3,425 ..... F4
66424 Everest 331 ..... G2
66425 Fairview 258 ..... G2
†66101 Fairway 4,619 ..... H2
67047 Fall River 173 ..... G4
66851 Florence 729 ..... E3
66026 Fontana 173 ..... H3
67842 Ford 272 ..... C4
66942 Formoso 166 ..... D2
67843 Fort Dodge 400 ..... C4
66027 Fort Leavenworth H2
66701 Fort Scott⊙ 8,893 ..... H4
67844 Fowler 592 ..... B4
66427 Frankfort 1,038 ..... F2
66735 Franklin 400 ..... H4
66736 Fredonia⊙ 3,047 ..... G4
67049 Freeport 12 ..... E4
66762 Frontenac 2,586 ..... H4
66738 Fulton 194 ..... H4
66739 Galena 3,587 ..... H4
66740 Galesburg 181 ..... G4
67443 Galva 651 ..... E3
67846 Garden City⊙ 18,256 ..... B4
67050 Garden Plain 775 ..... E4
66030 Gardner 2,392 ..... H3
67529 Garfield 277 ..... C3
66032 Garnett⊙ 3,310 ..... G3
66742 Gas 543 ..... G4
67638 Gaylord 203 ..... D2
67734 Gem 101 ..... B2
67444 Geneseo 496 ..... D3
67051 Geuda Springs 217 ..... E4
66743 Girard⊙ 2,888 ..... H4
67639 Glade 131 ..... C2
67445 Glasco 710 ..... E2
67446 Glen Elder 491 ..... D2
67052 Goddard 1,427 ..... E4
67053 Goessel 421 ..... E3
66428 Goff 196 ..... G2
67735 Goodland⊙ 5,708 ..... A2
67640 Gorham 355 ..... D3
67736 Gove⊙ 148 ..... B3
67737 Grainfield 417 ..... B2
†66441 Grandview Plaza 1,189 ..... F2
66429 Grantville 220 ..... G2
67530 Great Bend⊙ 16,608 ..... D3
66033 Greeley 405 ..... G3
67447 Green 155 ..... E2
66943 Greenleaf 462 ..... E2
67054 Greensburg⊙ 1,885 ..... C4
67346 Grenola 335 ..... F4
66852 Gridley 404 ..... G3
67738 Grinnell 410 ..... B2
67448 Gypsum 423 ..... E3
66944 Haddam 239 ..... E2
67056 Halstead 1,994 ..... E4
66853 Hamilton 363 ..... F4
66945 Hanover 802 ..... F2
67849 Hanston 257 ..... C3
67057 Hardtner 336 ..... D4
67058 Harper 1,823 ..... D4
66854 Hartford 551 ..... F3
66431 Harveyville 280 ..... F3
67347 Havana 169 ..... G4
67543 Haven 1,125 ..... E4
66432 Havensville 183 ..... F2
67059 Haviland 770 ..... C4

(continued on following page)

67601 Hays⊙ 16,301 ....C3	67545 Hudson 157 ....D3	66039 Kincaid 192 ....G3
67760 Haysville 8,006 ....E4	67951 Hugoton 3,165 ....A4	67068 Kingman⊙ 3,563 ....D4
67850 Healy 275 ....B3	66748 Humboldt 2,230 ....G4	67547 Kinsley⊙ 2,074 ....C4
67061 Hazelton 143 ....D4	67452 Hunter 135 ....C2	67070 Kiowa 1,409 ....D4
66746 Hepler 165 ....H4	66038 Huron 107 ....G2	67644 Kirwin 249 ....C2
67449 Herington 2,930 ....E3	67501 Hutchinson⊙ 40,284 ....D3	67859 Kismet 368 ....B4
67739 Herndon 220 ....B2	67301 Independence⊙ 10,598 ....G4	67350 Labette 123 ....G4
67062 Hesston 3,013 ....E3	67853 Ingalls 274 ....B4	67548 La Crosse⊙ 1,618 ....C3
66434 Hiawatha⊙ 3,702 ....G2	67546 Inman 947 ....D3	66040 La Cygne 1,025 ....H3
†67880 Hickock 68 ....A4	66749 Iola⊙ 6,938 ....G4	66751 La Harpe 687 ....G4
66035 Highland 954 ....G2	67065 Isabel 137 ....D4	67860 Lakin⊙ 1,823 ....A4
67642 Hill City⊙ 2,028 ....C2	67066 Iuka 235 ....D4	66041 Lancaster 274 ....G2
67063 Hillsboro 2,717 ....E3	66948 Jamestown 440 ....E2	66042 Lane 249 ....G3
67544 Hoisington 3,678 ....D3	67643 Jennings 194 ....B2	67549 Langdon 84 ....D4
67851 Holcomb 816 ....B3	67854 Jetmore⊙ 862 ....B3	67550 Larned⊙ 4,811 ....C3
66946 Hollenberg 57 ....F2	66949 Jewell 589 ....D2	67645 Latham 148 ....F4
66436 Holton⊙ 3,132 ....G2	67855 Johnson⊙ 1,244 ....A4	66044 Lawrence⊙ 52,738 ....G3
67450 Holyrood 567 ....D3	66441 Junction City⊙ 19,305 ....E3	Lawrence‡ 67,640 ....G3
67451 Hope 468 ....E3	67454 Kanopolis 729 ....D3	66048 Leavenworth⊙ 33,656 ....H2
†67879 Horace 137 ....A3	67741 Kanorado 217 ....A2	66206 Leawood 13,360 ....H3
66439 Horton 2,130 ....G2	*66101 Kansas City⊙ 161,148 ....H2	66952 Lebanon 440 ....D2
67349 Howard⊙ 965 ....F4	Kansas City‡ 1,327,020 ....H2	66856 Lebo 966 ....F3
67740 Hoxie⊙ 1,462 ....B2	67067 Kechi 288 ....E4	66050 Lecompton 576 ....G2
66440 Hoyt 536 ....G2	66951 Kensington 681 ....C2	

67073 Lehigh 189 ....E3	66053 Louisburg 1,744 ....H3	67745 McDonald 239 ....A2
66215 Lenexa 18,639 ....H2	66450 Louisville 207 ....F2	66501 McFarland 242 ....F2
67645 Lenora 444 ....C2	67648 Lucas 524 ....D2	66054 McLouth 700 ....G2
67074 Leon 667 ....F4	67649 Luray 295 ....D2	67460 McPherson⊙ 11,753 ....E3
66448 Leona 73 ....G2	66451 Lyndon⊙ 1,132 ....G3	67864 Meade⊙ 1,777 ....B4
66449 Leonardville 437 ....F2	67554 Lyons⊙ 4,134 ....D3	66510 Melvern 481 ....G3
67861 Leoti⊙ 1,869 ....A3	67557 Macksville 546 ....C4	67746 Menlo 42 ....B2
66857 Le Roy 701 ....G3	66860 Madison 1,099 ....F3	66512 Meriden 707 ....G2
67552 Lewis 551 ....C4	66955 Mahaska 119 ....E2	66203 Merriam 10,794 ....H3
67901 Liberal⊙ 14,911 ....B4	67101 Maize 1,294 ....E4	67105 Milan 135 ....E4
67351 Liberty 174 ....G4	67463 Manchester 98 ....E2	66055 Mildred 64 ....G3
67553 Liebenthal 163 ....C3	66502 Manhattan⊙ 32,644 ....F2	66514 Milford 465 ....E2
67455 Lincoln⊙ 1,599 ....D2	66956 Mankato⊙ 1,205 ....D2	67466 Miltonvale 588 ....E2
66858 Lincolnville 235 ....E3	67862 Manter 205 ....A4	67467 Minneapolis⊙ 2,075 ....E2
67456 Lindsborg 3,155 ....D3	66507 Maple Hill 381 ....F2	67865 Minneola 712 ....C4
66953 Linn 483 ....E2	66754 Mapleton 121 ....H3	66205 Mission 8,643 ....H3
67457 Little River 529 ....E3	66861 Marion⊙ 1,951 ....F3	67353 Moline 553 ....F4
67646 Logan 397 ....C2	67464 Marquette 639 ....E3	67867 Montezuma 730 ....B4
67647 Long Island 187 ....C2	66508 Marysville⊙ 3,670 ....F2	66755 Moran 643 ....G4
67352 Longford 109 ....E2	66862 Matfield Green 71 ....F3	67468 Morganville 261 ....E2
67647 Longton 396 ....F4	66509 Mayetta 287 ....G2	67650 Morland 273 ....B2
67459 Lorraine 157 ....D3	67103 Mayfield 128 ....E4	66515 Morrill 336 ....G2
67556 McCracken 292 ....C3		66958 Morrowville 180 ....E2
66050 Lecompton 576 ....G2	66859 Lost Springs 94 ....E3	66753 McCune 528 ....G4

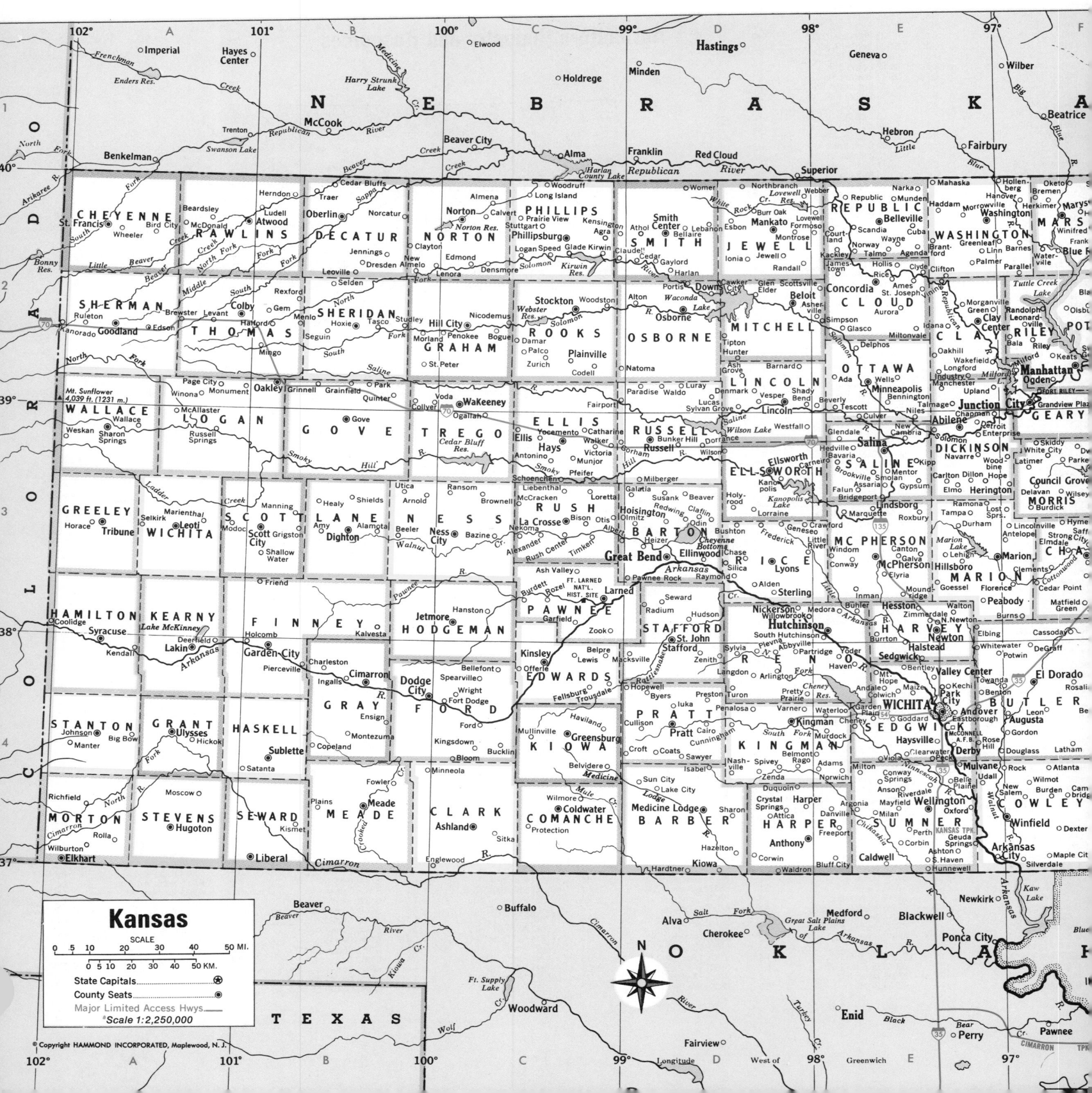

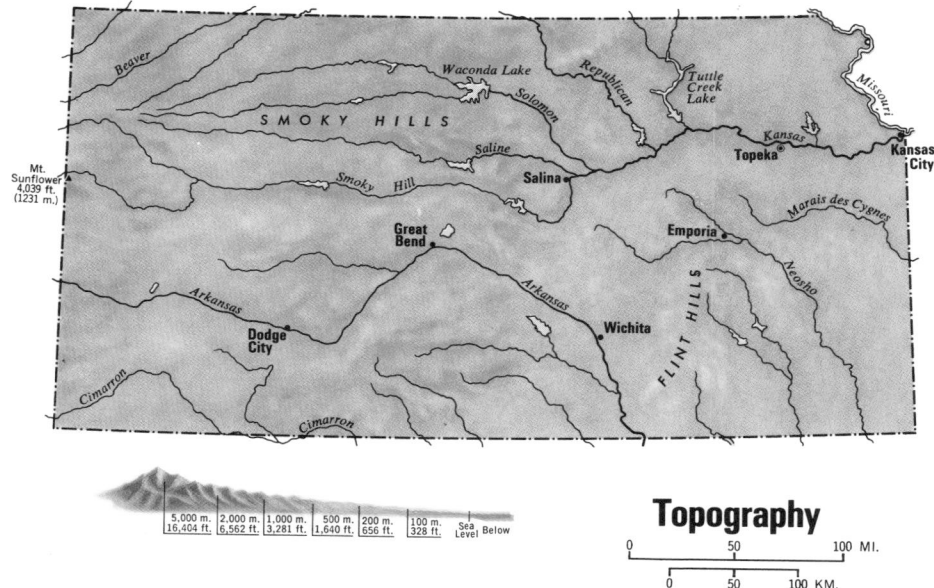

Topography

| 5,000 m. | 2,000 m. | 1,000 m. | 500 m. | 200 m. | 100 m. | Sea |
| 16,404 ft. | 6,562 ft. | 3,281 ft. | 1,640 ft. | 656 ft. | 328 ft. | Level Below |

67952 Moscow 228..............A4
66056 Mound City⊙ 755.......H3
67107 Moundridge 1,453......E3
67354 Mound Valley 381......G4
67108 Mount Hope 791........E4
66756 Mulberry 647...........H4
67109 Mullinville 339.......C4
67110 Mulvane 4,254.........E4
66959 Munden 152............E2
66058 Muscotah 248..........G2
66960 Narka 120.............E2
67112 Nashville 127.........D4
67651 Natoma 515............D2
66757 Neodesha 3,414........G4
67651 Neosho Falls 157......G3
66864 Neosho Rapids 289.....F3
67560 Ness City⊙ 1,769......C3
66516 Netawaka 218..........G2
66759 New Albany 78.........G4
67470 New Cambria 175.......E3
66839 New Strawn (Strawn) 457..G3
67114 Newton⊙ 16,332........E3
67561 Nickerson 1,292.......D3
67355 Niotaze 104...........F4

67653 Norcatur 226..........B2
67117 North Newton 1,222....E3
67654 Norton⊙ 3,400.........C2
66060 Nortonville 692.......G2
67118 Norwich 476...........E4
67472 Oakhill 35............E2
67748 Oakley⊙ 2,343.........B2
67749 Oberlin⊙ 2,387........B2
67563 Offerle 244...........C4
66517 Ogden 1,804...........F2
66518 Oketo 130.............F2
66061 Olathe⊙ 37,258........H3
67564 Olmitz 140............D3
66865 Olpe 477..............F3
66520 Olsburg 166...........F2
66521 Onaga 752.............F2
66522 Oneida 120............G2
66523 Osage City 2,667......G3
66064 Osawatomie 4,459......H3
67473 Osborne⊙ 2,120........D2
66066 Oskaloosa⊙ 1,092......G2
67356 Oswego⊙ 2,218.........G4
67565 Otis 410..............C3
66067 Ottawa⊙ 11,016........G3

66524 Overbrook 930..............G3
66204 Overland Park 81,784.......H3
67119 Oxford 1,125...............E4
66070 Ozawkie 472................G2
67657 Palco 329..................C2
66962 Palmer 149.................E2
66071 Paola⊙ 4,557...............H3
67658 Paradise 89................D2
67751 Park 183...................B2
66072 Parker 270.................H3
67357 Parsons 12,898.............G4
67566 Partridge 268..............D4
67567 Pawnee Rock 409............D3
66526 Paxico 168.................F2
66866 Peabody 1,474..............E3
67120 Peck 250...................E4
67121 Penalosa 31................D4
66073 Perry 907..................G2
67360 Peru 286...................F4
67661 Phillipsburg⊙ 3,229........C2
66762 Pittsburg 18,770...........H4
67869 Plains 1,044...............B4
67663 Plainville 2,458...........C2
66075 Pleasanton 1,303...........H3
67568 Plevna 115.................D4
66076 Pomona 868.................G3
67474 Portis 172.................D2
67123 Potwin 563.................F4
66527 Powhattan 95...............G2
67664 Prairie View 145...........C2
66208 Prairie Village 24,657.....H2
67124 Pratt⊙ 6,885...............D4
67767 Prescott 319...............H3
67569 Preston 227................D4
67570 Pretty Prairie 655.........D4
66078 Princeton 244..............G3
67127 Protection 684.............C4
66528 Quenemo 413................G3
67752 Quinter 951................B2
67571 Radium 47..................D3
67475 Ramona 116.................E3
66963 Randall 154................D2
66554 Randolph 131...............F2
67572 Ransom 448.................C3
66079 Rantoul 212................G3
67573 Raymond 132................D3
66868 Reading 244................F3
66769 Redfield 185...............H4
66964 Republic 223...............E2
66529 Reserve 105................G2
67753 Rexford 204................B2
67953 Richfield 81...............A4
66080 Richmond 510...............G3
66531 Riley 779..................F2
66770 Riverton 650...............H4
66532 Robinson 324...............G2
†66205 Roeland Park 7,962........H2
67954 Rolla 417..................A4
67133 Rose Hill 1,557............E4
†66773 Roseland 119..............H4
66533 Rossville 1,045............G2
67574 Rozel 219..................C3
67575 Rush Center 207............C3
67665 Russell⊙ 5,427.............D3
67755 Russell Springs 56.........A3
66534 Sabetha 2,286..............G2
67756 Saint Francis⊙ 1,610.......A2
66535 Saint George 309...........F2
67576 Saint John⊙ 1,501..........D3
66536 Saint Marys 1,598..........G2
66771 Saint Paul 746.............G4
67401 Salina⊙ 41,843.............E3
67870 Satanta 1,117..............B4
66772 Savonburg 113..............G4
67134 Sawyer 213.................D4
66773 Scammon 501................H4
66966 Scandia 480................E2
67667 Schoenchen 209.............C3
67871 Scott City⊙ 4,154..........B3
67477 Scottsville 56.............D2
66537 Scranton 664...............G3

67361 Sedan⊙ 1,579...............F4
67135 Sedgwick 1,471.............E4
67757 Selden 266.................B2
66538 Seneca⊙ 2,389..............F2
66081 Severance 134..............G2
67137 Severy 447.................F4
67577 Seward 88..................D3
67138 Sharon 283.................D4
67758 Sharon Springs⊙ 982........A3
*66202 Shawnee 29,653.............H2
66539 Silver Lake 1,350..........G2
67478 Simpson 123................E2
66967 Smith Center⊙ 2,240........D2
67479 Smolan 169.................E3
66540 Soldier 165................G2
67480 Solomon 1,018..............E3
67140 South Haven 439............E4
†67501 South Hutchinson 2,226....D3
67876 Spearville 693.............C4
67142 Spivey 83..................D4
66083 Spring Hill 2,005..........H3
67578 Stafford 1,425.............D4
67775 Stark 143..................G4
67579 Sterling 2,312.............D3
67669 Stockton⊙ 1,825............C2
66839 Strawn 457.................G3
66869 Strong City 675............F3
67877 Sublette⊙ 1,293............B4
66541 Summerfield 225............F2
67143 Sun City 85................D4
67481 Sylvan Grove 376...........D2
67581 Sylvia 353.................D4
67878 Syracuse⊙ 1,654............A3
67483 Tampa 113..................E3
67484 Tescott 331................E2
66542 Tecumseh 300...............G2
67485 Timken 99..................C3
67582 Tipton 321.................D2
66086 Tonganoxie 1,864...........G2
*66601 Topeka (cap.)⊙ 115,266.....G2
       Topeka‡ 185,442..........G2
66777 Toronto 466................G4
67144 Towanda 1,332..............E4
†66075 Trading Post 35...........H3
66778 Treece 194.................H4
67879 Tribune⊙ 955...............A3
66087 Troy⊙ 1,240................G2
67583 Turon 481..................D4
67364 Tyro 289...................G4
67146 Udall 891..................E4
67880 Ulysses⊙ 4,653.............A4
66779 Uniontown 371..............G4
67584 Utica 275..................B3
67147 Valley Center 3,300........E4
66088 Valley Falls 1,189.........G2
66544 Vermillion 191.............F2
67671 Victoria 1,328.............C3
†66937 Vining 85.................E2
67149 Viola 199..................E4
66870 Virgil 169.................F4
67672 WaKeeney⊙ 2,388............C2
67487 Wakefield 803..............E2
67673 Waldo 75...................D2
67150 Waldron 29.................D4
67761 Wallace 86.................A3
66780 Walnut 308.................G4
67151 Walton 269.................E3
66547 Wamego 3,159...............F2
66968 Washington⊙ 1,488..........E2
66548 Waterville 694.............F2
66090 Wathena 1,418..............H2
66871 Waverly 671................G3
67152 Wellington⊙ 8,212..........E4
66092 Wellsville 1,612...........G3
66782 West Mineral 229...........H4
66549 Westmoreland⊙ 598..........F2
66093 Westphalia 204.............G3
67869 West Plains
       (Plains) 1,044..........B4

66551 Wheaton 90.................F2
66872 White City 534.............F3
66094 White Cloud 234............G2
67154 Whitewater 751.............E4
66552 Whiting 270................G2
*67201 Wichita⊙ 279,835..........E4
       Wichita‡ 411,313.........E4
†66601 Willard 128...............G2
66095 Williamsburg 362...........G3
66435 Willis 85..................G2
†67501 Willowbrook 109...........D3
66873 Wilsey 179.................F3
67490 Wilson 978.................D3
66097 Winchester 570.............G2
67491 Windom 160.................E3
67156 Winfield⊙ 10,736...........E4
67764 Winona 258.................A2
67492 Woodbine 172...............E3
67675 Woodston 157...............C2
66783 Yates Center⊙ 1,998........G4
67159 Zenda 146..................D4
67676 Zurich 185.................C2

OTHER FEATURES

Arkansas (riv.).................D3
Beaver (creek).................A2
Big Blue (riv.)................F1
Cedar Bluff (res.).............C3
Cheney (res.)..................D4
Cheyenne Bottoms (lake)........D3
Chikaskia (riv.)...............E4
Cimarron (riv.)................A4
Cottonwood (riv.)..............F3
Council Grove (lake)...........F3
Crooked (creek)................B4
Elk (riv.).....................G4
Fall (riv.)....................G4
Fall River (lake)..............F4
Fort Larned Nat'l Hist. Site...C3
Fort Riley-Camp Whiteside 18,233..F2
John Redmond (res.)............G3
Hulah (lake)...................F5
Kanopolis (lake)...............D3
Kansas (riv.)..................F2
Kickapoo Ind. Res..............G2
Kirwin (res.)..................C2
Little Arkansas (riv.).........D3
Little Blue (riv.).............E1
Lovewell (res.)................D2
Marion (lake)..................E3
McConnell A.F.B................E4
McKinney (lake)................A3
Medicine Lodge (riv.)..........D4
Milford (lake).................E2
Missouri (riv.)................G1
Mule (creek)...................C4
Nemaha (riv.)..................G2
Neosho (riv.)..................G4
Ninnescah (riv.)...............D4
Norton (res.)..................C2
Olathe Nav. Air Sta............H3
Pawnee (riv.)..................B3
Perry (lake)...................G2
Pomona (lake)..................G3
Potawatomi Ind. Res............G2
Rattlesnake (creek)............D4
Republican (riv.)..............E2
Sac-Fox-Iowa Ind. Res..........G2
Saline (riv.)..................D3
Sappa (creek)..................C2
Smoky Hill (riv.)..............C3
Solomon (riv.).................D2
Sunflower (mt.)................A2
Toronto (lake).................G4
Tuttle Creek (lake)............F2
Verdigris (riv.)...............G5
Walnut (riv.)..................E4
Webster (res.).................C2
White Rock (creek).............D2
Wilson (lake)..................D3

⊙County seat.
‡Population of metropolitan area.
† Zip of nearest p.o.
* Multiple zips.

KENTUCKY

COUNTIES

Adair 15,233 L6
Allen 14,128 J7
Anderson 12,567 M5
Ballard 8,798 C6
Barren 34,009 K7
Bath 10,025 O4
Bell 34,330 O7
Boone 45,842 M3
Bourbon 19,405 N4
Boyd 55,513 R4
Boyle 25,066 M5
Bracken 7,738 N3
Breathitt 17,004 P5
Breckinridge 16,861 H5
Bullitt 43,346 K5
Butler 11,064 H6
Caldwell 13,473 F6
Calloway 30,031 E7
Campbell 83,317 N3
Carlisle 5,487 C7
Carroll 9,270 L3
Carter 25,060 P4
Casey 14,818 M6
Christian 66,878 F7
Clark 28,322 N4
Clay 22,752 O6
Clinton 9,321 L7
Crittenden 9,207 E6
Cumberland 7,289 L7
Daviess 85,949 G5
Edmonson 9,962 J6
Elliott 6,908 P4
Estill 14,495 O5
Fayette 204,165 N4
Fleming 12,323 O4
Floyd 48,764 R5
Franklin 41,830 M4
Fulton 8,971 C7
Gallatin 4,842 M3
Garrard 10,853 M5
Grant 13,308 M3
Graves 34,049 D7
Grayson 20,854 J5
Green 11,043 K6
Greenup 39,132 R3
Hancock 7,742 H5
Hardin 88,917 K5
Harlan 41,889 P7
Harrison 15,166 N4
Hart 15,402 K6
Henderson 40,849 F5
Henry 12,740 L4
Hickman 6,065 C7
Hopkins 46,174 F6
Jackson 11,996 N6
Jefferson 684,565 K4
Jessamine 26,065 M5
Johnson 24,432 R5
Kenton 137,058 M3
Knott 17,940 R6
Knox 30,239 O7
Larue 11,922 K5
Laurel 38,982 N6
Lawrence 14,121 R4
Lee 7,754 O5
Leslie 14,882 P6
Letcher 30,687 R6
Lewis 14,545 P3
Lincoln 19,053 M6
Livingston 9,219 E6
Logan 24,138 H7
Lyon 6,490 E6
Madison 53,352 N5
Magoffin 13,515 P5
Marion 17,910 L5
Marshall 25,637 E7
Martin 13,925 R5
Mason 17,765 O3
McCracken 61,310 D6
McCreary 15,634 N7
McLean 10,090 G5
Meade 22,854 J5
Menifee 5,117 O5
Mercer 19,011 M5
Metcalfe 9,484 K7
Monroe 12,353 K7
Montgomery 20,046 O4
Morgan 12,103 P5
Muhlenberg 32,238 G6
Nelson 27,584 K5
Nicholas 7,157 N4
Ohio 21,765 H6
Oldham 27,795 L4
Owen 8,924 M3
Owsley 5,709 O6
Pendleton 10,989 N3
Perry 33,763 P6
Pike 81,123 S6
Powell 11,101 O5
Pulaski 45,803 M6
Robertson 2,265 N3
Rockcastle 13,973 N6
Rowan 19,049 P4
Russell 13,708 L7
Scott 21,813 M4
Shelby 23,328 L4
Simpson 14,673 H7
Spencer 5,929 L4
Taylor 21,178 L6
Todd 11,874 G7
Trigg 9,384 F7
Trimble 6,253 L3
Union 17,821 F5
Warren 71,828 H6
Washington 10,764 L5
Wayne 17,022 M7
Webster 14,832 F5
Whitley 33,396 N7
Wolfe 6,698 O5
Woodford 17,778 M4

CITIES and TOWNS

Zip *Name/Pop.* *Key*

42202 Adairville 1,105 H7
42602 Albany⊙ 2,083 L7
41001 Alexandria⊙ 4,735 N3
41601 Allen 338 R5
42204 Allensville 170 G7
40223 Anchorage 1,726 L2
41101 Ashland 27,064 R4
Ashland-Huntington‡ 311,350 R4
42206 Auburn 1,467 H7
†40201 Audubon Park 1,571 J2
41002 Augusta 1,455 N3
41602 Auxier 900 R5
†40222 Bancroft 725 K1
41603 Banner 950 R5
†40201 Barbourmeade 1,038 K1
40906 Barbourville⊙ 3,333 O7
40004 Bardstown⊙ 6,155 L5
42023 Bardwell⊙ 988 D7
42024 Barlow 746 D6
41311 Beattyville⊙ 1,068 O5
42320 Beaver Dam 3,185 H6
40006 Bedford⊙ 835 L3
40359 Beechwood Village 1,462 K2
†40201 Bellemeade 918 L2
41073 Bellevue 7,678 S1
40807 Benham 936 R7
42025 Benton⊙ 3,700 E7
40403 Berea 8,226 N5
41003 Berry 287 N3
41605 Betsy Layne 975 R5
41124 Blaine 358 R4
40008 Bloomfield 954 L5
†40201 Blue Ridge Manor 465 L2
42713 Bonnieville 372 K6
†40403 Boone 300 N5
41314 Booneville⊙ 191 O6
42101 Bowling Green⊙ 40,450 H7
40009 Bradfordsville 331 L6
40108 Brandenburg⊙ 1,831 J4
†42025 Briensburg E7
†40201 Broadfields 311 K2
40409 Brodhead 686 N6
†41016 Bromley 844 S2
40109 Brooks 1,344 K4
41004 Brooksville⊙ 680 N3
†40201 Brownsboro Farm 790 L1
42210 Brownsville⊙ 674 J6
40218 Buechel 6,709 K2
40310 Burgin 1,008 M5
42717 Burkesville⊙ 2,051 L7
41005 Burlington⊙ 500 R2
42519 Burnside 775 M6
41006 Butler 663 N3
42211 Cadiz⊙ 1,661 F7
42327 Calhoun⊙ 1,080 G5
41007 California 135 N3
42029 Calvert City 2,388 E6
†40337 Camargo 1,301 K4
40011 Campbellsburg 714 L3
42718 Campbellsville⊙ 8,715 L6
41301 Campton⊙ 486 O5
42721 Caneyville 642 J6
40311 Carlisle⊙ 1,757 N4
41008 Carrollton⊙ 3,967 L3
42030 Carrsville 99 E6
†42459 Caseyville 43 E5
41129 Catlettsburg⊙ 3,005 R4
42127 Cave City 2,098 K6
†41522 Cedarville 81 S6
42328 Centertown 460 G6
42330 Central City 5,214 G6
42726 Clarkson 666 J6
42404 Clay 1,356 F6
40312 Clay City 1,276 O5
40313 Clearfield 1,250 P4
42031 Clinton⊙ 1,720 D7
40111 Cloverport 1,585 H5
†41501 Coal Run 348 R5
41076 Cold Spring 2,117 T2
42728 Columbia⊙ 3,710 L6
42032 Columbus 296 C7
41729 Combs 900 P6
41131 Concord 67 P3
40701 Corbin 8,075 N7
41010 Corinth 258 M3
42406 Corydon 874 F5
*41011 Covington 49,563 S2
40419 Crab Orchard 843 M6
†41016 Crescent Springs 1,951 R2
†41076 Crestview 528 S2
†41017 Crestview Hills 1,408 R2
40014 Crestwood 531 L4
41030 Crittenden 597 M3
42217 Crofton 823 G6
40823 Cumberland 3,712 R6
41031 Cynthiana⊙ 5,881 N4
40422 Danville⊙ 12,942 M5
42408 Dawson Springs 3,275 F6
41074 Dayton 6,979 T1
†40201 Devondale 1,164 K2
42036 Dexter E7
42409 Dixon⊙ 533 F5
†40243 Douglass Hills 4,384 L2
41034 Dover 305 O3
42337 Drakesboro 798 H6
41035 Dry Ridge 1,250 M3
42037 Dycusburg 64 E6
42410 Earlington 2,011 F6
42038 Eddyville⊙ 1,949 E6
†41017 Edgewood 7,230 S2
42129 Edmonton⊙ 1,401 K7
40117 Ekron 239 J5
42701 Elizabethtown⊙ 15,380 K5
41522 Elkhorn City 1,446 S6
42220 Elkton⊙ 1,815 G7
†41018 Elsmere 7,203 R2
40019 Eminence 2,260 L4
40826 Eolia 875 R6
41018 Erlanger 14,433 R2
40827 Essie 650 P6
42567 Eubank 207 M6
40828 Evarts 1,234 P7
41039 Ewing 144 O4
40118 Fairdale 7,315 K4
40020 Fairfield 169 L5
†41101 Fairview 198 S2
41040 Falmouth⊙ 2,482 N3
41524 Fedscreek 950 S6
42533 Ferguson 1,009 M6
†40222 Fincastle 804 L1
41139 Flatwoods 8,354 R4
41816 Fleming-Neon 1,195 R6
41041 Flemingsburg⊙ 2,835 O4
41042 Florence 15,586 R2
42343 Fordsville 561 H5
41527 Forest Hills 502 L2
40121 Fort Knox 31,055 K5
41017 Fort Mitchell 7,297 S2
41075 Fort Thomas 16,012 S2
†41011 Fort Wright 4,481 S2
41043 Foster 80 N3
42133 Fountain Run 340 K7
40601 Frankfort (cap.)⊙ 25,973 M4
42134 Franklin⊙ 7,738 J7
42411 Fredonia 535 E6
40322 Frenchburg⊙ 550 O5
†41175 Fullerton 950 P3
42041 Fulton 3,137 D7
42140 Gamaliel 456 K7
40324 Georgetown⊙ 10,972 M4
41044 Germantown 347 O3
41045 Ghent 439 L3
42044 Gilbertsville E7
42141 Glasgow⊙ 12,958 J7
41046 Glencoe 354 M3
†40222 Glenview 212 K1
†40222 Goose Creek 394 L1
42045 Grand Rivers 428 E7
†41005 Grant 150 M3
40327 Gratz 124 M4
†40201 Graymoor 1,167 K1
41143 Grayson⊙ 3,423 R4
42743 Greensburg⊙ 2,377 K6
41144 Greenup⊙ 1,386 R3
42345 Greenville⊙ 4,631 G6
42234 Guthrie 1,361 G7
42413 Hanson 485 G6
42048 Hardin 545 E7
40143 Hardinsburg⊙ 2,211 H5
41531 Hardy 900 S5
40831 Harlan⊙ 3,024 P7
40330 Harrodsburg⊙ 7,265 M5
42347 Hartford⊙ 2,512 H6
42348 Hawesville⊙ 1,036 H5
41701 Hazard⊙ 5,371 P6
42049 Hazel 465 E7
40949 Heidrick 400 O7
42420 Henderson⊙ 24,834 F5
42050 Hickman⊙ 2,894 C7
42051 Hickory D7
41076 Highland Heights 4,435 T2
41822 Hindman⊙ 876 R6
42152 Hiseville 349 K6
42748 Hodgenville⊙ 2,531 K5
†40228 Hollow Creek 1,023 K4
†41018 Hopeful Heights R2
42240 Hopkinsville⊙ 27,318 F7
42749 Horse Cave 2,045 K6
†40201 Houston Acres 608 K2
40437 Hustonville 339 M6
41749 Hyden⊙ 488 P6
41051 Independence⊙ 7,998 M3
†40201 Indian Hills 787 K1
41224 Inez⊙ 413 S5
40336 Irvine⊙ 2,889 O5
40146 Irvington 1,409 J5
42350 Island 532 G6
41642 Ivel 850 R5
41339 Jackson⊙ 2,651 P5
42629 Jamestown⊙ 1,441 L7
40299 Jeffersontown 15,795 L2
40337 Jeffersonville 1,528 O5
41537 Jenkins 3,271 R6
40440 Junction City 2,045 M5
40737 Keavy 900 N6
†41011 Kenton Vale 145 S2
42053 Kevil 382 D6
†40201 Kingsley 464 K2
42055 Kuttawa 560 E6
42056 La Center 1,044 C6
41643 Lackey R6
42254 La Fayette 160 F7
40031 La Grange⊙ 2,971 L4
†41017 Lakeside Park 3,038 R2
40444 Lancaster⊙ 3,365 M5
40342 Lawrenceburg⊙ 5,167 M4
40033 Lebanon⊙ 6,590 L5
40150 Lebanon Junction 1,581 K5
42754 Leitchfield⊙ 4,533 J6
42256 Lewisburg 972 G6
42351 Lewisport 1,832 H5
*40501 Lexington⊙ 204,165 N4
Lexington‡ 318,136 N4
42539 Liberty⊙ 2,206 M6
42352 Livermore 1,672 G5
40445 Livingston 334 N6
40036 Lockport 84 M4
40741 London⊙ 4,002 N6
42001 Lone Oak 443 D6
40037 Loretto 954 L5
41230 Louisa⊙ 1,832 R4
*40201 Louisville⊙ 298,840 J2
Louisville‡ 906,240 J2
40854 Loyall 1,210 P7
41016 Ludlow 4,959 S2
40855 Lynch 1,614 R7
†40201 Lynnview 1,157 K4
40040 Mackville 229 L5
42431 Madisonville⊙ 16,979 F6
40962 Manchester⊙ 1,838 O6
42064 Marion⊙ 3,392 E6
41649 Martin 827 R5
42066 Mayfield⊙ 10,705 D7
41056 Maysville⊙ 7,983 O3
41543 McAndrews 975 S5
42354 McHenry 582 H6
40447 McKee⊙ 759 O6
41835 McRoberts 1,106 R6
†40201 Meadow Vale 1,008 L1
41059 Melbourne 628 T2
†41060 Mentor 169 N3
40965 Middlesboro 12,251 O7
40243 Middletown 414 L2
40347 Midway 1,445 M4
40348 Millersburg 987 N4
40045 Milton 718 L3
†40201 Minor Lane Heights 1,882 K4
†40359 Monterey 186 M4
42633 Monticello⊙ 5,677 M7
†40223 Moorland 513 L2
40351 Morehead⊙ 7,789 P4
42437 Morganfield⊙ 3,781 E5
42261 Morgantown⊙ 2,000 H6
42440 Mortons Gap 1,201 F6
41064 Mount Olivet⊙ 346 N3
†40437 Mount Salem 50 M6
40353 Mount Sterling⊙ 5,820 N4
40456 Mount Vernon⊙ 2,334 N6
40047 Mount Washington 3,997 K4
41548 Mouthcard 900 S6
40155 Muldraugh 1,752 J5
42765 Munfordville⊙ 1,783 J6
42071 Murray⊙ 14,248 E7
42441 Nebo 269 F6
41840 Neon-Fleming 1,195 R6
40050 New Castle⊙ 832 L4
40051 New Haven 926 K5
*41071 Newport 21,587 S2
40356 Nicholasville⊙ 10,319 N5
†40201 Northfield 906 K1
40357 North Middletown 637 N4
42442 Nortonville 1,336 G6
42262 Oak Grove 2,088 G7
42159 Oakland 264 J6
41238 Oil Springs 900 P5
40219 Okolona 20,039 K4
41164 Olive Hill 2,539 P4
42301 Owensboro⊙ 54,450 G5
Owensboro‡ 85,949 G5
40359 Owenton⊙ 1,341 M3
40360 Owingsville⊙ 1,419 O4
42001 Paducah⊙ 29,315 D6
41240 Paintsville⊙ 3,815 R5
40361 Paris⊙ 7,935 N4
42160 Park City 614 J6
†41011 Park Hills 3,500 S2
†40201 Parkway Village 754 J2
42266 Pembroke 636 G7
40468 Perryville 841 M5
40056 Pewee Valley 982 L4
41553 Phelps 1,126 S6
41501 Pikeville⊙ 4,756 S6
42635 Pine Knot 1,389 M7
40977 Pineville⊙ 2,599 O7
†40201 Plantation 969 K1
40258 Pleasure Ridge Park 27,332 J4
40057 Pleasureville 837 L4
†42101 Plum Springs 393 J7
42367 Powderly 848 G6
41653 Prestonsburg⊙ 4,011 R5
†41008 Prestonville 205 L3
42445 Princeton⊙ 7,073 F6
40059 Prospect 1,981 K4
42450 Providence 4,434 F6
41169 Raceland 1,970 R3
40160 Radcliff 14,519 K5
40472 Ravenna 793 O5
40475 Richmond⊙ 21,705 N5
†40222 Riverwood 435 K1
42273 Rochester 289 H6

Agriculture, Industry and Resources

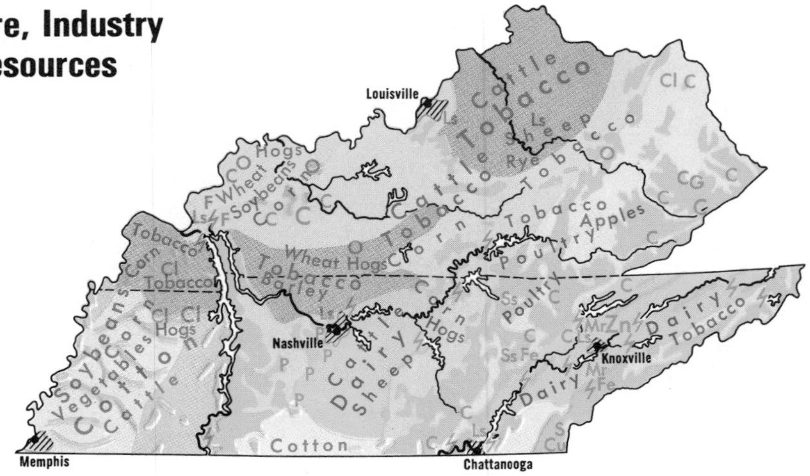

DOMINANT LAND USE

- Hogs, Soft Winter Wheat
- Tobacco, General Farming
- General Farming, Livestock, Tobacco
- General Farming, Livestock, Dairy
- General Farming, Livestock, Fruit, Tobacco
- Specialized Cotton
- Cotton, General Farming
- Cotton, Livestock
- Forests
- Swampland, Limited Agriculture

MAJOR MINERAL OCCURRENCES

C	Coal	G	Natural Gas	P	Phosphates
Cl	Clay	Ls	Limestone	S	Pyrites
Cu	Copper	Mr	Marble	Ss	Sandstone
F	Fluorspar	O	Petroleum	Zn	Zinc
Fe	Iron Ore				

ϟ Water Power

//// Major Industrial Areas

KENTUCKY

AREA 40,409 sq. mi. (104,659 sq. km.)
POPULATION 3,660,257
CAPITAL Frankfort
LARGEST CITY Louisville
HIGHEST POINT Black Mtn. 4,145 ft. (1263 m.)
SETTLED IN 1774
ADMITTED TO UNION June 1, 1792
POPULAR NAME Bluegrass State
STATE FLOWER Goldenrod
STATE BIRD Cardinal

TENNESSEE

AREA 42,144 sq. mi. (109,153 sq. km.)
POPULATION 4,591,120
CAPITAL Nashville
LARGEST CITY Memphis
HIGHEST POINT Clingmans Dome 6,643 ft. (2025 m.)
SETTLED IN 1757
ADMITTED TO UNION June 1, 1796
POPULAR NAME Volunteer State
STATE FLOWER Iris
STATE BIRD Mockingbird

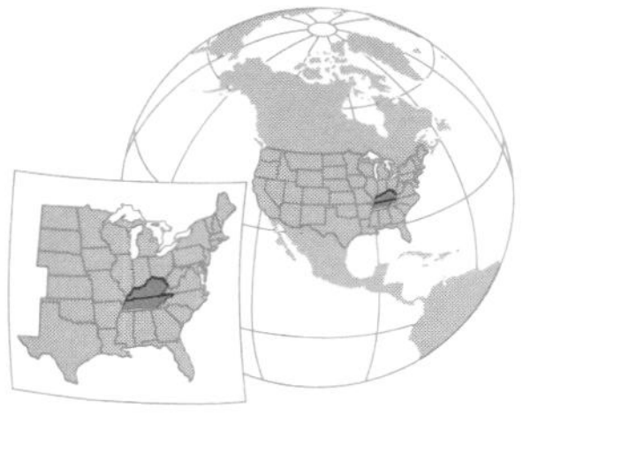

42369 Rockport 511 H6
†40201 Rolling Fields 731 K2
†40201 Rolling Hills 1,122 L1
41169 Russell 3,824 R3
42642 Russell Springs 1,831 L6
42276 Russellville⊙ 7,520 H7
†41015 Ryland Heights 252 M3
42372 Sacramento 538 G6
40370 Sadieville 253 M4
42453 Saint Charles 405 F6
40207 Saint Matthews 13,519 K2
†40201 Saint Regis Park 1,735 K2
42078 Salem 833 E6
40371 Salt Lick 347 O4
41465 Salyersville⊙ 1,352 P5
41083 Sanders 332 M3
41171 Sandy Hook⊙ 627 P4
41056 Sardis 198 O3
42553 Science Hill 655 M6
42164 Scottsville⊙ 4,278 J7
42455 Sebree 1,516 F5
†40201 Seneca Gardens 748 K2
40983 Sextons Creek 975 O6
40374 Sharpsburg 339 O4
40065 Shelbyville⊙ 5,329 L4
40165 Shepherdsville⊙ 4,454 K4
40216 Shively 16,819 K4
41085 Silver Grove 1,260 T2
40067 Simpsonville 642 L4
42456 Slaughters 269 F6
41764 Smilax 987 P6
40068 Smithfield 137 L4
42081 Smithland⊙ 512 E6
42171 Smiths Grove 767 J6
42501 Somerset⊙ 10,649 M6
42776 Sonora 416 K5
42374 South Carrollton 262 G6
41071 Southgate 2,833 T2
41174 South Portsmouth 900 P3
41175 South Shore 1,525 R3
25661 South Williamson 1,016 S5
41086 Sparta 192 M3
42458 Spottsville 914 G5
40069 Springfield⊙ 3,179 L5
†40201 Springlee 498 K2
40379 Stamping Ground 562 M4
40484 Stanford⊙ 2,764 M5
40380 Stanton⊙ 2,691 O5
42647 Stearns 1,557 N7
41567 Stone 900 S5
†40201 Strathmoor Village 466 J2
42459 Sturgis 2,293 F5
†41011 Taylor Mill 4,509 S2
40071 Taylorsville⊙ 801 L4
†40222 Thornhill 233 K1
41189 Tollesboro 808 O3
42167 Tompkinsville⊙ 4,366 K7
42286 Trenton 465 G7
41091 Union 601 M3
42461 Uniontown 1,169 F5
42784 Upton 731 K6
40272 Valley Station 24,474 K4
41179 Vanceburg⊙ 1,939 P3
41265 Van Lear 2,035 R5
†40828 Verda 1,133 P7
40383 Versailles⊙ 6,427 M4
41773 Vicco 456 P6
†41017 Villa Hills 4,402 R2
40175 Vine Grove 3,583 K5
†41063 Visalia 198 N3
40873 Wallins Creek 459 O7
41094 Walton 1,651 M3
41095 Warsaw⊙ 1,328 M3
41096 Washington 624 O3
42085 Water Valley 395 D7
42462 Waverly 434 F5
41666 Wayland 601 R6
41667 Weeksbury 850 R6
†40201 Wellington 653 K2
†40218 West Buechel 1,205 K2
41472 West Liberty⊙ 1,381 P5
40177 West Point 1,339 J4
†42501 West Somerset 850 M6
41101 Westwood 5,973 R4
†40207 Westwood 826 L1
42463 Wheatcroft 325 F5
41669 Wheelwright 865 R6
41390 Whick 280 P6
42464 White Plains 859 G6
41858 Whitesburg⊙ 1,525 R6
42378 Whitesville 788 H5
42653 Whitley City⊙ 1,683 N7
42087 Wickliffe⊙ 1,034 C7
†41071 Wilders 633 S2
40769 Williamsburg⊙ 5,560 N7
41097 Williamstown⊙ 2,502 M3
40078 Willisburg 235 L5
40390 Wilmore 3,787 M5
40391 Winchester⊙ 15,216 N5
†40201 Windy Hills 2,214 K1
42088 Wingo 606 D7
40771 Woodbine 900 N7
42170 Woodburn 330 J7
†40201 Woodland Hills 839 L2
†42001 Woodlawn-Oakdale 4,722 D6
†41071 Woodlawn 331 T2
†40201 Woodlawn Park 1,052 K2
41183 Worthington 1,948 R3
41098 Worthville 272 L3
41144 Wurtland 1,301 R3

OTHER FEATURES

Abraham Lincoln Birthplace Nat'l Hist. Site K5
Barkley (dam) E6
Barkley (lake) F7
Barren (riv.) H6
Barren River (lake) J7
Beech Fork (riv.) L5
Big Sandy (riv.) R4
Black (mt.) R7
Buckhorn (lake) O6
Chaplin (riv.) L5
Clarks, East Fork (riv.) E7
Cove Run (lake) O4
Cumberland (lake) M7
Cumberland (mt.) P7
Cumberland (riv.) K8
Cumberland Gap Nat'l Hist. Park P7
Dale Hollow (lake) L7
Dewey (lake) R5
Dix (riv.) M5
Drakes (creek) J7
Dry (creek) R3
Eagle (creek) M3
Fishtrap (lake) S6
Fort Campbell G7
Grayson (lake) P4
Green (riv.) G6
Green River (lake) L6
Herrington (lake) M5
Hinkston (creek) N4
Kentucky (dam) E7
Kentucky (lake) E8
Kentucky (riv.) M3
Land Between The Lakes Rec. Area E7
Laurel River (lake) N6
Lexington Blue Grass Army Depot N5
Licking (riv.) N3
Mammoth Cave Nat'l Park J6
Mayfield (creek) C7
Mississippi (riv.) A10
Mud (riv.) H7
Nolin (lake) K6
Nolin (riv.) J6
Obion (creek) C7
Ohio (riv.) F5
Paint Lick (riv.) M5
Panther (creek) G5
Pine (mt.) O7
Pond (riv.) G6
Red (riv.) O5
Red (riv.) G7
Rockcastle (riv.) N6
Rolling Fork (riv.) L5
Rough (riv.) H5
Rough River (lake) J5
Salt (riv.) K5
Tennessee (riv.) D6
Tradewater (riv.) F6
Tug Fork (riv.) S5

TENNESSEE

COUNTIES

Anderson 67,346 N8
Bedford 27,916 J9
Benton 14,901 E8
Bledsoe 9,478 L9
Blount 77,770 O9
Bradley 67,547 M10
Campbell 34,923 N8
Cannon 10,234 J9
Carroll 28,285 E9
Carter 50,205 S8
Cheatham 21,616 G8
Chester 12,727 D10
Claiborne 24,595 O8
Clay 7,676 K7
Cocke 28,792 P9
Coffee 38,311 J9
Crockett 14,941 C9
Cumberland 28,676 L9
Davidson 477,811 H8
Decatur 10,857 E9
De Kalb 13,589 K9
Dickson 30,037 G8
Dyer 34,663 C8
Fayette 25,305 C10
Fentress 14,826 M8
Franklin 31,983 J10
Gibson 49,467 D9
Giles 24,625 G10
Grainger 16,751 O8
Greene 54,422 R8
Grundy 13,787 K10
Hamblen 49,300 P8
Hamilton 287,740 L10
Hancock 6,887 P7
Hardeman 23,873 C10
Hardin 22,280 E10
Hawkins 43,751 P8
Haywood 20,318 C9
Henderson 21,390 E9
Henry 28,656 E8
Hickman 15,151 G9
Houston 6,871 F8
Humphreys 15,957 F8
Jackson 9,398 K8
Jefferson 31,284 P8
Johnson 13,745 T7
Knox 319,694 O9
Lake 7,455 B8
Lauderdale 24,555 B9
Lawrence 34,110 G10
Lewis 9,700 F9
Lincoln 26,483 H10
Loudon 28,553 N9
Macon 15,700 J7
Madison 74,546 D9
Marion 24,416 K10
Marshall 19,698 H10
Maury 51,095 G9
McMinn 41,878 M10
McNairy 22,525 D10
Meigs 7,431 M9
Monroe 28,700 N10
Montgomery 83,342 G8
Moore 4,510 J10
Morgan 16,604 M8
Obion 32,781 C8
Overton 17,575 L8
Perry 6,111 F9
Pickett 4,358 M7
Polk 13,602 N10
Putnam 47,690 K8
Rhea 24,235 M9
Roane 48,425 M9
Robertson 37,021 H7
Rutherford 84,058 J9
Scott 19,259 M8
Sequatchie 8,605 L10
Sevier 41,418 O9
Shelby 777,113 B10
Smith 14,935 J8
Stewart 8,665 F7
Sullivan 143,968 S7
Sumner 85,790 J8
Tipton 32,930 B9
Trousdale 6,137 J8
Unicoi 16,362 S8
Union 11,707 O8
Van Buren 4,728 L9
Warren 32,653 K9
Washington 88,755 R8
Wayne 13,946 F10
Weakley 32,896 D8
White 19,567 L9
Williamson 58,108 H9
Wilson 56,064 J8

CITIES and TOWNS

Zip Name/Pop. Key

†38301 Adair 70 D9
37010 Adams 600 G7
38310 Adamsville 1,453 E10
38001 Alamo⊙ 2,615 C9
37701 Alcoa 6,870 N9
37012 Alexandria 689 J8
38501 Algood 2,406 K8
38504 Allardt 654 M8
37301 Altamont⊙ 679 K10
38449 Ardmore 835 H10
38002 Arlington 1,778 B10
37015 Ashland City⊙ 2,329 G8
37303 Athens⊙ 12,080 M10
38004 Atoka 691 B10
38220 Atwood 1,143 D9
37016 Auburntown 204 J9
37743 Baileyton 333 R8
†37650 Banner Hill 2,913 R8
38134 Bartlett 17,170 B10
38544 Baxter 1,411 K8
37305 Beersheba Springs 643 K10
37020 Bell Buckle 450 J9
37205 Belle Meade 3,182 H8
38006 Bells 1,571 C9
37307 Benton⊙ 1,115 M10
†37201 Berry Hill 1,113 H8
†37027 Berry's Chapel 2,703 H9
38315 Bethel Springs 873 D10
38221 Big Sandy 650 E8
37709 Blaine 1,147 O8
37660 Bloomingdale 12,088 R7
37617 Blountville⊙ 2,554 S7
37618 Bluff City 1,121 S8
38008 Bolivar⊙ 6,597 C10
38010 Braden 293 B10
38316 Bradford 1,146 D8
37027 Brentwood 9,431 H8
37710 Briceville 850 N8
38011 Brighton 976 B10
37620 Bristol 23,986 S7
38012 Brownsville⊙ 9,307 C9
38317 Bruceton 1,579 E8
37711 Bulls Gap 821 P8
38015 Burlison 386 B9
37029 Burns 777 G8
38549 Byrdstown⊙ 884 L7
37309 Calhoun 590 M10
38320 Camden⊙ 3,279 E8
37030 Carthage⊙ 2,672 K8
37714 Caryville 2,039 N8
37032 Cedar Hill 420 H7
38551 Celina⊙ 1,580 K7
†37110 Centertown 300 K9
37033 Centerville⊙ 2,824 G9
37034 Chapel Hill 861 H9
37310 Charleston 756 M10
37036 Charlotte⊙ 788 G8
*37401 Chattanooga⊙ 169,558 K10
Chattanooga‡ 426,540 K10
37642 Church Hill 4,110 R7
38324 Clarksburg 400 E9
37040 Clarksville⊙ 54,777 G7
Clarksville‡ 150,220 G7
37311 Cleveland⊙ 26,415 M10
38425 Clifton 773 F10
37716 Clinton⊙ 5,245 N8
37313 Coalmont 625 K10
37315 Collegedale 4,607 M10
38017 Collierville 7,839 B10
38450 Collinwood 1,064 F10
37663 Colonial Heights 6,744 R8
38401 Columbia⊙ 26,571 G9
37720 Concord 8,569 N9
38501 Cookeville⊙ 20,535 L8
37317 Copperhill 418 N10
37047 Cornersville 722 H10
38224 Cottage Grove 117 E8
38326 Counce 975 E10
38019 Covington⊙ 6,065 B9
37318 Cowan 1,790 K10
37723 Crab Orchard 1,065 M9
37049 Cross Plains 655 H7
38555 Crossville⊙ 6,394 L9
37050 Cumberland City 276 F8
37724 Cumberland Gap 263 O8
37725 Dandridge⊙ 1,383 O8
37321 Dayton⊙ 5,913 L9
37322 Decatur⊙ 1,069 M9
38329 Decaturville⊙ 1,004 E9
37324 Decherd 2,233 J10
38391 Denmark 51 D9
37055 Dickson 7,040 G8
37058 Dover⊙ 1,197 F8
37059 Dowelltown 341 K8
38559 Doyle 344 K9
38225 Dresden⊙ 2,256 D8
37326 Ducktown 583 N10
37327 Dunlap⊙ 3,681 L10
38330 Dyer 2,419 D8
38024 Dyersburg⊙ 15,856 C8
†37801 Eagleton Village 5,331 O9
37060 Eagleville 444 H9
37412 East Ridge 21,236 L11
†38367 Eastview 552 D10
37643 Elizabethton⊙ 12,431 S8
38455 Elkton 540 H10
38029 Ellendale 850 B10
37329 Englewood 1,840 M10
38332 Enville 287 E10
37061 Erin⊙ 1,614 F8
37650 Erwin⊙ 4,739 S8
37330 Estill Springs 1,324 J10
38456 Ethridge 548 G10
37331 Etowah 3,758 M10
37062 Fairview 3,648 G9
37656 Fall Branch 1,340 R8
37334 Fayetteville⊙ 7,559 H10
38334 Finger 245 D10
38030 Finley 1,014 B8
†37201 Forest Hills 4,516 H8
37064 Franklin⊙ 12,407 H9
38034 Friendship 763 C9
37737 Friendsville 694 N9
38337 Gadsden 683 D9
38562 Gainesboro⊙ 1,119 K8
37066 Gallatin⊙ 17,191 H8
38036 Gallaway 804 B10
†38019 Garland 301 B9
38037 Gates 729 C9
37738 Gatlinburg 3,210 O9
38138 Germantown 21,482 B10
38338 Gibson 458 D9
†38015 Gilt Edge 142 B9
38229 Gleason 1,335 D8
37072 Goodlettsville 8,327 H8
38563 Gordonsville 893 K8
38039 Grand Junction 360 C10
37338 Graysville 1,380 L10
37742 Greenback 546 N9
37073 Greenbrier 3,180 H8
37743 Greeneville⊙ 14,097 R8
38230 Greenfield 2,109 D8
37339 Gruetli 910 K10
38040 Halls 2,444 C9
37658 Hampton 2,236 S8
37748 Harriman 8,303 M9
37341 Harrison 6,206 L10
37752 Harrogate-Shawanee 2,530 O8
37074 Hartsville⊙ 2,674 J8
38340 Henderson⊙ 4,449 D10
37075 Hendersonville 26,561 H8
38041 Henning 638 B9
38231 Henry 295 E8
38042 Hickory Valley 252 C10
38462 Hohenwald⊙ 3,922 F9
38342 Hollow Rock 955 E8
38232 Hornbeak 452 C8
38044 Hornsby 401 D10
38343 Humboldt 10,209 D9
38344 Huntingdon⊙ 3,962 E8
37345 Huntland 983 J10
37756 Huntsville⊙ 519 N8
37078 Hurricane Mills 850 F9
38463 Iron City 482 F10
37757 Jacksboro⊙ 1,722 N8
38301 Jackson⊙ 49,131 D9
38556 Jamestown⊙ 2,364 M8
37347 Jasper⊙ 2,633 K10
37760 Jefferson City 5,612 P8
37762 Jellico 2,798 N7
37601 Johnson City 39,753 S8
Johnson City-Kingspoint-Bristol‡ 433,638 S8
37659 Jonesboro⊙ 2,829 R8
37921 Karns 1,173 N9
38233 Kenton 1,551 C8
†37347 Kimball 1,220 K10
*37660 Kingsport 32,027 R7
37763 Kingston⊙ 4,441 N9
37082 Kingston Springs 1,017 G8
*37901 Knoxville⊙ 175,045 O9
Knoxville‡ 476,517 O9
37083 Lafayette⊙ 3,808 J7
37766 La Follette 8,198 N8
38046 La Grange 185 C10
37769 Lake City 2,335 N8
†38134 Lakeland 612 B10
†37379 Lakesite 651 L10
†37138 Lakewood 2,325 H8
37086 La Vergne 5,495 H9
38464 Lawrenceburg⊙ 10,184 G10
37087 Lebanon⊙ 11,872 J8
37771 Lenoir City 5,446 N9
37091 Lewisburg⊙ 8,760 H10
38351 Lexington⊙ 5,934 E9
37095 Liberty 365 K8
37096 Linden⊙ 1,087 F9
38570 Livingston⊙ 3,372 L8
37097 Lobelville 993 F9
37350 Lookout Mountain 1,886 L11
38469 Loretto 1,612 G10
37774 Loudon⊙ 3,943 N9
37779 Luttrell 962 O8
37352 Lynchburg⊙ 668 J10
38472 Lynnville 383 G10
37354 Madisonville⊙ 2,884 N9
37355 Manchester⊙ 7,250 J10
38237 Martin 8,898 D8
37801 Maryville⊙ 17,480 O9
37806 Mascot 2,203 O8
38049 Mason 471 B10
38050 Maury City 989 C9
37807 Maynardville⊙ 924 O8
37101 McEwen 1,352 F8
38201 McKenzie 5,405 E8
38235 McLemoresville 311 D9
37110 McMinnville⊙ 10,683 K9
38355 Medina 687 D9
38356 Medon 169 D10
*38101 Memphis⊙ 646,174 B10
Memphis‡ 912,887 B10
38357 Michie 530 E10
38052 Middleton 596 D10
38358 Milan 8,083 D9
38359 Milledgeville 392 E10
38053 Millington 20,236 B10
38473 Minor Hill 564 G10
37119 Mitchellville 209 J7
37356 Monteagle 1,126 K10
38574 Monterey 2,610 L8
37357 Morrison 587 K9
†37660 Morrison City 2,032 R7
37814 Morristown⊙ 19,683 P8
38057 Moscow 499 C10
37818 Mosheim 1,539 R8
37683 Mountain City⊙ 2,125 T8
37642 Mount Carmel 3,764 R8
37122 Mount Juliet 2,879 H8
38474 Mount Pleasant 3,375 G9
38058 Munford 2,336 B10
37130 Murfreesboro⊙ 32,845 J9
*37201 Nashville (cap.)⊙ 455,651 H8
Nashville-Davidson‡ 850,505 H8
38059 Newbern 2,794 C8
†37380 New Hope 681 K11
37134 New Johnsonville 1,824 E8
37820 New Market 1,216 O8
37821 Newport⊙ 7,580 P9
37825 New Tazewell 1,677 O8
37826 Niota 765 M9
37360 Normandy 118 J10
37828 Norris 1,374 N8
37829 Oakdale 323 M9
†37201 Oak Hill 4,609 H8
38060 Oakland 472 B10
37830 Oak Ridge 27,662 N8
38240 Obion 1,282 C8
37840 Oliver Springs 3,659 N8
37841 Oneida 3,717 N7
37363 Ooltewah 950 M10
†37660 Orebank 1,284 R7
37141 Orlinda 382 H7
35740 Orme 181 K10
37365 Palmer 1,027 K10
38242 Paris⊙ 10,728 E8
37843 Parrottsville 118 P8
38363 Parsons 2,422 E9
37143 Pegram 1,081 H8
37144 Petersburg 681 H10
37845 Petros 1,286 M8
37846 Philadelphia 507 M9
37863 Pigeon Forge 1,822 O9
37367 Pikeville⊙ 2,085 L9
†38017 Piperton 746 B10
†37738 Pittman Center 488 P9
38578 Pleasant Hill 371 L9
37148 Portland 4,030 H7
37849 Powell 7,220 N8
†37397 Powells Crossroads 918 L10
38478 Pulaski⊙ 7,184 G10
38251 Puryear 624 E8
38367 Ramer 429 D10
37415 Red Bank 13,299 L10
37150 Red Boiling Springs 1,173 K7
†37641 Rheatown R8
†37380 Richard City 87 K11
38080 Ridgely 1,932 B8
†37401 Ridgeside 417 L10
37152 Ridgetop 1,225 H8
38063 Ripley⊙ 6,366 B9
38253 Rives 386 C8
37687 Roan Mountain 1,108 S8
37853 Rockford 567 O9
37854 Rockwood 5,767 M9
37857 Rogersville⊙ 4,368 P8
38053 Rosemark 950 B10
38066 Rossville 379 B10
37860 Russellville 1,069 P8
38369 Rutherford 1,378 C8
37861 Rutledge⊙ 1,058 P8
38481 Saint Joseph 897 G10
37373 Sale Creek 900 L10
38370 Saltillo 434 E10
38254 Samburg 465 C8
38371 Sardis 301 E10
38067 Saulsbury 156 C10
38372 Savannah⊙ 6,992 E10
38374 Scotts Hill 668 E10
38375 Selmer⊙ 3,979 D10
37862 Sevierville⊙ 4,556 P9
37375 Sewanee 2,298 K10
38255 Sharon 1,134 D8
37160 Shelbyville⊙ 13,530 H10
37376 Sherwood 900 K10
37377 Signal Mountain 5,818 L10
38377 Silerton 100 D10
37165 Slayden 69 G8
37166 Smithville⊙ 3,839 K9
37167 Smyrna 8,839 H9
37869 Sneedville⊙ 1,110 P7
37319 Soddy-Daisy 8,388 L10
38068 Somerville⊙ 2,264 C10
†37030 South Carthage 1,004 K8
†37311 South Cleveland 4,360 M10
†37716 South Clinton 1,671 N8
†42041 South Fulton 2,735 D8
37380 South Pittsburg 3,636 K10
37171 Southside 800 G8
38583 Sparta⊙ 4,864 K9
38585 Spencer⊙ 1,126 L9
37381 Spring City 1,951 M9
37172 Springfield⊙ 10,814 H8
37174 Spring Hill 989 H9
38069 Stanton 540 C10
38379 Stantonville 271 E10
†37660 Sullivan Gardens 2,513 R8
38483 Summertown 850 G10
37873 Surgoinsville 1,536 R8
37874 Sweetwater 4,725 N9
37877 Talbott 975 P8
37879 Tazewell⊙ 2,090 O8
37385 Tellico Plains 698 N10
37178 Tennessee Ridge 1,325 F8
38079 Tiptonville⊙ 2,438 B8
38381 Toone 355 D10
37882 Townsend 351 O9
37387 Tracy City 1,356 K10
38382 Trenton⊙ 4,601 D9
38258 Trezevant 921 D8
38259 Trimble 722 C8
38260 Troy 1,093 C8
37388 Tullahoma 15,800 J10
37743 Tusculum 1,242 R8
38261 Union City⊙ 10,436 C8
37181 Vanleer 401 G8
†37397 Victoria 800 K10
37394 Viola 149 K9

(continued on following page)

37885 Vonore 528		N9
†37377 Walden 1,293		L10
37887 Wartburg⊙ 761		M8
37183 Wartrace 540		J9
37694 Watauga 376		S8
37184 Watertown 1,300		J8
37185 Waverly⊙ 4,405		F8
38485 Waynesboro⊙ 2,109		F10
38074 Western Institute 850		C10
37186 Westmoreland 1,754		J7
37187 White Bluff 2,055		G8
37188 White House 2,225		H8
37890 White Pine 1,900		P8
38075 Whiteville 1,270		C10
37397 Whitwell 1,783		K10
38076 Williston 395		C10
37398 Winchester⊙ 5,821		J10
37190 Woodbury⊙ 2,160		J9
38271 Woodland Mills 526		C8
38389 Yorkville 272		C8

OTHER FEATURES

Andrew Johnson Nat'l Hist. Site		R8
Appalachian (mts.)		M10
Bald (mts.)		R9
Barkley (lake)		F7
Big Sandy (riv.)		E9
Boone (lake)		S8
Buffalo (riv.)		F9
Caney Fork (riv.)		L9
Center Hill (lake)		K9
Cheatham (dam)		G8
Cheatham (lake)		H8
Cherokee (dam)		P8
Cherokee (lake)		P8
Chickamauga (dam)		L10
Chickamauga (lake)		L10
Chilhowee (mt.)		O9
Clinch (riv.)		N9
Clingmans Dome (mt.)		P10
Collins (riv.)		K9
Conasauga (riv.)		M11
Cordell Hull (res.)		K8
Cumberland (plat.)		L9
Cumberland (riv.)		K8
Cumberland Gap Nat'l Hist. Park		O7
Dale Hollow (lake)		L7
Douglas (lake)		P9
Duck (riv.)		F9
Elk (riv.)		H10
Emory (riv.)		M8
Forked Deer (riv.)		C9
Fort Campbell		G7
Fort Donelson Nat'l Mil. Park		F8
Fort Loudoun (lake)		N9
French Broad (riv.)		R9
Great Falls (dam)		K9
Great Smoky (mts.)		P9
Great Smoky Mountains Nat'l Park		P9
Green (riv.)		F10
Guyot (mt.)		P9
Harpeth (riv.)		G8
Hatchie (riv.)		B9
Hiwassee (riv.)		O10
Holston (riv.)		O8
Iron (mts.)		S8
Kentucky (lake)		E8
Land Between The Lakes Rec. Area		E7
Lick (creek)		R8
Little Tennessee (riv.)		N10
Loosahatchie (riv.)		B10
Melton Hill (lake)		N9
Memphis Naval Air Sta.		B10
Meriwether Lewis Park, Natchez Trace Pkwy		G10
Mississippi (riv.)		A10
Nolichucky (riv.)		R8
Norris (dam)		N8
Norris (lake)		O8
Obed (riv.)		M8
Obion (riv.)		C8
Ocoee (riv.)		M10
Old Hickory (dam)		H8
Old Hickory (lake)		J8
Pickwick (lake)		E11
Powell (riv.)		P8
Priest, J. Percy (lake)		J8
Red (riv.)		G7
Reelfoot (lake)		C8
Richland (creek)		G10
Rutherford Fork, Obion (riv.)		D8
Sequatchie (riv.)		L10
Sewart A.F.B.		J8
Shiloh Nat'l Mil. Park		E10
Shoal (creek)		F10
South Holston (lake)		S7
Stone (mts.)		T8
Stones (riv.)		H9
Stones River Nat'l Battlefield		H9
Sulphur Fork, Red (riv.)		H8
Tellico (riv.)		N10
Tennessee (riv.)		E10
Tims Ford (lake)		J10
Unaka (mts.)		S8
Unicoi (mts.)		N10
Watauga (lake)		T8
Watts Bar (dam)		M9
Watts Bar (lake)		M9
Whiteoak (creek)		F8
Wolf (riv.)		B10
Woods (res.)		J10
Yellow (creek)		F8

⊙County seat.
‡Population of metropolitan area.
† Zip of nearest p.o.
* Multiple zips.

Topography

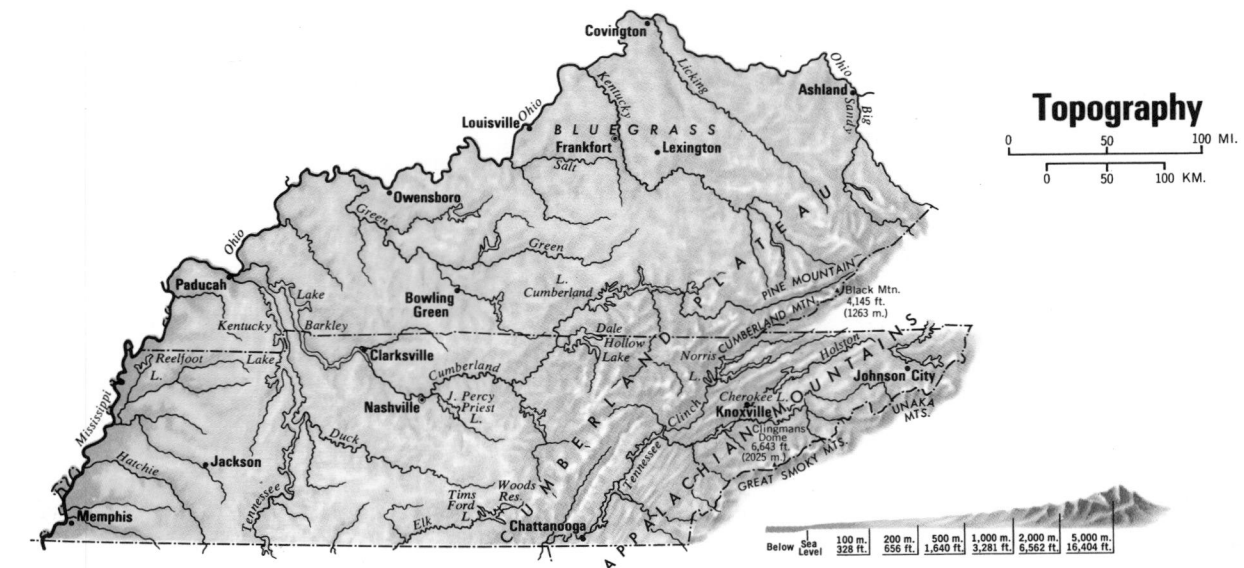

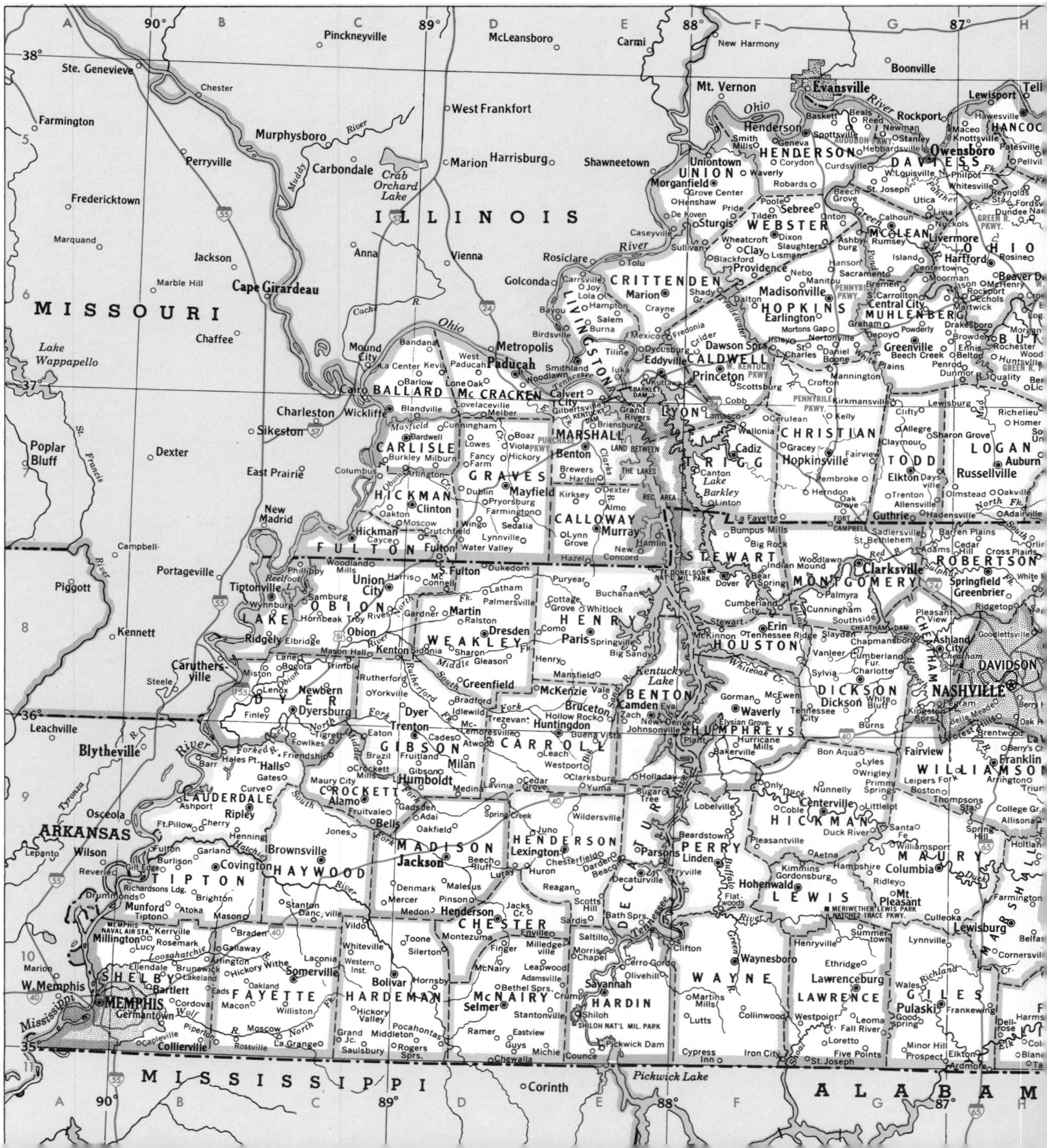

Kentucky and Tennessee

SCALE
0 5 10 20 30 40MI
0 5 10 20 30 40KM.

State Capitals ... ⊛
County Seats ... ◉
Major Limited Access Hwys. ———

Scale 1:1,970,000

© Copyright HAMMOND INCORPORATED, Maplewood, N.J.

Topography

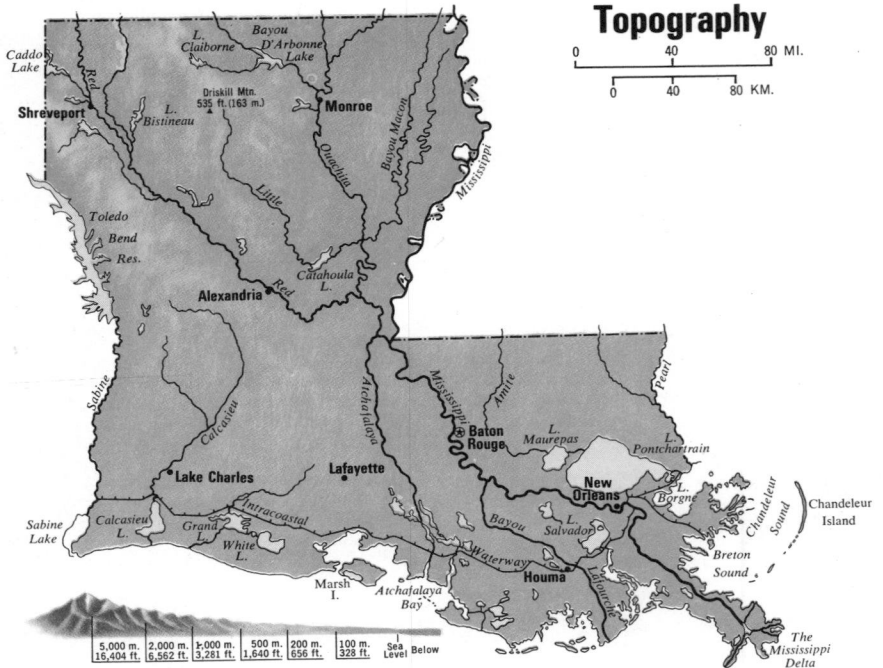

5,000 m. 2,000 m. 1,000 m. 500 m. 200 m. 100 m. Sea Below
16,404 ft. 6,562 ft. 3,281 ft. 1,640 ft. 656 ft. 328 ft. Level

PARISHES

Acadia 56,427	F6
Allen 21,390	E5
Ascension 50,068	J6
Assumption 22,084	H7
Avoyelles 41,393	G4
Beauregard 29,692	D5
Bienville 16,387	D2
Bossier 80,721	C1
Caddo 252,358	C1
Calcasieu 167,223	D6
Caldwell 10,761	F2
Cameron 9,336	D7
Catahoula 12,287	G3
Claiborne 17,095	D1
Concordia 22,981	G4
De Soto 25,727	C2
East Baton Rouge 366,191	K1
East Carroll 11,772	H1
East Feliciana 19,015	H5
Evangeline 33,343	F5
Franklin 24,141	G2
Grant 16,703	E3
Iberia 63,752	G7
Iberville 32,159	H6
Jackson 17,321	E2
Jefferson 454,592	K7
Jefferson Davis 32,168	E6
Lafayette 150,017	F6
Lafourche 82,483	K7
La Salle 17,004	F3
Lincoln 39,763	E1
Livingston 58,806	L2
Madison 15,975	H2
Morehouse 34,803	G1
Natchitoches 39,863	D3
Orleans 557,515	L6
Ouachita 139,241	F2
Plaquemines 26,049	L8
Pointe Coupee 24,045	G5
Rapides 135,282	E4
Red River 10,433	D2
Richland 22,187	G2
Sabine 25,280	C3
Saint Bernard 64,097	L7
Saint Charles 37,259	K7
Saint Helena 9,827	J5
Saint James 21,495	L3
Saint John the Baptist 31,924	M3
Saint Landry 84,128	F5
Saint Martin 40,214	G6
Saint Mary 64,253	H7
Saint Tammany 110,869	L6
Tangipahoa 80,698	K5
Tensas 8,525	H2
Terrebonne 94,393	J8
Union 21,167	F1
Vermilion 48,458	F7
Vernon 53,475	D4
Washington 44,207	K5
Webster 43,631	D1
West Baton Rouge 19,086	H6
West Carroll 12,922	H1
West Feliciana 12,186	H5
Winn 17,253	E3

CITIES and TOWNS

Zip	Name/Pop.	Key
70510	Abbeville⊙ 12,391	F7
70420	Abita Springs 1,072	L6
71316	Acme 235	G4
70710	Addis 1,320	J2
71401	Aimwell 55	G3
70421	Akers 150	N2

Zip	Name/Pop.	Key
70711	Albany 857	M1
71301	Alexandria⊙ 51,565	E4
	Alexandria‡ 151,985	E4
†70458	Alton 500	L6
70340	Amelia 3,617	H7
70422	Amite⊙ 4,301	K5
71403	Anacoco 820	D4
70426	Angie 311	L5
70712	Angola 600	G5
70032	Arabi 10,248	P4
71001	Arcadia⊙ 3,403	E1
71218	Archibald 425	G2
70512	Arnaudville 1,679	G6
71002	Ashland 307	D2
71003	Athens 419	E1
71404	Atlanta 127	E3
70513	Avery Island 500	G7
70714	Baker 12,865	K1
70514	Baldwin 2,644	H7
71405	Ball 3,405	F4
70036	Barataria 1,123	K7
70515	Basile 2,635	E5
71219	Baskin 286	G2
71220	Bastrop⊙ 15,527	G1
70715	Batchelor 500	G5
*70801	Baton Rouge (cap.)⊙ 219,419	K2
	Baton Rouge‡ 493,973	K2
†70360	Bayou Cane 15,723	J7
†70380	Bayou Vista 5,805	H7
71004	Belcher 436	C1
70630	Bell City 400	D6
70037	Belle Chasse 5,412	O4
71406	Belmont 350	C3
71407	Bentley 120	E3
71006	Benton⊙ 1,864	C1
†70558	Bermuda 50	D3
71222	Bernice 1,956	E1
70342	Berwick 4,466	H7
71007	Bethany 300	B2
71008	Bienville 249	D2
71009	Blanchard 1,128	C1
70427	Bogalusa 16,976	L5
71223	Bonita 503	G1
†71064	Bolinger 200	C1
71320	Bordelonville 350	G4
*71111	Bossier City 50,817	C1
70343	Bourg 2,073	J7
71409	Boyce 1,198	E4
70040	Braithwaite 350	P4
70516	Branch 200	F6
70517	Breaux Bridge 5,922	G6
70718	Brittany 475	L3
70518	Broussard 2,923	F6
70719	Brusly 1,762	J2
71220	Bryceland 94	E2
71321	Buckeye 280	F4
71322	Bunkie 5,364	F5
70041	Buras-Triumph 4,137	L8
70519	Cade 175	G6
71225	Calhoun 350	F2
71410	Calvin 263	E3
70631	Cameron⊙ 1,736	D7
71411	Campti 1,069	D3
†70584	Cankton 303	F6
70520	Carencro 3,712	G6
70042	Carlisle 975	L7
70721	Carville 1,037	K3
71015	Caspiana 50	C2
71016	Castor 195	D2
70522	Centerville 600	H7
70043	Chalmette⊙ 33,847	P4
†70767	Chamberlin 20	J1
71324	Chase 200	G2
70524	Chataignier 431	F5

Zip	Name/Pop.	Key
71226	Chatham 714	F2
70344	Chauvin 3,338	J8
71325	Cheneyville 865	F4
71412	Chopin 175	E4
71227	Choudrant 809	F1
70525	Church Point 4,599	F6
71414	Clarence 612	E3
71415	Clarks 931	F2
71326	Clayton 1,204	H3
70722	Clinton⊙ 1,919	J5
71416	Cloutierville 100	E3
71417	Colfax⊙ 1,680	E3
71229	Collinston 439	G1
71418	Columbia⊙ 687	F2
70723	Convent⊙ 400	L3
71419	Converse 449	C3
†71107	Cooper Road	C1
71327	Cottonport 1,911	F5
71018	Cotton Valley 1,445	D1
71019	Coushatta⊙ 2,084	D2
70433	Covington⊙ 7,892	K5
†70510	Cow Island 200	F7
†70656	Cravens 200	E5
71020	Creston 135	E3
70526	Crowley⊙ 16,036	F6
71230	Crowville 400	G2
71021	Cullen 1,869	D1
70345	Cut Off 5,049	K7
71420	Cypress 55	D3
70046	Davant 600	L7
70528	Delcambre 2,216	G7
71232	Delhi 3,290	H2
71233	Delta 295	J2
70726	Denham Springs 8,563	L2
70633	De Quincy 3,966	D6
70634	De Ridder⊙ 11,057	D5
71421	Derry 75	E3
70030	Des Allemands 2,920	N4
70047	Destrehan 2,382	N4
†71055	Dixie Inn 453	D1
71422	Dodson 469	E2
70346	Donaldsonville⊙ 7,901	K3
70352	Donner 500	J7
71234	Downsville 213	F1
71023	Doyline 801	D1
70637	Dry Creek 300	D5
71423	Dry Prong 526	E3
71235	Dubach 1,161	E1
71024	Dubberly 421	D2
70353	Dulac 675	J8
71236	Dunn 225	G2
70728	Duplessis 500	K2
70529	Duson 1,253	F6
70049	Edgard⊙ 400	M3
†71019	Edgefield 312	D2
71331	Effie 300	F4
70638	Elizabeth 454	E5
71424	Elmer 200	E4
71051	Elm Grove 100	C2
70532	Elton 1,450	E6
71425	Enterprise 375	G3
71332	Eola 47	F5
71237	Epps 672	G1
70533	Erath 2,133	F7
71238	Eros 158	F2
70534	Estherwood 691	F6
70730	Ethel 250	H5
70535	Eunice 12,479	F5
70639	Evans 500	D5
71333	Evergreen 272	F5
71240	Fairbanks 300	F1
71241	Farmerville⊙ 3,768	F1
70640	Fenton 491	E6

(continued)

Louisiana

SCALE
0 5 10 20 30 40 MI.
0 5 10 20 30 40 KM.

State Capitals ⊛
Parish Seats ⊙
Canals
Major Limited Access Hwys.

Scale 1:2,000,000

AREA 47,752 sq. mi. (123,678 sq. km.)
POPULATION 4,206,312
CAPITAL Baton Rouge
LARGEST CITY New Orleans
HIGHEST POINT Driskill Mtn. 535 ft. (163 m.)
SETTLED IN 1699
ADMITTED TO UNION April 30, 1812
POPULAR NAME Pelican State
STATE FLOWER Magnolia
STATE BIRD Eastern Brown Pelican

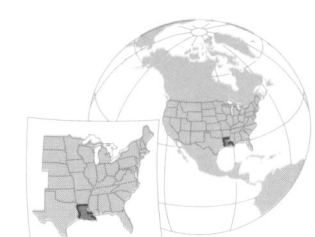

71334 Ferriday 4,472 .... G3
71426 Fisher 325 .... D4
71427 Flatwoods 360 .... E4
71428 Flora 300 .... D3
71429 Florien 964 .... D4
70436 Fluker 400 .... K5
70437 Folsom 319 .... K5
70732 Fordoche 676 .... G5
71242 Forest 299 .... H1
71430 Forest Hill 494 .... E4
70538 Franklin⊙ 9,584 .... G7
70438 Franklinton⊙ 4,119 .... K5
70733 French Settlement 761 .... L2
†71447 Galbraith 30 .... E4
70354 Galliano 5,159 .... K8
70540 Garden City 225 .... H7
70051 Garyville 2,856 .... M3
71432 Georgetown 381 .... F3
70355 Gheens 350 .... K7
71028 Gibsland 1,354 .... E1
71336 Gilbert 800 .... G2
71029 Gilliam 244 .... C1
71244 Girard 150 .... G2
71433 Glenmora 1,479 .... E5
71030 Gloster 780 .... C2
70736 Glynn 700 .... H5
70357 Golden Meadow 2,282 .... K8
71031 Goldonna 526 .... D2
70737 Gonzales 7,287 .... L2
†70079 Good Hope 500 .... N3
†71342 Good Pine-Trout 1,033 .... F3
71434 Gorum 150 .... E4
71338 Goudeau 25 .... G5
71245 Grambling 4,226 .... E1
70052 Gramercy 3,211 .... M3
71032 Grand Cane 252 .... C2
70541 Grand Coteau 1,165 .... G6
70358 Grand Isle 1,982 .... L8
70644 Grant 225 .... E5
71435 Grayson 564 .... F2
70441 Greensburg⊙ 662 .... J5
70739 Greenwell Springs 350 .... K1
71033 Greenwood 1,043 .... B2
70053 Gretna⊙ 20,615 .... O4
70740 Grosse Tete 749 .... G6
70542 Gueydan 1,695 .... E6
70057 Hahnville⊙ 2,947 .... N4
71034 Hall Summit 276 .... D2
70401 Hammond 15,043 .... N1
71035 Hanna 138 .... D3
70123 Harahan 11,384 .... O4
71340 Harrisonburg⊙ 610 .... G3
70058 Harvey 22,709 .... O4
71037 Haughton 1,510 .... C1
†71446 Hawthorn 400 .... D4
70646 Hayes 600 .... E6
71038 Haynesville 3,454 .... D1
71039 Heflin 279 .... D2
†70517 Henderson 1,560 .... G6
71341 Hessmer 743 .... F4
70743 Hester 250 .... L3
71437 Hicks 379 .... E4
71438 Hineston 400 .... E4
71247 Hodge 708 .... E2
70744 Holden 600 .... M1
71248 Holly Ridge 100 .... G2
71040 Homer⊙ 4,307 .... D1
71439 Hornbeck 470 .... D4
71043 Hosston 480 .... C1
70360 Houma⊙ 32,602 .... J7
70746 Iberville 367 .... K2
71044 Ida 306 .... C1
70443 Independence 1,684 .... M1
70747 Innis 200 .... G5
70543 Iota 1,326 .... E6
70647 Iowa 2,437 .... D6
†70427 Isabel 550 .... K5
70748 Jackson 3,133 .... H5
71045 Jamestown 131 .... D2
70544 Jeanerette 6,511 .... G7
†70067 Jean Lafitte 936 .... K7
70121 Jefferson 15,550 .... O4
71342 Jena⊙ 4,375 .... F3
70546 Jennings⊙ 12,401 .... E6
71249 Jigger 300 .... G2
71250 Jones 350 .... G1
71251 Jonesboro⊙ 5,061 .... E2
71343 Jonesville 2,828 .... G3
71749 Junction City 727 .... E1
70548 Kaplan 5,016 .... F6
71046 Keatchie 342 .... C2
71441 Kelly 325 .... F3
70062 Kenner 66,382 .... N4
70444 Kentwood 2,667 .... J5
71253 Kilbourne 286 .... H1
†70462 Killian 611 .... M2
70066 Killona 950 .... M3
70648 Kinder 2,603 .... E6
70371 Kraemer 350 .... M4
70750 Krotz Springs 1,374 .... G5
71443 Kurthwood 65 .... D4
70372 Labadieville 2,138 .... K4
71444 Lacamp 150 .... E4
70650 Lacassine 400 .... E6
70445 Lacombe 5,146 .... L6
*70501 Lafayette⊙ 81,961 .... F6
Lafayette‡ 150,017 .... F6
70067 Lafitte 1,312 .... K7
70549 Lake Arthur 3,615 .... E6
*70601 Lake Charles⊙ 75,226 .... D6
Lake Charles‡ 167,048 .... D6
70752 Lakeland 800 .... H5
71254 Lake Providence⊙ 6,361 .... H1
70068 La Place 16,112 .... N3
70373 Larose 5,234 .... K7
71344 Larto 500 .... G4
70550 Lawtell 1,014 .... F5
71445 Leander 145 .... E4
71345 Lebeau 200 .... F5
70651 Le Blanc 400 .... E5
71346 Lecompte 1,661 .... F4
71446 Leesville⊙ 9,054 .... D4
71447 Lena 300 .... E4
70551 Leonville 1,143 .... G6
†71008 Liberty Hill 50 .... E2
71348 Libuse 500 .... F4
71256 Lillie 172 .... E1

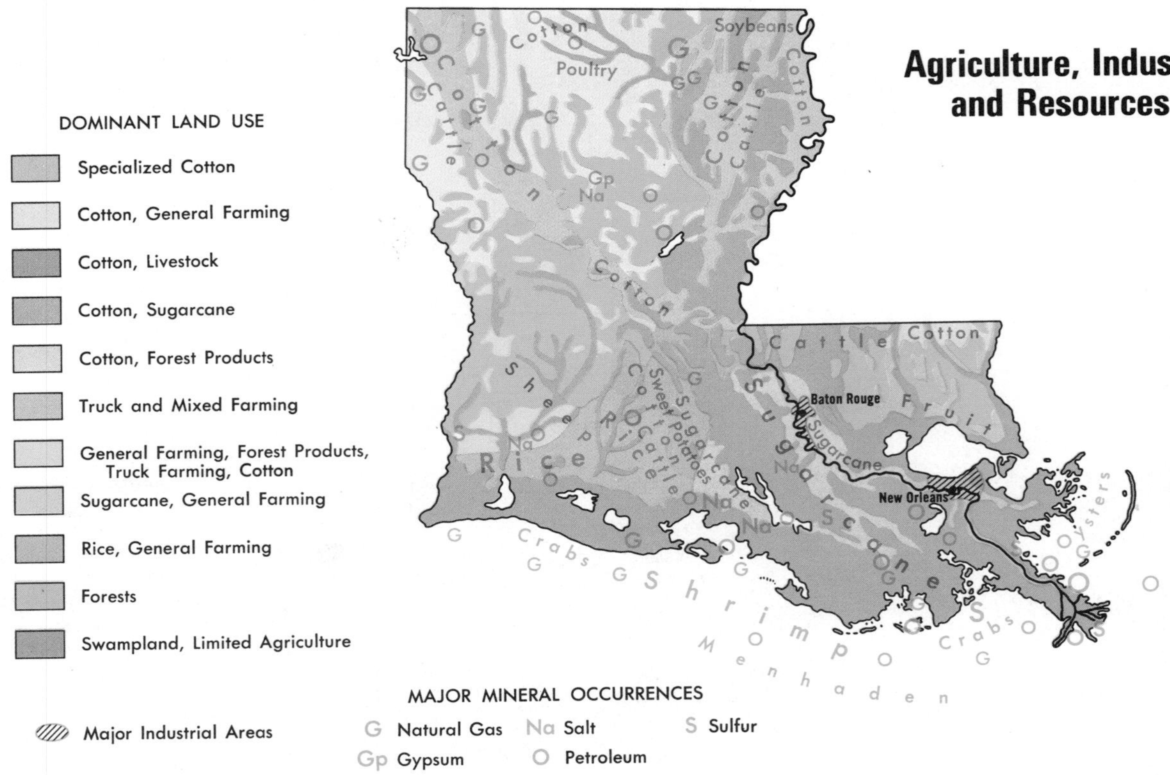
Agriculture, Industry and Resources
DOMINANT LAND USE
Specialized Cotton
Cotton, General Farming
Cotton, Livestock
Cotton, Sugarcane
Cotton, Forest Products
Truck and Mixed Farming
General Farming, Forest Products, Truck Farming, Cotton
Sugarcane, General Farming
Rice, General Farming
Forests
Swampland, Limited Agriculture
Major Industrial Areas
MAJOR MINERAL OCCURRENCES
G Natural Gas
Na Salt
S Sulfur
Gp Gypsum
O Petroleum
Soybeans
Cotton
Poultry
Cattle
Sheep
Rice
Sugarcane
Sweet Potatoes
Baton Rouge
Fruit
New Orleans
Oysters
Crabs
Shrimp
Menhaden

71257 Linville 150 .... F1
71048 Lisbon 138 .... E1
70754 Livingston⊙ 1,260 .... L1
70755 Livonia 980 .... G5
†70767 Lobdell 200 .... J1
70374 Lockport 2,424 .... K7
71049 Logansport 1,565 .... C3
71448 Longleaf 80 .... E4
71050 Longstreet 281 .... B2
70652 Longville 300 .... D5
70446 Loranger 250 .... N1
70552 Loreauville 860 .... G6
70756 Lottie 400 .... G5
†71008 Lucky 370 .... E2
70070 Luling 4,006 .... N4
70071 Lutcher 4,730 .... L3
70447 Madisonville 799 .... K6
70554 Mamou 3,194 .... F5
70448 Mandeville 6,076 .... L6
71259 Mangham 867 .... G2
71052 Mansfield⊙ 6,485 .... C2
71350 Mansura 2,074 .... G4
71449 Many⊙ 3,988 .... C3
70757 Maringouin 1,291 .... G6
71260 Marion 989 .... F1
71351 Marksville⊙ 5,113 .... G4
70072 Marrero 36,548 .... O4
†71019 Martin 584 .... D2
70555 Maurice 478 .... F6
†71433 McNary 240 .... E5
71346 Meeker 50 .... F4
71451 Melder 150 .... E4
71452 Melrose 500 .... E3
71353 Melville 1,764 .... G5
70556 Mermentau 771 .... E6
71261 Mer Rouge 802 .... G1
70653 Merryville 1,286 .... D5
*70001 Metairie 164,160 .... O4
70557 Midland 560 .... F6
70558 Milton 450 .... F6
†70070 Mimosa Park 3,737 .... N4
71055 Minden⊙ 15,074 .... D1
71059 Mira 354 .... C1
71453 Mitchell 155 .... C3
70376 Modeste 225 .... K3
*71201 Monroe⊙ 57,597 .... F1
Monroe‡ 139,241 .... F1
71454 Montgomery 843 .... E3
†70422 Montpelier 219 .... M1
71060 Mooringsport 911 .... B1
71455 Mora 427 .... E4
71355 Moreauville 853 .... G4
70380 Morgan City 16,114 .... H7
70759 Morganza 846 .... G5
71356 Morrow 600 .... F5
70559 Morse 835 .... F6
71262 Mound 40 .... H2
70450 Mount Hermon 170 .... K5
†71028 Mount Lebanon 105 .... D2
70390 Napoleonville⊙ 829 .... K4
70451 Natalbany 900 .... N1
71456 Natchez 527 .... D3
71457 Natchitoches⊙ 16,664 .... D3
71460 Negreet 400 .... C4
71357 Newellton 1,726 .... H2
70560 New Iberia⊙ 32,766 .... G6
71461 Newllano 2,213 .... D4
*70101 New Orleans⊙ 557,927 .... O4
New Orleans‡ 1,186,725 .... O4
70760 New Roads⊙ 3,924 .... G5

70078 New Sarpy 2,249 .... N4
71462 Noble 194 .... C3
70079 Norco 4,416 .... N3
†71247 North Hodge 573 .... E2
70761 Norwood 421 .... H5
71463 Oakdale 7,155 .... E5
71263 Oak Grove⊙ 2,214 .... H1
71264 Oak Ridge 257 .... G1
70655 Oberlin⊙ 1,764 .... E5
71061 Oil City 1,323 .... C1
71465 Olla 1,603 .... F3
70570 Opelousas⊙ 18,903 .... G5
70762 Oscar 650 .... H5
71466 Otis 400 .... E4
70391 Paincourtville 2,004 .... K3
71358 Palmetto 327 .... G5
70582 Parks 545 .... G6
70392 Patterson 4,693 .... H7
70452 Pearl River 1,693 .... L6
71063 Pelican 250 .... C3
70575 Perry 230 .... F7
70081 Pilottown 175 .... M8
70453 Pine Grove 570 .... J5
70576 Pine Prairie 734 .... E5
71360 Pineville 12,034 .... F4
71266 Pioneer 221 .... H1
70656 Pitkin 600 .... E5
71064 Plain Dealing 1,213 .... C1
70764 Plaquemine⊙ 7,521 .... J2
70393 Plattenville 205 .... K4
71362 Plaucheville 196 .... G5
71065 Pleasant Hill 776 .... C3
70082 Pointe a la Hache⊙ 750 .... L7
71467 Pollock 399 .... F3
70454 Ponchatoula 5,469 .... N2
70767 Port Allen⊙ 6,114 .... J2
70577 Port Barre 2,625 .... G5
70083 Port Sulphur 3,318 .... L8
†70726 Port Vincent 450 .... L2
71066 Powhatan 279 .... D3
71468 Provencal 695 .... D3
71268 Quitman 231 .... E2
70394 Raceland 6,302 .... J7
70578 Rayne 9,066 .... F6
71269 Rayville⊙ 4,610 .... G2
70580 Reddell 500 .... F5
70658 Reeves 199 .... D5
70084 Reserve 7,288 .... M3
†71282 Richmond 505 .... H2
†71201 Richwood 1,223 .... F2
†71334 Ridgecrest 895 .... G3
71068 Ringgold 1,655 .... D2
†70427 Rio 400 .... L5
70581 Roanoke 800 .... E6
71469 Robeline 238 .... D3
71069 Rodessa 337 .... B1
71364 Rosa 300 .... G5
70772 Rosedale 658 .... G6
70456 Roseland 1,346 .... J5
70659 Rosepine 953 .... D5
71365 Ruby 400 .... F4
71270 Ruston⊙ 20,585 .... E1
70457 Saint Benedict 190 .... K5
70775 Saint
Francisville⊙ 1,471 .... H5
71366 Saint Joseph⊙ 1,687 .... H3
71367 Saint Landry 550 .... F5
70582 Saint Martinville⊙ 7,965 .... G6
71471 Saint Maurice 560 .... E3
71070 Saline 293 .... E2

71071 Sarepta 831 .... D1
70807 Scotlandville 15,113 .... J1
70583 Scott 2,239 .... F6
†70764 Seymourville 2,891 .... J2
71072 Shongaloo 163 .... D1
*71101 Shreveport⊙ 205,820 .... C2
Shreveport‡ 376,646 .... C2
71073 Sibley 1,211 .... D1
71368 Sicily Island 691 .... G3
71472 Sieper 226 .... E4
71473 Sikes 226 .... F2
71369 Simmesport 2,293 .... G5
71474 Simpson 534 .... D4
71275 Simsboro 553 .... E1
70660 Singer 250 .... D5
71475 Slagle 650 .... D4
70777 Slaughter 729 .... H5
70458 Slidell 26,718 .... L6
71276 Sondheimer 225 .... H1
70778 Sorrento 1,197 .... L3
†71052 South Mansfield 1,463 .... C3
71277 Spearsville 181 .... E1
71278 Spencer 50 .... F1
70462 Springfield 424 .... M2
71075 Springhill 6,516 .... D1
†71049 Stanley 151 .... C3
71280 Sterlington 1,400 .... F1
71078 Stonewall 1,175 .... C2
70662 Sugartown 375 .... D5
70663 Sulphur 19,709 .... D6
70463 Sun 404 .... L5
70584 Sunset 2,300 .... F6
70464 Talisheek 315 .... L5
71282 Tallulah⊙ 11,634 .... H2
70465 Tangipahoa 493 .... J5
71080 Taylor 500 .... D1
71476 Temple 250 .... E4
71285 Terry 50 .... H1
†70053 Terry Town 23,548 .... O4
70397 Theriot 450 .... J8
70301 Thibodaux⊙ 15,810 .... J7
70466 Tickfaw 571 .... M1
71286 Transylvania 400 .... H1
71081 Trees 327 .... B1
†70041 Triumph-Buras 4,137 .... L8
71371 Trout-Good Pine 1,033 .... F3
71479 Tullos 776 .... F3
70782 Tunica 500 .... G5
70585 Turkey Creek 366 .... F5
71480 Urania 849 .... F3
70090 Vacherie 2,169 .... L3
70467 Varnado 249 .... L5
71481 Verda 100 .... E3
71373 Vidalia⊙ 5,936 .... G3
†71270 Vienna 519 .... E1
70586 Ville Platte⊙ 9,201 .... F5
70668 Vinton 3,631 .... C6
70092 Violet 11,678 .... P4
71082 Vivian 4,146 .... B1
†71418 Vixen 40 .... F2
70784 Wakefield 400 .... H5
†70433 Waldheim 25 .... L5
70785 Walker 2,957 .... L1
71289 Warden 130 .... H1
70589 Washington 1,266 .... G5
71375 Waterproof 1,339 .... H3
70786 Watson 800 .... L1
70591 Welsh 3,515 .... E6
70669 Westlake 5,246 .... D6
71291 West Monroe 14,993 .... F1

70094 Westwego 12,663 .... O4
70787 Weyanoke 500 .... H5
70788 White Castle 2,160 .... J3
†71371 White Sulphur Springs 50 .... F3
71376 Whiteville 150 .... F5
71377 Wildsville 800 .... G3
†70040 Wills Point 150 .... L7
70789 Wilson 656 .... H5
71483 Winnfield⊙ 7,311 .... E3
71295 Winnsboro⊙ 5,921 .... G2
71378 Wisner 1,424 .... G3
71485 Woodworth 412 .... E4
70592 Youngsville 1,053 .... G6
70791 Zachary 7,297 .... K1
†71371 Zenoria 76 .... F3
†71409 Zimmerman 20 .... E4
71486 Zwolle 2,602 .... C3

OTHER FEATURES

Allemands (lake) M4
Alligator (pt.) L6
Amite (riv.) L2
Anacoco (lake) D4
Atchafalaya (bay) H8
Atchafalaya (riv.) G6
Barataria (bay) L8
Barataria (passage) L8
Barksdale A.F.B. C2
Bayou D'Arbonne (lake) F1
Bird (isl.) M8
Bistineau (lake) D2
Black (lake) D3
Black Lake (bayou) D1
Boeuf (lake) J7
Boeuf (riv.) G1
Bonnet Carré Spillway and
Floodway N3
Borgne (lake) L7
Boudreau (bay) M7
Boudreaux (lake) J8
Breton (isls.) M8
Breton (sound) M7
Bundick (lake) D5
Caddo (lake) B1
Caillou (bay) J8
Calcasieu (lake) D7
Calcasieu (passage) D7
Calcasieu (riv.) E5
Catahoula (lake) F4
Cataouatche (lake) N4
Cat Island (chan.) M6
Cat Island (passage) J8
Chandeleur (isls.) N7
Chandeleur (sound) M7
Chenier (lake) F2
Chicot (pt.) M7
Claiborne (lake) E1
Clear (lake) D3
Cocodrie (lake) E5
Cotile (lake) E4
Cross (lake) C2
Curlew (isls.) M7
Dernieres (isls.) J8
Door (pt.) M6
Driskill (mt.) E2
Drum (bay) M7
East (bay) M8
East Cote Blanche (bay) G7
Edwards (lake) C2
Eloi (bay) M7
England A.F.B. E4
Fields (lake) J7
Fort Polk 14,142 .... D4
Free Mason (isls.) M7
Garden Island (bay) M8
Grand (lake) E7
Grand (lake) H8
Grand Terre (isls.) L8
Iatt (lake) E3
Jean Lafitte Nat'l Hist. Park P4
Lafourche (bayou) K8
Little (riv.) F3
Louisiana (pt.) C7
Macon (bayou) H1
Main (passage) M8
Manchac (passage) N2
Marsh (isl.) G7
Maurepas (lake) M2
Mermentau (riv.) E7
Mexico (gulf) F8
Mississippi (delta) M8
Mississippi (riv.) H3
Mississippi (sound) M6
Mississippi River Gulf Outlet
(canal) L7
Mozambique (pt.) M7
Mud (lake) D7
Naval Air Sta. O4
North (isls.) M7
North (pass) N8
North (pt.) M7
Northeast (pass) M8
Ouachita (riv.) F1
Palourde (lake) H7
Pearl (riv.) L5
Point au Fer (isl.) H8
Point au Fer (pt.) H8
Pontchartrain (lake) O3
Pontchartrain Causeway O3
Raccoon (pt.) H8
Red (riv.) G4
Sabine (lake) C7
Sabine (passage) C7
Sabine (riv.) C5
Saline (lake) E3
Salvador (lake) K7
Smithport (lake) C2
South (pass) M8
South (pt.) G8
Southeast (pass) M8
Southwest (pass) L8
Tangipahoa (riv.) N1
Tensas (riv.) G3
Terrebonne (bay) J8
Tickfaw (riv.) M1
Timbalier (bay) K8
Timbalier (isl.) K8
Toledo Bend (res.) C3
Turkey Creek (lake) G3
Vermilion (bay) F7
Vernon (lake) D4
Verret (lake) H7
Wallace (lake) C2
West (bay) M8
West Cote Blanche (bay) G7
White (lake) E7

⊙Parish seat.
‡Population of metropolitan area.
† Zip of nearest p.o. * Multiple zips.

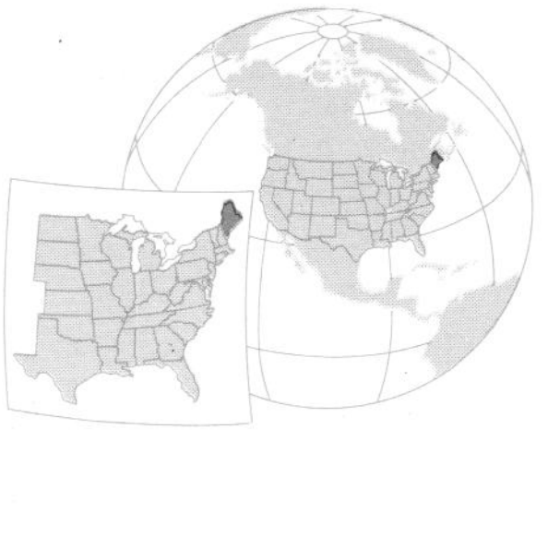

AREA 33,265 sq. mi. (86,156 sq. km.)
POPULATION 1,125,027
CAPITAL Augusta
LARGEST CITY Portland
HIGHEST POINT Katahdin 5,268 ft. (1606 m.)
SETTLED IN 1624
ADMITTED TO UNION March 15, 1820
POPULAR NAME Pine Tree State
STATE FLOWER White Pine Cone & Tassel
STATE BIRD Chickadee

COUNTIES

Androscoggin 99,657	C7
Aroostook 91,331	F2
Cumberland 215,789	C8
Franklin 27,098	B5
Hancock 41,781	G6
Kennebec 109,889	D7
Knox 32,941	E7
Lincoln 25,691	D7
Oxford 48,968	B7
Penobscot 137,015	F5
Piscataquis 17,634	E4
Sagadahoc 28,795	D7
Somerset 45,028	C4
Waldo 28,414	E6
Washington 34,963	H6
York 139,666	B9

CITIES and TOWNS

Zip	Name/Pop.	Key
04406	Abbot Village○ 576	D5
04001	Acton○ 1,228	B8
04606	Addison○ 1,061	H6
04910	Albion○ 1,551	E6
04002	Alfred○ 1,890	B9
†04610	Alexander○ 385	H5
†04774	Allagash○ 448	F1
†04938	Allens Mills 100	C6
04535	Alna○ 425	D7
†04468	Alton○ 468	F5
†04408	Amherst○ 203	G6
04216	Andover○ 850	B6
04911	Anson○ 2,226	D6
†04862	Appleton○ 818	E7
†04468	Argyle 225	F5
04732	Ashland○ 1,865	G2
04607	Ashville 36	G7
04912	Athens○ 802	D6
†04426	Atkinson○ 306	E5
04608	Atlantic 120	G7
04210	Auburn⊙ 23,128	C7
04330	Augusta (cap.)⊙ 21,819	D7
04408	Aurora○ 110	G6
04003	Bailey Island 500	D8
†04497	Bancroft○ 61	H4
04401	Bangor⊙ 31,643	F6
	Bangor‡ 83,919	F6
04609	Bar Harbor○ 4,124	G7
04609	Bar Harbor 2,685	G7
†04619	Baring○ 308	J5
04004	Bar Mills 800	C8
04653	Bass Harbor 450	G7
04530	Bath⊙ 10,246	D8
†04915	Bayside	F7
04611	Beals○ 695	H7
†04622	Beddington○ 36	H6
04915	Belfast⊙ 6,243	F7
04917	Belgrade○ 2,043	D7
†04915	Belmont○ 520	E7
†04733	Benedicta○ 225	G4
04937	Benton○ 2,188	D6
03901	Berwick○ 4,149	B9
03901	Berwick 2,378	B9
04217	Bethel○ 2,340	B7
04005	Biddeford 19,638	B9
04920	Bingham○ 1,184	D5
04920	Bingham 1,074	D5
†04613	Birch Harbor 300	H7
04734	Blaine○ 922	H2
04734	Blaine-Mars Hill 1,921	H2
04614	Blue Hill○ 1,644	F7
†04615	Blue Hill Falls 135	F7
04537	Boothbay○ 2,308	D8
04538	Boothbay Harbor 2,207	D8
04008	Bowdoinham○ 1,828	D7
†04481	Bowerbank○ 27	E5
04410	Bradford○ 888	F5
†04410	Bradford Center 105	F5
04411	Bradley○ 1,149	F6
04412	Brewer 9,017	F6
†04735	Bridgewater○ 742	H3
04009	Bridgton○ 3,528	B7
04009	Bridgton 1,639	B7
†04990	Brighton○ 74	D5
04539	Bristol○ 2,095	D8
†04616	Brooklin○ 619	F7
04921	Brooks○ 804	E6
04617	Brooksville○ 753	F7
04413	Brookton 175	H4
04010	Brownfield○ 767	B8
04414	Brownville○ 1,545	E5
04011	Brunswick○ 17,366	C8
04011	Brunswick 10,990	C8
04219	Bryant Pond 600	B7
04232	Buckfield○ 1,333	C7
04618	Bucks Harbor 300	J6
04416	Bucksport○ 4,345	F6
04416	Bucksport 2,853	F6
04540	Burkettville 120	E7
04417	Burlington○ 322	G5
04922	Burnham○ 951	E6
†04093	Buxton○ 5,775	C8
†04275	Byron○ 114	B6
04619	Calais 4,262	J5
04923	Cambridge○ 445	E5
04843	Camden○ 4,584	F7
04843	Camden 3,743	F7
04924	Canaan○ 1,189	D6
04221	Canton○ 831	C7
03902	Cape Neddick 850	B9
04014	Cape Porpoise 500	C9
04736	Caribou 9,916	G2
04419	Carmel○ 1,695	E6
†04947	Carrabassett Valley○ 107	C5
†04487	Carroll○ 175	G5
†04224	Carthage○ 438	C6
†04465	Cary○ 229	H4
04015	Casco○ 2,243	B7
04421	Castine○ 1,304	F7
†04941	Center Montville 16	E7
†04623	Centerville○ 28	H6
†04757	Chapman○ 406	G2
04422	Charleston○ 1,037	F5
†04666	Charlotte○ 300	J5
04017	Chebeague Island 900	C8
†04345	Chelsea○ 2,522	D7
04622	Cherryfield○ 983	H6
†04458	Chester○ 434	F5
†04938	Chesterville○ 869	C6
†04478	Chesuncook 6	D3
04926	China○ 2,918	E7
†04239	Chisholm 1,796	C7
†04428	Clifton○ 462	G6
04927	Clinton○ 2,696	D6
04927	Clinton 1,305	D6
†04623	Columbia○ 275	H6
04623	Columbia Falls○ 517	H6
†04638	Cooper○ 105	H6
04624	Corea 375	H7
04928	Corinna○ 1,887	E6
04020	Cornish○ 1,047	B8
†04976	Cornville○ 838	D6
04625	Cranberry Isles○ 198	G7
†04610	Crawford○ 86	H5
†04015	Crescent Lake 325	C7
†04851	Criehaven 5	F8
04738	Crouseville 450	G2
†04747	Crystal○ 349	G4
04021	Cumberland Center○ 5,284	C8
04021	Cumberland Center 2,015	C8
04563	Cushing○ 795	E7
04626	Cutler○ 726	J6
04543	Damariscotta○ 1,493	E7
04543	Damariscotta-Newcastle 1,411	E7
04424	Danforth○ 826	H4
†04622	Deblois○ 44	H6
†04429	Dedham○ 841	F6
04627	Deer Isle○ 1,492	F7
04022	Denmark○ 672	B8
04628	Dennysville○ 296	J6
04929	Detroit○ 744	E6
04930	Dexter○ 4,286	E5
04930	Dexter 3,118	E5
04224	Dixfield○ 2,389	C6
04224	Dixfield 1,725	C6
04932	Dixmont○ 812	E6
04426	Dover-Foxcroft○ 4,323	E5
04426	Dover-Foxcroft⊙ 2,974	E5
†04426	Dover South Mills 54	E5
04342	Dresden○ 998	D7
†04747	Dyer Brook○ 275	G3
04739	Eagle Lake○ 1,019	F1
04226	East Andover 250	B6
04544	East Boothbay 800	D8
04427	East Corinth 525	F5
04227	East Dixfield 250	C6
04429	East Holden 600	F6
04027	East Lebanon 950	B9
04228	East Livermore 500	C7
04630	East Machias○ 1,233	J6
04430	East Millinocket○ 2,372	F4
04430	East Millinocket 2,361	F4
04740	Easton○ 1,305	H2
04028	East Parsonfield 400	B8
04229	East Peru 200	C7
†04210	East Poland 200	C7
04631	Eastport 1,982	K6
04231	East Stoneham 300	B7
†04607	East Sullivan 496	G6
†04220	East Sumner 120	C7
†04862	East Union 75	E7
†04428	Eddington○ 1,769	F6
†04556	Edgecomb○ 841	D8
03903	Eliot○ 4,948	B9
04605	Ellsworth⊙ 5,179	F6
04031	Emery Mills 100	B8
04433	Enfield○ 1,397	F5
04434	Etna○ 758	E6
04936	Eustis○ 582	B5
04435	Exeter○ 823	E6
†04938	Fairbanks 400	C6
04937	Fairfield○ 6,113	D6
04937	Fairfield 3,169	D6
04105	Falmouth○ 6,853	C8
04105	Falmouth 1,655	C8
†04345	Farmingdale○ 2,535	D7
†04345	Farmingdale 2,014	D7
04938	Farmington○ 6,730	C6
04938	Farmington⊙ 3,583	C6
04940	Farmington Falls 500	C6
†04349	Fayette○ 812	C7
04546	Five Islands 225	D8
04742	Fort Fairfield○ 4,376	H2
04742	Fort Fairfield 2,282	H2
04743	Fort Kent○ 4,826	F1
04743	Fort Kent 2,375	F1
04744	Fort Kent Mills 200	F1
04438	Frankfort○ 783	F6
04634	Franklin○ 979	G6
04941	Freedom○ 458	E7
04032	Freeport○ 5,863	C8
04032	Freeport 1,906	C8
04635	Frenchboro○ 43	G7
04745	Frenchville○ 1,450	G1
04547	Friendship○ 1,000	E7
04037	Fryeburg○ 2,715	A7
04037	Fryeburg 1,644	A7
04345	Gardiner 6,485	D7
04939	Garland○ 718	E5
04548	Georgetown○ 735	D8
†04217	Gilead○ 191	B7
†04401	Glenburn○ 2,319	F6
04846	Glen Cove 250	E7
04038	Gorham○ 10,101	C8
04038	Gorham 4,052	C8
†04607	Gouldsboro○ 1,574	H7
04746	Grand Isle○ 719	G1
04637	Grand Lake Stream○ 198	H5
04039	Gray○ 4,344	C8
†04408	Great Pond○ 45	G6
04236	Greene○ 3,037	C7
04441	Greenville○ 1,839	D5
04441	Greenville 1,640	D5
04442	Greenville Junction 650	D5
04443	Guilford○ 1,793	E5
04443	Guilford 1,235	E5
04347	Hallowell 2,502	D7
†04785	Hamlin○ 340	H1
04444	Hampden○ 5,250	F6
04444	Hampden 3,538	F6
04445	Hampden Highlands 950	F6
04640	Hancock○ 1,409	G6
04237	Hanover○ 256	B7
04942	Harmony○ 755	D6
†04011	Harpswell○ 3,796	D8
04643	Harrington○ 859	H6
04040	Harrison○ 1,667	B7
†04221	Hartford○ 480	C7
04943	Hartland○ 1,669	D6
04943	Hartland 1,041	D6
04446	Haynesville○ 169	G4
04238	Hebron○ 665	C7
†04401	Hermon○ 3,170	F6
04944	Hinckley 140	D6
04041	Hiram○ 1,067	B8
†04730	Hodgdon○ 1,084	H3
04042	Hollis Center○ 2,892	B8
04847	Hope○ 730	E7
04730	Houlton○ 6,766	H3
04730	Houlton⊙ 5,730	H3
04448	Howland○ 1,602	F5
04448	Howland 1,502	F5
04449	Hudson○ 797	F5
04644	Hulls Cove 200	G7
04747	Island Falls○ 981	G3
04645	Isle Au Haut○ 57	F7
04848	Islesboro○ 521	F7
04945	Jackman○ 1,003	C4
†04630	Jacksonville 200	J6
04239	Jay○ 5,080	C7
04348	Jefferson○ 1,616	D7
04648	Jonesboro○ 553	J6
04649	Jonesport○ 1,512	H6
04649	Jonesport 1,050	H6
04450	Kenduskeag○ 1,210	E6
04043	Kennebunk○ 6,621	B9
04043	Kennebunk 3,294	B9
†04043	Kennebunk Beach 200	C9
04046	Kennebunkport○ 2,952	C9
04046	Kennebunkport 1,685	C9
04349	Kents Hill 300	D7
04947	Kingfield○ 1,083	C6
04451	Kingman 281	G4
†04990	Kingsbury○ 4	D5
03904	Kittery○ 9,314	B9
03904	Kittery 5,465	B9
03905	Kittery Point 1,260	B9
†04986	Knox○ 558	E6
04453	La Grange○ 509	F5
†04463	Lake View○ 20	F5
†04605	Lamoine○ 953	G7
04455	Lee○ 688	G5
†04263	Leeds○ 1,463	C7
04456	Levant○ 1,117	F6
04240	Lewiston 40,481	C7
	Lewiston-Auburn‡ 72,378	C7
04949	Liberty○ 694	E7
04749	Lille 300	G1
04048	Limerick○ 1,356	B8
04750	Limestone○ 8,719	H2
04750	Limestone 1,334	H2
04049	Limington○ 2,203	B8
04457	Lincoln○ 5,066	G5
04457	Lincoln 3,524	G5
04849	Lincolnville○ 1,414	E7
04850	Lincolnville Center 200	E7
†04730	Linneus○ 752	H3
04250	Lisbon○ 8,769	C7
04250	Lisbon-Lisbon Center 1,865	C7
04252	Lisbon Falls 4,370	D7
04350	Litchfield○ 1,954	D7
†04627	Little Deer Isle 475	F7
04082	Little Falls-South Windham 1,366	C8
†04760	Littleton○ 1,009	H3
04253	Livermore○ 1,826	C7
04254	Livermore Falls○ 3,572	C7
04254	Livermore Falls 2,441	C7
04255	Locke Mills 600	B7
04051	Lovell○ 767	B7
†04433	Lowell○ 194	F5
04652	Lubec○ 2,045	K6
†04730	Ludlow○ 403	G3
04654	Machias○ 2,458	J6
04654	Machias⊙ 1,277	J6
04655	Machiasport○ 1,108	H6
†04451	Macwahoc○ 126	G4
04756	Madawaska○ 5,282	G1
04756	Madawaska 4,165	G1
04950	Madison○ 4,367	D6
04950	Madison 2,788	D6
†04966	Madrid○ 178	B6
†04942	Mainstream 100	D6
04351	Manchester○ 1,949	D7
04757	Mapleton○ 1,895	G2
04758	Mars Hill○ 1,892	H2
04758	Mars Hill-Blaine 1,921	H2
04759	Masardis○ 328	G3
04851	Matinicus 66	F8
04459	Mattawamkeag○ 1,000	G5
04256	Mechanic Falls○ 2,616	C7
04256	Mechanic Falls 2,198	C7
04657	Meddybemps○ 110	J5
†04453	Medford○ 163	F5
†04453	Medford Center 100	F5
04460	Medway○ 1,871	G4
04957	Mercer○ 448	D6
04257	Mexico○ 3,698	B6
04257	Mexico 3,207	B6
†04216	Middledam 10	B6
04658	Milbridge○ 1,306	H6
04461	Milford○ 2,160	F6
04461	Milford 1,688	F6
04462	Millinocket○ 7,567	F4

(continued on following page)

Agriculture, Industry and Resources

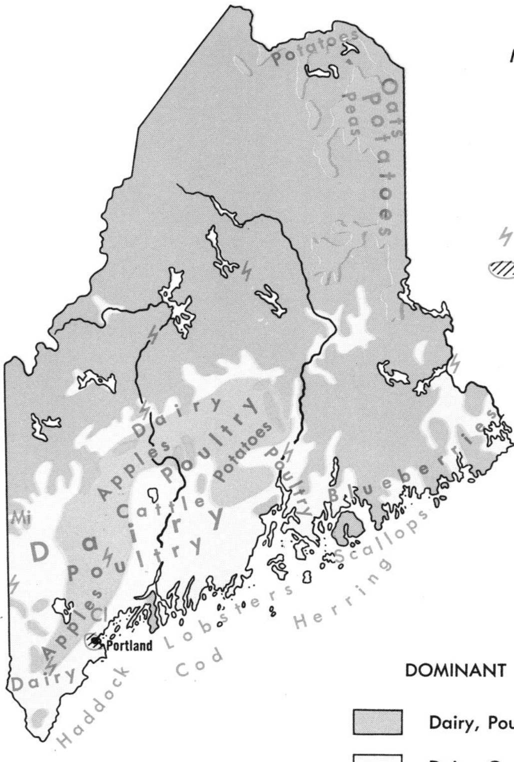

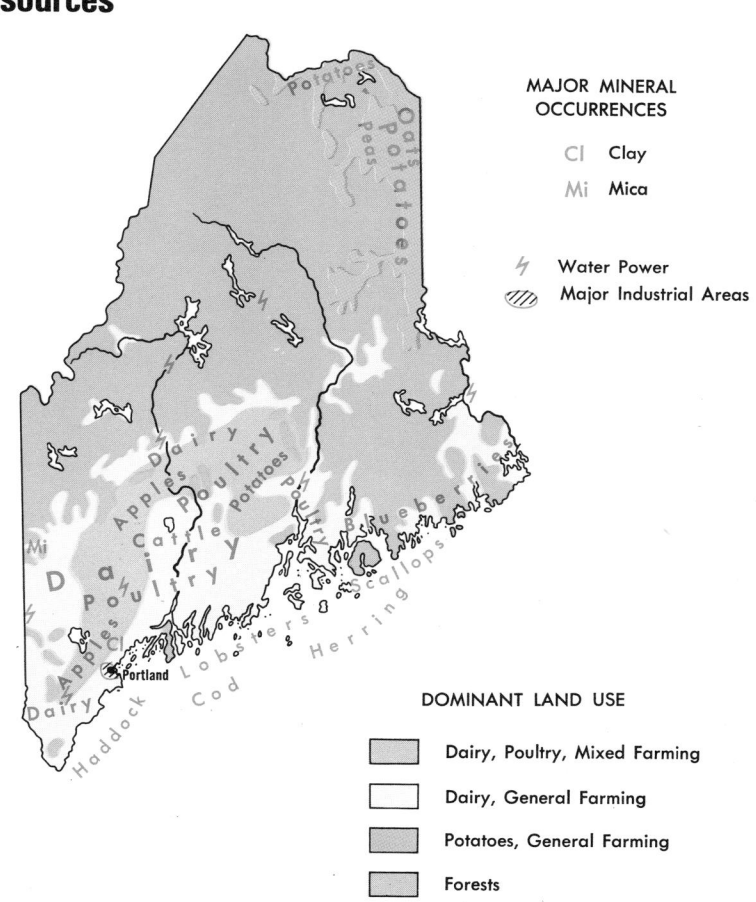

MAJOR MINERAL OCCURRENCES

Cl Clay
Mi Mica

Water Power
Major Industrial Areas

DOMINANT LAND USE

Dairy, Poultry, Mixed Farming
Dairy, General Farming
Potatoes, General Farming
Forests

04463 Milo○ 2,624 .... F5
04463 Milo 2,255 .... F5
04258 Minot○ 1,631 .... C7
04659 Minturn 150 .... G7
†04776 Monarda 100 .... G4
04852 Monhegan○ 109 .... E8
04259 Monmouth○ 2,888 .... D7
04951 Monroe○ 657 .... E6
04464 Monson○ 804 .... E5
04760 Monticello○ 950 .... H3
†04941 Montville○ 631 .... E7
04054 Moody 500 .... B9
†04478 Moosehead 6 .... D4
†04945 Moose River○ 252 .... C4
04952 Morrill○ 506 .... E7
04660 Mount Desert○ 2,063 .... G7
04352 Mount Vernon○ 1,021 .... D7
04055 Naples○ 1,833 .... B8
04552 Newagen 100 .... D8
†04445 Newburgh○ 1,228 .... F6
04553 Newcastle○ 1,227 .... D7
04553 Newcastle-Damariscotta 1,411 .... E7
04056 Newfield○ 644 .... B8
04260 New Gloucester○ 3,180 .... C8
04554 New Harbor 850 .... E8
04761 New Limerick○ 513 .... G3
04953 Newport○ 2,755 .... E6
04953 Newport 1,748 .... E6
04954 New Portland○ 651 .... C6
04261 Newry○ 235 .... B6
04955 New Sharon○ 969 .... C6
04762 New Sweden○ 737 .... G2
04956 New Vineyard○ 607 .... C6
04555 Nobleboro○ 1,154 .... D7
†04462 Norcross 13 .... F4
04957 Norridgewock○ 2,552 .... D6
04957 Norridgewock 1,318 .... D6
04958 North Anson 950 .... D6
03906 North Berwick○ 2,878 .... B9
03906 North Berwick 1,436 .... B9
04057 North Bridgton 300 .... B7
†04938 North Chesterville 50 .... C6
†04441 North East Carry 2 .... D4
04662 Northeast Harbor 800 .... G7
†04654 Northfield○ 88 .... H6
04853 North Haven○ 373 .... F7
04262 North Jay 800 .... C6
†04254 North Livermore 250 .... C7
04961 North New Portland 500 .... C6
†04476 North Penobscot 246 .... F7
†04849 Northport○ 958 .... E7
†04274 North Raymond 225 .... C8
04266 North Turner 350 .... C7
04962 North Vassalboro 950 .... D7
04267 North Waterford 390 .... B7
04062 North Windham 5,492 .... C8
†04219 North Woodstock 75 .... B7
†04096 North Yarmouth○ 1,919 .... C8
04268 Norway○ 4,042 .... B7
04268 Norway 2,653 .... B7
†04268 Norway Lake 75 .... B7
04763 Oakfield○ 847 .... G3
04963 Oakland○ 5,162 .... D6
04963 Oakland 3,387 .... D6
04063 Ocean Park 200 .... C9
03907 Ogunquit 1,492 .... B9
04064 Old Orchard Beach○ 6,291 .... C9
04064 Old Orchard Beach 6,023 .... C9
04468 Old Town 8,422 .... F6
04964 Oquossoc 150 .... B6
04471 Orient○ 97 .... H4
04472 Orland○ 1,645 .... F6
04473 Orono○ 10,578 .... F6
04473 Orono 9,891 .... F6
04474 Orrington○ 3,244 .... F6
04066 Orrs Island 600 .... D8
†04270 Otisfield○ 897 .... B7
04665 Otter Creek 260 .... G7
04854 Owls Head○ 1,633 .... F7
04764 Oxbow○ 84 .... G3
04270 Oxford○ 3,143 .... B7
04354 Palermo○ 760 .... E7
04965 Palmyra○ 1,485 .... E6
04271 Paris○ 4,168 .... B7
†04443 Parkman○ 621 .... D5
04475 Passadumkeag○ 430 .... F5
04765 Patten○ 1,368 .... F4
04765 Patten 1,057 .... F4
04558 Pemaquid 200 .... E8
04666 Pembroke○ 920 .... J6
04476 Penobscot○ 1,104 .... F7
04766 Perham○ 437 .... G2
04667 Perry○ 737 .... J6
04272 Peru○ 1,564 .... C6
04966 Phillips○ 1,092 .... C6
04562 Phippsburg○ 1,527 .... D8
04967 Pittsfield○ 4,125 .... E6
04967 Pittsfield 3,117 .... E6
†04345 Pittston○ 2,267 .... D7
04767 Plaisted 125 .... F1
†04925 Pleasant Pond 18 .... D5
04969 Plymouth○ 811 .... E6
04273 Poland○ 3,578 .... C7
04562 Popham Beach 40 .... D8
04768 Portage○ 562 .... G2
04855 Port Clyde 400 .... E8
04068 Porter○ 1,222 .... B8
*04101 Portland⊙ 61,572 .... C8
Portland‡ 183,625 .... C8
04069 Pownal○ 1,189 .... C8
†04487 Prentiss○ 205 .... G5
04769 Presque Isle 11,172 .... H2
04668 Princeton○ 994 .... H5
†04981 Prospect○ 511 .... F6
04669 Prospect Harbor 445 .... H7
04770 Quimby 50 .... F2
†04345 Randolph○ 1,834 .... D7
04970 Rangeley○ 1,023 .... B6
04071 Raymond○ 2,251 .... B8
04355 Readfield○ 1,943 .... D7
04357 Richmond○ 2,627 .... D7
04357 Richmond 1,578 .... D7
†04262 Riley 50 .... C6
†04930 Ripley○ 439 .... E5
04671 Robbinston○ 492 .... J5
†04734 Robinsons 160 .... H3
04841 Rockland⊙ 7,919 .... E7
04856 Rockport○ 2,749 .... F7
04478 Rockwood 265 .... D4
†04957 Rome○ 627 .... D6
†04654 Roque Bluffs○ 244 .... H6
04564 Round Pond 400 .... E8
04275 Roxbury○ 373 .... B6
04276 Rumford○ 8,240 .... B6
04276 Rumford 6,256 .... B6
04279 Rumford Point 320 .... B6
04280 Sabattus○ 3,081 .... C7
04280 Sabattus 1,234 .... C7
04072 Saco 12,921 .... C8
04772 Saint Agatha○ 1,035 .... G1
04971 Saint Albans○ 1,400 .... E6
04773 Saint David 915 .... G1
04774 Saint Francis○ 839 .... E1
04857 Saint George○ 1,948 .... E7
†04743 Saint John○ 322 .... F1
†04983 Salem 125 .... C6
†04009 Sandy Creek 132 .... B7
04972 Sandy Point 350 .... F7
04073 Sanford○ 18,020 .... B9
04073 Sanford 10,268 .... B9
04479 Sangerville○ 1,219 .... E5
†04417 Saponac 8 .... G5
04074 Scarborough○ 11,347 .... C8
04074 Scarborough 2,280 .... C8
04674 Seal Cove 215 .... G7
04675 Seal Harbor 500 .... G7
04973 Searsmont○ 782 .... E7
04974 Searsport○ 2,309 .... F7
04974 Searsport 1,348 .... F7
04075 Sebago Lake 800 .... B8
04481 Sebec○ 469 .... E5
04484 Seboeis○ 53 .... F5
†04478 Seboomook 3 .... D4
04676 Sedgwick○ 795 .... F7
04076 Shapleigh○ 1,370 .... B8
04975 Shawmut 500 .... D6
04775 Sheridan 300 .... F2
†04777 Sherman○ 1,021 .... G4
04777 Sherman Station 650 .... F4
04485 Shirley Mills○ 242 .... D5
†04330 Sidney○ 2,052 .... D7
04779 Sinclair 264 .... G1
04976 Skowhegan○ 8,098 .... D6
04976 Skowhegan⊙ 6,517 .... D6
04567 Small Point 22 .... D8
04978 Smithfield○ 748 .... D6
04780 Smyrna Mills○ 354 .... G3
04979 Solon○ 827 .... D6
†04341 Somerville○ 377 .... D7
†04660 Somesville (Mount Desert) 150 .... G7
04677 Sorrento○ 276 .... G7
03908 South Berwick○ 4,046 .... B9
†04009 South Bridgton 373 .... B8
04568 South Bristol○ 800 .... D8
04077 South Casco 750 .... B8
†03903 South Eliot 1,681 .... B9
†04928 South Exeter 100 .... E6
04080 South Hiram 350 .... B8
†04862 South Hope 200 .... E7
†04453 South La Grange 150 .... F5
†04259 South Monmouth 400 .... D7
04281 South Paris⊙ 2,128 .... C7
†04538 Southport○ 598 .... D8
04106 South Portland 22,712 .... C8
04858 South Thomaston○ 1,064 .... E7
†04864 South Union 50 .... E7
04081 South Waterford 300 .... B7
04679 Southwest Harbor○ 1,855 .... G7
04679 Southwest Harbor 1,052 .... G7
04082 South Windham (Little Falls-South Windham 1,366 .... C8
04487 Springfield○ 443 .... G5
04083 Springvale 2,940 .... B9
04782 Stacyville○ 554 .... F4
04084 Standish○ 5,946 .... B8
†04980 Starks○ 440 .... D6
04488 Stetson○ 618 .... E6
04680 Steuben○ 970 .... H6
04489 Stillwater 700 .... F6
04783 Stockholm○ 319 .... G1
04981 Stockton Springs○ 1,230 .... F7
04681 Stonington○ 1,273 .... F7
†04058 Stow○ 186 .... A7
04982 Stratton 600 .... B5
04983 Strong○ 1,506 .... C6
†04689 Sullivan○ 967 .... G6
†04292 Sumner○ 613 .... C7
04232 Sumner-East Sumner C7
04683 Sunset 165 .... F7
†04627 Sunshine 100 .... G7
04684 Surry○ 894 .... F7
04685 Swans Island○ 337 .... G7
†04915 Swanville○ 873 .... E6
†04040 Sweden○ 163 .... B7
04984 Temple○ 518 .... C6
04860 Tenants Harbor 900 .... E8
04861 Thomaston○ 2,900 .... E7
04861 Thomaston 2,348 .... E7
04986 Thorndike○ 603 .... E6
04490 Topsfield○ 240 .... H5
04086 Topsham○ 6,431 .... D8
04086 Topsham 4,657 .... D8
†04653 Tremont○ 1,222 .... G7
†04605 Trenton○ 718 .... G7
04571 Trevett 400 .... D8
04987 Troy○ 701 .... E6
04282 Turner○ 3,539 .... C7
04862 Union○ 1,569 .... E7
04988 Unity○ 1,431 .... E6
†04293 Upper Dam 2 .... B6
04784 Upper Frenchville 405 .... G1
04261 Upton○ 65 .... B6
04785 Van Buren○ 3,557 .... G1
04785 Van Buren 3,282 .... G1
04491 Vanceboro○ 256 .... J4
04989 Vassalboro○ 3,410 .... D7
04401 Veazie○ 1,610 .... F6
04360 Vienna○ 454 .... D6
04863 Vinalhaven○ 1,211 .... F7
04492 Waite○ 130 .... H5
†04915 Waldo○ 495 .... E7
04572 Waldoboro○ 3,985 .... E7
04572 Waldoboro 1,195 .... E7
†04605 Waltham○ 186 .... G6
04864 Warren○ 2,566 .... E7
04786 Washburn○ 2,028 .... G2
04786 Washburn 1,221 .... G2
04574 Washington○ 954 .... E7
04087 Waterboro○ 2,943 .... B8
04088 Waterford○ 951 .... B7
04901 Waterville 17,779 .... D6
04284 Wayne○ 680 .... D7
04285 Weld○ 435 .... C6
04990 Wellington○ 287 .... D5
04090 Wells○ 8,211 .... B9
04686 Wesley○ 140 .... H6
†04530 West Bath○ 1,309 .... D8
04092 Westbrook 14,976 .... C8
04493 West Enfield 609 .... F5
04787 Westfield○ 647 .... G2
04985 West Forks○ 72 .... D5
†04649 West Jonesport 400 .... H6
04094 West Kennebunk 750 .... B9
†04938 West Mills 75 .... C6
04288 West Minot 400 .... C7
04095 West Newfield 300 .... B8
04424 Weston○ 155 .... H4
04289 West Paris○ 1,390 .... B7
04290 West Peru 700 .... C7
04291 West Poland 250 .... C7
04074 West Scarborough 500 .... C8
04690 West Tremont 250 .... G7
04362 Whitefield○ 1,606 .... D7
04691 Whiting○ 335 .... J6
04692 Whitneyville○ 264 .... H6
†04443 Willimantic○ 164 .... E5
04293 Wilsons Mills 50 .... B6
04294 Wilton○ 4,382 .... C6
04294 Wilton 2,262 .... C6
04363 Windsor○ 1,702 .... D7
04495 Winn○ 503 .... G5
†04901 Winslow○ 8,057 .... D6
†04901 Winslow 5,903 .... D6
04693 Winter Harbor○ 1,120 .... G7
04496 Winterport○ 2,675 .... F6
04496 Winterport 1,126 .... F6
04788 Winterville○ 235 .... F2
04364 Winthrop○ 5,889 .... C7
04364 Winthrop 3,264 .... C7
04578 Wiscasset○ 2,832 .... D7
04694 Woodland○ 1,363 .... H5
04579 Woolwich○ 2,156 .... D8
04497 Wytopitlock 130 .... G4
04096 Yarmouth○ 6,585 .... C8
04096 Yarmouth 2,981 .... C8
03909 York○ 8,465 .... B9
03909 York 4,530 .... B9
03910 York Beach 900 .... B9
03911 York Harbor 950 .... B9

OTHER FEATURES

Abraham (mt.) C5
Acadia Nat'l Park G7
Allagash (lake) D3
Allagash (riv.) E2
Androscoggin (riv.) C7
Aroostook (riv.) G2
Attean (pond) C4
Baker (lake) D3
Baskahegan (lake) H5
Bear (riv.) B6
Big (brook) E2
Big (lake) H5
Big Black (riv.) D2
Bigelow (bight) C9
Big Spencer (mt.) E4
Black (pond) D3
Blue (mt.) C6
Blue Hill (bay) G7
Bog (lake) H6
Brassua (lake) D4
Casco (bay) C8
Cathance (lake) J6
Caucomgomoc (lake) D3
Center (pond) E5
Chamberlain (lake) E3
Chemquasabamticook (lake) D3
Chesuncook (lake) E3
Chiputneticook (lakes) H4
Clayton (lake) D2
Clifford (lake) H5
Cold Stream (pond) G5
Crawford (lake) H5
Cross (isl.) J6
Cross (lake) G1
Cupsuptic (riv.) B5
Dead (riv.) C5
Deer (isl.) F7
Duck (isls.) G7
Eagle (lake) E3
Eagle (lake) F1
East Machias (riv.) H6
East Musquash (lake) H5
Elizabeth (cape) C8
Ellis (pond) B6
Ellis (riv.) B6
Embden (pond) D6
Endless (lake) F5
Englishman (bay) J6
Eskutassis (pond) G5
Fifth (lake) H5
Fish (riv.) F2
Fish River (lake) F2
Flagstaff (lake) C5
Fourth (lake) H5
Frenchman (bay) G7
Gardner (lake) J6
Georges (isls.) E8
Graham (lake) G6
Grand (lake) H4
Grand Falls (lake) H5
Grand Lake Seboeis (lake) F3
Grand Manan (chan.) K6
Great Moose (lake) D6
Great Wass (isl.) J7
Green (isl.) F8
Harrington (lake) E4
Haut (isl.) G7
Indian Pond (lake) D4
Islesboro (isl.) F7
Jo-Mary (lakes) E4
Katahdin (mt.) F4
Kennebec (riv.) D7
Kezar (lake) B7
Kezar (pond) B7
Kingsbury (pond) D5
Little Black (riv.) E1
Little Madawaska (riv.) G2
Lobster (lake) E4
Long (lake) B7
Long (lake) E2
Long (lake) G1
Long (pond) C4
Long (pond) D6
Long (pond) E5
Long Falls (dam) C5
Longfellow (mts.) B6
Loon (lake) D3
Loring A.F.B. 6,572 .... H2
Lower Roach (pond) E4
Lower Sysladobsis (lake) G5
Machias (bay) J6
Machias (riv.) F2
Machias (riv.) H6
Machias Seal (isl.) J7
Madagascal (pond) G5
Marshall (isl.) G7
Matinicus Rock (isl.) F8
Mattamiscontis (lake) F4
Mattawamkeag (lake) G4
Mattawamkeag (riv.) G4
Meddybemps (lake) J5
Metinic (isl.) E8
Millinocket (lake) F4
Millinocket (lake) F3
Molunkus (lake) G4
Monhegan (isl.) E8
Moose (pond) B7
Moose (riv.) D4
Moosehead (lake) D4
Mooseleuk (stream) F2
Mooselookmeguntic (lake) B6
Mopang (lake) H6
Mount Desert (isl.) G7
Mount Desert Rock (isl.) G8
Moxie (lake) D5
Munsungan (lake) E3
Muscongus (bay) E8
Musquacook (lakes) E2
Nahmakanta (lake) E4
Nicatous (lake) G5
Nollesemic (lake) F4
Old (stream) H6
Onawa (lake) E5
Parlin (pond) C4
Parmachenee (lake) B5
Passamaquoddy (bay) J5
Passamaquoddy Ind. Res. J6
Pemadumcook (lake) E4
Penobscot (bay) F7
Penobscot (lake) C4
Penobscot (riv.) F5
Penobscot Ind. Res. F6
Pierce (pond) C5
Piscataqua (riv.) B9
Piscataquis (riv.) E5
Pleasant (lake) E3
Pleasant (lake) G3
Pleasant (lake) H5
Pleasant (riv.) H6
Pocomoonshine (lake) H5
Portage (lake) F2
Presque Isle A.F.B. G2
Priestly (lake) E2
Pushaw (lake) F6
Ragged (isl.) F8
Ragged (lake) E4
Rainbow (lake) E4
Rangeley (lake) B6
Richardson (lakes) B6
Rocky (lake) J6
Round (pond) E2
Rowe (lake) F2
Saco (riv.) B8
Saint Croix (riv.) J5
Saint Croix Isl. Nat'l Mon. J5
Saint Francis (riv.) E1
Saint Froid (lake) F2
Saint John (pond) D3
Saint John (riv.) G1
Salmon Falls (riv.) B9
Sandy (riv.) C6
Schoodic (lake) F5
Scraggly (lake) F3
Scraggly (lake) H5
Seal (isl.) F8
Sebago (lake) B8
Sebasticook (lake) E6
Seboeis (lake) F5
Seboeis (riv.) F3
Seboomook (lake) D4
Shallow (lake) E3
Small (cape) D8
Sourdnahunk (lake) F3
Spencer (pond) D4
Spencer (stream) C5
Spider (lake) E3
Squa Pan (lake) G2
Square (lake) G1
Sunday (riv.) B6
Swift (riv.) B6
Sysladobsis, Lower (lake) G5
Third (lake) H5
Twin (lakes) F4
Umbagog (lake) A6
Umcalcus (lake) G3
Umsaskis (lake) E2
Union, West Branch (riv.) G6
Vinalhaven (isl.) F7
Wassataquoik (stream) F4
Webb (lake) C6
Webster (brook) E3
West Grand (lake) H5
West Musquash (lake) H5
West Quoddy (head) K6
Wilson E5
Winnecook (lake) E6
Wooden Ball (isl.) F8
Wyman (lake) C5
Wytopitlock (lake) G4

⊙County seat.
‡Population of metropolitan area.
○Population of town or township.
† Zip of nearest p.o.
* Multiple zips.

Topography

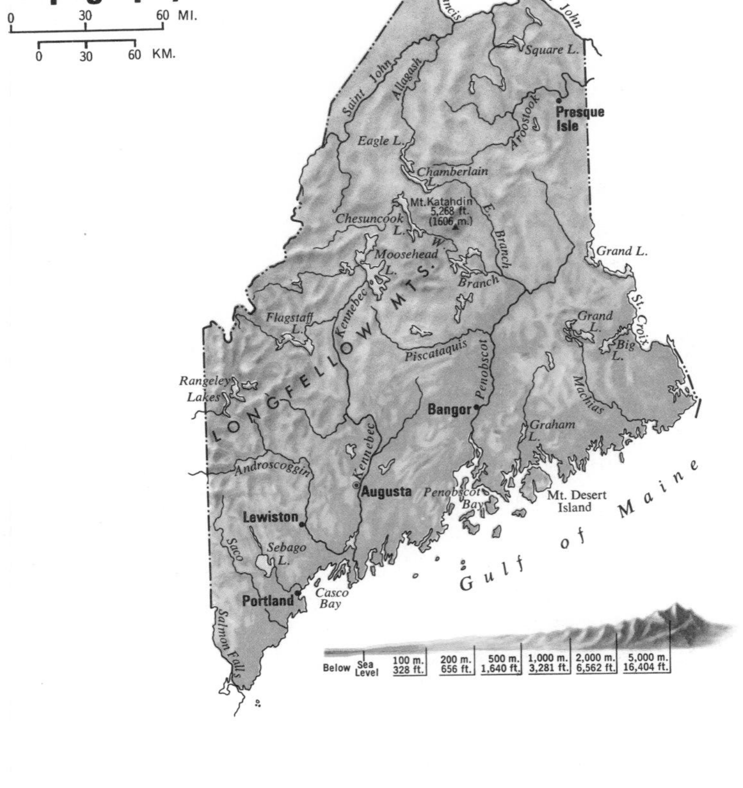
0 30 60 MI.
0 30 60 KM.
St. Francis
Saint John
Square L.
Saint John
Allagash
Aroostook
Presque Isle
Eagle L.
Chamberlain
Mt. Katahdin 5,268 ft. (1606 m.)
Chesuncook L.
E. Branch
W. Branch
Moosehead L.
Grand L.
Flagstaff L.
Kennebec
LONGFELLOW MTS.
Piscataquis
Penobscot
Grand L.
St. Croix
Big L.
Rangeley Lakes
Machias
Bangor
Graham L.
Androscoggin
Kennebec
Augusta
Penobscot Bay
Mt. Desert Island
Gulf of Maine
Lewiston
Saco
Sebago L.
Portland
Casco Bay
Salmon Falls
Below Sea Level
100 m. 328 ft.
200 m. 656 ft.
500 m. 1,640 ft.
1,000 m. 3,281 ft.
2,000 m. 6,562 ft.
5,000 m. 16,404 ft.

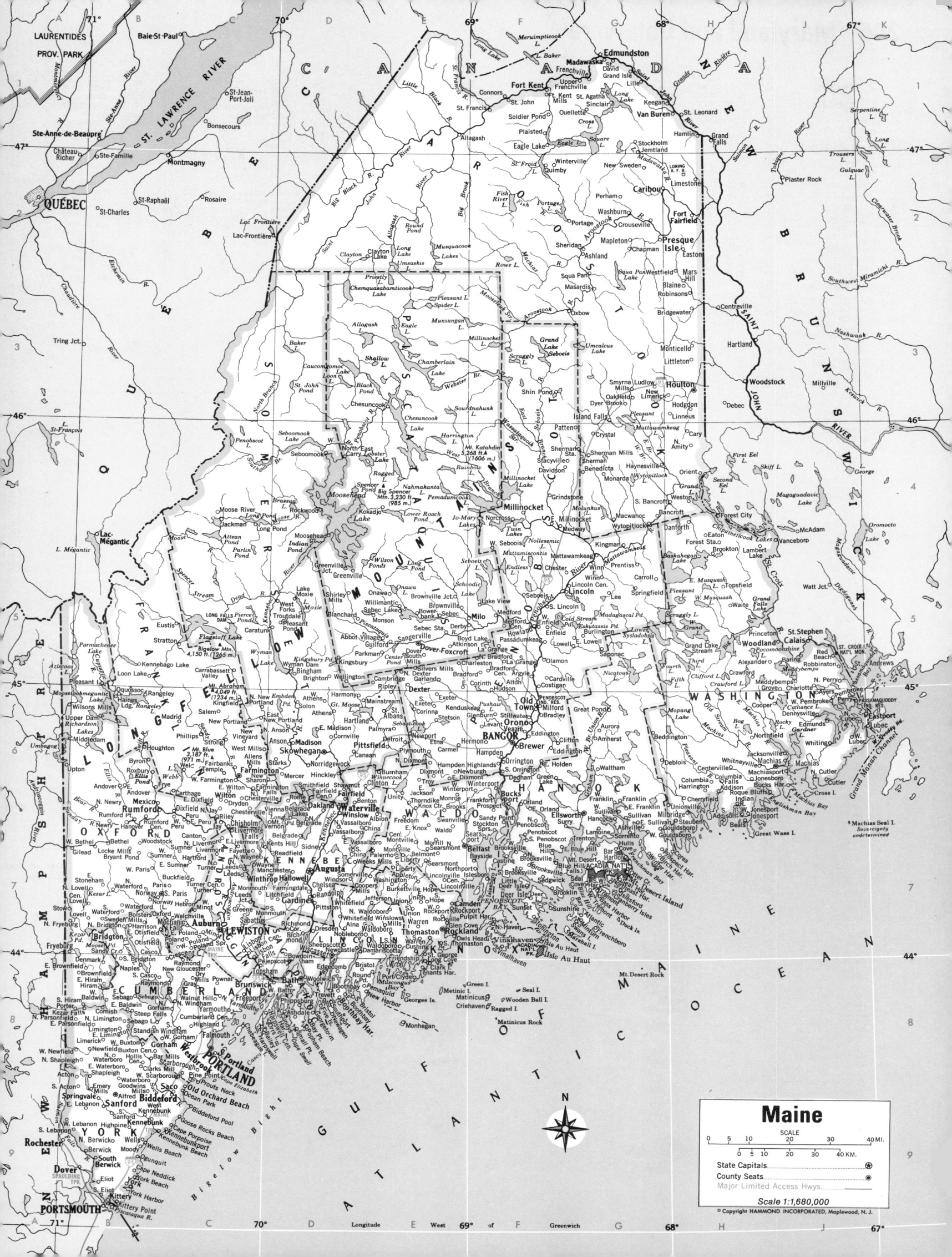

Maine

SCALE
0 5 10   20   30   40 MI.
0 5 10   20   30   40 KM.

State Capitals..................⊛
County Seats...................◉
Major Limited Access Hwys......

Scale 1:1,680,000

© Copyright HAMMOND INCORPORATED, Maplewood, N.J.

MARYLAND

COUNTIES

Allegany 80,548 ....C2
Anne Arundel 370,775 ....M4
Baltimore 655,615 ....M3
Baltimore (city county) 786,775 ...M3
Calvert 34,638 ....M6
Caroline 23,143 ....P5
Carroll 96,356 ....K2
Cecil 60,430 ....P2
Charles 72,751 ....K6
Dorchester 30,623 ....O7
Frederick 114,792 ....J4
Garrett 26,498 ....A2
Harford 145,930 ....N2
Howard 118,572 ....L4
Kent 16,695 ....O3
Montgomery 579,053 ....J4
Prince Georges 665,071 ......L5
Queen Annes 25,508 ....P4
Saint Marys 59,895 ....M7
Somerset 19,188 ....R8
Talbot 25,604 ....O5
Washington 113,086 ....G2
Wicomico 64,540 ....R7
Worcester 30,889 ....S8

CITIES and TOWNS

Zip   Name/Pop.   Key

21001 Aberdeen 11,533.............O2
21009 Abingdon 500 ....N3
21520 Accident 246 ....A2
20607 Accokeek 3,894 ....L6
*21401 Annapolis (cap.)⊙ 31,740 M5
20701 Annapolis Junction 775 ...M4
20608 Aquasco 950 ....L6
†21227 Arbutus 20,163............M4
†20785 Ardmore 500 ....G4
Aspen Hill 47,455 ....K4
*21201 Baltimore 786,775 ....M3
Baltimore‡ 2,174,023............M3
20610 Barstow 500 ....M6
21521 Barton 617 ....B2
21014 Bel Air⊙ 7,814 ....N2
20611 Bel Alton 800 ....L7
20705 Beltsville 12,760 ....G3
20612 Benedict 850 ....M6
21811 Berlin 2,162..............T7
†20740 Berwyn Heights 3,135 ....G4
21609 Bethlehem 500 ....P6
*20014 Bethesda 62,736 ....E4
21610 Betterton 356 ....O3
20710 Bladensburg 7,691 ....G4
21523 Bloomington 486 ....B3
21713 Boonsboro 1,908 ....H2
†20027 Boulevard Heights 500 ...F5
20715 Bowie 33,695 ....L4
21612 Bozman 700 ....N5
20613 Brandywine 1,319 ....L6
20722 Brentwood 2,988 ....F4
21225 Brooklyn 11,508 ....M4
†21659 Brookview 78 ....P6
21716 Brunswick 4,572 ....H3
21717 Buckeystown 400 ....J3
21718 Burkittsville 202 ....H3
20618 Bushwood 750 ....L7
20731 Cabin
     John-Brookmont 5,135 .E4
20619 California 5,770 ....M7
†20705 Calverton 7,649 ....L4
21613 Cambridge⊙ 11,703 ....O6
20748 Camp Springs 16,118 .....G6
21401 Cape Saint Claire 6,022 ..N4
20743 Capitol Heights 3,271 ....G5
21024 Cardiff 475 ....N2
†20028 Carmody Hills-Pepper Mill
     Village 5,571 ....G5
†21034 Castleton 750 ....N2
†21788 Catoctin Furnace 516 ....J2
21228 Catonsville 33,208 ....M3
21720 Cavetown 1,533 ....H2
21913 Ceciliton 508 ....P3
21617 Centreville⊙ 2,018 ....O4
21816 Chance 600 ....P8
21914 Charlestown 720 ....P2
20622 Charlotte Hall 1,901 .....M7
21027 Chase 900 ....N3
20623 Cheltenham 950 ....L6
20732 Chesapeake Beach 1,408 ..N6
21915 Chesapeake City 899 .....P2
21619 Chester 950 ....N5
21620 Chestertown⊙ 3,300 ......O4
20785 Cheverly 5,751 ....G4
20815 Chevy Chase 12,232 ....E4
†20015 Chevy Chase Section
     Four 3,189 ....E4
20783 Chillum 32,775 ....F4
21622 Church Creek 124 ....O6
21623 Church Hill 319 ....O4
21028 Churchville 500 ....N2
20734 Clarksburg 400 ....J4
21029 Clarksville 500 ....L4
21722 Clear Spring 477 ....G2
20624 Clements 800 ....L7
20735 Clinton 16,438...........G6
21030 Cockeysville 17,013 .....M3
20904 Colesville 14,359 ....K4
20740 College Park 23,614 .....G4
†20722 Colmar Manor 1,286 .....F4
20626 Coltons Point 600 ....M8
21043 Columbia 52,518..........L4
20627 Compton 500 ....M7
21723 Cooksville 497 ....K3
†20027 Coral Hills 11,602 ......G5
21524 Corriganville 1,020 .....C2
†20722 Cottage City 1,122 .....F4
†20611 Cox Station (Bel
     Alton) 800 ....L7
21502 Cresaptown 4,645 ....C2
21817 Crisfield 2,924 ....P9
21114 Crofton 12,009 ....M4
21032 Crownsville 900 ....M4
21502 Cumberland⊙ 25,933 ......D2
     Cumberland‡ 107,782 ....D2

20750 Damascus 4,129 ....K3
20628 Dameron 759 ....N8
21034 Darlington 850 ....N2
†20760 Darnestown 950 ....J4
20751 Deale 3,008 ....M5
21821 Deal Island 800 ....P8
21550 Deer Park 486 ....A3
†20784 Defense Heights G4
21875 Delmar 1,232 ....R7
21629 Denton⊙ 1,927 ....P5
20855 Derwood 413 ....K4
20753 Dickerson 850 ....J4
20747 District Heights 6,799 ..G5
20630 Drayden 400 ....N8
21222 Dundalk 71,293 ....N3
†20608 Eagle Harbor 45 ....M6
21631 East New Market 230 .....P6
21601 Easton⊙ 7,536 ....O5
21528 Eckhart Mines 1,333 .....C2
21822 Eden 800 ....R7
†21219 Edgemere 9,078 ....N4
21040 Edgewood 19,455 ....N3
†20781 Edmonston 1,109 ....F4
†21784 Eldersburg 4,959 ....L3

†21659 Eldorado 93 ....P6
21920 Elk Mills 550 ....P2
21034 Elk Neck 700 ....P2
21901 Elkton⊙ 6,468 ....P2
21529 Ellerslie 950 ....C2
21043 Ellicott City⊙ 21,784 ...L3
21727 Emmitsburg 1,552 ....J2
21221 Essex 39,614 ....N3
21824 Ewell 595 ....O9
†20027 Fairmount Heights 1,616 .G5
21047 Fallston 5,572 ....N2
21632 Federalsburg 1,952 ......P6
21061 Ferndale 14,314 ....M4
21048 Finksburg 450 ....L3
21634 Fishing Creek 595 ....N7
21530 Flintstone 400 ....D2
†20001 Forest Heights 2,999 ...F5
21050 Forest Hill 450 ....N2
†20028 Forestville 16,401 .....G5
†20022 Fort Foote 700 ....F6
20744 Fort Washington F6
†21740 Fountain Head 1,745 ....G2
†21760 Foxville 175 ....H2
21701 Frederick⊙ 28,086 ....J3

21053 Freeland 500 ....M2
20758 Friendship 600 ....M6
21531 Friendsville 511 ....A2
21532 Frostburg 7,715 ....C2
21826 Fruitland 2,694 ....R7
21734 Funkstown 1,103 ....H2
20760 Gaithersburg 26,424 .....K4
21635 Galena 374 ....P3
†19973 Galestown 142 ....M5
20765 Galesville 600 ....M5
21048 Gamber 500 ....L3
21054 Gambrills 460 ....M4
20766 Garrett Park 1,178 ......E3
21055 Garrison 950 ....L3
20767 Germantown 9,721 ....J4
†20801 Glenarden 4,993 ....G4
21061 Glen Burnie 37,263 ......M4
20768 Glen Echo 229 ....E4
21737 Glenelg 400 ....L3
21636 Goldsboro 188 ....P4
†21163 Granite 950 ....L3
21536 Grantsville 498 ....B2
21638 Grasonville 1,910 ....O5
20770 Greenbelt 17,332 ....G4

21122 Green Haven 6,577 ....M4
21639 Greensboro 1,253 ....P5
21740 Hagerstown⊙ 34,132 ......G2
     Hagerstown‡ 113,086 ....G2
†21740 Halfway 8,659 ....G2
21074 Hampstead 1,293 ....L2
21750 Hancock 1,887 ....F2
21201 Hanover 500 ....M4
21077 Harmans 400 ....M4
21078 Havre de Grace 8,763 ....O2
21830 Hebron 714 ....R7
21640 Henderson 156 ....P4
†21111 Hereford 680 ....M2
21753 Highfield-Cascade 1,096 .J2
21401 Highland Beach 8 ....M5
20903 Hillandale 9,686 ....F4
†20031 Hillcrest Heights 17,021 .F5
21641 Hillsboro 450 ....P5
21737 Hollywood 500 ....M7
20637 Hughesville 1,208 ....L6
20639 Huntingtown 450 ....M6
†21163 Hurlock 1,690 ....P6
20780 Hyattsville 12,709 ......F4
20770 Indian Head 1,381 ....K6

†20685 Island Creek 400 ....M7
21084 Jarrettsville 1,485 .....M2
†21085 Joppatowne 11,348 ......N3
21756 Keedysville 476 ....H3
†20901 Kemp Mill F3
20795 Kensington 1,822 ....E4
21087 Kingsville 2,824 ....N3
21538 Kitzmiller 387 ....B3
21758 Knoxville 500 ....H3
20785 Landover 5,374 ....G4
20784 Landover Hills 1,428 ....G4
20787 Langley Park 14,038 .....F4
20801 Lansdowne-Seabrook 15,814 .G4
21227 Lansdowne-Baltimore
     Highlands 16,759 ......M3
20646 La Plata⊙ 2,484 ....L6
20870 Largo 5,557 ....G5
*20810 Laurel 12,103 ....L4
21502 La Vale-Narrows
     Park 5,523 ....C2
20760 Laytonsville 195 ....K4
21761 Le Gore 500 ....J2
†21740 Leitersburg 350 ....H2
20650 Leonardtown⊙ 1,448 ......M7

(continued)

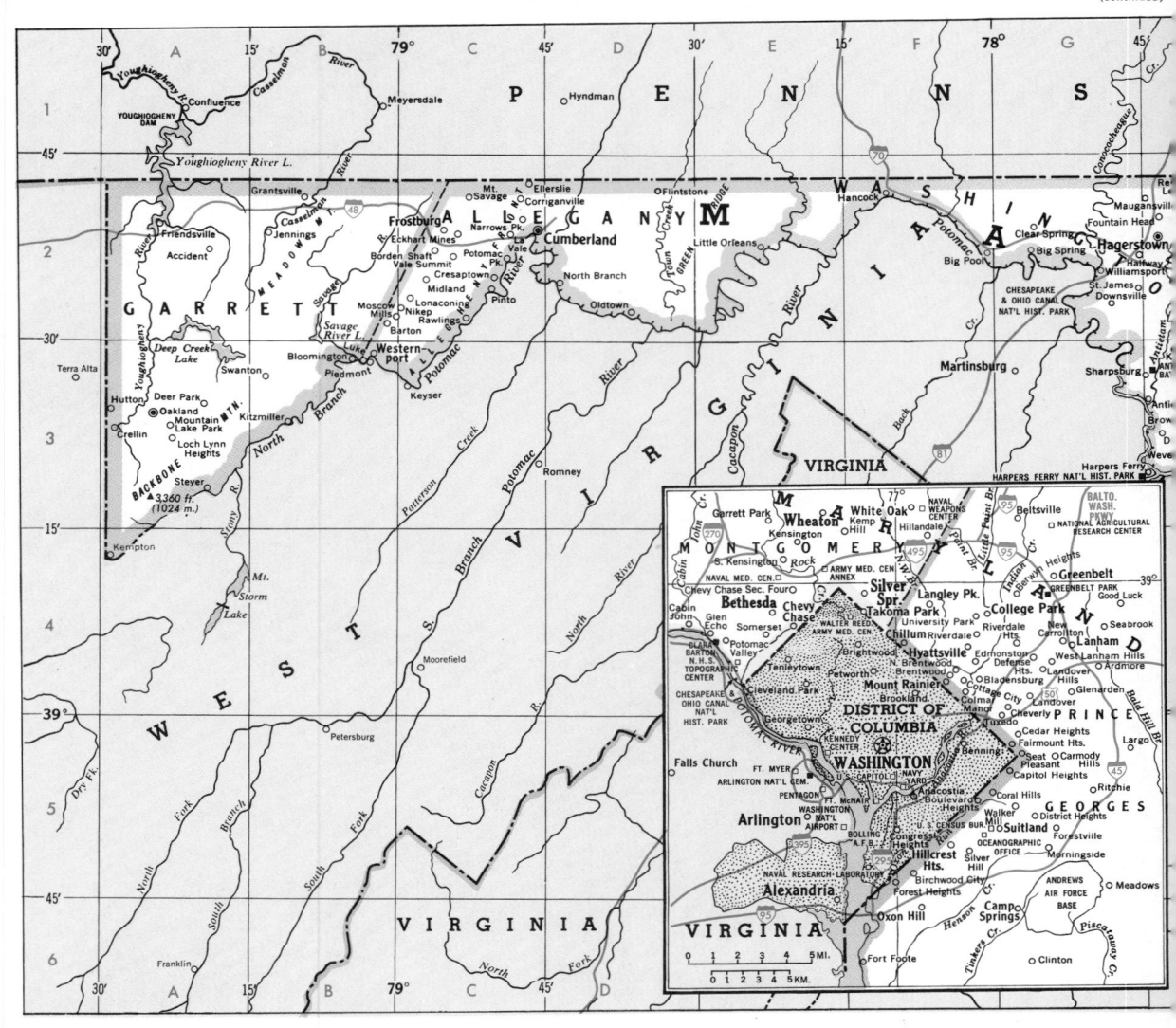

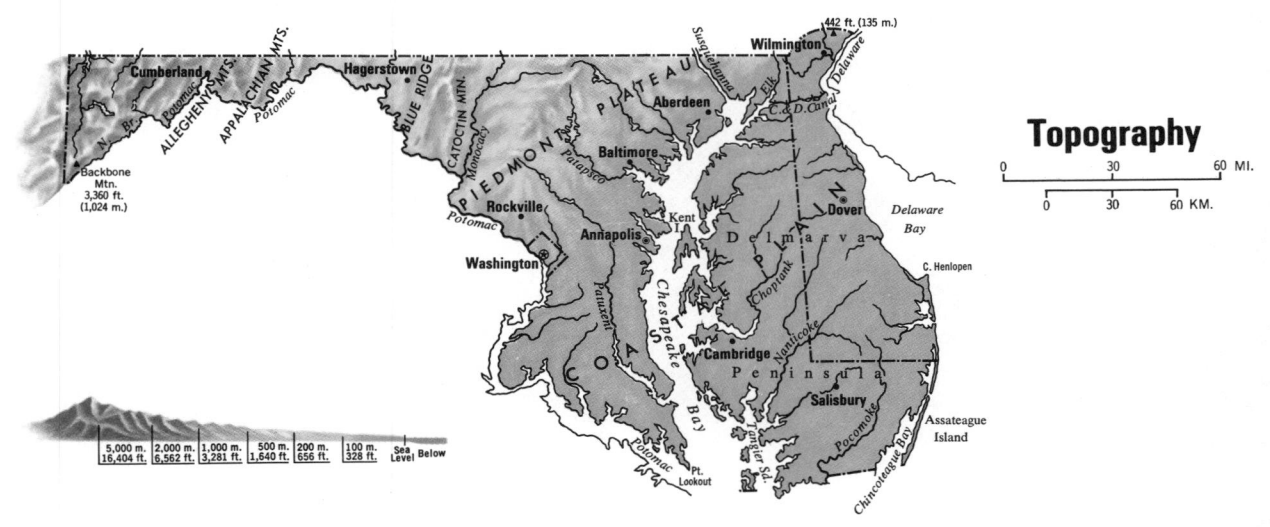

Topography

MARYLAND

AREA 10,460 sq. mi. (27,091 sq. km.)
POPULATION 4,216,975
CAPITAL Annapolis
LARGEST CITY Baltimore
HIGHEST POINT Backbone Mtn. 3,360 ft. (1024 m.)
SETTLED IN 1634
ADMITTED TO UNION April 28, 1788
POPULAR NAME Old Line State; Free State
STATE FLOWER Black-eyed Susan
STATE BIRD Baltimore Oriole

DELAWARE

AREA 2,044 sq. mi. (5,294 sq. km.)
POPULATION 594,317
CAPITAL Dover
LARGEST CITY Wilmington
HIGHEST POINT Ebright Road 442 ft. (135 m.)
SETTLED IN 1627
ADMITTED TO UNION December 7, 1787
POPULAR NAME First State; Diamond State
STATE FLOWER Peach Blossom
STATE BIRD Blue Hen Chicken

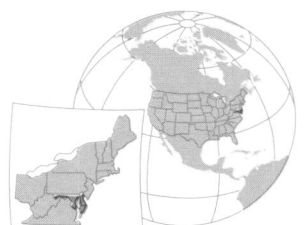

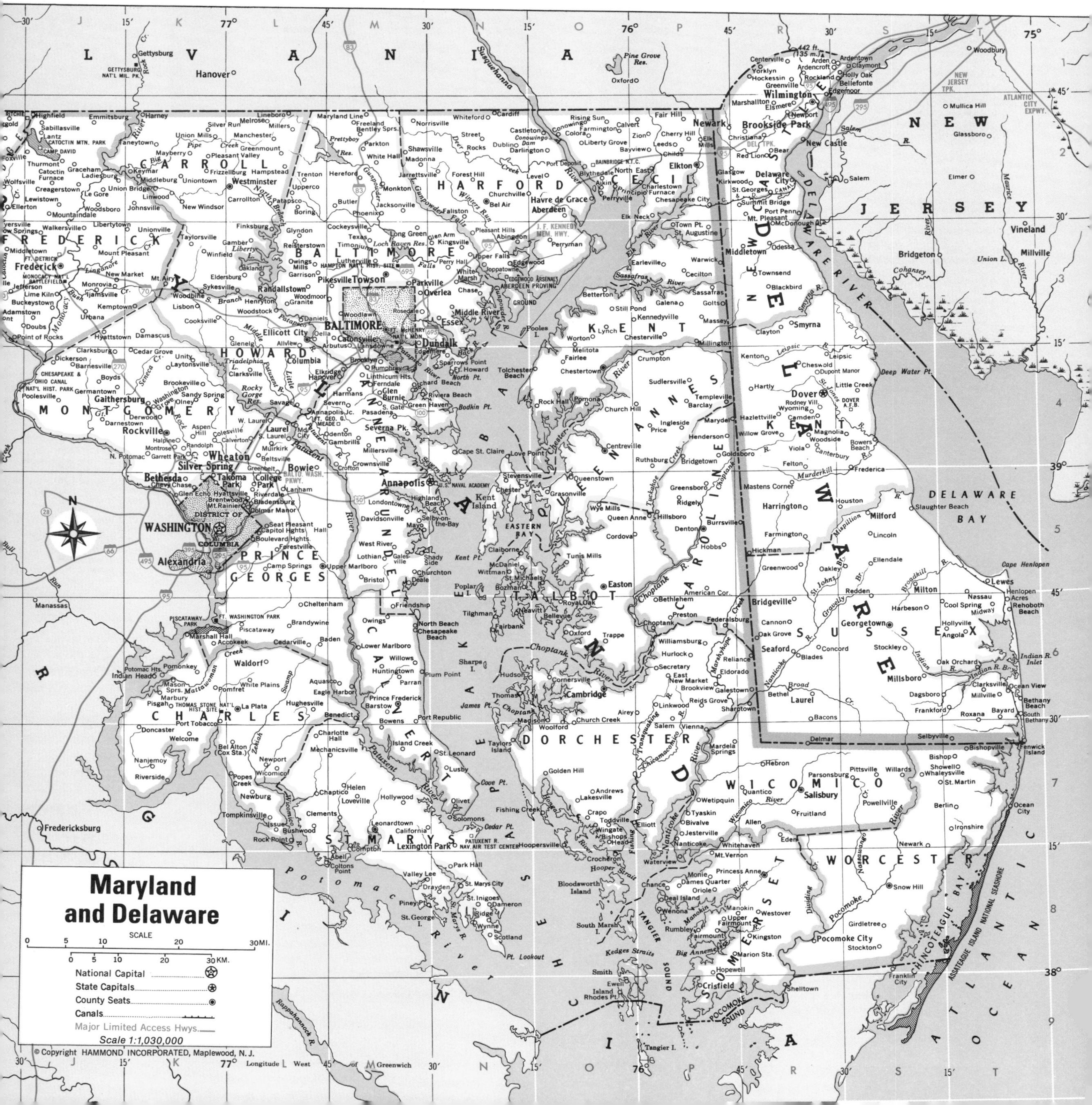

Maryland and Delaware

SCALE
0 5 10 20 30 MI.
0 5 10 20 30 KM.

National Capital .... ⊛
State Capitals .... ⊛
County Seats .... ◉
Canals
Major Limited Access Hwys.

Scale 1:1,030,000

© Copyright HAMMOND INCORPORATED, Maplewood, N.J.

(continued)

21701 Lewistown 600 .... J2
20653 Lexington Park 10,361..... M7
21762 Libertytown 400 .... J3
21090 Linthicum Heights 7,457 .. M4
21766 Little Orleans 600 .... E2
†21550 Loch Lynn Heights 503..... A3
21539 Lonaconing 1,420 .... C2
†21035 Londontowne 6,052 .... M4
21092 Long Green 1,626 .... M3
20656 Loveville 600................. M7
21540 Luke 329 .... B3
21093 Lutherville-Timonium 16,871 .... M3
21648 Madison 350.................. O6
21102 Manchester 1,830 .... L2
20658 Marbury 1,189 .... K6
21837 Mardela Springs 320 .... P7
21838 Marion Station 400 .... R8
†20616 Marshall Hall 325 .... K6
21649 Marydel 152 .... P4
†21113 Maryland City 6,949.......... L4
21767 Maugansville 1,707........... H2
21106 Mayo 2,795.................... M5
20659 Mechanicsville 784 .... M7
21220 Middle River 26,756 .... N3
21769 Middletown 1,748 .... J3
21542 Midland 601 .... C2
21108 Millersville 380 .... M4
21651 Millington 546 .... P3
†20028 Morningside 1,395 .... G5
†21701 Mountaindale 400............ J2
21550 Mountain Lake Park 1,597 . A3
21771 Mount Airy 2,450............ K3
†21701 Mount Pleasant 400 .... J3
20822 Mount Rainier 7,361 .... F4
21545 Mount Savage 1,640......... C2
†21853 Mount Vernon 900 .... P8
†20705 Muirkirk 950 .... L4
21773 Myersville 432................ H3
21840 Nanticoke 450 .... P7
†21502 Narrows Park-La Vale 5,523 .... C2
21841 Newark 900 .... S7
20664 Newburg 550 .... L7
20784 New Carrollton 12,632...... G4
21774 New Market 306 .... J3
21776 New Windsor 799 .... K2
20831 North Beach 1,504 .... N6
†20722 North Brentwood 580 .... F4
21901 North East 1,469.............. P2
†20854 North Potomac K4
21550 Oakland⊙ 1,994 .... A3
†21784 Oakland 2,242................. L3
21842 Ocean City 4,946 .... T7
21113 Odenton 13,270............... M4
†21228 Oella 600 .... L3
20832 Olney 13,026 .... K4
21206 Overlea 12,965 .... N3
20836 Owings 700 .... M6
21117 Owings Mills 9,526.......... L3
21654 Oxford 754 .... O6
20745 Oxon Hill 36,267............. F6
20667 Park Hall 775 .... N8
21234 Parkville 35,159 .... M3

21122 Pasadena 7,439 .... M4
21128 Perry Hall 13,455 .... N3
21130 Perryman 1,819.............. O3
21903 Perryville 2,018 .... O2
21208 Pikesville 22,555 .... M3
20674 Piney Point 950 .... M8
†20735 Piscataway 500............... L6
20640 Pisgah 650.................... K6
21850 Pittsville 519 .... S7
†21087 Pleasant Hills 2,790 .... N3
21851 Pocomoke City 3,558 .... R8
20675 Pomfret 600 .... L6
†20640 Pomonkey 410 .... K6
20837 Poolesville 3,428............. J4
21904 Port Deposit 664 .... O2
20677 Port Tobacco 40 .... K6
20640 Potomac Heights 2,456 .... K6
†21502 Potomac Park-Bowling Green 2,275 .... C2
21852 Powellville 400 .... S7
21655 Preston 498................... P6
20678 Prince Frederick⊙ 1,805 .. M6
21853 Princess Anne⊙ 1,499 ..... P8
†21090 Pumphrey 5,666 .... M4
21657 Queen Anne 259 .... O5
21658 Queenstown 491 .... O5
21133 Randallstown 25,927......... L3
21557 Rawlings 500 .... C2
21136 Reisterstown 19,385 .... L3
20680 Ridge 550 .... N8
21660 Ridgely 933 .... P5
21911 Rising Sun 1,160............. O2
†20027 Ritchie 950 .... G5
†20840 Riverdale Heights G4
†21061 Riviera Beach 8,812......... N4
21661 Rock Hall 1,511.............. O4
†21084 Rocks 450..................... N2
*20850 Rockville⊙ 43,811 .... K4
21779 Rohrersville 525.............. H3
21237 Rosedale 19,956.............. M3
†21758 Rosemont 300................ H3
21662 Royal Oak 600................. O6
21780 Sabillasville 450 .... J2
20684 Saint Inigoes 750 .... N8
21663 Saint Michaels 1,301......... N5
21801 Salisbury⊙ 16,429.......... R7
20860 Sandy Spring-Ashton 2,659 K4
20863 Savage-Guilford 2,928 ...... L4
20687 Scotland 475.................. N8
20801 Seabrook-Lanham 15,814 .. G4
20027 Seat Pleasant 5,217 .... G5
21664 Secretary 487................. P6
†21037 Selby-on-the-Bay 3,125..... N5
21144 Severn 20,147................ M4
21146 Severna Park 21,253 .... M4
20867 Shady Side 2,877............. M5
21782 Sharpsburg 721 .... G3
21861 Sharptown 654 .... R6
20023 Silver Hill-Suitland 32,164...... F5
†21157 Silver Run 350 .... K2
*20901 Silver Spring 72,893 .... F4
21783 Smithsburg 833 .... H2
21863 Snow Hill⊙ 2,192............ S8

†20015 Somerset 1,101 .... E4
†21113 South Gate 24,185 .... M4
†20795 South Kensington 9,344.... E4
†20810 South Laurel 18,034 .... L4
21219 Sparrows Point............... N4
21666 Stevensville 500.............. N5
21667 Still Pond 350 .... O3
21864 Stockton 400 .... S8
21668 Sudlersville 443 .... P4
†20746 Suitland-Silver Hill 32,164................. F5
21784 Sykesville 1,712.............. K3
20912 Takoma Park 16,231........ F4
21787 Taneytown 2,618 .... K2
21669 Taylors Island 400........... N7
21788 Thurmont 2,934.............. J2
21670 Templeville 96 .... P4
21671 Tilghman 979................. N6
21093 Timonium-Lutherville 16,871 .... M3
21672 Toddville 500 .... O7
21204 Towson⊙ 51,083............ M3
21673 Trappe 739.................... O6
†20780 Tuxedo 500 .... G5
21791 Union Bridge 927 .... K2
†20740 University Park 2,536 .... F4
21155 Upperco 500................... L2
21867 Upper Fairmount 500 .... P8
21156 Upper Falls 550 .... N3
20870 Upper Marlboro⊙ 828...... M5
20692 Valley Lee 600................ M8
21869 Vienna 300.................... P7
20601 Waldorf 9,782 .... L6
†20023 Walker Mill 10,651 .... F5
21793 Walkersville 2,212 .... J3
21912 Warwick 550.................. P3
20880 Washington Grove 527 ..... K4
20693 Welcome 438.................. K7
21562 Westernport 2,706........... B3
†20784 West Lanham Hills 350..... G4
21157 Westminster⊙ 8,808 .... L2
21871 Westover 450.................. R8
20902 Wheaton-Glenmont 48,598 E3
21160 Whiteford 500 .... N2
21161 White Hall 360 .... M2
21162 White Marsh 500 .... N3
†20901 White Oak 13,700 .... F3
20695 White Plains 5,167 .... L6
21874 Willards 540 .... S7
21795 Williamsport 2,153 .... G2
21676 Wittman 544.................. N5
21797 Woodbine 872................. K3
21798 Woodsboro 506............... J2
21163 Woodstock 700 .... L3
21677 Woolford 330................. O7
21679 Wye Mills 315................. O5
†20680 Wynne 450.................... N8
†21701 Yellow Springs 940.......... H3

OTHER FEATURES

Aberdeen Proving Ground 5,722.... N3
Allegheny Front (mts.) C2
Andrews A.F.B. 10,064............... G5
Antietam (creek)........................ H2
Antietam Nat'l Battlefield H3
Army Chemical Center O3
Back (riv.) N4
Backbone (mt.) A3
Bainbridge N.T.C. O2
Bald Hill Branch (riv.) G4
Big Annemessex (riv.)................. P8
Big Pipe (creek) K2
Bloodsworth (isl.) O8
Blue Ridge (mts.) H3
Bodkin (pt.) N4
Bush (creek) J3
Cabin John (creek)..................... E4
Camp David J2
Casselman (riv.)......................... B2
Catoctin (creek).......................... H3
Catoctin Mt. Park J2
Cedar (pt.)................................ N7
Census Bureau F5
Chesapeake (bay) N7
Chesapeake and Delaware (canal)...................... R2
Chesapeake and Ohio Canal Nat'l Hist. Park J4
Chester (riv.)............................. O4
Chicamacomico (riv.).................. P7
Chincoteague (bay) S8
Choptank (riv.)........................... O6
Clara Barton Nat'l Hist. Site E4
Conococheague (creek) G1
Conowingo (dam) O2
Cove (pt.)................................. N7
Deep Creek (lake) A3
Deer (creek) N2
Dividing (creek) R8
Eastern (bay) N5
Elk (riv.).................................. P3
Fishing (bay) O7
Fort Detrick J3
Fort George G. Meade 14,083....... L4
Fort McHenry Nat'l Mon. M3
Fort Ritchie 1,754........................ H2
Fort Washington Park L6
Great Seneca (creek) J4
Greenbelt Park G4
Green Ridge (mts.) E2
Gunpowder (riv.) N3
Gunpowder Falls (creek) M2
Hampton Nat'l Hist. Site M3
Harpers Ferry Nat'l Hist. Park G3
Henson (creek) F6
Honga (riv.) O7
Hooper (str.) O8
Indian (creek) G4
James (pt.) N6
Kedges (strs) O8
Kent (isl.) N5
Kent (pt.) N5
Liberty (lake) L3
Linganore (creek)........................ J3
Little Choptank (riv.).................. N6
Little Gunpowder Falls (creek)..................... M2

Little Paint Branch (riv.) F4
Little Patuxent (riv.).................... L4
Loch Raven (res.)....................... M3
Lookout (pt.)............................. N8
Manokin (riv.)............................ P8
Marshyhope (creek)..................... P6
Mattawoman (creek) K6
Meadow (mt.) B2
Middle Patuxent (riv.) L3
Monocacy (riv.).......................... J3
Monocacy Nat'l Battlefield J3
Nanticoke (riv.) P7
Nassawango (creek) S8
National Agricultural Research Center G3
Naval Academy, U.S. 5,367 .... N5
Naval Medical Center E4
Naval Weapons Center F3
North (pt.) N4
Oceanographic Office F5
Oxon Run (riv.) F5
Paint Branch (riv.) F4
Patapsco (riv.) M4
Patuxent (riv.) M7
Patuxent River Nav. Air Test Ctr. N7
Piscataway (creek) G6
Piscataway Park K6
Pocomoke (riv.) S8
Pocomoke (sound)...................... P9
Pooles (isl.) O3
Poplar (isl.)............................... N5
Potomac (riv.) M8
Prettyboy (res.) M2
Rock (creek) K4
Rocky Gorge (res.) L4
Saint George (isl.)...................... M8
Saint Marys (riv.) N8
Sassafras (riv.) P3
Savage (riv.) B2
Savage River (lake) B2
Severn (riv.).............................. N4
Sharps (isl.) N6
Smith (isl.) O8
South Marsh (isl.) O8
Susquehanna (riv.)..................... N1
Tangier (sound).......................... P8
Thomas Stone Nat'l Hist. Site K6
Tinkers (creek) F6
Topographic Center E4
Town (creek) E2
Transquaking (riv.) P7
Triadelphia (lake) L4
Tuckahoe (creek)........................ P5
Walter Reed Army Med. Ctr. Annex E4
Wicomico (riv.).......................... L7
Wicomico (riv.).......................... R7
Winters Run (creek) N2
Youghiogheny (riv.).................... A3
Youghiogheny River (lake) A2
Zekiah Swamp (riv.) L7

DELAWARE

COUNTIES

Kent 98,219 .... R
New Castle 398,115 .... R
Sussex 97,983............................ S

CITIES and TOWNS

Zip Name/Pop. *Ke*

†19801 Arden 516 .... R
†19810 Ardencroft 267 .... R
†19810 Ardentown 307 .... S
19809 Bellefonte 1,279.............. S
19930 Bethany Beach 330 .... T
19931 Bethel 197..................... R
†19973 Blades 664 .... R
†19962 Bowers Beach 198........... S
19993 Bridgeville 1,238 .... R
19711 Brookside 15,255............ R
19934 Camden 1,757................ R
†19801 Centerville 800............... R
19936 Cheswold 269 .... R
†19711 Christiana 500 .... R
19937 Clarksville 350 .... T
19703 Claymont 10,022............. S
19938 Clayton 1,216................. R
19930 Dagsboro 344 .... S
19706 Delaware City 1,858......... R
19940 Delmar 948 .... R
19901 Dover (cap.)⊙ 23,507...... R
†19901 Dupont Manor 1,059........ R
†19801 Edgemoor 7,397 .... S
19941 Ellendale 361 .... S
†19801 Elsmere 6,493................ R
19942 Farmington 141 .... R
19943 Felton 547..................... R
19944 Fenwick Island 114 .... T
19945 Frankford 828................. S
19946 Frederica 864 .... S
19947 Georgetown⊙ 1,710 .... S
†19711 Glasgow 350.................. R
19950 Greenwood 578 .... R
19952 Harrington 2,405............. R
†19971 Henlopen Acres 176......... T
19707 Hockessin 950................ R
†19801 Holly Oak .... S
19954 Houston 357 .... S
19955 Kenton 243.................... R
19708 Kirkwood 350................. R
19956 Laurel 3,052 .... R
†19901 Leipsic 228.................... S
19958 Lewes 2,197 .... T
19960 Lincoln 757 .... S
19961 Little Creek 230 .... S
19962 Magnolia 197 .... R
19709 Middletown 2,946 .... R
19963 Milford 5,366................. S
19966 Millsboro 1,233 .... S
19967 Millville 178 .... T
19968 Milton 1,359.................. S
19711 Newark 25,247 .... P
19720 New Castle 4,907 .... R
19804 Newport 1,167 .... R
†19966 Oak Orchard 350............. T
19970 Ocean View 495.............. T
19730 Odessa 384 .... R
19971 Rehoboth Beach 1,730 ..... T
19901 Rodney Village 1,753........ R
19733 Saint Georges 450............ R
19973 Seaford 5,256 .... R
19975 Selbyville 1,251 .... S
†19963 Slaughter Beach 121 .... S
19977 Smyrna 4,750 .... R
†19930 South Bethany 115 .... T
19734 Townsend 386................. R
19979 Viola 167 .... R
*19801 Wilmington⊙ 70,195 .... R
Wilmington‡ 524,108....... R
19980 Woodside 248................. R
19934 Wyoming 960 .... R
19736 Yorklyn 600................... R

OTHER FEATURES

Broad (creek) .... R
Broadkill (riv.)........................... S
Chesapeake and Delaware (canal) ... R
Choptank (riv.)........................... R
Deep Water (pt.) .... S
Delaware (bay)........................... S
Delaware (riv.) .... R
Dover A.F.B. 4,391 .... S
Henlopen (cape) .... T
Indian (riv.).............................. S
Indian River (bay) .... T
Indian River (inlet) .... T
Leipsic (riv.).............................. S
Mispillion (riv.) .... S
Murderkill (riv.) .... R
Nanticoke (riv.) .... R
Saint Jones (riv.) .... R
Smyrna (res.) .... R

DISTRICT OF COLUMBIA

CITIES and TOWNS

Zip Name/Pop. *K*

20007 Georgetown .... F
*20001 Washington, D.C. (cap.), U.S. 638,432 .... F
Washington‡ 3,060,240.... F

OTHER FEATURES

Anacostia (riv.) .... F
Bolling A.F.B. .... E
Fort Lesley J. McNair .... E
Kennedy Center .... A
Naval Yard .... F
U.S. Capitol .... F
Walter Reed Army Med. Ctr. .... E

⊙County seat.
‡Population of metropolitan area.
† Zip of nearest p.o.
* Multiple zips.

Agriculture, Industry and Resources

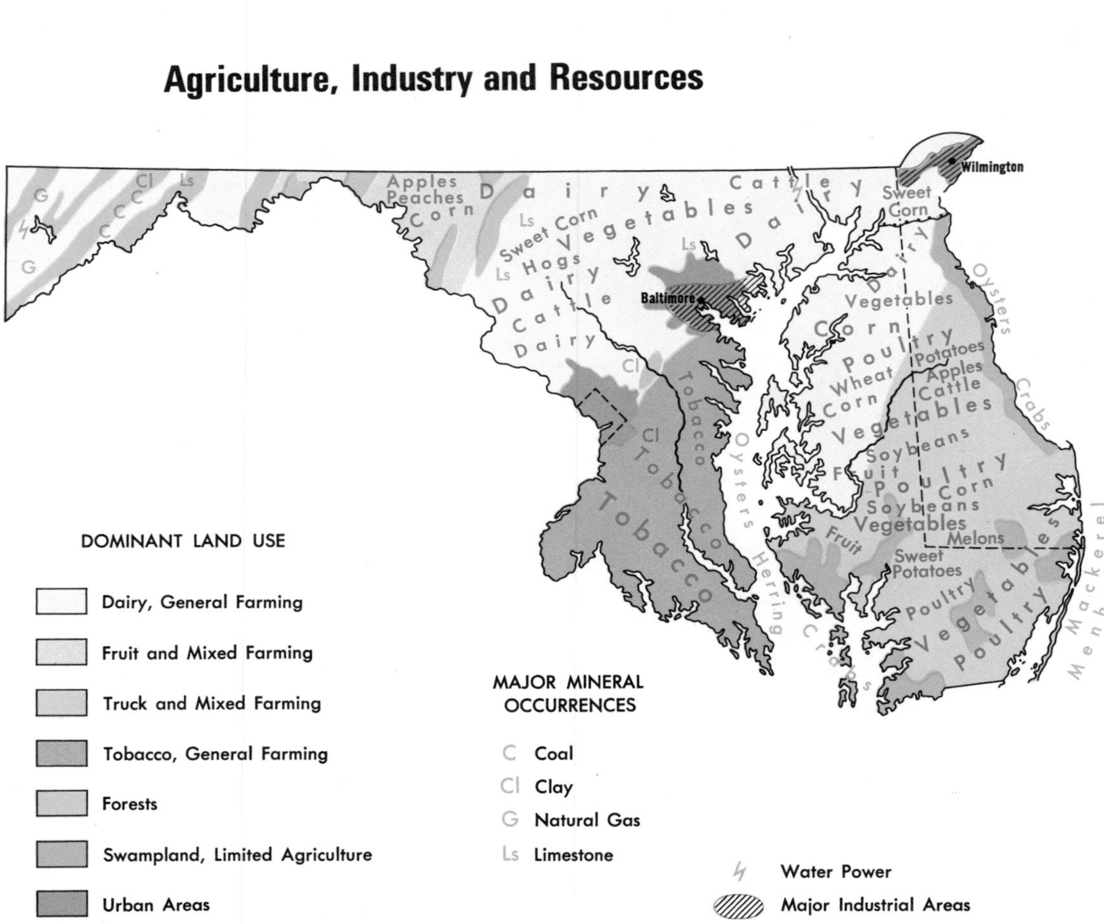
Wilmington
Baltimore
Apples
Peaches
Corn
Dairy
Cattle
Vegetables
Sweet Corn
Hogs
Poultry
Wheat
Potatoes
Tobacco
Soybeans
Fruit
Melons
Sweet Potatoes
Oysters
Crabs
Herring
Mackerel
Menhaden
DOMINANT LAND USE
Dairy, General Farming
Fruit and Mixed Farming
Truck and Mixed Farming
Tobacco, General Farming
Forests
Swampland, Limited Agriculture
Urban Areas
MAJOR MINERAL OCCURRENCES
C Coal
Cl Clay
G Natural Gas
Ls Limestone
Water Power
Major Industrial Areas

MASSACHUSETTS

AREA 8,284 sq. mi. (21,456 sq. km.)
POPULATION 5,737,037
CAPITAL Boston
LARGEST CITY Boston
HIGHEST POINT Mt. Greylock 3,491 ft. (1064 m.)
SETTLED IN 1620
ADMITTED TO UNION February 6, 1788
POPULAR NAME Bay State; Old Colony
STATE FLOWER Mayflower
STATE BIRD Chickadee

RHODE ISLAND

AREA 1,212 sq. mi. (3,139 sq. km.)
POPULATION 947,154
CAPITAL Providence
LARGEST CITY Providence
HIGHEST POINT Jerimoth Hill 812 ft. (247 m.)
SETTLED IN 1636
ADMITTED TO UNION May 29, 1790
POPULAR NAME Little Rhody; Ocean State
STATE FLOWER Violet
STATE BIRD Rhode Island Red

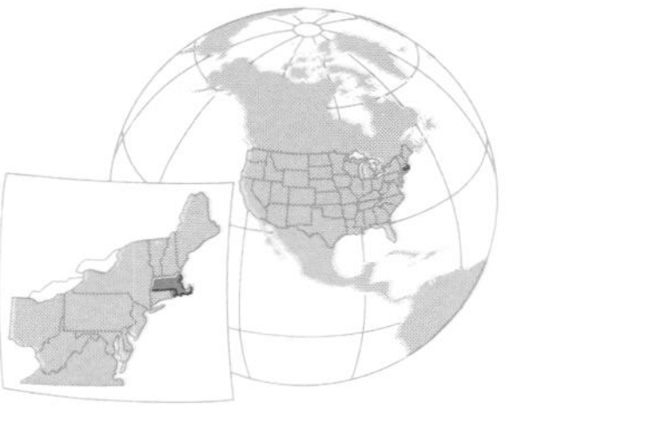

Agriculture, Industry and Resources

DOMINANT LAND USE

Specialized Dairy
Dairy, Poultry, Mixed Farming
Forests
Urban Areas

MAJOR MINERAL OCCURRENCES

Gn Granite

Water Power
Major Industrial Areas

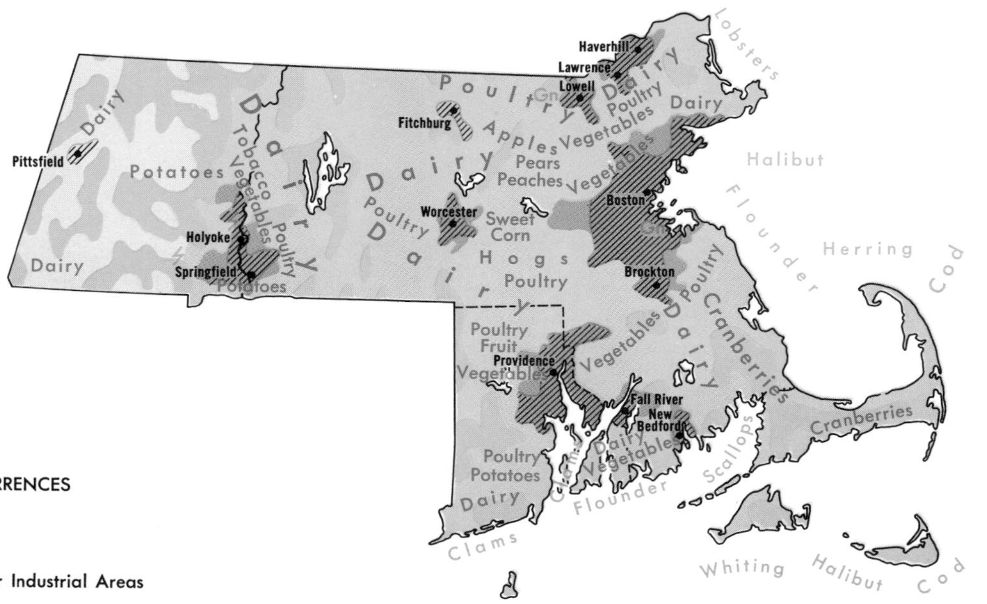

MASSACHUSETTS

COUNTIES

Barnstable 147,925 ..... N6
Berkshire 145,110 ..... B3
Bristol 474,641 ..... K5
Dukes 8,942 ..... M7
Essex 633,632 ..... L2
Franklin 64,317 ..... D2
Hampden 443,018 ..... D4
Hampshire 138,813 ..... D3
Middlesex 1,367,034 ..... J3
Nantucket 5,087 ..... O7
Norfolk 606,587 ..... K4
Plymouth 405,437 ..... L5
Suffolk 650,142 ..... K3
Worcester 646,352 ..... G3

CITIES and TOWNS

Zip	Name/Pop.	Key
02351	Abington○ 13,517	L4
01720	Acton○ 17,544	J3
02743	Acushnet○ 8,704	L6
01220	Adams○ 10,381	B2
01220	Adams 6,857	B2
01001	Agawam○ 26,271	D4
01261	Alford○ 394	A4
01913	Amesbury○ 13,971	L1
01913	Amesbury 12,236	L1
01002	Amherst○ 33,229	E3
01002	Amherst 17,773	E3
01810	Andover○ 26,370	K2
01810	Andover 8,445	K2
02174	Arlington○ 48,219	C6
01430	Ashburnham○ 4,075	G2
01430	Ashburnham 900	G2
01431	Ashby○ 2,311	G2
01330	Ashfield○ 1,458	C2
01721	Ashland○ 9,165	J3
01331	Athol○ 10,634	F2
01331	Athol 8,708	F2
02703	Attleboro 34,196	J5
01501	Auburn○ 14,845	G4
02322	Avon○ 5,026	K4
01432	Ayer○ 6,993	H2
01432	Ayer 3,165	H2
01436	Baldwinville 1,709	F2
02630	Barnstable○ 30,898	N6
02630	Barnstable⊙ 2,033	N6
01005	Barre○ 4,102	F3
01005	Barre 1,136	F3
01223	Becket○ 1,339	B3
01730	Bedford○ 13,067	B6
01007	Belchertown○ 8,339	E3
01007	Belchertown 2,531	E3
02019	Bellingham○ 14,300	J4
02019	Bellingham 4,454	J4
02178	Belmont○ 26,100	C6
02780	Berkley○ 2,731	K5
01503	Berlin○ 2,215	H3
01337	Bernardston○ 1,750	D2
01915	Beverly 37,655	E5
01821	Billerica○ 36,727	J2
01504	Blackstone○ 6,570	H4
01008	Blandford○ 1,038	C4
01740	Bolton○ 2,530	H3
01009	Bondsville 1,906	E4
*02101	Boston (cap.)⊙ 562,994	D7
	Boston‡ 2,763,357	D7
02532	Bourne○ 13,874	M6
02532	Bourne 2,678	M6
01719	Boxborough○ 3,126	H3
01921	Boxford○ 5,374	L2
01921	Boxford 1,841	L2
01505	Boylston○ 3,470	H3
02184	Braintree○ 36,337	D8
02020	Brant Rock-Ocean Bluff 4,055	M4
02631	Brewster○ 5,226	O5
02631	Brewster 1,744	O5
02324	Bridgewater○ 17,202	K5
02324	Bridgewater 6,781	K5
01010	Brimfield○ 2,318	F4
*C2401	Brockton 95,172	K4
	Brockton‡ 169,374	K4
01506	Brookfield○ 2,397	F4
01506	Brookfield 1,037	F4
02146	Brookline○ 55,062	C7
01338	Buckland○ 1,864	C2
01803	Burlington○ 23,486	C5
02532	Buzzards Bay 3,375	M5
02138	Cambridge⊙ 95,322	C7
02021	Canton○ 18,182	C8
01741	Carlisle○ 3,306	J2
02330	Carver○ 6,988	M5
02632	Centerville 3,640	N6
01339	Charlemont○ 1,149	C2
01507	Charlton○ 6,719	F4
02633	Chatham○ 6,071	P6
02633	Chatham 1,922	P6
01824	Chelmsford○ 31,174	J2
02150	Chelsea 25,431	D6
01225	Cheshire○ 3,124	B2
01011	Chester○ 1,123	C3
01012	Chesterfield○ 1,000	C3
*01013	Chicopee 55,112	D4
02535	Chilmark○ 489	M7
†02054	Clicquot-Millis 3,777	A8
01510	Clinton○ 12,771	H3
01778	Cochituate 6,126	A7
02025	Cohasset○ 7,174	F7
01340	Colrain○ 1,552	D2
01742	Concord○ 16,293	B6
01341	Conway○ 1,213	D2
†01772	Cordaville 1,384	H3
01026	Cummington○ 657	C3
01226	Dalton○ 6,797	B3
01923	Danvers○ 24,100	D5
02714	Dartmouth○ 23,966	K6
02026	Dedham⊙○ 25,298	C7
01342	Deerfield○ 4,517	D2
02638	Dennis○ 12,360	O5
02639	Dennis Port 2,570	O6
02715	Dighton○ 5,352	K5
†02122	Dorchester	D7
†01516	Douglas○ 3,730	H4
02030	Dover○ 4,703	B7
02030	Dover 2,051	B7
01826	Dracut○ 21,249	J2
01570	Dudley○ 8,717	G4
01827	Dunstable○ 1,671	J2
02332	Duxbury○ 11,807	M4
02332	Duxbury 1,685	M4
02333	East Bridgewater○ 9,945	L4
01515	East Brookfield○ 1,955	G4
01515	East Brookfield 1,443	G4
01516	East Douglas 1,683	G4
02536	East Falmouth (Teaticket) 5,181	M6
02642	Eastham○ 3,472	O5
01027	Easthampton○ 15,580	D3
01028	East Longmeadow○ 12,905	E4
02334	Easton○ 16,623	K4
01437	East Pepperell 2,212	H2
02539	Edgartown○ 2,204	M7
02539	Edgartown⊙ 1,138	M7
01344	Erving○ 1,326	E2
01929	Essex○ 2,998	L2
01929	Essex 1,490	L2
02149	Everett 37,195	D6
02719	Fairhaven○ 15,759	L6
*02720	Fall River 92,574	K6
	Fall River‡ 176,831	K6
*02540	Falmouth○ 23,640	M6
*02540	Falmouth 5,720	M6
01518	Fiskdale 1,859	F4
01420	Fitchburg⊙ 39,580	G2
	Fitchburg-Leominster‡ 99,957	G2
†01247	Florida○ 730	B2
02035	Foxboro○ 14,148	J4
02035	Foxboro 5,697	J4
01701	Framingham○ 65,113	A7
02038	Franklin○ 18,217	J4
02038	Franklin 9,296	J4
01440	Gardner 17,900	G2
†02535	Gay Head○ 220	L7
01833	Georgetown○ 5,687	L2
01031	Gilbertville 1,029	F3
†01376	Gill○ 1,259	D2
01930	Gloucester 27,768	M2
01032	Goshen○ 651	C3
01519	Grafton○ 11,238	H4
01033	Granby○ 5,380	E3
01033	Granby 1,302	E3
01034	Granville○ 1,204	C4
01230	Great Barrington○ 7,405	A4
01230	Great Barrington 3,150	A4
01301	Greenfield○ 18,436	D2
01301	Greenfield⊙ 14,198	D2
02041	Green Harbor 2,002	M4
01450	Groton○ 6,154	H2
01450	Groton 1,264	H2
01830	Groveland○ 5,040	L1
01035	Hadley○ 4,125	D3
02338	Halifax○ 5,513	L5
01936	Hamilton○ 6,960	L2
01036	Hampden○ 4,745	E4
01237	Hancock○ 643	A2
02339	Hanover○ 11,358	L4
02341	Hanson○ 8,617	L4
02341	Hanson 2,120	L4
01037	Hardwick○ 2,272	F3
01451	Harvard○ 12,170	H2
02645	Harwich○ 8,971	O6
02645	Harwich 4,399	O6
01038	Hatfield○ 3,045	D3
01038	Hatfield 1,251	D3
01830	Haverhill 46,865	K1
01346	Heath○ 482	C2
02043	Hingham○ 20,339	E8
02043	Hingham 5,742	E8
01235	Hinsdale○ 1,707	B3
02343	Holbrook○ 11,140	D8
01520	Holden○ 13,336	G3
†01550	Holland○ 1,589	F4
01746	Holliston○ 12,622	A8
01040	Holyoke 44,678	D4
01747	Hopedale○ 3,905	H4
01747	Hopedale 2,810	H4
01748	Hopkinton○ 7,114	J4
01748	Hopkinton 2,542	J4
01236	Housatonic 1,314	A3
01452	Hubbardston○ 1,797	F3
01749	Hudson○ 16,408	H3
01749	Hudson 14,156	H3
02045	Hull○ 9,714	E7
01050	Huntington○ 1,804	C4
02601	Hyannis 9,118	N6
01938	Ipswich○ 11,158	L2
01938	Ipswich 4,548	L2
02364	Kingston○ 7,362	M5
02364	Kingston 4,405	M5
02346	Lakeville○ 5,931	L5
02346	Lakeville 1,948	L5
01523	Lancaster○ 6,334	H3
01237	Lanesboro○ 3,131	A2
*01840	Lawrence⊙ 63,175	K2
	Lawrence-Haverhill‡ 281,981	K2
01238	Lee○ 6,247	B3
01238	Lee 2,140	B3
01524	Leicester○ 9,446	G4
01240	Lenox○ 6,523	A3
01240	Lenox 2,668	A3
01453	Leominster 34,508	G2
01054	Leverett○ 1,471	E3
02173	Lexington○ 29,479	B6
†01301	Leyden○ 498	D2
01773	Lincoln○ 7,098	B6
01460	Littleton○ 6,970	H2
†01460	Littleton Common 3,109	J2
01106	Longmeadow○ 16,301	D4
*01850	Lowell⊙ 92,418	J2
	Lowell‡ 233,410	J2
01056	Ludlow○ 18,150	E4
01462	Lunenburg○ 8,405	H2
01462	Lunenburg 1,789	H2
*01901	Lynn 78,471	D6
01940	Lynnfield○ 11,267	D5
02148	Malden 53,386	D6
01944	Manchester○ 5,424	F5
02048	Mansfield○ 13,453	J4
02048	Mansfield 6,786	J4
01945	Marblehead○ 20,126	E7
02738	Marion○ 3,932	L6
02738	Marion 1,438	L6
01752	Marlborough 30,617	H3
02050	Marshfield○ 20,916	M4
02050	Marshfield 4,421	M4
02051	Marshfield Hills 2,308	M4
02649	Mashpee○ 3,700	M6
02739	Mattapoisett○ 5,597	L6
02739	Mattapoisett 3,159	L6
01754	Maynard○ 9,590	J3
02052	Medfield○ 10,220	B8
02052	Medfield 6,108	B8
02155	Medford 58,076	C6
02053	Medway○ 8,447	J4
02176	Melrose 30,055	D6
01756	Mendon○ 3,108	H4
01860	Merrimac○ 4,451	L1
01844	Methuen○ 36,701	K2
02346	Middleboro○ 16,404	L5
02346	Middleboro 7,012	L5
01243	Middlefield○ 385	B3
01949	Middleton○ 4,135	K2
01757	Milford○ 23,390	H4
01757	Milford 21,730	H4
01527	Millbury○ 11,808	H4
01349	Millers Falls 1,101	E2
02054	Millis○ 6,908	A8
02054	Millis-Clicquot 3,777	A8
01529	Millville○ 1,693	H4
02186	Milton○ 25,860	D7
01057	Monson○ 7,315	E4
01057	Monson 2,167	E4
01351	Montague○ 8,011	E2
01245	Monterey○ 818	B4
†12517	Mount Washington○ 93	A4
01908	Nahant○ 3,947	E6
02554	Nantucket○ 5,087	O7
02554	Nantucket⊙ 3,229	O7
01760	Natick○ 29,461	A7
02192	Needham○ 27,901	B7
*02740	New Bedford⊙ 98,478	K6
	New Bedford‡ 169,425	K6
01531	New Braintree○ 671	F3
01950	Newbury○ 4,529	L1
01950	Newburyport⊙ 15,900	L1
†01230	New Marlborough○ 1,160	B4
01355	New Salem○ 688	E2
†02158	Newton 83,622	C7
02056	Norfolk○ 6,363	J4
01247	North Adams 18,063	B2
01059	North Amherst 5,616	E3
01060	Northampton⊙ 29,286	D3
01845	North Andover○ 20,129	K2
*02760	North Attleboro○ 21,095	J5
01532	Northborough○ 10,568	H3
01532	Northborough 5,670	H3
01534	Northbridge○ 12,246	H4
01535	North Brookfield○ 4,150	F3
01535	North Brookfield 2,543	F3
02764	North Dighton 1,174	K5
02651	North Eastham 1,318	O5
01360	Northfield○ 2,386	E2
01360	Northfield 1,182	E2
02358	North Pembroke 2,215	M4
02360	North Plymouth 3,250	L5
01864	North Reading○ 11,455	C5
02060	North Scituate 5,221	F8
02766	Norton○ 12,690	K5
02766	Norton 2,035	K5
02061	Norwell○ 9,182	F8
02062	Norwood○ 29,711	B8
02557	Oak Bluffs○ 1,984	M7
02557	Oak Bluffs 1,124	M7
01068	Oakham○ 994	F3
02065	Ocean Bluff-Brant Rock 4,055	M4
†01566	Old Sturbridge Village 500	F4
02558	Onset 1,493	M6
01364	Orange○ 6,844	E2
01364	Orange 3,942	E2
02653	Orleans○ 5,306	O5
02653	Orleans 1,811	O5
02655	Osterville 1,799	N6
01253	Otis○ 963	B4
01540	Oxford○ 11,680	G4
01540	Oxford 6,369	G4
01069	Palmer○ 11,389	E4
01069	Palmer 3,854	E4
01612	Paxton○ 3,762	G3
01960	Peabody 45,976	E5
†01002	Pelham○ 1,112	E3
02359	Pembroke○ 13,487	L4
01463	Pepperell○ 8,061	H2
01463	Pepperell 2,076	H2
01366	Petersham○ 1,024	F3
†01331	Phillipston○ 953	F2
01866	Pinehurst 6,588	B5
01201	Pittsfield⊙ 51,974	A3
	Pittsfield‡ 90,505	A3
01070	Plainfield○ 425	C2
02762	Plainville○ 5,857	J4
02360	Plymouth○ 35,913	M5
02360	Plymouth⊙ 7,232	M5
02367	Plympton○ 1,974	L5
01541	Princeton○ 2,425	G3
02657	Provincetown○ 3,536	O4
02657	Provincetown 3,372	O4
02169	Quincy 84,743	D7
02368	Randolph○ 28,218	D8
02767	Raynham○ 9,085	K5
02768	Raynham Center 3,776	K5
01867	Reading○ 22,678	C5
02769	Rehoboth○ 7,570	K5
02151	Revere 42,423	D6

(continued on following page)

ZIP	Place	Pop.	Grid
01254	Richmond○	1,659	A3
01542	Rochdale○	1,105	G4
02770	Rochester○	3,205	L6
02370	Rockland○	15,695	K4
01966	Rockport	6,345	M2
01367	Rowe○	336	C2
01969	Rowley○	3,867	L2
01969	Rowley	1,321	L2
†02119	Roxbury		C7
01368	Royalston○	955	F2
01071	Russell○	1,570	C4
01543	Rutland	4,334	G3
01543	Rutland	2,312	G3
02561	Sagamore○	1,152	M5
01970	Salem⊙	38,220	E5
01950	Salisbury○	5,973	L1
01950	Salisbury	3,265	L1
01255	Sandisfield○	720	B4
02563	Sandwich○	8,727	N5
02563	Sandwich	1,784	N5
01906	Saugus○	24,746	D6
01256	Savoy○	644	B2
02066	Scituate○	17,317	F8
02066	Scituate	5,351	F8
02771	Seekonk○	12,269	J5
02067	Sharon○	13,601	K4
02067	Sharon	5,976	K4
01257	Sheffield○	2,743	A4
01370	Shelburne Falls	2,046	D2
01770	Sherborn○	4,049	A8
01464	Shirley○	5,124	H2
01464	Shirley	1,630	H2
01545	Shrewsbury	22,674	H3
01072	Shutesbury○	1,049	E3
02725	Somerset○	18,813	K5
02143	Somerville	77,372	C6
†01002	South Amherst	4,861	E3
01073	Southampton○	4,137	C4
01466	South Ashburnham	1,123	G2
01772	Southborough○	6,193	H3
01550	Southbridge○	16,665	G4
01550	Southbridge	12,882	G4
01373	South Deerfield	1,926	D3
†02332	South Duxbury	2,985	M4
†01075	South Hadley	16,399	D4
01561	South Lancaster	2,329	H3
01077	Southwick○	7,382	C4
02664	South Yarmouth	7,525	O6
01562	Spencer○	10,774	F3
01562	Spencer	6,350	F3
*01101	Springfield⊙	152,319	D4
	Springfield-Chicopee-Holyoke‡	530,668	D4
01564	Sterling○	5,440	G3
01262	Stockbridge○	2,328	A3
01262	Stockbridge	1,109	A3
02180	Stoneham	21,424	C6
02072	Stoughton○	26,710	K4
01775	Stow○	5,144	H3
01566	Sturbridge	5,976	F4
01566	Sturbridge○	1,891	F4
01776	Sudbury○	14,027	A6
01375	Sunderland○	2,929	D3
01907	Swampscott	13,837	E6
02777	Swansea○	15,461	K5
02780	Taunton⊙	45,001	K5
02536	Teaticket	5,181	M6
01468	Templeton○	6,070	F2
01080	Three Rivers	3,322	E4
†01034	Tolland○	235	B4
01983	Topsfield○	5,709	L2
01983	Topsfield	2,647	L2
01469	Townsend○	7,201	H2
01469	Townsend	1,266	H2
02666	Truro○	1,486	O5
01376	Turners Falls	4,711	D2
01879	Tyngsboro○	5,683	J2
01264	Tyringham○	344	A4
01568	Upton○	3,886	H4
01568	Upton-West Upton	2,184	H4
01569	Uxbridge○	8,374	H4
02568	Vineyard Haven	1,704	M7
01880	Wakefield○	24,895	C5
01081	Wales○	1,177	F4
02081	Walpole○	18,859	B8
02081	Walpole	5,274	B8
02154	Waltham	58,200	B6
01082	Ware○	8,953	E3
01082	Ware	6,806	E3
02571	Wareham○	18,457	L5
02571	Wareham	2,493	L5
01083	Warren○	3,777	F4
01083	Warren	1,548	F4
01364	Warwick○	603	E2
02172	Watertown	34,384	C6
01778	Wayland○	12,170	A7
01570	Webster○	14,480	G4
01570	Webster	11,175	G4
02181	Wellesley	27,209	B7
02667	Wellfleet○	2,209	O5
01379	Wendell○	694	E2
01984	Wenham○	3,897	L2
01581	Westborough○	13,619	H3
01581	Westborough	4,238	H3
01583	West Boylston○	6,204	G3
02379	West Bridgewater	6,359	K4
01585	West Brookfield○	3,026	F4
01585	West Brookfield	1,423	F4
02669	West Chatham	1,398	O6
†01742	West Concord	5,331	A6
02670	West Dennis	2,023	O6
01085	Westfield	36,465	C4
01886	Westford○	13,434	J2
†01027	Westhampton○	1,137	C3
01473	Westminster○	5,139	G2
01985	West Newbury○	2,861	L1
02193	Weston○	11,169	B6
02790	Westport○	13,763	K6
01089	West Springfield	27,042	D4

Zip	Name/Pop.	Key
01266	West Stockbridge 1,280	A3
02575	West Tisbury○ 1,010	M7
01587	West Upton-Upton 2,184	H4
02576	West Wareham 1,837	L5
02090	Westwood○	B8
02673	West Yarmouth 3,852	N6
02188	Weymouth 55,601	D8
01093	Whately○ 1,341	D3
01588	Whitinsville 5,379	H4
02382	Whitman○ 13,534	L4
01095	Wilbraham 12,053	E4
01095	Wilbraham 3,379	E4
01096	Williamsburg○ 2,237	C3
01267	Williamstown 8,741	B2
01887	Wilmington 17,471	C5
01475	Winchendon 7,019	F2
01475	Winchendon 4,030	F2
01890	Winchester 20,701	C6
01270	Windsor○ 598	B2
02152	Winthrop 19,294	D6
01801	Woburn 36,626	C6
02543	Woods Hole 1,080	M6
*01601	Worcester⊙ 161,799	H3
	Worcester‡ 372,940	H3
01098	Worthington○ 932	C3
02093	Wrentham 7,580	J4
	Yarmouth○ 18,449	O6
02675	Yarmouth Port 2,490	N6

OTHER FEATURES

Adams Nat'l Hist. Site	D7
Agawam (riv.)	M5
Allerton (pt.)	E7
Ann (cape)	M2
Ashmere (lake)	B3
Assabet (riv.)	H3
Assawompset (pond)	L5
Bachelor (brook)	D3
Berkshire (hills)	B4
Big (pond)	B4
Bigelow (bight)	M1
Blackstone (riv.)	G3
Blue (hill)	C8
Boston (bay)	E6
Boston (harb.)	D7
Boston Nat'l Hist. Park	D6
Brewster (isls.)	E7
Buel (lake)	A4
Buzzards (bay)	L7
Cambridge (res.)	B6
Cape Cod (bay)	N5
Cape Cod (canal)	N5
Cape Cod Nat'l Seashore	P5
Chappaquiddick (isl.)	N7
Charles (riv.)	C7
Chicopee (riv.)	D4
Cobble Mountain (res.)	C4
Cochituate (lake)	A7
Cod (cape)	O4
Concord (riv.)	J2
Congamond (lkes.)	D4
Connecticut (riv.)	D2
Cuttyhunk (isl.)	L7
Deer (isl.)	E7
Deerfield (riv.)	C2
East (isl.)	E6
East Chop (pt.)	M7
Eastern (pt.)	M2
Elizabeth (isls.)	L7
Everett (mt.)	A4
Falls (riv.)	D2
Fort Devens	H2
Fort Rodman	L6
Fresh (pond)	C6
Gammon (pt.)	N6
Gay Head (prom.)	L7
Grace (mt.)	E2
Great (pt.)	O7
Green (riv.)	B2
Greylock (mt.)	B2
Gurnet (pt.)	M4
Hingham (bay)	E7
Holyoke (range)	D3
Hoosac (mts.)	B2
Hoosic (riv.)	A1
Housatonic (riv.)	A4
Ipswich (riv.)	L2
John F. Kennedy Nat'l Hist. Site	C7
Knightville (res.)	C3
Laurence G. Hanscom Field	B6
Little (riv.)	C4
Logan Internat'l Airport	D7
Long (isl.)	E7
Long (isl.)	O4
Long (pond)	L5
Lowell Nat'l Hist. Park	J2
Maine (gulf)	M2
Manhan (riv.)	D4
Manomet (pt.)	N5
Marblehead (neck)	F6
Martha's Vineyard (isl.)	M7
Massachusetts (bay)	M4
Merrimack (riv.)	K1
Mill (riv.)	C3
Mill (riv.)	D3
Millers (riv.)	E2
Minute Man Nat'l Hist. Park	B6
Mishaum (pt.)	L6
Monomonac (lake)	G2
Monomoy (isl.)	O6
Monomoy (isl.)	O6
Mount Hope (bay)	K6
Muskeget (chan.)	N7
Muskeget (isl.)	N7
Mystic (lake)	C6
Mystic (riv.)	C6
Nahant (bay)	E6
Nantucket (isl.)	O8
Nantucket (sound)	N6
Nashawena (isl.)	L7
Nashua (riv.)	H3
Naushon (isl.)	L7
Neponset (riv.)	C8
Nomans Land (isl.)	L7
Nonamesset (isl.)	M6
North (riv.)	D2
North (riv.)	L4
Onota (lake)	A3
Otis (res.)	B4
Otis A.F.B.	M6
Pasque (isl.)	L7
Plum (isl.)	L2
Plymouth (bay)	M5
Poge (cape)	N7
Pontoosuc (lake)	A3
Quabbin (res.)	E3
Quaboag (riv.)	F4
Quincy (bay)	D7
Quinebaug (riv.)	F4
Race (riv.)	N4
Salem Maritime Nat'l Hist. Site	E5
Saugus Iron Works Nat'l Hist. Site	D6
Shawshine (riv.)	K2
Silver (lake)	L4
South (riv.)	D2
Springfield Armory Nat'l Hist. Site	D4
Squibnocket (pt.)	M7
Stillwater (riv.)	G3
Sudbury (riv.)	A6
Sudbury (res.)	H3
Swift (riv.)	E4
Taconic (mts.)	A2
Taunton (riv.)	K5
Thompson (isl.)	D7
Toby (mt.)	E3
Tom (mt.)	D4
Tuckernuck (isl.)	N7
Vineyard (sound)	L7
Wachusett (mt.)	G3
Wachusett (res.)	H3
Walden (pond)	A6
Ware (riv.)	F3
Watuppa (pond)	K6
Webster (lake)	G4
Wellfleet (harb.)	O5
West (riv.)	H4
West Branch, Farmington (riv.)	B4
West Chop (pt.)	M7
Westfield (riv.)	C3
Westover A.F.B.	D4
Weweantic (riv.)	L5
Whitman (riv.)	G2
Winter I. Coast Guard Air Sta.	E5

RHODE ISLAND

COUNTIES

Bristol 46,942	J6
Kent 154,163	H6
Newport 81,383	K6
Providence 571,349	H5
Washington 93,317	H7

CITIES and TOWNS

Zip	Name/Pop.	Key
02804	Ashaway 1,747	G7
02806	Barrington○ 16,174	J6
02807	Block Island 620	H8
02808	Bradford 1,354	H7
02809	Bristol○ 20,128	J6
02863	Central Falls 16,995	J5
02816	Coventry○ 27,065	H6
02910	Cranston 71,992	J5
02818	East Greenwich○ 10,211	H6
02914	East Providence 50,980	J5
02822	Exeter○ 4,453	H6
02825	Foster○ 3,370	H5
02828	Greenville 7,516	H5
02830	Harrisville 1,224	H5
02832	Hope Valley 1,414	H6
02833	Hopkinton○ 6,406	H7
02835	Jamestown 4,040	J6
02835	Jamestown 2,156	J7
02881	Kingston 5,479	J7
02837	Little Compton○ 3,085	K6
02840	Middletown 17,216	J6
02882	Narragansett 12,088	J7
02882	Narragansett 3,342	J7
02840	Newport○ 29,259	J7
†02807	New Shoreham (Block Island)○ 620	H8
02852	North Kingstown○ 21,938	J6
02908	North Providence○ 29,188	J5
02859	Pascoag 3,807	H5
*02860	Pawtucket 71,204	J5
02883	Peace Dale-Wakefield 6,474	J7
02871	Portsmouth○ 14,257	J6
*02901	Providence (cap.)⊙ 156,804	H5
	Providence-Warwick-Pawtucket‡ 919,216	H5
02878	Tiverton○ 13,526	K6
02878	Tiverton 7,653	K6
†02864	Valley Falls 10,892	J5
*02879	Wakefield-Peace Dale 6,474	J7
02885	Warren○ 10,640	J6
*02886	Warwick 87,123	J6
02891	Westerly○ 18,580	G7
02891	Westerly 14,093	G7
02893	West Warwick 27,026	H6
02895	Woonsocket 45,914	J4

OTHER FEATURES

Black Rock (pt.)	H8
Block (isl.)	H8
Block Island (sound)	H8
Brenton (pt.)	J7
Conanicut (isl.)	J6
Dickens (pt.)	H8
Durfee (hill)	G5
Grace (pt.)	H8
Jerimoth (hill)	G5
Judith (pt.)	J7
Mount Hope (bay)	K6
Narragansett (bay)	J6
Noyes (pt.)	H7
Pawcatuck (riv.)	G7
Prudence (isl.)	J6
Rhode Island (isl.)	J6
Rhode Island (sound)	J7
Roger Williams Nat'l Mem.	J5
Sakonnet (pt.)	K7
Sakonnet (riv.)	K7
Sandy (pt.)	H8
Scituate (res.)	H5
Stillwater (res.)	C2
Touro Synagogue Nat'l Hist. Site	J7
Watch Hill (pt.)	G7

⊙County seat (Shire town).
‡Population of metropolitan area.
○Population of town or township.
† Zip of nearest p.o.   * Multiple zips.

Massachusetts and Rhode Island

SCALE

State Capitals
County Seats (Shire Towns)
Canals
Major Limited Access Hwys.

Scale 1:970,000

© Copyright HAMMOND INCORPORATED, Maplewood, N.J.

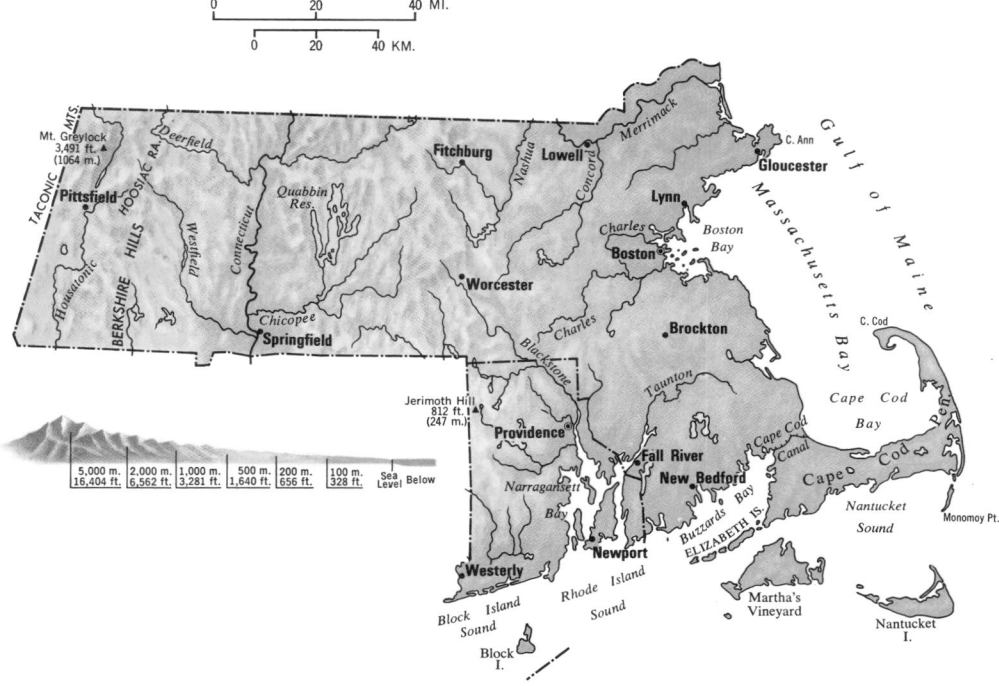

Topography

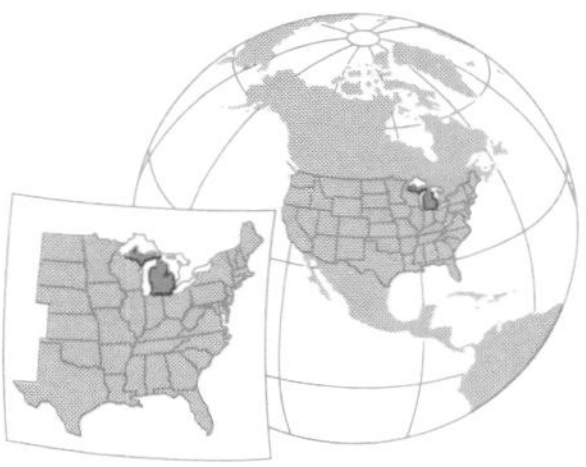

AREA 58,527 sq. mi. (151,585 sq. km.)
POPULATION 9,262,078
CAPITAL Lansing
LARGEST CITY Detroit
HIGHEST POINT Mt. Curwood 1,980 ft. (604 m.)
SETTLED IN 1650
ADMITTED TO UNION January 26, 1837
POPULAR NAME Wolverine State
STATE FLOWER Apple Blossom
STATE BIRD Robin

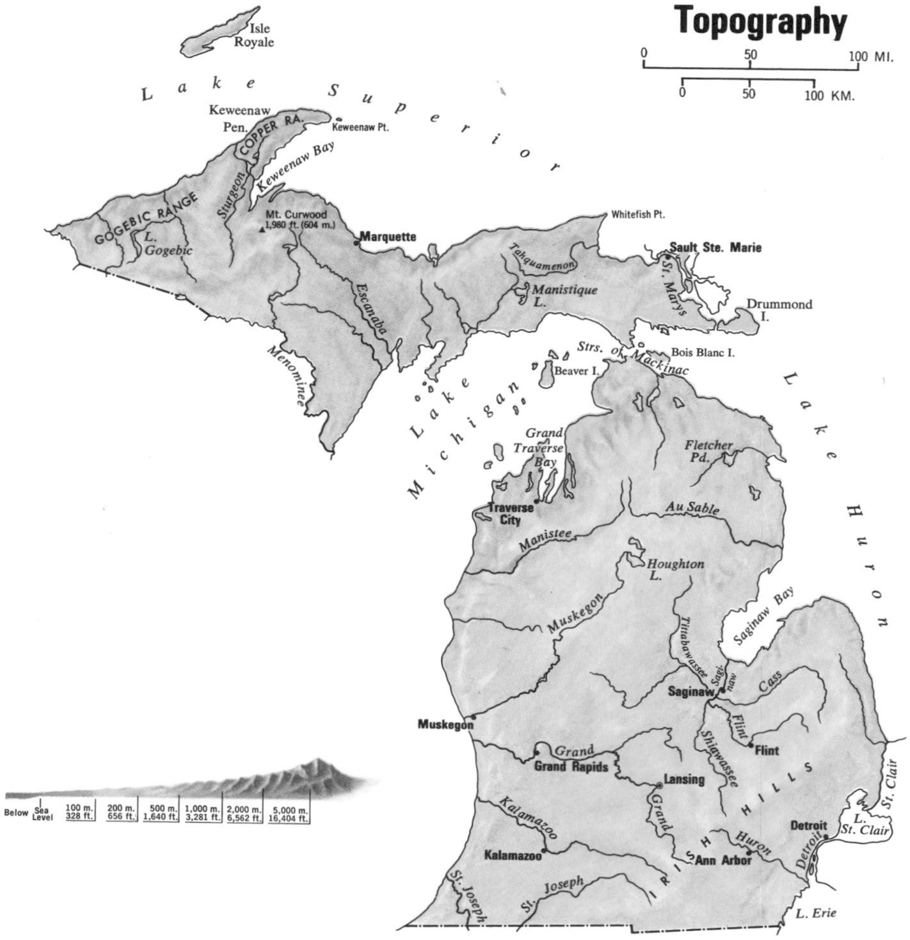

COUNTIES

Alcona 9,740 .... F4
Alger 9,225 .... C2
Allegan 81,555 .... D6
Alpena 32,315 .... F4
Antrim 16,194 .... D3
Arenac 14,706 .... F4
Baraga 8,484 .... A2
Barry 45,781 .... D6
Bay 119,881 .... E5
Benzie 11,205 .... C4
Berrien 171,276 .... C7
Branch 40,188 .... D7
Calhoun 141,557 .... D6
Cass 49,499 .... C7
Charlevoix 19,907 .... D3
Cheboygan 20,649 .... E3
Chippewa 29,029 .... E2
Clare 23,822 .... E5
Clinton 55,893 .... E6
Crawford 9,465 .... E4
Delta 38,947 .... C2
Dickinson 25,341 .... B2
Eaton 88,337 .... E6
Emmet 22,992 .... E3
Genesee 450,449 .... F5
Gladwin 19,957 .... E4
Gogebic 19,686 .... F2
Grand Traverse 54,899 .... D4
Gratiot 40,448 .... E5
Hillsdale 42,071 .... E7
Houghton 37,872 .... G1
Huron 36,459 .... F5
Ingham 275,520 .... E6
Ionia 51,815 .... D6
Iosco 28,349 .... F4
Iron 13,635 .... G2
Isabella 54,110 .... E5
Jackson 151,495 .... E6
Kalamazoo 212,378 .... D6
Kalkaska 10,952 .... D4
Kent 444,506 .... D5
Keweenaw 1,963 .... A1
Lake 7,711 .... D5
Lapeer 70,038 .... F5
Leelanau 14,007 .... D4
Lenawee 89,948 .... E7
Livingston 100,289 .... F6
Luce 6,659 .... D2
Mackinac 10,178 .... D2
Macomb 694,600 .... G6
Manistee 23,019 .... C4
Marquette 74,101 .... B2
Mason 26,365 .... C4
Mecosta 36,961 .... D5
Menominee 26,201 .... B3
Midland 73,578 .... E5
Missaukee 10,009 .... D4
Monroe 134,659 .... F7
Montcalm 47,555 .... D5
Montmorency 7,492 .... E3
Muskegon 157,589 .... C5
Newaygo 34,917 .... D5
Oakland 1,011,793 .... F6
Oceana 22,002 .... C5
Ogemaw 16,436 .... E4
Ontonagon 9,861 .... F1
Osceola 18,928 .... D5
Oscoda 6,858 .... E4
Otsego 14,993 .... E3
Ottawa 157,174 .... C6
Presque Isle 14,267 .... F3
Roscommon 16,374 .... E4
Saginaw 228,059 .... E5
Saint Clair 138,802 .... G6
Saint Joseph 56,083 .... D7
Sanilac 40,789 .... G5
Schoolcraft 8,575 .... C2
Shiawassee 71,140 .... E6
Tuscola 56,961 .... F5
Van Buren 66,814 .... C6
Washtenaw 264,748 .... F6
Wayne 2,337,891 .... F6
Wexford 25,102 .... D4

CITIES and TOWNS

Zip	Name/Pop.	Key
49220	Addison 655	E7
49221	Adrian⊙ 21,186	F7
48701	Akron 538	F5
†48763	Alabaster 46	F4
49224	Albion 11,059	E6
48001	Algonac 4,412	G6
49010	Allegan⊙ 4,576	D6
48101	Allen Park 34,196	B7
48801	Alma 9,652	E5
48003	Almont 1,857	F6
49707	Alpena⊙ 12,214	F3
*48103	Ann Arbor⊙ 107,966	F6
	Ann Arbor‡ 264,748	F6
48005	Armada 1,392	G6
48806	Ashley 570	E5
49011	Athens 960	D6
49709	Atlanta⊙ 475	E3
48611	Auburn 1,921	F5
48703	Au Gres 768	F4
49012	Augusta 913	D6
†48750	Au Sable 1,240	F4
48413	Bad Axe⊙ 3,184	G5
49304	Baldwin⊙ 674	D5
48414	Bancroft 618	E6
49013	Bangor 2,001	C6
49908	Baraga 1,055	G1
49101	Baroda 627	C7
*49014	Battle Creek 35,724	D6
	Battle Creek‡ 187,338	D6
48706	Bay City⊙ 41,593	F5
	Bay City‡ 119,881	F5
48612	Beaverton 1,025	E5
†49423	Beechwood 2,333	C6
48809	Belding 5,634	D5
49615	Bellaire⊙ 1,063	D4
48111	Belleville 3,366	F6
49021	Bellevue 1,289	E6
49022	Benton Harbor 14,707	C6
	Benton Harbor‡ 171,276	C6
†49022	Benton Heights 6,787	C6
48072	Berkley 18,637	B6
49103	Berrien Springs 2,042	C7
49911	Bessemer⊙ 2,553	F2
49617	Beulah⊙ 454	C4
†48010	Beverly Hills 11,598	B6
49307	Big Rapids⊙ 14,361	D5
48415	Birch Run 1,196	F5
*48008	Birmingham 21,689	B6
49228	Blissfield 3,107	F7
48013	Bloomfield Hills 3,985	B6
49026	Bloomingdale 537	C6
49712	Boyne City 3,348	E3
48615	Breckenridge 1,495	E5
49106	Bridgman 2,235	C7
48116	Brighton 4,268	F6
49229	Britton 693	F6
49028	Bronson 2,271	D7
49230	Brooklyn 1,110	E6
48416	Brown City 1,163	G5
49107	Buchanan 5,142	C7
49030	Burr Oak 853	D7
48507	Burton 29,976	F6
48418	Byron 689	E6
49601	Cadillac⊙ 10,199	D4
49316	Caledonia 722	D6
49913	Calumet 1,013	A1
48014	Capac 1,377	G5
48117	Carleton 2,786	F6
48723	Caro⊙ 4,317	F5
48724	Carrollton 7,482	E5
48811	Carson City 1,229	E5
48419	Carsonville 622	G5
48725	Caseville 851	F5
49915	Caspian 1,038	G2
48726	Cass City 2,258	F5
49031	Cassopolis⊙ 1,933	C7
49319	Cedar Springs 2,615	D5
49233	Cement City 539	E6
48015	Center Line 9,293	B6
49622	Central Lake 895	D3
49032	Centreville⊙ 1,202	D7
49720	Charlevoix⊙ 3,296	D3
48813	Charlotte⊙ 8,251	E6
49721	Cheboygan⊙ 5,106	E3
48118	Chelsea 3,816	E6
48616	Chesaning 2,656	E5
48617	Clare 3,300	E5
48016	Clarkston 968	F6
48017	Clawson 15,103	B6
49034	Climax 619	D6
49236	Clinton 2,342	F6
48420	Clio 2,669	F5
49036	Coldwater⊙ 9,461	D7
48618	Coleman 1,429	E5
49038	Coloma 1,833	C6
49040	Colon 1,190	D7
48421	Columbiaville 953	F5
49041	Comstock○ 11,162	D6
49237	Concord 900	E6
49042	Constantine 1,680	D7
49404	Coopersville 2,889	C5
48817	Corunna⊙ 3,206	E6
48422	Croswell 2,073	G5
49920	Crystal Falls⊙ 1,965	A2
49508	Cutlerville 8,256	D6
48423	Davison 6,087	F5
*48120	Dearborn 90,660	B7
48127	Dearborn Heights 67,706	B7
49045	Decatur 1,915	C6
48427	Deckerville 887	G5
49238	Deerfield 957	F7
*48201	Detroit⊙ 1,203,339	B7
	Detroit‡ 4,352,762	B7
†48161	Detroit Beach 2,112	F7
48820	De Witt 3,165	E6
48130	Dexter 1,524	F6
48821	Dimondale 1,008	E6
49406	Douglas 948	C6
49047	Dowagiac 6,307	D6
48020	Drayton Plains	F6
49726	Drummond Island○ 746	F3
48428	Dryden 650	F6
48131	Dundee 2,575	F7
48429	Durand 4,241	E6
49924	Eagle River⊙ 20	A1
48021	East Detroit 38,280	B6
†49506	East Grand Rapids 10,914	D6
49727	East Jordan 2,185	D3
†49801	East Kingsford	A3
48823	East Lansing 51,392	E6
48730	East Tawas 2,584	F4
†49001	Eastwood 7,186	D6
48827	Eaton Rapids 4,510	E6
49111	Eau Claire 573	C6
48229	Ecorse 14,447	B7
48829	Edmore 1,176	E5
49112	Edwardsburg 1,135	C7
49628	Elberta 556	C4
49629	Elk Rapids 1,504	D4
48731	Elkton 953	F5
48831	Elsie 1,022	E5
49829	Escanaba⊙ 14,355	C3
48732	Essexville 4,378	F5
49631	Evart 1,945	D5
48733	Fairgrove 691	F5
49022	Fair Plain 8,289	C6
*48024	Farmington 11,022	F6
48024	Farmington Hills 58,056	F6
48622	Farwell 804	E5
49408	Fennville 934	C6
48430	Fenton 8,098	F6
48220	Ferndale 26,227	B6
49409	Ferrysburg 2,440	C5
48134	Flat Rock 6,853	F6
*48501	Flint⊙ 159,611	F5
	Flint‡ 521,589	F5
48433	Flushing 8,624	F5
48835	Fowler 1,021	E5
48836	Fowlerville 2,289	F6
48734	Frankenmuth 3,753	F5
49635	Frankfort 1,603	C4
48025	Franklin 2,864	B6
48026	Fraser 14,560	B6
48623	Freeland 1,364	E5
49412	Fremont 3,672	D5
49415	Fruitport 1,143	C5
49053	Galesburg 1,822	D6
49113	Galien 692	C7
48135	Garden City 35,640	F6
49735	Gaylord⊙ 3,011	E3
48173	Gibraltar 4,458	F6
49837	Gladstone 4,533	C3
48624	Gladwin⊙ 2,479	E5
49055	Gobles 816	D6
48438	Goodrich 795	F6
48439	Grand Blanc 6,848	F6
49417	Grand Haven⊙ 11,763	C5
48837	Grand Ledge 6,920	E6
*49501	Grand Rapids⊙ 181,843	D5
	Grand Rapids‡ 601,680	D5
49418	Grandville 12,412	D6
49327	Grant 683	D5
49240	Grass Lake 962	E6
49738	Grayling⊙ 1,792	E4
48838	Greenville 8,019	D5
48138	Grosse Ile 9,320	B7
48236	Grosse Pointe 5,901	B7
†48236	Grosse Pointe Farms 10,551	B6
†48236	Grosse Pointe Park 13,639	B7
†48236	Grosse Pointe Shores 3,122	B6
†48236	Grosse Pointe Woods 18,886	B6
49841	Gwinn 1,408	B2
48212	Hamtramck 21,300	B6
49930	Hancock 5,122	G1
48441	Harbor Beach 2,000	G5
49740	Harbor Springs 1,567	D3
48225	Harper Woods 16,361	B6
48625	Harrison⊙ 1,700	E4
48740	Harrisville⊙ 559	F4
49420	Hart⊙ 1,888	C5
49057	Hartford 2,493	C6
48840	Haslett 7,025	E6
49058	Hastings⊙ 6,418	D6
48030	Hazel Park 20,914	B6
48626	Hemlock 1,362	E5
49421	Hesperia 876	D5
48203	Highland Park 27,909	B6
49242	Hillsdale⊙ 7,432	E7
49423	Holland 26,281	C6
48842	Holt 10,097	E6
49245	Homer 1,791	E6
49931	Houghton⊙ 7,512	G1
48629	Houghton Lake 2,449	E4
48630	Houghton Lake Heights	E4
49329	Howard City 1,118	D5
48843	Howell⊙ 6,976	E6
49934	Hubbell 1,278	A1
49247	Hudson 2,545	E7
49426	Hudsonville 4,844	D6
48444	Imlay City 2,495	F5
48141	Inkster 35,190	B7
49643	Interlochen 600	D4
48846	Ionia⊙ 5,920	D6
49801	Iron Mountain⊙ 8,341	B3
49935	Iron River 2,426	G2
49938	Ironwood 7,741	F2
49849	Ishpeming 7,538	B2
48847	Ithaca⊙ 2,950	E5
*49201	Jackson⊙ 39,739	E6
	Jackson‡ 151,495	E6
49428	Jenison 16,330	D6
49250	Jonesville 2,172	E6
*49001	Kalamazoo⊙ 79,722	D6
	Kalamazoo-Portage‡ 279,192	D6
49646	Kalkaska⊙ 1,654	D4
48030	Keego Harbor 3,083	F6
49330	Kent City 860	D5
49508	Kentwood 30,438	D6
48445	Kinde 600	G5
49801	Kingsford 5,290	A3
49649	Kingsley 664	D4
48848	Laingsburg 1,145	E6
49651	Lake City⊙ 843	D4
49945	Lake Linden 1,181	A1
†49039	Lake Michigan Beach 2,001	C6
48849	Lake Odessa 2,171	D6
48035	Lake Orion 2,907	F6
48850	Lakeview 1,139	D5
†49440	Lakewood Club 695	C5
48144	Lambertville 6,341	F7
49946	L'Anse⊙ 2,500	G1
*48901	Lansing (cap.) 130,414	E6
	Lansing-East Lansing‡ 468,482	E6
48446	Lapeer⊙ 6,198	F5
49913	Laurium 2,678	A1
49064	Lawrence 903	C6
49065	Lawton 1,558	D6
49654	Leland⊙ 776	D3
48449	Lennon 600	E5
49251	Leslie 2,110	E6
48450	Lexington 765	G5
48742	Lincoln 361	F4
48146	Lincoln Park 45,105	B7
48451	Linden 2,174	F6
49252	Litchfield 1,353	E6
*48150	Livonia 104,814	F6
49331	Lowell 3,707	D6
49431	Ludington⊙ 8,937	C5

(continued on following page)

48157 Luna Pier 1,443 F7
48851 Lyons 708 E6
49757 Mackinac Island 479 E3
49701 Mackinaw City 820 E3
48071 Madison Heights 35,375 B6
49659 Mancelona 1,432 E4
48158 Manchester 1,686 E6
49660 Manistee⊙ 7,566 C4
49854 Manistique⊙ 3,962 C3
49663 Manton 1,212 D4
48853 Maple Rapids 683 E5
49067 Marcellus 1,134 D6
48039 Marine City 4,414 G6
49665 Marion 816 D4
48453 Marlette 1,761 G5
49855 Marquette⊙ 23,288 B2
49068 Marshall⊙ 7,201 E6
49070 Martin 447 D6
48040 Marysville 7,345 G6
48854 Mason⊙ 6,019 E6
49071 Mattawan 2,143 D6
48744 Mayville 958 F5
49657 McBain 519 D4
48122 Melvindale 12,322 B7
48041 Memphis 1,171 G6
49072 Mendon 951 D7
49858 Menominee⊙ 10,099 B3
48637 Merrill 851 E5
48455 Metamora 552 F6
49254 Michigan Center 5,244 E6
49333 Middleville 1,797 D6
48640 Midland⊙ 37,250 E5
48160 Milan 4,182 F6
48042 Milford 5,041 F6
48746 Millington 1,237 F5
48647 Mio⊙ 975 E4
48161 Monroe⊙ 23,531 F7
49437 Montague 2,332 C5
48457 Montrose 1,706 F5
49256 Morenci 2,110 E7
49336 Morley 507 D5
48857 Morrice 733 E6
48043 Mount Clemens⊙ 18,806 G6
48458 Mount Morris 3,246 F5
48858 Mount Pleasant⊙ 23,746 E5
48860 Muir 698 D5
48861 Mulliken 550 E6
49862 Munising⊙ 3,083 C2
*49440 Muskegon⊙ 40,823 C5
Muskegon-Norton Shores-Muskegon Heights‡ 179,591 C5
49444 Muskegon Heights 14,611 C5
49261 Napoleon 1,400 E6
49073 Nashville 1,628 D6
49866 Negaunee 5,189 B2
49337 Newaygo 1,271 D5
48047 New Baltimore 5,439 G6
49868 Newberry⊙ 2,120 D2
48164 New Boston 1,200 F6
49117 New Buffalo 2,821 C7
48048 New Haven 1,871 G6
48460 New Lothrop 646 F5

49120 Niles 13,115 C7
49262 North Adams 565 E7
48461 North Branch 896 F5
49445 North Muskegon 4,024 C5
49670 Northport 611 D3
48167 Northville 5,698 F6
†49441 Norton Shores 22,025 C5
49870 Norway 2,919 B3
48050 Novi 22,525 F6
48237 Oak Park 31,537 B6
48864 Okemos 8,882 E6
49076 Olivet 1,604 E6
49765 Onaway 1,084 E3
49675 Onekama 582 C4
49265 Onsted 670 E6
49953 Ontonagon⊙ 2,182 F1
48033 Orchard Lake 1,798 F6
48462 Ortonville 1,190 F6
48750 Oscoda 2,431 F4
48463 Otisville 682 F5
49078 Otsego 3,802 D6
48866 Ovid 1,712 E5
48867 Owosso 16,455 E5
48051 Oxford 2,746 F6
49004 Parchment 1,817 D6
49269 Parma 873 E6
49079 Paw Paw⊙ 3,211 D6
†49038 Paw Paw Lake 4,193 C6
48052 Pearl Beach 3,430 G6
48466 Peck 606 G5
49769 Pellston 565 E3
49449 Pentwater 1,165 C5
48872 Perry 2,051 E6
49270 Petersburg 1,222 F7
49770 Petoskey⊙ 6,097 E3
48755 Pigeon 1,247 F5
48169 Pinckney 1,390 F6
48650 Pinconning 1,430 F5
49080 Plainwell 3,751 D6
48069 Pleasant Ridge 3,217 B6
*48170 Plymouth 9,986 F6
*48053 Pontiac⊙ 76,715 F6
49081 Portage 38,157 D6
48467 Port Austin 839 F4
48060 Port Huron⊙ 33,981 G6
48875 Portland 3,963 E6
48469 Port Sanilac 598 G5
49776 Posen 270 F3
48876 Potterville 1,502 E6
49082 Quincy 1,569 E7
49959 Ramsay F2
49451 Ravenna 951 D5
49274 Reading 1,203 E7
49677 Reed City⊙ 2,221 D5
48757 Reese 1,645 F5
48062 Richmond 3,536 G6
48218 River Rouge 12,912 B7
48192 Riverview 14,569 B7
48063 Rochester 7,203 F6
49341 Rockford 3,324 D5
48173 Rockwood 3,346 F6
49779 Rogers City⊙ 3,923 F3
48065 Romeo 3,509 F6

48174 Romulus 24,857 F6
49444 Roosevelt Park 4,015 C5
48653 Roscommon⊙ 834 E4
48654 Rose City 661 E4
48066 Roseville 54,311 B6
49452 Rothbury 522 C5
*48067 Royal Oak 70,893 B6
*48601 Saginaw⊙ 77,508 F5
Saginaw‡ 228,059 F5
48655 Saint Charles 2,276 E5
48079 Saint Clair 4,780 G6
*48080 Saint Clair Shores 76,210 B6
49781 Saint Ignace⊙ 2,632 E3
48879 Saint Johns⊙ 7,376 E6
49085 Saint Joseph⊙ 9,622 C6
48880 Saint Louis 4,107 E5
48176 Saline 6,483 F6
48471 Sandusky⊙ 2,216 G5
48657 Sanford 864 E5
48881 Saranac 1,421 D6
49453 Saugatuck 1,079 C6
49783 Sault Sainte Marie⊙ 14,448 E2
49087 Schoolcraft 1,359 D6
49454 Scottville 1,241 C5
48759 Sebewaing 2,046 F5
49455 Shelby 1,624 C5
48883 Shepherd 1,534 E5
48884 Sheridan 664 D5
†49085 Shoreham 742 C6
†49125 Shorewood 1,735 C7
*48034 Southfield 75,568 F6
48195 Southgate 32,058 F6
49090 South Haven 5,943 C6
48178 South Lyon 5,214 F6
†48161 South Monroe 4,232 F7
49963 South Range 861 G1
48179 South Rockwood 1,353 F7
†48060 Sparlingville 1,718 G6
49345 Sparta 3,373 D5
49283 Spring Arbor 2,101 E6
49015 Springfield 5,917 D6
49456 Spring Lake 2,731 C5
49284 Springport 675 E6
49964 Stambaugh 1,442 G2
48658 Standish⊙ 1,264 F5
48888 Stanton⊙ 1,315 D5
49887 Stephenson 967 B3
48659 Sterling 457 E4
48077 Sterling Heights 108,999 B6
49127 Stevensville 1,268 C6
49285 Stockbridge 1,213 E6
49091 Sturgis 9,468 D7
48890 Sunfield 591 D6
49682 Suttons Bay 504 D3
48473 Swartz Creek 5,013 F6
†48053 Sylvan Lake 1,949 F6
48763 Tawas City⊙ 1,967 F4
48180 Taylor 77,568 B7
49286 Tecumseh 7,320 E7
49092 Tekonsha 755 E6
49128 Three Oaks 1,774 C7
49093 Three Rivers 7,015 D7

49684 Traverse City⊙ 15,516 D4
48183 Trenton 22,762 B7
*48084 Troy 67,102 B6
48475 Ubly 862 G5
49094 Union City 1,667 D6
49129 Union Pier 1,039 C7
48767 Unionville 578 F5
*48087 Utica 5,282 F6
49095 Vandalia 447 D7
49795 Vanderbilt 525 E3
48768 Vassar 2,727 F5
49096 Vermontville 832 E6
48476 Vernon 1,008 F6
49097 Vicksburg 2,224 D6
49968 Wakefield 2,591 F2
49288 Waldron 570 E7
49504 Walker 15,088 D6
48088 Walled Lake 4,748 F6
*48089 Warren 161,134 B6
49098 Watervliet 1,867 C6
49348 Wayland 2,023 D6
48184 Wayne 21,159 F6
48892 Webberville 1,535 E6
49894 Wells B3
48661 West Branch⊙ 1,785 E4
48185 Westland 84,603 F6
48894 Westphalia 896 E6
49349 White Cloud⊙ 1,101 D5
49461 Whitehall 2,856 C5
49099 White Pigeon 1,478 D7
49971 White Pine 1,142 F1
48189 Whitmore Lake 2,920 F6
48770 Whittemore 438 F4
48895 Williamston 2,981 E6
48096 Wixom 6,705 F6
†49440 Wolf Lake 3,876 D5
49799 Wolverine 364 E3
†48183 Woodhaven 10,902 F6
48897 Woodland 431 D6
48192 Wyandotte 34,006 B7
49509 Wyoming 59,616 D6
48097 Yale 1,814 G5
48197 Ypsilanti 24,031 F6
49464 Zeeland 4,764 D6
†48601 Zilwaukee 2,201 F5

OTHER FEATURES

Abbaye (pt.) B2
Au Sable (pt.) C2
Au Sable (pt.) F4
Au Sable (riv.) E4
Au Train (bay) C2
Bad (riv.) E5
Barques (pt.) C3
Beaver (isl.) D3
Beaver (lake) F4
Belle (riv.) G6
Bete Grise (bay) B1
Betsy (riv.) D2
Big Bay (pt.) B2
Big Bay de Noc (bay) C3
Big Iron (riv.) F1
Big Sable (pt.) C4
Big Sable (riv.) C4
Big Star (lake) C5
Black (lake) E3
Black (riv.) E3
Black (riv.) G5
Blake (pt.) E1
Boardman (riv.) D4
Bois Blanc (isl.) E3
Bond Falls (res.) G2
Brevoort (lake) E3
Brule (riv.) A3
Burt (lake) E3
Cass (riv.) F5
Cedar (lake) F4
Charlevoix (lake) D3
Chippewa (riv.) E5
Crisp (pt.) D2
Crystal (lake) C4
Curwood (mt.) A2
Dead (riv.) B2
Deer (riv.) A2
De Tour (passage) E3
Detour (pt.) C3
Detroit (riv.) B7
Drummond (isl.) F2
Duck (lake) F4
Elk (lake) D4
Erie (lake) G7
Escanaba (riv.) B2
False Detour (chan.) F3
Fawn (riv.) D7
Fence (riv.) A2
Firesteel (riv.) G1
Fletcher (pond) F4
Flint (riv.) F5
Ford (riv.) B2
Forty Mile (pt.) F3
Fourteen Mile (pt.) F1
Garden (isl.) D3
Garden (pen.) C3
Glen (lake) C4
Gogebic (lake) F2
Good Harbor (bay) D3
Government (peak) F1
Grand (isl.) C2
Grand (lake) F3
Grand (riv.) D6
Grand Traverse (bay) D3
Granite (pt.) B2
Green (bay) B4
Gun (lake) D6
Hamlin (lake) C4
Higgins (lake) E4
High (isl.) D3
Hog (isl.) D3
Houghton (lake) E4
Hubbard (lake) F4
Huron (bay) A2
Huron (lake) G4
Huron (riv.) F6
Huron River (pt.) B2
Independence (lake) B2
Indian (lake) C2
Isle Royale Nat'l Park E1
Kalamazoo (riv.) C6
Keweenaw (bay) A1
Keweenaw (pt.) B1
K.I. Sawyer A.F.B. 7,345 B2
L'Anse Ind. Res. A2
Laughing Fish (pt.) B2
Leelanau (lake) D4
Light House (pt.) D3
Little Bay de Noc (bay) B3
Little Girl (pt.) E1
Little Sable (pt.) C5
Little Summer (isl.) C3
Little Traverse (bay) D3
Long (lake) F3
Lookingglass (riv.) E6
Mackinac (isl.) E3
Mackinac (str.) E3
Manistee (riv.) C4
Manistique (lake) D2
Manistique (riv.) C2
Manitou (isl.) B1
Maple (riv.) E5
Margrethe (lake) E4
Marquette (isl.) E3
Maumee (bay) F7
Menominee (riv.) B3
Michigamme (lake) A2
Michigamme (res.) B2
Michigamme (riv.) A2
Michigan (lake) B5
Mill (creek) G5
Millecoquins (lake) D2
Misery (bay) G1
Misery (riv.) G1
Montreal (riv.) F1
Mullett (lake) E3
Munuscong (lake) E2
Muskegon (riv.) C5
Neebish (isl.) E2
Net (riv.) G2
Ninemile (pt.) E3
North (chan.) F2
North (pt.) F3
North Fox (isl.) D3
North Manitou (isl.) C3
Oak (pt.) F5
Ontonagon (riv.) G1
Ontonagon Ind. Res. F1
Otsego (lake) E4
Paint (riv.) A2
Paradise (lake) E3
Passage (isl.) E1
Patterson (pt.) D3
Paw Paw (riv.) C6
Peninsula (pt.) C3
Perch (lake) G2
Perch (riv.) G2
Pere Marquette (riv.) D5
Pictured Rocks (cliff) C2
Pictured Rocks Nat'l Lakeshore C2
Pigeon (riv.) D7
Pigeon (riv.) E3
Pine (lake) F4
Pine (riv.) D4
Pine (riv.) E5
Platte (lake) C4
Porcupine (mts.) F1
Potagannissing (bay) F2
Poverty (isl.) C3
Prairie (riv.) D7
Presque Isle (riv.) F1
Rabbit (riv.) D6
Raisin (riv.) F7
Rapid (riv.) B2
Reedsburg (res.) E4
Rifle (riv.) E4
Royale (isl.) E1
Saginaw (bay) F5
Saginaw (riv.) F5
Saint Clair (lake) G6
Saint Clair (riv.) G6
Saint Joseph (riv.) C7
Saint Martin (bay) E3
Saint Martin (isl.) C3
Saint Marys (riv.) E2
Salt (pt.) E2
Sand (pt.) F5
Seul Choix (pt.) D3
Shiawassee (riv.) E5
Siskiwit (bay) E1
Sleeping Bear Dunes Nat'l Lakeshore C4
South (bay) C2
South (chan.) E3
South (pt.) F4
South Fox (isl.) D3
South Manitou (isl.) C3
Sturgeon (riv.) C2
Sugar (isl.) E2
Summer (isl.) C3
Superior (lake) C2
Tahquamenon (falls) D2
Tahquamenon (riv.) D2
Tawas (lake) F4
Tawas (pt.) F4
Thunder (bay) F4
Thunder Bay (riv.) F3
Tittabawassee (riv.) E5
Torch (lake) D3
Traverse (isl.) A1
Traverse (pt.) A1
Turtle (lake) F4
Two Hearted (riv.) D2
Vieux Desert (lake) G2
Walloon (lake) E3
White (riv.) C5
Whitefish (bay) E2
Whitefish (pt.) E2
Whitefish (riv.) C2
Wood (isl.) C2
Wurtsmith A.F.B. 5,166 F4
Yellow Dog (riv.) B2

⊙County seat.
‡Population of metropolitan area.
○Population of township.
† Zip of nearest p.o. * Multiple zip

Agriculture, Industry and Resources

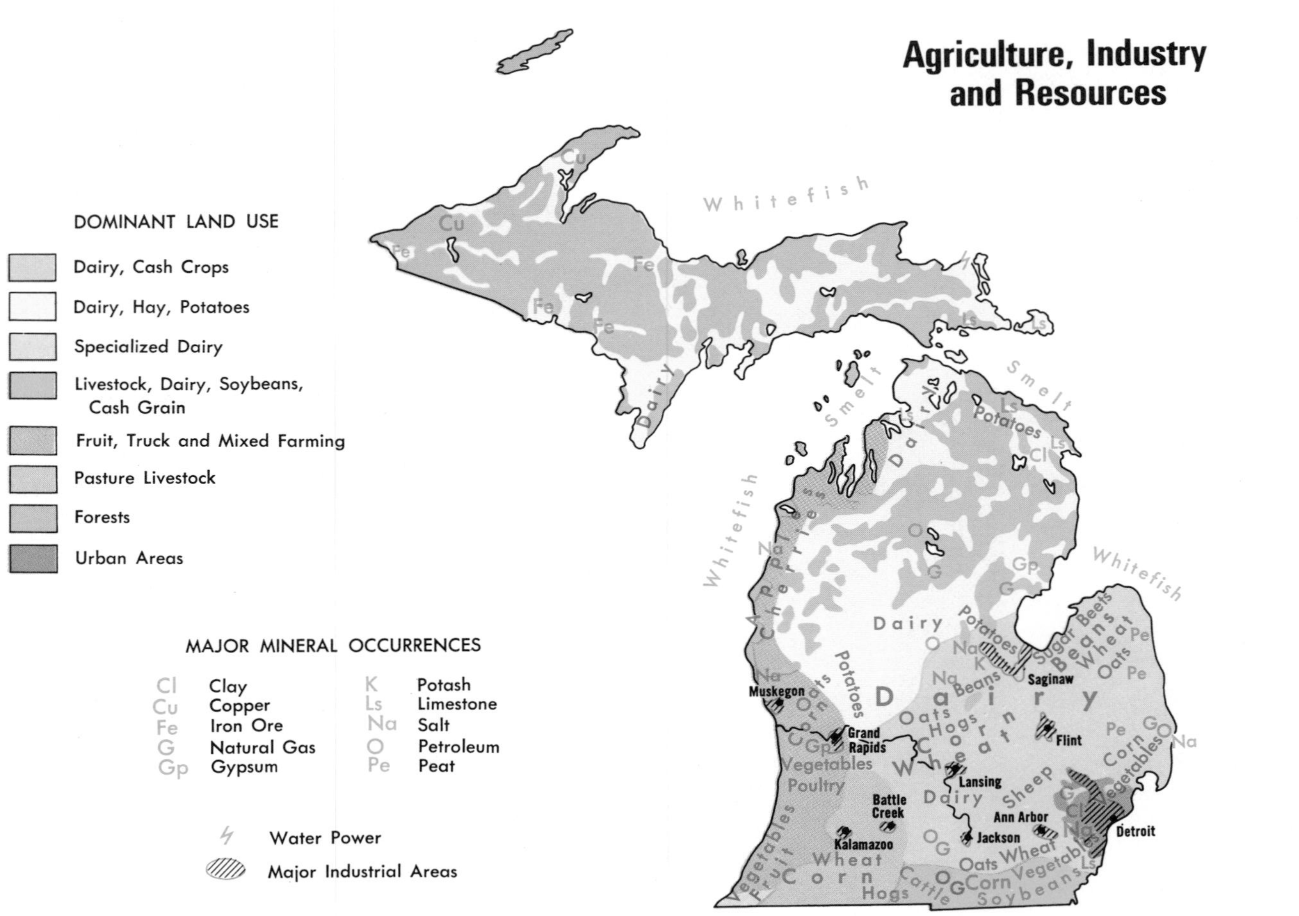
DOMINANT LAND USE
Dairy, Cash Crops
Dairy, Hay, Potatoes
Specialized Dairy
Livestock, Dairy, Soybeans, Cash Grain
Fruit, Truck and Mixed Farming
Pasture Livestock
Forests
Urban Areas
MAJOR MINERAL OCCURRENCES
Cl Clay
Cu Copper
Fe Iron Ore
G Natural Gas
Gp Gypsum
K Potash
Ls Limestone
Na Salt
O Petroleum
Pe Peat
Water Power
Major Industrial Areas
Whitefish
Smelt
Dairy
Potatoes
Sugar Beets
Beans
Wheat
Oats
Corn
Hogs
Sheep
Vegetables
Poultry
Fruit
Cattle
Soybeans
Muskegon
Grand Rapids
Saginaw
Flint
Lansing
Battle Creek
Kalamazoo
Jackson
Ann Arbor
Detroit

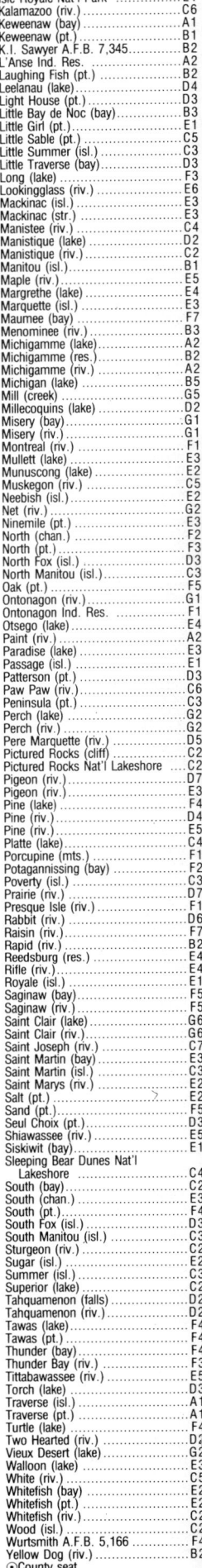

AREA 84,402 sq. mi. (218,601 sq. km.)
POPULATION 4,075,970
CAPITAL St. Paul
LARGEST CITY Minneapolis
HIGHEST POINT Eagle Mtn. 2,301 ft. (701 m.)
SETTLED IN 1805
ADMITTED TO UNION May 11, 1858
POPULAR NAME North Star State; Gopher State
STATE FLOWER Pink & White Lady's-Slipper
STATE BIRD Common Loon

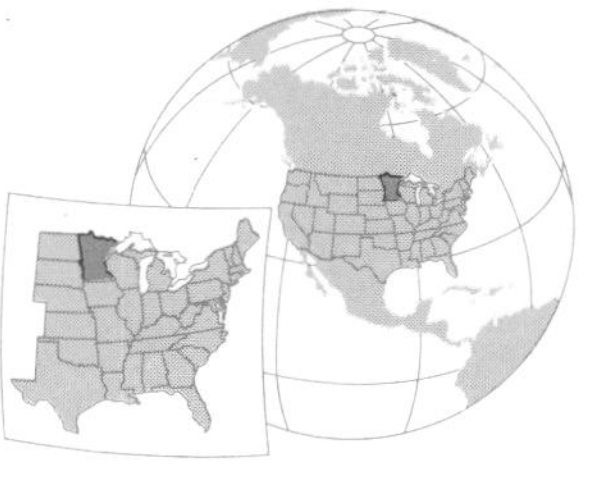

COUNTIES

Aitkin 13,404	E4
Anoka 195,998	E5
Becker 29,336	C4
Beltrami 30,982	C2
Benton 25,187	D5
Big Stone 7,716	B5
Blue Earth 52,314	D6
Brown 28,645	D6
Carlton 29,936	F4
Carver 37,046	E6
Cass 21,050	D4
Chippewa 14,941	C5
Chisago 25,717	F5
Clay 49,327	B4
Clearwater 8,761	C3
Cook 4,092	H3
Cottonwood 14,854	C6
Crow Wing 41,722	D4
Dakota 194,279	E6
Dodge 14,773	F7
Douglas 27,839	C5
Faribault 19,714	D7
Fillmore 21,930	F7
Freeborn 36,329	E7
Goodhue 38,749	F6
Grant 7,171	B5
Hennepin 941,411	E5
Houston 18,382	G7
Hubbard 14,098	D3
Isanti 23,600	E5
Itasca 43,069	E3
Jackson 13,690	C7
Kanabec 12,161	E5
Kandiyohi 36,763	C5
Kittson 6,672	B2
Koochiching 17,571	E2
Lac qui Parle 10,592	B6
Lake 13,043	G3
Lake of the Woods 3,764	D2
Le Sueur 23,434	E6
Lincoln 8,207	B6
Lyon 25,207	C6
Mahnomen 5,535	C3
Marshall 13,027	B2
Martin 24,687	D7
McLeod 29,657	D6
Meeker 20,594	D5
Mille Lacs 18,430	E5
Morrison 29,311	D4
Mower 40,390	F7
Murray 11,507	C6
Nicollet 26,929	D6
Nobles 21,840	C7
Norman 9,379	B3
Olmsted 92,006	F7
Otter Tail 51,937	C4
Pennington 15,258	B2
Pine 19,871	F4
Pipestone 11,690	B6
Polk 34,844	B3
Pope 11,657	C5
Ramsey 459,784	E5
Red Lake 5,471	B3
Redwood 19,341	C6
Renville 20,401	C6
Rice 46,087	E6
Rock 10,703	B7
Roseau 12,574	C2
Saint Louis 222,229	F3
Scott 43,784	E6
Sherburne 29,908	E5
Sibley 15,448	D6
Stearns 108,161	D5
Steele 30,328	E7
Stevens 11,322	B5
Swift 12,920	C5
Todd 24,991	D4
Traverse 5,542	B5
Wabasha 19,335	F6
Wadena 14,192	D4
Waseca 18,448	E6
Washington 113,571	F5
Watonwan 12,361	D7
Wilkin 8,454	B4
Winona 46,256	G6
Wright 58,681	D5
Yellow Medicine 13,653	B6

CITIES and TOWNS

Zip	Name/Pop.	Key
56510	Ada⊙ 1,971	B3
55909	Adams 797	F7
56110	Adrian 1,336	C7
55001	Afton 2,550	F6
56430	Ah-Gwah-Ching 400	D3
56431	Aitkin⊙ 1,770	E4
56433	Akeley 486	D3
56307	Albany 1,569	D5
56207	Alberta 145	B5
56007	Albert Lea⊙ 19,200	E7
55301	Albertville 564	E5
56009	Alden 687	E7
56308	Alexandria⊙ 7,608	C5
56111	Alpha 180	D7
55910	Altura 354	G6
56710	Alvarado 385	B2
56010	Amboy 606	D7
†55303	Andover 9,387	E5
55302	Annandale 1,568	D5
55303	Anoka⊙ 15,634	E5
56208	Appleton 1,842	C5
†55124	Apple Valley 21,818	G6
56713	Argyle 741	B2
55307	Arlington 1,779	D6
56309	Ashby 486	C4
55704	Askov 350	F4
56209	Atwater 1,128	D5
56511	Audubon 383	C4
55705	Aurora 2,670	F3
55912	Austin⊙ 23,020	E7
56114	Avoca 201	C7
56310	Avon 804	D5
55706	Babbitt 2,435	G3
56435	Backus 255	D4
56714	Badger 320	B2
56621	Bagley⊙ 1,321	C3
56115	Balaton 752	C6
56514	Barnesville 2,207	B4
55707	Barnum 464	F4
56311	Barrett 388	B5
56515	Battle Lake 708	C4
56623	Baudette⊙ 1,170	D2
†56401	Baxter 2,625	D4
55003	Bayport 2,932	F5
56211	Beardsley 344	B5
55601	Beaver Bay 283	G3
56116	Beaver Creek 260	B7
55308	Becker 601	E5
56312	Belgrade 805	C5
†55027	Bellechester 220	F6
56011	Belle Plaine 2,754	E6
56212	Bellingham 290	B5
56214	Belview 438	C6
56601	Bemidji⊙ 10,949	D3
56626	Bena 153	D3
56215	Benson⊙ 3,656	C5
56437	Bertha 510	C4
55005	Bethel 272	E5
56117	Bigelow 249	C7
56627	Big Falls 490	E2
56628	Bigfork 457	E3
55309	Big Lake 2,210	E5
56118	Bingham Lake 222	C7
55310	Bird Island 1,372	D6
55708	Biwabik 1,428	F3
56630	Blackduck 653	D3
†55433	Blaine 28,558	G5
56216	Blomkest 200	D6
55917	Blooming Prairie 1,969	E7
55420	Bloomington 81,831	G6
56013	Blue Earth⊙ 4,132	D7
56518	Bluffton 206	C4
56519	Borup 160	B3
55709	Bovey 813	E3
56314	Bowlus 276	D5
56218	Boyd 329	C6
55006	Braham 1,015	E5
56401	Brainerd⊙ 11,489	D4
†55056	Branch 1,866	F5
56315	Brandon 473	C5
56520	Breckenridge⊙ 3,909	B4
†56472	Breezy Point 384	D4
56119	Brewster 559	C7
56014	Bricelyn 487	E7
55429	Brooklyn Center 31,230	G5
†55444	Brooklyn Park 43,332	G5
56715	Brooks 173	B3
56316	Brooten 647	C5
56438	Browerville 693	D4
55918	Brownsdale 691	F7
56219	Browns Valley 887	B5
55919	Brownsville 418	G7
55312	Brownton 697	D6
56317	Buckman 171	D5
55313	Buffalo⊙ 4,560	E5
55314	Buffalo Lake 782	D6
55713	Buhl 1,284	F3
55337	Burnsville 35,674	E6
56318	Burtrum 177	D5
56120	Butterfield 634	D7
55920	Byron 1,715	F6
55921	Caledonia⊙ 2,691	G7
56521	Callaway 238	C3
55716	Calumet 469	E3
55008	Cambridge⊙ 3,287	E5
56522	Campbell 286	B4
56220	Canby 2,143	B6
55009	Cannon Falls 2,653	F6
55922	Canton 386	F7
56319	Carlos 364	C5
55718	Carlton⊙ 862	F4
55315	Carver 642	E6
56633	Cass Lake 1,001	D3
55012	Center City⊙ 458	F5
†55038	Centerville 734	E5
56121	Ceylon 543	D7
55316	Champlin 9,006	G5
56122	Chandler 344	C7
55317	Chanhassen 6,359	F6
55318	Chaska⊙ 8,346	F6
55923	Chatfield 2,055	F7
55013	Chisago City 1,634	E5
55719	Chisholm 5,930	E3
56221	Chokio 559	B5
55014	Circle Pines 3,321	G5
56222	Clara City 1,574	C6
55924	Claremont 591	E6
56440	Clarissa 663	C4
56223	Clarkfield 1,171	C6
56016	Clarks Grove 620	E7
56634	Clearbrook 579	C3
55319	Clear Lake 266	E5
55320	Clearwater 379	D5
56224	Clements 227	D6
56017	Cleveland 699	E6
56523	Climax 273	B3
56225	Clinton 622	B5
56226	Clontarf 196	C5
55720	Cloquet 11,142	F4
†55068	Coates 207	E6
55321	Cokato 2,056	D5
56320	Cold Spring 2,294	D5
55722	Coleraine 1,116	E3
55322	Cologne 545	E6
55421	Columbia Heights 20,029	G5
56019	Comfrey 548	D6
56020	Conger 183	E7
55723	Cook 800	F3
55433	Coon Rapids 35,826	G5
†55340	Corcoran 4,252	F5
56228	Cosmos 571	D6
55016	Cottage Grove 18,994	F6
56229	Cottonwood 924	C6
56021	Courtland 399	D6
55726	Cromwell 229	F4
56716	Crookston⊙ 8,628	B3
56441	Crosby 2,218	D4
56442	Crosslake 1,064	E4
†55428	Crystal 25,543	G5
55323	Crystal Bay (Orono) 6,845	F5
56123	Currie 359	C6
56323	Cyrus 334	C5
55925	Dakota 350	G7
56324	Dalton 248	C4
56230	Danube 590	C6
56231	Danvers 152	C5
56022	Darfur 139	D6
55324	Darwin 282	D5
55325	Dassel 1,066	D5
56232	Dawson 1,901	B6
55327	Dayton 4,070	E5
55391	Deephaven 3,716	G5
56527	Deer Creek 392	C4
56636	Deer River 907	E3
56444	Deerwood 580	E4
56233	De Graff 179	C5
55328	Delano 2,480	E5
56023	Delavan 262	D7
†55110	Dellwood 751	F5
56528	Dent 167	C4
56501	Detroit Lakes⊙ 7,106	C4
55926	Dexter 279	F7
56529	Dilworth 2,585	B4
55927	Dodge Center 1,816	F6
56235	Donnelly 317	B5
55929	Dover 312	F7
*55801	Duluth⊙ 92,811	F4
	Duluth-Superior‡ 266,650	F4
56236	Dumont 173	B5
55019	Dundas 422	E6
56127	Dunnell 216	D7
55111	Eagan 20,700	G6
56446	Eagle Bend 593	D4
56024	Eagle Lake 1,470	E6
†55005	East Bethel 6,626	E5
56721	East Grand Forks 8,537	B3
†56401	East Gull Lake 586	D4
56025	Easton 283	E7
56237	Echo 334	C6
55344	Eden Prairie 16,263	G6
55329	Eden Valley 763	D5
56128	Edgerton 1,123	B7
55424	Edina 46,073	G5
55931	Eitzen 226	G7
†55910	Elba 198	F6
56531	Elbow Lake⊙ 1,358	B5
55932	Elgin 667	F6
56533	Elizabeth 195	B4
55020	Elko 274	E6
55330	Elk River⊙ 6,785	E5
56026	Ellendale 555	E7
56129	Ellsworth 629	C7
56027	Elmore 882	D7
56325	Elrosa 214	C5
55731	Ely 4,820	G3
56028	Elysian 454	E6
56447	Emily 588	E4
56029	Emmons 465	E7
56534	Erhard 194	B4
56535	Erskine 585	B3
56326	Evansville 571	C4
55734	Eveleth 5,042	F3
55331	Excelsior 2,523	E6
55934	Eyota 1,244	F7
55332	Fairfax 1,405	D6
56031	Fairmont⊙ 11,506	D7
55113	Falcon Heights 5,291	G5
55021	Faribault⊙ 16,241	E6
55024	Farmington 4,370	E6
56641	Federal Dam 192	D3
56536	Felton 264	B3
56537	Fergus Falls⊙ 12,519	B4
56540	Fertile 869	B3
56448	Fifty Lakes 263	D4
55735	Finlayson 202	F4
56723	Fisher 453	B3
56328	Flensburg 256	D5
55736	Floodwood 648	E4
56329	Foley⊙ 1,606	D5
†56308	Forada 191	C5
55025	Forest Lake 4,596	F5
56330	Foreston 283	E5
56542	Fosston 1,599	C3
55935	Fountain 327	F7
56543	Foxhome 161	B4
55333	Franklin 512	D6
56544	Frazee 1,284	C4
56032	Freeborn 323	E7
56331	Freeport 563	D5
55432	Fridley 30,228	G5
56033	Frost 293	D7
56131	Fulda 1,308	C7
56332	Garfield 284	C5
56450	Garrison 174	E4
56132	Garvin 172	C6
56545	Gary 241	B3
55334	Gaylord⊙ 1,933	D6
56035	Geneva 417	E7
56239	Ghent 356	C6
55335	Gibbon 787	D6
55741	Gilbert 2,721	F3
56333	Gilman 156	E5
55336	Glencoe⊙ 4,396	D6
56036	Glenville 851	E7
56334	Glenwood⊙ 2,523	C5
56547	Glyndon 882	B4
55427	Golden Valley 22,775	G5
56644	Gonvick 362	C3
55027	Goodhue 657	F6
56725	Goodridge 191	C2
56037	Good Thunder 560	D6
55027	Goodview 2,567	G6
56240	Graceville 780	B5
56039	Granada 377	D7
55604	Grand Marais⊙ 1,289	G2
55936	Grand Meadow 965	F7
55744	Grand Rapids⊙ 7,934	E3
56241	Granite Falls⊙ 3,451	C6
55030	Grasston 123	E5
56726	Greenbush 817	B2
†55373	Greenfield 1,391	F5
55338	Green Isle 357	E6
56335	Greenwald 259	D5
56336	Grey Eagle 338	D5
56243	Grove City 596	D5
56727	Grygla 216	C2
56452	Hackensack 285	D4
56728	Hallock⊙ 1,405	A2
56548	Halstad 690	B3
55339	Hamburg 475	D6
55340	Hamel 2,623	F5
55304	Ham Lake 7,832	E5
55938	Hammond 178	F6
55031	Hampton 299	E6
56244	Hancock 877	C5
56245	Hanley Falls 265	C6
55341	Hanover 647	E5
56041	Hanska 429	D6
56134	Hardwick 279	B7
55939	Harmony 1,133	F7
55032	Harris 678	F5
56042	Hartland 322	E7
55033	Hastings⊙ 12,827	F6
56549	Hawley 1,634	B4
55940	Hayfield 1,243	F7
56043	Hayward 294	E7
55342	Hector 1,252	D6
56044	Henderson 739	E6
56136	Hendricks 737	B6
56550	Hendrum 336	B3
56551	Henning 832	C4
56248	Herman 600	B5
†55811	Hermantown 6,759	F4
56137	Heron Lake 783	C7
56453	Hewitt 299	C4
55746	Hibbing 21,193	F3
55748	Hill City 533	E4
56138	Hills 598	B7
55037	Hinckley 963	E4
56552	Hitterdal 253	B4

(continued on following page)

Agriculture, Industry and Resources

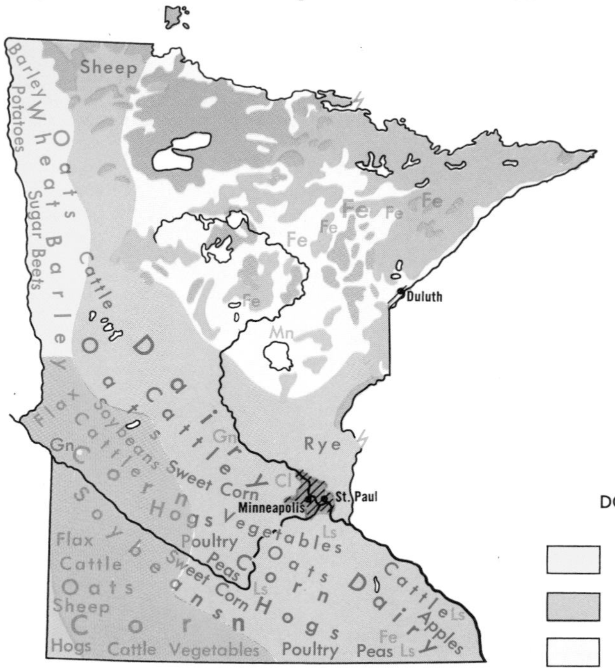

MAJOR MINERAL OCCURRENCES

Cl Clay
Fe Iron Ore
Gn Granite
Ls Limestone
Mn Manganese

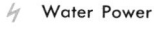

Water Power
Major Industrial Areas

DOMINANT LAND USE

Wheat, General Farming
Dairy, Livestock
Dairy, Hay, Potatoes
Cattle Feed, Hogs
Livestock, Cash Grain
Forests
Swampland, Limited Agriculture
Urban Areas

56339 Hoffman 631 C5
55941 Hokah 686 G7
56340 Holdingford 635 D5
56139 Holland 234 B6
56045 Hollandale 290 E7
56249 Holloway 142 C5
55343 Hopkins 15,336 G5
55943 Houston 1,057 G7
55349 Howard Lake 1,240 D5
55750 Hoyt Lakes 3,186 F3
55038 Hugo 3,771 E5
55350 Hutchinson 9,244 D6
†55359 Independence 2,640 F5
56649 International Falls⊙ 5,611 E2
55075 Inver Grove Heights 17,171 E6
56141 Iona 248 C7
56455 Ironton 537 D4
55040 Isanti 858 E5
56342 Isle 573 E4
56142 Ivanhoe⊙ 761 B6
56143 Jackson⊙ 3,797 C7
56048 Janesville 1,897 E6
56144 Jasper 731 B7
56145 Jeffers 437 C6
56456 Jenkins 219 D4
55352 Jordan 2,663 E6
56251 Kandiyohi 447 D5
56732 Karlstad 934 B2
56050 Kasota 739 D6
55944 Kasson 2,827 F6
55753 Keewatin 1,443 E3
56650 Kelliher 324 D3
55945 Kellogg 440 G6
55754 Kelly Lake 900 F3
56733 Kennedy 405 B2
56343 Kensington 331 C5
55946 Kenyon 1,529 E6
56252 Kerkhoven 761 C5
56051 Kiester 670 E7
56052 Kilkenny 177 E6
55353 Kimball 651 D5
55758 Kinney 447 F3
55947 La Crescent 3,674 G7
56054 Lafayette 507 D6
56149 Lake Benton 869 B6
56734 Lake Bronson 298 B2
55041 Lake City 4,505 F6
56055 Lake Crystal 2,078 D6
55042 Lake Elmo 5,296 F6
56150 Lakefield 1,845 C7
†55398 Lake Fremont (Zimmerman) 1,074 E5
55043 Lakeland 1,812 F6
56253 Lake Lillian 329 C6
56554 Lake Park 716 B4
†55043 Lake Saint Croix Beach 1,176 F6
†56401 Lake Shore 583 D4
55044 Lakeville 14,790 E6
56151 Lake Wilson 380 B7
56152 Lamberton 1,032 C6
56735 Lancaster 368 B2
55949 Lanesboro 923 G7
56461 Laporte 160 D3
†55744 La Prairie 536 E3
56344 Lastrup 150 D4
†55101 Lauderdale 1,985 G5
56057 Le Center⊙ 1,967 E6
55951 Le Roy 930 F7
55354 Lester Prairie 1,229 D6
56058 Le Sueur 3,763 E6
55952 Lewiston 1,226 G7
56060 Lewisville 273 D7
†55014 Lexington 2,150 G5
†55050 Lilydale 417 G5
55045 Lindstrom 1,972 F5
†55038 Lino Lakes 4,966 G5
56155 Lismore 276 B7
55355 Litchfield⊙ 5,904 D5
56345 Little Falls⊙ 7,250 D5
56653 Littlefork 918 E2
†56334 Long Beach 263 C5
55356 Long Lake 1,747 F5
56347 Long Prairie⊙ 2,859 D5
56655 Longville 191 D4
55046 Lonsdale 1,160 E6
55357 Loretto 297 F5
56349 Lowry 283 C5
56255 Lucan 262 C6
56156 Luverne⊙ 4,568 B7
55953 Lyle 576 F7
56157 Lynd 304 C6
55954 Mabel 861 G7
56062 Madelia 2,130 D6
56256 Madison⊙ 2,212 B5
56063 Madison Lake 592 E6
56158 Magnolia 234 B7
56557 Mahnomen⊙ 1,283 C3
55115 Mahtomedi 3,851 F5
56001 Mankato⊙ 28,651 E6
55955 Mantorville⊙ 705 F6
†55369 Maple Grove 20,525 G5
55358 Maple Lake 1,132 D5
55359 Maple Plain 1,421 F5
56065 Mapleton 1,516 E7
†55912 Mapleview 253 E7
55109 Maplewood 26,990 G5
55764 Marble 757 E3
56257 Marietta 279 B5
55047 Marine on Saint Croix 543 F5
56258 Marshall⊙ 11,161 C6
55360 Mayer 388 E6
56260 Maynard 428 C6
55956 Mazeppa 680 F6
55760 McGregor 447 E4
56556 McIntosh 681 C3
55761 McKinley 230 F3
55049 Medford 775 E6
55441 Medicine Lake 419 G5
†55340 Medina (Hamel) 2,623 F5
†56352 Meire Grove 174 C5
56352 Melrose 2,409 D5
56464 Menahga 980 C4
55050 Mendota 219 G5
†55050 Mendota Heights 7,288 G6
56736 Mentor 219 B3
56737 Middle River 349 B2
†55033 Miesville 179 F6
56262 Milan 417 C5
55957 Millville 186 F6
56263 Milroy 242 C6
56354 Miltona 187 C4
*55401 Minneapolis⊙ 370,951 G5
Minneapolis-Saint Paul‡ 2,114,256 G5
56264 Minneota 1,470 C6
55959 Minnesota City 265 G6
56068 Minnesota Lake 744 E7
55343 Minnetonka 38,683 G5
†55364 Minnetrista 3,236 F5
56265 Montevideo⊙ 5,845 C6
56069 Montgomery 2,349 E6
55362 Monticello 2,830 E5
55363 Montrose 762 E5
56560 Moorhead⊙ 29,998 B4
Moorhead-Fargo‡ 137,574 B4
55767 Moose Lake 1,408 F4
55051 Mora⊙ 2,890 E5
56266 Morgan 975 D6
56267 Morris⊙ 5,367 C5
55052 Morristown 639 E6
56270 Morton 549 C6
56466 Motley 444 D4
55364 Mound 9,280 E6
†55112 Mounds View 12,593 G5
55768 Mountain Iron 4,134 F3
56159 Mountain Lake 2,277 D7
56271 Murdock 343 C5
55769 Nashwauk 1,419 E3
56355 Nelson 209 C5
55053 Nerstrand 255 E6
56467 Nevis 332 D4
55366 New Auburn 331 D6
55112 New Brighton 23,269 G5
56738 Newfolden 384 B2
55367 New Germany 347 E6
†55428 New Hope 23,087 G5
56273 New London 812 C5
55054 New Market 286 E6
56356 New Munich 302 D5
55055 Newport 3,323 F6
56071 New Prague 2,952 E6
56072 New Richland 1,263 E7
56073 New Ulm⊙ 13,755 D6
56567 New York Mills 972 C4
56074 Nicollet 709 D6
56568 Nielsville 145 B3
56468 Nisswa 1,407 D4
55056 North Branch 1,597 F5
55057 Northfield 12,562 E6
56001 North Mankato 9,145 D6
†55101 North Oaks 2,846 G5
56661 Northome 312 D3
56275 North Redwood 206 D6
56075 Northrop 269 D7
55109 North Saint Paul 11,921 G5
55368 Norwood 1,219 E6
†55109 Oakdale 12,123 E5
56276 Odessa 177 B5
56160 Odin 134 D7
56569 Ogema 215 C3
56358 Ogilvie 423 E5
56161 Okabena 263 C7
56742 Oklee 536 C3
56277 Olivia⊙ 2,802 C6
56359 Onamia 691 E4
56162 Ormsby 181 D7
†55323 Orono 6,845 F5
55960 Oronoco 574 F6
55771 Orr 294 F2
56278 Ortonville⊙ 2,550 B5
56360 Osakis 1,355 C5
56744 Oslo 379 A2
55369 Osseo 2,974 G5
55961 Ostrander 293 F7
56571 Ottertail 239 C4
55060 Owatonna⊙ 18,632 E6
56469 Palisade 155 E4
56361 Parkers Prairie 917 C4
56470 Park Rapids⊙ 2,976 D4
56362 Paynesville 2,140 D5
56363 Pease 174 E5
†56472 Pelican Lakes (Breezy Point) 384 D4
56572 Pelican Rapids 1,867 B4
56078 Pemberton 208 E7
56279 Pennock 410 C5
56472 Pequot Lakes 681 D4
56573 Perham 2,086 C4
55962 Peterson 291 G7
†56364 Pierz 1,018 D5
56473 Pillager 341 D4
55063 Pine City⊙ 2,489 F5
55963 Pine Island 1,986 F6
56474 Pine River 881 D4
56164 Pipestone⊙ 4,887 B7
55964 Plainview 2,416 F6
55370 Plato 390 D6
56748 Plummer 353 B3
†55441 Plymouth 31,615 G5
56280 Porter 211 B6
55965 Preston⊙ 1,478 F7
55371 Princeton 3,146 E5
56281 Prinsburg 557 C6
55372 Prior Lake 7,284 E6
55810 Proctor 3,180 F4
55967 Racine 285 F7
56475 Randall 527 D4
55065 Randolph 351 E6
56668 Ranier 237 E2
56282 Raymond 723 C5
56750 Red Lake Falls⊙ 1,732 B3
55066 Red Wing⊙ 13,736 F6
56283 Redwood Falls⊙ 5,210 C6
56672 Remer 396 E3
56284 Renville 1,493 C6
56166 Revere 158 C6
56367 Rice 499 D5
55423 Richfield 37,851 G6
56368 Richmond 867 D5
55422 Robbinsdale 14,422 G5
55901 Rochester⊙ 57,890 F6
Rochester‡ 91,971 F6
55067 Rock Creek 890 F5
55373 Rockford 2,408 F5
56369 Rockville 597 D5
55374 Rogers 652 E5
55969 Rollingstone 528 G6
56371 Roscoe 154 D5
56751 Roseau⊙ 2,272 C2
55970 Rose Creek 371 F7
55068 Rosemount 5,083 E6
55113 Roseville 35,820 G5
56579 Rothsay 476 B4
56167 Round Lake 480 C7
56373 Royalton 660 D5
55069 Rush City 1,198 F5
55971 Rushford 1,478 G7
56168 Rushmore 387 C7
56169 Russell 412 C6
56170 Ruthton 328 B6
55778 Rutledge 185 F4
56580 Sabin 446 B4
56285 Sacred Heart 666 C6
55414 Saint Anthony 7,981 G5
55375 Saint Bonifacius 857 F5
55972 Saint Charles 2,184 F7
56080 Saint Clair 655 E6
56301 Saint Cloud⊙ 42,566 D5
Saint Cloud‡ 163,256 D5
55070 Saint Francis 1,184 E5
56554 Saint Hilaire 388 B2
56081 Saint James⊙ 4,346 D7
56374 Saint Joseph 2,994 D5
55426 Saint Louis Park 42,931 G5
56376 Saint Martin 220 D5
55376 Saint Michael 1,519 E5
*55101 Saint Paul (cap.)⊙ 270,230 G6
Saint Paul-Minneapolis‡ 2,114,256 G5
55071 Saint Paul Park 4,864 G6
56082 Saint Peter⊙ 9,056 E6
56375 Saint Stephen 453 D5
56755 Saint Vincent 141 A2
56083 Sanborn 518 C6
55072 Sandstone 1,594 F4
56377 Sartell 3,427 D5
56378 Sauk Centre 3,709 C5
56379 Sauk Rapids 5,793 D5
55337 Savage 3,954 G6
†55720 Scanlon 1,050 F4
56477 Sebeka 774 C4
55074 Shafer 180 F5
55379 Shakopee⊙ 9,941 F6
56581 Shelly 276 B3
56171 Sherburn 1,275 D7
56676 Shevlin 193 C3
†55112 Shoreview 17,300 G5
†55331 Shorewood 4,646 F5
55614 Silver Bay 2,917 G3
55381 Silver Lake 698 D6
†56001 Skyline 399 D6
56172 Slayton⊙ 2,420 C7
56085 Sleepy Eye 3,581 D6
†56345 Sobieski 219 D5
55382 South Haven 205 D5
56679 South International Falls 2,806 E2
55075 South Saint Paul 21,235 G6
56288 Spicer 909 C5
56087 Springfield 2,303 C6
55974 Spring Grove 1,275 G7
†55432 Spring Lake Park 6,477 E5
55384 Spring Park 1,465 F5
55975 Spring Valley 2,616 F7
56681 Squaw Lake 162 D3
55079 Stacy 996 E5
56479 Staples 2,887 D4
56381 Starbuck 1,224 C5
56173 Steen 153 B7
56757 Stephen 898 A2
55385 Stewart 616 D6
55976 Stewartville 3,925 F7
55082 Stillwater⊙ 12,290 F5
55988 Stockton 517 G6
56174 Storden 341 C6
56758 Strandquist 136 B2
55783 Sturgeon Lake 222 F4
†55075 Sunfish Lake 344 E6
56382 Swanville 295 D5
55786 Taconite 331 E3
56291 Taunton 177 B6
55084 Taylors Falls 623 F5
56683 Tenstrike 159 D3
56701 Thief River Falls⊙ 9,105 B2
†56319 Thomson 152 F4
†55331 Tonka Bay 1,354 F5
55790 Tower 640 F3
56175 Tracy 2,478 C6
56176 Trimont 805 D7
56088 Truman 1,392 D7
56089 Twin Lakes 210 E7
56584 Twin Valley 907 B3
55616 Two Harbors⊙ 4,039 G3
56178 Tyler 1,353 B6
56585 Ulen 514 B3
56586 Underwood 332 C4
56384 Upsala 400 D5
55979 Utica 249 G7
†55101 Vadnais Heights 5,111 G5
56587 Vergas 287 C4
55085 Vermillion 438 F6
56481 Verndale 504 C4
56090 Vernon Center 365 D7
56292 Vesta 360 C6
55386 Victoria 1,425 F6
56385 Villard 275 C5
55792 Virginia 11,056 F3
55981 Wabasha⊙ 2,372 G6
56293 Wabasso 745 C6
55387 Waconia 2,638 E6
56482 Wadena⊙ 4,699 C4
56386 Wahkon 271 E4
56387 Waite Park 3,496 D5
56091 Waldorf 249 E7
56484 Walker⊙ 970 D3
56180 Walnut Grove 753 C6
55982 Waltham 176 F7
55983 Wanamingo 717 F6
55743 Warba 150 E3
56762 Warren⊙ 2,105 B2
56763 Warroad 1,216 C2
56093 Waseca⊙ 8,219 E6
55388 Watertown 1,818 E6
56096 Waterville 1,717 E6
55389 Watkins 757 D5
56295 Watson 238 C5
56589 Waubun 390 C3
55390 Waverly 470 E5
55391 Wayzata 3,621 G5
56181 Welcome 855 D7
56097 Wells 2,777 E7
56590 Wendell 216 B4
56183 Westbrook 978 C6
55985 West Concord 762 F6
55118 West Saint Paul 18,527 G5
56296 Wheaton⊙ 1,969 B5
55110 White Bear Lake 22,538 G5
55090 Willernie 654 G5
56686 Williams 217 D2
56201 Willmar⊙ 15,895 C5
55795 Willow River 303 F4
56185 Wilmont 380 C7
56687 Wilton 176 C3
56101 Windom⊙ 4,666 C7
56592 Winger 200 B3
56098 Winnebago 1,869 D7
55987 Winona⊙ 25,075 G6
55395 Winsted 1,522 D6
55396 Winthrop 1,376 D6
55796 Winton 276 G3
56594 Wolverton 177 B4
†55798 Woodbury 10,297 F6
56297 Wood Lake 420 C6
56186 Woodstock 180 B7
56187 Worthington⊙ 10,243 C7
55797 Wrenshall 333 F4
55798 Wright 162 E4
55990 Wykoff 482 F7
55092 Wyoming 1,559 F5
55397 Young America 1,237 E6
55398 Zimmerman 1,074 E5
55991 Zumbro Falls 208 F6
55992 Zumbrota 2,129 F6

OTHER FEATURES

Ash (riv.) F2
Bald Eagle (lake) G3
Basswood (lake) G2
Battle (riv.) D3
Baudette (riv.) D2
Bear (riv.) E3
Bemidji (lake) D3
Benton (lake) B6
Big Fork (riv.) E2
Big Sandy (lake) E4
Big Stone (lake) B5
Birch (lake) G3
Black (riv.) D2
Blue Earth (riv.) D7
Bois de Sioux (riv.) B4
Bowstring (lake) E3
Buffalo (riv.) B4
Burntside (lake) F3
Cass (lake) D3
Cedar (riv.) F7
Chippewa (riv.) C5
Christina (lake) C4
Clearwater (riv.) C3
Cloquet (riv.) F4
Cobb (riv.) E7
Cottonwood (riv.) C6
Crooked (creek) F4
Crooked (lake) G2
Crow (riv.) F5
Crow Wing (riv.) D4
Cuyuna (range) D4
Dead (lake) C4
Deer (lake) E3
Des Moines (riv.) C7
Eagle (mt.) G2
East Swan (riv.) F3
Elbow (lake) C3
Emily (lake) C5
Fond du Lac Ind. Res. F4
Grand Portage Ind. Res. G2
Grand Portage Nat'l Mon. G2
Green (lake) D5
Greenwood (lake) G3
Gull (lake) D4
Heron (lake) C7
Hill (riv.) C3
Independence (lake) F5
Isabella (lake) G3
Itasca (lake) C3
Kabetogama (lake) E2
Kanaranzi (creek) C7
Kettle (riv.) F4
Knife (lake) G2
La Croix (lake) F2
Lac qui Parle (lake) C5
Lac qui Parle (riv.) B6
Lake of the Woods (lake) D1
Leaf (riv.) C4
Leech (lake) D3
Leech Lake Ind. Res. D3
Lida (lake) C4
Little Fork (riv.) E2
Little Rock (creek) C7
Long (lake) D4
Long (lake) F3
Long Prairie (riv.) D4
Lost (riv.) C3
Lower Red (lake) C3
Maple (lake) B3
Maple (riv.) E7
Marsh (lake) B5
Mary (lake) C5
Mesabi (range) E3
Middle (riv.) B2
Mille Lac Ind. Res. E4
Mille Lacs (lake) E4
Miltona (lake) C4
Minneapolis-Saint Paul Airport G5
Minnesota (riv.) E6
Minnetonka (lake) F5
Minnewaska (lake) C5
Misquah (hills) F2
Mississippi (riv.) D4
Moose (riv.) C2
Mud (lake) C2
Mud (riv.) C2
Muskeg (bay) C2
Mustinka (riv.) B5
Nemadji (riv.) F4
Nett (lake) E2
Nett Lake Ind. Res. E2
North (lake) F1
Otter Tail (lake) C4
Otter Tail (riv.) B4
Partridge (riv.) G3
Pelican (lake) C4
Pelican (lake) D4
Pelican (lake) F2
Pelican (riv.) B4
Pelican (riv.) F2
Pepin (lake) F6
Pigeon (riv.) G2
Pike (riv.) F3
Pipestone Nat'l Mon. B6
Pokegama (lake) E3
Pomme de Terre (riv.) C5
Poplar (riv.) C3
Prairie (riv.) E3
Rainy (lake) E2
Rainy (riv.) D2
Rapid (riv.) D2
Redeye (riv.) D4
Red Lake (riv.) B2
Red Lake Ind. Res. C2
Red River of the North (riv.) A2
Redwood (riv.) C6
Reno (lake) C5
Rice (lake) E4
Rock (riv.) B7
Root (riv.) G7
Roseau (riv.) B2
Rum (riv.) E5
Saganaga (lake) H2
Saint Croix (riv.) F5
Saint Louis (riv.) F4
Sand (creek) F5
Sand Hill (riv.) B3
Sarah (lake) F5
Schoolcraft (riv.) C3
Shakopee (creek) C5
Shell (riv.) C4
Shetek (lake) C6
Sleepy Eye (creek) C6
Snake (riv.) A2
Snake (riv.) E4
South Fowl (lake) G1
Star (lake) C4
Sturgeon (riv.) F3
Superior (lake) G3
Swan (lake) D6
Tamarac (riv.) A2
Tamarack (riv.) D2
Thief (lake) C2
Thief (riv.) B2
Traverse (lake) B5
Trout (lake) F2
Two Rivers (riv.) A1
Upper Red (lake) D2
Vermilion (lake) F3
Vermilion (range) F3
Vermilion (riv.) F2
Voyageurs Nat'l Park F2
Wabatawangang (lake) D3
West Swan (riv.) F3
White Earth Ind. Res. C3
Whiteface (riv.) F3
Whitefish (lake) D4
White Iron (lake) G3
Wild Rice (lake) F4
Wild Rice (riv.) B3
Willow (riv.) E4
Winnibigoshish (lake) D3
Woods (lake) D1
Zumbro (riv.) F6

⊙County seat.
‡Population of metropolitan area.
† Zip of nearest p.o. * Multiple zips.

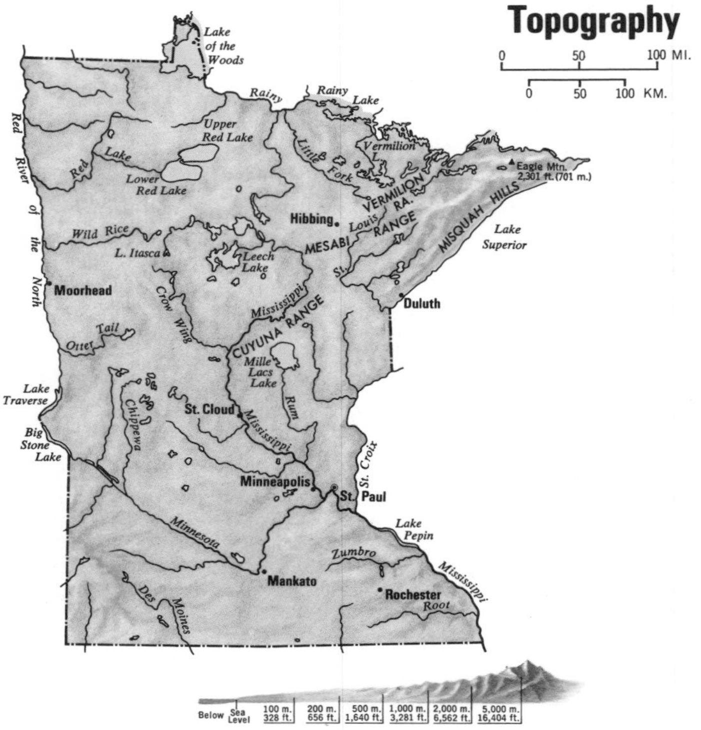
Topography
0 50 100 MI.
0 50 100 KM.
Lake of the Woods
Rainy
Rainy Lake
Red River of the North
Upper Red Lake
Red Lake
Lower Red Lake
Vermilion
Little Fork
Eagle Mtn. 2,301 ft. (701 m.)
VERMILION RA.
Louis
MISQUAH HILLS
Lake Superior
Hibbing
MESABI RANGE
Wild Rice
L. Itasca
Leech Lake
St.
Moorhead
Crow Wing
Mississippi
CUYUNA RANGE
Duluth
Otter Tail
Mille Lacs Lake
Lake Traverse
Chippewa
St. Cloud
Rum
Big Stone Lake
Minneapolis
St. Paul
St. Croix
Lake Pepin
Minnesota
Zumbro
Mississippi
Mankato
Rochester
Root
Des Moines
Below Sea Level
100 m. 328 ft.
200 m. 656 ft.
500 m. 1,640 ft.
1,000 m. 3,281 ft.
2,000 m. 6,562 ft.
5,000 m. 16,404 ft.

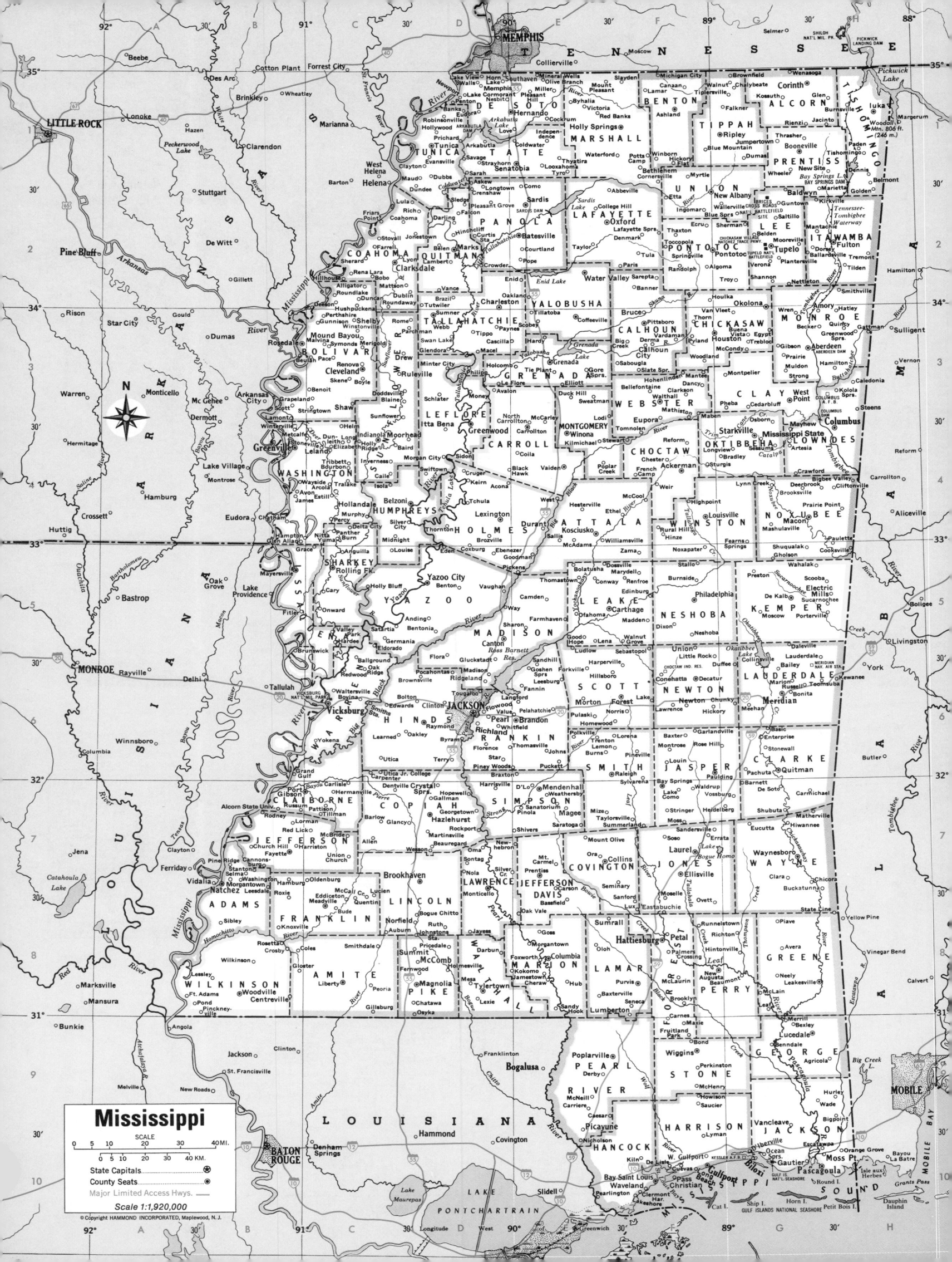

Mississippi

SCALE

0 5 10   20   30   40 MI.

0 5 10   20   30   40 KM.

State Capitals .... ⊛

County Seats .... ⊛

Major Limited Access Hwys. ————

Scale 1:1,920,000

© Copyright HAMMOND INCORPORATED, Maplewood, N.J.

AREA 47,689 sq. mi. (123,515 sq. km.)
POPULATION 2,520,638
CAPITAL Jackson
LARGEST CITY Jackson
HIGHEST POINT Woodall Mtn. 806 ft. (246 m.)
SETTLED IN 1716
ADMITTED TO UNION December 10, 1817
POPULAR NAME Magnolia State
STATE FLOWER Magnolia
STATE BIRD Mockingbird

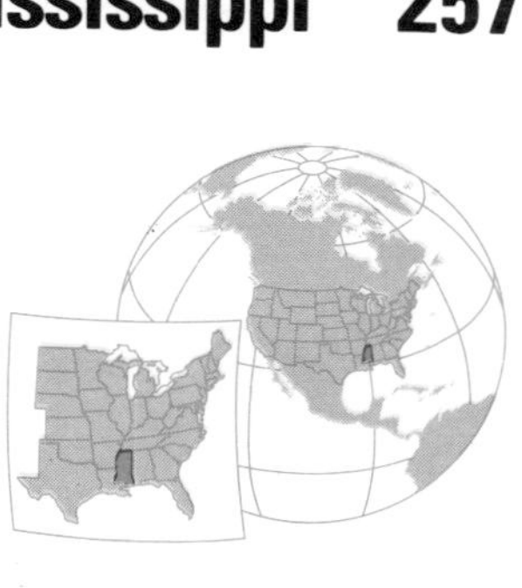

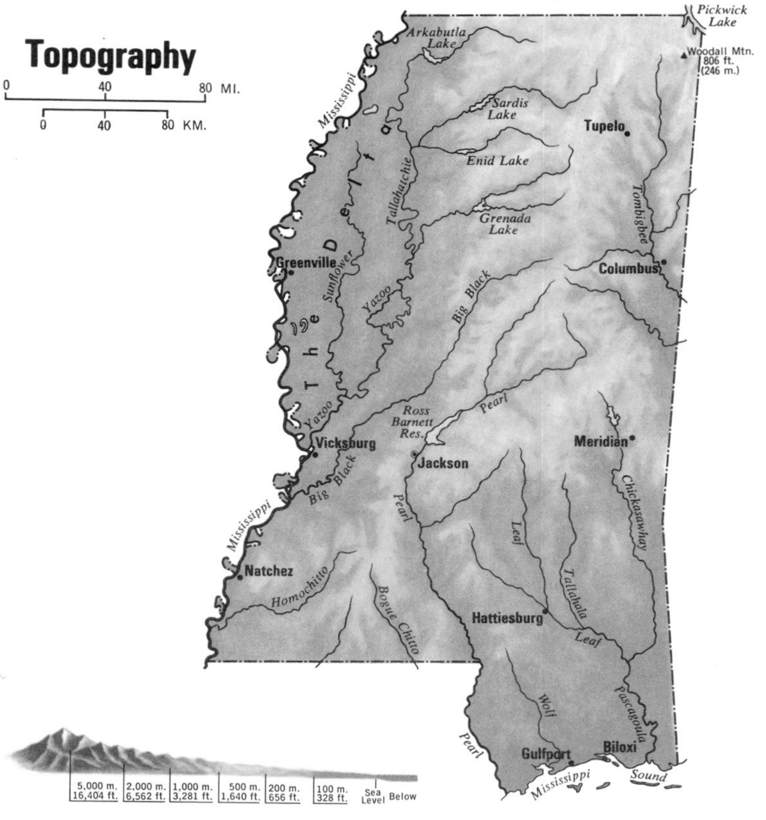

COUNTIES

Adams 38,035 B8
Alcorn 33,036 G1
Amite 13,369 C8
Attala 19,865 E4
Benton 8,153 F1
Bolivar 45,965 C3
Calhoun 15,664 F3
Carroll 9,776 E4
Chickasaw 17,853 G3
Choctaw 8,996 F4
Claiborne 12,279 C7
Clarke 16,945 G6
Clay 21,082 G3
Coahoma 36,918 C2
Copiah 26,503 D7
Covington 15,927 E7
De Soto 53,930 E1
Forrest 66,018 F8
Franklin 8,208 C8
George 15,297 G9
Greene 9,827 G8
Grenada 21,043 E3
Hancock 24,537 E10
Harrison 157,665 F10
Hinds 250,998 D6
Holmes 22,970 D4
Humphreys 13,931 C4
Issaquena 2,513 B5
Itawamba 20,518 H2
Jackson 118,015 G9
Jasper 17,265 F6
Jefferson 9,181 B7
Jefferson Davis 13,846 E7
Jones 61,912 F7
Kemper 10,148 G5
Lafayette 31,030 E2
Lamar 23,821 E8
Lauderdale 77,285 G6
Lawrence 12,518 D7
Leake 18,790 E5
Lee 57,061 G2
Leflore 41,525 D3
Lincoln 30,174 D8
Lowndes 57,304 H4
Madison 41,613 D5
Marion 25,708 E8
Marshall 29,296 E1
Monroe 36,404 H3
Montgomery 13,366 E4
Neshoba 23,789 F5
Newton 19,944 F6
Noxubee 13,212 G4
Oktibbeha 36,018 G4
Panola 28,164 E2
Pearl River 33,795 E9
Perry 9,864 G8
Pike 36,173 D8
Pontotoc 20,918 F2
Prentiss 24,025 G1
Quitman 12,636 D2
Rankin 69,427 E6
Scott 24,556 E6
Sharkey 7,964 C5
Simpson 23,441 E7
Smith 15,077 E6
Stone 9,716 F9
Sunflower 34,844 C3
Tallahatchie 17,157 D3
Tate 20,119 E1
Tippah 18,739 G1
Tishomingo 18,434 H1
Tunica 9,652 D1
Union 21,741 F2
Walthall 13,761 D8
Warren 51,627 C6
Washington 72,344 C4
Wayne 19,135 G7
Webster 10,300 F3
Wilkinson 10,021 B8
Winston 19,474 F4
Yalobusha 13,139 E2
Yazoo 27,349 D5

CITIES and TOWNS

Zip Name/Pop. Key

38601 Abbeville 448 F2
39730 Aberdeen⊙ 7,184 H3
39735 Ackerman⊙ 1,567 F4
39096 Alcorn State University B7
38820 Algoma 175 G2
†39083 Allen 15 C7
38720 Alligator 256 C2
38821 Amory 7,307 H3
38721 Anguilla 950 C5
38722 Arcola 588 C4
38602 Arkabutla 400 D1
39736 Artesia 526 G4
38603 Ashland⊙ 532 F1
38604 Askew 300 D1
†39664 Auburn 500 C8
38912 Avalon 100 D3
38723 Avon 400 B4
39320 Bailey 320 G6
38724 Baird 150 C4
38824 Baldwyn 3,427 G2
†39156 Ballground 30 C5
38913 Banner 120 F2
†39083 Barlow 20 C7
†39330 Basic 60 G6
39421 Bassfield 325 E8
38606 Batesville⊙ 4,692 E2
†39343 Baxter 75 F6
†39455 Baxterville 100 E8
39520 Bay Saint Louis⊙ 7,891 F10
39422 Bay Springs⊙ 1,884 F7
39423 Beaumont 1,112 G8
†39191 Beauregard 185 D7
38825 Becker 350 G3
38826 Belden 241 G2
38609 Belen 400 D2
39737 Bellefontaine 400 F3
38827 Belmont 1,420 H1
39038 Belzoni⊙ 2,982 C4
†39450 Benndale 500 G9
38725 Benoit 499 C3
39039 Benton 350 D5
39040 Bentonia 518 D5
†38659 Bethlehem 210 F1
38726 Beulah 431 B3
39738 Bigbee Valley 370 H4
38914 Big Creek 146 F3
†39567 Bigpoint 350 H9
*39530 Biloxi 49,311 G10
Biloxi-Gulfport‡ 191,918 G10
†38917 Black Hawk 41 E4
38727 Blaine 75 C3
38610 Blue Mountain 867 G1
38828 Blue Springs 131 G2
†38614 Bobo 200 C2
39629 Bogue Chitto 575 D8
39041 Bolton 664 D6
39550 Bond 350 F9
†39301 Bonita 300 G6
38829 Booneville⊙ 6,199 G1
†38756 Bourbon 200 C4
†39180 Bovina 50 C6
38730 Boyle 888 C3
39042 Brandon⊙ 9,626 E6
39044 Braxton 172 D6
38963 Brazil 229 D2
39601 Brookhaven⊙ 10,800 C7
39425 Brooklyn 450 F8
39739 Brooksville 1,038 G4
†38683 Brownfield 125 G1
38915 Bruce 2,208 F3
39322 Buckatunna 500 G7
39630 Bude 1,092 C8
38833 Burnsville 889 H1
38611 Byhalia 757 E1
†39205 Byram 250 D6
†38754 Caile 30 C4
39740 Caledonia 497 H3
38916 Calhoun City 2,033 F3
39045 Camden 150 E5
38612 Canaan 200 F1
39046 Canton⊙ 11,116 D5
39049 Carlisle 425 C7
†39360 Carmichael 75 G7
39050 Carpenter 200 C6
39426 Carriere 900 E9
38917 Carrollton⊙ 338 E4
39427 Carson 400 E7
39051 Carthage⊙ 3,453 E5
39054 Cary 470 C5
38920 Cascilla 230 D3
39741 Cedarbluff 175 G3
39631 Centreville 1,844 B8
38684 Chalybeate 350 G1
38921 Charleston⊙ 2,878 D2
39632 Chatawa 300 D8
38731 Chatham 150 B4
39323 Chunky 277 G6
39055 Church Hill 350 B7
39324 Clara 275 G7
38614 Clarksdale⊙ 21,137 D2
39551 Clermont Harbor 550 F10
38732 Cleveland⊙ 14,524 C3
39056 Clinton 14,660 D6
38617 Coahoma 350 C2
†38632 Cockrum 150 E1
38922 Coffeeville⊙ 1,129 E3
38923 Coila 75 E4
38618 Coldwater 1,505 E1
†39638 Coles 150 C8
†38655 College Hill 150 E2
39428 Collins⊙ 2,131 E7
39325 Collinsville 700 G6
39429 Columbia⊙ 7,733 E8
39701 Columbus⊙ 27,383 H3
38619 Como 1,378 E1
39057 Conehatta 200 F6
†39051 Conway 25 E5
38834 Corinth⊙ 13,839 G1
†38659 Cornersville 65 F1
38620 Courtland 381 E2
†39095 Coxburg 300 D5
39743 Crawford 495 G4
38621 Crenshaw 1,019 D2
39633 Crosby 349 B8
38622 Crowder 789 D2
38924 Cruger 540 D4
39059 Crystal Springs 4,902 D7
†38606 Curtis Station 350 D2
39326 Daleville 210 G5
†39643 Darbun 100 D8
38623 Darling 275 D2
39327 Decatur⊙ 1,148 F6
†39739 Deerbrook 30 G4
39328 De Kalb⊙ 1,159 G5
†39571 De Lisle 450 F10
39061 Delta City 310 C4
†38655 Denmark 40 F2
38838 Dennis 150 H1
†39059 Dentville 175 D7
†39470 Derby 298 E9
38839 Derma 793 F3
†39532 D'Iberville 13,369 G10
39062 D'Lo 463 E7
38736 Doddsville 232 C3
38737 Drew 2,528 C3
38739 Dublin 100 C2
38925 Duck Hill 706 E3
†39337 Duffee 175 G6
38625 Dumas 312 G1
38740 Duncan 501 C2
38626 Dundee 600 D1
39063 Durant 2,889 E4
39436 Eastabuchie 200 F8
39064 Ebenezer 200 D5
38841 Ecru 687 F2
39634 Eddiceton 65 C8
39065 Eden 150 D5
39066 Edwards 1,515 C6
†39156 Eldorado 20 C5
39329 Electric Mills 100 G5
38742 Elizabeth 500 C4
38926 Elliott 200 E3
39437 Ellisville⊙ 4,652 F7
38927 Enid E2
39330 Enterprise 607 G6
†39440 Errata 85 F7
39552 Escatawpa 5,367 G10
39067 Ethel 486 F4
38627 Etta 75 F2
39744 Eupora 2,048 F3
†38676 Evansville 60 D1
38628 Falcon 260 D2
38629 Falkner 251 G1
38630 Farrell 300 C2
39069 Fayette⊙ 2,033 B7
39635 Fernwood 500 D8
39070 Fitler 175 B5
39071 Flora 1,507 D5
39073 Florence 1,111 D6
†39201 Flowood 943 D6
39074 Forest⊙ 5,229 F6
39076 Forkville 185 E6
39636 Fort Adams 75 B8
39483 Foxworth 800 E8
39745 French Camp 306 F4
38631 Friars Point 1,400 C2
39577 Fruitland Park 75 F9
38843 Fulton⊙ 3,238 H2
39077 Gallman D7
38844 Gattman 151 H3
39553 Gautier 8,917 G10
39078 Georgetown 343 D7
†39354 Gholson 50 G5
†39083 Glancy 25 C7
38846 Glen 100 H1
38744 Glen Allan 650 B4
38928 Glendora 220 D3
39638 Gloster 1,726 B8
†39110 Gluckstadt 150 D5
38847 Golden 292 H2
39079 Goodman 1,285 E5
38929 Gore Springs 125 E3
38745 Grace 325 C5
†38725 Grapeland 200 B3
38701 Greenville⊙ 40,613 B4
38930 Greenwood⊙ 20,115 D4
38848 Greenwood Springs 170 H3
38901 Grenada⊙ 12,641 E3
*39501 Gulfport⊙ 39,676 F10
38746 Gunnison 708 C3
38849 Guntown 359 G2
†39661 Hamburg 150 B7
39746 Hamilton 500 H3
†38901 Hardy 45 E3
39080 Harperville 200 E6
39081 Harriston 500 C7
39082 Harrisville 500 D7
†38821 Hatley 497 H3
39401 Hattiesburg⊙ 40,829 F8
39083 Hazlehurst⊙ 4,437 D7
39439 Heidelberg 1,098 F7
39086 Hermanville 750 C7
38632 Hernando⊙ 2,969 E1
†39192 Hesterville 25 E4
39332 Hickory 670 F6
38633 Hickory Flat 458 F1
39087 Hillsboro 800 E6
†38646 Hinchcliff 60 D2
†39462 Hintonville 300 F8
†39108 Hinze 30 F4
†39751 Hohenlinden 96 F3
38940 Holcomb 50 D3
38748 Hollandale 4,336 C4
39088 Holly Bluff 700 C5
38749 Holly Ridge 350 C4
38635 Holly Springs⊙ 7,285 E1
†38676 Hollywood 80 D1
†39648 Holmesville 50 D8
38637 Horn Lake 4,326 D1
38850 Houlka 710 G2
38851 Houston⊙ 3,747 G3
†39574 Howison 300 F9
†39429 Hub 80 E8
39555 Hurley 500 H9
†38774 Hushpuckena 60 C2
38638 Independence 150 E1
38751 Indianola⊙ 8,221 C4
†38652 Ingomar 150 F2
38753 Inverness 1,034 C4
38754 Isola 834 C4
38941 Itta Bena 2,904 D4
38852 Iuka⊙ 2,846 H1
†38865 Jacinto 65 H1
*39201 Jackson (cap.)⊙ 202,895 D6
Jackson‡ 320,425 D6
39641 Jayess 200 D8
38639 Jonestown 1,231 D2
†38829 Jumpertown 472 G1
†38924 Keirn 3 D4
†39364 Kewanee 250 H6
39747 Kilmichael 906 E4
39556 Kiln 800 F10
†39661 Knoxville 65 B8
39643 Kokomo 250 E8
†39740 Kolola Springs 100 H3
39090 Kosciusko⊙ 7,415 E4
38834 Kossuth 190 G1
38640 Lafayette Springs 80 F2
39336 Lake 524 F6
38641 Lake Cormorant 300 D1
39558 Lakeshore 550 F10
38642 Lamar 200 F1
38643 Lambert 1,624 D2
38755 Lamont 400 B3
39335 Lauderdale 600 G5
39440 Laurel⊙ 21,897 F7
39336 Lawrence 250 F6
39451 Leaf 250 G8
39451 Leakesville⊙ 1,120 G8
39093 Learned 113 C6
38756 Leland 6,667 C4
39094 Lena 231 E5
†39667 Lexie 40 D8
39095 Lexington⊙ 2,628 D4
39645 Liberty⊙ 669 C8
39337 Little Rock 70 F5
39560 Long Beach 7,967 F10
†39759 Longview 800 G4
39096 Lorman 350 B7
39338 Louin 338 F6
39097 Louise 400 C5
39339 Louisville⊙ 7,323 G4
†38632 Love 50 D1
39452 Lucedale⊙ 2,429 G9
39646 Lucien 75 C7
39098 Ludlow 350 E5
38644 Lula 394 C2
39455 Lumberton 2,217 E8
†39501 Lyman 500 F10
†39739 Lynn Creek 20 G4
38645 Lyon 531 D2
39750 Maben 855 F3
39341 Macon⊙ 2,396 G4
39109 Madden 450 F5
39110 Madison 2,241 D6
39111 Magee 3,497 E7
39652 Magnolia⊙ 2,461 D8
†38769 Malvina 100 C3
38855 Mantachie 732 H2
39751 Mantee 158 F3
38856 Marietta 298 H2
39342 Marion 771 G6
38646 Marks⊙ 2,260 D2
†39083 Martinsville 30 D7
†39051 Marydell 99 F5
39752 Mathiston 632 F3
38758 Mattson 200 C2
†39458 Maxie 233 F9
39113 Mayersville⊙ 378 B5
39753 Mayhew 150 G4
39107 McAdams 350 E4
†39144 McBride 2 C7
39647 McCall Creek 250 C7
38943 McCarley 250 E3
39648 McComb 12,331 D8
38854 McCondy 150 G3
39108 McCool 203 F4
39561 McHenry 660 F9
39456 McLain 688 G8
39457 McNeill 800 E9
39653 Meadville⊙ 575 C8
39114 Mendenhall⊙ 2,533 E7
39301 Meridian⊙ 46,577 G6
38759 Merigold 574 C3
†39667 Mesa 30 D8
38760 Metcalfe 952 B4
38647 Michigan City 350 F1
39115 Midnight 500 C4
38648 Mineral Wells 250 E1
38944 Minter City 150 D3
39762 Mississippi State G4
39116 Mize 363 E7
38945 Money 350 D3
39654 Monticello⊙ 1,834 D7
39754 Montpelier 175 G3
†39338 Montrose 120 F6
38857 Mooreville 200 G2
38761 Moorhead 2,358 C4
38946 Morgan City 319 D4
39484 Morgantown 325 E8
†39120 Morgantown 3,445 B7
39117 Morton 3,303 E6
†39328 Moscow 30 G5
39459 Moselle 525 F8
39460 Moss 65 F7
39563 Moss Point 18,998 G10
38762 Mound Bayou 2,917 C3
†39474 Mount Carmel 30 E7
39119 Mount Olive 993 E7
38649 Mount Pleasant 250 E1
38650 Myrtle 402 F1
39120 Natchez⊙ 22,015 B7
39461 Neely 270 G8
38651 Nesbit 366 D1
39365 Neshoba 250 F5
38858 Nettleton 1,911 G2
38652 New Albany⊙ 7,072 G2
39462 New Augusta⊙ 589 F8
39140 Newhebron 470 D7
38850 New Houlka (Houlka) 710 G2
38859 New Site 100 H1
39345 Newton 3,708 F6
39463 Nicholson 400 E10
38763 Nitta Yuma 150 C4
†39629 Norfield 75 C8
38947 North Carrollton 859 E3
39346 Noxapater 516 F5
38948 Oakland 540 E2
†39154 Oakley 133 D6
39656 Oak Vale E8
39564 Ocean Springs 14,504 G10
39141 Ofahoma 350 E5
38860 Okolona⊙ 3,409 G2
38654 Olive Branch 2,067 E1
†39482 Oloh 93 E8
†39654 Oma 200 D7
†39501 Orange Grove 13,476 H10
39657 Osyka 581 D8
39464 Ovett 600 F8
38655 Oxford⊙ 9,882 F2
38764 Pace 519 C3
39347 Pachuta 256 G6
38861 Paden 119 H1
†39401 Palmers Crossing 2,765 F8
38765 Panther Burn 300 C4
38738 Parchman 200 D3
38949 Paris 253 F2
39567 Pascagoula⊙ 29,318 G10
Pascagoula-Moss Point‡ 118,015 G10
39571 Pass Christian 5,014 F10
39144 Pattison 540 C7
39348 Paulding⊙ 630 F6
39349 Paulette 230 H4
†38920 Paynes 100 D3
39028 Pearl 18,580 D6
39572 Pearlington 500 E10
39145 Pelahatchie 1,445 E6
39573 Perkinston 950 F9
†38746 Perthshire 25 C3
39465 Petal 8,476 F8
39755 Pheba 280 G3
39350 Philadelphia⊙ 6,434 F5
38950 Philipp 975 D3
†39476 Piave 150 G8
39466 Picayune 10,361 E9
39146 Pickens 1,386 E5
39148 Piney Woods 450 D6
39149 Pinola E7

(continued on following page)

Mississippi-Missouri River System

MILES
0   100   200   300

Navigable Waterways over 9 feet deep............
Major River Ports.................⊙

©Copyright HAMMOND INCORPORATED.

Agriculture, Industry and Resources

DOMINANT LAND USE

- Specialized Cotton
- Cotton, Livestock
- Cotton, General Farming
- Cotton, Forest Products
- Truck and Mixed Farming
- Forests
- Swampland, Limited Agriculture

MAJOR MINERAL OCCURRENCES

- Cl   Clay
- Fe   Iron Ore
- G   Natural Gas
- O   Petroleum
- ⫽⫽   Major Industrial Areas

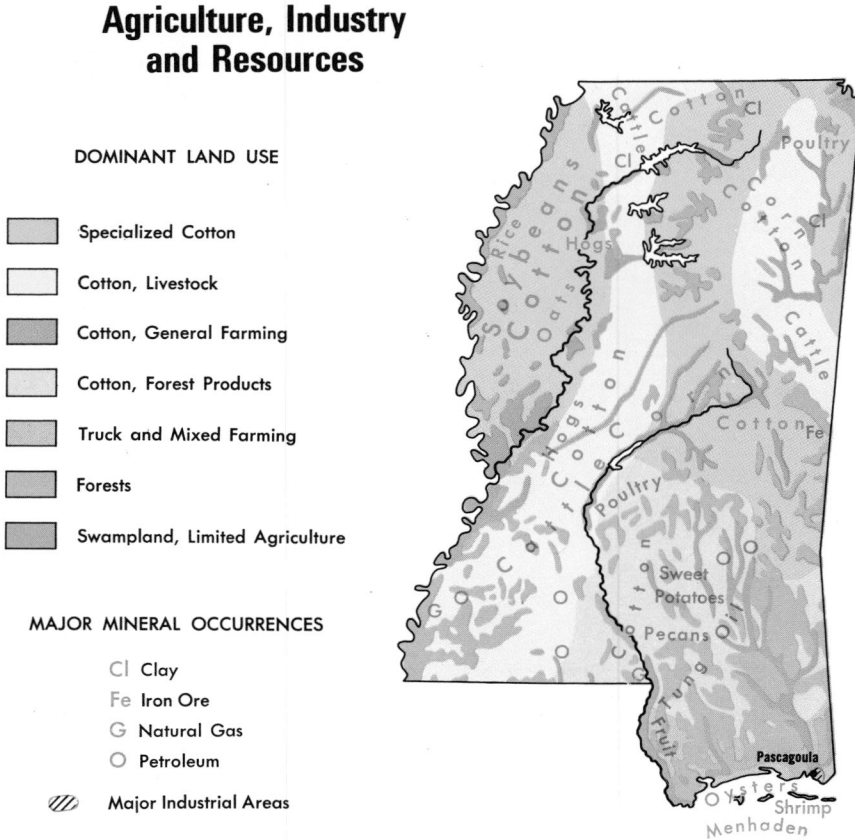

38951 Pittsboro⊙ 269	F3
38862 Plantersville 920	G2
38657 Pleasant Grove 100	D2
†38651 Pleasant Hill 400	E1
39072 Pocahontas 80	D6
39118 Polkville 129	E6
38863 Pontotoc⊙ 4,723	G2
38568 Pope 208	E2
39470 Poplarville⊙ 2,562	E9
39352 Porterville 150	G5
39150 Port Gibson⊙ 2,371	B7
38659 Potts Camp 525	F1
39756 Prairie	G3
39353 Prairie Point 150	H4
39474 Prentiss⊙ 1,465	E7
39354 Preston 500	G5
†39666 Pricedale 400	D8
†38676 Prichard 50	D1
39151 Puckett 279	E6
39152 Pulaski 108	E6
39475 Purvis⊙ 2,256	F8
39647 Quentin 40	C8
39355 Quitman⊙ 2,632	G6
39153 Raleigh⊙ 998	F6
38864 Randolph	F2
39154 Raymond⊙ 1,967	D6
38661 Red Banks 350	F1
†39096 Red Lick 100	B7
39156 Redwood 80	C6
39757 Reform 100	F4
38767 Rena Lara 350	C2
†39051 Renfroe 32	F5
†38732 Renova 659	C3
38662 Rich 72	D2
†39218 Richland 3,955	D6
39476 Richton 1,205	G8
39157 Ridgeland 5,461	D6
38865 Rienzi 423	G1
38663 Ripley⊙ 4,271	G1
38664 Robinsonville 285	D1
†39083 Rockport 30	D7
†39096 Rodney 100	B7
39159 Rolling Fork⊙ 2,590	C5
38768 Rome	C3
38769 Rosedale⊙ 2,793	B3
39356 Rose Hill 500	F6
†39633 Rosetta 120	B8
39661 Roxie 591	B8
38771 Ruleville 3,332	D3
†39108 Rural Hill 25	E4
†39150 Russum 200	B7
39662 Ruth 400	D8
39160 Sallis 211	E4
38866 Saltillo 1,271	G2
39112 Sanatorium 400	E7
39477 Sandersville 800	F7
39161 Sandhill 100	E5
39478 Sandy Hook 70	E8
*39479 Sanford 150	F8
38665 Sarah 150	D1
38666 Sardis⊙ 2,278	E2
38867 Sarepta 120	F2
39162 Satartia 73	C5
39574 Saucier 100	F9
38667 Savage 100	D1
38952 Schlater 429	D3
38953 Scobey 100	E3
39358 Scooba 511	G5
38772 Scott 400	B3
39359 Sebastopol 314	F5
39479 Seminary 327	E7
38668 Senatobia⊙ 5,013	E1
39758 Sessums 150	G4
38868 Shannon 680	G2
39163 Sharon 200	E5
38773 Shaw 2,461	C3
38774 Shelby 2,540	C3
38669 Sherard 150	C2
38869 Sherman 499	G2
39164 Shivers 100	E7
39360 Shubuta 626	G7
39361 Shuqualak 554	G5
39165 Sibley 350	B8
38954 Sidon 450	D4
39166 Silver City 378	C4
39663 Silver Creek 272	D7
38775 Skene 250	C3
38955 Slate Spring 102	F3
38670 Sledge 699	D2
39664 Smithdale 200	C8
38870 Smithville 866	H2
39665 Sontag 200	D7
39480 Soso 434	F7
38671 Southaven 16,071	E1
39167 Star 600	D6
39759 Starkville⊙ 15,169	G4
39362 State Line 484	G8
39766 Steens 125	H3
39767 Stewart 350	F4
38776 Stoneville 250	C4
39363 Stonewall 1,345	G6
38672 Stovall 50	C2
†38665 Strayhorn 275	D1
39481 Stringer 350	F7
38777 Stringtown 300	C3
39769 Sturgis 269	G4
39666 Summit 1,753	D8
38957 Sumner⊙ 452	D3
39482 Sumrall 1,197	E8
38778 Sunflower 1,027	C3
38958 Swan Lake 325	D3
38959 Swiftown 320	D4
39153 Sylvarena 102	F6
38673 Taylor 301	E2
39168 Taylorsville 1,387	F7
39169 Tchula 1,931	D4
39170 Terry 655	D6
38871 Thaxton 404	F2
39171 Thomastown 400	E5
†39073 Thomasville 50	E6
39172 Thornton 135	D4
†38829 Thrasher 100	G1
38960 Tie Plant 500	E3
38961 Tillatoba 106	E3
†39150 Tillman 65	C7
38674 Tiplersville 100	G1
38962 Tippo 200	D3

38873 Tishomingo 387	H1
38874 Toccopola 184	F2
39770 Tomnolen 200	F4
39364 Toomsuba 500	G6
39174 Tougaloo 800	D6
38757 Tralake 200	C4
38875 Trebloc 100	G3
38876 Tremont 379	H2
38779 Tribbett 100	C4
38675 Tula 140	F2
38676 Tunica⊙ 1,361	D1
38801 Tupelo⊙ 23,905	G2
38963 Tutwiler 1,174	C2
39667 Tylertown⊙ 1,976	D8
39365 Union 1,931	F5
39668 Union Church 75	C7
39178 Utica 865	C6
39175 Utica Junior College 40	C6
39176 Vaiden⊙ 924	E4
39177 Valley Park 400	C5
39178 Value 327	D6
38964 Vance 200	D2
†39564 Vancleave 1,330	G9
†38851 Van Vleet 400	G3
39179 Vaughan 210	D5
†39180 Vicksburg⊙ 25,434	C6
38679 Victoria 800	E1
39366 Vossburg 300	F7
†39567 Wade 800	G9
†39358 Wahalak 92	G5
38680 Walls 50	D1
38683 Walnut 513	G1
39189 Walnut Grove 439	F5
39771 Walthall⊙ 206	F3
39190 Washington 250	B7
38685 Waterford 400	E1
38965 Water Valley⊙ 4,147	E2
39576 Waveland 4,186	F10
39367 Waynesboro⊙ 5,349	G7
38780 Wayside 500	C4
†39114 Weathersby	E7
38966 Webb 782	D3
39772 Weir 553	F4
38834 Wenasoga 175	G1
39191 Wesson 1,313	D7
39192 West 253	E4
†39501 West Gulfport (North Gulfport) 6,660	F10
39773 West Point⊙ 8,811	G3
38880 Wheeler 600	G1
39193 Whitfield 900	E6
39577 Wiggins⊙ 3,205	F9
†38659 Winborn 70	F1
38967 Winona⊙ 6,177	E4
38781 Winstonville 486	C3
38782 Winterville 200	B4
39776 Woodland 135	F3
†39730 Wren 150	G3
39669 Woodville⊙ 1,512	B8
39194 Yazoo City⊙ 12,092	D5
†39090 Zama 100	F5

OTHER FEATURES

Amite (riv.)	C9
Arkabutla (lake)	D1
Big Black (riv.)	C6
Black (creek)	F8
Bogue Chitto (riv.)	D8
Bogue Homo (lake)	F7
Bowie (creek)	E7
Brices Cross Roads Nat'l Battlefield Site	G2
Buttahatchee (riv.)	H3
Cat (isl.)	F10
Catalpa (creek)	G4
Chickasaw Village, Natchez Trace Pkwy.	G2
Chickasawhay (riv.)	G7
Coldwater (riv.)	D1
Columbus A.F.B. 3,650	H3
Deer (creek)	C4
Enid (lake)	E2
Grenada (lake)	E3
Gulf Islands Nat'l Seashore	G10
Homochitto (riv.)	C8
Horn (isl.)	G10
Keesler A.F.B.	G10
Leaf (riv.)	F8
Little Tallahatchie (riv.)	D2
Meridian Naval Air Sta.	G5
Mississippi (riv.)	A8
Mississippi (sound)	G10
Noxubee (riv.)	G4
Okatibbee (lake)	G5
Pascagoula (riv.)	G9
Pearl (riv.)	D8
Petit Bois (isl.)	H10
Pickwick (lake)	H1
Pierre (bayou)	C7
Ross Barnett (res.)	E5
Round (isl.)	G10
Saint Louis (bay)	F10
Sardis (lake)	E2
Ship (isl.)	G10
Skuna (riv.)	F2
Strong (riv.)	D7
Sucarnoochee (creek)	G5
Sunflower (riv.)	C5
Tallahaga (creek)	F4
Tallahala (creek)	F6
Tallahatchie (riv.)	D3
Tchula (lake)	D4
Tennessee-Tombigbee Waterway	H2
Thompson (creek)	G8
Tombigbee (riv.)	H4
Trim Cane (creek)	G3
Tupelo Nat'l Battlefield	G2
Vicksburg Nat'l Mil. Park	C6
Wolf (riv.)	F9
Woodall (mt.)	H1
Yalobusha (riv.)	E3
Yazoo (riv.)	D3
Yockanookany (riv.)	E5

⊙County seat.
‡Population of metropolitan area.
† Zip of nearest p.o.   * Multiple zips

AREA 69,697 sq. mi. (180,515 sq. km.)
POPULATION 4,916,759
CAPITAL Jefferson City
LARGEST CITY St. Louis
HIGHEST POINT Taum Sauk Mtn. 1,772 ft. (540 m.)
SETTLED IN 1764
ADMITTED TO UNION August 10, 1821
POPULAR NAME Show Me State
STATE FLOWER Hawthorn
STATE BIRD Bluebird

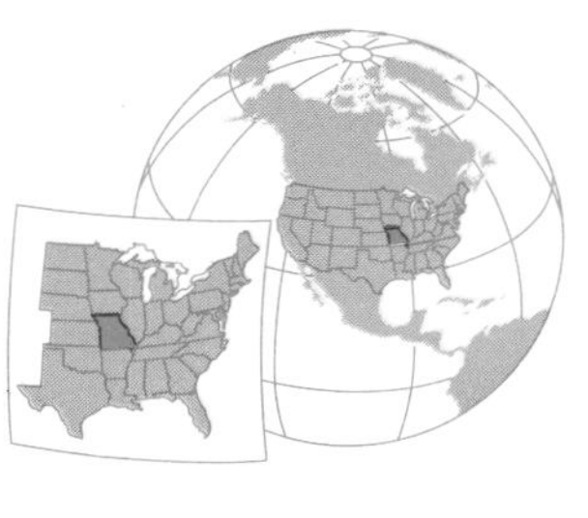

COUNTIES

Adair 24,870 G2
Andrew 13,980 C3
Atchison 8,605 B2
Audrain 26,458 J4
Barry 24,408 E9
Barton 11,292 D7
Bates 15,873 D6
Benton 12,183 F6
Bollinger 10,301 M8
Boone 100,376 H4
Buchanan 87,888 C3
Butler 37,693 M9
Caldwell 8,660 E3
Callaway 32,252 J5
Camden 20,017 G6
Cape Girardeau 58,837 N8
Carroll 12,131 F4
Carter 5,428 L9
Cass 51,029 D5
Cedar 11,894 E7
Chariton 10,489 F3
Christian 22,402 F9
Clark 8,493 J2
Clay 136,488 D4
Clinton 15,916 D3
Cole 56,663 H6
Cooper 14,643 G5
Crawford 18,300 K7
Dade 7,383 E8
Dallas 12,096 F7
Daviess 8,905 E3
De Kalb 8,222 D3
Dent 14,517 J7
Douglas 11,594 G9
Dunklin 36,324 M10
Franklin 71,233 K6
Gasconade 13,181 J6
Gentry 7,887 D2
Greene 185,302 F8
Grundy 11,959 E2
Harrison 9,890 E2
Henry 19,672 E6
Hickory 6,367 F7
Holt 6,882 B2
Howard 10,008 G4
Howell 28,807 J9
Iron 11,084 L7
Jackson 629,266 R5
Jasper 86,958 D8
Jefferson 146,183 L6
Johnson 39,059 E5
Knox 5,508 H2
Laclede 24,323 G7
Lafayette 29,925 E4
Lawrence 28,973 E8
Lewis 10,901 J2
Lincoln 22,193 L4
Linn 15,495 F3
Livingston 15,739 E3
Macon 16,313 G3
Madison 10,725 M8
Maries 7,551 J6
Marion 28,638 J3
McDonald 14,917 D9
Mercer 4,685 E2
Miller 18,532 H6
Mississippi 15,726 O9
Moniteau 12,068 G5
Monroe 9,716 H3
Montgomery 11,537 K5
Morgan 13,807 G6
New Madrid 22,945 N9
Newton 40,555 D9
Nodaway 21,996 C2
Oregon 10,238 K9
Osage 12,014 J6
Ozark 7,961 H9
Pemiscot 24,987 N10
Perry 16,784 N7
Pettis 36,378 F5
Phelps 33,633 J7
Pike 17,568 K4
Platte 46,341 C4
Polk 18,822 F7
Pulaski 42,011 H7
Putnam 6,092 F2
Ralls 8,984 J3
Randolph 25,460 G3
Ray 21,378 E4
Reynolds 7,230 L8
Ripley 12,458 L9
Saint Charles 144,107 M2
Saint Clair 8,622 E6
Sainte Genevieve 15,180 M7
Saint Francois 42,600 M7
Saint Louis 973,896 O3
Saint Louis (city county) 453,085 P3
Saline 24,919 F4
Schuyler 4,979 G2
Scotland 5,415 H2
Scott 39,647 N8
Shannon 7,885 K8
Shelby 7,826 H3
Stoddard 29,009 N9
Stone 15,587 F9
Sullivan 7,434 F2
Taney 20,467 F9
Texas 21,070 J8
Vernon 19,806 D7
Warren 14,900 K5
Washington 17,983 L7
Wayne 11,277 L8
Webster 20,414 G8
Worth 3,008 D2
Wright 16,188 H8

CITIES and TOWNS

Zip *Name/Pop.* *Key*

64720 Adrian 1,484 D6
63730 Advance 1,054 N8
63123 Affton 23,181 P4
64401 Agency 419 C3
64830 Alba 474 D8
64402 Albany⊙ 2,152 D2
63430 Alexandria 417 K2
64001 Alma 445 E4
65606 Alton⊙ 721 K9
64421 Amazonia 314 C3
64723 Amsterdam 231 D6
64831 Anderson 1,237 D9
63620 Annapolis 370 L8
63820 Anniston 320 O9
64724 Appleton City 1,257 D6
63821 Arbyrd 704 M10
63621 Arcadia 683 L7
64725 Archie 753 D5
65230 Armstrong 360 G4
63010 Arnold 19,141 M6
65604 Ash Grove 1,157 E8
65010 Ashland 1,021 H5
63530 Atlanta 441 H3
63332 Augusta 308 L5
65605 Aurora 6,437 E9
65231 Auxvasse 858 J4
65608 Ava⊙ 2,761 G9
64010 Avondale 612 P5
63011 Ballwin 12,656 N3
64011 Bates City 199 E5
†65619 Battlefield 1,227 F8
†63101 Bella Villa 758 R4
63735 Bell City 539 N8
65013 Belle 1,233 J6
†63137 Bellefontaine Neighbors 12,082 R2
63333 Bellflower 403 K4
†63101 Bel-Nor 2,047 P2
†63101 Bel-Ridge 3,682 P2
64012 Belton 12,708 C5
63736 Benton⊙ 674 O8
63134 Berkeley 15,922 P2
63822 Bernie 1,975 M9
63823 Bertrand 688 O9
64424 Bethany⊙ 3,095 E2
63532 Bevier 733 G3
65610 Billings 911 F8
65438 Birch Tree 622 K9
63624 Bismarck 1,625 L7
65321 Blackburn 314 F4
†63031 Black Jack 5,293 R1
65014 Bland 662 J6
63825 Bloomfield⊙ 1,795 M9
63627 Bloomsdale 397 M6
64015 Blue Springs 25,927 R6
†64101 Blue Summit R5
65613 Bolivar⊙ 5,919 F7
63628 Bonne Terre 3,797 L7
65233 Boonville⊙ 6,959 G5
64723 Bosworth 394 F4
65441 Bourbon 1,259 K6
63334 Bowling Green⊙ 3,022 K4
65616 Branson 2,550 F9
63533 Brashear 332 H2
64624 Braymer 986 E3
64625 Breckenridge 523 E3
†63114 Breckenridge Hills 5,666 O2
63144 Brentwood 8,209 P3
63044 Bridgeton 18,445 O2
†63044 Bridgeton Terrace 334 O2
64628 Brookfield 5,555 F3
64630 Browning 368 F2
65236 Brunswick 1,272 F4
64631 Bucklin 713 G3
64016 Buckner 2,848 R5
65622 Buffalo⊙ 2,217 F7
65237 Bunceton 419 G5
63629 Bunker 673 K8
64428 Burlington Junction 657 B2
64730 Butler⊙ 4,107 D6
65689 Cabool 2,090 H8
64632 Cainsville 496 E2
65239 Cairo 315 H4
65323 Calhoun 427 E6
65018 California⊙ 3,381 H5
63534 Callao 326 G3
†63101 Calverton Park 1,717 P2
65020 Camdenton⊙ 2,303 G6
64429 Cameron 4,519 D3
63933 Campbell 2,134 M9
63828 Canalou 369 N9
63435 Canton 2,435 J2
63701 Cape Girardeau 34,361 O8
63829 Cardwell 831 M10
64834 Carl Junction 3,937 C8
64633 Carrollton⊙ 4,700 E4
64835 Carterville 1,973 D8
64836 Carthage⊙ 11,104 D8
63830 Caruthersville⊙ 7,958 N10
65625 Cassville⊙ 2,091 E9
65022 Cedar City 427 H5
63436 Center 669 J3
65023 Centertown 304 H5
63633 Centerville⊙ 241 L8
65240 Centralia 3,537 H4
63740 Chaffee 3,241 N8
65024 Chamois 546 J5
†63101 Charlack 1,537 P2
63834 Charleston⊙ 5,230 O9
64733 Chilhowee 349 E5
64601 Chillicothe⊙ 9,089 E3
63437 Clarence 1,147 H3
65243 Clark 304 H4
65025 Clarksburg 352 G5
64430 Clarksdale 278 D3
†63017 Clarkson Valley 1,435 N3
63336 Clarksville 585 K4
63837 Clarkton 1,228 M10
†64119 Claycomo 1,671 P5
63105 Clayton⊙ 14,273 P3
64734 Cleveland 485 C5
65631 Clever 551 F8
64735 Clinton⊙ 8,366 E6
65325 Cole Camp 1,022 F6
65201 Columbia⊙ 62,061 H5
Columbia‡ 100,376 H5
†63128 Concord 20,896 P4
64020 Concordia 2,129 E5
65632 Conway 601 G7
†63101 Cool Valley 2,084 P2
63839 Cooter 479 N10
64021 Corder 483 E4
†64501 Country Club Village 1,234 C3
64437 Craig 379 B2
65633 Crane 1,185 E9
64739 Creighton 301 D6
†63126 Crestwood 12,815 O3
63141 Creve Coeur 11,757 O2
65452 Crocker 979 H7
63019 Crystal City 3,618 M6
†63101 Crystal Lake Park 496 O3
65453 Cuba 2,120 K6
63339 Curryville 323 K4
64439 Dearborn 547 C3
64740 Deepwater 475 E6
64440 De Kalb 245 C3
†63135 Dellwood 6,200 R2
63744 Delta 524 N8
63636 Des Arc 237 L8
63601 Desloge 3,481 M7
63020 De Soto 5,993 L6
63131 Des Peres 8,254 O3
63841 Dexter 7,043 N9
64840 Diamond 766 D9
65459 Dixon 1,402 H6
63935 Doniphan⊙ 1,921 L9
†65550 Doolittle 701 J7
63536 Downing 462 H2
64742 Drexel 908 C6
64841 Duenweg 703 D8
†64801 Duquesne 1,252 D8
64442 Eagleville 364 D2
64443 Easton 313 C3
63845 East Prairie 3,713 O9
64444 Edgerton 584 C3
63537 Edina⊙ 1,520 H2
†63101 Edmundson 1,374 O2
65026 Eldon 4,342 G6
64744 El Dorado Springs 3,868 E7
63638 Ellington 1,215 L8
†63011 Ellisville 6,233 M3
63937 Ellsinore 362 L9
63343 Elsberry 1,272 L4
63639 Elvins 1,548 L7
65466 Eminence⊙ 614 K8
63344 Eolia 401 L4
63846 Essex 545 N9
†63601 Esther 1,038 M7
63025 Eureka 3,862 M4
65646 Everton 317 E8
63440 Ewing 400 J2
64024 Excelsior Springs 10,424 R4
65647 Exeter 588 D9
64446 Fairfax 835 B2
65648 Fair Grove 863 F8
65649 Fair Play 384 E7
63345 Farber 503 J4
63640 Farmington⊙ 8,270 M7
65248 Fayette⊙ 2,983 G4
63026 Fenton 2,417 O4
63135 Ferguson 24,740 P2
64163 Ferrelview 447 O4
63028 Festus 7,574 M6
64449 Fillmore 265 C2
63940 Fisk 450 M9
63601 Flat River 4,443 M7
*63031 Florissant 55,372 P1
65652 Fordland 569 G8
64451 Forest City 387 B3
65653 Forsyth⊙ 1,010 F9
63441 Frankford 443 K4
63645 Fredericktown⊙ 4,036 M7
65035 Freeburg 554 J6
64746 Freeman 485 C5
†63101 Frontenac 3,654 O3
65251 Fulton⊙ 11,046 J5
65655 Gainesville⊙ 707 G9
65656 Galena⊙ 423 F9
64640 Gallatin⊙ 2,063 E3
64641 Galt 323 F2
64747 Garden City 1,021 D5
63037 Gerald 921 K6
63848 Gideon 1,240 N10
64642 Gilman City 414 D2
64118 Gladstone 24,990 P5
65254 Glasgow 1,336 G4
†64068 Glenaire 541 R5
63122 Glendale 6,035 P3
64748 Golden City 900 D8
63843 Goodman 1,030 C9
63543 Gorin H2
64454 Gower 1,276 C3
64029 Grain Valley 1,327 S6
64844 Granby 1,908 D9
64030 Grandview 24,502 P6
64456 Grant City⊙ 1,068 D2
†63155 Grantwood Village 1,002 O4
65037 Gravois Mills G6
63545 Green City 719 F2
65661 Greenfield⊙ 1,394 E8
65332 Green Ridge 488 F5
63546 Greentop 538 H2
63944 Greenville⊙ 393 M8
64034 Greenwood 1,315 R6
64643 Hale 529 F3
65255 Hallsville 624 H4
64644 Hamilton 1,582 E3
†63101 Hanley Hills 2,439 P2
63401 Hannibal 18,811 K3
64035 Hardin 688 E4
64701 Harrisonville⊙ 6,372 D5
65667 Hartville⊙ 576 G8
63945 Harviell M9
63349 Hawk Point 386 K5
63851 Hayti 3,964 N10
†63851 Hayti Heights 1,023 N10
†63736 Haywood City 425 N9
*63042 Hazelwood 12,935 P2
64036 Henrietta 424 E4
63048 Herculaneum 2,293 M6
65041 Hermann⊙ 2,695 K5
65668 Hermitage⊙ 384 F7
65257 Higbee 817 H4
64037 Higginsville 4,595 E4
63350 High Hill 254 K5
63050 Hillsboro⊙ 1,508 L6
†63101 Hillsdale 2,247 R2
63852 Holcomb 632 N10
64040 Holden 2,195 E5
63853 Holland 295 N10
65672 Hollister 1,439 F9
64048 Holt 276 D4
65043 Holts Summit 2,540 H5
†63879 Homestown 306 N10
64461 Hopkins 634 C1
63855 Hornersville 704 M10
65483 Houston⊙ 2,157 J8
65333 Houstonia 327 F5
†64152 Houston Lake 280 O5
†63869 Howardville 536 N9
65674 Humansville 907 E7
64752 Hume 315 C6
63443 Hunnewell 235 J3
†63101 Huntleigh 428 O3
65259 Huntsville⊙ 1,657 H4
63547 Hurdland 227 H2
65486 Iberia 852 H6
63754 Illmo 1,368 O8

(continued on following page)

Agriculture, Industry and Resources

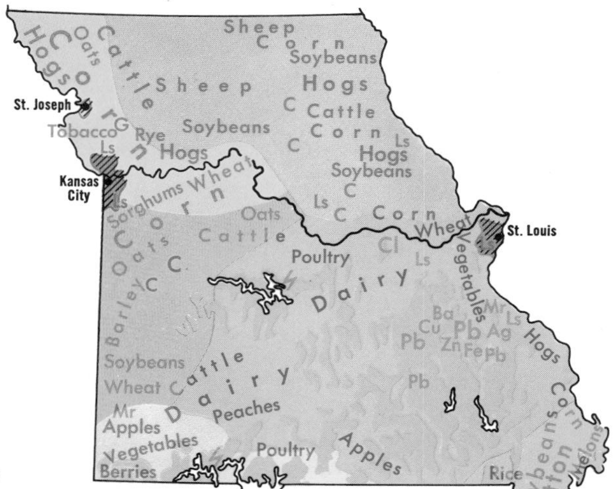

DOMINANT LAND USE

- Cattle Feed, Hogs
- Livestock, Cash Grain, Dairy
- Pasture Livestock
- Specialized Cotton
- General Farming, Dairy, Livestock, Poultry
- General Farming, Livestock, Truck Farming, Cotton
- Fruit and Mixed Farming
- Forests
- Urban Areas

MAJOR MINERAL OCCURRENCES

Ag	Silver	G	Natural Gas
Ba	Barite	Ls	Limestone
C	Coal	Mr	Marble
Cl	Clay	Pb	Lead
Cu	Copper	Zn	Zinc
Fe	Iron Ore		

Water Power — Major Industrial Areas

*64050 Independence⊙ 111,806 ... R5
63648 Irondale 349 ... L7
†64801 Iron Gates 314 ... C8
63650 Ironton⊙ 1,743 ... L7
63755 Jackson⊙ 7,827 ... N8
64648 Jamesport 651 ... E3
65046 Jamestown 317 ... G5
64755 Jasper 1,012 ... D8
65101 Jefferson City (cap.)⊙ 33,619 ... H5
63136 Jennings 17,026 ... R2
63351 Jonesburg 614 ... K5
64801 Joplin 39,023 ... C8
Joplin‡ 127,513 ... C8
†63645 Junction City 238 ... M7
63445 Kahoka⊙ 2,101 ... J2
*64101 Kansas City 448,159 ... P5
Kansas City‡ 1,327,020 ... P5
64060 Kearney 1,433 ... D4
63758 Kelso 455 ... O8
63857 Kennett⊙ 10,145 ... M10
65261 Keytesville⊙ 689 ... G4
64649 Kidder 265 ... D3
65686 Kimberling City 1,285 ... F9
64463 King City 1,063 ... D2
64650 Kingston⊙ 280 ... E3
64061 Kingsville 365 ... D5
63140 Kinloch 4,455 ... P2
63501 Kirksville⊙ 17,167 ... H2
63122 Kirkwood 27,987 ... O3
65336 Knob Noster 2,040 ... E5
63446 Knox City 281 ... H2
63447 La Belle 845 ... J2
64651 Laclede 445 ... F3
63352 Laddonia 726 ... J4
†63124 Ladue 9,376 ... P3
63448 La Grange 1,217 ... K2
64063 Lake Lotawana 1,875 ... R6
65049 Lake Ozark 427 ... G6
†63336 Lake Saint Louis 3,843 ... N2
†63101 Lakeshire 1,593 ... P4
†64015 Lake Tapawingo 925 ... R6
†64152 Lake Waukomis 1,050 ... P5
64034 Lake Winnebago 681 ... R6
64759 Lamar⊙ 4,053 ... D8
65337 La Monte 1,054 ... F5
64847 Lanagan 440 ... C9
63548 Lancaster⊙ 855 ... H1
63549 La Plata 1,423 ... H2
64652 Laredo 340 ... E2
64760 Latour 84 ... D5
64062 Lawson 1,688 ... D4
†63640 Leadington 238 ... M7
63653 Leadwood 1,371 ... L7
65535 Leasburg 304 ... K6
65536 Lebanon⊙ 9,507 ... G7
64063 Lee's Summit 28,741 ... R6
64761 Leeton 604 ... E5
63125 Lemay 35,424 ... R4
64066 Levasy 235 ... S5
63452 Lewistown 502 ... J2
64067 Lexington⊙ 5,063 ... E4
64762 Liberal 701 ... D7
64068 Liberty⊙ 16,251 ... R5
65542 Licking 1,272 ... J8
63862 Lilbourn 1,463 ... N9
65338 Lincoln 819 ... F6
65051 Linn⊙ 1,211 ... J5
65052 Linn Creek 242 ... G6
64653 Linneus⊙ 421 ... F3
65682 Lockwood 971 ... E8
64070 Lone Jack 420 ... S6
63353 Louisiana 4,261 ... K4
64763 Lowry City 676 ... E6
63762 Lutesville 865 ... M8
63552 Macon⊙ 5,680 ... H3
65263 Madison 656 ... H4
64466 Maitland 415 ... B2
63863 Malden 6,096 ... M9
65339 Malta Bend 292 ... F4
63011 Manchester 6,191 ... O3
65704 Mansfield 1,423 ... G8
63143 Maplewood 10,960 ... P3
63764 Marble Hill⊙ 601 ... N8
64658 Marceline 2,938 ... F3
65705 Marionville 1,920 ... E8
†63101 Marlborough 2,012 ... P3
63655 Marquand 397 ... M8
65340 Marshall⊙ 12,781 ... F4
65706 Marshfield⊙ 3,871 ... G8
63866 Marston 742 ... N9
63357 Marthasville 543 ... L5
65264 Martinsburg 309 ... J4
63043 Maryland Heights 5,676 ... O2
64468 Maryville⊙ 9,558 ... C2
63857 Matthews 547 ... N9
64469 Maysville⊙ 1,187 ... D3
64071 Mayview 291 ... E4
64659 Meadville 416 ... F3
63555 Memphis⊙ 2,105 ... H2
64660 Mendon 252 ... F3
64661 Mercer 442 ... F2
65058 Meta 336 ... H6
65265 Mexico⊙ 12,276 ... J4
63359 Middletown 268 ... J4
63556 Milan⊙ 1,947 ... F2
65707 Miller 795 ... E8
63952 Mill Spring 257 ... L8
64769 Mindenmines 318 ... C8
†63801 Miner 1,182 ... N9
63660 Mineral Point 358 ... L7
64072 Missouri City 343 ... R5
65270 Moberly 13,418 ... G4
65059 Mokane 293 ... J5
†63101 Moline Acres 2,774 ... R2
65708 Monett 6,148 ... E9
63456 Monroe City 2,557 ... J3
63361 Montgomery City⊙ 2,101 ... K5
63457 Monticello⊙ 134 ... J2
64770 Montrose 498 ... E6
63868 Morehouse 1,220 ... N9
63767 Morley 745 ... N8
65710 Morrisville 331 ... F8
64073 Mosby 284 ... R4
63362 Moscow Mills 484 ... K5
64470 Mound City 1,447 ... B2
65711 Mountain Grove 3,974 ... H8
65548 Mountain View 1,664 ... J8
64665 Mount Moriah 162 ... E2
65712 Mount Vernon⊙ 3,341 ... E8
†63088 Murphy 8,121 ... O4
64074 Napoleon 271 ... E4
63953 Naylor 602 ... L9
63954 Neelyville 474 ... M9
65347 Nelson 248 ... F4
64850 Neosho⊙ 9,493 ... D9
64772 Nevada⊙ 9,044 ... D7
65063 New Bloomfield 519 ... J5
65550 Newburg 743 ... J7
63558 New Cambria 246 ... G3
63363 New Florence 731 ... K5
65274 New Franklin 1,228 ... G4
†63736 New Hamburg ... O8
64471 New Hampton 358 ... D2
63068 New Haven 1,581 ... K5
63459 New London⊙ 1,161 ... K3
63869 New Madrid⊙ 3,204 ... O9
65713 Niangua 376 ... G8
65714 Nixa 2,662 ... F8
64854 Noel 1,161 ... D9
64668 Norborne 931 ... E4
63121 Normandy 5,174 ... R2
64116 North Kansas City 4,507 ... P5
†64152 Northmoor 506 ... P5
65717 Norwood 391 ... H8
63559 Novinger 626 ... G2
64075 Oak Grove 4,067 ... S6
†63080 Oak Grove 386 ... K6
†63101 Oakland 1,728 ... P3
63769 Oak Ridge 252 ... N7
†64116 Oakview 497 ... P5
63401 Oakwood 227 ... P5
64076 Odessa 3,088 ... E5
63366 O'Fallon 8,677 ... L5
63369 Old Monroe 272 ... L5
63124 Olivette 7,985 ... O2
63050 Olympian Village 774 ... M6
63771 Oran 1,266 ... N8
64473 Oregon⊙ 901 ... B2
64855 Oronogo 525 ... D8
64077 Orrick 922 ... D4
65065 Osage Beach 1,992 ... G6
64474 Osborn 381 ... D3
64776 Osceola⊙ 841 ... E6
65348 Otterville 472 ... G5
63114 Overland 19,620 ... O2
65066 Owensville 2,241 ... K6
65721 Ozark⊙ 2,980 ... F8
63069 Pacific 4,410 ... L5
†63101 Pagedale 4,542 ... P2
63461 Palmyra⊙ 3,469 ... J3
65275 Paris⊙ 1,598 ... J4
64152 Parkville 1,997 ... O5
64130 Parkway 254 ... L6
63870 Parma 1,081 ... N9
64670 Pattonsburg 502 ... D2
64078 Peculiar 1,571 ... D5
63462 Perry 836 ... J4
63775 Perryville⊙ 7,343 ... N7
63070 Pevely 2,732 ... M6
64476 Pickering 215 ... C2
63957 Piedmont 2,359 ... L8
65723 Pierce City 1,391 ... E8
65276 Pilot Grove 745 ... G5
63663 Pilot Knob 722 ... L7
†63120 Pine Lawn 6,662 ... R2
64856 Pineville⊙ 504 ... D9
64079 Platte City⊙ 2,114 ... C4
†64152 Platte Woods 467 ... O5
64477 Plattsburg⊙ 2,095 ... D3
64080 Pleasant Hill 3,301 ... D5
65725 Pleasant Hope 354 ... F8
†64836 Pleasant Valley 1,545 ... R5
64671 Polo 583 ... D3
63901 Poplar Bluff⊙ 17,139 ... L9
63373 Portage Des Sioux 488 ... M5
63873 Portageville 3,470 ... N10
63664 Potosi⊙ 2,528 ... L7
65068 Prairie Home 279 ... G5
64673 Princeton⊙ 1,264 ... E2
64857 Purcell 322 ... D8
64674 Purdin 243 ... F3
65734 Purdy 928 ... E9
63960 Puxico 833 ... M9
63561 Queen City 783 ... H2
63961 Qulin 545 ... M9
†64101 Randolph 91 ... P5
64479 Ravenwood 436 ... C2
65555 Raymondville 388 ... J8
64083 Raymore 3,154 ... D5
64133 Raytown 31,759 ... P6
65737 Reeds Spring 461 ... F9
65738 Republic 4,485 ... E8
64779 Rich Hill 1,471 ... D6
65556 Richland 1,922 ... H7
64085 Richmond⊙ 5,499 ... D4
63117 Richmond Heights 11,516 ... P3
64481 Ridgeway 516 ... D2
63874 Risco 446 ... N9
†63601 Rivermines 414 ... L7
64168 Riverside 3,206 ... O5
†63101 Riverview 3,367 ... R2
65279 Rocheport 272 ... H5
65740 Rockaway Beach 292 ... F9
†63119 Rock Hill 5,702 ... P3
64482 Rock Port⊙ 1,511 ... B2
64780 Rockville 281 ... D6
65742 Rogersville 741 ... G8
65401 Rolla⊙ 13,303 ... J7
63091 Rosebud 326 ... K6
64483 Rosendale 223 ... C2
64484 Rushville 271 ... B3
65074 Russellville 667 ... H6
64864 Saginaw 293 ... C8
63074 Saint Ann 15,523 ... O2
63301 Saint Charles⊙ 37,379 ... N1
63077 Saint Clair 3,485 ... K6
63670 Sainte Genevieve⊙ 4,481 ... M6
65075 Saint Elizabeth 312 ... H6
†63101 Saint George 1,545 ... P4
65559 Saint James 3,328 ... J6
63114 Saint John 7,854 ... P2
*64501 Saint Joseph⊙ 76,691 ... C3
Saint Joseph‡ 101,868 ... C3
*63101 Saint Louis⊙ 453,085 ... R3
Saint Louis‡ 2,355,276 ... R3
†65101 Saint Martins 739 ... H5
63673 Saint Marys 565 ... M7
63366 Saint Paul 607 ... L5
63376 Saint Peters 15,700 ... M1
65583 Saint Robert 1,735 ... H7
65560 Salem⊙ 4,454 ... J7
65281 Salisbury 1,975 ... G4
63126 Sappington 11,388 ... O4
64862 Sarcoxie 1,381 ... D8
64485 Savannah⊙ 4,184 ... C3
64783 Schell City 327 ... D6
63780 Scott City 3,262 ... O8
65301 Sedalia⊙ 20,927 ... F5
65745 Seligman 508 ... D9
63876 Senath 1,728 ... M10
64865 Seneca 1,853 ... C9
65746 Seymour 1,535 ... G8
63468 Shelbina 2,169 ... H3
63469 Shelbyville⊙ 645 ... H3
64784 Sheldon 491 ... D7
†63101 Shrewsbury 5,077 ... P3
64088 Sibley 382 ... S5
63801 Sikeston 17,431 ... N9
63377 Silex 287 ... K4
64487 Skidmore 437 ... B2
65349 Slater 2,492 ... G4
65350 Smithton 559 ... F5
64089 Smithville 1,873 ... D4
64863 South West City 516 ... D9
†63138 Spanish Lake 20,632 ... R1
65753 Sparta 743 ... F9
64679 Spickard 389 ... F2

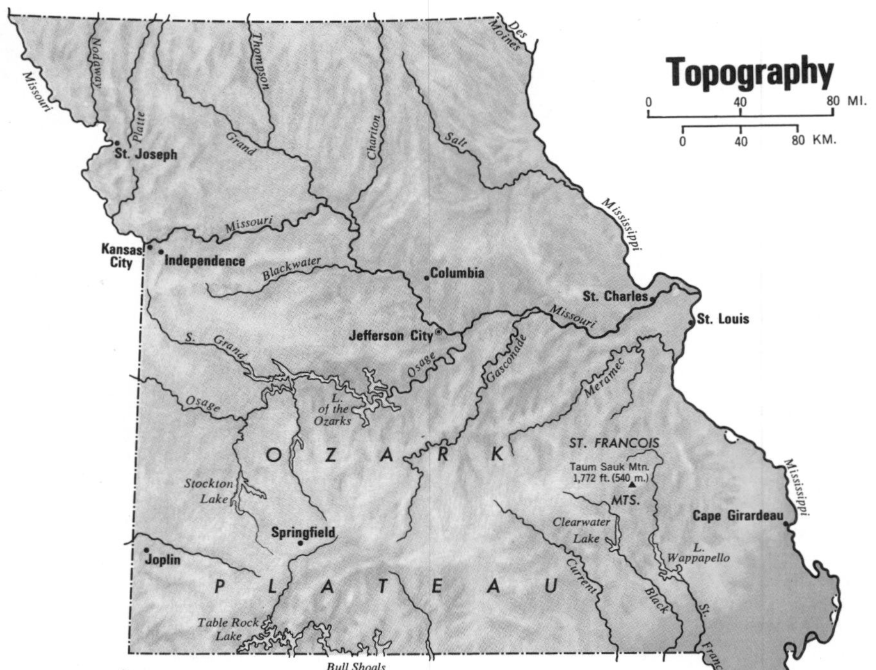
Topography
0 40 80 MI.
0 40 80 KM.
St. Joseph
Kansas City
Independence
Columbia
St. Charles
St. Louis
Jefferson City
O Z A R K
P L A T E A U
ST. FRANCOIS MTS.
Taum Sauk Mtn. 1,772 ft. (540 m.)
Springfield
Joplin
Cape Girardeau
Stockton Lake
Table Rock Lake
Bull Shoals Lake
L. of the Ozarks
L. Wappapello
Clearwater Lake
Missouri
Mississippi
Grand
Chariton
Salt
Thompson
Nodaway
Platte
Des Moines
Blackwater
S. Grand
Osage
Gasconade
Meramec
Current
Black
St. Francis
5,000 m. 16,404 ft.
2,000 m. 6,562 ft.
1,000 m. 3,281 ft.
500 m. 1,640 ft.
200 m. 656 ft.
100 m. 328 ft.
Sea Level
Below

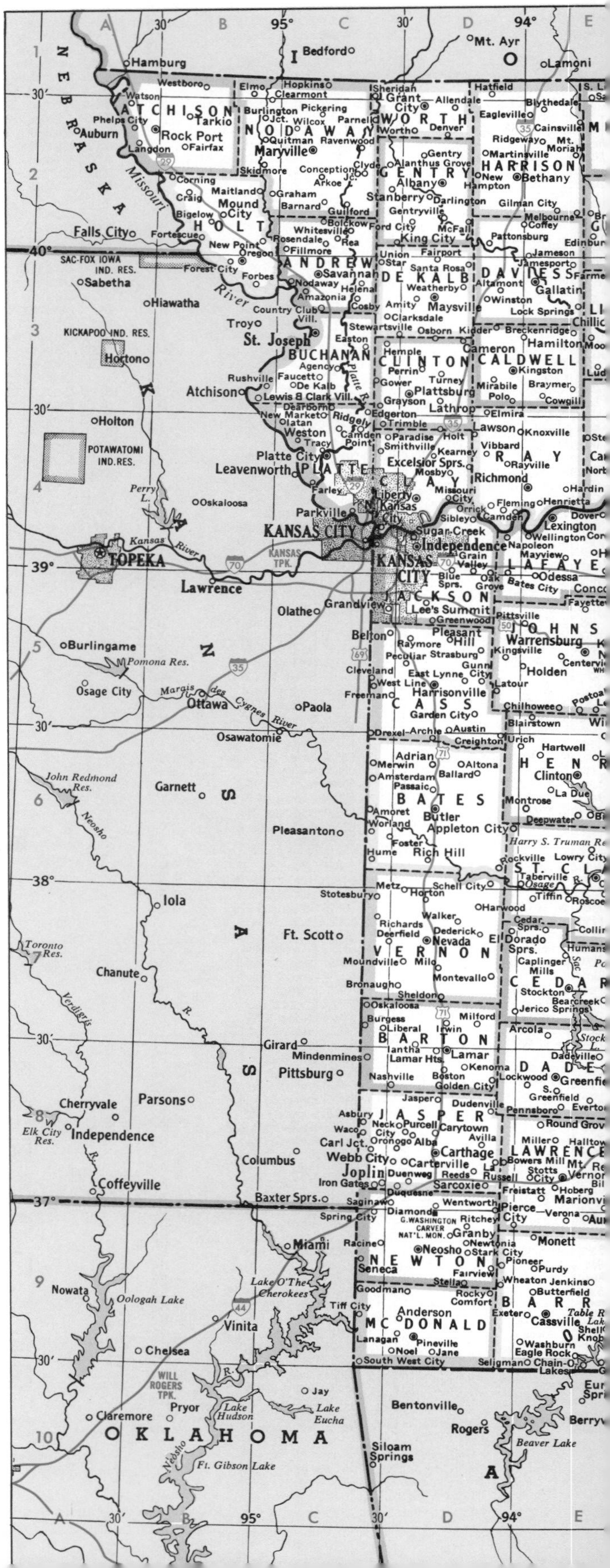
NEBRASKA
KANSAS
OKLAHOMA
TOPEKA
KANSAS CITY
Lawrence
Leavenworth
Atchison
St. Joseph
Independence
Maryville
Bedford
Hamburg
Mt. Ayr
Lamoni
Joplin
Carthage
Neosho
Nevada
Coffeyville
Independence
Parsons
Pittsburg
Chanute
Iola
Ft. Scott
Ottawa
Paola
Olathe
Osawatomie
Miami
Vinita
Pryor
Claremore
Rogers
Bentonville
Siloam Springs
Grand Lake O' The Cherokees
Oologah Lake
Lake Hudson
Lake Eucha
Beaver Lake
Ft. Gibson Lake
Pomona Res.
John Redmond Res.
Toronto Res.
Elk City Res.
POTAWATOMI IND. RES.
KICKAPOO IND. RES.
SAC-FOX IOWA IND. RES.
ATCHISON
NODAWAY
WORTH
HARRISON
GENTRY
HOLT
ANDREW
DE KALB
DAVIESS
BUCHANAN
CLINTON
CALDWELL
PLATTE
CLAY
RAY
JACKSON
LAFAYETTE
CASS
JOHNSON
BATES
HENRY
VERNON
CEDAR
BARTON
DADE
JASPER
LAWRENCE
NEWTON
McDONALD
BARRY
95°
94°
40°
39°
38°
37°

*65801 Springfield⊙ 133,116.....F8
   Springfield‡ 207,704.....F8
64489 Stanberry 1,387.....C2
63877 Steele 2,419.....N10
65565 Steelville⊙ 1,470.....K7
64490 Stewartsville 832.....C3
65785 Stockton⊙ 1,432.....E7
65567 Stoutland 286.....G7
65078 Stover 1,041.....G6
65757 Strafford 1,121.....F8
65284 Sturgeon 901.....H4
64054 Sugar Creek 4,305.....R5
63080 Sullivan 5,461.....K6
65571 Summersville 551.....J8
†63101 Sunset Hills 4,363.....O4
65351 Sweet Springs 1,694.....F5
65759 Taneyville 300.....F9
†65101 Taos 759.....H5
64491 Tarkio 2,375.....B2
†64063 Tarsney Lakes 329.....R6
65791 Thayer 2,211.....J9
†63025 Times Beach 2,041.....N4

65081 Tipton 2,155.....G5
†63101 Town and Country 3,187...O3
64079 Tracy 310.....C4
64683 Trenton⊙ 6,811.....E2
63379 Troy⊙ 2,624.....L5
65082 Tuscumbia⊙ 241.....H6
†63088 Twin Oaks 426.....N3
63084 Union⊙ 5,506.....L6
64494 Union Star 423.....C2
63565 Unionville⊙ 2,178.....D2
63130 University City 42,738.....P3
65767 Urbana 329.....F7
64788 Urich 509.....E6
63088 Valley Park 3,232.....O3
63382 Vandalia 3,170.....J4
63784 Vanduser 320.....N9
†63101 Velda 1,988.....P2
65769 Verona 592.....E9
65084 Versailles⊙ 2,406.....G6
65566 Viburnum 836.....K7
†63020 Victoria 375.....M6

65582 Vienna⊙ 514.....H6
†63101 Vinita Park 2,283.....P2
64790 Walker 325.....D7
65770 Walnut Grove 504.....F8
64093 Wardsville 535.....H6
64093 Warrensburg⊙ 13,807.....E5
63383 Warrenton⊙ 3,219.....K5
65355 Warsaw⊙ 1,494.....F6
†63101 Warson Woods 2,127.....O3
64096 Waverly 941.....E4
63472 Waynesville⊙ 2,879.....H7
65583 Waynesville⊙ 2,879.....H7
†64152 Weatherby Lake 1,446.....O5
65774 Weaubleau 464.....F7
64870 Webb City 7,309.....C8
63119 Webster Groves 23,097.....P3
64097 Wellington 780.....E4
63112 Wellston 4,495.....R2
63384 Wellsville 1,546.....K4
63385 Wentzville 3,193.....L5
64498 Westboro 188.....B1

64098 Weston 1,440.....C4
65775 West Plains⊙
   7,741.....J9
65779 Westwood 319.....O3
65779 Wheatland 364.....F7
64874 Wheaton 548.....E9
64688 Wheeling 379.....E3
†63101 Wilbur Park 564.....P3
65781 Willard 1,799.....F8
63977 Williamsville 418.....L9
65793 Willow Springs 2,215.....H9
63834 Wilson City 309.....O9
†63435 Winchester 2,237.....N3
65360 Windsor 3,058.....E5
63389 Winfield 592.....L5
65588 Winona 1,050.....K8
64689 Winston 246.....D3
†64024 Woods Heights 747.....S4
†63101 Woodson Terrace 4,788.....P2
63390 Wright City 1,179.....K5
63474 Wyaconda 359.....J2
63882 Wyatt 441.....O9

OTHER FEATURES

Bagnell (dam).....G6
Big (riv.).....L6
Black (riv.).....L10
Bull Shoals (lake).....G10
Chariton (riv.).....G1
Clearwater (lake).....L8
Cuivre (riv.).....N2
Current (riv.).....K8
Des Moines (riv.).....J1
Fort Leonard Wood 21,262.....H7
Gasconade (riv.).....H7
George Washington Carver Nat'l
   Mon......D9
Grand (riv.).....F3
Jefferson Nat'l Expansion Mem. Nat'l
   His.....R3
Lake City Arsenal.....R5
Meramec (riv.).....N3
Mississippi (riv.).....L4
Missouri (riv.).....H5

Norfork (lake).....H10
Osage (riv.).....E6
Ozark (plat.).....F9
Ozark Nat'l Scenic Riverways.....K8
Ozarks, Lake of the (lake).....F6
Platte (riv.).....C3
Pomme de Terre (lake).....E7
Richards Gebaur A.F.B. 4,305.....P6
Sac (riv.).....E7
Saint Francis (riv.).....M9
Salt (riv.).....J3
Stockton (lake).....E7
Table Rock (res.).....E9
Taneycomo (lake).....F9
Taum Sauk (mt.).....L7
Wappapello (lake).....K9
White (riv.).....G10
Whiteman A.F.B......E5
Wilson's Creek Nat'l Battlefield.....F8
⊙County seat.
‡Population of metropolitan area.
† Zip of nearest p.o. * Multiple zips.

Agriculture, Industry and Resources

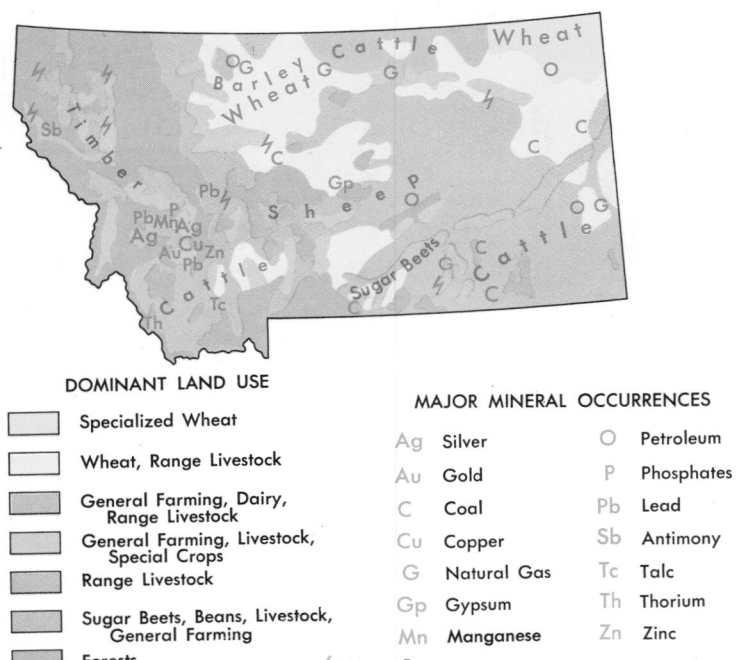

DOMINANT LAND USE

	Specialized Wheat
	Wheat, Range Livestock
	General Farming, Dairy, Range Livestock
	General Farming, Livestock, Special Crops
	Range Livestock
	Sugar Beets, Beans, Livestock, General Farming
	Forests

MAJOR MINERAL OCCURRENCES

Ag	Silver	O	Petroleum
Au	Gold	P	Phosphates
C	Coal	Pb	Lead
Cu	Copper	Sb	Antimony
G	Natural Gas	Tc	Talc
Gp	Gypsum	Th	Thorium
Mn	Manganese	Zn	Zinc

⚡ Water Power

COUNTIES

Beaverhead 8,186 .... C5
Big Horn 11,096 .... J5
Blaine 6,999 .... G2
Broadwater 3,267 .... E4
Carbon 8,099 .... G5
Carter 1,799 .... M5
Cascade 80,696 .... E3
Chouteau 6,092 .... F3
Custer 13,109 .... L4
Daniels 2,835 .... L2
Dawson 11,805 .... M3
Deer Lodge 12,518 .... C5
Fallon 3,763 .... M4
Fergus 13,076 .... G3
Flathead 51,966 .... B2
Gallatin 42,865 .... E5
Garfield 1,656 .... J3
Glacier 10,628 .... C2
Golden Valley 1,026 .... G4
Granite 2,700 .... C4
Hill 17,985 .... F2
Jefferson 7,029 .... D4
Judith Basin 2,646 .... F4
Lake 19,056 .... B3
Lewis and Clark 43,039 .... D3
Liberty 2,329 .... E2
Lincoln 17,752 .... A2
Madison 5,448 .... D5
McCone 2,702 .... L3
Meagher 2,154 .... F4
Mineral 3,675 .... B3
Missoula 76,016 .... C3
Musselshell 4,428 .... H4
Park 12,869 .... F5
Petroleum 655 .... H3
Phillips 5,367 .... J2
Pondera 6,731 .... D2
Powder River 2,520 .... L5
Powell 6,958 .... D4
Prairie 1,836 .... L4
Ravalli 22,493 .... B4
Richland 12,243 .... M3
Roosevelt 10,467 .... L2
Rosebud 9,899 .... K4
Sanders 8,675 .... A3
Sheridan 5,414 .... M2
Silver Bow 38,092 .... D5
Stillwater 5,598 .... G5
Sweet Grass 3,216 .... G5
Teton 6,491 .... D3
Toole 5,559 .... E2
Treasure 981 .... J4
Valley 10,250 .... K2
Wheatland 2,359 .... G4
Wibaux 1,476 .... M4
Yellowstone 108,035 .... H4
Yellowstone Nat'l Park 275 .... F6

CITIES and TOWNS

Zip	Name/Pop.	Key
59001	Absarokee 830	G5
59820	Alberton 368	B3
59710	Alder 120	D5
†59741	Amsterdam 130	E5
59711	Anaconda-Deer Lodge County⊙ 12,518	C4
59312	Angela 50	K4
59211	Antelope 83	M2
59821	Arlee 200	B3
59003	Ashland 600	K5
59410	Augusta 497	D3
59713	Avon 125	D4
59411	Babb 150	C2
59212	Bainville 245	M2
59313	Baker⊙ 2,354	M4
59006	Ballantine 380	J5
†59725	Bannack 2	D5
59613	Basin 350	D4
59007	Bearcreek 61	G5
59008	Belfry 300	H5
59714	Belgrade 2,336	E5
59412	Belt 825	E3
59314	Biddle 28	L5
59910	Big Arm 250	B3
59911	Bigfork 1,080	C2
59520	Big Sandy 835	G2
*59101	Billings⊙ 66,842	H5
	Billings‡ 108,035	H5
59012	Birney 100	K5
59414	Black Eagle 1,500	E3
59415	Blackfoot 100	D2
59823	Bonner-West Riverside 1,742	C4
59632	Boulder⊙ 1,441	E4
59521	Box Elder 300	F2
59715	Bozeman⊙ 21,645	E5
59416	Brady 450	E2
59014	Bridger 724	H5
59317	Broadus⊙ 712	L5
59015	Broadview 120	H4
59213	Brockton 374	M2
59417	Browning 1,226	C2
59016	Busby 700	J5
59701	Butte-Silver Bow County⊙ 37,205	D5
59720	Cameron 150	E5
59633	Canyon Creek 100	D4
†59347	Cartersville 115	K4
59421	Cascade 773	E3
59824	Charlo 250	B3
59522	Chester⊙ 963	E2
59523	Chinook⊙ 1,660	G2
59422	Choteau⊙ 1,798	D3
59215	Circle⊙ 931	L3
59634	Clancy 550	D4
59018	Clyde Park 283	F5
†59351	Coalwood 2	L5
59322	Cohagen 12	K3
59323	Colstrip 1,476	K5
59912	Columbia Falls 3,112	B2
59019	Columbus⊙ 1,439	G5
59826	Condon 300	C3
59827	Conner 420	B4
59425	Conrad⊙ 3,074	D2
59020	Cooke City 120	G5
59913	Coram 450	B2
59828	Corvallis 500	C4
59217	Crane 163	M3
59022	Crow Agency 975	J5
59218	Culbertson 887	M2
59024	Custer 300	J4
59427	Cut Bank⊙ 3,688	D2
59829	Darby 581	B4
59914	Dayton 140	B3
59830	De Borgia 300	A3
59025	Decker 150	K5
59722	Deer Lodge⊙ 4,023	C4
59430	Denton 356	G3

Montana

SCALE

0 5 10 20 40 60 MI.

0 5 10 20 40 60 KM.

State Capitals .... ⊛
County Seats .... ⊙
Major Limited Access Hwys.

Scale 1:3,450,000

© Copyright HAMMOND INCORPORATED, Maplewood, N.J.

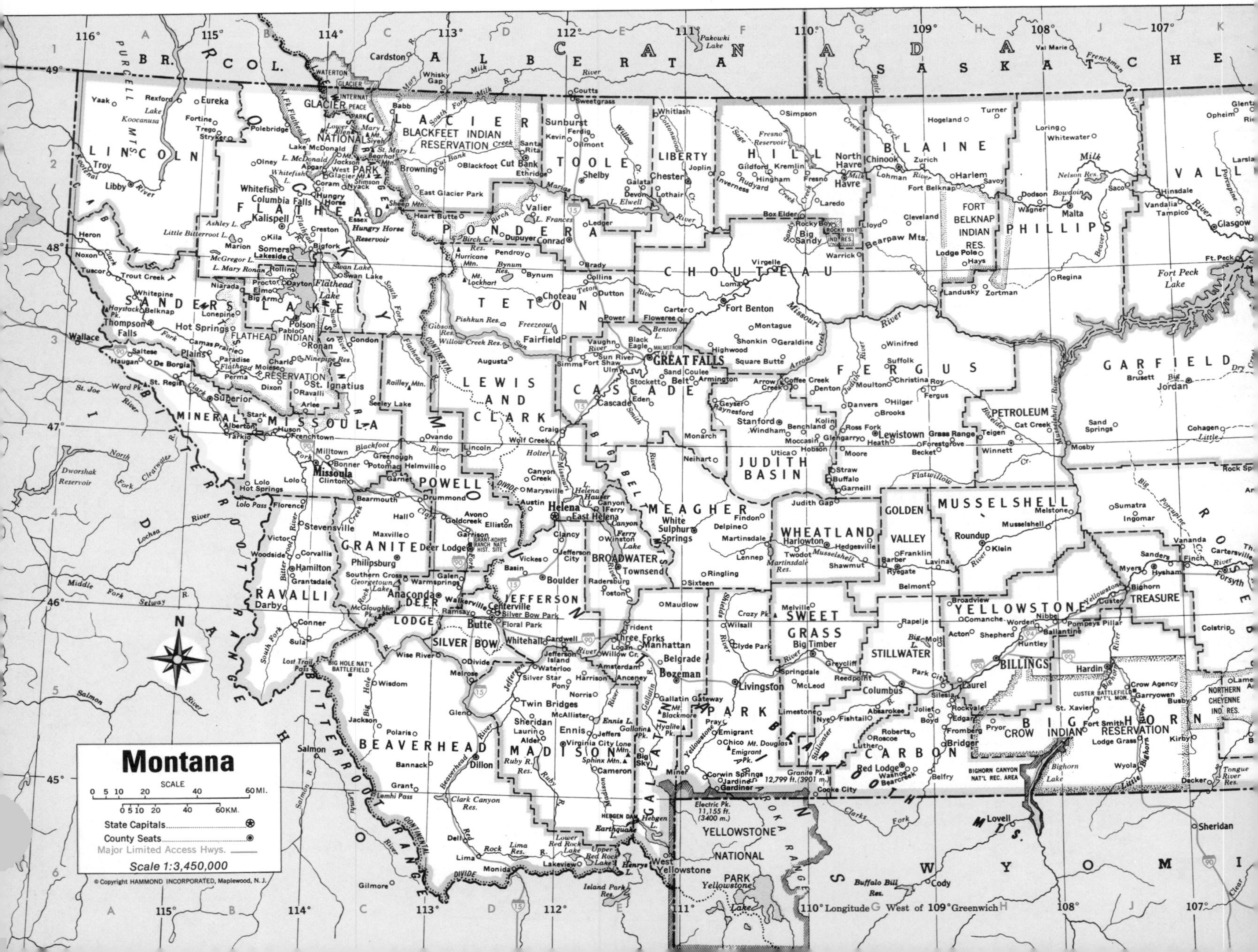

Topography

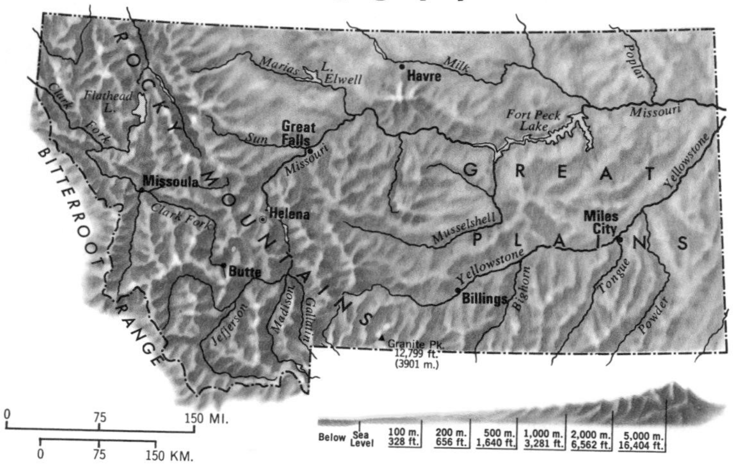
ROCKY MOUNTAINS
BITTERROOT RANGE
GREAT PLAINS
Marias
L. Elwell
Havre
Milk
Poplar
Flathead L.
Clark Fork
Sun
Great Falls
Missouri
Fort Peck Lake
Missoula
Helena
Clark Fork
Musselshell
Miles City
Yellowstone
Butte
Jefferson
Madison
Gallatin
Billings
Bighorn
Tongue
Powder
Granite Pk. 12,799 ft. (3901 m.)
0
75
150 MI.
0
75
150 KM.
Below Sea Level
100 m. 328 ft.
200 m. 656 ft.
500 m. 1,640 ft.
1,000 m. 3,281 ft.
2,000 m. 6,562 ft.
5,000 m. 16,404 ft.

AREA 147,046 sq. mi. (380,849 sq. km.)
POPULATION 786,690
CAPITAL Helena
LARGEST CITY Billings
HIGHEST POINT Granite Pk. 12,799 ft. (3901 m.)
SETTLED IN 1809
ADMITTED TO UNION November 8, 1889
POPULAR NAME Treasure State; Big Sky Country
STATE FLOWER Bitterroot
STATE BIRD Western Meadowlark

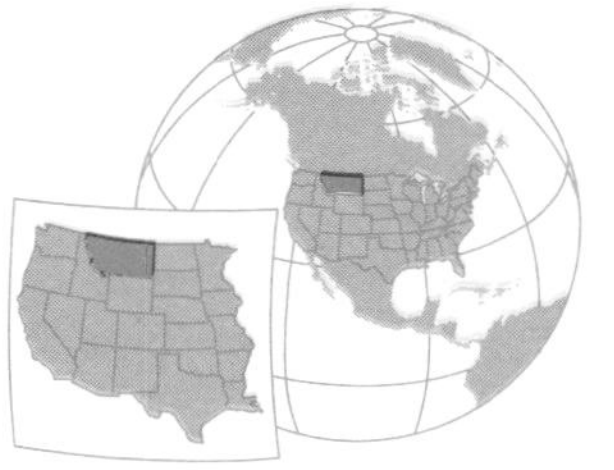

59725 Dillon⊙ 3,976 .... D5
59727 Divide 275 .... D5
59831 Dixon 550 .... B3
59524 Dodson 158 .... H2
59832 Drummond 414 .... D4
59432 Dupuyer 105 .... D2
59433 Dutton 359 .... E3
59434 East Glacier Park 475 .... C2
59635 East Helena 1,647 .... E4
59026 Edgar 220 .... H5
59324 Ekalaka⊙ 620 .... M5
59728 Elliston 250 .... D4
59915 Elmo 250 .... B3
59729 Ennis 660 .... E5
59917 Eureka 1,119 .... B2
59436 Fairfield 650 .... D3
59221 Fairview 1,366 .... M3
59326 Fallon 225 .... L4
59222 Flaxville 142 .... L2
59833 Florence 700 .... B4
59441 Forestgrove 100 .... H3
59327 Forsyth⊙ 2,553 .... K4
†59526 Fort Belknap 185 .... H2
59442 Fort Benton⊙ 1,693 .... F3
59918 Fortine 250 .... A2
59223 Fort Peck 456 .... K2
59443 Fort Shaw 200 .... E3
†59075 Fort Smith 300 .... J5
59225 Frazer 200 .... K2
59834 Frenchtown 300 .... B3
59226 Froid 323 .... M2
59029 Fromberg 469 .... H5
59444 Galata 100 .... E2
59730 Gallatin Gateway 600 .... E5
59030 Gardiner 600 .... F5
59731 Garrison 300 .... D4
59031 Garryowen 200 .... J5
59446 Geraldine 305 .... F3
59447 Geyser 125 .... F3
59525 Gildford 250 .... F2
59230 Glasgow⊙ 4,455 .... K2
59330 Glendive⊙ 5,978 .... M3
59733 Goldcreek 100 .... D4
59835 Grantsdale 500 .... B4
59032 Grass Range 139 .... H3
59401 Great Falls⊙ 56,725 .... E3
Great Falls‡ 80,696 .... E3
59836 Greenough 120 .... C4
59837 Hall 130 .... C4
59840 Hamilton⊙ 2,661 .... B4
59034 Hardin⊙ 3,300 .... J5
59526 Harlem 1,023 .... H2
59036 Harlowton⊙ 1,181 .... F4
59735 Harrison 94 .... E5
59842 Haugan 90 .... A3
59501 Havre⊙ 10,891 .... G2
59527 Hays 400 .... H2
59448 Heart Butte 300 .... C2
59601 Helena (cap.)⊙ 23,938 .... E4
59843 Helmville 250 .... C4
59450 Highwood 150 .... F3
59528 Hingham 186 .... F2
59241 Hinsdale 260 .... K2
59452 Hobson 261 .... G4
59845 Hot Springs 601 .... B3
59919 Hungry Horse 700 .... C2
59037 Huntley 250 .... H5
59846 Huson 97 .... B3
59038 Hysham⊙ 449 .... J4
59530 Inverness 150 .... F2
59336 Ismay 31 .... M4
59736 Jackson 210 .... C5
59638 Jefferson City 162 .... E4
59041 Joliet 580 .... G5
59531 Joplin 300 .... F2
59337 Jordan⊙ 485 .... J3
59453 Judith Gap 213 .... G4
59901 Kalispell⊙ 10,648 .... B2
59454 Kevin 208 .... D2
59920 Kila 350 .... B2
59338 Kinsey 100 .... L4
†59072 Klein 250 .... H4
59532 Kremlin 304 .... F2
59922 Lakeside 663 .... B2
59243 Lambert 203 .... M3
59043 Lame Deer 460 .... K5
59044 Laurel 5,481 .... H5
59046 Lavina 164 .... H4
59457 Lewistown⊙ 7,104 .... G3
59923 Libby⊙ 2,748 .... A2
59739 Lima 272 .... D6
59639 Lincoln 473 .... D4
59047 Livingston⊙ 6,994 .... F5
59050 Lodge Grass 771 .... J5
†59524 Lodge Pole 292 .... H2
59847 Lolo 2,418 .... B4
†59847 Lolo Hot Springs 25 .... B4
59460 Loma 200 .... F3
59225 Lustre 25 .... K2
59538 Malta⊙ 2,367 .... J2
59741 Manhattan 988 .... E5
59925 Marion 450 .... B2
59052 McLeod 150 .... G5
59247 Medicine Lake 408 .... M2
59743 Melrose 350 .... D5
59054 Melstone 238 .... H4
59055 Melville 100 .... F4
59301 Miles City⊙ 9,602 .... L4
59851 Milltown 300 .... C4
*59801 Missoula⊙ 33,388 .... C4
59463 Monarch 120 .... F3
59464 Moore 229 .... G4
59059 Musselshell 117 .... H4
59248 Nashua 495 .... K2
59465 Neihart 91 .... F4
†59501 North Havre 1,230 .... G2
59853 Noxon 800 .... A3
59927 Olney 200 .... B2
59250 Opheim 210 .... K2
59252 Outlook 122 .... M2
59854 Ovando 300 .... C3
59855 Pablo 500 .... B3
59856 Paradise 400 .... B3
59063 Park City 800 .... H5
59253 Peerless 110 .... L2
59467 Pendroy 100 .... D2
59858 Philipsburg⊙ 1,138 .... C4
59859 Plains 1,116 .... B3
59254 Plentywood⊙ 2,476 .... M2
59344 Plevna 191 .... M4
59860 Polson⊙ 2,798 .... B3
59064 Pompeys Pillar 300 .... J5
59747 Pony 130 .... E5
59255 Poplar 995 .... L2
59468 Power 159 .... E3
59929 Proctor 150 .... B3
59066 Pryor 146 .... H5
59641 Radersburg 104 .... E4
59863 Ravalli 150 .... B3
59068 Red Lodge⊙ 1,896 .... G5
59069 Reedpoint 160 .... G5
59258 Reserve 80 .... M2
59930 Rexford 130 .... A2
59259 Richey 417 .... L3
59642 Ringling 102 .... F4
59070 Roberts 312 .... G5
59931 Rollins 200 .... B3
59864 Ronan 1,530 .... C3
59347 Rosebud 259 .... K4
59072 Roundup⊙ 2,119 .... H4
59471 Roy 200 .... H3
59540 Rudyard 450 .... F2
59074 Ryegate⊙ 273 .... G4
59261 Saco 252 .... J2
59865 Saint Ignatius 877 .... C3
59866 Saint Regis 500 .... A3
59075 Saint Xavier 200 .... J5
59867 Saltese 90 .... A3
59472 Sand Coulee 600 .... E3
59473 Santa Rita 120 .... D2
59262 Savage 300 .... M3
59263 Scobey⊙ 1,382 .... L2
59868 Seeley Lake 900 .... C3
59474 Shelby⊙ 3,142 .... E2
59079 Shepherd 200 .... H5
59749 Sheridan 646 .... D5
59270 Sidney⊙ 5,726 .... M3
59751 Silver Star 125 .... D5
59477 Simms 200 .... E3
59932 Somers 700 .... B2
59479 Stanford⊙ 595 .... F3
59870 Stevensville 1,207 .... C4
59480 Stockett 500 .... E3
59933 Stryker 96 .... B2
59871 Sula 200 .... B5
59482 Sunburst 476 .... E2
59483 Sun River 300 .... E3
59872 Superior⊙ 1,054 .... B3
59911 Swan Lake 100 .... C3
59484 Sweetgrass 250 .... E2
59349 Terry⊙ 929 .... L4
59873 Thompson Falls⊙ 1,478 .... A3
59752 Three Forks 1,247 .... E5
59644 Townsend⊙ 1,587 .... E4
59874 Trout Creek 300 .... A3
59935 Troy 1,088 .... A2
59542 Turner 150 .... H2
59754 Twin Bridges 437 .... D5
59085 Twodot 285 .... F4
59485 Ulm 450 .... E3
59486 Valier 640 .... D2
59487 Vaughn 2,270 .... E3
59875 Victor 700 .... B4
59755 Virginia City⊙ 192 .... E5
59351 Volborg 125 .... L5
59701 Walkerville 887 .... D4
59756 Warmsprings 500 .... D4
59275 Westby 291 .... M2
59936 West Glacier 150 .... C2
59758 West Yellowstone 735 .... E6
59937 Whitefish 3,703 .... B2
59759 Whitehall 1,030 .... D5
59645 White Sulphur Springs⊙ 1,302 .... E4
59276 Whitetail 150 .... L2
59544 Whitewater 100 .... J2
59353 Wibaux⊙ 782 .... M3
59760 Willow Creek 150 .... E5
59086 Wilsall 250 .... F5
59489 Winifred 155 .... G3
59087 Winnett⊙ 207 .... H4
59647 Winston 120 .... E4
59761 Wisdom 140 .... C5
59762 Wise River 150 .... C5
59648 Wolf Creek 500 .... D3
59201 Wolf Point⊙ 3,074 .... L2
59088 Worden 600 .... H5
59089 Wyola 350 .... J5

OTHER FEATURES

Absaroka (range) F5
Allen (mt.) C2
Arrow (creek) F3
Ashley (lake) B2
Battle (creek) G1
Bearhat (mt.) C2
Bearpaw (mts.) G2
Beartooth (mts.) G5
Beaver (creek) J2
Beaverhead (riv.) D5
Benton (lake) E3
Big (lake) G5
Big Belt (mts.) E4
Big Dry (creek) K3
Big Hole (riv.) C5
Big Hole Nat'l Battlefield C5
Bighorn (lake) H5
Bighorn (riv.) J5
Bighorn Canyon Nat'l Rec. Area H5
Big Muddy (riv.) M2
Big Porcupine (creek) J4
Birch (creek) D2
Birch Creek (res.) D2
Bitterroot (range) B4
Bitterroot (riv.) B4
Blackfeet Ind. Res. D2
Blackfoot (riv.) C4
Blackmore (mt.) F5
Bowdoin (lake) J2
Boxelder (creek) H3
Boxelder (creek) M5
Bynum (res.) D2
Cabinet (mts.) A2
Canyon Ferry (lake) E4
Clark Canyon (res.) D6
Clark Fork (riv.) A3
Clarks Fork, Yellowstone (riv.) G6
Cottonwood (creek) E2
Cow (creek) G2
Crazy (peak) F4
Crow Ind. Res. H5
Custer Battlefield Nat'l Mon. J5
Cut Bank (creek) D2
Douglas (mt.) F5
Earthquake (lake) E6
Electric (peak) F6
Elwell (lake) E2
Emigrant (peak) F5
Ennis (lake) E5
Flathead (lake) C3
Flathead (riv.) B2
Flathead, North Fork (riv.) B2
Flathead, South Fork (riv.) C3
Flathead Ind. Res. B3
Flatwillow (creek) H4
Fort Belknap Ind. Res. H2
Fort Peck (lake) K3
Fort Union Trading Post Nat'l Hist. Site N2
Frances (lake) D2
Freezeout (lake) D3
Frenchman (riv.) J1
Fresno (res.) F2
Gallatin (peak) E5
Gallatin (riv.) E5
Georgetown (lake) C4
Gibson (res.) D3
Glacier Nat'l Park C2
Granite (peak) F5
Grant-Kohrs Ranch Nat'l Hist. Site D4
Hauser (lake) E4
Haystack (peak) A3
Hebgen (lake) E6
Helena (lake) E4
Holter (lake) D4
Hungry Horse (res.) C2
Hurricane (mt.) D2
Hyalite (peak) E5
Jackson (mt.) C2
Jefferson (riv.) D5
Judith (riv.) G3
Koocanusa (lake) A2
Kootenai (riv.) A2
Lemhi (pass) C6
Lewis (range) C2
Lima (res.) D6
Little Bighorn (riv.) J5
Little Bitterroot (lake) B2
Little Dry (creek) K3
Little Missouri (riv.) M5
Lockhart (mt.) D3
Lodge (creek) G1
Lolo (pass) B4
Lone (mt.) E5
Lost Trail (pass) B5
Lower Red Rock (lake) E6
Lower Saint Mary (lake) C2
Madison (riv.) E5
Malmstrom A.F.B. 6,675 .... E3
Marias (riv.) D2
Martinsdale (res.) F4
Mary Ronan (lake) B3
McDonald (lake) B2
McGloughlin (peak) C4
McGregor (lake) B3
Medicine (lake) M2
Milk (riv.) J2
Mission (range) C3
Missouri (riv.) L3
Musselshell (riv.) J3
Nelson (res.) J2
Ninepipe (res.) C3
Northern Cheyenne Indian Reservation K5
O'Fallon (creek) L4
Pishkun (res.) D3
Poplar (riv.) L2
Porcupine (creek) K2
Powder (riv.) L4
Purcell (mts.) A2
Railley (mt.) C3
Red Rock (lkes) E6
Red Rock (riv.) D6
Redwater (riv.) L3
Rock (creek) C4
Rocky (mts.) D4
Rocky Boy's Ind. Res. G2
Rosebud (creek) K4
Ruby (riv.) D5
Ruby River (res.) D5
Sage (creek) F2
Saint Mary (lake) C2
Saint Mary (riv.) C1
Sandy (creek) F2
Sheep (mt.) C2
Shields (riv.) F4
Siyeh (mt.) C2
Smith (riv.) E3
Sphinx (mt.) E5
Stillwater (riv.) G5
Stimson (mt.) C2
Sun (riv.) D3
Swan (lake) C3
Teton (riv.) E3
Tongue (riv.) K5
Upper Red Rock (lake) E6
Ward (peak) A3
Waterton-Glacier Int'l Peace Park C2
Whitefish (lake) B2
Willow (creek) E2
Willow Creek (res.) D3
Yellowstone (riv.) M3
Yellowstone National Park F6

⊙County seat.
‡Population of metropolitan area.
† Zip oт nearest p.o. * Multiple zips.

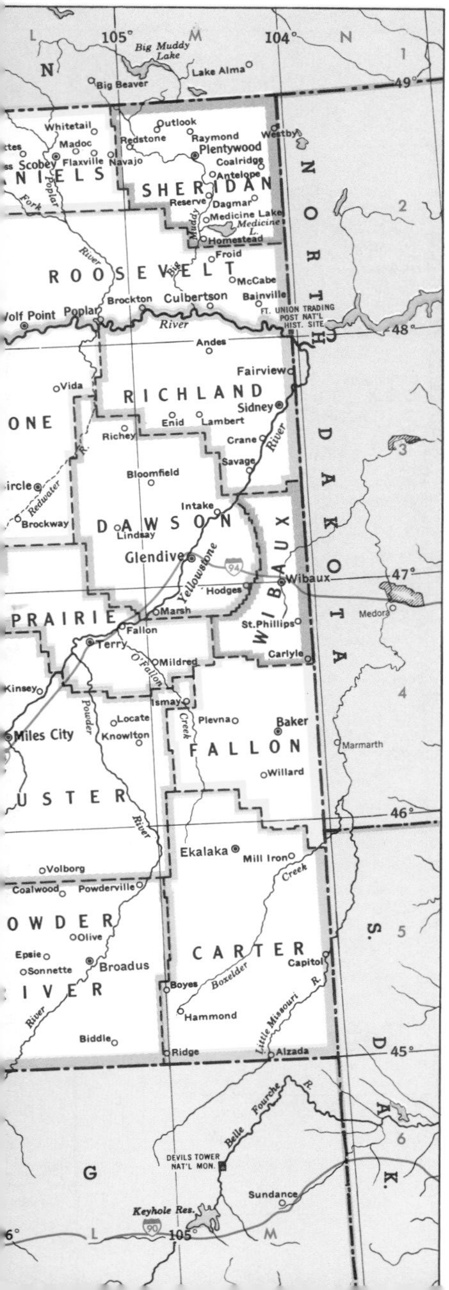
105°
104°
Big Muddy Lake
Lake Alma
Big Beaver
SHERIDAN
Plentywood
Medicine Lake
Homestead
ROOSEVELT
Culbertson
Poplar
Wolf Point
RICHLAND
Sidney
Fairview
DAWSON
Glendive
WIBAUX
Wibaux
PRAIRIE
Terry
Miles City
FALLON
Baker
CUSTER
Ekalaka
CARTER
POWDER RIVER
Broadus
NORTH DAKOTA
S. DAK.
Medora
Marmarth
Sundance
Keyhole Res.
DEVILS TOWER NAT'L MON.
FT. UNION TRADING POST NAT'L HIST. SITE

COUNTIES

Adams 30,656	F4
Antelope 8,675	F2
Arthur 513	C3
Banner 918	A3
Blaine 867	E3
Boone 7,391	F3
Box Butte 13,696	A2
Boyd 3,331	F2
Brown 4,377	E2
Buffalo 34,797	E4
Burt 8,813	H3
Butler 9,330	G3
Cass 20,297	H4
Cedar 11,375	G2
Chase 4,758	C4
Cherry 6,758	C2
Cheyenne 10,057	A3
Clay 8,106	F4
Colfax 9,890	G3
Cuming 11,664	H3
Custer 13,877	E3
Dakota 16,573	H2
Dawes 9,609	A2
Dawson 22,304	E4
Deuel 2,462	B3
Dixon 7,137	H2
Dodge 35,847	H3
Douglas 397,038	H3
Dundy 2,861	C4
Fillmore 7,920	G4
Franklin 4,377	F4
Frontier 3,647	D4
Furnas 6,486	E4
Gage 24,456	H4
Garden 2,802	B3
Garfield 2,363	F3
Gosper 2,140	E4
Grant 877	C3
Greeley 3,462	F3
Hall 47,690	F4
Hamilton 9,301	F4
Harlan 4,292	E4
Hayes 1,356	C4
Hitchcock 4,079	C4
Holt 13,552	F2
Hooker 990	C3
Howard 6,773	F3
Jefferson 9,817	G4
Johnson 5,285	H4
Kearney 7,053	F4
Keith 9,364	C3
Keya Paha 1,301	E2
Kimball 4,882	A3
Knox 11,457	G2
Lancaster 192,884	H4
Lincoln 36,455	D4
Logan 983	D3
Loup 859	E3
Madison 31,382	G3
McPherson 593	C3
Merrick 8,945	F3
Morrill 6,085	A3
Nance 4,740	F3
Nemaha 8,367	J4
Nuckolls 6,726	F4
Otoe 15,183	H4
Pawnee 3,937	H4
Perkins 3,637	C4
Phelps 9,769	E4
Pierce 8,481	G2
Platte 28,852	G3
Polk 6,320	G3
Red Willow 12,615	D4
Richardson 11,315	J4
Rock 2,383	E2
Saline 13,131	G4
Sarpy 86,015	H3
Saunders 18,716	H3
Scotts Bluff 38,344	A3
Seward 15,789	G4
Sheridan 7,544	B2
Sherman 4,226	F3
Sioux 1,845	A2
Stanton 6,549	G3
Thayer 7,582	G4
Thomas 973	D3
Thurston 7,186	H2
Valley 5,633	E3
Washington 15,508	H3
Wayne 9,858	G2
Webster 4,858	F4
Wheeler 1,060	F3
York 14,798	G4

CITIES and TOWNS

Zip	*Name/Pop.*	*Key*
68301	Adams 395	H4
69210	Ainsworth⊙ 2,256	D2
68620	Albion⊙ 1,997	F3
68810	Alda 601	F4
68710	Allen 390	H2
69301	Alliance⊙ 9,920	A2
68920	Alma⊙ 1,369	E4
68304	Alvo 144	H4
68812	Amherst 269	E4
68814	Ansley 644	E3
68922	Arapahoe 1,107	E4
68815	Arcadia 412	F3
68002	Arlington 1,117	H3
69120	Arnold 813	D3
69121	Arthur⊙ 124	C3
68003	Ashland 2,274	H3
68713	Atkinson 1,521	E2
68305	Auburn⊙ 3,482	J4
68818	Aurora⊙ 3,717	F4
68924	Axtell 602	E4
68004	Bancroft 552	H2
68622	Bartlett⊙ 144	F3
69020	Bartley 342	D4
68714	Bassett⊙ 1,009	E2
68715	Battle Creek 948	G3
69334	Bayard 1,435	A3
68310	Beatrice⊙ 12,891	H4
68926	Beaver City⊙ 775	E4
68313	Beaver Crossing 458	G4
68716	Beemer 853	H3
68005	Bellevue 21,813	J3
68624	Bellwood 407	G3
69021	Benkelman⊙ 1,235	C4
68317	Bennet 523	H4
68007	Bennington 631	H3
68927	Bertrand 775	E4
69122	Big Springs 505	B3
68928	Bladen 298	F4
68008	Blair⊙ 6,418	H3
68718	Bloomfield 1,393	G2
68930	Blue Hill 883	F4
68318	Blue Springs 521	H4
68010	Boys Town 622	H3
68319	Bradshaw 373	G4
69123	Brady 377	D3
68821	Brewster⊙ 46	D3
69336	Bridgeport⊙ 1,668	A3
68822	Broken Bow⊙ 3,979	E3
69127	Brule 438	C3
68322	Bruning 330	G4
68823	Burwell⊙ 1,383	E3
68722	Butte⊙ 529	F2
68824	Cairo 737	F3
68825	Callaway 579	D3
69022	Cambridge 1,206	D4
68932	Campbell 441	F4
68015	Cedar Bluffs 632	H3
68016	Cedar Creek 311	H3
68627	Cedar Rapids 447	F3
68724	Center⊙ 123	G2
68826	Central City⊙ 3,083	F3
68017	Ceresco 836	H3
69337	Chadron⊙ 5,933	B2
68725	Chambers 390	F2
68827	Chapman 349	F3
69129	Chappell⊙ 1,095	B3
68327	Chester 435	G4
68628	Clarks 445	G3
68629	Clarkson 817	G3
68328	Clatonia 273	H4
68933	Clay Center⊙ 962	F4
68726	Clearwater 409	F2
†69343	Clinton 80	B2
68727	Coleridge 673	G2
68601	Columbus⊙ 17,328	G3
68329	Cook 341	H4
68331	Cortland 403	H4
69130	Cozad 4,453	E4
69339	Crawford 1,315	A2
68729	Creighton 1,341	G2
68333	Crete 4,872	G4
68730	Crofton 948	G2
69024	Culbertson 767	C4
69025	Curtis 1,014	D4
68731	Dakota City⊙ 1,440	H2
69131	Dalton 345	B3
68831	Dannebrog 356	F3
68335	Davenport 445	G4
68632	David City⊙ 2,514	G3
68020	Decatur 723	H2
68340	Deshler 997	G4
68341	De Witt 642	G4
68342	Diller 311	H4
69133	Dix 275	A3
68633	Dodge 815	H3
68832	Doniphan 696	F4
68343	Dorchester 611	G4
68634	Duncan 410	G3
68347	Eagle 832	H4
68935	Edgar 705	F4
68636	Elgin 807	F3
68022	Elkhorn 1,344	H3
68836	Elm Creek 862	E4
68349	Elmwood 598	H4
68937	Elwood⊙ 716	E4
68733	Emerson 874	H2
68350	Endicott 198	G4
69028	Eustis 460	D4
68735	Ewing 520	F2
68351	Exeter 807	G4
68352	Fairbury⊙ 4,885	G4
68938	Fairfield 543	G4
68354	Fairmont 767	G4
68355	Falls City⊙ 5,374	J4
69029	Farnam 268	D4
68358	Firth 384	H4
68023	Fort Calhoun 641	J3
68939	Franklin⊙ 1,167	E4
68025	Fremont⊙ 23,979	H3
68359	Friend 1,079	G4
68638	Fullerton⊙ 1,506	F3
68361	Geneva⊙ 2,400	G4
68640	Genoa 1,090	G3
69341	Gering⊙ 7,760	A3
68840	Gibbon 1,531	F4
68841	Giltner 400	F4
68941	Glenvil 363	F4
69343	Gordon 2,167	B2
69138	Gothenburg 3,479	D4
68801	Grand Island⊙ 33,180	F4
69140	Grant⊙ 1,270	C4
68842	Greeley⊙ 597	F3
68366	Greenwood 587	H3
68367	Gresham 320	G3
68028	Gretna 1,609	H3
68942	Guide Rock 344	F4
68738	Hadar 286	G2
68368	Hallam 290	H4
68843	Hampton 419	G4
69346	Harrison⊙ 361	A2
68739	Hartington⊙ 1,730	G2
68944	Harvard 1,217	F4
68901	Hastings⊙ 23,045	F4
69032	Hayes Center⊙ 231	C4
69347	Hay Springs 794	B2
68370	Hebron⊙ 1,906	G4
69348	Hemingford 1,023	A2
68371	Henderson 1,072	G4
68029	Herman 340	H3
69143	Hershey 633	D3
68372	Hickman 687	H4
68947	Hildreth 394	E4
68948	Holbrook 297	D4
68949	Holdrege⊙ 5,624	E4
68030	Homer 564	H2
68031	Hooper 932	H3
68740	Hoskins 306	G2
68641	Howells 677	H3
68376	Humboldt 1,176	J4
68642	Humphrey 799	G3
69350	Hyannis⊙ 336	C3
69033	Imperial⊙ 1,941	C4
69034	Indianola 856	D4
68743	Jackson 287	H2
68378	Johnson 341	J4
68955	Juniata 703	F4
68847	Kearney⊙ 21,158	E4
68956	Kenesaw 854	F4
68034	Kennard 372	H3
69145	Kimball⊙ 3,120	A3
69035	Lamar 60	C4
68745	Laurel 1,031	G2
†68046	La Vista 9,588	J3
68957	Lawrence 350	F4
68643	Leigh 509	G3
69147	Lewellen 368	B3
68850	Lexington⊙ 7,040	E4
*68501	Lincoln (cap.)⊙ 171,932	H4
	Lincoln‡ 192,884	H4
68644	Lindsay 383	G3
69149	Lodgepole 413	B3
69217	Long Pine 521	E2
68958	Loomis 447	E4
68037	Louisville 1,022	H3
68853	Loup City⊙ 1,368	E3
69352	Lyman 551	A3
68746	Lynch 357	F2
68038	Lyons 1,214	H3
68748	Madison⊙ 1,950	G3
69150	Madrid 284	C4
68402	Malcolm 355	H4
68854	Marquette 303	G4
69151	Maxwell 410	D3
69038	Maywood 332	D4
69001	McCook⊙ 8,404	D4
68401	McCool Junction 404	G4
68041	Mead 506	H3
68752	Meadow Grove 400	G2
68856	Merna 389	E3
68405	Milford 2,108	H4
68406	Milligan 332	G4
69356	Minatare 969	A3
68959	Minden⊙ 2,939	F4
69357	Mitchell 1,956	A3
68647	Monroe 294	G3
69358	Morrill 1,097	A3
69152	Mullen⊙ 720	C2
68409	Murray 465	J4
68410	Nebraska City⊙ 7,127	J4
68413	Nehawka 270	H4
68756	Neligh⊙ 1,893	G2
68961	Nelson⊙ 733	F4
68757	Newcastle 348	H2
68758	Newman Grove 930	G3
68760	Niobrara 419	G2
68962	Nora 24	G4
68701	Norfolk 19,449	G2
68649	North Bend 1,368	H3
68859	North Loup 405	F3
69101	North Platte⊙ 24,509	D3
68761	Oakdale 410	F2
68045	Oakland 1,393	H3
68415	Odell 322	H4
69153	Ogallala⊙ 5,638	C3
*68101	Omaha⊙ 313,911	J3
	Omaha‡ 570,399	J3
68763	O'Neill⊙ 4,049	F2
68764	Orchard 482	F2
68862	Ord⊙ 2,658	F3
68966	Orleans 527	E4
68651	Osceola⊙ 975	G3
69154	Oshkosh⊙ 1,057	B3
68765	Osmond 871	G2
68863	Overton 633	E4
68967	Oxford 1,109	E4
69040	Palisade 401	C4
68864	Palmer 487	F3
68418	Palmyra 512	H4
68046	Papillion⊙ 6,399	J3
68420	Pawnee City⊙ 1,156	H4
69155	Paxton 568	C3
68047	Pender⊙ 1,318	H2
68421	Peru 998	J4
68652	Petersburg 381	G3
68865	Phillips 405	F4
68767	Pierce⊙ 1,535	G2
68768	Pilger 400	G2
68769	Plainview 1,483	G2
68653	Platte Center 367	G3
68048	Plattsmouth⊙ 6,295	J3
68866	Pleasanton 349	E4
68424	Plymouth 506	G4
68654	Polk 440	G3
68770	Ponca⊙ 1,057	H2
68867	Poole	F4
69156	Potter 369	A3
68050	Prague 285	H3
68127	Ralston 5,143	J3
68771	Randolph 1,106	G2
68869	Ravenna 1,296	F4
68970	Red Cloud⊙ 1,300	F4
68658	Rising City 392	G3
69360	Rushville⊙ 1,217	B2
68660	Saint Edward 891	G3
68873	Saint Paul⊙ 2,094	F3
†68760	Santee 388	G2
68874	Sargent 828	E3
68661	Schuyler⊙ 4,151	G3
68875	Scotia 349	F3
69361	Scottsbluff 14,156	A3
68057	Scribner 1,011	H3
68434	Seward⊙ 5,713	H4
68662	Shelby 724	G3
68876	Shelton 1,046	F4
68436	Shickley 413	G4
69162	Sidney⊙ 6,010	B3
68663	Silver Creek 496	G3
68664	Snyder 387	H3
68776	South Sioux City 9,339	H2
68665	Spalding 645	F3
68777	Spencer 596	F2
68059	Springfield 782	H3
68778	Springview⊙ 326	E2
68779	Stanton⊙ 1,603	G3
68439	Staplehurst 306	G4
69163	Stapleton⊙ 340	D3
68442	Stella 289	J4
68443	Sterling 526	H4
69042	Stockville⊙ 45	D4
69043	Stratton 499	C4
68666	Stromsburg 1,290	G3
68780	Stuart 641	E2
68978	Superior 2,502	F4
69165	Sutherland 1,238	C3
68979	Sutton 1,416	G4
68446	Syracuse 1,638	H4
68447	Table Rock 393	H4

Agriculture, Industry and Resources

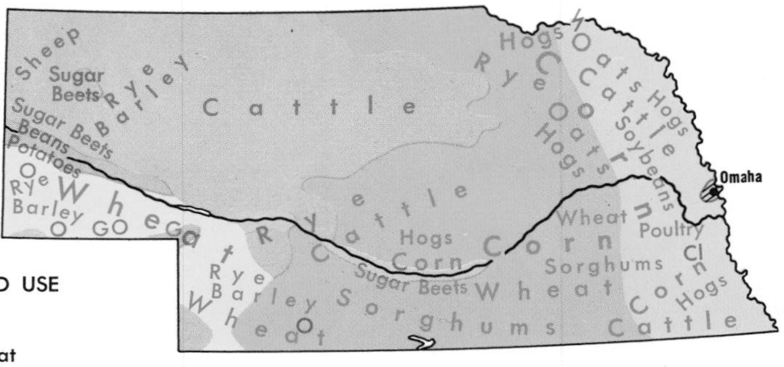

Sheep
Sugar Beets
Rye
Barley
Sugar Beets
Beans
Potatoes
Cattle
Hogs
Rye
Oats
Corn
Cattle
Soybeans
Hogs
Omaha
Wheat
Poultry
Cl
Hogs
Corn
Sorghums
Sugar Beets
Wheat
Rye
Barley
Cattle

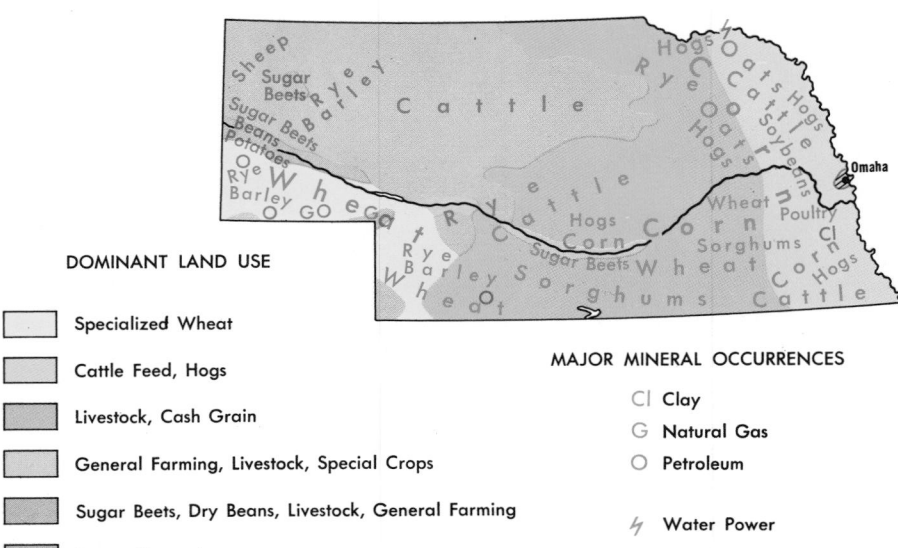

DOMINANT LAND USE

- Specialized Wheat
- Cattle Feed, Hogs
- Livestock, Cash Grain
- General Farming, Livestock, Special Crops
- Sugar Beets, Dry Beans, Livestock, General Farming
- Range Livestock

MAJOR MINERAL OCCURRENCES

- Cl Clay
- G Natural Gas
- O Petroleum
- Water Power
- Major Industrial Areas

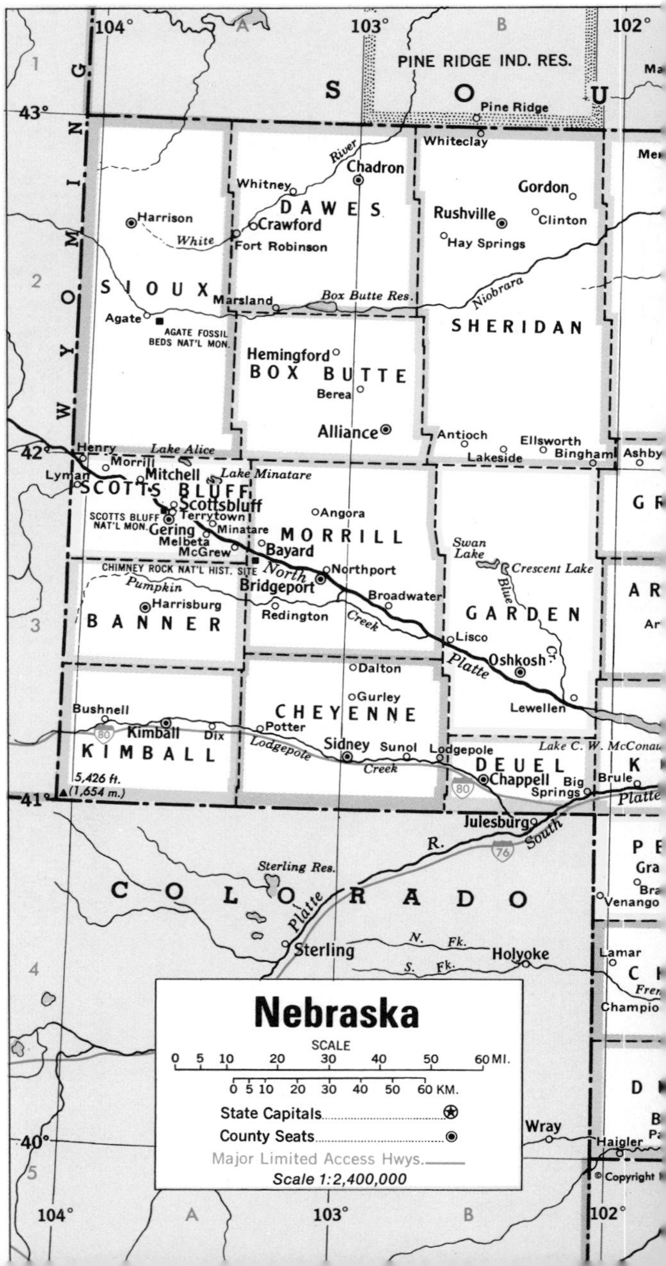

104°
103°
102°
A
B
PINE RIDGE IND. RES.
S O U
Pine Ridge
43°
Whiteclay
Chadron
Whitney
Gordon
Harrison
DAWES
Crawford
Rushville
Clinton
White
Fort Robinson
Hay Springs
SIOUX
Marsland
Box Butte Res.
Niobrara
Agate
AGATE FOSSIL BEDS NAT'L MON.
SHERIDAN
Hemingford
BOX BUTTE
Berea
Alliance
Antioch
Ellsworth
Lakeside
Bingham
Ashby
42°
Henry
Lake Alice
Morrill
Mitchell
Lake Minatare
Lyman
SCOTTS BLUFF
Scottsbluff
SCOTTS BLUFF NAT'L MON.
Terrytown
Gering
Angora
Minatare
Melbeta
MORRILL
McGrew
Bayard
Swan Lake
CHIMNEY ROCK NAT'L HIST. SITE
North
Northport
Crescent Lake
Pumpkin
Bridgeport
Broadwater
Harrisburg
BANNER
Redington
Creek
GARDEN
Lisco
Platte
Oshkosh
Dalton
Gurley
Bushnell
CHEYENNE
Lewellen
Kimball
Dix
Potter
Sidney
Sunol
Lodgepole
Lake C. W. McConaughy
KIMBALL
Lodgepole
Creek
DEUEL
5,426 ft. (1,654 m.)
Chappell
Big Springs
Brule
41°
Julesburg
South
R.
Sterling Res.
COLORADO
Platte
Sterling
N. Fk.
Holyoke
S. Fk.
Lamar
Venango
Champion
Wray
Haigler
40°
1
2
3
4
5
Nebraska
SCALE
0 5 10 20 30 40 50 60 MI.
0 5 10 20 30 40 50 60 KM.
State Capitals
County Seats
Major Limited Access Hwys.
Scale 1:2,400,000
© Copyright

68879 Taylor⊙ 278 ....E3	Colamus (riv.) E2
68450 Tecumseh⊙ 1,926 ....H4	Crescent (lake) B3
68061 Tekamah⊙ 1,886 ....H3	Dads (lake) D2
†69341 Terrytown 727 ....A3	Dismal (riv.) C3
69166 Thedford⊙ 313 ....D3	Elkhorn (riv.) G3
68781 Tilden 1,012 ....G2	Enders (res.) C4
69044 Trenton⊙ 796 ....D4	Frenchman (creek) C4
69167 Tryon⊙ 139 ....C3	Gavins Point (dam) G2
68063 Uehling 273 ....H3	Harlan County (lake) E5
68669 Ulysses 270 ....G3	Harry Strunk (lake) D4
68454 Unadilla 291 ....H4	Homestead Nat'l Mon. H4
68455 Union 307 ....J4	Hugh Butler (lake) D4
68456 Utica 689 ....G4	Jeffrey (res.) D4
69201 Valentine⊙ 2,829 ....D2	Johnson (lake) D4
68064 Valley 1,716 ....H3	Keya Paha (riv.) D1
68065 Valparaiso 484 ....H3	Kingsley (dam) C3
68783 Verdigre 617 ....F2	Lewis and Clark (lake) G2
68457 Verdon 278 ....J4	Little Blue (riv.) H5
68066 Wahoo⊙ 3,555 ....H3	Lodgepole (creek) A3
68784 Wakefield 1,125 ....H2	Logan (creek) H2
69169 Wallace 349 ....C4	Loup (riv.) F3
68067 Walthill 847 ....H2	Maloney (res.) D3
68069 Waterloo 450 ....H3	McConaughy, C. W. (lake) C3
69045 Wauneta 746 ....C4	Medicine (lake) D4
68786 Wausa 647 ....G2	Medicine Creek (dam) D4
68462 Waverly 1,726 ....H4	Merritt (res.) D2
68787 Wayne⊙ 5,240 ....G2	Middle Loup (riv.) D3
68463 Weeping Water 1,109 ......J4	Minatare (lake) A3
68464 Western 336 ....G4	Missouri (riv.) H3
68072 Weston 286 ....H3	Moon (lake) E2
68788 West Point⊙ 3,609 ....H3	Niobrara (riv.) E2
68465 Wilber⊙ 1,624 ....G4	North Loup (riv.) E3
68982 Wilcox 379 ....E4	North Platte (riv.) B3
68071 Winnebago 902 ....H2	Offutt A.F.B. 8,787 ....J3
68790 Winside 439 ....G2	Omaha Ind. Res. H3
68791 Wisner 1,335 ....H3	Pelican (lake) D2
68882 Wolbach 301 ....F3	Platte (riv.) E4
68883 Wood River 1,334 ....F4	Pumpkin (creek) A3
68466 Wymore 1,841 ....H4	Republican (riv.) A3
68467 York⊙ 7,723 ....G4	Santee Ind. Res. G2
68073 Yutan 631 ....H3	Scotts Bluff Nat'l Mon. A3
	Sherman (res.) E3
OTHER FEATURES	Snake (riv.) C2
	South Loup (riv.) E3
Agate Fossil Beds Nat'l Mon. A2	South Platte (riv.) C3
Alice (lake) A2	Sutherland (res.) C3
Beaver (creek) D5	Swan (lake) B3
Big Blue (riv.) H4	Swanson (lake) D4
Blue (creek) B3	White (riv.) A2
Box Butte (res.) A2	Winnebago Ind. Res. H2
Cedar (riv.) F3	⊙County seat.
Chimney Rock Nat'l Hist. Site A3	‡Population of metropolitan area.
	† Zip of nearest p.o.
	* Multiple zips.

AREA 77,355 sq. mi. (200,349 sq. km.)
POPULATION 1,569,825
CAPITAL Lincoln
LARGEST CITY Omaha
HIGHEST POINT (Kimball Co.) 5,246 ft. (1654 m.)
SETTLED IN 1847
ADMITTED TO UNION March 1, 1867
POPULAR NAME Cornhusker State
STATE FLOWER Goldenrod
STATE BIRD Western Meadowlark

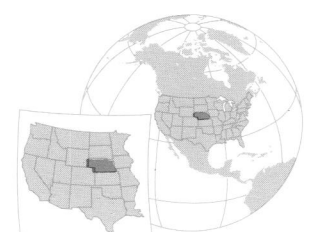

Topography

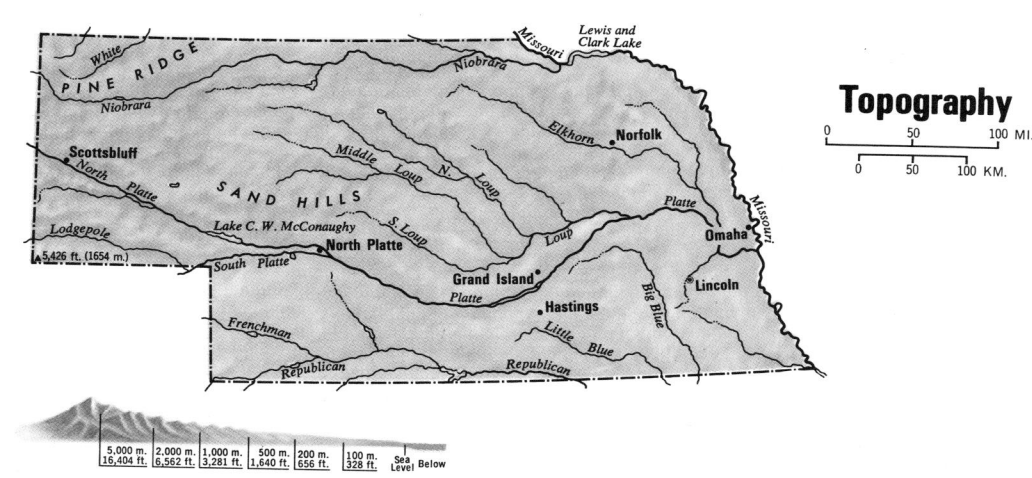

5,000 m. / 16,404 ft.   2,000 m. / 6,562 ft.   1,000 m. / 3,281 ft.   500 m. / 1,640 ft.   200 m. / 656 ft.   100 m. / 328 ft.   Sea Level   Below

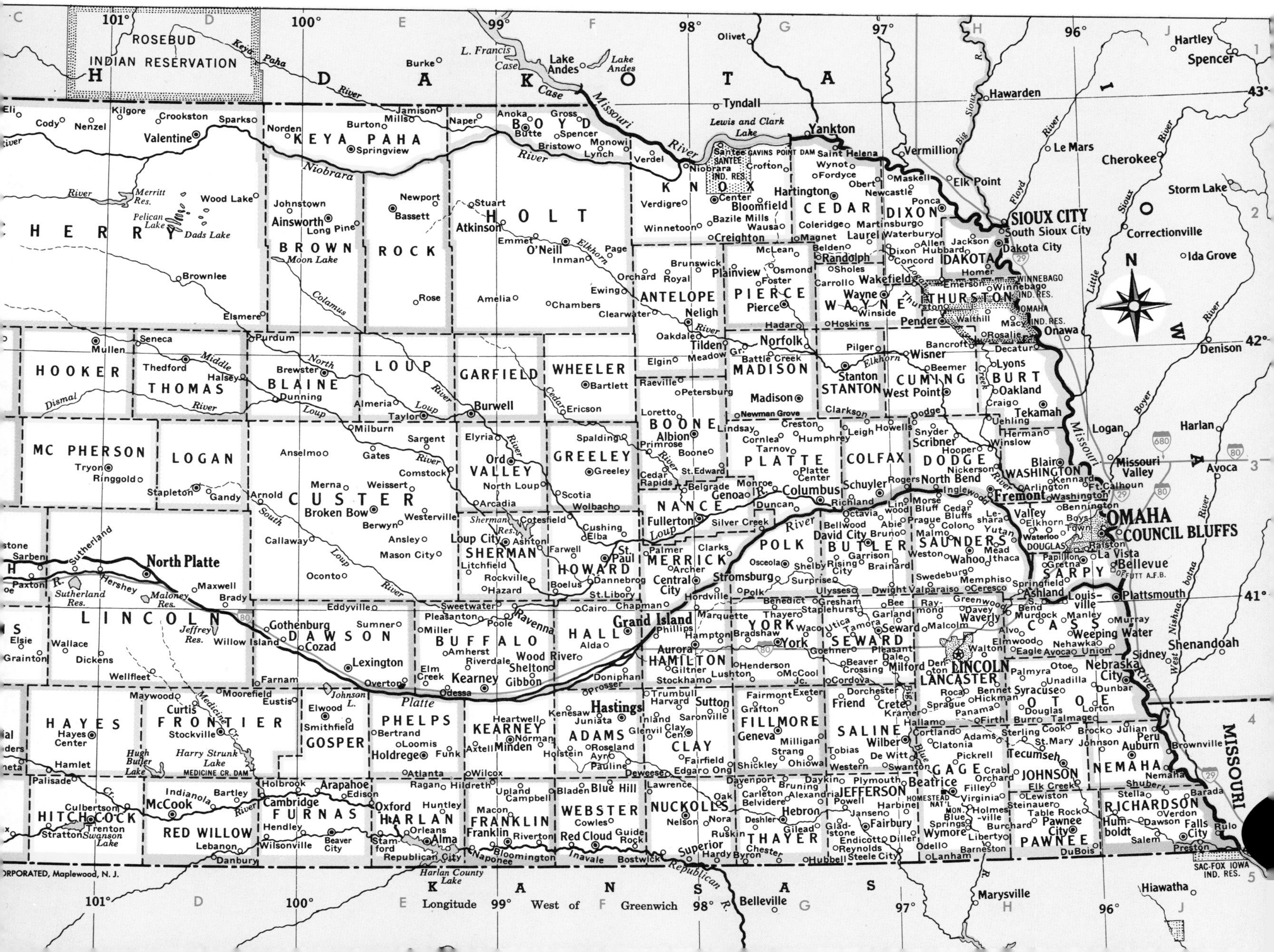

Copyright © Hammond Incorporated, Maplewood, N. J.

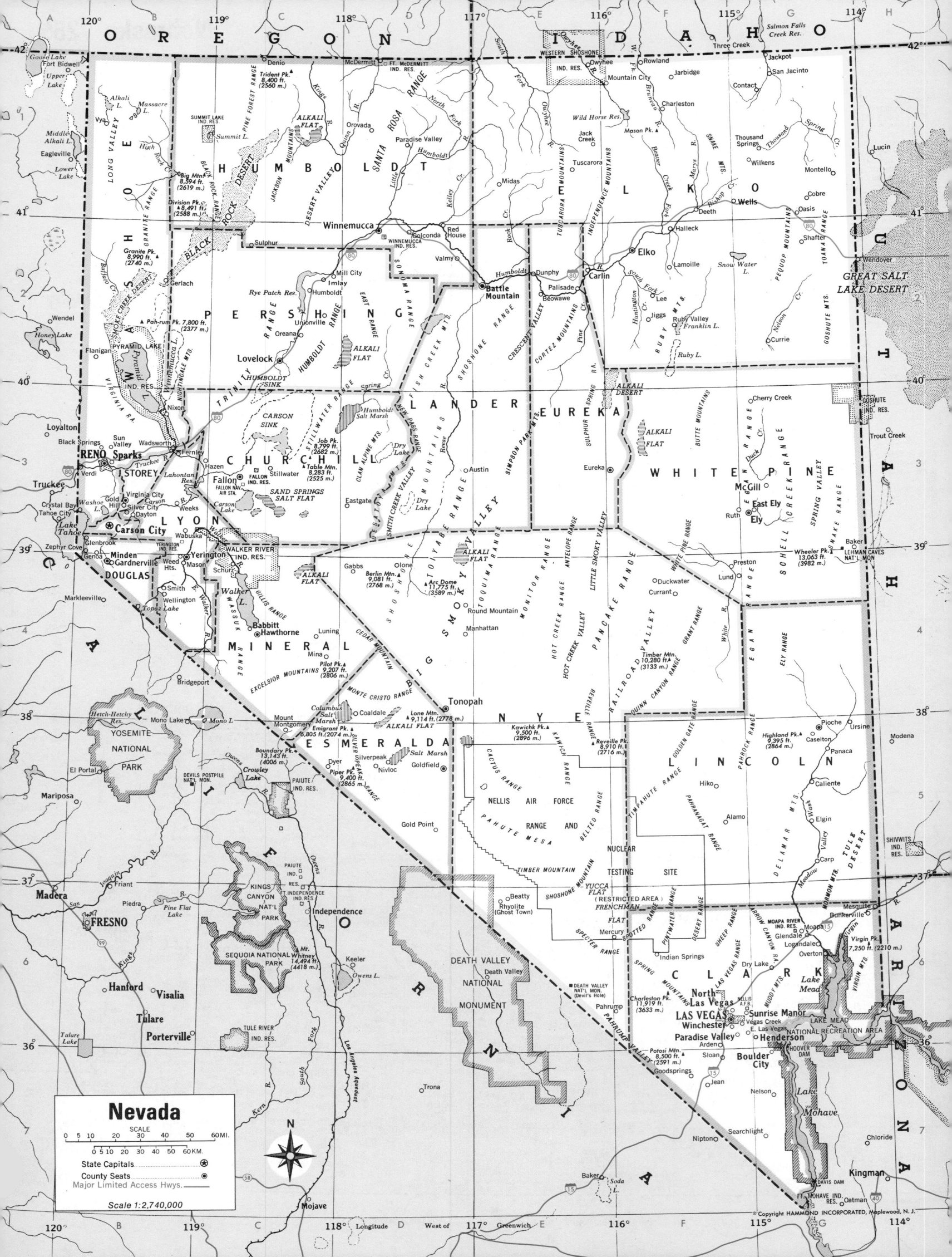

Nevada

SCALE
0 5 10 20 30 40 50 60MI.
0 5 10 20 30 40 50 60 KM.

State Capitals ✪
County Seats ◉
Major Limited Access Hwys. ▬▬▬

Scale 1:2,740,000

© Copyright HAMMOND INCORPORATED, Maplewood, N. J.

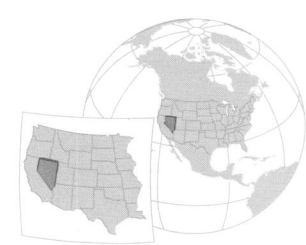

AREA 110,561 sq. mi. (286,353 sq. km.)
POPULATION 800,493
CAPITAL Carson City
LARGEST CITY Las Vegas
HIGHEST POINT Boundary Pk. 13,143 ft. (4006 m.)
SETTLED IN 1850
ADMITTED TO UNION October 31, 1864
POPULAR NAME Silver State; Sagebrush State
STATE FLOWER Sagebrush
STATE BIRD Mountain Bluebird

Agriculture, Industry and Resources

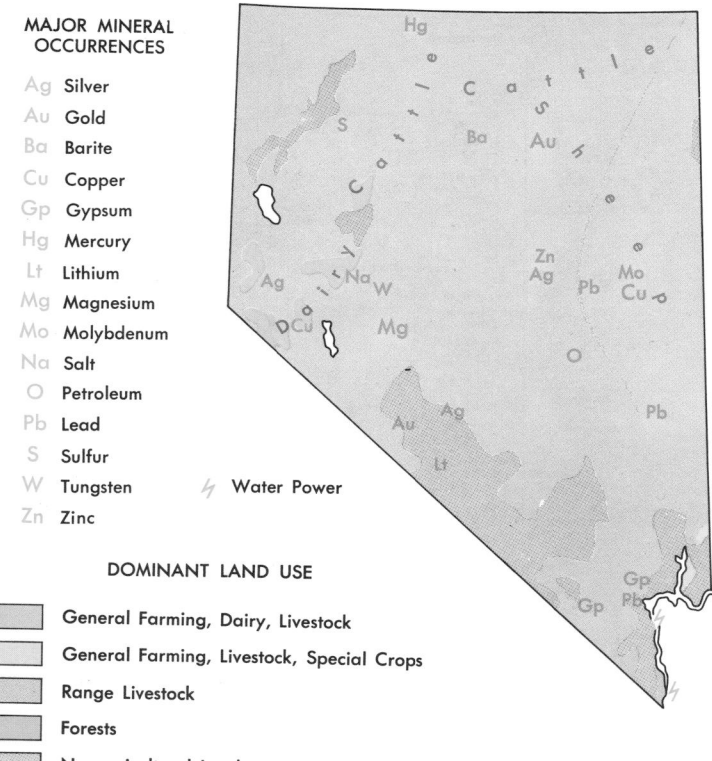

MAJOR MINERAL OCCURRENCES

Ag	Silver
Au	Gold
Ba	Barite
Cu	Copper
Gp	Gypsum
Hg	Mercury
Lt	Lithium
Mg	Magnesium
Mo	Molybdenum
Na	Salt
O	Petroleum
Pb	Lead
S	Sulfur
W	Tungsten
Zn	Zinc
⚡	Water Power

DOMINANT LAND USE

- General Farming, Dairy, Livestock
- General Farming, Livestock, Special Crops
- Range Livestock
- Forests
- Nonagricultural Land

Topography

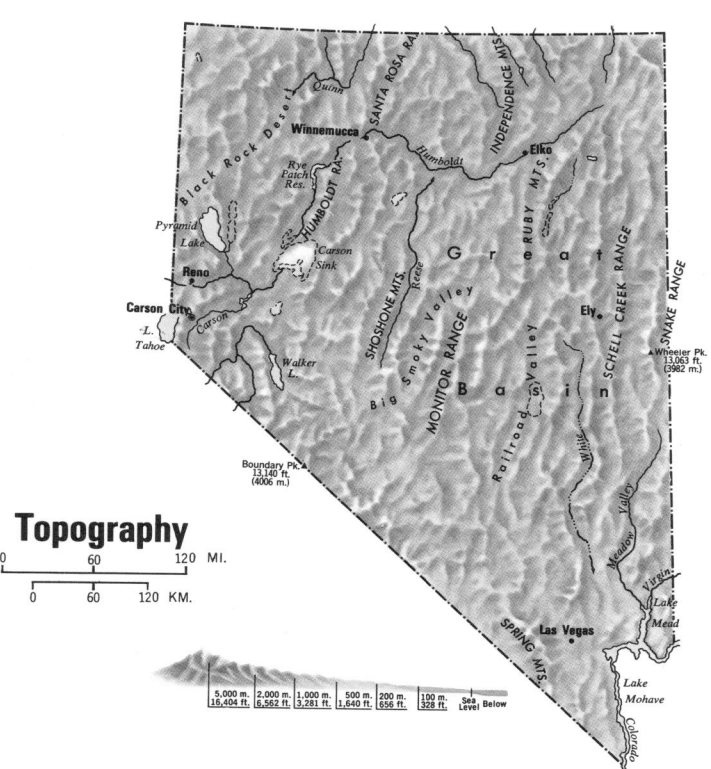

0   60   120 MI.

0   60   120 KM.

5,000 m. 16,404 ft. | 2,000 m. 6,562 ft. | 1,000 m. 3,281 ft. | 500 m. 1,640 ft. | 200 m. 656 ft. | 100 m. 328 ft. | Sea Level | Below

COUNTIES

Carson City (city) 32,022	B3
Churchill 13,917	C3
Clark 463,087	F6
Douglas 19,421	B4
Elko 17,269	F1
Esmeralda 777	D5
Eureka 1,198	E3
Humboldt 9,434	C1
Lander 4,076	D3
Lincoln 3,732	F5
Lyon 13,594	B3
Mineral 6,217	C4
Nye 9,048	E4
Pershing 3,408	C2
Storey 1,503	B3
Washoe 193,623	B2
White Pine 8,167	F3

CITIES and TOWNS

Zip	Name/Pop.	Key
89001	Alamo 300	F5
89310	Austin 300	E3
89416	Babbitt	C4
89311	Baker 140	G3
89820	Battle Mountain⊙ 2,749	E2
89003	Beatty 600	E6
89821	Beowawe 77	E2
†89508	Black Springs 180	B3
89005	Boulder City 9,590	G7
89007	Bunkerville 300	G6
89008	Caliente 982	G5
89822	Carlin 1,232	E2
†89008	Carp 30	G5
89701	Carson City (cap.) 32,022	B3
†89043	Caselton	G5
†89301	Cherry Creek 80	G3
89402	Crystal Bay 6,225	A3
89403	Dayton 350	B3
89823	Deeth 125	F1
89404	Denio 35	C1
89314	Duckwater 80	F4
89010	Dyer 56	C5
89315	East Ely	G3
89112	East Las Vegas 6,449	F6
89801	Elko⊙ 8,758	F2
89301	Ely⊙ 4,882	G3
89316	Eureka⊙ 300	E3
89406	Fallon⊙ 4,262	C3
89408	Fernley 750	B3
89409	Gabbs 811	D4
89410	Gardnerville 1,610	B4
89411	Genoa 254	B4
89412	Gerlach 400	B2
89413	Glenbrook 800	B3
89414	Golconda 275	D2
89013	Goldfield⊙ 500	D5
89019	Goodsprings 80	F7
89824	Halleck 68	F2
89415	Hawthorne⊙ 3,741	C4
89417	Hazen 76	C3
89015	Henderson 24,363	G6
89017	Hiko 210	F5
†89418	Humboldt 14	C2
89418	Imlay 250	C2
89018	Indian Springs 500	F6
†89310	Ione 20	D4
†89834	Jack Creek	E1
89825	Jackpot 400	G1
89826	Jarbidge 11	F1
89019	Jean 125	F7
89828	Lamoille 100	F2
*89101	Las Vegas⊙ 164,674	F6
	Las Vegas‡ 461,816	F6
89829	Lee 125	F2
89021	Logandale 410	G6
89419	Lovelock⊙ 1,680	C2
89317	Lund 380	F4
89420	Luning 90	C4
89022	Manhattan 93	E4
†89447	Mason 200	B4
89421	McDermitt 240	D1
89318	McGill 1,419	G3
89023	Mercury 500	E6
89024	Mesquite 500	G6
89422	Mina 450	C4
89423	Minden⊙ 1,029	B4
89025	Moapa 275	G6
89830	Montello 100	G1
89831	Mountain City 100	F1
†89046	Nelson 75	G7
89424	Nixon 400	B3
89030	North Las Vegas 42,739	F6
89425	Orovada 200	D1
89040	Overton 1,111	G6
89041	Pahrump 400	E6
89042	Panaca 650	G5
89119	Paradise Valley 84,818	F6
89426	Paradise Valley 115	D1
89043	Pioche⊙ 850	G5
*89501	Reno⊙ 100,756	B3
	Reno‡ 193,623	B3
†89003	Rhyolite (Ghost Town) 8	E6
89045	Round Mountain 400	E4
89833	Ruby Valley 150	F2
89319	Ruth 455	F3
89427	Schurz 800	C4
89046	Searchlight 500	F7
89428	Silver City 150	B3
89047	Silverpeak 100	D5
89430	Smith 200	B4
89431	Sparks 40,780	F6
†89406	Stillwater 150	C3
†89445	Sulphur	C2
†89110	Sunrise Manor 44,155	F6
†89431	Sun Valley 8,822	F6
†89835	Thousand Springs	G1
89049	Tonopah⊙ 1,952	D4
89834	Tuscarora 24	E1
89438	Valmy 200	D2
89121	Vegas Creek	G6
89440	Virginia City⊙ 750	B3
89442	Wadsworth 400	B3
89443	Weed Heights 8	B4
89444	Wellington 505	B4
89835	Wells 1,218	G1
†89109	Winchester 19,728	F6
89445	Winnemucca⊙ 4,140	D2
89447	Yerington⊙ 2,021	B4
89448	Zephyr Cove 1,316	A3

OTHER FEATURES

Alkali (lake)	B1
Antelope (range)	E3
Arc Dome (mt.)	D4
Arrow Canyon (range)	G6
Beaver Creek Fork, Humboldt (riv.)	F1
Belted (range)	E5
Berlin (mt.)	D4
Big (mt.)	B1
Big Smoky (valley)	D4
Bishop (creek)	F1
Black Rock (des.)	B2
Black Rock (range)	B1
Boundary (peak)	C5
Buffalo (creek)	B2
Butte (mts.)	F3
Cactus (range)	E5
Carson (lake)	C3
Carson (riv.)	B3
Carson (sink)	C3
Cedar (mt.)	D4
Charleston (peak)	F6
Clan Alpine (mts.)	D3
Columbus Salt (marsh)	C4
Cortez (mts.)	E2
Crescent (valley)	E2
Davis (dam)	G7
Death Valley Nat'l Mon.	E6
Delamar (mts.)	G5
Desatoya (mts.)	D3
Desert (range)	F6
Desert (valley)	C1
Devil's Hole (Death Valley Nat'l Mon.)	E6
Division (peak)	B1
Duck (creek)	G3
East (range)	D2
East Walker (riv.)	B4
Egan (range)	G4
Ely (range)	G4
Emigrant (peak)	C5
Excelsior (mts.)	C4
Fallon Ind. Res.	C3
Fallon Nav. Air Sta.	C3
Fish Creek (mts.)	D2
Fort McDermitt Ind. Res.	D1
Fort Mohave Ind. Res.	G7
Franklin (lake)	F2
Frenchman Flat (basin)	F6
Gillis (range)	C4
Golden Gate (range)	F5
Goshute (mts.)	G2
Goshute Ind. Res.	G3
Granite (peak)	B2
Granite (range)	B2
Grant (range)	F4
Great Salt Lake (des.)	H2
High Rock (creek)	B1
Highland (peak)	G5
Hoover (dam)	G7
Hot Creek (range)	E4
Hot Creek (valley)	E4
Humboldt (range)	C2
Humboldt (riv.)	C2
Humboldt (sink)	C2
Humboldt Salt (marsh)	D3
Huntington (creek)	F2
Independence (mts.)	E1
Jackson (mts.)	C1
Job (peak)	C3
Kawich (peak)	E5
Kawich (range)	E5
Kelley (creek)	D1
Kings (riv.)	C1
Lahontan (res.)	B3
Lake Mead Nat'l Rec. Area	G6
Las Vegas (range)	F6
Lehman Caves Nat'l Mon.	G4
Little Humboldt (riv.)	D1
Little Smoky (valley)	E4
Lone (mt.)	D4
Long (valley)	B1
Marys (riv.)	F1
Mason (peak)	F1
Massacre (lake)	B1
Mead (lake)	G6
Meadow Valley Wash (riv.)	G5
Moapa River Ind. Res.	G6
Mohave (lake)	G7
Monitor (range)	E4
Monte Cristo (range)	D4
Mormon (mts.)	G5
Muddy (mts.)	G6
Nellis A.F.B. 7,476	F6
Nellis Air Force Range and Nuclear Testing Site	E5
Nelson (creek)	G2
New Pass (range)	D3
Nightingale (mts.)	B2
Owyhee (riv.)	E1
Pahranagat (range)	F5
Pahrock (range)	F5
Pah-rum (peak)	B2
Pahrump (valley)	F6
Pahute (mesa)	E5
Pancake (range)	F4
Pequop (mts.)	G2
Pilot (peak)	C4
Pine (creek)	E2
Pine Forest (range)	C1
Pintwater (range)	F6
Piper (peak)	D5
Potosi (mt.)	F7
Pyramid (lake)	B2
Pyramid Lake Ind. Res.	B2
Quinn (riv.)	D1
Quinn Canyon (range)	F4
Railroad (valley)	F4
Reese (riv.)	D3
Reveille (peak)	E5
Reveille (range)	E4
Ruby (lake)	F2
Ruby (mts.)	F2
Rye Patch (res.)	C2
Sand Springs (salt flat)	C3
Santa Rosa (range)	D1
Schell Creek (range)	G3
Sheep (range)	F6
Shoshone (mt.)	E6
Shoshone (mts.)	D3
Shoshone (range)	E2
Silver Peak (range)	D5
Simpson Park (mts.)	E3
Smith Creek (valley)	D3
Smoke Creek (des.)	B2
Snake (mts.)	F1
Snake (range)	G3
Snow Water (lake)	G2
Sonoma (range)	D2
Specter (range)	E6
Spotted (range)	F6
Spring (creek)	D2
Spring (mts.)	F6
Spring (valley)	G3
Stillwater (range)	C3
Sulphur Spring (range)	E3
Summit (lake)	C1
Summit Lake Ind. Res.	B1
Table (mt.)	C3
Tahoe (lake)	A3
Thousand Spring (creek)	G1
Timber (mt.)	F4
Timber (mt.)	E5
Timpahute (range)	F5
Toana (range)	G1
Toiyabe (range)	D3
Topaz (lake)	B4
Toquima (range)	E4
Trident (peak)	C1
Trinity (range)	C2
Truckee (riv.)	B3
Tule (des.)	G5
Tuscarora (mts.)	E1
Virgin (mts.)	G6
Virgin (peak)	G6
Virgin (riv.)	G6
Virginia (range)	B3
Walker (lake)	C4
Walker (riv.)	C3
Walker River Ind. Res.	C3
Washoe (lake)	B3
Wassuk (range)	C4
Western Shoshone Ind. Res.	E1
Wheeler (peak)	G4
White (riv.)	F4
White Pine (range)	F3
Wild Horse (res.)	E1
Winnemucca (lake)	B2
Winnemucca Ind. Res.	D2
Yerington Ind. Res.	B3
Yucca Flat (basin)	E6

⊙County seat.
‡Population of metropolitan area.
† Zip of nearest p.o.
* Multiple zips.

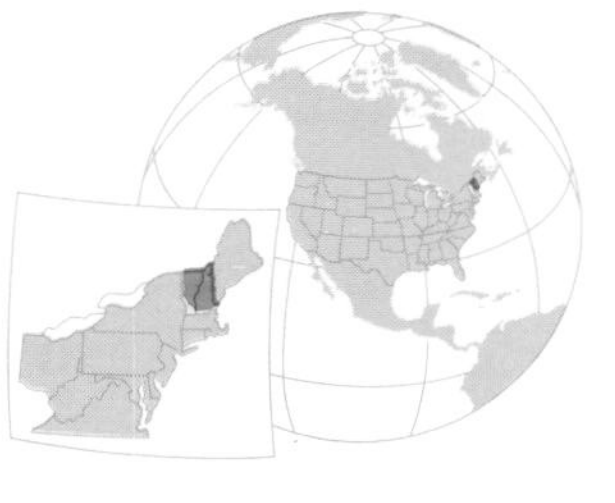

NEW HAMPSHIRE

AREA 9,279 sq. mi. (24,033 sq. km.)
POPULATION 920,610
CAPITAL Concord
LARGEST CITY Manchester
HIGHEST POINT Mt. Washington 6,288 ft. (1917 m.)
SETTLED IN 1623
ADMITTED TO UNION June 21, 1788
POPULAR NAME Granite State
STATE FLOWER Purple Lilac
STATE BIRD Purple Finch

VERMONT

AREA 9,614 sq. mi. (24,900 sq. km.)
POPULATION 511,456
CAPITAL Montpelier
LARGEST CITY Burlington
HIGHEST POINT Mt. Mansfield 4,393 ft. (1339 m.)
SETTLED IN 1764
ADMITTED TO UNION March 4, 1791
POPULAR NAME Green Mountain State
STATE FLOWER Red Clover
STATE BIRD Hermit Thrush

NEW HAMPSHIRE

COUNTIES

Belknap 42,884 .... D4
Carroll 27,931 .... E4
Cheshire 62,116 .... C6
Coos 35,147 .... E2
Grafton 65,806 .... D4
Hillsborough 276,608 .... D6
Merrimack 98,302 .... D5
Rockingham 190,345 .... E5
Strafford 85,408 .... E5
Sullivan 36,063 .... C5

CITIES and TOWNS

Zip Name/Pop. Key

03601 Acworth○ 590 .... C5
†03864 Albany○ 383 .... E4
†03222 Alexandria○ 706 .... D4
†03275 Allenstown○ 4,398 .... E5
03602 Alstead○ 1,461 .... C5
03809 Alton○ 2,440 .... E5
03810 Alton Bay 500 .... E5
03031 Amherst○ 8,243 .... D6
03216 Andover○ 1,587 .... D5
03440 Antrim○ 2,208 .... D5
03440 Antrim 1,142 .... D5
03217 Ashland○ 1,807 .... D4
03217 Ashland 1,479 .... D4
03441 Ashuelot 810 .... C6
03811 Atkinson○ 4,397 .... E6
03032 Auburn○ 2,883 .... E5
03218 Barnstead○ 2,292 .... E5
†03825 Barrington○ 4,404 .... F5
03812 Bartlett○ 1,566 .... E3
03740 Bath○ 761 .... D3
03102 Bedford○ 9,481 .... D6
03220 Belmont○ 4,026 .... E5
03442 Bennington○ 890 .... D5
†03785 Benton○ 333 .... D3
03570 Berlin 13,084 .... E3
03574 Bethlehem○ 1,784 .... D3
03301 Boscawen○ 3,435 .... D5
03221 Bradford○ 1,115 .... D5
†03833 Brentwood○ 2,004 .... E6
†03222 Bridgewater○ 606 .... D4
03222 Bristol○ 2,198 .... D4
03222 Bristol 1,258 .... D4
†03872 Brookfield○ 385 .... E4
03033 Brookline○ 1,766 .... D6
03223 Campton○ 1,694 .... D4
03741 Canaan○ 2,456 .... C4
03034 Candia○ 2,989 .... E5
03224 Canterbury○ 1,410 .... D5
†03595 Carroll○ 647 .... D3
03813 Center Conway 558 .... E4
03226 Center Harbor○ 808 .... E4
03814 Center Ossipee 800 .... E4
03603 Charlestown○ 4,417 .... C5
03603 Charlestown 1,294 .... C5
†04037 Chatham○ 189 .... E3
03036 Chester○ 2,006 .... E6
03443 Chesterfield○ 2,561 .... C6
†03258 Chichester○ 1,492 .... E5
03817 Chocorua 575 .... E4
03743 Claremont 14,557 .... C5
†05902 Clarksville○ 262 .... E1
03576 Colebrook○ 2,459 .... E2
03576 Colebrook 1,131 .... E2
03301 Concord (cap.)⊙ 30,400 .... D5
03229 Contoocook 1,499 .... D5
03818 Conway○ 7,158 .... E4
03818 Conway 1,781 .... E4
03746 Cornish Flat 450 .... C4
†03753 Croydon○ 457 .... C5
†03598 Dalton○ 672 .... D3
03230 Danbury○ 680 .... D4
03819 Danville○ 1,318 .... E6
03037 Deerfield○ 1,979 .... E5
†03244 Deering○ 1,041 .... D5
03038 Derry○ 18,875 .... E6
03038 Derry 12,248 .... E6
†03266 Dorchester○ 244 .... D4
03820 Dover⊙ 22,377 .... F5
03444 Dublin○ 1,303 .... C6
†03588 Dummer○ 390 .... E2
†03301 Dunbarton○ 1,174 .... D5
03824 Durham○ 10,652 .... F5
03824 Durham 8,448 .... F5
03231 East Andover 500 .... D5
03826 East Hampstead 900 .... E6
03827 East Kingston○ 1,135 .... F6
†03580 Easton○ 124 .... D3
03446 East Swanzey 500 .... C6
03832 Eaton (Eaton Center)○ 256 .... E4
†03264 Ellsworth○ 53 .... D4
03748 Enfield○ 3,175 .... C4
03748 Enfield 1,581 .... C4
03042 Epping○ 3,460 .... E5
03042 Epping 1,384 .... E5
03234 Epsom○ 2,743 .... E5
03579 Errol○ 313 .... E2
03750 Etna 550 .... C4
03833 Exeter○ 11,024 .... F6
03833 Exeter⊙ 8,947 .... F6
03835 Farmington○ 4,630 .... E5
03835 Farmington 3,284 .... E5
03447 Fitzwilliam○ 1,795 .... C6
03043 Francestown○ 830 .... D6
03580 Franconia○ 743 .... D3
03235 Franklin 7,901 .... D5
03836 Freedom○ 720 .... E4
03044 Fremont○ 1,333 .... E6
†03246 Gilford○ 4,841 .... E4
03237 Gilmanton○ 1,941 .... E5
03448 Gilsum○ 652 .... C5
03838 Glen 600 .... E3
03045 Goffstown○ 11,315 .... D5
03581 Gorham○ 3,322 .... E3
03581 Gorham 2,180 .... E3
03752 Goshen○ 549 .... C5
03240 Grafton○ 739 .... D4
03753 Grantham○ 704 .... C5
03047 Greenfield○ 972 .... D6
03840 Greenland○ 2,129 .... F5
03048 Greenville○ 1,988 .... D6
03048 Greenville 1,447 .... D6
†03241 Groton○ 255 .... D4
03582 Groveton 1,389 .... D2
03754 Guild 500 .... C5
03841 Hampstead○ 3,785 .... E6
03842 Hampton○ 10,493 .... F6
03842 Hampton 6,779 .... F6
03844 Hampton Falls○ 1,372 .... F6
03449 Hancock○ 1,193 .... C6
03755 Hanover○ 9,119 .... C4
03755 Hanover 6,861 .... C4
03450 Harrisville○ 860 .... C6
03765 Haverhill○ 3,445 .... C3
03241 Hebron○ 349 .... D4
03242 Henniker○ 3,246 .... D5
03242 Henniker 1,538 .... D5
03243 Hill○ 736 .... D4
03244 Hillsboro○ 3,437 .... D5
03244 Hillsboro 1,797 .... D5
03451 Hinsdale○ 3,631 .... C6
03451 Hinsdale 1,546 .... C6
03245 Holderness○ 1,586 .... D4
03049 Hollis○ 4,679 .... D6
03106 Hooksett○ 7,303 .... E5
03106 Hooksett 1,868 .... E5
03301 Hopkinton○ 3,861 .... D5
03051 Hudson○ 14,022 .... E6
03051 Hudson 6,248 .... E6
03845 Intervale 725 .... E3
03846 Jackson○ 642 .... E3
03452 Jaffrey○ 4,349 .... C6
03452 Jaffrey 2,684 .... C6
03583 Jefferson○ 803 .... D3
03431 Keene⊙ 21,449 .... C6
03848 Kingston○ 4,111 .... E6
03246 Laconia⊙ 15,575 .... E4
03584 Lancaster○ 3,401 .... D3
03584 Lancaster⊙ 2,134 .... D3
†03585 Landaff○ 266 .... D3
†03602 Langdon○ 437 .... C5
03766 Lebanon 11,134 .... C4
†03857 Lee○ 2,111 .... F5
03606 Lempster○ 637 .... C5
03251 Lincoln○ 1,313 .... D3
03585 Lisbon○ 1,517 .... D3
03585 Lisbon 1,151 .... D3
†03051 Litchfield○ 4,150 .... E6
03561 Littleton○ 5,558 .... D3
03561 Littleton 4,480 .... D3
03053 Londonderry○ 13,598 .... E6
03301 Loudon○ 2,454 .... E5
†03585 Lyman○ 281 .... D3
03768 Lyme○ 1,289 .... C4
†03082 Lyndeborough○ 1,070 .... D6
†03820 Madbury○ 987 .... F5
03849 Madison○ 1,051 .... E4
*03101 Manchester 90,936 .... E6
Manchester‡ 160,767 .... E6
03455 Marlborough○ 1,846 .... C6
03455 Marlborough 1,184 .... C6
03456 Marlow○ 542 .... C5
03850 Melvin Village 450 .... E4
03253 Meredith○ 4,646 .... D4
03253 Meredith 1,202 .... D4
03770 Meriden 800 .... C4
03054 Merrimack○ 15,406 .... D6
†03887 Middleton○ 734 .... E5
03588 Milan○ 1,013 .... E2
03055 Milford○ 8,685 .... D6
03055 Milford 6,269 .... D6
03851 Milton○ 2,438 .... F5
03852 Milton Mills 450 .... F4
03771 Monroe○ 619 .... C3
03057 Mont Vernon○ 1,444 .... D6
03254 Moultonboro○ 2,206 .... E4
03060 Nashua⊙ 67,865 .... D6
Nashua‡ 114,221 .... D6
†03457 Nelson○ 442 .... C5
03070 New Boston○ 1,928 .... D6
03255 Newbury○ 961 .... C5
03854 New Castle○ 936 .... F5
03855 New Durham○ 1,183 .... E5
03856 Newfields○ 817 .... F5
03256 New Hampton○ 1,249 .... D4
†03801 Newington○ 716 .... F5
03071 New Ipswich○ 2,433 .... D6
03257 New London○ 2,935 .... D5
03257 New London 1,335 .... D5
03857 Newmarket○ 4,290 .... F5
03857 Newmarket 3,749 .... F5
03773 Newport○ 6,229 .... C5
03773 Newport⊙ 4,388 .... C5
03858 Newton○ 3,068 .... E6
03859 Newton Junction 450 .... E6
03860 North Conway 2,104 .... E3
†03276 Northfield○ 3,051 .... D5
†03276 Northfield-Tilton 2,574 .... D5
03862 North Hampton○ 3,425 .... F6
03590 North Stratford 600 .... D2
†03582 Northumberland○ 2,520 .... D2
03261 Northwood○ 2,175 .... E5
03262 North Woodstock 750 .... D3
03290 Nottingham○ 1,952 .... E5
†03741 Orange○ 197 .... D4
03777 Orford○ 928 .... C4
03864 Ossipee 2,465 .... E4
03076 Pelham○ 8,090 .... E6
†03275 Pembroke○ 4,861 .... E5
03458 Peterborough○ 4,895 .... D6
03458 Peterborough 2,568 .... D6
03779 Piermont○ 507 .... C4
03592 Pittsburg○ 780 .... E1
03263 Pittsfield○ 2,889 .... E5
03263 Pittsfield 1,584 .... E5
03781 Plainfield○ 1,749 .... C4
03865 Plaistow○ 5,609 .... E6
03264 Plymouth○ 5,094 .... D4
03264 Plymouth 3,628 .... D4
03801 Portsmouth 26,254 .... F5
Portsmouth-Dover-Rochester‡ 163,880 .... F5
03593 Randolph○ 274 .... E3
03077 Raymond○ 5,453 .... E5
03077 Raymond 1,192 .... E5
†03470 Richmond○ 518 .... C6
03461 Rindge○ 3,375 .... C6
03867 Rochester 21,560 .... E5
†03431 Roxbury○ 190 .... C6
03266 Rumney○ 1,212 .... D4
03870 Rye○ 4,508 .... F5
03871 Rye Beach 600 .... F6
03079 Salem○ 24,124 .... E6
03268 Salisbury○ 781 .... D5
03269 Sanbornton○ 1,679 .... D5
03872 Sanbornville 750 .... F4
03873 Sandown○ 2,057 .... E6
03270 Sandwich○ 905 .... E4
03874 Seabrook○ 5,917 .... F6
†03458 Sharon○ 184 .... D6
†03581 Shelburne○ 318 .... E3
03878 Somersworth 10,350 .... F5
†01913 South Hampton○ 660 .... F6
03462 Spofford 750 .... C6
†03284 Springfield○ 532 .... C4
†03582 Stark○ 470 .... E2
†03576 Stewartstown○ 943 .... E2
03464 Stoddard○ 482 .... C5
03884 Strafford○ 1,663 .... E5
†03590 Stratford○ 989 .... D2
03885 Stratham○ 2,507 .... F5
03585 Sugar Hill○ 397 .... D3
†03445 Sullivan○ 585 .... C5
03782 Sunapee○ 2,312 .... C5
03275 Suncook 4,698 .... D5
03431 Surry○ 656 .... C5
†03260 Sutton○ 1,091 .... D5
†03431 Swanzey○ 5,183 .... C6
03886 Tamworth○ 1,672 .... E4
03084 Temple○ 692 .... D6
†03285 Thornton○ 952 .... D4
03276 Tilton○ 3,387 .... D5
03276 Tilton-Northfield 2,574 .... D5
03465 Troy○ 2,131 .... C6
03465 Troy 1,318 .... C6
†03816 Tuftonboro○ 1,500 .... E4
03595 Twin Mountain 500 .... D3
†03743 Unity○ 1,092 .... C5
†03872 Wakefield○ 2,237 .... F4
03608 Walpole○ 3,188 .... C5
03278 Warner○ 1,963 .... D5
03279 Warren○ 650 .... D4
03280 Washington○ 411 .... C5
03223 Waterville Valley○ 180 .... D4
03281 Weare○ 3,232 .... D5
†03301 Webster○ 1,095 .... D5
03282 Wentworth○ 527 .... D4
†03579 Wentworths Location○ 49 .... E2
†03242 West Henniker 500 .... D5
03784 West Lebanon C4
03467 Westmoreland○ 1,452 .... C6
03597 West Stewartstown 700 .... E2
03469 West Swanzey 1,022 .... C6
03865 Westville 750 .... E6
03598 Whitefield○ 1,681 .... D3
03598 Whitefield 1,005 .... D3
†03287 Wilmot○ 725 .... D5
03287 Wilmot Flat 450 .... D5
03086 Wilton○ 2,669 .... D6
03086 Wilton 1,310 .... D6
03470 Winchester○ 3,465 .... C6
03470 Winchester 1,732 .... C6
03087 Windham○ 5,664 .... E6
03289 Winnisquam 500 .... E5
03894 Wolfeboro○ 3,968 .... E4
03894 Wolfeboro 2,271 .... E4
03896 Wolfeboro Falls 600 .... E4
03293 Woodstock○ 1,008 .... D4
03785 Woodsville⊙ 1,195 .... C3

Topography

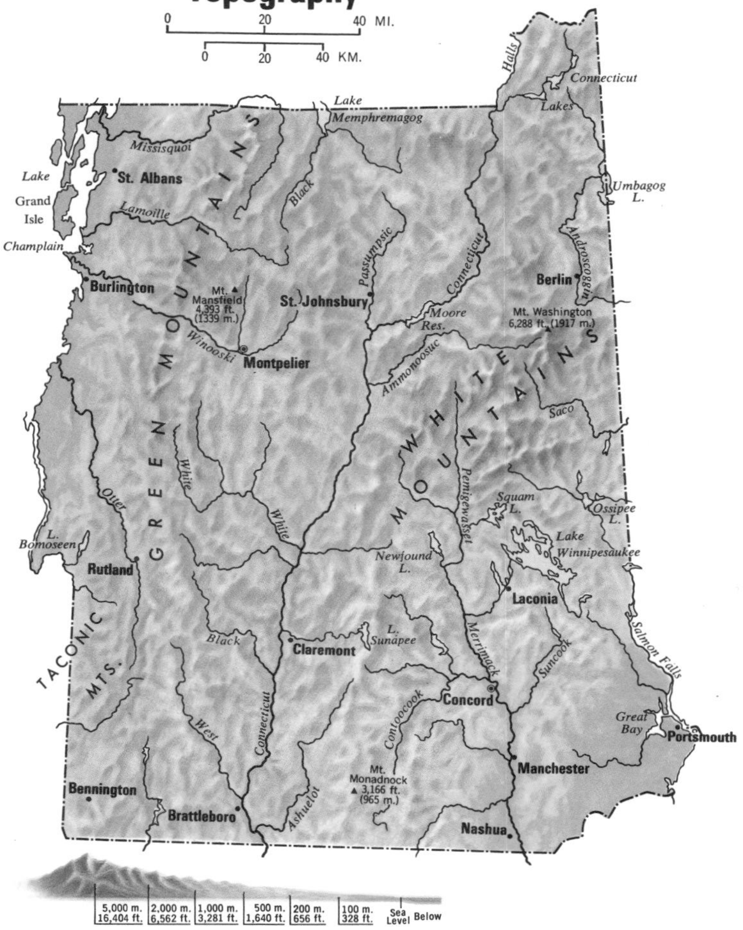

0 20 40 MI.
0 20 40 KM.
Halls
Connecticut Lakes
Lake Memphremagog
Missisquoi
Lake Champlain
Grand Isle
St. Albans
Umbagog L.
Lamoille
Black
Passumpsic
Androscoggin
Berlin
Burlington
Mt. Mansfield 4,393 ft. (1339 m.)
Winooski
Montpelier
St. Johnsbury
Moore Res.
Mt. Washington 6,288 ft. (1917 m.)
Ammonoosuc
Connecticut
GREEN MOUNTAINS
WHITE MOUNTAINS
Saco
White
Otter
Pemigewasset
Squam L.
Ossipee L.
L. Bomoseen
Lake Winnipesaukee
Newfound L.
Rutland
Laconia
TACONIC MTS.
Black
L. Sunapee
Claremont
Merrimack
Suncook
Salmon Falls
West
Contoocook
Concord
Great Bay
Portsmouth
Mt. Monadnock 3,166 ft. (965 m.)
Manchester
Bennington
Brattleboro
Ashuelot
Nashua
5,000 m. 16,404 ft.
2,000 m. 6,562 ft.
1,000 m. 3,281 ft.
500 m. 1,640 ft.
200 m. 656 ft.
100 m. 328 ft.
Sea Level
Below

OTHER FEATURES

Adams (mt.) E3
Ammonoosuc (riv.) D3
Androscoggin (riv.) E2
Ashuelot (riv.) C6
Back (lake) E1
Baker (riv.) D4
Bearcamp (riv.) E4
Beaver (brook) E6
Belknap (mt.) E5
Blackwater (res.) D5
Blue (mt.) E2
Bond (mt.) E3
Bow (lake) E5
Cabot (mt.) E3
Cannon (mt.) D3
Cardigan (mt.) D4
Carrigain (mt.) E3
Carter Dome (mt.) E3
Chocorua (mt.) E4
Cocheco (riv.) E5
Cold (riv.) C5
Comerford (dam) D3
Connecticut (riv.) B6
Contoocook (riv.) D6
Conway (lake) E4
Crawford Notch (pass) E3
Croydon (peak) C5
Croydon Branch, Sugar (riv.) C5
Crystal (lake) E5
Cube (mt.) D4
Dixville (peak) E2
Dixville Notch (pass) E2
Edward MacDowell (res.) D6
Ellis (riv.) E3
Everett (dam) D5
Exeter (riv.) E6
First Connecticut (lake) E1
Francis (lake) E1
Franconia Notch (pass) D3
Franklin Falls (res.) D4
Gale (riv.) D3
Great (bay) F5
Halls (stream) E1
Hancock (mt.) D3
Highland (lake) C5
Hutchins (mt.) E2
Indian (stream) E1
Jefferson (mt.) E3
Kearsarge (mt.) D5
Kinsman (mt.) D3
Kinsman Notch (pass) D3
Lafayette (mt.) D3
Lamprey (riv.) E5
Liberty (mt.) D3
Lincoln (mt.) D3
Long (mt.) E2
Mad (riv.) D4
Madison (mt.) E3
Mascoma (lake) C4
Massabesic (lake) E6
Merrimack (riv.) D5
Merrymeeting (lake) E5
Mohawk (riv.) E2
Monadnock (mt.) C6
Monroe (mt.) E3
Moore (dam) D3
Moore (res.) D3
Moosilauke (mt.) D3
Nash (stream) E2
Newfound (lake) D4
North Carter (mt.) E3
North Twin (mt.) D3
Nubanusit (lake) C5
Osceola (mt.) E3
Ossipee (lake) E4
Ossipee (mts.) E4
Ossipee (riv.) F4
Passaconaway (mt.) E4
Pawtuckaway (pond) E5
Pease A.F.B. F5
Pemigewasset (riv.) D4
Perry (stream) E1
Pine (riv.) E4
Pinkham Notch (pass) E3
Piscataqua (riv.) F5
Piscataquog (riv.) D5
Presidential (range) E3
Rice (mt.) E2
Saco (riv.) E3
Saint-Gaudens Nat'l Hist. Site B4
Salmon Falls (riv.) F5

(continued on following page)

Sandwich (mt.) E4
Sandwich (range) E4
Second (lake) E1
Shaw (mt.) E4
Shoals (isls.) F6
Smarts (mt.) C4
Souhegan (riv.) D6
South Twin (mt.) D3
Squam (lake) E4
Starr King (mt.) E3
Stub Hill (mt.) E1
Sugar (riv.) C5
Sunapee (lake) C5
Suncook (lkes) E5
Suncook (riv.) E5
Surry Mountain (lake) C5
Tarleton (lake) D4
Tecumseh (mt.) D4
Third (lake) E1
Tom (mt.) E3
Umbagog (lake) E2
Upper Ammonoosuc (riv.) E2
Warner (riv.) D5
Washington (mt.) E3
Waumbek (mt.) E3
Wentworth (lake) E4
White (isl.) F6
White (mts.) E3
Whiteface (mt.) E4
Wild Ammonoosuc (riv.) D3
Wilder (dam) C4
Winnipesaukee (lake) E4
Winnipesaukee (riv.) D5
Winnisquam (lake) D4

Agriculture, Industry and Resources

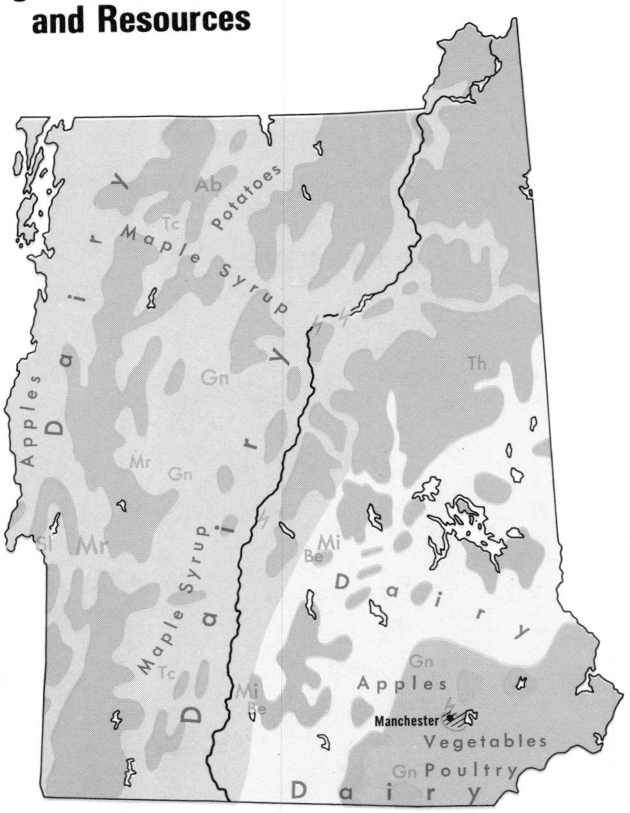

DOMINANT LAND USE

- Specialized Dairy
- Dairy, General Farming
- Dairy, Poultry, Mixed Farming
- Forests

- Water Power
- Major Industrial Areas

MAJOR MINERAL OCCURRENCES

Ab	Asbestos	Mr	Marble
Be	Beryl	Sl	Slate
Gn	Granite	Tc	Talc
Mi	Mica	Th	Thorium

VERMONT

COUNTIES

Addison 29,406 .... A3
Bennington 33,345 .... A6
Caledonia 25,808 .... C2
Chittenden 115,534 .... A3
Essex 6,313 .... D2
Franklin 34,788 .... B2
Grand Isle 4,613 .... A2
Lamoille 16,767 .... B2
Orange 22,739 .... C3
Orleans 23,440 .... C2
Rutland 58,347 .... A4
Washington 52,393 .... B3
Windham 36,933 .... B5
Windsor 51,030 .... B4

CITIES and TOWNS

Zip *Name/Pop.* *Key*

05820 Albany○ 705 .... C2
05440 Alburg○ 1,352 .... A2
05440 Alburg 496 .... A2
†05143 Andover○ 350 .... B5
05250 Arlington○ 2,184 .... A5
05250 Arlington 1,309 .... A5
05441 Bakersfield○ 852 .... B2
05031 Barnard○ 790 .... B4
05821 Barnet○ 1,338 .... C3
05641 Barre 9,824 .... C3
05641 Barre○ 7,090 .... C3
05822 Barton○ 2,990 .... C2
05822 Barton 1,062 .... C2
05823 Beebe Plain 500 .... C2
05902 Beecher Falls 950 .... D2
05101 Bellows Falls 3,456 .... C5
05442 Belvidere○ 218 .... B2
05201 Bennington○ 15,815 .... A6
05201 Bennington⊙ 9,349 .... A6
05731 Benson○ 739 .... A4
†05476 Berkshire○ 1,116 .... B2
05032 Bethel○ 1,715 .... B4
05032 Bethel 1,016 .... B4
†03590 Bloomfield○ 188 .... D2
†05466 Bolton○ 715 .... B3
05732 Bomoseen 700 .... A4
05340 Bondville 500 .... B5
05033 Bradford○ 2,191 .... C3
05033 Bradford 831 .... C3
†05669 Braintree○ 1,065 .... B4
05733 Brandon○ 4,194 .... A4
05733 Brandon 1,925 .... A4
05301 Brattleboro○ 11,886 .... B6
05301 Brattleboro 8,596 .... B6
05034 Bridgewater○ 867 .... B4
05734 Bridport○ 997 .... A4
05443 Bristol○ 3,293 .... A3
05443 Bristol 1,793 .... A3
05036 Brookfield○ 959 .... B3
†05345 Brookline○ 310 .... B5
†05860 Brownington○ 708 .... C2
†05871 Burke○ 1,385 .... D2
05401 Burlington⊙ 37,712 .... A3
Burlington‡ 114,070 .... A3
05647 Cabot○ 958 .... C3
05647 Cabot 259 .... C3
05648 Calais○ 1,207 .... B3
05444 Cambridge○ 2,019 .... B2
05444 Cambridge 217 .... B2
05903 Canaan○ 1,196 .... D2
05735 Castleton○ 3,637 .... A4
05142 Cavendish○ 1,355 .... B5
05736 Center Rutland 465 .... A4
05445 Charlotte○ 2,561 .... A3
05038 Chelsea 1,091 .... C4
05143 Chester○ 2,791 .... B5
05143 Chester-Chester Depot 1,267 .... B5
05737 Chittenden○ 927 .... B4
†05759 Clarendon○ 2,372 .... A4
05446 Colchester○ 12,629 .... A2
05824 Concord○ 1,125 .... D3
05039 Corinth○ 904 .... C3
†05753 Cornwall○ 993 .... A4
05825 Coventry○ 674 .... C2
05826 Craftsbury○ 844 .... C2
05739 Danby○ 992 .... A5
05828 Danville○ 1,705 .... C3
05829 Derby○ 4,222 .... C2
05829 Derby (Derby Center) 598 .... C2
05830 Derby Line 874 .... C2
05251 Dorset○ 1,648 .... A5
†05676 Duxbury○ 877 .... B3
05252 East Arlington 600 .... A5
05649 East Barre-Graniteville 2,172 .... C3
05253 East Dorset 550 .... A5
05837 East Haven○ 280 .... D2
05740 East Middlebury 550 .... A4
05651 East Montpelier○ 2,205 .... B3
05741 East Poultney 450 .... A4
05742 East Wallingford 500 .... B5
05652 Eden○ 612 .... B2
05450 Enosburg Falls 1,207 .... B2
05451 Essex○ 14,392 .... A2
05452 Essex Junction 7,033 .... A3
05454 Fairfax○ 1,805 .... B2
05455 Fairfield○ 1,493 .... B2
05743 Fair Haven○ 2,819 .... A4
05743 Fair Haven 2,363 .... A4
05045 Fairlee○ 770 .... C4
05456 Ferrisburg○ 2,117 .... A3
†05444 Fletcher○ 626 .... B2
05745 Forest Dale 500 .... A4
05457 Franklin○ 1,006 .... B2
†05478 Georgia○ 2,818 .... A2
05904 Gilman 600 .... D3
05839 Glover○ 843 .... C2
05146 Grafton○ 604 .... B5
05840 Granby○ 70 .... D2
05458 Grand Isle○ 1,238 .... A2
05654 Graniteville-East Barre 2,172 .... C3
05747 Granville○ 288 .... B4
05841 Greensboro○ 677 .... C2
05046 Groton○ 667 .... C3
05905 Guildhall 202 .... D2
†05301 Guilford○ 1,532 .... B6
†05358 Halifax○ 488 .... B6
05748 Hancock○ 334 .... B4
05843 Hardwick○ 2,613 .... C2
05843 Hardwick 1,476 .... C2
05047 Hartford○ 7,963 .... C4
05048 Hartland○ 2,396 .... C4
†05459 Highgate○ 2,493 .... B2
05461 Hinesburg○ 2,690 .... A3
†05830 Holland○ 473 .... D2
05749 Hubbardton○ 490 .... A4
05462 Huntington○ 1,161 .... B3
05655 Hyde Park○ 2,021 .... B2
05655 Hyde Park⊙ 475 .... B2
05750 Hydeville 500 .... A4
†05777 Ira○ 354 .... A4
05845 Irasburg○ 870 .... C2
05846 Island Pond 1,216 .... D2
05463 Isle La Motte○ 393 .... A2
05342 Jacksonville 252 .... B6
05343 Jamaica○ 681 .... B5
†05859 Jay○ 302 .... C2
05464 Jeffersonville 491 .... B2
05465 Jericho○ 3,575 .... A2
05465 Jericho 1,340 .... A2
05656 Johnson○ 2,581 .... B2
05656 Johnson 1,393 .... B2
05751 Killington 700 .... B4
†05752 Leicester○ 803 .... A4
†03576 Lemington○ 108 .... D2
†05443 Lincoln○ 870 .... B3
05148 Londonderry○ 1,510 .... B5
05847 Lowell○ 573 .... C2
05149 Ludlow○ 2,414 .... B5
05149 Ludlow 1,352 .... B5
05906 Lunenburg○ 1,138 .... D3
05849 Lyndon○ 4,924 .... C2
05850 Lyndon Center C2
05851 Lyndonville 1,401 .... D2
†05905 Maidstone○ 100 .... D2
05254 Manchester○ 3,261 .... A5
05254 Manchester⊙ 563 .... A5
05255 Manchester Center 1,719 .... A5
05344 Marlboro○ 695 .... B6
05658 Marshfield○ 1,267 .... C3
05658 Marshfield 301 .... C3
†05701 Mendon○ 1,056 .... B4
05753 Middlebury○ 7,574 .... A3
05753 Middlebury⊙ 5,591 .... A3
†05602 Middlesex○ 1,235 .... B3
05757 Middletown Springs○ 603 .... A5
05468 Milton○ 6,829 .... A2
05468 Milton 1,411 .... A2
05469 Monkton○ 1,201 .... A3
05470 Montgomery○ 681 .... B2
05471 Montgomery Center 400 .... B2
05602 Montpelier (cap.)⊙ 8,241 .... B3
05660 Moretown○ 1,221 .... B3
05853 Morgan○ 460 .... D2
†05661 Morristown○ 4,448 .... B2
05661 Morrisville 2,074 .... B2
05758 Mount Holly○ 938 .... B5
†05739 Mount Tabor○ 211 .... B5
†05871 Newark○ 280 .... D2
05051 Newbury○ 1,699 .... C3
05051 Newbury 425 .... C3
05345 Newfane○ 1,129 .... B6
05345 Newfane⊙ 119 .... B6
05472 New Haven○ 1,217 .... A3
05855 Newport○ 1,319 .... C2
05855 Newport⊙ 4,756 .... C2
05257 North Bennington 1,685 .... A6
05663 Northfield○ 5,435 .... B3
05663 Northfield 2,033 .... B3
05664 Northfield Falls 600 .... B3
05052 North Hartland 500 .... C4
05474 North Hero 442 .... A2
05665 North Hyde Park 450 .... B2
05053 North Pomfret 400 .... B4
05260 North Pownal 700 .... A6
05150 North Springfield B5
05859 North Troy 717 .... C2
†05101 North Westminster 310 .... B5
05907 Norton○ 184 .... D2
05055 Norwich○ 2,398 .... C4
†05201 Old Bennington 353 .... A6
†05649 Orange○ 752 .... C3
05860 Orleans 983 .... C2
05760 Orwell○ 901 .... A4
†05491 Panton○ 537 .... A3
05761 Pawlet○ 1,244 .... A5
05862 Peacham○ 531 .... C3
05151 Perkinsville 187 .... B5
05152 Peru○ 312 .... B5
05762 Pittsfield○ 396 .... B4
05763 Pittsford○ 2,590 .... A4
05763 Pittsford 666 .... A4
05667 Plainfield○ 1,249 .... C3
05667 Plainfield 599 .... C3
05056 Plymouth○ 405 .... B4
†05067 Pomfret○ 856 .... B4
05058 Post Mills 500 .... C4
05764 Poultney○ 3,196 .... A4
05764 Poultney 1,554 .... A4
05261 Pownal○ 3,269 .... A6
05765 Proctor○ 1,998 .... A4
05153 Proctorsville 481 .... B5
05346 Putney○ 1,850 .... B6
05059 Quechee 900 .... C4
05060 Randolph○ 4,689 .... B4
05060 Randolph 2,217 .... B4
05062 Reading○ 647 .... B5
05350 Readsboro○ 638 .... B6
05350 Readsboro 402 .... B6
05476 Richford○ 2,206 .... B2
05476 Richford 1,471 .... B2
05477 Richmond○ 3,159 .... A3
05477 Richmond 865 .... A3
05766 Ripton○ 327 .... A4
05767 Rochester○ 1,054 .... B4
†05101 Rockingham○ 5,538 .... B5
05669 Roxbury○ 452 .... B3
†05068 Royalton○ 2,100 .... B4
05768 Rupert○ 605 .... A5
05701 Rutland○ 3,300 .... B4
05701 Rutland⊙ 18,436 .... B4
05042 Ryegate○ 1,000 .... C3
05478 Saint Albans○ 3,555 .... A2
05478 Saint Albans⊙ 7,308 .... A2
†05401 Saint George○ 677 .... A2
05819 Saint Johnsbury○ 7,938 .... D3
05819 Saint Johnsbury⊙ 7,150 .... D3
05863 Saint Johnsbury Center 400 .... D3
05769 Salisbury○ 881 .... A4
†05250 Sandgate○ 234 .... A5
05154 Saxtons River 593 .... B5
†05363 Searsburg○ 72 .... A6
05262 Shaftsbury○ 3,001 .... A6
05065 Sharon○ 828 .... C4
05866 Sheffield○ 435 .... C2
05482 Shelburne○ 5,000 .... A3
05483 Sheldon○ 1,618 .... B2
05770 Shoreham○ 972 .... A4
†05738 Shrewsbury○ 866 .... B4
05670 South Barre 1,301 .... B3
05401 South Burlington 10,679 .... A3
05486 South Hero○ 1,188 .... A2
05155 South Londonderry 500 .... B5
05068 South Royalton 700 .... C4
05069 South Ryegate 400 .... C3
05156 Springfield○ 10,190 .... B5
05156 Springfield 5,603 .... B5
05352 Stamford○ 773 .... A6
05487 Starksboro○ 1,336 .... A3
05772 Stockbridge○ 508 .... B4
05672 Stowe○ 2,991 .... B3
05672 Stowe 531 .... B3
05072 Strafford○ 731 .... C4
†05360 Stratton○ 122 .... B5
†05733 Sudbury○ 380 .... A4
†05250 Sunderland○ 768 .... A5
05867 Sutton○ 667 .... C2
05488 Swanton○ 5,141 .... A2
05488 Swanton 2,520 .... A2
05074 Thetford○ 2,188 .... C4
†05773 Tinmouth○ 406 .... A5
05076 Topsham○ 767 .... C3
05353 Townshend○ 849 .... B5
05868 Troy○ 1,498 .... C2
05077 Tunbridge○ 925 .... C4
05489 Underhill○ 2,172 .... B2
05490 Underhill Center 575 .... B2
05491 Vergennes 2,273 .... A3
05354 Vernon○ 1,175 .... B6
05079 Vershire○ 442 .... C4
05673 Waitsfield○ 1,300 .... B3
†05873 Walden○ 575 .... C3
05773 Wallingford○ 1,893 .... B5
05773 Wallingford 1,141 .... B5
†05491 Waltham○ 394 .... A3
05355 Wardsboro○ 505 .... B5
05674 Warren○ 956 .... B3
05675 Washington○ 855 .... C3
05676 Waterbury○ 4,465 .... B3
05676 Waterbury 1,892 .... B3
05492 Waterville○ 470 .... B2
05678 Websterville 700 .... B3
05774 Wells○ 815 .... A5
05081 Wells River 396 .... C3
05301 West Brattleboro 2,795 .... B6
05871 West Burke 338 .... C2
05356 West Dover 550 .... B6
05083 West Fairlee○ 427 .... C4
05874 Westfield○ 418 .... C2
05494 Westford○ 1,413 .... A2
05875 West Glover C2
†05743 West Haven○ 253 .... A4
05158 Westminster○ 2,493 .... C5
05158 Westminster 319 .... C5
†05860 Westmore○ 257 .... C2
05161 Weston○ 627 .... B5
05777 West Rutland○ 2,351 .... A4
05777 West Rutland 2,169 .... A4
05359 West Townshend 500 .... B5
†05753 Weybridge○ 667 .... A3
†05851 Wheelock○ 444 .... C2
05001 White River Junction 2,582 .... C4
05778 Whiting○ 379 .... A4
05361 Whitingham○ 1,043 .... B6
05088 Wilder 1,461 .... C4
05679 Williamstown○ 2,284 .... B3
05495 Williston○ 3,843 .... A3
05363 Wilmington○ 1,808 .... B6
†05359 Windham○ 223 .... B5
05089 Windsor○ 4,084 .... C5
05089 Windsor 3,478 .... C5
05404 Winooski 6,318 .... A2
05680 Wolcott○ 986 .... C2
05681 Woodbury○ 573 .... C3
†05201 Woodford○ 314 .... A6
05091 Woodstock○ 3,214 .... B4
05091 Woodstock⊙ 1,178 .... B4
05682 Worcester○ 727 .... B3

OTHER FEATURES

Abraham (mt.) B3
Arrowhead Mountain (lake) A2
Ascutney (mt.) C5
Bald (mt.) D2
Barton (riv.) C2
Batten Kill (riv.) A5
Belvidere (mt.) B2
Black (riv.) B5
Black (riv.) C2
Bloodroot (mt.) B4
Bolton (mt.) B3
Bomoseen (lake) A4
Brandon Gap (pass) B4
Bread Loaf (mt.) A3
Bromley (mt.) B5
Brown's (riv.) A2
Burke (mt.) D2
Camels Hump (mt.) B3
Carmi (lake) B2
Caspian (lake) C2
Champlain (lake) A2
Chittenden (res.) B4
Clyde (riv.) C2
Comerford (dam) D3
Connecticut (riv.) C4
Crystal (lake) C2
Dorset (peak) A5
Dunmore (lake) A4
Echo (lake) D2
Ellen (mt.) B3
Equinox (mt.) A5
Fairfield (pond) A2
Glastenbury (mt.) A6
Gore (mt.) D2
Green (mts.) B4
Green River (res.) B2
Groton (lake) C3
Hardwick (lake) C2
Harriman (res.) B6
Harveys (lake) C3
Haystack (mt.) B6
Hoosic (riv.) A6
Hortonia (lake) A4
Hunger (mt.) B3
Iroquois (lake) A3
Island (pond) D2
Jay (peak) B2
Joes (brook) C3
Killington (peak) B4
Lamoille (riv.) A2
Lewis (creek) A3
Lincoln Gap (pass) B3
Little (riv.) B3
Mad (riv.) B3
Maidstone (lake) D2
Mansfield (mt.) B2
Memphremagog (lake) C1
Mettawee (riv.) A5
Middlebury Gap (pass) B4
Mill (riv.) B4
Missisquoi (riv.) B2
Mollys Falls (pond) C3
Moore (dam) D3
Moore (res.) D3
Moose (riv.) D2
Norton (pond) D2
Nulhegan (riv.) D2
Ottauquechee (riv.) B4
Otter (creek) A3
Passumpsic (riv.) D2
Pico (peak) B4
Poultney (riv.) A4
Saint Catherine (lake) A5
Salem (lake) C2
Seymour (lake) D2
Shelburne (pond) A3
Smugglers Notch (pass) B2
Snow (mt.) B6
Somerset (res.) A5
Spruce (mt.) C3
Stratton (mt.) B5
Tabor (mt.) B5
Trout (riv.) B2
Waits (riv.) C3
Waterbury (res.) B3
Wells (riv.) C3
West (riv.) B5
White (riv.) C4
White Face (mt.) B2
Wilder (dam) C4
Willoughby (lake) D2
Winooski (riv.) B3

⊙County seat.
‡Population of metropolitan area.
○Population of town or township.
† Zip of nearest p.o. * Multiple zips

AREA   7,787 sq. mi. (20,168 sq. km.)
POPULATION   7,364,823
CAPITAL   Trenton
LARGEST CITY   Newark
HIGHEST POINT   High Point 1,803 ft. (550 m.)
SETTLED IN   1617
ADMITTED TO UNION   December 18, 1787
POPULAR NAME   Garden State
STATE FLOWER   Purple Violet
STATE BIRD   Eastern Goldfinch

Agriculture, Industry and Resources

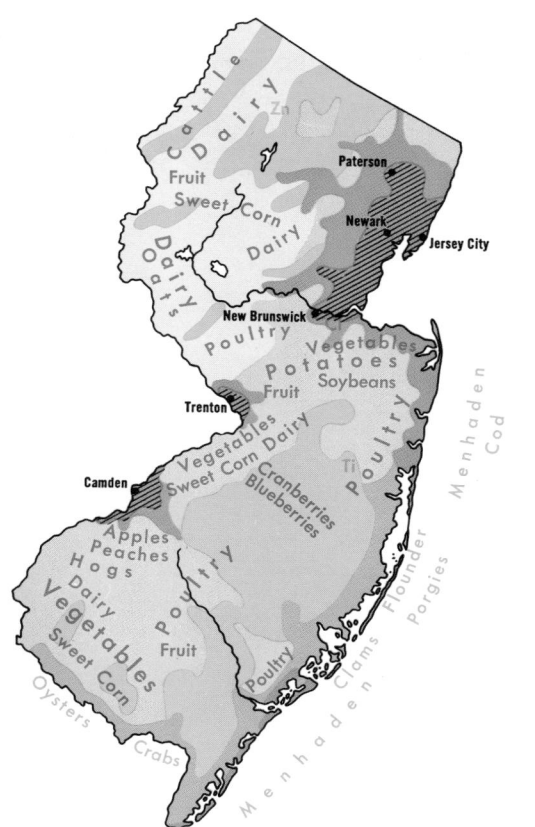

DOMINANT LAND USE

- Specialized Dairy
- Truck and Mixed Farming
- Forests
- Swampland, Limited Agriculture
- Urban Areas

MAJOR MINERAL OCCURRENCES

- Cl   Clay
- Ti   Titanium
- Zn   Zinc

 Major Industrial Areas

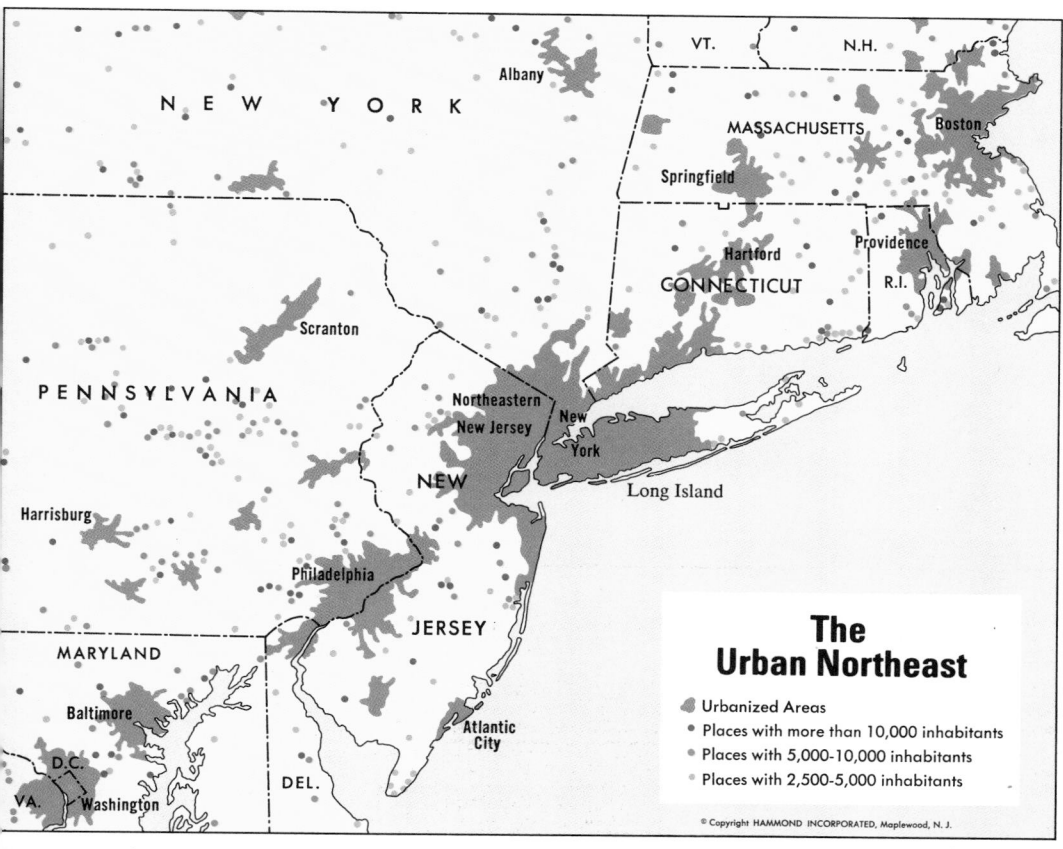

The Urban Northeast

- Urbanized Areas
- Places with more than 10,000 inhabitants
- Places with 5,000-10,000 inhabitants
- Places with 2,500-5,000 inhabitants

© Copyright HAMMOND INCORPORATED, Maplewood, N.J.

COUNTIES

Name	Key
Atlantic 194,119	D5
Bergen 845,385	E2
Burlington 362,542	D4
Camden 471,650	D4
Cape May 82,266	D5
Cumberland 132,866	C5
Essex 851,116	E2
Gloucester 199,917	C4
Hudson 556,972	E2
Hunterdon 87,361	D2
Mercer 307,863	D3
Middlesex 595,893	E3
Monmouth 503,173	E3
Morris 407,630	D2
Ocean 346,038	E4
Passaic 447,585	E1
Salem 64,676	C4
Somerset 203,129	D2
Sussex 116,119	D1
Union 504,094	E2
Warren 84,429	C2

CITIES and TOWNS

Zip	Name/Pop.	Key
08201	Absecon 6,859	D5
07820	Allamuchy 600	D2
07401	Allendale 5,901	B1
07711	Allenhurst 912	F3
08501	Allentown 1,962	D3
08720	Allenwood	E3
08001	Alloway 1,370	C4
08865	Alpha 2,644	C2
07620	Alpine 1,549	C1
07821	Andover 892	D2
08801	Annandale 1,040	D2
07712	Asbury Park 17,015	F3
	Asbury Park-Long Branch‡ 503,173	F3
†08033	Ashland	B3
08004	Atco	D4
*08401	Atlantic City 40,199	E5
	Atlantic City‡ 194,119	E5
07716	Atlantic Highlands 4,950	F3
08106	Audubon 9,533	B3
†08106	Audubon Park 1,274	B3
08202	Avalon 2,162	D5
07001	Avenel	E2
07717	Avon By The Sea 2,337	E3
08005	Barnegat 1,012	E4
08006	Barnegat Light 619	E4
08007	Barrington 7,418	B3
07920	Basking Ridge	D2
08742	Bay Head 1,340	E3
07002	Bayonne 65,047	B2
08008	Beach Haven 1,714	E4
08722	Beachwood 7,687	E4
07921	Bedminster 2,469	D2
08502	Belle Mead	D3
07109	Belleville 35,367	B2
08031	Bellmawr 13,721	B3
07719	Belmar 6,771	E3
07823	Belvidere⊙ 2,475	C2
07621	Bergenfield 25,568	C1
07922	Berkeley Heights○ 12,549	E2
08009	Berlin 5,786	D4
07924	Bernardsville 6,715	D2
08010	Beverly 2,919	D3
08012	Blackwood 5,219	C4
07825	Blairstown○ 4,360	C2
07003	Bloomfield 47,792	B2
07403	Bloomingdale 7,867	E1
08804	Bloomsbury 864	C2
07603	Bogota 8,344	B2
07005	Boonton 8,620	E2
08505	Bordentown 4,441	D3
08805	Bound Brook 9,710	D2
07720	Bradley Beach 4,772	F3
07826	Branchville 870	D1
08723	Breton Woods	E3
08723	Brick○ 53,629	E3
08014	Bridgeport 750	C4
08302	Bridgeton⊙ 18,795	C5
08807	Bridgewater○ 29,175	D2
08730	Brielle 4,068	E3
08203	Brigantine 8,318	E5
08030	Brooklawn 2,133	B3
08015	Browns Mills 10,568	D4
07828	Budd Lake 6,523	D2
08310	Buena 3,642	D4
08016	Burlington 10,246	D3
07405	Butler 7,616	E2
07006	Caldwell 7,624	B2
07830	Califon 1,023	D2
*08101	Camden⊙ 84,910	B3
†08701	Candlewood 6,750	E3
08204	Cape May 4,853	D6
08210	Cape May Court House⊙ 3,597	D5
07072	Carlstadt 6,166	B2
08069	Carneys Point 7,574	C4
07008	Carteret 20,598	E2
07009	Cedar Grove○ 12,600	B2
†08723	Cedarwood Park	E3
07928	Chatham 8,537	E2
08019	Chatsworth 700	D4
08879	Cheesequake	E3
*08034	Cherry Hill○ 68,785	B3
†08089	Chesilhurst 1,590	D4
07930	Chester 1,433	D2
†08505	Chesterfield 3,867	D3
†08077	Cinnaminson○ 16,072	B3
07066	Clark○ 16,699	A3
08020	Clarksboro	C4
08510	Clarksburg 800	E3
08312	Clayton 6,013	C4
08021	Clementon 5,764	D4
07010	Cliffside Park 21,464	C2
07721	Cliffwood	E3
*07011	Clifton 74,388	B2
08809	Clinton 1,910	D2
07624	Closter 8,164	C1
08108	Collingswood 15,838	B3
08213	Cologne 800	D4
07722	Colts Neck 950	E3
07832	Columbia 600	C2
08022	Columbus 800	D3
07961	Convent Station	E2
†08270	Corbin City 254	D5
†07821	Cranberry Lake 500	D2
08512	Cranbury 1,255	E3
07016	Cranford 24,573	E2
07626	Cresskill 7,609	C1
08515	Crosswicks 265	D3
07723	Deal 1,952	F3
08023	Deepwater 800	C4
08110	Delair	B3
08075	Delanco 3,730	D3
08075	Delran○ 14,811	B3
07627	Demarest 4,963	C1
08214	Dennisville 890	D5
07834	Denville○ 14,380	E2
08096	Deptford○ 23,473	B4
08317	Dorothy 900	D5
07801	Dover 14,681	D2
07628	Dumont 18,334	C1
08812	Dunellen 6,593	D2
08816	East Brunswick○ 37,711	E3
07936	East Hanover○ 9,319	E2
07734	East Keansburg	E3
08873	East Millstone 950	D3
†07100	East Newark 1,923	B2
*07017	East Orange 77,690	B2
07073	East Rutherford 7,849	B2
07724	Eatontown 12,703	E3
07020	Edgewater 4,628	C2
†08010	Edgewater Park○ 9,273	D3
*08817	Edison○ 70,193	E2
08215	Egg Harbor City 4,618	D4
07740	Elberon	F3
*07201	Elizabeth⊙ 106,201	B2
08318	Elmer 1,384	C4
†07407	Elmwood Park 18,377	B2
08217	Elwood 1,538	D4
07630	Emerson 7,793	B1
*07631	Englewood 23,701	C2
07632	Englewood Cliffs 5,698	C2
07726	Englishtown 976	E3
07021	Essex Fells 2,363	B2
08319	Estell Manor 848	D5
08025	Ewan 610	C4
07006	Fairfield○ 7,987	A2
07701	Fair Haven 5,679	E3
07410	Fair Lawn 32,229	B1
08320	Fairton 1,107	C5
07022	Fairview 10,519	C2
07023	Fanwood 7,767	E2
07931	Far Hills 677	D2
07727	Farmingdale 1,348	E3
†08505	Fieldsboro 597	D3
07836	Flanders	D2
08822	Flemington⊙ 4,132	D2
08518	Florence-Roebling 7,677	D3
07932	Florham Park 9,359	E2
†08037	Folsom 1,892	D4
08863	Fords	E2
08731	Forked River 900	E4
07024	Fort Lee 32,449	C2
07416	Franklin 4,486	D1
07417	Franklin Lakes 8,769	B1
†08823	Franklin Park○ 31,358	D3
08322	Franklinville	C4
07728	Freehold⊙ 10,020	E3
08825	Frenchtown 1,573	C2
07026	Garfield 26,803	B2
07027	Garwood 4,752	E2
08026	Gibbsboro 2,510	B4
08027	Gibbstown	C4
†08753	Gifford Park 6,528	E3
07933	Gillette	E2
08028	Glassboro 14,574	C4
08029	Glendora 5,632	B4
08826	Glen Gardner 834	D2
07028	Glen Ridge 7,855	B2
07452	Glen Rock 11,497	B1
08030	Gloucester City 13,121	B3
07435	Green Pond 800	E1
07935	Green Village 800	D2
08323	Greenwich○ 973	C5
08032	Grenloch 700	C4

(continued on following page)

07093 Guttenberg 7,340 C2
*07601 Hackensack⊙ 36,039 B2
07840 Hackettstown 8,850 D2
08033 Haddonfield 12,337 B3
08035 Haddon Heights 8,361 B3
08036 Hainesport○ 3,236 D4
07508 Haledon 6,607 B1
07419 Hamburg 1,832 D1
08690 Hamilton Square-Mercerville 25,446 D3
08037 Hammonton 12,298 D4
08827 Hampton 1,614 D2
07640 Harrington Park 4,532 C1
07029 Harrison 12,242 B2
†08057 Hartford 650 D4
08008 Harvey Cedars 363 E4
07604 Hasbrouck Heights 12,166 B2
07641 Haworth 3,509 C1
07507 Hawthorne 18,200 B2
07730 Hazlet 23,013 E3
08828 Helmetta 955 E3
07421 Hewitt 950 E1
08829 High Bridge 3,435 D2
07422 Highland Lakes 2,888 E1
08904 Highland Park 13,396 D2
07732 Highlands 5,187 F3
08520 Hightstown 4,581 D3
07642 Hillsdale 10,495 B1
07205 Hillside○ 21,440 B2
†08083 Hi-Nella 1,250 B4
07030 Hoboken 42,460 C2
07423 Ho Ho Kus 4,129 B1
07733 Holmdel○ 8,447 E3
07843 Hopatcong 15,531 D2
07844 Hope 310 D2
08525 Hopewell 2,001 D3
07731 Howell○ 25,065 E3
†07712 Interlaken 1,037 E3
07845 Ironia D2
07111 Irvington 61,493 B2
08830 Iselin E2
08732 Island Heights 1,575 E4
08527 Jackson○ 25,644 E3
08831 Jamesburg 4,114 E3
*07301 Jersey City⊙ 223,532 B2
Jersey City‡ 556,972 B2
07734 Keansburg 10,613 E3
07032 Kearny 35,735 B2
08824 Kendall Park 7,419 D3
07033 Kenilworth 8,221 E2
07735 Keyport 7,413 E3
08528 Kingston D3
07405 Kinnelon 7,770 E2
07848 Lafayette 900 D1
07034 Lake Hiawatha E2
07849 Lake Hopatcong D2
08733 Lakehurst 2,908 E3
†07871 Lake Mohawk 8,498 D1
08701 Lakewood 22,863 E3
08530 Lambertville 4,044 D3
07850 Landing D2
08734 Lanoka Harbor E4
08021 Laurel Springs 2,249 B4
08879 Laurence Harbor 6,737 E3
08735 Lavallette 2,072 E4
08045 Lawnside 3,042 B3
08648 Lawrenceville 19,724 D3
08833 Lebanon 820 D2
07852 Ledgewood D2
08327 Leesburg 700 D5
07737 Leonardo E3
07605 Leonia 8,027 C2
07938 Liberty Corner D2
07035 Lincoln Park 8,806 A1
07738 Lincroft E3
07036 Linden 37,836 A3
08021 Lindenwold 18,196 B4
08221 Linwood 6,144 D5
07424 Little Falls○ 11,496 B2
07643 Little Ferry 9,399 B2
07739 Little Silver 5,548 F3
07039 Livingston○ 28,040 E2
07644 Lodi 23,956 B2
07740 Long Branch 29,819 F3
Long Branch-Asbury Park‡ 503,173 F3
08403 Longport 1,249 D5
07853 Long Valley 1,682 D2
08048 Lumberton 600 D4
07071 Lyndhurst○ 20,326 B2
07939 Lyons D2
07940 Madison 15,357 E2
08049 Magnolia 4,881 B3
07430 Mahwah○ 12,127 E1
08328 Malaga 950 C4
08050 Manahawkin 1,469 E4
08736 Manasquan 5,354 E3
08738 Mantoloking 433 E3
08051 Mantua○ 9,193 C4
08835 Manville 11,278 D2
08052 Maple Shade○ 20,525 B3
07040 Maplewood○ 22,950 E2
08402 Margate City 9,179 E5
07746 Marlboro○ 17,560 E3
08053 Marlton 9,411 D4
08223 Marmora 650 D5
08836 Martinsville D2
07747 Matawan 8,837 E3
08330 Mays Landing⊙ 2,054 D5
07607 Maywood 9,895 B2
07428 McAfee 800 D1
†08232 McKee City 950 D5
08055 Medford D4
08055 Medford Lakes 4,958 D4
07945 Mendham 4,899 D2
08837 Menlo Park E2
08619 Mercerville-Hamilton Square 25,446 D3
08109 Merchantville 3,972 B3
08840 Metuchen 13,762 E2
08846 Middlesex 13,480 E2
07748 Middletown○ 62,574 E3
07432 Midland Park 7,381 B1
08848 Milford 1,368 C2
07041 Millburn○ 19,543 E2
07946 Millington 975 D2
†08876 Millstone 530 D2
08850 Milltown 7,136 E3
08332 Millville 24,815 C5
†07801 Mine Hill○ 3,325 D2
08342 Mizpah 900 D5
07750 Monmouth Beach 3,318 F3
08852 Monmouth Junction 2,579 D3
07434 Monroe○ 15,858 E3
*07042 Montclair 38,321 B2
07645 Montvale 7,318 B1
07045 Montville○ 14,290 E2
†07070 Moonachie 2,706 B2
08057 Moorestown 13,695 B3
07950 Morris Plains 5,305 D2
07960 Morristown⊙ 16,614 D2
07046 Mountain Lakes 4,153 E2
07092 Mountainside 7,118 E2
07856 Mount Arlington 4,251 D2
08059 Mount Ephraim 4,863 B3
07970 Mount Freedom D2
08060 Mount Holly 10,818 D4
†08054 Mount Laurel○ 17,614 D4
†07828 Mount Olive○ 18,748 D2
08061 Mount Royal 900 C4
08062 Mullica Hill 1,050 C4
08087 Mystic Islands 4,929 E4
08063 National Park 3,552 B3
07752 Navesink E3
07753 Neptune○ 28,366 E3
07753 Neptune City 5,276 E3
07857 Netcong 3,557 D2
*07101 Newark⊙ 329,248 B2
Newark‡ 1,965,304 B2
*08901 New Brunswick⊙ 41,442 E3
New Brunswick-Perth Amboy-Sayreville‡ 595,893 E3
08533 New Egypt 2,111 E3
08344 Newfield 1,563 D4
07435 Newfoundland 900 D1
08224 New Gretna 800 E4
07646 New Milford 16,876 B1
07974 New Providence 12,426 E2
07860 Newton⊙ 7,748 D1
08346 Newtonville 950 D4
07976 New Vernon D2
07032 North Arlington 16,587 B2
07047 North Bergen○ 47,019 B2
08876 North Branch 610 D2
08902 North Brunswick○ 22,220 D3
†07006 North Caldwell 5,832 B2
08204 North Cape May 4,029 C6
08225 Northfield 7,795 D5
07508 North Haledon 8,177 B1
07060 North Plainfield 19,108 E2
07647 Northvale 5,046 F1
08260 North Wildwood 4,714 D6
07648 Norwood 4,413 C1
07110 Nutley 28,998 B2
07755 Oakhurst E3
07436 Oakland 13,443 B1
08107 Oaklyn 4,223 B3
08226 Ocean City 13,949 D5
08740 Ocean Gate 1,385 E4
07756 Ocean Grove F3
07757 Oceanport 5,888 F3
07439 Ogdensburg 2,737 D1
08857 Old Bridge 21,815 E3
07675 Old Tappan 4,168 C1
07649 Oradell 8,658 B1
*07050 Orange 31,136 B2
08723 Osbornsville E3
07863 Oxford 1,587 C2
07470 Packanack Lake B1
07650 Palisades Park 13,732 C2
08065 Palmyra 7,085 B3
07652 Paramus 26,474 B1
07656 Park Ridge 8,515 B1
07054 Parsippany-Troy Hills○ 49,868 E2
07055 Passaic 52,463 E2
*07501 Paterson⊙ 137,970 B2
Paterson-Clifton-Passaic‡ 447,585 B2
08066 Paulsboro 6,944 C4
07977 Peapack-Gladstone 2,038 D2
08067 Pedricktown C4
08068 Pemberton 1,198 D4
08534 Pennington 2,109 D3
08110 Pennsauken○ 33,775 B3
08069 Penns Grove 5,760 C4
08070 Pennsville 12,467 C4
07440 Pequannock○ 13,776 B1
*08861 Perth Amboy 38,951 E2
08865 Phillipsburg 16,647 C2
08741 Pine Beach 1,796 E4
07058 Pine Brook E2
08021 Pine Hill 8,684 D4
08854 Piscataway○ 42,223 D2
08071 Pitman 9,744 C4
*07060 Plainfield 45,555 E2
08536 Plainsboro D3
08232 Pleasantville 13,435 D5
08742 Point Pleasant 17,747 E3
08742 Point Pleasant Beach 5,415 E3
08240 Pomona 2,358 D5
07442 Pompton Lakes 10,660 E1
07444 Pompton Plains B1
07758 Port Monmouth E3
†07850 Port Morris 616 D2
07865 Port Murray 250 D2
08349 Port Norris 1,730 C5
08241 Port Republic 837 D4
08540 Princeton 12,035 D3
08550 Princeton Junction 2,419 D3
†07885 Prospect Park 5,142 B1
08072 Quinton 750 C4
*07065 Rahway 26,723 E2
†08054 Ramblewood 6,475 D4
07446 Ramsey 12,899 B1
†07801 Randolph○ 17,828 D2
08869 Raritan 6,128 D2
07701 Red Bank 12,031 E3
07657 Ridgefield 10,294 B3
07660 Ridgefield Park 12,738 B2
*07450 Ridgewood 25,208 B1
08551 Ringoes 682 D3
07456 Ringwood 12,625 E1
08242 Rio Grande 2,016 D5
07457 Riverdale 2,530 A1
07661 River Edge 11,111 B1
08075 Riverside○ 7,941 B3
08077 Riverton 3,068 B3
07675 River Vale○ 9,489 B1
07662 Rochelle Park○ 5,603 B2
07866 Rockaway 6,852 D2
07647 Rockleigh 192 C1
08553 Rocky Hill 717 D3
08554 Roebling-Florence 7,677 D3
08555 Roosevelt 835 E3
07068 Roseland 5,330 A2
07203 Roselle 20,641 B2
07204 Roselle Park 13,377 A2
08352 Rosenhayn 950 C5
†07876 Roxbury○ 18,878 D2
07760 Rumson 7,623 F3
08078 Runnemede 9,461 B3
*07070 Rutherford 19,068 B2
07662 Saddle Brook○ 14,084 B1
07458 Saddle River 2,763 B1
08079 Salem⊙ 6,959 C4
08872 Sayreville 29,969 E3
07076 Scotch Plains○ 20,774 E2
07760 Sea Bright 1,812 F3
08302 Seabrook 1,411 C5
08750 Sea Girt 2,650 E3
08243 Sea Isle City 2,644 D5
08751 Seaside Heights 1,802 E4
08752 Seaside Park 1,795 E4
07094 Secaucus 13,719 B2
07077 Sewaren E2
08080 Sewell C4
08353 Shiloh 604 C5
08008 Ship Bottom 1,427 E4
07078 Short Hills E2
07701 Shrewsbury 2,962 E3
08081 Sicklerville D4
08558 Skillman D3
08201 Smithville 70 E5
08083 Somerdale 5,900 B4
08244 Somers Point 10,330 D5
08876 Somerville⊙ 11,973 D2
08879 South Amboy 8,322 E3
†07719 South Belmar 1,566 E3
08880 South Bound Brook 4,331 E2
†08852 South Brunswick○ 17,127 E3
07079 South Orange○ 15,864 A2
07080 South Plainfield 20,521 E2
08882 South River 14,361 E3
08753 South Toms River 3,954 E4
07871 Sparta○ 13,333 D1
08884 Spotswood 7,840 E3
07081 Springfield○ 13,955 E2
07762 Spring Lake 4,215 F3
†07762 Spring Lake Heights 5,424 E3
07874 Stanhope 3,638 D2
08886 Stewartsville 950 C2
07980 Stirling E2
07460 Stockholm D1
08559 Stockton 643 D3
08247 Stone Harbor 1,187 D5
08084 Stratford 8,005 B4
†07747 Strathmore E3
07876 Succasunna 10,931 D2
07901 Summit 21,071 E2
08008 Surf City 1,571 E4
07461 Sussex 2,418 D1
08085 Swedesboro 2,031 C4
07878 Tabor E2
07666 Teaneck○ 39,007 B2
07670 Tenafly 13,552 C1
07608 Teterboro 19 B2
08086 Thorofare B4
08887 Three Bridges 750 D2
07724 Tinton Falls 7,740 E3
08753 Toms River⊙ 7,465 E4
07512 Totowa 11,448 B1
07082 Towaco E2
*08601 Trenton (cap.)⊙ 92,124 D3
Trenton‡ 307,863 D3
08087 Tuckerton 2,472 E4
07083 Union○ 50,184 A2
07735 Union Beach 6,354 E3
07087 Union City 55,593 C2
†07421 Upper Greenwood Lake 2,734 E1
†07458 Upper Saddle River 7,958 B1
08406 Ventnor City 11,704 E5
07462 Vernon 800 E1
07044 Verona 14,166 B2
08251 Villas 5,909 D5
08088 Vincentown 900 D4
08360 Vineland 53,753 C5
Vineland-Millville-Bridgeton‡ 132,866 C5
†08043 Voorhees○ 12,919 B3
07463 Waldwick 10,802 B1
07719 Wall○ 18,952 E3
07057 Wallington 10,741 B2
†07712 Wanamassa E3
07465 Wanaque 10,025 B1
08758 Waretown 1,175 E4
†07060 Warren○ 9,805 D2
07882 Washington 6,429 D2
07060 Watchung 5,290 E2
07470 Wayne○ 46,474 A1
07087 Weehawken○ 13,168 C2
08090 Wenonah 2,303 C4
07006 West Caldwell 11,407 A2
†08204 West Cape May 1,091 D6
08092 West Creek 827 E4
†08086 West Deptford○ 18,002 B3
*07090 Westfield 30,447 E2
07764 West Long Branch 7,380 F3
07480 West Milford 950 E1
08108 Westmont 15,875 B3
07093 West New York 39,194 C2
07052 West Orange 39,510 A2
07424 West Paterson 11,293 B2
08628 West Trenton D3
08093 Westville 4,786 B3
†08260 West Wildwood 360 D6
07675 Westwood 10,714 B1
07885 Wharton 5,485 D2

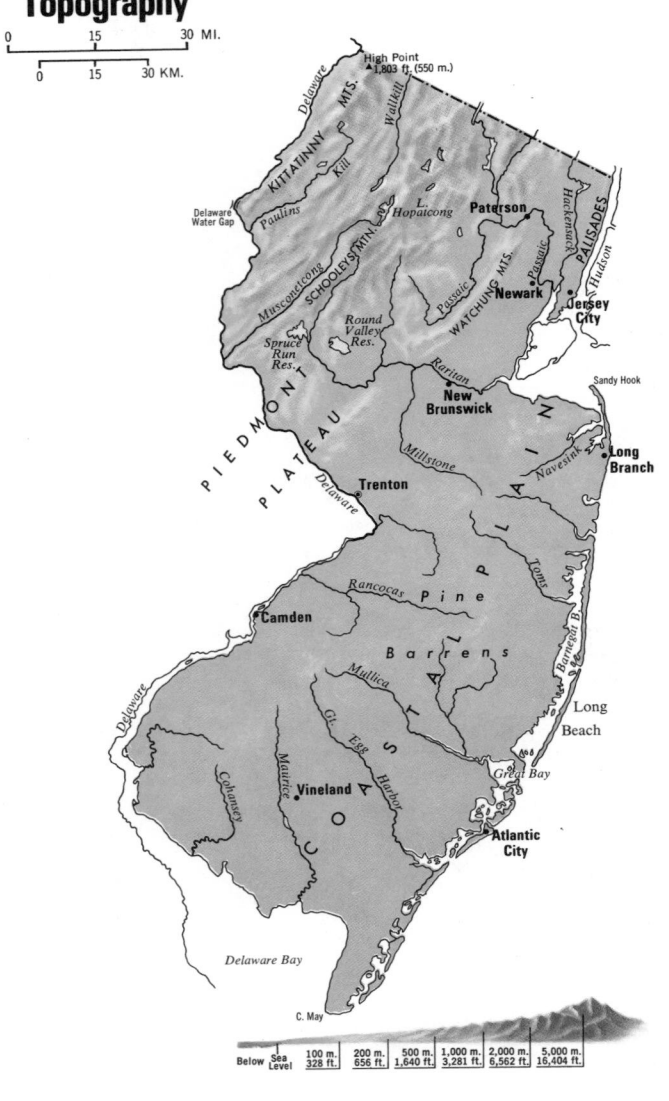

Topography

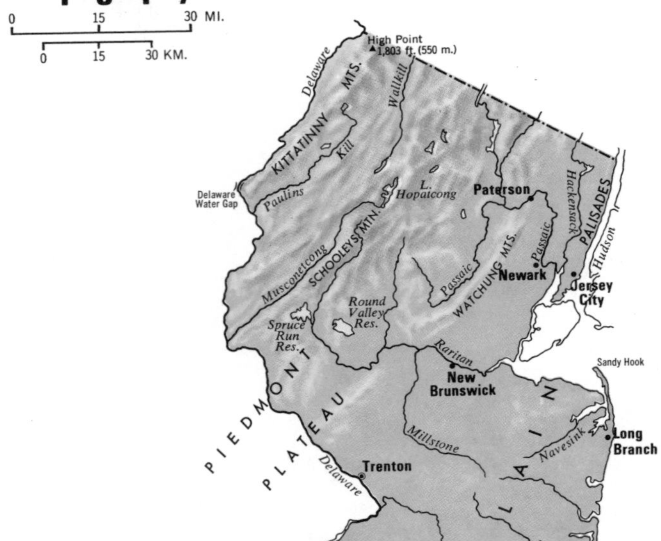
0 15 30 MI.
0 15 30 KM.
High Point 1,803 ft. (550 m.)
Delaware
KITTATINNY MTS.
Wallkill
Kill
Paulins
Delaware Water Gap
L. Hopatcong
Paterson
Hackensack
PALISADES
Hudson
Musconetcong
SCHOOLEYS MTN.
WATCHUNG MTS.
Passaic
Newark
Jersey City
Round Valley Res.
Spruce Run Res.
Raritan
Sandy Hook
New Brunswick
PIEDMONT PLATEAU
Millstone
Navesink
Long Branch
Trenton
Delaware
Toms
Rancocas
Pine
Camden
Barrens
Barnegat B.
Mullica
COASTAL PLAIN
Long Beach
Great Bay
Gt. Egg Harbor
Maurice
Cohansey
Vineland
Atlantic City
Delaware Bay
C. May
Below Sea Level
100 m. 328 ft.
200 m. 656 ft.
500 m. 1,640 ft.
1,000 m. 3,281 ft.
2,000 m. 6,562 ft.
5,000 m. 16,404 ft.

07981 Whippany E2
08889 White House Station D2
†07866 White Meadow Lake 8,429 D2
08252 Whitesboro 1,583 D5
07765 Wickatunk 950 E3
08260 Wildwood 4,913 D6
08260 Wildwood Crest 4,149 D6
08094 Williamstown 5,768 D4
08046 Willingboro○ 39,912 D3
08270 Woodbine 2,809 D5
07095 Woodbridge○ 90,074 E2
08096 Woodbury⊙ 10,353 B4
08097 Woodbury Heights 3,460 B4
07675 Woodcliff Lake 5,644 B1
†08107 Wood-Lynne 2,578 B3
†07885 Woodport D2
07075 Wood-Ridge 7,929 B2
08098 Woodstown 3,250 C4
08562 Wrightstown 3,031 D3
07481 Wyckoff○ 15,500 B1
08620 Yardville 9,414 D3

OTHER FEATURES

Absecon (inlet) E5
Alloways (creek) C4
Arthur Kill (str.) B3
Atlantic Highlands (ridge) E3
Barnegat (bay) E4
Batsto (riv.) D4
Bayonne Military Ocean Terminal B2
Beach Haven (inlet) E4
Beaver (brook) C2
Ben Davis (pt.) C5
Big Flat (brook) D1
Big Timber (creek) C4
Boonton (res.) E2
Brigantine (inlet) E5
Budd (lake) D2
Canistear (res.) E1
Cedar (creek) E4
Clinton (res.) E1
Cohansey (riv.) C5
Cold Spring (inlet) D6
Cooper (riv.) B3
Corson (inlet) D5
Crosswicks (creek) D3
Culvers (lake) D1
Delaware (bay) C5
Delaware (riv.) D3
Delaware Water Gap Nat'l Rec. Area C1
Earle Naval Weapons Sta. E3
Echo (lake) E1
Edison Nat'l Hist. Site A2
Egg Island (pt.) C5
Fort Dix 14,297 D3
Fort Hancock F3
Fort Monmouth E3
Gateway Nat'l Rec. Area E2
Great (bay) E4
Great Egg Harbor (inlet) E5
Greenwood (lake) E1
Hackensack (riv.) C1
Hereford (inlet) D5
High Point (mt.) D1
Hopatcong (lake) D2
Hudson (riv.) C1
Island (beach) E4
Kill Van Kull (str.) B2
Kittatinny (mts.) D1
Lakehurst Naval Air Engineering Center E3
Lamington (riv.) D2
Landing (creek) D4
Little Egg (harb.) E4
Lockatong (creek) C3
Long (beach) E4
Long Beach (isl.) E4
Lower New York (bay) E2
Manasquan (riv.) E3
Manumuskin (riv.) D5
Maurice (riv.) C4
May (cape) C6
McGuire A.F.B. 7,853 D3
Metedeconk (riv.) E3
Mill (creek) E4
Millstone (riv.) D3
Mohawk (lake) D1
Morristown Nat'l Hist. Park D2
Mullica (riv.) D4
Musconetcong (riv.) C2
Navesink (riv.) E3
Newark (bay) B2
Oak Ridge (res.) D1
Oldmans (creek) C4
Oradell (res.) B1
Oswego (riv.) E4
Owassa (lake) D1
Palisades C1
Passaic (riv.) E2
Paulins Kill (riv.) D1
Pennsauken (creek) B3
Pequest (riv.) D2
Picatinny Arsenal D2
Pohatcong (creek) C2
Pompton (lake) B1
Raccoon (creek) C4
Ramapo (riv.) E1
Rancocas (creek) D3
Raritan (bay) E3
Raritan (riv.) D2
Ridgeway Branch, Toms (riv.) E3
Round Valley (res.) D2
Saddle (riv.) B1
Salem (riv.) C4
Sandy Hook (spit) F3
Shoal Branch, Wading (riv.) D4
Spruce Run (res.) D2
Statue of Liberty Nat'l Mon. B2
Stony (brook) D3
Stow (creek) C5
Swartswood (lake) D1
Tappan (lake) C1
The Narrows (str.) E2
Toms (riv.) E3
Townsend (inlet) D5
Tuckahoe (riv.) D5
Union (lake) C5
Upper New York (bay) B2
Wading (riv.) D4
Wallkill (riv.) D1
Wanaque (res.) E1
Wawayanda (lake) E1

⊙County seat.
‡Population of metropolitan area.
○Population of town or township.
† Zip of nearest p.o. * Multiple zips.

New Jersey

SCALE

0   5   10   15   20 MI.

0   5   10   15   20 KM.

State Capitals .... ⊛

County Seats .... ⊛

Canals ....

Major Limited Access Hwys.

Scale 1:930,000

Copyright HAMMOND INCORPORATED, Maplewood, N.J.

COUNTIES

Bernalillo 419,700 .... C4
Catron 2,720 .... A4
Chaves 51,103 .... E5
Cibola B3
Colfax 13,667 .... E2
Curry 42,019 .... F4
De Baca 2,454 .... E4
Dona Ana 96,340 .... C6
Eddy 47,855 .... E6
Grant 26,204 .... A5
Guadalupe 4,496 .... E4
Harding 1,090 .... F3
Hidalgo 6,049 .... A7
Lea 55,993 .... F6
Lincoln 10,997 .... D5
Los Alamos 17,599 .... C3
Luna 15,585 .... B6
McKinley 56,449 .... A3
Mora 4,205 .... E3
Otero 44,665 .... D6
Quay 10,577 .... F3
Rio Arriba 29,282 .... B2
Roosevelt 15,695 .... F4
Sandoval 34,799 .... C3
San Juan 81,433 .... A2
San Miguel
   22,751 .... D3
Santa Fe 75,360 .... C3
Sierra 8,454 .... B5
Socorro 12,566 .... C5
Taos 19,456 .... D2
Torrance 7,491 .... D4
Union 4,725 .... F2
Valencia 61,115 .... C4

CITIES and TOWNS

Zip	Name/Pop.	Key
87510	Abiquiu 500	C2
†87034	Acoma 150	B4
†87034	Acomita (Pueblo of Acoma) 975	B3
88310	Alamogordo⊙ 24,024	C6
*87101	Albuquerque⊙ 331,767	C3
	Albuquerque‡ 454,499	C3
87511	Algodones 500	C2
87001	Algodones 195	C3
88312	Alto 285	D5
87512	Amalia 200	D2
88021	Anthony 3,285	C6
87711	Anton Chico 400	D3
87930	Arrey 367	B6
87513	Arroyo Hondo 400	D2
87514	Arroyo Seco 500	D2
88210	Artesia 10,385	E6
87410	Aztec⊙ 5,512	B2
88023	Bayard 3,036	A6
87002	Belen 5,617	C4
88314	Bent 294	D5
88024	Berino 600	C6
87004	Bernalillo⊙ 3,012	C3
87412	Blanco 200	B2
87413	Bloomfield 4,881	A2
87005	Bluewater 300	A3
87006	Bosque (Bosque Farms) 3,353	C4
87712	Buena Vista 178	D3
87714	Canjilon 380	C2
87515	Canones 300	C2
88316	Capitan 762	D5
88414	Capulin 100	F2
88220	Carlsbad⊙ 25,496	E6
88301	Carrizozo⊙ 1,222	D5
87007	Casa Blanca 560	B4
88113	Causey 81	F5
87518	Cebolla 100	C2
87008	Cedar Crest 600	C3
†87410	Cedar Hill 145	B2
88026	Central 1,968	A6
87010	Cerrillos 500	D3
87519	Cerro 400	D2
88713	Chacon 310	D2
87520	Chama 1,090	C2
88027	Chamberino 700	C6
87521	Chamisal 642	D2
87522	Chimayo 1,993	D3
88415	Cimarron 888	E2
87715	Cleveland 450	D2
88028	Cliff 600	A6
88317	Cloudcroft 521	D6
88101	Clovis⊙ 31,194	F4
†87041	Cochiti 983	C3
88029	Columbus 414	B7
88416	Conchas Dam 240	E3
87523	Cordova 750	D2
88318	Corona 236	D4
87048	Corrales 2,791	C3
87524	Costilla 400	D2
87313	Crownpoint 1,134	A3
†86504	Crystal 200	A3
87013	Cuba 609	B2
87014	Cubero 300	B4
87821	Datil 150	B5
88030	Deming⊙ 9,964	B6
87933	Derry 175	B6
88418	Des Moines 178	F2
88230	Dexter 882	E5
87527	Dixon 200	D2
88032	Dona Ana 800	C6

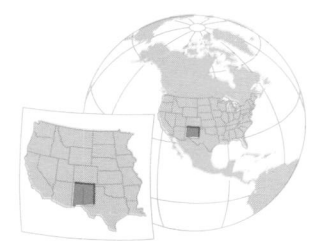

88115 Dora 168 ....F5
87528 Dulce 1,648 ....B2
87718 Eagle Nest 202 ...D2
88116 Elida 202 ....F5
87529 El Prado 200............D2
87530 El Rito 475 ....C2
87531 Embudo 400 ....C2
88321 Encino 155 ....D4
87532 Espanola 6,803 ....C2
87016 Estancia⊙ 830 ....D4
88231 Eunice 2,970.............C6
88033 Fairacres 700 ....C6
87401 Farmington 31,222 .....A2
†88041 Fierro 200 ....A6
87415 Flora Vista 500 ....A2
88118 Floyd 146.............F4
88419 Folsom 73 ....F2
88036 Fort Bayard 400 ....A6
88323 Fort Stanton 80 ....D5
88119 Fort Sumner⊙ 1,421 ..E4
87316 Fort Wingate 800 ....A3
87416 Fruitland 800 ....A2
†87540 Galisteo 125 ....D3
87017 Gallina 420 ....C2
87301 Gallup⊙ 18,167 ....A3
87317 Gamerco 800 ....A3
87936 Garfield 600 ....B6
88038 Gila 350.............A6
88324 Glencoe 125 ....D5
88039 Glenwood 220 ....A5
87535 Glorieta 300 ....D3
88120 Grady 122 ....F4
87020 Grants 11,439 ....B3
88424 Grenville 39 ....F2
87722 Guadalupita 300 ....D2
88232 Hagerman 936...........E5
88041 Hanover 300 ....A6
87937 Hatch 1,028...........B6
87537 Hernandez 500 ....C2
88325 High Rolls-Mountain
         Park 555 ....D5
88042 Hillsboro 175 ....B6
88240 Hobbs 29,153 ....F6
87723 Holman 400.............D2
88336 Hondo 425 ....D5
88250 Hope 111 ....E6
87901 Hot Springs (Truth or
         Consequences)⊙ 5,219.B5
88121 House 117 ....F4
88043 Hurley 1,616..........A6
87022 Isleta 1,246 ....C4
88252 Jal 2,675 ....F6
87023 Jarales 700 ....C4
87024 Jemez Pueblo 1,503 ....C3
87025 Jemez Springs 316 ......C3
87417 Kirtland 2,358 ....A2
87026 Laguna 900 ....B3
87027 La Jara 210 ....B2
88653 Lake Arthur 327 ....E5
88337 La Luz 1,194 ....C6
87539 La Madera 200 ....C2
88044 La Mesa 900 ....C6
87418 La Plata 150 ....A2
88001 Las Cruces⊙ 45,086 ....C6
         Las Cruces‡ 96,340 .....C6
87701 Las Vegas⊙ 14,322 ....D3
87725 Ledoux 300 ....D3
87823 Lemitar 800 ....B4
88338 Lincoln 100 ....D5
87543 Llano 325 ....D2
88255 Loco Hills 375 ....F6
88426 Logan 735.............F3
88045 Lordsburg⊙ 3,195 ......A6
87544 Los Alamos⊙ 11,039.......C3

87031 Los Lunas⊙ 3,525...........C4
†87101 Los Ranchos De
         Albuquerque 2,702......C3
88256 Loving 1,355 ....E6
88260 Lovington⊙ 9,727 ....F6
87547 Lumberton 175 ....C2
87824 Luna 200 ....A5
87825 Magdalena 1,022 ....B4
88263 Malaga 300 ....E6
88339 Mayhill 300 ....D6
†79901 Meadow Vista 3,377 ....C7
88124 Melrose 649 ....F4
87319 Mentmore 315............A3
88340 Mescalero 1,259 ....D5
88046 Mesilla 2,029 ....C6
88047 Mesilla Park 500 ....C6
88048 Mesquite 500 ....C6
87320 Mexican Springs 150 .....A3
87729 Miami 112.............E2
87021 Milan 3,747 ....B3
88049 Mimbres 300 ....B6
87731 Montezuma 250 ....D3
87939 Monticello 125 ....B5
88265 Monument 300 ....F6
87732 Mora⊙ 250............D3
87035 Moriarty 1,276 ....D4
87733 Mosquero⊙ 197 ....F3
87036 Mountainair 1,170 ....C4
†87501 Nambe 1,017...........D3
88430 Nara Visa 250 ....F3
87325 Newcomb 500 ....A2
87038 New Laguna 250 ....B4
88266 Oil Center 236 ....F6
87549 Ojo Caliente 600 ....D2
87735 Ojo Feliz 133 ....D2
87550 Ojo Sarco 380 ....D2
88052 Organ 300 ....C6
87040 Paguate 500 ....B3
87552 Pecos 885 ....D3
87041 Pena Blanca 700 ....C3
87553 Penasco 860 ....D2
87042 Peralta 400 ....C4
88343 Picacho 100 ....D5
88053 Pinos Altos 250 ....A6
87044 Ponderosa 300 ....C3
88130 Portales⊙ 9,940 ....E4
87045 Prewitt 300.............B3
88432 Puerto de Luna 175 .....D4
87829 Quemado 450 ....A4
87556 Questa 1,202 ....D2
88054 Radium Springs 150 .....B6
87736 Rainsville 350 ....D2
87321 Ramah 514............A3
87557 Ranches of Taos 1,411 ...D2
87740 Raton⊙ 8,225 ....E2
87558 Red River 332 ....D2
87322 Rehoboth 300 ....A3
87830 Reserve⊙ 439 ....A5
87560 Ribera 84 ....D3
87940 Rincon 300 ....C6
87124 Rio Rancho 9,985 ....C3
87561 Rodarte 650.............D2
88201 Roswell⊙ 39,676.........E5
87562 Rowe 290 ....D3
87743 Roy 381 ....E3
88345 Ruidoso 4,260 ....D5
88346 Ruidoso Downs 949 .....D5
87941 Salem 400 ....B6
87831 San Acacia 286.........B4
87832 San Antonio 359 ....B5
87564 San Cristobal 350 ....D2
87047 Sandia Park 450 ....C3

†87001 San Felipe Pueblo 1,465....C3
†87501 San Ildefonso 232 ....C3
88434 San Jon 341 ....F3
87565 San Jose 150 ....D3
87566 San Juan Pueblo 870 ....C2
88041 San Lorenzo 200 ....B6
88750 San Mateo 200 ....B3
88058 San Miguel 400 ....C6
88348 San Patricio 300 ....D5
87051 San Rafael 300 ....A3
87567 Santa Cruz 754 ....D2
87501 Santa Fe (cap.)⊙ 48,953 ..C3
†88041 Santa Rita 600...........B6
88435 Santa Rosa⊙ 2,469 .....E4
87052 Santo Domingo
         Pueblo 2,082...........C3
87053 San Ysidro 199...........C3
87745 Sapello 600 ....D3
87055 Seboyeta 125 ....B3
87568 Sena 150 ....D3
87569 Serafina 225 ....D3
87420 Shiprock 7,237..........A2
88061 Silver City⊙ 9,887 ......A6
87801 Socorro⊙ 7,173 ....C4
†87565 Soham 104.............D3
87747 Springer 1,657 ....E2
87057 Tajique 145 ....C4
87571 Taos⊙ 3,369 ....D2
†87571 Taos Pueblo 900 ....D2
88267 Tatum 896 ....F5
87574 Tesuque 1,014 ....C3
88135 Texico 958 ....F4
87323 Thoreau 1,099...........A3
87575 Tierra Amarilla⊙ 850 ....C2
87059 Tijeras 311 ....C3
87324 Toadlena 200 ....A2
87325 Tohatchi 1,011 ....A3
87060 Tome 500 ....C4
87577 Tres Piedras 200 ....D2
87578 Truchas 275 ....D2
†87701 Trujillo 148 ....E3
87901 Truth or
         Consequences⊙ 5,219..B5
88401 Tucumcari⊙ 6,765 ......F3
88352 Tularosa 2,536 ....C5
88003 University Park 4,353 ...C6
87579 Vadito 400 ....D2
88072 Vado 325 ....C6
87580 Valdez 300 ....D2
†87031 Valencia 500 ....C4
87581 Vallecitos 450...........C2
88073 Vanadium 150 ....A6
88353 Vaughn 737............D4
87582 Velarde 950 ....C2
87583 Villanueva 500 ....D3
†88055 Virden 246 ....A6
88752 Wagon Mound 416.......E2
87421 Waterflow 475 ....A2
87753 Watrous 175 ....D3
87544 White Rock 6,560 ....C3
88002 White Sands Missile
         Range 3,120...........C6
87063 Willard 166.............D4
87942 Williamsburg 433 ....B5
88136 Yeso 200 ....E4
87064 Youngsville 125 ....C2
†87053 Zia Pueblo 500 ....C3
87327 Zuni 5,551 ....A3

OTHER FEATURES

Abiquiu (res.) C2
Alamosa (riv.) B5
Animas (riv.) B1

Avalon (res.) E6
Aztec Ruins Nat'l Mon. A2
Baldy (peak) D3
Bandelier Nat'l Mon. C3
Big Burro (mts.) A6
Black (mt.) A6
Black (range) B5
Blanco (creek) F4
Bluewater (creek) B4
Bluewater (creek) D6
Bluewater (lake) A3
Boulder (lake) C2
Brazos (peak) C2
Burford (lake) C2
Caballo (res.) B6
Canadian (riv.) F3
Cannon A.F.B. 3,798 ....F4
Canyon Blanco (creek) B2
Capitan (mts.) D5
Capitan (peak) D5
Capulin Mountain Nat'l Mon. ..E2
Carlsbad Caverns Nat'l Park ..E6
Carrizo (creek) F2
Chaco (mesa) B3
Chaco (riv.) A2
Chaco Culture Nat'l Hist. Park .B2
Chico Arroyo (creek) B3
Chivato (mesa) B3
Chupadera (mesa) C5
Chuska (mts.) A2
Cimarron (riv.) E2
Colorado, Arroyo (riv.) B4
Compañero, Arroyo (creek) ...B4
Conchas (lake) E3
Conchas (riv.) E3
Cookes (range) B6
Corrumpa (creek) F2
Costilla (peak) D2
Cuchillo Negro (creek) B5
Cuervo (creek) E3
Dark Canyon (creek) E6
Datil (mts.) B4
Dry Cimarron (riv.) F2
Eagle Nest (lake) D2
Elephant Butte (res.) B5
El Morro Nat'l Mon. A3
El Rito (riv.) C2
Fifteenmile Arroyo (creek) D4
Florida (mts.) B7
Fort Bliss Mil. Res. C6
Fort Union Nat'l Mon. E3
Gallinas (mts.) B4
Gallinas (riv.) E3
Gila (riv.) A6

Gila Cliff Dwellings Nat'l Mon. A5
Grouse (mt.) A5
Guadalupe (mts.) D6
Hatchet (mts.) A7
Holloman A.F.B. 7,245 ....C6
Hueco (mts.) D6
Jemez (riv.) C3
Jemez Canyon (res.) C3
Jicarilla Ind. Res. B2
Jornada del Muerto (valley) ...C5
Kirtland A.F.B. C3
Ladron (mts.) B4
La Plata (riv.) A1
Largo, Cañon (creek) B2
Las Animas (creek) B5
Llano Estacado (Staked) (plain) .F5
Lucero (lake) C6
Macho, Arroyo del (creek) D5
Magdalena (mts.) B4
Manzano (mts.) C4
Manzano (peak) C4
McMillan (lake) E6
Mescalero (ridge) F5
Mescalero (valley) F5
Mescalero Apache Ind. Res. ...B6
Mimbres (mts.) B6
Mimbres (riv.) B6
Mogollon (mts.) A5
Mogollon Baldy (peak) A5
Montosa (mesa) B3
Mora (mts.) E3
Nacimiento (mts.) C2
Nacimiento (peak) C2
Navajo (riv.) B2
Navajo Ind. Res. A2
North Truchas (peak) D2
Ocate (creek) E2
O'Keeffe Nat'l Hist. Site C3
Oscura (mts.) C5
Osha (peak) C4
Padilla (creek) D5
Pajarito (creek) F3
Pecos (riv.) E5
Pecos Nat'l Mon. D3
Peloncillo (mts.) A6
Perro (mts.) D4
Pinos, Rio de los (riv.) B2
Pintada Arroyo (creek) E4
Playas (lake) A7
Potrillo (mts.) B7
Pueblo Ind. Res. B4
Pueblo Ind. Res. D3
Pueblo Ind. Res. C4
Pueblo Ind. Res. D2

Puerco (riv.) A3
Red Bluff (lake) E7
Revuelto (creek) F3
Rio Brazos (creek) C2
Rio Chama (riv.) C2
Rio Felix (riv.) E5
Rio Grande (riv.) C5
Rio Hondo (riv.) E5
Rio Penasco (riv.) E6
Rio Puerco (riv.) C4
Rio Salado (riv.) B4
Rocky (mts.) C1
Sacramento (mts.) D6
Salinas Nat'l Mon. C4
Salt (creek) E5
Salt (lake) F4
San Agustin (plains) B5
San Andres (mts.) C6
San Antonio (peak) C2
Sandia (peak) C3
San Francisco (riv.) A5
Sangre de Cristo (mts.) D3
San Jose (riv.) B3
San Juan (riv.) B2
San Mateo (mts.) B5
Seven Rivers (riv.) E6
Ship Rock (peak) A2
Sierra Blanca (peak) C5
Staked (Llano Estacado) (plain) .F5
Sumner (lake) E4
Taylor (mt.) B3
Tecolote (creek) D3
Tequesquite (creek) E2
Thompson (peak) D3
Tierra Blanca (creek) B6
Tramperos (creek) F2
Tularosa (valley) C6
Ute (creek) F3
Ute (peak) D2
Ute (res.) F3
Ute Mountain Ind. Res. A1
Vermejo (riv.) E2
Wheeler (peak) D2
White Sands (des.) C5
White Sands Missile Range C5
White Sands Nat'l Mon. C6
Whitewater Baldy (mt.) A5
Wingate Army Depot A3
Yeso (creek) E4
Zuni (mts.) A3
Zuni (riv.) A3
Zuni Ind. Res. A3

⊙County seat.
‡Population of metropolitan area.
† Zip of nearest p.o.   * Multiple zips.

STATE FACTS

AREA 121,593 sq. mi. (314,926 sq. km.)
POPULATION 1,302,981
CAPITAL Santa Fe
LARGEST CITY Albuquerque
HIGHEST POINT Wheeler Pk. 13,161 ft.
         (4011 m.)
SETTLED IN 1605
ADMITTED TO UNION January 6, 1912
POPULAR NAME Land of Enchantment
STATE FLOWER Yucca
STATE BIRD Road Runner

Topography

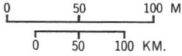

0   50   100 MI.
0   50   100 KM.

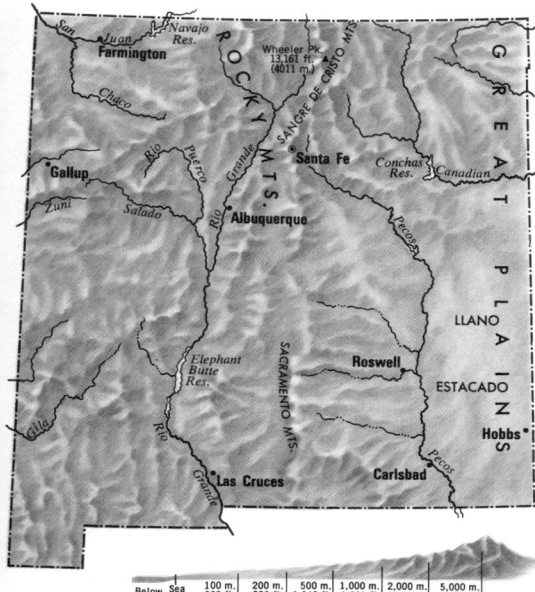

Below Sea Level | 100 m. 328 ft. | 200 m. 656 ft. | 500 m. 1,640 ft. | 1,000 m. 3,281 ft. | 2,000 m. 6,562 ft. | 5,000 m. 16,404 ft.

Agriculture, Industry and Resources

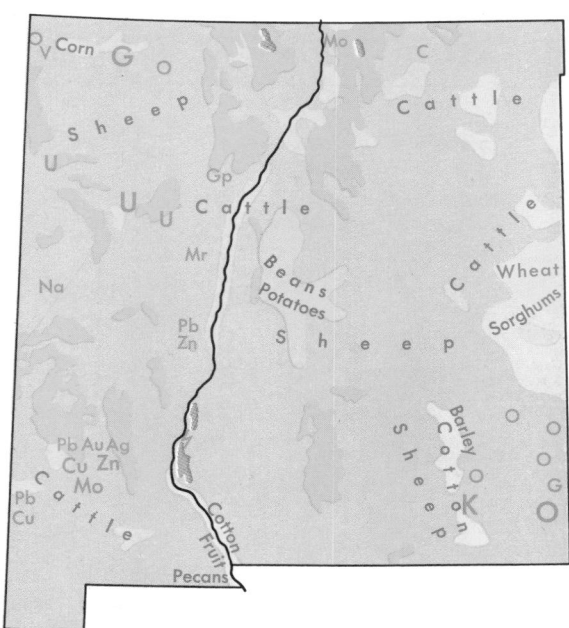

DOMINANT LAND USE

- Wheat, Grain Sorghums, Range Livestock
- General Farming, Livestock, Special Crops
- General Farming, Livestock, Cash Grain
- Dry Beans, General Farming
- Cotton, Forest Products
- Range Livestock
- Forests
- Nonagricultural Land

MAJOR MINERAL OCCURRENCES

Ag Silver
Au Gold
C Coal
Cu Copper
G Natural Gas
Gp Gypsum
K Potash
Mo Molybdenum
Mr Marble
Na Salt
O Petroleum
Pb Lead
U Uranium
V Vanadium
Zn Zinc
⚡ Water Power

New York

SCALE

0 5 10 20 30 40 MI.

0 5 10 20 30 40 KM.

State Capitals .... ⊛

County Seats .... ⊙

Canals ....

Major Limited Access Hwys. ——

Scale 1:1,920,000

COUNTIES

Albany 285,909	M5	
Allegany 51,742	D6	
Bronx 1,168,972	N9	
Broome 213,648	J6	
Cattaraugus 85,697	C6	
Cayuga 79,894	G4	
Chautauqua 146,925	B6	
Chemung 97,656	G6	
Chenango 49,344	J5	
Clinton 80,750	N1	
Columbia 59,487	N6	
Cortland 48,820	H5	
Delaware 46,824	K6	
Dutchess 245,055	N7	
Erie 1,015,472	C5	
Essex 36,176	N2	
Franklin 44,929	M1	
Fulton 55,153	M4	
Genesee 59,400	D4	
Greene 40,861	M6	
Hamilton 5,034	L3	
Herkimer 66,714	K4	
Jefferson 88,151	J2	
Kings 2,230,936	N9	
Lewis 25,035	K3	
Livingston 57,006	E5	
Madison 65,150	J5	
Monroe 702,238	E4	
Montgomery 53,439	M5	
Nassau 1,321,582	N9	
New York 1,428,285	N9	
Niagara 227,354	C4	
Oneida 253,466	J4	
Onondaga 463,920	H5	
Ontario 88,909	F5	

Orange 259,603	M8	
Orleans 38,496	D4	
Oswego 113,901	H4	
Otsego 59,075	K5	
Putnam 77,193	N8	
Queens 1,891,325	N9	
Rensselaer 151,966	O5	
Richmond 352,121	M9	
Rockland 259,530	M8	
Saint Lawrence 114,254	K2	
Saratoga 153,759	N4	
Schenectady 149,946	M5	
Schoharie 29,710	M5	
Schuyler 17,686	G6	
Seneca 33,733	G5	
Steuben 99,217	F6	
Suffolk 1,284,231	L7	
Sullivan 65,155	L7	
Tioga 49,812	H6	
Tompkins 87,085	H6	
Ulster 158,158	M7	
Warren 54,854	N3	
Washington 54,795	O4	
Wayne 84,581	F4	
Westchester 866,599	N8	
Wyoming 39,895	D5	
Yates 21,459	F5	

CITIES and TOWNS

Zip	Name/Pop.	Key
13605	Adams 1,701	J3
14801	Addison 2,028	F6
14001	Akron 2,971	C5
*12201	Albany (cap.) ⊙ 101,727	N5
	Albany-Schenectady-Troy‡	
	795,019	N5
14411	Albion ⊙ 4,897	D4

14004	Alden 2,488	C5
13607	Alexandria Bay 1,265	J2
14802	Alfred 4,967	E6
14706	Allegany 2,078	C6
12009	Altamont 1,292	M5
11930	Amagansett 2,188	R9
11701	Amityville 9,076	O9
12010	Amsterdam 21,872	M5
14006	Angola 2,292	C5
14009	Arcade 2,052	D5
10502	Ardsley 4,183	O6
12603	Arlington 11,305	N7
12015	Athens 1,738	N6
11509	Atlantic Beach 1,775	P7
14011	Attica 2,659	D5
13021	Auburn ⊙ 32,548	G5
13026	Aurora 926	G5
12018	Averill Park 1,337	O5
14414	Avon 3,006	E5
*11702	Babylon 12,388	O9
13733	Bainbridge 1,603	J6
11510	Baldwin 31,630	R7
13027	Baldwinsville 6,446	H4
12020	Ballston Spa ⊙ 4,711	N5
†12550	Balmville 2,919	M7
14020	Batavia ⊙ 16,703	D5
14810	Bath ⊙ 6,042	F6
11705	Bayport 9,282	O9
11706	Bay Shore 10,784	O9
12508	Beacon 12,937	N7
11710	Bellmore 18,106	R7
11713	Bellport 2,809	P9
14813	Belmont ⊙ 1,042	E6
11714	Bethpage 16,840	R7
14814	Big Flats 2,892	G6
*13901	Binghamton ⊙ 55,860	J6
	Binghamton‡ 301,336	J6

13612	Black River 1,384	J3
14219	Blasdell 3,288	C5
14715	Bolivar 1,345	D6
13309	Boonville 2,344	K4
13613	Brasher	
	Falls-Winthrop 1,454	L1
11717	Brentwood 44,321	O9
13029	Brewerton 2,472	H4
10509	Brewster 1,650	N8
11932	Bridgehampton 1,941	R9
†12524	Brinckerhoff 3,030	N7
12025	Broadalbin 1,415	M4
14420	Brockport 9,776	D4
14716	Brocton 1,416	B6
*10401	Bronx	
	(borough) ⊙ 1,168,972	N9
10708	Bronxville 6,267	O7
*11201	Brooklyn	
	(borough) ⊙ 2,230,936	N9
†11545	Brookville 3,290	R6
10511	Buchanan 2,041	N8
*14201	Buffalo ⊙ 357,870	B5
	Buffalo‡ 1,242,573	B5
12413	Cairo 1,281	M6
14423	Caledonia 2,188	E5
12816	Cambridge 1,820	O4
13316	Camden 2,667	J4
13031	Camillus 1,298	H4
13317	Canajoharie 2,412	L5
14424	Canandaigua ⊙ 10,419	F5
13032	Canastota 4,773	J4
14823	Canisteo 2,679	E6
13617	Canton ⊙ 7,055	K1
10512	Carmel ⊙ 27,948	N8
13619	Carthage 3,643	J3
12033	Castleton-on-Hudson 1,627	N5
12414	Catskill ⊙ 4,718	N6
†14850	Cayuga Heights 3,170	H6

13035	Cazenovia 2,599	J5
11516	Cedarhurst 6,162	P7
14720	Celoron 1,405	B6
11720	Centereach 30,136	O9
11934	Center Moriches 5,703	P9
11722	Central Islip 19,734	O9
13036	Central Square 1,418	H4
10917	Central Valley 1,705	M8
12919	Champlain 1,410	N1
12037	Chatham 2,001	N6
14225	Cheektowaga 92,145	C5
10918	Chester 1,910	M8
13037	Chittenango 4,290	J4
14428	Churchville 1,399	E4
14031	Clarence ⊙ 18,146	C5
13624	Clayton 1,816	H2
†12118	Clifton Park 23,989	N5
14432	Clifton Springs 2,039	F4
13323	Clinton 2,107	K4
14433	Clyde 2,491	G4
12043	Cobleskill 5,272	L5
12047	Cohoes 18,144	N5
10516	Cold Spring 2,161	N8
11724	Cold Spring Harbor 5,336	R6
*11201	Colonie 8,869	N5
13326	Cooperstown ⊙ 2,342	L5
11726	Copiague 20,132	O9
12822	Corinth 2,702	N4
14830	Corning 12,953	F6
12518	Cornwall On Hudson 3,164	M8
13045	Cortland ⊙ 20,138	H5
12051	Coxsackie 2,786	N6
10520	Croton-on-Hudson 6,889	N8
14727	Cuba 1,739	D6
11935	Cutchogue-New	
	Suffolk 2,788	P8
12929	Dannemora 3,770	N1

AREA 49,108 sq. mi. (127,190 sq. km.)
POPULATION 17,558,072
CAPITAL Albany
LARGEST CITY New York
HIGHEST POINT Mt. Marcy 5,344 ft. (1629 m.)
SETTLED IN 1614
ADMITTED TO UNION July 26, 1788
POPULAR NAME Empire State
STATE FLOWER Rose
STATE BIRD Bluebird

Topography

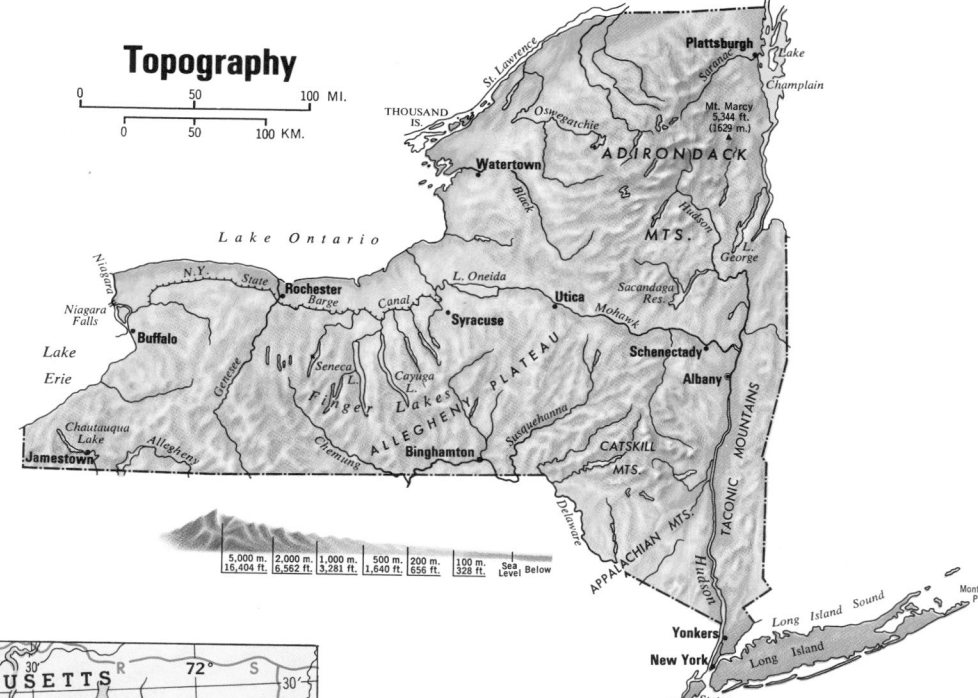

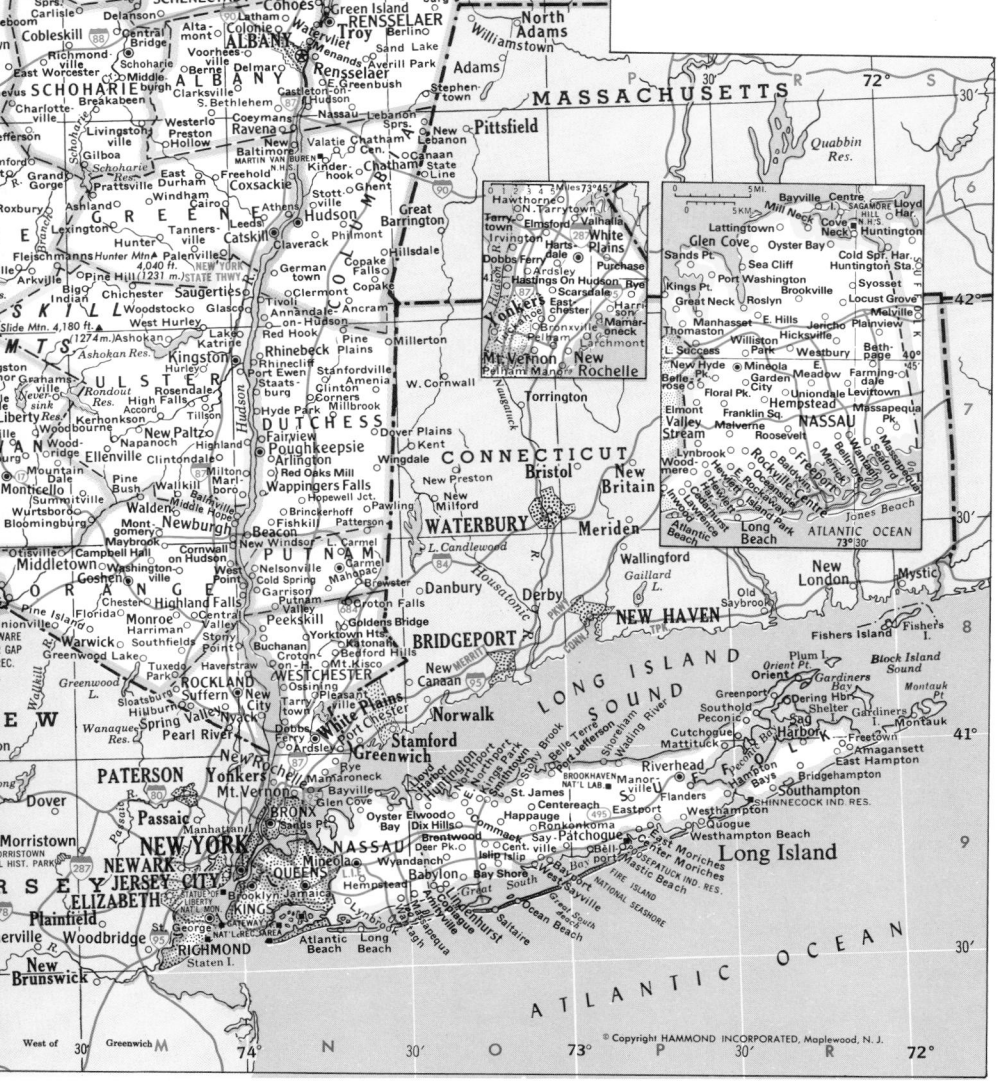

14437 Dansville 4,979	E5	
11729 Deer Park 30,394	O9	
13753 Delhi⊙ 3,374	L6	
12054 Delmar 8,423	N5	
14043 Depew 19,819	C5	
13754 Deposit 1,897	K6	
13214 DeWitt 9,024	H4	
11746 Dix Hills 26,693	O9	
10522 Dobbs Ferry 10,053	O6	
13329 Dolgeville 2,602	L4	
12522 Dover Plains 1,753	O7	
14837 Dundee 1,556	F5	
14048 Dunkirk 15,310	B5	
14052 East Aurora 6,803	C5	
10709 Eastchester 20,305	P6	
†11937 East Hampton 1,886	R9	
†11576 East Hills 7,160	R7	
11554 East Meadow 39,317	R7	
11731 East Northport 20,187	O9	
14445 East Rochester 7,596	F4	
11518 East Rockaway 10,917	R7	
13057 East Syracuse 3,412	H4	
14057 Eden 3,000	C5	
14058 Elba 750	D4	
12932 Elizabethtown⊙ 659	N2	
12428 Ellenville 4,405	M7	
14059 Elma 2,459	C5	
*14901 Elmira⊙ 35,327	G6	
Elmira‡ 97,656	G6	
14903 Elmira Heights 4,279	G6	
11003 Elmont 27,592	P7	
10523 Elmsford 3,361	O6	
11731 Elwood 11,847	O9	
13760 Endicott 14,457	H6	
13760 Endwell 13,745	H6	
14450 Fairport 5,970	F4	
†12601 Fairview 5,852	N7	
14733 Falconer 2,778	B6	
11735 Farmingdale 7,946	R7	
13066 Fayetteville 4,709	J4	
†12801 Fernwood 3,640	N4	
12524 Fishkill 1,555	N7	
†11901 Flanders-Riverside 5,400	P9	
*11001 Floral Park 16,805	P7	
10921 Florida 1,947	M8	
12068 Fonda⊙ 1,006	M5	
12937 Fort Covington 1,804	M1	
12828 Fort Edward 3,561	O4	
13339 Fort Plain 2,555	L5	
13340 Frankfort 2,995	K4	
11010 Franklin Square 29,051	R7	
14737 Franklinville 1,887	D6	
14063 Fredonia 11,126	B6	
11520 Freeport 38,272	R7	
14738 Frewsburg 1,908	B6	
14739 Friendship 1,461	D6	
13069 Fulton 13,312	H4	
11530 Garden City 22,927	R7	
14067 Gasport 1,339	C4	
14454 Geneseo⊙ 6,746	E5	
14456 Geneva 15,133	G5	
11542 Glen Cove 24,618	R6	
12801 Glens Falls 15,897	N4	
Glens Falls‡ 109,649	N4	
12078 Gloversville 17,836	M4	
10526 Golden's Bridge 1,367	N8	
10924 Goshen⊙ 4,874	M8	
13642 Gouverneur 4,285	K2	
14070 Gowanda 2,713	B6	
12832 Granville 2,696	O4	
*11020 Great Neck 9,168	P6	
14616 Greece 16,177	E4	
13778 Greene 1,747	J6	
12183 Green Island 2,696	N5	
11944 Greenport 2,273	P8	
12834 Greenwich 1,955	O4	
10925 Greenwood Lake 2,809	M8	
13073 Groton 2,313	H5	
12835 Hadley-Lake Luzerne 1,988	N4	
12086 Hagaman 1,331	M5	
14075 Hamburg 10,582	C5	
13346 Hamilton 3,725	J5	
11946 Hampton Bays 7,256	R9	
13783 Hancock 1,526	K7	
10528 Harrison 23,046	P6	
10530 Hartsdale 10,216	P6	
10706 Hastings On Hudson 8,573	O6	
11787 Hauppauge 20,960	O9	
10927 Haverstraw 8,800	M8	
10532 Hawthorne 5,010	O6	
*11550 Hempstead 40,404	R7	
13350 Herkimer⊙ 8,383	L4	
11557 Hewlett 6,986	P7	
†11557 Hewlett Harbor 1,331	P7	
*11801 Hicksville 43,245	R7	
12528 Highland 3,967	M7	
10928 Highland Falls 4,187	M8	
10931 Hillburn 926	M8	
†10977 Hillcrest 5,733	K8	
14468 Hilton 4,151	E4	
14080 Holland 1,347	C5	
14470 Holley 1,882	D4	
13077 Homer 3,635	H5	
14472 Honeoye Falls 2,410	F5	
12090 Hoosick Falls 3,609	O5	
12533 Hopewell Junction 1,754	N7	
14843 Hornell 10,234	E6	
14845 Horseheads 7,348	G6	
14744 Houghton 1,604	D6	
12534 Hudson⊙ 7,986	N6	
12839 Hudson Falls⊙ 7,419	O4	
11743 Huntington 21,727	R6	
11746 Huntington Station 28,769	R6	
12443 Hurley 4,892	M7	
12538 Hyde Park 2,550	N6	
13357 Ilion 9,450	K5	
11696 Inwood 8,228	P7	
14617 Irondequoit 57,648	E4	
10533 Irvington 5,774	O6	
11558 Island Park 4,847	R7	

(continued on following page)

(continued)

11751 Islip 13,438 .... O9
14850 Ithaca⊙ 28,732 .... G6
*11401 Jamaica⊙ N9
14701 Jamestown 35,775 .... B6
11753 Jericho 12,739 .... R6
13790 Johnson City 17,126 .... J6
12095 Johnstown⊙ 9,360 .... M4
13080 Jordan 1,371 .... H4
12944 Keeseville 2,025 .... O2
14271 Kenmore 18,474 .... C5
12446 Kerhonkson 1,646 .... M7
12106 Kinderhook 1,377 .... N6
11754 Kings Park 16,131 .... O9
11024 Kings Point 5,234 .... P6
12401 Kingston⊙ 24,481 .... M7
14218 Lackawanna 22,701 .... B5
10512 Lake Carmel 7,295 .... N8
†14006 Lake Erie Beach 4,625 .... B5
12845 Lake George⊙ 1,047 .... N4
12449 Lake Katrine 2,011 .... M7
12846 Lake Luzerne-Hadley 1,988 N4
12946 Lake Placid 2,490 .... N2
12108 Lake Pleasant⊙ 700 .... M4
11040 Lake Success 2,396 .... P7
14750 Lakewood 3,941 .... B6
14086 Lancaster 13,056 .... C5
14882 Lansing 3,039 .... H5
10538 Larchmont 6,308 .... P7
12110 Latham 11,182 .... N5
†11560 Lattingtown 1,749 .... R6
11559 Lawrence 6,175 .... P7
14482 Le Roy 4,900 .... E5
11756 Levittown 57,045 .... R7
14092 Lewiston 3,326 .... B4
12754 Liberty 4,293 .... L7
14485 Lima 2,025 .... E5
11757 Lindenhurst 26,919 .... O9
13365 Little Falls 6,156 .... L4
14755 Little Valley⊙ 1,203 .... C6
13088 Liverpool 2,849 .... H4
12758 Livingston Manor 1,436 .... L7
†11743 Lloyd Harbor 3,405 .... R6
14094 Lockport⊙ 24,844 .... C4
†11791 Locust Grove 9,670 .... R6
11561 Long Beach 34,073 .... R7
13367 Lowville⊙ 3,364 .... J3
11563 Lynbrook 20,424 .... P7
14489 Lyons⊙ 4,160 .... F4
14502 Macedon 1,400 .... F4
10541 Mahopac 7,681 .... N8
12953 Malone⊙ 7,668 .... M1
11565 Malverne 9,262 .... R7
10543 Mamaroneck 17,616 .... P7
14504 Manchester 1,698 .... F5
11030 Manhasset 8,485 .... P7
*10001 Manhattan
(borough) 1,428,285 ... M9
13104 Manlius 5,241 .... J5
13108 Marcellus 1,870 .... H5
12542 Marlboro 2,275 .... M7
11758 Massapequa 24,454 .... R7
11762 Massapequa Park 19,779 .. R7
13662 Massena 12,851 .... L1
11950 Mastic Beach 8,318 .... P9
11952 Mattituck 3,923 .... P9
12543 Maybrook 2,007 .... M8
14757 Mayville⊙ 1,626 .... A6
12118 Mechanicville 5,500 .... N5
14103 Medina 6,392 .... D4
†13021 Melrose Park 2,171 .... G5
11746 Melville 8,139 .... O9
†12201 Menands 4,012 .... N5
11566 Merrick 24,478 .... R7
13114 Mexico 1,621 .... H4
12122 Middleburgh 1,358 .... M5
12550 Middle Hope 3,229 .... M7
14105 Middleport 1,995 .... C4
10940 Middletown 21,454 .... L8
†12020 Milton 2,063 .... N4
11501 Mineola⊙ 20,757 .... R7
13115 Minetto 1,629 .... H4
12956 Mineville-Witherbee 1,925 .. O2
13116 Minoa 3,640 .... H4
13407 Mohawk 2,956 .... L4
10950 Monroe 5,996 .... M8
10952 Monsey 12,380 .... J8
12549 Montgomery 2,316 .... M7
12701 Monticello⊙ 6,306 .... L7
14865 Montour Falls 1,791 .... G6
13118 Moravia 1,582 .... H5
12962 Morrisonville 1,721 .... N1
13408 Morrisville 2,707 .... J5
10549 Mount Kisco 8,025 .... N8
14510 Mount Morris 3,039 .... E5
*10550 Mount Vernon 66,713 .... O7
10954 Nanuet 12,578 .... K8
12123 Nassau 1,285 .... N5
Nassau-Suffolk‡ 2,605,813 R7
14513 Newark 10,017 .... G4
13411 New Berlin 1,392 .... K5
12550 Newburgh 23,438 .... M7
Newburgh-Middletown‡
259,603 .... M7
10956 New City⊙ 35,859 .... K8
14108 Newfane 3,120 .... C4
13413 New Hartford 2,313 .... K4
11040 New Hyde Park 9,801 .... P7
12561 New Paltz 4,938 .... M7
*10801 New Rochelle 70,794 .... P7
†10901 New Square 1,750 .... K8
12550 New Windsor 7,812 .... N8
*10001 New York⊙ 7,071,639 .... M9
New York‡ 9,119,737 .... M9
13417 New York Mills 3,549 .... K4
*14301 Niagara Falls 71,384 .... C4
†12301 Niskayuna 5,223 .... N5
13667 Norfolk 1,599 .... K1
14110 North Boston 2,743 .... C5
14111 North Collins 1,496 .... C5
11768 Northport 7,651 .... O9
13212 North Syracuse 7,970 .... H4
10591 North Tarrytown 7,994 .... O6
14120 North Tonawanda 35,760 .. C4
12134 Northville 1,304 .... M4
13815 Norwich⊙ 8,082 .... J5
13668 Norwood 1,902 .... L1
10960 Nyack 6,428 .... K8
14125 Oakfield 1,791 .... D4
11572 Oceanside 33,639 .... R7
13669 Ogdensburg 12,375 .... K1
14126 Olcott 1,571 .... C4
14760 Olean 18,207 .... D6
13421 Oneida 10,810 .... J4
13820 Oneonta 14,933 .... K6
14127 Orchard Park 3,671 .... C5
13424 Oriskany 1,680 .... K4
10562 Ossining 20,196 .... N8
13126 Oswego⊙ 19,793 .... G4
14521 Ovid⊙ 666 .... G5
13827 Owego⊙ 4,364 .... H6
13830 Oxford 1,765 .... J6
11771 Oyster Bay 6,497 .... R6
14870 Painted Post 2,196 .... F6
14522 Palmyra 3,729 .... F4
11772 Patchogue 11,291 .... P9
12564 Pawling 1,996 .... N7
10965 Pearl River 15,893 .... K8
10566 Peekskill 18,236 .... N8
10803 Pelham 6,848 .... O7
†10803 Pelham Manor 6,130 .... O7
14527 Penn Yan⊙ 5,242 .... F5
14530 Perry 4,198 .... D5
12972 Peru 1,716 .... N1
14532 Phelps 2,004 .... F5
12565 Philmont 1,539 .... N6
13135 Phoenix 2,357 .... H4
10968 Piermont 2,269 .... K8
12567 Pine Plains 1,303 .... N7
14534 Pittsford 1,568 .... E4
11803 Plainview 28,037 .... R7
12901 Plattsburgh⊙ 21,057 .... O1
10570 Pleasantville 6,749 .... N8
13140 Port Byron 1,400 .... G4
10573 Port Chester 23,565 .... P7
†13901 Port Dickinson 1,974 .... J6
12466 Port Ewen 2,813 .... N7
12974 Port Henry 1,450 .... O2
11777 Port Jefferson 6,731 .... P9
12771 Port Jervis 8,699 .... L8
11050 Port Washington 14,521 .. R6
13676 Potsdam 10,635 .... K1
*12601 Poughkeepsie⊙ 29,757 .... N7
Poughkeepsie‡ 245,055 .... N7
14873 Prattsburg○ 1,657 .... F5
13142 Pulaski 2,415 .... H3
10579 Putnam Valley○ 8,994 .... N8
*11101 Queens (borough)
1,891,325 .... N9
14772 Randolph 1,398 .... C6
14131 Ransomville 1,401 .... C4
12143 Ravena 3,091 .... N6
12571 Red Hook 1,692 .... N7
†12601 Red Oaks Mill 5,236 .... N7
12144 Rensselaer 9,047 .... N5
12572 Rhinebeck 2,542 .... N7
13439 Richfield Springs 1,561 .... K5
*10301 Richmond (Staten Island)
(borough) 352,121 .... M9
11901 Riverhead⊙ 6,339 .... P9
*14601 Rochester⊙ 241,741 .... E4
Rochester‡ 971,879 .... E4
*11570 Rockville Centre 25,412 .... R7
13440 Rome 43,826 .... J4
11575 Roosevelt 14,109 .... R7
11576 Roslyn 2,134 .... R6
12979 Rouses Point 2,266 .... O1
10580 Rye 15,083 .... P6
11963 Sag Harbor 2,581 .... R8
11780 Saint James 12,122 .... O9
13452 Saint Johnsville 1,974 .... L5
14779 Salamanca 6,890 .... C6
†13132 Sand Ridge 1,293 .... H4
†11050 Sands Point 2,742 .... P6
12983 Saranac Lake 5,578 .... M2
12866 Saratoga Springs 23,906 .... N4
12477 Saugerties 3,882 .... M6
13146 Savannah○ 1,905 .... G4
11782 Sayville 12,013 .... O9
10583 Scarsdale 17,650 .... P6
*12301 Schenectady⊙ 67,972 .... M5
12157 Schoharie⊙ 1,016 .... M5
12871 Schuylerville 1,256 .... N4
12302 Scotia 7,280 .... N5
14546 Scottsville 1,789 .... E4
11579 Sea Cliff 5,364 .... R6
11783 Seaford 16,117 .... R7
13148 Seneca Falls 7,466 .... G5
13460 Sherburne 1,561 .... K5
13461 Sherrill 2,830 .... J4
14548 Shortsville 1,669 .... F5
13838 Sidney 4,861 .... K6
14136 Silver Creek 3,088 .... B5
13152 Skaneateles 2,789 .... H5
†14201 Sloan 4,529 .... C5
10974 Sloatsburg 3,154 .... M8
11787 Smithtown 30,906 .... O9
14551 Sodus 1,790 .... G4
14555 Sodus Point 1,334 .... G4
13209 Solvay 7,140 .... H4
11968 Southampton 4,000 .... R9
12779 South Fallsburg 2,196 .... L7
†12801 South Glens Falls 3,714 .... N4
†10960 South Nyack 3,602 .... K8
11971 Southold 4,770 .... P8
†14901 Southport 8,329 .... G6
14559 Spencerport 3,424 .... E4
10977 Spring Valley 20,537 .... K8
14141 Springville 4,285 .... C5
*10301 Staten Island
(borough) 352,121 .... M9
12170 Stillwater 1,572 .... N5
11790 Stony Brook 16,155 .... O9
10980 Stony Point 8,686 .... M8
12172 Stottville 1,387 .... N6
10901 Suffern 10,794 .... J8
11791 Syosset 9,818 .... R6
*13201 Syracuse⊙ 170,105 .... H4
Syracuse‡ 642,375 .... H4
10983 Tappan 8,267 .... K8
10591 Tarrytown 10,648 .... O6
†11020 Thomaston 2,684 .... P7
12883 Ticonderoga 2,938 .... N3
12486 Tillson 1,529 .... M7
14150 Tonawanda 18,693 .... B4
*12180 Troy⊙ 56,638 .... N5
14886 Trumansburg 1,722 .... G5
10707 Tuckahoe 6,076 .... O7
12986 Tupper Lake 4,478 .... M2
13849 Unadilla 1,367 .... K6
11553 Uniondale 20,016 .... R7
*13501 Utica⊙ 75,632 .... K4
Utica-Rome‡ 320,180 .... K4
12184 Valatie 1,492 .... N6
10989 Valley Cottage 8,214 .... K8
*11580 Valley Stream 35,769 .... P7
13850 Vestal○ 27,238 .... H6
14564 Victor 2,370 .... F5
†10901 Viola 5,340 .... J8
12186 Voorheesville 3,320 .... M5
12586 Walden 5,659 .... M7
12589 Wallkill 2,064 .... M7
13856 Walton 3,329 .... K6
13163 Wampsville⊙ 569 .... J4
11793 Wantagh 19,817 .... R7
12590 Wappingers Falls 5,110 .... N7
12885 Warrensburg 2,834 .... N3
14569 Warsaw⊙ 3,619 .... D5
10990 Warwick 4,320 .... M8
10992 Washingtonville 2,380 .... M8
12188 Waterford 2,405 .... N5
13165 Waterloo⊙ 5,303 .... G5
13601 Watertown⊙ 27,861 .... J3
13480 Waterville 1,672 .... K5
12189 Watervliet 11,354 .... N5
14891 Watkins Glen⊙ 2,440 .... G6
14892 Waverly 4,738 .... G7
14572 Wayland 1,846 .... E5
14580 Webster 5,499 .... F4
13166 Weedsport 1,952 .... G4
14895 Wellsville 5,769 .... E6
11590 Westbury 13,871 .... R7
†13619 West Carthage 1,824 .... J3
†14901 West Elmira 5,485 .... G6
14787 Westfield 3,446 .... A6
†12801 West Glens Falls 5,331 .... N4
11977 Westhampton 2,774 .... P9
11978 Westhampton Beach 1,629 P9
12491 West Hurley 2,382 .... M6
10994 West Nyack 8,553 .... K8
14788 Westons Mills 1,837 .... D6
10996 West Point 8,105 .... M8
11796 West Sayville 8,185 .... O9
14224 West Seneca 51,210 .... C5
12887 Whitehall 3,241 .... O3
*10601 White Plains⊙ 46,999 .... P6
13492 Whitesboro 4,460 .... K4
14588 Willard 1,339 .... G5
14589 Williamson 1,768 .... F4
14221 Williamsville 6,017 .... C5
11596 Williston Park 8,216 .... R7
13865 Windsor 1,155 .... J6
13697 Winthrop-Brasher
Falls 1,454 .... L1
12998 Witherbee-Mineville 1,925 .. N2
14590 Wolcott 1,496 .... G4
11598 Woodmere 17,205 .... P7
12498 Woodstock 2,280 .... M6
12790 Wurtsboro 1,128 .... L7
11798 Wyandanch 13,215 .... N9
*10701 Yonkers 195,351 .... O6
10598 Yorktown Heights 7,696 .... N8
13495 Yorkville 3,115 .... K4
14174 Youngstown 2,191 .... C4

OTHER FEATURES

Adirondack (mts.) M3
Algonquin (peak) M2
Allegany Ind. Res. 1,243 .... C6
Allegheny (res.) C7
Allegheny (riv.) C6
Ashokan (res.) M7
Ausable (riv.) N2
Batten Kill (riv.) O4
Beaver (riv.) K3
Big Moose (lake) L3
Black (lake) J1
Black (riv.) K3
Block Island (sound) S8
Blue Mountain (lake) M3
Bonaparte (lake) K2
Brandreth (lake) L3
Brant (lake) N3
Brookhaven Nat'l Lab. P9
Butterfield (lake) J2
Canandaigua (lake) F5
Canisteo (riv.) F6
Cannonsville (res.) K6
Catskill (mts.) L6
Cattaraugus (creek) C6
Cattaraugus Ind. Res. 1,994 .... C5
Cayuga (lake) G5
Champlain (lake) O1
Chateaugay, Upper (lake) M1
Chautauqua (lake) A6
Chazy (lake) N1
Chenango (riv.) J6
Cohocton (riv.) F6
Conesus (lake) E5
Conewango (creek) B6
Cranberry (lake) L2
Deer (riv.) J3
Deer (riv.) L1
Delaware (riv.) K7
East (riv.) N9
Erie (lake) A5
Fire Island Nat'l Seashore P9
Fishers (isl.) S8
Forked (lake) L3
Fort Drum J2
Fort Niagara C4
Fort Stanwix Nat'l Mon. J4
Fulton Chain (lkes) K3
Galloo (isl.) H3
Gardiners (bay) R8
Gardiners (isl.) R8
Gateway Nat'l Rec. Area M9
Genesee (riv.) E5
George (lake) N4
Grand (isl.) B5
Grass (riv.) K1
Great Sacandaga (lake) M4
Great South (bay) O9
Great South (beach) O9
Greenwood (lake) M8
Grenadier (isl.) H2
Griffiss A.F.B. K4
Haystack (mt.) N2
Hemlock (lake) E5
Hinckley (res.) K4
Honeoye (lake) F5
Honnedaga (lake) L3
Hudson (riv.) N7
Hunter (mt.) M6
Indian (lake) M3
Jones (beach) R7
Keuka (lake) F5
Lila (lake) L2
Little Tupper (lake) L2
Long (isl.) P9
Long (lake) M2
Long Island (sound) P9
Manhattan (isl.) M9
Marcy (mt.) N2
Martin Van Buren Nat'l Hist.
Site N6
Meacham (lake) M1
Mohawk (riv.) L5
Montauk (pt.) S8
Moose (riv.) K3
Neversink (res.) L7
New York State Barge (canal) C4
Niagara (riv.) B4
Oil Spring Ind. Res. 6 .... D6
Oneida (lake) J4
Onondaga Ind. Res. 596 .... H5
Ontario (lake) F3
Orient (pt.) R8
Oswegatchie (riv.) K2
Oswego (riv.) H4
Otisco (lake) H5
Otsego (lake) L5
Otselic (riv.) J5
Owasco (lake) G5
Peconic (bay) R9
Peninsula (pt.) H3
Pepacton (res.) L6
Piseco (lake) M4
Placid (lake) N2
Plattsburgh A.F.B. 5,905 .... N1
Pleasant (lake) M4
Plum (isl.) R8
Poosepatuck Ind. Res. 203 .... P9
Raquette (lake) L3
Rondout (res.) M7
Round (lake) L2
Sacandaga (lake) L3
Sackets (harb.) H3
Sagamore Hill Nat'l Hist. Site R6
Saint Lawrence (lake) K1
Saint Lawrence (riv.) J2
Saint Regis (riv.) L1
Saint Regis Ind. Res. 1,802 .... M1
Salmon (res.) J3
Salmon (riv.) H3
Salmon (riv.) M1
Saranac (lkes) M2
Saranac (riv.) N1
Saratoga (lake) N4
Saratoga Nat'l Hist. Park N4
Schoharie (res.) M6
Schroon (lake) N3
Seneca (lake) G5
Seneca (riv.) G5
Shelter (isl.) R8
Shinnecock Ind. Res. 297 .... R9
Silver (lake) N1
Skaneateles (lake) H5
Skylight (mt.) M2
Slide (mt.) L6
Staten (isl.) M9
Statue of Liberty Nat'l Mon. M9
Stony (isl.) H3
Stony (pt.) H3
Susquehanna (riv.) H6
Thousand (isls.) H2
Tioughnioga (riv.) H6
Titus (lake) M1
Tomhannock (res.) O5
Tonawanda Ind. Res. 467 .... D4
Toronto (res.) L7
Tupper (lake) M2
Tuscarora Ind. Res. 921 .... B4
Unadilla (riv.) K5
Upper Chateaugay (lake) M1
Valcour (isl.) N1
Wallkill (riv.) L8
Whiteface (mt.) N2
Whitney Point (lake) J6
Woodhull (lake) L3

⊙County seat.
‡Population of metropolitan area.
○Population of town or township.
† Zip of nearest p.o. * Multiple zips

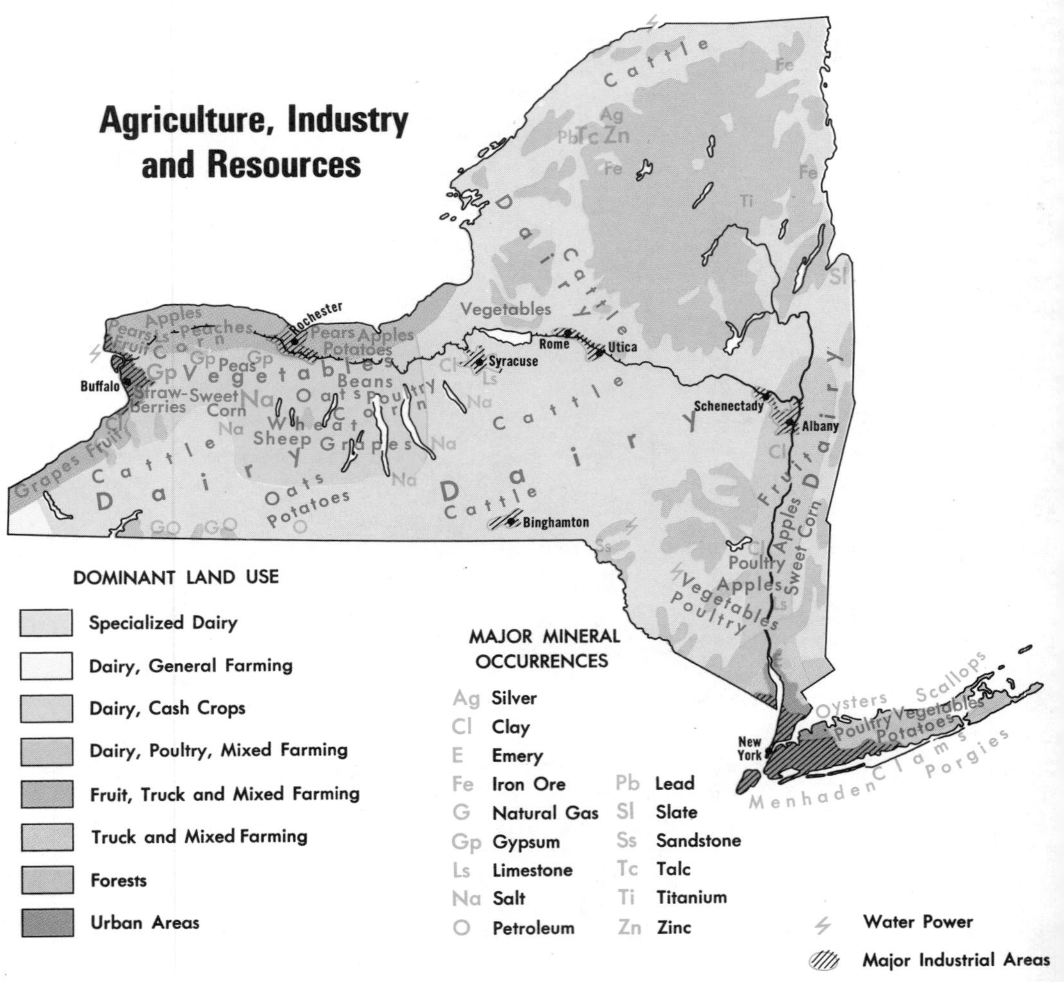
Agriculture, Industry and Resources
Cattle
Dairy
Vegetables
Apples
Pears
Peaches
Fruit
Corn
Peas
Straw-Sweet Berries Corn
Beans
Oats
Wheat
Poultry
Sheep
Grapes
Potatoes
Grapes Fruit
Sweet Corn
Vegetables
Poultry
Oysters
Scallops
Clams
Porgies
Menhaden
Rochester
Syracuse
Rome
Utica
Buffalo
Schenectady
Albany
Binghamton
New York
DOMINANT LAND USE
Specialized Dairy
Dairy, General Farming
Dairy, Cash Crops
Dairy, Poultry, Mixed Farming
Fruit, Truck and Mixed Farming
Truck and Mixed Farming
Forests
Urban Areas
MAJOR MINERAL OCCURRENCES
Ag Silver
Cl Clay
E Emery
Fe Iron Ore
G Natural Gas
Gp Gypsum
Ls Limestone
Na Salt
O Petroleum
Pb Lead
Sl Slate
Ss Sandstone
Tc Talc
Ti Titanium
Zn Zinc
Water Power
Major Industrial Areas

AREA 52,669 sq. mi. (136,413 sq. km.)
POPULATION 5,881,813
CAPITAL Raleigh
LARGEST CITY Charlotte
HIGHEST POINT Mt. Mitchell 6,684 ft. (2037 m.)
SETTLED IN 1650
ADMITTED TO UNION November 21, 1789
POPULAR NAME Tarheel State
STATE FLOWER Flowering Dogwood
STATE BIRD Cardinal

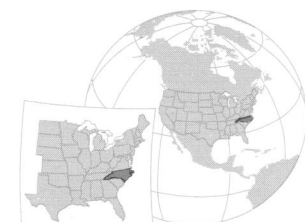

COUNTIES

Alamance 99,319	L3	
Alexander 24,999	G3	
Alleghany 9,587	G1	
Anson 25,649	J4	
Ashe 22,325	F2	
Avery 14,409	F2	
Beaufort 40,355	R4	
Bertie 21,024	P2	
Bladen 30,491	M5	
Brunswick 35,777	N6	
Buncombe 160,934	D3	
Burke 72,504	F3	
Cabarrus 85,895	H4	
Caldwell 67,746	F3	
Camden 5,829	S2	
Carteret 41,092	R5	
Caswell 20,705	L2	
Catawba 105,208	G3	
Chatham 33,415	L3	
Cherokee 18,933	A4	
Chowan 12,558	R2	
Clay 6,619	B4	
Cleveland 83,435	F4	
Columbus 51,037	M6	
Craven 71,043	P4	
Cumberland 247,160	M4	
Currituck 11,089	S2	
Dare 13,377	T3	
Davidson 113,162	J3	
Davie 24,599	H3	
Duplin 40,952	O5	
Durham 152,785	M3	
Edgecombe 55,988	O3	
Forsyth 243,683	J2	
Franklin 30,055	N2	
Gaston 162,568	G4	
Gates 8,875	R2	
Graham 7,217	B4	
Granville 34,043	M2	
Greene 16,117	O3	
Guilford 317,154	K3	
Halifax 55,286	O2	
Harnett 59,570	M4	
Haywood 46,495	C3	
Henderson 58,580	D4	
Hertford 23,368	P2	
Hoke 20,383	L4	
Hyde 5,873	S3	
Iredell 82,538	H3	
Jackson 25,811	C4	
Johnston 70,599	N4	
Jones 9,705	P4	
Lee 36,718	L4	
Lenoir 59,819	O4	
Lincoln 42,372	G3	
Macon 20,178	B4	
Madison 16,827	D3	
Martin 25,948	P3	
McDowell 35,135	E3	
Mecklenburg 404,270	H4	
Mitchell 14,428	E2	
Montgomery 22,469	K4	
Moore 50,505	L4	
Nash 67,153	O2	
New Hanover 103,471	O6	
Northampton 22,584	P2	
Onslow 112,784	P5	
Orange 77,055	L2	
Pamlico 10,398	R4	
Pasquotank 28,462	S2	
Pender 22,215	O5	
Perquimans 9,486	S2	
Person 29,164	M2	
Pitt 90,146	P3	
Polk 12,984	E4	
Randolph 91,728	K3	
Richmond 45,481	K4	
Robeson 101,610	L5	
Rockingham 83,426	K2	
Rowan 99,186	H3	
Rutherford 53,787	E4	
Sampson 49,687	N4	
Scotland 32,273	L5	
Stanly 48,517	J4	
Stokes 33,086	J2	
Surry 59,449	H2	
Swain 10,283	B3	
Transylvania 23,417	D4	
Tyrrell 3,975	S3	
Union 70,380	H4	
Vance 36,748	N2	
Wake 301,327	M3	
Warren 16,232	N2	
Washington 14,801	R3	
Watauga 31,666	F2	
Wayne 97,054	N4	
Wilkes 58,657	G2	
Wilson 63,132	O3	
Yadkin 28,439	H2	
Yancey 14,934	E3	

CITIES and TOWNS

Zip   Name/Pop.   Key

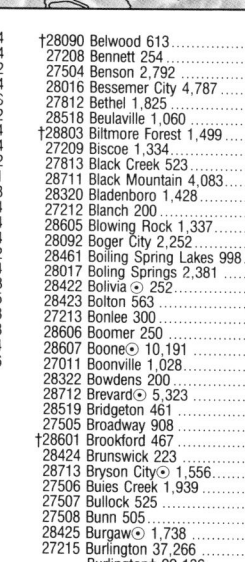

Great Smoky Mountains

28315 Aberdeen 1,945	L4	
27910 Ahoskie 4,887	P2	
27201 Alamance 320	K2	
28001 Albemarle⊙ 15,110	J4	
†28043 Alexander Mills 643	F4	
28509 Alliance 616	R4	
28702 Almond 140	B4	
28901 Andrews 1,621	B4	
28007 Ansonville 794	J4	
27501 Angier 1,709	M4	
27502 Apex 2,847	M3	
28510 Arapahoe 467	R4	
27263 Archdale 5,326	K3	
†28642 Arlington 872	H2	
28420 Ash 150	N6	
27203 Asheboro⊙ 15,252	K3	
*28801 Asheville⊙ 53,583	D3	
Asheville‡ 177,761	D3	
†27983 Askewville 227	R2	
28421 Atkinson 298	N5	
28512 Atlantic Beach 941	R5	
27805 Aulander 1,214	P2	
27806 Aurora 698	R4	
28318 Autryville 228	M4	
27915 Avon 500	U4	
28513 Ayden 4,361	P4	
27916 Aydlett 205	T2	
28009 Badin 1,514	J4	
27807 Bailey 685	N5	
28705 Bakersville⊙ 373	E2	
28706 Balfour 1,772	D4	
28707 Balsam 200	C4	
28604 Banner Elk 1,087	F2	
†27030 Bannertown 1,028	H1	
27008 Barber 155	H3	
†28739 Barker Heights 1,267	D4	
28710 Bat Cave 450	E4	
27808 Bath 207	R4	
27809 Battleboro 632	O2	
28515 Bayboro⊙ 759	R4	
†27892 Beargrass 82	P3	
28516 Beaufort⊙ 3,826	R5	
27810 Belhaven 2,430	R3	
28012 Belmont 4,607	H4	
†28451 Belville 102	N6	
†28090 Belwood 613	F4	
27208 Bennett 254	K3	
27504 Benson 2,792	N4	
28016 Bessemer City 4,787	G4	
27812 Bethel 1,825	P3	
28518 Beulaville 1,060	O5	
†28803 Biltmore Forest 1,499	E3	
27209 Biscoe 1,334	K4	
27813 Black Creek 523	O3	
28711 Black Mountain 4,083	E3	
28320 Bladenboro 1,428	M5	
27212 Blanch 200	L2	
28605 Blowing Rock 1,337	F2	
28092 Boger City 2,252	G4	
28461 Boiling Spring Lakes 998	N7	
28017 Boiling Springs 2,381	F4	
28422 Bolivia 752	N6	
28423 Bolton 563	N6	
27213 Bonlee 300	L3	
28606 Boomer 250	G2	
28607 Boone⊙ 10,191	F2	
27011 Boonville 1,028	H2	
28322 Bowdens 200	N4	
28712 Brevard⊙ 5,323	D4	
28519 Bridgeton 461	R4	
27505 Broadway 908	L4	
†28601 Brookford 467	G3	
28424 Brunswick 359	M6	
28713 Bryson City⊙ 1,556	C4	
27506 Buies Creek 1,939	M4	
27507 Bullock 525	M2	
27508 Bunn 505	N3	
28425 Burgaw⊙ 1,738	N5	
27215 Burlington 37,266	K2	
Burlington‡ 99,136	F2	
28714 Burnsville⊙ 1,452	E3	
27509 Butner 4,240	M2	
27312 Bynum 350	L3	
†29566 Calabash 128	M7	
28325 Calypso 689	N4	
27921 Camden⊙ 300	S2	
28326 Cameron 225	L4	
28229 Candor 868	K4	
28716 Canton 4,631	D3	
†28584 Cape Carteret 944	P5	
28428 Carolina Beach 2,000	O6	
27510 Carrboro 7,336	L3	
28327 Carthage⊙ 925	K4	
27511 Cary 21,763	M3	
28020 Casar 346	F3	
28717 Cashiers 553	C4	
28016 Castalia 358	O2	
28429 Castle Hayne 1,087	O6	
†28461 Caswell Beach 110	N7	
28609 Catawba 509	G3	
27230 Cedar Falls 400	K3	
27231 Cedar Grove 250	L2	
28520 Cedar Island 310	S5	
†27525 Centerville 135	N2	
28430 Cerro Gordo 295	M6	
28431 Chadbourn 2,000	M6	
†28445 Chadwick Acres 15	P6	
27514 Chapel Hill 32,421	L3	
*28201 Charlotte⊙ 314,447	H4	
Charlotte-Gastonia‡ 637,218	H4	
28021 Cherryville 4,844	G4	
28023 China Grove 2,081	H3	
28521 Chinquapin 280	O5	
27817 Chocowinity 644	P4	
28610 Claremont 880	G3	
28433 Clarkton 664	M6	
27520 Clayton 4,091	N3	
27012 Clemmons 7,401	J2	
27013 Cleveland 595	H3	
28328 Clinton⊙ 7,552	N5	
28721 Clyde 1,008	D3	
27521 Coats 1,385	M4	
27922 Cofield 465	R2	
27924 Colerain 284	R2	
27925 Columbia⊙ 758	S3	
28722 Columbus⊙ 727	E4	
28522 Comfort 325	O5	
27818 Como 89	P1	
28025 Concord⊙ 16,942	H4	
27819 Conetoe 215	O3	
28613 Conover 4,245	G3	
27820 Conway 678	P2	
27014 Cooleemee 1,448	H3	
28031 Cornelius 1,460	H4	
27927 Corolla 158	T2	
28523 Cove City 500	P4	
28032 Cramerton 1,869	G4	
27522 Creedmoor 1,641	M2	
28616 Crossnore 297	F2	
27852 Crisp 435	O3	
28331 Cumberland 400	M5	
27237 Cumnock 200	L3	
27929 Currituck⊙ 700	T2	
28034 Dallas 3,340	G4	
27016 Danbury⊙ 140	J2	
28036 Davidson 3,241	H4	
28524 Davis 612	R5	
27239 Denton 949	J3	
28725 Dillsboro 179	C4	
27017 Dobson⊙ 1,222	H2	
†27801 Dortches 885	O2	
28526 Dover 600	P4	
28619 Drexel 1,392	F3	
28332 Dublin 477	M5	
28334 Dunn 8,962	M4	
*27701 Durham⊙ 100,538	M2	
Durham-Raleigh‡ 530,673	M2	
27242 Eagle Springs 280	K4	
28038 Earl 206	F4	
†28434 East Arcadia 461	N6	
27018 East Bend 602	H2	
28726 East Flat Rock 3,365	E4	
†28723 East Laport 150	C4	
28352 East Laurinburg 536	L5	
†28752 East Marion 1,851	F3	
28039 East Spencer 2,150	J3	
27288 Eden 15,672	K1	
27932 Edenton⊙ 5,357	R2	
27909 Elizabeth City⊙ 14,004	S2	
28337 Elizabethtown⊙ 3,551	M5	
28621 Elkin 2,858	H2	
28622 Elk Park 535	E2	
28040 Ellenboro 560	F4	
28338 Ellerbe 1,415	K4	
27822 Elm City 1,561	O3	
27244 Elon College 2,873	L2	
†28557 Emerald Isle 865	P5	
27823 Enfield 2,995	O2	
28728 Enka 5,567	D3	
28339 Erwin 2,828	M4	
27247 Ether 425	K4	
27935 Eure 300	R2	
27830 Eureka 303	O3	
27825 Everetts 213	P3	
28438 Evergreen 310	M6	
28439 Fair Bluff 1,095	M6	
27826 Fairfield 900	S3	
28340 Fairmont 2,658	L6	
28730 Fairview 1,122	D3	
28341 Faison 636	N4	
28041 Faith 552	J3	

(continued on following page)

Agriculture, Industry and Resources

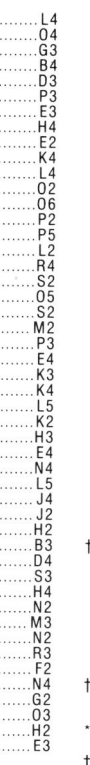

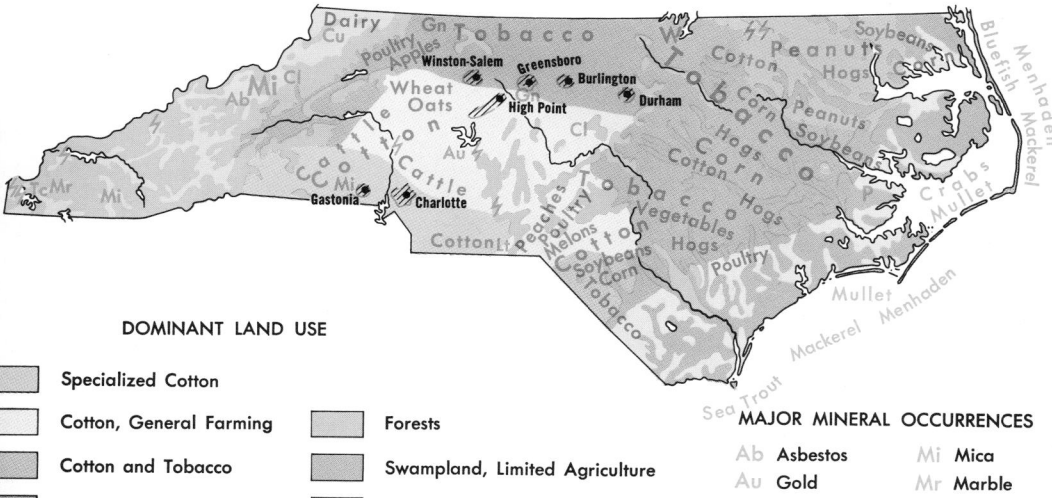

DOMINANT LAND USE

- Specialized Cotton
- Cotton, General Farming
- Cotton and Tobacco
- Tobacco, General Farming
- Peanuts, General Farming
- General Farming, Livestock, Fruit, Tobacco
- General Farming, Truck Farming, Tobacco, Livestock
- Forests
- Swampland, Limited Agriculture
- Nonagricultural Land
- ⚡ Water Power
- ▨ Major Industrial Areas

MAJOR MINERAL OCCURRENCES

Ab	Asbestos		Mi	Mica
Au	Gold		Mr	Marble
Cl	Clay		P	Phosphates
Cu	Copper		Tc	Talc
Gn	Granite		W	Tungsten
Lt	Lithium			

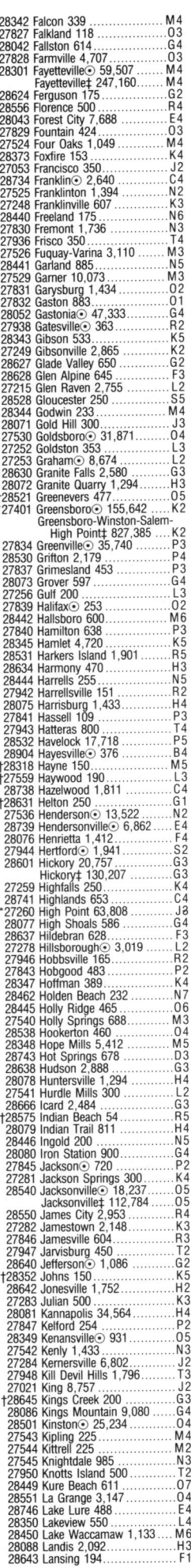

28342 Falcon 339 .... M4
27827 Falkland 118 .... O3
28042 Fallston 614 .... G4
27828 Farmville 4,707 .... O3
*28301 Fayetteville◉ 59,507 .... M4
  Fayetteville‡ 247,160 .... M4
28624 Ferguson 175 .... G2
†28556 Florence 500 .... R4
28043 Forest City 7,688 .... E4
27829 Fountain 424 .... O3
27524 Four Oaks 1,049 .... M4
†28734 Foxfire 153 .... K4
†27053 Francisco 350 .... J2
28734 Franklin◉ 2,640 .... C4
27525 Franklinton 1,394 .... N2
27248 Franklinville 607 .... K3
28440 Freeland 175 .... N6
27830 Fremont 1,736 .... N3
27936 Frisco 350 .... T4
27526 Fuquay-Varina 3,110 .... M3
28441 Garland 885 .... N5
27529 Garner 10,073 .... M3
27831 Garysburg 1,434 .... O2
27832 Gaston 883 .... O1
28052 Gastonia◉ 47,333 .... G4
27938 Gatesville◉ 363 .... R2
28343 Gibson 533 .... K5
27249 Gibsonville 2,865 .... K2
28627 Glade Valley 650 .... G2
28628 Glen Alpine 645 .... F3
27215 Glen Raven 2,755 .... L2
28528 Gloucester 250 .... S5
28344 Godwin 233 .... M4
28071 Gold Hill 300 .... J3
27530 Goldsboro◉ 31,871 .... O4
27252 Goldston 353 .... L3
27253 Graham◉ 8,674 .... L2
28630 Granite Falls 2,580 .... G3
28072 Granite Quarry 1,294 .... H3
†28521 Greenevers 477 .... O5
*27401 Greensboro◉ 155,642 .... K2
  Greensboro-Winston-Salem-
    High Point‡ 827,385 .... K2
27834 Greenville◉ 35,740 .... P3
28530 Grifton 2,179 .... P4
27837 Grimesland 453 .... P3
28073 Grover 597 .... G4
27256 Gulf 200 .... L3
27839 Halifax◉ 253 .... O2
28442 Hallsboro 600 .... M6
28740 Hamilton 638 .... P3
28345 Hamlet 4,720 .... K5
28531 Harkers Island 1,901 .... R5
28634 Harmony 470 .... H3
28444 Harrells 255 .... N5
27942 Harrellsville 151 .... R2
28075 Harrisburg 1,433 .... H4
27841 Hassell 109 .... P3
27943 Hatteras 800 .... T4
28532 Havelock 17,718 .... P5
28904 Hayesville◉ 376 .... B4
†28318 Hayne 150 .... M5
†27559 Haywood 190 .... L3
28738 Hazelwood 1,811 .... C4
†28631 Helton 250 .... G1
27536 Henderson◉ 13,522 .... N2
28739 Hendersonville◉ 6,862 .... E4
28076 Henrietta 1,412 .... F4
27944 Hertford◉ 1,941 .... S2
28601 Hickory 20,757 .... G3
  Hickory‡ 130,207 .... G3
27259 Highfalls 250 .... K4
28741 Highlands 653 .... C4
*27260 High Point 63,808 .... J3
28077 High Shoals 586 .... G4
28637 Hildebran 628 .... F3
27278 Hillsborough◉ 3,019 .... L2
27946 Hobbsville 165 .... R2
28843 Hobgood 483 .... P2
28347 Hoffman 389 .... K4
28462 Holden Beach 232 .... N7
28445 Holly Ridge 465 .... O6
27540 Holly Springs 688 .... M3
28538 Hookerton 460 .... O4
28348 Hope Mills 5,412 .... M5
28743 Hot Springs 678 .... D3
28638 Hudson 2,888 .... G3
28078 Huntersville 1,294 .... H4
27541 Hurdle Mills 300 .... L2
28666 Icard 2,484 .... G3
†28575 Indian Beach 54 .... R5
28079 Indian Trail 811 .... H4
28446 Ingold 200 .... N5
28080 Iron Station 900 .... G4
27845 Jackson◉ 720 .... P2
28281 Jackson Springs 300 .... K4
28540 Jacksonville◉ 18,237 .... O5
  Jacksonville‡ 112,784 .... O5
28550 James City 2,953 .... P4
27282 Jamestown 2,148 .... K3
27846 Jamesville 604 .... R3
27947 Jarvisburg 450 .... T2
28640 Jefferson◉ 1,086 .... G2
†28352 Johns 150 .... K5
28642 Jonesville 1,752 .... H2
27283 Julian 500 .... K3
28081 Kannapolis 34,564 .... H4
27847 Kelford 254 .... P2
28349 Kenansville◉ 931 .... O5
27542 Kenly 1,433 .... N3
27284 Kernersville 6,802 .... J2
27948 Kill Devil Hills 1,796 .... T3
27021 King 8,757 .... J2
†28645 Kings Creek 200 .... G3
28086 Kings Mountain 9,080 .... G4
28501 Kinston◉ 25,234 .... O4
27543 Kipling 225 .... M4
27544 Kittrell 225 .... M2
27545 Knightdale 985 .... N3
27950 Knotts Island 500 .... T2
28449 Kure Beach 611 .... O7
28551 La Grange 3,147 .... O4
28746 Lake Lure 488 .... E4
28350 Lakeview 550 .... L4
28450 Lake Waccamaw 1,133 .... M6
28088 Landis 2,092 .... H3
28643 Lansing 194 .... F1

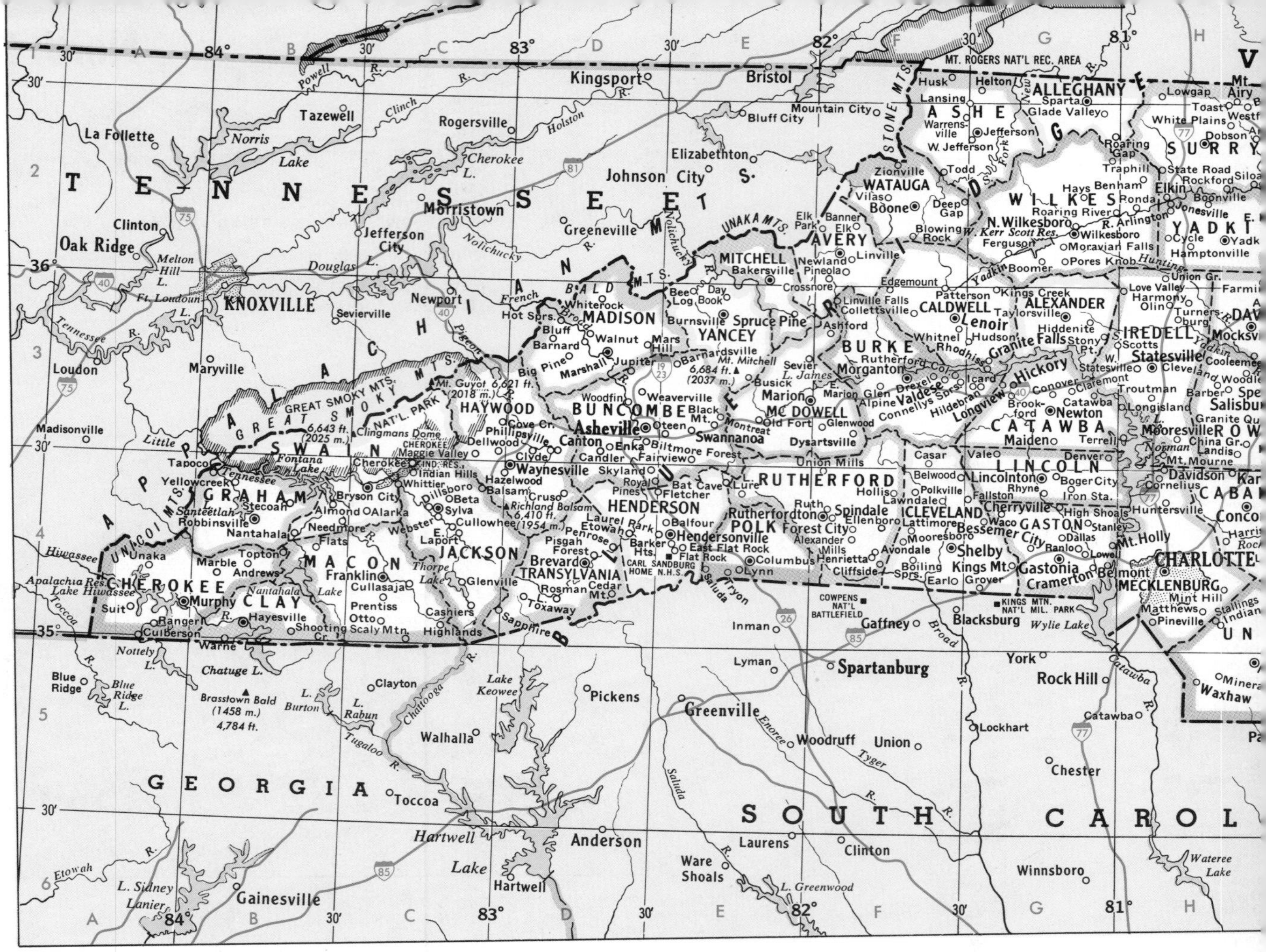

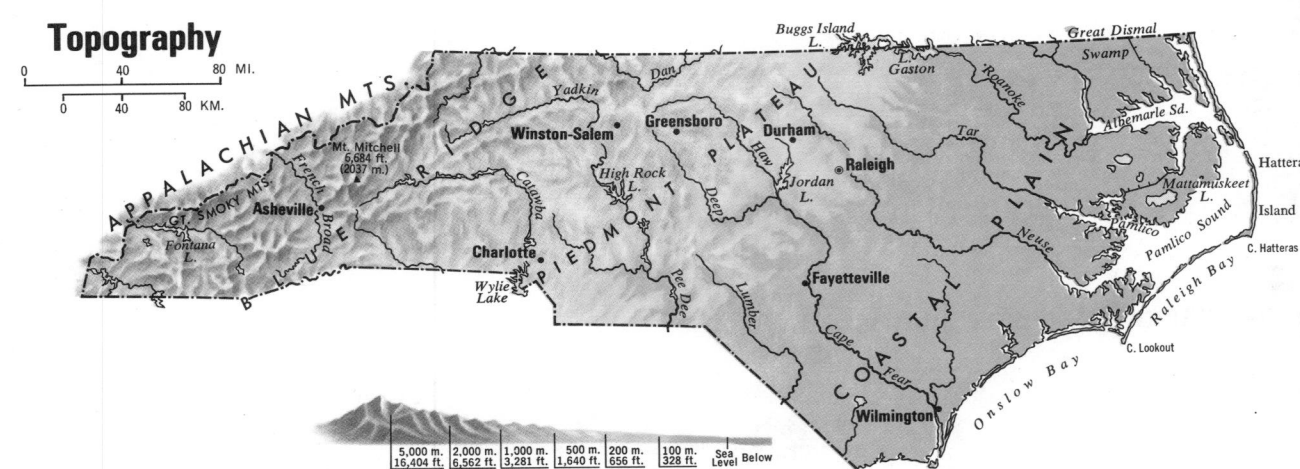

Topography

0   40   80 MI.

0   40   80 KM.

5,000 m. / 16,404 ft. — 2,000 m. / 6,562 ft. — 1,000 m. / 3,281 ft. — 500 m. / 1,640 ft. — 200 m. / 656 ft. — 100 m. / 328 ft. — Sea Level — Below

27848 Lasker 96 .... P2
28089 Lattimore 237 .... F4
28351 Laurel Hill 2,314 .... K5
†28739 Laurel Park 764 .... D4
28352 Laurinburg◉ 11,480 .... K5
28090 Lawndale 469 .... F4
28355 Leggett 99 .... O3
28645 Lenoir◉ 13,748 .... G3
27849 Lewiston 459 .... P2
27292 Lexington◉ 15,711 .... J3
27298 Liberty 1,997 .... K3
28091 Lilesville 588 .... K5
27546 Lillington◉ 1,948 .... M4
28092 Lincolnton◉ 4,879 .... G4
28356 Linden 365 .... M4
28646 Linville 244 .... F2
27850 Littleton 820 .... O2
28097 Locust 1,590 .... J4
28461 Long Beach 1,844 .... N7
28648 Longisland 200 .... H3
28601 Longview 3,587 .... F3
28452 Longwood 800 .... M7
27549 Louisburg◉ 3,238 .... N2
28677 Love Valley 55 .... H3
28098 Lowell 2,917 .... G4
28552 Lowland 600 .... S4
27851 Lucama 1,070 .... N3
28357 Lumber Bridge 171 .... L5
28358 Lumberton◉ 18,241 .... L5
28750 Lynn 650 .... E4
27852 Macclesfield 504 .... O3

27551 Macon 153 .... N2
27025 Madison 2,806 .... J2
28751 Maggie Valley 202 .... C3
28453 Magnolia 592 .... O5
28650 Maiden 2,574 .... G3
27552 Mamers 300 .... L4
†28387 Manly 500 .... L4
27953 Manns Harbor 550 .... T3
27954 Manteo◉ 902 .... T3
28752 Marion◉ 3,684 .... E3
28753 Marshall◉ 809 .... D3
28553 Marshallberg 400 .... S5
28754 Mars Hill 2,126 .... D3
28103 Marshville 2,011 .... J4
28105 Matthews 1,648 .... H4
28364 Maxton 2,711 .... L5
27027 Mayodan 2,627 .... K2
28555 Maysville 877 .... P5
28361 McCain 700 .... L4
28340 McDonald 117 .... L5
28102 McFarlan 133 .... J5
27302 Mebane 2,782 .... L2
†28515 Mesic 390 .... R4
27555 Micro 438 .... N3
27556 Middleburg 185 .... N2
27557 Middlesex 837 .... N3
27555 Milton 235 .... L1
†28510 Minnesott Beach 171 .... R5
28212 Mint Hill 7,915 .... H4
28753 Mocksville◉ 2,637 .... H3
27559 Moncure 700 .... L3
28110 Monroe◉ 12,639 .... J5

28757 Montreat 741 .... E3
28114 Mooresboro 405 .... F4
28115 Mooresville 8,575 .... H3
28654 Moravian Falls 1,552 .... G2
28557 Morehead City 4,359 .... R5
28119 Morven 765 .... J5
27030 Mount Airy 6,862 .... H1
27306 Mount Gilead 1,423 .... K4
28120 Mount Holly 4,530 .... H4
28365 Mount Olive 4,876 .... O4
28124 Mount Pleasant 1,210 .... J4
27855 Murfreesboro 3,007 .... R2
28906 Murphy◉ 2,070 .... B4
27959 Nags Head 1,020 .... T3
28056 Nashville◉ 3,033 .... N3
†28404 Navassa 439 .... O6
28560 New Bern◉ 14,557 .... P4
28885 New Holland 150 .... S4
28657 Newland◉ 722 .... F2
28127 New London 454 .... J4
28570 Newport 1,883 .... R5
28540 New River 5,401 .... O5
28658 Newton◉ 7,624 .... G3
28366 Newton Grove 564 .... N4
27563 Norlina 901 .... N2
28387 Norman 252 .... K4
28659 North Wilkesboro 3,260 .... G2
28128 Norwood 1,818 .... J4
28129 Oakboro 587 .... J4

27857 Oak City 475 .... P3
27310 Oak Ridge 675 .... K2
28459 Ocean Isle Beach 143 .... N7
27960 Ocracoke 500 .... T4
28762 Old Fort 752 .... E3
27974 Old Trap 500 .... T2
†28463 Olyphic 200 .... M7
28571 Oriental 536 .... R4
28369 Orrum 167 .... L6
27565 Oxford◉ 7,603 .... M2
28760 Otto 800 .... C4
28366 Pantego 185 .... R3
28371 Parkton 564 .... M5
27707 Parkwood 3,420 .... M3
28861 Parmele 484 .... P3
28133 Peachland 506 .... J4
28372 Pembroke 2,698 .... L5
28377 Penrose 500 .... D4
†28716 Phillipsville 1,642 .... D3
27863 Pikeville 660 .... N4
27041 Pilot Mountain 1,090 .... J2
28373 Pinebluff 935 .... K4
27042 Pine Hall 912 .... K2
28374 Pinehurst 3,421 .... K4
†28512 Pine Knoll Shores 646 .... R5
27568 Pine Level 953 .... N4
28662 Pineola 500 .... F2
27864 Pinetops 1,465 .... O3
28134 Pineville 1,525 .... H4
28572 Pink Hill 644 .... O4
28768 Pisgah Forest 1,899 .... D4
27312 Pittsboro◉ 1,332 .... L3

27962 Plymouth◉ 4,571 .... R3
27964 Point Harbor 250 .... T2
28135 Polkton 762 .... J4
28136 Polkville 528 .... F4
28573 Pollocksville 318 .... P5
†27810 Ponzer 225 .... R3
27966 Powells Point 750 .... T2
27967 Powellsville 320 .... R2
27569 Princeton 1,034 .... N4
†27886 Princeville 1,508 .... P3
28375 Proctorville 205 .... M6
28376 Raeford◉ 3,630 .... L5
*27601 Raleigh (cap.)◉ 150,255 .... M3
  Raleigh-Durham‡ 530,673 .... M3
27316 Ramseur 1,162 .... K3
27317 Randleman 2,156 .... K3
28052 Ranlo 1,774 .... G4
†28340 Raynham 83 .... L5
27868 Red Oak 314 .... N2
28377 Red Springs 3,642 .... L5
27320 Reidsville 12,492 .... L2
28386 Rennert 178 .... L5
28667 Rhodhiss 727 .... G3
28137 Richfield 373 .... J4
28574 Richlands 825 .... O5
27869 Rich Square 1,057 .... P2
27570 Ridgeway 100 .... N2
27870 Roanoke Rapids 14,702 .... O2
28669 Roaring River 287 .... G2
27325 Robbins 1,256 .... K4
28771 Robbinsville◉ 1,370 .... B4
27871 Robersonville 1,981 .... P3

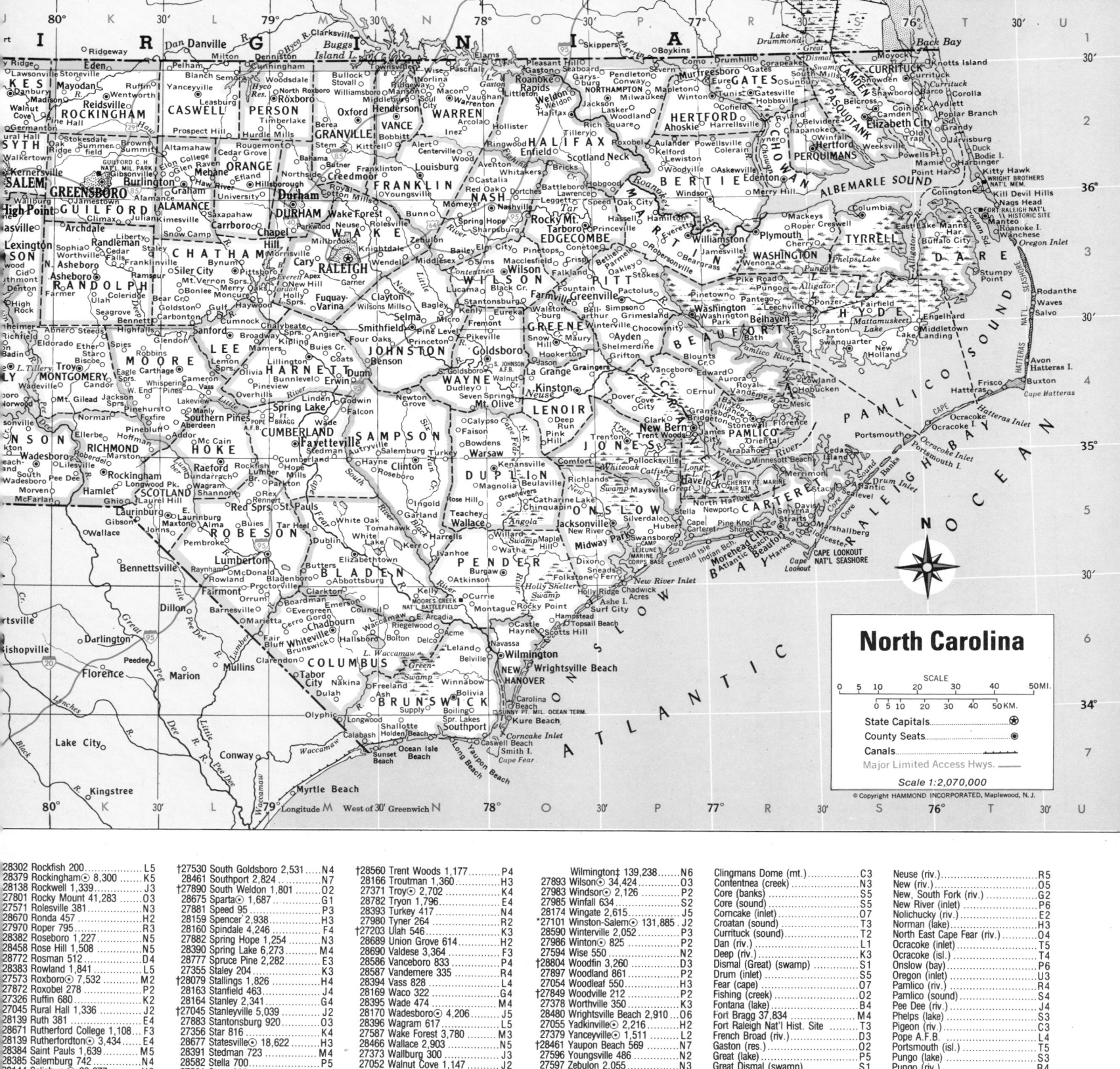

North Carolina

SCALE
0 5 10 20 30 40 50 MI.
0 5 10 20 30 40 50 KM.

State Capitals .... ⊛
County Seats .... ◎
Canals ....
Major Limited Access Hwys. ....

Scale 1:2,070,000

© Copyright HAMMOND INCORPORATED, Maplewood, N.J.

28302 Rockfish 200 ....L5
28379 Rockingham◎ 8,300 ....K5
28138 Rockwell 1,339 ....J3
27801 Rocky Mount 41,283 ....O3
27571 Rolesville 381 ....N3
28670 Ronda 457 ....H2
27970 Roper 795 ....R3
28382 Roseboro 1,227 ....N5
28458 Rose Hill 1,508 ....N5
28772 Rosman 512 ....D4
28383 Rowland 1,841 ....L5
27573 Roxboro◎ 7,532 ....M2
27872 Roxobel 278 ....P2
27326 Ruffin 680 ....K2
27045 Rural Hall 1,336 ....E3
28139 Ruth 381 ....E4
28671 Rutherford College 1,108..F3
28139 Rutherfordton◎ 3,434 ....E4
28384 Saint Pauls 1,639 ....M5
28385 Salemburg 742 ....N4
28144 Salisbury◎ 22,677 ....H3
Salisbury-Concord‡
185,081 ....H3
28773 Saluda 607 ....E4
27972 Salvo 150 ....U3
27330 Sanford◎ 14,773 ....L4
28774 Sapphire 350 ....D4
28775 Scaly Mountain 250 ....C4
27874 Scotland Neck 2,834 ......P2
28699 Scotts 500 ....H3
27535 Scranton 250 ....S4
27875 Seaboard 687 ....O1
27341 Seagrove 294 ....K4
27576 Selma 4,762 ....N3
27343 Semora 500 ....L2
28578 Seven Springs 166 ....P4
28459 Severn 309 ....P2
27878 Sharpsburg 997 ....O3
27973 Shawboro 300 ....S2
28150 Shelby◎ 15,310 ....G4
27344 Siler City 4,446 ....J4
27879 Simpson 407 ....P3
27880 Sims 192 ....N3
28579 Smyrna 291 ....R5
28580 Snow Hill◎ 1,374 ....O4
27350 Sophia 350 ....K3
28387 Southern Pines 8,620 ....L4

†27530 South Goldsboro 2,531 ....N4
28461 Southport 2,824 ....N7
†27890 South Weldon 1,801 .....O2
28675 Sparta◎ 1,687 ....G1
27881 Speed 95 ....P3
28159 Spencer 2,938 ....H3
28160 Spindale 4,246 ....F4
27882 Spring Hope 1,254 ....N3
28390 Spring Lake 6,273 ....M4
28777 Spruce Pine 2,282 ....E3
27355 Staley 204 ....K3
†28079 Stallings 1,826 ....H4
28163 Stanfield 463 ....J4
28164 Stanley 2,341 ....G4
†27045 Stanleyville 5,039 ....J2
27883 Stantonsburg 920 ....O3
27356 Star 816 ....K4
28677 Statesville◎ 18,622 ......H3
28391 Stedman 723 ....M4
28582 Stella 700 ....P5
27581 Stem 222 ....M2
27884 Stokes 450 ....P3
27357 Stokesdale 1,070 ....K2
27048 Stoneville 1,054 ....K2
28583 Stonewall 360 ....R4
28678 Stony Point 1,150 ....G3
27582 Stovall 417 ....M2
†28579 Straits 151 ....R5
27978 Stumpy Point 250 ....T3
28906 Suit 350 ....A4
27358 Summerfield 1,680 ....K2
27979 Sunbury 400 ....R2
28459 Sunset Beach 304 ....N7
28445 Surf City 421 ....O6
28778 Swannanoa 5,586 ....E3
27885 Swanquarter◎ 550 ....S4
28584 Swansboro 976 ....P5
28779 Sylva◎ 1,699 ....C4
28463 Tabor City 2,710 ....M6
27886 Tarboro◎ 8,634 ....O3
28392 Tar Heel 118 ....M5
28681 Taylorsville◎ 1,103 ......G3
28464 Teachey 373 ....N5
27360 Thomasville 14,144 ....J3
27887 Tillery 400 ....O2
27583 Timberlake 500 ....M2
27049 Toast 2,339 ....H2
28445 Topsail Beach 264 ....O6
28685 Traphill 550 ....H2
28585 Trenton◎ 407 ....P4

†28560 Trent Woods 1,177 ....P4
28166 Troutman 1,360 ....H3
27371 Troy◎ 2,702 ....K4
28782 Tryon 1,796 ....E4
28393 Turkey 417 ....N4
27980 Tyner 264 ....R2
†27203 Ulah 546 ....K3
28689 Union Grove 614 ....H2
28690 Valdese 3,364 ....F3
28586 Vanceboro 833 ....P4
28587 Vandemere 335 ....R4
28394 Vass 828 ....L4
28169 Waco 322 ....G4
28395 Wade 474 ....M4
28170 Wadesboro◎ 4,206 ......J5
28396 Wagram 617 ....L5
27587 Wake Forest 3,780 ....M3
28466 Wallace 2,903 ....N5
27373 Wallburg 300 ....J3
27052 Walnut Cove 1,147 ....J2
†27530 Walnut Creek 343 ....O4
27888 Walstonburg 181 ....O3
27981 Wanchese 1,105 ....T3
28909 Warne 200 ....B5
27589 Warrenton◎ 908 ....N2
28398 Warsaw 2,910 ....N4
27889 Washington◎ 8,418 ......P3
†28389 Washington Park 514 ....P3
28471 Watha 196 ....O5
28173 Waxhaw 1,208 ....H5
28786 Waynesville◎ 6,765 .....C4
28787 Weaverville 1,495 ....D3
28788 Webster 200 ....C4
27909 Weeksville 500 ....S2
27374 Welcome 3,243 ....J3
27890 Weldon 1,844 ....O2
27591 Wendell 2,222 ....N3
27375 Wentworth◎ 150 ....K2
27053 Westfield 450 ....J2
28694 West Jefferson 822 ....F2
†28389 Whispering Pines 1,160 ..L4
27891 Whitakers 924 ....O2
28337 White Lake 968 ....M5
27031 White Plains 200 ....H2
28472 Whiteville◎ 5,565 ....M6
28789 Whittier 200 ....C4
28697 Wilkesboro◎ 2,335 ......H2
†27536 Williamston◎ 59 ....O1
27892 Williamston◎ 6,159 .....P3
28401 Wilmington◎ 44,000 ....N6

Wilmington‡ 139,238 ....N6
27893 Wilson◎ 34,424 ....O3
27983 Windsor◎ 2,126 ....P2
27985 Winfall 634 ....S2
28174 Wingate 2,615 ....J5
*27101 Winston-Salem 131,885 .J2
28590 Winterville 2,052 ....P3
27986 Winton◎ 825 ....P2
27594 Wise 550 ....N2
†28804 Woodfin 3,260 ....D3
27897 Woodland 861 ....P2
27054 Woodleaf 550 ....H3
27894 Woodville 212 ....P2
27378 Worthville 350 ....K3
28480 Wrightsville Beach 2,910..O6
27055 Yadkinville◎ 2,216 ....H2
27379 Yanceyville◎ 1,511 ......L2
28461 Yaupon Beach 569 ....N7
27596 Youngsville 486 ....N2
27597 Zebulon 2,055 ....N3
28698 Zionville 525 ....F2

OTHER FEATURES

Albemarle (sound) S2
Alligator (lake) S3
Alligator (riv.) S3
Angola (swamp) O5
Apalachia (res.) A4
Appalachian (mts.) D2
Ashe (isl.) P6
Bald (mts.) D3
Black (mts.) N5
Blue Ridge (mts.) E3
Bodie (isl.) T2
Broad (riv.) E4
Buggs Island (lake) M1
Camp Lejeune Marine Corps
Base 30,764 ....P5
Cape Fear (riv.) M5
Cape Hatteras Nat'l Seashore ..T4
Carl Sandburg Home Nat'l Hist.
Site D4
Catawba (lake) G4
Catawba (riv.) H5
Catfish (lake) P5
Chatuge (lake) B5
Cherokee Ind. Res. C3
Cherry Point Marine Air Sta. ..R5
Chowan (riv.) R2

Clingmans Dome (mt.) C3
Contentnea (creek) N3
Core (banks) S5
Core (sound) S5
Corncake (inlet) O7
Croatan (sound) T3
Currituck (sound) T2
Dan (riv.) L1
Deep (riv.) K3
Dismal (Great) (swamp) S1
Drum (inlet) S5
Fear (cape) O7
Fishing (creek) O2
Fontana (lake) B4
Fort Bragg 37,834 ....M4
Fort Raleigh Nat'l Hist. Site ..T3
French Broad (riv.) D3
Gaston (res.) O2
Great (lake) P5
Great Dismal (swamp) S1
Great Smoky (mts.) B3
Great Smoky Mts. Nat'l Park ..B3
Green (swamp) N6
Guyot (mt.) D3
Hatteras (cape) U4
Hatteras (inlet) T4
Hatteras (isl.) T4
Haw (riv.) K2
High Rock (lake) J3
Hiwassee (lake) A4
Hiwassee (riv.) A4
Holly Shelter (swamp) O6
Hunting (riv.) N5
Hyco (riv.) L2
James (riv.) M1
Jordan, B. Everett (lake) M3
Kerr, W. Scott (res.) G2
Lanes (creek) J5
Little (riv.) N3
Little (riv.) L6
Little Pee Dee (riv.) L6
Little Tennessee (riv.) B4
Long (lake) P5
Lookout (cape) S5
Lumber (riv.) L5
Mattamuskeet (lake) S4
Meherrin (riv.) P1
Mitchell (mt.) E3
Moores Creek Nat'l Battlefield ..N5
Nantahala (lake) B4

Neuse (riv.) R5
New (riv.) O5
New, South Fork (riv.) G2
New River (inlet) P6
Nolichucky (riv.) E2
Norman (lake) H3
North East Cape Fear (riv.) ...O4
Ocracoke (inlet) T5
Ocracoke (isl.) T4
Onslow (bay) P6
Oregon (inlet) U3
Pamlico (riv.) R4
Pamlico (sound) R4
Pee Dee (riv.) J4
Phelps (lake) S3
Pigeon (riv.) C3
Pope A.F.B. L4
Portsmouth (isl.) T5
Pungo (lake) S3
Pungo (riv.) R4
Raleigh (bay) S5
Richland Balsam (mt.) D4
Roanoke (isl.) T3
Roanoke (riv.) O2
Rocky (riv.) H4
Santeetlah (lake) B4
Seymour Johnson A.F.B. O4
Six Run (creek) N4
Smith (isl.) O7
South (riv.) M5
South Yadkin (riv.) H3
Stone (mts.) F2
Sunny Point Mil. Ocean Term. ..O6
Tar (riv.) O3
Thorpe (lake) C4
Tillery (lake) J4
Trent (riv.) P4
Unaka (mts.) E2
Unicoi (mts.) A4
Waccamaw (lake) N6
Waccamaw (riv.) M7
Whiteoak (swamp) P5
W. Scott Kerr (res.) G2
Wright Brothers Nat'l Mem. ..T2
Yadkin (riv.) J3

◎ County seat.
‡ Population of metropolitan area.
† Zip of nearest p.o.
* Multiple zips.

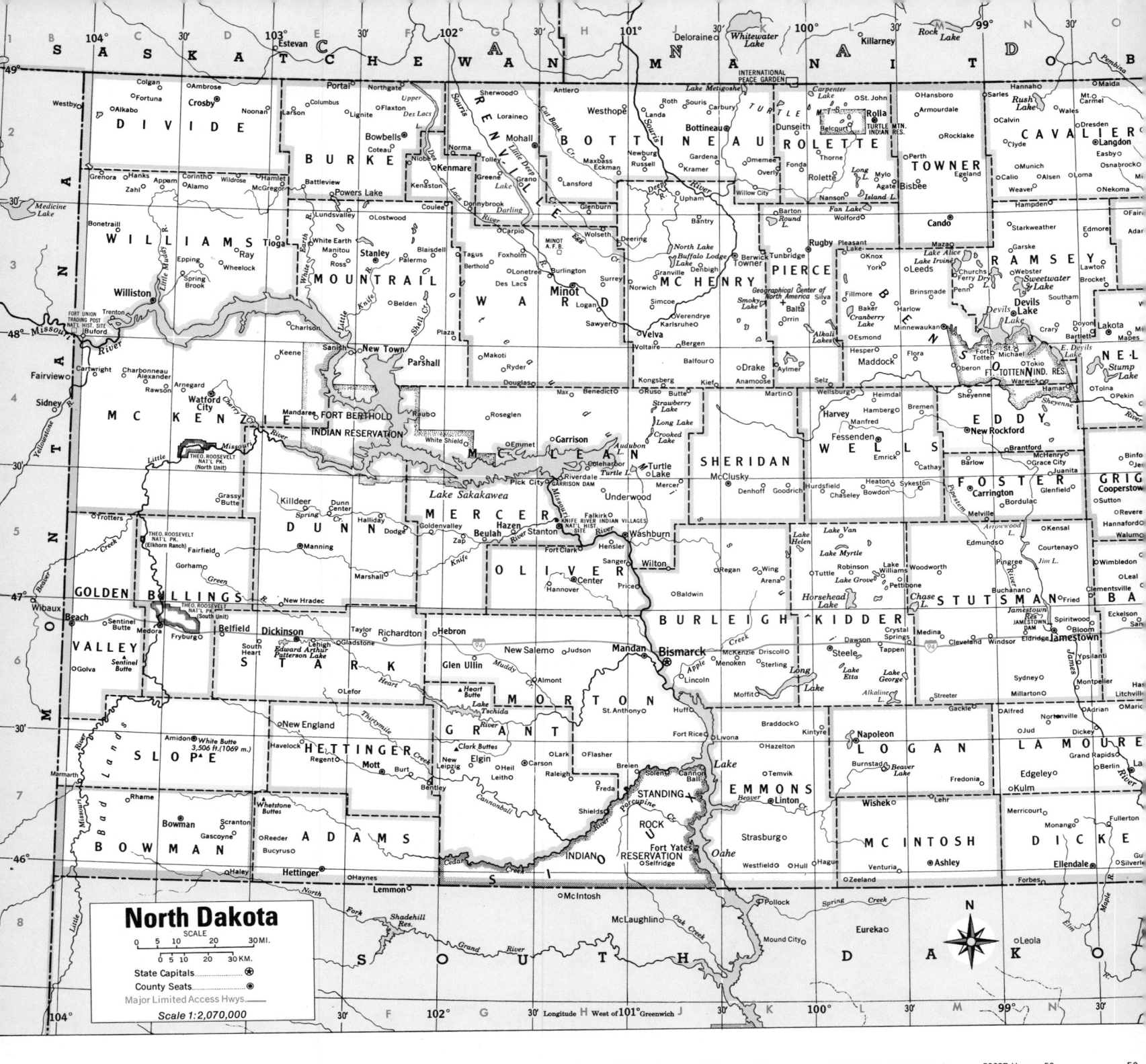

North Dakota

SCALE
0 5 10 20 30 MI.
0 5 10 20   30 KM.

State Capitals .... ⊛
County Seats .... ◉
Major Limited Access Hwys. ———

Scale 1:2,070,000

COUNTIES

Adams 3,584	F7
Barnes 13,960	O5
Benson 7,944	M3
Billings 1,138	D5
Bottineau 9,239	J2
Bowman 4,229	C7
Burke 3,822	E2
Burleigh 54,811	J6
Cass 88,247	R5
Cavalier 7,636	N2
Dickey 7,207	N7
Divide 3,494	C2
Dunn 4,627	E5
Eddy 3,554	N4
Emmons 5,877	K7
Foster 4,611	N5
Golden Valley 2,391	C5
Grand Forks 66,100	P3
Grant 4,274	G6
Griggs 3,714	O5
Hettinger 4,275	E7
Kidder 3,833	L6
LaMoure 6,473	N7
Logan 3,493	L7
McHenry 7,858	J3
McIntosh 4,800	L7
McKenzie 7,132	D4
McLean 12,383	G5
Mercer 9,404	G5
Morton 25,177	H6
Mountrail 7,679	E3

Nelson 5,233	O4
Oliver 2,495	H5
Pembina 10,399	P2
Pierce 6,166	K3
Ramsey 13,048	N3
Ransom 6,698	P7
Renville 3,608	G2
Richland 19,207	R7
Rolette 12,177	L2
Sargent 5,512	P7
Sheridan 2,819	H5
Sioux 3,620	H7
Slope 1,157	C7
Stark 23,697	E6
Steele 2,258	O5
Stutsman 24,154	M5
Towner 4,052	M2
Traill 9,624	R5
Walsh 15,371	P3
Ward 58,392	H3
Wells 6,979	L4
Williams 22,237	C3

CITIES and TOWNS

Zip	Name/Pop.	Key
58001	Abercrombie 260	S7
58210	Adams 303	O3
58831	Alexander 358	C4
58003	Alice 62	P6
58833	Ambrose 60	D2
58004	Amenia 93	R6
58620	Amidon◉ 43	D7

Zip	Name/Pop.	Key
58710	Anamoose 355	K4
58212	Aneta 341	P4
58213	Ardoch 78	R3
58835	Arnegard 193	D4
58006	Arthur 445	R5
58413	Ashley◉ 1,192	M7
58007	Ayr 42	P5
58712	Balfour 51	J4
58008	Barney 70	S7
58216	Bathgate 67	P2
58621	Beach◉ 1,381	C6
58316	Belcourt 1,803	L2
58622	Belfield 1,274	D6
58716	Benedict 68	H4
58415	Berlin 57	O7
58718	Berthold 485	G3
58523	Beulah 2,908	G5
58416	Binford 293	O4
58317	Bisbee 257	M2
58501	Bismarck (cap.)◉ 44,485	J6
58318	Bottineau◉ 2,829	J2
58721	Bowbells◉ 587	F2
58623	Bowman◉ 2,071	D7
58524	Braddock 86	K6
58320	Brinsmade 54	M3
58321	Brocket 74	O3
58722	Burlington 762	H3
58218	Buxton 336	R4
58322	Calio 60	N2
58323	Calvin 61	N2
58324	Cando◉ 1,496	M3
†58241	Canton (Hensel) 68	P2
58725	Carpio 244	G3

Zip	Name/Pop.	Key
58421	Carrington◉ 2,641	M5
58529	Carson◉ 469	H7
58012	Casselton 1,661	R6
58422	Cathay 66	M4
58220	Cavalier◉ 1,505	P2
58013	Cayuga 75	R7
58530	Center◉ 900	H5
58016	Clifford 51	R5
58017	Cogswell 227	P7
58727	Columbus 325	E2
58425	Cooperstown◉ 1,308	O5
58730	Crosby◉ 1,469	D2
58222	Crystal 256	P2
58021	Davenport 195	R6
58731	Deering 85	J3
58301	Devils Lake◉ 7,442	N3
58431	Dickey 74	N6
58601	Dickinson◉ 15,924	E6
58736	Drake 479	K4
58225	Drayton 1,082	R2
58329	Dunseith 625	K2
58024	Dwight 72	S7
58433	Edgeley 843	N7
58227	Edinburg 300	P3
58330	Edmore 416	O3
58533	Elgin 930	G7
58436	Ellendale◉ 1,967	N7
58228	Emerado 596	R4
58027	Enderlin 1,151	P6
58332	Esmond 337	L3
58229	Fairdale 97	O3
58030	Fairmount 480	S7
58102	Fargo 61,383	S6

Zip	Name/Pop.	Key
	Fargo-Moorhead‡ 137,574	S6
58438	Fessenden◉ 761	L4
58230	Finley◉ 718	P4
58535	Flasher 410	H7
58439	Forbes 84	N8
58231	Fordville 326	P3
58032	Forman◉ 629	P7
58033	Fort Ransom 99	P6
58844	Fortuna 98	C2
58538	Fort Yates◉ 771	J7
58440	Fredonia 82	M7
58442	Gackle 456	M6
58739	Gardena 66	J2
58036	Gardner 94	R5
58540	Garrison 1,830	H4
58235	Gilby 283	R3
58630	Gladstone 317	F6
58740	Glenburn 454	H2
58631	Glen Ullin 1,125	G6
58541	Goldenvalley 287	F5
58444	Goodrich 288	K5
58237	Grafton◉ 5,293	R3
58201	Grand Forks◉ 43,765	R4
	Grand Forks‡ 100,944	R4
58741	Granville 281	J3
58845	Grenora 362	C2
58040	Gwinner 725	P7
58636	Halliday 355	F5
58041	Hankinson 1,158	S7
58239	Hannah 90	N2
58341	Harvey 2,527	L4
58042	Harwood 326	R6
58240	Hatton 787	R4

Zip	Name/Pop.	Key
58637	Haynes 58	F8
58544	Hazelton 266	K7
58545	Hazen 2,365	G5
58638	Hebron 1,078	E6
58639	Hettinger◉ 1,739	E8
58045	Hillsboro◉ 1,600	S5
58243	Hoople 350	P3
58046	Hope 406	P5
58047	Horace 494	S6
58048	Hunter 369	R5
58244	Inkster 135	P3
58401	Jamestown◉ 16,280	N6
58049	Kathryn 95	P6
58746	Kenmare 1,456	F2
58640	Killdeer 790	E5
58051	Kindred 568	R6
58343	Knox 69	L3
58748	Kramer 84	J2
58456	Kulm 570	N7
58344	Lakota◉ 963	O3
58458	LaMoure◉ 1,077	O7
58749	Landa 62	J2
58249	Langdon◉ 2,335	O2
58251	Lansford 294	H2
58459	Larimore 1,524	Q4
58459	Leal 45	O5
58346	Leeds 678	M3
58460	Lehr 254	M7
58551	Leith 59	G6
58052	Leonard 289	R6
58053	Lidgerwood 971	R7
58752	Lignite 332	F2

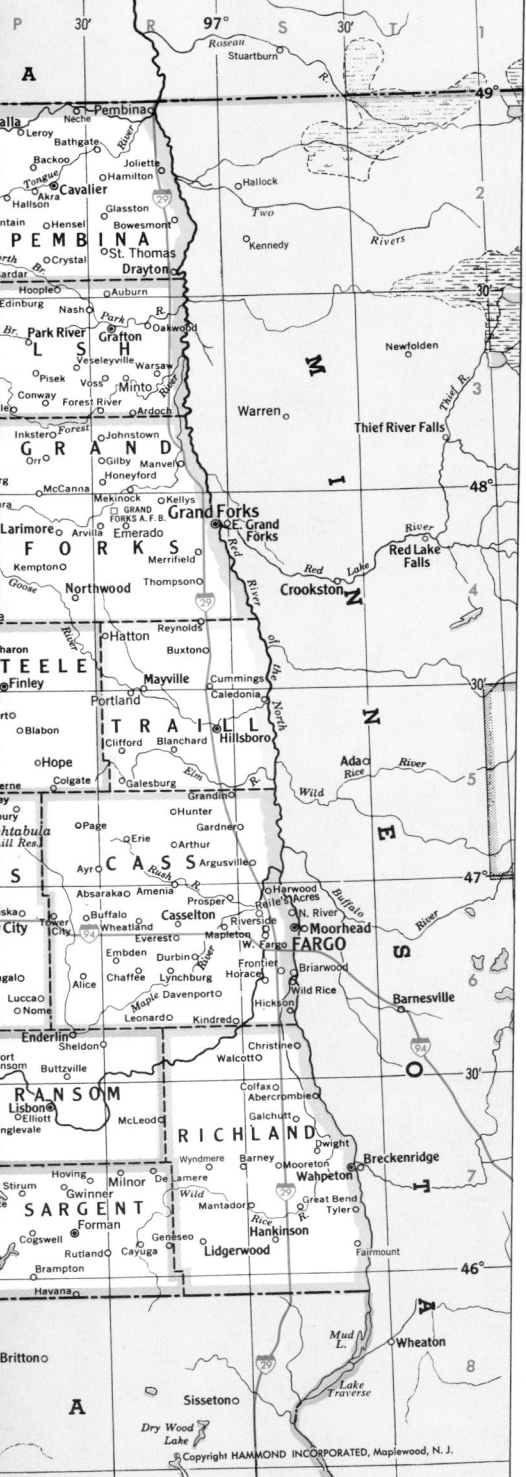

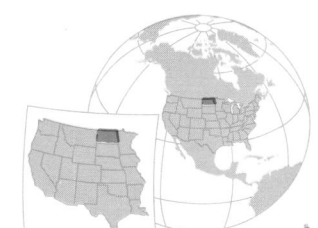

58276 Saint Thomas 528	R2
58780 Sanish	E4
58781 Sawyer 417	H3
58653 Scranton 415	D7
58568 Selfridge 273	J7
58654 Sentinel Butte 86	C6
58068 Sheldon 173	P6
58782 Sherwood 294	G2
58374 Sheyenne 307	M4
58655 South Heart 294	D6
58850 Spring Brook 52	D3
58784 Stanley⊙ 1,631	F3
58571 Stanton⊙ 623	H5
58482 Steele⊙ 796	L6
58573 Strasburg 623	K7
58483 Streeter 264	M6
58785 Surrey 999	H3
58487 Tappen 271	L6
58656 Taylor 239	F6
58852 Tioga 1,597	E3
58380 Tolna 241	O4
58071 Tower City 293	P6
58788 Towner⊙ 867	K3
58575 Turtle Lake 802	J4
58576 Underwood 1,329	H5
58072 Valley City⊙ 7,774	P6
58790 Velva 1,101	H4
58792 Voltaire 65	J3
58075 Wahpeton⊙ 9,064	S7
58281 Wales 74	N2
58282 Walhalla 1,429	P2
58577 Washburn⊙ 1,767	J5
58854 Watford City⊙ 2,119	D4
58078 West Fargo 10,099	S6
58793 Westhope 741	H2
58794 White Earth 98	E3
58795 Wildrose 214	D2
58801 Williston⊙ 13,336	C3
58384 Willow City 329	K2
58579 Wilton 950	J5
58492 Wimbledon 330	O5
58495 Wishek 1,345	L7
58385 Wolford 76	L3
58081 Wyndmere 550	R7
58386 York 69	L3
58580 Zap 511	G5
58581 Zeeland 253	L8

AREA 70,702 sq. mi. (183,118 sq. km.)
POPULATION 652,717
CAPITAL Bismarck
LARGEST CITY Fargo
HIGHEST POINT White Butte 3,506 ft. (1069 m.)
SETTLED IN 1780
ADMITTED TO UNION November 2, 1889
POPULAR NAME Flickertail State; Sioux State
STATE FLOWER Wild Prairie Rose
STATE BIRD Western Meadowlark

Topography

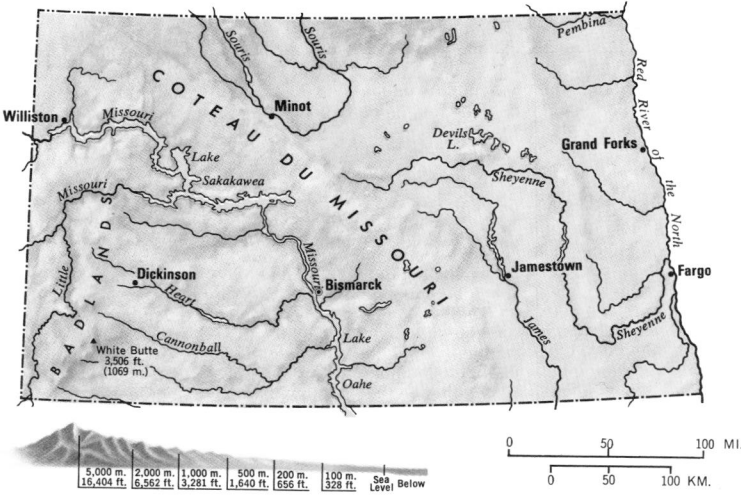

OTHER FEATURES

Alkali (lke)	L3
Alkaline (lake)	L6
Apple (creek)	J6
Arrowwood (lake)	N5
Ashtabula (lake)	P5
Audubon (lake)	H4
Bad Lands (reg.)	C7
Baldhill (Ashtabula) (res.)	P5
Bear (creek)	O7
Beaver (creek)	B5
Beaver (creek)	K7
Beaver (lake)	L7
Buffalo Lodge (lake)	J3
Cannonball (riv.)	G7
Carpenter (lake)	L2
Cedar (creek)	G7
Chase (lake)	M5
Cherry (creek)	D4
Clark (buttes)	G7
Coteau du Missouri (plain)	G3
Cranberry (lake)	L3
Crooked (lake)	J4
Cut Bank (creek)	H2
Darling (lake)	G2
Deep (riv.)	J1
Des Lacs (riv.)	G3
Devils (lake)	N3
Dry (lake)	M3
East Devils (lake)	N4
Egg (creek)	H3
Elm (riv.)	N8
Elm (riv.)	R5
Etta (lake)	L6
Fan (lake)	L2
Forest (riv.)	P3
Fort Berthold Ind. Res.	E4
Fort Totten Ind. Res.	N4
Fort Union Trading Post Nat'l Hist. Site	B3
Garrison (dam)	H5
George (lake)	L6
Goose (riv.)	P4
Grand, North Fork (riv.)	E8
Grand Forks A.F.B. 9,390	R4
Green (riv.)	D5
Grove (lake)	L5
Heart (butte)	G6
Heart (riv.)	H6
Helen (lake)	K5
Horsehead (lake)	L5
International Peace Garden	K1
Irvine (lake)	M3
Island (lake)	L2
James (riv.)	N6
Jamestown (res.)	N6
Jim (riv.)	N5
Knife (riv.)	G5
Knife R. Indian Villages Nat'l Hist. Site	H5
Little Deep (creek)	G2
Little Knife (riv.)	F3
Little Missouri (riv.)	D4
Little Muddy (riv.)	C3
Long (lake)	J4
Long (lake)	K6
Long (lake)	L2
Maple (riv.)	O8
Maple (riv.)	R6
Metigoshe (lake)	K2
Minot A.F.B. 9,880	H3
Missouri (riv.)	H5
Muddy (creek)	G6
Myrtle (lake)	L5
North (lake)	J3
Oahe (lake)	J7
Oak (creek)	J8
Park (riv.)	R3
Patterson, Edward A. (lake)	E6
Pembina (riv.)	O1
Pipestem (riv.)	M5
Porcupine (creek)	J7
Red River of the North (riv.)	S4
Round (lake)	K3
Rush (lake)	N2
Rush (riv.)	R5
Sakakawea (lake)	G5
Sentinel (butte)	C6
Shell (creek)	F3
Sheyenne (riv.)	O6
Smoky (lake)	K3
Souris (riv.)	J2
Spring (creek)	E5
Standing Rock Ind. Res.	J7
Strawberry (lake)	J4
Stump (lake)	O4
Sweetwater (lake)	N3
Theodore Roosevelt Nat'l Mem. Park	C5, D4,D6
Thirty Mile (creek)	F6
Tongue (riv.)	P2
Tschida (lake)	G6
Turtle (lake)	H4
Turtle (mts.)	K2
Turtle Mountain Ind. Res.	L2
Upper Des Lacs (lake)	F2
Van (lake)	L5
Whetstone (buttes)	E7
White (butte)	D7
White Butte (mt.)	D7
White Earth (riv.)	E3
Wild Rice (riv.)	R7
Yellowstone (riv.)	B4

⊙County seat.
‡Population of metropolitan area.
† Zip of nearest p.o.
* Multiple zips.

© Copyright HAMMOND INCORPORATED, Maplewood, N.J.

†58501 Lincoln 656	J6	58563 New Salem 1,081	G6
58552 Linton⊙ 1,561	K7	58763 New Town 1,335	F4
58054 Lisbon⊙ 2,283	P7	58266 Niagara 76	P4
58461 Litchville 251	O6	58062 Nome 67	P6
58056 Luverne 65	P5	58765 Noonan 283	D2
58348 Maddock 677	L4	†58102 North River 65	S6
58554 Mandan⊙ 15,513	J6	58267 Northwood 1,240	P4
58642 Manning⊙ 75	E5	58474 Oakes 2,112	O7
58058 Mantador 76	R7	58063 Oriska 125	P6
58256 Manvel 308	R3	58064 Page 329	P5
58059 Mapleton 306	R6	58769 Palermo 97	F3
58643 Marmarth 190	B7	58270 Park River 1,844	P3
58759 Max 317	H4	58770 Parshall 1,059	F4
58257 Mayville 2,255	R4	58271 Pembina 673	R2
58463 McClusky⊙ 658	K4	58476 Pingree 88	N5
58254 McVille 626	O4	58772 Portal 238	E2
58467 Medina 521	M6	58274 Portland 627	R5
58645 Medora⊙ 94	C6	58773 Powers Lake 466	E2
58259 Michigan 502	O3	58849 Ray 766	D3
58060 Milnor 716	R7	58649 Reeder 355	E7
58351 Minnewaukan⊙ 461	M3	58477 Regan 71	K5
58701 Minot⊙ 32,843	H3	58650 Regent 297	E7
58261 Minto 592	R3	58275 Reynolds 309	R4
58761 Mohall⊙ 1,049	G2	58651 Rhame 222	C7
58471 Monango 59	N7	58652 Richardton 699	F6
58472 Montpelier 96	N6	†58078 Riverside 465	S6
58646 Mott⊙ 1,315	F7	58365 Rocklake 287	M2
58352 Munich 300	N2	58479 Rogers 68	O5
58561 Napoleon⊙ 1,103	L6	58366 Rolette 667	L2
58265 Neche 471	P2	58367 Rolla⊙ 1,538	L2
58647 New England 825	E6	58368 Rugby⊙ 3,335	L3
58562 New Leipzig 352	G7	58067 Rutland 250	P7
58356 New Rockford⊙ 1,791	N4	58369 Saint John 401	L2

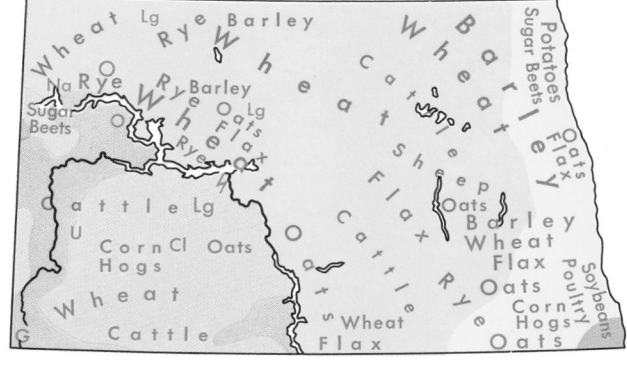

DOMINANT LAND USE

☐ Specialized Wheat

☐ Wheat, General Farming

☐ Wheat, Range Livestock

☐ Livestock, Cash Grain

☐ Sugar Beets, Dry Beans, Livestock, General Farming

☐ Range Livestock

⚡ Water Power

Agriculture, Industry and Resources

MAJOR MINERAL OCCURRENCES

Cl	Clay
G	Natural Gas
Lg	Lignite
Na	Salt
O	Petroleum
U	Uranium

Ohio

SCALE

0 5 10 20 30 40 MI.

0 5 10 20 30 40 KM.

State Capitals ⊛

County Seats ⊛

Major Limited Access Hwys. ────

Scale 1:1,800,000

© Copyright HAMMOND INCORPORATED, Maplewood, N.J.

AREA 41,330 sq. mi. (107,045 sq. km.)
POPULATION 10,797,624
CAPITAL Columbus
LARGEST CITY Cleveland
HIGHEST POINT Campbell Hill 1,550 ft. (472 m.)
SETTLED IN 1788
ADMITTED TO UNION March 1, 1803
POPULAR NAME Buckeye State
STATE FLOWER Scarlet Carnation
STATE BIRD Cardinal

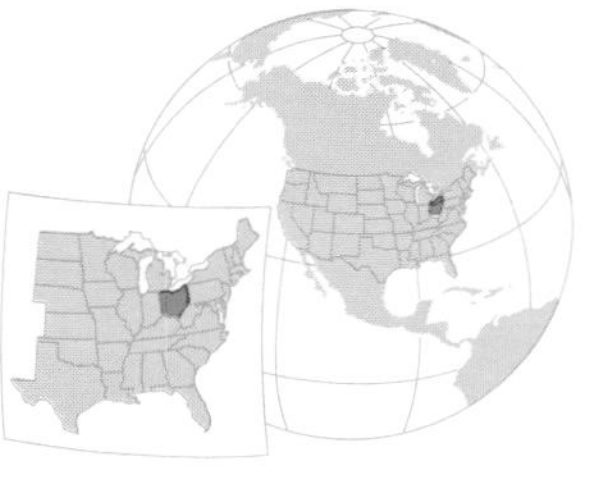

Topography

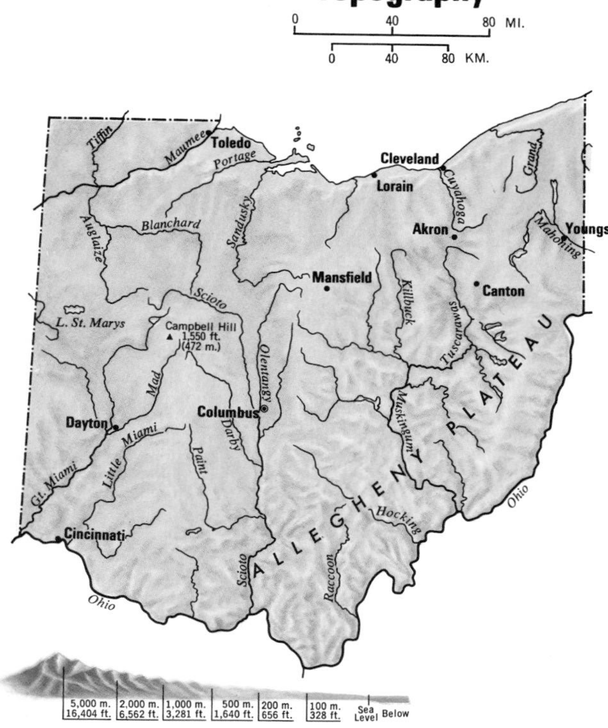
0 40 80 MI.
0 40 80 KM.
Toledo
Maumee
Tiffin
Portage
Cleveland
Lorain
Cuyahoga
Grand
Akron
Youngstown
Mahoning
Blanchard
Auglaize
Sandusky
Mansfield
Canton
Scioto
Kilbuck
Tuscarawas
L. St. Marys
Campbell Hill 1,550 ft. (472 m.)
Mad
Olentangy
Muskingum
PLATEAU
Columbus
Darby
Dayton
Miami
Little Miami
Paint
Gt. Miami
ALLEGHENY
Hocking
Ohio
Cincinnati
Raccoon
5,000 m. 16,404 ft.
2,000 m. 6,562 ft.
1,000 m. 3,281 ft.
500 m. 1,640 ft.
200 m. 656 ft.
100 m. 328 ft.
Sea Level
Below

COUNTIES

ams 24,328 ... D8
en 112,241 ... B4
hland 46,178 ... F4
htabula 104,215 ... J2
hens 56,399 ... F7
glaize 42,554 ... B4
lmont 82,569 ... J5
own 31,920 ... C8
tler 258,787 ... A7
rroll 25,598 ... H4
ampaign 33,649 ... C5
rk 150,236 ... C6
ermont 128,483 ... B7
nton 34,603 ... C7
lumbiana 113,572 ... J4
shocton 36,024 ... G5
awford 50,075 ... E4
yahoga 1,498,400 ... G3
rke 55,096 ... A5
fiance 39,987 ... A3
laware 53,840 ... D5
e 79,655 ... E3
rfield 93,678 ... E6
yette 27,467 ... D6
nklin 869,126 ... E5
ton 37,751 ... B2
llia 30,098 ... F8
auga 74,474 ... H3
eene 129,769 ... C6
ernsey 42,024 ... H5
milton 873,224 ... A7
ncock 64,581 ... C3
rdin 32,719 ... C4
rrison 18,152 ... H5
nry 28,383 ... B3
hland 33,477 ... C7
cking 24,304 ... F6
lmes 29,416 ... G4
ron 54,608 ... E3
kson 30,592 ... E7
ferson 91,564 ... J5
ox 46,304 ... F5
ke 212,801 ... H2
wrence 63,849 ... E8
king 120,981 ... F5
gan 39,155 ... C5
rain 274,909 ... F3
cas 471,741 ... C2
dison 33,004 ... D6
honing 289,487 ... J4
rion 67,974 ... D4
dina 113,150 ... G3
igs 23,641 ... F7
rcer 38,334 ... A4
ami 90,381 ... B5
nroe 17,382 ... H6
ntgomery 571,697 ... B6
rgan 14,241 ... G6
rrow 26,480 ... E4
skingum 83,340 ... G5
ble 11,310 ... G6
awa 40,076 ... D2
ulding 21,302 ... A3
ry 31,032 ... F6
Pickaway 43,662 ... D6
Pike 22,802 ... D7
Portage 135,856 ... H3
Preble 38,223 ... A6
Putnam 32,991 ... B3
Richland 131,205 ... E4
Ross 65,004 ... D7
Sandusky 63,267 ... D3
Scioto 84,545 ... D8
Seneca 61,901 ... D3
Shelby 43,089 ... B5
Stark 378,823 ... H4
Summit 524,472 ... G3
Trumbull 241,863 ... J3
Tuscarawas 84,614 ... H5
Union 29,536 ... D5
Van Wert 30,458 ... A4
Vinton 11,584 ... E7
Warren 99,276 ... B7
Washington 64,266 ... H7
Wayne 97,408 ... G4
Williams 36,369 ... A2
Wood 107,372 ... C3
Wyandot 22,651 ... D4

CITIES and TOWNS

Zip *Name/Pop.* *Key*

45101 Aberdeen 1,566 ... C8
45810 Ada 5,669 ... C4
45001 Addyston 1,195 ... B9
43101 Adelphi 472 ... E7
43901 Adena 1,062 ... J5
*44301 Akron⊙ 237,177 ... G3
Akron‡ 660,328 ... G3
45710 Albany 905 ... F7
43001 Alexandria 489 ... E5
45812 Alger 992 ... C4
44601 Alliance 24,315 ... H4
43102 Amanda 720 ... E6
†45201 Amberley 3,442 ... C9
45102 Amelia 1,108 ... D10
44001 Amherst 10,638 ... F3
43903 Amsterdam 783 ... J5
44003 Andover 1,205 ... J2
45302 Anna 1,038 ... B5
45303 Ansonia 1,267 ... A5
45813 Antwerp 1,765 ... A3
44606 Apple Creek 741 ... G4
44804 Arcadia 580 ... D3
45304 Arcanum 2,002 ... A6
43502 Archbold 3,318 ... B2
45814 Arlington 1,187 ... C4
†45201 Arlington Heights 1,082 ... C9
44805 Ashland⊙ 20,326 ... F4
43003 Ashley 1,057 ... E5
44004 Ashtabula 23,449 ... J2
43103 Ashville 2,046 ... E6
45701 Athens⊙ 19,743 ... F7
44807 Attica 865 ... E3
44201 Atwater 975 ... H3
44202 Aurora 8,177 ... H3
44010 Austinburg 900 ... J2
44515 Austintown 33,636 ... J3
44011 Avon 7,241 ... F3
44012 Avon Lake 13,222 ... F2
†43512 Ayersville 950 ... B3
†44805 Bailey Lakes 397 ... F4
45612 Bainbridge 1,042 ... D7
43804 Baltic 563 ... G5
43105 Baltimore 2,689 ... E6
44203 Barberton 29,751 ... G4
43713 Barnesville 4,633 ... H6
43905 Barton 1,039 ... J5
45103 Batavia⊙ 1,896 ... B7
†44870 Bay View 804 ... E3
44140 Bay Village 17,846 ... G9
44608 Beach City 1,083 ... G4
44122 Beachwood 9,983 ... J9
43716 Beallsville 601 ... J6
45808 Beaverdam 492 ... C4
44146 Bedford 15,056 ... H9
†44146 Bedford Heights 13,214 ... J9
43906 Bellaire 8,241 ... J5
45305 Bellbrook 5,174 ... C6
43310 Belle Center 930 ... C4
43311 Bellefontaine⊙ 11,888 ... C5
44811 Bellevue 8,187 ... E3
44813 Bellville 1,714 ... E4
43718 Belmont 714 ... J5
44609 Beloit 1,093 ... J4
45714 Belpre 7,193 ... G7
44017 Berea 19,567 ... G10
43908 Bergholz 914 ... J4
44814 Berlin Heights 756 ... F3
45106 Bethel 2,231 ... B8
43719 Bethesda 1,429 ... H5
44815 Bettsville 752 ... D3
45715 Beverly 1,471 ... G6
43209 Bexley 13,405 ... E6
45107 Blanchester 3,202 ... B7
44817 Bloomdale 744 ... D3
43106 Bloomingburg 869 ... D6
44818 Bloomville 1,019 ... D3
†45242 Blue Ash 9,506 ... C9
45817 Bluffton 3,310 ... C4
44512 Boardman 39,161 ... J3
44612 Bolivar 989 ... G4
†44264 Boston Heights 781 ... J10
45306 Botkins 1,372 ... B5
44695 Bowerston 487 ... H5
43402 Bowling Green⊙ 25,728 ... C3
45308 Bradford 2,166 ... B5
43406 Bradner 1,175 ... C3
44211 Brady Lake 470 ... H3
†44101 Bratenahl 1,485 ... H9
44141 Brecksville 10,132 ... H10
43107 Bremen 1,432 ... F6
44613 Brewster 2,321 ... G4
43912 Bridgeport 2,642 ... J5
†45211 Bridgetown 11,460 ... B9
43913 Brilliant 1,751 ... J5
†44240 Brimfield 3,161 ... H3
44402 Bristolville 900 ... J3
†44141 Broadview Heights 10,920 ... H10
44403 Brookfield 1,527 ... J3
44144 Brooklyn 12,342 ... H9
†44131 Brooklyn Heights 1,653 ... H9
44142 Brook Park 26,195 ... G9
†43912 Brookside 887 ... J5
45309 Brookville 4,322 ... B6
44212 Brunswick 28,104 ... G3
43506 Bryan⊙ 7,879 ... A3
45716 Buchtel 585 ... F7
43008 Buckeye Lake ... F6
44820 Bucyrus⊙ 13,433 ... E4
†45680 Burlington 900 ... F9
44021 Burton 1,401 ... H3
44822 Butler 991 ... F4
43723 Byesville 2,572 ... G6
43907 Cadiz⊙ 4,058 ... J5
45820 Cairo 596 ... B4
43920 Calcutta 1,121 ... J4
43724 Caldwell⊙ 1,935 ... G6
43314 Caledonia 759 ... D4
43725 Cambridge⊙ 13,573 ... G5
45311 Camden 1,971 ... A6
44405 Campbell 11,619 ... J3
45111 Camp Dennison 625 ... D9
44614 Canal Fulton 3,481 ... H4
43110 Canal Winchester 2,749 ... E6
44406 Canfield 5,535 ... J3
*44701 Canton⊙ 93,077 ... H4
Canton‡ 404,421 ... H4
43315 Cardington 1,665 ... E5
43316 Carey 3,674 ... D4
45005 Carlisle 4,276 ... B6
43112 Carroll 641 ... E6
44615 Carrollton⊙ 3,065 ... J4
44824 Castalia 973 ... E3
45314 Cedarville 2,799 ... C6
45822 Celina⊙ 9,137 ... A4
43011 Centerburg 1,275 ... E5
45459 Centerville 18,886 ... B6
44022 Chagrin Falls 4,335 ... J9
†45631 Chambersburg ... F8
44024 Chardon⊙ 4,434 ... H2
45719 Chauncey 1,050 ... F7
†45202 Cherry Grove 850 ... C10
45619 Chesapeake 1,370 ... E9
44026 Chesterland 2,301 ... H2
†45211 Cheviot 9,888 ... B9
45601 Chillicothe⊙ 23,420 ... E7
45389 Christiansburg 593 ... C5
*45201 Cincinnati⊙ 385,457 ... B9
Cincinnati‡ 1,401,403 ... B9
43113 Circleville⊙ 11,700 ... D6
43915 Clarington 558 ... J6
43115 Clarksburg 483 ... D7
45113 Clarksville 525 ... C7
45315 Clayton 752 ... B6
*44101 Cleveland⊙ 573,822 ... H9
Cleveland‡ 1,898,720 ... H9
44118 Cleveland Heights 56,438 ... H9
45002 Cleves 2,094 ... B9
44216 Clinton 1,277 ... G4
43410 Clyde 5,489 ... E3
†45638 Coal Grove 2,602 ... E9
45621 Coalton 639 ... E7
45828 Coldwater 4,220 ... A5
†44034 Colebrook 700 ... J2
44028 Columbia Station 518 ... G10
44408 Columbiana 4,987 ... J4
*43201 Columbus (cap.)⊙ 565,032 ... E6
Columbus‡ 1,093,293 ... E6
45830 Columbus Grove 2,313 ... B4
43811 Conesville 451 ... G5
44030 Conneaut 13,835 ... J2
45831 Continental 1,179 ... B3
45832 Convoy 1,140 ... A4
45723 Coolville 649 ... G7
43730 Corning 789 ... F6
44410 Cortland 5,011 ... J3
43812 Coshocton⊙ 13,405 ... G5
†45238 Covedale 5,830 ... B10
45318 Covington 2,610 ... B5
†44429 Craig Beach 1,657 ... H3
44827 Crestline 5,406 ... E4
44217 Creston 1,828 ... G3
45806 Cridersville 1,843 ... B4
43731 Crooksville 2,766 ... F6
45623 Crown City 513 ... F8
†45341 Crystal Lakes 1,463 ... C6
*44221 Cuyahoga Falls 43,890 ... G3
†44101 Cuyahoga Heights 739 ... H9
43413 Cygnet 646 ... C3
44618 Dalton 1,357 ... G4
43014 Danville 1,127 ... F5
†43123 Darbydale 825 ... D6
*45401 Dayton⊙ 193,444 ... B6
Dayton‡ 830,070 ... B6
44411 Deerfield 800 ... H3
45236 Deer Park 6,745 ... C9
43512 Defiance⊙ 16,810 ... B3
43318 Degraff 1,358 ... C5
43015 Delaware⊙ 18,780 ... E5
45833 Delphos 7,314 ... B4
43515 Delta 2,831 ... B2
44621 Dennison 3,398 ... H5
†45202 Dent 800 ... B9
43516 Deshler 1,870 ... C3
45750 Devola 2,708 ... H7
43917 Dillonvale 912 ... J5
44622 Dover 11,782 ... G4
44230 Doylestown 2,493 ... G4
43821 Dresden 1,646 ... G5
43017 Dublin 3,855 ... D5
43734 Duncan Falls 900 ... G6
45836 Dunkirk 954 ... C4
44730 East Canton 1,721 ... H4
44112 East Cleveland 36,957 ... H9
†44094 Eastlake 22,104 ... J8
43920 East Liverpool 16,687 ... J4
44413 East Palestine 5,306 ... J4
44626 East Sparta 868 ... H4
45320 Eaton⊙ 6,839 ... A6
†44035 Eaton Estates 1,806 ... G3
43517 Edgerton 1,813 ... A3
†44004 Edgewood 3,099 ... J2
43320 Edison 504 ... E4
43518 Edon 947 ... A2
45321 Eldorado 509 ... A6
45807 Elida 1,349 ... B4
43416 Elmore 1,271 ... D3
45216 Elmwood Place 2,840 ... B9
*44035 Elyria⊙ 57,538 ... F3
45322 Englewood 11,329 ... B6
45323 Enon 2,597 ... C6
44117 Euclid 59,999 ... J9
†45201 Evendale 1,954 ... C9
45042 Excello 900 ... B7
45324 Fairborn 29,702 ... B6
†45201 Fairfax 2,222 ... C9
45014 Fairfield 30,777 ... A7
44313 Fairlawn 6,100 ... G3
44077 Fairport Harbor 3,357 ... H2
44126 Fairview Park 19,311 ... G9
45325 Farmersville 950 ... A6
43521 Fayette 1,222 ... B2
45120 Felicity 929 ... B8
45840 Findlay⊙ 35,594 ... C3
45326 Fletcher 498 ... B5
43977 Flushing 1,266 ... J5
45843 Forest 1,633 ... C4
45405 Forest Park 18,675 ... B9
45230 Forestville 950 ... C10
45844 Fort Jennings 538 ... B4
45845 Fort Loramie 977 ... B5
†45426 Fort McKinley ... B6
45846 Fort Recovery 1,370 ... A5
†45801 Fort Shawnee 4,541 ... B4
44830 Fostoria 15,743 ... D3
45628 Frankfort 1,008 ... D7
45005 Franklin 10,711 ... B6
45629 Franklin Furnace 1,093 ... E8
43822 Frazeysburg 1,025 ... F5
44627 Fredericksburg 511 ... G4
43019 Fredericktown 2,299 ... F5
43973 Freeport 525 ... H5
43420 Fremont⊙ 17,834 ... D3
45630 Friendship 900 ... D8
43230 Gahanna 18,001 ... E5
44833 Galion 12,391 ... E4
45631 Gallipolis⊙ 5,576 ... F8
43022 Gambier 2,056 ... F5
44125 Garfield Heights 34,938 ... J9
44231 Garrettsville 1,769 ... H3
44040 Gates Mills 2,236 ... J9
44041 Geneva 6,655 ... J2
44043 Geneva-on-the-Lake 1,634 ... H2
43430 Genoa 2,213 ... D2
45121 Georgetown⊙ 3,467 ... C8
45327 Germantown 5,015 ... B6
45328 Gettysburg 545 ... A5
43431 Gibsonburg 2,479 ... D3
44420 Girard 12,517 ... J3
45848 Glandorf 746 ... B3
45246 Glendale 2,368 ... C9
†44139 Glenwillow 492 ... J10
45732 Glouster 2,211 ... F6
44629 Gnadenhutten 1,320 ... G5
†45201 Golf Manor 4,317 ... C9
45122 Goshen ... B7
44044 Grafton 2,231 ... F3
43522 Grand Rapids 962 ... C3
44045 Grand River 412 ... H2
†43212 Grandview Heights 7,420 ... D6
43023 Granville 3,851 ... E5
45330 Gratis 809 ... A6
43322 Green Camp 475 ... D4
45123 Greenfield 5,150 ... D7
45218 Greenhills 4,927 ... B9
44232 Greensburg 950 ... G4
44836 Green Springs 1,568 ... E3
44630 Greentown 300 ... H4
45331 Greenville⊙ 12,999 ... A5
44837 Greenwich 1,458 ... E3
43123 Grove City 16,816 ... D6
43125 Groveport 3,286 ... E6
45849 Grover Hill 486 ... B3
45634 Hamden 1,010 ... F7
45130 Hamersville 688 ... C8
*45011 Hamilton⊙ 63,189 ... A7
Hamilton-Middletown‡ 258,787 ... A7
43524 Hamler 625 ... B3
43931 Hannibal 550 ... J6
†43055 Hanover 926 ... F5
43126 Harrisburg 363 ... D6
45030 Harrison 5,855 ... A9
45850 Harrod 506 ... C4
†44085 Hartsgrove 200 ... J2
44632 Hartville 1,772 ... H4
43525 Haskins 568 ... C3
43127 Haydenville 395 ... F7
44838 Hayesville 518 ... F4
43055 Heath 6,969 ... F5
43025 Hebron 2,035 ... E6
43526 Hicksville 3,929 ... A3
†44143 Highland Heights 5,739 ... J9
43026 Hilliard 8,008 ... D5
45133 Hillsboro⊙ 6,356 ... C7
44234 Hiram 1,360 ... H3
43527 Holgate 1,315 ... B3
43528 Holland 1,048 ... C2
45033 Hooven 550 ... A9
43976 Hopedale 857 ... J5
44425 Hubbard 9,245 ... J3
45424 Huber Heights 35,480 ... B6
44236 Hudson 4,615 ... H3
†44022 Hunting Valley 786 ... J9
44839 Huron 7,123 ... E3
44131 Independence 6,607 ... H9
†45201 Indian Hill 5,521 ... C9
43932 Irondale 535 ... J4
45638 Ironton⊙ 14,290 ... E8
45640 Jackson⊙ 6,675 ... E7
45334 Jackson Center 1,310 ... B5
45740 Jacksonville 651 ... F7
45335 Jamestown 1,702 ... C6
44047 Jefferson⊙ 2,952 ... J2
†43162 Jefferson (West Jefferson) 4,448 ... D6
43128 Jeffersonville 1,252 ... C6
44840 Jeromesville 582 ... F4
43437 Jerry City 512 ... C3
43986 Jewett 972 ... H5
43031 Johnstown 3,158 ... E5
43748 Junction City 754 ... F6
45853 Kalida 1,019 ... B4
44240 Kent 26,164 ... H3
43326 Kenton⊙ 8,605 ... C4
45429 Kettering 61,186 ... B6
44637 Killbuck 937 ... G5
45034 Kings Mills 500 ... B7
45644 Kingston 1,208 ... E7
44048 Kingsville ... J2
44428 Kinsman 900 ... J3
43033 Kirkersville 626 ... E6
†44094 Kirtland 5,969 ... H2
43951 Lafferty 855 ... H5
44050 Lagrange 1,258 ... F3
44250 Lakemore 2,744 ... H3
43440 Lakeside 850 ... E2
43331 Lakeview 1,089 ... C4
44107 Lakewood 61,963 ... G9
43130 Lancaster⊙ 34,953 ... E6
43934 Lansing 950 ... J5
43332 La Rue 861 ... D4
43135 Laurelville 591 ... E7
†45501 Lawrenceville 307 ... C6
45036 Lebanon⊙ 9,636 ... B7
45135 Leesburg 1,019 ... D7
44431 Leetonia 2,121 ... J4
45856 Leipsic 2,171 ... C3
45338 Lewisburg 1,450 ... A6
44904 Lexington 3,823 ... E4
43532 Liberty Center 1,111 ... B3
*45801 Lima⊙ 47,381 ... B4
Lima‡ 218,244 ... B4
†45201 Lincoln Heights 5,259 ... C9
43442 Lindsey 571 ... D3
44432 Lisbon⊙ 3,159 ... J4
44253 Litchfield 650 ... F3
43136 Lithopolis 652 ... E6
45742 Little Hocking 800 ... G7
45215 Lockland 4,292 ... C9
44254 Lodi 2,942 ... F3
43138 Logan⊙ 6,557 ... F6
43140 London⊙ 6,958 ... C6
*44052 Lorain 75,416 ... F3
Lorain-Elyria‡ 274,909 ... F3
†44481 Lordstown 3,280 ... J3
44842 Loudonville 2,945 ... F4
44641 Louisville 7,996 ... H4
45140 Loveland 9,106 ... D9
45744 Lowell 729 ... H6
44436 Lowellville 1,558 ... J3
44843 Lucas 753 ... F4
45648 Lucasville 3,349 ... E8
43443 Luckey 895 ... D3
45142 Lynchburg 1,205 ... C7
44124 Lyndhurst 18,092 ... J9
43533 Lyons 596 ... B2
44056 Macedonia 6,571 ... J10
†45202 Mack ... B9
45243 Madeira 9,341 ... C9
44057 Madison 2,291 ... H2
44643 Magnolia 986 ... H4
43758 Malta 956 ... G6
44644 Malvern 1,032 ... H4
45144 Manchester 2,313 ... C8
*44901 Mansfield⊙ 53,927 ... F4
Mansfield‡ 131,205 ... F4
44255 Mantua 1,041 ... H3
44137 Maple Heights 29,735 ... H9
†43440 Marblehead 679 ... E2
45860 Maria Stein 950 ... A5

(continued on following page)

Agriculture, Industry and Resources

DOMINANT LAND USE

- Hogs, Soft Winter Wheat
- Livestock, Dairy, Soybeans, Cash Grain
- Dairy, General Farming
- General Farming, Livestock, Tobacco
- Fruit, Truck and Mixed Farming
- Forests
- Urban Areas

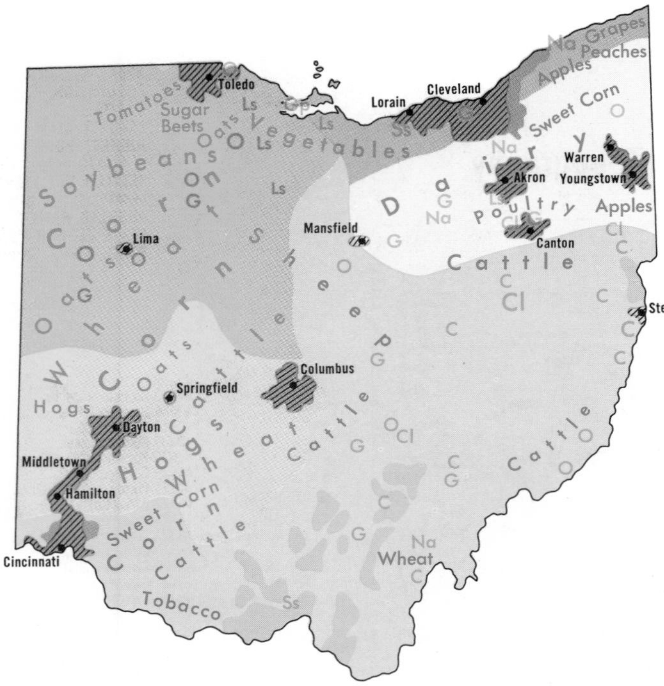

MAJOR MINERAL OCCURRENCES

- C Coal
- Cl Clay
- G Natural Gas
- Gp Gypsum
- Ls Limestone
- Na Salt
- O Petroleum
- Ss Sandstone
- Major Industrial Are[a]

45227 Mariemont 3,295 .... C9
45750 Marietta⊙ 16,467 .... G7
43302 Marion⊙ 37,040 .... D4
44645 Marshallville 788 .... G4
43935 Martins Ferry 9,331 .... J5
45146 Martinsville 539 .... C7
43040 Marysville⊙ 7,414 .... D5
45040 Mason 8,692 .... B7
44646 Massillon 30,557 .... H4
44438 Masury 1,836 .... J3
†45069 Maud 800 .... B7
43537 Maumee 15,747 .... C2
44124 Mayfield 3,577 .... J9
44124 Mayfield Heights 21,550 .... J9
45651 McArthur⊙ 1,912 .... F7
43534 McClure 694 .... C3
45858 McComb 1,608 .... C3
43756 McConnelsville⊙ 2,018 .... G6
44437 McDonald 3,744 .... J3
45859 McGuffey 646 .... C4
43044 Mechanicsburg 1,792 .... D5
44256 Medina⊙ 15,268 .... G3
45862 Mendon 749 .... A4
44060 Mentor 42,065 .... H2
44060 Mentor-on-the-Lake 7,919 .... G2
43540 Metamora 556 .... C2
45342 Miamisburg 15,304 .... B6
45041 Miamitown 800 .... A9
44652 Middlebranch 300 .... H4
†44017 Middleburg Heights 16,218 .... G10
44062 Middlefield 1,997 .... H3
45863 Middle Point 709 .... B4
45760 Middleport 2,971 .... F7
45042 Middletown 43,719 .... A6
44653 Midvale 654 .... H5
44846 Milan 1,569 .... E3
45150 Milford 5,232 .... D9
43045 Milford Center 764 .... D5
43447 Millbury 955 .... D2
44654 Millersburg⊙ 3,247 .... F4
43046 Millersport 844 .... E6
†45011 Millville 809 .... A7
44656 Mineral City 884 .... H4
44657 Minerva 4,549 .... H4
†43201 Minerva Park 1,618 .... E5
43938 Mingo Junction 4,834 .... J5
45865 Minster 2,557 .... B5
44260 Mogadore 4,190 .... H3
45050 Monroe 4,256 .... B7
44847 Monroeville 1,329 .... E3
45242 Montgomery 10,088 .... C9
43543 Montpelier 4,431 .... A2
†45439 Moraine 5,325 .... B6
†44022 Moreland Hills 3,083 .... J9
45152 Morrow 1,254 .... B7
43338 Mount Gilead⊙ 2,911 .... E4
45231 Mount Healthy 7,562 .... B9
45154 Mount Orab 1,573 .... C7
43939 Mount Pleasant 616 .... J5
43143 Mount Sterling 1,623 .... D6
43050 Mount Vernon⊙ 14,323 .... E5
43340 Mount Victory 667 .... D4
44262 Munroe Falls 4,731 .... H3
43144 Murray City 579 .... F6
43545 Napoleon⊙ 8,614 .... B3
44662 Navarre 1,343 .... H4
43940 Neffs 1,106 .... J5
44441 Negley 917 .... J4
45764 Nelsonville 4,567 .... F7
44849 Nevada 945 .... D4
43055 Newark⊙ 41,200 .... F5
Newark‡ 120,981 .... F5
45662 New Boston 3,188 .... E8
45869 New Bremen 2,393 .... B5
†44101 Newburgh Heights 2,678 .... H9
†45201 New Burlington 900 .... B9
45344 New Carlisle 6,498 .... C6
43832 Newcomerstown 3,986 .... G5
43762 New Concord 1,860 .... G6
43145 New Holland 783 .... D6
45871 New Knoxville 760 .... B5
45345 New Lebanon 4,501 .... B6
43764 New Lexington⊙ 5,179 .... F6
44851 New London 2,449 .... F3
45346 New Madison 1,008 .... A6
45767 New Matamoras J6
45011 New Miami 2,980 .... A7
44442 New Middletown 2,195 .... J4
45347 New Paris 1,709 .... A6
44663 New Philadelphia⊙ 16,883 .... G5
45768 Newport 975 .... H7
45157 New Richmond 2,769 .... B8
43766 New Straitsville 937 .... F6
44444 Newton Falls 4,960 .... J3
45244 Newtown 1,817 .... C10
45159 New Vienna 1,133 .... C7
44854 New Washington 1,213 .... E4
44445 New Waterford 1,314 .... J4
44446 Niles 23,088 .... J3
45872 North Baltimore 3,127 .... C3
45052 North Bend 546 .... B9
44450 North Bloomfield 650 .... J3
44720 North Canton 14,228 .... H4
45239 North College Hill 11,114 .... B9
44855 North Fairfield 525 .... E3
44067 Northfield 3,913 .... J10
44707 North Industry H4
44068 North Kingsville 2,939 .... J2
43060 North Lewisburg 1,072 .... C5
44452 North Lima 800 .... J4
†44057 North Madison 8,741 .... H2
44070 North Olmsted 36,486 .... G9
†44081 North Perry 897 .... H2
†44101 North Randall 1,054 .... H9
45414 Northridge 9,720 .... B6
44039 North Ridgeville 21,522 .... F3
44133 North Royalton 17,671 .... H10
†43619 Northwood 5,495 .... D2
†43701 North Zanesville 2,166 .... G6
44203 Norton 12,242 .... G3
44857 Norwalk⊙ 14,358 .... E3
45212 Norwood 26,342 .... C9
43449 Oak Harbor 2,678 .... D2
45656 Oak Hill 1,713 .... E8
†45419 Oakwood 9,372 .... B6
†44146 Oakwood 3,786 .... H9
45873 Oakwood 886 .... B3
44074 Oberlin 8,660 .... F3
†43201 Obetz 3,095 .... E6
45874 Ohio City 881 .... A4
44138 Olmsted Falls 5,868 .... G9
44862 Ontario 4,123 .... E4
†44101 Orange 2,376 .... J9
43616 Oregon 18,675 .... D2
44667 Orrville 7,511 .... G4
44076 Orwell 1,067 .... J2
45875 Ottawa⊙ 3,874 .... B3
†43601 Ottawa Hills 4,065 .... C2
45876 Ottoville 833 .... B4
45160 Owensville 858 .... B7
45056 Oxford 17,655 .... A6
44077 Painesville⊙ 16,391 .... H2
45877 Pandora 977 .... C4
44080 Parkman 600 .... H3
44129 Parma 92,548 .... H9
†44130 Parma Heights 23,112 .... G9
43062 Pataskala 2,284 .... E5
45879 Paulding⊙ 2,754 .... A3
45880 Payne 1,399 .... A3
45660 Peebles 1,790 .... D8
43450 Pemberville 1,321 .... C3
44264 Peninsula 604 .... G3
†44124 Pepper Pike 6,177 .... J9
44081 Perry 961 .... H2
43551 Perrysburg 10,215 .... C2
44864 Perrysville 836 .... F4
45354 Phillipsburg 705 .... B6
43771 Philo 799 .... G6
43147 Pickerington 3,917 .... E6
45661 Piketon 1,726 .... E7
43554 Pioneer 1,133 .... A2
45356 Piqua 20,480 .... B5
43064 Plain City 2,102 .... D5
43772 Pleasant City 481 .... G6
45359 Pleasant Hill 1,051 .... B5
43148 Pleasantville 780 .... F6
44865 Plymouth 1,939 .... E4
44514 Poland 3,084 .... J3
45769 Pomeroy⊙ 2,728 .... G7
43452 Port Clinton⊙ 7,223 .... E2
45770 Portland 150 .... G7
45662 Portsmouth⊙ 25,943 .... D8
43837 Port Washington 622 .... G5
43942 Powhatan Point 2,181 .... J6
45669 Proctorville 975 .... F9
43342 Prospect 1,159 .... D5
43456 Put-in-Bay 146 .... E2
43773 Quaker City 698 .... H6
43343 Quincy 633 .... C5
45771 Racine 908 .... G8
44265 Randolph 900 .... H3
44266 Ravenna⊙ 11,987 .... H3
43943 Rayland 566 .... J5
45215 Reading 12,843 .... C9
†44202 Reminderville 1,960 .... J10
†45202 Remington 600 .... C9
45773 Reno 576 .... H7
†43412 Reno Beach D2
44867 Republic 656 .... D3
43068 Reynoldsburg 20,661 .... E6
44286 Richfield 3,437 .... G3
43944 Richmond 624 .... J5
†44045 Richmond (Grand River) 412 .... H2
45673 Richmond Dale 950 .... E7
44143 Richmond Heights 10,095 .... H9
43344 Richwood 2,181 .... D5
45674 Rio Grande 864 .... F8
45167 Ripley 2,174 .... C8
43457 Risingsun 698 .... C3
44270 Rittman 6,063 .... G4
†43085 Riverlea 528 .... D5
44670 Robertsville 600 .... H4
44084 Rock Creek 652 .... J2
45882 Rockford 1,245 .... A4
44116 Rocky River 21,084 .... G9
44085 Rome 210 .... J2
44272 Rootstown 900 .... H3
†45662 Rosemount 1,747 .... D8
43777 Roseville 1,915 .... F6
45061 Ross 2,767 .... B9
43460 Rossford 5,978 .... C2
45236 Rossmoyne C9
†43943 Rush Run 560 .... J5
43347 Rushsylvania 610 .... C5
43348 Russells Point 1,156 .... C5
45775 Rutland 635 .... F7
45169 Sabina 2,799 .... C7
†44067 Sagamore Hills J10
45217 Saint Bernard 5,396 .... B9
43950 Saint Clairsville⊙ 5,452 .... J5
45883 Saint Henry 1,596 .... A5
45885 Saint Marys 8,414 .... B4
43072 Saint Paris 1,742 .... C5
44460 Salem 12,869 .... J4
43945 Salineville 1,629 .... J4
44870 Sandusky⊙ 31,360 .... E3
44671 Sandyville 500 .... H4
45171 Sardinia 826 .... C7
43946 Sardis 865 .... J6
43988 Scio 1,003 .... H5
†45662 Sciotodale 1,191 .... E8
45679 Seaman 1,039 .... C8
44672 Sebring 5,078 .... H4
†44131 Seven Hills 13,650 .... H9
45062 Seven Mile 841 .... A7
44273 Seville 1,568 .... G3
43947 Shadyside 4,315 .... J6
44120 Shaker Heights 32,487 .... H9
45241 Sharonville 10,108 .... C9
43782 Shawnee 924 .... F6
†44052 Sheffield 1,886 .... F3
44054 Sheffield Lake 10,484 .... F3
44875 Shelby 9,646 .... E4
43556 Sherwood 915 .... A3
44878 Shiloh 857 .... E4
44676 Shreve 1,608 .... F4
45365 Sidney⊙ 17,657 .... B5
†44221 Silver Lake 2,915 .... G3
†45201 Silverton 6,172 .... C9
43948 Smithfield 1,308 .... J5
44677 Smithville 1,467 .... G4
44139 Solon 14,341 .... J9
43783 Somerset 1,432 .... F6
†44001 South Amherst 1,848 .... F3
†43103 South Bloomfield 934 .... D6
45368 South Charleston 1,682 .... C6
44121 South Euclid 25,713 .... H9
45065 South Lebanon 2,700 .... B7
45680 South Point 3,918 .... E9
†44022 South Russell 2,784 .... H3
45369 South Vienna 464 .... C6
45682 South Webster 886 .... E8
43701 South Zanesville 1,739 .... F6
44275 Spencer 764 .... F3
45887 Spencerville 2,184 .... B4
45066 Springboro 4,962 .... B6
45246 Springdale 10,111 .... B9
*45501 Springfield⊙ 72,563 .... C6
Springfield‡ 183,885 .... C6
45370 Spring Valley 541 .... C6
44276 Sterling 600 .... G4
43952 Steubenville⊙ 26,400 .... J5
Steubenville-Weirton‡ 163,099 .... J5
43787 Stockport 558 .... G6
43154 Stoutsville 537 .... E6
44224 Stow 25,303 .... H3
44680 Strasburg 2,091 .... G4
44240 Streetsboro 9,055 .... H3
44136 Strongsville 28,577 .... G10
44471 Struthers 13,624 .... J3
43557 Stryker 1,423 .... B3
†44260 Suffield 650 .... H3
44681 Sugarcreek 1,966 .... G5
43074 Sunbury 2,101 .... E5
43558 Swanton 3,424 .... C2
44882 Sycamore 1,059 .... D4
43560 Sylvania 15,527 .... C2
45779 Syracuse 946 .... G7
44278 Tallmadge 15,269 .... H3
†43771 Taylorsville (Philo) 799 .... G6
45174 Terrace Park 2,044 .... D9
45780 The Plains 2,044 .... F7
43076 Thornville 838 .... F6
44883 Tiffin⊙ 19,549 .... D3
43963 Tiltonsville 1,750 .... J5
†44094 Timberlake 885 .... J8
45371 Tipp City 5,595 .... B6
†45245 Tobasco 950 .... C10
*43601 Toledo⊙ 354,635 .... D2
Toledo‡ 791,599 .... D2
43964 Toronto 6,934 .... J5
45067 Trenton 6,401 .... B7
45782 Trimble 579 .... F7
45426 Trotwood 7,802 .... B6
45373 Troy⊙ 19,086 .... B5
44682 Tuscarawas 917 .... H5
44087 Twinsburg 7,632 .... J10
44683 Uhrichsville 6,130 .... H5
45322 Union 5,219 .... B6
†47390 Union City 1,716 .... A5
44685 Uniontown 875 .... H4
44118 University Heights 15,401 .... H9
43221 Upper Arlington 35,648 .... D6
43351 Upper Sandusky⊙ 5,967 .... D4
43078 Urbana⊙ 10,762 .... C5
†43123 Urbancrest 880 .... D6
43080 Utica 2,238 .... F5
†43201 Valley View 730 .... D6
†44101 Valleyview 1,576 .... H9
45377 Vandalia 13,161 .... B6
45890 Vanlue 390 .... C4
45891 Van Wert⊙ 11,035 .... A4
44089 Vermilion 11,012 .... F3
45378 Verona 571 .... A6
45380 Versailles 2,384 .... A5
44473 Vienna 900 .... J3
44281 Wadsworth 15,166 .... G3
†44094 Waite Hill 529 .... H2
45687 Wakefield 300 .... E8
44889 Wakeman 906 .... F3
43465 Walbridge 2,900 .... C2
44687 Walnut Creek 550 .... G4
†44146 Walton Hills 2,199 .... J10
45895 Wapakoneta⊙ 8,402 .... B4
45785 Warner 250 .... H6
*44481 Warren⊙ 56,629 .... J3
44128 Warrensville Heights 16,565 .... H9
43844 Warsaw 765 .... G5
43160 Washington Court House⊙ 12,682 .... D6
44490 Washingtonville 865 .... J4
45786 Waterford 600 .... G6
43566 Waterville 3,884 .... C3
43567 Wauseon⊙ 6,173 .... B2
45690 Waverly⊙ 4,603 .... D7
43466 Wayne 894 .... C3
44688 Waynesburg 1,160 .... H4
45896 Waynesfield 826 .... C4
45068 Waynesville 1,796 .... B6
44090 Wellington 4,146 .... F3
45692 Wellston 6,016 .... F7
43968 Wellsville 5,095 .... J4
45381 West Alexandria 1,313 .... A6
45449 West Carrollton 13,148 .... B6
43081 Westerville 23,414 .... D5
44491 West Farmington 563 .... J3
44251 Westfield Center 791 .... G3
43162 West Jefferson 4,448 .... D6
43845 West Lafayette 2,225 .... G5
44145 Westlake 19,483 .... G9
43357 West Liberty 1,653 .... C5
43358 West Mansfield 716 .... C5
45383 West Milton 4,119 .... B6
43569 Weston 1,708 .... C3
†45662 West Portsmouth 4,095 .... D8
44287 West Salem 1,357 .... F4
45693 West Union⊙ 2,791 .... C8
43570 West Unity 1,639 .... B2
45694 Wheelersburg 4,796 .... E8
43213 Whitehall 21,299 .... E6
43571 Whitehouse 2,137 .... C2
44092 Wickliffe 16,790 .... J9
44890 Willard 5,720 .... E3
45176 Williamsburg 1,952 .... B7
44093 Williamsfield 950 .... J2
43164 Williamsport 792 .... D6
44094 Willoughby 19,329 .... J8
†44094 Willoughby Hills 8,612 .... J9
44094 Willowick 17,834 .... J8
45898 Willshire 564 .... A4
45177 Wilmington⊙ 10,431 .... C7
45697 Winchester 1,080 .... C8
44288 Windham 3,721 .... H3
43952 Wintersville 4,724 .... J5
45245 Withamsville 975 .... C10
†45201 Woodlawn 2,715 .... C9
†44101 Woodmere 877 .... J9
43793 Woodsfield⊙ 3,145 .... H6
43469 Woodville 2,050 .... D3
44691 Wooster⊙ 19,289 .... G4
43085 Worthington 15,016 .... E5
45215 Wyoming 8,282 .... C9
45385 Xenia⊙ 24,653 .... C6
45387 Yellow Springs 4,077 .... C6
43971 Yorkville 1,447 .... J5
*44501 Youngstown⊙ 115,436 .... J3
Youngstown-Warren‡ 531,350 .... J3
43701 Zanesville⊙ 28,655 .... G6
44697 Zoar 264 .... H4
44698 Zoarville 125 .... H4

OTHER FEATURES

Atwood (lake) H4
Auglaize (riv.) B4
Berlin (lake) H4
Big Walnut (creek) E5
Black (riv.) F3
Black Fork, Mohican (riv.) F4
Blanchard (riv.) C4
Blennerhassett (isl.) G7
Buckeye (lake) F6
Campbell (hill) C5
Captina (creek) J6
Cedar (pt.) ....
Chagrin (riv.) ....
Clear Fork (res.) ....
Clear Fork, Mohican (riv.) ....
Clendening (lake) ....
Cleveland-Hopkins Mun. Airport ....
Cuyahoga (riv.) ....
Darby (creek) ....
Deer (creek) ....
Delaware (lake) ....
Dillon (lake) ....
Dover (lake) ....
Duck (creek) ....
Erie (lake) ....
Eufaula (res.) ....
Grand (riv.) ....
Great Miami (riv.) ....
Hocking (riv.) ....
Hoover (res.) ....
Huron (riv.) ....
Indian (lake) ....
James A. Garfield Nat'l Hist. Site ....
Kelleys (isl.) ....
Keystone (res.) ....
Killbuck (creek) ....
Kokosing (riv.) ....
Leesville (lake) ....
Licking (riv.) ....
Little Beaver (creek) ....
Little Miami (riv.) ....
Little Miami, East Fork (riv.) ....
Little Muskingum (riv.) ....
Loramie (lake) ....
Mad (riv.) ....
Maumee (bay) ....
Maumee (riv.) ....
Middle Bass (isl.) ....
Mohican (riv.) ....
Mosquito Creek (lake) ....
Mound City Group Nat'l Mon. ....
Muskingum (riv.) ....
North Bass (isl.) ....
Ohio (riv.) ....
Ohio Brush (creek) ....
Olentangy (riv.) ....
Paint (creek) ....
Perry's Victory and Int'l Peace Mem. ....
Piedmont (lake) ....
Portage (riv.) ....
Pymatuning (res.) ....
Raccoon (creek) ....
Rattlesnake (creek) ....
Rickenbacker Air Force Base 1,763 ....
Rocky (riv.) ....
Rocky, West Branch (riv.) ....
Rocky Fork (lake) ....
Saint Joseph (riv.) ....
Saint Marys (lake) ....
Saint Marys (riv.) ....
Salt Fork (creek) ....
Sandusky (bay) ....
Sandusky (riv.) ....
Scioto (riv.) ....
Senecaville (lake) ....
Sevenmile (creek) ....
South Bass (isl.) ....
Stillwater (riv.) ....
Symmes (creek) ....
Tappan (lake) ....
Tiffin (riv.) ....
Tuscarawas (riv.) ....
Vermilion (riv.) ....
Wabash (riv.) ....
West Sister (isl.) ....
Whiteoak (creek) ....
William H. Taft Nat'l Hist. Site ....
Wills (creek) ....
Wills Creek (lake) ....
Wright-Patterson Air Force Base 9,155 ....
Yellow (creek) ....

⊙County seat.
‡Population of metropolitan area.
† Zip of nearest p.o. * Multiple z

AREA 69,956 sq. mi. (181,186 sq. km.)
POPULATION 3,025,290
CAPITAL Oklahoma City
LARGEST CITY Oklahoma City
HIGHEST POINT Black Mesa 4,973 ft. (1516 m.)
SETTLED IN 1889
ADMITTED TO UNION November 16, 1907
POPULAR NAME Sooner State
STATE FLOWER Mistletoe
STATE BIRD Scissor-tailed Flycatcher

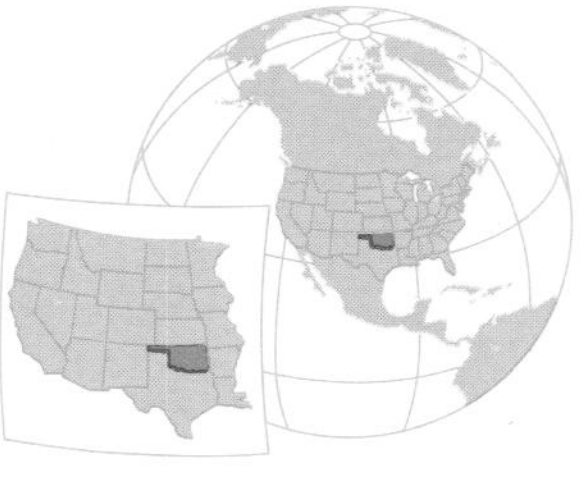

COUNTIES

lair 18,575 .... S3
falfa 7,077 .... K1
oka 12,748 .... O6
aver 6,806 .... E1
ckham 19,243 .... G4
aine 13,443 .... K3
yan 30,535 .... O7
ddo 30,905 .... K4
nadian 56,452 .... K3
rter 43,610 .... M6
erokee 30,684 .... R3
octaw 17,203 .... P6
narron 3,648 .... A1
eveland 133,173 .... M4
al 6,041 .... O5
manche 112,456 .... K5
tton 7,338 .... K6
aig 15,014 .... R1
eek 59,016 .... O3
ster 25,995 .... H3
laware 23,946 .... S2
wey 5,922 .... H2
s 5,596 .... G2
rfield 62,820 .... L2
rvin 27,856 .... M5
ady 39,490 .... L5
ant 6,518 .... L1
eer 7,028 .... G5
rmon 4,519 .... G5
rper 4,715 .... G1
skell 11,010 .... R4
ghes 14,338 .... O4
ckson 30,356 .... H5
ferson 8,183 .... L6
nnston 10,356 .... N6
y 49,852 .... M1
gfisher 14,187 .... L3
wa 12,711 .... J5
imer 9,840 .... R5
Flore 40,698 .... S5
coln 26,601 .... N3
jan 26,881 .... M3
ve 7,469 .... M7
jor 8,772 .... K2
rshall 10,550 .... N6
yes 32,261 .... R2
Clain 20,291 .... L5
McCurtain 36,151 .... S6
McIntosh 15,562 .... P4
Murray 12,147 .... M6
Muskogee 66,939 .... R3
Noble 11,573 .... M2
Nowata 11,486 .... P1
Okfuskee 11,125 .... O3
Oklahoma 568,933 .... M3
Okmulgee 39,169 .... P3
Osage 39,327 .... O1
Ottawa 32,870 .... S1
Pawnee 15,310 .... N2
Payne 62,435 .... N2
Pittsburg 40,524 .... P5
Pontotoc 32,598 .... N5
Pottawatomie 55,239 .... N4
Pushmataha 11,773 .... R6
Roger Mills 4,799 .... G3
Rogers 46,436 .... P2
Seminole 27,473 .... N4
Sequoyah 30,749 .... S3
Stephens 43,419 .... L6
Texas 17,727 .... C1
Tillman 12,398 .... J6
Tulsa 470,593 .... P2
Wagoner 41,801 .... P3
Washington 48,113 .... P1
Washita 13,798 .... J4
Woods 10,923 .... J1
Woodward 21,172 .... H2

CITIES and TOWNS

Zip *Name/Pop.* *Key*

74720 Achille 480 .... O7
74820 Ada⊙ 15,902 .... N5
74330 Adair 508 .... R2
73901 Adams 150 .... D1
73520 Addington 141 .... L6
74331 Afton 1,174 .... S1
74824 Agra 354 .... N3
74721 Albany 65 .... O7
73001 Albert 100 .... K4
74521 Albion 165 .... R5
74522 Alderson 366 .... P5
73002 Alex 769 .... L5
73716 Aline 313 .... K1
74825 Allen 998 .... O5
73521 Altus⊙ 23,101 .... H5
73717 Alva⊙ 6,416 .... J1
73004 Amber 416 .... L4
73718 Ames 314 .... K2
73719 Amorita 66 .... K1
73005 Anadarko⊙ 6,378 .... K4
74523 Antlers⊙ 2,989 .... P6
73006 Apache 1,560 .... K5
73620 Arapaho⊙ 851 .... H3
73401 Ardmore⊙ 23,689 .... M6
74901 Arkoma 2,175 .... T4
73832 Arnett⊙ 714 .... G2
74826 Asher 659 .... N5
74524 Ashland 72 .... O5
74525 Atoka⊙ 3,409 .... O6
74827 Atwood 225 .... O5
74001 Avant 461 .... O2
†73860 Avard 51 .... J1
73930 Baker 70 .... D1
74002 Barnsdall 1,501 .... O1
†74965 Baron 300 .... S3
74003 Bartlesville⊙ 34,568 .... O1
74722 Battiest 250 .... S6
73932 Beaver⊙ 1,939 .... F1
74421 Beggs 1,428 .... P3
†74966 Bengal 300 .... R5
74723 Bennington 302 .... P7
74331 Bernice 318 .... S1
73622 Bessie 245 .... H4
73008 Bethany 22,130 .... L3
74724 Bethel 350 .... S6
†74801 Bethel Acres 2,314 .... M4
74332 Big Cabin 252 .... R1
74630 Billings 632 .... M1
73009 Binger 791 .... K4
73720 Bison 103 .... L2
74008 Bixby 6,969 .... P3
74058 Blackburn 114 .... N2
74631 Blackwell 8,400 .... M1
73526 Blair 1,092 .... H5
73010 Blanchard 1,688 .... L4
74528 Blanco 215 .... P5
74529 Blocker 135 .... P4
†74701 Blue 150 .... O7
74333 Bluejacket 247 .... R1
73933 Boise City⊙ 1,761 .... B1
74726 Bokchito 628 .... O6
74930 Bokoshe 556 .... S4
74829 Boley 423 .... O4
74727 Boswell 702 .... P6
74830 Bowlegs 522 .... N4
74009 Bowring 115 .... O1
74422 Boynton 518 .... P3
73011 Bradley 284 .... L5
74423 Braggs 351 .... R3
74632 Braman 355 .... M1
73012 Bray 591 .... L5
73721 Breckinridge 261 .... L2
†73047 Bridgeport 115 .... K3
74010 Bristow 4,702 .... O3
74012 Broken Arrow 35,761 .... P2
74728 Broken Bow 3,965 .... S7
74530 Bromide 180 .... N6
†74873 Brooksville 46 .... M4
†74437 Bryant 74 .... P4
73834 Buffalo⊙ 1,381 .... G1
74931 Bunch 64 .... S3
74633 Burbank 161 .... N1
73722 Burlington 206 .... K1
73430 Burneyville 150 .... M7
73624 Burns Flat 2,431 .... H4
73625 Butler 388 .... H3
74831 Byars 353 .... N5
†74820 Byng 833 .... N5
73723 Byron 67 .... K1
73527 Cache 1,661 .... J5
74729 Caddo 923 .... O6
74730 Calera 1,390 .... O7
73014 Calumet 469 .... K3
74531 Calvin 315 .... O5
73835 Camargo 264 .... H2
74932 Cameron 365 .... T4
74425 Canadian 279 .... P4
74533 Caney 147 .... O6
73724 Canton 854 .... J2
73626 Canute 676 .... H4
73725 Capron 54 .... J1
74335 Cardin 500 .... S1
73726 Carmen 516 .... J1
73015 Carnegie 2,016 .... J4
74832 Carney 622 .... N3
73727 Carrier 259 .... K2
73627 Carter 367 .... H4
74934 Cartersville 79 .... S4
73016 Cashion 547 .... L3
74833 Castle 130 .... O4
74015 Catoosa 1,561 .... P2
73017 Cement 884 .... K5
74534 Centrahoma 166 .... O5
74834 Chandler⊙ 2,926 .... N3
73528 Chattanooga 403 .... J6
74426 Checotah 3,454 .... R4
74016 Chelsea 1,754 .... P1
73728 Cherokee⊙ 2,105 .... K1
73838 Chester 104 .... J2
73628 Cheyenne⊙ 1,207 .... G3
73018 Chickasha⊙ 15,828 .... L4
74635 Chilocco 400 .... M1
73020 Choctaw 7,520 .... M3
74337 Chouteau 1,559 .... R2
†74965 Christie 375 .... S3
73111 Cimarron L3
74017 Claremore⊙ 12,085 .... R2
74535 Clarita 72 .... O6
74536 Clayton 833 .... R5
74835 Clearview 250 .... O4
73729 Cleo Springs 514 .... K2
74020 Cleveland 2,972 .... O2
73601 Clinton 8,796 .... H3
74538 Coalgate⊙ 2,001 .... O5
74733 Colbert 1,122 .... O7
74338 Colcord 530 .... S2
†73010 Cole 309 .... L5
73432 Coleman 200 .... O6
74021 Collinsville 3,556 .... P2
73021 Colony 185 .... J4
73529 Comanche 1,937 .... L6
74339 Commerce 2,556 .... R1
73022 Concho 300 .... L3
†73041 Cooperton 31 .... J5
74022 Copan 960 .... P1
73632 Cordell⊙ 3,301 .... H4
73024 Corn 542 .... J4
†73456 Cornish 115 .... L6
74428 Council Hill 141 .... P3
73025 Countyline 550 .... L6
73730 Covington 715 .... L2
74429 Coweta 4,554 .... P3
†74934 Cowlington 546 .... S4
73027 Coyle 345 .... M3
73638 Crawford 53 .... G3
73028 Crescent 1,651 .... L3
74837 Cromwell 337 .... N4
74430 Crowder 431 .... P4
†73446 Cumberland 100 .... N6
74023 Cushing 7,720 .... N3
73639 Custer City 530 .... J3
73029 Cyril 1,220 .... K5
73731 Dacoma 226 .... J1
74838 Dale 160 .... M4
74026 Davenport 974 .... N3
73530 Davidson 501 .... J6
73030 Davis 2,782 .... M5
74636 Deer Creek 174 .... L1
74027 Delaware 544 .... P1
73115 Del City 28,523 .... L4
74028 Depew 682 .... O3
73531 Devol 186 .... J6
74431 Dewar 1,048 .... P4
74029 Dewey 3,545 .... P1
73031 Dibble 348 .... L4
†73401 Dickson 996 .... M6
73641 Dill City 649 .... H4
74340 Disney 464 .... S2
73032 Dougherty 210 .... M6
73733 Douglas 89 .... L2
74341 Douthat 30 .... S1
73734 Dover 570 .... L3
73735 Drummond 482 .... L2
74030 Drumright 3,162 .... N3
73533 Duncan⊙ 22,517 .... L5
74701 Durant⊙ 11,972 .... O6
73642 Durham 30 .... G3
74839 Dustin 498 .... O4
74734 Eagletown 650 .... S6
73033 Eakly 452 .... K4
74840 Earlsboro 266 .... N4
†73532 East Duke 484 .... H5
73034 Edmond 34,637 .... M3
73537 Eldorado 688 .... G6
73538 Elgin 1,003 .... K5
73644 Elk City 9,579 .... G4
73539 Elmer 131 .... H6
73035 Elmore City 582 .... M5
73935 Elmwood 300 .... F1
73036 El Reno⊙ 15,486 .... K3
†73529 Empire City 13 .... L6
73701 Enid⊙ 50,363 .... L2
73645 Erick 1,375 .... G4
74342 Eucha 210 .... S2
74432 Eufaula⊙ 3,159 .... P4
74637 Fairfax 1,949 .... N1
74343 Fairland 1,073 .... S1
73736 Fairmont 419 .... L2
†74080 Fair Oaks 346 .... P2
73737 Fairview⊙ 3,370 .... J2
†74881 Fallis 22 .... M3
74935 Fanshawe 416 .... S5
73840 Fargo 409 .... G2
73540 Faxon 140 .... J6
73646 Fay 140 .... J3
73937 Felt 120 .... A1
74543 Finley 350 .... R6
74842 Fittstown 500 .... N5
74843 Fitzhugh 150 .... N5
†73569 Fleetwood 12 .... L7
73541 Fletcher 1,074 .... K5
74652 Foraker 34 .... Q1
†73101 Forest Park 1,148 .... M3
73938 Forgan 611 .... E1
73038 Fort Cobb 760 .... K4
74434 Fort Gibson 2,477 .... R3
73841 Fort Supply 559 .... G1
74735 Fort Towson 789 .... R7
73647 Foss 188 .... H4
73039 Foster 100 .... M5
73435 Fox 400 .... M6
74031 Foyil 191 .... R2
74844 Francis 365 .... N5
73542 Frederick⊙ 6,153 .... H6
73842 Freedom 339 .... H1
73843 Gage 667 .... G2
74936 Gans 346 .... S4
73738 Garber 1,215 .... M2
74736 Garvin 162 .... S7
73844 Gate 146 .... F1
73040 Geary 1,700 .... K3
73436 Gene Autry 178 .... N6
73543 Geronimo 726 .... K6
†74531 Gerty 149 .... O5
74032 Glencoe 490 .... M2
74033 Glenpool 2,706 .... P3
74737 Golden 300 .... S6
†73093 Goldsby 603 .... L4
73739 Goltry 305 .... K1
†74740 Goodwater 240 .... S7
73939 Goodwell 1,186 .... C1
74435 Gore 445 .... R3
73041 Gotebo 457 .... J4
73544 Gould 318 .... G5
74545 Gowen 75 .... R5
73042 Gracemont 503 .... K4
73545 Grady 85 .... L6
73437 Graham 200 .... M6
†74652 Grainola 67 .... N1
73546 Grandfield 1,445 .... J6
†74349 Grand Lake Towne 36 .... S1
73547 Granite 1,617 .... H5
†74437 Grayson 150 .... P3
73043 Greenfield 233 .... K3
74344 Grove 3,378 .... S1
73044 Guthrie⊙ 10,312 .... M3
73942 Guymon⊙ 8,492 .... D1
74546 Haileyville 832 .... P5
74034 Hallett 186 .... N2
†73069 Hall Park 577 .... M4
73650 Hammon 866 .... H3
74845 Hanna 157 .... P4
74846 Harden City 250 .... N5
73944 Hardesty 243 .... D1
73832 Harmon 27 .... G2
73045 Harrah 2,897 .... M4
†74740 Harris 192 .... S7
74547 Hartshorne 2,380 .... R5
74436 Haskell 1,953 .... P3
73548 Hastings 246 .... K6
74740 Haworth 341 .... S7
73549 Headrick 223 .... H5
73438 Healdton 3,769 .... M6
74937 Heavener 2,776 .... S5
73741 Helena 710 .... K1
74741 Hendrix 106 .... O7
73046 Hennepin 300 .... M5
73742 Hennessey 2,287 .... L2
74437 Henryetta 6,432 .... O4
†73086 Hickory 95 .... N5
73743 Hillsdale 110 .... K1
73047 Hinton 1,432 .... K4
73744 Hitchcock 172 .... K3
74438 Hitchita 126 .... P3
73651 Hobart⊙ 4,735 .... J5
74439 Hoffman 407 .... P4
74848 Holdenville⊙ 5,469 .... O4
73550 Hollis⊙ 2,958 .... G5
73551 Hollister 82 .... J6
74035 Hominy 3,130 .... O2
74549 Honobia 80 .... R5
73945 Hooker 1,788 .... D1
†74366 Hoot Owl 3 .... R2
73746 Hopeton 42 .... J1
74940 Howe 562 .... S5
74440 Hoyt 160 .... R4
74743 Hugo⊙ 7,172 .... P7
74441 Hulbert 633 .... R3
74640 Hunter 276 .... L1
73048 Hydro 938 .... J3
74745 Idabel⊙ 7,622 .... S7
73552 Indiahoma 364 .... J5
74442 Indianola 254 .... P4
74036 Inola 1,550 .... P2
73747 Isabella 113 .... K2
74346 Jay⊙ 2,100 .... S2
†73759 Jefferson 92 .... L1
74037 Jenks 5,876 .... P2
74038 Jennings 395 .... N2
73749 Jet 352 .... K1
73049 Jones 2,270 .... M3
74347 Kansas 491 .... S2
74641 Kaw City 283 .... N1
74039 Kellyville 960 .... O3

(continued on following page)

Agriculture, Industry and Resources

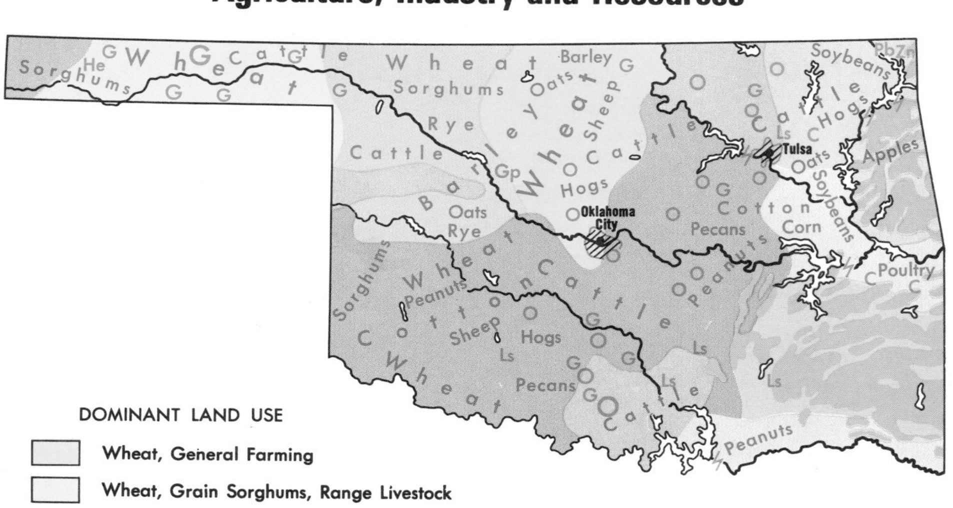

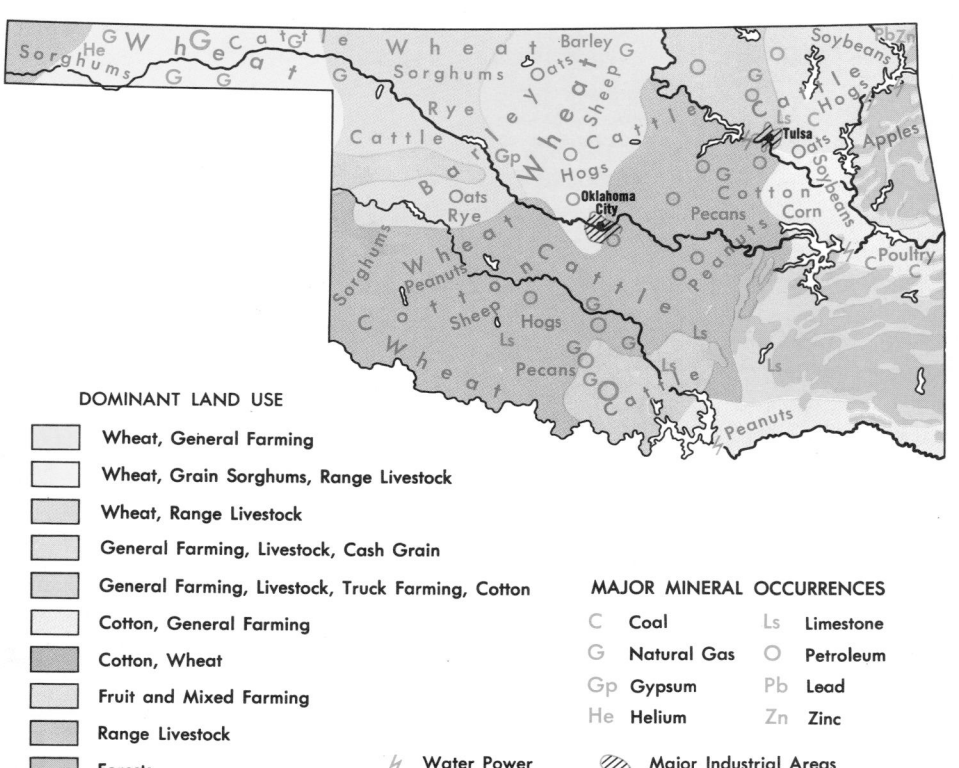

DOMINANT LAND USE

- Wheat, General Farming
- Wheat, Grain Sorghums, Range Livestock
- Wheat, Range Livestock
- General Farming, Livestock, Cash Grain
- General Farming, Livestock, Truck Farming, Cotton
- Cotton, General Farming
- Cotton, Wheat
- Fruit and Mixed Farming
- Range Livestock
- Forests

Water Power

MAJOR MINERAL OCCURRENCES

C	Coal	Ls	Limestone
G	Natural Gas	O	Petroleum
Gp	Gypsum	Pb	Lead
He	Helium	Zn	Zinc

Major Industrial Areas

74747 Kemp 178 .... O7
†74741 Kemp City (Hendrix) 106 ... O7
74040 Kendrick 132 .... N3
74748 Kenefic 140 .... O6
†74365 Kenwood 400 .... S2
74941 Keota 661 .... S4
74349 Ketchum 326 .... R1
73947 Keyes 557 .... B1
74041 Kiefer 912 .... O3
74601 Kildare 112 .... M1
73750 Kingfisher◉ 4,245 .... L3
73439 Kingston 1,171 .... N7
74552 Kinta 303 .... R4
74553 Kiowa 866 .... P5
73847 Knowles 44 .... F1
74849 Konawa 1,711 .... N5
74554 Krebs 1,754 .... P5
73753 Kremlin 301 .... L1
73754 Lahoma 537 .... K2
74850 Lamar 121 .... O4
†73728 Lambert 20 .... J1
74643 Lamont 571 .... L1
74350 Langley 582 .... R2
73050 Langston 443 .... M3
73848 Laverne 1,563 .... G1
73501 Lawton◉ 80,054 .... K5
   Lawton† 112,456 .... K5
74351 Leach 350 .... S2
73440 Lebanon 382 .... N7
73654 Leedey 499 .... H3
74942 Leflore 322 .... S5
74556 Lehigh 284 .... O6
74042 Lenapah 350 .... P1
73441 Leon 120 .... M7
74043 Leonard 400 .... P3
74943 Lequire 250 .... R4
73051 Lexington 1,731 .... M4
†74858 Lima (New Lima) 256 .... D4
73052 Lindsay 3,454 .... L5
73442 Loco 215 .... L6
74352 Locust Grove 1,179 .... R2
73849 Logan 18 .... F1
73443 Lone Grove 3,369 .... M6
73655 Lone Wolf 613 .... H5
73755 Longdale 405 .... K2
73053 Lookeba 221 .... K4
73842 Lookout 3 .... H1
†74063 Lotsee 7 .... O2
73553 Loveland 21 .... J6
73756 Loyal 112 .... K3
73757 Lucien 350 .... M2
73054 Luther 1,159 .... M3
†74578 Lutie 100 .... R5
74852 Macomb 58 .... M4
73446 Madill◉ 3,173 .... N6
73758 Manchester 146 .... L1
73554 Mangum◉ 3,833 .... G5
73555 Manitou 322 .... J5
74044 Mannford 1,610 .... O2
73447 Mannsville 568 .... N6
74045 Maramec 101 .... N2
74945 Marble City 294 .... S3
73448 Marietta◉ 2,494 .... M7
74644 Marland 340 .... M1
73055 Marlow 5,017 .... K5
73056 Marshall 372 .... L2
73556 Martha 219 .... H5
74854 Maud 1,444 .... N4
73851 May 89 .... G1
73656 Mayfield 17 .... G4
73057 Maysville 1,396 .... M5
74353 Mazie 118 .... R2
74501 McAlester◉ 17,255 .... P5
†74441 McBride 91 .... N7
74944 McCurtain 549 .... R4
74851 McLoud 4,061 .... M4
73445 McMillan 50 .... M6
73449 Mead 143 .... O7
73759 Medford◉ 1,419 .... L1
73557 Medicine Park 437 .... J5
74855 Meeker 1,032 .... N4
73760 Meno 171 .... K2
73058 Meridian 78 .... M3
74354 Miami◉ 14,237 .... S1
73110 Midwest City 49,559 .... M4
73450 Milburn 376 .... O6
74046 Milfay 200 .... N3
74856 Mill Creek 431 .... N6
74750 Millerton 262 .... S7
73451 Milo 25 .... M6
73059 Minco 1,489 .... L4
74946 Moffett 269 .... S4

74947 Monroe 150 .... S4
74444 Moodys 250 .... S2
73160 Moore 35,063 .... M4
73852 Mooreland 1,383 .... H2
74445 Morris 1,288 .... P3
73061 Morrison 671 .... M2
74047 Mounds 1,086 .... O3
73559 Mountain Park 557 .... J5
73062 Mountain View 1,189 .... J4
74557 Moyers 312 .... P6
74948 Muldrow 2,538 .... S4
73063 Mulhall 301 .... M2
74949 Muse 350 .... S5
74401 Muskogee◉ 40,011 .... R3
73064 Mustang 7,496 .... L4
73853 Mutual 135 .... H2
74354 Narcissa 100 .... S1
74646 Nardin 98 .... M1
73761 Nash 301 .... K1
74558 Nashoba 50 .... R6
74049 New Alluwe 129 .... R1
73065 Newcastle 3,076 .... L4
†73632 New Cordell
   (Cordell)◉ 3,301 .... H4
74647 Newkirk◉ 2,413 .... N1
74884 New Lima 256 .... O4
†74060 New Prue (Prue) 554 .... O2
†74080 New Tulsa 252 .... P2
†73116 Nichols Hills 4,171 .... L3
73066 Nicoma Park 2,588 .... M4
73068 Noble 3,497 .... M4
73018 Norge 87 .... K4
†73069 Norman◉ 68,020 .... M4
†73701 North Enid 992 .... L2

74358 North Miami 544 .... R1
74048 Nowata◉ 4,270 .... P1
73452 Oakland 485 .... N6
74359 Oaks 591 .... S2
73658 Oakwood 140 .... J3
74051 Ochelata 480 .... P1
74958 Octavia 30 .... S5
74052 Oilton 1,244 .... N2
74446 Okay 554 .... R3
73763 Okeene 1,601 .... K2
74859 Okemah◉ 3,381 .... O4
*73101 Oklahoma City
   (cap.)◉ 403,136 .... L4
   Oklahoma City† 834,088 .... L4
74447 Okmulgee◉ 16,263 .... O3
73450 Oktaha 376 .... R3
†74538 Olney 125 .... O6
73560 Olustee 721 .... H5
73764 Omega 50 .... K3
73948 Optima 133 .... D1
73765 Orienta 25 .... J2
73073 Orlando 218 .... M2
74054 Osage 243 .... O2
73561 Oscar 60 .... L7
73453 Overbrook 443 .... M6
74055 Owasso 6,149 .... P2
74860 Paden 448 .... N3
74951 Panama 1,425 .... S4
74759 Panola 75 .... R5
73074 Paoli 573 .... K4
†74435 Paradise Hill 154 .... R3
74451 Park Hill 200 .... R3

73075 Pauls Valley◉ 5,664 .... M5
74056 Pawhuska◉ 4,771 .... O1
74058 Pawnee◉ 1,688 .... N2
†74301 Pensacola 82 .... R2
†66713 Peoria 165 .... S1
74059 Perkins 1,762 .... M3
73076 Pernell 110 .... M5
73077 Perry◉ 5,796 .... M2
74862 Pharoah 100 .... O4
74360 Picher 2,180 .... S1
74752 Pickens 525 .... R4
73078 Piedmont 2,016 .... L3
†74873 Pink 911 .... M4
74560 Pittsburg 305 .... P5
73079 Pocasset 220 .... L4
74902 Pocola 3,268 .... T4
74601 Ponca City 26,238 .... M1
73766 Pond Creek 949 .... L1
74454 Porter 642 .... R3
74455 Porum 668 .... R4
74953 Poteau◉ 7,089 .... S4
74864 Prague 2,208 .... N4
74456 Preston 350 .... P3
74060 Prue 554 .... O2
74361 Pryor◉ 8,483 .... R2
73080 Purcell◉ 4,638 .... M4
73659 Putnam 74 .... J3
74363 Quapaw 1,097 .... S1
†74085 Quay 50 .... N2
73852 Quinlan 64 .... J2
74561 Quinton 1,228 .... R4
74650 Ralston 495 .... N2
74061 Ramona 567 .... P1

†73160 Ranchwood Manor 296 ... L4
73562 Randlett 461 .... K6
73081 Ratliff City 350 .... M6
74562 Rattan 332 .... R6
73455 Ravia 487 .... N6
74458 Redbird 199 .... P3
74563 Red Oak 676 .... R5
74651 Red Rock 376 .... M2
73563 Reed 48 .... G5
74801 Remus N4
†73759 Renfrow 27 .... L1
74459 Rentiesville 78 .... R4
73660 Reydon 252 .... G3
73456 Ringling 1,561 .... L6
74754 Ringold 200 .... R6
73768 Ringwood 389 .... K2
74062 Ripley 451 .... N2
†74932 Rock Island 160 .... T4
73661 Rocky 242 .... J4
74865 Roff 729 .... N5
74954 Roland 1,472 .... S4
73564 Roosevelt 396 .... J4
74364 Rose 100 .... R2
†74831 Rosedale 400 .... M5
73457 Rosston 66 .... G1
73457 Rubottom 35 .... M7
74755 Rufe 150 .... R6
73082 Rush Springs 1,451 .... L5
73565 Ryan 1,083 .... L6
74866 Saint Louis 109 .... N4
74365 Salina 1,115 .... R2
74955 Sallisaw◉ 6,403 .... S4
†73449 Sand Point 179 .... N7
74063 Sand Springs 13,121 .... O2

74066 Sapulpa◉ 15,853 .... O3
74867 Sasakwa 335 .... N5
74565 Savanna 828 .... P5
74756 Sawyer 200 .... R7
73662 Sayre◉ 3,177 .... G4
74460 Schulter 600 .... P3
73663 Seiling 1,103 .... J2
73856 Selman 25 .... H1
74868 Seminole◉ 8,590 .... N4
74068 Shamrock 218 .... N3
73857 Sharon 171 .... H2
73858 Shattuck 1,759 .... H1
74801 Shawnee◉ 26,506 .... N4
74652 Shidler 708 .... N1
†74701 Silo 43 .... N7
74069 Skedee 117 .... N2
74070 Skiatook 3,596 .... O2
†73051 Slaughterville 1,953 .... M4
74071 Slick 187 .... N3
74957 Smithville 133 .... S6
74567 Snow 200 .... R5
73566 Snyder 1,848 .... J5
74759 Soper 465 .... R7
74072 South Coffeyville 873 .... P1
74869 Sparks 772 .... N3
74366 Spavinaw 623 .... R2
73084 Spencer 4,064 .... M3
74760 Spencerville 275 .... R6
74073 Sperry 1,276 .... P2
74959 Spiro 2,221 .... S4
73458 Springer 679 .... M6
73567 Sterling 702 .... K5

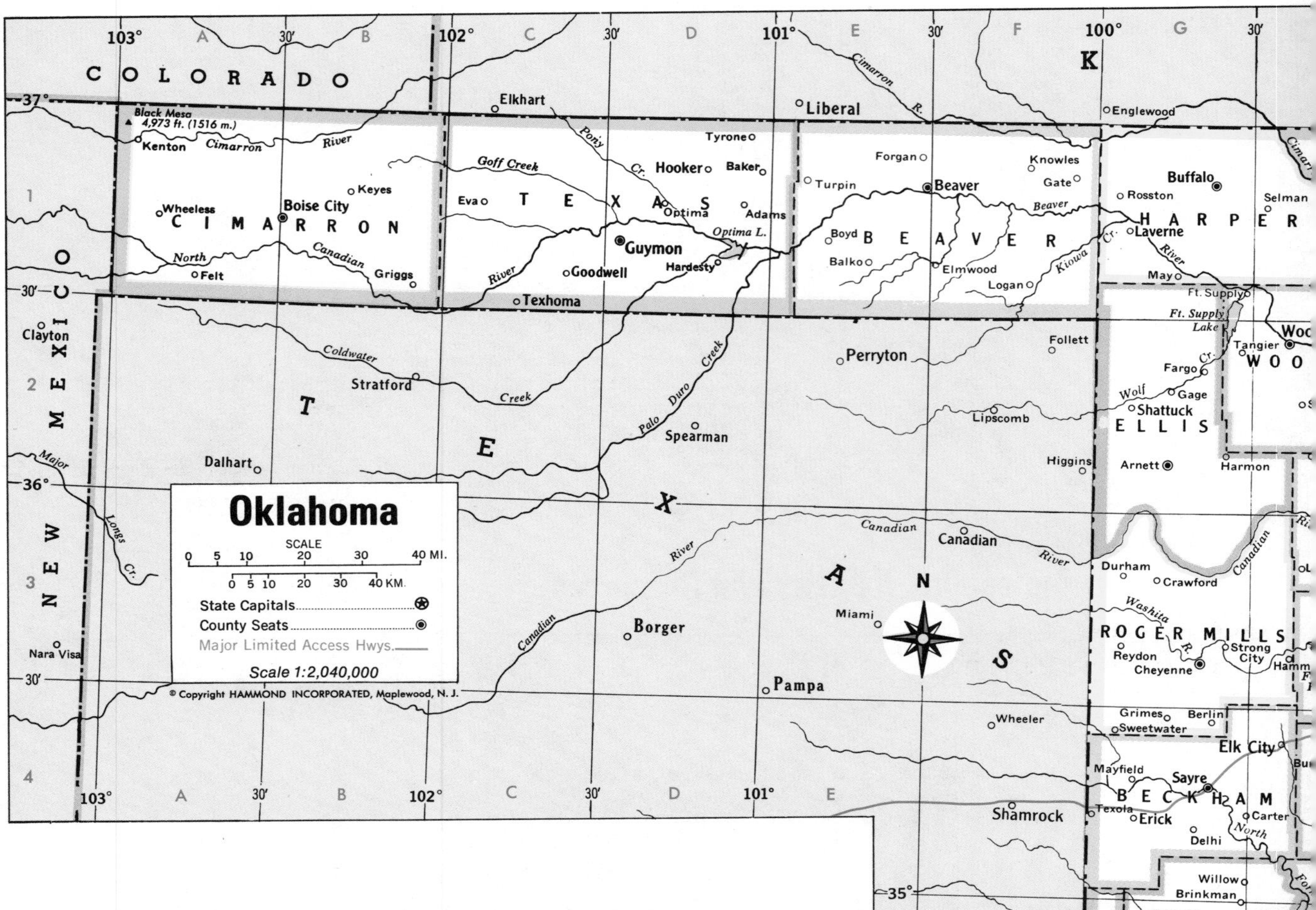

Oklahoma

SCALE
0 5 10   20   30   40 MI.
0 5 10 20   30   40 KM.

State Capitals .... ⊛
County Seats .... ◉
Major Limited Access Hwys. _____

Scale 1:2,040,000

© Copyright HAMMOND INCORPORATED, Maplewood, N.J.

Topography

0   50   100 MI.
0   50   100 KM.

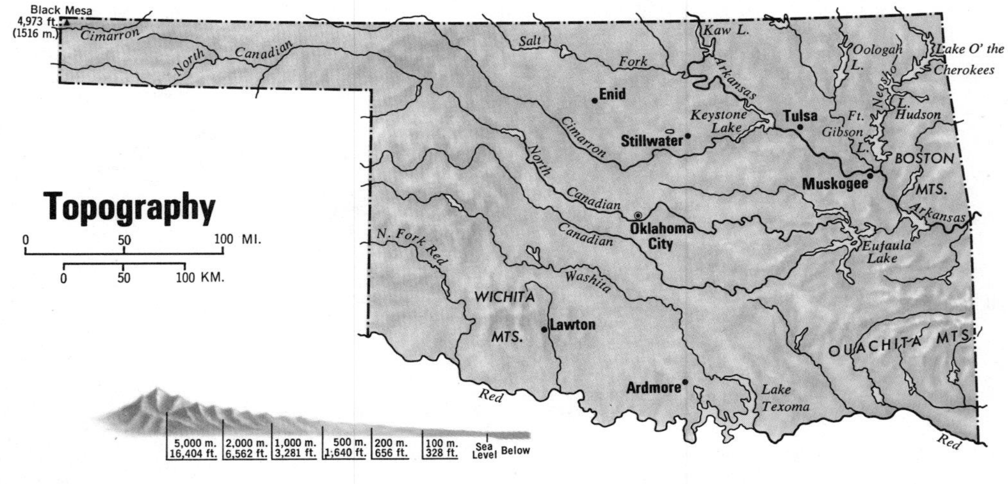

Black Mesa
4,973 ft.
(1516 m.)

5,000 m. | 2,000 m. | 1,000 m. | 500 m. | 200 m. | 100 m. | Sea Level | Below
16,404 ft. | 6,562 ft. | 3,281 ft. | 1,640 ft. | 656 ft. | 328 ft.

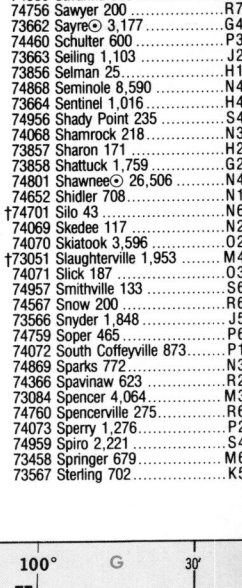

74461 Stidham 60	P4	74653 Tonkawa 3,524	M1
74462 Stigler⊙ 2,630	R4	†74852 Tribbey 215	M4
74074 Stillwater⊙ 38,268	N2	74856 Troy 92	N6
74960 Stilwell⊙ 2,369	S3	74875 Tryon 435	N3
74871 Stonewall 672	O5	74466 Tullahassee 145	P3
74367 Strang 126	R2	*74101 Tulsa⊙ 360,919	O2
74872 Stratford 1,459	M5	Tulsa‡ 689,628	O2
74569 Stringtown 1,047	P6	74572 Tupelo 542	O5
73665 Strong City 56	G3	74950 Turpin 450	E1
74079 Stroud 3,148	N3	74573 Tushka 358	O6
74570 Stuart 235	O5	74574 Tuskahoma 168	R5
†73565 Sugden 76		73088 Tussy 150	L6
73086 Sulphur⊙ 5,516	N5	73089 Tuttle 3,051	L4
74966 Summerfield 150	S5	73951 Tyrone 928	D1
73666 Sweetwater 85	G4	74763 Utica 38	O7
74463 Taft 489	R3	†73101 Valley Brook 921	M4
74464 Tahlequah⊙ 9,708	R3	74764 Valliant 927	R6
74080 Talala 191	P1	†74820 Vanoss 130	N5
74571 Talihina 1,387	S5	73091 Velma 831	L6
73667 Taloga⊙ 446	J2	74082 Vera 182	P2
†74462 Tamaha 145	S4	73092 Verden 625	K4
73087 Tatums 281	M6	74017 Verdigris 150	P2
74873 Tecumseh 5,123	N4	74877 Vernon 100	P4
73568 Temple 1,339	L6	74962 Vian 1,521	S4
74081 Terlton 155	O2	73859 Vici 845	H2
73569 Terral 604	L7	74301 Vinita⊙ 6,740	R1
73949 Texhoma 785	C1	73571 Vinson 42	G5
73668 Texola 106	G4	73572 Walters⊙ 2,778	K6
73459 Thackerville 431	M7	74878 Wanette 473	M5
73120 The Village 11,049	L3	74083 Wann 156	P1
73665 Thomas 1,515	J3		
†74017 Tiawah 125	P2		
73570 Tipton 1,475	H6		
73460 Tishomingo⊙ 3,212	N6		

73461 Wapanucka 472	N6	73801 Woodward⊙ 13,610	H2
74469 Warner 1,310	R4	74766 Wright City 1,168	R6
†74834 Warr Acres 9,940	L3	74470 Wyandotte 336	S1
†74834 Warwick 167	M3	73098 Wynnewood 2,615	M5
73093 Washington 477	L4	74084 Wynona 780	O1
73094 Washita 180	K4	74085 Yale 1,652	N2
73772 Watonga⊙ 4,139	K3	†74574 Yanush 123	R5
74964 Watts 316	S2	†74848 Yeager 138	O4
73773 Waukomis 1,551	K2	73099 Yukon 17,112	L3
73573 Waurika⊙ 2,258	L6		
73095 Wayne 621	M5	**OTHER FEATURES**	
73860 Waynoka 1,377	J1		
73096 Weatherford 9,640	J4	Altus (res.)	H5
74654 Webb City 157	N1	Altus A.F.B.	H5
74470 Webbers Falls 461	R3	Arbuckle Nat'l Rec. Area	N6
74369 Welch 697	R1	Arbuckles, Lake of the (lake)	M6
74880 Weleetka 1,195	O4	Arkansas (riv.)	S4
74471 Welling 115	S3	Atoka (lake)	P5
74881 Wellston 802	M3	Beaver (creek)	K6
74882 Welty 80	O3	Beaver (riv.)	F1
†74020 Westport 265	M4	Bird (creek)	O1
†72761 West Siloam Springs 431	S2	Black Bear (creek)	M2
74965 Westville 1,049	S2	Black Mesa (mt.)	A1
74883 Wetumka 1,725	O4	Blue (riv.)	O6
74884 Wewoka⊙ 5,480	O4	Bluestem (lake)	O1
74472 Whitefield 240	R4	Boston (mts.)	S3
74577 Whitesboro 450	S5	Broken Bow (lake)	S6
74578 Wilburton⊙ 2,996	R5	Cache (creek)	K6
†74932 Williams 110	T4	Canadian (riv.)	O4
73673 Willow 162	G4	Caney (riv.)	O1
73463 Wilson 1,585	M6	Canton (lake)	J2
74966 Wister 982	R5	Carl Blackwell (lake)	M2
†74868 Wolf 200	N4	Cherokees, Lake O'The (lake)	S1
73466 Woodville 94	N7	Chickasha (lake)	K4

Cimarron (riv.)	N2	North Canadian (riv.)	K3
Clear Boggy (creek)	O6	North Carrizo (riv.)	A1
Deep Fork, North Canadian (riv.)	N3	Oologah (lake)	P1
Denison (dam)	O7	Optima (lake)	D1
Elk (creek)	H3	Osage Ind. Res.	O1
Ellsworth (lake)	K6	Ouachita (lake)	R5
Eucha (lake)	S2	Pine Creek (lake)	R6
Eufaula (lake)	S3	Platt Nat'l Park	N6
Fort Cobb (res.)	J4	Poteau (riv.)	S5
Fort Gibson (lake)	R2	Prairie Dog Town Fork, Red	
Fort Sill 15,924	K5	(riv.)	F5
Fort Supply (lake)	G1	Red (riv.)	R7
Foss (lake)	H3	Red, North Fork (riv.)	G5
Great Salt Plains (lake)	K1	Salt Fork, Arkansas (riv.)	J1
Heyburn (res.)	O3	Salt Fork, Red (riv.)	G5
Hudson (lake)	R2	Sans Bois (mts.)	R4
Hugo (lake)	O1	Scott (mt.)	K5
Hulah (lake)	O1	Spavinaw (lake)	S2
Illinois (riv.)	S3	Tenkiller Ferry (lake)	S3
Jackfork (mt.)	P5	Texoma (lake)	N7
Kaw (lake)	O1	Thunderbird (lake)	M4
Kerr, Robert S. (res.)	S4	Tinker A.F.B.	M4
Keystone (lake)	O2	Tom Steed (lake)	J5
Kiamichi (mts.)	R5	Vance A.F.B.	K2
Kiamichi (riv.)	R5	Verdigris (riv.)	P2
Kiowa (creek)	F1	Washita (riv.)	M5
Lawtonka (lake)	K5	Waurika (lake)	K6
Little (riv.)	R6	Webbers Falls (res.)	R3
McAlester (lake)	P4	Wichita (mts.)	J5
Mountain Fork (riv.)	S6	Wildhorse (creek)	L5
Mud (creek)	L6	Wister (lake)	S5
Muddy Boggy (creek)	O5	Wolf (lake)	G2
Murray (lake)	M6	⊙County seat.	
Neosho (riv.)	R1	‡Population of metropolitan area.	
		† Zip of nearest p.o. * Multiple zips.	

COUNTIES

Baker 16,134 .... K3
Benton 68,211 .... D3
Clackamas 241,911 ...... E2
Clatsop 32,489 .... D1
Columbia 35,646 .... D2
Coos 64,047 .... C4
Crook 13,091 .... G3
Curry 16,992 .... C5
Deschutes 62,142 .... F4
Douglas 93,748 .... D4
Gilliam 2,057 .... J2
Grant 8,210 .... J3
Harney 8,314 .... H4
Hood River 15,835 ...... F2
Jackson 132,456 .... E5
Jefferson 11,599 .... F3
Josephine 58,855 .... D5
Klamath 59,117 .... F5
Lake 7,532 .... G5
Lane 275,226 .... E4
Lincoln 35,264 .... D3
Linn 89,495 .... E3

Malheur 26,896 .... K4
Marion 204,692 .... E3
Morrow 7,519 .... H2
Multnomah 562,640 ..... E2
Polk 45,203 .... D3
Sherman 2,172 .... G2
Tillamook 21,164 .... D2
Umatilla 58,861 .... J2
Union 23,921 .... J2
Wallowa 7,273 .... K2
Wasco 21,732 .... F2
Washington 245,860 .... G3
Wheeler 1,513 .... H3
Yamhill 55,332 .... D2

CITIES and TOWNS

Zip	Name/Pop.	Key
†97330	Adair Village 589	D3
†97810	Adams 240	J2
97620	Adel 24	H5
97901	Adrian 162	K4
†97365	Agate Beach 975	C3
97406	Agness 150	C5
97321	Albany ⊙ 26,678	D3
97407	Allegany 300	D4
97005	Aloha 28,353	A2
97324	Alsea 125	D3
†97601	Altamont 19,805	F5
97409	Alvadore 800	D3
97101	Amity 1,092	D2
97001	Antelope 39	G3
97530	Applegate 150	C4
97458	Arago 200	C4
97812	Arlington 521	H2
97520	Ashland 14,943	E5
97103	Astoria ⊙ 9,998	D1
97813	Athena 965	J2
97325	Aumsville 1,432	E3
97002	Aurora 523	B2
†97817	Austin 19	J3
97814	Baker ⊙ 9,471	K3
†97378	Ballston 120	D2
97411	Bandon 2,311	C4
97106	Banks 489	A1
†97013	Barlow 105	B2
97009	Barton 100	B2
97136	Bar View 170	C2
†97420	Barview 1,462	C4
97817	Bates 56	J3
97107	Bay City 986	D2
97621	Beatty 350	F5
97108	Beaver 350	D2
97004	Beavercreek 708	B2
97005	Beaverton 30,582	A2
97701	Bend ⊙ 17,263	F3
97522	Butte Falls 428	F3
†97058	Biggs 50	G2
97412	Blachly 80	D3
†97108	Blaine 38	D2
97326	Blodgett 250	D3
97413	Blue River 318	E3
97622	Bly 800	F5
97818	Boardman 1,261	H2
97623	Bonanza 270	F5
97008	Bonneville 80	E2
97009	Boring 150	E2
97010	Bridal Veil 20	E2
97458	Bridge 200	D4
†97136	Brighton 150	C2
97001	Brightwood 200	E2
97414	Broadbent 400	C4
97903	Brogan 130	K3
97415	Brookings 3,384	C5
97305	Brooks 490	A3
†97524	Brownsboro 150	E5
97327	Brownsville 1,261	E3
97351	Buena Vista 130	D3
†97420	Bunker Hill 1,555	C4
97720	Burns ⊙ 3,579	H4
†97002	Butteville 20	A2
97109	Buxton 450	D2
97416	Camas Valley 750	D4
97730	Camp Sherman 350	F3
†97493	Canary 23	D4
97013	Canby 7,659	B2
97110	Cannon Beach 1,187	D2
97820	Canyon City ⊙ 639	J3
97417	Canyonville 1,288	D5
97111	Carlton 1,302	D2
97014	Cascade Locks 838	E2
97329	Cascadia 250	E3
97523	Cave Junction 1,023	D5
97821	Cayuse 200	J2
97225	Cedar Hills 9,619	A2
†97005	Cedar Mill 900	A2
†97058	Celilo 50	G2
97502	Central Point 6,357	D5
97420	Charleston 500	C4
97306	Chemawa 400	A3
97731	Chemult 800	F4
†97058	Chenoweth 2,820	C2
97119	Cherry Grove 350	A2
97055	Cherryville 75	E2
97419	Cheshire 300	D3
97624	Chiloquin 778	F5
97015	Clackamas	D2
97016	Clatskanie 1,648	D1
97112	Cloverdale 300	D2
97401	Coburg 699	E3
97017	Colton 305	B3
97018	Columbia City 678	E2
97823	Condon ⊙ 783	G2
97420	Coos Bay 14,424	C4
97423	Coquille ⊙ 4,481	C4
97113	Cornelius 4,462	A2
97330	Corvallis ⊙ 40,960	D3
97424	Cottage Grove 7,148	D4

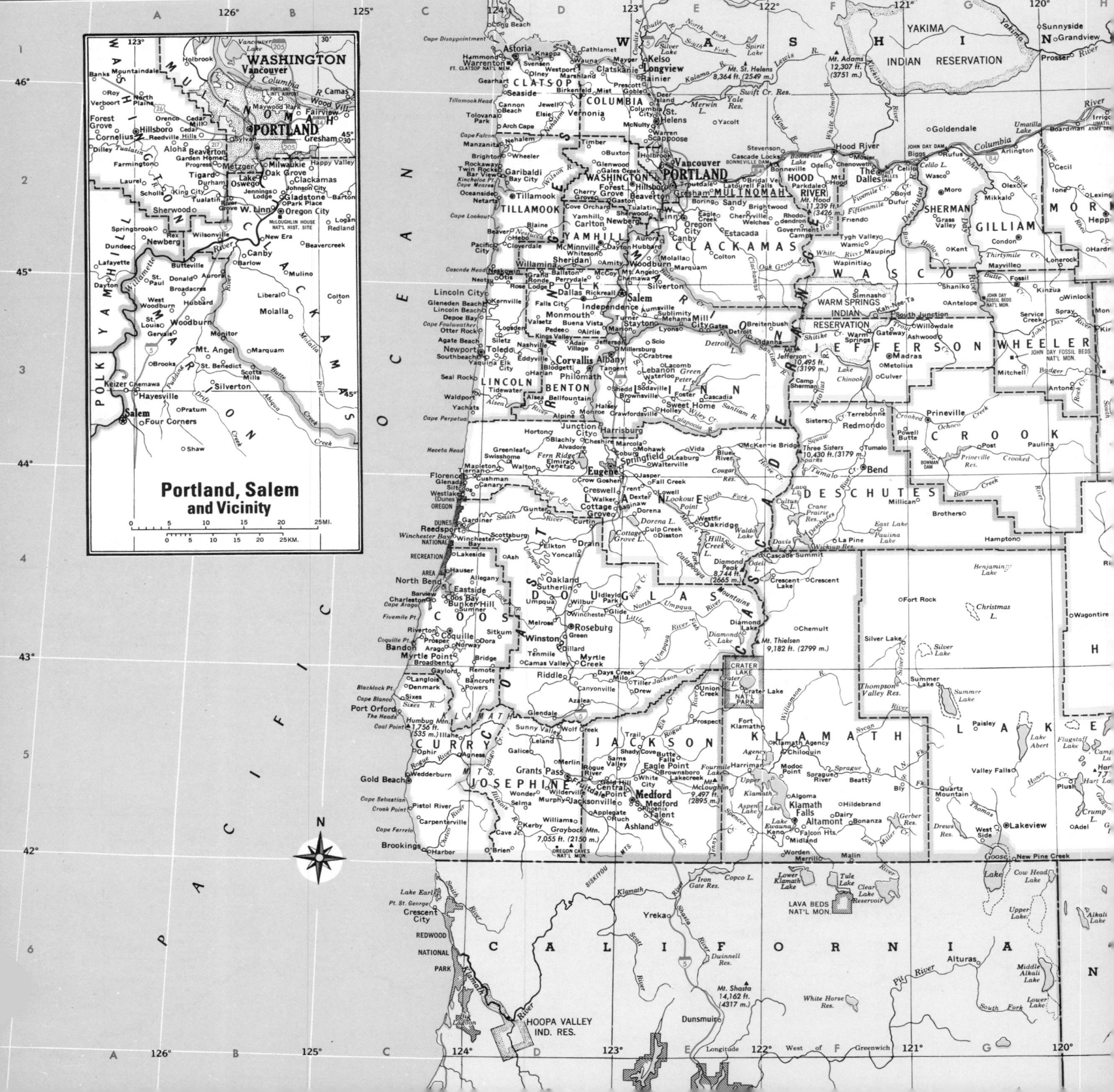

97824 Cove 451 ...K2
97335 Crabtree 200 ...E3
97732 Crane 84 ...J4
97336 Crawfordsville 350 ...E3
97733 Crescent 750 ...F4
97425 Crescent Lake 120 ...F4
97426 Creswell 1,770 ...D4
†97401 Crow 200 ...D4
97427 Culp Creek 600 ...E4
97734 Culver 514 ...F3
97428 Curtin 350 ...E4
†97439 Cushman 175 ...D4
97625 Dairy 80 ...F5
97338 Dallas⊙ 8,530 ...D3
97058 Dalles, The⊙ 10,820 ...F2
97429 Days Creek 550 ...D5
97114 Dayton 1,409 ...A3
97825 Dayville 199 ...H3
97054 Deer Island 225 ...E2
97341 Depoe Bay 723 ...C3
97342 Detroit 367 ...E3
97431 Dexter 500 ...E4
97432 Dillard 602 ...D4
†97116 Dilley 250 ...A2

†97427 Disston 123 ...E4
97200 Donald 267 ...A3
97434 Dorena 200 ...E4
97435 Drain 1,148 ...D4
97021 Dufur 560 ...F2
97115 Dundee 1,223 ...A2
†97493 Dunes (Westlake) 1,124 ...C4
†97233 Durham 707 ...B2
97905 Durkee 158 ...K3
97022 Eagle Creek 250 ...E2
97524 Eagle Point 2,764 ...E5
97420 Eastside 1,601 ...C4
97826 Echo 624 ...H2
97343 Eddyville 564 ...D3
97827 Elgin 1,701 ...K2
97436 Elkton 155 ...D4
97437 Elmira 900 ...D3
97828 Enterprise⊙ 2,003 ...K2
97023 Estacada 1,419 ...E2
*97401 Eugene⊙ 105,624 ...D3
   Eugene-Springfield‡
      275,226 ...D3
97024 Fairview 1,749 ...B2
†97601 Falcon Heights ...F5

97344 Falls City 804 ...D3
97710 Fields 150 ...J5
97439 Florence 4,411 ...C4
97116 Forest Grove 11,499 ...A2
97626 Fort Klamath 200 ...E5
97735 Fort Rock 150 ...G4
97830 Fossil⊙ 535 ...G2
97345 Foster 850 ...E3
97301 Four Corners 11,331 ...A3
97831 Fox 30 ...H3
†97526 Fruitdale-Harbeck 4,733 ...D5
97117 Gales Creek 150 ...A2
97223 Garden Home-
   Whitford 6,926 ...A2
97441 Gardiner 750 ...C4
97118 Garibaldi 999 ...D2
97119 Gaston 471 ...D2
97346 Gates 455 ...E3
†97741 Gateway 108 ...F3
97458 Gaylord 80 ...C5
97138 Gearhart 967 ...C1
97026 Gervais 799 ...A3
†97810 Gibbon 100 ...J2
97027 Gladstone 9,500 ...B2

AREA 97,073 sq. mi. (251,419 sq. km.)
POPULATION 2,633,149
CAPITAL Salem
LARGEST CITY Portland
HIGHEST POINT Mt. Hood 11,239 ft.
   (3426 m.)
SETTLED IN 1810
ADMITTED TO UNION February 14, 1859
POPULAR NAME Beaver State
STATE FLOWER Oregon Grape
STATE BIRD Western Meadowlark

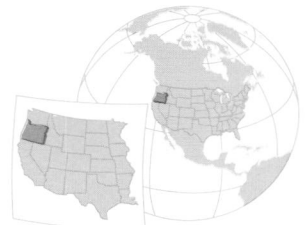

Topography

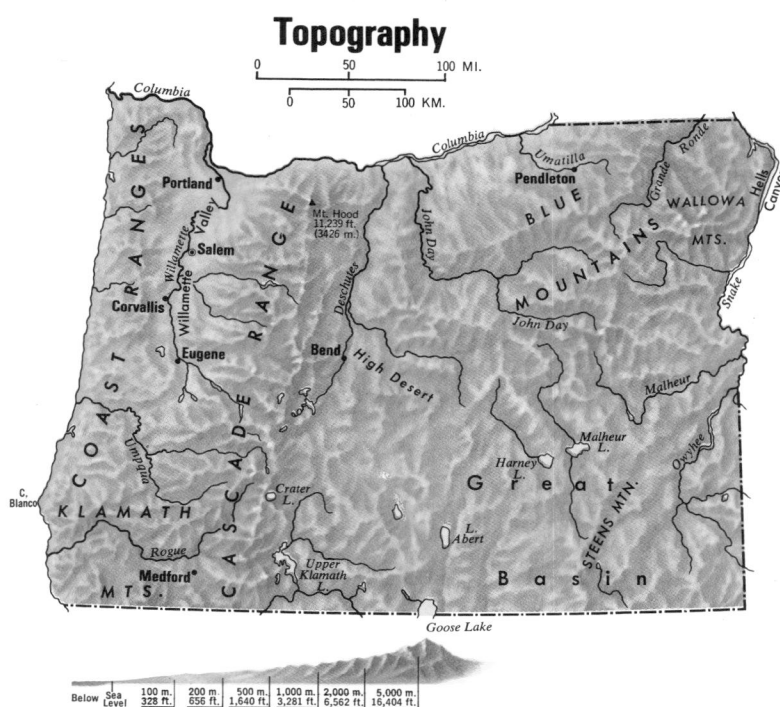

| Below Sea Level | 100 m. 328 ft. | 200 m. 656 ft. | 500 m. 1,640 ft. | 1,000 m. 3,281 ft. | 2,000 m. 6,562 ft. | 5,000 m. 16,404 ft. |

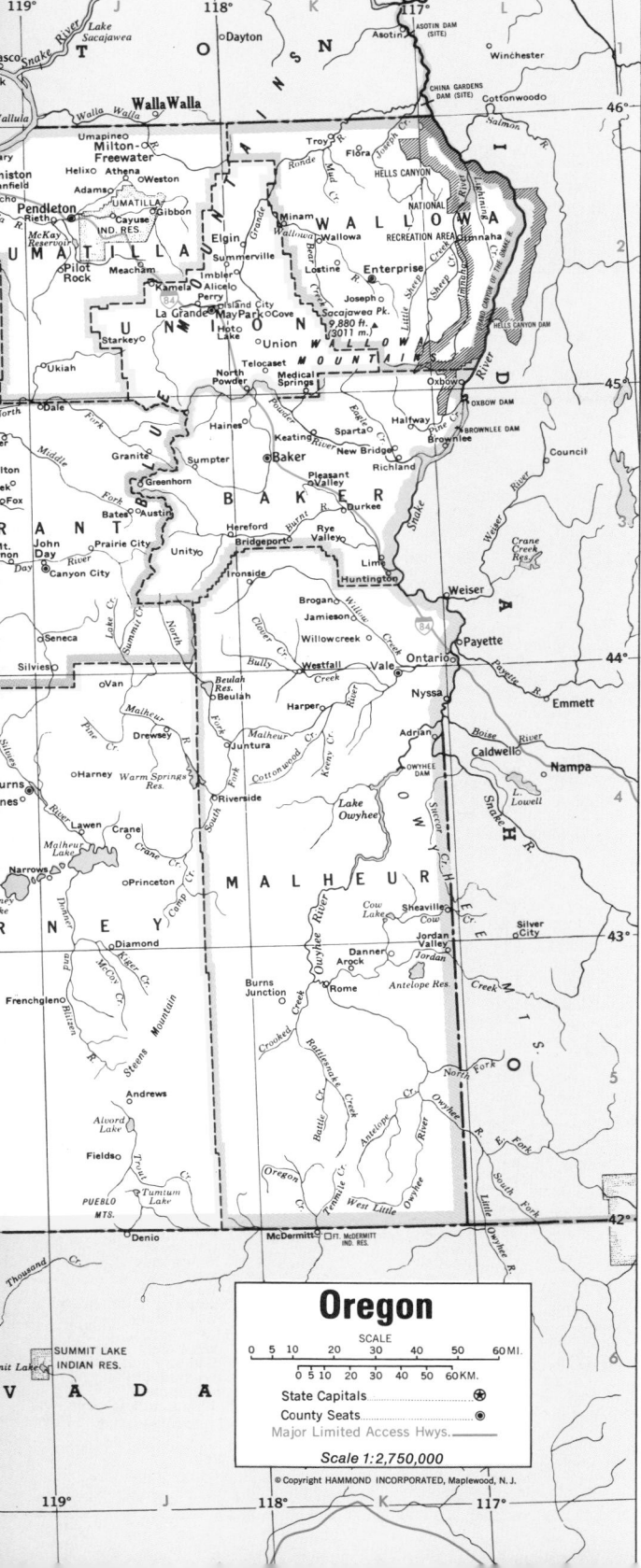

Oregon

SCALE
0 5 10 20 30 40 50 60 MI.
0 5 10 20 30 40 50 60 KM.

State Capitals ... ⊛
County Seats ... ⊙
Major Limited Access Hwys.

Scale 1:2,750,000

© Copyright HAMMOND INCORPORATED, Maplewood, N.J.

†97439 Glenada 300 ...C4
97442 Glendale 712 ...D5
97388 Gleneden Beach 400 ...C3
97120 Glenwood 225 ...D2
97443 Glide 470 ...D4
97048 Goble 108 ...E1
97444 Gold Beach⊙ 1,515 ...C5
97525 Gold Hill 904 ...D5
97401 Goshen 200 ...D4
97028 Government Camp 230 ...F2
97347 Grand Ronde 289 ...D2
†97877 Granite 17 ...J3
97526 Grants Pass⊙ 15,032 ...D5
97029 Grass Valley 164 ...G2
†97470 Green 3,897 ...D4
97030 Gresham 33,005 ...B2
97833 Haines 341 ...J3
97834 Halfway 310 ...K3
97348 Halsey 693 ...D3
97121 Hammond 516 ...C1
†97222 Happy Valley 1,499 ...B2
97415 Harbor 2,856 ...C5
97906 Harper 400 ...K4
†97601 Harriman 250 ...E5
97446 Harrisburg 1,881 ...D3
†97459 Hauser 400 ...C4
†97301 Hayesville 9,213 ...A3
97122 Hebo 400 ...D2
97835 Helix 155 ...J2
97836 Heppner⊙ 1,498 ...H2
97837 Hereford 128 ...K3
97838 Hermiston 9,408 ...H2
97123 Hillsboro⊙ 27,664 ...A2
97738 Hines 1,632 ...H4
†97208 Holbrook 494 ...A1
†97386 Holley 75 ...E3
97031 Hood River⊙ 4,329 ...F2
97448 Horton 175 ...D3
†97850 Hot Lake 4 ...K2
97032 Hubbard 1,640 ...A3
97907 Huntington 539 ...K3
97350 Idanha 319 ...E3
97447 Idleyld Park 300 ...D4
97841 Imbler 292 ...J2
97351 Independence 4,024 ...D3
97843 Ione 345 ...H2
97844 Irrigon 700 ...H2
97851 Island City 477 ...J2
97530 Jacksonville 2,030 ...D5
97909 Jamieson 120 ...K3
97401 Jasper 231 ...E3
97352 Jefferson 1,702 ...D3
†97845 John Day 2,012 ...J3
97027 Johnson City 378 ...B2
97910 Jordan Valley 473 ...K5
97846 Joseph 999 ...K2
97448 Junction City 3,320 ...D3
97911 Juntura 5 ...K4
97303 Keizer 18,592 ...A3

97627 Keno 500 ...F5
97033 Kent 200 ...G2
97531 Kerby 650 ...D5
97223 King City 1,853 ...A2
97630 Kings Valley 50 ...D3
97849 Kinzua 2 ...H3
97601 Klamath Falls⊙ 16,661 ...F5
†97103 Knappa 950 ...D1
†97355 Lacomb 425 ...E3
97127 Lafayette 1,215 ...A2
97850 La Grande⊙ 11,354 ...J2
†97524 Lakecreek 160 ...E5
97034 Lake Oswego 22,527 ...B2
97449 Lakeside 1,453 ...C4
97630 Lakeview⊙ 2,770 ...G5
97450 Langlois 150 ...C5
97739 La Pine 850 ...F4
97401 Leaburg 150 ...E3
97355 Lebanon 10,413 ...E3
97839 Lexington 307 ...H2
97042 Liberal 300 ...B3
†97341 Lincoln Beach 275 ...C3
97367 Lincoln City 5,469 ...C3
†97405 Logan 450 ...B2
†97823 Lonerock 26 ...H2
97856 Long Creek 252 ...H3
97857 Lostine 250 ...K2
97452 Lowell 661 ...E4
97358 Lyons 877 ...E3
97741 Madras⊙ 2,235 ...F3
97632 Malin 539 ...F5
97130 Manzanita 443 ...C2
97453 Mapleton 950 ...D3
97454 Marcola 900 ...E3
97359 Marion 300 ...D3
97037 Maupin 495 ...F2
†97850 May Park ...J2
97220 Maywood Park 1,083 ...B2
97401 McKenzie Bridge 500 ...E3
97128 McMinnville⊙ 14,080 ...D2
97858 McNary 330 ...H2
†97053 McNulty 1,805 ...E2
97859 Meacham 150 ...J2
97501 Medford⊙ 39,603 ...E5
   Medford‡ 132,456 ...E5
97384 Mehama 250 ...E3
97532 Merlin 500 ...D5
97633 Merrill 809 ...F5
97741 Metolius 451 ...F3
†97223 Metzger 5,544 ...A2
97634 Midland 520 ...F5
97360 Mill City 1,565 ...E3
97321 Millersburg 562 ...D3
†97417 Milo 600 ...E5
97862 Milton-Freewater 5,086 ...J2
97222 Milwaukie 17,931 ...B2
97038 Molalla 2,992 ...B3
97361 Monmouth 5,594 ...D3
97456 Monroe 412 ...D3

97864 Monument 192 ...H3
97039 Moro⊙ 336 ...G2
97040 Mosier 340 ...F2
97362 Mount Angel 2,876 ...B3
97041 Mount Hood 200 ...F2
97865 Mount Vernon 569 ...H3
97042 Mulino 720 ...B2
97533 Murphy 500 ...D5
97457 Myrtle Creek 3,365 ...D4
97458 Myrtle Point 2,859 ...C4
97131 Nehalem 258 ...D2
97364 Neotsu 300 ...C2
97149 Neskowin 250 ...D2
97143 Netarts 975 ...C2
97132 Newberg 10,394 ...A2
97635 New Pine Creek 400 ...G5
97365 Newport⊙ 7,519 ...C3
97459 North Bend 9,779 ...C4
97133 North Plains 715 ...A2
97867 North Powder 430 ...K2
97460 Norway 150 ...C4
97913 Nyssa 2,862 ...K4
97268 Oak Grove 11,640 ...B2
97462 Oakland 886 ...D4
97463 Oakridge 3,729 ...E4
97534 O'Brien 850 ...D5
97134 Oceanside 300 ...C2
97044 Odell 450 ...F2
97914 Ontario 8,814 ...K3
97464 Ophir 275 ...C5
97045 Oregon City⊙ 14,673 ...B2
†97123 Orenco 220 ...A2
97368 Otis 200 ...D2
97369 Otter Rock 450 ...C3
97840 Oxbow 100 ...L2
97135 Pacific City 500 ...C2
97636 Paisley 343 ...G5
97041 Parkdale 350 ...F2
†97045 Park Place 500 ...B2
97801 Pendleton⊙ 14,521 ...J2
†97101 Perrydale 200 ...D2
97370 Philomath 2,673 ...D3
97535 Phoenix 2,309 ...E5
97868 Pilot Rock 1,630 ...J2
*97201 Portland⊙ 366,383 ...B2
   Portland‡ 1,242,187 ...B2
97465 Port Orford 1,061 ...C5
97753 Powell Butte 350 ...F3
97466 Powers 819 ...D5
97869 Prairie City 1,106 ...J3
†97048 Prescott 73 ...E1
97721 Princeton 5 ...J4
97754 Prineville⊙ 5,276 ...G3
†97233 Progress 100 ...A2
97536 Prospect 200 ...E5
†97411 Prosper 110 ...C4
97048 Rainier 1,655 ...E1
†97045 Redland 700 ...B2
97756 Redmond 6,452 ...F3
97467 Reedsport 4,984 ...C4

(continued on following page)

Agriculture, Industry and Resources

DOMINANT LAND USE

- Specialized Wheat
- Wheat, Peas
- Specialized Dairy
- Dairy, Poultry, Mixed Farming
- Fruit and Mixed Farming
- Potatoes, General Farming
- General Farming, Dairy, Hay, Sugar Beets
- General Farming, Livestock, Special Crops
- Range Livestock
- Forests
- Nonagricultural Land

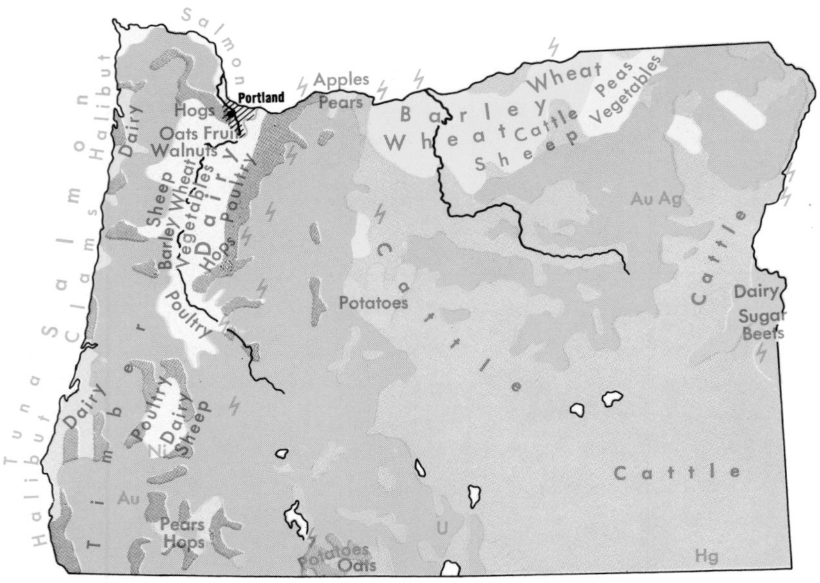

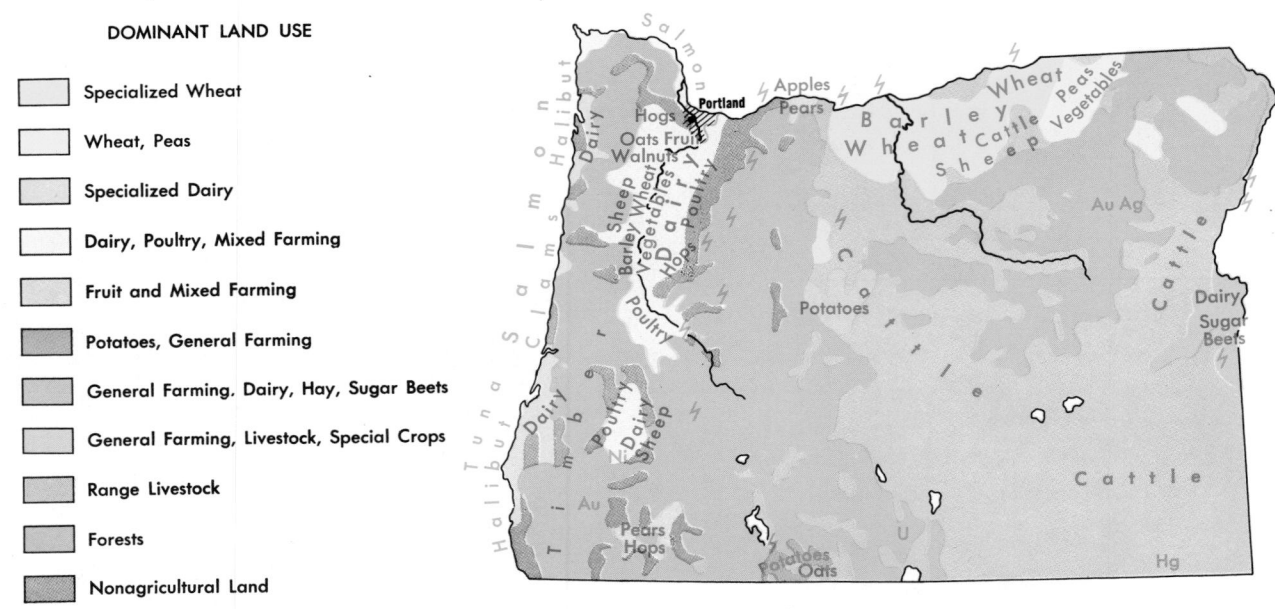

MAJOR MINERAL OCCURRENCES

Ag Silver
Au Gold
Hg Mercury
Ni Nickel
U Uranium

Water Power
Major Industrial Areas

†97005 Reedville 850 ..... A2
97870 Richland 181 ..... K3
97371 Rickreall 700 ..... D3
97469 Riddle 1,265 ..... D5
†97801 Rieth 300 ..... J2
97758 Riley 100 ..... H4
†97223 River Grove 314 ..... B2
†97423 Riverton 150 ..... C4
97136 Rockaway 906 ..... C2
97537 Rogue River 1,308 ..... D5
97470 Roseburg⊙ 16,644 ..... D4
97372 Rose Lodge 300 ..... D3
†97106 Roy 200 ..... A2
97050 Rufus 352 ..... G2
97472 Saginaw 150 ..... E4
97051 Saint Helens⊙ 7,064 ..... E2
†97026 Saint Louis 102 ..... A3
97137 Saint Paul 312 ..... A3
*97301 Salem (cap.)⊙ 89,233 ..... A3
Salem‡ 249,895 ..... A3
†97525 Sams Valley 100 ..... E5
97055 Sandy 2,905 ..... E2
97056 Scappoose 3,213 ..... E2
97374 Scio 579 ..... E3
97473 Scottsburg 300 ..... D4
97375 Scotts Mills 249 ..... B3
97376 Seal Rock 430 ..... C3
97138 Seaside 5,193 ..... D2
97538 Selma 150 ..... D5
97873 Seneca 285 ..... J3
97539 Shady Cove 1,097 ..... E5
97057 Shaniko 30 ..... G3
†97325 Shaw 800 ..... A3
97377 Shedd 850 ..... D3
97378 Sheridan 2,249 ..... D2
97140 Sherwood 2,386 ..... A2
97380 Siletz 1,001 ..... D3
97638 Silver Lake 200 ..... F4
97381 Silverton 5,168 ..... B3
97759 Sisters 696 ..... F3
97476 Sixes 300 ..... C5
†97355 Sodaville 171 ..... E3
97366 Southbeach 300 ..... C3
†97501 South Medford 2,898 ..... E5
97639 Sprague River 200 ..... F5
97874 Spray 155 ..... H3
†97132 Springbrook 500 ..... A2
97477 Springfield 41,621 ..... E3
97875 Stanfield 1,568 ..... H2
97383 Stayton 4,396 ..... E3
97385 Sublimity 1,077 ..... E3
97876 Summerville 143 ..... K2
†97420 Sumner 100 ..... C4
97877 Sumpter 133 ..... J3
97478 Sunny Valley 159 ..... D5
97479 Sutherlin 4,560 ..... D4
†97103 Svensen 950 ..... D1
97386 Sweet Home 6,921 ..... E3
97480 Swisshome 350 ..... D3
†97201 Sylvan B2
97540 Talent 2,577 ..... E5
97389 Tangent 478 ..... D3
97481 Tenmile 500 ..... D4
97760 Terrebonne 521 ..... F3
97058 The Dalles⊙ 10,820 ..... F2
97223 Tigard 14,286 ..... A2
97141 Tillamook⊙ 3,981 ..... D2
97484 Tiller 300 ..... E5
97144 Timber 175 ..... D2
97391 Toledo 3,151 ..... D3
97145 Tolovana Park 165 ..... C2
97541 Trail 350 ..... E5
†97431 Trent 100 ..... E4
97060 Troutdale 5,908 ..... E2
97062 Tualatin 7,483 ..... A2
†97701 Tumalo 500 ..... F3
97392 Turner 1,116 ..... E3
†97136 Twin Rocks 450 ..... C2
97063 Tygh Valley 663 ..... F2
97880 Ukiah 249 ..... J2
97881 Umapine 100 ..... J2
97882 Umatilla 3,199 ..... H2
97486 Umpqua 705 ..... D4
97883 Union 2,062 ..... K2
97884 Unity 115 ..... J3
97918 Vale⊙ 1,558 ..... K4
97393 Valsetz 320 ..... D3
97487 Veneta 2,449 ..... D3
†97116 Verboort 280 ..... A2
97064 Vernonia 1,785 ..... D2
97488 Vida 300 ..... E3
97394 Waldport 1,274 ..... C3
97885 Wallowa 847 ..... K2
97489 Walterville 250 ..... E3
97490 Walton 300 ..... D3
97063 Wamic 255 ..... F2
97761 Warm Springs 550 ..... F3
97053 Warren 750 ..... E2
97146 Warrenton 2,493 ..... C1
97065 Wasco 415 ..... G2
†97355 Waterloo 221 ..... E3
97491 Wedderburn 700 ..... C5
†97067 Welches 100 ..... E2
97492 Westfir 312 ..... E4
97493 Westlake 1,124 ..... C4
97068 West Linn 12,956 ..... B2
97886 Weston 719 ..... J2
97016 Westport 400 ..... D1
†97071 West Woodburn 600 ..... A3
97147 Wheeler 319 ..... D2
97503 White City 5,445 ..... E5
†97128 Whiteson 100 ..... D2
97494 Wilbur 476 ..... D4
97543 Wilderville 600 ..... D5
97396 Willamina 1,749 ..... D2
97544 Williams 750 ..... D5
97070 Wilsonville 2,920 ..... A2
97495 Winchester 900 ..... D4
97467 Winchester Bay 535 ..... C4
97496 Winston 3,359 ..... D4
97497 Wolf Creek 500 ..... D5
97071 Woodburn 11,196 ..... A3
†97060 Wood Village 2,253 ..... B2
97498 Yachats 482 ..... C3
97148 Yamhill 690 ..... D2
†97365 Yaquina 175 ..... C3
97499 Yoncalla 805 ..... D4

OTHER FEATURES

Abert (lake) G5
Abiqua (creek) B3
Agency (lake) E5
Alsea (riv.) D3
Alvord (lake) J5
Antelope (creek) K5
Arago (cape) C4
Aspen (lake) E5
Badger (creek) H3
Battle (creek) K5
Bear (creek) E5
Bear (creek) K2
Bear (creek) G4
Benjamin (lake) G4
Beulah (res.) J4
Blacklock (pt.) C5
Blanco (cape) C5
Blue (mts.) J3
Bonneville (dam) E2
Brownlee (dam) L3
Buck Hollow (creek) G2
Bully (creek) K3
Burnt (riv.) K3
Butte (creek) G2
Butte (creek) B3
Butter (creek) H2
Calapooia (riv.) E3
Calapooya (mts.) E4
Camp (creek) J4
Campbells (lake) H5
Cascade (head) C2
Cascade (range) E4
Celilo (lake) G2
Chetco (riv.) C5
Clackamas (riv.) E2
Clover (creek) K3
Coal (pt.) C5
Coast (ranges) D5
Columbia (riv.) G2
Coos (riv.) D4
Coquille (pt.) C4
Cottage Grove (lake) E4
Cottonwood (creek) K4
Cougar (res.) E3
Cow (creek) K4
Crane (creek) J4
Crane Prairie (res.) F4
Crater (lake) E5
Crater Lake Nat'l Park E5
Crook (pt.) C5
Crooked (creek) K5
Crooked (riv.) G3
Cultus (lake) F4
Dalles, The (dam) F2
Davis (lake) F4
Deschutes (riv.) G2
Detroit (lake) E3
Diamond (lake) E4
Donner and Blitzen (riv.) J4
Dorena (lake) E4
Drews (res.) G5
Drift (creek) B3
Eagle (creek) K3
East (lake) F4
Elk (creek) E5
Ewauna (lake) F5
Falcon (cape) C2
Fern Ridge (lake) D3
Ferrelo (cape) C5
Fifteenmile (creek) F2
Fish (creek) E4
Fivemile (creek) F2
Fivemile (pt.) C4
Flagstaff (lake) H5
Fort Clatsop Nat'l Mem. C1
Foulweather (cape) C3
Fourmile (lake) E5
Gerber (res.) F5
Goose (lake) G5
Grand Canyon, Snake R. (canyon) L2
Grande Ronde (riv.) K2
Green Peter (lake) E3
Guano (creek) H5
Guano (lake) H5
Harney (lake) H4
Hart (lake) H5
Hart (mt.) H5
Heads, The (prom.) C5
Heceta (head) C3
Hells Canyon (dam) L2
Hells Canyon Nat'l Rec. Area K2
Hills Creek (lake) E4
Honey (creek) G5
Hood (mt.) F2
Hood (riv.) F2
Horse (creek) F3
Illinois (riv.) D5
Imnaha (riv.) L2
Indigo (creek) D5
Jackson (creek) E5
Jefferson (mt.) F3
Jenny (creek) E5
John Day (dam) G2
John Day (riv.) G2
John Day Fossil Beds Nat'l Mon. G3
Jordan (creek) K5
Joseph (creek) K2
Keeny (creek) K4
Kiger (creek) J5
Kincheloe (pt.) C2
Klamath (mts.) C5
Lake (creek) J3
Lava (lake) F4
Lightning (creek) L2
Little (riv.) E4
Little Butter (creek) H2
Little Sheep (creek) K2
Lookout (cape) C2
Lookout Point (lake) E4
Lost (riv.) F5
Malheur (lake) J4
Malheur (riv.) J4
McCoy (creek) J5
McKay (res.) J2
McKenzie, South Fork (riv.) E3
McLoughlin (mt.) E5
McLoughlin House Nat'l Hist. Site B2
McNary (dam) H2
Meares (cape) C2
Metolius (riv.) F3
Miller (creek) F5
Molalla (riv.) B3
Mud (creek) K2
Murderers (creek) H3
Nehalem (riv.) D2
Nestucca (riv.) D2
North Santiam (riv.) E3
North Umpqua (riv.) E4
Oak Grove Fork, Clackamas (riv.) F2
Ochoco (creek) G3
Odell (lake) E4
Oregon (creek) K5
Oregon Caves Nat'l Mon. D5
Oregon Dunes Nat'l Rec. Area C4
Owyhee (dam) K4
Owyhee (lake) K4
Owyhee (mts.) K4
Owyhee (riv.) K5
Owyhee, North Fork (riv.) K5
Oxbow (dam) L3
Paulina (lake) F4
Perpetua (cape) C3
Pine (creek) L3
Pine (creek) J4
Portland Int'l Airport B2
Powder (riv.) K3
Prineville (res.) G3
Pudding (riv.) A3
Pueblo (mts.) J5
Rattlesnake (creek) K5
Rhea (creek) H2
Rock (creek) E4
Rock (creek) H3
Rock (creek) G2
Rogue (riv.) C5
Salt (creek) E4
Sebastian (cape) C5
Sheep (creek) L2
Shitike (creek) F3
Silver (creek) F4
Silver (creek) H4
Silver (lake) G4
Silver (lake) H4
Silvies (riv.) H4
Siskiyou (mts.) D6
Siuslaw (riv.) D4
Sixes (riv.) C5
Smith (riv.) D4
Snake (riv.) K3
South Santiam (riv.) E3
South Umpqua (riv.) E4
Sparks (lake) F3
Spencer (creek) E5
Sprague (riv.) F5
Squaw (creek) F3
Steens (mt.) J5
Succor (creek) K4
Summer (lake) G5
Summit (creek) J3
Sycan (riv.) F5
Tenmile (creek) K5
The Dalles (dam) F2
Thielsen (mt.) F4
Thirtymile (creek) G2
Thomas (creek) G5
Three Sisters (mt.) F3
Tillamook (head) C2
Trout (creek) J5
Trout (creek) F3
Tualatin (riv.) A2
Tumalo (creek) F3
Tumtum (lake) J5
Umatilla (lake) G2
Umatilla (riv.) H2
Umatilla Army Depot H2
Umatilla Ind. Res. J2
Umpqua (riv.) D4
Upper Klamath (lake) E5
Waldo (lake) E4
Walla Walla (riv.) J1
Wallowa (mts.) K2
Wallowa (riv.) K2
Wallula (lake) H1
Warm Springs (res.) J4
Warm Springs Ind. Res. F3
White (riv.) F2
Wickiup (res.) F4
Wiley (creek) E3
Willamette (riv.) A3
Willamette, Middle Fork (riv.) E4
Williamson (riv.) F5
Willow (creek) H2
Willow (creek) K3
Wilson (riv.) D2
Winchester (bay) C4

⊙County seat.
‡Population of metropolitan area.
† Zip of nearest p.o. * Multiple zips

AREA 45,308 sq. mi. (117,348 sq. km.)
POPULATION 11,863,895
CAPITAL Harrisburg
LARGEST CITY Philadelphia
HIGHEST POINT Mt. Davis 3,213 ft. (979 m.)
SETTLED IN 1682
ADMITTED TO UNION December 12, 1787
POPULAR NAME Keystone State
STATE FLOWER Mountain Laurel
STATE BIRD Ruffed Grouse

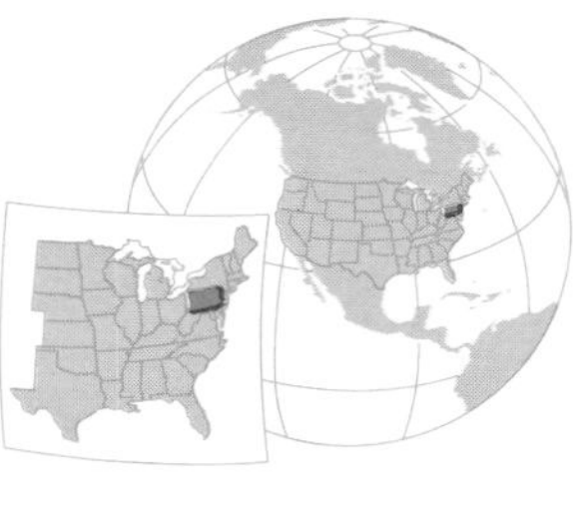

DOMINANT LAND USE

- Specialized Dairy
- Dairy, General Farming
- Fruit and Mixed Farming
- Fruit, Truck and Mixed Farming
- General Farming, Livestock, Tobacco
- General Farming, Livestock, Fruit, Tobacco
- Forests
- Urban Areas

MAJOR MINERAL OCCURRENCES

- C Coal
- Cl Clay
- Co Cobalt
- Fe Iron Ore
- G Natural Gas
- Ls Limestone
- O Petroleum
- Sl Slate
- Ss Sandstone
- Zn Zinc

Water Power
Major Industrial Areas

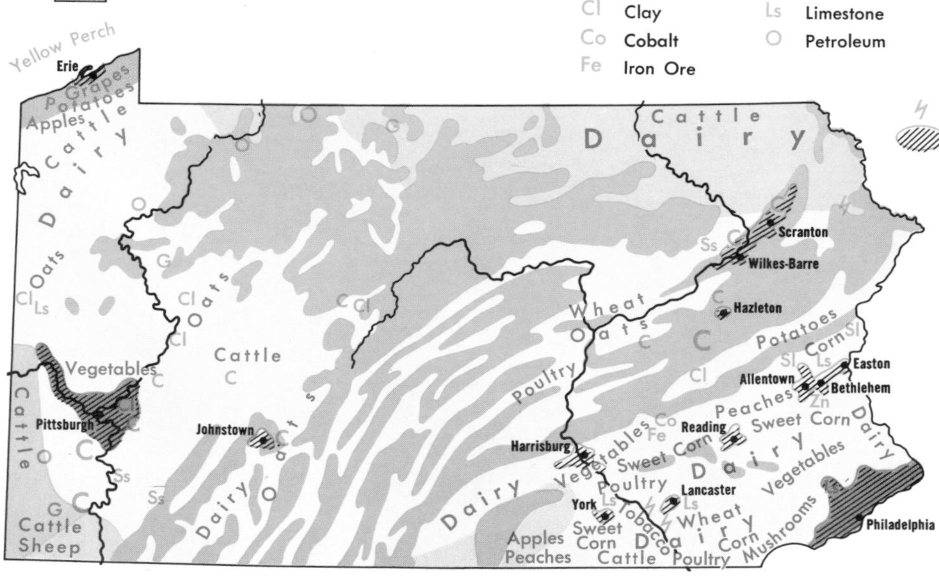

Agriculture, Industry and Resources

COUNTIES

Adams 68,292 ... H6
Allegheny 1,450,085 ... B5
Armstrong 77,768 ... D4
Beaver 204,441 ... B4
Bedford 46,784 ... E6
Berks 312,509 ... K5
Blair 136,621 ... F4
Bradford 62,919 ... J2
Bucks 479,211 ... M5
Butler 147,912 ... C4
Cambria 183,263 ... E4
Cameron 6,674 ... F3
Carbon 53,285 ... L4
Centre 112,760 ... G4
Chester 316,660 ... L6
Clarion 43,362 ... D3
Clearfield 83,578 ... F3
Clinton 38,971 ... G3
Columbia 61,967 ... K3
Crawford 88,869 ... B2
Cumberland 178,541 ... H5
Dauphin 232,317 ... J5
Delaware 555,007 ... M6
Elk 38,338 ... E3
Erie 279,780 ... B2
Fayette 159,417 ... C6
Forest 5,072 ... D2
Franklin 113,629 ... G6
Fulton 12,842 ... F6
Greene 40,476 ... B6
Huntingdon 42,253 ... F5
Indiana 92,281 ... D4
Jefferson 48,303 ... D3
Juniata 19,188 ... H4
Lackawanna 227,908 ... L3
Lancaster 362,346 ... K5
Lawrence 107,150 ... B4
Lebanon 108,582 ... K5
Lehigh 272,349 ... L4
Luzerne 343,079 ... L3
Lycoming 118,416 ... H3
McKean 50,635 ... E2
Mercer 128,299 ... B3
Mifflin 46,908 ... G4
Monroe 69,409 ... M3
Montgomery 643,621 ... M5
Montour 16,675 ... J3
Northampton 225,418 ... M4
Northumberland 100,381 ... J4
Perry 35,718 ... H5
Philadelphia (city county) 1,688,210 ... M6
Pike 18,271 ... M3
Potter 17,726 ... G2
Schuylkill 160,630 ... K4
Snyder 33,584 ... H4
Somerset 81,243 ... D6
Sullivan 6,349 ... J3
Susquehanna 37,876 ... L2
Tioga 40,973 ... H2
Union 32,870 ... H4
Venango 64,444 ... C3
Warren 47,449 ... D2
Washington 217,074 ... B5
Wayne 35,237 ... M2
Westmoreland 392,294 ... D5
Wyoming 26,433 ... K2
York 312,963 ... J6

CITIES and TOWNS

Zip Name/Pop. Key

19001 Abington○ 59,084 ... M5
19501 Adamstown 1,119 ... K5
17501 Akron 3,471 ... K5
16401 Albion 1,818 ... B2
18011 Alburtis 1,428 ... L5
†19018 Aldan 4,671 ... M7
15001 Aliquippa 17,094 ... B4
*18101 Allentown⊙ 103,758 ... L4
Allentown-Bethlehem-Easton‡ 636,714 ... L4
15101 Allison Park 10,000 ... C4
*16601 Altoona 57,078 ... F4
Altoona‡ 136,621 ... F4
19002 Ambler 6,628 ... M5
15003 Ambridge 9,575 ... B4
17003 Annville 4,493 ... J5
15613 Apollo 2,212 ... C4
18403 Archbald 6,295 ... F6
19003 Ardmore ... M6
15068 Arnold 6,853 ... C4
17921 Ashland 4,235 ... K4
18706 Ashley 3,512 ... E7
15215 Aspinwall 3,284 ... C6
18810 Athens 3,622 ... K2
17851 Atlas 1,162 ... K4
15202 Avalon 6,240 ... B6
15312 Avella 900 ... B5
17721 Avis 1,718 ... H3
18641 Avoca 3,536 ... F7
19311 Avondale 891 ... L6
15618 Avonmore 1,234 ... C4
15005 Baden 5,318 ... B4
19004 Bala-Cynwyd ... N6
†15208 Baldwin 24,598 ... B7
19503 Bally 1,051 ... L5
18013 Bangor 5,006 ... M4
15714 Barnesboro 2,741 ... E4
18014 Bath 1,953 ... M4
15009 Beaver⊙ 5,441 ... B4
15921 Beaverdale 1,187 ... E5
15010 Beaver Falls 12,525 ... B4
18216 Beaver Meadows 1,078 ... L4
15522 Bedford⊙ 3,326 ... F5
16823 Bellefonte⊙ 6,300 ... G4
15012 Belle Vernon 1,489 ... C5
17004 Belleville 1,689 ... G4
15202 Bellevue 10,128 ... B6
16617 Bellwood 2,114 ... F4
†15202 Ben Avon 2,314 ... B6
15314 Bentleyville 2,525 ... B5
15530 Berlin 1,999 ... E6
19506 Bernville 798 ... K5
18603 Berwick 11,850 ... K3
19312 Berwyn 5,246 ... L5
16112 Bessemer 1,293 ... B4
15102 Bethel Park 34,755 ... B7
*18015 Bethlehem 70,419 ... M4
19508 Birdsboro 3,481 ... L5
15716 Black Lick 1,313 ... D4
15717 Blairsville 4,166 ... D5
18447 Blakely 7,438 ... F6
15238 Blawnox 1,653 ... C6
17068 Bloomfield (New Bloomfield)⊙ 1,109 ... H5
17815 Bloomsburg⊙ 11,717 ... J3
16912 Blossburg 1,757 ... H2
16827 Boalsburg 2,295 ... G4
15315 Bobtown 1,008 ... B6
17007 Boiling Springs 2,223 ... H5
15923 Bolivar 706 ... D5
15531 Boswell 1,480 ... E5
18030 Bowmanstown 1,078 ... L4
19512 Boyertown 3,979 ... L5
15014 Brackenridge 4,297 ... C4
15104 Braddock 5,634 ... C7
16701 Bradford 11,211 ... E2
15227 Brentwood 11,907 ... B7
19405 Bridgeport 4,843 ... M5
15017 Bridgeville 6,154 ... B5
19007 Bristol 10,867 ... N5
19007 Bristol○ 58,733 ... N5
15824 Brockway 2,376 ... E3
19015 Brookhaven 7,912 ... M7
15825 Brookville⊙ 4,568 ... D3
19008 Broomall ... M6
15417 Brownsville 4,043 ... C5
19010 Bryn Mawr ... M5
15021 Burgettstown 1,867 ... A5
17009 Burnham 2,457 ... H4
16001 Butler⊙ 17,026 ... C4
15924 Cairnbrook 1,081 ... E5
15419 California 5,703 ... C5
16403 Cambridge Springs 2,102 ... C2
17011 Camp Hill 8,422 ... H5
15317 Canonsburg 10,459 ... B5
17724 Canton 1,959 ... J2
18407 Carbondale 11,255 ... L2
17013 Carlisle⊙ 18,314 ... H5
15106 Carnegie 10,099 ... B7
15722 Carrolltown 1,395 ... E4
15234 Castle Shannon 10,164 ... B7
18032 Catasauqua 6,711 ... M4
17820 Catawissa 1,568 ... K4
16404 Centerville 4,207 ... B6
15926 Central City 1,496 ... E5
17927 Centralia 1,017 ... K4
16828 Centre Hall 1,233 ... G4
18914 Chalfont 2,802 ... M5
17201 Chambersburg⊙ 16,174 ... G6
15022 Charleroi 5,717 ... C5
19012 Cheltenham○ 35,509 ... M5
*19013 Chester 45,794 ... L7
19017 Chester Heights 1,302 ... L7
†16866 Chester Hill 1,054 ... F4
15024 Cheswick 2,336 ... C6
16025 Chicora 1,192 ... C4
17509 Christiana 1,183 ... K6
†15235 Churchill 4,285 ... C7
15025 Clairton 12,188 ... C7
16214 Clarion⊙ 6,664 ... D3
†18411 Clarks Green 1,862 ... F6
18411 Clarks Summit 5,272 ... F6
16625 Claysburg 1,346 ... F5
15323 Claysville 1,029 ... B5
16830 Clearfield⊙ 7,580 ... F3
19018 Clifton Heights 7,320 ... M7
15728 Clymer 1,761 ... E4
18218 Coaldale 2,762 ... L4
19320 Coatesville 10,698 ... L5
16314 Cochranton 1,240 ... B2
19426 Collegeville 3,406 ... M5
19023 Collingdale 9,539 ... N7
17512 Columbia 10,466 ... K5
15927 Colver 1,165 ... E4
†19023 Colwyn 2,851 ... N7
15425 Connellsville 10,319 ... C5
19428 Conshohocken 8,475 ... M5
15027 Conway 2,747 ... B4
18219 Conyngham 2,242 ... K3
18036 Coopersburg 2,595 ... M5
18037 Coplay 3,130 ... L4
15108 Coraopolis 7,308 ... B4
17016 Cornwall 2,653 ... K5
16407 Corry 7,149 ... C2
16915 Coudersport⊙ 2,791 ... G2
15624 Crabtree 900 ... D5
15205 Crafton 7,623 ... B7
16630 Cresson 2,184 ... E5
17929 Cressona 1,810 ... K4
16833 Curwensville 3,116 ... E4
†15901 Dale 1,906 ... E5
18612 Dallas 2,679 ... E7
17313 Dallastown 3,949 ... J6
18414 Dalton 1,383 ... L2
17821 Danville⊙ 5,239 ... J4
19023 Darby 11,513 ... M7
18327 Delaware Water Gap 597 ... M4
15626 Delmont 2,159 ... D5
17517 Denver 2,018 ... K5
15627 Derry 3,072 ... D5
18519 Dickson City 6,699 ... F7
17019 Dillsburg 1,733 ... J5
15033 Donora 7,524 ... C5
15216 Dormont 11,275 ... B7
17315 Dover 1,910 ... J6
19335 Downingtown 7,650 ... L5
18901 Doylestown⊙ 8,717 ... M5
15034 Dravosburg 2,511 ... C7
19026 Drexel Hill ... M6
18221 Drifton 1,786 ... L3
18917 Dublin 1,565 ... M5
15801 DuBois 9,290 ... E3
†17701 Duboistown 1,218 ... H3
15431 Dunbar 1,369 ... C6
17020 Duncannon 1,645 ... H5
16635 Duncansville 1,355 ... F5
18512 Dunmore 16,781 ... F7
18641 Dupont 3,460 ... F7
15110 Duquesne 10,094 ... C7
18642 Duryea 5,415 ... F7
17316 East Berlin 1,054 ... J6
†18603 East Berwick 2,324 ... K3
16028 East Brady 1,153 ... C3
15909 East Conemaugh 2,128 ... E5
†17701 East Faxon 3,951 ... J3
18041 East Greenville 2,456 ... L5
†19050 East Lansdowne 2,806 ... M7
18042 Easton⊙ 26,027 ... M4
17520 East Petersburg 3,600 ... K5
18301 East Stroudsburg 8,039 ... M4
†15301 East Washington 2,241 ... B5
15931 Ebensburg⊙ 4,096 ... E5
†15005 Economy 9,538 ... B4
†19013 Eddystone 2,555 ... M7
†15218 Edgewood 4,382 ... B7
†15143 Edgeworth 1,738 ... B4
16412 Edinboro 6,324 ... B2
18704 Edwardsville 5,729 ... E7
16731 Eldred 965 ... F2
15037 Elizabeth 1,892 ... C5
17022 Elizabethtown 8,233 ... J5
17023 Elizabethville 1,531 ... J4
16920 Elkland 1,974 ... H1
15331 Ellsworth 1,228 ... B5
16117 Ellwood City 9,998 ... B4
17824 Elysburg 1,447 ... K4
17318 Emigsville 2,413 ... J5
16373 Emlenton 807 ... C3
18049 Emmaus 11,001 ... M4
15834 Emporium⊙ 2,837 ... F2
15202 Emsworth 3,074 ... B6
17025 Enola ... J5
17522 Ephrata 11,095 ... K5
*16501 Erie⊙ 119,123 ... B1
Erie‡ 279,780 ... B1
17815 Espy 1,571 ... K4
15223 Etna 4,534 ... B6
16033 Evans City 2,299 ... B4
15537 Everett 1,828 ... F5
15631 Everson 1,032 ... C5
15632 Export 1,143 ... C5
15436 Fairchance 2,106 ... C6
19030 Fairless Hills 16,000 ... N5
16415 Fairview 1,855 ... B1
15840 Falls Creek 1,208 ... E3
16121 Farrell 8,645 ... A3
17222 Fayetteville 3,202 ... G6
18921 Ferndale 2,204 ... E5
19522 Fleetwood 3,422 ... L5
†17745 Flemington 1,416 ... G3
19032 Folcroft 8,231 ... M7
16226 Ford City 3,923 ... D4
18421 Forest City 1,924 ... L2
†15221 Forest Hills 8,198 ... C7
18704 Forty Fort 5,590 ... F7
†18015 Fountain Hill 4,805 ... L4
†15238 Fox Chapel 5,049 ... C6
17931 Frackville 5,308 ... K4
16323 Franklin⊙ 8,146 ... C3
†16335 Fredericksburg 1,202 ... B2
15333 Fredericktown 1,052 ... C6
15042 Freedom 2,272 ... B4
18224 Freeland 4,285 ... L3
†18017 Freemansburg 1,879 ... M4
16229 Freeport 2,381 ... C4
16922 Galeton 1,462 ... G2
16641 Gallitzin 2,315 ... E4
†17701 Garden View 2,777 ... H3
15904 Geistown 3,304 ... E5
17325 Gettysburg⊙ 7,194 ... H6
17934 Gilberton 1,096 ... K4
16417 Girard 2,615 ... B2
17935 Girardville 2,268 ... K4
15045 Glassport 6,242 ... C7
18617 Glen Lyon 2,352 ... E7
19036 Glenolden 7,633 ... M7
17327 Glen Rock 1,662 ... J6
19038 Glenside ... M5
15634 Grapeville ... C5
18821 Great Bend 740 ... L2
17225 Greencastle 3,679 ... G6
15601 Greensburg⊙ 17,558 ... D5
15242 Greentree 5,722 ... B7
16125 Greenville 7,730 ... B3
16127 Grove City 8,162 ... B3
17032 Halifax 909 ... J5
17406 Hallam 1,428 ... J6
18822 Hallstead 1,280 ... L2
19526 Hamburg 4,011 ... L4
17331 Hanover 14,890 ... J6
16037 Harmony 1,334 ... B4
*17101 Harrisburg (cap.)⊙ 53,264 ... H5
Harrisburg‡ 446,072 ... H5
16038 Harrisville 1,033 ... B3
18618 Harveys Lake 2,318 ... E7
16646 Hastings 1,574 ... E4
19040 Hatboro 7,579 ... M5
19440 Hatfield 2,533 ... M5
19041 Haverford○ 52,349 ... M6
19083 Havertown ... M6
16840 Hawk Run 1,960 ... F4
18428 Hawley 1,181 ... M3
18201 Hazleton 27,318 ... L4
15106 Heidelberg 1,606 ... B7
17406 Hellam (Hallam) 1,428 ... J6
18055 Hellertown 6,025 ... M4
17033 Hershey 13,249 ... J5
†17044 Highland Park 1,879 ... H4
17034 Highspire 2,959 ... J5
16648 Hollidaysburg⊙ 5,892 ... F5
15748 Homer City 2,248 ... D4
15120 Homestead 5,092 ... B7
18431 Honesdale⊙ 5,128 ... M2
19344 Honey Brook 1,164 ... L5
15936 Hooversville 863 ... E5
15445 Hopwood 2,420 ... C6
15342 Houston 1,568 ... B5
16651 Houtzdale 1,222 ... F4
†18640 Hughestown 1,783 ... F7
17737 Hughesville 2,174 ... J3
17036 Hummelstown 4,267 ... J5
16652 Huntingdon⊙ 7,042 ... G5
16843 Hyde 1,791 ... F4
15545 Hyndman 1,106 ... E6
15126 Imperial 3,207 ... B5
15701 Indiana⊙ 16,051 ... D4
15052 Industry 2,417 ... B4
†15205 Ingram 4,346 ... B7
15642 Irwin 4,995 ... C5
17407 Jacobus 1,396 ... J6
15644 Jeannette 13,106 ... C5
†15025 Jefferson 8,643 ... B7
19046 Jenkintown 4,942 ... M5
18433 Jermyn 2,411 ... L2
15937 Jerome 1,196 ... D5
17740 Jersey Shore 4,631 ... H3
18434 Jessup 4,974 ... F6
18229 Jim Thorpe⊙ 5,263 ... L4
15845 Johnsonburg 3,938 ... E3
*15901 Johnstown 35,496 ... D5
Johnstown‡ 264,506 ... D5
16735 Kane 4,916 ... E2
†19607 Kenhorst 3,187 ... L5
19348 Kennett Square 4,715 ... L6
18704 Kingston 15,681 ... F7
16201 Kittanning⊙ 5,432 ... D4
16232 Knox 1,364 ... C3
16136 Koppel 1,146 ... B4
17834 Kulpmont 3,675 ... J4
19530 Kutztown 4,040 ... L4
16423 Lake City 2,384 ... B1
*17601 Lancaster⊙ 54,725 ... K5
Lancaster‡ 362,346 ... K5

(continued on following page)

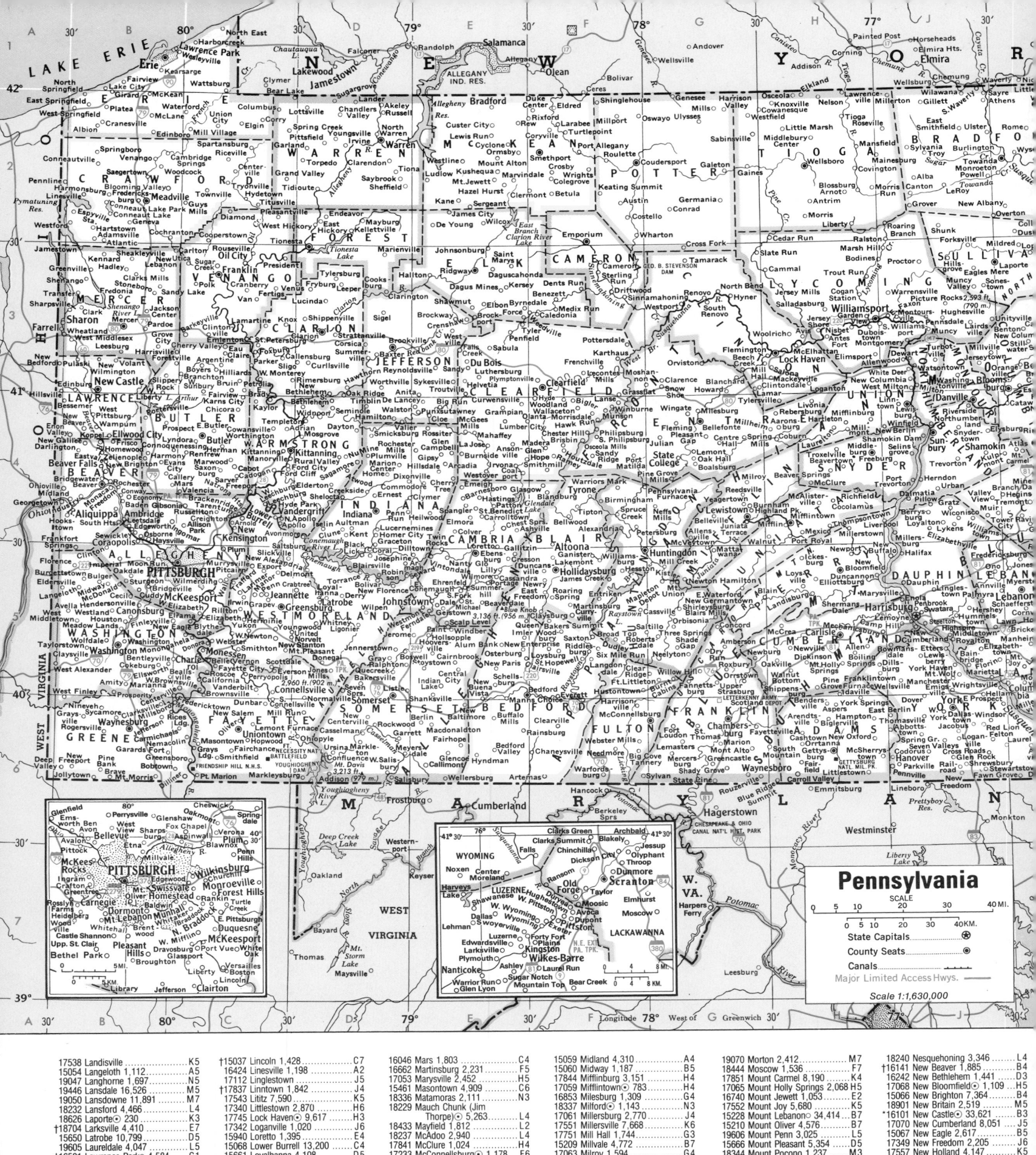

Pennsylvania

SCALE

0 5 10   20   30   40 MI.

0 5 10   20   30   40KM.

State Capitals .... ⊛
County Seats .... ◉
Canals .... ·—·—·—·
Major Limited Access Hwys. ———

Scale 1:1,630,000

17538 Landisville K5	†15037 Lincoln 1,428 ....C7	16046 Mars 1,803 ....C4	15059 Midland 4,310 ....A4	19070 Morton 2,412 ....M7	18240 Nesquehoning 3,346 ....L4
15054 Langeloth 1,112 ....A5	16624 Linesville 1,198 ....A2	16662 Martinsburg 2,231 ....F5	15060 Midway 1,187 ....B5	18444 Moscow 1,536 ....F7	†16141 New Beaver 1,885 ....B4
19047 Langhorne 1,697 ....N5	17112 Linglestown J5	17053 Marysville 2,452 ....H5	17844 Mifflinburg 3,151 ....H4	17851 Mount Carmel 8,190 ....K4	16242 New Bethlehem 1,441 ....C3
19050 Lansdale 16,526 ....M5	17543 Lititz 7,590 ....K5	15461 Masontown 4,909 ....C6	17059 Mifflintown⊙ 783 ....H4	17065 Mount Holly Springs 2,068 H5	17068 New Bloomfield⊙ 1,109 ....H5
19050 Lansdowne 11,891 ....M7	18336 Matamoras 2,111 ....N3	16853 Milesburg 1,309 ....G4	16740 Mount Jewett 1,053 ....E2	15066 New Brighton 7,364 ....B4
18232 Lansford 4,466 ....L4	17340 Littlestown 2,870 ....H6	18229 Mauch Chunk (Jim	18337 Milford⊙ 1,143 ....N3	17552 Mount Joy 5,680 ....K5	18901 New Britain 2,519 ....M5
18626 Laporte⊙ 230 ....K3	17745 Lock Haven⊙ 9,617 ....H3	Thorpe)⊙ 5,263 ....L4	17061 Millersburg 2,770 ....J4	15228 Mount Lebanon○ 34,414 ....B7	*16101 New Castle○ 33,621 ....A3
†18704 Larksville 4,410 ....E7	17342 Loganville 1,020 ....J6	18433 Mayfield 1,812 ....L2	17551 Millersville 7,668 ....K5	15210 Mount Oliver 4,576 ....B7	17070 New Cumberland 8,051 ....J5
†16501 Lawrence Park⊙ 4,584 ....C1	15940 Loretto 1,395 ....E4	18237 McAdoo 2,940 ....L4	17751 Mill Hall 1,744 ....H3	15209 Millvale 4,772 ....B7	15067 New Eagle 2,617 ....B5
17042 Lebanon○ 25,711 ....K5	15068 Lower Burrell 13,200 ....C4	17841 McClure 1,024 ....H4	17063 Milroy 1,594 ....G4	15666 Mount Pleasant 5,354 ....D5	17349 New Freedom 2,205 ....J6
15656 Leechburg 2,682 ....C4	15661 Loyalhanna 4,108 ....D5	17233 McConnellsburg⊙ 1,178 ....F6	17847 Milton 6,730 ....J3	18344 Mount Pocono 1,329 ....M3	17557 New Holland 4,147 ....K5
19533 Leesport 1,258 ....K5	15754 Lucernemines 1,195 ....C4	15057 McDonald 2,772 ....B5	17954 Minersville 5,635 ....K4	17066 Mount Union 3,101 ....G5	18938 New Hope 1,473 ....N5
15056 Leetsdale 1,604 ....B4	17048 Lykens 2,181 ....J4	*15130 McKeesport 31,012 ....C7	19540 Mohnton 2,156 ....L5	17554 Mountville 1,505 ....K5	15068 New Kensington 17,660 ....C4
18235 Lehighton 5,826 ....L4	18062 Macungie 1,899 ....L5	15136 McKees Rocks 8,742 ....B7	15061 Monaca 7,661 ....B4	17347 Mount Wolf 1,517 ....J5	17350 New Oxford 1,921 ....H6
16851 Lemont 2,613 ....G4	17948 Mahanoy City 6,167 ....K4	17344 McSherrystown 2,764 ....H6	15062 Monessen 11,928 ....C5	17756 Muncy 2,700 ....J3	17959 New Philadelphia 1,341 ....L4
17043 Lemoyne 4,178 ....J5	19355 Malvern 2,999 ....L5	16335 Meadville○ 15,544 ....B2	15063 Monongahela 5,950 ....B5	15120 Munhall 14,532 ....C7	17074 Newport 1,600 ....H5
*19053 Levittown 70,000 ....N5	*19063 Media⊙ 6,119 ....L7	17055 Mechanicsburg 9,487 ....H5	15146 Monroeville 30,977 ....C7	15668 Murrysville 16,036 ....K5	15468 New Salem 1,628 ....C6
17837 Lewisburg⊙ 5,407 ....J4	17345 Manchester 2,027 ....J5	*19063 Media⊙ 6,119 ....L7	17067 Myerstown 3,131 ....K5	†15626 New Salem (Delmont) 2,159 D5
17044 Lewistown○ 9,830 ....G4	17545 Manheim 5,015 ....K5	16137 Mercer⊙ 2,532 ....B3	17752 Montgomery 1,653 ....H3	18634 Nanticoke 13,044 ....E7	15672 New Stanton 2,600 ....C5
*15100 Liberty 3,112 ....C7	16933 Mansfield 3,322 ....H2	17236 Mercersburg 1,617 ....G6	18801 Montrose⊙ 1,980 ....L2	15943 Nanty Glo 3,936 ....E4	18940 Newtown 2,519 ....N5
15658 Ligonier 1,917 ....D5	19061 Marcus Hook 2,638 ....L7	19066 Merion Station M6	19072 Narberth 4,496 ....M6	17241 Newville 1,370 ....H5
15938 Lilly 1,462 ....E5	17547 Marietta 2,740 ....J5	15552 Meyersdale 2,581 ....D6	18507 Moosic 6,068 ....F7	18064 Nazareth 5,443 ....M4	16142 New Wilmington 2,774 ....B3
		17842 Middleburg⊙ 1,357 ....H4	19067 Morrisville 9,845 ....N5	15351 Nemacolin 1,235 ....B6	*19401 Norristown○ 34,684 ....M5
		17057 Middletown 10,122 ....J5	18635 Nescopeck 1,768 ....K3	

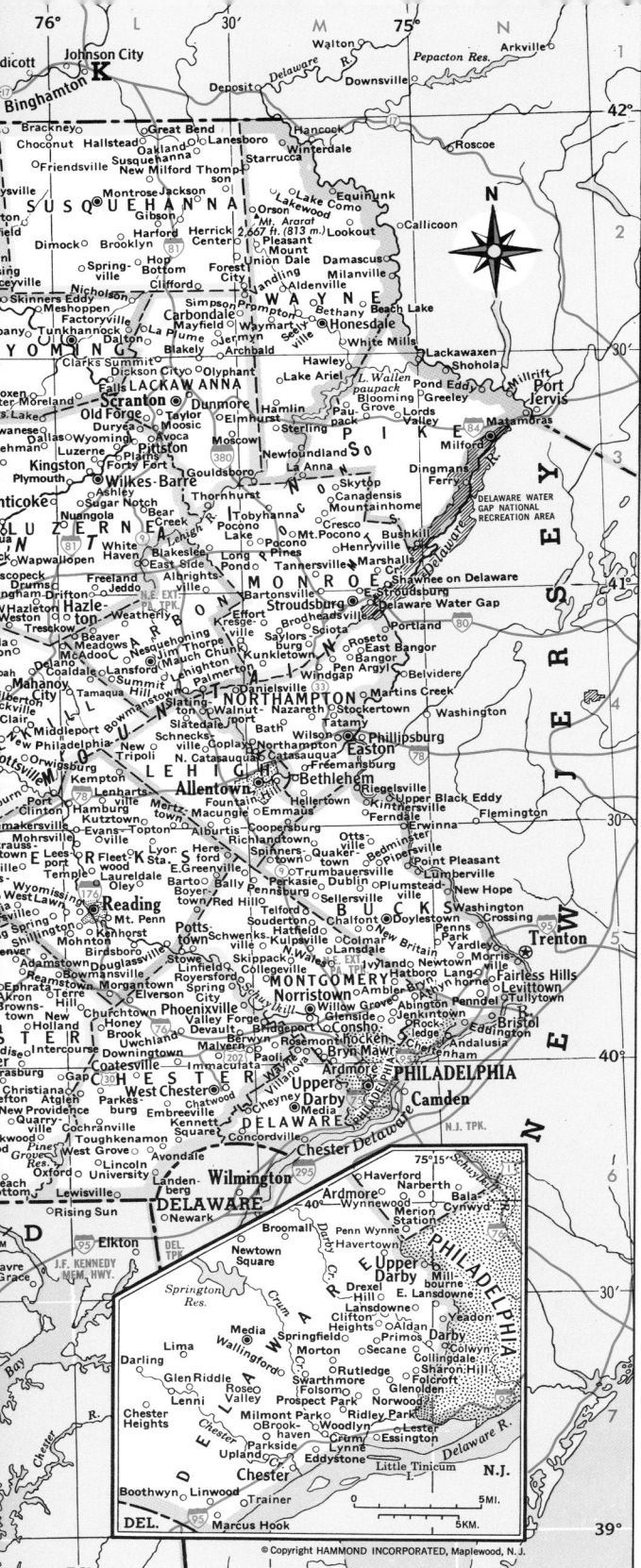

16823 Pleasant Gap 1,859 ....G4
15236 Pleasant Hills 9,676 ....B7
16341 Pleasantville 1,099 ....C2
15239 Plum 25,390 ....C5
18651 Plymouth 7,605 ....E7
15474 Point Marion 1,642 ....C6
16342 Polk 1,884 ....C3
15946 Portage 3,510 ....E5
16743 Port Allegany 2,593 ....F2
17965 Port Carbon 2,576 ....K4
††15133 Port Vue 5,316 ....C7
19464 Pottstown 22,729 ....L5
17901 Pottsville⊙ 18,195 ....K4
19076 Prospect Park 6,593 ....M7
15767 Punxsutawney 7,479 ....E4
18951 Quakertown 8,867 ....M5
17566 Quarryville 1,558 ....K6
*15104 Rankin 2,892 ....C7
*19601 Reading⊙ 78,686 ....L5
   Reading‡ 312,509 ....L5
17567 Reamstown 1,308 ....K5
18076 Red Hill 1,727 ....L5
17356 Red Lion 5,824 ....J6
17084 Reedsville 1,023 ....G4
17764 Renovo 1,812 ....G3
17087 Richland 1,470 ....K5
18955 Richlandtown 1,180 ....M5
15853 Ridgway⊙ 5,604 ....E3
19078 Ridley Park 7,889 ....M7
18077 Riegelsville 993 ....M4
16248 Rimersburg 1,096 ....D3
17868 Riverside 2,266 ....J4
16673 Roaring Spring 2,962 ....F5
19551 Robesonia 1,748 ....K5
15074 Rochester 4,759 ....B4
††19101 Rockledge 2,538 ....M5
15557 Rockwood 1,058 ....D6
15477 Roscoe 1,123 ....C5
18013 Roseto 1,812 ....M4
††19065 Rose Valley 1,038 ....L7
17250 Rouzerville 1,371 ....G6
19468 Royersford 4,243 ....L5
16249 Rural Valley 1,033 ....D4
15076 Russellton 1,878 ....C4
17970 Saint Clair 4,037 ....K4
15857 Saint Marys 6,417 ....E3
15951 Saint Michael 1,445 ....E5
15681 Saltsburg 964 ....C4
16056 Saxonburg 1,336 ....C4
18840 Sayre 6,951 ....K2
*15963 Scalp Level 1,186 ....E5
17972 Schuylkill Haven 5,977 ....K4
19473 Schwenksville 1,041 ....L5
15683 Scottdale 5,833 ....C5
*18501 Scranton⊙ 88,117 ....F7
   Scranton (Northeast
   Pa.)‡ 640,396 ....F7
17870 Selinsgrove 5,227 ....J4
18960 Sellersville 3,143 ....M5
15143 Sewickley 4,778 ....B4
17872 Shamokin 10,357 ....J4
17876 Shamokin Dam 1,622 ....J4
16146 Sharon 19,057 ....B3
   Sharon‡ 128,299 ....B3
19079 Sharon Hill 6,221 ....N7
15215 Sharpsburg 4,351 ....B6
16150 Sharpsville 5,375 ....A3
16347 Sheffield 1,471 ....D2
17976 Shenandoah 7,589 ....K4
18655 Shickshinny 1,192 ....K3
19607 Shillington 5,601 ....K5
16748 Shinglehouse 1,310 ....F2
17257 Shippensburg 5,261 ....H5
19555 Shoemakersville 1,391 ....K4
17361 Shrewsbury 2,688 ....J6
19608 Sinking Spring 2,617 ....K5
18080 Slatington 4,277 ....L4
15684 Slickville 1,178 ....C5
16057 Slippery Rock 3,047 ....B3
16749 Smethport⊙ 1,797 ....F2
15478 Smithfield 1,084 ....C6
15501 Somerset⊙ 6,474 ....D6
18964 Souderton 6,657 ....M5
15425 South Connellsville 2,296 ....C6
15956 South Fork 1,401 ....E5
††18840 South Waverly 1,176 ....J2

17701 South Williamsport 6,581 .. J3
15775 Spangler 2,399 ....E4
19475 Spring City 3,389 ....L5
15144 Springdale 4,418 ....C6
19064 Springfield 25,326 ....M7
17362 Spring Grove 1,832 ....J6
16801 State College 36,130 ....G4
   State College‡ 112,760 ....G4
17263 State Line 1,253 ....G6
17113 Steelton 6,484 ....J5
17363 Stewartstown 1,072 ....K6
16153 Stoneboro 1,177 ....B3
19464 Stowe 3,860 ....L5
17579 Strasburg 1,999 ....K6
18360 Stroudsburg⊙ 5,148 ....M4
15082 Sturgeon 1,312 ....B5
*16323 Sugar Creek 5,954 ....C3
18706 Sugar Notch 1,191 ....E7
18250 Summit Hill 3,418 ....L4
17801 Sunbury⊙ 12,292 ....J4
18847 Susquehanna 1,994 ....L2
19081 Swarthmore 5,950 ....M7
18704 Swoyersville 5,795 ....E7
15865 Sykesville 1,537 ....E3
18252 Tamaqua 8,843 ....L4
15084 Tarentum 6,419 ....C4
18517 Taylor 7,246 ....F7
18969 Telford 3,507 ....M5
19560 Temple 1,486 ....L5
17581 Terre Hill 1,217 ....L5
18512 Throop 4,166 ....F7
16351 Tidioute 844 ....D2
16353 Tionesta⊙ 659 ....C2
16684 Tipton 1,348 ....F4
16354 Titusville 6,884 ....C2
19562 Topton 1,818 ....L5
19374 Toughkenamon 1,111 ....L6
18848 Towanda⊙ 3,526 ....J2
17980 Tower City 1,667 ....J4
15085 Trafford 3,662 ....C5
*19013 Trainer 2,056 ....L7
17981 Tremont 1,796 ....J4
18254 Tresckow 1,128 ....K4
17881 Trevorton 2,192 ....J4
16947 Troy 1,381 ....J2
19007 Tullytown 2,277 ....N5
18657 Tunkhannock⊙ 2,144 ....L2
15145 Turtle Creek 6,959 ....C7
16686 Tyrone 6,346 ....F4
16438 Union City 3,623 ....C2
15401 Uniontown⊙ 14,510 ....C6
*19082 Upper Darby 84,054 ....M6
15241 Upper Saint Claire⊙ 19,023 B7
19481 Valley Forge 400 ....L5
17983 Valley View 1,722 ....J4
15690 Vandergrift 6,823 ....C4
15147 Verona 3,179 ....C6
15132 Versailles 2,150 ....C7
19085 Villanova M6
18088 Walnutport 2,007 ....L4
16365 Warren⊙ 12,146 ....D2
16441 Waterford 1,568 ....B2
17777 Watsontown 2,366 ....J3
19087 Wayne M6
17268 Waynesboro 9,726 ....G6
15370 Waynesburg⊙ 4,482 ....B6
18255 Weatherly 2,891 ....L4
16901 Wellsboro⊙ 3,805 ....H2
19565 Wernersville 1,811 ....K5
16510 Wesleyville 3,998 ....C1
15417 West Brownsville 1,433 ....C5
17435 West Chester⊙ 17,435 ....L6
16950 Westfield 1,268 ....H2
19390 West Grove 1,820 ....L6
18201 West Hazleton 4,871 ....K4
*16201 West Kittanning 1,591 ....C4
*15656 West Leechburg 1,395 ....C4
16159 West Middlesex 1,064 ....B3
15122 West Mifflin 26,279 ....C7
16160 West Pittsburg 1,133 ....B4
15089 West Newton 3,387 ....C5
*15905 West Newton 3,387 ....C5
*15905 Westmont 6,113 ....E5
18643 West Pittston 5,980 ....F7
15229 West View 7,648 ....B6

18644 West Wyoming 3,288 ....E7
†17401 West York 4,526 ....J6
15120 Whitaker 1,615 ....C7
*15234 Whitehall 15,206 ....B7
18661 White Haven 1,921 ....L3
15131 White Oak 9,480 ....C7
17097 Wiconisco 1,321 ....J4
*18701 Wilkes-Barre⊙ 51,551 ....F7
15221 Wilkinsburg 23,669 ....C7
16693 Williamsburg 1,400 ....F5
17701 Williamsport⊙ 33,401 ....H3
   Williamsport‡ 118,416 ....H3
17098 Williamstown 1,664 ....J4
19090 Willow Grove M5
15148 Wilmerding 2,421 ....C5
15025 Wilson 7,564 ....M4
15963 Windber 5,585 ....E5
18091 Windgap 2,651 ....M4
19567 Womelsdorf 1,827 ....K5
19094 Woodlyn M7
17368 Wrightsville 2,365 ....J5
18644 Wyoming 3,655 ....E7
19610 Wyomissing 6,551 ....K5
19067 Yardley 2,533 ....N5
19050 Yeadon 11,727 ....N7
17099 Yeagertown 1,305 ....G4
*17401 York⊙ 44,619 ....J6
   York‡ 381,255 ....J6
16371 Youngsville 2,006 ....D2
15697 Youngwood 3,749 ....D5
16063 Zelienople 3,502 ....B4

OTHER FEATURES

Allegheny (res.) E2
Allegheny (riv.) D2
Allegheny Front (mts.) E5
Appalachian (mts.) H4
Ararat (mt.) M2
Arthur (lake) C4
Beaver (riv.) B4
Blue (mt.) G5
Blue Knob (mt.) E5
Casselman (riv.) D6
Clarion (riv.) D3
Conemaugh (riv.) D5
Conemaugh River (lake) D4
Conewango (creek) D1
Davis (mt.) D6
Delaware (riv.) N3
Delaware Water Gap Nat'l Rec.
   Area N3
Glendale (lake) B1
Fort Necessity Nat'l
   Battlefield C6
George B. Stevenson (dam) G3
Gettysburg Nat'l Mil. Park H6
Glendale (lake) F4
Juniata (riv.) G5
Laurel Hill (mt.) D5
Lehigh (riv.) L3
Letterkenny Army Depot G6
Licking (creek) F6
Little Tinicum (isl.) M7
Lycoming (creek) H3
Monongahela (riv.) C6
North (mt.) K3
Ohio (riv.) A4
Oil (creek) C2
Pine (creek) H2
Pine Grove (res.) K6
Pocono (mts.) M3
Pymatuning (res.) A2
Redbank (creek) E3
Schuylkill (riv.) M5
Shenango River (lake) B3
Sinnemahoning (creek) F3
South (mt.) H6
Susquehanna (riv.) K6
Tioga (riv.) H1
Tionesta Creek (lake) D3
Towanda (creek) J2
Tuscarora (riv.) G5
Wallenpaupack (lake) M3
Youghiogheny River (lake) D6
⊙County seat.
‡Population of metropolitan area.
⊙Population of town or township.
† Zip of nearest p.o. * Multiple zips.

Topography

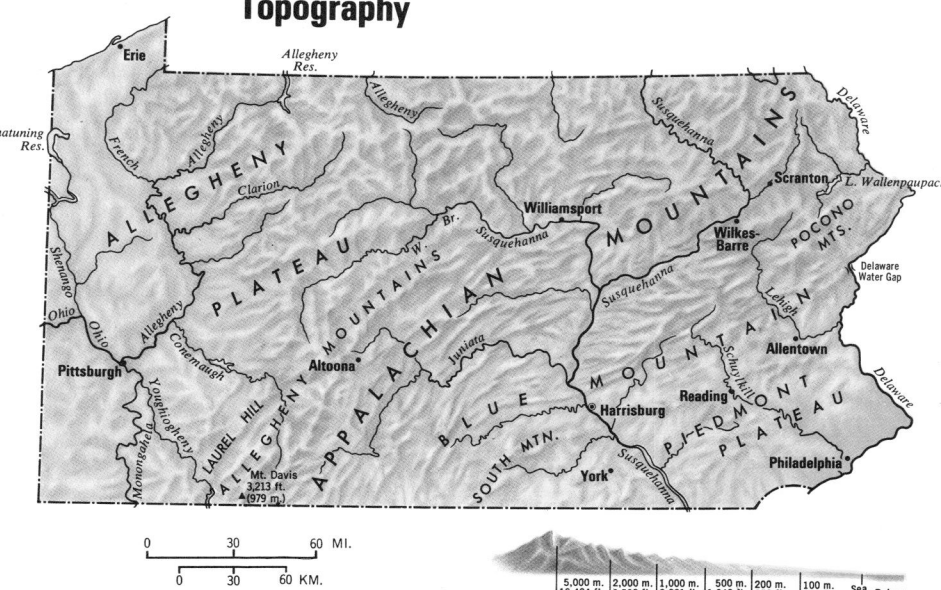

18067 Northampton 8,240 ....M4
15673 North Apollo 1,487 ....D4
15104 North Braddock 8,711 ....C7
†18032 North Catasauqua 2,554 ....L4
16428 North East 4,568 ....C1
17857 Northumberland 3,636 ....J4
19454 North Wales 3,391 ....M5
††16365 North Warren 1,232 ....D2
15674 Norvelt 2,541 ....C5
19074 Norwood 6,647 ....M7
15071 Oakdale 1,955 ....B5
15046 Oakmont 7,039 ....C6
††15059 Ohioville 4,217 ....A4
16301 Oil City 13,881 ....C3
18518 Old Forge 9,304 ....F7
15572 Oliver 3,777 ....C6
18447 Olyphant 5,204 ....F7
17961 Orwigsburg 2,700 ....K4
16666 Osceola Mills 1,466 ....F4
19363 Oxford 3,633 ....K6
††15963 Paint 1,177 ....E5
18071 Palmerton 5,455 ....L4
17078 Palmyra 7,228 ....J5
19301 Paoli 5,277 ....M5

17562 Paradise 1,107 ....K5
19365 Parkesburg 2,578 ....L6
††19013 Parkside 2,464 ....M7
†17331 Parkville 5,009 ....J6
16668 Patton 2,441 ....E4
18072 Pen Argyl 3,388 ....M4
17103 Penbrook 3,006 ....J5
19047 Penndel 2,703 ....N5
18073 Pennsburg 2,339 ....M5
†17331 Pennville 1,398 ....J6
††19151 Penn Wynne M6
18944 Perkasie 5,241 ....M5
15473 Perryopolis 2,139 ....C5
*19101 Philadelphia⊙ 1,688,210 ....N6
   Philadelphia‡ 4,716,818 ....N6
16866 Philipsburg 3,533 ....F4
19460 Phoenixville 14,165 ....L5
17963 Pine Grove 2,244 ....K4
16868 Pine Grove Mills 1,030 ....G4
15140 Pitcairn 4,175 ....C5
*15201 Pittsburgh⊙ 423,938 ....B7
   Pittsburgh‡ 2,263,894 ....B7
*18640 Pittston 9,930 ....F7
*18701 Plains 5,455 ....F7

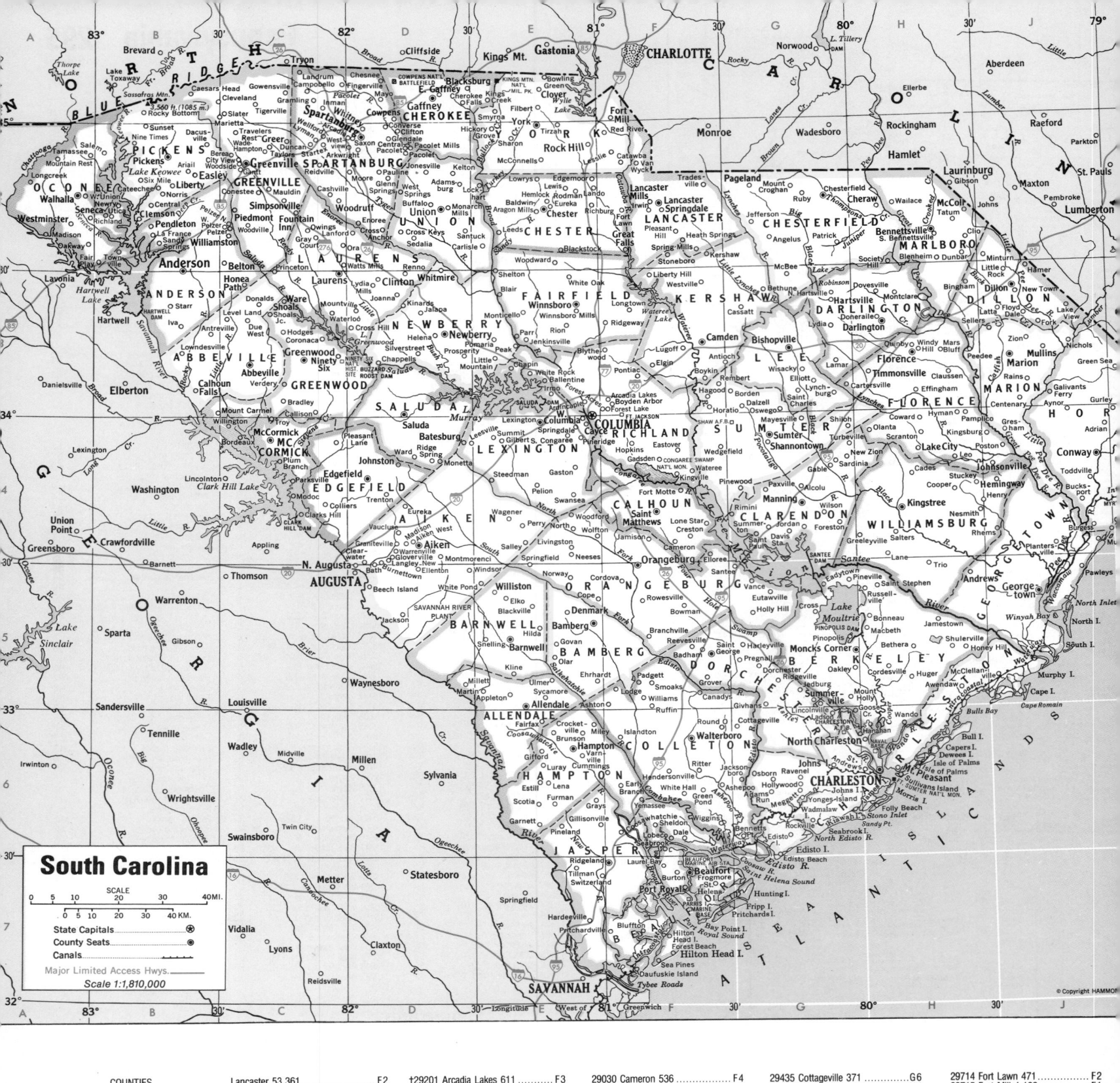

South Carolina

SCALE
0 5 10   20   30   40 MI.
0 5 10   20   30   40 KM.

State Capitals .... ⊛
County Seats .... ⊙
Canals .... ┴┴

Major Limited Access Hwys. _____

Scale 1:1,810,000

© Copyright HAMMOND

COUNTIES

Name/Pop.	Key
Abbeville 22,627	B3
Aiken 105,625	D4
Allendale 10,700	E6
Anderson 133,235	B2
Bamberg 18,118	E5
Barnwell 19,868	E5
Beaufort 65,364	F7
Berkeley 94,727	G5
Calhoun 12,206	F4
Charleston 276,974	H6
Cherokee 40,983	E2
Chester 30,148	E2
Chesterfield 38,161	G2
Clarendon 27,464	G4
Colleton 31,776	F5
Darlington 62,717	H3
Dillon 31,083	J3
Dorchester 58,761	G5
Edgefield 17,528	D4
Fairfield 20,700	F3
Florence 110,163	H3
Georgetown 42,461	J5
Greenville 287,913	C2
Greenwood 57,847	C3
Hampton 18,159	E6
Horry 101,419	J4
Jasper 14,504	E6
Kershaw 39,015	F3

Name/Pop.	Key
Lancaster 53,361	F2
Laurens 52,214	D2
Lee 18,929	G3
Lexington 140,353	E4
Marion 34,179	J3
Marlboro 31,634	H2
McCormick 7,797	C4
Newberry 31,242	D3
Oconee 48,611	A2
Orangeburg 82,276	F5
Pickens 79,292	B2
Richland 269,735	F4
Saluda 16,150	D3
Spartanburg 201,861	D2
Sumter 88,243	G4
Union 30,764	H4
Williamsburg 38,226	H4
York 106,720	E2

CITIES and TOWNS

Zip	Name/Pop.	Key
29620	Abbeville⊙ 5,833	C3
29801	Aiken⊙ 14,978	D4
†29801	Aiken West 3,083	D4
29810	Allendale⊙ 4,400	E5
*29621	Anderson⊙ 27,965	B2
	Anderson‡ 133,235	B2
29510	Andrews 3,129	H5
29320	Arcadia 2,088	C2

Zip	Name/Pop.	Key
†29201	Arcadia Lakes 611	F3
†29640	Ariail 2,419	B2
†29301	Arkwright 2,623	C2
†29582	Atlantic Beach 289	K4
29511	Aynor 643	J3
29003	Bamberg⊙ 3,672	E5
29812	Barnwell⊙ 5,572	E5
29006	Batesburg 4,023	D4
29816	Bath 2,242	D5
29902	Beaufort⊙ 8,634	F7
29627	Belton 5,312	C2
29512	Bennettsville⊙ 8,774	H2
29611	Berea 13,164	C2
29009	Bethune 481	G3
29010	Bishopville⊙ 3,429	G3
29702	Blacksburg 1,873	D1
29817	Blackville 2,840	E5
29516	Blenheim 202	H2
29910	Bluffton 541	F7
29016	Blythewood 92	E3
29431	Bonneau 401	G5
29018	Bowman 1,137	F5
29432	Branchville 1,769	F5
29911	Brunson 590	E6
29527	Bucksport 1,125	J4
29321	Buffalo 1,641	D2
†29844	Burnettown 359	D5
29902	Burton 3,619	F7
29628	Calhoun Falls 2,491	B3
29020	Camden⊙ 7,462	F3

Zip	Name/Pop.	Key
29030	Cameron 536	F4
29322	Campobello 472	C1
29031	Carlisle 503	D2
29169	Cayce 11,701	E4
29519	Centenary 700	J3
29630	Central 1,914	B2
†29372	Central Pacolet 315	D2
29036	Chapin 311	E3
29037	Chappells 109	D3
*29401	Charleston⊙ 69,510	G6
	Charleston-North	
	Charleston‡ 430,301	G6
29520	Cheraw 5,654	H2
29323	Chesnee 1,069	D1
29706	Chester⊙ 6,820	E2
29709	Chesterfield⊙ 1,432	G2
29611	City View 1,662	C2
29822	Clearwater 3,967	D4
29631	Clemson 8,118	B2
29635	Cleveland 800	C1
29324	Clifton 950	C2
29325	Clinton 8,596	D3
29526	Clio 1,031	H2
29710	Clover 3,451	E1
29048	Eutawville 615	G5
29527	Columbia (cap.)⊙ 100,385	F4
	Columbia‡ 408,176	F4
29329	Converse 1,173	D2
29526	Conway⊙ 10,240	J4
29038	Cope 167	E5
29039	Cordova 202	F5

Zip	Name/Pop.	Key
29435	Cottageville 371	G6
29530	Coward 428	H4
29330	Cowpens 2,023	D1
29332	Cross Hill 604	D3
29532	Darlington⊙ 7,989	H3
29042	Denmark 4,434	E5
29536	Dillon⊙ 7,060	J3
29638	Donalds 366	C3
†29532	Doraville 1,276	H3
29639	Due West 1,366	C3
29334	Duncan 1,259	C2
29640	Easley⊙ 15,195	B2
29044	Eastover 899	F4
†29340	East Gaffney 4,092	D1
29824	Edgefield⊙ 2,713	C4
†29438	Edisto Beach 193	G7
29438	Edisto Island 900	G6
29081	Ehrhardt 353	E5
29045	Elgin 595	F3
29826	Elko 329	E5
29047	Elloree 909	F4
29335	Enoree 1,107	D2
29918	Estill 2,308	E6
†29706	Eureka 1,627	E2
29048	Eutawville 615	G5
29827	Fairfax 2,154	E6
29501	Florence⊙ 29,176	H3
	Florence‡ 110,163	H3
29439	Folly Beach 1,478	H6
29206	Forest Acres 6,071	E3

Zip	Name/Pop.	Key
29714	Fort Lawn 471	F2
29715	Fort Mill 4,162	F1
29050	Fort Motte 700	F4
29644	Fountain Inn 4,226	C2
29921	Furman 348	E6
29340	Gaffney⊙ 13,453	D1
†29609	Gantt 13,719	C2
29053	Gaston 960	E4
29440	Georgetown⊙ 10,144	J5
29923	Gifford 385	E6
29054	Gilbert 211	E4
29346	Glendale 1,049	D2
29828	Gloverville 2,619	D4
29445	Goose Creek 17,811	H6
†29843	Govan 109	E5
29829	Graniteville 1,158	D4
29645	Gray Court 988	C2
29055	Great Falls 2,601	F2
29056	Greeleyville 593	H4
*29601	Greenville⊙ 58,242	C2
	Greenville-Spartanburg‡	
	568,758	C2
29646	Greenwood⊙ 21,613	C3
29651	Greer 10,525	C2
29924	Hampton⊙ 3,143	E6
29410	Hanahan 13,224	H6
29927	Hardeeville 1,250	E7
29448	Harleyville 606	G5
29550	Hartsville 7,631	G3
29058	Heath Springs 979	F3

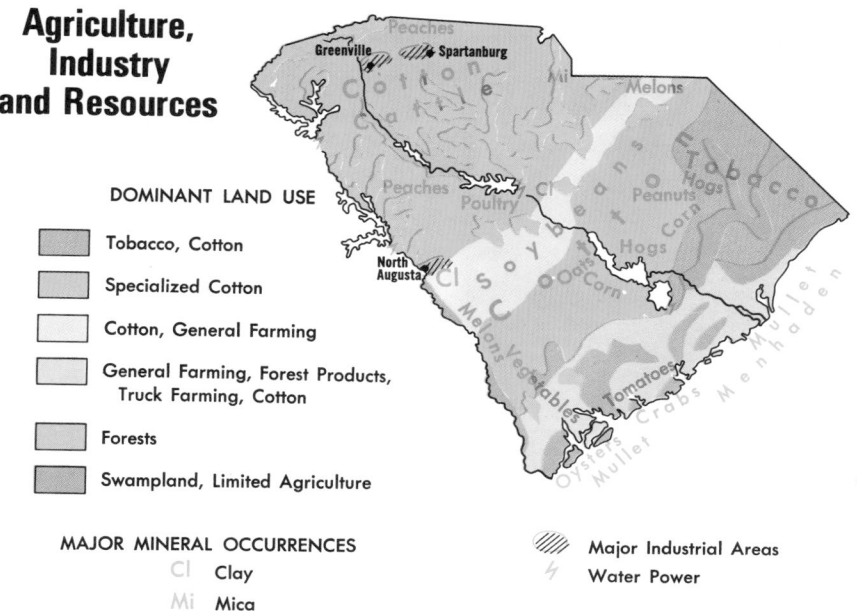

Agriculture, Industry and Resources

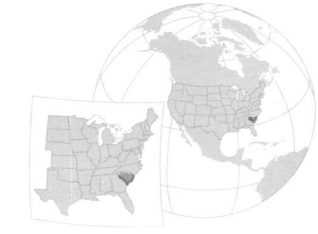

DOMINANT LAND USE

- Tobacco, Cotton
- Specialized Cotton
- Cotton, General Farming
- General Farming, Forest Products, Truck Farming, Cotton
- Forests
- Swampland, Limited Agriculture

MAJOR MINERAL OCCURRENCES

Cl Clay
Mi Mica

/// Major Industrial Areas
⚡ Water Power

AREA 31,113 sq. mi. (80,583 sq. km.)
POPULATION 3,121,833
CAPITAL Columbia
LARGEST CITY Columbia
HIGHEST POINT Sassafras Mtn. 3,560 ft. (1085 m.)
SETTLED IN 1670
ADMITTED TO UNION May 23, 1788
POPULAR NAME Palmetto State
STATE FLOWER Carolina (Yellow) Jessamine
STATE BIRD Carolina Wren

---

†29720 Lancaster Mills 2,096 ...... F2
29356 Landrum 2,141 ....C1
29564 Lane 554 ....H5
29834 Langley 1,714 ....D4
29565 Latta 1,804 ....J3
29902 Laurel Bay 5,238 ....F7
29360 Laurens⊙ 10,587 ......C3
29070 Leesville 2,296 ....E4
†29730 Lesslie 1,102 ....E2
29072 Lexington⊙ 2,131 ......E4
29657 Liberty 3,167 ....B2
†29483 Lincolnville 808 ......G6
29075 Little Mountain 282 ...E3
29076 Livingston 166 ....E4
29364 Lockhart 85 ....E2
29082 Lodge 145 ....F5
29569 Loris 2,193 ....K3
29659 Lowndesville 197 ......B3
†29706 Lowrys 225 ....E2
29078 Lugoff 2,939 ....F3
29932 Luray 149 ....E6
29325 Lydia Mills 925 ....D3
29365 Lyman 1,067 ....C2
29080 Lynchburg 534 ....G3
†29829 Madison 1,150 ....D4
29102 Manning⊙ 4,746 ....G4
29661 Marietta-Slater 1,834 ..C1
29571 Marion⊙ 7,700 ....J3
29662 Mauldin 8,143 ....C2
29104 Mayesville 663 ....G4
29101 McBee 774 ....G3
29458 McClellanville 436 ....H5
29570 McColl 2,677 ....H2
29726 McConnells 171 ....E2
29835 McCormick⊙ 1,725 ....C4
29460 Meggett 249 ....G6
†29379 Monarch Mills 2,353 ..D2
29461 Moncks Corner⊙ 3,699 .G5
29105 Monetta 167 ....D4
29840 Mount Carmel 182 .....C3
29727 Mount Croghan 146 ....G2
29464 Mount Pleasant 14,209 .H6
29574 Mullins 6,068 ....J3
29576 Murrells Inlet 2,410 ...K4
29577 Myrtle Beach 18,446 ...K4
29107 Neeses 557 ....E4
29108 Newberry⊙ 9,866 .....D3
29809 New Ellenton 2,628 ....D5
†29536 New Town 950 ....J3
29581 Nichols 606 ....J3
29666 Ninety Six 2,249 ......C3
29667 Norris 903 ....B2
29112 North 1,304 ....E4
29841 North Augusta 13,593 ..C5
29406 North Charleston 62,534 .G6
†29550 North Hartsville 2,650 .G3
29582 North Myrtle Beach 3,960 .K4
29113 Norway 518 ....E5
29114 Olanta 699 ....H4
29843 Olar 381 ....E5
29115 Orangeburg⊙ 14,933 ...F4
29372 Pacolet 1,556 ....D2
29373 Pacolet Mills 1,051 ...D2
29728 Pageland 2,720 ....G2
29583 Pamplico 1,213 ....H4
29584 Patrick 375 ....G2
29102 Paxville 244 ....G4
29671 Pickens⊙ 3,199 ......B2
29673 Piedmont 2,992 ....C2
29934 Pineland 800 ....E6
†29169 Pineridge 1,287 ......E4
29468 Pineville 900 ....H5
29125 Pinewood 689 ....G4
29469 Pinopolis 788 ....G5
29845 Plum Branch 73 ....C4
29126 Pomaria 271 ....D3
29935 Port Royal 2,977 .....F7
29127 Prosperity 803 ....D3
†29501 Quinby 952 ....H3
29470 Ravenel 1,655 ....G6

29471 Reevesville 241 ....F5
29729 Richburg 269 ....E2
29936 Ridgeland⊙ 1,143 ....E7
29129 Ridge Spring 969 ....D4
29472 Ridgeville 603 ....G5
29130 Ridgeway 343 ....F3
29730 Rock Hill 35,344 ....E2
      Rock Hill‡ 106,720 .....E2
29133 Rowesville 388 ....F5
29741 Ruby 256 ....G2
29407 Saint Andrews 9,908 .....G6
29477 Saint George⊙ 2,134 ....F5
29135 Saint Matthews⊙ 2,496 ..F4
29479 Saint Stephen 1,850 .....H5
29676 Salem 194 ....A2
29137 Salley 584 ....E4
29138 Saluda⊙ 2,752 ....D4
29142 Santee 612 ....F5
†29301 Saxon 4,383 ....D2
29939 Scotia 72 ....E6
29591 Scranton 861 ....H4
29592 Sellers 388 ....H3
29678 Seneca 7,436 ....A2
29742 Sharon 323 ....E2
29145 Silverstreet 200 ....D3
29681 Simpsonville 9,037 ......C2
29682 Six Mile 470 ....B2
29683 Slater-Marietta 1,834 ...C1
29481 Smoaks 165 ....F5
29743 Smyrna 47 ....E1
†29812 Snelling 111 ....D5
29593 Society Hill 848 ....H2
†29512 South Bennettsville 1,065 .H2
29169 South Congaree 2,113 ....E4
*29301 Spartanburg⊙ 43,826 ....C1
29169 Springdale 2,985 ....E4
†29720 Springdale 2,570 ....F2
29146 Springfield 604 ....E4
†29067 Spring Mills 1,419 .....F2
29684 Starr 241 ....B3
29377 Startex 1,006 ....C2
29554 Stuckey 222 ....H4
29482 Sullivans Island 1,867 ..H6
29148 Summerton 1,173 ....G4
29483 Summerville 6,706 ....G5
†29054 Summit 172 ....E4
29150 Sumter⊙ 24,890 ....G4
29577 Surfside Beach 2,522 ....K4
29160 Swansea 888 ....E4
29846 Sycamore 261 ....E5
29594 Tatum 101 ....H2
29687 Taylors 15,801 ....C2
29688 Tigerville 95 ....C1
29161 Timmonsville 2,112 ......H3
29690 Travelers Rest 3,017 ....C2
29847 Trenton 404 ....D4
29848 Troy 705 ....C4
29162 Turbeville 549 ....G4
29849 Ulmer 91 ....E6
29379 Union⊙ 10,523 ....D2
†29678 Utica 1,501 ....B2
29163 Vance 89 ....G5
29944 Varnville 1,948 ....E6
†29607 Wade-Hampton 20,180 ....C2
29164 Wagener 903 ....E4
29691 Walhalla⊙ 3,977 ....A2
29488 Walterboro⊙ 6,209 ......F6
29166 Ward 98 ....D4
29692 Ware Shoals 2,370 ....C3
29851 Warrenville 1,029 ....D4
29384 Waterloo 200 ....C3
†29360 Watts Mills 1,324 ......D2
29385 Wellford 2,143 ....C2
29169 West Columbia 10,409 ....E4
29693 Westminster 3,114 ....A2
29669 West Pelzer 944 ....B2
29696 West Union 300 ....A2
†29301 Westview 1,999 ....C2
29178 Whitmire 2,083 ....D3
29303 Whitney 4,052 ....D1
29493 Williams 205 ....F5
29697 Williamston 4,310 ....B2
29853 Williston 3,173 ....E5
29856 Windsor 55 ....E5
†29501 Windy Hill 1,622 ....H3
29180 Winnsboro⊙ 2,919 ....E3

†29180 Winnsboro Mills 1,890 ....E3
†29112 Woodford 206 ....E4
29388 Woodruff 5,171 ....D2
29945 Yemassee 789 ....F6
29745 York⊙ 6,412 ....E1

OTHER FEATURES

Ashepoo (riv.) F6
Ashley (riv.) G6
Bay Point (isl.) F7
Beaufort Marine Air Sta. F7
Big Black (creek) G2
Black (riv.) H4
Blue Ridge (mts.) B1
Broad (riv.) E2
Broad (riv.) F7
Buck (creek) J3
Bull (riv.) H6
Bullock (creek) E2
Bulls (bay) H6
Bush (riv.) D3
Buzzard Roost (dam) D3
Cape (isl.) J5
Capers (isl.) H6
Catawba (riv.) F2
Catfish (creek) J3
Charleston A.F.B. G6
Chattooga (riv.) A2
Clark Hill (dam) C4
Clark Hill (lake) C4
Combahee (riv.) F6
Congaree (riv.) F4
Congaree Nat'l Mon. F4
Cooper (riv.) H6
Coosaw (riv.) G7
Coosawhatchie (riv.) E6
Cowpens Nat'l Battlefield D1
Crooked (creek) H2
Deep (creek) B2
Dewees (isl.) H6
Donaldson A.F.B. C2

Edisto (isl.) G6
Edisto (riv.) G7
Enoree (riv.) C2
Fort Jackson F4
Fort Sumter Nat'l Mon. H6
Four Hole Swamp (creek) F5
Fripp (isl.) G7
Great Pee Dee (riv.) J4
Greenwood (lake) D3
Hartwell (dam) B3
Hartwell (lake) A3
Hilton Head (isl.) F7
Hunting (isl.) G7
Intracoastal Waterway H5
James (riv.) H6
Johns (isl.) G6
Juniper (creek) H2
Keowee (lake) B2
Keowee (riv.) B2
Kiawah (isl.) G6
Kings Mountain Nat'l Mil. Park .E1
Little (riv.) C3
Little (riv.) L4
Little Lynches (riv.) G3
Little Pee Dee (riv.) J4
Little River (inlet) L4
Lumber (riv.) J3
Lynches (riv.) H3
Marion (lake) H5
Morris (isl.) H6
Moultrie (lake) H5
Murphy (isl.) J5
Murray (lake) D4
Myrtle Beach A.F.B. K4
Naval Base H6
New (riv.) E6
Ninety Six Nat'l Hist. Site ..C3
North (inlet) J5
North (riv.) J5
North Edisto (riv.) G6
Pacolet (riv.) D1
Palms, Isle of (isl.) H6

Parris Island Marine Base F7
Pee Dee (riv.) H2
Pinopolis (dam) G5
Pocotaligo (riv.) G4
Port Royal (sound) F7
Pritchards (isl.) G7
Reedy (riv.) C2
Robinson (lake) G3
Romain (cape) J6
Saint Helena (isl.) F7
Saint Helena (sound) G7
Salkehatchie (riv.) E5
Saluda (riv.) D3
Sandy (pt.) H6
Sandy (riv.) E2
Santee (dam) G4
Santee (riv.) H5
Sassafras (mt.) B1
Savannah (riv.) E6
Savannah River Plant D5
Sea (isls.) G7
Seabrook (isl.) G6
Seneca (riv.) B2
Shaw A.F.B. 6,939 ....F4
South (isl.) J5
Stevens (creek) C4
Stono (inlet) H6
Thompsons (creek) G2
Tugaloo (riv.) A2
Turkey (creek) E2
Tybee Roads (chan.) F7
Tyger (riv.) D2
Waccamaw (riv.) J5
Wadmalaw (isl.) G6
Wando (riv.) H6
Wateree (lake) F3
Wateree (riv.) F3
Winyah (bay) J5
Wylie (lake) E1

⊙County seat.
‡Population of metropolitan area.
† Zip of nearest p.o.   * Multiple zips.

29554 Hemingway 853 ....J4
†29706 Hemlock (Eureka) 1,627 ..E2
29717 Hickory Grove 344 ....E2
29813 Hilda 355 ....E5
29928 Hilton Head Island 11,344 .F7
29653 Hodges 154 ....C3
29059 Holly Hill 1,785 ....G5
29449 Hollywood 729 ....G6
29349 Honea Path 4,114 ....C3
29349 Inman 1,554 ....C1
29063 Irmo 3,957 ....E3
†29720 Irwin 1,373 ....F2
29451 Isle of Palms 3,421 ......H6
29655 Iva 1,369 ....B3
29831 Jackson 1,771 ....D5
29453 Jamestown 193 ....H5
†29483 Jedburg 900 ....G5
29718 Jefferson 651 ....G2
29351 Joanna 1,839 ....D3
29555 Johnsonville 1,421 ....J4
29832 Johnston 2,624 ....D4
29353 Jonesville 1,201 ....D2
29067 Kershaw 1,993 ....G2
29556 Kingstree⊙ 4,147 ....H4
29814 Kline 315 ....E5
29456 Ladson 13,246 ....G6
29560 Lake City 6,731 ....H4
29563 Lake View 939 ....J3
29069 Lamar 1,333 ....G3
29720 Lancaster⊙ 9,703 ....F2

Topography

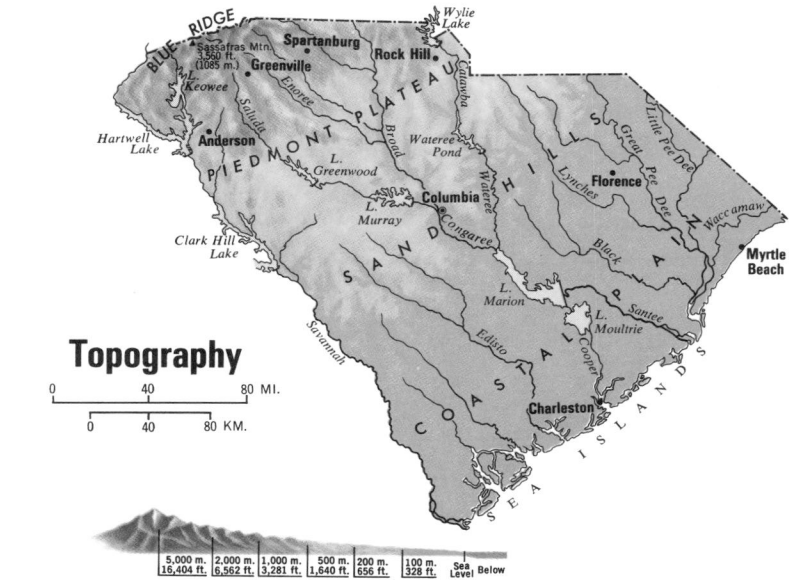

| 5,000 m. 16,404 ft. | 2,000 m. 6,562 ft. | 1,000 m. 3,281 ft. | 500 m. 1,640 ft. | 200 m. 656 ft. | 100 m. 328 ft. | Sea Level | Below |

0   40   80 MI.
0   40   80 KM.

COUNTIES

Aurora 3,628	M6
Beadle 19,195	N5
Bennett 3,044	F7
Bon Homme 8,059	O7
Brookings 24,332	R5
Brown 36,962	N2
Brule 5,245	L6
Buffalo 1,795	L5
Butte 8,372	B4
Campbell 2,243	J2
Charles Mix 9,680	M7
Clark 4,894	O4
Clay 13,689	P8
Codington 20,885	P4
Corson 5,196	G2
Custer 6,000	B6
Davison 17,820	N6
Day 8,133	O3

Deuel 5,289	R4
Dewey 5,366	G3
Douglas 4,181	N7
Edmunds 5,159	L3
Fall River 8,439	B7
Faulk 3,327	L3
Grant 9,013	R3
Gregory 6,015	L7
Haakon 2,794	F5
Hamlin 5,261	P4
Hand 4,948	L4
Hanson 3,415	O6
Harding 1,700	B2
Hughes 14,220	J5
Hutchinson 9,350	O7
Hyde 2,069	K4
Jackson 3,437	F6
Jerauld 2,929	M5
Jones 1,463	H6
Kingsbury 6,679	O5
Lake 10,724	P5
Lawrence 18,339	B5
Lincoln 13,942	R7

Lyman 3,864	J6
Marshall 5,404	O2
McCook 6,444	P6
McPherson 4,027	L2
Meade 20,717	D5
Mellette 2,249	H6
Miner 3,739	O5
Minnehaha 109,435	R6
Moody 6,692	R5
Pennington 70,361	C6
Perkins 4,700	D3
Potter 3,674	J3
Roberts 10,911	P2
Sanborn 3,213	N5
Shannon 11,323	D7
Spink 9,201	N4
Stanley 2,533	H5
Sully 1,990	J4
Todd 7,328	H7
Tripp 7,268	K7
Turner 9,255	P7
Union 10,938	R8
Walworth 7,011	J3

Yankton 18,952	P7
Ziebach 2,308	F4

CITIES and TOWNS

Zip	Name/Pop.	Key
57401 Aberdeen⊙ 25,851	M3
57310 Academy 10	M7
57520 Agar 139	J4
57420 Akaska 49	J3
57210 Albee 23	S3
57001 Alcester 885	R7
57311 Alexandria⊙ 588	O6
57714 Allen 300	F7
57312 Alpena 288	N5
57211 Altamont 58	R4
57421 Amherst 75	O2
57422 Armour⊙ 819	N7
57715 Ardmore 16	B7
57212 Arlington 991	P5
57313 Armour⊙ 819	N7
57423 Artas 43	K2
57314 Artesian 227	O6

57424 Ashton 154	N3
57213 Astoria 154	S4
57425 Athol 38	M3
57002 Aurora 507	R5
57315 Avon 576	N8
57214 Badger 99	P5
57003 Baltic 679	R6
57316 Bancroft 41	O4
57426 Barnard 65	N2
57716 Batesland 163	E7
57427 Bath 175	N3
57717 Belle Fourche⊙ 4,692	B4
57521 Belvidere 80	G6
57215 Bemis 37	F7
57004 Beresford 1,865	R7
†57310 Bijou Hills 12	L6
57620 Bison⊙ 457	E2
57718 Black Hawk 1,608	C5
57522 Blunt 424	J4
57317 Bonesteel 358	M7
57428 Bowdle 644	K3
57719 Box Elder 3,186	D5

57217 Bradley 135	O3
57005 Brandon 2,589	R6
57218 Brandt 129	R4
57429 Brentford 91	N3
57319 Bridgewater 653	P6
57219 Bristol 445	O3
57430 Britton⊙ 1,590	O2
57006 Brookings⊙ 14,951	R5
57220 Bruce 254	R5
57221 Bryant 388	P4
57720 Buffalo⊙ 453	B2
57722 Buffalo Gap 186	C6
57621 Bullhead 400	G2
57010 Burbank 92	R8
57523 Burke⊙ 859	L7
†57276 Bushnell 76	R5
57222 Butler 22	O3
57724 Camp Crook 100	B2
57012 Canistota 626	P6
57321 Canova 194	O6
57013 Canton⊙ 2,886	R7
57725 Caputa 50	D5

South Dakota

SCALE
0 5 10   20   40   60 MI.
0 5 10 20   40   60 KM.

State Capitals ⊛
County Seats ⊙
Major Limited Access Hwys.

Scale 1:2,220,000

© Copyright HAMMOND INCORPORATED, Maplewood, N.J.

57322 Carpenter 30 ....O4	57434 Conde 259 ....N3	57623 Dupree⊙ 562 ....F3
57526 Carter 7 ....J7	57227 Corona 126 ....R3	57625 Eagle Butte 435 ....G4
57323 Carthage 274 ....O5	57328 Corsica 644 ....N7	57232 Eden 142 ....P2
57223 Castlewood 557 ....R4	57019 Corson 125 ....R6	57735 Edgemont 1,468 ....B7
57324 Cavour 117 ....N5	57775 Cottonwood 4 ....F6	57024 Egan 248 ....R6
57527 Cedarbutte 10 ....H6	57729 Creighton 53 ....E5	57025 Elk Point⊙ 1,661 ....R8
†57058 Center 10 ....P6	57435 Cresbard 221 ....M3	57026 Elkton 632 ....S5
57014 Centerville 892 ....R7	57229 Crocker 70 ....O3	57736 Elm Springs 2 ....D5
†57754 Central City 232 ....B5	57020 Crooks 594 ....R6	57332 Emery 399 ....O6
57325 Chamberlain⊙ 2,258 ....L6	57730 Custer⊙ 1,830 ....B6	57737 Enning 25 ....E4
57015 Chancellor 257 ....R7	57529 Dallas 199 ....K7	57233 Erwin 66 ....P5
57431 Chelsea 41 ....M3	57329 Dante 83 ....N7	†57353 Esmond O5
57622 Cherry Creek 500 ....F4	57021 Davis 100 ....P7	57234 Estelline 719 ....R4
57016 Chester 375 ....R6	57732 Deadwood⊙ 2,035 ....B5	57334 Ethan 351 ....N6
57224 Claire City 87 ....P2	57022 Dell Rapids 2,389 ....R6	57437 Eureka 1,360 ....K2
57432 Claremont 180 ....N2	57330 Delmont 290 ....N7	57738 Fairburn 41 ....C6
57225 Clark⊙ 1,351 ....O4	57230 Dempster 62 ....R4	57335 Fairfax 225 ....M7
57581 Clearfield 12 ....K7	57231 De Smet⊙ 1,237 ....O5	57027 Fairview 90 ....R7
57226 Clear Lake⊙ 1,310 ....R4	57331 Dimock 140 ....O7	57626 Faith 576 ....E4
57017 Colman 501 ....R6	57530 Dixon 125 ....L7	57336 Farmer 27 ....O6
57528 Colome 361 ....K7	57436 Doland 381 ....N4	57438 Faulkton⊙ 981 ....L3
57018 Colton 757 ....P6	57023 Dolton 47 ....P7	57337 Fedora 80 ....O5
57433 Columbia 161 ....N2	57531 Draper 138 ....J6	57028 Flandreau⊙ 2,114 ....R5

(continued on following page)

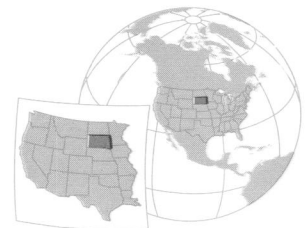

AREA   77,116 sq. mi. (199,730 sq. km.)
POPULATION   690,768
CAPITAL   Pierre
LARGEST CITY   Sioux Falls
HIGHEST POINT   Harney Pk. 7,242 ft.
     (2207 m.)
SETTLED IN   1856
ADMITTED TO UNION   November 2, 1889
POPULAR NAME   Coyote State; Sunshine
     State
STATE FLOWER   Pasqueflower
STATE BIRD   Ring-necked Pheasant

Topography

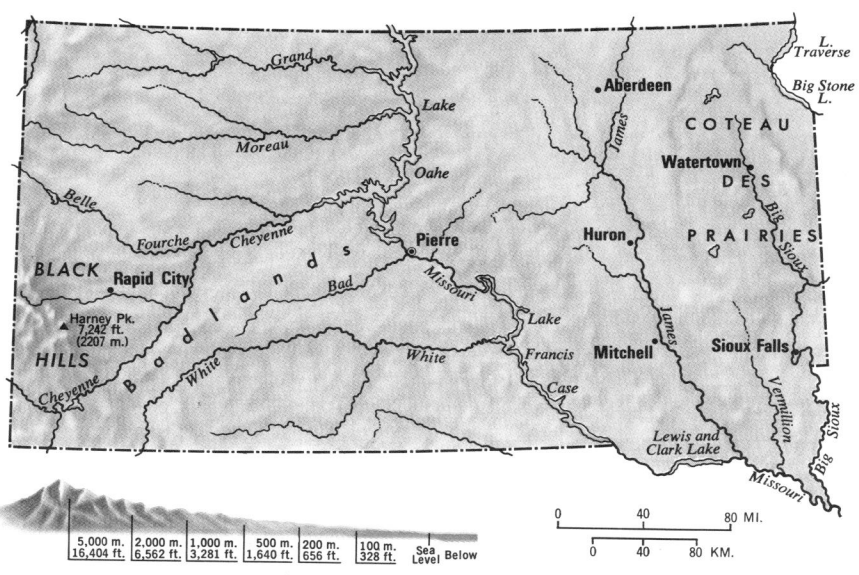

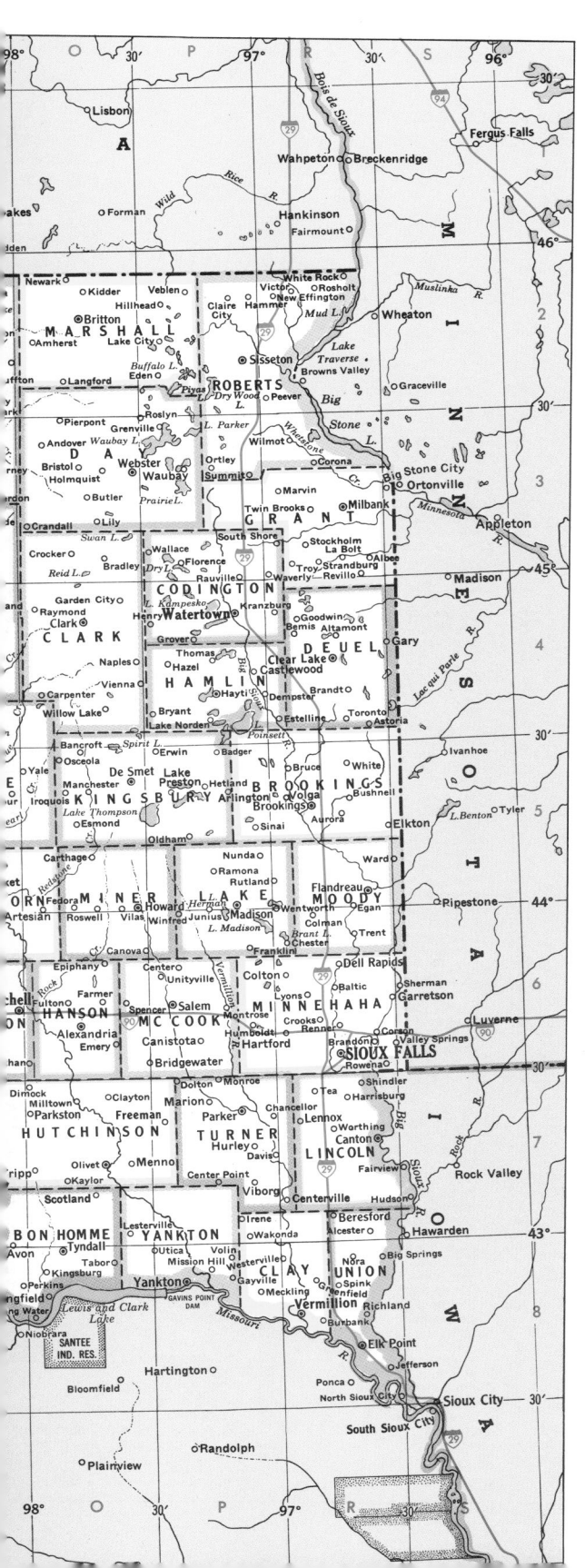

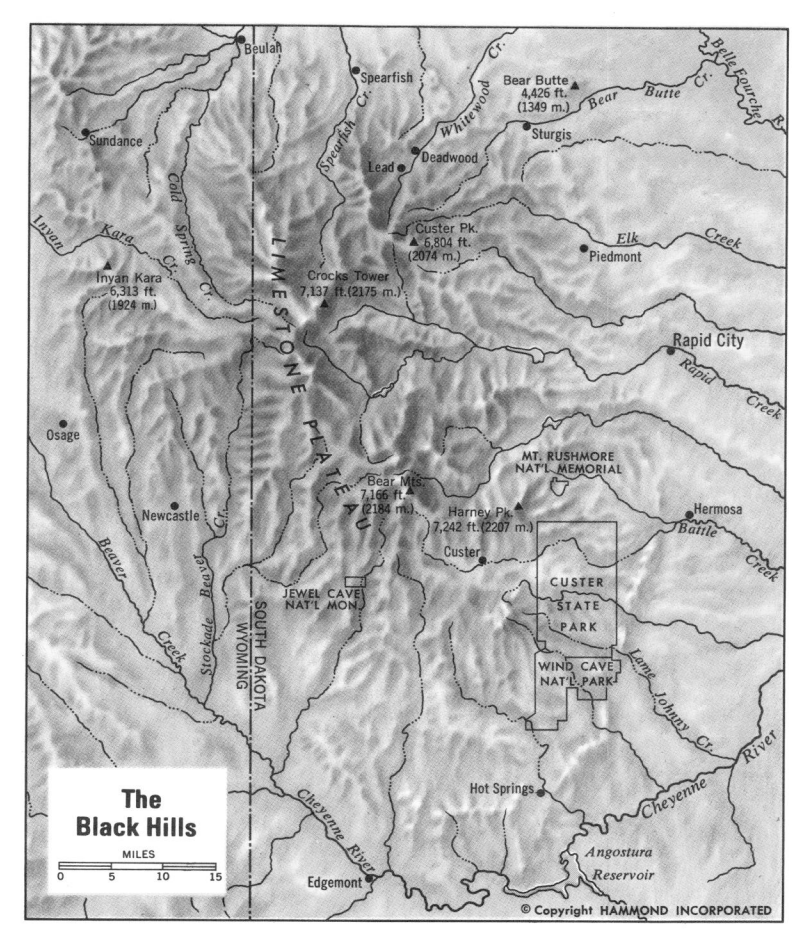

The Black Hills

© Copyright HAMMOND INCORPORATED

Agriculture, Industry and Resources

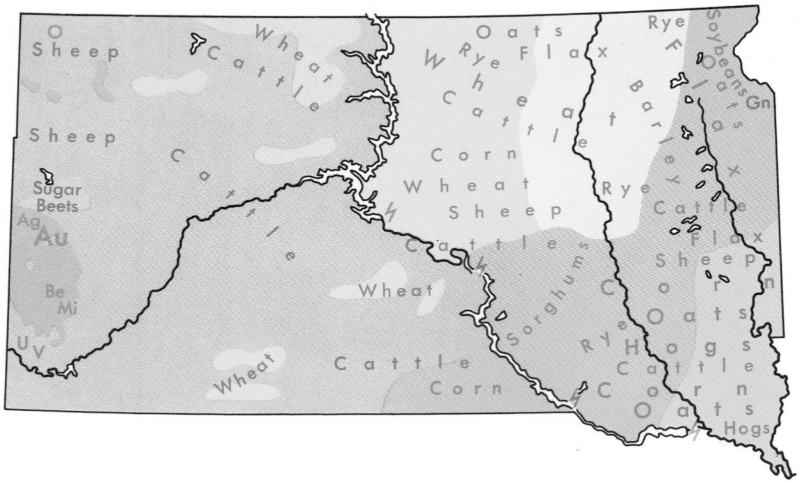

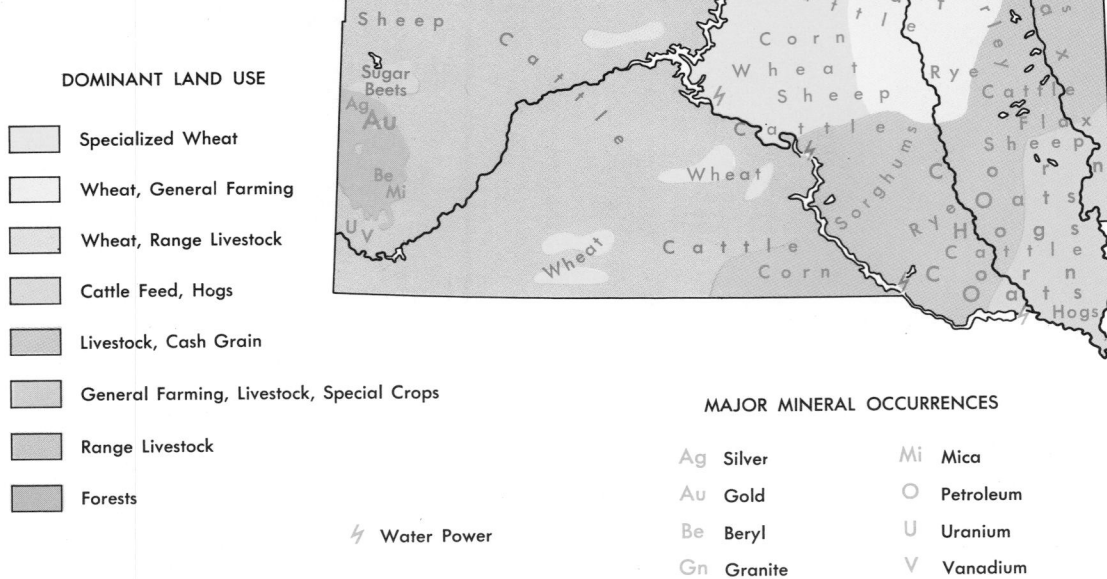

DOMINANT LAND USE

- Specialized Wheat
- Wheat, General Farming
- Wheat, Range Livestock
- Cattle Feed, Hogs
- Livestock, Cash Grain
- General Farming, Livestock, Special Crops
- Range Livestock
- Forests

Water Power

MAJOR MINERAL OCCURRENCES

- Ag Silver
- Au Gold
- Be Beryl
- Gn Granite
- Mi Mica
- O Petroleum
- U Uranium
- V Vanadium

57235 Florence 190 P3
57338 Forestburg 100 N5
57532 Fort Pierre⊙ 1,789 H5
57339 Fort Thompson 750 L5
57440 Frankfort 209 N4
57441 Frederick 307 N2
57029 Freeman 1,462 O7
57742 Fruitdale 88 B4
57340 Fulton 108 O6
57341 Gannvalley⊙ 70 L5
57236 Garden City 104 O4
57030 Garretson 963 S6
57237 Gary 354 S4
57031 Gayville 407 P8
57342 Geddes 303 M7
57442 Gettysburg⊙ 1,623 K3
57629 Glad Valley 75 F3
57630 Glencross 150 H3
57631 Glenham 169 J2
57238 Goodwin 139 R4
57533 Gregory 1,503 L7
57239 Grenville 119 O3
57445 Groton 1,230 N3
57534 Hamill 25 K6
57032 Harrisburg 558 R7
57344 Harrison 89 M7
57536 Harrold 196 K4
57033 Hartford 1,207 P6
57537 Hayes 25 H5
57241 Hayti⊙ 371 P4
57242 Hazel 94 P4
57446 Hecla 435 N2
57243 Henry 217 P4
57743 Hereford 50 D5
57744 Hermosa 251 C6
57632 Herreid 570 K2
57538 Herrick 115 L7
57244 Hetland 66 P5
57345 Highmore⊙ 1,055 L4
57745 Hill City 535 B6
†57437 Hillsview 9 L2
57348 Hitchcock 132 M4
57540 Holabird 30 K4
†57274 Holmquist 25 O3
57448 Hosmer 385 L2
57747 Hot Springs⊙ 4,742 C7
57449 Houghton 80 N2
57450 Hoven 615 K3
57349 Howard⊙ 1,169 P5
57748 Howes 4 E4
57034 Hudson 388 R7
57035 Humboldt 487 P6
57036 Hurley 419 P7
57350 Huron⊙ 13,000 N5
57541 Ideal 250 K6
57750 Interior 62 F6
57542 Iona 4 L6
57451 Ipswich⊙ 1,153 L3
57037 Irene 523 P7
57353 Iroquois 348 O5
57633 Isabel 332 G3
57452 Java 261 K3
57038 Jefferson 592 S8
57543 Kadoka⊙ 832 F6
57354 Kaylor 120 O7
57634 Keldron 17 F2
57544 Kennebec⊙ 334 K6
57545 Keyapaha 4 J7
57751 Keystone 295 C6
57355 Kimball 752 M6
57245 Kranzburg 136 R4
57752 Kyle 600 E7
57246 La Bolt 94 R3
57356 Lake Andes⊙ 1,029 M7
57247 Lake City 46 O2
57248 Lake Norden 417 P4
57249 Lake Preston 789 P5
57358 Lane 83 N5
57454 Langford 307 O2
57636 Lantry 200 G3
57754 Lead 4,330 B5
57455 Lebanon 129 K3
57638 Lemmon 1,871 E2
57039 Lennox 1,827 R7
57456 Leola⊙ 645 M2
57040 Lesterville 156 O7
57359 Letcher 221 N6
57250 Lily 38 O3
57639 Little Eagle 150 H2
57640 Lodgepole 20 D2
57457 Longlake 117 L2
57547 Longvalley 15 F7
57360 Loomis 55 N6
†57472 Lowry 22 K3
†57471 Loyalton 6 L3
57755 Ludlow 10 C2
57041 Lyons 100 R6
57042 Madison⊙ 6,210 P6
57643 Mahto 9 H2
57756 Manderson 450 D7
57460 Mansfield 120 N3
57757 Marcus 5 E4
57043 Marion 830 P7
57551 Martin⊙ 1,018 F7
57361 Marty 250 N8
57251 Marvin 52 R3
57641 McIntosh⊙ 418 G2
57642 McLaughlin 754 H2
57644 Meadow 21 E2
57044 Meckling 108 R8
57461 Mellette 192 N3
57045 Menno 793 P7
57552 Midland 277 G5
57252 Milbank⊙ 4,120 R3
57553 Milesville 6 F5
57554 Millboro 12 K7
57362 Miller⊙ 1,931 L4
57462 Mina 29 M3
57463 Miranda 30 M4
57555 Mission 748 H7
57046 Mission Hill 197 P8
57557 Mission Ridge 46 H4
57301 Mitchell⊙ 13,916 N6
57601 Mobridge 4,174 J2
57047 Monroe 170 P7
57048 Montrose 396 P6
57645 Morristown 127 F2
57558 Mosher 9 J7
57646 Mound City⊙ 111 K2
57363 Mount Vernon 402 N6
57758 Mud Butte 3 D4
57559 Murdo⊙ 723 H6
†57271 Naples 45 O4
57759 Nemo 42 B5
57255 New Effington 261 R2
57760 Newell 638 C4
57364 New Holland 125 M7
57761 New Underwood 517 D5
†57584 New Witten 134 K7
57762 Nisland 216 C4
57560 Norris 25 G7
†57625 North Eagle Butte 1,354 G3
57049 North Sioux City 1,992 R8
57465 Northville 138 M3
57050 Nunda 60 P5
57365 Oacoma 289 L6
57763 Oelrichs 124 C7
57764 Oglala 475 D7
57562 Okaton 30 H6
57563 Okreek 500 J7
57051 Oldham 222 P5
57052 Olivet⊙ 96 O7
57466 Onaka 70 L3
57564 Onida⊙ 851 K4
57765 Opal 5 D4
57766 Oral 60 C7
57467 Orient 87 L4
57256 Ortley 80 P3
57565 Ottumwa 3 G5
57767 Owanka 18 D5
57647 Parade 2 G3
57053 Parker⊙ 999 P7
57366 Parkston 1,545 O7
57566 Parmelee 600 G7
57257 Peever 232 R2
57567 Philip⊙ 1,088 F5
57367 Pickstown 225 M7
57769 Piedmont 500 C5
57468 Pierpont 184 O3
57501 Pierre (cap.)⊙ 11,973 J5
57770 Pine Ridge 3,059 E7
57771 Plainview 2 E4
57368 Plankinton⊙ 644 N6
57369 Platte 1,334 M7
57648 Pollock 355 J2
57772 Porcupine 260 E7
57649 Prairie City 50 D2
57568 Presho 760 J6
57773 Pringle 105 B6
57774 Provo 60 B7
57370 Pukwana 234 L6
57775 Quinn 80 E5
57650 Ralph 12 C2
57054 Ramona 241 P5
57701 Rapid City⊙ 46,492 C5
Rapid City‡ 90,850 C5
57357 Ravinia 88 N7
57258 Raymond 106 O4
57469 Redfield⊙ 3,027 N4
57776 Redig 50 C3
57777 Redowl 10 D4
57371 Ree Heights 88 L4
57569 Reliance 190 K6
57055 Renner 320 R6
57651 Reva 8 C2
57259 Revillo 158 R3
57652 Ridgeview 75 H3
†57701 Rockerville 28 C6
57470 Rockham 52 M4
57471 Roscoe 370 L3
57570 Rosebud 900 H7
57260 Rosholt 446 R2
57261 Roslyn 261 P2
57372 Roswell 19 O6
57056 Rowena 100 R6
57057 Rutland 30 P5
57571 Saint Charles 25 L7
57572 Saint Francis 766 H7
57373 Saint Lawrence 223 M4
57779 Saint Onge 250 B4
57058 Salem⊙ 1,486 P6
57780 Scenic 26 D6
57059 Scotland 1,022 O7
57472 Selby⊙ 884 J3
57473 Seneca 103 L3
57060 Sherman 100 S6
57781 Silver City 31 B5
57061 Sinai 129 P5
*57101 Sioux Falls⊙ 81,343 R6
Sioux Falls‡ 109,435 R6
57262 Sisseton⊙ 2,789 R2
57782 Smithwick 50 C7
57654 Sorum 2 D3
57263 South Shore 241 P3
57783 Spearfish 5,251 B5
57374 Spencer 380 O6
†57010 Spink 75 R8
57062 Springfield 1,377 N8
57346 Stephan 30 K5
57375 Stickney 409 M6
57264 Stockholm 95 R3
57784 Stoneville 20 D4
†57359 Storla 19 M6
57265 Strandburg 79 R3
57474 Stratford 82 N3
57785 Sturgis⊙ 5,184 B5
57266 Summit 290 P3
57063 Tabor 460 O8
†57433 Tacoma Park 20 N2
57064 Tea 729 R7
†57242 Thomas 12 P4
†57638 Thunder Hawk 26 F2
†57769 Tilford 75 C5
57656 Timber Lake⊙ 660 H3
57475 Tolstoy 97 K3
57268 Toronto 236 R4
57657 Trail City 68 H3
57065 Trent 197 R6
57376 Tripp 804 N7
†57754 Trojan 40 B5
†57265 Troy 18 R3
57476 Tulare 238 N4
57477 Turton 101 N3
57574 Tuthill 75 G7
57269 Twin Brooks 87 R3
57066 Tyndall⊙ 1,253 O8
57787 Union Center 63 D4
†57058 Unityville 20 P6
57067 Utica 100 P8
57788 Vale 160 C4
57068 Valley Springs 801 S6
†57381 Vayland 3 M5
57270 Veblen 368 P2
57478 Verdon 7 N3
57069 Vermillion⊙ 10,136 R8
57575 Vetal 19 G7
57070 Viborg 812 P7
†57260 Victor 9 R2
57271 Vienna 90 O4
†57349 Vilas 28 O6
†57701 Villa Ranchaero 1,666 C5
57379 Virgil 37 N5
57576 Vivian 95 J6
57071 Volga 1,221 R5
57072 Volin 156 P8
57380 Wagner 1,453 N7
57073 Wakonda 383 P7
57658 Wakpala 500 H2
57659 Walker 12 G2
57790 Wall 770 E6
57272 Wallace 90 P3
57577 Wanblee 550 F6
57074 Ward 43 R5
57479 Warner 322 M3
57791 Wasta 99 D5
57660 Watauga 50 F2
57201 Watertown⊙ 15,649 P4
57273 Waubay 675 P3
57202 Waverly 30 R3
57274 Webster⊙ 2,417 P3
57480 Wecota 30 L3
57075 Wentworth 193 R6
57381 Wessington 327 M5
57382 Wessington Springs⊙ 1,203 M5
†57069 Westerville 21 P8
57481 Westport 122 M2
57482 Wetonka 22 M2
57578 Wewela 6 K7
57276 White 474 R5
†57638 White Butte 21 E2
57661 Whitehorse 196 H3
57383 White Lake 414 M6
57792 White Owl 6 E4
57579 White River⊙ 561 H6
†57260 White Rock 10 R2
57793 Whitewood 821 B5
57278 Willow Lake 375 O4
57279 Wilmot 507 R3
57076 Winfred 81 P6
57580 Winner⊙ 3,472 K7
57584 Witten 134 J7
57384 Wolsey 437 N5
57585 Wood 134 J6
57385 Woonsocket⊙ 799 N5
57077 Worthing 388 R7
57794 Wounded Knee 376 D7
57386 Yale 136 O5
57078 Yankton⊙ 12,011 P8
57483 Zell 60 M4
57795 Zeona 2 D3

OTHER FEATURES

Aeber (creek) G4
Andes (lake) N7
Angostura (res.) B7
Antelope (creek) D3
Bad (riv.) G5
Badlands Nat'l Mon. E6
Battle (creek) C6
Bear in the Lodge (creek) F6
Beaver (creek) A6
Belle Fourche (res.) B4
Belle Fourche (riv.) C4
Big Bend (dam) K5
Big Sioux (riv.) S7
Big Stone (lake) R3
Black Hills (mts.) B5
Black Pine (creek) G6
Bois de Sioux (riv.) R1
Boxelder (creek) D5
Brant (lake) R6
Buffalo (creek) F6
Buffalo (lake) P2
Bull (creek) C2
Bull (creek) K6
Byron (lake) N4
Cain (creek) N5
Cherry (creek) F4
Cherry (creek) F5
Cheyenne (riv.) F4
Cheyenne River Ind. Res. F4
Choteau (creek) N7
Columbia Road (res.) N2
Cottonwood (creek) E5
Cottonwood (lake) M4
Crazy Horse Mon. B6
Crow (creek) A4
Crow Creek Ind. Res. L5
Dog Ear (creek) K6
Dry (creek) G4
Dry (lake) P3
Dry Wood (lake) P2
Elk (creek) C5
Ellsworth A.F.B. 4,766 C5
Elm (creek) D4
Elm (riv.) M2
Firesteel (creek) N6
Flint Rock (creek) E3
Fort Randall (dam) N7
Foster (creek) N4
Francis Case (lake) L7
French (creek) C6
Gavins Point (dam) P8
Geographical Center of U.S. B4
Grand (riv.) F2
Harney (peak) B6
Hat (creek) B7
Hell Canyon (creek) B6
Herman (lake) P5
Horsehead (creek) C7
Indian (creek) B4
James (riv.) N5
Jewel Cave Nat'l Mon. B6
Kampeska (lake) P4
Keya Paha (riv.) K7
Lame Johnny (creek) C6
Lewis and Clark (lake) O8
Little Missouri (riv.) B1
Little Moreau (riv.) G3
Little White (riv.) H7
Long (lake) L2
Lower Brule Ind. Res. K5
Madison (lake) P6
Maple (riv.) M1
Medicine (creek) J6
Medicine Knoll (creek) J5
Minnechaduza (creek) H7
Minnesota (riv.) S3
Missouri (riv.) P8
Mitchell (creek) G5
Moreau (riv.) G3
Mount Rushmore Nat'l Mem. B6
Mud (creek) N3
Mud (lake) R2
Mud Lake (res.) N2
Nasty (creek) C2
Oahe (dam) J5
Oahe (lake) J1
Oak (creek) H2
Oak (creek) J6
Okobojo (creek) J4
Old Lodge (creek) K6
Owl (creek) B4
Parker (lake) P3
Pearl (creek) N5
Pine Ridge Ind. Res. D7
Piyas (lake) P2
Platte (lake) M6
Pleasant Valley (creek) B6
Poinsett (lake) P4
Ponca (creek) L7
Prairie (lake) P3
Rabbit (creek) E3
Red (lake) L6
Red Owl (creek) E4
Red Scaffold (creek) F4
Redstone (creek) O5
Redwater (creek) A4
Reid (lake) O3
Rock (creek) O6
Rosebud Ind. Res. H7
Sand (creek) C2
Sand (creek) M5
Shadehill (res.) E2
Sharpe (lake) J5
Shue (creek) N5
Smith (creek) L6
Snake (creek) F4
Snake (creek) F5
Snake (creek) M3
Spirit (lake) O4
Spring (creek) C6
Spring (creek) J2
Squaw (creek) B3
Sulphur (creek) D4
Swan (creek) J3
Swan (lake) O3
Swan (lake) K3
Thompson (lake) O5
Thunder (creek) N4
Thunder Butte (creek) E3
Traverse (lake) R2
Turtle (creek) M4
Vermillion (riv.) P6
Virgin (creek) H3
Waubay (lake) O3
Whetstone (creek) R3
White (lake) M6
White (riv.) D7
Whitewood (creek) B4
Willow (creek) C4
Wind Cave Nat'l Park B6
Wolf (creek) L4
Wounded Knee (creek) E7

⊙County seat.
‡Population of metropolitan area.
† Zip of nearest p.o. * Multiple zips.

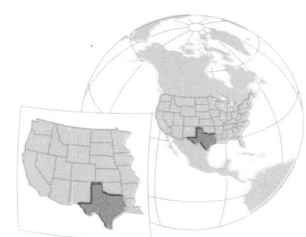

COUNTIES

Anderson 38,381 .... J6
Andrews 13,323 .... B5
Angelina 64,172 .... K6
Aransas 14,260 .... H10
Archer 7,266 .... F4
Armstrong 1,994 .... C3
Atascosa 25,055 .... F9
Austin 17,726 .... H8
Bailey 8,168 .... B3
Bandera 7,084 .... E8
Bastrop 24,726 .... G7
Baylor 4,919 .... E4
Bee 26,030 .... G9
Bell 157,820 .... G6
Bexar 988,798 .... F8
Blanco 4,681 .... F8
Borden 859 .... C5
Bosque 13,401 .... G6
Bowie 75,301 .... K4
Brazoria 169,587 .... J8
Brazos 93,588 .... H7
Brewster 7,573 .... A8
Briscoe 2,579 .... C3
Brooks 8,428 .... F11
Brown 33,057 .... F6
Burleson 12,313 .... H7
Burnet 17,803 .... F7
Caldwell 23,637 .... G8
Calhoun 19,574 .... H9
Callahan 10,992 .... E5
Cameron 209,727 .... G11
Camp 9,275 .... K5
Carson 6,672 .... C2
Cass 29,430 .... K4
Castro 10,556 .... B3
Chambers 18,538 .... K8
Cherokee 38,127 .... J6
Childress 6,950 .... D3
Clay 9,582 .... F4
Cochran 4,825 .... B4
Coke 3,196 .... D6
Coleman 10,439 .... E6
Collin 144,576 .... H4
Collingsworth 4,648 .... D3
Colorado 18,823 .... H8
Comal 36,446 .... F8
Comanche 12,617 .... F5
Concho 2,915 .... E6
Cooke 27,656 .... G4
Coryell 56,767 .... G6
Cottle 2,947 .... D3
Crane 4,600 .... B6
Crockett 4,608 .... C7
Crosby 8,859 .... C4
Culberson 3,315 .... C11
Dallam 6,531 .... B1

Dallas 1,556,390 .... H5
Dawson 16,184 .... C5
Deaf Smith 21,165 .... B3
Delta 4,839 .... J4
Denton 143,126 .... H4
De Witt 18,903 .... G9
Dickens 3,539 .... D4
Dimmit 11,367 .... E9
Donley 4,075 .... D2
Duval 12,517 .... F10
Eastland 19,480 .... F5
Ector 115,374 .... B6
Edwards 2,033 .... D7
El Paso 479,899 .... A10
Erath 22,560 .... F5
Falls 17,946 .... H6
Fannin 24,285 .... H4
Fayette 18,832 .... H8
Fisher 5,891 .... D5
Floyd 9,834 .... C3
Foard 2,158 .... E3
Fort Bend 130,846 .... J8
Franklin 6,893 .... J4
Freestone 14,830 .... H6
Frio 13,785 .... E9
Gaines 13,150 .... B5
Galveston 195,940 .... K8
Garza 5,336 .... C4
Gillespie 13,532 .... F7
Glasscock 1,304 .... C6
Goliad 5,193 .... G9
Gonzales 16,949 .... G8
Gray 26,386 .... D2
Grayson 89,796 .... H4
Gregg 99,495 .... K5
Grimes 13,580 .... J7
Guadalupe 46,708 .... G8
Hale 37,592 .... C3
Hall 5,594 .... D3
Hamilton 8,297 .... F6
Hansford 6,209 .... C1
Hardeman 6,368 .... E3
Hardin 40,721 .... K7
Harris 2,409,547 ...... J8
Harrison 52,265 ...... K5
Hartley 3,987 .... B2
Haskell 7,725 .... E4
Hays 40,594 .... F7
Hemphill 5,304 ...... D2
Henderson 42,606 .... J5
Hidalgo 283,323 .... F11
Hill 25,024 .... G5
Hockley 23,230 .... B4
Hood 17,714 ...... G5
Hopkins 25,247 .... J4
Houston 22,299 .... J6
Howard 33,142 .... C5

Hudspeth 2,728 .... B10
Hunt 55,248 .... H4
Hutchinson 26,304 .... C2
Irion 1,386 .... C6
Jack 7,408 .... F4
Jackson 13,352 .... H9
Jasper 30,781 .... K7
Jeff Davis 1,647 .... C11
Jefferson 250,938 ..... K8
Jim Hogg 5,168 .... F11
Jim Wells 36,498 ..... F10
Johnson 67,649 ...... G5
Jones 17,268 .... E5
Karnes 13,593 ...... G9
Kaufman 39,029 .... H5
Kendall 10,635 .... F8
Kenedy 543 .... G11
Kent 1,145 .... D4
Kerr 28,780 ...... E7
Kimble 4,063 ..... E7
King 425 .... D4
Kinney 2,279 .... D8
Kleberg 33,358 .. G10
Knox 5,329 ..... E4
Lamar 42,156 ... J4
Lamb 18,669 ... B3
Lampasas 12,005 . F6
La Salle 5,514 .. E9
Lavaca 19,004 .. H8
Lee 10,952 .... H7
Leon 9,594 .... J6
Liberty 47,088 . K7
Limestone 20,224 . H6
Lipscomb 3,766 .. D1
Live Oak 9,606 .. F9
Llano 10,144 .. F7
Loving 91 ..... A6
Lubbock 211,651 . C4
Lynn 8,605 ... C4
Madison 10,649 . J6
Marion 10,360 . K5
Martin 4,684 .. C5
Mason 3,683 .. E7
Matagorda 37,828 . H9
Maverick 31,398 .. D9
McCulloch 8,735 .. E6
McLennan 170,755 . G6
McMullen 789 ... F9
Medina 23,164 .. E8
Menard 2,346 .. E7
Midland 82,636 . B6
Milam 22,732 .. H7
Mills 4,477 ... F6
Mitchell 9,088 . D5
Montague 17,410 . G4
Montgomery 128,487 . J7
Moore 16,575 .. C2
Morris 14,629 .. K4

Motley 1,950 .... D3
Nacogdoches 46,786 ...... K6
Navarro 35,323 .... H5
Newton 13,254 .... L7
Nolan 17,359 .... D5
Nueces 268,215 .... G10
Ochiltree 9,588 ...... D1
Oldham 2,283 .... B2
Orange 83,838 ...... L7
Palo Pinto 24,062 .. F5
Panola 20,724 ..... K5
Parker 44,609 .... G5
Parmer 11,038 ... B3
Pecos 14,618 .... B7
Polk 24,407 .... K7
Potter 98,637 ... C2
Presidio 5,188 . C12
Rains 4,839 ... J5
Randall 75,062 . C2
Reagan 4,135 . C6
Real 2,469 ... E8
Red River 16,101 . J4
Reeves 15,801 . D11
Refugio 9,289 . G9
Roberts 1,187 . D2
Robertson 14,653 . H6
Rockwall 14,528 . H5
Runnels 11,872 .. E6
Rusk 41,382 ... K5
Sabine 8,702 .. L6
San Augustine 8,785 . K6
San Jacinto 11,434 . J7
San Patricio 58,013 . G10
San Saba 6,204 .. F6
Schleicher 2,820 . D7
Scurry 18,192 .. D5
Shackelford 3,915 . E5
Shelby 23,084 ... K6
Sherman 3,174 .. C1
Smith 128,366 ... J5
Somervell 4,154 . G5
Starr 27,266 ... F11
Stephens 9,926 . F5
Sterling 1,206 . C6
Stonewall 2,406 . D4
Sutton 5,130 .. D7
Swisher 9,723 . C3
Tarrant 860,880 . G5
Taylor 110,932 . E5
Terrell 1,595 .. B7
Terry 14,581 .. B4
Throckmorton 2,053 . E4
Titus 21,442 ... K4
Tom Green 84,784 .. D6

Travis 419,573 .... G7
Trinity 9,450 .... J6
Tyler 16,223 .... K7
Upshur 28,595 .... K5
Upton 4,619 .... B6
Uvalde 22,441 .... E8
Val Verde 35,910 ...... C8
Van Zandt 31,426 ..... J5
Victoria 68,807 ...... H9
Walker 41,789 ...... J7
Waller 19,798 ..... J8
Ward 13,976 ..... A6
Washington 21,998 . H7
Webb 99,258 ... E10
Wharton 40,242 . H8
Wheeler 7,137 . D2
Wichita 121,082 . F3
Wilbarger 15,931 . E3
Willacy 17,495 . G11
Williamson 76,507 . G7
Wilson 16,756 . F8
Winkler 9,944 . A6
Wise 26,575 . G4
Wood 24,697 . J5
Yoakum 8,299 . B4
Young 19,083 . F4
Zapata 6,628 . E11
Zavala 11,666 . E9

CITIES and TOWNS

Zip	Name/Pop.	Key
*79601	Abilene⊙ 98,315	E5
	Abilene‡ 139,192	E5
78516	Alamo 5,831	F11
78209	Alamo Heights 6,252	K10
76430	Albany⊙ 2,450	E5
78332	Alice⊙ 20,961	F10
75002	Allen 8,314	H1
79830	Alpine⊙ 5,465	D12
77511	Alvin 16,515	J3
*79101	Amarillo⊙ 149,230	C2
	Amarillo‡ 173,699	C2
77514	Anahuac⊙ 1,840	K8
77830	Anderson⊙ 500	J7
79714	Andrews⊙ 11,061	B5
77515	Angleton⊙ 13,929	J8
79501	Anson⊙ 2,831	E5
78336	Aransas Pass 7,173	G10
76351	Archer City⊙ 1,862	F4
*76010	Arlington 160,123	F2
79502	Aspermont⊙ 1,357	D4

75751 Athens⊙ 10,197 .... J5
75551 Atlanta 6,272 .... K4
*78701 Austin (cap.)⊙ 345,496 ... G7
Austin‡ 536,450 .... G7
76020 Azle 5,822 .... E2
77518 Bacliff 4,851 .... K2
79504 Baird⊙ 1,696 .... E5
75180 Balch Springs 13,746 .. H2
†78201 Balcones Heights 2,511 . J10
76821 Ballinger⊙ 4,207 ...... E6
78003 Bandera⊙ 947 .... F8
77532 Barrett 3,183 .... K1
78602 Bastrop⊙ 3,789 ...... G7
77414 Bay City⊙ 17,837 .... H9
77520 Baytown 56,923 ...... L2
*77701 Beaumont⊙ 118,102 .. K7
Beaumont-Port
Arthur-Orange‡ 375,497 K7
76021 Bedford 20,821 ...... F2
78102 Beeville⊙ 14,574 ..... G9
77401 Bellaire 14,950 ...... J2
76704 Bellmead 7,569 ...... H6
77418 Bellville⊙ 2,860 ..... H8
76513 Belton⊙ 10,660 ...... G7
76126 Benbrook 13,579 .... E2
79505 Benjamin⊙ 257 ..... E4
76932 Big Lake⊙ 3,404 .... C6
79720 Big Spring⊙ 24,804 . C5
78006 Boerne⊙ 3,229 ..... J10
75418 Bonham⊙ 7,338 .... H4
79007 Borger 15,837 ..... C2
75557 Boston⊙ 400 ..... K4
76230 Bowie 5,610 ..... G4
78832 Brackettville⊙ 1,676 . D8
76825 Brady⊙ 5,969 .... E6
77422 Brazoria 3,025 .... J9
76024 Breckenridge⊙ 6,921 . F5
77833 Brenham⊙ 10,966 .. H7
77611 Bridge City 7,667 .. L7
79316 Brownfield⊙ 10,387 . B4
*78520 Brownsville⊙ 84,997 . G12
Brownsville-Harlingen-San
Benito‡ 209,680 .. G12
76801 Brownwood⊙ 19,396 . F6
77801 Bryan⊙ 44,337 .... H7
Bryan-College
Station‡ 93,588 ... H7
76354 Burkburnett 10,668 . F3
76028 Burleson 11,734 .. F3
78611 Burnet⊙ 3,410 .... E5
77836 Caldwell⊙ 2,953 .. H7
76520 Cameron⊙ 5,721 .. H7
79014 Canadian⊙ 3,491 . D2
75103 Canton⊙ 2,845 .. J5
79015 Canyon⊙ 10,724 . C3
78834 Carrizo Springs⊙ 6,886 . E9
*75006 Carrollton 40,595 . G2
75633 Carthage⊙ 6,447 . K5
†78213 Castle Hills 4,773 . J10
75104 Cedar Hill 6,849 . G3
75935 Center⊙ 5,827 .. K6
75833 Centerville⊙ 799 . H6
77530 Channelview 17,471 . K1
79018 Channing⊙ 304 .. B2
79201 Childress⊙ 5,817 . D3
76437 Cisco 4,517 .... E5
79226 Clarendon⊙ 2,220 . C3
75426 Clarksville⊙ 4,917 . K4
79019 Claude⊙ 1,112 .. C2
†77565 Clear Lake Shores 755 . K2
76031 Cleburne⊙ 19,218 . G5
77327 Cleveland 5,977 . K7
77531 Clute 9,577 .... J9
77331 Coldspring⊙ 569 . J7
76834 Coleman⊙ 5,960 . E6
77840 College Station 37,272 . H7
76034 Colleyville 6,700 . F2
79512 Colorado City⊙ 5,405 . C5
78934 Columbus⊙ 3,923 . H8
76442 Comanche⊙ 4,075 . F6
75428 Commerce 8,136 . J4
*77301 Conroe⊙ 18,034 . J7
78109 Converse 5,150 . K11
75432 Cooper⊙ 2,338 . J4
76522 Copperas Cove 19,469 . G6
*78401 Corpus Christi⊙ 231,999 G10
Corpus Christi‡ 326,228 . G10
75110 Corsicana⊙ 21,712 . H5
78014 Cotulla⊙ 3,912 . E9
79731 Crane⊙ 3,622 . B6
75835 Crockett⊙ 7,405 . J6
79322 Crosbyton⊙ 2,289 . C4
79227 Crowell⊙ 1,509 . E4
76036 Crowley 5,852 . E3
78839 Crystal City⊙ 8,334 . E9
77954 Cuero⊙ 7,124 . G8
75638 Daingerfield⊙ 3,030 . K4
79022 Dalhart⊙ 6,854 . B1
*75201 Dallas⊙ 904,078 . G2
Dallas-Ft. Worth‡
2,974,878 .... G2
77535 Dayton 4,908 . J7
76234 Decatur⊙ 4,104 . G4
77536 Deer Park 22,648 . K2
76444 De Leon 2,478 . F5
78840 Del Rio⊙ 30,034 . D8
75020 Denison 23,884 . H4
76201 Denton⊙ 48,063 . G4

(continued on following page)

DOMINANT LAND USE

Wheat, Grain Sorghums, Range Livestock
Cotton, Wheat
Specialized Cotton
Cotton, General Farming
Cotton, Forest Products
Cotton, Range Livestock
Rice, General Farming
Peanuts, General Farming
General Farming, Livestock, Cash Grain
General Farming, Forest Products, Truck Farming, Cotton
Fruit, Truck and Mixed Farming
Range Livestock
Forests
Swampland, Limited Agriculture
Nonagricultural Land
Urban Areas

MAJOR MINERAL OCCURRENCES

At	Asphalt	He	Helium
Cl	Clay	Ls	Limestone
Fe	Iron Ore	Na	Salt
G	Natural Gas	O	Petroleum
Gn	Granite	S	Sulfur
Gp	Gypsum	Tc	Talc
Gr	Graphite	U	Uranium

⚡ Water Power
/// Major Industrial Areas

Agriculture, Industry and Resources

AREA 266,807 sq. mi. (691,030 sq. km.)
POPULATION 14,229,288
CAPITAL Austin
LARGEST CITY Houston
HIGHEST POINT Guadalupe Pk. 8,749 ft. (2667 m.)
SETTLED IN 1686
ADMITTED TO UNION December 29, 1845
POPULAR NAME Lone Star State
STATE FLOWER Bluebonnet
STATE BIRD Mockingbird

(continued)

79323 Denver City 4,704 ... B4
75115 De Soto 15,538 ... G3
78016 Devine 3,756 ... E8
75941 Diboll 5,227 ... K6
79229 Dickens⊙ 409 ... D4
77539 Dickinson 7,505 ... K3
79027 Dimmitt⊙ 5,019 ... B3
78537 Donna 9,952 ... F11
79029 Dumas⊙ 12,194 ... C2
75116 Duncanville 27,781 ... G3
78852 Eagle Pass⊙ 21,407 ... D9
76448 Eastland⊙ 3,747 ... F5
78539 Edinburg⊙ 24,075 ... F11
77957 Edna⊙ 5,650 ... H9
77437 El Campo 10,462 ... H8
76936 Eldorado⊙ 2,061 ... D7
78621 Elgin 4,535 ... G7
*79901 El Paso⊙ 425,259 ... A10
El Paso‡ 479,899 ... A10
78543 Elsa 5,061 ... G11
75440 Emory⊙ 813 ... J5
75119 Ennis 12,110 ... H5
76039 Euless 24,002 ... F2
76140 Everman 5,387 ... F3
79838 Fabens 4,285 ... B10
75840 Fairfield⊙ 3,505 ... H6
78355 Falfurrias⊙ 6,103 ... F10
75234 Farmers Branch 24,863 ... G2
79325 Farwell⊙ 1,354 ... A3
78114 Floresville⊙ 4,381 ... K11
†75067 Flower Mound 4,402 ... F1
79235 Floydada⊙ 4,193 ... C3
†76119 Forest Hill 11,684 ... F2
79734 Fort Davis⊙ 900 ... D11
79735 Fort Stockton⊙ 8,688 ... A7
*76101 Fort Worth⊙ 385,164 ... F2
77856 Franklin⊙ 1,349 ... H7
78624 Fredericksburg⊙ 6,412 ... E7
76842 Fredonia 50 ... E7
77541 Freeport 13,444 ... J9
77546 Friendswood 10,719 ... J2
79035 Friona 3,809 ... B3
75034 Frisco 3,499 ... H4
79738 Gail⊙ 171 ... C5
76240 Gainesville⊙ 14,081 ... G4
77547 Galena Park 9,879 ... J1
*77550 Galveston⊙ 61,902 ... L3
Galveston-Texas City‡ 195,940 ... L3
79739 Garden City⊙ 350 ... C6
*75040 Garland 138,857 ... H2
76528 Gatesville⊙ 6,260 ... G6
78626 Georgetown⊙ 9,468 ... G7
78022 George West⊙ 2,627 ... F9
78942 Giddings⊙ 3,950 ... H7
75644 Gilmer⊙ 5,167 ... J5
75647 Gladewater 6,548 ... K5
76043 Glen Rose⊙ 2,075 ... G5
76844 Goldthwaite⊙ 1,783 ... F6
77963 Goliad⊙ 1,990 ... G9
78629 Gonzales⊙ 7,152 ... G8
76046 Graham⊙ 9,170 ... F4
76048 Granbury⊙ 3,332 ... G5
*75050 Grand Prairie 71,462 ... G2
76051 Grapevine 11,801 ... F2
75401 Greenville⊙ 22,161 ... H4
76642 Groesbeck⊙ 3,373 ... H6
77619 Groves 17,090 ... L8
75845 Groveton⊙ 1,262 ... J7
79236 Guthrie⊙ 170 ... D4
77964 Hallettsville⊙ 2,865 ... G8
76117 Haltom City 29,014 ... F2
76531 Hamilton⊙ 3,189 ... G6
78550 Harlingen 43,543 ... G11
79521 Haskell⊙ 3,782 ... E4
77859 Hearne 5,418 ... H7
78361 Hebbronville⊙ 4,684 ... F10
75948 Hemphill⊙ 1,353 ... L6
77445 Hempstead⊙ 3,456 ... J7
75652 Henderson⊙ 11,473 ... K5
76365 Henrietta⊙ 3,149 ... F4
79045 Hereford⊙ 15,853 ... B3
†75201 Highland Park 8,909 ... G2
77562 Highlands 6,467 ... K1
76645 Hillsboro⊙ 7,397 ... G5
77563 Hitchcock 6,655 ... K3
78861 Hondo⊙ 6,057 ... E8
*77001 Houston⊙ 1,595,138 ... J2
Houston‡ 2,905,350 ... J2
*77338 Humble 6,729 ... J7
†77001 Hunters Creek Village 4,215 ... J1
77340 Huntsville⊙ 23,936 ... J7
76053 Hurst 31,420 ... F2
76367 Iowa Park 6,184 ... F4
*75061 Irving 109,943 ... G2
77029 Jacinto City 8,953 ... J1
76056 Jacksboro⊙ 4,000 ... F4
75766 Jacksonville 12,264 ... J5
75951 Jasper⊙ 6,959 ... L7
79528 Jayton⊙ 638 ... D4
75657 Jefferson⊙ 2,643 ... K5
†77001 Jersey Village 4,084 ... J1
78636 Johnson City⊙ 872 ... F7
78026 Jourdanton⊙ 2,743 ... F9
76849 Junction⊙ 2,593 ... E7
78118 Karnes City⊙ 3,296 ... G9
77450 Katy 5,660 ... J8
75142 Kaufman⊙ 4,658 ... H5
76248 Keller 4,156 ... F2
78119 Kenedy 4,356 ... G9
79745 Kermit⊙ 8,015 ... B6
78028 Kerrville⊙ 15,276 ... E7
75662 Kilgore 11,006 ... K5
76541 Killeen 46,296 ... G6
Killeen-Temple‡ 214,656 ... G6
78363 Kingsville⊙ 28,808 ... G10
†78109 Kirby 6,435 ... K11
77625 Kountze⊙ 2,716 ... K7
78945 La Grange⊙ 3,768 ... G8
77566 Lake Jackson 19,102 ... J8
76135 Lake Worth 4,394 ... E2
77568 La Marque 15,372 ... K3
79331 Lamesa⊙ 11,790 ... C5
76550 Lampasas⊙ 6,165 ... F6
*75146 Lancaster 14,807 ... G3
77571 La Porte 14,062 ... K2
*78040 Laredo⊙ 91,449 ... E10
Laredo‡ 99,258 ... E10
77573 League City 16,578 ... K2
78873 Leakey⊙ 468 ... E8
†78201 Leon Valley 9,088 ... J10
79336 Levelland⊙ 13,809 ... B4
*75067 Lewisville 24,273 ... G1
77575 Liberty⊙ 7,945 ... K7
75563 Linden⊙ 2,443 ... K4
79056 Lipscomb⊙ 52 ... D1
79339 Littlefield⊙ 7,409 ... B4
†78201 Live Oak 8,183 ... K10
77351 Livingston⊙ 4,928 ... K7
78643 Llano⊙ 3,071 ... F7
78644 Lockhart⊙ 7,953 ... G8
79241 Lockney 2,334 ... C3
*75601 Longview⊙ 62,762 ... K5
Longview-Marshall‡ 151,752 ... K5
*79401 Lubbock⊙ 173,979 ... C4
Lubbock‡ 211,651 ... C4
75901 Lufkin⊙ 28,562 ... K6
78648 Luling 5,039 ... G8
77864 Madisonville⊙ 3,660 ... J7
76063 Mansfield 8,092 ... F3
77578 Manvel 3,549 ... J3
79843 Marfa⊙ 2,466 ... C12
76661 Marlin⊙ 7,099 ... H6
75670 Marshall⊙ 24,921 ... K5
76856 Mason⊙ 2,153 ... E7
79244 Matador⊙ 1,052 ... D3
78368 Mathis 5,667 ... G9
78501 McAllen 66,281 ... F11
McAllen-Pharr-Edinburg‡ 283,229 ... F11
76657 McGregor 4,513 ... G6
75069 McKinney⊙ 16,256 ... H4
†77520 McNair ... K1
79245 Memphis⊙ 3,352 ... D3
76859 Menard⊙ 1,697 ... E7
79754 Mentone⊙ 50 ... D10
78570 Mercedes 11,851 ... F12
76665 Meridian⊙ 1,330 ... G6
76941 Mertzon⊙ 687 ... C6
*75149 Mesquite 67,053 ... H2
76667 Mexia 7,094 ... H6
79059 Miami⊙ 813 ... D2
*79701 Midland⊙ 70,525 ... C6
Midland‡ 82,636 ... C6
76065 Midlothian 3,219 ... G5
75773 Mineola 4,346 ... J5
76067 Mineral Wells 14,468 ... F5
78572 Mission 22,653 ... F11
77459 Missouri City 24,533 ... J2
79756 Monahans⊙ 8,397 ... B6
76251 Montague⊙ 1,253 ... G4
79346 Morton⊙ 2,674 ... B4
75455 Mount Pleasant⊙ 11,003 ... K4
75457 Mount Vernon⊙ 2,025 ... J4
79347 Muleshoe⊙ 4,842 ... B3
75961 Nacogdoches⊙ 27,149 ... J6
†77598 Nassau Bay 4,526 ... K2
77868 Navasota 5,971 ... J7
77627 Nederland 16,855 ... K8
75570 New Boston 4,628 ... K4
78130 New Braunfels⊙ 22,402 ... K10
75966 Newton⊙ 1,620 ... L7
76118 North Richland Hills 30,592 ... F2
79760 Odessa⊙ 90,027 ... B6
Odessa‡ 115,374 ... B6
76374 Olney 4,060 ... F4
77630 Orange⊙ 23,628 ... L7
76943 Ozona⊙ 3,766 ... C7
79248 Paducah⊙ 2,216 ... D4
76866 Paint Rock⊙ 256 ... E6
77465 Palacios 4,667 ... H9
75801 Palestine⊙ 15,948 ... J6
76072 Palo Pinto⊙ 350 ... F5
79065 Pampa⊙ 21,396 ... D2
79068 Panhandle⊙ 2,226 ... C2
75460 Paris⊙ 25,498 ... J4
*77501 Pasadena 112,560 ... J2
77581 Pearland 13,248 ... J2
78061 Pearsall⊙ 7,383 ... E9
79772 Pecos⊙ 12,855 ... D10
79070 Perryton⊙ 7,991 ... D1
78577 Pharr 21,381 ... F11
75686 Pittsburg⊙ 4,245 ... J4
79355 Plains⊙ 1,457 ... B4
79072 Plainview⊙ 22,187 ... C3
75074 Plano 72,331 ... G1
78064 Pleasanton 6,346 ... F9
77640 Port Arthur 61,251 ... K8
78578 Port Isabel 3,769 ... G11
78374 Portland 12,023 ... G10
77979 Port Lavaca⊙ 10,911 ... H9
77651 Port Neches 13,944 ... K7
79356 Post⊙ 3,961 ... C4
78065 Poteet 3,086 ... F8
77445 Prairie View 3,993 ... J7
79845 Presidio 1,723 ... C12
79252 Quanah⊙ 3,890 ... E3
76470 Ranger 3,142 ... F5
79778 Rankin⊙ 1,216 ... B6
78580 Raymondville⊙ 9,493 ... G11
78377 Refugio⊙ 3,898 ... G9
75080 Richardson 72,496 ... G2
76118 Richland Hills 7,977 ... F2
77469 Richmond⊙ 9,692 ... J8
78582 Rio Grande City⊙ 8,930 ... F11
77019 River Oaks 6,890 ... E2
76945 Robert Lee⊙ 1,202 ... D6
78380 Robstown 12,100 ... G10
79543 Roby⊙ 814 ... D5
76567 Rockdale 5,611 ... G7
78382 Rockport⊙ 3,686 ... H9
78880 Rocksprings⊙ 1,317 ... D8
75087 Rockwall⊙ 5,939 ... H5
78584 Roma-Los Saenz 3,384 ... E11
77471 Rosenberg 17,995 ... J8
78664 Round Rock 12,740 ... G7
75088 Rowlett 7,522 ... H2
75785 Rusk⊙ 4,681 ... J6
76179 Saginaw 5,736 ... E2
*76901 San Angelo⊙ 73,240 ... D6
San Angelo‡ 84,784 ... D6

Topography

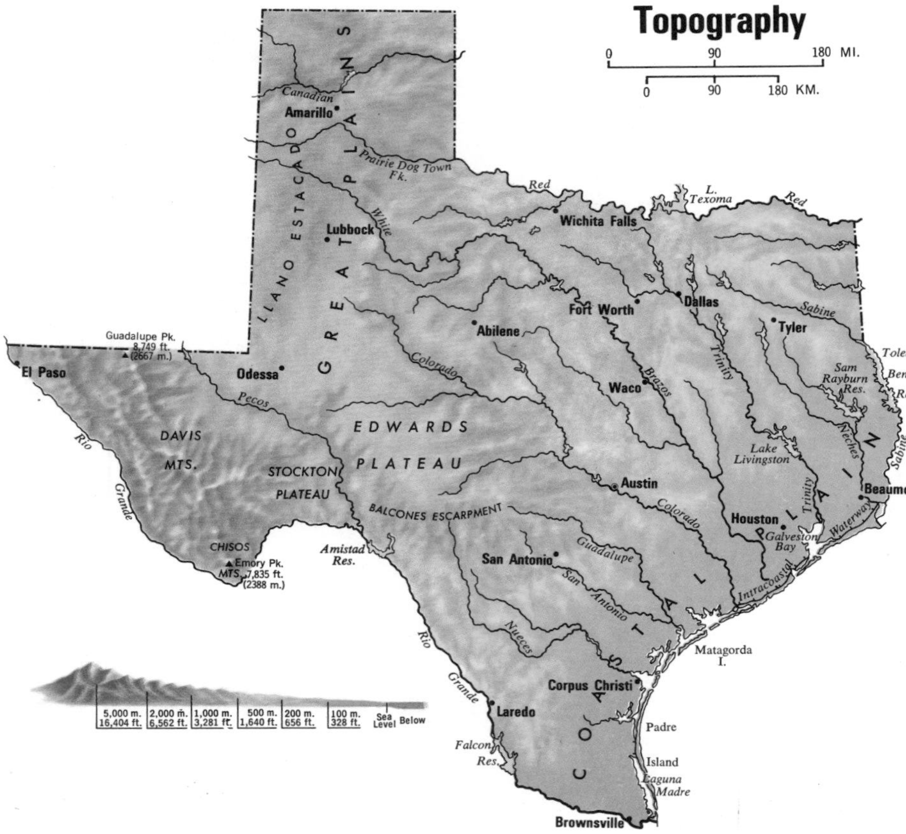

*78201 San Antonio⊙ 786,023 ... J11
San Antonio‡ 1,071,954 ... J11
75972 San Augustine⊙ 2,930 ... K6
78586 San Benito 17,988 ... G12
79848 Sanderson⊙ 1,241 ... B7
78384 San Diego⊙ 5,225 ... F10
76266 Sanger 2,574 ... G4
78589 San Juan 7,608 ... F11
78666 San Marcos⊙ 23,420 ... F8
76877 San Saba⊙ 2,847 ... F6
†76101 Sansom Park Village 3,921 ... E2
†77510 Santa Fe 6,172 ... K3
78385 Sarita⊙ 200 ... G10
78154 Schertz 7,262 ... K10
77586 Seabrook 4,670 ... K2
75159 Seagoville 7,304 ... H3
77474 Sealy 3,875 ... H8
78155 Seguin⊙ 17,854 ... G8
79360 Seminole⊙ 6,080 ... B5
†78357 Seven Sisters 2 ... F9
76380 Seymour⊙ 3,657 ... E4
75090 Sherman⊙ 30,413 ... H4
Sherman-Denison‡ 89,796 ... H4
79851 Sierra Blanca⊙ 800 ... B11
77656 Silsbee 7,684 ... K7
79257 Silverton⊙ 918 ... C3
78387 Sinton⊙ 6,044 ... G9
79364 Slaton 6,804 ... C4
78957 Smithville 3,470 ... G7
79549 Snyder⊙ 12,705 ... D5
76950 Sonora⊙ 3,856 ... D7
77587 South Houston 13,293 ... J2
79081 Spearman⊙ 3,413 ... C1
*77373 Spring ... J7
†77001 Spring Valley 3,353 ... J1
77477 Stafford 4,755 ... J2
79553 Stamford 4,542 ... E5
79782 Stanton⊙ 2,314 ... C5
76401 Stephenville⊙ 11,881 ... F5
76951 Sterling City⊙ 915 ... D6
79083 Stinnett⊙ 2,222 ... C2
79084 Stratford⊙ 1,917 ... C1
77478 Sugar Land 8,826 ... J8
75482 Sulphur Springs⊙ 12,804 ... J4
77480 Sweeny 3,538 ... J8
79556 Sweetwater⊙ 12,242 ... D5
78390 Taft 3,686 ... G9
79373 Tahoka⊙ 3,262 ... C4
76574 Taylor 10,619 ... G7
†77586 Taylor Lake Village 3,669 ... K2
75860 Teague 3,390 ... H6
76501 Temple 42,354 ... G6
79852 Terlingua 100 ... D12
75160 Terrell 13,269 ... H5
†78201 Terrell Hills 4,644 ... K11
*75501 Texarkana 31,271 ... L4
Texarkana, Tex.-Texarkana, Ark.‡ 27,019 ... L4
77590 Texas City 41,403 ... K3
73949 Texhoma 358 ... C1
The Colony 11,586 ... G1
76083 Throckmorton⊙ 1,174 ... F4
78072 Tilden⊙ 450 ... F9
77375 Tomball 3,996 ... J7
75862 Trinity 2,620 ... J7
79088 Tulia⊙ 5,033 ... C3
*75701 Tyler⊙ 70,508 ... J5
Tyler‡ 128,366 ... J5
78148 Universal City 10,720 ... K10
†75205 University Park 22,254 ... G2
78801 Uvalde⊙ 14,178 ... E8
75095 Van Alstyne 1,860 ... H4
79855 Van Horn⊙ 2,772 ... C11
79092 Vega⊙ 900 ... B2
76384 Vernon⊙ 12,695 ... E3
77901 Victoria⊙ 50,695 ... H9
Victoria‡ 68,807 ... H9
77662 Vidor 11,834 ... L7
*76701 Waco⊙ 101,261 ... G6
Waco‡ 170,755 ... G6
75501 Wake Village 3,865 ... K4
75165 Waxahachie⊙ 14,624 ... H5
76086 Weatherford⊙ 12,049 ... G5
79095 Wellington⊙ 3,043 ... D3
78596 Weslaco 19,331 ... F11
77486 West Columbia 4,109 ... J8
77630 West Orange 4,610 ... L7
†77005 West University Place 12,010 ... J2
†76101 Westworth 3,651 ... E2
77488 Wharton⊙ 9,033 ... J8
79096 Wheeler⊙ 1,584 ... D2
75693 White Oak 4,415 ... K5
76273 Whitesboro 3,197 ... H4
76108 White Settlement 13,508 ... E2
*76301 Wichita Falls⊙ 94,201 ... F4
Wichita Falls‡ 130,664 ... F4
†78201 Windcrest 5,332 ... K11
75494 Winnsboro 3,458 ... J5
79567 Winters 3,061 ... E6
75979 Woodville⊙ 2,821 ... K7
75098 Wylie 3,152 ... H1
78076 Zapata⊙ 3,831 ... E11

OTHER FEATURES

Amistad (res.) ... C8
Amistad Nat'l Rec. Area ... D8
Angelina (riv.) ... K6
Apache (mts.) ... C11
Aransas (passage) ... H10
Arlington (lake) ... F2
Baffin (bay) ... G10
Balcones Escarpment (plat.) ... E8
Beals (creek) ... C5
Benbrook (lake) ... E2
Bergstrom A.F.B. ... G7
Big Bend Nat'l Park ... A8
Bolivar (pen.) ... K8
Brazos (riv.) ... H7
Brownwood (lake) ... E6
Buchanan (lake) ... F7
Buck (creek) ... D3
Caddo (lake) ... L5
Canadian (riv.) ... D1
Carrizo (creek) ... A1
Carswell A.F.B. ... E2
Cathedral (mt.) ... D12
Cavallo (passage) ... H9
Cedar (lake) ... B5
Cerro Alto (mt.) ... B10
Chamizal Nat'l Mem. ... A10
Chase N.A.S. ... G9
Chinati (mts.) ... C12
Chinati (peak) ... C12
Chisos (mts.) ... A8
Cibolo (creek) ... K11
Clear Fork, Brazos (riv.) ... D5
Coldwater (creek) ... B1
Colorado (riv.) ... F7
Copano (bay) ... G9
Corpus Christi (lake) ... F9
Corpus Christi N.A.S. ... G10
Cottonwood Draw (dry riv.) ... C10
Davis (mts.) ... C11
Deep (creek) ... C5
Delaware (creek) ... C10
Delaware (mts.) ... C10
Denison (dam) ... H4
Devils (riv.) ... D7
Diablo, Sierra (mts.) ... C10
Double Mountain Fork, Brazos (riv.) ... C4
Dyess A.F.B. ... D5
Eagle (peak) ... C11
Eagle Mountain (lake) ... E2
Edwards (plat.) ... C7
Elephant (mt.) ... D12
Ellington A.F.B. ... K2
Elm Fork, Trinity (riv.) ... G2
Emory (peak) ... A8
Falcon (res.) ... E11
Finlay (mts.) ... B10
Fort Bliss 12,687 ... A10
Fort Davis Nat'l Hist. Site ... D11
Fort Hood 31,250 ... G6
Frio (riv.) ... E8
Galveston (bay) ... L2
Galveston (isl.) ... K8
Glass (mts.) ... A7
Goodfellow A.F.B. ... D6
Grapevine (lake) ... F2
Guadalupe (mts.) ... C10
Guadalupe (peak) ... B10
Guadalupe (riv.) ... G8
Guadalupe Mts. Nat'l Park ... C10
Houston (lake) ... J8
Houston Ship (chan.) ... K2
Howard (creek) ... C7
Hubbard Creek (lake) ... F5
Hueco (mts.) ... B10
Intracoastal Waterway ... J9
Johnson Draw (dry riv.) ... C7
Kelly A.F.B. ... J11
Kemp (lake) ... E4
Kingsville N.A.S. ... G10
Kiowa (creek) ... D1
Lackland A.F.B. 14,459 ... J11
Lake Meredith Nat'l Rec. Area ... C2
Lampasas (riv.) ... G6
Laughlin A.F.B. 2,994 ... D8
Lavon (lake) ... H1
Leon (riv.) ... F6
Livermore (mt.) ... C11
Livingston (lake) ... K7
Llano (riv.) ... D7
Llano Estacado (plain) ... B4
Locke (mt.) ... D11
Los Olmos (creek) ... F10
Los Olmos (creek) ... F11
Lyndon B. Johnson Nat'l Hist. Site ... F7
Lyndon B. Johnson Space Ctr. ... K2
Madre (lag.) ... G11
Maravillas (creek) ... A7
Matagorda (bay) ... H9
Matagorda (isl.) ... H9
Matagorda (pen.) ... J9
Matagorda Isl. Bombing and Gunnery Range ... H9
Medina (lake) ... E8
Medina (riv.) ... J11
Mexico (gulf) ... K9
Middle Concho (riv.) ... C6
Mountain Creek (lake) ... G2
Mustang (creek) ... A1
Mustang (isl.) ... G10
Mustang Draw (dry riv.) ... B5
Navasota (riv.) ... H7
Navidad (riv.) ... H8
Neches (riv.) ... K6
North Concho (riv.) ... C6
North Pease (riv.) ... D3
Nueces (riv.) ... F9
Padre (isl.) ... G10
Padre Island Nat'l Seashore ... G11
Palo Duro (creek) ... B2
Palo Duro (creek) ... C1
Pease (riv.) ... D3
Pecos (riv.) ... C7
Pedernales (riv.) ... F7
Possum Kingdom (lake) ... F5
Prairie Dog Town Fork, Red (riv.) ... C3
Quitman (mts.) ... B11
Red (riv.) ... F3
Red Bluff (lake) ... A6
Reese A.F.B. 1,934 ... B4
Rio Grande (riv.) ... D9
Rita Blanca (creek) ... B2
Sabine (riv.) ... L7
Salt Fork, Red (riv.) ... D3
Sam Rayburn (res.) ... K6
San Antonio (bay) ... H9
San Antonio (mt.) ... B10
San Antonio Missions Nat'l Hist. Park ... J11
San Francisco (creek) ... B8
San Luis (passage) ... K8
San Martine Draw (dry riv.) ... C11
San Saba (riv.) ... D7
Santa Isabel (creek) ... E1
Santiago (mts.) ... A8
Santiago (peak) ... D12
Sheppard A.F.B. ... F3
Sierra Diablo (mts.) ... C10
Sierra Vieja (mts.) ... C11
Staked (Llano Estacado) (plain) ... B4
Stamford (lake) ... E4
Stockton (plat.) ... B7
Sulphur (riv.) ... J4
Sulphur Draw (dry riv.) ... B4
Sulphur Springs (creek) ... B4
Tenmile (creek) ... G3
Terlingua (creek) ... D12
Texoma (lake) ... H3
Tierra Blanca (creek) ... B3
Toledo Bend (res.) ... L6
Toyah (creek) ... D11
Toyah (lake) ... A6
Travis (lake) ... G7
Trinity (bay) ... L2
Trinity (riv.) ... H5
Trinity, West Fork (riv.) ... G2
Trujillo (creek) ... A2
Vieja, Sierra (mts.) ... C11
Walnut (creek) ... F1
Washita (riv.) ... D1
West (bay) ... K3
White (riv.) ... C4
White River (lake) ... C4
White Rock (creek) ... G1
Wichita (riv.) ... F4
Wolf (creek) ... D1
Worth (lake) ... E2

⊙County seat.
‡Population of metropolitan area.
† Zip of nearest p.o. * Multiple zips

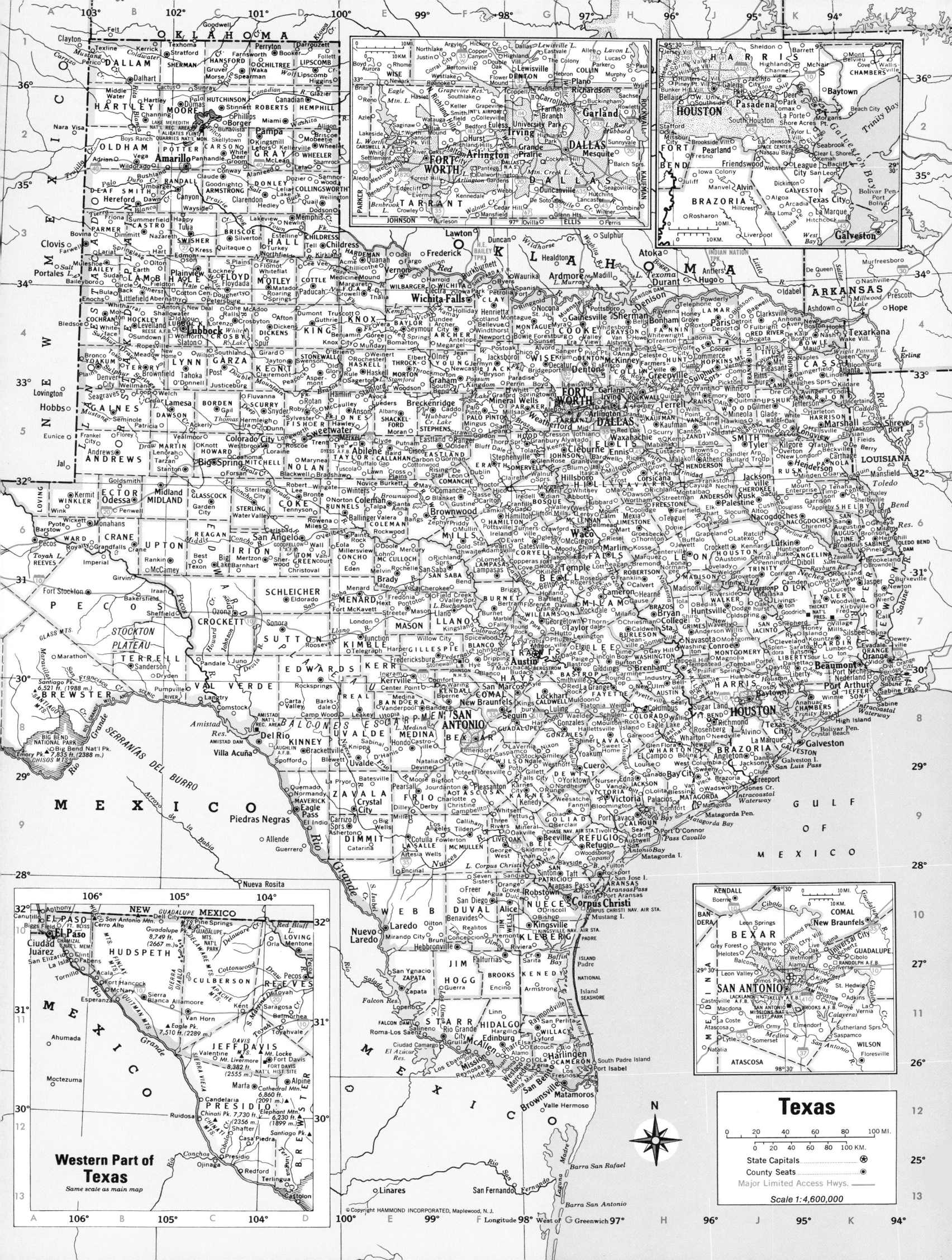

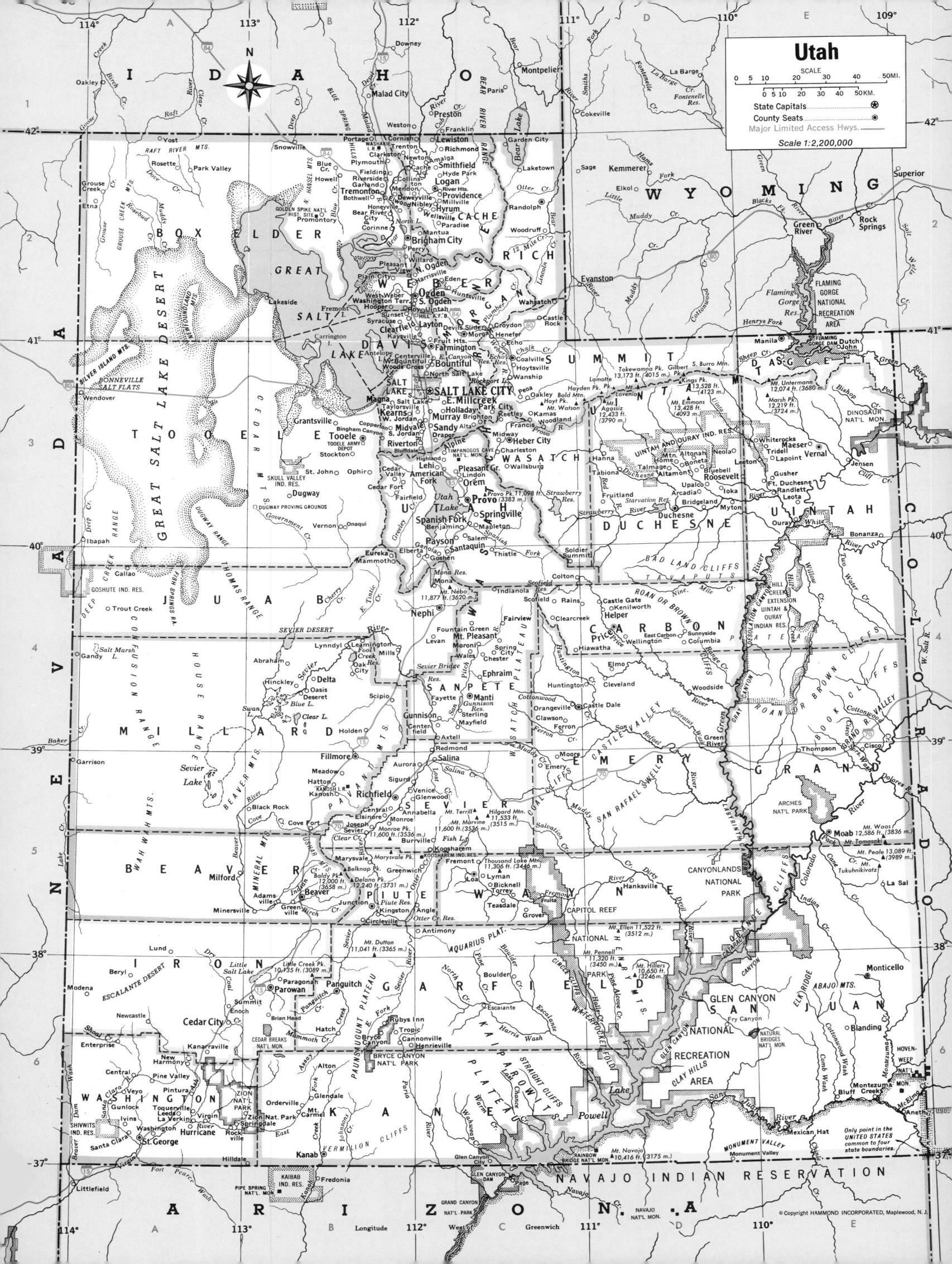

AREA 84,899 sq. mi. (219,888 sq. km.)
POPULATION 1,461,037
CAPITAL Salt Lake City
LARGEST CITY Salt Lake City
HIGHEST POINT Kings Pk. 13,528 ft. (4123 m.)
SETTLED IN 1847
ADMITTED TO UNION January 4, 1896
POPULAR NAME Beehive State
STATE FLOWER Sego Lily
STATE BIRD Sea Gull

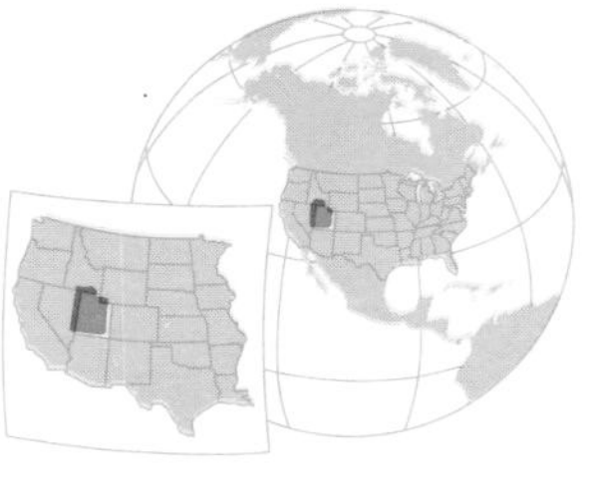

COUNTIES

Beaver 4,378 ..... A5
Box Elder 33,222 ..... A2
Cache 57,176 ..... C2
Carbon 22,179 ..... D4
Daggett 769 ..... E3
Davis 146,540 ..... B3
Duchesne 12,565 ..... D3
Emery 11,451 ..... D4
Garfield 3,673 ..... C6
Grand 8,241 ..... E5
Iron 17,349 ..... A6
Juab 5,530 ..... A4
Kane 4,024 ..... B6
Millard 8,970 ..... A4
Morgan 4,917 ..... C2
Piute 1,329 ..... B5
Rich 2,100 ..... C2
Salt Lake 619,066 ..... B3
San Juan 12,253 ..... E6
Sanpete 14,620 ..... C4
Sevier 14,727 ..... C5
Summit 10,198 ..... D3
Tooele 26,033 ..... A3
Uintah 20,506 ..... E3
Utah 218,106 ..... C3
Wasatch 8,523 ..... C3
Washington 26,065 ..... A6
Wayne 1,911 ..... C5
Weber 144,616 ..... B2

CITIES and TOWNS

Zip *Name/Pop.* *Key*

†84003 Alpine 2,649 ..... C3
84003 American Fork 12,693 ..... C3
84713 Beaver⊙ 1,792 ..... B5
84511 Blanding 3,118 ..... E6
†84065 Bluffdale 1,300 ..... B3
84010 Bountiful 32,877 ..... C3
84302 Brigham City⊙ 15,596 ..... C2
†84101 Brighton 150 ..... C3
84513 Castle Dale⊙ 1,910 ..... D4
84720 Cedar City 10,972 ..... A6
84014 Centerville 8,069 ..... C3
84015 Clearfield 17,982 ..... B2
84017 Coalville⊙ 1,031 ..... C3
84624 Delta 1,930 ..... B4
84020 Draper 5,521 ..... C3
84021 Duchesne⊙ 1,677 ..... D3
84022 Dugway 1,646 ..... B3
84520 East Carbon 1,942 ..... D4
84109 East Millcreek 24,150 ..... C3
84627 Ephraim 2,810 ..... C4
84025 Farmington⊙ 4,691 ..... C3
84523 Ferron 1,718 ..... C4
84631 Fillmore⊙ 2,083 ..... B5
†84037 Fruit Heights 2,728 ..... C2
84312 Garland 1,405 ..... B2
84029 Grantsville 4,419 ..... B3
84525 Green River 1,048 ..... D4
84634 Gunnison 1,255 ..... C4
†84401 Harrisville 1,371 ..... C2
84032 Heber City⊙ 4,362 ..... C3
84526 Helper 2,724 ..... D4
†84043 Highland 2,435 ..... C3
†84767 Hilldale 1,009 ..... A6
84117 Holladay 22,189 ..... C3
84528 Huntington 2,316 ..... C4
84737 Hurricane 2,361 ..... A6
84318 Hyde Park 1,495 ..... C2
84319 Hyrum 3,952 ..... C2
84740 Junction⊙ 151 ..... B5
84036 Kamas 1,064 ..... C3
84741 Kanab⊙ 2,148 ..... B6
84037 Kaysville 9,811 ..... B2
84118 Kearns 21,353 ..... B3
84745 La Verkin 1,174 ..... A6
84041 Layton 22,862 ..... C2
84043 Lehi 6,848 ..... C3
84320 Lewiston 1,438 ..... C2
†84062 Lindon 2,796 ..... C3
84747 Loa⊙ 364 ..... C5
84321 Logan⊙ 26,844 ..... C2
†84078 Maeser 2,216 ..... E3
84044 Magna 13,138 ..... B3
84046 Manila⊙ 272 ..... E3
84642 Manti⊙ 2,080 ..... C4
†84663 Mapleton 2,726 ..... C3
84531 Mexican Hat 250 ..... E6
84047 Midvale 10,146 ..... B3
84049 Midway 1,194 ..... C3
84751 Milford 1,293 ..... A5
84532 Moab⊙ 5,333 ..... E5
84754 Monroe 1,476 ..... B5
84535 Monticello⊙ 1,929 ..... E6
84050 Morgan⊙ 1,896 ..... C2
84646 Moroni 1,086 ..... C4
84647 Mount Pleasant 2,049 ..... C4
84107 Murray 25,750 ..... C3
84648 Nephi⊙ 3,285 ..... C4
†84321 Nibley 1,036 ..... C2
†84404 North Ogden 9,309 ..... C2
†84010 North Salt Lake 5,548 ..... C3
*84401 Ogden⊙ 64,407 ..... C2
Ogden-Salt Lake City‡ 936,255 ..... C4
84537 Orangeville 1,309 ..... C4
84057 Orem 52,399 ..... C3
Orem-Provo‡ 218,106 ..... C3
84759 Panguitch⊙ 1,343 ..... B6
84060 Park City 2,823 ..... C3
84761 Parowan⊙ 1,836 ..... B6
84651 Payson 8,246 ..... C3
†84302 Perry 1,084 ..... C2
†84401 Plain City 2,379 ..... B2
84062 Pleasant Grove 10,833 ..... C3
†84401 Pleasant View 3,983 ..... B2
84501 Price⊙ 9,086 ..... D4
84332 Providence 2,675 ..... C2
84601 Provo⊙ 74,108 ..... C3
Provo-Orem‡ 218,106 ..... C3
84064 Randolph⊙ 659 ..... C2
84701 Richfield⊙ 5,482 ..... B5
84333 Richmond 1,705 ..... C2
†84321 River Heights 1,211 ..... C2
84065 Riverton 7,293 ..... B3
84066 Roosevelt 3,842 ..... D3
84067 Roy 19,694 ..... C2
Saint George⊙ 11,350 ..... A6
84653 Salem 2,233 ..... C3
84654 Salina 1,992 ..... C5
*84101 Salt Lake City (cap)⊙ 163,697 ..... C3
Salt Lake City-Ogden‡ 936,255 ..... C3
*84070 Sandy 52,210 ..... C3
84765 Santa Clara 1,091 ..... A6
84655 Santaquin 2,175 ..... C4
84335 Smithfield 4,993 ..... C2
†84065 South Jordan 7,492 ..... B3
†84403 South Ogden 11,366 ..... C2
84115 South Salt Lake 9,884 ..... C3
84660 Spanish Fork 9,825 ..... C3
84663 Springville 12,101 ..... C3
†84015 Sunset 5,733 ..... B2
†84041 Syracuse 3,702 ..... B2
†84101 Taylorsville 17,448 ..... B3
84074 Tooele⊙ 14,335 ..... B3
84337 Tremonton 3,464 ..... B2
84078 Vernal⊙ 6,600 ..... E3
84780 Washington 3,092 ..... A6
†84403 Washington Terrace 8,212 ..... B2
84542 Wellington 1,406 ..... D4
84339 Wellsville 1,952 ..... C2
84083 Wendover 1,099 ..... A3
†84087 West Bountiful 3,556 ..... B3
84084 West Jordan 27,192 ..... B3
84340 Willard 1,241 ..... C2
84087 Woods Cross 4,263 ..... B3

OTHER FEATURES

Abajo (mts.) E6
Agassiz (mt.) D3
Antelope (isl.) B3
Aquarius (plat.) C5
Arches Nat'l Park E5
Assay (creek) B6
Bad Land (cliffs) D4
Baldy (peak) B5
Bear (lake) C2
Bear (riv.) B2
Beaver (mts.) A5
Beaver (riv.) A5
Beaver Dam Wash (creek) A6
Birch (creek) B5
Blue (creek) B2
Bonneville (salt flats) A3
Book (cliffs) E4
Brown (Roan) (cliffs) E4
Bryce Canyon Nat'l Park B6
Canyonlands Nat'l Park D5
Capitol Reef Nat'l Park C5
Castle (valley) D4
Cedar (mts.) B3
Cedar Breaks Nat'l Mon. B6
Chalk (creek) C3
Chinle (creek) E6
Clear (lake) B4
Cliff (creek) E3
Coal (cliffs) C5
Colorado (riv.) E5
Confusion (range) A4
Cottonwood (creek) C4
Cub (creek) C1
Deep (creek) B1
Deep Creek (range) A4
Delano (peak) B5
Desolation (canyon) E4
Dinosaur Nat'l Mon. E3
Dirty Devil (riv.) D5
Dolores (riv.) E5
Dry Coal (creek) A6
Duchesne (riv.) D3
Dugway (range) A3
Dugway Proving Grounds B3
Dutton (mt.) B5
East Canyon (res.) C3
Echo (res.) C3
Elk (ridge) E6
Ellen (mt.) D5
Emmons (mt.) D3
Escalante (des.) A6
Escalante (riv.) C6
Fish (lake) C5
Fish Springs (range) A4
Flaming Gorge (res.) E3
Flaming Gorge Nat'l Rec. Area E2
Fool Creek (res.) B4
Fremont (isl.) B2
Fremont (riv.) C5
Glen Canyon Nat'l Rec. Area D6
Golden Spike Nat'l Hist. Site B2
Goshute Ind. Res. A4
Government (creek) B3
Gray (canyon) D4
Great Salt (lake) B2
Great Salt Lake (des.) A3
Greeley (creek) B3
Green (riv.) D4
Grouse (creek) A2
Grouse Creek (mts.) A2
Gunnison (res.) C4
Henry (mts.) D6
Hilgard (mt.) C5
Hill (creek) E4
Hill A.F.B. C2
Hill Creek Ext., Uintah and Ouray Ind. Res. E4
Hillers (mt.) D6
House (range) A4
Hovenweep Nat'l Mon. E6
Huntington (creek) C4
Indian (creek) B5
Jordan (riv.) C3
Kaiparowits (plat.) C6
Kanab (creek) B7
Kanosh Ind. Res. B5
Kings (peak) D3
Koosharem Ind. Res. C5
Little Creek (peak) B6
Little Salt (lake) A6
Malad (riv.) B1
Marsh (peak) E3
Marvine (mt.) C5
Mineral (mts.) B5
Mona (res.) C4
Monroe (peak) B5
Montezuma (creek) E6
Monument (valley) D6
Muddy (creek) C4
Natural Bridges Nat'l Mon. E6
Navajo (mt.) D6
Navajo Ind. Res. D7
Nebo (mt.) C4
Newfoundland (mts.) A2
Nine Mile (creek) D4
North (lake) B2
Orange (cliffs) D5
Otter (creek) C5
Otter Creek (res.) C5
Paria (riv.) B6
Paunsaugunt (plat.) B6
Pavant (mts.) B5
Peale (mt.) E5
Pennell (mt.) D6
Piute (res.) B5
Plumber (creek) C2
Powell (lake) D6
Price (riv.) D4
Provo (peak) C3
Provo (riv.) C3
Raft River (mts.) A2
Rainbow Bridge Nat'l Mon. C6
Roan (cliffs) E4
Rockport (lake) C3
Salvation (creek) C5
San Juan (riv.) D6
San Pitch (riv.) C4
San Rafael (riv.) D4
San Rafael Swell (mts.) D5
Santa Clara (riv.) A6
Sevier (des.) B4
Sevier (lake) A5
Sevier (riv.) B4
Sevier Bridge (res.) C4
Shivwits Ind. Res. A6
Silver Island (mts.) A3
Skull Valley Ind. Res. B3
Spanish Fork (riv.) C3
Strait (cliffs) C6
Strawberry (res.) C3
Strawberry (riv.) D3
Swan (lake) B4
Tavaputs (plat.) D4
Thomas (range) A4
Thousand Lake (mt.) C5
Timpanogos Cave Nat'l Mon. C3
Tokewamna (peak) D3
Tooele Army Depot B3
Two Water (creek) E4
Uinta (mts.) D3
Uinta (riv.) D3
Uintah and Ouray Ind. Res. D3
Utah (lake) C3
Virgin (riv.) A6
Waas (mt.) E5
Wah Wah (mts.) A5
Wahweap (creek) C6
Wasatch (range) C3
Washakie Ind. Res. B2
Waterpocket Fold (cliffs) D6
Weber (riv.) C3
White (riv.) E3
Willow (creek) E4
Zion Nat'l Park A6

⊙County seat.
‡Population of metropolitan area.
† Zip of nearest p.o.
* Multiple zips.

Agriculture, Industry and Resources

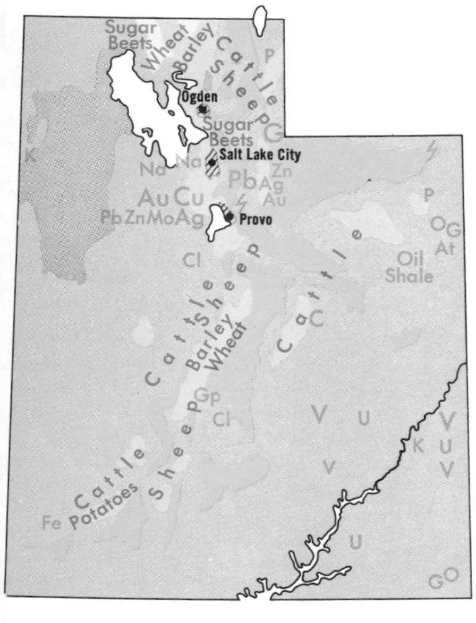

Sugar Beets
Wheat
Barley
Cattle
Sheep
Ogden
Salt Lake City
Provo
Oil Shale
Potatoes

DOMINANT LAND USE

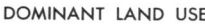

- Wheat, General Farming
- General Farming, Livestock, Special Crops
- Range Livestock
- Forests
- Nonagricultural Land

MAJOR MINERAL OCCURRENCES

Ag	Silver	Fe	Iron Ore	O	Petroleum
At	Asphalt	G	Natural Gas	P	Phosphates
Au	Gold	Gp	Gypsum	Pb	Lead
C	Coal	K	Potash	U	Uranium
Cl	Clay	Mo	Molybdenum	V	Vanadium
Cu	Copper	Na	Salt	Zn	Zinc

Water Power
Major Industrial Areas

Topography

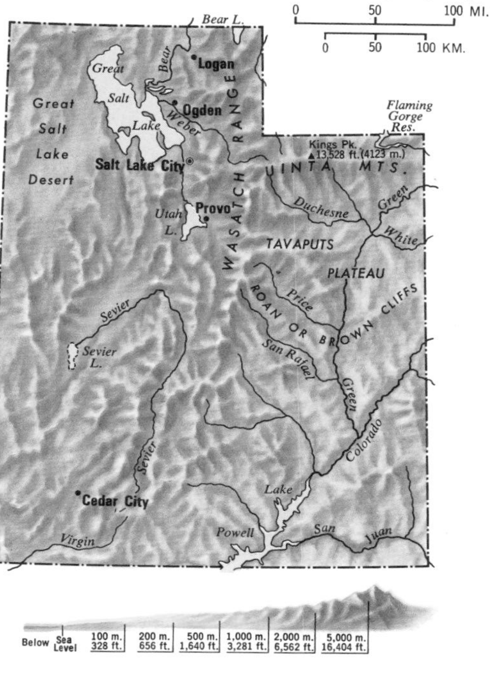

0 50 100 MI.
0 50 100 KM.
Bear L.
Logan
Great Salt Lake
Bear
Ogden
Weber
Great Salt Lake Desert
Salt Lake City
Kings Pk. 13,528 ft. (4123 m.)
UINTA MTS.
Flaming Gorge Res.
Utah L.
Provo
WASATCH RANGE
Duchesne
Green
White
TAVAPUTS PLATEAU
Price
ROAN OR BROWN CLIFFS
Sevier
Sevier L.
San Rafael
Green
Colorado
Sevier
Cedar City
Lake Powell
San Juan
Virgin
Below Sea Level
100 m. 328 ft.
200 m. 656 ft.
500 m. 1,640 ft.
1,000 m. 3,281 ft.
2,000 m. 6,562 ft.
5,000 m. 16,404 ft.

Topography

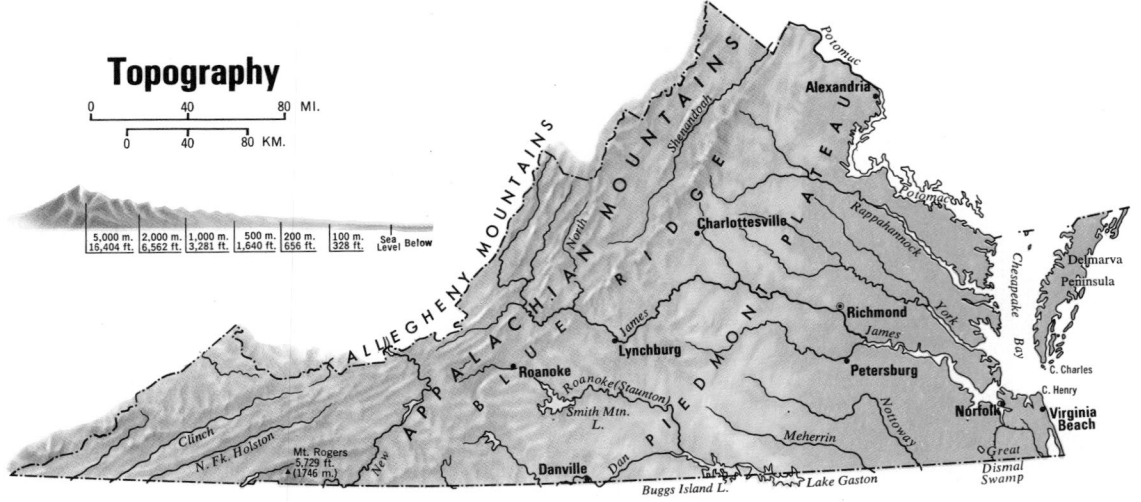

0   40   80 MI.

0   40   80 KM.

5,000 m. 2,000 m. 1,000 m. 500 m. 200 m. 100 m. Sea Below
16,404 ft. 6,562 ft. 3,281 ft. 1,640 ft. 656 ft. 328 ft. Level

COUNTIES

Accomack 31,268 ....S5
Albemarle 55,783 ....L5
Alleghany 14,333 ....H5
Amelia 8,405 ....M6
Amherst 29,122 ....K5
Appomattox 11,971 ....L6
Arlington 152,599 ....S2
Augusta 53,732 ....J4
Bath 5,860 ....J4
Bedford 34,927 ....J6
Bland 6,349 ....F6
Botetourt 23,270 ....J5
Brunswick 15,632 ....N7
Buchanan 37,989 ....D6
Buckingham 11,751 ....L5
Campbell 45,424 ....K6
Caroline 17,904 ....O4
Carroll 27,270 ....G7
Charles City 6,692 ....O6
Charlotte 12,266 ....L6
Chesterfield 141,372 ....M6
Clarke 9,965 ....M2
Craig 3,948 ....H6
Culpeper 22,620 ....M3
Cumberland 7,881 ....M6
Dickenson 19,806 ....D6
Dinwiddie 22,602 ....N6
Essex 8,864 ....P5
Fairfax 596,901 ....O3
Fauquier 35,889 ....N3
Floyd 11,563 ....H7
Fluvanna 10,244 ....M5
Franklin 35,740 ....J6
Frederick 34,150 ....M2
Giles 17,810 ....G6
Gloucester 20,107 ....P6
Goochland 11,761 ....N5
Grayson 16,579 ....F7
Greene 7,625 ....M4
Greensville 10,903 ....N7
Halifax 30,599 ....L7
Hanover 50,398 ....N5
Henrico 180,735 ....O6
Henry 57,654 ....J7
Highland 2,937 ....J4
Isle of Wight 21,603 ....P7
James City 22,763 ....P6
King and Queen 5,968 ....P5
King George 10,543 ....O4
King William 9,334 ....O5
Lancaster 10,129 ....R5
Lee 25,956 ....B7
Loudoun 57,427 ....N2
Louisa 17,825 ....N5
Lunenburg 12,124 ....M7
Madison 10,232 ....M4
Mathews 7,995 ....R6
Mecklenburg 29,444 ....M7
Middlesex 7,719 ....R5
Montgomery 63,516 ....H6
Nelson 12,204 ....L5
New Kent 8,781 ....P5
Northampton 14,625 ....S6
Northumberland 9,828 ....R5
Nottoway 14,666 ....M6
Orange 18,063 ....M4
Page 19,401 ....M3
Patrick 17,647 ....H7
Pittsylvania 66,147 ....K7
Powhatan 13,062 ....N5
Prince Edward 16,456 ....M6
Prince George 25,733 ....O6
Prince William 144,703 ....O3
Pulaski 35,229 ....G6
Rappahannock 6,093 ....M3
Richmond 6,952 ....P5
Roanoke 72,945 ....H6
Rockbridge 17,911 ....K5
Rockingham 57,038 ....L4
Russell 31,761 ....D7
Scott 25,068 ....C7
Shenandoah 27,559 ....L3
Smyth 33,366 ....E7
Southampton 18,731 ....O7
Spotsylvania 34,435 ....N4
Stafford 40,470 ....O4
Surry 6,046 ....P6
Sussex 10,874 ....O7
Tazewell 50,511 ....E6
Warren 21,200 ....M3

Washington 46,487 ....D7
Westmoreland 14,041 ....P4
Wise 43,863 ....C6
Wythe 25,522 ....F7
York 35,463 ....P6

CITIES and TOWNS

Zip   Name/Pop.   Key

24210 Abingdon⊙ 4,318 ....D7
23301 Accomac⊙ 522 ....S5
23001 Achilles 525 ....R6
22920 Afton 350 ....L4
23821 Alberta 394 ....N7
*22301 Alexandria (I.C.) 103,217 ..S3
24310 Allisonia 325 ....G7
24517 Altavista 3,849 ....K6
24520 Alton 500 ....K7
23002 Amelia Court House⊙ 500.N6
24521 Amherst⊙ 1,135 ....K5
24601 Amonate 350 ....E6
22003 Annandale 49,524 ....O3
24216 Appalachia 2,418 ....C7
24522 Appomattox⊙ 1,345 ....L6
24053 Ararat 500 ....G7
*22201 Arlington⊙ 152,599 ....T3
22922 Arrington 500 ....L5
23004 Arvonia 500 ....M5
22011 Ashburn 345 ....O2
23005 Ashland 4,640 ....N5
24311 Atkins 1,352 ....F7
24411 Augusta Springs 600 ......K4
24312 Austinville 750 ....F7
24054 Axton 540 ....J7
22041 Bailey's
      Crossroads 12,564 ....S3
24230 Banner 327 ....D7
22923 Barboursville 600 ....M4
24055 Bassett 2,034 ....J7
24314 Bastian 600 ....F6
22924 Batesville 575 ....L5
23015 Beaverdam 500 ....N5
24523 Bedford (I.C.)⊙ 5,991 ....J6
23306 Belle Haven 589 ....S5
24218 Ben Hur 400 ....B7
22610 Bentonville 500 ....M3
22611 Berryville⊙ 1,752 ....M2
24526 Big Island 500 ....K5
24603 Big Rock 900 ....D6
24219 Big Stone Gap 4,748 ......C7
24220 Birchleaf 650 ....D6
23307 Birdsnest 736 ....S6
24604 Bishop 600 ....E6
24060 Blacksburg 30,638 ....H6

23824 Blackstone 3,624 ....N6
24527 Blairs 500 ....K7
24315 Bland⊙ 950 ....F6
23308 Bloxom 407 ....S5
24605 Bluefield 5,946 ....F6
24064 Blue Ridge 2,347 ....J6
24606 Boissevain 975 ....F6
23235 Bon Air 16,224 ....N5
24065 Boones Mill 344 ....J6
22713 Boston 400 ....M3
22427 Bowling Green⊙ 665 ....O4
22620 Boyce 401 ....M2
23917 Boydton⊙ 486 ....M7
23827 Boykins 791 ....O7
22714 Brandy Station 400 ....N4
24607 Breaks 550 ....D6
22812 Bridgewater 3,289 ....K4
24201 Bristol (I.C.) 19,042 ....D7
24316 Broadford 500 ....E7
22815 Broadway 1,234 ....L3
23920 Brodnax 492 ....N7
22430 Brooke 245 ....O4
24528 Brookneal 1,454 ....L6
24415 Brownsburg 300 ....K5
22610 Browntown 300 ....M3
22622 Brucetown 250 ....M2
23923 Charlotte Court
      House⊙ 568 ....L6

23921 Buckingham⊙ 200 ....L5
24416 Buena Vista
      (I.C.) 6,717 ....K5
24529 Buffalo Junction 300 .....L7
22015 Burke 33,835 ....R3
24608 Burkes Garden 267 ....F6
23922 Burkeville 606 ....M6
22435 Callao 500 ....P5
24067 Callaway 225 ....H7
22016 Calverton 500 ....N3
23310 Cape Charles 1,512 ....R6
23313 Capeville 325 ....R6
22829 Capron 238 ....O7
23315 Carrsville 300 ....P7
23830 Carson 500 ....N6
22017 Casanova 370 ....N3
24069 Cascade 835 ....J7
24224 Castlewood 2,420 ....D7
24070 Catawba 350 ....H6
22019 Catlett 500 ....N3
24609 Cedar Bluff 1,550 ....E6
22437 Center Cross 360 ....P5
23030 Charles City⊙ 5 ....O6

23923 Charlotte Court

*22901 Charlottesville
      (I.C.)⊙ 39,916 ....M4
      Charlottesville‡ 113,568 ..M4
23924 Chase City 2,749 ....M7
24531 Chatham⊙ 1,390 ....K7
23316 Cheriton 695 ....R6
*23320 Chesapeake (I.C.)
      114,486 ....R7
23831 Chester 11,728 ....O6
23832 Chesterfield⊙ 950 ....N6
22623 Chester Gap 400 ....M3
24319 Chilhowie 1,269 ....E7
23336 Chincoteague 1,607 ....T5
24073 Christiansburg⊙ 10,345 ...H6
23032 Church View 200 ....P6
23899 Claremont 380 ....P6
23927 Clarksville 1,468 ....L7
†23061 Clay Bank 200 ....P6
†23139 Clayville 200 ....N6
22624 Clear Brook 300 ....M2
24225 Cleveland 360 ....D7
24422 Clifton Forge
      (I.C.) 5,046 ....J5
22321 Clinchburg 250 ....E7
24226 Clincho 900 ....D6
24244 Clinchport 89 ....C7
24228 Clintwood⊙ 1,369 ....D6
24534 Clover 215 ....L7
24077 Cloverdale 850 ....J6
24535 Cluster Springs 350 ......L7
23035 Cobbs Creek 700 ....R6
24230 Coeburn 2,625 ....C7
24536 Coleman Falls 250 ....K6
†24450 Collierstown 300 ....K5
24078 Collinsville 7,517 ....J7
22443 Colonial Beach 2,474 .....P4
23834 Colonial Heights
      (I.C.) 16,509 ....O6
23038 Columbia 111 ....M5
24538 Concord 500 ....K6
24837 Courtland⊙ 976 ....O7
22932 Covesville 475 ....L5
24426 Covington (I.C.)⊙
      9,063 ....H5
24430 Craigsville 845 ....J4
23930 Crewe 2,325 ....M6
24431 Crimora 450 ....L4
24322 Cripple Creek 200 ....F7
24323 Crockett 300 ....F7
22932 Crozet 2,553 ....L4
23039 Crozier 300 ....N5
24539 Crystal Hill 475 ....L7
23934 Cullen 725 ....L6
22701 Culpeper⊙ 6,621 ....M4
23040 Cumberland⊙ 300 ....M6
22448 Dahlgren 950 ....O4
22193 Dale City 33,127 ....O3
24083 Daleville 450 ....J6
24236 Damascus 1,330 ....E7
24237 Dante 1,083 ....D7
*24540 Danville (I.C.) 45,642 ...J7
      Danville‡ 111,789 ....J7

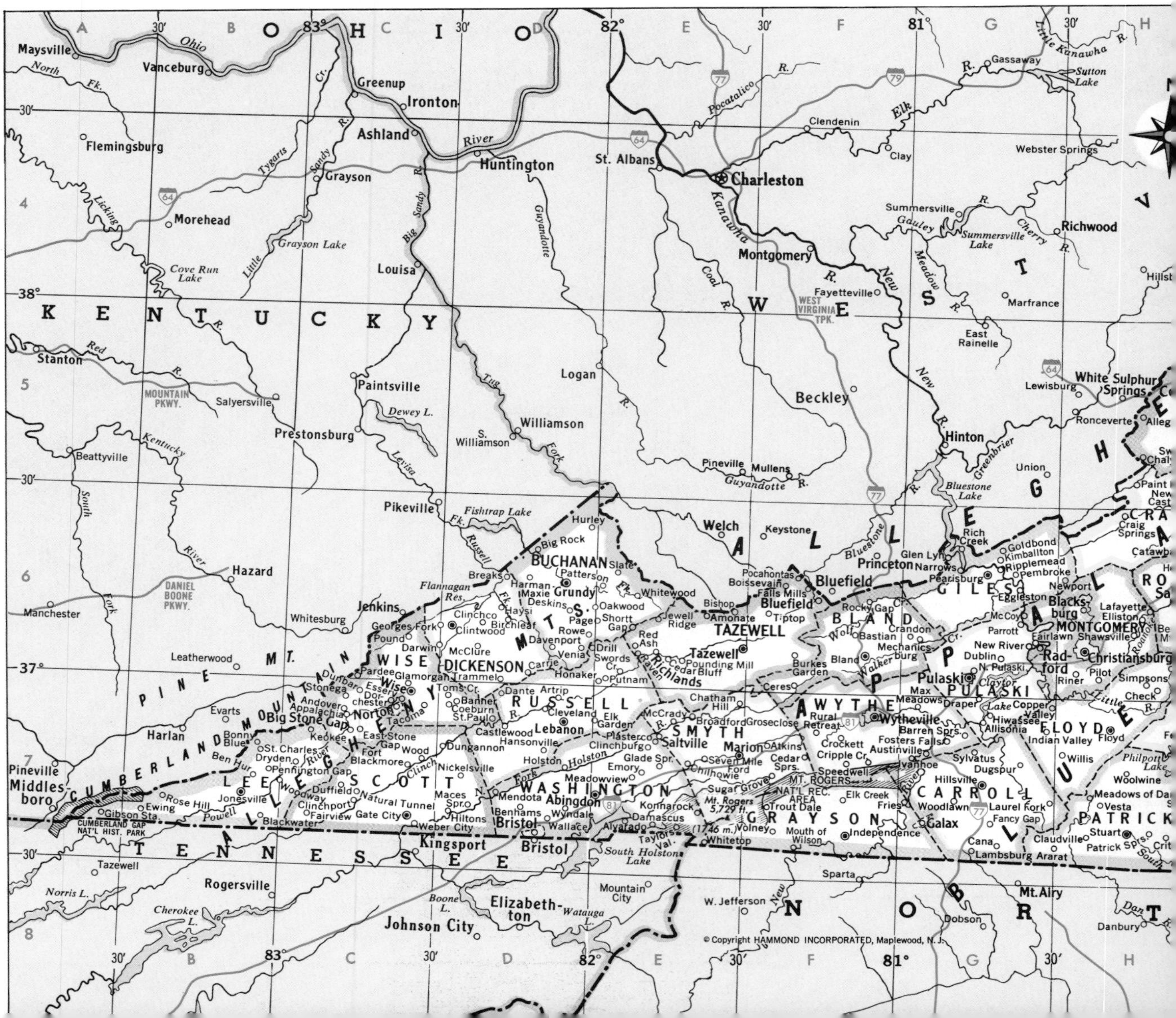

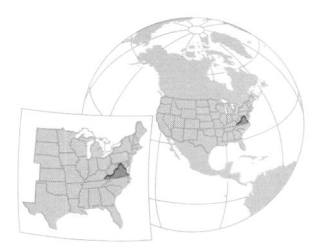

24239 Davenport 230	..........D6	
22821 Dayton 1,017	..........L4	
24432 Deerfield 500	..........K4	
23043 Deltaville 1,082	..........R5	
23839 Dendron 307	..........P6	
23936 Dillwyn 637	..........M5	
23841 Dinwiddie ⊙ 500	..........N6	
23842 Disputanta 800	..........O6	
23937 Drakes Branch 617	..........L7	
23844 Drewryville 200	..........O7	
24243 Dryden 400	..........B7	
24549 Dry Fork 200	..........K7	
24084 Dublin 2,368	..........G6	
24244 Duffield 148	..........C7	
22026 Dumfries 3,214	..........O3	
23938 Dundas 200	..........M7	
24245 Dungannon 339	..........D7	
22027 Dunn Loring 6,077	..........S2	
22454 Dunnsville 800	..........P5	
24085 Eagle Rock 750	..........J5	
22936 Earlysville 210	..........M4	
24246 East Stone Gap 240	..........C7	
23347 Eastville ⊙ 238	..........R6	
23845 Ebony 400	..........N7	
22824 Edinburg 752	..........M3	
24086 Eggleston 350	..........G6	
22827 Elkton 1,520	..........L4	
24087 Elliston-Lafayette 1,172	..........H6	
24327 Emory-Meadowview 2,292	..........E7	
23847 Emporia (I.C.) ⊙ 4,840	..........N7	
22937 Esmont 950	..........L5	
24274 Esserville 750	..........G2	
23803 Ettrick 4,890	..........O6	
23939 Evergreen 300	..........L6	
24248 Ewing 800	..........B7	
23350 Exmore 1,300	..........S5	
22030 Fairfax (I.C.) ⊙ 19,390	..........R3	
22039 Fairfax Station	..........R3	
24435 Fairfield 465	..........K5	
24141 Fairlawn 1,794	..........G6	
*22040 Falls Church (I.C.) 9,515	..........S2	
24613 Falls Mills 800	..........F6	
22401 Falmouth 3,271	..........O4	
24328 Fancy Gap 200	..........G7	
23901 Farmville ⊙ 6,067	..........M6	
24088 Ferrum 200	..........H7	
24089 Fieldale 1,190	..........H7	
24090 Fincastle ⊙ 200	..........J6	
22939 Fishersville 975	..........K4	
22627 Flint Hill 750	..........M3	
24091 Floyd ⊙ 411	..........H7	
24551 Forest 497	..........K6	
23055 Fork Union 350	..........M5	
24437 Fort Defiance 600	..........L4	
22578 Foxwells 400	..........R5	
22310 Franconia 8,476	..........S3	
23851 Franklin (I.C.) ⊙ 7,308	..........P7	
23354 Franktown 500	..........S6	
*22401 Fredericksburg (I.C.) 15,322	..........N4	
24330 Fries 758	..........F7	
22630 Front Royal ⊙ 11,126	..........M3	
22065 Gainesville 600	..........N3	
24333 Galax (I.C.) 6,524	..........G7	
22463 Garrisonville 200	..........N4	
24251 Gate City ⊙ 2,494	..........C7	
†24228 Georges Fork 200	..........C6	
24340 Glade Spring 1,722	..........E7	
24554 Gladys 500	..........K6	
24555 Glasgow 1,259	..........K5	
23060 Glen Allen 6,202	..........N5	
24093 Glen Lyn 235	..........G6	
†24541 Glenwood 2,276	..........K7	
23061 Gloucester ⊙ 1,545	..........P6	
23062 Gloucester Point 5,841	..........R6	
24094 Goldbond 250	..........G6	
22720 Goldvein 500	..........N4	
23063 Goochland ⊙ 800	..........N5	
24556 Goode 200	..........K6	
22942 Gordonsville 1,421	..........M4	
22637 Gore 500	..........M2	
24439 Goshen 134	..........K5	
23692 Grafton 500	..........P6	
23356 Greenbackville 300	..........T5	
23942 Green Bay 500	..........M6	
23357 Greenbush 200	..........S5	
24440 Greenville 400	..........K5	
22943 Greenwood 800	..........L4	
24557 Gretna 1,255	..........K7	
24441 Grottoes 1,369	..........L4	
†22306 Groveton 18,860	..........T3	
22614 Grundy ⊙ 1,699	..........D6	
23066 Gwynn 205	..........R5	
22469 Hague 425	..........P4	
24558 Halifax ⊙ 772	..........L7	
23359 Hallwood 243	..........S5	
22068 Hamilton 598	..........N2	
23943 Hampden-Sydney 1,011	..........L6	
*23601 Hampton (I.C.) 122,617	..........R6	
23069 Hanover ⊙ 500	..........O5	
23389 Harborton 200	..........S5	
24101 Hardy 325	..........J6	
24618 Harman-Maxie 650	..........D6	
22801 Harrisonburg ⊙ 19,671	..........K4	
23071 Hartfield 700	..........R5	
22069 Haymarket 230	..........N3	
22472 Haynesville 500	..........P5	
24256 Haysi 371	..........D6	
22473 Heathsville ⊙ 300	..........P5	
24102 Henry 300	..........J7	
*22070 Herndon 11,449	..........O3	
23075 Highland Springs 12,146	..........O5	
22132 Hillsboro 115	..........N2	
24343 Hillsville ⊙ 2,123	..........G7	
24258 Hiltons 300	..........D7	
24347 Hiwassee 250	..........G7	
24020 Hollins College 12,295	..........H6	
24260 Honaker 1,475	..........D6	
23860 Hopewell (I.C.) ⊙ 23,397	..........O6	
23395 Horntown 400	..........T5	
24445 Hot Springs 300	..........J4	
24104 Huddleston 200	..........K6	

(continued on following page)

AREA 40,767 sq. mi. (105,587 sq. km.)
POPULATION 5,346,818
CAPITAL Richmond
LARGEST CITY Norfolk
HIGHEST POINT Mt. Rogers 5,729 ft. (1746 m.)
SETTLED IN 1607
ADMITTED TO UNION June 26, 1788
POPULAR NAME Old Dominion
STATE FLOWER Dogwood
STATE BIRD Cardinal

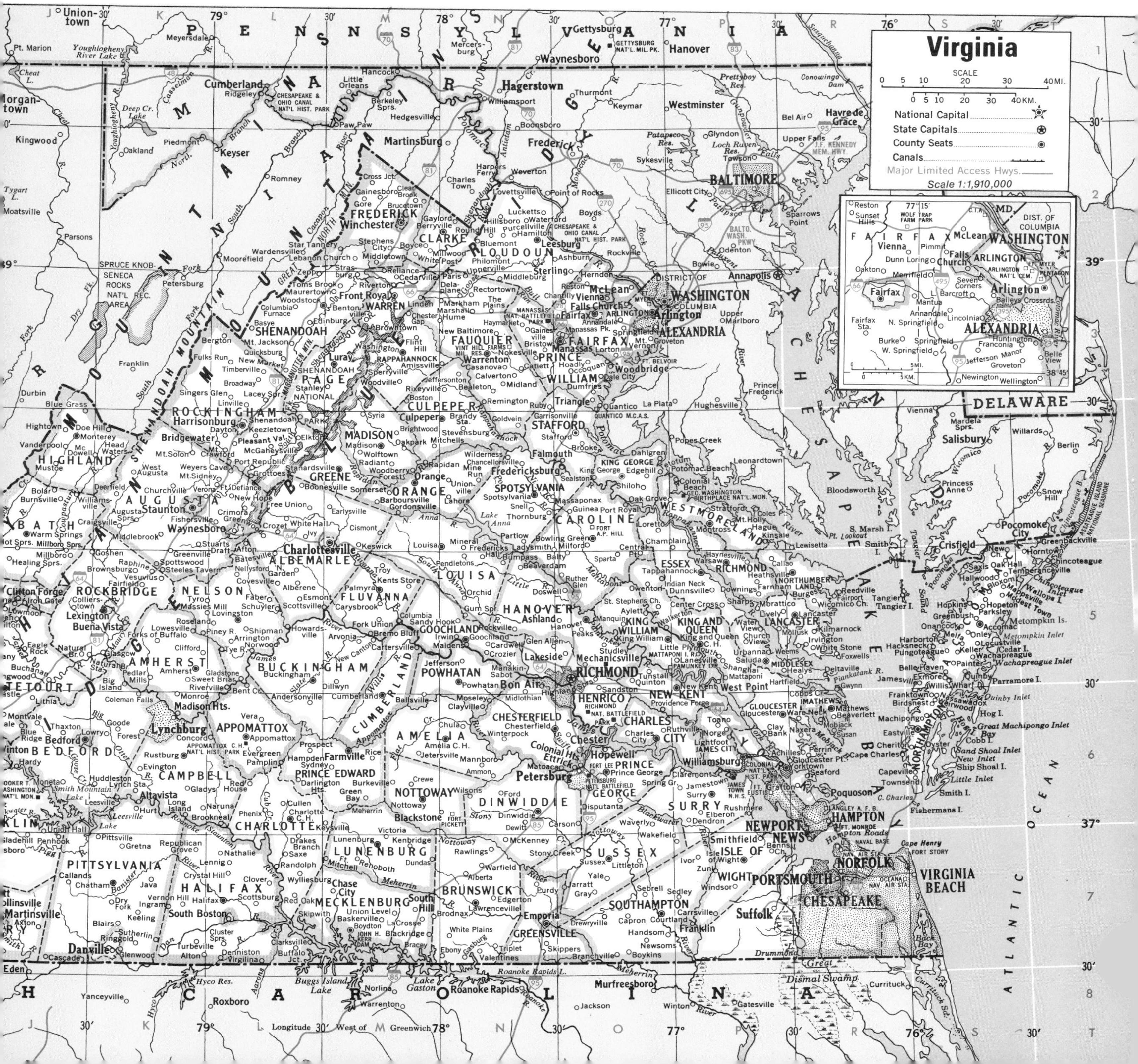

Agriculture, Industry and Resources

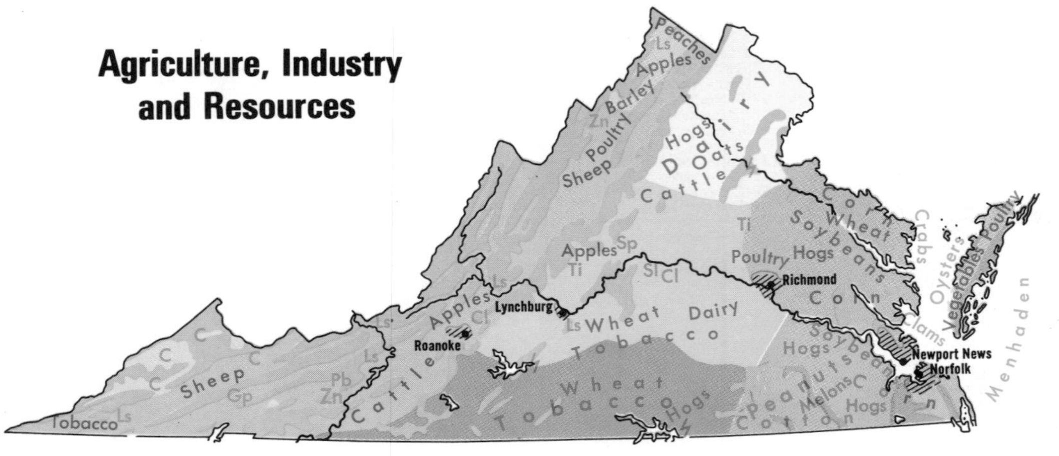

DOMINANT LAND USE

- Dairy, General Farming
- General Farming, Livestock, Dairy
- General Farming, Livestock, Tobacco
- General Farming, Livestock, Fruit, Tobacco
- General Farming, Truck Farming, Tobacco, Livestock
- Tobacco, General Farming
- Peanuts, General Farming
- Fruit and Mixed Farming
- Truck and Mixed Farming
- Forests
- Swampland, Limited Agriculture

MAJOR MINERAL OCCURRENCES

C	Coal	Sl	Slate		Water Power
Cl	Clay	Sp	Soapstone		Major Industrial Areas
Gp	Gypsum	Ti	Titanium		
Ls	Limestone	Zn	Zinc		
Pb	Lead				

22639 Hume 350 .... N3
†22301 Huntington 5,813 .... S3
24620 Hurley 850 .... D6
24563 Hurt 1,481 .... K6
24348 Independence⊙ 1,112 .... F7
24105 Indian Valley 300 .... G7
24448 Iron Gate 620 .... J5
22480 Irvington 567 .... R5
23397 Isle of Wight⊙ 185 .... P7
24350 Ivanhoe 900 .... G7
23866 Ivor 403 .... P7
22945 Ivy 900 .... L4
23081 Jamestown 12 .... P6
23398 Jamesville 500 .... S5
23867 Jarratt 614 .... O7
22303 Jefferson Manor S3
22724 Jeffersonton 300 .... N3
24622 Jewell Ridge 600 .... E6
24263 Jonesville⊙ 874 .... B7
24566 Keeling 680 .... K7
22832 Keezletown 975 .... L4
23401 Keller 236 .... S5
23944 Kenbridge 1,352 .... M7
24265 Keokee 300 .... C7
22947 Keswick 300 .... M4
23947 Keysville 704 .... M6
22482 Kilmarnock 945 .... R5
23085 King and Queen Court House⊙ 500 .... P5
22485 King George⊙ 575 .... O4
23086 King William⊙ 100 .... O5
22488 Kinsale 250 .... P4
23950 La Crosse 734 .... M7
22501 Ladysmith 360 .... N4
24108 Lafayette-Elliston 1,172 .... H6
†22041 Lake Barcroft 8,725 .... S3
23228 Lakeside 12,289 .... N5
24351 Lambsburg 800 .... G7
22503 Lancaster⊙ 110 .... R5
24352 Laurel Fork 300 .... G7
23868 Lawrenceville⊙ 1,484 .... N7
24266 Lebanon⊙ 3,206 .... D7
22641 Lebanon Church 300 .... L2
22075 Leesburg⊙ 8,357 .... N2
24450 Lexington (I.C.)⊙ 7,292 .... J5
†22313 Lincolnia 10,350 .... S3
22642 Linden 320 .... M3
22834 Linville 500 .... L3
22507 Lively 400 .... P5
22079 Lorton 5,813 .... O3
23093 Louisa⊙ 932 .... M4
22080 Lovettsville 613 .... N2
22949 Lovingston⊙ 600 .... L5
22951 Lowesville 500 .... K5
24457 Lowmoor 700 .... J5
†22075 Lucketts 500 .... N2
23952 Lunenburg⊙ 13 .... M7
22835 Luray⊙ 3,584 .... M3
*24501 Lynchburg (I.C.) 66,743 .... K6
Lynchburg‡ 153,260 .... K6
24571 Lynch Station 500 .... K6
23405 Machipongo 400 .... S6
22727 Madison⊙ 267 .... M4
24572 Madison Heights 14,146 .... K6
23103 Manakin-Sabot 200 .... N5
22110 Manassas (I.C.)⊙ 15,438 .... O3
22110 Manassas Park (I.C.) 6,524 .... O3
23106 Manquin 576 .... O5
†22030 Mantua 6,523 .... S3
23407 Mappsville 700 .... T5
24354 Marion⊙ 7,029 .... E7
22643 Markham 300 .... N3
22115 Marshall 800 .... N3
24112 Martinsville (I.C.)⊙ 18,149 .... J7
22954 Massies Mill 225 .... K5
23109 Mathews⊙ 500 .... R6
23803 Matoaca 1,967 .... N6
23110 Mattaponi 300 .... P5
24360 Max Meadows 782 .... G6
24269 McClure 300 .... D6
24111 McCoy 600 .... G6
22840 McGaheysville 600 .... L4
23872 McKenney 473 .... N7
*22101 McLean 35,664 .... S2
24361 Meadowview-Emory 2,292 .... D7
†24315 Mechanicsburg 350 .... G6
23111 Mechanicsville 9,269 .... O5
23954 Meherrin 400 .... M6
23410 Melfa 391 .... S5
24270 Mendota 375 .... D7
22116 Merrifield 7,525 .... S3
22117 Middleburg 619 .... N3
22645 Middletown 841 .... M2
22728 Midland 600 .... N3
23113 Midlothian 950 .... N6
22514 Milford 650 .... O4
24460 Millboro 400 .... J5
24460 Millboro Springs 200 .... J4
22646 Millwood 400 .... N2
23117 Mineral 399 .... N4
22568 Mine Run 450 .... N4
23118 Mobjack 210 .... R6
23412 Modest Town 225 .... T5
22517 Mollusk 800 .... P5
24121 Moneta 300 .... J6
24574 Monroe 500 .... K6
24465 Monterey⊙ 247 .... K4
22520 Montross⊙ 456 .... P4
24122 Montvale 900 .... J6
22523 Morattico 225 .... P5
23120 Moseley 210 .... N6
22841 Mount Crawford 315 .... L4
22524 Mount Holly 200 .... P4
22842 Mount Jackson 1,419 .... L3
24467 Mount Sidney 500 .... K4
22121 Mount Vernon 24,058 .... O3
24363 Mouth of Wilson 400 .... F7
24124 Narrows 2,516 .... G6
23413 Nassawadox 630 .... S6
24577 Nathalie 200 .... L7
24578 Natural Bridge 200 .... J5
24579 Natural Bridge Sta. 450 .... K5
23122 Naxera 300 .... R6
22958 Nellysford 290 .... L5
24127 New Castle⊙ 213 .... H5
23415 New Church 427 .... S5
24469 New Hope 200 .... L4
22122 Newington 8,313 .... S3
23124 New Kent⊙ 25 .... P5
22844 New Market 1,118 .... L3
24128 Newport 600 .... H6
*23601 Newport News (I.C.) 144,903 .... P6
Newport News-Hampton‡ 364,449 .... P6
24129 New River 500 .... G6
23874 Newsoms 368 .... O7
24271 Nickelsville 464 .... D7
22123 Nokesville 520 .... N3
24272 Nora 550 .... D6
*23501 Norfolk (I.C.) 266,979 .... R7
Norfolk-Virginia Beach-Portsmouth ‡ 806,691 .... R7
23127 Norge 750 .... P6
22959 North Garden 200 .... L5
†24301 North Pulaski 1,405 .... G6
22151 North Springfield 9,538 .... S3
24273 Norton (I.C.) 4,757 .... C7
23955 Nottoway⊙ 170 .... M6
23416 Oak Hall 221 .... S5
22124 Oakton 19,150 .... R3
24631 Oakwood 715 .... E6
22125 Occoquan 241 .... O3
23417 Onancock 1,461 .... S5
23418 Onley 526 .... S5
22960 Orange⊙ 2,631 .... M4
23419 Oyster 200 .... S6
24131 Paint Bank 235 .... H5
23420 Painter 321 .... S5
22963 Palmyra⊙ 250 .... M5
23958 Pamplin 273 .... L6
23421 Parksley 979 .... S5
24132 Parrott 750 .... G6
24133 Patrick Springs 800 .... H7
†23069 Peaks 500 .... O5
24134 Pearisburg⊙ 2,128 .... G6
24136 Pembroke 1,302 .... G6
24137 Penhook 500 .... J7
24277 Pennington Gap 1,716 .... C7
23803 Petersburg (I.C.) 41,055 .... N6
Petersburg-Colonial Heights-Hopewell‡ 129,296 .... N6
23959 Phenix 250 .... L6
22131 Philomont 265 .... N2
24138 Pilot 360 .... H6
22043 Pimmit 6,658 .... S2
22964 Piney River 778 .... L5
24139 Pittsville 600 .... K7
24635 Pocahontas 708 .... F6
23662 Poquoson (I.C.) 8,726 .... R6
22535 Port Royal 291 .... O4
*23701 Portsmouth (I.C.) 104,577 .... R7
24279 Pound 1,086 .... C6
24637 Pounding Mill 399 .... E6
23139 Powhatan⊙ 600 .... N5
23875 Prince George⊙ 150 .... O6
23960 Prospect 275 .... L6
23140 Providence Forge 500 .... P6
24301 Pulaski⊙ 10,106 .... G6
23422 Pungoteague 500 .... S5
22132 Purcellville 1,567 .... N2
†23847 Purdy 350 .... N7
22134 Quantico 621 .... O3
23423 Quinby 350 .... S5
24141 Radford (I.C.) 13,225 .... G6
22732 Radiant 250 .... M4
24472 Raphine 500 .... K5
24639 Raven 4,000 .... E6
23876 Rawlings 200 .... N7
22140 Rectortown 225 .... N3
24640 Red Ash 300 .... E6
23964 Red Oak 250 .... L7
22539 Reedville 400 .... R5
22734 Remington 425 .... N3
22090 Reston 36,407 .... R2
24147 Rich Creek 746 .... G6
24641 Richlands 5,796 .... E6
*23201 Richmond (cap.) (I.C.)⊙ 219,214 .... O5
Richmond‡ 632,015 .... O5
24148 Ridgeway 858 .... J7
24149 Riner 360 .... H6
24150 Ripplemead 600 .... G6
22651 Riverton 500 .... M3
*24001 Roanoke (I.C.) 100,220 .... H6
Roanoke‡ 224,341 .... H6
23146 Rockville 290 .... N5
24366 Rocky Gap 200 .... F6
24151 Rocky Mount⊙ 4,198 .... J7
24280 Rosedale 760 .... E7
24281 Rose Hill 700 .... B7
22967 Roseland 300 .... K5
22141 Round Hill 510 .... N2
24368 Rural Retreat 1,083 .... F7
†23430 Rushmere 1,070 .... P6
24588 Rustburg⊙ 650 .... K6
22546 Ruther Glen 200 .... O5
23147 Ruthville 300 .... P6
24282 Saint Charles 241 .... B7
24283 Saint Paul 973 .... D7
23148 St. Stephens Church 500 .... O5
24153 Salem (I.C.)⊙ 23,958 .... H6
24370 Saltville 2,376 .... E7
23149 Saluda⊙ 150 .... P5
23150 Sandston O5
23153 Sandy Hook 700 .... M5
23427 Saxis 415 .... S5
22969 Schuyler 250 .... L5
24589 Scottsburg 335 .... L7
24590 Scottsville 250 .... L5
23696 Seaford R6
22547 Sealston 200 .... O4
23878 Sedley 523 .... P7
24474 Selma 500 .... J5
22044 Seven Corners 6,058 .... S3
24373 Seven Mile Ford 425 .... E7
24162 Shawsville 950 .... H6
22849 Shenandoah 1,861 .... L4
22971 Shipman 350 .... L5
23430 Smithfield 3,718 .... P7
22553 Snell 300 .... N4
22972 Somerset 200 .... M4
24592 South Boston (I.C.) 7,093 .... L7
23970 South Hill 4,347 .... M7
22552 Sparta 485 .... O4
24374 Speedwell 650 .... F8
24165 Spencer 500 .... J7
22740 Sperryville 500 .... M3
22553 Spotsylvania⊙ 350 .... N4
*22150 Springfield 21,435 .... S3
22554 Stafford⊙ 750 .... O4
22973 Stanardsville⊙ 284 .... L4
22851 Stanley 1,204 .... L3
24168 Stanleytown 1,761 .... H7
22654 Star Tannery 500 .... L2
24401 Staunton (I.C.)⊙ 21,857 .... K4
24476 Steeles Tavern 200 .... K5
22655 Stephens City 1,179 .... M2
22170 Sterling 16,080 .... O2
24285 Stonega 275 .... C7
23882 Stony Creek 329 .... N7
22657 Strasburg 2,311 .... M3
24171 Stuart⊙ 1,131 .... H7
24477 Stuarts Draft 1,776 .... L4
23162 Studley 500 .... O5
*23432 Suffolk (I.C.) 47,621 .... P7
24375 Sugar Grove 1,027 .... E7
22090 Sunset Hills R2
23883 Surry⊙ 237 .... P6
23163 Susan 500 .... R6
23884 Sussex⊙ 75 .... O7
24595 Sweet Briar 900 .... K5
24649 Swords Creek 315 .... E6
†24343 Sylvatus 200 .... G7
23602 Tabb R6
23440 Tangier 771 .... R5
22560 Tappahannock⊙ 1,821 .... O5
24651 Tazewell⊙ 4,468 .... E6
23442 Temperanceville 400 .... T5
24174 Thaxton 450 .... K6
22171 The Plains 382 .... N3
22853 Timberville 1,510 .... L3
23168 Toano 950 .... P6
22660 Toms Brook 226 .... L3
23443 Townsend 525 .... R6
24289 Trammel 450 .... D6
22172 Triangle 4,770 .... O3
23886 Triplet 300 .... N7
24378 Trout Dale 248 .... F7
24175 Troutville 496 .... J6
22567 Unionville 500 .... N4
22176 Upperville 250 .... N2
23175 Urbanna 518 .... P5
23887 Valentines 400 .... N7
24656 Vansant 2,708 .... D6
24597 Vernon Hill 250 .... K7
24482 Verona 2,782 .... K4
24177 Vesta 350 .... H7
24483 Vesuvius 500 .... K5
23974 Victoria 2,004 .... M6
22180 Vienna 15,469 .... R2
24179 Vinton 8,027 .... J6
*23450 Virginia Beach (I.C.) 262,199 .... S7
23480 Wachapreague 404 .... S5
23888 Wakefield 1,355 .... O7
23177 Walkerton 985 .... O5
24484 Warm Springs⊙ 325 .... J4
22186 Warrenton⊙ 3,907 .... N3
22572 Warsaw⊙ 771 .... P5
22747 Washington⊙ 247 .... M3
22190 Waterford 350 .... N2
23180 Water View 265 .... P5
23890 Waverly 2,284 .... O6
22980 Waynesboro (I.C.) 15,329 .... K4
24251 Weber City 1,543 .... C7
22576 Weems 500 .... P5
23484 Weirwood 300 .... S6
24485 West Augusta 325 .... K4
23181 West Point 2,726 .... P5
22153 West Springfield 25,012 .... S3
24486 Weyers Cave 300 .... L4
22987 White Hall 250 .... L4
22578 White Stone 409 .... R5
24292 Whitetop 860 .... E7
24657 Whitewood 350 .... E6
22579 Wicomico Church 500 .... R5
†22553 Wilderness 200 .... N4
23185 Williamsburg (I.C.)⊙ 9,870 .... P6
23486 Willis Wharf 360 .... S5
22601 Winchester (I.C.)⊙ 20,217 .... M2
23487 Windsor 985 .... P7
24184 Wirtz 500 .... J6
24293 Wise⊙ 3,894 .... C7
22748 Wolftown 350 .... M4
22989 Woodberry Forest 450 .... M4
*22191 Woodbridge 24,004 .... O3
24381 Woodlawn 1,689 .... G7
22664 Woodstock⊙ 2,627 .... L3
†24277 Woodway 400 .... C7
23976 Wylliesburg 213 .... L7
24382 Wytheville⊙ 7,135 .... G7
23690 Yorktown⊙ 550 .... R6
23898 Zuni 300 .... P7

OTHER FEATURES

Aarons (creek) L7
Allegheny (mts.) H5
Anna (lake) N4
Appalachian (mts.) J5
Appomattox (riv.) M6
Appomattox Court House Nat'l Hist. Park K6
Arlington Nat'l Cemetery T3
Assateague Island Nat'l Seashore T4
Back (bay) S7
Back (creek) J4
Banister (riv.) K7
Big Otter (riv.) K6
Blackwater (riv.) J6
Blackwater (riv.) O6
Blue Ridge (mts.) J6
Bluestone (lake) G5
Booker T. Washington Nat'l Mon. J6
Buggs Island (lake) L8
Bull Run (creek) N3
Cedar (isl.) S5
Central Intelligence Agency (C.I.A.) S2
Charles (cape) R6
Chesapeake (bay) R5
Chesapeake and Ohio Canal Nat'l Mon. O2
Chincoteague (bay) T4
Chincoteague (inlet) T5
Claytor (lake) G6
Clinch (riv.) C7
Cobb (isl.) S6
Colonial Nat'l Hist. Park P6
Cowpasture (riv.) J4
Craig (creek) H5
Cub (creek) L6
Cumberland (mt.) B7
Cumberland Gap Nat'l Hist. Pk. A7
Dan (riv.) K7
Drummond (lake) P7
Fishermans (isl.) S6
Flannagan (res.) C6
Flat (creek) M6
Fort Belvoir 7,726 .... O3
Fort Eustis P6
Fort A.P. Hill O4
Fort Lee 9,784 .... O6
Fort Monroe R6
Fort Myer T2
Fort Pickett N6
Fort Story S7
Gaston (lake) M8
George Washington Birthplace Nat'l Mon. P4
Goose (creek) J6
Goose (creek) N3
Great Machipongo (inlet) S6
Great North (mt.) L2
Hampton Roads (est.) R7
Henry (cape) R7
Hog (isl.) S6
Hog Island (bay) S6
Holston, North Fork (riv.) D7
Hyco (riv.) K8
Jackson (riv.) J4
James (riv.) O6
Jamestown Nat'l Hist. Site .... P
John H. Kerr (dam) M7
Langley A.F.B. .... R
Leesville (lake) .... K
Levisa Fork (riv.) C
Little (inlet) .... S
Little (riv.) .... N
Little (riv.) .... H
Manassas Nat'l Battlefield Pk. .... K
Massanutten (mt.) L
Mattaponi (riv.) .... O
Mattaponi Ind. Res. .... P
Maury (riv.) .... K
Meherrin (riv.) .... M
Metompkin (inlet) .... T
Metompkin (isl.) .... T
Mobjack (bay) .... R
Mount Rogers Nat'l Rec. Area .... F
Naval Air Station .... R
New (inlet) .... S
New (riv.) .... F
Ni (riv.) .... N
North Anna (riv.) .... M
Nottoway (riv.) .... O
Oceana N.A.S. .... S
Pamunkey (riv.) .... O
Pamunkey Ind. Res. .... P
Parramore (isl.) .... S
Pentagon .... T
Petersburg Nat'l Battlefield .... O
Philpott (lake) .... H
Piankatank (riv.) .... R
Pigg (riv.) .... J
Po (riv.) .... N
Pocomoke (sound) .... S
Potomac (riv.) .... O
Powell (riv.) .... B
Quantico Marine Corps Air Sta. 7,121 .... O
Quinby (inlet) .... S
Rapidan (riv.) .... M
Rappahannock (riv.) .... P
Richmond Nat'l Battlefield Pk. .... O
Rivanna (riv.) .... M
Roanoke (riv.) .... N
Rogers (mt.) .... E
Russell Fork (riv.) C
Sand Shoal (inlet) .... S
Shenandoah (mt.) .... K
Shenandoah (riv.) .... N
Shenandoah Nat'l Park L
Ship Shoal (isl.) .... S
Slate (riv.) L
Smith (isl.) .... S
Smith (riv.) .... J
Smith Mountain (lake) .... J
South Anna (riv.) .... N
South Holston (lake) .... E
South Mayo (riv.) .... H
Staunton (Roanoke) (riv.) .... K
Stony (creek) .... N
Swift (creek) .... O
Tangier (isl.) .... R
Tangier (sound) .... S
Tug Fork (riv.) .... D
U.S. Naval Base .... R
Vint Hill Farms Mil. Res. .... N
Wachapreague (inlet) .... T
Walker (creek) .... F
Wallops (isl.) .... T
Willis (riv.) .... M
Wolf (creek) .... F
Wolf Trap Farm Park .... S
York (riv.) .... F

I.C. Independent City.
⊙County seat.
‡Population of metropolitan area.
† Zip of nearest p.o * Multiple zip

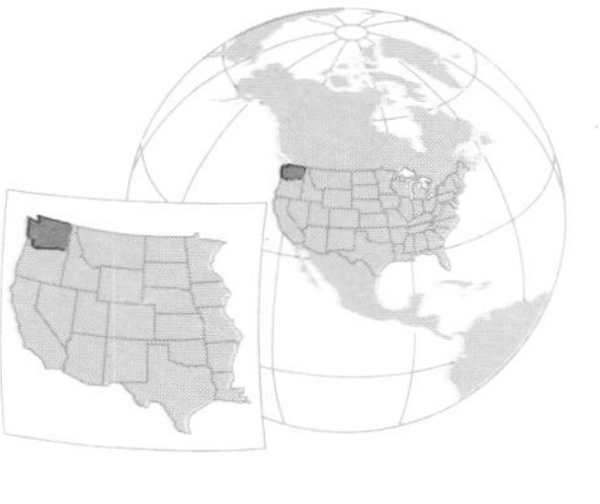

AREA 68,139 sq. mi. (176,480 sq. km.)
POPULATION 4,132,180
CAPITAL Olympia
LARGEST CITY Seattle
HIGHEST POINT Mt. Rainier 14,410 ft. (4392 m.)
SETTLED IN 1811
ADMITTED TO UNION November 11, 1889
POPULAR NAME Evergreen State
STATE FLOWER Western Rhododendron
STATE BIRD Willow Goldfinch

COUNTIES

Adams 13,267 .... G3
Asotin 16,823 .... H4
Benton 109,444 .... F4
Chelan 45,061 .... E3
Clallam 51,648 .... B2
Clark 192,227 .... C5
Columbia 4,057 .... H4
Cowlitz 79,548 .... C4
Douglas 22,144 .... F3
Ferry 5,811 .... G2
Franklin 35,025 .... G4
Garfield 2,468 .... H4
Grant 48,522 .... F3
Grays Harbor 66,314 .... B3
Island 44,048 .... C2
Jefferson 15,965 .... B3
King 1,269,749 .... D3
Kitsap 147,152 .... C3
Kittitas 24,877 .... E3
Klickitat 15,822 .... E5
Lewis 56,028 .... C4
Lincoln 9,604 .... G3
Mason 31,184 .... B3
Okanogan 30,639 .... F2
Pacific 17,237 .... B4
Pend Oreille 8,580 .... H2
Pierce 485,667 .... C3
San Juan 7,838 .... C2
Skagit 64,138 .... D2
Skamania 7,919 .... D5
Snohomish 337,720 .... D2
Spokane 341,835 .... H3
Stevens 28,979 .... H2
Thurston 124,264 .... C4
Wahkiakum 3,832 .... B4
Walla Walla 47,435 .... G4
Whatcom 106,701 .... D2
Whitman 40,103 .... H4
Yakima 172,508 .... E4

CITIES and TOWNS

Zip *Name/Pop.* *Key*

98520 Aberdeen 18,739 .... B3
98220 Acme 500 .... C2
99001 Airway Heights 1,730 .... H3
99102 Albion 631 .... H4
†98328 Alder 300 .... C4
98002 Algona 1,467 .... C3
98524 Allyn 850 .... C3
99103 Almira 330 .... G3
98526 Amanda Park 495 .... A3
98601 Amboy 480 .... C5
98221 Anacortes 9,013 .... C2
98603 Ariel 386 .... C5
98223 Arlington 3,282 .... C2
98304 Ashford 300 .... C4
99402 Asotin⊙ 943 .... H4
98002 Auburn 26,417 .... C3
98110 Bainbridge Island-Winslow (Winslow) 2,196 .... A2
98604 Battle Ground 2,774 .... C5
†98004 Beaux Arts Village 328 .... B2
98305 Beaver 450 .... A2
98528 Belfair 500 .... C3
*98004 Bellevue 73,903 .... B2
98225 Bellingham⊙ 45,794 .... C2
Bellingham‡ 106,701 .... C2
99320 Benton City 1,980 .... F4
98605 Bingen 644 .... D5
98010 Black Diamond 1,170 .... D3
98230 Blaine 2,363 .... C2
†98390 Bonney Lake 5,328 .... C3
98011 Bothell 7,943 .... B1
98310 Bremerton 36,208 .... A2
Bremerton‡ 146,609 .... A2
98812 Brewster 1,337 .... F2
98813 Bridgeport 1,174 .... F3
†98036 Brier 2,915 .... C3
98320 Brinnon 500 .... B3
†98101 Bryn Mawr-Skyway 11,754 B2
98321 Buckley 3,143 .... C3
98530 Bucoda 519 .... C4
98921 Buena 590 .... E4
98166 Burien 23,189 .... A2
98233 Burlington 3,894 .... C2
98013 Burton 650 .... C3
98607 Camas 5,681 .... C5
98323 Carbonado 456 .... D3
98324 Carlsborg 500 .... B2
98814 Carlton 410 .... F2
98014 Carnation 913 .... D3
98610 Carson 500 .... D5
98815 Cashmere 2,240 .... E3
98611 Castle Rock 2,162 .... B4
98612 Cathlamet⊙ 635 .... B4
98531 Centralia 11,555 .... C4
98520 Central Park 2,709 .... B3
98532 Chehalis⊙ 6,100 .... C4
98816 Chelan 2,802 .... E3
99004 Cheney 7,630 .... H3
99109 Chewelah 1,888 .... H2
98614 Chinook 928 .... B4
98326 Clallam Bay 600 .... A2
99403 Clarkston 6,903 .... H4
98235 Clearlake 750 .... C2
98922 Cle Elum 1,773 .... E3
98236 Clinton 900 .... C3
†98004 Clyde Hill 3,229 .... B2
†98055 Coalfield 500 .... B2
99111 Colfax⊙ 2,780 .... H4
99324 College Place 5,771 .... G4
99113 Colton 307 .... H4
†98632 Columbia Heights 2,515 .... C4
99114 Colville⊙ 4,510 .... H2
98819 Conconully 157 .... F2
98237 Concrete 592 .... D2
99326 Connell 1,981 .... G4
98535 Copalis Beach 600 .... A3
98536 Copalis Crossing 500 .... B3
98537 Cosmopolis 1,575 .... B4
99115 Coulee City 510 .... F3
99116 Coulee Dam 1,412 .... G3
98239 Coupeville⊙ 1,006 .... C2
99117 Creston 309 .... G3
99119 Cusick 246 .... H2
98240 Custer 300 .... C2
98617 Dallesport 600 .... D5
98241 Darrington 1,064 .... D2
99122 Davenport⊙ 1,559 .... G3
99328 Dayton⊙ 2,565 .... H4
98243 Deer Harbor 400 .... B2
99006 Deer Park 2,140 .... H3
98188 Des Moines 7,378 .... B2
99213 Dishman 10,169 .... H3
99329 Dixie 210 .... G4
98821 Dryden 500 .... E3
†98382 Dungeness 675 .... B2
98327 Du Pont 559 .... C3
98019 Duvall 729 .... D3
98245 Eastsound 800 .... B2
98801 East Wenatchee 1,640 .... E3
98328 Eatonville 998 .... C4
98020 Edmonds 27,679 .... C3
99123 Electric City 927 .... F3
98926 Ellensburg⊙ 11,752 .... E3
98541 Elma 2,720 .... B4
99124 Elmer City 312 .... G2
99125 Endicott 290 .... H4
†98310 Enetai 2,638 .... A2
98822 Entiat 445 .... E3
98022 Enumclaw 5,427 .... D3
98823 Ephrata⊙ 5,359 .... F3
†98310 Erlands Point 1,254 .... A2
*98201 Everett⊙ 54,413 .... C3
98247 Everson 898 .... C2
99012 Fairfield 582 .... H3
†98901 Fairview-Sumach 2,788 .... E4
98024 Fall City 1,528 .... D3
99128 Farmington 176 .... H3
98248 Ferndale 3,855 .... C2
98424 Fife 1,823 .... C3
98466 Fircrest 5,477 .... C3
†98531 Fords Prairie 2,582 .... B4
98331 Forks 3,060 .... A3
99014 Four Lakes 500 .... H3
98250 Friday Harbor⊙ 1,200 .... B2
†98901 Fruitvale 3,967 .... E4
99130 Garfield 599 .... H3
†99362 Garrett 1,134 .... G4
98824 George 261 .... F3
98335 Gig Harbor 2,429 .... C3
98336 Glenoma 500 .... C4
98619 Glenwood 626 .... D4
98251 Gold Bar 794 .... D3
98620 Goldendale⊙ 3,575 .... E5
98337 Gorst 750 .... C3
99133 Grand Coulee 1,180 .... G3
98930 Grandview 5,615 .... F4
98932 Granger 1,812 .... E4
98252 Granite Falls 911 .... D2
98547 Grayland 750 .... A4
98621 Grays River 350 .... B4
98253 Greenbank 600 .... C2
98339 Hadlock-Irondale 1,752 .... C2
98255 Hamilton 268 .... D2
†98366 Harper 300 .... A2
98933 Harrah 343 .... E4
99134 Harrington 507 .... G3
99135 Hartline 165 .... F3
99332 Hatton 81 .... G4
98025 Hobart 500 .... D3
98548 Hoodsport 500 .... B3
98550 Hoquiam 9,719 .... A3
†98004 Hunts Point 480 .... B2
98624 Ilwaco 604 .... A4
98256 Index 147 .... D3
98342 Indianola 800 .... A1
99139 Ione 594 .... H2
98027 Issaquah 5,536 .... C3
98343 Joyce 375 .... B2
98033 Juanita 17,232 .... B1
99335 Kahlotus 203 .... G4
98625 Kalama 1,216 .... C4
98344 Kapowsin 500 .... C4
98626 Kelso⊙ 11,129 .... C4
98028 Kenmore 7,281 .... B1
99336 Kennewick 34,397 .... F4
98031 Kent 23,152 .... C3
99141 Kettle Falls 1,087 .... H2
98345 Keyport 900 .... A2
98346 Kingston 950 .... C3
98033 Kirkland 18,779 .... B2
98934 Kittitas 782 .... E4
98628 Klickitat 750 .... D5
†98832 Krupp (Marlin) 83 .... F3
98629 La Center 439 .... C5
98503 Lacey 13,940 .... C3
98257 La Conner 633 .... C2
99143 Lacrosse 373 .... H4
†98101 Lake Forest Park 2,485 .... B1
98258 Lake Stevens 1,660 .... D3
99017 Lamont 101 .... H3
98260 Langley 650 .... C2
98350 La Push 500 .... A3
99018 Latah 155 .... H3
98826 Leavenworth 1,522 .... E3
99019 Liberty Lake 1,599 .... J3
98555 Lilliwaup 75 .... B3
99341 Lind 567 .... G4
98556 Littlerock 850 .... B4
98631 Long Beach 1,199 .... A4
98351 Longbranch 640 .... C3
98632 Longview 31,052 .... B4
99148 Loon Lake 500 .... H2
98262 Lummi Island 675 .... C2
98635 Lyle 580 .... D5
98263 Lyman 285 .... D2
98264 Lynden 4,022 .... C2
98036 Lynnwood 22,641 .... C3
98935 Mabton 1,248 .... E4
99149 Malden 200 .... H3
98829 Malott 350 .... F2
98353 Manchester 400 .... A2
98830 Mansfield 315 .... F3
98266 Maple Falls 300 .... D2
98038 Maple Valley 900 .... C3
99151 Marcus 174 .... H2
98268 Marietta-Alderwood 2,324 .. C2
98832 Marlin 83 .... F3
98270 Marysville 5,080 .... C2
99344 Mattawa 299 .... F4
98557 McCleary 1,419 .... B3
99022 Medical Lake 3,600 .... H3
98039 Medina 3,220 .... B2
98040 Mercer Island (city) 21,522 .... B2
99343 Mesa 278 .... G4
99152 Metaline 190 .... H2
99153 Metaline Falls 296 .... H2
†99210 Millwood 1,717 .... H3
98354 Milton 3,162 .... C3
98355 Mineral 550 .... C4
98562 Moclips 500 .... A3
98836 Monitor 650 .... E3
98272 Monroe 2,869 .... D3
98563 Montesano⊙ 3,247 .... B4
98356 Morton 1,264 .... C4
98837 Moses Lake 10,629 .... F3
98564 Mossyrock 463 .... C4
98043 Mountlake Terrace 16,534 . B1
98273 Mount Vernon⊙ 13,009 ... C2
98936 Moxee City 687 .... E4
98275 Mukilteo 1,426 .... C3
98937 Naches 644 .... E4
98565 Napavine 611 .... C4
98638 Naselle 500 .... B4
†98310 Navy Yard City 2,594 .... A2
98357 Neah Bay 800 .... A2
99155 Nespelem 284 .... G2
†98283 Newhalem 350 .... D2
99156 Newport⊙ 1,665 .... H2
†98501 Nisqually 500 .... C3
98276 Nooksack 429 .... C2
98358 Nordland 706 .... C2
†98100 Normandy Park 4,268 .... A2
98045 North Bend 1,701 .... D3
98639 North Bonneville 394 .... C5
99157 Northport 368 .... H2
99158 Oakesdale 444 .... H3
98277 Oak Harbor 12,271 .... C2
98568 Oakville 537 .... B4
98569 Ocean City 350 .... A3
98640 Ocean Park 918 .... A4
98551 Ocean Shores 1,692 .... A3
†98520 Ocosta 369 .... B4
99159 Odessa 1,009 .... G3
98840 Okanogan⊙ 2,302 .... F2
98359 Olalla 500 .... A2
*98501 Olympia (cap.)⊙ 27,447 ... C3
Olympia‡ 124,264 .... C3
98841 Omak 4,007 .... F2
98570 Onalaska 600 .... C4
99214 Opportunity 21,241 .... H3
98662 Orchards 8,828 .... C5
98844 Oroville 1,483 .... F2
98360 Orting 1,787 .... C3
99344 Othello 4,454 .... F4
99027 Otis Orchards-East Farms 4,597 .... H3
98938 Outlook 300 .... E4
98047 Pacific 2,261 .... C3
98571 Pacific Beach 900 .... A3
98361 Packwood 800 .... D4
99161 Palouse 1,005 .... H4
98939 Parker 500 .... E4
98444 Parkland 23,355 .... C3
99301 Pasco⊙ 18,425 .... F4
98846 Pateros 555 .... E2
98572 Pe Ell 617 .... B4
98847 Peshastin 500 .... E3
98281 Point Roberts 500 .... B2
99347 Pomeroy⊙ 1,716 .... H4
98362 Port Angeles⊙ 17,311 .... B2
†98101 Port Blakely 600 .... A2
98366 Port Orchard⊙ 4,787 .... A2
98368 Port Townsend⊙ 6,067 .... C2
†98584 Potlach 100 .... B3
98370 Poulsbo 3,453 .... A1
99348 Prescott 341 .... G4
98050 Preston 500 .... D3
99350 Prosser⊙ 3,896 .... F4
99163 Pullman 23,579 .... H4
98371 Puyallup 18,251 .... C3
98376 Quilcene 900 .... B3
98575 Quinault 450 .... B3
98848 Quincy 3,525 .... F3
98576 Rainier 891 .... C4

(continued on following page)

Agriculture, Industry and Resources

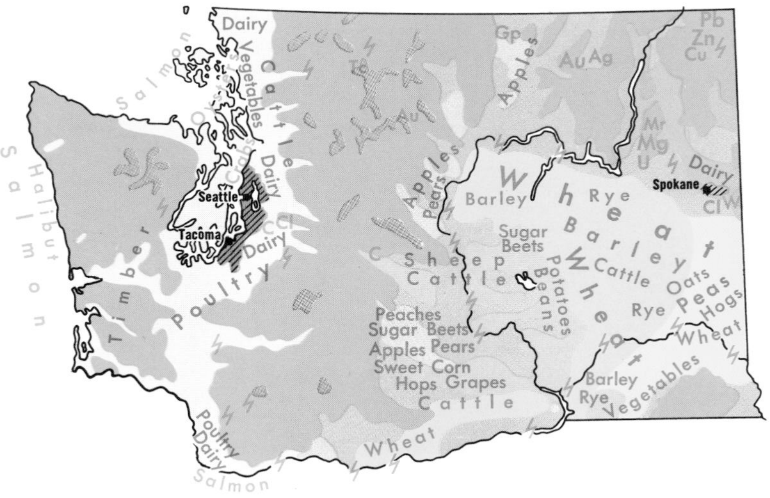

DOMINANT LAND USE

- Specialized Wheat
- Wheat, Peas
- Dairy, Poultry, Mixed Farming
- Fruit and Mixed Farming
- General Farming, Dairy, Range Livestock
- General Farming, Livestock, Special Crops
- Range Livestock
- Forests
- Urban Areas
- Nonagricultural Land

MAJOR MINERAL OCCURRENCES

Ag Silver
Au Gold
C Coal
Cl Clay
Cu Copper
Gp Gypsum
Mg Magnesium
Mr Marble
Pb Lead
Tc Talc
U Uranium
W Tungsten
Zn Zinc

Water Power
Major Industrial Areas

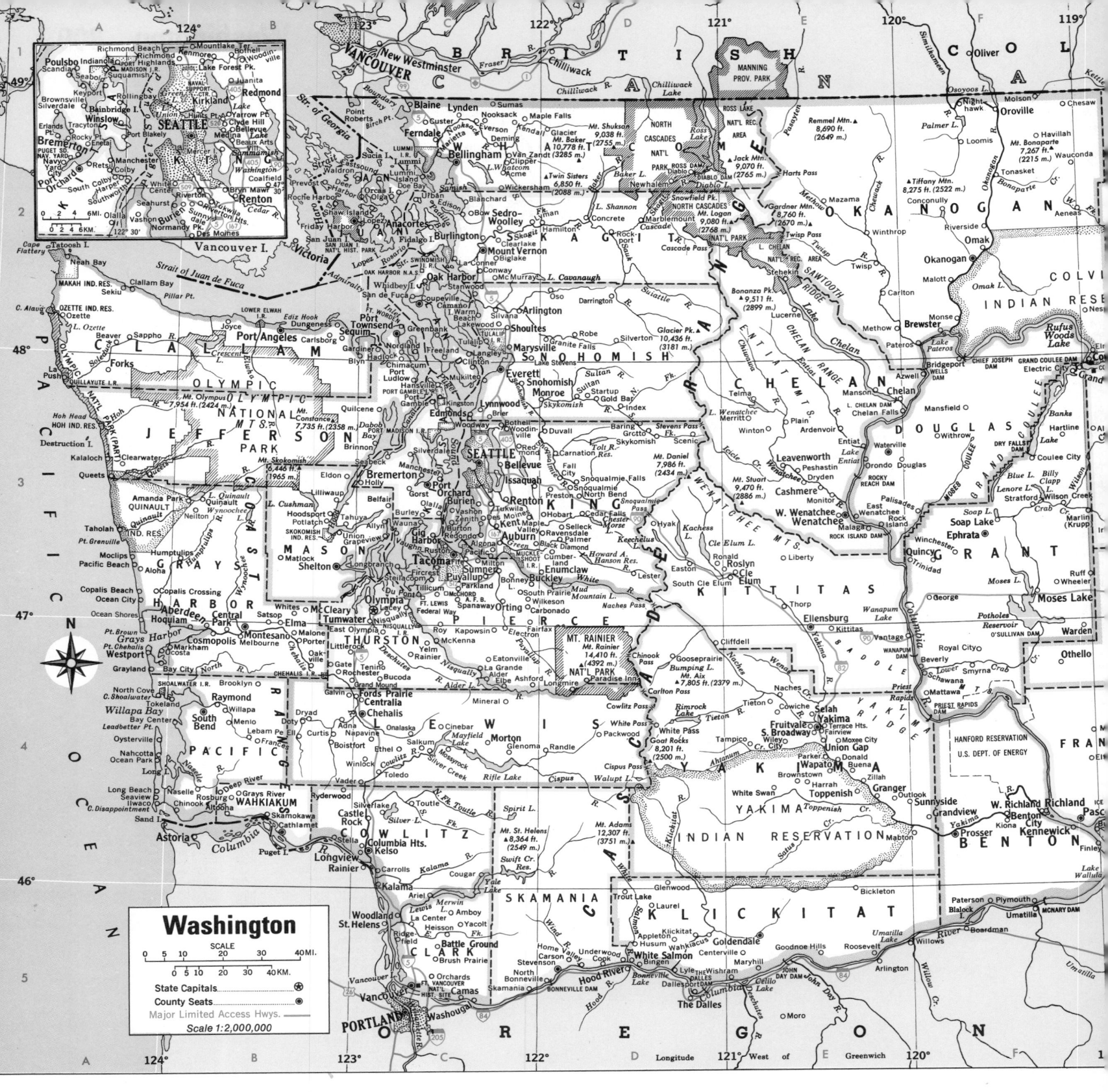

Washington

SCALE
0 5 10 20 30 40 MI.
0 5 10 20 30 40 KM.

⊛ State Capitals
⊙ County Seats
— Major Limited Access Hwys.
Scale 1:2,000,000

98377 Randle 950 ....D4
98051 Ravensdale 400 ....D3
98577 Raymond 2,991 ....B4
99029 Reardan 498 ....H3
98052 Redmond 23,318 ....B1
98054 Redondo 950 ....C3
98055 Renton 30,612 ....B2
99166 Republic⊙ 1,018 ....G2
98378 Retsil 1,524 ....A2
99352 Richland 33,578 ....F4
   Richland-Kennewick‡
   144,469 ....F4
98160 Richmond Beach-Innis
   Arden 6,700 ....A1
†98133 Richmond Highlands
   24,463 ....A1
98642 Ridgefield 1,062 ....C5
99169 Ritzville⊙ 1,800 ....H3
†98188 Riverton 14,182 ....B3
98579 Rochester 325 ....B3
99030 Rockford 442 ....H3
†98801 Rock Island 491 ....E3
†98626 Rocky Point 1,495 ....A2

98061 Rollingbay 950 ....A2
99170 Rosalia 572 ....H3
98643 Rosburg 419 ....B4
98941 Roslyn 938 ....E3
98580 Roy 417 ....C4
99357 Royal City 676 ....F4
†98401 Ruston 612 ....C3
98581 Ryderwood 367 ....B4
99171 Saint John 529 ....H3
98582 Salmon 390 ....C4
98583 Satsop 300 ....B3
*98101 Seattle⊙ 493,846 ....A2
   Seattle-Everett‡ 1,606,765..A2
98644 Seaview 500 ....A4
98284 Sedro-Woolley 6,110 ......C2
98381 Sekiu 328 ....A2
98942 Selah 4,500 ....E4
98382 Sequim 3,013 ....B2
98584 Shelton⊙ 7,629 ....B3
98383 Silverdale 950 ....B2
99347 Skamokawa 450 ....B4
98288 Skykomish 209 ....D3
98290 Snohomish 5,294 ....D3

†98065 Snoqualmie 1,370 ....D3
98851 Soap Lake 1,196 ....F3
98586 South Bend⊙ 1,686 ......B4
98901 South Broadway 3,500 .....C3
98943 South Cle Elum 449 ....D3
98384 South Colby 500 ....A2
98385 South Prairie 202 ....D3
98387 Spanaway 8,868 ....C3
99031 Spangle 276 ....H3
*98201 Spokane⊙ 171,300 ....H3
   Spokane‡ 341,835 ....H3
99032 Sprague 473 ....H3
99173 Springdale 281 ....H2
98292 Stanwood 1,646 ....C2
99293 Starbuck 198 ....G4
98852 Startup 350 ....D3
98283 Stehekin 98 ....E2
98388 Steilacoom 4,886 ....C3
98648 Stevenson⊙ 1,172 ....C5
98294 Sultan 1,578 ....D3
98295 Sumas 712 ....C2
98390 Sumner 4,936 ....C3
98944 Sunnyside 9,225 ....F4

98392 Suquamish 1,498 ....A1
*98401 Tacoma⊙ 158,501 ....C3
   Tacoma‡ 485,643 ....C3
98587 Taholah 550 ....A3
99033 Tekoa 854 ....H3
98589 Tenino 1,280 ....C4
98901 Terrace Heights 3,199 ....E4
98946 Thorp 500 ....E3
98947 Tieton 528 ....E4
98590 Tokeland 500 ....A4
98855 Tonasket 985 ....F2
98591 Toledo 637 ....C4
98948 Toppenish 6,517 ....E4
99371 Touchet 485 ....G4
99360 Toutle 825 ....C4
98649 Toutle 825 ....C4
†99218 Town and Country 5,578..H3
98393 Tracyton 2,304 ....A1
98650 Trout Lake 900 ....D5
98188 Tukwila 3,578 ....B3
98501 Tumwater 6,705 ....B3
98856 Twisp 911 ....E2
98651 Underwood 640 ....D5
98592 Union 380 ....B3

98903 Union Gap 3,184 ....E4
99179 Uniontown 286 ....H4
98593 Vader 406 ....B4
*98660 Vancouver⊙ 42,834 .....C5
98950 Vantage 300 ....E3
98070 Vashon 350 ....A2
98394 Vaughn 600 ....C3
99037 Veradale 7,256 ....H3
99361 Waitsburg 1,035 ....G4
99362 Walla Walla⊙ 25,618 ...G4
98951 Wapato 3,307 ....E4
98857 Warden 1,479 ....F3
99371 Washtucna 266 ....G4
98671 Washougal 3,834 ....C5
98395 Wawawai 300 ....H4
99039 Waverly 99 ....H3
98801 Wenatchee⊙ 17,257 .....E3
98595 Westport 1,954 ....A4
†98801 West Richland 2,938 ....F4
98801 West Wenatchee 2,187 ....E3
98146 White Center-Shorewood
   19,362.................A2

98672 White Salmon 1,853 ......D5
99185 Wilbur 1,122 ....G3
99396 Wilkeson 321 ....D3
†98577 Willapa 300 ....B4
98860 Wilson Creek 222 ....F3
98596 Winlock 1,052 ....C4
†98110 Winslow (Bainbridge
   Island-Winslow) 2,196 ...A2
98862 Winthrop 413 ....E2
98673 Wishram 575 ....D5
98674 Woodland 2,341 ....C5
98020 Woodway 832 ....A2
98675 Yacolt 544 ....C5
*98901 Yakima⊙ 49,826 ....E4
   Yakima‡ 172,508 ....E4
99004 Yarrow Point 1,064 ......B2
†98104 Yelm 1,294 ....C4
†98101 Zenith-Saltwater 8,982 .C3
98953 Zillah 1,599 ....E4

OTHER FEATURES

Abercrombie (mt.) E1

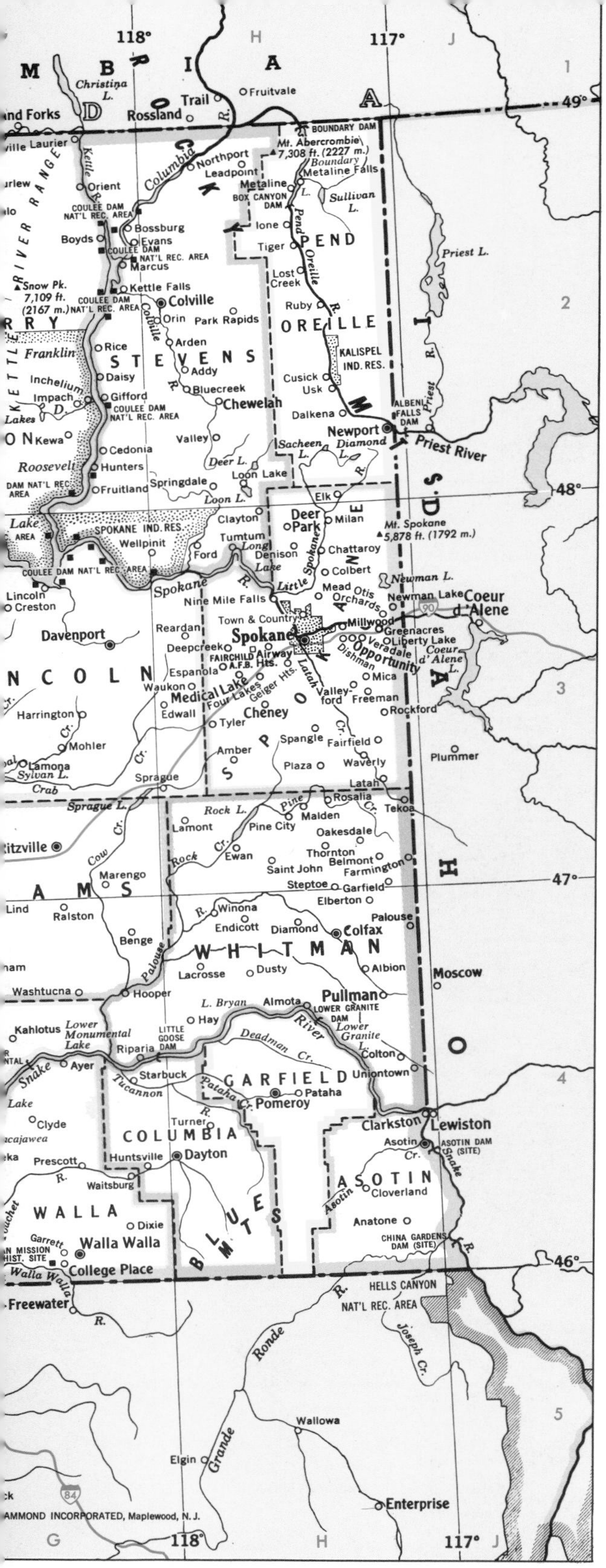
STEVENS
PEND OREILLE
SPOKANE
LINCOLN
ADAMS
WHITMAN
GARFIELD
COLUMBIA
ASOTIN
WALLA WALLA
Colville
Spokane
Cheney
Pullman
Colfax
Pomeroy
Dayton
Walla Walla
College Place
Clarkston
Lewiston
Moscow
Coeur d'Alene
Newport
Priest River
Davenport
Ritzville
Mt. Spokane 5,878 ft. (1792 m.)
Mt. Abercrombie 7,308 ft. (2227 m.)
Snow Pk. 7,109 ft. (2167 m.)
49°
48°
47°
46°
118°
117°
HAMMOND INCORPORATED, Maplewood, N.J.

Topography

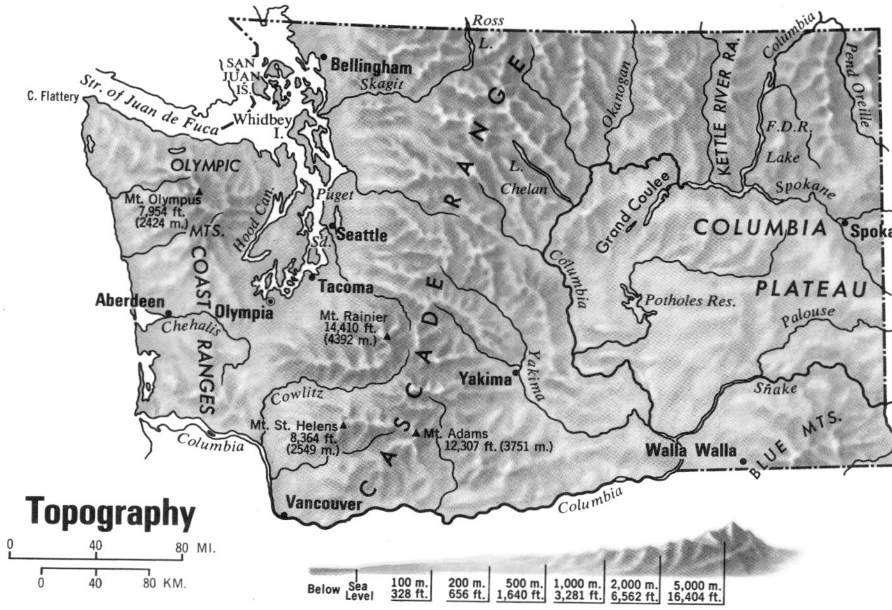
Bellingham
Seattle
Tacoma
Olympia
Aberdeen
Vancouver
Yakima
Walla Walla
Spokane
C. Flattery
Str. of Juan de Fuca
OLYMPIC MTS.
COAST RANGES
CASCADE RANGE
COLUMBIA PLATEAU
BLUE MTS.
KETTLE RIVER RA.
Mt. Olympus 7,954 ft. (2424 m.)
Mt. Rainier 14,410 ft. (4392 m.)
Mt. St. Helens 8,364 ft. (2549 m.)
Mt. Adams 12,307 ft. (3751 m.)
Potholes Res.
Grand Coulee
F.D.R. Lake
0 40 80 MI.
0 40 80 KM.
Below Sea Level
100 m. 328 ft.
200 m. 656 ft.
500 m. 1,640 ft.
1,000 m. 3,281 ft.
2,000 m. 6,562 ft.
5,000 m. 16,404 ft.

Adams (mt.) D4
Admiralty (inlet) B2
Ahtanum (creek) D4
Aix (mt.) D4
Alava (cape) A2
Alder (lake) C4
Asotin (creek) H4
Asotin (dam) J4
Bainbridge (isl.) A2
Baker (lake) D2
Baker (mt.) D2
Baker (riv.) D2
Banks (lake) F3
Birch (pt.) C2
Blalock (isl.) F5
Blue (lake) F3
Blue (mts.) H4
Bonanza (peak) E2
Bonaparte (creek) F2
Bonaparte (mt.) F2
Bonneville (dam) D5
Bonneville (lake) D5
Boundary (bay) C1
Boundary (dam) H2
Boundary (lake) H2
Box Canyon (dam) H2
Brown (pt.) A4
Bumping (lake) D4
Camano (isl.) C2
Carlton (pass) D4
Cascade (pass) D2
Cascade (range) D4
Cascade (riv.) D2
Cavanaugh (lake) D2
Cedar (riv.) B2
Celilo (lake) E5
Chehalis (pt.) A4
Chehalis (riv.) B4
Chehalis Ind. Res. B4
Chelan (lake) E2
Chelan (range) E2
Chester Morse (lake) D3
Chewack (riv.) E2
Chief Joseph (dam) F3
China Gardens (dam) J4
Chinook (pass) D4
Chiwawa (riv.) E2
Cispus (pass) D4
Cispus (riv.) D4
Cle Elum (lake) E3
Coal (creek) G3
Coast (ranges) B3
Columbia (riv.) B4
Colville (riv.) H2
Colville Ind. Res. G2
Constance (mt.) B3
Coulee Dam Nat'l Rec. Area G2
Cow (creek) G3
Cowlitz (pass) D4
Cowlitz (riv.) C4
Crab (creek) F3
Crescent (lake) B2
Curlew (lake) G2
Cushman (lake) B3
Dabob (bay) C3
Dalles, The (dam) D5
Daniel (mt.) D3
Deadman (creek) H4
Deer (lake) H2
Deschutes (riv.) C4
Destruction (isl.) A3
Diablo (lake) D2
Diamond (lake) H2
Disappointment (cape) A4
Dry Falls (dam) F3
Ediz Hook (pen.) B2
Elwha (riv.) B3
Entiat (lake) E3
Entiat (mts.) E2
Entiat (riv.) E3
Fairchild A.F.B. 5,353 H3
Fidalgo (isl.) C2
Flattery (cape) A2
Fort Lewis 23,761 C3
Fort Vancouver Nat'l Hist. Site C5
Fort Worden C2
Franklin D. Roosevelt (lake) G2
Gardner (mt.) E2
Georgia (str.) B2
Glacier (peak) D2
Goat Rocks (mt.) D4
Grand Coulee (canyon) F3
Grand Coulee (dam) F3
Grande Ronde (riv.) H5
Grays (harb.) A4
Green (lake) A2
Green (riv.) C3
Grenville (pt.) A3
Hanford Reservation-U.S. Dept. of Energy F4
Haro (str.) B2
Harts (pass) E2
Hells Canyon Nat'l Rec. Area H5
Hoh (head) A3
Hoh (riv.) A3
Hoh Ind. Res. A3
Hood (canal) B3
Howard A. Hanson (res.) D3
Humptulips (riv.) B3
Ice Harbor (dam) G4
Icicle (creek) E3
Jack (mt.) E2
John Day (dam) E5
Juan de Fuca (str.) A2
Kachess (lake) D3
Kalama (riv.) C4
Kalispel Ind. Res. H2
Keechelus (lake) D3
Kettle (riv.) G2
Kettle River (range) G2
Klickitat (riv.) D4
Lake (creek) G3
Lake Chelan Nat'l Rec. Area E2
Latah (creek) H3
Leadbetter (pt.) A4
Lenore (lake) F3
Lewis (riv.) C5
Little Goose (dam) G4
Little Spokane (riv.) H3
Logan (mt.) E2
Long (isl.) A4
Long (lake) H3
Loon (lake) H2
Lopez (isl.) C2
Lower Crab (creek) F4
Lower Elwah Ind. Res. B2
Lower Granite (lake) H4
Lower Monumental (lake) G4
Lummi (isl.) C2
Lummi Ind. Res. C2
Makah Ind. Res. A2
Mayfield (lake) C4
McChord A.F.B. 5,746 C3
McNary (dam) F5
Merwin (lake) C5
Methow (riv.) E2
Moses (lake) F3
Moses Coulee (canyon) F3
Mount Rainier Nat'l Park D4
Muckleshoot Ind. Res. C3
Mud Mountain (lake) D3
Naches (pass) D3
Naches (riv.) E4
Naselle (riv.) B4
Naval Support Ctr. B1
Newman (lake) H3
Nisqually (riv.) C4
Nisqually Ind. Res. C4
Nooksack (riv.) C2
North (riv.) B4
North Cascades Nat'l Park D2
Oak Harbor Naval Air Sta. C2
Okanogan (riv.) F2
Olympic (mts.) B3
Olympic Nat'l Park B3
Olympus (mt.) B3
Omak (lake) F2
Orcas (isl.) C2
Osoyoos (lake) F1
O'Sullivan (dam) F4
Ozette (lake) A2
Ozette Ind. Res. A2
Padilla (bay) C2
Palmer (lake) F2
Palouse (riv.) G4
Pasayten (riv.) E2
Pataha (creek) H4
Pateros (lake) F2
Pend Oreille (riv.) H2
Pillar (pt.) A2
Pine (creek) H3
Port Angeles Ind. Res. B2
Port Gamble Ind. Res. C3
Port Madison Ind. Res. A1
Potholes (res.) F3
Priest Rapids (lake) E4
Puget (isl.) B4
Puget (sound) C3
Puget Sound Navy Yard A2
Puyallup (riv.) C4
Queets (riv.) A3
Quillayute Ind. Res. A3
Quinault (lake) B3
Quinault (riv.) A3
Quinault Ind. Res. A3
Rainier (mt.) D4
Remmel (mt.) E2
Rifle (lake) C4
Rimrock (lake) D4
Rock (creek) H3
Rock (lake) H3
Rock Island (dam) E3
Rocky (mts.) H2
Rocky Reach (dam) E3
Rosario (str.) C2
Ross (dam) D2
Ross (lake) D2
Ross Lake Nat'l Rec. Area E2
Rufus Woods (lake) F2
Sacajawea (lake) G4
Sacheen (lake) H2
Saddle (mts.) E4
Saint Helens (mt.) C4
Samish (lake) C2
Sammamish (lake) B2
Sand (isl.) A4
San Juan (isl.) B2
San Juan Island Nat'l Hist. Park B2
Sanpoil (riv.) G2
Satus (creek) E4
Sauk (riv.) D2
Sawtooth (ridge) E2
Shannon (lake) D2
Shoalwater (cape) A4
Shoalwater Ind. Res. B4
Shuksan (mt.) D2
Silver (lake) C4
Similkameen (riv.) F1
Skagit (riv.) C2
Skokomish (mt.) B3
Skokomish Ind. Res. B3
Skykomish (riv.) D3
Snake (riv.) G4
Snohomish (riv.) C3
Snoqualmie (pass) D3
Snoqualmie (riv.) D3
Snow (peak) G2
Snowfield (peak) D2
Soap (lake) F3
Soleduck (riv.) A3
Spirit (lake) C4
Spokane (mt.) H3
Spokane (riv.) H3
Spokane Ind. Res. G3
Sprague (lake) G3
Stevens (pass) D3
Stuart (mt.) E3
Sucia (isl.) C2
Suiattle (riv.) D2
Sullivan (lake) H2
Sultan (riv.) D3
Swift Creek (res.) C4
Swinomish Ind. Res. C2
Sylvan (lake) G3
Tatoosh (isl.) A2
The Dalles (dam) D5
Tieton (riv.) D4
Tiffany (mt.) F2
Tolt River (res.) D3
Toppenish (creek) E4
Touchet (riv.) G4
Toutle, North Fork (riv.) C4
Toutle, South Fork (riv.) C4
Tucannon (riv.) G4
Tulalip Ind. Res. C2
Tule (lake) G3
Twin (lkes) G2
Twin Sisters (mt.) D2
Twisp (pass) E2
Twisp (riv.) E2
Umatilla (lake) E5
Union (lake) B2
Vancouver (lake) C5
Walla Walla (riv.) G4
Wallula (lake) F4
Walupt (lake) D4
Wanapum (lake) E3
Washington (lake) B2
Wells (dam) F3
Wenas (creek) E4
Wenatchee (lake) E3
Wenatchee (mts.) E3
Wenatchee (riv.) E3
Whatcom (lake) C2
Whidbey (isl.) C2
White (pass) D4
White (riv.) D3
White Salmon (riv.) D4
Whitman Mission Nat'l Hist. Site G4
Willapa (bay) A4
Wilson (creek) F3
Wind (riv.) D5
Wynoochee (lake) B3
Wynoochee (riv.) B3
Yakima (ridge) E4
Yakima (riv.) F4
Yakima Ind. Res. E4

⊙County seat.
‡Population of metropolitan area.
† Zip of nearest p.o. * Multiple zips.

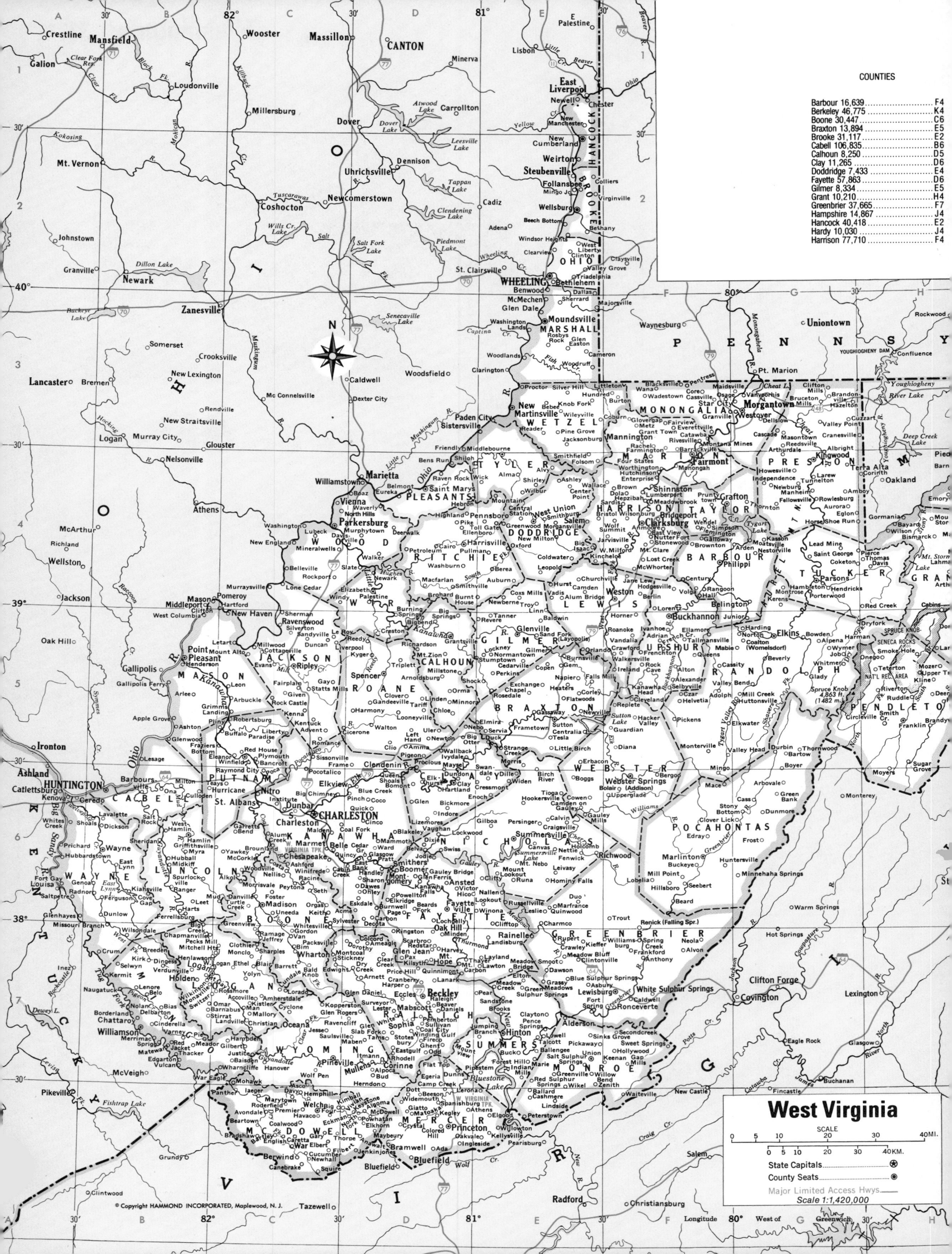

Jackson 25,794 ..... C5
Jefferson 30,302 ..... L4
Kanawha 231,414 ..... C6
Lewis 18,813 ..... E4
Lincoln 23,675 ..... B6
Logan 50,679 ..... C7
Marion 65,789 ..... F4
Marshall 41,608 ..... E3
Mason 27,045 ..... B5
McDowell 49,899 ..... C8
Mercer 73,942 ..... D8
Mineral 27,234 ..... J4
Mingo 37,336 ..... B7
Monongalia 75,024 ..... F3
Monroe 12,873 ..... E7
Morgan 10,711 ..... K3
Nicholas 28,126 ..... E6
Ohio 61,389 ..... E2
Pendleton 7,910 ..... H5
Pleasants 8,236 ..... D4
Pocahontas 9,919 ..... F6
Preston 30,460 ..... G4
Putnam 38,181 ..... C6
Raleigh 86,821 ..... D7
Randolph 28,734 ..... G5
Ritchie 11,442 ..... D4
Roane 15,952 ..... D5
Summers 15,875 ..... E7
Taylor 16,584 ..... F4
Tucker 8,675 ..... F4
Tyler 11,320 ..... E4
Upshur 23,427 ..... F5
Wayne 46,021 ..... B6
Webster 12,245 ..... F6
Wetzel 21,874 ..... E3
Wirt 4,922 ..... D4
Wood 93,648 ..... D4
Wyoming 35,993 ..... C7

CITIES and TOWNS

Zip *Name/Pop.* *Key*

25606 Accoville 975 ..... C7
†26288 Addison (Webster Springs)⊙ 939 ..... F6
26210 Adrian 510 ..... F5
26519 Albright 357 ..... G3
24910 Alderson 1,375 ..... E7
24807 Algoma 200 ..... D8
25501 Alkol 500 ..... C6
26320 Alma 197 ..... E4
24710 Alpoca 200 ..... D7
26321 Alum Bridge 150 ..... E4
25003 Alum Creek 900 ..... C6
26322 Alvy 150 ..... E4
25004 Ameagle 230 ..... D7
25607 Amherstdale 1,075 ..... C7
25005 Amma 200 ..... D5
24808 Anawalt 652 ..... D8

AREA 24,231 sq. mi. (62,758 sq. km.)
POPULATION 1,950,279
CAPITAL Charleston
LARGEST CITY Charleston
HIGHEST POINT Spruce Knob 4,863 ft. (1482 m.)
SETTLED IN 1774
ADMITTED TO UNION June 20, 1863
POPULAR NAME Mountain State
STATE FLOWER Big Rhododendron
STATE BIRD Cardinal

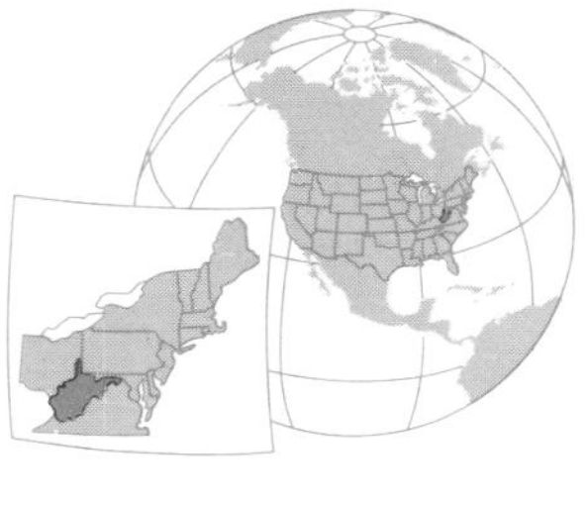

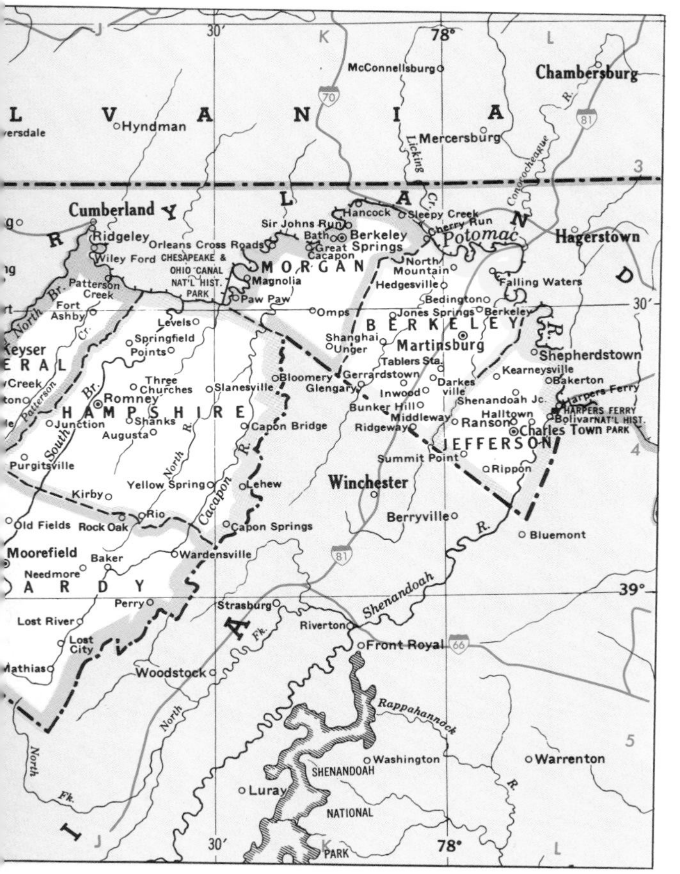

Chambersburg
McConnellsburg
Hyndman
Mercersburg
Cumberland
Hagerstown
Berkeley Springs
Great Cacapon
MORGAN
Hedgesville
Falling Waters
Paw Paw
Martinsburg
BERKELEY
Shepherdstown
Romney
HAMPSHIRE
Capon Bridge
Harpers Ferry
Charles Town
JEFFERSON
Winchester
Berryville
Bluemont
Moorefield
Wardensville
Capon Springs
Strasburg
Front Royal
Woodstock
Washington
Warrenton
Luray
SHENANDOAH NATIONAL PARK
Shenandoah
Rappahannock
Potomac

Topography

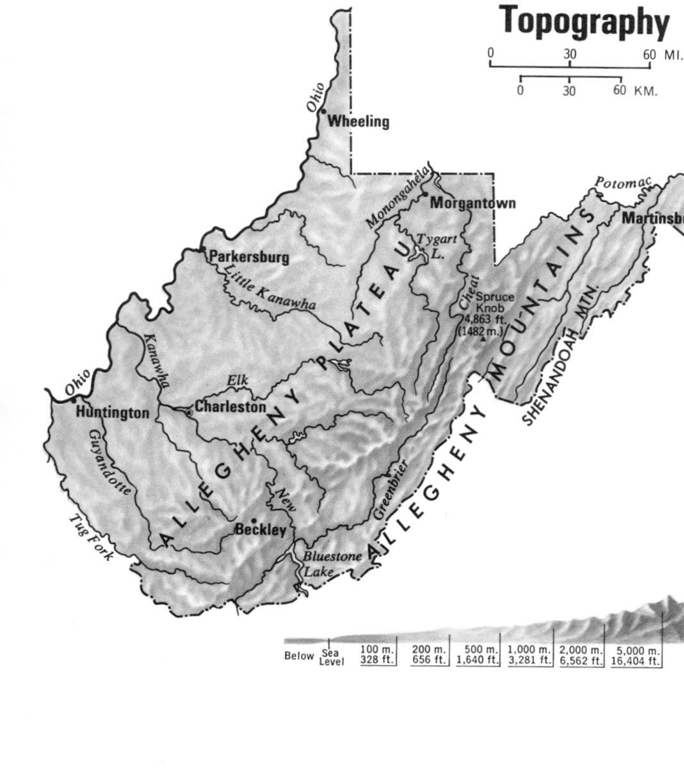

0 30 60 MI.
0 30 60 KM.
Ohio
Wheeling
Monongahela
Morgantown
Potomac
Martinsburg
Tygart L.
Parkersburg
Little Kanawha
Cheat
Spruce Knob 4,863 ft. (1482 m.)
ALLEGHENY PLATEAU
ALLEGHENY MOUNTAINS
SHENANDOAH MTN.
Kanawha
Elk
Charleston
Huntington
Guyandotte
New
Greenbrier
Beckley
Tug Fork
Bluestone Lake
ALLEGHENY
Below Sea Level
100 m. 328 ft.
200 m. 656 ft.
500 m. 1,640 ft.
1,000 m. 3,281 ft.
2,000 m. 6,562 ft.
5,000 m. 16,404 ft.

26323 Anmoore 865 ..... F4
25812 Ansted 1,952 ..... D6
25502 Apple Grove 900 ..... B5
24915 Arbovale 610 ..... G6
26816 Arthur 350 ..... H4
26520 Arthurdale 1,063 ..... G3
24916 Asbury 280 ..... E7
24809 Asco 175 ..... C8
25009 Ashford 400 ..... C6
25503 Ashton 259 ..... B5
24712 Athens 1,147 ..... E8
26325 Auburn 116 ..... E4
26704 Augusta 750 ..... J4
26705 Aurora 250 ..... G4
24811 Avondale 250 ..... C8
25608 Baisden 500 ..... C7
26801 Baker 200 ..... J4
25410 Bakerton 125 ..... L4
25010 Bald Knob 356 ..... C7
26326 Baldwin 92 ..... E5
25011 Bancroft 528 ..... C5
25504 Barboursville 2,871 ..... B6
25609 Barnabus 750 ..... C7
26559 Barrackville 1,815 ..... F3
25013 Barrett 950 ..... C7
24813 Bartley 900 ..... C8
24920 Bartow 500 ..... G5
†25411 Bath (Berkeley Springs) 789 ..... K3
26707 Bayard 540 ..... H4
25014 Beards Fork 400 ..... D6
25813 Beaver (Glen Hedrick) 1,122 ..... D7
25801 Beckley⊙ 20,492 ..... D7
26030 Beech Bottom 507 ..... E2
24714 Beeson 300 ..... D8
26250 Belington 2,038 ..... F4
25015 Belle 1,621 ..... C6
26133 Belleville 105 ..... C4
26134 Belmont 887 ..... D4
26656 Belva 275 ..... D6
26135 Bens Run 85 ..... D4
26031 Benwood 1,994 ..... E2
26298 Bergoo 220 ..... F6
25411 Berkeley Springs (Bath)⊙ 789 ..... K3
24815 Berwind 615 ..... C8
26032 Bethany 1,336 ..... E2
†26003 Bethlehem 3,045 ..... E2
26253 Beverly 475 ..... G5
25019 Bickmore 300 ..... D6
26136 Bigbend 120 ..... D5
25302 Big Chimney 450 ..... C6
25505 Big Creek 500 ..... B7
26137 Big Springs 485 ..... D5
25021 Bim 500 ..... C7
26610 Birch River 650 ..... E6
26521 Blacksville 248 ..... F3
25022 Blair 800 ..... C7
26817 Bloomery 200 ..... K4
25026 Blue Creek 650 ..... D6
24701 Bluefield 16,060 ..... D8
26288 Bolair 450 ..... F6
†25425 Bolivar 672 ..... L4
25030 Bomont 170 ..... D6
25031 Boomer 1,051 ..... D6
24817 Bradshaw 1,002 ..... C8
24715 Bramwell 989 ..... D8
26523 Brandonville 92 ..... G3
26802 Brandywine 300 ..... H5
25666 Breeden 600 ..... B7
26330 Bridgeport 6,604 ..... F4
26138 Brohard 80 ..... D4
25957 Brooks 196 ..... E7
26334 Brownton 400 ..... F4
26525 Bruceton Mills 296 ..... G3
24924 Buckeye 125 ..... F6
26201 Buckhannon⊙ 6,820 ..... F5
24716 Bud 400 ..... D7
25033 Buffalo 1,034 ..... C5
25413 Bunker Hill 600 ..... K4
26710 Burlington 300 ..... J4
26335 Burnsville 531 ..... E5
26336 Burnt House 175 ..... D4
26562 Burton 200 ..... F3
25035 Cabin Creek 900 ..... C6
26337 Cairo 428 ..... D4
24925 Caldwell 795 ..... F7
26660 Calvin 400 ..... E6
26208 Camden on Gauley 236 ..... E6
26033 Cameron 1,474 ..... E3
24819 Canebrake 300 ..... C8
26662 Canvas 300 ..... E6
26711 Capon Bridge 191 ..... K4
26823 Capon Springs 580 ..... K4
25037 Carbon 300 ..... D6
24821 Caretta 650 ..... C8
24927 Cass 148 ..... G6
26527 Cassville 800 ..... F3
25039 Cedar Grove 1,479 ..... D6
26339 Center Point 250 ..... E4
26612 Centralia 100 ..... E5
26340 Central Station 200 ..... E4
26214 Century 250 ..... F4
25507 Ceredo 2,255 ..... B6
25508 Chapmanville 1,164 ..... B7
*25301 Charleston (cap.)⊙ 63,968 ..... C6
Charleston‡ 269,595 ..... C6
25414 Charles Town⊙ 2,857 ..... L4
25958 Charmco 800 ..... E6
25667 Chattaroy 1,383 ..... B7
25418 Cherry Run 120 ..... L3
†25301 Chesapeake 2,364 ..... C6
26034 Chester 3,297 ..... E1
26301 Clarksburg⊙ 22,371 ..... F4
25043 Clay⊙ 940 ..... D6
25044 Clear Creek 300 ..... D7
†26003 Clearview 740 ..... E2
25045 Clendenin 1,373 ..... D5
26215 Cleveland 74 ..... F5
25822 Clifftop 100 ..... E6
25237 Clifton 325 ..... B5
24928 Clintonville 250 ..... E7
25046 Clio 300 ..... D5
25047 Clothier 900 ..... C7
25823 Coal City 2,324 ..... D7
25306 Coal Fork 2,775 ..... D6
26257 Coalton 306 ..... G5
24824 Coalwood 650 ..... C8
25048 Colcord 600 ..... D7
26035 Colliers 864 ..... E2
26615 Copen 50 ..... E5
25826 Corinne 900 ..... D7
25051 Costa 250 ..... C6
25239 Cottageville 300 ..... C5
25509 Cove Gap 650 ..... B6
26206 Cowen 723 ..... E6
26342 Coxs Mills 275 ..... E4
26205 Craigsville 1,562 ..... E6
25828 Cranberry 315 ..... D7
24931 Crawley 395 ..... E7
25669 Crum 500 ..... B7
24826 Cucumber 274 ..... C8
25510 Culloden 2,931 ..... B6
24827 Cyclone 500 ..... C7
26036 Dallas 450 ..... E2
25832 Daniels 1,959 ..... D7
25053 Danville 727 ..... C6
†25428 Darkesville 150 ..... L4
26260 Davis 979 ..... H4
24828 Davy 882 ..... C8
25054 Dawes 800 ..... D6
24932 Dawson 300 ..... E7
25670 Delbarton 981 ..... B7
26531 Dellslow 300 ..... G3
26217 Diana 300 ..... F5
26617 Dille 300 ..... E6
25671 Dingess 600 ..... B7
25059 Dixie 985 ..... D6
25060 Dorothy 400 ..... D7
24721 Dott 100 ..... D8
25062 Dry Creek 441 ..... D7
26263 Dryfork 425 ..... H5
25063 Duck 500 ..... E5
25064 Dunbar 9,285 ..... C6
24934 Dunmore 280 ..... G6
26264 Durbin 379 ..... G5
25067 East Bank 1,155 ..... D6
25835 Eastgulf 300 ..... D7
25512 East Lynn 150 ..... B6
†26301 East View 1,222 ..... F4
25836 Eccles 1,162 ..... D7
24829 Eckman 750 ..... C8
25672 Edgarton 415 ..... B7
26716 Eglon 70 ..... G4
24830 Elbert 400 ..... C8
25070 Eleanor 1,282 ..... C5
26143 Elizabeth⊙ 856 ..... D4
26717 Elk Garden 291 ..... H4
26241 Elkins⊙ 8,536 ..... G5
25071 Elkview 1,161 ..... C6
26267 Ellamore 250 ..... F5
26346 Ellenboro 357 ..... D4
25965 Elton 200 ..... E7
24832 English 500 ..... C8
26568 Enterprise 1,110 ..... F4
25075 Eskdale 400 ..... D6
25076 Ethel 450 ..... C7
26144 Eureka 125 ..... D4
25241 Evans 400 ..... C5
26533 Everettville 175 ..... F3
26554 Fairmont 23,863 ..... F4
26570 Fairview 759 ..... F3
†24966 Falling Spring (Renick) 240 ..... F6
26571 Farmington 583 ..... F3
25840 Fayetteville⊙ 2,366 ..... D6
26202 Fenwick 500 ..... E6
24835 Filbert 130 ..... D8
26818 Fisher 500 ..... H4
25841 Flat Top 550 ..... D7
26621 Flatwoods 405 ..... E5
26347 Flemington 452 ..... F4
26037 Follansbee 3,994 ..... E2
26348 Folsom 360 ..... E4
24935 Forest Hill 314 ..... E7
26719 Fort Ashby 1,205 ..... J4
25514 Fort Gay 886 ..... A6
26806 Fort Seybert 200 ..... H5
24936 Fort Spring 250 ..... E7
25081 Foster 500 ..... C6
26572 Four States 500 ..... F4
25071 Frame 76 ..... C5
26623 Frametown 150 ..... E5
26807 Franklin⊙ 780 ..... H5
25082 Fraziers Bottom 250 ..... B5
26219 Frenchton 102 ..... F5
26146 Friendly 242 ..... D3
25515 Gallipolis Ferry 325 ..... B5
26349 Galloway 500 ..... F4
25243 Gandeeville 150 ..... D5
24941 Gap Mills 300 ..... F7
24836 Gary 2,233 ..... C8
26624 Gassaway 1,225 ..... E5
25085 Gauley Bridge 1,177 ..... D6
26240 Gauley Mills 165 ..... E6
25244 Gay 300 ..... C5
25420 Gerrardstown 240 ..... K4
25843 Ghent 500 ..... D7
25621 Gilbert 757 ..... C7
26671 Gilboa 500 ..... E6
26350 Gilmer 110 ..... E5
26268 Glady 175 ..... G5
25086 Glasgow 1,031 ..... D6
25088 Glen 175 ..... D6
26038 Glen Dale 1,875 ..... E3
26039 Glen Easton 100 ..... E3
25090 Glen Ferris 200 ..... D6
25421 Glengary 250 ..... K4
†25813 Glen Hedrick (Beaver) 1,122 ..... D7
25846 Glen Jean D7
25848 Glen Rogers 500 ..... D7
26351 Glenville⊙ 2,155 ..... E5
25849 Glen White 300 ..... D7
25520 Glenwood 400 ..... B5
†26585 Glovergap 100 ..... F3
25093 Gordon 300 ..... C7
26720 Gormania 100 ..... H4
26354 Grafton⊙ 6,845 ..... G4
26147 Grantsville⊙ 788 ..... D5
26574 Grant Town 987 ..... F3
26534 Granville 992 ..... F3
24943 Grassy Meadows 100 ..... E7
25422 Great Cacapon 750 ..... K3
24944 Green Bank 115 ..... G6
25966 Green Sulphur Springs 225 ..... E7
24945 Greenville 125 ..... E7
26360 Greenwood 750 ..... E4
25095 Grimms Landing 350 ..... B5
26221 Guardian 175 ..... F5
26222 Hacker Valley 440 ..... F5
25423 Halltown 375 ..... L4
26269 Hambleton 403 ..... G4
25523 Hamlin⊙ 1,219 ..... B6
25623 Hampden 300 ..... C7
25424 Hancock 175 ..... K3
25102 Handley 633 ..... D6
†26250 Harding 100 ..... G5
26270 Harman 181 ..... G5
25246 Harmony 600 ..... D5
25851 Harper 400 ..... D7
25425 Harpers Ferry 361 ..... L4
26362 Harrisville⊙ 1,673 ..... E4
25247 Hartford 556 ..... C4
25524 Harts 400 ..... B6
25852 Harvey 300 ..... D7
24841 Havaco 350 ..... C8
26627 Heaters 440 ..... E5
25427 Hedgesville 217 ..... K3
26224 Helvetia 130 ..... F5
24842 Hemphill 700 ..... C8
25106 Henderson 604 ..... B5
26271 Hendricks 390 ..... G4
25624 Henlawson 900 ..... B7
26369 Hepzibah 600 ..... F4
24726 Herndon 500 ..... D7
25854 Hico 750 ..... D6
24946 Hillsboro 276 ..... F6
25951 Hinton⊙ 4,622 ..... E7
25625 Holden 2,036 ..... B7
26372 Horner 125 ..... F5
26769 Horse Shoe Run 500 ..... G4
†25506 Hubball 145 ..... B6
26575 Hundred 485 ..... E3
*25701 Huntington⊙ 63,684 ..... A6
Huntington-Ashland‡ 311,350 ..... A6
25526 Hurricane 3,751 ..... C6
26273 Huttonsville 242 ..... G5
24844 Iaeger 833 ..... C8
26374 Independence 200 ..... G4
24949 Indian Mills 150 ..... E7
25111 Indore 300 ..... D6
25112 Institute C6
25428 Inwood 1,159 ..... K4
24847 Itmann 600 ..... D7
25113 Ivydale 800 ..... D5
26377 Jacksonburg 400 ..... E3
26378 Jane Lew 406 ..... F4
25114 Jeffrey 900 ..... C7
24848 Jenkinjones 750 ..... D8
24849 Jesse 400 ..... C7
26674 Jodie 440 ..... D6
25969 Jumping Branch 700 ..... E7
26824 Junction 75 ..... J4

(continued on following page)

Agriculture, Industry and Resources

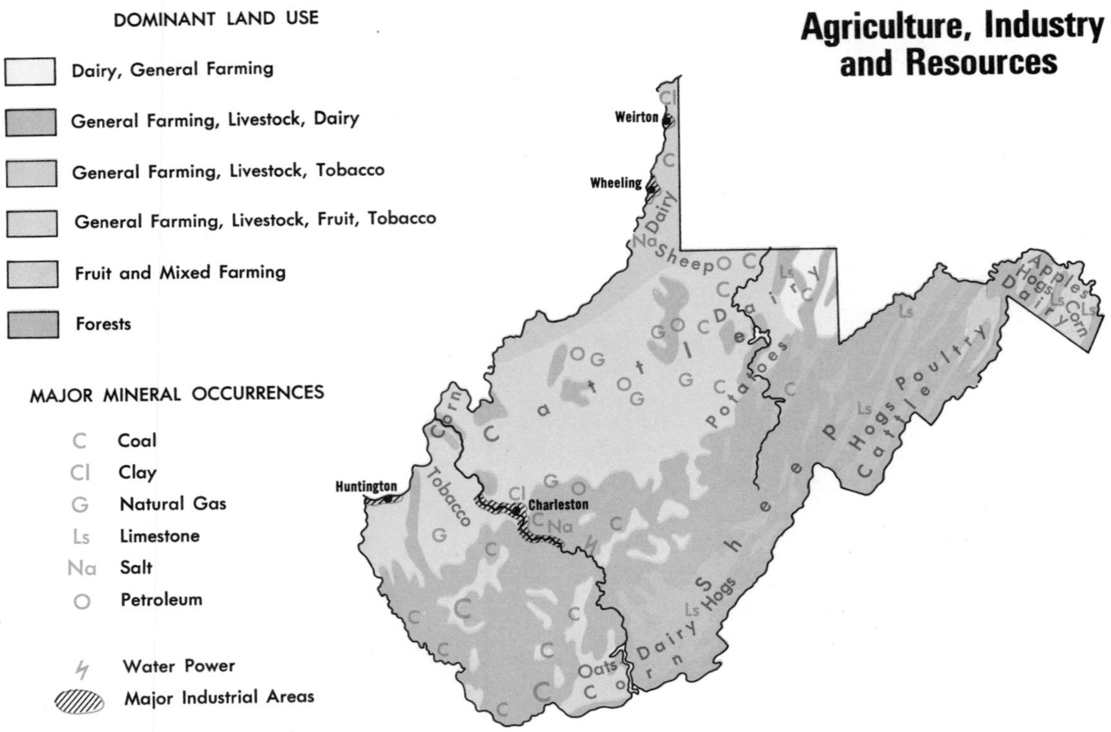

DOMINANT LAND USE
Dairy, General Farming
General Farming, Livestock, Dairy
General Farming, Livestock, Tobacco
General Farming, Livestock, Fruit, Tobacco
Fruit and Mixed Farming
Forests
MAJOR MINERAL OCCURRENCES
C Coal
Cl Clay
G Natural Gas
Ls Limestone
Na Salt
O Petroleum
Water Power
Major Industrial Areas
Weirton
Wheeling
Huntington
Charleston
Dairy
Sheep
Cattle
Potatoes
Corn
Tobacco
Hogs
Poultry
Apples
Oats

26275 Junior 591 .... G5
24851 Justice 600 .... C7
25115 Kanawha Falls 105 .... D6
25430 Kearneysville 250 .... L4
24731 Kegley 900 .... D8
24732 Kellysville 165 .... E8
25248 Kenna 150 .... C5
25530 Kenova 4,454 .... A6
25249 Kentuck 200 .... C5
25674 Kermit 705 .... B7
26726 Keyser⊙ 6,569 .... J4
24852 Keystone 902 .... D8
24950 Kieffer 135 .... E7
25859 Kilsyth 200 .... D7
24853 Kimball 871 .... C8
25120 Kingston 189 .... D7
26537 Kingwood⊙ 2,877 .... G4
26729 Kirby 110 .... J4
25628 Kistler 200 .... C7
26579 Knob Fork 106 .... E3
24854 Kopperston 700 .... C7
26731 Lahmansville 200 .... H4
25860 Lanark 559 .... D7
25629 Landville 400 .... C7
25535 Lavalette 600 .... B6
25863 Lawton 100 .... E7
25864 Layland 500 .... E7
†26430 Layopolis (Sand Fork) 280 .... E5
25251 Left Hand 700 .... D5
26676 Leivasy 200 .... E6
25676 Lenore 800 .... B7
25123 Leon 228 .... C5
25971 Lerona 550 .... D8
25537 Lesage 600 .... B5
25972 Leslie 350 .... E6
25865 Lester 626 .... D7
25253 Letart 350 .... C5
25431 Levels 180 .... J4
24901 Lewisburg⊙ 3,065 .... E7
26384 Linn 165 .... E4
26629 Little Birch 400 .... E5
26581 Littleton 335 .... F3
25125 Lizemores 400 .... D6
25866 Lochgelly 250 .... D6
25258 Lockney 190 .... E5
25601 Logan⊙ 3,029 .... B7
25630 Lorado 400 .... C7
†26201 Lorentz 200 .... F4
26810 Lost City 130 .... J5
26385 Lost Creek 604 .... F4
26811 Lost River 500 .... J5
†26101 Lubeck 1,356 .... C4
26386 Lumberport 939 .... F4
25631 Lundale 525 .... C7
25870 Maben 450 .... D7
26278 Mabie 550 .... F5
25871 Mabscott 1,668 .... D7
26148 Macfarlan 436 .... D4
25130 Madison⊙ 3,228 .... C6
26541 Maidsville 500 .... F3
25306 Malden 900 .... C6
25634 Mallory 1,330 .... C7
25132 Mammoth 563 .... D6
25635 Man 1,333 .... C7
26582 Mannington 3,036 .... F3
25975 Marfrance 225 .... E6
24954 Marlinton⊙ 1,352 .... F6
25315 Marmet 2,196 .... C6
25401 Martinsburg⊙ 13,063 .... K4
25260 Mason 1,432 .... B4
26542 Masontown 1,052 .... G3
25678 Matewan 822 .... B7
24736 Matoaka 613 .... D8
24861 Maybeury 300 .... D8
26833 Maysville 150 .... H4
24858 McDowell 500 .... D8
26040 McMechen 2,402 .... E3
26401 McWhorter 150 .... F4
24958 Meadow Bluff 250 .... E7
25976 Meadow Bridge 530 .... E7
26404 Meadowbrook 500 .... F4
25977 Meadow Creek 300 .... E7
26585 Metz 150 .... F3
26149 Middlebourne⊙ 941 .... E3
25540 Midkiff 650 .... B6
26280 Mill Creek 801 .... G5
24959 Mill Point 148 .... F6
25261 Millstone 850 .... D5
25262 Millwood 800 .... C5
25541 Milton 2,178 .... B6
25879 Minden 800 .... D7
26150 Mineralwells 325 .... C4
26281 Mingo 350 .... F5
25263 Minnora 500 .... D5
26405 Moatsville 150 .... G4
25636 Monaville 950 .... B7
26554 Monongah 1,132 .... F4
26586 Montana Mines 200 .... F3
25135 Montcoal 150 .... D7
26282 Monterville 250 .... F5
25136 Montgomery 3,104 .... D6
26283 Montrose 129 .... G4
26836 Moorefield⊙ 2,257 .... J4
26505 Morgantown⊙ 27,605 .... G3
25542 Morrisvale 450 .... C6
26041 Moundsville⊙ 12,419 .... E3
26407 Mountain 200 .... E4
25264 Mount Alto 200 .... C5
25139 Mount Carbon 450 .... D7
26408 Mount Clare 950 .... F4
25637 Mount Gay 4,366 .... C7
25880 Mount Hope 1,849 .... D7
26678 Mount Lookout 500 .... E6
26679 Mount Nebo 535 .... E6
26739 Mount Storm 500 .... H4
25882 Mullens 2,919 .... D7
26680 Nallen 250 .... E6
26631 Napier 158 .... E5
25685 Naugatuck 500 .... B7
25141 Nebo 200 .... D5
25142 Nellis 600 .... C6
24961 Neola 300 .... F7
26681 Nettie 500 .... E6
26410 Newburg 418 .... G4
26047 New Cumberland⊙ 1,752 .... E2
26050 Newell 2,032 .... E1
26154 New England 335 .... C4
24866 Newhall 400 .... C8
25265 New Haven 1,723 .... C5
26056 New Manchester 800 .... E1
26155 New Martinsville⊙ 7,109 .... E3
25266 Newton 390 .... D5
26632 Newville 160 .... E5
25143 Nitro 8,074 .... C6
25687 Nolan 250 .... B7
25267 Normantown 112 .... E5
24868 Northfork 1,105 .... D8
†26101 North Hills 940 .... D4
26285 Norton 400 .... G5
26301 Nutter Fort 2,078 .... F4
25901 Oak Hill 7,120 .... D6
24739 Oakvale 208 .... D8
24870 Oceana 2,143 .... C7
25902 Odd 500 .... D7
25147 Ohley 450 .... D6
25638 Omar 900 .... C7
26886 Onego 400 .... H5
25148 Orgas 500 .... C6
26412 Orlando 700 .... E5
25268 Orma D5
26543 Osage 285 .... F3
25151 Packsville 225 .... C7
26159 Paden City 3,671 .... D3
25152 Page 600 .... D6
26160 Palestine 110 .... D4
24872 Panther 450 .... C8
26101 Parkersburg⊙ 39,967 .... D4
Parkersburg-Marietta‡ 162,836 .... D4
26287 Parsons⊙ 1,937 .... G4
26746 Patterson Creek 157 .... J3
25434 Paw Paw 644 .... K3
25904 Pax 274 .... D7
†25955 Pear 100 .... E7
25547 Pecks Mill 350 .... B7
25905 Pemberton 300 .... D7
24962 Pence Springs 300 .... E7
26415 Pennsboro 1,652 .... E4
26544 Pentress 250 .... F3
26847 Petersburg⊙ 2,084 .... H5
24963 Peterstown 648 .... E8
25154 Peytona 175 .... C6
26416 Philippi⊙ 3,194 .... G4
24964 Pickaway 225 .... E7
26230 Pickens 240 .... F5
25689 Pie 250 .... B7
26750 Piedmont 1,491 .... H4
25156 Pinch 800 .... D6
26419 Pine Grove 767 .... E3
24874 Pineville⊙ 1,140 .... C7
25158 Pliny 900 .... B5
25159 Poca 1,142 .... C6
†25301 Pocotalico 2,420 .... C6
25550 Point Pleasant⊙ 5,682 .... B5
25437 Points 250 .... J4
25161 Powellton 1,339 .... D6
24877 Powhatan 400 .... D8
25162 Pratt 821 .... D6
24878 Premier 400 .... C8
†25880 Price Hill 175 .... D7
25555 Prichard 500 .... A6
24740 Princeton⊙ 7,493 .... D8
25164 Procious 600 .... D5
26055 Proctor 350 .... E3
26421 Pullman 196 .... D4
26852 Purgitsville 450 .... J4
25045 Quick 400 .... D6
†25015 Quincy 150 .... C6
25981 Quinwood 460 .... E6
26587 Rachel 550 .... F3
25165 Racine 725 .... C6
25556 Radnor 300 .... A6
25962 Rainelle 1,983 .... E7
25911 Raleigh 900 .... D7
25166 Ramage 350 .... C7
25557 Ranger 300 .... B6
25438 Ranson 2,471 .... L4
25913 Ravencliff 350 .... C7
26164 Ravenswood 4,126 .... C5
26167 Reader 950 .... E3
26289 Red Creek 125 .... H4
25168 Red House 600 .... C5
25692 Red Jacket 850 .... B7
26547 Reedsville 564 .... G3
25270 Reedy 338 .... D5
24966 Renick 240 .... F6
25915 Rhodell 472 .... D7
26261 Richwood 3,568 .... F6
26753 Ridgeley 994 .... J3
25440 Ridgeway 200 .... K4
26755 Rio 140 .... J4
25271 Ripley⊙ 3,464 .... C5
25441 Rippon 500 .... L4
26588 Rivesville 1,327 .... F3
26234 Rock Cave 400 .... F5
24881 Roderfield 900 .... C8
26757 Romney⊙ 2,094 .... J4
24970 Ronceverte 2,312 .... F7
26636 Rosedale 400 .... E5
25643 Rossmore 200 .... C7
26425 Rowlesburg 966 .... G4
26688 Runa 150 .... E6
25984 Rupert 1,276 .... E7
26689 Russellville 280 .... E6
25177 Saint Albans 12,402 .... C6
26290 Saint George 150 .... G4
26170 Saint Marys⊙ 2,219 .... D4
26426 Salem 2,706 .... E4
25559 Salt Rock 350 .... B6
26430 Sand Fork 280 .... E5
25985 Sandstone 300 .... E7
25275 Sandyville 500 .... C5
25876 Saulsville 250 .... C7
25917 Scarbro 800 .... D7
24975 Seebert 100 .... F6
25181 Seth 950 .... C6
26761 Shanks 500 .... J4
25182 Sharon 450 .... D6
25183 Sharples 250 .... C7
25443 Shepherdstown 1,791 .... L4
26173 Sherman 104 .... C5
26431 Shinnston 3,059 .... F4
26434 Shirley 275 .... E4
25562 Shoals 150 .... B6
26638 Shock 200 .... D5
†26164 Silverton 250 .... C5
26435 Simpson 250 .... F4
24976 Sinks Grove 156 .... F7
25320 Sissonville 450 .... C5
26175 Sistersville 2,367 .... D3
25920 Slab Fork 210 .... D7
25444 Slanesville 250 .... K4
26436 Smithburg 130 .... E4
25186 Smithers 1,482 .... D6
26437 Smithfield 278 .... E4
26178 Smithville 200 .... D4
24977 Smoot 300 .... E7
25921 Sophia 1,216 .... D7
25303 South Charleston 15,968 .... C6
25922 Spanishburg 550 .... D8
25276 Spencer⊙ 2,799 .... D5
25693 Sprigg 225 .... B7
26763 Springfield 250 .... J4
25565 Spurlockville 250 .... B6
24884 Squire 900 .... C8
26505 Star City 1,464 .... F3
25279 Statts Mills 400 .... C5
25188 Stickney 150 .... D7
25645 Stirrat 250 .... C7
26301 Stonewood 2,058 .... F4
24979 Stony Bottom 50 .... F6
25280 Stumptown 125 .... E5
26651 Summersville⊙ 2,972 .... E6
25446 Summit Point 455 .... K4
25932 Surveyor 300 .... D7
26601 Sutton⊙ 1,192 .... E5
26690 Swiss 500 .... D6
25647 Switzer 1,034 .... B7
25193 Sylvester 256 .... C6
24981 Talcott 800 .... E7
26237 Tallmansville 140 .... F5
26179 Tanner 375 .... E5
26764 Terra Alta 1,946 .... H4
26640 Tesla 300 .... E5
25694 Thacker 525 .... B7
26292 Thomas 747 .... H4
26440 Thornton 200 .... G4
24888 Thorpe 600 .... D8
26765 Three Churches 350 .... J4
25936 Thurmond 67 .... D7
26691 Tioga 825 .... E6
26059 Triadelphia 1,461 .... E2
26443 Troy 110 .... E4
26444 Tunnelton 510 .... G4
25203 Turtle Creek 566 .... C6
25205 Uneeda 700 .... C6
25447 Unger 300 .... K4
24983 Union⊙ 743 .... E7
26266 Upperglade 750 .... F6
26866 Upper Tract 155 .... H5
26445 Vadis 130 .... E4
26293 Valley Bend 950 .... F5
26060 Valley Grove 597 .... E2
26294 Valley Head 900 .... G5
25206 Van 800 .... C7
25696 Varney 750 .... B7
25649 Verdunville 950 .... B7
25938 Victor 500 .... D6
26105 Vienna 11,618 .... D4
24891 Vivian 500 .... D8
26238 Volga 125 .... F4
25697 Vulcan 130 .... B7
26589 Wadestown 300 .... F3
24984 Waiteville 230 .... F8
26180 Walker 100 .... D4
26448 Wallace 325 .... E4
25286 Walton 550 .... D5
26590 Wana 150 .... F3
24892 War 2,158 .... C8
26851 Wardensville 241 .... J4
26181 Washington 450 .... C4
26184 Waverly 500 .... D4
25570 Wayne⊙ 1,495 .... B6
26288 Webster Springs⊙ 939 .... F6
26062 Weirton 25,371 .... E2
Weirton-Steubenville‡ 163,099 .... E2
24801 Welch⊙ 3,885 .... C8
26070 Wellsburg⊙ 3,963 .... E2
25287 West Columbia 245 .... B5
25571 West Hamlin 643 .... B6
26074 West Liberty 744 .... E2
25601 West Logan 630 .... C7
26451 West Milford 510 .... F4
26452 Weston⊙ 6,250 .... F4
26505 Westover 4,884 .... G3
26456 West Union⊙ 1,090 .... E4
25651 Wharncliffe 900 .... C7
25208 Wharton 450 .... C7
26003 Wheeling⊙ 43,070 .... E2
Wheeling‡ 185,566 .... E2
24986 White Sulphur Springs 3,371 .... F7
25209 Whitesville 689 .... C6
26296 Whitmer 400 .... G5
25211 Widen 230 .... E6
26767 Wiley Ford 1,224 .... J3
26186 Wileyville 175 .... E3
25653 Wilkinson 975 .... B7
24991 Williamsburg 350 .... F7
25661 Williamson⊙ 5,219 .... B7
26187 Williamstown 3,095 .... C4
26461 Wilsonburg 350 .... F4
25699 Wilsondale 250 .... B7
26075 Windsor Heights 800 .... E2
25213 Winfield⊙ 329 .... C5
25214 Winifrede 750 .... C6
25942 Winona 250 .... E6
26462 Wolf Summit 750 .... F4
†26257 Womelsdorf (Coalton) 306 .... G5
25572 Woodville 300 .... C6
26591 Worthington 329 .... F4
25573 Yawkey 985 .... C6
26865 Yellow Spring 280 .... J4
25654 Yolyn 400 .... C7

OTHER FEATURES

Big Sandy (riv.) A6
Bluestone (lake) E7
Buckhannon (riv.) F5
Cacapon (riv.) J4
Cheat (riv.) G3
Cherry (riv.) E6
Chesapeake and Ohio Canal Nat'l Hist. Pa J3
Clear Fork, Guyandotte (riv.) C7
Coal (riv.) C6
Dry Fork (riv.) C8
Dry Fork (riv.) G5
East Lynn (lake) B6
Elk (riv.) D6
Fish (creek) E3
Gauley (riv.) D6
Greenbrier (riv.) F6
Guyandotte (riv.) B6
Harpers Ferry Nat'l Hist. Park L4
Hughes (riv.) D4
Kanawha (riv.) C5
Little Kanawha (riv.) D5
Meadow (riv.) E6
Mill (creek) C5
Monongahela (riv.) G3
Mount Storm (lake) H4
Mud (riv.) B6
New (riv.) E7
North (riv.) J4
Ohio (riv.) B5
Patterson (creek) J4
Pigeon (creek) B7
Pocatalico (riv.) C5
Pond Fork (riv.) C6
Potomac (riv.) L
Potts (creek) .... F
Reedy (creek) .... D
Shavers Fork (riv.) .... G
Shenandoah (riv.) .... K
Spruce Knob (mt.) .... G
Spruce Knob-Seneca Rocks Nat'l Rec. Area .... H
Stony (riv.) .... H
Summersville (lake) .... E
Sutton (lake) .... F
Tug Fork (riv.) .... B
Twelvepole (creek) .... A
Tygart (lake) .... G
Tygart Valley (riv.) .... F
West Fork (riv.) .... E
Williams (riv.) .... F

⊙County seat.
‡Population of metropolitan area.
† Zip of nearest p.o. * Multiple zip

AREA 56,153 sq. mi. (145,436 sq. km.)
POPULATION 4,705,521
CAPITAL Madison
LARGEST CITY Milwaukee
HIGHEST POINT Timms Hill 1,951 ft. (595 m.)
SETTLED IN 1670
ADMITTED TO UNION May 29, 1848
POPULAR NAME Badger State
STATE FLOWER Wood Violet
STATE BIRD Robin

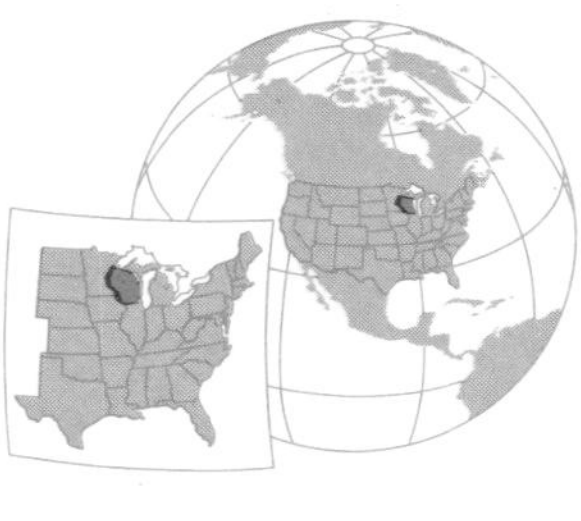

COUNTIES

Adams 13,457 ... G7
Ashland 16,783 ... E3
Barron 38,730 ... C5
Bayfield 13,822 ... D3
Brown 175,280 ... L7
Buffalo 14,309 ... C7
Burnett 12,340 ... B4
Calumet 30,867 ... K7
Chippewa 52,127 ... D5
Clark 32,910 ... E6
Columbia 43,222 ... H9
Crawford 16,556 ... E9
Dane 323,545 ... H9
Dodge 75,064 ... J9
Door 25,029 ... M6
Douglas 44,421 ... C3
Dunn 34,314 ... C6
Eau Claire 78,805 ... D6
Florence 4,172 ... K4
Fond du Lac 88,964 ... K8
Forest 9,044 ... J4
Grant 51,736 ... E10
Green 30,012 ... G10
Green Lake 18,370 ... H8
Iowa 19,802 ... F9
Iron 6,730 ... F3
Jackson 16,831 ... E7
Jefferson 66,152 ... J9
Juneau 21,039 ... F8
Kenosha 123,137 ... K10
Kewaunee 19,539 ... L6
La Crosse 91,056 ... D8
Lafayette 17,412 ... F10
Langlade 19,978 ... H5
Lincoln 26,555 ... G5
Manitowoc 82,918 ... L7
Marathon 111,270 ... G6
Marinette 39,314 ... K5
Marquette 11,672 ... H8
Menominee 3,373 ... J5
Milwaukee 964,988 ... L9
Monroe 35,074 ... E8
Oconto 28,947 ... K6
Oneida 31,216 ... G4
Outagamie 128,799 ... K7
Ozaukee 66,981 ... L9
Pepin 7,477 ... C6
Pierce 31,149 ... B6
Polk 32,351 ... B5
Portage 57,420 ... G6
Price 15,788 ... F4
Racine 173,132 ... K10
Richland 17,476 ... F9
Rock 139,420 ... H10
Rusk 15,589 ... D5
Saint Croix 43,262 ... B5
Sauk 43,469 ... G9
Sawyer 12,843 ... D4
Shawano 35,928 ... J6
Sheboygan 100,935 ... L8
Taylor 18,817 ... E5
Trempealeau 26,158 ... D7
Vernon 25,642 ... E8
Vilas 16,535 ... G3
Walworth 71,507 ... J10
Washburn 13,174 ... C4
Washington 84,848 ... K9
Waukesha 280,080 ... K9
Waupaca 42,831 ... J6
Waushara 18,526 ... H7
Winnebago 131,722 ... J8
Wood 72,799 ... F7

CITIES and TOWNS

Zip *Name/Pop.* *Key*

54405 Abbotsford 1,901 ... F6
53910 Adams 1,744 ... G8
53001 Adell 545 ... L8
53501 Afton 225 ... H10
53502 Albany 1,051 ... G10
†53534 Albion 300 ... H10
54201 Algoma 3,656 ... M6
53002 Allenton 915 ... K9
†54301 Allouez 14,882 ... L7
54610 Alma⊙ 876 ... C7
54611 Alma Center 454 ... E7
54805 Almena 526 ... B5
54909 Almond 477 ... G7
54720 Altoona 4,393 ... C6
54102 Amberg 875 ... K5
54001 Amery 2,404 ... B5
54406 Amherst 701 ... H7
54407 Amherst Junction 225 ... H7
54408 Aniwa 273 ... H6
54409 Antigo⊙ 8,653 ... H5
54911 Appleton⊙ 58,913 ... J7
Appleton-Oshkosh‡ 291,325 J7
†54568 Arbor Vitae 900 ... G4
54612 Arcadia 2,109 ... D7
53503 Arena 451 ... G9
54511 Argonne 600 ... J4
53504 Argyle 720 ... G10
54721 Arkansaw 400 ... B6
53911 Arlington 440 ... H9
54103 Armstrong Creek 615 ... K4
54410 Arpin 361 ... G6
53003 Ashippun 750 ... H1
54806 Ashland⊙ 9,115 ... E2
54304 Ashwaubenon 14,486 ... K7
54411 Athens 988 ... G5
54412 Auburndale 641 ... F6
54722 Augusta 1,560 ... D6
53506 Avoca 505 ... F9
†53520 Avon 120 ... H10
54413 Babcock 250 ... F7
53801 Bagley 317 ... D10
54202 Baileys Harbor 250 ... M5
54002 Baldwin 1,620 ... B6
54810 Balsam Lake⊙ 749 ... B5
54921 Bancroft 355 ... G7
54614 Bangor 1,012 ... E8
53913 Baraboo⊙ 8,081 ... G9
†54873 Barnes 225 ... D3
53507 Barneveld 579 ... F10
54812 Barron⊙ 2,595 ... C5
†53001 Batavia 125 ... K8
54723 Bay City 543 ... B6
54814 Bayfield 778 ... E2
†53201 Bayside 4,724 ... M1
54922 Bear Creek 454 ... J6
53916 Beaver Dam 14,149 ... J9
53802 Beetown 150 ... E10
53004 Belgium 892 ... L8
†54631 Bell Center 124 ... E9
53508 Belleville 1,302 ... G10
53510 Belmont 826 ... F10
53511 Beloit 35,207 ... H10
53803 Benton 983 ... F10
54923 Berlin 5,478 ... H8
†54410 Bethel 210 ... F6
†54440 Bevent 200 ... H6
53103 Big Bend 1,345 ... K2
54926 Big Falls 107 ... H6
54817 Birchwood 437 ... C4
54414 Birnamwood 688 ... H6
†54494 Biron 698 ... G7
54106 Black Creek 1,097 ... K7
53515 Black Earth 1,145 ... G9
54615 Black River Falls⊙ 3,434 ... E7
†54541 Blackwell 550 ... J4
54616 Blair 1,142 ... D7
53516 Blanchardville 803 ... G10
54617 Bloom City 167 ... E8
54724 Bloomer 3,342 ... D5
53804 Bloomington 743 ... E10
53517 Blue Mounds 387 ... G9
53518 Blue River 412 ... E9
†53581 Boaz 161 ... E9
†53105 Bohners Lake 1,507 ... K10
54107 Bonduel 1,160 ... K6
53805 Boscobel 2,662 ... E9
54512 Boulder Junction 780 ... G3
54416 Bowler 339 ... J6
54725 Boyceville 862 ... C5
54726 Boyd 660 ... E6
54203 Branch 300 ... L7
53919 Brandon 862 ... J8
54513 Brantwood 500 ... F4
53920 Briggsville 250 ... H8
54110 Brillion 2,907 ... L7
53520 Brodhead 3,153 ... G10
54417 Brokaw 298 ... G5
53005 Brookfield 34,035 ... K1
53521 Brooklyn 627 ... H10
53209 Brown Deer 12,921 ... L1
†53105 Brown's Lake 1,648 ... K3
53006 Brownsville 433 ... J8
53522 Browntown 284 ... G10
54819 Bruce 905 ... D5
54820 Brule 335 ... C2
54204 Brussels 500 ... L6
†54622 Buffalo 894 ... C7
53105 Burlington 8,385 ... K10
53922 Burnett 260 ... J9
53007 Butler 2,059 ... K1
54514 Butternut 438 ... E3
53009 Byron 40 ... K8
54821 Cable 227 ... D3
54727 Cadott 1,247 ... D6
53923 Cambria 680 ... H8
53523 Cambridge 844 ... H9
54822 Cameron 1,115 ... C5
†53019 Campbellsport 1,740 ... K8
54618 Camp Douglas 589 ... F8
53109 Camp Lake 2,060 ... K10
54823 Canton 100 ... C5
54928 Caroline 450 ... J6
53011 Cascade 615 ... K8
54205 Casco 484 ... L6
54619 Cashton 827 ... E8
53806 Cassville 1,270 ... E10
54620 Cataract 200 ... E7
54515 Catawba 205 ... E4
54206 Cato 85 ... L7
53924 Cazenovia 259 ... F8
54111 Cecil 445 ... K6
53012 Cedarburg 9,005 ... L9
53013 Cedar Grove 1,420 ... L8
54824 Centuria 711 ... A5
54621 Chaseburg 279 ... D8
54419 Chelsea 120 ... F5
†53029 Chenequa 532 ... J1
54728 Chetek 1,931 ... C5
54420 Chili 185 ... F6
53014 Chilton⊙ 2,965 ... K7
54729 Chippewa Falls⊙ 12,270 ... D6
54004 Clayton 425 ... B5
54005 Clear Lake 899 ... B5
53015 Cleveland 1,270 ... L8
53525 Clinton 1,751 ... J10
54929 Clintonville 4,567 ... J6
53016 Clyman 317 ... J9
53526 Cobb 409 ... F10
54622 Cochrane 512 ... C7
54421 Colby 1,496 ... F6
54112 Coleman 852 ... L5
54730 Colfax 1,149 ... C6
54930 Coloma 367 ... H7
53925 Columbus 4,049 ... H9
54113 Combined Locks 2,573 ... K7
†53147 Como 1,376 ... K10
54519 Conover 480 ... H3
54731 Conrath 86 ... E5
54623 Coon Valley 758 ... E8
54732 Cornell 1,583 ... D5
54827 Cornucopia 250 ... D2
54520 Crandon⊙ 1,969 ... H4
54114 Crivitz 1,041 ... L5
53528 Cross Plains 2,156 ... G9
53807 Cuba City 2,129 ... F10
53110 Cudahy 19,547 ... M2
54829 Cumberland 1,983 ... C4
54422 Curtiss 127 ... F6
54006 Cushing 150 ... A4
54931 Dale 410 ... J7
54733 Dallas 477 ... C5
53926 Dalton 300 ... H8
53529 Dane 518 ... G9
53114 Darien 1,152 ... J10
53530 Darlington⊙ 2,300 ... F10
53531 Deerfield 1,466 ... H9
54007 Deer Park 232 ... B5
53532 De Forest 3,367 ... H9
53018 Delafield 4,083 ... J1
53115 Delavan 5,684 ... J10
†53115 Delavan Lake 2,082 ... J10
†54856 Delta 35 ... D3
54208 Denmark 1,475 ... L7
54115 De Pere 14,892 ... K7
†54663 De Soto 318 ... D9
†54014 Diamond Bluff 100 ... A6
53808 Dickeyville 1,156 ... E10
54625 Dodge 185 ... D7
53533 Dodgeville⊙ 3,458 ... F10
54425 Dorchester 613 ... F5
53118 Dousman 1,153 ... J1
54734 Downing 242 ... B5
54735 Downsville 200 ... C6
53928 Doylestown 294 ... H9
54009 Dresser 670 ... A5
54832 Drummond 200 ... D3
54736 Durand⊙ 2,047 ... C6
53119 Eagle 1,008 ... H2
54521 Eagle River⊙ 1,326 ... H4
54626 Eastman 371 ... D9
53120 East Troy 2,385 ... J2
54701 Eau Claire⊙ 51,509 ... D6
Eau Claire‡ 130,507 ... D6
53019 Eden 534 ... K8
54426 Edgar 1,194 ... G6
53534 Edgerton 4,335 ... H10
54209 Egg Harbor 238 ... M5
54427 Eland 230 ... H6
54428 Elcho 500 ... H5
54429 Elderon 191 ... H6
54932 Eldorado 200 ... J8
54738 Eleva 593 ... D6
53020 Elkhart Lake 1,054 ... L8
53121 Elkhorn⊙ 4,605 ... J10
54739 Elk Mound 737 ... C6
54210 Ellison Bay 112 ... M5
54011 Ellsworth⊙ 2,143 ... A6
53122 Elm Grove 6,735 ... K1
54740 Elmwood 885 ... B6
†53401 Elmwood Park 483 ... M3
53929 Elroy 1,504 ... F8
54430 Elton 150 ... J5
54933 Embarrass 496 ... J6
53930 Endeavor 335 ... G8
54211 Ephraim 319 ... M5
54627 Ettrick 462 ... D7
53536 Evansville 2,835 ... H10
54835 Exeland 219 ... D4
54741 Fairchild 577 ... D6
53931 Fair Water 310 ... J8
54742 Fall Creek 1,148 ... D6
53932 Fall River 850 ... H9
†54840 Falun 95 ... A4
54120 Fence 200 ... K4
53809 Fennimore 2,212 ... E9
54431 Fenwood 165 ... F6
54628 Ferryville 227 ... D9
54524 Fifield 310 ... F4
54212 Fish Creek 119 ... M5
54121 Florence⊙ 780 ... K4
54935 Fond du Lac⊙ 35,863 ... K8
53125 Fontana 1,764 ... J10
53537 Footville 794 ... H10
54123 Forest Junction 140 ... K7
54213 Forestville 455 ... L6
53538 Fort Atkinson 9,785 ... J10
54629 Fountain City 963 ... C7
54836 Foxboro 360 ... B2
53933 Fox Lake 1,373 ... J8
†53117 Fox Point 7,649 ... M1
54214 Francis Creek 589 ... L7
53132 Franklin 16,871 ... L2
54837 Frederic 1,039 ... B4
53021 Fredonia 1,437 ... L8
54940 Fremont 510 ... J7
53934 Friendship⊙ 744 ... G8
53935 Friesland 267 ... H8
54630 Galesville 1,239 ... D7
54631 Gays Mills 627 ... E9
53127 Genesee Depot 350 ... J2
54632 Genoa 283 ... D8
53128 Genoa City 1,202 ... K11
53022 Germantown 10,729 ... K1
54124 Gillett 1,356 ... K6
54433 Gilman 436 ... E5
54743 Gilmanton 300 ... C7
54435 Gleason 200 ... G5
53023 Glenbeulah 423 ... L8
†53209 Glendale 13,882 ... M1
54526 Glen Flora 83 ... E4
53810 Glen Haven 160 ... E10
54013 Glenwood City 950 ... B5
54527 Glidden 940 ... E3
54125 Goodman 875 ... K4
54838 Gordon 600 ... C3
53540 Gotham 250 ... F9
53024 Grafton 8,381 ... L9
53936 Grand Marsh 725 ... G8
54839 Grand View 447 ... D3
54436 Granton 399 ... E6
54840 Grantsburg⊙ 1,153 ... A4
53541 Gratiot 280 ... F10
*54301 Green Bay⊙ 87,899 ... K6
Green Bay‡ 175,280 ... K6
53129 Greendale 16,928 ... L2
53220 Greenfield 31,467 ... L2
54941 Green Lake⊙ 1,208 ... H8
54126 Greenleaf 300 ... L7
54942 Greenville 900 ... J7
54437 Greenwood 1,124 ... E6
54128 Gresham 534 ... J6
54014 Hager City 110 ... A6
53130 Hales Corners 7,110 ... K2
54015 Hammond 991 ... A6
54943 Hancock 419 ... G7
54529 Harshaw 87 ... G4
53027 Hartford 7,046 ... K9
53029 Hartland 5,559 ... J1
54440 Hatley 300 ... H6
54841 Haugen 251 ... C4
54530 Hawkins 407 ... E4
54842 Hawthorne 200 ... C3
54843 Hayward⊙ 1,698 ... D3
53811 Hazel Green 1,282 ... F11
54531 Hazelhurst 630 ... G4
†53538 Hebron 450 ... J10
53137 Helenville 300 ... J10
54844 Herbster 100 ... D2
54441 Hewitt 470 ... F6
53543 Highland 860 ... F9
54129 Hilbert 1,176 ... K7
†54511 Hiles 350 ... J4

(continued on following page)

Agriculture, Industry and Resources

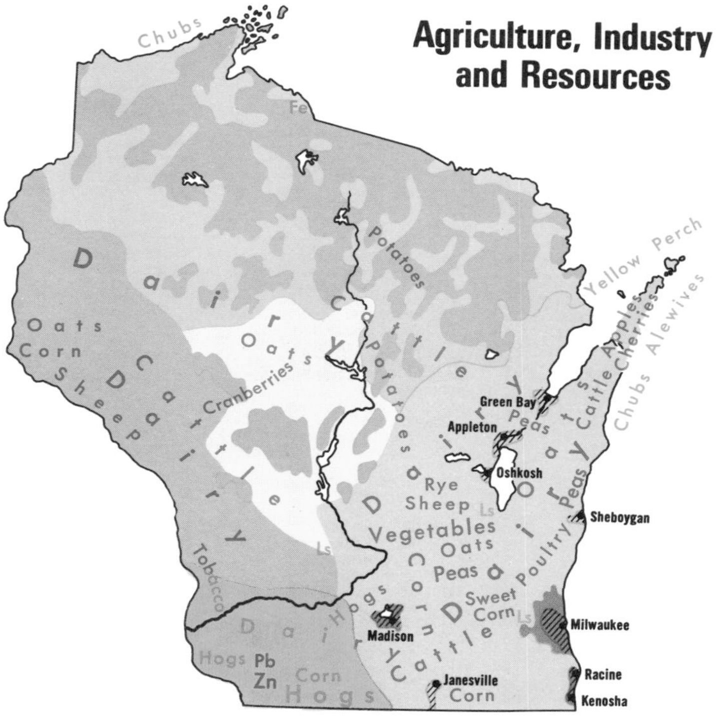

DOMINANT LAND USE

Specialized Dairy
Dairy, Hay, Potatoes
Dairy, General Farming
Hogs, Dairy
Dairy, Livestock
Forests
Urban Areas

MAJOR MINERAL OCCURRENCES

Fe Iron Ore
Pb Lead
Ls Limestone
Zn Zinc

Major Industrial Areas

54634 Hillsboro 1,263 ... F8
53031 Hingham 250 ... K8
54635 Hixton 364 ... E7
54745 Holcombe 200 ... D5
53544 Hollandale 271 ... G10
54636 Holmen 2,411 ... D8
53138 Honey Creek 300 ... J3
53032 Horicon 3,584 ... J9
54944 Hortonville 2,016 ... J7
†55082 Houlton 915 ... A5
54303 Howard 8,240 ... K6
53081 Howards
Grove-Millersville 1,838 . L8
53033 Hubertus 600 ... K1
54016 Hudson⊙ 5,434 ... A6
54746 Humbird 190 ... E6
54534 Hurley⊙ 2,015 ... F3
53034 Hustisford 874 ... J9
54637 Hustler 170 ... F8
54747 Independence 1,180 ... D7
54945 Iola 957 ... H6
54536 Iron Belt 300 ... F3
53035 Iron Ridge 766 ... K9
54847 Iron River 878 ... D2
†53941 Ironton 206 ... F8
53036 Ixonia 525 ... H1
53037 Jackson 1,817 ... K9
†54235 Jacksonport 150 ... M6
53545 Janesville⊙ 51,071 ... H10
Janesville-Beloit‡ 139,420 H10
53549 Jefferson⊙ 5,647 ... J10
54748 Jim Falls 100 ... D5
53038 Johnson Creek 1,136 ... J9
53550 Juda 500 ... H10
54443 Junction City 523 ... G6
53039 Juneau⊙ 2,045 ... J9
53139 Kansasville 150 ... L3
54130 Kaukauna 11,310 ... K7
†53050 Kekoskee 224 ... J8
54215 Kellnersville 369 ... L7
54638 Kendall 486 ... F8
54537 Kennan 194 ... F5
*53140 Kenosha⊙ 77,685 ... M3
Kenosha‡ 123,137 ... M3
54135 Keshena⊙ 980 ... J6
53040 Kewaskum 2,381 ... K8
54216 Kewaunee⊙ 2,801 ... M7
53042 Kiel 3,083 ... L8
53812 Kieler 800 ... E10
54136 Kimberly 5,881 ... K7
53939 Kingston 328 ... H8
54749 Knapp 419 ... B6
†54455 Knowlton 127 ... G6
53044 Kohler 1,651 ... L8
53147 Krakow 345 ... K6
54538 Lac du Flambeau 500 ... G4
†53066 Lac La Belle 289 ... H1
54601 La Crosse⊙ 48,347 ... D8
La Crosse‡ 91,056 ... D8
54848 Ladysmith⊙ 3,826 ... D5
54639 La Farge 746 ... E8
53940 Lake Delton 1,158 ... G8
53147 Lake Geneva 5,612 ... K10
53551 Lake Mills 3,670 ... H9
54849 Lake Nebagamon 780 ... C3
54539 Lake Tomahawk 600 ... H4
†54494 Lake Wazeecha 2,176 ... G7
†54729 Lake Wissota 1,788 ... D6
54138 Lakewood 425 ... K5
53813 Lancaster⊙ 4,076 ... E10
54540 Land O'Lakes 786 ... H3
53046 Lannon 987 ... K1
53941 La Valle 412 ... F8
53047 Lebanon 250 ... H1
54139 Lena 585 ... K6
†54656 Leon 100 ... E8
54948 Leopolis 200 ... J6
54851 Lewis 200 ... B4
53942 Limeridge 191 ... F9
53553 Linden 395 ... F10
54140 Little Chute 7,907 ... K7
53554 Livingston 642 ... E10
53555 Lodi 1,959 ... G9
53943 Loganville 239 ... F9
†54970 Lohrville 336 ... H7
53048 Lomira 1,446 ... J8
53556 Lone Rock 577 ... F9
54542 Long Lake 150 ... J4
53557 Lowell 326 ... J9
54446 Loyal 1,252 ... E6
54447 Lublin 142 ... E5
54853 Luck 997 ... B4
54217 Luxemburg 1,040 ... L6
53944 Lyndon Station 375 ... F8
54640 Lynxville 174 ... D9
53148 Lyons 550 ... K10
*53701 Madison (cap.)⊙ 170,616 . H9
Madison‡ 323,545 ... H9
54750 Maiden Rock 172 ... B6
54949 Manawa 1,205 ... J7
54220 Manitowoc⊙ 32,547 ... L7
54226 Maplewood 200 ... M6
54448 Marathon 1,552 ... G6
54855 Marengo 130 ... E3
54227 Maribel 363 ... L7
54143 Marinette⊙ 11,965 ... L5
54950 Marion 1,348 ... J6
53946 Markesan 1,446 ... J8
53947 Marquette 204 ... H8
53559 Marshall 2,363 ... H9
54449 Marshfield 18,290 ... F6
54856 Mason 102 ... D3
54450 Mattoon 382 ... J5
53948 Mauston⊙ 3,284 ... F8
53050 Mayville 4,333 ... K9
53560 Mazomanie 1,248 ... G9
53558 McFarland 3,783 ... H10
54543 McNaughton 450 ... H4
54451 Medford⊙ 4,035 ... F5
54546 Mellen 1,046 ... E3
54642 Melrose 507 ... E7
54619 Melvina 117 ... E8
54952 Menasha 14,728 ... J7
53051 Menomonee Falls 27,845 . K1
54751 Menomonie⊙ 12,769 ... C6
53092 Mequon 16,193 ... L1
54452 Merrill⊙ 9,578 ... G5
54754 Merrillan 587 ... E7
53561 Merrimac 365 ... G9
53056 Merton 1,045 ... K1
53562 Middleton 11,848 ... G9
54857 Mikana 200 ... C4
54453 Milan 153 ... F6
†53038 Milford 35 ... J9
54454 Milladore 250 ... G6
54643 Millston 110 ... E7
54858 Milltown 732 ... B4
53563 Milton 4,092 ... J10
*53201 Milwaukee⊙ 636,236 ... M1
Milwaukee‡ 1,397,143 ... M1
54644 Mindoro 200 ... D7
53565 Mineral Point 2,259 ... F10
54548 Minocqua 950 ... G4
54859 Minong 557 ... C3
54228 Mishicot 1,503 ... L7
54755 Mondovi 2,545 ... C6
54549 Monico 250 ... H4
53716 Monona 8,809 ... H9
53566 Monroe⊙ 10,027 ... G10
53949 Montello⊙ 1,273 ... H8
53569 Montfort 616 ... E10
53570 Monticello 1,021 ... G10
54550 Montreal 887 ... F3
53571 Morrisonville 375 ... G9
54455 Mosinee 3,015 ... G6
54149 Mountain 250 ... K5
53057 Mount Calvary 585 ... K8
53816 Mount Hope 197 ... D10
53572 Mount Horeb 3,251 ... G10
54645 Mount Sterling 223 ... D9
†53572 Mount Vernon 138 ... G10
53149 Mukwonago 4,014 ... J2
53573 Muscoda 1,331 ... F9
53150 Muskego 15,277 ... K2
53058 Nashotah 513 ... J1
54646 Necedah 773 ... F7
54956 Neenah 22,432 ... J7
54456 Neillsville⊙ 2,780 ... E6
54457 Nekoosa 2,519 ... G7
54756 Nelson 389 ... C7
54458 Nelsonville 199 ... H7
54150 Neopit 1,065 ... J6
53059 Neosho 575 ... J9
54960 Neshkoro 386 ... H8
54551 Newald 375 ... J4
54757 New Auburn 466 ... D5
53151 New Berlin 30,529 ... K2
53060 Newburg 783 ... K9
†61075 New Diggings 65 ... F10
54229 New Franken 150 ... L6
53574 New Glarus 1,763 ... G10
53061 New Holstein 3,412 ... K8
53950 New Lisbon 1,390 ... F8
54961 New London 6,210 ... J7
54017 New Richmond 4,306 ... A5
54151 Niagara 2,079 ... K4
54152 Nichols 267 ... K6
†53401 North Bay 219 ... M3
†54935 North Fond du Lac 3,844 . J8
53951 North Freedom 616 ... G9
†54016 North Hudson 2,218 ... A5
53064 North Lake 400 ... J1
53217 North Shore 14,930 ... M1
54648 Norwalk 517 ... E8
53154 Oak Creek 16,932 ... M2
54649 Oakdale 150 ... F8
53065 Oakfield 990 ... J8
53066 Oconomowoc 9,909 ... H1
†53066 Oconomowoc Lake 524 ... H1
54153 Oconto⊙ 4,505 ... L6
54154 Oconto Falls 2,500 ... K6
54962 Ogdensburg 214 ... J7
54459 Ogema 238 ... F5
53069 Okauchee 3,958 ... J1
†53555 Okee 250 ... H9
†54880 Oliver 253 ... B2
54963 Omro 2,763 ... J7
54650 Onalaska 9,249 ... D8
54155 Oneida 900 ... K7
54651 Ontario 398 ... E8
53070 Oostburg 1,647 ... L8
53575 Oregon 3,876 ... H10
53576 Orfordville 1,143 ... H10
54020 Osceola 1,581 ... A5
54901 Oshkosh⊙ 49,620 ... J8
54758 Osseo 1,474 ... D6
54460 Owen 998 ... F6
53952 Oxford 432 ... H8
53953 Packwaukee 271 ... G8
†53168 Paddock Lake 2,207 ... K10
53156 Palmyra 1,515 ... H2
53954 Pardeeville 1,594 ... H8
54552 Park Falls 3,192 ... F4
†54481 Park Ridge 643 ... H6
53817 Patch Grove 259 ... D10
53157 Pell Lake 1,826 ... K10
54553 Pence 234 ... F3
54759 Pepin 890 ... B7
54157 Peshtigo 2,807 ... L5
53072 Pewaukee 4,637 ... K1
54554 Phelps 950 ... H3
54555 Phillips⊙ 1,522 ... E4
54464 Phlox 150 ... J5
54465 Pickerel 107 ... J5
54760 Pigeon Falls 338 ... D7
54466 Pittsville 810 ... F7
53577 Plain 676 ... F9
54966 Plainfield 813 ... G7
†53017 Plat 120 ... K1
53818 Platteville 9,580 ... F10
53158 Pleasant Prairie 950 ... L10
54467 Plover 5,310 ... G7
54761 Plum City 505 ... B6
53073 Plymouth 6,027 ... L8
†54423 Polonia 200 ... H6
54864 Poplar 569 ... C2
53901 Portage⊙ 7,896 ... G8
54469 Port Edwards 2,077 ... G7
53074 Port Washington⊙ 8,612 .. L9
54865 Port Wing 290 ... D2
53820 Potosi 736 ... E10
54160 Potter 330 ... K7
54161 Pound 407 ... L5
53955 Poynette 1,447 ... G9
54967 Poy Sippi 425 ... J7
53821 Prairie du Chien⊙ 5,859 ... D9
53578 Prairie du Sac 2,145 ... G9
54762 Prairie Farm 387 ... C5
54556 Prentice 605 ... F4
54021 Prescott 2,654 ... A6
54968 Princeton 1,479 ... H8
54162 Pulaski 1,875 ... K6
54164 Pulcifer 35 ... K6
*53401 Racine⊙ 85,725 ... M3
Racine‡ 173,132 ... M3
54867 Radisson 280 ... D4
53956 Randolph 1,691 ... H8
53075 Random Lake 1,287 ... K8
†53126 Raymond 300 ... L2
54652 Readstown 396 ... E9
54970 Redgranite 976 ... J7
53959 Reedsburg 5,038 ... G8
54230 Reedsville 1,134 ... L7
53579 Reeseville 649 ... J9
53580 Rewey 233 ... F10
54501 Rhinelander⊙ 7,873 ... H4
54470 Rib Lake 945 ... F5
54868 Rice Lake 7,691 ... C5
53581 Richland Center⊙ 4,997 ... F9
54763 Ridgeland 300 ... B5
53582 Ridgeway 503 ... F10
53960 Rio 785 ... H9
54971 Ripon 7,111 ... J8
54022 River Falls 9,019 ... A6
†53201 River Hills 1,642 ... M1
54023 Roberts 833 ... A6
53167 Rochester 746 ... K3
†53523 Rockdale 200 ... J10
53077 Rockfield 200 ... L1
54653 Rockland 383 ... D8
53961 Rock Springs 426 ... F8
†53178 Rome 200 ... H1
54974 Rosendale 725 ... J8
54473 Rosholt 520 ... H6
54474 Rothschild 3,338 ... G6
†53583 Roxbury 260 ... G9
54475 Rudolph 392 ... G7
†54751 Rusk 40 ... C6
53079 Saint Cloud 560 ... K8
54024 Saint Croix Falls 1,497 ... A5
†53207 Saint Francis 10,042 ... M2
†54601 Saint Joseph Ridge 450 ... D8
54232 Saint Nazianz 738 ... L7
54765 Sand Creek 225 ... C5
53583 Sauk City 2,703 ... G9
53080 Saukville 3,494 ... L9
54559 Saxon 375 ... F3
54977 Scandinavia 292 ... H7
54476 Schofield 2,226 ... H6
†54843 Seeley 68 ... D3
54654 Seneca 235 ... E9
53584 Sextonville 225 ... F9
54165 Seymour 2,530 ... K6
53585 Sharon 1,280 ... J11
54166 Shawano⊙ 7,013 ... J6
53081 Sheboygan⊙ 48,085 ... L8
Sheboygan‡ 100,935 ... L8
53085 Sheboygan Falls 5,253 ... L8
54766 Sheldon 292 ... D5
54871 Shell Lake⊙ 1,135 ... C4
54169 Sherwood 372 ... K7
54170 Shiocton 805 ... K7
53211 Shorewood 14,327 ... M1
†53701 Shorewood Hills 1,837 ... G9
53586 Shullsburg 1,484 ... F10
53170 Silver Lake 1,598 ... K10
54872 Siren 896 ... B4
54234 Sister Bay 564 ... M5
53086 Slinger 1,612 ... K9
54655 Soldiers Grove 622 ... E9
54873 Solon Springs 590 ... C3
54025 Somerset 860 ... A5
53172 South Milwaukee 21,069 .. M2
53587 South Wayne 495 ... G10
54656 Sparta⊙ 6,934 ... E8
54479 Spencer 1,754 ... F6
54801 Spooner 2,365 ... B4
53588 Spring Green 1,265 ... G9
54767 Spring Valley 982 ... B6
54768 Stanley 2,095 ... E6
54026 Star Prairie 420 ... A5
54480 Stetsonville 487 ... F5
54657 Steuben 175 ... E9
54481 Stevens Point⊙ 22,970 ... G7
54172 Stiles 300 ... L6
53825 Stitzer 190 ... E10
53088 Stockbridge 567 ... K7
54769 Stockholm 104 ... B7
54658 Stoddard 762 ... D8
54876 Stone Lake 210 ... C4
53589 Stoughton 7,589 ... H10
54484 Stratford 1,385 ... F6
54770 Strum 944 ... D6
54235 Sturgeon Bay⊙ 8,847 ... M6
53177 Sturtevant 4,130 ... M3
54173 Suamico 900 ... K6
53178 Sullivan 434 ... H1
54485 Summit Lake 250 ... H5
53590 Sun Prairie 12,931 ... H9
54880 Superior⊙ 29,571 ... C2
Superior-Duluth‡ 266,650 . C2
†54880 Superior Village 580 ... B2
54174 Suring 581 ... K5
53089 Sussex 3,482 ... K1
53090 Taycheedah 350 ... K8
54659 Taylor 411 ... E7
†53820 Tennyson 476 ... E10
53091 Theresa 766 ... K8
53092 Thiensville 3,341 ... L1
54771 Thorp 1,635 ... E6
54562 Three Lakes 950 ... H4
54486 Tigerton 865 ... H6
54240 Tisch Mills 315 ... L7
54660 Tomah 7,204 ... F8
54487 Tomahawk 3,527 ... G5
54563 Tony 146 ... E5
54888 Trego 280 ... C4
54661 Trempealeau 956 ... C8
54662 Tunnel City 200 ... E7
54889 Turtle Lake 762 ... B5
53181 Twin Lakes 3,474 ... K11

Topography

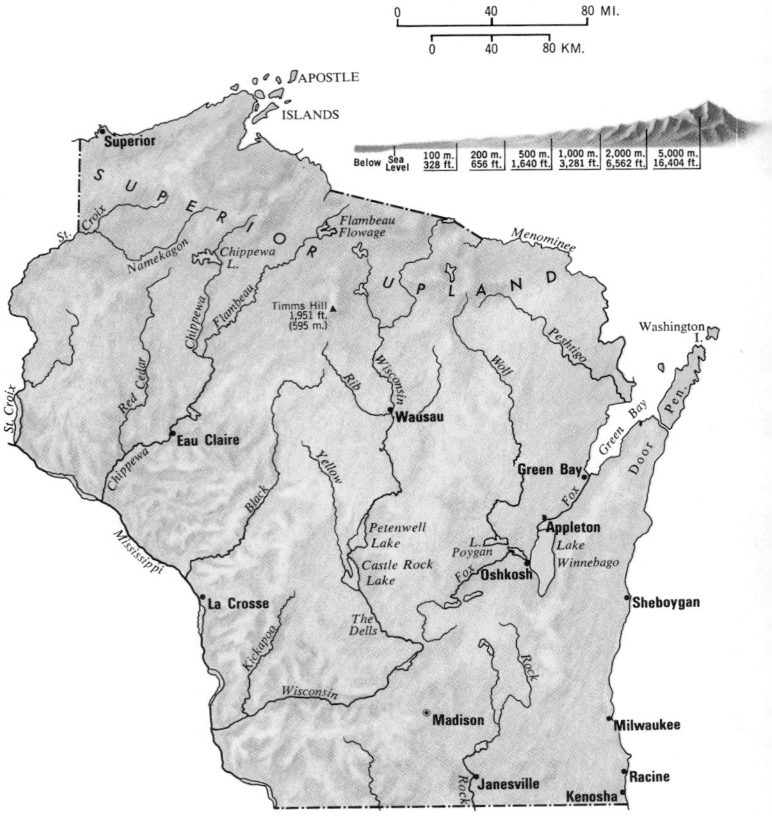
0 40 80 MI.
0 40 80 KM.
Below Sea Level
100 m. 328 ft.
200 m. 656 ft.
500 m. 1,640 ft.
1,000 m. 3,281 ft.
2,000 m. 6,562 ft.
5,000 m. 16,404 ft.
APOSTLE ISLANDS
Superior
SUPERIOR UPLAND
St. Croix
Namekagon
Chippewa L.
Flambeau Flowage
Menominee
Timms Hill 1,951 ft. (595 m.)
Washington I.
Peshtigo
Wolf
Rib
Wisconsin
Wausau
Green Bay
Door Pen.
Eau Claire
Red Cedar
Chippewa
Flambeau
St. Croix
Yellow
Black
Mississippi
Petenwell Lake
Castle Rock Lake
L. Poygan
Fox
Oshkosh
Appleton
Lake Winnebago
Sheboygan
La Crosse
The Dells
Kickapoo
Rock
Wisconsin
Madison
Milwaukee
Janesville
Racine
Kenosha

54241 Two Rivers 13,354 ... M7
53962 Union Center 216 ... F8
53182 Union Grove 3,517 ... L3
54488 Unity 418 ... F6
54245 Valders 984 ... L7
53593 Verona 3,336 ... G9
54489 Vesper 554 ... F7
54664 Viola 696 ... E8
54665 Viroqua⊙ 3,716 ... D8
54566 Wabeno 800 ... J5
53093 Waldo 416 ... L8
53183 Wales 1,992 ... J1
53184 Walworth 1,607 ... J10
54666 Warrens 300 ... E7
54890 Wascott 70 ... C3
54891 Washburn⊙ 2,080 ... D2
54246 Washington Island 550 ... M5
53185 Waterford 2,051 ... K3
53594 Waterloo 2,393 ... J9
53094 Watertown 18,113 ... J9
53021 Waubeka 450 ... L9
53186 Waukesha⊙ 50,365 ... K1
53597 Waunakee 3,866 ... G9
54981 Waupaca⊙ 4,472 ... H7
53963 Waupun 8,132 ... J8
54401 Wausau⊙ 32,426 ... G6
Wausau‡ 111,270 ... G6
54177 Wausaukee 648 ... K5
54982 Wautoma⊙ 1,629 ... H7
53226 Wauwatosa 51,308 ... L1
53826 Wauzeka 580 ... E9
†54126 Wayside 140 ... L7
54893 Webster 610 ... B4
53214 West Allis 63,982 ... L1
†53913 West Baraboo 846 ... G9
53095 West Bend⊙ 21,484 ... K9
54490 Westboro 750 ... F5
54667 Westby 1,797 ... E8
53964 Westfield 1,033 ... H8
†53201 West Milwaukee 3,535 ... L1
†54476 Weston 8,775 ... G6
54669 West Salem 3,276 ... D8
54983 Weyauwega 1,549 ... H7
54895 Weyerhaeuser 313 ... D5
54896 Wheeler 231 ... C5
54773 Whitehall⊙ 1,530 ... D7
54491 White Lake 309 ... J5
54247 Whitelaw 649 ... L7
53190 Whitewater 11,520 ... J10
†54481 Whiting 2,050 ... H7
54984 Wild Rose 741 ... H7
53191 Williams Bay 1,763 ... J10
54027 Wilson 155 ... B6
54670 Wilton 465 ... F8
54567 Winchester 300 ... G3
53185 Wind Lake 900 ... K2
†53401 Wind Point 1,695 ... M2
53598 Windsor 827 ... H9
54985 Winnebago 1,433 ... J8
54986 Winneconne 1,935 ... J7
54896 Winter 376 ... E4
53965 Wisconsin Dells 2,521 ... G8
54494 Wisconsin Rapids⊙ 17,995 G7
54498 Withee 509 ... E6
54499 Wittenberg 997 ... H6
53968 Wonewoc 842 ... F8
53827 Woodman 116 ... E9
54568 Woodruff 850 ... G4
54028 Woodville 725 ... B6
54180 Wrightstown 1,169 ... K7
54671 Wyeville 163 ... F7
53969 Wyocena 548 ... H9
54182 Zachow 135 ... K6

OTHER FEATURES

Apostle (isls.) ... F2
Apostle Islands Nat'l Lakeshore ... E1
Apple (riv.) ... A5
Bad River Ind. Res. ... E2
Bardon (lake) ... C3
Bear (isl.) ... E1
Beaver Dam (lake) ... J9
Beulah (lake) ... J2
Big Eau Pleine (res.) ... G6
Big Muskego (lake) ... L2
Big Rib (riv.) ... G5
Black (riv.) ... E7
Butternut (lake) ... J4
Castle Rock (lake) ... G8
Cat (isl.) ... E1
Chambers (isl.) ... M5
Chequamegon (bay) ... E2
Chetac (lake) ... D4
Chippewa (lake) ... D4
Chippewa (riv.) ... B7
Clam (lake) ... B4
Clam (riv.) ... A4
Dells, The (valley) ... G8
Denoon (lake) ... K2
Du Bay (lake) ... G6
Eagle (lake) ... H3
Eagle (lake) ... K3
Eau Claire (riv.) ... D6
Flambeau (riv.) ... E4
Flambeau Flowage (res.) ... F3
Fox (riv.) ... K2
Fox (riv.) ... K7
General Mitchell Field ... M2
Geneva (lake) ... K10
Golden (lake) ... H1
Green (bay) ... L6
Grindstone (lake) ... C4
Holcombe Flowage (res.) ... D5
Jump (riv.) ... E5
Kegonsa (lake) ... H10
Kickapoo (riv.) ... E9
Koshkonong (lake) ... H10
La Belle (lake) ... H1
Lac Court Oreilles Ind. Res. ... D4
Lac du Flambeau Ind. Res. ... G3
Long (lake) ... C4
Madeline (isl.) ... E2
Mendota (lake) ... H9
Menominee (riv.) ... L5
Metonga (lake) ... J4
Michigan (isl.) ... F2
Michigan (lake) ... M9
Mississippi (riv.) ... D10
Montreal (riv.) ... F2
Moose (lake) ... E3
Moose (lake) ... F3
Nagawicka (lake) ... J1
Namekagon (lake) ... D3
Namekagon (riv.) ... C3
North (lake) ... J1
Oak (isl.) ... E2
Oconomowoc (lake) ... H1
Oconto (riv.) ... K5
Okauchee (lake) ... J1
Outer (isl.) ... F1
Owen (lake) ... D3
Pecatonica (riv.) ... H11
Pelican (lake) ... H4
Pepin (lake) ... B7
Peshtigo (riv.) ... K5
Petenwell (lake) ... G7
Pewaukee (lake) ... K1
Phantom (lake) ... J2
Pine (lake) ... J1
Porte des Morts (str.) ... N5
Poygan (lake) ... J7
Puckaway (lake) ... H8
Red Cedar (riv.) ... C5
Red Cliff Ind. Res. ... E2
Rib (mt.) ... G6
Rock (riv.) ... J9
Round (lake) ... F4
Round (lake) ... D3
Saint Croix (lake) ... A6
Saint Croix (riv.) ... A4
Saint Croix Flowage (res.) ... C3
Saint Louis (riv.) ... A2
Sand (isl.) ... D2
Shawano (lake) ... K6
Shell (lake) ... C4
Spider (lake) ... D3
Stockbridge Ind. Res. ... J6
Stockton (isl.) ... F2
Sugar (riv.) ... H10
Sugarbush Hill (mt.) ... J4
Superior (lake) ... F1
Thunder (lake) ... H4
Tichigan (lake) ... K2
Timms Hill (mt.) ... F5
Trempealeau (riv.) ... C7
Trout (lake) ... G3
Vieux Desert (lake) ... G3
Washington (isl.) ... M5
Willow (res.) ... F4
Wind (lake) ... K2
Winnebago (lake) ... K7
Wisconsin (riv.) ... E9
Wolf (riv.) ... J5
Yellow (lake) ... B4
Yellow (riv.) ... F7

⊙County seat.
‡Population of metropolitan area.
† Zip of nearest p.o. * Multiple zips.

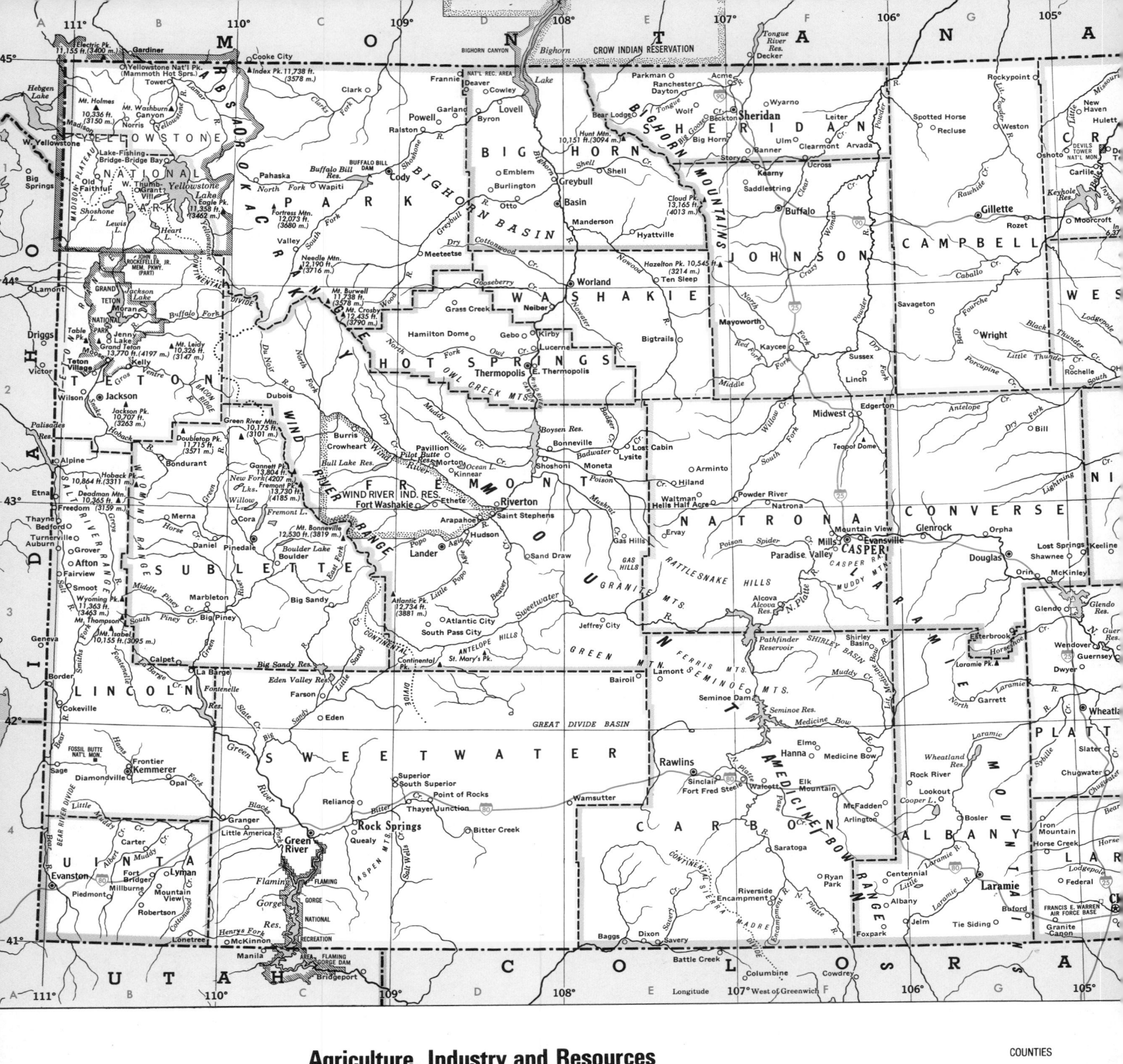

Agriculture, Industry and Resources

DOMINANT LAND USE

- Specialized Wheat
- Specialized Dairy
- General Farming, Livestock, Special Crops
- Sugar Beets, Dry Beans, Livestock, General Farming
- Range Livestock
- Forests
- Nonagricultural Land

MAJOR MINERAL OCCURRENCES

C	Coal	G	Natural Gas	So	Soda Ash
Cl	Clay	O	Petroleum	U	Uranium
Fe	Iron Ore	P	Phosphates	V	Vanadium
⚡	Water Power				

COUNTIES

Albany 29,062	G4
Big Horn 11,896	E1
Campbell 24,367	G1
Carbon 21,896	F4
Converse 14,069	G3
Crook 5,308	H1
Fremont 38,992	D2
Goshen 12,040	H4
Hot Springs 5,710	D2
Johnson 6,700	F1
Laramie 68,649	H4
Lincoln 12,177	B3
Natrona 71,856	F3
Niobrara 2,924	H2
Park 21,639	C1
Platte 11,975	H4
Sheridan 25,048	F1
Sublette 4,548	C3
Sweetwater 41,723	D4
Teton 9,355	B2
Uinta 13,021	E2
Washakie 9,496	E2
Weston 7,106	H2

CITIES and TOWNS

Zip	Name/Pop.	Key
83110	Afton 1,481	B3
82050	Albin 128	H4
82620	Alcova 275	F3

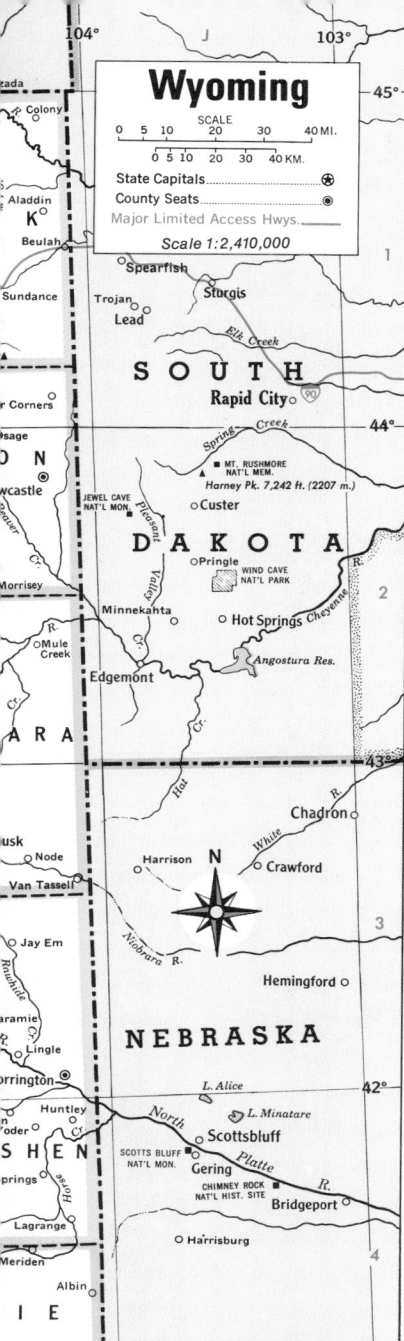

Wyoming

SCALE
0 5 10 20 30 40 MI.
0 5 10 20 30 40 KM.
State Capitals......................⊛
County Seats.......................◉
Major Limited Access Hwys._____

Scale 1:2,410,000

AREA 97,809 sq. mi. (253,325 sq. km.)
POPULATION 469,557
CAPITAL Cheyenne
LARGEST CITY Casper
HIGHEST POINT Gannett Pk. 13,804 ft. (4207 m.)
SETTLED IN 1834
ADMITTED TO UNION July 10, 1890
POPULAR NAME Equality State
STATE FLOWER Indian Paintbrush
STATE BIRD Meadowlark

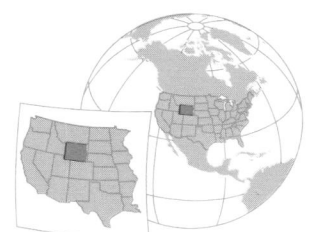

Topography

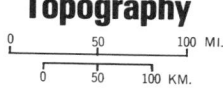

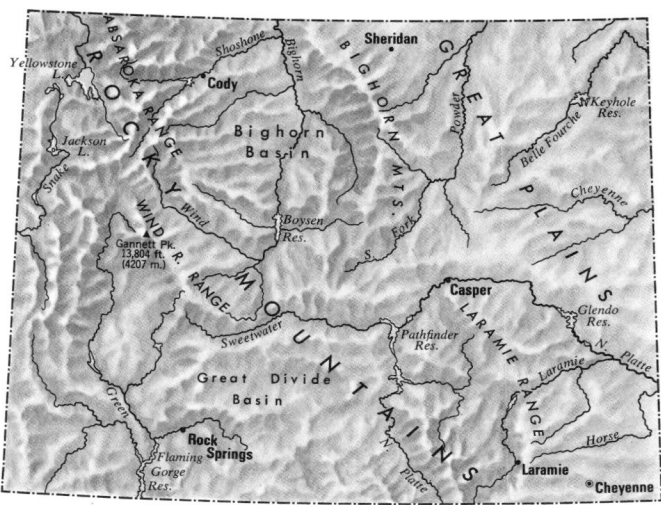

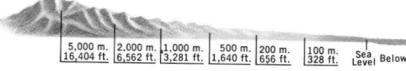

5,000 m. / 16,404 ft. | 2,000 m. / 6,562 ft. | 1,000 m. / 3,281 ft. | 500 m. / 1,640 ft. | 200 m. / 656 ft. | 100 m. / 328 ft. | Sea Level | Below

82510 Arapahoe 682.............D3	82926 Eden 198.............C3
83111 Auburn 360.............A3	82635 Edgerton 510.............F2
82321 Baggs 433.............E4	82324 Elk Mountain 338.............F4
82322 Bairoil 300.............E3	82325 Encampment 611.............F4
82410 Basin⊙ 1,349.............E1	83118 Etna 200.............A2
†82801 Beckton 110.............E1	82930 Evanston⊙ 6,421.............B4
83112 Bedford 350.............A3	82636 Evansville 2,335.............F3
82712 Beulah 184.............H1	83119 Fairview 150.............B3
82833 Big Horn 350.............E1	82932 Farson 350.............C3
83113 Big Piney 530.............B3	82933 Fort Bridger 300.............B4
82051 Bosler 195.............G4	82212 Fort Laramie 356.............H3
82834 Buffalo⊙ 3,799.............F1	82514 Fort Washakie 400.............C2
82411 Burlington 300.............D1	†82001 Fox Farm 2,850.............H4
82053 Burns 268.............H4	82423 Frannie 138.............D1
82412 Byron 633.............D1	83120 Freedom 400.............B3
82601 Casper⊙ 51,016.............F3	83121 Frontier 150.............B4
82055 Centennial 140.............F4	82501 Gas Hills 150.............E3
82001 Cheyenne (cap.)⊙ 47,283.............H4	82716 Gillette⊙ 12,134.............G1
82210 Chugwater 282.............G3	82213 Glendo 367.............G3
82835 Clearmont 191.............F1	82637 Glenrock 2,736.............G3
82414 Cody⊙ 6,790.............D1	82934 Granger 177.............C4
83114 Cokeville 515.............B3	82425 Grass Creek 152.............D2
82420 Cowley 455.............D1	82935 Green River⊙ 12,807.............C4
82512 Crowheart 200.............C2	82426 Greybull 2,277.............E1
83115 Daniel 130.............B3	83122 Grover 425.............B3
82836 Dayton 701.............E1	82214 Guernsey 1,512.............H3
82421 Deaver 178.............D1	82327 Hanna 2,288.............F4
83116 Diamondville 1,000.............B4	82215 Hartville 149.............H3
82323 Dixon 82.............E4	82060 Hillsdale 160.............H4
82633 Douglas⊙ 6,030.............G3	82061 Horse Creek 225.............G4
82513 Dubois 1,067.............C2	82515 Hudson 514.............D3
82443 East Thermopolis 359.............D2	82720 Hulett 291.............H1

83001 Jackson⊙ 4,511.............B2	82842 Story 637.............F1
82310 Jeffrey City 1,882.............E3	82729 Sundance⊙ 1,087.............H1
82639 Kaycee 271.............F2	82945 Superior 500.............D4
83011 Kelly 100.............B2	82442 Ten Sleep 407.............E1
83101 Kemmerer⊙ 3,273.............B4	83127 Thayne 256.............A3
82516 Kinnear 145.............D2	82443 Thermopolis⊙ 3,852.............D2
82430 Kirby 129.............D2	82240 Torrington⊙ 5,441.............H3
83123 La Barge 302.............B3	82730 Upton 1,193.............H1
82221 Lagrange 232.............H4	82242 Van Tassell 10.............H3
82520 Lander⊙ 7,867.............D3	82335 Walcott 200.............F4
82070 Laramie⊙ 24,410.............G4	82336 Wamsutter 681.............E4
82640 Linch 187.............F2	82201 Wheatland⊙ 5,816.............H3
82223 Lingle 475.............H3	83014 Wilson 480.............B2
82929 Little America 175.............C4	82401 Worland⊙ 6,391.............E1
†82642 Lost Cabin 25.............E2	82732 Wright 1,117.............G2
82224 Lost Springs 9.............G3	82190 Yellowstone Nat'l Pk. 350.............B1
82431 Lovell 2,447.............D1	82244 Yoder 110.............H4
†82443 Lucerne 240.............D2	
82225 Lusk⊙ 1,650.............H3	OTHER FEATURES
82937 Lyman 2,284.............B4	
82642 Lysite 175.............E2	Absaroka (range).............C1
†82190 Mammoth Hot Springs	Antelope (creek).............G2
(Yellowstone Nat'l Park 350.............B1	Antelope (hills).............D3
82432 Manderson 174.............E1	Aspen (mts.).............C4
82227 Manville 94.............H3	Atlantic (peak).............D3
†83113 Marbleton 537.............B3	Badwater (creek).............E2
82938 McKinnon 135.............C4	Bear (creek).............H4
82329 Medicine Bow 953.............F4	Bear (riv.).............B4
82433 Meeteetse 512.............D1	Bear Lodge (mts.).............H1
82643 Midwest 638.............F2	Bear River Divide (mts.).............B4
82644 Mills 2,139.............F3	Beaver (creek).............D3
82721 Moorcroft 1,014.............H1	Beaver (creek).............H2
83012 Moose 150.............B2	Belle Fourche (riv.).............H1
83013 Moran 200.............B2	Big Goose (creek).............E1
†82601 Mountain View.............F3	Bighorn (basin).............D1
82939 Mountain View 628.............B4	Bighorn (lake).............D1
82701 Newcastle⊙ 3,596.............H2	Bighorn (mts.).............E1
82190 Old Faithful 75.............B1	Bighorn (riv.).............D1
†82001 Orchard Valley 3,327.............H4	Bighorn Canyon Nat'l Rec. Area.............D1
82723 Osage 500.............H2	Big Sandy (riv.).............C3
†82601 Paradise Valley.............F3	Bitter (creek).............C4
82523 Pavillion 287.............D2	Blacks Fork, Green (riv.).............C4
82082 Pine Bluffs 1,077.............H4	Black Thunder (creek).............G2
82941 Pinedale⊙ 1,066.............C3	Bonneville (mt.).............C3
82942 Point of Rocks 425.............D4	Boysen (res.).............D2
82435 Powell 5,310.............C1	Buffalo Bill (dam).............C1
82839 Ranchester 655.............E1	Buffalo Bill (res.).............C1
82301 Rawlins⊙ 11,547.............E4	Buffalo Fork, Snake (riv.).............B2
82725 Recluse 225.............G1	Burwell (mt.).............C2
82943 Reliance 325.............C4	Caballo (creek).............G1
†82325 Riverside 55.............F4	Casper (range).............F3
82501 Riverton 9,247.............D2	Cheyenne (riv.).............H2
82944 Robertson 142.............B4	Chugwater (creek).............H4
82083 Rock River 415.............G4	Clarks Fork (riv.).............C1
82901 Rock Springs 19,458.............C4	Clear (creek).............F1
82331 Saratoga 2,410.............F4	Cloud (peak).............E1
82801 Sheridan⊙ 15,146.............F1	Cottonwood (creek).............B4
82615 Shirley Basin 400.............F3	Crazy Woman (creek).............F1
82649 Shoshoni 879.............D2	Crosby (mt.).............C2
82334 Sinclair 586.............E4	Crow (creek).............H4
83126 Smoot 310.............B3	Deadman (mt.).............B2
†82945 South Superior 586.............D4	Devils Tower Nat'l Mon.............H1

Doubletop (peak).............B2	Lodgepole (creek).............H2
Dry (creek).............C2	Lodgepole (creek).............H4
Dry Cottonwood (creek).............D1	Madison (plat.).............B1
Eagle (peak).............B1	Medicine Bow (range).............F4
Fivemile (creek).............D2	Medicine Bow (riv.).............F3
Flaming Gorge (res.).............C4	Middle Piney (creek).............B3
Flaming Gorge Nat'l Rec. Area.............C4	Muddy (creek).............D2
Fontenelle (creek).............B3	Muskrat (creek).............E2
Fontenelle (res.).............B3	Needle (mt.).............C1
Fort Laramie Nat'l Hist. Site.............H3	Niobrara (riv.).............J3
Fortress (mt.).............C1	North Laramie (riv.).............G3
Fossil Butte Nat'l Mon.............B4	North Platte (riv.).............H3
Francis E. Warren A.F.B. 3,627.............G4	Nowater (creek).............E2
Fremont (lake).............C3	Nowood (riv.).............E1
Fremont (peak).............C2	Owl, North Fork (creek).............D2
Gannett (peak).............C2	Owl Creek (mts.).............D2
Gas (hills).............E3	Palisades (res.).............A2
Glendo (res.).............H3	Pass (creek).............F4
Gooseberry (creek).............D1	Pathfinder (res.).............F3
Grand Teton (mt.).............B2	Poison (creek).............E2
Grand Teton Nat'l Park.............B2	Poison Spider (creek).............F3
Granite (mts.).............E3	Popo Agie (riv.).............D3
Great Divide (basin).............E3	Powder (riv.).............F2
Green (mt.).............E3	Rattlesnake (range).............E3
Green (riv.).............C4	Rawhide (creek).............G1
Green, East Fork (riv.).............C3	Rawhide (creek).............H3
Green River (mt.).............C2	Rocky (mts.).............C1
Greybull (riv.).............D1	Salt (riv.).............B3
Greys (riv.).............B3	Salt River (range).............B3
Gros Ventre (riv.).............B2	Salt Wells (creek).............D4
Guernsey (res.).............H3	Seminoe (mts.).............E3
Hams Fork (riv.).............B4	Seminoe (res.).............F3
Hazelton (peak).............E1	Shell (creek).............E1
Henrys Fork, Green (riv.).............C4	Shirley (mts.).............F3
Hoback (peak).............B2	Shoshone (lake).............B1
Hoback (riv.).............B2	Shoshone (riv.).............D1
Holmes (mt.).............B1	Sierra Madre (mts.).............E4
Horse (creek).............H4	Slate (creek).............C3
Horseshoe (creek).............G3	Smiths Fork (riv.).............B3
Hunt (mt.).............E1	Snake (riv.).............B2
Index (peak).............C1	South Cheyenne (riv.).............H2
Inyan Kara (creek).............H1	South Piney (creek).............F1
Inyan Kara (mt.).............H1	Sweetwater (riv.).............D3
Isabel (mt.).............B3	Sybille (creek).............G4
Jackson (lake).............B2	Teapot Dome (mt.).............F2
Jackson (peak).............B2	Teton (range).............B2
John D. Rockefeller, Jr., Mem.	Tongue (riv.).............E1
Pkwy.............B1	Washburn (mt.).............B1
Keyhole (res.).............H1	Wheatland (res.).............G4
Lamar (riv.).............B1	Willow (creek).............F2
Lance (creek).............G1	Wind (riv.).............D2
Laramie (mts.).............G3	Wind River (canyon).............D2
Laramie (peak).............G3	Wind River (range).............C2
Laramie (riv.).............G4	Wind River Ind. Res.............C2
Leidy (mt.).............B1	Wood (riv.).............C2
Lewis (lake).............B1	Wyoming (peak).............B3
Lightning (creek).............G1	Wyoming (range).............B3
Little Missouri (riv.).............H1	Yellowstone (lake).............B1
Little Muddy (creek).............B4	Yellowstone (riv.).............B1
Little Powder (riv.).............G1	Yellowstone Nat'l Park.............B1
Little Sandy (creek).............C3	⊙County seat.
Little Thunder (creek).............G2	† Zip of nearest p.o. * Multiple zips.

© Copyright HAMMOND INCORPORATED, Maplewood, N.J.

Acquisitions of Territory

WASHINGTON
OREGON COUNTRY
OREGON 1846 Treaty with Great Britain
MONTANA
NORTH DAKOTA
RED RIVER Title Established 1818
MINNESOTA
MICHIGAN
MAINE
VT. N.H.
NEW YORK
MASS.
CONN. R.I.
IDAHO
SOUTH DAKOTA
WISCONSIN
PENNSYLVANIA
NEW JERSEY
WYOMING
NEBRASKA
IOWA
OHIO
WEST VIRGINIA
MD. DEL.
NEVADA
UTAH
COLORADO
KANSAS
MISSOURI
ILLINOIS
INDIANA
UNITED STATES
VIRGINIA
KENTUCKY
CALIFORNIA
MEXICAN CESSION 1848
L O U I S I A N A Purchased from France 1803
KENTUCKY
TENNESSEE 1783
NORTH CAROLINA
ARIZONA
NEW MEXICO
OKLAHOMA
ARKANSAS
SOUTH CAROLINA
GADSDEN PURCHASE 1853
T E X A S Annexed in 1845
TEXAS
MISSISSIPPI
ALABAMA
GEORGIA
LOUISIANA
F L O R I D A 1819 Treaty with Spain

ALASKA 1867 Purchased from Russia

H A W A I I
Annexed in 1898

The United States in 1783 comprised the thirteen original states and included lands acquired by conquest during the Revolution and by the Treaty of 1783.

Rank by Area

Rank by Population

© Copyright
HAMMOND INCORPORATED

YEAR OF ADMISSION TO THE UNION

DELAWARE ☆	1787
PENNSYLVANIA ☆	
NEW JERSEY ☆	
GEORGIA ☆	1788
CONNECTICUT ☆	
MASSACHUSETTS ☆	
MARYLAND ☆	
SOUTH CAROLINA ☆	
NEW HAMPSHIRE ☆	
VIRGINIA ☆	
NEW YORK ☆	
NORTH CAROLINA ☆	1789
RHODE ISLAND ☆	1790
VERMONT ☆	1791
KENTUCKY ☆	1792
TENNESSEE ☆	1796
OHIO ☆	1803
LOUISIANA ☆	1812
INDIANA ☆	1816
MISSISSIPPI ☆	1817
ILLINOIS ☆	1818
ALABAMA ☆	1819
MAINE ☆	1820
MISSOURI ☆	1821

1959 ☆HAWAII ☆ALASKA
1912 ☆ARIZONA ☆NEW MEXICO
1907 ☆OKLAHOMA
☆UTAH ☆IDAHO ☆MONTANA ☆NORTH DAKOTA ☆SOUTH DAKOTA ☆WEST VIRGINIA ☆MINNESOTA ☆CALIFORNIA ☆WISCONSIN ☆IOWA ☆FLORIDA ☆ARKANSAS ☆MICHIGAN
☆WYOMING ☆WASHINGTON COLORADO ☆ NEBRASKA ☆ NEVADA ☆ KANSAS ☆ OREGON ☆ TEXAS ☆
1896 1889 1876 1867 1864 1861 1858 1848 1845 1836
1890 1863 1859 1850 1846 1837

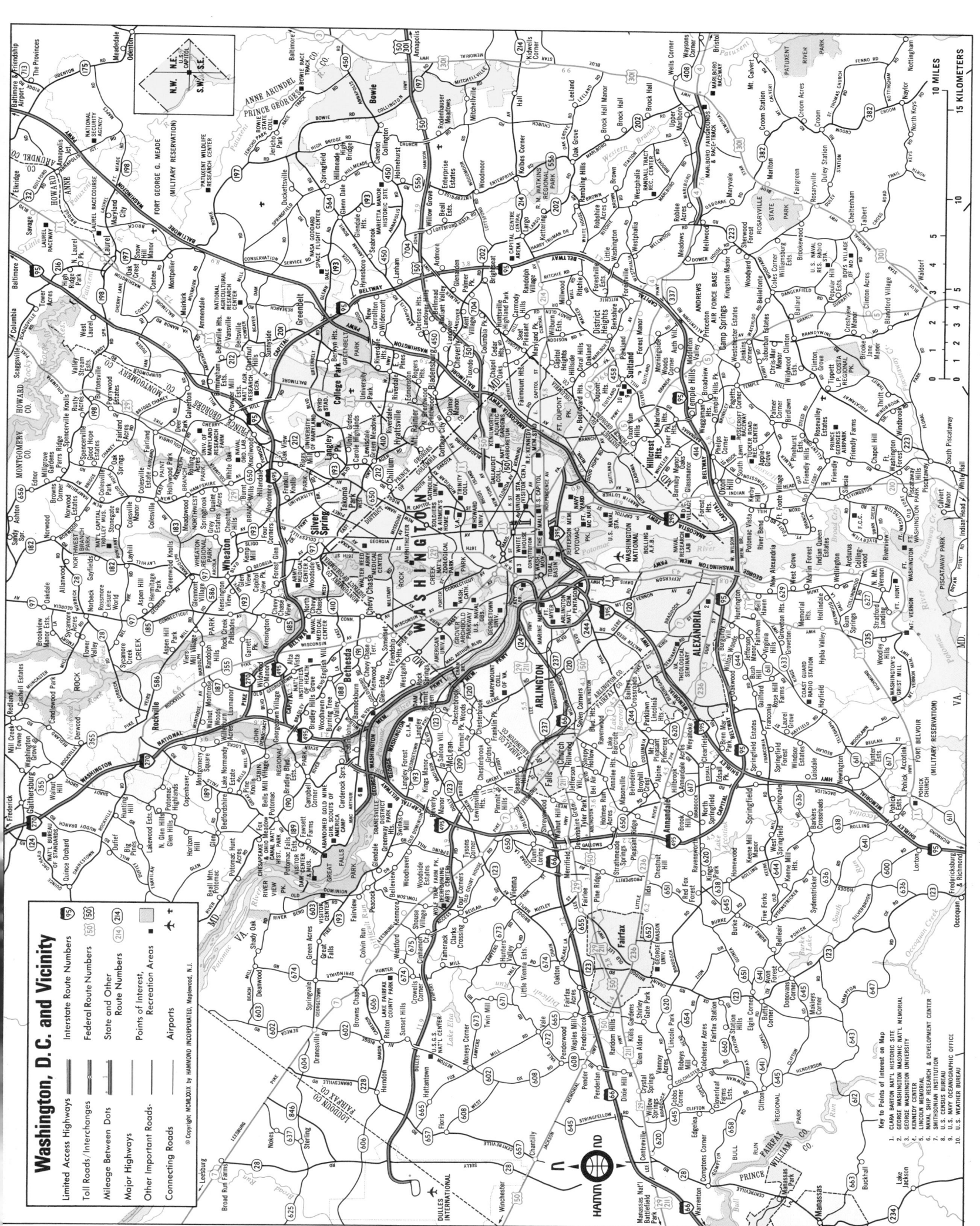

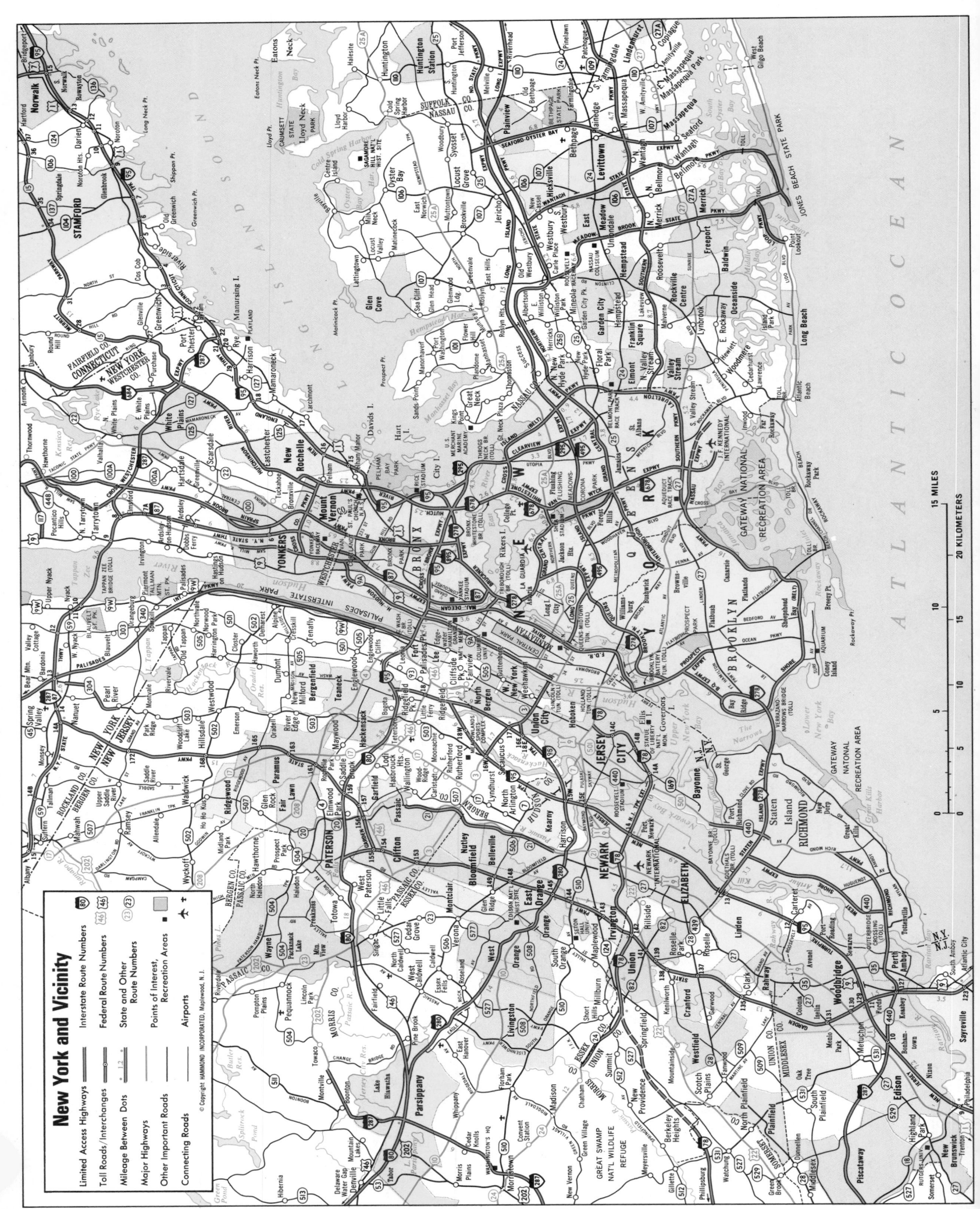

New York and Vicinity

Limited Access Highways	Interstate Route Numbers
Toll Roads/Interchanges	Federal Route Numbers
Mileage Between Dots	State and Other Route Numbers
Major Highways	Points of Interest, Recreation Areas
Other Important Roads	Airports
Connecting Roads	

© Copyright HAMMOND INCORPORATED, Maplewood, N.J.

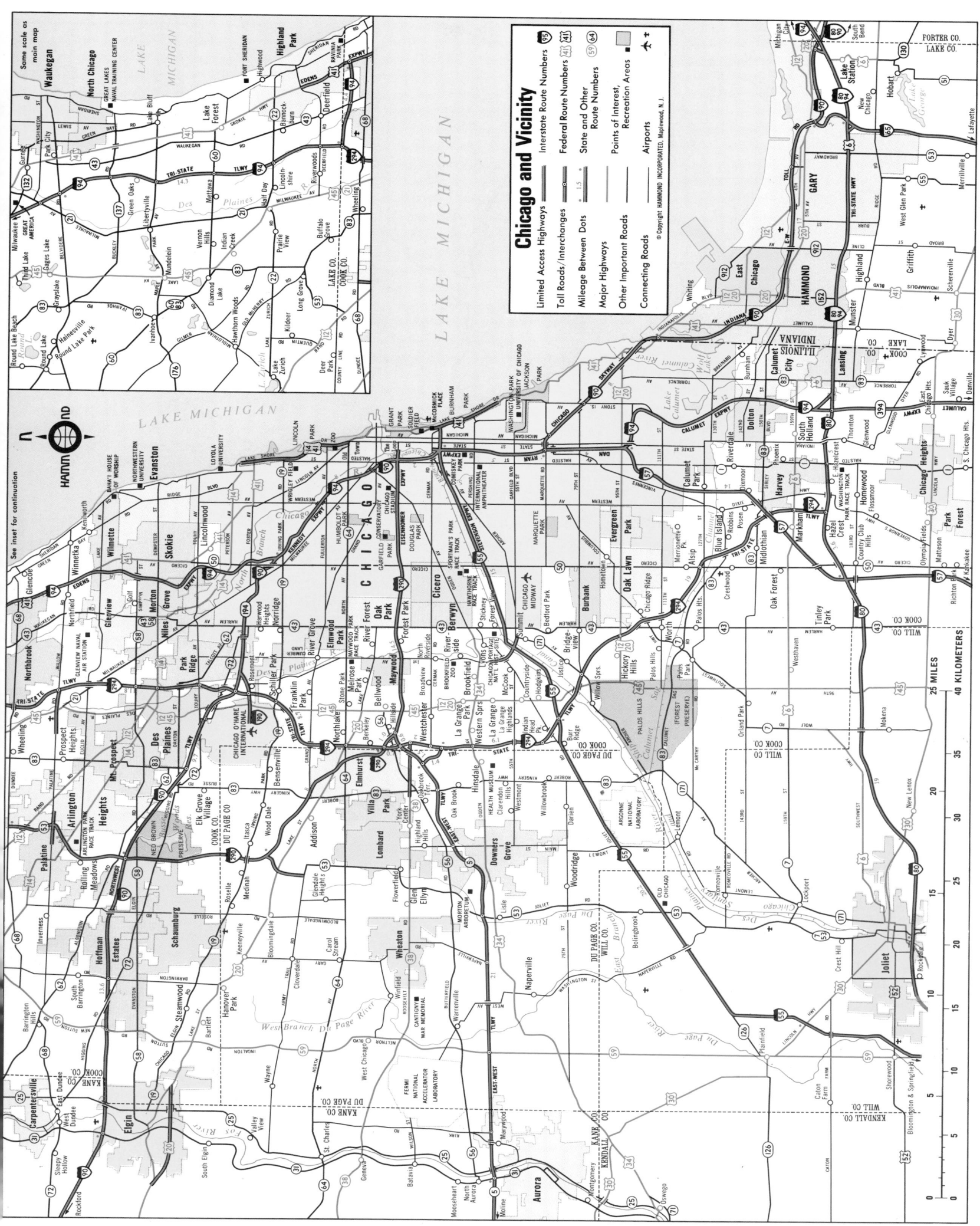

Chicago and Vicinity

Limited Access Highways — Interstate Route Numbers
Toll Roads/Interchanges — Federal Route Numbers
Mileage Between Dots — State and Other Route Numbers
Major Highways — Points of Interest, Recreation Areas
Other Important Roads — Airports
Connecting Roads

© Copyright HAMMOND INCORPORATED, Maplewood, N.J.

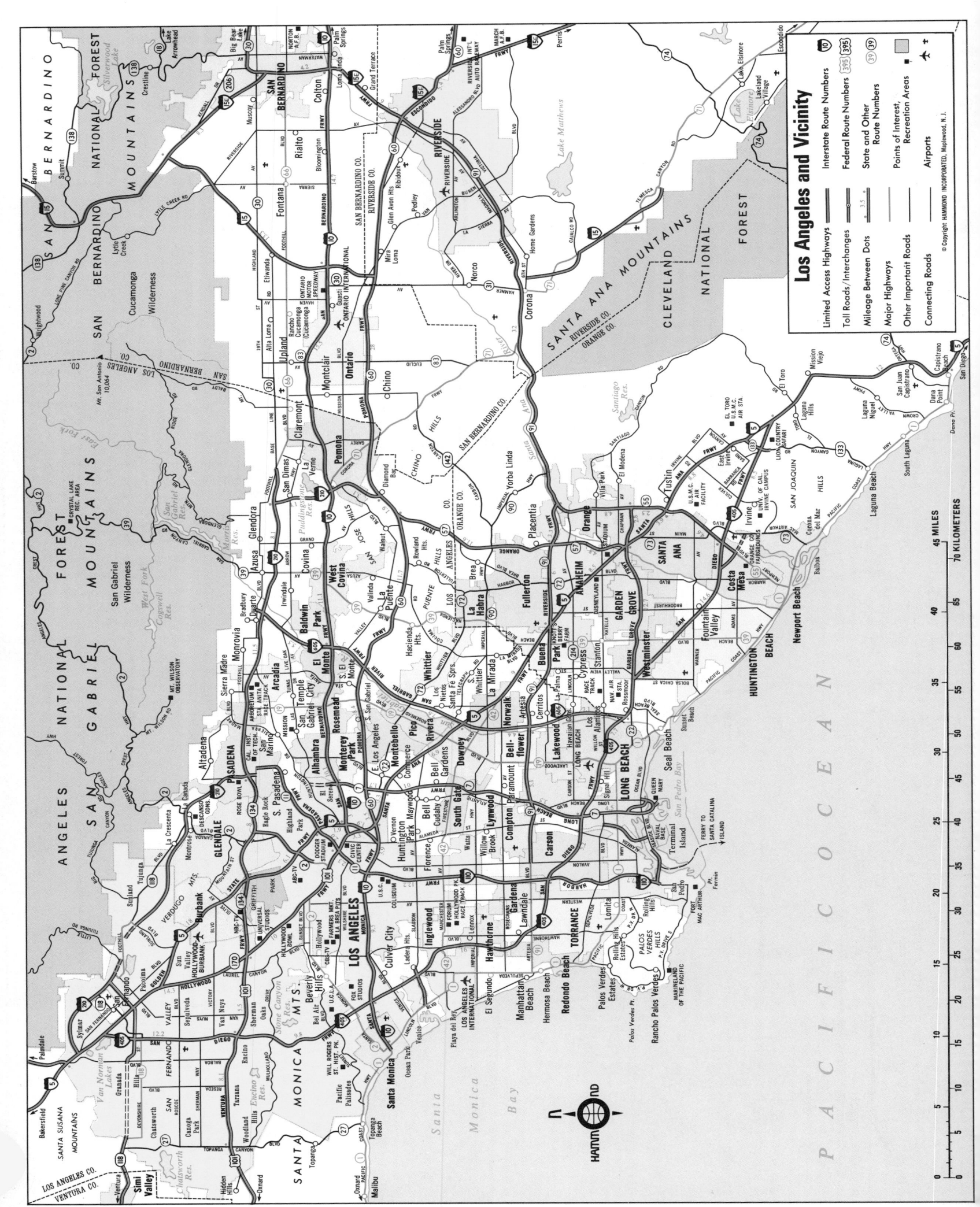

INDEX OF THE WORLD

Introduction

This index contains a complete alphabetical listing of more than one hundred thousand names shown on all the maps included in this atlas. Names not found in the individual indexes accompanying the maps appear here. The user who is unfamiliar with the location of a country, town, or physical feature, or who is in doubt as to which country, state or province a place belongs will find the answers to his questions in this index. Entries are indexed to all maps or insets showing the place.

The name of the feature sought will be found in its proper alphabetical sequence, followed by the name of the political division in which it is located, the page number of the map on which it will be found, and the key reference necessary for finding its location on the map. After noting the key reference letter-number combination for the place name, turn to the page number indicated. The place name will be found within the square formed by the two lines of latitude and the two lines of longitude which enclose the coordinates—i.e., the marginal letters and numbers. An open circle (○) after the name signifies a township — better known as a town — in the northeastern U.S.

All index entries for cities and towns in the United States are followed by a five-digit postal ZIP code number applying to the community. This useful feature permits the reader to address his mail so that it will be routed and delivered more efficiently and quickly by the U.S. Postal Service. A dagger (†) designates those places that do not possess a post office. The ZIP code number listed in such cases refers to that of the nearest post office. An asterisk (*) marks those larger cities which are divided into multiple ZIP code areas. Using the single ZIP code number listed in such cases will direct your letter to the proper city with dispatch. However, if the precise ZIP code number of the address within the city is needed, it is suggested that the reader refer to the latest National ZIP Code Directory at his local post office. This detailed guide lists every street in a multiple ZIP code city with the proper ZIP code for the street.

Because of limitations of space on the map, place names do not always appear in their complete form on the map. The complete forms are, however, given in the index. Variant spellings of names and alternate names are also given in this index. The alternate form or spelling of the name appears first, followed in parentheses by the name as it appears on the map. Physical features are usually listed under their proper names and not according to their generic terms; that is to say, Rio Negro will be found under Negro and not under Rio Negro. Exceptions are familiar names such as Rio Grande.

The abbreviations for the political division names and geographical features are explained on page XVI of the atlas. In addition, reference can be made to the Gazetteer-Index appearing on pages IX through XIII in which area, population, capital, map reference and population source data may be found for all major political and physical divisions of the world. Population figures for most entries are also included in the comprehensive individual indexes accompanying each map.

A

Aa (riv.), Switzerland 39/F3
Aachen, W. Germany 22/B3
Aadorf, Switzerland 39/G2
Aalen, W. Germany 22/D4
Aalsmeer, Netherlands 27/F4
Aalst, Belgium 27/D7
Aalten, Netherlands 27/K5
Aalter, Belgium 27/C6
Äänekoski, Finland 18/O5
Aarau, Switzerland 39/F2
Aarberg, Switzerland 39/D2
Aarburg, Switzerland 39/E2
Aardenburg, Netherlands 27/C6
Aare (riv.), Switzerland 39/E3
Aargau (canton), Switzerland 39/F2
Aarlen (Arlon), Belgium 27/H9
Aarons (creek), Va. 307/L7
Aaronsburg, Pa. (16820) 294/H4
Aarschot, Belgium 27/F7
Aat (Ath), Belgium 27/D7
Aba, China 77/F5
Aba, Hungary 41/E3
Aba, Nigeria 106/F7
Aba, Nigeria 102/C4
Aba, Zaire 115/F3
Aba as Sa'ud, Saudi Arabia 59/D6
Abacaxis (riv.), Brazil 132/B4
Abadan, Iran 54/F6
Abadan, Iran 66/F5
Abadan, Iran 59/E3
Abadeh, Iran 66/H5
Abadeh, Iran 59/F3
Abadla, Algeria 106/D2
Abádszalók, Hungary 41/F3
Abaeté, Brazil 132/E7
Abaetetuba, Brazil 132/D3
Abaetetuba, Brazil 120/E3
Abagnar (Silinhot), China 77/J3
Abaí, Paraguay 144/E4
Abaiang (atoll), Kiribati 87/H5
Abaila, Saudi Arabia 59/F5
Abajo (mts.), Utah 304/E6
Abakan, U.S.S.R. 54/L4
Abakan, U.S.S.R. 48/K4
Abala, Congo 115/C4
Abalos (pt.), Cuba 158/A2
Abana, Turkey 63/F2
Abancay, Peru 120/B4
Abancay, Peru 128/F9
Abapó, Bolivia 136/D6
Abaq, China 77/J3
Abarqu, Iran 59/F3
Abarqu, Iran 66/H5
Abasan, Gaza Strip 65/A5
Abashiri, Japan 81/M1
Abashiri (riv.), Japan 81/M1
Abau, Papua N.G. 85/C7
Abaújszántó, Hungary 41/F2
Abay (riv.), Ethiopia 111/G5
Abay, U.S.S.R. 48/H5
Abaya (lake), Ethiopia 111/G6
Abaza, U.S.S.R. 48/J4
Abbaye (pt.), Mich. 250/B2
Abbe (lake), Djibouti 111/H5
Abbeville, Ala. (36310) 195/H7
Abbeville, France 28/D2
Abbeville, Georgia (31001) 217/F7
Abbeville, La. (70510) 238/F7
Abbeville, Miss. (38601) 256/F2
Abbeville (co.), S.C. 296/B3
Abbeville, S.C. (29620) 296/C3
Abbey, Sask. 181/C5
Abbey (head), Scotland 15/E6
Abbeydorney, Ireland 17/B7
Abbeyfeale, Ireland 10/B4
Abbeyfeale, Ireland 17/C7
Abbeylara, Ireland 17/F4
Abbeyleix, Ireland 17/G6
Abbotsford, Br. Col. 184/L3
Abbotsford, Wis. (54405) 317/F6
Abbott, Ark. (†72944) 202/B3
Abbott, N. Mex. (†87747) 274/E2
Abbott, Texas (76621) 303/G6
Abbottabad, Pakistan 68/C2
Abbottabad, Pakistan 59/K3
Abbottsburg, N.C. (28321) 281/M5
Abbottsford, Georgia (†30240) 217/B4
Abbottstown, Pa. (17301) 294/J6
Abbot Village○, Maine (04406) 243/D5
Abbyville, Kansas (67510) 232/D4
'Abdul 'Aziz, Jebel (mts.), Syria 63/J4
Abdulino, U.S.S.R. 52/H4
Abéché, Chad 102/D3
Abéché, Chad 111/D5
Abee, Alberta 182/D2
Abell, Md. (20606) 245/M8
Abemama (atoll), Kiribati 87/H5
Abengourou, Ivory Coast 106/D7
Abengourou, Ivory Coast 102/B4
Åbenrå, Denmark 18/F9
Åbenrå, Denmark 21/C7
Abeokuta, Niger 106/E7
Abeokuta, Nigeria 102/C4
Aberaeron, Wales 13/C5
Aberaeron, Wales 10/D4
Abercarn, Wales 13/B6
Aberchirder, Scotland 15/F3
Abercorn, Québec 172/E4
Abercorn (Mbala), Zambia 115/F5
Abercrombie, N. Dak. (58001) 282/S7
Abercrombie, Nova Scotia 168/F3
Abercrombie (mt.), Wash. 310/H2
Aberdare, Wales 13/A6
Aberdare, Wales 10/E5
Aberdaron, Wales 13/C5
Aberdeen, Idaho (83210) 220/F7
Aberdeen, Ky. (42201) 237/H6
Aberdeen, Md. (21001) 245/O2
Aberdeen, Miss. (39730) 256/H3
Aberdeen (dam), Miss. 256/H3
Aberdeen○, N.J. (†07747) 273/E3
Aberdeen, N.S. Wales 97/F3
Aberdeen, N.C. (28315) 281/L4
Aberdeen (lake), N.W. Terrs. 187/J3
Aberdeen, Ohio (45101) 284/C8
Aberdeen, Sask. 181/E3
Aberdeen, Scotland 7/D3
Aberdeen, Scotland 15/F3
Aberdeen, Scotland 10/F2
Aberdeen (trad. co.), Scotland 15/B5
Aberdeen, S. Africa 118/C6
Aberdeen, S. Dak. 146/J5
Aberdeen, S. Dak. 188/G1
Aberdeen, S. Dak. (57401) 298/M3
Aberdeen, Wash. 188/B1
Aberdeen, Wash. (98520) 310/B3
Aberdeen Proving Ground, Md. 245/N3
Aberdour, Scotland 15/D1
Aberfeldy, Sask. 181/B2
Aberfeldy, Scotland 10/D2
Aberfeldy, Scotland 15/E4
Aberfoyle, Scotland 15/D4
Abergavenny, Wales 13/B6
Abergavenny, Wales 10/E5
Abergele, Wales 13/D4
Aberlady, Scotland 15/F4
Aberlour, Scotland 15/E3
Abernant, Ala. (35440) 195/D4
Abernathy, Texas (79311) 303/B4
Abernethy, Sask. 181/H5
Abernethy, Scotland 15/E4
Aberporth, Wales 13/C5
Abert (lake), Oreg. 188/C2
Abert (lake), Oreg. 291/G5
Abertillery, Wales 13/B6
Abertillery, Wales 10/E5
Aberystwyth, Wales 13/C5
Aberystwyth, Wales 10/D4
Abez', U.S.S.R. 52/K1
Abha, Saudi Arabia 59/D6
Abha, Saudi Arabia 54/F8
Abhar, Iran 66/F2
Abiad, Ras el (Blanc) (cape), Tunisia 106/G1
Abibe, Serranía de (mts.), Colombia 126/B3
'Abidiya, Sudan 59/B6
Abidjan (cap.), Ivory Coast 2/J5
Abidjan (cap.), Ivory Coast 102/B4
Abidjan (cap.), Ivory Coast 106/D7
Abie, Nebr. (68001) 264/H3
Abilene, Kansas (67410) 232/E3
Abilene, Texas 146/J6
Abilene, Texas (*79601) 303/E5
Abilene, Texas 188/G4
Abingdon, England 10/F5
Abingdon, England 13/F6
Abingdon, Ill. (61410) 222/C3
Abingdon, Iowa (†52533) 229/J6
Abingdon, Md. (21009) 245/N3
Abingdon, Va. (24210) 307/D7
Abingdon Downs, Queensland 95/B3
Abington, Conn. (06230) 210/G1
Abington, Ind. (†47330) 227/H5
Abington○, Mass. (02351) 249/L4
Abington, Pa. (19001) 294/M5
Abington, Scotland 15/E5
Abiqua (creek), Oreg. 291/B3
Abiquiu, N. Mex. (87510) 274/C2
Abiquiu (res.), N. Mex. 274/C2
Abita Springs, La. (70420) 238/L6
Abitibi (riv.), Ont. 162/H5
Abitibi (lake), Ont. 162/H6
Abitibi (lake), Ontario 175/E3
Abitibi (riv.), Ontario 175/D2
Abitibi (riv.), Ontario 177/J5
Abitibi (terr.), Québec 174/B3
Abitibi (county), Québec 174/B2
Abkhaz A.S.S.R., U.S.S.R. 48/E5
Abkhaz A.S.S.R., U.S.S.R. 52/F6
Abminga, S. Australia 94/D2
Abner, N.C. (†27371) 281/K4
Abnûb, Egypt 111/J4
Åbo (Turku), Finland 18/N6
Aboisso, Ivory Coast 106/D7
Aboite, Ind. (†46783) 227/G3
Abomey, Benin 106/E7
Abong-Mbang, Cameroon 115/B3
Abony, Hungary 41/E3
Abor (hills), India 68/G3
Aborlan, Philippines 82/B6
Abou Deïa, Chad 111/C5
Aboyne, Scotland 15/F3
Abqaiq, Saudi Arabia 59/E4
Abra (prov.), Philippines 82/C2
Abra (riv.), Philippines 82/C2
Abraham (lake), Alberta 182/B3
Abraham (mt.), Maine 243/C5
Abraham, Utah (†84635) 304/B4
Abraham (mt.), Vt. 268/B3
Abraham Lincoln Birthplace Nat'l Hist. Site, Ky. 237/K5
Abrams, Wis. (54101) 317/L6
Abrantes, Portugal 33/B3
Abra Pampa, Argentina 143/C1
Abreus, Cuba 158/D2
'Abri, Sudan 111/F3
Abricots, Haiti 158/A6
Abruzzi (reg.), Italy 34/D3
Absaraka, N. Dak. (58002) 282/P6
Absaroka (range), Mont. 262/F5
Absaroka (range), Wyo. 319/C1
Absarokee, Mont. (59001) 262/G5
Absecon, N.J. (08201) 273/D5
Absecon (inlet), N.J. 273/E5
Abu, India 68/C4
Abu 'Arish, Saudi Arabia 59/D6
Abu Dara, Ras (cape), Sudan 59/C5
Abu Dara, Ras (cape), Sudan 111/G3
Abu Deleiq, Sudan 59/B6
Abu Dhabi (cap.), U.A.E. 54/G7
Abu Dhabi (cap.), U.A.E. 59/F5
Abu ed Duhur, Syria 63/G5
Abu Habl, Wadi (dry riv.), Sudan 111/F5
Abu Hadriya, Saudi Arabia 59/E4
Abu Hamed, Sudan 111/F4
Abu Hamed, Sudan 59/B6
Abuja, Niger 106/F7
Abu Kemal, Syria 59/D3
Abu Kemal, Syria 63/J5
Abukuma (riv.), Japan 81/K4
Abu-Mad, Ras (cape), Saudi Arabia 59/C5
Abu Matariq, Sudan 111/E5
Abumombazi, Zaire 115/D3
Abuná (riv.), Bolivia 136/B2
Abunã, Brazil 132/H10
Abunã (riv.), Brazil 132/G10
Abu Qir (bay), Egypt 111/J2
Abu Qurqâs, Egypt 111/J4
Abu Road, India 68/C4
Abu Rujmein, Jebel (mts.), Syria 63/H5
Abu Shagara, Ras (cape), Sudan 111/G3
Abu Shagara, Ras (cape), Sudan 59/C5
Abut (head), N. Zealand 100/B5
Abu Tabari (well), Sudan 111/E4
Abuyog, Philippines 82/E5
Abu Zabad, Sudan 59/A7
Abu Zabad, Sudan 111/E5
Abwong, Sudan 111/F6
Aby (lag.), Ivory Coast 106/D8
Åbybro, Denmark 21/C3
Abydos (ruins), Egypt 111/F2
Abydos (ruins), Turkey 63/B6
Abyei, Sudan 111/E6
Acacías, Colombia 126/D6
Acaciaville, Nova Scotia 168/C4
Academy, S. Dak. (57310) 298/M7
Acadia (par.), La. 238/F6
Acadia Nat'l Park, Maine 243/G7
Acadia Valley, Alberta 182/E4
Acadie Siding, New Bruns. 170/E2
Acadieville, New Bruns. 170/E2
Acahay, Paraguay 144/B5
Acajutla, El Salvador 154/B4
Acala, Mexico 150/N8
Acala, Texas (†79839) 303/B10
Acámbaro, Mexico 150/J7
Acampo, Calif. (95220) 204/C9
Acandí, Colombia 126/B3
Acaponeta, Mexico 150/G5
Acapulco de Juárez, Mexico 146/H8
Acapulco de Juárez, Mexico 150/K8
Acaraí, Serra do (range), Brazil 132/B2
Acarai (mts.), Guyana 131/B5
Acaraú, Brazil 132/F3
Acaray (riv.), Paraguay 144/E4
Acarí, Peru 128/E10
Acarí (riv.), Peru 128/E10
Acarigua, Venezuela 124/D3
Acatlán de Osorio, Mexico 150/K7
Acatzingo de Hidalgo, Mexico 150/N2
Acayucan, Mexico 150/M8
Acchilla, Bolivia 136/C7
Accident, Md. (21520) 245/A2
Accokeek, Md. (20607) 245/L6
Accomac, Va. (23301) 307/S5
Accomack (co.), Va. 307/S5
Accord, Mass. (02018) 249/E8
Accord, N.Y. (12404) 276/M7
Accoville, W. Va. (25606) 312/C7
Accra (cap.), Ghana 102/B4
Accra (cap.), Ghana 106/D7
Accra (cap.), Ghana 2/J5
Accrington, England 10/G1
Accrington, England 13/H1
Aceguá, Uruguay 145/E2
Acequia, Idaho (83350) 220/E7
Acevedo, Argentina 143/F6
Achacachi, Bolivia 136/A5
Achaguas, Venezuela 124/D4
Achalpur, India 68/D4
Achao, Chile 138/D4
Achar, Uruguay 145/C3
Acharacle, Scotland 15/C4
Achégour (well), Niger 106/G5
Achenkirch, Austria 41/A3
Achill (head), Ireland 10/A4
Achill (head), Ireland 17/A4
Achill (isl.), Ireland 10/A4
Achill (isl.), Ireland 17/A4
Achille, Okla. (74720) 288/O7
Achilles, Va. (23001) 307/R6
Achill Sound, Ireland 17/B4
Achiltibuie, Scotland 15/C3
Achinsk, U.S.S.R. 48/K4
Achnasheen, Scotland 10/D2
Achnasheen, Scotland 15/C3
Achourat (well), Mali 106/D4
A'Chralaig (mt.), Scotland 15/C3
Acı (lake), Turkey 63/C4
Acıgöl, Turkey 63/F3
Acıpayam, Turkey 63/C4
Acireale, Italy 34/E6
Ackerly, Texas (79713) 303/C5
Ackerman, Miss. (39735) 256/F4
Ackerville, Ala. (†36778) 195/D6
Ackley, Iowa (50601) 229/G3
Acklins (isl.), Bahamas 146/L7
Acklins (isl.), Bahamas 156/C2
Ackworth, Iowa (50001) 229/G6
Aclare, Ireland 17/D3
Acle, England 13/J5
Acme, Alberta 182/D4
Acme, La. (71316) 238/G4
Acme, Mich. (49610) 250/D4
Acme, N.C. (†28456) 281/N6
Acme, Texas (†79252) 303/E3
Acme, Wash. (98220) 310/C2
Acme, W. Va. (†25122) 312/D6
Acme, Wyo. (82839) 319/E1
Acoaxet, Mass. (†02837) 249/K7
Acobamba, Peru 128/E9
Acolla, Peru 128/E8
Acoma, N. Mex. (†87034) 274/B4
Acomayo, Cusco, Peru 128/G9
Acomayo, Huánuco, Peru 128/E7
Acomita (Pueblo of Acoma), N. Mex. (†87034) 274/B3
Acona, Miss. (†39095) 256/D4
Aconcagua (mt.) 120/C6
Aconcagua, Cerro (mt.), Argentina 143/C3
Aconcagua, Chile 138/A9
Aconcagua (riv.), Chile 138/F2
Aconchi, Mexico 150/D2
Acopiara, Brazil 132/G4
Acora, Peru 128/H11
Acorizal, Brazil 132/C6
Acoyapa, Nicaragua 154/E5
Acqui Terme, Italy 34/B2
Acraman (lake), S. Australia 94/D5
Acre (state), Brazil 132/G10
Acre (riv.), Brazil 132/G10
Acre, Israel 65/C2

Acree, Georgia (†31791) 217/D7
Acri, Italy 34/F5
Ács, Hungary 41/E3
Actinolite, Ontario 177/G3
Acton, Maine (04001) 243/B8
Acton○, Maine (04001) 243/B8
Acton○, Mass. (01720) 249/J3
Acton, Mont. (59002) 262/H5
Acton Vale, Québec 172/E4
Actopan, Hidalgo, Mexico 150/K6
Actopan, Veracruz, Mexico 150/Q1
Açu, Brazil 120/F3
Açu, Brazil 132/G4
Aculeo, Chile 138/G4
Aculeo (lag.), Chile 138/G4
Acuña, Argentina 143/G5
Acuracay, Peru 128/F5
Acushnet○, Mass. (02743) 249/L6
Acworth, Georgia (30101) 217/C2
Acworth○, N.H. (03601) 268/C5
Acy, La. (†70774) 238/L3
Ada, Ghana 106/E7
Ada (co.), Idaho 220/B6
Ada, Kansas (67414) 232/E2
Ada, Minn. (56510) 255/B3
Ada, Ohio (45810) 284/C4
Ada, Okla. 188/G4
Ada, Okla. (74820) 288/N5
Ada, W. Va. (†24701) 312/D8
Adadle, Somalia 115/H2
Adafer (reg.), Mauritania 106/B5
Adair, Ill. (61411) 222/C3
Adair (co.), Iowa 229/E6
Adair, Iowa (50002) 229/D6
Adair (co.), Ky. 237/L6
Adair (co.), Mo. 261/G2
Adair (cape), N.W. Terrs. 187/L2
Adair (co.), Okla. 288/S3
Adair, Okla. (74330) 288/R2
Adair, Tenn. (†38301) 237/D9
Adairsville, Georgia (30103) 217/C2
Adair Village, Oreg. (†97330) 291/D3
Adairville, Ky. (42202) 237/H7
Adaja (riv.), Spain 33/D2
Adak (isl.), Alaska 196/L4
Adak (str.), Alaska 196/K4
Adak Naval Air Station, Alaska 196/L4
Adalar (isl.), Turkey 63/D6
Adalia (Antalya), Turkey 63/D4
Adam, Oman 59/G5
Adamawa (reg.), Cameroon 115/B2
Adamawa (reg.), Nigeria 106/G7
Adaminaby, N.S. Wales 97/E5
Adams (lake), Br. Col. 184/H4
Adams (riv.), Br. Col. 184/H4
Adams (co.), Colo. 208/L3
Adams (mt.), Colo. 208/H6
Adams (co.), Idaho 220/B5
Adams (co.), Ill. 222/B4
Adams (co.), Ind. 227/H3
Adams, Ind. (47240) 227/F6
Adams (co.), Iowa 229/D6
Adams, Kansas (†67128) 232/E4
Adams, Ky. (41201) 237/R4
Adams, Mass. (01220) 249/B2
Adams○, Mass. (01220) 249/B2
Adams, Minn. (55909) 255/F7
Adams (co.), Miss. 256/B8
Adams (co.), Nebr. 264/F4
Adams, Nebr. (68301) 264/H4
Adams (mt.), N.H. 268/E3
Adams, N.Y. (13605) 276/J3
Adams (co.), N. Dak. 282/F7
Adams, N. Dak. (58210) 282/O3
Adams (co.), Ohio 284/D8
Adams, Okla. (73901) 288/D1
Adams, Oreg. (97810) 291/J2
Adams (co.), Pa. 294/H6
Adam's (peak), Sri Lanka 68/E7
Adams, Tenn. (37010) 237/G7
Adams (co.), Wash. 310/G3
Adams (mt.), Wash. 310/D4
Adams (co.), Wis. 317/G7
Adams, Wis. (53910) 317/G8
Adamsboro, Ind. (†46947) 227/E3
Adam's Bridge (sound), India 68/D7
Adam's Bridge (shoals), Sri Lanka 68/D7
Adamsburg, S.C. (29380) 296/D2
Adams Center, N.Y. (13606) 276/H3
Adams Lake, Br. Col. 184/G5
Adams Mills, Ohio (43801) 284/G5
Adams Nat'l Hist. Site, Mass. 249/D7
Adamson, Okla. (†74547) 288/P5
Adams Run, S.C. (29426) 296/G6
Adamstown, Md. (21710) 245/H3
Adamstown, Pa. (19501) 294/K5
Adamstown (cap.), Pitcairn Is. 87/N8
Adamsville, Ala. (35005) 195/D3
Adamsville, New Bruns. 170/E2
Adamsville, Ohio (43802) 284/G5
Adamsville, Pa. (16110) 294/B2
Adamsville, Québec 172/E4
Adamsville, R.I. (02801) 249/K6
Adamsville, Tenn. (38310) 237/E10
Adamsville, Texas (76510) 303/F6
Adamsville, Utah (†84713) 304/B5
Adana (prov.), Turkey 63/F4
Adana, Turkey 59/C2
Adana, Turkey 63/F4
Adana, Turkey 54/E6
Adanac, Sask. 181/B3
Adapazarı, Turkey 63/D2
Adapazarı, Turkey 59/B1
Adarama, Sudan 111/G4
Adarama, Sudan 59/C6
Adare (cape), Ant. 2/T10
Adare (cape) 5/B9
Adare, Ireland 17/D6
Adar Qaga, Jebel (mt.), Sudan 59/C5
Adaut, Indonesia 85/J7
Adavale, Queensland 88/G5
Adavale, Queensland 95/C5
Adaza, Iowa (†50050) 229/E4
Adda (riv.), Italy 34/B2
Adda (riv.), Sudan 111/D6
Addanki, India 68/D5
Addieville, Ill. (62214) 222/D5
Addington, Okla. (73520) 288/L6
Addis, La. (70710) 238/J2
Addis Ababa (cap.), Ethiopia 102/F4
Addis Ababa (cap.), Ethiopia 2/L5
Addis Ababa (cap.), Ethiopia 111/G6
Addis Alam, Ethiopia 111/G6
Addison, Ala. (35540) 195/D2
Addison, Conn. (†06033) 210/E2
Addison, Ill. (60101) 222/B5
Addison, Maine (04606) 243/H6
Addison○, Maine (04606) 243/H6
Addison, Mich. (49220) 250/E7
Addison, N.Y. (14801) 276/F6
Addison, Ohio (45610) 284/F8
Addison, Pa. (15411) 294/D6
Addison, Texas (75001) 303/G2
Addison (co.), Vt. 268/A3
Addison, Vt. (†05491) 268/A3
Addison (Webster Springs), W. Va. (†26288) 312/F6
Addi Ugri, Ethiopia 59/C7
Ad Diwaniya, Iraq 66/D5
Addo Nat'l Park, S. Africa 118/D6
Addor, N.C. (†28373) 281/L4
Addy, Wash. (99101) 310/H2
Addyston, Ohio (45001) 284/B9
Ade, Ind. (†47922) 227/C3
Adel, Georgia (31620) 217/F8
Adel, Iowa (50003) 229/E5
Adel, Oreg. (97620) 291/H5
Adelaide (isl.) 5/C15
Adelaide, Australia 87/D9
Adelaide, Australia 2/S7
Adelaide (pen.), N.W. Terrs. 187/J3
Adelaide, S. Africa 118/D6
Adelaide (cap.), S. Australia 88/D8
Adelaide Airport 88/D8
Adelaide River, North. Terr. 88/E2
Adelaide River, North. Terr. 93/B2
Adelanto, Calif. (92301) 204/H9
Adelboden, Switzerland 39/E3
Adele (isl.), W. Australia 88/C3
Adele (isl.), W. Australia 92/C1
Adélie Coast (reg.) 5/C7
Adeline, Ill. (†61047) 222/D1
Adeline, La. (†70544) 238/G7
Adell, Wis. (53001) 317/L8
Adelong, N.S. Wales 97/D4
Adelphi, Jamaica 158/H5
Adelphi, Ohio (43101) 284/E7
Adelphia, N.J. (07710) 273/E3
Aden (gulf) 54/F8
Aden (gulf) 2/M5
Aden (gulf) 102/G3
Aden, Alberta 182/E5
Aden (gulf), Djibouti 111/J5
Aden (cap.), P.D.R. Yemen 54/F8
Aden (cap.), P.D.R. Yemen 59/E7
Aden (gulf), Somalia 115/J1
Adena, Ohio (43901) 284/J5
Adenau, W. Germany 22/B3
Adger, Ala. (35006) 195/D4
Adhaim (riv.), Iraq 66/D3
Adi (isl.), Indonesia 85/J6
Adicora, Venezuela 124/D2
Adige (riv.), Italy 34/C2
Adigrat, Ethiopia 111/G5
Adilabad, India 68/D5
Adilcevaz, Turkey 63/K3
Adin, Calif. (96006) 204/E2
Adirondack, N.Y. (12808) 276/N3
Adirondack (mts.), N.Y. 276/M3
Adi Ugri, Ethiopia 111/G5
Adıyaman (prov.), Turkey 63/H4
Adıyaman, Turkey 63/H4
Adjuntas, P. Rico 161/B2
Adjuntas, P. Rico 156/F1
Adkins, Texas (78101) 303/K11
Adlatok (bay), Newf. 166/B2
Adlavik (isls.), Newf. 166/C2
Adliswil, Switzerland 39/F2
Admiral, Sask. 181/C6
Admiral's Beach, Newf. 166/D2
Admiral's Cove, Newf. 166/D2
Admiralty (inlet) 162/H1
Admiralty (isl.), Alaska 196/M1
Admiralty (isls.), N.S. Wales 97/J1
Admiralty (inlet), N.W. Terrs. 187/K2
Admiralty, Papua N.G. 86/A1
Admiralty (isls.), Papua N.G. 87/E6
Admiralty (inlet), Wash. 310/B2
Admiralty (gulf), W. Australia 88/D2
Admiralty (gulf), W. Australia 92/D1
Admiralty Island Nat'l Mon., Alaska 196/M1
Admire, Kansas (66830) 232/F3
Admont, Austria 41/C3
Adna, Wash. (98522) 310/B4
Ado, Nigeria 106/E7
Adobe Creek (res.), Colo. 208/N6
Adok, Sudan 111/F6
Adolfo Alsina, Argentina 143/D4
Adolph, Minn. (55701) 255/F4
Adolph, W. Va. (†26280) 312/F5
Adolphus, Ky. (42120) 237/J7
Adona, Ark. (72001) 202/E3
Adonara (isl.), Indonesia 85/G7
Adoni, India 68/D5
Adorf, E. Germany 22/E3
Adour (riv.), France 28/C6
Adra, Spain 33/E4
Adrano, Italy 34/E6
Adrar, Algeria 102/B2
Adrar, Algeria 106/xD3
Adrar (reg.), Mauritania 106/B4
Adrar des Iforas (plat.), Algeria 106/E5
Adrar des Iforas (plat.), Mali 106/E5
Adré, Chad 111/D5
Adria, Italy 34/D2
Adrian, Georgia (31002) 217/G5
Adrian, Ill. (62310) 222/B3
Adrian, Mich. (49221) 250/F7
Adrian, Minn. (56110) 255/C7
Adrian, Mo. (64720) 261/D6
Adrian, N. Dak. (58410) 282/O6
Adrian, Ohio (44801) 284/D3
Adrian, Oreg. (97901) 291/K4
Adrian, Pa. (16210) 294/D4
Adrian, S.C. (†29526) 296/J4
Adrian, Texas (79001) 303/B2
Adrian, W. Va. (26210) 312/F5
Adriatic (sea) 7/F4
Adriatic (sea), Albania 45/B4
Adriatic (sea), Italy 34/E3
Adriatic (sea), Yugoslavia 45/B4
Aduwa, Ethiopia 59/C7
Advance, Ind. (46102) 227/D5
Advance, Mich. (†49712) 250/D3
Advance, Mo. (63730) 261/N8
Advance, N.C. (27006) 281/J3
Advent, W. Va. (25231) 312/C5
Adventure (sound), 143/E7
Adventure (bay), Chile 138/D5
Adventure, Guyana 131/B2
Adventure Bay, Tasmania 99/D5
Advocate (bay), Nova Scotia 168/D3
Advocate Harbour, Nova Scotia 168/D3
Adwa, Ethiopia 111/G5
Adwick le Street, England 13/K2
Adygey Aut. Obl., U.S.S.R. 52/F6
Adygey Aut. Obl., U.S.S.R. 48/D5
Adzhar A.S.S.R., U.S.S.R. 52/F6
Adzhar A.S.S.R., U.S.S.R. 48/E5
Ae, Scotland 15/E5
Aeber (creek), S. Dak. 298/G4
Aegean (sea), Greece 7/G5
Aegean (sea), Greece 45/G6
Aegean (sea), Turkey 63/A3
Aegean Islands (reg.), Greece 45/G6
Aeneas, Wash. (†98855) 310/F2
Aerial, Alberta 182/D4
AErø (isl.), Denmark 21/D8
AErøskøbing, Denmark 21/D8
Aeschi bei Spiez, Switzerland 39/E3
Aetna, Alberta 182/D5
Aetna, Tenn. (†37033) 237/G9
'Afaq, Iraq 66/D4
Afareaitu, Fr. Poly. 86/S13
Afayan (pt.), Guam 86/K7
Afdem, Ethiopia 111/H6
Afollé (reg.), Mauritania 106/B5
Affoltern am Albis, Switzerland 39/F2
Affoltern im Emmental, Switzerland 39/E2
Affric, Loch (lake), Scotland 15/D3
Affton, Mo. (63123) 261/P4
Afghanistan 2/N4
Afghanistan 54/H6
AFGHANISTAN 68/A2
Afgoi, Somalia 102/G4
Afgoi, Somalia 115/J3
Afif, Saudi Arabia 59/D5
Afikpo, Nigeria 106/F7
Afiqim, Israel 65/D2
Åfjord, Norway 18/G5
Aflex, Ky. (41510) 237/S5
Afmadu, Somalia 115/H3
Afognak (isl.), Alaska 196/H3
Africa 2/K4
AFRICA 102
Afşin, Turkey 63/G3
Afton, Iowa (50830) 229/E6
Afton, La. (†71282) 238/H2
Afton, Mich. (49705) 250/E3
Afton, Minn. (55001) 255/F6
Afton, N.Y. (13730) 276/J6
Afton, Okla. (74331) 288/S1
Afton, Tenn. (37616) 237/R8
Afton, Texas (79220) 303/D4
Afton, Va. (22920) 307/L4
Afton, Wis. (53501) 317/H10
Afton, Wyo. (83110) 319/B3
Afuá, Brazil 132/D3
'Afula, Israel 65/C2
Afyonkarahisar (prov.), Turkey 63/D3
Afyonkarahisar, Turkey 63/D3
Afyonkarahisar, Turkey 59/A2
Agadem (well), Niger 106/G5
Agadès, Niger 106/F5
Agadès, Niger 102/C3
Agadir, Morocco 106/C2
Agadir, Morocco 102/A1
Agaña (cap.), Guam 87/E4
Agaña (cap.), Guam 86/K7
Agano (riv.), Japan 81/J4
Agar, S. Dak. (57520) 298/J4
Agartala, India 68/G4
Agassiz (peak), Ariz. 198/D3
Agassiz (mt.), Utah 304/D3
Agat (bay), Guam 86/K7
Agata, U.S.S.R. 48/K3
Agate, Colo. (80101) 208/M4
Agate, Nebr. (†69346) 264/A2
Agate, N. Dak. (58310) 282/L2
Agate Beach, Oreg. (†97365) 291/C3
Agate Fossil Beds Nat'l Mon., Nebr. 264/A2
Agats, Indonesia 85/K7
Agatti (isl.), India 68/C6
Agattu (isl.), Alaska 196/J3
Agattu (str.), Alaska 196/J3
Agawa Bay, Ontario 177/J5
Agawa Bay, Ontario 175/D3
Agawam○, Mass. (01001) 249/D4
Agawam (riv.), Mass. 249/M5
Agboville, Ivory Coast 102/B4
Agboville, Ivory Coast 106/D7
Agdam, U.S.S.R. 52/G6
Agde, France 28/E6
Agen, France 28/D5
Agency, Iowa (52530) 229/J7
Agency, Mo. (64401) 261/C3
Agency (lake), Oreg. 291/E5
Agenda, Kansas (66930) 232/E2
Ageo, Japan 81/O2
Agerbaek, Denmark 21/B6
Ägerisee (lake), Switzerland 39/G2
Ages, Ky. (40801) 237/P7
Aghada-Farsid-Rostellan, Ireland 17/E8
Aghadoe, Ireland 17/B7
Aghagower, Ireland 17/C4
Agha Jari, Iran 66/F5
Aginsk Buryat Aut. Okr., U.S.S.R. 48/M4
Aginskoye, U.S.S.R. 48/M4
Agiobampo (bay), Mexico 150/E3
Agira, Italy 34/E6
Ağlasun, Turkey 63/D4
Ağlı, Turkey 63/E2
Agness, Oreg. (97406) 291/C5
Agno, Philippines 82/B2
Agnone, Italy 34/E4
Agnos, Ark. (72510) 202/G1
Agoo, Philippines 82/C2
Agordat, Ethiopia 111/G4
Agordat, Ethiopia 59/C6
Agra, India 54/J7
Agra, India 68/D3
Agra, Kansas (67621) 232/C2
Agra, Okla. (74824) 288/N3
Agraciada, Uruguay 145/A4
Agrado, Colombia 126/C6
Agramonte, Cuba 158/D1
Agreda, Spain 33/E2
Ağrı (prov.), Turkey 63/K3
Ağrı, Büyük (Ararat) (mt.), Turkey 63/L3
Ağrı (Karaköse), Turkey 63/K3
Agricola, Kansas (†66871) 232/G3
Agricola, Miss. (†39452) 256/G9
Agrigento (prov.), Italy 34/D6
Agrigento, Italy 34/D6
Agrihan (isl.), No. Marianas 87/E4
Agrinion, Greece 45/E6
Agropoli, Italy 34/E4
Agryz, U.S.S.R. 52/H3
Agua Caliente, Ariz. (†85333) 198/B6
Agua Caliente Ind. Res., Calif. 204/J10
Aguachica, Colombia 126/D3
Aguada, P. Rico 161/A1
Aguada de Pasajeros, Cuba 158/D2
Aguada Grande, Venezuela 124/D2
Aguadas, Colombia 126/C5
Agua de Dios, Colombia 126/C5
Aguadilla (dist.), P. Rico 161/A1
Aguadilla, P. Rico 156/F1
Aguadilla, P. Rico 161/A1
Aguadilla (bay), P. Rico 161/A1
Agua Dulce, Mexico 150/M7
Aguadulce, Panama 154/G6
Agua Dulce, Texas (78330) 303/F10
Agua Fria (riv.), Ariz. 198/C5
Agua Fría, Venezuela 124/D2
Agualeguas, Mexico 150/J3
Agua Linda, Venezuela 124/E5
Aguán (riv.), Honduras 154/D3
Aguanaval (riv.), Mexico 150/H4
Aguanish, Québec 174/E2
Aguanus (riv.), Newf. 166/B3
Aguanus (riv.), Québec 174/E2
Agua Prieta, Mexico 150/E1
Aguaray, Argentina 143/D1
Aguarico (riv.), Colombia 126/B7
Aguarico (riv.), Ecuador 128/D3
Aguasay, Venezuela 124/G3
Aguas Blancas, Chile 138/B4
Aguas Buenas, P. Rico 161/E2
Aguas Calientes, Cerro (mt.), Chile 138/C4
Aguascalientes (state), Mexico 150/H6
Aguascalientes, Mexico 150/H6
Aguascalientes, Mexico 146/H7
Aguas Corrientes, Uruguay 145/A6
Agua Vermelha (res.), Brazil 135/B1
Aguaytía (riv.), Peru 128/E7
Agudos, Brazil 135/B3
Águeda, Portugal 33/B2
Águeda (riv.), Spain 33/C2
Aguedaueda (riv.), Portugal 33/C2
Agueraktem (well), Mali 106/C4
Agueraktem (well), Mauritania 106/C4
Aguila, Ariz. (85320) 198/B5
Aguilar, Colo. (81020) 208/K8
Aguilar, Spain 33/D4
Aguilares, Argentina 143/C2
Aguilas, Spain 33/F4
Aguililla, Mexico 150/H7
Aguja, La (cape), Colombia 126/C2
Aguja (pt.), Peru 2/E6
Aguja (pt.), Peru 120/A3
Aguja (pt.), Peru 128/B5
Agulhas (cape), S. Africa 102/D8
Agulhas (cape), S. Africa 118/B6
Agusan (riv.), Philippines 82/E6
Agusan Canyon, Philippines 82/E6
Agusan del Norte (prov.), Philippines 82/E6
Agusan del Sur (prov.), Philippines 82/E6
Agustín Codazzi, Colombia 126/D3
Agutaya, Philippines 82/C5
Agutaya (isl.), Philippines 82/C5
Ahaggar (range), Algeria 102/C2
Ahaggar (range), Algeria 106/F4
Ahangaran, Afghanistan 59/J3
Ahar, Iran 66/E1
Ahascragh, Ireland 17/E5
Ahau, Fiji 87/H7
Ahaura, N. Zealand 100/C5
Ahaus, W. Germany 22/B2
Aherlow (riv.), Ireland 17/E7
Ah-Gwah-Ching, Minn. (56430) 255/D3
Ahipara, N. Zealand 100/D1
Ahlat, Turkey 63/K3
Ahlbeck, E. Germany 22/F2
Ahlen, W. Germany 22/B3
Ahmadabad, India 2/N4
Ahmadabad, India 68/C4
Ahmadabad, India 54/J7
Ahmadnagar, India 68/C5
Ahmadpur East, Pakistan 68/C3
Ahmeek, Mich. (49901) 250/A1
Ahmic (lake), Ontario 177/E2
Ahoghill, N. Ireland 17/J2
Ahome, Mexico 150/E4
Ahoskie, N.C. (27910) 281/P2
Ahousat, Br. Col. 184/D5
Ahrensburg, W. Germany 22/D2
Ahsahka, Idaho (83520) 220/B3
Ahtanum (creek), Wash. 310/D4
Ähtäri, Finland 18/N5
Ahua (pt.), Hawaii 218/B4
Ahuacatitlán, Mexico 150/L1
Ahuacatlán, Mexico 150/G6
Ahuachapán, El Salvador 154/B4
Ahuás, Honduras 154/E3
Ahumada, Mexico 150/F1
Ahurei, Fr. Poly. 87/M8
Åhus, Sweden 18/J9
Ahuzzam, Israel 65/B4
Ahvaz, Iran 54/F6
Ahvaz, Iran 59/E3
Ahvaz (Ahwaz), Iran 66/F5
Ahvenanmaa (prov.), Finland 18/L6
Ahwahnee, Calif. (93601) 204/F6
Ahwar, P.D.R. Yemen 59/E7
Ai-Ais, Namibia 118/B5
Aialik (bay), Alaska 196/C1
Aiama (lake), Brazil 132/H9
Aibonito, P. Rico 161/D2
Aichi (pref.), Japan 81/H6
Aid, Mo. (†63825) 261/M9
Aid, Ohio (†45645) 284/F8
Aiea, Hawaii (96701) 218/B3
Aigen im Mühlkreis, Austria 41/B2
Aigle (riv.), Québec 172/A3
Aigle, Switzerland 39/C4
Aiguá, Uruguay 145/E5
Aiguá (riv.), Uruguay 145/E4
Aigues-Mortes, France 28/F6
Aiguille d'Argentière (mt.), Switzerland 39/C5
Aihui (Aigun) (Heihe), China 77/L1
Aija, Peru 128/D7
Aikawa, Japan 81/H4
Aiken (co.), S.C. 296/D4
Aiken, S.C. (29801) 296/D4
Aikens (lake), Manitoba 179/G3
Aiken West, S.C. (†29801) 296/D4
Aikin, Md. (†21903) 245/O2
Aileron, North. Terr. 93/C7
Ailey, Georgia (30410) 217/G6
Ailinglapalap (atoll), Marshall Is. 87/G5
Aillon (lake), Québec 172/C2
Ailsa Craig, Ontario 177/C4
Ailsa Craig (isl.), Scotland 15/C5
Ailuk (atoll), Marshall Is. 87/H4
Aimogasta, Argentina 143/C2
Aimwell, La. (71401) 238/G3
Ain (dept.), France 28/F4
Ain (riv.), France 28/F4
Ainabo, Somalia 115/J2
Aina Haina, Hawaii (96821) 218/F2
'Ain al Mubarrak, Saudi Arabia 59/C5
Ainaži, U.S.S.R. 53/C2
Aïn Beïda, Algeria 106/F1
Aïn ben Tili (well), Mauritania 106/C3
'Ain el 'Arab, Syria 63/H4
Ain-Galakka, Chad 111/C4
Aïn Sefra, Algeria 106/D2
Ainslie (lake), Nova Scotia 168/G2
Ainsworth, Br. Col. 184/J5
Ainsworth, Iowa (52201) 229/K6
Ainsworth, Nebr. (69210) 264/D2
Aïn Temouchent, Algeria 106/D1
Ain Zueiya (well), Libya 111/D3
Aïoun el Atrous, Mauritania 106/C5
Aïoun el Atrous, Mauritania 102/B3
Aipe, Colombia 126/C6
Aiquile, Bolivia 136/C6
Aiquina, Chile 138/B3
Aïr (mts.), Niger 106/F5
Airdrie, Alberta 182/C4
Airdrie, Scotland 10/B1
Airdrie, Scotland 15/C2
Aire (riv.), England 10/F4
Aire (riv.), England 13/F4
Aire-sur-l'Adour, France 28/C6
Airey, Md. (†21613) 245/O6
Air Force (isl.), N.W. Terrs. 187/L3
Air Force Academy, Colo. 208/K5
Airlie, Oreg. (†97361) 291/D3
Airolo, Switzerland 39/G3
Air Ronge, Sask. 181/M3
Airville, Pa. (17302) 294/K6
Airway Heights, Wash. (99001) 310/H3
Aisén del General Carlos Ibáñez del Campo (reg.), Chile 138/E6
Aišiškes, U.S.S.R. 53/C3
Aisne (dept.), France 28/E3
Aisne (riv.), France 28/E3
Aitape, Papua N.G. 85/B6
Aith, Scotland 15/G2
Aitkin (co.), Minn. 255/E4
Aitkin, Minn. (56431) 255/E4
Aitutaki (atoll), Cook Is. 87/K7
Aiud, Romania 45/F2
Aix, France 7/E4
Aix (mt.), Wash. 310/D4
Aix-en-Provence, France 28/F6
Aix-les-Bains, France 28/G5
Aiyina, Greece 45/F7
Aiyion, Greece 45/F6
Aizpute, U.S.S.R. 53/A2
Aizuwakamatsu, Japan 81/J5
Aizwal, India 68/G4
Ajaccio, France 7/E4
Ajaccio, France 28/B7
Ajaccio (gulf), France 28/B7
Ajalpan, Mexico 150/L7
Ajana, W. Australia 92/A5
Ajanta, India 68/D4
Ajax, La. (†71450) 238/D3
Ajax, Ontario 177/E4
Ajedabia, Libya 111/D1
Aji Chai (riv.), Iran 66/E1
Ajigasawa, Japan 81/J3
'Ajja, West Bank 65/C3
Ajka, Hungary 41/D3
'Ajlun (dist.), Jordan 65/D3
'Ajlun, Jordan 65/D3
'Ajlun (range), Jordan 65/D3
'Ajman, U.A.E. 59/G4
Ajmer, India 68/C3
Ajmer, India 54/J7
Ajo, Ariz. (85321) 198/C6
Ajoewa, Suriname 131/C4
Ajoupa-Bouillon, Martinique 161/C5
Akan National Park, Japan 81/M2
Akaroa, N. Zealand 100/D5
Akasha, Sudan 111/F3
Akashi, Japan 81/H8
Akaska, S. Dak. (57420) 298/J3
Akbaba Tepesi (mt.), Turkey 63/H3
Akçaabat, Turkey 63/H2
Akçadağ, Turkey 63/G3
Akçakale, Turkey 63/H4
Akçakoca, Turkey 63/D2
Akçay, Turkey 63/C4
Akçay (riv.), Turkey 63/C4
Akdağ (mt.), Turkey 59/A2
Akdağ (mt.), Turkey 63/E4
Akdağ (mt.), Turkey 63/C4
Akdağmadeni, Turkey 63/F3
Akeley, Minn. (56433) 255/D3
Akeley, Pa. (†16345) 294/D2
Aken, E. Germany 22/D3
Akers, La. (70421) 238/N2
Akershus (co.), Norway 18/G6
Aketi, Zaire 102/E4
Aketi, Zaire 115/D3
Akhaltsikhe, U.S.S.R. 52/F6
Akhdar, Jebel (mts.), Libya 111/D1
Akhdar, Jebel (range), Oman 59/G5
Akhdar, Saudi Arabia 59/C4
Akhiok, Alaska (99615) 196/H3
Akhisar, Turkey 63/B3
Akhmim, Egypt 111/F2
Akhmim, Egypt 59/B4
Akhtopol, Bulgaria 45/H4
Akhtubinsk, U.S.S.R. 52/G5
Akhty, U.S.S.R. 52/G6
Akhtyrka, U.S.S.R. 52/E4
Aki, Japan 81/F7
Akiachak, Alaska (99551) 196/F2
Akiak, Alaska (99552) 196/F2
Akimiski (isl.), N.W.T. 162/H5
Akin, Ill. (62805) 222/E6
Akins, Okla. (†74955) 288/S3
Åkirkeby, Denmark 21/F9
Akita (pref.), Japan 81/J4
Akita, Japan 81/J4
Akita, Japan 54/P6
Akitio, N. Zealand 100/F4
Akjoujt, Mauritania 106/B5
Akkerman (Belgorod-Dnestrovskiy), U.S.S.R. 52/D5
Akkeshi, Japan 81/M2
Akko (Acre), Israel 65/C2
Akkrum, Netherlands 27/H2
Aklan (prov.), Philippines 82/D5
Aklavik, Canada 4/C16
Aklavik, N.W.T. 146/E3
Aklavik, N.W.T. 162/C2
Aklavik, N.W. Terrs. 187/E3
Akmolinsk (Tselinograd), U.S.S.R. 48/H4
Akobo (riv.), Ethiopia 111/F6
Akobo, Sudan 111/F6
Akobo (riv.), Sudan 111/F6
Akola, India 68/D4
Akolmiut (Kasigluk), Alaska (†99609) 196/F2
Akpatok (isl.), N.W.T. 162/K3
Akpatok (isl.), N.W. Terrs. 187/M3
Akpazar, Turkey 63/H3
Akpınar, Turkey 63/D5
Akqi, China 77/A3
Akra, N. Dak. (†58220) 282/P2
Akranes, Iceland 21/B1
Akreïjit, Mauritania 106/C5
Akritas (cape), Greece 45/E7
Akron, Ala. (35441) 195/C5
Akron, Colo. (80720) 208/N2
Akron, Ind. (46910) 227/E2
Akron, Iowa (51001) 229/A3
Akron, Mich. (48701) 250/F5
Akron, N.Y. (14001) 276/C4
Akron, Ohio 188/K2
Akron, Ohio (*44301) 284/G3
Akron, Ohio 146/K5
Akron, Pa. (17501) 294/K5
Aksai Chin (reg.), Pakistan 68/D2
Aksaray, Turkey 63/F3
Aksay, China 77/D4
Aksay, U.S.S.R. 48/F4
Akşehir, Turkey 63/D3
Akşehir, Turkey 59/B2
Akşehir (lake), Turkey 63/D3
Akseki, Turkey 63/D4
Aksu (Aqsu), China 77/B3
Aksu, China 54/K5
Aksu (riv.), Turkey 63/D4
Aksum, Ethiopia 102/F3
Aksum, Ethiopia 111/G5
Aksum, Ethiopia 59/C7
Aktas, U.S.S.R. 48/G5
Aktash, U.S.S.R. 48/J4
Aktí (pen.), Greece 45/G5
Aktyubinsk, U.S.S.R. 54/G4
Aktyubinsk, U.S.S.R. 48/F4
Aku, Nigeria 106/F7
Akun (isl.), Alaska 196/E4
Akune, Japan 81/E7
Akure, Nigeria 106/F7
Akureyri, Ice. 4/C10
Akureyri, Iceland 21/C1
Akureyri, Iceland 7/C2
Akuse, Ghana 106/E7
Akutan, Alaska (99553) 196/E4
Akutan (isl.), Alaska 196/E4
Akutan (passage), Alaska 196/E4
Akviran, Turkey 63/E4
Akyab (Sittwe), Burma 72/B2
Akyazı, Turkey 63/D2
Ål, Norway 18/F6
Alabam, Ark. (†72740) 202/C1
Alabama 188/J4
ALABAMA 195
Alabama (riv.), Ala. 188/J4
Alabama (riv.), Ala. 195/C8
Alabama (state), U.S. 146/K6
Alabaster, Ala. (35007) 195/F4
Alabaster, Mich. (†48763) 250/F4
Alabat, Philippines 82/D3

Alabat (isl.), Philippines 82/D3
Alaca, Turkey 63/F2
Alacahan, Turkey 63/G3
Alaçam, Turkey 63/F2
Alachua (co.), Fla. 212/D2
Alachua, Fla. (32615) 212/D2
Alacrán (reef), Mexico 150/P5
Alacranes, Cuba 158/D1
Aladağ (mt.), Turkey 63/F4
Aladagh, Kuh-i- (mt.), Iran 59/G2
'Aladagh, Kuh-e (mts.), Iran 66/K2
Aladdin, Wyo. (82710) 319/H1
Alaejos, Spain 33/D2
Alagir, U.S.S.R. 52/F6
Alagoa Grande, Brazil 132/H4
Alagoas (state), Brazil 132/G5
Alagoinhas, Brazil 120/F4
Alagoinhas, Brazil 132/G6
Alagón, Spain 33/F2
Alagón (riv.), Spain 33/C2
Alah (riv.), Philippines 82/E7
Al Ahqaf (Bahr es Safi) (des.), Saudi Arabia 59/E6
Al 'Ain, Saudi Arabia 59/C4
Alajuela, C. Rica 154/E6
Alakanuk, Alaska (99554) 196/E2
Alakol' (lake), U.S.S.R. 48/J5
Al 'Ala, Saudi Arabia 59/C4
Alalakeiki (chan.), Hawaii 218/J3
Alalapadu, Suriname 131/C4
Alamagan (isl.), No. Marianas 87/E4
Alamance (co.), N.C. 281/L3
Alamance, N.C. (27201) 281/K2
Alameda (co.), Calif. 204/D6
Alameda, Calif. (94501) 204/J2
Alameda (creek), Calif. 204/K3
Alameda, N. Mex. (87114) 274/C3
Alameda, Sask. 181/J6
Alamikamba, Nicaragua 154/E4
Alamo (lake), Ariz. 198/B4
Alamo (riv.), Calif. 204/K10
Alamo, Georgia (30411) 217/G6
Alamo, Ind. (47916) 227/C5
Alamo, Mexico 150/L6
Alamo, Nev. (89001) 266/F5
Alamo, N. Dak. (58830) 282/D2
Alamo, Tenn. (38001) 237/C9
Alamo, Texas (78516) 303/F11
Ala Moana, Hawaii 218/C4
Alamo-Danville, Calif. (94507) 204/K2
Alamogordo, N. Mex. 188/E4
Alamogordo, N. Mex. (88310) 274/C6
Alamo Heights, Texas (78209) 303/K10
Álamos, Mexico 150/E3
Alamosa (co.), Colo. 208/H7
Alamosa, Colo. (81101) 208/H8
Alamosa (creek), Colo. 208/G8
Alamosa (riv.), N. Mex. 274/B5
Alamota, Kansas (67830) 232/B3
Åland (Ahvenanmaa) (prov.), Finland 18/L6
Åland (isls.), Finland 7/F2
Åland (isls.), Finland 18/L6
Alanje, Panama 154/F6
Alanreed, Texas (79002) 303/D2
Alanson, Mich. (49706) 250/E3
Alanthus Grove, Mo. (†64489) 261/D2
Alanya, Turkey 59/B2
Alanya, Turkey 63/D4
Alaotra (lake), Madagascar 118/H3
Alapaha (riv.), Fla. 212/C1
Alapaha, Georgia (31622) 217/F8
Alapaha (riv.), Georgia 217/F7
Alaqua (creek), Fla. 212/C6
Alarcón (res.), Spain 33/E3
Alarka, N.C. (†28713) 281/C4
Alas (str.), Indonesia 85/F7
Alaşehir, Turkey 63/C3
Alashtar, Iran 66/E4
Alaska (reg.) 4/C17
Alaska 188/C5
Alaska (gulf) 146/D4
ALASKA 196
Alaska (gulf), Alaska 188/D6
Alaska (pen.), Alaska 188/C6
Alaska (range), Alaska 188/C6
Alaska (range), Alaska 146/C3
Alaska (pen.), Alaska 146/C4
Alaska (gulf), Alaska 196/K3
Alaska (pen.), Alaska 196/G3
Alaska (range), Alaska 196/H2
Alaska, Mich. (†49316) 250/D6
Alaska (state), U.S. 2/B2
Alaska (state), U.S. 146/C3
Alaska (range), U.S. 4/C17
Alaska (pen.), U.S. 4/D18
Alaska (gulf), U.S. 4/D17
Alaska Highway, Yukon 187/E3
Alassio, Italy 34/A2
Alatna, Alaska (†99720) 196/H1
Alatna (riv.), Alaska 196/H1
Alatri, Italy 34/D4
Alatyr', U.S.S.R. 52/G4
Al 'Auda, Saudi Arabia 59/E4
Alausí, Ecuador 128/C4
Álava (prov.), Spain 33/E1
Alava (cape), Wash. 188/A1
Alava (cape), Wash. 310/A2
Alaverdi, U.S.S.R. 52/F6
Alavus, Finland 18/N5
Alayor, Spain 33/J3
Al 'Azair, Iraq 66/E5
Alazeya (riv.), U.S.S.R. 48/Q3
Al 'Aziziya, Iraq 59/E3
Al 'Aziziya, Iraq 66/D4
Alba, Italy 34/B2
Alba, Mich. (49611) 250/E4
Alba, Mo. (64830) 261/D8
Alba, Pa. (16910) 294/J2
Alba, Texas (75410) 303/J5
Albacete (prov.), Spain 33/E3
Albacete, Spain 7/D5
Albacete, Spain 33/F3
Alba de Tormes, Spain 33/D2
Albaida, Spain 33/F3
Alba Iulia, Romania 45/F2
Albalate del Arzobispo, Spain 33/F2
Alban, Ontario 177/D1
Albanel, Québec 172/E1
Albanel (lake), Québec 174/C2
Albania 2/K3
Albania 7/G4
ALBANIA 45/E5
Albano (lake), Italy 34/F7
Albano Laziale, Italy 34/F7
Albany, Australia 87/B9
Albany, Calif. (94706) 204/J2
Albany, Ga. 146/K6
Albany, Georgia (*31701) 217/D7
Albany, Ill. (61230) 222/C2
Albany, Ind. (47320) 227/G4
Albany, Jamaica 158/J6
Albany, Ky. (42602) 237/L7
Albany, La. (70711) 238/M1
Albany, Minn. (56307) 255/D5
Albany, Mo. (64402) 261/D2
Albany○, N.H. (†03864) 268/E4
Albany (cap.), N.Y. 188/M2
Albany (cap.), N.Y. 146/L5
Albany (co.), N.Y. 276/M5
Albany (cap.), N.Y. (*12201) 276/N5
Albany, N. Zealand 100/B1
Albany, Nova Scotia 168/C4
Albany, Ohio (45710) 284/F7
Albany, Okla. (74721) 288/O7
Albany (riv.), Ont. 146/K4
Albany (riv.), Ont. 162/H5
Albany (riv.), Ontario 175/C2
Albany, Oreg. 188/B2
Albany, Oreg. (97321) 291/D3
Albany, Pr. Edward I. 168/E2
Albany, Texas (76430) 303/E5
Albany, Vt. (05820) 268/C2
Albany○, Vt. (05820) 268/C2
Albany, W. Australia 88/B6
Albany, W. Australia 92/B6
Albany, Wis. (53502) 317/G10
Albany (co.), Wyo. 319/G4
Albany, Wyo. (†82055) 319/F4
Albany Creek, Queensland 88/J2
Albardon, Argentina 143/C3
Albarracín, Spain 33/F2
Albatross (pt.), N. Zealand 100/E3
Albatross (bay), Queensland 88/G2
Albatross (bay), Queensland 95/B2
Albay (prov.), Philippines 82/D4
Albay (gulf), Philippines 82/D4
Albee, S. Dak. (57210) 298/S3
Albemarle (pt.), Ecuador 128/B9
Albemarle (sound), N.C. 188/L3
Albemarle, N.C. (28001) 281/J4
Albemarle (sound), N.C. 281/S2
Albemarle (co.), Va. 307/L5
Albenga, Italy 34/B3
Albeni Falls (dam), Idaho 220/B1
Alberdi, Paraguay 144/D5
Alberene, Va. (†22959) 307/L5
Alberga, S. Australia 94/D2
Alberga, The (riv.), S. Australia 94/D2
Alberga, The (riv.), S. Australia 88/E5
Alberhill, Calif. (†92330) 204/E11
Alberni (inlet), Br. Col. 184/H3
Albers, Ill. (62215) 222/D5
Albert (canal), Belgium 27/F6
Albert, France 28/E2
Albert, Kansas (67511) 232/C3
Albert (co.), New Bruns. 170/F3
Albert, N. Mex. (87733) 274/F3
Albert, N.S. Wales 97/D3
Albert, Okla. (73001) 288/K4
Albert (lake), Québec 172/C3
Albert (Mobuto Sese Seko) (lake), Uganda 115/F3
Albert (creek), Wyo. 319/B4
Albert (Mobutu Sese Seko) (lake), Zaire 115/F3
Alberta (prov.) 162/E5
Alberta (prov.) Canada 146/G4
Alberta, Ala. (36720) 195/D6
ALBERTA 182
Alberta (mt.), Alberta 182/B3
Alberta (mt.), Alta. 162/E5
Alberta, La. (†71016) 238/D2
Alberta, Minn. (56207) 255/B5
Alberta, Va. (23821) 307/N7
Alberta Beach, Alberta 182/C3
Albert City, Iowa (50510) 229/C3
Albert Edward (bay), N.W. Terrs. 187/H3
Albert Head, Br. Col. 184/J4
Alberti, Argentina 143/G7
Albertirsa, Hungary 41/E3
Albert Lea, Minn. (56007) 255/E7
Albert Mines, New Bruns. 170/F3
Alberton, Mont. (59820) 262/B3
Alberton, Pr. Edward I. 168/E2
Alberton, S. Africa 118/H6
Albert Town, Jamaica 158/H6
Albertville, Ala. (35950) 195/F2
Albertville, France 28/G5
Albertville, Minn. (55301) 255/E5
Albertville, Sask. 181/F2
Albeuve, Switzerland 39/D3
Albi, France 28/E6
Albia, Iowa (52531) 229/H6
Albin, Wyo. (82050) 319/H4
Albina, Suriname 131/D3
Albino, Italy 34/B2
Albion, Calif. (95410) 204/B4
Albion, Idaho (83311) 220/E7
Albion (mts.), Idaho 220/E7
Albion, Ill. (62806) 222/E5
Albion, Ind. (46701) 227/G2
Albion, Iowa (50005) 229/H4
Albion○, Maine (04910) 243/E6
Albion, Mich. (49224) 250/E6
Albion, Nebr. (68620) 264/F3
Albion, N.Y. (14411) 276/D4
Albion, Okla. (74521) 288/R5
Albion, Pa. (16401) 294/B2
Albion, R.I. (02802) 249/H5
Albion, Wash. (99102) 310/H4
Albion, Wis. (†53534) 317/H10
Al Birk, Saudi Arabia 59/D6
Albocácer, Spain 33/F2
Alborán (isl.), Spain 7/D5
Alborán (isl.), Spain 33/E5
Ålborg, Denmark 7/F3
Ålborg, Denmark 18/G8
Ålborg (bay), Denmark 21/D4
Alborn, Minn. (55702) 255/F4
Albox, Spain 33/E4
Albreda, Br. Col. 184/H4
Albright, W. Va. (26519) 312/G3
Albrightsville, Pa. (18210) 294/L3
Albristhorn (mt.), Switzerland 39/D4
Albufeira, Portugal 33/B4
Albuñol, Spain 33/E4
Albuquerque (cays), Colombia 126/A10
Albuquerque, N. Mex. 146/H6
Albuquerque, N. Mex. 188/E3
Albuquerque, N. Mex. (*87101) 274/C3
Alburg, Vt. (05440) 268/A2
Alburg○, Vt. (05440) 268/A2
Alburnett, Iowa (52202) 229/K4
Alburquerque, Spain 33/C3
Alburtis, Pa. (18011) 294/L5
Albury, Australia 87/E9
Albury, N. S. Wales 88/H7
Albury, N.S. Wales 97/D5
Albury, N. Zealand 100/C6
Alca, Peru 128/F10
Alcácer do Sal, Portugal 33/B3
Alcalá, Bolivia 136/C6
Alcalá de Chivert, Spain 33/G2
Alcalá de Guadaira, Spain 33/D4
Alcalá de Henares, Spain 33/G4
Alcalá de los Gazules, Spain 33/D4
Alcalá la Real, Spain 33/E4
Alcalde, N. Mex. (87511) 274/C2
Alcamo, Italy 34/D6
Alcanar, Spain 33/G2
Alcañices, Spain 33/C2
Alcañiz, Spain 33/F2
Alcàntara, Portugal 33/A1
Alcántara, Spain 33/C3
Alcántara (res.), Spain 33/C3
Alcántaratara (res.), Portugal 33/C3
Alcantarilla, Spain 33/F4
Alcaraz, Argentina 143/G5
Alcaraz, Spain 33/E3
Alcaraz, Sierra de (range), Spain 33/E3
Alcatraz (isl.), Calif. 204/J2
Alcaudete, Spain 33/E4
Alcázar de San Juan, Spain 33/E3
Alcester, S. Dak. (57001) 298/R7
Alcida, New Bruns. 170/E1
Alcira, Spain 33/F3
Alco, Ark. (72610) 202/F2
Alco, La. (71402) 238/D4
Alcoa, Tenn. (37701) 237/N9
Alcobaça, Brazil 132/G7
Alcobaça, Portugal 33/B3
Alcolu, S.C. (29001) 296/G4
Alcomdale, Alberta 182/C3
Alcona (co.), Mich. 250/F4
Alcona Beach, Ontario 177/E3
Alcones, Chile 138/F5
Alcony, Ohio (†45373) 284/B5
Alcora, Spain 33/F2
Alcorisa, Spain 33/F2
Alcorn, Ky. (†40447) 237/O5
Alcorn (co.), Miss. 256/G1
Alcorn State University, Miss. (39096) 256/B7
Alcorta, Argentina 143/F6
Alcoutim, Portugal 33/C4
Alcova, Wyo. (82620) 319/F3
Alcova (res.), Wyo. 319/F3
Alcoy, Spain 33/F3
Alcudia (bay), Spain 33/H3
Alda, Nebr. (68810) 264/F4
Aldabra (isls.), Seychelles 102/G5
Aldabra (isls.), Seychelles 118/H1
Aldama, Chihuahua, Mexico 150/G2
Aldama, Tamaulipas, Mexico 150/L5
Aldan, Pa. (†19018) 294/M7
Aldan, U.S.S.R. 54/O4
Aldan (riv.), U.S.S.R. 54/P3
Aldan, U.S.S.R. 48/N4
Aldan (plat.), U.S.S.R. 48/N4
Aldan (riv.), U.S.S.R. 48/O3
Aldeburgh, England 13/J5
Aldeburgh, England 10/G4
Aldeia Carajá, Brazil 132/D6
Aldeia Nova de São Bento, Portugal 33/C4
Alden, Ill. (60001) 222/E1
Alden, Iowa (50006) 229/G4
Alden, Kansas (67512) 232/D3
Alden, Mich. (49612) 250/D4
Alden, Minn. (56009) 255/E7
Alden, N.Y. (14004) 276/C5
Alden Bridge, La. (†71006) 238/C1
Aldenville, Pa. (18401) 294/M2
Alder, Mont. (59710) 262/D5
Alder, Wash. (†98328) 310/C4
Alder (lake), Wash. 310/C4
Alder Creek, N.Y. (13301) 276/K4
Alder Flats, Alberta 182/C3
Alderley, Wis. (†53066) 317/J1
Alderney (isl.), Chan. Is. 10/E6
Alderney (isl.), Chan. Is. 13/E8
Alderpoint, Calif. (95411) 204/B3
Alder Point, Nova Scotia 168/H2
Aldershot, England 10/F5
Aldershot, England 13/G8
Aldershot, Nova Scotia 168/D3
Alderson, Okla. (74522) 288/P5
Alderson, W. Va. (24910) 312/E7
Aldersyde, Alberta 182/C4
Aldine, Ind. (†46366) 227/D2
Aldora, Georgia (†30204) 217/D4
Aldouane, New Bruns. 170/E2
Aldrich, Ala. (†35115) 195/E4
Aldrich, Minn. (56434) 255/C4
Aldrich, Mo. (65601) 261/F7
Aldridge Brownhills, England 10/G3
Aldridge Brownhills, England 13/E5
Aledo, Ill. (61231) 222/C2
Aledo, Texas (76008) 303/E2
Aleg, Mauritania 106/B5
Alegre, Brazil 135/F2
Alegre, Brazil 132/F8
Alegrete, Brazil 132/B10
Alegrete, Brazil 120/D5
'Aleih, Lebanon 63/F6
Alejandra, Argentina 143/F5
Alejandría, Bolivia 136/C3
Alejandro Selkirk (isl.), Chile 120/A6
Aleknagik, Alaska (99555) 196/G3
Aleksandriya, U.S.S.R. 52/D5
Aleksandrov Gay, U.S.S.R. 52/G4
Aleksandrovsk, U.S.S.R. 52/J3
Aleksandrovsk-Sakhalinsky, U.S.S.R. 54/R4
Aleksandrovsk-Sakhalinskiy, U.S.S.R. 48/P5
Aleksandrów Kujawski, Poland 47/D2
Aleksandrów Łódzki, Poland 47/D3
Alekseyevka, U.S.S.R. 48/H4
Alekseyevka, U.S.S.R. 52/E4
Aleksin, U.S.S.R. 52/E4
Aleksinac, Yugoslavia 45/E4
Além Paraíba, Brazil 135/E2
Alençon, France 28/D3
Alenquer, Brazil 132/C3
Alenquer, Brazil 120/D3
Alenuihaha (chan.), Hawaii 218/E7
Aleppo (prov.), Syria 63/G4
Aleppo, Syria 54/E6
Aleppo, Syria 59/C2
Aleppo, Syria 63/G4
Aléria, France 28/B6
Alert, Canada 4/A12
Alert, Ind. (†47283) 227/F6
Alert, N.C. (†27589) 281/N2
Alert, N.W.T. 162/N3
Alert, N.W. Terrs. 187/M1
Alert (pt.), N.W. Terrs. 187/K1
Alert Bay, Br. Col. 184/D5
Alès, France 28/E5
Alessandria (prov.), Italy 34/B2
Alessandria, Italy 34/B2
Ålestrup, Denmark 21/C4
Ålesund, Norway 7/E2
Ålesund, Norway 18/D5
Aletschhorn (mt.), Switzerland 39/F4
Aleutian (isls.), Alaska 188/D6
Aleutian (isls.), Alaska 196/J4
Aleutian (range), Alaska 196/G3
Aleutian (isls.), U.S. 4/D18
Aleutian (isls.), U.S. 2/A3
Alex, Okla. (73002) 288/L5
Alexander (arch.), Alaska 146/E4
Alexander (arch.), Alaska 196/L1
Alexander (isl.) 5/B15
Alexander, Ark. (72002) 202/F4
Alexander (lake), Conn. 210/H1
Alexander, Georgia (30801) 217/J4
Alexander (co.), Ill. 222/D6
Alexander, Ill. (62601) 222/D4
Alexander, Iowa (50420) 229/G3
Alexander, Kansas (67513) 232/C3
Alexander○, Maine (†04610) 243/H5
Alexander, Manitoba 179/B5
Alexander, N.Y. (14005) 276/D5
Alexander (co.), N.C. 281/G3
Alexander, N. Dak. (58831) 282/C4
Alexander (cape), Solomon Is. 86/D2
Alexander (arch.), U.S. 4/D16
Alexander, W. Va. (26218) 312/F5
Alexander Bay, S. Africa 102/D7
Alexander Bay, S. Africa 118/B5
Alexander City, Ala. (35010) 195/G5
Alexander Mills, N.C. (†28043) 281/F4
Alexandra, N. Zealand 100/B6
Alexandra, S. Africa 118/H6
Alexandra, Victoria 97/C5
Alexandra Land (isl.), U.S.S.R. 52/A8
Alexandra Land (isl.), U.S.S.R. 48/E1
Alexandretta (Iskenderun), Turkey 63/G4
Alexandretta (gulf), Turkey 63/F4
Alexandria, Ala. (36250) 195/G3
Alexandria, Br. Col. 184/F4
Alexandria, Egypt 2/L4
Alexandria, Egypt 102/E1
Alexandria, Egypt 59/A3
Alexandria, Egypt 111/J2
Alexandria, Ind. (46001) 227/F4
Alexandria, Jamaica 158/J6
Alexandria, Ky. (41001) 237/N3
Alexandria, La. 146/J6
Alexandria, La. 188/H4
Alexandria, La. (71301) 238/E4
Alexandria, Minn. (56308) 255/C5
Alexandria, Mo. (63430) 261/K2
Alexandria, Nebr. (68303) 264/G4
Alexandria○, N.H. (†03222) 268/D4
Alexandria, North. Terr. 93/E5
Alexandria, Ohio (43001) 284/E5
Alexandria, Ontario 177/K2
Alexandria, Pa. (16611) 294/F4
Alexandria, Romania 45/G3
Alexandria, Scotland 15/A1
Alexandria, Scotland 10/A1
Alexandria, S. Dak. (57311) 298/O6
Alexandria, Tenn. (37012) 237/J8
Alexandria, Va. 188/L3
Alexandria (I.C.), Va. (*22301) 307/S3
Alexandria Bay, N.Y. (13607) 276/J2
Alexandrina (lake), S. Australia 94/F6
Alexandroúpolis, Greece 45/H5
Alexis, Ill. (61412) 222/C2
Alexis (riv.), Newf. 166/C3
Alexis Creek, Br. Col. 184/F4
Aleysk, U.S.S.R. 48/J4
Aleza Lake, Br. Col. 184/G3
Alfalfa (co.), Okla. 288/K1
Alfalfa, Okla. (†73015) 288/J4
Al Falluja, Iraq 59/D3
Al Falluja, Iraq 66/C4
Alfaro, Spain 33/F1
Alfatar, Bulgaria 45/H4
Al Fatha, Iraq 59/D2
Al Fatha, Iraq 66/C3
Alfeld, W. Germany 22/C2
Alfenas, Brazil 135/D2
Alférez (riv.), Uruguay 145/E5
Alford, England 13/H4
Alford, Fla. (32420) 212/D6
Alford○, Mass. (†01261) 249/A4
Alford, Scotland 15/F3
Alford, Scotland 10/E2
Alfordsville, Ind. (†47553) 227/C7
Alfred, Maine (04002) 243/B9
Alfred○, Maine (04002) 243/B9
Alfred, N.Y. (14802) 276/E6
Alfred, N. Dak. (58411) 282/N6
Alfred, Ontario 177/K2
Alfredton, N. Zealand 100/F4
Alfreton, England 13/F4
Alga, U.S.S.R. 48/F5
Ålgård, Norway 18/D7
Algarrobo, Chile 138/F3
Algarrobo (pt.), P. Rico 161/A2
Algarrobo del Águila, Argentina 143/C4
Algeciras, Colombia 126/C6
Algeciras, Spain 33/D4
Algemesí, Spain 33/F3
Alger (co.), Mich. 250/C2
Alger, Mich. (48610) 250/E4
Alger, Ohio (45812) 284/C4
Algeria 2/J4
Algeria 102/C2
ALGERIA 106/D3
Algés, Portugal 33/A1
Algete, Spain 33/G4
Alghero, Italy 34/B4
Algiers (cap.), Algeria 102/C1
Algiers (cap.), Algeria 106/E1
Algiers (cap.), Algeria 2/K4
Algiers, Ind. (†47567) 227/C7
Algoa, Ark. (†72112) 202/H3
Algoa (bay), S. Africa 118/D6
Algoa, Texas (†77511) 303/K3
Algodones, N. Mex. (87001) 274/C3
Algoma, Miss. (38820) 256/G2
Algoma (terr. dist.), Ontario 177/J5
Algoma (terr. dist.), Ontario 175/D3
Algoma, Oreg. (†97601) 291/F5
Algoma, W. Va. (24807) 312/D8
Algoma, Wis. (54201) 317/M6
Algoma Mills, Ontario 177/B1
Algona, Iowa (50511) 229/E2
Algona, Wash. (98002) 310/C3
Algonac, Mich. (48001) 250/G6
Algonquin, Ill. (60102) 222/E1
Algonquin (peak), N.Y. 276/M2
Algonquin Park, Ontario 177/F2
Algonquin Prov. Park, Ontario 177/F2
Algonquin Prov. Park, Ontario 175/E3
Algood, Tenn. (38501) 237/K8
Algorta, Uruguay 145/B3
Algrove, Sask. 181/H3
Alhama de Granada, Spain 33/E4
Alhama de Murcia, Spain 33/F4
Alhambra, Alberta 182/C3
Alhambra, Calif. (*91801) 204/C10
Alhambra, Ill. (62001) 222/D5
Al Hawtah, P.D.R. Yemen 59/E6
Al Hilla, Saudi Arabia 59/E5
Al Hoceima, Morocco 106/D1
Alhos Vedros, Portugal 33/B3
Alhué, Estero de (riv.), Chile 138/F4
Alia, Spain 33/D3
'Aliabad, Kuh-e (mt.), Iran 59/F3
'Aliabad, Kuh-e (mt.), Iran 66/G3
Aliağa, Turkey 63/B3
Alibag, India 68/C5
Alibates Flint Quarries Nat'l Mon., Texas 303/C2
Ali-Bayramly, U.S.S.R. 52/G7
Alibeyköyü, Turkey 63/D6
Alicante (prov.), Spain 33/F3
Alicante, Spain 33/F3
Alicante, Spain 7/D5
Alice (lake), Nebr. 264/A2
Alice, N. Dak. (58003) 282/P6
Alice (lake), N. Dak. 282/M3
Alice, Ontario 177/G2
Alice (chan.), Philippines 82/B8
Alice (riv.), Queensland 95/C4
Alice, Texas (78332) 303/F10
Alice Arm, Br. Col. 184/C2
Alicel, Oreg. (†97824) 291/J2
Alice Springs, Australia 87/D8
Alice Springs, North. Terr. 88/E4
Alice Springs, North. Terr. 93/D7
Aliceville, Ala. (35442) 195/B4
Aliceville (dam), Ala. 195/B4
Aliceville, Kansas (66832) 232/G3
Alicia, Ark. (72410) 202/H2
Alicia (bank), Colombia 126/B8
Alicudi (isl.), Italy 34/E5
Alida, Minn. (†56676) 255/C3
Alida, Sask. 181/K6
Aligarh, India 68/D3
'Ali Gharbi, Iraq 66/E4
Alijó, Portugal 33/C2
Alima (riv.), Congo 115/B4
Alimodian, Philippines 82/D5
Alindao, Cent. Afr. Rep. 115/D2
Aline, Georgia (†30420) 217/H6
Aline, Okla. (73716) 288/K1
Alingly, Sask. 181/E2
Alingsås, Sweden 18/H7
Alipore, India 68/F2
Aliquippa, Pa. (15001) 294/B4
Ali Sabieh, Djibouti 111/H5
'Ali Sharqi, Iraq 66/E4
Aliskerovo, U.S.S.R. 48/R3
Alivérion, Greece 45/G6
Aliwal North, S. Africa 118/D6
Alix, Alberta 182/D3
Alix, Ark. (72820) 202/C3
Aljezur, Portugal 33/B4
Aljojuca, Mexico 150/O1
Aljustrel, Portugal 33/B4
Alkabo, N. Dak. (58832) 282/C2
Alkali (lakes), Calif. 204/E2
Alkali (lake), Nev. 266/B1
Alkali (lakes), N. Dak. 282/L3
Alkali Lake, Br. Col. 184/F4
Alkaline (lake), N. Dak. 282/L6
Alken, Belgium 27/G7
Alkmaar, Netherlands 27/F3
Alkmaardermeer (lake), Netherlands 27/F3
Alkol, W. Va. (25501) 312/C6
Al Kufa, Iraq 66/D4
Al Kumait, Iraq 66/E4
Al Kuwait (cap.), Kuwait 59/E4
Al Kuwait (cap.), Kuwait 54/F7
Allagash○, Maine (†04774) 243/F1
Allagash (lake), Maine 243/D3
Allagash (riv.), Maine 243/E2
Allahabad, India 68/E3
Allahabad, India 54/K7
Allaine (riv.), Switzerland 39/D2
Allaire, N.J. (†07727) 273/E3
Allakaket, Alaska (99720) 196/H1
Allakh-Yun', U.S.S.R. 48/O3
Allamakee (co.), Iowa 229/L2
Allaman, Switzerland 39/B4
All American (canal), Calif. 204/K11
Allamoore, Texas (†79855) 303/C11
Allamuchy, N.J. (07820) 273/D2
Allan (mt.), Idaho 220/D4
Allan, Sask. 181/E4
Allan (hills), Sask. 181/E4
Allanmyo, Burma 72/B3
Allanwater, Ontario 175/C2
Allanwater, Ontario 177/G4
'Allaqi, Wadi (dry riv.), Egypt 111/F3
Allard (lake), Québec 174/E2
Allardt, Tenn. (38504) 237/M8
Allardville, New Bruns. 170/E1
Allariz, Spain 33/C1
Allatoona (lake), Georgia 217/C2
Alle, Switzerland 39/D2
Alleene, Ark. (71820) 202/B6
Allegan (co.), Mich. 250/D6
Allegan, Mich. (49010) 250/D6
Allegany (co.), Md. 245/C2
Allegany (co.), N.Y. 276/D6
Allegany, N.Y. (14706) 276/C6
Allegany, Oreg. (97407) 291/D4
Allegany Ind. Res., N.Y. 276/C6
Alleghany, Calif. (95910) 204/E4
Alleghany (co.), N.C. 281/G1
Alleghany (co.), Va. 307/H5
Alleghany, Va. (†24426) 307/H5
Allegheny (res.), N.Y. 276/C7
Allegheny (riv.), N.Y. 276/C6
Allegheny (co.), Pa. 294/B5
Allegheny (res.), Pa. 294/E2
Allegheny (riv.), Pa. 294/D2
Allegheny (mts.), Va. 307/H5
Allegheny Front (mts.), Md. 245/C2
Allegheny Front (mts.), Pa. 294/E5
Allègre (pt.), Guadeloupe 161/A6
Allegre, Ky. (42203) 237/G7
Alleman, Iowa (50007) 229/F5
Allemands (lake), La. 238/M4
Allen, Ala. (36419) 195/C7
Allen, Argentina 143/C4
Allen (co.), Ind. 227/G2
Allen, Lough (lake), Ireland 10/C3
Allen (lake), Ireland 17/E3
Allen, Bog of (marsh), Ireland 17/H5
Allen (co.), Kansas 232/G4
Allen, Kansas (66833) 232/F3
Allen (co.), Ky. 237/J7
Allen, Ky. (41601) 237/R5
Allen (par.), La. 238/E5
Allen, La. (†71469) 238/D3
Allen, Md. (21810) 245/R7
Allen, Mich. (49227) 250/E7
Allen, Miss. (†39083) 256/C7
Allen (mt.), Mont. 262/C2
Allen, Nebr. (68710) 264/H2
Allen (co.), Ohio 284/B4
Allen, Okla. (74825) 288/O5
Allen, Pa. (†17007) 294/H5
Allen, S. Dak. (57714) 298/F7
Allen, Texas (75002) 303/H1
Allendale, England 13/E3
Allendale, Ill. (62410) 222/F5
Allendale, Mo. (†64456) 261/D2
Allendale, N.J. (07401) 273/B1
Allendale (co.), S.C. 296/E6
Allendale, S.C. (29810) 296/E5
Allende, Coahuila, Mexico 150/J2
Allende, Nuevo León, Mexico 150/J4
Allendorf, Iowa (51330) 229/B2
Allenford, Ontario 177/C3
Allenhurst, Georgia (31301) 217/J7
Allenhurst, N.J. (07711) 273/F3
Allen Park, Mich. (48101) 250/B7
Allens Mills, Maine (†04938) 243/C6
Allenspark, Colo. (80510) 208/J2
Allen Springs, Ky. (†42122) 237/J7
Allenstein (Olsztyn), Poland 47/E2
Allenstown○, N.H. (†03275) 268/E5
Allensville, Ky. (42204) 237/G7
Allensville, Ohio (45611) 284/E7
Allensville, Pa. (17002) 294/G4
Allenton, Mo. (63001) 261/M4
Allenton, R.I. (†02852) 249/H6
Allenton, Wis. (53002) 317/K9
Allentown, Georgia (31003) 217/F5
Allentown, N.J. (08501) 273/D3
Allentown, N.Y. (14707) 276/E6
Allentown, Ohio (†45801) 284/B4
Allentown, Pa. 188/L2
Allentown, Pa. (*18101) 294/L4
Allentsteig, Austria 41/C2
Allenville, Ill. (†61951) 222/E4
Allenville, Mo. (†63740) 261/N8
Allenwood, N.J. (08720) 273/E3
Allenwood, Pa. (17810) 294/H3
Alleppey-Cochin, India 68/D7
Aller (riv.), W. Germany 22/C2
Allerton, Ill. (61810) 222/F4
Allerton, Iowa (50008) 229/G7
Allerton, Mass. (02045) 249/E7
Allerton (pt.), Mass. 249/E7
Alley, Jamaica 158/J7
Alley Spring, Mo. (†65466) 261/J8
Allgäu (reg.), W. Germany 22/D5

Allgäu Alps (mts.), Austria 41/A3
Allgood, Ala. (35013) 195/F3
Alliance, Alberta 182/E3
Alliance, Nebr. (69301) 264/A2
Alliance, N.C. (28509) 281/R4
Alliance, Ohio (44601) 284/H4
Al Lidam, Saudi Arabia 59/D5
Allier (dept.), France 28/E4
Allier (riv.), France 28/E5
Alligator (lake), Fla. 212/E3
Alligator (pt.), La. 238/L6
Alligator, Miss. (38720) 256/C2
Alligator (lake), N.C. 281/S3
Alligator (riv.), N.C. 281/S3
Alligator Pond, Jamaica 158/H6
Allingåbro, Denmark 21/D5
Allinge-Sandvig, Denmark 18/J9
Allinge-Sandvig, Denmark 21/F8
Allingham, Alberta 182/D4
Allingtown, Conn. (†06516) 210/D3
Allison, Iowa (50602) 229/H3
Allison, N. Mex. (†87301) 274/A3
Allison, Texas (79003) 303/D2
Allisona, Tenn. (†37046) 237/H9
Allisonia, Va. (24310) 307/G7
Allison Park, Pa. (15101) 294/C4
Alliston, Ontario 177/E3
Al Lith, Saudi Arabia 59/C5
Alloa, Scotland 10/B1
Alloa, Scotland 15/C1
Allock, Ky. (41710) 237/P6
Allons, Tenn. (38541) 237/L8
Allouez, Mich. (49805) 250/A1
Allouez, Wis. (†54301) 317/L7
Allow (riv.), Ireland 17/D7
Alloway, N.J. (08001) 273/C4
Alloways (creek), N.J. 273/C4
Allport, Ark. (†72046) 202/G4
Allred, Tenn. (38542) 237/L8
All Saints, Ant. & Bar. 161/E11
All Saints Village, Mo. (†63376) 261/M2
Allsboro, Ala. (†35616) 195/B1
Allsbrook, S.C. (†29569) 296/K3
Allschwil, Switzerland 39/D1
Allview, Md. (†21043) 245/L3
Allyn, Wash. (98524) 310/C3
Alma, Ala. (36501) 195/C8
Alma, Ark. (72921) 202/B3
Alma, Colo. (80420) 208/G4
Alma, Georgia (31510) 217/G7
Alma, Ill. (62807) 222/E5
Alma, Kansas (66401) 232/F2
Alma, Mich. (48801) 250/E5
Alma, Mo. (64001) 261/E4
Alma, Nebr. (68920) 264/E4
Alma, New Bruns. 170/F3
Alma, N. Mex. (†88039) 274/A5
Alma, N.C. (†28364) 281⅔5
Alma, Ontario 177/D4
Alma, Québec 174/C3
Alma, Québec 172/F1
Alma (isl.), Québec 172/F1
Alma, W. Va. (26320) 312/E4
Alma, Wis. (54610) 317/C7
Alma-Ata, U.S.S.R. 2/N3
Alma-Ata, U.S.S.R. 54/J5
Alma-Ata, U.S.S.R. 48/H5
Alma Center, Wis. (54611) 317/E7
Alma City, Minn. (†56048) 255/E6
Almada, Portugal 33/A1
Almadén, Spain 33/D3
Almagro, Spain 33/E3
Almaguer, Colombia 126/B7
Almanor (lake), Calif. 204/D3
Almansa, Spain 33/F3
Almanza, Spain 33/D1
Almanzor (mt.), Spain 33/D2
Almanzora (riv.), Spain 33/F4
Almartha, Mo. (†65773) 261/H9
Almazán, Spain 33/E2
Almaznyy, U.S.S.R. 48/M3
Almeida, Sierra (mts.), Chile 138/C4
Almeida, Portugal 33/C2
Almeirim, Portugal 33/B3
Almelo, Netherlands 27/K4
Almelund, Minn. (55002) 255/F5
Almena, Kansas (67622) 232/C2
Almena, Wis. (54805) 317/B5
Almenara, Brazil 120/E4
Almendra (res.), Spain 33/C2
Almendralejo, Spain 33/C3
Almere, Netherlands 27/G4
Almeria, Nebr. (68811) 264/E3
Almería (prov.), Spain 33/E4
Almería, Spain 7/D5
Almería, Spain 33/E4
Almería (gulf), Spain 33/E4
Al'met'yevsk, U.S.S.R. 52/H3
Älmhult, Sweden 18/H8
Almira, Wash. (99103) 310/G3
Almirantazgo (bay), Chile 138/F11
Almirante, Panama 154/F6
Almirante Montt (gulf), Chile 138/D9
Almirós, Greece 45/F6
Almo, Idaho (83312) 220/E7
Almo, Ky. (42020) 237/E7
Almodóvar, Portugal 33/B4
Almodóvar del Campo, Spain 33/D3
Almoharín, Spain 33/D3
Almoloya del Río, Mexico 150/K1
Almon, Georgia (†30209) 217/E3
Almond, N.Y. (14804) 276/E6
Almond, N.C. (28702) 281/B4
Almond (riv.), Scotland 15/E4
Almond, Wis. (54909) 317/G7
Almont, Colo. (81210) 208/of5
Almont, Mich. (48003) 250/F6
Almont, N. Dak. (58520) 282/H6
Almonte, Ontario 177/H2
Almonte, Spain 33/C4
Almora, India 68/D3
Almora, Minn. (†56551) 255/C4
Almorox, Spain 33/D2
Almota, Wash. (†99111) 310/H4
Al Muadhdam, Saudi Arabia 59/C4
Almudévar, Spain 33/F1
Almuñécar, Spain 33/E4
Almus, Turkey 63/G2
Al Musaiyib, Iraq 59/D3
Al Musaiyib, Iraq 66/D4
Almyra, Ark. (72003) 202/H5
Alna○, Maine (04535) 243/D7
Alness, Scotland 15/D3
Alness (riv.), Scotland 15/D3
Alnwick, England 10/F3
Alnwick, England 13/F2
Alofi (cap.), Niue 87/K7
Aloha, Oreg. (97005) 291/A2
Aloha, Wash. (98550) 310/A3
Alon, Burma 72/B2
Along, India 68/G3
Alonsa, Manitoba 179/C4
Alonso Rojas, Cuba 158/B2
Alor (isl.), Indonesia 85/G7
Álora, Spain 33/D4
Alor Gajah, Malaysia 72/D7
Alor Setar, Malaysia 72/D6
Alorton, Ill. (62207) 222/B2
Alost (Aalst), Belgium 27/D7
Alotau, Papua N.G. 87/E7
Aloysius (mt.), W. Australia 92/E4
Alpachiri, Argentina 143/D4
Alpaugh, Calif. (93201) 204/F8
Alpena, Ark. (72611) 202/D1
Alpena (co.), Mich. 250/F4
Alpena, Mich. (49707) 250/F3
Alpena, S. Dak. (57312) 298/N5
Alpena, W. Va. (†26254) 312/G5
Alpen Siding, Alberta 182/D2
Alpera, Spain 33/F3
Alpes-de-Haute-Provence (dept.), France 28/G5
Alpes-Maritimes (dept.), France 28/G6
Alpha, Ill. (61413) 222/C2
Alpha, Iowa (52130) 229/K3
Alpha, Ky. (42603) 237/L7
Alpha, Mich. (49902) 250/A2
Alpha, Minn. (56111) 255/D7
Alpha, N.J. (08865) 273/C2
Alpha, Queensland 88/H4
Alpharetta, Georgia (30201) 217/D2
Alphen aan de Rijn, Netherlands 27/F4
Alpiarça, Portugal 33/B3
Alpine, Ala. (35014) 195/F4
Alpine, Ariz. (85920) 198/F5
Alpine, Ark. (†71921) 202/D5
Alpine (co.), Calif. 204/F5
Alpine, Calif. (92001) 204/J11
Alpine, Ind. (†47331) 227/G5
Alpine, Ky. (42519) 237/M7
Alpine, N.J. (07620) 273/C1
Alpine, N.Y. (14805) 276/G6
Alpine, Oreg. (97456) 291/D3
Alpine, Tenn. (38543) 237/L8
Alpine, Texas (79830) 303/D12
Alpine, Utah (†84003) 304/C3
Alpine, Wyo. (83128) 319/B2
Alpirsbach, W. Germany 22/C4
Alpnach, Switzerland 39/F3
Alpoca, W. Va. (24710) 312/D7
Alportel, Portugal 33/C4
Alps (mts.) 7/E4
Alpu, Turkey 63/D3
Al Q'aim, Iraq 66/B3
Al Qaiyara, Iraq 66/C3
Al Qosh, Iraq 66/C2
Alquina, Ind. (†47331) 227/G5
Alquízar, Cuba 158/C1
Al Qurna, Iraq 66/E5
Al Qurna, Iraq 59/E3
Alroy Downs, North. Terr. 93/E5
Als (isl.), Denmark 21/C8
Alsace (trad. prov.), France, 29
Alsager, England 13/E4
Alsager, England 10/G2
Alsask, Sask. 181/B4
Alsatia, La. (†71276) 238/H1
Alsdorf, W. Germany 22/B3
Alsea, Oreg. (97324) 291/D3
Alsea (riv.), Oreg. 291/D3
Alsek (riv.), Alaska 196/L3
Alsek (riv.), Br. Col. 184/H1
Alsek (riv.), Yukon 187/E3
Alsen, N. Dak. (58311) 282/N2
Alsey, Ill. (62610) 222/C4
Alsfeld, W. Germany 22/C3
Alsip, Ill. (†60658) 222/B6
Alsózsolca, Hungary 41/F2
Alstead○, N.H. (03602) 268/C5
Alsten (isl.), Norway 18/H4
Alstfjorden (fjord), Norway 18/G3
Alston, England 13/E3
Alston, Georgia (30412) 217/H6
Alston, Mich. (†49958) 250/G1
Alstonville, N.S. Wales 97/G1
Alta, Iowa (51002) 229/C3
Alta, Norway 18/N2
Alta, Utah (84070) 304/C3
Altadena, Calif. (91001) 204/C10
Altaelv (riv.), Norway 18/N2
Alta Gracia, Argentina 143/D3
Altagracia, Cuba 158/G3
Altagracia, Venezuela 124/C2
Altagracia de Orituco, Venezuela 124/E3
Altai (mts.) 54/K5
Altai (mts.), Mongolia 77/C2
Alta Loma, Calif. (91701) 204/E10
Alta Loma, Texas (77510) 303/K3
Altamachi (riv.), Bolivia 136/B5
Altamaha (riv.), Ga. 188/K4
Altamaha (riv.), Georgia 217/H7
Altamaha (sound), Georgia 217/K8
Altamahaw, N.C. (27202) 281/L2
Altamira, Brazil 132/C3
Altamira, Chile 138/B5
Altamira, Dom. Rep. 158/D5
Altamira, Mexico 150/L5
Altamont, Ill. (62411) 222/E4
Altamont, Kansas (67330) 232/G4
Altamont, Manitoba 179/D5
Altamont, Mo. (64620) 261/D3
Altamont, N.Y. (12009) 276/M5
Altamont, Oreg. (†97601) 291/F5
Altamont, S. Dak. (57211) 298/R4
Altamont , Tenn. (37301) 237/K10
Altamont, Utah (84001) 304/D3
Altamonte Springs, Fla. (32701) 212/E3
Altamura, Italy 34/F4
Altar, Mexico 150/D1
Altario, Alberta 182/E4
Altata, Mexico 150/E4
Alt Aussee, Austria 41/B3
Alta Vista, Iowa (50603) 229/J2
Alta Vista, Kansas (66834) 232/F3
Altavista, Va. (24517) 307/K6
Altay, China 77/C2
Altay, Mongolia 77/E2
Altay (mts.), U.S.S.R. 48/J5
Altdorf, Switzerland 39/G3
Altea, Spain 33/G3
Altena, W. Germany 22/B3
Altenburg, E. Germany 22/E3
Altenburg, Mo. (63732) 261/O7
Altepexi, Mexico 150/L7
Alter do Chão, Portugal 33/C3
Altevatn (lake), Norway 18/L2
Altha, Fla. (32421) 212/A1
Altheim, Austria 41/B2
Altheimer, Ark. (72004) 202/G5
Althofen, Austria 41/C3
Alticane, Sask. 181/D3
Altındağ, Turkey 63/E2
Altınova, Turkey 63/B3
Altınözü, Turkey 63/G4
Altıntaş, Turkey 63/C3
Altiplano (plat.) 120/C4
Altkirch, France 28/G4
Altmar, N.Y. (13302) 276/J3
Altmark (reg.), E. Germany 22/D2
Altmühl (riv.), W. Germany 22/D4
Altnaharra, Scotland 15/D2
Alto, Georgia (30510) 217/E2
Alto, La. (71216) 238/G2
Alto, Mich. (49302) 250/D6
Alto, N. Mex. (88312) 274/D5
Alto, Tenn. (†37324) 237/K10
Alto, Texas (75925) 303/J6
Alto, Wis. (†53963) 317/J8
Alto Araguaia, Brazil 132/C7
Alto Chicapa, Angola 115/C6
Alto Cuale, Angola 115/C5
Alto Cuale, Angola 102/D5
Alto de Cantillana (mt.), Chile 138/G4
Alto Lucero, Mexico 150/P1
Alto Molócuè, Mozambique 118/F3
Alton, Ala. (†35210) 195/E3
Alton, Calif. (†95540) 204/A3
Alton, England 13/G6
Alton, England 10/F5
Alton, Ill. 188/J3
Alton, Ill. (62002) 222/A2
Alton, Ind. (†47137) 227/E8
Alton, Iowa (51003) 229/A3
Alton, Kansas (67623) 232/D2
Alton, Ky. (†40342) 237/M4
Alton, La. (†70458) 238/L6
Alton○, Maine (†04468) 243/F5
Alton, Mo. (65606) 261/K9
Alton○, N.H. (03809) 268/E5
Alton, N.Y. (14413) 276/G4
Alton, Nova Scotia 168/E3
Alton, R.I. (†02894) 249/G7
Alton, Utah (84710) 304/B6
Alton, Va. (24520) 307/K7
Alton, W. Va. (†26210) 312/F5
Altona, Ill. (61414) 222/C2
Altona, Ind. (†46738) 227/G2
Altona, Manitoba 179/E5
Altona, Mich. (†49336) 250/D5
Altona, Mo. (†64720) 261/D6
Altona, N.Y. (12910) 276/N1
Altona, Victoria 97/H5
Altona, Victoria 88/K7
Altona (bay), Victoria 97/H5
Altona (bay), Victoria 88/K7
Altona (lag.), Virgin Is. (U.S.) 161/F4
Altona, W. Germany 22/C2
Altonah, Utah (84002) 304/D3
Alton Bay, N.H. (03810) 268/E5
Alton Downs, S. Australia 94/F2
Alto Nevado, Cerro (mt.), Chile 138/E5
Altoona, Ala. (35952) 195/F2
Altoona, Fla. (32702) 212/E3
Altoona, Iowa (50009) 229/G5
Altoona, Kansas (66710) 232/G4
Altoona, Pa. 188/L2
Altoona, Pa. (*16601) 294/F4
Altoona, Wash. (†98643) 310/B4
Altoona, Wis. (54720) 317/C6
Alto Paraguay (dept.), Paraguay 144/C2
Alto Paraná (dept), Paraguay 144/E4
Alto Paraná (riv.), Paraguay 144/D5
Alto Park, Georgia (†30161) 217/B2
Alto Parnaíba, Brazil 132/E5
Alto Pass, Ill. (62905) 222/D6
Alto Ritacuva (mt.), Colombia 120/B2
Alto Ritacuva (mt.), Colombia 126/D4
Altos, Brazil 132/F4
Altos, Paraguay 144/B4
Alto Seco, Bolivia 136/D6
Alto Songo, Cuba 158/J4
Altotonga, Mexico 150/P1
Altötting, W. Germany 22/E4
Alto Velo (chan.), Dom. Rep. 158/C7
Alto Velo (isl.), Dom. Rep. 158/D7
Altrincham, England 13/H2
Altrincham, England 10/G2
Altro, Ky. (41306) 237/P6
Altstätten, Switzerland 39/J2
Altun Ha, Belize 154/C2
Altun Shan (range), China 54/K6
Altun Shan (range), China 77/C4
Altura, Minn. (55910) 255/G6
Alturas, Calif. (96101) 204/E2
Alturas, Fla. (33820) 212/E4
Altus, Ark. (72821) 202/C3
Altus, Okla. (73521) 288/H5
Altus (res.), Okla. 288/H5
Altus A.F.B., Okla. 288/H5
Alubijid, Philippines 82/E6
Alucra, Turkey 63/H2
Alūksne, U.S.S.R. 53/D2
Alula, Somalia 115/K1
Alula, Somalia 102/H3
Alum Bank, Pa. (15521) 294/E5
Alum Bridge, W. Va. (26321) 312/E4
Alum Creek, W. Va. (25003) 312/C6
Aluminé, Argentina 143/B4
Alum Rock, Calif. (†95116) 204/L3
Alus, Iraq 66/C3
Alushta, U.S.S.R. 52/D6
Alva, Fla. (33920) 212/E5
Alva, Ky. (†40977) 237/P7
Alva (lake), New Bruns. 170/D3
Alva, Okla. (73717) 288/J1
Alva, Scotland 10/B1
Alva, Scotland 15/C1
Alva, Wyo. (82711) 319/H1
Alva B. Adams (tunnel), Colo. 208/H2
Alvada, Ohio (44802) 284/D3
Alvadore, Oreg. (97409) 291/D3
Alvalade, Portugal 33/B4
Alvaneu, Switzerland 39/J3
Alvarado, Mexico 150/M7
Alvarado, Minn. (56710) 255/B2
Alvarado, Texas (76009) 303/G5
Álvaro S. Lima (res.), Brazil 135/B3
Alvaton, Georgia (30202) 217/C4
Alvaton, Ky. (42122) 237/J7
Alvdal, Norway 18/G5
Älvdalen, Sweden 18/J6
Alvear, Argentina 143/E2
Alvena, Sask. 181/E3
Alvesta, Sweden 18/J8
Alvin, Br. Col. 184/L2
Alvin, Ill. (61811) 222/F3
Alvin, Texas (77511) 303/J3
Alvin, Wis. (49936) 317/J4
Alvinston, Ontario 177/B5
Alvito, Portugal 33/B3
Älvkarleby, Sweden 18/K6
Alvo, Nebr. (68304) 264/H4
Alvon, W. Va. (†24986) 312/F7
Alvord, Iowa (51230) 229/A2
Alvord (lake), Oreg. 291/J5
Alvord, Texas (76225) 303/G4
Alvordton, Ohio (43501) 284/A2
Älvsborg (co.), Sweden 18/H7
Älvsbyn, Sweden 18/M4
Alvwood, Minn. (†56630) 255/D3
Alvy, W. Va. (26322) 312/E4
Alwar, India 68/D3
Alxa Shamo (des.), China 77/F4
Alxa Youqi, China 77/F4
Alxa Zuoqi, China 77/F4
Aly, Ark. (†72860) 202/D4
Alyangula, North. Terr. 88/F2
Alyangula, North. Terr. 93/E2
Alyth, Scotland 10/E2
Alyth, Scotland 15/E4
Alytus, U.S.S.R. 53/C3
Alz (riv.), W. Germany 22/E4
Alzada, Mont. (59311) 262/M5
Alzette (riv.), Luxembourg 27/J9
Alzey, W. Germany 22/C4
Amacuro (riv.), Venezuela 124/H4
Amadeus (lake), Australia 87/D8
Amadeus (lake), North. Terr. 88/E4
Amadeus (lake), North. Terr. 93/B8
Amadi, Sudan 111/F6
'Amadiya, Iraq 59/D2
'Amadiya, Iraq 66/C2
Amadjuak, N.W.T. 162/J3
Amadjuak (lake), N.W.T. 162/K3
Amadjuak, N.W. Terrs. 187/L3
Amadjuak (lake), N.W. Terrs. 187/L3
Amado, Ariz. (85640) 198/D7
Amador (co.), Calif. 204/E5
Amadora, Portugal 33/A1
Amador City, Calif. (95601) 204/C9
Amagansett, N.Y. (11930) 276/R9
Amagasaki, Japan 81/H8
Amager (isl.), Denmark 21/F6
Amagi, Japan 81/E7
Amagon, Ark. (72005) 202/H2
Amahai, Indonesia 85/H6
Amak (isl.), Alaska 196/F3
Amakura (riv.), Guyana 131/A2
Amakusa (isls.), Japan 81/D7
Åmål, Sweden 18/H7
Amalfi, Colombia 126/C4
Amalfi, Italy 34/E4
Amalga, Utah (†84335) 304/C2
Amalia, N. Mex. (87512) 274/D2
Amaliás, Greece 45/E7
Amalner, India 68/C4
Amambaí, Brazil 132/C8
Amambaí, Serra de (range), Brazil 132/C7
Amambay, Paraguay 144/D3
Amambay, Cordillera de (mts.), Paraguay 144/D-E3
Amami (isls.), Japan 54/P7
Amami (isls.), Japan 81/N5
Amami-O-Shima (isl.), Japan 81/N5
Amana, Iowa (52203) 229/K5
Amanavén, Colombia 126/G6
Amanda, Ohio (43102) 284/E6
Amanda Park, Wash. (98526) 310/A3
Amanos (mts.), Turkey 63/G4
Amantea, Italy 34/F5
Amanu (atoll), Fr. Poly. 87/N7
Amapá (terr.), Brazil 132/D2
Amapá, Brazil 120/D2
Amapá, Brazil 132/D2
Amapala, Honduras 154/D4
Amapari (riv.), Brazil 132/C2
'Amara, Iraq 66/E5
'Amara, Iraq 59/E3
Amarante, Brazil 132/F4
Amarante, Portugal 33/B2
Amaranth, Manitoba 179/D4
Amarapura, Burma 72/B2
Amareleja, Portugal 33/C3
Amarete, Bolivia 136/A4
Amargosa, Brazil 132/F6
Amargosa (range), Calif. 204/J7
Amargosa (riv.), Calif. 204/J7
Amarillas, Cuba 158/D2
Amarillo, Texas 188/F3
Amarillo, Texas 146/H6
Amarillo, Texas (*79101) 303/C2
Amasa, Mich. (49903) 250/G2
Amasra, Turkey 63/E2
Amasya (prov.), Turkey 63/F2
Amasya, Turkey 59/C1
Amasya, Turkey 63/G2
Amatignak (isl.), Alaska 196/K4
Amatitlán, Guatemala 154/B3
Amatlán de los Reyes, Mexico 150/P2
Amay, Belgium 27/G7
Amazon (riv.) 2/G6
Amazon (riv.), Brazil 120/D3
Amazon (riv.), Brazil 132/C3
Amazon (riv.), Colombia 126/E9
Amazon (riv.), Peru 128/F4
Amazon, Sask. 181/F4
Amazonas (state), Brazil 132/G9
Amazonas (comm.), Colombia 126/D8
Amazonas, Cuba 158/F2
Amazonas (dept.), Peru 128/C5
Amazonas (terr.), Venezuela 124/E5
Amazonia, Mo. (64421) 261/C3
Ambala, India 68/D2
Ambalavao, Madagascar 118/H4
Ambam, Cameroon 115/B3
Ambanja, Madagascar 118/H2
Ambarchik, U.S.S.R. 4/B1
Ambarchik, U.S.S.R. 48/R3
Ambato, Ecuador 120/B3
Ambato, Ecuador 128/C3
Ambato Boeny, Madagascar 118/H3
Ambatofinandrahana, Madagascar 118/H4
Ambatolampy, Madagascar 102/G6
Ambatolampy, Madagascar 118/H3
Ambatomainty, Madagascar 118/H3
Ambatondrazaka, Madagascar 118/H3
Ambatondrazaka, Madagascar 102/G6
Ambelau (isl.), Indonesia 85/H6
Amber, Iowa (†52205) 229/L4
Amber (Bobaomby) (cape), Madagascar 102/G6
Amber (Bobaomby) (cape), Madagascar 118/H2
Amber, Okla. (73004) 288/L4
Amber, Wash. (†99004) 310/H3
Amberg, W. Germany 22/D4
Amberg, Wis. (54102) 317/K5
Ambergris (cay), Belize 154/D1
Ambergris (cay), Turks & Caicos 156/D2
Ambérieu-en-Bugey, France 28/F5
Amberley, N. Zealand 100/D5
Amberley, Ohio (†45201) 284/C9
Amberson, Pa. (17210) 294/G5
Ambert, France 28/E5
Ambia, Ind. (47917) 227/C4
Ambikapur, India 68/E4
Ambil (isl.), Philippines 82/C4
Ambilobe, Madagascar 118/H2
Amble, England 13/F2
Amble, Mich. (†49329) 250/D5
Ambler, Alaska (99786) 196/G1
Ambler, Pa. (19002) 294/M5
Ambo, Peru 128/D8
Amboasary, Madagascar 118/H4
Ambodifototra, Madagascar 118/J3
Ambohimahasoa, Madagascar 118/H4
Amboise, France 28/D4
Ambon (Amboina), Indonesia 85/H6
Ambon, Indonesia 54/O10
Ambositra, Madagascar 102/G7
Ambositra, Madagascar 118/H4
Ambovombe, Madagascar 118/H5
Amboy, Calif. (92304) 204/K9
Amboy, Georgia (†31714) 217/E7
Amboy, Ill. (61310) 222/D2
Amboy, Ind. (46911) 227/F3
Amboy, Minn. (56010) 255/D7
Amboy, Wash. (98601) 310/C5
Amboy, W. Va. (26701) 312/G4
Amboyna (cay), Philippines 85/E4
Ambridge, Pa. (15003) 294/B4
Ambriz, Angola 115/B5
Ambrose, Georgia (31512) 217/G7
Ambrose, N. Dak. (58833) 282/D2
Ambrym (isl.), Vanuatu 87/G7
Ambunti, Papua N.G. 85/B6
Amburgey, Ky. (41801) 237/R6
Amchitka (isl.), Alaska 188/D6
Amchitka (isl.), Alaska 196/K4
Amchitka (passage), Alaska 196/K4
Am-Dam, Chad 111/D5
Amderma, U.S.S.R. 52/K1
Amderma, U.S.S.R. 48/F3
Amdo, China 77/D5
Ameagle, W. Va. (25004) 312/D7
Amealco, Mexico 150/K6
Ameca, Mexico 150/H6
Amecameca de Juárez, Mexico 150/L1
Ameghino, Argentina 143/D3
Amel, Belgium 27/J8
Ameland (isl.), Netherlands 27/H2
Amelia (isl.), Fla. 212/E1
Amelia, Italy 34/D3
Amelia, La. (70340) 238/H7
Amelia, Nebr. (68711) 264/F2
Amelia, Ohio (45102) 284/D10
Amelia (co.), Va. 307/M6
Amelia City, Fla. (†32034) 212/E1
Amelia Court House, Va. (23002) 307/N6
Amenia, N.Y. (12501) 276/N7
Amenia, N. Dak. (58004) 282/R6
America, Ill. (†62996) 222/D6
American (highlands), Ant. 2/N10
American (riv.), Calif. 204/C8
Americana, Brazil 135/C3
American Corner, Md. (†21632) 245/P5
American Falls, Idaho (83211) 220/E7
American Falls (res.), Idaho 188/D2
American Falls (res.), Idaho 220/F6
American Fork, Utah (84003) 304/C3
American Highland 5/B4
American Samoa 2/A6
AMERICAN SAMOA 86/N9
American Samoa 87/J7
Americus, Georgia (31709) 217/D6
Americus, Ind. (†47901) 227/D3
Americus, Kansas (66835) 232/F3
Americus, Mo. (†65069) 261/J5
Amersfoort, Netherlands 27/G4
Amersham, England 13/G7
Amery, Man. 162/G4
Amery, Manitoba 179/J2
Amery, Wis. (54001) 317/B5
Amery Ice Shelf, Ant. 2/N9
Amery Ice Shelf 5/C4
Ames, Iowa (50010) 229/F4
Ames, Kansas (66931) 232/E2
Ames, N.Y. (13317) 276/L5
Ames, Okla. (73718) 288/K2
Ames, Texas (†77575) 303/K7
Amesbury, England 13/F6
Amesbury, Mass. (01913) 249/L1
Amesbury○, Mass. (01913) 249/L1
Amesville, Conn. (†06031) 210/B1
Amesville, Ohio (45711) 284/F7
Amet (sound), Nova Scotia 168/E3
Amfilokhía, Greece 45/E6
Ámfissa, Greece 45/F6
Amga (riv.), U.S.S.R. 48/O3
Amguid, Algeria 106/F3
Amgun' (riv.), U.S.S.R. 48/O4
Amherst, Burma 72/C3
Amherst, Colo. (80721) 208/P1
Amherst○, Maine (†04408) 243/G6
Amherst, Mass. (01002) 249/E3
Amherst○, Mass. (01002) 249/E3
Amherst, Nebr. (68812) 264/E4
Amherst○, N.H. (03031) 268/D6
Amherst, N.Y. (†14226) 276/C4
Amherst, N.S. 162/K6
Amherst, Nova Scotia 168/D3
Amherst, Ohio (44001) 284/F3
Amherst (isl.), Ontario 177/H3
Amherst, S. Dak. (57421) 298/O2
Amherst, Texas (79312) 303/B4
Amherst (co.), Va. 307/K5
Amherst, Va. (24521) 307/K5
Amherst (mt.), W. Australia 92/D2
Amherst, Wis. (54406) 317/H7
Amherstburg, Ontario 177/A5
Amherstdale, W. Va. (25607) 312/C7
Amherst Junction, Wis. (54407) 317/H7
Amherst View, Ontario 177/H3
Amidon, N. Dak. (58620) 282/D7
Amiens, France 7/E4
Amiens, France 28/D3
Amindiri (isl.), India 68/C6
Amindivi (isls.), India 68/C6
Aminga, Argentina 143/C2
Amini (Amindiri) (isl.), India 68/C6
'Amir, Ras (cape), Libya 111/D1
Amiret, Minn. (56112) 255/C6
Amisk, Alberta 182/E3
Amisk (lake), Sask. 181/M4
Amissville, Va. (22002) 307/M3
Amistad (res.) 188/F5
Amistad (res.), Mexico 150/J2
Amistad, N. Mex. (88410) 274/F3
Amistad (dam), Texas 303/C8
Amistad (res.), Texas 303/C8
Amistad Nat'l Rec. Area, Texas 303/D8
Amite, La. (70422) 238/K5
Amite (riv.), La. 238/L2
Amite (co.), Miss. 256/C8
Amite (riv.), Miss. 256/C9
Amity, Ark. (71921) 202/D5
Amity, Georgia (†30817) 217/G3
Amity, Ind. (†46131) 227/E6
Amity, Mo. (†64469) 261/D3
Amity, Oreg. (97101) 291/D2
Amity, Pa. (15311) 294/B5
Amityville, N.Y. (11701) 276/O9
Åmli, Norway 18/F7
Amlia (isl.), Alaska 196/L4
Amlia (passage), Alaska 196/L4
Amlwch, Wales 10/D4
Amlwch, Wales 13/C4
Amma, W. Va. (25005) 312/D5
Amman (dist.), Jordan 65/D4
Amman (cap.), Jordan 54/E6
Amman (cap.), Jordan 65/D4
Amman (cap.), Jordan 59/C3
Ammanford, Wales 13/C6
Ammersee (lake), W. Germany 22/D4
Ammie, Ky. (41801) 237/O6
Ammon, Idaho (†83401) 220/G6
Ammon, Va. (23822) 307/N6
Ammonoosuc (riv.), N.H. 268/D3
Amnat, Thailand 72/E4
Amo, Ind. (46103) 227/D5
Amol, Iran 59/F2
Amol, Iran 66/H2
Amonate, Va. (24601) 307/E6
Amor, Minn. (†56515) 255/C4
Amora, Portugal 33/A1
Amorbach, W. Germany 22/C4
Amoret, Mo. (64722) 261/C6
Amorgós (isl.), Greece 45/G7
Amorita, Okla. (73719) 288/K1
Amory, Miss. (38821) 256/H3
Amos, Que. 162/J6
Amos, Québec 174/B3
Åmotfors, Sweden 18/H7
Amoy (Xiamen), China 77/J7
Amozoc de Mota, Mexico 150/N2
Ampana, Indonesia 85/G6
Ampanihy, Madagascar 118/G4
Amparo, Brazil 135/C3
Amper (riv.), W. Germany 22/D4
Amphitrite (isls.), China 85/E2
Amposta, Spain 33/G2
Ampthill, England 13/G5
Amqui, Québec 172/B2
'Amran, Yemen Arab Rep. 59/D6
Amravati, India 68/D4
Amreli, India 68/C4
Amriswil, Switzerland 39/H1
'Amrit (ruins), Syria 63/F5
Amritsar, India 54/J6

Amritsar, India 68/C2
Amrum (isl.), W. Germany 22/C1
Amsden, Ohio (44803) 284/D3
Amstelveen, Netherlands 27/B5
Amsterdam (isl.) 2/N7
Amsterdam, Georgia (31734) 217/D9
Amsterdam, Mo. (64723) 261/D6
Amsterdam, Mont. (†59741) 262/E5
Amsterdam (cap.), Netherlands 27/B4
Amsterdam (cap.), Netherlands 7/E3
Amsterdam, N.Y. (12010) 276/M5
Amsterdam, Ohio (43903) 284/J5
Amsterdam, Sask. 181/J4
Amstetten, Austria 41/C2
Amston, Conn. (06231) 210/F2
Am-Timan, Chad 111/D5
Amuay, Venezuela 124/C2
Amudar'ya (riv.) 2/N3
Amudar'ya (riv.), U.S.S.R. 54/H5
Amudar'ya (riv.), U.S.S.R. 48/G5
Amukta (isl.), Alaska 196/D4
Amukta (passage), Alaska 196/D4
Amuku (mts.), Guyana 131/B4
Amulet, Sask. 181/G6
Amund Ringnes (isl.), N.W.T. 162/M3
Amund Ringnes (isl.), N.W. Terrs. 187/J2
Amundsen (sea) 2/D10
Amundsen (bay) 5/C3
Amundsen (sea) 5/B13
Amundsen (gulf), Canada 4/B16
Amundsen (gulf), N.W.T. 162/D1
Amundsen (gulf), N.W.T. 146/F2
Amundsen (gulf), N.W. Terrs. 187/F2
Amundsen-Scott Station 5/A14
Amuntai, Indonesia 85/F6
Amur (riv.) 2/R3
Amur (riv.) 54/P5
Amur (Heilong Jiang) (riv.), China 77/L2
'Amur, Wadi (dry riv.), Sudan 111/G4
Amur (riv.), U.S.S.R. 48/O4
Amurang, Indonesia 85/G5
Amursk, U.S.S.R. 48/O4
Amy, Kansas (†67850) 232/B3
Amya (pass), Burma 72/C4
Amya (pass), Thailand 72/C4
Amyun, Lebanon 63/F5
An, Burma 72/B3
'Ana, Iraq 66/B3
'Ana, Iraq 59/D3
Anaa (atoll), Fr. Poly. 87/M7
Anabar (riv.), U.S.S.R. 48/M2
Anabel, Mo. (63431) 261/H3
Ana Branch, Darling (riv.), N.S. Wales 97/A3
'Anabta, West Bank 65/C3
Anacapa (isl.), Calif. 204/F10
Anaco, Venezuela 124/F3
Anacoco, La. (71403) 238/D4
Anacoco (lake), La. 238/D4
Anaconda, Mont. 188/D1
Anaconda-Deer Lodge County, Mont. (59711) 262/C4
Anacortes, Wash. (98221) 310/C2
Anacostia, D.C. (20020) 245/F5
Anacostia (riv.), D.C. 245/F5
Anadarko, Okla. (73005) 288/K4
Anadia, Portugal 33/B2
Anadolufeneri, Turkey 63/D5
Anadoluhisari, Turkey 63/D6
Anadyr', U.S.S.R 2/T2
Anadyr', U.S.S.R. 4/C1
Anadyr' (gulf), U.S.S.R. 4/C18
Anadyr' (gulf), U.S.S.R. 54/V3
Anadyr' (riv.), U.S.S.R. 54/U3
Anadyr' (riv.), U.S.S.R. 4/C1
Anadyr', U.S.S.R. 48/S3
Anadyr' (gulf), U.S.S.R. 48/T3
Anadyr' (range), U.S.S.R. 48/S3
Anadyr' (riv.), U.S.S.R. 48/S3
Anadyr' U.S.S.R. 54/U3
Anáfi (isl.), Greece 45/G7
Anagance, New Bruns. 170/E3
Anaheim, Calif. 188/C4
Anaheim, Calif. (*92801) 204/D11
Anahim Lake, Br. Col. 184/E4
Anahola, Hawaii (96703) 218/C1
Anáhuac, Chihuahua, Mexico 150/F2
Anáhuac, Nuevo León, Mexico 150/J3
Anahuac, Texas (77514) 303/K8
Anaï (well), Algeria 106/G4
Anai Mudi (mt.), India 68/D6
'Anaiza, Saudi Arabia 59/D4
'Anaiza, Saudi Arabia 54/F7
Anak, N. Korea 81/B4
Anakapalle, India 68/E5
Anaktalik Brook (riv.), Newf. 166/B2
Anaktuvuk Pass, Alaska (99721) 196/H1
Analalava, Madagascar 118/H2
Ana María (gulf), Cuba 158/F3
Anambas (isls.), Indonesia 85/D5
Anambra (state), Nigeria 106/F7
Anamoose, N. Dak. (58710) 282/K4
Anamosa, Iowa (52205) 229/L4
Anamur, Turkey 63/E4
Anamur (cape), Turkey 59/B2
Anamur (cape), Turkey 63/E5
Anan, Japan 81/G7
Anandale, La. (†71301) 238/F4
Ananea, Bolivia 136/A4
Anantapur, India 68/D6
Anantnag, India 68/D2
Anapa, U.S.S.R. 52/E6
Anápolis, Brazil 132/D7
Anápolis, Brazil 120/E4
Anar, Iran 66/J5
Anar, Iran 59/G3
Anarak, Iran 66/H4
Anarak, Iran 59/F3
Anar Darreh, Afghanistan 59/H3
Anar Darreh, Afghanistan 68/A2
Añasco, P. Rico 161/A1
Añasco, P. Rico 156/F1
Añasco (bay), P. Rico 161/A1
Anastasia (isl.), Fla. 212/E2
Anatahan (isl.), No. Marianas 87/E4
Anatolia (reg.), Turkey 63/D3
Anatone, Wash. (99401) 310/H4
Añatuya, Argentina 143/D2
Anauá (riv.), Brazil 132/B2
Anawalt, W. Va. (24808) 312/D8
Anaye (well), Niger 106/G5
Anbar (gov.), Iraq 66/B4
Ancash (dept.), Peru 128/D7
Ancaster, Ontario 177/D4
Anceney, Mont. (†59741) 262/E5
Ancenis, France 28/C4
Anchieta, Brazil 132/F8
Ancho, N. Mex. (†88301) 274/D5
Anchor, Ill. (61720) 222/E3
Anchorage, Alaska 188/D6
Anchorage, Alaska 146/D3
Anchorage, Alaska (*99501) 196/B1
Anchorage, Ky. (40223) 237/L2
Anchorage, U.S. 2/B2
Anchorage, U.S. 4/D17
Anchorena, Argentina 143/C4
Anchor Point, Alaska (99556) 196/B2
Anchor Point, Newf. 166/C3
Anchorville, Mich. (48004) 250/G6
Anchovy, Jamaica 158/H5
Ancienne-Lorette, Québec 172/H3
Ancitas (cay), Cuba 158/F3
Anclote (keys), Fla. 212/D3
Anco, Ky. (41711) 237/P6
Ancohuma (mt.), Bolivia 120/C4
Ancohuma, Nevada (mt.), Bolivia 136/A4
Ancón, Peru 128/D8
Ancona, Ill. (61311) 222/E2
Ancona (prov.), Italy 34/D3
Ancona, Italy 34/D3
Ancona, Italy 7/F4
Ancón de Sardinas (bay), Colombia 126/A7
Ancón de Sardinas (bay), Ecuador 128/C2
Ancoraimes, Bolivia 136/A4
Ancram, N.Y. (12502) 276/N6
Ancroft, England 13/F2
Ancrum, Scotland 15/F5
Ancud, Chile 120/B7
Ancud, Chile 138/D4
Ancud (gulf), Chile 138/D4
Anda (Anta), China 77/L2
Andacollo, Argentina 143/B4
Andacollo, Chile 138/A8
Andado, North. Terr. 93/D8
Andahuaylas, Peru 128/F9
Andale, Kansas (67001) 232/E4
Andalgalá, Argentina 143/C2
Åndalsnes, Norway 18/F5
Andalusia, Ala. (36420) 195/E8
Andalusia, Ill. (61232) 222/C2
Andalusia, Pa. (†19020) 294/N5
Andalusia (reg.), Spain 33/C4
Andaman (sea) 54/L8
Andaman (sea), Burma 72/B4
Andaman (isls.), India 2/P5
Andaman (isls.), India 54/L8
Andaman (isls.), India 68/G6
Andaman (sea), India 68/G6
Andaman and Nicobar Isls. (terr.), India 68/G6
Andamarca, Bolivia 136/B6
Andamarca, Peru 128/E8
Andamooka, S. Australia 94/E4
Andapa, Madagascar 118/H2
Andaraí, Brazil 132/F6
Andau, Austria 41/D3
Andeer, Switzerland 39/H3
Andelfingen, Switzerland 39/G1
Andenne, Belgium 27/G8
Anderlecht, Belgium 27/B9
Anderlues, Belgium 27/E8
Andermatt, Switzerland 39/G3
Andernach, W. Germany 22/B3
Anderson, Ala. (35610) 195/D1
Anderson, Alaska (†99760) 196/H2
Anderson, Argentina 143/F7
Anderson, Calif. (96007) 204/C3
Anderson, Ind. 188/J2
Anderson, Ind. (*46011) 227/F4
Anderson (riv.), Ind. 227/D8
Anderson, Iowa (†51652) 229/B7
Anderson (co.), Kansas 232/G3
Anderson (co.), Ky. 237/M5
Anderson (lake), Manitoba 179/D2
Anderson (pt.), Manitoba 179/F3
Anderson, Mo. (64831) 261/D9
Anderson (riv.), N.W.T. 162/D2
Anderson (riv.), N.W. Terrs. 187/F3
Anderson, S.C. 188/K4
Anderson (co.), S.C. 296/B2
Anderson, S.C. (*29621) 296/B2
Anderson (bay), Tasmania 99/D2
Anderson (co.), Tenn. 237/N8
Anderson, Tenn. (†37376) 237/K10
Anderson (co.), Texas 303/J6
Anderson, Texas (77830) 303/J7
Anderson Ranch (res.), Idaho 220/C6
Andersonville, Georgia (31711) 217/D6
Andersonville, Ind. (†47024) 227/G5
Andersonville, Tenn. (37705) 237/O8
Andersonville, Va. (23911) 307/L6
Andersonville Nat'l Hist. Site, Georgia 217/D6
Andes (range), 120/B2-6
Andes, Cordillera de los (mts.), Argentina 143/C2
Andes, Cordillera de los (mts.), Chile 138/C5,E
Andes, Colombia 126/C5
Andes, Mont. (†59218) 262/M3
Andes, N.Y. (13731) 276/L6
Andes, Cordillera de los (mts.), Peru 128/F10
Andes (lake), S. Dak. 298/N7
Andheri, India 68/B7
Andhra Pradesh (state), India 68/D5
Andijk, Netherlands 27/G3
Andikíthira (isl.), Greece 45/F8
Andilamena, Madagascar 118/H3
Andimeshk, Iran 66/F4
Anding, Miss. (†39040) 256/D5
Andırın, Turkey 63/G4
Ándissa, Greece 45/H6
Andizhan, U.S.S.R. 54/J5
Andizhan, U.S.S.R. 48/H5
Andkhvoy, Afghanistan 68/A1
Andkhvoy, Afghanistan 59/H2
Andoas Nuevo, Ecuador 128/D4
Andoma, Zaire 115/E3
Andong, S. Korea 81/D5
Andorra 7/E4
ANDORRA 33/G1
Andorra, Spain 33/F2
Andorra la Vella (cap.), Andorra 33/G1
Andover○, Conn. (06232) 210/F2
Andover, England 10/F5
Andover, England 13/F6
Andover, Ill. (61233) 222/C2
Andover, Iowa (52701) 229/N5
Andover, Kansas (67002) 232/E4
Andover, Maine (04216) 243/B6
Andover○, Maine (04216) 243/B6
Andover, Mass. (01810) 249/K2
Andover○, Mass. (01810) 249/K2
Andover, Minn. (†55303) 255/E5
Andover○, N.H. (03216) 268/D5
Andover, N.J. (07821) 273/D2
Andover, N.Y. (14806) 276/E6
Andover, Ohio (44003) 284/J2
Andover, S. Dak. (57422) 298/O3
Andover○, Vt. (†05143) 268/B5
Andover, Va. (24215) 307/C7
Andøya (isl.), Norway 18/J2
Andradas, Brazil 135/C3
Andradina, Brazil 132/D8
Andraitx, Spain 33/H3
Andravídha, Greece 45/E6
Andre (lake), Newf. 166/A3
Andreafski (Saint Marys), Alaska (†99658) 196/F2
Andreanof (isls.), Alaska 196/L4
Andreas (cape), Cyprus 63/F5
Andrelândia, Brazil 135/D2
Andrés, Nicaragua 154/F3
Andrespol, Poland 47/D3
Andrew, Alberta 182/D3
Andrew, Iowa (52030) 229/M4
Andrew, La. (†70548) 238/F6
Andrew (co.), Mo. 261/C3
Andrew (isl.), Nova Scotia 168/H3
Andrew Johnson Nat'l Hist. Site, Tenn. 237/R8
Andrews, (†21626) 245/O7
Andrews, Ind. (46702) 227/F3
Andrews, N.C. (28901) 281/B4
Andrews, Oreg. (†97732) 291/J5
Andrews, S.C. (29510) 296/H5
Andrews (co.), Texas 303/B5
Andrews, Texas (79714) 303/B5
Andrews A.F.B., Md. 245/G5
Andreyevka, U.S.S.R. 52/H4
Andria, Italy 34/F4
Androka, Madagascar 118/G5
Andros (isl.), Bahamas 146/L7
Andros (isl.), Bahamas 156/B1
Ándros, Greece 45/G7
Ándros (isl.), Greece 45/G7
Androscoggin (co.), Maine 243/C7
Androscoggin (riv.), Maine 243/C7
Androscoggin (riv.), N.H. 268/E2
Androth (isl.), India 68/C6
Andrychów, Poland 47/D4
Andsfjorden (fjord), Norway 18/K2
Andújar, Spain 33/D3
Andul, India 68/F2
Andulo, Angola 102/D6
Andulo, Angola 115/C6
Anéfis, Mali 106/E5
Anegada (isl.), Virgin Is. (Br.) 156/H1
Anegada (passage), Virgin Is. (Br.) 156/F3
Aného (Anécho), Togo 106/E7
Aneityum (Anatom) (isl.), Vanuatu 87/H8
'Aneiza, Jebel (mt.), Iraq 66/A4
'Aneiza, Jebel (mt.), Iraq 59/C3
'Aneiza, Jebel (mt.), Jordan 59/C3
'Aneiza, Jebel (mt.), Saudi Arabia 59/C3
Añelo, Argentina 143/C4
Anerley, Sask. 181/D4
Aneroid, Sask. 181/D6
Aneta, N. Dak. (58212) 282/P4
Aneth, Utah (84510) 304/E6
Aneto (peak), Spain 33/G1
Angaki (Quirino), Philippines 82/C2
Angamos (isl.), Chile 138/D8
Angamos (pt.), Chile 138/A4
Angara (riv.), U.S.S.R. 54/L4
Angara (riv.), U.S.S.R. 48/K4
Angarsk, U.S.S.R. 54/M4
Angarsk, U.S.S.R. 48/L4
Angas Downs, North. Terr. 93/C8
Angaston, S. Australia 94/F6
Angaur (isl.), Belau 87/D5
Ånge, Sweden 18/J5
Ange-Gardien, Québec 172/E4
Angel (isl.), Calif. 204/J2
Angel (falls), Venezuela 120/C2
Angel (fall), Venezuela 124/G5
Angela, Mont. (59312) 262/K4
Ángel de la Guarda (isl.), Mexico 150/C2
Angeles, Philippines 82/C3
Ángeles, P. Rico 161/B2
Ängelholm, Sweden 18/H8
Angélica, Argentina 143/E5
Angelica, N.Y. (14709) 276/E6
Angelica, Wis. (†54162) 317/K6
Angelina (co.), Texas 303/K6
Angelina (riv.), Texas 303/K6
Angelo, Wis. (†54656) 317/E8
Angels Camp, Calif. (95222) 204/E5
Angelus, S.C. (†29718) 296/G2
Angerman (riv.), Sweden 7/F2
Ångermanälven (riv.), Sweden 18/K5
Angermünde, E. Germany 22/E2
Angers, France 7/D4
Angers, France 28/C4
Angers, Québec 172/B4
Angicos, Brazil 132/G4
Angie, La. (70426) 238/L5
Angier, N.C. (27501) 281/M4
Angijak (isl.), N.W. Terrs. 187/M3
Angikuni (lake), N.W. Terrs. 187/J3
Angkor Wat (ruins), Cambodia 72/E4
Angle, Utah (†84712) 304/C5
Angle Inlet, Minn. (56711) 255/C1
Anglem (mt.), N. Zealand 100/A7
Anglesey (isl.), Wales 13/C4
Anglesey (isl.), Wales 10/D4
Angleton, Texas (77515) 303/J8
Anglia, Sask. 181/C4
Angliers, Québec 174/B3
Angmagssalik, Greenl. 4/C11
Angmagssalik, Greenland 146/Q3
Ango, Zaire 115/E3
Angoche, Mozambique 118/G3
Angoche, Mozambique 102/G6
Angoche (isl.), Mozambique 118/G3
Angol, Chile 138/D1
Angola 2/K6
Angola 102/D6
ANGOLA 115/C6
Angola, Del. (†19966) 245/T6
Angola, Ind. (46703) 227/G1
Angola, Kansas (67331) 232/G4
Angola, La. (70712) 238/G5
Angola, N.Y. (14006) 276/C5
Angola (swamp), N.C. 281/O5
Angola on the Lake, N.Y. (†14006) 276/B5
Angoon, Alaska (99820) 196/M1
Angora, Minn. (55703) 255/F3
Angora, Nebr. (69331) 264/A3
Angoram, Papua N.G. 85/B6
Angostura (falls), Colombia 120/B2
Angostura (falls), Colombia 126/E6
Angostura, Mexico 150/E4
Angostura (res.), S. Dak. 298/B7
Angoulême, France 28/D5
Angoumois (trad. prov.), France, 29
Angra do Heroísmo (dist.), Portugal 33/C1
Angra do Heroísmo, Portugal 33/C1
Angra dos Reis, Brazil 135/D3
Angren, U.S.S.R. 48/H5
Ang Thong, Thailand 72/C4
Anguil, Argentina 143/D4
Anguilla (isl.) 146/M8
ANGUILLA 156
Anguilla, Anguilla 156/F3
Anguilla, Miss. (38721) 256/C5
Anguillara Sabazia, Italy 34/F6
Anguille (cape), Newf. 166/C4
Angurugu, North. Terr. 93/E3
Angus, Iowa (†50220) 229/E5
Angus, Minn. (56712) 255/B2
Angus, Ontario 177/E3
Angus (trad. prov.), Scotland, 15/B5
Angusville, Manitoba 179/A4
Angwin, Calif. (94508) 204/C5
Anhée, Belgium 27/F8
Anholt, Denmark 21/E4
Anholt (isl.), Denmark 21/E4
Anholt (isl.), Denmark 18/G8
Anhua, China 77/H6
Anhui (Anhwei), China 77/J5
Aniak, Alaska (99557) 196/G2
Aniakchak (vol.), Alaska 196/G3
Aniakchak Nat'l Mon., Alaska 196/G3
Aniakchak Nat'l Preserve, Alaska 196/G3
Anicuns, Brazil 132/D7
Aniene (riv.), Italy 34/F6
Animas (riv.), Colo. 208/D8
Animas, N. Mex. (88020) 274/A7
Animas (riv.), N. Mex. 274/B1
Añimbo, Bolivia 136/C7
Anin, Burma 72/C4
Anin, West Bank 65/C2
Anina, Romania 45/E3
Anita, Iowa (50020) 229/D6
Anita, Pa. (15711) 294/D3
Aniva (cape), U.S.S.R. 48/P5
Aniwa, Wis. (54408) 317/H6
'Anjara, Jordan 65/D3
Anjidiv (Angediva) (isl.), India 68/C6
Anjou (trad. prov.), France, 29
Anjou, Québec 172/H4
Anjouan (isl.), Comoros 102/G6
Anjouan (isl.), Comoros 118/G2
Anju, N. Korea 81/B4
Anjum, Netherlands 27/J2
Ankang, China 77/G5
Ankara (prov.), Turkey 63/E3
Ankara (cap.), Turkey 63/E3
Ankara (cap.), Turkey 2/L4
Ankara (cap.), Turkey 54/E5
Ankara (cap.), Turkey 59/B2
Ankara (riv.), Turkey 63/D3
Ankazoabo, Madagascar 118/G4
Ankeny, Iowa (50021) 229/F5
Anker (riv.), England 10/G3
Ankerton, Alberta 182/D3
Ankhor, Somalia 115/J1
Anking (Anqing), China 77/J5
Anklam, E. Germany 22/E2
Ankober, Ethiopia 111/H6
Ankona, Fla. (†33450) 212/F4
Ankoro, Zaire 115/E5
An Loc (Binh Long), Vietnam 72/E5
Anlu, China 77/H5
Anmoore, W. Va. (26323) 312/F4
Ann (cape), Mass. 249/M2
Anna, Ill. (62906) 222/D6
Anna, Ky. (†42270) 237/J6
Anna, Ohio (45302) 284/B5
Anna, Texas (75003) 303/H4
Anna (lke), Va. 307/N4
Annaba, Algeria 102/C1
Annaba, Algeria 106/F1
Annabella, Utah (84711) 304/B5
Annaberg-Buchholz, E. Germany 22/E3
Anna Creek, S. Australia 94/D3
Annada, Mo. (63330) 261/L4
Annadel, Tenn. (†37770) 237/M8
Annagry, Ireland 17/E1
Annaheim, Sask. 181/G3
Annai, Guyana 131/B4
An Najaf (gov.), Iraq 66/C5
An Najaf, Iraq 59/D3
An Najaf, Iraq 66/D5
An Najaf, Iraq 54/F6
Annalee (riv.), Ireland 17/G3
Annalong, N. Ireland 17/K3
Annaly (bay), Virgin Is. (U.S.) 161/E3
Anna Maria, Fla. (33501) 212/D4
Annan, Scotland 15/E6
Annan, Scotland 10/E3
Annan (riv.), Scotland 15/E5
Annandale, Minn. (55302) 255/D5
Annandale, N.J. (08801) 273/D2
Annandale, Va. (22003) 307/S3
Annandale-on-Hudson, N.Y. (12504) 276/N6
Anna Plains, W. Australia 92/C2
Annapolis, Calif. (95412) 204/B5
Annapolis, Ill. (62413) 222/F4
Annapolis (cap.), Md. (*21401) 245/M5
Annapolis (cap.), Md. 188/L3
Annapolis, Mo. (63620) 261/L8
Annapolis (co.), Nova Scotia 168/C4
Annapolis (basin), Nova Scotia 168/C4
Annapolis (riv.), Nova Scotia 168/C4
Annapolis Junction, Md. (20701) 245/M4
Annapolis Royal, Nova Scotia 168/C4
Annapurna (mt.), Nepal 68/E3
Ann Arbor, Mich. 188/K2
Ann Arbor, Mich. (*48103) 250/F6
Anna Regina, Guyana 131/B2
Annascaul, Ireland 17/B7
An Nasiriya, Iraq 59/E3
An Nasiriya, Iraq 66/D5
Annat, Scotland 15/C3
Annaville, Québec 172/E3
Annawan, Ill. (61234) 222/C2
Annbank Station, Scotland 15/D5
Anne (mt.), Tasmania 99/C4
Anne Arundel (co.), Md. 245/M4
Annecy, France 28/G5
Annemanie, Ala. (36721) 195/D6
Anner (riv.), Ireland 17/F7
Anneta, Ky. (†42754) 237/J6
Annette, Alaska (99920) 196/N2
An Nhon, Vietnam 72/F4
Annieopscotch (mts.), Newf. 166/C4
Anniston, Ala. 188/J4
Anniston, Ala. (36201) 195/G3
Anniston, Mo. (63820) 261/O9
Anniston Army Depot, Ala. 195/F3
Annobon (isl.), Equat. Guinea 102/C5
Annona, Texas (75550) 303/K4
Annonay, France 28/F5
Annotto Bay, Jamaica 156/C3
Annotto Bay, Jamaica 158/K6
Annville, Ky. (40402) 237/O6
Annville, Pa. (17003) 294/J5
Annweiler am Trifels, W. Germany 22/B4
Anoka (co.), Minn. 255/E5
Anoka, Minn. (55303) 255/E5
Anoka, Nebr. (†68722) 264/F2
Anola, Manitoba 179/F5
Ano Nuevo (pt.), Calif. 204/J4
Áno Viánnos, Greece 45/G8
Anóyia, Greece 45/G8
Anqing (Anking), China 77/J5
Ans, Belgium 27/H7
Ansager, Denmark 21/B6
Ansai, China 77/G4
Ansbach, W. Germany 22/D4
Anse à Galets, Haiti 158/B6
Anse-à-Pitre, Haiti 158/C6
Anse-au-Griffon, Québec 172/D1
Anse-aux-Gascons, Québec 172/D2
Anse-à-Veau, Haiti 158/B6
Anse-Bertrand, Guadeloupe 161/A5
Anse-Bleue, New Bruns. 170/E1
Anse Boileau, Seychelles 118/H5
Anse-d'Hainault, Haiti 158/A6
Anse la Raye, St. Lucia 161/F6
Anselmo, Nebr. (68813) 264/E3
Anser Group (isls.), Tasmania 99/C1
Anserma, Colombia 126/B5
Anse Rouge, Haiti 158/B5
Anse Royale, Seychelles 118/H5
Anshan, China 77/K3
Anshan, China 54/O5
Anshun, China 77/G6
Ansley, Ala. (36001) 195/F7
Ansley, La. (†71228) 238/E2
Ansley, Nebr. (68814) 264/E3
Anson, Kansas (†67103) 232/E4
Anson, Maine (04911) 243/D6
Anson○, Maine (04911) 243/D6
Anson (pt.), Norfolk I. 88/K5
Anson (bay), Norfolk I. 88/K5
Anson (co.), N.C. 281/J4
Anson (bay), North. Terr. 88/D2
Anson, Texas (79501) 303/E5
Ansŏng, S. Korea 81/C5
Ansongo, Mali 106/E5
Ansonia, Conn. (06401) 210/C3
Ansonia, Ohio (45303) 284/A5
Ansonville, N.C. (28007) 281/J4
Ansonville, Pa. (†16656) 294/E4
Ansted, W. Va. (25812) 312/D6
Anta, Peru 128/F9
Antabamba, Peru 128/F10
Antakya, Turkey 59/C2
Antakya, Turkey 63/G4
Antalaha, Madagascar 118/J2
Antalaha, Madagascar 102/H6
Antalya (prov.), Turkey 63/D4
Antalya, Turkey 54/D6
Antalya, Turkey 63/D4
Antalya, Turkey 59/B2
Antalya (gulf), Turkey 63/D4
Antalya (gulf), Turkey 59/B2
Antananarivo (prov.), Madagascar 118/H3
Antananarivo (cap.), Madagascar 2/M6
Antananarivo (cap.), Madagascar 102/G6
Antananarivo (cap.), Madagascar 118/H3
Antarctic (pen.), Ant. 2/G9
Antarctic (pen.) 5/C15
Antarctica 2/E11
ANTARCTICA 5
Antarctic Circle 2/A9
An Teallach (mt.), Scotland 15/C3
Antelope (creek), Idaho 220/E6
Antelope, Kansas (66836) 232/F3
Antelope, Mont. (59211) 262/M2
Antelope (co.), Nebr. 264/F2
Antelope (range), Nev. 266/E3
Antelope, Oreg. (97001) 291/G3
Antelope (creek), Oreg. 291/K5
Antelope (res.), Oreg. 291/K5
Antelope, Sask. 181/C5
Antelope (lake), Sask. 181/C5
Antelope (creek), S. Dak. 298/D3
Antelope, Texas (76350) 303/F4
Antelope (isl.), Utah 304/B3
Antelope (creek), Wyo. 319/G2
Antelope (hills), Wyo. 319/D3
Antequera, Paraguay 144/D4
Antequera, Spain 33/D4
Antero (mt.), Colo. 208/G5
Antero (res.), Colo. 208/H5
Antes Fort, Pa. (17720) 294/H3
Anthon, Iowa (51004) 229/B4
Anthony, Fla. (32617) 212/D2
Anthony, Ind. (†47302) 227/G4
Anthony, Kansas (67003) 232/D4
Anthony, N. Mex. (88021) 274/C6
Anthony, R.I. (†02816) 249/H6
Anthony, Texas (88021) 303/A10
Anthony, W. Va. (24914) 312/F7
Anthony Lagoon, North. Terr. 88/E3
Anthony Lagoon, North. Terr. 93/D4
Anthracite, Alberta 182/C4
Anti-Atlas (ranges), Morocco 106/C3
Antibes, France 28/G6
Anticosti (isl.), Que. 146/M5
Anticosti (isl.), Que. 162/K6
Anticosti (isl.), Québec 174/E3
Antietam, Md. (†21782) 245/H3
Antietam (creek), Md. 245/H2
Antietam Nat'l Battlefield, Md. 245/H3
Antigo, Wis. (54409) 317/H5
Antigonish (co.), Nova Scotia 168/F3
Antigonish, Nova Scotia 168/F3
Antigonish (harb.), Nova Scotia 168/G3
Antigua (isl.) 146/M8
ANTIGUA & BARBUDA 156
ANTIGUA & BARBUDA 161
Antigua (isl.), Ant. & Bar. 161/E11
Antigua (isl.), Ant. & Bar. 156/G3
Antigua, Guatemala 154/B3
Antigua (riv.), Mexico 150/Q1
Antigua, Spain 33/B4
Antigues (pt.), Guadeloupe 161/A5
Antiguo Morelos, Mexico 150/K5
Antilla, Cuba 156/C2
Antilla, Cuba 158/J3
Antilles, Greater (isls.), W. Indies 156/B2
Antilles, Lesser (isls.), W. Indies 156/E4
Antimony, Utah (84712) 304/C5
Antioch, Calif. (94509) 204/L1
Antioch, Georgia (†30240) 217/B4
Antioch, Ill. (60002) 222/E1
Antioch, Nebr. (69340) 264/B2
Antioch, Ohio (43710) 284/H6
Antioch, S.C. (†29020) 296/F3
Antioch (Antakya), Turkey 63/G4
Antioch, W. Va. (†26743) 312/H4
Antioquia (dept.), Colombia 126/B4
Antioquia, Colombia 126/B4
Antique (prov.), Philippines 82/D5
Antiquity, Ohio (†45771) 284/G8
Antisana (mt.), Ecuador 128/D3
Anti-Taurus (mts.), Turkey 63/G3
Antler, N. Dak. (58711) 282/H2
Antler, Sask. 181/K6
Antler (riv.), Sask. 181/K6
Antler Lake, Alberta 182/D3
Antlers, Okla. (74523) 288/P6
Antofagasta (reg.), Chile 138/B4
Antofagasta, Chile 120/B5
Antofagasta, Chile 2/F7
Antofagasta, Chile 138/A4
Antofagasta de la Sierra, Argentina 143/C2
Antoine, Ark. (71922) 202/D5
Antoing, Belgium 27/C7
Anton, Colo. (80801) 208/N3
Antón, Panama 154/G6
Anton, Texas (79313) 303/B4
Anton Chico, N. Mex. (87711) 274/D3
Antone, Oreg. (†97750) 291/H3
Antongil (bay), Madagascar 118/J3
Antonina, Brazil 135/B4
Antonino, Kansas (67624) 232/C3
Antonito, Colo. (81120) 208/H8
Antony, France 28/B2
Antora (peak), Colo. 208/G6
Antreville, S.C. (†29620) 296/B3
Antrim (co.), Mich. 250/D3
Antrim, Mich. (†49659) 250/D4
Antrim, N.H. (03440) 268/D5
Antrim○, N.H. (03440) 268/D5
Antrim (dist.), N. Ireland 17/J2
Antrim, N. Ireland 10/C3
Antrim, N. Ireland 17/J2
Antrim, Ohio (†43973) 284/H5
Antrim, Pa. (†16901) 294/H2
Antsalova, Madagascar 118/G3
Antsirabe, Madagascar 102/G7
Antsirabe, Madagascar 118/H3
Antsiranana (prov.), Madagascar 118/H2
Antsiranana, Madagascar 118/H2
Antsiranana, Madagascar 102/G6

Antsia, U.S.S.R. 53/D2
Antsohihy, Madagascar 118/H2
Antu, China 77/L3
An Tuc (An Khe), Vietnam 72/F4
Antwerp (prov.), Belgium 27/F6
Antwerp, Belgium 7/E3
Antwerp, Belgium 27/E6
Antwerp, N.Y. (13608) 276/J2
Antwerp, Ohio (45813) 284/A3
Antwerpen (Antwerp), Belgium 27/E6
An Uaimh, Ireland 10/C4
An Uaimh, Ireland 17/H4
Anuenue (Sand) (isl.), Hawaii 218/C4
Anuradhapura, Sri Lanka 68/E7
Anutt, Mo. (†65401) 261/J7
Anvik, Alaska (99558) 196/F2
Anvil (peak), Alaska 196/K4
Anxi, China 77/E3
Anxious (bay), S. Australia 94/D5
Anyang, China 77/H4
A'nyêmaqên Shan (mts.), China 77/E5
Anykščiai, U.S.S.R. 53/C3
Anzá, Colombia 126/C4
'Anza, West Bank 65/C3
Anzac, Alberta 182/E1
Anzaldo, Bolivia 136/C5
Anzhero-Sudzhensk, U.S.S.R. 54/K4
Anzhero-Sudzhensk, U.S.S.R. 48/J4
Anzio, Italy 34/D4
Anzoátegui (state), Venezuela 124/F3
Aoiz, Spain 33/F1
Aoji-ri, N. Korea 81/E2
Aomori (pref.), Japan 81/K3
Aomori, Japan 54/R5
Aomori, Japan 81/K3
Ao Paray (riv.), Paraguay 144/A5
Aosta (reg.), Italy 34/A2
Aosta (prov.), Italy 34/A2
Aosta, Italy 34/A2
Aouara, Fr. Guiana 131/E3
Aouinet Bel Egrâ (well), Algeria 106/C3
Aoulef, Algeria 106/E3
Aozou, Chad 111/C3
Apa (riv.), Paraguay 144/D3
Apache (co.), Ariz. 198/F3
Apache (lake), Ariz. 198/D5
Apache, Okla. (73006) 288/K5
Apache (mts.), Texas 303/C11
Apache Creek, N. Mex. (†87830) 274/A5
Apache Junction, Ariz. (85220) 198/D5
Apalachee (bay), Fla. 188/K5
Apalachee (bay), Fla. 212/B2
Apalachee, Georgia (†30650) 217/E3
Apalachia (res.), N.C. 281/A4
Apalachicola, Fla. (32320) 212/A2
Apalachicola (bay), Fla. 212/B2
Apalachicola (riv.), Fla. 212/A1
Apalachin, N.Y. (13732) 276/H6
Apalona, Ind. (†47576) 227/D8
Apan, Mexico 150/M1
Apaporis (riv.), Colombia 126/F8
Aparecida, Brazil 135/D3
Aparri, Philippines 82/C1
Aparri, Philippines 85/G2
Aparurén, Venezuela 124/G5
Apataki (atoll), Fr. Poly. 87/M7
Apatin, Yugoslavia 45/D3
Apatity, U.S.S.R. 52/D1
Apatzingán de la Constitución, Mexico 150/H7
Ape, U.S.S.R. 53/D2
Apeldoorn, Netherlands 27/H4
Apennines (mts.), Italy 7/F4
Apennines, Central (range), Italy 34/D3
Apennines, Northern (range), Italy 34/B2
Apennines, Southern (range), Italy 34/E4
Apere (riv.), Bolivia 136/C4
Apex, N.C. (27502) 281/M3
Apgar, Mont. (†59936) 262/B2
Apia (cap.), W. Samoa 2/A6
Apia (cap.), W. Samoa 87/J7
Apia (cap.), W. Samoa 86/M8
Apiaí, Brazil 135/B4
Apishapa (riv.), Colo. 208/L8
Apison, Tenn. (37302) 237/L10
Ap Iwan, Cerro (mt.), Chile 138/E6
Apizaco, Mexico 150/N1
Aplao, Peru 128/F11
Aplin, Ark. (†72126) 202/E4
Aplington, Iowa (50604) 229/H3
Ap Long Ha, Vietnam 72/F5
Apo (vol.), Philippines 82/E7
Apohaqui, New Bruns. 170/E3
Apolda, E. Germany 22/D3
Apolima (str.), W. Samoa 86/L8
Apollo, Georgia (†31024) 217/F4
Apollo, Pa. (15613) 294/C4
Apollo Bay, Victoria 97/B6
Apollo Beach, Fla. (33570) 212/C3
Apolo, Bolivia 136/A4
Aponguao (riv.), Venezuela 124/H5
Apopka, Fla. (32703) 212/E3
Apopka (lake), Fla. 212/E3
Aporé (riv.), Brazil 132/D7
Apostle (isls.), Wis. 317/F2
Apostle Islands Nat'l Lakeshore, Wis. 317/E1
Apóstoles, Argentina 143/E2
Apoteri, Guyana 131/B3
Appalachia, Va. (24216) 307/C7
Appalachian (mts.) 188/K3
Appalachian (mts.), N.C. 281/D2
Appalachian (mts.), Pa. 294/H4
Appalachian (mts.), Tenn. 237/M10
Appalachian (mts.), U.S. 146/K6
Appalachian (mts.), Va. 307/J5
Appam, N. Dak. (†58830) 282/C2
Appanoose (co.), Iowa 229/H7
Appelscha, Netherlands 27/J3
Appenzell, Ausser Rhoden (canton), 39/H2
Appenzell, Inner Rhoden (canton), Switzerland 39/H2
Appenzell, Switzerland 39/H2
Apperson, Okla. (†74633) 288/N1
Appin, Ontario 177/C5
Appin (dist.), Scotland 15/C4
Appingedam, Netherlands 27/K2
Apple (creek), Ill. 222/C4
Apple (riv.), Ill. 222/C1
Apple (creek), N. Dak. 282/J6
Apple (riv.), Wis. 317/A5
Appleby, England 13/E3
Appleby, England 10/E3
Appleby, Texas (75961) 303/K6
Apple Creek, Ohio (44606) 284/G4
Applecross, Scotland 15/C3
Appledale, Br. Col. 184/J5
Applegate, Calif. (95703) 204/E5
Applegate, Mich. (48401) 250/G5
Applegate, Oreg. (97530) 291/D5
Apple Grove, W. Va. (25502) 312/B5
Apple Hill, Ontario 177/K2
Apple River, Ill. (61001) 222/C1
Apple River, Nova Scotia 168/D3
Apples, Switzerland 39/B3
Appleton, Ark. (72822) 202/E3
Appleton, Maine (†04540) 243/E7
Appleton○, Maine (†04862) 243/E7
Appleton, Minn. (56208) 255/C5
Appleton (Old Appleton), Mo. (†63770) 261/N7
Appleton, N.Y. (14008) 276/C4
Appleton, Ontario 177/H2
Appleton, S.C. (†29836) 296/E5
Appleton, Wash. (98602) 310/D5
Appleton, Wis. 188/J2
Appleton, Wis. (54911) 317/J7
Appleton City, Mo. (64724) 261/D6
Apple Valley, Calif. (92307) 204/H9
Apple Valley, Minn. (†55124) 255/G6
Appling (co.), Georgia 217/H7
Appling, Georgia (30802) 217/H3
Appomattox (co.), Va. 307/L6
Appomattox, Va. (24522) 307/L6
Appomattox (riv.), Va. 307/M6
Appomattox Court House Nat'l Hist. Park, Va. 307/K6
Apponaug, R.I. (†02887) 249/J6
Approuague (riv.), Fr. Guiana 131/E4
Apra (harb.), Guam 86/K7
Aprilia, Italy 34/D4
Apsheron (pen.), U.S.S.R. 52/H6
Apsheronsk, U.S.S.R. 52/F6
Apsley, Ontario 177/F3
Apsley, Victoria 97/A5
Apt, France 28/F6
Aptos, Calif. (95003) 204/K4
Apua (pt.), Hawaii 218/J6
Apulia (Puglia) (reg.), Italy 34/F4
Apulia Station, N.Y. (13020) 276/H5
Apure (state), Venezuela 124/D4
Apure (riv.), Venezuela 124/E4
Apurímac (dept.), Peru 128/F10
Apurímac (riv.), Peru 120/B4
Apurímac (riv.), Peru 128/F9
Apurito, Venezuela 124/D4
Ap Vinh Hao, Vietnam 72/F5
Aqaba (gulf) 54/E7
'Aqaba (gulf), Egypt 111/G2
'Aqaba (gulf), Egypt 59/C4
'Aqaba (gulf), Israel 65/D6
'Aqaba, Jordan 65/D6
'Aqaba, Jordan 59/C4
Aqaba (gulf), Jordan 65/D6
'Aqaba (gulf), Saudi Arabia 59/C4
Aqcheh, Afghanistan 68/B1
Aqcheh, Afghanistan 59/J2
Aq Darband, Iran 66/M2
'Aqiq, Sudan 111/G4
'Aqqaba, West Bank 65/C3
Aqqikkol Hu (lake), China 77/C4
'Aqra, Iraq 66/D2
'Aqraba, West Bank 65/C3
Aqsu (Aksu), China 77/B3
Aquades Beach, Sask. 181/C2
Aquaforte, Newf. 166/D2
Aqua Park, Okla. (†74435) 288/R3
Aquarius (range), Ariz. 198/B4
Aquarius (plat.), Utah 304/C5
Aquasco, Md. (20608) 245/L6
Aquia, Peru 128/D8
Aquidabán (riv.), Paraguay 144/D3
Aquidauana, Brazil 120/D5
Aquidauana, Brazil 132/C8
Aquila, Mexico 150/H7
Aquila, Switzerland 39/G3
Aquiles Serdán, Mexico 150/G2
Aquilla, Ohio (†44024) 284/H2
Aquin, Haiti 158/B6
Ara (riv.), Japan 81/O2
Arab, Ala. (35016) 195/E2
'Arab, Shatt-al- (riv.), Iran 59/E4
'Arab, Shatt-al- (riv.), Iran 66/F5
'Arab, Shatt-al- (riv.), Iraq 59/E4
'Arab, Shatt-al- (riv.), Iraq 66/F5
Arab, Mo. (63733) 261/M8
'Araba, Wadi (valley), Israel 65/D5
'Araba, Wadi (valley), Jordan 65/D5
Arabela, N. Mex. (†88351) 274/D5
Arabella, Sask. 181/K3
Arabi, Georgia (31712) 217/E7
'Arabi (isl.), Iran 66/G7
Arabi, La. (70032) 238/P4
Arabia, Ky. (†40437) 237/M6
Arabia, Ohio (†45659) 284/F8
Arabian (sea) 54/H8
Arabian (sea) 2/N5
Arabian (des.), Egypt 111/F2
Arabian (des.), Egypt 59/B4
Arabian (sea), India 68/B5
Arabian (sea), Pakistan 68/B5
Arabian (sea), P.D.R. Yemen 59/H5
Arabopó, Venezuela 124/H5
Araç, Turkey 63/E2
Araca, Bolivia 136/B5
Aracaju, Brazil 120/F4
Aracaju, Brazil 132/G5
Aracataca, Colombia 126/D2
Aracati, Brazil 132/G4
Araçatuba, Brazil 132/D8
Araçatuba, Brazil 135/A2
Araceli, Philippines 82/C5
Aracena, Spain 33/C4
Araçuaí, Brazil 132/F7
Arad, Israel 65/C5
Arad, Romania 7/G4
Arad, Romania 45/E2
Arada, Chad 111/D4
'Aradah, U.A.E. 59/F5
Aradan, Iran 66/H3
Arafat, Jebel (mt.), Saudi Arabia 59/D5
Arafura (sea) 87/D6
Arafura (sea) 2/R6
Arafura (sea) 88/E2
Arafura (sea), Indonesia 85/J8
Arafura (sea), North. Terr. 93/D1
Arago, Minn. (†56470) 255/C3
Arago, Oreg. (97458) 291/C4
Arago (cape), Oreg. 291/C4
Aragon, Georgia (30104) 217/B2
Aragon, N. Mex. (87820) 274/A5
Aragón (reg.), Spain 33/F2
Aragón (riv.), Spain 33/F1
Aragona, Italy 34/D6
Aragua (state), Venezuela 124/E3
Araguacema, Brazil 132/D5
Aragua de Barcelona, Venezuela 124/F3
Aragua de Maturín, Venezuela 124/G3
Araguaia (riv.), Brazil 120/E3
Araguaia (riv.), Brazil 132/D4
Araguaína, Brazil 132/C6
Araguaína, Brazil 120/E3
Araguari, Brazil 120/E4
Araguari, Brazil 132/D7
Araguari (riv.), Brazil 132/D2
Araioses, Brazil 132/F3
Arak, Algeria 106/E3
Arak, Iran 54/G6
Arak, Iran 59/E3
Arak, Iran 66/F3
Arakan (state), Burma 72/B3
Arakan Yoma (mts.), Burma 72/B3
Araks (riv.) 54/F6
Araks (riv.), Iran 59/E2
Araks (Aras) (riv.), Iran 66/E1
Araks (riv.), Turkey 63/K2
Araks (riv.), U.S.S.R. 7/J5
Araks (riv.), U.S.S.R. 52/G7
Aral (sea), U.S.S.R. 54/G5
Aral (sea), U.S.S.R. 48/F5
Aralık, Turkey 63/L3
Aral Sea (lake), U.S.S.R. 2/M3
Aral'sk, U.S.S.R. 54/H5
Aral'sk, U.S.S.R. 48/G5
Aramac, Queensland 95/C4
Aramberri, Mexico 150/J5
Arampampa, Bolivia 136/B5
Aran (isl.), Ireland 10/B3
Aran (isl.), Ireland 17/D2
Aran (isls.), Ireland 17/B5
Aran (isls.), Ireland 10/B4
Aranda de Duero, Spain 33/E2
Arandas, Mexico 150/H6
Aran Fawddwy (mt.), Wales 13/C5
Arani, Bolivia 136/C5
Aranjuez, Spain 33/E2
Aransas (co.), Texas 303/H10
Aransas (passage), Texas 303/H10
Aransas Pass, Texas (78336) 303/G10
Araoua (mts.), Fr. Guiana 131/E4
Araouane, Mali 106/D5
Araouane, Mali 102/B3
Arapaho, Okla. (73620) 288/H3
Arapahoe (co.), Colo. 208/L3
Arapahoe, Colo. (80802) 208/P5
Arapahoe (peak), Colo. 208/H2
Arapahoe, Nebr. (68922) 264/E4
Arapahoe, N.C. (28510) 281/R4
Arapahoe, Wyo. (82510) 319/D3
Arapaho Nat'l Rec. Area, Colo. 208/G2
Arapey, Uruguay 145/B1
Arapey Chico (riv.), Uruguay 145/B1
Arapey Grande (riv.), Uruguay 145/B2
Arapicos, Ecuador 128/D3
Arapiraca, Brazil 120/F3
Arapkir, Turkey 63/H3
Arapkir, Turkey 59/C2
'Ar'ar, Wadi (dry riv.), Iraq 66/B5
'Ar'ar, Wadi (dry riv.), Iraq 59/D3
'Ar'ar, Wadi (dry riv.), Saudi Arabia 59/D3
Araracuara, Colombia 126/E8
Araracuara, Cerros de (mts.), Colombia 126/E7
Araranguá, Brazil 132/D10
Araraquara, Brazil 132/E8
Araraquara, Brazil 135/B2
Araras, Brazil 135/C3
Ararat, Ala. (†36921) 195/B7
Ararat, N.C. (27007) 281/H2
Ararat (mt.), Pa. 294/M2
Ararat (mt.), Turkey 54/F6
Ararat (mt.), Turkey 63/L3
Ararat (mt.), Turkey 59/D2
Ararat, Victoria 88/G7
Ararat, Victoria 97/B5
Ararat, Va. (24053) 307/G7
Arari, Brazil 132/E3
Araruama (lake), Brazil 135/E3
Aras (Araks) (riv.), Iran 66/E1
Aras (Araks) (riv.), Iran 59/E2
Aratürük (Yiwu), China 77/D3
Arauca (riv.) 120/C2
Arauca (inten.), Colombia 126/E4
Arauca, Colombia 120/B2
Arauca, Colombia 126/E4
Arauca (riv.), Colombia 126/E4
Arauca (riv.), Venezuela 124/E4
Arauco, Chile 138/D1
Arauco (gulf), Chile 138/D1
Arauquita, Colombia 126/E4
Araure, Venezuela 124/D3
Aravaca, Spain 33/F4
Aravaipa (creek), Ariz. 198/E6
Arawa, Papua N.G. 86/C2
Arawe, Papua N.G. 86/B2
Arax (Araks)(riv), Asia 59/E2
Araxá, Brazil 132/E7
Araya, Venezuela 124/F2
Arba, Ind. (†47355) 227/H4
Arba Mench, Ethiopia 111/G6
Arba Mench, Ethiopia 102/F4
Arbeca, Spain 33/G2
Arbedo-Castione, Switzerland 39/G4
Arbela (Erbil), Iraq 59/D2
Arbela (Erbil), Iraq 66/D2
Arbela, Mo. (63432) 261/H2
Arboga, Sweden 18/J7
Arbois, France 28/F4
Arbon, Idaho (83212) 220/F7
Arbon, Switzerland 39/H1
Arborea, Italy 34/B5
Arborfield, Sask. 181/H2
Arborg, Manitoba 179/E4
Arbor Vitae, Wis. (†54568) 317/G4
Arbovale, W. Va. (24915) 312/G6
Arbrå, Sweden 18/K6
Arbroath, Scotland 15/F4
Arbroath, Scotland 10/E2
Arbroth, La. (†70736) 238/H5
Arbucias, Spain 33/H2
Arbuckle, Calif. (95912) 204/C4
Arbuckle (lake), Fla. 212/E4
Arbuckle, W. Va. (25006) 312/C5
Arbuckles, Lake of the (lake), Okla. 288/M6
Arbuthnot, Sask. 181/E6
Arbutus, Md. (†21227) 245/M4
Arbyrd, Mo. (63821) 261/M10
Arcachon, France 28/C5
Arcachon (bay), France 28/C5
Arcade, Georgia (†30549) 217/E2
Arcade, N.Y. (14009) 276/D5
Arcadia, Calif. (91006) 204/C10
Arcadia, Fla. (33821) 212/E4
Arcadia, Ind. (46030) 227/E4
Arcadia, Iowa (51430) 229/C4
Arcadia, Kansas (66711) 232/H4
Arcadia, La. (71001) 238/E1
Arcadia, Mich. (49613) 250/C4
Arcadia, Mo. (63621) 261/L7
Arcadia, Nebr. (68815) 264/F3
Arcadia, Nova Scotia 168/B5
Arcadia, Ohio (44804) 284/D3
Arcadia, Okla. (73007) 288/M3
Arcadia, Pa. (15712) 294/E4
Arcadia, R.I. (†02832) 249/H6
Arcadia, S.C. (29320) 296/C2
Arcadia, Texas (77517) 303/K3
Arcadia, Utah (†84012) 304/D3
Arcadia, Wis. (54612) 317/D7
Arcadia Lakes, S.C. (†29201) 296/F3
Arcahaie, Haiti 158/C6
Arcanum, Ohio (45304) 284/A6
Arcas (cay), Mexico 150/N6
Arcata, Calif. (95521) 204/A3
Arc Dome (mt.), Nev. 266/D4
Arcelia, Mexico 150/J7
Arch, N. Mex. (†88130) 274/F4
Archambault (lake), Québec 172/C3
Archangel, U.S.S.R. 4/C7
Archangel, U.S.S.R. 2/M2
Archangel, U.S.S.R. 7/J2
Archangel (Arkhangel'sk), U.S.S.R. 48/E3
Archangel (Arkhangel'sk), U.S.S.R. 52/F2
Archbald, Pa. (18403) 294/F6
Archbold, Ohio (43502) 284/B2
Arch Cape, Oreg. (97102) 291/D2
Archdale, N.C. (27263) 281/K3
Archena, Spain 33/F3
Archer, Fla. (32618) 212/D2
Archer, Iowa (51231) 229/B2
Archer, Nebr. (68816) 264/F3
Archer (fiord), N.W. Terrs. 187/M1
Archer (riv.), Queensland 95/B2
Archer (co.), Texas 303/F4
Archer City, Texas (76351) 303/F4
Archerfield, Queensland 88/K3
Archerfield, Queensland 95/D3
Archerwill, Sask. 181/H3
Arches Nat'l Park, Utah 304/E5
Archibald, La. (71218) 238/G2
Archidona, Ecuador 128/D3
Archidona, Spain 33/D4
Archie, La. (†71343) 238/G3
Archie, Mo. (64725) 261/D5
Archiestown, Scotland 15/E3
Archuleta (co.), Colo. 208/E8
Archydal, Sask. 181/F5
Arcis-sur-Aube, France 28/F3
Arckaringa (creek), S. Australia 94/D2
Arco, Georgia (†31520) 217/J8
Arco, Idaho (83213) 220/E6
Arco, Idaho 188/D2
Arco, Minn. (56113) 255/B6
Arcola, Ill. (61910) 222/E4
Arcola, Ind. (46704) 227/G2
Arcola, La. (†70456) 238/K5
Arcola, Miss. (38722) 256/C4
Arcola, Mo. (65603) 261/E7
Arcola, N.C. (27589) 281/N2
Arcola, Sask. 181/J6
Arcoponga, Bolivia 136/B5
Arcosanti, Ariz. (†86333) 198/C4
Arcos de Jalón, Spain 33/E2
Arcos de la Frontera, Spain 33/D4
Arcos de Valdevez, Portugal 33/B2
Arcot, India 68/D6
Arcoverde, Brazil 132/G5
Arctic (ocean) 54/C1
Arctic (ocean) 146/B2
Arctic (plain), Alaska 196/G1
Arctic, R.I. (†02893) 249/J6
Arctic Bay, Canada 4/B14
Arctic Bay, N.W.T. 162/H1
Arctic Bay, N.W. Terrs. 187/K2
Arctic Circle 2/J2
Arctic Ocean 2/B2
Arctic Ocean 4/A15
Arctic Ocean, U.S.S.R. 48/K1
Arctic Red (riv.), N.W. Terrs. 187/E3
Arctic Red River, N.W.T. 162/C2
Arctic Red River, N.W. Terrs. 187/E3
Arctic Village, Alaska (99722) 196/K1
Arda (riv.), Greece 45/G5
Ardabil, Iran 54/F6
Ardabil, Iran 59/E2
Ardabil, Iran 66/F1
Ardagh, Limerick, Ireland 17/C7
Ardagh, Longford, Ireland 17/F4
Ardahan, Turkey 59/D1
Ardahan, Turkey 63/K2
Ardal, Iran 66/G4
Årdal, Norway 18/E7
Årdalstangen, Norway 18/F6
Ardanuç, Turkey 63/K2
Ardara, Ireland 17/E2
Ardath, Sask. 181/D4
Ardavasar, Scotland 15/B3
Ardbeg, Ontario 177/D2
Ardèche (dept.), France 28/F5
Ardee, Ireland 17/H4
Ardee, Ireland 10/C4
Arden, Ark. (†71822) 202/B6
Arden, Del. (†19801) 245/R1
Arden, Denmark 21/C4
Arden, Manitoba 179/C4
Arden, Ontario 177/G3
Arden, Wash. (†99114) 310/H2
Arden, W. Va. (†26405) 312/G4
Arden-Arcade, Calif. (95825) 204/B8
Ardencroft, Del. (†19810) 245/R1
Ardennes (for.), Belgium 27/F9
Ardennes (dept.), France 28/F3
Ardenode, Alberta 182/D4
Ardentown, Del. (†19810) 245/S1
Ardenvoir, Wash. (98811) 310/E3
Ardersier, Scotland 15/E3
Ardeşen, Turkey 63/J2
Ardestan, Iran 59/F3
Ardestan, Iran 66/H4
Ardez, Switzerland 39/K3
Ardfert, Ireland 17/B7
Ardfinnan, Ireland 17/F7
Ardgay, Scotland 15/D3
Ardglass, N. Ireland 17/K3
Ardgour (dist.), Scotland 15/C4
Ardhéa, Greece 45/F5
Ardila (riv.), Spain 33/C3
Ardill, Sask. 181/F6
Ardino, Bulgaria 45/G5
Ardivachar (pt.), Scotland 15/A3
Ardle (riv.), Scotland 15/E4
Ardlethan, N.S. Wales 97/D4
Ardmore, Ala. (35805) 195/E1
Ardmore, Ind. (†46624) 227/E1
Ardmore, Ireland 17/F8
Ardmore, Md. (†20785) 245/G4
Ardmore, Mo. (†65247) 261/H3
Ardmore, Okla. 188/G4
Ardmore, Okla. (73401) 288/M6
Ardmore, Pa. (19003) 294/M6
Ardmore, S. Dak. (57715) 298/B7
Ardmore, Tenn. (38449) 237/H10
Ardnamurchan (pen.), Scotland 15/B4
Ardnamurchan (pt.), Scotland 15/B4
Ardoch, N. Dak. (58213) 282/R3
Ardon, Switzerland 39/D4
Ardooie, Belgium 27/C7
Ardrahan, Ireland 17/D5
Ardrishaig, Scotland 15/C4
Ardrossan, Alberta 182/D3
Ardrossan, Scotland 10/D3
Ardrossan, Scotland 15/D5
Ards (dist.), N. Ireland 17/K2
Ardsley, N.Y. (10502) 276/O6
Åre, Sweden 18/H5
Arecibo (dist.), P. Rico 161/C1
Arecibo, P. Rico 156/G1
Arecibo, P. Rico 161/B1
Aredale, Iowa (50605) 229/H3
Areguá, Paraguay 144/B4
Areia Branca, Brazil 132/G4
Arelee, Sask. 181/D3
Arena (pt.), Calif. 188/B3
Arena (pt.), Calif. 204/B5
Arena (pt.), Mexico 150/E5
Arena, N. Dak. (58412) 282/K5
Arena (isl.), Philippines 82/C6
Arena, Wis. (53503) 317/G9
Arenac (co.), Mich. 250/F4
Arenales, Cerro (mt.), Chile 138/D7
Arenas (pt.), Argentina 143/C7
Arenas (cay), Mexico 150/O5
Arenas (pt.), P. Rico 161/F2
Arenas de San Pedro, Spain 33/D2
Arendal, Norway 18/F7
Arendjelovac, Yugoslavia 45/E3
Arendonk, Belgium 27/G6
Arendtsville, Pa. (17303) 294/H6
Arenillas, Ecuador 128/B4
Arenys de Mar, Spain 33/H2
Arenzville, Ill. (62611) 222/C4
Areópolis, Greece 45/F7
Arequipa (dept.), Peru 128/F10
Arequipa, Peru 128/G11
Arequipa, Peru 120/B4
Arequipa, Peru 2/F6
Arequipa, Peru 128/72
Aresji, Neth. Ant. 161/D9
Areuse (riv.), Switzerland 39/C3
Arévalo, Spain 33/D2
Areyonga, North. Terr. 88/E4
Areyonga, North. Terr. 93/C8
Arezzo (prov.), Italy 34/C3
Arezzo, Italy 34/C3
Arfa Deh, Iran 66/H3
Arga (riv.), Spain 33/F1
Argadargada, North. Terr. 93/E6
Argalant, Mongolia 77/G3
Argalasti, Greece 45/F6
Argamasilla de Alba, Spain 33/E3
Arganda, Spain 33/G4
Argao, Philippines 82/D6
Argenta, Br. Col. 184/J5
Argenta, Ill. (62501) 222/E4
Argenta, Italy 34/D2
Argentan, France 28/D3
Argentat, France 28/D5
Argenteuil, France 28/A1
Argenteuil (co.), Québec 172/C4
Argentia, Newf. 166/C2
Argentina 2/F7
Argentina 120/C6
ARGENTINA 143
Argentine, Pa. (†16040) 294/C3
Argentino (lake), Argentina 143/B7
Argenton-sur-Creuse, France 28/D4
Argeş (riv.), Romania 45/G3
Argo, Ala. (†35173) 195/E3
Argo, Sudan 111/F4
Argo, Sudan 59/B6
Argolis (gulf), Greece 45/F7
Argonia, Kansas (67004) 232/E4
Argonne, Wis. (54511) 317/J4
Argonne Nat'l Laboratory, Ill. 222/B6
Árgos, Greece 45/F7
Argos, Ind. (46501) 227/E2
Argos (cape), Nova Scotia 168/G3
Argostólion, Greece 45/E6
Arguello (pt.), Calif. 204/E9
Arguin (bay), Mauritania 106/A4
Argun (riv.) 54/N4
Argun' (Ergun He) (riv.), China 77/K1
Argun' (riv.), U.S.S.R. 48/M4
Argungu, Nigeria 106/E6
Argus (range), Calif. 204/H7
Argusville, N. Dak. (58005) 282/R5
Arguvan, Turkey 63/H3
Argyle, Fla. (32422) 212/C6
Argyle, Georgia (31623) 217/G8
Argyle, Iowa (52619) 229/K7
Argyle, Maine (†04468) 243/F5
Argyle, Manitoba 179/E4
Argyle, Mich. (48410) 250/G5
Argyle, Minn. (56713) 255/B2
Argyle, Mo. (65001) 261/J6
Argyle, New Bruns. 170/C2
Argyle, N.Y. (12809) 276/O4
Argyle, Texas (76226) 303/F1
Argyle (lake), W. Australia 88/D3
Argyle (lake), W. Australia 92/E2
Argyle, Wis. (53504) 317/G10
Argyle Downs, W. Australia 92/E2
Argyll (dist.), Scotland 15/C4
Argyll (trad. co.), Scotland 15/B5
Arhangay, Mongolia 77/F2
Arhavi, Turkey 63/J2
Arhli (Arlit), Niger 106/F4
Ar Horqin, China 77/K3
Århus (co.), Denmark 21/D5
Århus, Denmark 7/E3
Århus, Denmark 21/D5
Århus, Denmark 18/F8
Aria, N. Zealand 100/E3
Ariah Park, N.S. Wales 97/D4
Ariail, S.C. (†29640) 296/B2
Ariano Irpino, Italy 34/E4
Ariari (riv.), Colombia 126/D6
Aribinda, Upper Volta 106/D6
Arica, Chile 120/B4
Arica, Chile 138/A1
Arica, Colombia 126/E9
Aricagua, Venezuela 124/C3
Ariccia, Italy 34/F7
Arichat, Nova Scotia 168/H3
Arichuna, Venezuela 124/E4
Arichuna (riv.), Venezuela 124/D4
Arid (cape), W. Australia 88/C6
Arid (cape), W. Australia 92/C6
Ariège (dept.), France 28/D6
Ariel, Wash. (98603) 310/C5
Ariguaní (riv.), Colombia 126/D3
Ariha (Jericho), West Bank 65/C4
Arikaree (riv.), Colo. 208/O3
Arima, Trin. & Tob. 156/G5
Arima, Trin. & Tob. 161/B10
Arimo, Idaho (83214) 220/F7
Arinagour, Scotland 15/B4
Aringa, Uganda 115/F3
Arinos (riv.), Brazil 132/B5
Ario de Rosales, Mexico 150/J7
Arion, Iowa (51520) 229/B5
Aripao, Venezuela 124/F4
Aripeka, Fla. (33502) 212/D3
Aripine, Ariz. (†85901) 198/E4
Aripo, El Cerro del (mt.), Trin. & Tob. 161/B10
Ariporo (riv.), Colombia 126/E4
Aripuanã, Brazil 120/C3
Aripuanã, Brazil 132/A5
Aripuanã (riv.), Brazil 120/D3
Aripuanã (riv.), Brazil 132/A4
Arisaig, Scotland 15/C4
Arisaig (sound), Scotland 15/C4
Arismendi, Venezuela 124/D3
Arispe, Iowa (50831) 229/E7
Aristazabal (isl.), Br. Col. 184/C4
Aritao, Philippines 82/C2
Ariton, Ala. (36311) 195/G7
Arivaca, Ariz. (85601) 198/D7
Arivonimamo, Madagascar 118/H3
Arixang (Wenquan), China 77/B3
Ariza, Spain 33/E2
Arizaro, Salar de (salt dep.), Argentina 143/C2
Arizona 188/D4
ARIZONA 198
Arizona (state), U.S. 146/G6
Arizona City, Ariz. (85223) 198/D6
Arizona Sunsites, Ariz. (85625) 198/F7
Arizpe, Mexico 150/D1
Årjäng, Sweden 18/H7
Arjay, Ky. (40902) 237/O7
Arjeplog, Sweden 18/L3
Arjona, Colombia 126/C2
Arkabutla, Miss. (38602) 256/D1
Arkabutla (dam), Miss. 256/D1
Arkabutla (lake), Miss. 256/D1
Arkadelphia, Ala. (35033) 195/E3
Arkadelphia, Ark. (71923) 202/D5
Arkaig, Loch (lake), Scotland 15/C4
Arkaig, Loch (lake), Scotland 10/D2
Arkalyk, U.S.S.R. 48/G4
Arkansas 188/H3
Arkansas (riv.) 188/H3
ARKANSAS 202
Arkansas (co.), Ark. 202/H5

Arkansas (riv.), Ark. 202/G5
Arkansas (riv.), Colo. 208/P6
Arkansas (riv.), Kansas 232/D3
Arkansas (riv.), Okla. 288/S4
Arkansas (state), U.S. 146/J6
Arkansas (riv.), U.S. 2/E4
Arkansas (riv.), U.S. 146/J6
Arkansas City, Ark. (71630) 202/H6
Arkansas City, Kans. 188/G3
Arkansas City, Kansas (67005) 232/E4
Arkansas Divide (mts.), Colo. 208/L4
Arkansas Post Nat'l Mem., Ark. 202/H5
Arkansaw, Wis. (54721) 317/B6
Arkdale, Wis. (54613) 317/G7
Arkhángelos, Greece 45/J7
Arkhipo-Osipovka, U.S.S.R. 52/E6
Arkinda, Ark. (71821) 202/B6
Arklow, Ireland 10/C4
Arklow, Ireland 17/J6
Arklow (bank), Ireland 17/K6
Arkoe, Mo. (†64468) 261/C2
Arkoma, Okla. (74901) 288/T4
Arkona (cape), E. Germany 22/E1
Arkona, Ontario 177/C4
Arkport, N.Y. (14807) 276/E6
Arkticheskiy Institut (isls.), U.S.S.R. 48/H2
Arkville, N.Y. (12406) 276/L6
Arkwright, S.C. (†29301) 296/C2
Arlee, Mont. (59821) 262/B3
Arlee, W. Va. (†25106) 312/B5
Arles, France 28/F6
Arley, Ala. (35541) 195/D2
Arlington, Ala. (36722) 195/C6
Arlington, Ariz. (85322) 198/C5
Arlington, Colo. (81021) 208/N6
Arlington, Georgia (31713) 217/C8
Arlington, Ill. (61312) 222/D2
Arlington, Ind. (46104) 227/F5
Arlington, Iowa (50606) 229/K3
Arlington, Kansas (67514) 232/D4
Arlington, Ky. (42021) 237/D7
Arlington○, Mass. (02174) 249/C6
Arlington, Minn. (55307) 255/D6
Arlington, Nebr. (68002) 264/H3
Arlington, N.Y. (12603) 276/N7
Arlington, N.C. (†28642) 281/H2
Arlington, Ohio (45814) 284/C4
Arlington, Oreg. (97812) 291/G2
Arlington, S. Dak. (57212) 298/P5
Arlington, Tenn. (38002) 237/B10
Arlington, Tex. 188/G4
Arlington, Texas (*76010) 303/F2
Arlington (lake), Texas 303/F2
Arlington, Vt. (05250) 268/A5
Arlington○, Vt. (05250) 268/A5
Arlington (co.), Va. 307/S2
Arlington, Va. (*22201) 307/T3
Arlington, Wash. (98223) 310/C2
Arlington, Wis. (53911) 317/H9
Arlington, Wyo. (†82080) 319/F4
Arlington Beach, Sask. 181/F4
Arlington Heights, Ill. (*60004) 222/B5
Arlington Heights, Ohio (†45201) 284/C9
Arlington Nat'l Cemetery, Va. 307/T3
Arlit (Arhli), Niger 106/F4
Arló, Hungary 41/F2
Arlon, Belgium 27/H9
Arltunga, North. Terr. 93/D7
Arm (riv.), Sask. 181/F5
Arma, Kansas (66712) 232/H4
Arma (plat.), Saudi Arabia 59/E4
Armada, Alberta 182/D4
Armada, Mich. (48005) 250/G6
Armadale, Scotland 15/C2
Armadale, Scotland 10/B1
Armagh (dist.), N. Ireland 17/H3
Armagh, N. Ireland 10/C3
Armagh, N. Ireland 17/H3
Armagh, Pa. (15920) 294/E5
Armagh, Québec 172/G3
Armathwaite, Tenn. (38506) 237/M8
Armavir, U.S.S.R. 7/J4
Armavir, U.S.S.R. 48/E5
Armavir, U.S.S.R. 52/F5
Armena, Alberta 182/D3
Armenia, Colombia 120/B2
Armenia, Colombia 126/B5
Armenian S.S.R., U.S.S.R. 7/J4
Armenian S.S.R., U.S.S.R. 52/F6
Armenian S.S.R., U.S.S.R. 48/E6
Armentières, France 28/E2
Armería, Mexico 150/G7
Armero, Colombia 126/C5
Armidale, Australia 87/F9
Armidale, N. S. Wales 88/J6
Armidale, N.S. Wales 97/F2
Armington, Ill. (61721) 222/D3
Armington, Mont. (†59412) 262/F3
Arminto, Wyo. (82630) 319/E2
Armistead, La. (†71019) 238/D3
Armit (lake), Manitoba 179/A2
Armley, Sask. 181/G2
Armona, Calif. (93202) 204/F7
Armorel, Ark. (72310) 202/L2
Armour, S. Dak. (57313) 298/N7
Armourdale, N. Dak. (†58365) 282/M2
Armoy, N. Ireland 17/J1
Armstrong, Br. Col. 184/H5
Armstrong, Ill. (61812) 222/F3
Armstrong, Ind. (†47708) 227/B8
Armstrong, Iowa (50514) 229/D2
Armstrong, Mo. (65230) 261/G4
Armstrong, Ont. 162/H5
Armstrong, Ontario 175/C2
Armstrong, Ontario 177/H4
Armstrong (co.), Pa. 294/D4
Armstrong (co.), Texas 303/C3
Armstrong, Texas (78338) 303/G11
Armstrong Brook, New Bruns. 170/E1
Armstrong Creek, Wis. (54103) 317/K4
Armstrongs Mills, Ohio (43904) 284/J6
Armuchee, Georgia (30105) 217/B2
Army Chemical Center, Md. 245/O3
Army Med. Ctr. Annex (Walter Reed), Md. 245/E4
Arnaía, Greece 45/F5
Arnaud, Manitoba 179/E5
Arnaud (riv.), Québec 174/F1
Arnaudville, La. (70512) 238/G6
Arnauti (cape), Cyprus 59/B2
Arnauti (cape), Cyprus 63/E5
Arnavutköy, Turkey 63/D6
Arnedo, Spain 33/E1
Arnegard, N. Dak. (58835) 282/D4
Årnes, Norway 18/G6
Arnett, Okla. (73832) 288/G2
Arnett, W. Va. (25007) 312/D7
Arney (riv.), N. Ireland 17/F3
Arnheim, Mich. (†49958) 250/G1
Arnhem (cape), Australia 87/D7
Arnhem, Netherlands 27/H4
Arnhem (cape), North. Terr. 88/F2
Arnhem (cape), North. Terr. 93/E2
Arnhem Land (reg.), Australia 87/D7
Arnhem Land (reg.), North. Terr. 88/E2
Arnhem Land (reg.), North. Terr. 93/D2
Arnhem Land Aboriginal Reserve, North. Terr. 88/E2
Arnhem Land Aboriginal Res., North. Terr. 93/C2
Arno (riv.), Italy 34/C3
Arno (atoll), Marshall Is. 87/H5
Arnold, Calif. (95223) 204/E5
Arnold, England 13/F4
Arnold, Kansas (67515) 232/B3
Arnold, Mich. (49819) 250/B2
Arnold, Minn. (†55801) 255/F4
Arnold, Mo. (63010) 261/M6
Arnold, Nebr. (69120) 264/D3
Arnold (riv.), North. Terr. 93/D3
Arnold, Pa. (15068) 294/C4
Arnold Mills, R.I. (†02864) 249/J5
Arnoldsburg, W. Va. (25234) 312/D5
Arnold's Cove, Newf. 166/C2
Arnolds Park, Iowa (51331) 229/C2
Arnoldstein, Austria 41/B3
Arnoldsville, Georgia (30619) 217/F3
Arnot, Pa. (16911) 294/H2
Arnøya (isl.), Norway 18/M1
Arnprior, Ontario 177/H2
Arnsberg, W. Germany 22/C3
Arnstadt, E. Germany 22/D3
Årø (isl.), Denmark 21/C7
Aro (riv.), Venezuela 124/F4
Aroa, Venezuela 124/D2
Aroab, Namibia 118/B5
Aroche, Spain 33/C4
Arock, Oreg. (97902) 291/K5
Aroland, Ontario 177/H4
Aroland, Ontario 175/C2
Arolla, Switzerland 39/E4
Arolsen, W. Germany 22/C3
Aroma, Bolivia 136/B6
Aroma, Sudan 111/G4
Aroma Park, Ill. (60910) 222/F2
Aromas, Calif. (95004) 204/D7
Aroostook (co.), Maine 243/F2
Aroostook (riv.), Maine 243/G2
Aroostook, New Bruns. 170/C2
Arorae (atoll), Kiribati 87/H6
Aroroy, Philippines 82/D4
Arosa, Ria de (est.), Spain 33/B1
Arosa, Switzerland 39/J3
Aroser Rothorn (mt.), Switzerland 39/J3
Årøsund, Denmark 21/C7
Arouca, Trin. & Tob. 161/B10
Arp, Georgia (†31783) 217/F7
Arp, Texas (75750) 303/J5
Arpa (riv.), Turkey 63/K2
Arpaçay, Turkey 63/K2
Arpin, Wis. (54410) 317/G6
Arque, Bolivia 136/B5
'Arraba, West Bank 65/C3
'Arrabe, Israel 65/C2
Arrah, India 68/E3
Ar Rahhaliya, Iraq 66/C4
Ar Rahhaliya, Iraq 59/D3
Arraias, Brazil 132/E6
Arran, Fla. (†32327) 212/B1
Arran, Sask. 181/K4
Arran (isl.), Scotland 15/C5
Arran (isl.), Scotland 10/D3
Arras, Br. Col. 184/G2
Arras, France 28/E2
Arrecifal, Colombia 126/F6
Arrecife, Spain 106/B3
Arrecife, Spain 33/C4
Arrecife de la Media Luna (reefs), Honduras 154/F3
Arrecifes, Argentina 143/F7
Arrecifes (riv.), Argentina 143/G6
Arrey, N. Mex. (87930) 274/B6
Arriaga, Mexico 150/N8
Arriba, Colo. (80804) 208/N4
Arribeños, Argentina 143/F7
Arrington, Kansas (†66436) 232/G2
Arrington, Tenn. (37014) 237/H9
Arrington, Va. (22922) 307/L5
Arriola, Colo. (†81323) 208/B8
Arrochar, Scotland 15/D4
Arronches, Portugal 33/C3
Arrow (lake), Ireland 17/E3
Arrow (creek), Mont. 262/F3
Arrow Canyon (range), Nev. 266/G6
Arrow Creek, Mont. (†59424) 262/F3
Arrowhead Mountain (lake), Vt. 268/A2
Arrow River, Manitoba 179/B4
Arrowrock (res.), Idaho 220/C6
Arrow Rock, Mo. (65320) 261/F4
Arrowsmith, Ill. (61722) 222/E3
Arrowtown, N. Zealand 100/B6
Arrowwood, Alberta 182/D4
Arrowwood (lake), N. Dak. 282/N5
Arroyas, Los (lake), Bolivia 136/C3
Arroyo, P. Rico 161/E3
Arroyo, P. Rico 156/G1
Arroyo Blanco, Cuba 158/F2
Arroyo de la Luz, Spain 33/C3
Arroyo del Valle (dry riv.), Calif. 204/L3
Arroyo Grande, Bolivia 136/A2
Arroyo Grande, Calif. (93420) 204/E8
Arroyo Hondo (dry riv.), Calif. 204/L3
Arroyo Hondo, N. Mex. (87513) 274/D2
Arroyo Mocho (dry riv.), Calif. 204/L2
Arroyo Seco, Argentina 143/F6
Arroyo Seco (dry riv.), Calif. 204/K10
Arroyo Seco, N. Mex. (87514) 274/D2
Arroyos y Esteros, Paraguay 144/B4
Ar Rumaila, Iraq 66/E5
Års, Denmark 21/C4
Ars-en-Ré, France 28/C4
Arsen'yev, U.S.S.R. 48/O5
Arsin, Turkey 63/H2
Arslanköy, Turkey 63/F4
Árta, Greece 45/E6
Artá, Spain 33/H3
Artas, S. Dak. (57423) 298/K2
Artawiya, Saudi Arabia 59/E4
Arteaga, Mexico 150/H7
Artem, U.S.S.R. 48/O5
Artemas, Pa. (17211) 294/E6
Artemisa, Cuba 158/B1
Artemisa, Cuba 156/A2
Artemovskiy, U.S.S.R. 48/M4
Artemus, Ky. (40903) 237/O7
Artena, Italy 34/F7
Artesia, Calif. (90701) 204/C11
Artesia, Miss. (39736) 256/G4
Artesia, N. Mex. 188/F4
Artesia, N. Mex. (88210) 274/E6
Artesian, S. Dak. (57314) 298/O6
Artesia Wells, Texas (78001) 303/E9
Arth, Switzerland 39/F2
Arthabaska (co.), Québec 172/E4
Arthabaska, Québec 172/F3
Arthur, Ill. (61911) 222/E4
Arthur, Ind. (†47598) 227/C8
Arthur, Iowa (51431) 229/C4
Arthur (co.), Nebr. 264/C3
Arthur, Nebr. (69121) 264/C3
Arthur (range), N. Zealand 100/D4
Arthur, N. Dak. (58006) 282/R5
Arthur, Ontario 177/D4
Arthur (lake), Pa. 294/C4
Arthur (lake), Tasmania 99/D4
Arthur (range), Tasmania 99/C5
Arthur (riv.), Tasmania 99/B3
Arthur, Tenn. (37707) 237/O7
Arthur (riv.), W. Australia 92/B2
Arthur, W. Va. (26816) 312/H4
Arthurdale, W. Va. (26520) 312/G3
Arthuret, England 13/E2
Arthurette, New Bruns. 170/C2
Arthur Kill (str.), N.J. 273/B3
Arthur's (pass), N. Zealand 100/C5
Arthurstown, Ireland 17/H7
Artibonite (dept.), Haiti 158/C5
Artibonite (riv.), Haiti 158/C5
Artigas (dept.), Uruguay 145/B1
Artigas, Uruguay 145/C1
Artillery (lake), N.W. Terrs. 187/H3
Artland, Sask. 181/B3
Artois, Calif. (95913) 204/C4
Artois (trad. prov.), France 29
Artova, Turkey 63/G2
Artux (Atushi), China 77/A4
Artvin (prov.), Turkey 63/J2
Artvin, Turkey 59/D1
Artvin, Turkey 63/J2
Aru (isls.), Indonesia 85/K7
Aru, Zaire 115/F3
Arua, Uganda 115/F3
Aruba (isl.), Neth. Ant. 161/E9
Aruba (isl.), Neth. Ant. 156/E4
Arucas, Spain 33/B5
Arunachal Pradesh (terr.), India 68/G3
Arundel, England 13/G7
Arundel, England 10/F5
Arundel, Québec 172/C4
Årup, Denmark 21/D7
Aruppukkottai, India 68/D7
'Arura, West Bank 65/C3
Arus, P. Rico 161/C3
Arusha (reg.), Tanzania 115/G4
Arusha, Tanzania 102/F5
Arusha, Tanzania 115/G4
Arusi (prov.), Ethiopia 111/G6
Aruwimi (riv.), Zaire 115/E3
Arva, Ireland 17/F4
Arva, Ontario 177/C4
Arvada, Colo. (*80001) 208/J3
Arvada, Wyo. (82831) 319/F1
Arvayheer, Mongolia 77/F2
Arvel, Ky. (†40447) 237/O5
Arvi, India 68/D4
Arvida, Québec 172/F1
Arvidsjaur, Sweden 18/L4
Arvika, Sweden 18/H7
Arvilla, N. Dak. (58214) 282/P4
Arvin, Calif. (93203) 204/G8
Arvonia, Va. (23004) 307/M5
Arwad (Ruad) (isl.), Syria 63/F5
Arxan, China 77/K2
Arys', U.S.S.R. 48/G5
Arzamas, U.S.S.R. 48/E4
Arzamas, U.S.S.R. 52/F3
Arzúa, Spain 33/B1
As, Belgium 27/H6
Aš, Czech. 41/B1
Aså, Denmark 21/D3
Asaba, Nigeria 106/F7
Asadabad, Iran 66/F3
Asahan (riv.), Indonesia 85/B5
Asahi, Japan 81/K6
Asahi (mt.), Japan 81/J4
Asahikawa, Japan 81/L2
Asahikawa, Japan 54/P5
Asama (mt.), Japan 81/J5
Asansol, India 68/F4
Åsarna, Sweden 18/J5
Asau, W. Samoa 86/L8
Asbest, U.S.S.R. 48/G4
Asbestos, Québec 172/F4
Asbury, Iowa (†52001) 229/M4
Asbury, Mo. (64832) 261/C8
Asbury, N.J. (08802) 273/C2
Asbury, W. Va. (24916) 312/E7
Asbury Park, N.J. (07712) 273/F3
As Busaiya, Iraq 66/E5
Ascención (Añez), Bolivia 136/D4
Ascensión, Argentina 143/F7
Ascension (par.), La. 238/J6
Ascensión, Mexico 150/E1
Ascension, Neth. Ant. 161/F8
Ascension (isl.), St. Helena 102/A5
Ascension (isl.), St. Helena 2/J6
Aschaffenburg, W. Germany 22/C4
Aschendorf, W. Germany 22/B2
Aschersleben, E. Germany 22/D3
Asco, W. Va. (24809) 312/C8
Ascog, Scotland 15/A2
Ascoli Piceno (prov.), Italy 34/D3
Ascoli Piceno, Italy 34/D3
Ascona, Switzerland 39/G4
Ascope, Peru 128/C6
Ascot, Queensland 88/K2
Ascot, Queensland 95/E2
Ascotán, Chile 138/B3
Ascotán, Salar de (salt dep.), Chile 138/B3
Ascot Corner, Québec 172/F4
Ascrib (isls.), Scotland 15/B3
Ascutney, Vt. (05030) 268/C5
Ascutney (mt.), Vt. 268/C5
Åseda, Sweden 18/J8
Åsele, Sweden 18/K4
Asenovgrad, Bulgaria 45/G5
Asèr, Ras (cape), Somalia 2/M5
Asèr, Ras (cape), Somalia 115/K1
Ash (riv.), Minn. 255/F2
Ash, N.C. (28420) 281/N6
Ash, Oreg. (†97473) 291/D4
Ash (creek), Utah 304/A6
'Ashaira, Saudi Arabia 59/D5
Ashanti (reg.), Ghana 102/B4
Ashanti (reg.), Ghana 106/D7
Ashaway, R.I. (02804) 249/G7
Ashboro, Ind. (†47840) 227/C6
Ashburn, Georgia (31714) 217/E7
Ashburn, Mo. (63433) 261/K3
Ashburn, Va. (22011) 307/O2
Ashburnham, Mass. (01430) 249/G2
Ashburnham○, Mass. (01430) 249/G2
Ashburton (riv.), Australia 87/B8
Ashburton, England 13/D7
Ashburton, N. Zealand 100/C5
Ashburton (riv.), W. Australia 88/B4
Ashburton (riv.), W. Australia 92/A3
Ashburton Downs, W. Australia 88/B4
Ashby, Ala. (†35035) 195/E4
Ashby○, Mass. (01431) 249/G2
Ashby, Minn. (56309) 255/C4
Ashby, Nebr. (69333) 264/C2
Ashbyburg, Ky. (†42456) 237/G5
Ash Creek, Minn. (†56173) 255/B7
Ashcroft, Br. Col. 184/G5
Ashdale, Maine (†04565) 243/D8
Ashdod, Israel 65/B4
Ashdot Ya'aqov, Israel 65/D2
Ashdown, Ark. (71822) 202/B6
Ashe (co.), N.C. 281/F2
Ashe (isl.), N.C. 281/P6
Asheboro, N.C. (27203) 281/K3
Ashepoo, S.C. (†29446) 296/G6
Ashepoo (riv.), S.C. 296/F6
Asher, Okla. (74826) 288/N5
Ashern, Manitoba 179/D3
Asherton, Texas (78827) 303/E9
Asherville, Ind. (†47834) 227/C6
Asherville, Kansas (67420) 232/D2
Asheville, N.C. 188/K3
Asheville, N.C. (*28801) 281/D3
Asheweig (riv.), Ontario 175/C2
Ashfield○, Mass. (01330) 249/C2
Ashfield, N. S. Wales 88/K4
Ashfield, N.S. Wales 97/J3
Ash Flat, Ark. (72513) 202/G1
Ashford, Ala. (36312) 195/H8
Ashford○, Conn. (06278) 210/G1
Ashford, England 10/G5
Ashford, England 13/H6
Ashford, Ireland 17/J5
Ashford, N.S. Wales 97/F1
Ashford, N.C. (†28752) 281/F3
Ashford, Wash. (98304) 310/C4
Ashford, W. Va. (25009) 312/C6
Ashford P.O. (Warrenville), Conn. (06278) 210/G1
Ash Fork, Ariz. (86320) 198/C3
Ash Grove, Kansas (†67455) 232/D2
Ash Grove, Mo. (65604) 261/E8
Ashgrove, Queensland 88/K2
Ashhurst, N. Zealand 100/E4
Ashibetsu, Japan 81/L2
Ashikaga, Japan 81/J5
Ashington, England 13/F2
Ashington, England 10/F3
Ashippun, Wis. (53003) 317/H1
Ashiya, Japan 81/H8
Ashizuri (cape), Japan 81/F7
Ashkhabad, U.S.S.R. 54/G6
Ashkhabad, U.S.S.R. 48/F6
Ashkum, Ill. (60911) 222/E3
Ash Lake, Minn. (†55771) 255/F2
Ashland, Ala. (36251) 195/G4
Ashland, Calif. (†94577) 204/K2
Ashland, Georgia (†30521) 217/F2
Ashland, Ill. (62612) 222/C4
Ashland, Kansas (67831) 232/C4
Ashland, Ky. (41101) 237/R4
Ashland, Ky. 188/K3
Ashland, La. (71002) 238/D2
Ashland, Maine (04732) 243/G2
Ashland○, Maine (04732) 243/G2
Ashland○, Mass. (01721) 249/J3
Ashland, Miss. (38603) 256/F1
Ashland, Mo. (65010) 261/H5
Ashland, Mont. (59003) 262/K5
Ashland, Nebr. (68003) 264/H3
Ashland, N.H. (03217) 268/D4
Ashland○, N.H. (03217) 268/D4
Ashland, N.J. (†08033) 273/B3
Ashland, N.Y. (12407) 276/M6
Ashland (co.), Ohio 284/F4
Ashland, Ohio (44805) 284/F4
Ashland, Okla. (74524) 288/O5
Ashland, Oreg. (97520) 291/E5
Ashland, Pa. (17921) 294/K4
Ashland, Va. (23005) 307/N5
Ashland (co.), Wis. 317/E3
Ashland, Wis. (54806) 317/E2
Ashland City, Tenn. (37015) 237/G8
Ashley (co.), Ark. 202/G7
Ashley, Ill. (62808) 222/D5
Ashley, Ind. (46705) 227/G1
Ashley, Mich. (48806) 250/E5
Ashley, Mo. (†63334) 261/K4
Ashley (lake), Mont. 262/B2
Ashley, N.S. Wales 97/E1
Ashley, N. Dak. (58413) 282/M7
Ashley, Ohio (43003) 284/E5
Ashley, Pa. (18706) 294/E7
Ashley (riv.), S.C. 296/G6
Ashley, W. Va. (†26339) 312/E4
Ashley Falls, Mass. (01222) 249/A4
Ashmere (lake), Mass. 249/B3
Ashmont, Alberta 182/E2
Ashmore, Ill. (61912) 222/F4
Ashmore, Nova Scotia 168/C4
Ashmore (isls.), Terr. of Ashmore and Cartier Is. 88/C2
Ashmore and Cartier Is., Terr. of, 88/C2
Ashokan, N.Y. (†12491) 276/M7
Ashokan (res.), N.Y. 276/M7
Ashport, Tenn. (†38063) 237/B9
Ashqelon, Israel 65/A4
Ash Shabicha, Iraq 66/C5
Ashtabula (lake), N. Dak. 282/P5
Ashtabula (co.), Ohio 284/J2
Ashtabula, Ohio (44004) 284/J2
Ashton, Idaho (83420) 220/G5
Ashton, Ill. (61006) 222/D2
Ashton, Iowa (51232) 229/B2
Ashton, Kansas (†67051) 232/E4
Ashton, Mich. (†49677) 250/D5
Ashton, Nebr. (68817) 264/F3
Ashton, R.I. (02864) 249/J5
Ashton, S.C. (†29082) 296/E5
Ashton, S. Dak. (57424) 298/N3
Ashton, W. Va. (25503) 312/B5
Ashton Creek, Br. Col. 184/H5
Ashton-under-Lyne, England 13/H2
Ashton-under-Lyne, England 10/G2
Ashuanipi (lake), Newf. 166/A3
Ashuanipi (riv.), Newf. 166/A3
Ashuanpi, Newf. 166/A3
Ashuelot, N.H. (03441) 268/C6
Ashuelot (riv.), N.H. 268/C6
Ash Valley, Kansas (†67550) 232/C3
Ashville, Ala. (35953) 195/F3
Ashville, Maine (04607) 243/G7
Ashville, Manitoba 179/B3
Ashville, Ohio (43103) 284/E6
Ashville, Pa. (16613) 294/F4
Ashwaubenon, Wis. (54304) 317/K7
Ashwood, Oreg. (97711) 291/G3
'Asi (Orontes) (riv.), Syria 63/G5
Asia 2/P3
Asia (isls.), Indonesia 85/J5
Asid (gulf), Philippines 82/D4
Asidonhoppo, Suriname 131/D4
Asilah, Morocco 106/C1
Asinara (gulf), Italy 34/B4
Asinara (isl.), Italy 34/B4
Asino, U.S.S.R. 48/J4
'Asir (reg.), Saudi Arabia 59/D6
Aşkale, Turkey 63/J3
Askeaton, Ireland 17/D6
Askew, Miss. (38604) 256/D1
Askewville, N.C. (†27983) 281/R2
Askim, Norway 18/E4
Askim, Sweden 18/G8
Aski Mosul, Iraq 66/C2
Askival (mt.), Scotland 15/B4
Askov, Denmark 21/C7
Askov, Minn. (55704) 255/F4
Askvoll, Norway 18/D6
Asmara, Ethiopia 111/G4
Asmara, Ethiopia 59/C6
Asmara, Ethiopia 102/F3
Asnaes, Denmark 21/E6
Åsnen (lake), Sweden 18/J8
Asnières-sur-Seine, France 28/A1
Aso (mt.), Japan 81/E7
Aso National Park, Japan 81/E7
Asosa, Ethiopia 111/F5
Asoteriba, Jebel (mt.), Sudan 111/G3
Asotin (co.), Wash. 310/H4
Asotin, Wash. (99402) 310/H4
Asotin (creek), Wash. 310/H4
Asotin (dam), Wash. 310/J4
Aspang Markt, Austria 41/D3
Aspatria, England 13/D3
Aspe, Spain 33/F3
Aspelund, Minn. (†55946) 255/F6
Aspen, Colo. (81611) 208/F4
Aspen, Nova Scotia 168/F3
Aspen (lake), Oreg. 291/E5
Aspen (mts.), Wyo. 319/C4
Aspen Grove, Br. Col. 184/G5
Aspen Hill, Md. 245/K4
Aspermont, Texas (79502) 303/D4
Aspers, Pa. (17304) 294/H6
Aspetuck, Conn. (†06880) 210/B4
Aspetuck (res.), Conn. 210/B4
Aspetuck (riv.), Conn. 210/B3
Aspinwall, Iowa (51432) 229/C5
Aspinwall, Pa. (15215) 294/C6
Aspiring (mt.), N. Zealand 100/B6
Aspley, Queensland 88/K2
Aspy (bay), Nova Scotia 168/H2
Asquith, Sask. 181/D3
Assab, Ethiopia 59/D7
Assab, Ethiopia 111/H5
Assaba (reg.), Mauritania 106/B5
Assabet (riv.), Mass. 249/H3
Assad, Bahrat (lake), Syria 63/H4
Assakarai (dry riv.), Niger 106/F5
Assale (lake), Ethiopia 111/H5
As Salman, Iraq 59/E3
As Salman, Iraq 66/D5
Assam (state), India 68/G3
Assapan (riv.), Manitoba 179/G2
Assaria, Kansas (67416) 232/E3
Assateague Island Nat'l Seashore, Va. 307/T4
Assawompset (pond), Mass. 249/L5
Assay (creek), Utah 304/B6
Asse, Belgium 27/E7
Asselar (well), Mali 106/D5
Asselle, Ethiopia 102/F4
Asselle, Ethiopia 111/G6
Assen, Netherlands 27/K3
Assenede, Belgium 27/D6
Assens, Århus, Denmark 21/D4
Assens, Fyn, Denmark 21/D7
Assesse, Belgium 27/G8
Assigny (lake), Newf. 166/A3
Assiniboia, Sask. 181/E6
Assiniboine (mt.), Alberta 182/C4
Assiniboine (mt.), Br. Col. 184/K5
Assiniboine (riv.), Manitoba 179/C5
Assiniboine (riv.), Sask. 181/J3
Assinica (lake), Québec 174/C3
Assinika (lake), Manitoba 179/G2
Assinika (riv.), Manitoba 179/G2
Assinippi, Mass. (02339) 249/E8
Assis, Brazil 132/D8
Assis, Brazil 135/A3
Assisi, Italy 34/D3
Assonet, Mass. (02702) 249/K5
Assumption, Ill. (62510) 222/E4
Assumption (par.), La. 238/H7
Assumption, Ohio (†43540) 284/B2
Assumption (isl.), Seychelles 118/H1
Assynt (dist.), Scotland 15/C2
Assynt, Loch (lake), Scotland 15/D2
Assyria, Mich. (†49021) 250/D6
Astara, U.S.S.R. 52/G7
Astatula, Fla. (32705) 212/E3
Asten, Netherlands 27/H6
Asterabad (Gorgan), Iran 59/F2
Asterabad (Gorgan), Iran 66/J2
Asti, Calif. (95413) 204/C5
Asti (prov.), Italy 34/B2
Asti, Italy 34/B2
Astillero, Peru 128/H9
Astipálaia, Greece 45/H7
Astipálaia (isl.), Greece 45/H7
Astle, New Bruns. 170/D2
Aston (bay), N.W. Terrs. 187/J2
Aston-Jonction, Québec 172/E3
Astor, Fla. (32002) 212/E2
Astorga, Spain 33/C1
Astoria, Ill. (61501) 222/C3
Astoria, Oreg. 188/B1
Astoria, Oreg. (97103) 291/D1
Astoria, S. Dak. (57213) 298/S4
Astorville, Ontario 177/E1
Astove (isl.), Seychelles 102/G6
Astove (isl.), Seychelles 118/H2
Astra, Argentina 143/C6
Astrakhan', U.S.S.R. 7/J4
Astrakhan', U.S.S.R. 52/G5
Astrakhan', U.S.S.R. 48/E5
Astray (lake), Newf. 166/A3
Astudillo, Spain 33/D1
Asturias (reg.), Spain 33/C1
Asunción, Bolivia 136/B2
Asuncion (isl.), No. Marianas 87/E4
Asunción, Paraguay 144/A4
Asunción (cap.), Paraguay 2/F7
Asunción (dept.), Paraguay 144/A4
Asunción (cap.), Paraguay 120/D5
Asuncion (passage), Philippines 82/D5
Asunción Mita, Guatemala 154/C3
Asunción Nochixtlán, Mexico 150/L8
Asunta, Bolivia 136/B5
Aswad, Ras al (cape), Saudi Arabia 59/C5
Aswân, Egypt 111/F3
Aswân, Egypt 59/B5
Aswân, Egypt 102/F2
Aswân (dam), Egypt 59/B5
Aswân (dam), Egypt 111/F3
Aswân High (dam), Egypt 102/F2
Aswân High (dam), Egypt 111/F3
Asyût, Egypt 111/J4
Asyût, Egypt 102/F2
Asyût, Egypt 59/B4
Aszód, Hungary 41/E3
Atabapo (riv.), Colombia 126/G6
Atabapo (riv.), Venezuela 124/E6
Atacama, Puna de (reg.), Argentina 143/C2
Atacama (reg.), Chile 138/B6
Atacama (des.), Chile 120/C5
Atacama (des.), Chile 138/B4
Atacama, Salar de (salt dep.), Chile 138/C4
Atafu (atoll), Tokelau Is. 87/J6
Atahona, Uruguay 145/B4
Atakora (mts.), Benin 106/E6
Atakpamé, Togo 106/E7
Atalándi, Greece 45/F6
Atalaya, Peru 128/E8
Atalissa, Iowa (52720) 229/L5
Atambua, Indonesia 85/G7
Atami, Japan 81/J6
Atapirire, Venezuela 124/F3
Atar, Mauritania 106/B4
Atar, Mauritania 102/A2
Ataran (riv.), Burma 72/C4
Atascadero, Calif. (93422) 204/E8
Atascosa (co.), Texas 303/F9
Atascosa, Texas (78002) 303/J11
Atbara (riv.), Ethiopia 111/G4
Atbara, Sudan 111/F4
Atbara, Sudan 59/B6
Atbara, Sudan 102/F3
Atbara (riv.), Sudan 59/C6
Atbara (riv.), Sudan 111/G4
Atbasar, U.S.S.R. 48/G4
Atchafalaya (bay), La. 238/H8
Atchafalaya (riv.), La. 238/G6
Atchison, Kans. 188/G3

Atchison (co.), Kansas 232/G2
Atchison, Kansas (66002) 232/G2
Atchison (co.), Mo. 261/B2
Atco, N.J. (08004) 273/D4
Ateca, Spain 33/F2
Atén, Bolivia 136/A4
Atenas, C. Rica 154/E6
Atessa, Italy 34/E3
Atglen, Pa. (19310) 294/K6
Ath, Belgium 27/D7
Athabasca (lake) 162/F4
Athabasca, Alberta 182/D2
Athabasca (lake), Alberta 182/C5
Athabasca (riv.), Alberta 182/D1
Athabasca, Alta. 162/E5
Athabasca (lake), Alta. 162/E4
Athabasca (riv.), Alta. 162/E2
Athabasca (riv.), Alta. 146/L2
Athabasca (lake), Canada 146/H4
Athabasca (lake), Sask. 181/L2
Athalia, Ohio (†45669) 284/F8
Athalmer, Br. Col. 184/K5
Athboy, Ireland 17/H4
Athea, Ireland 17/H4
Athelstan, Iowa (†50836) 229/D7
Athelstan, Québec 172/C4
Athelstane, Wis. (54104) 317/K5
Athena, Oreg. (97813) 291/J2
Athenry, Ireland 17/D5
Athens, Ala. (35611) 195/E1
Athens, (†71943) 202/C5
Athens, Ga. 188/K4
Athens, Georgia (*30601) 217/F3
Athens (cap.), Greece 7/G5
Athens (cap.), Greece 45/F7
Athens (cap.), Greece 2/K4
Athens, Greater, Greece 45/F7
Athens, Ill. (62613) 222/D4
Athens, Ind. (46912) 227/E2
Athens, La. (71003) 238/E1
Athens○, Maine (04912) 243/D6
Athens○, Maine (04912) 243/D6
Athens, Mich. (49011) 250/D6
Athens, N.Y. (12015) 276/N6
Athens (co.), Ohio 284/F7
Athens, Ohio (45701) 284/F7
Athens, Ontario 177/J3
Athens, Pa. (18810) 294/K2
Athens, Tenn. (37303) 237/M10
Athens, Texas (75751) 303/J5
Athens, W. Va. (24712) 312/E8
Athens, Wis. (54411) 317/G5
Athensville, Ill. (†62082) 222/C4
Atherley, Ontario 177/K5
Atherton, Calif. (94025) 204/K3
Atherton, Mo. (†64050) 261/R5
Atherton, Queensland 99/H3
Atherton, Queensland 88/H5
Athertonville, Ky. (†42748) 237/K5
Athleague, Ireland 17/E4
Athlone, Ireland 10/C4
Athlone, Ireland 17/F5
Athok, Burma 72/B3
Athol, Idaho (83801) 220/B2
Athol, Kansas (66932) 232/D2
Athol, Mass. (01331) 249/F2
Athol○, Mass. (01331) 249/F2
Athol, N.Y. (12810) 276/N4
Athol, N. Zealand 100/B6
Athol, Nova Scotia 168/D3
Athol (dist.), Scotland 15/D4
Athol, S. Dak. (57425) 298/M3
Atholville, New Bruns. 170/F1
Áthos (mt.), Greece 45/G5
Athy, Ireland 17/H6
Athy, Ireland 10/C4
Ati, Chad 111/C5
Ati, Chad 102/D3
Atibaia, Brazil 135/C3
Atienza, Spain 33/E2
Atico, Peru 128/F11
Atikameg, Alberta 182/C2
Atikokan, Ont. 162/G6
Atikokan, Ontario 177/G5
Atikokan, Ontario 175/B3
Atikonak (lake), Newf. 166/B3
Atim (lake), Manitoba 179/C2
Atiquizaya, El Salvador 154/C3
Atitlán (lake), Guatemala 154/B3
Atitlán (vol.), Guatemala 154/B3
Atiu, Cook Is. 87/L8
Atka, Alaska (99502) 196/D4
Atka, Alaska 188/D6
Atka (isl.), Alaska 196/L4
Atka, U.S.S.R. 48/Q3
Atkarsk, U.S.S.R. 52/G4
Atkins, Ark. (72823) 202/E3
Atkins, Iowa (52206) 229/K4
Atkins, Va. (24311) 307/F7
Atkinson (co.), Georgia 217/G8
Atkinson, Georgia (†31543) 217/J8
Atkinson, Ill. (61235) 222/C4
Atkinson○, Maine (†04426) 243/E5
Atkinson, Minn. (†55718) 255/F4
Atkinson, Nebr. (68713) 264/E2
Atkinson○, N.H. (03811) 268/E6
Atkinson, N.C. (28421) 281/N5
Atkinson (pt.), N.W. Terrs. 187/E2
Atkinson Field, Guyana 131/G2
Atlanta, Ark. (†71740) 202/D7
Atlanta, C. Rica 154/F6
Atlanta (cap.), Ga. 188/K4
Atlanta (cap.), Ga. 188/K4
Atlanta (cap.), Georgia (*30301) 217/K1
Atlanta, Idaho (83601) 220/C6
Atlanta, Ill. (61723) 222/D3
Atlanta, Ind. (46031) 227/E4
Atlanta, Kansas (67008) 232/F4
Atlanta, La. (71404) 238/E3
Atlanta, Mich. (49709) 250/E3
Atlanta, Mo. (63530) 261/H3
Atlanta, Nebr. (68923) 264/E4
Atlanta, N.Y. (14808) 276/F5
Atlanta, Ohio (43104) 284/D6
Atlanta, Texas (75551) 303/K4
Atlanta, (cap.) 111/K1
Atlanta Nav. Air Sta., Georgia 217/J1
Atlantic (ocean) 102/B5

Atlantic (ocean) 146/M6
Atlantic, Iowa (50022) 229/D6
Atlantic, Maine (04608) 243/G7
Atlantic (co.), N.J. 273/D5
Atlantic, N.C. (28511) 281/S5
Atlantic, Pa. (16111) 294/B3
Atlantic (peak), Wyo. 319/D3
Atlantic Beach, Fla. (32233) 212/E1
Atlantic Beach, N.Y. (11509) 276/P7
Atlantic Beach, N.C. (28512) 281/R5
Atlantic Beach, S.C. (†29582) 296/K4
Atlantic City, N.J. 188/M4
Atlantic City, N.J. (*08401) 273/E5
Atlantic City, Wyo. (†82520) 319/D3
Atlantic Highlands, N.J. (07716) 273/F3
Atlantic Highlands, N.J. 273/E3
Atlantic Highlands (ridge), N.J. 273/E3
Atlantic Mine, Mich. (49905) 250/G1
Atlantic Ocean 4/D11
Atlantic Ocean 7/C4
Atlantic Ocean, England 13/A7
Atlantic Ocean, Scotland 15/C2
Atlantic Ocean Ocean, Portugal 33/A3
Atlántida, Uruguay 145/B6
Atlas (mts.) 102/B1
Atlas (mts.), Algeria 106/E2
Atlas (mts.), Morocco 106/C2
Atlas, Pa. (17851) 294/K4
Atlas, Wis. (†54853) 317/H4
Atlin, Br. Col. 184/J1
Atlin (lake), Br. Col. 184/J1
Atlit, Israel 65/B2
Atlixco, Mexico 150/M2
Atmore, Ala. (36503) 195/C8
Atmore, Alberta 182/D2
Atnarko, Br. Col. 184/E4
Atocha, Bolivia 136/B7
Atoka (co.), Okla. 288/O6
Atoka, Okla. (74525) 288/P6
Atoka (res.), Okla. 288/P5
Atoka, Tenn. (38004) 237/B10
Atomic City, Idaho (83215) 220/F6
Atotonilco el Alto, Mexico 150/H6
Atoui, Wadi (dry riv.), Mauritania 106/A2
Atoui, Wadi (dry riv.), Western Sahara 106/A2
Atoyac (riv.), Mexico 150/N2
Atoyac (riv.), Mexico 150/Q2
Atoyac de Álvarez, Mexico 150/J8
Atrak (Atrek) (riv.), Iran 66/J2
Atrato (riv.), Colombia 126/B4
Atrek (riv.), Iran 59/G2
Atrek (Atrak), Iran 66/J2
Atrek (riv.), U.S.S.R. 48/F6
Atri, Italy 34/D3
Atsugi, Japan 81/J6
Atsumi (bay), Japan 81/H6
Attachie, Br. Col. 184/G2
Attala (co.), Miss. 256/E4
Attalens, Switzerland 39/C3
Attalla, Ala. (35954) 195/F2
Attapu, Laos 72/E4
Attapulgus, Georgia (31715) 217/D9
Attawapiskat (riv.), Ont. 146/K4
Attawapiskat (riv.), Ont. 162/H5
Attawapiskat (riv.), Ontario 175/D2
Attawapiskat (lake), Ontario 175/C2
Attawapiskat (riv.), Ontario 175/C2
Attawaugan, Conn. (†06241) 210/H1
Atteam (pond), Maine 243/C4
Attebubu, Ghana 106/D6
Atterberry, Ill. (†62675) 222/D3
Atter See (lake), Austria 41/B3
Attert, Belgium 27/F8
Attica, Ind. (47918) 227/C4
Attica, Iowa (†50138) 229/G6
Attica, Kansas (67009) 232/D4
Attica, Mich. (48412) 250/F5
Attica, N.Y. (14011) 276/E5
Attica, Ohio (44807) 284/E3
Attikamagen (lake), Newf. 166/A3
'Attil, West Bank 65/C3
Attleboro, Mass. (02703) 249/J5
Attleboro Falls, Mass. (02763) 249/J5
Attnang-Puchheim, Austria 41/B2
Attock, Pakistan 68/C3
Attu (isl.), Alaska 188/D6
Attu (isl.), Alaska 196/J3
Attu (isl.), U.S. 4/D1
Attunga, N.S. Wales 97/F2
Atuel (riv.), Argentina 143/C4
Atuntaqui, Ecuador 128/C2
Atuona, Fr. Poly. 87/M7
Atura, Uganda 115/F3
Atushi (Artux), China 77/A4
Atwater, Calif. (95301) 204/E6
Atwater, Minn. (56209) 255/D5
Atwater, Ohio (44201) 284/H3
Atwater, Sask. 181/J5
Atwood (Samana) (cay), Bahamas 156/D2
Atwood, Colo. (80722) 208/N1
Atwood, Ill. (61913) 222/E4
Atwood, Ind. (46502) 227/F2
Atwood, Kansas (67730) 232/B2
Atwood, Mich. (†49729) 250/D3
Atwood (lake), Ohio 284/H4
Atwood, Okla. (74827) 288/O5
Atwood, Ontario 177/J5
Atwood, Pa. (†16249) 294/D4
Atwood, Tenn. (38220) 237/F3
Atwoodville, Conn. (†06250) 210/G1
Atyrá, Paraguay 144/B4
Au, Switzerland 39/J2
Auasbila, Honduras 154/E3
Auau (chan.), Hawaii 218/H2
Aubagne, France 28/F6
Aubange, Belgium 27/H9
Aube (dept.), France 28/E4
Aube (riv.), France 28/E3
Aubenas, France 28/F5
Auberry, Calif. (93602) 204/F6
Aubervilliers, France 28/B1

Aubigny, Manitoba 179/E5
Aubonne, Switzerland 39/B4
Aubrey (cliffs), Ariz. 198/J3
Aubrey, Ark. (72311) 202/A4
Auburn, Ala. (36830) 195/H5
Auburn, Calif. (95603) 204/C8
Auburn, Ga. (30203) 217/E2
Auburn, Ill. (62615) 222/D4
Auburn, Ind. (46706) 227/G2
Auburn, Iowa (51433) 229/D4
Auburn, Kansas (66642) 232/F3
Auburn, Ky. (42206) 237/H7
Auburn, Maine (04210) 243/C7
Auburn, Maine 188/M2
Auburn○, Mass. (01501) 249/G4
Auburn, Mich. (48611) 250/F5
Auburn, Miss. (†39664) 256/C8
Auburn, N.S. Wales 88/K4
Auburn, N.Y. 188/L2
Auburn, N.Y. (13021) 276/G5
Auburn, N. Dak. (†58237) 282/R2
Auburn, Nova Scotia 168/C4
Auburn, Ohio 177/C4
Auburn, Pa. (17922) 294/K4
Auburn, Wash. (98002) 310/C3
Auburn, W. Va. (26325) 312/E4
Auburn, Wyo. (83111) 319/A3
Auburndale, Fla. (33823) 212/E3
Auburndale, Mass. (†02166) 249/B7
Auburndale, Wis. (54412) 317/F6
Auburn Heights, Mich. (48057) 250/F6
Auburntown, Tenn. (37016) 237/J9
Aubusson, France 28/E4
Aucanquilcha, Cerro (mt.), Chile 138/B3
Auce, U.S.S.R. 53/B2
Auch, France 28/D6
Auchenblae, Scotland 15/F4
Auchencairn, Scotland 15/E6
Auchinleck, Scotland 15/D6
Auchterarder, Scotland 10/D2
Auchterarder, Scotland 15/E4
Auchtermuchty, Scotland 15/E4
Aucilla (riv.), Fla. 212/C1
Aucilla (riv.), Fla. 212/C1
Auckland (isls.), N. Zealand 2/S8
Auckland, N. Zealand 2/T7
Auckland, N. Zealand 100/B1
Auckland, N. Zealand 87/B1
Auclair, Québec 172/J2
Aude (dept.), France 28/E6
Audegle, Somalia 115/J3
Auden, Ontario 177/H4
Auden, Ontario 175/C2
Audenarde (Oudenaarde), Belgium 27/D7
Auderghem, Belgium 27/C9
Audet, Québec 172/G4
Audincourt, France 28/G4
Audrain (co.), Mo. 261/J4
Audubon (co.), Iowa 229/D5
Audubon, Iowa (50025) 229/D5
Audubon, Minn. (56511) 255/C4
Audubon, N.J. (08106) 273/B3
Audubon (lake), N. Dak. 282/H4
Audubon Park, Ky. (†40201) 237/J2
Audubon Park, N.J. (†08106) 273/B3
Aue, E. Germany 22/E3
Auerbach, E. Germany 22/E3
Augathella, Queensland 95/H5
Augathella, Queensland 88/H5
Auger, Calif. (94550) 204/D1
Augher, N. Ireland 17/G3
Aughnacloy, N. Ireland 17/H3
Aughrabies (King George's) (falls), S. Africa 118/B5
Aughrim, Ireland 17/J6
Auglaize (co.), Ohio 284/B4
Auglaize (riv.), Ohio 284/B4
Au Gres, Mich. (48703) 250/F4
Augsburg, W. Germany 7/F4
Augsburg, W. Germany 22/D4
Augusta 2/R7
Augusta 87/C8
AUSTRALIA 88
Australia Aboriginal Reserve, W. Australia 88/D5
Australia Aboriginal Res., W. Australia 92/E4
Australian, Br. Col. 184/F4
Australian Alps (mts.), N.S. Wales 97/J3
Australian Alps (mts.), Victoria 97/J3
Australian Alps (mts.), Victoria 88/H7
Australian Capital Territory, /H7
Australian Capital Terr., Australia 87/F9
AUSTRALIAN CAPITAL TERRITORY 97/E4
Australind, W. Australia 92/A2
Austria 2/K4
Austria 7/F4
AUSTRIA 41
Austwell, Texas (77950) 303/H9
Autauga (co.), Ala. 195/E5
Autaugaville, Ala. (36003) 195/E6
Autlán de Navarro, Mexico 150/G7
Au Train, Mich. (49806) 250/C2
Au Train (bay), Mich. 250/C2
Autreyville, Georgia (†31768) 217/E8
Autryville, N.C. (28318) 281/M4
Autun, France 28/E4
Auvelais, Belgium 27/F8
Auvergne, Ark. (†72112) 202/H2
Auvergne (mts.), France 28/E5
Auvergne (trad. prov.) France 29
Auvergne, North. Terr. 93/B3
Auvergne, Québec 172/F3
Auxerre, France 28/E4
Auxier, Ky. (41602) 237/R5
Auxonne, France 28/F4
Auxvasse, Mo. (65231) 261/J4
Auyuittuq Nat'l Park, N.W. Terrs. /M3
Auyuittuq Nat'l Park, Que. 162/K2
Ava, Ill. (62907) 222/D6
Ava, Mo. (65608) 261/G9

Aultman, Pa. (15713) 294/D4
Aumsville, Oreg. (97325) 291/E3
Auning, Denmark 21/D5
Aunis (trad. prov.), France 29
Aur, Pulau (isl.), Malaysia 72/E7
Aura, Mich. (49906) 250/A2
Aura, N.J. (†08028) 273/C4
Aurangabad, Bihar, India 68/E4
Aurangabad, Maharashtra, India 68/D5
Auraria, Georgia (†30534) 217/E1
Auray, France 28/B4
Aurelia, Iowa (51005) 229/C3
Aurès (lag.), Algeria 106/F1
Aurich, W. Germany 22/B2
Aurignac, France 28/D6
Aurillac, France 28/E5
Aurland, Norway 18/E6
Aurora (cap.), Cook Is. 87/L8
Aurora, Colo. 188/F3
Aurora, Colo. (*80010) 208/K3
Aurora, Guyana 131/G3
Aurora, Ill. (*60504) 222/E2
Aurora, Ind. (47001) 227/H6
Aurora, Iowa (50607) 229/K3
Aurora, Kansas (67417) 232/E2
Aurora○, Maine (04408) 243/G6
Aurora, Minn. (55705) 255/F3
Aurora, Mo. (65605) 261/F9
Aurora, Nebr. (68818) 264/F4
Aurora, N.Y. (13026) 276/G5
Aurora, N.C. (28046) 281/R4
Aurora, Ohio (44202) 284/H3
Aurora, Ontario 177/J3
Aurora, Oreg. (97002) 291/B2
Aurora, Philippines 82/D4
Aurora (co.), S. Dak. 298/M6
Aurora, S. Dak. (57002) 298/R5
Aurora, Texas (76078) 303/E1
Aurora, Utah (84620) 304/B5
Aurora, W. Va. (26705) 312/G4
Aurora Lodge, Alaska (†99701) 196/J2
Auroraville, Wis. (†54923) 317/H7
Aus, Namibia 118/B3
Ausable (pt.), Mich. 250/F4
Au Sable (pt.), Mich. 250/F4
Au Sable (pt.), Mich. 250/C2
Au Sable (riv.), Mich. 250/F4
Ausable (riv.), N.Y. 276/N4
Au Sable Forks, N.Y. (12912) 276/N2
Auschwitz (Oświęcim), Poland 47/D3
Ausert (well), Western Sahara 106/B4
Auskerry (isl.), Scotland 15/F1
Aust-Agder (co.), Norway 18/E7
Austell, Georgia (30001) 217/J1
Austerlitz (Slavkov), Czech. 41/D2
Austin, Ark. (72007) 202/G4
Austin, Colo. (81410) 208/D5
Austin, Ind. (47102) 227/F7
Austin, Ky. (42123) 237/K7
Austin, Manitoba 179/D5
Austin, Minn. 188/H2
Austin, Minn. (55912) 255/E7
Austin, Mo. (†64725) 261/D5
Austin, Nev. 188/C3
Austin, Nev. (89310) 266/E3
Austin, Oreg. (97871) 291/J3
Austin, Pa. (16720) 294/F2
Austin (co.), Texas 303/H8
Austin (cap.), Texas 146/J6
Austin (cap.), Texas 188/G4
Austin (cap.), Texas (*78701) 303/G7
Austin (lake), W. Australia 88/B5
Austin (lake), W. Australia 92/B4
Austinburg, Ohio (44010) 284/J2
Austintown, Ohio (44515) 284/J3
Austinville, Iowa (50608) 229/H3
Austinville, Va. (24312) 307/F7
Austonio, Texas (†75835) 303/J6
Austral (isls.), Fr. Polynesia 2/B7
Austral (isls.), Fr. Poly. 87/L8
Australia 2/R7
Australia 87/C8
AUSTRALIA 88
Australia Aboriginal Reserve, W. Australia 88/D5
Australia Aboriginal Res., W. Australia 92/E4
Australian, Br. Col. 184/F4
Australian Alps (mts.), N.S. Wales 97/J3
Australian Alps (mts.), Victoria 97/J3
Australian Alps (mts.), Victoria 88/H7
Australian Capital Territory, /H7
Australian Capital Terr., Australia 87/F9
AUSTRALIAN CAPITAL TERRITORY 97/E4
Australind, W. Australia 92/A2
Austria 2/K4
Austria 7/F4
AUSTRIA 41
Austwell, Texas (77950) 303/H9
Autauga (co.), Ala. 195/E5
Autaugaville, Ala. (36003) 195/E6
Autlán de Navarro, Mexico 150/G7
Au Train, Mich. (49806) 250/C2
Au Train (bay), Mich. 250/C2
Autreyville, Georgia (†31768) 217/E8
Autryville, N.C. (28318) 281/M4
Autun, France 28/E4
Auvelais, Belgium 27/F8
Auvergne, Ark. (†72112) 202/H2
Auvergne (mts.), France 28/E5
Auvergne (trad. prov.) France 29
Auvergne, North. Terr. 93/B3
Auvergne, Québec 172/F3
Auxerre, France 28/E4
Auxier, Ky. (41602) 237/R5
Auxonne, France 28/F4
Auxvasse, Mo. (65231) 261/J4
Auyuittuq Nat'l Park, N.W. Terrs. /M3
Auyuittuq Nat'l Park, Que. 162/K2
Ava, Ill. (62907) 222/D6
Ava, Mo. (65608) 261/G9

Ava, N.Y. (13303) 276/K4
Ava, Ohio (43711) 284/G6
Avallon, Nova Scotia 168/D3
Avalon 28/E4
Avalon, Calif. (90704) 204/G10
Avalon, Miss. (38912) 256/D3
Avalon, N.J. (08202) 273/D5
Avalon (pen.), Newf. 166/D2
Avalon, N.J. (08202) 273/D5
Avalon (res.), N. Mex. 274/E6
Avalon, Pa. (15202) 294/B6
Avanos, Turkey 63/F3
Avans, Georgia (†30752) 217/A1
Avant, Okla. (74001) 288/O2
Avard, Okla. (†73860) 288/J1
Avaré, Brazil 132/D8
Avaré, Brazil 135/B3
Avarua (cap.), Cook Is. 87/L8
Avayalik (isls.), Newf. 166/B1
Avaz, Iran 66/M4
Aveiro (dist.), Portugal 33/B2
Aveiro, Portugal 33/B2
Avej, Iran 66/F3
Avella, Pa. (15312) 294/B5
Avellaneda, Argentina 143/G7
Avellino (prov.), Italy 34/E4
Avellino, Italy 34/E4
Avenal, Calif. (93204) 204/E8
Avenches, Switzerland 39/D3
Avenel, N.J. (07001) 273/E2
Aventon, N.C. (†27891) 281/O2
Avera, Georgia (30803) 217/G4
Avera, Miss. (†39456) 256/D8
Averías, Uruguay 145/B4
Averill, Mich. (†48640) 250/E5
Averill, Minn. (†56547) 255/B4
Averill○, Vt. (05901) 268/D2
Aversa, Italy 34/E4
Avery, Idaho (83802) 220/C2
Avery, Iowa (†52531) 229/H6
Avery (co.), N.C. 281/F2
Avery, Ohio (†44846) 284/E3
Avery, Okla. (†74023) 288/N3
Avery, Texas (75554) 303/K4
Avery Island, La. (70513) 238/G7
Aves (Bird) (isl.), Venezuela 156/F4
Avesnes-sur-Helpe, France 28/E3
Avesta, Sweden 18/J6
Aveyron (dept.), France 28/E5
Avezzano, Italy 34/D3
Aviemore, Scotland 15/E3
Avigliano, Italy 34/E4
Avignon, France 28/F6
Avignon, France 7/F4
Avihayil, Israel 65/B3
Ávila (prov.), Spain 33/D2
Ávila de los Caballeros, Spain 33/D2
Avilés, Spain 33/C1
Avilla, Ind. (46710) 227/G2
Avilla, Mo. (64833) 261/D8
Avinger, Texas (75630) 303/K5
Avion, France 28/E2
Avis, Pa. (17721) 294/H3
Avis, Portugal 33/C3
Aviston, Ill. (62216) 222/D5
Avize, France 28/E3
Avlum, Denmark 21/B5
Avoca, Ark. (72711) 202/B1
Avoca, Ind. (†47598) 227/C8
Avoca, Iowa (51521) 229/C6
Avoca, Ireland 17/J6
Avoca, Mich. (48006) 250/G5
Avoca, Minn. (56114) 255/C7
Avoca, Nebr. (68307) 264/H4
Avoca, N.Y. (14809) 276/F6
Avoca, Pa. (18641) 294/F7
Avoca, Tasmania 99/D3
Avoca, Texas (79503) 303/E5
Avoca, Victoria 97/B5
Avoca (riv.), Victoria 97/B5
Avoca, Wis. (53506) 317/F9
Avoch, Scotland 15/D3
Avola, Br. Col. 184/H4
Avola, Italy 34/E6
Avon, Ala. (†36312) 195/H8
Avon, Colo. (81620) 208/F3
Avon, Conn. (06001) 210/D1
Avon○, Conn. (06001) 210/D1
Avon (co.), England 13/E6
Avon (riv.), England 13/F7
Avon (riv.), England 13/F5
Avon (riv.), England 13/F5
Avon, Idaho (†83823) 220/B3
Avon, Ill. (61415) 222/C3
Avon○, Mass. (02322) 249/K4
Avon, Minn. (56310) 255/D5
Avon, Miss. (38723) 256/B4
Avon, Mont. (59713) 262/D4
Avon, N.Y. (14414) 276/E5
Avon, N.C. (27915) 281/U4
Avon (riv.), Nova Scotia 168/D4
Avon, Ohio (44011) 284/F3
Avon (riv.), Scotland 15/C1
Avon (riv.), Scotland 15/E3
Avon, S. Dak. (57315) 298/N8
Avon (riv.), W. Australia 88/B6
Avon (riv.), W. Australia 92/A1
Avon, Wis. (53530) 317/H10
Avonby The Sea, N.J. (07717) 273/E4
Avondale, Ariz. (85323) 198/E5
Avondale, Colo. (81022) 208/L6
Avondale, Mich. (†49631) 250/D4
Avondale, Mo. (64010) 261/P5
Avondale, N.S. Wales 97/F4
Avondale, Pa. (19311) 294/K6
Avondale, N.C. (128076) 281/F4
Avondale, W. Va. (24811) 312/C8
Avondale Estates, Georgia (30002) 217/L1
Avon Downs, North. Terr. 88/F4
Avon Downs, North. Terr. 93/E5
Avonhurst, Sask. 181/G5
Avon Lake, Ohio (44012) 284/F2
Avonlea, Sask. 181/G5
Avonmore, Ontario 177/K2
Avonmore, Pa. (15618) 294/C4

Avon Park, Fla. (33825) 212/E4
Avonport, Nova Scotia 168/D3
Avon Water (riv.), Scotland 15/D5
Avoyelles (par.), La. 238/G4
Avranches, France 28/C3
Awa (isl.), Japan 81/J4
Awaji, Japan 81/H8
Awaji (isl.), Japan 81/H8
Awanui, N. Zealand 100/D1
Awareh, Ethiopia 111/H6
Awarua (bay), N. Zealand 100/A6
Awash, Ethiopia 111/H6
Awash (riv.), Ethiopia 111/H5
Awaso, Ghana 106/D7
Awat, China 77/A3
Awatere (riv.), N. Zealand 100/D5
Awbeg (riv.), Ireland 17/D7
Awe, Loch (lake), Scotland 10/D2
Awe, Loch (lake), Scotland 15/C4
Aweil, Sudan 111/E6
Awendaw, S.C. (29429) 296/H5
Awosting, N.J. (†07421) 273/E1
Axe Edge (mt.), England 13/H2
Axel, Netherlands 27/D6
Axel Heiberg (isl.), Canada 4/A14
Axel Heiberg (isl.), N.W.T. 146/J1
Axel Heiberg (isl.), N.W. Terrs. 187/J2
Axel Heiburg (isl.), N.W.T. 162/N3
Axim, Ghana 106/D8
Axis, Ala. (36505) 195/B9
Ax-les-Thermes, France 28/D6
Axminster, England 13/D7
Axminster, England 10/E5
Axochiapan, Mexico 150/M2
Axson, Georgia (31624) 217/G8
Axtell, Kansas (66403) 232/F2
Axtell, Nebr. (68924) 264/F4
Axtell, Utah (84621) 304/C4
Axton, Va. (24054) 307/J7
Axum (Aksum), Ethiopia 111/G5
Ayabaca, Peru 128/C5
Ayabe, Japan 81/G6
Ayacucho, Argentina 143/E4
Ayacucho, Bolivia 136/D5
Ayacucho (dept.), Peru 128/E9
Ayacucho, Peru 128/F9
Ayacucho, Peru 128/F9
Ayaguz, U.S.S.R. 54/K5
Ayaguz, U.S.S.R. 48/J5
Ayakkum Hu (lake), China 77/C4
Ayamonte, Spain 33/C4
Ayan, U.S.S.R. 54/P4
Ayan, U.S.S.R. 48/O4
Ayancık, Turkey 63/F1
Ayapel, Colombia 126/C3
Ayapel, Serranía de (mts.), Colombia 126/C4
Ayaş, Turkey 63/E2
Ayata, Bolivia 136/A4
Ayaviri, Peru 128/G10
Aybak, Afghanistan 68/B1
Aybak, Afghanistan 59/J2
Aybastı, Turkey 63/G2
Aycliffe, England 13/F3
Ayden, N.C. (28513) 281/P4
Aydın (prov.), Turkey 63/B4
Aydın, Turkey 59/A2
Aydın, Turkey 63/B4
Aydıncık, Turkey 63/E4
Aydlett, N.C. (27916) 281/T2
Aydyrlinskiy, U.S.S.R. 52/K4
Ayer, Mass. (*01432) 249/H1
Ayer○, Mass. (*01432) 249/H2
Ayer, Switzerland 39/E4
Ayer, Wash. (†99348) 310/G4
Ayers, Maine (†04666) 243/J6
Ayer's Cliff, Québec 172/E4
Ayers Rock, Mt. Olga Nat'l Park, North. Terr. 88/E5
Ayers Rock (mt.), North. Terr. 88/E5
Ayers Rock Nat'l Park, North. Terr. 93/B8
Ayersville, Ohio (†43512) 284/B3
Ayiá, Greece 45/G3
Áyion Óros (aut. dist.), Greece 45/G5
Áyios Evstrátios (isl.), Greece 45/G5
Áyios Kírikos, Greece 45/H7
Áyios Matthaíos, Greece 45/F6
Áyios Nikólaos, Greece 45/G8
Áyios Yeóryios (cape), Greece 45/G5
Aykhal, U.S.S.R. 48/M3
Aylen (lake), Ontario 177/G2
Aylesbury, England 13/G7
Aylesbury, England 10/F5
Aylesbury, Sask. 181/G5
Aylesford, England 13/J8
Aylesford, Nova Scotia 168/D3
Aylett, Va. (23009) 307/O5
Ayllón, Spain 33/E2
Aylmer, N. Dak. (†58710) 282/K4
Aylmer (lake), N.W. Terrs. 187/H3
Aylmer, Ontario 177/C5
Aylmer, Québec 172/B4
Aylmer (lake), Québec 172/F4
Aylsham, England 13/J5
Aylsham, Sask. 181/H2
Aynor, S.C. (29511) 296/J3
Ayod, Sudan 111/F6
Ayolas, Paraguay 144/D5
Ayon (isl.), U.S.S.R. 48/R2
Ayora, Spain 33/F3
Ayr, Nebr. (68925) 264/F4
Ayr, N. Dak. (58007) 282/P5
Ayr, Ontario 177/C5
Ayr, Queensland 95/J4
Ayr, Queensland 88/H3
Ayr, Scotland 15/D5
Ayr, Scotland 10/D3
Ayr, Heads of (cape), Scotland 15/D5
Ayr (trad. co.) Scotland 15/A5
Ayr (riv.), Scotland 15/D5
Ayranci, Turkey 63/E4
Ayre (pt.), I. of Man 13/C3
Ayre (pt.), I. of Man 10/D3
Ayrshire, Iowa (50515) 229/D2
Ayton, Ontario 177/D3
Ayton, Scotland 15/F5

Aytos, Bulgaria 45/H4
Ayu (isls.), Indonesia 85/J5
Ayutla de los Libres, Mexico 150/K8
Ayutthaya (Phra Nakhon Si Ayutthaya), Thailand 72/D4
Ayvacık, Turkey 63/B3
Ayvalık, Turkey 59/A2
Ayvalık, Turkey 63/B3
Aywaille, Belgium 27/H8
Azalea, Oreg. (97410) 291/D5
Azalea Park, Fla. (32807) 212/E3
Azalia, Mich. (48110) 250/F6
Azamgarh, India 68/E3
Azángaro, Peru 128/H10
Azángaro (riv.), Peru 128/G10
Azaoua (reg.), Niger 106/F5
Azaouad (reg.), Mali 106/D5
Azaouak (reg.), Mali 106/E5
Azapa, Chile 138/A1
Azapa, Quebrada (riv.), Chile 138/B1
Azare, Nigeria 106/G6
Azaz, Syria 63/G4
Azbine (Air) (mts.), Niger 106/F5
Azcapotzalco, Mexico 150/L1
Azdavay, Turkey 63/E2
Azemmour, Morocco 106/C2
Azerbaidzhan S.S.R., U.S.S.R. 7/J4
Azerbaidzhan S.S.R., U.S.S.R. 48/E5
Azerbaidzhan S.S.R., U.S.S.R. 52/G6
Azerbaijan, East (prov.), Iran 66/E1
Azerbaijan, West (prov.), Iran 66/D1
Azerbaijan (reg.), Iran 66/D1
Aziscoos (lake), Maine 243/A5
Azle, Texas (76020) 303/E2
Azogues, Ecuador 128/C4
AZORES 33
Azores (isls.), Portugal 2/H4
Azores (isls.), Portugal 33/A2
Azoum, Bahr, Chad 111/C5
Azov, U.S.S.R. 7/H4
Azov, U.S.S.R. 52/E5
Azov (sea), U.S.S.R. 52/E5
Azov (sea), U.S.S.R. 48/D5
Azoyú, Mexico 150/K8
Azpeitia, Spain 33/E1
Azrou, Morocco 106/C2
Aztec, Ariz. (†85333) 198/B6
Aztec, N. Mex. (87410) 274/B2
Aztec Ruins Nat'l Mon., N. Mex. 274/A2
Azua (prov.), Dom. Rep. 158/D6
Azua, Dom. Rep. 156/D3
Azua, Dom. Rep. 158/D6
Azuaga, Spain 33/D3
Azuara, Spain 33/F2
Azuay (prov.), Ecuador 128/C4
Azuero (pen.), Panama 154/G7
Azul, Argentina 143/E4
Azul, Argentina 143/E4
Azul (riv.), Guatemala 154/C2
Azul, Cordillera (mts.), Peru 128/E7
Azurduy, Bolivia 136/D6
Azure (lake), Br. Col. 184/G4
Azusa, Calif. (91702) 204/D10
Azwell, Wash. (†98846) 310/F3
Azzel Mati, Sebkha (lake), Algeria 106/E3
Az Zubair, Iraq 66/E5

B

Ba, Fiji 86/P10
Baa, Indonesia 85/G8
Baaba (isl.), New Caled. 86/G4
Ba'albek, Lebanon 63/G5
Baan Baa, N.S. Wales 97/E2
Baar, Switzerland 39/F2
Baarle-Nassau, Netherlands 27/F6
Baarn, Netherlands 27/G4
Baatsagaan, Mongolia 77/E2
Baba, Ecuador 128/C3
Baba (cape), Turkey 63/D2
Baba (cape), Turkey 63/A3
Babadag, Romania 45/J3
Babadağ, Turkey 63/C4
Babaeski, Turkey 63/B2
Babahoyo, Ecuador 128/C3
Babanusa, Sudan 111/E5
Babar (isl.), Indonesia 85/H7
Babar (isls.), Indonesia 85/H7
Babati, Tanzania 115/G4
Babayevo, U.S.S.R. 52/E3
Babb, Mont. (59411) 262/C2
Babbie, Ala. (†36420) 195/F8
Babbitt, Minn. (55706) 255/G3
Babbitt, Nev. (89416) 266/C4
Babcock, Wis. (54413) 317/F7
Babel, Iran 59/F2
Babel (isl.), Tasmania 99/E1
Bab el Mandeb (str.) 102/G3
Bab el Mandeb (str.), Djibouti 111/H5
Babia, Mexico 150/J2
Babine (lake), Br. Col. 162/D5
Babine, Br. Col. 184/E3
Babine (lake), Br. Col. 184/E3
Babine (riv.), Br. Col. 184/D3
Babo, Indonesia 85/K7
Babol, Iran 66/G6
Babol, Iran 59/F2
Babol Sar, Iran 66/H2
Baboquivari (mts.), Ariz. 198/D7
Baboua, Cent. Afr. Rep. 115/C2
Babson Park, Fla. (33827) 212/E4
Babuyan (isls.), Philippines 54/O8
Babuyan (chan.), Philippines 82/A3
Babuyan (isl.), Philippines 82/B2
Babuyan (isl.), Philippines 85/G2
Babuyan (isl.), Philippines 82/A2
Babylon (ruins), Iraq 66/D4
Babylon, N.Y. (*11702) 276/O9
Baca (co.), Colo. 208/O8

Bacabal, Brazil 120/E3
Bacabal, Maranhão, Brazil 132/E4
Bacabal, Pará, Brazil 132/B4
Bacadéhuachi, Mexico 150/E2
Bacalar, Mexico 150/P7
Bacalar (lake), Mexico 150/P7
Bacan (isls.), Indonesia 85/H6
Bacanora, Mexico 150/E2
Bacarra, Philippines 82/C1
Bacău, Romania 7/G3
Bacău, Romania 45/H2
Baccalieu (isl.), Newf. 166/D2
Bac Can, Vietnam 72/E2
Baccaro (pt.), Nova Scotia 168/C5
Bacchus Marsh, Victoria 97/C5
Bacerac, Mexico 150/E1
Bac Giang, Vietnam 72/E2
Bach, Mich. (†48759) 250/F5
Bachaquero, Venezuela 124/C3
Bache (pen.), N.W. Terrs. 187/L2
Bache, Okla. (74526) 288/P5
Bachelor (brook), Mass. 249/D3
Bachíniva, Mexico 150/F2
Bach Long Vi, Dao (isl.), Vietnam 72/F2
Bachu (Maralwexi), China 77/A4
Back (bay), India 68/B7
Back (riv.), Md. 245/N4
Back (lake), N.H. 268/E1
Back (riv.), N.W.T. 146/H3
Back (riv.), N.W.T. 162/J2
Back (riv.), N.W. Terrs. 187/J3
Back (bay), Va. 307/S7
Back (creek), Va. 307/S7
Bačka Topola, Yugoslavia 45/D3
Back Bay, New Bruns. 170/C3
Backbone (mts.), Md. 245/A3
Backnang, W. Germany 22/C4
Backoo, N. Dak. (58215) 282/P2
Backus, Minn. (56435) 255/D4
Backway, The (inlet), Newf. 166/C2
Bac Lieu, Vietnam 72/E5
Bacliff, Texas (77518) 303/K2
Bac Ninh, Vietnam 72/E2
Baco (mt.), Philippines 82/C4
Bacolod, Philippines 85/G3
Bacolod, Philippines 54/O8
Bacolod, Philippines 82/C4
Bacon (co.), Georgia 217/G7
Bacone, Okla. (†74401) 288/R3
Bacon Ridge (mts.), Wyo. 319/B2
Bacons, Del. (†19940) 245/R6
Baconton, Georgia (31716) 217/D8
Bácsalmás, Hungary 41/E3
Bács-Kiskun (co.), Hungary 41/E3
Bacuna, Neth. Ant. 161/E8
Bacup, England 13/H1
Bacup, England 10/G1
Bad (riv.), Mich. 250/E5
Bad (hills), Sask. 181/C4
Bad (lake), Sask. 181/C4
Bad (riv.), S. Dak. 298/G5
Badacsonytomaj, Hungary 41/D3
Badagara, India 68/D6
Bad Aibling, W. Germany 22/D5
Badajoz (prov.), Spain 33/D3
Badajoz, Spain 33/D3
Badalona, Spain 33/H2
Bad Aussee, Austria 41/B3
Bad Axe, Mich. (48413) 250/G5
Bad Berleburg, W. Germany 22/C3
Bad Berneck, W. Germany 22/D3
Bad Bramstedt, W. Germany 22/C2
Bad Brückenau, W. Germany 22/C3
Baddeck, Nova Scotia 168/H2
Baddeck, Nova Scotia 168/H2
Bad Doberan, E. Germany 22/D1
Bad Driburg, W. Germany 22/C3
Bad Dürkheim, W. Germany 22/C4
Bad Dürrenberg, E. Germany 22/D3
Bad Ems, W. Germany 22/B3
Bad Gandersheim, W. Germany 22/D3
Bad Gastein, Austria 41/B3
Badger (peak), Idaho 220/E7
Badger, Iowa (50516) 229/E3
Badger, Minn. (56714) 255/B2
Badger, Newf. 166/C4
Badger (creek), Oreg. 291/H3
Badger, S. Dak. (57214) 298/P5
Badger (creek), Wyo. 319/E2
Badger's Quay, Newf. 166/D4
Bad Goisern, Austria 41/B3
Badham, S.C. (†29471) 296/F5
Bad Harzburg, W. Germany 22/D3
Bad Hersfeld, W. Germany 22/C3
Bad Hoevedorp, Netherlands 27/B5
Bad Hofgastein, Austria 41/B3
Bad Homburg vor der Höhe, W. Germany 22/C3
Bad Honnef, W. Germany 22/B3
Badian, Philippines 82/D6
Badin, N.C. (28009) 281/J4
Badin, Pakistan 68/B4
Badiraguato, Mexico 150/F4
Bad Ischl, Austria 41/B3
Bad Kissingen, W. Germany 22/C3
Bad Kreuznach, W. Germany 22/B4
Bad Land (butte), Utah 304/D4
Bad Lands (reg.), Idaho 220/E7
Bad Langensalza, E. Germany 22/D3
Bad Lauterberg im Harz, W. Germany 22/D3
Bad Leonfelden, Austria 41/C2

Bad Liebenwerda, E. Germany 22/E3
Bad Lippspringe, W. Germany 22/C3
Bad Mergentheim, W. Germany 22/C4
Bad Münster-Ebernburg, W. Germany 22/B4
Bad Münstereifel, W. Germany 22/B3
Bad Muskau, E. Germany 22/F3
Bad Nauheim, W. Germany 22/C3
Bad Neuenahr-Ahrweiler, W. Germany 22/B3
Bad Neustadt an der Saale, W. Germany 22/D3
Bado, Mo. (†65447) 261/H8
Bad Oldesloe, W. Germany 22/D2
Ba Don, Vietnam 72/E3
Bad Orb, W. Germany 22/C3
Bad Pyrmont, W. Germany 22/C3
Badr, Saudi Arabia 59/C5
Badra, Iraq 66/D4
Bad Ragaz, Switzerland 39/H2
Bad Reichenhall, W. Germany 22/E5
Bad River Ind. Res., Wis. 317/E2
Bad Sachsa, W. Germany 22/D3
Bad Salzschlirf, W. Germany 22/C3
Bad Salzuflen, W. Germany 22/C3
Bad Salzungen, E. Germany 22/D3
Bad Sankt-Leonhard im Lavanttal, Austria 41/C3
Bad Schwartau, W. Germany 22/D2
Bad Segeberg, W. Germany 22/D2
Bad Tölz, W. Germany 22/D5
Baduen, Somalia 115/J2
Badulla, Sri Lanka 68/E7
Bad Vilbel, W. Germany 22/C3
Bad Waldsee, W. Germany 22/C5
Badwater (creek), Wyo. 319/E2
Bad Wildungen, W. Germany 22/C3
Bad Wimpfen, W. Germany 22/C4
Baelum, Denmark 21/D4
Baena, Spain 33/D4
Baerle-Hertog, Belgium 27/F6
Báez, Cuba 158/G2
Baeza, Ecuador 128/D3
Baeza, Spain 33/E4
Bafa (lake), Turkey 63/B4
Bafang, Cameroon 115/B3
Bafatá, Guinea-Biss. 106/B6
Baffin (bay) 146/M2
Baffin (bay), Canada 2/F2
Baffin (isl.), Canada 2/F2
Baffin (bay), Canada 4/C13
Baffin (bay), N.W.T. 146/L2
Baffin (isl.), N.W.T. 146/L3
Baffin (isl.), N.W.T. 162/J1
Baffin (dist.), N.W. Terrs. 187/K2
Baffin (bay), N.W. Terrs. 187/M2
Baffin (isl.), N.W. Terrs. 187/L2
Baffin (bay), Texas 303/G10
Bafia, Cameroon 115/B3
Bafing (riv.), Guinea 106/B6
Bafing (riv.), Mali 106/B6
Bafoulabé, Mali 106/B6
Bafoussam, Cameroon 115/B2
Bafq, Iran 59/G3
Bafq, Iran 66/J5
Bafra, Turkey 59/C1
Bafra, Turkey 63/F2
Bafra (cape), Turkey 59/C1
Bafra (cape), Turkey 63/G2
Baft, Iran 66/K6
Baft, Iran 59/G4
Baga, Nigeria 106/G6
Bagabag, Philippines 82/C2
Bagac, Philippines 82/C3
Bagaces, C. Rica 154/E5
Bagadó, Colombia 126/B5
Bagalkot, India 68/D5
Bagam (well), Niger 106/F5
Bagamoyo, Tanzania 115/G5
Baganga, Philippines 82/F7
Baganian (pen.), Philippines 82/D7
Bagansiapiapi, Indonesia 85/C5
Bagata, Zaire 115/C4
Bagdad, Ariz. (86321) 198/B4
Bagdad, Fla. (32530) 212/B8
Bagdad, Ky. (40003) 237/L4
Bagdad (cap.), Iraq 59/E3
Bagdad (cap.), Iraq 54/P8
Bagdad (cap.), Iraq 2/M4
Bagdad (cap.), Iraq 66/D4
Bagdarin, U.S.S.R. 48/M4
Bagé, Brazil 120/D6
Bagé, Brazil 132/C10
Bagenalstown, Ireland 10/C4
Baile Átha Cliath (Dublin) (cap.), Ireland 17/K5
Baile Átha Cliath (Dublin) (cap.), Ireland 10/C4
Băile Herculane, Romania 45/F3
Bailén, Spain 33/E3
Băileşti, Romania 45/F3
Bagenkop, Denmark 21/D8
Baggs, Wyo. (82321) 319/E4
Baghbaghu, Iran 66/M3
Baghdad (heads), Iraq 66/D4
Baghdad (cap.), Iraq 59/E3
Baghdad (cap.), Iraq 54/P8
Baghdad (cap.), Iraq 2/M4
Baghdad (cap.), Iraq 66/D4
Bagheria, Italy 34/D5
Baghlan, Afghanistan 54/H6
Baghlan (prov.), Afghanistan 9/J2
Baghlan, Afghanistan 68/B1
Baghu, Iran 66/H5
Bağırpaşa Daği (mt.), Turkey 59/D2
Bağırpaşa Daği (mt.), Turkey 63/J3
Bagley, Iowa (50026) 229/E5
Bagley, Minn. (56621) 255/C3
Bagley, N.C. (†27542) 281/N3
Bagley, Wis. (53801) 317/D10
Bagnell (dam), Mo. 261/G6
Bagnell, Mo. (†65026) 261/G6
Bagnères-de-Bigorre, France 28/D6
Bagnères-de-Luchon, France 28/D6
Bagnolet, France 28/B2
Bagnols-sur-Cèze, France 28/F5
Bågø (isl.), Denmark 21/C7
Bago, Philippines 82/D5
Bagoé (riv.), Ivory Coast 106/C6
Bagoé (riv.), Mali 106/C6
Bagot, Manitoba 179/D5
Bagot (co.), Québec 172/E4
Bagrax (Bosten Hu) (lake), China 77/C3
Bagua, Peru 128/C5
Báguanos, Cuba 158/J3
Baguio, Philippines 54/N8

Baguio, Philippines 85/G2
Baguio, Philippines 82/C2
Baguirmi (reg.), Chad 111/C5
Bagwell, Texas (75412) 303/J4
Bagzane, Mont. N.C. (27503) 281/M2
Bahamas 2/F4
Bahamas 146/L3
BAHAMAS 156/C1
Bahariya (oasis), Egypt 111/E2
Bahariya (oasis), Egypt 59/A4
Bahawalnagar, Pakistan 68/C2
Bahawalpur, Pakistan 54/J7
Bahawalpur, Pakistan 68/C3
Bahawalpur, Pakistan 59/K4
Bahçe, Turkey 63/G4
Bahçesaray, Turkey 63/K3
Bahia (state), Brazil 132/F6
Bahía (Salvador), Brazil 132/G6
Bahía (isls.), Honduras 154/D2
Bahía Blanca, Argentina 2/F7
Bahía Blanca, Argentina 143/D4
Bahía Blanca, Argentina 120/C6
Bahía Bustamante, Argentina 143/C6
Bahía Honda, Cuba 158/B1
Bahía de Caráquez, Ecuador 128/B3
Bahía Kino, Mexico 150/C2
Bahía San Blas, Argentina 143/D5
Bahía Thetis, Argentina 143/C7
Bahía Tortugas, Mexico 150/B3
Bahir Dar, Ethiopia 111/G5
Bahomamey, P. Rico 161/A1
Bahoruco (prov.), Dom. Rep. 158/D6
Bahoruco, Sierra de (mts.), Dom. Rep. 158/D6
Bahraich, India 68/E3
BAHRAIN 59/F4
Bahrain 54/F7
Bahramabad (Rafsanjan), Iran 66/K5
Bahr el 'Arab (riv.), Sudan 111/D5
Bahr el Ghazal (dry riv.), Chad 111/C5
Bahr El Ghazal (prov.), Sudan 111/E6
Bahr es Safi (des.), Saudi Arabia 59/E6
Bahr ez Zeraf (riv.), Sudan 111/F6
Bahr Yusef (stream), Egypt 111/J4
Baia de Aramă, Romania 45/F3
Baia dos Tigres, Angola 115/B7
Baia Farta, Angola 115/B6
Baia Mare, Romania 45/F2
Baião, Brazil 132/B4
Baibiene, Argentina 143/D4
Baibokoum, Chad 111/C6
Bai Bung, Mui (Ca Mau) (pt.), Vietnam 72/E5
Baicheng (Bay), Xinjiang Uygur, China 77/B3
Baicheng, Jilin, China 77/K2
Baida, Libya 102/E1
Baida, Libya 111/D1
Baie-Comeau, Québec 172/A1
Baie-Comeau, Québec 174/D3
Baie de Henne, Haiti 158/D5
Baie-des-Bacons, Québec 172/H1
Baie-des-Moutons, Québec 174/F2
Baie-des-Rochers, Québec 172/H2
Baie-des-Sables, Québec 172/A1
Baie-du-Poste, Québec 174/C2
Baie-du-Vieux-Fort, Québec 174/F2
Baie-Johan-Beetz, Québec 174/E2
Baie-Mahault, Guadeloupe 161/A6
Baie-Sainte-Anne, New Bruns. 170/F1
Baie-Sainte-Catherine, Québec 172/H1
Baie-Saint-Paul, Que. 162/J6
Baie-Saint-Paul, Québec 174/C3
Baie-Trinité, Québec 172/B1
Baie-Verte, New Bruns. 170/F2
Baie Verte, Newf. 166/C4
Baieville, Québec 172/E3
Baigorrita, Argentina 143/F7
Baiji, Iraq 66/C3
Bailadon, Sask. 181/F5
Baileyboro, Texas (†79371) 303/B3
Bailey, Colo. (80421) 208/H4
Bailey, Iowa (†50455) 229/H2
Bailey, Mich. (49303) 250/D5
Bailey, Miss. (39320) 256/G6
Bailey, N.C. (27807) 281/N3
Bailey (co.), Texas 303/B3
Baileyboro, Texas (†79371) 303/B3
Bailey Island, Maine (04003) 243/D8
Bailey Lakes, Ohio (†44805) 284/F4
Bailey's Crossroads, Va. (22041) 307/S3
Baileys Harbor, Wis. (54202) 317/M5
Baileyton, Ala. (35019) 195/E2
Baileyton, Tenn. (37743) 237/R8
Baileyville, Conn. (†06455) 210/E2
Baileyville, Ill. (61007) 222/D1
Baileyville, Kansas (66404) 232/F2
Bailieborough, Ireland 17/G4
Bailique (isl.), Brazil 132/C3
Bailivanish, Scotland 15/A3
Baillie (isls.), N.W. Terrs. 187/F2
Baillieston, Scotland 15/B2
Baillif, Guadeloupe 161/A7
Bailundo, Angola 115/B6
Baima, China 77/H5
Bainbridge (isl.), Alaska 196/C1
Bainbridge, Georgia (31717) 217/C9
Bainbridge, Ind. (46105) 227/D5
Bainbridge, N.Y. (13733) 276/J6
Bainbridge (dist.), N. Ireland 17/J2
Bainbridge, Ohio (45612) 284/D7
Bainbridge, Pa. (17502) 294/J5
Bainbridge (isl.), Wash. 310/A2
Bainbridge Island-Winslow (Winslow), Wash. (98110) 310/A2

Bainbridge N.T.C., Md. 245/O2
Bainet, Haiti 158/E6
Baingoin, China 77/D5
Bains, La. (70713) 238/H5
Bainville, Mont. (59212) 262/M2
Baird (inlet), Alaska 196/C1
Baird (mts.), Alaska 196/F1
Baird (pen.), N.W. Terrs. 187/L3
Baird, Texas (79504) 303/E5
Bairdstown, Ohio (†45872) 284/C3
Bairdsville, New Bruns. 170/C2
Baire, Cuba 158/H4
Bairiki (cap.), Kiribati 87/H5
Bairin Zuoqi, China 77/J3
Bairnsdale, Victoria 88/H7
Bairnsdale, Victoria 97/D5
Bairoil, Wyo. (82322) 319/E3
Bais, Philippines 82/D6
Baisden, W. Va. (25608) 312/C7
Baïse (riv.), France 28/D6
Baisha, China 77/G8
Baïtadi, Nepal 68/E3
Bait al Faqih, Yemen Arab Rep. 59/D7
Bai Thuong, Vietnam 72/E3
Baixa da Banheira, Portugal 33/B3
Baixoaixo (isl.), Portugal 33/B2
Baixo Guandu, Brazil 132/F7
Baja, Hungary 41/E3
Baja California (state), Mexico 150/B1
Baja California Sur (state), Mexico 150/C3
Bajadero, P. Rico 161/C1
Bajgiran, Iran 66/L2
Bajo Boquete, Panama 154/F6
Bajo Nuevo (shoal), Colombia 126/C8
Bajos de Haina, Dom. Rep. 158/E6
Bajram Curri, Albania 45/D4
Bakala, Cent. Afr. Rep. 115/D2
Bakar, Yugoslavia 45/B3
Bakel, Senegal 106/B6
Baker (isl.), Alaska 196/M2
Baker (isl.), Calif. 146/D4
Baker, Calif. (92309) 204/J8
Baker (riv.), Chile 138/D7
Baker (mt.), Colo. 208/H2
Baker (co.), Fla. 212/D1
Baker, Fla. (32531) 212/C5
Baker (co.), Georgia 217/D8
Baker, Idaho (†83467) 220/E4
Baker, La. (70714) 238/K1
Baker (lake), Maine 243/D4
Baker, Minn. (56513) 255/B4
Baker, Mo. (†63846) 261/N9
Baker, Mont. (59313) 262/M4
Baker, Nev. (89311) 266/G3
Baker, N. Dak. (†58386) 282/L3
Baker (lake), N.W. Terrs. 187/J3
Baker, Okla. (73930) 288/D1
Baker, Oreg. 188/G2
Baker (co.), Oreg. 291/K3
Baker, Oreg. (97814) 291/K3
Baker (isl.), Pacific 87/J5
Baker (creek), Utah 304/A4
Baker (lake), Wash. 310/D2
Baker (mt.), Wash. 310/D2
Baker (riv.), Wash. 310/D2
Baker, W. Va. (26801) 312/J4
Baker Brook, New Bruns. 170/B1
Baker Butte (mt.), Ariz. 198/D4
Baker Hill, Ala. (36004) 195/H7
Baker Lake, N.W.T. 162/G3
Baker Lake, N.W. Terrs. 187/J3
Bakers (isls.), Mass. 249/F5
Bakersfield, Calif. 146/D4
Bakersfield, Calif. 188/D3
Bakersfield, Mo. (65609) 261/H9
Bakersfield, Texas (†79752) 303/B7
Bakersfield○, Vt. (05441) 268/B2
Bakers Summit, Pa. (16614) 294/F5
Bakersville, Conn. (†06057) 210/C1
Bakersville, N.C. (28705) 281/E2
Bakersville, Ohio (43803) 284/G5
Bakersville, Pa. (15501) 294/D5
Bakerton, Ky. (42771) 237/L7
Bakerton, W. Va. (25410) 312/L4
Bakerville, Tenn. (†37185) 237/F9
Bakewell, England 10/G2
Bakewell, England 13/J2
Bakewell, Tenn. (37304) 237/L10
Bakharz, Kuhha-ye (mt.), Iran 66/M3
Bakhchisaray, U.S.S.R. 52/D6
Bakhmach, U.S.S.R. 52/D4
Bakhtegan (lake), Iran 66/J6
Bakhtiari (gov.), Iran 66/F4
Bakhun, Kuh-e (mt.), Iran 66/K6
Bakhuys (mts.), Suriname 131/C3
Bakia, Cent. Afr. Rep. 115/E2
Bakırköy, Turkey 63/B2
Baklan, Turkey 63/C4
Bako, Ethiopia 111/G6
Bakony (mts.), Hungary 41/D3
Bakool (prov.), Somalia 115/H3
Bakouma, Cent. Afr. Rep. 115/D2
Bakoy (riv.), Guinea 106/B6
Bakoy (riv.), Mali 106/B6
Bakraband, Kuh-e (mts.), Iran 66/M7
Baktalórántháza, Hungary 41/G2
Baktu (Paektu) (mt.), N. Korea 81/C3
Baku, U.S.S.R. 2/M3
Baku, U.S.S.R. 7/J4
Baku, U.S.S.R. 52/F4
Baku, U.S.S.R. 52/H6
Bakus, Kansas (†66531) 232/F2
Bala, Ontario 177/D2
Bală, Turkey 63/E3
Bala, Wales 13/D5
Bala, Wales 10/E4
Balabac, Philippines 82/A7
Balabac (isl.), Philippines 85/F4
Balabac (isl.), Philippines 82/A7
Balabac (str.), Philippines 85/F4
Balabac (str.), Philippines 82/A7
Balabagan (isls.), Indonesia 85/F6
Balabio (isl.), New Caled. 86/G4
Balaclava, Jamaica 158/H6

Bala-Cynwyd, Pa. (19004) 294/N6
Balad, Somalia 115/J3
Balaghat, India 68/E4
Balaguer, Spain 33/G2
Balaïtous (mt.), Spain 33/F1
Balakai (mesa), Ariz. 198/F3
Balakhna, U.S.S.R. 52/F4
Balaklava, S. Australia 94/F6
Balaklava, U.S.S.R. 52/D6
Balakovo, U.S.S.R. 7/J3
Balakovo, U.S.S.R. 48/E4
Balakovo, U.S.S.R. 52/G4
Balallan, Scotland 15/B2
Bal'ama, Jordan 65/E3
Balambangan (isl.), Malaysia 85/F4
Balancán de Domínguez, Mexico 150/O8
Balandra (pt.), Dom. Rep. 158/G2
Balanga, Philippines 82/C3
Balangala, Zaire 115/D3
Balangiga, Philippines 82/E5
Ba Lang An, Mui (cape), Vietnam 72/F4
Balao, Ecuador 128/C4
Balasore, India 68/F4
Balassagyarmat, Hungary 41/E2
Balaton (lake), Hungary 7/F4
Balaton (lake), Hungary 41/D3
Balaton, Minn. (56115) 255/C6
Balatonfüred, Hungary 41/D3
Balatonszentgyörgy, Hungary 41/D3
Balayan (bay), Philippines 82/C4
Balbi (mt.), Papua N.G. 86/C2
Balboa, Panama 154/H6
Balboa Heights, Panama 154/H6
Balbriggan, Ireland 17/J4
Balbriggan, Ireland 10/C4
Balcarce, Argentina 143/E4
Balcarres, Sask. 181/H5
Balchik, Bulgaria 45/H4
Balch Springs, Texas (75180) 303/H2
Balclutha, N. Zealand 100/B7
Balcones Escarpment (plat.), Texas 303/E8
Balcones Heights, Texas (†78201) 303/J10
Bald (mt.), Colo. 208/H4
Bald (hill), Conn. 210/E4
Bald (mt.), Idaho 220/D5
Bald (mt.), New Bruns. 170/C1
Bald (mts.), N.C. 281/D3
Bald (mts.), Tenn. 237/R9
Bald (mt.), Utah 304/C3
Bald (mt.), Vt. 268/C2
Bald (head), W. Australia 88/B7
Bald (head), W. Australia 92/B6
Bald Eagle (lake), Minn. 255/S6
Baldeggersee (lake), Switzerland 39/F2
Baldhill (Ashtabula) (res.), N. Dak. 282/P5
Bald Hill Branch (riv.), Md. 245/G4
Bald Hills, Queensland 88/K2
Bald Knob, Ark. (72010) 202/G3
Bald Knob, W. Va. (25010) 312/C7
Baldonnel, Br. Col. 184/G2
Baldur, Manitoba 179/C5
Baldwin (riv.), Ala. 195/C9
Baldwin, Fla. (32234) 212/E1
Baldwin (co.), Georgia 217/F4
Baldwin, Georgia (30511) 217/E2
Baldwin, Ill. (62217) 222/D5
Baldwin, Iowa (52207) 229/M4
Baldwin, La. (70514) 238/H7
Baldwin, Mich. (49304) 250/D5
Baldwin, N.Y. (11510) 276/R7
Baldwin, N. Dak. (58521) 282/J5
Baldwin, Pa. (†15208) 294/B7
Baldwin, Va. (26326) 312/E5
Baldwin, Wis. (54002) 317/B6
Baldwin-Aragon Mills, S.C. (†29706) 296/E4
Baldwin City, Kansas (66006) 232/G3
Baldwin Park, Calif. (91706) 204/D10
Baldwinsville, N.Y. (13027) 276/H4
Baldwinton, Sask. 181/B3
Baldwinville, Mass. (01436) 249/F2
Baldwyn, Miss. (38824) 256/G2
Baldy (peak), Ariz. 198/F5
Baldy (mt.), Manitoba 179/B3
Baldy (peak), N. Mex. 274/D2
Baldy (peak), Utah 304/B5
Bale (prov.), Ethiopia 111/H6
Bale (mt.), Ethiopia 111/G6
Baleares (prov.), Spain 33/H3
Balearic (isls.), Spain 7/E5
Balearic (Baleares) (isls.), Spain 33/H3
Baleine, Grande R. de la (riv.), Que. 162/J4
Baleine, Grand Rivière de la (riv.), Québec 174/D1
Baleine, Petite Rivière de la (riv.), Québec 174/D1
Baleine (riv.), Québec 174/D1
Baleine, R. à la (riv.), Que. 162/K4
Balen, Belgium 27/G6
Baler, Philippines 82/C3
Baler (bay), Philippines 82/C3
Balerna, Switzerland 39/G5
Balerno, Scotland 15/D2
Baleshare (isl.), Scotland 15/A3
Balestrand, Norway 18/E6
Baley, U.S.S.R. 48/M4
Balfate, Honduras 154/D3
Balfour, Br. Col. 184/J5
Balfour, N.C. (28706) 281/E4
Balfron, Scotland 15/B3
Balgonie, Sask. 181/G5
Balhaf, P.D.R. Yemen 59/E7
Bal Harbour, Fla. (33101) 212/C4
Bali, Cameroon 115/A2
Bali (isl.), Indonesia 54/N10
Bali (isl.), Indonesia 85/F7
Bali (sea), Indonesia 85/F7
Bali (str.), Indonesia 85/E7

Baliangao, Philippines 82/D6
Balicuatro (isls.), Philippines 82/E4
Balige, Indonesia 85/B5
Balıkesir (prov.), Turkey 63/B3
Balıkesir, Turkey 63/B3
Balıkesir, Turkey 59/A2
Balikpapan, Indonesia 54/N10
Balikpapan, Indonesia 85/F6
Balık-Uzun (lake), Turkey 63/G2
Balimbing (Bato-Bato), Philippines 82/C8
Baling, Malaysia 72/D6
Balingasag, Philippines 82/E6
Balingen, W. Germany 22/C4
Balintang (chan.), Philippines 82/A2
Balintang (isls.), Philippines 82/A2
Baljennie, Sask. 181/C3
Balk, Netherlands 27/H3
Balkan (mts.) 7/G4
Balkan (mts.), Bulgaria 45/G4
Balkan, Ky. (40804) 237/O7
Balkány, Hungary 41/G3
Balkbrug, Netherlands 27/J3
Balkh, Afghanistan 68/B1
Balkh, Afghanistan 59/J2
Balkhash, U.S.S.R. 54/J5
Balkhash (lake), U.S.S.R. 2/N3
Balkhash (lake), U.S.S.R. 54/J5
Balkhash, U.S.S.R. 48/H5
Balkhash (lake), U.S.S.R. 48/H5
Balko, Okla. (73931) 288/E1
Ball (mt.), Conn. 210/C1
Ball (pond), Conn. 210/A3
Ball, La. (71405) 238/F4
Ball (bay), Norfolk I. 88/L6
Balla, Ireland 17/C4
Balladonia, W. Australia 92/D6
Ballaghaderreen, Ireland 17/E4
Ballaigues, Switzerland 39/B3
Ballantine, Mont. (59006) 262/J5
Ballantrae, Scotland 15/C5
Ballantyne (str.), N.W. Terrs. 187/G2
Ballarat, Australia 87/E9
Ballarat, Victoria 88/G7
Ballarat, Victoria 97/C5
Ballard (co.), Ky. 237/C6
Ballard, Mo. (†64730) 261/D6
Ballard (cape), Newf. 166/D2
Ballard (lake), W. Australia 88/B5
Ballard, W. Va. (24918) 312/E8
Ballardsville, Miss. (†38801) 256/H2
Ballardvale, Mass. (01810) 249/K2
Ballater, Scotland 10/E2
Ballater, Scotland 15/F3
Ball Club, Minn. (†56636) 255/E3
Ballenas (bay), Mexico 150/C3
Ballenero (chan.), Chile 138/E11
Ballengee, W. Va. (†24981) 312/E7
Ballens, Switzerland 39/B3
Ballenstedt, E. Germany 22/D3
Ballentine, S.C. (29002) 296/E3
Balleny (isls.), Ant. 2/S9
Balleny (isls.) 5/C9
Ballerup, Denmark 21/F6
Ballesteros, Philippines 82/C1
Balleza, Mexico 150/F3
Ball Ground, Georgia (30107) 217/D2
Ballground, Miss. (†39156) 256/C5
Ballia, India 68/E3
Ballidu, W. Australia 92/B5
Ballina, Mayo, Ireland 17/C3
Ballina, Tipperary, Ireland 17/E6
Ballina, Ireland 10/B3
Ballina, N.S. Wales 97/G1
Ballinagh, Ireland 17/G4
Ballinakill, Ireland 17/G6
Ballinamore, Ireland 17/F3
Ballinasloe, Ireland 10/B4
Ballinasloe, Ireland 17/E5
Ballincollig-Carrigrohane, Ireland 17/D8
Ballindine, Ireland 17/C4
Ballineen, Ireland 17/D8
Ballingarry, Limerick, Ireland 17/D7
Ballingarry, Tipperary, Ireland 17/F6
Ballinger, Texas (76821) 303/E6
Ballingry, Scotland 15/D1
Ballinlough, Ireland 17/D4
Ballinluig, Scotland 15/E4
Ballinrobe, Ireland 10/B4
Ballinrobe, Ireland 17/C4
Ballinskelligs (bay), Ireland 17/A8
Ballintober, Ireland 17/E4
Ballintra, Ireland 17/E2
Ballisodare, Ireland 17/E3
Ballivor, Ireland 17/H4
Balloch, Highland, Scotland 15/D3
Balloch, Strathclyde, Scotland 15/B1
Ballouville, Conn. (06233) 210/H1
Ballston, Oreg. (†97378) 291/D2
Ballston Spa, N.Y. (12020) 276/N5
Ballsville, Va. (†29139) 307/M6
Balltown, Iowa (†52073) 229/M3
Ballville, Ohio (†43420) 284/D3
Ballwin, Mo. (63011) 261/N3
Bally, India 68/F1
Bally, Pa. (19503) 294/L5
Ballybay, Ireland 17/G3
Ballybofey-Stranorlar, Ireland 17/F2
Ballybunio, Ireland 10/B4
Ballybunion, Ireland 17/B7
Ballycanew, Ireland 17/J6
Ballycarney, Ireland 17/J6
Ballycarry, N. Ireland 17/K2
Ballycastle, Ireland 17/C3
Ballycastle, N. Ireland 10/C3
Ballycastle, N. Ireland 17/J1
Ballyclare, N. Ireland 17/J2
Ballyconnell, Ireland 17/F3
Ballycotton, Ireland 17/F8
Ballycotton (bay), Ireland 17/F8
Ballydehob, Ireland 17/C8
Ballyduff, Ireland 17/B7
Ballygally, N. Ireland 17/K2
Ballygar, Ireland 17/E4
Ballygawley, N. Ireland 17/G3
Ballygeary, Ireland 17/J7
Ballygrant, Scotland 15/B5
Ballyhaise, Ireland 17/G3
Ballyhaunis, Ireland 17/D4
Ballyheige (bay), Ireland 17/B7
Ballyheigue, Ireland 17/B7
Ballyhoura (hills), Ireland 17/E7
Ballyjamesduff, Ireland 17/G4
Ballykelly, N. Ireland 17/G1
Ballylanders, Ireland 17/E7
Ballylongford, Ireland 17/B6
Ballymahon, Ireland 17/F4
Ballymakeery, Ireland 17/C8
Ballymena (dist.), N. Ireland 17/J2
Ballymena, N. Ireland 17/J2
Ballymena, N. Ireland 10/C3
Ballymoney (dist.), N. Ireland 17/J1
Ballymoney, N. Ireland 10/C3
Ballymoney, N. Ireland 17/J1
Ballymore, Ireland 17/F5
Ballymore Eustace, Ireland 17/J5
Ballymote, Ireland 17/D3
Ballymote, Ireland 10/B3
Ballynahinch, N. Ireland 17/J3
Ballynakill (harb.), Ireland 17/A4
Ballyporeen, Ireland 17/E7
Ballyragget, Ireland 17/G6
Ballyroan, Ireland 17/G6
Ballysadare (bay), Ireland 17/D3
Ballyshannon, Ireland 10/B3
Ballyshannon, Ireland 17/E3
Ballyteige (bay), Ireland 17/H7
Ballytore, Ireland 17/H5
Ballywalter, N. Ireland 17/K2
Balmaceda, Chile 138/E6
Balmat, N.Y. (13609) 276/K2
Balmazújváros, Hungary 41/F3
Balmertown, Ontario 175/B2
Balmhorn (mt.), Switzerland 39/E4
Balmoral, Manitoba 179/E4
Balmoral, New Bruns. 170/D1
Balmoral, Queensland 88/K2
Balmoral, Queensland 95/E2
Balmoral, Victoria 97/A5
Balmoral Castle, Scotland 10/
Balmoral Castle, Scotland 15/E3
Balmorhea, Texas (79718) 303/D11
Balmville, N.Y. (†12550) 276/M7
Balnearia, Argentina 143/D3
Balneario El Tesoro, Uruguay 145/E5
Balneario La Barra, Uruguay 145/E5
Balneario Solís, Uruguay 145/D5
Balombo, Angola 115/B6
Balonne (riv.), Queensland 88/H5
Balonne (riv.), Queensland 95/D6
Balotra, India 68/C3
Baloy (mt.), Philippines 82/D5
Balpunga, N.S. Wales 97/A3
Balrampur, India 68/E3
Balranald, N.S. Wales 88/G6
Balranald, N.S. Wales 97/B4
Balș, Romania 45/G3
Balsam, N.C. (28707) 281/C4
Balsam (lake), Ontario 177/F3
Balsam Creek, Ontario 177/E1
Balsam Lake, Wis. (54810) 317/B5
Balsapuerto, Peru 128/D5
Balsas, Brazil 120/E3
Balsas, Brazil 132/E4
Balsas (riv.), Brazil 132/E5
Balsas (riv.), Mexico 146/H8
Balsas (riv.), Mexico 150/J7
Bålsta, Sweden 18/G1
Balsthal, Switzerland 39/E2
Balta, N. Dak. (58313) 282/K3
Baltanás, Spain 33/D2
Baltasar Brum, Uruguay 145/B1
Baltasound, Scotland 15/G2
Baltic (sea) 2/K3
Baltic (sea) 7/F3
Baltic, Conn. (06330) 210/G2
Baltic (sea), Denmark 21/E9
Baltic (sea), E. Germany 22/E1
Baltic (sea), Finland 18/K9
Baltic, Mich. (†49905) 250/G1
Baltic, Ohio (43804) 284/G5
Baltic (sea), Poland 47/B1
Baltic, S. Dak. (57003) 298/R6
Baltic (sea), Sweden 18/K9
Baltic (sea), U.S.S.R. 52/B3
Baltic (sea), U.S.S.R. 48/B4
Baltimore, Ireland 10/B5
Baltimore, Ireland 17/C9
Baltimore (city county), Md. 245/M3
Baltimore (co.), Md. 245/M3
Baltimore, Md. (*21201) 245/M3
Baltimore, Md. 188/L3
Baltimore, Md. 146/L6
Baltimore, Ohio (43105) 284/E6
Baltimore, Ontario 177/F3
Baltinglass, Ireland 17/H6
Baltistan (reg.), Pakistan 68/D1
Baltit, Pakistan 68/C1
Baltiysk, U.S.S.R. 52/A4
Baltra (isl.), Ecuador 128/B9
Baltray, Ireland 17/J4
Baltrum (isl.), W. Germany 22/B2
Baly, Va. (†22546) 307/O5
Baluchistan (reg.), Iran 66/M7
Baluchistan (prov.), Pakistan 68/B3
Baluchistan (reg.), Pakistan 59/J4
Balurghat, India 68/F3
Balvi, U.S.S.R. 53/D2
Balwina Aboriginal Reserve, W. Australia 88/D4
Balwina Aboriginal Res., W. Australia 92/E3
Balya, Turkey 63/B3
Balykshi, U.S.S.R. 48/F5
Balzac, Alberta 182/C4
Balzar, Ecuador 128/C3
Bam, Iran 54/G7
Bam, Iran 66/L6
Bam, Iran 59/G4
Bam, U.S.S.R. 48/N4
Bama, Nigeria 106/G6
Bamako (cap.), Mali 2/J5
Bamako (cap.), Mali 106/C6
Bamako (cap.), Mali 102/B3
Bamba, Mali 106/D5
Bambamarca, Peru 128/C6
Bamban, Philippines 82/C3
Bambari, Cent. Afr. Rep. 102/E4
Bambari, Cent. Afr. Rep. 115/D2
Bamberg (co.), S.C. 296/E5
Bamberg, S.C. (29003) 296/E5
Bamberg, W. Germany 22/D4
Bambesa, Zaire 115/E3
Bambili, Zaire 115/E3
Bambio, Cent. Afr. Rep. 115/C3
Bamble, Norway 18/F7
Bamboo, Jamaica 158/J6
Bamboo Creek, W. Australia 92/C3
Bambuí, Brazil 132/E8
Bambuí, Brazil 135/C2
Bamenda, Cameroon 115/B2
Bamfield, Br. Col. 184/E6
Bamian, Afghanistan 59/J3
Bamian, Afghanistan 68/B2
Bamingui, Cent. Afr. Rep. 115/D2
Bamingui (riv.), Cent. Afr. Rep. 115/C2
Bamoa, Mexico 150/E4
Bampur, Iran 59/H4
Bampur, Iran 66/M7
Bampur (riv.), Iran 66/M7
Bamyili-Beswick, North. Terr. 93/C3
Banaba (isl.), Kiribati 87/G6
Bañado de Medina, Uruguay 145/E3
Bañado de Rocha, Uruguay 145/C2
Banagher, Ireland 17/F5
Banagüises, Cuba 158/D1
Banahao (mt.), Philippines 82/C3
Banalia, Zaire 115/E3
Banam, Cambodia 72/E5
Banamba, Mali 106/C6
Banamba, Mali 102/B3
Banamichi, Mexico 150/D2
Banana (riv.), Fla. 212/F3
Banana, Zaire 115/B5
Bananal (isl.), Brazil 120/D4
Bananal (isl.), Brazil 132/D5
Bananier, Guadeloupe 161/A7
Banao, Cuba 158/F2
Ban Aranyaprathet, Thailand 72/D4
Bânâs, Ras (cape), Egypt 111/G3
Bânâs, Ras (cape), Egypt 59/C5
Banas (riv.), India 68/D3
Banaz, Turkey 63/C3
Banaz (riv.), Turkey 63/C3
Banbar, China 77/E5
Ban Boun Tai, Laos 72/D2
Banbridge, N. Ireland 17/J3
Banbury, England 10/F4
Banbury, England 13/F5
Bancalan (isl.), Philippines 82/A6
Bancannia (lake), N.S. Wales 97/A2
Banchory, Scotland 10/E2
Banchory, Scotland 15/F3
Bancoran (isl.), Philippines 82/B7
Bancroft, Idaho (83217) 220/G7
Bancroft, Iowa (50517) 229/E2
Bancroft, Kansas (†66428) 232/G2
Bancroft, Ky. (†40222) 237/K1
Bancroft, La. (†70653) 238/C5
Bancroft, Maine (†04497) 243/H4
Bancroft○, Maine (†04497) 243/H4
Bancroft, Mich. (48414) 250/E6
Bancroft, Nebr. (68004) 264/H2
Bancroft, Ontario 177/G2
Bancroft, Oreg. (†97458) 291/D5
Bancroft, S. Dak. (57316) 298/O4
Bancroft, W. Va. (25011) 312/C5
Bancroft, Wis. (54921) 317/G7
Bancroft (Chililabombwe), Zambia 115/E6
Banda, Gabon 115/B4
Banda, India 68/D3
Banda (sea), Indonesia 54/O10
Banda (isls.), Indonesia 85/H6
Banda (sea), Indonesia 85/H7
Banda Aceh, Indonesia 85/A4
Banda Aceh, Indonesia 54/L9
Bandai (mt.), Japan 81/K5
Bandai-Asahi National Park, Japan 81/J4
Bandama (riv.), Ivory Coast 106/C7
Bandana, Ky. (42022) 237/D6
Bandanaira, Indonesia 85/H6
Bandar (Machilipatnam), India 68/E5
Bandar 'Abbas, Iran 66/J7
Bandar 'Abbas, Iran 54/G7
Bandar 'Abbas, Iran 59/G4
Bandar-e Deylam, Iran 66/G5
Bandar-e Lengeh, Iran 66/J7
Bandar-e Lengeh, Iran 59/F4
Bandar-e Ma'shur, Iran 66/F5
Bandar-e Pahlavi (Enzeli), Iran 59/E2
Bandar-e Pahlavi (Enzeli), Iran 66/F2
Bandar-e Rig, Iran 59/F4
Bandar-e Rig, Iran 66/G6
Bandar-e Torkaman, Iran 66/H2
Bandar-e Torkaman, Iran 59/F2
Bandar Khomeini, Iran 66/F5
Bandar Khomeini, Iran 59/E3
Bandar Maharani (Muar), Malaysia 72/D7
Bandar Penggaram (Batu Pahat), Malaysia 72/D7
Bandar Seri Begawan, Brunei 85/E4
Bandar Seri Begawan (cap.), Brunei 54/N9
Bandar Shahpur, Iran 66/F5
Bandawe, Malawi 115/F6
Bande, Spain 33/B1
Bandeira (mt.), Brazil 120/E5
Bandeira, Pico da (mt.), Brazil 132/F8
Bandeira (mt.), Brazil 135/E2
Bandelier Nat'l Mon., N. Mex. 274/C3
Bandera, Argentina 143/D2
Bandera (co.), Texas 303/E8
Bandera, Texas (78003) 303/F8
Banderas (bay), Mexico 150/G6
Banderilla, Mexico 150/P1
Bandholm, Denmark 21/E8
Bandiagara, Mali 106/D6
Bandırma, Turkey 59/A1
Bandırma, Turkey 63/B2
Bandon, Ireland 10/B5
Bandon, Ireland 17/D8
Bandon (riv.), Ireland 17/D8
Bandon, Oreg. (97411) 291/C4
Bandra, India 68/B7
Bandundu (prov.), Zaire 115/C4
Bandundu, Zaire 102/D5
Bandundu, Zaire 115/C4
Bandung, Indonesia 54/M10
Bandung, Indonesia 85/H2
Bandy, Va. (24602) 307/E6
Bandya, W. Australia 92/C4
Banes, Cuba 156/C2
Banes, Cuba 158/J3
Banff, Alberta 182/C4
Banff, Alta. 162/E5
Banff, Scotland 15/F3
Banff, Scotland 10/E2
Banff (trad. co.), Scotland 15/A5
Banff Nat'l Park, Alberta 182/B4
Banff Nat'l Park, Alta. 162/E5
Banfora, Upper Volta 106/D6
Bangalore, India 2/N5
Bangalore, India 54/J8
Bangalore, India 68/D6
Bangalow, N.S. Wales 97/G1
Bangar, Philippines 82/C2
Bangassou, Centr. Afr. Rep. 102/E4
Bangassou, Cent. Afr. Rep. 115/D3
Banggai, Indonesia 85/G6
Banggai (arch.), Indonesia 85/G6
Banggi (isl.), Malaysia 85/F4
Bangil, Indonesia 85/K2
Bangka (isl.), Indonesia 54/M10
Bangka (isl.), Indonesia 85/D6
Bangka (str.), Indonesia 85/C6
Bangkalan, Indonesia 85/K2
Bangkok (cap.), Thailand 2/P5
Bangkok (cap.), Thailand 72/D4
Bangkok (cap.), Thailand 54/M8
Bangladesh 2/P4
Bangladesh 54/L7
BANGLADESH 68/G4
Bang Lamung, Thailand 72/D4
Bangong Co (lake), China 77/A5
Bangor, Calif. (95914) 204/D4
Bangor, Maine 146/M5
Bangor, Maine (04401) 243/F6
Bangor, Maine 188/N2
Bangor, Mich. (49013) 250/C6
Bangor, N.Y. (12966) 276/M1
Bangor, N. Ireland 17/K2
Bangor, Pa. (18013) 294/M4
Bangor, Sask. 181/J5
Bangor, Wales 13/C4
Bangor, Wales 10/D4
Bangor, Wis. (54614) 317/E8
Bangs, Texas (76823) 303/E6
Bang Saphan, Thailand 72/C5
Bangued, Philippines 85/G2
Bangued, Philippines 82/C2
Banguezane (mt.), Niger 106/F5
Bangui (cap.), Cent. Afr. Rep. 102/D4
Bangui (cap.), Centr. Afr. Rep. 2/K5
Bangui (cap.), Cent. Afr. Rep. 115/C3
Bangui, Philippines 85/G2
Bangui, Philippines 82/C1
Bangui (bay), Philippines 82/C1
Bangweulu (lake), Zambia 115/F6
Ban Houayxay, Laos 72/D2
Baní, Dom. Rep. 158/E6
Baní, Dom. Rep. 156/D3
Bani (riv.), Mali 106/C6
Bani, Jebel (mts.), Morocco 106/C3
Bani, Philippines 82/B2
Bania, Cent. Afr. Rep. 115/C3
Baniara, Papua N.G. 85/C7
Bánica, Dom. Rep. 156/D3
Bánica, Dom. Rep. 158/D5
Banida, Idaho (†83263) 220/G7
Banister (riv.), Va. 307/K7
Bani Suheila, Gaza Strip 65/A5
Baniyas, Syria 63/F5
Banja Luka, Yugoslavia 7/F4
Banja Luka, Yugoslavia 45/C3
Banjarmasin, Indonesia 54/N10
Banjarmasin, Indonesia 85/E6
Banjul (cap.), Gambia 102/A3
Banjul (cap.), Gambia 106/A6
Banka Banka, North. Terr. 93/C5
Ban Kantang, Thailand 72/C6
Ban Kapong, Thailand 72/C5
Bankass, Mali 106/D6
Bankend, Sask. 181/H4
Ban Kèngkok, Laos 72/E3
Bankfoot, Scotland 15/E4
Bankhead (lake), Ala. 195/D4
Bankhead, Scotland 15/F3
Ban Khlong Yai, Thailand 72/D5
Ban Khon, Laos 72/E4
Banks, Ala. (36005) 195/G7
Banks (pt.), Alaska 196/H3
Banks, Ark. (71631) 202/F6
Banks (isl.), Br. Col. 184/B3
Banks (isl.), Canada 2/C2
Banks (isl.), Canada 4/B16
Banks (bay), Ecuador 128/B9
Banks (co.), Georgia 217/E2
Banks (lake), Georgia 217/F9
Banks, Idaho (83602) 220/B5
Banks, Miss. (†38664) 256/D1
Banks (cape), N. S. Wales 88/L4
Banks (cape), N.S. Wales 97/K4
Banks (pen.), N. Zealand 100/D5
Banks (isl.), N.W.T. 146/F2
Banks (isl.), N.W.T. 162/D1
Banks (isl.), N.W. Terrs. 187/F2
Banks, Oreg. (97106) 291/A1
Banks (isl.), Queensland 88/G2
Banks (isl.), Queensland 95/B1
Banks (str.), Tasmania 88/H8
Banks (str.), Tasmania 99/D2
Banks (isls.), Vanuatu 87/G7
Banks (lake), Wash. 310/F3
Bankston, Ala. (35542) 195/C3
Bankston, Iowa (†52045) 229/L3
Bankstown, N. S. Wales 88/K4
Bankstown, N.S. Wales 97/J3
Ban Kui Nua, Thailand 72/D4
Bankura, India 68/F4
Ban Lahanam, Laos 72/E3
Ban Me Thuot, Vietnam 72/F4
Bann (riv.), Ireland 17/J6
Bann (riv.), N. Ireland 17/H2
Bannack, Mont. (†59725) 262/C5
Banner, Ill. (†61520) 222/D3
Banner, Ky. (41603) 237/R5
Banner, Miss. (38913) 256/F2
Banner, Mo. (†63623) 261/L7
Banner (co.), Nebr. 264/A3
Banner, Va. (24230) 307/D7
Banner, Wyo. (82832) 319/F1
Banner Elk, N.C. (28604) 281/F2
Banner Hill, Tenn. (†37650) 237/R8
Banner Springs, Tenn. (†38556) 237/M8
Bannertown, N.C. (†27030) 281/H1
Ban Ngon, Thailand 72/D3
Banning, Calif. (92220) 204/J10
Banning, Georgia (†30185) 217/C3
Bannister, Mich. (48807) 250/E5
Bannock (co.), Idaho 220/F7
Bannock (creek), Idaho 220/F7
Bannock (peak), Idaho 220/F7
Bannock (range), Idaho 220/F7
Bannockburn, Ill. (†60015) 222/B5
Bannockburn, Ontario 177/G3
Bannockburn, Scotland 15/C1
Bannockburn, Scotland 10/B1
Bannow, Ireland 17/H7
Bannu, Pakistan 59/K3
Bannu, Pakistan 68/C2
Bañolas, Spain 33/H1
Bánovce nad Bebravou, Czech. 41/E2
Ban Pak Phanang, Thailand 72/D5
Banphot Phisai, Thailand 72/C3
Ban Pua, Thailand 72/D3
Banquo, Ind. (†46940) 227/F3
Ban Sattahip, Thailand 72/D4
Bansberia, India 68/F1
Bansha, Ireland 17/E7
Banská Bystrica, Czech. 41/E2
Banská Štiavnica, Czech. 41/E2
Bansko, Bulgaria 45/F5
Banstead, England 13/H8
Banstead, England 10/B6
Bansud, Philippines 82/C4
Banswara, India 68/C4
Bantam, Conn. (06750) 210/B2
Bantam (lake), Conn. 210/C2
Bantam (riv.), Conn. 210/B2
Bantayan, Philippines 82/D5
Bantayan (isl.), Philippines 82/D5
Ban Tha Uthen, Thailand 72/D3
Banton, Philippines 82/D4
Banton (isl.), Philippines 82/D4
Bantry, Ireland 10/B5
Bantry, Ireland 17/C8
Bantry (bay), Ireland 10/A5
Bantry (bay), Ireland 17/B8
Bantry, N. Dak. (58713) 282/J3
Bantul, Indonesia 85/J2
Bañuelo (mt.), Spain 33/D3
Banyak (isls.), Indonesia 85/B5
Banyo, Cameroon 115/B2
Banyo, Queensland 88/K2
Banyumas, Indonesia 85/H2
Banyuwangi, Indonesia 85/L2
Banzare Coast (reg.) 5/C7
Baode, China 77/H4
Baoding (Paoting), China 77/J4
Bao Ha, Vietnam 72/D2
Baoji (Paoki), China 77/G5
Baoji, China 54/M6
Bao Lac, Vietnam 72/E2
Baoshan, China 77/E7
Baoting, China 77/G8
Baotou (Paotow), China 77/G3
Baotou, China 54/M5
Baoulé (riv.), Ivory Coast 106/C6
Baoulé (dry riv.), Mali 106/C6
Baoulé (riv.), Mali 106/C6
Bapaume, Sask. 181/D2
Bapchule, Ariz. (85221) 198/D5
Bapsfontein, S. Africa 118/J6
Baptist, La. (†70401) 238/M1
Baptiste (lake), Ontario 177/G2
Baptistown, N.J. (08803) 273/D2
Baqên, China 77/D5
Ba'quba, Iraq 59/D3
Ba'quba, Iraq 66/D4
Baquedano, Chile 138/A4
Baquerizo Moreno, Ecuador 128/C9
Baqura, Jordan 65/D2
Bar, Yugoslavia 45/D4
Bara, Sudan 111/F5
Bara, Sudan 59/B7
Barabai, Indonesia 85/F6
Barabinsk, U.S.S.R. 48/H4
Baraboo, Wis. (53913) 317/G9
Baracaldo, Spain 33/E1
Barachois, New Bruns. 170/F2
Barachois (pt.), Nova Scotia 168/G4
Barachois, Québec 172/D1
Barachois Pond Prov. Park, Newf. 166/C4
Baracoa, Cuba 158/K4
Baracoa, Cuba 156/C2
Barada, Nebr. (†68457) 264/J4
Baradères, Haiti 158/B6
Baradères (bay), Haiti 158/B6
Baradero, Argentina 143/G6
Baradine, N.S. Wales 97/E2
Baradine (creek), N.S. Wales 97/E2
Baraga (co.), Mich. 250/G2
Baraga, Mich. (49908) 250/G1
Baragoi, Kenya 115/G3
Baraguá, Cuba 158/F2
Baragua, Venezuela 124/D2
Barahona (prov.), Dom. Rep. 158/D6
Barahona, Dom. Rep. 158/D6
Barahona, Dom. Rep. 156/D3
Barajas, Spain 33/F4
Barak, Turkey 63/G4
Baraka (riv.), Ethiopia 111/G4
Baraka (riv.), Sudan 111/G4
Baraka (riv.), Sudan 59/C6
Baraka, Zaire 115/E4
Baraki Barak, Afghanistan 59/J3
Baraki Barak, Afghanistan 68/B2
Baralzon (lake), Manitoba 179/J1
Barama (riv.), Guyana 131/A2
Baramanni, Guyana 131/B2
Baramati, India 68/C5
Baramita, Guyana 131/A2
Baramula, India 68/C2
Baranagar, India 68/F1
Barankwa, Sudan 59/B7
Baranoa, Colombia 126/C2
Baranof (isl.), Alaska 196/M1
Baranovichi, U.S.S.R. 7/G3
Baranovichi, U.S.S.R. 48/C4
Baranovichi, U.S.S.R. 52/C4
Baranya (co.), Hungary 41/E4
Barão de Cocais, Brazil 135/E1
Baras, Philippines 82/E4
Barasat, India 68/F1
Baratang (isl.), India 68/G6
Barataria, La. (70036) 238/K7
Barataria (bay), La. 238/L8
Barataria (passage), La. 238/L8
Barawa (Brava), Somalia 115/H3
Baraya, Colombia 126/C6
Barbacena, Brazil 120/E5
Barbacena, Brazil 135/E2
Barbacena, Brazil 132/F8
Barbacoas, Colombia 126/A7
Barbacoas, Venezuela 124/E3
Barbados 2/G5
Barbados 146/N8
BARBADOS 156/G4
BARBADOS 161/B8
Barbar (isls.), Indonesia 85/J7
Barbas (cape), Western Sahara 106/A4
Barbastro, Spain 33/F1
Barbate (riv.), Spain 33/D4
Barbeau, Mich. (49710) 250/E2
Barbeau (peak), N.W. Terrs. 187/L1
Barber, Ark. (72922) 202/B3
Barber (co.), Kansas 232/D4
Barber, Mont. (†59074) 262/G4
Barber, N.C. (27008) 281/H3
Barbers (pt.), Hawaii 218/E2
Barbers Point, Hawaii (†96706) 218/E2
Barbers Point Nav. Air Sta., Hawaii 218/E2
Barberton, Ohio (4r203) 284/G4
Barberton, S. Africa 118/E5
Barberville, Fla. (32005) 212/E2
Barbezieux-St-Hilaire, France 28/C5
Barbil, India 68/F4
Barbizon, France 28/E3
Barbosa, Colombia 126/D5
Barbour (co.), Ala. 195/H7
Barbour (co.), W. Va. 312/F4
Barbourmeade, Ky. (†40201) 237/K1
Barboursville, Va. (22923) 307/M4
Barboursville, W. Va. (25504) 312/B6
Barbourville, Ky. (40906) 237/O7
Barbuda (isl.) 146/M8
Barbuda (isl.), Ant. & Bar. 156/G3
Barcaldine, Queensland 88/G4
Barcaldine, Queensland 95/C4
Barcaldine, Scotland 15/C4
Barcarrota, Spain 33/C3
Barce (El Marj), Libya 111/D1
Barcellona Pozzo di Gotto, Italy 34/E5
Barcelona (prov.), Spain 33/G2
Barcelona, Spain 7/E4
Barcelona, Spain 33/H2
Barcelona, Venezuela 124/F2
Barcelona, Venezuela 120/C2
Barceloneta, P. Rico 161/C1
Barcelonnette, France 28/G5
Barcelos, Brazil 120/C3
Barcelos, Brazil 132/H9
Barcelos, Portugal 33/B2
Barclay, Md. (21607) 245/P4
Barco, N.C. (27917) 281/T2
Barcoo (creek), Queensland 88/G4
Barcoo (creek), Queensland 95/B5
Barcoo (creek), S. Australia 88/F5
Barcoo (creek), S. Australia 94/F3
Barcos (pt.), Cuba 158/B2
Barcs, Hungary 41/D4
Barczewo, Poland 47/E2
Bard, Calif. (92222) 204/L11
Bard, N. Mex. (88411) 274/F3
Bardai, Chad 111/C3
Bardai, Chad 102/D2
Bardejov, Czech. 41/F2
Bardera, Somalia 115/H3
Bardera, Somalia 102/G4
Bardney, England 13/G4
Bardolph, Ill. (61416) 222/C3
Bardon (lake), Wis. 317/C3
Bardonia, N.Y. (†10954) 276/K8
Bardsey (isl.), Wales 13/C5
Bardstown, Ky. (40004) 237/L5
Barduelv (riv.), Norway 18/L2
Bardwell, Ky. (42023) 237/D7
Bardwell, Texas (75101) 303/H5
Bareilly, India 54/K7
Bareilly, India 68/D3
Barellan, N.S. Wales 97/D4
Bärenhorn (mt.), Switzerland 39/H3
Barents (sea) 2/L2
Barents (sea) 7/J1
Barents (sea) 4/B8
Barents (sea), U.S.S.R. 48/D2
Barents (sea), U.S.S.R. 52/E1
Barentsburg, Norway 18/C2
Barentsøya (isl.), Norway 18/D2
Bäretswil, Switzerland 39/G2
Barfield, Ark. (†72315) 202/L2
Barfleur, France 28/C3
Barfleur (pt.), France 28/C3
Barga, China 77/B5
Bargal, Somalia 115/K1
Bargamin (creek), Idaho 220/C4
Bargersville, Ind. (46106) 227/E5
Bargo, N.S. Wales 97/F4
Bargrax (Bohu), China 77/C3

Barham, N.S. Wales 97/C4
Bar Harbor, Maine (04609) 243/G7
Bar Harbor○, Maine (04609) 243/G7
Bari (prov.), Italy 34/F4
Bari, Italy 34/F4
Bari, Italy 7/F4
Bari (prov.), Somalia 115/J1
Baria (riv.), Venezuela 124/E7
Barich, Alberta 182/D2
Barichara, Colombia 126/D4
Barida, Ras (cape), Saudi Arabia 59/C5
Barima (riv.), Guyana 131/B2
Barinas (state), Venezuela 124/D3
Barinas, Venezuela 124/C3
Barinas, Venzuela 120/C2
Baring, Maine (†04619) 243/J5
Baring○, Maine (†04619) 243/J5
Baring, Mo. (63531) 261/H2
Baring (head), N. Zealand 100/B3
Baring (cape), N.W. Terrs. 187/G3
Baring, Sask. 181/J5
Baring, Wash. (98224) 310/D3
Barinitas, Venezuela 124/C3
Baripada, India 68/F4
Bariri, Brazil 135/B3
Bariri (res.), Brazil 135/B3
Bâris, Egypt 111/F3
Barisal, Bangladesh 68/G4
Barisan (mts.), Indonesia 85/C6
Baritbog (riv.), New Bruns. 170/E1
Barito (riv.), Indonesia 85/E6
Bark (lake), Ontario 177/G2
Barkam, China 77/F5
Barker, N.Y. (14012) 276/C4
Barker Heights, N.C. (†28739) 281/D4
Barkeyville, Pa. (†16038) 294/C3
Barkhamsted○, Conn. (†06063) 210/D1
Barkhamsted (res.), Conn. 210/D1
Barkhan, Pakistan 68/B3
Barkhan, Pakistan 59/J4
Barking, England 10/C5
Barking, England 13/H8
Barkley (sound), Br. Col. 184/E6
Barkley (dam), Ky. 237/E6
Barkley (lake), Ky. 237/F7
Barkley (lake), Tenn. 237/F7
Barkly Downs, Queensland 95/A4
Barkly East, S. Africa 118/D6
Barkly Tableland (plat.), Australia 87/D7
Barkly Tableland, North. Terr. 88/F3
Barkly Tableland, North. Terr. 93/D4
Barkly Tableland, Queensland 95/A4
Barkmere, Québec 172/C3
Barkol, China 77/D3
Bark River, Mich. (49807) 250/B3
Barksdale, Texas (78828) 303/D8
Barksdale A.F.B., La. 238/C2
Barlby, England 13/G4
Bar-le-Duc, France 28/F3
Barlee (lake), Australia 87/B8
Barlee (lake), W. Australia 88/B5
Barlee (lake), W. Australia 92/B5
Barletta, Italy 34/F4
Barlinek, Poland 47/B2
Barling, Ark. (72923) 202/B3
Barlow, Br. Col. 184/F3
Barlow, Ky. (42024) 237/D6
Barlow, Miss. (†39083) 256/C7
Barlow, N. Dak. (†58421) 282/M4
Barlow, Ohio (45612) 284/G7
Barlow, Oreg. (†97013) 291/B2
Barlow Bend, Ala. (†36545) 195/C8
Barmedman, N.S. Wales 97/D4
Barmer, India 68/C3
Barmera, S. Australia 94/G6
Bar Mills, Maine (04004) 243/C8
Barmouth, Wales 10/D4
Barmouth, Wales 13/C5
Barna, Ireland 17/C5
Barnabus, W. Va. (25609) 312/C7
Barnaby (riv.), New Bruns. 170/E2
Barnaby River, New Bruns. 170/E2
Barnard, Kansas (67418) 232/D2
Barnard, Mo. (64423) 261/C2
Barnard, N.C. (†28753) 281/D3
Barnard, S. Dak. (57426) 298/N2
Barnard○, Vt. (05031) 268/B4
Barnard Castle, England 13/E3
Barnardsville, N.C. (28709) 281/E3
Barnaul, U.S.S.R. 54/K4
Barnaul, U.S.S.R. 48/J4
Barn Bluff (mt.), Tasmania 99/B3
Barnegat, Alberta 182/E2
Barnegat, N.J. (08005) 273/E4
Barnegat (bay), N.J. 273/E4
Barnegat (inlet), N.J. 273/E4
Barnegat Light, N.J. (08006) 273/E4
Barnes (sound), Fla. 212/F6
Barnes, Kansas (66933) 232/F2
Barnes (co.), N. Dak. 282/O5
Barnes, Wis. (†54873) 317/D3
Barnesboro, Pa. (15714) 294/E4
Barnes City, Iowa (50027) 229/H6
Barnes Corners, N.Y. (13610) 276/J3
Barneston, Nebr. (68309) 264/H4
Barnesville, Colo. (†80624) 208/L2
Barnesville, Georgia (30204) 217/D4
Barnesville, Md. (20703) 245/J4
Barnesville, Minn. (56514) 255/B4
Barnesville, N.C. (28319) 281/L6
Barnesville, Ohio (43713) 284/H6
Barnet, England 13/H7
Barnet, England 10/B5
Barnet○, Vt. (05821) 268/C3
Barnett, Georgia (†30821) 217/G3
Barnett, Miss. (†39347) 256/G7
Barnett, Mo. (65011) 261/G6
Barnettville, New Bruns. 170/E2
Barneveld, Netherlands 27/H4
Barneveld, N.Y. (13304) 276/K4
Barneveld, Wis. (53507) 317/F10
Barneville-Carteret, France 28/C3
Barney, Georgia (31625) 217/E8
Barney, N. Dak. (58008) 282/S7
Barnhart, Texas (76930) 303/C6
Barnhill, Ohio (†44663) 284/H5
Barnoldswick, England 13/H1
Barnrock, Ky. (†41219) 237/R5
Barnsdall, Okla. (74002) 288/O1
Barnsley, England 13/J2
Barnsley, England 10/F4
Barnstable (co.), Mass. 249/N6
Barnstable, Mass. (02630) 249/N6
Barnstable○, Mass. (02630) 249/N6
Barnstaple, England 10/E5
Barnstaple, England 13/D6
Barnstaple (bay), England 10/D5
Barnstaple (bay), England 13/C6
Barnstead○, N.H. (03218) 268/E5
Barnum, Iowa (50518) 229/E3
Barnum, Minn. (55707) 255/F4
Barnum, W. Va. (†26726) 312/H4
Barnum, Wis. (†54631) 317/E9
Barnwell, Ala. (†36532) 195/C10
Barnwell, Alberta 182/D5
Barnwell (co.), S.C. 296/E5
Barnwell, S.C. (29812) 296/E5
Baro (riv.), Ethiopia 111/G6
Baro, Nigeria 106/F7
Baroda (Vadodara), India 68/C4
Baroda, India 54/J7
Baroda, Mich. (49101) 250/C7
Baroghil (pass), Afghanistan 68/C1
Baroghil (pass), Pakistan 68/C1
Baron, Okla. (†74965) 288/S3
Baron Bluff (prom.), Virgin Is. (U.S.) 161/E3
Barons, Alberta 182/D4
Barooga, N.S. Wales 97/C4
Barossa (res.), S. Australia 94/C6
Barotseland (reg.), Zambia 115/D7
Barpeta, India 68/G3
Barqa (Cyrenaica) (reg.), Libya 111/D1
Barques (pt.), Mich. 250/C3
Barquisimeto, Venezuela 124/D2
Barquisimeto, Venezuela 120/C2
Barr, Scotland 15/D5
Barr, Tenn. (†38040) 237/B9
Barra, Brazil 132/F5
Barra (head), Scotland 10/C2
Barra (head), Scotland 15/A4
Barra (isl.), Scotland 15/A4
Barra (isl.), Scotland 10/C2
Barra (isls.), Scotland 10/C2
Barra (sound), Scotland 15/A3
Barraba, N.S. Wales 97/F2
Barra Bonita (res.), Brazil 135/B3
Barrackpore, India 68/F1
Barrackville, W. Va. (26559) 312/F3
Barra de Río Grande, Nicaragua 154/F4
Barra do Bugres, Brazil 132/B6
Barra do Corda, Brazil 132/E4
Barra do Piraí, Brazil 132/E8
Barra do Piraí, Brazil 135/E3
Barra Isles (isls.), Scotland 15/A4
Barra Mansa, Brazil 135/D3
Barranca, Lima, Peru 128/C8
Barranca, Loreto, Peru 128/D5
Barrancabermeja, Colombia 126/C4
Barranca de Upía, Colombia 126/D5
Barrancas, Argentina 143/F6
Barrancas (riv.), Argentina 143/G5
Barrancas, Chile 138/G3
Barrancas, Colombia 126/D2
Barrancas, Barinas, Venezuela 124/C3
Barrancas, Monagas, Venezuela 124/G3
Barranco de Loba, Colombia 126/C3
Barrancos, Cerro (mt.), Chile 138/D7
Barrancos, Portugal 33/C3
Barranqueras, Argentina 143/E2
Barranquilla, Colombia 120/B1
Barranquilla, Colombia 126/C2
Barranquitas, P. Rico 161/D2
Barras (riv.), Bolivia 136/B6
Barras, Brazil 132/F4
Barras, Colombia 126/D8
Barraute, Québec 174/B3
Barre, Mass. (01005) 249/F3
Barre○, Mass. (01005) 249/F3
Barre, Québec 172/G3
Barre, Vt. (05641) 268/C3
Barre○, Vt. (05641) 268/C3
Barreal, Argentina 143/C3
Barreau (pt.), New Bruns. 170/F1
Barre Center, N.Y. (†14411) 276/D4
Barreiras, Brazil 120/E4
Barreiras, Brazil 132/E6
Barreirinha, Brazil 132/B3
Barreirinhas, Brazil 132/F3
Barreiro, Portugal 33/B1
Barreiros, Brazil 132/H5
Barren (isls.), Alaska 196/B2
Barren (isl.), India 68/G6
Barren (co.), Ky. 237/K7
Barren (riv.), Ky. 237/H6
Barren (isls.), Madagascar 118/G3
Barren (isl.), Nova Scotia 168/G4
Barren (cape), Tasmania 99/E2
Barren Plains, Tenn. (†37172) 237/H7
Barren River (lake), Ky. 237/J7
Barren Springs, Va. (24313) 307/G7
Barre Plains, Mass. (†01005) 249/F3
Barrera, Bolivia 136/B3
Barretos, Brazil 132/D8
Barretos, Brazil 135/B2
Barrett, Minn. (56311) 255/B5
Barrett, Texas (77532) 303/K1
Barrett, W. Va. (25013) 312/C7
Barretts, Georgia (†31601) 217/F8
Barrhead, Alberta 182/C2
Barrhead, Scotland 10/A1
Barrhead, Scotland 15/B2
Barrhill, Scotland 15/D5
Barrie, Ontario 177/E3
Barrie (isl.), Ontario 177/B1
Barrière, Br. Col. 184/H4
Barrineau Park, Fla. (†32533) 212/B6
Barrington, Ill. (60010) 222/A5
Barrington○, N.H. (†03825) 268/F5
Barrington, N.J. (08007) 273/B3
Barrington, Nova Scotia 168/C5
Barrington (bay), Nova Scotia 168/C5
Barrington○, R.I. (02806) 249/J6
Barrington, Tasmania 99/C3
Barrington Hills, Ill. (†60010) 222/A5
Barrington P.O. (East Barrington), N.H. (03825) 268/F5
Barrington Passage, Nova Scotia 168/C5
Barrington Tops (mt.), N.S. Wales 97/F2
Barringun, N.S. Wales 97/C1
Barron (co.), Wis. 317/C5
Barron, Wis. (54812) 317/C5
Barronett, Wis. (54813) 317/B4
Barrouallie, St. Vin. & Grens. 161/A9
Barroui, Dominica 161/E6
Barrow, Alaska (99723) 196/G1
Barrow, Alaska 188/C5
Barrow, Alaska 146/C2
Barrow (pt.), Alaska 146/C2
Barrow (pt.), Alaska 196/G1
Barrow (isl.), Australia 87/B8
Barrow (co.), Georgia 217/E2
Barrow (riv.), Ireland 17/H7
Barrow (riv.), Ireland 10/C4
Barrow (str.), N.W.T. 162/G1
Barrow (str.), N.W. Terrs. 187/J2
Barrow (bay), Ontario 177/C2
Barrow, U.S. 4/B17
Barrow (pt.), U.S. 2/B2
Barrow (pt.), U.S. 4/B18
Barrow (isl.), W. Australia 88/A4
Barrow (isl.), W. Australia 92/A3
Barrow Creek, North. Terr. 93/D6
Barrow-in-Furness, England 10/E3
Barrow-in-Furness, England 13/D3
Barrows, Manitoba 179/A2
Barrowsville, Mass. (†02766) 249/K5
Barr Smith (mt.) 5/C5
Barruelo de Santullán, Spain 33/D1
Barry, Ill. (62312) 222/B4
Barry (co.), Mich. 250/D6
Barry, Minn. (56210) 255/B5
Barry (co.), Mo. 261/E9
Barry (mts.), Victoria 97/D5
Barry, Wales 13/B7
Barry, Wales 10/E5
Barry's Bay, Ontario 177/G2
Barryton, Mich. (49305) 250/D5
Barryville, N.Y. (12719) 276/L8
Barsi, India 68/D5
Barsinghausen, W. Germany 22/C2
Barss Corners, Nova Scotia 168/D4
Barstow, Calif. (92311) 204/H9
Barstow, Md. (20610) 245/M6
Barstow, Texas (79719) 303/A6
Bar-sur-Aube, France 28/F3
Bar-sur-Seine, France 28/F3
Bartelso, Ill. (62218) 222/D5
Barterville, Ky. (†40311) 237/N4
Barth, E. Germany 22/E1
Barth, Fla. (†32533) 212/B6
Barthel, Sask. 181/B2
Bartholomew (bayou), Ark. 202/G6
Bartholomew (co.), Ind. 227/F6
Bartibog Bridge, New Bruns. 170/E1
Bartica, Guyana 120/D2
Bartica, Guyana 131/B2
Bartın, Turkey 63/E2
Bartle, Cuba 158/H3
Bartle Frere (mt.), Queensland 88/H3
Bartle Frere (mt.), Queensland 95/C3
Bartlesville, Okla. (74003) 288/O1
Bartlett (dam), Ariz. 198/D5
Bartlett (res.), Ariz. 198/D5
Bartlett, Ill. (60103) 222/A5
Bartlett, Iowa (51655) 229/B7
Bartlett, Kansas (67332) 232/G4
Bartlett, Nebr. (68622) 264/F3
Bartlett○, N.H. (03812) 268/E3
Bartlett, N. Dak. (†58344) 282/N3
Bartlett, Ohio (45713) 284/G7
Bartlett, Tenn. (38134) 237/B10
Bartlett, Texas (76511) 303/G7
Bartlett Deep, Cayman Is. 156/B3
Bartletts Ferry (dam), Ala. 195/H5
Bartletts Ferry (dam), Georgia 217/B5
Bartley, Nebr. (69020) 264/D4
Bartley, W. Va. (24813) 312/C8
Barto, Pa. (19504) 294/L5
Bartolomeu Dias, Mozambique 118/F4
Barton, Ala. (†35616) 195/C1
Barton, Ark. (72312) 202/J4
Barton (co.), Kansas 232/D3
Barton, Md. (21521) 245/B2
Barton (co.), Mo. 261/D7
Barton, N. Dak. (58315) 282/K2
Barton, Ohio (43905) 284/J5
Barton, Oreg. (†97009) 291/B2
Barton, Vt. (05822) 268/C2
Barton○, Vt. (05822) 268/C2
Barton (riv.), Vt. 268/C2
Barton City, Mich. (48705) 250/F4
Barton Hills, Mich. (48105) 250/F6
Bartonsville, Pa. (18321) 294/M4
Bartonsville, Vt. (†05143) 268/B5
Barton-upon-Humber, England 13/G4
Barton-upon-Humber, England 10/F4
Bartonville, Ill. (61607) 222/D3
Bartonville, Texas (†76226) 303/F1
Bartoszyce, Poland 47/E1
Bartow, Fla. (33830) 212/E4
Bartow (co.), Georgia 217/C2
Bartow, Georgia (30413) 217/G5
Bartow, W. Va. (24920) 312/G5
Bartra Antiguo, Peru 128/E4
Bartra Nuevo, Peru 128/E4
Barú (isl.), Colombia 126/C2
Barú (vol.), Panama 154/F6
Baruipur, India 68/F2
Barus, Indonesia 85/B5
Barut, Tanjong (cape), Malaysia 85/E5
Baruun-Urt, Mongolia 77/H2
Barvas, Scotland 10/C1
Barvas, Scotland 15/B2
Barview, Oreg. (†97420) 291/C4
Bar View, Oreg. (†97136) 291/C2
Barville, Québec 174/B3
Barwani, India 68/D4
Barwick, Georgia (31720) 217/E9
Barwick, Ontario 175/B3
Barwick, Ontario 177/F5
Barwon (riv.) 88/H5
Barwon (riv.), N.S. Wales 97/D2
Barysh, U.S.S.R. 52/G4
Baryulgil, N.S. Wales 97/G1
Basalt, Colo. (81621) 208/E4
Basalt, Idaho (83218) 220/F6
Basankusu, Zaire 115/C3
Basavilbaso, Argentina 143/G6
Bas-Caraquet, New Bruns. 170/F1
Bascharage, Luxembourg 27/H9
Basco, Ill. (62313) 222/B3
Basco, Philippines 82/A2
Bascom, Fla. (32423) 212/A1
Bascom, Ohio (44809) 284/D3
Bascuñán (cape), Chile 138/A7
Basehor, Kansas (66007) 232/G2
Basel, Switzerland 39/E1
Basel, Switzerland 7/E4
Baselland (canton), Switzerland 39/E2
Baselstadt (canton), Switzerland 39/E1
Basey, Philippines 82/E5
Bashan, Conn. (†06423) 210/F2
Bashan (lake), Conn. 210/F3
Bashaw, Alberta 182/D3
Bashi, Ala. (†36784) 195/C7
Bashi (chan.), China 77/K7
Bashi (chan.), Philippines 82/A1
Bashkir A.S.S.R., U.S.S.R. 48/F4
Bashkir A.S.S.R., U.S.S.R. 52/J4
Basht, Iran 66/G5
Basic, Miss. (†39330) 256/G6
Basilan (prov.), Philippines 82/D7
Basilan, Philippines 82/C7
Basilan (isl.), Philippines 85/G4
Basilan (isl.), Philippines 82/D7
Basilan (str.), Philippines 82/C7
Basildon, England 13/J8
Basildon, England 10/G5
Basile, La. (70515) 238/E5
Basilicata (reg.), Italy 34/F4
Basim, India 68/D4
Basin, Mont. (59613) 262/D4
Basin (lake), Sask. 181/F3
Basin, Wyo. (82410) 319/E1
Basinger, Fla. (†33472) 212/F4
Basingstoke, England 10/F5
Basingstoke, England 13/F6
Basirhat, India 68/F4
Basit (cape), Syria 63/F5
Baskahegan (lake), Maine 243/H5
Başkale, Turkey 63/K3
Baskatong (res.), Que. 162/J6
Baskatong (res.), Québec 172/B3
Baskerville, Va. (23915) 307/M7
Basket (lake), Manitoba 179/C3
Baskett, Ky. (42402) 237/F5
Baskil, Turkey 63/H3
Baskin, La. (71219) 238/G2
Basking Ridge, N.J. (07920) 273/D2
Başmakçı, Turkey 63/C4
Basodino (peak), Switzerland 39/G4
Basoko, Zaire 115/D3
Basom, N.Y. (14013) 276/D4
Basongo, Zaire 115/D4
Basora (pt.), Neth. Ant. 161/E10
Basra (gov.), Iraq 66/E5
Basra, Iraq 2/M4
Basra, Iraq 66/E5
Basra, Iraq 59/E3
Basra, Iraq 54/F6
Bas-Rhin (dept.), France 28/G3
Bass (str.) 88/H7
Bass (str.), Australia 87/E9
Bass (isls.), Fr. Poly. 87/M8
Bass (lake), Ind. 227/D2
Bass (str.), Tasmania 99/C1
Bassano, Alberta 182/D4
Bassano del Grappa, Italy 34/C2
Bassas da India (isl.), Réunion 102/F7
Bassas da India (isl.), Réunion 118/F4
Bassecourt, Switzerland 39/D2
Bassein, Burma 54/L8
Bassein, Burma 72/B3
Bassein, India 68/C5
Basse-Pointe, Martinique 161/C5
Basse-Sambre, Belgium 27/F8
Basse Santa Su, Gambia 106/B6
Basse-Terre (cap.), Guadeloupe 161/A7
Basse-Terre (cap.), Guadeloupe 156/F4
Basse-Terre (isl.), Guadeloupe 161/A6
Basseterre (cap.), St. Chris.-Nevis 161/C10
Basseterre (cap.), St. Chris.-Nevis 156/F3
Basse Terre, Trin. & Tob. 161/B11
Bassett, Ark. (72313) 202/K2
Bassett, Iowa (†50645) 229/J2
Bassett, Kansas (†66749) 232/G4
Bassett, Nebr. (68714) 264/E2
Bassett, Va. (24055) 307/J7
Bassfield, Miss. (39421) 256/E8
Bass Harbor, Maine (04653) 243/G7
Bassikounou, Mauritania 106/C5
Bassin Bleu, Haiti 158/B5
Bass River, New Bruns. 170/E2
Bass River, Nova Scotia 168/E3
Bassum, W. Germany 22/C2
Basswood, Manitoba 179/B4
Basswood (lake), Minn. 255/G2
Basswood (lake), Ontario 175/B3
Båstad, Sweden 18/H8
Bastak, Iran 66/J7
Bastam, Iran 66/J2
Bastar, India 68/E5
Bastelica, France 28/B6
Bastenaken (Bastogne), Belgium 27/H9
Bastia, France 7/E4
Bastia, France 28/B6
Bastian, Va. (24314) 307/F6
Bastimentos (isl.), Panama 154/G6
Bastogne, Belgium 27/H9
Bastrop, La. (71220) 238/G1
Bastrop (co.), Texas 303/G7
Bastrop, Texas (78602) 303/G7
Bastuträsk, Sweden 18/L4
Basye, Va. (22810) 307/L3
Bas-Zaïre (prov.), Zaire 115/B4
Bata, Equat. Guinea 102/C4
Bata, Equat. Guinea 115/B3
Bataan (prov.), Philippines 82/C3
Batabanó (gulf), Cuba 158/C2
Batabanó (gulf), Cuba 156/A2
Batag (isl.), Philippines 82/E4
Batagay, U.S.S.R. 48/O3
Batala, India 68/D2
Batalha, Brazil 132/F3
Batalha, Portugal 33/B3
Batan (isls.), Philippines 54/O7
Batan, Albay (isl.), Philippines 82/E4
Batan, Batanes (isl.), Philippines 82/B2
Batan (isls.), Philippines 85/G1
Batan (isls.), Philippines 82/A2
Batanes (prov.), Philippines 82/A2
Batang, China 77/E5
Batang, China 54/L6
Batang, Indonesia 85/J2
Batangafo, Cent. Afr. Rep. 115/C2
Batangas (prov.), Philippines 82/C4
Batangas, Philippines 82/C4
Batangas, Philippines 85/G3
Batas (isl.), Philippines 82/B5
Bátaszék, Hungary 41/E3
Batatais, Brazil 135/C2
Batavia, Ill. (60510) 222/E2
Batavia (Jakarta) (cap.), Indonesia 85/H1
Batavia, Iowa (52533) 229/J7
Batavia, Mich. (†49036) 250/D7
Batavia, N.Y. (14020) 276/D5
Batavia, Ohio (45103) 284/B7
Batavia, Wis. (†53001) 317/K8
Batawa, Ontario 177/G3
Bataysk, U.S.S.R. 52/E5
Bat Cave, N.C. (28710) 281/E4
Batchelor, La. (70715) 238/G5
Batchelor, North. Terr. 93/B2
Batchtown, Ill. (62006) 222/C4
Batchwana Bay, Ontario 177/J5
Batdambang, Cambodia 54/M8
Batdambang (Battambang), Cambodia 72/D4
Bateman, Sask. 181/E5
Batemans Bay, N.S. Wales 97/F4
Bates, Ark. (72924) 202/B4
Bates, Mich. (†49690) 250/D4
Bates (co.), Mo. 261/D6
Bates (mt.), Norfolk I. 88/L5
Bates, Oreg. (97817) 291/J3
Batesburg, S.C. (29006) 296/D4
Bates City, Mo. (64011) 261/E5
Batesland, S. Dak. (57716) 298/E7
Batesville, Ala. (†36018) 195/H6
Batesville, Ark. (72501) 202/G2
Batesville, Ind. (47006) 227/G6
Batesville, Miss. (38606) 256/E2
Batesville, Ohio (43715) 284/H6
Batesville, Texas (78829) 303/E9
Batesville, Va. (22924) 307/L5
Bath, England 13/E6
Bath, England 10/E5
Bath, Ill. (62617) 222/C3
Bath, Ind. (47010) 227/H5
Bath, Jamaica 158/K6
Bath (co.), Ky. 237/O4
Bath, Maine (04530) 243/D8
Bath, Mich. (48808) 250/E6
Bath, Netherlands 27/E6
Bath, New Bruns. 170/C2
Bath○, N.H. (03740) 268/D3
Bath, N.Y. (14810) 276/F6
Bath, N.C. (27808) 281/R4
Bath, Ontario 177/H3
Bath, Pa. (18014) 294/M4
Bath, S.C. (29816) 296/D5
Bath, S. Dak. (57427) 298/N3
Bath (co.), Va. 307/J4
Bath (Berkeley Springs), W. Va. (†25411) 312/K3
Batha (riv.), Chad 111/C5
Bathgate, N. Dak. (58216) 282/P2
Bathgate, Scotland 15/C2
Bathgate, Scotland 10/C1
Bathsheba, Barbados 161/B8
Bath Springs, Tenn. (38311) 237/E10
Bathurst (isl.), Australia 87/C7
Bathurst (isl.), Canada 4/B14
Bathurst (Banjul)(cap.), Gambia 106/A6
Bathurst, N. Br. 162/K6
Bathurst, New Bruns. 170/E1
Bathurst, N. S. Wales 88/H6
Bathurst, N.S. Wales 97/E3
Bathurst (isl.), North Terr. 88/D2
Bathurst (isl.), North Terr. 93/A1
Bathurst (isl.), N.W.T. 162/F1
Bathurst (isl.), N.W.T. 146/H2
Bathurst (cape), N.W.T. 162/D1
Bathurst (cape), N.W.Terrs. 187/F2
Bathurst (inlet), N.W. Terrs. 187/H3
Bathurst (isl.), N.W. Terrs. 187/H2
Bathurst (harb.), Tasmania 99/C5
Bathurst Inlet, N.W. Terrs. 187/H3
Bathurst Island, North. Terr. 93/B1
Bathurst Island Mission, North. Terr. 88/E2
Bathurst Mines, New Bruns. 170/E1
Batié, Upper Volta 106/D7
Batı Fırat (riv.), Turkey 63/H3
Batin, Wadi al (dry riv.), Iraq 59/E4
Batin, Wadi al (dry riv.), Iraq 66/E6
Batin, Wadi al (dry riv.), Saudi Arabia 59/E4
Batina (reg.), Oman 59/G5
Batini (mt.), Fiji 86/Q10
Batiscan, Québec 172/E3
Batiscan (lake), Québec 172/E2
Batiscan (riv.), Québec 172/E2
Batley, England 13/J1
Batlow, N.S. Wales 97/E4
Batman, Turkey 63/J4
Batna, Algeria 102/C1
Batna, Algeria 106/F1
Bato, Catanduanes, Philippines 82/E4
Bato, Leyte, Philippines 82/E5
Bato-Bato, Philippines 82/C8
Batobato, Philippines 82/E7
Batoche, Sask. 181/E3
Batoche Nat'l Hist. Site, Sask. 181/E3
Baton Rouge (cap.), La. 146/J6
Baton Rouge (cap.), La. 188/H4
Baton Rouge (cap.), La. (*70801) 238/K2
Batopilas, Mexico 150/F3
Batouri, Cameroon 115/B3
Batovi, Uruguay 145/D2
Batrun, Lebanon 63/F5
Bat Shelomo, Israel 65/B2
Batson, Texas (77519) 303/K7
Batsto, N.J. (08037) 273/D4
Batsto (riv.), N.J. 273/D4
Batten Kill (riv.), N.Y. 276/O4
Batten Kill (riv.), Vt. 268/A5
Batterbee (cape), Ant. 2/N9
Batterbee (cape) 5/C3
Bätterkinden, Switzerland 39/E2
Battersea, Ontario 177/H3
Batticaloa, Sri Lanka 68/E7
Battiest, Okla. (74722) 288/S6
Batti Malv (isl.), India 68/G7
Battle (riv.) 162/E5
Battle (riv.), Alberta 182/D3
Battle, England 13/H7
Battle, England 10/G5
Battle (creek), Idaho 220/B7
Battle (riv.), Minn. 255/D3
Battle (creek), Mont. 262/G1
Battle (creek), Oreg. 291/K5
Battle (creek), Sask. 181/B6
Battle (riv.), Sask. 181/B3
Battle (creek), S. Dak. 298/C6
Battleboro, N.C. (27809) 281/O2
Battle Creek, Iowa (51006) 229/B4
Battle Creek, Mich. 188/J2
Battle Creek, Mich. (*49014) 250/D6
Battle Creek, Nebr. (68715) 264/G3
Battlefield, Mo. (†65619) 261/F8
Battleford, Sask. 162/E5
Battleford, Sask. 181/C3
Battle Ground, Ind. (47920) 227/D3
Battle Ground, Wash. (98604) 310/C5
Battle Harbour, Newf. 166/C3
Battle Harbour, Newf. 162/L5
Battle Lake, Alberta 182/C3
Battle Lake, Minn. (56515) 255/C4
Battle Mountain, Nev. (89820) 266/E2
Battles Wharf, Ala. (†36532) 195/C10
Battletown, Ky. (40104) 237/J4
Battleview, N. Dak. (58714) 282/E2
Battock (mt.), Scotland 15/F4
Battonya, Hungary 41/F3
Battrum, Sask. 181/C5
Batu (isls.), Indonesia 85/B6
Batuco, Chile 138/G3
Batu Gajah, Malaysia 72/D6
Batulaki, Philippines 82/E8
Batumi, U.S.S.R. 48/E5
Batumi, U.S.S.R. 52/F6
Batu Pahat, Malaysia 72/D7
Baturaja, Indonesia 85/C6
Baturité, Brazil 132/G4
Batusangkar, Indonesia 85/C6
Bat Yam, Israel 65/B3
Bauang, Philippines 82/C2
Baubau, Indonesia 85/G7
Bauchi (state), Nigeria 106/F6
Bauchi, Nigeria 106/F6
Baudette, Minn. (56623) 255/D2
Baudette (riv.), Minn. 255/D2
Baudh, India 68/E4
Baudó, Serranía de (mts.), Colombia 126/B5
Baudó (riv.), Colombia 126/B5
Baugé, France 28/D4
Bauld (cape), Newf. 166/C3
Bauld (cape), Newf. 162/L5
Bauline, Newf. 166/D2
Baulkham Hills, N. S. Wales 88/K4
Baulkham Hills, N.S. Wales 97/H3
Baulmes, Switzerland 39/C3
Bauma, Switzerland 39/G2
Baumann (fjord), N.W. Terrs. 187/K2
Baume-les-Dames, France 28/G4
Baures, Bolivia 136/D3
Baures (riv.), Bolivia 136/D3
Bauria, India 68/E2
Baurtregaum (mt.), Ireland 17/A7
Bauru, Brazil 120/E5
Bauru, Brazil 135/B3
Bauru, Brazil 132/D8
Bauska, U.S.S.R. 53/B2
Bauta, Cuba 158/C1
Bauta (riv.), P. Rico 161/C2
Bautzen, E. Germany 22/F3
Bauxite, Ark. (72011) 202/F4
Bavaria, Kansas (67419) 232/E3
Bavaria (state), W. Germany 22/D4
Bavarian (for.), W. Germany 22/E4
Bavarian Alps (mts.), Austria 41/A3
Bavarian Alps (range), W. Germany 22/D5
Baviácora, Mexico 150/E2
Bavispe, Mexico 150/E1
Bavispe, Río de (riv.), Mexico 150/E1
Bawean (isl.), Indonesia 85/K1
Bawku, Ghana 106/D6
Bawlf, Alberta 182/D3
Ba Xian, China 77/J4
Baxley, Georgia (31513) 217/H7
Baxoi, China 77/E5
Baxter (co.), Ark. 202/F1
Baxter, Iowa (50028) 229/G5
Baxter, Minn. (†56401) 255/D4
Baxter, Miss. (†39343) 256/F6
Baxter, Pa. (†15829) 294/D3
Baxter, Tenn. (38544) 237/K8
Baxter Springs, Kansas (66713) 232/H4
Baxterville, Miss. (†39455) 256/E8

Bay, Ark. (72411) 202/J2
Bay (Baicheng), China 77/B3
Bay (co.), Fla. 212/C6
Bay (co.), Mich. 250/E5
Bay, Mo. (65041) 261/J5
Bay, Laguna de (lake), Philippines 82/C3
Bay (prov.), Somalia 115/H3
Bayag (Calanasan), Philippines 82/C1
Bayaguana, Dom. Rep. 158/E6
Bayamhongor, Mongolia 77/E2
Bayamo, Cuba 156/C2
Bayamo, Cuba 158/H4
Bayamón (dist.), P. Rico 161/D1
Bayamón, P. Rico 161/D1
Bayamón, P. Rico 156/G1
Bayamón (riv.), P. Rico 161/D1
Bayanbaraat, Mongolia 77/G2
Bayandalay, Mongolia 77/F3
Bayan Dobo Suma, Mongolia 77/G3
Bayang, Philippines 82/E7
Bayangovĭ, Mongolia 77/F3
Bayan Har Shan (range), China 77/E5
Bayanhongor, Mongolia 77/E2
Bayan Mod, China 77/F3
Bayan Obo, China 77/G3
Bayan-Ölgiy, Mongolia 77/C2
Bayan-Öndör, Mongolia 77/E3
Bayan-Uul, Mongolia 77/E2
Bayard, Del. (†19945) 245/T6
Bayard, Iowa (50029) 229/D5
Bayard, Nebr. (69334) 264/A3
Bayard, N. Mex. (88023) 274/A6
Bayard, Sask. 181/F5
Bayard, W. Va. (26707) 312/H4
Bayat, Turkey 63/F2
Baybay, Philippines 82/E5
Baybay, Philippines 85/H3
Bayble, Scotland 15/B2
Bayboro, N.C. (28515) 281/R4
Bay Bulls, Newf. 166/D2
Bayburt, Turkey 59/D1
Bayburt, Turkey 63/J2
Bay Center, Wash. (98527) 310/A4
Bay Chimo, N.W. Terrs. 187/H3
Bay City, Mich. 188/K2
Bay City, Mich. (48706) 250/F5
Bay City, Oreg. (97107) 291/D2
Bay City, Texas (77414) 303/H9
Bay City, Wash. (†98520) 310/B4
Bay City, Wis. (54723) 317/B6
Baydarata (bay), U.S.S.R. 52/L1
Bay de Verde, Newf. 166/D2
Baydhabo, Somalia 115/H3
Baydhabo, Somalia 102/G4
Baydrag, Mongolia 77/E2
Bay du Vin (riv.), New Bruns. 170/E2
Bayerischer Wald Nat'l Park, W. Germany 22/E4
Bayeux, France 28/C3
Bayfield, Colo. (81122) 208/D8
Bayfield, New Bruns. 170/G2
Bayfield, Ontario 177/C4
Bayfield (sound), Ontario 177/B2
Bayfield (co.), Wis. 317/D3
Bayfield, Wis. (54814) 317/E2
Bayham, Ontario 177/D5
Bay Harbor Islands, Fla. (†33101) 212/B4
Bay Head, N.J. (08742) 273/E3
Bayhead, Nova Scotia 168/E3
Bayındır, Turkey 63/B3
Bayırköy, Turkey 63/B6
Baykal (lake), U.S.S.R. 54/N4
Baykal (lake), U.S.S.R. 2/Q3
Baykal (lake), U.S.S.R. 48/L4
Baykal (mts.), U.S.S.R. 48/L4
Baykan, Turkey 63/J3
Baykit, U.S.S.R. 48/K3
Baykonyr, U.S.S.R. 48/G5
Bay Lake, Fla. (†32786) 212/E3
Bay Lake, Minn. (†56444) 255/E4
Bay L'Argent, Newf. 166/D4
Baylis, Ill. (62314) 222/C4
Baylor (co.), Texas 303/E4
Bay Minette, Ala. (36507) 195/C9
Bay Mission, W. Australia 88/C3
Baynes Lake, Br. Col. 184/K5
Bayombong, Philippines 82/C2
Bayombong, Philippines 85/G2
Bayonne, France 28/C6
Bayonne, N.J. (07002) 273/B2
Bayonne Military Ocean Terminal, N.J. 273/B2
Bayou, Ky. (†42081) 237/E6
Bayou Barbary, La. (†70754) 238/M2
Bayou Bodcau (res.), Ark. 202/C7
Bayou Cane, La. (†70360) 238/J7
Bayou Chicot, La. (†70586) 238/F5
Bayou Current, La. (†71353) 238/G5
Bayou D'Arbonne (lake), La. 238/F1
Bayou Des Arc (riv.), Ark. 202/G3
Bayou Goula, La. (70716) 238/J3
Bayou La Batre, Ala. (36509) 195/B10
Bayou Meto, Ark. (†72160) 202/H5
Bayou Vista, La. (†70380) 238/H7
Bayóvar, Peru 128/B5
Bay Pines, Fla. (33504) 212/B3
Bay Point, Maine (†04548) 243/D8
Bay Point (isl.), S.C. 296/F7
Bayport, Fla. (†33512) 212/D3
Bay Port, Mich. (48720) 250/F5
Bayport, Minn. (55003) 255/F5
Bayport, N.Y. (11705) 276/O9
Bayram-Ali, U.S.S.R. 48/G6
Bayramiç, Turkey 63/B3
Bayreuth, W. Germany 22/D4
Bayrischzell, W. Germany 22/E5
Bay Roberts, Newf. 166/D2
Bays, Ky. (41310) 237/P5
Bays (lake), Ontario 177/F2
Bay Saint Lawrence, Nova Scotia 168/H1
Bay Saint Louis, Miss. (39520) 256/F10
Bayshore, Fla. (†33902) 212/E5
Bayshore, Mich. (49711) 250/D3
Bay Shore, N.Y. (11706) 276/O9
Bayshore Gardens, Fla. (33507) 212/D4
Bayside, Calif. (95524) 204/B3
Bayside, Maine (†04915) 243/F7
Bayside, New Bruns. 170/C3
Bayside, Ontario 177/G3
Bayside, Texas (78340) 303/G9
Bayside, Wis. (†53201) 317/M1
Bay Springs, Fla. (†36502) 212/B6
Bay Springs, Miss. (39422) 256/F7
Bay Springs (dam), Miss. 256/H1
Bay Springs (lake), Miss. 256/H1
Bayston Hill, England 13/E5
Baysville, Ontario 177/E2
Baytown, Texas (77520) 303/L2
Bay Tree, Alberta 182/A2
Bayuca, Spain 33/B1
Bayview, Calif. (†95501) 204/A3
Bayview, Idaho (83803) 220/B2
Bayview, Md. (†21901) 245/P2
Bay View, Mich. (49770) 250/E3
Bay View, N. Zealand 100/F3
Bay View, Ohio (†44870) 284/E3
Bay Village, Ohio (44140) 284/G9
Bayville, N.J. (08721) 273/E4
Bayville, N.Y. (11709) 276/R6
Baywood, La. (†70739) 238/K1
Baywood Park-Los Osos, Calif. (†93402) 204/E8
Baza, Spain 33/E4
Bazaar, Kansas (†66845) 232/F3
Bazaruto, Ilha do (isl.), Mozambique 118/F4
Bazas, France 28/C5
Bazhong, China 77/G5
Bazile Mills, Nebr. (†68729) 264/G2
Bazine, Kansas (67516) 232/C3
Bazman, Iran 66/M7
Bazman, Kuh-e (mt.), Iran 66/H4
Bazman, Kuh-e (mt.), Iran 59/H4
Beach (pond), Conn. 210/H2
Beach, Georgia (†31554) 217/G8
Beach, N. Dak. (58621) 282/C6
Beachburg, Ontario 177/H2
Beach City, Ohio (44608) 284/G4
Beach City, Texas (†77520) 303/L2
Beach Haven, N.J. (08008) 273/E4
Beach Haven (inlet), N.J. 273/E4
Beach Haven Crest, N.J. (†08008) 273/E4
Beach Haven Terrace, N.J. (†08008) 273/E4
Beach Lake, Pa. (18405) 294/M2
Beach Meadows, Nova Scotia 168/D4
Beachport, S. Australia 94/F7
Beachton, Georgia (†31792) 217/D9
Beachville, Ontario 177/D4
Beachwood, N.J. (08722) 273/E4
Beachwood, Ohio (44122) 284/J9
Beachy (head), England 10/G5
Beachy (head), England 13/H7
Beacon, Iowa (52534) 229/H6
Beacon, N.Y. (12508) 276/N7
Beacon, Tenn. (†38363) 237/E9
Beacon Falls○, Conn. (06403) 210/C3
Beaconia, Manitoba 179/F4
Beaconsfield, England 13/G8
Beaconsfield, Iowa (50030) 229/E7
Beaconsfield, Québec 172/H4
Beaconsfield, Tasmania 99/C3
Beadle, Sask. 181/B4
Beadle (co.), S. Dak. 298/N5
Beagle (chan.), Chile 138/E11
Beagle, Kansas (†66064) 232/G3
Beagle (gulf), North. Terr. 93/A2
Beagle, W. Australia 88/C3
Beaglebay Aboriginal Res., W. Australia 92/C2
Beagle Bay Mission, W. Australia 92/C2
Beal (range), Queensland 95/B5
Bealanana, Madagascar 118/H2
Beal City, Mich. (†48858) 250/D5
Beale (cape), Br. Col. 184/E6
Beale A.F.B., Calif. 204/D4
Bealeton, Va. (22712) 307/N3
Beallsville, Ohio (43716) 284/J6
Beallsville, Pa. (15313) 294/C5
Beals, Ky. (†42451) 237/G5
Beals○, Maine (04611) 243/H7
Beals (creek), Texas 303/C5
Beaman, Iowa (50609) 229/H4
Beaminster, England 13/E7
Beanblossom, Ind. (†46160) 227/E6
Beanblossom (creek), Ind. 227/D6
Bean City, Fla. (†33459) 212/F5
Bean Station, Tenn. (37708) 237/P8
Bear (mt.), Alaska 196/K2
Bear (lake), Alberta 182/A2
Bear (lake), Br. Col. 184/D2
Bear (creek), Colo. 208/P8
Bear (hill), Conn. 210/B2
Bear (mt.), Conn. 210/B1
Bear, Del. (19701) 245/R2
Bear, Idaho (83612) 220/B4
Bear (lake), Idaho 220/G7
Bear (riv.), Idaho 220/G7
Bear (isl.), Ireland 17/B8
Bear (riv.), Maine 243/B6
Bear (riv.), Minn. 255/E3
Bear (creek), N. Dak. 282/O7
Bear (isl.), Norway 4/B9
Bear (creek), Oreg. 291/K2
Bear (creek), Oreg. 291/E5
Bear (creek), Oreg. 291/G4
Bear (hills), Sask. 181/C4
Bear (isls.), U.S.S.R. 4/B1
Bear (lake), Utah 304/C2
Bear (riv.), Utah 304/B2
Bear (isl.), Wis. 317/E1
Bear (creek), Wyo. 319/H4
Bear (riv.), Wyo. 319/B4
Bear Branch, Ind. (†47018) 227/G7
Bearcamp (riv.), N.H. 268/E4
Bear Canyon, Alberta 182/A1
Bear Creek, Ala. (35543) 195/C2
Bearcreek, Mo. (†65649) 261/E7
Bearcreek, Mont. (59007) 262/G5
Bear Creek, N.C. (27207) 281/L3
Bear Creek, Pa. (18602) 294/F7
Bear Creek, Sask. 181/K5
Bear Creek, Wis. (54922) 317/J6
Beard, Ind. (†46041) 227/E4
Beard, W. Va. (†24946) 312/F6
Bearden, Ark. (71720) 202/E6
Bearden, Okla. (74859) 288/O4
Beardmore (glac.) 5/A8
Beardmore, Ontario 177/H5
Beardmore, Ontario 175/C3
Beards Fork, W. Va. (25014) 312/D6
Beardsley, Kansas (†67745) 232/A2
Beardsley, Minn. (56211) 255/B5
Beardstown, Ill. (62618) 222/C3
Beardstown, Ind. (†46996) 227/D2
Beardstown, Tenn. (†37097) 237/F9
Beargrass, N.C. (†27892) 281/P3
Bearhat (mt.), Mont. 262/C2
Bear in the Lodge (creek), S. Dak. 298/F6
Bear Island, Ontario 177/K5
Bear Lake, Br. Col. 184/F3
Bear Lake (co.), Idaho 220/G7
Bear Lake, Mich. (49614) 250/C4
Bear Lake, Pa. (16402) 294/C1
Bear Lodge, Wyo. (†82836) 319/E1
Bear Lodge (mts.), Wyo. 319/H1
Bearmouth, Mont. (†59832) 262/C4
Béarn (trad. prov.), France 29
Bearpaw (mts.), Mont. 262/G2
Bear River, Minn. (†55723) 255/E3
Bear River, Nova Scotia 168/C4
Bear River, Pr. Edward I. 168/F2
Bear River (range), Utah 304/C1
Bear River City, Utah (84301) 304/B2
Bear River Divide (mts.), Wyo. 319/B4
Bearsden, Scotland 15/B2
Bearskin Lake, Ontario 175/B2
Bear Spring, Tenn. (†37058) 237/F8
Beartooth (mts.), Mont. 262/G5
Beartown, W. Va. (†24817) 312/C8
Beas de Segura, Spain 33/E3
Beason, Ill. (62512) 222/D3
Beata (cape), Dom. Rep. 158/D7
Beata (cape), Dom. Rep. 156/D3
Beata (chan.), Dom. Rep. 158/C7
Beata (isl.), Dom. Rep. 158/C7
Beata (isl.), Dom. Rep. 156/D3
Beatenberg, Switzerland 39/E3
Beaton, Br. Col. 184/J5
Beatrice, Ala. (36425) 195/D7
Beatrice, Nebr. 188/G2
Beatrice, Nebr. (68310) 264/H4
Beatrice (cape), North. Terr. 88/F2
Beatrice (cape), North. Terr. 93/E3
Beattie, Kansas (66406) 232/F2
Beattock, Scotland 15/E5
Beatton (riv.), Br. Col. 184/G1
Beatton River, Br. Col. 184/G1
Beatty, Nev. (89003) 266/E6
Beatty, Oreg. (97621) 291/F5
Beatty, Sask. 181/G3
Beattyville, Ky. (41311) 237/O5
Beau (lake), Québec 172/H2
Beaubier, Sask. 181/G6
Beaubois, New Bruns. 170/E1
Beaucaire, France 28/F6
Beauce (co.), Québec 172/G3
Beauceville, Québec 172/G3
Beaucoup, Ill. (†62263) 222/D5
Beaudesert, Queensland 95/E6
Beauford, Minn. (†56037) 255/D7
Beaufort (sea) 4/B16
Beaufort (sea) 146/D2
Beaufort (sea), Alaska 196/K1
Beaufort, Malaysia 85/F4
Beaufort, Mo. (65013) 261/K6
Beaufort (co.), N.C. 281/R4
Beaufort, N.C. (28516) 281/R5
Beaufort (sea), N.W.T. 162/C1
Beaufort (sea), N.W. Terrs. 187/D2
Beaufort (co.), S.C. 296/F7
Beaufort, S.C. (29902) 296/F7
Beaufort, Victoria 97/B5
Beaufort (sea), Yukon 187/E2
Beaufort Marine Air Sta., S.C. 296/F7
Beaufort West, S. Africa 118/C6
Beauharnois (co.), Québec 172/C4
Beauharnois, Québec 172/D4
Beaulac, Québec 172/F4
Beaulieu, Minn. (†56557) 255/C3
Beauly, Scotland 10/D2
Beauly, Scotland 15/D3
Beauly (riv.), Scotland 15/D3
Beaumaris (bay), Victoria 97/J6
Beaumaris (bay), Victoria 88/L8
Beaumaris, Wales 13/C4
Beaumaris, Wales 10/D4
Beaumont, Alberta 182/D3
Beaumont, Belgium 27/E8
Beaumont, Calif. (92223) 204/J10
Beaumont, Kansas (67012) 232/F4
Beaumont, Miss. (39423) 256/G8
Beaumont, N. Zealand 100/B6
Beaumont, Québec 172/F3
Beaumont (pt.), St. Lucia 161/F6
Beaumont, Texas 146/J6
Beaumont, Texas 188/H4
Beaumont, Texas (*77701) 303/K7
Beaune, France 28/F4
Beauport, Québec 172/J3
Beaupré, Québec 172/G2
Beauraing, Belgium 27/F8
Beauregard (par.), La. 238/D5
Beauregard, Miss. (†39191) 256/D7
Beauséjour, Manitoba 179/F4
Beauty, Ky. (41203) 237/S5
Beauty Point, Tasmania 99/C3
Beauvais, France 28/E3
Beauval, Sask. 181/L3
Beauvallon, Alberta 182/E3
Beaux Arts Village, Wash. (†98004) 310/B2
Beaver (riv.) 162/F5
Beaver, Alaska (99724) 196/J1
Beaver (creek), Alaska 196/J1
Beaver (riv.), Alberta 182/E2
Beaver, Ark. (72613) 202/C1
Beaver (lake), Ark. 202/C1
Beaver (creek), Colo. 208/M3
Beaver (creek), Idaho 220/F5
Beaver, Iowa (50031) 229/E4
Beaver, Kansas (67517) 232/D3
Beaver (creek), Kansas 232/A2
Beaver, La. (†71463) 238/E5
Beaver (isl.), Mich. 250/D3
Beaver (lake), Mich. 250/F4
Beaver (creek), Mont. 262/J2
Beaver (creek), Nebr. 264/D5
Beaver (riv.), Newf. 166/B3
Beaver (brook), N.H. 268/E6
Beaver (brook), N.J. 273/C2
Beaver (riv.), N.Y. 276/K3
Beaver (creek), N. Dak. 282/K7
Beaver (creek), N. Dak. 282/B5
Beaver (lake), N. Dak. 282/L7
Beaver, Ohio (45613) 284/E7
Beaver (co.), Okla. 288/E1
Beaver, Okla. (73932) 288/F1
Beaver (creek), Okla. 288/K6
Beaver (riv.), Okla. 288/F1
Beaver, Oreg. (97108) 291/D2
Beaver (co.), Pa. 294/B4
Beaver, Pa. (15009) 294/B4
Beaver (riv.), Pa. 294/B4
Beaver (hills), Sask. 181/H4
Beaver (riv.), Sask. 181/L4
Beaver (creek), S. Dak. 298/A6
Beaver (co.), Utah 304/A5
Beaver, Utah (84713) 304/B5
Beaver (mts.), Utah 304/A5
Beaver (riv.), Utah 304/A5
Beaver, Wash. (98305) 310/A2
Beaver (Glen Hedrick), W. Va. (25813) 312/D7
Beaver, Wis. (54105) 317/K5
Beaver (creek), Wyo. 319/H2
Beaver (creek), Wyo. 319/D3
Beaverbank, Nova Scotia 168/E4
Beaver Bay, Minn. (55601) 255/G3
Beaver Brook Station, New Bruns. 170/E1
Beaver City, Nebr. (68926) 264/E4
Beaver Cove, Br. Col. 184/D5
Beaver Creek, Md. (†21740) 245/H2
Beaver Creek, Minn. (56116) 255/B7
Beavercreek, Ohio (†45690) 284/C6
Beavercreek, Oreg. (97004) 291/B2
Beaver Creek, Yukon 187/D3
Beaver Creek Fork, Humboldt (riv.), Nev. 266/F1
Beaver Crossing, Nebr. (68313) 264/G4
Beaverdale, Pa. (15921) 294/E5
Beaverdam, Alberta 182/E2
Beaver Dam, Ky. (42320) 237/H6
Beaverdam, Ohio (45808) 284/C4
Beaverdam, Va. (23015) 307/N5
Beaver Dam, Wis. (53916) 317/J9
Beaver Dam (lake), Wis. 317/J9
Beaver Dams, N.Y. (14812) 276/F6
Beaver Dam Wash (creek), Utah 304/A6
Beaverdell, Br. Col. 184/H5
Beaver Falls, N.Y. (13305) 276/K3
Beaver Falls, Pa. (15010) 294/B4
Beaver Harbour, New Bruns. 170/D3
Beaverhead (mts.), Idaho 220/E4
Beaverhead (co.), Mont. 262/C5
Beaverhead (riv.), Mont. 262/D5
Beaverhill (lake), Alberta 182/D3
Beaverhill (lake), Manitoba 179/J3
Beaver Lake, Alberta 182/E2
Beaver Lake, N.J. (†07416) 273/D1
Beaverlett, Va. (23016) 307/R6
Beaverlodge, Alberta 182/A2
Beaverlodge (lake), Sask. 181/L2
Beaver Meadows, Pa. (18216) 294/L4
Beaver Mines, Alberta 182/C5
Beaver Park, Sask. 181/J6
Beaver River, N.Y. (13306) 276/L3
Beaver River Flow (lake), N.Y. 276/K3
Beaver Springs, Pa. (17812) 294/H4
Beaverton, Ala. (35544) 195/B3
Beaverton, Mich. (48612) 250/E5
Beaverton, Ontario 177/E3
Beaverton, Oreg. (97005) 291/A2
Beavertown, Pa. (17813) 294/H4
Beaverville, Ill. (60912) 222/F3
Beawar, India 68/C3
Beazer, Alberta 182/D5
Beazley, Argentina 143/C3
Bebedouro, Brazil 132/D8
Bebedouro, Brazil 135/B2
Bebee, W. Va. (†26155) 312/E3
Bebington, England 10/F2
Bebington, England 13/G2
Bebra, W. Germany 22/C3
Bécancour, Québec 172/E3
Bécancour (riv.), Québec 172/F3
Beccles, England 13/J5
Beccles, England 10/G4
Bečej, Yugoslavia 45/E3
Becerreá, Spain 33/C1
Béchar, Algeria 102/B1
Béchar, Algeria 106/D2
Bechar (pt.), W. Australia 88/A3
Bechard, Sask. 181/G5
Becharof (lake), Alaska 196/G3
Bechyn, Minn. (†56275) 255/C6
Bechyně, Czech. 41/C2
Becida, Minn. (56625) 255/C3
Beckemeyer, Ill. (62219) 222/D5
Becker (co.), Minn. 255/C4
Becker, Minn. (55308) 255/E5
Becker, Miss. (38825) 256/G3
Becket○, Mass. (01223) 249/B3
Becket, Mont. (†59441) 262/G4
Beckham (co.), Okla. 288/G4
Beckley, W. Va. (25801) 312/D7
Beckton, Wyo. (†82801) 319/E1
Beckville, Texas (75631) 303/K5
Beckwourth, Calif. (96129) 204/E4
Bečva (riv.), Czech. 41/E2
Bedale, England 13/F3
Bédarieux, France 28/E6
Beddington○, Maine (†04622) 243/H6
Beddouza, Ras (cape), Morocco 106/C2
Bedele, Ethiopia 111/G6
Bedeque (bay), Pr. Edward I. 168/E2
Bedessa, Ethiopia 111/H6
Bedford, England 10/F4
Bedford, England 13/G5
Bedford (pt.), Grenada 161/D8
Bedford, Ind. (47421) 227/E7
Bedford, Iowa (50833) 229/D7
Bedford, Ky. (40006) 237/L3
Bedford○, Mass. (01730) 249/B6
Bedford, Mich. (49020) 250/D6
Bedford, Mo. (†64643) 261/F3
Bedford○, N.H. (03102) 268/D6
Bedford (basin), Nova Scotia 168/E4
Bedford, Ohio (44146) 284/H9
Bedford (co.), Pa. 294/E6
Bedford, Pa. (15522) 294/F5
Bedford, Québec 172/E4
Bedford (co.), Tenn. 237/J9
Bedford, Texas (76021) 303/F2
Bedford (co.), Va. 307/J6
Bedford (I.C.), Va. (24523) 307/J6
Bedford, Wyo. (83112) 319/A3
Bedford Heights, Ohio (†44146) 284/J9
Bedford Hills, N.Y. (10507) 276/N8
Bedford Park, Ill. (†60601) 222/B6
Bedfordshire (co.), England 13/G5
Bedford Valley, Pa. (†15522) 294/E6
Bedias, Texas (77831) 303/J7
Bedington, W. Va. (†25401) 312/L3
Bedlington, England 13/F2
Bedlington, England 10/E3
Bedminster○, N.J. (07921) 273/D2
Bedminster, Pa. (18910) 294/M5
Bedouaram (well), Niger 106/G5
Bedourie, Queensland 95/A5
Bedourie, Queensland 88/F4
Bedretto, Switzerland 39/G4
Bedrock, Colo. (81411) 208/B6
Bedsted, Denmark 21/B4
Bedwas and Machen, Wales 13/B6
Bedwellty, Wales 13/B6
Bedworth, England 13/F5
Bedworth, England 10/F4
Będzin, Poland 47/B3
Bee, Nebr. (68314) 264/H3
Bee (co.), Texas 303/G9
Beebe, Ark. (72012) 202/G3
Beebe Plain, Québec 172/E4
Beebe Plain, Vt. (05823) 268/C2
Beebe River, N.H. (03219) 268/D4
Bee Branch, Ark. (72013) 202/F3
Beech Bluff, Tenn. (38313) 237/D9
Beech Bottom, W. Va. (26030) 312/E2
Beech Creek, Ky. (42321) 237/G6
Beech Creek, Pa. (16822) 294/G3
Beecher, Ill. (60401) 222/F2
Beecher City, Ill. (62414) 222/E4
Beecher Falls, Vt. (05902) 268/D2
Beecher Island, Colo. (†80758) 208/P3
Beech Fork (riv.), Ky. 237/L5
Beech Grove, Ind. (46107) 227/E5
Beech Grove, Ky. (42322) 237/G5
Beechgrove, Tenn. (37018) 237/J9
Beech Island, S.C. (29842) 296/D5
Beechwood, Mass. (02025) 249/F8
Beechwood, Mich. (†49423) 250/C6
Beechwood, New Bruns. 170/C2
Beechwood, N.S. Wales 97/G2
Beechwood Village, Ky. (40359) 237/K2
Beechworth, Victoria 97/D5
Beechy, Sask. 181/D5
Beechy Point, Alaska (†99723) 196/H1
Beedeville, Ark. (72014) 202/H3
Beekman, La. (†71220) 238/G1
Beeler, Kansas (67518) 232/B3
Bee Log, N.C. (†28714) 281/E3
Beemer, Nebr. (68716) 264/H3
Beenleigh, Queensland 88/J5
Beer Ef'e (well), Israel 65/C5
Be'eri, Israel 65/A5
Bee Ridge, Fla. (†33578) 212/D4
Be'er Menuha, Israel 65/D5
Beernem, Belgium 27/C6
Be'er Ora, Israel 65/D5
Beersheba (Be'er Sheva), Israel 65/B5
Beersheba Springs, Tenn. (37305) 237/K10
Beer Sheva' (dry riv.), Israel 65/B5
Beersville, New Bruns. 170/E2
Be'er Tuveya, Israel 65/B4
Beeskow, E. Germany 22/F2
Beesleys Point, N.J. (†08226) 273/D5
Beeson, W. Va. (24714) 312/D8
Bee Spring, Ky. (42207) 237/J6
Beeston and Stapleford, England 13/F5
Beeton, Ontario 177/E3
Beetown, Wis. (53802) 317/E10
Beeville, Texas (78102) 303/G9
Befale, Zaire 115/D3
Befandriana, Madagascar 118/H3
Beg (lake), N. Ireland 17/J2
Bega, N. S. Wales 88/J7
Bega, N.S. Wales 97/E5
Begemdir (prov.), Ethiopia 111/G5
Beger, Mongolia 77/E2
Beggs, Okla. (74421) 288/P3
Begnins, Switzerland 39/B4
Béhague (pt.), Fr. Guiana 131/F3
Behan (lake), Alberta 182/E2
Behbehan, Iran 66/G5
Behistun (ruins), Iran 66/E3
Behm Canal (inlet), Alaska 196/N2
Behshahr, Iran 66/H2
Bei'an (Pehan), China 77/L2
Beica, Ethiopia 111/F6
Beihai (Pakhoi), China 77/G7
Beijing (Peking) (cap.), Peoples Rep. of China 54/N5
Beijing (Peking) (cap.), Peoples Rep. of China 77/J3
Beilen, Netherlands 27/K3
Beinn a Ghlo (mt.), Scotland 15/E4
Beinn Bhan (mt.), Scotland 15/C3
Beinn Bheigeir (mt.), Scotland 15/B5
Beinn Dearg (mt.), Scotland 15/D3
Beinn Dearg (mt.), Scotland 15/E4
Beinn Dhorain (mt.), Scotland 15/E2
Beinn Dorain (mt.), Scotland 15/D4
Beinn Eighe (mt.), Scotland 15/C3
Beinwil am See, Switzerland 39/F2
Beira, Mozambique 118/F3
Beira, Mozambique 102/F7
Beira, Somalia 115/J2
Beirne, Ark. (71721) 202/D6
Beirut (cap.), Lebanon 54/E6
Beirut (cap.), Lebanon 59/C3
Beirut (cap.), Lebanon 63/F6
Beiseker, Alberta 182/D4
Beishan, China 77/E3
Beitbridge, Zimbabwe 118/E4
Beit Fajjar, West Bank 65/C4
Beit Guvrin, Israel 65/B4
Beith, Scotland 10/A1
Beith, Scotland 15/D5
Beit Hanina, West Bank 65/C4
Beit Hanun, Gaza Strip 65/A4
Beit Jala, West Bank 65/C4
Beit Lahm (Bethlehem), West Bank 65/C4
Beit Nuba, West Bank 65/C4
Beit Sahur, West Bank 65/C4
Beiuş, Romania 45/F2
Beja (dist.), Portugal 33/C3
Beja, Portugal 33/C3
Béja, Tunisia 106/F1
Bejaïa, Algeria 106/F1
Bejaïa, Algeria 102/C1
Béjar, Spain 33/D2
Bejestan, Iran 59/G3
Bejestan, Iran 66/K3
Bejhi (riv.), Pakistan 68/B3
Bejou, Minn. (56516) 255/B3
Bejucal, Cuba 158/C1
Bejucal, Cuba 156/A2
Bekasi, Indonesia 85/H2
Békés (co.), Hungary 41/F3
Békés, Hungary 41/F3
Békéscsaba, Hungary 41/F3
Bekily, Madagascar 118/G4
Bekwai, Ghana 106/D7
Bel, La. (†70658) 238/D6
Bela, Pakistan 68/B3
Bela, Pakistan 59/J4
Bélabo, Cameroon 115/B3
Bela Crkva, Yugoslavia 45/E3
Bélair, Manitoba 179/F4
Bel Air, Md. (21014) 245/N2
Bélair, Québec 172/H3
Bel Alton, Md. (20611) 245/L7
Belas, Portugal 33/A1
Belau (Palau) 87/D5
Bela Vista, Angola 115/C6
Bela Vista, Brazil 120/D5
Bela Vista, Mato Grosso, Brazil 132/C8
Bela Vista, Rondônia, Brazil 132/H10
Bela Vista, Mozambique 118/E5
Bela Vista de Goiás, Brazil 132/D7
Belawan, Indonesia 85/B5
Belaya (riv.), U.S.S.R. 7/K3
Belaya (riv.), U.S.S.R. 52/H3
Belaya Tserkov', U.S.S.R. 52/C5
Belbeck, Sask. 181/F5
Belbutte, Sask. 181/D2
Bełchatów, Poland 47/D3
Belcher, Ky. (41513) 237/S6
Belcher, La. (71004) 238/C1
Belcher (isls.), N.W.T. 146/K4
Belcher (isl.), N.W.T. 162/H4
Belcher (chan.), N.W. Terrs. 187/J2
Belcheragh, Afghanistan 68/B1
Belcheragh, Afghanistan 59/J2
Belchertown, Mass. (01007) 249/E3
Belchertown○, Mass. (01007) 249/E3
Belchite, Spain 33/F2
Belcourt, N. Dak. (58316) 282/L2
Belcross, N.C. (†27921) 281/S2
Belden, Calif. (95915) 204/D3
Belden, Miss. (38826) 256/G2
Belden, Nebr. 264/G2
Belden, N. Dak. (58715) 282/F3
Beldenville, Wis. (54003) 317/A6
Belding, Mich. (48809) 250/D5
Belebey, U.S.S.R. 52/H4
Belém, Brazil 2/G6
Belém, Brazil 120/E3
Belém, Brazil 132/E3
Belém, Portugal 33/A1
Belén, Argentina 143/C2
Belén, Chile 138/B1
Belén, Honduras 154/C3
Belen, Miss. (38609) 256/D2
Belen, N. Mex. (87002) 274/C4
Belén, Panama 154/G6
Belén, Paraguay 144/D3
Belén, Uruguay 145/B1
Belén (range), Uruguay 145/C1
Belén de los Andaquíes, Colombia 126/C7
Belep (isls.), New Caled. 87/G7
Belet Weyne, Somalia 115/J3
Belet Weyne, Somalia 102/G4
Belev, U.S.S.R. 52/E4
Belfair, Wash. (98528) 310/C3
Belfast, Maine (04915) 243/F7
Belfast, N.Y. (14711) 276/D6
Belfast (cap.), N. Ireland 7/D3
Belfast (dist.), N. Ireland 17/J2
Belfast (cap.), N. Ireland 10/D3
Belfast (cap.), N. Ireland 17/J2
Belfast (inlet), N. Ireland 17/K2
Belfast, Tenn. (37019) 237/H10
Belfast Lough (inlet), N. Ireland 10/D3
Belfaux, Switzerland 39/D3
Belfield, N. Dak. (58622) 282/D6
Belford, England 13/F2
Belford, N.J. (07718) 273/E3
Belfort (terr.), France 28/G4
Belfort, France 28/G4
Belfry, Ky. (41514) 237/S5
Belfry, Mont. (59008) 262/H5
Belgaum, India 68/C5
Belgique, Mo. (†63775) 261/N7
Belgium 2/K3
Belgium 7/E3

BELGIUM 27
Belgium, Ill. (†61883) 222/F3
Belgium, Wis. (53004) 317/L8
Belgorod, U.S.S.R. 7/H3
Belgorod, U.S.S.R. 52/E4
Belgorod, U.S.S.R. 48/D4
Belgorod-Dnestrovskiy, U.S.S.R. 52/D5
Belgrade, Maine (04917) 243/D7
Belgrade○, Maine (04917) 243/D7
Belgrade, Minn. (56312) 255/C5
Belgrade, Mo. (63622) 261/L7
Belgrade, Mont. (59714) 262/E5
Belgrade, Nebr. (68623) 264/G3
Belgrade (cap.), Yugoslavia 7/G4
Belgrade (cap.), Yugoslavia 2/K3
Belgrade (cap.), Yugoslavia 45/E3
Belgrade Lakes, Maine (04918) 243/D6
Belgrave, Ontario 177/C4
Belgrave Heights, Victoria 97/J5
Belgrave South, Victoria 97/K5
Belgreen, Ala. (†35653) 195/C2
Belhaven, N.C. (27810) 281/R3
Belic, Cuba 158/G4
Belice (riv.), Italy 34/D6
Beli Manastir, Yugoslavia 45/D3
Belington, W. Va. (26250) 312/F4
Belitung (Billiton) (isl.), Indonesia 85/D6
Belize 2/E5
Belize 146/K8
BELIZE 154/C2
Belize (riv.), Belize 154/C2
Belize City, Belize 154/C2
Bélizon, Fr. Guiana 131/E3
Belk, Ala. (35545) 195/C3
Belknap, Ill. (†62995) 222/E6
Belknap, Iowa (†52537) 229/J7
Belknap, Mont. (†59874) 262/A3
Belknap (co.), N.H. 268/D4
Belknap (mt.), N.H. 268/E5
Belknap (peak), Utah 304/B5
Belkofski, Alaska (†99612) 196/F3
Bell, Calif. (90201) 204/C11
Bell, Fla. (32619) 212/D2
Bell (co.), Ky. 237/O7
Bell (isl.), Newf. 166/D2
Bell (isl.), Newf. 166/C3
Bell (isl.), Newf. 162/L5
Bell (pen.), N.W. Terrs. 187/K3
Bell (riv.), Que. 162/J6
Bell (riv.), Québec 174/B3
Bell (co.), Texas 303/G6
Bella Bella, Br. Col. 184/D4
Bellac, France 28/D4
Bellaco, Uruguay 145/B3
Bella Coola, Br. Col. 184/D4
Bella Coola (riv.), Br. Col. 184/D4
Belladère, Haiti 158/C6
Bella Flor, Bolivia 136/A2
Bellaghy, N. Ireland 17/H2
Bellagio, Italy 34/B2
Bellaire, Kansas (66934) 232/D2
Bellaire, Mich. (49615) 250/D4
Bellaire, Ohio (43906) 284/J5
Bellaire, Texas (77401) 303/J2
Bellamy, Ala. (36901) 195/B6
Bellarmin, Québec 172/G4
Bellarthur, N.C. (27811) 281/O3
Bellary, India 68/D5
Bellata, N.S. Wales 97/E1
Bella Unión, Uruguay 145/B1
Bella Villa, Mo. (†63101) 261/R4
Bella Vista, Corrientes, Argentina 143/E2
Bella Vista, Tucumán, Argentina 143/D2
Bella Vista, Ark. (†72712) 202/B1
Bella Vista, Bolivia 136/E3
Bella Vista, Salar de (salt dep.), Chile 138/B3
Bella Vista, Paraguay 144/D3
Bella Vista, Paraguay 144/E5
Bellavista, Peru 128/C5
Bell Bay, Tasmania 99/C3
Bellbird-Cessnock, N.S. Wales 97/F3
Bellbrook, Ohio (45305) 284/C6
Bell Buckle, Tenn. (37020) 237/J9
Bellburns, Newf. 166/C3
Bell Center, Wis. (†54631) 317/E9
Bell City, La. (70630) 238/D6
Bell City, Mo. (63735) 261/N8
Belle (riv.), Mich. 250/G6
Belle, Mo. (65013) 261/J6
Belle, W. Va. (25015) 312/C6
Belleair, Fla. (33540) 212/B2
Belleair Beach, Fla. (†33540) 212/B2
Belleair Bluffs, Fla. (33540) 212/B3
Belleair Shores, Fla. (†33540) 212/B3
Belle-Anse, Haiti 158/C6
Belle Center, Ohio (43310) 284/C4
Belle Chasse, La. (70037) 238/O4
Bellechasse (co.), Québec 172/G3
Bellechester, Minn. (†55027) 255/F6
Belle Côte, Nova Scotia 168/G2
Belle D'Eau, La. (†71330) 238/F4
Belledune, New Bruns. 170/E1
Belleek, N. Ireland 17/E3
Bellefleur, New Bruns. 170/C1
Bellefond, New Bruns. 170/E1
Bellefont, Kansas (†67876) 232/C4
Bellefontaine, Martinique 161/C6
Bellefontaine, Miss. (39737) 256/F3
Bellefontaine, Mo. (†63017) 261/N2
Bellefontaine, Ohio (43311) 284/C5
Bellefontaine Neighbors, Mo. (†63137) 261/R2
Bellefonte, Ark. (†72601) 202/D1
Bellefonte, Del. (19809) 245/S1
Bellefonte, Pa. (16823) 294/G4
Belle Fourche (riv.) 188/F2
Belle Fourche, S. Dak. (57717) 298/B4
Belle Fourche (res.), S. Dak. 298/B4
Belle Fourche (riv.), S. Dak. 298/C4
Belle Fourche (riv.), Wyo. 319/H1
Bellegarde, Sask. 181/K6
Belle Glade, Fla. (33430) 212/F5
Belle Glade Camp, Fla. (†33430) 212/F5
Belle Haven, Va. (23306) 307/S5
Belle-Île (isl.), France 28/B4
Belle Isle (str.), Canada 146/N5
Belle Isle (str.), Canada 2/G3
Belle Isle, Fla. (†32801) 212/E3
Belleisle (bay), New Bruns. 170/E3
Belle Isle (isl.), Newf. 166/C3
Belle Isle (str.), Newf. 166/C3
Belle Isle (str.), Newf. 162/L5
Belleisle Creek, New Bruns. 170/E3
Belle-Marche, Nova Scotia 168/H2
Belle Mead, N.J. (08502) 273/D3
Bellemeade, Ky. (†40201) 237/K2
Belle Meade, Tenn. (37205) 237/H8
Belle Mina, Ala. (35615) 195/E1
Bellemont, Ariz. (86015) 198/D3
Belleoram, Newf. 166/C4
Belleplain, N.J. (†08270) 273/D5
Belleplaine, Barbados 161/B8
Belle Plaine, Iowa (52208) 229/J5
Belle Plaine, Kansas (67013) 232/E4
Belle Plaine, Minn. (56011) 255/E6
Belle Plaine, Sask. 181/F5
Belle Prairie City, Ill. (†62828) 222/E5
Belle Rive, Ill. (62810) 222/E5
Belle River, Minn. (†56319) 255/C5
Belle River, Ontario 177/B5
Belle Rose, La. (70341) 238/K3
Bellerose, N.Y. (11426) 276/P7
Belle Terre, N.Y. (†11777) 276/O9
Belleterre, Québec 174/B3
Belle Union, Ind. (†46121) 227/D5
Belle Valley, Ohio (43717) 284/G6
Belle Vernon, Pa. (15012) 294/C5
Belleview, Fla. (32620) 212/D2
Belleview, Manitoba 179/B5
Belleview, Mo. (63623) 261/L7
Belle View, Va. (22307) 307/T3
Belleville, Ark. (72824) 202/D3
Belleville, Fla. (†31636) 212/C1
Belleville, France 28/F4
Belleville, Ill. 188/J3
Belleville, Ill. (*62220) 222/B3
Belleville, Kansas (66935) 232/E2
Belleville, Mich. (48111) 250/F6
Belleville, N.J. (07109) 273/B2
Belleville, N.Y. (13611) 276/H3
Belleville, Ontario 177/G3
Belleville, Pa. (17004) 294/G4
Belleville, W. Va. (26133) 312/C4
Belleville, Wis. (53508) 317/G10
Bellevue, Alberta 182/C5
Bellevue, Idaho (83313) 220/D6
Bellevue, Iowa (52031) 229/M4
Bellevue, Ky. (41073) 237/S1
Bellevue, Md. (†21662) 245/O6
Bellevue, Mich. (49021) 250/E6
Bellevue, Nebr. (68005) 264/J3
Bellevue, Newf. 166/D2
Bellevue, Ohio (44811) 284/E3
Bellevue, Pa. (15202) 294/B6
Bellevue, Sask. 181/F3
Bellevue, Texas (76228) 303/F4
Bellevue, Wash. (*98004) 310/B2
Belley, France 28/F5
Bell Farm, Ky. (†42647) 237/M7
Bellflower, Calif. (90706) 204/C11
Bellflower, Ill. (61724) 222/E3
Bellflower, Mo. (63333) 261/K4
Bellfountain, Oreg. (†97456) 291/D3
Bell Gardens, Calif. (90201) 204/C11
Bellin, Que. 162/J3
Bellin, Québec 174/F1
Bellingen, N.S. Wales 97/G2
Bellingham, England 13/E2
Bellingham, Mass. (02019) 249/J4
Bellingham○, Mass. (02019) 249/J4
Bellingham, Minn. (56212) 255/B5
Bellingham, Wash. 188/B1
Bellingham, Wash. (98225) 310/C2
Bellingham, Wash. 310/80
Bellingshausen (sea), Ant. 2/E9
Bellingshausen (sea) 5/C14
Bellinzona, Switzerland 39/H4
Bell-Irving (riv.), Br. Col. 184/C2
Bellis, Alberta 182/D2
Belliveau Cove, Nova Scotia 168/B4
Bellmawr, N.J. (08031) 273/B3
Bellmead, Texas (76704) 303/H6
Bellmont, Ill. (62811) 222/F5
Bellmore, Ind. (47830) 227/C5
Bellmore, N.Y. (11710) 276/R7
Bello, Colombia 126/C4
Bello, Colombia 120/B2
Bellona (reefs), New Caled. 87/G8
Bellona (isl.), Solomon Is. 86/D3
Bellot (str.), N.W.T. 162/G1
Bellot (str.), N.W. Terrs. 187/J2
Bellows Falls, Vt. (05101) 268/C5
Belloy, Alberta 182/A2
Bellport, N.Y. (11713) 276/P9
Bell Rock, N.W. Terrs. 187/G3
Bell Rock (isl.), Scotland 15/F4
Bells, Tenn. (38006) 237/C9
Bells, Texas (75414) 303/H4
Bellsbank, Scotland 15/D5
Bellshill, Scotland 15/C2
Bellsite, Manitoba 179/B2
Bellsund, Norway 18/C2
Belluno (prov.), Italy 34/D1
Belluno, Italy 34/D1
Bellview, Ala. (†36452) 195/D7
Bellview, N. Mex. (88111) 274/F4
Bell Ville, Argentina 143/D3
Bell Ville, Argentina 120/C6
Bellville, Georgia (30414) 217/H6
Bellville, Ohio (44813) 284/E4
Bellville, S. Africa 118/F6
Bellville, Texas (77418) 303/H8
Bellvue, Colo. (80512) 208/J1
Bellwald, Switzerland 39/F4
Bellwood, Ala. (36313) 195/G8
Bellwood, Ill. (60104) 222/B5
Bellwood, La. (†71468) 238/D3
Bellwood, Nebr. (68624) 264/G3
Bellwood, Pa. (16617) 294/F4
Belly (riv.), Alberta 182/D5
Belmar, N.J. (07719) 273/E3
Bélmez, Spain 33/D3
Belmond, Iowa (50421) 229/F3
Belmont, Ala. (†35450) 195/C5
Belmont, Calif. (94002) 204/J3
Belmont, Georgia (†30501) 217/E2
Belmont, Kansas (67014) 232/D4
Belmont, Ky. (40105) 237/K5
Belmont, La. (71406) 238/C3
Belmont○, Maine (†04915) 243/E7
Belmont, Manitoba 179/C5
Belmont○, Mass. (02178) 249/C6
Belmont, Miss. (38827) 256/H1
Belmont, Mont. (†59046) 262/G4
Belmont○, N.H. (03220) 268/E5
Belmont, N.Y. (14813) 276/E6
Belmont, N. Zealand 100/B2
Belmont, N.C. (28012) 281/H4
Belmont, Nova Scotia 168/E3
Belmont (co.), Ohio 284/J5
Belmont, Ohio (43718) 284/J5
Belmont, Ontario 177/C5
Belmont, Wash. (99104) 310/H3
Belmont, W. Va. (26134) 312/D4
Belmont, Wis. (53510) 317/F10
Belmonte, Brazil 132/G6
Belmonte, Portugal 33/C2
Belmonte, Spain 33/E3
Belmopan (cap.), Belize 146/K8
Belmopan (cap.), Belize 154/C2
Belmore, N.S. Wales 97/J3
Belmore, Ohio (45815) 284/B3
Belmullet, Ireland 17/B3
Bel-Nor, Mo. (†63101) 261/P2
Belo, W. Va. (†25661) 312/B7
Beloeil, Belgium 27/D7
Beloeil, Québec 172/D4
Belogorsk, U.S.S.R. 54/O4
Belogorsk, U.S.S.R. 48/N4
Belogradchik, Bulgaria 45/F4
Belo Horizonte, Brazil 2/G6
Belo Horizonte, Brazil 120/E4
Belo Horizonte, Brazil 132/F7
Belo Horizonte, Brazil 135/D1
Beloit, Ala. (†36759) 195/D6
Beloit, Kansas (67420) 232/D2
Beloit, Ohio (44609) 284/J4
Beloit, Wis. 188/J2
Beloit, Wis. (53511) 317/H10
Belomorsk, U.S.S.R. 48/D3
Belomorsk, U.S.S.R. 52/D2
Belorado, Spain 33/E1
Belorechensk, U.S.S.R. 52/E6
Beloretsk, U.S.S.R. 52/J4
Beloretsk, U.S.S.R. 48/F4
Belo-Tsiribihina, Madagascar 118/G3
Belovo, U.S.S.R. 48/J4
Beloye (lake), U.S.S.R. 48/D3
Beloye (lake), U.S.S.R. 52/E2
Belozersk, U.S.S.R. 52/E3
Belp, Switzerland 39/D3
Belpre, Kansas (67519) 232/C4
Belpre, Ohio (45714) 284/G7
Bel-Ridge, Mo. (†63101) 261/P2
Belshaw, Ind. (†46356) 227/C2
Belt, Mont. (59412) 262/E3
Belted (range), Nev. 266/E5
Belterra, Brazil 132/C3
Belton, Ky. (42324) 237/H6
Belton, Mo. (64012) 261/C5
Belton, S.C. (29627) 296/C2
Belton, Texas (76513) 303/G7
Beltra (lake), Ireland 17/C4
Beltrami (co.), Minn. 255/C2
Beltrami, Minn. (56517) 255/B3
Beltsville, Md. (20705) 245/G3
Bel'tsy, U.S.S.R. 7/G4
Bel'tsy, U.S.S.R. 52/C5
Belturbet, Ireland 17/G3
Beluga (lake), Alaska 196/B1
Belumut, Gunong (mt.), Malaysia 72/D7
Belush'ya Guba, U.S.S.R. 4/B7
Belush'ya Guba, U.S.S.R. 52/H1
Belva, W. Va. (26656) 312/D6
Belvedere, Calif. (94920) 204/H2
Belvedere, Georgia (†30032) 217/L1
Belvidere, Ill. (61008) 222/E1
Belvidere, Kansas (67015) 232/C4
Belvidere, Nebr. (68315) 264/G4
Belvidere, N.J. (07823) 273/C2
Belvidere, N.C. (27919) 281/S2
Belvidere, S. Dak. (57521) 298/G6
Belvidere, Tenn. (37306) 237/J10
Belvidere○, Vt. (05442) 268/B2
Belvidere (mt.), Vt. 268/B2
Belvidere Center, Vt. (05442) 268/B2
Belvidere Junction, Vt. (†05492) 268/B2
Belview, Minn. (56214) 255/C6
Belville, N.C. (†28451) 281/N6
Belvue, Kansas (66407) 232/F2
Belwood, N.C. (†28090) 281/F4
Belwood, Ontario 177/D4
Belyando (riv.), Queensland 88/H4
Belyando (riv.), Queensland 95/C4
Belyy (isl.), U.S.S.R. 4/B6
Belyy (isl.), U.S.S.R. 48/G2
Belzoni, Miss. (39038) 256/C4
Belzoni, Okla. (†74523) 288/R6
Belzyce, Poland 47/F3
Bem, Mo. (†65066) 261/K6
Bembe, Angola 115/B5
Bemboka, N.S. Wales 97/E5
Bement, Ill. (61813) 222/E4
Bemersyde, Sask. 181/J5
Bemidji, Minn. (56601) 255/D3
Bemidji (lake), Minn. 255/D3
Bemis, S. Dak. (57215) 298/R4
Bemiss, Georgia (†31601) 217/F9
Bemmel, Netherlands 27/H5
Bemus Point, N.Y. (14712) 276/B6
Bena, Minn. (56626) 255/D3
Benabarre, Spain 33/G1
Bena-Dibele, Zaire 115/D4
Ben Alder (mt.), Scotland 15/D4
Benalla, Victoria 97/D5
Benalto, Alberta 182/C3
Benanee, N.S. Wales 97/B4
Benares (Varanasi), India 68/E3
Benavente, Spain 33/D1
Benavides, Texas (78341) 303/F10
Ben Avon, Pa. (†15202) 294/B6
Ben Avon (mt.), Scotland 15/E3
Ben Barvas (mt.), Scotland 15/B2
Benbecula (isl.), Scotland 15/A3
Benbecula (isl.), Scotland 10/C2
Benbrook, Texas (76126) 303/E2
Benbrook (lake), Texas 303/E3
Benchland, Mont. (†59462) 262/F3
Ben Cruachan (mt.), Scotland 15/C4
Bencubbin, W. Australia 92/B5
Bend, Oreg. 188/B2
Bend, Oreg. (97701) 291/F3
Bend, Texas (76824) 303/F7
Ben Dash (hill), Ireland 17/C6
Ben Davis (pt.), N.J. 273/C5
Bendel (state), Nigeria 106/F7
Bendemeer, N.S. Wales 97/F2
Bendena, Kansas (66008) 232/G2
Bender Beila, Somalia 115/K2
Bender Beila, Somalia 102/G4
Bender Cassim (Bosaso), Somalia 115/J1
Bendersville, Pa. (17306) 294/H6
Bendery, U.S.S.R. 52/C5
Bendigo, Australia 87/E9
Bendigo, Victoria 88/G7
Bendigo, Victoria 97/C5
Bendoc, Victoria 97/E5
Bendon, Mich. (†49643) 250/D4
Bendorf, W. Germany 22/B3
Bene Beraq, Israel 65/B3
Benedict (pond), Conn. 210/C1
Benedict, Kansas (66714) 232/G4
Benedict, Md. (20612) 245/M6
Benedict, Minn. (56436) 255/D3
Benedict, Nebr. (68316) 264/G3
Benedict (mt.), Newf. 166/C3
Benedict, N. Dak. (58716) 282/H4
Benedicta○, Maine (04733) 243/G4
Beneditinos, Brazil 132/F4
Benenitra, Madagascar 118/H4
Benešov, Czech. 41/C2
Beneveian, Loch (lake), Scotland 15/D3
Benevento (prov.), Italy 34/E4
Benevento, Italy 34/E4
Benevolence, Georgia (31721) 217/C7
Benewah (co.), Idaho 220/B2
Benezett, Pa. (15821) 294/F3
Benfica, Portugal 33/A1
Benfleet, England 13/J8
Benga, Mozambique 118/E3
Bengal (bay) 54/K8
Bengal, Bay of (sea), Bangladesh 68/F5
Bengal, Bay of (sea) 2/P5
Bengal, Bay of (sea), Burma 72/B3
Bengal, Bay of (sea), India 68/F5
Bengal, Ind. (†46131) 227/F6
Bengal, Okla. (†74966) 288/R5
Ben Gardane, Tunisia 106/G2
Bengbis, Cameroon 115/B3
Bengbu (Pengpu), China 77/J5
Benge, Wash. (99105) 310/G4
Benggala (str.), Indonesia 85/A4
Benghazi, Libya 2/L4
Benghazi, Libya 102/D1
Benghazi (cap.), Libya 111/C1
Ben Ghnema, Jebel (mts.), Libya 111/C2
Bengkalis, Indonesia 85/C5
Bengkayang, Indonesia 85/E5
Bengkulu, Indonesia 85/C6
Bengo (dist.), Angola 115/B5
Bengough, Sask. 181/F6
Ben Griam More (mt.), Scotland 15/D2
Bengtsfors, Sweden 18/H7
Benguela (dist.), Angola 115/B6
Benguela, Angola 115/B6
Benguela, Angola 102/D6
Ben Guerdane (well), Mauritania 106/B3
Benguet (prov.), Philippines 82/C2
Benha, Egypt 111/J3
Benham, Ky. (40807) 237/R7
Benham, N.C. (†28621) 281/G2
Benhams, Va. (†24201) 307/D7
Ben Hee (mt.), Scotland 15/D2
Ben Hill (co.), Georgia 217/F7
Ben Hope (mt.), Scotland 15/D2
Ben Horn (mt.), Scotland 15/D2
Ben Hur, Va. (24218) 307/B7
Beni, El (dept.), Bolivia 136/C3
Beni (riv.), Bolivia 120/C4
Beni (riv.), Bolivia 136/B2
Beni, Zaire 115/E3
Beni Abbès, Algeria 106/D2
Benicarló, Spain 33/G2
Benicia, Calif. (94510) 204/K1
Benicito (riv.), Bolivia 136/C3
Beni Mazar, Egypt 111/J4
Beni Mellal, Morocco 106/C2
Beni Mellal, Morocco 102/B1
Benin 2/K5
Benin 102/C4
BENIN 106/E7
Benin (bight), Benin 106/E8
Benin (bight), Ghana 106/E8
Benin (bight), Nigeria 106/E8
Benin (bight), Togo 106/E8
Benin City, Nigeria 106/F7
Benin City, Nigeria 102/C4
Beni Ounif, Algeria 106/D2
Beni Saf, Algeria 106/D1
Beni Suef, Egypt 111/J3
Beni Suef, Egypt 102/E2
Beni Suef, Egypt 59/B4
Benito, Manitoba 179/A3
Beni Ulid, Libya 111/B1
Benjamín (isl.), Chile 138/D5
Benjamin, New Bruns. 170/D1
Benjamin (lake), Oreg. 291/G4
Benjamin, Texas (79505) 303/E4
Benjamin, Utah (†84660) 304/C3
Benjamín Aceval, Paraguay 144/C4
Benjamin Constant, Brazil 132/G9
Benjamin Constant, Brazil 120/B3
Benjamín Hill, Mexico 150/D1
Benkelman, Nebr. (69021) 264/C4
Ben Kilbreck (mt.), Scotland 15/D2
Ben Lawers (mt.), Scotland 15/D4
Benld, Ill. (62009) 222/D4
Ben Lomond, Ark. (71823) 202/B6
Ben Lomond, Calif. (95005) 204/K4
Ben Lomond, New Bruns. 170/E3
Ben Lomond (mt.), Scotland 15/D4
Ben Loyal (mt.), Scotland 15/D2
Ben Lui (mt.), Scotland 15/D4
Ben Macdhui (mt.), Scotland 15/E3
Ben Mhor (mt.), Scotland 15/A3
Ben More (mt.), Scotland 15/D4
Ben More (mt.), Scotland 15/B4
Ben More Assynt (mt.), Scotland 15/D2
Bennan (head), Scotland 15/C5
Bennane (head), Scotland 15/C5
Benndale, Miss. (†39450) 256/G9
Bennet, Nebr. (68317) 264/H4
Bennett, Br. Col. 184/J1
Bennett, W.A.C. (dam), Br. Col. 184/F2
Bennett, Colo. (80102) 208/L3
Bennett (peak), Colo. 208/G7
Bennett (creek), Idaho 220/C6
Bennett, Iowa (52721) 229/L5
Bennett, N.C. (27208) 281/K3
Bennett (lake), North. Terr. 93/B7
Bennett (co.), S. Dak. 298/F7
Bennett, Wis. (54815) 317/C3
Bennettsbridge, Ireland 17/G6
Bennetts Point, S.C. (†29446) 296/G6
Bennetts Switch, Ind. (†46901) 227/E3
Bennettsville, S.C. (29512) 296/H2
Bennettville, Minn. (†56431) 255/E4
Ben Nevis (mt.), Scotland 7/D3
Ben Nevis (mt.), Scotland 15/D4
Ben Nevis (mt.), Scotland 10/D2
Benning, D.C. (20019) 245/F5
Bennington, Idaho (†83254) 220/G7
Bennington, Ind. (47011) 227/G7
Bennington, Kansas (67422) 232/E2
Bennington, Nebr. (68007) 264/H3
Bennington○, N.H. (03442) 268/D5
Bennington, Okla. (74723) 288/P7
Bennington (co.), Vt. 268/A6
Bennington, Vt. (05201) 268/A6
Bennington○, Vt. (05201) 268/A6
Benns Church, Va. (†23430) 307/P7
Benoit, Miss. (38725) 256/C3
Benoit, Wis. (54816) 317/D3
Benom, Gunong (mt.), Malaysia 72/D7
Benoni, S. Africa 118/J6
Bensenville, Ill. (60106) 222/B5
Bensheim, W. Germany 22/C4
Benson, Ariz. (85602) 198/E7
Benson, Ill. (61516) 222/D3
Benson, La. (†71419) 238/C3
Benson, Minn. (56215) 255/C5
Benson, N.C. (27504) 281/N4
Benson (co.), N. Dak. 282/M3
Benson (Hollsopple), Pa. (15935) 294/E5
Benson, Sask. 181/J6
Benson○, Vt. (05731) 268/A4
Benson Landing, Vt. (†05731) 268/A4
Bens Run, W. Va. (26135) 312/D4
Bent (co.), Colo. 208/N7
Bent, N. Mex. (88314) 274/D5
Bent Creek, Va. (†24553) 307/L5
Benteng, Indonesia 85/F7
Bentham, England 13/E3
Bentheim, W. Germany 22/B2
Bentinck (isl.), Burma 72/C5
Bentinck (isl.), Queensland 88/F3
Bentinck (isl.), Queensland 95/A3
Bentiu, Sudan 111/E6
Bentley, Alberta 182/C3
Bentley, Ill. (†62321) 222/B3
Bentley, Iowa (†51559) 229/B6
Bentley, Kansas (67016) 232/E4
Bentley, La. (71407) 238/E3
Bentley, Mich. (48613) 250/E5
Bentley, N. Dak. (58522) 282/F7
Bentley, Okla. (†74525) 288/O6
Bentley Springs, Md. (21019) 245/M2
Bentleyville, Ohio (†44022) 284/J9
Bentleyville, Pa. (15314) 294/B5
Bentley with Arksey, England 13/F4
Bent Mountain, Va. (24059) 307/H6
Bento Gonçalves, Brazil 132/C10
Benton, Ala. (36785) 195/E6
Benton, Alberta 182/E4
Benton (co.), Ark. 202/B1
Benton, Ark. (72015) 202/E4
Benton, Calif. (93512) 204/G6
Benton, Ill. (62812) 222/E6
Benton (co.), Ind. 227/C3
Benton, Ind. (†46526) 227/F2
Benton (co.), Iowa 229/J4
Benton, Iowa (50835) 229/E7
Benton, Kansas (67017) 232/E4
Benton, Ky. (42025) 237/E7
Benton, La. (71006) 238/C1
Benton○, Maine (†04937) 243/D6
Benton (co.), Minn. 255/D5
Benton, Minn. (†55322) 255/F6
Benton (lake), Minn. 255/B6
Benton (co.), Miss. 256/F1
Benton, Miss. (39039) 256/D5
Benton (co.), Mo. 261/F6
Benton, Mo. (63736) 261/O8
Benton (lake), Mont. 262/E3
Benton, New Bruns. 170/C3
Benton○, N.H. (†03785) 268/D3
Benton, Ohio (44654) 284/G4
Benton (co.), Oreg. 291/D3
Benton, Pa. (17814) 294/K3
Benton (co.), Tenn. 237/E8
Benton, Tenn. (37307) 237/M10
Benton (co.), Wash. 310/F4
Benton, Wis. (53803) 317/F10
Benton City, Mo. (65232) 261/J4
Benton City, Wash. (99320) 310/F4
Bentong, Malaysia 72/D7
Benton Harbor, Mich. (49022) 250/C6
Benton Heights, Mich. (†49022) 250/C6
Bentonia, Miss. (39040) 256/D5
Benton Ridge, Ohio (45816) 284/C4
Bentonsport, Iowa (†52565) 229/K7
Bentonville, Ark. (72712) 202/B1
Bentonville, Ind. (47322) 227/G5
Bentonville, Ohio (45105) 284/C8
Bentonville, Va. (22610) 307/M3
Bent's Old Fort Nat'l Hist. Site, Colo. 208/M6
Benué (riv.), Cameroon 115/A2
Benue (state), Nigeria 106/F7
Benue (riv.), Nigeria 102/C4
Benue (riv.), Nigeria 106/F7
Ben Vorlich (mt.), Scotland 15/D4
Ben Vrackie (mt.), Scotland 15/E4
Benwee (head), Ireland 17/B3
Benwood, W. Va. (26031) 312/E2
Ben Wyvis (mt.), Scotland 15/D3
Benxi (Penki), China 77/K3
Benxi, China 54/O5
Benzie (co.), Mich. 250/C4
Benzonia, Mich. (49616) 250/D4
Beo, Indonesia 85/H5
Beograd (Belgrade) (cap.), Yugoslavia 45/E3
Beowawe, Nev. (89821) 266/E2
Beppu, Japan 81/E7
Bequia (isl.), St. Vin. & Grens. 156/G4
Beragh, N. Ireland 17/G2
Berar (reg.), India 68/D4
Berat, Albania 45/D5
Berau (bay), Indonesia 85/J6
Berber, Sudan 111/F4
Berber, Sudan 59/B6
Berber, Sudan 102/F3
Berbera, Somalia 115/J1
Berbera, Somalia 102/G3
Berberati, Cent. Afr. Rep. 102/D4
Berberati, Cent. Afr. Rep. 115/C3
Berbice (riv.), Guyana 131/B3
Berchem, Belgium 27/F6
Berchem-Sainte-Agathe, Belgium 27/B9
Bercher, Switzerland 39/C3
Berchtesgaden, W. Germany 22/E5
Berck, France 28/D2
Berclair, Texas (78107) 303/G9
Berdichev, U.S.S.R. 48/C5
Berdichev, U.S.S.R. 52/C5
Berdsk, U.S.S.R. 48/J4
Berdyansk, U.S.S.R. 7/H4
Berdyansk, U.S.S.R. 52/E5
Berea, Ky. (40403) 237/N5
Berea, Nebr. (†69301) 264/A2
Berea, N.C. (†27565) 281/M2
Berea, Ohio (44017) 284/G10
Berea, S.C. (29611) 296/C2
Berea, Spain 33/C1
Berea, W. Va. (26327) 312/E4
Bereda, Somalia 115/K1
Bereda, Somalia 102/H3
Beregovo, U.S.S.R. 52/B5
Berekum, Ghana 106/D7
Berenguela, Bolivia 136/A5
Berenice (ruins), Egypt 111/F3
Berens (riv.), Man. 162/G5
Berens (isl.), Manitoba 179/E2
Berens (riv.), Manitoba 179/F2
Berens (riv.), Ontario 175/A2
Berens River, Man. 162/G5
Berens River, Manitoba 179/F2
Beresford (lake), Fla. 212/E3
Beresford, New Bruns. 170/E1
Beresford, S. Dak. (57004) 298/R7
Beresford Lake, Manitoba 179/G4
Bereşti Tîrg, Romania 45/H2
Berettyó (riv.), Hungary 41/F3
Berettyóújfalu, Hungary 41/F3
Berezina (riv.), U.S.S.R. 52/C4
Bereznik, U.S.S.R. 52/F2
Berezniki, U.S.S.R. 7/K3
Berezniki, U.S.S.R. 48/F4
Berezniki, U.S.S.R. 52/J3
Berezovo, U.S.S.R. 48/G3
Berg, Norway 18/K2
Berg, Switzerland 39/H1
Berga, Algeria 106/E3
Berga, Spain 33/G1
Bergama, Turkey 63/B3
Bergama, Turkey 59/A2
Bergamo (prov.), Italy 34/B2
Bergamo, Italy 34/B2
Bergeijk, Netherlands 27/G6
Bergen (Mons), Belgium 27/E8
Bergen, E. Germany 22/E1
Bergen, Minn. (†56101) 255/D7
Bergen, Netherlands 27/F3
Bergen (co.), N.J. 273/E2
Bergen, N.Y. (14416) 276/E4
Bergen, N. Dak. (58792) 282/J3
Bergen, Norway 18/D6
Bergen, Norway 7/E2
Berg en Dal, Suriname 131/D3
Bergenfield, N.J. (07621) 273/C1
Bergen op Zoom, Netherlands 27/E5
Berger, Mo. (63014) 261/K5
Bergerac, France 28/D5
Bergholz, Ohio (43908) 284/J4
Bergisch Gladbach, W. Germany 22/B3
Bergland, Mich. (49910) 250/F1
Bergman, Ark. (72615) 202/E1
Bergoo, W. Va. (26298) 312/F6
Bergos (riv.), Turkey 63/C6
Bergshamra, Sweden 18/L7
Bergsjö, Sweden 18/K5
Bergstrom A.F.B., Texas 303/G7
Bergton, Va. (22811) 307/L3
Berguent, Morocco 106/D2
Bergum, Netherlands 27/H2
Bergumermeer (lake), Netherlands 27/J2
Bergün-Bravuogn, Switzerland 39/J3
Berhala (str.), Indonesia 85/C6
Berhampore, India 68/F4
Berhampur, India 68/F5
Berhida, Hungary 41/E3
Bering (sea) 2/A3

Bering (str.) 4/C18
Bering (strait) 54/W3
Bering (sea) 54/V4
Bering (sea) 146/A3
Bering (str.) 146/B3
Bering (str.) 146/B3
Bering (str.), Alaska 188/C5
Bering (sea), Alaska 188/C6
Bering (glac.), Alaska 196/K2
Bering (sea), Alaska 196/D2
Bering (str.), Alaska 196/E1
Bering (isl.), U.S.S.R. 48/R4
Bering (str.), U.S.S.R. 48/S4
Bering (str.), U.S.S.R. 48/U3
Beringen, Belgium 27/G6
Bering Land Bridge Nat'l Preserve, Alaska 196/F1
Beringovskiy, U.S.S.R. 48/T3
Berino, N. Mex. (88024) 274/C6
Berja, Spain 33/E4
Berkåk, Norway 18/G5
Berkel, Netherlands 27/F5
Berkeley, Calif. 188/B3
Berkeley, Calif. (*94701) 204/J2
Berkeley, Ill. (60162) 222/B5
Berkeley, Mo. (63134) 261/P2
Berkeley (co.), S.C. 296/G5
Berkeley (co.), W. Va. 312/K4
Berkeley, W. Va. (*12541) 312/L4
Berkeley Heights○, N.J. (07922) 273/E2
Berkeley Lake, Georgia (†30136) 217/D3
Berkeley Springs (Bath), W. Va. (25411) 312/K3
Berken, Libya 111/B2
Berkey, Ohio (43504) 284/C2
Berkhamsted, England 13/G7
Berkhout, Netherlands 27/F3
Berkley, Iowa (†50220) 229/F5
Berkley, Mich. (48072) 250/B6
Berkner (isl.), Ant. 2/G10
Berkner (isl.), Ant. 2/G10
Berkner (isl.), 5/B16
Berkovitsa, Bulgaria 45/F4
Berks (co.), Pa. 294/K5
Berkshire, Conn. (†06482) 210/B3
Berkshire (co.), England 13/F6
Berkshire (co.), Mass. 249/B3
Berkshire (hills), Mass. 249/B4
Berkshire, N.Y. (13736) 276/H6
Berkshire○, Vt. (†05476) 268/C2
Berland (riv.), Alberta 182/A3
Berlanga de Duero, Spain 33/E2
Berleburg (Bad Berleburg), W. Germany 22/C3
Berlevåg, Norway 18/Q1
Berlin○, Conn. (06037) 210/E2
Berlin (cap.), E. Germany 7/F3
Berlin (cap.), E. Germany 2/K3
Berlin (dist.), E. Germany 22/F4
Berlin, East (cap.), E. Germany 22/F4
Berlin, Georgia (31722) 217/E8
Berlin, Ill. (†62670) 222/D4
Berlin, Md. (21811) 245/T7
Berlin○, Mass. (01503) 249/H3
Berlin (riv.), Nev. 266/D4
Berlin, N.H. 188/M2
Berlin, N.H. (03570) 268/E3
Berlin, N.J. (08009) 273/D4
Berlin, N.Y. (12022) 276/O5
Berlin, N. Dak. (58415) 282/O7
Berlin, Ohio (44610) 284/C4
Berlin (center) Ohio (44401) 284/H4
Berlin, Okla. (†73662) 288/G4
Berlin, Pa. (15530) 294/E6
Berlin (pond), Vt. 268/C3
Berlin, W. Germany 7/F3
Berlin (West) (free city), W. Germany 22/E4
Berlin (West), W. Germany 22/E4
Berlin, W. Va. (†26452) 312/F4
Berlin, Wis. (54923) 317/H4
Berlin Center, Ohio (44401) 284/H3
Berlin Heights, Ohio (44814) 284/F2
Bermagui, N.S. Wales 97/F5
Bermeo, Spain 33/E1
Bermillo de Sayago, Spain 33/D2
Bermuda (isls.) 146/M6
Bermuda (isl.), (Br.) 2/F4
Bermuda, Ala. (†36438) 195/D8
BERMUDA 156/G3
Bermuda (isl.), Bermuda 156/H3
Bermuda, La. (†70558) 238/B3
Bern, Idaho (83220) 220/G7
Bern, Kansas (66408) 232/F2
Bern (canton), Switzerland 39/D2
Bern (cap.), Switzerland 7/E4
Bern (cap.), Switzerland 39/D3
Bernabé Rivera, Uruguay 145/B1
Bernadotte, Minn. (†56054) 255/D6
Bernal○, N. Mex. (87004) 274/C4
Bernalillo, N. Mex. (87004) 274/C4
Bernalillo (co.), N. Mex. 274/A6
Bernard, Iowa (52032) 229/M4
Bernardino de Sahagún, Mexico 150/M1
Bernardo, N. Mex. (†87028) 274/C4
Bernardo de Irigoyen, Argentina 143/F2
Bernardston○, Mass. (01337) 249/H1
Bernardsville, N.J. (07924) 273/D2
Bernasconi, Argentina 143/D4
Bernau bei Berlin, E. Germany 22/E2
Bernay, France 28/D3
Bernburg, E. Germany 22/D3
Berndorf, Austria 41/C3
Berne, Ind. (46711) 227/H3
Berne, N.Y. (12023) 276/M5
Berneray (isl.), Scotland 15/A4
Berneray (isl.), Scotland 15/A3
Bernese Oberland (reg.), Switzerland 39/E3
Bernhards Bay, N.Y. (13028) 276/J4
Bernic (lake), Manitoba 179/G4
Bernice, La. (71222) 238/E1
Bernice, Okla. (74331) 288/S1

Bernic Lake, Manitoba 179/G4
Bernie, Mo. (63822) 261/M9
Bernier (bay), N.W. Terrs. 187/K2
Bernier (isl.), W. Australia 88/A4
Bernier (isl.), W. Australia 92/A4
Bernierville, Québec 172/G1
Bernina (pass), Italy 34/C1
Bernina, Piz (peak), Italy 34/B1
Bernina (mts.), Switzerland 39/J4
Bernina (pass), Switzerland 39/K4
Bernina (peak), Switzerland 39/J4
Bernina (riv.), Switzerland 39/J4
Bernkastel-Kues, W. Germany 22/B4
Bernstadt, Ky. (†40741) 237/N6
Bernville, Pa. (19506) 294/K5
Bero (riv.), Angola 115/B5
Beromünster, Switzerland 39/F2
Beroroha, Madagascar 118/G4
Beroun, Czech. 41/B2
Beroun, Minn. (55004) 255/F5
Berounka (riv.), Czech. 41/B2
Berovo, Yugoslavia 45/F5
Berre (lag.), France 28/F4
Berri, S. Australia 88/G6
Berri, S. Australia 94/G6
Berridale, N.S. Wales 97/E5
Berriedale, Scotland 15/E2
Berrien (co.), Georgia 217/F8
Berrien (co.), Mich. 250/C7
Berrien Springs, Mich. (49103) 250/C7
Berrigan, N.S. Wales 97/C4
Berrondo, Uruguay 145/C5
Berry, Ala. (35546) 195/C3
Berry (creek), Alberta 182/E4
Berry (isls.), Bahamas 156/B1
Berry (head), England 13/D7
Berry (trad. reg.) France 29
Berry, Ky. (41003) 237/N3
Berry, N.S. Wales 97/F4
Berry (head), Nova Scotia 168/G3
Berryessa (lake), Calif. 204/D5
Berry Hill, Tenn. (†37216) 237/H8
Berry Mills, New Bruns. 170/E2
Berrymoor, Alberta 182/C3
Berrysburg, Pa. (17005) 294/J4
Berry's Chapel, Tenn. (†72501) 237/H9
Berryton, Georgia (30718) 217/B2
Berryton, Kansas (66409) 232/G3
Berryville, Ark. (72616) 202/C1
Berryville, Va. (22611) 307/M2
Bersaba, Namibia 118/B5
Bertha, Minn. (56437) 255/C4
Berthier (co.), Québec 172/E3
Berthier (county), Québec 174/B3
Berthier-en-Bas, Québec 172/G3
Berthierville, Québec 172/D3
Berthold, N. Dak. (58718) 282/G3
Berthoud, Colo. (80513) 208/J2
Berthoud Pass, Colo. (†80438) 208/H3
Bertie (co.), N.C. 281/P2
Bertogne, Belgium 27/H8
Bertolínia, Brazil 132/F4
Bertoua, Cameroon 115/B3
Bertraghboy (bay), Ireland 17/A5
Bertram, Iowa (†52401) 229/K5
Bertram, Texas (78605) 303/F7
Bertrand, Cerro (mt.), Chile 138/D8
Bertrand, Mo. (63823) 261/O9
Bertrand, Nebr. (68927) 264/F4
Bertrand (riv.), Mo. 261/O1
Bertrandville, La. (†70040) 238/L7
Bertrix, Belgium 27/G9
Bertwell, Sask. 181/J3
Bervie, Ontario 177/G3
Berwick, Ill. (61417) 222/C3
Berwick, Iowa (50032) 229/G5
Berwick, Kansas (†66534) 232/G2
Berwick, La. (70342) 238/H7
Berwick, Maine (03901) 243/B9
Berwick○, Maine (03901) 243/B9
Berwick, New Bruns. 170/E3
Berwick, N. Dak. (†58788) 282/K3
Berwick, Nova Scotia 168/D4
Berwick, Pa. (18603) 294/K3
Berwick (riv.), Sask. 181/G6
Berwick, Victoria 97/K6
Berwick-upon-Tweed, England (52722) 229/N5
Berwick-upon-Tweed, England 10/F3
Berwind, W. Va. (24815) 312/C8
Berwyn, Alberta 182/B1
Berwyn, Ill. (60402) 222/B6
Berwyn, Nebr. (68819) 264/F3
Berwyn, Pa. (19312) 294/L5
Berwyn (mts.), Wales 13/D5
Berwyn Heights, Md. (†20740) 245/G4
Beryl, Utah (84714) 304/A6
Beryl, W. Va. (†26726) 312/H4
Berzence, Hungary 41/D3
Besalampy, Madagascar 118/G3
Besançon, France 28/G4
Besançon, France 7/E4
Betzdorf, W. Germany 22/B3
Beşiktaş, Turkey 63/D6
Beşiri, Turkey 63/J4
Beskids, West (mts.), Czech. 41/E2
Beskids, East (mts.), Czech. 41/F1
Beskids (range), Poland 47/E4
Beslan, U.S.S.R. 52/F6
Besni, U.S.S.R. 52/F2
Besor (riv.), Israel 65/B5
Bessbrook, N. Ireland 17/J3
Bessèges, France 28/F5
Bessemer, Ala. (35020) 195/D4
Bessemer, Ala. (35020) 195/D4
Bessemer, Mich. (49911) 250/F2
Bessemer City, N.C. (28016) 281/H4
Bessie, Texas (76931) 303/C6
Bessmay, Texas (76931) 303/C6
Best, Texas (76931) 303/C6
Beswick, North. Terr. 88/E2
Beswick Aboriginal Reserve, North. Terr. 88/E2
Beswick Aboriginal Res., North. Terr. 93/C3
Beta, N.C. (†28779) 281/C4
Betanzos, Bolivia 136/C5
Betanzos, Spain 33/B1

Bétaré-Oya, Cameroon 115/B2
Bete Grise (bay), Mich. 250/B1
Bethalto, Ill. (62010) 222/B2
Bethanie, Namibia 118/B5
Bethany (res.), Calif. 204/L2
Bethany, N.J. (08010) 273/D3
Bethany, Ill. (61914) 222/E4
Bethany, Ind. (†46111) 227/E5
Bethany, La. (71007) 238/B2
Bethany, Manitoba 179/G4
Bethany, Minn. (†55910) 255/F6
Bethany, Mo. (64424) 261/E2
Bethany, Ohio (†45042) 284/B7
Bethany, Okla. (73008) 288/L3
Bethany, Ontario 177/F3
Bethany, Pa. (†18431) 294/M2
Bethany, Sask. 181/G4
Bethany, W. Va. (26032) 312/E2
Bethany Beach, Del. (19930) 245/T6
Bethel, Alaska 188/C6
Bethel, Alaska (99559) 196/F2
Bethel○, Conn. (06801) 210/B3
Bethel, Del. (19931) 245/R6
Bethel, Ky. (40306) 237/O4
Bethel, Maine (04217) 243/B7
Bethel○, Maine (04217) 243/B7
Bethel, Minn. (55005) 255/E5
Bethel, N.C. (63434) 261/J3
Bethel, N.Y. (12720) 276/L7
Bethel, N.C. (27812) 281/P3
Bethel, Ohio (45106) 284/B8
Bethel, Ohio (74724) 288/S1
Bethel, Vt. (05032) 268/B4
Bethel, Vt. (05032) 268/B4
Bethel, Wis. (†54410) 317/F6
Bethel Acres, Okla. (†74801) 288/M4
Bethel Heights, Ark. (†72764) 202/B1
Bethel Island, Calif. (94511) 204/L1
Bethel Park, Pa. (15102) 294/B7
Bethel Springs, Tenn. (38315) 237/D10
Bethel Town, Jamaica 158/G6
Bethera, S.C. (29430) 296/H5
Bethesda, Md. (*20014) 245/E4
Bethesda, Ohio (43719) 284/H5
Bethesda, Wales 13/C4
Bethlehem, Conn. (06751) 210/C2
Bethlehem○, Conn. (06751) 210/C2
Bethlehem, Georgia (30620) 217/E3
Bethlehem, Ind. (47104) 227/F5
Bethlehem, Ky. (40007) 237/L4
Bethlehem, Iowa (†50238) 229/G7
Bethlehem, Md. (21609) 245/P6
Bethlehem, Miss. (†38659) 256/F1
Bethlehem○, N.Y. (†12036) 276/M6
Bethlehem, Pa. (*18015) 294/M4
Bethlehem, S. Africa 102/E7
Bethlehem, S. Africa 118/D5
Bethlehem, Virgin Is. (\) 161/E4
Bethlehem, West Bank 65/C4
Bethlehem, W. Va. (†26003) 312/E2
Bethpage, N.Y. (11714) 276/R7
Bethpage, Tenn. (37022) 237/J7
Bethulie, S. Africa 118/D6
Bethune, Colo. (80805) 208/P4
Béthune, France 28/E2
Bethune, Sask. 181/F5
Bethune, S.C. (29009) 296/G3
Betijoque, Venezuela 124/C3
Betim, Brazil 135/D2
Betioky, Madagascar 118/G4
Betoota, Queensland 95/B5
Bet-Pak-Dala (des.), U.S.S.R. 48/H5
Betroka, Madagascar 118/H4
Bet She'an, Israel 65/D3
Bet Shemesh, Israel 65/B4
Betsiamites, Québec 174/D3
Betsiamites, Québec 174/D3
Betsiamites (riv.), Québec 174/C2
Betsiboka (riv.), Madagascar 118/H3
Betsy (riv.), Mich. 250/D2
Betsy Layne, Ky. (41605) 237/R5
Bette (peak), Libya 102/D4
Bette (peak), Libya 111/C3
Bettembert, Iowa (52722) 229/N5
Betteravia, Calif. (†93454) 204/E9
Betterton, Md. (21610) 245/O3
Bettiah, India 68/F3
Bettlach, Switzerland 39/D2
Bettles, Alaska (†99726) 196/H1
Bettles Field, Alaska (99726) 196/H1
Bettsville, Ohio (44815) 284/D3
Bettyhill, Scotland 15/D2
Betul, India 68/D4
Betula, India (†16749) 294/F2
Betwa (riv.), India 68/D4
Between, Georgia (†30655) 217/E3
Betws-y-Coed, Wales 13/D4
Betzdorf, W. Germany 22/B3
Beulah, Ala. (†36872) 195/H5
Beulah, Colo. (81023) 208/K6
Beulah, Manitoba 179/A4
Beulah, Mich. (49617) 250/C4
Beulah, Miss. (38726) 256/B3
Beulah, Mo. (65436) 261/J7
Beulah, N. Dak. (58523) 282/G5
Beulah, Oreg. (†97911) 291/J4
Beulah (res.), Oreg. 291/J4
Beulah, Victoria 97/B4
Beulah Valley, Colo. 208/J6
Beulah, Wis. 317/J2
Belah, Wyo. (82712) 319/H1
Beulaker Wijde (lake), Netherlands 27/H3
Beulaville, N.C. (28518) 281/O5
Beutthen (Bytom), Poland 47/A3
Bevans, N.J. (†07851) 273/D1
Bevent, Wis. (†54440) 317/H6
Bevercotes, England 13/F6
Beverin (peak), Switzerland 39/H3
Beverley, England 10/F4
Beverley, England 13/G4
Beverley, Sask. 181/C5
Beverley, W. Australia 92/B4
Beverly, C. Rica 154/F6

Beverly, Kansas (67423) 232/E2
Beverly, Ky. (40913) 237/P7
Beverly, Mass. (01915) 249/E5
Beverly, Mo. (†64079) 261/O4
Beverly, N.J. (08010) 273/D3
Beverly, Ohio (45715) 284/G5
Beverly, Wash. (99321) 310/F4
Beverly, W. Va. (26253) 312/G5
Beverly○, Conn. (†06501) 210/C3
Beverly (lake), N.W. Terrs. 187/H3
Beverly Beach, Fla. (†32036) 212/E2
Beverly Hills, Calif. (*90210) 204/B10
Beverly Hills, Mich. (48010) 250/B6
Beverly Shores, Ind. (46301) 227/C1
Bevier, Mo. (63532) 261/G3
Bevington, Iowa (50033) 229/F6
Bewdley, England 13/E5
Bewdley, England 10/E3
Bewdley, Ontario 177/F3
Bex, Switzerland 39/D4
Bexar, Ala. (†35570) 195/B2
Bexar (co.), Texas 303/F8
Bexhill, England 10/G5
Bexhill, England 13/H7
Bexley, England 13/H8
Bexley, England 13/H8
Bexley, Miss. (†39452) 256/G9
Bexley, Ohio (43209) 284/F6
Bey (mts.), Turkey 63/D4
Bey el Kebir, Wadi (dry riv.), Libya 111/B1
Beykoz, Turkey 63/D5
Beyla, Guinea 106/C7
Beylerbeyi, Turkey 63/F3
Beynon, Alberta 182/D4
Beyoğlu, Turkey 63/D6
Beypazarı, Turkey 59/B1
Beypazarı, Turkey 63/D4
Beyşehir, Turkey 59/B2
Beyşehir, Turkey 63/D4
Beyşehir (lake), Turkey 63/D4
Beytüşşebap, Turkey 63/K4
Bezanson, Alberta 182/A2
Bezhetsk, U.S.S.R. 52/E3
Béziers, France 7/E4
Béziers, France 28/E6
Bhadrak, India 68/F4
Bhadravati, India 68/D5
Bhadreswar, India 68/F1
Bhag, Pakistan 68/B3
Bhag, Pakistan 59/J4
Bhagalpur, India 68/F4
Bhaktapur, Nepal 68/F3
Bhaktapur, Nepal 68/F3
Bhamo, Burma 72/C1
Bhandara, India 68/E4
Bhanjanagar, India 68/E4
Bharatpur, India 68/D3
Bharuch, India 68/C4
Bhatapara, India 68/E4
Bhatinda, India 68/C2
Bhatkal, India 68/C6
Bhatpara, India 68/F1
Bhavani, India 68/D6
Bhavnagar, India 68/C4
Bhavnagar, India 54/J7
Bhawanipatna, India 68/E5
Bhera, Pakistan 68/C2
Bheri (riv.), Nepal 68/E3
Bhilai, India 68/E4
Bhilwara, India 68/C3
Bhima (riv.), India 68/D5
Bhimavaram, India 68/E5
Bhimunipatnam, India 68/E5
Bhind, India 68/D3
Bhinmal, India 68/C3
Bhir (Bir), India 68/D5
Bhiwandi, India 68/B7
Bhiwani, India 68/D3
Bhojpur, Nepal 68/F3
Bhopal, India 54/J7
Bhopal, India 68/D4
Bhor, India 68/C5
Bhubaneswar, India 68/F4
Bhuj, India 68/B4
Bhusawal, India 68/D4
Bhutan 2/P4
Bhutan 54/L7
BHUTAN 68/G3
Biafra (bight), Cameroon 115/A3
Biafra (bight), Equat. Guinea 115/A3
Biafra (bight), Nigeria 106/F8
Biak, Indonesia 85/K6
Biak (isl.), Indonesia 85/K6
Biała Podlaska, Poland 47/F3
Biała Podlaska (prov.), Poland 47/F3
Białogard, Poland 47/C1
Białystok (prov.), Poland 47/F2
Białystok, Poland 47/F2
Białystok, Poland 7/G3
Biancavilla, Italy 34/E6
Biarritz, France 28/C6
Bias, W. Va. (†25661) 312/B7
Biasca, Switzerland 39/H4
Biba, Egypt 111/J4
Bibai, Japan 81/L2
Bibala, Angola 115/B6
Bibb (co.), Ala. 195/D5
Bibb (co.), Georgia 217/E5
Bibb City, Georgia (†31901) 217/B5
Bibbenluke, N.S. Wales 97/E5
Biberach an der Riss, W. Germany 22/C4
Biberist, Switzerland 39/D2
Bible Grove, Ill. (62813) 222/E5
Bible Hill, Nova Scotia 168/E3
Bic, Québec 172/J1
Bic (isl.), Québec 172/J1
Bicas, Brazil 135/E2
Bicaz, Romania 45/H2
Bicester, England 13/F6
Biche (lake), Alberta 182/E2
Biche, Trin. & Tob. 161/B10
Bicheno, Tasmania 99/E3
Bickerdike, Alberta 182/B3
Bickerton (isl.), North. Terr. 93/E2
Bickerton West, Nova Scotia 168/G3

Bickleigh, Sask. 181/C4
Bickleton, Wash. (99322) 310/E5
Bickmore, W. Va. (25019) 312/D6
Bicknell, Ind. (47512) 227/C7
Bicknell, Utah (84715) 304/C5
Bicske, Hungary 41/E3
Bida, Nigeria 106/F7
Bidar, India 68/D5
Biddeford, Maine (04005) 243/B9
Biddeford, Maine 188/N2
Biddeford Pool, Maine (04006) 243/C9
Biddinghuizen, Netherlands 27/H4
Biddle, Mont. (59313) 262/L5
Biddu, West Bank 65/C4
Biddulph, England 13/H2
Bidean nam Bian (mt.), Scotland 15/D4
Bide Arm, Newf. 166/C3
Bideford, England 13/C6
Bideford, England 10/D5
Bidokht, Iran 66/J3
Bidon 5 (Poste Maurice Cordier), Algeria 106/C4
Bidwell, Ohio (45614) 284/F8
Bidyadhari (riv.), India 68/G1
Bié (dist.), Angola 115/C6
Bié, Angola 102/D6
Bié, Angola 115/C6
Bieber, Calif. (96009) 204/D2
Biebrza (riv.), Poland 47/F2
Bielawa, Poland 47/C3
Bield, Manitoba 179/A3
Bieldside, Scotland 15/F3
Bielefeld, W. Germany 22/C2
Bieler (lake), N.W. Terrs. 187/K3
Bielersee (lake), Switzerland 39/D2
Biella, Italy 34/B2
Bielsko (prov.), Poland 47/D4
Bielsko-Biała, Poland 47/D4
Bielsk Podlaski, Poland 47/F2
Biencourt, Québec 172/J2
Bienfait, Sask. 181/J6
Bien Hoa, Vietnam 72/E5
Bienvenue, Fr. Guiana 131/E4
Bienvenue, Loc (lake), Que. 162/J4
Bienville (par.), La. 238/D2
Bienville, La. (71008) 238/D2
Bienville (lake), Québec 174/C2
Bière, Switzerland 39/B3
Bietigheim-Bissingen, W. Germany 22/C4
Bietschhorn (mt.), Switzerland 39/E4
Bièvre, France 27/F9
Bièvres, France 28/A2
Big (isl.), Alberta 182/B5
Big (creek), Idaho 220/G4
Big (creek), Ind. 227/B8
Big (brook), Maine 243/H2
Big (lake), Maine 243/H5
Big (pond), Mass. 249/B4
Big (riv.), Mo. 261/L6
Big (lake), Mont. 262/G5
Big (mt.), Nev. 266/B1
Big (bay), Newf. 166/B2
Big (isl.), Newf. 166/C3
Big (riv.), Newf. 166/C3
Big (isl.), N.W. Terrs. 187/L3
Biga, Turkey 63/B2
Bigadiç, Turkey 63/C3
Bigalı, Turkey 63/B6
Big Annemessex (riv.), Md. 245/P8
Big Arm, Mont. (59910) 262/B3
Big Bald (mt.), New Bruns. 170/D1
Big Bar, Calif. (96010) 204/B3
Big Bar Creek, Br. Col. 184/F4
Big Basin, Calif. (†95006) 204/J4
Big Bay, Mich. (49808) 250/B2
Big Bay (pt.), Mich. 250/B2
Big Bay De Noc (bay), Mich. 250/C3
Big Bear City, Calif. (92314) 204/J9
Big Bear Lake, Calif. (92315) 204/J9
Big Beaver, Sask. 181/G6
Bigbee, Ala. (36510) 195/B7
Bigbee Valley, Miss. (39738) 256/H4
Big Bell, W. Australia 92/B4
Big Belt (mts.), Mont. 262/E4
Big Bend (dam), S. Dak. 298/K5
Big Bend, Wis. (53103) 317/K2
Big Bend City, Minn. (†56262) 255/C5
Big Bend National Park, Texas (79834) 303/A8
Big Bend Nat'l Park, Texas 303/A8
Big Bend City, Minn. (†56262) 255/C5
Bigbend, Oreg. 291/J4
Bigbend, Mont. 262/J4
Big Black (riv.), Maine 243/G3
Big Black (riv.), Miss. 256/C6
Big Black (creek), S.C. 296/G2
Big Black River, Manitoba 179/E1
Big Blue (riv.), Ind. 227/E5
Big Blue (riv.), Kansas 232/F1
Big Blue (riv.), Nebr. 264/H4
Big Boulder (creek), Idaho 220/B7
Big Bow, Kansas (67855) 232/A4
Big Bras d'Or, Nova Scotia 168/H2
Big Bureau (riv.), Ill. 222/D4
Big Burro (mts.), N. Mex. 274/A6
Bigbury (bay), England 13/C7
Big Cabin, Okla. (74332) 288/R1
Big Canoe (creek), Ala. 195/F3
Big Chimney, W. Va. (25302) 312/C6
Big Chino Wash (dry riv.), Ariz. 198/C4
Big Clifty, Ky. (42712) 237/J5
Big Coulee, Alberta 182/D2
Big Cove Tannery, Pa. (17212) 294/F6
Big Creek, Calif. (93605) 204/F6
Big Creek, Idaho (†83677) 220/E4
Big Creek, Ky. (40914) 237/O6
Big Creek, Miss. (38914) 256/F3
Big Creek, W. Va. (25505) 312/B7
Big Cypress (swamp), Fla. 212/E5
Big Cypress Nat'l Preserve, Fla. 212/F5

Bigelow (brook), Conn. 210/G1
Bigelow (bight), Maine 243/C9
Bigelow (mt.), Maine 243/C5
Bigelow (bight), Mass. 249/M1
Bigelow, Minn. (56117) 255/C7
Bigelow, Mo. (64425) 261/B2
Big Falls, Minn. (56627) 255/E2
Big Falls, Wis. (54926) 317/H6
Big Flat, Ark. (72617) 202/F1
Big Flats, N.Y. (14814) 276/G6
Bigfoot, Texas (78005) 303/F9
Big Fork, Ark. (71928) 202/B5
Big Fork (riv.), Minn. 255/E2
Bigfork, Mont. (56628) 255/E5
Bigfork, Mont. (59911) 262/C2
Big Four, W. Va. (†24853) 312/C8
Bigga, N.S. Wales 97/E4
Biggar, Sask. 162/F5
Biggar, Sask. 181/C3
Biggar, Scotland 10/E3
Biggar, Scotland 15/E5
Bigge (range), Queensland 95/D5
Bigge (isl.), W. Australia 92/D1
Biggers, Ark. (72413) 202/J1
Biggleswade, England 10/F4
Biggleswade, England 13/G5
Big Goose (creek), Wyo. 319/E1
Big Grizzly (creek), Colo. 208/G1
Biggs, Calif. (95917) 204/D4
Biggs, Oreg. (†97058) 291/G2
Biggs Field, Texas (†79908) 303/A10
Biggsville, Ill. (61418) 222/C3
Big Hole (mts.), Idaho 220/G6
Big Hole (riv.), Mont. 262/C5
Big Hole Nat'l Battlefield, Mont. 262/C5
Bighorn (riv.) 188/E2
Big Horn (dam), Alberta 182/B3
Big Horn (range), Alberta 182/B3
Big Horn (mts.), Ariz. 198/B5
Bighorn (lake), Mont. 262/H5
Bighorn, Mont. (59010) 262/J4
Bighorn (lake), Mont. 262/H5
Bighorn, Mont. 262/J5
Big Horn (co.), Wyo. 319/E1
Big Horn, Wyo. (82833) 319/E1
Bighorn (basin), Wyo. 319/D1
Bighorn (lake), Wyo. 319/D1
Bighorn (mts.), Wyo. 319/D1
Bighorn (riv.), Wyo. 319/D1
Bighorn Canyon Nat'l Rec. Area, Mont. 262/H5
Bighorn Canyon Nat'l Rec. Area, Wyo. 319/D1
Big Indian, N.Y. (12410) 276/M6
Big Iron (riv.), Mich. 250/F1
Big Isaac, W. Va. (†26426) 312/E4
Big Island, Ontario 177/G3
Big Island, Va. (24526) 307/K5
Big Lake, Alaska (†99687) 196/J6
Big Lake, Alaska (†99716) 196/J1
Big Lake, Minn. (55309) 255/E5
Big Lake, Texas (76932) 303/C6
Big Lake, Wash. (†98273) 310/C2
Big Lake Ranch, Br. Col. 184/G4
Bigler, Pa. (16825) 294/F4
Biglerville, Pa. (17307) 294/H6
Big Lick, Tenn. (†38555) 237/L9
Big Lost (riv.), Idaho 220/G6
Big Moose, N.Y. (†13331) 276/L3
Big Moose (lake), N.Y. 276/L3
Big Muddy (riv.), Ill. 222/D6
Big Muddy (riv.), Mont. 262/M2
Big Muddy (lake), Sask. 181/G6
Big Muskego (lake), Wis. 317/L2
Bignona, Senegal 106/A6
Big Oak Flat, Calif. (95305) 204/F6
Big Otter (riv.), Va. 307/K6
Big Otter, W. Va. (25113) 312/D5
Big Pine, Calif. (93513) 204/G6
Big Pine (key), Fla. 212/E7
Big Pine (creek), Ind. 227/C3
Big Pine, N.C. (†28753) 281/D3
Big Piney, Mo. (65437) 261/H7
Big Piney, Wyo. (83113) 319/B3
Big Pipe (creek), Md. 245/K2
Big Plain, Ohio (†43140) 284/D6
Bigpoint, Miss. (39567) 256/H9
Big Pond, Newf. 166/D2
Big Pond, Nova Scotia 168/H3
Big Pool, Md. (21711) 245/F2
Big Porcupine (creek), Mont. 262/J4
Big Prairie, Ohio (44611) 284/F4
Big Raccoon (riv.), Ind. 227/C5
Big Rapids, Mich. (49307) 250/D5
Big Rib (riv.), Wis. 317/G5
Big Rideau (lake), Ontario 177/H3
Big River, Sask. 162/F5
Big River, Sask. 181/D2
Big Rock, Ill. (60511) 222/E2
Big Rock, Iowa (52725) 229/M5
Big Rock, Tenn. (37023) 237/F7
Big Rock, Va. (24603) 307/F6
Big Run, Pa. (15715) 294/E4
Big Sable (pt.), Mich. 250/C4
Big Sable (riv.), Mich. 250/C4
Big Sage (res.), Calif. 204/F2
Big Salmon (riv.), New Bruns. 170/E3
Big Sand (lake), Manitoba 179/H2
Big Sandy (riv.), Ariz. 198/B4
Big Sandy (creek), Colo. 208/N4
Big Sandy, Ky. 237/R4
Big Sandy (lake), Minn. 255/E4
Big Sandy, Mont. (59520) 262/G2
Big Sandy, Tenn. (38221) 237/E8
Big Sandy, Texas (75755) 303/J5
Big Sandy, W. Va. 312/A6
Big Sandy, Wyo. (†82923) 319/C3
Big Sandy (res.) Wyo. 319/C3
Big Sandy (riv.), Wyo. 319/C3
Big Sioux (riv.), Iowa 229/A3
Big Sioux (riv.), S. Dak. 188/G2
Big Sioux (riv.), S. Dak. 298/S7
Big Sky, Mont. (59716) 262/F5
Big Smoky (valley), Nev. 266/D4
Big Southern (butte), Idaho 220/E6

Big Spencer (mt.), Maine 243/E4
Big Spring, Georgia (†30240) 217/C5
Big Spring, Ky. (40106) 237/J5
Big Spring, Md. (21722) 245/G2
Big Spring, Tenn. (37323) 237/M10
Big Spring, Texas 188/F4
Big Spring, Texas (79720) 303/C5
Big Springs, Nebr. (69122) 264/B3
Big Springs, S. Dak. (†57001) 298/S8
Big Springs, W. Va. (26137) 312/D5
Big Star (lake), Mich. 250/C5
Bigstick (lake), Sask. 181/B5
Big Stone, Alberta 182/E4
Bigstone (lake), Manitoba 179/J3
Bigstone (pt.), Manitoba 179/E2
Bigstone (riv.), Manitoba 179/J3
Big Stone (co.), Minn. 255/B5
Big Stone (lake), Minn. 255/B5
Big Stone (lake), S. Dak. 298/R3
Big Stone City, S. Dak. (57216) 298/S3
Big Stone Gap, Va. (24219) 307/C7
Big Sur, Calif. (93920) 204/D7
Big Thicket Nat'l Preserve, Texas 303/K7
Big Thompson (riv.), Colo. 208/H2
Big Timber, Mont. (59011) 262/G5
Big Timber (creek), N.J. 273/C4
Big Tracadie (riv.), New Bruns. 170/E1
Bigtrails, Wyo. (†82442) 319/E2
Big Trout (lake), Ontario 177/F2
Big Trout (lake), Ontario 175/B2
Big Trout Lake, Ontario 175/C2
Big Valley, Alberta 182/D3
Big Walnut (creek), Ind. 227/D5
Big Walnut (creek), Ohio 284/E5
Big Wells, Texas (78830) 303/E9
Big Whiteshell Lake, Manitoba 179/G4
Big Wood (riv.), Idaho 220/D6
Bihać, Yugoslavia 45/B3
Bihar (state), India 68/F4
Bihar, India 68/F3
Biharamulo, Tanzania 115/F4
Biharkeresztes, Hungary 41/F3
Biharnagybajom, Hungary 41/F3
Bijagós (isls.), Guinea-Biss. 106/A6
Bijagós (isls.), Guinea-Biss. 102/A3
Bijapur, Karnataka, India 68/D5
Bijapur, Madhya Pradesh, India 68/E5
Bijar, Iran 66/E3
Bijeljina, Yugoslavia 45/D3
Bijelo Polje, Yugoslavia 45/D4
Bijiang, China 77/E6
Bijie, China 77/G6
Bijnor, India 68/D3
Bijou (creek), Colo. 208/L3
Bijou Hills, S. Dak. (†57310) 298/L6
Bikaner, India 54/J7
Bikaner, India 68/C3
Bikar (atoll), Marshall Is. 87/H4
Bikin, U.S.S.R. 48/O5
Bikini (atoll), Marshall Is. 87/G4
Bikoro, Zaire 115/C4
Bikoro, Zaire 102/D5
Bilaspur, India 68/E4
Bilauktaung (range), Burma 72/C4
Bilauktaung (range), Thailand 72/C4
Bilbao, Spain 33/E1
Bilbao, Spain 7/D4
Bileća, Yugoslavia 45/D4
Bilecik (prov.), Turkey 63/D2
Bilecik, Turkey 59/A1
Bilecik, Turkey 63/D2
Biłgoraj, Poland 47/F3
Bilibino, U.S.S.R. 4/C1
Bilibino, U.S.S.R. 48/R3
Bilin, Burma 72/C3
Bílina, Czech. 41/B1
Biliran (isl.), Philippines 82/E5
Bill, Wyo. (82631) 319/G2
Billate (riv.), Ethiopia 111/G6
Billerica◯, Mass. (01821) 249/J2
Billings (lake), Conn. 210/H2
Billings, Mo. (65610) 261/F8
Billings, Mont. 146/H5
Billings, Mont. 188/E1
Billings, Mont. (*59101) 262/H5
Billings (co.), N. Dak. 282/D5
Billings, Okla. (74630) 288/M1
Billingsgate (isl.), Mass. 249/O5
Billingshurst, England 13/G6
Billingsley, Ala. (36006) 195/E5
Billiton (isl.), Indonesia 54/M10
Billiton (isl.), Indonesia 85/D6
Bill Williams (riv.), Ariz. 198/B4
Billy Clapp (lake), Wash. 310/F3
Bilma, Niger 102/D3
Bilma, Niger 106/G5
Biloela, Queensland 88/J4
Biloela, Queensland 95/D5
Biloku, Guyana 131/B5
Biloxi, Miss. 146/K6
Biloxi, Miss. 188/J4
Biloxi, Miss. (*39530) 256/G10
Biltine, Chad 111/D5
Biltine, Chad 102/D3
Biltmore Forest, N.C. (†28803) 281/E3
Bilwaskarma, Nicaragua 154/F3
Bilzen, Belgium 27/G7
Bim, W. Va. (25021) 312/C7
Biminis, The (isls.), Bahamas 156/B1
Bina-Itawa, India 68/D4
Binalbagan, Philippines 82/D5
Binalong, N.S. Wales 97/E4
Binboğa (mts.), Turkey 63/G3
Binbrook, Ontario 177/E4
Binche, Belgium 27/E8
Binda, N.S. Wales 97/E4
Bindloss, Alberta 182/E4
Bindoon, W. Australia 92/B1
Bindura, Zimbabwe 118/E3
Binéfar, Spain 33/G2
Binevenagh (mt.), N. Ireland 17/H1
Binford, N. Dak. (58416) 282/O4
Binga (mt.), Mozambique 118/E3
Bingara, N.S. Wales 97/F1
Bingen, Wash. (98605) 310/D5
Bingen, W. Germany 22/B4
Binger, Okla. (73009) 288/K4
Bingerville, Ivory Coast 106/D7
Bingham (co.), Idaho 220/F6
Bingham, Ill. (62011) 222/D4
Bingham, Maine (04920) 243/D5
Bingham◯, Maine (04920) 243/D5
Bingham, Nebr. (69335) 264/B2
Bingham, N. Mex. (87815) 274/C5
Bingham, S.C. (†29565) 296/H3
Bingham Lake, Minn. (56118) 255/C7
Binghamton, N.Y. 188/L2
Binghamton, N.Y. (*13901) 276/J6
Bingöl (prov.), Turkey 63/J3
Bingöl (Çapakçur), Turkey 63/J3
Bingöl Dağları (mts.), Turkey 63/J3
Binhai, China 77/K5
Binh Long (An Loc), Vietnam 72/E5
Binh Son, Vietnam 72/F4
Binjai, Indonesia 85/B5
Binn, Switzerland 39/F4
Binnaway, N.S. Wales 97/E2
Binningen, Switzerland 39/D1
Binongko (isl.), Indonesia 85/G7
Binscarth, Manitoba 179/A4
Bintan (isl.), Indonesia 85/C5
Bintuhan, Indonesia 85/C6
Bintulu, Malaysia 85/E5
Binyamina, Israel 65/B2
Binyang, China 77/G7
Bíobío (reg.), Chile 138/E1
Bío-Bío (riv.), Chile 138/E2
Biograd, Yugoslavia 45/B4
Bioko (isl.), Equat. Guinea 102/C4
Bioko (terr.), Equat. Guinea 115/A3
Bioko (isl.), Equat. Guinea 115/A3
Biola, Calif. (93606) 204/E7
Bippus, Ind. (46713) 227/F3
Bir, India 68/D5
Bira, U.S.S.R. 48/O5
Birag, Kuh-e (mts.), Iran 66/M7
Bir 'Ali, P.D.R. Yemen 59/E7
Birama (pt.), Cuba 158/G4
Birao, Cent. Afr. Rep. 115/D1
Biratnagar, Nepal 68/F3
Biratori, Japan 81/L2
Bir Bala, Iran 66/L8
Bir Bala, Iran 59/G4
Bircao, Somalia 115/H4
Birch (creek), Alaska 196/J1
Birch (hills), Alberta 182/A2
Birch (lake), Alberta 182/E3
Birch (mts.), Alberta 182/B5
Birch (riv.), Alberta 182/B5
Birch (creek), Idaho 220/F5
Birch (isl.), Manitoba 179/C2
Birch (lake), Minn. 255/G3
Birch (creek), Mont. 262/D2
Birch (lake), Sask. 181/C2
Birch (creek), Utah 304/B5
Birch (pt.), Wash. 310/C2
Birch Creek, Alaska (†99740) 196/J1
Birch Creek (valley), Idaho 220/E5
Birch Creek (res.), Mont. 262/D2
Birchdale, Minn. (56629) 255/D2
Birch Harbor, Maine (04613) 243/H7
Birch Hills, Sask. 181/F3
Birchip, Victoria 97/B4
Birch Island, Br. Col. 184/H4
Birchleaf, Va. (24220) 307/D6
Birch River, Manitoba 179/A2
Birch River, W. Va. (26610) 312/E6
Birch Run, Mich. (48415) 250/F5
Birchtown, Nova Scotia 168/C5
Birch Tree, Mo. (65438) 261/K9
Birchwood, Md. (†20021) 245/F5
Birchwood, Tenn. (37308) 237/M10
Birchwood, Wis. (54817) 317/C4
Birchy Bay, Newf. 166/D4
Bird (isl.), La. 238/M8
Bird (creek), Okla. 288/O1
Bird City, Kansas (67731) 232/A2
Bird Cove, Newf. 166/C3
Bird Island, Minn. (55310) 255/D6
Birds, Ill. (62415) 222/F5
Birdsboro, Pa. (19508) 294/L5
Birdseye, Ind. (47513) 227/D8
Birds Hill, Manitoba 179/F4
Birdsnest, Va. (23307) 307/S6
Birdsong, Ark. (†72386) 202/K3
Birdsville, Ky. (†42081) 237/D6
Birdsville, Queensland 88/F5
Birdsville, Queensland 95/A5
Birdtail, Manitoba 179/B4
Birdum, North. Terr. 93/C3
Birdwood, S. Australia 94/C7
Birecik, Turkey 63/H4
Bir el Khzaim (well), Mauritania 106/C4
Bireuen, Indonesia 85/B4
Bir Ganduz (well), Western Sahara 106/A4
Birganj, Nepal 68/F3
Bir Hakeim (ruins), Libya 111/D1
Birigui, Brazil 135/A2
Birjand, Iran 66/L4
Birjand, Iran 59/G3
Birjand, Iran 54/G6
Birken, Br. Col. 184/F5
Birkenfeld, Oreg. (97016) 291/D1
Birkenfeld, W. Germany 22/B4
Birkenhead, England 13/G2
Birkenhead, England 10/F2
Birkenhead, N. Zealand 100/B1
Birkenhead Lake Prov. Park, Br. Col. 184/F5
Birkerød, Denmark 21/F6
Birket Qârûn (lake), Egypt 111/J3
Bir Ksaib Ounane (well), Mali 106/C4
Birksgate (range), S. Australia 94/A2
Bîrlad, Romania 45/H2
Bîrlad (riv.), Romania 45/H2
Birmingham, Ala. 146/K6
Birmingham, Ala. 188/J4
Birmingham, Ala. 195/D3
Birmingham, England 7/D3
Birmingham, England 10/G3
Birmingham, England 13/F5
Birmingham, Iowa (52535) 229/K7
Birmingham, Mich. (*48008) 250/B6
Birmingham, Mo. (†64068) 261/R5
Birmingham, N.J. (08011) 273/D4
Birmingham, Ohio (44816) 284/F3
Birmingham, Pa. (†16686) 294/F4
Birmingham, Sask. 181/H5
Birmitrapur, India 68/E4
Bir Mogreïn, Mauritania 106/B3
Birnam, Scotland 15/E4
Birnamwood, Wis. (54414) 317/H6
Birney, Mont. (59012) 262/K5
Birnie, Manitoba 179/C4
Birnin Kebbi, Nigeria 106/E6
Birni-N'Konni, Niger 106/E6
Birni-N'Konni, Niger 102/C3
Bir Nzaran (well), Western Sahara 106/B4
Birobidzhan, U.S.S.R. 54/O5
Birobidzhan, U.S.S.R. 48/O5
Biron, Wis. (†54494) 317/G7
Bir Ounane (well), Mali 106/D4
Birqin, West Bank 65/C3
Birr, Ireland 17/F5
Birr, Ireland 10/B4
Birregurra, Victoria 97/B6
Birrie (riv.), N. S. Wales 88/H5
Birrie (riv.), N.S. Wales 97/D1
Birrimbah, North. Terr. 93/C3
Birrindudu, North. Terr. 93/A5
Birriwa, N.S. Wales 97/E3
Birs (riv.), Switzerland 39/D2
Birsay, Sask. 181/D4
Birsk, U.S.S.R. 52/J3
Birta, Ark. (†72853) 202/D3
Bir Taba, Egypt 59/B4
Bir Taba (well), Egypt 111/F2
Birtle, Manitoba 179/B4
Biru, China 77/D5
Biruaca, Venezuela 124/E4
Biruni, U.S.S.R. 48/G5
Biržai, U.S.S.R. 53/C2
Bir Zeit, West Bank 65/C4
Bisbee, Ariz. 188/E4
Bisbee, Ariz. (85603) 198/F7
Bisbee, N. Dak. (58317) 282/M2
Biscarrosse (lake), France 28/C5
Biscay (bay) 2/J3
Biscay (bay) 7/D4
Biscay (bay), France 28/B5
Biscay, Minn. (†55336) 255/D6
Biscay (bay), Spain 33/E1
Biscay Bay (riv.), Newf. 166/D2
Biscayne (bay), Fla. 212/F6
Biscayne (key), Fla. 212/B5
Biscayne Nat'l Park, Fla. 212/F6
Biscayne Park, Fla. (33152) 212/B4
Bisceglie, Italy 34/F4
Bischofshofen, Austria 41/B3
Bischofswerda, E. Germany 22/F3
Bischofszell, Switzerland 39/H1
Biscoe (isls.) 5/C15
Biscoe, Ark. (72017) 202/H4
Biscoe, N.C. (27209) 281/K4
Biscotasing, Ontario 177/J5
Biscotasing, Ontario 175/D3
Biscucuy, Venezuela 124/D3
Bisha, Saudi Arabia 59/D5
Bisha, Wadi (dry riv.), Saudi Arabia 59/D5
Bishiara (well), Libya 111/D3
Bisho (cap.), Ciskei, S. Africa 102/E8
Bishop, Calif. (93514) 204/G6
Bishop, Georgia (30621) 217/F3
Bishop, Md. (†21813) 245/S7
Bishop (creek), Nev. 266/F1
Bishop, Texas (78343) 303/G10
Bishop (creek), Utah 304/E3
Bishop, Va. (24604) 307/E6
Bishop Auckland, England 10/E3
Bishop Auckland, England 13/E3
Bishopbriggs, Scotland 15/B2
Bishop Hill, Ill. (61419) 222/C2
Bishopric, Sask. 181/F5
Bishop's Falls, Newf. 166/C4
Bishops Head, Md. (21611) 245/O7
Bishops Mitre (mt.), Newf. 166/B2
Bishop's Stortford, England 10/G5
Bishop's Stortford, England 13/H6
Bishopton, Québec 172/F4
Bishopton, Scotland 15/B2
Bishopville, Md. (21813) 245/T7
Bishopville, S.C. (29010) 296/G3
Bishri, Jebel el (mts.), Syria 63/H5
Biskra, Algeria 106/F2
Biskra, Algeria 102/C1
Biskupiec, Poland 47/E2
Bislig, Philippines 85/H4
Bislig, Philippines 82/F6
Bismarck, Ark. (71929) 202/D5
Bismarck, Ill. (61814) 222/F3
Bismarck, Mo. (63624) 261/L7
Bismarck (cap.), N. Dak. 146/H5
Bismarck (cap.), N. Dak. 188/G1
Bismarck (cap.), N. Dak. (58501) 282/J6
Bismarck (arch.), Papua N.G. 87/E6
Bismarck (arch.), Papua N.G. 86/B1
Bismarck (sea), Papua N.G. 86/B1
Bismarck (arch.), Papua N.G. 2/S6
Bismarck, W. Va. (†26739) 312/H4
Bismil, Turkey 63/J4
Bison (lake), Alberta 182/B1
Bison, Kansas (67520) 232/C3
Bison, Okla. (73720) 288/L2
Bison, S. Dak. (57620) 298/E2
Bispgården, Sweden 18/K5
Bissau (cap.), Guinea-Biss. 106/A6
Bissau (cap.), Guinea-Biss. 102/A3
Bissett, Manitoba 179/G4
Bistineau (lake), La. 238/D2
Bistrița, Romania 45/G2
Bita (riv.), Colombia 126/F5
Bitagron, Suriname 131/C3
Bitam, Gabon 115/B3
Bitburg, W. Germany 22/B4
Bitely, Mich. (49309) 250/D5
Bithlo, Fla. (†32801) 212/E3
Bitkine, Chad 111/C5
Bitlis (prov.), Turkey 63/J3
Bitlis, Turkey 63/J3
Bitlis, Turkey 59/D2
Bitola, Yugoslavia 45/E5
Bitola, Yugoslavia 7/G4
Bitonto, Italy 34/F4
Bitter (lakes), Egypt 111/K3
Bitter (lake), Sask. 181/B5
Bitter (creek), Wyo. 319/C4
Bitter Creek, Wyo. (†82901) 319/D4
Bitterfeld, E. Germany 22/E3
Bitterfontein, S. Africa 118/B6
Bittern (lake), Alberta 182/D3
Bittern Lake, Alberta 182/D3
Bitterroot (range) 188/D1
Bitterroot (range), Idaho 220/D3
Bitterroot (range), Mont. 262/B4
Bitterroot (riv.), Mont. 262/B4
Bitterroot (range), U.S. 146/G5
Bitti, Italy 34/B4
Bitumount, Alberta 182/E1
Bitung, Indonesia 85/H5
Biu, Nigeria 106/G6
Biu (plat.), Nigeria 106/G6
Bivalve, Md. (21814) 245/P7
Bivalve, N.J. (08301) 273/C5
Bivolari, Romania 45/H2
Biwa (lake), Japan 81/H6
Biwabik, Minn. (55708) 255/F3
Bixby, Minn. (55916) 255/E7
Bixby, Mo. (65439) 261/K7
Bixby, Okla. (74008) 288/P3
Biyang, China 77/H5
Biysk, U.S.S.R. 54/K4
Biysk, U.S.S.R. 48/J4
Bizcocho, Uruguay 145/B4
Bizerte, Tunisia 106/F1
Bizerte, Tunisia 102/C1
Bjargtangar (pt.), Iceland 21/A1
Bjelovar, Yugoslavia 45/C3
Bjerringbro, Denmark 21/C5
Bjorkdale, Sask. 181/H3
Bjørnafjorden (fjord), Norway 18/D6
Bjorne (pen.), N.W. Terrs. 187/K2
Bjornøya (isl.), Norway 18/D3
Blabon, N. Dak. (†58046) 282/P5
Blachly, Oreg. (97412) 291/D3
Black (sea) 2/L3
Black (sea) 54/E5
Black (sea) 7/H4
Black, Ala. (36314) 195/G8
Black (riv.), Alaska 196/K1
Black (mesa), Ariz. 198/E2
Black (mts.), Ariz. 198/A3
Black (riv.), Ariz. 198/E5
Black (riv.), Ark. 202/H2
Black (sea), Bulgaria 45/J4
Black (pond), Conn. 210/G1
Black (pt.), Conn. 210/G3
Black (mts.), England 13/D6
Black (creek), Fla. 212/E1
Black (head), Ireland 17/C5
Black (riv.), Jamaica 158/H6
Black (mt.), Ky. 237/R7
Black (lake), La. 238/D3
Black (pond), Maine 243/D3
Black (isl.), Manitoba 179/F3
Black (riv.), Manitoba 179/F4
Black (lake), Mich. 250/E3
Black (riv.), Mich. 250/E3
Black (riv.), Mich. 250/G5
Black (riv.), Minn. 255/D2
Black (creek), Miss. 256/F8
Black, Mo. (63625) 261/L7
Black (riv.), Mo. 261/L10
Black (mt.), N. Mex. 274/A6
Black (range), N. Mex. 274/B5
Black (lake), N.Y. 276/J1
Black (riv.), N.Y. 276/K3
Black (riv.), N.C. 281/N5
Black (riv.), Ohio 284/F3
Black (riv.), Ontario 177/E3
Black (sea), Romania 45/J4
Black (lake), Sask. 181/M2
Black (riv.), S.C. 296/H4
Black (sea), Turkey 63/E1
Black (sea), U.S.S.R. 48/D5
Black (sea), U.S.S.R. 52/D6
Black (creek), Vt. 268/B2
Black (riv.), Vt. 268/C2
Black (riv.), Vt. 268/B5
Black (riv.), Vietnam 72/D2
Black (mts.), Wales 13/D6
Black (for.), W. Germany 22/C4
Black (riv.), Wis. 317/E7
Black (riv.), Yukon 187/D3
Blackall, Australia 87/E8
Blackall, Queensland 88/H4
Blackall, Queensland 95/C5
Black Bear (creek), Okla. 288/M2
Blackberry (riv.), Conn. 210/B1
Blackbird, Del. (†19734) 245/R3
Blackbourne (pt.), Norfolk I. 88/L6
Black Branch, Nulhegan (riv.), Vt. 268/D2
Blackburn (mt.), Alaska 196/K2
Blackburn, England 13/H1
Blackburn, England 10/G1
Blackburn, La. (†71038) 238/D1
Blackburn, Mo. (65321) 261/F4
Blackburn, Okla. (74058) 288/N2
Blackburn, Ontario 177/J2
Blackburn, Scotland 15/C2
Black Butte (lake), Calif. 204/C4
Black Canyon City, Ariz. (85324) 198/C4
Black Canyon of the Gunnison Nat'l Mon., Colo. 208/D5
Black Creek, Br. Col. 184/E5
Black Creek, N.C. (27813) 281/O3
Black Creek, Wis. (54106) 317/K7
Black Diamond, Alberta 182/C4
Black Diamond, Wash. (98010) 310/D3
Blackduck, Minn. (56630) 255/D3
Black Duck (riv.), Ontario 175/C1
Black Eagle, Mont. (59414) 262/E3
Black Earth, Wis. (53515) 317/G9
Black Elster (riv.), E. Germany 22/E3
Blackey, Ky. (41804) 237/R6
Blackfalds, Alberta 182/D3
Blackfeet Ind. Res., Mont. 262/D2
Blackfoot, Alberta 182/E3
Blackfoot, Idaho (83221) 220/F6
Blackfoot (res.), Idaho 220/G7
Blackfoot (riv.), Idaho 220/G6
Blackfoot, Mont. (59415) 262/D2
Blackfoot (riv.), Mont. 262/C4
Blackford (co.), Ind. 227/G4
Blackford, Ky. (42403) 237/F6
Blackford, Scotland 15/E4
Black Forest, Colo. (80908) 208/K4
Blackfork, Ohio (45615) 284/E8
Black Fork, Mohican (riv.), Ohio 284/F4
Blackgum, Okla. (†74962) 288/S3
Black Hall, Conn. (†06371) 210/F3
Black Hawk, Colo. (80422) 208/J3
Blackhawk, Ind. (†47866) 227/C6
Black Hawk (co.), Iowa 229/J4
Black Hawk, Miss. (†38917) 256/E4
Black Hawk, S. Dak. (57718) 298/C5
Blackhead (bay), Newf. 166/D2
Blackhead Road, Newf. 166/D2
Black Hills (mts.) 188/F2
Black Hills (mts.), S. Dak. 298/B5
Blackie, Alberta 182/D4
Black Isle (pen.), Scotland 15/D3
Black Jack, Mo. (†63031) 261/R1
Black Lake (bayou), La. 238/D1
Black Lake, Québec 172/F3
Black Lake, Sask. 181/M2
Blackledge (riv.), Conn. 210/F2
Black Lick, Pa. (15716) 294/D4
Blacklock (pt.), Oreg. 291/C5
Black Mesa (mt.), Okla. 288/A1
Blackmore (mt.), Mont. 262/F5
Black Mountain, N.C. (28711) 281/E3
Black Oak, Ark. (72414) 202/K2
Black Oak, Ind. (†46406) 227/C1
Black Pine (mts.), Idaho 220/E7
Black Pine (peak), Idaho 220/E7
Black Pine (creek), S. Dak. 298/G6
Black Point, Calif. (†94947) 204/J1
Black Point, Conn. (†06357) 210/G3
Black Point, New Bruns. 170/D1
Blackpool, England 13/G1
Blackpool, England 10/F1
Blackridge, Va. (23916) 307/M7
Black River, Jamaica 158/H6
Black River, Jamaica 156/B3
Black River (bay), Jamaica 158/G6
Black River, Mich. (48721) 250/F4
Black River, New Bruns. 170/E3
Black River (pond), Newf. 166/C2
Black River, N.Y. (13612) 276/J3
Black River Bridge, New Bruns. 170/E2
Black River Falls, Wis. (54615) 317/E7
Black Rock, Ark. (72415) 202/H1
Black Rock (des.), Nev. 266/B2
Black Rock (range), Nev. 266/B1
Black Rock (pt.), R.I. 249/H8
Black Rock, Utah (†84751) 304/B5
Blacksburg, S.C. (29702) 296/D1
Blacksburg, Va. (24060) 307/H6
Blacks Fork, Green (riv.), Wyo. 319/C4
Blacks Harbour, New Bruns. 170/D3
Blackshear, Georgia (31516) 217/H8
Blackshear (lake), Georgia 217/E7
Blacksher, Ala. (†36507) 195/C8
Blacksod (bay), Ireland 17/A3
Black Springs, Ark. (†71960) 202/C5
Black Springs, Nev. (†89508) 266/B3
Black Squirrel (creek), Colo. 208/L5
Blackstairs (mt.), Ireland 17/H6
Blackstock, Ontario 177/F3
Blackstock, S.C. (29014) 296/E2
Blackstone◯, Mass. (01504) 249/H4
Blackstone (riv.), Mass. 249/G3
Blackstone, Va. (23824) 307/N6
Blacksville, W. Va. (26521) 312/F3
Black Thunder (creek), Wyo. 319/G2
Black Tickle, Newf. 166/C3
Blackton, Ark. (†72069) 202/H4
Blacktown, N.S. Wales 88/K4
Blacktown, N.S. Wales 97/H3
Blackville, New Bruns. 170/E2
Blackville, S.C. (29817) 296/E5
Black Volta (riv.) 102/B3
Black Volta (riv.), Ghana 106/D6
Black Volta (riv.), Ivory Coast 106/D6
Black Volta (riv.), Upper Volta 106/D6
Black Warrior (riv.), Ala. 195/C5
Blackwater (riv.), England 13/H6
Blackwater (riv.), Fla. 212/B6
Blackwater, Ireland 17/J7
Blackwater (riv.), Ireland 10/B4
Blackwater (riv.), Ireland 17/D7
Blackwater (riv.), Ireland 17/H4
Blackwater, Mo. (65322) 261/G5
Blackwater (res.), N.H. 268/D5
Blackwater (riv.), N. Ireland 17/H3
Blackwater, Queensland 95/D4
Blackwater, Queensland 88/H4
Blackwater (res.), Scotland 15/D4
Blackwater, Va. (24221) 307/B7
Blackwater (riv.), Va. 307/J6
Blackwater (riv.), Va. 307/O6
Blackwell, Ark. (72019) 202/E3
Blackwell (brook), Conn. 210/H1
Blackwell, Okla. (74631) 288/M1
Blackwell, Texas (79506) 303/D5
Blackwell, Wis. (†54511) 317/J4
Blackwood (Ngunju) (cape), Indonesia 85/F8
Blackwood, N.J. (08012) 273/C4
Blackwood Terrace, N.J. (†08096) 273/C4
Bladen, Nebr. (68928) 264/F4
Bladen (co.), N.C. 281/M5
Bladenboro, N.C. (28320) 281/M5
Bladensburg, Md. (20710) 245/G4
Bladensburg, Ohio (43005) 284/F5
Blades, Del. (†19973) 245/R6
Bladon Springs, Ala. (36902) 195/B7
Bladworth, Sask. 181/E4
Blaeberry, Br. Col. 184/J4
Blaenavon, Wales 13/B6
Blagodarnoye, U.S.S.R. 52/F5
Blagoevgrad, Bulgaria 45/F5
Blagoveshchensk, U.S.S.R. 54/O4
Blagoveshchensk, U.S.S.R. 48/N4
Blagoveshchensk, U.S.S.R. 52/J4
Blain, France 28/C4
Blain, Pa. (17006) 294/H5
Blaine, Georgia (†30175) 217/C1
Blaine (co.), Idaho 220/D6
Blaine, Kansas (66410) 232/F2
Blaine, Ky. (41124) 237/R4
Blaine◯, Maine (04734) 243/H2
Blaine, Mich. (†48032) 250/G5
Blaine, Minn. (†55433) 255/G5
Blaine, Miss. (38727) 256/C3
Blaine (co.), Mont. 262/G2
Blaine (co.), Nebr. 264/E3
Blaine, Ohio (43909) 284/J5
Blaine (co.), Okla. 288/K3
Blaine, Oreg. (†97108) 291/D2
Blaine, Tenn. (37709) 237/O8
Blaine, Wash. (98230) 310/C2
Blaine Lake, Sask. 181/D3
Blaine-Mars Hill, Maine (04734) 243/H2
Blainville, Québec 172/H4
Blair, Kansas (†66090) 232/H2
Blair, Nebr. (68008) 264/H3
Blair, Okla. (73526) 288/H5
Blair (co.), Pa. 294/F4
Blair, S.C. (29015) 296/E3
Blair, W. Va. (25022) 312/C7
Blair, Wis. (54616) 317/D7
Blair Athol, Queensland 95/C4
Blair Atholl, Scotland 10/E2
Blair Atholl, Scotland 15/E4
Blairgowrie and Rattray, Scotland 15/E4
Blairgowrie and Rattray, Scotland 10/E2
Blairmore, Alberta 182/C5
Blairs, Va. (24527) 307/K7
Blairsburg, Iowa (50034) 229/F4
Blairsden, Calif. (96103) 204/E4
Blairs Mills, Ky. (41402) 237/P4
Blairs Mills, Pa. (17213) 294/G5
Blairstown, Iowa (52209) 229/J5
Blairstown, Mo. (64726) 261/E5
Blairstown◯, N.J. (07825) 273/C2
Blairsville, Georgia (30512) 217/E1
Blairsville, Pa. (15717) 294/D5
Blaisdell, N. Dak. (58720) 282/F3
Blaj, Romania 45/F2
Blake (pt.), Mich. 250/E1
Blakeley, Minn. (†56011) 255/E6
Blakeley, W. Va. (†25160) 312/D6
Blakely, Georgia (31723) 217/C8
Blakely, Pa. (18447) 294/F6
Blakesburg, Iowa (52536) 229/H7
Blakeslee, Ohio (43505) 284/A2
Blakeslee, Pa. (18610) 294/L3
Blaketown, Newf. 166/D2
Blalock, Ala. (†36773) 195/D6
Blalock, Georgia (†30525) 217/E1
Blalock (isl.), Wash. 310/F5
Blanc (cape) 2/J4
Blanc (mt.), France 7/E4
Blanc (mt.), France 28/G5
Blanc (mt.), Italy 34/A2
Blanc (cape), Mauritania 102/A2
Blanc (cape), Mauritania 106/A4
Blanc (cape), Tunisia 106/G1
Blanc (cape), Western Sahara 106/A4
Blanca (bay), Argentina 120/C6
Blanca (bay), Argentina 143/D4
Blanca (lag.), Chile 138/E10
Blanca (peak), Colo. 188/F3
Blanca, Colo. (81123) 208/H8
Blanca (peak), Colo. 208/H7
Blanca (pt.), C. Rica 154/F5
Blanca, Cordillera (mts.), Peru 128/D7
Blanch, N.C. (27212) 281/L2
Blanchard, Idaho (83804) 220/A1
Blanchard, Iowa (51630) 229/C7
Blanchard, La. (71009) 238/C1
Blanchard◯, Maine (†04406) 243/D5
Blanchard, Mich. (49310) 250/D5
Blanchard, N. Dak. (58009) 282/R5
Blanchard (riv.), Ohio 284/C4
Blanchard, Okla. (73010) 288/L4
Blanchard, Pa. (16826) 294/G3
Blanchard, Wash. (†98232) 310/C2
Blanchardstown, Ireland 17/H5
Blanchardville, Wis. (53516) 317/G10
Blanche, Ky. (†40902) 237/O7
Blanche (riv.), Québec 172/E2
Blanche (lake), S. Australia 88/F5
Blanche (lake), S. Australia 94/F3
Blanche, Tenn. (†38488) 237/H10
Blanche (lake), W. Australia 88/C4
Blanche Marie (fall), Suriname 131/C3
Blanchester, Ohio (45107) 284/B7
Blanchisseuse, Trin. & Tob. 161/B10
Blanco (riv.), Argentina 143/C2
Blanco (riv.), Bolivia 136/D4
Blanco (lake), Chile 138/F10
Blanco (cape), C. Rica 154/E6
Blanco (peak), C. Rica 154/F6
Blanco (riv.), Mexico 150/Q2
Blanco, N. Mex. (87412) 274/B2
Blanco (creek), N. Mex. 274/F4
Blanco, Okla. (74528) 288/P5
Blanco (cape), Oreg. 188/A2
Blanco (cape), Oreg. 291/C5
Blanco (cape), Peru 128/B5
Blanco (riv.), Peru 128/F6
Blanco (co.), Texas 303/F8
Blanco, Texas (78606) 303/F7
Blanc-Sablon, Québec 174/F2
Bland, Mo. (65014) 261/J6
Bland (co.), Va. 307/F6
Bland, Va. (24315) 307/F6
Blandburg, Pa. (16619) 294/F4

Blandford○, Mass. (01008) 249/C4
Blandford, Nova Scotia 168/D4
Blandford Forum, England 13/E7
Blandford Forum, England 10/E5
Blanding, Utah (84511) 304/E6
Blandinsville, Ill. (61420) 222/C3
Blandville, Ky. (42026) 237/D7
Blanefield, Scotland 15/B1
Blanes, Spain 33/H2
Blaney Park, Mich. (†49836) 250/D2
Blanford, Ind. (47831) 227/B5
Blankenberge, Belgium 27/C6
Blankenburg am Harz, E. Germany 22/D3
Blanket, Texas (76432) 303/F6
Blanquillo, Uruguay 145/D3
Blansko, Czech. 41/D2
Blanton, Ala. (†36872) 195/H5
Blanton, Fla. (†33525) 212/D3
Blantyre, Malawi 115/F7
Blantyre, Malawi 102/F6
Blantyre, Scotland 15/B2
Blarney, Ireland 10/B5
Blarney, Ireland 17/D8
Blas (peak), Switzerland 39/G3
Blasdell, N.Y. (14219) 276/C5
Blasket (isls.), Ireland 10/A4
Blasket (isls.), Ireland 17/A7
Blatná, Czech. 41/B2
Blato, Yugoslavia 45/C4
Blatten, Switzerland 39/E4
Blaubeuren, W. Germany 22/C4
Blauvelt, N.Y. (10913) 276/K8
Blåvands Huk (pt.), Denmark 21/A6
Blawenburg, N.J. (08504) 273/D3
Blawnox, Pa. (15238) 294/C6
Blaydon, England 10/F3
Blaydon, England 13/H3
Blaye, France 28/C5
Blayney, N.S. Wales 97/E3
Blaze (pt.), North. Terr. 88/D2
Blaze (pt.), North. Terr. 93/A2
Bleckley (co.), Georgia 217/F6
Bled, Yugoslavia 45/A2
Bledsoe (co.), Tenn. 237/L9
Bledsoe, Texas (79314) 303/A4
Bleecker, Ala. (†36874) 195/H5
Blekinge (co.), Sweden 18/J8
Blencoe, Iowa (51523) 229/A5
Blenheim, N. Zealand 100/D4
Blenheim, Ontario 177/C5
Blenheim, S.C. (29516) 296/H2
Blenker, Wis. (54415) 317/F6
Blennerhassett (isl.), Ohio 284/G7
Blerick, Netherlands 27/J6
Blesbok (riv.), S. Africa 118/J7
Blessing, Texas (77419) 303/H9
Blessington, Ireland 17/J5
Blevins, Ark. (71825) 202/C6
Blewett, Texas (†78801) 303/D8
Blida, Algeria 106/E1
Blida, Algeria 102/C1
Bligh (sound), N. Zealand 100/A6
Bligh Water (bay), Fiji 86/P10
Blind Channel, Br. Col. 184/E5
Blind River, Ont. 162/H6
Blind River, Ontario 177/J5
Blind River, Ontario 175/D3
Blinman, S. Australia 88/F6
Blinman, S. Australia 94/F4
Blinnenhorn (mt.), Switzerland 39/F4
Bliss, Idaho (83314) 220/D7
Bliss, N.Y. (14024) 276/D5
Blissfield, Mich. (49228) 250/F7
Blissfield, New Bruns. 170/D2
Blissfield, Ohio (43805) 284/G5
Blitar, Indonesia 85/K2
Blitchton, Georgia (†31308) 217/J6
Blocher, Ind. (47138) 227/F7
Block (isl.), R.I. 249/H8
Blocker, Okla. (74529) 288/P4
Block House, Nova Scotia 168/D4
Block Island (sound), N.Y. 276/S8
Block Island, R.I. (02807) 249/H8
Block Island (sound), R.I. 249/H8
Blockton, Iowa (50836) 229/D7
Blodgett, Mo. (63824) 261/O8
Blodgett, Oreg. (97326) 291/D3
Blodgett Landing, N.H. (†03255) 268/D5
Bloemendaal, Netherlands 27/E4
Bloemfontein, S. Africa 102/E7
Bloemfontein, S. Africa 118/C5
Blois, France 28/D4
Blokzijl, Netherlands 27/H3
Blomkest, Minn. (56216) 255/D6
Błonie, Poland 47/E2
Bloodroot (mt.), Vt. 268/B4
Bloodsworth (isl.), Md. 245/O8
Bloodvein (riv.), Manitoba 179/F3
Bloodvein (riv.), Ontario 175/A2
Bloodvein River, Manitoba 179/F3
Bloody Foreland (prom.), Ireland 17/E1
Bloody Foreland (prom.), Ireland 10/B3
Bloom, Kansas (67833) 232/C4
Bloom, N. Dak. (†58401) 282/N6
Bloomburg, Texas (75556) 303/L4
Bloom City, Wis. (54617) 317/E8
Bloomdale, Ohio (44817) 284/D3
Bloomer, Ark. (†72933) 202/B3
Bloomer, Wis. (54724) 317/D5
Bloomery, W. Va. (26817) 312/K4
Bloomfield, Sierra (mts.), Bolivia 136/D4
Bloomfield○, Conn. (06002) 210/E1
Bloomfield, Ind. (47424) 227/D6
Bloomfield, Iowa (52537) 229/J7
Bloomfield, Ky. (40008) 237/L5
Bloomfield, Mo. (63825) 261/M9
Bloomfield, Mont. (59315) 262/M3
Bloomfield, Nebr. (68718) 264/G2
Bloomfield, Newf. 166/D2
Bloomfield, N.J. (07003) 273/B2
Bloomfield, N. Mex. (87413) 274/A2
Bloomfield, Ontario 177/G4
Bloomfield (New Bloomfield), Pa. (17068) 294/H5
Bloomfield○, Vt. (†03590) 268/D2
Bloomfield Hills, Mich. (48013) 250/B6
Bloomfield Ridge, New Bruns. 170/D2
Bloomfield Station, New Bruns. 170/E3
Bloomingburg, N.Y. (12721) 276/L7
Bloomingburg, Ohio (43106) 284/D6
Bloomingdale, Georgia (31302) 217/K6
Bloomingdale, Ill. (60108) 222/A5
Bloomingdale, Ind. (47832) 227/C5
Bloomingdale, Mich. (49026) 250/C6
Bloomingdale, N.J. (07403) 273/E1
Bloomingdale, N.Y. (12913) 276/M2
Bloomingdale, Ohio (43910) 284/J5
Bloomingdale, Tenn. (37660) 237/R7
Bloomingdale, Wis. (†54667) 317/E8
Blooming Grove, Ind. (†47012) 227/G5
Blooming Grove, Pa. (†18428) 294/M3
Blooming Grove, Texas (76626) 303/H5
Bloomingport, Ind. (†47355) 227/G5
Blooming Prairie, Minn. (55917) 255/E7
Bloomington, Calif. (92316) 204/E10
Bloomington, Idaho (83223) 220/G7
Bloomington, Ill. 188/J2
Bloomington, Ill. (61701) 222/D3
Bloomington, Ind. (47401) 227/D6
Bloomington, Md. (21523) 245/B3
Bloomington, Minn. (55420) 255/G6
Bloomington, Nebr. (68929) 264/F4
Bloomington, Texas (77951) 303/H9
Bloomington, Wis. (53804) 317/E10
Bloomington Springs, Tenn. (38545) 237/K8
Blooming Valley, Pa. (†16335) 294/B2
Bloomsburg, Pa. (17815) 294/J3
Bloomsbury, Alberta 182/C2
Bloomsbury, N.J. (08804) 273/C2
Bloomsdale, Mo. (63627) 261/M6
Bloomville, N.Y. (13739) 276/L6
Bloomville, Ohio (44818) 284/D3
Blora, Indonesia 85/K2
Blossburg, Pa. (16912) 294/H2
Blossom, Texas (75416) 303/J4
Bloubergstrand, S. Africa 118/E6
Blount (co.), Ala. 195/E2
Blount (co.), Tenn. 237/O9
Blounts Creek, N.C. (27814) 281/P4
Blount Springs, Ala. (†35079) 195/E3
Blountstown, Fla. (32424) 212/A1
Blountsville, Ala. (35031) 195/E2
Blountsville, Ind. (†47360) 227/G4
Blountville, Tenn. (37617) 237/S7
Blowering (res.), N.S. Wales 97/E4
Blowing Rock, N.C. (28605) 281/F2
Bloxom, Va. (23308) 307/S5
Bludenz, Austria 41/A3
Blue, Ariz. (85922) 198/F5
Blue (riv.), Ariz. 198/F5
Blue (mt.), Colo. 208/B2
Blue (riv.), Colo. 208/G3
Blue (riv.), Ind. 227/E8
Blue (mts.), Jamaica 158/J6
Blue (mt.), Maine 243/C6
Blue (hills), Mass. 249/C8
Blue (creek), Nebr. 264/B3
Blue (mt.), New Bruns. 170/D1
Blue (mt.), N.H. 268/E2
Blue (mts.), N. S. Wales 88/H6
Blue (mts.), N.S. Wales 97/F3
Blue, Okla. (†74701) 288/O7
Blue (riv.), Okla. 288/O6
Blue (mts.), Oreg. 291/J3
Blue (mt.), Pa. 294/G5
Blue (creek), Utah 304/B2
Blue (lake), Utah 304/B4
Blue (lake), Wash. 310/F3
Blue (mts.), Wash. 310/H4
Blue Ash, Ohio (†45242) 284/C9
Blue Ball, Ark. (†72866) 202/C4
Blue Bell, S. Dak. (†57773) 298/C6
Bluebell, Utah (84007) 304/D3
Blueberry Creek, Br. Col. 184/J5
Blueberry Mountain, Alberta 182/A2
Blue Creek, Ohio (45616) 284/D8
Blue Creek, Utah (†84337) 304/B2
Bluecreek, Wash. (†99109) 310/H2
Blue Creek, W. Va. (25026) 312/D6
Blue Cypress (lake), Fla. 212/F4
Blue Diamond, Ky. (41718) 237/P6
Blue Earth (co.), Minn. 255/D6
Blue Earth, Minn. (56013) 255/D7
Blue Earth (riv.), Minn. 255/D7
Blue Eye, Ark. (65611) 202/D1
Blue Eye, Mo. (65611) 261/F9
Bluefield, Va. (24605) 307/F6
Bluefield, W. Va. 188/K3
Bluefield, W. Va. (24701) 312/D8
Bluefields, Jamaica 158/G6
Bluefields, Nicaragua 154/F4
Bluegrass, Ind. (†46939) 227/E3
Blue Grass, Iowa (52726) 229/M5
Blue Grass, Minn. (†56477) 255/C4
Blue Grass, Va. (24413) 307/J3
Blue Heron, Sask. 181/E2
Blue Hill, Maine (04614) 243/F7
Blue Hill○, Maine (04614) 243/F7
Blue Hill (bay), Maine 243/G7
Blue Hill, Nebr. (68930) 264/F4
Blue Hill Falls, Maine (04615) 243/F7
Blue Hills, Conn. (06112) 210/E1
Blue Island, Ill. (60406) 222/B6
Bluejacket, Okla. (74333) 288/R1
Blue Jay, Calif. (92317) 204/H9
Blue Joint (lake), Oreg. 291/H5
Blue Knob (mt.), Pa. 294/E5
Blue Lake, Calif. (95525) 204/A3
Blue Mesa (res.), Colo. 208/E5
Bluemont, Va. (22012) 307/N2
Blue Mound, Ill. (62513) 222/D4
Blue Mound, Kansas (66010) 232/H3
Blue Mound, Texas (†76101) 303/E2
Blue Mounds, Wis. (53517) 317/G9
Blue Mountain, Ala. (36201) 195/G3
Blue Mountain, Ark. (72826) 202/C3
Blue Mountain (lake), Ark. 202/C3
Blue Mountain (peak), Jamaica 158/K6
Blue Mountain (peak), Jamaica 156/C3
Blue Mountain, Miss. (38610) 256/G1
Blue Mountain (lake), N.Y. 276/M3
Blue Mountain Lake, N.Y. (12812) 276/M3
Blue Mountains, Australia 87/E9
Blue Mountains, N. S. Wales 88/J6
Blue Mountains, N.S. Wales 97/F3
Blue Nile (riv.) 102/F3
Blue Nile (Abay) (riv.), Ethiopia 111/G5
Blue Nile (prov.), Sudan 111/F5
Blue Nile (riv.), Sudan 59/B6
Blue Nile (riv.), Sudan 111/F5
Blue Nose (mt.), Idaho 220/D4
Bluenose (lake), N.W. Terrs. 187/G3
Blue Rapids, Kansas (66411) 232/F2
Blue Ridge, Alberta 182/C2
Blue Ridge, Georgia (30513) 217/D1
Blue Ridge (lake), Georgia 217/D1
Blue Ridge (mts.), Georgia 217/D1
Blue Ridge, Ind. (†46176) 227/F5
Blue Ridge (mts.), Md. 245/H3
Blue Ridge (mts.), N.C. 281/E3
Blue Ridge (mts.), S.C. 296/B1
Blue Ridge, Va. (24064) 307/J6
Blue Ridge (mts.), Va. 307/J6
Blue Ridge Manor, Ky. (†40201) 237/L2
Blue Ridge Summit, Pa. (17214) 294/G6
Blue River, Br. Col. 184/H4
Blue River, Colo. (†80424) 208/G4
Blue River, Oreg. (97413) 291/E3
Blue River, Wis. (53518) 317/E9
Blue Rock, Nova Scotia 168/D4
Blue Rock (Gaysport), Ohio (43720) 284/G6
Blue Sea Lake, Québec 172/A3
Bluesky, Alberta 182/A1
Blue Spring (hills), Utah 304/B1
Blue Springs, Ala. (†36017) 195/G7
Blue Springs, Miss. (38828) 256/G2
Blue Springs, Mo. (64015) 261/R6
Blue Springs, Nebr. (68318) 264/H4
Blue Stack (mts.), Ireland 17/E2
Bluestem (lake), Okla. 288/O1
Bluestone (lake), W. Va. 307/G5
Bluestone (lake), W. Va. 312/E7
Blue Sulphur Springs, W. Va. (†25545) 312/E7
Blue Summit, Mo. (†64101) 261/R5
Bluevale, Ontario 177/C4
Bluewater, N. Mex. (87005) 274/A3
Bluewater (creek), N. Mex. 274/B4
Bluewater (creek), N. Mex. 274/D6
Bluewater (lake), N. Mex. 274/A3
Bluff (cape), Newf. 166/C3
Bluff, N. Zealand 100/B7
Bluff, N.C. (†28743) 281/D3
Bluff, Utah (84512) 304/E6
Bluff City, Ark. (71722) 202/D6
Bluff City, Ill. (†62624) 222/E5
Bluff City, Kansas (67018) 232/E4
Bluff City, Tenn. (37618) 237/S8
Bluff Dale, Texas (76433) 303/F5
Bluffdale, Utah (†84065) 304/B3
Bluff Knoll (mt.), W. Australia 92/B6
Bluff Park, Ala. (35226) 195/E4
Bluffs, Ill. (62621) 222/C4
Bluffsprings, Fla. (†32535) 212/B5
Bluffton, Alberta 182/C3
Bluffton, Ark. (72827) 202/C4
Bluffton, Georgia (31724) 217/C7
Bluffton, Ind. (46714) 227/G3
Bluffton, Minn. (56518) 255/C4
Bluffton, Ohio (45817) 284/C4
Bluffton, S.C. (29910) 296/F7
Bluford, Ill. (62814) 222/E5
Blum, Texas (76627) 303/G5
Blumenau, Brazil 132/D9
Blumenau, Brazil 120/E5
Blumenfeld, Manitoba 179/D5
Blumenheim, Sask. 181/E3
Blumenhof, Sask. 181/D5
Blumenort, Manitoba 179/E5
Blumenort, Manitoba 179/F5
Blumenort, Sask. 181/D6
Blumenstein, Switzerland 39/E3
Blumenthal, Sask. 181/E3
Blümlisalp (mt.), Switzerland 39/E3
Blunt, S. Dak. (57522) 298/J4
Bly, Oreg. (97622) 291/F5
Blying (sound), Alaska 196/C1
Blyn, Wash. (†98382) 310/B3
Blyth, England 13/F2
Blyth, England 10/F3
Blyth, Ontario 177/C4
Blyth Bridge, Scotland 15/E5
Blythe, Calif. (92225) 204/L10
Blythe, Georgia (30805) 217/H4
Blythedale, Md. (†21904) 245/O2
Blythedale, Mo. (64426) 261/E2
Blythedale, Pa. (†15018) 294/C5
Blytheswood, Ontario 177/B5
Blytheville, Ark. 188/H3
Blytheville, Ark. (72315) 202/L2
Blytheville A.F.B., Ark. 202/K2
Blythewood, S.C. (29016) 296/E3
Bo, S. Leone 102/A4
Bo, S. Leone 106/B7
Boac, Philippines 82/C4
Boaco, Nicaragua 154/E4
Boa Esperança, Brazil 135/D2
Boalsburg, Pa. (16827) 294/G4
Boano (isl.), Indonesia 85/H6
Boa Nova, Brazil 132/B5
Board Camp, Ark. (71932) 202/B4
Boardman (riv.), Mich. 250/D4
Boardman, N.C. (†28438) 281/M6
Boardman, Ohio (44512) 284/J3
Boardman, Oreg. (97818) 291/H2
Boardman, Wis. (†54016) 317/A5
Boardmans Bridge, Conn. (†06776) 210/B2
Boas (riv.), N.W. Terrs. 187/K3
Boat Basin, Br. Col. 184/D5
Boat Harbour, Tasmania 99/B2
Boat of Garten, Scotland 15/E3
Boa Vista, Brazil 120/C2
Boa Vista, Brazil 132/H8
Boa Vista (isl.), C. Verde 106/B8
Boayan (isl.), Philippines 82/B5
Boaz, Ala. (35957) 195/F2
Boaz, Ky. (42027) 237/D7
Boaz, Mo. (†65631) 261/F8
Boaz, W. Va. (†26187) 312/D4
Boaz, Wis. (†53581) 317/E9
Bobadah, N.S. Wales 97/D3
Bobai, China 77/H7
Bobaomby (cape), Madagascar 102/G6
Bobaomby (Amber) (cape), Madagascar 118/H2
Bobare, Venezuela 124/D2
Bobbili, India 68/E5
Bobbitt, N.C. (†27544) 281/N2
Bobcaygeon, Ontario 177/F3
Bobigny, France 28/B1
Böblingen, W. Germany 22/C4
Bobo, Miss. (†38614) 256/C2
Bobo Dioulasso, Upper Volta 106/D6
Bobo Dioulasso, Upper Volta 102/B3
Bobon, Philippines 82/F7
Bobonaza (riv.), Ecuador 128/D3
Bobonong, Botswana 118/D4
Bobotov Kuk (mt.), Yugoslavia 45/D4
Bobr (riv.), Poland 47/B3
Bobrov, U.S.S.R. 52/F4
Bobruysk, U.S.S.R. 7/G3
Bobruysk, U.S.S.R. 52/C4
Bobruysk, U.S.S.R. 48/C4
Bobs (lake), Ontario 177/H3
Bobtown, Pa. (15315) 294/B6
Bobures, Venezuela 124/C3
Boby, Pic (mt.), Madagascar 118/H4
Bocabec, New Bruns. 170/C3
Boca Chica, Dom. Rep. 158/E6
Boca Chica (key), Fla. 212/E7
Boca Ciega (bay), Fla. 212/B3
Boca de Aroa, Venezuela 124/D2
Boca del Mangle, Venezuela 124/D2
Boca del Pao, Venezuela 124/F3
Boca del Pepé, Colombia 126/B5
Boca del Río, Mexico 150/Q2
Boca del Soco, Dom. Rep. 158/F6
Boca do Acre, Brazil 132/G10
Boca Grande, Fla. (33921) 212/D5
Boca Grande (key), Fla. 212/D7
Boca Grande (passage), Trin. & Tob. 161/A10
Boca Grande (gulf), Venezuela 124/H3
Bocaiúva, Brazil 132/E7
Bocaranga, Cent. Afr. Rep. 115/C2
Boca Raton, Fla. (*33432) 212/F5
Bocas del Toro, Panama 154/F6
Bocay, Nicaragua 154/E3
Bochnia, Poland 47/E4
Bocholt, Belgium 27/H6
Bocholt, W. Germany 22/B3
Bochov, Czech. 41/B1
Bochum, W. Germany 22/B3
Bock, Minn. (56313) 255/E5
Boco, Chile 138/F2
Boconó, Venezuela 124/C3
Boda, Cent. Afr. Rep. 115/C3
Bodalla, N.S. Wales 97/F5
Bodaybo, U.S.S.R. 54/N4
Bodaybo, U.S.S.R. 48/M4
Bodcaw, Ark. (†71858) 202/D6
Boddam, Scotland 15/G3
Boddington, W. Australia 92/B2
Bode, Iowa (50519) 229/E3
Bodega (bay), Calif. 204/B5
Bodega (head), Calif. 204/B5
Bodega Bay, Calif. (94923) 204/B5
Bodegraven, Netherlands 27/F4
Bodélé (depr.), Chad 102/D3
Bodélé (depr.), Chad 111/C4
Boden, Sweden 18/M4
Bodensee (Constance) (lake), Austria 41/A3
Bodensee (Constance) (lake), Switzerland 39/H1
Bodensee (Constance) (lake), W. Germany 22/C5
Boderg (lake), Ireland 17/E4
Bodfish, Calif. (93205) 204/G8
Bodhan, India 68/D5
Bodie (isl.), N.C. 281/T2
Bodinayakkanur, India 68/D6
Bodines, Pa. (17722) 294/H3
Bodio, Switzerland 39/G4
Bodkin (pt.), Md. 245/N4
Bodmin, England 13/C7
Bodmin, England 10/D5
Bodmin, Sask. 181/D2
Bodo, Alberta 182/E3
Bodø, Norway 18/J3
Bodø, Norway 7/F2
Bodrum, Turkey 63/B4
Bodrum, Turkey 59/A2
Bo Duc, Vietnam 72/E4
Bódvaszilas, Hungary 41/F2
Boelus, Nebr. (68820) 264/F3
Boende, Zaire 115/D4
Boerne, Texas (78006) 303/J10
Boeuf (lake), La. 238/J7
Boeuf (riv.), La. 238/G1
Boffa, Guinea 106/B6
Bog (lake), Maine 243/H6
Bogalusa, La. 188/H4
Bogalusa, La. (70427) 238/L5
Bogan (riv.), N.S. Wales 97/D2
Bogandé, Upper Volta 106/E6
Bogan Gate, N.S. Wales 97/D3
Bogantungan, Queensland 95/C4
Bogard, Mo. (64622) 261/E4
Bogart, Georgia (30622) 217/E3
Bogata, Texas (75417) 303/J4
Bogatynia, Poland 47/B3
Boğazlıyan, Turkey 63/F3
Bogen, W. Germany 22/E4
Bogenfels, Namibia 118/B5
Bogense, Denmark 21/D6
Boger City, N.C. (28092) 281/G4
Boggabilla, N.S. Wales 97/F1
Boggabri, N.S. Wales 97/F2
Boggeragh (mts.), Ireland 17/D7
Boggs, W. Va. (26299) 312/E6
Boggstown, Ind. (46110) 227/F5
Boggy (peak), Ant. & Bar. 161/D11
Boggy Creek, Manitoba 179/A3
Boggy Depot, Okla. (†74525) 288/O6
Bogia, Papua N.G. 85/B6
Bogie (riv.), Scotland 15/F3
Bognor Regis, England 13/G7
Bognor Regis, England 10/F5
Bogny-sur-Meuse, France 28/F3
Bogo, Philippines 82/E5
Bogon (riv.), N. S. Wales 88/H6
Bogong (mt.), Victoria 97/D5
Bogor, Indonesia 54/M10
Bogor, Indonesia 85/H2
Bogoslof (isl.), Alaska 196/E4
Bogotá (cap.), Colombia 126/D5
Bogotá (cap.), Colombia 120/B2
Bogotá (cap.), Colombia 2/F5
Bogota, Ill. (†62448) 222/E5
Bogota, N.J. (07603) 273/B2
Bogota, Tenn. (38007) 237/C8
Bogra, Bangladesh 68/F4
Boguchar, U.S.S.R. 52/F5
Bogue, Kansas (67625) 232/C2
Bogué, Mauritania 106/B5
Bogue Chitto, Miss. (39629) 256/D8
Bogue Chitto (riv.), Miss. 256/D8
Bogue Homo (lake), Miss. 256/F7
Boguszów-Gorce, Poland 47/B3
Bog Walk, Jamaica 158/J6
Bo Hai (gulf), China 77/J4
Boharm, Sask. 181/F5
Bohemian (for.), Czech. 41/B2
Bohemian (for.), W. Germany 22/E4
Bohemian-Moravian Heights (hills), Czech. 41/C2
Boherbue, Ireland 17/C7
Bohners Lake, Wis. (†53105) 317/K10
Bohodleh, Somalia 115/J2
Bohol (prov.), Philippines 82/E6
Bohol (isl.), Philippines 85/G4
Bohol (isl.), Philippines 82/E6
Bohol (sea), Philippines 82/E6
Bohol (str.), Philippines 82/D6
Böhönye, Hungary 41/D3
Bohu (Bagrax), China 77/C3
Boicourt, Kansas (†66075) 232/H3
Boiestown, New Bruns. 170/D2
Boiling Spring Lakes, N.C. (28461) 281/N7
Boiling Springs, N.C. (28017) 281/F4
Boiling Springs, Pa. (17007) 294/H5
Bois Blanc (isl.), Mich. 250/E3
Boischatel, Québec /J3
Boisdale, Nova Scotia 168/H2
Boisdale, Loch (inlet), Scotland 15/A3
Bois D'Arc, Mo. (65612) 261/F8
Bois-des-Filion, Québec 172/H4
Bois de Sioux (riv.), Minn. 255/B4
Bois de Sioux (riv.), S. Dak. 298/R1
Boise (co.), Idaho 220/C6
Boise (cap.), Idaho 146/G5
Boise (cap.), Idaho (*83701) 220/B6
Boise (cap.), Idaho 188/C2
Boise (mts.), Idaho 220/B6
Boise (riv.), Idaho 220/B6
Boise City, Okla. (73933) 288/B1
Boissevain, Man. 162/G6
Boissevain, Manitoba 179/C5
Boissevain, Va. (24606) 307/F6
Boistfort, Wash. (†98532) 310/B4
Boisvert (pt), Québec 172/J1
Boizenburg an der Elbe, E. Germany 22/D2
Bojador (cape), W. Sahara 102/A2
Bojador (cape), Western Sahara 106/B3
Bojeador (cape), Philippines 82/C1
Bojnurd, Iran 66/K2
Bojnurd, Iran 59/G2
Bojonegoro, Indonesia 85/J2
Bokchito, Okla. (74726) 288/O6
Boké, Guinea 106/B6
Bokeelia, Fla. (33922) 212/D5
Bokel (cay), Belize 154/D2
Bokhara (riv.), N.S. Wales 97/D1
Bokhoma, Okla. (†71821) 288/S7
Boknafjord (fjord), Norway 18/D7
Boko, Congo 115/B4
Bokoro, Chad 111/C5
Bokoshe, Okla. (74930) 288/S4
Bokote, Zaire 115/D4
Bokpyin, Burma 72/C5
Boksburg, S. Africa 118/J6
Bokungu, Zaire 115/D4
Bol, Chad 111/B5
Bol, Chad 102/D3
Bolair, W. Va. (26288) 312/F6
Bolama, Guinea-Biss. 106/A6
Bolan (pass), Pakistan 68/B3
Bolangir, India 68/E4
Bolar, Va. (24414) 307/J4
Bolatusha, Miss. (†39160) 256/E5
Bolayır, Turkey 63/C5
Bolbec, France 28/D3
Bolckow, Mo. (64427) 261/C2
Bolderslev, Denmark 21/C8
Bolding, Ark. (†71747) 202/F7
Boldman, Ky. (†41501) 237/R5
Boldon, England 13/J3
Bolduc, Québec 172/G4
Bole, China 77/B3
Bole, Ghana 106/D7
Boles, Ark. (72926) 202/B4
Bolesławiec, Poland 47/B3
Boley, Okla. (74829) 288/O4
Bolgatanga, Ghana 106/D6
Boli, China 77/M2
Boligee, Ala. (35443) 195/C5
Bolinao, Philippines 82/B2
Bolinao (cape), Philippines 82/B2
Bolinas, Calif. (94924) 204/H1
Boling, Texas (77420) 303/H8
Bolingbroke, Georgia (31004) 217/E5
Bolingbrook, Ill. (60439) 222/A6
Bolinger, Ala. (36903) 195/B7
Bolinger, La. (†71064) 238/C1
Bolívar, Argentina 143/D4
Bolívar, Bolivia 136/B3
Bolívar (dept.), Colombia 126/C3
Bolívar, Antioquia, Colombia 126/C5
Bolívar, Cauca, Colombia 126/B7
Bolívar (prov.), Ecuador 128/C3
Bolívar, Ecuador 128/C2
Bolivar (co.), Miss. 256/C3
Bolivar, Mo. (65613) 261/F7
Bolivar, N.Y. (14715) 276/D6
Bolivar, Ohio (44612) 284/G4
Bolivar, Pa. (15923) 294/D5
Bolívar, Peru 128/D6
Bolivar, Tenn. (38008) 237/C10
Bolivar (pen.), Texas 303/K8
Bolívar (state), Venezuela 124/F7
Bolívar, Cerro (mt.), Venezuela 124/G4
Bolívar, Pico (peak), Venezuela 124/C3
Bolivar, W. Va. (†25425) 312/L4
Bolivia 2/F6
Bolivia 120/C4
BOLIVIA 136
Bolivia, N.C. (28422) 281/N6
Bolkar (mts.), Turkey 63/F4
Bolkhov, U.S.S.R. 52/E4
Bolligen, Switzerland 39/E3
Bolling, Ala. (36007) 195/E7
Bolling A.F.B., D.C. 245/E5
Bollinger (co.), Mo. 261/M8
Bollington, England 10/G2
Bollington, England 13/H2
Bollnäs, Sweden 18/K6
Bollon, Queensland 95/C6
Bollon, Queensland 88/H5
Bollstabruk, Sweden 18/L5
Bolmen (lake), Sweden 18/H8
Bolobo, Zaire 115/C4
Bologna (prov.), Italy 34/C2
Bologna, Italy 34/C2
Bologna, Italy 7/F4
Bolognesi, Peru 128/F8
Bolognesi, Peru 128/F6
Bologoye, U.S.S.R. 52/D3
Bolomba, Zaire 115/C3
Bolonchén de Rejón, Mexico 150/O7
Bolondrón, Cuba 156/B2
Bolondrón, Cuba 158/D1
Bolovens (plat.), Laos 72/E4
Bolpebra, Bolivia 136/A2
Bolsena (lake), Italy 34/C3
Bol'shevik (isl.), U.S.S.R. 54/N2
Bol'shevik (isl.), U.S.S.R. 4/A4
Bol'shevik (isl.), U.S.S.R. 48/K2
Bol'shoy Lyakhov (isl.), U.S.S.R. 54/R2
Bol'shoy Lyakhovskiy (isl.), U.S.S.R. 48/P2
Bolsover, England 13/J2
Bolsters Mills, Maine (†04040) 243/B7
Bolsward, Netherlands 27/H2
Boltaña, Spain 33/F1
Boltigen, Switzerland 39/D3
Bolton○, Conn. (06040) 210/F1
Bolton, England 10/G2
Bolton, England 13/H2
Bolton○, Mass. (01740) 249/H3
Bolton, Miss. (39041) 256/D6
Bolton, N.C. (28423) 281/N6
Bolton○, Vt. (†05466) 268/B3
Bolton (mt.), Vt. 268/B3
Bolton Landing, N.Y. (12814) 276/N3
Bolu (prov.), Turkey 63/D2
Bolu, Turkey 59/B1
Bolu, Turkey 63/D2
Bolus (head), Ireland 17/A8
Bolvadin, Turkey 63/D3
Bolvanskiy Nos (cape), U.S.S.R. 52/K1
Bolvanskiy Nos (cape), U.S.S.R. 48/G2
Bolzano (Bolzen), Italy 34/C1
Bolzano, Italy 7/F4
Bolzano-Bozen (prov.), Italy 34/C1
Bolzen (Bolzano), Italy 34/C1
Boma, Zaire 102/D5
Boma, Zaire 115/B5
Bomaderry-Nowra, N.S. Wales 97/F4
Bomarton, Texas (†76380) 303/E4
Bomba (gulf), Libya 111/D1
Bombala, N.S. Wales 97/E5
Bombardopolis, Haiti 158/B5
Bombay, India 54/J8
Bombay, India 2/N5
Bombay (harb.), India 68/B7
Bombay, Minn. (†55946) 255/F6
Bombay, N.Y. (12914) 276/M1
Bomboma, Zaire 115/C3
Bom Conselho, Brazil 132/G5
Bom Despacho, Brazil 135/D1
Bom Despacho, Brazil 132/E7
Bomdila, India 68/G3
Bom Futuro, Brazil 120/C4
Bom Futuro, Brazil 132/A5
Bomi, China 77/E6
Bom Jesus, Brazil 132/E5
Bom Jesus da Lapa, Brazil 120/E4
Bom Jesus da Lapa, Brazil 132/F6
Bom Jesus do Itabapoana, Brazil 135/F2
Bomongo, Zaire 115/C3
Bomont, W. Va. (25030) 312/D6
Bomoseen, Vt. (05732) 268/A4
Bomoseen (lake), Vt. 268/A4
Bom Retiro, Brazil 132/D10
Bom Sucesso, Brazil 135/D2
Bomu (riv.), Cent. Afr. Rep. 102/E4
Bomu (riv.), Cent. Afr. Rep. 115/D3
Bomu (riv.), Zaire 115/D3
Bon (cape), Tunisia 102/D1
Bon (cape), Tunisia 106/G1
Bona (mt.), Alaska 196/K2
Bonabéri, Cameroon 115/A3
Bonacca (Guanaja) (isl.), Honduras 154/E2
Bon Accord, Alberta 182/D3
Bonaduz, Switzerland 39/H3
Bon Air, Ala. (35032) 195/F4
Bonair, Iowa (†52155) 229/J2
Bon Air, Tenn. (†38583) 237/L9
Bon Air, Va. (23235) 307/N5
Bonaire (isl.), Neth. Ant. 156/E4
Bonaire (isl.), Neth. Ant. 161/E9
Bonalbo, N.S. Wales 97/G1

Bonanza, Alberta 182/A2
Bonanza, Ark. (†72901) 202/B3
Bonanza, Colo. (†81155) 208/G6
Bonanza, Nicaragua 154/E4
Bonanza, Oreg. (97623) 291/F5
Bonanza, Utah (84008) 304/E3
Bonanza (peak), Wash. 310/E2
Bonao, Dom. Rep. 158/E6
Bonaparte, Iowa (52620) 229/K7
Bonaparte (lake), N.Y. 276/K2
Bonaparte (creek), Wash. 310/F2
Bonaparte (mt.), Wash. 310/F2
Bonaparte (arch.), W. Australia 88/C2
Bonaparte (arch.), W. Australia 92/D1
Bon Aqua, Tenn. (37025) 237/G9
Bonar Bridge, Scotland 15/D3
Bonaventure (cape), Newf. 166/D2
Bonaventure (co.), Québec 172/C1
Bonaventure (county), Québec 174/D3
Bonaventure, Québec 172/C2
Bonaventure (isl.), Québec 172/D1
Bonaventure (riv.), Québec 172/C1
Bonavista, Newf. 166/D2
Bonavista (bay), Newf. 166/D1
Bonavista (cape), Newf. 166/D1
Bonavista, Newf. 162/L6
Boncarbo, Colo. (81024) 208/K8
Bonchester Bridge, Scotland 15/F5
Boncourt, Switzerland 39/C2
Bond, Colo. (80423) 208/F3
Bond (co.), Ill. 222/D5
Bond, Ky. (40407) 237/N6
Bond, Miss. (39550) 256/F9
Bond (mt.), N.H. 268/E3
Bondeno, Italy 34/C2
Bond Falls (res.), Mich. 250/G2
Bondi (beach), N.S. Wales 97/K3
Bondiss, Alberta 182/E2
Bondo, Zaire 115/D3
Bondoukou, Ivory Coast 106/D7
Bondsville, Mass. (01009) 249/E4
Bonduel, Wis. (54180) 317/K6
Bondurant, Iowa (50035) 229/G5
Bondurant, Wyo. (82922) 319/B2
Bondville, Ill. (61815) 222/E3
Bondville, Ky. (40308) 237/M5
Bondville, Vt. (05340) 268/B5
Bondy, France 28/B1
Bône (Annaba), Algeria 106/F1
Bone, Idaho (†83401) 220/G6
Bone (gulf), Indonesia 54/O10
Bone (gulf), Indonesia 85/G7
Bone Cave, Tenn. (†38581) 237/L9
Bone Gap, Ill. (62815) 222/F5
Bo'ness, Scotland 10/C1
Bo'ness, Scotland 15/C1
Bonesteel, S. Dak. (57317) 298/M7
Bonet (riv.), Ireland 17/E3
Boneta, Utah (†58801) 304/D3
Bonetraill, N. Dak. (†58801) 282/C3
Boneville, Georgia (30806) 217/G4
Bonfield, Ill. (60913) 222/E2
Bonfield, Ontario 177/E1
Bonfol, Switzerland 39/D2
Bong (range), Liberia 106/B7
Bongabong, Philippines 82/C4
Bongandanga, Zaire 115/D3
Bonggaw, Philippines 82/B8
Bongo (isl.), Philippines 82/D7
Bongor, Chad 111/C5
Bongor, Chad 102/D3
Bong Son (Hoai Nhon), Vietnam 72/F4
Bonham, Texas (75418) 303/H4
Bonhill, Scotland 15/B1
Bon Homme (co.), S. Dak. 298/O7
Bonifacio, France 28/B7
Bonifacio (str.), France 28/B7
Bonifacio (str.), Italy 34/B4
Bonifay, Fla. (32425) 212/C5
Bönigen, Switzerland 39/E3
Bonilla, S. Dak. (†57348) 298/N4
Bonin (isls.), Japan 2/S4
Bonin (isls.), Japan 54/R7
Bonin (isls.), Japan 87/E3
Bonin (isls.), Japan 81/M3
Bonita, Ariz. (†85643) 198/E6
Bonita (pt.), Calif. 204/H2
Bonita, La. (71223) 238/G1
Bonita, Texas (†39301) 256/E1
Bonita Springs, Fla. (33923) 212/E5
Bonlee, N.C. (27213) 281/L3
Bonn (cap.), W. Germany 7/E3
Bonn (cap.), W. Germany 22/B3
Bonne (bay), Newf. 166/C4
Bonneau, S.C. (29431) 296/H5
Bonner (co.), Idaho 220/B1
Bonners Ferry, Idaho (83805) 220/B1
Bonner Springs, Kansas (66012) 232/N2
Bonner-West Riverside, Mont. (59823) 262/C4
Bonnet (lake), Manitoba 179/G4
Bonnétable, France 28/D3
Bonnet Carré Spillway and Floodway, La. 238/N3
Bonne Terre, Mo. (63628) 261/L7
Bonnet Plume (riv.), Yukon 187/E3
Bonneville, France 28/G4
Bonneville, Idaho 220/G6
Bonneville, Oreg. (97008) 291/F2
Bonneville (dam), Oreg. 291/F2
Bonneville (salt flats), Utah 304/A3
Bonneville (dam), Wash. 310/D5
Bonneville (lake), Wash. 310/D5
Bonneville, Wyo. (†82649) 319/E2
Bonneville (mt.), Wyo. 319/C3
Bonney Lake, Wash. (†98390) 310/C3
Bonnie, Ill. (62816) 222/E5
Bonnieville, Ky. (42713) 237/K6
Bonnots Mill, Mo. (65016) 261/J5
Bonny (res.), Colo. 208/P3
Bonny, Nigeria 106/F8
Bonny (bight), Nigeria 106/F8
Bonnybridge, Scotland 15/C1
Bonnyman, Ky. (41719) 237/P6
Bonnyrigg, N. S. Wales 88/K4

Bonnyrigg, N.S. Wales 97/H3
Bonnyrigg and Lasswade, Scotland 10/C1
Bonnyrigg and Lasswade, Scotland 15/D2
Bonny River, New Bruns. 170/D3
Bonnyville, Alberta 182/E2
Bono, Ark. (72416) 202/J2
Bono, Ohio (†43445) 284/D2
Bonorva, Italy 34/B4
Bonpas (creek), Ill. 222/F5
Bonpland (mt.), N. Zealand 100/A6
Bon Secour, Ala. (36511) 195/C10
Bon Secour (bay), Ala. 195/C10
Bonsecours, Québec 172/E4
Bonshaw, Pr. Edward I. 168/E2
Bonthain, Indonesia 85/F7
Bonthe, S. Leone 106/B7
Bontoc, Philippines 85/G2
Bontoc, Philippines 82/C2
Bon Wier, Texas (75928) 303/L7
Bonyhád, Hungary 41/E3
Boody, Ill. (62514) 222/D4
Book (cliffs), Utah 304/E4
Booker, Texas (79005) 303/D1
Booker T. Washington Nat'l Mon., Va. 307/J6
Boolaloo, W. Australia 92/B3
Booligal, N.S. Wales 97/K3
Boom, Belgium 27/E6
Boom, Tenn. (†38573) 237/L7
Boomer, N.C. (28606) 281/G2
Boomer, W. Va. (25031) 312/D6
Boomi, N. S. Wales 88/K8
Boomi, N.S. Wales 97/E1
Boon (pt.), Ant. & Bar. 161/E11
Boon, Mich. (49618) 250/D4
Boondall, Queensland 88/K2
Boone (co.), Ark. 202/C1
Boone, Colo. (81025) 208/L6
Boone (co.), Ill. 222/E1
Boone (co.), Ind. 227/D3
Boone (co.), Iowa 229/F5
Boone, Iowa (50036) 229/F4
Boone (co.), Ky. 237/M3
Boone (co.), Mo. 261/H4
Boone, Nebr. (68625) 264/F3
Boone (co.), Nebr. 264/F3
Boone, N.C. (28607) 281/F2
Boone (lake), Tenn. 237/S6
Boone (co.), W. Va. 312/C6
Boone Grove, Ind. (46302) 227/C2
Boonesboro, Mo. (†65250) 261/H4
Booneville, Va. (†22935) 307/L4
Booneville, Ark. (72927) 202/C3
Booneville, Ky. (41314) 237/P6
Booneville, Miss. (38829) 256/G1
Boonsboro, Md. (21713) 245/H2
Boonton, N.J. (07005) 273/E2
Boonton (res.), N.J. 273/E2
Boonville, Calif. (95415) 204/B5
Boonville, Ind. (47601) 227/C8
Boonville, Mo. (65233) 261/G5
Boonville, N.Y. (13309) 276/K4
Boonville, N.C. (27011) 281/H2
Boopi (riv.), Bolivia 136/B4
Boorooban, N.S. Wales 97/C4
Boorowa, N.S. Wales 97/E4
Boort, Victoria 97/B5
Booth, Ala. (36008) 195/E6
Boothbay, Maine (04537) 243/D8
Boothbay○, Maine (†46577) 243/D8
Boothbay Harbor, Maine (04538) 243/D8
Boothia (pen.), Canada 4/B14
Boothia (gulf), Canada 4/B14
Boothia (pen.), N.W.T. 146/J2
Boothia (gulf), N.W.T. 146/J2
Boothia (pen.), N.W.T. 162/G1
Boothia (gulf), N.W.T. 162/G1
Boothia (isthmus), N.W.T. 162/G2
Boothia (pen.), U.S.S.R. 52/D3
Boothia (gulf), N.W. Terrs. 187/K3
Boothia (pen.), N.W. Terrs. 187/J2
Boothville, La. (70038) 238/M8
Boothwyn, Pa. (19061) 294/L7
Bootle, England 10/F2
Bootle, England 13/G2
Booué, Gabon 115/B3
Bophuthatswana (bantustan), S. Africa 102/E7
Bophuthatswana (rep.), S. Africa 118/D5
Boppard, W. Germany 22/B3
Boquerón, Cuba 158/K4
Boquerón (bay), Paraguay 144/B3
Boquerón, Cuba 156/K4
Boquerón, El (pass), Peru 128/E7
Boquerón, P. Rico 156/F1
Boquerón, P. Rico 161/A3
Boquerón (bay), P. Rico 161/A3
Boquilla del Carmen, Mexico 150/H2
Bor, Czech. 41/B2
Bor, Sudan 111/F6
Borzya, U.S.S.R. 48/M4
Bor, U.S.S.R. 52/B5
Bor, Turkey 63/F4
Bor, Yugoslavia 45/E3
Bora-Bora (isl.), Fr. Poly. 87/L7
Borah (peak), Idaho 220/E5
Borah (peak), Idaho 220/E5
Borama, Somalia 115/H1
Borås, Sweden 7/F3
Borås, Sweden 18/H8
Borazjan, Iran 66/G6
Borazjan, Iran 59/E4
Borba, Brazil 120/D3
Borba, Brazil 132/H9
Borba, Portugal 33/C3
Borbón, Venezuela 124/F4
Borça, Turkey 63/J2
Borculo, Netherlands 27/J4
Bordeaux, France 28/C5
Bordeaux, France 7/D4
Bordeaux, S.C. (†29835) 296/C4
Bordeaux (mt.), Virgin Is. (U.S.) 161/C4
Bordelonville, La. (71320) 238/G4
Borden (isl.), Canada 4/B15

Borden, Ind. (47106) 227/F8
Borden (isl.), N.W. Terrs. 187/G2
Borden (pen.), N.W. Terrs. 187/K2
Borden, Pr. Edward I. 168/E2
Borden, Sask. 181/D3
Borden, S.C. (29017) 296/G3
Borden, W. Australia 92/B6
Borden Shaft, Md. (†21532) 245/B2
Borden Springs, Ala. (†36262) 195/H3
Bordentown, N.J. (08505) 273/D3
Border, Minn. (†56623) 255/C2
Border, Wyo. (83114) 319/B3
Borderland, W. Va. (25665) 312/B7
Borders (reg.), Scotland 15/E5
Bordertown, S. Australia 88/F7
Bordertown, S. Australia 94/G7
Bordighera, Italy 34/A3
Bordj Bou Arreridj, Algeria 106/E1
Bordj Fly Sainte Marie, Algeria 106/D3
Bordj Omar Driss, Algeria 106/F3
Bordj Omar Driss, Algeria 102/C2
Bordulac, N. Dak. (58417) 282/N5
Boreing, Ky. (†40740) 237/N6
Boreray (isl.), Scotland 15/A2
Boreray (isl.), Scotland 15/A3
Borgå, Finland 18/N6
Borge, Norway 18/H2
Borger, Netherlands 27/K3
Borger, Texas (79007) 303/C2
Borger, Texas 188/F3
Borgerhout, Belgium 27/E6
Borgholm, Sweden 18/K8
Borghorst, W. Germany 22/B2
Borgloon, Belgium 27/G7
Borgne (lake), La. 238/L7
Borgne (riv.), Switzerland 39/D4
Borgo, Italy 34/C1
Borgomanero, Italy 34/C2
Borgo San Lorenzo, Italy 34/C2
Borgworm (Waremme), Belgium 27/G7
Borikan, Laos 72/D3
Borislav, U.S.S.R. 52/B5
Borisoglebsk, U.S.S.R. 48/E4
Borisoglebsk, U.S.S.R. 52/F4
Borisov, U.S.S.R. 52/C4
Borisovka, U.S.S.R. 52/E4
Bo River Post, Sudan 111/E6
Borja, Peru 128/C3
Borja, Spain 33/F2
Borjas Blancas, Spain 33/G2
Borken, W. Germany 22/B3
Børkop, Denmark 21/C6
Borku, Chad 111/C4
Borkum, W. Germany 22/B2
Borkum (isl.), W. Germany 22/B2
Borlänge, Sweden 18/J6
Borna, E. Germany 22/E3
Borndiep (chan.), Netherlands 27/H2
Borne, Netherlands 27/K4
Borneo (isl.) 2/Q6
Borneo (isl.) 54/N9
Borneo (isl.), Indonesia 85/E5
Borneo (isl.), Malaysia 85/E5
Bornheim, W. Germany 22/B3
Bornholm (co.), Denmark 21/F9
Bornholm (isl.), Denmark 7/F3
Bornholm (isl.), Denmark 18/J9
Bornholm (isl.), Denmark 21/F9
Borno (state), Nigeria 106/G6
Bornova, Turkey 63/B3
Boracay (isl.), Philippines 82/D5
Borojó, Venezuela 124/C2
Boron, Calif. (93516) 204/H8
Borongan, Philippines 82/E5
Borot Kidod (well), Israel 65/C5
Borovichi, U.S.S.R. 52/D3
Borradaile, Alberta 182/E3
Borre, Norway 18/H4
Borrego Springs, Calif. (92004) 204/J10
Borris, Ireland 17/H6
Borris-in-Ossory, Ireland 17/F6
Borrisokane, Ireland 17/F6
Borrisoleigh, Ireland 17/F6
Borroloola, North. Terr. 88/F3
Borroloola, North. Terr. 93/E4
Borşa, Romania 45/G2
Borsod-Abaúj-Zemplén (co.), Hungary 41/F2
Bortala (Bole), China 77/B3
Borth, Wales 13/C5
Bort-les-Orgues, France 28/E5
Boruca, C. Rica 154/F6
Borujerd, Iran 59/E3
Borujerd, Iran 66/F4
Borup, Denmark 21/E7
Borup, Minn. (56519) 255/B3
Börzsöny (mts.), Hungary 41/E3
Borzya, U.S.S.R. 48/M4
Bosa, Italy 34/B4
Bosanska Dubica, Yugoslavia 45/C3
Bosanska Gradiška, Yugoslavia 45/C3
Bosanska Kostajnica, Yugoslavia 45/B3
Bosanska Krupa, Yugoslavia 45/C3
Bosanski Brod, Yugoslavia 45/D3
Bosanski Novi, Yugoslavia 45/C3
Bosanski Petrovac, Yugoslavia 45/C3
Bosanski Šamac, Yugoslavia 45/D3
Bosaso, Somalia 115/J1
Bosaso, Somalia 102/G3
Boscawen○, N.H. (03301) 268/D5
Bosch, van den (cape), Indonesia 85/J6
Bosco, La. (†71201) 238/F2
Boscobel, Wis. (53805) 317/E9
Boshan, China 77/J4
Boshof, S. Africa 102/E6
Boshrūyeh, Iran 66/H4
Bosilegrad, Yugoslavia 45/F4
Bosiljevo, Yugoslavia 45/B3
Boskoop, Netherlands 27/F4
Boskovice, Czech. 41/D2
Bosler, Wyo. (82051) 319/G4
Bosna (riv.), Yugoslavia 45/D3
Bosnia and Hercegovina (rep.), Yugoslavia 45/C3

Boso (pen.), Japan 81/K6
Bosobolo, Zaire 115/D3
Bosporus (str.), Turkey 7/G4
Bosporus (str.), Turkey 59/A1
Bosporus (str.), Turkey 63/C1
Bosque (Bosque Farms), N. Mex. (87006) 274/C4
Bosque (co.), Texas 303/G6
Boss, Mo. (65440) 261/K7
Bossangoa, Cent. Afr. Rep. 102/D4
Bossangoa, Cent. Afr. Rep. 115/C2
Bossé, New Bruns. 170/B1
Bossembele, Cent. Afr. Rep. 115/C2
Bossier (par.), La. 238/C1
Bossier City, La. (*71111) 238/C1
Bosso, Niger 106/G6
Bostan, Iran 66/F5
Bostan, Pakistan 68/B2
Bostanabad-e-Bala, Iran 66/E2
Bosten (Bagrax) Hu (lake), China 77/C3
Boston, England 13/G5
Boston, England 10/F4
Boston, Georgia (31626) 217/E9
Boston, Ind. (47324) 227/H5
Boston, Ky. (40107) 237/K5
Boston (cap.), Mass. 146/L5
Boston (cap.), Mass. 188/M2
Boston (cap.), Mass. (*02101) 249/D7
Boston (bay), Mass. 249/E6
Boston (harb.), Mass. 249/D7
Boston, Mo. (†64759) 261/D8
Boston, N.Y. (14025) 276/C5
Boston (mts.), Okla. 288/S3
Boston, Pa. (15135) 294/M5
Boston, Tenn. (†37064) 237/G9
Boston, Texas 75557) 303/K4
Boston, U.S. 2/F3
Boston, Va. (22713) 307/M3
Boston Bar, Br. Col. 184/G3
Boston Heights, Ohio (†44264) 284/J10
Bostonnais (isl.), Québec 172/E2
Bostonnais, Grand Lac (lake), Québec 172/E2
Bostonnais (riv.), Québec 172/E2
Boston Nat'l Hist. Park, Mass. 249/D6
Bostwick, Fla. (32007) 212/E2
Bostwick, Georgia (30623) 217/E3
Bostwick, Nebr. (†68978) 264/F4
Boswell, Ark. (72516) 202/F1
Boswell, Br. Col. 184/J4
Boswell, Ind. (47921) 227/C3
Boswell, Okla. (74727) 288/P6
Boswell, Pa. (15531) 294/H5
Boswell Bay, Alaska (†99574) 196/J2
Bosworth, Mo. (64541) 261/F4
Bot (riv.), S. Africa 118/G7
Botany, N. S. Wales 88/L4
Botany (bay), N. S. Wales 88/L4
Botany, N.S. Wales 97/J4
Botany (bay), N.S. Wales 97/J4
Botene, Laos 72/D3
Botetourt (co.), Va. 307/J5
Botevgrad, Bulgaria 45/F4
Botha, Alberta 182/D3
Botha (riv.), Alberta 182/B1
Bothell, Wash. (98011) 310/B1
Bothnia (gulf) 7/G2
Bothnia (gulf), Finland 18/M5
Bothnia (gulf), Sweden 18/N4
Bothwell, Ontario 177/C5
Bothwell, Tasmania 99/C4
Bothwell, Utah (†84337) 304/B2
Botkins, Ohio (45306) 284/B5
Botna, Iowa (†51454) 229/C5
Botoşani, Romania 45/H2
Botrange (mt.), Belgium 27/J8
Botrivier, S. Africa 118/F7
Botsford, Conn. (06404) 210/C3
Botswana 2/L7
BOTSWANA 118/C4
Bottineau (co.), N. Dak. 282/J2
Bottineau, N. Dak. (58318) 282/J2
Bottrel, Alberta 182/C4
Bottrop, W. Germany 22/B3
Botucatu, Brazil 135/B3
Botucatu, Brazil 132/D8
Botwood, Newf. 166/C4
Bouaflé, Ivory Coast 106/C7
Bouaké, Ivory Coast 102/B4
Bouaké, Ivory Coast 106/D7
Bouali, Cent. Afr. Rep. 115/C3
Bouar, Cent. Afr. Rep. 102/D4
Bouar, Cent. Afr. Rep. 115/C2
Bou Arfa, Morocco 106/D2
Bouca, Cent. Afr. Rep. 115/C2
Boucaut (bay), North. Terr. 93/D1
Boucherville, Québec 172/J4
Boucherville (isl.), Québec 172/J4
Bouches-du-Rhône (dept.), France 28/F6
Bouchette, Québec 172/A3
Bouckville, N.Y. (13310) 276/J5
Bou Djebeha, Mali 106/D5
Boudreau (bay), La. 238/M7
Boudreaux, La. (†70353) 238/J8
Boudreaux (lake), La. 238/J8
Boudry, Switzerland 39/C3
Boufarik, Algeria 106/E1
Bougainville (reef), 95/C2
Bougainville (reef), Coral Sea Is. Terr. 88/H3
Bougainville (isl.), Papua N.G. 87/F6
Bougainville (isl.), Papua N.G. 86/C3
Bougainville (str.), Papua N.G. 86/D2
Bougainville (str.), Solomon Is. 86/D2
Bougainville (cape), W. Australia 88/D2
Bougainville (cape), W. Australia 92/D1
Bougaroun (cape), Algeria 106/F1
Boughton (isl.), Pr. Edward I. 168/F2

Bougie (Béjaïa), Algeria 106/F1
Bougouni, Mali 106/C6
Bouillante, Guadeloupe 161/A6
Bouillon, Belgium 27/G9
Bou Izakarn, Morocco 106/C3
Boujad, Morocco 106/C2
Boula, Cent. Afr. Rep. 115/C3
Boulanger, Québec 172/E1
Boularderie (isl.), Nova Scotia 168/J2
Boulder, Australia 87/C9
Boulder, Colo. 188/D3
Boulder, Colo. 146/H6
Boulder (co.), Colo. 208/J2
Boulder, Colo. (*80301) 208/J2
Boulder (mts.), Idaho 220/D6
Boulder, Mont. (59632) 262/E4
Boulder (lake), N. Mex. 274/C3
Boulder, Utah (84716) 304/C6
Boulder (creek), Utah 304/C6
Boulder, W. Australia 88/C6
Boulder, Wyo. (82923) 319/C3
Boulder (lake), Wyo. 319/C3
Boulder City, Nev. (89005) 266/G7
Boulder Creek, Calif. (95006) 204/J4
Boulder Junction, Wis. (54512) 317/G3
Boulder-Kalgoorlie, W. Australia 88/H3
Boulder-Kalgoorlie, W. Australia 92/C5
Boulevard, Calif. (92005) 204/J11
Boulevard Heights, Md. (†20027) 245/F5
Boulia, Queensland 95/A4
Boulia, Queensland 88/F4
Boulogne, Fla. (†32046) 212/E1
Boulogne-Billancourt, France 28/A2
Boulogne-sur-Mer, France 28/D2
Bouna, Ivory Coast 106/D6
Boundary, Alaska (†99732) 196/K2
Boundary (co.), Idaho 220/B1
Boundary (peak), Nev. 266/C5
Boundary (plat.), Sask. 181/B6
Boundary (bay), Wash. 310/C1
Boundary (dam), Wash. 310/H2
Boundary (lake), Wash. 310/H2
Boundary Bend, Victoria 97/B4
Bound Brook, N.J. (08805) 273/D2
Boundiali, Ivory Coast 106/C6
Boundji, Congo 115/C4
Boun Nua, Laos 72/D2
Bountiful, Utah (84010) 304/C3
Bounty (isl.), N. Zealand 87/H10
Bounty, Sask. 181/D4
Bourail, New Caled. 87/G8
Bourail, New Caled. 86/G4
Bourbon, Ill. (†61953) 222/E4
Bourbon, Ind. (46504) 227/E2
Bourbon (co.), Kansas 232/H4
Bourbon (co.), Ky. 237/N4
Bourbon, Miss. (†38756) 256/C4
Bourbon, Mo. (65441) 261/K6
Bourbonnais (trad. prov.), France 29
Bourbonnais, Ill. (60914) 222/F2
Bourem, Mali 106/E5
Bourg, La. (70343) 238/J7
Bourganeuf, France 28/D5
Bourg-des-Saintes, Guadeloupe 161/A7
Bourg-en-Bresse, France 28/F4
Bourgeois, New Bruns. 170/F2
Bourges, France 28/E4
Bourget, Ontario 177/J2
Bourg-Léopold (Leopoldsburg), Belgium 27/G6
Bourgoin-Jallieu, France 28/F5
Bourg Saint-Pierre, Switzerland 39/D5
Bourke, N. S. Wales 88/H6
Bourke, N. S. Wales 97/D2
Bourne, England 13/G5
Bourne, Mass. (02532) 249/M6
Bourne○, Mass. (02532) 249/M6
Bournedale, Mass. (†02532) 249/M5
Bournemouth, England 13/F6
Bournemouth, England 10/F5
Bourneville, Ohio (45617) 284/D7
Bourscheid, Luxembourg 27/J8
Bouse, Ariz. (85325) 198/A5
Bouse Wash (dry riv.), Ariz. 198/A4
Boussac, France 28/D4
Bousso, Chad 111/C5
Boussu, Belgium 27/D8
Boutilimit, Mauritania 106/B5
Boutilimit, Mauritania 102/A3
Bouton, Iowa (50039) 229/E5
Bouvard (cape), W. Australia 92/A3
Bouvet (isl.) 5/D1
Bouvetøya (Bouvet) (isl.) 5/D1
Boven Bolivia, Neth. Ant. 161/E8
Boves, Italy 34/A2
Bovey, Minn. (55709) 255/E3
Bovey Tracey, England 13/D7
Bovill, Idaho (83806) 220/B3
Bovina, Miss. (†39180) 256/C6
Bovina, Texas (79009) 303/A3
Bovril, Argentina 143/G5
Bow (riv.), Alberta 182/D4
Bow (riv.), Alta. 162/E5
Bow (lake), N.H. 268/E5
Bow, Wash. (98232) 310/C2
Bowbells, N. Dak. (58721) 282/F2
Bow City, Alberta 182/D4
Bowden, Alberta 182/C4
Bowden, Jamaica 158/B2
Bowden, W. Va. (26254) 312/G5
Bowdens, N.C. (28322) 281/N4
Bowdle, S. Dak. (57428) 298/N3
Bowdoin (lake), Mont. 262/J2
Bowdoinham○, Maine (04008) 243/D7
Bowdon, N. Dak. (58418) 282/L5
Bowdon, Georgia (30108) 217/B3
Bowdon Junction, Georgia (30109) 217/B3
Bowell, Alberta 182/E4
Bowen, Australia 87/F3
Bowen, Ill. (62316) 222/B3
Bowen, Ky. (40309) 237/O5
Bowen, Queensland 95/D3
Bowen, Queensland 88/H3
Bowen Island, Br. Col. 184/K3
Bowenville, Queensland 88/M2
Bowenville, Queensland 95/D1

Bowerbank○, Maine (†04481) 243/E5
Bowers, Ind. (†47940) 227/D4
Bowers Beach, Del. (†19962) 245/S4
Bowers Mill, Mo. (†64848) 261/E8
Bowerston, Ohio (44695) 284/H5
Bowersville, Georgia (30516) 217/G2
Bowersville, Ohio (44695) 284/C6
Bowes, England 13/F3
Bowesmont, N. Dak. (58217) 282/R2
Bowie, Ariz. (85605) 198/F6
Bowie (co.), Texas 303/K4
Bowie, Md. (20715) 245/L4
Bowie, Md. (81428) 208/D5
Bowie (creek), Miss. 256/E7
Bowie (co.), Texas 303/K4
Bowie, Texas (76230) 303/G4
Bow Island, Alberta 182/E5
Bowkan, Iran 66/E2
Bowlegs, Okla. (74830) 288/N4
Bowler, Wis. (54416) 317/J6
Bowling Green, Fla. (33834) 212/E4
Bowling Green, Ind. (47833) 227/D6
Bowling Green, Ky. (42101) 237/H7
Bowling Green, Ky. 188/L3
Bowling Green, Mo. (63334) 261/K4
Bowling Green, Ohio (43402) 284/C3
Bowling Green (cape), Queensland 88/H3
Bowling Green (cape), Queensland 95/C3
Bowling Green, S.C. (29703) 296/E1
Bowling Green, Va. (22427) 307/O4
Bowlus, Minn. (56314) 255/D5
Bowman, Calif. (95604) 204/C8
Bowman, Georgia (30624) 217/G2
Bowman (co.), N. Dak. 282/C7
Bowman, N. Dak. (58623) 282/D7
Bowman (bay), N.W.T. 162/J2
Bowman (bay), N.W. Terrs. 187/L3
Bowman (dam), Oreg. 291/G3
Bowman, S.C. (29018) 296/F5
Bowmansdale, Pa. (†17008) 294/J5
Bowmanstown, Pa. (18030) 294/L4
Bowmansville, Pa. (17507) 294/L5
Bow Mills, N.H. (†03301) 268/D5
Bowmont, Idaho (†83651) 220/B6
Bowmore, Scotland 15/B5
Bowmore, Scotland 10/C3
Bowral, N.S. Wales 97/J4
Bowraville, N.S. Wales 97/G2
Bowring, Okla. (74009) 288/O1
Bowron Lake Prov. Park, Br. Col. 184/G3
Bowser, Br. Col. 184/C2
Bowser (lake), Br. Col. 184/C2
Bowsman, Manitoba 179/A2
Bowstring, Minn. (56631) 255/E3
Bowstring (lake), Minn. 255/E3
Boxborough○, Mass. (01719) 249/H3
Box Butte (co.), Nebr. 264/A2
Box Butte (res.), Nebr. 264/A2
Box Canyon (dam), Wash. 310/H2
Box Elder (creek), Colo. 208/K4
Box Elder, Mont. (59521) 262/F2
Boxelder (creek), Mont. 262/M5
Boxelder (creek), Mont. 262/H3
Box Elder, S. Dak. (57719) 298/D5
Boxelder (creek), S. Dak. 298/D5
Box Elder (co.), Utah 304/A2
Boxford, Mass. (01921) 249/L2
Boxford○, Mass. (01921) 249/L2
Box Hill, Victoria 97/J5
Box Hill, Victoria 88/L7
Boxholm, Iowa (50040) 229/E4
Bo Xian (Pohsien), China 77/J5
Boxley, Ark. (†72742) 202/D2
Boxmeer, Netherlands 27/H5
Box Springs, Georgia (31801) 217/C5
Boxtel, Netherlands 27/G5
Boyabat, Turkey 63/F2
Boyacá (dept.), Colombia 126/D5
Boyama (Stanley) (falls), Zaire 102/E3
Boyama (Stanley) (falls), Zaire 115/D3
Boyanup, W. Australia 92/A2
Boyce, La. (71409) 238/E4
Boyce, Va. (22620) 307/M2
Boyceville, Wis. (54725) 317/C5
Boyd, Ala. (†35470) 195/B5
Boyd, Fla. (†32347) 212/C1
Boyd (co.), Ky. 237/R4
Boyd, Minn. (56218) 255/C6
Boyd, Mont. (59013) 262/G5
Boyd (co.), Nebr. 264/F2
Boyd, Okla. (†73931) 288/E1
Boyd, Oreg. (†97021) 291/F2
Boyd, Texas (76023) 303/E1
Boyd, Wis. (54726) 317/E6
Boydell, Ark. (†71658) 202/H7
Boyden, Iowa (51234) 229/B2
Boyden Arbor, S.C. (†29128) 296/F5
Boyd Lake, Maine (†04463) 243/F5
Boyds, Md. (20720) 245/J4
Boyds, Wash. (99107) 310/G2
Boydton, Va. (23917) 307/M7
Boyer (riv.), Alberta 182/A5
Boyer, Iowa (†51448) 229/C4
Boyer (riv.), Iowa 229/B5
Boyer, W. Va. (†24915) 312/G5
Boyer Ahmediyeh and Kohkiluyeh (gov.), Iran 66/G5
Boyero, Colo. (80821) 208/N5
Boyers, Pa. (16020) 294/C4
Boyertown, Pa. (19512) 294/L5
Boykin, Ala. (†31737) 217/C8
Boykin, S.C. (†29128) 296/F3
Boykins, Va. (23827) 307/O7
Boyle, Alberta 182/E2
Boyle, Ireland 17/E4
Boyle, Ireland 10/B3
Boyle (co.), Ky. 237/M5
Boyle, Miss. (38730) 256/C3
Boyleston, Ind. (†46057) 227/C4
Boylston○, Mass. (01505) 249/H3
Boylston, Nova Scotia 168/J2
Boyne (riv.), Ireland 17/J4
Boyne City, Mich. (49712) 250/E3

Boyne Falls, Mich. (49713) 250/E3
Boyne Lake, Alberta 182/E5
Boynton, Okla. (74422) 288/P3
Boynton Beach, Fla. (*33435) 212/F5
Boy River, Minn. (56632) 255/D3
Boysen (res.), Wyo. 319/D2
Boysen Bay, N.Y. (†13212) 276/H4
Boys Ranch, Texas (79010) 303/B2
Boys Town, Nebr. (68010) 264/H3
Boyuibe, Bolivia 136/D7
Bozcaada (isl.), Turkey 63/A3
Bozdoğan, Turkey 63/C4
Bozeman, Mont. 188/D1
Bozeman, Mont. (59715) 262/E5
Bozkır, Turkey 63/E4
Bozkurt, Turkey 63/F2
Bozman, Md. (21612) 245/N5
Bozoum, Cent. Afr. Rep. 115/C2
Bozova, Turkey 63/H4
Bozqush, Kuh-e (mts.), Iran 66/E2
Bozüyük, Turkey 59/B2
Bozüyük, Turkey 63/C3
Bra, Italy 34/A2
Brabant (prov.), Belgium 27/F7
Brabant Lake, Sask. 181/M3
Brač (isl.), Yugoslavia 45/C4
Bracadale, Loch (inlet), Scotland 15/B3
Bracciano, Italy 34/C3
Bracciano (lake), Italy 34/D3
Bracebridge, Ontario 177/E2
Braceville, Ill. (60407) 222/E2
Bracey, Va. (23919) 307/M7
Bräcke, Sweden 18/J5
Bracken (co.), Ky. 237/N3
Bracken, Sask. 181/C6
Brackendale, Br. Col. 184/F5
Brackenridge, Pa. (15014) 294/C4
Brackett, Wis. (†54742) 317/D6
Brackettville (canton), Texas 303/D8
Brackley, England 10/F4
Brackley, England 13/F5
Bracknell, England 13/G8
Bracknell, Tasmania 99/C3
Brackney, Pa. (18812) 294/K2
Brackwede, W. Germany 22/C3
Braço Maior do Araguaia (riv.), Brazil 132/D5
Braço Menor do Araguaia (riv.), Brazil 132/D6
Brad, Romania 45/F2
Bradbury, Calif. (91010) 204/D10
Braddock, N. Dak. (58524) 282/K6
Braddock, Pa. (15104) 294/C7
Braddock, Sask. 181/D5
Braddyville, Iowa (51631) 229/D7
Braden, Okla. (†74959) 288/S4
Braden, Tenn. (38010) 237/B10
Bradenton, Fla. (*33506) 212/D4
Bradenton Beach, Fla. (33510) 212/D4
Bradford, Ark. (72020) 202/G3
Bradford, England 13/J1
Bradford, England 10/H1
Bradford (co.), Fla. 212/D2
Bradford, Ill. (61421) 222/D2
Bradford, Ind. (47107) 227/E8
Bradford, Iowa (50041) 229/G3
Bradford, Ky. (†41043) 237/N3
Bradford, Maine (04410) 243/F5
Bradford○, Maine (04410) 243/F5
Bradford○, N.H. (03221) 268/D5
Bradford, Ohio (45308) 284/B5
Bradford, Ontario 177/E3
Bradford (co.), Pa. 294/J2
Bradford, Pa. (16701) 294/E2
Bradford, R.I. (02808) 249/H7
Bradford, Tenn. (38316) 237/D8
Bradford, Vt. (05033) 268/C3
Bradford○, Vt. (05033) 268/C3
Bradford Center, Maine (†04410) 243/F5
Bradford-on-Avon, England 13/E6
Bradfordsville, Ky. (40009) 237/L6
Bradgate, Iowa (50520) 229/E3
Bradley (co.), Ark. 202/F7
Bradley, Ark. (71826) 202/C7
Bradley, Calif. (93426) 204/E8
Bradley, Fla. (33835) 212/D4
Bradley, Georgia (†31032) 217/E4
Bradley, Ill. (60915) 222/F2
Bradley○, Maine (04411) 243/F6
Bradley, Miss. (†39759) 256/G4
Bradley, Ohio (†43917) 284/J5
Bradley, Okla. (73011) 288/L5
Bradley, S.C. (29819) 296/C3
Bradley, S. Dak. (57217) 298/O3
Bradley (co.), Tenn. 237/M10
Bradley, Wis. (†54487) 317/G4
Bradley Beach, N.J. (07720) 273/F3
Bradleyton, Ala. (†36041) 195/F7
Bradleyville, Mo. (65614) 261/F9
Bradner, Ohio (43406) 284/C3
Bradshaw, Nebr. (68319) 264/G4
Bradshaw, Texas (†79567) 303/D5
Bradshaw, W. Va. (24817) 312/C8
Bradwardine, Manitoba 179/B5
Bradwell, Sask. 181/E4
Brady (glac.), Alaska 196/M1
Brady, Mont. (59416) 262/E2
Brady, Nebr. (69123) 264/D3
Brady (mt.), S. Australia 94/D3
Brady, Texas (76825) 303/E6
Brady Lake, Ohio (44211) 284/H3
Bradyville, Tenn. (37026) 237/J9
Brae, Scotland 15/G2
Braedstrup, Denmark 21/C6
Braemar, Scotland 15/E3
Braemar, Scotland 10/E2
Braemar (dist.), Scotland 15/E3
Braemar, Tenn. (37658) 237/S8
Braeside, Ontario 177/H2
Braeside, W. Australia 92/C3
Braga (dist.), Portugal 33/B2
Braga, Portugal 7/D4
Braga, Portugal 33/B2
Bragado, Argentina 143/F7
Bragança, Brazil 120/E3
Bragança, Brazil 132/E3
Bragança (dist.), Portugal 33/C2
Bragança, Portugal 33/C2
Bragança Paulista, Brazil 135/C3
Bragança Paulista, Brazil 132/E8
Braggadocio, Mo. (63826) 261/N10
Bragg City, Mo. (63827) 261/N10
Bragg Creek, Alberta 182/C4
Braggs, Ala. (†36761) 195/E6
Braggs, Okla. (74423) 288/R3
Bragman's Bluff (Puerto Cabezas), Nicaragua 154/F3
Braham, Minn. (55006) 255/E5
Brahmaputra (riv.) 54/L7
Brahmaputra (riv.), Bangladesh 68/G3
Brahmaputra (riv.), India 68/G3
Braich-y-Pwll (prom.), Wales 10/D4
Braich-y-Pwll (prom.), Wales 13/C5
Braidwood, Ill. (60408) 222/E2
Braidwood, N.S. Wales 97/E4
Brăila, Romania 7/G4
Brăila, Romania 45/H3
Brăila (marshes), Romania 45/H3
Brainard, Nebr. (68626) 264/G3
Brainards, N.J. (08865) 273/C2
Braine-l'Alleud, Belgium 27/E7
Braine-le-Comte, Belgium 27/D7
Brainerd, Minn. 188/H1
Brainerd, Minn. (56401) 255/D4
Braintree○, Mass. (02184) 249/D8
Braintree (West Braintree), Vt. 268/B4
Braintree○, Vt. (†05669) 268/B4
Braintree and Bocking, England 13/H6
Braintree and Bocking, England 10/G5
Braithwaite, La. (70040) 238/P4
Brak, Libya 102/D2
Brak, Libya 111/B2
Brake, W. Germany 22/C2
Brakna (reg.), Mauritania 106/B5
Brakpan, S. Africa 118/J6
Bralorne, Br. Col. 184/F5
Braman, Okla. (74632) 288/M1
Bramber, Nova Scotia 168/D3
Bramberg am Wildkogel, Austria 41/B3
Bramble (bay), Queensland 95/E2
Bramming, Denmark 21/B7
Bramon, Venezuela 124/B4
Brampton, England 13/E3
Brampton, Mich. (49810) 250/B3
Brampton, N. Dak. (58010) 282/P7
Brampton, Ontario 177/J4
Bramsche, W. Germany 22/B2
Bramwell, W. Va. (24715) 312/D8
Bran (riv.), Scotland 15/D3
Brancepeth, Sask. 181/F2
Branch, Ark. (72928) 202/C3
Branch, La. (70516) 238/F6
Branch (co.), Mich. 250/D7
Branch, Mich. (49402) 250/D5
Branch, Minn. (†55056) 255/F5
Branch, Mo. (†65786) 261/F7
Branch, Newf. 166/D2
Branch (riv.), Newf. 166/C2
Branch, Wis. (54203) 317/L7
Branch Dale, Pa. (17923) 294/K4
Branchport, N.Y. (14418) 276/F5
Branchton, Pa. (16021) 294/C3
Branchville, Ala. (†35120) 195/F3
Branchville, Conn. (†06829) 210/B3
Branchville, Ind. (47514) 227/D8
Branchville, N.J. (07826) 273/D1
Branchville, S.C. (29432) 296/F5
Branchville, Va. (23828) 307/O7
Branco (riv.), Brazil 120/C2
Branco (riv.), Brazil 132/H8
Brandberg (mt.), Namibia 118/A4
Brande, Denmark 21/B6
Brandenburg, E. Germany 22/E2
Brandenburg (reg.), E. Germany 22/E2
Brandenburg, Ky. (40108) 237/J4
Brandon, Colo. (81026) 208/P6
Brandon, England 13/H5
Brandon, Fla. (33511) 212/D4
Brandon, Iowa (52210) 229/K4
Brandon (bay), Ireland 17/A7
Brandon (head), Ireland 17/A7
Brandon (mt.), Ireland 17/A7
Brandon, Man. 146/H4
Brandon, Man. 162/F6
Brandon, Manitoba 179/C5
Brandon, Minn. (56315) 255/C5
Brandon, Miss. (39042) 256/E6
Brandon, Nebr. (69102) 264/C4
Brandon, Ohio (†43050) 284/F5
Brandon, S. Dak. (57005) 298/R6
Brandon, Vt. (05733) 268/A4
Brandon○, Vt. (05733) 268/A4
Brandon, Wis. (53919) 317/J8
Brandon Gap (pass), Vt. 268/B4
Brandonville, W. Va. (26523) 312/G3
Brandreth (lake), N.Y. 276/L3
Brandsville, Mo. (65688) 261/J9
Brandt, Ohio (†45371) 284/B6
Brandt, S. Dak. (57218) 298/R4
Brandvlei, S. Africa 118/B6
Brandýs nad Labem-Stará Boleslavv, Czech. 41/C1
Brandy Station, Va. (22714) 307/N4
Brandywine, Md. (20613) 245/L6
Brandywine, W. Va. (26802) 312/H5
Branford, Conn. (06405) 210/D3
Branford○, Conn. (06405) 210/D3
Branford (harb.), Conn. 210/D4
Branford (riv.), Conn. 210/D3
Branford, Fla. (32008) 212/D2
Braniewo, Poland 47/D1
Brannock (isls.), Ireland 17/A5
Bransfield (str.) 5/C16
Branson, Colo. (81027) 208/M8
Branson, Mo. (65616) 261/F9
Brant, Alberta 182/D4
Brant, Mich. (48614) 250/E5
Brant, N.Y. (14027) 276/B5
Brant (lake), N.Y. 276/N3
Brant (county), Ontario 177/D4
Brant (lake), S. Dak. 298/R6
Brant Beach, N.J. (†08008) 273/E4
Brantford, Kansas (†66938) 232/E2
Brantford, N. Dak. (†58356) 282/N4
Brantford, Ontario 177/D4
Brant Lake, N.Y. (12815) 276/N3
Brantley, Ala. (36009) 195/F7
Brantley (co.), Georgia 217/J8
Brant Rock-Ocean Bluff, Mass. (02020) 249/M4
Brantville, New Bruns. 170/E1
Brantwood, Wis. (54513) 317/F4
Branxholm, Tasmania 99/D3
Branxholme, Victoria 97/A5
Branxton-Greta, N.S. Wales 97/F3
Bras d'Or, Nova Scotia 168/H2
Bras d'Or (lake), Nova Scotia 168/H3
Braselton, Georgia (30517) 217/E2
Brasfield, Ark. (†72017) 202/H4
Brashear, Mo. (63533) 261/H2
Brasher, Mo. (†63830) 261/N10
Brasher Falls-Winthrop, N.Y. (13613) 276/L1
Brasiléia, Brazil 132/G10
Brasília (cap.), Brazil 2/G6
Brasília (cap.), Brazil 120/E4
Brasília (cap.), Brazil 132/E6
Brasília de Minas, Brazil 132/F7
Brașov, Romania 45/G3
Brașov, Romania 7/G4
Brass, Nigeria 106/F8
Brass (isls.), Virgin Is. (U.S.) 161/A4
Brassey (range), W. Australia 92/C4
Brasstown Bald (mt.), Georgia 217/E1
Brassua (lake), Maine 243/D4
Braswell, Georgia (†30153) 217/C3
Brate, Norway 18/G7
Bratenahl, Ohio (†44101) 284/H9
Bratislava, Czech. 7/F4
Bratislava (city), Czech. 41/D2
Bratislava, Czech. 41/D2
Bratsk, U.S.S.R. 54/M4
Bratsk, U.S.S.R. 48/L4
Bratsk (res.), U.S.S.R. 48/L4
Brattleboro, Vt. (05301) 268/B6
Brattleboro○, Vt. (05301) 268/B6
Bratton, Sask. 181/D4
Braunau am Inn, Austria 41/B2
Braunlage, W. Germany 22/D3
Braunschweig (Brunswick), W. Germany 22/D2
Braunton, England 13/C6
Brava (isl.), C. Verde 106/B8
Brava, Somalia 115/H3
Brava, Somalia 102/G4
Brava (pt.), Uruguay 145/B7
Brave, Pa. (15316) 294/B6
Bravo (riv.), Chile 138/D7
Bravo (Grande) (riv.), Mexico 150/G2
Brawley, Calif. 188/C4
Brawley, Calif. (92227) 204/K11
Braxton, Miss. (39044) 256/D6
Braxton (co.), W. Va. 312/E5
Bray, Ireland 17/K5
Bray, Ireland 10/C4
Bray (head), Ireland 17/A8
Bray (isl.), N.W. Terrs. 187/L3
Bray, Okla. (73012) 288/L5
Braymer, Mo. (64624) 261/E3
Brayton, Iowa (50042) 229/D5
Brazeau (dam), Alberta 182/C3
Brazeau (mt.), Alberta 182/B3
Brazeau (riv.), Alberta 182/B3
Brazil 2/F6
Brazil 120/D4
BRAZIL 132, 135
Brazil, Ind. (47834) 227/C5
Brazil, Miss. (38963) 256/D2
Brazil, Tenn. (†38382) 237/C9
Brazilian Highlands (plat.), Brazil 120/E4
Brazilton, Kansas (†66743) 232/H4
Brazito, Mo. (†65101) 261/H6
Brazoria (co.), Texas 303/J8
Brazoria, Texas (77422) 303/J9
Brazos (peak), N. Mex. 274/C2
Brazos (co.), Texas 303/H7
Brazos (riv.), Texas 188/G4
Brazos (riv.), Texas 146/J6
Brazos (riv.), Texas 303/H7
Brazo Sur, Pilcomayo (riv.), Argentina 143/E1
Brazzaville (cap.), Congo 115/C4
Brazzaville (cap.), Congo 2/K6
Brazzaville (cap.), Congo 102/D5
Brčko, Yugoslavia 45/D3
Brda (riv.), Poland 47/C2
Brea, Calif. (92621) 204/D11
Breadalbane (dist.), Scotland 15/D4
Bread Loaf, Vt. (05753) 268/B4
Bread Loaf (mt.), Vt. 268/B4
Breakabeen, N.Y. (†12122) 276/M5
Breakeyville, Québec 172/J3
Breaks, Va. (24607) 307/D6
Breaksea (sound), N. Zealand 100/A6
Bream (bay), N. Zealand 100/E1
Breasclete, Scotland 15/B2
Breathitt (co.), Ky. 237/P5
Breau-Village, New Bruns. 170/F2
Breaux Bridge, La. (70517) 238/G6
Brebes, Indonesia 85/H2
Brébeuf, Québec 172/C3
Brébeuf (lake), Québec 172/G1
Brechin, Ontario 177/E3
Brechin, Scotland 10/E2
Brechin, Scotland 15/F4
Brecht, Belgium 27/F6
Breckenridge, Colo. (80424) 208/G4
Breckenridge, Mich. (48615) 250/E5
Breckenridge, Minn. (56520) 255/B4
Breckenridge, Mo. (64625) 261/E3
Breckenridge, Texas (76024) 303/F5
Breckenridge Hills, Mo. (†63114) 261/O2
Breckinridge (co.), Ky. 237/H5
Breckinridge, Okla. (73721) 288/L2
Brecknock (Brecon), Wales 13/D6
Brecksville, Ohio (44141) 284/H10
Břeclav, Czech. 41/D2
Brecon, Wales 13/D6
Brecon, Wales 10/E5
Brecon Beacons (mt.), Wales 13/D6
Brecon Beacons National Park, Wales 13/D6
Breda, Iowa (51436) 229/C4
Breda, Netherlands 27/F5
Bredasdorp, S. Africa 118/B6
Bredasdorp Nat'l Park, S. Africa 118/C6
Bredbo, N.S. Wales 97/E4
Bredbyn, Sweden 18/L5
Bredebro, Denmark 21/B7
Bredenbury, Sask. 181/K5
Bredene, Belgium 27/B6
Bredstedt, W. Germany 22/C1
Bree, Belgium 27/H6
Breed, Wis. (†54174) 317/K5
Breeden, W. Va. (25666) 312/B7
Breeding, Ky. (42715) 237/L7
Breedsville, Mich. (49027) 250/C6
Breese, Ill. (62230) 222/D5
Breesport, N.Y. (14816) 276/G6
Breezand, Netherlands 27/F3
Breezy Point, Minn. (†56472) 255/D4
Bregenz, Austria 41/A3
Bregovo, Bulgaria 45/F3
Breidhafjördhur (fjord), Iceland 7/B2
Breidhafjördhur (fjord), Iceland 21/B1
Breien, N. Dak. (58525) 282/H7
Breil-Brigels, Switzerland 39/H3
Breil-sur-Roya, France 28/G6
Breisach am Rhein, W. Germany 22/B4
Breisgau (reg.), W. Germany 22/B5
Breitenbach, Switzerland 39/E2
Breitenbush, Oreg. (†97342) 291/F3
Breithorn (mt.), Switzerland 39/E5
Breithorn (mt.), Switzerland 39/E4
Brejo, Brazil 132/F3
Bremanger (isl.), Norway 18/D6
Bremen, Ala. (35033) 195/E3
Bremen, Georgia (30110) 217/B3
Bremen, Ill. (†62233) 222/D6
Bremen, Ind. (46506) 227/E2
Bremen, Kansas (66412) 232/F2
Bremen, Ky. (42325) 237/G6
Bremen, N. Dak. (58319) 282/M4
Bremen, Ohio (43107) 284/F6
Bremen, Sask. 181/F3
Bremen, W. Germany 7/E3
Bremen (state), W. Germany 22/C2
Bremen, W. Germany 22/C2
Bremer (co.), Iowa 229/J3
Bremer, Iowa (†50677) 229/J3
Bremerhaven, W. Germany 22/C2
Bremerton, Wash. 188/B1
Bremerton, Wash. (98310) 310/A2
Bremervörde, W. Germany 22/C2
Bremgarten, Switzerland 39/F2
Bremo Bluff, Va. (23022) 307/M5
Bremond, Texas (76629) 303/H6
Brenham, Texas (77833) 303/H7
Brenner (pass), Austria 41/A3
Brenner (pass), Italy 34/C1
Brent, Ala. (35034) 195/D5
Brent, England 13/H8
Brent, England 10/B5
Brent, Ontario 177/F1
Brentford, S. Dak. (57429) 298/N3
Brenton (pt.), R.I. 249/J7
Brentwood, Ark. (†72959) 202/B2
Brentwood, Calif. (94513) 204/L2
Brentwood, England 10/C5
Brentwood, England 13/J8
Brentwood, Md. (20722) 245/F4
Brentwood, Mo. (63144) 261/P3
Brentwood○, N.H. (†03833) 268/E6
Brentwood, N.Y. (11717) 276/O9
Brentwood, Pa. (15227) 294/B7
Brentwood, Tenn. (37027) 237/H8
Brentwood Park, S. Africa 118/J6
Brereton Lake, Manitoba 179/G5
Bresaylor, Sask. 181/C3
Brescia (prov.), Italy 34/C2
Brescia, Italy 7/E4
Brescia, Italy 34/C2
Breskens, Netherlands 27/C6
Breslau (Wrocław), Poland 47/C3
Bressanone, Italy 34/C1
Bressay (isl.), Scotland 15/G2
Bressay (isl.), Scotland 10/G1
Bressuire, France 28/C4
Brest, France 7/D4
Brest, France 28/A3
Brest, Georgia (†31716) 217/D8
Brest, New Bruns. 170/E2
Brest, U.S.S.R. 7/G3
Brest, U.S.S.R. 48/C4
Brest, U.S.S.R. 52/B4
Bretaña, Peru 128/E5
Brethren, Mich. (49619) 250/D4
Breton, Alberta 182/C3
Breton (isls.), La. 238/M8
Breton (sound), La. 238/M7
Breton (cape), Nova Scotia 168/J3
Breton Cove, Nova Scotia 168/H2
Breton Woods, N.J. (08723) 273/E3
Brett (cape), N. Zealand 100/E1
Bretten, W. Germany 22/C4
Bretton Woods, N.H. (03575) 268/E3
Brevard (co.), Fla. 212/F3
Brevard, N.C. (28712) 281/D4
Breves, Brazil 132/D3
Brevig Mission, Alaska (99785) 196/E1
Brevik, Minn. (†56655) 255/D3
Brevoort (lake), Mich. 250/D3
Brevoort (isl.), N.W. Terrs. 187/M3
Brevort, Mich. (†49760) 250/E2
Brewarrina, N. S. Wales 88/H5
Brewarrina, N.S. Wales 97/D1
Brewer, Maine (04412) 243/F6
Brewer, Mo. (†63775) 261/N7
Brewers, Ky. (†42025) 237/E7
Brewers Mills, New Bruns. 170/C2
Brewersville, Ind. (†47265) 227/F6
Brewerton, N.Y. (13029) 276/H4
Brewster (pond), Conn. 210/F2
Brewster, Kansas (67732) 232/A2
Brewster, Mass. (02631) 249/O5
Brewster○, Mass. (02631) 249/O5
Brewster (isls.), Mass. 249/E7
Brewster, Minn. (56119) 255/C7
Brewster, Nebr. (68821) 264/D3
Brewster (lake), N.S. Wales 97/D3
Brewster, N.Y. (10509) 276/N8
Brewster, Ohio (44613) 284/G4
Brewster, Cerro (mt.), Panama 154/H6
Brewster (co.), Texas 303/A8
Brewster, Wash. (98812) 310/F2
Brewton, Ala. (36426) 195/D8
Breynat, Alberta 182/D2
Brežice, Yugoslavia 45/B3
Brezina, Algeria 106/E2
Březnice, Czech. 41/B2
Breznik, Bulgaria 45/F4
Brezno, Czech. 41/E2
Bria, Cent. Afr. Rep. 102/E4
Bria, Cent. Afr. Rep. 115/D2
Briançon, France 28/G5
Brian Head, Utah (84719) 304/B6
Briar, Texas (†76023) 303/E1
Briar Creek, Pa. (†18603) 294/K3
Briare, France 28/E4
Briartown, Okla. (†74455) 288/R4
Briarwood, (†58102) 282/S6
Bribbaree, N.S. Wales 97/D3
Brice, Ohio (43109) 284/E6
Bricelyn, Minn. (56014) 255/E7
Brices Cross Roads Nat'l Battlefield Site, Miss.256/G2
Briceville, Tenn. (37710) 237/N8
Brí Chualann (Bray), Ireland 17/K5
Brick○, N.J. (08723) 273/E3
Brickaville (Vohibinany), Madagascar 118/H3
Brickerville, Pa. (†17543) 294/K5
Brickeys, Ark. (72320) 202/J4
Bricks, N.C. (†27891) 281/O2
Brickton, Nova Scotia 168/C4
Bridal Veil, Oreg. (97010) 291/E2
Bride (riv.), Ireland 17/E7
Bridesville, Br. Col. 184/H6
Bridge, Idaho (†83342) 220/E7
Bridge, Oreg. (†97458) 291/D4
Bridgeboro, Georgia (31705) 217/E8
Bridge City, Texas (77611) 303/L7
Bridgedale, New Bruns. 170/F3
Bridgeford, Sask. 181/E5
Bridgehampton, N.Y. (11932) 276/R9
Bridge Lake, Br. Col. 184/G4
Bridgeland, Utah (84012) 304/D3
Bridgend, Wales 13/A7
Bridgenorth, Ontario 177/F3
Bridge of Allan, Scotland 10/B1
Bridge of Allan, Scotland 15/C1
Bridge of Don, Scotland 15/F3
Bridge of Weir, Scotland 15/A2
Bridgeport, Ala. (35740) 195/G1
Bridgeport, Calif. (93517) 204/F5
Bridgeport, Conn. 188/M2
Bridgeport, Conn. (*06601) 210/C4
Bridgeport, Ill. (62417) 222/F5
Bridgeport, Kansas (67416) 232/E3
Bridgeport, Mich. (48722) 250/F5
Bridgeport, Nebr. (69336) 264/A3
Bridgeport, N.J. (08014) 273/C4
Bridgeport, N.Y. (13030) 276/J4
Bridgeport, Ohio (43912) 284/J5
Bridgeport, Okla. (†73047) 288/K3
Bridgeport, Oreg. (97819) 291/K3
Bridgeport, Pa. (19405) 294/M5
Bridgeport, Texas (76026) 303/G4
Bridgeport, Wash. (98813) 310/F3
Bridgeport, W. Va. (26330) 312/F4
Bridgeport, Wis. (†53821) 317/D9
Bridger, Mont. (59014) 262/H5
Bridgeton, Ind. (47836) 227/C5
Bridgeton, Mich. (†49327) 250/D5
Bridgeton, Mo. (63044) 261/O2
Bridgeton, N.J. (08302) 273/C5
Bridgeton, N.C. (28519) 281/R4
Bridgeton Terrace, Mo. (†63044) 261/O2
Bridgetown (cap.), Barbados 156/G4
Bridgetown (cap.), Barbados 161/B9
Bridgetown, Md. (†21640) 245/P4
Bridgetown, Nova Scotia 168/C4
Bridgetown, Ohio (†45211) 284/B9
Bridgetown, W. Australia 88/B6
Bridgetown, W. Australia 92/B6
Bridgeview, Ill. (60455) 222/B6
Bridgeville, Calif. (95526) 204/B3
Bridgeville, Del. (19933) 245/R6
Bridgeville, Nova Scotia 168/F3
Bridgeville, Pa. (15017) 294/B5
Bridgeville, Québec 172/D1
Bridgewater○, Conn. (06752) 210/B2
Bridgewater, Iowa (50837) 229/D6
Bridgewater○, Maine (04735) 243/H3
Bridgewater, Mass. (02324) 249/K5
Bridgewater○, Mass. (02324) 249/K5
Bridgewater○, N.H. (†03222) 268/D4
Bridgewater○, N.J. (08807) 273/D2
Bridgewater, N.Y. (13313) 276/K5
Bridgewater, N.S. 162/K7
Bridgewater, Nova Scotia 168/D4
Bridgewater, Pa. (15009) 294/B4
Bridgewater, S. Dak. (57319) 298/P6
Bridgewater, Tasmania 99/D4
Bridgewater○, Vt. (05034) 268/B4
Bridgewater (cape), Victoria 97/A6
Bridgewater, Va. (22812) 307/K4
Bridgewater Center, Vt. (†05034) 268/B4
Bridgewater Corners, Vt. (05035) 268/B4
Bridgman, Mich. (49106) 250/C7
Bridgnorth, England 13/E5
Bridgnorth, England 10/E4
Bridgton, Maine (04009) 243/B7
Bridgton○, Maine (04009) 243/B7
Bridgwater, England 10/E5
Bridgwater, England 13/E6
Bridlington, England 13/G3
Bridlington, England 10/F3
Bridlington (bay), England 13/G3
Bridport, England 13/E7
Bridport, England 10/E5
Bridport, Tasmania 99/D3
Bridport○, Vt. (05734) 268/A4
Brieg (Brzeg), Poland 47/C3
Brielle, Netherlands 27/E5
Brielle, N.J. (08730) 273/E3
Briensburg, Ky. (†42025) 237/E7
Brienz, Switzerland 39/F3
Brienzer Rothorn (mt.), Switzerland 39/F3
Brienzersee (lake), Switzerland 39/F3
Brier (isl.), Nova Scotia 168/B4
Brier, Wash. (†98036) 310/C3
Briercrest, Sask. 181/F5
Brierfield, Ala. (35035) 195/E4
Brier Hill, N.Y. (13614) 276/J1
Brig, Switzerland 39/F4
Brigantine, N.J. (08203) 273/E5
Brigantine (inlet), N.J. 273/E5
Brigden, Ontario 177/B5
Brigg, England 13/G4
Briggs, Texas (78608) 303/F7
Briggs Corner, New Bruns. 170/E2
Briggsdale, Colo. (80611) 208/L1
Briggsville, Ark. (72828) 202/C4
Briggsville, Wis. (53920) 317/H8
Brigham City, Utah 188/D2
Brigham City, Utah (84302) 304/C2
Brighowe, England 13/J1
Bright, Ind. (†45030) 227/H6
Bright, Victoria 97/D5
Brightlingsea, England 13/J6
Brightlingsea, England 10/G5
Brighton, Ala. (35020) 195/D4
Brighton, Colo. (80601) 208/K3
Brighton, England 10/F5
Brighton, England 13/G7
Brighton, Fla. (†33472) 212/E4
Brighton, Ill. (62012) 222/C4
Brighton, Ind. (†46746) 227/G1
Brighton, Iowa (52540) 229/K6
Brighton○, Maine (†04990) 243/D5
Brighton, Mich. (48116) 250/F6
Brighton, Mo. (65617) 261/F8
Brighton, Nova Scotia 168/C4
Brighton, Ohio (†44090) 284/F3
Brighton, Ontario 177/G3
Brighton, Oreg. (†97136) 291/C2
Brighton, S. Australia 88/D8
Brighton, S. Australia 94/A8
Brighton, Tasmania 99/D4
Brighton, Tenn. (38011) 237/B10
Brighton, Utah (84101) 304/C3
Brighton, Victoria 97/J5
Brighton, Victoria 88/L7
Brighton, Wis. (†53139) 317/K3
Brightons, Scotland 15/C1
Brightsand (lake), Sask. 181/B2
Brights Grove, Ontario 177/B4
Brightshade, Ky. (40962) 237/O7
Brightstar, Ark. (†75556) 202/C7
Brightwood, D.C. (20011) 245/F4
Brightwood, Oreg. (97001) 291/E2
Brightwood, Va. (22715) 307/M4
Brignoles, France 28/G6
Brigus, Newf. 166/D2
Brihuega, Spain 33/E2
Brikama, Gambia 106/A6
Brill, Wis. (54818) 317/C4
Brilliant, Ala. (35548) 195/C2
Brilliant, Ohio (43913) 284/J5
Brillion, Wis. (54110) 317/L7
Brilon, W. Germany 22/C3
Brimfield, Ill. (61517) 222/D3
Brimfield, Ind. (46720) 227/G2
Brimfield○, Mass. (01010) 249/F4
Brimfield, Ohio (†44240) 284/H3
Brimley, Mich. (49715) 250/E2
Brimson, Minn. (55602) 255/F3
Brimson, Mo. (64642) 261/E2
Brimstone (hill), St. Chris.-Nevis 161/C10
Brinckerhoff, N.Y. (†12524) 276/N7
Brindakit, U.S.S.R. 48/O4
Brindisi (prov.), Italy 34/G4
Brindisi, Italy 7/F4
Brindisi, Italy 34/G4
Bringhurst, Ind. (46913) 227/E3
Brinkhaven, Ohio (43006) 284/F5
Brinkley, Ark. (72021) 202/H4
Brinkman, Okla. (†73673) 288/G4
Brinktown, Mo. (65443) 261/J6
Brinnon, Wash. (98320) 310/B3
Brinsmade, N. Dak. (58320) 282/M3
Brinson, Georgia (31725) 217/C9
Briny Breezes, Fla. (†33435) 212/G5
Brione, Switzerland 39/G4
Brioude, France 28/E5
Brisbane, Australia 2/S7
Brisbane, Calif. (94005) 204/J2
Brisbane (cap.), Queensland 95/D2
Brisbane (cap.), Queensland 88/K3
Brisbane (riv.), Queensland 88/J3
Brisbane (riv.), Queensland 95/D2
Brisbane Airport, Queensland 95/E2
Brisbane International Airport, Queensland 88/K2
Brisbane Water, N. S. Wales 88/J6
Brisbane Water, N.S. Wales 97/F3
Brisbin, Pa. (16620) 294/F4
Brisco, Br. Col. 184/J5
Briscoe (co.), Texas 303/C3
Briscoe, Texas (79011) 303/D2
Brisighella, Italy 34/C2
Brissago, Switzerland 39/G4
Bristol (bay), Alaska 188/C6
Bristol (bay), Alaska 146/B4
Bristol (bay), Alaska 196/F3
Bristol (lake), Calif. 204/K9
Bristol, Colo. (81028) 208/P6
Bristol, Conn. (06010) 210/D2
Bristol, England 13/E6
Bristol, England 7/D3
Bristol, England 10/E5
Bristol (chan.), England 13/C6
Bristol (chan.), England 10/E5
Bristol, Fla. (32321) 212/B1
Bristol, Georgia (31518) 217/H8
Bristol, Ind. (46507) 227/F1

Bristol, Maine (04539) 243/D8
Bristol○, Maine (04539) 243/D8
Bristol (co.), Mass. 249/K5
Bristol, Mich. (†49688) 250/D4
Bristol, New Bruns. 170/C2
Bristol (co.), Ohio 284/J3
Bristol, N.H. (03222) 268/D4
Bristol○, N.H. (03222) 268/D4
Bristol, Pa. (19007) 294/N5
Bristol○, Pa. (19007) 294/N5
Bristol (co.), R.I. 249/J6
Bristol, R.I. (02809) 249/J6
Bristol, S. Dak. (57219) 298/O3
Bristol, Tenn. 188/K3
Bristol, Tenn. (37620) 237/S7
Bristol (bay), U.S. 4/D18
Bristol, Va. 188/K3
Bristol, Vt. (05443) 268/A3
Bristol○, Vt. (05443) 268/A3
Bristol (I.C.), Va. (24201) 307/D7
Bristol (chan.), Wales 13/C6
Bristol (chan.), Wales 10/E5
Bristol, W. Va. (26332) 312/F4
Bristolville, Ohio (44402) 284/J3
Bristow, Ohio (47515) 227/D8
Bristow, Iowa (50611) 229/H3
Bristow, Nebr. (68719) 264/F2
Bristow, Okla. (74010) 288/O3
Bristow, Va. (22013) 307/N3
Britannia Beach, Br. Col. 184/K2
British (mts.), Alaska 196/K1
British (mts.), Yukon 187/D3
British Columbia (prov.) 162/D4
BRITISH COLUMBIA 184
British Columbia (prov.), Canada 146/F4
British Indian Ocean Territory 2/N6
British Indian Ocean Territory 54/J10
British Isles 7/D3
Brits, S. Africa 118/D5
Britstown, S. Africa 118/C6
Britt, Iowa (50423) 229/F2
Britt, Minn. (55710) 255/F3
Britt, Ontario 177/D2
Brittany (trad. prov.), France 29
Brittany, La. (70718) 238/L3
Brittnau, Switzerland 39/E2
Britton, Mich. (49229) 250/F6
Britton, S. Dak. (57430) 298/O2
Brive-la-Gaillarde, France 28/D5
Briviesca, Spain 33/E1
Brno, Czech. 7/F4
Brno, Czech. 41/D2
Broa (inlet), Cuba 158/C1
Broach (Bharuch), India 68/C4
Broad (brook), Conn. 210/H2
Broad (creek), Del. 245/P7
Broad (riv.), N.C. 281/E4
Broad (riv.), S.C. 296/F7
Broad (sound), Queensland 88/H4
Broad (sound), Queensland 95/D4
Broad (bay), Scotland 15/B2
Broad (riv.), S.C. 296/E2
Broadacres, Oreg. (†97032) 291/A3
Broadacres, Sask. 181/B3
Broadalbin, N.Y. (12025) 276/M4
Broad Arrow, W. Australia 88/C6
Broad Arrow, W. Australia 92/C5
Broadback (riv.), Québec 174/B2
Broadbent, Oreg. (97414) 291/A4
Broad Brook, Conn. (06016) 210/E1
Broad Cove, Newf. 166/D2
Broad Cove, Nova Scotia 168/D4
Broaddus, Texas (75929) 303/K6
Broadfields, Ky. (†40201) 237/K2
Broadford, Ireland 17/C7
Broadford, Scotland 15/C3
Broadford, Victoria 97/C5
Broadford, Va. (24316) 307/E7
Broad Haven (harb.), Ireland 17/B3
Broadhurst, Georgia (†31545) 217/J8
Broadkill (riv.), Del. 245/S5
Broadland, S. Dak. (†57350) 298/N4
Broadlands, Ill. (61816) 222/E4
Broad Law (mt.), Scotland 15/E5
Broadmeadows, Victoria 88/K8
Broadmeadows, Victoria 97/H4
Broadstairs and Saint Peter's, England 13/J6
Broad Top, Pa. (16621) 294/F5
Broadus, Mont. (59315) 262/L3
Broad Valley, Manitoba 179/E4
Broadview, Ill. (60153) 222/B6
Broadview, Mont. (59015) 262/H4
Broadview, N. Mex. (88112) 274/F4
Broadview, Sask. 181/J4
Broadview Heights, Ohio (†44141) 284/H10
Broadview Park, Fla. (†33314) 212/B4
Broadwater (co.), Mont. 262/E4
Broadwater, Nebr. (69125) 264/B3
Broadway, N.J. (08808) 273/C2
Broadway, N.C. (27505) 281/L4
Broadway, Ohio (43007) 284/C5
Broadway, Va. (22851) 307/L3
Broadwell, Ill. (62623) 222/D3
Broager, Denmark 21/C5
Broc, Switzerland 39/D3
Brochet, Man. 162/F4
Brochet, Manitoba 179/H2
Brock (isl.), N.W.T. 162/M3
Brock, Nebr. (68320) 264/H4
Brock (isl.), N. W. Terrs. 187/G2
Brock, Sask. 181/C4
Brockdell, Tenn. (†37367) 237/L10
Brocken (mt.), E. Germany 22/D3
Brocket, Alberta 182/D5
Brocket, N. Dak. (58321) 282/O3
Brockington, Sask. 181/G2
Brockport, N.Y. (14420) 276/H4
Brockport, Pa. (15823) 294/E3
Brockton, Mass. (*02401) 249/K4
Brockton, Mont. (59213) 262/M2
Brockville, Ontario 177/J3
Brockway, Mont. (59214) 262/L3
Brockway, New Bruns. 170/C3
Brockway, Pa. (15824) 294/E3
Brocton, Ill. (61917) 222/F4

Brocton, N.Y. (14716) 276/B6
Broderick, Sask. 181/E4
Broderick-Bryte, Calif. (95605) 204/B8
Brodeur (pen.), Canada 4/B14
Brodeur (pen.), N.W.T. 146/K2
Brodeur (pen.), N.W.T. 162/H1
Brodeur (pen.), N. W. Terrs. 187/K2
Brodhead, Ky. (40409) 237/N6
Brodhead, Wis. (53520) 317/G10
Brodheadsville, Pa. (18322) 294/M4
Brodick, Scotland 15/C5
Brodick, Scotland 10/D3
Brodnax, Va. (23920) 307/N7
Broek in Waterland, Netherlands 27/C4
Brogan, Oreg. (97903) 291/K3
Brohard, W. Va. (26138) 312/D4
Brohman, Mich. (49312) 250/D5
Brokaw, Wis. (54417) 317/G5
Broken (bay), N.S. Wales 97/F3
Broken Arrow, Okla. (74012) 288/P2
Broken Bow, Nebr. (68822) 264/E3
Broken Bow, Okla. (74728) 288/S7
Broken Bow (lake), Okla. 288/S6
Broken Hill, Australia 87/E9
Broken Hill, N. S. Wales 88/G6
Broken Hill, N. S. Wales 97/A3
Broken Hill (Kabwe), Zambia 115/E6
Brokensword, Ohio (†44820) 284/E4
Brokopondo (dist.), Suriname 131/D4
Brokopondo, Suriname 131/D3
Brome (co.), Québec 172/E4
Brome, Québec 172/E4
Brome (lake), Québec 172/E4
Bromer, Ind. (†47452) 227/E7
Bromhead, Sask. 181/H6
Bromide, Okla. (74530) 288/N4
Bromley, England 13/H8
Bromley, England 10/C5
Bromley, Ky. (†41016) 237/S2
Bromley (mt.), Vt. 268/E5
Brompton (lake), Québec 172/E4
Bromptonville, Québec 172/E4
Bromsgrove, England 13/E5
Bromyard, England 13/E5
Bronaugh, Mo. (64728) 261/C7
Bronco, Texas (†79355) 303/B6
Brønderslev, Denmark 18/F8
Brønderslev, Denmark 21/C3
Bronkton, W. Australia 92/B7
Bronnøysund, Norway 18/G4
Brøns, Denmark 21/B7
Bronson, Fla. (32621) 212/D2
Bronson, Iowa (51007) 229/A4
Bronson, Kansas (66716) 232/H4
Bronson, Mich. (49028) 250/D7
Bronson (lake), Sask. 181/B1
Bronson, Texas (75930) 303/L6
Bronston, Ky. (42515) 237/M7
Bronte, Italy 34/E6
Bronte, Texas (76933) 303/D6
Bronwood, Georgia (31726) 217/D7
Bronx (co.), N.Y. 276/N9
Bronx (borough), N.Y. (*10401) 276/N9
Bronxville, N.Y. (10708) 276/O7
Brook, Ind. (47922) 227/C3
Brookdale, Calif. (95007) 204/J4
Brookdale, Manitoba 179/C4
Brookdale, Nova Scotia 168/D3
Brooke, Va. (22430) 307/O4
Brooke (co.), W. Va. 312/E2
Brookeborough, N. Ireland 17/G3
Brookeland, Texas (75931) 303/L6
Brooker, Fla. (32622) 212/D2
Brooke's Point, Philippines 82/A6
Brookeville, Md. (20729) 245/K4
Brookfield○, Conn. (06804) 210/B3
Brookfield, Georgia (31727) 217/F8
Brookfield, Ill. (60513) 222/B6
Brookfield, Mass. (01506) 249/F4
Brookfield○, Mass. (01506) 249/F4
Brookfield, Mo. (64628) 261/F3
Brookfield○, N.H. (03872) 268/E4
Brookfield, N.Y. (13314) 276/K5
Brookfield, Nova Scotia 168/E3
Brookfield, Ohio (44403) 284/J3
Brookfield○, Vt. (05036) 268/B3
Brookfield, Wis. (53005) 317/K1
Brookfield Center, Conn. (06805) 210/B3
Brookford, N.C. (†28601) 281/G3
Brookhaven, Georgia (†30304) 217/K1
Brookhaven, Miss. (39601) 256/C7
Brookhaven, Pa. (19015) 294/M7
Brookhaven Nat'l Lab., N.Y. 276/P9
Brookings (co.), S. Dak. 298/R5
Brookings, Oreg. (97415) 291/A4
Brookings, S. Dak. (57006) 298/R5
Brookland, Ark. (72417) 202/J2
Brookland, D.C. (20017) 245/F4
Brooklawn, N.J. (08030) 273/B3
Brooklet, Georgia (30415) 217/J6
Brookley Air Force Base, Ala. 195/B9
Brooklin○, Maine (04616) 243/F7
Brookline, Mass. (02146) 249/C17
Brookline○, Mass. 249/C17
Brookline○, N.H. (03033) 268/D6
Brookline○, Vt. (†05345) 268/B5
Brookline Station (Brookline), Mo. (65619) 261/F4
Brooklyn, Ala. (36429) 195/E8
Brooklyn○, Conn. (06234) 210/H1
Brooklyn, Georgia (31814) 217/C6
Brooklyn (Lovejoy), Ill. (†62059) 222/A2
Brooklyn, Ill. (62367) 222/C3
Brooklyn, Ind. (46111) 227/E5
Brooklyn, Iowa (52211) 229/J5
Brooklyn, Ky. (42209) 237/H6
Brooklyn, Md. (21225) 245/M4
Brooklyn, Mich. (49230) 250/E6
Brooklyn, Miss. (39425) 256/F4
Brooklyn, Newf. 166/D2
Brooklyn (borough), N.Y. (*11201) 276/N9
Brooklyn, Nova Scotia 168/D4
Brooklyn, Ohio (44144) 284/H9
Brooklyn, Wash. (†98537) 310/B4

Browning, Mo. (64630) 261/F2
Browning, Mont. (59417) 262/C2
Browning, N.S. Wales 97/E4
Browning, Sask. 181/J6
Browning Entrance (str.), Br. Col. 184/B3
Browningson, Idaho (†64740) 261/E6
Browningson, Vt. (†05860) 268/C2
Brownlee (dam), Idaho 220/B5
Brownlee, Nebr. (69126) 264/D2
Brownlee, Oreg. (†97840) 291/L3
Brownlee (dam), Oreg. 291/L3
Brownlee, Sask. 181/F5
Browns (pen.), Br. Col. 184/D5
Browns, Calif. (95606) 204/C5
Browns (co.), Georgia 217/E9
Browns, Iowa (50841) 229/D7
Browns, Ill. (62818) 222/F5
Brown's (riv.), Vt. 268/A2
Brownsboro, Ala. (35741) 195/F1
Brownsboro, Oreg. (†97524) 291/B4
Brownsboro, Texas (75656) 303/J5
Brownsboro Farm, Ky. (†40201) 237/L1
Brownsburg, Ind. (46112) 227/E4
Brownsburg, Québec 172/C4
Brownsburg, Va. (24415) 307/K5
Brownsdale, Minn. (55918) 255/F7
Brownsdale, Newf. 166/D2
Browns Flat, New Bruns. 170/D3
Brown's Lake, Wis. (53105) 317/K3
Browns Mills, N.J. (08015) 273/D4
Browns Spring, Mo. (†65610) 261/F9
Browns Summit, N.C. (27214) 281/K2
Brownstown, Ill. (62418) 222/E5
Brownstown, Ind. (47220) 227/F7
Browns Town, Jamaica 158/J6
Brownstown, Pa. (17508) 294/K5
Brownstown, Wash. (98920) 310/E4
Browns Valley, Ind. (†47933) 227/D5
Browns Valley, Minn. (56219) 255/B5
Browns Village, Fla. (†33142) 212/B4
Brownsville, Ind. (47325) 227/H5
Brownsville, Ky. (42210) 237/J6
Brownsville, Md. (21715) 245/H3
Brownsville, Minn. (55919) 255/G7
Brownsville, Miss. (†39140) 256/D6
Brownsville, Oreg. (97327) 291/B2
Brownsville, Pa. (15417) 294/C5
Brownsville, Tenn. (38012) 237/C9
Brownsville, Texas (*78520) 303/G12
Brownsville, Texas 188/G5
Brownsville, Texas 146/J7
Brownsville, Vt. (†05037) 268/B5
Brownsville, Wash. (†98310) 310/A2
Brownsville, Wis. (53006) 317/J7
Brownton, Minn. (55312) 255/D6
Brownton, W. Va. (26334) 312/F4
Browntown, Va. (22610) 307/M3
Browntown, Wis. (53522) 317/G10
Brownvale, Alberta 182/B1
Brownville, Ala. (†35476) 195/C4
Brownville, Fla. (†33821) 212/D3
Brownville, Maine (04414) 243/E5
Brownville○, Maine (04414) 243/E5
Brownville, Nebr. (68321) 264/J4
Brownville, N.Y. (13615) 276/H3
Brownville Junction, Maine (04415) 243/E5
Brown Willy (mt.), England 13/C7
Brownwood, Texas (76801) 303/F6
Brownwood (lake), Texas 303/E6
Browse (isl.), W. Australia 88/C2
Browse (isl.), W. Australia 92/C1
Broxburn, Scotland 15/D1
Broxton, Georgia (31519) 217/G7
Broye (riv.), Switzerland 39/C3
Broyle (cape), Newf. 166/D2
Brozas, Spain 33/C3
Brozville, Miss. (†39095) 256/D4
Brtnice, Czech. 41/C2
Bruay-en-Artois, France 28/E2
Bruce, Alberta 182/E3
Bruce, Fla. (32455) 212/C6
Bruce, Miss. (38915) 256/F3
Bruce (mts.), N. W. Terrs. 187/L2
Bruce (county), Ontario 177/C2
Bruce, S. Dak. (57220) 298/R5
Bruce (mt.), W. Australia 88/B4
Bruce (mt.), W. Australia 92/B3
Bruce, Wis. (54819) 317/D5
Bruce Crossing, Mich. (49912) 250/A3
Brucefield, Ontario 177/C4
Bruce Lake, Ontario 175/B2
Bruce Mines, Ontario 177/J5
Bruce Mines, Ontario 175/D3
Bruce Rock, W. Australia 88/B5
Bruce Rock, W. Australia 92/B5
Bruceton, Tenn. (38317) 237/E8
Bruceton Mills, W. Va. (26525) 312/G3
Brucetown, Va. (22622) 307/M2
Bruceville, Ind. (†47516) 227/C7
Bruchsal, W. Germany 22/C4
Bruck an der Leitha, Austria 41/D2
Bruck an der Mur, Austria 41/C3
Bruderheim, Alberta 182/D3
Bruff, Ireland 17/D7
Bruges, Belgium 27/C6
Brugg, Switzerland 39/F1
Brugge (Bruges), Belgium 27/C6
Brühl, W. Germany 22/B3
Bruin, Ky. (41125) 237/P4
Bruin (cape), New Bruns. 170/G2
Bruin, Pa. (16022) 294/C3
Bruins, Ark. (†72348) 202/K4
Brule (riv.), Mich. 250/A3
Brûlé (lake), Québec 172/C3
Brûlé (lake), Québec 172/B2
Brule (co.), S. Dak. 298/L6
Brule (riv.), Switzerland 39/D4
Brule, Wis. (54820) 317/C2
Brumado, Brazil 120/E4
Brumado, Brazil 132/F4
Brumley, Mo. (65017) 261/H6
Brummen, Netherlands 27/J4
Brundidge, Ala. (36010) 195/G7
Bruneau, Idaho (83604) 220/C7

Bruneau (riv.), Idaho 220/C7
Brunei 2/Q5
Brunei 54/N9
BRUNEI 85/E4
Bruner, Mo. (65620) 261/F8
Brunete, Spain 33/F4
Brunette (isl.), Newf. 166/C4
Bruni, Texas (78344) 303/F10
Brunico, Italy 34/D1
Bruning, Nebr. (68322) 264/G4
Brunkild, Manitoba 179/E5
Brunner, N. Zealand 100/C5
Brunner (lake), N. Zealand 100/C5
Brunnen, Switzerland 39/F2
Brunot, Mo. (†63636) 261/M8
Brunsbüttel, W. Germany 22/C2
Brunson, S.C. (29911) 296/E6
Brunssum, Netherlands 27/J7
Brunswick, Ga. 188/K4
Brunswick, Ohio (51008) 229/A3
Brunswick (pen.), Chile 138/E10
Brunswick, Ga. (31520) 217/K8
Brunswick, Maine (04011) 243/C8
Brunswick○, Maine (04011) 243/C8
Brunswick, Md. (21716) 245/H3
Brunswick, Miss. (†55051) 255/E5
Brunswick, Mo. (65236) 261/F4
Brunswick, Nebr. (68720) 264/G2
Brunswick (co.), N.C. 281/N6
Brunswick, N.C. (28424) 281/M6
Brunswick, Ohio (44212) 284/G3
Brunswick, Tenn. (38014) 237/B10
Brunswick (co.), Va. 307/N7
Brunswick (bay), W. Australia 88/C3
Brunswick, W. Germany 7/E3
Brunswick (bay), W. Australia 92/D1
Brunswick, W. Germany 22/D2
Brunswick Heads, N.S. Wales 97/G1
Brunswick Junction, W. Australia 92/A2
Bruntál, Czech. 41/D2
Bruree, Ireland 17/D7
Brus (lag.), Honduras 154/E2
Brusett, Mont. (59318) 262/J3
Brush, Colo. (80723) 208/M2
Brush Creek, Minn. (†56014) 255/E7
Brush Creek, Mo. (†65536) 261/G7
Brush Creek, Tenn. (38547) 237/J8
Brush Prairie, Wash. (98606) 310/C5
Brushton, N.Y. (12916) 276/L1
Brushy Prairie, Ind. (†46761) 227/G1
Brusio, Switzerland 39/K4
Brus Laguna, Honduras 154/E3
Brusly, La. (70719) 238/J2
Brusque, Brazil 132/D9
Brussels (cap.), Belgium 7/E3
Brussels, Ill. (62013) 222/C5
Brussels, Ontario 177/C4
Brussels, Wis. (54204) 317/L6
Bruthen, Victoria 97/D5
Brutus, Mich. (49716) 250/E3
Bruxelles, Manitoba 179/D5
Bruzual, Venezuela 124/D3
Bryan (co.), Georgia 217/K6
Bryan, Ohio (43506) 284/A3
Bryan (co.), Okla. 288/O7
Bryan, Texas (77801) 303/H7
Bryan (lake), Wash. 310/H4
Bryans, U.S.S.R. 7/H3
Bryansk, U.S.S.R. 52/D4
Bryansk, U.S.S.R. 48/D4
Bryanston, Ontario 177/C4
Bryant, Ala. (35958) 195/G1
Bryant, Ark. (72022) 202/E4
Bryant, Fla. (33439) 212/F5
Bryant (lake), Fla. 212/E2
Bryant, Ill. (61519) 222/C3
Bryant, Ind. (47326) 227/G3
Bryant, Iowa (52727) 229/N5
Bryant, Okla. (†74437) 288/P4
Bryant, S. Dak. (57221) 298/P4
Bryant, Wis. (54418) 317/G5
Bryant Pond, Maine (04219) 243/B7
Bryantsburg, Ind. (†47250) 227/G7
Bryantsville, Ky. (40410) 237/M5
Bryantville, Mass. (02327) 249/L4
Bryce (mt.), Br. Col. 184/J4
Bryce Canyon, Utah (84717) 304/B6
Bryce Canyon Nat'l Park, Utah 304/B6
Bryceland, La. (71014) 238/E2
Bryceville, Fla. (32009) 212/D1
Bryn Athyn, Pa. (19009) 294/M5
Brynica (riv.), Poland 47/B4
Bryn Mawr, Pa. (19010) 294/M5
Bryn Mawr-Skyway, Wash. (†98101) 310/B2
Brynmawr, Wales 10/E5
Brynmawr, Wales 13/B6
Bryn Mawr-Skyway, Wash. (†98101) 310/B2
Bryrup, Denmark 21/C5
Bryson, Texas (76027) 303/F4
Bryson City, N.C. (28713) 281/C4
Bryte-Broderick, Calif. (95605) 204/B8
Brzeg, Poland 47/D3
Brzeg Dolny, Poland 47/C3
Brzesko, Poland 47/F4
Brzozów, Poland 47/F4
B-Say-Tah, Sask. 181/G5
Bua Chum, Thailand 72/D4
Buad (isl.), Philippines 82/F5
Buba, Guinea-Biss: 106/B6
Bubali, Neth. Ant. 161/D10
Bubaque, Guinea-Biss. 106/A6
Bubendorf, Switzerland 39/E1
Bubikon, Switzerland 39/G2
Bubiyan (isl.), Kuwait 59/E4
Bucak, 63/D4
Bucaramanga, Colombia 120/B2
Bucaramanga, Colombia 126/D3
Bucareli (bay), Alaska 196/M2
Bucas Grande (isl.), Philippines 82/F6

Bucasia, Queensland 95/D4
Buccaneer (arch.), W. Australia 88/C3
Buccaneer (arch.), W. Australia 92/C2
Buchan (gulf), N.W. Terrs. 187/L2
Buchan (dist.), Scotland 15/F3
Buchan, Georgia (30113) 217/B3
Buchanan (co.), Iowa 229/K4
Buchanan, Iowa (†52772) 229/L5
Buchanan, Ky. (†41129) 237/R4
Buchanan, Liberia 106/B7
Buchanan, Liberia 102/A4
Buchanan, Mich. (49107) 250/C7
Buchanan (co.), Mo. 261/C3
Buchanan, N.Y. (10511) 276/N8
Buchanan, N. Dak. (58420) 282/N5
Buchanan, Sask. 181/J4
Buchanan, Tenn. (38222) 237/E8
Buchanan (lake), Texas 303/F7
Buchanan (co.), Va. 307/D6
Buchanan, Va. (24066) 307/J5
Buchan Ness (prom.), Scotland 15/G3
Buchans, Newf. 166/C4
Bucharest (cap.), Romania 7/G4
Bucharest (cap.), Romania 2/L3
Bucharest (Bucureşti) (cap.), Romania 45/G3
Buchegg (mts.), Switzerland 39/D2
Buchholz in der Nordheide, W. Germany 22/C2
Buchlyvie, Scotland 15/B1
Buchon (pt.), Calif. 204/D8
Buchs, Switzerland 39/H2
Buchtel, Ohio (45716) 284/F6
Buck (creek), Ind. 227/E8
Buck (creek), S.C. 296/J3
Buck (creek), Texas 303/D3
Buck (isl.), Virgin Is. (U.S.) 161/G3
Buck, W. Va. (†24935) 312/E7
Buckatunna, Miss. (39322) 256/G7
Buck Creek, Alberta 182/C3
Buck Creek, Ind. (47924) 227/D4
Buckeye, Ariz. (85326) 198/C5
Buckeye, Iowa (50043) 229/G4
Buckeye, La. (71321) 238/F4
Buckeye, N. Mex. (†88260) 274/F6
Buckeye, W. Va. (24924) 312/F6
Buckeye Lake, Ohio (43008) 284/F5
Buckeystown, Md. (21717) 245/J3
Buckfastleigh, England 13/C7
Buckfield○, Maine (†04232) 243/C7
Buck Grove, Iowa (†51442) 229/C5
Buckhannon, W. Va. (26201) 312/F5
Buckhannon (riv.), W. Va. 312/F5
Buckhaven and Methil, Scotland 15/F4
Buckhaven and Methil, Scotland 10/E2
Buckhead, Georgia (30625) 217/F5
Buck Hollow (creek), Oreg. 291/G2
Buckholts, Texas (76518) 303/H7
Buckhorn, Ky. 237/O6
Buckhorn, Mo. (†63636) 261/M8
Buckhorn (lake), Ontario 177/F3
Buckie, Scotland 15/E3
Buckie, Scotland 10/E2
Buckingham, Colo. (†80742) 208/L1
Buckingham, Conn. (†06033) 210/E2
Buckingham, England 13/G6
Buckingham, England 10/F5
Buckingham, Ill. (60917) 222/E2
Buckingham, Iowa (50612) 229/J4
Buckingham, Québec 172/B4
Buckingham, Texas (†75080) 303/H2
Buckingham (co.), Va. 307/L5
Buckingham, Va. (23921) 307/L5
Buckinghamshire (c.), England 13/G6
Buck Island (chan.), Virgin Is. (U.S.) 161/F3
Buck Island Reef Nat'l Mon., Virgin Is. (U.S.) 161/G3
Buck Lake, Alberta 182/C3
Buckland, Alaska (99727) 196/F1
Buckland, Conn. (06040) 210/E1
Buckland○, Mass. (01338) 249/C2
Buckland, Ohio (45819) 284/B4
Buckland, Québec 172/G4
Buckley, Ill. (60918) 222/F3
Buckley, Mich. (49620) 250/D4
Buckley, Wales 13/G2
Buckley, Wash. (98321) 310/C3
Bucklin, Kansas (67834) 232/C4
Bucklin, Mo. (64631) 261/G3
Buckman, Minn. (56317) 255/E5
Buckner, Ark. (71827) 202/D7
Buckner, Ill. (62819) 222/E6
Buckner, Ky. (40010) 237/L4
Buckner, Mo. (64016) 261/R5
Bucks, Ala. (36512) 195/B8
Bucks (co.), Pa. 294/M5
Bucksburn, Scotland 15/F3
Bucks Harbor, Maine (04618) 243/A6
Buckskin (mts.), Ariz. 198/B4
Buckskin, Ind. (47647) 227/C8
Bucksport, Maine (04416) 243/F6
Bucksport○, Maine (04416) 243/F6
Bucksport, S.C. (29527) 296/J4
Buckville, Ark. (71934) 202/D4
Bucoda, Mo. (†63876) 261/M10
Bucoda, Wash. (98530) 310/B4
Buco-Zau, Angola 111/B4
Buctouche (harb.), New Bruns. 170/F2
Buctouche, New Bruns. 170/F2
Buctouche (riv.), New Bruns. 170/F2
Bucureşti (Bucharest) (cap.), Romania 45/G3
Bucyrus, Kansas (66013) 232/H3
Bucyrus, Mo. (65444) 261/H8
Bucyrus, N. Dak. (58624) 282/E7
Bucyrus, Ohio (44820) 284/E4
Bud, Ind. (†24716) 312/D7
Bud, W. Va. (24716) 312/D7
Buda, Ill. (61314) 222/D2
Buda, Texas (78610) 303/G7
Budafok, Hungary 41/E3
Budakeszi, Hungary 41/E3
Budaörs, Hungary 41/E3

Budapest (city), Hungary 41/E3
Budapest (cap.), Hungary 41/E3
Budapest (cap.), Hungary 7/F4
Budaun, India 68/D3
Budd (lake), N.J. 273/D2
Budd Coast (reg.) 5/C6
Budd Lake, N.J. (07828) 273/D2
Buddon Ness (prom.), Scotland 15/F4
Bude (bay), England 13/C7
Bude, Miss. (39630) 256/C8
Bude-Stratton, England 13/C7
Budge-Budge, India 68/F2
Budgewoi Lake, N.S. Wales 97/F3
Budia, Spain 33/E2
Büdingen, W. Germany 22/C3
Budišov, Czech. 41/D2
Budjala, Zaire 115/C3
Budleigh Salterton, England 13/D7
Budrio, Italy 34/C2
Budva, Yugoslavia 45/D4
Buea, Cameroon 115/A3
Buechel, Ky. (40218) 237/K2
Buel (lake), Mass. 249/A4
Buellton, Calif. (93427) 204/E9
Buena, N.J. (08310) 273/D4
Buena, Wash. (98921) 310/E4
Buena Esperanza, Argentina 143/C3
Buena Park, Calif. (*90622) 204/D11
Buenaventura, Colombia 126/B6
Buenaventura, Colombia 120/B2
Buenaventura (bay), Colombia 126/B6
Buenaventura, Cuba 158/H3
Buenaventura, Mexico 150/F2
Buena Vista, Ala. (†36425) 195/D7
Buena Vista, Ark. (†71764) 202/D7
Buena Vista, Bolivia 136/D5
Buena Vista (lake), Calif. 204/F8
Buena Vista, Colo. (81211) 208/G5
Buenavista, Cuba 158/F2
Buenavista (bay), Cuba 158/F2
Buena Vista, Georgia (31803) 217/C6
Buena Vista (co.), Iowa 229/C3
Buena Vista, Miss. (†38851) 256/G3
Buena Vista, N. Mex. (87712) 274/D3
Buena Vista, Ohio (†45684) 284/D8
Buena Vista, Oreg. (†97351) 291/D3
Buena Vista, Paraguay 144/D5
Buenavista, Philippines 82/E6
Buena Vista, Sask. 181/F5
Buena Vista, Tenn. (38318) 237/E9
Buena Vista, Uruguay 145/E3
Buena Vista, Anzoátegui, Venezuela 124/F3
Buena Vista, Apure, Venezuela 124/D4
Buena Vista, Falcón, Venezuela 124/D2
Buena Vista (I.C.), Va. (24416) 307/K5
Buendía (res.), Spain 33/E2
Bueno (riv.), Chile 138/D3
Buenos Aires (lake) 120/B7
Buenos Aires (prov.), Argentina 143/D4
Buenos Aires (cap.), Argentina 120/C6
Buenos Aires (cap.), Argentina 143/H7
Buenos Aires (cap.), Argentina 2/F7
Buenos Aires (lake), Argentina 143/B6
Buenos Aires (lake), Chile 138/E6
Buenos Aires, Amazonas, Colombia 126/F9
Buenos Aires, Caquetá, Colombia 126/D7
Buenos Aires, C. Rica 154/F6
Buesaco, Colombia 126/B7
Buey Arriba, Cuba 158/H4
Bueyeros, N. Mex. (88412) 274/F3
Buffalo, Ala. (†36862) 195/H5
Buffalo, Alberta 182/E4
Buffalo (lake), Alberta 182/D3
Buffalo (riv.), Ark. 202/E2
Buffalo, Ill. (62515) 222/D4
Buffalo, Ind. (47925) 227/D3
Buffalo, Iowa (52728) 229/M6
Buffalo, Kansas (66717) 232/G4
Buffalo, Ky. (42716) 237/K6
Buffalo (bay), Manitoba 179/G5
Buffalo, Minn. (55313) 255/E5
Buffalo (riv.), Minn. 255/B4
Buffalo, Mo. (65622) 261/F7
Buffalo, Mont. (59418) 262/G4
Buffalo (co.), Nebr. 264/E4
Buffalo (creek), Nev. 266/B2
Buffalo, N.Y. 146/L5
Buffalo, N.Y. 188/L2
Buffalo, N.Y. (*14201) 276/B5
Buffalo, N. Dak. (58011) 282/R6
Buffalo, Ohio (43722) 284/G6
Buffalo, Okla. (73834) 288/G1
Buffalo, S.C. (29321) 296/D2
Buffalo (co.), S. Dak. 298/L5
Buffalo, S. Dak. (57720) 298/B2
Buffalo (creek), S. Dak. 298/F6
Buffalo (lake), S. Dak. 298/P2
Buffalo (riv.), Tenn. 237/F9
Buffalo, Texas (75831) 303/J6
Buffalo, W. Va. (25033) 312/C5
Buffalo (co.), Wis. 317/C7
Buffalo, Wis. (†54622) 317/C7
Buffalo, Wyo. (82834) 319/F1
Buffalo Bill (dam), Wyo. 319/C1
Buffalo Bill (res.), Wyo. 319/C1
Buffalo Center, Iowa (50424) 229/F2
Buffalo City, Ark. (†72653) 202/E1
Buffalo City, N.C. (†27931) 281/T3
Buffalo Creek, Br. Col. 184/G4
Buffalo Creek, Colo. (80425) 208/J4
Buffalo Fork, Snake (riv.), Wyo. 319/B2
Buffalo Gap, Sask. 181/F6
Buffalo Gap, S. Dak. (57722) 298/C6
Buffalo Gap, Texas (79508) 303/E5
Buffalo Grove, Ill. (†60090) 222/B5
Buffalo Head (hills), Alberta 182/B5
Buffalo Junction, Va. (24529) 307/L7
Buffalo Lake, Minn. (55314) 255/D6
Buffalo Lodge (lake), N. Dak. 282/J3
Buffalo Mills, Pa. (15534) 294/E6
Buffalo Narrows, Sask. 181/L3
Buffalo Nat'l River, Ark. 202/E1
Buffalo Pound Prov. Park, Sask. 181/F5
Buffalo River Junction, N.W. Terrs. 187/G3
Buffalo Valley, Tenn. (38548) 237/K8
Buffaloville, Ind. (47518) 227/D8
Buff Bay, Jamaica 158/K6
Buford, Alberta 182/D3
Buford, Georgia (30518) 217/D2
Buford, N. Dak. (†58853) 282/C3
Buford, Ohio (45110) 284/C7
Buford, Wyo. (82052) 319/G4
Bug (riv.) 7/G3
Bug (riv.), Poland 47/F2
Bug (riv.), U.S.S.R. 7/G4
Bug (riv.), U.S.S.R. 52/B4
Bug (riv.), U.S.S.R. 52/D5
Buga, Colombia 126/B6
Bugaboo Glacier Prov. Park, Br. Col. 184/J5
Bugak, Hungary 41/E3
Bugaldie, N.S. Wales 97/E2
Bugasong, Philippines 82/C5
Buggs Island (lake), N.C. 281/M1
Buggs Island (lake), Va. 307/L8
Bugiougio (isl.), Portugal 33/B2
Bugojno, Yugoslavia 45/C3
Bugrino, U.S.S.R. 52/G1
Bugsuk (isl.), Philippines 85/F4
Bugsuk (isl.), Philippines 82/A6
Bugt, China 77/K2
Bugui (pt.), Philippines 82/D4
Bugul'ma, U.S.S.R. 7/K3
Bugul'ma, U.S.S.R. 52/H4
Bugul'ma, U.S.S.R. 48/F4
Buguruslan, U.S.S.R. 52/H4
Buhl, Ala. (35446) 195/C4
Buhl, Idaho (83316) 220/D7
Buhl, Minn. (55713) 255/F3
Bühl, W. Germany 22/C4
Buhler, Kansas (67522) 232/E3
Buhuşi, Romania 45/H2
Buie, Loch (inlet), Scotland 15/C4
Buies, N.C. (†28377) 281/L5
Buies Creek, N.C. (27506) 281/M4
Buiksloot, Netherlands 27/C4
Builth Wells, Wales 13/D5
Builth Wells, Wales 10/E4
Buin, Chile 138/G4
Buin, Papua N.G. 86/C2
Buin (peak), Switzerland 39/K3
Buinsk, U.S.S.R. 52/G4
Bujalance, Spain 33/D4
Bujumbura (cap.), Burundi 102/F5
Bujumbura (cap.), Burundi 115/E4
Buka, Papua N.G. 87/F6
Buka (isl.), Papua N.G. 86/C2
Buka (passage), Papua N.G. 86/C2
Bukachacha, U.S.S.R. 48/M4
Bukama, Zaire 115/E5
Buka Passage, Papua N.G. 86/C2
Bukavu, Zaire 102/E5
Bukavu, Zaire 115/E4
Bukene, Tanzania 115/F4
Bukhara, U.S.S.R. 54/H5
Bukhara, U.S.S.R. 48/G5
Bukidnon (prov.), Philippines 82/E6
Bukittinggi, Indonesia 85/B6
Bükk (mts.), Hungary 41/F2
Bukoba, Tanzania 115/F4
Bukowno, Poland 47/C4
Bul, Kuh-e (mt.), Iran 66/H5
Bula, Indonesia 85/J6
Bula, Texas (79320) 303/B4
Bulacan (prov.), Philippines 82/C3
Bülach, Switzerland 39/G1
Bulahdelah, N.S. Wales 97/G3
Bulak (Bole), China 77/B3
Bulalakao, Philippines 82/C4
Bulan, Ky. (41722) 237/P6
Bulan, Philippines 82/D4
Bulancak, Turkey 63/H2
Bulanık, Turkey 63/K3
Bûlâq, Egypt 111/F2
Bûlâq, Egypt 59/B4
Bulawayo, Zimbabwe 118/D3
Bulawayo, Zimbabwe 102/E7
Buldan, Turkey 59/A2
Buldan, Turkey 63/C3
Buldibuyo, Peru 128/D7
Buldir (isl.), Alaska 196/J3
Bulgan, Mongolia 77/F2
Bulgan, Ömnögovĭ, Mongolia 77/F3
Bulgan, Hovd, Mongolia 77/D2
Bulgan, Bulgan, Mongolia 77/F2
Bulgaria 2/L3
Bulgaria 7/G4
BULGARIA 45/G4
Bulger, Pa. (15019) 294/B5
Bulgroo, Queensland 95/B5
Bulgroo, Queensland 88/G5
Bulhar, Somalia 115/H1
Buli, Indonesia 85/H5
Buliluyan (cape), Philippines 85/F4
Buliluyan (cape), Philippines 82/A6
Bulimba (creek), Queensland 95/E3
Bulkley (riv.), Br. Col. 184/D2
Bull, The (isl.), Ireland 17/A8
Bull (isl.), Newf. 166/D2
Bull (isl.), S.C. 296/H6
Bull (creek), S. Dak. 298/K6
Bull (creek), S. Dak. 298/C2
Bullard, Georgia (†31020) 217/F5
Bullard, Texas (75757) 303/J5
Bull Arm (inlet), Newf. 166/D2
Bullas, Spain 33/F4
Bulldog (isl.), Newf. 166/C3
Bulle, Switzerland 39/D3
Bullen (bay), Neth. Ant. 161/F8
Buller (riv.), N. Zealand 100/D4
Buller (mt.), Victoria 97/D5
Bullfinch, W. Australia 92/B5
Bull Harbour, Br. Col. 184/C5
Bullhead, S. Dak. (57621) 298/G2
Bullhead City-Riviera, Ariz. (86430) 198/A3
Bullitt (co.), Ky. 237/K5
Bull Lake (res.), Wyo. 319/C2
Bulloch (co.), Georgia 217/J6
Bullock (co.), Ala. 195/G6
Bullock, N.C. (27507) 281/M2
Bullock (creek), S.C. 296/E2
Bulloo (riv.), Queensland 88/G5
Bulloo (lake), Queensland 95/B6
Bulloo (riv.), Queensland 95/B6
Bulloo Downs, Queensland 88/G5
Bull Run (creek), Va. 307/N3
Bulls, N. Zealand 100/E4
Bulls (bay), S.C. 296/H6
Bull Savanna-Junction, Jamaica 158/H6
Bulls Bridge, Conn. (†06785) 210/B2
Bulls Gap, Tenn. (37711) 237/P8
Bull Shoals, Ark. (72619) 202/E1
Bull Shoals (lake), Ark. 202/E1
Bull Shoals (lake), Mo. 261/G10
Bully (creek), Oreg. 291/K3
Bulnes, Chile 138/E1
Bulo Burti, Somalia 115/J3
Bulolo, Papua N.G. 85/B7
Bulpitt, Ill. (62517) 222/D4
Buluan (lake), Philippines 82/E7
Bulukumba, Indonesia 85/G7
Bulun, U.S.S.R. 4/B3
Bulun, U.S.S.R. 48/N2
Bulungu, Zaire 115/C4
Bulusan, Philippines 82/E4
Bulusan (vol.), Philippines 82/D4
Bulyea, Sask. 181/G5
Bumba, Zaire 115/D3
Bumble Bee, Ariz. (*86301) 198/C4
Bumiayu, Indonesia 85/H2
Bumpass, Va. (23024) 307/N5
Bumping (lake), Wash. 310/D4
Bumpus Mills, Tenn. (37028) 237/F7
Bumthang, Bhutan 68/G3
Buna, Kenya 115/G3
Buna, Papua N.G. 85/C7
Buna, Texas (77612) 303/L7
Bunavista, Texas (†79007) 303/C2
Bunawan, Philippines 82/E6
Bunbeg-Derrybeg, Ireland 17/E1
Bunbury, Australia 87/B9
Bunbury, Pr. Edward I. 168/F2
Bunbury, W. Australia 88/B6
Bunbury, W. Australia 92/A2
Bunceton, Mo. (65237) 261/G5
Bunch, Iowa (†52552) 229/H7
Bunch, Okla. (74931) 288/S3
Bunche Park, Fla. (†33054) 212/B4
Bunclody-Carrickduff, Ireland 17/H6
Buncombe, Ill. (62912) 222/E6
Buncombe (co.), N.C. 281/D3
Buncrana, Ireland 17/G1
Buncrana, Ireland 10/C3
Bundaberg, Australia 87/E8
Bundaberg, Queensland 88/J4
Bundaberg, Queensland 95/D5
Bundanoon, N.S. Wales 97/F4
Bundarra, N.S. Wales 97/F2
Bünde, W. Germany 22/C2
Bundi, India 68/D3
Bundick (lake), La. 238/D5
Bundooma, North. Terr. 93/D8
Bundoora, Victoria 97/J4
Bundoran, Ireland 17/E3
Bunessan, Scotland 15/B4
Bunga (pt.), Philippines 82/E4
Bungalaut (chan.), Indonesia 85/B6
Bungay, England 10/G4
Bungay, England 13/J5
Bungee (brook), Conn. 210/G1
Bu Ngem, Libya 111/C1
Bungendore, N.S. Wales 97/E4
Bungo (str.), Japan 81/F7
Bunguran (Great Natuna) (isl.), Indonesia 85/D5
Bunguran (Natuna) (isls.), Indonesia 85/D5
Bunia, Zaire 115/E3
Bunia, Zaire 102/E4
Bunji, Pakistan 68/C1
Bunker, Mo. (63629) 261/K8
Bunker Group (isls.), Queensland 95/E4
Bunker Hill, Ill. (62014) 222/D4
Bunker Hill, Ind. (46914) 227/E3
Bunker Hill, Kansas (67626) 232/D3
Bunker Hill, Oreg. (†97420) 291/C4
Bunker Hill, W. Va. (25413) 312/K4
Bunker Hill Village, Texas (†77001) 303/J1
Bunkerville, Nev. (89007) 266/G6
Bunkeya, Zaire 115/E6
Bunkie, La. (71322) 238/F5
Bunn, N.C. (27508) 281/N3
Bunnell, Fla. (32010) 212/E2
Bunnlevel, N.C. (28323) 281/M4
Buntok, Indonesia 85/E6
Bunyala, Kenya 115/F3
Bünyan, Turkey 63/G3
Bunyan's Cove, Newf. 166/C2
Bunyu (isl.), Indonesia 85/F5
Buochs, Switzerland 39/F3
Buol, Indonesia 85/G5
Buq, Iran 66/M6
Bura, Kenya 115/H4
Bur Acaba, Somalia 115/H3
Bur Acaba, Somalia 102/G4
Buraida, Saudi Arabia 59/D4
Buraimi, Oman 59/G5
Buraimi, U.A.E. 59/G5
Buram, Sudan 111/E5
Burang, China 77/B5
Burao, Somalia 102/G4
Burao, Somalia 115/J2
Buras-Triumph, La. (70041) 238/L8
Burauen, Philippines 82/E5
Buraz, Turkey 63/B6
Burbank, Calif. (*91501) 204/C10
Burbank, Ill. (60459) 222/B6
Burbank, Ohio (44214) 284/F4
Burbank, Okla. (74633) 288/N1
Burbank, S. Dak. (57010) 298/R8
Burbank, Wash. (99323) 310/G4
Burchard, Minn. (†56115) 255/C6
Burchard, Nebr. (68323) 264/H4
Burcher, N.S. Wales 97/D3
Burchinal, Iowa (†50469) 229/G2
Burdekin (riv.), Queensland 88/H3
Burdekin (riv.), Queensland 95/C3
Burden, Kansas (67019) 232/F4
Burdett, Alberta 182/E5
Burdett, Kansas (67523) 232/C3
Burdett, N.Y. (14818) 276/G6
Burdette, Ark. (72321) 202/L2
Burdette, S. Dak. (†57476) 298/M4
Burdick, Kansas (66838) 232/F3
Burdur (prov.), Turkey 63/D4
Burdur, Turkey 59/A2
Burdur, Turkey 63/D4
Burdur (lake), Turkey 63/D4
Burdwan, India 68/F4
Bureå, Sweden 18/M4
Bureau (co.), Ill. 222/D2
Bureau, Ill. (61315) 222/D2
Burei, Ethiopia 111/G6
Büren, W. Germany 22/C3
Büren an der Aare, Switzerland 39/D2
Bürentsogt, Mongolia 77/H2
Burford (lake), N. Mex. 274/C2
Burford, Ontario 177/D4
Burfordville, Mo. (63739) 261/N8
Burgas, Bulgaria 45/H4
Burgas, Bulgaria 7/G4
Burg auf Fehmarn, W. Germany 22/D1
Burgaw, N.C. (28425) 281/N5
Burgaz (isl.), Turkey 63/D6
Burg bei Magdeburg, E. Germany 22/D2
Burgdorf, Switzerland 39/E2
Burgenland (prov.), Austria 41/D3
Burgeo, Newf. 166/C4
Burgersdorp, S. Africa 118/D6
Burgess, Mo. (†66756) 261/C7
Burgess, S.C. (†29576) 296/J4
Burgess, Va. (22432) 307/R5
Burgess (mt.), Yukon 187/D3
Burgess Hill, England 13/G7
Burgessville, Ontario 177/D4
Burgettstown, Pa. (15021) 294/A5
Burghausen, W. Germany 22/E4
Burghead, Scotland 10/E2
Burghead, Scotland 15/E3
Burghill, Ohio (44404) 284/J3
Burgin, Ky. (40310) 237/M5
Burgis, Sask. 181/J4
Bürglen, Thurgau, Switzerland 39/H1
Bürglen, Uri, Switzerland 39/G3
Burglengenfeld, W. Germany 22/D4
Burgoon, Ohio (43407) 284/D3
Burgos, Mexico 150/K4
Burgos (prov.), Spain 33/E1
Burgos, Spain 7/D4
Burgos, Spain 33/E1
Burg Stargard, E. Germany 22/E2
Burgsteinfurt, W. Germany 22/B2
Burgsvik, Sweden 18/K8
Burgundy (trad. prov.), France 29
Burhaniye, Turkey 63/B3
Burhanpur, India 68/D4
Buri, Brazil 135/B3
Buri (pen.), Ethiopia 111/H4
Burias (isl.), Philippines 82/D4
Burias (passage), Philippines 82/D4
Buribay, U.S.S.R. 52/J4
Burica (pt.), C. Rica 154/F6
Burica, Punta (cape), Panama 154/F6
Burien, Wash. (98166) 310/A2
Burin, Newf. 166/C4
Burin (pen.), Newf. 166/C4
Buriram, Thailand 72/D4
Buriti, Brazil 132/F3
Buriti Alegre, Brazil 132/D7
Buriti dos Lopes, Brazil 132/F3
Burj al Hattaba, Tunisia 106/F2
Burkburnett, Texas (76354) 303/F3
Burke (chan.), Br. Col. 184/D4
Burke (co.), Georgia 217/J4
Burke, Idaho (†83873) 220/C2
Burke, N.Y. (12917) 276/M1
Burke (co.), N.C. 281/F3
Burke (co.), N. Dak. 282/E2
Burke, S. Dak. (57523) 298/L7
Burke, Texas (75941) 303/K6
Burke○, Vt. (†05871) 268/D2
Burke (mt.), Vt. 268/D2
Burke, Va. (22015) 307/R3
Bürkelkopf (mt.), Switzerland 39/K3
Burkes Garden, Va. (24608) 307/F6
Burkesville, Ky. (42717) 237/L7
Burket, Ind. (46508) 227/F2
Burketown, Queensland 95/A3
Burketown, Queensland 88/F3
Burkett, Texas (76828) 303/E5
Burkettsville, Ohio (45310) 284/A5
Burkettville, Maine (04540) 243/E7
Burkeville, Texas (75932) 303/L7
Burkeville, Va. (23922) 307/M6
Burkittsville, Md. (21718) 245/H3
Burkley, Ky. (†42021) 237/C7
Burk's Falls, Ontario 177/E2
Burkville, Ala. (36725) 195/E6
Burleigh (co.), N. Dak. 282/J6
Burleson (co.), Texas 303/H7
Burleson, Texas (76028) 303/F3
Burley, Idaho (83318) 220/E7
Burley, Wash. (98322) 310/C3
Burlingame, Calif. (94010) 204/J2
Burlingame, Kansas (66413) 232/G3
Burlington, Colo. (80807) 208/P4
Burlington○, Conn. (06013) 210/D1
Burlington, Ill. (60109) 222/E1
Burlington, Ind. (46915) 227/E4
Burlington, Iowa (52601) 229/L7
Burlington, Iowa 188/H2
Burlington, Kansas (66839) 232/G3
Burlington, Ky. (41005) 237/R2
Burlington○, Maine (04417) 243/G5
Burlington○, Mass. (01803) 249/C5
Burlington, Mich. (49029) 250/D6
Burlington (co.), N.J. 273/D4
Burlington, N.J. (08016) 273/D3
Burlington, N.C. (27215) 281/K2
Burlington, N. Dak. (58722) 282/H3
Burlington, Ohio (†45680) 284/E9
Burlington, Okla. (73722) 288/K1
Burlington, Ontario 177/E4
Burlington, Pa. (18814) 294/J2
Burlington, Vt. (05401) 268/A3
Burlington, Vt. 146/L5
Burlington, Vt. 188/M2
Burlington, Wash. (98233) 310/C2
Burlington, W. Va. (26710) 312/J4
Burlington, Wis. (53105) 317/K10
Burlington, Wyo. (82411) 319/D1
Burlington Flats, N.Y. (13315) 276/K5
Burlington Junction, Mo. (64428) 261/B2
Burlison, Tenn. (38015) 237/B9
Burma 2/P4
Burma 54/L7
BURMA 72
Burmis, Alberta 182/C5
Burna, Ky. (42028) 237/E6
Burnaby, Br. Col. 184/K3
Burnaby (isl.), Br. Col. 184/B4
Burnet (co.), Texas 303/F7
Burnet, Texas (78611) 303/F7
Burnett, Minn. (†55727) 255/F4
Burnett (co.), Wis. 317/B4
Burnett, Wis. (53922) 317/J9
Burnettown, S.C. (†29834) 296/D5
Burnettsville, Ind. (47926) 227/D3
Burney, Calif. (96013) 204/D3
Burney (mt.), Chile 138/D9
Burney, Ind. (47222) 227/F6
Burneyville, Okla. (73430) 288/M7
Burnham, Ill. (†60601) 222/C6
Burnham○, Maine (04922) 243/E6
Burnham, Mo. (†65793) 261/J9
Burnham, Pa. (17009) 294/H4
Burnham-on-Crouch, England 13/H6
Burnham-on-Sea, England 13/D6
Burnie, Tasmania 99/B3
Burnie-Somerset, Tasmania 88/H8
Burning Springs, Ky. (40922) 237/O6
Burning Springs, W. Va. (26139) 312/D5
Burnips, Mich. (49314) 250/D6
Burnley, England 10/G1
Burnley, England 13/H1
Burnmouth, Scotland 15/F5
Burns, Colo. (80426) 208/F3
Burns, Kansas (66840) 232/F3
Burns, Miss. (†39153) 256/E6
Burns, N.S. Wales 97/A3
Burns, Oreg. (97720) 291/H4
Burns, Tenn. (37029) 237/G8
Burns, Wyo. (82053) 319/H4
Burns City, Ind. (47553) 227/D7
Burns Flat, Okla. (73624) 288/H4
Burns Harbor, Ind. (†46401) 227/C1
Burnside, Conn. (†06108) 210/E1
Burnside, Ill. (62318) 222/B3
Burnside, Iowa (50521) 229/E4
Burnside, Ky. (42519) 237/M6
Burnside, La. (70738) 238/L3
Burnside, Miss. (†39350) 256/F5
Burnside (riv.), N.W. Terrs. 187/G3
Burnside, Pa. (15721) 294/E4
Burnside, S. Australia 88/E8
Burnside, S. Australia 94/B8
Burnside, Suriname 131/C2
Burns Junction, Oreg. (†97902) 291/K5
Burns Lake, Br. Col. 162/D5
Burns Lake, Br. Col. 184/D3
Burnstad, N. Dak. (58526) 282/L7
Burnsville, Ala. (†36701) 195/E6
Burnsville, Ind. (†47201) 227/F6
Burnsville, Minn. (55337) 255/E6
Burnsville, Miss. (38833) 256/H1
Burnsville, New Bruns. 170/E1
Burnsville, N.C. (28714) 281/E3
Burnsville, Va. (24420) 307/J4
Burnsville, W. Va. (26335) 312/E5
Burnt (lakes), Alberta 182/C1
Burnt (lake), Newf. 166/B3
Burnt (riv.), Ontario 177/F3
Burnt (riv.), Oreg. 291/K3
Burnt (lake), Québec 174/E2
Burnt Cabins, Pa. (17215) 294/G5
Burnt Corn, Ala. (36431) 195/D7
Burnt House, W. Va. (26336) 312/D4
Burnt Island (lake), Ontario 177/F2
Burntisland, Scotland 15/D1
Burntisland, Scotland 10/C1
Burnt Islands, Newf. 166/C4
Burnt Point, Newf. 166/D2
Burnt Prairie, Ill. (62820) 222/E5
Burnt River, Ontario 177/F3
Burntroot (lake), Ontario 177/F2
Burntside (lake), Minn. 255/F3
Burntwood, England 13/F5
Burntwood, England 10/G2
Burntwood (riv.), Manitoba 179/J2
Burnwell, W. Va. (25034) 312/D6
Burqa, West Bank 65/C3
Burqin, China 77/C2
Burr (pond), Conn. 210/C1
Burr, Minn. (†56220) 255/B6
Burr, Nebr. (68324) 264/H4
Burr, Sask. 181/F3
Burra, S. Australia 94/F5
Burraboi, N.S. Wales 97/C4
Burramurra, North. Terr. 93/E6
Burray (isl.), Scotland 15/F2
Burrel, Albania 45/D5
Burren Junction, N.S. Wales 97/E2
Burriana, Spain 33/G3
Burringbar, N.S. Wales 97/G1
Burrinjuck (res.), N.S. Wales 97/E4
Burris, Wyo. (†82501) 319/C2
Burro (creek), Ariz. 198/B4
Burro (mts.), Mexico 150/J2
Burr Oak, Ind. (46509) 227/E2
Burr Oak, Iowa (52131) 229/K2
Burr Oak, Kansas (66936) 232/D2
Burr Oak, Mich. (49030) 250/D7
Burro-Burro (riv.), Guyana 131/B3
Burrow (head), Scotland 15/D6
Burrows, Ind. (46916) 227/E3
Burrows, Sask. 181/K5
Burr Ridge, Ill. (†60558) 222/B6
Burrsville, Md. (†21629) 245/P5
Burrton, Kansas (67020) 232/E3
Burrville, Conn. (†06790) 210/C1
Burrville, Tenn. (†37872) 237/M8
Burrville, Utah (†84701) 304/C5
Burrwood, La. (†70091) 238/M8
Burry Port, Wales 13/C6
Bursa (prov.), Turkey 63/C2
Bursa, Turkey 63/C2
Bursa, Turkey 59/A1
Bursa, Turkey 54/D5
Bur Sa'id (Port Said), Egypt 111/K2
Bur Said (Port Said), Egypt 59/B3
Burstall, Sask. 181/B5
Burt, Iowa (50522) 229/E2
Burt, Mich. (48417) 250/F5
Burt (lake), Mich. 250/E3
Burt (co.), Nebr. 264/H3
Burt, N.Y. (14028) 276/C4
Burt, N. Dak. (†58646) 282/F7
Burta, N.S. Wales 97/A3
Burt Lake, Mich. (49717) 250/E3
Burton, Br. Col. 184/H5
Burton (lake), Georgia 217/E1
Burton, Mich. (48507) 250/F6
Burton, Nebr. (†68778) 264/E2
Burton, New Bruns. 170/D3
Burton, Ohio (44021) 284/H3
Burton, S.C. (29902) 296/F7
Burton, Texas (77835) 303/H7
Burton, Wash. (98013) 310/C3
Burton, W. Va. (26562) 312/F3
Burtonport, Ireland 17/E2
Burtonport, Ireland 10/B3
Burton upon Trent, England 13/F5
Burton upon Trent, England 10/G2
Burtonville, Ky. (†41179) 237/P4
Burträsk, Sweden 18/M4
Burtrum, Minn. (56318) 255/D5
Burtts Corner, New Bruns. 170/D2
Buru (isl.), Indonesia 54/O10
Buru (isl.), Indonesia 85/H6
Buru (sea), Indonesia 85/H6
Burultokay (Fuhai), China 77/C2
Burundi 2/L6
Burundi 102/F5
BURUNDI 115/E4
Bururi, Burundi 115/F4
Burutu, Nigeria 106/F7
Burwash Landing, Yukon 187/D3
Burwell, Nebr. (68823) 264/E3
Burwell (mt.), Wyo. 319/C2
Burwood, N.S. Wales 88/K4
Burwood, N.S. Wales 97/J3
Bury, England 13/H2
Bury, England 10/G2
Bury, Québec 172/F4
Buryat A.S.S.R., U.S.S.R. 48/M4
Burye, Ethiopia 102/F3
Burye, Ethiopia 111/G5
Bury Saint Edmunds, England 13/H5
Bury Saint Edmunds, England 10/G4
Busby, Alberta 182/C3
Busby, Mont. (59016) 262/J5
Buseno, Switzerland 39/H4
Bush, Ill. (†62924) 222/D6
Bush, Ky. (40724) 237/O6
Bush, La. (70431) 238/L5
Bush (creek), Md. 245/J3
Bush (riv.), N. Ireland 17/H1
Bush (riv.), S.C. 296/D3
Bush City, Kansas (†66032) 232/G3
Bushehr (prov.), Iran 66/G6
Bushehr (Bushire), Iran 66/G6
Bushehr, Iran 59/F4
Bushell, Sask. 181/L2
Bushey, England 10/B5
Bushey, England 13/H7
Bushire, Iran 54/G7
Bushiribana, Neth. Ant. 161/E10
Bush Island, Nova Scotia 168/D4
Bushkill, Pa. (18324) 294/M3
Bushland, Texas (79012) 303/B2
Bushmills, N. Ireland 17/J1
Bushnell, Fla. (33513) 212/D3
Bushnell, Ill. (61422) 222/C3
Bushnell, Nebr. (69128) 264/A3
Bushnell, S. Dak. (57276) 298/R5
Bushong, Kansas (66841) 232/F3
Bushton, Ill. (†61920) 222/E4
Bushton, Kansas (67427) 232/D3
Bushwood, Md. (20618) 245/L7
Bushyhead, Okla. (†74016) 288/P2
Bushy Park, Tasmania 99/C4
Busick, N.C. (28714) 281/E3
Businga, Zaire 115/D3
Buskerud (co.), Norway 18/F6
Buskirk, Ky. (41406) 237/P5
Busko Zdrój, Poland 47/E3
Busra, Syria 63/G6
Busselton, W. Australia 88/A6
Busselton, W. Australia 92/A6
Busseron (creek), Ind. 227/C7
Busse Woods (res.), Ill. 222/B5
Bussey, Iowa (50044) 229/H6
Bussigny-près-Lausanne, Switzerland 39/B3
Bussum, Netherlands 27/G4
Bustard (isls.), Ontario 177/C2
Busti, N.Y. (†14701) 276/B6
Bustinza, Argentina 143/F6
Busto Arsizio, Italy 34/B2
Busuanga (isl.), Philippines 82/B4
Busuanga (isl.), Philippines 85/F3
Busu-Djanoa, Zaire 115/D3
Büsum, W. Germany 22/C1
Buta, Zaire 102/E4
Buta, Zaire 115/D3
Buta-Ranquil, Argentina 143/C4
Butare, Rwanda 115/E4
Butaritari (atoll), Kiribati 87/H5
Butcher (isl.), India 68/B7
Bute (inlet), Br. Col. 184/E5
Bute (isl.), Scotland 15/C5
Bute (trad. co.), Scotland 15/A5
Bute (sound), Scotland 15/C5
Butedale, Br. Col. 184/C3
Butembo, Zaire 115/E3
Butembo, Zaire 102/E5
Butha, China 77/K2
Butiaba, Uganda 115/F3

Butler (co.), Ala. 195/E7
Butler, Ala. (36904) 195/B6
Butler, Georgia (31006) 217/D5
Butler, Ill. (62015) 222/D4
Butler, Ind. (46721) 227/H2
Butler (co.), Iowa 229/H3
Butler (co.), Kansas 232/F4
Butler (co.), Ky. 237/H6
Butler, Ky. (41006) 237/N3
Butler, Md. (21023) 245/M2
Butler, Minn. (†56567) 255/C4
Butler (co.), Mo. 261/M9
Butler, Mo. (64730) 261/D6
Butler (co.), Nebr. 264/G3
Butler, N.J. (07405) 273/E2
Butler (co.), Ohio 284/A7
Butler, Ohio (44822) 284/F4
Butler, Okla. (73625) 288/H3
Butler (co.), Pa. 294/C4
Butler, Pa. (16001) 294/C4
Butler, S. Dak. (57222) 298/O3
Butler, Tenn. (37640) 237/T8
Butler (bay), Virgin Is. (U.S.) 161/E4
Butler, Wis. (53007) 317/K1
Butler Springs, Ala. (†36030) 195/E7
Butlerville, Ark. (†72176) 202/G4
Butlerville, Ind. (47223) 227/F6
Butlerville, Ohio (†45162) 284/B7
Butner, N.C. (27509) 281/M2
Bütschelegg (mt.), Switzerland 39/D3
Bütschwil, Switzerland 39/H2
Buttahatchee (riv.), Ala. 195/B3
Buttahatchee (riv.), Miss. 256/H3
Butte (co.), Calif. 204/D4
Butte (co.), Idaho 220/E6
Butte, Mont. 146/G5
Butte, Mont. 188/D1
Butte, Nebr. (68722) 264/F2
Butte (mts.), Nev. 266/F3
Butte, N. Dak. (58723) 282/J4
Butte (creek), Oreg. 291/G2
Butte (creek), Oreg. 291/B3
Butte (co.), S. Dak. 298/B3
Butte City, Calif. (95920) 204/C4
Butte City, Idaho (83213) 220/E6
Butte Des Morts, Wis. (†54901) 317/J7
Butte Falls, Oreg. (97522) 291/E5
Butter (creek), Oreg. 291/H2
Butterfield, Ark. (†72104) 202/E5
Butterfield, Minn. (56120) 255/D7
Butterfield, Mo. (65623) 261/E9
Butterfield (lake), N.Y. 276/J2
Butternut, Mich. (†48811) 250/E5
Butternut, Wis. (54514) 317/E3
Butternut (lake), Wis. 317/J4
Butter Pot Prov. Park, Newf. 166/D2
Butters, N.C. (28324) 281/M5
Butterworth, Malaysia 72/D6
Butterworth (Gcuwa), S. Africa 118/D6
Buttes, Switzerland 39/C3
Butte-Silver Bow County, Mont. (59701) 262/D5
Buttevant, Ireland 17/D7
Butteville, Oreg. (†97002) 291/A2
Butt of Lewis (prom.), Scotland 15/B2
Button (isls.), N.W. Terrs. 187/M3
Buttonwillow, Calif. (93206) 204/F8
Butts (co.), Georgia 217/E4
Buttzville, N.J. (07829) 273/D2
Buttzville, N. Dak. (†58054) 282/P6
Butuan, Philippines 82/E6
Butuan, Philippines 85/H4
Butuan, Philippines 54/O9
Butuan (bay), Philippines 82/E6
Butumi, U.S.S.R. 7/J4
Butung (isl.), Indonesia 54/O10
Butung (isl.), Indonesia 85/G6
Buturlinovka, U.S.S.R. 52/F4
Butzbach, W. Germany 22/C3
Bützow, E. Germany 22/E2
Buxtehude, W. Germany 22/C2
Buxton, England 10/G2
Buxton, England 13/J2
Buxton○, Maine (†04093) 243/C8
Buxton, N.C. (27920) 281/U4
Buxton, N. Dak. (58218) 282/R4
Buxton, Oreg. (97109) 291/D2
Buxton Center, Maine (†04093) 243/B8
Buy, U.S.S.R. 52/F3
Buyck, Minn. (55771) 255/F2
Buynaksk, U.S.S.R. 52/G6
Büyükada, Turkey 63/D6
Büyük Ağrı (Ararat) (mt.), Turkey 63/L3
Büyük Ağrı (Ararat) (mt.), Turkey 59/D2
Büyükanafarta, Turkey 63/B6
Büyükdere, Turkey 63/D5
Büyük Hasan Dağı, Turkey 63/E3
Büyük Menderes (riv.), Turkey 59/A2
Buzău, Romania 45/H3
Buzău (riv.), Romania 45/H3
Buzeima (well), Libya 111/D3
Buziaş, Romania 45/E3
Buzios (cape), Brazil 135/F3
Buzuluk, U.S.S.R. 52/H4
Buzuluk, U.S.S.R. 48/F4
Buzzard Roost (dam), S.C. 296/D3
Buzzards (bay), Mass. 249/L6
Buzzards Bay, Mass. (02532) 249/M5
Byala, Bulgaria 45/G4
Byala Slatina, Bulgaria 45/F4
Byam Martin (chan.), N.W. Terrs. 187/H2
Byam Martin (isl.), N.W. Terrs. 187/H2
Byars, Okla. (74831) 288/N5
Bybee, Tenn. (37713) 237/P8
Bydgoszcz (prov.), Poland 47/C2
Bydgoszcz, Poland 47/C2
Bydgoszcz, Poland 7/F3
Byemoor, Alberta 182/D4
Byers, Colo. (80103) 208/L3
Byers, Kansas (67021) 232/D4
Byers, Texas (76357) 303/F3
Byesville, Ohio (43723) 284/G6
Byfield, Mass. (01922) 249/L1
Byford, W. Australia 88/B2
Bygland, Minn. (†56723) 255/B3
Bygland, Norway 18/F7
Byhalia, Miss. (38611) 256/E1
Bykhov, U.S.S.R. 52/C4
Bylas, Ariz. (85530) 198/E5
Bylot (isl.), N.W.T. 146/L2
Bylot (isl.), N.W.T. 162/J1
Bylot (isl.), N.W. Terrs. 187/L2
Byng, Okla. (†74820) 288/N5
Byng Inlet, Ontario 177/D2
Byng Inlet, Ontario 175/D3
Bynum, Mont. (59419) 262/D3
Bynum (res.), Mont. 262/D2
Bynum, N.C. (27312) 281/L3
Bynumville, Mo. (†65281) 261/G3
Byram, Conn. (06830) 210/A4
Byram (pt.), Conn. 210/A4
Byram (riv.), Conn. 210/A4
Byram, Miss. (†39205) 256/D6
Byrd Station 5/A12
Byrdstown, Tenn. (38549) 237/L7
Byrnedale, Pa. (15827) 294/E3
Byrock, N.S. Wales 97/D2
Byromville, Georgia (31007) 217/E6
Byron, Calif. (94514) 204/L2
Byron (isl.), Chile 138/D7
Byron, Georgia (31008) 217/E5
Byron, Ill. (61010) 222/D1
Byron, Ind. (†46371) 227/C5
Byron, Maine (†04275) 243/B6
Byron○, Maine (†04275) 243/B6
Byron, Mich. (48418) 250/E6
Byron, Minn. (55920) 255/F6
Byron, Nebr. (68325) 264/G4
Byron (bay), Newf. 166/C3
Byron (cape), N. S. Wales 88/J5
Byron (cape), N.S. Wales 97/G1
Byron, N.Y. (14422) 276/D4
Byron, Okla. (73723) 288/K1
Byron (lake), S. Dak. 298/N4
Byron, Wis. (53009) 317/K8
Byron, Wyo. (82412) 319/D1
Byron Bay, N.S. Wales 97/G1
Byron Center, Mich. (49315) 250/D6
Byrum, Denmark 21/E3
Byskeälv (riv.), Sweden 18/L4
Bystřice nad Pernštejnem, Czech. 41/D2
Bystřice pod Hostýnem, Czech. 41/D2
Bystrzyca Kłodzka, Poland 47/C3
Bytča, Czech. 41/E2
Bytom, Poland 47/A3
Bytów, Poland 47/C1

C

Caacupé, Paraguay 144/B5
Caaguazú (dept.), Paraguay 144/D-E4
Caaguazú, Paraguay 144/D4
Caála, Angola 115/C6
Caamaño (sound), Br. Col. 184/C4
Caapucú, Paraguay 144/D5
Caatingas (for.), Brazil 120/E3
Caazapá (dept.), Paraguay 144/D-E5
Caazapá, Paraguay 144/D5
Caba, Philippines 82/C2
Cabadbaran, Philippines 82/E6
Cabaiguán, Cuba 158/E2
Cabalasan (mt.), Philippines 82/E5
Caballero, Paraguay 144/B5
Caballo, N. Mex. (87931) 274/B6
Caballo (res.), N. Mex. 274/B6
Caballo (creek), Wyo. 319/G1
Caballococha, Peru 128/G4
Caballones (chan.), Cuba 158/F3
Cabana, Peru 128/C7
Cabañaquinta, Spain 33/D1
Cabañas, Cuba 158/B1
Cabanatuan, Philippines 54/O8
Cabanatuan, Philippines 82/C3
Cabanatuan, Philippines 85/G2
Cabanes, Spain 33/F2
Cabano, Québec 172/J2
Cabarroguis, Philippines 82/C2
Cabarrus (co.), N.C. 281/H4
Cabazon, Calif. (92230) 204/J10
Cabbage Tree (creek), Queensland 95/D2
Cabedelo, Brazil 132/H4
Cabell (co.), W. Va. 312/B6
Cabery, Ill. (60919) 222/E3
Cabet, Pitons du (mt.), Martinique 161/C6
Cabeza del Buey, Spain 33/D3
Cabezas, Bolivia 136/D6
Cabezas, Cuba 158/D1
Cabildo, Chile 138/A9
Cabimas, Venezuela 120/B1
Cabimas, Venezuela 124/C2
Cabin Creek, W. Va. (25035) 312/C6
Cabinda (dist.), Angola 115/B5
Cabinda, Angola 115/B5
Cabinda, Angola 102/D5
Cabinet (mts.), Mont. 262/A2
Cabin John (creek), Md. 245/E4
Cabin John-Brookmont, Md. (20731) 245/E4
Cabins, W. Va. (26855) 312/H4
Cable, Minn. (†56301) 255/D5
Cable, Ohio (43009) 284/C5
Cable, Wis. (54821) 317/D3
Cabo Blanco, Peru 128/B5
Cabo Delgado (prov.), Mozambique 118/F2
Cabo Frio, Brazil 132/F8
Cabo Frio, Brazil 135/F3
Cabo Gracias a Dios, Nicaragua 154/F3
Cabonga (res.), Québec 174/B3
Cabool, Mo. (65689) 261/H8
Cabora Bassa (dam), Mozambique 118/E3
Caborn, Ind. (†47620) 227/B9
Cabo Rojo, P. Rico 161/A2
Cabo San Lucas, Mexico 150/E5
Cabot (str.) 162/K6
Cabot, Ark. (72023) 202/F4
Cabot (str.), Canada 146/N5
Cabot (lake), Newf. 166/B2
Cabot (str.), Newf. 166/B4
Cabot (mt.), N.H. 268/E2
Cabot (head), Ontario 177/C2
Cabot, Pa. (16023) 294/C4
Cabot, Vt. (05647) 268/C3
Cabot○, Vt. (05647) 268/C3
Cabo Vírgenes, Argentina 143/C7
Cabra, Spain 33/D4
Cabra de Santo Cristo, Spain 33/E4
Cabral, Dom. Rep. 158/D6
Cabral (lag.), Paraguay 144/A5
Cabrera, Dom. Rep. 158/E5
Cabrera (isl.), Spain 33/H3
Cabri, Sask. 181/C5
Cabri (lake), Sask. 181/B4
Cabrillo Nat'l Mon., Calif. 204/H11
Cabrits (isl.), Martinique 161/D7
Cabrón (cape), Dom. Rep. 158/F5
Cabruta, Venezuela 124/E4
Cabudare, Venezuela 124/D3
Cabugao, Philippines 82/C2
Cabulauan (isls.), Philippines 82/C5
Cabullones (pt.), P. Rico 161/C3
Caburai (mt.), Guyana 131/A3
Cabure, Venezuela 124/D2
Caçador, Brazil 132/D9
Cacahoatán, Mexico 150/N9
Čačak, Yugoslavia 45/E4
Caçapava, Brazil 135/D3
Caçapava do Sul, Brazil 132/C10
Cacapon (riv.), W. Va. 312/J4
Cáceres (lag.), Bolivia 136/G6
Cáceres, Brazil 132/B7
Cáceres, Brazil 120/D4
Cáceres, Colombia 126/C4
Cáceres (prov.), Spain 33/C3
Cáceres, Spain 33/C3
Cáceres, Spain 7/D5
Cachapoal (riv.), Chile 138/G5
Cache (riv.), Ark. 202/H3
Cache (riv.), Ill. 222/D6
Cache, Okla. (73527) 288/J5
Cache (creek), Okla. 288/K6
Cache (co.), Utah 304/C2
Cache Bay, Ontario 177/D1
Cache Creek, Br. Col. 184/G5
Cache Junction, Utah (84304) 304/C2
Cache la Poudre (riv.), Colo. 208/H1
Cacheu, Guinea-Biss. 106/A6
Cachi, Argentina 143/C2
Cachina, Quebrada (riv.), Chile 138/A5
Cachipo, Venezuela 124/G3
Cachoeira, Brazil 132/G6
Cachoeira de Itapemirim, Brazil 120/E5
Cachoeira do Arari, Brazil 132/D3
Cachoeira do Sul, Brazil 132/C10
Cachoeira do Sul, Brazil 120/D6
Cachoeiro de Itapemirim, Brazil 132/G8
Cachorras, Colombia 126/D8
Cachos (pt.), Chile 138/A6
Cachuela Esperanza, Bolivia 136/C2
Cachuma (lake), Calif. 204/F9
Cacocum, Cuba 158/H3
Cacocum, Cuba 156/C2
Cacolo, Angola 115/C6
Caconda, Angola 115/B6
Cacouna, Québec 172/H2
Cactus (range), Nev. 266/E5
Cactus (hills), Sask. 181/F5
Cactus, Texas (79013) 303/B1
Cactus Lake, Sask. 181/B3
Cacuri, Venezuela 124/F5
Cacuso, Angola 115/C5
Čadca, Czech. 41/E2
Caddo (riv.), Ark. 202/D5
Caddo (par.), La. 238/C1
Caddo (lake), La. 238/B1
Caddo (co.), Okla. 288/K4
Caddo, Okla. (74729) 288/O6
Caddo, Texas (76029) 303/F5
Caddo (lake), Texas 303/L5
Caddo Gap, Ark. (71935) 202/C5
Caddo Valley, Ark. (†71923) 202/D5
Caddy Lake, Manitoba 179/G5
Cade, La. (70519) 238/G6
Cadereyta Jiménez, Mexico 150/K4
Cades, S.C. (29518) 296/H4
Cades, Tenn. (†38358) 237/D9
Cades Cove, Tenn. (†37882) 237/O9
Cadet, Mo. (63630) 261/L6
Cadibarrawirracanna (lake), S. Australia 94/D3
Cadillac, Mich. (49601) 250/D4
Cadillac, Québec 174/B3
Cadillac, Sask. 181/D6
Cadiz, Calif. (92319) 204/K9
Cadiz (lake), Calif. 204/K9
Cadiz, Ind. (†47362) 227/G5
Cadiz, Ky. (42211) 237/F7
Cadiz, Ohio (43907) 284/J5
Cadiz, Philippines 82/D5
Cádiz (prov.), Spain 33/D4
Cádiz, Spain 33/C4
Cádiz, Spain 7/D5
Cádiz (gulf), Spain 33/C4
Cádizadiz (gulf), Portugal 33/C4
Cadogan, Alberta 182/E3
Cadogan○, Pa. (16212) 294/C4
Cadomin, Alberta 182/B3
Cadott, Wis. (54727) 317/D6
Cadotte (lake), Alberta 182/B1
Cadotte (riv.), Alberta 182/B1
Cadotte Lake, Alberta 182/B1
Cadron (creek), Ark. 202/F3
Caduruan (pt.), Philippines 82/D5
Cadwell, Georgia (31009) 217/G6
Cadyville, N.Y. (12918) 276/N1
Caen, France 28/C3
Caen, France 7/D4
Caerleon, Wales 13/B6
Caernarfon, Wales 13/C4
Caernarfon, Wales 10/D4
Caernarfon (bay), Wales 13/C4
Caernarfon (bay), Wales 10/D4
Caerphilly, Wales 13/B6
Caerphilly, Wales 10/E5
Caesar, Miss. (†39466) 256/E9
Caesarea, Ontario 177/F3
Caesars Head, S.C. (†29635) 296/B1
Caeté, Brazil 135/E1
Caetité, Brazil 132/F6
Cafayate, Argentina 143/C2
Cafelândia, Brazil 135/B2
Cagayan (prov.), Philippines 82/C1
Cagayan (isls.), Philippines 82/C6
Cagayan (isls.), Philippines 85/F4
Cagayan (riv.), Philippines 82/C2
Cagayancillo, Philippines 82/C6
Cagayan de Oro, Philippines 82/E6
Cagayan de Oro, Philippines 85/G4
Cagayan Sulu (isl.), Philippines 85/F4
Cagayan Sulu (isl.), Philippines 82/B7
Cagle, Tenn. (†37327) 237/L10
Cagles Mill (lake), Ind. 227/D6
Cagli, Italy 34/D3
Cagliari (prov.), Italy 34/B5
Cagliari, Italy 7/E5
Cagliari, Italy 34/B5
Cagliari (gulf), Italy 34/B5
Cagua (vol.), Philippines 82/D1
Cagua, Venezuela 124/E2
Caguán (riv.), Colombia 126/C7
Caguas, P. Rico 161/E2
Caguas, P. Rico 156/G1
Caha (mts.), Ireland 17/B8
Cahaba, Ala. (†36767) 195/D6
Cahaba (riv.), Ala. 195/D5
Cahabón, Guatemala 154/C3
Cahir, Ireland 10/B4
Cahir, Ireland 17/F7
Cahirciveen, Ireland 17/A8
Cahirciveen, Ireland 10/A5
Cahokia, Ill. (62206) 222/A3
Cahone, Colo. (81320) 208/B7
Cahore (pt.), Ireland 17/J6
Cahors, France 28/D5
Cahuapanas, Peru 128/D5
Cahuilla Ind. Res., Calif. 204/J10
Cahuinari (riv.), Colombia 126/E8
Cahuita (pt.), C. Rica 154/F6
Caiapônia, Brazil 132/C7
Caibarién, Cuba 158/F2
Caibarién, Cuba 156/B2
Caibiran, Philippines 82/E5
Caicara, Venezuela 124/G3
Caicara de Orinoco, Venezuela 124/E4
Caicedonia, Colombia 126/C5
Caicó, Brazil 120/F3
Caicó, Brazil 132/G4
Caicos (passage), Bahamas 156/D2
Caicos (bank), Turks & Caicos 156/D2
Caicos (isls.), Turks & Caicos 156/D2
Caicos (passage), Turks & Caicos 156/D2
Caille, Miss. (†38754) 256/C4
Cailloma, Peru 128/G10
Caillou (bay), La. 238/J8
Caimanera, Cuba 158/J4
Caimanera, Cuba 156/C3
Cain (creek), S. Dak. 298/N5
Cainde, Angola 115/B7
Cains (riv.), New Bruns. 170/D2
Cains Store, Ky. (42520) 237/M6
Cainsville, Mo. (64632) 261/E2
Cainsville, Tenn. (†37085) 237/J9
Caird Coast (reg.) 5/B17
Cairnbaan, Scotland 15/C4
Cairnbrook, Pa. (15924) 294/E5
Cairndow, Scotland 15/D4
Cairn Gorm (mt.), Scotland 15/E3
Cairngorm (mts.), Scotland 15/E3
Cairnryan, Scotland 15/D6
Cairns, Australia 87/E7
Cairns, Queensland 95/C3
Cairns, Queensland 88/H3
Cairnsmore (mt.), Scotland 15/D5
Cairn Toul (mt.), Scotland 15/E3
Cairo (gap.), Egypt 102/F2
Cairo (cap.), Egypt 111/J3
Cairo (cap.), Egypt 2/L4
Cairo, Egypt 59/B4
Cairo, Georgia (31728) 217/D9
Cairo, Ill. 188/J3
Cairo, Ill. (62914) 222/D6
Cairo, Kansas (†67035) 232/D4
Cairo, Mo. (65239) 261/H4
Cairo, Nebr. (68824) 264/F3
Cairo, N.Y. (12413) 276/M6
Cairo, Ohio (45820) 284/B4
Cairo, Okla. (†74538) 288/O5
Cairo, W. Va. (26337) 312/D4
Caissie (pt.), New Bruns. 170/F2
Caister-on-Sea, England 13/J5
Caistor, England 13/G4
Caithness (trad. co.), Scotland 15/B4
Caiundo, Angola 102/D6
Caiundo, Angola 115/D7
Caiza, Bolivia 136/C7
Cajabamba, Ecuador 128/C3
Cajabamba, Peru 128/C6
Cajacay, Peru 128/D8
Caja de Muertos (isl.), P. Rico 161/C3
Cajamarca (dept.), Peru 128/C6
Cajamarca, Peru 128/C6
Cajamarca, Peru 120/B3
Cajatambo, Peru 128/D8
Cajazeiras, Brazil 132/G4
Cajidiocan, Philippines 82/D4
Cajuata, Bolivia 136/B5
Cajuru, Brazil 135/C2
Čakovec, Yugoslavia 45/C2
Çal, Turkey 63/C3
Çala, Turkey 63/K2
Calabar, Nigeria 102/D4
Calabar, Nigeria 106/F7
Calabash, N.C. (†29566) 281/M7
Calabazar de Sagua, Cuba 158/E1
Calabogie, Ontario 177/H2
Calabozo, Venezuela 124/E3
Calabria (reg.), Italy 34/F5
Cala Burras (pt.), Spain 33/D4
Calaceite, Spain 33/G2
Calacoto, Bolivia 136/A5
Caladesi (isl.), Fla. 212/B2
Calafat, Romania 45/F3
Calafate, Argentina 143/B7
Calafquén (lake), Chile 138/E3
Calagnaan (isl.), Philippines 82/D5
Calagua (isls.), Philippines 82/D3
Calahoo, Alberta 182/D3
Calahorra, Spain 33/E1
Calais, Alberta 182/B2
Calais, France 28/D2
Calais, France 7/E3
Calais (Dover) (str.), France 28/D2
Calais, Maine 188/N1
Calais, Maine (04619) 243/J5
Calais○, Vt. (05648) 268/B3
Calama, Brazil 132/H10
Calama, Chile 120/C5
Calama, Chile 138/B3
Calamar, Bolívar, Colombia 126/C2
Calamar, Vaupés, Colombia 126/D7
Calamarca, Bolivia 136/A5
Calamba, Laguna, Philippines 82/C3
Calamba, Misamis Occ., Philippines 82/D6
Calamian Group (isls.), Philippines 85/F3
Calamian Group (isls.), Philippines 82/B4
Calamine, Ark. (72418) 202/H1
Calamocha, Spain 33/F2
Calamus, Iowa (52729) 229/M5
Calanasan, Philippines 82/C1
Calancasca (riv.), Switzerland 39/H4
Calanda, Spain 33/F2
Calang, Indonesia 85/B5
Calansho, Serir (des.), Libya 111/D2
Calansho Sand Sea (des.), Libya 111/D2
Calapan, Philippines 82/C4
Calapan, Philippines 85/G3
Calapooia (riv.), Oreg. 291/E3
Calapooya (mts.), Oreg. 291/E4
Călăraşi, Romania 45/H3
Calarcá, Colombia 126/C5
Calasparra, Spain 33/F3
Calatayud, Spain 33/F2
Calatorao, Spain 33/F2
Calauag, Philippines 82/D4
Calaveras (co.), Calif. 204/E5
Calaveras (res.), Calif. 204/L3
Calaveras (lake), Texas 303/K11
Calavite (cape), Philippines 82/C4
Calayan, Philippines 82/A2
Calayan (isl.), Philippines 82/A2
Calbayog, Philippines 82/E4
Calbe, E. Germany 22/D3
Calbuco, Chile 138/D4
Calca, Peru 128/G9
Calcasieu (par.), La. 238/D6
Calcasieu, La. (71433) 238/E4
Calcasieu (lake), La. 238/D7
Calcasieu (passage), La. 238/D7
Calcasieu (riv.), La. 238/E5
Calceta, Ecuador 128/C3
Calchaquí, Argentina 143/F5
Calcis, Ala. (†35178) 195/F4
Calcutta, India 68/F2
Calcutta, India 54/K7
Calcutta, India 2/P4
Calcutta, Ohio (43920) 284/J4
Calcutta, Suriname 131/C3
Caldas (dept.), Colombia 126/C5
Caldas da Rainha, Portugal 33/B3
Caldas Novas, Brazil 132/D7
Calder, Idaho (83808) 220/B2
Calder, Sask. 181/K4
Calder, Loch (lake), Scotland 15/E2
Caldera, Chile 120/B5
Caldera, Chile 138/A6
Calderas (bay), Dom. Rep. 158/D6
Calderas, Venezuela 124/C3
Calderwood, Tenn. (†37801) 237/N9
Caldicot, Wales 13/E6
Çaldıran, Turkey 63/K3
Caldwell, Ark. (72322) 202/J3
Caldwell, Idaho (83605) 220/B6
Caldwell, Idaho 188/C2
Caldwell, Kansas (67022) 232/E4
Caldwell (co.), Ky. 237/F6
Caldwell (par.), La. 238/F2
Caldwell (co.), Mo. 261/E3
Caldwell, N.J. (07006) 273/B2
Caldwell (co.), N.C. 281/F3
Caldwell, Ohio (43724) 284/G6
Caldwell (co.), Texas 303/G8
Caldwell, Texas (77836) 303/H7
Caldwell, W. Va. (24925) 312/F7
Caldwell, Wis. (†53149) 317/J2
Caldy (isl.), Wales 13/C6
Cale, Ark. (71828) 202/D6
Cale, Ind. (†47544) 227/D7
Caledon, N. Ireland 17/H3
Caledon, Ontario 177/E4
Caledon, S. Africa 118/G7
Caledonia, Ill. (61011) 222/E1
Caledonia, Mich. (49316) 250/D6
Caledonia, Minn. (55921) 255/G7
Caledonia, Miss. (39740) 256/H3
Caledonia, Mo. (63631) 261/L7
Caledonia, N.Y. (14423) 276/E5
Caledonia, N. Dak. (58219) 282/S5
Caledonia, Guysborough, Nova Scotia 168/F3
Caledonia, Queens, Nova Scotia 168/C4
Caledonia, Ohio (43314) 284/D4
Caledonia, Pa. (†15868) 294/F3
Caledonia (co.), Vt. 268/C2
Caledonia, Wis. (53108) 317/L2
Caledonian (canal), Scotland 15/D3
Calella, Spain 33/H2
Calenzana, France 28/B6
Calera, Ala. (35040) 195/E4
Calera, Okla. (74730) 288/O7
Calera de Tango, Chile 138/G4
Caleta Barquito, Chile 138/A6
Caleta Clarencia, Chile 138/E10
Caleta Olivia, Argentina 143/C6
Caleta Olivia, Argentina 120/C7
Caleta Pan de Azúcar, Chile 138/A5
Caleu, Chile 138/G2
Caleufú, Argentina 143/C4
Calexico, Calif. (92231) 204/K11
Calf of Man (isl.), I. of Man 13/C3
Calfsound, Scotland 15/F1
Calgary, Alberta 182/C4
Calgary, Alta. 162/E5
Calgary (cap.), Alta.. 146/G4
Calgary, Canada 2/D3
Calhan, Colo. (80808) 208/L4
Calheta, Portugal 33/A2
Calhoun (co.), Ala. 195/G3
Calhoun, Ala. (†36047) 195/F6
Calhoun (co.), Ark. 202/E6
Calhoun (co.), Fla. 212/D6
Calhoun (co.), Georgia 217/C7
Calhoun, Georgia (30701) 217/C1
Calhoun (co.), Ill. 222/C4
Calhoun, Ill. (62419) 222/E5
Calhoun (co.), Iowa 229/D4
Calhoun, Ky. (42327) 237/G5
Calhoun, La. (71225) 238/F2
Calhoun (co.), Mich. 250/D6
Calhoun (co.), Miss. 256/F3
Calhoun, Mo. (65323) 261/E6
Calhoun (co.), S.C. 296/F4
Calhoun, Tenn. (37309) 237/M10
Calhoun (co.), Texas 303/H9
Calhoun (co.), W. Va. 312/D5
Calhoun City, Miss. (38916) 256/F3
Calhoun Falls, S.C. (29628) 296/B3
Cali, Colombia 126/B6
Cali, Colombia 120/B2
Calicito, Cuba 158/H4
Calicoan (isl.), Philippines 82/E5
Calico Rock, Ark. (72519) 202/F1
Calicut (Kozhikode), India 68/D6
Caliente, Nev. (89008) 266/G5
Califon, N.J. (07830) 273/D2
California 188/B3
CALIFORNIA 204
California, Ky. (41007) 237/N3
California, Md. (20619) 245/M7
California (gulf), Mexico 146/G7
California (gulf), Mexico 150/D3
California, Mo. (65018) 261/H5
California, Pa. (15419) 294/C5
California, Trin. & Tob. 161/A11
California (state), U.S. 146/G6
California Aqueduct, Calif. 204/E7
California City, Calif. (93505) 204/H8
California Hot Springs, Calif. (93207) 204/G8
California Junction, Iowa (†51555) 229/B5
Calimete, Cuba 158/D1
Calio, N. Dak. (58322) 282/N2
Calion, Ark. (71724) 202/E7
Calipatria, Calif. (92233) 204/K10
Calistoga, Calif. (94515) 204/C5
Calixa-Lavallée, Québec 172/J4
Calkini, Mexico 150/O6
Çalköy, Turkey 63/C3
Call, Texas (75933) 303/L7
Callabonna (lake), S. Australia 88/G5
Callabonna (lake), S. Australia 94/F3
Callafo, Ethiopia 111/H6
Callahan, Calif. (96014) 204/C2
Callahan, Fla. (32011) 212/E1
Callahan (co.), Texas 303/E5
Callalli, Peru 128/G10
Callan, Ireland 17/G7
Callan, Ireland 10/C4
Callander, Ont. 162/H6
Callander, Ontario 177/E1
Callander, Scotland 10/D2
Callander, Scotland 15/D4
Callands, Va. (24530) 307/J7
Callantsoog, Netherlands 27/F3
Callao, Mo. (63534) 261/G3
Callao (prov.), Peru 128/D9
Callao, Peru 128/D9
Callao, Peru 2/F6
Callao, Peru 120/B4
Callao, Utah (†84034) 304/A4
Callao, Va. (22435) 307/P5
Callapa, Bolivia 136/A5
Callaway, Minn. (56521) 255/C3
Callaway (co.), Mo. 261/J5
Callaway, Nebr. (68825) 264/D3
Callaway, Va. (24067) 307/H7
Calle Larga, Chile 138/G2
Callender, Iowa (50523) 229/E4
Callensburg, Pa. (16213) 294/D3
Callery, Pa. (16024) 294/C4
Calleuque, Chile 138/F5
Calliaqua, St. Vin. & Grens. 161/A9
Callicoon, N.Y. (12723) 276/K7
Callicoon Center, N.Y. (12724) 276/L7
Calliham, Texas (78007) 303/F9
Callimont, Pa. (†15552) 294/E6
Calling (lake), Alberta 182/D2
Callis, Somalia 115/J2
Callison, S.C. (29819) 296/C3
Callosa de Ensarriá, Spain 33/G3
Calloway, Fla. (32401) 212/D6
Calloway (co.), Ky. 237/E7
Calmar, Alberta 182/D3
Calmar, Iowa (52132) 229/K2
Calmer, Ark. (†71665) 202/F6
Calnali, Mexico 150/K6
Calne, England 13/F6
Calobre, Panama 154/G6
Caloosahatchee (riv.), Fla. 212/E5
Caloundra, Queensland 88/J5
Caloundra, Queensland 95/E5
Čalovo, Czech. 41/D3
Calpella, Calif. (95418) 204/B4
Calpet, Wyo. (†83123) 319/B3
Calpulálpan, Mexico 150/M1
Calstock, England 13/C7
Caltagirone, Italy 34/E6

Caltanissetta (prov.), Italy 34/D6
Caltanissetta, Italy 34/D6
Caluire-et-Cuire, France 28/F5
Calulo, Angola 115/C6
Calumet (lake), Ill. 222/C6
Calumet, Iowa (51009) 229/B3
Calumet, La. (†70538) 238/H7
Calumet, Mich. 188/J1
Calumet, Mich. (49913) 250/A1
Calumet, Minn. (55716) 255/E3
Calumet, Okla. (73014) 288/K3
Calumet, Québec 172/C4
Calumet (co.), Wis. 317/K7
Calumet City, Ill. (60409) 222/C6
Calumet Park, Ill. (†60643) 222/C6
Calumetville, Wis. (†53049) 317/K8
Caluquembe, Angola 102/D6
Caluquembe, Angola 115/B6
Calva, Ariz. (†85530) 198/E5
Calvados (dept.), France 28/C3
Calvary, Georgia (31729) 217/D9
Calvary, Ky. (†40033) 237/L6
Calvert, Ala. (36513) 195/B8
Calvert (isl.), Br. Col. 184/C4
Calvert, Kansas (†67622) 232/C2
Calvert (co.), Md. 245/M6
Calvert, Md. (†21911) 245/O2
Calvert, Newf. 166/D2
Calvert, Texas (77837) 303/H7
Calvert City, Ky. (42029) 237/E6
Calvert Hills, North. Terr. 93/E4
Calverton, Md. (†20705) 245/L4
Calverton, Va. (22016) 307/N3
Calverton Park, Mo. (†63101) 261/P2
Calvertville, Ind. (†47424) 227/D6
Calvi, France 28/B6
Calvillo, Mexico 150/H6
Calvin, Ky. (40813) 237/O7
Calvin, La. (71410) 238/E3
Calvin, N. Dak. (58323) 282/N2
Calvin, Okla. (74531) 288/O5
Calvin, W. Va. (26660) 312/E6
Calvinia, S. Africa 102/E8
Calvinia, S. Africa 118/B6
Calwa, Calif. (93745) 204/F7
Calypso, N.C. (28325) 281/N4
Calzada de Calatrava, Spain 33/E3
Camabatela, Angola 115/C5
Camacho, Bolivia 136/C7
Camacupa, Angola 115/C6
Camaguán, Venezuela 124/E3
Camagüey (prov.), Cuba 158/G2
Camagüey, Cuba 158/G3
Camagüey, Cuba 146/L7
Camagüey, Cuba 156/B2
Camagüey (arch.), Cuba 158/G2
Camaiore, Italy 34/C3
Camajuaní, Cuba 158/E2
Camak, Georgia (30807) 217/G4
Camaná, Peru 128/F11
Camanche (res.), Calif. 204/C9
Camanche, Iowa (52730) 229/N5
Camano (isl.), Wash. 310/C2
Camanongue, Angola 115/D6
Camanongue, Angola 102/E6
Camaquã, Brazil 132/C10
Câmara de Lobos, Portugal 33/A2
Çamardı, Turkey 63/F4
Camargo, Bolivia 136/C7
Camargo, Ill. (61919) 222/E4
Camargo, Ky. (†40337) 237/K4
Camargo, Okla. (73835) 288/H2
Camarillo, Calif. (93010) 204/F9
Camarines Norte (prov.), Philippines 82/D3
Camarines Sur (prov.), Philippines 82/D4
Camarón (cape), Honduras 154/E2
Camarones, Argentina 143/C5
Camarones, Chile 138/B2
Camarones (riv.), Chile 138/A2
Camas (co.), Idaho 220/D6
Camas (creek), Idaho 220/D6
Camas (creek), Idaho 220/F5
Camas (creek), Idaho 220/D5
Camas, Wash. (98607) 310/C5
Camas Prairie, Mont. (†59857) 262/B3
Camas Valley, Oreg. (97416) 291/D4
Camatagua, Venezuela 124/E3
Camatindi, Bolivia 136/D7
Ca Mau (Mui Bai Bung) (pt.), Vietnam 72/E5
Cambará, Brazil 135/A3
Cambará, Brazil 132/D8
Cambay, India 68/C4
Cambay (gulf), India 54/J7
Cambay (gulf), India 68/C4
Camberwell, Victoria 88/L7
Camberwell, Victoria 97/J5
Cambodia 2/Q5
Cambodia 54/M8
CAMBODIA (KAMPUCHEA) 72
Camborne-Redruth, England 10/D5
Camborne-Redruth, England 13/B7
Cambra, Pa. (18611) 294/K3
Cambrai, France 28/E2
Cambria, Alberta 182/D4
Cambria, Calif. (93428) 204/D8
Cambria, Ill. (62915) 222/D6
Cambria, Ind. (†46041) 227/D4
Cambria, Iowa (†50060) 229/G7
Cambria, Mich. (†49242) 250/E7
Cambria, Minn. (†56073) 255/D6
Cambria (co.), Pa. 294/E4
Cambria, Wis. (53923) 317/H8
Cambrian (mts.), Wales 13/D5
Cambridge, England 13/G5
Cambridge, England 10/G4
Cambridge, Idaho (83610) 220/B5
Cambridge, Ill. (61238) 222/C2
Cambridge, Iowa (50046) 229/G5
Cambridge, Jamaica 158/H6
Cambridge, Kansas (67023) 232/F4
Cambridge○, Maine (04923) 243/E5
Cambridge, Md. (21613) 245/O6
Cambridge, Mass. (02138) 249/C7
Cambridge (res.), Mass. 249/B6
Cambridge, Minn. (55008) 255/E5
Cambridge, Nebr. (69022) 264/D4
Cambridge, N.Y. (12816) 276/O4
Cambridge, N. Zealand 100/E2
Cambridge, Ohio (43725) 284/G5
Cambridge, Ontario 177/D4
Cambridge, Tasmania 99/D4
Cambridge, Vt. (05444) 268/B2
Cambridge○, Vt. (05444) 268/B2
Cambridge, Wis. (53523) 317/H9
Cambridge Bay, Canada 4/B15
Cambridge Bay, N.W.T. 162/F2
Cambridge Bay, N.W. Terrs. 187/H3
Cambridge City, Ind. (47327) 227/G5
Cambridge-Narrows, New Bruns. 170/E3
Cambridgeport, Vt. (05141) 268/B5
Cambridgeshire (co.), England 13/G5
Cambridge Springs, Pa. (16403) 294/C2
Cambridge Station, Nova Scotia 168/D3
Cambuí, Brazil 135/C3
Cambulo, Angola 115/D5
Cambulo, Angola 102/E5
Cambuslang, Scotland 15/B2
Camden, Ala. (36726) 195/D7
Camden (bay), Alaska 196/K1
Camden, Ark. (71701) 202/E6
Camden, Del. (19934) 245/R4
Camden, England 13/H8
Camden, England 10/B5
Camden (co.), Georgia 217/J9
Camden, Ill. (62319) 222/C3
Camden, Ind. (46917) 227/D3
Camden, Maine (04843) 243/F7
Camden○, Maine (04843) 243/F7
Camden, Mich. (49232) 250/E7
Camden, Miss. (39045) 256/E5
Camden (co.), Mo. 261/G6
Camden, Mo. (64017) 261/D4
Camden, N.J. 188/M3
Camden (co.), N.J. 273/D4
Camden, N.J. (*08101) 273/B3
Camden, N.S. Wales 97/F4
Camden, N.Y. (13316) 276/J4
Camden (co.), N.C. 281/S2
Camden, N.C. (27921) 281/S2
Camden, Ohio (45311) 284/A6
Camden, S.C. (29020) 296/F3
Camden, Tenn. (38320) 237/E8
Camden, Texas (75934) 303/K7
Camden, W. Va. (26338) 312/E4
Camden Haven, N.S. Wales 97/G2
Camden on Gauley, W. Va. (26208) 312/E6
Camden Park, St. Vin. & Grens. 161/A9
Camden Point, Mo. (64018) 261/C4
Camdenton, Mo. (65020) 261/G6
Cameia, Angola 115/D6
Camelford, England 13/C7
Çameli, Turkey 63/C4
Camels Hump (mt.), Vt. 268/B3
Camerino, Italy 34/D3
Cameron, Ariz. (86020) 198/D3
Cameron (peak), Colo. 208/H1
Cameron, Ill. (61423) 222/C3
Cameron (par.), La. 238/D7
Cameron, La. (70631) 238/D7
Cameron, Mo. (64429) 261/D3
Cameron, Mont. (59720) 262/E5
Cameron, N.Y. (14819) 276/F6
Cameron (mts.), N. Zealand 100/A7
Cameron, N.C. (28326) 281/L4
Cameron (isl.), N.W. Terrs. 187/H2
Cameron, Ohio (43914) 284/J6
Cameron, Okla. (74932) 288/T4
Cameron (co.), Pa. 294/F3
Cameron, Pa. (†15834) 294/F3
Cameron, S.C. (29030) 296/F4
Cameron (co.), Texas 303/G11
Cameron, Texas (76520) 303/H7
Cameron, W. Va. (26033) 312/E3
Cameron, Wis. (54822) 317/C5
Cameron Falls, Ontario 177/H5
Cameron Highlands, Malaysia 72/D6
Cameroon 2/K5
Cameroon 102/D4
CAMEROON 115/B2
Cameroon (mt.), Cameroon 102/C4
Cameroon (mt.), Cameroon 115/A3
Camerota, Italy 34/E4
Cametá, Brazil 132/D3
Camiguin (prov.), Philippines 82/E6
Camiguin, Cagayan (isl.), Philippines 82/B3
Camiguin, Camiguin (isl.), Philippines 82/E6
Camiling, Philippines 82/C3
Camilla, Georgia (31730) 217/D8
Camillus, N.Y. (13031) 276/H4
Camiña, Chile 138/B2
Camiña, Quebrada (riv.), Chile 138/B2
Caminha, Portugal 33/B2
Camino, Calif. (95709) 204/E5
Camiri, Bolivia 120/C5
Camiri, Bolivia 136/D7
Camlachie, Ontario 177/B4
Çamlıdere, Turkey 63/E2
Cammack, Ind. (†47302) 227/G4
Cammack Village, Ark. (†72201) 202/E4
Cammal, Pa. (17723) 294/H3
Camoapa, Nicaragua 154/E4
Camocim, Brazil 132/F3
Camolin, Ireland 17/J6
Camooweal, Queensland 88/F3
Camooweal, Queensland 95/A3
Camopi, Fr. Guiana 131/E4
Camopi (riv.), Fr. Guiana 131/E4
Camorta (isl.), India 68/G7
Camoruco, Colombia 126/E4
Camotes (isls.), Philippines 82/E5
Camotes (sea), Philippines 82/E5
Camp (creek), Georgia 217/J2
Camp (creek), Ind. 227/E6
Camp (creek), Oreg. 291/J4
Camp (co.), Texas 303/K5
Campaign, Tenn. (38550) 237/K9
Campamento, Uruguay 145/C1
Campana, Argentina 143/G6
Campana (isl.), Chile 120/B7
Campana (isl.), Chile 138/D7
Campanario, Cerro (mt.), Argentina 143/C4
Campanario, Cerro (mt.), Chile 138/A10
Campanario, Spain 33/D3
Campanha, Brazil 135/D2
Campania, Georgia (†30814) 217/H4
Campania (reg.), Italy 34/E4
Campaspe (riv.), Victoria 97/C5
Campbell, Ala. (36727) 195/C7
Campbell, Alaska (†99901) 196/M2
Campbell, Calif. (95008) 204/K3
Campbell (co.), Ky. 237/N3
Campbell, Minn. (56522) 255/B4
Campbell, Mo. (63933) 261/M9
Campbell, Nebr. (68932) 264/F4
Campbell, N.Y. (14821) 276/F6
Campbell (cape), N. Zealand 100/E4
Campbell, Ohio (44405) 284/J3
Campbell (hill), Ohio 284/C5
Campbell (co.), S. Dak. 298/J2
Campbell (co.), Tenn. 237/N8
Campbell (co.), Va. 307/K6
Campbell (co.), Wyo. 319/G1
Campbell (mt.), Yukon 187/E3
Campbellford, Ontario 177/G3
Campbell Hall, N.Y. (10916) 276/M8
Campbell Hill, Ill. (62916) 222/D6
Campbell Island, Br. Col. 184/C4
Campbellpore, Pakistan 68/C2
Campbell River, Br. Col. 184/E5
Campbells (lake), Oreg. 291/H5
Campbellsburg, Ind. (47108) 227/E7
Campbellsburg, Ky. (40011) 237/L3
Campbellsport, Wis. (†53019) 317/K8
Campbell Station, Ark. (†72473) 202/H2
Campbellsville, Ky. (42718) 237/L6
Campbellton, Fla. (32426) 212/D5
Campbellton, Mo. (†63068) 261/K5
Campbellton, N. Br. 162/K6
Campbellton, N. Br. 146/M5
Campbellton, New Bruns. 170/D1
Campbellton, Newf. 166/D4
Campbellton, Pr. Edward I. 168/D2
Campbellton, Texas (78008) 303/F9
Campbelltown, N.S. Wales 97/F4
Campbelltown, S. Australia 88/E8
Campbelltown, S. Australia 94/B7
Campbell Town, Tasmania 99/D3
Campbeltown, Scotland 15/C5
Campbeltown, Scotland 10/C3
Camp Creek, Alberta 182/C2
Camp Creek, W. Va. (25820) 312/D7
Camp Crook, S. Dak. (57724) 298/B2
Camp David, Md. 245/J2
Camp Dennison, Ohio (45111) 284/D9
Camp Dix, Ky. (41127) 237/P3
Camp Douglas, Wis. (54618) 317/F8
Campeche (state), Mexico 150/O7
Campeche, Mexico 146/J8
Campeche, Mexico 150/O7
Campeche (bay), Mexico 146/J7
Campeche (bank), Mexico 150/O6
Campeche (bay), Mexico 150/N7
Campechuela, Cuba 158/G4
Camper, Manitoba 179/D3
Camperdown, Victoria 97/C6
Camperville, Manitoba 179/B2
Camp Grove, Ill. (61424) 222/D2
Camp Hale, Colo. 208/G4
Camp Hill, Ala. (36850) 195/G5
Camp Hill, Pa. (17011) 294/H5
Camp Hill, Queensland 88/K3
Camp Hill, Queensland 95/E3
Campiglia Marittima, Italy 34/C3
Campile, Ireland 17/H7
Campillo de Altobuey, Spain 33/F3
Campillos, Spain 33/D4
Campina Grande, Brazil 120/F3
Campina Grande, Brazil 132/G4
Campinas, Brazil 120/E5
Campinas, Brazil 135/C3
Campinas, Brazil 132/E8
Campina Verde, Brazil 135/B1
Campina Verde, Brazil 132/D7
Camp Lake, Wis. (53109) 317/K10
Camp Lejeune Marine Corps Base, N.C. 281/P5
Campli, Italy 34/D3
Camp Morton, Manitoba 179/E4
Camp Nelson, Calif. (93208) 204/G7
Campo, Calif. (92006) 204/J11
Campo, Cameroon 115/B3
Campo, Colo. (81029) 208/O8
Campoalegre, Colombia 126/C6
Campobasso (prov.), Italy 34/E4
Campobasso, Italy 34/E4
Campobello (isl.), New Bruns. 170/D4
Campobello, S.C. (29322) 296/C1
Campo Belo, Brazil 132/E8
Campo Belo, Brazil 135/D2
Campo Claro, Venezuela 124/G2
Campo de Criptana, Spain 33/E3
Campo de la Cruz, Colombia 126/C2
Campo Florido, Brazil 135/B1
Campo Formoso, Brazil 132/F5
Campo Grande, Brazil 120/D5
Campo Grande, Brazil 132/8
Campo Ind. Res., Calif. 204/J11
Campo Largo, Brazil 135/B4
Campo Maior, Brazil 132/F4
Campo Maior, Portugal 33/C3
Campos, Brazil 132/F8
Campos, Brazil 135/F2
Campos, Brazil 120/E5
Campos (reg.), Brazil 120/E4
Campos Altos, Brazil 135/C1
Campo Seco, Calif. (95226) 204/D9
Campo Tencia (peak), Switzerland 39/G4
Campo Tures, Italy 34/C1
Camp Pendleton, Calif. 204/H10
Camp Perrin, Haiti 158/A6
Camp Point, Ill. (62320) 222/B3
Camp Robinson, Ontario 177/G4
Camp Robinson, Ontario 175/B2
Camp Sherman, Oreg. (97730) 291/F3
Camp Springs, Md. (20748) 245/G6
Campti, La. (71411) 238/D3
Campton, Georgia (†30655) 217/E3
Campton, Ky. (41301) 237/O5
Campton○, N.H. (03223) 268/D4
Camptown, Pa. (18815) 294/K2
Campus, Ill. (60920) 222/E2
Camp Verde, Ariz. (86322) 198/D4
Campville, Fla. (†32640) 212/D2
Camp Wood, Texas (78833) 303/D8
Cam Ranh, Vietnam 72/F5
Cam Ranh, Vinh (bay), Vietnam 72/F5
Camrose, Alberta 182/D3
Camrose, Alta. 162/E5
Camsell (riv.), N.W. Terrs. 187/G3
Camsell Portage, Sask. 181/L2
Camuy, P. Rico 161/B1
Camuy, P. Rico 156/F1
Camuy (riv.), P. Rico 161/B1
Çan, Turkey 63/B2
Caña, Czech. 41/F2
Cana (pt.), Dom. Rep. 158/F6
Cana, Sask. 181/J5
Cana, Va. (24317) 307/G7
Canaan, Conn. (06018) 210/B1
Canaan○, Conn. (06018) 210/B1
Canaan (mt.), Conn. 210/B1
Canaan, Ind. (47224) 227/G7
Canaan○, Maine (04924) 243/D6
Canaan, Miss. (38612) 256/F1
Canaan, New Bruns. 170/E2
Canaan (riv.), New Bruns. 170/E2
Canaan○, N.H. (03741) 268/C4
Canaan, N.Y. (12029) 276/O6
Canaan○, Vt. (05903) 268/D2
Canaan Center, N.H. (†03741) 268/C4
Canaan Forks, 170/E2
Canaan Road, New Bruns. 170/E2
Canada 2/D3
Canada 4/C13
Canada 146/G4
CANADA, 163
Cañada, La (mt.), Cuba 158/B2
Canada, Ky. (41519) 237/S5
Canada (bay), Newf. 166/C3
Cañada de Gómez, Argentina 143/F6
Cañada Nieto, Uruguay 145/B4
Canadensis, Pa. (18325) 294/M3
Canadian (riv.) 188/F3
Canadian (riv.), N. Mex. 274/F3
Canadian (co.), Okla. 288/K3
Canadian, Okla. (74425) 288/P4
Canadian (riv.), Okla. 288/O4
Canadian, Texas (79014) 303/D2
Canadian (riv.), Texas 303/D1
Canadian (riv.), U.S. 146/H6
Canadian City, Okla. (†73064) 288/L4
Canadice (lake), N.Y. 276/F5
Canadys, S.C. (29433) 296/F5
Canagua (riv.), Venezuela 124/C3
Canajoharie, N.Y. (13317) 276/L5
Çanakkale (prov.), Turkey 63/B2
Çanakkale, Turkey 63/B6
Çanakkale, Turkey 59/A2
Çanakkale Boğazı (Dardanelles) (str.), Turkey 63/B6
Çanakkale Boğazı (str.), Turkey 59/A2
Canal (creek), Alberta 182/E5
Canala, New Caled. 86/G4
Canala (bay), New Caled. 86/H4
Canal Flats, Br. Col. 184/K5
Canal Fulton, Ohio (44614) 284/H4
Canalou, Mo. (63828) 261/N9
Canal Point, Fla. (33438) 212/F5
Canals, Argentina 143/D3
Canal Winchester, Ohio (43110) 284/E6
Canandaigua, N.Y. (14424) 276/F5
Canandaigua (lake), N.Y. 276/F5
Cananea, Mexico 150/D1
Cananéia, Brazil 132/E9
Cananéia, Brazil 135/C4
Canancova, Cuba 158/K3
Cañar (prov.), Ecuador 128/C4
Cañar, Ecuador 128/C4
Canaries, St. Lucia 161/G6
Canaries, Piton (mt.), St. Lucia 161/G6
Canarreos, Los (arch.), Cuba 158/C2
Canary, Oreg. (†97493) 291/D4
Canary (isls.), Spain 102/A2
Canary (isls.), Spain 2/H4
Canary (isls.), Spain 33/B4
Canary (isls.), Spain 106/A3
Cañas, C. Rica 154/E5
Cañas, Cuba 158/B1
Cañas (range), Uruguay 145/C2
Canaseraga, N.Y. (14822) 276/E6
Cañasgordas, Colombia 126/B4
Canasí, Cuba 158/C1
Canastota, N.Y. (13032) 276/J4
Canatlán, Mexico 150/G4
Canaveral (cape), Fla. 146/L7
Canaveral (Kennedy) (cape), Fla. 188/L5
Canaveral (cape), U.S. 2/F4
Canavieiras, Brazil 132/G6
Cañazas, Panama 154/G6
Canbelego, N.S. Wales 97/D2
Canberra (cap.), Australia 87/F9
Canberra (cap.), Australia 2/S7
Canberra (cap.), Australia, Aust. Cap. Terr. 97/E4
Canby, Calif. (96015) 204/E2
Canby, Minn. (56220) 255/B6
Canby, Oreg. (97013) 291/B2
Cancún, Mexico 150/Q6
Candala, Somalia 115/J1
Candarave, Peru 128/G11
Çandarlı (gulf), Turkey 63/B3
Candás, Spain 33/D1
Candéias, Brazil 120/F4
Candela, Mexico 150/J3
Candelaria, Bolivia 136/F5
Candelaria (riv.), Bolivia 136/F5
Candelaria, Cuba 158/B1
Candelaria, Mexico 150/O7
Candelaria (riv.), Mexico 150/O8
Candelaria, Philippines 82/B3
Candelaria, Texas (†79843) 303/C12
Candelaria, Venezuela 124/F4
Candeleda, Spain 33/D2
Candelero (pt.), P. Rico 161/F2
Candelo, N.S. Wales 97/E5
Candia (Iráklion), Greece 45/G8
Candia○, N.H. (03034) 268/E5
Candiac, Québec 172/J4
Candiac, Sask. 181/H5
Cândido Mendes, Brazil 132/E3
Çandır, Turkey 63/F3
Candle (lake), Sask. 181/F2
Candle Lake, Sask. 181/F2
Candler, Fla. (32624) 212/E2
Candler (co.), Georgia 217/H6
Candler, N.C. (28715) 281/D3
Candlewood (lake), Conn. 210/A2
Candlewood, N.J. (†08701) 273/E3
Cando, N. Dak. (58324) 282/M3
Cando, Sask. 181/B3
Candon, Philippines 82/C2
Candor, N.Y. (13743) 276/H6
Candor, N.C. (27229) 281/K4
Cane (creek), Utah 304/E5
Canea (Khaniá), Greece 45/G8
Caneadea, N.Y. (14717) 276/D6
Canebay, Virgin Is. (U.S.) 161/E3
Cane Beds, Ariz. (†86022) 198/B2
Canebrake, W. Va. (24819) 312/C8
Caneel (bay), Virgin Is. (U.S.) 161/B4
Canehill, Ark. (72717) 202/B2
Canelones (dept.), Uruguay 145/D5
Canelones, Uruguay 120/D6
Canelones, Uruguay 145/B6
Canelos, Ecuador 128/D3
Canendiyu (dept.), Paraguay 144/E4
Cañete (riv.), 128/D9
Cañete, Chile 138/D2
Cañete, Spain 33/F2
Cane Valley, Ky. (42720) 237/L6
Caney, Kansas (67333) 232/G4
Caney, Ky. (41407) 237/P5
Caney, Okla. (74533) 288/O6
Caney (riv.), Okla. 288/O1
Caney Fork (riv.), Tenn. 237/L9
Caneyville, Ky. (42721) 237/J6
Canfield, Ark. (71829) 202/C7
Canfield, Ohio (44406) 284/J3
Canford, Br. Col. 184/G5
Cangallo, Peru 128/E9
Cangamba, Angola 115/C6
Cangas, Spain 33/B1
Cangas de Narcea, Spain 33/C1
Cangas de Onís, Spain 33/D1
Canguaretama, Brazil 132/H4
Cangyuan, China 77/E7
Cangzhou (Tsangchow), China 77/J4
Caniapiscau (riv.), Que. 162/K4
Caniapiscau, Québec 174/D1
Caniapiscau (res.), Québec 174/D2
Caniapiscau (riv.), Québec 174/D1
Canicattì, Italy 34/E6
Canigao (chan.), Philippines 82/E5
Canik (mts.), Turkey 63/G2
Caniles, Spain 33/E4
Canim (lake), Br. Col. 184/G4
Canim Lake, Br. Col. 184/G4
Canindé, Brazil 132/G4
Canisteo (res.), N.J. 273/E1
Canisteo, N.Y. (14823) 276/E6
Canisteo (riv.), N.Y. 276/F6
Canistota, S. Dak. (57012) 298/P6
Cañitas de Felipe Pescador, Mexico 150/H5
Canjáyar, Spain 33/E4
Canje (riv.), Guyana 131/C2
Canjilon, N. Mex. (87515) 274/C2
Çankaya, Turkey 63/E3
Çankırı (prov.), Turkey 63/E2
Çankırı, Turkey 63/E2
Çankırı, Turkey 59/B1
Cankton, La. (†70584) 238/F6
Canlaon, Philippines 82/D5
Canlaon (peak), Philippines 82/D5
Canmer, Ky. (42722) 237/K6
Canmore, Alberta 182/C4
Canna (isl.), Scotland 10/C2
Canna (isl.), Scotland 15/B3
Canna (sound), Scotland 15/B3
Cannalville, Ohio (†43777) 284/F6
Cannanore, India 68/C6
Cannelburg, Ind. (47519) 227/C7
Cannel City, Ky. (41408) 237/P5
Cannelles (pt.), St. Lucia 161/G7
Cannelles (riv.), St. Lucia 161/G6
Cannelton, Ind. (47520) 227/D9
Cannes, France 28/G6
Cannich, Scotland 15/D3
Canning (riv.), Alaska 196/J1
Canning, Nova Scotia 168/D3
Canning, S. Dak. (†57501) 298/K5
Canning (riv.), W. Australia 88/B3
Canning (res.), W. Australia 88/B2
Canning, W. Australia 92/A1
Cannington, Ontario 177/E3
Cannington Manon Hist. Park, Sask. 181/J6
Cannock, England 10/G2
Cannock, England 13/E5
Cannon, Del. (19935) 245/R6
Cannon (mt.), N.H. 268/D3
Cannon (co.), Tenn. 237/J9
Cannon A.F.B., N. Mex. 274/F4
Cannon Ball, N. Dak. (58528) 282/J7
Cannonball (riv.), N. Dak. 282/G7
Cannon Beach, Oreg. (97110) 291/D2
Cannondale, Conn. (†06897) 210/B4
Cannon Falls, Minn. (55009) 255/F6
Cannonsburg, Miss. (†39120) 256/B7
Cannonsville (res.), N.Y. 276/K6
Cannonville, Utah (84718) 304/B6
Cann River, Victoria 97/E5
Caño (isl.), C. Rica 154/F6
Canoas, Brazil 132/D10
Canoas, Brazil 120/D5
Canobie Lake, N.H. (†03079) 268/E6
Caño Capure (riv.), Venezuela 124/H3
Canoe, Ala. (†36502) 195/D8
Canoe (riv.), Br. Col. 184/H4
Canoe (lake), Sask. 181/L3
Canoe Lake, Sask. 181/L3
Canoe River, Br. Col. 184/H4
Canoga Park, Calif. (*91303) 204/B10
Canoinhas, Brazil 132/D9
Caño Macareo (riv.), Venezuela 124/H3
Caño Mánamo (riv.), Venezuela 124/G3
Canon, Georgia (30520) 217/F2
Canonbie, Scotland 15/F5
Canonchet, R.I. (†02833) 249/H7
Canon City, Colo. (81212) 208/J6
Canones, N. Mex. (87516) 274/C2
Canonsburg, Pa. (15317) 294/B5
Canoochee, Georgia (30416) 217/H5
Canoose Flowage (lake), New Bruns. 170/C3
Canora, Sask. 181/J4
Canosa di Puglia, Italy 34/E4
Canouan (isl.), St. Vin. & Grens. 156/G4
Canova, S. Dak. (57321) 298/O6
Canovanas (riv.), P. Rico 161/E1
Canowindra, N.S. Wales 97/E3
Canquella, Bolivia 136/A7
Cansado, Mauritania 106/A4
Canso, Nova Scotia 168/H3
Canso (cape), Nova Scotia 168/H3
Canso (str.), Nova Scotia 168/G3
Canta, Peru 128/D8
Cantabrian (range), Spain 33/C1
Cantagalo, Brazil 135/E3
Cantal (dept.), France 28/E5
Cantal, Sask. 181/K6
Cantalejo, Spain 33/D2
Cantanhede, Portugal 33/B2
Cantaura, Venezuela 124/F3
Canterbury○, Conn. (06331) 210/H2
Canterbury, Del. (†19943) 245/R4
Canterbury, England 10/G5
Canterbury, England 13/H6
Canterbury, New Bruns. 170/C3
Canterbury○, N.H. (03224) 268/D5
Canterbury, N.S. Wales 88/K4
Canterbury, N.S. Wales 97/J3
Canterbury (bight), N. Zealand 100/D6
Can Tho, Vietnam 72/E5
Can Tho, Vietnam 54/M9
Cantil, Calif. (93519) 204/H8
Cantiles (cay), Cuba 158/C3
Cantillana, Alto de (mt.), Chile 138/G4
Cantley, Québec 172/B4
Canto del Agua, Chile 138/A7
Canto do Buriti, Brazil 132/F5
Canton (Guangzhou), China 77/H7
Canton, China 54/N7
Canton, China 2/Q4
Canton, Conn. (06019) 210/D1
Canton○, Conn. (06019) 210/D1
Canton, Georgia (30114) 217/C2
Canton, Ill. (61520) 222/C3
Canton, Ind. (†47167) 227/E7
Canton, Kansas (67428) 232/E3
Canton, Ky. (42212) 237/F7
Canton (isl.), Kiribati 87/J6
Canton○, Maine (04221) 243/C7
Canton○, Mass. (02021) 249/C8
Canton, Minn. (55922) 255/F7
Canton, Miss. (39046) 256/D5
Canton, Mo. (63435) 261/J2
Canton, N.J. (†08079) 273/C5
Canton, N.Y. (13617) 276/K1
Canton, N.C. (28716) 281/D3
Canton (Hensel), N. Dak. (†58241) 282/P2
Canton, Ohio 188/K2
Canton, Ohio (*44701) 284/H4
Canton, Okla. (73724) 288/J2
Canton (lake), Okla. 288/J2
Canton, Pa. (17724) 294/J2
Canton, S. Dak. (57013) 298/R7
Canton, Texas (75103) 303/J5
Canton, Wis. (54823) 317/C5
Canton-Bégin, Québec 172/F1
Canton Bend, Ala. (†36726) 195/D6
Canton Center, Conn. (06020) 210/D1
Cantonment, Fla. (32533) 212/B6
Canton-Patapédia, Québec 172/B2
Cantoria, Spain 33/E4
Cantrall, Ill. (62625) 222/D4
Cantril, Iowa (52542) 229/J7
Cantù, Italy 34/B2
Cantuar, Sask. 181/C5
Cantwell, Alaska (99729) 196/J2
Canuck, Sask. 181/C6
Cañuelas, Argentina 143/G7
Canumã (riv.), Brazil 132/B4
Canutama, Brazil 132/G9
Canute, Okla. (73626) 288/H4
Canutillo, Texas (79835) 303/A10
Canvas, W. Va. (26662) 312/E6
Canvey Island, England 13/J8
Canvey Island, England 10/G5
Canwood, Sask. 181/E2
Canyon (lake), Ariz. 198/D5
Canyon, Br. Col. 184/J5
Canyon, Calif. (94516) 204/K2
Canyon (co.), Idaho 220/B6
Canyon (creek), Idaho 220/C6
Canyon, Minn. (55717) 255/F3
Canyon, Texas (79015) 303/C3
Canyon, Wyo. (82190) 319/B1
Canyon Blanco (creek), N. Mex. 274/B2
Canyon City, Oreg. (97820) 291/J3
Canyon Creek, Alberta 182/C2
Canyon Creek, Mont. (59633) 262/D4
Canyon de Chelly Nat'l Mon., Ariz. 198/F2
Canyon Ferry, Mont. (†59601) 262/E4
Canyon Ferry (lake), Mont. 262/E4
Canyonlands Nat'l Park, Utah 304/D5
Canyonville, Oreg. (97417) 291/D5
Cao Bang, Vietnam 72/E2
Caol, Scotland 15/C4
Cao Lanh, Vietnam 72/E5
Caonao, Cuba 158/E2
Caonillas (lake), P. Rico 161/C2

Cap (isl.), Philippines 82/C8
Cap (pt.), St. Lucia 161/G5
Capa, S. Dak. (57525) 298/H5
Capac, Mich. (48014) 250/G5
Capachica, Peru 128/H10
Çapakçur, Turkey 59/D2
Çapakçur, Turkey 63/J3
Cap-à-l'Aigle, Québec 172/G2
Capalonga, Philippines 82/D3
Capanaparo (riv.), Venezuela 124/E4
Capanema, Brazil 132/E3
Capannori, Italy 34/C3
Capão Bonito, Brazil 132/D9
Capão Bonito, Brazil 135/B4
Caparica, Portugal 33/A1
Caparo (riv.), Venezuela 124/C4
Capasin, Sask. 181/D2
Capatárida, Venezuela 124/C2
Capay, Calif. (95607) 204/C5
Cap-Bateau, New Bruns. 170/F1
Cap-Chat, Que. 162/K6
Cap-Chat, Québec 172/B1
Cap-Chat, Québec 174/D3
Cap-de-la-Madeleine, Québec 172/E3
Cap-des-Rosiers, Québec 172/D1
Cap d'Or (cape), Nova Scotia 168/D3
Cape (pen.), S. Africa 118/E7
Cape (pt.), S. Africa 118/F7
Cape (isl.), S.C. 296/J5
Cape Barren (isl.), Tasmania 88/H8
Cape Barren (isl.), Tasmania 99/E2
Cape Breton (isl.), N.S. 146/N5
Cape Breton (isl.), N.S. 162/K6
Cape Breton (co.), Nova Scotia 168/H3
Cape Breton (isl.), Nova Scotia 168/J2
Cape Breton Highlands Nat'l Park, Nova Scotia 168/H2
Cape Broyle, Newf. 166/D2
Cape Canaveral, Fla. (32920) 212/F3
Cape Carteret, N.C. (†28584) 281/P5
Cape Charles, Newf. 166/C3
Cape Charles, Va. (23310) 307/R6
Cape Coast, Ghana 106/D7
Cape Coast, Ghana 102/B4
Cape Cod (bay), Mass. 249/N5
Cape Cod (canal), Mass. 249/N5
Cape Cod Nat'l Seashore, Mass. 249/P5
Cape Coral, Fla. (33904) 212/E5
Cape Dorset, N.W.T. 162/J3
Cape Dorset, N.W. Terrs. 187/L3
Cape Dyer, N.W. Terrs. 187/M3
Cape Fanshaw, Alaska (†99833) 196/N1
Cape Fear (riv.), N.C. 188/L4
Cape Fear (riv.), N.C. 281/M5
Cape George, Nova Scotia 168/F3
Cape Girardeau (co.), Mo. 261/N8
Cape Girardeau, Mo. (63701) 261/O8
Cape Girardeau, Mo. 188/H3
Cape Hatteras Nat'l Seashore, N.C. 281/T4
Cape Horn (mt.), Idaho 220/C5
Cape Krusenstern Nat'l Mon., Alaska 196/F1
Capel, W. Australia 92/A2
Capela, Brazil 132/G5
Cape Lisburne, Alaska (†99766) 196/E1
Capella, Queensland 95/D4
Capella (isls.), Virgin Is. (U.S.) 161/B5
Capelle, Netherlands 27/F5
Capelongo, Angola 115/C6
Cape Lookout Nat'l Seashore, N.C. 281/S5
Cape May (co.), N.J. 273/D5
Cape May, N.J. (08204) 273/D6
Cape May Coastguard Ctr., N.J. 273/D6
Cape May Court House, N.J. (08210) 273/D5
Cape May Point, N.J. (08212) 273/D6
Capenda-Camulemba, Angola 115/C5
Capenda-Camulemba, Angola 102/D5
Cape Neddick, Maine (03902) 243/B9
Cape Negro (isl.), Nova Scotia 168/C5
Cape North, Nova Scotia 168/H2
Cape of Good Hope (prov.), S. Africa 102/E8
Cape of Good Hope (prov.), S. Africa 118/C6
Cape Pole, Alaska (†99901) 196/M2
Cape Porpoise, Maine (04014) 243/C9
Cape Ray, Newf. 166/C4
Capers (isl.), S.C. 296/H6
Cape Sable (isl.), Nova Scotia 168/C5
Cape Saint Claire, Md. (21401) 245/N4
Cape Smith, N.W. Terrs. 187/L3
Capesterre, Basse-Terre, Guadeloupe 161/A7
Capesterre, Marie-Galante, Guadeloupe 161/B7
Cape Tormentine, New Bruns. 170/G2
Cape Town (cap.), S. Africa 2/L7
Cape Town (cap.), S. Africa 102/D8
Cape Town (cap.), S. Africa 118/E6
Cape Verde 2/H5
CAPE VERDE 106/A8
Capeville, Va. (23313) 307/R6
Cape Vincent, N.Y. (13618) 276/H2
Cape Yakataga, Alaska (99560) 196/K2
Cape York (pen.), Australia 87/E7
Cape York, Queensland 95/B1
Cape York (pen.), Queensland 88/G2
Cape York (pen.), Queensland 95/B2
Cap-Haïtien, Haiti 158/C5
Cap-Haïtien, Haiti 156/D3
Capiatá, Paraguay 144/B4
Capibara, Venezuela 124/E6
Capilla de Farruco, Uruguay 145/D3
Capim (riv.), Brazil 132/D3
Capinota, Bolivia 136/B5
Capira, Panama 154/G6
Capirenda, Bolivia 136/D7
Capistrano Beach, Calif. (92624) 204/H10
Capitan, N. Mex. (88316) 274/D5
Capitan (mts.), N. Mex. 274/D5
Capitan (peak), N. Mex. 274/D5
Capitán Aracena (isl.), Chile 138/E10
Capitán Bado, Paraguay 144/E3
Capitan Grande Ind. Res., Calif. 204/J11
Capitán Meza, Paraguay 144/E5
Capitán Pastene, Chile 138/D2
Capitán Ustarés, Cerro (mt.), Bolivia 136/E6
Capitol, Mont. (59319) 262/M5
Capitola, Calif. (95010) 204/K4
Capitola, Fla. (†32302) 212/B1
Capitol Heights, Md. (20743) 245/G5
Capitol Hill (cap.), No. Marianas 87/E4
Capitol Reef Nat'l Park, Utah 304/C5
Capiz (prov.), Philippines 82/D5
Caplan, Québec 172/C2
Capleville, Tenn. (†38101) 237/B10
Caplin Cove, Newf. 166/D2
Caplinger Mills, Mo. (65607) 261/E7
Čapljina, Yugoslavia 45/C4
Cap Lumière, New Bruns. 170/F2
Capon Bridge, W. Va. (26711) 312/K4
Capon Springs, W. Va. (26823) 312/K4
Capotoan (mt.), Philippines 82/E4
Cappahayden, Newf. 166/D2
Cappamore, Ireland 17/E6
Cappawhite, Ireland 17/E6
Cap-Pelé, New Bruns. 170/F2
Cappoquin, Ireland 17/F7
Capps, Ala. (†36353) 195/H8
Capraia (isl.), Italy 34/B3
Capreol, Ontario 175/D3
Capreol, Ontario 177/K5
Capri (isl.), Italy 34/E4
Capricorn (chan.), Queensland 95/D4
Capricorn Group (isls.), Queensland 88/J4
Capricorn Group (isls.), Queensland 95/E4
Caprivi Strip (reg.), Namibia 102/E6
Caprivi Strip (reg.), Namibia 118/C3
Caprock, N. Mex. (88213) 274/F5
Capron, Ill. (61012) 222/E1
Capron, Okla. (73725) 288/J1
Capron, Va. (23829) 307/O7
Cap-Rouge, Québec 172/H3
Cap-Saint-Ignace, Québec 172/G2
Cap-Santé, Québec 172/F3
Cap-Seize, Québec 172/C1
Capshaw, Ala. (35742) 195/E1
Capstan (cape), Nova Scotia 168/D3
Capstick, Nova Scotia 168/H1
Captain Bermúdez, Argentina 143/F6
Captain Cook, Hawaii (96704) 218/G5
Captains Flat, N.S. Wales 97/E4
Captieux, France 28/C5
Captina (creek), Ohio 284/J6
Captiva, Fla. (33924) 212/D5
Captiva (isl.), Fla. 212/D5
Capua, Italy 34/E4
Capuchin (cape), Dominica 161/E5
Capulhuac de Mirafuentes, Mexico 150/K1
Capulin, Colo. (81124) 208/G8
Capulin, N. Mex. (88414) 274/F2
Capulin Mountain Nat'l Mon., N. Mex. 274/E2
Caputa, S. Dak. (57725) 298/D5
Caquetá (inten.), Colombia 126/C7
Caquetá (riv.), Colombia 120/B2
Caquetá (riv.), Colombia 126/E8
Caquiaviri, Bolivia 136/A5
Carabao (isl.), Philippines 82/D4
Carabelas, Argentina 143/F6
Carabobo (state), Venezuela 124/D2
Carabobo, Bolívar, Venezuela 124/H4
Carabobo, Carabobo, Venezuela 124/D3
Carabuco, Bolivia 136/A4
Caracal, Romania 45/G3
Caracaraí, Brazil 120/C2
Caracas (bay), Neth. Ant. 161/G9
Caracas (cap.), Venezuela 120/C2
Caracas (cap.), Venezuela 124/E2
Caracas (cap.), Venezuela 2/F5
Carache, Venezuela 124/C3
Caracollo, Bolivia 136/B5
Caraga, Philippines 82/F7
Caragabal, N.S. Wales 97/D3
Caraguatá, Uruguay 145/D3
Caraguatá (riv.), Uruguay 145/D3
Caraguatatuba, Brazil 135/D3
Caraguatay, Paraguay 144/B4
Carahue, Chile 138/D2
Carajás, Serra dos (range), Brazil 132/D4
Caramat, Ontario 177/H5
Caramat, Ontario 175/C3
Caramoan, Philippines 82/D4
Caranavi, Bolivia 136/B4
Carandaí, Brazil 135/E2
Carandaiti, Bolivia 136/D7
Carandotta, Queensland 95/A4
Carangola, Brazil 135/E2
Caransebeş, Romania 45/F3
Carapa (riv.), Paraguay 144/E4
Carapa, Venezuela 124/G3
Caraparaná (riv.), Colombia 126/D8
Caraparí, Bolivia 136/D7
Carapeguá, Paraguay 144/B5
Carapichaima, Trin. & Tob. 161/B10
Caraquet, New Bruns. 170/E1
Caraquet (isl.), New Bruns. 170/F1
Carás, Peru 128/D7
Caratasca, Honduras 154/F3
Caratasca (cays), Honduras 154/F2
Caratasca (lag.), Honduras 154/F3
Caratinga, Brazil 132/F7
Caratinga, Brazil 135/E1
Caratunk, Maine (04925) 243/C5
Caratunk○, Maine (04925) 243/C5
Carauari, Brazil 120/C3
Carauari, Brazil 132/G9
Caraúbas, Brazil 132/G4
Caravaca de la Cruz, Spain 33/E3
Caravaggio, Italy 34/B2
Caravelas, Brazil 132/G7
Caraveli, Peru 128/F10
Caravelle (pen.), Martinique 161/D6
Caraway, Ark. (72419) 202/K2
Carayaó, Paraguay 144/C4
Carazinho, Brazil 132/C10
Carballino, Spain 33/B1
Carballo, Spain 33/B1
Carberry, Manitoba 179/C5
Carbo, Mexico 150/D2
Carbon, Alberta 182/D4
Carbon (peak), Colo. 208/E5
Carbon, Ind. (47837) 227/C5
Carbon, Iowa (50839) 229/D6
Carbon (co.), Mont. 262/G5
Carbon (co.), Pa. 294/L4
Carbon, Texas (76435) 303/F5
Carbon (co.), Utah 304/D4
Carbon, W. Va. (25037) 312/D6
Carbon (co.), Wyo. 319/F4
Carbonado, Wash. (98323) 310/D3
Carbonara (cape), Italy 34/B5
Carbon Cliff, Ill. (61239) 222/C2
Carbondale, Alberta 182/D3
Carbondale, Colo. (81623) 208/E4
Carbondale, Ill. (62901) 222/D6
Carbondale, Kansas (66414) 232/G3
Carbondale, Ohio (45717) 284/F7
Carbondale, Pa. (18407) 294/L2
Carbonear, Newf. 166/D2
Carbon Hill, Ala. (35549) 195/D3
Carbon Hill, Ill. (†60416) 222/E2
Carbon Hill, Ohio (43111) 284/F7
Carbonia, Italy 34/B5
Carbonton, N.C. (†27330) 281/L3
Carbost, Scotland 15/B3
Carbury, Ireland 17/H5
Carbury, N. Dak. (58724) 282/J2
Carcagente, Spain 33/F3
Carcans (lake), France 28/C5
Carcarañá, Argentina 143/F6
Carcarañá (riv.), Argentina 143/F6
Carcassonne, France 28/D6
Carchi (prov.), Ecuador 128/C2
Carcoar, N.S. Wales 97/E3
Carcross, Yukon 187/E3
Çardak, Turkey 63/C6
Cardal, Uruguay 145/C5
Cardale, Manitoba 179/B4
Cárdenas, Cuba 158/D1
Cárdenas, Cuba 156/B2
Cárdenas (bay), Cuba 158/D1
Cárdenas, San Luis Potosí, Mexico 150/K6
Cárdenas, Tabasco, Mexico 150/N8
Cardenden, Scotland 15/D1
Cardiel (lake), Argentina 143/B6
Cardiff, Ala. (35041) 195/E3
Cardiff, Md. (21024) 245/N2
Cardiff, Wales 7/D3
Cardiff, Wales 13/B7
Cardiff, Wales 10/E5
Cardiff-by-the-Sea, Calif. (92007) 204/H10
Cardigan (mt.), N.H. 268/D4
Cardigan, Pr. Edward I. 168/F2
Cardigan (bay), Pr. Edward I. 168/F2
Cardigan, Wales 13/C5
Cardigan, Wales 10/D4
Cardigan (bay), Wales 10/D4
Cardigan (bay), Wales 13/C5
Cardin, Okla. (74335) 288/S1
Cardinal (lake), Alberta 182/B1
Cardinal, Manitoba 179/D5
Cardinal, Ontario 177/J3
Cardington, Ohio (43315) 284/E5
Cardona, Uruguay 145/B4
Cardoso (isl.), Brazil 135/C4
Cardozo, Uruguay 145/C3
Cardross, Sask. 181/F6
Cardston, Alberta 182/D5
Cardston, Alta. 162/E6
Cardville, Maine (04418) 243/F5
Cardwell, Mo. (63829) 261/M10
Cardwell, Mont. (59721) 262/E5
Cardwell, Queensland 95/C3
Cardwell, Va. (†23039) 307/N5
Carefree, Ariz. (85331) 198/C5
Carefree, Ind. (†47137) 227/E8
Carei, Romania 45/F2
Carén, Chile 138/A8
Carencro, La. (70520) 238/G6
Carentan, France 28/C3
Caretta, W. Va. (24821) 312/C8
Carey, Idaho (83320) 220/E6
Carey, Ohio (43316) 284/D4
Carey (lake), W. Australia 88/C5
Carey (lake), W. Australia 92/C5
Careywood, Idaho (83809) 220/B1
Cargill, Ontario 177/C3
Carhuás, Peru 128/D7
Cariaco, Venezuela 124/G2
Cariamanga, Ecuador 128/C5
Caribbean (sea) 2/F5
Caribbean (sea) 146/K8
Caribbean (sea), 156/B4
Caribbean (sea), Ant. & Bar. 156/B4
Caribbean (sea), Cayman Is. 156/B4
Caribbean (sea), Cuba 156/B4
Caribbean (sea), Dominica 156/B4
Caribbean (sea), Dom. Rep. 156/B4
Caribbean (sea), Grenada 156/B4
Caribbean (sea), Guadeloupe 156/B4
Caribbean (sea), Haiti 156/B4
Caribbean (sea), Jamaica 156/B4
Caribbean (sea), Martinique 156/B4
Caribbean (sea), Neth. Ant. 156/B4
Caribbean (sea), P. Rico 156/B4
Caribbean (sea), St. Chris.-Nevis 156/B4
Caribbean (sea), St. Lucia 156/B4
Caribbean (sea), St. Vin. & Grens. 156/B4
Caribbean (sea), Virgin Is. (Br.) 156/B4
Caribbean (sea), Virgin Is. (U.S.) 156/B4
Caribén, Venezuela 124/E4
Cariboo (mts.), Br. Col. 184/G3
Caribou (mts.), Alberta 182/B5
Caribou (co.), Idaho 220/G7
Caribou (mt.), Idaho 220/G6
Caribou (range), Idaho 220/G6
Caribou, Maine 188/N1
Caribou, Maine (04736) 243/G2
Caribou (riv.), Manitoba 179/J1
Caribou, Nova Scotia 168/F3
Caribou (isl.), Nova Scotia 168/F3
Caribou (isl.), Ontario 175/C3
Caribou (lake), Ontario 177/H4
Caribou (lake), Québec 172/D3
Caribou River, Nova Scotia 168/F3
Carib Reserve, Dominica 161/F6
Caribrod (Dimitrovgrad), Yugoslavia 45/F4
Carichic, Mexico 150/F2
Carievale, Sask. 181/K6
Carigara, Philippines 82/E5
Carignan, Québec 172/J4
Carignan (lake), Québec 172/E2
Carillon, Québec 172/C4
Carina, Queensland 88/K3
Carinda, N.S. Wales 97/D2
Cariñena, Spain 33/F2
Carinhanha, Brazil 132/E6
Carini, Italy 34/D5
Carinthia (prov.), Austria 41/B3
Caripe, Venezuela 124/G2
Caripito, Venezuela 124/G2
Cariquima, Chile 138/B2
Carirubana, Venezuela 124/C2
Carite (lake), P. Rico 161/E2
Cark (mt.), Ireland 17/F2
Carl, Georgia (30203) 217/E3
Carl Blackwell (lake), Okla. 288/M2
Carlea, Sask. 181/H2
Carleton, Mich. (48117) 250/F6
Carleton, Nebr. (68326) 264/G4
Carleton (co.), New Bruns. 170/C2
Carleton (mt.), New Bruns. 170/D1
Carleton, Nova Scotia 168/C4
Carleton (riv.), Nova Scotia 168/C4
Carleton, Québec 172/C2
Carleton Place, Ontario 177/H2
Carlile, Wyo. (82713) 319/H1
Carlin, Nev. (89822) 266/E2
Carlingford, Ireland 17/J3
Carlingford (inlet), Ireland 17/J3
Carlingford (mt.), Ireland 17/J3
Carlingford, New Bruns. 170/C2
Carlinville, Ill. (62626) 222/D4
Carlisle, Ark. (72024) 202/G4
Carlisle (bay), Barbados 161/B9
Carlisle, England 13/D3
Carlisle, England 10/E3
Carlisle, Ind. (47838) 227/C7
Carlisle, Iowa (50047) 229/G6
Carlisle (co.), Ky. 237/C7
Carlisle, Ky. (40311) 237/N4
Carlisle, La. (70042) 238/L7
Carlisle○, Mass. (01741) 249/J2
Carlisle, Minn. (56538) 255/B4
Carlisle, Miss. (39049) 256/C7
Carlisle, New Bruns. 170/C2
Carlisle, N.Y. (12031) 276/L5
Carlisle, Ohio (45005) 284/B6
Carlisle, Ontario 177/D4
Carlisle, Pa. (17013) 294/H5
Carlisle, S.C. (29031) 296/D2
Carl Junction, Mo. (64834) 261/C8
Carlock, Ill. (61725) 222/D3
Carlock, S. Dak. (†57533) 298/L7
Carloforte, Italy 34/B5
Carlos, Ind. (†47355) 227/G4
Carlos, Minn. (56319) 255/C5
Carlos Casares, Argentina 143/F7
Carlos Reyles, Uruguay 145/C4
Carlos Tejedor, Argentina 143/D4
Carlow (co.), Ireland 17/H6
Carlow, Ireland 17/H6
Carlow, Ireland 10/C4
Carloway, Scotland 15/B2
Carlowrie, Manitoba 179/E5
Carlowville, Ala. (†36761) 195/D6
Carl Sandburg Home Nat'l Hist. Site, N.C. 281/D4
Carlsbad, (92008) 204/H10
Carlsbad, N. Mex. 188/F4
Carlsbad, N. Mexico 146/H6
Carlsbad, N. Mex. (88220) 274/E6
Carlsbad, Texas (76934) 303/D6
Carlsbad Caverns Nat'l Park, N. Mex. 274/E6
Carlsbad Springs, Ontario 177/J2
Carlsborg, Wash. (98324) 310/B2
Carlshend, Mich. (49811) 250/B2
Carlstadt, N.J. (07072) 273/B2
Carlton, Ala. (36515) 195/C8
Carlton, Barbados 161/B8
Carlton, England 13/F5
Carlton, Georgia (30627) 217/F2
Carlton, Kansas (67429) 232/E3
Carlton (co.), Minn. 255/F4
Carlton, Minn. (55718) 255/F4
Carlton, N.Y. (†14411) 276/D4
Carlton, Oreg. (97111) 291/D2
Carlton, Pa. (16311) 294/C3
Carlton, Sask. 181/E3
Carlton, Texas (76436) 303/F6
Carlton, Wash. (98814) 310/F2
Carlton (pass), Wash. 310/D4
Carltonville, S. Africa 118/H7
Carluke, Scotland 15/E5
Carluke, Scotland 10/B1
Carlyle, Ill. (62231) 222/D5
Carlyle (lake), Ill. 222/D5
Carlyle, Kansas (66718) 232/G4
Carlyle, Mont. (59320) 262/M4
Carlyle, Sask. 181/J6
Carlyle Lake Resort, Sask. 181/J6
Carmacks, Yukon 187/E3
Carmagnola, Italy 34/A2
Carman, Ill. (61425) 222/B3
Carman, Man. 162/G6
Carman, Manitoba 179/D5
Carmangay, Alberta 182/D4
Carmanville, Newf. 166/D4
Carmarthen, Wales 13/C6
Carmarthen, Wales 10/D5
Carmarthen (bay), Wales 10/D5
Carmarthen (bay), Wales 13/C6
Carmaux, France 28/E5
Carmel, Calif. (93923) 204/D7
Carmel, Ind. (46032) 227/E5
Carmel (cape), Israel 65/B2
Carmel (mt.), Israel 65/C2
Carmel○, Maine (04419) 243/E6
Carmel○, N.Y. (10512) 276/N8
Carmel, Sask. 181/F3
Carmel (head), Wales 13/C4
Carmelo, Uruguay 145/A4
Carmelo, Venezuela 124/C2
Carmel Valley, Calif. (93924) 204/D7
Carmen, Ariz. (†85640) 198/D7
Carmen, Bolivia 136/B2
Carmen (riv.), Chile 138/B7
Carmen, C. Rica 154/F5
Carmen, Idaho (83462) 220/E4
Carmen (isl.), Mexico 150/D3
Carmen, Okla. (73726) 288/J1
Carmen, Bohol, Philippines 82/E6
Carmen, North Cotabato, Philippines 82/E7
Carmen, Uruguay 145/D4
Carmen de Areco, Argentina 143/F7
Carmen del Paraná, Paraguay 144/D5
Carmen de Patagones, Argentina 143/D5
Carmensa, Argentina 143/C4
Carmi, Br. Col. 184/H5
Carmi, Ill. (62821) 222/E5
Carmi (lake), Vt. 268/B2
Carmichael, Calif. (95608) 204/C8
Carmichael, Miss. (†39360) 256/G7
Carmichael, Sask. 181/C5
Carmichaels, Pa. (15320) 294/B6
Carmiel, Israel 65/C2
Carmila, Queensland 95/D4
Carmine, Texas (78932) 303/H7
Carmody Hills-Pepper Mill Village, Md. (†20028) 245/G5
Carmona, Spain 33/D4
Carnac, France 28/B4
Carnadero (creek), Calif. 204/L4
Carnamah, W. Australia 92/A5
Carnarvon, Australia 87/B8
Carnarvon, Iowa (51437) 229/C4
Carnarvon, Ontario 177/F2
Carnarvon (range), Queensland 95/D5
Carnarvon, S. Africa 118/C6
Carnarvon, W. Australia 88/A4
Carnarvon, W. Australia 92/A4
Carnation, Wash. (98014) 310/D3
Carnaxide, Portugal 33/A1
Carn Ban (mt.), Scotland 15/D3
Carndonagh, Ireland 17/G1
Carnduff, Sask. 181/K6
Carnegie, Georgia (†31740) 217/C7
Carnegie, Okla. (73015) 288/J4
Carnegie, Pa. (15106) 294/B7
Carnegie (lake), W. Australia 88/C5
Carnegie (lake), W. Australia 92/C4
Carn Eige (mt.), Scotland 15/C3
Carneiro, Kansas (†67425) 232/D3
Carnes, Miss. (†39455) 256/F8
Carnesville, Georgia (30521) 217/F2
Carnew, Ireland 17/H6
Carney, Mich. (49812) 250/B3
Carney, Okla. (74832) 288/N3
Carneys Point, N.J. (08069) 273/C4
Carnic Alps (mts.), Austria 41/B3
Carnic Alps (range), Italy 34/D1
Car Nicobar (isl.), India 68/G7
Carnlough, N. Ireland 17/K2
Carn More (mt.), Scotland 15/E3
Carnot, Cent. Afr. Rep. 115/C3
Carnoustie, Scotland 15/F4
Carnoustie, Scotland 10/E2
Carnsore (pt.), Ireland 10/C4
Carnsore (pt.), Ireland 17/J7
Carnwath (riv.), N.W. Terrs. 187/F3
Carnwath, Scotland 15/E5
Carnwood, Alberta 182/C3
Caro, Mich. (48723) 250/F5
Caroga Lake, N.Y. (12032) 276/L4
Carol City, Fla. (33055) 212/B4
Carolina, Ala. (†36420) 195/E8
Carolina, Brazil 132/E4
Carolina, P. Rico 161/E1
Carolina, R.I. (02812) 249/H7
Carolina Beach, N.C. (28428) 281/O6
Caroline, Alberta 182/C3
Caroline (isl.), Kiribati 87/M7
Caroline (co.), Md. 245/P5
Caroline, N.Y. (†14817) 276/H6
Caroline (isls.), Pac. Is. Terr. 87/E5
Caroline (isls.), Pacific Is. Terr. 2/S5
Caroline (co.), Va. 307/O4
Caroline, Wis. (54928) 317/J6
Carol Stream, Ill. (†60187) 222/A5
Caron, Sask. 181/F5
Caron Brook, New Bruns. 170/B1
Carondelet, Ecuador 128/C2
Caroni, Trin. & Tob. 161/B10
Caroni (riv.), Venezuela 120/C2
Caroní (riv.), Venezuela 124/G4
Carora, Venezuela 124/C2
Carouge, Switzerland 39/B4
Carp, Ind. (†47460) 227/D6
Carp, Minn. (†56623) 255/D2
Carp, Nev. (†89008) 266/G5
Carp, Ontario 177/H2
Carpathian (mts.) 7/G4
Carpathian (mts.), Romania 45/G2
Carpentaria (gulf) 88/F2
Carpentaria (gulf), Australia 87/D7
Carpentaria (gulf), North. Terr. 93/E3
Carpentaria (gulf), Queensland 95/A2
Carpenter (lake), Br. Col. 184/F5
Carpenter, Iowa (50426) 229/H2
Carpenter, Ky. (†40769) 237/O7
Carpenter, Miss. (39050) 256/C6
Carpenter (lake), N. Dak. 282/L2
Carpenter, Ohio (†47518) 284/F7
Carpenter, S. Dak. (57322) 298/O4
Carpenter, Wyo. (82054) 319/H4
Carpentersville, Ill. (60110) 222/E1
Carpentersville, N.J. (†08865) 273/C2
Carpenterville, Oreg. (†97415) 291/C5
Carpentras, France 28/F5
Carpertee (riv.), N.S. Wales 97/F3
Carpi, Italy 34/C2
Carpinteria, Calif. (93013) 204/F9
Carpio, N. Dak. (58725) 282/G3
Carp Lake, Mich. (49718) 250/E3
Carp Lake Prov. Park, Br. Col. 184/F3
Carr, Colo. (80612) 208/K1
Carra (lake), Ireland 17/C4
Carrabassett Valley○, Maine (†04947) 243/C5
Carrabelle, Fla. 212/B2
Carradale, Scotland 15/C5
Carragana, Sask. 181/J3
Carraguao (pt.), Cuba 158/B2
Carraipía, Colombia 126/D2
Carraízo (lake), P. Rico 161/E1
Carranglan, Philippines 82/C3
Carrantuohill (mt.), Ireland 10/B5
Carrantuohill (mt.), Ireland 17/B7
Carranza, Venustiano (res.), Mexico 150/J3
Carrao (riv.), Venezuela 124/G5
Carrara, Italy 34/C2
Carrasco, Uruguay 145/B7
Carrasquero, Venezuela 124/B2
Carrathool, N. S. Wales 88/G6
Carrathool, N.S. Wales 97/C4
Carrboro, N.C. (27510) 281/L3
Carrbridge, Scotland 15/E3
Carrera de Yeguas, Dom. Rep. 158/D6
Carrera Pinto, Chile 138/B6
Carreta (pt.), C. Rica 154/F6
Carreto, Panama 154/J6
Carriacou (isl.), Grenada 156/G4
Carrick, Manitoba 179/F5
Carrick (dist.), Scotland 15/D5
Carrickfergus (dist.), N. Ireland 17/K2
Carrickfergus, N. Ireland 10/D3
Carrickfergus, N. Ireland 17/K2
Carrickmacross, Ireland 10/C3
Carrickmacross, Ireland 17/H4
Carrick-on-Shannon, Ireland 17/F4
Carrick-on-Shannon, Ireland 10/C4
Carrick-on-Suir, Ireland 17/F7
Carrick-on-Suir, Ireland 10/C4
Carrier, Okla. (73727) 288/K2
Carriere, Miss. (39426) 256/E9
Carrier Mills, Ill. (62917) 222/E6
Carrigaholt, Ireland 17/B6
Carrigain (mt.), N.H. 268/E3
Carrigaline, Ireland 17/E8
Carrigallen, Ireland 17/F4
Carrigan (head), Ireland 17/D2
Carrigart, Ireland 17/F1
Carrigtwohill, Ireland 17/E8
Carrington, Mo. (†65251) 261/H5
Carrington, N. Dak. (58421) 282/M5
Carrión de los Condes, Spain 33/D1
Carrizal, Colombia 126/D1
Carrizal Bajo, Chile 138/A7
Carrizo (creek), Ariz. 198/E4
Carrizo (mts.), Ariz. 198/G2
Carrizo (creek), N. Mex. 274/F2
Carrizo (creek), Texas 303/A1
Carrizo Springs, Texas (78834) 303/E9
Carrizozo, N. Mex. (88301) 274/D5
Carroll (co.), Ark. 202/C1
Carroll (co.), Georgia 217/B3
Carroll (co.), Ill. 222/D1
Carroll (co.), Ind. 227/D3
Carroll (co.), Iowa 229/D4
Carroll, Iowa (51401) 229/D4
Carroll (co.), Ky. 237/L3
Carroll○, Maine (†04487) 243/G5
Carroll, Manitoba 179/B5
Carroll (lake), Manitoba 179/G3
Carroll (co.), Md. 245/K2
Carroll (co.), Miss. 256/E4
Carroll (co.), Mo. 261/F4
Carroll, Nebr. (68723) 264/G2
Carroll (co.), N.H. 268/E4
Carroll○, N.H. (†03595) 268/D3
Carroll, N.S. Wales 97/F2
Carroll (co.), Ohio 284/H4
Carroll, Ohio (43112) 284/E6
Carroll (co.), Tenn. 237/E9
Carroll (co.), Va. 307/G7
Carrolls, Wash. (98609) 310/C4
Carrolls Crossing, New Bruns. 170/D2
Carrollton, Ala. (35447) 195/B4
Carrollton, Georgia (30117) 217/C3
Carrollton, Ill. (62016) 222/C4
Carrollton, Iowa (†51440) 229/D5
Carrollton, Ky. (41008) 237/L3
Carrollton, Md. (†21157) 245/L2
Carrollton, Mich. (48724) 250/E5
Carrollton, Miss. (38917) 256/E4
Carrollton, Mo. (64633) 261/E4
Carrollton, Ohio (44615) 284/J4
Carrollton, Texas (*75006) 303/G2
Carrolltown, Pa. (15722) 294/E4
Carroll Valley, Pa. (†17320) 294/H6
Carron, Scotland 15/C1
Carron (riv.), Scotland 15/D3
Carron (riv.), Scotland 15/C3
Carron (riv.), Scotland 15/C1
Carron Valley (res.), Scotland 15/B1
Carrot (riv.), Sask. 181/J2
Carrot Creek, Alberta 182/B3
Carrothers, Ohio (44823) 284/D3
Carrot River, Sask. 181/H2
Carrowdore, N. Ireland 17/K2
Carrowkeel, Ireland 17/G1
Carrowmore (lake), Ireland 17/B3
Carrsville, Ky. (42030) 237/E6
Carrsville, Va. (23315) 307/P7
Carruthers, Sask. 181/B3
Carrville, Ala. (†36023) 195/G5
Carryduff, N. Ireland 17/K2
Carryville, Ark. (†72454) 202/K1
Çarşamba, Turkey 63/G2
Carseland, Alberta 182/D4

Carson, Ala. (†36548) 195/C8
Carson, Calif. (90745) 204/C11
Carson, Iowa (51525) 229/C6
Carson, Miss. (39427) 256/E7
Carson (lake), Nev. 266/C3
Carson (riv.), Nev. 266/B3
Carson (sink), Nev. 266/C3
Carson, N. Mex. (87517) 274/D2
Carson, N. Dak. (58529) 282/H7
Carson (co.), Texas 303/C2
Carson, Va. (23830) 307/O6
Carson, Wash. (98610) 310/D5
Carson City, Mich. (48811) 250/E5
Carson City (cap.), Nev. 146/G6
Carson City (cap.), Nev. 188/C3
Carson City (co.), Nev. 266/B3
Carson City (cap.), Nev. (89701) 266/B3
Carson Lake, Ark. (†72370) 202/K2
Carson Sink (depr.), Nev. 188/C3
Carsonville, Georgia (†31827) 217/D5
Carsonville, Mich. (48419) 250/G5
Carsphairn, Scotland 15/D5
Carstairs, Alberta 182/D4
Carswell A.F.B., Texas 303/E2
Cartagena, Chile 138/F3
Cartagena, Colombia 120/B1
Cartagena, Colombia 126/C2
Cartagena, Cuba 158/D2
Cartagena, Spain 7/D5
Cartagena, Spain 33/F4
Cartago, Calif. (93549) 204/G7
Cartago, Colombia 126/B5
Cartago, C. Rica 154/F6
Carta Valley, Texas (78835) 303/D8
Cartaxo, Portugal 33/B3
Cartecay, Georgia (†30540) 217/D1
Carter (co.), Ky. 237/P4
Carter, Ky. (41128) 237/P4
Carter (co.), Mo. 261/L9
Carter (co.), Mont. 262/M5
Carter, Mont. (59420) 262/E3
Carter (co.), Okla. 288/M6
Carter, Okla. (73627) 288/H4
Carter, S. Dak. (57526) 298/J7
Carter (co.), Tenn. 237/S8
Carter, Tenn. (†37643) 237/S8
Carter, Wis. (†54566) 317/J5
Carter, Wyo. (†82937) 319/B4
Carter Dome (mt.), N.H. 268/E3
Carteret, N.J. (07008) 273/E2
Carteret (co.), N.C. 281/R5
Carter Lake, Iowa (†68101) 229/B6
Carter Nine, Okla. (†74633) 288/N1
Carters, Georgia (30704) 217/C1
Carters (lake), Georgia 217/C1
Cartersburg, Ind. (46114) 227/E5
Cartersville, Georgia (30120) 217/C2
Cartersville, Iowa (†50469) 229/G2
Cartersville, Mont. (†59347) 262/K4
Cartersville, Okla. (74934) 288/S4
Cartersville, S.C. (†29161) 296/H3
Cartersville, Va. (23027) 307/M5
Carterton, N. Zealand 100/E4
Carterton and Black Bourton, England 13/F6
Carterville, Ill. (62918) 222/D6
Carterville, Mo. (64835) 261/D8
Carthage, Ark. (71725) 202/E5
Carthage, Ill. (62321) 222/B3
Carthage, Ind. (46115) 227/F5
Carthage, Maine (†04224) 243/C6
Carthage○, Maine (†04224) 243/C6
Carthage, Miss. (39051) 256/E5
Carthage, Mo. (64836) 261/D8
Carthage, N.Y. (13619) 276/J3
Carthage, N.C. (28327) 281/K4
Carthage, S. Dak. (57323) 298/O5
Carthage, Tenn. (37030) 237/K8
Carthage, Texas (75633) 303/K5
Cartier, Ontario 177/J5
Cartier (isl.), Terr. Ashmore and Cartier Is. 88/C2
Cartwright, Newf. 146/N4
Cartwright, Manitoba 179/C5
Cartwright, Newf. 166/C3
Cartwright, Newf. 162/L5
Cartwright, N. Dak. (58838) 282/C4
Caruaí (riv.), Venezuela 124/H5
Caruaru, Brazil 132/G5
Carumás, Peru 128/G11
Carúpano, Venezuela 120/C1
Carúpano, Venezuela 124/G2
Carurú, Colombia 126/E7
Carutapera, Brazil 132/E3
Caruth, Mo. (†63857) 261/N10
Caruthers, Calif. (93609) 204/E7
Caruthersville, Mo. (63830) 261/N10
Carver○, Mass. (02330) 249/M5
Carver (co.), Minn. 255/E6
Carver, Minn. (55315) 255/E6
Carville, La. (70721) 238/K3
Carvoeiro (cape), Portugal 33/B3
Cary, Ill. (60013) 222/E1
Cary○, Maine (†04465) 243/H4
Cary, Miss. (39054) 256/C5
Cary, N.C. (27511) 281/M3
Caryapundy (swamp), N.S. Wales 97/B1
Caryapundy (swamp), Queensland 95/B6
Carysbrook, Va. (23055) 307/M5
Carytown, Mo. (†64836) 261/D8
Caryville, Fla. (32427) 212/C6
Caryville, Mass. (02024) 249/J4
Caryville, Tenn. (37714) 237/N8
Casa, Ark. (72025) 202/D3
Casa Agapito, Colombia 126/D6
Casablanca, Chile 138/F3
Casablanca, Estero de (riv.), Chile 138/F3
Casablanca, Morocco 102/B1
Casablanca, Morocco 106/C2
Casablanca, Morocco 2/J4
Casa Blanca, N. Mex. (87007) 274/B4
Casa Branca, Brazil 135/C2
Casa Cruz (cape), Trin. & Tob. 161/B11
Casa Grande, Ariz. 188/D4
Casa Grande, Ariz. (85222) 198/D6
Casa Grande Nat'l Mon., Ariz. 198/D6
Casale Monferrato, Italy 34/B2
Casalmaggiore, Italy 34/C2
Casamance (riv.), Senegal 106/A6
Casanare (inten.), Colombia 126/B3
Casanare (riv.), Colombia 126/E4
Casanay, Venezuela 124/G2
Casa Nova, Brazil 132/F5
Casanova, Va. (22017) 307/N3
Casa Piedra, Texas (†79843) 303/C12
Casar, N.C. (28020) 281/F3
Casar de Cáceres, Spain 33/C3
Casas Grandes (riv.), Mexico 150/F1
Casas-Ibáñez, Spain 33/F3
Cascade (range) 188/B1
Cascade (range), Calif. 204/D1
Cascade, Colo. (80909) 208/K5
Cascade, Idaho (83611) 220/C5
Cascade (res.), Idaho 220/C5
Cascade, Iowa (52033) 229/L4
Cascade, Jamaica 158/G6
Cascade (co.), Mont. 262/E3
Cascade, Mont. (59421) 262/E3
Cascade, N.H. (†03581) 268/E3
Cascade (pt.), N. Zealand 100/B6
Cascade, Norfolk I. 88/L5
Cascade (bay), Norfolk I. 88/L5
Cascade (head), Oreg. 291/C2
Cascade (range), Oreg. 291/E4
Cascade, Seychelles 118/H5
Cascade (range), U.S. 146/F5
Cascade, Va. (24069) 307/J7
Cascade (pass), Wash. 310/D2
Cascade (range), Wash. 310/D4
Cascade (riv.), Wash. 310/D2
Cascade, W. Va. (26526) 312/G3
Cascade, Wis. (53011) 317/K8
Cascade Locks, Oreg. (97014) 291/E2
Cascade Summit, Oreg. (97425) 291/F4
Cascadia, Oreg. (97329) 291/E3
Cascais, Portugal 33/B3
Cascajal, Cuba 158/E1
Cascapédia (riv.), Québec 172/C1
Cascas, Peru 128/C6
Cascavel, Brazil 132/G4
Cascilla, Miss. (38920) 256/D3
Cascina-Navacchio, Italy 34/C3
Casco, Maine (04015) 243/B7
Casco○, Maine (04015) 243/B7
Casco (bay), Maine 243/C8
Casco, Wis. (54205) 317/L6
Cascorro, Cuba 158/H3
Cascumpeque (bay), Pr. Edward I. 168/E2
Caselton, Nev. (†89043) 266/G5
Case-Pilote, Martinique 161/C6
Caserta (prov.), Italy 34/E4
Caserta, Italy 34/E4
Caseville, Mich. (48725) 250/F5
Casey (key), Fla. 212/D4
Casey, Ill. (62420) 222/F4
Casey, Iowa (50048) 229/D5
Casey (co.), Ky. 237/M6
Casey, Québec 174/C3
Casey Creek, Ky. (42723) 237/L6
Caseyville, Ill. (62232) 222/B2
Caseyville, Ky. (†42459) 237/E5
Cash, Ark. (72421) 202/J2
Cashel, Ireland 17/F7
Cashel, Ireland 10/C4
Cashiers, N.C. (28717) 281/C4
Cashion, Ariz. (85329) 198/C5
Cashion, Okla. (73016) 288/L3
Cashmere, Wash. (98815) 310/E3
Cashmere, W. Va. (†24918) 312/E8
Cashton, Wis. (54619) 317/E8
Cashtown, Pa. (17310) 294/H6
Cashville, S.C. (†29388) 296/C2
Casigua, Falcón, Venezuela 124/C2
Casigua, Zulia, Venezuela 124/B3
Casiguran, Philippines 82/D2
Casiguran (sound), Philippines 82/C2
Casilda, Argentina 143/F6
Casilda, Cuba 158/E2
Casilda (pt.), Cuba 158/E2
Casino, N. S. Wales 88/J5
Casino, N.S. Wales 97/G1
Casiquiare, Brazo (riv.), Venezuela 124/E6
Casitas Springs, Calif. (†93001) 204/F9
Caslan, Alberta 182/D2
Čáslav, Czech. 41/C2
Casma, Peru 128/C7
Casma (riv.), Peru 128/C7
Casmalia, Calif. (93429) 204/E9
Casnovia, Mich. (49318) 250/D5
Caspar, Calif. (95420) 204/B4
Caspe, Spain 33/G2
Casper, Wyo. 188/E2
Casper, Wyo. 146/H5
Casper, Wyo. (82601) 319/F3
Casper (range), Wyo. 319/F3
Caspian (sea) 54/G5
Caspian (sea) 2/M3
Caspian (sea), Iran 66/G1
Caspian, Mich. (49915) 250/G2
Caspian (sea), U.S.S.R. 7/J4
Caspian (sea), U.S.S.R. 52/G6
Caspian (sea), U.S.S.R. 48/F6
Caspian (lake), Vt. 268/C2
Caspiana, La. (71015) 238/C2
Cass, Ark. (†72949) 202/C2
Cass (co.), Ill. 222/C4
Cass (co.), Ind. 227/E3
Cass, Ind. (†47882) 227/C6
Cass (co.), Iowa 229/D6
Cass (co.), Mich. 250/C7
Cass (riv.), Mich. 250/F5
Cass (co.), Minn. 255/D4
Cass (lake), Minn. 255/D3
Cass (co.), Mo. 261/D5
Cass (co.), Nebr. 264/H4
Cass (co.), N. Dak. 282/R5
Cass (co.), Texas 303/K4
Cass, W. Va. (24927) 312/G6
Cassadaga, Fla. (32706) 212/E3
Cassadaga, N.Y. (14718) 276/B6
Cassà de la Selva, Spain 33/H2
Cassai, Angola 115/D6
Cassamba, Angola 115/D6
Cassandra, Georgia (†30727) 217/B1
Cassandra, Pa. (15925) 294/E5
Cassano allo Ionio, Italy 34/F5
Cassatt, S.C. (29032) 296/G3
Cass City, Mich. (48726) 250/F5
Casscoe, Ark. (72026) 202/H4
Casselberry, Fla. (32707) 212/E3
Casselman (riv.), Md. 245/B2
Casselman, Ontario 177/J2
Casselman, Pa. (†15557) 294/D6
Casselman (riv.), Pa. 294/D6
Casselton, N. Dak. (58012) 282/R6
Cássia, Brazil 135/C2
Cassia (co.), Idaho 220/E7
Cassiar, Br. Col. 184/K2
Cassiar (mts.), Br. Col. 184/K2
Cassiar (mts.), Yukon 187/E3
Cassidy, Br. Col. 184/J3
Cassilis, N.S. Wales 97/E3
Cassils, Alberta 182/D4
Cassino, Italy 34/D4
Cassiporé (cape), Brazil 132/D2
Cassity, W. Va. (†26278) 312/F5
Cass Lake, Minn. (56633) 255/D3
Cassoday, Kansas (66842) 232/F3
Cassopolis, Mich. (49031) 250/C7
Casstown, Ohio (45312) 284/B5
Cassville, Georgia (30123) 217/C2
Cassville, Ind. (†46901) 227/E3
Cassville, Mo. (65625) 261/E9
Cassville, Pa. (16623) 294/G5
Cassville, W. Va. (26527) 312/F3
Cassville, Wis. (53806) 317/E10
Castagnola, Switzerland 39/G4
Castaic, Calif. (91310) 204/G9
Castalia, Iowa (52133) 229/K2
Castalia, New Bruns. 170/D4
Castalia, N.C. (27816) 281/O2
Castalia, Ohio (44824) 284/E3
Castalian Springs, Tenn. (37031) 237/J8
Castana, Iowa (51010) 229/B4
Castanhal, Brazil 120/E3
Castanhal, Brazil 132/E3
Castaños, Mexico 150/J3
Castelfranco Veneto, Italy 34/D2
Castel Gandolfo, Italy 34/F7
Casteljaloux, France 28/D5
Castella, Calif. (96017) 204/C2
Castellammare (gulf), Italy 34/D5
Castellammare del Golfo, Italy 34/D5
Castellammare di Stabia, Italy 34/E4
Castellane, France 28/G6
Castellanos, Uruguay 145/B6
Castelli, Buenos Aires, Argentina 143/H7
Castelli, Chaco, Argentina 143/D2
Castellón (prov.), Spain 33/G2
Castellón de la Plana, Spain 33/G2
Castellote, Spain 33/F2
Castelnaudary, France 28/E6
Castelo, Brazil 132/F8
Castelo Branco (dist.), Portugal 33/C3
Castelo Branco, Portugal 33/C3
Castelo de Vide, Portugal 33/C3
Castelo do Piauí, Brazil 132/F4
Castel San Pietro Terme, Italy 34/C2
Castelsarrasin, France 28/D6
Castelvetrano, Italy 34/D6
Casterton, Victoria 97/A5
Castiglione del Lago, Italy 34/C3
Castiglion Fiorentino, Italy 34/C3
Castile, N.Y. (14427) 276/D5
Castilletes, Venezuela 124/C2
Castillo, Cerro (mt.), Chile 138/E6
Castillo, Dom. Rep. 158/E5
Castillo de San Marcos Nat'l Mon., Fla. 212/E2
Castillos, Uruguay 145/F5
Castillos (lag.), Uruguay 145/F5
Castine○, Maine (04421) 243/F7
Castine, Ohio (45313) 284/A6
Castle (harb.), Bermuda 156/H2
Castle (mt.), Br. Col. 184/A2
Castle (peak), Colo. 208/F5
Castle (creek), Idaho 220/B7
Castle (peak), Idaho 220/D5
Castle (pt.), N. Zealand 100/F4
Castle, Okla. (74833) 288/O4
Castle (valley), Utah 304/D4
Castle A.F.B., Calif. 204/E6
Castlebar, Ireland 17/C4
Castlebar, Ireland 10/B4
Castlebay, Scotland 15/A4
Castlebellingham, Ireland 17/J4
Castleberry, Ala. (36432) 195/D8
Castleblayney, Ireland 10/C3
Castleblayney, Ireland 17/H3
Castlebridge, Ireland 17/J7
Castle Bruce, Dominica 161/F6
Castlecomer-Donaguile, Ireland 17/G6
Castlecomer-Donaguile, Ireland 10/C4
Castle Dale, Utah (84513) 304/D4
Castle Danger, Minn. (†56616) 255/G3
Castledawson, N. Ireland 17/H2
Castlederg, N. Ireland 17/F2
Castledermot, Ireland 17/H6
Castle Dome (mts.), Ariz. 198/A5
Castle Douglas, Scotland 15/E6
Castle Douglas, Scotland 10/D3
Castlefin, Ireland 17/F2
Castleford, Idaho (83321) 220/C7
Castlegar, Br. Col. 184/J5
Castlegregory, Ireland 10/B4
Castlegregory, Ireland 17/A7
Castle Hayne, N.C. (28429) 281/O6
Castle Hills, Texas (†78213) 303/J10
Castle Hot Springs, Ariz. (†85342) 198/C5
Castleisland, Ireland 17/B7
Castle Kennedy, Scotland 15/D6
Castlemaine, Victoria 97/C5
Castlemartyr, Ireland 17/E8
Castle Park, Mich. (†49423) 250/C6
Castlepollard, Ireland 17/G4
Castlerea, Ireland 17/D4
Castlerea, Ireland 10/B4
Castlereagh (riv.), N.S. Wales 97/E2
Castlereagh (dist.), N. Ireland 17/K2
Castle Rock, Colo. (80104) 208/K4
Castle Rock, Minn. (55010) 255/E6
Castle Rock, S. Dak. (†57760) 298/C4
Castle Rock, Utah (†82930) 304/C2
Castle Rock, Wash. (98611) 310/B4
Castle Rock (lake), Wis. 317/G8
Castle Shannon, Pa. (15234) 294/B7
Castleton, Ill. (61426) 222/D2
Castleton, Jamaica 158/J6
Castleton, Md. (†21034) 245/N2
Castleton, Ontario 177/F3
Castleton○, Vt. (05735) 268/A4
Castleton-on-Hudson, N.Y. (12033) 276/N5
Castletown, Ireland 17/F6
Castletown, I. of Man 13/C3
Castletown, Scotland 15/E2
Castletownbere, Ireland 17/B8
Castletownroche, Ireland 17/D7
Castletownshend, Ireland 17/C9
Castlewellan, N. Ireland 17/K3
Castlewood, S. Dak. (56623) 298/R4
Castlewood, Va. (24224) 307/D7
Castolon, Texas (†79852) 303/D12
Castor, Alberta 182/D3
Castor, La. (71016) 238/D2
Castorland, N.Y. (13620) 276/J3
Castres, France 28/E6
Castries (cap.), St. Lucia 156/G4
Castries (cap.), St. Lucia 161/G6
Castro, Brazil 135/B4
Castro, Brazil 132/D9
Castro, Chile 138/D4
Castro (co.), Texas 303/B3
Castro Alves, Brazil 132/G6
Castro Daire, Portugal 33/C2
Castro del Río, Spain 33/D4
Castrojeriz, Spain 33/E1
Castro Marim, Portugal 33/C4
Castropol, Spain 33/C1
Castrop-Rauxel, W. Germany 22/B3
Castroreale, Italy 34/E5
Castro-Urdiales, Spain 33/E1
Castro Valley, Calif. (94546) 204/K2
Castro Verde, Portugal 33/B4
Castrovillari, Italy 34/F5
Castroville, Calif. (95012) 204/D7
Castroville, Texas (78009) 303/J11
Castrovirreyna, Peru 128/E9
Castuera, Spain 33/D3
Casuarito, Colombia 126/G5
Casupá, Uruguay 145/D5
Caswell, Alaska (†99688) 196/B1
Caswell (co.), N.C. 281/L2
Caswell Beach, N.C. (†28461) 281/N7
Cat (isl.), Bahamas 146/L7
Cat (isl.), Bahamas 156/C1
Cat (isl.), Miss. 256/F10
Çat, Turkey 63/J3
Cat (isl.), Wis. 317/E1
Catacamas, Honduras 154/E3
Catacaos, Peru 128/B5
Catacocha, Ecuador 128/C5
Catadupa, Jamaica 158/H6
Cataguases, Brazil 135/E2
Catahoula (par.), La. 238/G3
Catahoula (lake), La. 238/F4
Cataingan, Philippines 82/E5
Çatak, Turkey 63/K4
Catalão, Brazil 132/D7
Çatalca, Turkey 63/C2
Cataldo, Idaho (83810) 220/B2
Catalina, Chile 138/B5
Catalina (pt.), Chile 138/F10
Catalina (isl.), Dom. Rep. 158/F6
Catalina, Newf. 166/D2
Catalone, Nova Scotia 168/H3
Catalonia (reg.), Spain 33/G2
Catalpa (creek), Miss. 256/G4
Çatalzeytin, Turkey 63/F1
Catamarca (prov.), Argentina 143/C2
Catamarca, Argentina 143/C2
Catamarca, Argentina 120/C5
Catamayo, Ecuador 128/C4
Catanauan, Philippines 82/D4
Catandica, Mozambique 118/E3
Catanduanes (prov.), Philippines 82/E4
Catanduanes (isl.), Philippines 82/E4
Catanduanes (isl.), Philippines 85/H3
Catanduva, Brazil 135/B2
Catanduva, Brazil 132/D8
Catania (prov.), Italy 34/E6
Catania, Italy 34/E6
Catania, Italy 7/F5
Cataño, P. Rico 156/G1
Cataño, P. Rico 161/D1
Catanzaro (prov.), Italy 34/F5
Catanzaro, Italy 34/F5
Cataouatche (lake), La. 238/N4
Cataract (creek), Ariz. 198/C3
Cataract, Ind. (†47460) 227/D6
Cataract (canyon), Utah 304/D5
Cataract, Wis. (54620) 317/E7
Catarama, Ecuador 128/C3
Cataricahua, Bolivia 136/B6
Catarina, Texas (78836) 303/E9
Catarman, Philippines 82/E4
Catarman (pt.), Philippines 82/F7
Catasauqua, Pa. (18032) 294/M4
Catastrophe (cape), S. Australia 88/F7
Catastrophe (cape), S. Australia 94/D6
Catatumbo (riv.), Colombia 126/D3
Catatumbo (riv.), Venezuela 124/B3
Cataula, Georgia (31804) 217/C5
Cataumet, Mass. (02534) 249/M6
Catawba (co.), N.C. 281/G3
Catawba, N.C. (28609) 281/G3
Catawba (lake), N.C. 281/G4
Catawba (riv.), N.C. 281/H5
Catawba, Ohio (43010) 284/C6
Catawba, S.C. (29704) 296/F2
Catawba (riv.), S.C. 296/F2
Catawba, Va. (24070) 307/H6
Catawba, W. Va. (†26554) 312/F3
Catawba, Wis. (54515) 317/E4
Catawba Island, Ohio (†43452) 284/E2
Catawissa, Mo. (63015) 261/L6
Catawissa, Pa. (17820) 294/K4
Cat Ba, Dao (isl.), Vietnam 72/E2
Catbalogan, Philippines 85/H3
Catbalogan, Philippines 82/E5
Cat Creek, Mont. (59017) 262/H3
Cateechee, S.C. (29629) 296/B2
Cateel, Philippines 85/H4
Cateel, Philippines 82/F7
Catemaco, Mexico 150/M7
Catemu, Chile 138/G2
Cater, Sask. 181/C2
Caterham and Warlingham, England 13/H8
Caterham and Warlingham, England 10/B6
Cates, Ind. (47927) 227/C4
Catete, Angola 115/B5
Catfish (lake), N.C. 281/P5
Catfish (lake), Ontario 177/F2
Catfish (creek), S.C. 296/J3
Cathance (lake), Maine 243/J6
Catharine, Kansas (67627) 232/C3
Catharine Lake, N.C. (†28754) 281/O5
Cathay, N. Dak. (58422) 282/M4
Cathedral (mt.), Texas 303/D12
Cathedral City, Calif. (92234) 204/J10
Cathedral Prov. Park, Br. Col. 184/H5
Catherine, Ala. (36728) 195/D6
Catherine (lake), Ark. 202/E5
Catheys Valley, Calif. (95306) 204/E6
Cathlamet, Wash. (98612) 310/B4
Cat Island (chan.), La. 238/M6
Cat Island (passage), La. 238/J8
Catlett, Va. (22019) 307/N3
Catlettsburg, Ky. (41129) 237/R4
Catlin, Ill. (61817) 222/F3
Catlin, Ind. (†47872) 227/C5
Catmon, Philippines 82/E5
Cato, Ark. (†72076) 202/F4
Cato (isl.), Australia 87/F8
Cato, Ind. (†47598) 227/C8
Cato, N.Y. (13033) 276/G4
Cato, Wis. (54206) 317/L7
Catoche (cape), Mexico 150/Q6
Catoctin (creek), Md. 245/H3
Catoctin Furnace, Md. (†21788) 245/J2
Catoctin Mt. Park, Md. 245/J2
Catolé do Rocha, Brazil 132/G4
Catonsville, Md. (21228) 245/M3
Catoosa (co.), Georgia 217/B1
Catoosa, Okla. (74015) 288/P2
Catriló, Argentina 143/D4
Catrimani, Brazil 132/H9
Catrine, Scotland 15/D5
Catron, Mo. (63833) 261/N9
Catron (co.), N. Mex. 274/A4
Catskill, N.Y. (12414) 276/N6
Catskill (mts.), N.Y. 276/L6
Cattaraugus (co.), N.Y. 276/C6
Cattaraugus, N.Y. (14719) 276/C6
Cattaraugus (creek), N.Y. 276/C6
Cattaraugus Ind. Res., N.Y. 276/C5
Catumbela, Angola 115/B6
Cauayan, Isabela, Philippines 82/C2
Cauayan, Negros Occ., Philippines 82/D6
Cauca (dept.), Colombia 126/B6
Cauca (riv.), Colombia 120/B2
Cauca (riv.), Colombia 126/C4
Caucagua, Venezuela 124/E2
Caucasia, Colombia 126/C4
Caucasus (mts.), U.S.S.R. 7/J4
Caucasus (mts.), U.S.S.R. 52/F6
Caucasus (mts.), U.S.S.R. 48/E5
Caucedo (cape), Dom. Rep. 158/E6
Caucete, Argentina 143/C3
Caucomgomoc (lake), Maine 243/D3
Caudete, Spain 33/F3
Caughnawaga, Québec 172/H4
Cauit (pt.), Philippines 82/F6
Cauldcleuch Head (mt.), Scotland 15/F5
Caulfield, Mo. (65626) 261/H9
Caulfield, Victoria 88/L7
Caulfield, Victoria 97/J5
Caulksville, Ark. (†72951) 202/C3
Caulonia, Italy 34/F5
Caúngula, Angola 115/C5
Cauquenes, Chile 138/A11
Caura (riv.), Venezuela 120/D2
Caura (riv.), Venezuela 124/F5
Causapscal, Québec 172/B2
Causeway, Ireland 17/B7
Causey, N. Mex. (88113) 274/F5
Causses (reg.), France 28/D5
Cauterets, France 28/C6
Cauthron, Ark. (†72958) 202/B4
Cauto (riv.), Cuba 158/H3
Cauto del Embarcadero, Cuba 158/H4
Cauto el Cristo, Cuba 158/J3
Cava de'Tirreni, Italy 34/E4
Cavaillon, France 28/F6
Cavaillon, Haiti 158/A6
Cavalier (co.), N. Dak. 282/N2
Cavalier, N. Dak. (58220) 282/P2
Cavalla (riv.), Liberia 106/C7
Cavalli (isls.), N. Zealand 100/E1
Cavallo (passage), Texas 303/H9
Cavally (riv.), Ivory Coast 106/C7
Cavan (co.), Ireland 17/G4
Cavan, Ireland 10/C4
Cavan, Ireland 17/G3
Cavan, Ontario 177/F3
Cavanaugh (lake), Wash. 310/D2
Cavari, Bolivia 136/B5
Cavarzere, Italy 34/D2
Cave City, Ark. (72521) 202/G2
Cave City, Ky. (42127) 237/K6
Cave Creek, Ariz. (85331) 198/D5
Cave Hill, Barbados 161/B9
Cave in Rock, Ill. (62919) 222/E6
Cave Junction, Oreg. (97523) 291/D5
Cavell, Sask. 181/C3
Cavendish, Alberta 182/E4
Cavendish, Idaho (†83550) 220/B3
Cavendish, Newf. 166/D2
Cavendish○, Vt. (05142) 268/B5
Cavergno, Switzerland 39/G4
Cave Spring, Georgia (30124) 217/B2
Cave Springs, Ark. (72718) 202/B1
Cavetown, Md. (21720) 245/H2
Caviana (isl.), Brazil 120/E2
Caviana (isl.), Brazil 132/D2
Cavili (isl.), Philippines 82/C6
Cavinas, Bolivia 136/B3
Cavite (prov.), Philippines 82/C3
Cavite, Philippines 82/C3
Cavite, Philippines 85/G3
Cavour, S. Dak. (57324) 298/N5
Cavour, Wis. (54516) 317/J4
Cawdor, Scotland 15/E3
Cawker City, Kansas (67430) 232/D2
Cawndilla (lake), N.S. Wales 97/A3
Cawnpore (Kanpur), India 68/E3
Cawood, Ky. (40815) 237/P7
Cawston, Br. Col. 184/H5
Cawston, England 13/J5
Caxambu, Brazil 135/D2
Caxias, Brazil 120/E3
Caxias, Brazil 132/F4
Caxias do Sul, Brazil 120/D5
Caxias do Sul, Brazil 132/D10
Caxito, Angola 115/B5
Çay, Turkey 63/D3
Cayacoa, Dom. Rep. 158/E6
Cayamas (cays), Cuba 158/C2
Cayambe, Ecuador 128/D3
Cayambe (mt.), Ecuador 128/D3
Cayasta, Argentina 143/F5
Cayastacito, Argentina 143/F5
Cayce, Ky. (†42041) 237/C7
Cayce, S.C. (29169) 296/E4
Çaycuma, Turkey 63/E2
Caycuse, Br. Col. 184/J3
Çayeli, Turkey 63/J2
Cayenne (cap.), Fr. Guiana 120/D2
Cayenne (cap.), Fr. Guiana 2/G5
Cayenne (dist.), Fr. Guiana 131/E3
Cayenne (cap.), Fr. Guiana 131/E3
Cayer, Manitoba 179/D3
Cayes Jacmel, Haiti 158/C6
Cayeux-sur-Mer, France 28/D2
Cayey, P. Rico 156/G1
Cayey, P. Rico 161/D2
Cayey, Sierra de (mts.), P. Rico 161/D2
Çayıralan, Turkey 63/F3
Çayırlı, Turkey 63/J3
Cayley, Alberta 182/D4
Cayman (isls.) 146/K8
Cayman Brac (isl.), Cayman Is. 156/B3
CAYMAN ISLANDS 156/B3
Cayo Costa (isl.), Fla. 212/D5
Cayon, St. Chris.-Nevis 161/C10
Cay Sal (bank), Bahamas 156/B2
Cayucos, Calif. (93430) 204/E8
Cayuga, Ind. (47928) 227/C5
Cayuga (co.), N.Y. 276/G4
Cayuga, N.Y. (13034) 276/G5
Cayuga (lake), N.Y. 276/G5
Cayuga, N. Dak. (58013) 282/R7
Cayuga, Texas (75832) 303/J6
Cayuga Heights, N.Y. (†14850) 276/H6
Cayuse, Oreg. (97821) 291/J2
Cayuta (creek), N.Y. 276/G6
Cazadero, Calif. (95421) 204/B5
Cazalla de la Sierra, Spain 33/D4
Cazaux (lake), France 28/C5
Cazenovia, N.Y. (13035) 276/J5
Cazenovia, Wis. (53924) 317/F8
Cazin, Yugoslavia 45/B3
Cazis, Switzerland 39/H3
Cazma (riv.), Yugoslavia 45/C3
Cazombo, Angola 115/D6
Cazones (gulf), Cuba 158/C2
Cazorla, Spain 33/E4
Cazorla, Venezuela 124/E3
Cazot, Uruguay 145/B6
Cazueleja, Cerro (mt.), Colombia 126/C6
Ceanannus Mór, Ireland 17/G4
Ceanannus Mór, Ireland 10/C4
Ceará (state), Brazil 132/G4
Ceará (Fortaleza), Brazil 132/G3
Ceará-Mirim, Brazil 132/H4
Ceará-Mirim, Brazil 120/F3
Cébaco (isl.), Panama 154/G7
Ceballos, Mexico 150/H3
Cebeci, Turkey 63/E3
Cebolla (creek), Colo. 208/E6
Cebolla, N. Mex. (87518) 274/C2
Cebollatí, Uruguay 145/F4
Cebollatí (riv.), Uruguay 145/F4
Cebreros, Spain 33/D2
Cebu (prov.), Philippines 82/D5
Cebu, Philippines 82/D5
Cebu, Philippines 85/G3
Cebu, Philippines 54/O8
Cebu, Philippines 2/R5
Cebu (isl.), Philippines 85/G3
Cebu (isl.), Philippines 82/D5
Cecelia, La. (70521) 238/G6
Cecil, Ala. (36013) 195/F6
Cecil, Ark. (72930) 202/C3
Cecil, Georgia (31627) 217/F8
Cecil (co.), Md. 245/P2
Cecil, Ohio (45821) 284/A3
Cecil, Oreg. (†97843) 291/H2
Cecil, Pa. (15321) 294/B5
Cecil, Wis. (54111) 317/K6
Cecil Field Naval Air Sta., Fla. 212/E1
Cecilia, Ky. (42724) 237/K5
Cecil Lake, Br. Col. 184/G2
Cecil M. Harden (lake), Ind. 227/C5
Cecilton, Md. (21913) 245/P3
Cecilville, Calif. (†96027) 204/B2
Cecina, Italy 34/C3

Ceclavín, Spain 33/C3
Cedar (pt.), Ala. 195/B10
Cedar, Br. Col. 184/J3
Cedar (creek), Colo. 208/M1
Cedar (lake), Conn. 210/E3
Cedar (creek), Ind. 227/G2
Cedar (co.), Iowa 229/L5
Cedar, Iowa (52543) 229/H6
Cedar (riv.), Iowa 188/H2
Cedar (riv.), Iowa 229/K4
Cedar, Kansas (67628) 232/D2
Cedar (lake), Manitoba 179/B1
Cedar (pt.), Md. 245/N7
Cedar, Mich. (49621) 250/D4
Cedar (lake), Mich. 250/F4
Cedar (riv.), Minn. 255/F7
Cedar (co.), Mo. 261/E7
Cedar (co.), Nebr. 264/G2
Cedar (riv.), Nebr. 264/F3
Cedar (mt.), Nev. 266/D4
Cedar (creek), N.J. 273/E4
Cedar (creek), N. Dak. 282/G7
Cedar (pt.), Ohio 284/D2
Cedar (lake), Ontario 177/F1
Cedar (lake), Texas 303/B5
Cedar (mts.), Utah 304/B3
Cedar (isl.), Va. 307/S5
Cedar (riv.), Wash. 310/B2
Cedar Bluff, Ala. (35959) 195/G2
Cedar Bluff, Iowa (†52772) 229/L5
Cedar Bluff (res.), Kansas 232/C3
Cedarbluff, Miss. (39741) 256/G3
Cedar Bluff, Va. (24609) 307/E6
Cedar Bluffs, Kansas (†67749) 232/B2
Cedar Bluffs, Nebr. (68015) 264/H3
Cedar Breaks Nat'l Mon., Utah 304/B6
Cedar Brook, N.J. (08018) 273/D4
Cedarburg, Wis. (53012) 317/L9
Cedarbutte, S. Dak. (57527) 298/H6
Cedar City, Mo. (65022) 261/H5
Cedar City, Utah 188/D3
Cedar City, Utah (84720) 304/A6
Cedar Cove, Ala. (†35453) 195/D4
Cedar Creek, Ark. (†72950) 202/C4
Cedar Creek (peak), Idaho 220/E7
Cedar Creek (res.), Idaho 220/D7
Cedarcreek, Mo. (†65680) 261/G9
Cedar Creek, Nebr. (68016) 264/H3
Cedar Crest, N. Mex. (87008) 274/C3
Cedaredge, Colo. (81413) 208/D5
Cedar Falls, Iowa (50613) 229/H3
Cedar Falls, N.C. (27230) 281/K3
Cedar Falls, Wash. (†98045) 310/D3
Cedar Falls, Wis. (†54751) 317/C6
Cedar Fort, Utah (†84013) 304/B3
Cedar Gap, Mo. (†65746) 261/G8
Cedar Grove, Ant. & Bar. 161/E11
Cedar Grove, Fla. (†32401) 212/D6
Cedar Grove, Georgia (†30727) 217/L2
Cedar Grove, Ind. (47016) 227/H6
Cedar Grove, Md. (†20767) 245/K4
Cedar Grove○, N.J. (07009) 273/B2
Cedar Grove, N.C. (27231) 281/L2
Cedar Grove, Tenn. (38321) 237/D9
Cedar Grove, W. Va. (25039) 312/D6
Cedar Grove, Wis. (53013) 317/L8
Cedar Heights, Md. (†20027) 245/G5
Cedar Hill, N. Mex. (†87410) 274/B2
Cedar Hill, Tenn. (37032) 237/H7
Cedar Hill, Texas (75104) 303/G3
Cedar Hill Lakes, Mo. (63016) 261/L6
Cedar Hills, Oreg. (97225) 291/A2
Cedarhurst, N.Y. (11516) 276/P7
Cedar Island, N.C. (28520) 281/S5
Cedar Key, Fla. (32625) 212/C2
Cedar Knolls, N.J. (07927) 273/E2
Cedar Lake, Ind. (46303) 227/C2
Cedar Lake, Minn. (†56431) 255/E4
Cedar Mill, Oreg. (†97005) 291/A2
Cedar Mills, Minn. (55351) 255/D6
Cedar Mountain, N.C. (28718) 281/D4
Cedar Park, Texas (78613) 303/G7
Cedar Point, Ill. (61316) 222/D2
Cedar Point, Kansas (66843) 232/F3
Cedar Rapids, Iowa 188/H2
Cedar Rapids, Iowa (*52401) 229/K5
Cedar Rapids, Iowa 146/J5
Cedar Rapids, Nebr. (68627) 264/F3
Cedar River, Mich. (49813) 250/B3
Cedar Run, N.J. (†08092) 273/E4
Cedar Run, Pa. (17727) 294/H2
Cedar Springs, Georgia (31732) 217/C8
Cedar Springs, Mich. (49319) 250/D5
Cedar Springs, Mo. (†64744) 261/E7
Cedar Springs, Ontario 177/B5
Cedar Springs, Va. (†24368) 307/F7
Cedar Swamp (pond), Conn. 210/G2
Cedartown, Georgia (30125) 217/B2
Cedarvale, Br. Col. 184/C2
Cedar Vale, Kansas (67024) 232/F4
Cedarvale, N. Mex. (87009) 274/D4
Cedar Valley, Utah (84013) 304/B3
Cedarville, Ark. (72932) 202/B2
Cedarville, Calif. (96104) 204/E2
Cedarville, Ill. (61013) 222/D1
Cedarville, Ind. (†46741) 227/G2
Cedarville, Ky. (†41522) 237/S6
Cedarville, Md. (†20613) 245/L6
Cedarville, Mich. (49719) 250/E2
Cedarville, N.J. (08311) 273/C5
Cedarville, N.Y. (†13357) 276/K5
Cedarville, Ohio (45314) 284/C6
Cedarville, Va. (†22630) 307/M3
Cedarville, W. Va. (26611) 312/E5
Cedarwood Park, N.J. (†08723) 273/E3
Cedonia, Wash. (99137) 310/G2
Cedoux, Sask. 181/H6
Cedral, Mexico 150/J5
Cedros, Honduras 154/D3
Cedros (isl.), Mexico 146/G7
Cedros (isl.), Mexico 150/B2
Cedros, Trin. & Tob. 161/A11
Ceduna, S. Australia 88/E6
Ceduna, S. Australia 94/D5
Cee Vee, Texas (79223) 303/D3
Cefalù, Italy 34/E5
Cegléd, Hungary 41/E3
Ceglie Messapico, Italy 34/F4
Cehegín, Spain 33/F3
Ceiba, P. Rico 161/F2
Çekerek, Turkey 63/F2
Çekerek (riv.), Turkey 63/F3
Cela, Angola 115/C6
Celada Cué, Paraguay 144/D3
Celano, Italy 34/D3
Celanova, Spain 33/B1
Celaya, Mexico 150/J6
Celbridge, Ireland 17/H5
Celebes (sea) 54/O9
Celebes (isl.), Indonesia 54/N10
Celebes (isl.), Indonesia 2/R6
Celebes (Sulawesi) (isl.), Indonesia 85/G5
Celebes (sea), Indonesia 85/G5
Celebes (sea), Philippines 82/D8
Celendín, Peru 128/D6
Celerigna-Schlarigna, Switzerland 39/J3
Celeste, Texas (75423) 303/H4
Celestine, Ind. (47521) 227/D8
Celestún, Mexico 150/O6
Celica, Ecuador 128/B4
Céligny, Switzerland 39/B4
Çelikhan, Turkey 63/H3
Celilo, Oreg. (†97058) 291/G2
Celilo (lake), Oreg. 291/G2
Celilo (lake), Wash. 310/E5
Celina, Minn. (†55788) 255/E3
Celina, Ohio (45822) 284/A4
Celina, Tenn. (38551) 237/K7
Celina, Texas (75009) 303/H4
Celista, Br. Col. 184/H5
Celje, Yugoslavia 45/B2
Cella, Spain 33/F2
Cellar (head), Scotland 15/B2
Celldömölk, Hungary 41/D3
Celle, W. Germany 22/D2
Celorico da Beira, Portugal 33/C2
Celoron, N.Y. (14720) 276/B6
Cement, Okla. (73017) 288/K5
Cement City, Mich. (49233) 250/E6
Çemişkezek, Turkey 63/H3
Cemmaes (head), Wales 13/C5
Cenderawasih (bay), Indonesia 85/K6
Ceneri (mt.), Switzerland 39/G4
Cenia, Spain 33/G2
Census Bureau, Md. 245/F5
Centenary, Ind. (†47842) 227/B5
Centenary, S.C. (29519) 296/J3
Centennial (mts.), Idaho 220/F5
Centennial, Wyo. (82055) 319/F4
Centennial Wash (dry riv.), Ariz. 198/B5
Center, Colo. (81125) 208/G7
Center, Georgia (†30601) 217/F2
Center, Ind. (†46901) 227/E4
Center, Ky. (42214) 237/K6
Center (pond), Maine 243/E5
Center, Mo. (63436) 261/J3
Center, Nebr. (68724) 264/G2
Center, N. Dak. (58530) 282/H5
Center, Okla. (†74820) 288/N5
Center, S. Dak. (†57058) 298/P6
Center, Texas (75935) 303/K6
Center Barnstead, N.H. (03225) 268/E5
Center Belpre, Ohio (†45714) 284/G7
Centerbrook, Conn. (06409) 210/F3
Centerburg, Ohio (43011) 284/E5
Center City, Minn. (55012) 255/F5
Center Conway, N.H. (03813) 268/E4
Center Cross, Va. (22437) 307/P5
Centerdale, R.I. (02911) 249/H5
Centereach, N.Y. (11720) 276/O9
Centerfield, Utah (84622) 304/C4
Center Groton, Conn. (†06340) 210/G3
Center Harbor○, N.H. (03226) 268/E4
Center Hill, Ark. (72143) 202/G3
Center Hill, Fla. (33514) 212/D3
Center Hill (lake), Tenn. 237/K9
Center Junction, Iowa (52212) 229/L4
Center Line, Mich. (48015) 250/B6
Center Lovell, Maine (04016) 243/B7
Center Montville, Maine (†04941) 243/E7
Center Moreland, Pa. (18657) 294/E7
Center Moriches, N.Y. (11934) 276/P9
Center Ossipee, N.H. (03814) 268/E4
Center Point, Ark. (71830) 202/C5
Centerpoint, Ind. (47840) 227/C6
Center Point, Iowa (52213) 229/K4
Center Point, La. (71323) 238/F4
Center Point, S. Dak. (†57070) 298/P7
Center Point, Texas (78010) 303/E8
Center Point, W. Va. (26339) 312/E4
Center Ridge, Ark. (72027) 202/E3
Center Rutland, Vt. (05736) 268/A4
Center Sandwich, N.H. (03227) 268/D4
Center Square, Ind. (†47043) 227/H7
Center Strafford, N.H. (03815) 268/E5
Centerton, Ark. (72719) 202/B1
Centerton, Ind. (46116) 227/E5
Centerton, N.J. (†08318) 273/C4
Centertown, Ky. (42328) 237/G6
Centertown, Mo. (65023) 261/H5
Centertown, Tenn. (†37110) 237/K9
Center Tuftonboro, N.H. (03816) 268/E4
Centerview, Mo. (64019) 261/E5
Center Village, Ohio (†43021) 284/E5
Centerville, Ark. (72829) 202/D3
Centerville, Del. (†19801) 245/R1
Centerville, Georgia (31028) 217/E5
Centerville, Ind. (47330) 227/H5
Centerville, Iowa (52544) 229/H7
Centerville, Kansas (66014) 232/H3
Centerville, Ky. (†41522) 237/S6
Centerville, La. (70522) 238/H7
Centerville○, Maine (†04623) 243/H6
Centerville, Mass. (02632) 249/N6
Centerville, Minn. (†55038) 255/E5
Centerville, Mo. (63633) 261/L8
Centerville, N.C. (†27549) 281/N2
Centerville, Ohio (45459) 284/B6
Centerville, Pa. (15417) 294/B6
Centerville, Pa. (16404) 294/C2
Centerville, S. Dak. (57014) 298/R7
Centerville, Tenn. (37033) 237/G9
Centerville, Texas (75833) 303/H6
Centerville, Utah (84014) 304/C3
Centerville, Wash. (98613) 310/D5
Centrahoma, Okla. (74534) 288/O5
Central, Ala. (36014) 195/F5
Central (sen. dist.), Alaska 196/H2
Central, Alaska (99730) 196/J1
Central, Ariz. (85531) 198/F6
Central, Cordillera (range), Bolivia 136/C6
Central, Cordillera (range), Colombia 126/C5
Central, Cordillera (range), Dom. Rep. 158/D5
Central, Idaho (†83241) 220/G7
Central, Ind. (47110) 227/E8
Central (Markazi) (prov.), Iran 66/G3
Central (dist.), Israel 65/B3
Central (prov.), Kenya 115/G4
Central, La. (†70723) 238/L3
Central, N. Mex. (88026) 274/A6
Central (dept.), Paraguay 144/D4
Central (Baganga), Philippines 82/F7
Central, Cordillera (range), P. Rico 161/C2
Central (reg.), Scotland 15/D4
Central, S.C. (29630) 296/B2
Central, Utah (†84701) 304/A6
Central, Utah (84722) 304/B5
Central Aboriginal Reserve, W. Australia 88/D4
Central Aboriginal Res., W. Australia 92/E3
Central African Republic 2/K5
Central African Republic 102/D4
CENTRAL AFRICAN REPUBLIC 115/C2
Central Aguirre, P. Rico 161/D3
Central Amancio Rodríguez, Cuba 158/G3
Central America 2/E5
Central América, Cuba 158/J4
Central Bedeque, Pr. Edward I. 168/E2
Central Blissville, New Bruns. 170/D3
Central Bolivia, Cuba 158/G2
Central Brasil, Cuba 158/G2
Central Bridge, N.Y. (12035) 276/M5
Central Butte, Sask. 181/E5
Central Cándido González, Cuba 158/G3
Central City, Ark. (†72923) 202/B3
Central City, Colo. (80427) 208/J3
Central City, Ill. (†62801) 222/D5
Central City, Iowa (52214) 229/K4
Central City, Ky. (42330) 237/G6
Central City, Nebr. (68826) 264/F3
Central City, Pa. (15926) 294/E5
Central City, S. Dak. (†57754) 298/B5
Central Colombia, Cuba 158/G3
Central Falls, R.I. (02863) 249/J5
Central Frank País, Cuba 158/K3
Central Greece and Euboea (reg.), Greece 45/F6
Central Guatemala, Cuba 158/J3
Central Haití, Cuba 158/G3
Centralhatchee, Georgia (†30217) 217/B4
Central Heights-Midland City, Ariz. (†85501)198/E5
Centralia, Ill. (62801) 222/D5
Centralia, Iowa (†52068) 229/M4
Centralia, Kansas (66415) 232/F2
Centralia, Mo. (65240) 261/H4
Centralia, Okla. (74336) 288/R1
Centralia, Pa. (17927) 294/K4
Centralia, Texas (75834) 303/K6
Centralia, Wash. 188/B1
Centralia, Wash. (98531) 310/C4
Centralia, W. Va. (26612) 312/E5
Central Intelligence Agency (C.I.A.), Va. 307/S2
Central Islip, N.Y. (11722) 276/O9
Central Lake, Mich. (49622) 250/D3
Central Los Reynaldos, Cuba 158/J4
Central Loynaz Echevarría, Cuba 158/J3
Central Manuel Tames, Cuba 158/K4
Central Niágara, Cuba 158/B1
Central Pacolet, S.C. (†29372) 296/D2
Central Park, Wash. (98520) 310/B3
Central Patricia, Ontario 175/B2
Central Point, Oreg. (97502) 291/D5
Central Point, Va. (†22427) 307/O4
Central Saanich, Br. Col. 184/K3
Central Square, N.Y. (13036) 276/H4
Central Station, W. Va. (26340) 312/E4
Central Ural (mts.), U.S.S.R. 52/J2
Central Valley, Calif. (96019) 204/C3
Central Valley, N.Y. (10917) 276/M8
Central Village, Conn. (06332) 210/H2
Central Village, Mass. (02790) 249/K6
Central Wedge (mt.), North. Terr. 93/C7
Centre, Ala. (35960) 195/G2
Centre (co.), Pa. 294/G4
Centre Hall, Pa. (16828) 294/G4
Centre Island, N.Y. (†11771) 276/R6
Centre-Saint-Simon, New Bruns. 170/E1
Centreville, Ala. (35042) 195/D5
Centreville, Ill. (62206) 222/B3
Centreville, Md. (21617) 245/O4
Centreville, Mich. (49032) 250/D7
Centreville, Miss. (39631) 256/B8
Centreville, New Bruns. 170/C2
Centreville, Digby, Nova Scotia 168/B4
Centreville, Kings, Nova Scotia 168/D3
Centreville (Thurman), Ohio (†45685) 284/F8
Centuria, Wis. (54824) 317/A5
Centurión, Uruguay 145/F3
Century, Fla. (32535) 212/B5
Century, W. Va. (26414) 312/F4
Cephalonia (Kefallinía) (isl.), Greece 45/E6
Ceram (isl.), Indonesia 54/P10
Ceram (isl.), Indonesia 85/H6
Cerbat (mts.), Ariz. 198/A3
Cercal, Portugal 33/B4
Cerca la Source, Haiti 158/C5
Cereal, Alberta 182/E4
Ceredo, W. Va. (25507) 312/B6
Ceres, Argentina 143/D2
Ceres, Brazil 132/D6
Ceres, Brazil 120/E4
Ceres, Calif. (95307) 204/D6
Ceres, N.Y. (14721) 276/D6
Ceres, S. Africa 118/B6
Ceres, Va. (24318) 307/F6
Ceresco, Nebr. (68017) 264/H3
Céret, France 28/E6
Cereté, Colombia 126/C3
Cerf (lake), Québec 172/B3
Cerf (isl.), Seychelles 118/H5
Cerfontaine, Belgium 27/E8
Cerignola, Italy 34/E4
Çerkeş, Turkey 63/E2
Çerkezköy, Turkey 63/C2
Çermik, Turkey 63/H3
Cernavodă, Romania 45/J3
Cernier, Switzerland 39/C2
Cernobbio, Italy 34/B2
Cerralvo (isl.), Mexico 150/E4
Cerrillos, N. Mex. (87010) 274/D3
Cerrillos, Uruguay 145/A6
Cerrito, Paraguay 144/D5
Cerritos, Calif. (†90701) 204/C11
Cerritos, Mexico 150/J5
Cerro, N. Mex. (87519) 274/D2
Cerro Aconcagua (mt.) 120/C6
Cerro Alto (mt.), Texas 303/B10
Cêrro Azul, Brazil 135/B4
Cerro Azul, Mexico 150/L6
Cerro Azul, Peru 128/D9
Cerro Castillo, Chile 138/E9
Cerro Chato, Cerro Largo, Uruguay 145/F3
Cerro Chato, Rivera, Uruguay 145/D2
Cerro Chato, Treinta y Tres, Uruguay 145/D4
Cerro Colorado, Uruguay 145/D4
Cerro Corá, Paraguay 144/E3
Cerro de las Armas, Uruguay 145/B5
Cerro de las Cuentas, Uruguay 145/E3
Cerro de Pasco, Peru 120/B4
Cerro de Pasco, Peru 128/D8
Cerro de San Antonio, Colombia 126/C
Cerro Gordo, Ill. (61818) 222/E4
Cerro Gordo (co.), Iowa 229/G2
Cerro Gordo, N.C. (28430) 281/M6
Cerro Gordo (pt.), P. Rico 161/D1
Cerro Gordo, Tenn. (38322) 237/E10
Cerro Largo (dept.), Uruguay 145/E3
Cerro Manantiales, Chile 138/F10
Cerulean, Ky. (42215) 237/F7
Cervera, Spain 33/G2
Cervera del Río Alhama, Spain 33/E1
Cervera de Pisuerga, Spain 33/D1
Cerveteri, Italy 34/E6
Cervione, France 28/B6
Cesano, Italy 34/F6
César (dept.), Colombia 126/D3
César (riv.), Colombia 126/D2
Cesena, Italy 34/D2
Cesenatico, Italy 34/D2
Cēsis, U.S.S.R. 53/C2
Cēsis, U.S.S.R. 52/C3
Česká Kamenice, Czech. 41/C1
Česká Lípa, Czech. 41/C1
Česká Třebová, Czech. 41/D2
České Budějovice, Czech. 41/C2
Český Brod, Czech. 41/C1
Český Krumlov, Czech. 41/C2
Český Těšín, Czech. 41/E2
Çeşme, Turkey 63/B3
Céspedes, Cuba 158/G2
Cessford, Alberta 182/E4
Cessnock-Bellbird, N.S. Wales 88/J6
Cessnock-Bellbird, N.S. Wales 97/F3
Cestos (riv.), Liberia 106/C7
Cetinje, Yugoslavia 45/D4
Çetinkaya, Turkey 63/G3
Ceuta, Spain 106/C1
Ceuta, Spain 7/D5
Ceuta, Spain 102/B1
Ceuta, Spain 33/D5
Cévennes (mts.), France 28/E5
Cevio, Switzerland 39/G4
Cevizli, Turkey 63/D4
Ceyhan, Turkey 63/F4
Ceyhan (riv.), Turkey 63/F4
Ceyhan (riv.), Turkey 59/C2
Ceylânpınar, Turkey 63/H4
Ceylon (Sri Lanka) 54/K9
Ceylon, Minn. (56121) 255/D7
Ceylon, Sask. 181/G6
Chabás, Argentina 143/F6
Chaca, Chile 138/B1
Chacabuco, Argentina 143/F7
Chacabuco, Chile 138/G2
Chacachacare (isl.), Trin. & Tob. 161/A10
Chacahoula, La. (†70395) 238/J7
Chacalluta, Chile 138/A1
Chachacomani, Bolivia 136/A6
Chachapoyas, Peru 128/D6
Chachapoyas, Peru 120/B3
Chachoengsao, Thailand 72/D4
Chachro, Pakistan 68/C3
Chaco (prov.), Argentina 143/D2
Chaco (mesa), N. Mex. 274/B3
Chaco (riv.), N. Mex. 274/A2
Chaco (dept.), Paraguay 144/B-C2
Chaco Austral (reg.), Argentina 143/D2
Chaco Boreal (reg.), Paraguay 144/B2-3
Chaco Central (reg.), Argentina 143/D1
Chaco Culture Nat'l Hist. Park, N. Mex. 274/B2
Chacoma, Bolivia 136/A6
Chacon (cape), Alaska 196/N2
Chacon, N. Mex. (87713) 274/D2
Chacuaco (creek), Colo. 208/M8
Chad 2/K5
Chad 102/D3
Chad (lake) 102/D3
CHAD 111/C4
Chad (lake), Chad 111/C5
Chad (lake), Niger 106/G6
Chad (lake), Nigeria 106/G6
Chadan, U.S.S.R. 48/K4
Chadbourn, N.C. (28431) 281/M6
Chadron, Nebr. (69337) 264/B2
Chadwick, Ill. (61014) 222/D1
Chadwick, Mo. (65629) 261/G9
Chadwick Acres, N.C. (†28445) 281/P6
Chadwicks, N.Y. (13319) 276/K4
Chadyr-Lunga, U.S.S.R. 52/C5
Chaffee (co.), Colo. 208/G5
Chaffee, Mo. (63740) 261/N8
Chaffee, N.Y. (14030) 276/C5
Chaffee, N. Dak. (58014) 282/R6
Chaffers (isl.), Chile 138/D5
Chafurray, Colombia 126/D6
Chagai (hills), Afghanistan 68/A3
Chagai, Pakistan 59/H4
Chagai, Pakistan 68/A3
Chagai (hills), Pakistan 68/A3
Chagai (hills), Pakistan 59/H4
Chagda, U.S.S.R. 48/O4
Chaghcharan, Afghanistan 68/B2
Chagoda, U.S.S.R. 52/E3
Chagoness, Sask. 181/G3
Chagos (arch.), Br. Ind. Ocean Terr. 2/N6
Chagos (arch.), Br. Ind. Ocean Terr. 54/J10
Chagrin (riv.), Ohio 284/J8
Chagrin Falls, Ohio (44022) 284/J9
Chaguanas, Trin. & Tob. 161/B10
Chaguaramas, Trin. & Tob. 161/A10
Chaguaramas, Venezuela 124/E3
Chaguaya, Bolivia 136/C7
Chagulak (isl.), Alaska 196/D4
Chahal, Guatemala 154/C3
Chahar Borjak, Afghanistan 59/H3
Chahar Borjak, Afghanistan 68/A2
Chah Bahar, Afghanistan 59/H4
Chah Bahar, Iran 66/M8
Chahuites, Mexico 150/M8
Chai Badan, Thailand 72/D4
Chaibasa, India 68/F4
Chai Buri, Thailand 72/D3
Chainat, Thailand 72/D4
Chain-O-Lakes, Mo. (†65641) 261/E9
Chaira, Laguna (lake), Colombia 126/C7
Chaires, Fla. (†32302) 212/B1
Chaitén, Chile 138/E4
Chaiya, Thailand 72/C5
Chaiyaphum, Thailand 72/D4
Chajarí, Argentina 143/G5
Chajul, Guatemala 154/B3
Chake Chake, Tanzania 115/H5
Chala, Peru 128/E10
Chalais, Switzerland 39/E4
Chalatenango, El Salvador 154/C3
Chalchihuites, Mexico 150/G5
Chalco de Díaz Covarrubias, Mexico 150/M1
Chaleur (bay), New Bruns. 170/E1
Chaleur (bay), Québec 172/C2
Chaleur (bay), Québec 174/D3
Chalfont, Pa. (18914) 294/M5
Chalhuanca, Peru 128/F10
Chaling, China 77/H6
Chalk (creek), Utah 304/C3
Chalk River, Ontario 175/E3
Chalk River, Ontario 177/G1
Chalkyitsik, Alaska (99788) 196/K1
Challacollo, Bolivia 136/B6
Challana, Bolivia 136/A4
Challapata, Bolivia 136/B6
Challenger (mts.), N.W. Terrs. 187/L1
Challis, Idaho (83226) 220/D5
Challviri (salt dep.), Bolivia 136/B8
Chalmers, Ind. (47929) 227/D3
Chalmette, La. (70043) 238/P4
Chalna Port, Bangladesh 68/F4
Chalonnes-sur-Loire, France 28/C4
Châlons-sur-Marne, France 28/F3
Chalon-sur-Saône, France 28/F4
Chaltel, Cerro (mt.), Chile 138/E8
Chalus, Iran 59/F2
Chalus, Iran 66/G2
Chalybeate, Miss. (38684) 256/G1
Chalybeate Springs, Georgia (†31816) 217/C5
Chalybeate Springs, N.C. (†27526) 281/M3
Cham, Switzerland 39/F2
Cham, W. Germany 22/E4
Chama, Colo. (81126) 208/J8
Chama, N. Mex. (87520) 274/C2
Chaman, Pakistan 68/B2
Chaman, Pakistan 59/J3
Chamba, India 68/D2
Chambal (riv.), India 68/D3
Chambas, Cuba 158/F2
Chamberino, N. Mex. (88027) 274/C6
Chamberlain (creek), Idaho 220/C4
Chamberlain (lake), Maine 243/E3
Chamberlain, Sask. 181/F5
Chamberlain, S. Dak. (57325) 298/L6
Chamberlain, Uruguay 145/C3
Chamberlin, La. (†70767) 238/J1
Chambers (co.), Ala. 195/H5
Chambers, Ariz. (86502) 198/F3
Chambers, Nebr. (68725) 264/F2
Chambers (co.), Texas 303/K8
Chambers (isl.), Wis. 317/M5
Chambersburg, Ill. (62323) 222/C4
Chambersburg, Ind. (†47454) 227/E7
Chambersburg, Ohio (†45631) 284/F8
Chambersburg, Pa. (17201) 294/G6
Chambéry, France 28/F5
Chambeshi (riv.), Zambia 115/F6
Chambeyron (mt.), France 28/G5
Chambeyron (mt.), Italy 34/A2
Chambi, Jebel (mt.), Tunisia 106/F2
Chamblee, Georgia (30341) 217/K1
Chambly (co.), Québec 172/J4
Chambly, Québec 172/J4
Chambord, France 28/D4
Chambord, Québec 172/E1
Chamdo (Qamdo), China 77/E5
Chame (pt.), Panama 154/H6
Chamela (bay), Mexico 150/G7
Chamical, Argentina 143/C3
Chamisal, N. Mex. (†87521) 274/D2
Chamizal Nat'l Mem., Texas 303/A10
Chamizo, Uruguay 145/D5
Chamo (lake), Ethiopia 111/G6
Chamois, Mo. (65024) 261/J5
Chamonix-Mont-Blanc, France 28/G5
Chamoson, Switzerland 39/D4
Champ, Mo. (†63042) 261/O2
Champagne (trad. prov.), France 29
Champagne, Yukon 187/E3
Champaign, Ill. 188/J2
Champaign (co.), Ill. 222/E3
Champaign, Ill. (61820) 222/E3
Champaign (co.), Ohio 284/C5
Champasak, Laos 72/E4
Champdani, India 68/F1
Champerico, Guatemala 154/A3
Champéry, Switzerland 39/C4
Champex, Switzerland 39/D4
Champigny-sur-Marne, France 28/C2
Champion, Alberta 182/D4
Champion, Mich. (49814) 250/B2
Champion, Nebr. (69023) 264/C4
Champlain (lake) 188/M2
Champlain, N.Y. (12919) 276/N1
Champlain (lake), N.Y. 276/O1
Champlain (county), Québec 174/C3
Champlain (co.), Québec 172/E2
Champlain, Québec 172/E3
Champlain (lake), Québec 172/D4
Champlain (lake), Vt. 268/A2
Champlain, Va. (22438) 307/O4
Champlain Park, N.Y. (†12901) 276/O1
Champlin, Minn. (55316) 255/G5
Champney's West, Newf. 166/D2
Champotón, Mexico 150/O7
Chamusa, Sierra (mts.), Colombia 126/C6
Chamusca, Portugal 33/B3
Chan, Ko (isl.), Thailand 72/C5
Chana, Ill. (61015) 222/D2
Chañaral, Chile 120/B5
Chañaral, Chile 138/A6
Chañaral (isl.), Chile 138/A7
Chancay, Peru 128/D8
Chance, Ala. (36729) 195/C7
Chance, Ky. (†42728) 237/L7
Chance, Md. (21816) 245/P8
Chance Cove, Newf. 166/D2
Chance Cove (cape), Newf. 166/D2
Chance Harbour, New Bruns. 170/D3
Chancellor, Ala. (36316) 195/G8
Chancellor, Alberta 182/D4
Chancellor, S. Dak. (57015) 298/R7
Chancellorsville, Va. (†22401) 307/N4
Chanco, Chile 138/A11
Chancy, Switzerland 39/A4
Chandalar, Alaska (†99726) 196/J1
Chandalar (riv.), Alaska 196/J1
Chandalar, East Fork (riv.), Alaska 196/J1
Chandeleur (isls.), La. 238/N7
Chandeleur (sound), La. 238/M7
Chanderi, India 68/D4
Chandernagore, India 68/F1
Chandigarh (terr.), India 68/D2
Chandigarh, India 68/D2
Chandler, Ariz. (85224) 198/D5
Chandler, Ind. (47610) 227/C8
Chandler, Minn. (56122) 255/C7
Chandler, Okla. (74834) 288/N3
Chandler, Que. 162/K6
Chandler, Québec 174/E3
Chandler, Québec 172/D2
Chandler, Texas (75758) 303/J5
Chandler Springs, Ala. (†35160) 195/F4
Chandlers Valley, Pa. (16312) 294/D2
Chandlersville, Ohio (43727) 284/G6
Chandlerville, Ill. (62627) 222/C3
Chandmanĭ, Mongolia 77/E2
Chandolin, Switzerland 39/E4
Chandos (lake), Ontario 177/G3
Chandrapur, India 68/D5
Chaneysville, Pa. (†21530) 294/F6
Chang, Ko (isl.), Thailand 72/D4
Changane (riv.), Mozambique 118/E4
Changbaek-sanmaek (mts.), N. Korea 81/D2
Changchih (Changzhi), China 77/H4
Changchow (Zhangzhou), China 77/J7
Changchow (Changzhou), China 77/J5
Changchun, China 77/K3
Changchun, China 54/O5
Changchun, China 2/R3
Changde (Changteh), China 77/H6
Changde, China 54/N7
Change Islands, Newf. 166/D4
Changewater, N.J. (07831) 273/D2
Changhua, China 77/K7
Changhŭng, S. Korea 81/C6
Changji, China 77/C3
Changjiang, China 77/G8
Chang Jiang (Yangtze) (riv.), China 2/Q4
Chang Jiang (Yangtze) (riv.), China 54/N6
Chang Jiang (Yangtze) (riv.), China 77/K5
Changjin (res.), N. Korea 81/C3
Chang Khoeng, Thailand 72/C3
Changling, China 77/K3
Changsha, China 77/H6
Changsha, China 2/Q4
Changsha, China 54/N7
Changshun, China 77/G6
Changsŏng, S. Korea 81/C6
Changteh (Changde), China 77/H6
Changuinola, Panama 154/F6
Changwu, China 77/G4
Changyang, China 77/H5
Changyeh (Zhangye), China 77/F4
Changyŏn, N. Korea 81/B4

Changzhi (Changchih), China 77/H4
Changzhi, China 54/N6
Changzhou (Changchow), China 77/K5
Chanhassen, Minn. (55317) 255/F6
Chankiang (Zhanjiang), China 77/H7
Channahon, Ill. (60410) 222/E2
Channel (isls.) 7/D4
Channel (isl.), Manitoba 179/B2
CHANNEL ISLANDS 10/E6
CHANNEL ISLANDS 13/E8
Channel Islands Nat'l Park, Calif. 204/E11
Channel-Port aux Basques, Newf. 166/C4
Channel-Port aux Basques, Newf. 162/L6
Channelview, Texas (77530) 303/K1
Channing, Mich. (49815) 250/B2
Channing, Texas (79018) 303/B2
Chantada, Spain 33/C1
Chanthaburi, Thailand 72/D4
Chantilly, France 28/E3
Chantilly, Va. (22021) 307/O3
Chantonnay, France 28/C4
Chantrey (inlet), N.W. Terrs. 187/J3
Chanute, Kansas (66720) 232/G4
Chanute A.F.B., Ill. 222/E3
Chao, Peru 128/C7
Chao'an (Chaochow), China 77/J7
ChaoChao (isl.), Portugal 33/B2
Chao Phraya, Mae Nam (riv.), Thailand 72/D4
Chaotung (Zhaotung), China 77/F6
Chaoyang, Guangdong, China 77/J7
Chaoyang, Liaoning, China 77/J3
Chapa, Vietnam 72/E2
Chapacura, Bolivia 136/A2
Chapais, Québec 174/B3
Chapala (lake), Mexico 150/H6
Chapanoke, N.C. (†27944) 281/S2
Chaparé (riv.), Bolivia 136/C5
Chaparra, Cuba 158/H3
Chaparral, Colombia 126/C6
Chapayevsk, U.S.S.R. 48/F4
Chapayevsk, U.S.S.R. 52/G4
Chapecó, Brazil 132/C9
Chapel, W. Va. (†26624) 312/E5
Chapel Arm, Newf. 166/D2
Chapel en le Frith, England 13/J2
Chapel Hill, Ark. (†71832) 202/B5
Chapel Hill, Ind. (†47436) 227/E6
Chapel Hill, N.C. (27514) 281/L3
Chapel Hill, Tenn. (37034) 237/H9
Chapelton, Jamaica 158/J6
Chapicuy, Uruguay 145/B2
Chapin, Ill. (62628) 222/C4
Chapin, Iowa (50427) 229/G3
Chapin, S.C. (29036) 296/E3
Chapleau, Ont. 162/H6
Chapleau, Ontario 175/D3
Chapleau, Ontario 177/J5
Chaplin○, Conn. (06235) 210/G1
Chaplin, Ky. (40012) 237/L5
Chaplin (riv.), Ky. 237/L5
Chaplin, Sask. 181/E5
Chaplin (lake), Sask. 181/E5
Chapman, Ala. (36015) 195/E7
Chapman (pt.), Conn. 210/F3
Chapman, Kansas (67431) 232/E3
Chapman○, Maine (†04757) 243/G2
Chapman, Nebr. (68827) 264/F3
Chapmansboro, Tenn. (37035) 237/G8
Chapmanville, W. Va. (25508) 312/B7
Chappaquiddick (isl.), Mass. 249/N7
Chappell, Ky. (40816) 237/P7
Chappell, Nebr. (69129) 264/B3
Chappell (isls.), Tasmania 99/D2
Chappell Hill, Texas (77426) 303/H7
Chappells, S.C. (29037) 296/D3
Chapra, India 68/F3
Chaptico, Md. (20621) 245/M7
Chapultepec, Mexico 150/A1
Chaqui, Bolivia 136/C6
Chara, U.S.S.R. 48/M4
Charadai, Argentina 143/D2
Charagua, Bolivia 136/D6
Charagua, Sierra de (mts.), Bolivia 136/D6
Charagua, Paraguay 144/D3
Charak, Iran 66/J7
Charambirá (pt.), Colombia 126/B5
Charaña, Bolivia 136/A5
Charata, Argentina 143/D2
Charbon, N.S. Wales 97/F3
Charbonneau, N. Dak. (†58831) 282/C4
Charcas, Mexico 150/J5
Charcot (isl.) 5/C15
Chard, Alberta 182/E2
Chard, England 13/E7
Chard, England 10/E5
Chardon, Ohio (44024) 284/H2
Chardonnière, Haiti 158/A6
Chardzhou, U.S.S.R. 54/H6
Chardzhou, U.S.S.R. 48/G6
Charente (dept.), France 28/D5
Charente (riv.), France 28/C5
Charente-Maritime (dept.), France 28/C5
Charenton, La. (70523) 238/H7
Charenton-le-Pont, France 28/B2
Charette, Québec 172/D3
Charikar, Afghanistan 68/B1
Charikar, Afghanistan 59/J2
Charing, Georgia (†31058) 217/D6
Charing Cross, Ontario 177/B5
Chariton, Iowa (50049) 229/G6
Chariton (riv.), Iowa 229/G7
Chariton (co.), Mo. 261/F3
Chariton (riv.), Mo. 261/G1
Charity, Guyana 131/B2
Charity, Mo. (†65644) 261/G7
Charkhlia (Ruoqiang), China 77/C4
Charlack, Mo. (†63101) 261/P2
Charlemagne, Québec 172/H4
Charlemont○, Mass. (01339) 249/C2
Charleroi, Belgium 27/E8
Charleroi, Pa. (15022) 294/C5
Charles, Georgia (†30474) 217/H6
Charles (co.), Md. 245/K6
Charles (riv.), Mass. 249/C7
Charles (isl.), N.W. Terrs. 187/L3
Charles (cape), Va. 188/L3
Charles (cape), Va. 307/R6
Charlesbourg, Québec 172/J3
Charles City, Iowa (50616) 229/H2
Charles City (co.), Va. 307/O6
Charles City, Va. (23030) 307/O6
Charles Mix (co.), S. Dak. 298/M7
Charles Mound (hill), Ill. 222/C1
Charleston, Ark. (72933) 202/B3
Charleston, Ill. (61920) 222/E4
Charleston, Kansas (†67853) 232/B4
Charleston○, Maine (04422) 243/F5
Charleston, Miss. (38921) 256/D2
Charleston, Mo. (63834) 261/O9
Charleston, Nev. (†89801) 266/F1
Charleston (peak), Nev. 266/F6
Charleston (lake), Ontario 177/J3
Charleston, Oreg. (97420) 291/C4
Charleston, S.C. 146/L6
Charleston, S.C. 188/L4
Charleston (co.), S.C. 296/H6
Charleston, S.C. (*29401) 296/G6
Charleston, Tenn. (37310) 237/M10
Charleston, Utah (†84032) 304/C3
Charleston (cap.), W. Va. 188/K3
Charleston (cap.), W. Va. 146/K6
Charleston (cap.), W. Va. (*25301) 312/C6
Charleston A.F.B., S.C. 296/G6
Charlestown, Ind. (47111) 227/F8
Charlestown, Md. (21914) 245/P2
Charlestown, N.H. (03603) 268/C5
Charlestown○, N.H. (03603) 268/C5
Charlestown, R.I. (02813) 249/H7
Charlestown, St. Chris.-Nevis 161/C11
Charlestown (cap.), Nevis, St. Chris.-Nevis 156/F3
Charles Town, W. Va. (25414) 312/L4
Charlestown-Bellahy, Ireland 17/D4
Charleville, Australia 87/E8
Charleville (Rathluirc), Ireland 17/D7
Charleville, Queensland 95/C5
Charleville, Queensland 88/H5
Charleville-Mézières, France 28/F3
Charlevoix (co.), Mich. 250/D3
Charlevoix, Mich. (49720) 250/D3
Charlevoix (lake), Mich. 250/D3
Charlevoix-Est (co.), Québec 172/G2
Charlevoix-Est (county), Québec 174/C3
Charlevoix-Ouest (co.), Québec 172/G2
Charlevoix-Ouest (county), Québec 174/C3
Charley, Ky. (†41230) 237/R5
Charlie Lake, Br. Col. 184/G2
Charlo, Mont. (59824) 262/B3
Charlo, New Bruns. 170/D1
Charlo (riv.), New Bruns. 170/D1
Charlotte, Ark. (72522) 202/H2
Charlotte (lake), Br. Col. 184/E4
Charlotte (harb.), Fla. 188/K5
Charlotte (co.), Fla. 212/E5
Charlotte (harb.), Fla. 212/D5
Charlotte, Iowa (52731) 229/M5
Charlotte○, Maine (†04666) 243/J5
Charlotte, Mich. (48813) 250/E6
Charlotte (co.), New Bruns. 170/C3
Charlotte, N.C. (*28201) 281/H4
Charlotte, N.C. 188/L3
Charlotte, N.C. 146/K6
Charlotte (lake), Nova Scotia 168/F4
Charlotte, Tenn. (37036) 237/G8
Charlotte, Texas (78011) 303/F9
Charlotte○, Vt. (05445) 268/A3
Charlotte (co.), Va. 307/L6
Charlotte Amalie (cap.), Virgin Is. (U.S.) 156/H1
Charlotte Amalie (cap.), Virgin Is. (U.S.) 161/B4
Charlotte Court House, Va. (23923) 307/L6
Charlotte Hall, Md. (20622) 245/M7
Charlotte Harbor, Fla. (†33950) 212/E5
Charlottenberg, Sweden 18/H6
Charlottenburg, W. Germany 22/E4
Charlottesville, Ind. (46117) 227/F5
Charlottesville, Va. 188/L3
Charlottesville (I.C.), Va. (*22901) 307/M4
Charlottetown, Newf. 166/D2
Charlottetown, Newf. 166/C3
Charlottetown (cap.), P.E.I. 146/M5
Charlottetown (cap.), P.E.I. 162/K6
Charlottetown (cap.), Pr. Edward I. 168/E2
Charlotteville, N.Y. (12036) 276/L5
Charlotte Waters, North. Terr. 93/D8
Charlson, N. Dak. (58726) 282/E3
Charlton (co.), Georgia 217/H9
Charlton○, Mass. (01507) 249/F4
Charlton (isl.), N.W.T. 162/H5
Charlton, Ontario 177/K5
Charlton, Ontario 175/D3
Charlton, Victoria 97/B5
Charlton City, Mass. (01508) 249/F4
Charlton Depot, Mass. (01509) 249/F4
Charlton Kings, England 13/F6
Charmco, W. Va. (25958) 312/E6
Charmey, Switzerland 39/D3
Charny, Québec 172/J3
Charolles, France 28/F4
Charouine, Algeria 106/D3
Charqueada Aguas de São Pedro, Brazil 135/B3
Charron (lake), Manitoba 179/G2
Charsk, U.S.S.R. 48/J5
Charter Oak, Iowa (51439) 229/C4
Charters, Ky. (†41179) 237/P3
Charters Towers, Australia 87/E7
Charters Towers, Queensland 95/C4
Charters Towers, Queensland 88/H4
Chartierville, Québec 172/F4
Chartley, Mass. (02712) 249/K5
Chartres, France 28/D3
Chascomús, Argentina 143/H7
Chase, Ala. (†35811) 195/E1
Chase, Br. Col. 184/H5
Chase (co.), Kansas 232/F3
Chase, Kansas (67524) 232/D3
Chase, La. (71324) 238/G2
Chase, Md. (21027) 245/N3
Chase, Mich. (49623) 250/D5
Chase (co.), Nebr. 264/C4
Chase (lake), N. Dak. 282/M5
Chaseburg, Wis. (54621) 317/D8
Chase City, Va. (23924) 307/M7
Chaseley, N. Dak. (58423) 282/L5
Chase Mills, N.Y. (13621) 276/K1
Chase N.A.S., Texas 303/G9
Chaska, Minn. (55318) 255/F6
Chaska, Tenn. (†37729) 237/N7
Chasm, Br. Col. 184/G4
Chasŏng, N. Korea 81/C3
Chasseron (mt.), Switzerland 39/C3
Chastang, Ala. (36517) 195/B8
Chastre, Belgium 27/F7
Chaswood, Nova Scotia 168/E3
Chataignier, La. (70524) 238/F5
Chatanika, Alaska (99731) 196/J1
Chatawa, Miss. (39632) 256/D8
Chatcolet, Idaho (†83851) 220/B2
Chateaubelair, St. Vin. & Grens. 161/A8
Châteaubriant, France 28/C4
Château-Chinon, France 28/E4
Château-d'Oex, Switzerland 39/D4
Château-du-Loir, France 28/D4
Châteaudun, France 28/D3
Châteaugay, N.Y. (12920) 276/N1
Châteaugay, Upper (lake), N.Y. 276/M1
Château-Gontier, France 28/C4
Châteauguay (co.), Québec 172/D4
Châteauguay, Québec 172/H4
Châteauguay-Centre, Québec 172/H4
Châteauneuf-sur-Loire, France 28/E4
Château-Renault, France 28/D4
Château-Richer, Québec 172/F3
Châteauroux, France 28/D4
Château-Salins, France 28/G3
Château-Thierry, France 28/E3
Châteaux (pt.), Guadeloupe 161/B6
Chateh, Alberta 182/A5
Châtelet, Belgium 27/F8
Châtellerault, France 28/D4
Châtel-Saint-Denis, Switzerland 39/C3
Chater, Manitoba 179/C5
Chatfield, Ark. (†72348) 202/K3
Chatfield, Manitoba 179/E4
Chatfield, Minn. (55923) 255/F7
Chatfield, Ohio (44825) 284/E4
Chatham (str.), Alaska 196/M1
Chatham (sound), Br. Col. 184/B3
Chatham (isl.), Chile 138/D9
Chatham, England 13/J8
Chatham, England 10/G5
Chatham (co.), Georgia 217/K6
Chatham, Ill. (62629) 222/D4
Chatham, La. (71226) 238/F2
Chatham, Mass. (02633) 249/P6
Chatham○, Mass. (02633) 249/P6
Chatham, Mich. (49816) 250/B2
Chatham, Miss. (38731) 256/B4
Chatham, New Bruns. 162/K6
Chatham, New Bruns. 170/E1
Chatham○, N.H. (†04037) 268/E3
Chatham, N.J. (07928) 273/E2
Chatham, N.Y. (12037) 276/N6
Chatham (isls.), N. Zealand 87/J10
Chatham (isl.), N. Zealand 100/D7
Chatham (isls.), N. Zealand 100/D7
Chatham (co.), N.C. 281/L3
Chatham, Ontario 177/B5
Chatham, Va. (24531) 307/K7
Chatham Center, N.Y. (†12037) 276/N6
Chatham Head, New Bruns. 170/E2
Chatham Port, Mass. (†02650) 249/P6
Châtillon, France 28/B2
Châtillon-sur-Indre, France 28/D4
Châtillon-sur-Seine, France 28/F4
Chato, Cerro (mt.), Argentina 143/B5
Chato, Cerro (mt.), Chile 138/E4
Chatom, Ala. (36518) 195/B8
Chatou, France 28/A1
Chatrapur, India 68/F5
Chatsworth, Calif. (91311) 204/B10
Chatsworth, Georgia (30705) 217/C1
Chatsworth, Ill. (60921) 222/E3
Chatsworth, Iowa (51011) 229/A3
Chatsworth, N.J. (08019) 273/D4
Chatsworth, Ontario 177/D3
Chattahoochee (riv.) 188/K4
Chattahoochee (riv.), Ala. 195/H8
Chattahoochee, Fla. (32324) 212/B1
Chattahoochee (riv.), Fla. 212/B1
Chattahoochee (co.), Georgia 217/C6
Chattahoochee (riv.), Georgia 217/B8
Chattahoochee River Nat'l Rec. Area, Georgia 217/K1
Chattanooga, Ohio (†45882) 284/A4
Chattanooga, Okla. (73528) 288/J6
Chattanooga, Tenn. 188/J3
Chattanooga, Tenn. 146/K6
Chattanooga, Tenn. (*37401) 237/K10
Chattaroy, Wash. (99003) 310/H3
Chattaroy, W. Va. (25667) 312/B7
Chatteris, England 13/H5
Chattooga (riv.), Ala. 195/H2
Chattooga (co.), Georgia 217/B1
Chattooga (riv.), Georgia 217/A2
Chattooga (riv.), Georgia 217/F1
Chattooga (riv.), S.C. 296/A2
Chatuge (lake), Georgia 217/E1
Chatuge (lake), N.C. 281/B5
Chatwood, Pa. (†19380) 294/L6
Chaud (lake), Québec 172/C3
Chaudière (riv.), Québec 172/G4
Chauk, Burma 72/B2
Chauk, Burma 54/L7
Chaukan (pass), Burma 72/C1
Chaumont, France 28/F3
Chaumont, N.Y. (13622) 276/H2
Chauncey, Georgia (31011) 217/F6
Chauncey, Ohio (45719) 284/F7
Chauny, France 28/E3
Chau Phu, Vietnam 72/E5
Chauques (isls.), Chile 138/D4
Chautauqua, Ill. (†62028) 222/C5
Chautauqua (co.), Kansas 232/F4
Chautauqua, Kansas (67334) 232/F4
Chautauqua (co.), N.Y. 276/B6
Chautauqua, N.Y. (14722) 276/A6
Chautauqua (lake), N.Y. 276/A6
Chauvin, Alberta 182/E3
Chauvin, La. (70344) 238/J8
Chavantes, Serra dos (range), Brazil 132/D5
Chaves, Brazil 132/D3
Chaves (Santa Cruz) (isl.), Ecuador 128/C9
Chaves (co.), N. Mex. 274/E5
Chaves, Portugal 33/C2
Chavies, Ky. (41727) 237/P6
Chavornay, Switzerland 39/C3
Chayanta, Bolivia 136/B6
Chaykovskiy, U.S.S.R. 52/H3
Chazelles-sur-Lyon, France 28/F5
Chazy, N.Y. (12921) 276/N1
Chazy (lake), N.Y. 276/N1
Cheadle, Alberta 182/D4
Cheadle, England 13/E5
Cheadle and Gatley, England 13/H2
Cheadle and Gatley, England 10/G2
Cheaha (mt.), Ala. 195/G4
Cheam View, Br. Col. 184/M3
Cheap (chan.), Chile 138/D7
Cheat (lake), W. Va. 312/G3
Cheat (riv.), W. Va. 312/G3
Cheatham (co.), Tenn. 237/G8
Cheatham (dam), Tenn. 237/G8
Cheatham (lake), Tenn. 237/H8
Cheb, Czech. 41/B1
Chebanse, Ill. (60922) 222/F3
Chebeague Island, Maine (04017) 243/C8
Chebogue (harb.), Nova Scotia 168/B5
Cheboksary, U.S.S.R. 7/J3
Cheboksary, U.S.S.R. 52/G3
Cheboksary, U.S.S.R. 48/E4
Cheboygan, Mich. 188/K1
Cheboygan (co.), Mich. 250/E3
Cheboygan, Mich. (49721) 250/E3
Chech, Erg (des.), Algeria 106/D3
Chech, Erg (des.) 102/B2
Chech, Erg (des.), Mali 106/D4
Chechaouene, Morocco 106/D1
Chechen-Ingush A.S.S.R., U.S.S.R. 48/E5
Chechen-Ingush A.S.S.R., U.S.S.R. 52/G6
Chech'ŏn, S. Korea 81/D5
Check, Va. (24072) 307/H6
Checker Hall, Barbados 161/B8
Checotah, Okla. (74426) 288/R4
Chedabucto (bay), Nova Scotia 168/G3
Cheduba (isl.), Burma 72/B3
Cheektowaga, N.Y. (14225) 276/C5
Cheekye, Br. Col. 184/F5
Cheesequake, N.J. (08879) 273/E3
Cheesman (lake), Colo. 208/J4
Chefoo (Yantai), China 77/K4
Chefornak, Alaska (99561) 196/F2
Chegdomyn, U.S.S.R. 48/O4
Chegga (well), Mauritania 106/C3
Chehalis (lake), Br. Col. 184/L3
Chehalis, Wash. (98532) 310/C4
Chehalis (pt.), Wash. 310/A4
Chehalis (riv.), Wash. 310/B4
Chehalis Ind. Res., Wash. 310/B4
Chehar Deh, Iran 66/K4
Cheju (isl.), S. Korea 54/O6
Cheju, S. Korea 81/C7
Cheju (isl.), S. Korea 81/C7
Cheju (str.), S. Korea 81/C7
Chekiang (Zhejiang), China 77/K6
Chelan, Sask. 181/H3
Chelan (lake), Wash. 188/B1
Chelan (co.), Wash. 310/E3
Chelan, Wash. (98816) 310/E3
Chelan (dam), Wash. 310/E3
Chelan (lake), Wash. 310/E2
Chelan (range), Wash. 310/E2
Chelan Falls, Wash. (98817) 310/E3
Cheleken, U.S.S.R. 48/F6
Chelia (mt.), Algeria 106/F1
Chelif (riv.), Algeria 106/E1
Chelkar, U.S.S.R. 48/F5
Chelles, France 28/C1
Chelm (prov.), Poland 47/F3
Chelm, Poland 47/F3
Chelmno, Poland 47/D2
Chelmsford, England 13/J7
Chelmsford, England 10/G5
Chelmsford○, Mass. (01824) 249/J2
Chelmza, Poland 47/D2
Chelsea, Ala. (35043) 195/E4
Chelsea, Ind. (†47138) 227/F7
Chelsea, Iowa (52215) 229/J5
Chelsea○, Maine (†04345) 243/D7
Chelsea, Mass. (02150) 249/D6
Chelsea, Mich. (48118) 250/E6
Chelsea, Okla. (74016) 288/P1
Chelsea, S. Dak. (57431) 298/M3
Chelsea, Vt. (05038) 268/C4
Chelsea, Victoria 88/L8
Chelsea, Victoria 97/J6
Chelsea, Wis. (54419) 317/F5
Cheltenham, England 13/E6
Cheltenham, England 10/F5
Cheltenham, Md. (20623) 245/L6
Cheltenham○, Pa. (19012) 294/M5
Chelva, Spain 33/F3
Chelyabinsk, U.S.S.R. 2/N3
Chelyabinsk, U.S.S.R. 54/H4
Chelyabinsk, U.S.S.R. 48/G4
Chelyuskin (cape), U.S.S.R. 54/N2
Chelyuskin (cape), U.S.S.R. 4/B4
Chelyuskin (cape), U.S.S.R. 48/M2
Chemainus, Br. Col. 184/J3
Chemawa, Oreg. (97306) 291/A3
Chemba, Mozambique 118/E3
Chembur, India 68/B7
Chemehuevi Valley Ind. Res., Calif. 204/L9
Chemeketa Park-Redwood Estates, Calif. (95044) 204/K4
Chemnitz (Karl-Marx-Stadt), E. Germany 22/E3
Chemquasabamticook (lake), Maine 243/D3
Chemult, Oreg. (97731) 291/F4
Chemung, Ill. (†60033) 222/E1
Chemung (co.), N.Y. 276/G6
Chemung, N.Y. (14825) 276/G6
Chenab (riv.), India 68/C2
Chenab (riv.), Pakistan 68/C2
Chenab (riv.), Pakistan 59/K4
Chenachane, Algeria 106/D3
Chena Hot Springs, Alaska (†99701) 196/J1
Chenango (co.), N.Y. 276/J6
Chenango (riv.), N.Y. 276/J6
Chenango Bridge, N.Y. (13745) 276/J6
Chenango Forks, N.Y. (13746) 276/J6
Chen Barag, China 77/J2
Chêne-Bougeries, Switzerland 39/B4
Chenequa, Wis. (†53029) 317/J1
Chénéville, Québec 172/B4
Cheney, Kansas (67025) 232/E4
Cheney (res.), Kansas 232/E4
Cheney, Wash. (99004) 310/H3
Cheneyville, Ill. (†60942) 222/F3
Cheneyville, La. (71325) 238/F4
Chengchow (Zhengzhou), China 77/H5
Chengde (Chengteh), China 77/J3
Chengdu (Chengtu), China 77/F5
Chengdu, China 54/M6
Chengdu, China 2/P4
Chengkou, China 77/G5
Chengteh (Chengde), China 77/J3
Chengtu (Chengdu), China 77/F5
Chenier (lake), La. 238/F2
Chenoa, Ill. (61726) 222/E3
Chenoa, Ky. (40925) 237/O7
Chenoweth, Oreg. (†97058) 291/F2
Chen Xian, China 77/H6
Chepachet, R.I. (02814) 249/H5
Chepén, Peru 128/C6
Chépénéhé, New Caled. 86/H4
Chepes, Argentina 143/C3
Chépica, Chile 138/A10
Chepo, Panama 154/H6
Chepo (riv.), Panama 154/H6
Chepstow, Ontario 177/C3
Chepstow, Wales 10/E5
Chepstow, Wales 13/E6
Chequamegon (bay), Wis. 317/E2
Cher (dept.), France 28/E4
Cher (riv.), France 28/D4
Cheraw, Colo. (81030) 208/N6
Cheraw, Miss. (†39483) 256/E8
Cheraw, S.C. (29520) 296/H2
Cherbourg, France 28/C3
Cherbourg, France 7/D4
Cherbourg, Queensland 95/D5
Cherchell, Algeria 106/E1
Cherchen (Qiemo), China 77/C4
Cherdyn', U.S.S.R. 52/J2
Cheremkhovo, U.S.S.R. 54/L4
Cheremkhovo, U.S.S.R. 48/L4
Cherepovets, U.S.S.R. 7/H3
Cherepovets, U.S.S.R. 52/E3
Cherepovets, U.S.S.R. 48/D4
Chergui, Chott Ech (salt lake), Algeria 106/E2
Cherhill, Alberta 182/C3
Cherial (riv.), India 68/F2
Cheriton, Va. (23316) 307/R6
Cherkassy, U.S.S.R. 7/H4
Cherkassy, U.S.S.R. 52/D5
Cherkessk, U.S.S.R. 48/E5
Cherkessk, U.S.S.R. 52/F6
Chermside, Queensland 88/K2
Chermside, Queensland 95/D2
Chernigov, U.S.S.R. 7/H3
Chernigov, U.S.S.R. 48/D4
Chernigov, U.S.S.R. 52/D4
Chernogorsk, U.S.S.R. 48/K4
Chernorechenskiy, U.S.S.R. 52/H2
Chernovtsy, U.S.S.R. 7/G4
Chernovtsy, U.S.S.R. 52/C5
Chernovtsy, U.S.S.R. 48/C5
Chernushka, U.S.S.R. 52/J3
Chernyshevsk, U.S.S.R. 48/M4
Chernyshevskiy, U.S.S.R. 48/M3
Cherokee (co.), Ala. 195/G2
Cherokee, Ala. (35616) 195/C1
Cherokee (co.), Georgia 217/D2
Cherokee (co.), Iowa 229/B3
Cherokee, Iowa (51012) 229/B3
Cherokee (co.), Kansas 232/H4
Cherokee, Kansas (66724) 232/H4
Cherokee, Ky. (†41180) 237/R4
Cherokee (co.), N.C. 281/A4
Cherokee, N.C. (28719) 281/C4
Cherokee (co.), Okla. 288/R3
Cherokee, Okla. (73728) 288/K1
Cherokee (co.), S.C. 296/D1
Cherokee (dam), Tenn. 237/P8
Cherokee (lake), Tenn. 237/P8
Cherokee (co.), Texas 303/J6
Cherokee, Texas (76832) 303/F7
Cherokee City, Ark. (†72734) 202/A1
Cherokee Falls, S.C. (29705) 296/D1
Cherokee Ind. Res., N.C. 281/C3
Cherokees, Lake O'The (lake), Okla. 288/S1
Cherokee Village, Ark. (†72525) 202/G1
Cherrapunji, India 68/G3
Cherry, Ariz. (†86327) 198/C4
Cherry (creek), Ariz. 198/E4
Cherry (brook), Conn. 210/D1
Cherry, Ill. (61317) 222/D2
Cherry (co.), Nebr. 264/C2
Cherry, N.C. (†27928) 281/R3
Cherry (creek), N. Dak. 282/D4
Cherry (creek), S. Dak. 298/F5
Cherry (creek), S. Dak. 298/F4
Cherry, Tenn. (†38041) 237/B9
Cherry (creek), Utah 304/B4
Cherry (riv.), W. Va. 312/E6
Cherry Creek, Br. Col. 184/G5
Cherry Creek, Nev. (†89301) 266/G3
Cherry Creek, N.Y. (14723) 276/B6
Cherry Creek, S. Dak. (57622) 298/F4
Cherryfield○, Maine (04622) 243/H6
Cherry Fork, Ohio (45618) 284/C8
Cherry Grove, Alberta 182/E2
Cherry Grove, Minn. (†55975) 255/F7
Cherry Grove, Ohio (†45202) 284/C10
Cherry Grove, Oreg. (†97119) 291/D2
Cherry Hill, Ark. (†71953) 202/B4
Cherry Hill, Md. (†21921) 245/P2
Cherry Hill○, N.J. (*08034) 273/B3
Cherry Lake Farms, Fla. (†32350) 212/C1
Cherryland, Calif. (†94541) 204/K2
Cherrylog, Georgia (30522) 217/D1
Cherry Point, Alberta 182/A1
Cherry Point Marine Air Sta., N.C. 281/R5
Cherry Run, W. Va. (25418) 312/L3
Cherry Tree, Pa. (15724) 294/E4
Cherryvale, Kansas (67335) 232/G4
Cherry Valley, Ark. (72324) 202/J3
Cherry Valley, Ill. (61016) 222/D1
Cherry Valley, Mass. (01611) 249/G3
Cherry Valley, N.Y. (13320) 276/L5
Cherry Valley, Ontario 177/G4
Cherry Valley, Pa. (†16030) 294/C3
Cherryville, Br. Col. 184/H5
Cherryville, Mo. (65446) 261/K7
Cherryville, N.C. (28021) 281/G4
Cherryville, Oreg. (†97055) 291/E2
Cherskiy, U.S.S.R. 54/T3
Cherskiy, U.S.S.R. 4/C1
Cherskiy (range), U.S.S.R. 54/R3
Cherskiy (range), U.S.S.R. 48/Q3
Cherskiy (range), U.S.S.R. 48/P3
Cherta, Spain 33/G2
Chertsey, England 13/G8
Chertsey, England 10/B6
Chervonograd, U.S.S.R. 52/B4
Chesaning, Mich. (48616) 250/E5
Chesapeake (bay) 188/L3
Chesapeake (bay), Md. 245/N7
Chesapeake, Ohio (45619) 284/E9
Chesapeake (bay), U.S. 146/L6
Chesapeake (I.C.), Va. (*23320) 307/R7
Chesapeake (bay), Va. 307/R5
Chesapeake, W. Va. (†25301) 312/C6
Chesapeake and Delaware (canal), Del. 245/R2
Chesapeake and Delaware (canal), Md. 245/R2
Chesapeake and Ohio Canal Nat'l Hist. Park, Md. 245/J4
Chesapeake and Ohio Canal Nat'l Mon., Va. 307/O2
Chesapeake and Ohio Canal Nat'l Hist. Park, W. Va. 312/J3
Chesapeake Beach, Md. (20732) 245/N6
Chesapeake City, Md. (21915) 245/P2
Chesaw, Wash. (98818) 310/G2
Chésery, Pointe de (mt.), Switzerland 39/C4
Chesham, England 10/F5
Chesham, England 13/G7
Chesham, N.H. (†03455) 268/C6
Cheshire, Conn. (06410) 210/D2
Cheshire○, Conn. (06410) 210/D2
Cheshire (co.), England 13/E4
Cheshire○, Mass. (01225) 249/B2
Cheshire (res.), Mass. 249/A2
Cheshire (co.), N.H. 268/C6
Cheshire, Ohio (45620) 284/F8
Cheshire, Oreg. (97419) 291/D3
Cheshskaya (bay), U.S.S.R. 7/J2
Cheshskaya (bay), U.S.S.R. 52/G1
Cheshunt, England 13/H7
Cheshunt, England 10/B5
Chesilhurst, N.J. (†08089) 273/D4
Cheslatta, Br. Col. 184/E3
Chesley, Ontario 177/C3
Chesnaye, Manitoba 179/K2
Chesnee, S.C. (29323) 296/D1
Chester, Ark. (72934) 202/B2
Chester, Calif. (96020) 204/D3
Chester, Conn. (06412) 210/F3
Chester○, Conn. (06412) 210/F3
Chester, England 13/E2
Chester, England 13/G2
Chester, Georgia (31012) 217/F6
Chester, Idaho (83421) 220/G5
Chester, Ill. (62233) 222/D6
Chester, Ind. (†47374) 227/H5
Chester, Iowa (52134) 229/J2
Chester○, Maine (†04458) 243/F5
Chester, Md. (21619) 245/N5
Chester (riv.), Md. 245/O4
Chester○, Mass. (01011) 249/C3
Chester, Minn. (55904) 255/F6
Chester, Miss. (†39735) 256/F4
Chester, Mont. (59522) 262/E2
Chester, Nebr. (68327) 264/G4
Chester○, N.H. (03036) 268/E6
Chester, N.J. (07930) 273/D2
Chester, N.Y. (10918) 276/M8
Chester, Nova Scotia 168/D4
Chester, Ohio (45720) 284/G7
Chester, Okla. (73838) 288/J2
Chester (co.), Pa. 294/L6
Chester, Pa. (*19013) 294/L7
Chester (creek), Pa. 294/L7
Chester (co.), S.C. 296/E2
Chester, S.C. (29706) 296/E2
Chester, S. Dak. (57016) 298/R6
Chester (co.), Tenn. 237/D10
Chester, Texas (75936) 303/K7
Chester, Utah (84623) 304/C4
Chester○, Vt. (05143) 268/B5
Chester, Va. (23831) 307/O6
Chester, W. Va. (26034) 312/E1
Chester Basin, Nova Scotia 168/D4
Chester-Chester Depot, Vt. (05143) 268/B5
Chesterfield, Conn. (†06370) 210/G3
Chesterfield, England 13/J2

Chesterfield, England 10/F4
Chesterfield, Idaho (†83217) 220/G7
Chesterfield, Ill. (62630) 222/D4
Chesterfield, Ind. (46017) 227/F4
Chesterfield (isl.), Madagascar 118/G3
Chesterfield○, Mass. (01012) 249/C3
Chesterfield, Mo. (63017) 261/N2
Chesterfield (isls.), New Caled. 87/F7
Chesterfield○, N.H. (03443) 268/C6
Chesterfield○, N.J. (†08505) 273/D3
Chesterfield (inlet), N.W.T. 146/J3
Chesterfield (inlet), N.W.T. 162/G2
Chesterfield (inlet), N.W. Terrs. 187/J3
Chesterfield (co.), S.C. 296/G2
Chesterfield, S.C. (29709) 296/G2
Chesterfield, Tenn. (†38351) 237/E9
Chesterfield (co.), Va. 307/N6
Chesterfield, Va. (23832) 307/N6
Chesterfield Inlet, N.W.T. 162/G3
Chesterfield Inlet, N.W. Terrs. 187/K3
Chester Gap, Va. (22623) 307/M3
Chester Heights, Pa. (19017) 294/L7
Chesterhill, Ohio (43728) 284/G6
Chester Hill, Pa. (†16866) 294/F4
Chesterland, Ohio (44026) 284/H2
Chester-le-Street, England 10/E3
Chester-le-Street, England 13/J3
Chester Morse (lake), Wash. 310/D3
Chesterton, Ind. (46304) 227/D1
Chestertown, Md. (21620) 245/O4
Chestertown, N.Y. (12817) 276/N3
Chesterville, Ill. (†61911) 222/E4
Chesterville, Maine (†04938) 243/C6
Chesterville○, Maine (†04938) 243/C6
Chesterville, Md. (†21651) 245/P3
Chesterville, Ohio (43317) 284/E5
Chesterville, Ontario 177/J2
Chesterville, Québec 172/F4
Chestnut, Ala. (†36425) 195/D7
Chestnut, Ill. (62518) 222/D3
Chestnut, La. (71017) 238/D2
Chestnut Hill, Conn. (†06249) 210/G2
Chestnut Mound, Tenn. (38552) 237/K8
Chest Springs, Pa. (16624) 294/E4
Chesuncook, Maine (†04478) 243/D3
Chesuncook (lake), Maine 243/E3
Cheswick, Pa. (15024) 294/C6
Cheswold, Del. (19936) 245/R4
Chetac (lake), Wis. 317/D4
Chetco (riv.), Oreg. 291/C5
Chetek, Wis. (54728) 317/C5
Chéticamp, Nova Scotia 168/G2
Chéticamp (isl.), Nova Scotia 168/G2
Chetlat (isl.), India 68/C6
Chetopa, Kansas (67336) 232/G4
Chetumal, Mexico 150/Q7
Chetumal (bay), Mexico 150/P8
Chetwynd, Br. Col. 184/G2
Chevak, Alaska (99563) 196/E2
Cheval Blanc (pt.), Haiti 158/B5
Chevelon (creek), Ariz. 198/E4
Cheverly, Md. (20785) 245/G4
Cheville (pass), Switzerland 39/D4
Cheviot (hills), England 13/E2
Cheviot, The (mt.), England 13/E2
Cheviot, N. Zealand 100/D5
Cheviot, Ohio (†45211) 284/B9
Cheviot (hills), Scotland 15/F5
Cheviot, The (mt.), Scotland 15/F5
Chevrolet, Ky. (40817) 237/P7
Chevy Chase, Md. (20815) 245/E4
Chevy Chase Section Four, Md. (†20015) 245/E4
Chewack (riv.), Wash. 310/E2
Chewalla, Tenn. (38393) 237/D10
Chewelah, Wash. (99109) 310/H2
Chewsville, Md. (21721) 245/H2
Chexbres, Switzerland 39/C3
Cheyenne (riv.) 188/F2
Cheyenne (co.), Colo. 208/O5
Cheyenne (co.), Kansas 232/A2
Cheyenne (co.), Nebr. 264/A3
Cheyenne, Okla. (73628) 288/G3
Cheyenne (riv.), S. Dak. 298/F4
Cheyenne (riv.), U.S. 146/H5
Cheyenne (cap.), Wyo. 146/H5
Cheyenne (cap.), Wyo. 188/E2
Cheyenne (cap.), Wyo. (82001) 319/H4
Cheyenne (riv.), Wyo. 319/H2
Cheyenne Bottoms (lake), Kansas 232/D3
Cheyenne River Ind. Res., S. Dak. 298/F4
Cheyenne Wells, Colo. (80810) 208/P5
Cheyne (bay), W. Australia 92/B6
Cheyney, Pa. (19319) 294/M6
Cheyres, Switzerland 39/C3
Chezacut, Br. Col. 184/F4
Chhatarpur, India 68/D4
Chhindwara, India 68/D4
Chi, Mae Nam (riv.), Thailand 72/D3
Chiai, China 77/K7
Chiambone, Somalia 115/H4
Chiang Dao, Thailand 72/C3
Chiange, Angola 115/B7
Chiang Khan, Thailand 72/D3
Chiang Mai, Thailand 72/C3
Chiang Mai, Thailand 54/L8
Chiang Rai, Thailand 72/C3
Chiang Saen, Thailand 72/C2
Chiapa de Corzo, Mexico 150/N8
Chiapas (state), Mexico 150/N8
Chiari, Italy 34/C2
Chiasso, Switzerland 39/G5
Chiatura, U.S.S.R. 52/F6
Chiautempan, Mexico 150/N1
Chiavari, Italy 34/B2
Chiba (pref.), Japan 81/P2
Chiba, Japan 81/P2
Chibabava, Mozambique 118/E4
Chibia, Angola 115/B7
Chiblow (lake), Ontario 177/A1
Chibougamau, Québec 174/C3
Chibougamu, Que. 162/J6
Chibukak (cape), Alaska 196/D2
Chibuto, Mozambique 118/E4
Chibwe, Zambia 115/E6
Chicago, Ill. 146/K5
Chicago, Ill. 188/J2
Chicago, Ill. (*60601) 222/C5
Chicago, North Branch (riv.), Ill. 222/B5
Chicago, U.S. 2/E3
Chicago Heights, Ill. (60411) 222/C6
Chicago Portage Nat'l Hist. Site, Ill. 222/B6
Chicago Ridge, Ill. (60415) 222/B6
Chicama, Peru 128/C6
Chicamacomico (riv.), Md. 245/P7
Chicamocha (riv.), Colombia 126/D4
Chicanán (riv.), Venezuela 124/H4
Chicapa (riv.), Angola 115/D5
Chicapa (riv.), Zaire 115/D5
Chic-Chocs (mts.), Québec 172/C1
Chichagof (isl.) Alaska 188/D6
Chichagof (isl.), Alaska 196/M1
Chichén-Itzá (ruin), Mexico 150/P6
Chichester, England 13/G7
Chichester, England 10/F5
Chichester○, N.H. (†03258) 268/E5
Chichester, N.Y. (12416) 276/M6
Chichi (isl.), Japan 87/E3
Chichi (isl.), Japan 81/M3
Chichibu, Japan 81/J5
Chichibu-Tama National Park, Japan 81/J6
Chichicaste, Honduras 154/E3
Chichicastenango, Guatemala 154/B3
Chichigalpa, Nicaragua 154/D4
Chichiriviche, Venezuela 124/D2
Chickalah, Ark. (†72833) 202/D3
Chickaloon, Alaska (†99645) 196/C1
Chickamauga, Georgia (30707) 217/B1
Chickamauga (lake), Tenn. 188/J3
Chickamauga (dam), Tenn. 237/L10
Chickamauga (lake), Tenn. 237/L10
Chickamauga and Chattanooga Nat'l Mil. Park, Georgia 217/B1
Chickamaw Beach, Minn. (†56474) 255/D4
Chickasaw, Ala. (36611) 195/B9
Chickasaw (co.), Iowa 229/J2
Chickasaw (co.), Miss. 256/G3
Chickasaw, Ohio (45826) 284/A5
Chickasawhay (riv.), Miss. 256/G7
Chickasaw Village, Natchez Trace Pkwy., Miss. 256/G2
Chickasha, Okla. 188/G4
Chickasha, Okla. (73018) 288/L4
Chickasha (lake), Okla. 288/K4
Chicken, Alaska (99732) 196/K2
Chiclana de la Frontera, Spain 33/C4
Chiclayo, Peru 128/C6
Chiclayo, Peru 120/B3
Chico (riv.), Argentina 120/C7
Chico (riv.), Argentina 143/C6
Chico (riv.), Argentina 143/C5
Chico, Calif. (95926) 204/D4
Chico, Mont. (†59027) 262/F5
Chico (riv.), Philippines 82/C2
Chico, Texas (76030) 303/G4
Chicoana, Argentina 143/C2
Chico Arroyo (creek), N. Mex. 274/B3
Chicopee, Kansas (†66762) 232/H4
Chicopee, Mass. (*01013) 249/D4
Chicopee (riv.), Mass. 249/D4
Chicora, Miss. (†39322) 256/G7
Chicora, Pa. (16025) 294/C4
Chicot (co.), Ark. 202/H7
Chicot, Ark. (†71640) 202/H7
Chicot (pt.), La. 238/M7
Chicoutimi, Que. 162/J6
Chicoutimi, Que. 146/L5
Chicoutimi (county), Québec 174/C2
Chicoutimi (co.), Québec 172/G1
Chicoutimi, Québec 172/G1
Chicoutimi, Québec 174/C3
Chicoutimi (riv.), Québec 172/F2
Chicoutimi-Nord, Québec 172/F1
Chicualacuala, Mozambique 118/E4
Chidambaram, India 68/E6
Chidester, Ark. (71726) 202/D6
Chidley (cape), Canada 146/M3
Chidley (cape), Newf. 166/B1
Chidley (cape), Newf. 162/K3
Chidley (cape), N.W. Terrs. 187/M3
Chidlow, W. Australia 88/B2
Chief Joseph (dam), Wash. 310/F3
Chiefland, Fla. (32626) 212/D2
Chiefs (pt.), Ontario 177/C3
Chiemsee (lake), W. Germany 22/E5
Chienti (riv.), Italy 34/D3
Chieri, Italy 34/A2
Chieti (prov.), Italy 34/E3
Chieti, Italy 34/E3
Chietla, Mexico 150/M2
Chièvres, Belgium 27/D7
Chifeng, China 77/J3
Chigasaki, Japan 81/O3
Chiginagak (mt.), Alaska 196/G3
Chignahuapan, Mexico 150/N1
Chignecto (bay), New Bruns. 170/F3
Chignecto (bay), Nova Scotia 168/D3
Chignecto (cape), Nova Scotia 168/C3
Chignecto (isth.), Nova Scotia 168/D3
Chignik, Alaska (†99613) 196/G3
Chignik (bay), Alaska 196/G3
Chignik Lagoon, Alaska (99565) 196/G3
Chignik Lake, Alaska (†99613) 196/G3
Chiguana, Bolivia 136/A7
Chigubo, Mozambique 118/E4
Chigwell, England 13/H8
Chigwell, England 10/C5
Chihuahua (state), Mexico 150/F2
Chihuahua, Mexico 150/F2
Chihuahua, Mexico 146/H7
Chikaskia (riv.), Kansas 232/E4
Chik Ballapur, India 68/D6
Chikmagalur, India 68/D6
Chilanga, Zambia 115/E7
Chilanko Forks, Br. Col. 184/E4
Chilapa de Álvarez, Mexico 150/K8
Chilas, Pakistan 68/D1
Chilas, Pakistan 59/K2
Chilca (Pucusana), Peru 128/D9
Chilcoot, Calif. (96105) 204/E4
Chilcotin (riv.), Br. Col. 184/E4
Childers, Queensland 88/J5
Childersburg, Ala. (35044) 195/F4
Childress (co.), Texas 303/D3
Childress, Texas (79201) 303/D3
Childs (lake), Manitoba 179/A3
Childs, Md. (21916) 245/P2
Childwold, N.Y. (12922) 276/L2
Chile 2/F7
Chile 120/B5
CHILE 138
Chile Chico, Chile 138/E6
Chilecito, Argentina 143/C2
Chiles, Colombia 126/A7
Chilete, Peru 128/C6
Chilga, Ethiopia 111/G5
Chilhowee, Mo. (64733) 261/E5
Chilhowee (mt.), Tenn. 237/O9
Chilhowie, Va. (24319) 307/E7
Chili, Ind. (46926) 227/F3
Chili, Wis. (54420) 317/F6
Chililabombwe, Zambia 115/E6
Chililabombwe, Zambia 102/E6
Chilili, N. Mex. (†87059) 274/C4
Chilka (lake), India 68/F5
Chilko (lake), Br. Col. 184/F4
Chilko (riv.), Br. Col. 184/E4
Chilkoot (pass), Alaska 196/M1
Chilkoot (pass), Br. Col. 184/J1
Chillán, Chile 120/B6
Chillán, Chile 138/A11
Chillicothe, Ill. (61523) 222/D3
Chillicothe, Iowa (52548) 229/J6
Chillicothe, Mo. (64601) 261/E3
Chillicothe, Ohio (45601) 284/E7
Chillicothe, Texas (79225) 303/E3
Chilliwack, Br. Col. 162/D6
Chilliwack, Br. Col. 184/M3
Chillum, Md. (20783) 245/F4
Chilmark○, Mass. (02535) 249/M7
Chilo, Ohio (45112) 284/B8
Chilocco, Okla. (74635) 288/M1
Chiloé (isl.), Chile 120/B7
Chiloé (isl.), Chile 138/D4
Chiloquin, Oreg. (97624) 291/F5
Chilpancingo de los Bravos, Mexico 150/K8
Chiltern (hills), England 13/G6
Chilton (co.), Ala. 195/E5
Chilton, Texas (76632) 303/G6
Chilton, Wis. (53014) 317/K7
Chiltonville, Mass. (†02360) 249/M5
Chilumba, Malawi 115/F6
Chilwa (lake), Malawi 115/G7
Chilwa (lake), Mozambique 118/F3
Chimacum, Wash. (98325) 310/C3
Chimaltenango, Guatemala 154/B3
Chimán, Panama 154/H6
Chimantá-tepuí (mt.), Venezuela 124/G5
Chimay, Belgium 27/E8
Chimayo, N. Mex. (87522) 274/D3
Chimbarongo, Chile 138/A10
Chimbay, U.S.S.R. 48/F5
Chimborazo (prov.), Ecuador 128/C3
Chimborazo (mt.), Ecuador 120/B3
Chimborazo (mt.), Ecuador 128/C3
Chimbote, Peru 120/B3
Chimbote, Peru 128/C7
Chimbote (bay), Peru 128/C7
Chimichaqua, Colombia 126/D3
Chimkent, U.S.S.R. 54/H5
Chimkent, U.S.S.R. 48/H5
Chimney Point, Vt. (†05491) 268/A3
Chimney Rock, Colo. (81127) 208/E8
Chimney Rock Nat'l Hist. Site, Nebr. 264/A3
Chimoio, Mozambique 118/E3
Chimoio, Mozambique 102/F6
Chin, Alberta 182/D5
Chin (state), Burma 72/B2
Chin (hills), Burma 72/B2
Chin (cape), Ontario 177/C2
China 2/P4
China 54/L6
CHINA 85
CHINA 77
China, Maine (04926) 243/E7
China○, Maine (04926) 243/E7
China, Campeche, Mexico 150/O7
China, Nuevo León, Mexico 150/K4
Chinácota, Colombia 126/D4
China Gardens (dam), Wash. 310/J4
China Grove, Ala. (†36081) 195/G7
China Grove, N.C. (28023) 281/H3
China Grove, Texas (†78201) 303/K11
China Lake, Calif. (†93555) 204/H8
China Lake Naval Weapons Center, Calif. 204/H8
Chinameca, El Salvador 154/C4
Chinandega, Nicaragua 154/D4
Chinati (mts.), Texas 303/C12
Chinati (peak), Texas 303/C12
Chincha (isls.), Peru 128/D9
Chincha Alta, Peru 128/D9
Chincha Alta, Peru 120/B4
Chinchaga (riv.), Alberta 182/A5
Chinchilla, Pa. (18410) 294/F6
Chinchilla, Queensland 88/J5
Chinchilla de Monte-Aragón, Spain 33/F3
Chinchiná, Colombia 126/C5
Chinchón, Spain 33/G5
Chinchoua, Gabon 115/A4
Chinchow (Jinzhou), China 77/K3
Chincoteague (bay), Md. 245/S8
Chincoteague, Va. (23336) 307/T5
Chincoteague (bay), Va. 307/T5
Chincoteague (inlet), Va. 307/T5
Chinde, Mozambique 118/F3
Chinde, Mozambique 102/F6
Chindu, China 77/E5
Chindwin (riv.), Burma 72/B2
Chinese Camp, Calif. (95309) 204/E6
Chingleput, India 68/E6
Chingola, Zambia 115/E6
Chingola, Zambia 102/E6
Chinguar, Angola 115/C6
Chinguetti, Mauritania 106/B4
Chinhae, S. Korea 81/D6
Chiniak (cape), Alaska 196/H3
Chiñijo, Bolivia 136/A4
Chiniot, Pakistan 68/C2
Chinipas, Mexico 150/E3
Chinju, S. Korea 81/D6
Chinkapin Knob (mt.), Ark. 202/E2
Chinkiang (Zhenjiang), China 77/J5
Chinle, Ariz. (86503) 198/F2
Chinle (creek), Ariz. 198/F2
Chinle (valley), Ariz. 198/F2
Chinle (creek), Utah 304/E6
Chinle Wash (dry riv.), Ariz. 198/F2
Chino (valley), Ariz. 198/C3
Chino, Calif. (91710) 204/D10
Chinon, France 28/D4
Chinook, Alberta 182/E4
Chinook, Mont. (59523) 262/G2
Chinook (lake), Oreg. 291/F3
Chinook, Wash. (98614) 310/B4
Chinook (pass), Wash. 310/D4
Chinook Valley, Alberta 182/B1
Chino Valley, Ariz. (86323) 198I/C4
Chinquapin, N.C. (28521) 281/O5
Chinsali, Zambia 115/F6
Chinsi (Jinxi), China 77/K3
Chintheche, Malawi 115/F6
Chinú, Colombia 126/C3
Chinwangtao (Qinhuangdao), China 77/K4
Chiny, Belgium 27/G9
Chioggia, Italy 34/D2
Chip (lake), Alberta 182/C3
Chipamanu (riv.), Bolivia 136/A2
Chipata, Zambia 115/F6
Chipata, Zambia 102/F6
Chipewyan (lake), Alberta 182/D1
Chipewyan (riv.), Alberta 182/D1
Chipewyan Lake, Alberta 182/D1
Chipindo, Angola 102/D6
Chipindo, Angola 115/C6
Chipinga, Zimbabwe 118/E4
Chipley, Fla. (32428) 212/D6
Chiplun, India 68/C5
Chipman, Alberta 182/D3
Chipman, New Bruns. 170/E2
Chipman (riv.), Sask. 181/M2
Chipoka, Malawi 115/F6
Chipola (riv.), Fla. 212/D6
Chipola, La. (†70441) 238/J5
Chippenham, England 13/E6
Chippenham, England 10/F5
Chippewa (co.), Mich. 250/E2
Chippewa (riv.), Mich. 250/E5
Chippewa (co.), Minn. 255/C5
Chippewa (riv.), Minn. 255/C5
Chippewa (riv.), Wis. 188/H1
Chippewa (co.), Wis. 317/D5
Chippewa (lake), Wis. 317/D4
Chippewa (riv.), Wis. 317/B7
Chippewa Falls, Wis. (54729) 317/D6
Chippewa Lake, Mich. (49320) 250/D5
Chipping Norton, England 13/F6
Chippis, Switzerland 39/E4
Chiputneticook (lakes), Maine 243/H4
Chiputneticook (lakes), New Bruns. 170/C3
Chiquián, Peru 128/D8
Chiquimula, Guatemala 154/C3
Chiquinquirá, Colombia 126/C5
Chiquita (lake), Argentina 120/C6
Chir (riv.), U.S.S.R. 52/F5
Chira (riv.) 120/A3
Chira (riv.), Ecuador 128/B5
Chirala, India 68/E5
Chirchik, U.S.S.R. 48/H5
Chireno, Texas (75937) 303/K6
Chirfa, Niger 106/G4
Chiri (mt.), S. Korea 81/C6
Chiribiquete, Sierra de (mts.), Colombia 126/D7
Chiricahua (mts.), Ariz. 198/F6
Chiricahua Nat'l Mon., Ariz. 198/F6
Chiriguaná, Colombia 126/D3
Chirikof (isl.), Alaska 196/G3
Chirinos, Peru 128/C5
Chiriquí (gulf), Panama 154/F7
Chiriquí (lag.), Panama 154/G6
Chiriquí Grande, Panama 154/F6
Chirk, Wales 13/D5
Chirnside, Scotland 15/F5
Chiromo, Malawi 115/F7
Chironico, Switzerland 39/G4
Chirpan, Bulgaria 45/G4
Chirripó Grande (mt.), C. Rica 154/F6
Chirundu, Zimbabwe 118/D3
Chisago (co.), Minn. 255/F5
Chisago City, Minn. (55013) 255/E5
Chisamba, Zambia 115/E6
Chisana, Alaska (†99566) 196/K2
Chisec, Guatemala 154/B3
Chisholm, Maine (†04239) 243/C7
Chisholm, Minn. (55719) 255/E3
Chisholm Mills, Alberta 182/C2
Chishui, China 77/G6
Chisimayu, Somalia 115/H4
Chisimayu, Somalia 102/G5
Chişineu Criş, Romania 45/E2
Chismville, Ark. (†72943) 202/C3
Chisos (mts.), Texas 303/A8
Chistochina, Alaska (†99586) 196/K2
Chistopol', U.S.S.R. 52/H3
Chiswick, Ontario 177/E1
Chita, U.S.S.R. 54/N4
Chita, U.S.S.R. 48/M4
Chitado, Angola 115/B7
Chitado, Namibia 118/A3
Chitek (lake), Manitoba 179/C2
Chitek (lake), Sask. 181/D2
Chitek Lake, Sask. 181/D2
Chitembo, Angola 115/C6
Chitina, Alaska (99566) 196/K2
Chitina (riv.), Alaska 196/K2
Chitipa, Malawi 115/F5
Chitorgarh, India 68/C4
Chitose, Japan 81/K2
Chitradurga, India 68/D6
Chitral, Pakistan 59/K2
Chitral, Pakistan 68/C1
Chitré, Panama 154/G7
Chittagong, Bangladesh 54/L7
Chittagong, Bangladesh 68/G4
Chittenango, N.Y. (13037) 276/J4
Chittenden (co.), Vt. 268/A3
Chittenden○, Vt. (05737) 268/B4
Chittenden (res.), Vt. 268/B4
Chittering, W. Australia 88/B6
Chittoor, India 68/D6
Chiumbe (riv.), Angola 115/D5
Chiva, Spain 33/F3
Chivacoa, Venezuela 124/D2
Chivapure (riv.), Venezuela 124/E4
Chivasso, Italy 34/A2
Chivato (mesa), N. Mex. 274/B3
Chivay, Peru 128/G10
Chive, Bolivia 136/A3
Chivilcoy, Argentina 143/F7
Chivilcoy, Argentina 120/C6
Chivington, Colo. (81031) 208/O6
Chiwawa (riv.), Wash. 310/E2
Chixoy (riv.), Guatemala 154/B2
Chizha, U.S.S.R. 52/F1
Chloe, W. Va. (25235) 312/D5
Chloride, Ariz. (86431) 198/A3
Chloride, Mo. (†63646) 261/L8
Chlumec, Czech. 41/C1
Choam Khsant, Cambodia 72/E4
Choapa, Chile 138/A9
Choapa (riv.), Chile 138/A9
Choate, Br. Col. 184/M3
Chobe (riv.), Botswana 118/C3
Chobe (riv.), Namibia 118/C3
Chobe Nat'l Park, Botswana 118/D3
Choc, St. Lucia 161/G5
Choc (bay), St. Lucia 161/G5
Chocalán, Chile 138/F4
Chocaya, Bolivia 136/B7
Choccolocco, Ala. (36254) 195/G3
Choceń, Czech. 41/C1
Choch'iwŏn, S. Korea 81/C5
Chocó (dept.), Colombia 126/B4
Chocó (bay), Colombia 126/B6
Chocolate (mts.), Ariz. 198/A5
Chocolate (mts.), Calif. 204/K10
Chocomán, Mexico 150/P2
Choconut, Pa. (†18818) 294/K2
Chocorua, N.H. (03817) 268/E4
Chocorua (mt.), N.H. 268/E4
Chocowinity, N.C. (27817) 281/P4
Choctaw (co.), Ala. 195/B6
Choctaw, Ala. (36905) 195/B6
Choctaw, Ark. (72028) 202/F2
Choctaw (co.), Miss. 256/F4
Choctaw (co.), Okla. 288/P6
Choctaw, Okla. (73020) 288/M3
Choctaw Bluff, Ala. (†36545) 195/C8
Choctawhatchee (riv.), Ala. 195/H8
Choctawhatchee (bay), Fla. 212/C6
Choctawhatchee (riv.), Fla. 212/C6
Choctaw Ind. Res., Miss. 256/F6
Chodov, Czech. 41/B1
Chodziez, Poland 47/C2
Choele-Choel, Argentina 143/C4
Choele-Choel, Argentina 120/C6
Choestoe, Georgia (†30512) 217/E1
Chofu, Japan 81/O2
Choiceland, Sask. 181/G2
Choiseul (county), St. Lucia 161/F7
Choiseul, St. Lucia 161/F7
Choiseul (isl.), Solomon Is. 87/F6
Choiseul (isl.), Solomon Is. 86/D2
Choisy-le-Roi, France 28/B2
Choix, Mexico 150/E3
Chojna, Poland 47/B2
Chojnice, Poland 47/C2
Chojnów, Poland 47/B3
Chokai (mt.), Japan 81/J4
Chokio, Minn. (56221) 255/B5
Chokoloskee, Fla. (33925) 212/E6
Chokurdakh, U.S.S.R. 4/B2
Chokurdakh, U.S.S.R. 54/R2
Chokurdakh, U.S.S.R. 48/P2
Cholame, Calif. (93431) 204/E8
Cholet, France 28/C4
Choloma, Honduras 154/C3
Cholula de Rivadavia, Mexico 150/M1
Choluteca, Honduras 154/D4
Choluteca (riv.), Honduras 154/D4
Choma, Zambia 115/E7
Choma, Zambia 102/E6
Chomes, C. Rica 154/E5
Chomo Lhari (mt.), Bhutan 68/F3
Chomutov, Czech. 41/B1
Ch'ŏnan, S. Korea 81/C5
Chon Buri, Thailand 72/D4
Chonchi, Chile 138/D4
Ch'ŏnch'ŏn, N. Korea 81/C3
Chone, Ecuador 128/B3
Ch'ŏngjin, N. Korea 54/P5
Ch'ŏngjin, N. Korea 81/E3
Chŏngju, N. Korea 81/B4
Ch'ŏngju, S. Korea 81/C5
Chongqing (Chungking), China 77/G6
Chongqing, China 2/P4
Chongqing, China 54/M7
Chŏngŭp, S. Korea 81/C6
Chongyang, China 77/H6
Chongzuo, China 77/G7
Chŏnju, S. Korea 81/C6
Chon May, Vung (bay), Vietnam 72/F3
Chonos (arch.), Chile 120/B7
Chonos (arch.), Chile 138/D6
Chopin, La. (71412) 238/E4
Chopin (lake), Québec 172/D2
Choptank (riv.), Del. 245/P5
Choptank, Md. (†21655) 245/P6
Choptank (riv.), Md. 245/O6
Choquecota, Bolivia 136/A6
Chorley, England 10/G2
Chorley, England 13/G4
Chorleywood, England 13/G7
Chorleywood, England 10/A5
Choroní, Venezuela 124/E2
Choros (cape), Chile 138/A7
Choros, Los (riv.), Chile 138/A7
Chortitz, Sask. 181/D5
Chortkov, U.S.S.R. 52/B5
Ch'ŏrwŏn, S. Korea 81/C4
Chorzele, Poland 47/E2
Chorzów, Poland 47/B4
Ch'osan, N. Korea 81/C3
Choshi, Japan 81/K6
Chosica, Peru 128/D8
Chos-Malal, Argentina 143/C4
Choszczno, Poland 47/B2
Chota, Peru 128/C6
Choteau, Mont. (59422) 262/D3
Choteau (creek), S. Dak. 298/N7
Chotěboř, Czech. 41/C2
Choudrant, La. (71227) 238/F1
Chouteau (co.), Mont. 262/F3
Chouteau, Okla. (74337) 288/R2
Chovoreca, Cerro (mt.), Bolivia 136/F6
Chovoreca (mt.), Paraguay 144/C1
Chowan (co.), N.C. 281/R2
Chowan (riv.), N.C. 281/R2
Chowchilla, Calif. (93610) 204/E6
Choybalsan, Mongolia 54/N5
Choybalsan, Mongolia 77/J2
Chrastava, Czech. 41/C1
Chriesman, Texas (77838) 303/H7
Chrisman, Ill. (61924) 222/F4
Chrisney, Ind. (47611) 227/C8
Christchurch, England 10/F5
Christchurch, England 13/F7
Christchurch, N. Zealand 2/T8
Christchurch, N. Zealand 100/D5
Christian (sound), Alaska 196/M2
Christian (co.), Ill. 222/D4
Christian (co.), Ky. 237/F7
Christian (co.), Mo. 261/F9
Christian (cape), N.W. Terrs. 187/M2
Christian (isl.), Ontario 177/D3
Christian, W. Va. (†25135) 312/C7
Christiana, Del. (†19711) 245/R2
Christiana, Jamaica 158/H6
Christiana, Pa. (17509) 294/K6
Christiana, S. Africa 118/D5
Christiana, Tenn. (37037) 237/J9
Christiansburg, Ohio (45389) 284/C5
Christiansburg, Va. (24073) 307/H6
Christiansfeld, Denmark 21/C7
Christiansted, Virgin Is. (U.S.) 156/H2
Christiansted, Virgin Is. (U.S.) 161/F4
Christiansted Nat'l Hist. Site, Virgin Is. (U.S.) 161/F4
Christie, Okla. (†74965) 288/S3
Christina (lake), Alberta 182/E2
Christina (riv.), Alberta 182/E1
Christina (lake), Minn. 255/C4
Christina, Mont. (59423) 262/G3
Christina Lake, Br. Col. 184/H5
Christine, N. Dak. (58015) 282/S6
Christine, Texas (78012) 303/F9
Christmas, Ariz. (†85292) 198/E5
Christmas (isl.), Australia 54/M11
Christmas (isl.), Australia 2/Q6
Christmas, Fla. (32709) 212/E3
Christmas (isl.), Kiribati 87/L5
Christmas, Mich. (49862) 250/C2
Christmas (lake), Oreg. 291/G4
Christmas (riv.), W. Australia 88/D3
Christmas Creek, W. Australia 92/D2
Christmas Island, Nova Scotia 168/H3
Christopher, Ill. (62822) 222/D6
Christopher Lake, Sask. 181/F2
Christoval, Texas (76935) 303/D6
Chromo, Colo. (81128) 208/F8
Chrudim, Czech. 41/C2
Chrudimka (riv.), Czech. 41/C2
Chrysler, Ala. (†36550) 195/C8
Chryston, Scotland 15/C2
Chrzanów, Poland 47/B4
Chu (riv.), U.S.S.R. 48/H5
Chualar, Calif. (93925) 204/D7
Chüanchow (Quanzhou), China 77/J7
Chuathbaluk, Alaska (†99559) 196/G2
Chubbuck, Idaho (83202) 220/F7
Chubu-Sangaku National Park, Japan 81/H5
Chubut (prov.), Argentina 143/C5
Chubut (riv.), Argentina 120/C7
Chubut (riv.), Argentina 143/C5
Chuchi (lake), Br. Col. 184/E2
Chuchow (Zhuzhou), China 77/H6
Chuckey, Tenn. (37641) 237/R8
Chucunaque (riv.), Panama 154/J6
Chudleigh, Tasmania 99/C3
Chudovo, U.S.S.R. 52/D3
Chugach (isls.), Alaska 196/B2
Chugash (isls.), Alaska 196/C1
Chugiak, Alaska (99567) 196/C1
Chuginadak (isl.), Alaska 196/D4
Chuguchak (Tacheng), China 77/B2
Chugwater, Wyo. (82210) 319/H4
Chugwater (creek), Wyo. 319/H4
Chukai, Malaysia 72/D6
Chukchi (sea) 4/C18
Chukchi (sea) 54/W3
Chukchi (sea), Alaska 196/E1
Chukchi (pen.), U.S.S.R. 54/V3
Chukchi (pen.), U.S.S.R. 4/C18
Chukchi (pen.), U.S.S.R. 48/T3
Chukchi (sea), U.S.S.R. 48/T2
Chukchi Aut. Okr., U.S.S.R. 48/R3
Chukhloma, U.S.S.R. 52/F3
Chula, Ark. (†72857) 202/C4
Chula, Georgia (31733) 217/E7
Chula, Mo. (64635) 261/F3
Chula, Va. (†23002) 307/N6
Chu Lai, Vietnam 72/F4
Chula Vista, Calif. (*92010) 204/J11
Chulucanas, Peru 128/B5
Chulumani, Bolivia 136/B5
Chulym (riv.), U.S.S.R. 48/J4
Chuma, Bolivia 136/A4
Chumatien (Zhumadian), China 77/H5
Chumbicha, Argentina 143/C2

Chumikan, U.S.S.R. 48/O4
Chumphon, Thailand 72/C5
Chuna (riv.), U.S.S.R. 48/K4
Chunchi, Ecuador 128/C4
Ch'unch'ŏn, S. Korea 81/D5
Chunchula, Ala. (36521) 195/B9
Ch'ungju, S. Korea 81/C5
Chungking (Chongqing), China 77/G6
Chŭngsan, N. Korea 81/B4
Chungshan (Zhongshan), China 77/H7
Chunky, Miss. (39323) 256/G6
Chunya, Tanzania 115/F5
Chunya (riv.), U.S.S.R. 48/K3
Chupaca, Peru 128/E9
Chupadera (mesa), N. Mex. 274/C5
Chupara (pt.), Trin. & Tob. 161/B10
Chuquibamba, Peru 128/F10
Chuquibambilla, Peru 128/F9
Chuquicamata, Chile 138/B3
Chuquichambi, Bolivia 136/B5
Chuquisaca (dept.), Bolivia 136/C6
Chur, Switzerland 39/J3
Churachandpur, India 68/G4
Church, Iowa (†52151) 229/L2
Churchbridge, Sask. 181/J5
Church Creek, Md. (21622) 245/O6
Church Hill, Md. (21623) 245/O4
Church Hill, Miss. (39055) 256/B7
Church Hill, Tenn. (37642) 237/R7
Churchill (riv.) 162/G4
Churchill (pk.), Br. Col. 162/D4
Churchill (peak), Br. Col. 184/L2
Churchill (riv.), Canada 146/J4
Churchill, Man. 146/J4
Churchill, Man. 162/G4
Churchill (cape), Man. 162/G4
Churchill, Manitoba 179/K2
Churchill (cape), Manitoba 179/K2
Churchill (riv.), Manitoba 179/J2
Churchill (co.), Nev. 266/C3
Churchill (falls), Newf. 166/B3
Churchill (riv.), Newf. 166/B3
Churchill, Pa. (†15235) 294/C7
Churchill (riv.), Que. 162/K5
Churchill (riv.), Sask. 181/M3
Churchill, Victoria 97/D6
Churchill Falls, Newf. 166/B3
Churchman (mt.), W. Australia 92/B5
Church Point, La. (70525) 238/F6
Church Point, Nova Scotia 168/B4
Church's Ferry, N. Dak. (58325) 282/M3
Church Stretton, England 13/E5
Churchton, Md. (20733) 245/N5
Churchtown, Pa. (†17555) 294/L5
Church View, Va. (23032) 307/P5
Churchville, Md. (21028) 245/N2
Churchville, N.Y. (14428) 276/E4
Churchville, Va. (24421) 307/K4
Churchville, W. Va. (†26338) 312/E4
Churdan, Iowa (50050) 229/D4
Churfirsten (mt.), Switzerland 39/H2
Churin, Peru 128/D8
Churu, India 68/D3
Churubusco, Ind. (46723) 227/G2
Churubusco, N.Y. (12923) 276/N1
Churuguara, Venezuela 124/D2
Churwalden, Switzerland 39/J3
Chushul, India 68/D2
Chuska (mts.), N. Mex. 274/A2
Chusovoy, U.S.S.R. 52/J3
Chute-à-Blondeau, Ontario 177/K2
Chute-aux-Outardes, Québec 172/A1
Chute-des-Passes, Québec 174/C3
Chute-Saint-Philippe, Québec 172/B3
Chuvash A.S.S.R., U.S.S.R. 48/E4
Chuvash A.S.S.R., U.S.S.R. 52/G3
Chu Xian, China 77/J5
Chuxiong, China 77/F7
Chuy, Uruguay 145/F4
Chvalšiny, Czech. 41/C2
Ciales, P. Rico 161/C1
Ciamis, Indonesia 85/H2
Ciampino, Italy 34/F7
Cianjur, Indonesia 85/H2
Cibecue, Ariz. (85911) 198/E4
Cibola (co.), N.Mex. 274/B3
Cibolo, Texas (78108) 303/K10
Cibolo (creek), Texas 303/K11
Çiçekdağı, Turkey 63/F3
Cicero, Ill. (60650) 222/B5
Cicero, Ind. (46034) 227/E4
Cicero Dantas, Brazil 132/G5
Cicerone, W. Va. (†25243) 312/D5
Ciconsine (lake), Québec 172/D2
Cid, N.C. (†27292) 281/J3
Cide, Turkey 63/E2
Cidlina (riv.), Czech. 41/C1
Cidra, Cuba 158/D1
Cidra, P. Rico 161/D2
Ciechanów (prov.), Poland 47/E2
Ciechanów, Poland 47/E2
Ciechocinek, Poland 47/D2
Ciego de Ávila (prov.), Cuba 158/F2
Ciego de Ávila, Cuba 156/B2
Ciego de Ávila, Cuba 158/F2
Ciempozuelos, Spain 33/F5
Ciénaga, Colombia 120/B1
Ciénaga, Colombia 126/C2
Ciénaga de Oro, Colombia 126/C3
Cienfuegos (prov.), Cuba 158/E2
Cienfuegos, Cuba 156/B2
Cienfuegos, Cuba 146/K7
Cienfuegos, Cuba 158/D2
Cienfuegos (bay), Cuba 158/D2
Cieplice Śląskie-Zdrój, Poland 47/B3
Čierny Balog, Czech. 41/E2
Cieszyn, Poland 47/D4
Cieza, Spain 33/F3
Çifteler, Turkey 63/D3
Cifuentes, Spain 33/E2
Cigánd, Hungary 41/F2
Cihanbeyli, Turkey 63/E3
Cihuatlán, Mexico 150/G7
Cijara (res.), Spain 33/D3
Cijulang, Indonesia 85/H2
Cilacap, Indonesia 85/H2
Çıldır, Turkey 63/K2
Çıldır (lake), Turkey 63/K2
Cilleros, Spain 33/C2
Cilo Dağı (mt.), Turkey 63/K4
Cima, Calif. (92323) 204/K8
Cimahi, Indonesia 85/H2
Cimarron (riv.) 188/G3
Cimarron, Colo. (81220) 208/D6
Cimarron, Kansas (67835) 232/B4
Cimarron (riv.), Kansas 232/B4
Cimarron, N. Mex. (87714) 274/E2
Cimarron (riv.), N. Mex. 274/E2
Cimarron (co.), Okla. 288/A1
Cimarron, Okla. (73111) 288/L3
Cimarron (riv.), Okla. 288/N2
Cimin, Turkey 63/H3
Cimone (mt.), Italy 34/C2
Cîmpeni, Romania 45/F2
Cîmpia Turzii, Romania 45/F2
Cîmpina, Romania 45/H3
Cîmpulung, Romania 45/G3
Cîmpulung Moldovenesc, Romania 45/G2
Cinaruco (riv.), Colombia 126/F4
Cinaruco (riv.), Venezuela 124/D4
Cinca (riv.), Spain 33/G2
Cincinnati, Ark. (†72769) 202/B1
Cincinnati, Iowa (52549) 229/G7
Cincinnati, Ohio 146/K6
Cincinnati, Ohio (*45201) 284/B9
Cincinnati, Ohio 188/K3
Cincinnatus, N.Y. (13040) 276/H5
Cinclare, La. (†70767) 238/J2
Cinco, W. Va. (†25301) 312/D6
Cinco Balas (cays), Cuba 158/E3
Cinco Bayou, Fla. (†32548) 212/B6
Cinco Saltos, Argentina 143/C4
Cinderella, W. Va. (†25661) 312/B7
Cinderford, England 13/E6
Çine, Turkey 63/B4
Cinebar, Wash. (98533) 310/C4
Ciney, Belgium 27/G8
Cinnaminson○, N.J. (†08077) 273/B3
Cintalapa de Figueroa, Mexico 150/N8
Cinto (mt.), France 28/B6
Cipolletti, Argentina 143/C4
Circeo (cape), Italy 34/D4
Circle, Alaska (99733) 196/K1
Circle, Mont. (59215) 262/L3
Circle (butte), Utah 304/C6
Circle Back, Texas (†79371) 303/B3
Circle City, Mo. (†63846) 261/N9
Circle Pines, Minn. (55014) 255/G5
Circle Springs, Alaska (†99730) 196/K1
Circleville, Kansas (66416) 232/G2
Circleville, Ohio (43113) 284/D6
Circleville, Utah (84723) 304/B5
Circleville, W. Va. (26804) 312/H5
Circular (head), Tasmania 99/B2
Cirebon, Indonesia 85/H2
Cirebon, Indonesia 54/M10
Ciremay (mt.), Indonesia 85/H2
Cirencester, England 10/F5
Cirencester, England 13/E6
Cirque (mt.), Newf. 166/B2
Cisco, Georgia (30708) 217/C1
Cisco, Ill. (61830) 222/E3
Cisco, Texas (76437) 303/E5
Cisco, Utah (84515) 304/E5
Cisco Springs Wash (creek), Utah 304/E4
Ciskei (bantustan), S. Africa 102/E8
Cismont, Va. (†22947) 307/M4
Cisnădie, Romania 45/G3
Cisne (isls.), Honduras 154/F2
Cisne, Ill. (62823) 222/E5
Cisneros, Colombia 126/C4
Cisnes (riv.), Chile 138/E5
Cispus (pass), Wash. 310/D4
Cispus (riv.), Wash. 310/D4
Cissna Park, Ill. (60924) 222/F3
Citlaltépetl (mt.), Mexico 150/O2
Citra, Fla. (32627) 212/D2
Citronelle, Ala. (36522) 195/B8
Citrus (co.), Fla. 212/D3
Citrus Center, Fla. (†33471) 212/E5
Citrus Heights, Calif. (95610) 204/C8
Cittadella, Italy 34/C2
Città di Castello, Italy 34/C3
Cittanova, Italy 34/F5
City Mills, Mass. (†02056) 249/J4
City Point, Fla. (†32922) 212/F3
City Point, Wis. (†54466) 317/F7
City View, Ontario 177/J2
City View, S.C. (29611) 296/C2
Ciudad Acuña (Villa Acuña), Mexico 150/J2
Ciudad Altamirano, Mexico 150/J7
Ciudad Bolívar, Venezuela 120/C2
Ciudad Bolívar, Venezuela 124/G3
Ciudad Bolivia, Venezuela 124/C3
Ciudad Camargo, Chihuahua, Mexico 150/G3
Ciudad Camargo, Tamaulipas, Mexico 150/K3
Ciudad Darío, Nicaragua 154/D4
Ciudad del Carmen, Mexico 150/N7
Ciudad Delicias, Mexico 150/G2
Ciudad del Maíz, Mexico 150/K5
Ciudad de Nutrias, Venezuela 124/D3
Ciudad de Río Grande, Mexico 150/H5
Ciudadela, Spain 33/H2
Ciudad Guayana, Venezuela 120/C2
Ciudad Guayana, Venezuela 124/G3
Ciudad Guerrero, Mexico 150/F2
Ciudad Guzmán, Mexico 150/H7
Ciudad Hidalgo, Chiapas, Mexico 150/N9
Ciudad Hidalgo, Michoacán, Mexico 150/J7
Ciudad Juárez, Mexico 146/H6
Ciudad Juárez, Mexico 150/F1
Ciudad Lerdo, Mexico 150/H4
Ciudad Madero, Mexico 150/L5
Ciudad Mante, Mexico 150/K5
Ciudad Mendoza, Mexico 150/O2
Ciudad Miguel Alemán, Mexico 150/K3
Ciudad Obregón, Mexico 146/H7
Ciudad Obregón, Mexico 150/E3
Ciudad Ojeda, Venezuela 120/B2
Ciudad Ojeda, Venezuela 124/C2
Ciudad Piar (co.), Venezuela 124/G4
Ciudad Quesada, C. Rica 154/E5
Ciudad Real (prov.), Spain 33/D3
Ciudad Real, Spain 33/D3
Ciudad Río Bravo, Mexico 150/K4
Ciudad-Rodrigo, Spain 33/C2
Ciudad Satélite, Mexico 150/L1
Ciudad Serdán, Mexico 150/O2
Ciudad Valles, Mexico 150/K5
Ciudad Victoria, Mexico 150/K5
Civa (cape), Turkey 63/G2
Cividale del Friuli, Italy 34/D1
Civitavecchia, Italy 34/C3
Civitella del Tronto, Italy 34/D3
Civray, France 28/D4
Çivril, Turkey 63/C3
Cizre, Turkey 63/K4
Clachan, Scotland 15/C5
Clackamas (co.), Oreg. 291/E2
Clackamas, Oreg. (97015) 291/B2
Clackamas (riv.), Oreg. 291/E2
Clackmannan, Scotland 10/B1
Clackmannan, Scotland 15/C1
Clackmannan (trad. co.), Scotland 15/A5
Clacton, England 13/J6
Clacton, England 10/G5
Claflin, Kansas (67525) 232/D3
Claiborne, Ala. (36434) 195/D7
Claiborne (par.), La. 238/D1
Claiborne (lake), La. 238/E1
Claiborne, Md. (21624) 245/N5
Claiborne (co.), Miss. 256/C7
Claiborne (co.), Tenn. 237/O8
Clair, New Bruns. 170/B1
Clair, Sask. 181/G3
Claire (lake), Alberta 182/B5
Claire (lake), Alta. 162/E4
Claire City, S. Dak. (57224) 298/P2
Clairemont, Texas (†79518) 303/D4
Clair Engle (lake), Calif. 204/C3
Clairette, Texas (†76457) 303/F5
Clairfield, Tenn. (37715) 237/O7
Clairmont, Alberta 182/A2
Clairmont Springs, Ala. (†35160) 195/G4
Clairton, Pa. (15025) 294/C7
Clallam (co.), Wash. 310/B2
Clallam Bay, Wash. (98326) 310/A2
Clam (bay), Nova Scotia 168/F4
Clam (lake), Wis. 317/B4
Clam (riv.), Wis. 317/A4
Clamart, France 28/A2
Clamecy, France 28/E4
Clam Falls, Wis. (54825) 317/B4
Clam Gulch, Alaska (99568) 196/B1
Clam Lake, Wis. (54517) 317/E3
Clan Alpine (mts.), Nev. 266/D3
Clancy, Mont. (59634) 262/E4
Clandeboye, Manitoba 179/E4
Clandeboye, Ontario 177/C4
Clandonald, Alberta 182/E3
Clanton, Ala. (35045) 195/E5
Clanwilliam, Manitoba 179/C4
Clanwilliam, S. Africa 118/B6
Clapperton (isl.), Ontario 177/B1
Clara, Ireland 10/C4
Clara, Ireland 17/F5
Clara, Miss. (39324) 256/G7
Clara, Uruguay 145/D3
Clara Barton Nat'l Hist. Site, Md. 245/E4
Clara City, Minn. (56222) 255/C6
Claravale, North. Terr. 93/B3
Clare, Ill. (60111) 222/E1
Clare, Ind. (†46060) 227/F4
Clare, Iowa (50524) 229/E3
Clare (co.), Ireland 17/D6
Clare (isl.), Ireland 10/A4
Clare (isls.), Ireland 17/A4
Clare (riv.), Ireland 17/D5
Clare (co.), Mich. 250/E5
Clare, Mich. (48617) 250/E5
Clare, N.S. Wales 97/B3
Clare, S. Australia 94/F5
Claregalway, Ireland 17/D5
Claremont, Calif. (91711) 204/D10
Claremont, Ill. (62421) 222/F5
Claremont, Jamaica 158/J6
Claremont, Minn. (55924) 255/E6
Claremont, N.H. (03743) 268/C5
Claremont, N.C. (28610) 281/G3
Claremont, S. Dak. (57432) 298/N2
Claremont, Va. (23899) 307/P6
Claremore, Okla. (74017) 288/R2
Claremorris, Ireland 17/C4
Claremorris, Ireland 10/B4
Clarence (str.), Alaska 196/N2
Clarence (isl.), Chile 120/B8
Clarence (isl.), Chile 138/E10
Clarence, Iowa (52216) 229/M5
Clarence, La. (71414) 238/E3
Clarence, Mo. (63437) 261/H3
Clarence (riv.), N. S. Wales 88/J5
Clarence (riv.), N.S. Wales 97/G1
Clarence○, N.Y. (14031) 276/C5
Clarence (riv.), N. Zealand 100/E5
Clarence (str.), North. Terr. 88/E2
Clarence (str.), North. Terr. 93/B2
Clarence (cape), N.W. Terrs. 187/K2
Clarence (head), N.W. Terrs. 187/L2
Clarence, Pa. (16829) 294/G3
Clarence Bridge, N. Zealand 100/E5
Clarence Creek, Ontario 177/J2
Clarenceville, Québec 172/D4
Clarendon, Ark. (72029) 202/H4
Clarendon, New Bruns. 170/D3
Clarendon, N.C. (28432) 281/M6
Clarendon (lake), Ontario 177/G3
Clarendon, Pa. (16313) 294/D2
Clarendon (co.), S.C. 296/G4
Clarendon, Texas (79226) 303/C3
Clarendon○, Vt. (†05759) 268/A4
Clarendon Hills, Ill. (60514) 222/B6
Clarens, Switzerland 39/C4
Clarenville, Newf. 166/C2
Claresholm, Alberta 182/D4
Clariden (mt.), Switzerland 39/G3
Clarie Coast (reg.) 5/C7
Clarinda, Iowa (51632) 229/C7
Clarines, Venezuela 124/F3
Clarington, Ohio (43915) 284/J6
Clarington, Pa. (15828) 294/D3
Clarion, Iowa (50525) 229/F3
Clarión (isl.), Mexico 150/B7
Clarion, Mich. (†49796) 250/E3
Clarion (co.), Pa. 294/D3
Clarion, Pa. (16214) 294/D3
Clarion (riv.), Pa. 294/D3
Clarion River, East Branch (lake), Pa. 294/E2
Clarissa, Minn. (56440) 255/C4
Clarita, Okla. (74535) 288/O6
Clark (lake), Alaska 196/H2
Clark (co.), Ark. 202/D5
Clark, Colo. (80428) 208/F1
Clark (co.), Idaho 220/F5
Clark (co.), Ill. 222/F4
Clark (co.), Ind. 227/F8
Clark (co.), Kansas 232/C4
Clark (co.), Ky. 237/N4
Clark (co.), Mo. 261/J2
Clark, Mo. (65243) 261/H4
Clark (co.), Nev. 266/F6
Clark○, N.J. (07066) 273/A3
Clark, N.C. (†28560) 281/P4
Clark (buttes), N. Dak. 282/G7
Clark (co.), Ohio 284/C6
Clark, Ohio (43810) 284/G5
Clark (pt.), Ontario 177/C3
Clark, Pa. (16113) 294/B3
Clark (co.), S. Dak. 298/O4
Clark, S. Dak. (57225) 298/O4
Clark (co.), Wash. 310/C5
Clark (co.), Wis. 317/E6
Clark, Wyo. (†59008) 319/C1
Clark Canyon (res.), Mont. 262/D6
Clark Center, Ill. (†62441) 222/F4
Clarkdale, Ariz. (86324) 198/C4
Clarke (co.), Ala. 195/C7
Clarke (co.), Georgia 217/F3
Clarke (co.), Iowa 229/F6
Clarke (co.), Miss. 256/G6
Clarke (range), Queensland 95/C4
Clarke (isl.), Tasmania 99/E2
Clarke (co.), Va. 307/M2
Clarke City, Québec 174/D2
Clarkedale, Ark. (72325) 202/K3
Clarke's Beach, Newf. 166/D2
Clarkesville, Georgia (30523) 217/F1
Clarkfield, Minn. (56223) 255/C6
Clark Fork, Idaho (83811) 220/B1
Clark Fork (riv.), Mont. 188/D1
Clark Fork (riv.), Mont. 262/A3
Clark Hill (lake), Georgia 217/H3
Clark Hill (dam), S.C. 296/C4
Clark Hill (lake), S.C. 296/C4
Clarkia, Idaho (83812) 220/B2
Clark Island, Maine (†04859) 243/E8
Clarklake, Mich. (49234) 250/E6
Clarkrange, Tenn. (38553) 237/L8
Clarks, East Fork (riv.), Ky. 237/E7
Clarks, La. (71415) 238/F2
Clarks, Nebr. (68628) 264/G3
Clarksboro, N.J. (08020) 273/C4
Clarksburg, Calif. (95612) 204/B9
Clarksburg, Ind. (47225) 227/G6
Clarksburg, Md. (20734) 245/J4
Clarksburg, Mo. (65025) 261/G5
Clarksburg, N.J. (08510) 273/E3
Clarksburg, Ohio (43115) 284/D7
Clarksburg, Ontario 177/D3
Clarksburg, Tenn. (38324) 237/E9
Clarksburg, W. Va. 188/K3
Clarksburg, W. Va. (26301) 312/F4
Clarks Corner, Conn. (†06256) 210/G1
Clarksdale, Miss. 188/J4
Clarksdale, Miss. (38614) 256/D2
Clarksdale, Mo. (64430) 261/D3
Clarks Falls, Conn. (†06359) 210/H3
Clarks Fork, Yellowstone (riv.), Mont. 262/G6
Clarks Fork (riv.), Wyo. 319/C1
Clarks Green, Pa. (†18411) 294/F6
Clarks Grove, Minn. (56016) 255/E7
Clark's Harbour, Nova Scotia 168/C5
Clarks Hill, Ind. (47930) 227/D4
Clarks Hill, S.C. (29821) 296/C4
Clarks Mill, Maine (†04847) 243/B8
Clarks Mills, Pa. (16114) 294/B3
Clarkson, Ky. (42726) 237/J6
Clarkson, Miss. (†39752) 256/F3
Clarkson, Nebr. (68629) 264/G3
Clarkson, N.Y. (14430) 276/E4
Clarkson Valley, Mo. (†63017) 261/N3
Clarks Point, Alaska (99569) 196/G3
Clarks Summit, Pa. (18411) 294/F6
Clarkston, Georgia (30021) 217/L1
Clarkston, Mich. (48016) 250/F6
Clarkston, Scotland 15/B2
Clarkston, Utah (84305) 304/B2
Clarkston, Wash. (99403) 310/H4
Clark's Town, Jamaica 158/H6
Clarksville, Ark. (72830) 202/D3
Clarksville, Del. (19937) 245/T6
Clarksville, Fla. (32430) 212/D6
Clarksville, Ind. (47130) 227/F8
Clarksville, Iowa (50619) 229/H3
Clarksville, Md. (21029) 245/L4
Clarksville, Mich. (48815) 250/D6
Clarksville, Mo. (63336) 261/K4
Clarksville○, N.H. (†05902) 268/E1
Clarksville, N.Y. (12041) 276/M5
Clarksville, Ohio (45113) 284/C7
Clarksville (Clark), Pa. (16113) 294/B3
Clarksille, Pa. (15322) 294/B6
Clarksville, Tenn. 188/J3
Clarksville, Tenn. (37040) 237/G7
Clarksville, Texas (75426) 303/K4
Clarksville, Va. (23927) 307/L7
Clarkton, Mo. (63837) 261/M10
Clarkton, N.C. (28433) 281/M6
Clarno, Wis. (†53566) 317/G10
Claro (riv.), Bolivia 136/A3
Claro (riv.), Brazil 132/D7
Claro (riv.), Chile 138/G5
Claro, Switzerland 39/G4
Clashmoor, Sask. 181/H3
Clashmore, Ireland 17/F8
Clatonia, Nebr. (68328) 264/H4
Clatskanie, Oreg. (97016) 291/D1
Clatsop (co.), Oreg. 291/D1
Claud, Ala. (†36024) 195/F5
Claude, Texas (79019) 303/C2
Claudell, Kansas (†66951) 232/C2
Claudville, Va. (24076) 307/H7
Claudy, N. Ireland 17/G2
Claunch, N. Mex. (87011) 274/C4
Claussen, S.C. (†29501) 296/H3
Clausthal-Zellerfeld, W. Germany 22/D3
Claverack-Red Mills, N.Y. (12513) 276/N6
Claveria, Philippines 82/C1
Clavet, Sask. 181/E4
Clawson, Mich. (48017) 250/B6
Clawson, Utah (84516) 304/C4
Claxton, Georgia (30417) 217/J6
Clay (co.), Ala. 195/G4
Clay (co.), Ark. 202/K1
Clay, Calif. (†95638) 204/C9
Clay (creek), Colo. 208/O7
Clay (co.), Fla. 212/E2
Clay (co.), Georgia 217/B7
Clay (co.), Ill. 222/E5
Clay (co.), Ind. 227/C6
Clay (co.), Iowa 229/C2
Clay (co.), Kansas 232/E2
Clay (co.), Ky. 237/O6
Clay, Ky. (42404) 237/F6
Clay, La. (†71270) 238/E2
Clay (co.), Minn. 255/B4
Clay (co.), Miss. 256/G3
Clay (co.), Mo. 261/D4
Clay (co.), Nebr. 264/F4
Clay (co.), N.C. 281/B4
Clay (co.), S. Dak. 298/P8
Clay (co.), Tenn. 237/K5
Clay (co.), Texas 303/F4
Clay (hills), Utah 304/D6
Clay (co.), W. Va. 312/D6
Clay, W. Va. (25043) 312/D6
Claybank, Sask. 181/F5
Clay Bank, Va. (†23061) 307/P6
Clay Center, Kansas (67432) 232/E2
Clay Center, Nebr. (68933) 264/F4
Clay Center, Ohio (43408) 284/D2
Clay City, Ill. (62824) 222/E5
Clay City, Ind. (47841) 227/C6
Clay City, Ky. (40312) 237/O5
Claycomo, Mo. (†64119) 261/P5
Clay Cross, England 13/J2
Claydon, Sask. 181/B6
Clayhatchee, Ala. (†36322) 195/G8
Claymont, Del. (19703) 245/S1
Claymour, Ky. (†42220) 237/G7
Clayoquot (sound), Br. Col. 184/D5
Claypool, Ariz. (85532) 198/E5
Claypool, Ind. (46510) 227/F2
Claypool, Ky. (†42101) 237/J7
Claysburg, Pa. (16625) 294/F5
Clay Springs, Ariz. (†85934) 198/E4
Claysville, Ohio (43729) 284/G6
Claysville, Pa. (15323) 294/B5
Clayton, Ala. (36015) 195/G7
Clayton, Calif. (94517) 204/K2
Clayton, Del. (19938) 245/R3
Clayton (co.), Georgia 217/D3
Clayton, Georgia (30525) 217/F1
Clayton, Idaho (83227) 220/D5
Clayton, Ill. (62324) 222/B3
Clayton, Ind. (46118) 227/D5
Clayton (co.), Iowa 229/L3
Clayton, Iowa (†52049) 229/L3
Clayton, Kansas (67629) 232/B2
Clayton, La. (71326) 238/H3
Clayton (lake), Maine 243/D2
Clayton, Mich. (49235) 250/E7
Clayton, Miss. (†38626) 256/D1
Clayton, Mo. (63105) 261/P3
Clayton, N.J. (08312) 273/C4
Clayton, N. Mex. (88415) 274/F2
Clayton, N.Y. (13624) 276/H2
Clayton, N.C. (27520) 281/N3
Clayton, Ohio (45315) 284/B6
Clayton, Okla. (74536) 288/R5
Clayton, S. Dak. (†57332) 298/O7
Clayton, Victoria 97/J5
Clayton, Wash. (99110) 310/H3
Clayton, W. Va. (†24910) 312/E7
Clayton, Wis. (54004) 317/B5
Clayton Lake, Maine (04018) 243/E2
Claytonville, Ill. (60926) 222/F3
Claytor (lake), Va. 307/G6
Clayville, N.Y. (13322) 276/K5
Clayville, R.I. (02815) 249/H5
Clayville, S. Africa 118/H6
Clayville, Va. (†23139) 307/N6
Clear, Alaska (99704) 196/J2
Clear (cape), Alaska 196/D1
Clear (hills), Alberta 182/A1
Clear (creek), Ariz. 198/D4
Clear (lake), Calif. 188/B3
Clear (lake), Calif. 204/C4
Clear (lake), Iowa 229/G2
Clear (cape), Ireland 7/C3
Clear (cape), Ireland 17/B9
Clear (cape), Ireland 10/B5
Clear (isl.), Ireland 17/C9
Clear (lake), La. 238/D3
Clear (lake), Manitoba 179/C4
Clear (lake), Ontario 177/F3
Clear (lake), Ontario 177/G2
Clear (creek), Utah 304/B5
Clear (lake), Utah 304/B4
Clear (creek), Wyo. 319/F1
Clear Boggy (creek), Okla. 288/O6
Clearbrook, Br. Col. 184/L3
Clearbrook, Minn. (56634) 255/C3
Clear Brook, Va. (22624) 307/M2
Clear Creek, Calif. (†96039) 204/B2
Clear Creek (co.), Colo. 208/H3
Clear Creek, Ind. (47426) 227/E6
Clearcreek, Utah (†84538) 304/C4
Clear Creek, W. Va. (25044) 312/D7
Clearfield, Iowa (50840) 229/D7
Clearfield, Ky. (40313) 237/P4
Clearfield (co.), Pa. 294/F3
Clearfield, Pa. (16830) 294/F3
Clearfield, S. Dak. (57581) 298/K7
Clearfield, Utah (84015) 304/B2
Clear Fork (res.), Ohio 284/E4
Clear Fork, Mohican (riv.), Ohio 284/F4
Clear Fork, Brazos (riv.), Texas 303/D5
Clear Fork, Guyandotte (riv.), W. Va. 312/C7
Clear Hills, Alberta 182/B1
Clearlake, Calif. (95422) 204/C5
Clear Lake (res.), Calif. 204/D2
Clear Lake, Ind. (†46737) 227/H1
Clear Lake, Iowa (50428) 229/G2
Clear Lake, La. (†71414) 238/E3
Clear Lake, Minn. (55319) 255/E5
Clear Lake, S. Dak. (57226) 298/R4
Clearlake, Wash. (98235) 310/C2
Clear Lake, Wis. (54005) 317/B5
Clearlake Oaks, Calif. (95423) 204/C4
Clear Lake Shores, Texas (†77565) 303/K2
Clearmont, Mo. (64431) 261/C1
Clearmont, Wyo. (82835) 319/F1
Clear Ridge, Pa. (†17229) 294/F5
Clear Spring, Ind. (†47220) 227/E7
Clear Spring, Md. (21722) 245/G2
Clearview, Okla. (74835) 288/O4
Clearview, W. Va. (†26003) 312/E2
Clearview City, Kansas (66019) 232/G3
Clearville, Pa. (15535) 294/F6
Clearwater (riv.), Alberta 182/C4
Clearwater (riv.), Alberta 182/E1
Clearwater, Br. Col. 184/G4
Clearwater (lake), Br. Col. 184/G4
Clearwater (riv.), Br. Col. 184/G4
Clearwater, Fla. 188/K5
Clearwater, Fla. (*33515) 212/B2
Clearwater (co.), Idaho 220/C3
Clearwater, Idaho (83521) 220/C3
Clearwater (mts.), Idaho 220/C3
Clearwater (riv.), Idaho 220/B3
Clearwater, Kansas (67026) 232/E4
Clearwater, Manitoba 179/D5
Clearwater (co.), Minn. 255/C3
Clearwater, Minn. (55320) 255/D5
Clearwater (riv.), Minn. 255/C3
Clearwater (lake), Mo. 261/L8
Clearwater, Nebr. (68726) 264/F2
Clearwater (brook), New Bruns. 170/D2
Clearwater (riv.), Sask. 181/L3
Clearwater, S.C. (29822) 296/D4
Clearwater, Wash. (†98331) 310/A3
Clearwater Beach (isl.), Fla. 212/B2
Clearwater Lake, Manitoba 179/H3
Clearwater Lake, Wis. (54518) 317/H4
Clearwater Lake Beach, Sask. 181/D5
Clearwater Lake Prov. Park, Manitoba 179/H3
Cleator Moor, England 13/D3
Cleburne (co.), Ala. 195/G3
Cleburne (co.), Ark. 202/F2
Cleburne, Texas 188/G4
Cleburne, Texas (76031) 303/G5
Cle Elum, Wash. (98922) 310/E3
Cle Elum (lake), Wash. 310/E3
Cleethorpes, England 13/H4
Cleethorpes, England 10/F4
Cleeves, Sask. 181/C2
Cleghorn, Iowa (51014) 229/B3
Cleghorn, Wis. (†54738) 317/C6
Clem, Georgia (†30117) 217/B3
Clemenceau, Ariz. (†86326) 198/C4
Clemenceau, Sask. 181/J3
Clément, Fr. Guiana 131/E4
Clemente (isl.), Chile 138/D6
Clementon, N.J. (08021) 273/D4
Clements, Calif. (95227) 204/C9
Clements, Kansas (66844) 232/F3
Clements, Md. (20624) 245/L7
Clements, Minn. (56224) 255/D6
Clementson, Minn. (56623) 255/D2
Clementsport, Nova Scotia 168/C4
Clementsvale, Nova Scotia 168/C4
Clementsville, N. Dak. (†58492) 282/O5
Clemmons, N.C. (27012) 281/J2
Clemons, Iowa (50051) 229/G4
Clemscot, Okla. (†73437) 288/L6
Clemson, S.C. (29631) 296/B2
Clendenin, W. Va. (25045) 312/D5
Clendening (lake), Ohio 284/H5
Cleopatra Needle (mt.), Philippines 82/B5
Cleora, Okla. (†74331) 288/S1
Cleo Springs, Okla. (73729) 288/K2
Clerf (riv.), Luxembourg 27/J8
Clermont, Fla. (32711) 212/E3
Clermont, France 28/E3
Clermont, Georgia (30527) 217/E2
Clermont, Iowa (52135) 229/K3
Clermont, Ky. (40110) 237/K5
Clermont, N.Y. (†12526) 276/N6
Clermont (co.), Ohio 284/B7
Clermont, Pa. (†16740) 294/E2
Clermont, Québec 172/G2
Clermont, Queensland 88/H4
Clermont, Queensland 95/C4
Clermont-Ferrand, France 7/E4
Clermont-Ferrand, France 28/E5
Clermont Harbor, Miss. (39551) 256/F10
Clervaux, Luxembourg 27/J8
Cleve, S. Australia 88/F6
Cleve, S. Australia 94/E5
Clevedon, England 13/D6
Cleveland, Ala. (35049) 195/E3
Cleveland (co.), Ark. 202/F6
Cleveland, Ark. (72030) 202/E3
Cleveland (co.), England 13/F3
Cleveland (hills), England 13/F3
Cleveland, Fla. (†33950) 212/E5
Cleveland, Georgia (30528) 217/E1
Cleveland, Minn. (56017) 255/E6
Cleveland, Miss. (38732) 256/C3

Cleveland, Mo. (64734) 261/C5
Cleveland, Mont. (†59523) 262/G2
Cleveland, N. Mex. (87715) 274/D2
Cleveland, N.Y. (13042) 276/J4
Cleveland (co.), N.C. 281/F4
Cleveland, N.C. (27013) 281/H3
Cleveland, N. Dak. (58424) 282/M6
Cleveland, Ohio 188/K2
Cleveland, Ohio (*44101) 284/H9
Cleveland, Ohio 146/K5
Cleveland (co.), Okla. 288/M4
Cleveland, Okla. (74020) 288/O2
Cleveland, S.C. (29635) 296/C1
Cleveland, Tenn. (37311) 237/M10
Cleveland, Texas (77327) 303/K7
Cleveland, Utah (84518) 304/D4
Cleveland, Va. (24225) 307/D7
Cleveland, W. Va. (26215) 312/F5
Cleveland, Wis. (53015) 317/L8
Cleveland Heights, Ohio (44118) 284/H9
Cleveland-Hopkins Mun. Airport, Ohio 284/G9
Clevelândia do Norte, Brazil 132/D2
Cleveland Park, D.C. (20008) 245/E4
Clever, Mo. (65631) 261/F8
Cleves, Iowa (†50601) 229/G4
Cleves, Ohio (45002) 284/B9
Clew (bay), Ireland 17/B4
Clew (bay), Ireland 10/B4
Clewiston, Fla. (33440) 212/E5
Clichy, France 28/B1
Clicquot-Millis, Mass. (†02054) 249/A8
Clifden, Ireland 10/B4
Clifden, Ireland 17/B5
Cliff, N. Mex. (88028) 274/A6
Cliff (cape), Nova Scotia 168/E3
Cliff (creek), Utah 304/E3
Cliffdell, Wash. (†98937) 310/E4
Clifford, Ind. (47226) 227/F6
Clifford, Ky. (41208) 237/S4
Clifford (lake), Maine 243/H5
Clifford, Mich. (48727) 250/F5
Clifford, N. Dak. (58016) 282/R5
Clifford, Ontario 177/D4
Clifford, Pa. (18413) 294/L2
Clifford, Va. (24533) 307/K5
Clifford, Wis. (†54564) 317/F4
Cliffordvale, New Bruns. 170/C2
Cliffside, N.C. (28024) 281/F4
Cliffside Park, N.J. (07010) 273/C2
Clifftop, W. Va. (25822) 312/E6
Cliffwood, N.J. (07721) 273/E3
Clifton, Ariz. (85533) 198/F5
Clifton, Colo. (81520) 208/C4
Clifton, Idaho (83228) 220/F7
Clifton, Ill. (60927) 222/F3
Clifton, Kansas (66937) 232/E2
Clifton, La. (†70438) 238/K5
Clifton○, Maine (†04428) 243/G6
Clifton, New Bruns. 170/E1
Clifton, N.J. (*07011) 273/B2
Clifton, S.C. (29324) 296/D2
Clifton, Tenn. (38425) 237/F10
Clifton, Texas (76634) 303/G6
Clifton, W. Va. (25237) 312/B5
Clifton, Wis. (†54618) 317/F8
Clifton City, Mo. (†65348) 261/G5
Clifton Dartmouth Hardness, England 10/E5
Clifton Dartmouth Hardness, England 13/D7
Clifton Forge (I.C.), Va. (24422) 307/J5
Clifton Heights, Pa. (19018) 294/M7
Clifton Hill, Mo. (65244) 261/G4
Clifton Hills, S. Australia 94/F2
Clifton Mills, W. Va. (†26525) 312/G3
Clifton Park○, N.Y. (†12118) 276/N5
Clifton Springs, N.Y. (14432) 276/F4
Cliftonville, Miss. (†39739) 256/H4
Clifty, Ark. (†72756) 202/C1
Clifty (creek), Ind. 227/F6
Clifty, Ky. (42216) 237/G7
Clifty, Tenn. (†38583) 237/L9
Clifty, W. Va. (†25854) 312/E6
Climax, Colo. (80429) 208/G4
Climax, Georgia (31734) 217/D9
Climax, Kansas (67027) 232/F4
Climax, Ky. (40413) 237/N6
Climax, Mich. (49034) 250/D6
Climax, Minn. (56523) 255/B3
Climax, N.C. (27233) 281/K3
Climax, Sask. 181/C6
Climax Springs, Mo. (65324) 261/G6
Climbing Hill, Iowa (51015) 229/B4
Clinch (co.), Georgia 217/G9
Clinch (riv.), Tenn. 237/N9
Clinch (riv.), Va. 307/C7
Clinchburg, Va. (24321) 307/E7
Clinchco, Va. (24226) 307/D6
Clinchfield, Georgia (31013) 217/E6
Clinchmore, Tenn. (†37714) 237/N8
Clinchport, Va. (24244) 307/C7
Cline Settlement, Alberta 182/B3
Clingmans Dome (mt.), N.C. 281/C3
Clingmans Dome (mt.), Tenn. 237/P10
Clint, Texas (79836) 303/B10
Clinton, Ala. (35448) 195/C5
Clinton, Ark. (72031) 202/F2
Clinton, Br. Col. 184/G4
Clinton, Conn. (06413) 210/E3
Clinton○, Conn. (06413) 210/E3
Clinton (co.), Ill. 222/D5
Clinton, Ill. (61727) 222/E3
Clinton (co.), Ind. 227/E4
Clinton, Ind. (47842) 227/C5
Clinton (co.), Iowa 229/M5
Clinton, Iowa 188/J2
Clinton, Iowa (52732) 229/N5
Clinton (co.), Ky. 237/L7
Clinton, Ky. (42031) 237/D7
Clinton, La. (70722) 238/J5
Clinton, Maine (04927) 243/D6
Clinton○, Maine (04927) 243/D6
Clinton, Md. (20735) 245/G6
Clinton○, Mass. (01510) 249/H3
Clinton (co.), Mich. 250/E6
Clinton, Mich. (49236) 250/F6
Clinton, Minn. (56225) 255/B5
Clinton, Miss. (39056) 256/D6
Clinton (co.), Mo. 261/D3
Clinton, Mo. (64735) 261/E6
Clinton, Mont. (59825) 262/C4
Clinton, Nebr. (†69343) 264/B2
Clinton, N.J. (08809) 273/D2
Clinton (res.), N.J. 273/E1
Clinton (co.), N.Y. 276/N1
Clinton, N.Y. (13323) 276/K4
Clinton, N. Zealand 100/B7
Clinton, N.C. (28328) 281/N5
Clinton (co.), Ohio 284/C7
Clinton, Ohio (44216) 284/G4
Clinton, Okla. (73601) 288/H3
Clinton, Ontario 177/C4
Clinton (co.), Pa. 294/G3
Clinton, Pa. (15026) 294/B5
Clinton, S.C. (29325) 296/D3
Clinton, Tenn. (37716) 237/N8
Clinton, Wash. (98236) 310/C3
Clinton, W. Va. (†26058) 312/E2
Clinton, Wis. (53525) 317/J10
Clinton-Colden (lake), N.W. Terrs. 187/H3
Clinton Corners, N.Y. (12514) 276/N7
Clinton Creek, Yukon 187/D3
Clintondale, N.Y. (12515) 276/M7
Clintondale, Pa. (†17751) 294/H3
Clinton Falls, Minn. (†55060) 255/E6
Clintonville, Conn. (†06473) 210/D3
Clintonville, Ky. (†40361) 237/N4
Clintonville, Pa. (16372) 294/C3
Clintonville, W. Va. (24928) 312/E7
Clintonville, Wis. (54929) 317/J6
Clintwood, Va. (24228) 307/D6
Clio, Ala. (36017) 195/G7
Clio, Iowa (50052) 229/G7
Clio, La. (†70462) 238/M2
Clio, Mich. (48420) 250/F5
Clio, S.C. (29525) 296/H2
Clio, W. Va. (25046) 312/D5
Clipper, Wash. (98244) 310/C2
Clipperton (isl.) 146/H8
Clipperton (isl.) 2/D5
Clisham (mt.), Scotland 15/B3
Clitherall, Minn. (56524) 255/C4
Clitheroe, England 13/H1
Clitheroe, England 10/G1
Clive, Alberta 182/D3
Clive, Iowa (50318) 229/F5
Clive, N. Zealand 100/F3
Cliza, Bolivia 136/C5
Cloan, Sask. 181/C3
Cloates (pt.), W. Australia 92/A3
Clode (sound), Newf. 166/D2
Cloe, Pa. (†15767) 294/E4
Cloghan, Ireland 17/F5
Clogh-Chatsworth, Ireland 17/G6
Clogheen, Ireland 17/F7
Clogher, N. Ireland 17/G3
Clogherhead, Ireland 17/J4
Cloghy, N. Ireland 17/K3
Clonakilty, Ireland 10/B5
Clonakilty, Ireland 17/D8
Clonakilty (bay), Ireland 17/D8
Clonaslee, Ireland 17/F5
Cloncurry, Australia 87/E8
Cloncurry, Queensland 95/B4
Cloncurry, Queensland 88/G4
Cloncurry (riv.), Queensland 95/B4
Clondalkin, Ireland 17/J5
Clonegal, Ireland 17/H6
Clones, Ireland 10/C3
Clones, Ireland 17/G3
Clonfert, Ireland 17/E5
Clonmany, Ireland 17/G1
Clonmel, Ireland 17/F7
Clonmel, Ireland 10/C4
Clonmellon, Ireland 17/H4
Clonroche, Ireland 17/H7
Clontarf, Minn. (56226) 255/C5
Clontuskert, Ireland 17/E4
Cloone, Ireland 17/F4
Cloppenburg, W. Germany 22/B2
Clopton, Ala. (36317) 195/G7
Cloquet, Minn. (55720) 255/F4
Cloquet (riv.), Minn. 255/F4
Cloridorme, Québec 172/D1
Clorinda, Argentina 143/E2
Closeburn, Scotland 15/E5
Closplint, Ky. (40927) 237/P7
Closter, N.J. (07624) 273/C1
Clothier, W. Va. (25047) 312/C7
Clotho, Minn. (†56347) 255/C4
Cloud (co.), Kansas 232/E2
Cloud (peak), Wyo. 319/E1
Cloud Chief, Okla. (†73632) 288/J4
Cloudcroft, N. Mex. (88317) 274/D6
Cloudland, Georgia (30709) 217/A1
Cloudy (bay), N. Zealand 100/E4
Cloudy, Okla. (74537) 288/R6
Cloughjordan, Ireland 17/E6
Cloughmills, N. Ireland 17/J2
Cloutierville, La. (71416) 238/E3
Clova, Québec 174/B3
Clover (creek), Oreg. 291/K3
Clover, S.C. (29710) 296/E1
Clover, Va. (24534) 307/L7
Clover, W. Va. (†25276) 312/D5
Clover Bar, Alberta 182/D3
Clover Bend, Ark. (†72433) 202/H2
Clover Bottom, Ky. (40447) 237/N5
Cloverdale, Ala. (35617) 195/C1
Cloverdale, Calif. (95425) 204/B5
Cloverdale, Ind. (46120) 227/D5
Cloverdale, Minn. (†55037) 255/F4
Cloverdale, Ohio (45827) 284/B3
Cloverdale, Oreg. (97112) 291/D2
Cloverdale, Va. (24077) 307/J6
Cloverland, Ind. (†47834) 227/C6
Cloverland, Wash. (†99402) 310/H4
Cloverleaf, Manitoba 179/F5
Clover Lick, W. Va. (†24979) 312/F6
Clover Pass, Alaska (†99901) 196/N2
Cloverport, Ky. (40111) 237/H5
Cloverton, Minn. (†55048) 255/F4
Clovis, Calif. (93612) 204/F7
Clovis, N.Mex. 188/F4
Clovis, N. Mex. (88101) 274/F4
Clovulin, Scotland 15/C4
Cloyne, Ireland 17/E8
Cloyne, Ontario 177/G3
Cluanie, Loch (lake), Scotland 15/C3
Club (isl.), Ontario 177/C2
Cluff Lake, Sask. 181/L2
Cluj-Napoca, Romania 45/F2
Cluj-Napoca, Romania 7/G4
Clun, England 10/E4
Clun, England 13/D6
Clune, Pa. (15727) 294/D4
Clunes, Victoria 97/B5
Cluny, Alberta 182/D4
Cluny, France 28/F4
Cluses, France 28/G4
Clusone-Fiorine, Italy 34/C2
Cluster Springs, Va. (24535) 307/L7
Clute, Texas (77531) 303/J9
Clutha (riv.), N. Zealand 100/B6
Clutier, Iowa (52217) 229/J4
Clwyd (co.), Wales 13/D4
Clyattville, Georgia (31604) 217/F9
Clyde, Alberta 182/D2
Clyde (lake), Alberta 182/E2
Clyde, Canada 4/B13
Clyde (riv.), Dominica 161/F6
Clyde, Kansas (66938) 232/E2
Clyde, Mo. (64432) 261/C2
Clyde, N.Y. (14433) 276/G4
Clyde, N.C. (28721) 281/D3
Clyde, N. Dak. (†58352) 282/N2
Clyde, N.W.T. 162/J1
Clyde (inlet), N.W.T. 162/K1
Clyde, N.W. Terrs. 187/M2
Clyde (inlet), N.W. Terrs. 187/M2
Clyde (riv.), Nova Scotia 168/C5
Clyde, Ohio (43410) 284/E3
Clyde (firth), Scotland 15/D5
Clyde (firth), Scotland 10/D3
Clyde (riv.), Scotland 10/E3
Clyde (riv.), Scotland 15/D5
Clyde (riv.), Tasmania 99/D4
Clyde, Texas (79510) 303/E5
Clyde (riv.), Vt. 268/C2
Clyde, Wash. (†99348) 310/G4
Clydebank, Scotland 15/B2
Clydebank, Scotland 10/A1
Clyde Hill, Wash. (†98004) 310/B2
Clyde Park, Mont. (59018) 262/F5
Clyde River, Nova Scotia 168/C5
Clyman, Wis. (53016) 317/J9
Clymer, N.Y. (14724) 276/A6
Clymer, Pa. (15728) 294/E4
Clymers, Ind. (†46947) 227/E3
Clyo, Georgia (31303) 217/K6
Cnoc May (mt.), Scotland 15/C5
Coachella, Calif. (92236) 204/J10
Coachella (canal), Calif. 204/K10
Coachford, Ireland 17/D8
Coahoma (co.), Miss. 256/C2
Coahoma, Miss. (38617) 256/C2
Coahoma, Texas (79511) 303/C5
Coahuila (state), Mexico 150/H3
Coakley, Ky. (†42782) 237/K6
Coal (creek), Ind. 227/C4
Coal, Mo. (†64735) 261/E6
Coal (co.), Okla. 288/O5
Coal (pt.), Oreg. 291/C5
Coal (butte), Utah 304/C5
Coal (creek), Wash. 310/G3
Coal (riv.), W. Va. 312/C6
Coal Bluff, Ind. (†47874) 227/C5
Coal Branch, New Bruns. 170/E2
Coalburn, Scotland 15/E5
Coal City, Ill. (60416) 222/E2
Coal City, Ind. (47427) 227/D6
Coal City, W. Va. (25823) 312/D7
Coalcomán de Matamoros, Mexico 150/H7
Coal Creek, Alaska (†99701) 196/K1
Coal Creek, Colo. (81221) 208/J6
Coal Creek, Ind. (†47932) 227/C4
Coal Creek, New Bruns. 170/E2
Coaldale, Alberta 182/D5
Coaldale, Colo. (81222) 208/H6
Coaldale, Nev. (†89049) 266/D4
Coaldale, Pa. (18218) 294/L4
Coaldale (Six Mile Run), Pa. (16679) 294/F5
Coalfield, Tenn. (37719) 237/N8
Coalfield, Wash. (†98055) 310/B2
Coal Fork, W. Va. (25306) 312/D6
Coalgate, Okla. (74538) 288/O5
Coal Grove, Ohio (†45638) 284/E9
Coal Harbour, Br. Col. 184/D5
Coal Hill, Ark. (72832) 202/C3
Coalhurst, Alberta 182/D5
Coaling, Ala. (35449) 195/D4
Coalinga, Calif. (93210) 204/E7
Coalisland, N. Ireland 17/H2
Coalmont, Br. Col. 184/G5
Coalmont, Colo. (80430) 208/F1
Coalmont, Ind. (47845) 227/C6
Coalmont, Tenn. (37313) 237/K10
Coalport, Pa. (16627) 294/E4
Coalridge, Mont. (†59219) 262/M2
Coal Run, Ky. (†41501) 237/R5
Coalspur, Alberta 182/B3
Coalton, Ill. (†62075) 222/D4
Coalton, Ohio (45621) 284/E7
Coalton, W. Va. (26257) 312/G5
Coal Valley, Alberta 182/B3
Coal Valley, Ill. (61240) 222/C2
Coalville, England 13/F5
Coalville, Iowa (†50501) 229/E4
Coalville, Utah (84017) 304/C3
Coalwood, Mont. (†59351) 262/L5
Coalwood, W. Va. (24824) 312/C8
Coambo, Angola 115/C5
Coambo, Angola 102/D5
Coamo, P. Rico 161/D2
Coamo, P. Rico 156/G1
Coamo (res.), P. Rico 161/D3
Coamo (riv.), P. Rico 161/D3
Coari, Brazil 120/C3
Coari, Brazil 132/H9
Coarsegold, Calif. (93614) 204/F6
Coast (mts.) 162/C4
Coast (ranges) 188/B2
Coast (mts.), Alaska 196/N1
Coast (mts.), Br. Col. 146/E4
Coast (mts.), Br. Col. 184/D3
Coast (ranges), Calif. 204/D7
Coast (prov.), Kenya 115/G4
Coast (ranges), Oreg. 291/D5
Coast (ranges), U.S. 146/F5
Coast (ranges), Wash. 310/B3
Coast Guard Academy, Conn. 210/G3
Coatbridge, Scotland 10/B1
Coatbridge, Scotland 15/C2
Coatepec, Mexico 150/P1
Coatepeque, Guatemala 154/A3
Coates, Minn. (†55068) 255/E6
Coatesville, Ind. (46121) 227/D5
Coatesville, Pa. (19320) 294/L5
Coatetelco, Mexico 150/L2
Coaticook, Québec 172/F4
Coatopa, Ala. (35450) 195/B6
Coats, Kansas (67028) 232/D4
Coats, N.C. (27521) 281/M4
Coats (isl.), N.W.T. 162/H3
Coats (isl.), N.W.T. 146/K3
Coats (isl.), N.W. Terrs. 187/K3
Coatsburg, Ill. (62325) 222/B3
Coats Land (reg.), Ant. 2/H10
Coats Land (reg.) 5/B17
Coatsville, Mo. (63535) 261/G1
Coatzacoalcos, Mexico 146/J8
Coatzacoalcos, Mexico 150/M7
Coatzingo, Mexico 150/N2
Cobalt, Conn. (06414) 210/E2
Cobalt, Idaho (83229) 220/D4
Cobalt, Ont. 162/H6
Cobalt, Ontario 177/K5
Cobalt, Ontario 175/D3
Cobalt City, Mo. (†63645) 261/M7
Cobán, Guatemala 154/B3
Cobar, N. S. Wales 88/H6
Cobar, N.S. Wales 97/C2
Cobargo, N.S. Wales 97/E5
Cobb (co.), Georgia 217/C3
Cobb, Georgia (31735) 217/E7
Cobb, Ky. (42405) 237/F6
Cobb (riv.), Minn. 255/E7
Cobb (isl.), Va. 307/S6
Cobb, Wis. (53526) 317/F10
Cobbadah, N.S. Wales 97/F2
Cobble Hill, Br. Col. 184/K3
Cobble Mountain (res.), Mass. 249/C4
Cobbs Creek, Va. (23035) 307/R6
Cobbtown, Georgia (30420) 217/H6
Cobden, Ill. (62920) 222/D6
Cobden, Minn. (†56085) 255/D6
Cobden, Ontario 177/H2
Cobden, Victoria 97/B6
Cobequid (bay), Nova Scotia 168/E3
Cóbh, Ireland 10/B5
Cóbh, Ireland 17/E8
Cobham (riv.), Manitoba 179/G1
Cobham (riv.), Ontario 175/A2
Cobija, Bolivia 136/A2
Cobija, Brazil 120/C4
Coble, Tenn. (†37033) 237/F9
Cobleskill, N.Y. (12043) 276/L5
Coboconk, Ontario 177/F3
Cobourg (pen.), North. Terr. 88/E2
Cobourg (pen.), North. Terr. 93/C1
Cobourg, Ontario 177/F4
Cobquecura, Chile 138/D1
Cobram, Victoria 97/C4
Cobre, Nev. (†89830) 266/G1
Cóbuè, Mozambique 118/F2
Coburg, Iowa (†51566) 229/C7
Coburg (isl.), N.W. Terrs. 187/L2
Coburg, Oreg. (97401) 291/E3
Coburg, Victoria 88/K7
Coburg, Victoria 97/H5
Coburg, W. Germany 22/D3
Coburn, Pa. (16832) 294/H4
Coburn, W. Va. (26562) 312/F3
Coca, Ecuador 128/D3
Cocachacra, Peru 128/G11
Cocagne, New Bruns. 170/F2
Cocagne (isl.), New Bruns. 170/F2
Cocagne (riv.), New Bruns. 170/F2
Cocagne Cape, New Bruns. 170/F2
Cocama, Peru 128/G8
Cocanada (Kakinada), India 68/E5
Cocani, Bolivia 136/B7
Cocapata, Bolivia 136/B5
Cocentaina, Spain 33/F3
Cochabamba (dept.), Bolivia 136/C5
Cochabamba, Bolivia 120/C4
Cochabamba, Bolivia 136/C5
Cochamó, Chile 138/E3
Coche (isl.), Venezuela 124/F2
Cocheco (riv.), N.H. 268/E5
Cochecton, N.Y. (†12726) 276/K7
Cochem, W. Germany 22/B3
Cochenour, Ontario 175/B2
Cochetopa (creek), Colo. 208/F6
Cochimbo, Serra do (mts.), Brazil 132/C5
Cochin, Sask. 181/C2
Cochin-Alleppey, India 68/D6
Cochinos (bay), Cuba 158/D2
Cochise (co.), Ariz. 198/F7
Cochise, Ariz. (85606) 198/F6
Cochiti, N. Mex. (†87041) 274/C3
Cochituate, Mass. (01778) 249/A7
Cochituate (lake), Mass. 249/A7
Cochran, Georgia (31014) 217/F6
Cochran (co.), Texas 303/B4
Cochrane, Ala. (†35442) 195/B4
Cochrane, Alberta 182/C4
Cochrane (lake), Chile 138/E7
Cochrane, Cerro (mt.), Chile 138/E7
Cochrane (riv.), Manitoba 179/H2
Cochrane, Ont. 146/K5
Cochrane, Ont. 162/H6
Cochrane (terr. dist.), Ontario 177/J4
Cochrane (terr. dist.), Ontario 175/D2
Cochrane, Ontario 177/K5
Cochrane, Ontario 175/D3
Cochrane (riv.), Sask. 181/N2
Cochrane, Wis. (54622) 317/C7
Cochranton, Pa. (16314) 294/B2
Cochranville, Pa. (19330) 294/L6
Cockburn (chan.), Chile 138/E11
Cockburn (isl.), Ontario 177/A2
Cockburn, S. Australia 94/G5
Cockburn (sound), W. Australia 88/B2
Cockburn Harbour, Turks & Caicos 156/D2
Cockburnspath, Scotland 15/F5
Cocke (co.), Tenn. 237/P9
Cockenoe (isl.), Conn. 210/B4
Cockenzie and Port Seton, Scotland 15/D1
Cockermouth, England 13/D3
Cockermouth, England 10/E3
Cockeysville, Md. (21030) 245/M3
Cockrell Hill, Texas (75211) 303/G2
Cockrum, Miss. (†38632) 256/E1
Coclé del Norte, Panama 154/G6
Coco (chan.), Burma 72/B4
Coco (cay), Cuba 158/G1
Coco (riv.), Honduras 154/E3
Coco (chan.), India 68/G6
Coco (riv.), Nicaragua 154/E3
Coco, W. Va. (†25071) 312/D6
Cocoa, Fla. (32922) 212/F3
Cocoa Beach, Fla. (32931) 212/F3
Cocobeach, Gabon 115/B3
Cocodrie (lake), La. 238/E5
Cocolamus, Pa. (17014) 294/H4
Coconino (co.), Ariz. 198/C3
Coconino (plat.), Ariz. 198/C3
Coconut Creek, Fla. (†33060) 212/F5
Cocopah Ind. Res., Ariz. 198/A6
Cocorit, Mexico 150/E3
Cocos (isls.), Australia 2/P6
Cocos (isls.), Australia 54/L11
Cocos (isl.), C. Rica 146/K9
Cocos (isl.), Guam 86/K7
Cocos (bay), Trin. & Tob. 161/B10
Cocuy, Sierra Nevada del (mts.), Colombia 126/D4
Cod (cape), Mass. 146/M5
Cod (cape), Mass. 188/N2
Cod (cape), Mass. 249/O4
Cod (isl.), Newf. 166/B2
Codajás, Brazil 120/C3
Codajás, Brazil 132/H9
Coddle (harb.), Nova Scotia 168/G3
Codegua, Chile 138/G4
Codell, Kansas (67630) 232/C2
Coden, Ala. (36523) 195/B10
Codera (cape), Venezuela 124/F2
Coderre, Sask. 181/E5
Codesa, Alberta 182/B2
Codes Corner, Ontario 177/H3
Codette, Sask. 181/H2
Codfish (isl.), N. Zealand 100/A7
Codigua, Chile 138/F4
Codington (co.), S. Dak. 298/P4
Codó, Brazil 120/E3
Codó, Brazil 132/E4
Codorus, Pa. (17311) 294/J6
Codpa, Chile 138/B1
Codrington, Ant. & Bar. 156/G3
Codrington, Barbados 161/B8
Codroipo, Italy 34/D2
Codroy, Newf. 166/C4
Cody, Nebr. (69211) 264/C2
Cody, Wyo. (82414) 319/D1
Codys, New Bruns. 170/E3
Coe, Ind. (†47598) 227/C8
Coeburn, Va. (24230) 307/D7
Coe Hill, Ontario 177/G3
Coelemu, Chile 138/D1
Coello, Ill. (62825) 222/D6
Coen, Queensland 88/G2
Coen, Queensland 95/B2
Coeroeni (riv.), Suriname 131/C4
Coesfeld, W. Germany 22/B3
Coesse, Ind. (†46725) 227/G2
Coeur d'Alene, Idaho (83814) 220/B2
Coeur d'Alene, Idaho 188/C1
Coeur d'Alene (lake), Idaho 220/B2
Coeur d'Alene (mts.), Idaho 220/C2
Coeur d'Alene (riv.), Idaho 220/B2
Coevorden, Netherlands 27/K3
Coeymans, N.Y. (12045) 276/N6
Coffee (co.), Ala. 195/G8
Coffee (co.), Georgia 217/G8
Coffee (co.), Tenn. 237/J9
Coffee Creek, Mont. (59424) 262/F3
Coffeen, Ill. (62017) 222/D4
Coffee Springs, Ala. (36318) 195/G8
Coffeeville, Ala. (36524) 195/B7
Coffeeville (dam), Ala. 195/B7
Coffeeville, Miss. (38922) 256/E3
Coffey (co.), Kansas 232/G3
Coffey, Mo. (64636) 261/E2
Coffeyville, Kans. 188/G3
Coffeyville, Kansas (67337) 232/G4
Coffin (bay), S. Australia 94/D6
Coffin Bay (pen.), S. Australia 94/D6
Coffs Harbour, N. S. Wales 88/J6
Coffs Harbour, N.S. Wales 97/G2
Cofield, N.C. (27922) 281/R2
Cogan Station, Pa. (†17728) 294/H3
Cogar, Okla. (†73059) 288/K4
Cogdell, Georgia (31628) 217/G8
Cogealac, Romania 45/J3
Coggon, Iowa (52218) 229/L4
Coghinas (riv.), Italy 34/B4
Coglians (Hohe Warte) (mt.), Austria 41/B3
Cognac, France 28/C5
Cogolludo, Spain 33/E2
Cogotí, Chile 138/A8
Cogswell, N. Dak. (58017) 282/P7
Cogton, Philippines 82/E6
Cohagen, Mont. (59322) 262/K3
Cohansey (riv.), N.J. 273/C5
Cohasset, Ala. (†36474) 195/E8
Cohasset○, Mass. (02025) 249/F7
Cohasset, Minn. (55721) 255/E3
Cohoctah, Mich. (48816) 250/F6
Cohocton, N.Y. (14826) 276/F5
Cohocton (riv.), N.Y. 276/F6
Cohoe, Alaska (†99669) 196/B1
Cohoes, N.Y. (12047) 276/N5
Cohoni, Bolivia 136/B5
Cohuna, Victoria 97/C4
Cohutta, Georgia (30710) 217/C1
Coiba, Isla de (isl.), Panama 154/F7
Coihaique, Chile 138/E6
Coihaique Alto, Chile 138/E6
Coihueco, Chile 138/A11
Coila, Miss. (38923) 256/E4
Coill Dubh, Ireland 17/H5
Coimbatore, India 54/J8
Coimbatore, India 68/D6
Coimbra (dist.), Portugal 33/B2
Coimbra, Portugal 7/D4
Coimbra, Portugal 33/B2
Coin, Iowa (51636) 229/C7
Coín, Spain 33/D4
Coinco, Chile 138/G5
Coinjock, N.C. (27923) 281/S2
Coipasa, Bolivia 136/A6
Coipasa (lake), Bolivia 136/B6
Coipasa (salt dep.), Bolivia 136/A6
Coire, Loch (lake), Scotland 15/D2
Cojata, Peru 128/H10
Cojedes (state), Venezuela 124/D3
Cojedes (riv.), Venezuela 124/D3
Cojimíes, Ecuador 128/B2
Cojoro, Venezuela 124/C2
Cojutepeque, El Salvador 154/C4
Cokato, Minn. (55321) 255/D5
Coke (co.), Texas 303/D6
Cokeburg, Pa. (15324) 294/B5
Cokedale, Colo. (81032) 208/K8
Coker, Ala. (35452) 195/C4
Cokercreek, Tenn. (37314) 237/N10
Coketon, W. Va. (†26292) 312/G4
Cokeville, Wyo. (83114) 319/B3
Colaba (pt.), India 68/B7
Colac, Victoria 88/G7
Colac, Victoria 97/B6
Colachel, India 68/D7
Colair (lake), India 68/E5
Colamus (riv.), Nebr. 264/E2
Colasay, Peru 128/C5
Colatina, Brazil 120/E4
Colatina, Brazil 132/F7
Colbeck (cape) 5/B10
Colbert (co.), Ala. 195/C1
Colbert, Georgia (30628) 217/F2
Colbert, Okla. (74733) 288/O7
Colbert, Wash. (99005) 310/H3
Colborne, Ontario 177/G4
Colbún, Chile 138/A11
Colburn, Idaho (83865) 220/B1
Colburn, Ind. (47931) 227/D3
Colby, Kansas (67701) 232/A2
Colby, Wash. (†98366) 310/A2
Colby, Wis. (54421) 317/F6
Colcamar, Peru 128/D6
Colchester, Conn. (06415) 210/F2
Colchester○, Conn. (06415) 210/F2
Colchester, England 13/H6
Colchester, England 10/G5
Colchester, Ill. (62326) 222/C3
Colchester (co.), Nova Scotia 168/E3
Colchester, Ontario 177/B6
Colchester○, Vt. (05446) 268/A2
Colcord, Okla. (74338) 288/S2
Colcord, W. Va. (25048) 312/D7
Cold (bay), Alaska 196/F4
Cold (lake), Alberta 182/E2
Cold (riv.), N.H. 268/C5
Cold Bay, Alaska (99571) 196/F3
Cold Brook, N.Y. (13324) 276/L4
Coldbrook Station, Nova Scotia 168/D3
Colden, N.Y. (14033) 276/C5
Coldingham, Scotland 15/F5
Cold Lake, Alberta 182/E2
Cold Spring, Ky. (41076) 237/T2
Cold Spring, Minn. (56320) 255/D5
Cold Spring, N.J. (†08204) 273/D6
Cold Spring (inlet), N.J. 273/D6
Cold Spring, N.Y. (10516) 276/N8
Coldspring (head), Nova Scotia 168/E3
Coldspring, Texas (77331) 303/J7
Cold Spring Harbor, N.Y. (11724) 276/R6
Cold Springs, Okla. (†73564) 288/J5
Coldstream, Br. Col. 184/H5
Cold Stream (pond), Maine 243/G5
Coldstream, New Bruns. 170/C2
Coldstream, Scotland 10/E3
Coldstream, Scotland 15/F5
Coldstream, Victoria 97/K4
Coldwater, Kansas (67029) 232/C4
Coldwater, Mich. (49036) 250/D7
Coldwater, Miss. (38618) 256/E1
Coldwater (riv.), Miss. 256/D1
Coldwater, Ohio (45828) 284/A5
Coldwater, Ontario 177/E3
Coldwater, Tenn. (†37334) 237/H10
Coldwater (creek), Texas 303/B1
Coldwater, W. Va. (†26411) 312/E4
Cole (co.), Mo. 261/H6
Cole (harb.), Nova Scotia 168/E4
Cole, Okla. (†73010) 288/L5
Colebrook, Conn. (06021) 210/C1
Colebrook○, Conn. (06021) 210/C1
Colebrook, N.H. (03576) 268/E2
Colebrook○, N.H. (03576) 268/E2
Colebrook, Ohio (†44034) 284/J2
Colebrook, Tasmania 99/D4
Colebrook River (lake), Conn. 210/C1
Cole Camp, Mo. (65325) 261/F6
Coleen (riv.), Alaska 196/K1
Colegrove, Pa. (†16749) 294/F2
Coleharbor, N. Dak. (58531) 282/H4
Coleman, Alberta 182/C5
Coleman, Alta. 162/E6
Coleman, Fla. (33521) 212/D3
Coleman, Georgia (31736) 217/C7
Coleman, Mich. (48618) 250/E5
Coleman, Okla. (73432) 288/O6
Coleman (riv.), Queensland 95/B2
Coleman (co.), Texas 303/E6
Coleman, Texas (76834) 303/E6

Coleman, Wis. (54112) 317/L5
Coleman Falls, Va. (24536) 307/K6
Colemans Lake, Georgia (†30441) 217/H5
Çölemerik, Turkey 63/K4
Colerain, N.C. (27924) 281/R2
Coleraine, Minn. (55722) 255/E3
Coleraine (dist.), N. Ireland 17/H1
Coleraine, N. Ireland 17/H1
Coleraine, N. Ireland 10/C3
Coleraine, Québec 172/F4
Coleraine, Victoria 97/A5
Coleridge, Nebr. (68727) 264/G2
Coleridge (lake), N. Zealand 100/C5
Coleridge, N.C. (27234) 281/K3
Coles (pt.), Peru 128/G11
Coles, Miss. (†39638) 256/C8
Coles (co.), Ill. 222/E4
Coles, Ky. (†41222) 237/R5
Colesberg, S. Africa 118/D6
Colesburg, Georgia (†31569) 217/J9
Colesburg, Iowa (52035) 229/L3
Colesburg, Ky. (†40150) 237/K5
Coles Island, New Bruns. 170/E3
Colesville, Md. (20904) 245/K4
Colesville, N.J. (†07461) 273/D1
Coleta, Ill. (61017) 222/D2
Coleville, Calif. (96107) 204/F5
Coleville, Sask. 181/B4
Colfax, Calif. (95713) 204/E4
Colfax, Ill. (61728) 222/E3
Colfax, Ind. (46035) 227/D4
Colfax, Iowa (50054) 229/H5
Colfax, La. (71417) 238/E3
Colfax (co.), Nebr. 264/G3
Colfax (co.), N. Mex. 274/E2
Colfax, N. Dak. (58018) 282/S7
Colfax, Sask. 181/H6
Colfax, Wash. (99111) 310/H4
Colfax, Wis. (54730) 317/C6
Colgan, N. Dak. (†58844) 282/C2
Colgate, N. Dak. (†58046) 282/P5
Colgate (cape), N.W. Terrs. 187/J1
Colgate, Sask. 181/H6
Colgate, Wis. (53017) 317/K1
Colhué Huapi (lake), Argentina 143/C6
Co Lieu, Vietnam 72/E3
Colignan, Victoria 97/B4
Colijnsplaat, Netherlands 27/D5
Colima (state), Mexico 150/G7
Colima, Mexico 150/H7
Colina, Chile 138/G3
Colina (riv.), Chile 138/G3
Colinas, Brazil 132/F4
Colinet, Newf. 166/D2
Colington, N.C. (†27949) 281/T3
Colinton, Alberta 182/D2
Coll, Scotland 15/B2
Coll (isl.), Scotland 15/B4
Coll (isl.), Scotland 10/C2
Collaguasi, Chile 138/C5
Collamer, Ind. (†46787) 227/F2
Collarenebri, N.S. Wales 97/E1
Collbran, Colo. (81624) 208/C4
Colle di Val d'Elsa, Italy 34/C3
College, Alaska (99701) 196/J1
College Bridge, New Bruns. 170/F3
College City, Ark. 202/J1
Collegedale, Tenn. (37315) 237/M10
College Grove, Tenn. (37046) 237/H9
College Heights, Alberta 182/D3
College Hill, Ky. (40416) 237/N5
College Hill, Miss. (†38655) 256/E2
College Mound, Mo. (†65247) 261/G3
College Park, Georgia (30337) 217/K2
College Park, Md. (20740) 245/G4
College Place, Wash. (99324) 310/G4
College Springs, Iowa (51637) 229/C7
College Station, Texas (77840) 303/H7
Collegeville, Ind. (47978) 227/C3
Collegeville, Minn. (†56321) 255/D5
Collegeville, Pa. (19426) 294/M5
Colle Sestriere, Italy 34/A2
Colleton (co.), S.C. 296/F6
Collett, Ind. (†47371) 227/H4
Collette, New Bruns. 170/E2
Collettsville, N.C. (28611) 281/F3
Colley, Pa. (†18614) 294/K2
Colleyville, Texas (76034) 303/F2
Collie, Australia 87/B9
Collie, N.S. Wales 97/E2
Collie, W. Australia 88/B6
Collie, W. Australia 92/B2
Collier (co.), Fla. 212/E5
Collier (bay), W. Australia 88/C3
Collier (bay), W. Australia 92/C1
Colliers, Newf. 166/D2
Colliers, S.C. (29838) 296/C4
Colliers, W. Va. (26035) 312/E2
Collierstown, Va. (†24450) 307/J5
Collierville, Tenn. (38017) 237/B10
Colliguay, Chile 138/F3
Collin (co.), Texas 303/H4
Collin (co.), Ill. 222/E4
Collingdale, Pa. (19023) 294/N7
Collingswood, N.J. (08108) 273/B3
Collingsworth (co.), Texas 303/D3
Collingwood, N. Zealand 100/D4
Collingwood, Ontario 177/D3
Collingwood, Victoria 97/J5
Collingwood, Victoria 88/L7
Collingwood Corner, Nova Scotia 168/E3
Collins, Ark. (71634) 202/G6
Collins, Georgia (30421) 217/H6
Collins, Iowa (50055) 229/G5
Collins, Miss. (39428) 256/C5
Collins, Mo. (64738) 261/E7
Collins, Mont. (†59433) 262/E3
Collins, N.Y. (14034) 276/C6
Collins (head), Norfolk I. 88/L6
Collins, Ohio (44826) 284/E3
Collins, Ontario 177/G4
Collins (co.), Tenn. 237/K9
Collins Bay, Ontario 177/H3
Collins Bay, Sask. 181/M2
Collins Center, N.Y. (14035) 276/C6
Collinston, La. (71229) 238/G1
Collinston, Utah (84306) 304/B2

Collinsville, Ala. (35961) 195/G2
Collinsville, Calif. (†94585) 204/L1
Collinsville, Conn. (06022) 210/D1
Collinsville, Ill. (62234) 222/B2
Collinsville, Mass. (†01826) 249/J2
Collinsville, Miss. (39325) 256/G6
Collinsville, Ohio (45004) 284/A6
Collinsville, Okla. (74021) 288/P2
Collinsville, Queensland 88/H4
Collinsville, Queensland 95/C4
Collinsville, Va. (24078) 307/J7
Collinwood, Tenn. (38450) 237/F10
Collipulli, Chile 138/E2
Collirene, Ala. (†36785) 195/E6
Collis, Minn. (†56236) 255/B5
Collison, Ill. (61831) 222/F3
Collista, Ky. (†41222) 237/R5
Collombey-Muraz, Switzerland 39/C4
Collon, Ireland 17/J4
Collon (mt.), Switzerland 39/D5
Collonge-Bellerive, Switzerland 39/B4
Collooney, Ireland 17/E3
Collpa, Bolivia 136/C6
Collyer, Kansas (67631) 232/B2
Colma, Calif. (94014) 204/J2
Colman, S. Dak. (57017) 298/R6
Colmar, France 28/G3
Colmar, Pa. (18915) 294/M5
Colmar Manor, Md. (†20722) 245/F4
Colmenar de Oreja, Spain 33/G5
Colmenar, Spain 33/F4
Colmenar Viejo, Spain 33/F4
Colmesneil, Texas (75938) 303/K7
Colmonell, Scotland 15/D5
Colne, England 10/G1
Colne, England 13/H1
Colne (riv.), England 10/B5
Colne (riv.), England 13/G8
Colne Valley, England 10/G2
Colne Valley, England 13/J2
Colo, Iowa (50056) 229/G4
Colo (riv.), N.S. Wales 97/F3
Cologne, Minn. (55322) 255/E6
Cologne, N.J. (08213) 273/D4
Cologne, W. Germany 7/E3
Cologne, W. Germany 22/B3
Coloma, Calif. (95613) 204/D5
Coloma, Mich. (49038) 250/C6
Coloma, Wis. (54930) 317/H7
Colombes, France 28/A1
Colombia 2/F5
Colombia 120/B2
COLOMBIA 126
Colombia, Colombia 126/C6
Colombo (cap.), Sri Lanka 54/J9
Colombo (cap.), Sri Lanka 2/N5
Colombo (cap.), Sri Lanka 68/D7
Colome, S. Dak. (57528) 298/K7
Colón, Buenos Aires, Argentina 143/F6
Colón, Entre Ríos, Argentina 143/G6
Colón, Ark. 202/D7
Colón, Colombia 126/B1
Colón, Colombia 126/B7
Colón, Cuba 158/D1
Colón, Cuba 156/B2
Colón, Archipiélago de (terr.), Ecuador 128/C8
Colón (mts.), Honduras 154/E3
Colón, Mexico 150/K6
Colon, Mich. (49040) 250/D7
Colon, Nebr. (68018) 264/H3
Colón, Pan. 146/K9
Colón, Panama 154/H1
Colón, Isla de (isl.), Panama 154/G6
Colón, Lavalleja, Uruguay 145/C4
Colón, Montevideo, Uruguay 145/B7
Colón, Venezuela 124/E6
Colona, Colo. (†81401) 208/D6
Colona, Ill. (61241) 222/C2
Colonarie, St. Vin. & Grens. 161/A9
Colonarie (pt.), St. Vin. & Grens. 161/A9
Colonel Light Gardens, S. Australia 88/D8
Colonel Light Gardens, S. Australia 94/A8
Colonia, N.J. (07067) 273/E2
Colonia (dept.), Uruguay 145/B5
Colonia, Uruguay 145/B5
Colonia (cap.), S.C. 146/K6
Colonia Agraciada, Uruguay 145/A4
Colonia Arrué, Uruguay 145/B5
Colonia Artigas, Uruguay 145/B1
Colonia Concordia, Uruguay 145/A4
Colonia Elisa, Argentina 143/E2
Colonia Itacumbú, Uruguay 145/B1
Colonia Josefa, Argentina 143/D4
Colonia Las Heras, Argentina 143/C6
Colonia Lavalleja, Uruguay 145/C2
Colonial Beach, Va. (22443) 307/P4
Colonial Heights, Tenn. (37663) 237/R8
Colonial Heights (I.C.), Va. (23834) 307/O6
Colonia Neuland, Paraguay 144/B3
Colonia Palma, Uruguay 145/B1
Colonia Pte. Stroessner, Paraguay 144/D3
Colonia Rossel y Rius, Uruguay 145/D4
Colonias, N. Mex. (†88435) 274/E3
Colonia San Alfredo, Paraguay 144/B3
Colonia Sgto. José E. López, Paraguay 144/D3
Colonia Valdense, Uruguay 145/B5
Colonia Yby Yu, Paraguay 144/E3
Colonie, N.Y. (†12201) 276/N5
Colonne (cape), Italy 34/F5
Colonsay, Sask. 181/F4
Colonsay (isl.), Scotland 10/C2
Colonsay (isl.), Scotland 15/B4
Colony, Kansas (66015) 232/G3
Colony, Okla. (†63563) 261/H2
Colony, Okla. (73021) 288/J4
Colony, Wyo. (†57717) 319/H1
Colora, Md. (21917) 245/O2
Colorado (lag.), Bolivia 136/A8
Colorado 188/E3
COLORADO 208
Colorado (riv.), 188/D4
Colorado (riv.), Argentina 2/F5

Colorado (riv.), Argentina 120/C6
Colorado (riv.), Argentina 143/D4
Colorado (riv.), Ariz. 198/A5
Colorado (riv.), Calif. 204/L8
Colorado (riv.), Colo. 208/A5
Colorado, Arroyo (riv.), N. Mex. 274/B4
Colorado (riv.), Texas 303/H8
Colorado (riv.), Texas 188/G4
Colorado (riv.), Texas 146/H4
Colorado (riv.), Texas 303/F7
Colorado (state), U.S. 146/H6
Colorado (riv.), U.S. 2/L4
Colorado (riv.), Utah 304/E5
Colorado City, Ariz. (86021) 198/B2
Colorado City, Colo. (81019) 208/K6
Colorado City, Texas (79512) 303/C5
Colorado Nat'l Mon., Colo. 208/B4
Colorado River Aqueduct, Calif. 204/K10
Colorado River Ind. Res., Ariz. 198/A5
Colorado River Ind. Res., Calif. 204/L10
Colorados, Los (arch.), Cuba 158/A1
Colorado Springs, Colo. 146/H6
Colorado Springs, Colo. 188/F3
Colorado Springs, Colo. (*80901) 208/K5
Colored Hill, W. Va. (†24740) 312/D8
Colotlán, Mexico 150/H5
Colp, Ill. (62921) 222/D6
Colpoy (bay), Ontario 177/C3
Colpoys Bay, Ontario 177/C3
Colquechaca, Bolivia 136/B6
Colquiri, Bolivia 136/B5
Colquitt, Georgia (31737) 217/C8
Colquitt (co.), Georgia 217/E8
Colrain○, Mass. (01340) 249/C2
Colson (riv.), Brazil 120/B2
Colson, Ky. (†41858) 237/R6
Colstrip, Mont. (59323) 262/K5
Colton, Calif. (92324) 204/E10
Colton, N.Y. (13625) 276/L1
Colton, Ohio (43510) 284/C3
Colton, Oreg. (97017) 291/B3
Colton, S. Dak. (57018) 298/P6
Colton, Utah (†84601) 304/C4
Colton, Wash. (99113) 310/H4
Coltons Point, Md. (20626) 245/M8
Colts Neck, N.J. (07722) 273/E3
Coluene (riv.), Brazil 120/C3
Columbia (riv.) 188/B1
Columbia, Ala. (36319) 195/H8
Columbia (glac.), Alaska 196/C1
Columbia (mt.), Alberta 182/B3
Columbia (co.), Ark. 202/D7
Columbia (lake), Br. Col. 184/K5
Columbia (mt.), Br. Col. 184/J4
Columbia, Calif. (95310) 204/E5
Columbia (co.), Fla. 212/D1
Columbia, Fla. (†32055) 212/D1
Columbia (co.), Georgia 217/H3
Columbia, Ill. (62236) 222/C5
Columbia, Iowa (50057) 229/G6
Columbia, Ky. (42728) 237/L6
Columbia, La. (71418) 238/F2
Columbia○, Maine (†04623) 243/H6
Columbia, Md. (21043) 245/L4
Columbia, Miss. (39429) 256/E8
Columbia, Mo. (65201) 261/H5
Columbia, Mo. 188/H3
Columbia, N.J. (07832) 273/C2
Columbia (co.), N.Y. 276/N6
Columbia, N.C. (27925) 281/S3
Columbia (cape), N.W.T. 162/N3
Columbia (cape), N.W. Terrs. 187/M1
Columbia (co.), Oreg. 291/B2
Columbia (riv.), Oreg. 291/B2
Columbia (riv.), Pa. 294/K3
Columbia, Pa. (17512) 294/K5
Columbia (cap.), S.C. 188/K4
Columbia (cap.), S.C. (*29201) 296/F4
Columbia, S. Dak. (57433) 298/N2
Columbia, Tenn. 188/J3
Columbia, Tenn. (38401) 237/G9
Columbia (riv.), U.S. 146/F5
Columbia, Utah (†84501) 304/C4
Columbia, Va. (23038) 307/M5
Columbia (co.), Wash. 310/H4
Columbia (riv.), Wash. 310/B4
Columbia (co.), Wis. 317/H9
Columbia City, Ind. (46725) 227/G2
Columbia City, Oreg. (97018) 291/E2
Columbia Falls○, Maine (04623) 243/H6
Columbia Falls, Mont. (59912) 262/B2
Columbia Furnace, Va. (†22824) 307/L3
Columbia Heights, Minn. (55421) 255/G6
Columbia Heights, Wash. (†98632) 310/C4
Columbiana, Ala. (35051) 195/E4
Columbiana○, Ohio 284/J4
Columbiana, Ohio (44408) 284/J4
Columbia Road (res.), S. Dak. 298/N2
Columbia Station, Ohio (44028) 284/G10
Columbiaville, Mich. (48421) 250/F5
Columbine, Colo. (†80428) 208/E1
Columbretes (isls.), Spain 33/G3
Columbus (dam), 256/H3
Columbus, Ark. (71831) 202/C6
Columbus, Ga. 188/K4
Columbus, Ga. 146/K6
Columbus, Georgia (*31901) 217/C6
Columbus, Ill. (62328) 222/B4
Columbus, Ind. (47201) 227/E6
Columbus, Kansas (66725) 232/H4
Columbus, Ky. (42032) 237/C7
Columbus, Miss. 188/J4
Columbus, Miss. (39701) 256/H3
Columbus, Mont. (59019) 262/G5
Columbus, Nebr. (68601) 264/G3

Columbus, N.J. (08022) 273/D3
Columbus, N. Mex. (88029) 274/B7
Columbus (co.), N.C. 281/M6
Columbus, N.C. (28722) 281/E4
Columbus, N. Dak. (58727) 282/E2
Columbus (cap.), Ohio 188/K3
Columbus (cap.), Ohio 146/K6
Columbus (cap.), Ohio (*43201) 284/E6
Columbus, Pa. (16405) 294/F1
Columbus, Texas (78934) 303/H8
Columbus, Wis. (53925) 317/H9
Columbus A.F.B., Miss. 256/H3
Columbus City, Iowa (52737) 229/L6
Columbus Grove, Ohio (45830) 284/B4
Columbus Junction, Iowa (52738) 229/L6
Columbus Salt (marsh), Nev. 266/C4
Colusa (co.), Calif. 204/C4
Colusa, Calif. (95932) 204/C4
Colusa (co.), S. Dak. 298/S2
Colusa (co.), S. Dak. 222/B3
Colver, Pa. (15927) 294/E4
Colville (riv.), Alaska 188/C5
Colville (riv.), Alaska 196/G1
Colville (cape), N. Zealand 100/E2
Colville (bay), N.W. Terrs. 187/F3
Colville (lake), N.W. Terrs. 187/F3
Colville, Wash. (99114) 310/H2
Colville (riv.), Wash. 310/H2
Colville Ind. Res., Wash. 310/G2
Colville Lake, N.W. Terrs. 187/F3
Colwell, Iowa (50620) 229/H2
Colwich, Kansas (67030) 232/E4
Colwyn, Pa. (†19023) 294/N7
Colwyn Bay, Wales 10/E4
Colwyn Bay, Wales 13/D4
Comacchio, Italy 34/D2
Comal (co.), Texas 303/F8
Comala, Mexico 150/H7
Comalapa, Guatemala 154/B3
Comalapa, Nicaragua 154/E4
Comalcalco, Mexico 150/N7
Comanche (co.), Kansas 232/C4
Comanche, Mont. (†59015) 262/H4
Comanche, Okla. (73529) 288/L6
Comanche (co.), Okla. 288/K5
Comanche, Okla. 288/K5
Comanche (co.), Texas 303/F6
Comanche, Texas (76442) 303/F6
Comandante Fontana, Argentina 143/D2
Comandante Luis Piedrabuena, Argentina 143/C6
Comănești, Romania 45/H2
Comarapa, Bolivia 136/C5
Comayagua, Honduras 154/D3
Combahee (riv.), S.C. 296/F6
Combarbalá, Chile 138/A8
Comber, N. Ireland 17/K2
Comber, Ontario 177/B5
Combermere (bay), Burma 72/B3
Combermere, Ontario 177/G2
Combine, Texas (†75159) 303/H3
Combined Locks, Wis. (54113) 317/K7
Comblain-au-Pont, Belgium 27/G8
Combourg, France 28/C3
Comboyne, N.S. Wales 97/G2
Combs, Ark. (72721) 202/C2
Combs, Ky. (41729) 237/P6
Comb Wash (creek), Utah 304/E6
Comeauville, Nova Scotia 168/B4
Come By Chance, Newf. 166/D2
Come-by-Chance, N.S. Wales 97/E2
Comendador, Dom. Rep. 158/C2
Comer, Ala. (†36053) 195/H6
Comer, Georgia (30629) 217/F2
Comeragh (mts.), Ireland 17/F7
Comerford (dam), N.H. 268/B3
Comerford (dam), Vt. 268/B3
Comerío, P. Rico 161/D2
Comet (riv.), Queensland 88/H4
Comet (riv.), Queensland 95/D5
Comfort, N.C. (28522) 281/O5
Comfort (cape), N.W. Terrs. 187/K3
Comfort, Texas (78013) 303/F7
Comfrey, Minn. (56019) 255/D6
Comilla, Bangladesh 68/G4
Comines, Belgium 27/B7
Comino (isl.), Malta 34/E7
Comitán de Domínguez, Mexico 150/O8
Comite (riv.), La. 238/K1
Comitán de Domínguez, Mexico 150/O8
Commack, N.Y. (11725) 276/O9
Commentry, France 28/E4
Commerce, Calif. (90040) 204/C10
Commerce, Georgia (30529) 217/E2
Commerce, Mo. (63742) 261/O8
Commerce, Okla. (74339) 288/R1
Commerce, Texas (75428) 303/J4
Commerce City, Colo. (80022) 208/K3
Commercial Point, Ohio (43116) 284/E6
Commercy, France 28/F3
Commewijne (dist.), Suriname 131/D3
Commewijne (riv.), Suriname 131/D3
Commiskey, Ind. (47227) 227/F7
Commissaries (lake), Queensland 172/E1
Commissioner (isl.), Manitoba 179/G2
Committee (bay), N.W.T. 162/H2
Committee (bay), N.W. Terrs. 187/K3
Commodore, Pa. (15729) 294/D4
Commodore (reef), Philippines 85/F4
Commonwealth, Wis. (†54121) 317/K4
Commonwealth (mt.), U.S.S.R. 48/H6
Communism (peak), U.S.S.R. 48/H6
Como, Colo. (80432) 208/H4
Como (prov.), Italy 34/B2
Como, Italy 34/B1
Como (lake), Italy 34/B2
Como, La. (†71295) 238/G2
Como, Miss. (38619) 256/E1
Como, N.C. (27818) 281/P1
Como, Tenn. (38223) 237/E8
Como, Texas (75431) 303/J4
Como, Wis. (†53147) 317/K10
Comodoro Rivadavia, Argentina 143/C6
Comodoro Rivadavia, Argentina 120/C7
Comoé (riv.), Ivory Coast 106/D7
Comoé (riv.), Upper Volta 106/D7
Comorin (cape), India 54/J9

Comorin (cape), India 2/N5
Comorin (cape), India 68/D7
Comoros 2/M6
Comoros 102/G6
COMOROS 118/D2
Comox, Br. Col. 184/H2
Compañero, Arroyo (creek), N. Mex. 274/B2
Compass Lake, Fla. (32448) 212/D6
Compeer, Alberta 182/E4
Compiègne, France 28/E3
Compostela, Mexico 150/G6
Comprida (isl.), Brazil 135/C4
Comptche, Calif. (95427) 204/B4
Compton, Calif. (*90220) 204/C11
Compton, Ill. (61318) 222/D2
Compton, Md. (20627) 245/M7
Compton, Québec 172/F4
Compton, Québec 172/F4
Comrie, Scotland 15/E4
Comstock, Mich. (49041) 250/D6
Comstock, Minn. (56525) 255/B4
Comstock, Nebr. (68828) 264/E3
Comstock, N.Y. (12821) 276/O4
Comstock, Texas (78837) 303/C8
Comstock, Wis. (54826) 317/C5
Comstock Bridge, Conn. (†06424) 210/F2
Comté (riv.), Fr. Guiana 131/E3
Comunidad, Venezuela 124/E6
Cona, Peru 77/D6
Conakry (cap.), Guinea 102/A4
Conara Junction, Tasmania 99/D3
Conargo, N.S. Wales 97/C4
Concarneau, France 28/A4
Conceição da Barra, Brazil 132/F7
Conceição do Araguaia, Brazil 132/D5
Conceição do Araguaia, Brazil 120/D3
Concepción, Corrientes, Argentina 143/E2
Concepción, El Beni, Bolivia 136/B2
Concepción, Santa Cruz, Bolivia 136/D5
Concepción (lag.), Bolivia 136/E5
Concepción, Chile 138/D1
Concepción, Chile 120/B6
Concepción (chan.), Chile 138/D9
Concepción (bay), Mexico 150/D3
Concepción (dept.), Paraguay 144/D3
Concepción, Paraguay 144/D3
Concepción, Paraguay 120/D5
Concepción, Peru 128/E8
Concepción, Texas (78349) 303/F10
Concepción de la Sierra, Argentina 143/E2
Concepción del Oro, Mexico 150/J4
Concepción del Uruguay, Argentina 143/G6
Concepción de María, Honduras 154/D4
Conception (pt.), Calif. 188/B4
Conception (pt.), Calif. 204/E9
Conception (bay), Newf. 166/D2
Conception Harbour, Newf. 166/D2
Conception Junction, Mo. (64434) 261/C2
Conchas (res.), N. Mex. 188/F3
Conchas (dam), N. Mex. 274/E3
Conchas (lake), N. Mex. 274/E3
Conchas Dam, N. Mex. (88416) 274/E3
Conche, Newf. 166/C3
Conchi, Chile 138/B3
Conchillas, Uruguay 145/B5
Conchi Viejo, Chile 138/B3
Concho, Ariz. (85924) 198/F4
Concho, Okla. (73022) 288/L3
Concho (co.), Texas 303/E6
Concho (riv.), Texas 303/E6
Conchos (riv.), Mexico 146/H7
Conchos (riv.), Mexico 150/G3
Concise, Switzerland 39/C3
Concón, Chile 138/F2
Conconully, Wash. (98819) 310/F2
Concord, Ark. (72523) 202/G2
Concord, Calif. (*94520) 204/K1
Concord, Del. (†19973) 245/R8
Concord, Fla. (†32333) 212/B1
Concord, Georgia (30206) 217/D4
Concord, Ill. (62631) 222/C4
Concord, Ky. (41131) 237/P3
Concord, Mich. (49237) 250/E6
Concord, Mo. (†63128) 261/P4
Concord○, Mass. (01742) 249/B6
Concord (riv.), Mass. 249/J2
Concord, Mich. (49237) 250/E6
Concord, Mo. (†63128) 261/P4
Concord, Nebr. (68728) 264/H2
Concord, N.C. (28025) 281/H4
Concord (cap.), N.H. 146/L5
Concord (cap.), N.H. 188/M2
Concord (cap.), N.H. (†03301) 268/C5
Concord, N. S. Wales 88/K4
Concord, N.S. Wales 97/J3
Concord, Tenn. (37720) 237/N9
Concord○, Vt. (05824) 268/D3
Concord, Va. (24538) 307/K6
Concord, Wis. (†53066) 317/H1
Concordia, Argentina 120/D6
Concordia, Brazil 132/D9
Concordia, Honduras 154/D3
Concordia, Kansas (66901) 232/E2
Concordia, Ky. (†40157) 237/J4
Concordia (par.), La. 238/G4
Concordia, Mexico 150/G5
Concordia, Mo. (64020) 261/E5
Concordia, Peru 128/E5
Concordville, Pa. (19331) 294/M6
Concrete, N. Dak. (58221) 282/P2
Concrete, Wash. (98237) 310/D2
Con Cuong, Vietnam 72/E3
Conda, Idaho (83230) 220/G7
Condado, Cuba 158/E2

Condar, Colombia 126/D8
Conde, Brazil 132/G5
Conde, S. Dak. (57434) 298/N3
Condega, Nicaragua 154/D4
Condit, Ohio (†43074) 284/E5
Condo, Bolivia 136/B6
Condoblin, N.S. Wales 97/D3
Condoblin, N.S. Wales 88/H6
Condom, France 28/D5
Condon, Mont. (59826) 262/C3
Condon, Oreg. (97823) 291/G2
Condor, Alberta 182/C3
Cóndor, Cordillera del (range), Ecuador 128/C5
Cóndor, Cordillera del (range), Peru 128/C5
Condoto, Colombia 126/B5
Cone, Texas (79321) 303/C4
Conecuh (co.), Ala. 195/E8
Conecuh (riv.), Ala. 195/D8
Conegliano, Italy 34/D2
Conehatta, Miss. (39057) 256/F6
Conejos (co.), Colo. 208/G8
Conejos, Colo. (81129) 208/G8
Conejos (peak), Colo. 208/G8
Conejos (riv.), Colo. 208/G8
Conemaugh (riv.), Pa. 294/D5
Conemaugh River (lake), Pa. 294/D4
Conemaugh, Pa. (15929) 294/C2
Conestee, S.C. (29636) 296/C2
Conestoga, Pa. (17516) 294/K6
Conesus, N.Y. (14435) 276/E5
Conesville, Iowa (52739) 229/L6
Conesville, Ohio (43811) 284/G5
Conetoe, N.C. (27819) 281/O3
Conewango, N.Y. (†14726) 276/C6
Conewango (creek), N.Y. 276/B6
Conewango (creek), Pa. 294/D1
Confidence, Iowa (†52569) 229/G7
Confluence, Ky. (41730) 237/P6
Confluence, Pa. (15424) 294/D6
Confolens, France 28/D4
Confusion (range), Utah 304/A4
Confuso (riv.), Paraguay 144/C4
Cong, Ireland 17/C4
Congamond (lakes), Conn. 210/E1
Congamond (lakes), Mass. 249/D4
Congaree (riv.), S.C. 296/F4
Congaree Nat'l Mon., S.C. 296/F4
Conger, Minn. (56020) 255/F6
Conger (range), N.W. Terrs. 187/K1
Congerville, Ill. (61729) 222/D3
Conghua, China 77/H7
Congleton, England 13/H2
Congo 2/K6
Congo 102/D5
Congo (riv.), 102/D5
Congo (riv.), 2/L5
Congo (riv.), Angola 115/C4
CONGO 115/C4
Congo (riv.), Congo 115/C4
Congo (riv.), Zaire 115/C4
Congonhas, Brazil 135/E2
Congress, Ariz. (85332) 198/C4
Congress, Ohio (†44287) 284/F4
Congress, Sask. 181/E6
Congress Heights, D.C. (20032) 245/F4
Conical, Cerro (mt.), Argentina 143/B5
Cónico, Cerro (mt.), Chile 138/E4
Conimicut, R.I. (02889) 249/J6
Coningsby, England 13/G4
Coniston, North. Terr. 93/C7
Conjuror (bay), N.W. Terrs. 187/G3
Conklin, Alberta 182/E2
Conklin, Mich. (49403) 250/D5
Conley, Georgia (30027) 217/K2
Conn, Lough (lake), Ireland 10/B3
Conn (lake), Ireland 17/C3
Conn (lake), N.W. Terrs. 187/L2
Connacht (riv.), Ireland 17/F2
Connacht (trad. prov.), Ireland 17
Connah's Quay, Wales 13/G2
Connaught Heights, Sask. 181/A9
Conneaut, Ohio (44030) 284/J2
Conneaut Lake, Pa. (16316) 294/B2
Conneaut Lake Park, Pa. (16316) 294/B2
Conneautville, Pa. (16406) 294/B2
Connecticut 188/M2
Connecticut (riv.), 188/M2
CONNECTICUT 210
Connecticut (riv.), Conn. 210/E2
Connecticut (riv.), Mass. 249/E1
Connecticut (riv.), N.H. 268/B6
Connecticut (state), U.S. 146/L5
Connecticut (riv.), Vt. 268/C4
Connell, Scotland 15/C4
Connell, New Bruns. 170/C2
Connell, Wash. (99326) 310/G4
Connell Creek, Sask. 181/H2
Connellsville, Pa. (15425) 294/C5
Connelly Springs, N.C. (28612) 281/F3
Connelsville, Pa. (†63559) 261/G2
Connemara (dist.), Ireland 17/B5
Conner, Mont. (59827) 262/B5
Conner (mt.), North. Terr. 93/B8
Connersville, Ind. (47331) 227/G5
Connerville, Okla. (74836) 288/N6
Connétable (isls.), Fr. Guiana 131/F3
Connoquenessing, Pa. (16027) 294/B4
Connors, New Bruns. 170/B1
Conoble, N.S. Wales 97/C3
Conococheague (creek), Md. 245/G1
Cononbridge, Scotland 15/D3
Conover, N.C. (28613) 281/G3
Conover, Ohio (45317) 284/B5
Conover, Wis. (54519) 317/H3
Conowingo, Md. (21918) 245/O2
Conowingo (dam), Md. 245/O2
Conquerall Bank, Nova Scotia 168/D4
Conquest, Sask. 181/D4
Conquista, Brazil 135/B2
Conrad, Alberta 182/D5
Conrad, Iowa (50631) 229/H4
Conrad, Mont. (59425) 262/F2
Conrad, Pa. (†16720) 294/G2
Conran, Mo. (63873) 261/N10
Conrath, Wis. (54731) 317/E5

Conroe, Texas (*77301) 303/J7
Conroy, Iowa (52220) 229/J5
Consecon, Ontario 177/G3
Conselheiro Lafaiete, Brazil 135/E2
Conselheiro Lafaiete, Brazil 132/E8
Consett, England 13/E3
Consolación del Norte, Cuba 158/B1
Consolación del Sur, Cuba 158/B2
Consolación del Sur, Cuba 156/A2
Consort, Alberta 182/E3
Constable, N.Y. (12926) 276/M1
Constableville, N.Y. (13325) 276/K3
Constance (lake), Austria 41/A3
Constance, Ky. (41009) 237/R2
Constance, Sask. 181/F6
Constance (lake), Switzerland 39/H1
Constance (mt.), Wash. 310/B3
Constance (lake), W. Germany 22/C5
Constancia, Uruguay 145/A2
Constant, Morne (hill), Guadeloupe 161/B7
Constanţa, Romania 7/G4
Constanţa, Romania 45/J3
Constantia, N.Y. (13044) 276/H4
Constantia, S. Africa 118/E6
Constantina, Spain 33/D4
Constantine (cape), Alaska 196/G3
Constantine, Algeria 102/C1
Constantine, Algeria 106/F1
Constantine, Mich. (49042) 250/D7
Constanza, Argentina 143/G6
Constanza, Dom. Rep. 158/D6
Constitución, Chile 138/A11
Constitución, Uruguay 145/A2
Constitution, Georgia 217/K2
Constitution, Ohio (45722) 284/G7
Consuegra, Spain 33/E3
Consul, Ala. (†36783) 195/C6
Consul, Sask. 181/B6
Contact, Nev. (†89825) 266/G1
Contamana, Peru 128/E6
Contas (riv.), Brazil 132/F6
Contentnea (creek), N.C. 281/N3
Conthey, Switzerland 39/D4
Continental, Ariz. (†85640) 198/D7
Continental, Ohio (45831) 284/B4
Continental (peak), Wyo. 319/D3
Contla, Mexico 150/N1
Contoocook, N.H. (03229) 268/D5
Contoocook (riv.), N.H. 268/D6
Contra Costa (co.), Calif. 204/D6
Contramaestre, Cuba 158/G3
Contratación, Colombia 126/D4
Contrecoeur, Québec 172/D4
Contreras (isl.), Chile 138/D9
Contreras (isls.), Panama 154/F7
Controller (bay), Alaska 196/J3
Contulmo, Chile 138/D2
Contumazá, Peru 128/C6
Contwoyto (lake), N.W.T. 162/E2
Contwoyto (lake), N.W. Terrs. 187/H3
Convención, Colombia 126/D3
Convent, La. (70723) 238/L3
Convent Station, N.J. (†07961) 273/E2
Conversano, Italy 34/F4
Converse (hill), Conn. 210/F1
Converse, Ind. (46919) 227/F3
Converse, La. (71419) 238/C3
Converse, S.C. (29329) 296/D2
Converse, Texas (78109) 303/K11
Converse (co.), Wyo. 319/G3
Convoy, Ireland 17/F2
Convoy, Ohio (45832) 284/A4
Conway (co.), Ark. 202/E3
Conway, Ark. (72032) 202/F3
Conway (lake), Ark. 202/F3
Conway, Iowa (50834) 229/D7
Conway, Kansas (67434) 232/E3
Conway, Ky. (40417) 237/N6
Conway◯, Mass. (01341) 249/D2
Conway, Mich. (49722) 250/E4
Conway, Miss. (†39051) 256/E5
Conway, Mo. (65632) 261/G7
Conway, N.H. (03818) 268/E4
Conway◯, N.H. (03818) 268/E4
Conway (lake), N.H. 268/E4
Conway, N.C. (27820) 281/P2
Conway, N. Dak. (†58233) 282/P3
Conway, Nova Scotia 168/C4
Conway, Pa. (15027) 294/B4
Conway, S.C. (29526) 296/J4
Conway, Texas (79068) 303/C4
Conway, Wash. (98238) 310/C2
Conway Springs, Kansas (67031) 232/E4
Conwy, Wales 10/E4
Conwy, Wales 13/D4
Conwy (bay), Wales 13/C4
Conyers, Georgia (*30207) 217/D3
Conyngham, Pa. (18219) 294/K3
Coober Pedy, S. Australia 88/E5
Coober Pedy, S. Australia 94/D3
Cooch Behar, India 68/F3
Coogee, N.S. Wales 97/K3
Cook (isls.) 87/K7
Cook (inlet), Alaska 196/B1
Cook (inlet), Alaska 196/K2
Cook (cape), Br. Col. 184/C5
Cook (bay), Chile 138/E11
Cook (co.), Georgia 217/F8
Cook (co.), Ill. 222/F2
Cook (co.), Minn. 255/H3
Cook, Minn. (55723) 255/F3
Cook, Nebr. (68329) 264/H4
Cook (isls.), N. Zealand 2/B6
Cook (mt.), N. Zealand 87/G10
Cook (str.), N. Zealand 87/H10
Cook (str.), N. Zealand 100/C5
Cook (str.), N. Zealand 100/E4
Cook, S. Australia 88/E6
Cook, S. Australia 94/B4
Cook (inlet), U.S. 4/D17
Cook (pt.), Victoria 97/H5
Cook (pt.), Victoria 88/K7
Cook, Wash. 310/D5
Cooke (co.), Texas 303/G4
Cooke City, Mont. (59020) 262/G5
Cookes (range), N. Mex. 274/B6

Cookeville, Tenn. (38501) 237/L8
Cooking Lake, Alberta 182/D3
Cook's (Paopao) (bay), Fr. Poly. 86/S12
Cooks, Mich. (49817) 250/C3
Cooksburg, Pa. (16217) 294/D3
Cooks Falls, N.Y. (12728) 276/K7
Cook's Harbour, Newf. 166/C3
Cookshire, Québec 173/H1
Cooks Mills, Ill. (†61931) 222/E4
Cook Station, Mo. (65449) 261/K7
Cookstown, N.J. (08511) 273/D3
Cookstown, N. Ireland 17/H2
Cookstown, N. Ireland 10/C3
Cookstown, N. Ireland 17/H2
Cookstown, Ontario 177/E3
Cooksville, Ill. (61730) 222/E3
Cooksville, Md. (21723) 245/K3
Cooksville, Miss. (†39341) 256/H5
Cooktown, Australia 87/F4
Cooktown, Queensland 95/C2
Cooktown, Queensland 88/H4
Cool, Texas (†76086) 303/G5
Coolabah, N.S. Wales 95/C5
Cooladdi, Queensland 95/C5
Cooladdi, Queensland 88/H5
Coolah, N.S. Wales 97/E3
Coolamon, N.S. Wales 97/D4
Coolaney, Ireland 17/D3
Coolatai, N.S. Wales 97/F1
Cooleemee, N.C. (27014) 281/H3
Cooley, Minn. (†55769) 255/E3
Coolgardie, W. Australia 88/C6
Coolgardie, W. Australia 92/C5
Coolgreany, Ireland 17/J6
Coolibah, North. Terr. 93/B3
Coolidge, Ariz. (85228) 198/D6
Coolidge (dam), Ariz. 198/E5
Coolidge, Georgia (31738) 217/E8
Coolidge, Kansas (67836) 232/A3
Coolidge, Texas (76635) 303/H6
Coolidge Dam, Ariz. (†85542) 198/E5
Coolin, Idaho (83821) 220/B1
Cool Spring, Del. (†19951) 245/T6
Cool Valley, Mo. (†63101) 261/P2
Coolville, Ohio (45723) 284/F6
Cooma, N.S. Wales 88/H7
Cooma, N.S. Wales 97/E5
Coombs, Br. Col. 184/H3
Coonabarabran, N.S. Wales 97/E2
Coonamble, N.S. Wales 88/H6
Coonamble, N.S. Wales 97/E2
Coondapoor, India 68/C6
Coon Rapids, Iowa (50058) 229/D5
Coon Rapids, Minn. (55433) 255/G6
Coon Valley, Wis. (54623) 317/E8
Cooper, Ala. (†35045) 195/E6
Cooper (pt.), Calif. 204/D7
Cooper, Iowa (50059) 229/D5
Cooper, Ky. (†42633) 237/M7
Cooper, Maine (†04638) 243/H6
Cooper◯, Maine (†04638) 243/H6
Cooper (co.), Mo. 261/G5
Cooper (riv.), N.J. 273/B3
Cooper, S.C. (†29560) 296/H4
Cooper (riv.), S.C. 296/H6
Cooper, Texas (75432) 303/J4
Cooper (lake), Wyo. 319/F4
Co-Operative, Ky. (42610) 237/M7
Cooper City, Fla. (33328) 212/B4
Cooperdale, Ohio (†43842) 284/F5
Cooper Landing, Alaska (99572) 196/C1
Coopers (Barcoo) (creek), Queensland 95/B5
Coopers (Barcoo) (creek), S. Australia 88/G5
Coopers (Barcoo) (creek), S. Australia 94/F3
Coopersburg, Pa. (18036) 294/M5
Coopers Mills, Maine (04341) 243/E7
Coopers Plains, N.Y. (14827) 276/F6
Coopers Plains, Queensland 95/B5
Coopers Plains, Queensland 88/K3
Cooperstown, N.Y. (13326) 276/L5
Cooperstown, N. Dak. (58425) 282/O5
Cooperstown, Pa. (16371) 294/C2
Coopersville, Ky. (42611) 237/M7
Coopersville, Mich. (49404) 250/C5
Cooperton, Okla. (†73041) 288/J5
Coorabie, S. Australia 88/E6
Coorabie, S. Australia 94/B4
Coorong, The (lag.), S. Australia 94/F6
Coorow, W. Australia 92/B5
Coos (co.), N.H. 268/E2
Coos (co.), Oreg. 291/C4
Coos (bay), Oreg. 291/B4
Coosa (co.), Ala. 195/F5
Coosa (riv.), Ala. 195/F5
Coosa, Georgia (30129) 217/B2
Coosa (riv.), Georgia 217/A2
Coosada, Ala. (36020) 195/F5
Coosaw (riv.), S.C. 296/F7
Coosawattee (riv.), Georgia 217/C1
Coosawatchie, S.C. (29912) 296/F6
Coosawhatchie (riv.), S.C. 296/F6
Coos Bay, Oreg. 188/C4
Coos Bay, Oreg. (97420) 291/C4
Cootamundra, N.S. Wales 88/H6
Cootamundra, N.S. Wales 97/D4
Cootehill, Ireland 17/G3
Cootehill, Ireland 10/C3
Cooter, Mo. (63839) 261/N10
Copacabana, Argentina 143/C2
Copacabana, Bolivia 136/A5
Copake, N.Y. (12516) 276/N6
Copake Falls, N.Y. (12517) 276/N6
Copala, Mexico 150/K8
Copalis Beach, Wash. (98535) 310/A3
Copalis Crossing, Wash. (98536) 310/B3
Copan, Okla. (74022) 288/F1
Copano (bay), Texas 303/G9
Copco (lake), Calif. 204/C2
Cope, Colo. (80812) 208/O3
Cope, Ind. (†46151) 227/E6
Cope, S.C. (29038) 296/E5
Cope (cape), Spain 33/F4
Copeland, Ala. (†36558) 195/B7

Copeland, Fla. (33926) 212/E6
Copeland, Idaho (†83805) 220/B1
Copeland, Kansas (67837) 232/B4
Copeland (isl.), N. Ireland 17/K2
Copemish, Mich. (49625) 250/D4
Copen, W. Va. (26615) 312/E5
Copenhagen (commune), Denmark 21/F6
Copenhagen (cap.), Denmark 7/F3
Copenhagen (cap.), Denmark 21/F6
Copenhagen (cap.), Denmark 18/G9
Copenhagen, N.Y. (13626) 276/J3
Copere, Bolivia 136/D6
Copiague, N.Y. (11726) 276/O9
Copiah (co.), Miss. 256/D7
Copiapó, Chile 120/B5
Copiapó, Chile 138/B6
Copiapó (bay), Chile 138/A6
Copiapó (riv.), Chile 138/A6
Copinsay (isl.), Scotland 15/F2
Coplay, Pa. (18037) 294/L4
Copley, S. Australia 94/F4
Copmanhurst, N.S. Wales 97/G1
Copoolo, Venezuela 124/H3
Coporolo (riv.), Angola 118/L6
Coppell, Texas (75019) 303/G2
Copper (riv.), Alaska 196/J2
Copper (mts.), Ariz. 198/B6
Copperas Cove, Texas (76522) 303/G6
Copper Canyon, Texas (†76226) 303/F1
Copper Center, Alaska (99573) 196/J2
Copper City, Mich. (49917) 250/A1
Copperfield, W. Australia 88/C6
Copperfield, W. Australia 92/B5
Copper Harbor, Mich. (49918) 250/B1
Copperhill, Tenn. (37317) 237/N10
Copper Hill, Va. (24079) 307/H6
Coppermine, Canada 4/C15
Coppermine, N.W.T. 162/E2
Coppermine (riv.), N.W.T. 162/E2
Coppermine, N. Terrs. 187/G3
Coppermine (riv.), N.W. Terrs. 187/G3
Copper Mountain, Br. Col. 184/G5
Copperton, Utah (†84006) 304/B3
Copper Valley, Va. (†24141) 307/G7
Coppet, Switzerland 39/B4
Coppock, Iowa (†52654) 229/K6
Coqên, China 71/E5
Coquet (riv.), England 13/F2
Coquí, P. Rico 161/D3
Coquille, Oreg. (97423) 291/C4
Coquille (pt.), Oreg. 291/C4
Coquimatlán, Mexico 150/G7
Coquimbo (reg.), Chile 138/A8
Coquimbo, Chile 120/B6
Coquimbo, Chile 138/B6
Coquitlam, Br. Col. 184/K3
Cora, Ill. (†62280) 222/D6
Cora, Wyo. (82925) 319/C3
Corabia, Romania 45/G4
Coracora, Peru 128/F10
Corail, Haiti 158/A6
Coraki, N.S. Wales 97/G1
Coral (sea) 87/F7
Coral (sea) 88/H2
Coral (sea) 2/S6
Coral, Mich. (49322) 250/D5
Coral (sea), New Caled. 86/G4
Coral (sea), Papua N.G. 85/B7
Coral, Pa. (15731) 294/D5
Coral (sea), Philippines 82/A6
Coral (sea), Queensland 95/C1
Coral (bay), Philippines 82/A6
Coral (bay), Virgin Is. (U.S.) 161/C4
Coral Cove, Fla. (†33559) 212/D4
Coral Gables, Fla. (33134) 212/B5
Coral Harbour, N.W.T. 162/H2
Coral Harbour, N.W. Terrs. 187/K3
Coral Hills, Md. (†20027) 245/G5
Coral Sea Islands (terr.), Australia 87/E7
CORAL SEA ISLANDS TERR. 95/C2
Coral Sea Islands Territory, /J3
Coral Springs, Fla. (33060) 212/F5
Coralville, Iowa (52241) 229/K5
Coralville (lake), Iowa 229/K5
Coram, Mont. (59913) 262/C2
Coramba, N.S. Wales 97/G2
Corangamite (lake), Victoria 97/B6
Corantijn (riv.), Suriname 131/C3
Corapolis, N.C. (27829) 281/N1
Corapeake, N.C. (27926) 281/R1
Corato, Italy 34/F4
Corbeil, Ontario 177/E1
Corberrie, Nova Scotia 168/C4
Corbigny, France 28/E5
Corbin, Kansas (67032) 232/E4
Corbin, Ky. (40701) 237/N7
Corbin City, N.J. (†08270) 273/D5
Corbridge, England 13/E3
Corby, England 13/G5
Corcelles-près-Payerne, Switzerland 39/C3
Corcoran, Calif. (93212) 204/F7
Corcoran, Minn. (†55340) 255/F5
Corcovado (gulf), Chile 120/B7
Corcovado (gulf), Chile 138/A4
Corcovado (vol.), Chile 138/D5
Corcubión, Spain 33/B1
Cord, Ark. (72524) 202/H2
Cordaville, Mass. (†01772) 249/H3
Cordele, Georgia (31015) 217/F7
Cordelia, Calif. (†94585) 204/K1
Cordell, Okla. (73632) 288/J4
Cordell Hull (res.), Tenn. 237/K8
Corder, Mo. (64021) 261/F4
Cordesville, S.C. (29434) 296/H5
Cordillera (dept.), Paraguay 144/D2
Cordillo Grounds, S. Australia 94/G2
Córdoba (prov.), Argentina 143/D3
Córdoba, Argentina 2/F7
Córdoba, Argentina 143/D3
Córdoba, Argentina 120/C6
Córdoba (dept.), Colombia 126/C3
Córdoba, Mexico 150/P2
Córdoba (prov.), Spain 33/D3
Córdoba, Spain 33/D4
Córdoba, Spain 7/D5
Cordobés (riv.), Uruguay 145/D3

Cordova, Ala. (35550) 195/D3
Cordova, Alaska 4/D4
Cordova, Alaska 188/D6
Cordova (bay), Alaska 196/M2
Cordova, Ill. (61242) 222/C2
Cordova, Manitoba 179/C4
Cordova, Md. (21625) 245/O5
Cordova, N. Mex. (87523) 274/D2
Córdova, Peru 128/E7
Cordova, Tenn. (38018) 237/B10
Cordova, U.S. 4/C17
Cordova Mines, Ontario 177/G3
Core (banks), N.C. 281/S5
Core (sound), N.C. 281/S5
Core, W. Va. (26529) 312/F3
Corea, Maine (04624) 243/H7
Corella, Spain 33/F1
Corey, La. (†71201) 238/F2
Corfield, Queensland 88/G4
Corfield, Queensland 95/B4
Corfu (Kérkira) (isl.), Greece 45/D6
Corfu, N.Y. (14036) 276/D5
Corgémont, Switzerland 39/D2
Cori, Italy 34/F7
Coria, Spain 33/C3
Coria del Río, Spain 33/C4
Corigliano Calabro, Italy 34/F5
Corinda, Queensland 88/K3
Corinda, Queensland 95/A3
Corinda, Queensland 95/D3
Coringa (islets), Australia 87/F7
Coringa (isls.), Coral Sea Is. Terr. 88/H3
Corinna◯, Maine (04928) 243/E6
Corinne, Okla. (†74751) 288/R6
Corinne, Sask. 181/G5
Corinne, Utah (84307) 304/B2
Corinne, W. Va. (25826) 312/D7
Corinth, Ark. (†72833) 202/C3
Corinth, Georgia (†30203) 217/B4
Corinth, Greece 45/F7
Corinth, Ky. (41010) 237/M3
Corinth, Miss. (38834) 256/G1
Corinth, N.Y. (12822) 276/N4
Corinth, N. Dak. (†58830) 282/D2
Corinth, W. Va. (26713) 312/H4
Corinth◯, Vt. (05039) 268/C3
Corinto, Brazil 132/E7
Corinto, Colombia 126/B6
Corinto, Nicaragua 154/D4
Coripata, Bolivia 136/B5
Corisco (isl.), Equat. Guinea 115/A3
Cork (co.), Ireland 17/D7
Cork, Ireland 7/D3
Cork, Ireland 10/B5
Cork, Ireland 17/E8
Cork (harb.), Ireland 17/E8
Cork (harb.), Ireland 10/B5
Cork, New Bruns. 170/D3
Corker (cay), Belize 154/D2
Corleone, Italy 34/D6
Corley, W. Va. (26616) 312/E5
Cormack, Manitoba 179/J3
Cormorant (lake), Manitoba 179/H3
Cormorant, Minn. (†56572) 255/B4
Corn (creek), Ariz. 198/E3
Corn, Okla. (73024) 288/J4
Cornaca, Bolivia 136/D7
Corncake (inlet), N.C. 281/O7
Cornelia, Georgia (30531) 217/E1
Cornélio Procópio, Brazil 132/D8
Cornelius, N.C. (28031) 281/H4
Cornelius, Oreg. (97113) 291/A4
Cornell, Ill. (61319) 222/E3
Cornell, Mich. (49818) 250/B3
Cornell, Wis. (54732) 317/D5
Corner (inlet), Victoria 97/D6
Corner Brook, Newf. 166/C3
Corner Brook, Newf. 162/K6
Cornerstone, Ark. (†72004) 202/G5
Cornersville, Md. (†21613) 245/O6
Cornersville, Tenn. (†38659) 256/F1
Cornersville, Tenn. (37047) 237/H10
Cornerville, Ark. (†71667) 202/G6
Cornettes de Bise (mts.), Switzerland 39/C4
Cornfield (pt.), Conn. 210/F3
Cornfields, Ariz. (†86505) 198/F3
Cornhill, New Bruns. 170/E4
Cornhill, Scotland 15/F3
Corning, Ark. (72442) 202/J1
Corning, Calif. (96021) 204/C4
Corning, Iowa (50841) 229/D7
Corning, Kansas (66417) 232/F2
Corning, Mo. (64435) 261/B2
Corning, N.Y. (14830) 276/F6
Corning, Ohio (43730) 284/F6
Corning, Sask. 181/J6
Cornish Colo. (†80611) 208/L2
Cornish◯, Maine (04020) 243/B8
Cornish, Okla. (†73456) 288/L6
Cornish, Utah (84308) 304/B2
Cornish Flat, N.H. (03746) 268/C4
Cornishville, Ky. (40314) 237/M5
Cornland, Ill. (62519) 222/D3
Cornlea, Nebr. (68630) 264/G3
Corno (mt.), Italy 34/F3
Cornucopia, Wis. (54827) 317/D2
Cornville, Ariz. (86325) 198/D4
Cornville◯, Maine (†04976) 243/D6
Cornwall◯, Conn. (06753) 210/B1
Cornwall (cape), England 13/B7
Cornwall (isl.), N.W.T. 162/F1
Cornwall (isl.), N.W. Terrs. 187/J2
Cornwall, Ont. 162/J7
Cornwall, Ontario 177/K2
Cornwall, Pa. (17016) 294/K5
Cornwall, Pr. Edward I. 168/E2

Corso, Mo. (63377) 261/K4
Corson (inlet), N.J. 273/D5
Corson (co.), S. Dak. 298/G2
Corson, S. Dak. (57019) 298/R6
Cortaro, Ariz. (85230) 198/D6
Corte, France 28/B6
Corte Madera, Calif. (94925) 204/J2
Cortés, Cuba 158/A2
Cortés (inlet), Cuba 158/B2
Cortez, Colo. (81321) 208/B8
Cortez, Fla. (33522) 212/D4
Cortez (mts.), Nev. 266/E2
Cortina d'Ampezzo, Italy 34/D1
Cortland, Ill. (60112) 222/E1
Cortland, Ind. (47228) 227/F7
Cortland, Nebr. (68331) 264/H4
Cortland (co.), N.Y. 276/H5
Cortland, N.Y. (13045) 276/H5
Cortland, Ohio (44410) 284/J3
Cortona, Italy 34/C3
Coruche, Portugal 33/B3
Çoruh (riv.), Turkey 59/D1
Çoruh (riv.), Turkey 63/J2
Corum (prov.), Turkey 63/F2
Çorum, Turkey 59/B1
Çorum, Turkey 63/F2
Çorum (riv.), Turkey 63/F2
Corumbá, Brazil 120/D4
Corumbá, Brazil 132/B7
Corunna, Ind. (46730) 227/G2
Corunna, Mich. (48817) 250/E6
Corunna, Ontario 177/B5
Corvallis, Mont. (59828) 262/C4
Corvallis, Oreg. 188/B2
Corvallis, Oreg. (97330) 291/D3
Corvo (isl.), Portugal 33/A1
Corvuso, Minn. (†56228) 255/D6
Corwen, Wales 10/E4
Corwen, Wales 13/D5
Corwin, Kansas (†67061) 232/D4
Corwin, Ohio (†45068) 284/B6
Corwith, Iowa (50430) 229/F3
Corwin Springs, Mont. (59021) 262/F5
Cory, Ind. (47846) 227/C6
Corydon, Ind. (47112) 227/E8
Corydon, Iowa (50060) 229/G7
Corydon, Ky. (42406) 237/F5
Coryell (co.), Texas 303/G6
Coryville, Pa. (†16731) 294/F2
Corzoneso, Switzerland 39/G4
Cosalá, Mexico 150/E5
Cosamaloapan de Carpio, Mexico 150/M7
Cosapa, Bolivia 136/A6
Cosautlán de Carvajal, Mexico 150/P1
Cosby, Mo. (64436) 261/C3
Cosby, Tenn. (37722) 237/P9
Cos Cob, Conn. (06807) 210/A4
Coscomatepec de Bravo, Mexico 150/P2
Cosegüina (pt.), Nicaragua 154/D4
Cosenza (prov.), Italy 34/F5
Cosenza, Italy 34/F5
Cosenza, Italy 7/F5
Coshocton (co.), Ohio 284/G5
Coshocton, Ohio (43812) 284/G5
Cosine, Sask. 181/A3
Cosío, Mexico 150/H5
Cosmoledo (isls.), Seychelles 102/G5
Cosmoledo (isls.), Seychelles 118/H1
Cosmo Newbery Aboriginal Reserve, W. Australia 88/C5
Cosmo Newbery Aboriginal Res., W. Australia 92/C5
Cosmopolis, Wash. (98537) 310/B4
Cosmos, Minn. (56228) 255/D6
Cosne-Cours-sur-Loire, France 28/E4
Cosperville, Ind. (†46794) 227/F1
Cosquín, Argentina 143/D3
Cossonay, Switzerland 39/B3
Costa, W. Va. (25051) 312/C6
Costa Azul, Uruguay 145/E4
Costa Brava (reg.), Spain 33/H2
Costa da Caparica, Portugal 33/A1
Costa de Sola (Costa del Sol) (reg.), Spain 33/D4
Costa Mesa, Calif. (*92626) 204/D11
Costa Rica 2/E5
Costa Rica, Bolivia 136/A2
COSTA RICA 154/E5
Costa Rica, Mexico 150/E5
Costa Smeralda (reg.), Italy 34/B4
Costa Verde (reg.), Italy 34/B5
Costello, Pa. (†16720) 294/G2
Costessey, England 13/J5
Costeşti, Romania 45/G3
Costigan, Maine (04423) 243/F5
Costilla (co.), Colo. 208/J8
Costilla, N. Mex. (87524) 274/D2
Costilla (peak), N. Mex. 274/D2
Cosumnes (riv.), Calif. 204/C9
Coswig, Dresden, E. Germany 22/E3
Coswig, Halle, E. Germany 22/D3
Cotabato, Philippines 85/G4
Cotabato, Philippines 82/D7
Cotacajes (riv.), Bolivia 136/B5
Cotagaita, Bolivia 136/C7
Cotahuasi, Peru 128/F10
Cotati, Calif. (94928) 204/C5
Coteau, N. Dak. (58728) 282/F2
Coteau (hills), Sask. 181/B6
Coteau-du-Lac, Québec 172/C4
Coteau du Missouri (plain), N. Dak. 282/G3
Coteau-Landing, Québec 172/C4
Coteaux, Haiti 158/A6
Côte-d'Or (dept.), France 28/F4
Côte-d'Or (mts.), France 28/F4
Cotentin (pen.), France 28/C3
Corse (cape), France 28/B6
Corse (isl.), France 28/B6
Corse du Sud (dept.), France 28/B6
Corserine (mt.), Scotland 15/D5
Corsewall (pt.), Scotland 15/C5
Corsham, England 13/E6
Corsica (isl.), France 28/B6
Corsica (isl.), France 7/B5
Corsica, Pa. (15829) 294/D3
Corsica, S. Dak. (57328) 298/N7
Corsicana, Texas 188/G4
Corsicana, Texas (75110) 303/H5

Côtes de Fer, Haiti 158/B6
Côtes-du-Nord (dept.), France 28/B3
Cotesfield, Nebr. (68829) 264/F3
Cotija de la Paz, Mexico 150/H7
Cotile (lake), La. 238/E4
Coto, Argentina 143/D2
Cotoca, Bolivia 136/D5
Coto Laurel, P. Rico 161/C2
Cotonou, Benin 102/C4

Cotonou, Benin 106/E7
Cotopaxi, Colo. (81223) 208/H6
Cotopaxi (prov.), Ecuador 128/C3
Cotopaxi (mt.), Ecuador 128/C3
Cotswold (hills), England 13/E6
Cottage City, Md. (†20722) 245/F4
Cottage Grove, Ala. (†35089) 195/F5
Cottage Grove, Ind. (†47353) 227/H5
Cottage Grove, Minn. (55016) 255/F6
Cottage Grove, Oreg. (97424) 291/D4
Cottage Grove (lake), Oreg. 291/E4
Cottage Grove, Tenn. (38224) 237/E8
Cottagehill, Fla. (32533) 212/B6
Cottage Hills, Ill. (62018) 222/B2
Cottageville, S.C. (29435) 296/G6
Cottageville, W. Va. (25239) 312/C5
Cottam, Ontario 177/B5
Cottbus (dist.), E. Germany 22/F3
Cottbus, E. Germany 22/F3
Cotter, Ark. (72626) 202/E1
Cotter, Iowa (52221) 229/L6
Cottesloe, W. Australia 88/B2
Cottian Alps (range), France 28/G5
Cottian Alps (range), Italy 34/A2
Cottica, Suriname 131/D4
Cottica (riv.), Suriname 131/D3
Cottle, Ky. (41412) 237/P5
Cottle (co.), Texas 303/D3
Cottleville, Mo. (63338) 261/M2
Cotton, Georgia (31739) 217/D8
Cotton, Minn. (55724) 255/F3
Cotton (co.), Okla. 288/K6
Cottonburg, Ky. (†40475) 237/N5
Cotton Center, Texas (79021) 303/C4
Cottondale, Ala. (35453) 195/D4
Cottondale, Fla. (32431) 212/D6
Cotton Ground, St. Chris.-Nevis 161/C11
Cotton Plant, Ark. (72036) 202/H3
Cottonport, La. (71327) 238/F5
Cottonton, Ala. (36851) 195/H6
Cottontown, Tenn. (37048) 237/H8
Cotton Valley, La. (71018) 238/D1
Cottonwood, Ala. (36320) 195/H8
Cottonwood, Ariz. (86326) 198/D4
Cottonwood (cliffs), Ariz. 198/B3
Cottonwood, Br. Col. 184/G3
Cottonwood, Calif. (96022) 204/C3
Cottonwood (creek), Calif. 204/C3
Cottonwood, Idaho (83522) 220/B3
Cottonwood (butte), Idaho 220/C4
Cottonwood (riv.), Kansas 232/F3
Cottonwood (co.), Minn. 255/C6
Cottonwood, Minn. (56229) 255/C6
Cottonwood (riv.), Minn. 255/C6
Cottonwood (creek), Mont. 262/E2
Cottonwood (creek), Oreg. 291/K4
Cottonwood, S. Dak. (57775) 298/F6
Cottonwood (creek), S. Dak. 298/E5
Cottonwood (lake), S. Dak. 298/M4
Cottonwood, Texas (†79504) 303/D5
Cottonwood (creek), Utah 304/C4
Cottonwood (creek), Utah 304/E4
Cottonwood (creek), Wyo. 319/B4
Cottonwood Draw (dry riv.), Texas 303/C10
Cottonwood Falls, Kansas (66845) 232/F3
Cottonwood Wash (dry riv.), Ariz. 198/E4
Cottonwood Wash (creek), Utah 304/E6
Cotuí, Dom. Rep. 158/E5
Cotuit, Mass. (02635) 249/N6
Cotulla, Texas (78014) 303/E9
Couch, Mo. (65690) 261/K9
Couchiching (lake), Ontario 177/E3
Couchwood, La. (†71018) 238/D1
Coudekerque-Branche, France 28/E2
Couderay, Wis. (54828) 317/D4
Coudersport, Pa. (16915) 294/G2
Coudres (isl.), Québec 172/G2
Cougar (res.), Oreg. 291/E3
Cougar, Wash. (98616) 310/C4
Coughlan, New Bruns. 170/E2
Coulee, N. Dak. (†58746) 282/F2
Coulee City, Wash. (99115) 310/F3
Coulee Dam, Wash. (99116) 310/G3
Coulee Dam Nat'l Rec. Area, Wash. 310/G2
Coulihaut, Dominica 161/E6
Coulommiers, France 28/E3
Coulter, Iowa (50431) 229/G3
Coulter, Manitoba 179/B5
Coulterville, Calif. (95311) 204/E6
Coulterville, Ill. (62237) 222/D5
Counamama, Fr. Guiana 131/E3
Counce, Tenn. (38326) 237/E10
Council, Alaska (†99784) 196/F2
Council, Georgia (†31631) 217/G9
Council, Idaho (83612) 220/B5
Council, N.C. (28434) 281/M6
Council Bluffs, Iowa (51501) 229/B6
Council Bluffs, Iowa 188/G2
Council Grove, Kansas (66846) 232/F3
Council Grove (lake), Kansas 232/F3
Council Hill, Okla. (74428) 288/P3
Countess, Alberta 182/D4
Country (harb.), Nova Scotia 168/G3
Country Club Hills, Ill. (60477) 222/B6
Country Club Village, Mo. (†64501) 261/C3
Country Harbour Mines, Nova Scotia 168/G3
Country Life Acres, Mo. (†63101) 261/N3
Countryside, Ill. (†60525) 222/B6
County Line, Ala. (†35172) 195/E3
County Line, Ala. (†36467) 195/F8
Countyline, Okla. (73025) 288/L6
Coupar Angus, Scotland 10/E2
Coupar Angus, Scotland 15/E4
Coupeville, Wash. (98239) 310/C2
Courantyne (riv.) 120/D2
Courantyne (riv.), Guyana 131/C3
Courbevoie, France 28/A1
Courcelles, Belgium 27/E8
Courcelles, Québec 172/G4
Courgenay, Switzerland 39/D2
Courmayeur, Italy 34/A2
Courrendlin, Switzerland 39/D2
Courroux, Switzerland 39/D2
Courtelary, Switzerland 39/C2
Courtenay, Br. Col. 162/D6
Courtenay, Br. Col. 184/E5
Courtenay, N. Dak. (58426) 282/N5
Courtételle, Switzerland 39/D2
Courtland, Ala. (35618) 195/D1
Courtland, Calif. (95615) 204/B9
Courtland, Kansas (66939) 232/E2
Courtland, Minn. (56021) 255/D6
Courtland, Miss. (38620) 256/E2
Courtland, Ontario 177/D5
Courtland, Va. (23837) 307/O7
Courtmacsherry, Ireland 17/D8
Courtmacsherry (bay), Ireland 17/D8
Courtney, Mo. (†64051) 261/R5
Courtney, Okla. (†73456) 288/L7
Courtois, Mo. (65451) 261/K7
Courtown (Este Sudeste) (cays), Colombia 126/A10
Courtown Harbour, Ireland 17/J6
Courtrai (Kortrijk), Belgium 27/C7
Courtright, Ontario 177/B5
Courval, Sask. 181/E5
Courville, Québec 172/J3
Coushatta, La. (71019) 238/D2
Coutances, France 28/C3
Coutras, France 28/C5
Coutts, Alberta 182/D5
Coutts (inlet), N.W. Terrs. 187/L2
Couva, Trin. & Tob. 161/B10
Couvet, Switzerland 39/C3
Couvin, Belgium 27/F8
Cova da Piedade, Portugal 33/A1
Cove, Ark. (71937) 202/B5
Cove (pt.), Md. 245/N7
Cove, Minn. (†56359) 255/E4
Cove, Ohio (†45640) 284/E8
Cove (isl.), Ontario 177/C2
Cove, Oreg. (97824) 291/K2
Cove, Texas (†77580) 303/L1
Cove (creek), Utah 304/B5
Cove and Kilcreggan, Scotland 15/A1
Cove Bay, Scotland 15/F3
Cove City, N.C. (28523) 281/P4
Cove Creek, N.C. (†28786) 281/D3
Covedale, Ohio (†45238) 284/B10
Cove Fort, Utah (†84713) 304/B5
Cove Gap, W. Va. (25509) 312/B6
Covelo, Calif. (95428) 204/B4
Covena, Georgia (30422) 217/H6
Covendo, Bolivia 136/B4
Cove Neck, N.Y. (†11771) 276/R6
Coventry○, Conn. (06238) 210/F1
Coventry, England 13/F5
Coventry, England 10/F4
Coventry○, R.I. (02816) 249/H6
Coventry○, Vt. (05825) 268/C2
Coventry Center, R.I. (02816) 249/H6
Cove Orchard, Oreg. (†97148) 291/D2
Coverdale, Georgia (†31714) 217/E7
Coverdale, Ontario 177/F4
Covert, Mich. (49043) 250/C6
Cove Run (lake), Ky. 237/O4
Covesville, Va. (22931) 307/L5
Covilhã, Portugal 33/C2
Covin, Ala. (†35555) 195/C3
Covina, Calif. (*91722) 204/D10
Covington (co.), Ala. 195/F8
Covington, Georgia (30209) 217/E3
Covington, Ind. (47932) 227/C4
Covington, Iowa (†52324) 229/K5
Covington, Ky. (*41011) 237/S2
Covington, Ky. 188/J3
Covington, La. (70433) 238/K5
Covington, Mich. (49919) 250/G2
Covington (co.), Miss. 256/E7
Covington, Ohio (45318) 284/B5
Covington, Okla. (73730) 288/L2
Covington, Pa. (16917) 294/J2
Covington, Tenn. (38019) 237/B9
Covington (I.C.), Va. (24426) 307/H5
Cow (creek), Mont. 262/G2
Cow (creek), Oreg. 291/K4
Cow (lake), Oreg. 291/K4
Cow (creek), Wash. 310/G3
Cowal (lake), N.S. Wales 97/D3
Cowal (dist.), Scotland 15/C4
Cowan, Ind. (†47302) 227/G4
Cowan, Ky. (†41039) 237/O4
Cowan, Manitoba 179/B2
Cowan (lake), Sask. 181/D2
Cowan, Tenn. (37318) 237/K10
Cowan (lake), W. Australia 88/C6
Cowan (lake), W. Australia 92/C5
Cowanesque, Pa. (16918) 294/H2
Cowangie, Victoria 97/A4
Cowansville, Pa. (16218) 294/C4
Cowansville, Québec 172/E4
Cowaramup, W. Australia 92/A6
Coward, S.C. (29530) 296/H4
Coward Springs, S. Australia 94/E3
Cowarie, S. Australia 94/F2
Cowarts, Ala. (36321) 195/H8
Cow Bay, Nova Scotia 168/E4
Cowbridge, Wales 13/A7
Cowcreek, Ky. (†41314) 237/O6
Cowden, Ill. (62422) 222/E4
Cowdenbeath, Scotland 10/C1
Cowdenbeath, Scotland 15/D1
Cowdrey, Colo. (80434) 208/G1
Cowell, S. Australia 88/F6
Cowell, S. Australia 94/E5
Cowen, W. Va. (26206) 312/E6
Cowes, England 10/F5
Cowes, England 13/F7
Coweta (co.), Georgia 217/C4
Coweta, Okla. (74429) 288/P3
Cowgill, Mo. (64637) 261/E3
Cow Head, Newf. 166/C4
Cowichan (lake), Br. Col. 184/J3
Cowiche, Wash. (98923) 310/E4
Cowie, Scotland 15/C1
Cowikee, North Fork (creek), Ala. 195/H6
Cow Island, La. (†70510) 238/F7
Cowles, Nebr. (†68930) 264/F4
Cowles, N. Mex. (†87535) 274/D3
Cowlesville, N.Y. (14037) 276/D5
Cowley, Alberta 182/D5
Cowley (co.), Kansas 232/F4
Cowley, Wyo. (82420) 319/D1
Cowley, Yukon 187/E3
Cowlington, Okla. (†74934) 288/S4
Cowlitz (co.), Wash. 310/C4
Cowlitz (pass), Wash. 310/D4
Cowlitz (riv.), Wash. 310/C4
Cowpasture (riv.), Va. 307/J4
Cowpens, S.C. (29330) 296/D1
Cowpens Nat'l Battlefield, S.C. 296/D1
Cowra, N. S. Wales 88/H6
Cowra, N.S. Wales 97/E3
Cox (bight), Tasmania 99/C5
Coxburg, Miss. (†39095) 256/D5
Cox City, Okla. (†73082) 288/L5
Coxim, Brazil 132/C7
Coxsackie, N.Y. (12051) 276/N6
Cox's Bazar (Maheshkhali), Bangladesh 68/G4
Cox's Cove, Newf. 166/C4
Coxs Mills, W. Va. (26342) 312/E4
Cox Station (Bel Alton), Md. (†20611) 245/L7
Coxton, Ky. (40831) 237/P7
Coy, Ala. (36435) 195/D7
Coy, Ark. (72037) 202/G4
Coyame, Mexico 150/G2
Coyle (riv.), Argentina 143/B7
Coyle, Okla. (73027) 288/M3
Coyoacán, Mexico 150/L1
Coyote (creek), Calif. 204/L3
Coyote (res.), Calif. 204/L4
Coyote, N. Mex. (87012) 274/C2
Coyotepec, Mexico 150/L1
Coyuca (riv.), Mexico 150/O1
Coyuca de Benítez, Mexico 150/J8
Coyuca de Catalán, Mexico 150/J7
Coyutla, Mexico 150/L6
Coyville, Kansas (66727) 232/G4
Cozad, Nebr. (69130) 264/E4
Cozumel, Mexico 150/Q6
Cozumel (isl.), Mexico 150/Q6
Crab (creek), Wash. 310/F3
Crab Hill, Barbados 161/B8
Crab Orchard (lake), Ill. 222/E6
Crab Orchard, Ky. (40419) 237/M6
Crab Orchard, Nebr. (68332) 264/H4
Crab Orchard, Tenn. (37723) 237/M9
Crabtree, Oreg. (97335) 291/E3
Crabtree, Pa. (15624) 294/D5
Crabtree, Québec 172/D4
Cracow (city), Poland 47/E4
Cracow (Kraków) (prov.), Poland 47/E4
Cracow, Poland 47/E4
Cracow, Poland 7/F3
Cradle (mt.), Tasmania 99/B3
Cradle Mt. Lake St. Clair Nat'l Park, Tasmania 99/B3
Cradock, S. Africa 102/E8
Cradock, S. Africa 118/D6
Crafters-Bridgewater, S. Australia 88/E8
Crafters-Bridgewater, S. Australia 94/B8
Crafton, Pa. (15205) 294/B7
Craftsbury○, Vt. (05826) 268/C2
Craftsbury Common, Vt. (05827) 268/C2
Cragford, Ala. (36255) 195/G4
Craig, Alaska (99921) 196/M2
Craig, Colo. (81625) 208/D2
Craig (mts.), Idaho 220/B4
Craig, Iowa (51017) 229/A3
Craig, Mo. (64437) 261/B2
Craig, Mont. (†59648) 262/D3
Craig, Nebr. (68019) 264/H3
Craig (co.), Okla. 288/R1
Craig (co.), Va. 307/H6
Craig (creek), Va. 307/H5
Craigavon (dist.), N. Ireland 17/J3
Craigavon, N. Ireland 10/C3
Craigavon, N. Ireland 17/J3
Craig Beach, Ohio (†44429) 284/H3
Craigellachie, Scotland 15/E3
Craighead (co.), Ark. 202/J2
Craighouse, Scotland 15/C5
Craigieburn, Victoria 97/C5
Craigleath, Ontario 177/D3
Craigmont, Idaho (83523) 220/B3
Craigmyle, Alberta 182/D4
Craignish (hills), Nova Scotia 168/G3
Craignure, Scotland 15/C4
Craigs (Sainte Rita), Manitoba 179/F5
Craig Springs, Va. (†24127) 307/H6
Craigsville, Va. (24430) 307/J4
Craigsville, W. Va. (26205) 312/E6
Craigville, Ind. (46731) 227/G3
Craigville, Minn. (†56639) 255/E3
Craik, Sask. 181/F4
Crail, Scotland 15/F4
Crail, Scotland 10/E2
Crailsheim, W. Germany 22/D4
Craiova, Romania 45/F3
Craiova, Romania 7/G4
Cramerton, N.C. (28032) 281/G4
Cramond (isl.), Scotland 15/D1
Cranberry (lake), N.Y. 276/L2
Cranberry (lake), N. Dak. 282/L3
Cranberry, Pa. (16319) 294/C3
Cranberry, W. Va. (25828) 312/D7
Cranberry Isles, Maine (04625) 243/G7
Cranberry Isles○, Maine (04625) 243/G7
Cranberry Lake, N.J. (†07821) 273/D2
Cranberry Lake, N.Y. (12927) 276/L2
Cranberry Portage, Manitoba 179/H3
Cranbourne, Victoria 97/C6
Cranbourne, Victoria 88/M8
Cranbrook, Br. Col. 162/E6
Cranbrook, Br. Col. 184/K5
Cranbrook, Tasmania 99/D4
Cranbrook, W. Australia 92/B6
Cranbury, Conn. (†06856) 210/B4
Cranbury, N.J. (08512) 273/E3
Crandall, Georgia (30711) 217/C1
Crandall, Ind. (47114) 227/E8
Crandall, Manitoba 179/B4
Crandall, S. Dak. (†57434) 298/O3
Crandall, Texas (75114) 303/H5
Crandon, Va. (†24315) 307/G6
Crandon, Wis. (†54520) 317/H4
Crane, Barbados 161/C9
Crane, Ind. (47522) 227/D7
Crane, Mo. (65633) 261/E9
Crane, Mont. (59217) 262/M3
Crane, Oreg. (97732) 291/J4
Crane (creek), Oreg. 291/J4
Crane (lake), Sask. 181/B5
Crane (co.), Texas 303/B6
Crane, Texas (79731) 303/B6
Crane Creek (res.), Idaho 220/B5
Crane Hill, Ala. (35053) 195/D2
Crane Lake, Minn. (55725) 255/F2
Crane Nest, Ky. (40928) 237/O7
Crane Prairie (res.), Oreg. 291/F4
Crane River, Manitoba 179/C3
Cranesville, Pa. (16410) 294/B2
Cranesville, W. Va. (†26764) 312/G3
Crane Valley, Sask. 181/F6
Cranfills Gap, Texas (76637) 303/G6
Cranford○, N.J. (07016) 273/E2
Cranleigh, England 13/G6
Cransac, France 28/E5
Cranston, Iowa (52741) 229/L6
Cranston, R.I. (02910) 249/J5
Crapaud, Pr. Edward I. 168/E2
Crapo, Md. (21626) 245/O7
Crary, N. Dak. (58327) 282/N3
Craster, England 13/F2
Crater (lake), Oreg. 291/E5
Crater Lake, Oreg. (97604) 291/E5
Crater Lake Nat'l Park, Oreg. 291/E5
Craters of the Moon Nat'l Mon., Idaho 220/E6
Crateús, Brazil 132/F4
Crateús, Brazil 120/E3
Crati (riv.), Italy 34/F5
Crato, Brazil 132/G4
Crato, Brazil 120/E3
Crato e Mártires, Portugal 33/C3
Crauford (cape), N.W. Terrs. 187/K2
Craven (co.), N.C. 281/P4
Craven, Sask. 181/G5
Cravens, La. (†70656) 238/E5
Cravo Norte, Colombia 126/F4
Cravo Norte (riv.), Colombia 126/E4
Cravo Sur (riv.), Colombia 126/E5
Crawford (co.), Ark. 202/B2
Crawford, Colo. (81415) 208/D5
Crawford (co.), Georgia 217/E5
Crawford, Georgia (30630) 217/F3
Crawford (co.), Ill. 222/F4
Crawford (co.), Ind. 227/E8
Crawford (co.), Iowa 229/C4
Crawford (co.), Kansas 232/H4
Crawford, Kansas (†67444) 232/E3
Crawford○, Maine (†04610) 243/H5
Crawford (lake), Maine 243/H5
Crawford (co.), Mich. 250/E4
Crawford, Miss. (39743) 256/G4
Crawford (co.), Mo. 261/K7
Crawford, Nebr. (69339) 264/A2
Crawford (co.), Ohio 284/E4
Crawford, Okla. (73638) 288/G3
Crawford (co.), Pa. 294/B2
Crawford, Scotland 15/E5
Crawford, Tenn. (38554) 237/L8
Crawford, Texas (76638) 303/G6
Crawford, W. Va. (26343) 312/F5
Crawford (co.), Wis. 317/E9
Crawford Bay, Br. Col. 184/J5
Crawford House, N.H. (†03595) 268/E3
Crawford Notch (pass), N.H. 268/E3
Crawfordsville, Ark. (72327) 202/K3
Crawfordsville, Ind. 227/D4
Crawfordsville, Iowa (52621) 229/K6
Crawfordsville, Oreg. (97336) 291/E3
Crawfordville, Fla. (32327) 212/B1
Crawfordville, Georgia (30631) 217/G3
Crawley, England 13/G6
Crawley, W. Va. (24931) 312/E7
Crayne, Ky. (42033) 237/E6
Crazy (peak), Mont. 262/F4
Crazy Horse Mon., S. Dak. 298/B6
Crazy Woman (creek), Wyo. 319/F1
Creach Bheinn (mt.), Scotland 15/C4
Creagerstown, Md. (†21788) 245/J2
Creag Meagaidh (mt.), Scotland 15/D4
Creal Springs, Ill. (62922) 222/E6
Cream (hill), Conn. 210/B1
Cream, Wis. (†54610) 317/C7
Creamridge, N.J. (08514) 273/E3
Crean (lake), Sask. 181/E1
Creciente (isl.), Mexico 150/D5
Credenhill, England 13/E5
Crediton, England 13/D7
Crediton, England 10/E5
Crediton, Ontario 177/C4
Cree (lake), Sask. 162/F4
Cree (lake), Sask. 181/L3
Cree (riv.), Sask. 181/M2
Cree (riv.), Scotland 15/D5
Creede, Colo. (81130) 208/E7
Creedmoor, N.C. (27522) 281/M2
Creek (co.), Okla. 288/O3
Creekside, Pa. (15732) 294/D4
Creek Stand, Ala. (†36070) 195/G6
Creekville, Ky. (†40963) 237/P6
Creel, Mexico 150/E3
Cree Lake, Sask. 181/L3
Creelman, Sask. 181/H6
Creelsboro, Ky. (†42629) 237/L7
Creemore, Ontario 177/D3
Creeslough, Ireland 17/F1
Creetown, Scotland 15/D6
Creighton, Mo. (64739) 261/D6
Creighton, Nebr. (68729) 264/G2
Creighton, Pa. (15030) 294/C4
Creighton, Sask. 181/N4
Creighton, S. Dak. (57729) 298/E5
Creignish, Nova Scotia 168/G3
Creil, France 28/E3
Crellin, Md. (21525) 245/A3
Crema, Italy 34/B2
Cremona, Alberta 182/C4
Cremona (prov.), Italy 34/B2
Cremona, Italy 34/B2
Crenshaw (co.), Ala. 195/F7
Crenshaw, Miss. (38621) 256/D2
Crenshaw, Pa. (†15824) 294/E3
Creola, Ala. (36525) 195/B9
Creola, Ohio (45622) 284/E7
Creole, La. (70632) 238/D7
Crépy-en-Valois, France 28/E3
Creran, Loch (inlet), Scotland 15/C4
Cres (isl.), Yugoslavia 45/B3
Cresaptown, Md. (21502) 245/C2
Cresbard, S. Dak. (57435) 298/M3
Crescent (isls.), China 85/E2
Crescent (lake), Fla. 212/E2
Crescent, Iowa (51526) 229/B6
Crescent, La. (†70764) 238/J2
Crescent, Mo. (63018) 261/N4
Crescent (lake), Nebr. 264/B3
Crescent (valley), Nev. 266/E2
Crescent, Okla. (73028) 288/L3
Crescent, Oreg. (97733) 291/F4
Crescent (lake), Tasmania 99/D4
Crescent (lake), Wash. 310/B3
Crescent Beach, Conn. (†06537) 210/G3
Crescent City, Calif. (95531) 204/A2
Crescent City, Fla. (32012) 212/E2
Crescent City, Ill. (60928) 222/F3
Crescent Head, N.S. Wales 97/G2
Crescent Lake, Maine (†04015) 243/C7
Crescent Lake, Oreg. (97425) 291/F4
Crescent Lake, Sask. 181/J4
Crescent Mills, Calif. (95934) 204/E3
Crescent Springs, Ky. (†41016) 237/R2
Cresco, Iowa (52136) 229/J2
Cresco, Pa. (18326) 294/M3
Crespo, Argentina 143/F6
Cresskill, N.J. (07626) 273/C1
Cressmont, W. Va. (†25074) 312/E6
Cresson, Pa. (16630) 294/E5
Cresson, Texas (76035) 303/G5
Cressona, Pa. (17929) 294/K4
Cressy, Tasmania 99/C3
Crest, France 28/F5
Crest, Georgia (†30286) 217/D5
Crested Butte, Colo. (81224) 208/E5
Crest Hill, Ill. (60431) 222/E2
Crestline, Calif. (92325) 204/H9
Crestline, Kansas (66728) 232/H4
Crestline, Ohio (44827) 284/E4
Creston, Br. Col. 184/J5
Creston, Calif. (93432) 204/E8
Creston, Ill. (60113) 222/D2
Creston, Ind. (†46356) 227/C2
Creston, Iowa (50801) 229/E6
Creston, Ky. (42524) 237/L6
Creston, La. (71020) 238/E3
Creston, Mont. (59902) 262/C2
Creston, Nebr. (68631) 264/G3
Creston, Newf. 166/C4
Creston, Ohio (44217) 284/G3
Creston, S.C. (†29030) 296/F4
Creston, Wash. (99117) 310/G3
Creston, W. Va. (26141) 312/D5
Crestone, Colo. (81131) 208/H7
Crestone (peak), Colo. 208/H7
Crestview, Fla. (32536) 212/C6
Crestview, Ky. (†41076) 237/S2
Crestview Hills, Ky. (†41017) 237/R2
Crestwood, Ill. (†60445) 222/B6
Crestwood, Ky. (40014) 237/L4
Crestwood, Mo. (†63126) 261/O3
Crestwynd, Sask. 181/F5
Creswell, N.C. (27928) 281/S3
Creswell (bay), N.W. Terrs. 187/J2
Creswell, Oreg. (97426) 291/D4
Creswell Downs, North. Terr. 93/E4
Creswick, Victoria 97/B5
Crete (reg.), Greece 45/G8
Crete (isl.), Greece 7/G5
Crete (isl.), Greece 45/G8
Crete (sea), Greece 45/G7
Crete, Ill. (60417) 222/F2
Crete, Nebr. (68333) 264/G4
Crete, N. Dak. (58020) 282/P7
Créteil, France 28/B2
Cretin (cape), Papua N.G. 86/B2
Creus (cape), Spain 33/H1
Creuse (dept.), France 28/D4
Creuse (riv.), France 28/D4
Creve Coeur, Ill. (61611) 222/D3
Creve Coeur, Mo. (63141) 261/O2
Crevillente, Spain 33/F3
Crewe, Va. (23930) 307/M6
Crewe and Nantwich, England 13/E4
Crewe and Nantwich, England 10/F2
Crewkerne, England 13/E7
Crewkerne, England 10/E5
Crews, Ala. (†35586) 195/B3
Crianlarich, Scotland 15/D4
Criccieth, Wales 13/C5
Criccieth, Wales 10/D4
Crichton, Sask. 181/D6
Criciúma, Brazil 132/D10
Crider, Ky. (†42445) 237/F6
Cridersville, Ohio (45806) 284/B4
Crieff, Scotland 15/E4
Crieff, Scotland 10/E2
Criehaven, Maine (†04851) 243/F8
Crillon (mt.), Alaska 196/L1
Crimea (pen.), U.S.S.R. 7/H4
Crimea (pen.), U.S.S.R. 48/D5
Crimea (pen.), U.S.S.R. 52/D6
Crimean Oblast, U.S.S.R. 52/D6
Crimmitschau, E. Germany 22/E3
Crimond, Scotland 15/G3
Crimora, Va. (24431) 307/L4
Crinan, Scotland 15/C4
Cripple Creek, Colo. 188/F3
Cripple Creek, Colo. (80813) 208/J5
Cripple Creek, Va. (24322) 307/F7
Crisfield, Md. (21817) 245/P9
Crisp (co.), Georgia 217/E7
Crisp (pt.), Mich. 250/D2
Crisp, N.C. (27852) 281/O3
Crissolo, Italy 34/A2
Cristal, Sierra del (mts.), Cuba 158/J3
Cristalina, Brazil 132/E7
Cristóbal (mt.), Colombia 120/B1
Cristóbal (pt.), Ecuador 128/B9
Cristóbal, Panama 154/G6
Cristóbal Colón, Pico (peak), Colombia 126/D2
Crişul Alb (riv.), Romania 45/F2
Crişul Repede (riv.), Romania 45/F2
Crittenden (co.), Ark. 202/K3
Crittenden (co.), Ky. 237/E6
Crittenden, Ky. (41030) 237/M3
Critz, Va. (24082) 307/H7
Crivitz, Wis. (54114) 317/L5
Croagh Patrick (mt.), Ireland 17/C4
Croatan (sound), N.C. 281/T3
Croatia (rep.), Yugoslavia 45/C3
Croche (riv.), Québec 172/E2
Crocheron, Md. (21627) 245/O8
Crochu, Grenada 161/D8
Crocker, Mo. (65452) 261/H7
Crocker, S. Dak. (57229) 298/O3
Crocketford, Scotland 15/E5
Crockett, Calif. (94525) 204/J1
Crockett (co.), Tenn. 237/C9
Crockett (co.), Texas 303/C7
Crockett, Texas (75835) 303/J6
Crockett, Va. (24323) 307/F7
Crockett Mills, Tenn. (38021) 237/C9
Crocketts Bluff, Ark. (72038) 202/H5
Crocketville, S.C. (29913) 296/E6
Crocodile (riv.), S. Africa 118/H6
Croft, Kansas (†67028) 232/D4
Crofton, Br. Col. 184/J3
Crofton, Ky. (42217) 237/G6
Crofton, Md. (21114) 245/M4
Crofton, Nebr. (68730) 264/G2
Croghan, N.Y. (13327) 276/K3
Croix des Bouquets, Haiti 158/C6
Croker (cape), North. Terr. 88/E2
Croker (cape), North. Terr. 93/C1
Croker (bay), N.W. Terrs. 187/K2
Croker (cape), Ontario 177/D3
Croker Island Mission, North. Terr. 88/E2
Croker Island Mission, North. Terr. 93/C1
Cromarty, Scotland 15/E3
Cromarty, Scotland 10/D2
Cromarty (firth), Scotland 15/D3
Cromdale, Scotland 15/E3
Cromer, England 13/J5
Cromer, England 10/G4
Cromer, Manitoba 179/A5
Cromwell, Ala. (36906) 195/B6
Cromwell○, Conn. (06416) 210/E2
Cromwell, Ind. (46732) 227/F2
Cromwell, Iowa (50842) 229/E6
Cromwell, Ky. (42333) 237/H6
Cromwell, Minn. (55726) 255/F4
Cromwell, N. Zealand 100/B6
Cromwell, Okla. (74837) 288/N4
Cronulla, N. S. Wales 88/L5
Cronulla, N.S. Wales 97/J4
Crook, Colo. (80726) 208/O1
Crook (co.), Oreg. 291/G3
Crook (pt.), Oreg. 291/C5
Crook (co.), Wyo. 319/H1
Crook and Willington, England 13/E3
Crooked (isl.), Bahamas 156/D2
Crooked (creek), Ind. 227/D2
Crooked (creek), Kansas 232/B4
Crooked (creek), Minn. 255/F4
Crooked (lake), Minn. 255/G2
Crooked (lake), N. Dak. 282/J4
Crooked (creek), Oreg. 291/K5
Crooked (riv.), Oreg. 291/G3
Crooked (creek), S.C. 296/H2
Crooked Creek, Alaska (99575) 196/G2
Crooked Creek, Alberta 182/B2
Crooked Island (passage), Bahamas 156/C2
Crooked River, Sask. 181/H3
Crookhaven, Ireland 17/B9
Crooks, S. Dak. (57020) 298/R6
Crookston, Minn. 188/G1
Crookston, Minn. (56716) 255/B3
Crookston, Nebr. (69212) 264/D2
Crooksville, Ohio (43731) 284/F6
Crookwell, N.S. Wales 97/E4
Croom, Ireland 17/D6
Cropper, Ky. (40015) 237/L4
Cropsey, Ill. (61731) 222/E3
Crosby, Ala. (†36343) 195/H8
Crosby, England 13/G2
Crosby, England 10/G2
Crosby, Minn. (56441) 255/D4
Crosby, Miss. (39633) 256/B8
Crosby, N. Dak. (58730) 282/D2
Crosby, Pa. (16724) 294/F2
Crosby (co.), Texas 303/C4
Crosby, Texas (77532) 303/J8
Crosby (mt.), Wyo. 319/C2
Crosbyton, Texas (79322) 303/C4
Crosland, Georgia (†31771) 217/E8
Cross (sound), Alaska 196/L1
Cross (co.), Ark. 202/J3
Cross (riv.), Cameroon 115/A2
Cross (lake), La. 238/C2
Cross (isl.), Maine 243/J6
Cross (lake), Maine 243/G1
Cross (bay), Manitoba 179/C1
Cross (lake), Manitoba 179/J3
Cross (cape), Namibia 118/A4
Cross (riv.), Nigeria 106/F7
Cross (isl.), Nova Scotia 168/D4
Cross, S.C. (29436) 296/G5
Cross Anchor, S.C. (29331) 296/D2
Crossapol, Scotland 15/B4
Cross City, Fla. (32628) 212/C2
Cross Creek, New Bruns. 170/D2
Crossett, Ark. (71635) 202/G7
Crossfarnoge (pt.), Ireland 17/J7
Cross Fell (mt.), England 13/E3
Crossfield, Alberta 182/C4
Cross Fork, Pa. (17729) 294/G3

Crossgar, N. Ireland 17/K3
Crosshaven, Ireland 17/E8
Crosshill, Scotland 15/D5
Cross Hill, S.C. (29332) 296/D3
Cross Junction, Va. (22625) 307/M2
Cross Keys, S.C. (†29379) 296/D2
Cross Lake, Manitoba 179/J3
Crosslake, Minn. (56442) 255/E4
Crossley (mt.), N. Zealand 100/D5
Crossmaglen, N. Ireland 17/H3
Crossmichael, Scotland 15/D6
Crossmolina, Ireland 17/C3
Crossnore, N.C. (28616) 281/F2
Cross Plains, Ind. (47017) 227/G7
Cross Plains, Tenn. (37049) 237/H7
Cross Plains, Texas (76443) 303/E5
Cross Plains, Wis. (53528) 317/G9
Cross River (state), Nigeria 106/F7
Cross Roads, Ga. (†92242) 204/L9
Crossroads, N. Mex. (88114) 274/F5
Cross Roads, Pa. (†17322) 294/J6
Cross Timbers, Mo. (65634) 261/F6
Crosstown, Mo. (†63775) 261/N7
Cross Village, Mich. (49723) 250/D3
Crossville, Ala. (35962) 195/G2
Crossville, Ill. (62827) 222/F5
Crossville, Tenn. (38555) 237/L9
Crosswicks, N.J. (08515) 273/D3
Crosswicks (creek), N.J. 273/D3
Croswell, Mich. (48422) 250/G5
Crotch (lake), Ontario 177/H3
Crothersville, Ind. (47229) 227/F7
Croton (Hartford), Ohio (43013) 284/E5
Crotone, Italy 34/F5
Croton Falls, N.Y. (10519) 276/N8
Croton-on-Hudson, N.Y. (10520) 276/N8
Crouch, Idaho (†83622) 220/B5
Crouseville, Maine (04738) 243/G2
Crow (creek), Colo. 208/L1
Crow (riv.), Minn. 255/F5
Crow, Oreg. (†97401) 291/D4
Crow (creek), S. Dak. 298/A4
Crow (creek), Wyo. 319/H4
Crow Agency, Mont. (59022) 262/J5
Crowborough, England 13/H6
Crow Creek Ind. Res., S. Dak. 298/L5
Crowder, Miss. (38622) 256/D2
Crowder, Okla. (74430) 288/P4
Crowduck (lake), Manitoba 179/G4
Crowdy (head), N.S. Wales 97/G2
Crowell, Texas (79227) 303/E4
Crowfoot, Alberta 182/D4
Crowheart, Wyo. (82512) 319/C2
Crow Ind. Res., Mont. 262/H5
Crowl (creek), N.S. Wales 97/C2
Crow Lake, S. Dak. (†57382) 298/M6
Crowle, England 13/G4
Crowley (lake), Calif. 204/G6
Crowley (co.), Colo. 208/M6
Crowley, Colo. (81033) 208/M6
Crowley, La. (70526) 238/F6
Crowley, Texas (76036) 303/E3
Crowley Lake, Calif. (93546) 204/G6
Crowley's Ridge (mt.), Ark. 202/J2
Crown, Minn. (†55005) 255/E5
Crown (mt.), Virgin Is. (U.S.) 161/A4
Crown City, Ohio (45623) 284/F8
Crown King, Ariz. (86333) 198/C4
Crown Point, Ind. (46307) 227/C2
Crownpoint, N. Mex. (87313) 274/A3
Crown Point, N.Y. (12928) 276/N3
Crown Prince Frederik (isl.), N.W. Terrs. 187/K3
Crownsville, Md. (21032) 245/M4
Crows Landing, Calif. (95313) 204/D6
Crowsnest (pass), Alberta 182/C5
Crowsnest, Br. Col. 184/K5
Crowsnest (pass), Br. Col. 184/K5
Crowville, La. (71230) 238/G2
Crow Wing (co.), Minn. 255/D4
Crow Wing (riv.), Minn. 255/D4
Croydon, England 13/H8
Croydon, England 10/B6
Croydon○, N.H. (†03753) 268/C5
Croydon (peak), N.H. 268/C5
Croydon, Queensland 88/G3
Croydon, Queensland 95/B3
Croydon, Utah (84018) 304/C2
Croydon, Victoria 88/M7
Croydon, Victoria 97/K5
Croydon Branch, Sugar (riv.), N.H. 268/C5
Crozet (isls.) 2/M8
Crozet, Va. (22932) 307/L4
Crozier (chan.), N.W. Terrs. 187/G2
Crozier, Va. (23039) 307/N5
Cruces, Cuba 158/E2
Cruces, Cuba 156/B2
Cruden Bay, Scotland 15/G3
Cruger, Miss. (38924) 256/D4
Cruillas, Mexico 150/K4
Crum (creek), Pa. 294/M7
Crum, W. Va. (25669) 312/B7
Crumlin, N. Ireland 17/J2
Crum Lynne, Pa. (19022) 294/M7
Crummies, Ky. (40821) 237/P7
Crump, Mich. (†48634) 250/E5
Crump (lake), Oreg. 291/H5
Crump, Tenn. (38327) 237/E10
Crumpton (pt.), Dominica 161/F5
Crumpton, Md. (21628) 245/P4
Crumrod, Ark. (72328) 202/H5
Crumstown, Ind. (†46554) 227/E1
Crusheen, Ireland 17/D6
Cruso, N.C. (†28716) 281/D4
Cruta, Honduras 154/F3
Crutchfield, Ky. (42034) 237/D7
Crutwell, Sask. 181/E2
Cruz (cape), Cuba 156/C3
Cruz (cape), Cuba 158/G4
Cruz Alta, Brazil 120/D5
Cruz Alta, Brazil 132/C10
Cruz Bay, Virgin Is. (U.S.) 161/C4
Cruz del Eje, Argentina 143/C3
Cruz del Eje, Argentina 120/C6
Cruz de Piedra, Uruguay 145/E3
Cruz de San Pedro, Uruguay 145/E2
Cruzeiro, Brazil 135/D3
Cruzeiro do Sol, Brazil 120/B3
Cruzeiro do Sul, Brazil 132/G10
Cruz Grande, Chile 138/A7
Crysler, Ontario 177/J2
Crystal (mts.), Congo 115/B4
Crystal (lake), Conn. 210/F1
Crystal (pond), Conn. 210/G1
Crystal (bay), Fla. 212/D3
Crystal (mts.), Gabon 115/B4
Crystal, Ind. (†47527) 227/D8
Crystal○, Maine (†04747) 243/G4
Crystal, Mich. (48818) 250/E5
Crystal (lake), Mich. 250/C4
Crystal, Minn. (†55428) 255/G5
Crystal, N.H. (†03591) 268/E2
Crystal (lake), N.H. 268/E5
Crystal, N. Mex. (†86504) 274/A2
Crystal, N. Dak. (58222) 282/P2
Crystal (lake), Vt. 268/C2
Crystal, W. Va. (†24747) 312/D8
Crystal Bay (Orono), Minn. (55323) 255/F5
Crystal Bay, Nev. (89402) 266/A3
Crystal Beach, Texas (77650) 303/K8
Crystal Brook, S. Australia 94/E5
Crystal City, Manitoba 179/C5
Crystal City, Mo. (63019) 261/M6
Crystal City, Texas (78839) 303/E9
Crystal Falls, Mich. (49920) 250/A2
Crystal Falls, Ontario 177/E1
Crystal Hill, Va. (24539) 307/L7
Crystal Lake, Conn. (†06066) 210/F1
Crystal Lake, Fla. (†32463) 212/D6
Crystal Lake, Ill. (60014) 222/E1
Crystal Lake, Iowa (50432) 229/F2
Crystal Lake Park, Mo. (†63101) 261/O3
Crystal Lakes, Ohio (†45341) 284/C6
Crystal River, Fla. (32629) 212/D3
Crystal Springs, Ark. (†71968) 202/D5
Crystal Springs (res.), Calif. 204/J3
Crystal Springs, Fla. (33524) 212/D3
Crystal Springs, Georgia (†30105) 217/B2
Crystal Springs, Kansas (†67058) 232/D4
Crystal Springs, Miss. (39059) 256/D7
Crystal Springs, N. Dak. (58427) 282/L6
Crystal Springs, Sask. 181/F3
Crystal Valley, Mich. (†49420) 250/C5
Csabrendek, Hungary 41/D3
Csákvár, Hungary 41/E3
Csanádpalota, Hungary 41/F3
Csenger, Hungary 41/G3
Csepel, Hungary 41/E3
Csepelsziget (isl.), Hungary 41/E3
Csepreg, Hungary 41/D3
Csongrád (co.), Hungary 41/F3
Csongrád, Hungary 41/E3
Csorna, Hungary 41/D3
Csorvás, Hungary 41/F3
Csurgó, Hungary 41/D3
Ctesiphon (ruins), Iraq 66/D4
Cúa, Venezuela 124/E2
Cuadro Nacional, Argentina 143/C3
Cuamba, Mozambique 118/F2
Cuando (riv.), Angola 115/C7
Cuando (riv.), Zambia 115/D7
Cuando Cubango (dist.), Angola 115/C7
Cuangar, Angola 115/C7
Cuango (riv.) 102/D5
Cuango, Angola 115/C5
Cuango (riv.), Angola 115/C5
Cuanza (riv.), Angola 102/D6
Cuanza (riv.), Angola 115/C5
Cuanza-Norte (dist.), Angola 115/B5
Cuanza-Sul (dist.), Angola 115/C6
Cuao (riv.), Venezuela 124/E5
Cua Rao, Vietnam 72/E3
Cuareim (riv.), Uruguay 145/B1
Cuaró, Uruguay 145/D2
Cuatrociénagas de Carranza, Mexico 150/H3
Cuatro Compañeros, Cuba 158/G3
Cuatro Ojos, Bolivia 136/D5
Cuauhtémoc, Mexico 150/F2
Cuautepec de Hinojosa, Mexico 150/K6
Cuautitlán de Romero Rubio, Mexico 150/L1
Cuautla Morelos, Mexico 150/L2
Cub (creek), Utah 304/C1
Cub (creek), Va. 307/L6
Cuba 2/E4
Cuba 146/L7
Cuba, Ala. (36907) 195/B6
CUBA 156/B2
CUBA 158
Cuba, Ill. (61427) 222/C3
Cuba, Ind. (†47460) 227/D6
Cuba, Kansas (66940) 232/E2
Cuba, Mo. (65453) 261/K6
Cuba, N. Mex. (87013) 274/B2
Cuba, N.Y. (14727) 276/D6
Cuba (chan.), N. Zealand 100/D7
Cuba, Ohio (45114) 284/C7
Cuba, Portugal 33/C3
Cuba City, Wis. (53807) 317/F10
Cubage, Ky. (40822) 237/O7
Cubagua (isl.), Venezuela 124/F2
Cuballing, W. Australia 92/B2
Cubango (riv.), Angola 102/D6
Cubango (riv.), Angola 115/C7
Cubango (riv.), Namibia 118/B3
Cubatão, Brazil 135/C3
Cube (mt.), N.H. 268/D4
Cubero, N. Mex. (87014) 274/B3
Cubiro, Venezuela 124/D3
Cub Run, Ky. (42729) 237/J6
Çubuk, Turkey 63/E2
Cubulco, Guatemala 154/B3
Cuchara, Colo. (81055) 208/J8
Cuchi, Angola 115/C1
Cuchi, Angola 102/D6
Cuchillo, N. Mex. (87932) 274/B5
Cuchillo-Có, Argentina 143/D4
Cuchillo Negro (creek), N. Mex. 274/B5
Cuchivero, Venezuela 124/F4
Cuchivero (riv.), Venezuela 124/F4
Cuckfield, England 13/G6
Cuckfield, England 10/F5
Cucumber, W. Va. (24826) 312/C8
Cúcuta, Colombia 126/D4
Cúcuta, Colombia 120/B2
Cudahy, Calif. (90201) 204/C5
Cudahy, Wis. (53110) 317/M2
Cudal, N.S. Wales 97/E3
Cuddalore, India 68/E6
Cuddapah, India 68/D6
Cuddeback (lake), Calif. 204/H8
Cuddy, Pa. (15031) 294/B5
Cudgewa, Victoria 97/D5
Cudillero, Spain 33/C1
Cudjoe (key), Fla. 212/E7
Cudrefin, Switzerland 39/D3
Cudworth, Sask. 181/F3
Cue, W. Australia 88/B5
Cue, W. Australia 92/B4
Cuéllar, Spain 33/D2
Cuéllar-Baza, Spain 33/E4
Cuemaní (riv.), Colombia 126/D7
Cuenca, Ecuador 120/B3
Cuenca, Ecuador 128/C4
Cuenca (prov.), Spain 33/E2
Cuenca, Spain 33/E2
Cuenca, Sierra de (range), Spain 33/F3
Cuencamé de Ceniceros, Mexico 150/H4
Cuernavaca, Mexico 150/L2
Cuero, Texas (77954) 303/G8
Cuervo, N. Mex. (88417) 274/E3
Cuervo (creek), N. Mex. 274/E3
Cueto, Cuba 158/J3
Cuevas, Miss. (†39571) 256/F10
Cuevas del Almanzora, Spain 33/F4
Cuevas de Vinromá, Spain 33/F2
Cuevo, Bolivia 136/D7
Cufré, Uruguay 145/B5
Cuiabá, Brazil 120/D4
Cuiabá, Brazil 132/C6
Cuiabá (riv.), Brazil 132/B7
Cuicatlán, Mexico 150/L8
Cuicuina, Nicaragua 154/E4
Cuilapa, Guatemala 154/B3
Cuilapa Miravalles (vol.), C. Rica 154/E5
Cuilcagh (mt.), Ireland 17/F3
Cuilco, Guatemala 154/B3
Cuillin (hills), Scotland 15/B3
Cuillin (sound), Scotland 10/C2
Cuillin (sound), Scotland 15/B3
Cuilo, Angola 115/C5
Cuitlahuac, Mexico 150/P2
Cuito (riv.), Angola 115/C7
Cuito-Cuanavale, Angola 115/C7
Cuitzeo (lake), Mexico 150/J7
Cuivre (riv.), Mo. 261/N2
Cujmir, Romania 45/F3
Cukmantl, Czech. 41/D1
Çukur, Turkey 63/F3
Çukurca, Turkey 63/K4
Culaba, Philippines 82/E5
Cu Lao, Hon (isls.), Vietnam 72/F5
Culberson, N.C. (28903) 281/A4
Culberson (co.), Texas 303/C11
Culbertson, Mont. (59218) 262/M2
Culbertson, Nebr. (69024) 264/C4
Culcairn, N.S. Wales 97/D4
Culdaff, Ireland 17/G1
Culdaff (bay), Ireland 17/G1
Culdesac, Idaho (83524) 220/B3
Cul-de-Sac du Marin (bay), Martinique 161/D6
Culebra (creek), Colo. 208/H8
Culebra (peak), Colo. 208/J8
Culebra, P. Rico 161/G1
Culebra (isl.), P. Rico 161/G1
Culebra (isl.), P. Rico 156/G1
Culebras, Peru 128/C7
Culebrinas (riv.), P. Rico 161/A1
Culebrita (isl.), P. Rico 161/G2
Culemborg, Netherlands 27/G5
Culgoa (riv.), N.S. Wales 97/D1
Culgoa (riv.), Queensland 95/C6
Culiacán, Mexico 150/F4
Culiacán, Mexico 146/H7
Culion, Philippines 82/C5
Culion (isl.), Philippines 82/B5
Cullasaja, N.C. (†28734) 281/C4
Cullburra-Orient Point, N.S. Wales 97/F4
Cullen, La. (71021) 238/D1
Cullen, Sask. 181/J6
Cullen, Scotland 15/F3
Cullen, Va. (23934) 307/L6
Cullen Bullen, N.S. Wales 97/F3
Culleoka, Tenn. (38451) 237/G10
Cullera, Spain 33/F3
Cullin (lake), Ireland 17/C4
Cullison, Kansas (†67124) 232/D4
Cullman (co.), Ala. 195/E2
Cullman, Ala. (35055) 195/E2
Culloden, Georgia (31016) 217/D5
Culloden, W. Va. (25510) 312/B6
Cullom, Ill. (60929) 222/E3
Cullomburg, Ala. (36920) 195/B7
Cullompton, England 13/D7
Cullowhee, N.C. (28723) 281/C4
Cully, Switzerland 39/C4
Cullybackey, N. Ireland 17/J2
Culotte (lake), Québec 172/C2
Culp, Alberta 182/B2
Culp Creek, Oreg. (97427) 291/E4
Culpeper (co.), Va. 307/M3
Culpeper, Va. (22701) 307/M4
Culpepper (isl.), Ecuador 128/B8
Culpina, Bolivia 136/C7
Culross, Manitoba 179/E5
Culross, Scotland 10/B1
Culross, Scotland 15/C1
Culta, Bolivia 136/B6
Cults, Scotland 15/F3
Cultus (lake), Oreg. 291/F4
Cultus Lake, Br. Col. 184/M3
Culuene (riv.), Brazil 132/C6
Culver, Ind. (46511) 227/E2
Culver, Kansas (67435) 232/E3
Culver, Minn. (55727) 255/F4
Culver, Oreg. (97734) 291/F3
Culver (pt.), W. Australia 88/D6
Culver (pt.), W. Australia 92/D6
Culver City, Calif. (90230) 204/B10
Culverden, N. Zealand 100/D5
Culvers (lake), N.J. 273/D1
Culverton, Georgia (†31087) 217/G4
Cuma, Angola 115/B6
Cumaná, Venezuela 120/C2
Cumaná, Venezuela 124/F2
Cumanacoa, Venezuela 124/F2
Cumanayagua, Cuba 158/E2
Cumaria, Peru 128/F7
Cumback, Ind. (†47501) 227/C7
Cumbal, Colombia 126/B7
Cumberland (riv.) 188/J3
Cumberland (plat.), Ala. 195/F1
Cumberland, Br. Col. 184/E5
Cumberland (sound), Canada 4/C13
Cumberland (isl.), Georgia 217/K9
Cumberland (co.), Ill. 222/E4
Cumberland, Ind. (46229) 227/E5
Cumberland, Iowa (50843) 229/D6
Cumberland (co.), Ky. 237/L7
Cumberland, Ky. (40823) 237/R6
Cumberland (lake), Ky. 237/M7
Cumberland (mt.), Ky. 237/P7
Cumberland (riv.), Ky. 237/K8
Cumberland (co.), Maine 243/C8
Cumberland, Md. (21502) 245/D2
Cumberland, Md. 245/80
Cumberland, Md. 188/L3
Cumberland (basin), New Bruns. 170/F3
Cumberland (co.), N.J. 273/C5
Cumberland (co.), N.C. 281/M4
Cumberland, N.C. (28331) 281/M5
Cumberland (pen.), N.W.T. 162/K2
Cumberland (pen.), N.W. Terrs. 187/M3
Cumberland (sound), N.W.T. 146/N3
Cumberland (sound), N.W.T. 162/K2
Cumberland (sound), N.W. Terrs. 187/M3
Cumberland (co.), Nova Scotia 168/D3
Cumberland (basin), Nova Scotia 168/D3
Cumberland, Ohio (43732) 284/G6
Cumberland, Okla. (†73446) 288/N6
Cumberland, Ontario 177/J2
Cumberland (co.), Pa. 294/H5
Cumberland (isls.), Queensland 88/H4
Cumberland (isls.), Queensland 95/D4
Cumberland (bay), St. Vin. & Grens. 161/A8
Cumberland (lake), Sask. 181/J1
Cumberland (co.), Tenn. 237/L9
Cumberland (plat.), Tenn. 237/L9
Cumberland (riv.), Tenn. 237/K8
Cumberland (co.), Va. 307/M6
Cumberland, Va. (23040) 307/M6
Cumberland (mt.), Va. 307/B7
Cumberland, Wash. (†98022) 310/D3
Cumberland, Wis. (54829) 317/C4
Cumberland Bay, New Bruns. 170/E2
Cumberland Beach, Ontario 177/E3
Cumberland Center, Maine (04021) 243/C8
Cumberland Center○, Maine (04021) 243/C8
Cumberland City, Tenn. (37050) 237/F8
Cumberland Furnace, Tenn. (37051) 237/G8
Cumberland Gap, Tenn. (37724) 237/O8
Cumberland Gap Nat'l Hist. Park, Ky. 237/P7
Cumberland Gap Nat'l Hist. Park, Tenn. 237/O7
Cumberland Gap Nat'l Hist. Park, Va. 307/A7
Cumberland House, Sask. 181/J2
Cumberland Island Nat'l Seashore, Georgia 217/K9
Cumbernauld, Scotland 15/C1
Cumbre del Laudo (mt.), Argentina 143/C2
Cumbre Negra, Cerro (mt.), Argentina 143/C5
Cumbre Negra, Cerro (mt.), Chile 138/E5
Cumbria (co.), England 13/D3
Cumbrian (mts.), England 13/D3
Cumbum, India 68/D5
Cumby, Texas (75433) 303/J4
Cuming (co.), Nebr. 264/H3
Cummaquid, Mass. (02637) 249/N6
Cumming, Georgia (30130) 217/D2
Cumming, Iowa (50061) 229/F6
Cummings, Kansas (66016) 232/G2
Cummings, N. Dak. (58223) 282/S4
Cummings, S.C. (†29944) 296/E6
Cummingsville, Tenn. (†38583) 237/L9
Cummington○, Mass. (01026) 249/C3
Cummins, S. Australia 94/D6
Cumnock, N.S. Wales 97/E3
Cumnock, N.C. (27237) 281/L3
Cumnock and Holmhead, Scotland 10/D3
Cumnock and Holmhead, Scotland 15/D5
Cumpas, Mexico 150/E1
Çumra, Turkey 63/E4
Cuñapirú, Uruguay 145/D2
Cuñapirú, Arroyo (riv.), Uruguay 145/D2
Cunapo, Trin. & Tob. 161/B10
Cuñare, Colombia 126/D7
Cunaviche, Venezuela 124/E4
Cunco, Chile 138/E2
Cuncumén, Coquimbo, Chile 138/A9
Cuncumén, Santiago, Chile 138/F4
Cundeelee Aboriginal Reserve, W. Australia 88/C6
Cundeelee Aboriginal Res., W. Australia 92/C5
Cunderdin, W. Australia 92/B5
Cundiff, Ky. (42730) 237/L7
Cundinamarca (dept.), Colombia 126/C5
Cundiyo, N. Mex. (87522) 274/D3
Cunduacán, Mexico 150/N7
Cundys Harbor, Maine (04011) 243/D8
Cunene (riv.) 102/D6
Cunene (dist.), Angola 115/C7
Cunene (dam), Angola 115/B7
Cunene (riv.), Angola 115/B7
Cuneo (prov.), Italy 34/A2
Cuneo, Italy 34/A2
Çüngüş, Turkey 63/H3
Cunnamulla, Australia 87/E8
Cunnamulla, Queensland 95/C5
Cunnamulla, Queensland 88/H5
Cunningham, Kansas (67035) 232/D4
Cunningham, Ky. (42035) 237/D7
Cunningham, N.C. (†27343) 281/L1
Cunningham, Tenn. (37052) 237/G8
Cunningham, Wash. (99327) 310/G4
Cuorgnè, Italy 34/A2
Cupar, Sask. 181/G5
Cupar, Scotland 15/E4
Cupar, Scotland 10/E2
Cupertino, Calif. (95014) 204/K3
Cupica (gulf), Colombia 126/B4
Cupids, Newf. 166/D2
Ćuprija, Yugoslavia 45/E4
Cuprum, Idaho (†83612) 220/B4
Cupsuptic (riv.), Maine 243/B5
Cuquenán (riv.), Venezuela 124/H5
Cuquiari (riv.), Colombia 126/E7
Curaçá, Brazil 132/G5
Curaçao (isl.), Neth. Ant. 161/G7
Curaçao (isl.), Neth. Ant. 156/E4
Curacautín, Chile 138/E2
Curacaví, Chile 138/G3
Curahuara de Carangas, Bolivia 136/A5
Curahuara de Pacajes, Bolivia 136/A5
Curanilahue, Chile 138/D1
Curaray (riv.), Ecuador 128/D3
Curaumilla (pt.), Chile 138/E2
Curdsville, Ky. (42334) 237/G5
Curecanti Nat'l Rec. Area, Colo. 208/F6
Curepipe, Mauritius 118/G5
Curepto, Chile 138/A10
Curiapo, Venezuela 124/H3
Curiche, Bolivia 136/D6
Curicó, Chile 120/B6
Curicó, Chile 138/A10
Curieuse (isl.), Seychelles 118/H5
Curitiba, Brazil 132/D9
Curitiba, Brazil 120/F5
Curitiba, Brazil 135/B4
Curlew, Iowa (50527) 229/D3
Curlew (isls.), La. 238/M7
Curlew, Wash. (99118) 310/G2
Curlew (lake), Wash. 310/G2
Curlewis, N.S. Wales 97/F2
Curllsville, Pa. (16221) 294/D3
Curnamona, S. Australia 94/F4
Curragh, The, Ireland 17/H5
Curragh, The (racecourse), Ireland 10/C4
Currais Novos, Brazil 132/G4
Curran, Ill. (62632) 222/D4
Curran, Mich. (48728) 250/F4
Currant, Nev. (†89314) 266/F4
Currawilla, Queensland 95/B5
Current (riv.), Ark. 202/J1
Current (riv.), Mo. 261/K8
Currie, Minn. (56123) 255/C6
Currie, Nev. (†89301) 266/G2
Currie, N.C. (28435) 281/N6
Currie, Scotland 15/D2
Currie, Tasmania 99/A1
Currituck (co.), N.C. 281/S2
Currituck, N.C. (27929) 281/T2
Currituck (sound), N.C. 281/T2
Curry, Alaska (†99676) 196/J2
Curry (co.), N. Mex. 274/F4
Curry (co.), Oreg. 291/C5
Curryville, Mo. (63339) 261/K4
Curryville, Pa. (16631) 294/F5
Curtea de Argeș, Romania 45/G3
Curtice, Ohio (43412) 284/D2
Curtin, Oreg. (97428) 291/D4
Curtina, Uruguay 145/C3
Curtis, Ark. (71728) 202/D6
Curtis, La. (†71101) 238/C2
Curtis, Mich. (49820) 250/D2
Curtis, Nebr. (69025) 264/D4
Curtis, Okla. (†73852) 288/H2
Curtis (isl.), Queensland 88/J4
Curtis (isl.), Queensland 95/D4
Curtis, Wash. (98538) 310/B4
Curtis Group (isls.), Tasmania 99/C1
Curtiss, Wis. (54422) 317/F6
Curtis Station, Miss. (†38606) 256/D2
Curtisville, Ind. (†46036) 227/F4
Curuá (riv.), Brazil 132/C4
Curuçá, Brazil 132/E3
Curuguaty, Paraguay 144/E4
Curup, Indonesia 85/C6
Cururú, Bolivia 136/D4
Cururupu, Brazil 132/E3
Curutú (riv.), Venezuela 124/G5
Curuzú Cuatiá, Argentina 143/G5
Curuzú Cuatiá, Argentina 120/D5
Curve, Tenn. (†38063) 237/B9
Curvelo, Brazil 132/E7
Curwensville, Pa. (16833) 294/E4
Curwood (mt.), Mich. 250/A2
Cusachón (isl.), Colombia 126/D1
Cusco, Peru 120/B4
Cusco (dept.), Peru 128/F9
Cusco (Cuzco), Peru 128/F9
Cushendall, N. Ireland 17/J1
Cushing, Iowa (51018) 229/B4
Cushing○, Maine (04563) 243/E7
Cushing, Minn. (56443) 255/D4
Cushing, Nebr. (†68873) 264/F3
Cushing, Okla. (74023) 288/N3
Cushing, Texas (75760) 303/J6
Cushing, Wis. (54006) 317/A4
Cushman, Ark. (72526) 202/G2
Cushman, Mass. (01002) 249/D3
Cushman, Oreg. (†97439) 291/D4
Cushman (lake), Wash. 310/B3
Cusiana (riv.), Colombia 126/D5
Cusick, Wash. (99119) 310/H2
Cuslett, Newf. 166/C2
Cusset, France 28/E4
Cusseta, Ala. (36852) 195/H5
Cusseta, Georgia (31805) 217/C6
Cusson, Minn. (†55771) 255/F2
Custar, Ohio (43511) 284/C3
Custer (co.), Colo. 208/J6
Custer (co.), Idaho 220/D5
Custer, Ky. (40115) 237/J5
Custer, Mich. (49405) 250/C5
Custer (co.), Mont. 262/L4
Custer, Mont. (59024) 262/J4
Custer (co.), Nebr. 264/E3
Custer (co.), Okla. 288/H3
Custer (co.), S. Dak. 298/B6
Custer, S. Dak. (57730) 298/B6
Custer, Wash. (98240) 310/C2
Custer Battlefield Nat'l Mon., Mont. 262/J5
Custer City, Okla. (73639) 288/J3
Custer City, Pa. (16725) 294/E2
Custer Park, Ill. (60418) 222/E2
Cut Bank, Mont. (59427) 262/D2
Cut Bank (creek), Mont. 262/D2
Cut Bank (creek), N. Dak. 282/H2
Cutbank, Sask. 181/E4
Cutchogue-New Suffolk, N.Y. (11935) 276/P8
Cutervo, Peru 128/C6
Cuthbert, Georgia (31740) 217/C7
Cut Knife, Sask. 181/B3
Cutler, Calif. (93615) 204/F7
Cutler, Ill. (62238) 222/D5
Cutler, Ind. (46920) 227/D4
Cutler, Maine (04626) 243/J6
Cutler○, Maine (04626) 243/J6
Cutler, Ohio (45724) 284/G7
Cutler Ridge, Fla. (33157) 212/F6
Cutlerville, Mich. (49508) 250/D6
Cut Off, La. (70345) 238/K7
Cutra (lake), Ireland 17/D5
Cutral-Có, Argentina 143/C4
Cutshin, Ky. (41732) 237/P6
Cuttaburra (creek), N.S. Wales 97/C1
Cuttack, India 54/K7
Cuttack, India 68/F4
Cutten, Calif. (95534) 204/A3
Cuttingsville, Vt. (05738) 268/B4
Cuttyhunk, Mass. (02713) 249/L7
Cuttyhunk (isl.), Mass. 249/L7
Cuvier (isl.), N. Zealand 100/E2
Cuvier (cape), W. Australia 88/A4
Cuvier (cape), W. Australia 92/A4
Cuvo (riv.), Angola 115/B6
Cuxhaven, W. Germany 22/C2
Cuya, Chile 138/B2
Cuyabeno, Ecuador 128/E3
Cuyahoga (co.), Ohio 284/G3
Cuyahoga (riv.), Ohio 284/H10
Cuyahoga Falls, Ohio (*44221) 284/G3
Cuyahoga Heights, Ohio (†44101) 284/H9
Cuyama, Calif. (93214) 204/F9
Cuyama (riv.), Calif. 204/E8
Cuyapaipe Ind. Res., Calif. 204/J11
Cuyk, Netherlands 27/H5
Cuylerville, N.Y. (†14481) 276/E5
Cuyo, Philippines 82/C5
Cuyo (isl.), Philippines 82/C5
Cuyo (isls.), Philippines 82/C5
Cuyo (isls.), Philippines 85/G3
Cuyocuyo, Peru 128/H10
Cuyo East (passage), Philippines 82/C5
Cuyo West (passage), Philippines 82/C5
Cuyuna, Minn. (†56444) 255/E4
Cuyuna (range), Minn. 255/D4
Cuyuni (riv.) 120/C2
Cuyuni (riv.), Guyana 131/B2
Cuyuni (riv.), Venezuela 124/H4
Cuyu Tigni, Nicaragua 154/F3
Cuzco, Ind. (†47432) 227/D8
Cuzzart, W. Va. (26530) 312/H3
Čvrsnica (mt.), Yugoslavia 45/C4
Cwmamman, Wales 13/D6
Cwmbran, Wales 13/B6
Cyangugu, Rwanda 115/E4
Cyclades (isls.), Greece 45/G7
Cycle, N.C. (27015) 281/H2
Cyclone, Ind. (†46041) 227/E4
Cyclone, Pa. (16726) 294/E2
Cyclone, W. Va. (24827) 312/C7
Cygnet, Ohio (43413) 284/C3
Cygnet, Tasmania 99/C5
Cylinder, Iowa (50528) 229/D2
Cylon, Wis. (†54017) 317/B5
Cymric, Sask. 181/G4
Cynthia, Alberta 182/C3
Cynthiana, Ind. (47612) 227/B8
Cynthiana, Ky. (41031) 237/N4
Cynthiana, Ohio (45624) 284/D7
Cypert, Ark. (†72366) 202/J5
Cypress, Ala. (35454) 195/C5
Cypress (hills), Alberta 182/E5
Cypress (bayou), Ark. 202/F3
Cypress, Calif. (90630) 204/D11
Cypress, Fla. (32432) 212/A1
Cypress (lake), Fla. 212/E3
Cypress, Ill. (62923) 222/D6
Cypress, Ind. (†47708) 227/B9
Cypress (pond), Ind. 227/B8
Cypress, La. (71420) 238/D3
Cypress (hills), Sask. 181/B6
Cypress (lake), Sask. 181/B6
Cypress Gardens, Fla. (33880) 212/E3
Cypress Hills Prov. Park, Alberta 182/E5
Cypress Hills Prov. Park, Sask. 181/B6
Cypress Inn, Tenn. (38452) 237/F10
Cypress Prov. Park, Br. Col. 184/K3
Cypress Quarters, Fla. (†33472) 212/F4
Cypress River, Manitoba 179/D5
Cyprus 2/L4
Cyprus 54/E6
CYPRUS 59/B2
CYPRUS 63/E5
Cyrenaica (reg.), Libya 102/E1
Cyrenaica (reg.), Libya 111/D1

Cyrene (Shahat), Libya 111/D1
Cyrene, Mo. (†63334) 261/K4
Cyril, Okla. (73029) 288/K5
Cyrus, Minn. (56323) 255/C5
Czar, Alberta 182/E3
Czar, W. Va. (†26224) 312/F5
Czarna Białostocka, Poland 47/F2
Czarnków, Poland 47/C2
Czechoslovakia 2/K3
Czechoslovakia 7/F4
CZECHOSLOVAKIA 41
Czechowice-Dziedzice, Poland 47/D4
Czech Socialist Rep., Czech. 41/B1
Czeladź, Poland 47/B4
Czersk, Poland 47/D2
Częstochowa (prov.), Poland 47/D3
Częstochowa, Poland 47/D3
Częstochwa, Poland 7/F3
Człuchów, Poland 47/C2

D

Da'an (Talai), China 77/K2
Daaquam, Québec 172/H3
Dabajuro, Venezuela 124/C2
Dabakala, Ivory Coast 106/D7
Dabas, Hungary 41/E3
Daba Shan (range), China 77/G5
Dabeiba, Colombia 126/B4
Dabhoi, India 68/C4
Dabney, Ind. (†47023) 227/G6
Dabob (bay), Wash. 310/C3
Dabola, Guinea 106/B6
Dabou, Ivory Coast 106/D7
Daboya, Ghana 106/D7
Dąbrowa Górnicza, Poland 47/B3
Dąbrowa Tarnowska, Poland 47/E3
Dăbuleni, Romania 45/F4
Dacca (cap.), Bangladesh 54/L7
Dacca (cap.), Bangladesh 68/G4
Dachau, W. Germany 22/D4
Dačice, Czech. 41/C2
Dac Lac, Cao Nguyen (plat.), Vietnam 72/F4
Dacoma, Okla. (73731) 288/J1
Dacono, Colo. (80514) 208/K2
Dacre, Ontario 177/G2
Dacula, Georgia (30211) 217/E3
Dacusville, S.C. (†29640) 296/B2
Dadanawa, Guyana 131/B4
Daday, Turkey 63/E2
Dade (co.), Fla. 212/F6
Dade (co.), Georgia 217/A1
Dade (co.), Mo. 261/E8
Dade City, Fla. (33525) 212/D3
Dadeville, Ala. (36853) 195/G5
Dadeville, Mo. (65635) 261/E8
Dadra and Nagar Haveli (terr.), India 68/C4
Dads (lake), Nebr. 264/D2
Dadu, Pakistan 68/B3
Dadu, Pakistan 59/J4
Dăeni, Romania 45/J3
Daer (res.), Scotland 15/E5
Daet, Philippines 85/G3
Daet, Philippines 82/D3
Dafang, China 77/G6
Dafna, Israel 65/D1
Dafoe, Sask. 181/G4
Dafter, Mich. (49724) 250/E2
Dagabur, Ethiopia 111/H6
Dagana, Senegal 106/A5
Dagda, U.S.S.R. 53/D2
Dagelet (Ullŭng) (isl.), S. Korea 81/E5
Dagestan A.S.S.R., U.S.S.R. 48/E5
Dagestan A.S.S.R., U.S.S.R. 52/G6
Dagestanskiye Ogni, U.S.S.R. 52/G6
Daggett, Calif. (92327) 204/H9
Daggett, Mich. (49821) 250/B3
Daggett (co.), Utah 304/E3
Dagmar, Mont. (59219) 262/M2
Dagö (Hiiumaa) (isl.), U.S.S.R. 52/B3
Dagsboro, Del. (19930) 245/S6
Dagua, Colombia 126/B6
Daguan, China 77/F6
D'Aguilar (range), Tasmania 99/B4
Dagupan, Philippines 82/C2
Daguscahonda, Pa. (†15853) 294/E3
Dagus Mines, Pa. (15831) 294/E3
Dahab, Egypt 111/F2
Dahana (des.), Saudi Arabia 54/F7
Dahana (des.), Saudi Arabia 59/E4
Dahinda, Ill. (61428) 222/C3
Dahinda, Sask. 181/G6
Da Hingan Ling (Great Khingan) (range), China 54/O5
Da Hingan Ling (range), China 77/J3
Dahlak (arch.), Ethiopia 111/H4
Dahlak (arch.), Ethiopia 59/D6
Dahlak (isl.), Ethiopia 59/D6
Dahlak (isl.), Ethiopia 111/H4
Dahlem, W. Germany 22/E4
Dahlen, N. Dak. (58224) 282/P3
Dahlgren, Ill. (62828) 222/E5
Dahlgren, Va. (22448) 307/O4
Dahlia, N. Mex. (†87711) 274/D3
Dahlonega, Georgia (30533) 217/D1
Dahme, E. Germany 22/E3
Dai (mt.), Japan 81/F6
Dailekh, Nepal 68/E3
Dailey, Colo. (†80728) 208/O1
Dailly, Scotland 15/D5
Daimanji (mt.), Japan 81/F5
Daimiel, Spain 33/E3
Daingean, Ireland 17/G5
Daingerfield, Texas (75638) 303/K4
Daio (cape), Japan 81/H6
Daiquirí, Cuba 158/J4
Daireaux, Argentina 143/D4
Dairût, Egypt 111/J4
Dairy, Oreg. (97625) 291/F5
Dairy Flat-Redvale, N. Zealand 100/B1
Dairyland, Wis. (†54830) 317/B3
Daisen-Oki National Park, Japan 81/F6
Daisetsu (mt.), Japan 81/L2
Daisetsu-Zan National Park, Japan 81/L2
Daisetta, Texas (77533) 303/K7
Daisy, Ark. (†71950) 202/C5
Daisy, Georgia (30423) 217/J6
Daisy, Ky. (41733) 237/P6
Daisy, Mo. (63743) 261/N7
Daisy, Okla. (74540) 288/P5
Daisy, Wash. (†99167) 310/G2
Daito, Japan 81/J8
Daito (isls.), Japan 54/P7
Dajabón (prov.), Dom. Rep. 158/D5
Dajabón, Dom. Rep. 158/D5
Dajarra, Queensland 88/F4
Dajarra, Queensland 95/A4
Dakar (cap.), Senegal 2/J5
Dakar (cap.), Senegal 102/A3
Dakar (cap.), Senegal 106/A6
Dakhla (oasis), Egypt 111/E2
Dakhla (oasis), Egypt 59/A4
Dakhla, W. Sahara 102/A2
Dakhla, Western Sahara 106/A4
Dakoro, Niger 106/F6
Dakota, Georgia (†31714) 217/E7
Dakota, Ill. (61018) 222/D1
Dakota (co.), Minn. 255/E6
Dakota, Minn. (55925) 255/G7
Dakota (co.), Nebr. 264/H2
Dakota City, Iowa (50529) 229/E3
Dakota City, Nebr. (68731) 264/H2
Dal (riv.), Sweden 7/F2
Dala, Angola 115/D6
Dalaba, Guinea 106/B6
Dalälven (riv.), Sweden 18/K6
Dalaman (riv.), Turkey 63/C4
Dalandzadgad, Mongolia 77/G3
Dalanganem (isls.), Philippines 82/C5
Dalark, Ark. (†71923) 202/E5
Da Lat, Vietnam 72/F5
Dalavich, Scotland 15/C4
Dalbandin, Pakistan 68/A3
Dalbandin, Pakistan 59/H4
Dalbeattie, Scotland 10/E3
Dalbeattie, Scotland 15/E6
Dalbo, Minn. (55017) 255/E5
Dalby, Queensland 95/D5
Dalby, Queensland 88/J5
Dalby, Sweden 18/H6
Dalcahue, Chile 138/D4
Dalcour, La. (†70040) 238/P4
Dale (co.), Ala. 195/G8
Dale, Ill. (62829) 222/E6
Dale, Ind. (47523) 227/D8
Dale, Minn. (†56549) 255/B4
Dale, Norway 18/E6
Dale, Okla. (74838) 288/M4
Dale, Oreg. (97880) 291/J3
Dale, Pa. (†15901) 294/E5
Dale, S.C. (29914) 296/F6
Dale (mt.), W. Australia 88/B2
Dale (mt.), W. Australia 92/B1
Dale, Wis. (54931) 317/J7
Dale City, Va. (22193) 307/O3
Dale Hollow (lake), Ky. 237/L7
Dale Hollow (lake), Tenn. 237/L7
Dalemead, Alberta 182/D4
Dalen, Netherlands 27/K3
Daleside, S. Africa 118/H7
Daleville, Ala. (36322) 195/G8
Daleville, Ind. (47334) 227/F4
Daleville, Miss. (39326) 256/G5
Daleville, Va. (24083) 307/J6
Dale West, W. Australia 92/B2
Dalhart, Texas (79022) 303/B1
Dalhousie, New Bruns. 170/D1
Dalhousie (cape), N.W. Terrs. 187/E2
Dalhousie (mt.), Nova Scotia 168/E3
Dalhousie East, Nova Scotia 168/D4
Dalhousie Junction, New Bruns. 170/D1
Dalhousie West, Nova Scotia 168/C4
Dali, China 77/E6
Dalías, Spain 33/E4
Daliburgh, Scotland 15/A3
Dalizi, China 77/L3
Dalkeith, Ontario 177/K2
Dalkeith, Scotland 10/C1
Dalkeith, Scotland 15/D2
Dalkena, Wash. (†99156) 310/H2
Dall (isl.), Alaska 196/M2
Dall (mt.), Alaska 196/H2
Dallam (co.), Texas 303/B1
Dallas (co.), Ala. 195/D6
Dallas (co.), Ark. 202/E6
Dallas, Georgia (30132) 217/C3
Dallas (co.), Iowa 229/E5
Dallas, Iowa (50062) 229/G6
Dallas, Manitoba 179/E3
Dallas (co.), Mo. 261/F7
Dallas, N.C. (28034) 281/G4
Dallas, Oreg. (97338) 291/D3
Dallas, Pa. (18612) 294/E7
Dallas, Scotland 15/E3
Dallas, S. Dak. (57529) 298/K7
Dallas (co.), Texas 303/H5
Dallas, Texas (*75201) 303/G2
Dallas, Texas 188/G4
Dallas, Texas 146/J6
Dallas, U.S. 2/E4
Dallas, W. Va. (26036) 312/E2
Dallas, Wis. (54733) 317/C5
Dallas Center, Iowa (50063) 229/E5
Dallas City, Ill. (62330) 222/B3
Dallas Naval Air Sta., Texas 303/G2
Dallastown, Pa. (17313) 294/J6
Dalles, The, Oreg. (97058) 291/F2
Dalles, The (dam), Oreg. 291/F2
Dalles, The (dam), Wash. 310/D5
Dallesport, Wash. (98617) 310/D5
Dalloi, Ethiopia 111/G5
Dallol Bosso (dry riv.), Niger 106/E6
Dalmaj, Hor (lake), Iraq 66/D4
Dalmally, Scotland 10/D2
Dalmally, Scotland 15/D4
Dalmatia, Pa. (17017) 294/J4
Dalmatia (reg.), Yugoslavia 45/C4
Dalmellington, Scotland 15/D5
Dalmellington, Scotland 10/D3
Dalmeny, Sask. 181/E3
Dal'negorsk, U.S.S.R. 48/O5
Dal'nerechensk, U.S.S.R. 48/O5
Daloa, Ivory Coast 106/C7
Daloa, Ivory Coast 102/B4
Dalroy, Alberta 182/D4
Dalry, Scotland 10/A1
Dalry, Scotland 15/D5
Dalrymple, Scotland 15/D5
Dalton, Ark. (72423) 202/H1
Dalton, Georgia (30720) 217/C1
Dalton, Ky. (†42445) 237/F6
Dalton○, Mass. (01226) 249/B3
Dalton, Mich. (†49445) 250/C5
Dalton, Minn. (56324) 255/C4
Dalton, Mo. (65246) 261/F4
Dalton, Nebr. (69131) 264/B3
Dalton○, N.H. (†03598) 268/D3
Dalton, N.Y. (14836) 276/E5
Dalton, N.C. (†27043) 281/J2
Dalton, Ohio (44618) 284/G4
Dalton, Pa. (18414) 294/L2
Dalton, Wis. (53926) 317/H8
Dalton City, Ill. (61925) 222/E4
Daltonganj, India 68/E4
Dalton Gardens, Idaho (†83814) 220/B2
Dalton-in-Furness, England 13/D3
Dalupiri (isl.), Philippines 82/A3
Dalwallinu, W. Australia 88/B6
Dalwallinu, W. Australia 92/B5
Dalwhinnie, Scotland 15/D4
Dalworthington Gardens, Texas (†76101) 303/F2
Daly (cape) 5/C4
Daly (riv.), North. Terr. 88/E2
Daly (riv.), North. Terr. 93/B2
Daly (bay), N.W. Terrs. 187/K3
Dalyat al-Karmel, Israel 65/B2
Daly City, Calif. (*94014) 204/H2
Daly River, North. Terr. 88/E2
Daly River, North. Terr. 93/B2
Daly River Aboriginal Reserve, North. Terr. 88/D2
Daly River Aboriginal Res., North. Terr. 93/A2
Dalyup, W. Australia 92/C6
Daly Waters, Australia 87/D7
Daly Waters, North. Terr. 88/E3
Daly Waters, North. Terr. 93/C4
Dalzell, Ill. (61320) 222/D2
Dalzell, S.C. (29040) 296/G3
Dam, Saudi Arabia 59/D5
Daman (dist.), India 68/C4
Damanhur, Egypt 111/J3
Damanhur, Egypt 59/A3
Damar (isl.), Indonesia 85/H7
Damar (isls.), Indonesia 85/H7
Damar, Kansas (67632) 232/C2
Damara, Cent. Afr. Rep. 115/C2
Damaraland (reg.), Namibia 118/B4
Damariscotta○, Maine (04543) 243/E7
Damariscotta-Newcastle, Maine (04543) 243/E7
Damascus, Ark. (72039) 202/F3
Damascus, Georgia (31741) 217/C8
Damascus, Md. (20750) 245/K3
Damascus, Ohio (44619) 284/J4
Damascus, Pa. (18415) 294/M2
Damascus (prov.), Syria 63/G6
Damascus (cap.), Syria 59/C3
Damascus (cap.), Syria 54/E6
Damascus (cap.), Syria 63/G6
Damascus, Va. (24236) 307/E7
Damavand, Iran 66/H3
Damavand (mt.), Iran 54/G6
Damavand (mt.), Iran 59/F2
Damavend (Demavend) (mt.), Iran 66/G3
Damazin (Ed Damazin), Sudan 111/F5
Damba, Angola 115/B5
Dam Doi, Vietnam 72/E5
Dame Marie, Haiti 158/A6
Dame Marie (cape), Haiti 158/A6
Dame Marie (cape), Haiti 156/C3
Dameron, Md. (20628) 245/N8
Dames Ferry, Georgia (†31046) 217/E4
Dames Quarter, Md. (21820) 245/P8
Damghan, Iran 59/F2
Damghan, Iran 66/J2
Damh, Loch (lake), Scotland 15/C3
Damietta, Egypt 102/F1
Damietta, Egypt 111/J3
Damietta, Egypt 59/B3
Damiya, Jordan 65/D3
Dammam, Saudi Arabia 59/F4
Dammastock (mt.), Switzerland 39/F3
Damme, Belgium 27/C6
Damodar (riv.), India 68/F4
Damoh, India 68/D4
Damongo, Ghana 106/D7
Dampier (str.), Indonesia 85/J6
Dampier (str.), Papua N.G. 86/B2
Dampier (str.), Papua N.G. 85/C7
Dampier, W. Australia 88/B4
Dampier (arch.), W. Australia 88/B4
Dampier, W. Australia 92/B3
Dampier (arch.), W. Australia 92/B3
Dampier Downs, W. Australia 92/C2
Dampier Land (reg.), W. Australia 88/C3
Dampier Land (reg.), W. Australia 92/C2
Damqut, P.D.R. Yemen 59/F6
Damvant, Switzerland 39/C2
Dan, Israel 65/D1
Dan (riv.), N.C. 281/L1
Dan (riv.), Va. 307/K7
Dana, Ill. (61321) 222/E3
Dana, Ind. (47847) 227/C5
Dana, Iowa (50064) 229/E4
Dana, Jordan 65/E5
Dana, Sask. 181/F3
Danakil (reg.), Ethiopia 111/H5
Danané, Ivory Coast 106/C7
Da Nang, Vietnam 72/F3
Da Nang, Vietnam 54/M8
Da Nang, Mui (cape), Vietnam 72/F3
Danao, Philippines 82/D5
Dana Point, Calif. (92629) 204/H10
Danba, China 77/F5
Danburg, Georgia (30668) 217/G3
Danbury, Conn. (06810) 210/B3
Danbury, Iowa (51019) 229/B4
Danbury, Nebr. (69026) 264/D4
Danbury○, N.H. (03230) 268/D4
Danbury, N.C. (27016) 281/J2
Danbury, Sask. 181/J4
Danbury, Texas (77534) 303/J8
Danbury, Wis. (54830) 317/B3
Danbury P.O. (South Danbury), N.H. (03230) 268/D5
Danby (lake), Calif. 204/K9
Danby○, Vt. (05739) 268/A5
Dancing (pt.), Manitoba 179/D2
Dancy, Ala. (†35442) 195/B4
Dancy, Miss. (†39751) 256/F3
Dancy, Wis. (†54455) 317/G6
Dancyville, Tenn. (†38069) 237/C10
Dand, Manitoba 179/B5
Dandaragan, W. Australia 88/B6
Dandaragan, W. Australia 92/A5
Dandenong, Victoria 97/K5
Dandenong, Victoria 88/M7
Dandenong (creek), Victoria 97/K5
Dandenong (mt.), Victoria 97/K5
Dandenong (creek), Victoria 88/M7
Danderyd, Sweden 18/H1
Dandong (Tantung), China 77/K3
Dandong, China 54/O5
Dandridge, Tenn. (37725) 237/O8
Dane (riv.), England 13/H2
Dane (co.), Wis. 317/H9
Dane, Wis. (53529) 317/G9
Daneborg, Greenl. 4/B10
Danford Lake, Québec 172/A4
Danforth, Ill. (60930) 222/E3
Danforth, Maine (04424) 243/H4
Danforth○, Maine (04424) 243/H4
Danger (Pukapuka) (atoll), Cook Is. 87/K7
Dangila, Ethiopia 111/G5
Dangrek (mts.), Cambodia 72/D4
Dangrek (Dong Rak) (mts.), Thailand 72/D4
Dangriga (Stann Creek), Belize 153/C2
Dania, Fla. (33004) 212/B4
Daniel (mt.), Wash. 310/D3
Daniel, Wyo. (83115) 319/B3
Daniel Boone, Ky. (†42442) 237/G6
Daniel-Johnson (dam), Québec 174/D2
Daniels, Md. (†21043) 245/L3
Daniels (co.), Mont. 262/L2
Daniels, W. Va. (25832) 312/D7
Daniel's Harbour, Newf. 166/C3
Danielson, Conn. (06239) 210/H1
Danielson Prov. Park, Sask. 181/E4
Danielstown, Guyana 131/B2
Danielsville, Georgia (30633) 217/F2
Danielsville, Pa. (18038) 294/M4
Danilov, U.S.S.R. 52/E3
Dankov, U.S.S.R. 52/E4
Danlí, Honduras 154/D3
Danmarkshavn, Greenl. 4/B10
Dannebrog, Nebr. (68831) 264/F3
Dannelly (res.), Ala. 195/D6
Dannemora, N.Y. (12929) 276/N1
Dannemora, Sweden 18/K6
Dannenberg, W. Germany 22/D2
Danner, Oreg. (†97910) 291/K5
Dannevirke, N. Zealand 100/F4
Dan Sai, Thailand 72/D3
Dansville, Mich. (48819) 250/E6
Dansville, N.Y. (14437) 276/E5
Dante (Hafun), Somalia 115/K1
Dante, S. Dak. (57329) 298/N7
Dante, Va. (24237) 307/D7
Danube (riv.) 7/G4
Danube (riv.), Austria 41/C2
Danube (riv.), Bulgaria 45/H4
Danube (riv.), Czech. 41/C2
Danube (riv.), Hungary 41/E3
Danube, Minn. (56230) 255/C6
Danube (delta), Romania 45/J3
Danube (riv.), Romania 45/H4
Danube (riv.), W. Germany 22/C4
Danube (riv.), Yugoslavia 45/E3
Danubyu, Burma 72/B3
Danvers, Ill. (61732) 222/D3
Danvers○, Mass. (01923) 249/D5
Danvers, Minn. (56231) 255/C5
Danvers, Mont. (59429) 262/G3
Danversport, Mass. (†01923) 249/E5
Danville, Ala. (35619) 195/D2
Danville, Ark. (72833) 202/D3
Danville, Calif. (94526) 204/K2
Danville, Georgia (31017) 217/F5
Danville, Ill. 188/J3
Danville, Ill. (61832) 222/F3
Danville, Ind. (46122) 227/D5
Danville, Iowa (52623) 229/L7
Danville, Kansas (67036) 232/E4
Danville, Ky. (40422) 237/M5
Danville, La. (†71008) 238/E2
Danville, Mo. (†63361) 261/J5
Danville○, N.H. (03819) 268/E6
Danville, Ohio (43014) 284/F5
Danville, Pa. (17821) 294/J4
Danville, Québec 172/E4
Danville○, Vt. (05828) 268/C3
Danville, Va. 188/L3
Danville, Va. 146/L6
Danville (I.C.), Va. (*24540) 307/J7
Danville, Wash. (99121) 310/G2
Danville, W. Va. (25053) 312/C6
Danville, Wis. (†53925) 317/J9
Dan Xian, China 77/G8
Danzig (Gdańsk), Poland 47/D1
Danzig (Gdańsk) (gulf), Poland 47/D1
Daocheng, China 77/F6
Dao Xian, China 77/H6
Dapa, Philippines 82/E6
Dapaong, Togo 106/E6
Daphne, Ala. (36526) 195/C9
Daphne, Sask. 181/G3
Dapitan, Philippines 82/D6
Dapoli, India 68/C5
Dapp, Alberta 182/C2
Da Qaidam, China 77/E4
Darab, Iran 59/G4
Darab, Iran 66/J6
Darabani, Romania 45/H1
Dar al Hamra, Saudi Arabia 59/C4
Daram (isl.), Philippines 82/E5
Daran, Iran 66/G4
Darbandikhan (dam), Iraq 66/D3
Darbhanga, India 68/F3
Darbun, Miss. (†39643) 256/D8
Darby (cape), Alaska 196/F2
Darby, Mont. (59829) 262/B4
Darby (creek), Ohio 284/D5
Darby, Pa. (19023) 294/M7
Darby (creek), Pa. 294/M6
Darby, Victoria 97/D6
Darbydale, Ohio (†43123) 284/D6
Darbyville, Ohio (†43164) 284/D6
D'Arcy, Br. Col. 184/F5
D'Arcy, Sask. 181/C4
Dardanelle, Ark. (72834) 202/D3
Dardanelle (lake), Ark. 202/D3
Dardanelles (str.), Turkey 7/G5
Dardanelles (str.), Turkey 59/A2
Dardanelles (str.), Turkey 63/B6
Darden, Tenn. (38328) 237/E9
Dare (co.), N.C. 281/T3
Dar-el-Beida (Casablanca), Morocco 106/C2
Darende, Turkey 63/G3
Dar es Salaam (cap.), Tanzania 102/F5
Dar es Salaam (cap.), Tanzania 2/M6
Dar es Salaam (cap.), Tanzania 115/G5
Dareton, N.S. Wales 97/B4
Daretown, N.J. (†08318) 273/C4
Darfur, Minn. (56022) 255/D6
Darfur, Northern (prov.), Sudan 111/D5
Darfur, Southern (prov.), Sudan 111/E5
Dargan, Md. (†25425) 245/H3
Dargaville, N. Zealand 100/D1
Dar Hamid (reg.), Sudan 111/F5
Darham Mumingan Lianheqi, China 77/H3
Darhan (Darkhan), Mongolia 77/G2
Darien○, Conn. (06820) 210/B4
Darien, Georgia (31305) 217/K8
Darien, Ill. (†60559) 222/B6
Darien, N.Y. (†14040) 276/D5
Darién (mts.), Panama 154/J6
Darien, Wis. (53114) 317/J10
Darien Center, N.Y. (14040) 276/D5
Dariense, Cordillera (range), Nicaragua 154/E4
Darjeeling, India 68/F3
Dark (head), St. Vin. & Grens. 161/A8
Darkan, W. Australia 92/B2
Dark Canyon (creek), N. Mex. 274/E6
Dark Cove, Newf. 166/D4
Darke (co.), Ohio 284/A5
Darkesville, W. Va. (†25428) 312/L4
Darkin (riv.), W. Australia 88/B2
Darlag, China 77/E5
Darling (river), Australia 87/E9
Darling, Miss. (38623) 256/D2
Darling (riv.), N.S. Wales 88/G6
Darling (riv.), N.S. Wales 97/B3
Darling (lake), N. Dak. 282/G2
Darling, Pa. (†19063) 294/L7
Darling (range), W. Australia 88/B6
Darling (range), W. Australia 92/A1
Darling Downs, Queensland 95/D5
Darlingford, Manitoba 179/D5
Darlington, Ala. (36730) 195/D7
Darlington, England 10/F3
Darlington, England 13/F3
Darlington, Fla. (†32464) 212/C5
Darlington, Idaho (83231) 220/E6
Darlington, Ind. (47940) 227/D4
Darlington, La. (†70441) 238/J5
Darlington, Md. (21034) 245/N2
Darlington, Mo. (64438) 261/D2
Darlington, New Bruns. 170/D1
Darlington, Pa. (16115) 294/A4
Darlington (co.), S.C. 296/H3
Darlington, S.C. (29532) 296/H3
Darlington, Wis. (53530) 317/F10
Darlington Heights, Va. (23935) 307/L6
Darlington Point, N.S. Wales 97/C4
Darliston, Jamaica 158/H6
Darłowo, Poland 47/C1
Dar Masalit (reg.), Sudan 111/D5
Darmody, Sask. 181/E5
Darmstadt, Ill. (†62255) 222/D5
Darmstadt, Ind. (†47618) 227/B8
Darmstadt, W. Germany 22/C4
Darnell, La. (71231) 238/G1
Darnestown, Md. (†20760) 245/J4
Darnick, N.S. Wales 97/B3
Darnley (cape) 5/C4
Darnley (bay), N.W. Terrs. 187/F3
Daroca, Spain 33/F2
Darra, Queensland 88/K3
Darreh Gaz, Iran 66/L2
Darrington, Wash. (98241) 310/D2
Dar Rounga (reg.), Cent. Afr. Rep. 115/D2
Darrouzett, Texas (79024) 303/D1
Darrow, La. (70725) 238/K3
Darrtown, Ohio (†45056) 284/A7
Darsser Ort (pt.), E. Germany 22/E1
Dart (cape) 5/B12
Dart (riv.), England 13/D7
D'Artagnan, Québec 172/J3
Dartford, England 13/J8
Dartford, England 10/C5
Dartmoor, Victoria 97/A5
Dartmoor National Park, England 13/C7
Dartmouth (Clifton Dartmouth Hardness), England 10/E5
Dartmouth (Clifton Dartmouth Hardness), England 13/D7
Dartmouth○, Mass. (02714) 249/K6
Dartmouth, N.S. 162/K7
Dartmouth, Nova Scotia 168/E4
Dartmouth (riv.), Québec 172/D1
Darton, England 13/J2
Dartuch (cape), Spain 33/H3
Daru, Papua N.G. 87/E6
Daru, Papua N.G. 85/B7
Daruvar, Yugoslavia 45/C3
Darvel, Scotland 15/D5
Darwell, Alberta 182/C3
Darwen, England 10/G1
Darwen, England 13/H1
Darwin, Australia 2/R6
Darwin, Australia 87/D7
Darwin, Calif. (93522) 204/H7
Darwin (bay), Chile 138/D6
Darwin, Cordillera (mts.), Chile 138/D8
Darwin, Cordillera (mts.), Chile 138/E11
Darwin (Culpepper) (isl.), Ecuador 128/B8
Darwin, Ill. (†62477) 222/F4
Darwin, Minn. (55324) 255/D5
Darwin (cap.), North. Terr. 88/E2
Darwin (cap.), North. Terr. 93/B2
Darwin, Okla. (†74523) 288/P6
Das (isl.), U.A.E. 59/F4
Dash, Ben (hill), Ireland 17/C6
Dashan, Ras (mt.), Ethiopia 59/C7
Dashbalbar, Mongolia 77/H2
Dasher, Georgia (31601) 217/F9
Dashinchilen, Mongolia 77/F2
Dasht (riv.), Pakistan 68/A3
Dasht (riv.), Pakistan 59/H4
Dashtiari, Iran 66/M8
Dashtiari, Iran 59/H4
Dashwood, Br. Col. 184/H3
Dashwood, Ontario 177/C4
Dasol (bay), Philippines 82/B3
Dassel, Minn. (55325) 255/D5
Datça, Turkey 63/B4
Dateland, Ariz. (85333) 198/B6
Datia, India 68/D3
Datil, N. Mex. (87821) 274/B4
Datil (mts.), N. Mex. 274/B4
Datong, Qinghai, China 77/F4
Datong (Tatung), Shanxi, China 77/H3
Datto, Ark. (72424) 202/J1
Datu Piang, Philippines 82/E7
Daua (riv.), Kenya 115/H3
Daufuskie Island, S.C. (29915) 296/F7
Daugava (Western Dvina) (riv.), U.S.S.R. 53/D2
Daugavpils, U.S.S.R. 7/G3
Daugavpils, U.S.S.R. 53/D3
Daugavpils, U.S.S.R. 48/C4
Daugavpils, U.S.S.R. 52/C3
Daule, Ecuador 128/B3
Daulnay, New Bruns. 170/E1
Daun, W. Germany 22/B3
Daung Kyun (isl.), Burma 72/C4
Dauphin, Man. 162/F5
Dauphin, Manitoba 179/B3
Dauphin (lake), Manitoba 179/C3
Dauphin (riv.), Manitoba 179/D3
Dauphin (cape), Nova Scotia 168/H2
Dauphin (co.), Pa. 294/J5
Dauphin, Pa. (17018) 294/J5
Dauphin, St. Lucia 161/G5
Dauphiné (trad. prov.), France 29
Dauphin Island, Ala. (36528) 195/B10
Daus, Tenn. (†37327) 237/L10
Davangere, India 68/D6
Davant, La. (70046) 238/L7
Davao, Philippines 85/H4
Davao, Philippines 54/O9
Davao, Philippines 2/R5
Davao, Philippines 82/E7
Davao (gulf), Philippines 82/E7
Davao (gulf), Philippines 85/H4
Davao del Norte (prov.), Philippines 82/E7
Davao del Sur (prov.), Philippines 82/E7
Davao Oriental (prov.), Philippines 82/F7
Daveluyville, Québec 172/E3
Davenport, Calif. (95017) 204/K4
Davenport, Fla. (33837) 212/E3
Davenport, Iowa (*52801) 229/M5
Davenport, Iowa 188/H2
Davenport, Nebr. (68335) 264/G4
Davenport, N.Y. (13750) 276/L6
Davenport, N. Dak. (58021) 282/R6
Davenport (mt.), North. Terr. 93/B7
Davenport, Okla. (74026) 288/N3
Davenport, Va. (24239) 307/D6
Davenport, Wash. (99122) 310/G3
Daventry, England 13/F5
Davey, Nebr. (68336) 264/H4
Davey (riv.), Tasmania 99/B4
David (pt.), Grenada 161/D8
David, Ky. (41616) 237/R5
David, Panama 154/F6
David City, Nebr. (68632) 264/G3
Davidson (mts.), Alaska 196/K1
Davidson, Maine (†04782) 243/F4
Davidson (co.), N.C. 281/J3
Davidson, N.C. (28036) 281/H4
Davidson, Okla. (73530) 288/J6
Davidson, Sask. 181/E4
Davidson (co.), Tenn. 237/H8
Davidson (mts.), Yukon 187/D3
Davidsonville, Md. (21035) 245/M5
Davie, Fla. (33314) 212/B4
Davie (co.), N.C. 281/H3
Daviess (co.), Ind. 227/C7
Daviess (co.), Ky. 237/G5
Daviess (co.), Mo. 261/E3
Davik, Norway 18/D6
Davilla, Texas (76523) 303/G7
Davin, Sask. 181/H5
Daviot, Scotland 15/D3
Davis (str.) 2/G2
Davis (str.) 146/N3
Davis (str.) 4/C12
Davis (sea) 5/C5
Davis (dam), Ariz. 198/A3
Davis, Calif. (95616) 204/B8
Davis (isl.), Fla. 212/C3
Davis, Ill. (61019) 222/D1

Davis (co.), Iowa 229/J7
Davis (dam), Nev. 266/G7
Davis, N.C. (28524) 281/R5
Davis (str.), N.W.T. 162/K1
Davis (str.), N.W. Terrs. 187/M3
Davis, Okla. (73030) 288/M5
Davis (lake), Oreg. 291/F4
Davis (mt.), Pa. 294/D6
Davis, Sask. 181/F2
Davis, S. Dak. (57021) 298/P7
Davis (mts.), Texas 303/C11
Davis (co.), Utah 304/B3
Davis, W. Va. (26260) 312/H4
Davisboro, Georgia (31018) 217/G5
Davis City, Iowa (50065) 229/F7
Davis Cove, Newf. 166/D4
Davis Creek, Calif. (96108) 204/E2
Davis Dam, Ariz. (†86430) 198/A3
Davis Inlet, Newf. 166/B2
Davis Junction, Ill. (61020) 222/D1
Davis-Monthan A.F.B., Ariz. 198/E6
Davison, Mich. (48423) 250/F5
Davison (co.), S. Dak. 298/N6
Davis Station 5/C4
Davis Station, S.C. (29041) 296/G4
Daviston, Ala. (36256) 195/G4
Davisville, Mo. (65456) 261/K7
Davisville, R.I. (02854) 249/H6
Davisville, W. Va. (26142) 312/C4
Davlekanovo, U.S.S.R. 52/H4
Davos, Switzerland 39/J3
Davos (valley), Switzerland 39/J3
Davy, W. Va. (24828) 312/C8
Dawa (riv.), Ethiopia 111/G7
Dawasir, Hadhb (range), Saudi Arabia 59/D5
Dawasir, Wadi (dry riv.), Saudi Arabia 59/E5
Dawes (co.), Nebr. 264/A2
Dawes, W. Va. (25054) 312/D6
Dawlish, England 13/D7
Dawn, Mo. (64638) 261/E3
Dawn, Texas (79025) 303/B3
Dawna (range), Burma 72/C3
Dawson, Ala. (35963) 195/G2
Dawson, Canada 4/C16
Dawson (isl.), Chile 138/E10
Dawson (co.), Georgia 217/D2
Dawson, Georgia (31742) 217/D7
Dawson, Ill. (62520) 222/D4
Dawson, Iowa (50066) 229/E5
Dawson (bay), Manitoba 179/B2
Dawson, Minn. (56232) 255/B6
Dawson, Mo. (†65548) 261/H8
Dawson (co.), Mont. 262/M3
Dawson (co.), Nebr. 264/E4
Dawson, Nebr. (68337) 264/J4
Dawson, N. Dak. (58428) 282/L6
Dawson (inlet), N.W. Terrs. 187/J3
Dawson (riv.), Queensland 88/H4
Dawson (riv.), Queensland 95/D5
Dawson (co.), Texas 303/C5
Dawson, Texas (76639) 303/H6
Dawson, W. Va. (24932) 312/F7
Dawson, Yukon 146/E3
Dawson, Yukon 162/E3
Dawson, Yukon 187/E3
Dawson Bay, Manitoba 179/B2
Dawson Creek, Br. Col. 146/F4
Dawson Creek, Br. Col. 162/D4
Dawson Creek, Br. Col. 184/G2
Dawson Springs, Ky. (42408) 237/F6
Dawsonville, Georgia (30534) 217/D2
Dawsonville, New Bruns. 170/C1
Dawu, China 77/H5
Dawu, China 77/F5
Dax, France 28/C6
Da Xian, China 77/G5
Day, Fla. (32013) 212/C1
Day, Minn. (†55006) 255/E5
Day (co.), S. Dak. 298/O3
Day Book, N.C. (†28714) 281/E3
Daykin, Nebr. (68338) 264/G4
Daylesford, Victoria 97/C5
Daylight, Tenn. (†37110) 237/K9
Daymán, Uruguay 145/B2
Daymán (range), Uruguay 145/B2
Daymán (riv.), Uruguay 145/B2
Dayong, China 77/H6
Days Creek, Oreg. (97429) 291/D5
Daysland, Alberta 182/D3
Daysville, Ky. (†42276) 237/G7
Dayton, Ala. (36731) 195/C6
Dayton, Idaho (83232) 220/F7
Dayton, Ill. (†61350) 222/E2
Dayton, Ind. (47941) 227/D4
Dayton, Iowa (50530) 229/E4
Dayton, Ky. (41074) 237/T1
Dayton, Mich. (†49113) 250/C7
Dayton, Minn. (55327) 255/E5
Dayton, Mont. (59914) 262/B3
Dayton, Nev. (89403) 266/B3
Dayton, N.J. (08810) 273/D3
Dayton, N.Y. (4041) 276/C6
Dayton, Ohio (*45401) 284/B6
Dayton, Ohio 146/K6
Dayton, Ohio 188/K3
Dayton, Oreg. (97114) 291/A3
Dayton, Pa. (16222) 294/D4
Dayton, Tenn. (37321) 237/L9
Dayton, Texas (77535) 303/J7
Dayton, Va. (22821) 307/L4
Dayton, Wash. (99328) 310/H4
Dayton, Wis. (†53508) 317/H10
Dayton, Wyo. (82836) 319/E1
Daytona Beach, Fla. 188/K5
Daytona Beach, Fla. 146/K7
Daytona Beach, Fla. (*32014) 212/F2
Daytona Beach Shores, Fla. (32016) 212/F2
Dayu, China 77/H6
Dayville, Conn. (06241) 210/H1
Dayville, Oreg. (97825) 291/H3
Dazey, N. Dak. (58429) 282/O5
Dazhai, China 77/H4
Dazkırı, Turkey 63/D4
De Aar, S. Africa 118/C6
Dead (lake), Fla. 212/D6
Dead (sea), Israel 65/C4
Dead (sea), Israel 59/C3
Dead (sea), Jordan 59/C3
Dead (sea), Jordan 65/C4
Dead (riv.), Maine 243/C5
Dead (riv.), Mich. 250/B2
Dead (lake), Minn. 255/C4
Dead (sea), West Bank 59/C3
Deadhorse, Alaska (†99723) 196/J1
Deadman (creek), Wash. 310/H4
Deadman (mt.), Wyo. 319/B2
Deadwood, Alberta 182/B1
Deadwood (res.), Idaho 220/C5
Deadwood (riv.), Idaho 220/C5
Deadwood, S. Dak. (57732) 298/B5
Deaf Smith (co.), Texas 303/B3
Deal, England 13/J6
Deal, England 10/G5
Deal, N.J. (07723) 273/F3
Deal (isl.), Tasmania 99/D1
Deale, Md. (20751) 245/M5
Deal Island, Md. (21821) 245/P8
Dean (chan.), Br. Col. 184/D4
Dean (riv.), Br. Col. 184/D4
Dean, Nova Scotia 168/F3
Deán Funes, Argentina 143/D3
Deanville, Texas (77852) 303/H7
Dearborn (co.), Ind. 227/H6
Dearborn, Mich. (*48120) 250/B7
Dearborn, Mo. (64439) 261/C3
Dearborn Heights, Mich. (48127) 250/B7
Dearing, Georgia (30808) 217/H4
Dearing, Kansas (67340) 232/G4
De Armanville, Ala. (36257) 195/G3
Dearne, England 13/K2
Deary, Idaho (83823) 220/B3
Dease (inlet), Alaska 196/H1
Dease (lake), Br. Col. 184/K2
Dease (riv.), Br. Col. 184/K2
Dease (str.), N.W.T. 146/G3
Dease (str.), N.W.T. 162/F2
Dease (str.), N.W. Terrs. 187/H3
Dease Arm (inlet), N.W. Terrs. 187/F3
Death (valley), Calif. 204/H7
Death Valley (depr.), Calif. 188/C3
Death Valley, Calif. (92328) 204/J7
Death Valley Junction, Calif. (92328) 204/J7
Death Valley Nat'l Mon., Calif. 204/H7
Death Valley Nat'l Mon., Nev. 266/E6
Deatsville, Ala. (36022) 195/F5
Deauville, France 28/C3
Deauville, Québec 172/E4
Deaver, Wyo. (82421) 319/D1
Deavertown, Ohio (†43731) 284/G6
De Baca (co.), N. Mex. 274/E4
Deba Habe, Nigeria 106/G6
Debar, Yugoslavia 45/E5
De Bary, Fla. (32713) 212/E3
Debden, Sask. 181/D2
Débé, Trin. & Tob. 161/B11
Debec, New Bruns. 170/C2
De Beque, Colo. (81630) 208/C4
De Berry, Texas (75639) 303/L5
Debert, Nova Scotia 168/E3
Dębica, Poland 47/E3
De Bilt, Netherlands 27/G4
Dęblin, Poland 47/E3
Deblois○, Maine (†04622) 243/H6
Dębno, Poland 47/B2
Debo (lake), Mali 106/D5
Debolt, Alberta 182/B2
De Borgia, Mont. (59830) 262/A3
Debra Birhan, Ethiopia 111/G6
Debra Markos, Ethiopia 111/G5
Debra Markos, Ethiopia 102/F3
Debra Tabor, Ethiopia 111/G5
Debrecen, Hungary 41/F3
Debrecen, Hungary 7/G4
Decatur, Ala. (*35601) 195/D1
Decatur, Ark. (72722) 202/A1
Decatur (co.), Georgia 217/C9
Decatur, Georgia (*30030) 217/K1
Decatur, Ill. 188/J3
Decatur, Ill. 146/K6
Decatur, Ill. (*62521) 222/E4
Decatur (co.), Ind. 227/G6
Decatur, Ind. (46733) 227/H3
Decatur (co.), Iowa 229/F7
Decatur, Iowa (50067) 229/F7
Decatur (co.), Kansas 232/B2
Decatur, Mich. (49045) 250/C6
Decatur, Miss. (39327) 256/F6
Decatur, Nebr. (68020) 264/H2
Decatur, Ohio (45115) 284/C8
Decatur (co.), Tenn. 237/E9
Decatur, Tenn. (37322) 237/M9
Decatur, Texas (76234) 303/G4
Decaturville, Tenn. (38329) 237/E9
Decazeville, France 28/E5
Deccan (plat.), India 68/D6
Decherd, Tenn. (37324) 237/J10
Děčín, Czech. 41/C1
Decision (cape), Alaska 196/M2
Decize, France 28/E4
Decker, Ind. (47524) 227/B7
Decker, Manitoba 179/B4
Decker, Mich. (48426) 250/F5
Decker, Mont. (59025) 262/K5
Decker Lake, Br. Col. 184/E3
Deckers, Colo. (†80135) 208/J4
Deckerville, Ark. (†72386) 202/K3
Deckerville, Mich. (48427) 250/G5
Declo, Idaho (83323) 220/E7
Decorah, Iowa (52101) 229/K2
Decota, W. Va. (†25122) 312/D6
Decoy, Ky. (41321) 237/P5
Dededo, Guam 86/K6
Dedegül Dağı (mt.), Turkey 63/D4
Dedemsvaart, Netherlands 27/J3
Dederick, Mo. (†64744) 261/D7
Dedham, Iowa (51440) 229/D5
Dedham, Maine (†04429) 243/F6
Dedham○, Maine (†04429) 243/F6
Dedham, Mass. (02026) 249/C7
Dédougou, Upper Volta 106/D6
Dedza, Malawi 115/F6
Dee (riv.), England 13/D4
Dee (riv.), England 10/E4
Dee (riv.), Ireland 17/H4
Dee (riv.), Scotland 15/D5
Dee (riv.), Scotland 15/F3
Dee (riv.), Scotland 10/E2
Dee (riv.), Tasmania 99/C4
Dee (riv.), Wales 10/E4
Dee (riv.), Wales 13/D4
Deedsville, Ind. (46921) 227/E3
Deel (riv.), Ireland 17/C3
Deel (riv.), Ireland 17/G4
Deel (riv.), Ireland 17/D7
Deele (riv.), Ireland 17/F2
Deenwood, Georgia (†31501) 217/H8
Deep (creek), Idaho 220/F7
Deep (creek), Idaho 220/B7
Deep (inlet), Newf. 166/B2
Deep (riv.), N.C. 281/K3
Deep (riv.), N. Dak. 282/J1
Deep (creek), S.C. 296/B2
Deep (creek), Texas 303/C5
Deep (creek), Utah 304/B1
Deep (creek), Utah 304/A3
Deep Bight, Newf. 166/C2
Deep Brook, Nova Scotia 168/C4
Deep Creek (mts.), Idaho 220/F7
Deep Creek (lake), Md. 245/A3
Deep Creek (range), Utah 304/A4
Deepcreek, Wash. (†99010) 310/H3
Deepdale, Manitoba 179/A3
Deep Fork, North Canadian (riv.), Okla. 288/N3
Deep Gap, N.C. (28618) 281/F2
Deephaven, Minn. (55391) 255/G5
Deeping Saint James, England 13/G5
Deep River, Conn. (06417) 210/F3
Deep River○, Conn. (06417) 210/F3
Deep River (res.), Conn. 210/F2
Deep River, Iowa (52222) 229/J5
Deep River, Ontario 177/G1
Deep River, Ontario 175/E3
Deep River, Wash. (†98638) 310/B4
Deep Run, N.C. (28525) 281/O4
Deep Springs, Calif. (†93513) 204/H6
Deepstep, Georgia (31082) 217/G4
Deep Valley, Pa. (†15352) 294/A6
Deep Water (pt.), Del. 245/S4
Deepwater, Mo. (64740) 261/E6
Deepwater, N.J. (08023) 273/C4
Deepwater, N.S. Wales 97/F1
Deer (isl.), Alaska 196/F4
Deer, Ariz. (72628) 202/D2
Deer (creek), Ind. 227/D5
Deer (creek), Ind. 227/E3
Deer (isl.), Maine 243/F7
Deer (creek), Md. 245/N2
Deer (isl.), Mass. 249/E7
Deer (riv.), Mich. 250/A2
Deer (lake), Minn. 255/E3
Deer (creek), Miss. 256/C4
Deer (isl.), New Bruns. 170/D4
Deer (harb.), Newf. 166/D2
Deer (riv.), N.Y. 276/L1
Deer (riv.), N.Y. 276/J3
Deer (creek), Ohio 284/D6
Deer (lake), Wash. 310/H2
Deerbrook, Miss. (†39739) 256/G4
Deerbrook, Wis. (54424) 317/H5
Deer Creek, Ill. (61733) 222/D3
Deer Creek, Ind. (†46917) 227/E3
Deer Creek, Minn. (56527) 255/C4
Deer Creek (lake), Ohio 284/D6
Deer Creek, Okla. (74636) 288/L1
Deerfield, Ill. (60015) 222/B5
Deerfield, Ind. (†47380) 227/H4
Deerfield, Kansas (67838) 232/A4
Deerfield○, Mass. (01342) 249/D2
Deerfield (riv.), Mass. 249/C2
Deerfield, Mich. (49238) 250/F7
Deerfield, Mo. (64741) 261/D7
Deerfield○, N.H. (03037) 268/E5
Deerfield, Ohio (44411) 284/H3
Deerfield (riv.), Vt. 268/B6
Deerfield, Va. (24432) 307/K4
Deerfield, Wis. (53531) 317/H9
Deerfield Beach, Fla. (33441) 212/F5
Deerfield Street, N.J. (08313) 273/C4
Deerford, La. (†70791) 238/K1
Deer Grove, Ill. (61243) 222/D2
Deer Harbor, Wash. (98243) 310/B2
Deerhorn, Manitoba 179/E4
Deering, Alaska (99736) 196/F1
Deering○, N.H. (†03244) 268/D5
Deering, N. Dak. (58731) 282/J3
Deer Island, Oreg. (97054) 291/E2
Deer Isle, Maine (04627) 243/F7
Deer Isle○, Maine (04627) 243/F7
Deer Lake, Newf. 166/C4
Deer Lake, Ontario 175/B2
Deer Lodge (co.), Mont. 262/C5
Deer Lodge, Mont. (59722) 262/D4
Deer Lodge, Tenn. (37726) 237/M8
Deer Park, Ala. (36529) 195/B8
Deer Park, Calif. (94576) 204/C5
Deer Park, Fla. (†32901) 212/F3
Deer Park, Ill. (†60010) 222/A5
Deer Park, Md. (21550) 245/A3
Deer Park, N.Y. (11729) 276/O9
Deer Park, Ohio (45236) 284/C9
Deer Park, Texas (77536) 303/K2
Deer Park, Wash. (99006) 310/H3
Deer Park, Wis. (54007) 317/B5
Deer River, Minn. (56636) 255/E3
Deer River, N.Y. (13627) 276/J3
Deer Run, W. Va. (†26807) 312/H5
Deersville, Ohio (44693) 284/H5
Deerton, Mich. (49822) 250/B2
Deer Trail, Colo. (80105) 208/M3
Deerwalk, W. Va. (†26180) 312/D4
Deerwood, Minn. (56444) 255/E4
Deesa, India 68/C4
Deeson, Miss. (†38740) 256/C2
Deeth, Nev. (89823) 266/F1
Dee Why, N.S. Wales 88/L4
Dee Why, N.S. Wales 97/K3
Defense Heights, Md. (†20784) 245/G4
Deferiet, N.Y. (13628) 276/J2
Defiance (plat.), Ariz. 198/F3
Defiance, Iowa (51527) 229/C5
Defiance, Mo. (63341) 261/L5
Defiance (co.), Ohio 284/A3
Defiance, Ohio (43512) 284/B3
De Fluessen (lake), Netherlands 27/G3
Defoe, Ky. (40017) 237/L4
Deford, Mich. (48729) 250/F5
De Forest, Wis. (53532) 317/H9
Defoy, Québec 172/E3
De Funiak Springs, Fla. (32433) 212/C6
Dégelis, Québec 172/J2
Degema, Nigeria 106/F8
Degersheim, Switzerland 39/H2
Deggendorf, W. Germany 22/E4
De Graff, Kansas (†66840) 232/F4
De Graff, Minn. (56233) 255/C5
Degraff, Ohio (43318) 284/C5
Degrasse, N.Y. (13629) 276/L2
De Gray (lake), Ark. 202/D5
De Grey, W. Australia 88/B4
De Grey, W. Australia 92/B3
De Grey (riv.), W. Australia 88/C4
De Grey (riv.), W. Australia 92/B3
De Haan, Belgium 27/C6
Deh Bid, Iran 66/H5
Dehdez, Iran 66/G5
Deheq, Iran 66/G4
Dehiwala-Mt. Lavinia, Sri Lanka 68/D7
Dehkhvareqan, Iran 66/D2
Dehlco, La. (†71269) 238/G2
De Honte (bay), Netherlands 27/D6
Dehra Dun, India 68/D2
Dehua, China 77/J6
Deim Zubeir, Sudan 111/E6
Deinze, Belgium 27/C7
Deir Abu Sa'id, Jordan 65/D3
Deir Ballut, West Bank 65/C3
Deir el Balah, Gaza Strip 65/A5
Deir ez Zor (prov.), Syria 63/H5
Deir ez Zor, Syria 63/H5
Deir ez Zor, Syria 59/C2
Deir Sharaf, West Bank 65/C3
Dej, Romania 45/F2
De Kalb (co.), Ala. 195/G2
De Kalb (co.), Georgia 217/D3
De Kalb (co.), Ill. 222/E2
De Kalb, Ill. (60115) 222/E2
De Kalb (co.), Ind. 227/H2
De Kalb, Miss. (39328) 256/G5
De Kalb (co.), Mo. 261/D3
De Kalb, Mo. (64440) 261/C3
De Kalb (co.), Tenn. 237/K9
De Kalb, Texas (75559) 303/K4
De Kalb Junction, N.Y. (13630) 276/K2
Dekese, Zaire 115/D4
Dekoa, Cent. Afr. Rep. 115/C2
De Koog, Netherlands 27/F2
De Koven, Ky. (†42459) 237/E5
Dela, Okla. (†74523) 288/P6
Delacour, Alberta 182/D4
Delacroix, La. (†70085) 238/L7
Delafield, Ill. (†62859) 222/E5
Delafield, Wis. (53018) 317/J1
Delagoa (bay), Mozambique 118/E5
Delair, N.J. (08110) 273/B3
Delamar (mts.), Nev. 266/G5
De Lamere, N. Dak. (58022) 282/R7
DeLancey, Pa. (15733) 294/D4
Delanco○, N.J. (08075) 273/D3
De Land, Fla. (32720) 212/E2
De Land, Ill. (61839) 222/E3
Delaney, Ark. (†72727) 202/C2
Delano, Calif. (93215) 204/F8
Delano, Minn. (55328) 255/E5
Delano, Pa. (18220) 294/K4
Delano, Tenn. (37325) 237/M10
Delano (peak), Utah 304/B5
Delanson, N.Y. (12053) 276/M5
Delaplaine, Ark. (72425) 202/J1
Delaplane, Va. (22025) 307/N3
Delaram, Afghanistan 59/H3
Delaram, Afghanistan 68/A2
Delaronde (lake), Sask. 181/E1
Delavan, Ill. (61734) 222/D3
Delavan, Kansas (66847) 232/F3
Delavan, Minn. (56023) 255/D7
Delavan, Wis. (53115) 317/J10
Delavan Lake, Wis. (†53115) 317/J10
Delaware 188/L3
Delaware (bay) 188/M3
Delaware, Ark. (72835) 202/D3
DELAWARE 245
Delaware (bay), Del. 245/T5
Delaware (riv.), Del. 245/R3
Delaware (co.), Ind. 227/G4
Delaware, Ind. (†47037) 227/G6
Delaware (co.), Iowa 229/L4
Delaware, Iowa (52036) 229/L4
Delaware, N.J. (07833) 273/C2
Delaware (bay), N.J. 273/C5
Delaware (riv.), N.J. 273/D3
Delaware (co.), N.Y. 276/K6
Delaware (riv.), N.Y. 276/K7
Delaware (co.), Ohio 284/D5
Delaware, Ohio (43015) 284/E5
Delaware (lake), Ohio 284/E5
Delaware (co.), Okla. 288/S2
Delaware, Okla. (74027) 288/P1
Delaware, Ontario 177/C5
Delaware (co.), Pa. 294/M6
Delaware (riv.), Pa. 294/N3
Delaware (creek), Texas 303/C10
Delaware (mts.), Texas 303/C10
Delaware (state), U.S. 146/L6
Delaware City, Del. (19706) 245/R2
Delaware Water Gap Nat'l Rec. Area, N.J. 273/C1
Delaware Water Gap, Pa. (18327) 294/M4
Delaware Water Gap Nat'l Rec. Area, Pa. 294/N3
Delbarton, W. Va. (25670) 312/B7
Del Bonita, Alberta 182/D5
Delburne, Alberta 182/D3
Delcambre, La. (70528) 238/G7
Del City, Okla. (73115) 288/L4
Delco, N.C. (28436) 281/N6
Deldoul, Algeria 106/E3
Deleau, Manitoba 179/B5
Delegate, N.S. Wales 97/E5
Delémont, Switzerland 39/D2
De Leon, Texas (76444) 303/F5
De Leon Springs, Fla. (32028) 212/E2
Delevan, N.Y. (14042) 276/D6
Delfí, Greece 45/F6
Delft, Minn. (56124) 255/C7
Delft, Netherlands 27/E4
Delfzijl, Netherlands 27/K2
Delgada (pt.), Argentina 143/D5
Delgada (pt.), Calif. 204/A3
Delgada (pt.), Mexico 150/L7
Delgado (cape), Mozambique 102/G6
Delgado (cape), Mozambique 118/G2
Delgado Chalbaud, Cerro (mt.), Venezuela 124/G6
Delgertsogt, Mongolia 77/G2
Delgo, Sudan 111/F3
Delhi, Calif. (95315) 204/E6
Delhi, Colo. (†81059) 208/M7
Delhi, Ill. (†62052) 222/C4
Delhi (terr.), India 68/D3
Delhi, India 68/D3
Delhi, India 54/J7
Delhi, India 2/N4
Delhi, Iowa (52223) 229/L4
Delhi, La. (71232) 238/H2
Delhi, Minn. (56234) 255/C6
Delhi, N.Y. (13753) 276/L6
Delhi, Okla. (†73662) 288/G4
Delhi, Ontario 177/D5
Delia, Alberta 182/D4
Delia, Kansas (66418) 232/G2
Delice, Dominica 161/F7
Delice, Turkey 63/E3
Delice (riv.), Turkey 63/F3
Délices, Fr. Guiana 131/E3
Delicias, Cuba 158/H3
Delicias, Venezuela 124/B4
Delight, Ark. (71940) 202/C5
Delijan, Iran 66/G4
Delingha, China 77/E4
De Lisle, Miss. (†39571) 256/F10
Delisle, Québec 172/F1
Delisle, Sask. 181/D4
Delitzsch, E. Germany 22/E3
Dell, Ark. (72426) 202/K2
Dell, Mont. (59724) 262/D6
Dell City, Texas (79837) 303/C10
Dell Rapids, S. Dak. (57022) 298/R6
Dellrose, Tenn. (38453) 237/H10
Dellroy, Ohio (44620) 284/H4
Dells, The (valley), Wis. 317/G8
Dellslow, W. Va. (26531) 312/G3
Dellwood, Minn. (†55110) 255/F5
Dellwood, Mo. (†63135) 261/R2
Dellwood, N.C. (†28786) 281/C3
Dellwood, Wis. (53927) 317/G7
Dellys, Algeria 106/E1
Delmar, Ala. (35551) 195/C2
Del Mar, Calif. (92014) 204/H11
Delmar, Del. (19940) 245/R7
Delmar, Iowa (52037) 229/M4
Delmar, Md. (21875) 245/R7
Delmar, N.Y. (12054) 276/N5
Delmas, Sask. 181/C3
Delmas, S. Africa 118/J6
Delmenhorst, W. Germany 22/C2
Delmont, N.J. (08314) 273/C5
Delmont, Pa. (15626) 294/D5
Delmont, S. Dak. (57330) 298/N7
Del Norte (co.), Calif. 204/B2
Del Norte, Colo. (81132) 208/G7
Del Norte (peak), Colo. 208/F7
Deloit, Iowa (51441) 229/C4
DeLong (mts.), Alaska 196/F1
De Long, Ill. (†61436) 222/C3
Delong, Ind. (46922) 227/E2
Deloraine, Man. 162/G6
Deloraine, Manitoba 179/B5
Deloraine, Tasmania 99/C3
Delorme (lake), Québec 174/C2
Deloro, Ontario 177/G3
Delphi, Ind. (46923) 227/D3
Delphia, Ky. (41735) 237/P6
Delphos, Iowa (50844) 229/E7
Delphos, Kansas (67436) 232/E2
Delphos, Ohio (45833) 284/B4
Delpine, Mont. (†59053) 262/F4
Delran○, N.J. (08075) 273/B3
Delray Beach, Fla. (*33444) 212/F5
Del Rey Oaks, Calif. (93940) 204/D7
Del Rio, Tenn. (37727) 237/P9
Del Rio, Texas 188/F5
Del Rio, Texas 146/H7
Del Rio, Texas (78840) 303/D8
Del Rosa, Calif. (92404) 204/E10
Delson, Québec 172/H4
Delta, Ala. (36258) 195/G4
Delta, Br. Col. 184/K3
Delta (co.), Colo. 208/D5
Delta, Colo. (81416) 208/D5
Delta, Iowa (52550) 229/J6
Delta, La. (71233) 238/J2
Delta, Manitoba 179/D4
Delta (co.), Mich. 250/C2
Delta, Mo. (63744) 261/N8
Delta, Ohio (43515) 284/B2
Delta, Ontario 177/H3
Delta, Pa. (17314) 294/K6
Delta (co.), Texas 303/J4
Delta, Utah (84624) 304/B4
Delta, Wis. (†54856) 317/D3
Delta Amacuro (terr.), Venezuela 124/H3
Delta City, Miss. (39061) 256/C4
Delta Junction, Alaska (99737) 196/J2
Deltaville, Va. (23043) 307/R5
Delton, Mich. (49046) 250/D6
Deltona, Fla. (32725) 212/E3
Delungra, N.S. Wales 97/F1
Del Valle, Argentina 143/F7
Del Valle (lake), Calif. 204/L3
Delvin, Ireland 17/G4
Delvinákion, Greece 45/E6
Delvinë, Albania 45/D6
Delwin, Mich. (†48858) 250/E5
Demaine, Sask. 181/D5
Demak, Indonesia 85/J2
Demanda, Sierra de la (range), Spain 33/E1
Demarcation (pt.), Alaska 196/K1
Demarest, N.J. (07627) 273/C1
Demavend (Damavend) (mt.), Iran 66/G3
Demba, Zaire 115/D5
Dembidollo, Ethiopia 111/F6
Demchok, India 68/D2
Demerara (riv.), Guyana 131/B3
Demidov, U.S.S.R. 52/D3
Deming, N. Mex. (88030) 274/B6
Deming, Wash. (98244) 310/C2
Demini (riv.), Brazil 132/H8
Demirci, Turkey 63/C3
Demirkent, Turkey 63/E4
Demirköy, Turkey 63/B2
Demir Qapu, Syria 63/J4
Demmin, E. Germany 22/E2
Democracia, Venezuela 124/E6
Demopolis, Ala. (36732) 195/C6
Demopolis (dam), Ala. 195/C5
Demopolis (lake), Ala. 195/C5
Demorest, Georgia (30535) 217/F1
De Mossville, Ky. (41033) 237/N3
Demotte, Ind. (46310) 227/C2
Dempo (mt.), Indonesia 85/C6
Dempster, S. Dak. (57230) 298/R4
Demster, N.Y. (†13126) 276/H3
Demta, Indonesia 85/L6
Denain, France 28/E2
Denali, Alaska (†99729) 196/J2
Denali Nat'l Park, Alaska 196/H2
Denali Nat'l Preserve, Alaska 196/H2
Denare Beach, Sask. 181/M4
Denau, U.S.S.R. 48/G6
Denbigh (cape), Alaska 196/F2
Denbigh, N. Dak. (58732) 282/J3
Denbigh, Ontario 177/G2
Denbigh, Wales 13/D4
Denbigh, Wales 10/E4
Den Burg, Netherlands 27/F2
Denby, S. Dak. (57733) 298/E7
Denby Dale, England 13/J2
Den Chai, Thailand 72/C3
Dender (riv.), Belgium 27/D7
Denderleeuw, Belgium 27/E7
Dendermonde, Belgium 27/E6
Dendron, Va. (23839) 307/P6
Denekamp, Netherlands 27/L4
Denezhkin Kamen' (mt.), U.S.S.R. 52/J2
Dengkou, China 77/G3
Dêngqên, China 77/E5
Denham, Ind. (46925) 227/D2
Denham, Minn. (55728) 255/F4
Denham, W. Australia 92/A4
Denham Springs, La. (70726) 238/L2
Den Helder, Netherlands 27/F3
Denhoff, N. Dak. (58430) 282/K5
Denholm, Sask. 181/C3
Denholm, Scotland 15/F5
Denia, Spain 33/G3
Deniliquin, N.S. Wales 88/G7
Deniliquin, N.S. Wales 97/C4
Denio, Nev. (89404) 266/C1
Denison, Iowa (51442) 229/C4
Denison, Kansas (66419) 232/G2
Denison (dam), Okla. 288/O7
Denison (range), Tasmania 99/C4
Denison, Texas 188/G4
Denison, Texas (75020) 303/H4
Denison (dam), Texas 303/H4
Denison, Wash. (†99006) 310/H3
Denizli (prov.), Turkey 63/C4
Denizli, Turkey 63/C4
Denizli, Turkey 59/A2
Denman, N.S. Wales 97/F3
Denman Island, Br. Col. 184/H2
Denmark 2/K3
Denmark 7/E3
Denmark (strait) 4/C11
Denmark (str.) 146/S3
Denmark (str.) 7/B2
DENMARK 18/D9
DENMARK 21/E6
Denmark, Iowa (52624) 229/L7
Denmark, Kansas (†67455) 232/D2
Denmark○, Maine (04022) 243/B8
Denmark, Miss. (†38655) 256/F2
Denmark (bay), N.W. Terrs. 187/H2
Denmark, Oreg. (†97450) 291/C5
Denmark, S.C. (29042) 296/E5
Denmark, Tenn. (38391) 237/D9
Denmark, W. Australia 88/B7
Denmark, W. Australia 92/B6
Denmark, Wis. (54208) 317/L7
Dennard, Ark. (72629) 202/E2
Dennehotso, Ariz. (86535) 198/F2
Dennery, St. Lucia 161/G6
Denning, Ark. (†72821) 202/C3
Dennis (hill), Conn. 210/C1
Dennis, Kansas (67341) 232/G4
Dennis○, Mass. (02638) 249/O5
Dennis, Miss. (38838) 256/H1
Dennis (head), Scotland 15/F1
Dennison, Minn. (55018) 255/E6
Dennison, Ohio (44621) 284/H5
Dennis Port, Mass. (02639) 249/O6
Denniston, Va. (†24520) 307/L7
Dennisville, N.J. (08214) 273/D5
Dennisville, Sask. 181/K6
Denny and Dunipace, Scotland 10/B1
Denny and Dunipace, Scotland 15/C1
Dennysville○, Maine (04628) 243/J6
Den Oever, Netherlands 27/G3
Denoon (lake), Wis. 317/K2
Denpasar, Indonesia 85/E7
Densmore, Kansas (67633) 232/C2
Dent, Minn. (56528) 255/C4
Dent (co.), Mo. 261/J7
Dent, Ohio (†45202) 284/B9
Dent Blanche (mt.), Switzerland 39/E4
Dent de Lys (mt.), Switzerland 39/D4
Dent de Ruth (mt.), Switzerland 39/D3
Dent d'Hérens (mt.), Switzerland 39/E5

Denton, England 13/H2
Denton, Georgia (31532) 217/G7
Denton, Kansas (66017) 232/G2
Denton, Ky. (41132) 237/R4
Denton, Md. (21629) 245/P5
Denton, Mo. (†63877) 261/N10
Denton, Mont. (59430) 262/G3
Denton, Nebr. (68339) 264/H4
Denton, N.C. (27239) 281/J3
Denton (co.), Texas 303/G4
Denton, Texas 188/G4
Denton, Texas (76201) 303/G4
D'Entrecasteaux (isls.), Papua N.G. 87/F6
D'Entrecasteaux (isls.), Papua N.G. 85/C7
D'Entrecasteaux (chan.), Tasmania 99/D5
D'Entrecasteaux (pt.), W. Australia 88/B7
D'Entrecasteaux (pt.), W. Australia 92/A6
Dents du Midi (mt.), Switzerland 39/C4
Dents Run, Pa. (†15832) 294/F3
Dentville, Miss. (†39059) 256/C7
Denver, Ark. (†72632) 202/D1
Denver (co.), Colo. 208/K3
Denver (cap.), Colo. (*80201) 208/K3
Denver (cap.), Colo. 188/F3
Denver (cap.), Colo. 146/H6
Denver, Ill. (†62321) 222/B3
Denver, Ind. (46926) 227/E3
Denver, Iowa (50622) 229/J3
Denver, Mo. (64441) 261/D2
Denver, N.C. (28037) 281/G3
Denver, Pa. (17517) 294/K5
Denver, Tenn. (37054) 237/F8
Denver, U.S. 2/D3
Denver City, Texas (79323) 303/B4
Denville○, N.J. (07834) 273/E2
Denzil, Sask. 181/B3
Deogarh, India 68/E4
Deoghar, India 68/F4
Deolali, India 68/C5
Deora, Colo. (†81054) 208/O7
Deoria, India 68/E3
De Panne, Belgium 27/B6
Depauville, N.Y. (13632) 276/H2
Depauw, Ind. (47115) 227/E8
De Peel (reg.), Netherlands 27/H6
Dependencias Federales (terr.), Venezuela 124/E2
De Pere, Wis. (54115) 317/K7
Depew, N.Y. (14043) 276/C5
Depew, Okla. (74028) 288/O3
De Peyster, N.Y. (13633) 276/K1
Depoe Bay, Oreg. (97341) 291/C3
Deport, Texas (75435) 303/J4
Deposit, N.Y. (13754) 276/K6
Dépôt Lézard, Fr. Guiana 131/E3
Depoy, Ky. (42336) 237/G6
Deptford○, N.J. (08096) 273/B4
Depue, Ill. (61322) 222/D2
Deputy, Ind. (47230) 227/F7
Dêqên, China 77/E6
De Queen, Ark. (71832) 202/B5
De Quincy, La. (70633) 238/D6
Der'a (prov.), Syria 63/G6
Der'a, Syria 63/G6
Dera Bugti, Pakistan 68/B3
Dera Bugti, Pakistan 59/J4
Dérac, Haiti 158/C5
Dera Ghazi Khan, Pakistan 68/C3
Dera Ghazi Khan, Pakistan 59/J4
Dera Ismail Khan, Pakistan 59/K3
Dera Ismail Khan, Pakistan 68/C2
Derbent, U.S.S.R. 7/J4
Derbent, U.S.S.R. 52/G6
Derby, Australia 87/C7
Derby, Conn. (06418) 210/C3
Derby, England 13/F5
Derby, England 10/F4
Derby, Ind. (47525) 227/D8
Derby, Iowa (50068) 229/G7
Derby, Kansas (67037) 232/E4
Derby, Maine (04425) 243/E5
Derby, Miss. (†39470) 256/E9
Derby, N.Y. (14047) 276/B5
Derby, Ohio (43117) 284/D6
Derby, Tasmania 99/D3
Derby, Texas (†78017) 303/E9
Derby (Derby Center), Vt. (05829) 268/C2
Derby○, Vt. (05829) 268/C2
Derby, W. Australia 88/C3
Derby, W. Australia 92/C2
Derby Line, Vt. (05830) 268/C2
Derbyshire (co.), England 13/F5
Derecske, Hungary 41/F3
Dereli, Turkey 63/H2
Derendingen, Switzerland 39/E2
Derg (lake), Ireland 17/E6
Derg (lake), Ireland 17/F2
Derg, Lough (lake), Ireland 10/B4
Derg (riv.), N. Ireland 17/F2
De Ridder, La. (70634) 238/D5
Derik, Turkey 63/J4
Dering Harbor, N.Y. (†11964) 276/R8
Derinkuyu, Turkey 63/F3
Derj, Libya 111/B1
Derma, Miss. (38839) 256/F3
Dermott, Ark. (71638) 202/H7
Derna, Libya 102/E1
Derna, Libya 111/D1
Dernic, Sask. 181/J4
Dernieres (isls.), La. 238/J8
Deroche, Br. Col. 184/L3
Deronda, Wis. (54008) 317/B5
De Rossett, Tenn. (†38583) 237/L9
Déroute (passage), Chan. Is. 13/F8
Derravaragh (lake), Ireland 17/G4
Derrinallum, Victoria 97/B5
Derry, La. (71421) 238/E3
Derry, N.H. (03038) 268/E6
Derry○, N.H. (03038) 268/E6
Derry, N. Mex. (87933) 274/B6
Derry, Pa. (15627) 294/D5
Derrygonnelly, N. Ireland 17/F3
Derryveagh (mts.), Ireland 17/E2
Derudeb, Sudan 111/G4
De Ruyter, N.Y. (13052) 276/J5
Dervaig, Scotland 15/B4
Derventa, Yugoslavia 45/C3
Dervock, N. Ireland 17/J1
Derwent, Alberta 182/E3
Derwent (riv.), England 13/H3
Derwent (riv.), England 13/G3
Derwent (riv.), England 10/F3
Derwent (riv.), Tasmania 99/C4
Derwent Bridge, Tasmania 99/C4
Derwood, Md. (20855) 245/K4
Desaguadero (riv.), Argentina 143/C3
Desaguadero, Bolivia 136/A5
Desaguadero (riv.), Bolivia 136/B5
Desaguadero, Peru 128/H11
De Salis (bay), N.W. Terrs. 187/F2
Des Allemands, La. (70030) 238/N4
Des Arc, Ark. (72040) 202/G4
Des Arc (bayou), Ark. 202/G3
Des Arc, Mo. (63636) 261/L8
Desatoya (mts.), Nev. 266/D3
Desbiens, Québec 172/E1
Des Bois (lake), N.W. Terrs. 187/F3
Desboro, Ontario 177/C3
Desborough, England 13/G5
Descanso, Calif. (92016) 204/J11
Deschaillons-sur-Saint-Laurent, Québec 172/E3
Deschambault, Québec 172/E3
Deschambault Lake, Sask. 181/M3
Descharme Lake, Sask. 181/L3
Deschênes, Québec 172/B4
Deschênes (lake), Québec 172/A4
Deschutes (co.), Oreg. 291/F4
Deschutes (riv.), Oreg. 291/G2
Deschutes (riv.), Wash. 310/C4
Deseado (riv.), Argentina 120/C7
Deseado (riv.), Argentina 143/C6
Deseado (cape), Chile 138/D10
Desemboque Seris, Mexico 150/C2
Desengaño (pt.), Argentina 143/C6
Desenzano del Garda, Italy 34/C2
Deseret, Utah (†84624) 304/B4
Deseronto, Ontario 177/G3
Desert (range), Nev. 266/F6
Desert (valley), Nev. 266/C1
Deserta Grande (isl.), Portugal 33/B2
Desertas (isl.), Portugal 102/A1
Desertas (isls.), Portugal 106/A2
Desertas (isls.), Portugal 33/A2
Desert Center, Calif. (92239) 204/K10
Desert Hot Springs, Calif. (92240) 204/J9
Desert View Highlands, Calif. (†93550) 204/G9
Desha (co.), Ark. 202/H6
Deshaies, Guadeloupe 161/A6
Deshler, Nebr. (68340) 264/G4
Deshler, Ohio (43516) 284/C3
Désirade, La (isl.), Guadeloupe 161/B6
Des Lacs, N. Dak. (58733) 282/G3
Des Lacs (riv.), N. Dak. 282/G3
Desloge, Mo. (63601) 261/M7
Desmarais, Alberta 182/D2
Desmaraisville, Québec 174/B3
Desmet, Idaho (83824) 220/B2
De Smet, S. Dak. (57231) 298/O5
Desmochados, Paraguay 144/C5
Des Moines (riv.) 188/H2
Des Moines (co.), Iowa 229/L7
Des Moines (cap.), Iowa 146/J5
Des Moines (cap.), Iowa 188/H2
Des Moines (cap.), Iowa (*50301) 229/G5
Des Moines (riv.), Iowa 229/J7
Des Moines (riv.), Minn. 255/C7
Des Moines (riv.), Mo. 261/J1
Des Moines, N. Mex. (88418) 274/F2
Des Moines, Wash. (98188) 310/B2
Desna (riv.), U.S.S.R. 52/D4
Desolación (isl.), Chile 120/B8
Desolación (isl.), Chile 138/D10
Desolation (canyon), Utah 304/E4
De Soto (co.), Fla. 212/E4
De Soto, Georgia (31743) 217/D7
De Soto, Ill. (62924) 222/D6
Desoto, Ind. (47302) 227/G4
De Soto, Iowa (50069) 229/E5
De Soto, Kansas (66018) 232/H3
De Soto (par.), La. 238/C2
De Soto (co.), Miss. 256/E1
De Soto, Miss. (†39360) 256/G7
De Soto, Mo. (63020) 261/L6
De Soto, Texas (75115) 303/G3
De Soto, Wis. (†54663) 317/D9
De Soto Nat'l Mem., Fla. 212/D4
Des Peres, Mo. (63131) 261/O3
Des Plaines, Ill. (*60016) 222/B5
Des Plaines (riv.), Ill. 222/A6
Dessa, Niger 106/E6
Dessalines, Haiti 158/C5
Dessau, E. Germany 22/E3
Dessel, Belgium 27/G6
Dessye, Ethiopia 102/G3
Dessye, Ethiopia 111/G5
Destelbergen, Belgium 27/D6
Destin, Fla. (32541) 212/C6
Destrehan, La. (70047) 238/N4
Destruction (isl.), Wash. 310/A3
Destruction Bay, Yukon 187/E3
Deta, Romania 45/E3
Detah, N.W. Terrs. 187/G3
Detlor, Ontario 177/G2
Detmold, W. Germany 22/C3
De Tour (passage), Mich. 250/E3
Detour (pt.), Mich. 250/C3
De Tour Village, Mich. (49725) 250/E3
Detrital Wash (dry riv.), Ariz. 198/A3
Detroit, Ala. (35552) 195/B2
Detroit, Ill. (62332) 222/C4
Detroit, Kansas (†67410) 232/E3
Detroit○, Maine (04929) 243/E6
Detroit, Mich. 146/K5
Detroit, Mich. 188/K2
Detroit, Mich. (*48201) 250/B7
Detroit (riv.), Mich. 250/B7
Detroit, Oreg. (97342) 291/E3
Detroit (lake), Oreg. 291/E3
Detroit, Texas (75436) 303/J4
Detroit, U.S. 2/E3
Detroit Beach, Mich. (†48161) 250/F7
Detroit Lakes, Minn. (56501) 255/C4
Dett, Zimbabwe 118/D3
Detva, Czech. 41/E2
De Twente (reg.), Netherlands 27/K4
Deuel (co.), Nebr. 264/B3
Deuel (co.), S. Dak. 298/R4
Deûle (riv.), Belgium 27/B7
Deurne, Belgium 27/F6
Deurne, Netherlands 27/H6
Deustua, Peru 128/G10
Deutsch Feistritz, Austria 41/C3
Deutschkreutz, Austria 41/D3
Deutsch Landsberg, Austria 41/C3
Deutsch Wagram, Austria 41/D2
Deux Frères, Les (isls.), Vietnam 72/E5
Deux-Montagnes (co.), Québec 172/C4
Deux-Montagnes, Québec 172/H4
Deux Montagnes (lake), Québec 172/C4
Deux Rivières, Ontario 177/F1
Deux-Sèvres (dept.), France 28/C4
Deva, Romania 45/F3
De Valls Bluff, Ark. (72041) 202/H4
Devault, Pa. (19432) 294/L5
Dévaványa, Hungary 41/F3
Devecser, Hungary 41/D3
Develi, Turkey 59/C2
Develi, Turkey 63/F3
Deventer, Netherlands 27/J4
Devenyns (lake), Québec 172/D2
Devereux, Georgia (†31087) 217/F4
Deveron (riv.), Scotland 15/F3
De View (bayou), Ark. 202/J3
Deville, La. (71328) 238/F4
Devil River (peak), N. Zealand 100/D4
Devils (isl.), Fr. Guiana 120/D2
Devil's (isl.), Fr. Guiana 131/E3
Devils (lake), N. Dak. 282/N3
Devils (riv.), Texas 303/D7
Devilsbit (mt.), Ireland 17/F6
Devil's Hole (Death Valley Nat'l Mon.), Nev. 266/E6
Devils Lake, N. Dak. 188/G1
Devils Lake, N. Dak. (58301) 282/N3
Devils Paw (mt.), Alaska 196/N1
Devils Postpile Nat'l Mon., Calif. 204/F6
Devils Slide, Utah (†84050) 304/C2
Devils Thumb (mt.), Br. Col. 184/A1
Devils Tower, Wyo. (82714) 319/H1
Devils Tower Nat'l Mon., Wyo. 319/H1
Devin, Bulgaria 45/G5
Devine, Texas (78016) 303/E8
Devizes, England 13/F6
Devizes, England 10/F5
Devol, Okla. (73531) 288/J6
Devola, Ohio (45750) 284/H7
De Volet (pt.), St. Vin. & Grens. 161/A8
Devon, Alberta 182/D3
Devon (isl.), Canada 4/B14
Devon, Conn. (†06460) 210/C4
Devon (co.), England 13/D7
Devon, Jamaica 158/H6
Devon, Kansas (†66701) 232/H4
Devon, Mont. (†59474) 262/E2
Devon (isl.), N.W.T. 162/M3
Devon (isl.), N.W.T. 146/K2
Devon (isl.), N.W. Terrs. 187/K2
Devondale, Ky. (†40201) 237/K2
Devonia, Tenn. (†37710) 237/N8
Devonport, Australia 87/E10
Devonport, N. Zealand 100/C1
Devonport, Tasmania 88/H8
Devonport, Tasmania 99/C3
Devrek, Turkey 63/D2
Devrekâni, Turkey 63/E2
Devrez (riv.), Turkey 63/E2
Dewar, Iowa (50623) 229/J3
Dewar, Okla. (74431) 288/P4
Dewart, Pa. (17730) 294/J3
Dewas, India 68/D4
Dewberry, Alberta 182/E3
Dewees (isl.), S.C. 296/H6
De Weese (plat.), Colo. 208/J6
Deweese, Nebr. (68934) 264/F4
Dewey, Ariz. (86327) 198/C4
Dewey, Ill. (61840) 222/E3
Dewey (lake), Ky. 237/R5
Dewey (co.), Okla. 288/H2
Dewey, Okla. (74029) 288/P1
Dewey (Culebra), P. Rico 161/G1
Dewey (co.), S. Dak. 298/G3
Dewey Park, Ill. (†62025) 222/B2
Deweyville, Texas (77614) 303/L7
Deweyville, Utah (84309) 304/B2
De Wijk, Netherlands 27/J3
De Winton, Alberta 182/C4
De Witt, Ark. (72042) 202/H5
De Witt (co.), Ill. 222/E3
Dewitt, Ill. (61735) 222/E3
De Witt, Iowa (52742) 229/N5
Dewitt, Ky. (†40935) 237/O7
De Witt, Mich. (48820) 250/E6
De Witt, Mo. (64639) 261/F4
De Witt, Nebr. (68341) 264/G4
DeWitt, N.Y. (13214) 276/H4
De Witt (co.), Texas 303/G9
Dewitt, Va. (23840) 307/N6
Dewittville, Québec 172/C4
Dewsbury, England 10/H2
Dewsbury, England 13/J1
Dewy Rose, Georgia (30634) 217/G2
Dexter (lake), Fla. 212/E2
Dexter, Georgia (31019) 217/G6
Dexter, Iowa (50070) 229/E5
Dexter, Kansas (67038) 232/F4
Dexter, Ky. (42036) 237/E7
Dexter, Maine (04930) 243/E5
Dexter○, Maine (04930) 243/E5
Dexter, Mich. (48130) 250/F6
Dexter, Minn. (55926) 255/F7
Dexter, Mo. (63841) 261/N9
Dexter, N. Mex. (88230) 274/E5
Dexter, N.Y. (13634) 276/H2
Dexter, Oreg. (97431) 291/E4
Dexter City, Ohio (45727) 284/G6
Dexterville, Wis. (†54466) 317/F7
Deyang, China 77/F5
Dey Dey (lake), S. Australia 88/E5
Dey Dey (lake), S. Australia 94/B3
DeYoung, Pa. (16728) 294/E2
Dez (riv.), Iran 59/E3
Dez (riv.), Iran 66/F4
De Zaan (riv.), Netherlands 27/B4
Dezful, Iran 59/E3
Dezful, Iran 54/F6
Dezful, Iran 66/F4
Dezhnev (cape), U.S.S.R. 4/C18
Dezhnev (cape), U.S.S.R. 48/T3
Dezhou (Tehchow), China 77/J4
Dezh Shahpur, Iran 66/E3
Dezh Shahpur, Iran 59/E2
Dhaba, Saudi Arabia 59/C4
Dhahiriya, West Bank 65/B5
Dhahran, Saudi Arabia 54/F7
Dhahran, Saudi Arabia 59/E4
Dhali, Cyprus 63/E5
Dhamar, Yemen Arab Rep. 59/D7
Dhamtari, India 68/E4
Dhanbad, India 68/F4
Dhangarhi, Nepal 68/E3
D'Hanis, Texas (78850) 303/E8
Dhank, Oman 59/G5
Dhankuta, Nepal 68/F3
Dhar, India 68/C4
Dharma, Saudi Arabia 59/E5
Dharmsala, India 68/D2
Dharwar-Hubli, India 68/C5
Dhaulagiri (mt.), Nepal 68/E3
Dhenkanal, India 68/E4
Dhidhimótikhon, Greece 45/H5
Dhikaia, Greece 45/H5
Dhimitsána, Greece 45/F7
Dhi Qar (gov.), Iraq 66/E5
Dhira', Jordan 65/D5
Dhofar (reg.), Oman 59/F6
Dholpur, India 68/D3
Dhomokós, Greece 45/F6
Dhond, India 68/C5
Dhoraji, India 68/C4
Dhubri, India 68/G3
Dhulia, India 68/C4
Día (isl.), Greece 45/G8
Diable (pt.), Martinique 161/D5
Diablerets (mt.), Switzerland 39/D4
Diablo (canyon), Ariz. 198/D4
Diablo, Calif. (94528) 204/K2
Diablo, Sierra (mts.), Texas 303/C10
Diablo, Wash. (†98283) 310/D2
Diablo (dam), Wash. 310/D2
Diablo (lake), Wash. 310/D2
Diablotin, Morne (mt.), Dominica 161/E6
Diadema, Brazil 135/C3
Diagonal, Iowa (50845) 229/E7
Dial, Georgia (30536) 217/D1
Diamant, Rocher du (isl.), Martinique 161/C7
Diamante, Argentina 143/F6
Diamante (riv.), Argentina 143/C3
Diamantina, Brazil 132/F7
Diamantina (riv.), Queensland 88/G4
Diamantina (riv.), Queensland 95/B4
Diamantina Lakes, Queensland 95/B4
Diamantino, Brazil 132/B6
Diamond (lake), Conn. 210/F2
Diamond (head), Hawaii 218/C5
Diamond (peak), Idaho 220/E5
Diamond, Ind. (†47874) 227/C5
Diamond, La. (†70083) 238/L7
Diamond, Mo. (64840) 261/D9
Diamond, Ohio (44412) 284/H3
Diamond, Oreg. (97722) 291/J4
Diamond (lake), Oreg. 291/E4
Diamond (peak), Oreg. 291/E4
Diamond, Pa. (†16354) 294/C2
Diamond, Virgin Is. (U.S.) 161/F4
Diamond, Wash. (†99111) 310/H4
Diamond (lake), Wash. 310/H2
Diamond Bluff, Wis. (†54014) 317/A6
Diamond City, Alberta 182/D5
Diamond City, Ark. (72644) 202/E1
Diamond Coast (reg.), Namibia 118/A5
Diamond Lake, Oreg. (97731) 291/E4
Diamond Point, N.Y. (12824) 276/N4
Diamond Springs, Calif. (95619) 204/D8
Diamondville, Wyo. (83116) 319/B4
Diana, W. Va. (26217) 312/F5
Dian Chi (lake), China 77/F7
Dianjiang, China 77/G5
Diano Marina, Italy 34/B3
Dianópolis, Brazil 132/E5
Diapaga, Upper Volta 106/E6
Dias Creek, N.J. (†08210) 273/D5
Díaz, Argentina 143/F6
Diaz, Ark. (72043) 202/H2
Dibaya, Zaire 115/D5
Dibaya-Lubue, Zaire 115/C4
Dibble, Okla. (73031) 288/L4
Dibeng, S. Africa 118/C5
D'Iberville, Miss. (†39532) 256/G10
Dibete, Botswana 118/D4
Diboll, Texas (75941) 303/K6
Dibrugarh, India 68/G3
Dibulla, Colombia 126/D2
Dickens, Iowa (51333) 229/C2
Dickens, Nebr. (69132) 264/C4
Dickens (pt.), R.I. 249/H8
Dickens (co.), Texas 303/D4
Dickens, Texas (79229) 303/D4
Dickenson (co.), Va. 307/D6
Dickerson, Md. (20753) 245/J4
Dickey, Georgia (†31746) 217/C7
Dickey (co.), N. Dak. 282/N7
Dickey, N. Dak. (58431) 282/N6
Dickeyville, Wis. (53808) 317/E10
Dickinson, Ala. (36436) 195/C7
Dickinson (co.), Iowa 229/C2
Dickinson (co.), Kansas 232/E3
Dickinson (co.), Mich. 250/B2
Dickinson, N. Dak. 188/F1
Dickinson, N. Dak. (58601) 282/E6
Dickinson, Pa. (17218) 294/H5
Dickinson, Texas (77539) 303/K3
Dickinson Center, N.Y. (12930) 276/M1
Dickson, Alberta 182/C3
Dickson, Okla. (†73401) 288/M6
Dickson (lake), Ontario 177/F2
Dickson (co.), Tenn. 237/G8
Dickson, Tenn. (37055) 237/G8
Dickson, W. Va. (†25535) 312/B6
Dickson City, Pa. (18519) 294/F7
Dicle, Turkey 63/J3
Dicle (riv.), Turkey 63/J4
Didam, Netherlands 27/J5
Didcot, England 13/F6
Dido, La. (†70656) 238/E5
Didsbury, Alberta 182/C4
Didsbury, Alta. 162/E5
Didyme, Québec 172/E1
Die, France 28/F5
Diébougou, Upper Volta 106/D6
Diefenbaker (lake), Sask. 181/E4
Diego de Almagro (isl.), Chile 138/D9
Diego Garcia (isl.), Br. Ind. Ocean Terr. 54/J10
Diego Lamas, Uruguay 145/C1
Diego Pérez (cay), Cuba 158/C2
Diégo-Suarez (Antsiranana), Madagascar 118/H2
Diehlstadt, Mo. (†63834) 261/N9
Diekirch, Luxembourg 27/J9
Dielsdorf, Switzerland 39/F1
Diemen, Netherlands 27/C5
Diemtigen, Switzerland 39/D3
Dien Bien Phu, Vietnam 72/D2
Diep (riv.), S. Africa 118/F6
Diepholz, W. Germany 22/C2
Diepoldsau, Switzerland 39/J2
Dieppe, France 28/D3
Dieppe, New Bruns. 170/F2
Dieppe Bay, St. Chris.-Nevis 161/C10
Dieren, Netherlands 27/J4
Dierks, Ark. (71833) 202/B5
Diessenhofen, Switzerland 39/G1
Diest, Belgium 27/F7
Dieterich, Ill. (62424) 222/E4
Dietikon, Switzerland 39/F2
Dietrich, Idaho (83324) 220/D7
Diever, Netherlands 27/J3
Diez y Nueve (19) de Abril, Uruguay 145/E5
Diez y Ocho (18) de Julio, Uruguay 145/F4
Dif, Somalia 115/H3
Diffa, Niger 106/G6
Differdange, Luxembourg 27/H9
Difficult, Tenn. (†37145) 237/K8
Difficult (mt.), Victoria 97/B5
Digboi, India 68/H3
Digby (co.), Nova Scotia 168/C4
Digby, Nova Scotia 168/C4
Digby Gut (chan.), Nova Scotia 168/C4
Digby Neck (pen.), Nova Scotia 168/B4
Digdeguash (riv.), New Bruns. 170/C3
Digges (isls.), N.W. Terrs. 187/L3
Diggins, Mo. (65636) 261/G8
Dighton, Kansas (67839) 232/B3
Dighton○, Mass. (02715) 249/K5
Dighton, Mich. (†49688) 250/D4
Digne, France 28/G5
Digoin, France 28/F4
Digor, Turkey 63/K2
Digos, Philippines 82/E7
Digul (riv.), Indonesia 85/K7
Dijon, France 28/F4
Dijon, France 7/E4
Dike, Iowa (50624) 229/H4
Dikhil, Djibouti 111/H5
Dikili, Turkey 63/B3
Diksmuide, Belgium 27/B6
Dikson, U.S.S.R. 4/B5
Dikson, U.S.S.R. 48/J2
Dikwa, Nigeria 106/G6
Dikwa, Nigeria 102/D3
Dilam, Saudi Arabia 59/E5
Dilbeek, Belgium 27/B9
Dildo, Newf. 166/D2
Dili, Indonesia 54/O10
Dili, Indonesia 85/H7
Dilia, N. Mex. (†87711) 274/D3
Diligent River, Nova Scotia 168/D3
Di Linh, Vietnam 72/F5
Dilke, Sask. 181/F5
Dilkon, Ariz. (†86047) 198/E3
Dilla, Ethiopia 111/G6
Dillabough, Sask. 181/J3
Dillard, Georgia (30537) 217/F1
Dillard, Mo. (65458) 261/K7
Dillard, Oreg. (97432) 291/D4
Dill City, Okla. (73641) 288/H4
Dille, W. Va. (26617) 312/E6
Dillenburg, W. Germany 22/C3
Diller, Nebr. (68342) 264/H4
Dilley, Oreg. (†97116) 291/A2
Dilley, Texas (78017) 303/E9
Dillia (dry riv.), Niger 106/G5
Dilliner, Pa. (15327) 294/B6
Dilling, Sudan 111/E5
Dillingen, W. Germany 22/B4
Dillingen an der Donau, W. Germany 22/D4
Dillingham, Alaska 188/C6
Dillingham, Alaska (99576) 196/G3
Dillon (riv.), Alberta 182/E2
Dillon, Colo. (80435) 208/H3
Dillon, Kansas (†67451) 232/E3
Dillon, Mont. (59725) 262/D5
Dillon (lake), Ohio 284/F5
Dillon (co.), S.C. 296/J3
Dillon, S.C. (29536) 296/J3
Dillonvale, Ohio (43917) 284/J5
Dillsboro, Ind. (47018) 227/G6
Dillsboro, N.C. (28725) 281/C4
Dillsburg, Pa. (17019) 294/J5
Dilltown, Pa. (15929) 294/E5
Dillwyn, Va. (23936) 307/M5
Dilolo, Zaire 115/D6
Dilsen, Belgium 27/H6
Dilworth, Minn. (56529) 255/B4
Dimas, Cuba 158/A2
Dimas, Mexico 150/F5
Dimashq (Damascus) (cap.), Syria 63/G6
Dimashq (Damascus) (cap.), Syria 59/C3
Dimbelenge, Zaire 115/D5
Dimbokro, Ivory Coast 106/D7
Dimboola, Victoria 97/B5
Dime Box, Texas (77853) 303/H7
Dimitrovgrad, Bulgaria 45/G4
Dimitrovgrad, U.S.S.R. 52/G4
Dimitrovgrad, U.S.S.R. 48/F4
Dimitrovgrad, Yugoslavia 45/F4
Dimlang (mt.), Nigeria 106/G7
Dimmit (co.), Texas 303/E9
Dimmitt, Texas (79027) 303/B3
Dimock, Pa. (18816) 294/L2
Dimock, S. Dak. (57331) 298/O7
Dimona, Israel 65/D4
Dimona (mt.), Israel 65/C5
Dimondale, Mich. (48821) 250/E6
Dimsdale, Alberta 182/A2
Dinagat, Philippines 82/E5
Dinagat (isl.), Philippines 85/H3
Dinagat (isl.), Philippines 82/E5
Dinagat (sound), Philippines 82/E5
Dinajpur, Bangladesh 68/F3
Dinan, France 28/B3
Dinant, Belgium 27/G8
Dinar, Kuh-e (mts.), Iran 66/G5
Dinar, Turkey 63/D3
Dinard, France 28/B3
Dinaric Alps (mts.), Yugoslavia 45/B3
Dinas Powis, Wales 13/B7
Dinder (riv.), Ethiopia 111/F5
Dinder (riv.), Sudan 59/B7
Dinder (riv.), Sudan 111/F5
Dindigul, India 68/D6
Dingalan (bay), Philippines 82/C3
Dingbian, China 77/G4
Dingess, W. Va. (25671) 312/B7
Dinggye, China 77/C6
Dinghai, China 77/K5
Dingle, Idaho (83233) 220/G7
Dingle, Ireland 10/A4
Dingle, Ireland 17/A7
Dingle (bay), Ireland 10/A4
Dingle (bay), Ireland 17/A7
Dingmans Ferry, Pa. (18328) 294/N3
Dingolfing, W. Germany 22/E4
Dinguiraye, Guinea 106/B6
Dingwall, Nova Scotia 168/H2
Dingwall, Scotland 15/D3
Dingwall, Scotland 10/D2
Dingxi, China 77/F4
Dingxing, China 77/H4
Dinh, Mui (cape), Vietnam 72/F5
Dinkelsbühl, W. Germany 22/D4
Dinnebito Wash (dry riv.), Ariz. 198/E3
Dinokwe, Botswana 118/D4
Dinorwic, Ontario 177/G5
Dinorwic, Ontario 175/B3
Dinosaur, Colo. (81610) 208/B2
Dinosaur Nat'l Mon., Colo. 208/B2
Dinosaur Nat'l Mon., Utah 304/E3
Dinsdale, Iowa (†50669) 229/H4
Dinsmore, Sask. 181/D4
Dinsor, Somalia 115/H3
Dinuba, Calif. (93618) 204/F7
Dinwiddie (co.), Va. 307/N6
Dinwiddie, Va. (23841) 307/N6
Dinxperlo, Netherlands 27/K5
Diogo (isl.), Philippines 82/B2
Dioila, Mali 106/C6
Diomede, Alaska (†99762) 196/E1
Diourbel, Senegal 106/A6
Diphu, India 68/G3
Dipilto, Cordillera (range), Nicaragua 154/D4
Diplo, Pakistan 68/B4
Dipolog, Philippines 82/D6
Dipper Harbour, New Bruns. 170/D3
Dir, Pakistan 68/C1
Dir, Pakistan 59/K2
Dire, Mali 106/D5
Direction (cape), Queensland 88/G2
Direction (cape), Queensland 95/B2
Dire Dawa, Ethiopia 102/G4
Dire Dawa, Ethiopia 111/H6
Diriamba, Nicaragua 154/D5
Dirico, Angola 115/D7
Dirico, Namibia 118/C3
Dirk Hartogs (isl.), Australia 87/B8
Dirk Hartogs (isl.), W. Australia 88/A5
Dirk Hartogs (isl.), W. Australia 92/A4
Dirksland, Netherlands 27/E5
Dirmil, Turkey 63/C4
Dirranbandi, Queensland 88/H5
Dirranbandi, Queensland 95/D6
Dirty Devil (riv.), Utah 304/D5
Disappointment (lake), Australia 87/C8
Disappointment (creek), Colo. 208/B7
Disappointment (isls.), Fr. Poly. 87/N7
Disappointment (lake), Newf. 166/B3
Disappointment (cape), Wash. 188/A1
Disappointment (cape), Wash. 310/A4
Disappointment (lake), W. Australia 88/C4
Disappointment (lake), W. Australia 92/C3
Discovery (bay) 88/E7
Discovery (bay), Victoria 97/A6
Discovery Bay, Jamaica 158/J5
Disentis-Muster, Switzerland 39/G3
Dishman, Wash. (99213) 310/H3
Disko (isl.), Greenl. 4/C12

Disko (isl.), Greenland 146/N3
Disko, Ind. (†46982) 227/E2
Disley, Sask. 181/F5
Dismal (riv.), Nebr. 264/C3
Dismal (Great) (swamp), N.C. 281/S1
Disney, Okla. (74340) 288/S2
Dison, Belgium 27/H7
Dispur, India 68/G3
Disputanta, Va. (23842) 307/O6
Disraeli (bay), N.W. Terrs. 187/L1
Disraëli, Québec 172/F4
Diss, England 13/J5
Diss, England 10/G4
Disston (lake), Fla. 212/E2
Disston, Oreg. (†97427) 291/E4
District Heights, Md. (20747) 245/G5
District of Columbia 146/L6
District of Columbia 188/L3
DISTRICT OF COLUMBIA 245
Distrito Especial, Colombia 126/C5
Distrito Federal, Argentina 143/H7
Distrito Federal, Mexico 150/L1
Distrito Federal, Venezuela 124/E2
Distrito Nacional, Dom. Rep. 158/E6
Disûq, Egypt 111/J3
Dittmer, Mo. (63023) 261/L6
Ditton (riv.), Québec 172/F4
Diu, India 68/C4
Diu (dist.), India 68/C4
Diuata (mts.), Philippines 82/E6
Divernon, Ill. (62530) 222/D4
Divide, Colo. (80814) 208/J5
Divide, Mont. (59727) 262/D5
Divide (co.), N. Dak. 282/C2
Dividing (creek), Md. 245/R8
Dividing Creek, N.J. (08315) 273/C5
Divino, Brazil 135/E2
Divinópolis, Brazil 132/E8
Divinópolis, Brazil 120/E5
Divinópolis, Brazil 135/D2
Divis (mt.), N. Ireland 17/J2
Divisa Nova, Brazil 135/C2
Division (peak), Nev. 266/B1
Divo, Ivory Coast 106/C7
Divriği, Turkey 63/H3
Divriği, Turkey 59/C2
Dix, Ill. (62830) 222/E5
Dix (riv.), Ky. 237/M5
Dix, Nebr. (69133) 264/A3
Dixfield, Maine (04224) 243/C6
Dixfield○, Maine (04224) 243/C6
Dix Hills, N.Y. (11746) 276/O9
Dixie, Ala. (†36420) 195/E8
Dixie (co.), Fla. 212/D2
Dixie, Georgia (31629) 217/E9
Dixie, Idaho (83525) 220/C4
Dixie, La. (†71107) 238/C1
Dixie, Wash. (99329) 310/G4
Dixie, W. Va. (25059) 312/D6
Dixie Inn, La. (†71055) 238/D1
Dixmont, Maine (04932) 243/E6
Dixmont○, Maine (04932) 243/E6
Dixmoor, Ill. (†60469) 222/C6
Dixmude (Diksmuide), Belgium 27/B6
Dixon, Calif. (95620) 204/B9
Dixon, Ill. (61021) 222/D2
Dixon, Iowa (52745) 229/M5
Dixon, Ky. (42409) 237/F5
Dixon, Miss. (†39350) 256/F5
Dixon, Mo. (65459) 261/H6
Dixon, Mont. (59831) 262/B3
Dixon (co.), Nebr. 264/H2
Dixon, Nebr. (68732) 264/H2
Dixon, N. Mex. (87527) 274/D2
Dixon, N.C. (†28445) 281/O5
Dixon, Ohio (†46773) 284/A4
Dixon, S. Dak. (57530) 298/L7
Dixon, Wyo. (82323) 319/E4
Dixon Entrance (chan.) 146/E4
Dixon Entrance (chan.), Alaska 196/M2
Dixon Entrance (chan.), Br. Col. 184/A3
Dixons Mills, Ala. (36736) 195/C6
Dixon Springs, Ill. (†62911) 222/E6
Dixon Springs, Tenn. (37057) 237/J8
Dixonville, Ala. (†36426) 195/E8
Dixonville, Alberta 182/B1
Dixonville, Pa. (15734) 294/D4
Dixville (peak), N.H. 268/E2
Dixville, Québec 172/F4
Dixville Notch, N.H. (†03576) 268/E2
Dixville Notch (pass), N.H. 268/E2
Diyadin, Turkey 63/K3
Diyala (heads), Iraq 66/D4
Diyala (riv.), Iraq 66/D4
Diyarbakır (prov.), Turkey 63/H4
Diyarbakır, Turkey 54/F6
Diyarbakır, Turkey 63/H4
Diyarbakır, Turkey 59/C2
Dizful (Dezful), Iran 66/F4
Dja (riv.), Cameroon 115/B3
Dja (riv.), Congo 115/B3
Djado (plat.) 102/D2
Djado, Niger 102/D2
Djado, Niger 106/G4
Djado (plat.), Niger 106/G4
Djakarta (Jakarta) (cap.), Indonesia 85/H1
Djakovica, Yugoslavia 45/E4
Djakovo, Yugoslavia 45/D3
Djambala, Congo 115/B4
Djambi (Jambi), Indonesia 85/C6
Djanet, Algeria 106/F4
Djanet, Algeria 102/C2
Djelfa, Algeria 106/E2
Djema, Cent. Afr. Rep. 115/E2
Djemaa, Algeria 106/F2
Djenné, Mali 102/B3
Djenné, Mali 106/D6
Djerba (isl.), Tunisia 106/G2
Djerid, Shott el (salt lake), Tunisia 106/F2
Djibo, Upper Volta 106/D6
Djibouti 2/L5
Djibouti 102/G3
DJIBOUTI 111/H5
Djibouti (cap.), Djibouti 111/H5
Djibouti (cap.), Djibouti 102/G3
Djokjakarta (Yogyakarta), Indonesia 85/J2
Djolu, Zaire 115/D3
Djouf, El (des.), Mauritania 106/C4
Djougou, Benin 106/E7
Djoum, Cameroon 115/B3
Djugu, Zaire 115/F3
D'Lo, Miss. (39062) 256/E7
Dmitriya Lapteva (str.), U.S.S.R. 4/B2
Dmitriya Lapteva (str.), U.S.S.R. 48/O2
Dneprodzerzhinsk, U.S.S.R. 7/H4
Dneprodzerzhinsk, U.S.S.R. 52/D5
Dnepropetrovsk, U.S.S.R. 7/H4
Dnepropetrovsk, U.S.S.R. 48/D5
Dnepropetrovsk, U.S.S.R. 52/D5
Dnieper (riv.), U.S.S.R. 7/H3
Dnieper (riv.), U.S.S.R. 48/D5
Dnieper (riv.), U.S.S.R. 52/D5
Dniester (riv.), U.S.S.R. 7/G4
Dniester (riv.), U.S.S.R. 52/C5
Dniester (riv.), U.S.S.R. 48/C5
Dno, U.S.S.R. 52/D3
Doaghbeg, Ireland 17/F1
Doaktown, New Bruns. 170/D2
Doans, Ind. (†47424) 227/D7
Doba, Chad 111/C6
Doba, Chad 102/D4
Dobbie (mt.), North. Terr. 93/E7
Dobbin (bay), N.W. Terrs. 187/L2
Dobbins A.F.B., Georgia 217/J1
Dobbs Ferry, N.Y. (10522) 276/O6
Dobbyn, Queensland 95/A3
Dobele, U.S.S.R. 53/B2
Döbeln, E. Germany 22/E3
Doberai (pen.), Indonesia 85/J6
Dobiegniew, Poland 47/B2
Doblas, Argentina 143/D4
Dobo, Indonesia 85/J7
Doboj, Yugoslavia 45/C3
Doboy (sound), Georgia 217/K8
Dobřany, Czech. 41/B2
Dobre Miasto, Poland 47/E2
Dobrich (Tolbukhin), Bulgaria 45/H4
Dobříš, Czech. 41/C2
Dobrush, U.S.S.R. 52/D4
Dobruška, Czech. 41/D1
Dobryanka, U.S.S.R. 52/J3
Dobšiná, Czech. 41/F2
Dobson, N.C. (27017) 281/H2
Doce (riv.), Brazil 135/E2
Doce (riv.), Brazil 132/F7
Doce Leguas (cays), Cuba 158/F3
Docker River, North. Terr. 93/A8
Docking, England 13/H5
Dock Junction (Arco), Georgia (†31520) 217/J8
Doctor Arroyo, Mexico 150/K5
Doctor Cecilio Báez, Paraguay 144/D4
Doctor Juan L. Mallorquín, Paraguay 144/E4
Doctor Juan Manuel Frutos, Paraguay 144/E4
Doctor M. Irala, Paraguay 144/E4
Doctor Pedro P. Peña, Paraguay 144/A3
Doctors Inlet, Fla. (32030) 212/E1
Doctortown, Georgia (†31545) 217/J7
Doddridge, Ark. (71834) 202/Ci
Doddridge (co.), W. Va. 312/E4
Dodds, Alberta 182/D3
Doddsville, Miss. (38736) 256/C3
Dodecanese (isls.), Greece 45/H8
Dodge (co.), Georgia 217/F6
Dodge, Mass. (†01507) 249/G4
Dodge (co.), Minn. 255/F7
Dodge (co.), Nebr. 264/H3
Dodge, Nebr. (68633) 264/H3
Dodge, N. Dak. (58625) 282/F5
Dodge, Texas (77334) 303/J7
Dodge (co.), Wis. 317/J9
Dodge, Wis. (54625) 317/D7
Dodge Center, Minn. (55927) 255/F6
Dodge City, Kans. 188/F3
Dodge City, Kansas (67801) 232/B4
Dodgeville, Wis. (53533) 317/F10
Dodgingtown, Conn. (†06470) 210/B3
Dodman (pt.), England 13/C7
Dodoma (reg.), Tanzania 115/G5
Dodoma, Tanzania 102/F5
Dodoma, Tanzania 115/G5
Dodsland, Sask. 181/C4
Dodson, La. (71422) 238/E2
Dodson, Mont. (59524) 262/H2
Dodson, Texas (79230) 303/D3
Doe (lake), Ontario 177/E2
Doe (bay), Wash. 310/C2
Doe Bay, Wash. (†98279) 310/C2
Doe Hill, Va. (24433) 307/K4
Doering, Wis. (†54435) 317/G5
Doerun, Georgia (31744) 217/E8
Doe Run, Mo. (63637) 261/M7
Doesburg, Netherlands 27/J4
Doetinchem, Netherlands 27/J5
Dog (pond), Conn. 210/C1
Dog (isl.), Fla. 212/B2
Dog (lake), Manitoba 179/D3
Dog (isl.), Newf. 166/B2
Dog (lake), Ontario 177/G5
Dogai Coring (lake), China 77/C5
Doğanbey, Turkey 63/D4
Doğanhisar, Turkey 63/D3
Doğanşehir, Turkey 63/G3
Dog Creek, Br. Col. 184/G4
Dog Ear (creek), S. Dak. 298/K6
Döger, Turkey 63/D3
Dogo (isl.), Japan 81/F5
Dogondoutchi, Niger 106/E6
Dogondoutchi, Niger 102/C3
Dogpatch, Ark. (72648) 202/D1
Dog Pound, Alberta 182/C4
Dogskin (lake), Manitoba 179/G3
Doğubeyazit, Turkey 63/K3
Dogwood (pt.), St. Chris.-Nevis 161/D11
Doha (cap.), Qatar 54/G7
Doha (cap.), Qatar 59/F4
Dohad, India 68/C4
Doheny, Québec 172/E2
Dohuk (gov.), Iraq 66/C2
Dohuk, Iraq 66/C2
Doi Inthanon (mt.), Thailand 72/C3
Doilungdeqen, China 77/C6
Doi Pha Hom Pok (mt.), Thailand 72/C2
Doi Pia Fai (mt.), Thailand 72/D4
Doische, Belgium 27/F8
Dois Córregos, Brazil 135/B3
Dois Irmãos, Serra (range), Brazil 132/F5
Dokkum, Netherlands 27/H2
Doksy, Czech. 41/C1
Dokterstuin, Neth. Ant. 161/F8
Dola, Ohio (45835) 284/C4
Dola, W. Va. (†26386) 312/F4
Dolan, Ind. (†47401) 227/E6
Doland, S. Dak. (57436) 298/N4
Dolan Springs, Ariz. (86441) 198/A3
Dolavon, Argentina 143/C5
Dolbeau, Québec 174/C3
Dolbeau, Québec 172/E1
Doldenhorn (mt.), Switzerland 39/E4
Dole, France 28/F4
Dolega, Panama 154/F6
Dolent (mt.), Switzerland 39/C5
Doles, Georgia (†31791) 217/E7
Dolgellau, Wales 13/D5
Dolgellau, Wales 10/E4
Dolgeville, N.Y. (13329) 276/L4
Dolgiy (isl.), U.S.S.R. 52/J1
Dolinsk, U.S.S.R. 48/P5
Dollar, Scotland 10/B1
Dollar, Scotland 15/E4
Dollar Bay, Mich. (49922) 250/G1
Dollard (bay), Netherlands 27/L2
Dollard, Sask. 181/C6
Dollard-des-Ormeaux, Québec 172/H4
Dollart (est.), W. Germany 22/B2
Dollarville, Mich. (†49868) 250/D2
Dolliver, Iowa (50531) 229/D2
Dolný Kubín, Czech. 41/E2
Dolo, Ethiopia 111/H7
Dolomite, Ala. (35061) 195/D4
Dolomite Alps (range), Italy 34/C1
Dolores, Argentina 143/E4
Dolores, Argentina 120/D6
Dolores (co.), Colo. 208/C7
Dolores, Colo. (81323) 208/C8
Dolores (riv.), Colo. 208/B5
Dolores, Guatemala 154/C2
Dolores, Philippines 82/E4
Dolores, Spain 33/F3
Dolores, Uruguay 145/A4
Dolores (riv.), Utah 304/E5
Dolores, Venezuela 124/D3
Dolores Hidalgo de la Independencia Nacional, Mexico 150/J6
Dolphin and Union (str.), N.W. Terrs. 187/G3
Dölsach, Austria 41/B3
Dolton, Ill. (60419) 222/C6
Dolton, S. Dak. (57023) 298/P7
Dom (mt.), Switzerland 39/E4
Domain, Manitoba 179/E5
Domanic, Turkey 63/C3
Domar (dry riv.), Chad 111/C4
Domat-Ems, Switzerland 39/H3
Domažlice, Czech. 41/B2
Dombås, Norway 18/F5
Dombe Grande, Angola 115/B6
Dombóvár, Hungary 41/E3
Dombrád, Hungary 41/F2
Dombresson, Switzerland 39/C2
Domburg, Netherlands 27/C5
Domburg, Suriname 131/D3
Dome, Ariz. (†85364) 198/A6
Dome Creek, Br. Col. 184/G3
Domeiko, Chile 138/A4
Domeyko, Cordillera (mts.), Chile 138/B4
Domínguez, Argentina 143/G6
Dominica 2/F5
Dominica 146/M8
DOMINICA 156/G4
DOMINICA 161/E7
Dominica (passage), Dominica 161/E5
Dominican Republic 2/F4
Dominican Republic 146/L8
DOMINICAN REPUBLIC 156/D3
DOMINICAN REPUBLIC 158
Dominion (lake), Newf. 166/B3
Dominion (cape), N.W. Terrs. 187/L3
Dominion, Nova Scotia 168/J2
Dominion City, Manitoba 179/E5
Domino, Newf. 166/C3
Dömitz, E. Germany 22/D2
Domjor, India 68/F1
Domleschg (valley), Switzerland 39/E2
Dommel (riv.), Netherlands 27/H6
Domo, Ethiopia 111/J6
Domodossola, Italy 34/A1
Dom Pedrito, Brazil 132/C10
Dompu, Indonesia 85/F7
Domremy, Sask. 181/F3
Domrémy-la-Pucelle, France 28/F3
Dom Silvério, Brazil 135/E2
Dömsöd, Hungary 41/E3
Domuyo (vol.), Argentina 143/B4
Don (riv.), England 10/F4
Don (riv.), England 13/F4
Don (riv.), Ontario 177/J4
Don (riv.), Scotland 15/F3
Don (riv.), Scotland 10/E2
Don (riv.), U.S.S.R. 7/J4
Don (riv.), U.S.S.R. 48/E5
Don (riv.), U.S.S.R. 52/F5
Dona Ana (Mutarara), Mozambique 118/F3
Dona Ana (co.), N. Mex. 274/C6
Dona Ana, N. Mex. (88032) 274/C6
Donabate, Ireland 17/J5
Donaghadee, N. Ireland 17/K2
Donahue, Iowa (52746) 229/M5
Donald, Br. Col. 184/J4
Donald, Oreg. (97020) 291/A3
Donald, Victoria 97/B5
Donald, Wash. (†98951) 310/E4
Donald, Wis. (†54433) 317/E5
Donalda, Alberta 182/D3
Donalds, S.C. (29638) 296/C3
Donaldson, Ark. (71941) 202/E5
Donaldson, Ind. (46513) 227/E2
Donaldson, Minn. (56720) 255/B2
Donaldson A.F.B., S.C. 296/C2
Donaldsonville, La. (70346) 238/K3
Donalsonville, Georgia (31745) 217/C8
Donansburg, Ky. (†42743) 237/K6
Donath, Switzerland 39/H3
Donatville, Alberta 182/D2
Donau (Danube) (riv.), Austria 41/D2
Donau (Danube) (riv.), W. Germany 22/C4
Donaueschingen, W. Germany 22/C5
Donauwörth, W. Germany 22/D4
Donavon, Sask. 181/D4
Donbar, Queensland 95/B3
Don Benito, Spain 33/C3
Doncaster, England 13/F4
Doncaster, England 10/F4
Doncaster, Md. (†20646) 245/K7
Doncaster and Templestowe, Victoria 88/L7
Doncaster and Templestowe, Victoria 97/J5
Dondo, Angola 115/B5
Dondo, Mozambique 118/F3
Dondra (head), Sri Lanka 68/E7
Dondra Head (cape), Sri Lanka 54/K9
Donegal (co.), Ireland 17/K2
Donegal, Ireland 10/B3
Donegal, Ireland 17/F2
Donegal (bay), Ireland 17/D3
Donegal (bay), Ireland 10/B3
Donegal (harb.), Ireland 17/E2
Donegal (pt.), Ireland 17/B6
Donegal, Pa. (15628) 294/D5
Donel, Honduras 154/E3
Doneraile, Ireland 17/D7
Doneraile, S.C. (†29532) 296/H3
Donets (riv.), U.S.S.R. 7/H4
Donets (riv.), U.S.S.R. 48/D5
Donets (riv.), U.S.S.R. 52/E5
Donetsk, U.S.S.R. 7/H4
Donetsk, U.S.S.R. 48/D5
Donetsk, U.S.S.R. 52/E5
Donga (riv.), Cameroon 115/B2
Donga, Nigeria 106/G7
Donga (riv.), Nigeria 106/G7
Dongara, W. Australia 92/A5
Dongchuan, China 77/F6
Dongen, Netherlands 27/F5
Dongfang, China 77/G8
Dongfanghong, China 77/M2
Donggala, Indonesia 85/F6
Dönghén, Laos 72/E3
Dong Hoi, Vietnam 72/E3
Dongio, Switzerland 39/H4
Dongning, China 77/M3
Dongo, Zaire 115/C3
Dongola, Ill. (62926) 222/D6
Dongola, Sudan 102/E3
Dongola, Sudan 59/B6
Dongola, Sudan 111/F4
Dongou, Congo 115/C3
Dong Rak (mts.), Thailand 72/D4
Dongsha (isl.), China 77/J7
Dongsheng, China 77/H4
Dongtai, China 77/K5
Dongting (lake), China 54/N7
Dongting Hu (riv.), China 77/H6
Dong Ujimqin, China 77/J2
Dongwe (riv.), Zambia 115/D6
Donie, Texas (75838) 303/H6
Doñihue, Chile 138/G5
Doniphan (co.), Kansas 232/G2
Doniphan, Mo. (63935) 261/L9
Doniphan, Nebr. (68832) 264/F4
Donji Vakuf, Yugoslavia 45/C3
Donkin, Nova Scotia 168/J2
Donley (co.), Texas 303/D2
Dønna (isl.), Norway 18/H3
Donna, Texas (78537) 303/F11
Donnacona, Québec 172/F3
Donnan, Iowa (52139) 229/K3
Donnellson, Ill. (62019) 222/D4
Donnellson, Iowa (52625) 229/K7
Donnelly, Alberta 182/B2
Donnelly, Idaho (83615) 220/B5
Donnelly, Minn. (56235) 255/B5
Donner (pass), Calif. 204/E4
Donner, La. (70352) 238/J7
Donner and Blitzen (riv.), Oreg. 291/J4
Donnybrook, N. Dak. (58734) 282/G2
Donnybrook, Queensland 95/D5
Donnybrook, W. Australia 92/A2
Donora, Pa. (15033) 294/C5
Donovan, Georgia (†31096) 217/G5
Donovan, Ill. (60931) 222/F3
Donsol, Philippines 82/D4
Donwell, Sask. 181/J4
Donzère, France 28/F5
Dooagh-Keel, Ireland 17/A4
Doole, Texas (76836) 303/E6
Dooling, Georgia (†31063) 217/E6
Doolittle (pond), Conn. 210/C1
Doolittle, Mo. (†65550) 261/J7
Doolittle Mills, Ind. (†47118) 227/D8
Dooly (co.), Georgia 217/E6
Doon, Iowa (51235) 229/A2
Doon, Ireland 17/E6
Doon, Loch (lake), Scotland 15/D5
Doon (riv.), Scotland 15/D5
Doonerak (mt.), Alaska 196/H1
Doonside, Sask. 181/K6
Door (pt.), La. 238/M6
Door (co.), Wis. 317/M6
Door (pen.), Wis. 317/M6
Doorn, Netherlands 27/G4
Doornik (Tournai), Belgium 27/C7
Doqa, Saudi Arabia 59/D6
Dor, Israel 65/B2
Dora, Ala. (35062) 195/D3
Dora, Mo. (65637) 261/H9
Dora, N. Mex. (88115) 274/F5
Dora, Oreg. (†97458) 291/D4
Dora (lake), W. Australia 88/C4
Dora (lake), W. Australia 92/C3
Dora Baltea (riv.), Italy 34/A2
Dorado, P. Rico 161/D1
Dora Lake, Minn. (†56661) 255/D3
Doran, Minn. (56530) 255/B4
Dora Riparia (riv.), Italy 34/A2
Doraville, Georgia (30340) 217/K1
D'Orbigny, Bolivia 136/D7
Dorbiljin (Emin), China 77/B2
Dorbod, China 77/K2
Dorcas, W. Va. (26835) 312/H5
Dorchester, England 10/E5
Dorchester, England 13/E7
Dorchester, Georgia (†31317) 217/K7
Dorchester, Ill. (62020) 222/D4
Dorchester, Iowa (52140) 229/L2
Dorchester (co.), Md. 245/O7
Dorchester, Mass. (†02122) 249/D7
Dorchester, Nebr. (68343) 264/G4
Dorchester, New Bruns. 170/F3
Dorchester○, N.H. (†03266) 268/D4
Dorchester, N.J. (08316) 273/D5
Dorchester (cape), N.W. Terrs. 187/L3
Dorchester, Ontario 177/C5
Dorchester (co.), Québec 172/C3
Dorchester (co.), S.C. 296/G5
Dorchester, S.C. (29437) 296/G5
Dorchester, Wis. (54425) 317/F5
Dorchester Crossing, New Bruns. 170/F2
Dordogne (dept.), France 28/D5
Dordogne (riv.), France 7/E4
Dordogne (riv.), France 28/D5
Dordrecht, Netherlands 27/F5
Doré (lake), Ontario 177/G2
Doré (lake), Sask. 181/L3
Dore Alps (mts.), France 28/E5
Doré Lake, Sask. 181/L4
Dorena, Mo. (†63845) 261/O9
Dorena, Oreg. (97434) 291/E4
Dorena (lake), Oreg. 291/E4
Dorenlee, Alberta 182/D3
Dores, Scotland 15/D3
Dores do Indaiá, Brazil 132/E7
Dorgali, Italy 34/B4
Dörgön Nuur (lake), Mongolia 77/D2
Dori, Mali 102/B3
Dori, Upper Volta 106/D6
Doring (riv.), S. Africa 118/B6
Dorintosh, Sask. 181/L4
Dorion, Ontario 177/H5
Dorion, Québec 172/C4
Dorking, England 13/G8
Dorking, England 10/F5
Dormont, Pa. (15216) 294/B7
Dornach, Switzerland 39/E2
Dornbirn, Austria 41/A3
Dornie, Scotland 15/C3
Dornoch, Scotland 10/D2
Dornoch, Scotland 15/D3
Dornoch (firth), Scotland 15/E3
Dornoch (firth), Scotland 10/E2
Dornod, Mongolia 77/H2
Dornogovĭ, Mongolia 77/G3
Dorog, Hungary 41/E3
Dorohoi, Romania 45/H2
Dorotea, Sweden 18/K4
Dorothy, Alberta 182/D4
Dorothy, Minn. (†56750) 255/B3
Dorothy, N.J. (08317) 273/D5
Dorothy, W. Va. (25060) 312/D7
Dorr (lake), Fla. 212/E2
Dorr, Mich. (49323) 250/D6
Dorrance, Kansas (67634) 232/D3
Dorre (isl.), W. Australia 88/A5
Dorre (isl.), W. Australia 92/A4
Dorreen, Br. Col. 184/C3
Dorrigo, N.S. Wales 97/G2
Dorris, Calif. (96023) 204/D2
Dorset (co.), England 13/E7
Dorset, Minn. (†56470) 255/D4
Dorset, Ohio (44032) 284/J2
Dorset○, Vt. (05251) 268/A5
Dorset (peak), Vt. 268/A5
Dorset Heights (hills), England 13/E7
Dorsey, Miss. (†38801) 256/H2
Dorsten, W. Germany 22/B3
Dortches, N.C. (†27801) 281/O2
Dortmund, W. Germany 7/E3
Dortmund, W. Germany 22/B3
Dorton, Ky. (41520) 237/R6
Dörtyol, Turkey 63/F4
Doruma, Zaire 115/E3
Dorval, Québec 172/H4
Dory Point, N.W. Terrs. 187/G3
Dos Bahías (cape), Argentina 143/D5
Dos Cabezas, Ariz. (†85643) 198/F6
Dos Caminos, Cuba 158/J4
Dos de Mayo, Peru 128/E6
Dos Hermanas, Spain 33/D4
Dos Palos, Calif. (93620) 204/E6
Dosquet, Québec 172/F3
Dos Reyes (pt.), Chile 138/A5
Dos Ríos, Cuba 158/J4
Dosso, Niger 106/E6
Dossor, U.S.S.R. 48/F5
Dossville, Miss. (†39051) 256/F5
Doswell, Va. (23047) 307/N5
Dothan, Ala. 188/J4
Dothan, Ala. (*36303) 195/H8
Doti, Nepal 68/E3
Dot Klish (canyon), Ariz. 198/E2
Dot Lake, Alaska (99737) 196/K2
Dott, W. Va. (24721) 312/D8
Döttingen, Switzerland 39/F1
Doty, Wash. (98539) 310/B4
Douai, France 28/E2
Douala, Cameroon 115/B3
Douala, Cameroon 102/D4
Douarnenez, France 28/A3
Double Branches, Georgia (†30817) 217/H3
Double Mer (lake), Newf. 166/C3
Double Mountain Fork, Brazos (riv.), Texas 303/C4
Double Oak, Texas (†76226) 303/F1
Double Springs, Ala. (35553) 195/D2
Doubletop (peak), Wyo. 319/B2
Doubs (dept.), France 28/G4
Doubs (riv.), France 28/G4
Doubs, Md. (†21710) 245/J3
Doubs (riv.), Switzerland 39/C2
Doubtful (sound), N. Zealand 100/A6
Doubtless (bay), N. Zealand 100/D1
Doucette, Texas (79542) 303/K7
Douds, Iowa (52551) 229/J7
Doué-la-Fontaine, France 28/C4
Douentza, Mali 106/D6
Douentza, Mali 102/B3
Dougherty (co.), Georgia 217/D7
Dougherty, Iowa (50433) 229/G3
Dougherty, Okla. (73032) 288/M6
Dougherty, Texas (79231) 303/C4
Douglas, Ala. (35964) 195/F2
Douglas (mt.), Alaska 196/H3
Douglas, Ariz. 146/G6
Douglas, Ariz. 188/E4
Douglas, Ariz. (85607) 198/F7
Douglas (chan.), Br. Col. 184/C3
Douglas (co.), Colo. 208/K4
Douglas (creek), Colo. 208/B3
Douglas (bay), Dominica 161/E5
Douglas (co.), Georgia 217/C3
Douglas, Georgia (31533) 217/G7
Douglas (co.), Ill. 222/E4
Douglas, Ireland 17/D8
Douglas (cap.), I. of Man 13/C3
Douglas (cap.), I. of Man 10/D3
Douglas (co.), Kansas 232/G3
Douglas, Manitoba 179/C5
Douglas○, Mass. (†01516) 249/H4
Douglas, Mich. (49406) 250/C6
Douglas (co.), Minn. 255/C5
Douglas, Minn. (†55960) 255/F6
Douglas (co.), Mo. 261/G9
Douglas (mt.), Mont. 262/F5
Douglas (co.), Nebr. 264/H3
Douglas, Nebr. (68344) 264/H4
Douglas (co.), Nev. 266/B4
Douglas, N. Dak. (58735) 282/G4
Douglas, North. Terr. 93/B2
Douglas, Okla. (73733) 288/L2
Douglas, Ontario 177/H2
Douglas (pt.), Ontario 177/C3
Douglas (co.), Oreg. 291/D4
Douglas, Scotland 15/E5
Douglas, S. Africa 118/C5
Douglas (co.), S. Dak. 298/N7
Douglas (lake), Tenn. 237/P9
Douglas (co.), Wash. 310/F3
Douglas, Wash. (†98858) 310/F3
Douglas (co.), Wis. 317/C3
Douglas, Wyo. (82633) 319/G3
Douglas Harbour, New Bruns. 170/D3
Douglas Lake, Br. Col. 184/H5
Douglas Prov. Park, Sask. 181/E4
Douglass, Kansas (67039) 232/F4
Douglass, Texas (75943) 303/K6
Douglass Hills, Ky. (†40243) 237/L2
Douglassville, Pa. (19518) 294/L5
Douglastown, New Bruns. 170/E1
Douglastown, Québec 172/D1
Douglasville, Georgia (*30133) 217/C3
Doullens, France 28/E2
Doulus (head), Ireland 17/A8
Doumé, Cameroon 115/B3
Dounby, Scotland 15/E1
Doune, Scotland 15/D4
Dour, Belgium 27/D8
Dourados, Brazil 120/D5
Dourados, Brazil 132/C8
Douro (riv.), Portugal 7/D4
Douro (riv.), Portugal 33/B2
Douro (riv.), Spain 33/C2
Dousman, Wis. (53118) 317/J1
Douthat, Okla. (74341) 288/S1
Douville, Québec 172/D4
Dove (riv.), England 13/J2
Dove (creek), Utah 304/A2
Dove Creek, Colo. (81324) 208/A7
Dover, Ark. (72837) 202/D3
Dover (cap.), Del. 146/L6
Dover (cap.), Del. 188/L3
Dover (cap.), Del. (19901) 245/R4
Dover, England 7/E3
Dover, England 10/G5
Dover, England 13/J6
Dover (str.), England 13/J7
Dover (str.), England 10/G5
Dover, Fla. (33527) 212/D4
Dover, Georgia (30424) 217/J5
Dover, Idaho (83825) 220/B1
Dover, Ill. (61323) 222/D2
Dover, Ind. (†46052) 227/H6
Dover, Kansas (66420) 232/G3
Dover, Ky. (41034) 237/O3
Dover, Mass. (02030) 249/B7
Dover○, Mass. (02030) 249/B7
Dover, Minn. (55929) 255/F7
Dover, Mo. (64022) 261/E4
Dover, N.H. (03820) 268/F5
Dover, N.J. (07801) 273/D2
Dover, N.C. (28526) 281/P4
Dover, Ohio (44622) 284/G4
Dover (lake), Ohio 284/H4
Dover, Okla. (73734) 288/L3
Dover, Pa. (17315) 294/J6
Dover, Tasmania 99/C5
Dover, Tenn. (37058) 237/F8
Dover (pt.), W. Australia 88/D6
Dover (pt.), W. Australia 92/D6
Dover A.F.B., Del. 245/S4
Doverel, Georgia (†31742) 217/D7
Dover-Foxcroft, Maine (04426) 243/E5
Dover-Foxcroft○, Maine (04426) 243/E5
Dover Hill, Ind. (†47581) 227/D7
Dover Plains, N.Y. (12522) 276/O7
Dover South Mills, Maine (†04426) 243/E5
Dovesville, S.C. (29540) 296/H3
Dovey (riv.), Wales 10/D4
Dovey (riv.), Wales 13/D5
Dovns Klint (cliff), Denmark 21/D8
Dovray, Minn. (56125) 255/C6

Dovre, Norway 18/F6
Dovrefjell (hills), Norway 18/F5
Dow (Xau) (lake), Botswana 118/C4
Dow, Ill. (62022) 222/C4
Dow, Okla. (†74547) 288/P5
Dowa, Malawi 115/F6
Dowagiac, Mich. (49047) 250/D6
Dow City, Iowa (51528) 229/B5
Dowell, Ill. (62927) 222/D6
Dowelltown, Tenn. (37059) 237/K8
Dowlatabad, Afghanistan 59/H3
Dowlatabad, Afghanistan 68/A2
Dowlatabad, Kerman, Iran 66/K6
Dowlatabad, Khorasan, Iran 66/M2
Dowlat Yar, Afghanistan 59/J3
Dowlat Yar, Afghanistan 68/B2
Dowling, Alberta 182/E4
Dowling (lake), Alberta 182/D4
Dowling, Mich. (49050) 250/D6
Dowling Park, Fla. (32060) 212/C1
Down (dist.), N. Ireland 17/K3
Downe, Sask. 181/C4
Downer, Minn. (†56514) 255/B4
Downers Grove, Ill. (60515) 222/A6
Downey, Calif. (*90240) 204/C11
Downey, Idaho (83234) 220/F7
Downey, Iowa (†52358) 229/L5
Downfall (creek), Queensland 95/D2
Downham Market, England 13/H5
Downham Market, England 10/G4
Downieville, Calif. (95936) 204/E4
Downing, Mo. (63536) 261/H2
Downing, Wis. (54734) 317/B5
Downings, Va. (†22460) 307/P5
Downingtown, Pa. (19335) 294/L5
Downpatrick (head), Ireland 17/C3
Downpatrick, N. Ireland 10/C3
Downpatrick, N. Ireland 17/K3
Downs, Ill. (61736) 222/E3
Downs, Kansas (67437) 232/D2
Downsville, La. (71234) 238/F1
Downsville, Md. (†21795) 245/G2
Downsville, N.Y. (13755) 276/L6
Downsville, Wis. (54735) 317/C6
Downton, England 13/F6
Dows, Iowa (50071) 229/F3
Dowshi, Afghanistan 59/J2
Dowshi, Afghanistan 68/B1
Doyle, Calif. (96109) 204/E3
Doyle, Georgia (†31803) 217/D6
Doyle, Tenn. (38559) 237/K9
Doylestown, Ohio (44230) 284/G4
Doylestown, Pa. (18901) 294/M5
Doylestown, Wis. (53928) 317/H9
Doyleville, Colo. (†81239) 208/F6
Doyline, La. (71023) 238/D1
Doyon, N. Dak. (58328) 282/O3
Dozen (isls.), Japan 81/F5
Dozier, Ala. (36028) 195/F7
Dozier, Texas (†79079) 303/D2
Dozois (res.), Québec 174/B3
Dra, Wadi (dry riv.), Morocco 106/C3
Drachten, Netherlands 27/J2
Dracut○, Mass. (01826) 249/J2
Drăgănești Olt, Romania 45/G3
Drăgășani, Romania 45/F3
Dragonera (isl.), Spain 33/H3
Dragons Mouth (str.), Trin. & Tob. 156/F5
Dragons Mouth (str.), Trin. & Tob. 161/A10
Dragons Mouth (str.), Venezuela 124/H2
Dragoon, Ariz. (85609) 198/F6
Dragoon (mts.), Ariz. 198/F7
Draguignan, France 28/G6
Drain, Oreg. (97435) 291/D4
Drake (passage) 2/F8
Drake (passage) 5/C15
Drake (passage), Chile 138/E11
Drake, Colo. (80515) 208/J2
Drake, Mo. (†65066) 261/K6
Drake, N. Dak. (58736) 282/K4
Drake, Sask. 181/G4
Drakensberg (range), Lesotho 118/D6
Drakensberg (range), S. Africa 118/D6
Drakensberg (range), Swaziland 118/D6
Drakes (creek), Ky. 237/J7
Drakesboro, Ky. (42337) 237/H6
Drakes Branch, Va. (23937) 307/L7
Drakesville, Iowa (52552) 229/J7
Draketown, Georgia (†30179) 217/B3
Dráma, Greece 45/F5
Drammen, Norway 7/E3
Drammen, Norway 18/C4
Drance (riv.), Switzerland 39/D4
Drancy, France 28/B1
Drang, Ia (riv.), Cambodia 72/E4
Draper, S. Dak. (57531) 298/J6
Draper, Utah (84020) 304/C3
Draper, Va. (24324) 307/G7
Draper, Wis. (†54852) 317/E4
Draperstown, N. Ireland 17/H2
Draperstown, N. Ireland 10/C3
Drasco, Ark. (72530) 202/G2
Drau (riv.), Austria 41/C3
Drava (riv.) 7/F4
Dráva (riv.), Hungary 41/D3
Drava (riv.), Yugoslavia 45/C3
Dravosburg, Pa. (15034) 294/C7
Drawsko Pomorskie, Poland 47/B2
Drax Hall, Barbados 161/B8
Drayden, Md. (20630) 245/N8
Drayton, N. Dak. (58225) 282/R2
Drayton, Ontario 177/D4
Drayton Plains, Mich. (48020) 250/F6
Drayton Valley, Alberta 182/C3
Drenthe (prov.), Netherlands 27/K3
Dresbach, Minn. (55930) 255/G7
Dresden, E. Germany 7/F3
Dresden (dist.), E. Germany 22/E3
Dresden, E. Germany 22/E3
Dresden, Kansas (67635) 232/B2
Dresden○, Maine (04342) 243/D7
Dresden, Mo. (†65301) 261/F5
Dresden, N.Y. (14441) 276/F5
Dresden, N. Dak. (†58249) 282/O2
Dresden, Ohio (43821) 284/G5
Dresden, Ontario 177/B5
Dresden, Tenn. (38225) 237/D8
Dresden Station, N.Y. (†12887) 276/O3
Dresser, Wis. (54009) 317/A5
Dreux, France 28/D3
Drew (co.), Ark. 202/G6
Drew, Miss. (38737) 256/C3
Drew, Oreg. (†97484) 291/E5
Drewry, Ala. (†36460) 195/D8
Drewryville, Va. (23844) 307/O7
Drews (res.), Oreg. 291/G5
Drewsey, Oreg. (97904) 291/J4
Drewsville, N.H. (03604) 268/C5
Drexel, Mo. (64742) 261/C6
Drexel, N.C. (28619) 281/F3
Drexel Hill, Pa. (19026) 294/M6
Dreyfus, Ky. (40426) 237/N5
Drezdenko, Poland 47/B2
Driebergen, Netherlands 27/G4
Driffield, England 13/G3
Driffield, England 10/F4
Drift (creek), Oreg. 291/B3
Drifton, Pa. (18221) 294/L3
Driftwood, Okla. (†73722) 288/K1
Driftwood, Pa. (15832) 294/F3
Driggs, Ark. (†72943) 202/C3
Driggs, Idaho (83422) 220/G6
Drill, Va. (†24260) 307/E6
Drimoleague, Ireland 17/C8
Drin (riv.), Albania 45/E4
Drina (riv.), Yugoslavia 45/D3
Drinkwater, Sask. 181/F5
Dripping Springs, Texas (78620) 303/F7
Driscoll, N. Dak. (58532) 282/K6
Driscoll, Texas (78351) 303/G10
Drishane, Ireland 17/C7
Driskill (mt.), La. 238/E2
Drøbak, Norway 18/D4
Drobeta-Turnu Severin, Romania 45/F3
Drogenbos, Belgium 27/B10
Drogheda, Ireland 17/J4
Drogheda, Ireland 10/C4
Drogobych, U.S.S.R. 52/B5
Drogobych, U.S.S.R. 48/C5
Droichead Nua, Ireland 10/C4
Droichead Nua, Ireland 17/H5
Droitwich, England 13/E5
Dromahair, Ireland 17/E3
Drome (dept.), France 28/F5
Drome (riv.), France 28/F5
Dromore, Bainbridge, N. Ireland 17/J3
Dromore, Omagh, N. Ireland 17/G3
Dromore West, Ireland 17/D3
Dronfield, England 13/J2
Drongan, Scotland 15/D5
Dronne (riv.), France 28/D5
Dronninglund, Denmark 21/D3
Dronten (prov.), Netherlands 27/H4
Dronten, Netherlands 27/H3
Dropmore, Manitoba 179/A3
Drouin, Victoria 97/C6
Druid, Sask. 181/C4
Druif, Neth. Ant. 161/D10
Drum (hills), Ireland 17/F7
Drum (bay), La. 238/M7
Drum (inlet), N.C. 281/S5
Drumaness, N. Ireland 17/K3
Drumbeg, Scotland 15/C2
Drumbo, Ontario 177/D4
Drumcar, Ireland 17/J4
Drumconrath, Ireland 17/H4
Drumheller, Alberta 182/D4
Drumheller, Alta. 162/E5
Drumhill, N.C. (†27937) 281/R1
Drumkeerin, Ireland 17/E3
Drumlish, Ireland 17/F4
Drummond, Idaho (†83420) 220/G5
Drummond (isl.), Mich. 250/F2
Drummond, Mont. (59832) 262/D4
Drummond, New Bruns. 170/C1
Drummond (mt.), North. Terr. 93/E5
Drummond, Okla. (73735) 288/L2
Drummond (co.), Québec 172/E4
Drummond (range), Queensland 88/H4
Drummond (range), Queensland 95/C5
Drummond (lake), Va. 307/P7
Drummond, Wis. (54832) 317/D3
Drummond Island, Mich. (49726) 250/F3
Drummonds, Tenn. (38023) 237/A10
Drummondville, Québec 172/E4
Drummondville-Nord, Québec 172/E4
Drummondville-Sud, Québec 172/E4
Drummore, Scotland 15/D6
Drummoyne, N. S. Wales 88/K4
Drummoyne, N.S. Wales 97/J3
Drumnadrochit, Scotland 15/D3
Drumquin, N. Ireland 17/F2
Drumright, Okla. (74030) 288/N3
Drums, Pa. (18222) 294/K3
Drumshanbo, Ireland 17/E3
Drury, Mo. (65638) 261/H9
Druskininkai, U.S.S.R. 53/C3
Druten, Netherlands 27/H5
Druz, Jebel ed (mts.), Syria 63/G6
Druzhba, U.S.S.R. 48/J5
Druzhina, U.S.S.R. 48/P3
Drvar, Yugoslavia 45/C3
Dry (bay), Alaska 196/L3
Dry (creek), Ky. 237/R3
Dry (lake), N. Dak. 282/M3
Dry (riv.), North. Terr. 88/E3
Dry (riv.), North. Terr. 93/C3
Dry (creek), S. Dak. 298/G4
Dry (lake), S. Dak. 298/P3
Dry (creek), Wyo. 319/C2
Dryad, Wash. (†98532) 310/B4
Dryanovo, Bulgaria 45/G4
Dry Branch, Georgia (31020) 217/F5
Dry Cimarron (riv.), N. Mex. 274/F2
Dry Coal (creek), Utah 304/A6
Dry Cottonwood (creek), Wyo. 319/D1
Dry Creek, La. (70637) 238/D5
Dry Creek, W. Va. (25062) 312/vD7
Dryden, Ark. (†72401) 202/J2
Dryden, Maine (04225) 243/C6
Dryden, Mich. (48428) 250/F6
Dryden, N.Y. (13053) 276/H6
Dryden, Ontario 177/G4
Dryden, Ontario 175/B3
Dryden, Texas (78851) 303/C7
Dryden, Va. (24243) 307/B7
Dryden, Wash. (98821) 310/E3
Dry Falls (dam), Wash. 310/F3
Dry Fork, Va. (24549) 307/K7
Dryfork, W. Va. (26263) 312/H5
Dry Fork (riv.), W. Va. 312/G5
Dry Fork (riv.), W. Va. 312/C8
Dry Fork, Cheyenne (riv.), Wyo. 319/G2
Dry Fork, Powder (riv.), Wyo. 319/F2
Dry Lake, Nev. (†89040) 266/G6
Drymen, Scotland 15/B1
Dry Mills, Maine (†04039) 243/C8
Dry Prong, La. (71423) 238/E3
Dry Ridge, Ky. (41035) 237/M3
Dry Run, Pa. (17220) 294/G5
Drysdale (riv.), W. Australia 88/D3
Drysdale (riv.), W. Australia 92/D1
Dry Tortugas (keys), Fla. 212/D7
Drytown, Calif. (95699) 204/C8
Dry Wood (lake), S. Dak. 298/P2
Dschang, Cameroon 115/A2
Duaca, Venezuela 124/D2
Duaringa, Queensland 95/D4
Duart, Ontario 177/C5
Duarte, Calif. (91010) 204/D10
Duarte (prov.), Dom. Rep. 158/E5
Duarte (peak), Dom. Rep. 158/D5
Dubach, La. (71235) 238/E1
Dubai, U.A.E. 59/F4
Dubawnt (lake), N.W.T. 162/F3
Dubawnt (lake), N.W.T. 146/H3
Dubawnt (lake), N.W. Terrs. 187/H3
Dubawnt (riv.), N.W.T. 162/F3
Dubawnt (riv.), N.W. Terrs. 187/H3
Du Bay (lake), Wis. 317/G6
Dubberly, La. (71024) 238/D1
Dubbo, N. S. Wales 88/H6
Dubbo, N.S. Wales 97/E3
Dubbs, Miss. (†38626) 256/D1
Dübendorf, Switzerland 39/G2
Dublin, Calif. (94566) 204/K2
Dublin, Georgia (31021) 217/G5
Dublin, Ind. (47335) 227/G5
Dublin (co.), Ireland 17/J5
Dublin (cap.), Ireland 7/D3
Dublin (cap.), Ireland 17/K5
Dublin (cap.), Ireland 10/C4
Dublin (bay), Ireland 10/C4
Dublin (bay), Ireland 17/J5
Dublin, Ky. (†42039) 237/D7
Dublin, Md. (†21154) 245/N2
Dublin, Mich. (†49689) 250/D4
Dublin, Miss. (38739) 256/C2
Dublin○, N.H. (03444) 268/C6
Dublin, N.C. (28332) 281/M5
Dublin, Ohio (43017) 284/D5
Dublin, Ontario 177/C4
Dublin, Pa. (18917) 294/M5
Dublin, Texas (76446) 303/F5
Dublin, Va. (24084) 307/G6
Dubna, U.S.S.R. 52/E4
Dubna, U.S.S.R. 52/E3
Dubnica nad Váhom, Czech. 41/E2
Dubno, U.S.S.R. 52/C4
Dubois, Idaho (83423) 220/F5
Dubois, Ill. (62831) 222/D5
Dubois (co.), Ind. 227/D8
Dubois, Ind. (47525) 227/D8
Du Bois, Nebr. (68345) 264/H4
DuBois, Pa. (15801) 294/E3
Dubois, Wyo. (82513) 319/C2
Duboistown, Pa. (†17701) 294/H3
Dubréka, Guinea 106/B7
Dubreuilville, Ontario 177/J5
Dubrovnik, Yugoslavia 45/C4
Dubreuilville, Ontario 175/D3
Dubuc, Sask. 181/J5
Dubuque (co.), Iowa 229/M4
Dubuque, Iowa 188/H2
Dubuque, Iowa (52001) 229/M3
Duchcov, Czech. 41/B1
Duchesne (co.), Utah 304/D3
Duchesne, Utah (84021) 304/D3
Duchesne (riv.), Utah 304/D3
Duchess, Alberta 182/E4
Duchess, Queensland 88/F4
Duchess, Queensland 95/A4
Ducie (isl.), Pitcairn Is. 87/O8
Duck (isls.), Maine 243/G7
Duck (lake), Mich. 250/F4
Duck (creek), Nev. 266/G3
Duck, N.C. (†27949) 281/T2
Duck (creek), Ohio 284/H6
Duck (isl.), Ontario 177/H4
Duck (isls.), Ontario 177/A2
Duck (riv.), Tenn. 237/F9
Duck, W. Va. (25063) 312/E5
Duck Bay, Manitoba 179/B2
Duck Hill, Miss. (38925) 256/E3
Duck Lake, Sask. 181/E3
Duck Lake Hist. Park, Sask. 181/E3
Duck Lake Post, Manitoba 179/J2
Duck Mountain Prov. Park, Manitoba 179/B3
Duck Mountain Prov. Park, Sask. 181/K4
Duck River, Tenn. (38454) 237/G9
Ducktown, Georgia (†30130) 217/D2
Ducktown, Tenn. (37326) 237/N10
Duckwater, Nev. (89314) 266/F4
Duclos, Québec 172/A4
Ducor, Calif. (93218) 204/G8
Ducos, Martinique 161/D6
Dudelange, Luxembourg 27/J10
Dudenville, Mo. (†64748) 261/D8
Duderstadt, W. Germany 22/D3
Dudhi, India 68/E4
Dudignac, Argentina 143/F7
Düdingen, Switzerland 39/D3
Dudinka, U.S.S.R. 54/K3
Dudinka, U.S.S.R. 4/B5
Dudinka, U.S.S.R. 48/J3
Dudley, England 13/E5
Dudley, England 10/G3
Dudley, Georgia (31022) 217/F5
Dudley○, Mass. (01570) 249/G4
Dudley, Mo. (63936) 261/M9
Dudley, N.C. (28333) 281/N4
Dudley, Pa. (16634) 294/F5
Dudley (lake), Québec 172/B3
Dudleytown, Ind. (†47274) 227/F7
Dudváh (riv.), Czech. 41/D2
Dudweiler, W. Germany 22/B4
Dueñas, Spain 33/D2
Duenweg, Mo. (64841) 261/D8
Duero (Douro) (riv.), Spain 33/C2
Due West, S.C. (29639) 296/C3
Duff, Sask. 181/H5
Duff, Tenn. (37729) 237/N8
Duffee, Miss. (†39337) 256/G6
Duffel, Belgium 27/F6
Dufferin (county), Ontario 177/D3
Duffield, Alberta 182/C3
Duffield, Va. (24244) 307/C7
Dufftown, Scotland 10/E2
Dufftown, Scotland 15/E3
Dufourspitze (mt.), Switzerland 39/E5
Dufresne, Manitoba 179/F5
Dufrost, Manitoba 179/E5
Dufur, Oreg. (97021) 291/F2
Dugald, Manitoba 179/F5
Dugger, Ind. (47848) 227/C6
Dugi Otok (isl.), Yugoslavia 45/B3
Dugspur, Va. (24325) 307/G7
Duguayville, New Bruns. 170/E1
Du Gué (riv.), Québec 174/C1
Dugway, Utah (84022) 304/B3
Dugway (range), Utah 304/A3
Dugway Proving Grounds, Utah 304/B3
Duhamel, Alberta 182/D3
Duhamel, Québec 172/B3
Duich, Loch (inlet), Scotland 15/C3
Duida, Cerro (mt.), Venezuela 124/F6
Duifken (pt.), Queensland 88/G2
Duifken (pt.), Queensland 95/B2
Duiker (pt.), S. Africa 118/E6
Duinain (riv.), Scotland 15/D3
Duirinish (dist.), Scotland 15/B3
Duisburg, W. Germany 22/B3
Duitama, Colombia 126/D5
Duiveland (isl.), Netherlands 27/D5
Duivendrecht, Netherlands 27/C5
Duke, Ala. (†36279) 195/G3
Duke (isl.), Alaska 196/N2
Duke, Mo. (65461) 261/H7
Duke, Okla. (73532) 288/G5
Duke Center, Pa. (16729) 294/F2
Dukedom, Tenn. (38226) 237/D8
Duke of Gloucester (isls.), Fr. Poly. 87/M8
Dukes (co.), Mass. 249/M7
Dukes, Mich. (†49885) 250/B2
Dukhan, Qatar 59/F4
Duki, Pakistan 68/B2
Dukla (pass), Czech. 41/F2
Dukla (pass), Poland 47/E4
Dukou, China 77/F6
Dulac, La. (70353) 238/J8
Dulah, N.C. (†28463) 281/M6
Dulan, China 77/E4
Dulce (riv.), Argentina 143/D2
Dulce (gulf), C. Rica 154/F6
Dulce, N. Mex. (87528) 274/B2
Duleek, Ireland 17/J4
Dulgalakh (riv.), U.S.S.R. 48/O3
Dülmen, W. Germany 22/B3
Dulunguin (pt.), Philippines 82/C7
Duluth, Georgia (30136) 217/D2
Duluth, Kansas (66421) 232/F2
Duluth, Minn. 146/J5
Duluth, Minn. 188/H1
Duluth, Minn. (*55801) 255/F4
Dulverton, England 13/D6
Duma, Syria 63/G6
Duma, West Bank 65/C3
Dumagasa (pt.), Philippines 82/C7
Dumaguete, Philippines 82/D6
Dumaguete, Philippines 85/G4
Dumanquilas (bay), Philippines 82/D7
Dumaran (isl.), Philippines 85/G3
Dumaran (isl.), Philippines 82/C5
Dumaresq (riv.), N.S. Wales 97/F1
Dumas, Ark. (71639) 202/H6
Dumas, Miss. (38625) 256/G1
Dumas, Sask. 181/J6
Dumas, Texas (79029) 303/C2
Dumbarton, New Bruns. 170/C3
Dumbarton, Scotland 10/A1
Dumbarton, Scotland 15/B1
Dum Dum, India 68/F1
Dume (pt.), Calif. 204/G10
Dumeir, Syria 63/G6
Dumfoundling (bay), Fla. 212/C4
Dumfries, New Bruns. 170/C3
Dumfries, Scotland 15/E5
Dumfries, Scotland 10/E3
Dumfries (trad. co.), Scotland, 15/B5
Dumfries, Va. (22026) 307/O3
Dumfries and Galloway (reg.), Scotland 15/E5
Dumlu, Turkey 63/J2
Dummer○, N.H. (†03588) 268/E2
Dummer, Sask. 181/G6
Dümmersee (lake), W. Germany 22/C2
Dumont, Iowa (50625) 229/H3
Dumont, Minn. (56236) 255/B5
Dumont, N.J. (07628) 273/C1
Dumont, Texas (79232) 303/D4
Dumont d'Urville Station 5/C7
Dumyât (Damietta), Egypt 111/J3
Dumyat (Damietta), Egypt 59/B3
Dun (isl.), Scotland 15/A2
Duna (Danube) (riv.), Hungary 41/E3
Dunaff (head), Ireland 17/F1
Dunaföldvár, Hungary 41/E3
Dunaharaszti, Hungary 41/E3
Dunajec (riv.), Czech. 41/F2
Dunajec (riv.), Poland 47/E4
Dunajská Streda, Czech. 41/D3
Dunakeszi, Hungary 41/E3
Dunalley, Tasmania 99/D4
Dunany (pt.), Ireland 17/J4
Dunaszekcső, Hungary 41/E3
Dunaújváros, Hungary 41/E3
Dunav (Danube) (riv.), Bulgaria 45/H4
Dunavecse, Hungary 41/E3
Dunbar, Iowa (†50158) 229/H5
Dunbar, Nebr. (68346) 264/J4
Dunbar, Okla. (†74557) 288/P6
Dunbar, Pa. (15431) 294/C6
Dunbar, Scotland 10/E2
Dunbar, Scotland 15/F4
Dunbar, S.C. (†29525) 296/H2
Dunbar, W. Va. (25064) 312/C6
Dunbar, Wis. (54119) 317/K4
Dunbarton○, N.H. (†03301) 268/D5
Dunbarton (trad. co.), Scotland 15/A5
Dunbarton Center, N.H. (†03301) 268/D5
Dunbeath, Scotland 15/E2
Dunbeg, Scotland 15/C4
Dunblane, Sask. 181/D4
Dunblane, Scotland 15/E4
Dunblane, Scotland 10/D2
Dunbridge, Ohio (43414) 284/C3
Duncan, Ariz. (85534) 198/F6
Duncan, Br. Col. 184/J3
Duncan (riv.), Br. Col. 184/J5
Duncan (isls.), China 85/E2
Duncan, Ill. (†61559) 222/D3
Duncan (passage), India 68/G6
Duncan, Miss. (38740) 256/C2
Duncan, Nebr. (68634) 264/G3
Duncan, Okla. (73533) 288/L5
Duncan (lake), Québec 174/B2
Duncan, S.C. (29334) 296/C2
Duncan, W. Va. (25240) 312/C5
Duncan Falls, Ohio (43734) 284/G6
Duncannon, Ireland 17/H7
Duncannon, Pa. (17020) 294/H5
Duncans, Jamaica 158/H5
Duncans Bridge, Mo. (†63437) 261/H3
Duncansby (head), Scotland 15/F2
Duncansby (head), Scotland 10/E1
Duncansville, Pa. (16635) 294/F5
Duncanville, Ala. (35456) 195/D4
Duncanville, Texas (75116) 303/G3
Dunchurch, Ontario 177/E2
Duncombe, Iowa (50532) 229/E4
Duncombe (bay), Norfolk I. 88/L5
Dundaga, U.S.S.R. 53/B2
Dundalk, Ireland 17/H3
Dundalk, Ireland 10/C4
Dundalk (bay), Ireland 10/C4
Dundalk (bay), Ireland 17/J4
Dundalk, Md. (21222) 245/N3
Dundalk, Ontario 177/D3
Dundarrach, N.C. (†28386) 281/L5
Dundas (isl.), Br. Col. 184/B3
Dundas, Greenl. 4/B13
Dundas, Greenland 146/M2
Dundas, Ill. (62425) 222/E5
Dundas, Minn. (55019) 255/E6
Dundas (str.), North. Terr. 88/E2
Dundas (str.), North. Terr. 93/B1
Dundas (pen.), N.W. Terrs. 187/G2
Dundas, Ohio (45625) 284/E7
Dundas (county), Ontario 177/J2
Dundas, Ontario 177/E4
Dundas, Va. (23938) 307/M7
Dundas (lake), W. Australia 88/C6
Dundas (lake), W. Australia 92/C6
Dundee, Fla. (33838) 212/E3
Dundee (East and West Dundee), Ill. (60118) 222/E1
Dundee, Ind. (†47348) 227/F4
Dundee, Iowa (52038) 229/L3
Dundee, Ky. (42338) 237/H5
Dundee, Mich. (48131) 250/F7
Dundee, Minn. (56126) 255/C7
Dundee, Miss. (38626) 256/D1
Dundee, N.Y. (14837) 276/F5
Dundee, Oreg. (97115) 291/A2
Dundee, Scotland 7/D3
Dundee, Scotland 15/F4
Dundee, Scotland 10/E2
Dundee, S. Africa 118/E5
Dundee, Texas (76358) 303/F4
Dundgovĭ, Mongolia 77/G2
Dundon, W. Va. (†25043) 312/D6
Dundonald, Scotland 15/D5
Dundrum, N. Ireland 17/K3
Dundrum (bay), N. Ireland 17/K3
Dundurn, Sask. 181/E4
Dundy (co.), Nebr. 264/C4
Dune Acres, Ind. (†46304) 227/C1
Dunedin, Fla. (33528) 212/B2
Dunedin, N. Zealand 2/T8
Dunedin, N. Zealand 100/C6
Dunedoo, N.S. Wales 97/E3
Dunellen, N.J. (08812) 273/D2
Dunes (Westlake), Oreg. (†97493) 291/C4
Dunfanaghy, Ireland 17/F1
Dunfee, Ind. (†46802) 227/G2
Dunfermline, Ill. (61524) 222/D3
Dunfermline, Sask. 181/D3
Dunfermline, Scotland 15/D1
Dunfermline, Scotland 10/C1
Dungalear Station, N.S. Wales 97/D1
Dungannon (dist.), N. Ireland 17/H3
Dungannon, N. Ireland 17/H3
Dungannon, Ontario 177/C4
Dungannon, Va. (24245) 307/D7
Dungarpur, India 68/C4
Dungarvan, Ireland 10/C4
Dungarvan, Ireland 17/F7
Dungarvan (harb.), Ireland 10/C4
Dungarvan (harb.), Ireland 17/G7
Dungarvon (riv.), New Bruns. 170/D2
Dungeness (pt.), Argentina 143/C7
Dungeness (pt.), Chile 138/F10
Dungeness (prom.), England 13/J7
Dungeness (prom.), England 10/G5
Dungeness, Wash. (†98382) 310/B2
Dungiven, N. Ireland 17/H2
Dungloe, Ireland 17/E2
Dungog, N.S. Wales 97/F3
Dungu, Zaire 115/E3
Dungunab, Sudan 59/C5
Dungunab, Sudan 111/G3
Dunham, Québec 172/E4
Dunhua (Tunhwa), China 77/L3
Dunhuang, China 77/E3
Dunkeld, Queensland 95/D5
Dunkeld, Scotland 15/E4
Dunkeld, Victoria 97/B5
Dunkellin (riv.), Ireland 17/D5
Dunkerton, Iowa (50626) 229/J3
Dunkery (hill), England 13/D6
Dunkineely, Ireland 17/E2
Dunkirk (riv.), Alberta 182/D1
Dunkirk (Dunkerque), France 28/E2
Dunkirk, France 28/E2
Dunkirk, Ind. (47336) 227/G4
Dunkirk, N.Y. (14048) 276/B5
Dunkirk, Ohio (45836) 284/C4
Dunkley, Br. Col. 184/F3
Dunklin (co.), Mo. 261/M10
Dunkwa, Ghana 106/D7
Dún Laoghaire, Ireland 10/D4
Dún Laoghaire, Ireland 17/K5
Dunlap, Ill. (61525) 222/D3
Dunlap, Ind. (†46514) 227/F1
Dunlap, Iowa (51529) 229/B5
Dunlap, Kansas (66848) 232/F3
Dunlap, Tenn. (37327) 237/L10
Dunlavin, Ireland 17/H5
Dunleath, Sask. 181/K4
Dunleer, Ireland 17/J4
Dunleith, Miss. (†38756) 256/C4
Dunlow, W. Va. (25511) 312/B6
Dunloy, N. Ireland 17/J1
Dunmanus (bay), Ireland 17/B8
Dunmanway, Ireland 17/C8
Dunmanway, Ireland 10/B5
Dunmor, Ky. (42339) 237/G6
Dunmore, Alberta 182/E5
Dunmore, Ireland 17/D4
Dunmore, Pa. (18512) 294/F7
Dunmore (lake), Vt. 268/A4
Dunmore, W. Va. (24934) 312/G6
Dunmore East, Ireland 17/G7
Dunn, La. (71236) 238/G2
Dunn, N.C. (28334) 281/M4
Dunn (co.), N. Dak. 282/E5
Dunn, Texas (79516) 303/D5
Dunn (co.), Wis. 317/C6
Dunnamanagh, N. Ireland 17/G2
Dunn Center, N. Dak. (58626) 282/E5
Dunnegan, Mo. (65640) 261/E7
Dunnell, Minn. (56127) 255/D7
Dunnellon, Fla. (32630) 212/D2
Dunnet, Scotland 15/E2
Dunnet (bay), Scotland 15/E2
Dunnet (head), Scotland 10/E1
Dunnet (head), Scotland 15/E2
Dunnigan, Calif. (95937) 204/C5
Dunning, Nebr. (68833) 264/D3
Dunning, Scotland 15/E4
Dunn Loring, Va. (22027) 307/S2
Dunnottar, Manitoba 179/E4
Dunnottar, S. Africa 118/J6
Dunns, W. Va. (†25841) 312/D7
Dunnsville, Va. (22454) 307/P5
Dunnville, Ky. (42528) 237/M6
Dunnville, Ontario 177/E5
Du Noir (riv.), Wyo. 319/C2
Dunolly, Victoria 97/B5
Dunoon, Scotland 15/A2
Dunoon, Scotland 10/A1
Dunphy, Nev. (†89821) 266/E2
Dunragit, Scotland 15/D6
Dunrea, Manitoba 179/C5
Dunreith, Ind. (47337) 227/F5
Duns, Scotland 10/E3
Duns, Scotland 15/F5
Dunscore, Scotland 15/E5
Dunseith, N. Dak. (58329) 282/K2
Dunshaughlin, Ireland 17/H5
Dunsmuir, Calif. (96025) 204/C2
Dunstable, England 10/F5
Dunstable, England 13/G6
Dunstable○, Mass. (01827) 249/J2
Dunster, Br. Col. 184/G3
Duntochter, Scotland 15/B2
Dunure, Scotland 15/D5
Dunvegan, Nova Scotia 168/F4
Dunvegan, Scotland 15/B3
Dunvegan, Loch (inlet), Scotland 15/B3
Dunville, Newf. 166/D2
Dunwoody, Georgia (†30338) 217/K1
Duo, W. Va. (†25984) 312/E6
Duolun, China 77/J3
Duong Dong, Vietnam 72/D5
Du Page (co.), Ill. 222/E2
Du Page, East Branch (riv.), Ill. 222/A6
Du Page, West Branch (riv.), Ill. 222/A6
Du Page (riv.), Ill. 222/E2
Duparquet, Québec 174/B3
Duperow, Sask. 181/C4
Duplessis, La. (70728) 238/K2
Duplin (co.), N.C. 281/O5
Dupo, Ill. (62239) 222/A3
Du Pont, Georgia (31630) 217/G9
Dupont, Ind. (47231) 227/G7
Dupont, Ohio (45837) 284/B3
Dupont, Pa. (18641) 294/F7
Du Pont, Wash. (98327) 310/C3
Dupont Manor, Del. (†19901) 245/R4
Dupree, S. Dak. (57623) 298/F3
Dupuis Corner, New Bruns. 170/F2
Dupuy, Québec 174/B3
Dupuyer, Mont. (59432) 262/D2
Duque de Bragança, Angola 115/C5
Duque de Caxias, Brazil 135/E3
Duque de York (isl.), Chile 138/C9
Duquesne, Mo. (†64801) 261/D8
Duquesne, Pa. (15110) 294/C7
Duquette, Minn. (55729) 255/F4
Du Quoin, Ill. (62832) 222/D5
Duquoin, Kansas (†67058) 232/D4
Dura, West Bank 65/C4
Durack (range), W. Australia 88/D3
Durağan, Turkey 63/F2
Duran, N. Mex. (88319) 274/D4
Durance (riv.), France 28/F6

Durand, Georgia (†31830) 217/C5
Durand, Ill. (61024) 222/D1
Durand, Mich. (48429) 250/E6
Durand, Wis. (54736) 317/C6
Durango, Colo. 188/E3
Durango, Colo. (81301) 208/D8
Durango, Iowa (52039) 229/M3
Durango (state), Mexico 150/G4
Durango, Mexico 146/H7
Durango, Mexico 150/G4
Durango, Spain 33/E1
Duranillin, W. Australia 92/B2
Durant, Iowa (52747) 229/M5
Durant, Miss. (39063) 256/E4
Durant, Okla. 188/G4
Durant, Okla. (74701) 288/O6
Duratón (riv.), Spain 33/E2
Durazno (dept.), Uruguay 145/C3
Durazno, Uruguay 145/C4
Durazno, Grande del (range), Uruguay 145/D4
Durban, Manitoba 179/A3
Durban, S. Africa 2/L7
Durban, S. Africa 102/F7
Durban, S. Africa 118/E5
Durbanville, S. Africa 118/F6
Durbe, U.S.S.R. 53/B2
Durbin, Ind. (†46060) 227/F4
Durbin, N. Dak. (58023) 282/R6
Durbin, W. Va. (26264) 312/G5
Durbuy, Belgium 27/H8
Düren, W. Germany 22/B3
Durfee (hill), R.I. 249/G5
Durg, India 68/E4
Durgapur, India 68/F4
Durgerdam, Netherlands 27/C4
Durham, Ark. (†72701) 202/C2
Durham, Calif. (95938) 204/D4
Durham, Conn. (06422) 210/E3
Durham○, Conn. (06422) 210/E3
Durham (co.), England 13/F3
Durham, England 13/J3
Durham, England 10/F3
Durham, Kansas (67438) 232/E3
Durham, Mo. (63438) 261/J3
Durham, N.H. (03824) 268/F5
Durham○, N.H. (03824) 268/F5
Durham (pt.), N. Zealand 100/D7
Durham, N.C. 188/L3
Durham (co.), N.C. 281/M3
Durham, N.C. (*27701) 281/M2
Durham, Okla. (73642) 288/G3
Durham (reg. munic.), Ontario 177/F3
Durham, Ontario 177/D3
Durham, Oreg. (†97233) 291/A2
Durham Bridge, New Bruns. 170/D2
Durham Center, Conn. (†06422) 210/E3
Durham Downs, Queensland 95/B5
Durham-Sud, Québec 172/E4
Durhamville, N.Y. (13054) 276/J4
Duri, N.S. Wales 97/F2
Durkee, Oreg. (97905) 291/K3
Durness, Scotland 15/D2
Durnford (pt.), Western Sahara 106/A4
Dürnten, Switzerland 39/G2
Duror, Scotland 15/C4
Durrell, Newf. 166/D4
Dürrenroth, Switzerland 39/E2
Durrës (Durazzo), Albania 45/D5
Durrës, Albania 7/F4
Durrington, England 13/F6
Durrow, Laoighis, Ireland 17/G6
Durrow, Offaly, Ireland 17/F5
Dursey (isl.), Ireland 17/A8
Dursunbey, Turkey 63/C3
Duruh, Iran 59/H3
Duruh, Iran 66/M4
D'Urville (isl.), N. Zealand 100/D4
Duryea, Pa. (18642) 294/F7
Dusa Marreb, Somalia 115/J2
Dūsh, Egypt 59/B5
Dūsh, Egypt 111/F3
Dushan, China 77/G6
Dushanbe, U.S.S.R. 54/H6
Dushanbe, U.S.S.R. 2/N4
Dushanbe, U.S.S.R. 48/G6
Dushore, Pa. (18614) 294/K2
Dusky (sound), N. Zealand 100/A6
Duson, La. (70529) 238/F6
Düsseldorf, W. Germany 7/E3
Düsseldorf, W. Germany 22/B3
Dustin, Okla. (74839) 288/O4
Dusty, N. Mex. (87934) 274/B5
Dusty, Wash. (†99143) 310/H4
Dutch (creek), Ark. 202/C4
Dutch Cap (cay), Virgin Is. (U.S.) 161/A4
Dutchess (co.), N.Y. 276/N7
Dutch Flat, Calif. (95714) 204/E4
Dutch Harbor, Alaska (†99685) 196/E4
Dutch John, Utah (84023) 304/E3
Dutch Mills, Ark. (†72744) 202/B2
Dutch Neck, N.J. (†08550) 273/D3
Dutchtown, Mo. (63745) 261/N8
Dutton, Ala. (35744) 195/G1
Dutton, Ark. (†72760) 202/C2
Dutton (mt.), Conn. 210/C1
Dutton, Mont. (59433) 262/E3
Dutton, Ontario 177/C5
Dutton (mt.), Utah 304/B5
Duval (co.), Fla. 212/E1
Duval, Sask. 181/G4
Duval (co.), Texas 303/F10
Duvalierville, Haiti 158/C6
Duvall, Wash. (98019) 310/D3
Duvergé, Dom. Rep. 158/D6
Duvernay, Alberta 182/E3
Duwadami, Saudi Arabia 59/D5
Duxbury (pt.), Calif. 204/H2
Duxbury, Mass. (02332) 249/M4
Duxbury○, Mass. (02332) 249/M4
Duxbury○, Vt. (†05676) 268/B3
Duyun (Tuyün), China 77/G6
Düzce, Turkey 63/D2
Duzdab (Zahedan), Iran 66/M6
Dvina, (bay), U.S.S.R. 52/E2
Dvina, Northern (riv.), U.S.S.R. 4/C7
Dvina, Northern (riv.), U.S.S.R. 7/J2
Dvina, Northern (riv.), U.S.S.R. 48/E3
Dvina, Northern (riv.), U.S.S.R. 52/F2
Dvina, Western (riv.) U.S.S.R. 53/C2
Dvina, Western (riv.), U.S.S.R. 48/C4
Dvina, Western (riv.), U.S.S.R. 52/C3
Dvina, Western (riv.), U.S.S.R. 7/G3
Dvinsk (Daugavpils), U.S.S.R. 52/C3
Dvory nad Žitavou, Czech. 41/E3
Dvůr Králové nad Labem, Austria 41/C1
Dwale, Ky. (41621) 237/R5
Dwarka, India 68/B4
Dwellingup, W. Australia 92/B2
Dwight, Ill. (60420) 222/E2
Dwight, Kansas (66849) 232/F3
Dwight, Nebr. (68635) 264/G3
Dwight, N. Dak. (58024) 282/S7
Dwight, Ontario 177/F2
Dworshak (res.), Idaho 220/C3
Dwyer, N. Mex. (†88034) 274/B6
Dwyer, Wyo. (82211) 319/G3
Dyas, Ala. (†36507) 195/C9
Dyat'kovo, U.S.S.R. 52/D4
Dybvad, Denmark 21/D3
Dyce, Scotland 15/F3
Dyckesville, Wis. (†54217) 317/L6
Dycusburg, Ky. (42037) 237/E6
Dyer, Ark. (72935) 202/B3
Dyer, Ind. (46311) 227/C1
Dyer, Ky. (†40115) 237/J5
Dyer, Nev. (89010) 266/C5
Dyer (cape), N.W.T. 162/K2
Dyer (cape), N.W. Terrs. 187/M3
Dyer (co.), Tenn. 237/C8
Dyer, Tenn. (38330) 237/D8
Dyer Brook○, Maine (†04747) 243/G3
Dyersburg, Tenn. (38024) 237/C8
Dyersville, Iowa (52040) 229/L3
Dyess, Ark. (72330) 202/K2
Dyess A.F.B., Texas 303/D5
Dyfed, Wales 13/C6
Dyje (riv.), Czech. 41/D2
Dyke (lake), Newf. 166/A3
Dykh-Tau (mt.), U.S.S.R. 52/F6
Dyle (riv.), Belgium 27/F7
Dysart, Iowa (52224) 229/J4
Dysart, Sask. 181/H5
Dysartsville, N.C. (†28761) 281/F3
Dzamïn Üüd, Mongolia 77/H3
Dzaoudzi (cap.), Comoros 118/H2
Dzavhan, Mongolia 77/E2
Dzavhan Gol (riv.), Mongolia 77/D2
Dzerzhinsk, U.S.S.R. 7/J3
Dzerzhinsk, U.S.S.R. 48/E4
Dzerzhinsk, U.S.S.R. 52/F3
Dzhalal-Abad, U.S.S.R. 48/H5
Dzhalilabad, U.S.S.R. 52/G7
Dzhalinda, U.S.S.R. 48/N4
Dzhambul, U.S.S.R. 54/J5
Dzhambul, U.S.S.R. 48/H5
Dzhankoy, U.S.S.R. 52/D5
Dzhelinda, U.S.S.R. 48/M2
Dzhetygara, U.S.S.R. 48/G4
Dzhezkazgan, U.S.S.R. 54/H5
Dzhezkazgan, U.S.S.R. 48/G5
Dzhugdzhur (range), U.S.S.R. 54/P4
Dzhugdzhur (range), U.S.S.R. 48/O4
Dzhul'fa, U.S.S.R. 52/G7
Dzhusaly, U.S.S.R. 48/G5
Działdowo, Poland 47/E2
Dzibalchén, Mexico 150/P7
Dzibichaltún (ruin), Mexico 150/P6
Dzidzantún, Mexico 150/P6
Dzierzoniów, Poland 47/C3
Dzilam de Bravo, Mexico 150/P6
Dzitbalché, Mexico 150/P6
Dzurh, Mongolia 77/E2
Dzüünharaa, Mongolia 77/G2
Dzuunmod, Mongolia 77/G2

E

Eabamet (lake), Ontario 175/C2
Eads, Colo. (81036) 208/O6
Eads, Tenn. (38028) 237/B10
Eadytown, S.C. (†29468) 296/G5
Eagan, Minn. (55111) 255/G6
Eagan, Tenn. (37730) 237/O7
Eagar, Ariz. (85925) 198/F4
Eagarville, Ill. (†62033) 222/D4
Eagle, Alaska 188/D4
Eagle, Alaska (99738) 196/K2
Eagle (creek), Ariz. 198/F5
Eagle (lake), Calif. 204/E3
Eagle (peak), Calif. 204/E2
Eagle (co.), Colo. 208/F3
Eagle, Colo. (81631) 208/F3
Eagle (riv.), Colo. 208/E3
Eagle, Idaho (83616) 220/B6
Eagle (creek), Ind. 227/E4
Eagle (lake), Iowa 229/F2
Eagle (creek), Ky. 237/M3
Eagle (lake), Maine 243/F1
Eagle (lake), Maine 243/E3
Eagle, Mich. (48822) 250/E6
Eagle (mt.), Minn. 255/G2
Eagle, Nebr. (68347) 264/H4
Eagle (riv.), Newf. 166/C3
Eagle, Ontario 177/C5
Eagle (lake), Ontario 177/F5
Eagle (lake), Ontario 177/E2
Eagle (creek), Oreg. 291/K3
Eagle (hills), Sask. 181/C3
Eagle (peak), Texas 303/C11
Eagle (mt.), Virgin Is. (U.S.) 161/E4
Eagle, Wis. (53119) 317/H2
Eagle (lake), Wis. 317/H2
Eagle (lake), Wis. 317/K3
Eagle (peak), Wyo. 319/B1
Eagle Bay, N.Y. (13331) 276/L3
Eagle Bend, Minn. (56446) 255/D4
Eagle Bridge, N.Y. (12057) 276/O5
Eagle Butte, S. Dak. (57625) 298/G4
Eagle City, Okla. (73658) 288/J3
Eagle Crags (mt.), Calif. 204/J8
Eagle Creek, Oreg. (97022) 291/E2
Eagle Grove, Iowa (50533) 229/F3
Eagle Harbor, Md. (†20608) 245/M6
Eagle Harbor, Mich. (49951) 250/A1
Eaglehawk, Victoria 97/C5
Eaglehill (creek), Sask. 181/D4
Eagle Lake, Fla. (33839) 212/E4
Eagle Lake, Maine (04739) 243/F1
Eagle Lake○, Maine (04739) 243/F1
Eagle Lake, Minn. (56024) 255/E6
Eagle Lake, Ontario 177/F2
Eagle Lake, Texas (77434) 303/H8
Eagle Mills, Ark. (71729) 202/E6
Eagle Mountain, Calif. (92241) 204/K10
Eagle Mountain (lake), Texas 303/E2
Eagle Nest, N. Mex. (87718) 274/D2
Eagle Nest (lake), N. Mex. 274/D2
Eagle Pass, Texas (78852) 303/D9
Eagle Point, Oreg. (97524) 291/E5
Eagle River, Alaska (99577) 196/C1
Eagle River, Mich. (49924) 250/A1
Eagle River, Wis. (54521) 317/H4
Eagle Rock, Mo. (65641) 261/E9
Eagle Rock, Va. (24085) 307/J5
Eaglesfield, Scotland 15/E5
Eaglesham, Alberta 182/B2
Eaglesham, Scotland 15/D5
Eagles Mere, Pa. (17731) 294/J3
Eagle Springs, N.C. (27242) 281/K4
Eagleton Village, Tenn. (†37801) 237/O9
Eagletown, Ind. (†46074) 227/E4
Eagletown, Okla. (74734) 288/S6
Eagleville, Calif. (96110) 204/E2
Eagleville, Conn. (†06268) 210/F1
Eagleville, Mo. (64442) 261/D2
Eagleville, Tenn. (37060) 237/H9
Eakly, Okla. (73033) 288/K4
Ealing, England 13/H8
Ealing, England 10/B5
Ear (lake), Sask. 181/B3
Earby, England 13/H1
Eardley (lake), Manitoba 179/F2
Ear Falls, Ontario 175/B2
Earl (lake), Calif. 204/A2
Earl, N.C. (28038) 281/F4
Earl, Wis. (54833) 317/C4
Earle, Ark. (72331) 202/K3
Earle Naval Weapons Sta., N.J. 273/E3
Earleton, Fla. (32631) 212/D2
Earleville, Md. (21919) 245/P3
Earl Grey, Sask. 181/G5
Earlham, Iowa (50072) 229/E6
Earlimart, Calif. (93219) 204/F8
Earling, Iowa (51530) 229/C5
Earlington, Ky. (42410) 237/F6
Earl Park, Ind. (47942) 227/C3
Earlsboro, Okla. (74840) 288/N4
Earlston, Scotland 15/F5
Earlton, Kansas (†66720) 232/G4
Earlton, Ontario 177/K5
Earltown, Nova Scotia 168/E3
Earlville, Ill. (60518) 222/E2
Earlville, Iowa (52041) 229/L4
Earlville, N.Y. (13332) 276/J5
Early (co.), Georgia 217/C8
Early, Iowa (50535) 229/C4
Early Branch, S.C. (29916) 296/F6
Earlysville, Va. (22936) 307/M4
Earn, Loch (lake), Scotland 15/D4
Earn (riv.), Scotland 15/E4
Earnslaw (mt.), N. Zealand 100/B6
Earp, Calif. (92242) 204/L9
Earth, Texas (79031) 303/B3
Earthquake (lake), Mont. 262/E6
Easby, N. Dak. (†58249) 282/O2
Easington, England 13/J3
Easingwold, England 13/F3
Eask (lake), Ireland 17/E2
Easky, Ireland 17/D3
Easley, S.C. (29640) 296/B2
East (cape), Alaska 196/K4
East (riv.), Conn. 210/E3
East (pt.), Fla. 212/E6
East (bay), La. 238/M8
East (pt.), Mass. 249/E6
East (range), Nev. 266/D2
East (pt.), N.S. Wales 97/J2
East (riv.), N.Y. 276/N9
East (cape), N. Zealand 87/H9
East (cape), N. Zealand 100/G2
East (bay), Nova Scotia 168/H3
East (riv.), Nova Scotia 168/F3
East (lake), Oreg. 291/F4
East (pt.), Pr. Edward I. 168/G2
East (Dezhnev) (cape), U.S.S.R. 4/C18
East (pt.), Virgin Is. (U.S.) 161/G4
Eastaboga, Ala. (36260) 195/F3
Eastabuchie, Miss. (39436) 256/F8
East Albany, Vt. (†05820) 268/C2
East Alburg, Vt. (†05440) 268/A2
East Aldfield, Québec 172/A4
East Alligator (riv.), North. Terr. 93/C2
East Alton, Ill. (62024) 222/A2
East Andover, Maine (04226) 243/B6
East Andover, N.H. (03231) 268/D5
East Angus, Québec 172/F4
Eastanollee, Georgia (30538) 217/F1
East Arcadia, N.C. (†28434) 281/N6
East Arlington, Vt. (05252) 268/A5
East Arrow Park, Br. Col. 184/J5
East Aspetuck (riv.), Conn. 210/B2
East Aurora, N.Y. (14052) 276/C5
East Baldwin, Maine (04024) 243/B8
East Bangor, Pa. (18013) 294/M4
East Bank, W. Va. (25067) 312/D6
East Barnet, Vt. (†05821) 268/C3
East Barre-Graniteville, Vt. (05649) 268/C3
East Barrington, N.H. (03825) 268/F5
East Baton Rouge (par.), La. 238/K1
East Bay, Nova Scotia 168/H2
East Bay (hills), Nova Scotia 168/H3
East Bend, N.C. (27018) 281/H2
East Berbice-Corantyne (dist.), Guyana 131/C3
East Berkshire, Vt. (05447) 268/B2
East Berlin, Conn. (06023) 210/E2
East Berlin, Pa. (17316) 294/J6
East Bernard, Texas (77435) 303/H8
East Bernstadt, Ky. (40729) 237/N6
East Berwick, Pa. (†18603) 294/K3
East Bethany, N.Y. (14054) 276/D5
East Bethel, Minn. (†55005) 255/E5
East Bethel, Vt. (†05032) 268/B4
East Bloomfield, N.Y. (14443) 276/E5
East Blue Hill, Maine (04629) 243/G7
East Blythe, Calif. (†92225) 204/L10
East Boothbay, Maine (04544) 243/D8
Eastborough, Kansas (†67201) 232/E4
Eastbourne, England 13/H7
Eastbourne, England 10/G5
Eastbourne, N. Zealand 100/B3
East Brady, Pa. (16028) 294/C3
East Braintree, Manitoba 179/G5
East Braintree, Mass. (†02184) 249/D8
East Braintree, Vt. (†05060) 268/B3
East Branch, N.Y. (13756) 276/K7
East Branch, Rocky (riv.), Ohio 284/G10
East Brewster, Mass. (†02631) 249/O5
East Brewton, Ala. (36426) 195/E8
East Bridgewater○, Mass. (02333) 249/L4
East Brisbane, Queensland 88/K3
East Brisbane, Queensland 95/E3
East Brookfield, Mass. (01515) 249/G4
East Brookfield○, Mass. (01515) 249/G4
East Brookfield, Vt. (†05036) 268/C3
East Brooklyn, Conn. (†06239) 210/H1
East Broughton, Québec 172/F3
East Broughton Station, Québec 172/F3
East Brownfield, Maine (†04010) 243/B8
East Brunswick○, N.J. (08816) 273/E3
East Burke, Vt. (05832) 268/D2
East Butler, Pa. (16029) 294/C4
East Calais, Vt. (05650) 268/C3
East Calder, Scotland 15/D2
East Camden, Ark. (71701) 202/E6
East Canaan, Conn. (06024) 210/B1
East Candia, N.H. (03040) 268/E5
East Canton, Ohio (44730) 284/H4
East Canyon (res.), Utah 304/C3
East Cape Girardeau, Ill. (†62957) 222/D6
East Carbon, Utah (84520) 304/D4
East Carondelet, Ill. (62240) 222/A3
East Carroll (par.), La. 238/H1
East Chain, Minn. (†56031) 255/D7
East Charleston, Vt. (05833) 268/D2
Eastchester, N.Y. (10709) 276/P6
East Chester, Nova Scotia 168/D4
East Chevington, England 13/F2
East Chezzetcook, Nova Scotia 168/E4
East Chicago, Ind. (46312) 227/C1
East Chicago Heights, Ill. (†60411) 222/C6
East China (sea) 54/O7
East China (sea), China 77/L6
East China (sea), Japan 81/C8
East China (sea), S. Korea 81/C8
East Chop (pt.), Mass. 249/M7
East Claridon, Ohio (44033) 284/H2
East Cleveland, Ohio (44112) 284/H9
East Coast Bays, N. Zealand 100/B1
East Concord, Vt. (†05906) 268/D3
East Conemaugh, Pa. (15909) 294/E5
East Corinth, Maine (04427) 243/F5
East Corinth, Vt. (05040) 268/C3
East Cote Blanche (bay), La. 238/G7
East Coulée, Alberta 182/D4
East Craftsbury, Vt. (†05826) 268/C2
East Dedham, Mass. (02026) 249/C8
East Demerara-West Coast Berbice (dist.), Guyana 131/C2
East Dennis, Mass. (02641) 249/O5
East Dereham, England 13/H5
East Dereham, England 10/G4
East Derry, N.H. (03041) 268/E6
East Detroit, Mich. (48021) 250/B6
East Devils (lake), N. Dak. 282/N4
East Dixfield, Maine (04227) 243/C6
East Dixmont, Maine (†04932) 243/E6
East Dorset, Vt. (05253) 268/A5
East Douglas, Mass. (01516) 249/G4
East Dover, Vt. (05341) 268/B6
East Dublin, Georgia (31021) 217/G5
East Dubuque, Ill. (61025) 222/C1
East Duke, Okla. (†73532) 288/H5
East Dundee (Dundee), Ill. (†60118) 222/E1
East Durham, N.Y. (12423) 276/M6
East Eddington, Maine (04428) 243/F6
East Ellijay, Georgia (30539) 217/C1
East Ely, Nev. 266/G3
Eastend, Sask. 181/C6
Eastend, Virgin Is. (U.S.) 161/D4
East Enterprise, Ind. (47019) 227/H7
Easter (isl.), Chile 87/Q8
Easter (isl.), Chile 2/D7
Eastern (pt.), Conn. 210/G3
Eastern (Arabian) (des.), Egypt 111/F2
Eastern (prov.), Kenya 115/G4
Eastern (bay), Md. 245/N5
Eastern (pt.), Mass. 249/M2
Eastern (creek), N.S. Wales 97/H3
Eastern Channel (str.), Japan 81/D7
Eastern Ghats (mts.), India 68/D6
Eastern Samar (prov.), Philippines 82/E5
Eastern Scheldt (est.), Netherlands 27/D5
Eastern Taurus (mts.), Turkey 63/J3
Eastern Wolf (isl.), New Bruns. 170/D4
Easterville, Manitoba 179/C1
East Fairfield, Vt. (05448) 268/B2
East Falkland (isl.), 193/E7
East Falkland (isl.), Falk. Is. 120/D8
East Falmouth (Teaticket), Mass. (02536) 249/M6
East Farnham, Québec 172/E4
East Faxon, Pa. (†17701) 294/J3
East Feliciana (par.), La. 238/H5
East Ferry, Nova Scotia 168/B4
East Flanders (prov.), Belgium 27/D7
East Flat Rock, N.C. (28726) 281/E4
Eastford○, Conn. (06242) 210/G1
East Fork, Little Miami (riv.), Ohio 284/C7
East Fork, Green (riv.), Wyo. 319/C3
East Foxboro, Mass. (†02035) 249/K4
East Franklin, Maine (†04634) 243/G6
East Franklin, Vt. (†05457) 268/B2
East Freedom, Pa. (16637) 294/E5
East Freetown, Mass. (02717) 249/L5
East Friesland (reg.), W. Germany 22/B2
East Frisian (isls.), W. Germany 22/B2
East Gaffney, S.C. (†29340) 296/D1
East Galesburg, Ill. (61430) 222/C3
Eastgate, Nev. (†89406) 266/D3
East Georgia, Vt. (†05455) 268/A2
East Germantown (Pershing), Ind. (†47370) 227/G5
East Germany 7/F3
EAST GERMANY 22
East Gillespie, Ill. (†62033) 222/D4
East Glacier Park, Mont. (59434) 262/C2
East Glastonbury, Conn. (06025) 210/E2
East Grafton, N.H. (†03240) 268/D4
East Granby○, Conn. (06026) 210/E1
East Grand Forks, Minn. (56721) 255/B3
East Grand Rapids, Mich. (†49506) 250/D6
East Granville, Vt. (†05669) 268/B3
East Greenbush, N.Y. (12061) 276/N5
East Green Harbour, Nova Scotia 168/C5
East Greenville, Ohio (†44666) 284/G4
East Greenville, Pa. (18041) 294/L5
East Greenwich, R.I. (02818) 249/H6
East Grinstead, England 10/G5
East Grinstead, England 13/G6
Eastgulf, W. Va. (25835) 312/D7
East Gull Lake, Minn. (†56401) 255/D4
East Haddam○, Conn. (06423) 210/F3
Eastham○, Mass. (02642) 249/O5
East Hampstead, N.H. (03826) 268/E6
East Hampton, Conn. (06424) 210/E2
East Hampton○, Conn. (06424) 210/E2
Easthampton○, Mass. (01027) 249/D3
East Hampton, N.Y. (11937) 276/R9
East Hanover○, N.J. (07936) 273/E2
East Hardin, Ill. (†62031) 222/C4
East Hardwick, Vt. (05836) 268/C2
East Hartford○, Conn. (06108) 210/E1
East Hartland, Conn. (06027) 210/D1
East Harwich, Mass. (†02645) 249/O6
East Haven○, Conn. (06512) 210/D3
East Haven○, Vt. (05837) 268/D2
East Haverhill, N.H. (†03780) 268/D3
East Hazelcrest, Ill. (†60429) 222/C6
East Hebron, N.H. (03232) 268/D4
East Helena, Mont. (59635) 262/E4
East Hereford, Québec 172/F4
East Herkimer, N.Y. 276/L4
East Hickory, Pa. (16321) 294/D2
East Hills, N.Y. (†11576) 276/R7
East Hiram, Maine (†04041) 243/B8
East Hodge, La. (†71247) 238/E2
East Holden, Maine (04429) 243/F6
East Hope, Idaho (†83836) 220/B1
East Jackson, Maine (†04986) 243/E6
East Jamaica, Vt. (†05343) 268/B5
East Jordan, Mich. (49727) 250/D3
East Juliette, Georgia (†31046) 217/E4
East Keansburg, N.J. (07734) 273/E3
East Kelowna, Br. Col. 184/H5
East Kent, Conn. (†06785) 210/B2
East Kilbride, Scotland 15/B2
East Killingly, Conn. (06243) 210/H1
East Kingsford, Mich. (†49801) 250/A3
East Kingston○, N.H. (03827) 268/F6
East Knox, Maine (†04921) 243/E7
East Korea (bay), N. Korea 81/D4
East Lake, Mich. (49626) 250/C4
East Lake, Minn. (†55760) 255/E4
East Lake, N.C. (27931) 281/S3
Eastlake, Ohio (†44094) 284/J8
East Lake-Orient Park, Fla. (†33601) 212/C2
Eastland, Tenn. (†38583) 237/L9
Eastland (co.), Texas 303/F5
Eastland, Texas (76448) 303/F5
East Lansdowne, Pa. (†19050) 294/M7
East Lansing, Mich. (48823) 250/E6
East Laport, N.C. (†28723) 281/C4
East Las Vegas, Nev. (89112) 266/F6
East Laurinburg, N.C. (28352) 281/L5
East Lebanon, Maine (04027) 243/B9
East Lee, Mass. (†01238) 249/B3
Eastleigh, England 13/F7
Eastleigh, England 10/F5
East Lempster, N.H. (03605) 268/C5
East Limington, Maine (†04049) 243/B8
East Linton, Scotland 15/F5
East Litchfield, Conn. (06759) 210/C1
East Livermore, Maine (04228) 243/C7
East Liverpool, Ohio (43920) 284/J4
East Loch (inlet), Hawaii 218/B3
East Loch Tarbert (inlet), Scotland 15/B3
East London, S. Africa 102/E8
East London, S. Africa 118/D6
East Longmeadow○, Mass. (01028) 249/E4
East Los Angeles, Calif. (90022) 204/C10
East Lowell, Maine (†04433) 243/G5
East Lyme○, Conn. (06333) 210/G3
East Lynn, Ill. (60932) 222/F3
East Lynn, W. Va. (25512) 312/B6
East Lynn (lake), W. Va. 312/B6
East Lynne, Mo. (64743) 261/D5
East Machias, Maine (04630) 243/J6
East Machias○, Maine (04630) 243/J6
East Machias (riv.), Maine 243/H6
East Madison, Maine (†04950) 243/D6
East Madison, N.H. (†03849) 268/E4
Eastmain, Que. 162/J5
Eastmain (riv.), Que. 146/L4
Eastmain (riv.), Que. 162/J5
Eastmain, Québec 174/B2
Eastmain (riv.), Québec 174/B2
Eastman, Georgia (31023) 217/F6
Eastman, Québec 172/E4
Eastman, Wis. (54626) 317/D9
East Marion, N.C. (†28752) 281/F3
East Meadow, N.Y. (11554) 276/R7
East Meredith, N.Y. (13757) 276/L6
East Middlebury, Vt. (05740) 268/A4
East Millcreek, Utah (84109) 304/C3
East Millinocket, Maine (04430) 243/F4
East Millinocket○, Maine (04430) 243/F4
East Millstone, N.J. (08873) 273/D3
East Milton, Mass. (†02186) 249/D7
East Mines, Nova Scotia 168/E3
East Moline, Ill. (61244) 222/C2
East Montpelier○, Vt. (05651) 268/B3
East Moriches, N.Y. (11940) 276/P9
East Morris, Conn. (†06763) 210/C2
East Murton, England 13/J3
East Musquash (lake), Maine 243/H5
East Naples, Fla. (†33940) 212/E5
East Newark, N.J. (07100) 273/B2
East New Market, Md. (21631) 245/P6
East Newnan, Georgia (†30263) 217/C4
East New Portland, Maine (†04954) 243/D6
East Nishnabotna (riv.), Iowa 229/C6
East Northfield, Mass. (†01360) 249/E2
East Northport, N.Y. (11731) 276/O9
East Norton, Mass. (†02766) 249/K5
East Norwalk, Conn. (†06856) 210/B4
East Olympia, Wash. (98540) 310/B4
Easton, Calif. (93706) 204/F7
Easton○, Conn. (06612) 210/B4
Easton (res.), Conn. 210/B3
Easton, Ill. (62633) 222/D3
Easton, Kansas (66020) 232/G2
Easton, La. (†70586) 238/F5
Easton○, Maine (04740) 243/H2
Easton, Md. (21601) 245/O5
Easton○, Mass. (02334) 249/K4
Easton, Minn. (56025) 255/E7
Easton, Mo. (64443) 261/C3
Easton○, N.H. (†03580) 268/D3
Easton, Pa. (18042) 294/M4
Easton, Wash. (98925) 310/D3
Easton, Wis. (†53936) 317/G8
Eastondale, Mass. (†02375) 249/K4
East Orange, N.J. (*07017) 273/B2
East Orland, Maine (04431) 243/F6
East Orleans, Mass. (02643) 249/P5
East Otis, Mass. (01029) 249/B4
East Otisfield, Maine (†04270) 243/B7
East Otto, N.Y. (14729) 276/C6
Eastover, S.C. (29044) 296/F4
East Palatka, Fla. (32031) 212/E2
East Palestine, Ohio (44413) 284/J4
East Park (res.), Calif. 204/C4
East Parsonfield, Maine (04028) 243/B8
East Peacham, Vt. (†05821) 268/C3
East Pembroke, Mass. (†02359) 249/M4
East Pembroke, N.Y. (14056) 276/D5
East Peoria, Ill. (61611) 222/D3
East Pepperell, Mass. (01437) 249/H2
East Peru (Peru), Iowa (†50222) 229/F6
East Peru, Maine (04229) 243/C7
East Petersburg, Pa. (17520) 294/K5
East Pleasant Plain, Iowa (†52540) 229/K6
Eastpoint, Fla. (32328) 212/B2
East Point, Georgia (30344) 217/K2
East Point, Ky. (41216) 237/R5
East Point, La. (71025) 238/D2
East Poland, Maine (†04210) 243/C7
East Poplar, Sask. 181/F6
Eastport, Idaho (83826) 220/B1
Eastport, Maine 188/N2
Eastport, Maine (04631) 243/K6
Eastport, Mich. (49627) 250/D3
Eastport, Newf. 166/D1
Eastport, N.Y. (11941) 276/P9
East Poultney, Vt. (05741) 268/A4
East Prairie, Mo. (63845) 261/O9
East Preston, England 13/G7
East Prospect, Pa. (17317) 294/J6
East Providence, R.I. (02914) 249/J5
East Putnam, Conn. (†06260) 210/H1
East Randolph, N.Y. (14730) 276/C6
East Randolph, Vt. (05041) 268/B4
East Retford, England 13/G4
East Retford, England 10/F4
East Richford, Vt. (†05476) 268/B2
East Ridge, Tenn. (37412) 237/L11
Eastriggs, Scotland 15/E6
East Rindge, N.H. (†03461) 268/D6
East River, Conn. (†06443) 210/E3
East River Saint Marys, Nova Scotia 168/F3
East Riverside-Kingshurst, New Bruns. 170/E3
East Rochester, N.Y. (14445) 276/F4
East Rochester, Ohio (44625) 284/H4
East Rockaway, N.Y. (11518) 276/R7
East Rutherford, N.J. (07073) 273/B2
Eastry, England 13/J6
East Ryegate, Vt. (05042) 268/C3
East Saint Louis, Ill. 188/J3
East Saint Louis, Ill. (*62201) 222/A2
East Sandwich, Mass. (02537) 249/N6
East Saugus, Mass. (†01906) 249/D6
East Sebago, Maine (04029) 243/B8
East Selkirk, Manitoba 179/F4
East Shoal (lake), Manitoba 179/E4
East Siberian (sea), U.S.S.R. 4/B1
East Siberian (sea), U.S.S.R. 54/T2
East Siberian (sea), U.S.S.R. 48/S2
Eastside, Oreg. (97420) 291/C4
East Side, Pa. (18634) 294/L3
East Sister (peak), Idaho 220/C2
East Sister (isl.), Tasmania 99/E1
East Smithfield, Pa. (18817) 294/J2
Eastsound, Wash. (98245) 310/B2

East Sparta, Ohio (44626) 284/H4
East Spencer, N.C. (28039) 281/J3
East Springfield, N.Y. (13333) 276/L5
East Springfield, Pa. (16411) 294/A2
East Stone Gap, Va. (24246) 307/C7
East Stoneham, Maine (04231) 243/B7
East Stroudsburg, Pa. (18301) 294/M4
East Sullivan, Maine (†04607) 243/C6
East Sullivan, N. H. (03445) 268/C6
East Sumner, Maine (†04220) 243/C7
East Sussex (co.), England 13/H7
East Swan (riv.), Minn. 255/F3
East Swanzey, N.H. (03446) 268/C6
East Syracuse, N.Y. (13057) 276/H4
East Tawas, Mich. (48730) 250/F4
East Templeton, Mass. (01438) 249/G2
East Thermopolis, Wyo. (†82443) 319/D2
East Thetford, Vt. (05043) 268/C4
East Thompson, Conn. (†06255) 210/H1
East Tintic (creek), Utah 304/C3
East Tohopekaliga (lake), Fla. 212/E3
East Troy, Wis. (53120) 317/J2
East Union, Maine (†04862) 243/E7
Eastvale, Pa. (†15010) 294/B4
Eastvale, Texas (†75067) 303/G1
East Vassalboro, Maine (04935) 243/D7
Eastview, Tenn. (†38367) 237/D10
East View, W. Va. (†26301) 312/F4
East Village, Conn. (†06468) 210/C3
Eastville, Georgia (†30677) 217/E3
Eastville, Va. (23347) 307/R6
East Wakefield, N.H. (03830) 268/E4
East Walker (riv.), Nev. 266/B4
East Wallingford, Vt. (05742) 268/B5
East Walpole, Mass. (02032) 249/C8
East Wareham, Mass. (02538) 249/M5
East Washington, Pa. (†15301) 294/B4
East Waterboro, Maine (04030) 243/B8
East Waterford, Pa. (17021) 294/G5
East Wenatchee, Wash. (98801) 310/E3
East Weymouth, Mass. (02189) 249/E8
East Whately, Mass. (†01373) 249/G4
East Williamson, N.Y. (14449) 276/F4
East Willington, Conn. (†06279) 210/G1
East Wilton, Maine (04234) 243/C6
East Windsor○, Conn. (†06088) 210/E1
East Windsor Hill, Conn. (06028) 210/E1
East Winn, Maine (†04495) 243/G5
East Wolfeboro, N.H. (†03894) 268/E4
Eastwood, Mich. (†49001) 250/D6
Eastwood, N. S. Wales 88/K4
Eastwood, N. S. Wales 97/J3
Eastwood, Ontario 177/D4
East Woodstock, Conn. (06244) 210/H1
East Worcester, N.Y. (12064) 276/L5
East York, Ontario 177/J4
Eaton, Colo. (80615) 208/K1
Eaton, Ill. (†62454) 222/F4
Eaton, Ind. (47338) 227/G4
Eaton, Maine (†04424) 243/H4
Eaton (co.), Mich. 250/E6
Eaton (Eaton Center)○, N.H. (03832) 268/E4
Eaton, N.Y. (13334) 276/J5
Eaton, Ohio (45320) 284/A6
Eaton, Tenn. (38331) 237/C9
Eaton Center, N.H. (03832) 268/E4
Eaton Estates, Ohio (†44035) 284/G3
Eatonia, Sask. 181/B4
Eaton Rapids, Mich. (48827) 250/E6
Eatonton, Georgia (31024) 217/F4
Eatontown, N.J. (07724) 273/E3
Eatonville, Fla. (32751) 212/E3
Eatonville, Wash. (98328) 310/C4
Eau Claire, Mich. (49111) 250/C6
Eau Claire, Pa. (16030) 294/C3
Eau Claire, Lac à l' (lake), Que. 162/J4
Eau Claire (lake), Québec 174/C1
Eau Claire, Wis. 188/H2
Eau Claire (co.), Wis. 317/D6
Eau Claire, Wis. (54701) 317/D6
Eau Claire (riv.), Wis. 317/D6
Eau Galle, Wis. (54737) 317/B6
Eauripik (atoll), Micronesia 87/F5
Ebal (mt.), Jordan 65/C3
Ebano, Mexico 150/K5
Ebb, Fla. (†32331) 212/C1
Ebb and Flow (lake), Manitoba 179/C3
Ebbw Vale, Wales 13/B6
Ebbw Vale, Wales 10/E5
Ebeltoft, Denmark 21/D5
Ebeltoft, Denmark 18/G8
Ebenezer, Miss. (39064) 256/D5
Ebenezer, Sask. 181/J4
Ebenfurth, Austria 41/D3
Eben Junction, Mich. (49825) 250/B2
Ebensburg, Pa. (15931) 294/E5
Ebensee, Austria 41/B3
Eberbach, W. Germany 22/C4
Ebersbach, E. Germany 22/F3
Eberswalde-Finow, E. Germany 22/E2
Ebetsu, Japan 81/K2
Ebingen, W. Germany 22/C4
Ebinur Hu (lake), China 77/B2
Ebnat-Kappel, Switzerland 39/H2
Eboli, Italy 34/E4
Ebolowa, Cameroon 102/D4
Ebolowa, Cameroon 102/B3
Ebon (atoll), Marshall Is. 87/G5
Ebony, W. Va. (23845) 307/N7
Ebor, Manitoba 179/A5
Ebrach, W. Germany 22/D4
Ebrié (lag.), Ivory Coast 106/D8
Ebro, Fla. (32437) 212/C6
Ebro, Minn. (†56621) 255/C3
Ebro (riv.), Spain 7/D4
Ebro (riv.), Spain 33/G2
Ecatepec de Morelos, Mexico 150/L1
Ecaussines, Belgium 27/E7
Ecclefechan, Scotland 15/E5
Eccles, W. Va. (25836) 312/D7
Ecclesville, Trin. & Tob. 161/B11
Eceabat, Turkey 63/B6
Echallens, Switzerland 39/C3
Echarate, Peru 128/F9

Echeconnee, Georgia (†31008) 217/E5
Echmiadzin, U.S.S.R. 52/F6
Echo, Ala. (†36360) 195/G8
Echo (cliffs), Ariz. 198/D3
Echo, La. (71330) 238/F4
Echo, Minn. (56237) 255/D5
Echo (lake), N.J. 273/E1
Echo (lake), Oreg. (97826) 291/H2
Echo (lake), Tasmania 99/C4
Echo, Utah (84024) 304/C3
Echo (res.), Utah 304/C3
Echo, Wis. 268/D2
Echo Bay, Ontario 177/J5
Echo Bay, Ontario 175/D3
Echola, Ala. (35457) 195/C4
Echo Lake, N.J. (†07435) 273/E1
Echo Lake, Nova Scotia 168/E4
Echols (co.), Georgia 217/G9
Echols, Ky. (42340) 237/H6
Echo Valley Prov. Park, Sask. 181/G5
Echt, Netherlands 27/H6
Echternach, Luxembourg 27/J9
Echuca, Victoria 97/C5
Echuca, Victoria 88/C5
Écija, Spain 33/D4
Eck, Loch (lake), Scotland 15/A1
Eckelson, N. Dak. (58432) 282/O6
Eckerman, Mich. (49728) 250/E2
Eckernförde, W. Germany 22/D1
Eckerty, Ind. (47116) 227/D8
Eckhart Mines, Md. (21528) 245/C2
Eckley, Colo. (80727) 208/P2
Eckman, N. Dak. (†58760) 282/H2
Eckman, W. Va. (24829) 312/C8
Eckville, Alberta 182/C3
Eclectic, Ala. (36024) 195/F5
Eclipse (harb.), Newf. 166/B2
Eclipse (sound), N. W. Terrs. 187/L2
Economy, Ind. (47339) 227/G5
Economy, Nova Scotia 168/D3
Economy, Pa. (†15005) 294/B4
Écorce (lake), Québec 172/A2
Écorces (riv.), Québec 172/F1
Écorse, Mich. (48229) 250/P8
Écru, Miss. (38841) 256/F2
Ector (co.), Texas 303/B6
Ecuador 2/D6
Ecuador 120/B3
ECUADOR 128
Ecublens, Switzerland 39/B3
Ecum Secum, Nova Scotia 168/F3
Ecum Secum Bridge, Nova Scotia 168/F4
Edam, Sask. 181/C2
Edam-Volendam, Netherlands 27/G4
Eday (isl.), Scotland 10/E1
Eday (isl.), Scotland 15/F1
Edberg, Alberta 182/D3
Edcouch, Texas (78538) 303/G11
Edd, Ethiopia (†32801) 212/E3
Edd, Ethiopia 59/D7
Ed Da'ein, Sudan (†117) 111/H5
Ed Damazin, Sudan 111/F5
Ed Damer, Sudan 111/F4
Ed Damer, Sudan 59/B6
Ed Damer, Sudan 102/F3
Ed Debba, Sudan 111/F5
Ed Debba, Sudan 59/B6
Eddiceton, Miss. (39634) 256/C8
Eddington, Maine (†04428) 243/F6
Eddington○, Maine (†04428) 243/F6
Eddington, Pa. (19020) 294/N5
Eddleston, Scotland 15/E5
Eddontenajon, Br. Col. 184/K2
Eddrachillis (bay), Scotland 15/C2
Ed Dueim, Sudan 59/B7
Ed Dueim, Sudan 111/F5
Ed Dueim, Sudan 102/F3
Eddy (co.), N. Mex. 274/E6
Eddy (co.), N. Dak. 282/N4
Eddy, Texas (76524) 303/G6
Eddystone (rocks), England 13/C7
Eddystone (rocks), England 10/D5
Eddystone, Manitoba 179/C3
Eddystone, Pa. (19013) 294/M7
Eddystone (pt.), Tasmania 88/H8
Eddystone (pt.), Tasmania 99/E2
Eddyville, Ill. (62928) 222/E6
Eddyville, Iowa (52553) 229/H6
Eddyville, Ky. (42038) 237/E6
Eddyville, Nebr. (68834) 264/F4
Eddyville, Oreg. (97343) 291/D3
Ede, Netherlands 27/H4
Edéa, Cameroon 115/B3
Edelény, Hungary 41/F2
Edelstein, Ill. (61526) 222/D3
Eden, Ariz. (85535) 198/F6
Edén, Ecuador 128/E3
Eden (riv.), England 13/E3
Eden (riv.), England 10/E3
Eden, Georgia (31307) 217/K6
Eden, Idaho (83325) 220/D7
Eden, Ind. (†46140) 227/F5
Eden, Manitoba 179/C4
Eden, Md. (21822) 245/R7
Eden, Miss. (39065) 256/D5
Eden, Mont. (59401) 262/E3
Eden, N.S. Wales 97/E5
Eden, N.Y. (14057) 276/C5
Eden, N.C. (27288) 281/K1
Eden (riv.), Scotland 15/F4
Eden, S. Dak. (57232) 298/P2
Eden, Texas (76837) 303/E6
Eden, Utah (84310) 304/C2
Eden○, Vt. (05652) 268/C2
Eden, Wis. (53019) 317/K8
Eden, Wyo. (82929) 319/C3
Edenburg, Sask. 181/J3
Edenburg, S. Africa 118/D5
Edendale, S. Africa 118/D5
Edenderry, Ireland 10/C4
Edenderry, Ireland 17/G5
Edenhope, Victoria 97/A5
Eden Mills, Ontario 177/D4
Eden Mills, Vt. (05653) 268/C2
Eden Prairie, Minn. (55344) 255/G6

Edenton, N.C. (27932) 281/R2
Edenton, Ohio (†45122) 284/C7
Edenvale, S. Africa 118/H6
Eden Valley, Minn. (55329) 255/D5
Eden Valley (res.), Wyo. 319/C3
Edenville, Mich. (48620) 250/E5
Edenwold, Sask. 181/G5
Eder (res.), W. Germany 22/C3
Eder (riv.), W. Germany 22/C3
Ederney and Kesh, N. Ireland 17/F2
Edessa, Greece 45/F5
Edievale, N. Zealand 100/B6
Edgar (co.), Ill. 222/F4
Edgar, Mont. (59026) 262/H5
Edgar, Nebr. (68935) 264/F4
Edgar, Wis. (54426) 317/G6
Edgard, La. (70049) 238/M3
Edgar Springs, Mo. (65462) 261/J7
Edgartown, Mass. (02539) 249/M7
Edge (isl.), Norway 4/B8
Edgecliff, Texas (†76011) 303/E2
Edgecomb○, Maine (†04556) 243/D8
Edgecombe (co.), N.C. 281/R3
Edgecumbe (cape), Alaska 196/L1
Edgecumbe, N. Zealand 100/F2
Edgefield (co.), S.C. 296/D4
Edgefield, S.C. (29824) 296/C4
Edge Hill, Georgia (†30810) 217/G4
Edgehill, Va. (†22937) 307/O4
Edgeley, N. Dak. (58433) 282/N7
Edgemere, Idaho (†83856) 220/B1
Edgemere, Md. (†21219) 245/N4
Edgemont, Ark. (72044) 202/F2
Edgemont, S. Dak. (57735) 298/B7
Edgemoor, Del. (†19801) 245/S1
Edgemount, N.C. (28645) 281/F2
Edgerton, Ind. (†46797) 227/H2
Edgerton, Kansas (66021) 232/H3
Edgerton, Minn. (56128) 255/B7
Edgerton, Mo. (64444) 261/C3
Edgerton, Ohio (43517) 284/A3
Edgerton, Va. (23868) 307/N7
Edgerton, Wis. (53534) 317/H10
Edgerton, Wyo. (82635) 319/F2
Edgewater, Br. Col. 184/J5
Edgewater, Colo. (80214) 208/J3
Edgewater, Fla. (32032) 212/F3
Edgewater, N.J. (07020) 273/C2
Edgewater, Wis. (54834) 317/D4
Edgewater Park○, N.J. (†08010) 273/D3
Edgewood, Br. Col. 184/H5
Edgewood, Calif. (96094) 204/C2
Edgewood, Fla. (†32801) 212/E3
Edgewood, Ill. (62426) 222/E5
Edgewood, Ind. (†46012) 227/F5
Edgewood, Iowa (52042) 229/K3
Edgewood, Ky. (†41017) 237/S2
Edgewood, Md. (21040) 245/N3
Edgewood, N. Mex. (87015) 274/C3
Edgewood, Ohio (†44004) 284/J2
Edgewood, Pa. (†15218) 294/E7
Edgeworth, Pa. (†15143) 294/B4
Edgington, Ill. (†61284) 222/C2
Edhessa, Greece 45/F5
Edina, Minn. (55424) 255/G5
Edina, Mo. (63537) 261/H2
Edinboro, Pa. (16412) 294/B2
Edinburg, Ill. (62531) 222/D4
Edinburg, Miss. (†39051) 256/F5
Edinburg, Mo. (†64683) 261/E2
Edinburg, N. Dak. (58227) 282/P3
Edinburg, Pa. (16116) 294/B3
Edinburg, Texas (78539) 303/F11
Edinburg, Va. (22824) 307/M3
Edinburgh, Ind. (46124) 227/E6
Edinburgh (cap.), Scotland 7/D3
Edinburgh (cap.), Scotland 15/E5
Edinburgh (cap.), Scotland 10/C1
Edingen (Enghien), Belgium 27/D7
Edirne (prov.), Turkey 63/B2
Edirne, Turkey 7/H4
Edirne, Turkey 63/B2
Edirne, Turkey 59/A1
Edison, Calif. (93220) 204/G8
Edison, Georgia (31746) 217/C7
Edison (lake), Calif. 204/F6
Edison, Nebr. (68936) 264/E4
Edison○, N.J. (*08817) 273/E2
Edison, Ohio (43320) 284/E4
Edison Nat'l Hist. Site, N.J. 273/A2
Edison, Wash. (98246) 310/C3
Edisto (riv.), S.C. 296/F4
Edisto (riv.), S.C. 296/G7
Edisto Beach, S.C. (29438) 296/G7
Edisto Island, S.C. (29438) 296/G7
Edith, Georgia (†31631) 217/G9
Edithburgh, S. Australia 94/E6
Ediz Hook (pt.), Wash. 310/B2
Edjeleh, Algeria 106/F3
Edmeston, N.Y. (13335) 276/K5
Edmond, Kansas (67636) 232/C2
Edmond, Okla. (73034) 288/M3
Edmonds, Wash. (98020) 310/C3
Edmondson, Ark. (72332) 202/K3
Edmonson (co.), Ky. 237/J6
Edmonston, Md. (†20781) 245/F4
Edmonston (mt.), Alberta 182/D3
Edmonton, Alta. 146/G4
Edmonton, Canada 2/D3
Edmonton, Ky. (42129) 237/K7
Edmonton, Alta. 162/E5
Edmonton, Canada 2/D3
Edmore, Mich. (48829) 250/E6
Edmore, N. Dak. (58330) 282/O3
Edmund, Wis. (†53533) 317/F9
Edmunds, Maine (†04628) 243/J6
Edmunds, N. Dak. (†58746) 282/M5

Edmunds (co.), S. Dak. 298/L3
Edmundson, Mo. (†63101) 261/O2
Edmundston, New Bruns. 170/B1
Edna, Ala. (†36922) 195/B6
Edna, Iowa (†51246) 229/H2
Edna, Kansas (67342) 232/G4
Edna, Texas (77957) 303/H9
Edna Bay, Alaska (†99901) 196/M2
Edo (riv.), Japan 81/P2
Edolo, Italy 34/C2
Edom, Texas (†75656) 303/J5
Edon, Ohio (43518) 284/A2
Édouard (lake), Québec 172/E2
Edrans, Manitoba 179/C4
Edremit, Turkey 63/B3
Edremit, Turkey 59/A2
Edremit (gulf), Turkey 63/B3
Edri, Libya 111/B2
Edsbyn, Sweden 18/J6
Edson, Alberta 182/B3
Edson, Alta. 162/E5
Edson, Kansas (67733) 232/A2
Eduardo Castex, Argentina 143/D4
Edwall, Wash. (99008) 310/H3
Edwand, Alberta 182/D2
Edward (lake) 102/E5
Edward, N.C. (27821) 281/R4
Edward (lake), Uganda 115/E4
Edward (lake), Zaïre 115/E4
Edward, N.Y. (13635) 276/K2
Edward MacDowell (res.), N.H. 268/D6
Edwards, Colo. (81632) 208/F3
Edwards (co.), Ill. 222/E5
Edwards (riv.), Ill. 222/C2
Edwards (co.), Kansas 232/C4
Edwards (lake), La. 238/C2
Edwards, Miss. (39066) 256/C6
Edwards, N.Y. (13635) 276/K2
Edwards (plat.), Texas 303/C7
Edwards A.F.B., Calif. 204/H9
Edwards (co.), Texas 303/C7
Edwardsburg, Mich. (49112) 250/C7
Edwardsport, Ind. (47528) 227/C7
Edwardsville, Ala. (36261) 195/H3
Edwardsville, Ill. (62025) 222/B4
Edwardsville, Kansas (66113) 232/H2
Edwardsville, Pa. (18704) 294/E7
Edward VII (pen.) 5/B17
Edward VIII (bay) 5/C4
Edwight, W. Va. (†25189) 312/C7
Edwin, Ala. (†36317) 195/H7
Edwin, Manitoba 179/D5
Eefde, Netherlands 27/J4
Eek, Alaska (99578) 196/F2
Eeklo, Belgium 27/D6
Eel (riv.), Calif. 204/B4
Eel (riv.), Ind. 227/C6
Eel (riv.), Ind. 227/F3
Eel River Bridge, New Bruns. 170/D1
Eel River Crossing, New Bruns. 170/D1
Eems, Netherlands 27/K2
Eersterivier, S. Africa 118/B8
Efate (isl.), Vanuatu 87/G7
Eferding, Austria 41/B2
Effie, La. (71331) 238/F4
Effie, Minn. (56639) 255/F3
Effigy Mounds Nat'l Mon., Iowa 229/L2
Effingham (co.), Georgia 217/K6
Effingham (co.), Ill. 222/E4
Effingham, Ill. (62401) 222/E4
Effingham, Kansas (66023) 232/G4
Effingham, S.C. (29541) 296/H3
Effingham Falls, N.H. (†03814) 268/E4
Effort, Pa. (18330) 294/M4
Efland, N.C. (27243) 281/L2
Eflâni, Turkey 63/E2
Egadi (isls.), Italy 34/C6
Egan, Ill. (†61026) 222/D1
Egan (range), Nev. 266/G4
Egan, S. Dak. (57024) 298/R6
Egaña, Uruguay 145/B4
Eger, Hungary 41/F3
Egeria, W. Va. (†25902) 312/D7
Egernsund, Denmark 21/C8
Egerton (mt.), W. Australia 92/B4
Egerton, Ind. (†46072) 227/H4
Egg (isl.), Manitoba 179/E3
Egg (creek), N. Dak. 282/H3
Egg, Switzerland 39/G2
Eggenburg, Austria 41/C2
Eggertsville, N.Y. (†14226) 276/C5
Egg Harbor, Wis. (54209) 317/M5
Egg Harbor City, N.J. (08215) 273/D4
Egg Island (pt.), N.J. 273/C5
Eggiwil, Switzerland 39/E3
Egg Lagoon, Tasmania 99/A1
Eggleston, Va. (24086) 307/L6
Egham, England 13/G8
Egham, England 10/B5
Éghezée, Belgium 27/F7
Egilsay (isl.), Scotland 15/F1
Eglin A.F.B., Fla. 212/C6
Eglington (isl.), N.W.T. 162/L3
Eglinton (cape), N.W. Terrs. 187/M2
Eglisau, Switzerland 39/G1
Eglon, W. Va. (26716) 312/G4
Egmond aan Zee, Netherlands 27/E3
Egmont (key), Fla. 212/D4
Egmont (cape), N. Zealand 100/D3
Egmont (mt.), N. Zealand 100/D3
Egmont (cape), Nova Scotia 168/E4
Egmont (bay), Pr. Edward I. 168/D2
Egmont (cape), Pr. Edward I. 168/D2
Egnach, Switzerland 39/H1
Egnar, Colo. (81325) 208/B7
Egremont, Alberta 182/D2
Egremont, England 13/D3
Egremont, England 10/D3

Eğridir, Turkey 63/D4
Eğridir (lake), Turkey 59/B2
Eğridir (lake), Turkey 63/D4
Egtved, Denmark 21/C6
Egvekinot, U.S.S.R. 48/S3
Egyek, Hungary 41/F3
Egypt 2/L4
Egypt 102/E2
EGYPT 111/E2
EGYPT 59/A4
Egypt, Georgia (†31329) 217/K6
Egypt, Miss. (†38860) 256/G3
Egypt Lake, Fla. (†33614) 212/C2
Eha Amufu, Nigeria 106/F7
Ehime (pref.), Japan 81/F7
Ehingen, W. Germany 22/C4
Ehrenberg, Ariz. (85334) 198/A5
Ehrenberg (range), North. Terr. 93/B7
Ehrenfeld, Pa. (†15956) 294/E5
Ehrhardt, S.C. (29081) 296/E5
Ehrwald, Austria 22/D4
Eiao (isl.), Fr. Poly. 87/M6
Eibar, Spain 33/E1
Eichstätt, W. Germany 22/D4
Eider (riv.), W. Germany 22/C1
Eidfjord, Norway 18/E6
Eidsfoss, Norway 18/G6
Eidson, Tenn. (37731) 237/P7
Eidsvold, Queensland 95/D5
Eidsvoll, Norway 18/G6
Eielson A.F.B., Alaska 196/J2
Eigersund, Norway 18/D7
Eigg (mt.), Nova Scotia 168/F3
Eigg (isl.), Scotland 15/B4
Eigg (isl.), Scotland 10/C2
Eigg (sound), Scotland 15/B4
Eight Degree (chan.), India 68/C7
Eighteen Mile (peak), Idaho 220/E5
Eight Mile (brook), Conn. 210/C3
Eight Mile (riv.), Conn. 210/F3
Eights Coast (reg.) 5/B14
Eighty Mile, Ky. (42130) 237/K7
Eighty Mile (beach), W. Australia 88/C3
Eighty Mile (beach), W. Australia 92/C2
Eijerlandsche Gat (str.), Netherlands 27/F2
Eil, Loch (lake), Scotland 15/C4
Eil, Somalia 115/J2
Eildon, Victoria 97/C5
Eildon (lake), Victoria 97/C5
Eileen, Ill. (†60416) 222/E2
Eileen (lake), N.W.T. 162/F3
Eilerts de Haan (mts.), Suriname 131/C4
Eina, Norway 18/G6
Einbeck, W. Germany 22/D3
Eindhoven, Netherlands 27/G6
'Ein Gedi, Israel 65/C5
'Ein Harod, Israel 65/C3
'Ein Netafim (well), Israel 65/D5
Einsiedeln, Switzerland 39/G2
Eirunepé, Brazil 132/G10
Eirunepé, Brazil 120/B3
Eisenach, E. Germany 22/D3
Eisenberg, E. Germany 22/D3
Eisenerz, Austria 41/C3
Eisenhower (mt.), Alberta 182/C4
Eisenhüttenstadt, E. Germany 22/F2
Eisenkappel-Vellach, Austria 41/C3
Eisenstadt, Austria 41/D3
Eiserfeld, W. Germany 22/C3
Eishort, Loch (inlet), Scotland 15/B3
Eisleben, E. Germany 22/D3
Eisling (mts.), Luxembourg 27/H9
Eitzen, Minn. (55931) 255/G7
Ejby, Denmark 21/C7
Ejea de los Caballeros, Spain 33/F1
Ejido, Venezuela 124/C3
Ejin, China 77/F3
Ejin Horo, China 77/G4
Ejutla de Crespo, Mexico 150/L8
Ekalaka, Mont. (59324) 262/M5
Ekenäs, Finland 18/N6
Ekeren, Belgium 27/E6
Eketahuna, N. Zealand 100/E4
Ekibastuz, U.S.S.R. 48/H4
Ekibin, Queensland 88/K3
Ekimchan, U.S.S.R. 48/O4
Ekin, Ind. (†46072) 227/E4
Ekonk, Conn. (†06384) 210/H2
Ekron, Ky. (40117) 237/J5
Eksjö, Sweden 18/J8
Ekuk, Alaska (†99569) 196/G3
Ekwan (riv.), Ont. 162/H5
Ekwan (riv.), Ontario 175/C2
Ekwok, Alaska (99580) 196/G3
El Aaiún (Laayoune), Morocco 102/A2
El Aaiún (Laayoune), Western Sahara 106/B3
El Abbasiya, Sudan 111/F5
El Abiar, Libya 111/D1
El Abiod Sidi Cheikh, Algeria 106/E2
El Agheila, Libya 111/C1
Elaine, Ark. (72333) 202/J5
El 'Al, Jordan 65/D4
El 'Alamein, Egypt 111/E1
El 'Alamein, Egypt 59/A3
El Almacén, Venezuela 124/G4
El Amparo de Apure, Venezuela 124/C4
Elams, N.C. (†23919) 281/O1
Elamton, Ky. (41420) 237/P5
Elamville, Ala. (†36311) 195/G7
Eland, Wis. (54427) 317/H6
El Ángel, Ecuador 128/C2
El Arahal, Spain 33/D4
El 'Arish, Egypt 59/B3
El 'Arish, Egypt 111/F1
El Asiento, Bolivia 136/B6
El Asnam, Algeria 106/E1
El Asnam, Algeria 102/C1
Elassón, Greece 45/F5
Elat, Israel 65/D6

Elath (Elat), Israel 65/D6
Elath, Israel 59/B4
El Athale (Itala), Somalia 115/J3
Elato (atoll), Micronesia 87/E5
El 'Atrun (oasis), Sudan 111/E4
El 'Auja, Israel 65/D5
Elâziğ (prov.), Turkey 63/H3
Elâziğ, Turkey 59/C2
Elâziğ, Turkey 63/H3
El Azizia, Libya 111/B1
El Azúcar (res.), Mexico 150/K3
Elba, Ala. (36323) 195/F8
Elba (isl.), Italy 7/E4
Elba, Idaho (83326) 220/E7
Elba (isl.), Italy 34/C3
Elba, Minn. (†55910) 255/F6
Elba, Nebr. (68835) 264/F3
Elba, N.Y. (14058) 276/D4
Elba, Ohio (45728) 284/H6
El Bab, Syria 63/G4
El Balqa (dist.), Jordan 65/D4
El Banco, Colombia 126/D2
El Barco, Spain 33/C1
El Barco de Ávila, Spain 33/D2
El Bardi, Libya 111/D1
El Barkat, Libya 111/B3
Elbasan, Albania 45/E5
El Baúl, Venezuela 124/D3
El Bawiti, Egypt 111/E2
El Bawiti, Egypt 59/A4
El Bawiti, Egypt 111/E2
El Bayadh, Algeria 106/E2
El Bayadh, Algeria 102/C1
Elbe (riv.) 7/F3
Elbe, E. Germany 22/D2
Elbe (riv.), W. Germany 22/C2
Elbe, Wash. (98330) 310/C4
El Beida, Yemen Arab Rep. 59/E7
El Beni (dept.), Bolivia 136/C3
Elberfeld, Ind. (47613) 227/C8
Elberon, Iowa (52225) 229/J4
Elberon, N.J. (†07740) 273/F3
Elberon, Va. (23846) 307/P6
Elbert (co.), Colo. 208/L4
Elbert, Colo. (80106) 208/L4
Elbert (mt.), Colo. 208/G4
Elbert (co.), Georgia 117/G2
Elbert, W. Va. (24830) 312/C8
Elberta, Ala. (36530) 195/C10
Elberta, Georgia (†31093) 217/E5
Elberta, Mich. (49628) 250/C4
Elberta, Utah (84626) 304/B4
Elberton, Georgia (30635) 217/G2
Elberton, Wash. (†99130) 310/H4
Elbeuf, France 28/C2
Elbing, Kansas (67041) 232/E3
Elbing (Elblag), Poland 47/D1
El Bira, West Bank 65/C4
Elbistan, Turkey 63/G3
Elblag (prov.), Poland 47/D1
Elblag, Poland 47/D1
Elblag, Poland 7/F3
El Bolsón, Argentina 143/B5
Elbon, Pa. (†15823) 294/E3
El Bonillo, Spain 33/E3
El Boquerón (pass), Peru 128/E7
El Borma, Tunisia 106/F2
Elbow (riv.), Alberta 182/C4
Elbow (lake), Manitoba 179/A4
Elbow, Sask. 181/E4
Elbow (lake), Minn. 255/C5
Elbow Lake, Minn. (56531) 255/B5
Elbridge, N.Y. (13060) 276/G5
Elbridge, Tenn. (38227) 237/C8
El'brus (mt.), U.S.S.R. 7/J4
El'brus (mt.), U.S.S.R. 52/F6
El Buheyrat (prov.), Sudan 111/E6
El Bur, Somalia 115/J3
Elburn, Netherlands 27/H4
El Burgo de Osma, Spain 33/E2
Elburn, Ill. (60119) 222/F2
Elburz (mts.), Iran 59/F2
Elburz (mts.), Iran 66/G2
El Cajon, Calif. (*92020) 204/J11
El Callao, Venezuela 124/G4
El Calvario, Venezuela 124/E3
El Campo, Texas (77437) 303/H8
El Caney, Cuba 158/J4
El Carmen, El Beni, Bolivia 136/D3
El Carmen, Santa Cruz, Bolivia 136/F6
El Carmen, O'Higgins, Chile 138/F5
El Carmen, Ñuble, Chile 138/A11
El Carmen, Chocó, Colombia 126/C2
El Carmen, Nariño, Colombia 126/A6
El Carmen, Norte de Santander, Colombia 126/D3
El Carmen de Bolívar, Colombia 126/C3
El Carre, Ethiopia 111/H6
El Centro, Calif. 188/D4
El Centro, Calif. (92243) 204/K11
El Cerrito, Colombia 126/D4
El Cerrito, Calif. (94530) 204/J2
El Cerrito, Colombia 126/B6
El Cerro, Bolivia 136/E5
El Chaparro, Venezuela 124/F3
Elche, Spain 33/F3
Elche de la Sierra, Spain 33/E3
Elcho (isl.), North. Terr. 93/D1
Elcho, Wis. (54428) 317/H5
El Chocón, Argentina 143/C4
Elcho Island Mission, North. Terr. 93/D1
El Choro, Bolivia 136/B6
El Chorro, Argentina 143/D1
Elco, Ill. (62929) 222/D6
El Cobre, Chile 138/A4
El Cobre, Cuba 158/J4
El Cocuy, Colombia 126/D4
El Convento, Chile 138/F4
El Corazón, Ecuador 128/C3
El Cristo, Venezuela 124/C4
El Cuey, Dom. Rep. 158/F6
El Cuy, Argentina 143/C4
Elda, Spain 33/F3
El Dara, Ill. (†62312) 222/B4

Elde (riv.), E. Germany 22/D2
Elden (mt.), Ariz. 198/D3
El Der, Ethiopia 111/H6
Elderbank, Nova Scotia 168/E4
El Dere, Somalia 115/J3
El Dere, Somalia 115/J3
Elderon, Wis. (54429) 317/H6
Eldersburg, Md. (†21784) 245/L3
Eldersley, Sask. 181/H3
Elderslie, Scotland 15/B2
Elderslie, Pa. (15036) 294/A5
Elderton, Pa. (15736) 294/D4
El Diente (peak), Colo. 208/C7
El'dikan, U.S.S.R. 49/O3
Eldivan, Turkey 63/E2
El Diviso, Colombia 126/A7
El Djem, Tunisia 106/G1
El Djezair (Algiers) (cap.), Algeria 106/E1
El Djouf (des.) 102/B2
Eldon, Iowa (52554) 229/J7
Eldon, Mo. (65026) 261/G6
Eldon, Wash. (†98555) 310/B3
Eldora, Colo. (†80466) 208/H3
Eldora, Iowa (50627) 229/G4
Eldora, N.J. (†08270) 273/D5
Eldorado, Argentina 143/E3
El Dorado, Ark. 188/H4
El Dorado (†71730) 202/E7
El Dorado (co.), Calif. 204/E5
El Dorado, Calif. (95623) 204/C8
Eldorado (Fender), Georgia (†31794) 217/E8
Eldorado, Ill. (62930) 222/E6
Eldorado, Iowa (52175) 229/K2
El Dorado, Kansas (67042) 232/F4
Eldorado, Md. (†21659) 245/P6
Eldorado, Mexico 150/G5
Eldorado, Miss. (†39156) 256/C5
Eldorado, N.C. (†27371) 281/K4
Eldorado, Ohio (45321) 284/A6
Eldorado, Okla. (73537) 288/G6
Eldorado, Texas (76936) 303/D7
Eldorado, Venezuela 124/H4
Eldorado, Wis. (54932) 317/J8
El Dorado Hills, Calif. (95630) 204/C8
El Dorado Springs, Mo. (64744) 261/E7
Eldorendo, Georgia (†31737) 217/C8
Eldoret, Kenya 102/F4
Eldoret, Kenya 115/G3
Eldred, Ill. (62027) 222/C4
Eldred, Minn. (56532) 255/B3
Eldred, Pa. (16731) 294/F4
Eldridge, Ala. (35554) 195/C3
Eldridge, Iowa (52748) 229/M5
Eldridge, Mo. (65463) 261/G7
Eldridge, N. Dak. (58435) 282/N6
El Dulce Nombre, Honduras 154/E3
Eleanor, W. Va. (25070) 312/C5
Electra, Texas (76360) 303/F4
Electric (peak), Mont. 262/F6
Electric City, Wash. (99123) 310/F3
Electric Mills, Miss. (39327) 256/G5
Electron, Wash. (†98360) 310/C4
Eleele, Hawaii 96/H05) 218/C2
Elefantes (gulf), Chile 138/D6
Elek, Hungary 41/F3
Elektrostal', U.S.S.R. 7/J4
Elektrostal', U.S.S.R. 52/E3
El Empedrado, Venezuela 124/C3
Elena, Bulgaria 45/G4
Elephant (isl.) 5/D16
Elephant (riv.), Namibia 118/B5
Elephant (mt.), Texas 303/D12
Elephanta (isl.), India 68/B7
Elephant Butte, N. Mex. (87935) 274/B5
Elephant Butte (res.), N. Mex. 168/B4
Elephant Butte (res.), N. Mex. 274/B5
Eleroy, Ill. (61027) 222/D1
Eleşkirt, Turkey 63/K3
El Espinar, Spain 33/D2
El Estor, Guatemala 154/C3
Eleuthera (isl.), Bahamas 146/L7
Eleuthera (isl.), Bahamas 156/C1
Eleva, Wis. (54738) 317/D6
Eleven Mile Canyon (res.), Colo. 208/H5
Elevtheroúpolis, Greece 45/G5
El Faiyûm, Egypt 111/J3
El Faiyûm, Egypt 59/B4
El Faiyûm, Egypt 102/E2
El Fasher, Sudan 111/E5
El Fasher, Sudan 102/E3
El Fashn, Egypt 111/J4
El Ferrol, Spain 33/B1
El Fifi, Sudan 111/F5
Elfin Cove, Alaska (99825) 196/M1
El Fogaha, Libya 111/C2
Elfrida, Ariz. (†85617) 198/F7
Elfros, Sask. 181/J4
El Fuerte, Mexico 150/E3
El Furat (riv.), Syria 63/H4
El Gallo, Nicaragua 154/E4
El Gatrun, Libya 111/B3
El Gatrun, Libya 102/D2
El Geneina, Sudan 111/D5
El Geneina, Sudan 102/E3
El Geteina, Sudan 111/F5
El Geteina, Sudan 59/B7
El Gezira (prov.), Sudan 111/F5
El Gezira, Libya 111/D2
Elgg, Switzerland 39/G2
El Gheria esh Sherqia, Libya 111/B1
El Ghor (reg.), Jordan 65/C6
Elgin, Ariz. (†85637) 198/E7
Elgin, Ill. 188/J2
Elgin, Ill. (60120) 222/E1
Elgin, Iowa (52141) 229/K3
Elgin, Kansas (†67361) 232/F4
Elgin, Manitoba 179/B5
Elgin, Minn. (55932) 255/F6
Elgin, Nebr. (68636) 264/F3
Elgin, Nev. (†89009) 266/G5
Elgin, New Bruns. 170/E3
Elgin, N. Dak. (58533) 282/G7

Elgin, Ohio (45838) 284/A4
Elgin, Okla. (73538) 288/K5
Elgin (county), Ontario 177/C5
Elgin, Ontario 177/H3
Elgin, Oreg. (97827) 291/K2
Elgin, Pa. (16413) 294/C2
Elgin, Scotland 10/E2
Elgin, Scotland 15/E3
Elgin, S.C. (29045) 296/F3
Elgin, Tenn. (37732) 237/M8
Elgin, Texas (78621) 303/G7
El Golea, Algeria 102/C1
El Goléa, Algeria 106/E2
Elgon (mt.) 102/F4
Elgon (mt.), Kenya 115/F3
Elgon (mt.), Uganda 115/F3
Elgood, W. Va. (24723) 312/E8
El Granada, Calif. (94018) 204/H3
El Guapo, Venezuela 124/F2
El Guayabo, Dom. Rep. 158/E5
El Hamad (des.), Iraq 59/F2
El Hammam, Egypt 111/E1
El Hammam, Egypt 59/A3
El Hamurre, Somalia 115/J2
El Haseke, Syria 59/D2
El Haseke, Syria 63/J4
El Hilla, Sudan 111/E5
El Huecu, Argentina 143/B4
El Husn, Jordan 65/D3
Eli, Nebr. (69213) 264/C2
Elias, Ky. (58447) 237/O6
Elida, N. Mex. (88116) 274/F5
Elida, Ohio (45807) 284/B4
Elie, Manitoba 179/E5
Elie and Earlsferry, Scotland 15/F4
Elihu, Ky. (42530) 237/M6
Elijah, Mo. (†65626) 261/H9
Elila (riv.), Zaire 115/E4
Elim, Alaska (99739) 196/F2
Elimsport, Pa. (†17810) 294/H3
El Indio, Texas (78860) 303/D9
Eling, England 13/F7
Elin Pelin, Bulgaria 45/F4
Eliot○, Maine (03903) 243/B9
Eliot (mt.), Newf. 166/B2
Elisa, Argentina 143/D3
Eliska, Ala. (†36500) 195/C8
El Iskandariya (Alexandria), Egypt 59/A3
El Iskandariya (Alexandria), Egypt 111/J2
Elista, U.S.S.R. 7/J4
Elista, U.S.S.R. 48/E5
Elista, U.S.S.R. 52/E3
Elizabeth, Ark. (72531) 202/F1
Elizabeth, Colo. (80107) 208/K4
Elizabeth, Georgia (30060) 217/J1
Elizabeth, Ill. (61028) 222/C1
Elizabeth, Ind. (47117) 227/F8
Elizabeth, La. (70638) 238/E5
Elizabeth (cape), Maine 243/C8
Elizabeth (isls.), Mass. 249/L7
Elizabeth, Minn. (56533) 255/B4
Elizabeth, Miss. (38742) 256/C4
Elizabeth (mt.), New Bruns. 170/D1
Elizabeth, N.J. (*07201) 273/B2
Elizabeth, Pa. (15037) 294/C5
Elizabeth, S. Australia 88/E6
Elizabeth, S. Australia 94/B7
Elizabeth, W. Va. (26143) 312/D4
Elizabethton, Tenn. (37643) 237/S8
Elizabethtown, Ill. (62931) 222/E6
Elizabethtown, Ind. (47232) 227/F4
Elizabethtown, Ky. (42701) 237/K5
Elizabethtown, N.Y. (12932) 276/N2
Elizabethtown, N.C. (28337) 281/M5
Elizabethtown, Ohio (†45052) 284/A9
Elizabethtown, Pa. (17022) 294/J5
Elizabethville, Pa. (17023) 294/J4
Elizaville, Ind. (†46052) 227/E4
Elizaville, Ky. (41037) 237/O4
Elizondo, Spain 33/F1
El Jadida, Morocco 102/B1
El Jadida, Morocco 106/C2
El Jauf, Libya 102/B1
El Jauf, Libya 111/D3
El Jicaral, Nicaragua 154/D4
El Jícaro, Nicaragua 154/E4
Elk (riv.), Ala. 195/D1
Elk (riv.), Br. Col. 184/K5
Elk, Calif. (95432) 204/B4
Elk (riv.), Colo. 208/F1
Elk (co.), Kansas 232/F4
Elk (riv.), Kansas 232/F4
Elk (isl.), Manitoba 179/F4
Elk (riv.), Mar. 245/P3
Elk (lake), Mich. 250/D4
Elk, N. Mex. (†88210) 274/D6
Elk (creek), Okla. 288/H5
Elk (creek), Oreg. 291/E5
Elk (co.), Pa. 294/E3
Elk, Poland 47/F2
Elk (creek), S. Dak. 298/C5
Elk (riv.), Tenn. 237/H10
Elk (ridge), Utah 304/E6
Elk, Wash. (99009) 310/H2
Elkader, Iowa (52043) 229/L3
El Kamlin, Sudan 59/B7
El Karak (dist.), Jordan 65/E5
El Karak, Jordan 59/C3
El Karak, Jordan 65/E4
El Karnak, Egypt 111/F2
El Karnak, Egypt 59/B4
Elkatawa, Ky. (41322) 237/P5
Elk City, Idaho (83525) 220/C4
Elk City, Kansas (67344) 232/F4
Elk City (lake), Kansas 232/F4
Elk City, Okla. (73644) 288/H4
Elk City, Oreg. (†97391) 291/D3
Elk Creek, Calif. (95939) 204/B4
Elk Creek, Ky. (†40071) 237/L4
Elk Creek, Mo. (65464) 261/H8
Elk Creek, Nebr. (68348) 264/H4
Elk Creek, Va. (24326) 307/F7
Elk Creek, Wis. (†54747) 317/C7

El Kelaa des Srarhna, Morocco 106/C2
Elk Falls, Kansas (67345) 232/F4
Elkford, Br. Col. 184/K5
Elkfork, Ky. (41421) 237/P5
Elk Garden, Va. (†24266) 307/E7
Elk Garden, W. Va. (26717) 312/H4
Elk Grove, Calif. (95624) 204/B9
Elk Grove Village, Ill. (60007) 222/B5
El Khalil (Hebron), West Bank 65/C4
El Khandaq, Sudan 59/B6
El Khandaq, Sudan 111/E4
El Khârga, Egypt 111/F2
El Khârga, Egypt 102/E2
El Khârga, Egypt 59/B4
Elkhart, Ill. (62634) 222/D3
Elkhart, Ind. 188/J2
Elkhart (co.), Ind. 227/F1
Elkhart, Ind. (46514) 227/F1
Elkhart (riv.), Ind. 227/F1
Elkhart, Iowa (50073) 229/F5
Elkhart, Kansas (67950) 232/A4
Elkhart, Texas (75839) 303/J6
Elkhart Lake, Wis. (53020) 317/L8
Elkhorn, Nebr. (68022) 264/H3
Elkhorn (riv.), Nebr. 264/G3
Elkhorn, W. Va. (24831) 312/D8
Elkhorn, Wis. (53121) 317/J10
Elkhorn City, Ky. (41522) 237/S6
Elkhovo, Bulgaria 45/H4
Elkhurst, W. Va. (†25043) 312/D6
Elkin, N.C. (28621) 281/H2
Elkins, Ark. (72727) 202/C1
Elkins, N. Mex. (†88101) 274/E5
Elkins, W. Va. (26241) 312/G5
Elkinsville, Ind. (†47448) 227/E6
Elkland, Mo. (65644) 261/F8
Elkland, Pa. (16920) 294/H1
Elk Mills, Md. (21920) 245/P2
Elkmont, Ala. (35620) 195/E1
Elkmont, Tenn. (†37862) 237/O9
Elk Mound, Wis. (54739) 317/C6
Elk Mountain, Wyo. (82324) 319/F4
Elk Neck, Md. (†21901) 245/P2
Elko, Br. Col. 184/K5
Elko, Georgia (31025) 217/E6
Elko, Minn. (55020) 255/E6
Elko, Nev. 188/C3
Elko, Nev. 146/G6
Elko (co.), Nev. 266/F1
Elko, Nev. (89801) 266/F2
Elko, S.C. (29826) 296/E5
Elk Park, N.C. (28622) 281/E2
Elk Point, Alberta 182/E3
Elk Point, S. Dak. (57025) 298/R8
Elkport, Iowa (52044) 229/L3
Elkridge, Md. (21227) 245/M4
Elkridge, W. Va. (†24868) 312/D6
Elk River, Idaho (83827) 220/B3
Elk River, Minn. (55330) 255/E5
Elk Run Heights, Iowa (†50700) 229/J4
Elk Springs, Colo. (81633) 208/C2
Elkton, Fla. (32033) 212/E2
Elkton, Ky. (42220) 237/G7
Elkton, Md. (21921) 245/P2
Elkton, Mich. (48731) 250/F4
Elkton, Minn. (55933) 255/F7
Elkton, Mo. (†65650) 261/F7
Elkton, Oreg. (97436) 291/D4
Elkton, S. Dak. (57026) 298/S5
Elkton, Tenn. (38455) 237/H10
Elkton, Va. (22827) 307/L4
Elk Valley, Tenn. (†37848) 237/N7
Elkview, W. Va. (25071) 312/C6
Elkville, Ill. (62932) 222/D6
Elkwater, Alberta 182/F5
Elkwater, W. Va. (†26273) 312/G5
Ellabell, Georgia (31308) 217/K6
El Ladhiqiya (Latakia), Syria 59/C2
El Ladhiqiya (Latakia), Syria 63/F5
El Lago, Texas (†77586) 303/K2
Ellamar, Alaska (†99695) 196/D1
Ellamore, W. Va. (26267) 312/F5
Ellaville, Fla. (†32060) 212/C1
Ellaville, Georgia (31806) 217/D6
Ellef Ringnes (isl.), Canada 4/B15
Ellef Ringnes (isl.), N.W.T. 162/M3
Ellef Ringnes (isl.), N.W.T. 146/H2
Ellef Ringnes (isl.), N. W. Terrs. 187/H2
Ellen (mt.), Utah 304/D5
Ellen (mt.), Vt. 268/B3
Ellenboro, N.C. (28040) 281/F4
Ellenboro, W. Va. (26346) 312/D4
Ellenboro, Wis. (†53813) 317/E10
Ellenburg Center, N.Y. (†12934) 276/N1
Ellenburg Depot, N.Y. (12935) 276/N1
Ellendale, Del. (19941) 245/S5
Ellendale, Minn. (56026) 255/E7
Ellendale, N. Dak. (58436) 282/N7
Ellendale, Tasmania 99/C4
Ellendale, Tenn. (38029) 237/B10
Ellendale, W. Australia 92/D2
Ellensburg, Wash. (98926) 310/E3
Ellenton, Fla. (34222) 212/C4
Ellenton, Georgia (31747) 217/E8
Ellenville, N.Y. (12428) 276/M7
Ellenwood, Georgia (30049) 217/L2
Ellerbe, N.C. (28338) 281/K4
Ellershouse, Nova Scotia 168/D4
Ellerslie, Georgia (31807) 217/C5
Ellerslie, Md. (21529) 245/C2
Ellerslie, N. Zealand 100/C1
Ellerslie, Pr. Edward I. 168/F2
Ellerton, Barbados 161/B9
Ellerton, Md. (†34512) 245/H2
Ellery, Ill. (62833) 222/E5
Ellesmere (isl.), Canada 4/B14
Ellesmere (isl.), Canada 2/F1

Ellesmere, England 13/E5
Ellesmere (lake), N. Zealand 100/D5
Ellesmere (isl.), N.W.T. 146/K1
Ellesmere (isl.), N. W. Terrs. 187/K2
Ellesmere Port, England 13/G2
Ellesmere Port, England 10/F2
Ellettsville, Ind. (47429) 227/D6
Ellezelles, Belgium 27/D7
El Libertador General Bernardo O'Higgins (reg.), Chile 138/A10
Ellice (riv.), N. W. Terrs. 187/H3
Ellicott City, Md. (21043) 245/L3
Ellicottville, N.Y. (14731) 276/C6
Ellijay, Georgia (30540) 217/C1
El Limón, Nicaragua 154/E4
Ellington○, Conn. (06029) 210/F1
Ellington, Mo. (63638) 261/L8
Ellington, N.Y. (14732) 276/B6
Ellinwood, Kansas (67526) 232/D3
Elliot, Manitoba 179/G2
Elliot, S. Africa 118/D6
Elliot Lake, Ontario 177/B1
Elliot Lake, Ontario 175/D3
Elliott, Ark. (†71701) 202/E7
Elliott (key), Fla. 212/F6
Elliott, Iowa (51532) 229/C6
Elliott, Ill. (60933) 222/D3
Elliott, Ky. 237/P4
Elliott, Md. (21823) 245/P7
Elliott, Miss. (38926) 256/E3
Elliott, N. Dak. (58025) 282/P7
Elliott, North. Terr. 88/E3
Elliott, North. Terr. 93/C4
Elliott, S.C. (29046) 296/G3
Elliott, Tasmania 99/B3
Elliott (bay), Tasmania 99/B5
Elliottsburg, Pa. (17024) 294/H5
Elliottville, Ky. (40317) 237/P4
El Kitta, Jordan 65/D3
Ellis, Idaho (83235) 220/D5
Ellis (co.), Kansas 232/C3
Ellis, Kansas (67663) 232/C3
Ellis (pond), Maine 243/B6
Ellis (riv.), Maine 243/B6
Ellis (riv.), N.H. 268/E3
Ellis (co.), Okla. 288/G2
Ellis (co.), Texas 303/H5
El Lisan (pen.), Jordan 65/C5
Ellisburg, N.Y. (13636) 276/H3
Ellis Grove, Ill. (62241) 222/D5
Ellison Bay, Wis. (†54210) 317/M5
Elliston, Mont. (59728) 262/D4
Elliston, Newf. 166/D2
Elliston, Ohio (43432) 284/D2
Elliston, S. Africa 118/A6
Elliston, Va. (24087) 307/H6
Elliston-Lafayette, Va. (24087) 307/H6
Ellisville, Ill. (61431) 222/C3
Ellisville, Miss. (39437) 256/F7
Ellisville, Mo. (†63011) 261/M3
Ellisville, Wis. (†54217) 317/L7
Ellon, Scotland 15/G3
Ellon, Scotland 15/F3
Elloree, S.C. (29047) 296/F4
Elloughton, England 13/G4
Ellis (riv.), Alberta 182/D1
Ellport, Pa. (†16117) 294/B4
Ells (riv.), Alberta 182/A2
Ellsinore, Mo. (63937) 261/L9
Ellston, Iowa (50074) 229/F7
Ellsworth (hill), Conn. 210/B1
Ellsworth, Ill. (61737) 222/E3
Ellsworth, Iowa (50075) 229/F4
Ellsworth (co.), Kansas 232/D3
Ellsworth, Kansas (67439) 232/D3
Ellsworth, Maine (04605) 243/F6
Ellsworth, Mich. (49729) 250/D3
Ellsworth, Minn. (56129) 255/B7
Ellsworth, Nebr. (69340) 264/B2
Ellsworth○, N.H. (†03264) 268/D4
Ellsworth (lake), Okla. 288/K5
Ellsworth, Pa. (15331) 294/B5
Ellsworth, Wis. (54011) 317/A6
Ellsworth A.F.B., S. Dak. 298/C5
Ellsworth Land (reg.) 5/B14
Ellwangen, W. Germany 22/D4
Ellwood City, Pa. (16117) 294/B4
Elm (riv.), N. Dak. 282/N8
Elm (creek), S. Dak. 298/D5
Elm (riv.), S. Dak. 282/M2
Elm, Switzerland 39/H3
Elma, Iowa (50628) 229/J2
Elma, Manitoba 179/G5
Elma, N.Y. (14059) 276/C5
Elma, Wash. (98541) 310/B4
El Macao, Dom. Rep. 158/F6
El Madwar, Jordan 65/E3
El Mafraq, Jordan 65/E3
El Mahalla el Kubra, Egypt 111/J3
El Majdal, Jordan 65/D3
Elmalı, Turkey 63/C4
El Manaqil, Sudan 111/F5
El Mansûra, Egypt 111/K3
El Mansûra, Egypt 59/B3
El Manteco, Venezuela 124/G4
El Manzano, Chile 138/F5
El Ma'qil, Iraq 66/E5
El Marj, Libya 102/E1
El Marj, Libya 111/D1
El Marmol, Mexico 150/B2
Elm City, N.C. (27822) 281/O3
Elm Creek, Manitoba 179/E5
Elm Creek, Nebr. (68836) 264/E4
Elmdale, Minn. (†56314) 255/D5
Elmendorf, Texas (78112) 303/K11
Elmendorf A.F.B., Alaska 196/B1
El Pilar, Venezuela 124/G2
El Pintado, Argentina 143/D1
Elmer, Mo. (63538) 261/J3
Elmer, Minn. (†55765) 255/F3
Elmer, N.J. (08318) 273/C4
Elmer, Okla. (73539) 288/H6
Elmer City, Wash. (99124) 310/G2
Elm Fork, Trinity (riv.), Texas 303/G2
Elm Grove, La. (71051) 238/C2

Elm Grove, Wis. (53122) 317/K1
Elmhurst, Ill. (60126) 222/B5
Elmhurst, Pa. (18416) 294/P7
Elmina, Ghana 106/D8
El Minya, Egypt 102/E2
El Minya, Egypt 59/B4
El Minya, Egypt 111/J4
Elmira, Ill. (†61483) 222/D2
Elmira, Mich. (49730) 250/E3
Elmira, Mo. (†64062) 261/D3
Elmira, N.Y. 188/L2
Elmira, N.Y. (*14901) 276/G6
Elmira, Ontario 177/D4
Elmira, Oreg. (97431) 291/D3
Elmira, Pr. Edward I. 168/F2
Elmira, W. Va. (26618) 312/E5
Elmira Heights, N.Y. (†14903) 276/G6
Elmira, Ariz. (85335) 198/C5
El Mirage, Ariz. (85335) 198/C5
El Misti (mt.), Peru 128/G11
Elmo, Kansas (†67451) 232/E3
Elmo, Mo. (64445) 261/B1
Elmo, Mont. (59915) 262/B3
Elmo, Texas (75118) 303/H5
Elmo, Utah (84521) 304/D4
Elmo, Wyo. (82324) 319/F4
Elmodel, Georgia (31748) 217/D8
Elmont, Kansas (†66603) 232/G2
Elmont, N.Y. (11003) 276/F7
El Monte, Calif. (*91731) 204/D10
El Monte, Chile 138/E4
Elmora, Pa. (15737) 294/E4
Elmore (co.), Ala. 195/F5
Elmore, Ala. (36025) 195/F5
Elmore (co.), Idaho 220/C6
Elmore, Minn. (56027) 255/D7
Elmore, Ohio (43416) 284/D3
Elmore City, Okla. (73035) 288/M5
Elmore, N. Mex. (†87321) 274/A3
El Morro Nat'l Mon., N. Mex. 274/A3
Elm Park, La. (†70775) 238/H5
El Mraiti (well), Mali 106/D5
El Mrayer (well), Mauritania 106/C4
El Mreïti (well), Mauritania 106/C4
Elmrock, Ky. (41624) 237/P6
Elmsdale, Nova Scotia 168/E4
Elmsdale, Pr. Edward I. 168/D2
Elmsford, N.Y. (10523) 276/O6
Elmshorn, W. Germany 22/D2
Elm Springs, Ark. (72728) 202/B1
Elm Springs, S. Dak. (57736) 298/D5
Elmsvale, Nova Scotia 168/E4
Elmvale, Ontario 177/E3
Elmville, Conn. (†06239) 210/H1
Elmwood, Conn. (06110) 210/D2
Elmwood, Ill. (61529) 222/D3
Elmwood, Nebr. (68349) 264/H4
Elmwood, Okla. (73935) 288/F1
Elmwood, Ontario 177/C3
Elmwood, Wis. (54740) 317/B6
Elmwood Park, Ill. (60635) 222/B5
Elmwood Park, N.J. (†07407) 273/B2
Elmwood Park, Wis. (†53401) 317/M3
Elmwood Place, Ohio (45216) 284/B9
Elmworth, Alberta 182/A2
Elne, France 26/E6
El Nido, Calif. (95317) 204/E6
El Nido, Philippines 82/B5
El Ñilhue, Chile 138/G2
Elnora, Alberta 182/D3
Elnora, Ind. (47529) 227/C7
El Obeid, Sudan 102/E3
El Obeid, Sudan 59/B7
El Obeid, Sudan 111/E5
Elobey (isls.), Equat. Guinea 115/A3
El Odaiya, Sudan 111/E5
Eloff, S. Africa 118/A6
Eloi (bay), La. 238/M7
Elon, Ala. (†35760) 195/F4
Elon College, N.C. (27244) 281/L2
Elora, Ontario 177/D4
Elora, Tenn. (37328) 237/J10
El Oro (prov.), Ecuador 128/C4
Elortondo, Argentina 143/F6
El Oso, Venezuela 124/F3
El Oued, Algeria 106/F2
Eloy, Ariz. (85231) 198/D6
El Pájaro, Colombia 126/D2
El Palmar, Chuquisaca, Bolivia 136/D7
El Palmar, Santa Cruz, Bolivia 136/D7
El Palmar, Tarija, Bolivia 136/D7
El Palmar, Venezuela 124/G3
El Pao, Anzoátegui, Venezuela 124/F3
El Pao, Bolívar, Venezuela 124/G3
El Pao, Cojedes, Venezuela 124/D3
El Paraíso, Colombia 126/C7
El Paraíso, Copán, Honduras 154/C3
El Paraíso, El Paraíso, Honduras 154/D4
El Pardo, Spain 33/F4
El Paso, Ark. (72045) 202/F3
El Paso (co.), Colo. 208/K5
El Paso, Ill. (61738) 222/D3
El Paso (co.), Texas 303/A10
El Paso, Texas 146/H6
El Paso, Texas 188/F4
El Paso, Texas (*79901) 303/A10
El Paso, U.S. 2/D4
El Pato, Colombia 126/C6
El Perú, Bolivia 136/B3
El Perú, Venezuela 124/H4
Elphin, Ireland 17/E4
Elphinstone, Manitoba 179/B4
El Pico, Bolivia 136/C4
El Pilar, Venezuela 124/G2
El Pintado, Argentina 143/D1
El Piquete, Argentina 143/D1
El Portal, Calif. (95318) 204/F6
El Portón, Honduras 154/D4
El Portugués, Peru 128/C7
El Porvenir, Honduras 154/D4
El Porvenir, Mexico 150/G1
El Porvenir, N. Mex. (†87731) 274/D4
El Porvenir, Panama 154/H6

El Potosí, Mexico 150/J4
El Pozo, Dom. Rep. 158/F5
El Prado, N. Mex. (87529) 274/D2
El Progreso, Ecuador 128/C9
El Progreso, Guatemala 154/B3
El Progreso, Honduras 154/D3
El Puente, Santa Cruz, Bolivia 136/D5
El Puente, Tarija, Bolivia 136/C7
El Puerto de Santa María, Spain 33/C4
El Pun, Ecuador 128/D2
El Qâhira (Cairo) (cap.), Egypt 59/B4
El Qâhira (Cairo) (cap.), Egypt 111/J3
El Qantara, Egypt 111/K3
El Qantara, Egypt 59/B3
El Qasr, Egypt 111/E2
El Qasr, Egypt 59/A4
El Quadmus, Syria 63/F6
El Quebrachal, Argentina 143/D2
Elqui (riv.), Chile 138/A8
El Quisco, Chile 138/E3
El Quneitra (prov.), Syria 63/F6
El Quneitra, Syria 63/F6
El Quseir, Egypt 102/F2
El Quseir, Egypt 59/B4
El Quseir, Egypt 111/F2
El Quseir, Syria 63/G5
El Quwaira, Jordan 65/E5
Elrama, Pa. (15038) 294/C5
El Rashid, Syria 63/H5
El Rashid, Syria 59/C2
El Rastro, Venezuela 124/E3
El Real de Santa María, Panama 154/J6
El Realejo, Nicaragua 154/D4
El Reno, Okla. (73036) 288/K3
El Rio, Calif. (†93030) 204/F8
El Rito, N. Mex. (87530) 274/C2
El Rito, N. Mex. 274/C2
Elrod, Ala. (35458) 195/C4
Elrod, Ind. (†47018) 227/G6
El Roque, Venezuela 124/E2
Elrosa, Minn. (56325) 255/C5
El Rosario, Estero (riv.), Chile 138/F3
El Rosario, Mexico 150/B1
Elrose, Sask. 181/D4
Elroy, Wis. (53929) 317/F8
Elsa, Texas (78543) 303/G11
Elsa, Yukon 187/E3
Elsah, Ill. (62028) 222/C5
El Salado, Dom. Rep. 158/F6
El Salto, Mexico 150/G5
El Salvador 2/E5
El Salvador 146/J8
El Salvador, C. Rica 154/F5
EL SALVADOR 154/C4
El Samán de Apure, Venezuela 124/D4
El Santo, Cuba 158/D1
Elsas, Ontario 175/D4
El Sauce, Nicaragua 154/D4
Elsberry, Mo. (63343) 261/L4
Elsburg, S. Africa 118/A6
El Segundo, Calif. (90245) 204/B11
El Seibo (prov.), Dom. Rep. 158/F6
El Seibo, Dom. Rep. 158/F6
Elsey, Mo. (†65633) 261/E9
Elsie, Mich. (48831) 250/E5
Elsie, Nebr. (69134) 264/C4
Elsie, Oreg. (†97138) 291/D2
Elsiesrivier, S. Africa 118/A6
Elsinore (lake), Calif. 204/E11
Elsinore, Utah (84724) 304/B5
Elsmere, Del. (†19801) 245/R2
Elsmere, Ky. (†41018) 237/K3
Elsmere, Nebr. (69135) 264/D2
Elsmore, Kansas (66732) 232/G4
El Socorro, Venezuela 124/F3
El Sollum (gulf), Egypt 111/E1
El Sombrero, Venezuela 124/E3
Elst, Netherlands 27/H5
Elster, Black (riv.), E. Germany 22/J3
Elster, White (riv.), E. Germany 22/J3
Elston, Ind. (†47901) 227/D4
Elstow, Sask. 181/E4
El Tabo, Chile 138/F3
El Tambo, Colombia 126/B6
El Teleno (mt.), Spain 33/C1
Eltham, N. Zealand 100/E3
Eltham, Victoria 97/J4
Eltham, Victoria 88/L6
El Tiemblo, Spain 33/D2
El Tigre, Venezuela 120/C2
El Tigre, Venezuela 124/F3
El Tocuyo, Venezuela 124/C3
El Tofo, Chile 138/A7
Elton, La. (70532) 238/E6
El'ton, U.S.S.R. 52/G5
Elton, W. Va. (25965) 312/E7
Elton, Wis. (54430) 317/J5
Eltopia, Wash. (99330) 310/G4
El Toro, Calif. (92630) 204/E11
El Toro, Venezuela 124/H3
El Toro Marine Air Sta., Calif. 204/D11
El Tránsito, Chile 138/B7
El Triunfo, Honduras 154/D4
El Tucuche (mt.), Trin. & Tob. 161/B10
El Tûr, Egypt 111/F2
El Tur, Egypt 59/B4
Eluru, India 68/E5
El'Uweinat, Libya 111/B2
Elva, Manitoba 179/A5
Elva, U.S.S.R. 53/D1
El Vado, N. Mex. (†87575) 274/C2
Elvas, Portugal 33/C3
Elvaston, Ill. (62334) 222/B3
Elverson, Pa. (19520) 294/L5
Elverum, Norway 18/G6
El Viejo, Nicaragua 154/D4
El Vigía, Venezuela 124/C3
El Vínculo, Venezuela 124/D1
Elvins, Mo. (63639) 261/L7
Elvira, Iowa (†52732) 229/N5

Elvira (isl.), N.W. Terrs. 187/H2
El Volcán, Chile 138/B10
El Wak, Kenya 115/H3
El War (well), Niger 106/G4
El Wasta, Egypt 111/J3
Elwell, Mich. (48832) 250/E5
Elwell (lake), Mont. 262/E2
Elwha (riv.), Wash. 310/B3
Elwood, Ill. (60421) 222/E2
Elwood, Ind. (46036) 227/F4
Elwood, Iowa (52226) 229/M4
Elwood, Kansas (66024) 232/H2
Elwood, Nebr. (68937) 264/E4
Elwood, N.J. (08217) 273/D4
Elwood, N.Y. (11731) 276/O9
Elwood, Utah (†84337) 304/B2
Ely, England 13/H5
Ely, England 10/G4
Ely, Iowa (52227) 229/K5
Ely, Minn. (55731) 255/G3
Ely, Nev. 188/D3
Ely, Nev. (89301) 266/G3
Ely (range), Nev. 266/G4
Ely, Vt. (05044) 268/C4
Ely (riv.), Wales 13/B7
El Yaduda, Jordan 65/D4
El Yagual, Venezuela 124/D4
Elyakim, Israel 65/C2
Elyashiv, Israel 65/B3
Elyria, Kansas (†67460) 232/E3
Elyria, Nebr. (68837) 264/E3
Elyria, Ohio (*44035) 284/F3
Elysburg, Pa. (17824) 294/K4
Elysian, Minn. (56028) 255/E6
Elysian Fields, Texas (75642) 303/L5
Elysian Grove, Tenn. (†37185) 237/F8
El Yunque (mt.), P. Rico 161/F1
El Zacatón, Mexico 150/J5
Elze, W. Germany 22/C2
Emanguk (Emmonak), Alaska (99581) 196/E2
Emanuel (co.), Georgia 217/H5
Emba, U.S.S.R. 48/F5
Emba (riv.), U.S.S.R. 48/F5
Embar, Newf. 166/A3
Embarcación, Argentina 143/D1
Embarras (riv.), Ill. 222/E4
Embarras Airport, Alberta 182/C5
Embarrass, Minn. (55732) 255/F3
Embarrass, Wis. (54933) 317/J6
Embden (pond), Maine 243/D6
Embden, N. Dak. (†58079) 282/R6
Emblem, Wyo. (82422) 319/D1
Embo, Scotland 15/E3
Emboscada, Paraguay 144/B4
Embree, Newf. 166/C4
Embreeville, Pa. (†19320) 294/L6
Embreeville Junction, Tenn. (†37601) 237/R8
Embro, Ontario 177/C4
Embrun, France 28/G5
Embrun, Ontario 177/J2
Embu, Kenya 115/G4
Embudo, N. Mex. (87531) 274/C2
Emden, Ill. (62635) 222/D3
Emden, Mo. (63439) 261/J3
Emden, W. Germany 22/B2
Emeigh, Pa. (15738) 294/E4
Emelle, Ala. (35459) 195/B5
Emerado, N. Dak. (58228) 282/R4
Emerald (isl.), N.W. Terrs. 187/G2
Emerald, Queensland 88/H4
Emerald, Queensland 95/C4
Emerald, Wis. (54012) 317/B5
Emerald Isle, N.C. (†28557) 281/P5
Emero (riv.), Bolivia 136/B3
Emerson, Ark. (71740) 202/D7
Emerson, Georgia (30137) 217/C2
Emerson, Iowa (51533) 229/C6
Emerson, Man. 162/G6
Emerson, Manitoba 179/E5
Emerson, Nebr. (68733) 264/H2
Emerson, N.J. (07630) 273/B1
Emerson, N.C. (28433) 281/M6
Emery, S. Dak. (57332) 298/O6
Emery (co.), Utah 304/D4
Emery, Utah (84522) 304/C5
Emery Mills, Maine (04031) 243/B8
Emeryville, Calif. (94608) 204/J2
Emeryville, Ontario 177/B5
Emet, Turkey 63/C3
Emida, Idaho (†83861) 220/B2
Emigrant, Mont. (59027) 262/F5
Emigrant (peak), Mont. 262/F5
Emigrant (peak), Nev. 266/C5
Emigsville, Pa. (17318) 294/J5
Emi Koussi (mt.), Chad 102/D3
Emi Koussi (mt.), Chad 111/C4
Emilia-Romagna (reg.), Italy 34/C2
Emilio Ayarza, Argentina 143/F7
Emily, Minn. (56447) 255/E4
Emily (lake), Minn. 255/C5
Emin (Dorbiljin), China 77/B2
Emine (cape), Bulgaria 45/J4
Eminence, Ind. (46125) 227/D5
Eminence, Ky. (40019) 237/L4
Eminence, Mo. (65466) 261/K8
Emington, Ill. (60934) 222/E3
Emirau (isl.), Papua N.G. 86/B1
Emirdağ, Turkey 63/D3
Emir Dağı (mt.), Turkey 63/D3
Emir Dağı (mt.), Turkey 59/B2
Emirgazi, Turkey 63/E4
Emison, Ind. (47530) 227/C7
Emita, Tasmania 99/D2
Emlenton, Pa. (16373) 294/C3
Emlyn, Ky. (40730) 237/N7
Emma, Ill. (62834) 222/E6
Emma, Ind. (†46571) 227/F1
Emma, Ky. (41625) 237/R5
Emma, Mo. (65327) 261/F5
Emma (range), Suriname 131/C4
Emmaboda, Sweden 18/J8
Emmalane, Georgia (†30442) 217/H5
Emmastad, Neth. Ant. 161/F9
Emmaus, Pa. (18049) 294/M4
Emmaus, Virgin Is. (U.S.) 161/C4
Emmaville, N.S. Wales 97/F1
Emmeloord, Netherlands 27/H3
Emmen, Netherlands 27/K3
Emmen, Switzerland 39/F2
Emmendingen, W. Germany 22/B4
Emmental (riv.), Switzerland 39/E3
Emmerich, W. Germany 22/B3
Emmet, Ark. (71835) 202/D6
Emmet (co.), Iowa 229/D2
Emmet (co.), Mich. 250/E3
Emmet, Nebr. (68734) 264/F2
Emmet, N. Dak. (58534) 282/G4
Emmet, Queensland 88/G4
Emmet, Queensland 95/C5
Emmetsburg, Iowa (50536) 229/D2
Emmett, Idaho (83617) 220/B6
Emmett, Kansas (66422) 232/F2
Emmett, Mich. (48022) 250/G6
Emmitsburg, Md. (21727) 245/J2
Emmonak, Alaska (99581) 196/E2
Emmons, Minn. (56029) 255/E7
Emmons (co.), N. Dak. 282/K7
Emmons (mt.), Utah 304/D3
Emneth, England 13/H5
Emo, Ontario 175/B3
Emo, Ontario 177/F5
Emory (riv.), Tenn. 237/M8
Emory, Texas (75440) 303/J5
Emory (peak), Texas 303/A8
Emory Gap, Tenn. (37735) 237/M9
Emory-Meadowview, Va. (24327) 307/E7
Emoryville, W. Va. (†26717) 312/H4
Empalme, Mexico 150/D2
Empalme Olmos, Uruguay 145/B6
Empangeni, S. Africa 118/E5
Empedrado, Argentina 143/E2
Empedrado, Chile 138/A11
Empexa (salt dep.), Bolivia 136/A7
Empire, Ala. (35063) 195/D3
Empire, Calif. (95319) 204/D6
Empire, Colo. (80438) 208/H3
Empire (res.), Colo. 208/L2
Empire, Georgia (31026) 217/F6
Empire, La. (70050) 238/L8
Empire, Mich. (49630) 250/C4
Empire, Ohio (43926) 284/J5
Empire City, Okla. (†73529) 288/L6
Empire Landing, Ariz. (85344) 198/A4
Empoli, Italy 34/C3
Emporia, Fla. (†32080) 212/E2
Emporia, Ind. (†46056) 227/F5
Emporia, Kans. 188/G3
Emporia, Kansas (66801) 232/F3
Emporia (I.C.), Va. (23847) 307/N7
Emporium, Pa. (15834) 294/F2
Empress, Alberta 182/E4
Empress Augusta (bay), Papua N.G. 86/C2
'Emrani, Iran 66/L3
Emrick, N. Dak. (†58422) 282/L4
Ems (riv.), W. Germany 22/B2
Emsdale, Ontario 177/E2
Emsdetten, W. Germany 22/B2
Emsworth, Pa. (15202) 294/B6
Emyvale, Ireland 17/G3
Enarotali, Indonesia 85/K6
Enard (bay), Scotland 15/C2
Enaville, Idaho (83829) 220/B2
Encampment, Wyo. (82325) 319/F4
Encampment (riv.), Wyo. 319/F4
Encarnación, Paraguay 120/D5
Encarnación, Paraguay 144/E5
Encarnación de Díaz, Mexico 150/H6
Enchant, Alberta 182/D4
Enchi, Ghana 106/D7
Encinal, Texas (78019) 303/E9
Encinitas, Calif. (92024) 204/H10
Encino, Calif. (91316) 204/B10
Encino, N. Mex. (88321) 274/D4
Encino, Texas (78353) 303/F11
Enciso, Colombia 126/E7
Encontrados, Venezuela 124/B3
Encounter (bay), S. Australia 88/F7
Encounter (bay), S. Australia 94/F6
Encrucijada, Cuba 158/E1
Endako, Br. Col. 184/E3
Ende, Indonesia 85/G7
Endeavor, Pa. (16322) 294/D2
Endeavor (str.), Queensland 88/G2
Endeavor, Wis. (53930) 317/G8
Endeavour (str.), Queensland 95/B1
Endeavour, Sask. 181/J3
Endee, N. Mex. (†88411) 274/F3
Endelave (isl.), Denmark 21/D6
Enderbury (isl.), Kiribati 87/J6
Enderby, Br. Col. 184/H5
Enderby Land (reg.), Ant. 2/M9
Enderby Land (reg.) 5/B3
Enderlin, N. Dak. (58027) 282/P6
Enders, Nebr. (69027) 264/C4
Enders (res.), Nebr. 264/C4
Endiang, Alberta 182/D4
Endicott (mts.), Alaska 196/H1
Endicott, Nebr. (68350) 264/G4
Endicott, N.Y. (13760) 276/H6
Endicott, S. Africa 118/J6
Endicott, Wash. (99125) 310/H4
Endless (lake), Maine 243/F5
Endrick Water (riv.), Scotland 15/B1
Endrőd, Hungary 41/F3
Endwell, N.Y. (13760) 276/H6
Ene (riv.), Peru 128/E8
Energy, Ill. (62933) 222/E6
Enes, Hungary 41/F2
Enetai, Wash. (†98310) 310/A2
Enewetak (Eniwetok) (atoll), Marshall Is. 87/G4
Enez, Turkey 63/B2
Enfield, Conn. (06082) 210/E1
Enfield○, Conn. (06082) 210/E1
Enfield, England 13/H7
Enfield, England 10/B5
Enfield, Ill. (62835) 222/E5
Enfield, Maine (04433) 243/F5
Enfield○, Maine (04433) 243/F5
Enfield, Minn. (†55362) 255/E5
Enfield, N.H. (03748) 268/C4
Enfield○, N.H. (03748) 268/C4
Enfield, N.C. (27823) 281/O2
Enfield, Nova Scotia 168/E4
Enfield, S. Australia 88/D7
Enfield, S. Australia 94/B7
Enfield Center, N.H. (03749) 268/C4
Enfield P.O. (Thompsonville), Conn. (06082) 210/E1
Engadine, Mich. (49827) 250/D2
Engadine (valley), Switzerland 39/K3
Engaño (cape), Dom. Rep. 158/F6
Engaño (cape), Philippines 82/D1
Engaño (cape), Philippines 54/O8
Engelberg, Switzerland 39/F3
Engelhard, N.C. (27824) 281/T3
Engelhartszell, Austria 41/B2
Engel's, U.S.S.R. 7/J3
Engel's, U.S.S.R. 52/G4
Engel's, U.S.S.R. 48/E4
Engen, Br. Col. 184/E3
Enggano (isl.), Indonesia 85/C7
Enghien, Belgium 27/D7
Engi, Switzerland 39/H3
England, Ark. (72046) 202/G4
ENGLAND 13
ENGLAND 10/F5
England, U.K. 7/D3
England A.F.B., La. 238/E4
Engle, N. Mex. (†87935) 274/B5
Englee, Newf. 166/C3
Englefeld, Sask. 181/G3
Englehart, Ont. 162/H6
Englehart, Ontario 177/K5
Englehart, Ontario 175/E3
Englevale, Kansas (†66756) 232/H4
Englevale, N. Dak. (58028) 282/P7
Englewood, Colo. (†80110) 208/K3
Englewood, Fla. (33533) 212/D5
Englewood, Kansas (67840) 232/C4
Englewood, N.J. (*07631) 273/C2
Englewood, Ohio (45322) 284/B6
Englewood, Tenn. (37329) 237/M10
Englewood Cliffs, N.J. (07632) 273/C2
English (chan.) 7/D4
English (chan.), Chan. Is. 13/D8
English (chan.), England 13/D8
English (chan.), England 10/E6
English (chan.), France 28/B3
English, Ind. (47118) 227/E8
English, Ky. (†41008) 237/L3
English (riv.), Ontario 175/B2
English, W. Va. (24832) 312/C8
English Bay, Alaska (99603) 196/B2
English Bazar, India 68/F3
English Coast (reg.) 5/B15
English Creek, N.J. (†08330) 273/D5
English Harbour, Newf. 166/D2
English Harbour West, Newf. 166/C4
English Lake, Ind. (46366) 227/D2
Englishman (bay), Maine 243/J6
English River, Ontario 177/G5
English River, Ontario 175/B3
Englishtown, N.J. (07726) 273/E3
Englishtown, Nova Scotia 168/H2
Énguera, Spain 33/F3
Enid, Miss. (38927) 256/E2
Enid (lake), Miss. 256/E2
Enid, Mont. (59220) 262/M3
Enid, Okla. 188/G3
Enid, Okla. (73701) 288/L2
Enid (mt.), W. Australia 92/B3
Enigma, Georgia (31749) 217/F8
Enilda, Alberta 182/B2
Eniwa, Japan 81/K2
Enka, N.C. (28728) 281/D3
Enkeldoorn, Zimbabwe 118/E3
Enkhuizen, Netherlands 27/G3
Enköping, Sweden 18/G1
Enmore, Guyana 131/C2
Enna (prov.), Italy 34/E6
Enna, Italy 34/E6
Ennadai (lake), N.W. Terrs. 187/H3
En Nahud, Sudan 111/E5
En Nahud, Sudan 102/E3
En Naqura, Lebanon 63/F6
En Nebk, Syria 59/C3
En Nebk, Syria 63/G5
Ennedi (plat.), Chad 111/D4
Ennell (lake), Ireland 17/G5
Ennenda, Switzerland 39/H2
Ennery, Haiti 158/C5
Enngonia, N.S. Wales 97/C1
Enning, S. Dak. (57737) 298/E4
Ennis, Ireland 10/B4
Ennis, Ireland 17/D6
Ennis, Ky. (†42337) 237/H6
Ennis, Mont. (59729) 262/E5
Ennis (lake), Mont. 262/E5
Ennis, Texas (75119) 303/H5
Enniscorthy, Ireland 17/J7
Enniscorthy, Ireland 10/C4
Enniskerry, Ireland 17/J5
Enniskillen, New Bruns. 170/D3
Enniskillen, N. Ireland 10/C3
Enniskillen, N. Ireland 17/F3
Ennistymon, Ireland 10/B4
Ennistymon, Ireland 17/C6
Enns, Austria 41/C2
Enns (riv.), Austria 41/C3
Enoch, Utah (†84720) 304/A6
Enoch, W. Va. (†25043) 312/E6
Enochs, Texas (79324) 303/B4
Enoggera, Queensland 88/K2
Enoggera (creek), Queensland 95/D2
Enola, Ark. (72047) 202/F3
Enola, Pa. (17025) 294/J5
Enon, Mo. (†65074) 261/G6
Enon, Ohio (45323) 284/C6
Enontekiö, Finland 18/N2
Enon Valley, Pa. (16120) 294/B4
Enoree, S.C. (29335) 296/D2
Enoree (riv.), S.C. 296/C2
Enos, Ind. (†47963) 227/C2
Enosburg Falls, Vt. (05450) 268/B2
Enrage (cape), New Bruns. 170/F3
Enrile, Philippines 82/C2
Enrique Carbó, Argentina 143/G6
Enriquillo, Dom. Rep. 158/D7
Enriquillo, Dom. Rep. 156/D3
Enriquillo (lake), Dom. Rep. 158/C6
Ens, Sask. 181/F3
Enschede, Netherlands 27/K4
Ensenada, Argentina 143/H7
Ensenada, Mexico 150/A1
Ensenada, P. Rico 161/B3
Enshi, China 77/G5
Ensign, Alberta 182/D4
Ensign, Kansas (67841) 232/B4
Ensign, Mich. (†49878) 250/C3
Ensley, Fla. (32504) 212/B6
Ent A.F.B., Colo. 208/K5
Entebbe, Uganda 115/F4
Entebbe, Uganda 102/F5
Enterprise, Ala. (36330) 195/G8
Enterprise, Calif. (96001) 204/C3
Enterprise, Guyana 131/B2
Enterprise, Kansas (67441) 232/E3
Enterprise, La. (71425) 238/G3
Enterprise, Miss. (39330) 256/G6
Enterprise, N.W. Terrs. 187/G3
Enterprise, Ohio (†43138) 284/F6
Enterprise, Okla. (†74561) 288/R4
Enterprise, Ontario 177/H3
Enterprise, Oreg. (97828) 291/K2
Enterprise, Utah (84725) 304/A6
Enterprise, W. Va. (26568) 312/F4
Entiat, Wash. (98822) 310/E3
Entiat (lake), Wash. 310/E3
Entiat (mts.), Wash. 310/E2
Entiat (riv.), Wash. 310/E3
Entlebuch, Switzerland 39/F3
Entrance, Alberta 182/B3
Entrejo, Neth. Ant. 161/E8
Entrelacs, Québec 172/C3
Entrepeñas (res.), Spain 33/E2
Entre Ríos (prov.), Argentina 143/E3
Entre Ríos, Bolivia 136/C7
Entre Rios, Brazil 132/C4
Entre Rios de Minas, Brazil 135/D2
Entriken, Pa. (16638) 294/F5
Entwistle, Alberta 182/C3
Enugu, Nigeria 102/C4
Enugu, Nigeria 106/F7
Enumclaw, Wash. (98022) 310/D3
Envigado, Colombia 126/C4
Enville, Tenn. (38332) 237/E10
Enwarak (mt.), Guyana 131/B3
Enyellé, Congo 115/C3
Enying, Hungary 41/E3
Enzeli, Iran 66/F2
Enzeli, Iran 59/E2
Eola, La. (71332) 238/F5
Eola, Texas (76937) 303/E6
Eolia, Ky. (40826) 237/R6
Eolia, Mo. (63344) 261/L4
Eoline, Ala. (35042) 195/D4
Epe, Netherlands 27/H4
Epéna, Congo 115/C3
Epéna, Congo 102/D4
Epenarra, North. Terr. 93/D6
Épernay, France 28/E3
Epes, Ala. (35460) 195/B5
Ephesus, Georgia (†30217) 217/B4
Ephesus (ruins), Turkey 63/B3
Ephraim, Utah (84627) 304/C4
Ephraim, Wis. (54211) 317/M5
Ephrata, Pa. (17522) 294/K5
Ephrata, Wash. (98823) 310/F3
Ephratah, N.Y. (†13339) 276/L4
Épinal, France 28/G3
Épinay-sur-Seine, France 28/B1
Epiphany, S. Dak. (†57321) 298/O6
Epira, Guyana 131/C3
Epirus (reg.), Greece 45/E6
Episkopi, Cyprus 63/E5
Eport, Loch (inlet), Scotland 15/A3
Epoufette, Mich. (†49762) 250/D2
Epping, England 13/H7
Epping, England 10/C5
Epping, N.H. (03042) 268/E5
Epping○, N.H. (03042) 268/E5
Epping, N. Dak. (58843) 282/D3
Epps, La. (71237) 238/G1
Epsie, Mont. (†59317) 262/L5
Epsom, Ind. (†47568) 227/C7
Epsom○, N.H. (03234) 268/E5
Epsom and Ewell, England 13/G8
Epsom and Ewell, England 10/B6
Epukiro, Namibia 118/B4
Epworth, Georgia (30541) 217/D1
Epworth, Iowa (52045) 229/M4
Equality, Ala. (36026) 195/F5
Equality, Ill. (62947) 222/E6
Equateur (prov.), Zaire 115/D3
Equator 2/H5
Equatoria, Eastern (prov.), Sudan 111/F6
Equatoria, Western (prov.), Sudan 111/E6
Equatorial Guinea 2/K5
Equatorial Guinea 102/C4
EQUATORIAL GUINEA 115/A3
Equinox (mt.), Vt. 268/A5
Equinunk, Pa. (18417) 294/M2
Era, Ohio (†43143) 284/D6
Era, Texas (76238) 303/G4
Eran, Philippines 82/A6
Erath, La. (70533) 238/F7
Erath (co.), Texas 303/F5
Erbaa, Turkey 63/G2
Erbach, W. Germany 22/C4
Erbacon, W. Va. (26203) 312/E5
Erbil (gov.), Iraq 66/C3
Erbil, Iraq 59/D2
Erbil, Iraq 54/F6
Erbil, Iraq 66/D2
Erçek (lake), Turkey 63/K3
Ercilla, Chile 138/E2
Erciş, Turkey 59/D2
Erciş, Turkey 63/K3
Erciyas Dağı (mt.), Turkey 63/F3
Érd, Hungary 41/E3
Erdahl, Minn. (†56531) 255/C5
Erdek, Turkey 63/B2
Erdemli, Turkey 63/F4
Erdenetsagaan, Mongolia 77/J2
Erdőtelek, Hungary 41/F3
Erebato (riv.), Venezuela 124/F5
Ereen, Mongolia 77/H2
Ereğli, Turkey 63/D2
Ereğli, Turkey 59/B2
Ereğli, Turkey 63/F4
Erenhot, China 77/H3
Erenhot, China 54/N5
Erenköy, Turkey 63/D6
Eresma (riv.), Spain 33/D2
Erexim, Brazil 132/C9
Érezée, Belgium 27/G8
Erfoud, Morocco 106/D2
Erfurt, E. Germany 7/F3
Erfurt (dist.), E. Germany 22/D3
Erfurt, E. Germany 22/D3
Ergani, Turkey 63/H3
Ergene (riv.), Turkey 63/B2
Ergli, U.S.S.R. 53/C2
Ergun He (Argun') (riv.), China 77/K1
Ergun Youqi, China 77/K1
Ergun Zuoqi, China 77/K1
Ergun Zuoqi, China 54/O4
Er Hai (lake), China 77/F6
Erhard, Minn. (56534) 255/B4
Eriboll, Loch (inlet), Scotland 15/D2
Erica, Netherlands 27/K3
Erica, Victoria 97/D5
Erice, Italy 34/D5
Ericeira, Portugal 33/B3
Erichsen (lake), N.W. Terrs. 187/K2
Ericht, Loch (lake), Scotland 15/D4
Erick, Okla. (73645) 288/G4
Erickson, Br. Col. 184/J5
Erickson, Manitoba 179/C4
Ericsburg, Minn. (†56649) 255/E2
Ericson, Nebr. (68637) 264/F3
Erie (lake) 146/K5
Erie (lake) 188/K2
Erie (lake) 162/H7
Erie, Colo. (80516) 208/K2
Erie, Ill. (61250) 222/C2
Erie, Kansas (66733) 232/G4
Erie, Mich. (48133) 250/F7
Erie (lake), Mich. 250/G7
Erie (co.), N.Y. 276/C5
Erie (lake), N.Y. 276/A5
Erie, N. Dak. (58029) 282/R5
Erie (co.), Ohio 284/E3
Erie (lake), Ohio 284/H1
Erie (lake), Ontario 177/E5
Erie, Pa. 188/K2
Erie, Pa. 146/L5
Erie (co.), Pa. 294/B2
Erie, Pa. (*16501) 294/B1
Erie (lake), Pa. 294/B1
Erie, Tenn. (†37846) 237/M9
Erieau, Ontario 177/C5
Erie Beach, Ontario 177/B5
Erieville, N.Y. (13061) 276/J5
Erigabo, Somalia 115/J1
Eriksdale, Manitoba 179/D4
Erimo (cape), Japan 81/L3
Erin, Ontario 177/D4
Erin, Tenn. (37061) 237/F8
Erin (bay), Trin. & Tob. 161/A11
Erin (pt.), Trin. & Tob. 161/A11
Erinferry, Sask. 181/D2
Erinview, Manitoba 179/E4
Eriskay (isl.), Scotland 15/A3
Erisort, Loch (inlet), Scotland 15/B2
Erith, Alberta 182/B3
Eritrea (prov.), Ethiopia 111/G4
Eritrea (reg.), Ethiopia 102/F3
Eritrea (reg.), Ethiopia 59/C6
Erivan, U.S.S.R. 7/J4
Erivan, U.S.S.R. 48/E6
Erivan, U.S.S.R. 52/F6
Erkilet, Turkey 63/F3
Erkina (riv.), Ireland 17/G6
Erkowit, Sudan 59/C6
Erlach, Switzerland 39/D2
Erlands Point, Wash. (†98310) 310/A2
Erlangen, W. Germany 22/D4
Erlanger, Ky. (41018) 237/R2
Erldunda, North. Terr. 93/C8
Erlenbach im Simmental, Switzerland 39/E3
Erling (lake), Ark. 202/C7
Ermatingen, Switzerland 39/H1
Ermelo, Netherlands 27/H4
Ermelo, S. Africa 118/E5
Ermenak, Turkey 63/E4
Ermeran Station, N.S. Wales 97/D3
Ermoúpolis, Greece 45/G7
Ernabella, S. Australia 94/C2
Ernakulam, India 54/J8
Erne (riv.), Ireland 17/E3
Erne (lake), N. Ireland 17/F3
Erne, Lough (lake), N. Ireland 10/C3
Ernée, France 28/C3
Ernen, Switzerland 39/F4
Ernest, Pa. (15739) 294/D4
Ernfold, Sask. 181/D5
Ernul, N.C. (28527) 281/P4
Erode, India 68/D6
Eromanga, Queensland 88/G5
Eromanga, Queensland 95/B5
Eros, La. (71238) 238/F2
Erquelinnes, Belgium 27/E8
Err (peak), Switzerland 39/J3
Er Rafid, Jordan 65/D2
Er Rahad, Sudan 111/F5
Er Rahad, Sudan 59/B7
Er Ramtha, Jordan 65/E2
Er Ras, Saudi Arabia 59/D4
Errata, Miss. (†39440) 256/F7
Errego, Mozambique 118/F3
Er Rif (range), Morocco 106/D2
Errigal (mt.), Ireland 17/E1
Er Rihiya, West Bank 65/C5
Errington, Br. Col. 184/J3
Erris (head), Ireland 10/A3
Erris (head), Ireland 17/A3
Errol○, N.H. (03579) 268/E2
Errol, Scotland 15/E4
Erromango (isl.), Vanuatu 87/H7
Er Roseires, Sudan 111/F5
Er Rumman, Jordan 65/D3
Er Ruseifa, Jordan 65/E3
Ersekë, Albania 45/E5
Erskine, Alberta 182/D3
Erskine, Minn. (56535) 255/B3
Erstein, France 28/G3
Erstfeld, Switzerland 39/G3
Ertai, China 77/C2
Ertil', U.S.S.R. 52/F4
Eruh, Turkey 63/K4
Erval, Brazil 132/C11
Ervay, Wyo. (†82638) 319/E3
Erving○, Mass. (01344) 249/E2
Erwin, N.C. (28339) 281/M4
Erwin, S. Dak. (57233) 298/P5
Erwin, Tenn. (37650) 237/S8
Erwinna, Pa. (18920) 294/N5
Erwinville, La. (70729) 238/H5
Erwood, Sask. 181/J3
Eryuan, China 77/F6
Erzgebirge (mts.), Czech. 41/B1
Erzgebirge (mts.), E. Germany 22/E3
Erzin, Turkey 63/G4
Erzincan (prov.), Turkey 63/H3
Erzincan, Turkey 63/H3
Erzincan, Turkey 59/C2
Erzurum (prov.), Turkey 63/J3
Erzurum, Turkey 59/D2
Erzurum, Turkey 54/E5
Erzurum, Turkey 63/J3
Esan (pt.), Japan 81/K3
Esashi, Iwate, Japan 81/K4
Esashi, Hokkaido, Japan 81/J3
Esashi, Hokkaido, Japan 81/L1
Esbjerg, Denmark 21/B7
Esbjerg, Denmark 7/E3
Esbjerg, Denmark 18/F9
Esbon, Kansas (66941) 232/D2
Escabosa, N. Mex. (†87059) 274/C4
Escalante, Philippines 82/D5
Escalante, Utah (84726) 304/C6
Escalante (des.), Utah 304/A6
Escalante (riv.), Utah 304/C6
Escalon, Calif. (95320) 204/E6
Escalón, Mexico 150/G3
Escalona, Spain 33/D2
Escambia (co.), Ala. 195/D8
Escambia (creek), Ala. 195/D8
Escambia (riv.), Ala. 195/D8
Escambia (co.), Fla. 212/B6
Escambia (riv.), Fla. 212/B6
Escanaba, Mich. (49829) 250/C3
Escanaba (riv.), Mich. 250/B2
Escarcega, Mexico 150/O7
Escatawpa (riv.), 256/H8
Escatawpa, Ala. (†36584) 195/B8
Escatawpa (riv.), Ala. 195/B9
Escatawpa, Miss. (39552) 256/G10
Eschenbach, Switzerland 39/G2
Escholzmatt, Switzerland 39/E3
Esch-sur-Alzette, Luxembourg 27/J9
Esch-sur-Sûre, Luxembourg 27/H9
Eschwege, W. Germany 22/C3
Eschweiler, W. Germany 22/B3
Escobar, Argentina 143/G7
Escobar, Paraguay 144/B5
Escocesa (bay), Dom. Rep. 158/E5
Escoma, Bolivia 136/A4
Escondido, Calif. (*92025) 204/J10
Escondido (riv.), Nicaragua 154/F4
Escoumins, Québec 172/H1
Escudo de Veraguas (isl.), Panama 154/G6
Escuinapa de Hidalgo, Mexico 150/G5
Escuintla, Guatemala 154/B3
Escuintla, Mexico 150/N9
Escuminac, New Bruns. 170/F1
Escuminac (bay), New Bruns. 170/D1
Escuminac (pt.), New Bruns. 170/F1
Esdaile, Wis. (†54723) 317/A6
Eséka, Cameroon 115/B3
Eşen, Turkey 63/C4
Esfahan (Isfahan) (prov.), Iran 66/H4
Esfahan (Isfahan), Iran 66/G4
Esfandak, Iran 66/N7
Esh, England 13/H3
Esher, England 13/H8
Esher, England 10/B6
Eshowe, S. Africa 118/E5
Esh Shaubak, Jordan 65/D5
Esk (riv.), England 13/D2
Esk, Queensland 95/E5
Esk, Sask. 181/G4
Esk (riv.), Scotland 15/F5
Eska, Alaska (†99674) 196/B1
Eskdale, W. Va. (25075) 312/D6
Esker, Newf. 166/A3
Eskilstuna, Sweden 18/K7
Eskimalatya, Turkey 63/H3
Eskimo (lkes), N.W. Terrs. 187/E3
Eskimo Point, N.W.T. 162/G3
Eskimo Point, N.W. Terrs. 187/J3
Eskipazar, Turkey 63/E2
Eskişehir (prov.), Turkey 63/D3
Eskişehir, Turkey 54/D6
Eskişehir, Turkey 59/B2
Eskişehir, Turkey 63/D3
Esko, Minn. (55733) 255/F4
Eskridge, Kansas (66423) 232/F3
Eskutassis (pond), Maine 243/G5
Esla (riv.), Spain 33/D2
Eslöv, Sweden 18/H9
Esme, Sask. 181/D6
Eşme, Turkey 63/C3
Esmeralda (isl.), Chile 138/C8
Esmeralda, Cuba 158/G1
Esmeralda (co.), Nev. 266/D5
Esmeraldas (prov.), Ecuador 128/C2
Esmeraldas, Ecuador 120/A2
Esmeraldas, Ecuador 128/B2
Esmeraldas (riv.), Ecuador 128/C2
Esmond, Ill. (60129) 222/E1
Esmond, N. Dak. (58332) 282/L3
Esmond, R.I. (02917) 249/H5
Esmond, S. Dak. (†57353) 298/O5
Esmont, Va. (22937) 307/L5
Esmoraca, Bolivia 136/B7
Esneux, Belgium 27/H7
Esom Hill, Georgia (30138) 217/B3

Espada (pt.), Colombia 126/E1
Espada (pt.), Dom. Rep. 158/F6
Espagnol (pt.), St. Vin. & Grens. 161/A8
Espaillat (prov.), Dom. Rep. 158/E5
Espalion, France 28/E5
Española (isl.), Ecuador 128/C10
Espanola, Fla. (†32010) 212/E2
Espanola, N. Mex. (87532) 274/C3
Espanola, Ontario 177/J5
Espanola, Ontario 175/D3
Espanola, Wash. (†99022) 310/H3
Esparta, C. Rica 154/E5
Esparto, Calif. (95627) 204/C5
Espejo, Chile 138/G3
Espejo, Spain 33/D4
Espelkamp, W. Germany 22/C2
Espenberg (cape), Alaska 196/F1
Esperança, Brazil 132/G4
Esperance, Australia 87/C9
Esperance, N.Y. (12066) 276/M5
Esperance, W. Australia 88/C6
Esperance, W. Australia 92/C6
Esperance (bay), W. Australia 92/C6
Esperanza, Argentina 143/F5
Esperanza, Br. Col. 184/D5
Esperanza, Cuba 158/E2
Esperanza, Dom. Rep. 158/D5
Esperanza (mts.), Honduras 154/E3
Esperanza, Puebla, Mexico 150/O2
Esperanza, Sonora, Mexico 150/E3
Esperanza, Peru 128/G7
Esperanza, P. Rico 161/G2
Esperanza, Texas (†79841) 303/B11
Esperanza, Venezuela 124/E6
Espichel (cape), Portugal 33/B3
Espigão Mestre (Geral de Goiás) (range), Brazil 132/E6
Espinal, Colombia 126/C5
Espinhaço (mts.), Brazil 120/E4
Espinhaço, Serra do (range), Brazil 132/F7
Espinho, Portugal 33/B2
Espinillo, Argentina 143/E2
Espinillo (pt.), Uruguay 145/A7
Espino, Venezuela 124/F3
Espírito Santo (state), Brazil 132/F7
Espírito Santo (state), Brazil 135/F2
Espíritu Santo (isl.), Mexico 150/D4
Espiritu Santo (cape), Philippines 85/H3
Espiritu Santo (cape), Philippines 82/E4
Espíritu Santo (isl.), Vanuatu 87/G7
Espita, Mexico 150/Q6
Espiye, Turkey 63/H2
Esplanada, Brazil 132/G5
Espluga de Francolí, Spain 33/G2
Espoir (bay), Newf. 166/C4
Espoo, Finland 18/O6
Esposende, Portugal 33/B2
Esprit-Saint, Québec 172/J1
Espungabera, Mozambique 118/E4
Espy, Pa. (17815) 294/K3
Espyville Station, Pa. (16414) 294/B2
Esqueda, Mexico 150/E1
Esquel, Argentina 143/B5
Esquel, Argentina 120/B7
Esquimalt, Br. Col. 184/K4
Esquina, Argentina 143/G5
Esquipulas, Nicaragua 154/E4
Es Sahab, Jordan 65/E4
Es Salt, Jordan 65/D3
Es Salt, Jordan 59/C3
Essaouira, Morocco 106/B2
Essaouira, Morocco 102/A1
Essé, Cameroon 115/B3
Essen, Belgium 27/F6
Essen, W. Germany 7/E3
Essen, W. Germany 22/B3
Essendon, Victoria 88/K7
Essendon, Victoria 97/H5
Essequibo (riv.), Guyana 120/D2
Essequibo (riv.), Guyana 131/B3
Esserville, Va. (24274) 307/G2
Essex, Calif. (92332) 204/K9
Essex, Conn. (06426) 210/F3
Essex○, Conn. (06426) 210/F3
Essex (co.), England 13/H6
Essex, Ill. (60935) 222/E2
Essex, Iowa (51638) 229/C7
Essex, Md. (21221) 245/N3
Essex (co.), Mass. 249/L2
Essex, Mass. (01929) 249/L2
Essex○, Mass. (01929) 249/L2
Essex, Mo. (63846) 261/N9
Essex, Mont. (59916) 262/C2
Essex (co.), N.J. 273/E2
Essex (co.), N.Y. 276/N2
Essex, N.Y. (12936) 276/O2
Essex (county), Ontario 177/B5
Essex, Ontario 177/B5
Essex (co.), Vt. 268/D2
Essex○, Vt. (05451) 268/A2
Essex (co.), Va. 307/P5
Essex Fells, N.J. (07021) 273/B2
Essex Junction, Vt. (05452) 268/A3
Essexville, Mich. (48732) 250/F5
Es Sidr, Libya 111/C1
Essie, Ky. (40827) 237/P6
Essig, Minn. (56030) 255/D6
Essington, Pa. (19029) 294/M7
Esslingen am Neckar, W. Germany 22/C4
Essonne (dept.), France 28/E3
Es Sukhna, Jordan 65/E3
Es Sukhne, Syria 59/C3
Es Sukhne, Syria 63/H5
Es Suki, Sudan 59/B7
Es Sweida (prov.), Syria 63/G6
Es Suweida, Syria 59/C3
Es Suweida, Syria 63/G6
Est (pt.), Haiti 158/C4
Est (lake), Québec 172/H2
Estacada, Oreg. (97023) 291/E2
Estaca de Vares (pt.), Spain 33/C1
Estación Atlántida, Uruguay 145/B6
Estación Cuaró, Uruguay 145/C1
Estación J.J. Castro, Uruguay 145/C4
Estación José Ignacio, Uruguay 145/E5
Estación La Floresta, Uruguay 145/C7
Estación Lasala, Uruguay 145/C7
Estación Laureles, Uruguay 145/C2
Estación Margat, Uruguay 145/B6
Estación Migues, Uruguay 145/C6
Estación Pampa, Uruguay 145/C3
Estación Puma, Uruguay 145/D5
Estación Rincón, Uruguay 145/F3
Estación Sosa Díaz, Uruguay 145/C6
Estación Tapia, Uruguay 145/C6
Estación Villasboas, Uruguay 145/C4
Estación Yí, Uruguay 145/C4
Estados (isl.), Argentina 120/C8
Estados, Los (isl.), Argentina 143/D7
Estahbanat, Iran 59/F4
Estahbanat, Iran 66/J6
Estaire, Ontario 177/D1
Estampuis, Belgium 27/C7
Estância, Brazil 132/G5
Estância, Brazil 120/F4
Estancia, N. Mex. (87016) 274/D4
Estancia Caleta Josefina, Chile 138/F10
Estancia Laguna Blanca, Chile 138/E9
Estancia Morro Chico, Chile 138/E9
Estancia Punta Delgada, Chile 138/E9
Estancia San Gregorio, Chile 138/E9
Estancia Springhill (Cerro Manantiales), Chile 138/F10
Estanzuela, Uruguay 145/B5
Estanzuelas, El Salvador 154/C4
Estarca, Bolivia 136/C7
Estats (peak), Spain 33/G1
Estavayer-le-Lac, Switzerland 39/C3
Estcourt, S. Africa 118/D5
Este (pt.), Cuba 158/C3
Este, Italy 34/C2
Este (pt.), P. Rico 161/G2
Este (pt.), Uruguay 120/D6
Este (pt.), Uruguay 145/D6
Esteban Rams, Argentina 143/F5
Estelí, Nicaragua 154/D4
Estella, Spain 33/E1
Estelline, S. Dak. (57234) 298/R4
Estelline, Texas (79233) 303/D3
Estell Manor, N.J. (08319) 273/D5
Estepa, Spain 33/D4
Estepona, Spain 33/D4
Ester, Alaska (99725) 196/J2
Esterbrook, Wyo. (†82633) 319/G3
Estérel, Québec 172/C3
Esterhazy, Sask. 181/K5
Estero (bay), Calif. 204/D8
Estero (pt.), Calif. 204/D8
Estero, Fla. (33928) 212/E5
Estero (isl.), Fla. 212/E5
Estes Park, Colo. (80517) 208/J2
Este Sudeste (cays), Colombia 126/A10
Estevan, Sak. 162/F6
Estevan, Sask. 181/J6
Estevan Point, Br. Col. 184/D5
Estey, Mich. (†48652) 250/E5
Esther (isl.), Alaska 196/C1
Esther, Alberta 182/E4
Esther, La. (†70510) 238/F7
Esther, Mo. (†63601) 261/M7
Estherville, Iowa (51334) 229/D2
Estherwood, La. (70534) 238/F6
Estill (co.), Ky. 237/O5
Estill, Miss. (†38748) 256/C4
Estill, S.C. (29918) 296/E6
Estillfork, Ala. (35745) 195/F1
Estill Springs, Tenn. (37330) 237/J10
Estlin, Sask. 181/G5
Esto, Fla. (32425) 212/C5
Eston, England 13/F3
Eston, Sask. 162/F5
Eston, Sask. 181/C4
ESTONIA 53
Estonian S.S.R., U.S.S.R. 7/G3
Estonian S.S.R., U.S.S.R. 48/C4
Estonian S.S.R, U.S.S.R. 52/C3
Estoril, Portugal 33/B3
Estral Beach, Mich. (†48166) 250/F7
Estreito (res.), Brazil 135/C2
Estrela, Serra da (mts.), Portugal 33/C2
Estrella (riv.), Calif. 204/E8
Estremadura (reg.), Spain 33/C3
Estremoz, Portugal 33/C3
Estrondo, Serra do (range), Brazil 132/D4
Estuary, Sask. 181/B5
Esztergom, Hungary 41/E3
Etadunna, S. Australia 94/F3
Etalle, Belgium 27/H9
Étampes, France 28/E3
Étaples, France 28/D2
Etawah, India 68/D3
Etawney (lake), Manitoba 179/J2
Etchojoa, Mexico 150/E3
Ethan, S. Dak. (57334) 298/N6
Ethel, Ark. (72048) 202/H5
Ethel (mt.), Colo. 208/F1
Ethel, La. (70730) 238/H5
Ethel, Miss. (39067) 256/F4
Ethel, Mo. (63539) 261/G3
Ethel, Ontario 177/C4
Ethel, Wash. (98542) 310/C4
Ethel, W. Va. (25076) 312/C7
Ethelbert, Manitoba 179/B3
Ethel Creek, W. Australia 92/C3
Ethelsville, Ala. (35461) 195/B4
Ethelton, Sask. 181/G3
Ether, N.C. (27247) 281/K4
Ethete, Wyo. (82520) 319/D2
Ethiopia 2/L5
Ethiopia 102/F4
ETHIOPIA 59/C7
ETHIOPIA 111/G5
Ethridge, Mont. (59435) 262/D2
Ethridge, Tenn. (38456) 237/G10
Etive, Loch (inlet), Scotland 15/C4
Etiwanda, Calif. (91739) 204/E10
Etna, Calif. (96027) 204/C2
Etna, Ind. (†46725) 227/F2
Etna (vol.), Italy 7/F5
Etna (vol.), Italy 34/E6
Etna○, Maine (04434) 243/E6
Etna, N.H. (03750) 268/C4
Etna, Ohio (43018) 284/E6
Etna, Pa. (15223) 294/B6
Etna, Utah (†84313) 304/A2
Etna, Wyo. (83118) 319/A2
Etna Green, Ind. (46524) 227/E2
Etobicoke, Ontario 177/J4
Etoile, Ky. (42131) 237/K7
Etoile, Zaire 115/E6
Etolin (isl.), Alaska 196/N2
Etolin (str.), Alaska 196/E2
Etomami (riv.), Manitoba 179/F2
Etomami (riv.), Sask. 181/J3
Eton, England 10/F5
Eton, England 13/G8
Eton, Georgia (30724) 217/C1
Etosha Pan (salt pan), Namibia 118/B3
Etosha Salt Pan, Namibia 102/D6
Etoumbi, Congo 115/B3
Etowah (co.), Ala. 195/F2
Etowah, Ark. (72428) 202/K2
Etowah (riv.), Georgia 217/C2
Etowah, N.C. (28729) 281/D4
Etowah, Tenn. (37331) 237/M10
Étretat, France 28/D3
Etta, Miss. (38627) 256/F2
Etta (lake), N. Dak. 282/L6
Et Tafila, Jordan 65/E5
Et Taiyiba, Jordan 65/D2
Ettelbruck, Luxembourg 27/J9
Et Tell el Abyad, Syria 63/H4
Etten-Leur, Netherlands 27/F5
Etter, Minn. (†55033) 255/F6
Etterbeek, Belgium 27/B9
Etters, Pa. (17319) 294/J5
Etters Beach, Sask. 181/F4
Ettington, Sask. 181/F6
Ettlingen, W. Germany 22/C4
Ettrick, Scotland 15/E5
Ettrick, Va. (23803) 307/O6
Ettrick, Wis. (54627) 317/D7
Ettrick Pen (mt.), Scotland 15/E5
Etty, Ky. (41523) 237/R6
Etzatlán, Mexico 150/G6
Etzikom, Alberta 182/E5
Etzikom Coulee (riv.), Alberta 182/E5
Eu, France 28/D3
Euabalong, N.S. Wales 97/D3
Eubank, Ky. (42567) 237/M6
Euboea (Évvoia) (isl.), Greece 45/G6
Eucha, Okla. (74342) 288/S2
Eucha (lake), Okla. 288/S2
Eucla, W. Australia 92/E5
Euclid, Minn. (56722) 255/B3
Euclid, Ohio (44117) 284/J9
Eucumbene (lake), N.S. Wales 97/E5
Eucutta, Miss. (†39360) 256/G7
Eudora, Ark. (71640) 202/H7
Eudora, Kansas (66025) 232/G3
Eudora, Miss. (†38632) 256/D1
Eudora, Mo. (65645) 261/E7
Eufaula, Ala. (36027) 195/H7
Eufaula (Walter F. George Res.) (lake), Ala 195/H7
Eufaula (Walter F. George Res.) (lake), Georgia 217/B7
Eufaula (res.), Ohio 284/L4
Eufaula, Okla. (74432) 288/P4
Eufaula (lake), Okla. 288/P4
Eugene, Ind. (†47928) 227/B5
Eugene, Mo. (65032) 261/H6
Eugene, Oreg. 188/B2
Eugene, Oreg. 146/F5
Eugene, Oreg. (*97401) 291/D3
Eugene O'Neill Nat'l Hist. Site, Calif. 204/K2
Eugowra, N.S. Wales 97/E3
Euharlee, Georgia (†30120) 217/C2
Euless, Texas (76039) 303/F2
Eulo, Queensland 95/C6
Eulonia, Georgia (†31331) 217/K7
Eumungerie, N.S. Wales 97/E2
Eunice, La. (70535) 238/F6
Eunice, N. Mex. (88231) 274/F6
Eunola, Ala. (†36340) 195/G8
Eupen, Belgium 27/J7
Euphrates (riv.) 54/F6
Euphrates (riv.), Iran 59/E3
Euphrates (riv.), Iraq 59/E3
Euphrates (riv.), Iraq 66/D4
Euphrates (riv.), Syria 59/E3
Euphrates (El Furat) (riv.), Syria 63/H4
Euphrates (Fırat) (riv.), Turkey 63/G4
Eupora, Miss. (39744) 256/F3
Eure (dept.), France 28/D3
Eure (riv.), France 28/D3
Eure, N.C. (27935) 281/R2
Eure-et-Loir (dept.), France 28/D3
Eureka, Calif. 188/B2
Eureka, Calif. 146/F5
Eureka, Calif. (95501) 204/A3
Eureka, Canada 4/A14
Eureka, Colo. (†81433) 208/D7
Eureka (res.), Fla. 212/E2
Eureka, Ill. (61530) 222/D3
Eureka, Ind. (†47635) 227/C9
Eureka, Kansas (67045) 232/F4
Eureka, Mo. (63025) 261/M4
Eureka, Mont. (59917) 262/B2
Eureka (co.), Nev. 266/E3
Eureka, Nev. (89316) 266/E3
Eureka, N.C. (27830) 281/O3
Eureka, N.W.T. 146/K2
Eureka, N.W.T. 162/N3
Eureka, N.W. Terrs. 187/K2
Eureka (sound), N.W. Terrs. 187/K2
Eureka, Nova Scotia 168/F3
Eureka, S.C. (†29706) 296/E2
Eureka, S.C. (†29847) 296/D4
Eureka, S. Dak. (57437) 298/K2
Eureka, Utah (84628) 304/B4
Eureka, Wash. (†99348) 310/G4
Eureka, W. Va. (26144) 312/D4
Eureka Lodge, Alaska (†99588) 196/C1
Eureka Springs, Ark. (72632) 202/C1
Euroa, Victoria 97/C5
Europa (pt.), Gibraltar 33/D4
Europa (isl.), Réunion 102/G7
Europa (isl.), Réunion 118/G4
Europe 2/K3
Europoort, Netherlands 27/E5
Eusebio Ayala, Paraguay 144/B4
Euskirchen, W. Germany 22/B3
Eustace, Texas (75124) 303/H5
Eustis, Fla. (32726) 212/E3
Eustis, Maine (04936) 243/B5
Eustis○, Maine (04936) 243/B5
Eustis, Nebr. (69028) 264/D4
Euston, N.S. Wales 97/B4
Eutaw, Ala. (35462) 195/C5
Eutawville, S.C. (29048) 296/G5
Eutin, W. Germany 22/D1
Eutsuk (lake), Br. Col. 184/D3
Eva, Ala. (35621) 195/E2
Eva (lake), Alberta 182/B5
Eva, La. (†71354) 238/G4
Eva, Okla. (†73949) 288/C1
Eva, Tenn. (38333) 237/E8
Evadale, Texas (77615) 303/L7
Eva Downs, North. Terr. 93/D5
Évain, Québec 174/B3
Evan, Minn. (56238) 255/D6
Evan (lake), Québec 174/B2
Evandale, New Bruns. 170/D3
Evandale, Tasmania 99/D3
Evangeline (par.), La. 238/F5
Evangeline, La. (70537) 238/F6
Evangeline, New Bruns. 170/F1
Evans, Colo. (80620) 208/K2
Evans (mt.), Colo. 208/H3
Evans (co.), Georgia 217/J6
Evans, Georgia (30809) 217/H3
Evans, La. (70639) 238/D5
Evans (head), N.S. Wales 97/G1
Evans (str.), N.W.T. 162/H3
Evans (str.), N.W. Terrs. 187/K3
Evans, Wash. (99126) 310/H2
Evans, W. Va. (25241) 312/C5
Evansburg, Alberta 182/C3
Evans Center, N.Y. (†14006) 276/B5
Evans City, Pa. (16033) 294/B4
Evansdale, Iowa (50707) 229/J4
Evans Head, N.S. Wales 97/G1
Evans Mills, N.Y. (13637) 276/J2
Evansport, Ohio (43519) 284/B3
Evanston, Ill. (*60201) 222/B5
Evanston, Ind. (47531) 227/D8
Evanston, Wyo. 188/D2
Evanston, Wyo. (82930) 319/B4
Evansville (Bettles Field), Alaska (†99726) 196/H1
Evansville, Ark. (72729) 202/B2
Evansville, Ill. (62242) 222/D5
Evansville, Ind. 188/J3
Evansville, Ind. 146/K6
Evansville, Ind. (*47701) 227/C9
Evansville, Minn. (56326) 255/C4
Evansville, Miss. (†38676) 256/D1
Evansville, Pa. (19521) 294/L5
Evansville, Wis. (53536) 317/H10
Evansville, Wyo. (82636) 319/F3
Evant, Texas (76525) 303/G6
Evanton, Scotland 15/D3
Evart, Mich. (49631) 250/D5
Evarts, Ky. (40828) 237/P7
Evaton, S. Africa 118/H7
Evaz, Iran 66/J7
Eveleth, Minn. (55734) 255/F3
Evelyn, La. (†71052) 238/D3
Evendale, Ohio (†45201) 284/C9
Evening Shade, Ark. (72532) 202/G1
Evenki Aut. Okr., U.S.S.R. 48/K3
Evensk, U.S.S.R. 4/C1
Evensk, U.S.S.R. 48/Q3
Evensville, Tenn. (37332) 237/M9
Even Yehuda, Israel 65/B3
Everard (lake), S. Australia 88/E6
Everard (lake), S. Australia 94/D4
Everard (ranges), S. Australia 94/C2
Evere, Belgium 27/C9
Everest (mt.) 54/K7
Everest (mt.), China 77/C6
Everest, Kansas (66424) 232/G2
Everest (mt.), Nepal 68/F3
Everest, N. Dak. (†58023) 282/R6
Everett, Georgia (31536) 217/J8
Everett, Mass. (02149) 249/D6
Everett (mt.), Mass. 249/A4
Everett, New Bruns. 170/C1
Everett (dam), N.H. 268/D5
Everett (mts.), N.W. Terrs. 187/M3
Everett, Ontario 177/E3
Everett, Pa. (15537) 294/F5
Everett, Wash. 188/B1
Everett, Wash. (*98201) 310/C3
Everetts, N.C. (27825) 281/P3
Everettville, W. Va. (26533) 312/F3
Evergem, Belgium 27/D6
Everglades, The (swamp), Fla. 212/F6
Everglades, The (swamp), Fla. 188/K5
Everglades City, Fla. (33929) 212/E6
Everglades Nat'l Park, Fla. 212/F6
Evergreen, Ala. (36401) 195/E8
Evergreen, Colo. (80439) 208/J3
Evergreen, La. (71333) 238/F5
Evergreen, N.C. (28438) 281/M6
Evergreen, Va. (23939) 307/L6
Evergreen Park, Ill. (60642) 222/B6
Everly, Iowa (51338) 229/C2
Everman, Texas (76140) 303/F3
Everson, Pa. (15631) 294/C5
Everson, Wash. (98247) 310/C2
Eversonville, Mo. (†64688) 261/F3
Everton, Ark. (72633) 202/E1
Everton, Ind. (†47331) 227/G5
Everton, Mo. (65646) 261/E8
Evesham, England 10/E4
Evesham, England 13/F5
Evesham, Sask. 181/B3
Evington, Va. (24550) 307/K6
Evolène, Switzerland 39/D4
Évora (dist.), Portugal 33/C3
Évora, Portugal 7/D5
Évora, Portugal 33/C3
Évreux, France 28/D3
Évros (riv.), Greece 45/H5
Évry, France 28/E3
Évvoia (isl.), Greece 7/G5
Évvoia (isl.), Greece 45/G6
Ewa, Hawaii (96706) 218/A4
Ewab (Kai) (isls.), Indonesia 85/J7
Ewa Beach, Hawaii (96706) 218/A4
Ewan, N.J. (08025) 273/C4
Ewan, Wash. (99127) 310/H3
Ewaninga, North. Terr. 93/D7
Ewart, Iowa (†50171) 229/H5
Ewarton, Jamaica 156/C3
Ewarton, Jamaica 158/J6
Ewauna (lake), Oreg. 291/F5
Ewe, Loch (inlet), Scotland 15/C3
Ewell, Md. (21824) 245/O9
Ewen, Mich. (49925) 250/F2
Ewing, Ill. (62836) 222/E5
Ewing, Ky. (41039) 237/O4
Ewing, Mo. (63440) 261/J2
Ewing, Nebr. (68735) 264/F2
Ewing (mt.), North. Terr. 93/E7
Ewing, Va. (24248) 307/B7
Ewington, Ohio (45627) 284/F8
Ewo, Congo 115/B4
Exaltación, Bolivia 136/C3
Excel, Ala. (36439) 195/D8
Excel, Alberta 182/E4
Excello, Mo. (65247) 261/H3
Excello, Ohio (45042) 284/B7
Excelsior, Minn. (55331) 255/E6
Excelsior (mts.), Nev. 266/C4
Excelsior, Wis. (†53518) 317/E9
Excelsior Springs, Mo. (64024) 261/R4
Exchange, W. Va. (26619) 312/E5
Excursion Inlet, Alaska (†99826) 196/M1
Exe (riv.), England 13/D7
Exe (riv.), England 10/E5
Executive Committee (range) 5/B12
Exeland, Wis. (54835) 317/D4
Exeter, Calif. (93221) 204/F7
Exeter, Conn. (†06249) 210/F2
Exeter, England 13/D7
Exeter, England 10/E5
Exeter, Ill. (†62694) 222/C4
Exeter, Maine (04435) 243/E6
Exeter○, Maine (04435) 243/E6
Exeter, Mo. (65647) 261/D9
Exeter, Nebr. (68351) 264/G4
Exeter, N.H. (03833) 268/F6
Exeter○, N.H. (03833) 268/F6
Exeter (riv.), N.H. 268/E6
Exeter (sound), N.W. Terrs. 187/M3
Exeter, Ontario 177/C4
Exeter○, R.I. (02822) 249/H6
Exeter, Tasmania 99/C3
Exira, Iowa (50076) 229/D5
Exline, Iowa (52555) 229/H7
Exminster, England 13/D7
Exmoor National Park, England 13/D6
Exmore, Va. (23350) 307/S5
Exmouth, England 13/D7
Exmouth, England 10/E5
Exmouth, W. Australia 88/A4
Exmouth (gulf), W. Australia 88/A4
Exmouth, W. Australia 92/A3
Exmouth (gulf), W. Australia 92/A3
Expanse, Sask. 181/E6
Experiment, Georgia (30212) 217/D4
Exploits (riv.), Newf. 166/C4
Export, Pa. (15632) 294/C5
Exshaw, Alberta 182/C4
Extension, Br. Col. 184/J3
Extension, La. (71239) 238/G3
Exu, Brazil 132/G4
Exuma (cays), Bahamas 156/C1
Exuma (sound), Bahamas 156/C1
Eyasi (lake), Tanzania 115/F4
Eye, England 10/G4
Eye, England 13/J5
Eye (pen.), Scotland 15/B2
Eyebrow, Sask. 181/E5
Eyebrow (lake), Sask. 181/E5
Eyehill (creek), Sask. 181/B3
Eyemouth, Scotland 15/F5
Eyemouth, Scotland 10/F3
Eynesil, Turkey 63/H2
Eynhallow (sound), Scotland 15/E1
Eynort, Loch (inlet), Scotland 15/A3
Eyota, Minn. (55934) 255/F7
Eyre (lake), Australia 87/D8
Eyre (bay), Chile 138/D8
Eyre (mts.), N. Zealand 100/B6
Eyre (riv.), Queensland 88/F5
Eyre (lake), S. Australia 88/F5
Eyre (pen.), S. Australia 88/F6
Eyre (pen.), S. Australia 94/D5
Eyre, W. Australia 92/D6
Eyrecourt, Ireland 17/E5
Eyre North (lake), S. Australia 94/E3
Eyre South (lake), S. Australia 94/E3
Eysturoy (isl.), Denmark 21/B3
Eyüp, Turkey 63/D6
Ezel, Ky. (41425) 237/P5
Ezequiel Montes, Mexico 150/K6
Ezibider, Turkey 63/H2
Ezine, Turkey 63/B3
Ezna, Iran 66/F4
Ez Zababida, West Bank 65/C3
Ez Zarqa', Jordan 65/E3
Ez Zuetina, Libya 111/D1

F

Faaa, Fr. Poly. 86/S13
Fabens, Texas (79838) 303/B10
Faber (lake), N.W. Terrs. 187/G3
Faber, Va. (22938) 307/L5
Fabius, Ala. (35965) 195/G1
Fabius, N.Y. (13063) 276/J5
Fåborg, Denmark 21/D7
Fåborg, Denmark 18/G9
Fabriano, Italy 34/D3
Fabyan, Alberta 182/E3
Fabyan, Conn. (06245) 210/H1
Fabyan House, N.H. (†03595) 268/E3
Facatativá, Colombia 126/C5
Faceville, Georgia (†31717) 217/C9
Fachi, Niger 106/G5
Fackler, Ala. (35746) 195/G1
Factoryville, Pa. (18419) 294/L2
Facundo, Argentina 143/C6
Fada, Chad 111/D4
Fada-N'Gourma, Upper Volta 106/E6
Fadd, Hungary 41/E3
Faddeyevskiy (isl.), U.S.S.R. 4/B2
Faddeyevskiy (isl.), U.S.S.R. 48/P2
Faden, Newf. 166/A3
Faenza, Italy 34/D2
Faerøe (isls.), Den. 4/C10
Faerøe (isls.), Denmark 7/D2
Faerøe (isls.), Denmark 21/B2
FAERØE ISLANDS, Denmark 21/B2
Faerøe Islands, Denmark 21/B2
Fafan (riv.), Ethiopia 111/H6
Fafe, Portugal 33/B2
Fagan, Ky. (†40322) 237/O5
Făgăraș, Romania 45/G3
Fagernes, Norway 18/F6
Fagersta, Sweden 18/J6
Fagnano (lake), Argentina 143/C7
Fagnano (lake), Chile 138/F11
Faguibine (lake), Mali 106/D5
Fagundes, Brazil 132/G4
Fagus, Mo. (63938) 261/M9
Fahan, Ireland 17/G1
Fahrej (Iranshahr), Iran 66/M7
Fahrej (Iranshahr), Iran 59/H4
Faial (isl.), Portugal 33/B1
Faid, Saudi Arabia 59/D4
Faido, Switzerland 39/G4
Fainaven (mt.), Scotland 15/D2
Fair (head), N. Ireland 17/J1
Fair (isl.), Scotland 10/F1
Fairacres, N. Mex. (88033) 274/C6
Fairbank, Ariz. (85612) 198/E7
Fairbank, Iowa (50629) 229/K3
Fairbank, Md. (†21671) 245/N6
Fairbanks, Alaska 146/D3
Fairbanks, Alaska (99701) 196/J2
Fairbanks, Alaska 188/D5
Fairbanks, Fla. (†32601) 212/D2
Fairbanks, Ind. (47849) 227/B6
Fairbanks, La. (71240) 238/F1
Fairbanks, Maine (†04938) 243/C6
Fairbanks, Minn. (†55602) 255/G3
Fairbanks, U.S. 4/C17
Fairbanks, U.S. 2/C2
Fair Bluff, N.C. (28439) 281/M6
Fairborn, Ohio (45324) 284/B6
Fairburn, Georgia (30213) 217/J2
Fairburn, S. Dak. (57738) 298/C6
Fairbury, Ill. (61739) 222/E3
Fairbury, Nebr. (68352) 264/G4
Fairchance, Pa. (15436) 294/C6
Fairchild, Wis. (54741) 317/D6
Fairchild A.F.B., Wash. 310/H3
Fairdale, Ill. (†60146) 222/E1
Fairdale, Ky. (40118) 237/K4
Fairdale, N. Dak. (58229) 282/O3
Fairdealing, Mo. (63939) 261/L9
Fairfax, Ala. (36854) 195/H5
Fairfax, Calif. (94930) 204/H1
Fairfax, Iowa (52228) 229/K5
Fairfax, Manitoba 179/B5
Fairfax, Minn. (55332) 255/D6
Fairfax, Mo. (64446) 261/B2
Fairfax, Ohio (†45201) 284/C9
Fairfax, Okla. (74637) 288/N1
Fairfax, S.C. (29827) 296/E6
Fairfax, S. Dak. (57335) 298/M7
Fairfax○, Vt. (05454) 268/B2
Fairfax (co.), Va. 307/O3
Fairfax (I.C.), Va. (22030) 307/R3
Fairfax, Wash. (†98323) 310/C4
Fairfax Station, Va. (22039) 307/R3
Fairfield, Ala. (35064) 195/E4
Fairfield, Calif. (94533) 204/K1
Fairfield (co.), Conn. 210/B3
Fairfield○, Conn. (06430) 210/B4
Fairfield, Fla. (32634) 212/D2
Fairfield, Idaho (83327) 220/D6
Fairfield, Ill. (62837) 222/E5
Fairfield, Iowa (52556) 229/J6
Fairfield, Ky. (40020) 237/L5
Fairfield, Maine (04937) 243/D6
Fairfield○, Maine (04937) 243/D6
Fairfield, Mont. (59436) 262/D3
Fairfield, Nebr. (68938) 264/G4
Fairfield, New Bruns. 170/E3
Fairfield○, N.J. (07006) 273/A2
Fairfield, N.S. Wales 88/K4
Fairfield, N.S. Wales 97/H3
Fairfield, N. Zealand 100/C6
Fairfield, N.C. (27826) 281/S3
Fairfield, N. Dak. (58627) 282/D5
Fairfield (co.), Ohio 284/E6
Fairfield, Ohio (45014) 284/A7
Fairfield, Pa. (17320) 294/H6
Fairfield (co.), S.C. 296/E3
Fairfield, Tenn. (†37183) 237/J9
Fairfield, Texas (75840) 303/H6
Fairfield, Utah (†84013) 304/B3
Fairfield○, Vt. (05455) 268/B2
Fairfield (pond), Vt. 268/A2
Fairfield, Va. (24435) 307/K5
Fairfield, Wash. (99012) 310/H3
Fairfield Center, Maine (†04937) 243/D6
Fairford, Ala. (†36553) 195/B8
Fairford, Manitoba 179/D3
Fairgrange, Ill. (†61920) 222/E4
Fairgrove, Mich. (48733) 250/F5
Fair Grove, Mo. (65648) 261/F8W
Fair Harbour, Br. Col. 184/D5
Fairhaven○, Mass. (02719) 249/L6
Fair Haven, Mich. (48023) 250/G6
Fairhaven, Minn. (†55382) 255/D5
Fairhaven, New Bruns. 170/C4

Fair Haven, N.J. (07701) 273/E3
Fair Haven, N.Y. (13064) 276/G4
Fairhaven, Ohio (†45003) 284/A6
Fair Haven, Vt. (05743) 268/A4
Fair Haven○, Vt. (05743) 268/A4
Fair Hill, Md. (†21921) 245/P2
Fairholme, Sask. 181/C2
Fairhope, Ala. (36532) 195/C10
Fairhope, Pa. (15538) 294/E6
Fairisle, New Bruns. 170/E1
Fair Isle (isl.), Scotland 15/F3
Fairland, Ind. (46126) 227/F5
Fairland, Okla. (74343) 288/S1
Fair Lawn, N.J. (07410) 273/B1
Fairlawn, Ohio (44313) 284/G3
Fairlawn, Va. (24141) 307/G6
Fairlee, Md. (†21620) 245/O4
Fairlee○, Vt. (05045) 268/C4
Fairless Hills, Pa. (19030) 294/N5
Fairlie, N. Zealand 100/C6
Fairlie, Scotland 15/D5
Fairlight, Sask. 181/K6
Fairmead, Calif. (†93610) 204/E6
Fairmont, Ill. (†62002) 222/A2
Fairmont, Minn. (56031) 255/D7
Fairmont, Mo. (†63474) 261/J2
Fairmont, Nebr. (68354) 264/G4
Fairmont, N.C. (28340) 281/L6
Fairmont, Okla. (73736) 288/L2
Fairmont, W. Va. 188/K3
Fairmont, W. Va. (26554) 312/F4
Fairmont City, Ill. (†62201) 222/B2
Fairmont Hot Springs, Br. Col. 184/J5
Fairmount, Georgia (30139) 217/C2
Fairmount, Ill. (61841) 222/F3
Fairmount, Ind. (46928) 227/F4
Fairmount, Md. (†21871) 245/P8
Fairmount, N. Dak. (58030) 282/S7
Fairmount, Sask. 181/B4
Fairmount Heights, Md. (†20027) 245/G5
Fair Oaks, Ark. (72397) 202/J3
Fair Oaks, Calif. (95628) 204/C8
Fair Oaks, Georgia (†30060) 217/J1
Fair Oaks, Ind. (47943) 227/C2
Fair Oaks, Okla. (†74080) 288/P2
Fair Plain, Mich. (49022) 250/C6
Fairplain, W. Va. (†25271) 312/C5
Fairplay, Colo. (80440) 208/H4
Fairplay, Ky. (42735) 237/L7
Fair Play, Mo. (65649) 261/E7
Fair Play, S.C. (29643) 296/A2
Fairpoint, Ohio (43927) 284/J5
Fairpoint, S. Dak. (†57785) 298/D4
Fairport, Iowa (†52761) 229/M6
Fairport, Kansas (†67665) 232/C2
Fairport, Mo. (64447) 261/D2
Fairport, N.Y. (14450) 276/F4
Fairport, Va. (†22539) 307/R5
Fairport Harbor, Ohio (44077) 284/H2
Fairton, N.J. (08320) 273/C5
Fairvale, New Bruns. 170/E3
Fairview, Ala. (35208) 195/E2
Fairview, Alberta 182/A1
Fairview, Ill. (61432) 222/B3
Fairview, Ind. (†46127) 227/G5
Fairview, Ind. (†47018) 227/G7
Fairview, Kansas (66425) 232/G2
Fairview, Ky. (†41101) 237/S2
Fairview, Ky. (42221) 237/G7
Fairview, Mich. (48621) 250/F4
Fairview, Mo. (64842) 261/D9
Fairview, Mont. (59221) 262/M3
Fairview, N.J. (07022) 273/C2
Fairview, N.Y. (†12601) 276/N7
Fairview, N.C. (28730) 281/D3
Fairview, Ohio (43736) 284/H5
Fairview, Okla. (73737) 288/J2
Fairview, Oreg. (97024) 291/B2
Fairview, Pa. (16415) 294/B1
Fairview, Pa. (†16041) 294/C3
Fairview, S. Dak. (57027) 298/R7
Fairview, Tenn. (37062) 237/G9
Fairview, Utah (84629) 304/C4
Fairview, W. Va. (26570) 312/F3
Fairview, Wyo. (83119) 319/B3
Fairview Heights, Ill. (62208) 222/B3
Fairview Park, Ind. (†47842) 227/C5
Fairview Park, Ohio (44126) 284/G9
Fairview-Sumach, Wash. (†98901) 310/E4
Fair Water, Wis. (53931) 317/J8
Fairway, Kansas (†66101) 232/H2
Fairweather (cape), Alaska 196/L1
Fairweather (mt.), Alaska 196/L1
Fairweather (mt.), Br. Col. 184/H1
Fairy Glen, Sask. 181/G2
Fais (isl.), Micronesia 87/E5
Faisalabad, Pakistan 54/J6
Faisalabad, Pakistan 59/K3
Faisalabad, Pakistan 68/C2
Faison, N.C. (28341) 281/N4
Faith, Minn. (†56584) 255/B3
Faith, N.C. (28041) 281/J3
Faith, S. Dak. (57626) 298/E4
Faithorn, Mich. (†49892) 250/B3
Faizabad-cum-Ayodhya, India 68/E3
Fajami, Syria 63/J5
Fajardo, P. Rico 161/F1
Fajardo (riv.), P. Rico 161/F1
Fajou (isl.), Guadeloupe 161/A6
Fakaofo (atoll), Tokelau Is. 87/J6
Fakarava (atoll), Fr. Poly. 87/M7
Fakenham, England 13/H5
Fakfak, Indonesia 85/J6
Fakılı, Turkey 63/F3
Fakse, Denmark 21/F7
Fakse (bay), Denmark 21/F7
Fakse Ladeplads, Denmark 21/F7
Falaise, France 28/C3
Falam, Burma 72/B2
Falama, West Bank 65/C3
Falcarragh, Ireland 17/E1
Fălciu, Romania 45/J2
Falcon, Ky. (41426) 237/P5
Falcon (lake), Manitoba 179/G5
Falcón (res.), Mexico 150/K3
Falcon, Miss. (38628) 256/D2
Falcon, N.C. (28342) 281/M4
Falcon (cape), Oreg. 291/C2
Falcon (res.), Texas 188/G5
Falcon (dam), Texas 303/E11
Falcon (res.), Texas 303/E11
Falcón (state), Venezuela 124/D2
Falcone (cape), Italy 34/B4
Falconer, N.Y. (14733) 276/B6
Falcon Heights, Minn. (55113) 255/G5
Falcon Heights, Oreg. (†97601) 291/F5
Falcon Lake, Manitoba 179/G5
Falémé (riv.), Mali 106/B6
Falémé (riv.), Senegal 106/B6
Faleolo, W. Samoa 86/L8
Falfurrias, Texas (78355) 303/F10
Falher, Alberta 182/B2
Falkenberg, Sweden 18/H8
Falkensee, E. Germany 22/E3
Falkenstein, E. Germany 22/E3
Falkirk, N. Dak. (†58577) 282/H5
Falkirk, Scotland 10/B1
Falkirk, Scotland 15/C1
Falkland (isls.) 2/G8
Falkland (isls.), 143/D7
Falkland (sound), 143/D7
Falkland, Br. Col. 184/H5
Falkland, N.C. (27827) 281/O3
Falkland, Scotland 15/E4
FALKLAND ISLANDS 143
Falkland Islands 120/D8
Falkner, Miss. (38629) 256/G1
Falknov (Sokolov), Czech. 41/B1
Falköping, Sweden 18/H7
Falkville, Ala. (35622) 195/E2
Fall (riv.), Kansas 232/G4
Falla, Cuba 158/F2
Fall Branch, Tenn. (37656) 237/R8
Fallbrook, Calif. (92028) 204/H10
Fall City, Wash. (98024) 310/D3
Fall Creek, Oreg. (97438) 291/E4
Fall Creek, Wis. (54742) 317/D6
Fallin, Scotland 15/C1
Falling Spring (Renick), W. Va. (†24966) 312/F6
Falling Waters, W. Va. (25419) 312/L3
Fallis, Okla. (†74881) 288/M3
Fall Mills, Tenn. (†37345) 237/J10
Fallon (co.), Mont. 262/M4
Fallon, Mont. (59326) 262/L4
Fallon, Nev. (89406) 266/C3
Fallon Ind. Res., Nev. 266/C3
Fallon Nav. Air Sta., Nev. 266/C3
Fall River, Kansas (67047) 232/G4
Fall River (lake), Kansas 232/F4
Fall River, Mass. 188/M2
Fall River, Mass. (*02720) 249/K6
Fall River, Nova Scotia 168/E4
Fall River (co.), S. Dak. 298/B7
Fall River, Tenn. (†38468) 237/G10
Fall River, Wis. (53932) 317/H9
Fall River Mills, Calif. (96028) 204/D3
Falls (riv.), Mass. 249/D2
Falls, Pa. (18615) 294/E6
Falls (co.), Texas 303/H6
Falls Church (I.C.), Va. (*22040) 307/S2
Falls City, Nebr. (68355) 264/J4
Falls City, Oreg. (97344) 291/D3
Falls City, Texas (78113) 303/G9
Falls Creek, Pa. (15840) 294/E3
Falls Mill, W. Va. (26620) 312/E5
Falls Mills, Va. (24613) 307/F6
Falls of Rough, Ky. (40119) 237/J5
Fallston, Md. (21047) 245/N2
Fallston, N.C. (28042) 281/G4
Falls Village, Conn. (06031) 210/B1
Falls ville, Ark. (†72861) 202/D2
Falmouth, Ant. & Bar. 161/E11
Falmouth, Ant. & Bar. 156/F3
Falmouth, England 10/D5
Falmouth, England 13/B7
Falmouth (bay), England 13/B7
Falmouth, Ind. (46127) 227/G5
Falmouth, Jamaica 158/H5
Falmouth, Jamaica 156/C3
Falmouth, Ky. (41040) 237/N3
Falmouth, Maine (04105) 243/C8
Falmouth○, Maine (04105) 243/C8
Falmouth, Mass. (*02540) 249/M6
Falmouth○, Mass. (*02540) 249/M6
Falmouth, Mich. (49632) 250/E4
Falmouth, Nova Scotia 168/D3
Falmouth, Va. (22401) 307/O4
False (bay), S. Africa 118/F7
False Detour (chan.), Mich. 250/F3
False Divi (pt.), India 68/E5
False Pass, Alaska (99583) 196/F4
Falso (cape), Dom. Rep. 158/C7
Falso (cape), Honduras 154/F3
Falso (cape), Mexico 150/D5
Falster (isl.), Denmark 21/F8
Fălticeni, Romania 45/H2
Falun, Kansas (67442) 232/E3
Falun, Sweden 18/J6
Falun, Sweden 7/F2
Falun, Wis. (†54840) 317/A4
Famagusta, Cyprus 63/F5
Famagusta, Cyprus 59/B3
Famagusta (bay), Cyprus 63/F5
Famaka, Sudan 111/F5
Famatina, Argentina 143/C2
Famatina, Sierra de (mts.), Argentina 143/C2
Family (lake), Manitoba 179/G3
Famoso, Calif. (†93280) 204/F8
Fan (lake), N. Dak. 282/L2
Fanad (head), Ireland 17/F1
Fancy Farm, Ky. (42039) 237/D7
Fancy Gap, Va. (24328) 307/G7
Fancy Prairie, Ill. (62637) 222/D4
Fandriana, Madagascar 118/H4
Fangak, Sudan 111/F6
Fang Xian, China 77/G5
Fangzheng, China 77/L2
Fannettsburg, Pa. (17221) 294/G5
Fannich, Loch (lake), Scotland 15/D3
Fannin (co.), Georgia 217/D1
Fannin, Miss. (†39042) 256/E6
Fannin (co.), Texas 303/H4
Fannin, Texas (77960) 303/G9
Fanning (isl.), Kiribati 87/L5
Fanning (isl.), Kiribati 2/B5
Fanning Springs (Suwannee River), Fla. (†32693) 212/D2
Fanny Bay, Br. Col. 184/H2
Fannystelle, Manitoba 179/E5
Fanø (isl.), Denmark 21/B7
Fanø (isl.), Denmark 18/F9
Fano, Italy 34/D3
Fanshawe, Okla. (74935) 288/S5
Fan Si Pan (mt.), Vietnam 72/D2
Fantasque (pt.), Haiti 158/B6
Fanwood, N.J. (07023) 273/E2
Fao, Iraq 66/F6
Faradje, Zaire 115/E3
Faradofay, Madagascar 102/G7
Faradofay, Madagascar 118/H5
Farafangana, Madagascar 102/G7
Farafangana, Madagascar 118/H4
Farâfra (oasis), Egypt 111/E2
Farâfra (oasis), Egypt 59/A4
Farah, Afghanistan 68/A2
Farah, Afghanistan 54/H6
Farah, Afghanistan 59/H3
Farah Rud (riv.) 54/H6
Farah Rud (riv.), Afghanistan 59/H3
Farah Rud (riv.), Afghanistan 68/A2
Farallon (isls.), Calif. 204/B6
Farallon de Pajaros (isl.), No. Marianas 87/E3
Farallons, The (gulf), Calif. 204/H2
Faranah, Guinea 106/B6
Farasan (isls.), Saudi Arabia 59/D6
Faraulep (atoll), Micronesia 87/E5
Farber, Mo. (63345) 261/J4
Farciennes, Belgium 27/E8
Fareham, England 10/F5
Fareham, England 13/F7
Farewell, Alaska (†99629) 196/H2
Farewell (cape), Greenl. 4/D12
Farewell (cape), Greenland 146/P4
Farewell (cape), Greenland 2/G3
Farewell (cape), N. Zealand 100/D4
Farfa, Italy 34/D3
Farfán, Ecuador 128/D2
Fargo, Ark. (†72021) 202/H4
Fargo, Georgia (31631) 217/G9
Fargo, Mich. (†48006) 250/G5
Fargo, N. Dak. 146/J5
Fargo, N. Dak. 188/G1
Fargo, N. Dak. (58102) 282/S6
Fargo, Okla. (73840) 288/G2
Fargo, Texas (†76384) 303/E3
Far Hills, N.J. (07931) 273/D2
Faribault, Minn. 188/H2
Faribault (co.), Minn. 255/D7
Faribault, Minn. (55021) 255/E6
Faridabad, India 68/D3
Faridpur, Bangladesh 68/F4
Fariman, Iran 66/L3
Farina, Ill. (62838) 222/E5
Farisita, Colo. (81037) 208/J7
Fariston, Ky. (†40741) 237/N6
Färjestaden, Sweden 18/K8
Farler, Ky. (41742) 237/P6
Farley, Iowa (52046) 229/L4
Farley, Mo. (64028) 261/O4
Farley, N. Mex. (†87747) 274/E2
Farlin, Iowa (50077) 229/E4
Farlington, Kansas (66734) 232/H4
Farm (riv.), Conn. 210/D3
Farmdale, Ohio (44417) 284/J3
Farmer, N.C. (27203) 281/K3
Farmer (isl.), N.W. Terrs. 187/K4
Farmer, Ohio (43520) 284/A3
Farmer, S. Dak. (57366) 298/O6
Farmer City, Ill. (61842) 222/E3
Farmers, Ky. (40319) 237/P4
Farmers Branch, Texas (75234) 303/G2
Farmersburg, Ind. (47850) 227/C6
Farmersburg, Iowa (52047) 229/L3
Farmersville, Ala. (36737) 195/E6
Farmersville, Calif. (93223) 204/F7
Farmersville, Ill. (62533) 222/D4
Farmersville, Mo. (†64683) 261/E3
Farmersville, Ohio (45325) 284/A6
Farmersville, Texas (75031) 303/H4
Farmerville, La. (71241) 238/F1
Farmhaven, Miss. (†39046) 256/E5
Farmill (riv.), Conn. 210/C3
Farmingdale, Maine (†04345) 243/D7
Farmingdale○, Maine (†04345) 243/D7
Farmingdale, N.J. (07727) 273/E3
Farmingdale, N.Y. (11735) 276/R7
Farmingdale, S. Dak. (†57725) 298/D6
Farmington, Ark. (72730) 202/B1
Farmington, Br. Col. 184/G2
Farmington, Calif. (95230) 204/E6
Farmington○, Conn. (06032) 210/D2
Farmington (riv.), Conn. 210/D1
Farmington, Del. (19942) 245/R5
Farmington, Georgia (30638) 217/F3
Farmington, Ill. (61531) 222/C3
Farmington, Iowa (52626) 229/K7
Farmington, Ky. (42040) 237/D7
Farmington, Maine (04938) 243/C6
Farmington○, Maine (04938) 243/C6
Farmington, Md. (†21911) 245/O2
Farmington, Mich. (*48024) 250/F6
Farmington, Minn. (55024) 255/E6
Farmington, Mo. (63640) 261/M7
Farmington, N.H. (03835) 268/E5
Farmington○, N.H. (03835) 268/E5
Farmington, N.M. 188/E3
Farmington, N. Mex. (87401) 274/A2
Farmington, N.C. (†27028) 281/H3
Farmington, Oreg. (†97123) 291/A2
Farmington, Tenn. (†37091) 237/H9
Farmington, Utah (84025) 304/C3
Farmington, Wash. (99128) 310/H3
Farmington, W. Va. (26571) 312/F3
Farmington, Wis. (†53050) 317/A5
Farmington Falls, Maine (04940) 243/C6
Farmington Hills, Mich. (48024) 250/F6
Farmland, Ind. (47340) 227/G4
Farmville, N.C. (27828) 281/O3
Farmville, Va. (23901) 307/M6
Farnam, Nebr. (69029) 264/D4
Farnams, Mass. (†01225) 249/B2
Farnborough, England 13/G8
Farner, Tenn. (37333) 237/N10
Farnham, England 13/G8
Farnham, England 10/F5
Farnham, N.Y. (14061) 276/B6
Farnham, Québec 172/E4
Farnham, Va. (22460) 307/P5
Farnhamville, Iowa (50538) 229/D4
Farnsworth, Texas (79033) 303/C1
Farnworth, England 13/H2
Faro, Brazil 132/B3
Faro (dist.), Portugal 33/B4
Faro, Portugal 7/D5
Faro, Portugal 33/B4
Fårö (isl.), Sweden 18/L8
Faro, Yukon 187/E3
Faro, Yukon 162/C3
Fårösund, Sweden 18/L8
Farquhar (cape), W. Australia 88/A4
Farquhar (cape), W. Australia 92/A3
Farr (bay) 5/C5
Farragut, Iowa (51639) 229/C7
Farrar, Georgia (†31085) 217/E4
Farrar, Iowa (†50161) 229/G5
Farrar, Mo. (63746) 261/N7
Farrar (riv.), Scotland 15/D3
Farrashband, Iran 66/G6
Farrell (lake), Alberta 182/D4
Farrell, Miss. (38630) 256/C2
Farrell, Pa. (16121) 294/A3
Farrellton, Québec 172/A4
Farris, Okla. (74542) 288/P6
Farrow, Alberta 182/D4
Farrukhabad-cum-Fatehgarh, India 68/D3
Fars (prov.), Iran 66/H6
Fársala, Greece 45/F6
Farsi, Afghanistan 68/A2
Farsi, Afghanistan 59/H3
Farsi (isl.), Iran 66/G7
Farsø, Denmark 21/C4
Farson, Iowa (†52563) 229/J6
Farson, Wyo. (82932) 319/C3
Farsund, Norway 18/E7
Fartak, Ras (cape), P.D.R. Yemen 59/F6
Farum, Denmark 21/F6
Farwell, Mich. (48622) 250/E5
Farwell, Minn. (56327) 255/C5
Farwell, Nebr. (68838) 264/F3
Farwell, Texas (79325) 303/A3
Fasa, Iran 66/H6
Fasa, Iran 59/F4
Fasano, Italy 34/F4
Fashoda (Kodok), Sudan 111/F6
Fassett, Québec 172/C4
Fastnet Rock (isl.), Ireland 17/B9
Fastov, U.S.S.R. 52/C4
Fatagar Tuting (cape), Indonesia 85/J6
Fatehpur, Rajasthan, India 68/C3
Fatehpur, Uttar Pradesh, India 68/E3
Fatih, Turkey 63/D6
Fátima, Portugal 33/B3
Fatsa, Turkey 63/G2
Fatshan (Foshan), China 77/H7
Fatuhiva (isl.), Fr. Poly. 87/N7
Faubush, Ky. (42532) 237/M6
Faucett, Mo. (64448) 261/C3
Faucilles (mts.), France 28/G3
Fauldhouse, Scotland 10/C1
Fauldhouse, Scotland 15/C2
Faulk (co.), S. Dak. 298/L3
Faulkner (co.), Ark. 202/F3
Faulkner, Iowa (†50601) 229/G3
Faulkner, Manitoba 179/D3
Faulkton, S. Dak. (57438) 298/L3
Faunsdale, Ala. (36738) 195/C6
Fauquier, Ontario 177/J5
Fauquier, Ontario 175/D3
Fauquier (co.), Va. 307/N3
Faurei, 45/H3
Faust, Alberta 182/C2
Favara, Italy 34/D6
Faversham, England 13/H6
Faversham, England 10/G5
Favignana (isl.), Italy 34/D6
Fawcett, Alberta 182/C2
Fawn (riv.), Ind. 227/G1
Fawn (riv.), Mich. 250/D7
Fawn (lake), Ont. 162/H5
Fawn (riv.), Ontario 175/C2
Fawn Grove, Pa. (17321) 294/J6
Fawnskin, Calif. (92333) 204/J9
Faxaflói (bay), Iceland 21/B1
Faxon, Okla. (73540) 288/J6
Fay, Ill. (†61285) 222/C2
Fay, Okla. (73646) 288/J3
Faya-Largeau, Chad 102/D3
Faya-Largeau, Chad 111/C4
Fayette (co.), Ala. 195/C3
Fayette, Ala. (35555) 195/C3
Fayette (co.), Georgia 217/C4
Fayette (co.), Ill. 222/D4
Fayette (co.), Ind. 227/G5
Fayette, Ind. (†46052) 227/E5
Fayette (co.), Iowa 229/K3
Fayette, Iowa (52142) 229/K3
Fayette (co.), Ky. 237/N4
Fayette○, Maine (†04349) 243/C7
Fayette, Miss. (39069) 256/B7
Fayette, Mo. (65248) 261/G4
Fayette (co.), Ohio 284/D6
Fayette, Ohio (43521) 284/B2
Fayette (co.), Pa. 294/C6
Fayette (co.), Tenn. 237/C10
Fayette (co.), Texas 303/H8
Fayette, Utah (84630) 304/C4
Fayette (co.), W. Va. 312/D6
Fayette City, Pa. (15438) 294/C5
Fayetteville, Ala. (†35150) 195/F4
Fayetteville, Ark. 188/H3
Fayetteville, Ark. (72701) 202/B1
Fayetteville, Georgia (30214) 217/C4
Fayetteville, Ill. (†62258) 222/D5
Fayetteville, Ind. (†47421) 227/D7
Fayetteville, Mo. (†64093) 261/E5
Fayetteville, N.Y. (13066) 276/J4
Fayetteville, N.C. 188/L3
Fayetteville, N.C. (*28301) 281/M4
Fayetteville, Ohio (45118) 284/C7
Fayetteville, Pa. (17222) 294/G6
Fayetteville, Tenn. (37334) 237/H10
Fayetteville, Texas (78940) 303/H8
Fayetteville, W. Va. (25840) 312/D6
Fayville, Mass. (01745) 249/H3
Faywood, N. Mex. (88034) 274/B6
Fdérik (Fort-Gouraud), Mauritania 106/B4
Feakle, Ireland 17/D6
Feale (riv.), Ireland 17/C7
Fear (cape), N.C. 188/L4
Fear (cape), N.C. 281/O7
Fearer, Md. (†21531) 245/A2
Fearns Springs, Miss. (†39339) 256/G4
Feather (riv.), Calif. 204/D4
Feather Falls, Calif. (95940) 204/D4
Featherston, N. Zealand 100/E4
Featherston, Okla. (†74561) 288/P4
Fécamp, France 28/D3
Fédala (Mohammedia), Morocco 106/C2
Federación, Argentina 143/G5
Federal, Alberta 182/E3
Federal, Wyo. (†82001) 319/G4
Federal Dam, Minn. (56641) 255/D3
Federal District, Brazil 132/E6
Federal Heights, Colo. (†80221) 208/J3
Federalsburg, Md. (21632) 245/P6
Federal Territory (state), Laos 72/D7
Fedora, S. Dak. (57337) 298/O5
Fedscreek, Ky. (41524) 237/S6
Feeagh (lake), Ireland 17/B4
Feeding Hills, Mass. (01030) 249/D4
Feeny, N. Ireland 17/H2
Feesburg, Ohio (45119) 284/B8
Fegyvernek, Hungary 41/F3
Fehérgyarmat, Hungary 41/G3
Fehmarn (str.), Denmark 21/E8
Fehmarn (isl.), W. Germany 22/D1
Fehmarn (str.), W. Germany 22/D1
Feia (lake), Brazil 135/F3
Feijó, Brazil 132/G10
Feilding, N. Zealand 100/E4
Feio (riv.), Brazil 135/B2
Feira, Portugal 33/B2
Feira (Luangwa), Zambia 115/E7
Feira de Santana, Brazil 132/G5
Feira de Santana, Brazil 120/F4
Fejér (co.), Hungary 41/E3
Fejø (isl.), Denmark 21/E8
Feke, Turkey 63/G4
Felanitx, Spain 33/H3
Felch, Mich. (49831) 250/B3
Felchville (Reading), Vt. (†05062) 268/B5
Felda, Fla. (33930) 212/E5
Feldbach, Austria 41/C3
Feldberg (mt.), W. Germany 22/C5
Feldkirch, Austria 41/A3
Feldkirchen in Kärnten, Austria 41/B3
Feliciano (riv.), Argentina 143/G5
Felicité (isl.), Seychelles 118/J5
Felicity, Ohio (45120) 284/B8
Felipe Carillo Puerto, Mexico 150/P7
Felipe Matiauda, Paraguay 144/D4
Felipe Yofré, Argentina 143/G4
Felix (cape), N.W. Terrs. 187/J3
Felixstowe, England 10/G5
Felixstowe, England 13/J6
Fellbach, W. Germany 22/C4
Felletin, France 28/E5
Felling, England 13/J3
Fellows, Calif. (93224) 204/F8
Fellowsville, W. Va. (†26410) 312/G4
Fellsburg, Kansas (67048) 232/C4
Fellsmere, Fla. (32948) 212/F4
Fels am Wagram, Austria 41/C2
Felsberg, Switzerland 39/H3
Felsenthal, Ark. (†71747) 202/F7
Felt, Idaho (83424) 220/G6
Felt, Okla. (73937) 288/A1
Felton, Ark. (†72360) 202/J4
Felton, Calif. (95018) 204/K4
Felton, Del. (19943) 245/R4
Felton, Georgia (30140) 217/B3
Felton, Minn. (56536) 255/B3
Felton, Pa. (17322) 294/J6
Feltre, Italy 34/C1
Felts Mills, N.Y. (13638) 276/J3
Feltwell, England 13/H5
Felty, Ky. (†40962) 237/O6
Femø (isl.), Denmark 21/E8
Femundsjø (lake), Norway 18/G5
Fence (riv.), Mich. 250/A2
Fence, Wis. (54120) 317/K4
Fence Lake, N. Mex. (87315) 274/A4
Fender, Georgia (†31794) 217/E8
Fenelon Falls, Ontario 177/F3
Fengcheng, China 77/K3
Fengjie, China 77/G5
Fengning, China 77/J3
Fengqing, China 77/E7
Fengtai, China 77/J5
Feng Xian, China 77/G5
Fengxiang, China 77/G5
Fengyang, China 77/H4
Fengzhen, China 77/H3
Fen He (riv.), China 77/H4
Fenholloway (riv.), Fla. 212/C1
Feni (isls.), Papua N.G. 86/C2
Fenimore (passage), Alaska 196/L4
Fenit, Ireland 17/B7
Fenn, Alberta 182/D3
Fenn, Idaho (83531) 220/B4
Fenner, Calif. (†92332) 204/K9
Fennimore, Wis. (53809) 317/E9
Fennville, Mich. (49408) 250/C6
Fenoarivo, Fianarantsoa, Madagascar 118/H4
Fenoarivo, Toamasina, Madagascar 118/H3
Fens, The (reg.), England 13/G5
Fenton (riv.), Conn. 210/G1
Fenton, Ill. (61251) 222/C2
Fenton, Iowa (50539) 229/E2
Fenton, La. (70640) 238/E6
Fenton, Mich. (48430) 250/F6
Fenton, Mo. (63026) 261/O4
Fenton, Sask. 181/F2
Fentress (co.), Tenn. 237/M8
Fenwick, Conn. (†06475) 210/F3
Fenwick, Mich. (48834) 250/D5
Fenwick, Nova Scotia 168/D3
Fenwick, W. Va. (26202) 312/E6
Fenwick Island, Del. (19944) 245/T7
Fenwood, Sask. 181/H4
Fenwood, Wis. (54431) 317/F6
Feodosiya, U.S.S.R. 52/D5
Ferbane, Ireland 17/F5
Ferdig, Mont. (59437) 262/E2
Ferdinand, Idaho (83526) 220/B3
Ferdinand, Ind. (47532) 227/D8
Ferdows, Iran 59/G3
Ferdows, Iran 66/K3
Ferfer, Somalia 115/J2
Fergana, U.S.S.R. 54/J6
Fergana, U.S.S.R. 48/H5
Fergus (riv.), Ireland 17/D6
Fergus (co.), Mont. 262/G3
Fergus, Mont. (59451) 262/H3
Fergus, Ontario 177/D4
Fergus Falls, Minn. (56537) 255/B4
Ferguson, Br. Col. 184/J5
Ferguson, Iowa (50078) 229/H5
Ferguson, Ky. (42533) 237/M6
Ferguson, Mo. (63135) 261/P2
Ferguson, N.C. (28624) 281/G2
Ferguson, W. Va. (†25511) 312/B6
Ferintosh, Alberta 182/D3
Ferkessédougou, Ivory Coast 106/D7
Ferlach, Austria 41/C3
Ferland, Ontario 177/H4
Ferland, Ontario 175/C2
Ferland, Québec 172/G1
Ferland, Sask. 181/D6
Ferlo (reg.), Senegal 106/B6
Fermanagh (dist.), N. Ireland 17/F3
Ferme-Neuve, Québec 172/B3
Fermeuse, Newf. 166/D2
Fermo, Italy 34/D3
Fermont, Québec 174/D2
Fermoselle, Spain 33/C2
Fermoy, Ireland 17/E7
Fermoy, Ireland 10/B4
Fernald, Iowa (†50201) 229/G4
Fernández, Argentina 143/D2
Fernandina (isl.), Ecuador 128/B9
Fernandina Beach, Fla. (32034) 212/E1
Fernando de la Mora, Paraguay 144/B4
Fernando Po (Bioko) (isl.), Equat. Guinea 102/C4
Fernando Po (Bioko) (isl.), Equat. Guinea 115/A3
Fernandópolis, Brazil 135/A2
Fernan Lake, Idaho (†83814) 220/B2
Fernbank, Ala. (35558) 195/B3
Fernberg, Queensland 88/K3
Ferndale, Calif. (95536) 204/A3
Ferndale, Md. (21061) 245/M4
Ferndale, Mich. (48220) 250/B6
Ferndale, Pa. (18921) 294/M4
Ferndale, Pa. (15905) 294/E5
Ferndale, Wash. (98248) 310/C2
Fernelmont, Belgium 27/F7
Ferness, Scotland 15/E3
Ferney, S. Dak. (57439) 298/N3
Fernie, Br. Col. 162/E6
Fernie, Br. Col. 184/K5
Fernley, Nev. (89408) 266/B3
Fern Ridge (lake), Oreg. 291/D3
Ferns, Ireland 17/J6
Fern Tree Gully, Victoria 97/K5
Fernwood, Idaho (83830) 220/B2
Fernwood, Miss. (39635) 256/D8
Fernwood, N.Y. (†12801) 276/N4
Fernwood, N.Y. (†13142) 276/H4
Ferolle (pt.), Newf. 166/C3
Ferrandina, Italy 34/F4
Ferrara (prov.), Italy 34/C2
Ferrara, Italy 7/F4
Ferrara, Italy 34/C2
Ferré (cape), Martinique 161/E7
Ferreira do Alentejo, Portugal 33/B3
Ferreira Gomes, Brazil 132/D2
Ferrellsburg, W. Va. (25513) 312/B6
Ferrelo (cape), Oreg. 291/C5
Ferrelview, Mo. (64163) 261/O4
Ferreñafe, Peru 128/C6
Ferriday, La. (71334) 238/G3
Ferrier, Alberta 182/C3
Ferrières, Belgium 27/H8
Ferris, Ill. (62336) 222/B3
Ferris, Texas (75125) 303/H3
Ferris (mts.), Wyo. 319/E3
Ferrisburg○, Vt. (05456) 268/A3
Ferrol (pen.), Peru 128/C7
Ferrol del Caudillo, Spain 33/B1
Ferron, Utah (84523) 304/C4
Ferron (creek), Utah 304/C4
Ferros, Brazil 132/F7
Ferrum, Va. (24088) 307/H7
Ferry, Guadeloupe 161/A6
Ferry, Mich. (49455) 250/C5
Ferry (co.), Wash. 310/G2
Ferryden, Scotland 15/F4
Ferryland, Newf. 166/D2
Ferryland (cape), Newf. 166/D2
Ferry Road, New Bruns. 170/E1
Ferrysburg, Mich. (49409) 250/C5
Ferryville, Wis. (54628) 317/D9
Fertigs, Pa. (†16364) 294/C3
Fertile, Iowa (50434) 229/G2
Fertile, Minn. (56540) 255/B3
Fertile, Sask. 181/K6
Fertilia, Italy 34/B4
Fertö tó (Neusiedler See) (lake), Austria 41/C3
Fertö tó (Neusiedler See) (lake), Hungary 41/D3
Fès (Fez), Morocco 106/D2
Fès, Morocco 102/B1

Feshi, Zaire 115/C5
Fessenden, N. Dak. (58438) 282/L4
Fesserton, Ontario 177/E3
Festina, Iowa (52143) 229/K2
Festus, Mo. (63028) 261/M6
Fetești, Romania 45/H3
Fethard, Tipperary, Ireland 17/F7
Fethard, Wexford, Ireland 17/H7
Fethiye, Turkey 63/C4
Fethiye, Turkey 59/A2
Fetlar (isl.), Scotland 15/G2
Fetlar (isl.), Scotland 10/H1
Feudal, Sask. 181/D4
Feuerkogel (mt.), Austria 41/B3
Feuerthalen, Switzerland 39/G1
Feuilles (riv.), Que. 162/J4
Feuilles (riv.), Que. 146/L4
Feuilles (riv.), Québec 174/C1
Feversham, Ontario 177/D3
Fevzipaşa, Turkey 63/G4
Feyzabad, Afghanistan 54/H6
Feyzabad, Afghanistan 59/K2
Feyzabad, Afghanistan 68/C1
Fezzan (reg.), Libya 102/D2
Fezzan (reg.), Libya 111/B2
Ffestiniog, Wales 10/E4
Ffestiniog, Wales 13/D5
Fiambalá, Argentina 143/C2
Fianarantsoa (prov.), Madagascar 118/H4
Fianarantsoa, Madagascar 102/G7
Fianarantsoa, Madagascar 118/H4
Fianga, Chad 111/C6
Fiat, Ind. (†47326) 227/G3
Fiatt, Ill. (61433) 222/C3
Fichtelberg (mt.), E. Germany 22/E3
Fichtelgebirge (range), W. Germany 22/D3
Fickle, Ind. (†46035) 227/D4
Ficklin, Georgia (†30673) 217/G3
Ficksburg, S. Africa 118/D5
Fidalgo (isl.), Wash. 310/C2
Fidelity, Ill. (62030) 222/C4
Fidenza, Italy 34/B2
Fieberbrunn, Austria 41/B3
Field, Br. Col. 184/J4
Field, Ky. (40934) 237/O7
Field, Ontario 177/E1
Fieldale, Va. (24089) 307/H7
Field Creek, Texas (†76869) 303/F7
Fielding, New Bruns. 170/C2
Fielding, Sask. 181/D3
Fielding, Utah (84311) 304/B2
Fieldon, Ill. (62031) 222/C4
Fields, La. (70641) 238/C5
Fields (lake), La. 238/J7
Fields, Oreg. (97710) 291/J5
Fieldsboro, N.J. (†08505) 273/D3
Fieldton, Texas (79326) 303/B3
Fier, Albania 45/D5
Fierro, N. Mex. (†88041) 274/A6
Fiesch, Switzerland 39/F4
Fiesole, Italy 34/C3
Fife (lake), Sask. 181/E6
Fife (reg.), Scotland 15/E4
Fife (trad. co.), Scotland 15/B5
Fife, Wash. (98424) 310/C3
Fife Lake, Mich. (49633) 250/D4
Fife Lake, Sask. 181/F6
Fife Ness (prom.), Scotland 15/F4
Fifield, N.S. Wales 97/D3
Fifield, Wis. (54524) 317/F4
Fifteenmile (creek), Oreg. 291/F2
Fifteenmile Arroyo (creek), N. Mex. 274/D4
Fifth (lake), Maine 243/H5
Fifth Cataract, Sudan 111/F4
Fifth Cataract, Sudan 59/B6
Fifth Cataract (dam), Sudan 102/F3
Fifty Lakes, Minn. (56448) 255/D4
Fiftysix, Ark. (72533) 202/F2
Fig (riv.), Newf. 166/B3
Figeac, France 28/D5
Figueira da Foz, Portugal 33/B2
Figueras, Spain 33/H1
Figuig, Morocco 102/B1
Figuig, Morocco 106/D2
Figuras (pt.), P. Rico 161/E3
Fiji 2/A6
Fiji 87/H8
FIJI 86/P11
Filadelfia, Bolivia 136/A2
Filadelfia, C. Rica 154/E5
Filadelfia, Paraguay 144/B3
Fil'akovo, Czech. 41/E2
Filbert, S.C. (†29745) 296/E1
Filbert, W. Va. (24835) 312/D8
Filchner Ice Shelf, Ant. 2/H10
Filchner Ice Shelf, Ant. 5/B16
Fil di Remia (peak), Switzerland 39/H4
File (hills), Sask. 181/H5
Filer, Idaho (83328) 220/D7
Filer City, Mich. (49634) 250/C4
Filey, England 13/G3
Filey, England 10/F3
Filiátes, Greece 45/E6
Filiatrá, Greece 45/E7
Filicudi (isl.), Italy 34/E5
Filingué, Niger 106/E6
Filion, Mich. (48432) 250/G5
Filippiás, Greece 45/E6
Filipstad, Sweden 18/H7
Filisur, Switzerland 39/J3
Filley, Nebr. (68357) 264/H4
Fillmore, Calif. (93015) 204/G9
Fillmore, Ill. (62032) 222/D4
Fillmore, Ind. (46128) 227/D5
Fillmore (co.), Minn. 255/F7
Fillmore, Minn. (†55990) 255/F7
Fillmore, Mo. (64449) 261/C2
Fillmore (co.), Nebr. 264/G4
Fillmore, N.Y. (14735) 276/D6
Fillmore, N. Dak. (58333) 282/L3
Fillmore, Okla. (†73450) 288/N6
Fillmore, Sask. 181/H6
Fillmore, Utah (84631) 304/B5
Filtu, Ethiopia 111/H6
Filyos (riv.), Turkey 63/D2
Fimi (riv.), Zaire 115/C4
Finale Emilia, Italy 34/C2
Finale Ligure, Italy 34/B2
Fiñana, Spain 33/E4
Fincastle, Ind. (†46172) 227/D5
Fincastle, Ky. (†40222) 237/L1
Fincastle, Va. (24090) 307/J6
Finch, Mont. (†59076) 262/K4
Finch, Ontario 177/J2
Finchburg, Ala. (†36444) 195/D7
Finchville, Ky. (40022) 237/L4
Findhorn, Scotland 15/E3
Findhorn (riv.), Scotland 15/E3
Findıklı, Turkey 63/J2
Findlater, Sask. 181/F5
Findlay, Ill. (62534) 222/E4
Findlay, Ohio (45840) 284/C3
Findley Lake, N.Y. (14736) 276/A6
Findochty, Scotland 15/E3
Findon, Mont. (†59053) 262/F4
Fine, N.Y. (13639) 276/K2
Finesville, N.J. (†08865) 273/C2
Fingal, N. Dak. (58031) 282/P6
Fingal, Ontario 177/C5
Fingal, Tasmania 99/E3
Finger (lake), Ontario 175/B2
Finger, Tenn. (38334) 237/D10
Fingerville, S.C. (29338) 296/D1
Fingoè, Mozambique 118/E2
Finhaut, Switzerland 39/C4
Finike, Turkey 63/D4
Finistère (dept.), France 28/A3
Finisterre (cape), Spain 7/C4
Finisterre (cape), Spain 33/B1
Finke (riv.), North. Terr. 88/E5
Finke (riv.), North. Terr. 93/C8
Finke (riv.), S. Australia 94/C1
Finksburg, Md. (21048) 245/L3
Finland 2/L2
Finland 4/C8
Finland 7/G2
Finland (gulf) 7/G3
FINLAND 18
Finland (gulf), Finland 18/P7
Finland, Minn. (55603) 255/G3
Finland (gulf), U.S.S.R. 52/C3
Finland (gulf), U.S.S.R. 48/C4
Finland (gulf), U.S.S.R. 53/D1
Finlay (riv.), Br. Col. 162/D4
Finlay (riv.), Br. Col. 184/E1
Finlay (mts.), Texas 303/B10
Finlayson, Minn. (55735) 255/F4
Finley, Ky. (42736) 237/L6
Finley, N. S. Wales 88/H7
Finley, N.S. Wales 97/C4
Finley, N. Dak. (58230) 282/P4
Finley, Okla. (74543) 288/R6
Finley, Tenn. (38030) 237/B8
Finley, Wash. (†99336) 310/F4
Finleyson, Georgia (†31071) 217/F6
Finleyville, Pa. (15332) 294/B5
Finly, Ind. (46129) 227/F5
Finn (riv.), Ireland 17/F2
Finn (riv.), Ireland 17/G3
Finnegan, Alberta 182/E4
Finney (co.), Kansas 232/B3
Finnmark (co.), Norway 18/O2
Finschhafen, Papua N.G. 86/B2
Finschhafen, Papua N.G. 85/C7
Finspång, Sweden 18/J7
Finsteraarhorn (mt.), Switzerland 39/F3
Finstermünz (pass), Switzerland 39/K3
Finsterwalde, E. Germany 22/E3
Finstown, Scotland 15/E1
Fintona, N. Ireland 17/G3
Fintry, Scotland 15/B1
Fionn Loch (lake), Scotland 15/C3
Fir (riv.), Sask. 181/J2
Fırat (riv.), Turkey 63/G4
Fircrest, Wash. (98466) 310/C3
Fire (isl.), Alaska 196/B1
Firebag (riv.), Alberta 182/E1
Firebaugh, Calif. (93622) 204/E7
Firebrick, Ky. (41137) 237/P3
Fireco, W. Va. (†25856) 312/D7
Fire Island National Seashore, N.Y. 276/P9
Firenze (Florence), Italy 34/C3
Fires (bay), Tasmania 99/E3
Firesteel (riv.), Mich. 250/G1
Firesteel, S. Dak. (57628) 298/G3
Firesteel (creek), S. Dak. 298/N6
Firestone, Colo. (80520) 208/K2
Firgrove, S. Africa 118/F6
Firmat, Argentina 143/F6
Firminy, France 28/F5
Fir Mountain, Sask. 181/E6
Firozabad, India 68/D3
Firozpur, India 68/C2
First (lake), New Bruns. 170/B1
First Cataract, Egypt 59/B5
First Connecticut (lake), N.H. 268/E1
First Eel (lake), New Bruns. 170/C3
Firstview, Colo. (†80810) 208/O5
Firth (riv.), Alaska 196/K1
Firth, Idaho (83236) 220/F6
Firth, Nebr. (68358) 264/H4
Firth (riv.), Yukon 187/D3
Firuzabad, Iran 66/H6
Firuzkuh, Iran 66/H3
Fischot Islands, Newf. 166/C3
Fish (creek), Ind. 227/H2
Fish (riv.), Maine 243/F2
Fish (riv.), Namibia 118/B5
Fish (creek), Oreg. 291/E4
Fish (lake), Utah 304/C5
Fish (creek), W. Va. 312/E3
Fish Camp, Calif. (93623) 204/F6
Fish Creek (res.), Idaho 220/E6
Fish Creek (mts.), Nev. 266/D2
Fish Creek, Wis. (54212) 317/M5
Fisher, Ark. (72429) 202/J2
Fisher (isl.), Fla. 212/B5
Fisher, Ill. (61843) 222/E3
Fisher, La. (71426) 238/D4
Fisher (bay), Manitoba 179/E3
Fisher (riv.), Manitoba 179/E3
Fisher, Minn. (56723) 255/B3
Fisher (str.), N.W.T. 162/H3
Fisher (str.), N.W. Terrs. 187/K3
Fisher (lake), Nova Scotia 168/C4
Fisher, S. Australia 94/B4
Fisher (co.), Texas 303/D5
Fisher, W. Va. (26818) 312/H4
Fisher Bay, Manitoba 179/E3
Fisher Branch, Manitoba 179/E3
Fishermans (isl.), Va. 307/S6
Fishers, Ind. (46038) 227/E5
Fishers (isl.), N.Y. 276/S8
Fishers Island, N.Y. (06390) 276/R8
Fishersville, Va. (22939) 307/K4
Fisherville (South Grafton), Mass. (01560) 249/H4
Fishguard and Goodwick, Wales 13/B5
Fishguard and Goodwick, Wales 10/D4
Fish Haven, Idaho (83261) 220/G7
Fishing (lake), Manitoba 179/G2
Fishing (bay), Md. 245/O7
Fishing (creek), N.C. 281/O2
Fishing Creek, Md. (21634) 245/N7
Fishing Lake, Alberta 182/E3
Fishing Ships Harbour, Newf. 166/C3
Fishkill, N.Y. (12524) 276/N7
Fish River (lake), Maine 243/F2
Fishs Eddy, N.Y. (13774) 276/K8
Fish Springs (range), Utah 304/A4
Fishtail, Mont. (59028) 262/G5
Fishtoft, England 13/H5
Fishtrap, Ky. (41525) 237/S6
Fishtrap (lake), Ky. 237/S6
Fisk, Mo. (63940) 261/M9
Fiskdale, Mass. (01518) 249/F4
Fiske, Sask. 181/C4
Fiskeville, R.I. (02823) 249/H6
Fitch Bay, Québec 172/E4
Fitchburg, Mass. (01420) 249/G2
Fitchville, Conn. (06334) 210/G2
Fitchville, Ohio (†44851) 284/E3
Fithian, Ill. (61844) 222/F3
Fitler, Miss. (39070) 256/B5
Fittri (lake), Chad 111/C5
Fittstown, Okla. (74842) 288/N5
Fitzcarrald, Peru 128/G8
Fitzgerald, Alberta 182/C4
Fitzgerald, Georgia (31750) 217/F7
Fitzgerald, N.W.T. 162/E4
Fitzhugh (sound), Br. Col. 184/D4
Fitzhugh, Okla. (74843) 288/N5
Fitzmaurice (riv.), North. Terr. 93/B3
Fitzpatrick, Ala. (36029) 195/G6
Fitzpatrick, Georgia (†31044) 217/F5
Fitzroy (riv.), Australia 87/C7
Fitz Roy (Chaltel) (mt.), Chile 138/E8
Fitzroy, North. Terr. 93/B4
Fitzroy (riv.), Queensland 88/J4
Fitzroy (riv.), Queensland 95/D4
Fitzroy, Victoria 97/H5
Fitzroy, Victoria 88/L7
Fitzroy (riv.), W. Australia 88/C3
Fitzroy (riv.), W. Australia 92/D2
Fitzroy Crossing, W. Australia 88/D3
Fitzroy Crossing, W. Australia 92/D2
Fitzroy Harbour, Ontario 177/H2
Fitzwilliam○, N.H. (03447) 268/C6
Fitzwilliam (isl.), Ontario 177/C2
Fitzwilliam Depot, N.H. (03447) 268/C6
Fiume (Rijeka), Yugoslavia 45/B3
Fiumicino, Italy 34/F7
Five (isls.), Nova Scotia 168/D3
Five Fingers, New Bruns. 170/C1
Five Island (lake), Iowa 229/D2
Five Islands, Maine (04546) 243/D8
Five Islands, Nova Scotia 168/D3
Five Mile (riv.), Conn. 210/H1
Fivemile (creek), Oreg. 291/F2
Fivemile (pt.), Oreg. 291/C4
Fivemile (creek), Wyo. 319/D2
Fivemiletown, N. Ireland 17/G3
Five Points, Ala. (36855) 195/H4
Five Points, Fla. (32922) 212/D1
Five Points, Tenn. (38457) 237/G10
Five Stars, Guyana 131/A2
Fivizzano, Italy 34/B2
Fizi, Zaire 115/E4
Fjerritslev, Denmark 21/C3
Flagler, Colo. (80815) 208/N4
Flagler (co.), Fla. 212/F2
Flagler Beach, Fla. (32036) 212/E2
Flag Pond, Tenn. (37657) 237/R8
Flagstaff, Ariz. 146/G6
Flagstaff, Ariz. 188/D3
Flagstaff, Ariz. (86001) 198/D3
Flagstaff (lake), Maine 243/C5
Flagstaff (lake), Oreg. 291/H5
Flagtown, N.J. (08821) 273/D2
Flambeau (riv.), Wis. 317/E4
Flambeau Flowage (res.), Wis. 317/F3
Flamborough (head), England 13/G3
Flamborough (head), England 10/G3
Flamenco de San Pedro, Cuba 158/F3
Flaming Gorge (dam), Utah 304/E3
Flaming Gorge (res.), Utah 304/E3
Flaming Gorge (res.), Wyo. 319/C4
Flaming Gorge Nat'l Rec. Area, Utah 304/E2
Flaming Gorge Nat'l Rec. Area, Wyo. 319/C4
Flamingo (cay), Bahamas 156/C2
Flanagan, Ill. (61740) 222/E3
Flanagan (passage), Virgin Is. (Br.) 161/D4
Flanagan (passage), Virgin Is. (U.S.) 161/D4
Flanagin Town, Trin. & Tob. 161/B10
Flanders, Conn. (†06757) 210/B1
Flanders (trad. prov.) France 29
Flanders, N.J. (07836) 273/D2
Flanders, Ontario 175/B3
Flanders-Riverside, N.Y. (†11901) 276/P9
Flandreau, S. Dak. (57028) 298/R5
Flanigan, Nev. (†89501) 266/B2
Flannagan (res.), Va. 307/C6
Flannan (isls.), Scotland 15/A2
Flannan (isls.), Scotland 10/C1
Flasher, N. Dak. (58535) 282/H7
Flat, Ky. (41325) 237/O5
Flat, Mo. (†65550) 261/J7
Flat (isl.), Philippines 85/F3
Flat, Texas (76526) 303/G6
Flat (creek), Va. 307/M6
Flat (cays), Virgin Is. (U.S.) 161/A4
Flat Bay, Newf. 166/C4
Flatbrookville, N.J. (†07832) 273/D1
Flatbush, Alberta 182/C2
Flat Creek, Tenn. (†37160) 237/H10
Flat Creek-Wegra, Ala. (†35129) 195/D3
Flat Fork, Ky. (41427) 237/P5
Flatgap, Ky. (41219) 237/R5
Flathead (riv.), Br. Col. 184/K6
Flathead (co.), Mont. 262/B2
Flathead (lake), Mont. 188/D1
Flathead (lake), Mont. 262/C3
Flathead (riv.), Mont. 262/B2
Flathead, North Fork (riv.), Mont. 262/B2
Flathead, South Fork (riv.), Mont. 262/C3
Flathead Ind. Res., Mont. 262/B3
Flatlands, New Bruns. 170/D1
Flat Lick, Ky. (40935) 237/O7
Flatonia, Texas (78941) 303/G8
Flat River, Mo. (63601) 261/M7
Flat Rock, Ala. (35966) 195/G1
Flat Rock, Ill. (62427) 222/F5
Flat Rock, Ind. (47234) 227/F6
Flatrock (creek), Ind. 227/F5
Flat Rock, Ky. (†42634) 237/M7
Flat Rock, Mich. (48134) 250/F6
Flat Rock, Newf. 166/D2
Flat Rock, N.C. (†27043) 281/E4
Flat Rock, Ohio (44828) 284/E3
Flats, N.C. (†28781) 281/B4
Flattery (cape), Br. Col. 162/D6
Flattery (cape), Queensland 88/H2
Flattery (cape), Queensland 95/C2
Flattery (cape), Wash. 146/F5
Flattery (cape), Wash. 188/A1
Flattery (cape), Wash. 310/A2
Flat Top, W. Va. (25841) 312/D7
Flatwillow (creek), Mont. 262/H4
Flatwood, Ala. (†36739) 195/C6
Flatwoods, Ky. (41139) 237/R4
Flatwoods, La. (71427) 238/E4
Flatwoods, Tenn. (38458) 237/F9
Flatwoods, W. Va. (26621) 312/E5
Flawil, Switzerland 39/H2
Flaxcombe, Sask. 181/B4
Flaxman (isl.), Alaska 196/J1
Flaxton, N. Dak. (58737) 282/F2
Flaxville, Mont. (59222) 262/L2
Fleet, Alberta 182/E3
Fleet, England 13/G8
Fleet, Loch (inlet), Scotland 15/D3
Fleetwood, England 10/E4
Fleetwood, England 13/D4
Fleetwood, Okla. (†73569) 288/L7
Fleetwood, Pa. (19522) 294/L5
Fleischmanns, N.Y. (12430) 276/L6
Flekkefjord, Norway 18/E7
Flémalle, Belgium 27/G7
Fleming, Colo. (80728) 208/O1
Fleming, Georgia (31309) 217/K7
Fleming (co.), Ky. 237/O4
Fleming, Mo. (†64077) 261/D4
Fleming, Pa. (19375) 294/G4
Fleming, Sask. 181/K5
Fleming-Neon, Ky. (41816) 237/R6
Flemingsburg, Ky. (41041) 237/O4
Flemington, Georgia (†31313) 217/K7
Flemington, Mo. (65650) 261/F7
Flemington, N.J. (08822) 273/D2
Flemington, Pa. (†17745) 294/G3
Flemington, W. Va. (26347) 312/F4
Flen, Sweden 18/K7
Flensburg, Minn. (56328) 255/D5
Flensburg, W. Germany 22/C1
Flers, France 28/C3
Flesherton, Ontario 177/D3
Flesk (riv.), Ireland 17/C7
Fleta, Ala. (†36043) 195/F6
Fletcher (pond), Mich. 250/F4
Fletcher, Mo. (63030) 261/L6
Fletcher, N.C. (28732) 281/E4
Fletcher, Ohio (45326) 284/B5
Fletcher, Okla. (73541) 288/K5
Fletcher○, Vt. (†05444) 268/B2
Fletschhorn (mt.), Switzerland 39/F4
Fleurance, France 28/D6
Fleur de Lys, Newf. 166/C3
Fleur-de-May (lake), Newf. 166/B3
Fleurier, Switzerland 39/C3
Fleurus, Belgium 27/E8
Flevoland Polders, Netherlands 27/G4
Flims, Switzerland 39/H3
Flinders (reefs), 95/D3
Flinders (riv.), Australia 87/E7
Flinders (reef), Coral Sea Is. Terr. 88/H3
Flinders (riv.), Queensland 88/G3
Flinders (riv.), Queensland 95/B3
Flinders (range), S. Australia 88/F6
Flinders (range), S. Australia 94/F4
Flinders (isl.), Tasmania 88/H7
Flinders (isl.), Tasmania 99/D1
Flinders (bay), W. Australia 88/A6
Flinders (bay), W. Australia 92/A6
Flin Flon, Man. 146/H4
Flin Flon, Manitoba 179/H3
Flin Flon, Man.-Sask. 162/F4
Flin Flon, Sask. 181/N4
Flint (riv.), Ga. 188/K4
Flint, Georgia (†31716) 217/D8
Flint (riv.), Georgia 217/D8
Flint, Ind. (†46703) 227/G1
Flint (isl.), Kiribati 87/L7
Flint, Mich. 146/K5
Flint, Mich. 188/K2
Flint, Mich. (*48501) 250/F5
Flint (riv.), Mich. 250/F5
Flint (lake), N.W. Terrs. 187/L3
Flint, Wales 13/G2
Flint City, Ala. (†35601) 195/D1
Flinthill, Mo. (63346) 261/L5
Flint Hill, Va. (22627) 307/M3
Flinton, Ontario 177/G3
Flint Rock (creek), S. Dak. 298/E3
Flintstone, Georgia (30725) 217/B1
Flintstone (lake), Manitoba 179/G4
Flintstone, Md. (21530) 245/D2
Flintville, Tenn. (37335) 237/H10
Flippen, Georgia (30215) 217/D3
Flippin, Ark. (72634) 202/E1
Flippin, Ky. (42132) 237/K7
Flix, Spain 33/G2
Flom, Minn. (56541) 255/B3
Flomaton, Ala. (36441) 195/D8
Flomot, Texas (79234) 303/D3
Flood, Br. Col. 184/M3
Floodwood, Minn. (55736) 255/E4
Flora, Ill. (62839) 222/E5
Flora, Ind. (46929) 227/E3
Flora, La. (71428) 238/D3
Flora, Miss. (39071) 256/D5
Flora, N. Dak. (†58348) 282/M4
Flora (riv.), North. Terr. 93/B3
Flora, Norway 18/D6
Flora, Oreg. (†97828) 291/K2
Florac, France 28/E5
Florahome, Fla. (32635) 212/E2
Floral, Ark. (72534) 202/G2
Florala, Ala. (36442) 195/F8
Floral City, Fla. (32636) 212/D3
Floral Park, N.Y. (*11001) 276/P7
Floraville, Queensland 95/B3
Flora Vista, N. Mex. (87415) 274/A2
Floreana (Sta. María), Ecuador 128/B10
Floreana (Santa María) (isl.), Ecuador 128/B10
Florence, Ala. 188/J4
Florence, Ala. (*35630) 195/C1
Florence, Ariz. (85232) 198/D5
Florence, Ark. (†71655) 202/G6
Florence (lake), Calif. 204/G6
Florence, Colo. (81226) 208/J6
Florence, Ill. (†62363) 222/C4
Florence, Ind. (47020) 227/H7
Florence (prov.), Italy 34/C3
Florence, Italy 7/F4
Florence, Italy 34/C3
Florence, Kansas (66851) 232/E3
Florence, Ky. (41042) 237/R2
Florence, Minn. (56130) 255/B6
Florence, Miss. (39073) 256/D6
Florence, Mo. (65329) 261/G5
Florence, Mont. (59833) 262/B4
Florence, N.Y. (†13316) 276/J4
Florence, N.C. (†28556) 281/R4
Florence, Nova Scotia 168/H2
Florence, Ontario 177/B5
Florence, Oreg. (97439) 291/C4
Florence, Pa. (†15021) 294/A5
Florence, S.C. 188/L4
Florence (co.), S.C. 296/H3
Florence, S.C. (29501) 296/H3
Florence, S. Dak. (57235) 298/P3
Florence (riv.), Tasmania 99/C4
Florence, Tenn. (†37130) 237/H9
Florence, Texas (76527) 303/G7
Florence, Vt. (05744) 268/A4
Florence (co.), Wis. 317/K4
Florence, Wis. (54121) 317/K4
Florence Junction, Ariz. (†85220) 198/D5
Florence-Roebling, N.J. (08518) 273/D3
Florenceville, New Bruns. 170/C2
Florencia, Colombia 126/C7
Florencia, Colombia 120/B2
Florencia, Cuba 158/F2
Florennes, Belgium 27/F8
Florenton, Minn. (†55792) 255/F3
Florenville, Belgium 27/G9
Flores, Las (riv.), Argentina 143/G7
Flores, Brazil 132/G4
Flores (isl.), Br. Col. 184/D5
Flores, Guatemala 154/C2
Flores (isl.), Indonesia 54/O10
Flores (isl.), Indonesia 85/G7
Flores (sea), Indonesia 2/Q6
Flores (sea), Indonesia 54/N10
Flores (sea), Indonesia 85/F7
Flores (isl.), Portugal 33/A1
Flores (dept.), Uruguay 145/C4
Flores (isl.), Uruguay 145/D5
Floresville, Texas (78114) 303/K11
Florey, Texas (†79714) 303/B5
Florham Park, N.J. (07932) 273/E2
Floriano, Brazil 132/F4
Floriano, Brazil 120/E3
Florianópolis, Brazil 132/E9
Florianópolis, Brazil 120/E5
Florida 188/K5
FLORIDA 212
Florida (strs.) 146/K7
Florida, Bolivia 136/D6
Florida, Cuba 158/G3
Florida (str.), Cuba 156/B1
Florida (bay), Fla. 188/K6
Florida (bay), Fla. 212/F6
Florida (cape), Fla. 212/F6
Florida (keys), Fla. 188/K6
Florida (keys), Fla. 212/E7
Florida (strs.), Fla. 188/K6
Florida (strs.), Fla. 212/F7
Florida○, Mass. (†01247) 249/B2
Florida, Mo. (†65283) 261/J4
Florida (mts.), N. Mex. 274/B7
Florida, N.Y. (10921) 276/M8
Florida, Ohio (†43545) 284/B3
Florida, P. Rico 161/C1
Florida (isls.), Solomon Is. 86/E3
Florida (state), U.S. 146/K7
Florida (dept.), Uruguay 145/D4
Florida, Uruguay 145/C5
Florida City, Fla. (33034) 212/F6
Florida Ridge, Fla. (†32960) 212/F4
Floridia, Italy 34/E6
Florien, La. (71429) 238/D4
Florin, Calif. (95828) 204/B8
Florin, Pa. (†17552) 294/J5
Flórina, Greece 45/E5
Floris, Iowa (52560) 229/J7
Florissant, Colo. (80816) 208/J5
Florissant, Mo. (*63031) 261/P1
Florissant Fossil Beds Nat'l Mon., Colo. 208/J5
Flossmoor, Ill. (60422) 222/B6
Flovilla, Georgia (30216) 217/E4
Flowerdale, Tasmania 99/B2
Floweree, Mont. (59440) 262/E3
Flower Mound, Texas (†75067) 303/F1
Flowerpot (isl.), Ontario 177/C2
Flowers (bay), Newf. 166/B2
Flowers Cove, Newf. 166/C3
Flowery Branch, Georgia (30542) 217/E2
Flowood, Miss. (†39201) 256/D6
Floyd (co.), Georgia 217/B2
Floyd, Georgia (30059) 217/J1
Floyd (co.), Ind. 227/F8
Floyd (co.), Iowa 229/H2
Floyd, Iowa (50435) 229/H2
Floyd (riv.), Iowa 229/A3
Floyd (co.), Ky. 237/R5
Floyd, La. (†71266) 238/H1
Floyd, N. Mex. (88118) 274/F4
Floyd (co.), Texas 303/C3
Floyd (co.), Va. 307/H7
Floyd, Va. (24091) 307/H7
Floydada, Texas (79235) 303/C3
Floyd Dale, S.C. (29542) 296/J3
Floyds Knobs, Ind. (47119) 227/F8
Fluchthorn (mt.), Switzerland 39/K3
Flüela (pass), Switzerland 39/J3
Flüelen, Switzerland 39/G3
Fluhberg (mt.), Switzerland 39/G2
Fluker, La. (70436) 238/K5
Flums, Switzerland 39/H2
Flushing, Mich. (48433) 250/F5
Flushing, Netherlands 27/C6
Flushing, Ohio (43977) 284/J5
Fluvanna, Texas (79517) 303/D5
Fluvanna (co.), Va. 307/M5
Fly, Ohio (45730) 284/H6
Fly (riv.), Papua N.G. 87/E6
Fly (riv.), Papua N.G. 85/A7
Fly Creek, N.Y. (13337) 276/K5
Flying H, N. Mex. (88322) 274/E5
Flying Shot, Alberta 182/A2
Flynns Lick, Tenn. (†38562) 237/K8
Foam Lake, Sask. 181/H4
Foard (co.), Texas 303/E3
Foça, Turkey 63/B3
Foča, Yugoslavia 45/D4
Fochabers, Scotland 15/E3
Focşani, Romania 45/H3
Foge (isl.), Nigeria 106/E6
Foggia (prov.), Italy 34/E4
Foggia, Italy 7/F4
Foggia, Italy 34/E4
Fogo (isl.), C. Verde 106/B8
Fogo, Newf. 166/D4
Fogo (isl.), Newf. 166/D4
Fogo (isl.), Newf. 162/L6
Fohnsdorf, Austria 41/C3
Föhr (isl.), W. Germany 22/C1
Foisy, Alberta 182/E3
Foix, France 28/D6
Foix (trad. prov.) France 29
Folcroft, Pa. (19032) 294/M7
Folda (fjord), Norway 18/J3
Folda (fjord), Norway 18/G4
Földeák, Hungary 41/F3
Földes, Hungary 41/F2
Foley, Ala. (36535) 195/C10
Foley, Fla. (†32347) 212/C1
Foley, Minn. (56329) 255/D5
Foley, Mo. (63347) 261/L4
Foley (isl.), N.W. Terrs. 187/L3
Foleyet, Ontario 177/J5
Foleyet, Ontario 175/D3
Folgares, Angola 115/C6
Foligno, Italy 34/D3
Folkestone, England 13/J6
Folkestone, England 10/G5
Folkston, Georgia (31537) 217/H9
Folkstone, N.C. (†28445) 281/O5
Follansbee, W. Va. (26037) 312/E2
Follett, Texas (79034) 303/D1
Föllinge, Sweden 18/J5
Folly Beach, S.C. (29439) 296/H6
Folsom, Calif. (95630) 204/C8
Folsom (lake), Calif. 204/C8
Folsom, La. (70437) 238/K5
Folsom, N.J. (†08037) 273/D4
Folsom, N. Mex. (88419) 274/F2
Folsom, Pa. (19033) 294/M7
Folsom, W. Va. (26348) 312/E4
Folsomville, Ind. (47614) 227/C8
Folteşti, Romania 45/H3
Fomboni, Comoros 118/G2
Fomento, Cuba 158/E2
Fonda, Iowa (50540) 229/D3
Fonda, N.Y. (12068) 276/M5
Fonda, N. Dak. (†58366) 282/K2
Fond d'Or (bay), St. Lucia 161/G6
Fond-du-Lac (riv.), Sask. 162/F4
Fond du Lac, Sask. 181/L2
Fond du Lac (riv.), Sask. 181/M2
Fond du Lac, Wis. 188/J2
Fond du Lac (co.), Wis. 317/K8
Fond du Lac, Wis. (54935) 317/K8
Fond du Lac Ind. Res., Minn. 255/F4
Fonde, Ky. (40937) 237/O7
Fondi, Italy 34/D4
Fond-Lahaye, Martinique 161/C6
Fond-Saint-Denis, Martinique 161/C6
Fond Verrettes, Haiti 158/C6
Fonehill, Sask. 181/J4
Fongafale (cap.), Tuvalu 87/H6
Fonsagrada, Spain 33/C1
Fonseca, Colombia 126/D2
Fonseca (gulf), El Salvador 154/D4
Fonseca (gulf), Honduras 154/D4
Fonseca (gulf), Nicaragua 154/D4
Fontaine, New Bruns. 170/F2

Fontainebleau, France 28/E3
Fontainebleau, Québec 172/F4
Fontana, Calif. (92335) 204/E10
Fontana, Kansas (66026) 232/H3
Fontana (lake), N.C. 281/B4
Fontana, Wis. (53125) 317/J10
Fontanelle, Iowa (50846) 229/E6
Fontanet, Ind. (47851) 227/C5
Fontas (riv.), Br. Col. 184/M2
Fonte Boa, Brazil 132/G9
Fontein, Neth. Ant. 161/E8
Fontenay-le-Comte, France 28/C4
Fontenay-sous-Bois, France 28/C2
Fonteneau (lake), Newf. 166/B3
Fontenelle, Québec 172/D1
Fontenelle (creek), Wyo. 319/B3
Fontenelle (res.), Wyo. 319/B3
Fontibón, Colombia 126/C5
Fontur (prom.), Iceland 7/C2
Fontur (pt.), Iceland 21/D1
Fonyód, Hungary 41/D3
Foochow (Fuzhou), China 77/J6
Fool Creek (res.), Utah 304/B4
Foosland, Ill. (61845) 222/E3
Foothills, Alberta 182/B3
Footscray, Victoria 97/H5
Footscray, Victoria 88/K7
Footville, Ohio (†44084) 284/J2
Footville, Wis. (53537) 317/H10
Foping, China 77/G5
Forada, Minn. (†56308) 255/C5
Foraker (mt.), Alaska 196/H2
Foraker, Ind. (46525) 227/F1
Foraker, Ohio (†45812) 284/C4
Foraker, Okla. (74652) 288/O1
Forbach, France 28/G3
Forbes (mt.), Alberta 182/B4
Forbes (mt.), Br. Col. 184/J4
Forbes (isl.), Fla. 212/D6
Forbes, Minn. (55738) 255/F3
Forbes, Mo. (†64473) 261/B3
Forbes, N. S. Wales 88/H6
Forbes, N.S. Wales 97/E3
Forbes, N. Dak. (58439) 282/N8
Forbes (lake), Québec 172/C3
Forbing, La. (71106) 238/C2
Forbus, Tenn. (38561) 237/M7
Forcados, Nigeria 106/E7
Forcalquier, France 28/F6
Force, Pa. (15841) 294/E3
Forchheim, W. Germany 22/D4
Forchu (bay), Nova Scotia 168/H3
Forchu (cape), Nova Scotia 168/B5
Ford, England 13/F2
Ford (isl.), Hawaii 218/B3
Ford (co.), Ill. 222/E3
Ford (co.), Kansas 232/C4
Ford, Kansas (67842) 232/C4
Ford, Ky. (40320) 237/N5
Ford (riv.), Mich. 250/B2
Ford (cape), North. Terr. 88/D2
Ford (cape), North. Terr. 93/A2
Ford, Va. (23850) 307/N6
Ford, Wash. (99013) 310/H3
Ford City, Calif. (†93268) 204/F8
Ford City, Mo. (†64463) 261/C2
Ford City, Pa. (16226) 294/D4
Ford Cliff, Pa. (16228) 294/D4
Fordland, Mo. (65652) 261/G8
Fordoche, La. (70732) 238/G5
Ford Ranges (mts.) 5/B11
Fords, N.J. (08863) 273/E2
Ford's Bridge, N.S. Wales 97/C1
Fords Prairie, Wash. (†98531) 310/B4
Fordsville, Ky. (42343) 237/H5
Fordville, N. Dak. (58231) 282/P3
Fordwich, Ontario 177/C4
Fordyce, Ark. (71742) 202/F6
Fordyce, Nebr. (68736) 264/G2
Forécariah, Guinea 106/B7
Foreman, Ark. (71836) 202/B6
Foremost, Alberta 182/E5
Foresman, Ind. (†47922) 227/C3
Forest, Belgium 27/B9
Forest, Ind. (46039) 227/E4
Forest, La. (71242) 238/H1
Forest, Miss. (39074) 256/F6
Forest (riv.), N. Dak. 282/P3
Forest, Ohio (45843) 284/C4
Forest, Ontario 177/C4
Forest (co.), Pa. 294/D2
Forest, Va. (24551) 307/K6
Forest (co.), Wis. 317/J4
Forest Acres, S.C. (29206) 296/E3
Forest Beach, S.C. (†29928) 296/F7
Forestburg, Alberta 182/E3
Forestburg, S. Dak. (57338) 298/N5
Forestburg, Texas (76239) 303/G4
Forest City, Ill. (61532) 222/D3
Forest City, Iowa (50436) 229/F2
Forest City, Maine (04413) 243/H4
Forest City, Mo. (64451) 261/B3
Forest City, New Bruns. 170/C3
Forest City, N.C. (28043) 281/E4
Forest City, Pa. (18421) 294/L2
Forestdale, Ala. (35214) 195/E3
Forestdale, R.I. (02824) 249/H5
Forest Dale, Vt. (05745) 268/A4
Forester, Mich. (†48419) 250/G5
Foresters Falls, Ontario 177/H2
Forest Glen, Georgia (†31001) 217/F7
Forest Green, Mo. (†65281) 261/G4
Forest Grove, Br. Col. 184/G4
Forestgrove, Mont. (59441) 262/H3
Forest Grove, Oreg. (97116) 291/A2
Forest Heights, Md. (†20001) 245/F5
Foresthill, Calif. (†95703) 204/E4
Forest Hill, Ind. (†47240) 227/F6
Forest Hill, La. (71430) 238/E4
Forest Hill, Md. (21050) 245/N2
Forest Hill, N.S. Wales 97/D4
Forest Hill, Texas (†76119) 303/F2
Forest Hill, W. Va. (24935) 312/E7
Forest Hills, Ky. (41527) 237/L2
Forest Hills, Pa. (†15221) 294/C7
Forest Hills, Tenn. (†37201) 237/H8
Forest Home, Ala. (36030) 195/E7
Forest Homes, Ill. (†62018) 222/B2
Forestier (cape), Tasmania 99/E4
Forestier (pen.), Tasmania 99/E4
Forest Junction, Wis. (54123) 317/K7
Forest Knolls-Lagunitas, Calif. (94933) 204/H1
Forest Lake, Mich. (49832) 250/C2
Forest Lake, Minn. (55025) 255/F5
Foreston, Minn. (56330) 255/E5
Foreston, S.C. (†29102) 296/G4
Forest Park, Georgia (30050) 217/K2
Forest Park, Ill. (60130) 222/B5
Forest Park, Ohio (45405) 284/B9
Forest Park, Okla. (†73101) 288/M3
Forestport, N.Y. (13338) 276/K4
Forest River, N. Dak. (58233) 282/P3
Forest Station, Maine (†04413) 243/H4
Forest View, Ill. (†60402) 222/B6
Forestville, Conn. (†06010) 210/D2
Forestville, Md. (†20028) 245/G5
Forestville, Mich. (48434) 250/G5
Forestville, N.Y. (14062) 276/B6
Forestville, Ohio (45230) 284/C10
Forestville, Pa. (16035) 294/B3
Forestville, Québec 172/H1
Forestville, Québec 174/D3
Forestville, Wis. (54213) 317/L6
Forez (mts.), France 28/E5
Forfar, Scotland 10/E2
Forfar, Scotland 15/F4
Forgan, Okla. (73938) 288/E1
Forgan, Sask. 181/D4
Forget, Sask. 181/J6
Forge Village, Mass. (01828) 249/H2
Forillon Nat'l Park, Que. 162/K6
Forillon Nat'l Park, Québec 172/D1
Foristell, Mo. (63348) 261/L5
Fork, N.C. (†27028) 281/J3
Fork, S.C. (29543) 296/J3
Forked (lake), N.Y. 276/L3
Forked Deer (riv.), Tenn. 237/C9
Forked Deer, Middle Fork (riv.), Tenn. 237/C9
Forked Deer, North Fork (riv.), Tenn. 237/C8
Forked Deer, South Fork (riv.), Tenn. 237/C9
Forked River, N.J. (08731) 273/E4
Fork Lake, Alberta 182/E2
Forkland, Ala. (36740) 195/C5
Fork Mountain, Tenn. (†37728) 237/N8
Fork River, Manitoba 179/B3
Forks, Wash. (98331) 310/A3
Forks of Buffalo, Va. (†24521) 307/K5
Forks of Elkhorn, Ky. (†40601) 237/M4
Forks of Salmon, Calif. (96031) 204/B2
Forksville, Pa. (18616) 294/J3
Fork Union, Va. (23055) 307/M5
Forkville, Miss. (39076) 256/E6
Forlì (prov.), Italy 34/D2
Forlì, Italy 34/D2
Forman, N. Dak. (58032) 282/P7
Formartine (dist.), Scotland 15/F3
Formby, England 13/G2
Formby, England 10/F2
Formby (head), England 13/G2
Formentera (isl.), Spain 33/G3
Formentor (cape), Spain 33/H2
Formia, Italy 34/D4
Formiga, Brazil 135/D2
Formiga, Brazil 132/E8
Formosa (prov.), Argentina 143/D1
Formosa, Argentina 143/E2
Formosa, Argentina 120/D5
Formosa, Ark. (†72031) 202/E3
Formosa, Brazil 132/E6
Formosa, Serra (range), Brazil 132/C5
Formosa (Taiwan) (isl.), China 2/R4
Formosa (Taiwan) (isl.), China 77/K7
Formosa (Taiwan) (str.), China 77/J7
Formosa (bay), Kenya 115/H4
Formosa, Ontario 177/C3
Formosa, Paraguay 144/C5
Formoso, Kansas (66942) 232/D2
Forney, Ala. (†30124) 195/H2
Forney, Texas (75126) 303/H5
Forres, Scotland 10/E2
Forres, Scotland 15/E3
Forrest, Ill. (61741) 222/E3
Forrest (co.), Miss. 256/F8
Forrest, N. Mex. (†88401) 274/F4
Forrest (lake), Sask. 181/L3
Forrest, W. Australia 88/D6
Forrest (lakes), W. Australia 88/D5
Forrest, W. Australia 92/D5
Forrest City, Ark. (72335) 202/J3
Forreston, Ill. (61030) 222/D1
Forrest River Aboriginal Res., W. Australia 92/D1
Forrest River Mission, W. Australia 92/D1
Forrest Station, Manitoba 179/C5
Forsan, Texas (79733) 303/C5
Forsayth, Queensland 95/B3
Forsayth, Queensland 88/G3
Forshaga, Sweden 18/H7
Forssa, Finland 18/N6
Forst, E. Germany 22/F3
Forster-Tuncurry, N.S. Wales 97/G3
Forsyth (co.), Georgia 217/D2
Forsyth, Georgia (31029) 217/E4
Forsyth, Ill. (62535) 222/D4
Forsyth, Mo. (65653) 261/F9
Forsyth, Mont. (59327) 262/K4
Forsyth (co.), N.C. 281/J2
Fort (pt.), St. Chris.-Nevis 161/C11
Fort (mt.), Switzerland 39/D4
Fort A.P. Hill, Va. 307/O4
Fort Adams, Miss. (39336) 256/B8
Fort à la Corne, Sask. 181/G2
Fort Albany, Ont. 146/K4
Fort Albany, Ont. 162/H5
Fort Albany, Ontario 175/D2
Fort Alexander, Manitoba 179/F4
Fortaleza, Bolivia 136/C1
Fortaleza, Bolivia 136/B3
Fortaleza, Brazil 132/G3
Fortaleza, Brazil 120/F3
Fortaleza, Brazil 2/H6
Fortaleza de Santa Teresa, Uruguay 145/F5
Fort Ann, N.Y. (12827) 276/N4
Fort Apache, Ariz. (85926) 198/F5
Fort Apache Ind. Res., Ariz. 198/E5
Fort Ashby, W. Va. (26719) 312/J4
Fort Assiniboine, Alberta 182/C2
Fort Atkinson, Iowa (52144) 229/J2
Fort Atkinson, Wis. (53538) 317/J10
Fort Augustus, Scotland 10/D2
Fort Augustus, Scotland 15/D3
Fort Battleford Nat'l Hist. Park, Sask. 181/C3
Fort Bayard, N. Mex. (88036) 274/A6
Fort Beaufort, S. Africa 118/D6
Fort Beauséjour Nat'l Hist. Park, New Bruns 170/F3
Fort Belknap, Mont. (†59526) 262/H2
Fort Belknap Ind. Res., Mont. 262/H2
Fort Belvoir, Va. 307/O3
Fort Bend (co.), Texas 303/J8
Fort Benjamin Harrison, Ind. 227/E5
Fort Benning, Georgia 217/B6
Fort Benton, Mont. (59442) 262/F3
Fort Berthold Ind. Res., N. Dak. 282/E4
Fort Bidwell, Calif. (96112) 204/E2
Fort Bidwell Ind. Res., Calif. 204/E2
Fort Blackmore, Va. (24250) 307/C7
Fort Bliss, Texas 303/A10
Fort Bliss Mil. Res., N. Mex. 274/C6
Fort Bowie Nat'l Hist. Site, Ariz. 198/F6
Fort Bragg, Calif. (95437) 204/B4
Fort Bragg, N.C. 281/M4
Fort Branch, Ind. (47648) 227/B8
Fort Bridger, Wyo. (82933) 319/B4
Fort Calhoun, Nebr. (68023) 264/J3
Fort Campbell, Ky. 237/G7
Fort Campbell, Tenn. 237/G7
Fort Carlton Hist. Park, Sask. 181/E3
Fort Caroline Nat'l Mem., Fla. 212/E1
Fort Carson, Colo. 208/K5
Fort Chaffee, Ark. 202/B3
Fort Chambly Nat'l Hist. Park, Québec 172/J4
Fort-Chimo, Que. 162/K4
Fort-Chimo, Québec 174/F2
Fort Chipewyan, Alberta 182/C5
Fort Chipewyan, Alta 162/E4
Fort Chipewyan, Alta. 146/G4
Fort Churchill, Manitoba 179/K2
Fort Clark, N. Dak. (†58571) 282/H5
Fort Clatsop Nat'l Mem., Oreg. 291/C1
Fort Cobb, Okla. (73038) 288/K4
Fort Cobb (res.), Okla. 288/J4
Fort Collins, Colo. 146/H6
Fort Collins, Colo. 188/E2
Fort Collins, Colo. (80521) 208/J1
Fort Covington○, N.Y. (12937) 276/M1
Fort-Dauphin (Faradofay), Madagascar 118/H5
Fort Davis, Ala. (36031) 195/G6
Fort Davis, Alaska 196/E2
Fort Davis, Texas (79734) 303/D11
Fort Davis Nat'l Hist. Site, Texas 303/D11
Fort Defiance, Ariz. (86504) 198/F3
Fort Defiance, Va. (24437) 307/L4
Fort-de-France (cap.), Martinique 161/C6
Fort-de-France (cap.), Martinique 156/G4
Fort-de-France (bay), Martinique 161/C6
Fort Denaud, Fla. (†33935) 212/E5
Fort Deposit, Ala. (36032) 195/E7
Fort-Desaix, Martinique 161/D6
Fort Detrick, Md. 245/J3
Fort Devens, Mass. 249/H2
Fort Dick, Calif. (95531) 204/A2
Fort Dix, N.J. 273/D3
Fort Dodge, Iowa (50501) 229/E3
Fort Dodge, Iowa 188/H2
Fort Dodge, Kansas (67843) 232/C4
Fort Donelson Nat'l Mil. Park, Tenn. 237/F8
Fort Drum, Fla. (†33472) 212/F4
Fort Drum, N.Y. 276/J2
Fort Duchesne, Utah (84026) 304/E3
Forteau, Newf. 166/C3
Fort Edward, N.Y. (12828) 276/O4
Forte República, Angola 115/C5
Forte República, Angola 102/D5
Fort Erie, Ontario 177/E5
Fortescue, Mo. (64452) 261/B2
Fortescue, N.J. (08321) 273/C5
Fortescue (riv.), W. Australia 88/B4
Fortescue (riv.), W. Australia 92/B3
Fort Eustis, Va. 307/P6
Fort Fairfield, Maine (04742) 243/H2
Fort Fairfield○, Maine (04742) 243/H2
Fort Foote, Md. (†20022) 245/F6
Fort-Foureau (Kousseri), Cameroon 115/B1
Fort Frances, Ont. 162/G6
Fort Frances, Ontario 177/F5
Fort Frances, Ontario 175/B3
Fort Franklin, N.W.T. 162/D3
Fort Franklin, N.W. Terrs. 187/F3
Fort Fraser, Br. Col. 184/E3
Fort Frederica Nat'l Mon., Georgia 217/K8
Fort Fred Steele, Wyo. (†82301) 319/E4
Fort Gaines, Ala. 195/B10
Fort Gaines, Georgia (31751) 217/C7
Fort Garland, Colo. (81133) 208/J8
Fort Gay, W. Va. (25514) 312/A6
Fort-George, Que. 146/K4
Fort-George, Que. 162/J5
Fort-George, Québec 174/B2
Fort George G. Meade, Md. 245/L4
Fort Gibson, Okla. (74434) 288/R3
Fort Gibson (lake), Okla. 288/R2
Fort Good Hope, N.W.T. 162/D2
Fort Good Hope, N.W. Terrs. 187/F3
Fort Gordon, Georgia 217/H4
Fort-Gouraud (Fdérik), Mauritania 106/B4
Fort Grant, Ariz. (85643) 198/E6
Fort Greely, Alaska 196/J2
Fort Green, Fla. (33834) 212/E4
Forth, Scotland 15/C2
Forth (firth), Scotland 10/E2
Forth (firth), Scotland 15/F4
Forth (riv.), Scotland 15/B1
Forth (riv.), Scotland 10/C1
Forth, Tasmania 99/C3
Forth (riv.), Tasmania 99/C3
Fort Hall, Idaho (83203) 220/F6
Fort Hall, Kenya 115/G4
Fort Hall Ind. Res., Idaho 220/F6
Fort Hancock, N.J. 273/F3
Fort Hancock, Texas (79839) 303/B11
Forth and Clyde (canal), Scotland 10/B1
Forth and Clyde (canal), Scotland 15/B2
Fort Hertz (Putao), Burma 72/C1
Fort Hood, Texas 303/G6
Fort Howard, Md. (21052) 245/N4
Fort Huachuca, Ariz. 198/E7
Fort Hunter Liggett, Calif. 204/D8
Fortierville, Québec 172/F3
Fortín 10 de Octubre, Paraguay 144/A2
Fortín Ávalos Sánchez, Paraguay 144/C3
Fortín Boquerón, Paraguay 144/C3
Fortín Buenos Aires, Paraguay 144/B3
Fortín Campero, Bolivia 136/C8
Fortín Capitán Escobar, Paraguay 144/B3
Fortín Carlos Antonio López (Pitiantuta), Paraguay 144/C2
Fortín Casanillo, Paraguay 144/C3
Fortín Coronel Bogado, Paraguay 144/C2
Fortín Coronel Sánchez, Paraguay 144/C1
Fortín de las Flores, Mexico 150/P2
Fort Independence Ind. Res., Calif. 204/G7
Fortine, Mont. (59918) 262/A2
Fortín Falcon, Paraguay 144/C3
Fortín Florida, Paraguay 144/C2
Fortín Galpón, Paraguay 144/C1
Fortín General Bruguez, Paraguay 144/C4
Fortín General Caballero, Paraguay 144/C4
Fortín General Delgado, Paraguay 144/B4
Fortín General Díaz, Paraguay 144/B3
Fortín Guaraní, Paraguay 144/C3
Fortín Hernandarias, Paraguay 144/A2
Fortín Infante Rivarola, Paraguay 144/A2
Fortín Isla Poí, Paraguay 144/C3
Fortín Lagerenza i, Paraguay 144/B2
Fortín Madrejón, Paraguay 144/B2
Fortín Max Paredes, Bolivia 136/F6
Fortín Mayor Alberto Gardel, Paraguay 144/A3
Fortín Mayor Rodríguez, Paraguay 144/B3
Fortín Mutum, Bolivia 136/F6
Fortín Nueva Asunción (Picuiba), Paraguay 144/B2
Fortín Olmos, Argentina 143/F4
Fortín Palmar de las Islas, Paraguay 144/B1
Fortín Pilcomayo, Paraguay 144/B3
Fortín Presidente Ayala, Paraguay 144/C3
Fortín Presidente Cardozo, Paraguay 144/B5
Fortín Ravelo, Bolivia 136/E6
Fortín Santiago Rodríguez, Paraguay 144/A2
Fortín Suárez Arana, Bolivia 136/F6
Fortín Teniente Américo Picco, Paraguay 144/C1
Fortín Teniente E. Ochoa, Paraguay 144/B2
Fortín Teniente Esteban Martínez, Paraguay 144/C3
Fortín Teniente Gabino Mendoza, Paraguay 144/B2
Fortín Teniente Juan E. López, Paraguay 144/B2
Fortín Teniente Martínez, Paraguay 144/C2
Fortín Teniente Montenía, Paraguay 144/B3
Fortín Teniente Primero Anselmo Escobar, Paraguay 144/A3
Fortín Teniente Primero M. Cabello, Paraguay 144/B3
Fortín Teniente Primero Ramiro Espínola, Paraguay 144/B2
Fortín Toledo, Paraguay 144/B3
Fortín Torres, Paraguay 144/C2
Fortín Vanguardia, Bolivia 136/F6
Fortín Zalazar, Paraguay 144/C3
Fort Jackson, N.Y. (12938) 276/L1
Fort Jackson, S.C. 296/F4
Fort Jefferson Nat'l Mon., Fla. 212/C7
Fort Jennings, Ohio (45844) 284/B4
Fort Jesup, La. (†71449) 238/C3
Fort Johnson, N.Y. (12070) 276/M5
Fort Jones, Calif. (96032) 204/C2
Fort Kent, Alberta 182/E2
Fort Kent, Maine (04743) 243/F1
Fort Kent○, Maine (04743) 243/F1
Fort Kent Mills, Maine (04744) 243/F1
Fort Klamath, Oreg. (97626) 291/E5
Fort Knox, Ky. (40121) 237/K5
Fort Lallemand, Algeria 106/F2
Fort Lamar, Georgia (†30633) 217/F2
Fort Langley, Br. Col. 184/L3
Fort Laramie, Wyo. (82212) 319/H3
Fort Laramie Nat'l Hist. Site, Wyo. 319/H3
Fort Larned Nat'l Hist. Site, Kansas 232/C3
Fort Lauderdale, Fla. 188/K5
Fort Lauderdale, Fla. (*33301) 212/C4
Fort Lawn, S.C. (29714) 296/F2
Fort Lawrence, Nova Scotia 168/D3
Fort Leavenworth, Kansas (66027) 232/H2
Fort Lee, N.J. (07024) 273/C2
Fort Lee, Va. 307/O6
Fort Leonard Wood, Mo. 261/H7
Fort Lesley J. McNair, D.C. 245/E5
Fort Lewis, Wash. 310/C3
Fort Liard, N.W.T. 146/F3
Fort Liard, N.W.T. 162/D3
Fort Liard, N.W. Terrs. 187/F3
Fort Liberté, Haiti 156/D3
Fort Liberté, Haiti 158/C5
Fort Littleton, Pa. (17223) 294/F5
Fort Loramie, Ohio (45845) 284/B5
Fort Loudon, Pa. (17224) 294/G6
Fort Loudoun (lake), Tenn. 237/N9
Fort Lupton, Colo. (80621) 208/K2
Fort Lyon, Colo. (81038) 208/N6
Fort MacArthur, Calif. 204/C11
Fort Mackay, Alta. 162/E4
Fort Macleod, Alberta 182/D5
Fort Macleod, Alta. 162/E6
Fort MacMahon, Algeria 106/E3
Fort Madison, Iowa 188/H2
Fort Madison, Iowa (52627) 229/L7
Fort Matanzas Nat'l Mon., Fla. 212/E2
Fort McClellan Mil. Res., Ala. 195/G3
Fort McCoy, Fla. (32637) 212/E2
Fort McCoy, Wis. 317/E7
Fort McDermitt Ind. Res., Nev. 266/D1
Fort McDowell Ind. Res., Ariz. 198/D5
Fort McHenry Nat'l Mon., Md. 245/M3
Fort McKavett, Texas (76841) 303/E7
Fort McKay, Alberta 182/E1
Fort McKinley, Ohio (†45426) 284/B6
Fort McMurray, Alberta 182/E1
Fort McMurray, Alta. 146/G4
Fort McMurray, Alta. 162/E4
Fort McPherson, Georgia 217/K1
Fort McPherson, N.W.T. 146/E3
Fort McPherson, N.W.T. 162/C2
Fort McPherson, N.W. Terrs. 187/E3
Fort Meade, Fla. (33841) 212/E4
Fort Meade, Uruguay (57741) 298/C5
Fort Mill, S.C. (29715) 296/F1
Fort Miribel, Algeria 106/E3
Fort Mitchell, Ala. (36856) 195/H6
Fort Mitchell, Ky. (41017) 237/S2
Fort Mitchell, Va. (23941) 307/M7
Fort Mohave Ind. Res., Ariz. 198/A4
Fort Mohave Ind. Res., Calif. 204/L9
Fort Mohave Ind. Res., Nev. 266/G7
Fort Monmouth, N.J. 273/E3
Fort Monroe, Va. 307/R6
Fort Morgan, Ala. 195/C10
Fort Morgan, Colo. (80701) 208/M2
Fort Motte, S.C. (29050) 296/F4
Fort Myer, Va. 307/T2
Fort Myers, Fla. 188/K5
Fort Myers, Fla. (*33901) 212/E5
Fort Myers Beach, Fla. (33931) 212/E5
Fort Necessity, La. (71243) 238/G2
Fort Necessity Nat'l Battlefield, Pa. 294/C6
Fort Nelson, Br. Col. 146/F4
Fort Nelson, Br. Col. 162/D4
Fort Nelson, Br. Col. 184/M2
Fort Nelson (riv.), Br. Col. 184/M2
Fort Niagara, N.Y. 276/C4
Fort Norman, N.W.T. 162/D3
Fort Norman, N.W.T. 146/F3
Fort Norman, N.W. Terrs. 187/F3
Fort Ogden, Fla. (33842) 212/E4
Fort Oglethorpe, Georgia (30742) 217/B1
Fort Ord, Calif. 204/D7
Fort Payne, Ala. (35967) 195/G2
Fort Pearce Wash (dry riv.), Ariz. 198/B2
Fort Pearce Wash (creek), Utah 304/A6
Fort Peck, Mont. (59223) 262/K2
Fort Peck (dam), Mont. 262/K3
Fort Peck (lake), Mont. 262/K3
Fort Peck (lake), Mont. 146/H5
Fort Peck (lake), Mont. 188/E1
Fort Pickett, Va. 307/N6
Fort Pierce, Fla. 188/K5
Fort Pierce, Fla. (*33450) 212/F4
Fort Pierre, S. Dak. (57532) 298/H5
Fort Pillow, Tenn. (38032) 237/B9
Fort Pitt Hist. Park, Sask. 181/B2
Fort Plain, N.Y. (13339) 276/L5
Fort Point Nat'l Hist. Site, Calif. 204/J2
Fort Polk, La. 238/D4
Fort Portal, Uganda 115/F3
Fort Providence, N.W.T. 162/E3
Fort Providence, N.W. Terrs. 187/G3
Fort Pulaski Nat'l Mon., Georgia 217/L6
Fort Qu'Appelle, Sask. 181/H5
Fort Raleigh Nat'l Hist. Site, N.C. 281/T3
Fort Randall (dam), S. Dak. 298/N7
Fort Ransom, N. Dak. (58033) 282/P6
Fort Recovery, Ohio (45846) 284/A5
Fort Resolution, N.W.T. 146/G3
Fort Resolution, N.W.T. 162/E3
Fort Resolution, N.W. Terrs. 187/G3
Fortress (mt.), Wyo. 319/C1
Fort Rice, N. Dak. (58537) 282/J6
Fort Richardson, Alaska 196/C1
Fort Riley-Camp Whiteside, Kansas 232/F2
Fort Ripley, Minn. (56449) 255/D4
Fort Ritchie, Md. 245/H2
Fort Ritner, Ind. (47430) 227/E7
Fort Robinson, Nebr. (†69339) 264/A2
Fort Rock, Oreg. (97735) 291/G4
Fort Rodman, Mass. 249/L6
Fortrose, Scotland 10/D2
Fortrose, Scotland 15/D3
Fort Rosebery (Mansa), Zambia 115/E6
Fort Ross, Calif. (†95421) 204/B5
Fort Rucker, Ala. 195/G8
Fort-Rupert, Que. 146/L4
Fort Rupert (Rupert House), Québec 174/B2
Fort Saint James, Br. Col. 162/D5
Fort Saint James, Br. Col. 184/E3
Fort Saint John, Br. Col. 162/D4
Fort Saint John, Br. Col. 184/G2
Fort Sam Houston, Texas 303/K11
Fort San, Sask. 181/H5
Fort Sandeman, Pakistan 68/B2
Fort Sandeman, Pakistan 59/J3
Fort Saskatchewan, Alberta 182/D3
Fort Saskatchewan, Alta. 162/E5
Fort Scott, Kans. 188/H3
Fort Scott, Kansas (66701) 232/H4
Fort Seneca, Ohio (44829) 284/D3
Fort Severn, Ontario 175/C1
Fort Seward, Calif. (†95440) 204/B3
Fort Seybert, W. Va. (26806) 312/H5
Fort Shafter, Hawaii 218/C3
Fort Shaw, Mont. (59443) 262/E3
Fort Shawnee, Ohio (†45801) 284/B4
Fort Sheridan, Ill. 222/B5
Fort-Shevchenko, U.S.S.R. 48/F5
Fort Sill, Okla. 288/K5
Fort Simpson, Canada 4/C15
Fort Simpson, N.W.T. 146/F3
Fort Simpson, N.W.T. 162/D3
Fort Simpson, N.W. Terrs. 187/F3
Fort Smith, Ark. 188/H3
Fort Smith, Ark. 146/J6
Fort Smith, Ark. (*72901) 202/B3
Fort Smith, Mont. (†59075) 262/J5
Fort Smith, N.W.T. 146/G3
Fort Smith, N.W.T. 162/E3
Fort Smith (dist.), N.W. Terrs. 187/G3
Fort Smith, N.W. Terrs. 187/G4
Fort Smith Nat'l Hist. Site, Ark. 202/B3
Fort Spring, W. Va. (24936) 312/E7
Fort Stanton, N. Mex. (88323) 274/D5
Fort Stanwix Nat'l Mon., N.Y. 276/J4
Fort Steele, Br. Col. 184/K5
Fort Stewart, Georgia 217/J7
Fort Stewart, Ontario 177/G2
Fort Stockton, Texas (79735) 303/A7
Fort Story, Va. 307/S7
Fort Sumner, N. Mex. (88119) 274/E4
Fort Sumter Nat'l Mon., S.C. 296/H6
Fort Supply, Okla. (73841) 288/G1
Fort Supply (lake), Okla. 288/G1
Fort Tarat, Algeria 106/F3
Fort Thomas, Ariz. (85536) 198/E5
Fort Thomas, Ky. (41075) 237/S2
Fort Thompson, S. Dak. (57339) 298/L5
Fort Ticonderoga, N.Y. (†12883) 276/O3
Fort Totten, N. Dak. (58335) 282/M4
Fort Totten Ind. Res., N. Dak. 282/N4
Fort Towson, Okla. (74735) 288/R7
Fortuna, Calif. (95540) 204/A3
Fortuna, Mo. (65034) 261/G5
Fortuna, N. Dak. (58844) 282/C2
Fortuna, Spain 33/F3
Fortuna Ledge, Alaska (99585) 196/F2
Fortune, Newf. 166/C4
Fortune (bay), Newf. 166/C4
Fort Union Nat'l Mon., N. Mex. 274/E3
Fort Union Trading Post Nat'l Hist. Site, Mont. 262/N2
Fort Union Trading Post Nat'l Hist. Site, N. Dak 282/B3
Fort Valley, Georgia (31030) 217/E5
Fort Vancouver Nat'l Hist. Site, Wash. 310/C5
Fort Vermilion, Alberta 182/B5
Fort Vermilion, Alta. 162/E4
Fort Vermilion, Alta. 146/G4
Fort Victoria, Zimbabwe 102/F7
Fort Victoria, Zimbabwe 118/E4
Fortville, Ind. (46040) 227/F5
Fort Wainwright, Alaska 196/J1
Fort Walsh Nat'l Hist. Park, Sask. 181/A6
Fort Walton Beach, Fla. (32548) 212/C6
Fort Washakie, Wyo. (82514) 319/C2
Fort Washington, Md. (†20744) 245/L6
Fort Washington Park, Md. 245/L6
Fort Wayne, Ind. 188/J2
Fort Wayne, Ind. (*46801) 227/G2
Fort Wellington, Guyana 131/C2
Fort White, Fla. (32038) 212/D2
Fort William, Scotland 10/D2
Fort William, Scotland 15/C4
Fort Wingate, N. Mex. (87316) 274/A3
Fort Worden, Wash. 310/C2
Fort Worth, Texas 146/J6
Fort Worth, Texas (*76101) 303/F2
Fort Worth, Texas 188/G4
Fort Wright, Ky. (†41011) 237/S2
Fort Yates, N. Dak. (58538) 282/J7
Forty Fort, Pa. (18704) 294/F7
Forty Mile (pt.), Mich. 250/F3
Fort Yukon, Alaska 146/D3
Fort Yukon, Alaska (99740) 196/J1
Fort Yukon, Alaska 188/D5
Fort Yukon, U.S. 4/C17
Forum, Ark. (†72740) 202/C1
Fosforescente (bay), P. Rico 161/A3
Foshan (Fatshan), China 77/H7
Fosheim (pen.), N.W. Terrs. 187/K1
Foss, Okla. (73647) 288/H4
Foss (res.), Okla. 288/H3
Fossano, Italy 34/A2
Fosses-La-Ville, Belgium 27/F8
Fossil (creek), Ariz. 198/D4
Fossil, Oreg. (97830) 291/G2
Fossil Butte Nat'l Mon., Wyo. 319/B4
Fossombrone, Italy 34/D3
Fosston, Minn. (56542) 255/C3
Fosston, Sask. 181/H3
Foster, Ind. (†47932) 227/C4
Foster, Ky. (41043) 237/N3
Foster, Mo. (64745) 261/D6
Foster, Nebr. (68737) 264/G2
Foster (co.), N. Dak. 282/N5
Foster, Okla. (73039) 288/M5
Foster, Oreg. (97345) 291/E3
Foster○, R.I. (02825) 249/H5

Foster (riv.), Sask. 181/M3
Foster (creek), S. Dak. 298/N4
Foster, W. Va. (25081) 312/C6
Foster, Wis. (†54758) 317/D6
Foster Center (Foster P.O.), R.I. (†02825) 249/H5
Foster City, Calif. (94404) 204/J2
Foster City, Mich. (49834) 250/B3
Fosters, Ala. (35463) 195/C4
Fosters, Mich. (†48415) 250/F5
Fosters Falls, Va. (24329) 307/G7
Fosterton, Sask. 181/C5
Foster Village, Hawaii (†96701) 218/B3
Fosterville, New Bruns. 170/C3
Fosterville, Tenn. (37063) 237/J9
Fostoria, Ala. (†36737) 195/E6
Fostoria, Iowa (51340) 229/C2
Fostoria, Kansas (66426) 232/F2
Fostoria, Mich. (48435) 250/F5
Fostoria, Ohio (44830) 284/D3
Fougamou, Gabon 115/A4
Fougères, France 28/C3
Fouke, Ark. (71837) 202/C7
Foul (bay), Egypt 111/G3
Foul (sound), Ireland 17/B5
Foula (isl.), Scotland 15/F2
Foula (isl.), Scotland 10/G1
Foules, La. (†71326) 238/G3
Foulness Island (pen.), England 13/J6
Foulpointe, Madagascar 118/H3
Foulweather (cape), Oreg. 291/C3
Foulwind (cape), N. Zealand 100/C4
Foumban, Cameroon 115/B2
Foumban, Cameroon 102/D4
Fountain, Ala. (36460) 195/D7
Fountain, Colo. (80817) 208/K5
Fountain (creek), Colo. 208/K5
Fountain, Fla. (32438) 212/D6
Fountain (co.), Ind. 227/C4
Fountain, Ind. (†47918) 227/C4
Fountain, Mich. (49410) 250/C4
Fountain, Minn. (55935) 255/F7
Fountain, N.C. (27829) 281/O3
Fountain City, Ind. (47341) 227/H5
Fountain City, Wis. (54629) 317/C7
Fountain Green, Ill. (†62321) 222/C3
Fountain Green, Utah (84632) 304/C4
Fountain Head, Md. (†21740) 245/G2
Fountain Head, Tenn. (†37148) 237/J7
Fountain Hill, Ark. (71642) 202/G7
Fountain Hill, Pa. (†18015) 294/L4
Fountain Inn, S.C. (29644) 296/C2
Fountain Run, Ky. (42133) 237/K7
Fountaintown, Ind. (46130) 227/F5
Fountain Valley, Calif. (92708) 204/D11
Four (peaks), Newf. 166/B2
Four Buttes, Mont. (59224) 262/L2
Fourche, Ariz. (†72016) 202/E4
Fourche LaFave (riv.), Ark. 202/D4
Four Corners, Oreg. (97301) 291/A3
Four Corners, Wyo. (82715) 319/H1
Four Falls, New Bruns. 170/C2
Four Hole Swamp (creek), S.C. 296/F5
Four Lakes, Wash. (99014) 310/H3
Fourmies, France 28/F2
Four Mile (riv.), Conn. 210/F3
Fourmile (lake), Oreg. 291/E5
Four Mountains (isls.), Alaska 196/E4
Fournier (cape), N. Zealand 100/E7
Fournier, Ontario 177/K2
Four Oaks, N.C. (27524) 281/M4
Four Paths, Jamaica 158/J6
Four Peaks (mt.), Ariz. 198/D5
Four States, W. Va. (26572) 312/F4
Fourteen Mile (pt.), Mich. 250/F1
Fourth (lake), Maine 243/H5
Fourth Cataract, Sudan 111/F4
Fourth Cataract, Sudan 59/B6
Fourth Cataract (dam), Sudan 102/F3
Foveaux (str.), N. Zealand 87/G10
Foveaux (str.), N. Zealand 100/A7
Fowler, Calif. (93625) 204/F7
Fowler, Colo. (81039) 208/L6
Fowler, Ill. (62338) 222/B3
Fowler, Ind. (47944) 227/C3
Fowler, Kansas (67844) 232/B4
Fowler, Mich. (48835) 250/E5
Fowlerton, Ind. (46930) 227/F4
Fowlerton, Texas (78021) 303/F9
Fowlerville, Mich. (48836) 250/F6
Fowlkes, Tenn. (38033) 237/C9
Fowlstown, Georgia (31752) 217/D9
Fowman, Iran 66/F2
Fowyang (Fuyang), China 77/J5
Fox (isls.), Alaska 196/E4
Fox, Ark. (72051) 202/F2
Fox (lake), Ill. 222/A4
Fox (riv.), Ill. 222/E5
Fox (riv.), Ill. 222/E2
Fox (riv.), Manitoba 179/K2
Fox, Mich. (†49813) 250/B3
Fox (isl.), New Bruns. 170/F1
Fox, Okla. (73435) 288/M6
Fox, Oreg. (97831) 291/H3
Fox (riv.), Wis. 317/K7
Fox (riv.), Wis. 317/K2
Foxboro, Mass. (02035) 249/J4
Foxboro○, Mass. (02035) 249/J4
Foxboro, Ontario 177/G3
Foxboro, Wis. (54836) 317/B2
Foxburg, Pa. (16036) 294/C3
Fox Chapel, Pa. (†15238) 294/C6
Fox Creek, Alberta 182/B2
Foxe (basin), Canada 4/C13
Foxe (basin), N.W.T. 146/L3
Foxe (basin), N.W.T. 162/J2
Foxe (basin), N.W. Terrs. 187/L3
Foxe (chan.), N.W.T. 162/H2
Foxe (chan.), N.W.T. 146/K3
Foxe (chan.), N.W. Terrs. 187/K3
Foxe (pen.), N.W.T. 146/L3
Foxe (pen.), N.W.T. 162/J3
Foxe (pen.), N.W. Terrs. 187/L3
Fox Farm, Wyo. (†82001) 319/H4
Foxfire, N.C. (†28373) 281/K4
Foxford, Ireland 17/C4
Foxford, Sask. 181/F2
Fox Glacier, N. Zealand 100/B5
Fox Harbour, Newf. 166/D2
Fox Harbour, Newf. 166/C3
Foxholm, N. Dak. (58738) 282/G3
Foxhome, Minn. (56543) 255/B4
Fox Lake, Alberta 182/B5
Fox Lake, Ill. (60020) 222/A4
Fox Lake, Wis. (53933) 317/J8
Foxon, Conn. (†06512) 210/D3
Foxpark, Wyo. (82057) 319/F4
Fox Point, Wis. (†53117) 317/M1
Fox River, Nova Scotia 168/D3
Fox River Grove, Ill. (60021) 222/A5
Foxton, Colo. (80441) 208/J4
Foxton, N. Zealand 100/E4
Fox Valley, Sask. 181/B5
Foxville, Md. (†21760) 245/H2
Foxwarren, Manitoba 179/A4
Foxwells, Va. (22578) 307/R5
Foxworth, Miss. (39483) 256/E8
Foyers, Scotland 15/D3
Foyil, Okla. (74031) 288/R2
Foyle (inlet), Ireland 17/G1
Foyle, Lough (inlet), Ireland 10/C3
Foyle (riv.), Ireland 17/G2
Foyle, Lough (inlet), N. Ireland 10/C3
Foyle (inlet), N. Ireland 17/G1
Foyle (riv.), N. Ireland 17/G2
Foynes, Ireland 10/B4
Foynes, Ireland 17/C6
Foz do Breu, Brazil 132/F10
Foz do Cunene, Angola 115/B7
Foz do Iguaçu, Brazil 132/C9
Frackville, Pa. (17931) 294/K4
Fraga, Spain 33/G2
Fragoso (cay), Cuba 158/F1
Fraile Muerto, Uruguay 145/E3
Frailes, Los (isl.), Dom. Rep. 158/C7
Fram, Paraguay 144/E5
Framboise, Nova Scotia 168/H3
Framboise Cove (bay), Nova Scotia 168/H3
Frame, W. Va. (25071) 312/C5
Frameries, Belgium 27/D8
Frametown, W. Va. (26623) 312/E5
Framingham○, Mass. (01701) 249/A7
Framingham Center, Mass. (01701) 249/J3
Framlingham, England 13/J5
Frampton, Québec 172/G3
Franca, Brazil 135/C2
Franca, Brazil 132/E8
Francavilla Fontana, Italy 34/F4
France 2/J3
France 7/E4
FRANCE 28
Francés (cape), Cuba 158/B2
Francés (cape), Cuba 158/A2
Frances (lake), Mont. 262/D2
Frances, Wash. (†98577) 310/B4
Frances (lake), Yukon 187/E3
Francestown○, N.H. (03043) 268/D6
Francés Viejo (cape), Dom. Rep. 158/E5
Francesville, Ind. (47946) 227/D3
Franceville, Gabon 115/B4
Franche Comté (trad. prov.), France 29
Francia, Uruguay 145/C3
Francis (lake), N.H. 268/E1
Francis, Okla. (74844) 288/N5
Francis, Sask. 181/H5
Francis, Utah (†84036) 304/C3
Francis Case (lake), S. Dak. 188/F2
Francis Case (lake), S. Dak. 146/J5
Francis Case (lake), S. Dak. 298/L7
Francisco, Ala. (†37345) 195/F1
Francisco, Ind. (47649) 227/B8
Francisco, N.C. (†27053) 281/J2
Francisco de Orellana, Peru 128/F4
Francisco I. Madero, Mexico 150/H4
Francis Creek, Wis. (54214) 317/L7
Francis E. Warren A.F.B., Wyo. 319/G4
Francistown, Botswana 118/D4
Francoeur, Québec 172/F3
Francois (lake), Br. Col. 162/D5
François (lake), Br. Col. 184/D3
François, Newf. 166/C4
François Lake, Br. Col. 184/D3
Franconia○, N.H. (03580) 268/D3
Franconia, Va. (22310) 307/S3
Franconian Jura (range), W. Germany 22/D4
Franconia Notch (pass), N.H. 268/D3
Franeker, Netherlands 27/H2
Frank, Alberta 182/C5
Frankel City, Texas (79737) 303/B5
Frankenberg-Eder, W. Germany 22/C3
Frankenmarkt, Austria 41/B3
Frankenmuth, Mich. (48734) 250/F5
Frankenthal, W. Germany 22/C4
Frankenwald (for.), W. Germany 22/D3
Frankewing, Tenn. (38459) 237/H10
Frankfield, Jamaica 158/H6
Frankford, Del. (19945) 245/S6
Frankford (Kilcormac), Ireland 17/F5
Frankford, Mo. (63441) 261/K4
Frankford, Ontario 177/G3
Frankford, W. Va. (24938) 312/F7
Frankfort, Ala. (†35653) 195/C1
Frankfort, Ill. (60423) 222/B6
Frankfort, Ind. (46041) 227/E4
Frankfort, Kansas (66427) 232/F2
Frankfort (cap.), Ky. (40601) 237/M4
Frankfort (cap.), Ky. 188/K3
Frankfort (cap.), Ky. 146/K6
Frankfort○, Maine (04438) 243/F6
Frankfort, Mich. (49635) 250/C4
Frankfort, N.Y. (13340) 276/K4
Frankfort, Ohio (45628) 284/D7
Frankfort, S. Dak. (57440) 298/N4
Frankfort Springs, Pa. (†15050) 294/A4
Frankfurt (dist.), E. Germany 22/F2
Frankfurt, W. Germany 7/E3
Frankfurt am Main, W. Germany 22/C3
Frankfurt an der Oder, E. Germany 22/F2
Frankland (cape), Tasmania 99/D1
Frankland (range), Tasmania 99/B4
Franklin (co.), Ala. 195/C2
Franklin (pt.), Alaska 196/G1
Franklin, Ala. (36444) 195/G6
Franklin, Ariz. (85534) 198/F6
Franklin (co.), Ark. 202/C2
Franklin, Ark. (72536) 202/G1
Franklin○, Conn. (†06254) 210/G2
Franklin (co.), Fla. 212/B2
Franklin (co.), Georgia 217/F2
Franklin, Georgia (30217) 217/B4
Franklin (co.), Idaho 220/G7
Franklin, Idaho (83237) 220/G7
Franklin (co.), Ill. 222/E5
Franklin, Ill. (62638) 222/C4
Franklin (co.), Ind. 227/G6
Franklin, Ind. (46131) 227/E6
Franklin (co.), Iowa 229/G3
Franklin, Iowa (†52625) 229/L7
Franklin (co.), Kansas 232/G3
Franklin, Kansas (66735) 232/H4
Franklin (co.), Ky. 237/M4
Franklin, Ky. (42134) 237/J7
Franklin (par.), La. 238/G2
Franklin, La. (70538) 238/G7
Franklin (co.), Maine 243/B5
Franklin, Maine (04634) 243/G6
Franklin○, Maine (04634) 243/G6
Franklin, Manitoba 179/C4
Franklin (co.), Mass. 249/D2
Franklin, Mass. (02038) 249/J4
Franklin○, Mass. (02038) 249/J4
Franklin, Mich. (48025) 250/B6
Franklin, Minn. (†55792) 255/F3
Franklin, Minn. (55333) 255/D6
Franklin (co.), Miss. 256/C8
Franklin (co.), Mo. 261/K6
Franklin, Mo. (65250) 261/G4
Franklin, Mont. (†59074) 262/G4
Franklin (co.), Nebr. 264/F4
Franklin, Nebr. (68939) 264/E4
Franklin (lake), Nev. 266/F2
Franklin, N.H. (03235) 268/D5
Franklin, N.J. (07416) 273/D1
Franklin (co.), N.Y. 276/M1
Franklin, N.Y. (13775) 276/K6
Franklin (co.), N.C. 281/N2
Franklin, N.C. (28734) 281/C4
Franklin (dist.), N.W.T. 162/H1
Franklin (bay), N.W.Terrs. 187/F2
Franklin (lake), N.W. Terrs. 187/J3
Franklin (mts.), N.W. Terrs. 187/F3
Franklin (str.), N.W. T. 162/G1
Franklin (str.), N.W. Terrs. 187/J2
Franklin (co.), Ohio 284/E5
Franklin, Ohio (45005) 284/B6
Franklin (co.), Pa. 294/G6
Franklin, Pa. (16323) 294/C3
Franklin, S. Dak. (†57042) 298/P6
Franklin, Tasmania 99/C5
Franklin (riv.), Tasmania 99/B4
Franklin (co.), Tenn. 237/J10
Franklin, Tenn. (37064) 237/H9
Franklin (co.), Texas 303/J4
Franklin, Texas (77856) 303/H7
Franklin (co.), Vt. 268/B2
Franklin○, Vt. (05457) 268/B2
Franklin (co.), Va. 307/J6
Franklin (I.C.), Va. (23851) 307/P7
Franklin (co.), Wash. 310/G4
Franklin, W. Va. (26807) 312/H5
Franklin, Wis. (53132) 317/L2
Franklin D. Roosevelt (lake), Wash. 310/G2
Franklin Falls (res.), N.H. 268/D4
Franklin Furnace, Ohio (45629) 284/E8
Franklin Grove, Ill. (61031) 222/D2
Franklin Lakes, N.J. (07417) 273/B1
Franklin Park, Ill. (60131) 222/B5
Franklin Park○, N.J. (†08823) 273/D3
Franklin River, Br. Col. 184/H3
Franklin Springs, Georgia (30639) 217/F2
Franklin Square, N.Y. (11010) 276/R7
Franklinton, La. (70438) 238/K5
Franklinton, N.C. (27525) 281/N2
Franklintown, Pa. (17323) 294/H5
Franklinville, N.J. (08322) 273/C4
Franklinville, N.Y. (14737) 276/D6
Franklinville, N.C. (27248) 281/K3
Franks (pond), Newf. 166/D2
Frankslake, Sask. 181/G5
Frankston, Texas (75763) 303/J5
Franksville, Wis. (53126) 317/M3
Frankton, Ind. (46044) 227/F4
Franktown, Colo. (80116) 208/K4
Franktown, Ontario 177/H2
Franktown, Va. (23354) 307/S6
Frankville, Ala. (36538) 195/B7
Frankville, Iowa (†52162) 229/K2
Frankville, Nova Scotia 168/G3
Frankville, Ontario 177/J2
Frannie, Wyo. (82423) 319/D1
Franquelin, Québec 172/B1
Franquia, Uruguay 145/B1
Franschhoek, S. Africa 118/F6
Fransfontein, Namibia 118/A4
Františkovy Lázně, Czech. 41/B1
Franz, Ontario 177/J5
Franz, Ontario 175/D3
Franz Josef Land (isls.), U.S.S.R. 2/L1
Franz Josef Land (isls.), U.S.S.R. 4/A7
Franz Josef Land (isls.), U.S.S.R. 48/F1
Frascati, Italy 34/F7
Fraser (isl.), Australia 87/F8
Fraser (riv.), Br. Col. 146/F4
Fraser (riv.), Br. Col. 162/D5
Fraser (lake), Br. Col. 184/E3
Fraser (riv.), Br. Col. 184/F4
Fraser, Colo. (80442) 208/H3
Fraser, Iowa (†50036) 229/E4
Fraser, Mich. (48026) 250/B6
Fraser, Minn. (†55719) 255/F3
Fraser (riv.), Newf. 166/B2
Fraser (isl.), Queensland 88/J4
Fraser (isl.), Queensland 95/E5
Fraserburgh, Scotland 15/G3
Fraserburgh, Scotland 10/E2
Fraserdale, Ontario 175/D3
Fraserdale, Ontario 177/J5
Fraser Lake, Br. Col. 184/E3
Fraser Mills, Br. Col. 184/K3
Fraser Reach (chan.), Br. Col. 184/C3
Frasertown, N. Zealand 100/F3
Fraserwood, Manitoba 179/E4
Frasnes-lez Anvaing, Belgium 27/D7
Frauenfeld, Switzerland 39/G1
Frauenkirchen, Austria 41/D3
Fray Benito, Cuba 158/J3
Fray Bentos, Uruguay 145/A4
Fray Marcos, Uruguay 145/D5
Frazee, Minn. (56544) 255/C4
Frazer, Mont. (59225) 262/K2
Frazeysburg, Ohio (43822) 284/F5
Frazier Park, Calif. (93225) 204/F9
Fraziers Bottom, W. Va. (25082) 312/B5
Frechen, W. Germany 22/B3
Fred, Texas (77616) 303/K7
Freda, N. Dak. (†58569) 282/H7
Fredensborg, Denmark 21/F6
Fredensdal, Virgin Is. (U.S.) 161/F4
Frederic, Mich. (49733) 250/E4
Frederic, Wis. (54837) 317/B4
Frederica, Del. (19946) 245/S4
Fredericia, Denmark 21/C6
Fredericia, Denmark 18/F9
Frederick (sound), Alaska 196/N1
Frederick, Colo. (80530) 208/K2
Frederick, Ill. (62639) 222/C3
Frederick, Kansas (†67444) 232/D3
Frederick (co.), Md. 245/J3
Frederick, Md. (21701) 245/J3
Frederick, Okla. (73542) 288/H6
Frederick, S. Dak. (57441) 298/N2
Frederick (co.), Va. 307/M2
Fredericksburg, Ind. (47120) 227/E8
Fredericksburg, Iowa (50630) 229/J3
Fredericksburg, Ohio (44627) 284/G4
Fredericksburg, Pa. (17026) 294/J5
Fredericksburg, Pa. (†16335) 294/B2
Fredericksburg, Texas (78624) 303/E7
Fredericksburg (I.C.), Va. (*22401) 307/N4
Fredericks Hall, Va. (†23117) 307/N4
Frederickton, N.S. Wales 97/G2
Fredericktown, Mo. (63645) 261/M7
Fredericktown, Ohio (43019) 284/F5
Fredericktown, Pa. (15333) 294/C6
Fredericton, N. Br. 146/M5
Fredericton, N. Br. 162/K6
Fredericton (cap.), New Bruns. 170/D3
Fredericton Junction, New Bruns. 170/D3
Frederika, Iowa (50631) 229/J3
Frederik Hendrik (Kolepom) (isl.), Indonesia 85/K7
Frederiksberg (commune), Denmark 21/F6
Frederiksberg, Denmark 21/F6
Frederiksborg (co.), Denmark 21/E5
Frederikshåb, Greenl. 4/C12
Frederikshåb, Greenland 146/N3
Frederikshavn, Denmark 18/G8
Frederikshavn, Denmark 21/D3
Frederikssund, Denmark 21/E6
Frederiksted, Virgin Is. (U.S.) 161/E4
Frederiksted, Virgin Is. (U.S.) 156/G2
Frederiksvaerk, Denmark 21/E6
Frederiksvaerk, Denmark 18/G8
Frederik Willem IV (falls), Suriname 131/C4
Fredonia, Ala. (†31833) 195/H5
Fredonia, Ariz. (86022) 198/C2
Fredonia (Biscoe), Ark. (72017) 202/H4
Fredonia, Ind. (†47137) 227/E8
Fredonia, Iowa (†52738) 229/L6
Fredonia, Kansas (66736) 232/G4
Fredonia, Ky. (42411) 237/E6
Fredonia, N.Y. (14063) 276/B6
Fredonia, N. Dak. (58440) 282/M7
Fredonia, Pa. (16124) 294/B3
Fredonia, Texas (76842) 303/E7
Fredonia, Wis. (53021) 317/L8
Fredric, Iowa (†52531) 229/H6
Fredrika, Sweden 18/L4
Fredrikstad, Norway 18/D4
Freeborn (co.), Minn. 255/E7
Freeborn, Minn. (56032) 255/E7
Freeburg, Ill. (62243) 222/D5
Freeburg, Minn. (†55921) 255/G7
Freeburg, Mo. (65035) 261/J6
Freeburg, Pa. (17827) 294/H4
Freeburn, Ky. (41528) 237/S5
Freedhem, Minn. (†56345) 255/D4
Freedom, Calif. (95019) 204/L4
Freedom, Ind. (47431) 227/D6
Freedom, Ky. (†42157) 237/K7
Freedom○, Maine (04941) 243/E7
Freedom○, N.H. (03836) 268/E4
Freedom, Okla. (73842) 288/H1
Freedom, Pa. (15042) 294/B4
Freedom, Wyo. (83120) 319/B3
Freehold, N.J. (07728) 273/E3
Freehold, N.Y. (12431) 276/N6
Freel (peak), Calif. 204/F5
Freeland, Md. (21053) 245/M2
Freeland, Mich. (48623) 250/E5
Freeland, N.C. (28440) 281/N6
Freeland, Pa. (18224) 294/L3
Freeland, Wash. (98249) 310/C2
Freeland Park, Ind. (†47944) 227/C3
Freelandville, Ind. (47535) 227/C7
Freels (cape), Newf. 166/D3
Freelton, Ontario 177/D4
Freeman (riv.), Alberta 182/C2
Freeman, Ind. (†47460) 227/D6
Freeman (lake), Ind. 227/D3
Freeman, Mo. (64746) 261/C5
Freeman, S. Dak. (57029) 298/O7
Freeman, Wash. (99015) 310/H3
Freemansburg, Pa. (†18017) 294/M4
Freemanville, Ala. (†36502) 195/D8
Free Mason (isls.), La. 238/M7
Freemont, Calif. 188/B3
Freemont, Sask. 181/B3
Freeport, Bahamas 156/B1
Freeport, Fla. (32439) 212/C6
Freeport, Ill. 188/J2
Freeport, Ill. (61032) 222/D1
Freeport, Ind. (†46161) 227/F5
Freeport, Kansas (67049) 232/E4
Freeport, Maine (04032) 243/C8
Freeport○, Maine (04032) 243/C8
Freeport, Mich. (49325) 250/D6
Freeport, Minn. (56331) 255/D5
Freeport, N.Y. (11520) 276/R7
Freeport, Nova Scotia 168/B4
Freeport, Ohio (43973) 284/H5
Freeport, Pa. (16229) 294/C4
Freeport, Texas (77541) 303/J9
Freer, Texas (78357) 303/F10
Free Soil, Mich. (49411) 250/C4
Freestone (co.), Texas 303/H6
Freetown, Ant. & Bar. 161/E11
Freetown, Ind. (47235) 227/E7
Freetown, N.Y. (†11937) 276/R9
Freetown (cap.), S. Leone 102/A4
Freetown (cap.), S. Leone 106/B7
Free Union, Va. (22940) 307/L4
Freeville, N.Y. (13068) 276/H5
Freezeout (lake), Mont. 262/D3
Fregenal de la Sierra, Spain 33/C3
Fregene, Italy 34/F6
Freiberg, E. Germany 22/E3
Freiburg, W. Germany 7/E4
Freiburg im Breisgau, W. Germany 22/B5
Freidberg, Austria 41/D3
Freienbach, Switzerland 39/G2
Freire, Chile 138/E2
Freirina, Chile 138/A7
Freising, W. Germany 22/D4
Freistadt, Austria 41/C2
Freistatt, Mo. (65654) 261/E8
Freital, E. Germany 22/E3
Freixo de Espada à Cinta, Portugal 33/C2
Fréjus, France 28/G6
Fréjus (pass), France 28/G5
Fréjus (pass), Italy 34/A2
Frelighsburg, Québec 172/E4
Fremantle, Australia 2/Q7
Fremantle, Australia 87/B9
Fremantle, W. Australia 88/B6
Fremantle, W. Australia 92/A1
Fremington, England 13/C6
Fremont, Calif. (*94536) 204/K3
Fremont (peak), Calif. 204/H8
Fremont (co.), Colo. 208/J5
Fremont (co.), Idaho 220/G5
Fremont, Ind. (46737) 227/H1
Fremont (co.), Iowa 229/B7
Fremont, Iowa (52561) 229/H6
Fremont, Mich. (49412) 250/D5
Fremont, Mo. (63941) 261/K9
Fremont, Nebr. 188/G2
Fremont, Nebr. (68025) 264/H3
Fremont○, N.H. (03044) 268/E6
Fremont, N.C. (27830) 281/N3
Fremont, Ohio (43420) 284/D3
Fremont, Utah (84727) 304/C5
Fremont (isl.), Utah 304/B2
Fremont (riv.), Utah 304/C5
Fremont, Wis. (54940) 317/J7
Fremont (co.), Wyo. 319/D2
Fremont (lake), Wyo. 319/C3
Fremont (peak), Wyo. 319/C2
French, Argentina 143/F7
French (riv.), Conn. 210/H1
French (riv.), Ontario 177/D1
French (creek), Pa. 294/C2
French (creek), S. Dak. 298/C6
French (isl.), Victoria 97/C6
Frenchboro○, Maine (04635) 243/G7
French Broad (riv.), N.C. 281/D3
French Broad (riv.), Tenn. 237/R9
Frenchburg, Ky. (40322) 237/O5
French Camp, Miss. (39745) 256/F4
French Creek, W. Va. (26218) 312/F5
French Frigate (shoal), Hawaii 188/F6
French Frigate (shoals), Hawaii 87/K3
French Frigate (shoals), Hawaii 218/C6
Frenchglen, Oreg. (97736) 291/H5
French Guiana 2/G5
French Guiana 120/D2
FRENCH GUIANA 131/E3
French Lick, Ind. (47432) 227/D7
Frenchman (creek), Colo. 208/P1
Frenchman (bay), Maine 243/G7
Frenchman (riv.), Mont. 188/E1
Frenchman (riv.), Mont. 262/J1
Frenchman (creek), Nebr. 264/C4
Frenchman (riv.), Sask. 181/C6
Frenchman (cay), Virgin Is. (Br.) 161/C4
Frenchman Butte, Sask. 181/B2
Frenchman Flat (basin), Nev. 266/F6
Frenchmans Cap (mt.), Tasmania 99/B4
Frenchmans Island, Newf. 166/C3
Frenchpark, Ireland 17/E4
French Polynesia 87/L8
French River, Minn. (†55801) 255/G4
French River, Ontario 177/D1
French Settlement, La. (70733) 238/L2
Frenchton, W. Va. (26219) 312/F5
Frenchtown, Mont. (59834) 262/B3
Frenchtown, N.J. (08825) 273/C2
Frenchville, Maine (04745) 243/G1
Frenchville○, Maine (04745) 243/G1
Frenchville, Pa. (16836) 294/F3
Frenštát pod Radhoštěm, Czech. 41/E2
Fresco, Ivory Coast 106/C7
Fresh (pond), Mass. 249/C6
Freshford, Ireland 17/G6
Freshwater, Calif. (†95501) 204/B3
Freshwater (Guffey), Colo. (80820) 208/H5
Freshwater, England 13/F7
Freshwater, Newf. 166/D2
Fresia, Chile 138/D3
Fresillo, Mexico 146/H7
Fresnillo de González Echeverría, Mexico 150/H5
Fresno (co.), Calif. 204/E7
Fresno, Calif. 146/G6
Fresno, Calif. 188/C3
Fresno, Calif. (*93706) 204/F7
Fresno (riv.), Calif. 204/E7
Fresno, Colombia 126/C5
Fresno, Mont. (†59532) 262/G2
Fresno (res.), Mont. 262/F2
Fresno, Texas (77545) 303/J2
Freudenstadt, W. Germany 22/C4
Frew, Ky. (41744) 237/P6
Frewena, North. Terr. 93/D5
Frewsburg, N.Y. (14738) 276/B6
Freycinet (pen.), Tasmania 99/E4
Fria, Guinea 106/B6
Fria (cape), Namibia 102/D6
Fria (cape), Namibia 118/A3
Friant, Calif. (93626) 204/F7
Friant-Kern (canal), Calif. 204/F8
Friars Point, Miss. (38631) 256/C2
Frias, Argentina 143/D2
Fribourg (canton), Switzerland 39/D3
Fribourg, Switzerland 39/D3
Frick, Switzerland 39/E1
Friday Harbor, Wash. (98250) 310/B2
Fridley, Minn. (55432) 255/G5
Fried, N. Dak. (†58401) 282/N5
Friedberg, W. Germany 22/C3
Friedland, E. Germany 22/E2
Friedrichshafen, W. Germany 22/C5
Friedrichstadt, W. Germany 22/C1
Friend, Kansas (67845) 232/B3
Friend, Nebr. (68359) 264/G4
Friend, Oreg. (97021) 291/F2
Friendly, W. Va. (26146) 312/D3
Friendship, Ark. (71942) 202/E5
Friendship, Ind. (47021) 227/G7
Friendship, Maine (04547) 243/E7
Friendship○, Maine (04547) 243/E7
Friendship, Md. (20758) 245/M6
Friendship, N.Y. (14739) 276/D6
Friendship, Ohio (45630) 284/D8
Friendship, Tenn. (38034) 237/C9
Friendship, Wis. (53934) 317/G8
Friendship Hill Nat'l Hist. Site, Pa. 294/C6
Friendsville, Ill. (†62863) 222/F5
Friendsville, Md. (21531) 245/A2
Friendsville, Pa. (18818) 294/L2
Friendsville, Tenn. (37737) 237/N9
Friendswood, Texas (77546) 303/J2
Frienisberg (mt.), Switzerland 39/D2
Frierson, La. (71027) 238/C2
Fries, Va. (24330) 307/F7
Friesach, Austria 41/C3
Friesche Gat (chan.), Netherlands 27/J2
Friesland, Minn. (†55037) 255/E4
Friesland (prov.), Netherlands 27/H2
Friesland, Wis. (53935) 317/H8
Frigate (isl.), Seychelles 118/J5
Frigate Bay, St. Chris.-Nevis 161/C10
Frimley and Camberley, England 13/G8
Frink, Fla. (†32430) 212/D6
Frinton and Walton, England 10/G5
Frinton and Walton, England 13/J6
Frio (cape), Brazil 120/E5
Frio (cape), Brazil 135/F3
Frio (co.), Texas 303/E9
Frio (riv.), Texas 303/E8
Friockheim, Scotland 15/F4
Friol, Spain 33/C1
Friona, Texas (79035) 303/B3
Fripp (isl.), S.C. 296/G7
Frisches Haff (lag.), Poland 47/D1
Frisco, Colo. (80443) 208/G3
Frisco, N.C. (27936) 281/T4
Frisco, Pa. (†16117) 294/B4
Frisco, Texas (75034) 303/H4
Frisco City, Ala. (36445) 195/D8
Frisian (isls.) 7/E3
Frisian, North (isls.), Denmark 21/B7
Frisian, West (isls.), Netherlands 27/G2
Frisian, East (isls.), W. Germany 22/B2
Frisian, North (isls.), W. Germany 22/B1
Frissell (mt.), Conn. 210/B1
Fristoe, Mo. (†65355) 261/F6
Fritch, Texas (79036) 303/C2
Fritchton, Ind. (†47591) 227/C7
Fritz Creek, Alaska (†99603) 196/B2
Fritzlar, W. Germany 22/C3
Friuli-Venezia Giulia (reg.), Italy 34/D1
Frizzellburg, Md. (†21157) 245/K2
Frobisher (bay), N.W.T. 162/K3
Frobisher (bay), N.W. Terrs. 187/M3
Frobisher, Sask. 181/J6
Frobisher (lake), Sask. 181/L3
Frobisher Bay, N.W.T. 162/K3
Frobisher Bay, N.W. Terrs. 187/M3
Froelich, Iowa (†52047) 229/L2
Frog (lake), Alberta 182/E3
Frog Lake, Alberta 182/E3
Frogmore, S.C. (29920) 296/F7
Frogue, Ky. (†42714) 237/L7
Frohavet (bay), Norway 18/F5
Frohna, Mo. (63748) 261/N7
Frohnleiten, Austria 41/C3
Froid, Mont. (59226) 262/M2
Froidchapelle, Belgium 27/E8
Frolovo, U.S.S.R. 48/E5
Frolovo, U.S.S.R. 52/F5
Fromberg, Mont. (59029) 262/H5
Frome (lake), Australia 87/E9
Frome, England 10/E5
Frome, England 13/E6
Frome, Jamaica 158/G6
Frome (lake), S. Australia 88/G6
Frome (lake), S. Australia 94/G4
Front (range), Colo. 208/H1

Fronteira, Portugal 33/C3
Fronteiras, Brazil 132/F4
Frontenac, Kansas (66762) 232/H4
Frontenac, Minn. (55026) 255/F6
Frontenac, Mo. (†63101) 261/O3
Frontenac (county), Ontario 177/H3
Frontenac (co.), Québec 172/G4
Frontera, Mexico 150/N7
Frontier, Mich. (49239) 250/E7
Frontier (co.), Nebr. 264/D4
Frontier, N. Dak. (†58102) 282/S6
Frontier, Sask. 181/C6
Frontier, Wyo. (83121) 319/B4
Front Royal, Va. (22630) 307/M3
Frosinone (prov.), Italy 34/D4
Frosinone, Italy 34/D4
Frösö, Sweden 18/J5
Frost, La. (†70753) 238/L2
Frost, Minn. (56033) 255/D7
Frost, Texas (76641) 303/H5
Frost, W. Va. (†24954) 312/G6
Frostburg, Md. (21532) 245/C2
Frostproof, Fla. (33843) 212/E4
Froude, Sask. 181/H6
Frövi, Sweden 18/J7
Frøya (isl.), Norway 18/F5
Frozen (str.), N.W.T. 162/H2
Frozen (str.), N.W. Terrs. 187/K3
Fruita, Colo. (81521) 208/B4
Fruita, Utah (†84775) 304/C5
Fruitdale, Ala. (36539) 195/B8
Fruitdale, S. Dak. (57742) 298/B4
Fruitdale-Harbeck, Oreg. (†97526) 291/D5
Fruitgrove, Queensland 88/K3
Fruit Heights, Utah (†84037) 304/C2
Fruithurst, Ala. (36262) 195/G3
Fruitland, Idaho (83619) 220/B6
Fruitland, Iowa (52749) 229/L6
Fruitland, Md. (21826) 245/R7
Fruitland, Mo. (†63755) 261/N8
Fruitland, N. Mex. (87416) 274/A2
Fruitland, Tenn. (†38343) 237/D9
Fruitland, Utah (84027) 304/D3
Fruitland, Wash. (99129) 310/G2
Fruitland Park, Fla. (32731) 212/D3
Fruitland Park, Miss. (39577) 256/F9
Fruitport, Mich. (49415) 250/C5
Fruitvale, Br. Col. 184/J5
Fruitvale, Idaho (83620) 220/B5
Fruitvale, Tenn. (38336) 237/C9
Fruitvale, Wash. (†98901) 310/E4
Fruitville, Fla. (33578) 212/D4
Frunze, U.S.S.R. 54/J5
Frunze, U.S.S.R. 48/H5
Frutal, Brazil 135/B2
Frutigen, Switzerland 39/E3
Frutillar, Chile 138/D3
Fry, Georgia (37317) 217/D1
Fryburg, N. Dak. (†58622) 282/D6
Fryburg, Ohio (†45895) 284/B4
Fryburg, Pa. (16326) 294/D3
Fry Canyon, Utah (†84511) 304/D6
Frýdek-Místek, Czech. 41/E2
Frýdlant nad Ostravicí, Czech. 41/E2
Frýdlant v. Čechách, Czech. 41/C1
Frye, Maine (04235) 243/B6
Fryeburg, La. (†71039) 238/D2
Fryeburg, Maine (04037) 243/A7
Fryeburg○, Maine (04037) 243/A7
Fu'an, China 77/K6
Fuchu, Hiroshima, Japan 81/F6
Fuchu, Tokyo, Japan 81/O2
Fuding, China 77/K6
Fuengirola, Spain 33/D4
Fuensalida, Spain 33/D2
Fuente-Álamo, Spain 33/F4
Fuente de Cantos, Spain 33/C3
Fuentelapeña, Spain 33/D2
Fuente Obejuna, Spain 33/D3
Fuenterrabía, Spain 33/E1
Fuentesaúco, Spain 33/D2
Fuentes de Andalucía, Spain 33/D4
Fuentes de Oñoro, Spain 33/C2
Fuerte (isl.), Colombia 126/B3
Fuerte (riv.), Mexico 150/E3
Fuerte Bulnes, Chile 138/E10
Fuerte Olimpo, Argentina 120/D5
Fuerte Olimpo, Paraguay 144/C2
Fuerteventura (isl.), Spain 102/A2
Fuerteventura (isl.), Spain 106/B3
Fuerteventura (isl.), Spain 33/C4
Fuga (isl.), Philippines 82/A3
Fuglebjerg, Denmark 21/E7
Fugu, China 77/H4
Fuhai (Burultokay), China 77/C2
Fuik, Neth. Ant. 161/G9
Fujairah, U.A.E. 59/G4
Fuji, Japan 81/J6
Fuji (mt.), Japan 81/J6
Fuji (riv.), Japan 81/J6
Fujian (Fukien), China 77/J6
Fujieda, Japan 81/J6
Fuji-Hakone-Izu National Park, Japan 81/H6
Fujin, China 77/M2
Fujisawa, Japan 81/O3
Fukagawa, Japan 81/L2
Fukang, China 77/C3
Fukuchiyama, Japan 81/G6
Fukue, Japan 81/D7
Fukui (pref.), Japan 81/G5
Fukui, Japan 81/G5
Fukuoka (pref.), Japan 81/D7
Fukuoka, Japan 54/O6
Fukuoka, Japan 81/D7
Fukushima (pref.), Japan 81/K5
Fukushima, Japan 81/K5
Fukuyama, Japan 81/F6
Fulbourn, England 13/H5
Fulbright, Texas (75436) 303/J4
Fulda, Ind. (47536) 227/D8
Fulda, Minn. (56131) 255/C7
Fulda, Sask. 181/F3
Fulda, W. Germany 22/C3
Fulda (riv.), W. Germany 22/C3
Fulford, England 13/F4
Fulford Harbour, Br. Col. 184/K3
Fuling, China 77/G6
Fulks Run, Va. (22830) 307/L3
Fullarton, Trin. & Tob. 161/A11
Fullerton, Calif. (*92631) 204/D11
Fullerton, Ky. (†41175) 237/P3
Fullerton, La. (70642) 238/D4
Fullerton, Nebr. (68638) 264/F3
Fullerton, N. Dak. (58441) 282/O7
Fully, Switzerland 39/D4
Fulnek, Czech. 41/D2
Fulpmes, Austria 41/A3
Fulton, Ala. (36446) 195/C7
Fulton (co.), Ark. 202/G1
Fulton, Ark. (71838) 202/C6
Fulton (co.), Georgia 217/D3
Fulton (co.), Ill. 222/C3
Fulton, Ill. (61252) 222/C2
Fulton (co.), Ind. 227/E2
Fulton, Ind. (46931) 227/E3
Fulton, Iowa (†52060) 229/M4
Fulton, Kansas (66738) 232/H4
Fulton (co.), Ky. 237/C7
Fulton, Ky. (42041) 237/D7
Fulton, Mich. (49052) 250/D6
Fulton, Miss. (38843) 256/H2
Fulton, Mo. (65251) 261/J5
Fulton (co.), N.Y. 276/M4
Fulton, N.Y. (13069) 276/H4
Fulton (co.), Ohio 284/B2
Fulton, Ohio (43321) 284/E5
Fulton (co.), Pa. 294/F6
Fulton, S. Dak. (57340) 298/O6
Fulton, Tenn. (†38041) 237/B9
Fulton, Texas (78358) 303/H9
Fulton Chain (lakes), N.Y. 276/K3
Fultondale, Ala. (35068) 195/E3
Fultonham, Ohio (43738) 284/F6
Fultonville, N.Y. (12072) 276/M5
Fults, Ill. (62244) 222/C5
Fulwood, England 10/G1
Fulwood, England 13/G1
Funabashi, Japan 81/P2
Funafuti (atoll), Tuvalu 87/H6
Funchal (cap.), Madeira, Port. 102/A1
Funchal (dist.), Portugal 33/A2
Funchal (cap.), Madeira, Portugal 106/A2
Funchal, Portugal 33/A2
Fundación, Colombia 126/C2
Fundão, Portugal 33/C2
Fundy (bay) 162/K7
Fundy (bay), New Bruns. 170/E3
Fundy (bay), Nova Scotia 168/C3
Fundy Nat'l Park, New Bruns. 170/E3
Funhalouro, Mozambique 118/E4
Funing, China 77/K5
Funk, Nebr. (68940) 264/E4
Funk (isl.), Newf. 166/D4
Funkley, Minn. (†56630) 255/D3
Funkstown, Md. (21734) 245/H2
Funston, Georgia (31753) 217/E8
Funter, Alaska (†99801) 196/M1
Funtua, Nigeria 106/F6
Fuping, China 77/H4
Fuquay-Varina, N.C. (27526) 281/M3
Furancungo, Mozambique 118/E2
Furka (pass), Switzerland 39/F3
Furman, Ala. (36741) 195/E6
Furman, S.C. (29921) 296/E6
Furmanov, U.S.S.R. 52/F3
Furnace, Ky. (†40472) 237/O5
Furnace, Mass. (†01031) 249/F3
Furnace, Scotland 15/C4
Furnas (res.), Brazil 120/E5
Furnas (dam), Brazil 135/C2
Furnas (co.), Nebr. 264/E4
Furneaux Group (isls.), Australia 87/E9
Furneaux Group (isls.), Tasmania 88/H8
Furneaux Group (isls.), Tasmania 99/E1
Furnes (Veurne), Belgium 27/B6
Furness, Sask. 181/B2
Furry Creek, Br. Col. 184/K2
Fürstenberg, E. Germany 22/E2
Fürstenfeld, Austria 41/C3
Fürstenfeldbruck, W. Germany 22/D4
Fürstenwalde, E. Germany 22/F2
Fürth, W. Germany 22/D4
Furth im Wald, W. Germany 22/E4
Furukawa, Japan 81/K4
Fury and Hecla (str.), N.W.T. 162/H2
Fury and Hecla (str.), N.W. Terrs. 187/K3
Fusagasugá, Colombia 126/C5
Fushun, China 77/K3
Fushun, China 54/O5
Fusilier, Sask. 181/B4
Fusin (Fuxin), China 77/K3
Fusingchen (Simao), China 77/F7
Fusio, Switzerland 39/G4
Fusong, China 77/L3
Füssen, W. Germany 22/D5
Futa Jallon (mts.), Guinea 106/B6
Futaleufú, Chile 138/E4
Futrono, Chile 138/E3
Futuna (Hoorn) (isls.), Wallis and Futuna 87/J7
Fu Xian, Liaoning, China 77/K4
Fu Xian, Shaanxi, China 77/G4
Fuxin (Fusin), China 77/K3
Fuxin, China 54/O5
Fuyang (Fowyang), China 77/J5
Fuyu, Heilongjiang, China 77/K2
Fuyu, Jilin, China 77/L2
Fuyuan, Heilongjiang, China 77/M2
Fuyuan, Yunnan, China 77/F6
Fuyun, China 77/C2
Füzesabony, Hungary 41/F3
Füzesgyarmat, Hungary 41/F3
Fuzhou (Foochow), Fujian, China 77/J6
Fuzhou, Jiangxi, China 77/J6
Fuzhou, China 2/R4
Fuzhou, China 54/N7
Fyffe, Ala. (35971) 195/G2
Fylingdales, England 13/G3
Fyn (co.), Denmark 21/D7
Fyn (isl.), Denmark 21/D7
Fyn (isl.), Denmark 18/G9
Fyne, Loch (inlet), Scotland 10/D2
Fyne, Loch (inlet), Scotland 15/C5
Fyns Hoved (pt.), Denmark 21/D6
Fyvie, Scotland 15/F3
Fyzabad, Trin. & Tob. 161/A11

G

Gaastra, Mich. (49927) 250/G2
Gabarus, Nova Scotia 168/H3
Gabarus (bay), Nova Scotia 168/H3
Gabarus (cape), Nova Scotia 168/J3
Gabbettville, Georgia (†30240) 217/B5
Gabbs, Nev. (89409) 266/D4
Gabela, Angola 115/B6
Gabès, Tunisia 106/F2
Gabès, Tunisia 102/D1
Gabès (gulf), Tunisia 106/G2
Gabgaba, Wadi (dry riv.), Sudan 111/F3
Gable, S.C. (29051) 296/G4
Gabon 2/K6
Gabon 102/D4
GABON 115/B4
Gaborone (cap.), Botswana 2/L7
Gaborone (cap.), Botswana 118/D4
Gaborone (cap.), Botswana 102/E7
Gabras, Sudan 111/E5
Gabredarre, Ethiopia 111/H6
Gabriel (str.), N.W. Terrs. 187/M3
Gabrik (riv.), Iran 66/L7
Gabriola, Br. Col. 184/J3
Gabrovo, Bulgaria 45/G4
Gachalá, Colombia 126/D5
Gach Saran, Iran 59/F3
Gach Saran, Iran 66/G5
Gackle, N. Dak. (58442) 282/M6
Gacko, Yugoslavia 45/D4
Gadag-Betgeri, India 68/D5
Gäddede, Sweden 18/J4
Gadê, China 77/E5
Gadebusch, E. Germany 22/D2
Gadmen, Switzerland 39/F3
Gadsby, Alberta 182/D3
Gadsden, Ala. 188/J4
Gadsden, Ala. (*35901) 195/G2
Gadsden, Ariz. (85336) 198/A6
Gadsden (co.), Fla. 212/B1
Gadsden, S.C. (29052) 296/F4
Gadsden, Tenn. (38337) 237/D9
Gads Hill, Mo. (†63957) 261/L8
Gadston (pt.), Fla. 212/C3
Gadwal, India 68/D5
Gadyach, U.S.S.R. 52/D4
Găești, Romania 45/G3
Gaeta, Italy 34/D4
Gaeta (gulf), Italy 34/D4
Gaferut (isl.), Micronesia 87/E5
Gaffney, S.C. (29340) 296/D1
Gafsa, Tunisia 106/F2
Gagarin, U.S.S.R. 52/D3
Gage, Alberta 182/A1
Gage (co.), Nebr. 264/H4
Gage, N. Mex. (†88030) 274/A6
Gage, Okla. (73843) 288/G2
Gagetown, Mich. (48735) 250/F5
Gagetown, New Bruns. 170/D3
Gaggenau, W. Germany 22/C4
Gagnoa, Ivory Coast 102/B4
Gagnoa, Ivory Coast 106/C7
Gagnon, Que. 162/K5
Gagnon, Québec 174/D2
Gagnon (lake), Québec 172/B3
Gagny, France 28/C1
Gagra, U.S.S.R. 52/E6
Gahanna, Ohio (43230) 284/E5
Gaiba (lag.), Bolivia 136/F5
Gail, Saudi Arabia 59/E5
Gail, Texas (79738) 303/C5
Gaillac, France 28/D6
Gaillard (lake), Conn. 210/D3
Gaillard, Georgia (†31078) 217/D5
Gaima, Papua N.G. 85/B7
Gaiman, Argentina 143/C5
Gaines, Mich. (48436) 250/F6
Gaines, Pa. (16921) 294/G2
Gaines (co.), Texas 303/B5
Gainesboro, Tenn. (38562) 237/K8
Gainesboro, Va. (†22601) 307/M2
Gainestown, Ala. (36540) 195/C8
Gainesville, Ala. (35464) 195/B5
Gainesville (dam), Ala. 195/B5
Gainesville, Fla. 188/K5
Gainesville, Fla. (*32601) 212/D2
Gainesville, Georgia (30501) 217/E2
Gainesville, Mo. (65655) 261/G9
Gainesville, N.Y. (14066) 276/D5
Gainesville, Texas (76240) 303/G4
Gainesville, Va. (22065) 307/N3
Gainsborough, England 10/F4
Gainsborough, England 13/G4
Gainsborough, Sask. 181/K6
Gairdner (lake), Australia 87/D9
Gairdner (lake), S. Australia 88/F6
Gairdner (lake), S. Australia 94/D4
Gairloch, Scotland 15/C3
Gairloch, Loch (inlet), Scotland 15/C3
Gais, Switzerland 39/H2
Gaithersburg, Md. (20760) 245/K4
Gajdel, Czech. 41/E2
Gakona, Alaska (99586) 196/K2
Galadi, Ethiopia 111/J6
Galahad, Alberta 182/E3
Galana (riv.), Kenya 115/G4
Galand, Iran 66/J2
Galanta, Czech. 41/D2
Galápagos (isls.), Ecuador 2/E6
Galápagos (isls.), Ecuador 128/C8
Galashiels, Scotland 10/E3
Galashiels, Scotland 15/F5
Galata, Mont. (59444) 262/E2
Galata, Turkey 63/C6
Galați, Romania 7/G4
Galați, Romania 45/H3
Galatia, Ill. (62935) 222/E6
Galatia, Kansas (†67567) 232/D3
Galatina, Italy 34/G4
Galatone, Italy 34/F4
Galax (I.C.), Va. (24333) 307/G7
Galbally, Ireland 17/E7
Galbraith, La. (†71447) 238/E4
Galcaio, Somalia 102/G4
Galcaio, Somalia 115/J2
Galchutt, N. Dak. (58034) 282/S7
Gale, Ill. (62936) 222/D6
Gale (riv.), N.H. 268/D3
Galeana, Chihuahua, Mexico 150/F1
Galeana, Nuevo León, Mexico 150/J4
Galela, Indonesia 85/H5
Galen, Mont. (†59722) 262/D4
Galena, Alaska (99741) 196/G2
Galena, Ill. (61036) 222/C1
Galena, Ind. (†47119) 227/F8
Galena, Kansas (66739) 232/H4
Galena, Md. (21635) 245/P3
Galena, Mo. (65656) 261/F9
Galena, Ohio (43021) 284/E5
Galena Park, Texas (77547) 303/J1
Galeota (pt.), Trin. & Tob. 161/B11
Galera (pt.), Chile 138/D3
Galera (pt.), Ecuador 128/B2
Galera (pt.), Trin. & Tob. 161/C10
Galera (pt.), Trin. & Tob. 156/G5
Galesburg, Ill. 188/H2
Galesburg, Ill. (61401) 222/C3
Galesburg, Kansas (66740) 232/G4
Galesburg, Mich. (49053) 250/D6
Galesburg, N. Dak. (58035) 282/R5
Gales Creek, Oreg. (97117) 291/D2
Gales Ferry, Conn. (06335) 210/G3
Galestown, Md. (†19973) 245/P6
Galesville, Md. (20765) 245/M5
Galesville, Wis. (54630) 317/D7
Galeton, Colo. (80622) 208/K1
Galeton, Pa. (16922) 294/G2
Galetta, Ontario 177/H2
Galgenberg (hill), Netherlands 27/H4
Galguduud (prov.), Somalia 115/J2
Galiano, Br. Col. 184/K3
Galiano (isl.), Br. Col. 184/K3
Galice, Oreg. (†97532) 291/D5
Galich, U.S.S.R. 52/F3
Galicia (reg.), Spain 33/B1
Galien, Mich. (49113) 250/C7
Galilee, Sea of (lake), Israel 59/C3
Galilee, Sea of (Tiberias) (lake), Israel 65/D2
Galilee (reg.), Israel 65/C2
Galilee (lake), Queensland 95/C4
Galina (pt.), Jamaica 158/J6
Galion (bay), Martinique 161/D6
Galion, Ohio (44833) 284/E4
Galisteo, N. Mex. (†87540) 274/D3
Galiuro (mts.), Ariz. 198/E6
Galivants Ferry, S.C. (29544) 296/J3
Gallabat, Sudan 111/G5
Gallan (head), Scotland 15/A2
Gallant, Ala. (35972) 195/F2
Gallarate, Italy 34/B2
Gallatin (co.), Ill. 222/E6
Gallatin (co.), Ky. 237/M3
Gallatin, Mo. (64640) 261/E3
Gallatin (co.), Mont. 262/E5
Gallatin (peak), Mont. 262/E5
Gallatin (riv.), Mont. 262/E5
Gallatin, Tenn. (37066) 237/H8
Gallatin Gateway, Mont. (59730) 262/E5
Galloway, Tenn. (38036) 237/B10
Galle, Sri Lanka 54/J9
Galle, Sri Lanka 68/D7
Gallegos (riv.), Argentina 143/B7
Gallegos, N. Mex. (†87733) 274/F3
Galley (head), Ireland 17/D9
Gallia (co.), Ohio 284/F8
Galliano, La. (70354) 238/K8
Gallina, N. Mex. (87017) 274/C2
Gallinas (pt.), Colombia 120/B1
Gallinas (pt.), Colombia 126/E1
Gallinas (mts.), N. Mex. 274/B4
Gallinas (riv.), N. Mex. 274/E3
Gallion, Ala. (36742) 195/C6
Gallion, La. (†71223) 238/G1
Gallipoli, Italy 34/F4
Gallipoli, Turkey 59/A1
Gallipoli, Turkey 63/C5
Gallipolis, Ohio (45631) 284/F8
Gallipolis Ferry, W. Va. (25515) 312/B5
Gallitzin, Pa. (16641) 294/E4
Gällivare, Sweden 18/M3
Gallman, Miss. (39077) 256/D7
Gallo (pt.), Chile 138/E3
Gallo (mt.), Dom. Rep. 158/D5
Gällö, Sweden 18/J5
Galloo (isl.), N.Y. 276/H3
Galloway, Ark. (†72114) 202/F4
Galloway, Br. Col. 184/K5
Galloway (dist.), Scotland 15/D5
Galloway, Mull of (prom.), Scotland 15/D6
Galloway, Mull of (prom.), Scotland 10/D3
Galloway, W. Va. (26349) 312/F4
Galloway, Wis. (54432) 317/H6
Gallup, N. Mex. 188/E3
Gallup, N. Mex. (87301) 274/A3
Gallur, Spain 33/F2
Galole, Kenya 115/G4
Gal'on, Israel 65/B4
Galston, Scotland 10/D3
Galston, Scotland 15/D5
Galt, Calif. (95632) 204/C9
Galt, Iowa (50101) 229/F3
Galt, Mo. (64641) 261/F2
Galtee (mts.), Ireland 17/E7
Galtymore (mt.), Ireland 17/E7
Galva, Ill. (61434) 222/D2
Galva, Iowa (51020) 229/C3
Galva, Kansas (67443) 232/E3
Galván (mt.), Paraguay 144/C3
Galvarino, Chile 138/D2
Galveston, Ind. (46932) 227/E3
Galveston (co.), Texas 303/K8
Galveston, Texas 146/J7
Galveston, Texas 188/H5
Galveston, Texas (*77550) 303/L3
Galveston (bay), Texas 188/H5
Galveston (bay), Texas 303/L2
Galveston (isl.), Texas 303/K8
Gálvez, Argentina 143/F6
Galvez, La. (†70769) 238/L2
Gálvez, Spain 33/D3
Galvin, Wash. (98544) 310/B4
Galway (co.), Ireland 17/D5
Galway, Ireland 17/C5
Galway, Ireland 7/D3
Galway, Ireland 10/B4
Galway (bay), Ireland 17/C5
Galway (bay), Ireland 10/B4
Galway, N.Y. (12074) 276/N4
Gamaliel, Ky. (42140) 237/K7
Gamarra, Colombia 126/D3
Gamas Ab (riv.), Iran 66/E3
Gamay, Philippines 82/E3
Gamay (bay), Philippines 82/E4
Gamba, China 77/C6
Gambaga, Ghana 106/D6
Gambela, Ethiopia 111/F6
Gambell, Alaska (99742) 196/D2
Gamber, Md. (†21048) 245/L3
Gambia 2/J5
Gambia 102/A3
GAMBIA 106/A6
Gambia (riv.), Gambia 106/B6
Gambia (riv.), Senegal 106/B6
Gambier (isls.), Fr. Poly. 87/N8
Gambier, Ohio (43022) 284/F5
Gambo, Newf. 166/D4
Gamboma, Congo 115/C4
Gambos, Angola 115/B6
Gambrills, Md. (21054) 245/M4
Gamerco, N. Mex. (87317) 274/A3
Gaming, Austria 41/C3
Gamleby, Sweden 18/J8
Gammon (riv.), Manitoba 179/G3
Gammon (pt.), Mass. 249/N6
Gampel, Switzerland 39/E4
Gamu-Gofa (prov.), Ethiopia 111/G6
Gamvik, Norway 18/Q1
Ganado, Ariz. (86505) 198/F3
Ganado, Texas (77962) 303/H8
Ganale Dorya (riv.), Ethiopia 111/H6
Gananoque, Ontario 177/H3
Ganassi, Philippines 82/D7
Ganaveh, Iran 66/G6
Ganda, Angola 115/B6
Gandajika, Zaire 115/D5
Gandara, Philippines 82/E4
Gándara, Spain 33/C1
Gandava, Pakistan 68/B3
Gandava, Pakistan 59/J4
Gandeeville, W. Va. (25243) 312/D5
Gander, Newf. 166/D4
Gander (lake), Newf. 166/D4
Gander (riv.), Newf. 166/D4
Gander, Newf. 162/L6
Gandesa, Spain 33/G2
Gandhinagar, India 68/C4
Gandía, Spain 33/F3
Gandy, Nebr. (†69163) 264/D3
Gandy, Utah (†84728) 304/A4
Gandzha (Kirovabad), U.S.S.R. 52/G6
Ganga (Ganges) (riv.), India 68/F3
Gan Gan, Argentina 143/C5
Ganganagar, India 68/C3
Gangapur, India 68/D3
Gangara, Niger 106/F6
Gangaw, Burma 72/B2
Gangca, China 77/F4
Gangdisê Shan (range), China 77/B5
Ganges (riv.) 54/K7
Ganges (riv.) 2/P4
Ganges, Mouths of the (delta), Bangladesh 68/F3
Ganges (riv.), Bangladesh 68/F3
Ganges, Br. Col. 184/K3
Ganges, Mouths of the (delta), India 68/F4
Ganges (riv.), India 68/F3
Gangtok, India 68/F3
Gan He (riv.), China 77/K2
Gani, Indonesia 85/H6
Ganister, Pa. (†16693) 294/F5
Ganmain, N.S. Wales 97/D4
Gann (Brinkhaven), Ohio (†43006) 284/F5
Gannat, France 28/E4
Gannett, Idaho (†83313) 220/D6
Gannett (peak), Wyo. 188/D2
Gannett (peak), Wyo. 319/C2
Gannvalley, S. Dak. (57341) 298/L5
Ganquan, China 77/H4
Gans, Okla. (74936) 288/S4
Gänserndorf, Austria 41/D2
Gansevoort, N.Y. (12831) 276/N4
Ganshoren, Belgium 27/B9
Gansu (Kansu), China 77/E3
Gansville, La. (†71422) 238/E2
Gantt, Ala. (36038) 195/E8
Gantt, S.C. (†29609) 296/C2
Ganzhou (Kanchow), China 77/H6
Gao (mt.), Cent. Afr. Rep. 115/C2
Gao, Mali 102/C3
Gao, Mali 106/E5
Gao'an, China 77/H6
Gaolan, China 77/F4
Gaotai, China 77/F4
Gaoua, Upper Volta 106/D6
Gaoual, Guinea 106/B6
Gaoyou Hu (lake), China 77/J5
Gap, France 28/G5
Gap, Pa. (17527) 294/L6
Gap (creek), Sask. 181/B6
Gapan, Philippines 82/C3
Gapcreek, Ky. (†42603) 237/M7
Gap Mills, W. Va. (24941) 312/F7
Gar, China 77/B5
Gara (lake), Ireland 17/D4
Gara, Lough (lake), Ireland 10/B4
Garachiné, Panama 154/H6
Garad, Somalia 115/J2
Garadice (lake), Ireland 17/F3
Garah, N.S. Wales 97/E1
Garamba Nat'l Park, Zaire 115/E3
Garanhuns, Brazil 120/F3
Garanhuns, Brazil 132/G5
Garards Fort, Pa. (15334) 294/B6
Garbahaarrey, Somalia 115/H3
Garba Tula, Kenya 115/G3
Garber, Iowa (52048) 229/L3
Garber, Okla. (73738) 288/M2
Garberville, Calif. (95440) 204/B3
Garbosh, Kuh-e (mt.), Iran 66/G4
Garbsen, W. Germany 22/C2
Garça, Brazil 135/B3
Garcia, Colo. (81134) 208/J8
García de Sola (res.), Spain 33/D3
Garcitas, Venezuela 124/C3
Gard (dept.), France 28/F6
Gard (riv.), France 28/F5
Garda (lake), Italy 34/C2
Gardanne, France 28/F6
Gardar, N. Dak. (58234) 282/P2
Gardelegen, E. Germany 22/D2
Garden, Mich. (49835) 250/C3
Garden (isl.), Mich. 250/D3
Garden (pen.), Mich. 250/C3
Garden (co.), Nebr. 264/B3
Garden (isl.), W. Australia 88/A2
Garden (isl.), W. Australia 92/A1
Gardena, Calif. (*90747) 204/C11
Gardena, Idaho (†83629) 220/B5
Gardena, N. Dak. (58739) 282/J2
Garden City, Ala. (35070) 195/E2
Garden City, Georgia (31408) 217/K6
Garden City, Idaho (†83704) 220/B6
Garden City, Iowa (50102) 229/G4
Garden City, Kans. 188/F3
Garden City, Kansas (67846) 232/B4
Garden City, La. (70540) 238/H7
Garden City, Mich. (48135) 250/F6
Garden City, Minn. (56034) 255/D6
Garden City, Mo. (64747) 261/D5
Garden City, N.Y. (11530) 276/R7
Garden City, S. Dak. (57236) 298/O4
Garden City, Texas (79739) 303/C6
Garden City, Utah (84028) 304/C2
Garden City Beach, S.C. (29576) 296/K4
Gardendale, Ala. (35071) 195/E3
Garden Grove, Calif. (*92640) 204/D11
Garden Grove, Iowa (50103) 229/F7
Garden Home-Whitford, Oreg. (97223) 291/A2
Garden Island (bay), La. 238/M8
Garden Plain, Kansas (67050) 232/E4
Garden Prairie, Ill. (61038) 222/E1
Garden Reach, India 68/F2
Garden River, Alberta 182/B5
Gardenstown, Scotland 15/F3
Gardenton, Manitoba 179/F5
Garden Valley, Idaho (83622) 220/C5
Garden View, Pa. (†17701) 294/H3
Garden Village, Ontario 177/E1
Gardez, Afghanistan 59/J3
Gardez, Afghanistan 68/B2
Gardi, Georgia (†31545) 217/J7
Gardiner, Maine (04345) 243/D7
Gardiner, Mont. (59030) 262/F5
Gardiner, Oreg. (97441) 291/C4
Gardiner (dam), Sask. 181/D4
Gardiner, Wash. (98334) 310/B2
Gardiners (bay), N.Y. 276/R8
Gardiners (isl.), N.Y. 276/R8
Gardner (canal), Br. Col. 184/C3
Gardner, Colo. (81040) 208/J7
Gardner (lake), Conn. 210/G2
Gardner, Fla. (†33890) 212/E4
Gardner, Ill. (60424) 222/E2
Gardner, Kansas (66030) 232/H3
Gardner (isl.), Kiribati 87/J6
Gardner (lake), Maine 243/J6
Gardner, Mass. (01440) 249/G2
Gardner, N. Dak. (58036) 282/R5
Gardner, Tenn. (†38237) 237/D8
Gardner (mt.), Wash. 310/E2
Gardner Creek, New Bruns. 170/E3
Gardner Pinnacles (isls.), Hawaii 87/K3
Gardner Pinnacles (isls.), Hawaii 188/F6
Gardner Pinnacles (isls.), Hawaii 218/C6
Gardnerville, Nev. (89410) 266/B4
Gardo, Somalia 115/J2
Gardula, Ethiopia 111/G6
Gare Loch (inlet), Scotland 15/A1
Garelochhead, Scotland 15/A1
Garelochhead, Scotland 10/A1
Gareloi (isl.), Alaska 196/K4
Garessio, Italy 34/A2
Garfield, Ark. (72732) 202/C1
Garfield (co.), Colo. 208/C3
Garfield, Colo. (81227) 208/G5
Garfield, Georgia (30425) 217/H5
Garfield, Kansas (67529) 232/C3
Garfield, Ky. (40140) 237/J5
Garfield, Minn. (56332) 255/C5
Garfield (co.), Mont. 262/J3
Garfield (co.), Nebr. 264/F3
Garfield, N.J. (07026) 273/B2
Garfield, N. Mex. (87936) 274/B6
Garfield (co.), Okla. 288/L2
Garfield (co.), Utah 304/C6
Garfield (co.), Wash. 310/H4
Garfield, Wash. (99130) 310/H3
Garfield Heights, Ohio (44125) 284/J9
Gargaliánoi, Greece 45/E7
Gargunnock, Scotland 15/B1
Garibaldi, Br. Col. 184/F5
Garibaldi, Oreg. (97118) 291/D2
Garibaldi Prov. Park, Br. Col. 184/F5
Garies, S. Africa 118/B6
Garioch (dist.), Scotland 15/F3
Garissa, Kenya 115/G4
Garita, N. Mex. (88421) 274/E3
Garland, Ala. (†36456) 195/E7
Garland (co.), Ark. 202/D4

Garland, Ark. (71839) 202/C7
Garland, Kansas (66741) 232/H4
Garland, Maine (04939) 243/E5
Garland○, Maine (04939) 243/E5
Garland, Manitoba 179/B3
Garland, Nebr. (68360) 264/G4
Garland, N.C. (28441) 281/N5
Garland, Pa. (16416) 294/C2
Garland, Tenn. (†38019) 237/B9
Garland, Tex. 188/G4
Garland, Texas (*75040) 303/H2
Garland, Utah (84312) 304/B2
Garland, Wyo. (†82435) 319/D1
Garlandville, Miss. (†39345) 256/F6
Garlieston, Scotland 15/D6
Garlin, Ky. (†42728) 237/L6
Garmisch-Partenkirchen, W. Germany 22/D5
Garmouth, Scotland 15/E3
Garmsar, Iran 59/F2
Garmsar, Iran 66/H3
Garnavillo, Iowa (52049) 229/L3
Garneill, Mont. (59445) 262/G4
Garner, Ark. (72052) 202/G3
Garner, Iowa (50438) 229/F2
Garner (lake), Manitoba 179/G4
Garner, N.C. (27529) 281/M3
Garnet, Mich. (†49762) 250/D2
Garnet, Mont. (†59832) 262/C4
Garnet (bay), N.W. Terrs. 187/L3
Garnett, Kansas (66032) 232/G3
Garnett, S.C. (29922) 296/E6
Garnish, Newf. 166/C4
Garoe, Somalia 115/J2
Garonne (riv.), France 7/D4
Garonne (riv.), France 28/C5
Garoua, Cameroon 102/D4
Garoua, Cameroon 115/B2
Garrabost, Scotland 15/B2
Garrard (co.), Ky. 237/M5
Garretson, S. Dak. (57030) 298/S6
Garrett, Ill. (†61913) 222/E4
Garrett, Ind. (46738) 227/G2
Garrett, Ky. (41630) 237/R6
Garrett (co.), Md. 245/A2
Garrett, Pa. (15542) 294/D6
Garrett, Wash. (†99362) 310/G4
Garrett, Wyo. (82058) 319/G3
Garrett Park, Md. (20766) 245/E3
Garretts Bend, W. Va. (†25523) 312/C6
Garrettsville, Ohio (44231) 284/H3
Garrick, Sask. 181/G2
Garrison, Iowa (52229) 229/J4
Garrison, Ky. (41141) 237/P3
Garrison, Md. (21055) 245/L3
Garrison, Minn. (56450) 255/E4
Garrison, Mo. (65657) 261/F9
Garrison, Mont. (59731) 262/D4
Garrison, Nebr. (68632) 264/G3
Garrison, N.Y. (10524) 276/N8
Garrison, N. Dak. (58540) 282/H4
Garrison (dam), N. Dak. 282/H5
Garrison, Texas (75946) 303/K6
Garrison, Utah (84728) 304/A5
Garrisonville, Va. (22463) 307/N4
Garron (pt.), N. Ireland 17/K1
Garrovillas, Spain 33/C3
Garry (lake), Canada 4/C14
Garry (lake), N.W.T. 162/G2
Garry (lake), N.W. Terrs. 187/H3
Garry, Loch (lake), Scotland 15/D3
Garry (riv.), Scotland 15/D4
Garryowen, Mont. (59031) 262/J5
Garsen, Kenya 115/G4
Garske, N. Dak. (†58382) 282/N3
Garson (lake), Alberta 182/E1
Garson, Manitoba 179/F4
Garstang, England 13/G1
Gartan (lake), Ireland 17/F2
Gartmore, Scotland 15/B1
Garulia, India 68/F1
Garut, Indonesia 85/H2
Garvagh, N. Ireland 17/H2
Garvan (isls.), Ireland 17/G1
Garvin, Minn. (56132) 255/C6
Garvin (co.), Okla. 288/M5
Garvin, Okla. (74736) 288/S7
Garwin, Iowa (50632) 229/H4
Garwolin, Poland 47/E3
Garwood, Mo. (†63965) 261/L8
Garwood, N.J. (07027) 273/E2
Garwood, Texas (77442) 303/H8
Gary, Ind. 146/K5
Gary, Ind. 188/J2
Gary, Ind. (*46401) 227/C1
Gary, Minn. (56545) 255/B3
Gary, S. Dak. (57237) 298/S4
Gary, Texas (75643) 303/K5
Gary, W. Va. (24836) 312/C8
Garyarsa (Gartok), China 77/B5
Garyarsa, China 54/K6
Garysburg, N.C. (27831) 281/O2
Garyville, La. (70051) 238/M3
Garza (co.), Texas 303/C4
Garzê, China 77/F5
Garzón, Colombia 126/C6
Garzón, Uruguay 145/E5
Garzón (lag.), Uruguay 145/E5
Gas, Kansas (66742) 232/G4
Gas (hills), Wyo. 319/E3
Gasan-Kuli, U.S.S.R. 48/F6
Gasburg, Va. (23857) 307/N7
Gas City, Ind. (46933) 227/F4
Gasconade (co.), Mo. 261/J6
Gasconade, Mo. (65036) 261/J5
Gasconade (riv.), Mo. 261/H7
Gascony (trad. prov.), France 29
Gascoyne (riv.), Australia 87/B8
Gascoyne, N. Dak. (58629) 282/D7
Gascoyne (riv.), W. Australia 88/A4
Gascoyne (riv.), W. Australia 92/B4
Gascoyne Junction, W. Australia 92/A4
Gash (Mareb) (riv.), Ethiopia 59/C7
Gash (riv.), Sudan 59/C6
Gashaka, Nigeria 106/G7
Gas Hills, Wyo. (82501) 319/E3
Gash Mareb (riv.), Ethiopia 111/G5
Gasht, Iran 66/M7
Gasker (isl.), Scotland 15/A3
Gaskiers, Newf. 166/D2
Gasmata, Papua N.G. 86/B2
Gaspar, Cuba 158/F2
Gaspar Hernández, Dom. Rep. 158/E5
Gasparilla (isl.), Fla. 212/D5
Gaspé, Que. 162/K6
Gaspé, Québec 172/D1
Gaspé, Québec 174/E3
Gaspé (bay), Québec 172/D1
Gaspé (cape), Québec 172/D1
Gaspé (pen.), Québec 172/D2
Gaspé-Est (county), Québec 174/E3
Gaspé-Est (co.), Québec 172/D1
Gaspé-Ouest (co.), Québec 172/C1
Gaspé-Ouest (county), Québec 174/D3
Gaspereau (riv.), New Bruns. 170/D2
Gaspereau (lake), Nova Scotia 168/D4
Gaspésie Prov. Park, Québec 174/D3
Gaspésie Prov. Park, Québec 172/C1
Gasport, N.Y. (14067) 276/C4
Gasque, Ala. (†36542) 195/C10
Gassan (mt.), Japan 81/J4
Gassaway, Tenn. (†37095) 237/K9
Gassaway, W. Va. (26624) 312/E5
Gassetts, Vt. (†05144) 268/B5
Gassville, Ark. (72635) 202/F1
Gaston, Ind. (47342) 227/G4
Gaston (co.), N.C. 281/G4
Gaston, N.C. (27832) 281/O1
Gaston (res.), N.C. 281/O2
Gaston, Oreg. (97119) 291/D2
Gaston, S.C. (29053) 296/E4
Gaston (lake), Va. 307/M8
Gastonburg, Ala. (†36728) 195/C6
Gastonia, N.C. 188/K3
Gastonia, N.C. (28052) 281/G4
Gastre, Argentina 143/C5
Gat, Israel 65/B4
Gata (cape), Cyprus 59/B3
Gata (cape), Cyprus 63/E5
Gata (cape), Spain 33/F4
Gata (mts.), Spain 33/C2
Gatchel, Ind. (†47586) 227/D8
Gatchina, U.S.S.R. 52/C3
Gate, Okla. (73844) 288/F1
Gate, Wash. (†98579) 310/B4
Gate City, Va. (24251) 307/C7
Gatehouse of Fleet, Scotland 10/E3
Gatehouse of Fleet, Scotland 15/D6
Gates, Nebr. (68839) 264/E3
Gates (co.), N.C. 281/R2
Gates, N.C. (27937) 281/R2
Gates, Oreg. (97346) 291/E3
Gates, Tenn. (38037) 237/C9
Gateshead, England 10/F3
Gateshead, England 13/J3
Gateshead (isl.), N.W. Terrs. 187/J2
Gates Mills, Ohio (44040) 284/J9
Gates of the Arctic Nat'l Park, Alaska 196/H1
Gates of the Arctic Nat'l Preserve, Alaska 196/H1
Gatesville, N.C. (27938) 281/R2
Gatesville, Texas (76528) 303/G6
Gateswood, Ala. (†36507) 195/C9
Gateway, Ark. (72733) 202/B1
Gateway, Colo. (81522) 208/B5
Gateway, Oreg. (†97741) 291/F3
Gateway Nat'l Rec. Area, N.J. 273/E2
Gateway Nat'l Rec. Area, N.Y. 276/M9
Gatewood, Mo. (63942) 261/K9
Gatico, Chile 138/A4
Gatineau (co.), Québec 172/B3
Gatineau (county), Québec 174/B3
Gatineau, Québec 172/B4
Gatineau (riv.), Québec 172/B3
Gatliff, Ky. (†40769) 237/O7
Gatlinburg, Tenn. (37738) 237/O9
Gatooma, Zimbabwe 118/D3
Gatooma, Zimbabwe 102/E6
Gatow, W. Germany 22/E4
Gatteville-le-Phare, France 28/C3
Gattman, Miss. (38844) 256/H3
Gatton, Queensland 88/J5
Gatton, Queensland 95/E5
Gatun (lake), Panama 154/G6
Gatzke, Minn. (56724) 255/C2
Gaucín, Spain 33/D4
Gauhati, India 68/G3
Gauhati, India 54/L7
Gauja (riv.), U.S.S.R. 53/C2
Gauley (riv.), W. Va. 312/D6
Gauley Bridge, W. Va. (25085) 312/D6
Gauley Mills, W. Va. (26240) 312/E6
Gaultois, Newf. 166/C4
Gausdale, Ky. (40906) 237/N7
Gause, Texas (77857) 303/H7
Gaussberg (mt.) 5/C5
Gautier, Miss. (39553) 256/G10
Gavater, Iran 59/H5
Gavater, Iran 66/M8
Gávdhos (isl.), Greece 45/F8
Gave de Pau (riv.), France 28/C6
Gavião, Portugal 33/C3
Gavins Point (dam), Nebr. 264/G2
Gavins Point (dam), S. Dak. 298/P8
Gaviota, Calif. (†93017) 204/E9
Gavkhuni (lake), Iran 59/F3
Gavkhuni (marsh), Iran 66/H4
Gävle, Sweden 7/F2
Gävle, Sweden 18/K6
Gävleborg (co.), Sweden 18/K6
Gawai, Burma 72/C1
Gawler, S. Australia 88/F6
Gawler (ranges), S. Australia 88/F6
Gawler, S. Australia 94/B6
Gawler (ranges), S. Australia 94/E5
Gawler (riv.), S. Australia 94/B6
Gay, Georgia (30218) 217/C4
Gay, Mich. (49928) 250/A1
Gay, U.S.S.R. 52/J4
Gay, W. Va. (25244) 312/C5
Gaya, India 68/F4
Gaya, Niger 106/E6
Gay Head○, Mass. (†02535) 249/L7
Gay Head (prom.), Mass. 249/L7
Gay Hill, Texas (†77833) 303/H7
Gayle, Jamaica 158/J6
Gaylesville, Ala. (35973) 195/G2
Gaylord, Kansas (67638) 232/D2
Gaylord, Mich. (49735) 250/E3
Gaylord, Minn. (55334) 255/D6
Gaylord, Oreg. (97458) 291/C5
Gaylord, Va. (†22611) 307/M2
Gaylordsville, Conn. (06755) 210/A2
Gayndah, Queensland 95/D5
Gayndah, Queensland 88/J5
Gayny, U.S.S.R. 52/H2
Gays, Ill. (61928) 222/E4
Gaysin, U.S.S.R. 52/C5
Gays Mills, Wis. (54631) 317/E9
Gaysport, Ohio (†43720) 284/G6
Gaysville, Vt. (05746) 268/B4
Gayville, S. Dak. (57031) 298/P8
Gaza, Cent. Afr. Rep. 115/C3
Gaza, Egypt 59/B3
Gaza, Gaza Strip 65/A5
Gaza, Iowa (†51245) 229/B2
Gaza (prov.), Mozambique 118/E4
Gaza, N.H. (†03269) 268/D4
GAZA STRIP 59/B3
Gaza Strip 65/A5
Gazelle, Calif. (96034) 204/C2
Gazelle (pen.), Papua N.G. 86/B2
Gaziantep (prov.), Turkey 63/G4
Gaziantep, Turkey 54/E6
Gaziantep, Turkey 63/G4
Gaziantep, Turkey 59/C2
Gazik, Iran 66/L4
Gazipaşa, Turkey 63/E4
Gbarnga, Liberia 106/C7
Gbarnga, Liberia 102/B4
Gbogo, Nigeria 106/F7
Gcuwa, S. Africa 118/D6
Gdańsk (prov.), Poland 47/D1
Gdańsk, Poland 7/F3
Gdańsk, Poland 47/D1
Gdov, U.S.S.R. 52/C3
Gdynia, Poland 7/F3
Gdynia, Poland 47/D1
Gearhart, Oreg. (97138) 291/C1
Geary (co.), Kansas 232/F3
Geary, New Bruns. 170/D3
Geary, Okla. (73040) 288/K3
Geashill, Ireland 17/G5
Geauga (co.), Ohio 284/H3
Gebe (isl.), Indonesia 85/H6
Gebeit Mine, Sudan 111/G3
Gebo, Wyo. (†82430) 319/D2
Gebze, Turkey 63/C2
Gedaref, Sudan 111/G5
Gedaref, Sudan 59/C7
Gedaref, Sudan 102/F3
Geddes, S. Dak. (57342) 298/M7
Gede (mt.), Indonesia 85/H2
Gedera, Israel 65/B4
Gedi (ruins), Kenya 115/G4
Gedinne, Belgium 27/F9
Gediz, Turkey 63/C3
Gediz (riv.), Turkey 63/C3
Gedo, Ethiopia 111/G6
Gedo (prov.), Somalia 115/H3
Gedser, Denmark 21/F8
Gedser Odde (pt.), Denmark 21/E8
Gedsted, Denmark 21/C4
Geebung, Queensland 88/K2
Geebung, Queensland 95/E2
Geel, Belgium 27/F6
Geelong, Victoria 88/L7
Geelong, Victoria 97/C6
Geelong West, Victoria 88/G7
Geelong West, Victoria 97/C6
Geelvink (Cenderawasih) (bay), Indonesia 85/K6
Geelvink (chan.), W. Australia 88/A5
Geelvink (chan.), W. Australia 92/A5
Geertruidenberg, Netherlands 27/F5
Geesthacht, W. Germany 22/D2
Geetingsville, Ind. (†46041) 227/D4
Geeveston, Tasmania 99/C5
Geff, Ill. (62842) 222/E5
Gê'gyai, China 77/B5
Geh, Iran 59/H4
Geh, Iran 66/L7
Gehua, Papua N.G. 85/C8
Geidam, Nigeria 106/G6
Geiger, Ala. (†35459) 195/B5
Geiger Heights, Wash. (†99219) 310/H3
Geikie (riv.), Sask. 181/M3
Geilo, Norway 18/F6
Geiranger, Norway 18/E5
Geislingen an der Steige, W. Germany 22/C4
Geismar, La. (70734) 238/K3
Geist (res.), Ind. 227/F5
Geistown, Pa. (15904) 294/E5
Geita, Tanzania 115/F4
Gejiu (Kokiu), China 77/F7
Gejiu, China 54/M7
Gela, Italy 34/E6
Gelang, Tanjong (pt.), Malaysia 72/D6
Geldenaken (Jodoigne), Belgium 27/F7
Gelderland (prov.), Netherlands 27/H4
Geldermalsen, Netherlands 27/G5
Geldern, W. Germany 22/B3
Geldrop, Netherlands 27/H6
Geleen, Netherlands 27/H7
Gelendzhik, U.S.S.R. 52/E6
Gelgia (riv.), Switzerland 39/J3
Gelibolu (Gallipoli), Turkey 63/C5
Gelidonya (cape), Turkey 59/B2
Gelidonya (cape), Turkey 63/D4
Gelligaer, Wales 13/A6
Gelnhausen, W. Germany 22/C3
Gelnica, Czech. 41/F2
Gelså (riv.), Denmark 21/C7
Gelsenkirchen, W. Germany 22/B3
Gelsted, Denmark 21/C7
Gelterkinden, Switzerland 39/E2
Gem, Alberta 182/D4
Gem (co.), Idaho 220/B6
Gem, Idaho (†83873) 220/C2
Gem, Ind. (†46140) 227/F5
Gem, Kansas (67734) 232/B2
Gem (lake), Manitoba 179/G4
Gem, W. Va. (26625) 312/E5
Gemas, Malaysia 72/D7
Gembloux-sur-Orneau, Belgium 27/F7
Gemena, Zaire 115/D3
Gemena, Zaire 102/E4
Gemerek, Turkey 63/G3
Gemert, Netherlands 27/H5
Gemlik, Turkey 63/C2
Gemmell, Minn. (†56660) 255/D3
Gemona, Italy 34/D1
Gemsa, Egypt 111/F2
Genç, Turkey 63/J3
Gendringen, Netherlands 27/J5
Gene Autry, Okla. (73436) 288/N6
Genemuiden, Netherlands 27/H3
General Acha, Argentina 143/C4
General Alvear, Buenos Aires, Argentina 143/F7
General Alvear, Mendoza, Argentina 143/C3
General Arenales, Argentina 143/F7
General Artigas, Paraguay 144/D5
General Belgrano, Argentina 143/G7
General Bravo, Mexico 150/K4
General Campos, Argentina 143/G5
General Cepeda, Mexico 150/J4
General Conesa, Argentina 143/C5
General Elizardo Aquino, Paraguay 144/D4
General Enrique Martínez, Uruguay 145/F4
General Eugenio A. Garay, Paraguay 144/A2
General Galarza, Argentina 143/C6
General Grant Grove Section (King's Canyon), Calif. 204/G7
General Güemes, Argentina 143/D1
General Guido, Argentina 143/E4
General José de San Martín, Argentina 143/E2
General Juan Madariaga, Argentina 143/E4
General Lagos, Chile 138/B1
General La Madrid, Argentina 143/D4
General Las Heras, Argentina 143/G7
General Lavalle, Argentina 143/E4
General Manuel Belgrano, Cerro (mt.), Argentina 143/C2
General Mitchell Field, Wis. 317/M2
General O'Brien, Argentina 143/F7
General Paz, Argentina 143/E2
General Paz, Argentina 143/H7
General Paz (lake), Chile 138/E5
General Pico, Argentina 143/D4
General Ramírez, Argentina 143/F6
General Roca, Argentina 143/C4
General Saavedra, Bolivia 136/D5
General San Martín, Argentina 143/D4
General San Martín, Argentina 143/G4
General Santos, Philippines 82/E7
General Terán, Mexico 150/K4
General Tinio, Philippines 82/C3
General-Toshevo, Bulgaria 45/H4
General Viamonte, Argentina 143/F7
General Villegas, Argentina 143/D4
Generoso (mt.), Switzerland 39/H5
Genesee, Idaho (83832) 220/B3
Genesee (co.), Mich. 250/F5
Genesee (co.), N.Y. 276/D4
Genesee (riv.), N.Y. 276/E6
Genesee, Pa. (16923) 294/G2
Genesee, Wis. (†53127) 317/J2
Genesee Depot, Wis. (53127) 317/J2
Geneseo, Ill. (61254) 222/C2
Geneseo, Kansas (67444) 232/D3
Geneseo, N.Y. (14454) 276/E5
Geneseo, N. Dak. (58037) 282/R7
Geneva (lake) 7/E4
Geneva (co.), Ala. 195/G8
Geneva, Ala. (36340) 195/G8
Geneva, Fla. (32732) 212/E3
Geneva (lake), France 28/G4
Geneva, Georgia (31810) 217/C5
Geneva, Idaho (83238) 220/G7
Geneva, Ill. (60134) 222/E2
Geneva, Ind. (46740) 227/H3
Geneva, Iowa (50633) 229/G3
Geneva, Ky. (†42406) 237/F5
Geneva, Minn. (56035) 255/E7
Geneva, Nebr. (68361) 264/G4
Geneva, N.Y. (14456) 276/G5
Geneva, Ohio (44041) 284/J2
Geneva, Pa. (16316) 294/B2
Geneva (Genève) (canton), Switzerland 39/B4
Geneva (Genève), Switzerland 39/B4
Geneva, Switzerland 7/E4
Geneva (lake), Switzerland 39/C4
Geneva, Texas (75947) 303/L6
Geneva (lake), Wis. 317/K10
Geneva-on-the-Lake, Ohio (44043) 284/H2
Genezin, Turkey 63/F3
Genichesk, U.S.S.R. 52/E5
Genil (riv.), Spain 33/D4
Genk, Belgium 27/H7
Gennargentu, Monti del (mt.), Italy 34/B5
Gennep, Netherlands 27/H5
Gennevilliers, France 28/B1
Genoa, Ark. (71840) 202/C7
Genoa, Colo. (80818) 208/N4
Genoa, Ill. (60135) 222/E1
Genoa (prov.), Italy 34/B2
Genoa, Italy 7/E4
Genoa, Italy 34/B2
Genoa (gulf), Italy 34/B2
Genoa, Nebr. (68640) 264/G3
Genoa, Nev. (89411) 266/B4
Genoa, N.Y. (13071) 276/G5
Genoa, Ohio (43430) 284/D2
Genoa, W. Va. (25517) 312/A6
Genoa, Wis. (54632) 317/D8
Genoa City, Wis. (53128) 317/K11
Genola, Minn. (†56364) 255/D5
Genola, Utah (†84655) 304/C4
Genova (Genoa), Italy 34/B2
Genovesa (isl.), Ecuador 128/C9
Gent (Ghent), Belgium 27/D6
Genthin, E. Germany 22/E2
Gentilly, France 28/B2
Gentilly, Minn. (†56716) 255/B3
Genting, Indonesia 85/D5
Gentofte, Denmark 21/F6
Gentry, Ark. (72734) 202/A1
Gentry (co.), Mo. 261/D2
Gentry, Mo. (64453) 261/D2
Gentryville, Ind. (47537) 227/C8
Gentryville, Mo. (†64402) 261/D2
Genzano di Roma, Italy 34/F7
Geographe (chan.), W. Australia 88/A4
Geographe (bay), W. Australia 92/A6
Geographe (chan.), W. Australia 92/A4
Geographical Center of North America, N. Dak. 282/K3
Geographical Center of U.S., S. Dak. 298/B4
George (isl.), 143/E7
George (lake), Fla. 212/E2
George, Iowa (51237) 229/B2
George (isl.), Manitoba 179/E2
George (lake), Manitoba 179/G4
George (co.), Miss. 256/G9
George (isl.), Newf. 166/C3
George (lake), N.S. Wales 97/E4
George (lake), N.Y. 276/N4
George (sound), N. Zealand 100/A6
George (lake), N. Dak. 282/L6
George (cape), Nova Scotia 168/G3
George (lake), Nova Scotia 168/B5
George (riv.), Que. 162/K4
George (riv.), Que. 146/M4
George (riv.), Québec 174/F2
George, S. Africa 118/C6
George (lake), Uganda 115/F4
George, Wash. (98824) 310/F3
George A.F.B., Calif. 204/H9
George B. Stevenson (dam), Pa. 294/G3
George Land (isl.), U.S.S.R. 4/B7
George Land (isl.), U.S.S.R. 48/E1
George Rogers Clark Nat'l Hist. Park, Ind. 227/B7
Georges (isls.), Maine 243/E8
Georges (riv.), N. S. Wales 88/K4
Georges (riv.), N.S. Wales 97/H4
Georges Brook, Newf. 166/D2
George's Cove, Newf. 166/C3
Georges Fork, Va. (†24228) 307/C6
Georges Mills, N.H. (03751) 268/C5
Georgetown, Ark. (72054) 202/G3
Georgetown, Calif. (95634) 204/E5
George Town (cap.), Cayman Is. 156/B3
Georgetown, Colo. (80444) 208/H3
Georgetown, Conn. (06829) 210/B4
Georgetown, Del. (19947) 245/S6
Georgetown, D.C. (20007) 245/E5
Georgetown, Fla. (32039) 212/E2
Georgetown, Gambia 106/A6
Georgetown, Georgia (31754) 217/B7
Georgetown (cap.), Guyana 2/G5
Georgetown (cap.), Guyana 131/C2
Georgetown (cap.), Guyana 120/D2
Georgetown, Idaho (83239) 220/G7
Georgetown, Ill. (61846) 222/F4
Georgetown, Ind. (47122) 227/F8
Georgetown, Ky. (40324) 237/M4
Georgetown, La. (71432) 238/F3
Georgetown, Maine (04548) 243/D8
Georgetown○, Maine (04548) 243/D8
George Town (Pinang), Malaysia 72/C6
George Town, Malaysia 54/M9
Georgetown, Mass. (01833) 249/L2
Georgetown○, Mass. (01833) 249/L2
Georgetown, Minn. (56546) 255/B3
Georgetown, Miss. (39078) 256/D7
Georgetown (lake), Mont. 262/C4
Georgetown, Ohio (45121) 284/C8
Georgetown, Pa. (15043) 294/A4
Georgetown, Pr. Edward I. 168/F2
Georgetown, Queensland 88/G3
Georgetown, Queensland 95/B3
Georgetown, St. Vin. & Grens. 161/A8
Georgetown, St. Vin. & Grens. 156/G4
Georgetown (co.), S.C. 296/J5
Georgetown, S.C. 188/L4
Georgetown, S.C. (29440) 296/J5
George Town, Tasmania 99/C3
George Town, Tasmania 88/H8
Georgetown, Tenn. (37336) 237/L10
Georgetown, Texas (78626) 303/G7
George V Coast (reg.) 5/C8
Georgeville, Minn. (†56312) 255/C5
Georgeville, Nova Scotia 168/F3
George Washington Carver Nat'l Mon., Mo. 261/D9
George Washington Birthplace Nat'l Mon., Va. 307/P4
George West, Texas (78022) 303/F9
Georgia 188/K4
GEORGIA 217
Georgia (str.), Br. Col. 184/J3
Georgia (state), U.S. 146/K6
Georgia○, Vt. (†05478) 268/A2
Georgia (str.), Wash. 310/B2
Georgia Center, Vt. (†05478) 268/A2
Georgian (bay), Ont. 162/H6
Georgian (bay), Ontario 177/D2
Georgian (bay), Ontario 175/D3
Georgiana, Ala. (36033) 195/E7
Georgian Bay Is. Nat'l Park, Ontario 177/C2D3
Georgian S.S.R., U.S.S.R. 7/J4
Georgian S.S.R., U.S.S.R. 52/F6
Georgian S.S.R., U.S.S.R. 48/D5
Georgiaville, R.I. (†02917) 249/H5
Georgina (riv.), North. Terr. 93/E6
Georgina (isl.), Ontario 177/E3
Georgina (riv.), Queensland 88/F4
Georgina (riv.), Queensland 95/A4
Georgiu-Dezh, U.S.S.R. 52/E4
Georgsmarienhütte, W. Germany 22/B2
Gera (dist.), E. Germany 22/D3
Gera, E. Germany 22/E3
Geraardsbergen, Belgium 27/D7
Gerald, Mo. (63037) 261/K6
Gerald, Sask. 181/K5
Geral de Goiás, Serra (range), Brazil 132/E6
Geraldine, Ala. (35974) 195/G2
Geraldine, Mont. (59446) 262/F3
Geraldine, N. Zealand 100/C6
Geraldton, Australia 87/B8
Geraldton, Ont. 162/H6
Geraldton, Ontario 177/H5
Geraldton, Ontario 175/C3
Geraldton, W. Australia 88/A5
Geraldton, W. Australia 92/A5
Gerar (dry riv.), Israel 65/B5
Gerber, Calif. (96035) 204/C3
Gerber (res.), Oreg. 291/F5
Gercüş, Turkey 63/J4
Gerdine (mt.), Alaska 196/A1
Gerede, Turkey 63/E2
Gereshk, Afghanistan 59/H3
Gereshk, Afghanistan 68/A2
Geretsried, W. Germany 22/D5
Gérgal, Spain 33/E4
Gerger, Turkey 63/H3
Gerik, Malaysia 72/D6
Gering, Nebr. (69341) 264/A3
Gerlach, Nev. (89412) 266/B2
Gerlachovka (mt.), Czech. 41/E2
Gerlogubi, Ethiopia 111/H6
Germania, Miss. (†39162) 256/C5
Germania, Pa. (16922) 294/G2
Germania, Wis. (†54968) 317/H8
Germano, Ohio (†43825) 284/J5
Germansen (lake), Br. Col. 184/E2
Germansen Landing, Br. Col. 184/E2
Germanton, N.C. (27019) 281/J2
Germantown, Ill. (62245) 222/D5
Germantown, Ky. (41044) 237/O3
Germantown, Md. (20767) 245/J4
Germantown, New Bruns. 170/F3
Germantown, N.Y. (12526) 276/N6
Germantown, Ohio (45327) 284/B6
Germantown, Tenn. (38138) 237/B10
Germantown, Wis. (53022) 317/K1
German Valley, Ill. (61039) 222/D1
Germany (East) 2/K3
Germany (West) 2/K3
GERMANY, EAST 22/E2
GERMANY, WEST 22
Germencik, Turkey 63/B4
Germersheim, W. Germany 22/C4
Germfask, Mich. (49836) 250/C2
Germiston, S. Africa 102/E7
Germiston, S. Africa 118/H6
Gerofit, Israel 65/D5
Gerolstein, W. Germany 22/B3
Gerona (prov.), Spain 33/H1
Gerona, Spain 33/H2
Geronimo, Ariz. (†85536) 198/F5
Geronimo, Okla. (73543) 288/K6
Gerpinnes, Belgium 27/F8
Gerpir (cape), Iceland 21/D1
Gerra, Switzerland 39/G4
Gerrardstown, W. Va. (25420) 312/K4
Gerringong, N.S. Wales 97/F4
Gerrish, N.H. (†03301) 268/D5
Gerry, N.Y. (14740) 276/B6
Gers (dept.), France 28/D6
Gers (riv.), France 28/D6
Gersau, Switzerland 39/G2
Gersfeld, W. Germany 22/C3
Gerster, Mo. (†64776) 261/E7
Gerty, Okla. (†74531) 288/O5
Gervais, Oreg. (97026) 291/A3
Gervasio, Uruguay 145/F4
Gêrzê, China 77/B5
Gerze, Turkey 63/F2
Geser, Indonesia 85/J6
Gesher, Israel 65/C2
Gesher Haziv, Israel 65/C1
Gessie, Ind. (†47974) 227/C4
Getafe, Spain 33/F4
Getaway, Ohio (†45675) 284/F9
Gettysburg, Ohio (45328) 284/A5
Gettysburg, Pa. (17325) 294/H6
Gettysburg, S. Dak. (57442) 298/K3
Gettysburg Nat'l Mil. Park, Pa. 294/H6
Getulio Vargas, Uruguay 145/F3
Getz Ice Shelf 5/B12
Geuda Springs, Kansas (67051) 232/E4
Geurie, N.S. Wales 97/E3
Gevar'am, Israel 65/B4
Gevaş, Turkey 63/K3
Gevgelija, Yugoslavia 45/F5
Gex, France 28/G4
Geyser, Mont. (59447) 262/F3
Geyserville, Calif. (95441) 204/B5
Geyve, Turkey 63/D2
Gezira, El (reg.), Sudan 111/F5
Ghabaghib, Syria 63/H5
Ghadames, Libya 102/D2
Ghadames, Libya 111/A2
Ghaghra (riv.), India 68/E3
Ghaida, P.D.R. Yemen 59/F6
Ghalla, Wadi el (dry riv.), Sudan 111/E5
Ghana 2/J5
Ghana 102/B4
GHANA 106/D7
Ghanzi, Botswana 118/C4
Gharb Abu Muharik (des.), Egypt 111/J4
Ghardaïa, Algeria 106/E2
Ghardaïa, Algeria 102/C1
Gharian, Libya 102/D1
Gharian, Libya 111/B1
Gharib, Jebel (mt.), Egypt 59/B4
Ghat, Libya 102/D2
Ghat, Libya 111/B3
Ghat Kopar, India 68/B7
Ghazaouet, Algeria 106/D2
Ghaziabad, India 68/D3
Ghazipur, India 68/E3
Ghazni, Afghanistan 68/B2
Ghazni, Afghanistan 59/J3
Ghea (riv.), India 68/F1
Gheen, Minn. (55740) 255/F3
Gheens, La. (70355) 238/K7
Ghemines, Libya 111/C1

Ghenghis Khan Wall (ruin), China 77/H2
Ghenghis Khan Wall (ruins), Mongolia 77/H2
Ghent, Belgium 27/D6
Ghent, Ky. (41045) 237/L3
Ghent, Minn. (56239) 255/C6
Ghent, N.Y. (12075) 276/N6
Ghent, W. Va. (25843) 312/D7
Gheorghe Gheorghiu-Dej, Romania 45/H2
Gheorghieni, Romania 45/G2
Gherla, Romania 45/G2
Ghimbi, Ethiopia 111/G6
Ghio, N.C. (†28345) 281/K5
Ghisonaccia, France 28/B7
Ghizar, Pakistan 68/C1
Gholson, Miss. (†39354) 256/G5
Ghost Dam, Alberta 182/C4
Ghost Lake, Alberta 182/C4
Ghurian, Afghanistan 68/A2
Ghurian, Afghanistan 59/H3
Giacomo (pass), Switzerland 39/G4
Gia Dinh, Vietnam 72/E5
Giannutri (isl.), Italy 34/C3
Giant's Causeway, N. Ireland 17/H1
Giarre, Italy 34/E6
Giatto, W. Va. (†24736) 312/D8
Gibara, Cuba 156/C2
Gibara, Cuba 158/J3
Gibbon, Minn. (55335) 255/D6
Gibbon, Nebr. (68840) 264/F4
Gibbon, Oreg. (†97810) 291/J2
Gibbons, Alberta 182/D3
Gibbonsville, Idaho (83463) 220/E4
Gibb River, W. Australia 92/D2
Gibbs, Mo. (63540) 261/H2
Gibbs, Sask. 181/G5
Gibbsboro, N.J. (08026) 273/B4
Gibbs City, Mich. (†49935) 250/G2
Gibbstown, N.J. (08027) 273/C4
Gibbsville, Wis. (†53085) 317/L8
Gibeon, Namibia 118/B5
Gibloux (mt.), Switzerland 39/D3
Gibraltar 7/D5
Gibraltar (str.) 2/J4
Gibraltar (str.) 102/B1
Gibraltar (str.) 7/D5
Gibraltar, 2/J4
GIBRALTAR 33
Gibraltar, 33/D4
Gibraltar (pt.), England 13/H4
Gibraltar, Mich. (48173) 250/F6
Gibraltar (str.), Morocco 106/C1
Gibraltar (str.), Spain 33/D5
Gibsland, La. (71028) 238/E1
Gibson (desert), Australia 87/C8
Gibson, Georgia (30810) 217/G4
Gibson (co.), Ind. 227/B8
Gibson, Iowa (50104) 229/J6
Gibson, La. (70356) 238/J7
Gibson, Miss. (†39730) 256/G3
Gibson, Mo. (63847) 261/M10
Gibson (res.), Mont. 262/D3
Gibson, N.C. (28343) 281/K5
Gibson, Pa. (18820) 294/L2
Gibson (co.), Tenn. 237/D9
Gibson, Tenn. (38338) 237/D9
Gibson, W. Australia 88/C6
Gibson, W. Australia 92/C6
Gibson (des.), W. Australia 88/C4
Gibson (des.), W. Australia 92/D3
Gibsonburg, Ohio (43431) 284/D3
Gibson City, Ill. (60936) 222/E3
Gibsonia, Pa. (15044) 294/B4
Gibsons, Br. Col. 184/K3
Gibson Station, Va. (†24248) 307/A7
Gibsonton, Fla. (33534) 212/C3
Gibsonville, N.C. (27249) 281/K2
Giddings, Texas (78942) 303/H7
Gideälv (riv.), Sweden 18/L5
Gideon, Mo. (63848) 261/N10
Gideon, Okla. (†74464) 288/R2
Gien, France 28/E4
Giese, Minn. (†56350) 255/E4
Giessendam-Hardinxveld, Netherlands 27/F5
Giethoorn, Netherlands 27/J3
Gif, France 28/E3
Gifan, Iran 66/K2
Giffard, Québec 172/J3
Giffnock, Scotland 15/B2
Gifford, Fla. (32960) 212/F4
Gifford, Ill. (61847) 222/E3
Gifford, Iowa (50259) 229/G4
Gifford (riv.), N.W. Terrs. 187/K2
Gifford, Scotland 15/F5
Gifford, S.C. (29923) 296/E6
Gifford, Wash. (99131) 310/G2
Gifhorn, W. Germany 22/D2
Gift Lake, Alberta 182/C2
Gifu (pref.), Japan 81/H6
Gifu, Japan 81/H6
Giganta, Sierra de la (mts.), Mexico 150/D4
Gigante, Colombia 126/C6
Gigha (isl.), Scotland 15/C5
Gigha (sound), Scotland 15/C5
Gig Harbor, Wash. (98335) 310/C3
Giglio (isl.), Italy 34/C3
Gijón, Spain 7/D4
Gijón, Spain 33/D1
Gil (isl.), Br. Col. 184/C3
Gila (riv.) 188/D4
Gila (co.), Ariz. 198/E5
Gila (mts.), Ariz. 198/A6
Gila (mts.), Ariz. 198/F5
Gila (riv.), Ariz. 198/B6
Gila, N. Mex. (88038) 274/A6
Gila (riv.), N. Mex. 274/A6
Gila (riv.), U.S. 146/G6
Gila Bend, Ariz. (85337) 198/C6
Gila Bend (mts.), Ariz. 198/B5
Gila Bend Ind. Res., Ariz. 198/C6
Gila Cliff Dwellings Nat'l Mon., N. Mex. 274/A5
Gila Hot Springs, N. Mex. (†88061) 274/A5
Gilan (prov.), Iran 66/F2
Gilan (reg.), Iran 66/F2
Gila River Ind. Res., Ariz. 198/C5
Gilat, Israel 65/B5
Gilbert, Ariz. (85234) 198/D5
Gilbert, Ark. (72636) 202/E2
Gilbert, Iowa (50105) 229/F4
Gilbert (isls.), Kiribati 2/T5
Gilbert (isls.), Kiribati 87/H6
Gilbert, La. (71336) 238/G2
Gilbert, Minn. (55741) 255/F3
Gilbert (riv.), Newf. 166/C3
Gilbert (riv.), Queensland 88/G3
Gilbert (riv.), Queensland 95/B3
Gilbert, S.C. (29054) 296/E4
Gilbert (peak), Utah 304/D3
Gilbert, W. Va. (25621) 312/C7
Gilberton, Pa. (17934) 294/K4
Gilbertown, Ala. (36908) 195/B7
Gilbert Plains, Manitoba 179/B3
Gilberts, Ill. (60136) 222/E1
Gilbertsville, Ky. (42044) 237/E7
Gilbertsville, N.Y. (13776) 276/K6
Gilbertville, Iowa (50634) 229/J4
Gilbertville, Mass. (01031) 249/F3
Gilbjerg Hoved (pt.), Denmark 21/F5
Gilboa, N.Y. (12076) 276/M6
Gilboa, Ohio (45847) 284/C3
Gilboa, W. Va. (26671) 312/E6
Gilbués, Brazil 132/E5
Gilby, N. Dak. (58235) 282/R3
Gilchrist (co.), Fla. 212/D2
Gilchrist (creek), Manitoba 179/F2
Gilchrist (lake), Manitoba 179/G2
Gilcrest, Colo. (80623) 208/K2
Gildford, Mont. (59525) 262/F2
Gilé, Mozambique 118/F3
Gilead, Conn. (†06232) 210/F2
Gilead, Ind. (†46951) 227/E3
Gilead○, Maine (†04217) 243/B7
Gilead, Nebr. (68362) 264/G4
Giles (co.), Tenn. 237/G10
Giles (co.), Va. 307/G6
Gilf Kebir (plat.), Egypt 111/E3
Gilford, Mich. (48736) 250/F5
Gilford○, N.H. (†03246) 268/E4
Gilford, N. Ireland 17/J3
Gilford Park, N.J. (†08753) 273/E4
Gilgai, N.S. Wales 97/F1
Gilgandra, N. S. Wales 88/H6
Gilgandra, N.S. Wales 97/E2
Gilgil, Kenya 115/G4
Gilgit, Pakistan 68/C1
Gilgit, Pakistan 59/K2
Gilgunnia, N.S. Wales 97/C3
Gill, Colo. (80624) 208/L2
Gill (lake), Ireland 17/E3
Gill○, Mass. (†01376) 249/D2
Gillam, Manitoba 179/K2
Gilleleje, Denmark 21/F5
Gilles (lake), S. Australia 94/E5
Gillespie, Ill. (62033) 222/D4
Gillespie, New Bruns. 170/C2
Gillespie (co.), Texas 303/F7
Gillett, Ark. (72055) 202/H5
Gillett, Pa. (16925) 294/J2
Gillett, Texas (78116) 303/G8
Gillett, Wis. (54124) 317/K6
Gillette, N.J. (07933) 273/E2
Gillette, Wyo. (82716) 319/G1
Gillett Grove, Iowa (51341) 229/C2
Gilleyville, La. 238/G2
Gillham, Ark. (71841) 202/B5
Gilliam, La. (71029) 238/C1
Gilliam, Mo. (65330) 261/F4
Gilliam (co.), Oreg. 291/G2
Gillies Bay, Br. Col. 184/H2
Gillingham, Dorset, England 13/E6
Gillingham, Kent, England 13/J8
Gillis (range), Nev. 266/C4
Gillisonville, S.C. (†29936) 296/E6
Gillsburg, Miss. (†39657) 256/C8
Gillsville, Georgia (30543) 217/E2
Gilly, Switzerland 39/B4
Gilman, Colo. (81634) 208/G3
Gilman, Conn. (06336) 210/G2
Gilman, Ill. (60938) 222/E3
Gilman, Ind. (†46001) 227/F4
Gilman, Iowa (50106) 229/H5
Gilman, Minn. (56333) 255/E5
Gilman, Vt. (05904) 268/D3
Gilman, Wis. (54433) 317/E5
Gilman City, Mo. (64642) 261/D2
Gilmanton○, N.H. (03237) 268/E5
Gilmanton, Wis. (54743) 317/C7
Gilmanton Iron Works, N.H. (03837) 268/E5
Gilmer (co.), Georgia 217/D1
Gilmer, Texas (75644) 303/J5
Gilmer (co.), W. Va. 312/E5
Gilmer, W. Va. (26350) 312/E5
Gilmore, Ark. (72339) 202/K3
Gilmore, Georgia (†30080) 217/J1
Gilmore City, Iowa (50541) 229/D3
Gilpin (co.), Colo. 208/H3
Gilroy, Calif. (95020) 204/D6
Gilson, Ill. (61436) 222/C3
Gilsum○, N.H. (03448) 268/C5
Gilt Edge, Tenn. (†38015) 237/B9
Giltner, Nebr. (68841) 264/F4
Gilze, Netherlands 27/F5
Gimel, Switzerland 39/B3
Gimie (mt.), St. Lucia 161/G6
Gimlet, Ky. (†41164) 237/P4
Gimli, Manitoba 179/F4
Gimo, Sweden 18/K6
Gingerland, St. Chris.-Nevis 161/D11
Gingin, W. Australia 92/A1
Gingoog, Philippines 82/E6
Gingoog (bay), Philippines 82/E6
Gings, Ind. (†46173) 227/G5
Ginir, Ethiopia 111/H6
Ginnosar, Israel 65/D2
Ginzo de Limia, Spain 33/C1
Gio, Hon (isl.), Vietnam 72/E3
Giohar, Somalia 102/G4
Giohar, Somalia 115/J3
Gioia del Colle, Italy 34/F4
Gioiosa Ionica, Italy 34/F5
Giornico, Switzerland 39/G4
Giovinazzo, Italy 34/F4
Gi-Paraná (riv.), Brazil 132/H10
Gippsland (reg.), Victoria 97/D6
Gipsy (lake), Alberta 182/E1
Gipsy, Mo. (63750) 261/M8
Gipsy, Pa. (15741) 294/E4
Giraltovce, Czech. 41/F2
Girard, (62640) 222/D4
Girard, Georgia (30426) 217/J4
Girard, Kansas (66743) 232/H4
Girard, La. (71244) 238/G2
Girard, Mich. (†49036) 250/E6
Girard, Ohio (44420) 284/J3
Girard, Pa. (16417) 294/B2
Girard, Texas (79518) 303/D4
Girardot, Colombia 126/C5
Girardville, Pa. (17935) 294/K4
Girardville, Québec 172/E1
Girdle Ness (prom.), Scotland 15/G3
Girdler, Ky. (40943) 237/O7
Girdletree, Md. (21829) 245/S8
Giresun (prov.), Turkey 63/H2
Giresun, Turkey 59/C1
Giresun, Turkey 63/H2
Girga, Egypt 59/B4
Girga, Egypt 111/F2
Girga, Egypt 102/F2
Giri (riv.), Zaire 115/C3
Girilambone, N.S. Wales 97/D2
Girón, Ecuador 128/C4
Gironde (dept.), France 28/C5
Gironde (riv.), France 28/C5
Giroux, Manitoba 179/F5
Girouxville, Alberta 182/B2
Girvan, Scotland 15/D5
Girvan, Scotland 10/D3
Girvin, Sask. 181/F4
Girvin, Texas (79740) 303/B6
Gisborne, N. Zealand 100/G3
Giscome, Br. Col. 184/F3
Gisenyi, Rwanda 115/E4
Gislaved, Sweden 18/H8
Gisors, France 28/D3
Gistel, Belgium 27/B6
Giswil, Switzerland 39/F3
Gitega, Burundi 115/F4
Giuba (riv.), Somalia 115/H3
Giubiasco, Switzerland 39/H4
Giulianova, Italy 34/E3
Giurgiu, Romania 45/G3
Giv'atayim, Israel 65/B3
Giv'at Brenner, Israel 65/B4
Giv'at Hayyim, Israel 65/B3
Give, Denmark 21/C6
Given, W. Va. (25245) 312/C5
Givet, France 28/F2
Givhans, S.C. (†29472) 296/G5
Givors, France 28/F5
Giza, Egypt 111/J3
Giza, Egypt 59/B4
Gizab, Afghanistan 59/J3
Gizab, Afghanistan 68/B2
Gizhiga (bay), U.S.S.R. 48/Q3
Gizo, Solomon Is. 86/D3
Giżycko, Poland 47/E1
Gjerlev, Denmark 21/D4
Gjerrild Klint (cliff), Denmark 21/D5
Gjirokastër, Albania 45/D5
Gjoa Haven, N.W.T. 162/G2
Gjoa Haven, N.W. Terrs. 187/J3
Gjøvik, Norway 18/G6
Gjøvik, Norway 7/E2
Glace Bay, N.S. 162/L6
Glace Bay, Nova Scotia 168/J2
Glacier (bay), Alaska 196/M1
Glacier, Br. Col. 184/J4
Glacier (co.), Mont. 262/C2
Glacier, Wash. (98244) 310/D2
Glacier (peak), Wash. 310/D2
Glacier Bay Nat'l Park, Alaska 196/M1
Glacier Bay Nat'l Preserve, Alaska 196/L1
Glacier Nat'l Park, Br. Col. 184/J4
Glacier Nat'l Park, Mont. 188/D1
Glacier Nat'l Park, Mont. 262/C2
Glacier Nat'l Pk., Br. Col. 162/D5
Gladbrook, Iowa (50635) 229/H4
Glade, Kansas (67639) 232/C2
Glade, La. (†71374) 238/G4
Gladehill, Va. (24092) 307/J7
Glade Park, Colo. (81523) 208/B5
Glades (co.), Fla. 212/E5
Glade Spring, Va. (24340) 307/E7
Glade Valley, N.C. (28627) 281/G2
Gladeville, Tenn. (37071) 237/J8
Gladewater, Texas (75647) 303/K5
Gladmar, Sask. 181/G6
Gladstone, Ill. (61437) 222/B3
Gladstone, Manitoba 179/D4
Gladstone, Mich. (49837) 250/C3
Gladstone, Mo. (64118) 261/P5
Gladstone, Nebr. (68363) 264/G4
Gladstone, N. Mex. (88422) 274/F2
Gladstone, N. Dak. (58630) 282/F6
Gladstone, Oreg. (97027) 291/B2
Gladstone, Queensland 88/J4
Gladstone, Queensland 95/D4
Gladstone, S. Australia 94/F5
Gladstone, Tasmania 99/D2
Gladstone, Va. (24553) 307/L5
Glad Valley, S. Dak. (57629) 298/F3
Gladwin (co.), Mich. 250/E4
Gladwin, Mich. (48624) 250/E5
Glady, W. Va. (26268) 312/G5
Gladys, Va. (24554) 307/K6
Glåma (riv.), Norway 7/F2
Glåma (riv.), Norway 18/G6
Glamis, Calif. (†92227) 204/K11
Glamis, Sask. 181/D4
Glamis, Scotland 15/F4
Glamoč, Yugoslavia 45/C3
Glamsbjerg, Denmark 21/D7
Glan, Philippines 85/G4
Glan, Philippines 82/E8
Glancy, Miss. (†39083) 256/C7
Gland, Switzerland 39/B4
Glandore, Ireland 17/C8
Glandore (harb.), Ireland 17/C9
Glandorf, Ohio (45848) 284/B3
Glâne (riv.), Switzerland 39/C3
Glanmire-Riverstown, Ireland 17/E8
Glanworth, Ireland 17/E7
Glärnisch (mt.), Switzerland 39/H2
Glarus (canton), Switzerland 39/H3
Glarus, Switzerland 39/H2
Glarus Alps (mts.), Switzerland 39/H3
Glasco, Kansas (67445) 232/E2
Glasco, N.Y. (12432) 276/M6
Glascock (co.), Georgia 217/G4
Glasford, Ill. (61533) 222/D3
Glasgo, Conn. (06337) 210/H2
Glasgow, Del. (†19711) 245/R2
Glasgow, Ill. (†62610) 222/C4
Glasgow, Ky. (42141) 237/J7
Glasgow, Mo. (65254) 261/G4
Glasgow, Mont. (59230) 262/K2
Glasgow, Pa. (16644) 294/E4
Glasgow, Scotland 7/D3
Glasgow, Scotland 10/B1
Glasgow, Scotland 15/B2
Glasgow, Va. (24555) 307/K5
Glasgow, W. Va. (25086) 312/D6
Glasier (lake), New Bruns. 170/A1
Glaslyn, Sask. 181/C2
Glas Maol (mt.), Scotland 15/E4
Glasnevin, Sask. 181/F6
Glass, Manitoba 179/F5
Glass (riv.), Scotland 15/D3
Glass (mts.), Texas 303/A7
Glassboro, N.J. (08028) 273/C4
Glasscock (co.), Texas 303/C6
Glasser, N.J. (07837) 273/D2
Glassport, Pa. (15045) 294/C7
Glasston, N. Dak. (58236) 282/R2
Glassville, New Bruns. 170/C2
Glastenbury (mt.), Vt. 268/A6
Glastonbury, Conn. (06033) 210/E2
Glastonbury○, Conn. (06033) 210/E2
Glastonbury, England 13/E6
Glastonbury, England 10/E5
Glatt (riv.), Switzerland 39/G2
Glattfelden, Switzerland 39/F1
Glauchau, E. Germany 22/E3
Glazier, Texas (79037) 303/D2
Glazov, U.S.S.R. 7/K3
Glazov, U.S.S.R. 52/H3
Gleason, Tenn. (38229) 237/D8
Gleason, Wis. (54435) 317/G5
Gleasondale, Mass. (01749) 249/J3
Gleeson, Ariz. (†85617) 198/F7
Gleichen, Alberta 182/D4
Gleisdorf, Austria 41/C3
Gleiwitz (Gliwice), Poland 47/A4
Glen (lake), Ireland 17/F1
Glen (lake), Mich. 250/C4
Glen, Minn. (†56431) 255/E4
Glen, Miss. (38846) 256/H1
Glen, Mont. (†59725) 262/D5
Glen, N.H. (03838) 268/E3
Glen (canyon), Utah 304/D6
Glen, W. Va. (25088) 312/D6
Glenada, Oreg. (†97439) 291/C4
Glenaire, Mo. (†64068) 261/R5
Glen Alice, Tenn. (†37854) 237/M9
Glen Allan, Miss. (38744) 256/B4
Glen Allen, Ala. (35559) 195/C3
Glenallen, Mo. (63751) 261/M8
Glen Allen, Va. (23060) 307/N5
Glen Almond, Québec 172/B4
Glen Alpine, N.C. (28628) 281/F3
Glenamaddy, Ireland 17/D4
Glen Arbor, Mich. (49636) 250/C4
Glenarden, Md. (†20801) 245/G4
Glen Arm, Md. (21057) 245/N3
Glenarm, N. Ireland 17/J2
Glenavon, Sask. 181/J5
Glen Avon Heights, Calif. (92509) 204/E10
Glenavy, N. Zealand 100/C6
Glenavy, N. Ireland 17/J2
Glen Bain, Sask. 181/E6
Glenbarr, Scotland 15/C5
Glenbeigh, Ireland 17/B7
Glenbeulah, Wis. (53023) 317/L8
Glenboro, Manitoba 179/C5
Glenbrook, Nev. (89413) 266/B3
Glenburn○, Maine (†04401) 243/F6
Glenburn, N. Dak. (58740) 282/H2
Glen Burnie, Md. (21061) 245/M4
Glenbush, Sask. 181/D2
Glen Campbell, Pa. (15742) 294/E4
Glen Canyon (dam), Ariz. 198/D2
Glen Canyon City, Utah (84741) 304/C6
Glen Canyon Nat'l Rec. Area, Ariz. 198/D1
Glen Canyon Nat'l Rec. Area, Utah 304/D6
Glencaple, Scotland 15/E5
Glen Carbon, Ill. (62034) 222/B2
Glencliff, N.H. (03238) 268/D4
Glencoe, Ala. (35905) 195/G3
Glencoe, Ill. (60022) 222/B5
Glencoe, Ky. (41046) 237/M3
Glencoe, La. (†70538) 238/G7
Glencoe, Minn. (55336) 255/D6
Glencoe, Mo. (63038) 261/M3
Glencoe, New Bruns. 170/D1
Glencoe, N. Mex. (88324) 274/D5
Glencoe, Ohio (43928) 284/J6
Glencoe, Okla. (74032) 288/M2
Glencoe, Ontario 177/C5
Glencoe, Pa. (15543) 294/E6
Glencoe, Scotland 15/C4
Glencoe, S. Africa 118/E5
Glencolumbkille, Ireland 17/D2
Glen Cove, Maine (04846) 243/E7
Glen Cove, N.Y. (11542) 276/R6
Glencross, S. Dak. (57630) 298/H3
Glendale, Ariz. (*85301) 198/C5
Glendale, Calif. 188/C4
Glendale, Calif. (*91201) 204/C10
Glendale, Fla. (†32433) 212/C5
Glendale, Ind. (†47558) 227/C7
Glendale, Kansas (†67425) 232/E3
Glendale, Ky. (42740) 237/K5
Glendale, Mass. (01229) 249/A3
Glendale, Mo. (63122) 261/P3
Glendale, Nev. (†89025) 266/G6
Glendale, N.H. (†03246) 268/E4
Glendale, Nova Scotia 168/G3
Glendale, Ohio (45246) 284/C9
Glendale, Oreg. (97442) 291/D5
Glendale (lake), Pa. 294/F4
Glendale, R.I. (02826) 249/H5
Glendale, S.C. (29346) 296/D2
Glendale, Utah (84729) 304/B6
Glen Dale, W. Va. (26038) 312/E3
Glendale, Wis. (†53209) 317/M1
Glendale Heights, Ill. (†60118) 222/A5
Glen Daniel, W. Va. (25844) 312/D7
Glen Dean, Ky. (40141) 237/J5
Glendevey, Colo. (†80485) 208/H1
Glendive, Mont. (59330) 262/M3
Glendo, Wyo. (82213) 319/G3
Glendo (res.), Wyo. 319/H3
Glendon, Alberta 182/E2
Glendon, N.C. (27251) 281/L4
Glendora, Calif. (91740) 204/D10
Glendora, Miss. (38928) 256/D3
Glendora, N.J. (08029) 273/B4
Glen Easton, W. Va. (26039) 312/E3
Glen Echo, Md. (20768) 245/E4
Glen Eden, N. Zealand 100/B1
Gleneden Beach, Oreg. (97388) 291/C3
Glen Elder, Kansas (67446) 232/D2
Glenelg, Md. (21737) 245/L3
Glenelg, Scotland 10/D2
Glenelg, Scotland 15/C3
Glenelg, S. Australia 88/D8
Glenelg, S. Australia 94/A8
Glenelg (riv.), Victoria 97/A5
Glenella, Manitoba 179/C4
Glen Ellyn, Ill. (60137) 222/A5
Glenevis, Alberta 182/C3
Glen Ewen, Sask. 181/K6
Glen Ferris, W. Va. (25090) 312/D6
Glenfield, N.Y. (13343) 276/K3
Glenfield, N. Zealand 100/B1
Glenfield, N. Dak. (58443) 282/N5
Glenfield, Pa. (†15143) 294/B6
Glen Flora, Texas (77443) 303/H8
Glen Flora, Wis. (54526) 317/E4
Glenford, Ohio (43739) 284/F6
Glen Gardner, N.J. (08826) 273/D2
Glengarriff, Ireland 17/C8
Glengarry, Mont. (†59457) 262/G3
Glengarry (county), Ontario 177/K2
Glengary, W. Va. (25421) 312/K4
Glenham, S. Dak. (57631) 298/J2
Glen Harbour, Sask. 181/G5
Glen Haven, Colo. (80532) 208/H2
Glen Haven, Mich. (†49621) 250/C4
Glen Haven, Wis. (53810) 317/E10
Glenhayes, W. Va. (25519) 312/A6
Glen Hedrick (Beaver), W. Va. (†25813) 312/D7
Glen Hope, Pa. (16645) 294/F4
Glen Innes, N. S. Wales 88/J5
Glen Innes, N.S. Wales 97/F1
Glen Jean, W. Va. (25846) 312/D7
Glen Kerr, Sask. 181/D5
Glenlea, Manitoba 179/E5
Glenlivet, New Bruns. 170/D1
Glenluce, Scotland 15/D6
Glen Lyn, Va. (24093) 307/G6
Glen Lyon, Pa. (18617) 294/E7
Glenmary, Tenn. (37740) 237/M8
Glenmere, Sask. 181/J5
Glen Miller, Ontario 177/G3
Glenmont, Ohio (44628) 284/F4
Glenmora, La. (71433) 238/E5
Glen More (dist.), Scotland 15/D3
Glenmorgan, Queensland 95/D5
Glenn (co.), Calif. 204/C4
Glenn, Georgia (30219) 217/B4
Glenn, Mich. (49416) 250/C6
Glennallen, Alaska (99588) 196/D1
Glenn Heights, Texas (†75115) 303/G3
Glennie, Mich. (48737) 250/F4
Glenns Ferry, Idaho (83623) 220/C7
Glenn Springs, S.C. (29347) 296/D2
Glennville, Calif. (93226) 204/G8
Glennville, Georgia (30427) 217/J7
Glenolden, Pa. (19036) 294/M7
Glenoma, Wash. (98336) 310/C4
Glenora, Br. Col. 184/K2
Glenorchy, Tasmania 88/H8
Glenorchy, Tasmania 99/D4
Glenormiston, Queensland 95/A4
Glen Park, N.Y. (†13601) 276/J3
Glenpool, Okla. (74033) 288/P3
Glen Raven, N.C. (27215) 281/L2
Glenreagh, N.S. Wales 97/G2
Glen Riddle, Pa. (19037) 294/L7
Glen Ridge, N.J. (07028) 273/B2
Glenrio, N. Mex. (88423) 274/F3
Glen Robertson, Ontario 177/K2
Glen Rock, N.J. (07452) 273/B1
Glen Rock, Pa. (17327) 294/J6
Glenrock, Wyo. (82637) 319/G3
Glen Rogers, W. Va. (25848) 312/D7
Glen Rose, Texas (76043) 303/G5
Glenrothes, Scotland 15/E4
Glen Roy, Ohio (†45692) 284/E7
Glen Saint Mary, Fla. (32040) 212/D1
Glens Falls, N.Y. (12801) 276/N4
Glens Fork, Ky. (42741) 237/L6
Glenshaw, Pa. (15116) 294/C6
Glenside, Pa. (19038) 294/M5
Glenside, Sask. 181/E4
Glenside-Churton Park, N. Zealand 100/B2
Glensted, Mo. (†65084) 261/G5
Glentana, Mont. (59240) 262/K2
Glenties, Ireland 10/B3
Glenties, Ireland 17/E2
Glentrool, Scotland 15/D5
Glentworth, Sask. 181/E6
Glen Ullin, N. Dak. (58631) 282/G6
Glenview, Ill. (60025) 222/B5
Glenview, Ky. (†40222) 237/K1
Glenview Nav. Air. Sta., Ill. 222/B5
Glenvil, Nebr. (68941) 264/F4
Glenville, Conn. (†06830) 210/A4
Glenville, Ireland 17/D7
Glenville, Minn. (56036) 255/E7
Glenville, N.C. (28736) 281/C4
Glenville, W. Va. (26351) 312/E5
Glen Walter, Ontario 177/K2
Glen White, W. Va. (25849) 312/D7
Glenwillow, Ohio (†44139) 284/J10
Glen Wilton, Va. (24438) 307/J5
Glenwood, Ala. (36034) 195/F7
Glenwood, Alberta 182/D5
Glenwood, Ark. (71943) 202/C5
Glenwood, Fla. (32722) 212/E2
Glenwood, Georgia (30428) 217/L1
Glenwood, Ill. (60425) 222/C6
Glenwood, Ind. (46133) 227/G5
Glenwood, Iowa (51534) 229/B6
Glenwood, Mich. (†49047) 250/C6
Glenwood, Minn. (56334) 255/C5
Glenwood, Mo. (63541) 261/G1
Glenwood, Newf. 166/D4
Glenwood, N.J. (07418) 273/D1
Glenwood, N. Mex. (88039) 274/A5
Glenwood, N.C. (28737) 281/F3
Glenwood, Oreg. (97120) 291/D2
Glenwood, Utah (84730) 304/C5
Glenwood, W. Va. (†24541) 307/K7
Glenwood, Wash. (98619) 310/D4
Glenwood, W. Va. (25520) 312/B5
Glenwood City, Wis. (54013) 317/B5
Glenwood Springs, Colo. (81601) 208/E4
Glezen, Ind. (†47567) 227/C8
Glidden, Iowa (51443) 229/D4
Glidden, Sask. 181/B4
Glidden, Wis. (54527) 317/E3
Glide, Oreg. (97443) 291/D4
Glin, Ireland 17/C6
Glis, Switzerland 39/E4
Glittertinden (mt.), Norway 7/E2
Glittertinden (mt.), Norway 18/F6
Gliwice, Poland 47/A4
Globe, Ariz. 188/D4
Globe, Ariz. (85501) 198/E5
Gloggnitz, Austria 41/D3
Głogów (Glogau), Poland 47/C3
Glomawr, Ky. (†41701) 237/P6
Glomfjord, Norway 18/J3
Gloria (bay), Cuba 158/G2
Glorieta, N. Mex. (87535) 274/D3
Glorioso (isls.), Réunion 118/H2
Glory, Minn. (†56431) 255/E4
Glory of Russia (cape), Alaska 196/D2
Glossop, England 13/J2
Glossop, England 10/G2
Gloster, La. (71030) 238/C2
Gloster, Miss. (39638) 256/B8
Glostrup, Denmark 21/F6
Gloucester, England 13/E6
Gloucester, England 10/E5
Gloucester, Mass. (01930) 249/M2
Gloucester (co.), New Bruns. 170/E1
Gloucester (co.), N.J. 273/C4
Gloucester, N.S. Wales 97/F2
Gloucester, N.C. (28528) 281/S5
Gloucester (cape), Papua N.G. 86/B2
Gloucester (co.), Va. 307/P6
Gloucester, Va. (23061) 307/P6
Gloucester City, N.J. (08030) 273/B3
Gloucester Junction, New Bruns. 170/E1
Gloucester Point, Va. (23062) 307/R6
Gloucestershire (co.), England 13/E6
Glouster, Ohio (45732) 284/F6
Glover (reef), Belize 154/D2
Glover, Mo. (63646) 261/L8
Glover (isl.), Newf. 166/C4
Glover, N. Dak. (†58474) 282/O7
Glover, Okla. (†74728) 288/S6
Glover○, Vt. (05839) 268/C2
Glovergap, W. Va. (†26585) 312/F3
Gloversville, N.Y. (12078) 276/M4
Glovertown, Newf. 166/C1
Gloverville, S.C. (29828) 296/D4
Głowno, Poland 47/D2
Głubczyce, Poland 47/C3
Glubokoye, U.S.S.R. 52/C3
Głuchołazy, Poland 47/C3
Glücksburg, W. Germany 22/C1
Gluckstadt, Miss. (†39110) 256/D5
Glückstadt, W. Germany 22/C2
Glukhov, U.S.S.R. 52/D4
Glumsø, Denmark 21/E7
Glyde (riv.), Ireland 17/H4
Glynco, Georgia (†31520) 217/J8
Glyncorrwg, Wales 13/D6
Glyndon, Md. (21071) 245/L3
Glyndon, Minn. (56547) 255/B4
Glyngøre, Denmark 21/C4
Glynn (co.), Georgia 217/J8
Glynn, La. (70736) 238/H5
Glynn, N. Ireland 17/K2
Gmünd, Carinthia, Austria 41/B3
Gmünd, Lower Austria, Austria 41/C2
Gmunden, Austria 41/B3
Gnadenhutten, Ohio (44629) 284/G5
Gnaw Bone, Ind. (†47448) 227/E6
Gnesta, Sweden 18/G2
Gniew, Poland 47/D2
Gniewkowo, Poland 47/D2
Gniezno, Poland 47/C2
Gnjilane, Yugoslavia 45/E4
Gnowangerup, W. Australia 88/B6
Gnowangerup, W. Australia 92/B6
Goa, Daman and Diu (terr.), India 68/C4
Goa (dist.), India 68/C5
Goalpara, India 68/G3
Goascorán, Honduras 154/D4
Goat Fell (mt.), Scotland 15/C5
Goat River, Br. Col. 184/G3
Goat Rock (dam), Ala. 195/H5
Goat Rock (lake), Ala. 195/H5
Goat Rock (dam), Georgia 217/B5
Goat Rock (lake), Georgia 217/B5
Goat Rocks (mt.), Wash. 310/D4
Goba, Ethiopia 111/H6
Goba, Ethiopia 102/G4
Goba, Mozambique 118/E5

Gobabis, Namibia 118/B4
Gobabis, Namibia 102/D7
Gobernador Crespo, Argentina 143/F5
Gobernador Gregores, Argentina 143/C6
Gobernador Mansilla, Argentina 143/G6
Gobi (des.) 54/M5
Gobi (des.), China 77/G3
Gobi (des.), Mongolia 77/G3
Goble, Oreg. (97048) 291/E1
Gobler, Mo. (63849) 261/N10
Gobles, Mich. (49055) 250/D6
Gobo, Japan 81/G7
Gobwen, Somalia 115/H4
Goch, W. Germany 22/B3
Go Cong, Vietnam 72/E5
Godahl, Minn. (†56081) 255/D6
Godalming, England 13/G8
Godalming, England 10/F5
Godavari (riv.), India 54/J8
Godavari (riv.), India 68/D5
Godbout, Québec 174/D3
Godbout, Québec 172/B1
Goddard, Kansas (67052) 232/E4
Godech, Bulgaria 45/F4
Goderich, Ontario 177/C4
Godfrey, Georgia (30650) 217/F4
Godfrey, Ill. (62035) 222/A2
Godhavn, Greenl. 4/C12
Godhavn, Greenland 146/N3
Godhra, India 68/C4
Gödöllő, Hungary 41/E3
Godoy Cruz, Argentina 143/C3
Gods (lake), Man. 162/G5
Gods (lake), Manitoba 179/K3
Gods (riv.), Man. 162/G4
Gods (riv.), Manitoba 179/K3
Gods Mercy (bay), N.W. Terrs. 187/K3
Gods River, Manitoba 179/K3
Godthåb (Nûk) (cap.), Greenl. 4/C12
Godthåb (Nûk) (cap.), Greenland 2/G2
Godthåb (Nûk) (cap.), Greenland 146/N3
Godwin, N.C. (28344) 281/M4
Godwin Austen (K2) (mt.), Pakistan 68/D1
Godwinsville, Georgia (†31023) 217/F6
Goehner, Nebr. (68364) 264/G4
Goéland (lake), Québec 174/B3
Goélands (lake), Québec 174/E1
Goeree (isl.), Netherlands 27/D5
Goes, Netherlands 27/D6
Goessel, Kansas (67053) 232/E3
Goetzville, Mich. (49736) 250/E2
Goff, Kansas (66428) 232/G2
Goff (creek), Okla. 288/C1
Goffstown○, N.H. (03045) 268/D5
Gogama, Ontario 175/D3
Gogama, Ontario 177/J5
Gogebic (co.), Mich. 250/F2
Gogebic (lake), Mich. 250/F2
Göggingen, W. Germany 22/D4
Gogrial, Sudan 111/E6
Goi, Ben (bay), Vietnam 72/F4
Goiana, Brazil 132/H4
Goiandira, Brazil 132/E7
Goiânia, Brazil 132/D7
Goiânia, Brazil 120/D4
Goiás (state), Brazil 132/D6
Goiás, Brazil 132/D6
Goiás, Brazil 120/D4
Goil, Loch (lake), Scotland 15/A1
Goin, Tenn. (†37825) 237/O8
Goirle, Netherlands 27/G5
Góis, Portugal 33/B2
Gojjam (prov.), Ethiopia 111/G5
Gökçe, Turkey 63/B2
Gökçeada (isl.), Turkey 59/A1
Gökçeada (isl.), Turkey 63/A2
Gökırmak (riv.), Turkey 63/F2
Göksu (riv.), Turkey 63/E4
Göksun, Turkey 63/G3
Gokteik, Burma 72/C2
Gol, Norway 18/F6
Gola (isl.), Ireland 17/E1
Golan Heights (reg.), West Bank 65/D1
Gölbaşı, Turkey 63/G4
Golborne, England 13/G2
Gol'chikha, U.S.S.R. 48/J2
Golconda, Ill. (62938) 222/E6
Golconda (ruins), India 68/D5
Golconda, Nev. (89414) 266/D2
Gölcük, Turkey 63/C2
Golčův Jeníkov, Czech. 41/C2
Gold (riv.), Nova Scotia 168/D4
Goldap, Poland 47/F1
Gold Bar, Wash. (98251) 310/D3
Gold Beach, Oreg. (97444) 291/C5
Goldbond, Va. (24094) 307/G6
Goldboro, Nova Scotia 168/G3
Gold Bridge, Br. Col. 184/F5
Gold Coast (reg.), Ghana 106/D8
Gold Coast, Queensland 95/E6
Gold Coast, Queensland 88/J5
Goldcreek, Mont. (59733) 262/D4
Golden, Br. Col. 184/J4
Golden, Colo. (80401) 208/J3
Golden, Idaho (†83530) 220/C4
Golden, Ill. g)62339) 222/B3
Golden, Ireland 17/F7
Golden, Miss. (38847) 256/H2
Golden, Mo. (65658) 261/E9
Golden, N. Mex. (†87047) 274/C3
Golden (bay), N. Zealand 100/D4
Golden, Okla. (74737) 288/S6
Golden (lake), Ontario 177/G2
Golden (lake), Wis. 317/H1
Golden Beach, Fla. (†33160) 212/C4
Golden City, Mo. (64748) 261/D8
Goldendale, Wash. (98620) 310/E5
Golden Ears Prov. Park, Br. Col. 184/L2
Golden Gate (chan.), Calif. 188/B3
Golden Gate (chan.), Calif. 204/H2
Golden Gate, Fla. (33999) 212/E5
Golden Gate, Ill. (62843) 222/E5
Golden Gate (range), Nev. 266/F5
Golden Gate Nat'l Rec. Area, Calif. 204/H2
Golden Grove, Jamaica 158/K6
Golden Hill, Md. (†21622) 245/O7
Golden Lake, Ontario 177/G2
Golden Meadow, La. (70357) 238/K8
Golden Prairie, Sask. 181/B5
Golden Rock, St. Chris.-Nevis 161/C10
Golden's Bridge, N.Y. (10526) 276/N8
Golden Shores, Ariz. (†86436) 198/A4
Golden Spike Nat'l Hist. Site, Utah 304/B2
Golden Vale (plain), Ireland 17/E7
Golden Valley, Minn. (55427) 255/G5
Golden Valley (co.), Mont. 262/G4
Golden Valley (co.), N. Dak. 282/C5
Goldenvalley, N. Dak. (58541) 282/F5
Golden Valley, Ontario 177/E2
Goldfield, Iowa (50542) 229/F3
Goldfield, Nev. 188/C3
Goldfield, Nev. (89013) 266/D5
Gold Hill, Ala. (†36879) 195/G5
Gold Hill, Nev. (†89440) 266/B3
Gold Hill, N.C. (28071) 281/J3
Gold Hill, Oreg. (97525) 291/D5
Goldonna, La. (71031) 238/D2
Gold Point, Nev. (†89013) 266/D5
Gold River, Br. Col. 184/D5
Goldsberry, Mo. (†63539) 261/G3
Goldsboro, Md. (21636) 245/P4
Goldsboro, N.C. 188/L3
Goldsboro, N.C. (27530) 281/O4
Goldsboro (Etters), Pa. (17319) 294/J5
Goldsby, Okla. (†73093) 288/L4
Goldsmith, Ind. (46045) 227/E4
Goldsmith, Texas (79741) 303/B5
Goldston, N.C. (27252) 281/L3
Goldstone (mt.), Idaho 220/E4
Goldsworthy, W. Australia 88/C4
Goldsworthy, W. Australia 92/B3
Goldthwaite, Texas (76844) 303/F6
Goldvein, Va. (22720) 307/N4
Goldville, Ala. (†35010) 195/G4
Göle, Turkey 63/K2
Goleniów, Poland 47/B2
Goleta, Calif. (93117) 204/F9
Golf, Fla. (†33444) 212/F5
Golf, Ill. (60029) 222/B5
Golfito, C. Rica 154/F6
Golf Manor, Ohio (†45201) 284/C9
Golfo Santa Clara, Mexico 150/B1
Gölhisar, Turkey 63/C4
Goliad (co.), Texas 303/G9
Goliad, Texas (77963) 303/G9
Gölköy, Turkey 63/G2
Golling an der Salzach, Austria 41/B3
Gölmarmara, Turkey 63/C3
Golmud (Golmo), China 77/D4
Golmud, China 54/L6
Golo (riv.), France 28/B6
Golo (isl.), Philippines 82/C4
Golovin, Alaska (99762) 196/F2
Golpayegan, Iran 59/F3
Golpayegan, Iran 66/G4
Gölpazarı, Turkey 63/D2
Golshan (Tabas), Iran 66/K4
Golspie, Scotland 15/E3
Goltry, Okla. (73739) 288/K1
Golts, Md. (21637) 245/P3
Golub-Dobrzyn, Poland 47/D2
Golungo Alto, Angola 115/B5
Golva, N. Dak. (58632) 282/C6
Goma, Zaire 115/E4
Goma, Zaire 102/E5
Gombari, Zaire 115/E3
Gombe, Nigeria 106/G6
Gomel', U.S.S.R. 7/H3
Gomel', U.S.S.R. 48/D4
Gomel', U.S.S.R. 52/D4
Gomer, Ohio (45809) 284/B4
Gomera (isl.), Spain 106/A3
Gomera (isl.), Spain 33/B5
Gometra (isl.), Scotland 15/B4
Gomez, Fla. (†33455) 212/F4
Gómez Farías, Mexico 150/F2
Gómez Palacio, Mexico 150/G4
Gomishan, Iran 66/J2
Goms (valley), Switzerland 39/F4
Gona, Papua N.G. 85/C7
Gonabad, Iran 59/G3
Gonabad, Iran 66/L3
Gonaïves, Haiti 158/B5
Gonaïves, Haiti 156/D3
Gonâve (gulf), Haiti 158/B5
Gonâve (isl.), Haiti 158/B6
Gonâve (isl.), Haiti 156/D3
Gonbad-e Kavus, Iran 66/J2
Gonbadli, Iran 66/M2
Gönc, Hungary 41/F2
Gonda, India 68/E3
Gondal, India 68/C4
Gondar, Ethiopia 102/F3
Gondar, Ethiopia 59/C7
Gondar, Ethiopia 111/G5
Gondia, India 68/E4
Gondola Point, New Bruns. 170/D3
Gondomar, Portugal 33/B2
Gönen, Turkey 63/B2
Gonggar, China 77/D6
Gongga Shan (mt.), China 77/F6
Gonghe, China 77/E4
Gongliu, China 77/B3
Gongola (state), Nigeria 106/G7
Gongola (riv.), Nigeria 106/G6
Gongolgan, N.S. Wales 97/D2
Góngora (mt.), C. Rica 154/E5
Goñi, Uruguay 145/C4
Gonjo, China 77/E5
Gonvick, Minn. (56644) 255/C3
Gonzaga, Philippines 82/D1
Gonzales, Calif. (93926) 204/D7
Gonzales, La. (70737) 238/L2
Gonzales (co.), Texas 303/G8
Gonzales, Texas (78629) 303/G8
Gonzalez, Fla. (32560) 212/B6
González, Mexico 150/K5
González, Riacho (riv.), Paraguay 144/C3
Goobies, Newf. 166/D2
Goochland (co.), Va. 307/N5
Goochland, Va. (23063) 307/N5
Goodbee, La. (†70433) 238/K6
Goode (mt.), Alaska 196/C1
Goode, Va. (24556) 307/K6
Goodell, Iowa (50439) 229/F3
Goodenough (cape) 5/C7
Gooderham, Ontario 177/F3
Goodeve, Sask. 181/H4
Goodfare, Alberta 182/A2
Goodfellow A.F.B., Texas 303/D6
Goodfield, Ill. (61742) 222/D3
Good Harbor (bay), Mich. 250/D3
Good Hart, Mich. (49737) 250/D3
Good Hope, Ala. (†36024) 195/E2
Goodhope (bay), Alaska 196/F1
Good Hope, Georgia (30641) 217/E3
Good Hope, Ill. (61438) 222/C3
Good Hope, La. (†70079) 238/N3
Good Hope, Miss. (†39094) 256/E5
Good Hope, Ohio (43121) 284/D7
Good Hope (cape), S. Africa 102/D8
Good Hope (cape), S. Africa 2/K7
Good Hope (cape), S. Africa 118/E7
Goodhue (co.), Minn. 255/F6
Goodhue, Minn. (55027) 255/F6
Gooding (co.), Idaho 220/D6
Gooding, Idaho (83330) 220/D7
Goodland, Fla. (33933) 212/E6
Goodland, Ind. (47948) 227/C3
Goodland, Kansas (67735) 232/A2
Goodland, Minn. (55742) 255/E3
Goodlands, Manitoba 179/B5
Goodlettsville, Tenn. (37072) 237/H8
Goodlow, Br. Col. 184/G2
Good Luck, Md. (†20715) 245/G4
Goodman, Miss. (39079) 256/E5
Goodman, Mo. (63843) 261/C9
Goodman, Wis. (54125) 317/K4
Goodnews Bay, Alaska (99589) 196/F3
Goodnight, Texas (†79226) 303/C3
Goodnoe Hills, Wash. (†99356) 310/E5
Goodooga, N.S. Wales 97/D1
Good Pine, La. (†71342) 238/F3
Goodrich, Colo. (†80653) 208/M2
Goodrich, Mich. (48438) 250/F6
Goodrich, N. Dak. (58444) 282/K5
Goodrich, Texas (77335) 303/K7
Goodrich, Wis. (†54451) 317/G5
Goodridge, Alberta 182/E2
Goodridge, Minn. (56725) 255/C2
Goodsoil, Sask. 181/L4
Goodson, Mo. (65659) 261/F7
Good Spirit (lake), Sask. 181/J4
Goodspirit Lake Prov. Park, Sask. 181/J4
Goodspring, Tenn. (38460) 237/G10
Goodsprings, Ala. (35560) 195/D3
Goodsprings, Nev. (89019) 266/F7
Good Thunder, Minn. (56037) 255/D6
Goodview, Minn. (55027) 255/G6
Goodwater, Ala. (35072) 195/F4
Goodwater, Okla. (†74740) 288/S7
Goodwater, Sask. 181/H6
Goodway, Ala. (36449) 195/D8
Goodwell, Okla. (73939) 288/C1
Goodwin, Ark. (72340) 202/J4
Goodwin, S. Dak. (57238) 298/R4
Goodwins Mills, Maine (†04005) 243/B8
Goodwood, Ontario 177/E3
Goodwood, S. Africa 118/F6
Goodyear, Ariz. (85323) 198/C5
Gooik, Belgium 27/E7
Goole, England 10/F4
Goole, England 13/G4
Goolgowi, N.S. Wales 97/C3
Gooloogong, N.S. Wales 97/E3
Goomalling, W. Australia 88/B6
Goomalling, W. Australia 92/B1
Goombalie, N.S. Wales 97/C1
Goondiwindi, Queensland 88/H5
Goondiwindi, Queensland 95/D6
Goor, Netherlands 27/K4
Goose (lake) 188/B2
Goose (lake), Calif. 204/E1
Goose (creek), Idaho 220/E7
Goose (riv.), Newf. 166/B3
Goose (riv.), N. Dak. 282/P4
Goose (isl.), Nova Scotia 168/F4
Goose (isl.), Nova Scotia 168/G3
Goose (lake), Oreg. 291/G5
Goose (creek), Va. 307/J6
Goose (creek), Va. 307/N3
Goose Airport P.O. (Goose Bay), Newf. 162/K5
Goose Bay, Newf. 162/K5
Goose Bay, Newf. 146/M4
Goose Bay-Happy Valley, Newf. 166/B3
Gooseberry (creek), Wyo. 319/D1
Gooseberry Cove, Newf. 166/C2
Goose Cove, Newf. 166/C2
Goose Cove, Newf. 166/C3
Goose Cove, Nova Scotia 168/H2
Goose Creek (mts.), Idaho 220/E7
Goose Creek, Ky. (†40222) 237/L1
Goose Creek, S.C. (29445) 296/H6
Goose Lake, Iowa (52750) 229/N5
Gooseprairie, Wash. (†98937) 310/D4
Goose Rock, Ky. (40944) 237/O6
Goose Rocks Beach, Maine (†04046) 243/C9
Göppingen, W. Germany 22/C4
Góra, Poland 47/C3
Gorakhpur, India 68/E3
Gorchs, Argentina 143/G7
Gorda (pt.), Cuba 158/C2
Gorda (bank), Honduras 154/F3
Gorda (cay), Honduras 154/F3
Gorda (pt.), Nicaragua 154/F5
Gorda (pt.), Panama 154/H6
Gördes, Turkey 63/C3
Gordevio, Switzerland 39/G4
Gørding, Denmark 21/B7
Gordo, Ala. (35466) 195/C4
Gordola, Switzerland 39/G4
Gordon, Ala. (36343) 195/H8
Gordon (lake), Alberta 182/E1
Gordon (riv.), Br. Col. 184/H3
Gordon (isl.), Chile 138/E11
Gordon (co.), Georgia 217/C2
Gordon, Georgia (31031) 217/F5
Gordon, Kansas (†67010) 232/F4
Gordon, Nebr. (69343) 264/B2
Gordon, Ohio (45329) 284/B6
Gordon, Scotland 15/F5
Gordon, Tasmania 99/G5
Gordon (lake), Tasmania 99/C4
Gordon (riv.), Tasmania 99/B4
Gordon, Texas (76453) 303/F5
Gordon, W. Va. (25093) 312/C7
Gordon, Wis. (54838) 317/C3
Gordondale, Alberta 182/A2
Gordon Downs, W. Australia 92/E2
Gordon's Bay, S. Africa 118/F7
Gordonsburg, Tenn. (†38462) 237/F9
Gordonsville, Ala. (†36040) 195/E6
Gordonsville, Minn. (†56036) 255/E7
Gordonsville, Tenn. (38563) 237/K8
Gordonsville, Va. (22942) 307/M4
Gordonvale, Queensland 88/H3
Gordonvale, Queensland 95/C3
Gordonville, Mo. (63752) 261/N8
Gore (pt.), Alaska 196/C2
Goré, Chad 111/C6
Gore (range), Colo. 208/G3
Gore, Ethiopia 111/G6
Gore, Ethiopia 102/F4
Gore, N. Zealand 100/B7
Gore, Ohio (†43188) 284/F6
Gore, Okla. (74435) 288/R3
Gore (mt.), Vt. 268/D2
Gore, Va. (22637) 307/M2
Gore Bay, Ontario 177/B2
Gorebridge, Scotland 10/C1
Gorebridge, Scotland 15/D2
Goree, Texas (76363) 303/E4
Goregaon, India 68/B7
Görele, Turkey 63/H2
Gore Springs, Miss. (38929) 256/E3
Goreville, Ill. (62939) 222/E6
Gorey, Chan. Is. 13/F8
Gorey, Ireland 17/J6
Gorey, Ireland 10/C4
Gorgan, Iran 54/G6
Gorgan, Iran 59/F2
Gorgan (Gurgan), Iran 66/J2
Gorgan (riv.), Iran 66/J2
Gorgan (riv.), Iran 59/F2
Gorgas, Ala. (†35580) 195/D3
Gorgol (reg.), Mauritania 106/B5
Gorgona (isl.), Colombia 126/A6
Gorgona (isl.), Italy 34/B3
Gorham, Ill. (62940) 222/D6
Gorham, Kansas (67640) 232/D3
Gorham, Maine (04038) 243/C8
Gorham○, Maine (04038) 243/C8
Gorham, N.H. (03581) 268/E3
Gorham○, N.H. (03581) 268/E3
Gorham, N.Y. (14461) 276/F5
Gorham, N. Dak. (†58627) 282/D5
Gori, U.S.S.R. 52/F6
Gorin, Mo. (63543) 261/H2
Gorinchem, Netherlands 27/G5
Gorizia (prov.), Italy 34/D2
Gorizia, Italy 34/D2
Gorki, U.S.S.R. 52/D4
Gor'kiy, U.S.S.R. 2/M3
Gor'kiy, U.S.S.R. 7/J3
Gor'kiy, U.S.S.R. 48/E4
Gor'kiy, U.S.S.R. 52/F3
Gørlev, Denmark 21/E7
Gorlice, Poland 47/E4
Görlitz, E. Germany 22/F3
Görlitz, E. Germany 7/F3
Gorlitz, Sask. 181/J4
Gorlovka, U.S.S.R. 7/H4
Gorlovka, U.S.S.R. 52/E5
Gorman, Calif. (93243) 204/G9
Gorman, Tenn. (†37101) 237/F8
Gorman, Texas (76454) 303/F5
Gormania, W. Va. (26720) 312/H4
Gormanston, Ireland 17/J4
Gormanston, Tasmania 99/B4
Gorna Oryakhovitsa, Bulgaria 45/G4
Gornji Milanovac, Yugoslavia 45/D3
Gornji Vakuf, Yugoslavia 45/C4
Gorno-Altay Aut. Obl., U.S.S.R. 48/J4
Gorno-Altaysk, U.S.S.R. 48/J4
Gorno-Badakhshan Aut. Obl., U.S.S.R. 48/H6
Gornyak, U.S.S.R. 48/J4
Gorodets, U.S.S.R. 52/F3
Gorodok, U.S.S.R. 52/D3
Goroka, Papua N.G. 85/B7
Goroke, Victoria 97/A5
Gorong (isl.), Indonesia 85/J6
Gorong (isls.), Indonesia 85/J6
Gorongosa Nat'l Park, Mozambique 118/E3
Gorontalo, Indonesia 85/G5
Gorrahei, Ethiopia 111/H6
Gorredijk, Netherlands 27/J2
Gorrie, Ontario 177/C4
Gorst, Wash. (98337) 310/C3
Gort, Ireland 10/B4
Gort, Ireland 17/D5
Gortin, N. Ireland 17/G2
Gorum, La. (71434) 238/E4
Gorumna (isl.), Ireland 17/B5
Goryn' (riv.), U.S.S.R. 52/C4
Gorzów (prov.), Poland 47/B2
Gorzów Wielkopolski, Poland 47/B2
Gorzów Wielkopolski, Poland 7/F3
Göschenen, Switzerland 39/G3
Gose, Japan 81/J8
Gosen, Japan 81/J5
Goshen, Ala. (36035) 195/F7
Goshen, Ark. (72735) 202/C1
Goshen, Calif. (93227) 204/F7
Goshen○, Conn. (06756) 210/C1
Goshen (pt.), Conn. 210/G3
Goshen, Ind. (46526) 227/F1
Goshen, Ky. (40026) 237/K4
Goshen○, Mass. (01032) 249/C3
Goshen○, N.H. (03752) 268/C5
Goshen, N.J. (08218) 273/D5
Goshen, N.Y. (10924) 276/M8
Goshen, Nova Scotia 168/G3
Goshen, Ohio (45122) 284/B7
Goshen, Oreg. (97401) 291/D4
Goshen, Utah (84633) 304/C4
Goshen, Va. (24439) 307/K5
Goshen (co.), Wyo. 319/H4
Goshen Springs, Miss. (†39042) 256/E6
Goshogawara, Japan 81/K3
Goshute (mts.), Nev. 266/G2
Goshute Ind. Res., Nev. 266/G3
Goshute Ind. Res., Utah 304/A4
Gosier, Guadeloupe 161/B6
Goslar, W. Germany 22/D3
Gosnell, Ark. (†72315) 202/K2
Gosper (co.), Nebr. 264/E4
Gospić, Yugoslavia 45/B3
Gosport, Ala. (†36482) 195/C7
Gosport, England 13/F7
Gosport, England 10/F5
Gosport, Ind. (47433) 227/D6
Goss, Miss. (†39429) 256/E8
Gossau, Switzerland 39/H2
Gossville, N.H. (†03234) 268/E5
Gostivar, Yugoslavia 45/E5
Gostyń, Poland 47/C3
Gostynin, Poland 47/D2
Göta (canal), Sweden 18/J7
Göta (riv.), Sweden 18/H7
Gotebo, Okla. (73041) 288/J4
Göteborg, Sweden 7/F3
Göteborg, Sweden 18/G8
Göteborg och Bohus (co.), Sweden 18/G7
Gotha, E. Germany 22/D3
Gotham, Wis. (53540) 317/F9
Gothenburg, Nebr. (69138) 264/D4
Gothic (mesa), Ariz. 198/F2
Gotland (co.), Sweden 18/L8
Gotland (isl.), Sweden 7/F3
Gotland (isl.), Sweden 18/L8
Goto (isls.), Japan 81/D7
Goto (lake), Neth. Ant. 161/D8
Gotse Delchev, Bulgaria 45/F5
Gotska Sandön (isl.), Sweden 18/L7
Gotsu, Japan 81/F6
Göttingen, W. Germany 22/D3
Gottwaldov, Czech. 41/D2
Götzis, Austria 41/A3
Goubere, Cent. Afr. Rep. 115/E2
Gouda, Netherlands 27/F4
Goudeau, La. (71338) 238/G5
Gough (lake), Alberta 182/D3
Gough, Georgia (30811) 217/H4
Gough (isl.), St. Helena 2/J8
Gouin (res.), Que. 162/J6
Gouin (res.), Québec 174/C3
Goulburn, N.S. Wales 88/J6
Goulburn, N.S. Wales 97/E4
Goulburn (isls.), North. Terr. 88/E2
Goulburn (isls.), North. Terr. 93/C1
Goulburn (riv.), Victoria 97/C5
Goulburn Island, North. Terr. 93/C1
Gould, Ark. (71643) 202/G6
Gould, Colo. (†80480) 208/G2
Gould, Okla. (73544) 288/G5
Gould, Québec 172/F4
Gould City, Mich. (49838) 250/D2
Goulding, Fla. (†32502) 212/B6
Goulds, Fla. (33170) 212/F6
Goulds, Newf. 166/D2
Gouldsboro, Maine (†04607) 243/H7
Gouldsboro○, Maine (†04607) 243/H7
Gouldsboro, Pa. (18424) 294/L3
Gouldtown, Sask. 181/D5
Goulmima, Morocco 106/C2
Goumbou, Mali 106/C6
Gounamitz (riv.), New Bruns. 170/C1
Goundam, Mali 106/D5
Goundam, Mali 102/B3
Gourara (oasis), Algeria 106/E3
Gourbeyre, Guadeloupe 161/A7
Gourdon, France 28/D5
Gouré, Niger 106/G6
Gourma-Rharous, Mali 106/D5
Gournay-en-Bray, France 28/D3
Gouro, Chad 111/C4
Gourock, Scotland 10/A1
Gourock, Scotland 15/A1
Gouveia, Portugal 33/C2
Gouverneur, N.Y. (13642) 276/K2
Gouvy, Belgium 27/H8
Gouyave, Grenada 161/C8
Gouyave, Grenada 156/F4
Govan, Sask. 181/G4
Govan, S.C. (†29843) 296/E5
Gove (co.), Kansas 232/B3
Gove, Kansas (67736) 232/B3
Gove (Nhulunbuy), North. Terr. 93/E2
Govena (cape), U.S.S.R. 48/R4
Govenlock, Sask. 181/B6
Governador Valadares, Brazil 132/F7
Governador Valadares, Brazil 120/E4
Government (mt.), Ariz. 198/C3
Government (peak), Mich. 250/F1
Government (creek), Utah 304/B3
Government Camp, Oreg. (97028) 291/F2
Governor (lake), Nova Scotia 168/F3
Govi-Altay, Mongolia 77/E3
Gowan, Minn. (†55736) 255/F4
Gowanda, N.Y. (14070) 276/B6
Gowd-e Zerreh (depr.), Afghanistan 59/H4
Gowen, Mich. (49326) 250/D5
Gowen, Okla. (74545) 288/R5
Gowensville, S.C. (†29356) 296/C1
Gower, Mo. (64454) 261/C3
Gower (isl.), N.S. Wales 97/J2
Gower (mt.), N.S. Wales 97/J2
Gower (pen.), Wales 13/C6
Gowna (lake), Ireland 17/G4
Gowran, Ireland 17/G6
Gowrie, Iowa (50543) 229/E4
Gowrie Park, Tasmania 99/C3
Goya, Argentina 120/D5
Goya, Argentina 143/G4
Goyave, Guadeloupe 161/A6
Goyder (riv.), North. Terr. 93/D2
Goyders (lag.), S. Australia 94/F2
Göynücek, Turkey 63/F2
Göynük, Turkey 63/D2
Goz Beïda, Chad 111/D5
Gozo (isl.), Malta 34/E6
Goz Regeb, Sudan 111/G4
Graaff-Reinet, S. Africa 118/C6
Graal-Müritz, E. Germany 22/E1
Graauw, Netherlands 27/E6
Grabill, Ind. (46741) 227/H2
Grabouw, S. Africa 118/F7
Grabs, Switzerland 39/H2
Gračac, Yugoslavia 45/B3
Gračanica, Yugoslavia 45/D3
Grace, Idaho (83241) 220/G7
Grace (mt.), Mass. 249/E2
Grace, Miss. (38745) 256/C5
Grace (pt.), R.I. 249/H8
Grace City, N. Dak. (58445) 282/N4
Gracefield, Québec 172/A3
Graceham, Md. (†21788) 245/J2
Gracemont, Okla. (73042) 288/K4
Graceton, Minn. (†56686) 255/D2
Graceton, Pa. (15743) 294/D4
Graceville, Fla. (32440) 212/D5
Graceville, Minn. (56240) 255/B5
Graceville, Queensland 88/K3
Gracewood, Georgia (30812) 217/H4
Gracey, Ky. (42232) 237/F7
Grächen, Switzerland 39/E4
Gracias, Honduras 154/C3
Gracias a Dios (cape), Nic. 146/K8
Gracias a Dios (cape), Nicaragua 154/F3
Graciosa (isl.), Portugal 33/C1
Gradačac, Yugoslavia 45/D3
Gradaús, Brazil 120/D3
Gradaús, Brazil 132/D4
Gradaús, Serra dos (range), Brazil 132/D4
Grado, Spain 33/D1
Grady, Ala. (36036) 195/F7
Grady, Ark. (71644) 202/G5
Grady (co.), Georgia 217/D9
Grady (isl.), Newf. 166/C3
Grady, N. Mex. (88120) 274/F4
Grady (co.), Okla. 288/L5
Grady, Okla. (73545) 288/L6
Gradyville, Ky. (42742) 237/L6
Graeagle, Calif. (96103) 204/E4
Graested, Denmark 21/F5
Graettinger, Iowa (51342) 229/D2
Graf, Iowa (†52039) 229/M3
Grafenau, W. Germany 22/E4
Grafenwöhr, W. Germany 22/D4
Graff, Mo. (65660) 261/H8
Graff-Reinet, S. Africa 102/E8
Graford, Texas (76045) 303/F5
Grafton, Australia 87/F8
Grafton (isls.), Chile 138/D10
Grafton, Ill. (62037) 222/C5
Grafton, Ind. (†47620) 227/B9
Grafton, Iowa (50440) 229/G2
Grafton○, Mass. (01519) 249/H4
Grafton, Nebr. (68365) 264/G4
Grafton, New Bruns. 170/C2
Grafton (co.), N.H. 268/D4
Grafton○, N.H. (03240) 268/D4
Grafton, N. S. Wales 88/J5
Grafton, N.S. Wales 97/G1
Grafton, N.Y. (12082) 276/N5
Grafton, N. Dak. (58237) 282/R3
Grafton, Ohio (44044) 284/F3
Grafton, Ontario 177/G4
Grafton○, Vt. (05146) 268/B5
Grafton, Va. (23692) 307/P6
Grafton, W. Va. (26354) 312/G4
Grafton, Wis. (53024) 317/L9
Grafton Center, N.H. (†03240) 268/D4
Graham, Ala. (36263) 195/H4
Graham (lake), Alberta 182/C1
Graham (co.), Ariz. 198/E6
Graham (mt.), Ariz. 198/F6
Graham (isl.), Br. Col. 184/A3
Graham (peak), Colo. 208/E8
Graham, Fla. (32042) 212/D2
Graham, Georgia (†31513) 217/H7
Graham (creek), Ind. 227/F7
Graham (co.), Kansas 232/C2
Graham, Ky. (42344) 237/G6
Graham (lake), Maine 243/G6
Graham, Mo. (64455) 261/C2
Graham (co.), N.C. 281/B4
Graham, N.C. (27253) 281/L2
Graham (isl.), N.W.T. 162/M3
Graham (isl.), N.W. Terrs. 187/J2
Graham, Okla. (73437) 288/M6
Graham, Ontario 175/B3
Graham, Ontario 177/G5
Graham, Texas (76046) 303/F4
Graham Bell (isl.), U.S.S.R. 4/A6
Graham Bell (isl.), U.S.S.R. 48/G1
Grahamdale, Manitoba 179/D3
Graham Land (reg.), Ant. 2/G9
Graham Land (reg.) 5/C15
Graham Reach (chan.), Br. Col. 184/C3
Grahamstown, S. Africa 102/E8
Grahamstown, S. Africa 118/D6
Grahamsville, N.Y. (12740) 276/L7
Grahn, Ky. (41142) 237/P4
Graian Alps (range), France 28/G5
Graian Alps (range), Italy 34/A2
Graiguenamanagh-Tinnahinch, Ireland 17/H6
Grain Coast (reg.), Liberia 106/B8
Grainfield, Kansas (67737) 232/B2
Grainger (co.), Tenn. 237/O8
Graingers, N.C. (†28501) 281/O4
Grainola, Okla. (†74652) 288/N1
Grainton, Nebr. (69169) 264/C4
Grain Valley, Mo. (64029) 261/S6
Grajaú, Brazil 132/E4
Grajaú (riv.), Brazil 132/E4
Grajewo, Poland 47/F2
Gram, Denmark 21/C7
Gramalote, Colombia 126/D4
Gramat, France 28/D5
Grambling, La. (71245) 238/E1
Gramercy, La. (70052) 238/M3

Gramling, S.C. (29348) 296/C1
Grammer, Ind. (47236) 227/F6
Grammont (Geraardsbergen), Belgium 27/D7
Grampian, Pa. (16838) 294/E4
Grampian (reg.), Scotland 15/F3
Grampian (mts.), Scotland 15/D4
Gramsbergen, Netherlands 27/K3
Gran, Norway 18/G6
Granada, Colo. (81041) 208/P6
Granada, Minn. (56039) 255/D7
Granada, Nicaragua 154/E5
Granada (prov.), Spain 33/E4
Granada, Spain 33/E4
Granada, Spain 7/D5
Granada Hills, Calif. (91344) 204/B10
Granados, Mexico 150/E2
Granard, Ireland 17/F4
Granbury, Texas (76048) 303/G5
Granby, Colo. (80446) 208/H2
Granby (lake), Colo. 208/G2
Granby, Conn. (06035) 210/D1
Granby○, Conn. (06035) 210/D1
Granby, Mass. (01033) 249/E3
Granby○, Mass. (01033) 249/E3
Granby, Mo. (64844) 261/D9
Granby, Québec 172/E4
Granby○, Vt. (05840) 268/D2
Gran Canaria (isl.), Spain 33/B5
Gran Chaco (reg.) 120/C5
Gran Chaco (reg.), Argentina 143/D1
Gran Chaco (reg.), Paraguay 144/B2-3
Gran Couva, Trin. & Tob. 161/B11
Grand (canal), China 54/N6
Grand (canal), China 77/J4
Grand (co.), Colo. 208/G2
Grand (bay), Dominica 161/F7
Grand (canal), Ireland 17/G5
Grand (lake), La. 188/H4
Grand (lake), La. 238/H8
Grand (lake), La. 238/E7
Grand (lake), Maine 243/H4
Grand (isl.), Mich. 250/C2
Grand (lake), Mich. 250/F3
Grand (riv.), Mich. 250/D6
Grand (riv.), Mo. 261/F3
Grand (bay), New Bruns. 170/D3
Grand (lake), New Bruns. 170/D3
Grand (lake), New Bruns. 170/C3
Grand (lake), Newf. 166/B3
Grand (lake), Newf. 166/C4
Grand (isl.), N.Y. 276/B5
Grand (riv.), Ohio 284/H2
Grand (riv.), Ontario 177/D4
Grand (riv.), S. Dak. 298/F2
Grand (co.), Utah 304/E5
Grand Anse, Grenada 161/C9
Grand Bahama (isl.), Bahamas 146/L7
Grand Bahama (isl.), Bahamas 156/B1
Grand Bank, Newf. 166/C4
Grand-Bassam, Ivory Coast 106/D7
Grand Bay, Ala. (36541) 195/B10
Grand Bay, Dominica 161/F7
Grand Bay, New Bruns. 170/D3
Grand Bayou, La. (†71052) 238/C2
Grand Beach, Manitoba 179/F4
Grand Beach, Mich. (†49117) 250/C7
Grand Bend, Ontario 177/C4
Grand Blanc, Mich. (48439) 250/F6
Grand-Bourg, Guadeloupe 161/B7
Grand Bruit, Newf. 166/C4
Grand Caicos (isl.), Turks & Caicos 156/D2
Grand Caille (pt.), St. Lucia 161/F6
Grand Canary (isl.), Spain 102/A2
Grand Canary (isl.), Spain 106/A3
Grand Cane, La. (71032) 238/C2
Grand Canyon, Ariz. 188/D3
Grand Canyon, Ariz. (86023) 198/C2
Grand Canyon, Snake R. (canyon), Oreg. 291/L2
Grand Canyon Nat'l Mon., Ariz. 198/C2
Grand Canyon Nat'l Park, Ariz. 188/D3
Grand Canyon Nat'l Park, Ariz. 198/C2
Grand Canyon of the Snake River (canyon), Idaho 220/B4
Grand Cayman (isl.), Cayman Is. 156/B3
Grand Centre, Alberta 182/E2
Grand Cess, Liberia 106/C8
Grand Chain, Ill. (62941) 222/E6
Grand Chenier, La. (70643) 238/E7
Grand Combin (mt.), Switzerland 39/D5
Grand Comoro (isl.), Comoros 102/G6
Grand Comoro (isl.), Comoros 118/G2
Grand Coteau, La. (70541) 238/G6
Grand Coulee, Sask. 181/G5
Grand Coulee, Wash. (99133) 310/G3
Grand Coulee (canyon), Wash. 310/F3
Grand Coulee (dam), Wash. 310/F3
Grandcour, Switzerland 39/C3
Grand Cul de Sac (riv.), St. Lucia 161/G6
Grand Cul-de-Sac Marin (bay), Guadeloupe 161/A6
Grand Desert, Nova Scotia 168/E4
Grand Detour, Ill. (†61021) 222/D2
Grande (bay), Argentina 120/C8
Grande (bay), Argentina 143/C7
Grande (falls), Argentina 143/E3
Grande (riv.), Argentina 143/C4
Grande (marsh), Bolivia 136/F5
Grande (riv.), Bolivia 136/C6
Grande (riv.), Bolivia 136/C4
Grande (isl.), Brazil 132/C8
Grande (isl.), Brazil 135/D3
Grande (riv.), Brazil 120/E5
Grande (riv.), Brazil 132/D8
Grande (riv.), Brazil 132/E5
Grande (riv.), Brazil 135/B2
Grande (isl.), Chile 138/A6
Grande (riv.), Chile 138/F10
Grande, Salar (salt dep.), Chile 138/B3
Grande, Salto (falls), Colombia 126/D8
Grande (isl.), Colombia 126/B4
Grande (riv.), Guatemala 154/A3
Grande (riv.), Jamaica 158/K6
Grande (riv.), Mexico 150/G2
Grande (riv.), Mexico 150/N8
Grande (riv.), New Bruns. 170/C1
Grande (riv.), Nicaragua 154/E4
Grande (riv.), Peru 128/E10
Grande (range), Uruguay 145/D4
Grande, Arroyo (riv.), Uruguay 145/B4
Grande-Anse, New Bruns. 170/E1
Grande-Anse, Québec 172/E2
Grande Cache, Alberta 182/A3
Grande-Cascapédia, Québec 172/C2
Grande Cayemite (isl.), Haiti 158/B6
Grande-Clairière, Manitoba 179/B5
Grande de Añasco (riv.), P. Rico 161/B2
Grande de Arecibo (riv.), P. Rico 161/C1
Grande de Lípez (riv.), Bolivia 136/B7
Grande de Loíza (riv.), P. Rico 161/E1
Grande de Manatí (riv.), P. Rico 161/C1
Grande de Santiago (riv.), Mexico 150/G6
Grande de Tierra del Fuego (isl.), Argentina 143/C7
Grande de Tierra del Fuego (isl.), Chile 138/E11
Grande Dixence (dam), Switzerland 39/D4
Grande-Grève, Québec 172/D1
Grande Inferior (range), Uruguay 145/C4
Grande Pointe, Manitoba 179/F5
Grande Prairie, Alberta 182/A2
Grande Prairie, Alta. 162/E4
Grande-Prairie, Alta. 146/G4
Grande Prairie, Texas (*75050) 303/G2
Grand Erg Occidental (des.), Algeria 102/C1
Grand Erg Occidental (des.), Algeria 106/E2
Grand Erg Oriental (des.), Algeria 102/C1
Grand Erg Oriental (des.), Algeria 106/F2
Grand Erg Oriental (des.), Tunisia 106/F2
Grande' Rivière, Martinique 161/C5
Grande-Rivière, Québec 172/D2
Grande-Rivière, La (riv.), Que. 146/L4
Grande Rivière, La (riv.), Québec 174/B2
Grande Rivière, Trin. & Tob. 161/B10
Grande Rivière du Nord, Haiti 158/C5
Grande Ronde (riv.), Oreg. 291/K2
Grande Ronde (riv.), Wash. 310/H5
Grande Saline, Haiti 158/B5
Grandes-Bergeronnes, Québec 172/H1
Grandes-Piles, Québec 172/E3
Grand-Étang, Nova Scotia 168/G2
Grande-Terre, Guadeloupe 161/B6
Grande-Vallée, Québec 172/D1
Grande Vigie, Guadeloupe 161/B5
Grand Falls (lake), Maine 243/H5
Grand Falls, New Bruns. 170/C1
Grand Falls, Newf. 166/C4
Grand Falls, Newf. 146/N5
Grand Falls, Newf. 162/L6
Grandfalls, Texas (79742) 303/B6
Grand Falls Hill, New Bruns. 170/C1
Grandfield, Okla. (73546) 288/J6
Grand Forks, Br. Col. 184/H6
Grand Forks, N. Dak. 146/J5
Grand Forks, N. Dak. 188/G1
Grand Forks (co.), N. Dak. 282/P3
Grand Forks, N. Dak. (58201) 282/R4
Grand Forks A.F.B., N. Dak. 282/R4
Grand Glaise, Ark. (†72020) 202/G3
Grand Goâve, Haiti 158/B6
Grand Gorge, N.Y. (12434) 276/L6
Grand Gosier, Haiti 158/C6
Grand Gulf, Miss. (†39150) 256/B6
Grand Harbour, New Bruns. 170/D4
Grand Haven, Mich. (49417) 250/C5
Grand-Îlet (isl.), Guadeloupe 161/A7
Grandin, Fla. (32638) 212/E2
Grandin, Mo. (63943) 261/L9
Grandin, N. Dak. (58038) 282/R5
Grand Island, Nebr. 146/J5
Grand Island, Nebr. 188/G2
Grand Island, Nebr. (68801) 264/F4
Grand Island, N.Y. (14072) 276/B5
Grand Isle, La. (70358) 238/L8
Grand Isle, Maine (04746) 243/G1
Grand Isle, Maine 243/G18
Grand Isle (co.), Vt. 268/A2
Grand Isle○, Vt. (05458) 268/A2
Grand Junction, Colo. 146/H6
Grand Junction, Colo. 188/E3
Grand Junction, Colo. (81501) 208/B4
Grand Junction, Iowa (50107) 229/E4
Grand Junction, Mich. (49056) 250/C6
Grand Junction, Tenn. (38039) 237/C10
Grand-Lahou, Ivory Coast 106/C8
Grand Lake, Ark. (†71640) 202/H7
Grand Lake, Colo. (80447) 208/H2
Grand Lake, La. (†70601) 238/D6
Grand Lake, Nova Scotia 168/E4
Grand Lake Seboeis (lake), Maine 243/F3
Grand Lake Stream○, Maine (04637) 243/H5
Grand Lake Towne, Okla. (†74349) 288/S1
Grand Ledge, Mich. (48837) 250/E6
Grand-Lieu (lake), France 28/C4
Grand Manan (chan.), Maine 243/K6
Grand Manan (chan.), New Bruns. 170/C4
Grand Manan (isl.), New Bruns. 170/D4
Grand Marais, Manitoba 179/F4
Grand Marais, Mich. (49839) 250/D2
Grand Marais, Minn. (55604) 255/G2
Grand Marsh, Wis. (53936) 317/G8
Grand Meadow, Minn. (55936) 255/F7
Grand'Mère, Québec 172/E3
Grand Mound, Iowa (52751) 229/M5
Grand Mound, Wash. (†98531) 310/C4
Grand Muveran (mt.), Switzerland 39/D4
Grand Narrows, Nova Scotia 168/H3
Grândola, Portugal 33/B3
Grandora, Sask. 181/E3
Grand Pass, Mo. (65331) 261/F4
Grand-Popo, Benin 106/E7
Grand Portage, Minn. (55617) 255/G2
Grand Portage Ind. Res., Minn. 255/G2
Grand Portage Nat'l Mon., Minn. 255/G2
Grand Pré, Nova Scotia 168/D3
Grand Rapids, Manitoba 179/C1
Grand Rapids, Mich. 146/K5
Grand Rapids, Mich. 188/K2
Grand Rapids, Mich. (*49501) 250/D5
Grand Rapids, Minn. (55744) 255/E3
Grand Rapids, N. Dak. (†58458) 282/N7
Grand Rapids, Ohio (43522) 284/C3
Grand-Remous, Québec 172/B3
Grand Ridge, Fla. (32442) 212/A1
Grand Ridge, Ill. (61325) 222/E2
Grand River, Iowa (50108) 229/F7
Grand River, Nova Scotia 168/H3
Grand River, Ohio (44045) 284/H2
Grand River (valley), Utah 304/E4
Grand Rivers, Ky. (42045) 237/E7
Grand Ronde, Oreg. (97347) 291/D2
Grand Roy, Grenada 161/C8
Grand Saline, Texas (75140) 303/J5
Grand Santi, Fr. Guiana 131/D3
Grandson, Switzerland 39/C3
Grand Terrace, Calif. (92324) 204/E10
Grand Terre (isls.), La. 238/L8
Grand Teton (mt.), Wyo. 319/B2
Grand Teton Nat'l Park, Wyo. 319/B2
Grand Tower, Ill. (62942) 222/D6
Grand Traverse (co.), Mich. 250/D4
Grand Traverse (bay), Mich. 250/D3
Grand Turk (isl.), Turks & Caicos 156/D2
Grand Valley, Ontario 177/D4
Grand Valley, Pa. (16420) 294/C2
Grandview, Ark. (72637) 202/C1
Grand View, Idaho (83624) 220/B7
Grandview, Ill. (†62701) 222/D4
Grand View, Ill. (†61944) 222/F4
Grandview, Ind. (47615) 227/C9
Grandview, Iowa (52752) 229/L6
Grandview, Manitoba 179/B3
Grandview, Mo. (64030) 261/P6
Grandview, Ohio (†45767) 284/H7
Grandview, Tenn. (37337) 237/M9
Grandview, Texas (76050) 303/G5
Grandview, Wash. (98930) 310/F4
Grand View, Wis. (54839) 317/D3
Grandview Heights, Ohio (†43212) 284/D6
Grand View-on-Hudson, N.Y. (†10960) 276/K8
Grandview Plaza, Kansas (†66441) 232/F2
Grandville, Mich. (49418) 250/D6
Grand Wash (butte), Ariz. 198/B2
Grand Wash (riv.), Ariz. 198/B2
Grandy, Minn. (55029) 255/E5
Grandy, N.C. (27939) 281/T2
Graneros, Chile 138/G5
Grange, England 13/E3
Grange, N.H. (†03584) 268/E3
Grangeburg, Ala. (†36343) 195/H8
Grange City, Ky. (†41049) 237/O4
Grangemouth, Scotland 10/B1
Grangemouth, Scotland 15/C1
Granger, Ind. (46530) 227/E1
Granger, Iowa (50109) 229/F5
Granger, Minn. (55937) 255/F7
Granger, Mo. (63442) 261/H2
Granger, Texas (76530) 303/G7
Granger, Wash. (98932) 310/E4
Granger, Wyo. 188/D2
Granger, Wyo. (82934) 319/C4
Grangeville, Idaho (83530) 220/B4
Grangeville, La. (†70422) 238/J5
Granisle, Br. Col. 184/D3
Granite, Colo. (81228) 208/G4
Granite, Md. (†21163) 245/L3
Granite (pt.), Mich. 250/B2
Granite (co.), Mont. 262/C4
Granite (peak), Mont. 262/F5
Granite (peak), Nev. 266/B2
Granite (range), Nev. 266/B2
Granite, Okla. (73547) 288/H5
Granite, Oreg. (†97877) 291/J3
Granite (mts.), Wyo. 319/E3
Granite Bay, Br. Col. 184/E5
Granite Canon, Wyo. (82059) 319/G4
Granite City, Ill. (62040) 222/A2
Granite City Army Depot, Ill. 222/A2
Granite Falls, Minn. (56241) 255/C6
Granite Falls, N.C. (28630) 281/G3
Granite Falls, Wash. (98252) 310/D2
Granite Quarry, N.C. (28072) 281/H3
Graniteville, Mass. (01829) 249/J2
Graniteville, Mo. (†63650) 261/L7
Graniteville, S.C. (29829) 296/D4
Graniteville-East Barre, Vt. (05654) 268/C3
Granitique, Chaîne (range), Fr. Guiana 131/E4
Granity, N. Zealand 100/C4
Granja, Brazil 132/F3
Granma (prov.), Cuba 158/H4
Gränna, Sweden 18/J8
Grannis, Ark. (71944) 202/B5
Grano, N. Dak. (†58750) 282/G2
Granollers, Spain 33/H2
Gran Paradiso (mt.), Italy 34/A2
Gran Piedra (mt.), Cuba 158/J4
Gran Quivira, N. Mex. (†87036) 274/D4
Gran Sabana, La (plain), Venezuela 124/G5
Grant, Ala. (35747) 195/F1
Grant (co.), Ark. 202/F5
Grant, Colo. (80448) 208/H4
Grant, Fla. (32949) 212/F4
Grant (co.), Ind. 227/F3
Grant, Iowa (50847) 229/C6
Grant (co.), Kansas 232/A4
Grant (co.), Ky. 237/M3
Grant, Ky. (†41005) 237/M3
Grant (par.), La. 238/E3
Grant, La. (70644) 238/E5
Grant, Mich. (49327) 250/D5
Grant (co.), Minn. 255/B5
Grant, Mont. (†59725) 262/C5
Grant (co.), Nebr. 264/C3
Grant, Nebr. (69140) 264/C4
Grant (range), Nev. 266/F4
Grant (co.), N. Mex. 274/A5
Grant (co.), N. Dak. 282/G6
Grant (co.), Okla. 288/L1
Grant, Okla. (74738) 288/R7
Grant (co.), Oreg. 291/J3
Grant (co.), S. Dak. 298/R3
Grant (co.), Wash. 310/F3
Grant (co.), W. Va. 312/H4
Grant (co.), Wis. 317/E10
Grant Center, Iowa (†51026) 229/A4
Grant City, Mo. (64456) 261/D2
Grantfork, Ill. (†62249) 222/D5
Grantham, Alberta 182/E4
Grantham, England 13/G5
Grantham, England 10/F4
Grantham○, N.H. (03753) 268/C5
Granthams Landing, Br. Col. 184/J3
Grant-Kohrs Ranch Nat'l Hist. Site, Mont. 262/D4
Granton, Ontario 177/C4
Granton, Wis. (54436) 317/E6
Grantown-on-Spey, Scotland 10/E2
Grantown-on-Spey, Scotland 15/E3
Grant Park, Ill. (60940) 222/F2
Grants, N. Mex. (87020) 274/B3
Grantsboro, N.C. (28529) 281/R4
Grantsburg, Ill. (62943) 222/E6
Grantsburg, Ind. (47123) 227/E8
Grantsburg, Wis. (54840) 317/A4
Grantsdale, Mont. (59835) 262/B4
Grants Pass (chan.), Ala. 195/B10
Grants Pass, Oreg. 188/B2
Grants Pass, Oreg. (97526) 291/D5
Grantsville, Md. (21536) 245/B2
Grantsville, Utah (84029) 304/B3
Grantsville, W. Va. (26147) 312/D5
Grant Town, W. Va. (26574) 312/F3
Grantville, Alberta 182/D4
Grantville, Georgia (30220) 217/C4
Grantville, Kansas (66429) 232/G2
Grantwood Village, Mo. (†63155) 261/O4
Granum, Alberta 182/D5
Granville, France 28/C3
Granville, Ill. (61326) 222/D2
Granville, Iowa (51022) 229/B3
Granville (lake), Man. 162/G4
Granville (lake), Manitoba 179/H2
Granville○, Mass. (01034) 249/C4
Granville, Mo. (†65275) 261/H3
Granville, N.Y. (12832) 276/O4
Granville (co.), N.C. 281/M2
Granville, N. Dak. (58741) 282/J3
Granville, Ohio (43023) 284/E5
Granville, Tenn. (38564) 237/K8
Granville○, Vt. (05747) 268/B4
Granville, W. Va. (26534) 312/F3
Granville Centre, Nova Scotia 168/C4
Granville Ferry, Nova Scotia 168/C4
Granville Lake, Manitoba 179/H2
Grapeland, Miss. (†38725) 256/B3
Grapeland, Texas (75844) 303/J6
Grapeview, Wash. (98546) 310/C3
Grapeville, Pa. (15634) 294/C5
Grapevine, Ark. (72057) 202/F5
Grapevine (mts.), Calif. 204/H7
Grapevine, Texas (76051) 303/F2
Grapevine (lake), Texas 303/F2
Grasmere, Br. Col. 184/K5
Grasmere, N.H. (†03045) 268/D5
Grasmere, S. Africa 118/H7
Gräsö (isl.), Sweden 18/L6
Grasonville, Md. (21638) 245/O5
Grass (riv.), Manitoba 179/J3
Grass (riv.), N.Y. 276/K1
Grass (pt.), Virgin Is. (U.S.) 161/G4
Grass Creek, Ind. (46935) 227/E3
Grass Creek, Wyo. (82425) 319/D2
Grasse, France 28/G6
Grassflat, Pa. (16839) 294/F3
Grassington, England 13/F3
Grass Lake, Mich. (49240) 250/E6
Grassland, Alberta 182/D2
Grass Patch, W. Australia 92/C6
Grass Range, Mont. (59032) 262/H3
Grass River Prov. Park, Manitoba 179/H3
Grasston, Minn. (55030) 255/E5
Grass Valley, Calif. 188/B3
Grass Valley, Calif. (95945) 204/D4
Grass Valley, Oreg. (97029) 291/G2
Grassy (key), Fla. 212/F7
Grassy, Mo. (63753) 261/M8
Grassy Butte, N. Dak. (58634) 282/D5
Grassy Lake, Alberta 182/E5
Grassy Meadows, W. Va. (24943) 312/E7
Grassy Park, S. Africa 118/E6
Grassy Plains, Br. Col. 184/E3
Gråsten, Denmark 21/C8
Grates (pt.), Newf. 166/D2
Grates Cove, Newf. 166/D2
Gratiot (co.), Mich. 250/E5
Gratiot, Ohio (43740) 284/F6
Gratiot, Wis. (53541) 317/F10
Gratis, Ohio (4533) 284/A6
Graton, Calif. (95444) 204/C5
Gratwein, Austria 41/C3
Gratz, Ky. (40327) 237/M4
Gratz, Pa. (17030) 294/J4
Graubünden (Grisons) (canton), Switzerland 39/H3
Grauehörner (mts.), Switzerland 39/H3
Graulhet, France 28/E6
Graus, Spain 33/G1
Grauspitz (mt.), Liecht. 39/J2
Grave (pt.), France 28/C5
Grave, Netherlands 27/H5
Gravelbourg, Sask. 181/E6
Gravelly, Ark. (72838) 202/C4
Gravelly Beach, Tasmania 99/C3
Gravelly Branch, Nanticoke (riv.), Del. 245/R6
Gravel Switch, Ky. (40328) 237/L5
Gravelton, Ind. (†46542) 227/F2
Gravenhurst, Ontario 177/E3
Graves, Georgia (31755) 217/C7
Graves (co.), Ky. 237/D7
Gravesend, England 10/C5
Gravesend, England 13/J8
Gravette, Ark. (72736) 202/B1
Gravina (isl.), Alaska 196/N2
Gravina in Puglia, Italy 34/F4
Gravity, Iowa (50848) 229/D7
Gravois (pt.), Haiti 158/A7
Gravois Mills, Mo. (65037) 261/G6
Grawn, Mich. (49637) 250/D4
Gray, France 28/F4
Gray, Georgia (31032) 217/F4
Gray, Iowa (50110) 229/D5
Gray (co.), Kansas 232/B4
Gray, Ky. (40734) 237/O7
Gray, La. (70359) 238/J7
Gray, Maine (04039) 243/C8
Gray○, Maine (04039) 243/C8
Gray, Pa. (15544) 294/D5
Gray, Sask. 181/G5
Gray (co.), Texas 303/D2
Gray (canyon), Utah 304/D4
Gray, Va. (†23897) 307/O7
Grayback (mt.), Oreg. 291/D5
Gray Court, S.C. (29645) 296/C2
Gray Hawk, Ky. (40434) 237/N6
Gray Horse, Okla. (†74637) 288/N1
Grayland, Wash. (98547) 310/A4
Grayling, Alaska (99590) 196/G2
Grayling, Mich. (49738) 250/E4
Graymont, Ill. (61743) 222/E3
Graymoor, Ky. (†40201) 237/K1
Gray Rapids, New Bruns. 170/E2
Grayridge, Mo. (63850) 261/N9
Grays, Ark. (†72101) 202/H3
Grays (lake), Idaho 220/G6
Grays, S.C. (†29916) 296/E6
Grays (harb.), Wash. 310/A4
Graysbranch, Ky. (†41144) 237/R3
Grays Harbor (co.), Wash. 310/B3
Grayslake, Ill. (60030) 222/B4
Grays Lake Outlet (creek), Idaho 220/G6
Grays Landing, Pa. (15461) 294/C6
Grayson, Georgia (30221) 217/E3
Grayson (co.), Ky. 237/J5
Grayson, Ky. (41143) 237/R4
Grayson (lake), Ky. 237/P4
Grayson, La. (71435) 238/F2
Grayson, Mo. (†64492) 261/D3
Grayson, Okla. (†74437) 288/P3
Grayson, Sask. 181/J5
Grayson (co.), Texas 303/H4
Grayson (co.), Va. 307/F7
Grays River, Wash. (98621) 310/B4
Gray Summit, Mo. (63039) 261/L6
Graysville, Ala. (35073) 195/D3
Graysville, Georgia (30726) 217/B1
Graysville, Ind. (47852) 227/B6
Graysville, Manitoba 179/D5
Graysville, Ohio (45734) 284/H6
Graysville, Pa. (15337) 294/B6
Graysville, Tenn. (37338) 237/L10
Graytown, Ohio (43432) 284/D2
Grayville, Ill. (62844) 222/B4
Graz, Austria 7/F4
Graz, Austria 41/C3
Grazalema, Spain 33/D4
Greaney, Minn. (†55740) 255/F3
Great (lakes) 2/E3
Great (sound), Bermuda 156/G3
Great (fall), Guyana 131/B3
Great (chan.), India 68/G7
Great (riv.), Jamaica 158/H6
Great (pond), Mass. 249/D8
Great (pt.), Mass. 249/O7
Great (bay), N.H. 268/F5
Great (bay), N.J. 273/E4
Great (isl.), N. Zealand 100/D1
Great (lake), N.C. 281/P5
Great (lake), Tasmania 99/C3
Great (pond), Virgin Is. (U.S.) 161/F4
Great Abaco (isl.), Bahamas 146/L7
Great Abaco (isl.), Bahamas 156/C1
Great Alföld (plain), Hungary 41/F3
Great Australian (bight) 88/D6
Great Australian (bight), Australia 87/C9
Great Australian (bight), S. Australia 94/A5
Great Australian (bight), W. Australia 92/E6
Great Averill (pond), Vt. 268/D2
Great Bacolet (pt.), Grenada 161/D8
Great Baddow, England 13/J7
Great Bahama (bank), Bahamas 156/B1
Great Barrier (reef) 88/H2
Great Barrier (reef), 95/C2
Great Barrier (reef), Australia 87/E7
Great Barrier (isl.), N. Zealand 100/E2
Great Barrington, Mass. (01230) 249/A4
Great Barrington○, Mass. (01230) 249/A4
Great Barton, England 13/H5
Great Bear (lake), Canada 4/C16
Great Bear (lake), N.W.T. 162/D2
Great Bear (lake), N.W.T. 146/F3
Great Bear (lake), N.W. Terrs. 187/F3
Great Bear (riv.), N.W. Terrs. 187/F3
Great Bend, Kansas (67530) 232/D3
Great Bend, N.Y. (13643) 276/J2
Great Bend, N. Dak. (58039) 282/S7
Great Bend, Pa. (18821) 294/L2
Great Blasket (isl.), Ireland 17/A7
Great Bras d'Or (chan.), Nova Scotia 168/H2
Great Burnt (lake), Newf. 166/C4
Great Cacapon, W. Va. (25422) 312/K3
Great Coco (isl.), Burma 72/B4
Great Colinet (isl.), Newf. 166/D2
Great Corn (isl.), Nicaragua 154/F4
Great Cornard, England 13/H5
Great Cumbrae (isl.), Scotland 15/A2
Great Dismal (swamp), N.C. 281/S1
Great Dismal (swamp), Va. 307/R8
Great Divide (basin), Wyo. 319/E3
Great Dividing (range), N. S. Wales 88/J6
Great Dividing (range), N.S. Wales 97/E3
Great Dividing (range), Queensland 88/H4
Great Dividing (range), Queensland 95/C4
Great Egg Harbor (inlet), N.J. 273/E5
Great Egg Harbor (riv.), N.J. 273/D4
Greater Antilles (isls.) 146/K7
Greater London, England 13/H8
Greater Manchester (co.), England 13/H2
Greaterville, Ariz. (†85637) 198/E7
Great Exhibition (bay), N. Zealand 100/D1
Great Exuma (isl.), Bahamas 146/L7
Great Exuma (isl.), Bahamas 156/C2
Great Falls, Manitoba 179/F4
Great Falls, Mont. 188/D1
Great Falls, Mont. 146/G5
Great Falls, Mont. (59401) 262/E3
Great Falls, S.C. (29055) 296/F2
Great Falls (dam), Tenn. 237/K9
Great Fish (riv.), S. Africa 118/D6
Great Harwood, England 13/H1
Greathead (bay), St. Vin. & Grens. 161/A9
Great Inagua (isl.), Bahamas 146/L7
Great Inagua (isl.), Bahamas 156/D2
Great Indian (des.), India 68/C3
Great Isaac (isl.), Bahamas 156/B1
Great Kai (isl.), Indonesia 85/J7
Great Karoo (reg.), S. Africa 118/C6
Great Kei (riv.), S. Africa 118/D6
Great Khingan (range), China 54/O5
Great Lakes Nav. Trng. Ctr., Ill. 222/B4
Great Machipongo (inlet), Va. 307/S6
Great Meadows, N.J. (07838) 273/D2
Great Mercury (isl.), N. Zealand 100/F2
Great Miami (riv.), Ohio 284/A7
Great Misery (isl.), Mass. 249/F5
Great Moose (lake), Maine 243/D6
Great Namaland (reg.), Namibia 118/B4
Great Natuna (isl.), Indonesia 85/D5
Great Neck, N.Y. (*11020) 276/P6
Great Nicobar (isl.), India 68/G7
Great North (mt.), Va. 307/L2
Great Ormes (head), Wales 13/C4
Great Ouse (riv.), England 13/H5
Great Ouse (riv.), England 10/G4
Great Pedro Bluff (prom.), Jamaica 158/H6
Great Pee Dee (riv.), S.C. 296/J4
Great Pond○, Maine (†04408) 243/G6
Great Pond (bay), Virgin Is. (U.S.) 161/F4
Great Pubnico (lake), Nova Scotia 168/C5
Great Ruaha (riv.), Tanzania 115/G5
Great Sacandaga (lake), N.Y. 276/M4
Great Saint Bernard (pass), Italy 34/A2
Great Saint Bernard (mt.), Switzerland 39/D5
Great Saint Bernard (pass), Switzerland 39/D5
Great Saint Bernard (tunnel), Switzerland 39/D5
Great Salt (pond), St. Chris.-Nevis 161/D10
Great Salt (lake), Utah 188/D2
Great Salt (lake), Utah 146/G6
Great Salt (lake), Utah 304/B2
Great Salt Lake (des.), Nev. 266/H2
Great Salt Lake (des.), Utah 304/A3
Great Salt Plains (lake), Okla. 288/K1
Great Sand (hills), Sask. 181/B5
Great Sand Dunes Nat'l Mon., Colo. 208/H7
Great Sand Sea (des.), Egypt 111/D2
Great Sand Sea (des.), Libya 111/D2
Great Sandy (desert), Australia 87/C8
Great Sandy (Fraser) (isl.), Queensland 88/J5
Great Sandy (Fraser) (isl.), Queensland 95/E5
Great Sandy (des.), W. Australia 88/C4
Great Sandy (des.), W. Australia 92/C3
Great Seneca (creek), Md. 245/J4
Great Sitkin (isl.), Alaska 196/L4
Great Slave (lake), Canada 4/C15
Great Slave (lake), N.W.T. 162/E3
Great Slave (lake), N.W.T. 146/H3
Great Slave (lake), N.W. Terrs. 187/G3
Great Slave Lake Highway, N.W. Terrs. 187/G3
Great Smoky (mts.), N.C. 281/B3
Great Smoky (mts.), Tenn. 237/P9
Great Smoky Mountains Nat'l Park, Tenn. 237/P9
Great Smoky Mts. Nat'l Park, N.C. 281/B3
Great South (bay), N.Y. 276/O9
Great South (beach), N.Y. 276/O9
Great Tenasserim (riv.), Burma 72/C4
Great Thatch (isl.), Virgin Is. (Br.) 161/C4
Great Tobago (isl.), Virgin Is. (Br.) 161/B3

Great Torrington, England 10/D5
Great Torrington, England 13/C7
Great Valley, N.Y. (14741) 276/C6
Great Victoria (des.) 88/D5
Great Victoria (desert), Australia 87/C8
Great Victoria (des.), S. Australia 94/B3
Great Victoria (des.), W. Australia 92/D5
Great Village, Nova Scotia 168/E3
Great Wall (ruins), China 54/N5
Great Wall (ruins), China 77/G4,J
Great Wass (isl.), Maine 243/J7
Great Western Tiers (mts.), Tasmania 99/C3
Great Yarmouth, England 13/J5
Great Yarmouth, England 10/G4
Great Zab (riv.), Iraq 66/C2
Grecco, Uruguay 145/B3
Grecia, C. Rica 154/E5
Greco (cape), Cyprus 63/F5
Gredos, Sierra de (range), Spain 33/D2
Greece 2/L4
Greece 7/G5
GREECE 45/F6
Greece, N.Y. (14616) 276/E4
Greeley, Colo. 188/F2
Greeley, Colo. (80631) 208/K2
Greeley, Iowa (52050) 229/L3
Greeley (co.), Kansas 232/A3
Greeley, Kansas (66033) 232/G3
Greeley (co.), Nebr. 264/F3
Greeley, Nebr. (68842) 264/F3
Greeley, Pa. (18425) 294/N3
Greeley (creek), Utah 304/B3
Greeleyville, S.C. (29056) 296/H4
Greely (fjord), N.W. Terrs. 187/K1
Greely, Ontario 177/J2
Green (bay) 188/J1
Green (riv.) 188/D3
Green (isl.), Ant. & Bar. 161/E11
Green (riv.), Colo. 208/A2
Green (isl.), Grenada 161/D8
Green (riv.), Ill. 222/D2
Green, Kansas (67447) 232/E2
Green (co.), Ky. 237/K6
Green (riv.), Ky. 237/G6
Green (isl.), Maine 243/F8
Green (riv.), Mass. 249/B2
Green, Mich. (†49953) 250/F1
Green (bay), Mich. 250/B4
Green (lake), Minn. 255/D5
Green (riv.), New Bruns. 170/B1
Green (cape), N.S. Wales 97/F5
Green (swamp), N.C. 281/N6
Green (riv.), N. Dak. 282/D5
Green (pt.), Nova Scotia 168/C5
Green (isl.), Ontario 177/A2
Green, Oreg. (†97470) 291/D4
Green (isls.), Papua N.G. 86/C2
Green (lake), Sask. 181/D1
Green (riv.), Tenn. 237/F10
Green (riv.), U.S. 146/H6
Green (riv.), Utah 304/D4
Green (mts.), Vt. 268/B4
Green (cay), Virgin Is. (U.S.) 161/F4
Green (lake), Wash. 310/A2
Green (riv.), Wash. 310/C3
Green (co.), Wis. 317/G10
Green (bay), Wis. 317/L6
Green (mt.), Wyo. 319/E3
Green (riv.), Wyo. 319/C4
Greenacres, Calif. (93308) 204/F8
Green Acres, Sask. 181/F2
Greenacres, Wash. (99016) 310/J3
Greenacres City, Fla. (33463) 212/F5
Greenan, Sask. 181/C4
Greenback, Tenn. (37742) 237/N9
Greenbackville, Va. (23356) 307/T5
Green Bank, N.J. (†08215) 273/D4
Greenbank, Wash. (98253) 310/C2
Green Bank, W. Va. (24944) 312/G6
Green Bay, N. Zealand 100/B1
Green Bay, Va. (23942) 307/M6
Green Bay, Wis. 188/J2
Green Bay, Wis. 146/J5
Green Bay, Wis. (*54301) 317/K6
Greenbelt, Md. (20770) 245/G4
Greenbelt Park, Md. 245/G4
Greenbrier, Ala. (†35758) 195/E1
Greenbrier, Ark. (72058) 202/F3
Greenbrier, Mo. (†63730) 261/M8
Greenbrier, Tenn. (37073) 237/H8
Greenbrier (co.), W. Va. 312/F7
Greenbrier (riv.), W. Va. 312/F6
Green Brook, N.J. (08812) 273/D2
Greenbush, Mass. (02040) 249/F8
Greenbush, Mich. (48738) 250/F4
Greenbush, Minn. (56726) 255/B2
Greenbush, Va. (23357) 307/S5
Green Camp, Ohio (43322) 284/D4
Greencastle, Ind. (46135) 227/D5
Greencastle, Ireland 17/H1
Green Castle, Mo. (63544) 261/G2
Greencastle, Pa. (17225) 294/G6
Green Center, Ind. (†46701) 227/G2
Green City, Mo. (63545) 261/F2
Green Court, Alberta 182/C2
Green Cove Springs, Fla. (32043) 212/E2
Greencreek, Idaho (83533) 220/B3
Green Creek, N.J. (08219) 273/D5
Greendale, Ind. (†47025) 227/H6
Greendale, Wis. (53129) 317/L2
Greendell, N.J. (07839) 273/D2
Greene (co.), Ala. 195/C5
Greene (co.), Ark. 202/J1
Greene (co.), Georgia 217/F3
Greene (co.), Ill. 222/C4
Greene (co.), Ind. 227/D6
Greene (co.), Iowa 229/E5
Greene, Iowa (50636) 229/H3
Greene○, Maine (04236) 243/C7
Greene (co.), Miss. 256/G8
Greene (co.), Mo. 261/F8
Greene (co.), N.Y. 276/M6
Greene, N.Y. (13778) 276/J6
Greene (co.), N.C. 281/O3
Greene, N. Dak. (†58787) 282/G2
Greene (co.), Ohio 284/C6
Greene (co.), Pa. 294/B6
Greene, R.I. (02827) 249/G6
Greene (co.), Tenn. 237/R8
Greene (co.), Va. 307/M4
Greenevers, N.C. (†28521) 281/O5
Greeneville, Tenn. (37743) 237/R8
Greenfield, Calif. (93927) 204/D7
Greenfield, Ill. (62044) 222/C4
Greenfield, Ind. (46140) 227/F5
Greenfield, Iowa (50849) 229/D6
Greenfield, Mass. (01301) 249/D2
Greenfield○, Mass. (01301) 249/D2
Greenfield, Minn. (†55373) 255/F5
Greenfield, Mo. (65661) 261/E8
Greenfield○, N.H. (03047) 268/D6
Greenfield, Nova Scotia 168/D4
Greenfield, Ohio (45123) 284/D7
Greenfield, Okla. (73043) 288/K3
Greenfield, S. Dak. (†57010) 298/R8
Greenfield, Tenn. (38230) 237/D8
Greenfield, Wis. (53220) 317/L2
Greenfield Hill, Conn. (†06430) 210/B4
Greenfield Park, Québec 172/J4
Greenford, Ohio (44422) 284/J4
Green Forest, Ark. (72638) 202/D1
Green Hall, Ky. (41328) 237/O6
Green Harbor, Mass. (02041) 249/M4
Green Haven, Md. (21122) 245/M4
Greenhills, Ohio (45218) 284/B9
Greenhorn, Oreg. (†97877) 291/J3
Green Island, Iowa (52051) 229/N4
Green Island, Jamaica 158/G6
Green Island, N.Y. (12183) 276/N5
Green Island, N. Zealand 100/C7
Greenisland, N. Ireland 17/K2
Green Island (bay), Philippines 82/B5
Green Island Cove, Newf. 166/C3
Green Isle, Minn. (55338) 255/E6
Green Lake, Maine (†04429) 243/F6
Green Lake, Sask. 181/L4
Green Lake (co.), Wis. 317/H8
Green Lake, Wis. (54941) 317/H8
Greenland 2/G2
Greenland 4/B12
Greenland 146/P2
Greenland (sea) 146/T2
Greenland (sea) 4/B10
Greenland, Ark. (72737) 202/B1
Greenland, Barbados 161/B8
Greenland, Colo. (†80118) 208/K4
Greenland, Mich. (49929) 250/G1
Greenland○, N.H. (03840) 268/F5
Greenlaw, Scotland 15/F5
Greenleaf, Idaho (83626) 220/B6
Greenleaf, Kansas (66943) 232/E2
Greenleaf, Minn. (†55355) 255/D6
Greenleaf, Oreg. (97445) 291/D3
Greenleaf, Wis. (54126) 317/L7
Greenleafton, Minn. (†55965) 255/F7
Greenlee (co.), Ariz. 198/F5
Green Lowther (mt.), Scotland 15/E5
Greenmount, Ky. (†40741) 237/N6
Greenmount, Md. (†21074) 245/L2
Green Mountain (res.), Colo. 208/G3
Green Mountain, Iowa (50637) 229/H4
Green Mountain, N.C. (28740) 281/E3
Green Mountain Falls, Colo. (80819) 208/K5
Green Oaks, Ill. (†60048) 222/B4
Greenock, Scotland 10/A1
Greenock, Scotland 15/A2
Greenore, Ireland 17/J4
Greenore (pt.), Ireland 17/J7
Greenough (mt.), Alaska 196/K1
Greenough, Georgia (†31716) 217/D8
Greenough, Mont. (59836) 262/C4
Green Peter (lake), Oreg. 291/E3
Green Pond, Ala. (35074) 195/D4
Green Pond, N.J. (07435) 273/E1
Green Pond, S.C. (29446) 296/F6
Greenport, N.Y. (11944) 276/P8
Green Ridge (mts.), 245/E2
Green Ridge, Mo. (65332) 261/F5
Green River (lake), Ky. 237/L6
Green River, Utah (84525) 304/D4
Green River (res.), Vt. 268/B2
Green River, Wyo. 188/E2
Green River, Wyo. (82935) 319/C4
Green River (mt.), Wyo. 319/C2
Green Rock, Ill. (†61241) 222/C2
Greens (peak), Ariz. 198/F4
Greensboro, Ala. (36744) 195/C5
Greensboro, Fla. (32330) 212/B1
Greensboro, Georgia (30642) 217/F3
Greensboro, Ind. (47344) 227/G5
Greensboro, Md. (21639) 245/P5
Greensboro, N.C. 146/L6
Greensboro, N.C. 188/K3
Greensboro, N.C. (*27401) 281/K2
Greensboro, Pa. (15338) 294/B6
Greensboro○, Vt. (05841) 268/C2
Greensburg, Ind. (47240) 227/G6
Greensburg, Kansas (67054) 232/C4
Greensburg, Ky. (42743) 237/K6
Greensburg, La. (70441) 238/J5
Greensburg, Mo. (†63531) 261/H2
Greensburg, Ohio (44232) 284/G4
Greensburg, Pa. (15601) 294/D5
Green Sea, S.C. (29545) 296/J3
Greens Farms, Conn. (06436) 210/B4
Greens Fork, Ind. (47345) 227/H5
Green's Harbour, Newf. 166/D2
Greenshields, Alberta 182/E3
Greenslopes, Queensland 88/K3
Greenslopes, Queensland 95/E3
Greenspond, Newf. 166/D4
Green Springs, Ohio (44836) 284/E3
Greenstone (pt.), Scotland 15/C3
Greenstreet, Sask. 181/A2
Green Sulphur Springs, W. Va. (25966) 312/E7
Greensville (co.), Va. 307/N7
Greentop, Mo. (63546) 261/H2
Greentown, Ind. (46936) 227/E4
Greentown, Ohio (44630) 284/H4
Greentree, Pa. (15242) 294/B7
Greenup, Ill. (62428) 222/E4
Greenup (co.), Ky. 237/R3
Greenup, Ky. (41144) 237/R3
Greenvale, Queensland 95/C3
Green Valley, Ariz. (85614) 198/D7
Green Valley, Ill. (61534) 222/D3
Green Valley, Minn. (†56258) 255/C6
Green Valley, Ontario 177/K2
Green Valley, Wis. (54127) 317/K6
Greenview, Calif. (96037) 204/B2
Greenview, Ill. (62642) 222/D3
Greenview, W. Va. (†25166) 312/C6
Green Village, N.J. (07935) 273/D2
Greenville, Ala. (36037) 195/E7
Greenville, Calif. (95947) 204/E3
Greenville, Del. (19807) 245/R1
Greenville, Fla. (32331) 212/C1
Greenville, Georgia (30222) 217/C4
Greenville, Ill. (62246) 222/D5
Greenville, Ind. (47124) 227/F8
Greenville, Iowa (51343) 229/C3
Greenville, Ky. (42345) 237/G6
Greenville, Liberia 106/C8
Greenville, Maine (04441) 243/D5
Greenville○, Maine (04441) 243/D5
Greenville, Mich. (48838) 250/D5
Greenville, Miss. 146/J6
Greenville, Miss. 188/H4
Greenville, Miss. (38701) 256/B4
Greenville, Mo. (63944) 261/M8
Greenville, N.H. (03048) 268/D6
Greenville○, N.H. (03048) 268/D6
Greenville, N.C. (27834) 281/P3
Greenville, Ohio (45331) 284/A5
Greenville, Pa. (16125) 294/B3
Greenville, R.I. (02828) 249/H5
Greenville, S.C. 146/K6
Greenville, S.C. 188/K4
Greenville (co.), S.C. 296/C2
Greenville, S.C. (*29601) 296/C2
Greenville, Texas 188/G4
Greenville, Texas (75401) 303/H4
Greenville, Utah (84731) 304/B5
Greenville, Va. (24440) 307/K5
Greenville, W. Va. (24945) 312/E7
Greenville, Wis. (54942) 317/J7
Greenville Junction, Maine (04442) 243/D5
Greenwald, Minn. (56335) 255/D5
Greenwater Lake, Sask. 181/H3
Greenwater Lake Prov. Park, Sask. 181/H3
Greenway, Ark. (72430) 202/K1
Greenway, Manitoba 179/C5
Greenway, S. Dak. (†57437) 298/K2
Greenwell Springs, La. (70739) 238/K1
Greenwich○, Conn. (06830) 210/A4
Greenwich (pt.), Conn. 210/A4
Greenwich, England 13/H8
Greenwich, England 10/B5
Greenwich (Kapingamarangi) (atoll), Micronesia 87/F5
Greenwich○, N.J. (08323) 273/C5
Greenwich, N.Y. (12834) 276/O4
Greenwich, Ohio (44837) 284/E3
Greenwich, Utah (84732) 304/B5
Greenwood, Ark. (72936) 202/B3
Greenwood, Br. Col. 184/H5
Greenwood, Calif. (95635) 204/E5
Greenwood, Del. (19950) 245/R5
Greenwood, Fla. (32443) 212/A1
Greenwood, Ind. (46142) 227/E5
Greenwood (co.), Kansas 232/F4
Greenwood, Ky. (†42634) 237/N7
Greenwood, La. (71033) 238/B2
Greenwood, Mass. (01880) 249/D6
Greenwood (lake), Minn. 255/G3
Greenwood, Miss. (38930) 256/D4
Greenwood, Mo. (64034) 261/R6
Greenwood, Nebr. (68366) 264/H3
Greenwood (lake), N.J. 273/E1
Greenwood, N.Y. (14839) 276/E6
Greenwood (lake), N.Y. 276/M8
Greenwood, S.C. 188/K4
Greenwood (co.), S.C. 296/C3
Greenwood, S.C. (29646) 296/C3
Greenwood (lake), S.C. 296/D3
Greenwood, S. Dak. (†57380) 298/N8
Greenwood, Va. (22943) 307/L4
Greenwood, W. Va. (26360) 312/E4
Greenwood, Wis. (54437) 317/E6
Greenwood Lake, N.Y. (10925) 276/M8
Greenwood Springs, Miss. (38848) 256/H3
Greer, Ariz. (85927) 198/F4
Greer, Idaho (†83544) 220/B3
Greer, Mo. (†65606) 261/K9
Greer, Ohio (†44628) 284/F4
Greer (co.), Okla. 288/G5
Greer, S.C. (29651) 296/C2
Greers Ferry, Ark. (†72067) 202/F2
Greers Ferry (lake), Ark. 202/G2
Greeson (lake), Ark. 202/C5
Gregg (co.), Texas 303/K5
Greggs, Georgia (†31620) 217/F8
Gregory, Ark. (72059) 202/H3
Gregory, Mich. (48137) 250/E6
Gregory (range), Queensland 95/B3
Gregory (riv.), Queensland 88/F3
Gregory (riv.), Queensland 95/A3
Gregory (lake), S. Australia 88/F5
Gregory (lake), S. Australia 94/F3
Gregory (co.), S. Dak. 298/L7
Gregory, S. Dak. (57533) 298/L7
Gregory (lake), W. Australia 92/C4
Gregory Landing, Mo. (†63435) 261/K2
Gregory's (sound), Ireland 17/B5
Greian (head), Scotland 15/A3
Greifensee (lake), Switzerland 39/G2
Greifswald, E. Germany 22/E1
Grein, Austria 41/C2
Greina (pass), Switzerland 39/G3
Greiz, E. Germany 22/E3
Grelton, Ohio (43523) 284/C3
Gremikha, U.S.S.R. 52/E1
Gremyachinsk, U.S.S.R. 52/J3
Grenå, Denmark 21/D5
Grenå, Denmark 18/G8
Grenada 2/F5
Grenada 146/M8
Grenada, Calif. (96038) 204/C2
GRENADA 161/D9
GRENADA 156/G4
Grenada (isl.), Grenada 156/G4
Grenada (co.), Miss. 256/E3
Grenada, Miss. (38901) 256/E3
Grenada (lake), Miss. 256/E3
Grenadier (isl.), N.Y. 276/H2
Grenadines (isls.), Grenada 156/G4
Grenadines (isls.), St. Vin. & Grens. 156/G4
Grenchen, Switzerland 39/D2
Grenfell, N.S. Wales 97/E3
Grenfell, Sask. 181/J5
Grenloch, N.J. (08032) 273/C4
Grenoble, France 7/E4
Grenoble, France 28/F5
Grenola, Kansas (67346) 232/F4
Grenora, N. Dak. (58845) 282/C2
Grenville (chan.), Br. Col. 184/C3
Grenville, Grenada 161/D8
Grenville (bay), Grenada 161/D8
Grenville, N. Mex. (88424) 274/F2
Grenville (county), Ontario 177/J3
Grenville, Québec 172/C4
Grenville (cape), Queensland 88/G2
Grenville (cape), Queensland 95/B1
Grenville, S. Dak. (57239) 298/O3
Grenville (pt.), Wash. 310/A3
Gresham, Nebr. (68367) 264/G3
Gresham, Oreg. (97030) 291/B2
Gresham, S.C. (29546) 296/J4
Gresham, Wis. (54128) 317/J6
Greshamville, Georgia (†30650) 217/F3
Gresik, Indonesia 85/K2
Gresston, Georgia (†31023) 217/F6
Greta-Branxton, N.S. Wales 97/F3
Greta East, N.S. Wales 97/F3
Gretna, Fla. (32332) 212/B1
Gretna, La. (70053) 238/O4
Gretna, Manitoba 179/E5
Gretna, Nebr. (68028) 264/H3
Gretna, Scotland 10/E3
Gretna, Scotland 15/E5
Gretna, Tasmania 99/D4
Gretna, Va. (24557) 307/K7
Grevelingen (str.), Netherlands 27/E5
Greven, W. Germany 22/B2
Grevená, Greece 45/E5
Grevenbroich, W. Germany 22/B3
Grevenmacher, Luxembourg 27/J9
Grevesmühlen, E. Germany 22/D2
Greville (bay), Nova Scotia 168/D3
Grey (isls.), Newf. 166/C3
Grey (riv.), N. Zealand 100/C5
Grey (cape), North. Terr. 88/F2
Grey (cape), North. Terr. 93/E2
Grey (county), Ontario 177/D3
Grey (range), Queensland 88/G5
Grey (range), Queensland 95/B5
Grey Abbey, N. Ireland 17/K2
Greybull, Wyo. (82426) 319/E1
Greybull (riv.), Wyo. 319/D1
Greycliff, Mont. (59033) 262/G5
Grey Eagle, Minn. (56336) 255/D5
Grey Forest, Texas (†78201) 303/J10
Grey Islands, Newf. 166/C3
Greylock (mt.), Idaho 220/C6
Greylock (mt.), Mass. 249/B2
Greymouth, N. Zealand 100/C5
Greymouth, N. Zealand 87/G10
Grey River, Newf. 166/C4
Greys (riv.), Wyo. 319/B3
Greystone, Colo. (†81640) 208/B1
Greystone, Conn. (†06786) 210/C2
Greystone Park, N.J. (07950) 273/D2
Greystones-Delgany, Ireland 10/D4
Greystones-Delgany, Ireland 17/K5
Greytown, N. Zealand 100/E4
Greytown (San Juan del Norte), Nicaragua 154/F5
Greytown, S. Africa 118/E5
Grez-Doiceau, Belgium 27/F7
Gribbles Settlement, North. Terr. 93/B1
Gridley, Calif. (95948) 204/D4
Gridley (mt.), Conn. 210/B1
Gridley, Ill. (61744) 222/E3
Gridley, Kansas (66852) 232/G3
Gridone (mt.), Switzerland 39/G4
Griend (isl.), Netherlands 27/G2
Grier, N. Mex. (†88101) 274/F4
Gries am Brenner, Austria 41/A3
Griesheim, W. Germany 22/C4
Grieskirchen, Austria 41/B2
Griffin, Georgia (30223) 217/D4
Griffin, Ind. (47616) 227/B8
Griffin, Sask. 181/H6
Griffiss A.F.B., N.Y. 276/K4
Griffith, Ind. (46319) 227/C1
Griffith, N.S. 88/H6
Griffith, N.S. Wales 97/C4
Griffith (isl.), Ontario 177/D3
Griffithsville, W. Va. (25521) 312/B6
Griffithville, Ark. (72060) 202/G3
Grifton, N.C. (28530) 281/P4
Griggs (co.), N. Dak. 282/O5
Griggs, Okla. (†73949) 288/B1
Griggsville, Ill. (62340) 222/C4
Grigston, Kansas (†67871) 232/B3
Grijalva (riv.), Mexico 150/N7
Grim (cape), Tasmania 99/A2
Grimari, Cent. Afr. Rep. 115/C2
Grimbergen, Belgium 27/E7
Grimes, Ala. (†36350) 195/H8
Grimes, Calif. (95950) 204/C4
Grimes, Iowa (50111) 229/F5
Grimes, Okla. (†73628) 288/G4
Grimes (co.), Texas 303/J7
Grimesland, N.C. (27837) 281/P3
Griminish, Scotland 15/A3
Grimma, E. Germany 22/E3
Grimmen, E. Germany 22/E1
Grimms Landing, W. Va. (25095) 312/B5
Grimsby, England 13/G4
Grimsby, England 10/F4
Grimsby, Ontario 177/E4
Grimsel (pass), Switzerland 39/F3
Grimsey (isl.), Iceland 21/C1
Grimshaw, Alberta 182/B1
Grimsley, Tenn. (38565) 237/L2
Grimstad, Norway 18/F7
Grindelwald, Switzerland 39/E3
Grindrod, Br. Col. 184/H5
Grindsted, Denmark 21/B6
Grindstone, Maine (†04460) 243/F4
Grindstone (isl.), New Bruns. 170/F3
Grindstone (lake), Wis. 317/C4
Grind Stone City, Mich. (48467) 250/G4
Grindstone Prov. Rec. Park, Manitoba 179/F3
Grinnell, Iowa (50112) 229/H5
Grinnell, Kansas (67738) 232/B2
Grinnell (pen.), N.W. Terrs. 187/J2
Grippon, Guadeloupe 161/B6
Griqualand West (reg.), S. Africa 118/C5
Griquatown, S. Africa 118/C5
Grise Fiord, Canada 4/B13
Grise Fiord, N.W.T. 162/H1
Grise Fiord, N.W. Terrs. 187/K2
Gris-Nez (cape), France 28/D2
Grisons (Graubünden) (elec. div.), Switzerland 39/H3
Grissom A.F.B., Ind. 227/E3
Griswold, Iowa (51535) 229/C6
Griswold, Manitoba 179/B5
Griswoldville, Mass. (01345) 249/D2
Griva, U.S.S.R. 53/D3
Grizzly (bay), Calif. 204/K1
Grizzly Flats, Calif. (95636) 204/E5
Groais (isl.), Newf. 166/C3
Grobiņa, U.S.S.R. 53/A2
Grodno, U.S.S.R. 7/G3
Grodno, U.S.S.R. 48/C4
Grodno, U.S.S.R. 52/B4
Grodzisk Mazowiecki, Poland 47/E2
Grodzisk Wielkopolski, Poland 47/C2
Groenlo, Netherlands 27/K4
Groesbeck, Ohio (45239) 284/B9
Groesbeck, Texas (76642) 303/H6
Groesbeek, Netherlands 27/H5
Groix (isl.), France 28/B4
Grójec, Poland 47/E3
Grömitz, W. Germany 22/D1
Gronau, W. Germany 22/B2
Grondines, Québec 172/E3
Grong, Norway 18/H4
Grong Grong, N.S. Wales 97/D4
Groningen, Minn. (†55072) 255/E4
Groningen (prov.), Netherlands 27/K2
Groningen, Netherlands 27/K2
Groningen, Suriname 131/D2
Groninger Wad (sound), Netherlands 27/J2
Gronlid, Sask. 181/G2
Grønnedal, Greenl. 4/C12
Grono, Switzerland 39/H4
Groom, Texas (79039) 303/C2
Groomsport, N. Ireland 17/K2
Groot-Drakenstein, S. Africa 118/F6
Groote (is.), North. Terr. 88/F2
Groote (riv.), S. Africa 118/C6
Groote Eylandt (isl.), Australia 87/D7
Groote Eylandt (isl.), North. Terr. 93/E3
Groote IJ Polder, Netherlands 27/B4
Grootfontein, Namibia 118/B3
Groot Sint Joris, Neth. Ant. 161/G9
Gros (pt.), Grenada 161/C8
Gros Islet, St. Lucia 161/G5
Gros Islet (bay), St. Lucia 161/G5
Grosmont (Island Lake), Alberta 182/D2
Gros Morne, Haiti 158/B5
Gros-Morne, Martinique 161/D6
Gros Morne (mt.), Newf. 166/C4
Gros-Morne, Québec 172/C1
Gros Morne Nat'l Park, Newf. 166/C4
Gros Piton (mt.), St. Lucia 161/G6
Gross, Nebr. (†68719) 264/F2
Grosse Ile, Mich. (48138) 250/B7
Grosse Isle, Manitoba 179/E4
Gross Emme (riv.), Switzerland 39/E2
Grossenbrode, W. Germany 22/D1
Grossenhain, E. Germany 22/E3
Grosse Pointe, Mich. (48236) 250/B7
Grosse Pointe Farms, Mich. (†48236) 250/B6
Grosse Pointe Park, Mich. (†48236) 250/B7
Grosse Pointe Shores, Mich. (†48236) 250/B6
Grosse Pointe Woods, Mich. (†48236) 250/B6
Grosser Arber (mt.), W. Germany 22/E4
Grosser Peilstein (mt.), Austria 41/C2
Grosses Coques, Nova Scotia 168/B4
Grosses-Roches, Québec 172/B1
Grosse Tete, La. (70740) 238/G6
Grosseto (prov.), Italy 34/C3
Grosseto, Italy 34/C3
Grossglockner (mt.), Austria 41/B3
Gross Litzner (mt.), Switzerland 39/K3
Grossräschen, E. Germany 22/E3
Grosssiegharts, Austria 41/C2
Grosswangen, Switzerland 39/F2
Grosvenor Dale, Conn. (06246) 210/H1
Gros Ventre (riv.), Wyo. 319/B2
Groswater (bay), Newf. 166/C3
Groton, Conn. (06340) 210/G3
Groton○, Conn. (06340) 210/G3
Groton, Mass. (01450) 249/H2
Groton○, Mass. (01450) 249/H2
Groton○, N.H. (†03241) 268/D4
Groton, N.Y. (13073) 276/H5
Groton, S. Dak. (57445) 298/N3
Groton○, Vt. (05046) 268/C3
Groton (lake), Vt. 268/C3
Groton Long (pt.), Conn. 210/H3
Groton Long Point, Conn. (†06340) 210/G3
Grottaferrata, Italy 34/F7
Grottaglie, Italy 34/F4
Grotto, Wash. (98288) 310/D3
Grottoes, Va. (24441) 307/L4
Grouard, Alberta 182/B2
Grouard Mission, Alberta 182/C2
Grouard Mission, Alta. 162/E4
Groundhog (riv.), Ontario 175/D3
Grouse (Lost River), Idaho (†83255) 220/E6
Grouse (mt.), N. Mex. 274/A5
Grouse (creek), Utah 304/A2
Grouse Creek, Utah (84313) 304/A2
Grouse Creek (mts.), Utah 304/A2
Grouw, Netherlands 27/H2
Grovania, Georgia (†31036) 217/E6
Grove, Maine (04638) 243/J5
Grove (lake), N. Dak. 282/L5
Grove, Okla. (74344) 288/S1
Grove Beach, Conn. (†06413) 210/E3
Grove Center, Ky. (†42437) 237/F5
Grove City, Fla. (33533) 212/D5
Grove City, Minn. (56243) 255/D5
Grove City, Ohio (43123) 284/D6
Grove City, Pa. (16127) 294/B3
Grovedale, Alberta 182/A2
Grove Hill, Ala. (36451) 195/C7
Groveland, Calif. (95321) 204/E6
Groveland, Fla. (32736) 212/E3
Groveland, Georgia (†31321) 217/J6
Groveland, Ind. (†46121) 227/D5
Groveland○, Mass. (01830) 249/L1
Groveland, N.Y. (14462) 276/E5
Groveoak, Ala. (35975) 195/F2
Grove Place, Virgin Is. (U.S.) 161/E4
Groveport, Ohio (43125) 284/E6
Grover, Colo. (80729) 208/L1
Grover, Mo. (63040) 261/M3
Grover, N.C. (28073) 281/G4
Grover, Pa. (17735) 294/J2
Grover, S.C. (29447) 296/F5
Grover, S. Dak. (†57201) 298/P4
Grover, Utah (†84773) 304/C5
Grover, Wyo. (83122) 319/B3
Grover City, Calif. (93433) 204/E8
Grover Hill, Ohio (45849) 284/B3
Grovertown, Ind. (46531) 227/D2
Groves, Texas (77619) 303/L8
Grovespring, Mo. (65662) 261/G8
Groveton, N.H. (03582) 268/D2
Groveton, Texas (75845) 303/J7
Groveton, Va. (†22306) 307/T3
Grovetown, Georgia (30813) 217/H4
Groveville, N.J. (†08601) 273/D3
Growler (mts.), Ariz. 198/B6
Groznyy, U.S.S.R. 7/J4
Groznyy, U.S.S.R. 48/E5
Groznyy, U.S.S.R. 52/G6
Grubbs, Ark. (72431) 202/H2
Grubišno Polje, Yugoslavia 45/C3
Grudovo, Bulgaria 45/H4
Grudziądz, Poland 47/D2
Gruenthal, Sask. 181/E3
Gruetli, Tenn. (37339) 237/K10
Gruinard (bay), Scotland 15/C3
Grulla, Texas (78548) 303/F11
Grünberg (Zielona Góra), Poland 47/B3
Grünburg, Austria 41/C3
Grundy (co.), Ill. 222/E2
Grundy (co.), Iowa 229/H4
Grundy (co.), Mo. 261/E2
Grundy (co.), Tenn. 237/K10
Grundy, Va. (24614) 307/D6
Grundy Center, Iowa (50638) 229/H4
Grunthal, Manitoba 179/F5
Gruver, Iowa (51344) 229/D2
Gruver, Texas (79040) 303/C1
Gruyères, Switzerland 39/D3
Gryazi, U.S.S.R. 52/F4
Gryazovets, U.S.S.R. 52/F3
Gryfice, Poland 47/B2
Gryfino, Poland 47/B2
Grygla, Minn. (56727) 255/C2
Gryon, Switzerland 39/D4
Grytviken 5/D17
Gstaad, Switzerland 39/D4
Gsteig, Switzerland 39/D4
Guacamaya, Colombia 126/C6
Guacamayas, Colombia 126/F6
Guacanayabo (gulf), Cuba 156/C2
Guacanayabo (gulf), Cuba 158/G4
Guacara, Venezuela 124/D2
Guachara, Venezuela 124/D4
Gu Achi, Ariz. (†85634) 198/C6
Guácimo, C. Rica 154/F5
Guacuí, Brazil 135/F2
Guadalajara, Mexico 2/D5
Guadalajara, Mexico 150/H6
Guadalajara, Mexico 146/H7
Guadalajara (prov.), Spain 33/E2
Guadalajara, Spain 33/E2
Guadalcanal (isl.), Solomon Is. 87/F7
Guadalcanal (isl.), Solomon Is. 86/D3
Guadalcanal, Spain 33/D3
Guadalimar (riv.), Spain 33/E3
Guadaloupe (isl.), Mexico 146/G7
Guadalquivir (riv.), Spain 7/D5
Guadalquivir (riv.), Spain 33/C4
Guadalupe, Potosí, Bolivia 136/B7
Guadalupe, Santa Cruz, Bolivia 136/C6
Guadalupe, Calif. (93434) 204/E9
Guadalupe (riv.), Calif. 204/K3
Guadalupe, Nuevo León, Mexico 150/K4
Guadalupe, Zacatecas, Mexico 150/H5
Guadalupe (co.), N. Mex. 274/E4
Guadalupe (mts.), N. Mex. 274/D6
Guadalupe, Peru 128/E9
Guadalupe, Spain 33/D3
Guadalupe, Sierra de (range), Spain 33/D3
Guadalupe (co.), Texas 303/G8
Guadalupe (mts.), Texas 303/C10
Guadalupe (peak), Texas 303/B10
Guadalupe (riv.), Texas 303/G8
Guadalupe Bravo, Mexico 150/F1

Guadalupe Mts. Nat'l Park, Texas 303/C10
Guadalupe Victoria, Durango, Mexico 150/H4
Guadalupe Victoria, Puebla, Mexico 150/O1
Guadalupe y Calvo, Mexico 150/F3
Guadalupita, N. Mex. (87722) 274/D2
Guadarrama, Sierra de (range), Spain 33/E2
Guadarrama (riv.), Spain 33/F4
Guadarrama, Venezuela 124/D3
Guadeloupe (isl.) 146/M8
GUADELOUPE 161/A5
GUADELOUPE 156/F3
Guadeloupe (isl.), Guadeloupe 161/B6
Guadeloupe (isl.), Guadeloupe 156/F3
Guadeloupe (passage), Guadeloupe 161/A5
Guadeloupe Nat'l Park, Guadeloupe 161/A6
Guadiana (riv.) 7/D5
Guadiana (riv.), Portugal 33/C4
Guadiana (riv.), Spain 33/D3
Guadix, Spain 33/E4
Guafo (gulf), Chile 138/D5
Guafo (isl.), Chile 138/D5
Guage, Ky. (41329) 237/P5
Guaicanamar, Cuba 158/G3
Guaico, Trin. & Tob. 161/B10
Guaimaca, Honduras 154/D3
Guáimaro, Cuba 158/G3
Guaina, Venezuela 124/G5
Guainía (riv.) 120/C2
Guainía (comm.), Colombia 126/F6
Guainía (riv.), Colombia 126/F6
Guainía (riv.), Venezuela 124/E6
Guaira (dept.), Paraguay 144/D4
Guairá (falls), Paraguay 144/E4
Guaitecas (isls.), Chile 138/D5
Guajaba (cay), Cuba 158/G2
Guajará-Mirim, Brazil 132/H10
Guajará-Mirim, Brazil 120/C4
Guajataca (lake), P. Rico 161/B1
Guajira (pen.) 120/B1
Guajira, La (dept.), Colombia 126/D2
Guajira (pen.), Colombia 126/E1
Gualaca, Panama 154/F6
Gualaceo, Ecuador 128/C4
Gualala, Calif. (95445) 204/B5
Gualán, Guatemala 154/C3
Gualaquiza, Ecuador 128/C4
Guale, Ecuador 128/B3
Gualeguay, Argentina 143/G6
Gualeguay (riv.), Argentina 143/G5
Gualeguaychú, Argentina 143/G6
Gualpatanta, Honduras 154/E3
Guam (isl.) 87/E4
GUAM 86/K7
Guam (isl.), U.S. 2/S5
Guamal, Magdalena, Colombia 126/C3
Guamal, Meta, Colombia 126/D6
Guamblin (isl.), Chile 138/D5
Guamo, Cuba 158/H3
Guamote, Ecuador 128/C4
Guampí, Sierra de (mts.), Venezuela 124/F4
Guamúchil, Mexico 150/E4
Guana, Venezuela 124/G5
Guanabacoa, Cuba 158/C1
Guanabacoa, Cuba 156/B2
Guanabara (bay), Brazil 135/E3
Guanacevi, Mexico 150/G4
Guanahacabibes (gulf), Cuba 158/A2
Guanahacabibes (pen.), Cuba 158/A2
Guanaja, Honduras 154/E2
Guanaja (isl.), Honduras 154/E2
Guanajay, Cuba 158/B1
Guanajay, Cuba 156/A2
Guanajibo (pt.), P. Rico 161/A2
Guanajibo (riv.), P. Rico 161/A2
Guanajuato (state), Mexico 150/J6
Guanajuato, Mexico 150/J6
Guanambi, Brazil 120/E4
Guañape (isls.), Peru 128/C7
Guanare, Venezuela 124/D3
Guanare (riv.), Venezuela 124/D3
Guanare Viejo (riv.), Venezuela 124/D3
Guanarito, Venezuela 124/D3
Guandacol, Argentina 143/C2
Guane, Cuba 158/A2
Guane, Cuba 156/A2
Guangdong (Kwangtung) (prov.), China 77/H7
Guangnam, China 77/G7
Guangshan, China 77/J5
Guangxi Zhuangzu (Kwangsi Chuang Aut. Reg.), China 77/G7
Guangyuan, China 77/G5
Guangze, China 77/J6
Guangzhou (Canton), China 77/H7
Guangzhou (Canton), China 54/N7
Guánica, P. Rico 161/B3
Guánica, P. Rico 156/F1
Guánica (lake), P. Rico 161/B3
Guanipa (riv.), Venezuela 124/G3
Guaniquilla (pt.), P. Rico 161/A2
Guano, Ecuador 128/C3
Guano (creek), Oreg. 291/H5
Guano (lake), Oreg. 291/H5
Guanoco, Venezuela 124/G2
Guanta, Venezuela 124/F2
Guantánamo (prov.), Cuba 158/K4
Guantánamo, Cuba 146/L7
Guantánamo, Cuba 158/K4
Guantánamo, Cuba 156/C2
Guantánamo (bay), Cuba 158/J4
Guantánamo (bay), Cuba 156/C3
Guantánamo Bay U.S. Nav. Reserve, Cuba 158/K4
Guan Xian, China 77/F5
Guape, Colombia 126/D6
Guapí, Colombia 126/B6
Guapi (bay), Colombia 126/A6
Guápiles, C. Rica 154/F5
Guapo (bay), Trin. & Tob. 161/A11
Guaporé (riv.) 120/C4
Guaporé (riv.), Bolivia 136/C3
Guaporé (riv.), Brazil 132/H10
Guaqui, Bolivia 136/A5
Guarambaré, Paraguay 144/B5
Guaranda, Ecuador 128/C3
Guarapuava, Brazil 132/C9
Guaratinguetá, Brazil 135/D3
Guaratinguetá, Brazil 132/E8
Guarda (dist.), Portugal 33/C2
Guarda, Portugal 33/C2
Guardatinajas, Venezuela 124/E3
Guardiagrele, Italy 34/E3
Guardia Mitre, Argentina 143/D5
Guardian (bank), C. Rica 154/D6
Guardian, W. Va. (26221) 312/F5
Guareña, Spain 33/C3
Guarenésia, Brazil 135/C2
Guarero, Venezuela 124/B2
Guarico (pt.), Cuba 158/K3
Guárico (state), Venezuela 124/E3
Guarico, Venezuela 124/D3
Guárico (res.), Venezuela 124/E3
Guárico (riv.), Venezuela 124/E3
Guariquén, Venezuela 124/G2
Guarita, Honduras 154/C3
Guaro, Cuba 158/J3
Guarujá, Brazil 135/C4
Guarulhos, Brazil 135/C3
Guasave, Mexico 150/E4
Guasdualito, Venezuela 124/C4
Guasimal, Cuba 158/E2
Guasimal, Venezuela 124/D4
Guasipati, Venezuela 124/H4
Guastalla, Italy 34/C2
Guatemala 2/D5
Guatemala 146/J8
Guatemala (cap.), Guat. 146/J8
GUATEMALA 154/B3
Guatemala (cap.), Guatemala 154/B3
Guateque, Colombia 126/D5
Guatuaro (pt.), Trin. & Tob. 161/B11
Guaviare (riv.), Colombia 120/B2
Guaviare (riv.), Colombia 126/F6
Guaxupé, Brazil 135/C2
Guayabal, Cuba 158/G3
Guayabal (lake), P. Rico 161/C2
Guayabal, Amazonas, Venezuela 124/E6
Guayabal, Guárico, Venezuela 124/E3
Guayabero (riv.), Colombia 126/D6
Guayacán, Chile 138/A8
Guayaguayare, Trin. & Tob. 161/B11
Guayama (dist.), P. Rico 161/D2
Guayama, P. Rico 161/E3
Guayama, P. Rico 156/G1
Guayaneco (arch.), Chile 138/D7
Guayanés (pt.), P. Rico 161/F2
Guayanés (riv.), P. Rico 161/E2
Guayanilla, P. Rico 161/B3
Guayanilla, P. Rico 156/F1
Guayanilla (bay), P. Rico 161/B3
Guayape, Honduras 154/D3
Guayapo, Serranía (mts.), Venezuela 124/E5
Guayaquil, Ecuador 2/E6
Guayaquil, Ecuador 128/B4
Guayaquil, Ecuador 120/A3
Guayaquil (gulf), Ecuador 120/A3
Guayaquil (gulf), Ecuador 128/B4
Guayaquilaró (riv.), Argentina 143/G5
Guayaramerín, Bolivia 136/C2
Guayas (prov.), Ecuador 128/B4
Guayas (riv.), Ecuador 128/C4
Guaymas, Mexico 146/G7
Guaymas, Mexico 150/D3
Guaynabo, P. Rico 161/D1
Guayo (lake), P. Rico 161/B2
Guayos, Cuba 158/F2
Guayubín, Dom. Rep. 158/D5
Guazú-cuá, Paraguay 144/D5
Gubakha, U.S.S.R. 48/F4
Gubakha, U.S.S.R. 52/J3
Guban (reg.), Somalia 115/H1
Gubat, Philippines 82/E4
Gubbio, Italy 34/D3
Guben (Wilhelm-Pieck-Stadt), E. Germany 22/F3
Guben (Gubin), Poland 47/B3
Gubin, Poland 47/B3
Gubkin, U.S.S.R. 52/E4
Guckeen, Minn. (†56013) 255/D7
Gúdar, Sierra de (range), Spain 33/F2
Gudauta, U.S.S.R. 52/F6
Gudenå (riv.), Denmark 21/C5
Gudermes, U.S.S.R. 52/G6
Güdül, Turkey 63/E2
Gudur, India 68/D6
Guebwiller, France 28/G4
Guéckédou, Guinea 106/B7
Guelma, Algeria 106/F1
Guelph, N. Dak. (58447) 282/O7
Guelph, Ont. 162/H7
Guelph, Ontario 177/D4
Guelta de Zemmur (well), Western Sahara 106/B3
Guemar, Algeria 102/C1
Guemar, Algeria 106/F2
Güémez, Mexico 150/K5
Güeppí, Peru 128/E3
Guerara, Algeria 106/E2
Güere (riv.), Venezuela 124/F3
Guéréda, Chad 111/D5
Guéret, France 28/D4
Guerneville, Calif. (95446) 204/B5
Guernica y Luno, Spain 33/E1
Guernsey (isl.), Chan. Is. 13/E8
Guernsey (isl.), Chan. Is. 10/E6
Guernsey, Iowa (50172) 229/J5
Guernsey (co.), Ohio 284/H5
Guernsey, Sask. 181/F4
Guernsey, Wyo. (82214) 319/H3
Guernsey (res.), Wyo. 319/H3
Guerra, Texas (78360) 303/F11
Guerrero (state), Mexico 150/J8
Guerzim, Algeria 106/E3
Gueydan, La. (70542) 238/E6
Guffey, Colo. (80820) 208/H5
Gugerd, Kuh-e (mts.), Iran 66/H3
Guggisberg, Switzerland 39/D3
Gughe (mt.), Ethiopia 111/G6
Guiana (isl.), Ant. & Bar. 161/E11
Guiana Highlands (plat.) 120/C2
Guichi, China 77/J5
Guichón, Uruguay 145/B3
Guidder, Cameroon 115/B2
Guide, China 77/F4
Guide Rock, Nebr. (68942) 264/F4
Guidonia, Italy 34/F6
Guiglo, Ivory Coast 106/C7
Guihulngan, Philippines 82/D5
Güija (lake), El Salvador 154/C3
Güija (lake), Guatemala 154/C3
Guija, Mozambique 118/E4
Guijuelo, Spain 33/D2
Guilarte (mt.), P. Rico 161/B2
Guild, N.H. (03754) 268/C5
Guildford, England 13/G8
Guildford, England 10/F5
Guildford Junction, Tasmania 99/B3
Guildhall, Vt. (05905) 268/D2
Guilford, Conn. (06437) 210/E3
Guilford◯, Conn. (06437) 210/E3
Guilford, Ind. (47022) 227/H6
Guilford, Maine (04443) 243/E5
Guilford◯, Maine (04443) 243/E5
Guilford, Mo. (64457) 261/C2
Guilford, N.Y. (13780) 276/J6
Guilford (co.), N.C. 281/K3
Guilford◯, Vt. (†05301) 268/B6
Guilin (Kweilin), China 77/G6
Guilin, China 54/N7
Guillaume-Delisle (lake), Québec 174/B1
Guimarães, Brazil 132/E3
Guimarães, Portugal 33/B2
Guimaras (isl.), Philippines 82/D5
Guimaras (str.), Philippines 82/D5
Guimba, Philippines 82/C3
Guin, Ala. (35563) 195/C3
Guinan, China 77/F4
Guinda, Calif. (95637) 204/C5
Guinea 2/J5
Guinea 102/A3
GUINEA 106/B6
Guinea (gulf) 2/K5
Guinea (gulf) 102/C4
Guinea (gulf), Benin 106/E8
Guinea (gulf), Ghana 106/E8
Guinea (gulf), Guinea-Biss. 106/E8
Guinea (gulf), Ivory Coast 106/E8
Guinea (gulf), Nigeria 106/E8
Guinea (gulf), Togo 106/E8
Guinea, Va. (†22580) 307/O4
Guinea-Bissau 2/H5
Guinea-Bissau 102/A3
GUINEA-BISSAU 106/A6
Güines, Cuba 158/C1
Güines, Cuba 156/B2
Guingamp, France 28/B3
Guion, Ark. (72540) 202/G2
Guiones (pt.), C. Rica 154/E6
Guiping, China 77/G7
Guipúzcoa (prov.), Spain 33/E1
Güira de Melena, Cuba 158/C1
Guiratinga, Brazil 132/C7
Guir Hamada (des.), Algeria 106/D2
Güiria, Venezuela 124/G2
Guisa, Cuba 158/H4
Guisanbourg, Fr. Guiana 131/F3
Guisborough, England 13/F3
Guise, France 28/E3
Guitiriz, Spain 33/C1
Guiuan, Philippines 82/E5
Guixi, China 77/J6
Gui Xian, China 77/G7
Guiyang (Kweiyang), Guizhou, China 77/G6
Guiyang, Hunan, China 77/H6
Guiyang, China 54/M7
Guizhou (Kweichow) (prov.), China 77/G6
Gujarat (state), India 68/C4
Gujranwala, Pakistan 59/K3
Gujranwala, Pakistan 68/C2
Gujrat, Pakistan 59/K3
Gujrat, Pakistan 68/C2
Gukovo, U.S.S.R. 52/F5
Gulang, China 77/F4
Gulargambone, N.S. Wales 97/E2
Gulbarga, India 68/D5
Gulbene, U.S.S.R. 53/D2
Gulch (cape), Newf. 166/B2
Gulen, Norway 18/D6
Gulf (co.), Fla. 212/D7
Gulf, N.C. (27256) 281/L3
Gulf Breeze, Fla. (32561) 212/B6
Gulf Crest, Ala. (†36521) 195/B8
Gulf Hammock, Fla. (32639) 212/D2
Gulf Harbors, Fla. (†33552) 212/D3
Gulf Island Nat'l Seashore, Fla. 212/B6
Gulf Islands Nat'l Seashore, Miss. 256/G10
Gulfport, Fla. (33737) 212/B3
Gulf Port, Ill. (†52601) 222/B3
Gulfport, Miss. 188/J4
Gulfport, Miss. (*39501) 256/F10
Gulf Shores, Ala. (36542) 195/C10
Gulf Stream, Fla. (†33444) 212/F5
Gulgong, N.S. Wales 97/E3
Gulian, China 77/K1
Gulin, China 77/G6
Gulistan, 48/G5
Gulja (Yining), China 77/B3
Gulkana, Alaska (†99586) 196/J2
Gull (lake), Alberta 182/C3
Gull (lake), Minn. 255/D4
Gull (isl.), Newf. 166/D2
Gullane, Scotland 15/F4
Gull Bay, Ontario 177/H5
Gull Bay, Ontario 175/C3
Gullfoot (lake), Ontario 177/F3
Gull Island, Newf. 166/D2
Gull Island (pt.), Newf. 166/D2
Gulliver, Mich. (49840) 250/D2
Gull Lake, Alberta 182/D3
Gull Lake, Sask. 181/C5
Gully, Minn. (56646) 255/C3
Gülnar, Turkey 63/E4
Gulnare, Colo. (81042) 208/K8
Gulnare, Ky. (41530) 237/S5
Gulquac (lake), New Bruns. 170/D2
Gulquac (riv.), New Bruns. 170/C2
Gülşehir, Turkey 63/F3
Gulu, Uganda 115/F3
Gulvain (mt.), Scotland 15/C4
Guma (Pishan), China 77/A4
Gumaca, Philippines 82/D4
Gumare, Botswana 118/C3
Gumbranch, Georgia (†31313) 217/J7
Gumel, Nigeria 106/F6
Gumeracha, S. Australia 94/C7
Gumma (pref.), Japan 81/J5
Gummersbach, W. Germany 22/B3
Gummi, Nigeria 106/F6
Gum Spring, Va. (23065) 307/N5
Gum Springs, Ark. (†71923) 202/D5
Gümüş, Turkey 63/F2
Gümüşhacıköy, Turkey 63/F2
Gümüşhane (prov.), Turkey 63/H2
Gümüşhane, Turkey 59/C1
Gümüşhane, Turkey 63/H2
Gun (cay), Bahamas 156/B1
Gun (lake), Mich. 250/D6
Guna, India 68/D4
Gunbower, Victoria 97/C4
Gundagai, N.S. Wales 97/D4
Gunderbooka (ranges), N.S. Wales 97/C2
Gündoğmuş, Turkey 63/D4
Güney, Turkey 63/C3
Gunflint Trail, Minn. (†55604) 255/F1
Gungu, Zaire 115/C5
Gunisao (lake), Manitoba 179/J3
Gunlock, Utah (84733) 304/A6
Gunn, Alberta 182/C3
Gunna (isl.), Scotland 15/B4
Gunnbjørn (mt.), Greenl. 4/C11
Gunn City, Mo. (†64760) 261/D5
Gunnedah, N. S. Wales 88/H6
Gunnedah, N.S. Wales 97/F2
Gunning, N.S. Wales 97/E4
Gunnison (co.), Colo. 208/E5
Gunnison, Colo. (81230) 208/E5
Gunnison (riv.), Colo. 208/C5
Gunnison (tunnel), Colo. 208/D6
Gunnison, Miss. (38746) 256/C3
Gunnison, Utah (84634) 304/C4
Gunnison (res.), Utah 304/C4
Gunnworth, Sask. 181/C4
Gunpowder (riv.), Md. 245/N3
Gunpowder, Queensland 95/A3
Gunpowder, Queensland 88/F3
Gunpowder Falls (creek), Md. 245/M2
Guntakal, India 68/D5
Gunter, Ontario 177/G3
Gunter, Oreg. (†97436) 291/D4
Gunter Air Force Base, Ala. 195/F6
Guntersville, Ala. (35976) 195/F2
Guntersville (dam), Ala. 195/F2
Guntersville (lake), Ala. 195/F2
Gunton, Manitoba 179/E4
Guntown, Miss. (38849) 256/G2
Guntur, India 54/K8
Guntur, India 68/D5
Gunungapi (isl.), Indonesia 85/H7
Günzburg, W. Germany 22/D4
Gunzenhausen, W. Germany 22/D4
Gurabo, P. Rico 161/E2
Gurais, India 68/D2
Gurdon, Ark. (71743) 202/D6
Gurgan (Gorgan), Iran 66/J2
Gurguéia (riv.), Brazil 132/E5
Guri, Venezuela 124/G4
Guri (dam), Venezuela 120/C2
Guri (res.), Venezuela 120/C2
Guri (res.), Venezuela 124/G4
Gurk, Austria 41/C3
Gurla Mandhada (mt.), China 77/B5
Gurley, Ala. (35748) 195/F1
Gurley, La. (†70730) 238/H5
Gurley, Nebr. (69141) 264/B3
Gurley, N.S. Wales 97/E1
Gurley, S.C. (†29569) 296/J3
Gurleyville, Conn. (†06268) 210/G1
Gurnee, Ill. (60031) 222/B4
Gurnet (pt.), Mass. 249/M4
Gurney, Wis. (54528) 317/F3
Gurneyville, Alberta 182/E2
Guro, Mozambique 118/E3
Gürpınar, Turkey 63/K3
Gurteen, Ireland 17/D3
Gurtnellen, Switzerland 39/G3
Gürün, Turkey 63/G3
Gurupá, Brazil 132/D3
Gurupi, Brazil 132/D5
Gurupi, Brazil 120/E4
Gurupi, Serra do (range), Brazil 132/E4
Gurupi (riv.), Brazil 132/E3
Gur'yev, U.S.S.R. 54/G5
Gur'yev, U.S.S.R. 48/F5
Gusau, Nigeria 106/F6
Gusau, Nigeria 102/C3
Gusher, Utah (84030) 304/E3
Gusinje, Yugoslavia 45/D4
Gusinoozersk, U.S.S.R. 48/L4
Gus'-Khrustal'nyy, U.S.S.R. 52/F3
Güssing, Austria 41/D3
Gustavo Díaz Ordaz, Mexico 150/K3
Gustavus, Alaska (99826) 196/M1
Gustavus, Ohio (†44417) 284/J3
Gustine, Calif. (95322) 204/D6
Gustine, Texas (76455) 303/F6
Guston, Ky. (40142) 237/J5
Güstrow, E. Germany 22/E2
Gütersloh, W. Germany 22/C3
Guthrie, Ind. (†47421) 227/D7
Guthrie (co.), Iowa 229/D5
Guthrie, Ky. (42234) 237/G7
Guthrie, Minn. (56451) 255/D3
Guthrie, Mo. (†65063) 261/H5
Guthrie, Okla. 188/G3
Guthrie, Okla. (73044) 288/M3
Guthrie, Texas (79236) 303/D4
Guthrie Center, Iowa (50115) 229/D5
Gutiérrez Zamora, Mexico 150/L6
Guttannen, Switzerland 39/F3
Guttenberg, Iowa (52052) 229/L3
Guttenberg, N.J. (07093) 273/C2
Guttingen, Switzerland 39/H1
Gu-Win, Ala. (†35563) 195/C3
Guy, Alberta 182/B2
Guy, Ark. (72061) 202/F3
Guyana 2/G5
Guyana 120/D2
GUYANA 131/B3
Guyandotte (riv.), W. Va. 312/B6
Guyang, China 77/G3
Guymon, Okla. (73942) 288/D1
Guyot (glac.), Alaska 196/K2
Guyot (mt.), N.C. 281/C3
Guyot (mt.), Tenn. 237/P9
Guyra, N.S. Wales 97/F2
Guys, Tenn. (38339) 237/D10
Guysborough (co.), Nova Scotia 168/F3
Guysborough, Nova Scotia 168/G3
Guysborough (riv.), Nova Scotia 168/G3
Guys Mills, Pa. (16327) 294/C2
Guysville, Ohio (45735) 284/G7
Guyton, Georgia (31312) 217/K6
Guyuan, China 77/G4
Guzmán (lake), Mexico 150/F1
Guzmán Blanco, Venezuela 124/E6
Guzmanes (cays), Cuba 158/B2
Gwa, Burma 72/B3
Gwaai, Zimbabwe 118/D3
Gwabegar, N.S. Wales 97/E2
Gwadabawa, Nigeria 106/F6
Gwadar, Pakistan 59/H5
Gwadar, Pakistan 68/A4
Gwalior, India 54/J7
Gwalior, India 68/D3
Gwanda, Zimbabwe 118/D4
Gwda (riv.), Poland 47/C2
Gweebarra (bay), Ireland 17/D2
Gweebarra (riv.), Ireland 17/E2
Gwelo, (Gweru) Zimbabwe 118/D3
Gwelo, Zimbabwe 102/F6
Gwent, Wales 13/D6
Gwersyllt, Wales 13/G4
Gwinn, Mich. (49841) 250/B2
Gwinner, N. Dak. (58040) 282/P7
Gwinnett (co.), Georgia 217/D2
Gwydir (riv.), N.S. Wales 97/E1
Gwynedd, Wales 13/C4
Gwynn, Va. (23066) 307/R5
Gwynne, Alberta 182/D3
Gwynneville, Ind. (46144) 227/F5
Gyaca, China 77/D6
Gyangzê, China 77/C6
Gyaring Co (lake), China 77/C5
Gyaring Hu (lake), China 77/E5
Gyasikan, Ghana 106/D7
Gyda (pen.), U.S.S.R. 54/J2
Gyda (pen.), U.S.S.R. 4/C6
Gyda, U.S.S.R. 48/H2
Gyda (pen.), U.S.S.R. 48/H2
Gydan (Kolyma) (range), U.S.S.R. 48/Q3
Gyirong, China 77/B6
Gylling, Denmark 21/D6
Gympie, Australia 87/F8
Gympie, Queensland 88/J5
Gympie, Queensland 95/E5
Gyobingauk, Burma 72/C3
Gyoma, Hungary 41/F3
Gyöngyös, Hungary 41/E3
Gyönk, Hungary 41/E3
Győr, Hungary 7/F4
Győr, Hungary 41/D3
Győr-Sopron (co.), Hungary 41/D3
Gypsum, Colo. (81637) 208/F3
Gypsum, Kansas (67448) 232/E3
Gypsum (lake), Manitoba 179/D3
Gypsum, Ohio (43433) 284/E2
Gypsumville, Manitoba 179/D3
Gyrfalcon (isls.), N.W. Terrs. 187/M4
Gyula, Hungary 41/F3

H

Haacht, Belgium 27/F7
Haag, Austria 41/C2
Haakon (co.), S. Dak. 298/F5
Haamstede, Netherlands 27/D5
Ha'apai Group (isls.), Tonga 87/J8
Haapajärvi, Finland 18/O5
Haapamäki, Finland 18/O5
Haapsalu, U.S.S.R. 53/B1
Haar, W. Germany 22/D4
Haarlem, Netherlands 27/F4
Haarlemmermeer (Hoofddorp), Netherlands 27/F4
Haarlemmermeer Polder, Netherlands 27/B5
Haast, N. Zealand 100/B5
Haast (pass), N. Zealand 100/B6
Haast (riv.), N. Zealand 100/B5
Haasts Bluff, North. Terr. 88/E4
Haasts Bluff, North. Terr. 93/B7
Haasts Bluff Aboriginal Reserve, North. Terr. 88/E5
Haasts Bluff Aboriginal Res., North. Terr. 93/B7
Hab (riv.), Pakistan 68/B3
Hab (riv.), Pakistan 59/J4
Habahe, China 77/C2
Habana, La (Havana) (prov.), Cuba 158/C1
Habana, Cuba 158/C1
Habay, Alberta 182/A5
Habay, Belgium 27/H9
Habban, P.D.R. Yemen 59/E7
Habbaniya, Iraq 59/D3
Habbaniya, Iraq 66/C4
Habbaniya, Hor al (lake), Iraq 66/C4
Habersham (co.), Georgia 217/E1
Habersham, Georgia (30544) 217/F1
Habersham, Tenn. (†37766) 237/N8
Habiganj, Bangladesh 68/G4
Habikino, Japan 81/J8
Habomai (isls.), Japan 81/N2
Habonim, Israel 65/B2
Haboro, Japan 81/K1
Hachenburg, W. Germany 22/B3
Hachinohe, Japan 81/K3
Hachioji, Japan 81/O2
Hachiro (lag.), Japan 81/J3
Hachita, N. Mex. (88040) 274/A7
Hacıbektaş, Turkey 63/F3
Hacienda Village, Fla. (†33301) 212/B4
Hacılar, Turkey 63/F3
Hack (mt.), S. Australia 94/F4
Hackberry, Ariz. (86411) 198/B3
Hackberry, La. (70645) 238/D7
Hackensack, Minn. (56452) 255/D4
Hackensack, N.J. (*07601) 273/B2
Hackensack (riv.), N.J. 273/C1
Hacker Valley, W. Va. (26222) 312/F5
Hacketstown, Ireland 17/H6
Hackett, Ark. (72937) 202/B3
Hacketts Cove, Nova Scotia 168/E4
Hackettstown, N.J. (07840) 273/D2
Hackleburg, Ala. (35564) 195/C2
Hackleman, Ind. (†46928) 227/F4
Hackney, England 13/H8
Hackney, England 10/B5
Hacksneck, Va. (23358) 307/S5
Hacoda, Ala. (†36442) 195/F8
Hadano, Japan 81/O3
Hadar, Nebr. (68738) 264/G2
Hadarba, Ras (cape), Sudan 111/G3
Hadashville, Manitoba 179/G5
Hadd, Ras al (cape), Oman 59/G5
Hadd, Ras al (cape), Oman 54/H7
Haddam◯, Conn. (06438) 210/E3
Haddam, Kansas (66944) 232/E2
Haddam Neck, Conn. (†06424) 210/E2
Haddar, Saudi Arabia 59/E5
Haddington, Scotland 10/E3
Haddington, Scotland 15/F5
Haddix, Ky. (41331) 237/P6
Haddock, Georgia (31033) 217/F4
Haddonfield, N.J. (08033) 273/B3
Haddon Heights, N.J. (08035) 273/B3
Hadejia, Nigeria 106/G6
Hadejia (riv.), Nigeria 106/F6
Hadensville, Ky. (†42234) 237/G7
Hadera, Israel 65/B3
Hadera (dry riv.), Israel 65/B3
Haderslev, Denmark 21/C7
Haderslev, Denmark 18/F9
Hadhar, Iraq 66/C3
Hadhramaut (reg.), P.D.R. Yemen 54/F8
Hadhramaut (dist.), P.D.R. Yemen 59/E7
Hadhramaut, Wadi (dry riv.), P.D.R. Yemen 59/F7
Hadibu, P.D.R. Yemen 54/G8
Hadibu, P.D.R. Yemen 59/F7
Hadım, Turkey 63/E4
Haditha, Iraq 66/C3
Haditha, Iraq 59/D3
Hadiya, Saudi Arabia 59/C4
Hadleigh, England 13/H5
Hadley, Ind. (†46122) 227/D5
Hadley, Ky. (42235) 237/H6
Hadley◯, Mass. (01035) 249/D3
Hadley, Minn. (56133) 255/C7
Hadley (bay), N.W. Terrs. 187/H2
Hadley, Pa. (16130) 294/B3
Hadley-Lake Luzerne, N.Y. (12835) 276/N4
Hadlock-Irondale, Wash. (98339) 310/C2
Hadlyme, Conn. (06439) 210/F3
Hadselfjorden (fjord), Norway 18/J2
Hadspen, Tasmania 99/D3
Hadsten, Denmark 21/C5
Hadsund, Denmark 21/D4
Haedo (range), Uruguay 145/C2
Haeju, N. Korea 81/B4
Haena, Hawaii (†96714) 218/C1
Haena (pt.), Hawaii 218/C1
Hafar al Batin, Saudi Arabia 59/E4
Haffe, Syria 63/G5
Hafford, Sask. 181/D3
Hafik, Turkey 63/G3
Haflong, India 68/G3
Hafnarfjördhur, Iceland 21/B2
Haft Gel, Iran 66/F5
Hafun, Somalia 115/K1
Hafun, Ras (cape), Somalia 115/K1
Hagaman, N.Y. (12086) 276/M5
Hagan, Georgia (30429) 217/J6
Hagar, Ontario 177/D1
Hagari (riv.), India 68/D6
Hagarstown, Ill. (62247) 222/D5
Hagarville, Ark. (72839) 202/D2
Hagemeister (isl.), Alaska 196/F3
Hagen, Sask. 181/F3
Hagen, W. Germany 22/B3
Hagenow, E. Germany 22/D2
Hagensborg, Br. Col. 184/D4
Hager City, Wis. (54014) 317/A6
Hagerman, Idaho (83332) 220/D7
Hagerman, N. Mex. (88232) 274/E5
Hagerstown, Ind. (47346) 227/G5
Hagerstown, Md. (21740) 245/G2
Hagerstown, Md. 188/L3
Hagfors, Sweden 18/H6
Hagi, Japan 81/E6
Ha Giang, Vietnam 72/E2
Hagley, Tasmania 99/C3
Hagood, S.C. (†29128) 296/F3
Hags (head), Ireland 17/B6
Hague, Fla. (†32601) 212/D2
Hague (cape), France 28/C3
Hague, The (cap.), Netherlands 7/E3
Hague, The (cap.), Netherlands 27/E4
Hague, N.Y. (12836) 276/N3
Hague, N. Dak. (58542) 282/L7
Hague, Sask. 181/E3
Hague, Va. (22469) 307/P4
Haguenau, France 28/G3
Haha (isl.), Japan 87/E3
Haha (isl.), Japan 81/M3
Ha! Ha! (lake), Qué. 172/G1

Ha! Ha! (riv.) Qué. 172/G1
Hahatonka, Mo. (†65020) 261/G7
Hahira, Georgia (31632) 217/F9
Hahndorf, S. Australia 94/C8
Hahnville, La. (70057) 238/N4
Hai, Iraq 59/E3
Hai, Iraq 66/E4
Haifa (dist.), Israel 65/C2
Haifa, Israel 65/B2
Haifa, Israel 59/B3
Haifa (bay), Israel 65/C2
Haifeng, China 77/J7
Haig (lake), Alberta 182/B1
Haight, Alberta 182/D3
Haigler, Nebr. (69030) 264/C4
Haikang, China 77/G7
Haikou (Hoihow), China 77/H7
Haikou, China 54/N8
Haiku, Hawaii (96708) 218/J2
Hail, Saudi Arabia 54/F7
Hail, Saudi Arabia 59/D4
Hailar, China 77/J2
Hailar He (riv.), China 77/K2
Haile, La. (†71260) 238/F1
Hailesboro, N.Y. (13645) 276/K2
Hailey, Idaho (83333) 220/D6
Haileybury, Ontario 177/K5
Haileybury, Ontario 175/D3
Haileyville, Okla. (74546) 288/P5
Hailong, China 77/L3
Hailsham, England 13/H7
Hailun, China 77/L2
Hailuoto, Finland 18/O4
Hailuoto (isl.), Finland 18/O4
Haina, Hawaii (96727) 218/H3
Hainan (isl.), China 2/Q5
Hainan (isl.), China 54/N8
Hainan (isl.), China 77/H8
Hainaut (prov.), Belgium 27/D7
Hainburg an der Donau, Austria 41/D2
Haines, Alaska (99827) 196/M1
Haines, Oreg. (97833) 291/J3
Hainesburg, N.J. (†07832) 273/C2
Haines City, Fla. (33844) 212/E3
Haines Junction, Yukon 187/E3
Haines Landing, Maine (†04964) 243/B6
Hainesport○, N.J. (08036) 273/D4
Hainesville, Ill. (†60030) 222/A4
Hainesville, N.J. (†07826) 273/D1
Hainfeld, Austria 41/C2
Haiphong, Vietnam 54/M7
Hairy Hill, Alberta 182/D3
Haiti 2/F5
Haiti 146/L8
HAITI 158
HAITI 156/D3
Haiwee, Calif. 204/H7
Haiya Junction, Sudan 59/C6
Haiya Junction, Sudan 111/G4
Haiyan, China 77/F4
Haiyang, China 77/K4
Haiyuan, China 77/G4
Hajara, Al (plain), Iraq 66/D5
Hajarain, P.D.R. Yemen 59/E6
Hajdú-Bihar (co.), Hungary 41/F3
Hajdúböszörmény, Hungary 41/F3
Hajdúdorog, Hungary 41/F3
Hajdúhadház, Hungary 41/F3
Hajdúnánás, Hungary 41/F3
Hajdúsámson, Hungary 41/F3
Hajdúszoboszló, Hungary 41/F3
Haji Ibraham (mt.), Iraq 66/D2
Hajja, Yemen Arab Rep. 59/D6
Hajnowka, Poland 47/F2
Hajós, Hungary 41/E3
Haka, Burma 72/B2
Hakalau, Hawaii (96710) 218/J4
Hakkâri (prov.), Turkey 63/K4
Hakkâri (Çölemerik), Turkey 63/K4
Hakkâri (mts.), Turkey 63/K4
Hakken (mt.), Japan 81/H6
Hakodate, Japan 81/K3
Hakodate, Japan 54/R5
Haku (mt.), Japan 81/H5
Hakui, Japan 81/H5
Hakusan National Park, Japan 81/H5
Hal (Halle), Belgium 27/E7
Halabja, Iraq 66/D3
Halachó, Mexico 150/O6
Halaib, Sudan 59/C5
Halaib, Sudan 111/G3
Halalii (lake), Hawaii 218/A2
Halaula, Hawaii 188/G5
Halawa, Hawaii, Hawaii (†96711) 218/G3
Halawa, Molokai, Hawaii (†96748) 218/H1
Halawa (bay), Hawaii 218/H1
Halawa (cape), Hawaii 218/H1
Halawa (stream), Hawaii 218/B3
Halawa Heights, Hawaii (†96701) 218/B3
Halberstadt, E. Germany 22/D3
Halbrite, Sask. 181/H6
Halbur, Iowa (51444) 229/D4
Halcon (mt.), Philippines 82/C4
Halcyon Dale, Georgia (†30467) 217/J5
Haldane, Ill. (†61030) 222/D1
Haldeman, Ky. (40329) 237/P4
Halden, Norway 18/G7
Haldensleben, E. Germany 22/D2
Haldimand, Ontario 177/E5
Haldimand-Norfolk (reg. munic.), Ontario 177/E5
Hale (co.), Ala. 195/C5
Hale, Argentina 143/F7
Hale, Colo. (80730) 208/P3
Hale, Camp, Colo. 208/G4
Hale, England 13/H2
Hale, Iowa (52230) 229/L4
Hale, Mich. (48739) 250/F4
Hale, Mo. (64643) 261/F3
Hale (riv.), North. Terr. 93/D8
Hale (co.), Texas 303/C3
Hale (mt.), W. Australia 92/B4
Haleakala (crater), Hawaii 218/K2
Haleakala Nat'l Park, Hawaii 218/K2
Haleb (Aleppo), Syria 59/C2
Haleb (Aleppo), Syria 63/G4
Haleburg, Ala. (†36319) 195/H8
Hale Center, Texas (79041) 303/C3
Haledon, N.J. (07508) 273/B1
Haleiwa, Hawaii (96712) 218/E1
Halen, Belgium 27/G7
Hales Corners, Wis. (53130) 317/K2
Halesowen, England 13/E5
Halesowen, England 10/G3
Hales Point, Tenn. (†38040) 237/B9
Halesworth, England 13/J5
Haley, N. Dak. (†58629) 282/D8
Haley Station, Ontario 177/H2
Haleyville, Ala. (35565) 195/C2
Haleyville, N.J. (†08349) 273/C5
Half Assini, Ghana 106/D8
Halfeti, Turkey 63/H4
Half Island Cove, Nova Scotia 168/G3
Half Moon (cay), Belize 154/D2
Halfmoon Bay, Alberta 182/C3
Halfmoon Bay, Br. Col. 184/J2
Half Moon Bay, Calif. (94019) 204/H3
Half Moon Bay (Oban), N. Zealand 100/B7
Half Moon Lake, Alberta 182/D2
Halford, Kansas (†67701) 232/B2
Halfway (riv.), Br. Col. 184/F2
Halfway, Ky. (42150) 237/J7
Halfway, Md. (†21740) 245/G2
Half Way, Mo. (65663) 261/F7
Halfway, Oreg. (97834) 291/K3
Halfway House, Hawaii (†96718) 218/H6
Halfway House, S. Africa 118/H6
Halfweg, Netherlands 27/B4
Halhul, West Bank 65/C4
Haliburton (county), Ontario 177/F2
Haliburton, Ontario 177/F2
Haliburton (lake), Ontario 177/F2
Halieli, Turkey 63/B6
Halifax, Canada 2/F3
Halifax, England 13/J1
Halifax, England 10/G1
Halifax (harb.), Grenada 161/C8
Halifax○, Mass. (02338) 249/L5
Halifax (co.), N.C. 281/O2
Halifax, N.C. (27839) 281/O2
Halifax (co.), Nova Scotia 168/E4
Halifax (cap.), N.S. 162/K7
Halifax (cap.), N.S. 146/M5
Halifax (cap.), Nova Scotia 168/E4
Halifax (harb.), Nova Scotia 168/E4
Halifax, Pa. (17032) 294/J5
Halifax (bay), Queensland 88/H3
Halifax (bay), Queensland 95/C3
Halifax○, Vt. (†05358) 268/B6
Halifax (co.), Va. 307/L7
Halifax, Va. (24558) 307/L7
Halifax Center, Vt. (†05358) 268/B6
Haliimaile, Hawaii (96787) 218/J2
Halil (riv.), Iran 59/G4
Halin, Somalia 115/J2
Halkett (cape), Alaska 196/H1
Halkirk, Alberta 182/D3
Halkirk, Scotland 10/E1
Halkirk, Scotland 15/E2
Hall (isl.), Alaska 196/D2
Hall (co.), Georgia 217/E2
Hall, Ind. (†46157) 227/D5
Hall, Ky. (†41840) 237/R6
Hall, Md. (†20716) 245/L5
Hall (isls.), Micronesia 87/F5
Hall, Mont. (59837) 262/C4
Hall (co.), Nebr. 264/F4
Hall (basin), N.W. Terrs. 187/M1
Hall (lake), N.W. Terrs. 187/K3
Hall (pen.), N.W.T. 162/K3
Hall (pen.), N.W. Terrs. 187/M3
Hall (riv.), Québec 172/C2
Hall (co.), Texas 303/D3
Hall, W. Va. (†26201) 312/F4
Halla (mt.), S. Korea 81/C7
Hallam, Nebr. (68368) 264/H4
Hallam, Victoria 97/K5
Halland (co.), Sweden 18/H8
Hallandale, Fla. (33009) 212/B4
Hallandale (riv.), Scotland 15/E2
Hallaniya (isl.), P.D.R. Yemen 59/G6
Hallau, Switzerland 39/F1
Hall Beach, N.W. Terrs. 187/K3
Hallboro, Manitoba 179/C4
Halle, Belgium 27/E7
Halle, E. Germany 7/F3
Halle (dist.), E. Germany 22/D3
Halle, E. Germany 22/D3
Halleck, Nev. (89824) 266/F2
Hällefors, Sweden 18/J7
Hallein, Austria 41/B3
Halle-Neustadt, E. Germany 22/D3
Hallett, Okla. (74034) 288/N2
Hallettsville, Texas (77964) 303/G8
Halley, Ark. (†71638) 202/H6
Halliday, N. Dak. (58636) 282/F5
Hallie, Wis. (†54729) 317/D6
Halligen (isls.), W. Germany 22/C1
Hall Meadow (brook), Conn. 210/C1
Hallock, Minn. (56728) 255/A2
Hallonquist, Sask. 181/D5
Hallowell, Kansas (66744) 232/H4
Hallowell, Maine (04347) 243/D7
Hall Park, Okla. (†73069) 288/M4
Halls (stream), N.H. 268/E1
Halls, Tenn. (38040) 237/C9
Halls (creek), Utah 304/D6
Hallsberg, Sweden 18/J7
Hallsboro, N.C. (28442) 281/M6
Halls Creek, Australia 87/C7
Halls Creek, W. Australia 88/D3
Halls Creek, W. Australia 92/D2
Halls Crossroads, Tenn. (37918) 237/O8
Hallson, N. Dak. (†58220) 282/P2
Halls Summit, Kansas (†66871) 232/G3
Hallstahammar, Sweden 18/K7
Hallstatt, Austria 41/B3
Hallstavik, Sweden 18/L6
Hallstead, Pa. (18822) 294/L2
Hall Summit, La. (71034) 238/D2
Hallsville, Ill. (†61727) 222/D3
Hallsville, Mo. (65255) 261/H4
Hallsville, Ohio (45633) 284/E7
Hallsville, Texas (75650) 303/K5
Hallton, Pa. (†15860) 294/E3
Halltown, Mo. (65664) 261/E8
Halltown, W. Va. (25423) 312/L4
Hallum, Netherlands 27/H2
Hallwilersee (lake), Switzerland 39/F2
Hallwood, Va. (23359) 307/S5
Halma, Minn. (56729) 255/B2
Halmahera (isl.), Indonesia 54/O9
Halmahera (isl.), Indonesia 85/H5
Halmahera (sea), Indonesia 85/H5
Halmstad, Sweden 18/H8
Halpine, Md. (†20852) 245/K4
Halq el Oued, Tunisia 106/G1
Hals, Denmark 21/D3
Halsell, Ala. (†36912) 195/B6
Halsey, Nebr. (69142) 264/D3
Halsey, Oreg. (97348) 291/D3
Halstad, Minn. (56548) 255/B3
Halstead, England 13/H6
Halstead, England 10/G5
Halstead, Kansas (67056) 232/E4
Haltdalen, Norway 18/G5
Haltemprice, England 13/G4
Haltemprice, England 10/F4
Haltern, W. Germany 22/B3
Haltiatunturi (mt.), Finland 18/M2
Haltom City, Texas (76117) 303/F2
Halton (reg. munic.), Ontario 177/E4
Halton Hills, Ontario 177/E4
Haltwhistle, England 13/E2
Halulu (lake), Hawaii 218/A2
Ham, Chad 111/C5
Ham, France 28/E3
Hama (prov.), Syria 63/G5
Hama, Syria 63/G5
Hama, Syria 59/C2
Hamada, Jebel (mt.), Egypt 59/B5
Hamada, Japan 81/E6
Hamadan (gov.), Iran 66/F3
Hamadan, Iran 66/F3
Hamadan, Iran 59/E3
Hamadan, Iran 54/F6
Hamamatsu, Japan 54/P6
Hamamatsu, Japan 81/H6
Hamar, N. Dak. (58336) 282/N4
Hamar, Norway 18/G6
Hamar, Saudi Arabia 59/E5
Hambantota, Sri Lanka 68/E7
Hamberg, N. Dak. (58337) 282/L4
Hamber Prov. Park, Br. Col. 184/H4
Hamblen (co.), Tenn. 237/P8
Hambleton, W. Va. (26269) 312/G4
Hamburg, Ark. (71646) 202/G7
Hamburg, Conn. (†06371) 210/F3
Hamburg, Ill. (62045) 222/C4
Hamburg, Iowa (51640) 229/B7
Hamburg, Mich. (48139) 250/F6
Hamburg, Minn. (55339) 255/D6
Hamburg, Miss. (†39661) 256/B7
Hamburg, N.J. (07419) 273/D1
Hamburg, N.Y. (14075) 276/C5
Hamburg, Pa. (19526) 294/L4
Hamburg, W. Germany 7/F3
Hamburg (state), W. Germany 22/D2
Hamburg, W. Germany 22/D2
Hamburg, Wis. (54438) 317/G5
Hamda, Saudi Arabia 59/D6
Hamden○, Conn. (06514) 210/D3
Hamden, N.Y. (13782) 276/K6
Hamden, Ohio (45634) 284/F7
Häme (prov.), Finland 18/O6
Hämeenlinna, Finland 18/O6
Hamel, Ill. (62046) 222/B2
Hamel, Minn. (55340) 255/F5
Hamel, Québec 172/G3
Hamelin Pool, W. Australia 88/A5
Hamelin Pool, W. Australia 92/A4
Hameln, W. Germany 22/C2
Hamer, Idaho (83425) 220/F6
Hamer, S.C. (29547) 296/J3
Hamersley (range), W. Australia 88/B4
Hamersley (range), W. Australia 92/B3
Hamersville, Ohio (45130) 284/C8
Hamhŭng, N. Korea 81/C4
Hami (Kumul), China 77/D3
Hami, China 54/L5
Hamill, S. Dak. (57534) 298/K6
Hamilton, Ala. (35570) 195/C2
Hamilton (lake), Ark. 202/D5
Hamilton (cap.), Bermuda 156/G3
Hamilton (mt.), Calif. 204/L3
Hamilton, Colo. (81638) 208/D2
Hamilton (co.), Fla. 212/D1
Hamilton, Georgia (31811) 217/C5
Hamilton (co.), Ill. 222/E5
Hamilton, Ill. (62341) 222/B3
Hamilton (co.), Ind. 227/E4
Hamilton, Ind. (46742) 227/H1
Hamilton (co.), Iowa 229/F4
Hamilton, Iowa (50116) 229/H6
Hamilton (co.), Kansas 232/A3
Hamilton, Kansas (66853) 232/F4
Hamilton○, Mass. (01936) 249/L2
Hamilton, Mich. (49419) 250/C6
Hamilton, Miss. (39746) 256/H3
Hamilton, Mo. (64644) 261/E3
Hamilton, Mont. (59840) 262/B4
Hamilton (co.), Nebr. 264/F4
Hamilton (inlet), Newf. 166/C3
Hamilton (inlet), Newf. 146/N4
Hamilton (inlet), Newf. 162/L5
Hamilton (sound), Newf. 166/D4
Hamilton (co.), N.Y. 276/L3
Hamilton, N.Y. (13346) 276/J5
Hamilton, N. Zealand 100/E2
Hamilton, N.C. (27840) 281/P3
Hamilton, N. Dak. (58238) 282/R2
Hamilton (co.), Ohio 284/A7
Hamilton, Ohio 188/K3
Hamilton, Ohio (*45011) 284/A7
Hamilton, Ont. 146/K5
Hamilton, Ont. 162/H7
Hamilton, Ontario 177/E4
Hamilton, Oreg. (†97856) 291/H3
Hamilton, Pa. (15744) 294/D4
Hamilton (riv.), Queensland 95/B4
Hamilton, R.I. (†02852) 249/J6
Hamilton, Scotland 15/C2
Hamilton, Scotland 10/B1
Hamilton, The (riv.), S. Australia 94/D2
Hamilton, The (riv.), S. Australia 88/E5
Hamilton, Tasmania 99/C4
Hamilton (co.), Tenn. 237/L10
Hamilton (co.), Texas 303/F6
Hamilton, Texas (76531) 303/G6
Hamilton, Victoria 88/G7
Hamilton, Victoria 97/B5
Hamilton, Va. (22068) 307/N2
Hamilton, Wash. (98255) 310/D2
Hamilton City, Calif. (95951) 204/C4
Hamilton Dome, Wyo. (82427) 319/D2
Hamilton Square-Mercerville, N.J. (08690) 273/D3
Hamilton-Wentworth (reg. munic.), Ontario 177/D4
Hamina, Finland 18/P6
Hamiota, Manitoba 179/B4
Ham Lake, Minn. (55304) 255/E5
Hamler, Ohio (43524) 284/B3
Hamlet, Ind. (46532) 227/D2
Hamlet, Nebr. (69031) 264/C4
Hamlet, N.Y. (†14138) 276/B6
Hamlet, N.C. (28345) 281/K5
Hamlet, N. Dak. (†58795) 282/E2
Hamlet, Ohio (†45102) 284/B8
Hamletsburg, Ill. (62944) 222/E6
Hamlin, Alberta 182/D2
Hamlin, Iowa (50117) 229/D5
Hamlin, Kansas (†66434) 232/G2
Hamlin, Ky. (42046) 237/E7
Hamlin○, Maine (†04785) 243/H1
Hamlin (lake), Mich. 250/C4
Hamlin, N.Y. (14464) 276/E4
Hamlin, Pa. (18427) 294/M3
Hamlin, Sask. 181/C3
Hamlin (co.), S. Dak. 298/P4
Hamlin, Texas (79520) 303/E5
Hamlin, W. Va. (25523) 312/B6
Hamm, W. Germany 22/B3
Hammamet (gulf), Tunisia 106/G1
Hammar, Hor al (lake), Iraq 66/E5
Hammarstrand, Sweden 18/J5
Hamme, Belgium 27/E6
Hammel, Denmark 21/C5
Hammelburg, W. Germany 22/C3
Hammer, S. Dak. (†57255) 298/R2
Hammerdal, Sweden 18/J5
Hammerfest, Norway 4/B9
Hammerfest, Norway 18/N1
Hammerfest, Norway 7/G1
Hammersmith, England 10/B5
Hammersmith, England 13/H8
Hammerum, Denmark 21/C5
Hammett, Idaho (83627) 220/C7
Hammon, Okla. (73650) 288/H3
Hammonasset (pt.), Conn. 210/E3
Hammonasset (res.), Conn. 210/E3
Hammonasset (riv.), Conn. 210/E3
Hammond, Ill. (61929) 222/E4
Hammond, Ind. (*46320) 227/B1
Hammond, Ky. (†40935) 237/O7
Hammond, La. (70401) 238/N1
Hammond, Minn. (55938) 255/F6
Hammond, Mo. (†65762) 261/G9
Hammond, Mont. (59332) 262/M5
Hammond (riv.), New Bruns. 170/E3
Hammond, N.Y. (13646) 276/J2
Hammond, Oreg. (97121) 291/C1
Hammond, Wis. (54015) 317/A6
Hammondsport, N.Y. (14840) 276/F6
Hammondsville, Ohio (43930) 284/J4
Hammondvale, New Bruns. 170/E3
Hammondville, Ala. (†35989) 195/G1
Hammonton, N.J. (08037) 273/D4
Hamnavoe, Scotland 15/G2
Ham-Nord, Québec 172/F4
Hamoa, Hawaii (†96713) 218/K2
Hamois, Belgium 27/G8
Hamont-Achel, Belgium 27/H6
Hampden, Maine (04444) 243/F6
Hampden○, Maine (04444) 243/F6
Hampden (co.), Mass. 249/D4
Hampden○, Mass. (01036) 249/E4
Hampden, Newf. 166/C4
Hampden, N. Zealand 100/C6
Hampden, N. Dak. (58338) 282/N2
Hampden, W. Va. (25623) 312/C7
Hampden Highlands, Maine (04445) 243/F6
Hampden-Sydney, Va. (23943) 307/L6
Hampshire (co.), England 13/F6
Hampshire, Ill. (60140) 222/E1
Hampshire (co.), Mass. 249/D3
Hampshire, Tenn. (38461) 237/G9
Hampshire (co.), W. Va. 312/J4
Hampshire, Wyo. (†82701) 319/H2
Hampstead, Dominica 161/E5
Hampstead, Md. (21074) 245/L2
Hampstead, New Bruns. 170/D3
Hampstead○, N.H. (03841) 268/E6
Hampstead, N.C. (28443) 281/O6
Hampstead, Québec 172/H4
Hampton, Ark. (71744) 202/F6
Hampton○, Conn. (06247) 210/G1
Hampton, Fla. (32044) 212/D2
Hampton, Georgia (30228) 217/D4
Hampton, Ill. (61256) 222/C2
Hampton, Iowa (50441) 229/G3
Hampton, Ky. (42047) 237/E6
Hampton, Minn. (55031) 255/E6
Hampton, Miss. (†38744) 256/B4
Hampton, Nebr. (68843) 264/G4
Hampton, New Bruns. 170/E3
Hampton, N.H. (03842) 268/F6
Hampton○, N.H. (03842) 268/F6
Hampton, N.J. (08827) 273/D2
Hampton, N.Y. (12837) 276/O3
Hampton, Nova Scotia 168/C4
Hampton, Oreg. (†97712) 291/G4
Hampton, Pa. (†17350) 294/H6
Hampton (co.), S.C. 296/E6
Hampton, S.C. (29924) 296/E6
Hampton, Tenn. (37658) 237/S8
Hampton (I.C.), Va. (*23601) 307/R6
Hampton Bays, N.Y. (11946) 276/R9
Hampton Beach, N.H. (03842) 268/F6
Hampton Falls○, N.H. (03844) 268/F6
Hampton Nat'l Hist. Site, Md. 245/M3
Hampton Park, Victoria 97/K6
Hampton Park, Victoria 88/M8
Hampton Roads (est.), Va. 307/R7
Hampton Springs, Fla. (†32347) 212/C1
Hamptonville, N.C. (27020) 281/H2
Hamrat esh Sheikh, Sudan 111/E5
Hamrin, Jabal (mts.), Iraq 66/D3
Hams Bluff (prom.), Virgin Is. (U.S.) 161/E3
Hams Fork (riv.), Wyo. 319/B4
Ham-Sud, Québec 172/F4
Hamton, Sask. 181/J4
Hamtramck, Mich. (48212) 250/B6
Hamur, Turkey 63/K3
Han (riv.), China 54/N6
Han (riv.), S. Korea 81/C5
Hana, Hawaii 188/F5
Hana, Hawaii (96713) 218/K2
Hanac, Turkey 63/K2
Hanaford (Logan), Ill. (†62856) 222/E6
Hanagita (peak), Alaska 196/K2
Hanahan, S.C. (29410) 296/H6
Hanakiya, Saudi Arabia 59/D5
Hanalei, Hawaii (96714) 218/C1
Hanalei (bay), Hawaii 218/C1
Hanalei (riv.), Hawaii 218/C1
Hanamaki, Japan 81/K4
Hanamalo (pt.), Hawaii 218/F7
Hanamaulu, Hawaii (96715) 218/C1
Hanapepe, Hawaii (96716) 218/C2
Hanapepe (bay), Hawaii 218/C2
Hanau, W. Germany 22/C3
Hanbogd, Mongolia 77/G3
Hanceville, Ala. (35077) 195/E2
Hancheng, China 77/H4
Hanchung (Hanzhong), China 77/G5
Hancock, Conn. (†06786) 210/C2
Hancock (co.), Georgia 217/G4
Hancock (co.), Ill. 222/B3
Hancock (co.), Ind. 227/F5
Hancock (co.), Iowa 229/E2
Hancock, Iowa (51536) 229/C6
Hancock (co.), Ky. 237/H5
Hancock (co.), Maine 243/G6
Hancock○, Maine (04640) 243/G6
Hancock, Md. (21750) 245/F2
Hancock○, Mass. (01237) 249/A2
Hancock, Mich. (49930) 250/G1
Hancock, Minn. (56244) 255/C5
Hancock (co.), Miss. 256/E10
Hancock, Mo. (†65452) 261/H7
Hancock○, N.H. (03449) 268/C6
Hancock (mt.), N.H. 268/D3
Hancock, N.Y. (13783) 276/K7
Hancock (co.), Ohio 284/C3
Hancock (co.), Tenn. 237/P7
Hancock○, Vt. (05748) 268/B4
Hancock (co.), W. Va. 312/E2
Hancock, W. Va. (25424) 312/K3
Hancock, Wis. (54943) 317/G7
Hancocks Bridge, N.J. (08038) 273/C4
Hand (co.), S. Dak. 298/L4
Handa (isl.), Scotland 15/C2
Handan (Hantan), China 77/H4
Handan, China 54/N6
Handel, Sask. 181/C3
Handeni, Tanzania 115/G5
Handies (peak), Colo. 208/E7
Handley, W. Va. (25102) 312/D6
Handlová, Czech. 41/E2
Handsom, Va. (23859) 307/O7
Handsworth, Sask. 181/J6
Haney, Br. Col. 184/L3
Hanford, Calif. (93230) 204/F7
Hanford Reservation, Wash. 310/F4
Hangayn Nuruu (mts.), Mongolia 77/E2
Hangchow (Hangzhou), China 77/J5
Hanggin, China 77/G4
Hanging Rock, Ohio (45635) 284/E8
Hangklip (cape), S. Africa 118/F7
Hangö, Finland 18/N7
Hangöudd (prom.), Finland 18/N7
Hangzhou (Hangchow), China 77/J5
Hangzhou, China 54/N6
Hangzhou Wan (bay), China 77/K5
Hanh, Mongolia 77/F1
Hani, Turkey 63/J3
Haniqra, Rosh (cape), Israel 65/C1
Hanish (isls.), Yemen Arab Rep. 59/D7
Hankinson, N. Dak. (58041) 282/S7
Hanko (Hangö), Finland 18/N7
Hanks, N. Dak. (†58856) 282/C2
Hanksville, Utah (84734) 304/D5
Hanle, India 68/D2
Hanley, Sask. 181/E4
Hanley Falls, Minn. (56245) 255/C6
Hanley Hills, Mo. (†63101) 261/P2
Hanlontown, Iowa (50444) 229/G2
Hanmer, N. Zealand 100/D5
Hann (mt.), W. Australia 92/D1
Hanna, Alberta 182/E4
Hanna, Alta. 162/E5
Hanna, Ind. (46340) 227/D2
Hanna, La. (71035) 238/D3
Hanna, Okla. (74845) 288/P4
Hanna, Utah (84031) 304/D3
Hanna, Wyo. (82327) 319/F4
Hanna City, Ill. (61536) 222/D3
Hannaford, N. Dak. (58448) 282/O5
Hannah, N. Dak. (58239) 282/N2
Hannah (bay), Ontario 175/D2
Hannawa Falls, N.Y. (13647) 276/L1
Hannibal, Mo. (63401) 261/K3
Hannibal, Mo. 188/H3
Hannibal, N.Y. (13074) 276/G4
Hannibal, Ohio (43931) 284/J6
Hannibal, Wis. (54439) 317/E5
Hanno, Japan 81/O2
Hannover, N. Dak. (58543) 282/H5
Hannover, W. Germany 7/E3
Hannover, W. Germany 22/C2
Hannuit (Hannut), Belgium 27/G7
Hannut, Belgium 27/G7
Hanöbukten (bay), Sweden 18/J9
Hanoi (cap.), Vietnam 2/Q4
Hanoi (cap.), Vietnam 54/M7
Hanover (isl.), Chile 120/B8
Hanover (isl.), Chile 138/D9
Hanover, Conn. (06350) 210/G2
Hanover, Ill. (61041) 222/C1
Hanover, Ind. (47243) 227/F7
Hanover, Kansas (66945) 232/F2
Hanover○, Maine (04237) 243/B7
Hanover, Md. (21201) 245/M4
Hanover○, Mass. (02339) 249/L4
Hanover○, Mich. (49241) 250/E6
Hanover, Minn. (55341) 255/E5
Hanover, N.H. (03755) 268/C4
Hanover○, N.H. (03755) 268/C4
Hanover, N. Mex. (88041) 274/A6
Hanover, Ohio (†43055) 284/F5
Hanover, Ontario 177/C3
Hanover, Pa. (17331) 294/J6
Hanover (co.), Va. 307/N5
Hanover, Va. (23069) 307/O5
Hanover, W. Va. (24839) 312/C7
Hanover Park, Ill. (60103) 222/A5
Hanoverton, Ohio (44423) 284/J4
Hansboro, N. Dak. (58339) 282/M2
Hansell, Iowa (50640) 229/G3
Hansen, Idaho (83334) 220/D7
Hansford (co.), Texas 303/C1
Han Shui (riv.), China 77/H5
Hanska, Minn. (56041) 255/D6
Hans Lollik (isls.), Virgin Is. (U.S.) 161/B4
Hanson, Ky. (42413) 237/G6
Hanson, Mass. (02341) 249/L4
Hanson○, Mass. (02341) 249/L4
Hanson (bay), N. Zealand 100/E7
Hanson (riv.), North. Terr. 93/C6
Hanson, Okla. (†74955) 288/S4
Hanson (co.), S. Dak. 298/O6
Hansonville, Va. (†24266) 307/D7
Hanstholm, Denmark 21/B3
Hanston, Kansas (67849) 232/C3
Hansville, Wash. (98340) 310/C3
Hantan (Handan), China 77/H4
Hants (co.), Nova Scotia 168/D4
Hant's Harbour, Newf. 166/D2
Hantsport, Nova Scotia 168/D3
Hantzsch (riv.), N.W. Terrs. 187/L3
Hanumangarh, India 68/C3
Hanwood, N.S. Wales 97/C4
Hanyuan, China 77/F6
Hanzhong (Hanchung), China 77/G5
Hao (atoll), Fr. Poly. 87/N7
Haouach, Wadi (dry riv.), Chad 111/C4
Haparanda, Sweden 18/N4
Hapeville, Georgia (30354) 217/K2
Happy, Ky. (41746) 237/P6
Happy, Texas (79042) 303/C3
Happy Adventure, Newf. 166/D2
Happy Camp, Calif. (96039) 204/B2
Happy Jack, Ariz. (86024) 198/D4
Happy Jack, La. (†70083) 238/L7
Happy Valley, Oreg. (†97222) 291/B2
Happy Valley-Goose Bay, Newf. 166/B3
Haql, Saudi Arabia 59/C4
Harad, Saudi Arabia 59/E5
Harads, Sweden 18/M3
Harahan, La. (70123) 238/O4
Haraja, Saudi Arabia 59/D6
Haralson (co.), Georgia 217/B3
Haralson, Georgia (30229) 217/C4
Haramachi, Japan 81/K5
Harar (prov.), Ethiopia 111/H6
Harar, Ethiopia 111/H6
Harar, Ethiopia 102/G4
Harardera, Somalia 115/J3
Harare (Salisbury) (cap.), Zimbabwe 102/E6
Haraz, Chad 111/C5
Harbel, Liberia 106/B7
Harbeson, Del. (19951) 245/S6
Harbin, China 77/L2
Harbin, China 2/R3
Harbin, China 54/O5
Harbine, Nebr. (†68377) 264/G4
Harbinger, N.C. (27941) 281/T2
Harboør, Denmark 21/B4
Harbor, Oreg. (97415) 291/C5
Harbor Beach, Mich. (48441) 250/G5
Harbor City, Calif. (90710) 204/C11
Harborcreek, Pa. (16421) 294/C1
Harbor Springs, Mich. (49740) 250/D3
Harborton, Va. (23389) 307/S5
Harbor View, Ohio (43434) 284/C2
Harbour (isl.), Bahamas 156/C1
Harbour Breton, Newf. 166/C4
Harbour Deep, Newf. 166/C3
Harbour Grace, Newf. 166/D2
Harbour Grace, Newf. 162/L6
Harbour Main, Newf. 166/D2
Harbourton, N.J. (†08530) 273/D3
Harbourville, Nova Scotia 168/D3
Harburg-Wilhelmsburg, W. Germany 22/C2
Hårby, Denmark 21/D7
Harco, Ill. (†62935) 222/E6
Harcourt, Iowa (50544) 229/E4
Harcourt, New Bruns. 170/E2
Harcourt, Ontario 177/F2
Harcuvar (mts.), Ariz. 198/B5
Harda, India 68/D4
Hardangerfjord (fjord), Norway 18/D7
Hardangerfjorden (fjord), Norway 7/E3
Hardangervidda (plat.), Norway 18/E6
Hardaway, Ala. (36039) 195/G6
Hardburly, Ky. (41747) 237/P6
Hardee (co.), Fla. 212/E4
Hardee, Miss. (†39177) 256/C5
Hardeeville, S.C. (29927) 296/E7
Hardeman (co.), Tenn. 237/C10
Hardeman (co.), Texas 303/E3
Hardenberg, Netherlands 27/J3
Harden City, Okla. (74846) 288/N5
Harderwijk, Netherlands 27/H4
Hardesty, Okla. (73944) 288/D1

Hardieville, Alberta 182/D5
Hardin (co.), Ill. 222/E6
Hardin, Ill. (62047) 222/C4
Hardin (co.), Iowa 229/G4
Hardin (co.), Ky. 237/K5
Hardin, Ky. (42048) 237/E7
Hardin, Mo. (64035) 261/E4
Hardin, Mont. (59034) 262/J5
Hardin (co.), Ohio 284/C4
Hardin (co.), Tenn. 237/E10
Hardin (co.), Texas 303/K7
Harding (lake), Ala. 195/H5
Harding (lake), Georgia 217/B5
Harding, Manitoba 179/B5
Harding, Minn. (56364) 255/E4
Harding (co.), N. Mex. 274/F3
Harding (pt.), Nova Scotia 168/D5
Harding (co.), S. Dak. 298/B2
Harding, W. Va. (†26250) 312/G5
Harding Icefield, Alaska 196/C2
Hardingville, N.J. (†08343) 273/C4
Hardinsburg, Ind. (47125) 227/E8
Hardinsburg, Ky. (40143) 237/H5
Hardin Springs, Ky. (†42712) 237/J5
Hardinville, Ill. (†62449) 222/F5
Hardinxveld-Giessendam, Netherlands 27/G5
Hardisty, Alberta 182/E3
Hardisty (lake), N.W. Terrs. 187/G3
Hardman, Oreg. (†97836) 291/H2
Hardoi, India 68/E3
Hardshell, Ky. (41348) 237/P6
Hardt (mts.), W. Germany 22/C4
Hardtner, Kansas (67057) 232/D4
Hardwar, India 68/D2
Hardwick (Midway-Hardwick), Georgia (31034) 217/F4
Hardwick○, Mass. (01037) 249/F3
Hardwick, Minn. (56134) 255/B7
Hardwick, Vt. (05843) 268/C2
Hardwick○, Vt. (05843) 268/C2
Hardwick (lake), Vt. 268/C2
Hardwicke, New Bruns. 170/E1
Hardwicke Island, Br. Col. 184/E5
Hardwood Ridge, New Bruns. 170/D2
Hardy, Ark. (72542) 202/H1
Hardy (pen.), Chile 138/F11
Hardy, Iowa (50545) 229/E3
Hardy, Ky. (41531) 237/S5
Hardy, Miss. (†38901) 256/E3
Hardy, Nebr. (68943) 264/G4
Hardy, Okla. (†74641) 288/N1
Hardy, Sask. 181/G6
Hardy, Va. (24101) 307/J6
Hardy (co.), W. Va. 312/J4
Hardyville, Ky. (42746) 237/K6
Hare (bay), Newf. 166/C3
Hare (fjord), N.W. Terrs. 187/K1
Hare Bay, Newf. 166/D4
Harelbeke, Belgium 27/C7
Harfleur, France 28/D3
Harford (co.), Md. 245/N2
Harford, N.Y. (13784) 276/H6
Harford, Pa. (18823) 294/L2
Hargeysa, Somalia 115/H2
Hargeysa, Somalia 102/G4
Hargill, Texas (78549) 303/F11
Hargrave, Manitoba 179/A5
Hargwen, Alberta 182/B3
Har Hu (lake), China 77/E4
Hari (riv.), Indonesia 85/C6
Harib, Yemen Arab Rep. 59/E7
Haricha Hamada (des.), Mali 106/D4
Harim, Syria 63/G4
Harima (sea), Japan 81/G6
Harima, Jordan 65/D2
Haringey, England 10/B5
Haringey, England 13/H8
Haringvliet (str.), Netherlands 27/E5
Hariq, Saudi Arabia 59/E5
Harirud (riv.), Afghanistan 68/A1
Harirud (riv.), Afghanistan 59/H3
Hari Rud (riv.), Iran 66/M3
Haris, West Bank 65/C3
Harjavalta, Finland 18/M6
Harjo, Okla. (†74854) 288/N4
Harkaway, Victoria 97/K5
Harkers Island, N.C. (28531) 281/R5
Harkiko, Ethiopia 111/G4
Harlan, Ind. (46743) 227/H2
Harlan, Iowa (51537) 229/C5
Harlan, Kansas (67641) 232/D2
Harlan (co.), Ky. 237/P7
Harlan, Ky. (40831) 237/P7
Harlan (co.), Nebr. 264/E4
Harlan, Oreg. (†97343) 291/D3
Harlan County (lake), Nebr. 264/E5
Harlech, Wales 10/E4
Harlech, Wales 13/C5
Harlem, Fla. (33440) 212/F5
Harlem, Georgia (30814) 217/H4
Harlem, Mont. (59526) 262/H2
Harlem Springs, Ohio (44631) 284/J4
Harleston, England 13/J5
Harleton, Texas (75651) 303/K5
Hårlev, Denmark 21/F7
Harleyville, S.C. (29448) 296/G5
Harlingen, Netherlands 27/G2
Harlingen, N.J. (†08502) 273/D3
Harlingen, Texas (78550) 303/G11
Harlingen, Texas 188/G5
Harlow, England 13/H7
Harlow, N. Dak. (58340) 282/M3
Harlowton, Mont. (59036) 262/F4
Harman, W. Va. (26270) 312/G5
Harman-Maxie, Va. (24618) 307/D6
Harmans, Md. (21077) 245/M4
Harmattan, Alberta 182/C4
Harmon, Ill. (61042) 222/D2
Harmon (co.), Okla. 288/G5
Harmon, Okla. (73832) 288/G2
Harmonsburg, Pa. (16422) 294/B2
Harmony, Ark. (†72830) 202/D2
Harmony, Ind. (47853) 227/C5
Harmony, Maine (04942) 243/D6
Harmony○, Maine (04942) 243/D6
Harmony, Minn. (55939) 255/F7
Harmony, N.C. (28634) 281/H3
Harmony, Pa. (16037) 294/B4
Harmony, R.I. (02829) 249/H5
Harmony, W. Va. (25246) 312/D5
Harms, Tenn. (†37334) 237/H10
Harned, Ky. (40144) 237/J5
Harnett (co.), N.C. 281/M4
Harney (lake), Fla. 212/F3
Harney, Md. (†21787) 245/K2
Harney (co.), Oreg. 291/H4
Harney (lake), Oreg. 188/C2
Harney, Oreg. (†97720) 291/J4
Harney (lake), Oreg. 291/H4
Harney (peak), S. Dak. 298/B6
Härnösand, Sweden 18/L5
Haro, Spain 33/E1
Haro (str.), Wash. 310/B2
Harold, Fla. (32563) 212/B6
Harold, Ky. (41635) 237/R5
Harp (lake), Newf. 166/B2
Harper, Ill. (†61030) 222/D1
Harper, Iowa (52231) 229/J6
Harper (co.), Kansas 232/D4
Harper, Kansas (67058) 232/D4
Harper, Liberia 106/C8
Harper, Liberia 102/B4
Harper (co.), Okla. 288/G1
Harper, Oreg. (97906) 291/K4
Harper, Texas (78631) 303/E7
Harper, Wash. (†98366) 310/A2
Harper, W. Va. (25851) 312/D7
Harpers Ferry, Iowa (52146) 229/L2
Harpers Ferry, W. Va. (25425) 312/L4
Harpers Ferry Nat'l Hist. Park, Md. 245/G3
Harpers Ferry Nat'l Hist. Park, W. Va. 312/L4
Harpersville, Ala. (35078) 195/F4
Harperville, Miss. (39080) 256/E6
Harper Woods, Mich. (48225) 250/B6
Harpeth (riv.), Tenn. 237/G8
Harpster, Idaho (†83521) 220/C4
Harpster, Ohio (43323) 284/D4
Harpswell○, Maine (†04011) 243/D8
Harpswell Center, Maine (†04011) 243/D8
Harpursville, N.Y. (13787) 276/J6
Harput, Turkey 63/H3
Harquahala (mts.), Ariz. 198/B5
Harrah, Okla. (73045) 288/M4
Harrah, Wash. (98933) 310/E4
Harran, Turkey 63/H4
Harrell, Ark. (71745) 202/F7
Harrells, N.C. (28444) 281/N5
Harrellsville, N.C. (27942) 281/R2
Harricana (riv.), Québec 174/B3
Harriet, Ark. (72939) 202/E2
Harrietsfield, Nova Scotia 168/E4
Harrietta, Mich. (49638) 250/D4
Harriettsville, Ohio (†45745) 284/H6
Harrigan Cove, Nova Scotia 168/F4
Harriman, N.Y. (10926) 276/M8
Harriman, Oreg. (†97601) 291/E5
Harriman, Tenn. (37748) 237/M9
Harriman (res.), Vt. 268/B6
Harrington (sound), Bermuda 156/G3
Harrington, Del. (19952) 245/R5
Harrington○, Maine (04643) 243/H6
Harrington (lake), Maine 243/E4
Harrington, N.S. Wales 97/G2
Harrington, S. Dak. (†57551) 298/G7
Harrington, Wash. (99134) 310/G3
Harrington Harbour, Québec 174/F2
Harrington Park, N.J. (07640) 273/C1
Harris, Calif. (†95440) 204/B3
Harris (co.), Georgia 217/C5
Harris, Iowa (51345) 229/C2
Harris, Kansas (†66032) 232/G3
Harris, Mich. (49845) 250/B3
Harris, Minn. (55032) 255/F5
Harris, Mo. (64645) 261/F2
Harris, Okla. (†74740) 288/S7
Harris, Sask. 181/D4
Harris (dist.), Scotland 15/B3
Harris (dist.), Scotland 10/C2
Harris (sound), Scotland 15/A3
Harris (sound), Scotland 10/C2
Harris (lake), S. Australia 94/D4
Harris, Tenn. (†38261) 237/C8
Harris (co.), Texas 303/J8
Harrisburg, Ark. (72432) 202/J2
Harrisburg, Ill. (62946) 222/E6
Harrisburg, Ind. (†47331) 227/G5
Harrisburg, Mo. (65256) 261/H4
Harrisburg, Nebr. (69345) 264/A3
Harrisburg, N.C. (28075) 281/H4
Harrisburg, Ohio (43126) 284/D6
Harrisburg, Oreg. (97446) 291/D3
Harrisburg (cap.), Pa. 188/L2
Harrisburg (cap.), Pa. 146/L5
Harrisburg (cap.), Pa. (*17101) 294/H5
Harrisburg, S. Dak. (57032) 298/R7
Harrismith, S. Africa 118/D5
Harrison (bay), Alaska 196/H1
Harrison, Ark. (72601) 202/D1
Harrison (lake), Br. Col. 184/M2
Harrison, Georgia (31035) 217/G5
Harrison, Idaho (83833) 220/B2
Harrison, Ill. (†61072) 222/D1
Harrison (co.), Ind. 227/E8
Harrison (co.), Iowa 229/B5
Harrison (co.), Ky. 237/N4
Harrison○, Maine (04040) 243/B7
Harrison, Mich. (48625) 250/E4
Harrison (co.), Miss. 256/F10
Harrison (co.), Mo. 261/E2
Harrison, Mont. (59735) 262/E5
Harrison, Nebr. (69346) 264/A2
Harrison (cape), Newf. 166/C3
Harrison (cape), Newf. 162/L5
Harrison, N.J. (07029) 273/B2
Harrison, N.Y. (10528) 276/P6
Harrison (co.), Ohio 284/H5
Harrison, Ohio (45030) 284/A9
Harrison, S. Dak. (57344) 298/M7
Harrison, Tenn. (37341) 237/L10
Harrison (co.), Texas 303/K5
Harrison (co.), W. Va. 312/F4
Harrison, Wis. (†54435) 317/G5
Harrisonburg, La. (71340) 238/G3
Harrisonburg (I.C.), Va. (22801) 307/K4
Harrison Hot Springs, Br. Col. 184/M3
Harrison Valley, Pa. (16927) 294/G2
Harrisonville, Ill. (†62295) 222/C5
Harrisonville, Mo. (64701) 261/D5
Harrisonville, N.J. (08039) 273/C4
Harrisonville, Ohio (†45769) 284/F7
Harrisonville, Pa. (17228) 294/F6
Harriston, Miss. (39081) 256/C7
Harriston, Ontario 177/D4
Harristown, Ill. (62537) 222/D4
Harrisville, Ind. (†47390) 227/H4
Harrisville, Mich. (48740) 250/F4
Harrisville, Miss. (39082) 256/D7
Harrisville○, N.H. (03450) 268/C6
Harrisville, N.Y. (13648) 276/K2
Harrisville, Ohio (43974) 284/J5
Harrisville, Pa. (16038) 294/B3
Harrisville, R.I. (02830) 249/H5
Harrisville, Utah (†84401) 304/C2
Harrisville, W. Va. (26362) 312/E4
Harris Wash (creek), Utah 304/C6
Harrod, Ohio (45850) 284/C4
Harrodsburg, Ind. (47434) 227/D6
Harrodsburg, Ky. (40330) 237/M5
Harrods Creek, Ky. (40027) 237/K4
Harrogate, Br. Col. 184/J5
Harrogate, England 13/J1
Harrogate, England 10/F4
Harrogate-Shawanee, Tenn. (37752) 237/O8
Harrold, S. Dak. (57536) 298/K4
Harrold, Texas (76364) 303/F3
Harrop (lake), Manitoba 179/G2
Harrow, England 13/G8
Harrow, England 10/B5
Harrow, Ontario 177/B5
Harrow, Victoria 97/A5
Harrowby, Manitoba 179/A4
Harrowsmith, Ontario 177/H3
Harry Strunk (lake), Nebr. 264/D4
Harsens Island, Mich. (48028) 250/G6
Harshaw, Wis. (54529) 317/G4
Harstad, Norway 18/K2
Hart (lake), Fla. 212/E3
Hart (co.), Georgia 217/G2
Hart (co.), Ky. 237/K6
Hart, Mich. (49420) 250/C5
Hart (lake), Oreg. 291/H5
Hart (mt.), Oreg. 291/H5
Hart, Texas (79043) 303/B3
Hart (riv.), Yukon 187/E3
Hartbees (riv.), S. Africa 118/C5
Hartberg, Austria 41/C3
Harte, Manitoba 179/C4
Harte (mt.), Manitoba 179/A2
Hartell, Alberta 182/C4
Hartfield, Va. (23071) 307/R5
Hartford, Ala. (36344) 195/G8
Hartford, Ark. (72938) 202/B3
Hartford (co.), Conn. 210/D1
Hartford (cap.), Conn. 146/L5
Hartford (cap.), Conn. 188/M2
Hartford (cap.), Conn. (*06101) 210/E1
Hartford, Ill. (62048) 222/A2
Hartford, Iowa (50118) 229/G6
Hartford, Kansas (66854) 232/F3
Hartford, Ky. (42347) 237/H6
Hartford○, Maine (†04221) 243/C7
Hartford, Mich. (49057) 250/C6
Hartford, N.J. (†08057) 273/D4
Hartford, N.Y. (12838) 276/O4
Hartford, Ohio (†43013) 284/E5
Hartford, Ohio (44424) 284/J3
Hartford, S. Dak. (57033) 298/P6
Hartford, Tenn. (37753) 237/P9
Hartford○, Vt. (05047) 268/C4
Hartford, W. Va. (25247) 312/C4
Hartford, Wis. (53027) 317/K9
Hartford City, Ind. (47348) 227/G4
Harthill, Scotland 15/C2
Hartington, Nebr. (68739) 264/G2
Hartington, Ontario 177/H3
Hartland○, Conn. (†06091) 210/D1
Hartland, England 13/C7
Hartland (pt.), England 13/C6
Hartland (pt.), England 10/D5
Hartland, Maine (04943) 243/D6
Hartland○, Maine (04943) 243/D6
Hartland, Mich. (48029) 250/F6
Hartland, Minn. (56042) 255/E7
Hartland, New Bruns. 170/C2
Hartland○, Vt. (05048) 288/C4
Hartland, W. Va. (†25043) 312/D6
Hartland, Wis. (53029) 317/J1
Hartland Four Corners, Vt. (05049) 268/C4
Hartlepool, England 10/F3
Hartlepool, England 13/F3
Hartleton, Pa. (17829) 294/H4
Hartley, Iowa (51346) 229/C2
Hartley (co.), Texas 303/B2
Hartley, Texas (79044) 303/B2
Hartley, Zimbabwe 118/E3
Hartleyville, Alberta 182/D5
Hartline, Wash. (99135) 310/F3
Hartly, Del. (19953) 245/R4
Hartman, Ark. (72840) 202/C3
Hartman, Colo. (81043) 208/P6
Hartney, Manitoba 179/B5
Harts (pass), Wash. 310/E2
Harts, W. Va. (25524) 312/B6
Hartsburg, Ill. (62643) 222/D3
Hartsburg, Mo. (65039) 261/H5
Hartsdale, N.Y. (10530) 276/P6
Hartsel, Colo. (80449) 208/H4
Hartselle, Ala. (35640) 195/E2
Hartsfield, Georgia (31756) 217/E8
Hartsgrove, Ohio (†44085) 284/J2
Hartshorn, Mo. (65479) 261/J8
Hartshorne, Okla. (74547) 288/R5
Harts Range, North. Terr. 88/F4
Harts Range, North. Terr. 93/D7
Hartstown, Pa. (16131) 294/B2
Hartsville, Ind. (47244) 227/F6
Hartsville, Mass. (†01230) 249/B4
Hartsville, S.C. (29550) 296/G3
Hartsville, Tenn. (37074) 237/J8
Hartville, Mo. (65667) 261/G8
Hartville, Ohio (44632) 284/H4
Hartville, Wyo. (82215) 319/H3
Hartwell, Georgia (30643) 217/G2
Hartwell (dam), Georgia 217/G2
Hartwell (lake), Georgia 217/G2
Hartwell, Mo. (†64788) 261/E6
Hartwell (dam), S.C. 296/B3
Hartwell (lake), S.C. 296/A3
Hartwick, Iowa (52232) 229/J5
Hartwick, N.Y. (13348) 276/K5
Hartz (mt.), Tasmania 99/C5
Harug el Asued, El (mts.), Libya 111/C2
Haruniye, Turkey 63/G4
Har Us Nuur (lake), Mongolia 77/D2
Harvard (mt.), Colo. 208/G5
Harvard, Idaho (83834) 220/B3
Harvard, Ill. (60033) 222/E1
Harvard, Iowa (†50008) 229/G7
Harvard○, Mass. (01451) 249/H2
Harvard, Nebr. (68944) 264/F4
Harvel, Ill. (62538) 222/D4
Harvest, Ala. (35749) 195/E1
Harvester, Mo. (63303) 261/N2
Harvey, Ill. (60426) 222/B6
Harvey, Iowa (50119) 229/H6
Harvey (co.), Kansas 232/E3
Harvey, La. (70058) 238/O4
Harvey, Albert, New Bruns. 170/F3
Harvey, York, New Bruns. 170/D3
Harvey (lake), New Bruns. 170/C3
Harvey (mt.), New Bruns. 170/D3
Harvey, N. Dak. (58341) 282/L4
Harvey, W. Australia 88/B6
Harvey, W. Australia 92/A2
Harvey, W. Va. (25852) 312/D7
Harvey Cedars, N.J. (08008) 273/E4
Harveys (lake), Vt. 268/C3
Harveysburg, Ohio (45032) 284/C7
Harveys Lake, Pa. (18618) 294/E7
Harveyton, Ky. (†41718) 237/P6
Harveyville, Kansas (66431) 232/F3
Harviell, Mo. (63945) 261/M9
Harwich, England 13/J6
Harwich, England 10/G5
Harwich, Mass. (02645) 249/O6
Harwich○, Mass. (02645) 249/O6
Harwich Port, Mass. (02646) 249/O6
Harwinton, Conn. (06791) 210/C1
Harwinton○, Conn. (06791) 210/C1
Harwood, Mo. (64750) 261/D7
Harwood, N. Dak. (58042) 282/S6
Harwood, Ontario 177/F3
Harwood, Texas (78632) 303/G8
Harwood Heights, Ill. (60656) 222/B5
Harwood Island, N.S. Wales 97/G1
Harworth, England 13/F4
Haryana (state), India 68/D3
Harz (mts.), E. Germany 22/D3
Harz (mts.), W. Germany 22/D3
Harzgerode, E. Germany 22/D3
Hasa, Wadi el (dry riv.), Jordan 65/E5
Hasan Dağı, Büyük (mt.), Turkey 63/E3
Hasbrouck Heights, N.J. (07604) 273/B2
Hase (riv.), W. Germany 22/B2
Haseke (prov.), Syria 63/J4
Haselünne, W. Germany 22/B2
Hasenkamp, Argentina 143/F5
Hashtpar, Iran 66/F2
Haskeir (isl.), Scotland 15/A3
Haskell, Ark. (†72015) 202/E4
Haskell (co.), Kansas 232/B4
Haskell, N.J. (07420) 273/A1
Haskell (co.), Okla. 288/R4
Haskell, Okla. (74436) 288/P3
Haskell (co.), Texas 303/E4
Haskell, Texas (79521) 303/E4
Haskett, Manitoba 179/D5
Haskins, Iowa (†52201) 229/K6
Haskins, Ohio (43525) 284/C3
Haslach an der Mühl, Austria 41/C2
Hasle, Denmark 21/F8
Haslemere, England 13/G6
Haslemere, England 10/F5
Haslet, Texas (76052) 303/E2
Haslett, Mich. (48840) 250/E6
Haslev, Denmark 21/E7
Haslingden, England 13/H1
Hassa, Turkey 63/G4
Hassan, India 68/D6
Hassayampa (riv.), Ariz. 198/C5
Hasse, Texas (76456) 303/F6
Hassel (sound), N.W. Terrs. 187/J2
Hassel (isl.), Virgin Is. (U.S.) 161/B4
Hassell, N.C. (27841) 281/P3
Hasselt, Belgium 27/G7
Hasselt, Netherlands 27/J3
Hassfurt, W. Germany 22/D3
Hassi Messaoud, Algeria 106/F2
Hassi R'Mel, Algeria 106/E2
Hässleholm, Sweden 18/H8
Hassloch, W. Germany 22/C4
Haster, Scotland 15/E2
Hastière, Belgium 27/F8
Hastings, England 10/G5
Hastings, England 13/H7
Hastings, Fla. (32045) 212/E2
Hastings, Iowa (51540) 229/C6
Hastings, Mich. (49058) 250/D6
Hastings, Minn. (55033) 255/F6
Hastings, Nebr. 188/G2
Hastings, Nebr. (68901) 264/F4
Hastings, N. Zealand 100/F3
Hastings, N. Dak. (†58049) 282/O6
Hastings, Okla. (73548) 288/K6
Hastings (county), Ontario 177/G3
Hastings, Ontario 177/G3
Hastings, Pa. (16646) 294/E4
Hastings On Hudson, N.Y. (10706) 276/O6
Hasty, Ark. (72640) 202/D1
Hasty, Colo. (81044) 208/O6
Hasvik, Norway 18/M1
Haswell, Colo. (81045) 208/N6
Hat (peak), Calif. 204/E2
Hat (creek), S. Dak. 298/B7
Hatay (prov.), Turkey 63/G4
Hatay (Antakya), Turkey 63/G4
Hatboro, Pa. (19040) 294/M5
Hatch, N. Mex. (87937) 274/B6
Hatch, Utah (84735) 304/B6
Hatchechubbee, Ala. (36858) 195/H6
Hatcher, Georgia (†31754) 217/B7
Hatches Creek, North. Terr. 88/F4
Hatches Creek, North. Terr. 93/D6
Hatchet (mts.), N. Mex. 274/A7
Hatchett (pt.), Conn. 210/G3
Hatchie (riv.), Tenn. 237/B9
Hatchineha (lake), Fla. 212/E3
Hateg, Romania 45/F3
Hatfield, Ark. (71945) 202/B5
Hatfield, England 13/H7
Hatfield, Ind. (47617) 227/C9
Hatfield, Ky. (†41514) 237/S5
Hatfield, Mass. (01038) 249/D3
Hatfield○, Mass. (01038) 249/D3
Hatfield, Minn. (56135) 255/B7
Hatfield, Mo. (64458) 261/D1
Hatfield, N.S. Wales 97/B3
Hatfield, Pa. (19440) 294/M5
Hatfield, Sask. 181/F4
Hatfield, Wis. (†54754) 317/E7
Hatfield Point, New Bruns. 170/E3
Hatgal, Mongolia 77/E1
Hathaway, Mont. (59333) 262/K4
Hatherleigh, Sask. 181/C2
Hathras, India 68/D3
Hatiba, Ras (cape), Saudi Arabia 59/C5
Ha Tien, Vietnam 72/E5
Hatillo, P. Rico 161/B1
Ha Tinh, Vietnam 72/E3
Hatira (mt.), Israel 65/B6
Hatley, Miss. (†38821) 256/H3
Hatley, Québec 172/E4
Hatley, Wis. (54440) 317/H6
Hato, Neth. Ant. 161/G8
Hato del Volcán, Panama 154/F6
Hato Mayor, Dom. Rep. 158/F6
Hato Rey, P. Rico 161/E1
Hatseva, Israel 65/D5
Hattem, Netherlands 27/H4
Hatteras (cape), N.C. 146/L6
Hatteras (cape), N.C. 188/M3
Hatteras, N.C. (27943) 281/T4
Hatteras (cape), N.C. 281/U4
Hatteras (inlet), N.C. 281/T4
Hatteras (isl.), N.C. 281/U4
Hatteras (cape), U.S. 2/F4
Hattiesburg, Miss. 188/H4
Hattiesburg, Miss. (39401) 256/F8
Hattieville, Ark. (72063) 202/E3
Hattieville, Belize 154/C2
Hatton, Ala. (†35672) 195/D1
Hatton, N. Dak. (58240) 282/R4
Hatton, Sask. 181/B5
Hatton, Scotland 15/G3
Hatton, Utah (†84637) 304/B5
Hatton, Wash. (99332) 310/G4
Hatuey, Cuba 158/G3
Hatvan, Hungary 41/E3
Hat Yai, Thailand 72/C6
Hatzic, Br. Col. 184/L3
Hau Bon, Vietnam 72/E4
Haubstadt, Ind. (47639) 227/B8
Haud (reg.), Ethiopia 111/J6
Haud (plat.), Somalia 115/J2
Haugan, Mont. (59842) 262/A3
Hauge, Norway 18/E7
Haugen, Wis. (54841) 317/C4
Haugesund, Norway 7/E3
Haugesund, Norway 18/D7
Haughton, La. (71037) 238/C1
Hauhungaroa (range), N. Zealand 100/E3
Haukivesi (lake), Finland 18/Q5
Haultain (riv.), Sask. 181/L3
Haunstetten, W. Germany 22/D4
Hauppauge, N.Y. (11787) 276/O9
Haura, P.D.R. Yemen 59/E7
Hauraki (gulf), N. Zealand 100/C1
Hauran, Wadi (dry riv.), Iraq 59/D3
Hauran, Wadi (dry riv.), Iraq 66/D4
Hauroko (lake), N. Zealand 100/A6
Hauru (pt.), Fr. Poly. 86/S12
Hauser, Idaho (†83854) 220/A2
Hauser (lake), Mont. 262/E4
Hauser, Oreg. (†97459) 291/C4
Hausstock (mt.), Switzerland 39/H3
Haut (isl.), Maine 243/G7
Haute (isl.), Nova Scotia 168/C3
Haute-Corse (dept.), France 28/B6
Haute-Garonne (dept.), France 28/D6
Haute-Loire (dept.), France 28/E5
Haute-Marne (dept.), France 28/F3
Hauterive, Québec 172/A1
Hauterive, Québec 174/D3
Hautes-Alpes (dept.), France 28/G5
Haute-Saône (dept.), France 28/G4
Haute-Savoie (dept.), France 28/G5
Hautes-Pyrénées (dept.), France 28/D6
Haute-Vienne (dept.), France 28/D5
Hautmont, France 28/F2
Haut-Rhin (dept.), France 28/G4
Hauts-de-Seine (dept.), France 28/A2
Haut-Zaïre (prov.), Zaire 115/E3
Hauula, Hawaii (96717) 218/E1
Havaco, W. Va. (24841) 312/C8
Havana, Ala. (35467) 195/C5
Havana, Ark. (72842) 202/D3
Havana (cap.), Cuba 156/A2
Havana (cap.), Cuba 2/E4
Havana (cap.), Cuba 146/K7
Havana (cap.), Cuba 158/C1
Havana, Fla. (32333) 212/B1
Havana, Ill. (62644) 222/D3
Havana, Kansas (67347) 232/G4
Havana, Minn. (†55060) 255/E6
Havana, N. Dak. (58043) 282/P8
Havana, Ohio (†44890) 284/E3
Havannah (chan.), New Caled. 86/H5
Havant and Waterloo, England 13/G7
Havasu (lake) 188/D4
Havasu (lake), Ariz. 198/A4
Havasu (lake), Calif. 204/L9
Havasu (lake), U.S. 146/G6
Havasupai Ind. Res., Ariz. 198/C2
Havdrup, Denmark 21/F6
Havel (riv.), E. Germany 22/E2
Havelange, Belgium 27/G8
Havelberg, E. Germany 22/D2
Havelock, Iowa (50546) 229/D3
Havelock, New Bruns. 170/E3
Havelock, N. Zealand 100/D4
Havelock, N.C. (28532) 281/P5
Havelock, N. Dak. (†58647) 282/E7
Havelock, Ontario 177/G3
Havelock North, N. Zealand 100/F3
Haven, Kansas (67543) 232/E4
Havensville, Kansas (66432) 232/F2
Haverford○, Pa. (19041) 294/M6
Haverfordwest, Wales 10/D5
Haverfordwest, Wales 13/B6
Haverhill, England 13/H5
Haverhill, Iowa (50120) 229/H5
Haverhill, Mass. (01830) 249/K1
Haverhill○, N.H. (03765) 268/C3
Haverhill, Ohio (45636) 284/E8
Havering, England 10/C5
Havering, England 13/J8
Haverstraw, N.Y. (10927) 276/M8
Havertown, Pa. (19083) 294/M6
Haviland, Kansas (67059) 232/C4
Haviland, Ohio (45851) 284/A3
Havillah, Wash. (†98855) 310/F2
Havířov, Czech. 41/E2
Havlíčkův Brod, Czech. 41/C2
Havran, Turkey 63/B3
Havre, Mont. 146/G5
Havre, Mont. 188/E1
Havre, Mont. (59501) 262/G2
Havre Boucher, Nova Scotia 168/G3
Havre de Grace, Md. (21078) 245/O2
Havre-Saint-Pierre, Québec 174/E2
Havre-St-Pierre, Que. 162/K5
Havsa, Turkey 63/B2
Havza, Turkey 63/F2
Haw (riv.), N.C. 281/K2
Hawaii 188/F5
HAWAII 218
Hawaii (co.), Hawaii 218/K7
Hawaii (isl.), Hawaii 87/L4
Hawaii (isl.), Hawaii 188/F6
Hawaii (isl.), Hawaii 218/H5
Hawaii (state), U.S. 2/B4
Hawaii (state), U.S. 87/K4
Hawaiian (isls.) 87/J3
Hawaii Kai, Hawaii (96825) 218/F2
Hawaii Nat'l Park, Hawaii (96718) 218/J6
Hawaii Volcanoes Nat'l Park, Hawaii 218/H6
Hawara, Jordan 65/D2
Hawarden, Iowa (51023) 229/A2
Hawarden, N. Zealand 100/D5
Hawarden, Sask. 181/E4
Hawarden, Wales 13/G2
Hawea (lake), N. Zealand 100/B6
Hawera, N. Zealand 100/E3
Hawes, England 13/E3
Hawesville, Ky. (42348) 237/H5
Hawi, Hawaii (96719) 218/G3
Hawick, Minn. (56246) 255/D5
Hawick, Scotland 10/E3
Hawick, Scotland 15/F5
Hawk (hills), Alberta 182/B1
Hawke (hills), Newf. 166/D2
Hawke (isl.), Newf. 166/C3
Hawke (riv.), Newf. 166/C3
Hawke (bay), N. Zealand 100/F3
Hawker, S. Australia 88/F6
Hawker, S. Australia 94/F4
Hawke's Bay, Newf. 166/C3
Hawkesbury (isl.), Br. Col. 184/C3
Hawkesbury, Ontario 177/K2
Hawkestone, Ontario 177/E3
Hawkeye, Iowa (52147) 229/J3
Hawkins, Mich. (†49677) 250/D5
Hawkins (co.), Tenn. 237/P8
Hawkins, Texas (75765) 303/J5
Hawkins, Wis. (54530) 317/E4
Hawkinsville, Georgia (31036) 217/E6
Hawk Junction, Ontario 175/D3
Hawk Junction, Ontario 177/J5
Hawk Point, Mo. (63349) 261/K5
Hawk Run, Pa. (16840) 294/F4
Hawks, Mich. (49743) 250/F3
Hawk Springs, Wyo. (82217) 319/H4
Hawley, Minn. (56549) 255/B4
Hawley, Pa. (18428) 294/M3
Hawley, Texas (79525) 303/E5
Hawleyville, Conn. (06440) 210/B3
Haworth, N.J. (07641) 273/C1
Haworth, Okla. (74740) 288/S7
Hawston, S. Africa 118/G7
Hawthorn, La. (†71446) 238/D4
Hawthorn, Pa. (16230) 294/D3
Hawthorn, Victoria 97/J5
Hawthorne, Calif. (90250) 204/C11
Hawthorne, Fla. (32640) 212/D2
Hawthorne, Nev. (89415) 266/C4
Hawthorne, N.J. (07507) 273/B2
Hawthorne, N.Y. (10532) 276/O6
Hawthorne, Victoria 88/L7
Hawthorne, Wis. (54842) 317/C3
Hawthorn Woods, Ill. (†60047) 222/B5
Haxby, England 13/F3
Haxtun, Colo. (80731) 208/O1
Hay (riv.) 162/E4
Hay (lake), Alberta 182/A5
Hay (riv.), Alberta 182/A5
Hay (riv.), Canada 146/G4
Hay, N.S. Wales 88/H6
Hay, N.S. Wales 97/C4
Hay (dry riv.), North. Terr. 88/F4
Hay (cape), North. Terr. 93/A3
Hay (dry riv.), North. Terr. 93/E7
Hay (lake), Ontario 177/F2

Hay, Wales 10/E4
Hay, Wales 13/D5
Hay, Wash. (99136) 310/H4
Hayama, Japan 81/O3
Hayange, France 28/F3
Haycock, Alaska (†99762) 196/F1
Hayden, Ala. (35079) 195/E3
Hayden, Ariz. (85235) 198/E5
Hayden, Colo. (81639) 208/E2
Hayden, Idaho (†83835) 220/B2
Hayden (lake), Idaho 220/B2
Hayden, Ind. (47245) 227/F7
Hayden, Mo. (†65459) 261/H6
Hayden, N. Mex. (†88410) 274/F3
Hayden (peak), Utah 304/C3
Haydenburg, Tenn. (†38588) 237/K8
Hayden Lake, Idaho (83835) 220/B2
Haydenville, Mass. (01039) 249/C3
Haydenville, Ohio (43127) 284/F7
Hayes (mt.), Alaska 196/J2
Hayes (pen.), Greenl. 4/B13
Hayes, Jamaica 158/J6
Hayes, La. (70646) 238/E6
Hayes (riv.), Man. 162/G4
Hayes (riv.), Manitoba 179/K3
Hayes (co.), Nebr. 264/C4
Hayes (riv.), N.W. Terrs. 187/J3
Hayes, S. Dak. (57537) 298/H5
Hayes, Wis. (†54174) 317/J5
Hayes Center, Nebr. (69032) 264/C4
Hayesville, Iowa (52562) 229/J6
Hayesville, New Bruns. 170/D2
Hayesville, N.C. (28904) 281/B4
Hayesville, Ohio (44838) 284/F4
Hayesville, Oreg. (†97301) 291/A3
Hayfield, Iowa (50445) 229/F2
Hayfield, Minn. (55940) 255/F7
Hayfork, Calif. (96041) 204/B3
Hay Fork, Trinity (riv.), Calif. 204/B3
Hay Lakes, Alberta 182/D3
Hayle, England 13/B7
Haylow, Georgia (†31648) 217/G9
Haymana, Turkey 63/E3
Haymarket, Va. (22069) 307/N3
Hayne, N.C. (†28318) 281/M5
Haynes, Alberta 182/D3
Haynes, Ark. (72341) 202/J4
Haynes, N. Dak. (58637) 282/F8
Haynesville, La. (71038) 238/D1
Haynesville○, Maine (04446) 243/G4
Haynesville, Va. (22472) 307/P5
Hayneville, Ala. (36040) 195/E6
Hayrabolu, Turkey 63/B2
Hay River, N.W.T. 162/E3
Hay River, N.W.T. 146/G3
Hay River, N.W. Terrs. 187/G3
Hays, Alberta 182/E4
Hays, Kansas (67601) 232/C3
Hays, Mont. (59527) 262/H2
Hays, N.C. (28635) 281/G2
Hays (co.), Texas 303/F7
Haysi, Va. (24256) 307/D6
Hay Springs, Nebr. (69347) 264/B2
Haystack (mt.), Conn. 210/C1
Haystack (peak), Mont. 262/A3
Haystack (mt.), N.Y. 276/N2
Haystack (mt.), Vt. 268/B6
Haysville, Ind. (†47546) 227/D8
Haysville, Kansas (67060) 232/E4
Haysville, Pa. (†15143) 294/B4
Hayter, Alberta 182/E3
Hayti, Mo. (63851) 261/N10
Hayti, S. Dak. (57241) 298/P4
Hayti Heights, Mo. (†63851) 261/N10
Hayton, Wis. (†53014) 317/K7
Hayward, Calif. (*94541) 204/K2
Hayward (lake), Conn. 210/F2
Hayward, Minn. (56043) 255/E7
Hayward, Mo. (†63873) 261/N10
Hayward, Wis. (54843) 317/D3
Haywards-Manor Park, N. Zealand 100/B2
Haywood, Manitoba 179/D5
Haywood (co.), N.C. 281/C3
Haywood, N.C. (†27559) 281/L3
Haywood, Okla. (74548) 288/P5
Haywood (co.), Tenn. 237/C9
Haywood City, Mo. (†63736) 261/N9
Hazar (lake), Turkey 63/H3
Hazaran, Kuh-e (mt.), Iran 66/K6
Hazard, Ky. (41701) 237/P6
Hazard, Nebr. (68844) 264/F3
Hazardville, Conn. (06082) 210/E1
Hazaribagh, India 68/F4
Hazar Qadam, Afghanistan 59/J3
Hazar Qadam, Afghanistan 68/B2
Hazebrouck, France 28/E2
Hazel, Ky. (42049) 237/E7
Hazel, S. Dak. (57242) 298/P4
Hazel Cliffe, Sask. 181/J5
Hazel Crest, Ill. (60429) 222/B6
Hazeldean, New Bruns. 170/C2
Hazel Dell, Ill. (62430) 222/E4
Hazel Dell, Sask. 181/H4
Hazeldine, Alberta 182/E3
Hazel Green, Ala. (35750) 195/E1
Hazel Green, Ky. (41332) 237/O5
Hazelgreen, Mo. (†65556) 261/H7
Hazel Green, Wis. (53811) 317/F11
Hazel Grove and Bramhall, England 13/H2
Hazel Hill, Nova Scotia 168/G3
Hazelhurst, Ill. (†61064) 222/D2
Hazel Hurst, Pa. (16733) 294/E2
Hazelhurst, Wis. (54531) 317/G4
Hazel Park, Mich. (48030) 250/B6
Hazel Patch, Ky. (†40729) 237/N6
Hazelridge, Manitoba 179/F5
Hazelrigg, Ind. (†46052) 227/D4
Hazel Run, Minn. (56247) 255/C6
Hazelton, Br. Col. 162/D5
Hazelton, Br. Col. 184/D2
Hazelton (mts.), Br. Col. 184/C2
Hazelton, Idaho (83335) 220/E7
Hazelton, Kansas (67061) 232/D4
Hazelton, N. Dak. (58544) 282/K7
Hazelton, W. Va. (26535) 312*G3
Hazelton (peak), Wyo. 319/E1
Hazelwood, Ind. (†46118) 227/D5
Hazelwood, Mo. (*63042) 261/P2
Hazelwood, N.C. (28738) 281/C4
Hazen (bay), Alaska 196/E2
Hazen, Ark. (72064) 202/G4
Hazen, Nev. (89417) 266/C3
Hazen, N. Dak. (58545) 282/G5
Hazen (lake), N.W. Terrs. 187/L1
Hazen (str.), N.W. Terrs. 187/G2
Hazenmore, Sask. 181/D6
Hazerim, Israel 65/B5
Hazlehurst, Georgia (31539) 217/G7
Hazlehurst, Miss. (39083) 256/D7
Hazlet, N.J. (07730) 273/E3
Hazlet, Sask. 181/C5
Hazleton, Ind. (47640) 227/B8
Hazleton, Iowa (50641) 229/K3
Hazleton, Pa. (18201) 294/L4
Hazlett (lake), W. Australia 88/D4
Hazlettville, Del. (†19953) 245/R4
Hazor Hagelilit, Israel 65/D2
Hazro, Turkey 63/J3
Heacham, England 13/H5
Headford, Ireland 17/C5
Headland, Ala. (36345) 195/H8
Headlee, Ind. (†47960) 227/D3
Head of Amherst, Nova Scotia 168/E3
Head of Bay d'Espoir, Newf. 166/C4
Head of Bight (bay), S. Australia 94/B4
Head of Grassy, Ky. (41145) 237/P4
Head of Island, La. (†70462) 238/L2
Head of Jeddore, Nova Scotia 168/E4
Head of Millstream, New Bruns. 170/E3
Head of Saint Margarets Bay, Nova Scotia 168/E4
Headquarters, Idaho (83534) 220/C3
Headrick, Okla. (73549) 288/H5
Heads, The (prom.), Oreg. 291/C5
Heads of Ayr (cape), Scotland 15/D5
Head Waters, Va. (24442) 307/K4
Heafford Junction, Wis. (54532) 317/G4
Healdsburg, Calif. (95448) 204/B5
Healdton, Okla. (73438) 288/M6
Healdville, Vt. (05147) 268/B5
Healesville, Victoria 97/C5
Healing Springs, Ala. (†36558) 195/B7
Healing Springs, Va. (†24445) 307/J5
Healy, Alaska (99743) 196/H2
Healy, Kansas (67850) 232/B3
Healys, Va. (†23071) 307/R5
Heanor, England 13/F4
Heard (isl.), Australia 2/N8
Heard (co.), Georgia 217/B4
Hearne, Sask. 181/F5
Hearne, Texas (77859) 303/H7
Hearst (isl.) 5/B16
Hearst, Ont. 162/H6
Hearst, Ontario 177/J5
Hearst, Ontario 175/D3
Heart (lake), Alberta 182/E2
Heart (butte), N. Dak. 282/G6
Heart (riv.), N. Dak. 282/F6
Heart (lake), Wyo. 319/B1
Heart Butte, Mont. (59448) 262/C2
Heart River Settlement, Alberta 182/B2
Heart's Content, Newf. 166/D2
Heart's Delight, Newf. 166/D2
Heart's Desire, Newf. 166/D2
Hearts Hill, Sask. 181/B3
Heartwell, Nebr. (68945) 264/F4
Heartwellville, Vt. (†05350) 268/A6
Heaters, W. Va. (26627) 312/E5
Heath, Ala. (†36420) 195/F8
Heath, Alberta 182/E3
Heath (riv.), Bolivia 136/A3
Heath○, Mass. (01346) 249/C2
Heath, Mont. (†59457) 262/G3
Heath, Ohio (43055) 284/F5
Heath (riv.), Peru 128/H9
Heath (pt.), Québec 174/E3
Heathcote, Victoria 97/C5
Heatherton, Newf. 166/C4
Heatherton, Nova Scotia 168/G3
Heathhall, Scotland 15/E5
Heath Springs, S.C. (29058) 296/F2
Heath Steele, New Bruns. 170/D1
Heathsville, Va. (22473) 307/P5
Heaton, N. Dak. (58450) 282/L5
Heavener, Okla. (74937) 288/S5
Hebbardsville, Ky. (†42420) 237/G5
Hebbronville, Texas (78361) 303/F10
Hebbs Cross, Nova Scotia 168/D4
Hebburn, England 13/J3
Hebei (Hopei) (prov.), China 77/J4
Hebel, Queensland 95/C6
Heber, Ariz. (85928) 198/E4
Heber, Calif. (92249) 204/K11
Heber City, Utah (84032) 304/C3
Heber Springs, Ark. (72543) 202/G2
Hebert, La. (71436) 238/G2
Hébert (riv.), Nova Scotia 168/D3
Hébertville, Québec 172/F1
Hébertville-Station, Québec 172/F1
Hebgen (dam), Mont. 262/E6
Hebgen (lake), Mont. 262/E6
Hebi, China 77/H4
Hebo, Oreg. (97122) 291/D2
Hebrides, Inner (isls.), Scotland 10/C2
Hebrides, Inner (isls.), Scotland 15/B4
Hebrides, Outer (isls.), Scotland 10/C2
Hebrides, Outer (isls.), Scotland 15/A3
Hebrides (sea), Scotland 15/B3
Hebrides (sea), Scotland 10/C2
Hebron○, Conn. (06248) 210/F2
Hebron, Ill. (60034) 222/E1
Hebron, Ind. (46341) 227/C2
Hebron, Ky. (41048) 237/R2
Hebron○, Maine (04238) 243/C7
Hebron, Md. (21830) 245/R7
Hebron, Nebr. (68370) 264/G4
Hebron, Newf. 146/M4
Hebron, Newf. 162/K4
Hebron (fjord), Newf. 166/B2
Hebron○, N.H. (03241) 268/D4
Hebron, N. Dak. (58638) 282/G6
Hebron, Nova Scotia 168/B5
Hebron, Ohio (43025) 284/F6
Hebron, Texas (†75067) 303/G1
Hebron, West Bank 65/C4
Hebron, West Bank 59/C3
Hebron, W. Va. (26368) 312/D4
Hebron, Wis. (†53538) 317/J10
Hecate (str.), Br. Col. 162/C5
Hecate (str.), Br. Col. 146/E4
Hecate (str.), Br. Col. 184/B3
Hecelchakán, Mexico 150/O6
Heceta (isl.), Alaska 196/M2
Heceta (head), Oreg. 291/C3
Hechi, China 77/G7
Hechingen, W. Germany 22/C4
Hechuan (Hochwan), China 77/G5
Hecker, Ill. (62248) 222/D5
Hecla, Manitoba 179/F3
Hecla (isl.), Manitoba 179/F3
Hecla, S. Dak. (57446) 298/N2
Hecla Prov. Park, Manitoba 179/F3
Hector, Ark. (72843) 202/E3
Hector, Minn. (55342) 255/D6
Hector, N.Y. (14841) 276/G5
Hede, Sweden 18/H5
Hedemora, Sweden 18/K6
Hedenäset, Sweden 18/N3
Hedensted, Denmark 21/C6
Hedgesville, Mont. (†59078) 262/G4
Hedgesville, W. Va. (25427) 312/K3
Hedley, Br. Col. 184/G5
Hedley, Texas (79237) 303/D3
Hedmark (co.), Norway 18/G6
Hedon, England 13/G4
Hedrick, Iowa (52563) 229/J6
Hedville, Kansas (†67401) 232/E3
Hedwig Village, Texas (†77001) 303/H1
Heemskerk, Netherlands 27/F3
Heemstede, Netherlands 27/F4
Heer, Netherlands 27/H7
Heerde, Netherlands 27/H4
Heerenveen, Netherlands 27/H3
Heerhugowaard, Netherlands 27/F3
Heerlen, Netherlands 27/J7
Heesch, Netherlands 27/G5
Hefei (Hofei), China 77/J5
Hefei, China 54/N6
Heffley Creek, Br. Col. 184/G5
Heflin, Ala. (36264) 195/G3
Heflin, La. (71039) 238/D2
Hegang (Hokang), China 77/L2
Hegang, China 54/O5
Hegau (reg.), W. Germany 22/C5
Hegeler, Ill. (†61832) 222/F3
Hegins, Pa. (17938) 294/K4
Heiban, Sudan 111/F5
Heiberger, Ala. (†36756) 195/D5
Heide, W. Germany 22/C1
Heidelberg, Ky. (41333) 237/O5
Heidelberg, Minn. (†56071) 255/E6
Heidelberg, Miss. (39439) 256/F7
Heidelberg, Pa. (15106) 294/B7
Heidelberg, S. Africa 118/J7
Heidelberg, Victoria 97/J5
Heidelberg, Victoria 88/L7
Heidelberg, W. Germany 22/C4
Heiden, Switzerland 39/H2
Heidenau, E. Germany 22/E3
Heidenheim an der Brenz, W. Germany 22/D4
Heidenreichstein, Austria 41/C2
Heidrick, Ky. (40949) 237/O7
Heihe (Aihui) (Aigun), China 77/L1
Heijo (P'yŏngyang) (cap.), N. Korea 81/C4
Heil, N. Dak. (58546) 282/G7
Heilbron, S. Africa 118/D5
Heilbronn, W. Germany 22/C4
Heiligenblut, Austria 41/B3
Heiligenhafen, W. Germany 22/D1
Heiligenstadt, E. Germany 22/D3
Heilman, Ind. (†47523) 227/C8
Heilongjiang (Heilungkiang) (prov.), China 77/K2
Heilong Jiang (Amur) (riv.), China 77/L2
Heiloo, Netherlands 27/F3
Heilwood, Pa. (15745) 294/E4
Heimberg, Switzerland 39/E3
Heimdal, N. Dak. (58342) 282/L4
Heinola, Finland 18/P6
Heinola, Minn. (†56567) 255/C4
Heinsburg, Alberta 182/E3
Heinze Chaung (bay), Burma 72/C4
Heise, Idaho (†83443) 220/G6
Heiskell, Tenn. (37754) 237/O8
Heisler, Alberta 182/D3
Heislerville, N.J. (08324) 273/D5
Heisson, Wash. (98622) 310/C5
Heist-Knokke, Belgium 27/C6
Heist-op-den-Berg, Belgium 27/F6
Heizer, Kansas (†67530) 232/D3
Hejaz (reg.), Saudi Arabia 59/C4
Hejian, China 77/J4
Hejing, China 77/C3
Hekimhan, Turkey 63/G3
Hekla (mt.), Iceland 4/C11
Hekla (mt.), Iceland 7/C2
Hekla (vol.), Iceland 21/B1
Hekou, China 77/F7
Hel, Poland 47/D1
Hel (pen.), Poland 47/D1
Helan, China 77/G4
Heldens (pt.), St. Chris.-Nevis 161/C10
Helechawa, Ky. (41334) 237/P5
Helen, Georgia (30545) 217/E1
Helen, Md. (20635) 245/M7
Helen (lake), N. Dak. 282/K5
Helena, Ala. (35080) 195/E4
Helena, Ark. (72342) 202/J4
Helena, Calif. (96042) 204/B3
Helena, Georgia (31037) 217/G6
Helena, Mo. (64459) 261/C3
Helena (cap.), Mont. 146/G5
Helena (cap.), Mont. 188/D1
Helena (cap.), Mont. (59601) 262/E4
Helena (lake), Mont. 262/E4
Helena, N.Y. (13649) 276/L1
Helena, Ohio (43435) 284/D3
Helena, Okla. (73741) 288/K1
Helena, S.C. (†29108) 296/D3
Helensburgh, N.S. Wales 97/F4
Helensburgh, Scotland 10/A1
Helensburgh, Scotland 15/A1
Helen Springs, North. Terr. 93/C5
Helensville, N. Zealand 100/B1
Helenville, Wis. (53137) 317/J10
Helenwood, Tenn. (37755) 237/M8
Helez, Israel 65/B4
Helgoland (bay), W. Germany 22/C1
Helgoland (isl.), W. Germany 22/B1
Heliopolis, Egypt 111/J3
Helix, Oreg. (97835) 291/J2
Hellam, Pa. (17406) 294/J6
Hell Canyon (creek), S. Dak. 298/B6
Hellebaek, Denmark 21/F5
Hellendoorn, Netherlands 27/J4
Hellertown, Pa. (18055) 294/M4
Helles (cape), Turkey 63/B6
Hellevoetsluis, Netherlands 27/E5
Hellier, Ky. (41534) 237/S6
Hellín, Spain 33/F3
Hells (canyon), Idaho 220/B4
Hells Canyon (dam), Idaho 220/B4
Hells Canyon (dam), Oreg. 291/L2
Hells Canyon Nat'l Rec. Area, Idaho 220/B4
Hells Canyon Nat'l Rec. Area, Oreg. 291/K2
Hells Canyon Nat'l Rec. Area, Wash. 310/H5
Hells Half Acre, Wyo. (82601) 319/E2
Hell-Ville, Madagascar 102/G6
Hell-Ville, Madagascar 118/H2
Helm, Miss. (†38756) 256/C4
Helmand (riv.), Afghanistan 54/H6
Helmand (riv.), Afghanistan 59/J3
Helmand (riv.), Afghanistan 68/B2
Helmand (Sistan, Daryacheh-ye) (lake), Iran 66/M5
Helmer, Ind. (46744) 227/G1
Helmetta, N.J. (08828) 273/E3
Helmond, Netherlands 27/H6
Helmsburg, Ind. (47435) 227/E6
Helmsdale, Scotland 10/E1
Helmsdale, Scotland 15/E2
Helmsdale (riv.), Scotland 15/E2
Helmsley, England 13/G3
Helmstedt, W. Germany 22/D2
Helmville, Mont. (59843) 262/C4
Helotes, Texas (78023) 303/J10
Helper, Utah (84526) 304/D4
Helsenhorn (mt.), Switzerland 39/F4
Helsingborg, Sweden 18/H8
Helsingborg, Sweden 7/F3
Helsinge, Denmark 21/F6
Helsingør, Denmark 21/F6
Helsingør, Denmark 18/H8
Helsinki (cap.), Finland 18/O6
Helsinki (cap.), Finland 7/G2
Helsinki (cap.), Finland 2/L2
Helston, England 13/B7
Helston, England 10/D5
Helston, Manitoba 179/C4
Helton, Ky. (40840) 237/P7
Helton, N.C. (†28631) 281/G1
Heltonville, Ind. (47436) 227/E7
Helvecia, Argentina 143/F5
Helvetia, Pa. (†15848) 294/E3
Helvetia, W. Va. (26224) 312/F5
Helvick (head), Ireland 17/G7
Helwân, Egypt 59/B4
Helwân, Egypt 111/J3
Hemar (dry riv.), Israel 65/C5
Hemaruka, Alberta 182/E4
Hematite, Mo. (63047) 261/L6
Hemel Hempstead, England 10/F5
Hemel Hempstead, England 13/G7
Hemet, Calif. (92343) 204/H10
Hemford, Nova Scotia 168/D4
Hemingford, Nebr. (69348) 264/A2
Hemingway, S.C. (29554) 296/J4
Hemlock, Ind. (46937) 227/F4
Hemlock, Mich. (48626) 250/E5
Hemlock, N.Y. (14466) 276/E5
Hemlock (lake), N.Y. 276/E5
Hemlock, Ohio (43743) 284/F6
Hemlock (Eureka), S.C. (†29706) 296/E2
Hemlock Grove, Ohio (45738) 284/F7
Hemmingford, Québec 172/D4
Hemnes, Norway 18/J3
Hemphill (co.), Texas 303/D2
Hemphill, Texas (75948) 303/L6
Hemphill, W. Va. (24842) 312/C8
Hemple, Mo. (†64490) 261/D3
Hempstead (co.), Ark. 202/C6
Hempstead, N.Y. (*11550) 276/R7
Hempstead, Texas (77445) 303/J7
Hemse, Sweden 18/L8
Henagar, Ala. (35978) 195/G1
Henan (Honan) (prov.), China 77/H5
Henan, China 77/F5
Hen and Chickens (isls.), N. Zealand 100/E1
Henares (riv.), Spain 33/G4
Henbury, North. Terr. 93/C8
Hendaye, France 28/C6
Hendek, Turkey 63/D2
Henderson, Ala. (†36035) 195/F7
Henderson (co.), Ill. 222/C3
Henderson, Ill. (61439) 222/C2
Henderson (riv.), Ill. 222/C2
Henderson, Ind. (†46173) 227/F5
Henderson, Iowa (51541) 229/B6
Henderson (co.), Ky. 237/F5
Henderson, Ky. (42420) 237/F5
Henderson, La. (†70517) 238/G6
Henderson, Md. (21640) 245/P4
Henderson, Minn. (56044) 255/E6
Henderson, Nebr. (68371) 264/G4
Henderson, Nev. (89015) 266/G6
Henderson, N.Y. (13650) 276/H3
Henderson, N. Zealand 100/B1
Henderson (co.), N.C. 281/D4
Henderson, N.C. (27536) 281/N2
Henderson (isl.), Pitcairn Is. 87/O8
Henderson (co.), Tenn. 237/E9
Henderson, Tenn. (38340) 237/D10
Henderson (co.), Texas 303/J5
Henderson, Texas (75652) 303/K5
Henderson, W. Va. (25106) 312/B5
Hendersonville, N.C. (28739) 281/E4
Hendersonville, Pa. (15339) 294/B5
Hendersonville, S.C. (†29945) 296/F6
Hendersonville, Tenn. (37075) 237/H8
Hendley, Nebr. (68946) 264/D4
Hendon, Sask. 181/H3
Hendorabi (isl.), Iran 66/H7
Hendra, Queensland 88/K2
Hendricks (co.), Ind. 227/D5
Hendricks, Ky. (41441) 237/P5
Hendricks, Minn. (56136) 255/B6
Hendricks, W. Va. (26271) 312/G4
Hendrickson, Mo. (†63967) 261/M9
Hendrix, Okla. (74741) 288/O7
Hendrix Lake, Br. Col. 184/G4
Hendrum, Minn. (56550) 255/B3
Hendry (co.), Fla. 212/E5
Hendrysburg, Ohio (43744) 284/H5
Henefer, Utah (84033) 304/C2
Hengchun, China 77/K7
Hengduan Shan (mts.), China 77/E6
Hengelo, Gelderland, Netherlands 27/J4
Hengelo, Overijssel, Netherlands 27/K4
Hengshan, China 77/G4
Hengshui, China 77/J4
Heng Xian, China 77/G7
Hengyang, China 77/H6
Hengyang, China 54/N7
Henik (lakes), N.W. Terrs. 187/J3
Hénin-Beaumont, France 28/E2
Henjam (isl.), Iran 66/J7
Henlawson, W. Va. (25624) 312/B7
Henley, Mo. (65040) 261/H6
Henley and Grange, S. Australia 88/D8
Henley Harbour, Newf. 166/C3
Henley on Klip, S. Africa 118/H7
Henley-on-Thames, England 13/G8
Henlopen (cape), Del. 245/T5
Henlopen Acres, Del. (†19971) 245/T6
Henne, Denmark 21/B6
Hennebont, France 28/B4
Hennef, W. Germany 22/B3
Hennepin, Ill. (61327) 222/D2
Hennepin (co.), Minn. 255/E5
Hennepin, Okla. (73046) 288/M5
Hennessey, Okla. (73742) 288/L2
Hennigsdorf bei Berlin, E. Germany 22/E3
Henniker, N.H. (03242) 268/D5
Henniker○, N.H. (03242) 268/D5
Henning, Ill. (61848) 222/F3
Henning, Minn. (56551) 255/C4
Henning, Tenn. (38041) 237/B9
Henribourg, Sask. 181/F2
Henrico (co.), Va. 307/O6
Henrietta, Mo. (64036) 261/E4
Henrietta, N.Y. (14467) 276/E4
Henrietta, N.C. (28076) 281/E4
Henrietta, Texas (76365) 303/F4
Henrietta Maria (cape), Ont. 162/H4
Henrietta Maria (cape), Ontario 175/D1
Henriette, Minn. (55036) 255/E5
Henrieville, Utah (84736) 304/C6
Henry (co.), Ala. 195/H7
Henry (co.), Georgia 217/D4
Henry, Idaho (†83230) 220/G7
Henry (co.), Ill. 222/C2
Henry, Ill. (61537) 222/D2
Henry (co.), Ind. 227/G5
Henry (co.), Iowa 229/K6
Henry (co.), Ky. 237/L4
Henry (co.), Mo. 261/E6
Henry, Nebr. (69349) 264/A2
Henry (isl.), Nova Scotia 168/G3
Henry (co.), Ohio 284/B3
Henry, S.C. (†29554) 296/J4
Henry, S. Dak. (57243) 298/P4
Henry (co.), Tenn. 237/E8
Henry, Tenn. (38231) 237/E8
Henry (mts.), Utah 304/D6
Henry (co.), Va. 307/J7
Henry, Va. (24102) 307/J7
Henry (cape), Va. 307/R7
Henryetta, Okla. (74437) 288/O4
Henry House, Alberta 182/B3
Henry Kater (cape), N.W. Terrs. 187/M3
Henrys (lake), Idaho 220/G5
Henrys Fork, Snake (riv.), Idaho 220/G5
Henrys Fork, Green (riv.), Wyo. 319/C4
Henryton, Md. (21080) 245/L3
Henryville, Ind. (47126) 227/F7
Henryville, Pa. (18332) 294/M3
Henryville, Québec 172/D4
Henryville, Tenn. (†38483) 237/G10
Hensall, Ontario 177/C4
Hensel, N. Dak. (58241) 282/P2
Henshaw, Ky. (†42459) 237/F5
Hensies, Belgium 27/D8
Hensler, N. Dak. (58547) 282/H5
Hensley, Ark. (72065) 202/F4
Henson (creek), Md. 245/F6
Hentiy, Mongolia 77/H2
Henty, N.S. Wales 97/D4
Henzada, Burma 72/B3
Henzada, Burma 54/L8
Hepburn, Iowa (†51632) 229/C7
Hepburn, Ohio (†43326) 284/D4
Hepburn, Sask. 181/E3
Hephzibah, Georgia (30815) 217/H4
Hepler, Kansas (66746) 232/H4
Heppner, Oreg. (97836) 291/H2
Hepu (Hoppo), China 77/G7
Hepworth, Ontario 177/C3
Hepzibah, W. Va. (26369) 312/F4
Hequ, China 77/H4
Herald, Calif. (95638) 204/C9
Heralds (cays), 95/D3
Herat, Afghanistan 54/H6
Herat, Afghanistan 59/H3
Herat, Afghanistan 68/A2
Hérault (dept.), France 28/E6
Hérault (riv.), France 28/E6
Herbert, Ala. (†36401) 195/E8
Herbert (riv.), Queensland 88/H3
Herbert, Sask. 181/D5
Herbert Hoover Nat'l Hist. Site, Iowa 229/L5
Herbes (isl.), Ala. 195/B10
Herbeumont, Belgium 27/G9
Herb Lake, Manitoba 179/H3
Herborn, W. Germany 22/C3
Herbst, Ind. (46952) 227/F3
Herbster, Wis. (54844) 317/D2
Herceg Novi, Yugoslavia 45/D4
Herchmer, Man. 162/G4
Herchmer, Manitoba 179/K2
Herculaneum, Mo. (63048) 261/M6
Hercules, Calif. (94547) 204/J1
Herd, Okla. (†74056) 288/O1
Heredia, C. Rica 154/E5
Hereford, Ariz. (85615) 198/E7
Hereford, Colo. (80732) 208/L1
Hereford, England 13/E5
Hereford, England 10/E4
Hereford, Md. (†21111) 245/M2
Hereford (inlet), N.J. 273/D5
Hereford, Oreg. (97837) 291/K3
Hereford, Pa. (18056) 294/L5
Hereford, S. Dak. (57743) 298/D5
Hereford, Texas (79045) 303/B3
Hereford and Worcester (co.), England 13/E5
Hérémence, Switzerland 39/D4
Herencia, Spain 33/E3
Herentals, Belgium 27/F6
Heretaunga-Pinehaven, N. Zealand 100/C2
Herford, W. Germany 22/C2
Hergiswil, Switzerland 39/F3
Héricourt, France 28/G4
Heringsdorf, E. Germany 22/F1
Herington, Kansas (67449) 232/E3
Heriot Bay, Br. Col. 184/E5
Herisau, Switzerland 39/H2
Herkimer, Kansas (66433) 232/F2
Herkimer (co.), N.Y. 276/L4
Herkimer, N.Y. (13350) 276/L4
Herlen Gol (Kerulen) (riv.), Mongolia 77/H2
Herlong, Calif. (96113) 204/E3
Herm (isl.), Chan. Is. 13/E8
Hermagor-Preseggersee, Austria 41/B3
Herman, Mich. (†49946) 250/A2
Herman, Minn. (56248) 255/B5
Herman, Nebr. (68029) 264/H3
Herman, Pa. (16039) 294/C4
Herman (lake), S. Dak. 298/P5
Herma Ness (prom.), Scotland 15/G2
Hermann, Mo. (65041) 261/K5
Hermannsburg, North. Terr. 88/E4
Hermannsburg, North. Terr. 93/C7
Hermansverk, Norway 18/E6
Hermansville, Mich. (49847) 250/B3
Hermantown, Minn. (†55811) 255/F4
Hermanus, S. Africa 118/G7
Hermanville, Miss. (39086) 256/C7
Hermidale, N.S. Wales 97/D2
Hermil, Lebanon 63/G5
Herminie, Pa. (15637) 294/C5
Hermiston, Oreg. (97838) 291/H2
Hermitage, Ark. (71647) 202/F7
Hermitage, Grenada 161/D8
Hermitage, Mo. (65668) 261/F7
Hermitage, Newf. 166/C4
Hermitage (bay), Newf. 166/C4
Hermitage Springs, Tenn. (†37150) 237/K7
Hermite (isls.), Chile 138/F11
Hermleigh, Texas (79526) 303/D5
Hermon, Ill. (†61458) 222/C3
Hermon○, Maine (†04401) 243/F6
Hermon, N.Y. (13652) 276/K2
Hermon (mt.), Syria 63/F6
Hermosa (peak), Colo. 208/D7
Hermosa, S. Dak. (57744) 298/C6
Hermosa Beach, Calif. (90254) 204/B11
Hermosillo, Mexico 146/G7
Hermosillo, Mexico 150/D2
Hermsdorf, W. Germany 22/E3
Hernád (riv.), Hungary 41/F2
Hernandarias, Argentina 143/F5
Hernandarias, Paraguay 144/E4
Hernández, Argentina 143/F6
Hernandez, N. Mex. (87537) 274/C2
Hernando, Argentina 143/D3
Hernando (co.), Fla. 212/D3
Hernando, Fla. (32642) 212/D3
Hernando, Miss. (38632) 256/E1
Herndon, Georgia (†30442) 217/H5
Herndon, Iowa (†50128) 229/E5
Herndon, Kansas (67739) 232/B2
Herndon, Ky. (42236) 237/G7
Herndon, Pa. (17830) 294/J4
Herndon, Va. (*22070) 307/O3
Herndon, W. Va. (24726) 312/D7
Herne, Belgium 27/E7
Herne, W. Germany 22/B3
Herning, Denmark 18/F8
Herning, Denmark 21/B5
Herod, Georgia (†31742) 217/D7
Herod, Ill. (62947) 222/E6
Heroica Caborca, Mexico 150/C1
Heroica Nogales, Mexico 150/D1
Heron (lake), Minn. 255/C7
Heron, Mont. (59844) 262/A2
Heron (isl.), New Bruns. 170/D1
Heron Bay, Ontario 177/H5
Heron Bay, Ontario 175/C3
Heron Lake, Minn. (56137) 255/C7
Hérouxville, Québec 172/E3

Herowabad, Iran 66/F2
Herradura, Argentina 143/E2
Herradura, Cuba 158/B1
Herreid, S. Dak. (57632) 298/K2
Herrera, Argentina 143/D2
Herrera del Duque, Spain 33/D3
Herrera de Pisuerga, Spain 33/D1
Herrero (pt.), Mexico 150/Q7
Herrick, Ill. (62431) 222/D4
Herrick, S. Dak. (57538) 298/L7
Herrick, Tasmania 99/D3
Herrick Center, Pa. (18430) 294/L2
Herrin, Ill. (62948) 222/E6
Herring Cove, Nova Scotia 168/A4
Herrings, N.Y. (13653) 276/J2
Herrington (lake), Ky. 237/M5
Herron, Ill. (49744) 250/F3
Herronton, Alberta 182/D5
Hersbruck, W. Germany 22/D4
Herschel, Sask. 181/C4
Herschel (isl.), Yukon 187/E3
Herscher, Ill. (60941) 222/E2
Herselt, Belgium 27/F6
Hersey, Mich. (49639) 250/D5
Hersey, Wis. (†54027) 317/B6
Hershey, Nebr. (69143) 264/F3
Hershey, Pa. (17033) 294/J5
Hersman, Ill. (†62353) 222/C4
Herstal, Belgium 27/H7
Hertel, Wis. (54845) 317/B4
Hertford, England 13/H7
Hertford, England 10/G5
Hertford (co.), N.C. 281/P2
Hertford, N.C. (27944) 281/S2
Hertfordshire (co.), England 13/G6
Hervás, Spain 33/D2
Herve, Belgium 27/H7
Hervey Bay, Queensland 88/J5
Hervey Bay, Queensland 95/K5
Hervey (bay), Queensland 88/J4
Hervey (bay), Queensland 95/E5
Herzberg, E. Germany 22/E3
Herzeliyya, Israel 65/B3
Herzogenbuchsee, Switzerland 39/E2
Herzogenburg, Austria 41/C2
Heshui, China 77/G4
Hesketh, Alberta 182/D4
Hesper, Iowa (†52101) 229/K2
Hesper, N. Dak. (†58348) 282/L4
Hesperange, Luxembourg 27/J9
Hesperia, Calif. (92345) 204/H9
Hesperia, Mich. (49421) 250/D5
Hespero, Alberta 182/C3
Hesperus, Colo. (81326) 208/C8
Hesperus (mt.), Colo. 208/C8
Hess, Okla. (†73539) 288/H6
Hess (riv.), Yukon 187/E3
Hesse (state), W. Germany 22/C3
Hessel, Mich. (49745) 250/E2
Hessmer, La. (71341) 238/F4
Hesston, Kansas (67062) 232/E3
Hesston, Pa. (16647) 294/F5
Hester, La. (70743) 238/L3
Hester, Okla. (†73554) 288/H5
Hesterville, Miss. (†39192) 256/F4
Hetch Hetchy (res.), Calif. 204/F6
Heth, Ark. (72346) 202/K3
Hetland, S. Dak. (57244) 298/P5
Hettick, Ill. (62649) 222/C4
Hettinger (co.), N. Dak. 282/E7
Hettinger, N. Dak. (58639) 282/E8
Hetton, England 13/J3
Hettstedt, E. Germany 22/D3
Heusden, Netherlands 27/G5
Heuvelland, Belgium 27/B7
Heuvelton, N.Y. (13654) 276/K1
Heves (co.), Hungary 41/F3
Heves, Hungary 41/F3
Heward, Sask. 181/H6
Hewins, Kansas (67024) 232/F4
Hewitt, Minn. (56453) 255/C4
Hewitt, N.J. (07421) 273/E1
Hewitt, Wis. (54441) 317/F6
Hewlett, N.Y. (11557) 276/P7
Hewlett Harbor, N.Y. (†11557) 276/P7
Hexham, England 13/H3
Hexham, England 10/E3
He Xian, China 77/H7
Hexigten, China 77/J3
Hext, Texas (76848) 303/E7
Heybeli (isl.), Turkey 63/D6
Heybridge, Tasmania 99/C3
Heyburn, Idaho (83336) 220/E7
Heyburn (res.), Okla. 288/S1
Heyden, Ontario 177/D4
Heyfield, Victoria 97/D6
Heywood (chan.), Burma 72/B3
Heywood, England 13/H2
Heywood, Victoria 97/A6
Heyworth, Ill. (61745) 222/E3
Heze, China 77/J4
Hezuo, China 77/F5
Hialeah, Fla. 188/K5
Hialeah, Fla. (*33010) 212/B4
Hialeah Gardens, Fla. (†33010) 212/B4
Hiattville, Kansas (66747) 232/G4
Hiawassee, Georgia (30546) 217/E1
Hiawatha, Iowa (52233) 229/K4
Hiawatha, Kansas (66434) 232/G2
Hiawatha, Mich. (†49854) 250/C2
Hiawatha, Utah (84527) 304/D4
Hibbard, Ind. (†46511) 227/E2
Hibbing, Ind. 188/H1
Hibbing, Minn. (55746) 255/F3
Hibbs, Okla. (†35741) 288/J6
Hibbs (co.), Tasmania 99/B4
Hibernia, N.J. (07842) 273/E2
Hibuson (isl.), Philippines 82/E5
Hicacos (pen.), Cuba 158/D1
Hicacos (pt.), Cuba 158/D1
Hickam A.F.B., Hawaii 218/B4
Hickam Housing, Hawaii (96824) 218/B4
Hickman, Ark. (†72315) 202/L2
Hickman, Del. (†21629) 245/R5
Hickman (co.), Ky. 237/C7
Hickman, Ky. (42050) 237/C7

Hickman, Nebr. (68372) 264/H4
Hickman (co.), Tenn. 237/G9
Hickman, Tenn. (38567) 237/K8
Hickman's Harbour, Newf. 166/D2
Hickok, Kansas (†67880) 232/A4
Hickory, Ky. (42051) 237/D7
Hickory, Miss. (39332) 256/F6
Hickory (co.), Mo. 261/F7
Hickory, N.C. (28601) 281/G3
Hickory, Okla. (†73086) 288/N5
Hickory Corners, Mich. (49060) 250/D6
Hickory Creek, Texas (75423) 303/F1
Hickory Flat, Ala. (†36274) 195/H4
Hickory Flat, Miss. (38633) 256/F1
Hickory Grove, S.C. (29717) 296/E2
Hickory Hills, Ill. (60457) 222/B6
Hickory Plains, Ark. (72066) 202/G3
Hickory Ridge, Ark. (72347) 202/J3
Hickory Valley, Tenn. (38042) 237/C10
Hickory Withe, Tenn. (38043) 237/B10
Hickox, Georgia (†31553) 217/H8
Hicks, La. (71437) 238/E4
Hickson, N. Dak. (58047) 282/S6
Hickson, Ontario 177/D4
Hicksville, N.Y. (*11801) 276/R7
Hicksville, Ohio (43526) 284/A3
Hico, La. (†71235) 238/E4
Hico, Texas (76457) 303/F6
Hico, W. Va. (25854) 312/D6
Hida (riv.), Japan 81/H6
Hidalgo, Ill. (62432) 222/E4
Hidalgo, Ky. (42622) 237/M7
Hidalgo (state), Mexico 150/K6
Hidalgo, Coahuila, Mexico 150/K3
Hidalgo, Tamaulipas, Mexico 150/K4
Hidalgo (co.), N. Mex. 274/A7
Hidalgo (co.), Texas 303/F11
Hidalgo, Texas (78557) 303/F11
Hidalgo del Parral (Parral), Mexico 150/G3
Hidden Hills, Calif. (†91302) 204/B10
Hiddenite, N.C. (28636) 281/G3
Hiddensee (isl.), E. Germany 22/E1
Hijar, Spain 33/F2
Hiko, Nev. (89017) 266/F5
Hikone, Japan 81/H6
Hikueru (atoll), Fr. Poly. 87/M7
Hikurangi, N. Zealand 100/E1
Hikurangi (mt.), N. Zealand 100/G2
Hiland, Wyo. (82638) 319/F2
Hiland Park, Fla. (32405) 212/C6
Hilbert, Wis. (54129) 317/K7
Hilbre (isl.), England 13/G2
Hilda, Alberta 182/E4
Hilda, S.C. (29813) 296/E5
Hildburghausen, E. Germany 22/D3
Hildebran, N.C. (28637) 281/F3
Hildebrand, Oreg. (†97625) 291/F5
Hilden, Nova Scotia 168/E3
Hildesheim, W. Germany 22/D2
Hildreth, Nebr. (68947) 264/F4
Hiles, Wis. (†54511) 317/J4
Hilgard (mt.), Calif. 204/G7
Hilger, Mont. (59451) 262/G3
Hilham, Tenn. (38568) 237/L8
Hill (riv.), Minn. 255/C3
Hill (co.), Mont. 262/F2
Hill○, N.H. (03243) 268/D4
Hill (co.), Texas 303/G5
Hill (creek), Utah 304/E4
Hilla, Iraq 66/D4
Hilla, Iraq 59/D3
Hillaby (mt.), Barbados 161/B8
Hillandale, Md. (†20903) 245/F4
Hill Bank, Belize 154/C2
Hillburn, N.Y. (10931) 276/M8
Hill City, Idaho (83337) 220/D6
Hill City, Kansas (67642) 232/C2
Hill City, Minn. (55748) 255/E4
Hill City, S. Dak. (57745) 298/B6
Hill Country Village, Texas (†78232) 303/K10
Hill Creek Ext., Uintah and Ouray Ind. Res., Utah 304/E4
Hillcrest, Alberta 182/C5
Hillcrest, Ill. (†61244) 222/D2
Hillcrest, Ill. (*10977) 276/K8
Hillcrest, Texas (†75511) 303/J3
Hillcrest Heights, Fla. (†33827) 212/E4
Hillcrest Heights, Md. (†20031) 245/F5
Hilldale, Utah (84767) 304/A6
Hillegom, Netherlands 27/E4
Hillemann, Ark. (†72101) 202/H3
Hill End, N.S. Wales (†72101) 202/H3
Hillerød, Denmark 18/H9
Hillerød, Denmark 21/F6
Hillers (mt.), Utah 304/D6
Hillhead, S. Dak. (†57270) 298/O2
Hillhouse, Miss. (†38720) 256/C2
Hilliard, Alberta 182/D3
Hilliard, Fla. (32046) 212/E1
Hilliard, Ohio (43026) 284/D5
Hilliards, Pa. (16040) 294/C3
Hilliardville, Fla. (†32327) 212/B1
Hillingdon, England 10/B5
Hillisburg, Ind. (46046) 227/E4
Hill Island (lake), N.W. Terrs. 187/H3
Hillman, Mich. (49746) 250/F3
Hillman, Minn. (56338) 255/E4
Hillman, New Bruns. 170/C2
Hillmond, Sask. 181/B3
Hill of Fearn, Scotland 15/D3
Hillridge, Australia 179/C3
Hillrose, Colo. (80733) 208/N2
Hills (lake) (52235) 229/K5
Hills, Minn. (56138) 255/B7
Hillsboro, Ala. (35643) 195/D1
Hillsboro, Georgia (31038) 217/E4
Hillsboro, Ind. (47949) 227/C4
Hillsboro, Iowa (52630) 229/K7
Hillsboro, Kansas (67063) 232/E3
Hillsboro, Ky. (41049) 237/O4
Hillsboro, Md. (21641) 245/P5

Hillsboro, Miss. (39087) 256/E6
Highpoint, Miss. (†39339) 256/F4
Hillsboro, Mo. (63050) 261/L6
Hillsboro, N. Mex. (03244) 268/D5
Hillsboro○, N.H. (03244) 268/D5
Hillsboro, N. Mex. (88042) 274/B6
Hillsboro, N. Dak. (58045) 282/S5
Hillsboro, Ohio (45133) 284/C7
Hillsboro, Oreg. (97123) 291/A2
Hillsboro, Tenn. (37342) 237/K10
Hillsboro, Texas (76645) 303/G5
Hillsboro, Va. (22132) 307/N2
Hillsboro, W. Va. (24946) 312/F6
Hillsboro Beach, Fla. (†33060) 212/F5
Hillsboro Lower Village, N.H. (†03244) 268/D5
Hillsborough, Calif. (94010) 204/J2
Hillsborough (co.), Fla. 212/D4
Hillsborough (bay), Fla. 212/D5
Hillsborough (canal), Fla. 212/F5
Hillsborough (riv.), Fla. 212/D3
Hillsborough, New Bruns. 170/F2
Hillsborough (co.), N.H. 268/D6
Hillsboro, N.C. (27278) 281/L2
Hillsborough, N. Ireland 17/J3
Hillsborough (bay), Pr. Edward I. 168/E2
Hillsboro Upper Village, N.H. (†03244) 268/D5
Hillsburgh, Ontario 177/D4
Hillsburn, Nova Scotia 168/C4
Hills Creek (lake), Oreg. 291/E4
Hillsdale, Ind. (47854) 227/C5
Hillsdale, Kansas (66036) 232/H3
Hillsdale (co.), Mich. 250/E7
Hillsdale, Mich. (49242) 250/E7
Hillsdale, Mo. (†63101) 261/R2
Hillsdale, N.J. (07642) 273/B1
Hillsdale, N.Y. (12529) 276/O6
Hillsdale, Okla. (73743) 288/K1
Hillsdale, Ontario 177/E3
Hillsdale○, Mass. (01235) 249/B3
Hillsdale, Pa. (15746) 294/E4
Hillsdale, Wis. (54744) 317/C5
Hillsdale, Wyo. (82060) 319/H4
Hillsdale○, N.H. (03451) 268/C6
Hillsgrove, Pa. (18619) 294/J3
Hillsgrove, R.I. (†02887) 249/J6
Hillside, Ariz. (†86301) 198/B4
Hillside, Colo. (81232) 208/H6
Hillside, Ill. (60162) 222/B6
Hillside○, N.J. (07205) 273/B2
Hillside, S. Dak. (†57328) 298/N7
Hillside Beach, Manitoba 179/G4
Hillsport, Ontario 175/C3
Hill Spring, Alberta 182/D5
Hillston, N. Wales 88/G6
Hillston, N.S. Wales 97/C3
Hillsview, S. Dak. (†57437) 298/L2
Hillsville, Pa. (16132) 294/A4
Hillsville, Va. (24343) 307/G7
Hillswick, Scotland 15/G2
Hilltonia, Georgia (30467) 217/J5
Hilltop, Ariz. (†85632) 198/F6
Hilltown, N. Ireland 17/J3
Hillview, Ill. (62050) 222/C4
Hillview, Minn. (†56477) 255/C4
Hillview, Newf. 166/D2
Hilmar-Irwin, Calif. (95324) 204/E6
Hilo, Hawaii 87/L4
Hilo, Hawaii (96720) 218/J5
Hilo, Hawaii 188/G6
Hilo (bay), Hawaii 218/J5
Hilongos, Philippines 82/E5
Hilshire Village, Texas (†77001) 303/J1
Hilt, Calif. (†96044) 204/C2
Hilterfingen, Switzerland 39/E3
Hilton (inlet), Br. Col. 184/H4
Hilton, Ga. (†31723) 217/C8
Hilton, Manitoba 179/C5
Hilton, N.Y. (14468) 276/E4
Hilton Beach, Ontario 177/J5
Hilton Head (isl.), S.C. 296/F7
Hilton Head Island, S.C. (29928) 296/F7
Hiltons, Va. (24258) 307/D7
Hilvan, Turkey 63/H4
Hilvarenbeek, Netherlands 27/G6
Hilversum, Netherlands 27/G4
Hima, Ky. (40951) 237/O6
Himachal Pradesh (state), India 68/D2
Himalaya (mts.) 54/L7
Himalaya (mts.), Bhutan 68/E2
Himalaya (mts.), China 77/C6
Himalaya (mts.), India 68/D2
Himalaya (mts.), Nepal 68/D2
Himanka, Finland 18/N5
Himeji, Japan 81/G6
Himi, Japan 81/H5
Himlerville (Beauty), Ky. (†41203) 237/S5
Himrod, N.Y. (14842) 276/F5
Himyar, Ky. (40952) 237/O7
Hinatuan, Philippines 82/F6
Hinchcliff, Miss. (†38646) 256/D2
Hinche, Haiti 158/C5
Hinche, Haiti 156/D3
Hinchinbrook (isl.), Alaska 196/J3
Hinchinbrook (isl.), Queensland 88/H3
Hinchinbrook (isl.), Queensland 95/C3
Hinchinbrook Entrance (chan.), Alaska 196/J3
Hinchliffe, Sask. 181/J3
Hinckley, England 10/F4
Hinckley (lake), N.C. 281/A4
Hinckley, Ill. (60520) 222/E2
Hinckley, Maine (04944) 243/D6
Hinckley, Minn. (55037) 255/E4
Hinckley, N.Y. (13352) 276/K4
Hinckley (res.), N.Y. 276/K4
Hinckley, Ohio (44233) 284/G3
Hinckley, Utah (84635) 304/B4
Hindeloopen, Netherlands 27/G3
Hindenburg (Zabrze), Poland 47/A4
Hinderwell, England 13/J3
Hindiya, Iraq 66/C4

Hindman, Ky. (41822) 237/R6
Hindmarsh, S. Australia 88/D8
Hindmarsh, S. Australia 94/A7
Hindmarsh○ (lake), Victoria 97/A5
Hinds (co.), Miss. 256/D6
Hinds, N. Zealand 100/C6
Hindsboro, Ill. (61930) 222/E4
Hindsville, Ark. (72738) 202/C1
Hindu (lake), Afghanistan 68/B1
Hindu Kush (mts.), Afghanistan 68/B1
Hindu Kush (mts.), Afghanistan 59/J2
Hindu Kush (mts.), India 68/C1
Hindu Kush (mts.), Pakistan 68/B1
Hindupur, India 68/D6
Hi-Nella, N.J. (†08083) 273/B4
Hines, Minn. (56647) 255/D3
Hines, Oreg. (97738) 291/H4
Hinesburg○, Vt. (05461) 268/A3
Hines Creek, Alberta 182/A1
Hines Creek, Alta. 162/E4
Hineston, La. (71438) 238/E4
Hinesville, Georgia (31313) 217/J7
Hinganghat, India 68/D4
Hingham, Mass. (02043) 249/E8
Hingham○, Mass. (02043) 249/E8
Hingham (bay), Mass. 249/E7
Hingham, Mont. (59528) 262/F2
Hingham, Wis. (53031) 317/K8
Hingoli, India 68/D5
Hinigaran, Philippines 82/D5
Hinis, Turkey 63/J3
Hinkley, Calif. (92347) 204/H9
Hinkle, Nev. 188/F4
Hinkston (creek), Ky. 237/N4
Hinlopenstreten (str.), Norway 18/C1
Hinnera (prov.), Somalia 115/J3
Hinnøya (isl.), Norway 18/K2
Hino, Japan 81/G2
Hinojosa del Duque, Spain 33/D3
Hinsdale (co.), Colo. 208/E7
Hinsdale, Ill. (60521) 222/B6
Hinsdale (co.), Mont. 262/K2
Hinsdale, Mont. (59241) 262/K2
Hinsdale, N.H. (03451) 268/C6
Hinsdale, N.Y. (14743) 276/D6
Hinsdale○, N.H. (03451) 268/C6
Hinson, Fla. (†32333) 212/B1
Hinsonton, Georgia (†31779) 217/D8
Hinterrhein (riv.), Switzerland 39/H3
Hinton, Alberta 182/B3
Hinton, Iowa (51024) 229/B3
Hinton, Ky. (†65201) 261/H4
Hinton, Okla. (73047) 288/K4
Hinton, W. Va. (25951) 312/E7
Hintonville, Miss. (†39462) 256/F8
Hinwil, Switzerland 39/G2
Hinze, Miss. (39108) 256/F4
Hippolytushoef, Netherlands 27/G3
Hipswell, England 13/F3
Hiram, Georgia (30141) 217/C3
Hiram, Ky. (†40823) 237/P7
Hiram, Maine (04041) 243/B8
Hiram○, Maine (04041) 243/B8
Hiram, Mo. (63947) 261/M8
Hiram, Ohio (44234) 284/H3
Hirara, Japan 81/L7
Hirata, Japan 81/F6
Hiratsuka, Japan 81/J3
Hîrlău, Romania 45/H2
Hiroo, Japan 81/L2
Hirosaki, Japan 81/K3
Hiroshima (pref.), Japan 81/E6
Hiroshima, Japan 54/P6
Hiroshima, Japan 81/E6
Hirsch, Sask. 181/J6
Hirschberg (Jelenia Góra), Poland 47/B3
Hirson, France 28/F3
Hîrşova, Romania 45/J3
Hirtshals, Denmark 21/C2
Hisban, Jordan 65/D4
Hisega, S. Dak. (†57701) 298/C5
Hiseville, Ky. (42152) 237/K6
Hisle, S. Dak. (†57577) 298/F7
Hispaniola (isl.) 146/L7
Hispaniola (isl.), Dom. Rep. 156/D2
Hispaniola (isl.), Haiti 156/D2
Hispaniola (isl.), W. Indies 156/D2
Hissar, India 68/D3
Hissop, Ala. (35081) 195/F5
Hit, Iraq 66/C3
Hit, Iraq 59/D3
Hitachi, Japan 81/K5
Hitachiota, Japan 81/K5
Hitchcock (co.), Nebr. 264/C4
Hitchcock, Okla. (73744) 288/K3
Hitchcock, Sask. 181/J6
Hitchcock, S. Dak. (57348) 298/M4
Hitchcock, Texas (77563) 303/K3
Hitchin, England 13/G6
Hitchin, England 10/F5
Hitchins, Ky. (41146) 237/R4
Hitchita, Okla. (74438) 288/P3
Hiteman, Iowa (†52531) 229/H6
Hitoyoshi, Japan 81/E7
Hitra (isl.), Norway 18/J5
Hitt, Miss. (63555) 261/H1
Hitterdal, Minn. (56552) 255/B4
Hitzacker, W. Germany 22/D2
Hitzkirch, Switzerland 39/F2
Hivaoa (isl.), Fr. Poly. 87/N6
Hiwannee, Miss. (†39367) 256/G7
Hiwasse, Ark. (72739) 202/B1
Hiwassee (lake), N.C. 281/A4
Hiwassee (riv.), Tenn. 237/O10
Hiwassee, Va. (24347) 307/G7
Hixon, Br. Col. 184/F3
Hixson, Tenn. (37343) 237/L10
Hixton, Wis. (54635) 317/E7
Hizan, Turkey 63/K3
Hjallerup, Denmark 21/C3
Hjälmaren (lake), Sweden 18/J3
Hjerm, Denmark 21/B5
Hjerting, Denmark 21/B6

Hjo, Sweden 18/J7
Hjørring, Denmark 18/F8
Hjørring, Denmark 21/C3
Hka, Nam (riv.), Burma 72/C2
Hkakabo Razi (mt.), Burma 72/C1
Hlinsko, Czech. 41/D2
Hlohovec, Czech. 41/D3
Hlučín, Czech. 41/E2
Hmawbi, Burma 72/C3
Hnúšť'a-Likier, Czech. 41/E2
Ho, Ghana 106/E7
Hoa Binh, Vietnam 72/E2
Hoa Da, Vietnam 72/E3
Hoadley, Alberta 182/C3
Hoadly, Va. (†22191) 307/O3
Hoagland, Ind. (46745) 227/H3
Hoai Nhon, Vietnam 72/F4
Hoaksbergen, Netherlands 27/K4
Hoare (bay), N.W. Terrs. 187/M3
Hoback (peak), Wyo. 319/B2
Hoback, Wyo. 319/B2
Hobart, Australia 2/S8
Hobart, Ind. (46342) 227/C1
Hobart, N.Y. (13788) 276/L6
Hobart, Okla. (73651) 288/J5
Hobart (cap.), Tasmania 88/H8
Hobart (cap.), Tasmania 99/D4
Hobart (bay), Mass. 249/E7
Hobart, Wash. (98025) 310/D3
Hobbema, Alberta 182/D3
Hobbs, Ind. (46047) 227/F4
Hobbs (lake), Manitoba 179/G3
Hobbs, Md. (†21629) 245/P5
Hobbs, N. Mex. 188/F4
Hobbs, N. Mex. (88240) 274/F6
Hobbs Coast (reg.) 5/B12
Hobbs Island, Ala. (†35804) 195/F1
Hobbsville, N.C. (27946) 281/R2
Hoberg, Mo. (†65712) 261/E8
Hobe Sound, Fla. (33455) 212/F4
Hobgood, N.C. (27843) 281/P2
Hoboken, Belgium 27/E6
Hoboken, Georgia (31542) 217/H8
Hoboken, N.J. (07030) 273/C2
Hoboksar, China 77/C2
Hobro, Denmark 21/C4
Hobro, Denmark 18/F8
Hobson (lake), Br. Col. 184/H4
Hobson, Mont. (59452) 262/G4
Hobson City, Ala. (†36201) 195/G3
Hobson, Ohio. (†37460) 293/N4
Hobsons (bay), Victoria 97/H5
Hobsons (bay), Victoria 88/L7
Hobucken, N.C. (28537) 281/S4
Hoburgen (cliff), Sweden 18/L8
Hochdorf, Switzerland 39/F2
Hochfeld, Manitoba 179/F5
Hochgolling (mt.), Austria 41/B3
Hochwan (Hechuan), China 77/F5
Hochwang (mt.), Switzerland 39/J3
Hockaday, Mich. (†48624) 250/E4
Hockanum, Conn. (†06108) 210/E2
Hockanum (riv.), Conn. 210/E1
Hockenheim, W. Germany 22/C4
Hockerville, Okla. (†74363) 288/S1
Hockessin, Del. (19707) 245/R1
Hocking (co.), Ohio 284/F6
Hocking (riv.), Ohio 284/F7
Hockingport, Ohio (45739) 284/G7
Hockley (co.), Texas 303/B4
Hodaka (mt.), Japan 81/H5
Hodder (riv.), England 13/H1
Hoddesdon, England 13/H7
Hodeida, Yemen Arab Rep. 54/F8
Hodeida, Yemen Arab Rep. 59/D7
Hodgdon○, Maine (†04730) 243/H3
Hodge, La. (71247) 238/E2
Hodge, Mo. (†64096) 261/E4
Hodgeman (co.), Kansas 232/C3
Hodgen, Okla. (74939) 288/S5
Hodgenville, Ky. (42748) 237/K5
Hodges, Ala. (35571) 195/C2
Hodges, Mont. (†59353) 262/M4
Hodges, S.C. (29653) 296/C3
Hodge's Cove, Newf. 166/D2
Hodgeville, W. Va. (†26201) 312/F4
Hodgeville, Sask. 181/E5
Hodgkins, Ill. (60525) 222/B6
Hodgson, Manitoba 179/F3
Hodh (reg.), Mauritania 106/C5
Hod Hasharon, Israel 65/B3
Hodiyya, Israel 65/B4
Hódmezővásárhely, Hungary 41/F3
Hodonín, Czech. 41/D2
Hoehne, Colo. (81046) 208/L8
Hoei (Huy), Belgium 27/G8
Hoek, Netherlands 27/D6
Hoek van Holland (Hook of Holland), Netherlands 27/D4
Hoek van Holland (cape), Netherlands 27/D5
Hoensbroek, Netherlands 27/H7
Hoeryŏng, N. Korea 81/D2
Hoeselt, Belgium 27/G7
Hoey, Sask. 181/F3
Hof, W. Germany 22/D3
Hoffman, Ill. (62250) 222/D5
Hoffman, Minn. (56339) 255/C5
Hoffman, N.C. (28347) 281/K4
Hoffman, Okla. (74439) 288/P3
Hoffman Estates, Ill. (60195) 222/A5
Hofgeismar, W. Germany 22/C3
Hofors, Sweden 18/K6
Hofs (glac.), Iceland 21/C1
Hofu, Japan 81/E6
Hofuf, Saudi Arabia 59/E4
Hofuf, Saudi Arabia 54/F7
Hog (isl.), Mich. 250/D3
Hog (isl.), Pr. Edward I. 168/E2
Hog (isl.), Va. 307/S6
Hogan, Mo. (†63646) 261/L7
Hogan (lake), Ontario 177/F2
Hogan Group (isls.), Tasmania 99/D1
Hogansburg, N.Y. (13655) 276/L1
Hogansville, Georgia (30230) 217/C4

Hogarth (mt.), North. Terr. 93/E6
Hogatza, Alaska (99744) 196/G1
Hogeland, Mont. (59529) 262/H2
Hog Island (bay), Va. 307/S6
Hog River (Hogatza), Alaska (99744) 196/G1
Hogshead (pt.), Conn. 210/E3
Högyész, Hungary 41/E3
Hoh (head), Wash. 310/A3
Hoh (riv.), Wash. 310/A3
Hohenau, Paraguay 144/E5
Hohenau an der March, Austria 41/D2
Hohenberg, Austria 41/C3
Hohenems, Austria 41/A3
Hohenlinden, Miss. (†39751) 256/F3
Hohen Neuendorf, E. Germany 22/E2
Hohen Solms, La. (†70788) 238/K3
Hohenstollen (mt.), Switzerland 39/F3
Hohenwald, Tenn. (38462) 237/F9
Hohe Tauern (range), Austria 41/B3
Hohe Venn (plat.), Belgium 27/H8
Hohe Warte (mt.), Austria 41/B3
Hohhot (Huhehot), China 77/H3
Hohhot, China 54/N5
Hoh Ind. Res., Wash. 310/A3
Hohokam Pima Nat'l Mon., Ariz. 198/D5
Ho Ho Kus, N.J. (07423) 273/B1
Hoholitna (riv.), Alaska 196/G2
Hoh Xil Shan (mts.), China 77/C4
Hoi An, Vietnam 72/F4
Hoihow (Haikou), China 77/H7
Hoima, Uganda 115/F3
Hoisington, Kansas (67544) 232/D3
Hoi Xuan, Vietnam 72/E2
Højer, Denmark 21/B8
Højslev, Denmark 21/C4
Hokah, Minn. (55941) 255/G7
Hokang (Hegang), China 77/L2
Hoke (co.), N.C. 281/L4
Hokes Bluff, Ala. (35903) 195/G3
Hokianga (harb.), N. Zealand 100/D1
Hokitika, N. Zealand 100/C5
Hokkaido (pref.), Japan 81/K2
Hokkaido (isl.), Japan 2/S3
Hokkaido (isl.), Japan 54/R5
Hokkaido (isl.), Japan 81/L2
Holabird, S. Dak. (57540) 298/K4
Holbaek, Denmark 21/E6
Holbaek, Denmark 18/G9
Holbeach, England 10/F4
Holbeach, England 13/H5
Holbein, Sask. 181/E2
Holberg, Br. Col. 184/C5
Holbrook, Ariz. (86025) 198/E4
Holbrook, Idaho (83243) 220/F7
Holbrook, Iowa (†52325) 229/K5
Holbrook○, Mass. (02343) 249/D8
Holbrook, Nebr. (68948) 264/D4
Holbrook, N.S. Wales 97/D4
Holbrook, Oreg. (†97208) 291/A1
Holcomb, Ill. (61043) 222/D1
Holcomb, Kansas (67851) 232/B3
Holcomb, Miss. (38940) 256/D3
Holcomb, Mo. (63852) 261/N10
Holcomb, N.Y. (14469) 276/F5
Holcomb, W. Va. (26262) 312/E6
Holcombe, Wis. (54745) 317/D5
Holcombe Flowage (res.), Wis. 317/D5
Holden, Alberta 182/D3
Holden, La. (70744) 238/M1
Holden○, Mass. (01520) 249/G3
Holden, Mo. (64040) 261/E5
Holden, Utah (84636) 304/B4
Holden, W. Va. (25625) 312/B7
Holden Beach, N.C. (28462) 281/N7
Holdenville, Okla. (74848) 288/O4
Holder, Fla. (32645) 212/D3
Holderness (pen.), England 13/G4
Holderness○, N.H. (03245) 268/D4
Holdfast, Sask. 181/F5
Holdingford, Minn. (56340) 255/D5
Holdrege, Nebr. (68949) 264/E4
Holeby, Denmark 21/E8
Holen, Norway 18/D4
Holešov, Czech. 41/D2
Holetown, Barbados 161/B8
Holgate, Ohio (43527) 284/B3
Holguín (prov.), Cuba 158/J3
Holguín, Cuba 146/L7
Holguín, Cuba 158/J3
Holguín, Cuba 156/C2
Holíč, Czech. 41/D2
Holice, Czech. 41/D1
Holiday Hills, Ill. (†60050) 222/A4
Holijsloot, Netherlands 27/C4
Holitna (riv.), Alaska 196/G2
Hollabrunn, Austria 41/D2
Holladay, Tenn. (38341) 237/E9
Holladay, Utah (84117) 304/C3
Hollam's Bird (isl.), Namibia 118/A4
Holland, Georgia (†30730) 217/B2
Holland, Ind. (47541) 227/C8
Holland, Iowa (50642) 229/H4
Holland, Ky. (42153) 237/J7
Holland, Manitoba 179/D5
Holland○, Mass. (†01550) 249/F4
Holland, Mich. (49423) 250/C6
Holland, Minn. (56139) 255/B6
Holland, Mo. (63853) 261/N10
Holland, N.Y. (14080) 276/C5
Holland, Ohio (43528) 284/C2
Holland, Texas (76534) 303/G7
Holland○, Vt. (†05830) 268/D2
Hollandale, Minn. (56045) 255/E7
Hollandale, Miss. (38748) 256/C4
Hollandale, Wis. (53544) 317/G10
Holland Centre, Ontario 177/D3
Hollandia (Jayapura), Indonesia 85/K6
Holland Landing, Ontario 177/E3
Holland Park, Queensland 88/K3
Holland Park, Queensland 95/E3
Holland Patent, N.Y. (13354) 276/K4
Hollandsburg, Ind. (†47872) 227/C5
Hollandstoun, Scotland 15/F1
Hollansburg, Ohio (45332) 284/A5
Hollenbeck (riv.), Conn. 210/B1
Hollenberg, Kansas (66946) 232/F2
Holley, N.Y. (14470) 276/D4
Holley, Oreg. (†97386) 291/E3
Hollick-Kenyon (plat.) 5/B13
Holliday, Mo. (65258) 261/H3
Holliday, Texas (76366) 303/F4
Hollidaysburg, Pa. (16648) 294/F5
Hollins, Ala. (35082) 195/F4
Hollins College, Va. (24020) 307/H6
Hollis, Ark. (†72857) 202/D4
Hollis, Kansas (†66901) 232/E2
Hollis○, N.H. (03049) 268/D6
Hollis, N.C. (†28040) 281/F4
Hollis, Okla. (73550) 288/G5
Hollis Center○, Maine (04042) 243/B8
Hollister, Calif. (95023) 204/D7
Hollister, Fla. (32047) 212/E2
Hollister, Idaho (†83301) 220/D7
Hollister, Mo. (65672) 261/F9
Hollister, N.C. (27844) 281/O2
Hollister, Okla. (73551) 288/J6
Hollister, Wis. (†54491) 317/J5
Holliston○, Mass. (01746) 249/A8
Holloman A.F.B., N. Mex. 274/C6
Holloway, Minn. (56249) 255/C5
Holloway, Ohio (43985) 284/H5
Hollowayville, Ill. (†61356) 222/D2
Hollow Creek, Ky. (†40228) 237/K4
Hollow Rock, Tenn. (38342) 237/E8
Hollsopple, Pa. (15935) 294/E5
Hollum, Netherlands 27/H2
Holly, Colo. (81047) 208/P6
Holly, Mich. (48442) 250/F6
Holly, Wash. (†98310) 310/C3
Holly Bluff, Miss. (39088) 256/C5
Holly Grove, Ark. (72069) 202/H4
Holly Hill, Fla. (32017) 212/E2
Holly Hill, S.C. (29059) 296/G5
Holly Oak, Del. (†19801) 245/S1
Holly Pond, Ala. (35083) 195/E2
Holly Ridge, La. (71248) 238/G2
Holly Ridge, Miss. (38749) 256/C4
Holly Ridge, N.C. (28445) 281/O6
Holly Shelter (swamp), N.C. 281/O6
Holly Springs, Ark. (71746) 202/E6
Holly Springs, Georgia (30142) 217/D2
Holly Springs, Miss. (38635) 256/E1
Holly Springs, N.C. (27540) 281/M3
Hollytree, Ala. (35751) 195/F1
Hollyville, Del. (†19951) 245/T6
Hollywood, Ala. (35752) 195/G1
Hollywood, Ark. (†71923) 202/D5
Hollywood, Calif. (90028) 204/C10
Hollywood, Fla. 188/K5
Hollywood, Fla. (*33020) 212/B4
Hollywood, Georgia (30523) 217/E1
Hollywood, La. (†70663) 238/D6
Hollywood, Md. (20636) 245/M7
Hollywood, Miss. (†38676) 256/D1
Hollywood, Mo. (†63821) 261/O10
Hollywood, S.C. (29449) 296/G6
Hollywood, W. Va. (†24983) 312/F7
Hollywood Park, Texas (†78201) 303/K10
Holman, N. Mex. (87723) 274/D2
Holman (isl.), N.W.T. 162/D1
Holman Island, Canada 4/B15
Holman Island, N.W. Terrs. 187/G2
Holmdel○, N.J. (07733) 273/E3
Holmen, Wis. (54636) 317/D8
Holmes (reef), 95/C3
Holmes (reef), Coral Sea Is. Terr. 88/H3
Holmes (co.), Fla. 212/C5
Holmes (creek), Fla. 212/D5
Holmes, Iowa (†50525) 229/F3
Holmes (co.), Miss. 256/D4
Holmes (co.), Ohio 284/G4
Holmes (mt.), Wyo. 319/B1
Holmes Beach, Fla. (33509) 212/D4
Holmes City, Minn. (56341) 255/C5
Holmeson, N.J. (†08526) 273/E3
Holmestrand, Norway 18/C4
Holmesville, Miss. (†39648) 256/D8
Holmesville, Nebr. (68374) 264/H4
Holmesville, New Bruns. 170/C2
Holmesville, Ohio (44633) 284/G4
Holmesville, Ontario 177/C4
Holmfield, Manitoba 179/C5
Holmfirth, England 13/J2
Holmquist, S. Dak. (†57274) 298/O3
Holmsbu, Norway 18/D4
Holmsund, Sweden 18/M5
Holmwood, La. (†70647) 238/D6
Holon, Israel 65/B3
Holopaw, Fla. (†32901) 212/E3
Holroyd, N.S. Wales 97/H3
Holroyd (riv.), Queensland 95/B2
Holstebro, Denmark 18/F8
Holstebro, Denmark 21/B5
Holsted, Denmark 21/B6
Holstein, Iowa (51025) 229/B4
Holstein, Mo. (†63378) 261/K5
Holstein, Nebr. (68950) 264/F4
Holstein, Ontario 177/D3
Holsteinsborg, Greenl. 4/C12
Holston (riv.), Tenn. 237/O8
Holston, Va. (†24210) 307/D7
Holston, North Fork (riv.), Va. 307/D7
Holston Valley, Tenn. (†37620) 237/S7
Holsworthy, England 10/D5
Holsworthy, England 13/C7
Holt, Ala. (35404) 195/D4
Holt (dam), Ala. 195/D4
Holt, Calif. (95234) 204/D6
Holt, England 13/J5
Holt, Fla. (32564) 212/C6
Holt, Mich. (48842) 250/E6
Holt, Minn. (†56738) 255/B2
Holt (co.), Mo. 261/B2
Holt, Mo. (64048) 261/D4
Holt (co.), Nebr. 264/F2
Holte, Denmark 21/F6
Holter (lake), Mont. 262/D4
Holtland, Tenn. (†37034) 237/H9
Holton, Ind. (47023) 227/G6
Holton, Kansas (66436) 232/G2
Holton, La. (†70422) 238/K5
Holton, Mich. (49425) 250/C5
Holton, Newf. 166/C3
Holts Summit, Mo. (65043) 261/H5
Holtville, Ala. (†36022) 195/F5
Holtville, Calif. (92250) 204/K11
Holtville, New Bruns. 170/D2
Holtwood, Pa. (17532) 294/K6
Holualoa, Hawaii (96725) 218/G5
Holualoa, Hawaii 188/F6
Holwerd, Netherlands 27/H2
Holy (isl.), England 13/F2
Holy (isl.), England 10/F3
Holy (isl.), Scotland 15/C5
Holy (isl.), Wales 10/D4
Holy (isl.), Wales 13/C4
Holy City, Calif. (95026) 204/K4
Holy Cross (mt.), Colo. 208/F4
Holy Cross, Alaska (99602) 196/G2
Holy Cross, Iowa (52053) 229/L3
Holycross, Ireland 17/F6
Holy Cross, Wis. (†53004) 317/L9
Holyhead, Wales 10/D4
Holyhead, Wales 13/C4
Holy Loch (inlet), Scotland 15/A1
Holyoke, Colo. (80734) 208/P1
Holyoke, Mass. (01040) 249/D4
Holyoke (range), Mass. 249/D3
Holyoke, Minn. (55749) 255/F4
Holyrood, Kansas (67450) 232/D3
Holyrood, Newf. 166/D2
Holyrood (bay), Newf. 166/D2
Holyrood (pond), Newf. 166/D2
Holyroyd, N. S. Wales 88/K4
Holy Trinity, Ala. (36859) 195/H6
Holywell, Wales 13/G2
Holywood, N. Ireland 17/K2
Holzminden, W. Germany 22/C3
Homalin, Burma 72/B1
Homathko (riv.), Br. Col. 184/E4
Homayunshahr, Iran 66/G4
Hombori, Mali 106/D5
Hombori (mts.), Mali 106/D5
Homburg, W. Germany 22/B4
Home (bay) 162/K2
Home, Kansas (66438) 232/F2
Home (isl.), Newf. 166/B1
Home (bay), N.W. Terrs. 187/M3
Home, Pa. (15747) 294/D4
Homedale, Idaho (83628) 220/A6
Home Gardens, Calif. (†91720) 204/E11
Home Hill, Queensland 88/H3
Home Hill, Queensland 95/C3
Homeland, Calif. (92348) 204/H10
Homeland, Georgia (†31537) 217/H9
Home Place, La. (†70083) 238/L8
Homer, Alaska (99603) 196/B2
Homer, Georgia (30547) 217/F2
Homer, Ill. (61849) 222/F3
Homer, Ind. (46146) 227/F5
Homer, Ky. (†42276) 237/H7
Homer, La. (71040) 238/D1
Homer, Mich. (49245) 250/E6
Homer, Minn. (55942) 255/G6
Homer, Nebr. (68030) 264/H2
Homer, N.Y. (13077) 276/H5
Homer, Ohio (43027) 284/E5
Homer City, Pa. (15748) 294/D4
Homerville, Georgia (31634) 217/G8
Homerville, Ohio (44235) 284/F3
Homestead, Fla. (*33030) 212/F6
Homestead, Iowa (52236) 229/K5
Homestead, Mont. (59242) 262/M2
Homestead, Okla. (†73763) 288/K2
Homestead, Pa. (15120) 294/B7
Homestead, Queensland 95/C4
Homestead A.F.B., Fla. 212/F6
Homestead Nat'l Mon., Nebr. 264/H4
Homestown, Mo. (†63879) 261/N10
Hometown, Ill. (60456) 222/B6
Home Valley, Wash. (†98648) 310/D5
Homewood, Ala. (35209) 195/E4
Homewood, Calif. (95718) 204/E4
Homewood, Ill. (60430) 222/B6
Homewood, Kansas (†66095) 232/G3
Homewood, Manitoba 179/E5
Homewood, Miss. (†39152) 256/E6
Homewood, Pa. (15208) 294/B4
Homeworth, Ohio (44634) 284/J4
Hominy, Okla. (74035) 288/O2
Hominy Falls, W. Va. (†26651) 312/E6
Homochitto (riv.), Miss. 256/B8
Homoine, Mozambique 118/E4
Homonhon (isl.), Philippines 82/E5
Homosassa, Fla. (32646) 212/D3
Homosassa (isls.), Fla. 212/D3
Homosassa Springs, Fla. (32647) 212/D3
Homra, Hamada el (des.), Libya 111/B2
Homs, Libya 102/D1
Homs, Libya 111/B1
Homs (prov.), Syria 63/G5
Homs, Syria 59/C3
Homs, Syria 63/G5
Homs, Syria 54/E6
Hon, Ark. (†72958) 202/B4
Hon, Libya 102/D2
Hon, Libya 111/C2
Honaker, Va. (24260) 307/D6
Honan (Henan) (prov.), China 77/H5
Honaunau, Hawaii (96726) 218/G6
Honavar, India 68/C6
Honaz Dağı (mt.), Turkey 63/C4
Hon Chong, Vietnam 72/E5
Honda, Colombia 126/C5
Honda (bay), Cuba 158/B1
Honda (bay), Philippines 82/B6
Honda (bay), P. Rico 161/F2
Hondo (riv.), Belize 154/C1
Hondo, Japan 81/E7
Hondo (riv.), Mexico 150/P7
Hondo, N. Mex. (88336) 274/D5
Hondo, Texas (78861) 303/E8
Hondsrug (hills), Netherlands 27/K3
Honduras 2/E5
Honduras 146/K8
HONDURAS 154
Honduras (gulf) 146/K8
Honduras (gulf), Belize 154/D2
Honduras (gulf), Guatemala 154/D2
Honduras (cape), Honduras 154/E2
Honduras (gulf), Honduras 154/D2
Honea Path, S.C. (29654) 296/C3
Honegg (mt.), Switzerland 39/E3
Honeoye, N.Y. (14471) 276/F5
Honeoye (lake), N.Y. 276/F5
Honeoye Falls, N.Y. (14472) 276/F5
Honesdale, Pa. (18431) 294/M2
Honey (lake), Calif. 204/E3
Honey (creek), Ind. 227/C6
Honey (creek), Oreg. 291/G5
Honey Brook, Pa. (19344) 294/L5
Honey Creek, Ind. (†47356) 227/F4
Honey Creek, Iowa (51542) 229/B6
Honey Creek, Wis. (53138) 317/J3
Honeydale, New Bruns. 170/C3
Honeyford, N. Dak. (†58242) 282/R3
Honey Grove, Texas (75446) 303/J4
Honey Harbour, Ontario 177/E3
Honey Hill, S.C. (†29479) 296/H5
Honey Island, Texas (†77625) 303/K7
Honeymoon Bay, Br. Col. 184/J3
Honeyville, Utah (84314) 304/B2
Honeywood, Ontario 177/D3
Honfleur, France 28/D3
Honfleur, Québec 172/G3
Høng, Denmark 21/E7
Honga (riv.), Md. 245/O7
Hon Gai, Vietnam 72/E2
Hongch'ŏn, S. Korea 81/D5
Hong Kong 54/N7
Hong Kong, 2/Q4
HONG KONG 77
Hongliuhe, China 77/E3
Hongor, Mongolia 77/H2
Hongshui He (riv.), China 77/G7
Hongsŏng, S. Korea 81/C5
Hongtong, China 77/H4
Honguedo (passage), Québec 174/E3
Hongwŏn, N. Korea 81/C3
Hongze Hu (lake), China 77/J5
Honiara (cap.), Solomon Is. 86/D3
Honiara (cap.), Solomon Is. 87/F6
Honiton, England 10/E5
Honiton, England 13/D7
Honjo, Japan 81/J4
Honnedaga (lake), N.Y. 276/L3
Honnelles, Belgium 27/D8
Honningsvag, Norway 18/O1
Honobia, Okla. (74549) 288/R5
Honohina, Hawaii (†96710) 218/J4
Honokaa, Hawaii 188/G5
Honokaa, Hawaii (96727) 218/H4
Honokahau, Hawaii (†96761) 218/H1
Honokohau, Hawaii, Hawaii (†96740) 218/G5
Honokohau, Maui, Hawaii (†96725) 218/J1
Honolulu (co.), Hawaii 218/D3
Honolulu (cap.), Hawaii 87/L3
Honolulu (cap.), Hawaii 188/F5
Honolulu (cap.), Hawaii (*96801) 218/C4
Honolulu (harb.), Hawaii 218/C4
Honolulu, U.S. 2/B5
Honolulu Int'l Airport, Hawaii 218/B4
Honomu, Hawaii (96728) 218/J4
Honor, Mich. (49640) 250/D4
Honoraville, Ala. (36042) 195/F7
Honouliuli, Hawaii (†96706) 218/A3
Honshu (isl.), Japan 2/S4
Honshu (isl.), Japan 54/P6
Honshu (isl.), Japan 81/J5
Honuapo, Hawaii (†96772) 218/H7
Hood, Calif. (95639) 204/B9
Hood (riv.), N.W. Terrs. 187/G3
Hood (mt.), Oreg. 291/F2
Hood (riv.), Oreg. 291/F2
Hood (co.), Texas 303/G5
Hood (canal), Wash. 310/B3
Hood River (co.), Oreg. 291/F2
Hood River, Oreg. (97031) 291/F2
Hoodsport, Wash. (98548) 310/B3
Hoofddorp (Haarlemmermeer), Netherlands 27/F4
Hoogeveen, Netherlands 27/J3
Hoogezand-Sappemeer, Netherlands 27/K2
Hooghly (riv.), India 68/F2
Hooghly-Chinsura, India 68/F1
Hoogkarspel, Netherlands 27/G3
Hoogstraten, Belgium 27/F6
Hook (head), Ireland 17/H7
Hook (isl.), Queensland 88/H4
Hook (isl.), Queensland 95/D4
Hookena, Hawaii (†96704) 218/G6
Hooker (co.), Nebr. 264/C3
Hooker, Okla. (73945) 288/D1
Hooker Creek, North. Terr. 88/E3
Hooker Creek, North. Terr. 93/B5
Hooker Creek Aboriginal Reserve, North. Terr. 88/E3
Hookersville, W. Va. (†26651) 312/E6
Hookerton, N.C. (28538) 281/O4
Hook of Holland, Netherlands 27/D4
Hooks, Texas (75561) 303/K4
Hooksett, N.H. (03106) 268/E5
Hooksett○, N.H. (03106) 268/E5
Hookstown, Pa. (15050) 294/B4
Hoolehua, Hawaii 188/F5
Hoolehua, Hawaii (96729) 218/G1
Hoonah, Alaska (99829) 196/M1
Hoonah (sound), Alaska 196/M1
Hoopa, Calif. (95546) 204/B2
Hoopa Valley Ind. Res., Calif. 204/A2
Hooper, Colo. (81136) 208/H7
Hooper (str.), Md. 245/O8
Hooper, Nebr. (68031) 264/H3
Hooper, Utah (84315) 304/B2
Hooper, Wash. (99333) 310/G4
Hooper Bay, Alaska (99604) 196/E2
Hooper Bay, Alaska 188/C5
Hoopersville, Md. (21642) 245/O7
Hoopeston, Ill. (60942) 222/F3
Hoople, N. Dak. (58243) 282/P2
Hooppole, Ill. (61258) 222/D2
Hoorn, Netherlands 27/G3
Hoorn (isls.), Wallis and Futuna 87/J7
Hoosac (mts.), Mass. 249/B2
Hoosac Tunnel, Mass. (†01339) 249/C2
Hoosic (riv.), Mass. 249/A1
Hoosic (riv.), Vt. 268/A6
Hoosick Falls, N.Y. (12090) 276/O5
Hoosier, Sask. 181/B4
Hoot Owl, Okla. (†74356) 288/R2
Hooven, Ohio (45033) 284/A9
Hoover, Ala. (†35216) 195/E4
Hoover (dam), Nev. 266/G7
Hoover (res.), Ohio 284/E5
Hoover, S. Dak. (†57760) 298/C3
Hooversville, Pa. (15936) 294/E5
Hop (riv.), Conn. 210/F1
Hopa, Turkey 63/J2
Hopatcong, N.J. (07843) 273/D2
Hopatcong (lake), N.J. 273/D2
Hop Bottom, Pa. (18824) 294/L2
Hope, Alaska (99605) 196/C1
Hope (pt.), Alaska 196/E1
Hope (bay) 5/C16
Hope, Ark. (71801) 202/C6
Hope, Br. Col. 162/D6
Hope, Br. Col. 184/M3
Hope, Idaho (83836) 220/B1
Hope, Ind. (47246) 227/F6
Hope, Kansas (67451) 232/E3
Hope, Ky. (40334) 237/O4
Hope, Maine (04847) 243/E7
Hope○, Maine (04847) 243/E7
Hope, Mich. (48628) 250/E5
Hope, Minn. (56045) 255/E7
Hope, Mo. (†65061) 261/J5
Hope (lake), Newf. 166/B3
Hope, N.J. (07844) 273/D2
Hope, N. Mex. (88250) 274/E6
Hope, N. Dak. (58046) 282/P5
Hope (isl.), Norway 4/B8
Hope, R.I. (02831) 249/H6
Hope, Loch (lake), Scotland 15/D2
Hope Bay, Jamaica 158/K6
Hopedale, Ill. (61747) 222/D3
Hopedale, Mass. (01747) 249/H4
Hopedale○, Mass. (01747) 249/H4
Hopedale, Newf. 166/B2
Hopedale, Newf. 162/L4
Hopedale, Newf. 146/N4
Hopedale, Ohio (43976) 284/J5
Hopeful Heights, Ky. (†41018) 237/R2
Hope Hull, Ala. (36043) 195/F6
Hopei (Hebei) (prov.), China 77/J4
Hopeland, Pa. (17533) 294/K5
Hopelchén, Mexico 150/P7
Hopeman, Scotland 15/E3
Hope Mills, N.C. (28348) 281/M5
Hopen (isl.), Norway 18/E2
Hopes Advance (cape), Québec 174/F1
Hopeton, Okla. (73746) 288/J1
Hopeton, Va. (†23421) 307/S5
Hopetoun, Victoria 97/B4
Hopetoun, W. Australia 92/C6
Hopetown, Québec 172/D2
Hopetown, S. Africa 118/C5
Hopetown, W. Australia 88/C6
Hope Valley, R.I. (02832) 249/H6
Hope Valley (res.), S. Australia 88/E7
Hopeville, Iowa (†50174) 229/F7
Hopewell, Ala. (†36264) 195/H3
Hopewell, Jamaica 158/G5
Hopewell, Kansas (†67557) 232/D4
Hopewell, Md. (†21817) 245/P8
Hopewell, Miss. (†39059) 256/D7
Hopewell, N.J. (08525) 273/D3
Hopewell (isls.), N.W. Terrs. 187/L4
Hopewell, Nova Scotia 168/F3
Hopewell, Ohio (43746) 284/F6
Hopewell, Pa. (16650) 294/F5
Hopewell (I.C.), Va. (23860) 307/O6
Hopewell Cape, New Bruns. 170/F3
Hopewell Hill, New Bruns. 170/F3
Hopewell Junction, N.Y. (12533) 276/N7
Hopfgarten in Nordtirol, Austria 41/B3
Hopi (buttes), Ariz. 198/E3
Hopi Ind. Res., Ariz. 198/E2
Hopkins (co.), Ky. 237/F6
Hopkins, Mich. (49328) 250/D6
Hopkins, Minn. (55343) 255/G5
Hopkins, Mo. (64461) 261/C1
Hopkins (lake), North. Terr. 93/A8
Hopkins, S.C. (29061) 296/F4
Hopkins (co.), Texas 303/J4
Hopkins (riv.), Victoria 97/B5
Hopkins, Va. (†23421) 307/S5
Hopkins (lake), W. Australia 88/D4
Hopkins (lake), W. Australia 92/E4
Hopkinsville, Ky. (42240) 237/F7
Hopkinton, Iowa (52237) 229/L4
Hopkinton, Mass. (01748) 249/J4
Hopkinton○, Mass. (01748) 249/J4
Hopkinton○, N.H. (03301) 268/D5
Hopkinton, N.Y. (12940) 276/L1
Hopkinton○, R.I. (02833) 249/H7
Hopland, Calif. (95449) 204/B5
Hoppo (Hepu), China 77/G7
Hop River, Conn. (†06237) 210/F2
Hopwood, Pa. (15445) 294/C6
Hoquiam, Wash. 188/B1
Hoquiam, Wash. (98550) 310/A3
Horace, Kansas (†67879) 232/A3
Horace, N. Dak. (58047) 282/S6
Horasan, Turkey 63/K2
Horatio, Ark. (71842) 202/B3
Horatio, S.C. (29062) 296/F3
Horažďovice, Czech. 41/B2
Horche, Spain 33/E2
Horconcitos, Panama 154/F6
Hordaland (co.), Norway 18/E6
Horden, England 13/J4
Hordio, Somalia 115/K1
Hordville, Nebr. (68846) 264/G3
Horgen, Switzerland 39/G2
Hořice v Podkrkonoší, Czech. 41/C1
Horicon, Wis. (53032) 317/J9
Horine, Mo. (†63070) 261/M6
Horizon, Sask. 181/F6
Horley, England 13/H8
Hormigüeros, P. Rico 161/A2
Hormoz, Iran 66/J7
Hormoz (isl.), Iran 66/K7
Hormozgan (prov.), Iran 66/J7
Hormuz (str.), Iran 59/G4
Hormuz (str.), Iran 66/K7
Hormuz (str.), Oman 59/G4
Horn (cape) 2/F8
Horn, Austria 41/C2
Horn (cape), Chile 120/C8
Horn (cape), Chile 138/F11
Horn (cape), Iceland 7/B2
Horn (cape), Iceland 21/B1
Horn (head), Ireland 17/E1
Horn (isl.), Miss. 256/G10
Horn (mts.), N.W. Terrs. 187/G3
Horn (riv.), N.W. Terrs. 187/G3
Hornád (riv.), Czech. 41/F2
Hornaday (riv.), N.W. Terrs. 187/F3
Hornafjördhur (fjord), Iceland 21/D1
Horná Štubňa, Czech. 41/E2
Horn-Bad Meinberg, W. Germany 22/C3
Hornbeak, Tenn. (38232) 237/C8
Hornbeck, Alberta 182/B3
Hornbeck, La. (71439) 238/D4
Hornbrook, Calif. (96044) 204/C2
Hornby, N. Zealand 100/D5
Hornby (bay), N.W. Terrs. 187/G3
Hornby Island, Br. Col. 184/H2
Horncastle, England 13/G4
Horncastle, England 10/F4
Horndean, Manitoba 179/E5
Hörnefors, Sweden 18/L5
Hornell, N.Y. (14843) 276/E6
Hornepayne, Ontario 175/C3
Hornepayne, Ontario 177/J5
Horner, W. Va. (26372) 312/F5
Hornerstown, N.J. (†08514) 273/E3
Hornersville, Mo. (63855) 261/M10
Horn Hill, Ala. (†36467) 195/F8
Horní Benešov, Czech. 41/D2
Hornick, Iowa (51026) 229/A4
Horní Libina, Czech. 41/D2
Hornings Mills, Ontario 177/D3
Hornitos, Calif. (95325) 204/E6
Horn Lake, Miss. (38637) 256/D1
Hörnli (mt.), Switzerland 39/G2
Hornos, Falso (cape), Chile 138/F11
Hornsby, N. S. Wales 88/K3
Hornsby, N.S. Wales 97/J3
Hornsby, Tenn. (38044) 237/D10
Hornsea, England 13/G4
Hornsea, England 10/F4
Hornslandet (pen.), Sweden 18/K6
Hornslet, Denmark 21/D5
Horns Road, Nova Scotia 168/H2
Hornsund (bay), Norway 18/C2
Horntown, Va. (23395) 307/T5
Hořovice, Czech. 41/C2
Horqin Youyi Qianqi (Ulanhot), China 77/K2
Horqueta, Paraguay 144/D3
Horry (co.), S.C. 296/J4
Horse (lake), Calif. 204/E3
Horse (creek), Colo. 208/M5
Horse (creek), Fla. 212/E4
Horse (isls.), Newf. 166/C3
Horse (creek), Oreg. 291/F3
Horse (creek), Wyo. 319/H4
Horse (creek), Wyo. 319/B3
Horse Branch, Ky. (42349) 237/H6
Horse Cave, Ky. (42749) 237/K6
Horse Chops (head), Newf. 166/D2
Horse Creek, Calif. (96045) 204/C2
Horse Creek (res.), Colo. 208/N6
Horse Creek, Wyo. (82061) 319/G4
Horsefly, Br. Col. 184/G4
Horsefly (lake), Br. Col. 184/G4
Horsehead (lake), N. Dak. 282/L5
Horsehead (creek), S. Dak. 298/C7
Horseheads, N.Y. (14845) 276/G6
Horsens, Denmark 18/F9
Horsens, Denmark 21/C6
Horsens (fjord), Denmark 21/D6
Horseshoe (lake), Ariz. 198/D5
Horseshoe (pt.), Fla. 212/C2
Horseshoe (lake), Manitoba 179/G2
Horse Shoe (pt.), St. Chris.-Nevis 161/C11
Horseshoe (creek), Wyo. 319/G3
Horseshoe Beach, Fla. (32648) 212/C2
Horseshoe Bend, Ark. (72512) 202/G1
Horseshoe Bend, Idaho (83629) 220/B6
Horseshoe Bend Nat'l Mil. Park, Ala. 195/G5
Horseshoe Lake, Ontario 177/E2
Horse Shoe Run, W. Va. (26769) 312/G4
Horsetooth (res.), Colo. 208/J1
Horsham, England 10/F5
Horsham, England 13/G6
Horsham, Sask. 181/B5
Horsham, Victoria 97/B5
Horsham, Victoria 88/G7
Hørsholm, Denmark 21/F6
Horšovský Týn, Czech. 41/B2
Horst, Netherlands 27/H6
Horta (dist.), Portugal 33/A1
Horta, Portugal 33/B1
Hortaleza, Spain 33/G4
Horten, Norway 18/D4
Hortense, Georgia (31543) 217/J8
Hortensfjord (fjord), Norway 18/G4
Horton, Ala. (35980) 195/F2
Horton, Iowa (†50677) 229/J3
Horton, Kansas (66439) 232/G2
Horton, Mich. (49246) 250/E6
Horton, Mo. (64751) 261/D7
Horton (riv.), N.S. Wales 97/F2
Horton (riv.), N.W. Terrs. 187/F3
Horton, Oreg. (97448) 291/D3
Horton Bay, Mich. (†49712) 250/D3
Hortonia (lake), Vt. 268/A4
Hortonville, Ind. (†46069) 227/E4
Hortonville, Mass. (†02777) 249/K5
Hortonville, Wis. (54944) 317/J7
Hørve, Denmark 21/E6
Horwich, England 10/G2
Horwich, England 13/G2

Hoschton, Georgia (30548) 217/E2
Hoselaw, Alberta 182/E2
Hosenofu (well), Libya 111/D3
Hosford, Fla. (32334) 212/B1
Hoshab, Pakistan 68/A3
Hoshab, Pakistan 59/H4
Hoshangabad, India 68/D4
Hoskins, Nebr. (68740) 264/G2
Hosmer, Br. Col. 184/K5
Hosmer, S. Dak. (57448) 298/L2
Hospental, Switzerland 39/F3
Hospers, Iowa (51238) 229/B2
Hospet, India 68/D5
Hospital, Chile 138/G4
Hospital, Ireland 17/E7
Hospitalet, Spain 33/H2
Hosseina, Ethiopia 111/G6
Hosston, La. (71043) 238/C1
Hoste (isl.), Chile 120/B8
Hoste (isl.), Chile 138/F11
Hostinné, Czech. 41/C1
Hoswick, Scotland 15/G2
Hot, Thailand 72/C3
Hotan, China 77/B4
Hotan, China 54/K6
Hotan He (riv.), China 77/B4
Hotchkiss, Alberta 182/B1
Hotchkiss, Colo. (81419) 208/D5
Hotchkissville, Conn. (†06798) 210/C2
Hot Creek (range), Nev. 266/E4
Hot Creek (valley), Nev. 266/E4
Hotevilla, Ariz. (86030) 198/E3
Hotham (inlet), Alaska 196/F1
Hoting, Sweden 18/K4
Hot Lake, Oreg. (†97850) 291/K2
Hot Spring (co.), Ark. 202/E5
Hot Springs, Mont. (59845) 262/B3
Hot Springs (Truth or Consequences), N. Mex. (8790 274/B5
Hot Springs, N.C. (28743) 281/D3
Hot Springs, S. Dak. 188/F2
Hot Springs, S. Dak. (57747) 298/C7
Hot Springs, Va. (24445) 307/J4
Hot Springs (co.), Wyo. 319/D2
Hot Springs Cove, Br. Col. 184/D5
Hot Springs National Park, Ark. 188/H4
Hot Springs National Park, Ark. (71901) 202/D4
Hot Springs Nat'l Park, Ark. 202/D4
Hot Sulphur Springs, Colo. (80451) 208/H2
Hottah (lake), N.W.T. 162/E2
Hottah (lake), N.W. Terrs. 187/G3
Hottentot (bay), Namibia 118/A5
Hotton, Belgium 27/G8
Hou, Nam (riv.), Laos 72/D2
Houck, Ariz. (86506) 198/F3
Houcktown, Ohio (†45840) 284/C4
Houffalize, Belgium 27/H8
Houghton, Iowa (52631) 229/K7
Houghton, Maine (†04275) 243/B6
Houghton, Mich. 188/J1
Houghton (co.), Mich. 250/G1
Houghton, Mich. (49931) 250/G1
Houghton (lake), Mich. 250/E4
Houghton, N.Y. (14744) 276/D6
Houghton, S. Dak. (57449) 298/N2
Houghton Lake, Mich. (48629) 250/E4
Houghton Lake Heights, Mich. (48630) 250/E4
Houghton-le-Spring, England 13/J3
Houhoek, S. Africa 118/F7
Houlka, Miss. (38850) 256/G2
Houlton, Maine 188/N1
Houlton, Maine (04730) 243/H3
Houlton○, Maine (04730) 243/H3
Houlton, Wis. (†55082) 317/A5
Houma, China 77/H4
Houma, La. (70360) 238/J7
Houndé, Upper Volta 106/D6
Hounslow, England 13/G8
Hounslow, England 10/B5
Hourn, Loch (inlet), Scotland 15/C3
Housatonic (riv.), Conn. 210/C3
Housatonic, Mass. (01236) 249/A3
Housatonic (riv.), Mass. 249/A4
House (mt.), Alberta 182/C2
House (riv.), Alberta 182/D2
House, N. Mex. (88121) 274/F4
House (range), Utah 304/A4
House Springs, Mo. (63051) 261/L6
Houston (co.), Ala. 195/H8
Houston, Ala. (35572) 195/D2
Houston, Alaska (†99687) 196/B1
Houston, Ark. (72070) 202/E3
Houston, Br. Col. 184/D3
Houston, Del. (19954) 245/S5
Houston, Fla. (†32060) 212/D1
Houston (co.), Georgia 217/E6
Houston, Ind. (†47235) 227/E6
Houston (co.), Minn. 255/G7
Houston, Minn. (55943) 255/G7
Houston, Miss. (38851) 256/G3
Houston, Mo. (65483) 261/J8
Houston, Ohio (45333) 284/B5
Houston, Pa. (15342) 294/B5
Houston (co.), Tenn. 237/F8
Houston (co.), Texas 303/J6
Houston, Texas (*77001) 303/J2
Houston, Texas 188/G5
Houston, Texas 146/J7
Houston (lake), Texas 303/J8
Houston, U.S. 2/E4
Houston Acres, Ky. (†40201) 237/K2
Houstonia, Mo. (65333) 261/F5
Houston Lake, Mo. (†64152) 261/O5
Houston Ship (chan.), Texas 303/K2
Hout (bay), S. Africa 118/E6
Houtbaai, S. Africa 118/E6
Houtman Abrolhos (isls.), W. Australia 88/A5
Houtman Abrolhos (isls.), W. Australia 92/A5
Houtrak Polder, Netherlands 27/A4
Houtzdale, Pa. (16651) 294/F4
Hov, Denmark 21/D6
Hovd, Mongolia 77/D2
Hovd (Kobdo, Jirgalanta), Mongolia 77/D2
Hovd, Mongolia 54/L5
Hovd Gol (riv.), Mongolia 77/D2
Hove, England 13/G7
Hove, England 10/F5
Hoven, S. Dak. (57450) 298/K3
Hovenweep Nat'l Mon., Colo. 208/A8
Hovenweep Nat'l Mon., Utah 304/E6
Hoveyzeh, Iran 66/F5
Hoving, N. Dak. (†58060) 282/P7
Hovland, Minn. (55606) 255/G2
Hövsgöl, Mongolia 77/E1
Hövsgöl Nuur (lake), Mongolia 77/F1
Howar, Wadi (dry riv.), Sudan 111/E4
Howard (pass), Alaska 196/G1
Howard (co.), Ark. 202/C5
Howard, Colo. (81233) 208/H6
Howard, Georgia (31039) 217/D5
Howard (co.), Ind. 227/E4
Howard (co.), Iowa 229/J2
Howard, Kansas (67349) 232/F4
Howard (co.), Md. 245/L4
Howard (co.), Mo. 261/G4
Howard (co.), Nebr. 264/F3
Howard, New Bruns. 170/E2
Howard, Ohio (43028) 284/F5
Howard, Pa. (16841) 294/G3
Howard, S. Dak. (57349) 298/P5
Howard (co.), Texas 303/C5
Howard (creek), Texas 303/C7
Howard, Wis. (54303) 317/K6
Howard A. Hanson (res.), Wash. 310/D3
Howard City, Mich. (49329) 250/D5
Howard City (Boelus), Nebr. (68820) 264/F3
Howard Lake, Minn. (55349) 255/D5
Howards Grove-Millersville, Wis. (53081) 317/L8
Howards Ridge, Mo. (†65655) 261/H9
Howardstown, Ky. (40028) 237/K5
Howardsville, Va. (24562) 307/L5
Howardville, Mo. (†63869) 261/N9
Howden, England 13/G4
Howe (cape), Australia 87/F9
Howe (sound), Br. Col. 184/K2
Howe, Idaho (83244) 220/F6
Howe, Ind. (46746) 227/G1
Howe (cape), N. S. Wales 88/J7
Howe (cape), N. S. Wales 97/F5
Howe, Okla. (74940) 288/S5
Howe, Texas (75059) 303/H4
Howell, Ark. (72071) 202/H3
Howell, Georgia (†31636) 217/F9
Howell, Mich. (48843) 250/E6
Howell (co.), Mo. 261/J9
Howell○, N.J. (07731) 273/E3
Howell, Tenn. (†37334) 237/H10
Howell, Utah (84316) 304/B2
Howells, Nebr. (68641) 264/H3
Howes, S. Dak. (57748) 298/E4
Howesville, Ind. (†47438) 227/C6
Howesville, W. Va. (†26444) 312/G4
Howey In The Hills, Fla. (32737) 212/E3
Howick, N. Zealand 100/C1
Howick, Québec 172/D4
Howick, S. Africa 118/E5
Howison, Miss. (†39574) 256/F9
Howland, Maine (04448) 243/F5
Howland○, Maine (04448) 243/F5
Howland (isl.), Pacific 87/J5
Howland Ridge, New Bruns. 170/C2
Howley, Newf. 166/C4
Howlong, N.S. Wales 97/D4
Howrah, India 68/F2
Howrah, India 54/K7
Howser, Br. Col. 184/J5
Hoxeyville, Mich. (49641) 250/D4
Hoxie, Ark. (72433) 202/H1
Hoxie, Kansas (67740) 232/B2
Höxter, W. Germany 22/C3
Hoxud, China 77/C3
Hoy (isl.), Scotland 15/E2
Hoy (isl.), Scotland 10/E1
Hoy (sound), Scotland 15/E2
Hoyerswerda, E. Germany 22/F3
Hoylake, England 13/G2
Hoylake, England 10/F2
Hoyland Nether, England 13/J2
Hoyleton, Ill. (62803) 222/D5
Hoyos, Spain 33/C2
Hoyran (lake), Turkey 63/D3
Hoyt, Colo. (80641) 208/L2
Hoyt, Kansas (66440) 232/G2
Hoyt, New Bruns. 170/D3
Hoyt, Okla. (74440) 288/R4
Hoyt (peak), Utah 304/C3
Hoyt Lakes, Minn. (55750) 255/F3
Hoytsville, Utah (†84017) 304/C3
Hoytville, Ohio (43529) 284/C3
Hozat, Turkey 59/C2
Hozat, Turkey 63/H3
Hradec Králové, Czech. 41/C1
Hranice, Czech. 41/D2
Hrinova, Czech. 41/E2
Hron (riv.), Czech. 41/E2
Hronov, Czech. 41/D1
Hrubieszów, Poland 47/F3
Hrušovany, Czech. 41/D2
Hsewi, Burma 72/C2
Hsipaw, Burma 72/C2
Hsüchang (Xuchang), China 77/H5
Htawgaw, Burma 72/C1
Huacaraje, Bolivia 136/D3
Huacareta, Bolivia 136/C7
Huacaya, Bolivia 136/D7
Huachacalla, Bolivia 136/A6
Huachi, Bolivia 136/D4
Huachi, China 77/G4
Huachipato, Chile 138/D1
Huacho, Peru 120/B4
Huacho, Peru 128/D8
Huachuca (peak), Ariz. 198/E7
Huachuca City, Ariz. (85616) 198/E7
Huacrachuco, Peru 128/D7
Huade, China 77/H3
Huadian, China 77/L3
Hua Hin, Thailand 72/D4
Huahine (isl.), Fr. Poly. 87/L7
Huaibei, China 77/J5
Huaibin, China 77/H5
Huaide (Hwaiteh), China 77/K3
Huaiji, China 77/H7
Huainan, China 77/J5
Huainan, China 54/N6
Huairen, China 77/H4
Huajuapan de León, Mexico 150/L8
Hualaihué, Chile 138/E4
Hualalai (mt.), Hawaii 218/G5
Hualañé, Chile 138/A10
Hualapai (mts.), Ariz. 198/B4
Hualapai (peak), Ariz. 198/B3
Hualapai Ind. Res., Ariz. 198/B3
Hualgayoc, Peru 128/C6
Hualien, China 77/K7
Hualla, Peru 128/F9
Huallaga (riv.), Peru 120/B3
Huallaga (riv.), Peru 128/D5
Huallanca, Ancash, Peru 128/D7
Huallanca, Huánuco, Peru 128/D7
Huallen, Alberta 182/A2
Huamachuco, Peru 128/D6
Huamantla, Mexico 150/N1
Huambo (dist.), Angola 115/C6
Huambo, Angola 102/D6
Huambo, Angola 115/C6
Huambo, Angola 2/K6
Huanaqui, Bolivia 136/A7
Huanay, Bolivia 136/B4
Huancabamba, Peru 128/C5
Huancané, Bolivia 136/B6
Huancané, Peru 128/H10
Huancapi, Peru 128/E9
Huancavelica (dept.), Peru 128/E9
Huancavelica, Peru 120/B4
Huancavelica, Peru 128/E9
Huancayo, Peru 128/E9
Huancayo, Pru 120/B4
Huanchaca, Bolivia 136/B7
Huanchaca, Cerro (mt.), Bolivia 136/B7
Huanchaca, Serranía de (mts.), Bolivia 136/E4
Huanchaco, Peru 128/C7
Huanggang, China 77/J5
Huang He (riv.), China 2/Q4
Huang He (Hwang Ho) (riv.), China 54/N6
Huang He (Ma Qu) (riv.), China 77/F5
Huang He (Yellow) (riv.), China 77/J4
Huangling, China 77/G4
Huangliu, China 77/G8
Huangshi, China 77/J5
Huangzhong, China 77/F4
Huanqueros, Argentina 143/F5
Huanta, Peru 128/E9
Huánuco (dept.), Peru 128/D7
Huánuco, Peru 128/E7
Huánuco, Peru 120/B3
Huanuni, Bolivia 136/B6
Huanuni, Bolivia 120/C4
Huan Xian, China 77/G4
Huapai, N. Zealand 100/B1
Huapí (mts.), Nicaragua 154/E4
Huaquechula, Mexico 150/M2
Huara, Chile 138/B2
Huaral, Peru 128/D8
Huaráz, Peru 128/D7
Huaráz, Peru 120/B3
Huari, Bolivia 136/B6
Huari, Peru 128/D7
Huariaca, Peru 128/E8
Huarina, Bolivia 136/A5
Huarmey, Peru 128/C8
Huarochirí, Peru 128/D9
Huarocondo, Peru 128/F9
Huásabas, Mexico 150/E2
Huasaga (riv.), Peru 128/D4
Huascarán (mt.), Peru 120/B3
Huascarán (mt.), Peru 128/D7
Huasco, Chile 138/A7
Huasco (riv.), Chile 138/A7
Huatabampo, Mexico 150/D3
Huatunas (lag.), Bolivia 136/B3
Huatusco de Chicuellar, Mexico 150/P2
Huauchinango, Mexico 150/L6
Huaura, Peru 128/D8
Hueutla de Jiménez, Mexico 150/L7
Huayabamba (riv.), Peru 128/D6
Huaylas, Peru 128/C7
Huayllas, Bolivia 136/C6
Hub, Miss. (†39429) 256/E8
Hubball, W. Va. (†25506) 312/B6
Hubbard, Iowa (50122) 229/G4
Hubbard (lake), Mich. 250/F4
Hubbard (co.), Minn. 255/D3
Hubbard, Minn. (†56470) 255/C4
Hubbard, Nebr. (68741) 264/H2
Hubbard, Ohio (44425) 284/J3
Hubbard, Oreg. (97032) 291/A3
Hubbard, Sask. 181/H4
Hubbard, Texas (76648) 303/H6
Hubbard Creek (lake), Texas 303/F5
Hubbard Lake, Mich. (49747) 250/F4
Hubbards, Nova Scotia 168/D4
Hubbardston○, Mass. (01452) 249/F3
Hubbardston, Mich. (48845) 250/E5
Hubbardstown, W. Va. (†25555) 312/A6
Hubbardsville, N.Y. (13355) 276/J5
Hubbardton○, Vt. (05749) 268/A4
Hubbart (pt.), Manitoba 179/K2
Hubbell, Mich. (49934) 250/A1
Hubbell, Nebr. (68375) 264/G4
Hubbell Trading Post Nat'l Hist. Site, Ariz. 198/F3
Hub City, Wis. (†53581) 317/F9
Hubei (Hupei) (prov.), China 77/H5
Huberdeau, Québec 172/C4
Huber Heights, Ohio (45424) 284/B6
Hubert, N.C. (28539) 281/P5
Hubertus, Wis. (53033) 317/K1
Hubli-Dharwar, India 68/C5
Hubli-Dharwar, India 54/J8
Huch'ang, N. Korea 81/C3
Hückelhoven, W. Germany 22/B3
Hucknall, England 13/F4
Huddersfield, England 10/G2
Huddersfield, England 13/J2
Huddinge, Sweden 18/H1
Huddleston, Va. (24104) 307/K6
Huddy, Ky. (41535) 237/S5
Hudiksvall, Sweden 18/K6
Hudson (bay) 162/H3
Hudson (str.) 162/J3
Hudson (bay), Canada 2/E3
Hudson (str.), Canada 146/L3
Hudson (bay), Canada 146/K3
Hudson, Colo. (80642) 208/K2
Hudson, Fla. (33568) 212/D3
Hudson, Ill. (61748) 222/E3
Hudson, Ind. (46747) 227/G1
Hudson, Iowa (50643) 229/H4
Hudson, Kansas (67545) 232/D3
Hudson, Ky. (40145) 237/J5
Hudson○, Maine (04449) 243/F5
Hudson (bay), Manitoba 179/K2
Hudson, Md. (†21613) 245/N6
Hudson, Mass. (01749) 249/H3
Hudson○, Mass. (01749) 249/H3
Hudson, Mich. (49247) 250/E7
Hudson, N.H. (03051) 268/E6
Hudson○, N.H. (03051) 268/E6
Hudson (co.), N.J. 273/E2
Hudson (riv.), N.J. 273/C1
Hudson, N.Y. (12534) 276/N6
Hudson (riv.), N.Y. 276/N7
Hudson, N.C. (28638) 281/G3
Hudson (bay), N.W. Terrs. 187/K3
Hudson (str.), N.W. Terrs. 187/L3
Hudson, Ohio (44236) 284/H3
Hudson (lake), Okla. 288/R2
Hudson, Ontario 175/B2
Hudson, Ontario 177/G4
Hudson (bay), Ontario 175/D1
Hudson, Québec 172/C4
Hudson (bay), Québec 174/A1
Hudson (str.), Québec 174/F1
Hudson, S. Dak. (57034) 298/R7
Hudson, Wis. (54016) 317/A6
Hudson, Wyo. (82515) 319/D3
Hudson Bay, Sask. 181/J3
Hudson Falls, N.Y. (12839) 276/O4
Hudson Hope, Br. Col. 184/F2
Hudson Lake, Ind. (46552) 227/D1
Hudsons Bay, Alberta 182/E4
Hudsonville, Mich. (49426) 250/D6
Hudspeth (co.), Texas 303/B10
Hudwin (lake), Manitoba 179/G1
Hue, Vietnam 54/M8
Hue, Vietnam 72/E3
Hueco (mts.), N. Mex. 274/D6
Hueco (mts.), Texas 303/B10
Huedin, Romania 45/F2
Huehue, Hawaii (†96740) 218/G5
Huehuetenango, Guatemala 154/B3
Huehuetlán el Chico, Mexico 150/M2
Huejotzingo, Mexico 150/M1
Huejutla, Mexico 150/K6
Huelma, Spain 33/E4
Huelva (prov.), Spain 33/C4
Huelva, Spain 33/C4
Huelva, Spain 7/D5
Huelva (riv.), Spain 33/C4
Huentelauquén, Chile 138/A8
Huercal-Overa, Spain 33/F4
Huerfano (co.), Colo. 208/K7
Huerfano (riv.), Colo. 208/L7
Huesca (prov.), Spain 33/F1
Huesca, Spain 33/F1
Huéscar, Spain 33/E4
Huetamo, Mexico 150/J7
Huete, Spain 33/E2
Huetter, Idaho (†83854) 220/B2
Huey, Ill. (62252) 222/D5
Hueyotlipan de Hidalgo, Mexico 150/M1
Hueytown, Ala. (35020) 195/D4
Huff, N. Dak. (58555) 282/J6
Huffman, Ark. (†72315) 202/L2
Huffton, S. Dak. (†57432) 298/N2
Huger, S.C. (29450) 296/H5
Huggins, Mo. (65484) 261/H8
Hugh Butler (lake), Nebr. 264/D4
Hughenden, Alberta 182/E3
Hughenden, Australia 87/E8
Hughenden, Queensland 88/G4
Hughenden, Queensland 95/B4
Hughes, Alaska (99745) 196/H1
Hughes, Ark. (72348) 202/J4
Hughes (co.), Okla. 288/O4
Hughes, S. Australia 94/A4
Hughes (co.), S. Dak. 298/J5
Hughes (riv.), W. Va. 312/D4
Hughes Springs, Texas (75656) 303/K5
Hughestown, Pa. (†18640) 294/F7
Hughesville, Md. (20637) 245/L6
Hugheville, Mo. (65334) 261/F5
Hughesville, Pa. (17737) 294/J3
Hughson, Calif. (95326) 204/E6
Hughton, Sask. 181/D4
Hugh Town, England 13/A8
Hugo, Colo. (80821) 208/N4
Hugo, Minn. (55038) 255/E5
Hugo, Okla. (74743) 288/P7
Hugo (lake), Okla. 288/R6
Hugo Stroessner, Paraguay 144/C4
Hugoton, Kansas (67951) 232/A4
Huhehot (Hohhot), China 77/H3
Huiarau (range), N. Zealand 100/F3
Hŭich'ŏn, N. Korea 81/C3
Huila (dist.), Angola 115/B7
Huila (dept.), Colombia 126/C6
Huila (mt.), Colombia 120/B2
Huila, Nevado del (mt.), Colombia 126/C6
Huimanguillo, Mexico 150/N8
Huimin, China 77/J4
Huinca Renancó, Argentina 143/D3
Huining, China 77/G4
Huissen, Netherlands 27/H5
Huitzilán, Mexico 150/O1
Huitzuco de los Figueroa, Mexico 150/K7
Huixcolotla, Mexico 150/N2
Huixtepec, Mexico 150/L8
Huixtla, Mexico 150/N9
Huize, China 77/F6
Huizen, Netherlands 27/G4
Huizhou, China 77/H7
Hulaco, Ala. (†35087) 195/E2
Hulah (lake), Kansas 232/F5
Hulah, Okla. (†67333) 288/O1
Hulah (lake), Okla. 288/O1
Hulan, China 77/L2
Hulbert, Mich. (49748) 250/D2
Hulbert, Okla. (74441) 288/R3
Hulberton, N.Y. (14473) 276/D4
Hulett, Wyo. (82720) 319/H1
Hulin, China 77/M2
Hull, England 7/E3
Hull, England 10/F4
Hull, England 13/G4
Hull, Fla. (†33842) 212/E4
Hull, Georgia (30646) 217/F2
Hull, Ill. (62343) 222/B4
Hull, Iowa (51239) 229/A2
Hull (isl.), Kiribati 87/J6
Hull○, Mass. (02045) 249/E7
Hull, N. Dak. (†58542) 282/K7
Hull, Que. 162/J6
Hull (co.), Québec 172/B4
Hull, Québec 172/B4
Hulls Cove, Maine (04644) 243/G7
Hulopee (bay), Hawaii 218/G2
Hulopoe Bay, Hawaii (†96763) 218/H2
Hulst, Netherlands 27/E6
Hultsfred, Sweden 18/K8
Hulun Nur (lake), China 77/J2
Huma, China 77/L1
Humacao (dist.), P. Rico 161/E2
Humacao, P. Rico 161/E2
Humacao, P. Rico 156/G1
Humacao (riv.), P. Rico 161/E2
Huma He (riv.), China 77/K1
Humahuaca, Argentina 143/C1
Humaitá, Bolivia 136/B2
Humaitá, Brazil 132/H10
Humaitá, Brazil 120/C3
Humaitá, Brazil 132/G10
Humaitá, Paraguay 144/C5
Humansdorp, S. Africa 118/C6
Humansville, Mo. (65674) 261/E7
Humarock, Mass. (02047) 249/M4
Humber (riv.), England 13/G4
Humber (riv.), England 10/G4
Humber (riv.), Newf. 166/C4
Humber (riv.), Ontario 177/J3
Humberside (co.), England 13/G4
Humberto, Argentina 143/F5
Humbird, Wis. (54746) 317/E6
Humble, Texas (*77338) 303/J7
Humble City, N. Mex. (†88240) 274/F6
Humboldt, Ariz. (86329) 198/C4
Humboldt (co.), Calif. 204/B3
Humboldt (bay), Calif. 204/A3
Humboldt (bay), Colombia 126/B4
Humboldt, Ill. (61931) 222/E4
Humboldt (co.), Iowa 229/E3
Humboldt, Iowa (50548) 229/E3
Humboldt, Kansas (66748) 232/G4
Humboldt, Minn. (56731) 255/A2
Humboldt, Nebr. (68376) 264/J4
Humboldt (co.), Nev. 266/C1
Humboldt, Nev. (†89418) 266/C2
Humboldt (range), Nev. 266/C2
Humboldt (riv.), Nev. 188/C2
Humboldt (riv.), Nev. 266/E2
Humboldt (sink), Nev. 266/C2
Humboldt (mt.), New Caled. 86/H4
Humboldt, Sask. 162/F5
Humboldt, Sask. 181/F3
Humboldt, S. Dak. (57035) 298/P6
Humboldt, Tenn. (38343) 237/D9
Humboldt Salt (marsh), Nev. 266/D3
Humbug (mt.), Oreg. 291/C5
Hume, Ill. (61932) 222/F4
Hume, Mo. (64752) 261/C6
Hume (res.), N.S. Wales 97/D4
Hume, N.Y. (14745) 276/D5
Hume, Sask. 181/H6
Hume (lake), Victoria 97/D4
Hume, Va. (22639) 307/N3
Humenné, Czech. 41/G2
Humeston, Iowa (50123) 229/G7
Humlum, Denmark 21/B4
Hummelstown, Pa. (17036) 294/J5
Hummock (isl.), Tasmania 99/D2
Humnoke, Ark. (72072) 202/G4
Humphrey (pt.), Alaska 196/K1
Humphrey, Ark. (72073) 202/G5
Humphrey, Idaho (†83446) 220/F5
Humphrey, Nebr. (68642) 264/G3
Humphreys (peak), Ariz. 198/D3
Humphreys, La. (†70356) 238/J7
Humphreys (co.), Miss. 256/C4
Humphreys, Mo. (64646) 261/F2
Humphreys, Okla. (†73521) 288/H5
Humphreys (co.), Tenn. 237/F8
Humpolec, Czech. 41/C2
Humptulips, Wash. (98552) 310/A3
Humptulips (riv.), Wash. 310/B3
Humpty Doo, North. Terr. 93/B2
Húnaflói (bay), Iceland 7/B2
Húnaflói (bay), Iceland 21/B1
Hunan (prov.), China 77/H6
Hunchun, China 77/M3
Hundested, Denmark 21/E6
Hundred, W. Va. (26575) 312/E3
Hunedoara, Romania 7/G4
Hunedoara, Romania 45/F3
Hünfeld, W. Germany 22/C3
Hungary 2/K3
Hungary 7/F4
HUNGARY 41
Hunger (mt.), Vt. 268/B3
Hungerford, Queensland 95/B6
Hŭngnam, N. Korea 54/O6
Hungry Horse, Mont. (59919) 262/C2
Hungry Horse (res.), Mont. 262/C2
Hungtow (isl.), China 77/K7
Hunjiang, China 77/L3
Hunmanby, England 13/G3
Hunnewell, Kansas (†67140) 232/E4
Hunnewell, Mo. (63443) 261/J3
Hunse (riv.), Netherlands 27/K3
Hunsrück (mts.), W. Germany 22/B4
Hunstanton, England 13/H5
Hunstanton, England 10/G4
Hunt, Ill. (†62480) 222/E4
Hunt (co.), Texas 303/H4
Hunt (mt.), Wyo. 319/E1
Hunte (riv.), W. Germany 22/C2
Hunter, Ark. (72074) 202/H3
Hunter (isl.), Br. Col. 184/C4
Hunter (peak), Idaho 220/D3
Hunter, Kansas (67452) 232/D2
Hunter, Mo. (†63943) 261/L9
Hunter (riv.), N.S. Wales 97/F3
Hunter, N.Y. (12442) 276/M6
Hunter (mt.), N.Y. 276/M6
Hunter (mts.), N. Zealand 100/A6
Hunter, N. Dak. (58048) 282/R5
Hunter, Okla. (74640) 288/L1
Hunter (isls.), Tasmania 88/G8
Hunter (isl.), Tasmania 99/A2
Hunter (isls.), Tasmania 99/B2
Hunterdon (co.), N.J. 273/D2
Hunter River, Pr. Edward I. 168/E2
Hunters, Wash. (99137) 310/G2
Hunters Creek Village, Texas (†77001) 303/J1
Hunters Hill, N. S. Wales 88/K4
Hunters Hill, N.S. Wales 97/J3
Huntersville, Ky. (†42602) 237/L7
Huntersville, Minn. (†56464) 255/D4
Huntersville, N.C. (28078) 281/H4
Huntersville, W. Va. (†24954) 312/G6
Huntertown, Ind. (46748) 227/G2
Hunterville, N. Zealand 100/E3
Hunting (riv.), N.C. 281/H2
Hunting (isl.), S.C. 296/G7
Huntingburg, Ind. (47542) 227/D8
Huntingdon, Br. Col. 184/L3
Huntingdon (isl.), Newf. 166/C3
Huntingdon (co.), Pa. 294/F5
Huntingdon, Pa. (16652) 294/G5
Huntingdon (co.), Québec 172/C4
Huntingdon, Québec 172/C4
Huntingdon, Tenn. (38344) 237/E8
Huntingdon and Godmanchester, England 13/G5
Huntingdon and Godmanchester, England 10/F4
Huntington, Ark. (72940) 202/B3
Huntington, Conn. (†06484) 210/C3
Huntington, England 13/G3
Huntington (co.), Ind. 227/G3
Huntington, Ind. (46750) 227/G3
Huntington (lake), Ind. 227/F3
Huntington, Iowa (†51334) 229/D2
Huntington○, Mass. (01050) 249/C4
Huntington (creek), Nev. 266/F2
Huntington, N.J. (†08865) 273/C2
Huntington, N.Y. (11743) 276/R6
Huntington, Oreg. (97907) 291/K3
Huntington, Texas (75949) 303/K6
Huntington, Utah (84528) 304/C4
Huntington (creek), Utah 304/C4
Huntington○, Vt. (05462) 268/B3
Huntington, Va. (†22301) 307/S3
Huntington, W. Va. 188/K3
Huntington, W. Va. (*25701) 312/A6
Huntington Beach, Calif. (*92646) 204/C11
Huntington Center, Vt. (†05462) 268/B3
Huntington Park, Calif. (90255) 204/C11
Huntington Station, N.Y. (11746) 276/R6
Huntingtown, Md. (20639) 245/M6
Hunting Valley, Ohio (†44022) 284/J9
Huntland, Tenn. (37345) 237/J10
Huntleigh, Mo. (†63101) 261/O3
Huntley, Ill. (60142) 222/E1
Huntley, Minn. (56047) 255/D7
Huntley, Mont. (59037) 262/H5
Huntley, Nebr. (68951) 264/E4
Huntley, Wyo. (82218) 319/H4
Huntly, N. Zealand 100/E2
Huntly, Scotland 10/E2
Huntly, Scotland 15/F3
Huntoon, Sask. 181/H6
Huntsburg, Ohio (44046) 284/H2
Hunts Inlet, Br. Col. 184/B3
Hunts Point, Nova Scotia 168/D5
Hunts Point, Wash. (†98004) 310/B2
Huntsville, Ala. 188/J4
Huntsville, Ala. (*35801) 195/E1
Huntsville, Ark. (72740) 202/C1
Huntsville, Conn. (†06031) 210/B1
Huntsville, Ind. (†47358) 227/F4
Huntsville, Ind. (†46064) 227/G4
Huntsville, Ky. (42251) 237/H6
Huntsville, Mo. (65259) 261/H4
Huntsville, Ohio (43324) 284/C5
Huntsville, Ontario 177/E2
Huntsville, Ontario 175/E3
Huntsville, Tenn. (37756) 237/N8
Huntsville, Texas (77340) 303/J7
Huntsville, Utah (84317) 304/C2
Huntsville, Wash. (†99328) 310/G4
Hunucmá, Mexico 150/O6
Hunza (Baltit), Pakistan 68/C1
Huocheng, China 77/B3
Huon (isls.), New Caled. 87/G7
Huon (gulf), Papua N.G. 87/E6
Huon (gulf), Papua N.G. 85/C7
Huon (pen.), Papua N.G. 86/A2
Huon (riv.), Tasmania 99/C5
Huong Khe, Vietnam 72/E3
Huonville-Ranelagh, Tasmania 99/C5
Huoshan, China 77/J5
Huot, Minn. (†56716) 255/B3
Huo Xian, China 77/H4
Hupei (Hubei) (prov.), China 77/H5
Hurbanovo, Czech. 41/E3
Hurd (cape), Ontario 177/C2
Hurdland, Mo. (63547) 261/H2
Hurdle Mills, N.C. (27541) 281/L2
Hurdsfield, N. Dak. (58451) 282/L5

Hure, China 77/K3
Hureidha, P.D.R. Yemen 59/E6
Hurghada, Egypt 111/F2
Hurghada, Egypt 59/B4
Hurlburt, Fla. (†32548) 212/B6
Hurley, Miss. (39555) 256/H9
Hurley, Mo. (65675) 261/F9
Hurley, N. Mex. (88043) 274/A6
Hurley, N.Y. (12443) 276/M7
Hurley, S. Dak. (57036) 298/P7
Hurley, Va. (24620) 307/D6
Hurley, Wis. (54534) 317/F3
Hurleyville, N.Y. (12747) 276/L7
Hurlford, Scotland 15/D5
Hurlock, Md. (21643) 245/P6
Huron (lake) 146/K5
Huron (lake) 162/H7
Huron, Calif. (93234) 204/E7
Huron, Ind. (47437) 227/D7
Huron, Kansas (66038) 232/G2
Huron (lake), Mich. 188/K2
Huron (co.), Mich. 250/F5
Huron (bay), Mich. 250/A2
Huron (lake), Mich. 250/G4
Huron (riv.), Mich. 250/F6
Huron (co.), Ohio 284/E3
Huron, Ohio (44839) 284/E3
Huron (riv.), Ohio 284/E3
Huron (county), Ontario 177/C4
Huron (lake), Ontario 177/B3
Huron (lake), Ontario 175/D3
Huron, S. Dak. 188/G2
Huron, S. Dak. (57350) 298/N5
Huron, Tenn. (38345) 237/E9
Huron City, Mich. (†48467) 250/G4
Huron Mountain, Mich. (†49808) 250/B2
Huron Park, Ontario 177/C4
Huron River (pt.), Mich. 250/B2
Hurricane, Ala. (†36507) 195/C9
Hurricane (cliffs), Ariz. 198/B2
Hurricane (mt.), Utah 304/A6
Hurricane, Utah (84737) 304/A6
Hurricane, W. Va. (25526) 312/C6
Hurricane Deck, Mo. (†65079) 261/G6
Hurricane Mills, Tenn. (37078) 237/F9
Hurst, Georgia (†30560) 217/D1
Hurst, Ill. (62949) 222/D6
Hurst, Texas (76053) 303/F2
Hurst, W. Va. (†26445) 312/E4
Hurstville, Iowa (†52060) 229/M4
Hurstville, N. S. Wales 88/K4
Hurstville, N.S. Wales 97/J4
Hurt, Va. (24563) 307/K6
Hürth, W. Germany 22/B3
Hurtsboro, Ala. (36860) 195/H6
Hurunui (riv.), N. Zealand 100/D5
Hurup, Denmark 21/B4
Húsavík, Iceland 21/C1
Husher, Wis. (†53108) 317/L2
Hushpuckena, Miss. (†38774) 256/C2
Huşi, Romania 45/J2
Husk, N.C. (28639) 281/F1
Huskisson, N.S. Wales 97/F4
Huslia, Alaska (99746) 196/G1
Huson, Mont. (59846) 262/B3
Hussar, Alberta 182/D4
Hustisford, Wis. (53034) 317/J9
Hustler, Wis. (54637) 317/F8
Hustontown, Pa. (17229) 294/F5
Hustonville, Ky. (40437) 237/M6
Hustopeče, Czech. 41/D2
Husum, Sweden 18/L5
Husum, Wash. (98623) 310/D5
Husum, W. Germany 22/C1
Hutchins (mt.), N.H. 268/E2
Hutchins, Texas (75141) 303/G3
Hutchins, Wis. (†54450) 317/H6
Hutchinson, Kans. 188/G3
Hutchinson, Kansas (67501) 232/D3
Hutchinson, Kansas 146/J6
Hutchinson, Minn. (55350) 255/D6
Hutchinson (co.), S. Dak. 298/O7
Hutchinson (co.), Texas 303/C2
Hutchinson, W. Va. (†26591) 312/F4
Huth, Yemen Arab Rep. 59/D6
Hutsonville, Ill. (62433) 222/F4
Hutt (riv.), N. Zealand 100/C2
Hüttenberg, Austria 41/C3
Hüttental, W. Germany 22/C3
Hutte Sauvage (lake), Québec 174/E1
Huttig, Ark. (71747) 202/F7
Hutto, Texas (78634) 303/G7
Hutton, La. (†71402) 238/D4
Hutton, Md. (†21550) 245/A3
Huttonsville, W. Va. (26273) 312/G5
Hutton Valley, Mo. (†65793) 261/J9
Huttwil, Switzerland 39/E2
Hutubi, China 77/C3
Huumula, Hawaii (†96743) 218/H5
Huutokoski, Finland 18/P5
Huwelijkszorg, Suriname 131/C2
Huxford, Ala. (36543) 195/D8
Hu Xian, China 77/G5
Huxley, Alberta 182/D4
Huxley, Iowa (50124) 229/F5
Huxley, Texas (†75973) 303/L6
Huy, Belgium 27/G8
Huyton-with-Roby, England 13/G2
Hvannadalshnúkur (mt.), Iceland 21/C1
Hvar (isl.), Yugoslavia 45/C4
Hvidbjerg, Denmark 21/B4
Hvide Sande, Denmark 21/A6
Hviding, Denmark 21/B7
Hvítá (riv.), Iceland 21/B1
Hwainan (Huainan), China 77/J5
Hwaiteh (Huaide), China 77/K3
Hwange (Wankie), Zimbabwe 118/D3
Hwang Ho (riv.), China 54/N6
Hwangju, N. Korea 81/C4
Hwangshih (Huangshi), China 77/J5
Hyak, Wash. (98068) 310/D3
Hyalite (peak), Mont. 262/E5
Hyannis, Mass. (02601) 249/N6
Hyannis, Nebr. (69350) 264/C3
Hyannis Port, Mass. (02647) 249/N6
Hyargas, Mongolia 77/D2
Hyargas Nuur (lake), Mongolia 77/D2
Hyas, Sask. 181/J4
Hyattstown, Md. (20734) 245/J3
Hyattsville, Md. (*20780) 245/F4
Hyattville, Wyo. (82428) 319/E1
Hybart, Ala. (36452) 195/D7
Hybord, Manitoba 179/C1
Hyco (riv.), N.C. 281/L2
Hyco (riv.), Va. 307/K8
Hydaburg, Alaska (99922) 196/M2
Hyde, England 13/H2
Hyde, England 10/G2
Hyde, N. Zealand 100/C6
Hyde (co.), N.C. 281/S3
Hyde, Pa. (16843) 294/F4
Hyde (co.), S. Dak. 298/K4
Hyden, Ky. (41749) 237/P6
Hyden, W. Australia 92/B6
Hyde Park, Mass. (02136) 249/C7
Hyde Park, N.Y. (12538) 276/N6
Hyde Park, Ontario 177/C4
Hyde Park, Pa. (15641) 294/D4
Hyde Park, Utah (84318) 304/C2
Hyde Park, Vt. (05655) 268/B2
Hyde Park○, Vt. (05655) 268/B2
Hyder, Alaska (99923) 196/P2
Hyderabad, India 2/N5
Hyderabad, India 68/D5
Hyderabad, India 54/J8
Hyderabad, Pakistan 68/B3
Hyderabad, Pakistan 59/J4
Hyderabad, Pakistan 54/H7
Hydesville, Calif. (95547) 204/B3
Hydetown, Pa. (16328) 294/C2
Hydeville, Vt. (05750) 268/A4
Hydraulic, Br. Col. 184/F4
Hydro, Okla. (73048) 288/J3
Hye, Texas (78635) 303/F7
Hyères, France 28/G6
Hyères (isls.), France 28/G6
Hyesan, N. Korea 81/D3
Hygiene, Colo. (80533) 208/J2
Hyland (riv.), Yukon 187/F3
Hylo, Alberta 182/D2
Hyltebruk, Sweden 18/H8
Hyman, S.C. (†29583) 296/H4
Hymer, Kansas (†66869) 232/F3
Hymera, Ind. (47855) 227/C6
Hyndman (peak), Idaho 220/D6
Hyndman, Pa. (15545) 294/E6
Hyner, Pa. (17738) 294/G3
Hynish (bay), Scotland 15/B4
Hyogo (pref.), Japan 81/H7
Hypoluxo, Fla. (†33460) 212/F5
Hyrra Banda, Cent. Afr. Rep. 115/D2
Hyrum, Utah (84319) 304/C2
Hyrynsalmi, Finland 18/Q4
Hysham, Mont. (59038) 262/J4
Hythe, Alberta 182/A2
Hythe, Alta. 162/E4
Hythe, England 10/G5
Hythe, England 13/H6
Hythe, Tasmania 99/C5
Hytop, Ala. (35753) 195/F1
Hyuga, Japan 81/E7
Hyvinkää, Finland 18/O6

I

Ia Drang (riv.), Vietnam 72/E4
Iaeger, W. Va. (24844) 312/C8
Ialomiţa (marshes), Romania 45/J3
Ialomiţa (riv.), Romania 45/H3
Iamonia (lake), Fla. 212/B1
Iantha, Mo. (64753) 261/D8
Iar Connacht (dist.), Ireland 17/C5
Iaşi, Romania 7/G4
Iaşi, Romania 45/H2
Iatan, Mo. (†64098) 261/C4
Iatt (lake), La. 238/E3
Iba, Philippines 85/F4
Iba, Philippines 82/B3
Ibadan, Nigeria 2/K5
Ibadan, Nigeria 102/C4
Ibadan, Nigeria 106/E7
Ibagué, Colombia 126/C5
Ibagué, Colombia 120/B2
Ibaiti, Brazil 135/A3
Ibapah, Utah (84034) 304/A3
Ibar (riv.), Yugoslavia 45/E4
Ibaraki (pref.), Japan 81/K5
Ibaraki, Japan 81/J7
Ibarra, Ecuador 128/D2
Ibarra, Ecuador 120/B2
Ibarreta, Argentina 143/D2
Ibb, Yemen Arab Rep. 59/D7
Ibbenbüren, W. Germany 22/B2
'Ibbin, Jordan 65/D3
Iberia (par.), La. 238/G7
Iberia, Mo. (65486) 261/H6
Iberia, Ohio (43325) 284/E4
Iberia, Peru 128/F5
Iberville (par.), La. 238/H6
Iberville, La. (70746) 238/K2
Iberville (co.), Québec 172/D4
Iberville, Québec 172/D4
Iberville, D' (lake), Québec 174/C1
Ibi, Nigeria 106/F7
Ibiá, Brazil 132/E7
Ibibobo, Bolivia 136/D7
Ibicaraí, Brazil 132/G6
Ibicuí (riv.), Brazil 132/C10
Ibicuy, Argentina 143/G6
Ibipetuba, Brazil 132/F5
Ibitinga, Brazil 135/B2
Ibiza, Spain 33/G3
Ibiza (isl.), Spain 7/E5
Ibiza (isl.), Spain 33/G3
Ibo, Bolivia 136/D7
Ibo, Mozambique 118/G2
Ibounzi (mt.), Gabon 115/B4
Ibra, Oman 59/G5
Ibra, Wadi (dry riv.), Sudan 111/D5
Ibrány, Hungary 41/F2
'Ibri, Oman 59/G5
Ibusuki, Japan 81/E8
Içá (riv.), Brazil 120/C3
Içá (riv.), Brazil 132/G9
Ica (dept.), Peru 128/E10
Ica, Peru 128/E10
Ica, Peru 120/B4
Ica (riv.), Peru 128/E10
Icabarú, Venezuela 124/H5
Icabarú (riv.), Venezuela 124/G5
Icacos (pt.), Trin. & Tob. 161/A11
Icaño, Catamarca, Argentina 143/C2
Icaño, Santiago del Estero, Argentina 143/D2
Icard, N.C. (28666) 281/G3
Ice Harbor (dam), Wash. 310/G4
Içel (prov.), Turkey 63/F4
Içel (Mersin), Turkey 63/F4
Iceland 2/J2
Iceland 7/C2
Iceland 4/C10
ICELAND 21/B1
Ichang (Yichang), China 77/H5
Ichchapuram, India 68/F5
Ichhapur, India 68/F1
Ichihara, Japan 81/P3
Ichikawa, Japan 81/P2
Ichilo (riv.), Bolivia 136/C5
Ichinohe, Japan 81/K3
Ichinomiya, Japan 81/H6
Ichinoseki, Japan 81/K4
Ichnya, U.S.S.R. 52/D4
Ichoa (riv.), Bolivia 136/C4
Ichoca, Bolivia 136/B5
Ichtegem, Belgium 27/B6
Ichun (Yichun), China 77/L2
Ichuña, Peru 128/G11
Icicle (creek), Wash. 310/E3
Ickesburg, Pa. (17037) 294/H5
Icla, Bolivia 136/C6
Içme, Turkey 63/H3
Icó, Brazil 132/G4
Iconium, Mo. (†64776) 261/E6
Icy (bay), Alaska 196/K3
Icy (cape), Alaska 196/F1
Icy (cape), Alaska 196/K3
Icy (pt.), Alaska 196/L1
Icy (str.), Alaska 196/M1
Ida (co.), Iowa 229/C4
Ida, La. (71044) 238/C1
Ida, Mich. (48140) 250/F7
Idabel, Okla. (74745) 288/S7
Ida Grove, Iowa (51445) 229/B4
Idaho 188/D2
IDAHO 220
Idaho (co.), Idaho 220/C4
Idaho, 202
Idaho, Ohio (45661) 284/D7
Idaho (state), U.S. 146/G5
Idaho City, Idaho (83631) 220/C6
Idaho Falls, Idaho 146/G5
Idaho Falls, Idaho 188/D2
Idaho Falls, Idaho (*83401) 220/F6
Idaho Springs, Colo. (80452) 208/H3
Idahue, Chile 138/F5
Idalia, Colo. (80735) 208/P3
Idalou, Texas (79329) 303/C4
Idana, Kansas (67432) 232/E2
Idanha, Oreg. (97350) 291/E3
Idanha-a-Nova, Portugal 33/C3
Idar-Oberstein, W. Germany 22/B4
Idaville, Ind. (47950) 227/D3
Idaville, Pa. (17337) 294/H5
Iddan, Somalia 115/J2
Iddesleigh, Alberta 182/E4
Ide, Japan 81/J7
Ideal, Georgia (31041) 217/D6
Ideal, S. Dak. (57541) 298/K6
Idehan Murzuk (des.), Libya 111/B2
Idehan Ubari (des.), Libya 111/B2
Idelès, Algeria 106/F4
Ider, Ala. (35981) 195/G1
Ider Gol (riv.), Mongolia 77/E2
Idfu, Egypt 111/F3
Idfu, Egypt 59/B5
Ídhi (mt.), Greece 45/G8
Ídhra, Greece 45/F7
Idi, Indonesia 85/B4
Idil, Turkey 63/J4
Idiofa, Zaire 115/C4
Idlewild, Mich. (49642) 250/D5
Idlewild, Tenn. (38346) 237/D8
Idleyld Park, Oreg. (97447) 291/D4
Idlib (prov.), Syria 63/G5
Idlib, Syria 63/G5
Idna, West Bank 65/B4
Idrigill (pt.), Scotland 15/B3
Idyllwild-Pine Cove, Calif. (92349) 204/J10
Ie (isl.), Japan 81/N6
Ieper, Belgium 27/B7
Ierápetra, Greece 45/G8
Iet, Somalia 115/H3
Ifakara, Tanzania 115/G5
Ifalik (atoll), Micronesia 87/E5
Ifanadiana, Madagascar 118/H4
Ife, Nigeria 106/E7
Iférouane, Niger 102/C3
Iférouane, Niger 106/F5
Iffley, Sask. 181/C3
Ifni, Morocco 106/B3
Ifni, Morocco 102/B2
Ifugao (prov.), Philippines 82/C2
Igal, Hungary 41/D3
Igara-Paraná (riv.), Colombia 126/D8
Igarapava, Brazil 135/C2
Igarapé-Miri, Brazil 132/D3
Igarka, U.S.S.R. 4/C5
Igarka, U.S.S.R. 54/K3
Igarka, U.S.S.R. 48/J3
Iğdır, Turkey 63/K3
Iggesund, Sweden 18/K6
Igis, Switzerland 39/J3
Igiugig, Alaska (†99613) 196/G3
Iglesias, Italy 34/B5
Igli, Algeria 106/D2
Igloo, S. Dak. (†57774) 298/B7
Igloolik, Canada 4/B14
Igloolik, N.W.T. 162/H2
Igloolik, N.W. Terrs. 187/K3
Iglosiatik (isl.), Newf. 166/B2
Ignace, Ont. 162/G6
Ignace, Ontario 175/B3
Ignace, Ontario 177/G5
Ignacio, Calif. (94947) 204/H1
Ignacio, Colo. (81137) 208/D8
Ignacio Agramonte, Cuba 158/G3
Ignacio de la Llave, Mexico 150/Q2
İğneada (cape), Turkey 63/C2
Igoumenítsa, Greece 45/E6
Igra, U.S.S.R. 52/H3
Igrim, U.S.S.R. 48/G3
Iguaçu (riv.) 120/D5
Iguaçu (riv.), Brazil 132/C9
Igualada, Spain 33/G2
Iguala de la Independencia, Mexico 150/K7
Iguape, Brazil 135/C4
Iguassú (falls) 120/D5
Iguatu, Brazil 120/F3
Iguatu, Brazil 132/G4
Iguazú (falls), Argentina 143/F2
Iguazú (falls), Brazil 132/C9
Iguazú (falls), Paraguay 144/E4
Iguazú Nat'l Park, Argentina 143/E2
Iguéla, Gabon 115/A4
Iguidi, Erg (des.), 102/B2
Iguidi, Erg (des.), Algeria 106/C3
Iguidi, Erg (des.), Mauritania 106/C3
Iheya (isl.), Japan 81/N6
Ihlen, Minn. (56140) 255/B7
Ihosy, Madagascar 118/H4
Ihu, Papua N.G. 85/B7
Ii (riv.), Finland 7/G2
Iida, Japan 81/H6
Iijoki (riv.), Finland 18/O4
Iisalmi, Finland 18/P5
Iizuka, Japan 81/E7
Ijamsville, Md. (21754) 245/J3
Ijebu-Ode, Nigeria 106/E7
IJlst, Netherlands 27/H2
IJmeer (bay), Netherlands 27/C4
IJmuiden, Netherlands 27/E4
IJssel (riv.), Netherlands 27/J4
IJsselmeer (lake), Netherlands 27/G3
IJsselstein, Netherlands 27/F4
Ijuí, Brazil 132/C10
IJzendijke, Netherlands 27/D6
Ikaalinen, Finland 18/N6
Ikaría (isl.), Greece 45/G7
Ikast, Denmark 21/C5
Ikeda, Hokkaido, Japan 81/L2
Ikeda, Osaka, Japan 81/H7
Ikeja, Nigeria 106/E7
Ikela, Zaire 115/D4
Ikelemba, Congo 115/C3
Ikhtiman, Bulgaria 45/F4
Iki (isl.), Japan 81/D7
Ikom, Nigeria 106/F7
Ikoma, Japan 81/J8
Ikopa (riv.), Madagascar 118/H3
Ikpikpuk (riv.), Alaska 196/H1
Iksal, Israel 65/C2
Ikuno, Japan 81/G6
Ila, Georgia (30647) 217/F2
Ilagan, Philippines 82/C2
Ilam (gov.), Iran 66/E4
Ilam, Iran 66/E4
Ilam, Nepal 68/F3
Ilan, China 77/K7
Ilanskiy, U.S.S.R. 48/K4
Ilanz, Switzerland 39/H3
Ilaro, Nigeria 106/E7
Ilasco, Mo. (†63401) 261/K3
Ilava, Czech. 41/E2
Ilave, Peru 128/H11
Iława, Poland 47/D2
Ilderton, Ontario 177/C4
Île-à-la-Crosse, Sask. 181/L3
Île-à-la-Crosse (lake), Sask. 181/L3
Île-Bizard, Québec 172/H4
Ilebo, Zaire 115/D4
Île de France (trad. prov.), France, 29
Île-de-Montréal (co.), Québec 172/H4
Île des Chênes, Manitoba 179/F5
Île-Jésus (co.), Québec 172/H4
Ilek (riv.), U.S.S.R. 52/J4
Île-Perrot, Québec 172/G4
Îles (lake), Québec 172/B3
Ilesha, Nigeria 106/F7
Ilfeld, N. Mex. (87538) 274/D3
Ilfis (riv.), Switzerland 39/E3
Ilford, Manitoba 179/J2
Ilford, N.S. Wales 97/E3
Ilfracombe, England 10/D5
Ilfracombe, England 13/C6
Ilgaz, Turkey 63/E2
Ilgaz (mts.), Turkey 63/E2
Ilgın, Turkey 63/D3
Ilha Grande (bay), Brazil 135/D3
Ílhavo, Portugal 33/B2
Ilhéus, Brazil 120/F4
Ilhéus, Brazil 132/G6
Ili (riv.), U.S.S.R. 54/J5
Ili (riv.), U.S.S.R. 48/H5
Iliamna, Alaska (99606) 196/G3
Iliamna (lake), Alaska 188/C6
Iliamna (lake), Alaska 196/G3
Iliamna (vol.), Alaska 196/H2
Iliç, Turkey 63/H3
Ilıca, Turkey 63/J3
Iliff, Colo. (80736) 208/N1
Iligan, Philippines 82/E6
Iligan (bay), Philippines 82/E6
Ilin (isl.), Philippines 82/C4
Ilio (pt.), Hawaii 218/G1
Ilion, N.Y. (13357) 276/K5
Ilium (ruins), Turkey 63/B6
Ilkeston, England 13/F5
Ilkeston, England 10/F4
Illabo, N.S. Wales 97/D4
Illahe, Oreg. (†97406) 291/C5
Illampu, Nevado (mt.), Bolivia 136/A4
Illana (bay), Philippines 82/D7
Illana, Spain 33/E2
Illapel, Chile 138/A8
Ille-et-Vilaine (dept.), France 28/C3
Illéla, Niger 106/F6
Iller (riv.), W. Germany 22/D4
Illescas, Spain 33/D2
Illescas, Uruguay 145/D4
Ille-sur-Têt, France 28/E6
Illimani, Nevada (mt.), Bolivia 136/B5
Illinois 188/J3
ILLINOIS 222
Illinois (bayou), Ark. 202/D3
Illinois (riv.), Colo. 208/G1
Illinois (riv.), Ill. 188/H2
Illinois (riv.), Ill. 222/C4
Illinois (riv.), Okla. 288/S3
Illinois (riv.), Oreg. 291/D5
Illinois (state), U.S. 146/K6
Illinois - Mississippi (canal), Ill. 222/C2
Illiopolis, Ill. (62539) 222/D4
Illizi, Algeria 106/F3
Illizi, Algeria 102/C2
Illmo, Mo. (63754) 261/O8
Illnau, Switzerland 39/G2
Illora, Spain 33/E4
Illuka, N.S. Wales 97/G1
Il'men (lake), U.S.S.R. 7/H3
Il'men' (lake), U.S.S.R. 52/D3
Ilmenau, E. Germany 22/D3
Ilmenau (riv.), W. Germany 22/D2
Ilminster, England 13/D7
Ilo, Peru 128/G11
Ilobasco, El Salvador 154/C4
Ilocos Norte (prov.), Philippines 82/C1
Ilocos Sur (prov.), Philippines 82/C2
Iloilo (prov.), Philippines 82/D5
Iloilo, Philippines 85/G3
Iloilo, Philippines 82/D5
Iloilo, Philippines 54/O8
Iloilo (str.), Philippines 82/D5
Ilomantsi, Finland 18/R5
Ilorin, Nigeria 106/E7
Ilorin, Nigeria 102/C4
Ilpendam, Netherlands 27/C4
Ilsley, Ky. (†42408) 237/F6
Ilubabor (prov.), Ethiopia 111/F6
Ilükste, U.S.S.R. 53/D3
Ilwaco, Wash. (98624) 310/A4
Iłża, Poland 47/E3
Imabari, Japan 81/F6
Imandra (lake), U.S.S.R. 48/D3
Imandra (lake), U.S.S.R. 52/D1
Imari, Japan 81/D7
Imataca, Serranía (mts.), Venezuela 124/H4
Imatra, Finland 18/Q6
Imazu, Japan 81/G6
Imbâba, Egypt 111/J3
Imbabura (prov.), Ecuador 128/C2
Imbaimadai, Guyana 131/A3
Imbert, Dom. Rep. 158/D5
Imbituba, Brazil 132/D10
Imbituva, Brazil 135/A4
Imbler, Oreg. (97841) 291/J2
Imboden, Ark. (72434) 202/H1
Imeri, Sierra (mts.), Venezuela 124/F7
Imese, Zaire 115/C3
Imi, Ethiopia 111/H6
Imías, Cuba 158/K4
Imilac, Chile 138/B4
Imishli, U.S.S.R. 52/G7
Imlay, Nev. (89418) 266/C2
Imlay, S. Dak. (†57780) 298/E6
Imlay City, Mich. (48444) 250/F5
Imlaystown, N.J. (08526) 273/D3
Imler, Pa. (16655) 294/E5
Immaculata, Pa. (19345) 294/L6
Immenstadt im Allgäu, W. Germany 22/C5
Immingham, England 13/G4
Immokalee, Fla. (33934) 212/E5
Imnaha, Oreg. (97842) 291/L2
Imnaha (riv.), Oreg. 291/L2
Imo (state), Nigeria 106/F7
Imogene, Iowa (51645) 229/C7
Imola, Italy 34/C2
Impach, Wash. (†99138) 310/G2
Imperatriz, Brazil 120/E3
Imperatriz, Brazil 132/E4
Imperia (prov.), Italy 34/B3
Imperia, Italy 34/B3
Imperial (dam), Ariz. 198/A6
Imperial (res.), Ariz. 198/A6
Imperial (co.), Calif. 204/K10
Imperial, Calif. (92251) 204/K11
Imperial (dam), Calif. 204/L11
Imperial (res.), Calif. 204/L10
Imperial (valley), Calif. 204/K10
Imperial (riv.), Chile 138/D2
Imperial, Mo. (63052) 261/M6
Imperial, Nebr. (69033) 264/C4
Imperial, Pa. (15126) 294/B5
Imperial, Peru 128/D9
Imperial, Sask. 181/F4
Imperial, Texas (79743) 303/B6
Imperial Beach, Calif. (92032) 204/H11
Imperial Mills, Alberta 182/E2
Impfondo, Congo 115/C3
Imphal, India 54/L7
Imphal, India 68/G4
Impora, Bolivia 136/C7
İmralı (isl.), Turkey 63/C2
İmranlı, Turkey 63/H2
Imroz (Gökçeada) (isl.), Turkey 63/A2
Imst, Austria 41/A3
Imuris, Mexico 150/D1
Imuruan (bay), Philippines 82/B5
Imuruk (basin), Alaska 196/E1
'Imwas, West Bank 65/B4
Ina, Ill. (62846) 222/E5
Ina, Japan 81/H6
Ina (riv.), Japan 81/H7
Inaha, Georgia (†31790) 217/E7
Inala, Queensland 88/K3
Inala, Queensland 95/D3
Inambari, Peru 128/H9
Inambari (riv.), Peru 128/H9
In Amenas, Algeria 106/F3
In Amguel, Algeria 106/E4
Inangahua Junction, N. Zealand 100/C4
Iñapari, Peru 128/H8
Inarajan, Guam 86/K7
Inari, Finland 18/P2
Inari (lake), Finland 7/G2
Inari (lake), Finland 18/P2
Inavale, Nebr. (68952) 264/F4
Inawashiro (lake), Japan 81/K5
In Azaoua (well), Niger 106/F4
Inca, Spain 33/H3
Incacamachi, Cerro (mt.), Bolivia 136/A6
Inca de Oro, Chile 138/B6
Incaguasi, Nevada (mt.), Chile 138/C6
Incahuasi, Cerro de (mt.), Argentina 143/C2
Ince (cape), Turkey 63/F1
Incekum (cape), Turkey 63/F4
Incesu, Turkey 63/F3
Inchard, Loch (inlet), Scotland 15/C2
Inchcape (Bell Rock) (isl.), Scotland 15/F4
Inchelium, Wash. (99138) 310/G2
Inchigeelagh, Ireland 17/C8
Inchiri (reg.), Mauritania 106/A5
Inchkeith, Sask. 181/J5
Inchkeith (isl.), Scotland 15/D1
Inchnadamph, Scotland 15/D2
Inch'ŏn, S. Korea 54/O6
Inch'ŏn, S. Korea 81/C5
Indaal, Loch (inlet), Scotland 15/B5
In Dagouber (well), Mali 106/D4
Indalsälven (riv.), Sweden 18/H5
Indawgyi (lake), Burma 72/C1
Indé, Mexico 150/G4
Independence (co.), Ark. 202/G2
Independence, Belize 154/C2
Independence, Calif. (93526) 204/H7
Independence, Ind. (†47918) 227/C4
Independence, Iowa (50644) 229/K4
Independence, Kansas (67301) 232/G4
Independence, Ky. (41051) 237/M3
Independence, La. (70443) 238/M1
Independence (lake), Mich. 250/B2
Independence, Minn. (†55359) 255/F5
Independence (lake), Minn. 255/F5
Independence, Miss. (38638) 256/E1
Independence, Mo. (*64050) 261/R5
Independence (mts.), Nev. 266/E1
Independence, Ohio (44131) 284/H9
Independence, Oreg. (97351) 291/D3
Independence, Va. (24348) 307/F7
Independence, W. Va. (26374) 312/G4
Independence, Wis. (54747) 317/D7
Independencia, Bolivia 136/B5
Independencia, Brazil 132/C6
Independencia (prov.), Dom. Rep. 158/D6
Independencia (bay), Peru 128/D10
Independencia (isl.), Peru 128/D10
Independencia, Venezuela 124/B4
Index, Wash. (98256) 310/D3
Index (peak), Wyo. 319/C1
India 2/N4
India 54/J7
INDIA 68/D4
Indiahoma, Okla. (73552) 288/J5
Indialantic, Fla. (32903) 212/F3
India Muerta (riv.), Uruguay 145/E4
Indian (ocean) 54/H10
Indian (ocean) 102/G7
Indian (mt.), Conn. 210/B1
Indian (pond), Conn. 210/A1
Indian (riv.), Del. 245/S6
Indian (riv.), Fla. 212/F3
Indian (creek), Idaho 220/C5
Indian (creek), Ind. 227/D6
Indian (creek), Ind. 227/E8
Indian (creek), Md. 245/G4
Indian (lake), Mich. 250/C2
Indian (stream), N.H. 268/E1
Indian (lake), N.Y. 276/M3
Indian (harb.), Nova Scotia 168/G3
Indian (lake), Ohio 284/C5
Indian (creek), S. Dak. 298/B4
Indian (creek), Utah 304/B5
Indian (creek), Utah 304/E5
Indiana 188/J3
INDIANA 227
Indiana (co.), Pa. 294/D4
Indiana, Pa. (15701) 294/D4
Indiana (state), U.S. 146/K6
Indiana Dunes Nat'l Lakeshore, Ind. 227/C1
Indianapolis (cap.), Ind. 146/K5
Indianapolis (cap.), Ind. 188/J3
Indianapolis (cap.), Ind. (*46201) 227/E5
Indian Bay, Manitoba 179/G5
Indian Beach, N.C. (†28575) 281/R5
Indian Brook, Nova Scotia 168/H2
Indian Cabins, Alberta 182/B4
Indian Creek, Fla. (†33139) 212/B4
Indian Creek, Ill. (†60069) 222/B4
Indian Harbour Beach, Fla. (†32901) 212/F3
Indian Head, Md. (20640) 245/K6
Indian Head, Sask. 162/F5
Indian Head, Sask. 181/H5
Indian Hill, Ohio (†45201) 284/C9
Indian Hills, Ky. (†40201) 237/K1
Indian Hills, N.C. (†28719) 281/C4
Indian Lake, N.Y. (12842) 276/M3
Indian Lake, Pa. (†15560) 294/E5
Indian Mills, W. Va. (24949) 312/E7
Indian Mound, Tenn. (37079) 237/F7
Indian Neck, Conn. (†06405) 210/D3
Indian Neck, Va. (23077) 307/O5
Indian Ocean 2/N6
Indian Ocean 5/C3
Indian Ocean, Indonesia 85/E8
Indian Ocean, S. Australia 94/E7
Indian Ocean, Tasmania 99/A4
Indian Ocean, Victoria 97/B6
Indian Ocean, W. Australia 92/A5
Indianola, Ill. (61850) 222/F4
Indianola, Iowa (50125) 229/F6

Indianola, Miss. (38751) 256/C4
Indianola, Nebr. (69034) 264/D4
Indianola, Okla. (74442) 288/P4
Indianola, Utah (†84629) 304/C4
Indianola, Wash. (98342) 310/A1
Indian Pond (lake), Maine 243/D4
Indian River (bay), Del. 245/T6
Indian River (inlet), Del. 245/T6
Indian River (co.), Fla. 212/F4
Indian River, Maine (†04649) 243/H6
Indian River, Mich. (49749) 250/E3
Indian River, Ontario 177/F3
Indian River Shores, Fla. (32960) 212/F4
Indian Rocks Beach, Fla. (33535) 212/B3
Indian Shores, Fla. (†33535) 212/B3
Indian Springs, Georgia (30231) 217/E4
Indian Springs, Ind. (47544) 227/D7
Indian Springs, Nev. (89018) 266/F6
Indiantown, Fla. (33456) 212/F4
Indian Trail, N.C. (28079) 281/H4
Indian Valley, Idaho (83632) 220/B5
Indian Valley, Va. (24105) 307/G7
Indian Village, Ind. (†46601) 227/E1
Indian Village, La. (†70764) 238/H6
Indian Wells, Ariz. (86031) 198/E3
Indian Wells, Calif. (†92260) 204/J10
Indiga, U.S.S.R. 48/E3
Indiga, U.S.S.R. 52/G1
Indigirka, U.S.S.R. 54/R3
Indigirka (riv.), U.S.S.R. 4/C2
Indigirka (riv.), U.S.S.R. 48/P3
Indigo (creek), Oreg. 291/D5
Indio, Calif. (92201) 204/J10
Indios (chan.), Cuba 158/B2
Indispensable (str.), Solomon Is. 86/E3
Indochina (reg.), Vietnam 72/D2
Indonesia 2/Q6
Indonesia 54/M10
INDONESIA 85
Indooroopilly, Queensland 95/D3
Indooroopilly, Queensland 88/K3
Indore, India 68/D4
Indore, India 54/J7
Indore, W. Va. (25111) 312/D6
Indramayu, Indonesia 85/H2
Indramayu (pt.), Indonesia 85/H1
Indravati (riv.), India 68/E5
Indre (dept.), France 28/D4
Indre (riv.), France 28/D4
Indre-et-Loire (dept.), France 28/D4
Indus (riv.) 2/N4
Indus (riv.) 54/H7
Indus, Alberta 182/D4
Indus (riv.), India 68/B3
Indus, Minn. (†56629) 255/E2
Indus, Mouths of the (delta), Pakistan 68/B4
Indus (riv.), Pakistan 59/J4
Indus (riv.), Pakistan 68/B3
Industrial City, Georgia (†30705) 217/C1
Industry, Ill. (61440) 222/C3
Industry, Kansas (†67410) 232/E2
Industry, Pa. (15052) 294/B4
Industry, Texas (78944) 303/H7
Inebolu, Turkey 59/B1
Inebolu, Turkey 63/E2
Inegöl, Turkey 63/C2
In Eker, Algeria 106/F4
Ineu, Romania 45/E2
Inez, Ky. (41224) 237/S5
Inez, N.C. (27589) 281/N2
Inezgane, Morocco 106/C2
In Ezzane (well), Algeria 106/G4
Infanta, Philippines 82/C3
Infieles (pt.), Chile 138/A6
Infiesto, Spain 33/D1
In-Gall, Niger 106/F5
Ingalls, Ark. (71648) 202/F7
Ingalls (mt.), Calif. 204/E3
Ingalls, Ind. (46048) 227/F5
Ingalls, Kansas (67853) 232/B4
Ingalls, Mich. (49848) 250/B3
Ingavi, Bolivia 136/B2
Ingelmunster, Belgium 27/C7
Ingelow, Manitoba 179/C5
Ingenbohl, Switzerland 39/G2
Ingende, Zaire 115/C4
Ingende, Zaire 102/D5
Ingeniero Huergo, Argentina 143/C4
Ingeniero Jacobacci, Argentina 143/C5
Ingeniero Luiggi, Argentina 143/D4
Ingeniero Montero Hoyos (Tocomechi), Bolivia 136/D5
Ingersoll, Ontario 177/C4
Ingham (co.), Mich. 250/E6
Ingham, Queensland 88/H3
Ingham, Queensland 95/C3
Inglefield, Ind. (47618) 227/B8
Inglés (pt.), Cuba 158/G4
Inglesa (bay), Chile 138/A6
Ingleside, Md. (21644) 245/P4
Ingleside, Ontario 177/J2
Ingleside, W. Va. (†24740) 312/E8
Inglewood, Calif. (*90301) 204/B11
Inglewood, Nebr. (†68025) 264/H3
Inglewood, N. Zealand 100/E3
Inglewood, Victoria 97/B5
Inglis, Fla. (32649) 212/D2
Inglis, Manitoba 179/A4
Ingold, N.C. (28446) 281/N5
Ingoldsby, Ontario 177/F3
Ingolstadt, W. Germany 22/D4
Ingomar, Miss. (†38652) 256/F2
Ingomar, Mont. (59039) 262/J4
Ingomar, Nova Scotia 168/C5
Ingomar, Pa. (15127) 294/C4
Ingonish, Nova Scotia 168/H2
Ingonish Beach, Nova Scotia 168/H2
Ingonish North (bay), Nova Scotia 168/H2
Ingornachoix (bay), Newf. 166/C3
Ingraham (lake), Fla. 212/E6
Ingraham, Ill. (62434) 222/E5
Ingram, Pa. (†15205) 294/B7
Ingram, Texas (78025) 303/E7
Ingram, Va. (24564) 307/K7
Ingram, Wis. (†54530) 317/E5
Ingre, Bolivia 136/D7
In Guezzam, Algeria 106/F5
Ingwavuma, S. Africa 118/E5
Inhambane (prov.), Mozambique 118/E4
Inhambane, Mozambique 118/F4
Inhambane, Mozambique 102/F7
Inhaminga, Mozambique 118/F3
Inharrime, Mozambique 118/F4
Inharrime, Mozambique 102/F7
Inhumas, Brazil 132/D7
Iniesta, Spain 33/F3
Inini, Fr. Guiana 131/E4
Inini (riv.), Fr. Guiana 131/E4
Inírida, Colombia 120/C2
Inírida, Colombia 126/F6
Inírida (riv.), Colombia 126/F6
Inishannon, Ireland 17/D8
Inishbofin (isl.), Ireland 17/A4
Inishbofin (isl.), Ireland 10/A4
Inishbofin (isl.), Ireland 17/E1
Inisheer (isl.), Ireland 17/B5
Inishmaan (isl.), Ireland 17/C5
Inishmore (isl.), Ireland 17/B5
Inishmurray (isl.), Ireland 17/D3
Inishowen (head), Ireland 17/H1
Inishowen (pen.), Ireland 17/G1
Inishshark (isl.), Ireland 17/A4
Inishtrahull (isl.), Ireland 17/G1
Inishtrahull (sound), Ireland 17/G1
Inishturk (isl.), Ireland 10/A4
Inishturk (isls.), Ireland 17/A4
Inistioge, Ireland 17/G7
Injune, Queensland 95/D5
Injune, Queensland 88/H5
Inkerman, New Bruns. 170/F1
Inklin (riv.), Br. Col. 184/J2
Inkom, Idaho (83245) 220/F7
Inkster, Mich. (48141) 250/B7
Inkster, N. Dak. (58244) 282/P3
Inland (lake), Ala. 195/E3
Inland (lake), Alaska 196/G1
Inland (lake), Manitoba 179/C2
Inland, Nebr. (68954) 264/F4
Inle (lake), Burma 72/C2
Inlet, N.Y. (13360) 276/L3
Inman, Georgia (30232) 217/D4
Inman, Kansas (67546) 232/E3
Inman, Nebr. (68742) 264/F2
Inman, S.C. (29349) 296/C1
Inn (riv.), Austria 41/B2
Inn (riv.), Switzerland 39/K3
Inn (riv.), W. Germany 22/E4
Innamincka, S. Australia 94/G2
Innellan, Scotland 15/A2
Inner (sound), Scotland 10/D2
Inner (sound), Scotland 15/C3
Inner Hebrides (isls.), Scotland 15/B4
Innerkip, Ontario 177/D4
Innerleithen, Scotland 10/E3
Innerleithen, Scotland 15/E5
Inner Mongolia (reg.), China 54/N5
Inner Mongolia (reg.), China 77/H3
Inner Mongolian Aut. Reg. (Nei Monggol), China 77/H3
Innertkirchen, Switzerland 39/F3
Innes (lake), N.S. Wales 97/G2
Innis, La. (70747) 238/G5
Inniscrone, Ireland 17/C3
Innisfail, Alberta 182/D3
Innisfail, Queensland 95/C3
Innisfail, Queensland 88/H3
Innisfree, Alberta 182/E3
Innisville, Ontario 177/H2
Innoko (riv.), Alaska 196/G2
Innsbruck, Austria 7/F4
Innsbruck, Austria 41/A3
Inny (riv.), Ireland 17/F4
Inny (riv.), Ireland 17/A8
Inola, Okla. (74036) 288/P2
Inongo, Zaire 115/C4
İnönü, Turkey 63/D3
Inoucdjouac, Que. 162/H4
Inoucdjouac, Québec 174/E1
Inowrocław, Poland 47/D2
Inquisivi, Bolivia 136/B5
In Rhar, Algeria 106/E3
Ins, Switzerland 39/D2
In Salah, Algeria 106/E3
In Salah, Algeria 102/C2
Insch, Scotland 15/F3
Insein, Burma 72/C3
Inset, Norway 18/G5
Insinger, Sask. 181/H4
Inspiration, Ariz. (85537) 198/D5
Institute, W. Va. (25112) 312/C6
Instow, Sask. 181/C6
Inta, U.S.S.R. 7/K2
Inta, U.S.S.R. 52/K1
Inta, U.S.S.R. 48/G3
Intake, Mont. (†59330) 262/M3
Intelewa, Suriname 131/D4
Intendente Alvear, Argentina 143/D4
Intepe, Turkey 63/B6
Intercession City, Fla. (33848) 212/E3
Intercourse, Pa. (17534) 294/K5
Interior, S. Dak. (57750) 298/F6
Interlachen, Fla. (32048) 212/E2
Interlaken, Mass. (†01266) 249/A3
Interlaken, N.J. (†07712) 273/E3
Interlaken, N.Y. (14847) 276/G5
Interlaken, Switzerland 39/E3
Interlochen, Mich. (49643) 250/D4
International Airport, Georgia 217/K2
International Airport, Mo. 261/P2
International Airport (Dallas-Ft. Worth), Texas 303/F2
International Falls, Minn. 188/H1
International Falls, Minn. (56649) 255/E2
International Peace Garden, Manitoba 179/B5
International Peace Garden, N. Dak. 282/K1
Intervale, N.H. (03845) 268/E3
Interview (isl.), India 68/G6
Inthanon, Doi (mt.), Thailand 72/C3
Intipucá, El Salvador 154/D4
Intracoastal Waterway, S.C. 296/H5
Intracoastal Waterway, Texas 303/J9
Intragna, Switzerland 39/G4
Intutu, Peru 128/E4
Inubo (cape), Japan 81/K6
Inútil (bay), Chile 138/E10
Inuvik, Canada 4/C16
Inuvik, N.W.T. 146/E3
Inuvik, N.W.T. 162/C2
Inuvik (dist.), N.W. Terrs. 187/F3
Inuvik, N.W. Terrs. 187/E3
Inver (bay), Ireland 17/E2
Inveraray, Scotland 10/D2
Inveraray, Scotland 15/C4
Inverbervie, Scotland 10/E2
Inverbervie, Scotland 15/F4
Invercargill, N. Zealand 100/B7
Invercassley, Scotland 15/D3
Inverell, N. S. Wales 88/J5
Inverell, N.S. Wales 97/F1
Invergarry, Scotland 15/D3
Invergordon, Scotland 10/D2
Invergordon, Scotland 15/D3
Invergowrie, Scotland 15/E4
Inver Grove Heights, Minn. (55075) 255/E6
Inverhuron, Ontario 177/C3
Inverie, Scotland 15/C3
Inverkeilor, Scotland 15/F4
Inverkeithing, Scotland 15/D1
Inverkeithing, Scotland 10/C1
Inverloch, Victoria 97/C6
Invermay, Ontario 177/C3
Invermay, Sask. 181/J4
Invermere, Br. Col. 184/J5
Invermoriston, Scotland 15/D3
Inverness, Ala. (†36089) 195/G6
Inverness, Calif. (94937) 204/B5
Inverness, Fla. (32650) 212/D3
Inverness, Ill. (†60067) 222/A5
Inverness, Miss. (38753) 256/C4
Inverness, Mont. (59530) 262/F2
Inverness, N.S. 162/K6
Inverness (co.), Nova Scotia 168/G2
Inverness, Nova Scotia 168/G2
Inverness, Québec 172/F3
Inverness, Scotland 15/D3
Inverness, Scotland 7/D3
Inverness, Scotland 10/D2
Inverness, (trad. co.), Scotland, 15/A5
Inverurie, Scotland 15/F3
Inverurie, Scotland 10/E2
Inverway, North. Terr. 93/A4
Investigator (shoal), Philippines 85/E4
Investigator (str.), S. Australia 88/F7
Investigator (str.), S. Australia 94/E6
Investigator Group (isls.), S. Australia 88/E6
Investigator Group (isls.), S. Australia 94/D5
Inwood, Ind. (46533) 227/E2
Inwood, Iowa (51240) 229/A2
Inwood, Manitoba 179/E4
Inwood, N.Y. (11696) 276/P7
Inwood, Ontario 177/C5
Inwood, W. Va. (25428) 312/K4
Inyanga, Zimbabwe 118/E3
Inyan Kara (creek), Wyo. 319/H1
Inyan Kara (mt.), Wyo. 319/H1
Inyo (co.), Calif. 204/H7
Inyo (mts.), Calif. 204/G6
Inyokern, Calif. (93527) 204/H8
Inza, U.S.S.R. 52/G4
Inzana (lake), Br. Col. 184/E3
Ioánnina, Greece 7/G4
Ioánnina, Greece 45/E6
Ioco, Br. Col. 184/K3
Ioka, Utah (†84052) 304/D3
Iola, Ill. (62847) 222/E5
Iola, Kansas (66749) 232/G4
Iola, Texas (77861) 303/H7
Iola, Wis. (54945) 317/H6
Iolotan', U.S.S.R. 48/G6
Ioma, Papua N.G. 85/C7
Iona, Angola 115/B7
Iona, Idaho (83427) 220/G6
Iona, Minn. (56141) 255/C7
Iona (isls.), Newf. 166/C2
Iona, N.J. (†08322) 273/C4
Iona, Nova Scotia 168/H3
Iona, Ontario 177/C5
Iona (isl.), Scotland 15/B4
Iona (isl.), Scotland 10/C2
Iona, S. Dak. (57542) 298/L6
Ione, Ark. (†72927) 202/B3
Ione, Calif. (95640) 204/C9
Ione, Nev. (†89310) 266/D4
Ione, Oreg. (97843) 291/H2
Ione, Wash. (99139) 310/H2
Ionia, Iowa (50645) 229/J2
Ionia, Kansas (66947) 232/D2
Ionia (co.), Mich. 250/D6
Ionia, Mich. (48846) 250/D6
Ionia, Mo. (65335) 261/F6
Ionian (sea) 7/F5
Ionian (isls.), Greece 7/F5
Ionian (sea), Greece 45/D7
Ionian (sea), Italy 34/F6
Ionian Islands (reg.), Greece 45/D6
Íos, Greece 45/G7
Íos (isl.), Greece 45/G7
Iosco (co.), Mich. 250/F4
Iosegun (lake), Alberta 182/B2
Iosegun (riv.), Alberta 182/B2
Iota, La. (70543) 238/E6
Iowa 188/H2
IOWA 229
Iowa (co.), Iowa 229/J5
Iowa (riv.), Iowa 188/H2
Iowa (riv.), Iowa 229/H4
Iowa, La. (70647) 238/D6
Iowa (state), U.S. 146/J5
Iowa (co.), Wis. 317/F9
Iowa City, Iowa 188/H2
Iowa City, Iowa (52240) 229/L5
Iowa Colony, Texas (†77583) 303/J2
Iowa Falls, Iowa (50126) 229/G3
Iowa Park, Texas (76367) 303/F4
Iowa Point, Kansas (†66035) 232/G2
Ipala, Guatemala 154/C3
Ipameri, Brazil 132/E7
Ipanema, Brazil 135/F1
Iparia, Peru 128/E7
Ipatovo, U.S.S.R. 52/F5
Ipava, Ill. (61441) 222/C3
Ipel' (riv.), Czech. 41/E2
Iphigenia (bay), Alaska 196/M2
Ipiales, Colombia 126/B7
Ipil, Philippines 82/D7
Ipin (Yibin), China 77/F6
Ipoh, Malaysia 54/M9
Ipoh, Malaysia 72/D6
Ipoly (riv.), Hungary 41/E2
Iporá, Brazil 120/D4
Ipperwash Prov. Park, Ontario 177/C4
Ippy, Cent. Afr. Rep. 115/D2
Ipsala, Turkey 63/B2
Ipsile, Turkey 63/G2
Ipswich, Australia 87/F8
Ipswich, England 13/J5
Ipswich, England 10/G4
Ipswich, Mass. (01938) 249/L2
Ipswich○, Mass. (01938) 249/L2
Ipswich (riv.), Mass. 249/L2
Ipswich, Queensland 88/J3
Ipswich, Queensland 95/E5
Ipswich, S. Dak. (57451) 298/L3
Ipu, Brazil 132/F4
Iquique, Chile 120/B5
Iquique, Chile 138/A2
Iquitos, Peru 120/B3
Iquitos, Peru 128/F4
Ira, Iowa (50127) 229/G5
Ira, N.Y. (†13033) 276/G4
Ira, Texas (79527) 303/C5
Ira○, Vt. (†05777) 268/A4
Iraan, Texas (79744) 303/B7
Iracoubo, Fr. Guiana 131/E3
Iraël', U.S.S.R. 52/H2
Iráklion, Greece 45/G8
Iráklion, Greece 7/G5
Iran 2/M4
Iran 54/G7
IRAN 66
IRAN 59/F3
Iran (mts.), Malaysia 85/E5
Iranshahr, Iran 66/M7
Iranshahr, Iran 59/H4
Irapa, Venezuela 124/G2
Irapuato, Mexico 150/J6
Iraq 2/M4
Iraq 54/F6
IRAQ 59/D3
IRAQ 66
Irasburg○, Vt. (05845) 268/C2
Irati, Brazil 132/D9
Irati, Brazil 135/A4
Irawan, Philippines 82/B6
Irazú (mt.), C. Rica 154/F6
Irbid, Jordan 65/D2
Irby, Wash. (99159) 310/G3
Irecê, Brazil 120/E4
Iredell (co.), N.C. 281/H3
Iredell, Texas (76649) 303/G6
Ireland 2/J3
Ireland 7/D3
Ireland (isl.), Bermuda 156/G3
Ireland, Ind. (47545) 227/C8
IRELAND 17
IRELAND 10/B4
IRELAND, NORTHERN 10, 17
Ireland, Texas (76536) 303/F6
Ireland, W. Va. (26376) 312/F5
Ireland's Eye (isl.), Ireland 17/K5
Ireland's Eye (isl.), Newf. 166/D2
Irene, S. Africa 118/H6
Irene, S. Dak. (57037) 298/P7
Ireng (riv.), Guyana 131/B3
Ireton, Iowa (51027) 229/A3
Irharhar, Wadi (dry riv.), Algeria 106/F3
Iri, S. Korea 81/C6
Irian Jaya (reg.), Indonesia 2/R6
Irian Jaya (reg.), Indonesia 85/J6
Iriba, Chad 111/D4
Iriga, Philippines 82/D4
Iringa (reg.), Tanzania 115/G5
Iringa, Tanzania 102/F5
Iringa, Tanzania 115/G5
Iriomote (isl.), Japan 81/K7
Irion (co.), Texas 303/C6
Iriona, Honduras 154/E2
Iriri (riv.), Brazil 132/C4
Irish (sea) 7/D3
Irish (sea), England 13/B4
Irish (sea), England 10/D4
Irish (sea), Ireland 10/D4
Irish (sea), Ireland 17/K4
Irish (sea), I. of Man 13/B4
Irish (sea), Wales 13/B4
Irish (sea), Wales 10/D4
Irishtown, New Bruns. 170/F2
Irishtown, Tasmania 99/B2
Irish Vale, Nova Scotia 168/H3
Irkutsk, U.S.S.R. 2/Q3
Irkutsk, U.S.S.R. 54/M4
Irkutsk, U.S.S.R. 48/L4
Irma, Alberta 182/E3
Irma, Wis. (54442) 317/G5
Irmo, S.C. (29063) 296/E3
Iro (cape), Japan 81/J6
Irois (cape), Haiti 158/A6
Iron (mt.), Fla. 212/E4
Iron (co.), Mich. 250/G2
Iron, Minn. (55751) 255/F3
Iron (co.), Mo. 261/L7
Iron (mts.), Tenn. 237/S8
Iron (co.), Utah 304/A6
Iron (co.), Wis. 317/F3
Iron Belt, Wis. (54536) 317/F3
Ironbound (isls.), Newf. 166/C2
Iron Bridge, Ontario 177/A1
Iron City, Georgia (31759) 217/C8
Iron City, Tenn. (38463) 237/F10
Irondale, Ala. (35210) 195/E3
Irondale, Mo. (63648) 261/L7
Irondale, Ohio (43932) 284/J4
Irondequoit, N.Y. (14617) 276/E4
Iron Gate (res.), Calif. 204/C2
Iron Gate, Va. (24448) 307/J5
Iron Gates, Mo. (†64801) 261/C8
Ironia, N.J. (07845) 273/D2
Iron Knob, S. Australia 88/F6
Iron Knob, S. Australia 94/E5
Iron Mountain, Mich. (49801) 250/B3
Iron Mountain, Wyo. (82062) 319/G4
Iron Range, Queensland 95/B2
Iron Ridge, Wis. (53035) 317/K9
Iron River, Alberta 182/E2
Iron River, Mich. (49935) 250/G2
Iron River, Wis. (54847) 317/D2
Irons, Mich. (49644) 250/D4
Ironshire, Md. (†21811) 245/T7
Ironside, Oreg. (97908) 291/K3
Ironspring (creek), Sask. 181/G3
Iron Springs, Alberta 182/D5
Iron Springs, Ariz. (86330) 198/C4
Iron Station, N.C. (28080) 281/G4
Ironton, Minn. (56455) 255/D4
Ironton, Mo. (63650) 261/L7
Ironton, Ohio (45638) 284/E8
Ironton, Wis. (†53941) 317/F8
Ironwood, Mich. (49938) 250/F2
Iroquois (co.), Ill. 222/F3
Iroquois, Ill. (60945) 222/F3
Iroquois (riv.), Ill. 222/F3
Iroquois (riv.), Ind. 227/B3
Iroquois, Ontario 177/J3
Iroquois, S. Dak. (57353) 298/O5
Iroquois (lake), Vt. 268/A3
Iroquois Falls, Ont. 162/H6
Iroquois Falls, Ontario 175/D3
Iroquois Falls, Ontario 177/J5
Iroquois Point, Hawaii (†96706) 218/A4
'Irqa, P.D.R. Yemen 59/E7
Irrara (creek), N.S. Wales 97/C1
Irrawaddy (div.), Burma 72/B3
Irrawaddy, Mouths of the (delta), Burma 72/B4
Irrawaddy (riv.), Burma 54/L7
Irrawaddy (riv.), Burma 72/B3
Irricana, Alberta 182/D4
Irrigon, Oreg. (97844) 291/H2
Irtysh (riv.), U.S.S.R. 54/J4
Irtysh (riv.), U.S.S.R. 48/H4
Irumu, Zaire 115/E3
Irún, Spain 33/F1
Irupana, Bolivia 136/B5
Iruya, Argentina 143/D1
Irvine, Alberta 182/E5
Irvine, Calif. (92713) 204/D11
Irvine, Ky. (40336) 237/O5
Irvine (lake), N. Dak. 282/M3
Irvine, Pa. (16329) 294/D2
Irvine, Scotland 15/D5
Irvine, Scotland 10/D3
Irvinestown, N. Ireland 17/F3
Irving, Ill. (62051) 222/D4
Irving, Iowa (†52225) 229/J5
Irving, N.Y. (14081) 276/B5
Irving, Texas (*75061) 303/G2
Irvington, Ala. (36544) 195/B9
Irvington, Ill. (62848) 222/D5
Irvington, Iowa (50550) 229/E3
Irvington, Ky. (40146) 237/J5
Irvington, N.J. (07111) 273/B2
Irvington, N.Y. (10533) 276/O6
Irvington, Va. (22480) 307/R5
Irvin's (bay), Grenada 161/D8
Irvona, Pa. (16656) 294/E4
Irwin (co.), Georgia 217/F7
Irwin, Idaho (83428) 220/G6
Irwin, Iowa (51446) 229/C5
Irwin, Mo. (64754) 261/D7
Irwin, Ohio (43029) 284/D5
Irwin, Pa. (15642) 294/C5
Irwin, S.C. (†29720) 296/F2
Irwin, Va. (23063) 307/N5
Irwin, W. Australia 88/A5
Irwinton, Georgia (31042) 217/F5
Irwinville, Georgia (31760) 217/F7
Isa, Nigeria 106/F6
Isaac (lake), Br. Col. 184/G3
Isaacs (riv.), Queensland 88/H4
Isaacs (riv.), Queensland 95/D4
Isaac's Harbour North, Nova Scotia 168/G3
Isabel (bay), Ecuador 128/B9
Isabel, Kansas (67065) 232/D4
Isabel, La. (†70427) 238/K5
Isabel, S. Dak. (57633) 298/G3
Isabel (mt.), Wyo. 319/B3
Isabela (bay), Dom. Rep. 158/D5
Isabela (cape), Dom. Rep. 158/D5
Isabela (isl.), Ecuador 128/B9
Isabela (prov.), Philippines 82/C2
Isabela, Philippines 82/C7
Isabela, P. Rico 156/F1
Isabela, P. Rico 161/A1
Isabela de Sagua, Cuba 158/E1
Isabelia, Cordillera (range), Nicaragua 154/E4
Isabella (lake), Calif. 204/G8
Isabella, Manitoba 179/B4
Isabella (co.), Mich. 250/E5
Isabella, Mich. (†49878) 250/C3
Isabella, Minn. (55607) 255/G3
Isabella (lake), Minn. 255/G3
Isabella (riv.), Minn. 255/G3
Isabella (bay), N.W. Terrs. 187/M3
Isabella, Okla. (73747) 288/K2
Isabella, Tenn. (37346) 237/N10
Isabel Segunda, P. Rico 161/G2
Isaccea, Romania 45/J3
Isachsen, Canada 4/B15
Isachsen, N.W. Terrs. 187/H2
Isachsen (cape), N.W. Terrs. 187/H2
Isachsens, N.W.T. 146/H2
Ísafjördhardjúp (fjord), Iceland 21/A1
Ísafjördhur, Iceland 7/B2
Ísafjördhur, Iceland 21/B1
Isahaya, Japan 81/D7
Isana (riv.), Colombia 126/F7
Isangi, Zaire 115/D3
Isanti (co.), Minn. 255/E5
Isanti, Minn. (55040) 255/E5
Isar (riv.), W. Germany 22/E4
Isarog (mt.), Philippines 82/D4
Isbell, Ala. (†35653) 195/C2
Iscar, Spain 33/D2
Ischia (isl.), Italy 34/D4
Ischua, N.Y. (14746) 276/D6
Iscuandé, Colombia 126/A6
Ise, Japan 81/H6
Ise (bay), Japan 81/H6
Isefjord (fjord), Denmark 21/E6
Iselin, N.J. (08830) 273/E2
Iselin, Pa. (†15681) 294/D4
Isenthal, Switzerland 39/F3
Iseo (lake), Italy 34/C2
Isère (dept.), France 28/F5
Isère (riv.), France 28/F5
Iserlohn, W. Germany 22/B3
Isernia (prov.), Italy 34/E4
Isernia, Italy 34/E4
Ise-Shima National Park, Japan 81/H6
Iseyin, Nigeria 106/E7
Isfahan (prov.), Iran 66/H4
Isfahan, Iran 54/G6
Isfahan, Iran 66/G4
Isfahan, Iran 59/F3
Isfjorden (fjord), Norway 18/C2
Isham, Sask. 181/C4
Isherton, Guyana 131/B4
Ishigaki, Japan 81/L7
Ishigaki (isl.), Japan 81/L7
Ishige, Japan 81/P2
Ishikari (bay), Japan 81/K2
Ishikari (riv.), Japan 81/L2
Ishikawa (pref.), Japan 81/H5
Ishim (riv.), U.S.S.R. 54/H4
Ishim, U.S.S.R. 48/H4
Ishim (riv.), U.S.S.R. 48/G4
Ishimbay, U.S.S.R. 52/J4
Ishinomaki, Japan 81/K4
Ishioka, Japan 81/K5
Ishizuchi (mt.), Japan 81/F7
Ishpeming, Mich. (49849) 250/B2
Isiboro (riv.), Bolivia 136/C5
Isil'kul', U.S.S.R. 48/H4
Isimu, Indonesia 85/G5
Isiolo, Kenya 115/G3
Isiro, Zaire 102/E4
Isiro, Zaire 115/E3
Isisford, Queensland 88/G4
Isisford, Queensland 95/C5
Iskenderun, Turkey 59/C2
Iskenderun, Turkey 63/G4
Iskilip, Turkey 63/F2
Iskŭr (riv.), Bulgaria 45/G4
Iskut (riv.), Br. Col. 184/B2
Isla, Salar de la (salt dep.), Chile 138/B5
Isla, Veracruz, Mexico 150/M7
Isla (riv.), Scotland 15/E4
Isla Cristina, Spain 33/C4
Isla de Aguada, Mexico 150/O7
Isla de Maipo, Chile 138/G4
Islâhiye, Turkey 63/G4
Isla Holbox, Mexico 150/Q6
Islamabad (cap.), Pakistan 68/C2
Islamabad (cap.), Pakistan 2/N4
Islamabad (cap.), Pakistan 54/J6
Islamabad (cap.), Pakistan 59/K3
Islamabad District, Pakistan 68/C2
Islamorada, Fla. (33036) 212/F7
Isla Mujeres, Mexico 150/Q6
Island, Ky. (42350) 237/G6
Island (lake), Man. 162/G5
Island (lake), Manitoba 179/K3
Island (beach), N.J. 273/E4
Island (lake), N. Dak. 282/L2
Island (bay), Philippines 82/B6
Island (lag.), S. Australia 94/E4
Island (pond), Vt. 268/D2
Island (co.), Wash. 310/C2
Island City, Oreg. (97851) 291/K2
Island Creek, Md. (†20685) 245/M7
Island Falls○, Maine (04747) 243/G3
Island Falls, Ontario 175/D3
Island Grove, Fla. (32654) 212/D2
Island Heights, N.J. (08732) 273/E4
Islandia, Fla. (†33101) 212/F6
Island Lake, Ill. (60042) 222/A4
Island Lake, Manitoba 179/J3
Island Lake, Wis. (†54757) 317/D5
Island Park, Idaho (83429) 220/G5
Island Park (res.), Idaho 220/G5
Island Park, N.Y. (11558) 276/R7
Island Park, R.I. (†02871) 249/J6
Island Pond, Vt. (05846) 268/D2
Islands (bay), Newf. 166/C4
Islands (bay), Newf. 166/C2
Islands (bay), N. Zealand 100/E1
Islandton, S.C. (29929) 296/F6
Island View, Minn. (†56649) 255/E2
Island View, New Bruns. 170/D3
Isla Patrulla, Uruguay 145/E3
Isla Pucú, Paraguay 144/B4
Isla Umbú, Paraguay 144/C5
Isla Vista, Calif. (93117) 204/E9
Islay, Alberta 182/E3
Islay (isl.), Scotland 15/B5
Islay (isl.), Scotland 10/C3
Islay (sound), Scotland 15/C5
Isle (riv.), France 28/D5
Isle, Minn. (56342) 255/E4
Isle Au Haut○, Maine (04645) 243/F7
Isle-aux-Coudres, Québec 172/G2
Isle-aux-Grues, Québec 172/G2
Isle aux Morts, Newf. 166/C4
Isle La Motte○, Vt. (05463) 268/A2
Isle of Hope, Georgia (†31406) 217/K7
ISLE of MAN 13/C3
ISLE OF MAN 10/D3

Isle of Palms, S.C. (29451) 296/H6
Isle of Whithorn, Scotland 15/D6
Isle of Wight (co.), England 13/F7
Isle of Wight (co.), Va. 307/P7
Isle of Wight, Va. (23397) 307/P7
Isle Pierre, Br. Col. 184/F3
Isle Royale (isl.), Mich. 250/D1
Isle Royale National Park, Mich. (†55605) 250/E1
Isle Royale Nat'l Park, Mich. 250/E1
Islesboro, Maine (04848) 243/F7
Islesboro○, Maine (04848) 243/F7
Islesboro (isl.), Maine 243/F7
Islesford, Maine (04646) 243/G7
Isles of Scilly, England 13/A7
Isleta, N. Mex. (87022) 274/C4
Isleton, Calif. (95641) 204/L1
Islington, England 10/B5
Islington, England 13/H8
Islington, Mass. (02090) 249/C8
Islip, N.Y. (11751) 276/O9
Ismailia, Egypt 111/K3
Ismailia, Egypt 59/B3
Ismay, Mont. (59336) 262/M4
Isna, Egypt 111/F2
Isna, Egypt 59/B4
Isney, Ala. (†36919) 195/B7
Isny im Allgäu, W. Germany 22/D5
Isojoki, Finland 18/M5
Isoka, Zambia 115/F6
Isola, Miss. (38754) 256/C4
Isonville, Ky. (41149) 237/P4
Isparta (prov.), Turkey 63/D4
Isparta, Turkey 63/D4
Isparta, Turkey 59/B2
Isperikh, Bulgaria 45/H4
Ispir, Turkey 63/J2
Israel 2/L4
Israel 54/E6
ISRAEL 59/B3
ISRAEL 65
Issano, Guyana 131/B3
Issaouane Erg (des.), Algeria 106/F3
Issaquah, Wash. (98027) 310/C3
Issaquena (co.), Miss. 256/B5
Issia, Ivory Coast 106/C7
Issineru, Guyana 131/A2
Issoire, France 28/E5
Issoudun, France 28/D4
Issue, Md. (20645) 245/L7
Issyk-Kul' (lake), U.S.S.R. 54/J5
Issyk-Kul' (lake), U.S.S.R. 48/H5
Issy-les-Moulineaux, France 28/A2
Istanbul (prov.), Turkey 63/C2
Istanbul, Turkey 2/L3
Istanbul, Turkey 7/G4
Istanbul, Turkey 63/D6
Istanbul, Turkey 59/A1
Isthmus (bay), Ontario 177/C2
Istiaia, Greece 45/F6
Istmina, Colombia 126/B5
Istokpoga (lake), Fla. 212/E4
Istranca (mts.), Turkey 63/B2
Istres, France 28/F6
Istria (pen.), Yugoslavia 45/A3
Isulan, Philippines 82/E7
Itá, Paraguay 144/B5
Itabaiana, Paraíba, Brazil 132/H4
Itabaiana, Sergipe, Brazil 132/G5
Itaberaba, Brazil 132/F6
Itabira, Brazil 120/E4
Itabira, Brazil 132/F7
Itabirito, Brazil 135/E2
Itabuna, Brazil 132/G6
Itabuna, Brazil 120/F4
Itacoatiara, Brazil 132/B3
Itacoatiara, Brazil 120/D3
Itacurubí, Paraguay 144/B5
Itacurubí del Rosario, Paraguay 144/D4
Itaguara, Brazil 135/D2
Itaguatins, Brazil 132/D4
Itagüí, Colombia 126/C4
Itaí, Brazil 135/B3
Itaipu (dam) 120/D5
Itaipu (dam), Brazil 132/C9
Itaipu (dam), Paraguay 144/E4
Itaituba, Brazil 132/C4
Itajaí, Brazil 120/E5
Itajaí, Brazil 132/D9
Itajubá, Brazil 135/D3
Itajubá, Brazil 132/E8
Itala, Somalia 115/J3
Italy 2/K3
Italy 7/F4
ITALY 34
Italy, Texas (76651) 303/H5
Itamarandiba, Brazil 132/F7
Itami, Japan 81/H7
Itanagar, India 68/G3
Itanhaem, Brazil 135/C4
Itapagipe, Brazil 135/B1
Itapé, Paraguay 144/C5
Itapeby, Uruguay 145/B2
Itapecerica, Brazil 135/D2
Itapecuru (riv.), Brazil 132/F4
Itapecuru-Mirim, Brazil 132/F3
Itapemirim, Brazil 132/F8
Itaperuna, Brazil 135/F2
Itapetinga, Brazil 120/F4
Itapetinga, Brazil 132/G6
Itapetininga, Brazil 135/B3
Itapetininga, Brazil 132/D8
Itapeva, Brazil 132/D8
Itapeva, Brazil 135/B3
Itapi (riv.), Brazil 132/B3
Itapicuru, Brazil 132/G5
Itapicuru (riv.), Brazil 132/G5
Itapipoca, Brazil 132/G3
Itapira, Brazil 135/C3
Itapiranga, Brazil 132/B3
Itápolis, Brazil 132/D8
Itápolis, Brazil 135/B2
Itaporanga, Brazil 132/G4
Itapúa (dept.), Paraguay 144/E5
Itaqui, Brazil 132/B10
Itaquyry, Paraguay 144/E4
Itararé, Brazil 132/D9
Itararé, Brazil 135/B4
Itararé (riv.), Brazil 135/B3
Itariri, Brazil 135/C4
Itarsi, India 68/D4
Itasca, Ill. (60143) 222/B5
Itasca (co.), Minn. 255/E3
Itasca (lake), Minn. 255/C3
Itasca, Texas (76055) 303/G5
Itasy (lake), Madagascar 118/H3
Itata (riv.), Chile 138/A11
Itatí, Argentina 143/E2
Itatiba, Brazil 135/C3
Itaú, Bolivia 136/D7
Itauguá, Paraguay 144/B5
Itaúna, Brazil 135/D2
Itawamba (co.), Miss. 256/H2
Itbayat (isl.), Philippines 82/A2
Itchen (lake), N.W. Terrs. 187/G3
Iténez (Guaporé) (riv.), Bolivia 136/C3
Ithaca, Mich. (48847) 250/E5
Ithaca, Nebr. (68033) 264/H3
Ithaca, N.Y. (14850) 276/G6
Ithaca, Ohio (†45329) 284/A6
Ithaca, Queensland 88/K2
Ithaca, Wis. (†53581) 317/F9
Itháki, Greece 45/E6
Itháki (Ithaca) (isl.), Greece 45/E6
Itigi, Tanzania 115/F5
Itimbiri (riv.), Zaire 115/D3
Itkillik (riv.), Alaska 196/H1
Itmann, W. Va. (24847) 312/D7
Itnay (riv.), Fr. Guiana 120/D2
Ito, Japan 81/J6
Itoigawa, Japan 81/H5
Itoman, Japan 81/N6
Itonamas (riv.), Bolivia 136/C3
Itta Bena, Miss. (38941) 256/D4
Ittre, Belgium 27/E7
Itu, Brazil 135/C3
Ituaçu, Brazil 132/F6
Ituango, Colombia 126/C4
Ituberá, Brazil 132/G6
Ituiutaba, Brazil 132/D7
Itumbiara, Brazil 120/E4
Itumbiara, Brazil 132/D7
Ituna, Sask. 181/H4
Ituni, Guyana 131/B3
Itupiranga, Brazil 132/D4
Iturama, Brazil 135/A1
Iturbe, Paraguay 144/C5
Ituri (for.), Zaire 115/E3
Iturup (isl.), U.S.S.R. 54/R5
Iturup (isl.), U.S.S.R. 48/P5
Ituverava, Brazil 135/C2
Ituzaingó, Argentina 143/E2
Ituzaingó, Uruguay 145/A6
Itzehoe, W. Germany 22/C2
Iuka, Ill. (62849) 222/E5
Iuka, Kansas (67066) 232/D4
Iuka, Ky. (42052) 237/E6
Iuka, Miss. (38852) 256/H1
Iul'tin, U.S.S.R. 48/T3
Iva, S.C. (29655) 296/B3
Ivaí (riv.), Brazil 132/C8
Ivalo, Finland 18/P2
Ivalojoki (riv.), Finland 18/P2
Ivan, Ark. (71748) 202/F6
Ivan, La. (†71006) 238/C1
Ivančice, Czech. 41/D2
Ivangrad, Yugoslavia 45/E4
Ivanhoe, Calif. (93235) 204/F7
Ivanhoe, Minn. (56142) 255/B6
Ivanhoe, N. S. Wales 88/G6
Ivanhoe, N.S. Wales 97/C3
Ivanhoe, N.C. (28447) 281/N5
Ivanhoe, Va. (24350) 307/G7
Ivanhoe, W. Australia 92/E1
Ivanhoe, W. Va. (†26201) 312/F5
Ivanjica, Yugoslavia 45/E4
Ivanof, Alaska (†99502) 196/G3
Ivano-Frankovsk, U.S.S.R. 7/G4
Ivano-Frankovsk, U.S.S.R. 48/C5
Ivano-Frankovsk, U.S.S.R. 52/B5
Ivanovo, U.S.S.R. 7/J3
Ivanovo, U.S.S.R. 48/E4
Ivanovo, U.S.S.R. 52/E3
Ivarib, Namibia 118/B4
Ivaylovgrad, Bulgaria 45/H5
Ivdel, U.S.S.R. 48/G3
Ivel, Ky. (41642) 237/R5
Ives (mesa), Ariz. 198/E3
Ivesdale, Ill. (61851) 222/E4
Ivey, Georgia (†31031) 217/F5
Ivigtut, Greenl. 4/D12
Ivindo (riv.), Cameroon 115/B3
Ivindo (riv.), Congo 115/B3
Ivindo (riv.), Gabon 115/B3
Ivins, Utah (84738) 304/A6
Ivohibe, Madagascar 118/H4
Ivón, Bolivia 136/C2
Ivor, Va. (23866) 307/P7
Ivory Coast 2/J5
Ivory Coast 102/B4
IVORY COAST 106/C7
Ivory Coast (reg.), Ivory Coast 106/C8
Ivoryton, Conn. (06442) 210/F3
Ivrea, Italy 34/B2
Ivrindi, Turkey /B3
Ivry-sur-Seine, France 28/B2
Ivujivik, Que. 162/J3
Ivujivik, Québec 174/E1
Ivy, Va. (22945) 307/L4
Ivybridge, England 13/D7
Ivydale, W. Va. (25113) 312/D5
Ivyland, Pa. (†18974) 294/M5
Ivy Mountain (brook), Conn. 210/C1
Ivyton, Ky. (41444) 237/P5
Ivyton, Tenn. (†38543) 237/L8
Iwaizumi, Japan 81/K4
Iwaki, Japan 54/R6
Iwaki, Japan 81/K5
Iwaki (mt.), Japan 81/K3
Iwakuni, Japan 81/E6
Iwami, Japan 81/G6
Iwamizawa, Japan 81/L2
Iwanai, Japan 81/K2
Iwasaki, Japan 81/J3
Iwata, Japan 81/H6
Iwate (pref.), Japan 81/K4
Iwate (mt.), Japan 81/K4
Iwatsuki, Japan 81/O2
Iwilei, Hawaii (†96801) 218/C4
Iwo (isl.), Japan 87/E3
Iwo (isl.), Japan 81/M4
Iwo, Nigeria 106/E7
Iwŏn, N. Korea 81/D3
Ixelles, Belgium 27/C9
Ixiamas, Bolivia 136/A3
Ixmiquilpan, Mexico 150/K6
Ixonia, Wis. (53036) 317/H1
Ixtapa, Mexico 150/J8
Ixtapalapa, Mexico 150/L1
Ixtenco, Mexico 150/N1
Ixtepec, Mexico 150/M8
Ixtlán del Río, Mexico 150/G6
Iyo, Japan 81/F7
Iyo (sea), Japan 81/E7
Izabal (lake), Guatemala 154/C3
Izamal, Mexico 150/P6
Izard (co.), Ark. 202/G1
Izberbash, U.S.S.R. 52/G6
Izegem, Belgium 27/C7
Izeh, Iran 66/F5
Izhevsk, U.S.S.R. 7/K3
Izhevsk, U.S.S.R. 48/F4
Izhevsk, U.S.S.R. 52/H3
Izhma (riv.), U.S.S.R. 52/H2
Izigan (cape), Alaska 196/E4
Izmail, U.S.S.R. 48/C5
Izmail, U.S.S.R. 52/C5
Izmir (prov.), Turkey 63/B3
Izmir, Turkey 54/D6
Izmir, Turkey 63/B3
Izmir, Turkey 59/A2
Izmir (gulf), Turkey 63/B3
Izmit, Turkey 59/A1
Izmit, Turkey 63/D2
Iznájar, Spain 33/D4
Iznalloz, Spain 33/E4
Iznik, Turkey 63/C2
Iznik (lake), Turkey 63/C2
Izozog, Bolivia 136/D6
Izozog (swamp), Bolivia 136/E6
Izra, Syria 63/G6
Izsák, Hungary 41/E3
Izsófalva, Hungary 41/F2
Iztapa, Guatemala 154/B4
Izu (isls.), Japan 81/J6
Izu (pen.), Japan 81/J6
Izúcar de Matamoros, Mexico 150/M2
Izuhara, Japan 81/D6
Izumi, Japan 81/J8
Izumiotsu, Japan 81/J8
Izumisano, Japan 81/G6
Izumo, Japan 81/F6
Izyum, U.S.S.R. 52/E5

J

Jaba, West Bank 65/C3
Jabaliya, Gaza Strip 65/A4
Jabalpur, India 54/K7
Jabalpur, India 68/D4
Jaba Rud (riv.), Iran 66/L2
Jabbeke, Belgium 27/C6
Jabir, Jordan 65/E2
Jablonec nad Nisou, Czech. 41/C1
Jablonica, Czech. 41/D2
Jablunka (pass), Czech. 41/E2
Jablunkov, Czech. 41/E2
Jaboatão, Brazil 120/F3
Jaboatão, Brazil 132/H5
Jaboticabal, Brazil 135/B2
Jaboticabal, Brazil 132/D8
Jabrin, Saudi Arabia 59/E5
Jaca, Spain 33/F1
Jacaguas (riv.), P. Rico 161/C2
Jacaleapa, Honduras 154/D3
Jacaltenango, Guatemala 154/B3
Jacaréacanga, Brazil 132/B4
Jacareí, Brazil 132/E8
Jacareí, Brazil 135/D3
Jacarezinho, Brazil 135/A3
Jacarezinho, Brazil 132/D8
Jáchal, Argentina 143/C3
Jachin, Ala. (36910) 195/B6
Jáchymov, Czech. 41/B1
Jacinto, Miss. (†38865) 256/H1
Jacinto City, Texas (77029) 303/J1
Jack, Ala. (36346) 195/F7
Jack (creek), Minn. 255/C7
Jack (lake), Ontario 177/F3
Jack (co.), Texas 303/F4
Jack (mt.), Wash. 310/E2
Jack Creek, Nev. (†89834) 266/E1
Jackfish (riv.), Alberta 182/B5
Jackfish (lake), Sask. 181/C2
Jackfish Lake, Sask. 181/C2
Jackfork (mt.), Okla. 288/P5
Jackman, Maine (04945) 243/C4
Jackman○, Maine (04945) 243/C4
Jackpot, Nev. (89825) 266/G1
Jacksboro, Tenn. (37757) 237/N8
Jacksboro, Texas (76056) 303/F4
Jacks Creek, Tenn. (38347) 237/D10
Jacks Fork (riv.), Mo. 261/J8
Jackson (co.), Ala. 195/F1
Jackson, Ala. (36545) 195/C8
Jackson (co.), Ark. 202/H2
Jackson, Calif. (95642) 204/C9
Jackson (co.), Colo. 208/G1
Jackson (co.), Fla. 212/D5
Jackson (lake), Fla. 212/B1
Jackson (lake), Fla. 212/E4
Jackson (co.), Georgia 217/E2
Jackson, Georgia (30233) 217/E4
Jackson (lake), Georgia 217/E4
Jackson (co.), Ill. 222/D6
Jackson (co.), Ind. 227/E7
Jackson (co.), Iowa 229/M4
Jackson (co.), Kansas 232/G2
Jackson (co.), Ky. 237/N6
Jackson, Ky. (41339) 237/P5
Jackson (par.), La. 238/E2
Jackson, La. (70748) 238/H5
Jackson (co.), Mich. 250/E6
Jackson, Mich. 188/J2
Jackson, Mich. (*49201) 250/E6
Jackson (co.), Minn. 255/C7
Jackson, Minn. (56143) 255/C7
Jackson (co.), Miss. 256/G9
Jackson (cap.), Miss. 146/K6
Jackson (cap.), Miss. 188/J4
Jackson (cap.), Miss. (*39201) 256/D6
Jackson (co.), Mo. 261/R5
Jackson, Mo. (63755) 261/N8
Jackson, Mont. (59736) 262/C5
Jackson (mt.), Mont. 262/C2
Jackson, Nebr. (68743) 264/H2
Jackson (mts.), Nev. 266/C1
Jackson○, N.H. (03846) 268/E3
Jackson○, N.J. (08527) 273/E3
Jackson (bay), N. Zealand 100/B5
Jackson (co.), N.C. 281/C4
Jackson, N.C. (27845) 281/P2
Jackson (co.), Ohio 284/E7
Jackson, Ohio (45640) 284/E7
Jackson (co.), Okla. 288/H5
Jackson (co.), Oreg. 291/E5
Jackson (creek), Oreg. 291/E5
Jackson, Pa. (18825) 294/L2
Jackson, S.C. (29831) 296/D5
Jackson (co.), S. Dak. 298/F6
Jackson (co.), Tenn. 237/K8
Jackson, Tenn. 188/J3
Jackson, Tenn. (38301) 237/D9
Jackson (co.), Texas 303/H9
Jackson (riv.), Va. 307/J4
Jackson (co.), W. Va. 312/C5
Jackson (co.), Wis. 317/E7
Jackson, Wis. (53037) 317/K9
Jackson (lake), Wyo. 188/E2
Jackson, Wyo. (83001) 319/B2
Jackson (lake), Wyo. 319/B2
Jackson (peak), Wyo. 319/B2
Jacksonboro, S.C. (29452) 296/G6
Jacksonburg, Ind. (†47327) 227/G5
Jacksonburg, Ohio (†45067) 284/B6
Jacksonburg, W. Va. (26377) 312/E3
Jackson Center, Ohio (45334) 284/B5
Jackson Center, Pa. (16133) 294/B3
Jackson Junction, Iowa (52150) 229/K2
Jackson Lake (res.), Colo. 208/L2
Jacksonport, Ark. (72075) 202/H2
Jacksonport, Wis. (†54235) 317/M6
Jackson's Arm, Newf. 166/C4
Jacksons Gap, Ala. (36861) 195/G5
Jackson Springs, N.C. (27281) 281/K4
Jacksontown, Ohio (43030) 284/F6
Jacksonville, Ala. (36265) 195/G3
Jacksonville, Ark. (72076) 202/F4
Jacksonville, Fla. 146/K6
Jacksonville, Fla. 188/K4
Jacksonville, Fla. (*32201) 212/E1
Jacksonville, Georgia (31544) 217/G7
Jacksonville, Ill. (62650) 222/C4
Jacksonville, Maine (†04630) 243/J6
Jacksonville, Md. (†21131) 245/M2
Jacksonville, Mo. (65260) 261/G3
Jacksonville, New Bruns. 170/C2
Jacksonville, N.C. (28540) 281/O5
Jacksonville, Ohio (45740) 284/F7
Jacksonville, Oreg. (97530) 291/D5
Jacksonville (Kent), Pa. (15752) 294/D4
Jacksonville, Texas (75766) 303/J5
Jacksonville, Vt. (05342) 268/B6
Jacksonville Beach, Fla. (32250) 212/E1
Jacksonville Naval Air Sta., Fla. 212/E1
Jacmel, Haiti 158/C6
Jacmel, Haiti 156/D3
Jacobabad, Pakistan 59/J4
Jacobabad, Pakistan 68/B3
Jacobina, Brazil 120/E4
Jacobina, Brazil 132/F5
Jacob Lake, Ariz. (86051) 198/C2
Jacobson, Minn. (55752) 255/E4
Jacobstown, N.J. (†08562) 273/D3
Jacobsville, Mich. (†49945) 250/A1
Jacobus, Pa. (17407) 294/J6
Jacques-Cartier (lake), Québec 172/F2
Jacques-Cartier (mt.), Québec 172/C1
Jacques-Cartier (passage), Québec 174/E3
Jacques-Cartier (riv.), Québec 172/F2
Jacquet (riv.), New Bruns. 170/D1
Jacquet River, New Bruns. 170/E1
Jacquinot (bay), Papua N.G. 86/B2
Jaculpe (riv.), Brazil 132/F5
Jacumba, Calif. (92034) 204/J11
Jacupiranga, Brazil 135/B4
Jaddi, Ras (cape), Pakistan 59/H4
Jaddi, Ras (pt.), Pakistan 68/A4
Jade (bay), W. Germany 22/C2
Jadwin, Mo. (65501) 261/K8
Jaén, Peru 128/C5
Jaén (prov.), Spain 33/E4
Jaén, Spain 7/D5
Jaén, Spain 33/E4
Jaffa (cape), S. Australia 94/F7
Jaffna, Sri Lanka 68/E7
Jaffna, Sri Lanka 54/K9
Jaffray, Br. Col. 184/K5
Jaffrey, N.H. (03452) 268/C6
Jaffrey○, N.H. (03452) 268/C6
Jaffrey Center, N.H. (03454) 268/C6
Jafura (des.), Saudi Arabia 59/F5
Jagdalpur, India 68/E5
Jagdaqi, China 77/K1
Jagersfontein, S. Africa 118/D5
Jagfontein, S. Africa 118/G7
Jaghbub (Jarabub), Libya 111/D2
Jagin (riv.), Iran 66/L8
Jagna, Philippines 82/E6
Jagtial, India 68/D5
Jagua, Cuba 158/D2
Jaguaquara, Brazil 132/F6
Jaguara (res.), Brazil 135/C2
Jaguarão, Brazil 132/C11
Jaguariaíva, Brazil 132/D9
Jaguariaíva, Brazil 135/B4
Jaguaribe (riv.), Brazil 132/G4
Jagüey Grande, Cuba 158/D2
Jagüey Grande, Cuba 156/B2
Jahrom, Iran 59/F4
Jahrom, Iran 66/H6
Jaicoa, Cordillera (mts.), P. Rico 161/B1
Jaicós, Brazil 132/F4
Jailolo, Indonesia 85/H5
Jainca, China 77/F4
Jaipur, India 54/J7
Jaipur, India 68/D3
Jaisalmer, India 68/C3
Jajarm, Iran 66/K2
Jajce, Yugoslavia 45/C3
Jajpur, India 68/F4
Jakarta (cap.), Indonesia 2/Q6
Jakarta (cap.), Indonesia 54/M10
Jakarta (cap.), Indonesia 85/H1
Jakin, Georgia (31761) 217/C8
Jakobstad, Finland 18/N5
Jakubany, Czech. 41/F2
Jal, N. Mex. (88252) 274/F6
Jala, Mexico 150/G6
Jalacingo, Mexico 150/P1
Jalaid, China 77/K2
Jalalabad, Afghanistan 68/B2
Jalalabad, Afghanistan 59/K3
Jalama, West Bank 65/C3
Jalapa, Guatemala 154/B3
Jalapa, Ind. (†46952) 227/F3
Jalapa, Nicaragua 154/E4
Jalapa, S.C. (†29108) 296/D3
Jalapa Enríquez, Mexico 146/J8
Jalapa Enríquez, Mexico 150/P1
Jalbun, West Bank 65/C3
Jaleswar, Nepal 68/F3
Jalgaon, India 68/D4
Jalingo, Nigeria 106/G7
Jalisco (state), Mexico 150/H6
Jalkot, Pakistan 59/K2
Jalna, India 68/D4
Jalo, Libya 111/D2
Jalo (oasis), Libya 111/D2
Jalón (riv.), Spain 33/E2
Jalor, India 68/C3
Jalpa, Mexico 150/H6
Jalpa de Méndez, Mexico 150/N7
Jalpaiguri, India 68/F3
Jalpan, Mexico 150/K6
Jalq, Iran 66/N7
Jáltipan de Morelos, Mexico 150/M8
Jalud, West Bank 65/C3
Jaluit (atoll), Marshall Is. 87/G5
Jam, Iran 66/H7
Jama, Ecuador 128/B3
Jamaica 2/F5
Jamaica 146/L8
Jamaica, Cuba 158/K4
Jamaica (chan.), Haiti 156/C3
Jamaica, Iowa (50128) 229/E5
JAMAICA 158
JAMAICA 156/C3
Jamaica (chan.), Jamaica 156/C3
Jamaica, N.Y. (*11401) 276/N9
Jamaica○, Vt. (05343) 268/B5
Jamaica Plain, Mass. (02130) 249/C7
Jamaiké, Suriname 131/D4
Jamalpur, Bangladesh 68/F4
Jamalpur, India 68/F3
Jamama, Somalia 115/H3
Jamanota (mt.), Neth. Ant. 161/E10
Jamanxim (riv.), Brazil 132/C4
Jambi, Indonesia 85/C6
Jambi, Indonesia 54/M10
Jambuair (cape), Indonesia 85/B4
James (bay) 162/H5
James (riv.) 188/G2
James (bay), Canada 146/K4
James (isl.), Chile 138/D5
James (peak), Colo. 208/H3
James, Georgia (†31032) 217/E5
James, Iowa (†51101) 229/A3
James (pt.), Md. 245/N6
James, Miss. (†38748) 256/B4
James (lake), N.C. 281/E3
James (riv.), N. Dak. 282/N6
James (bay), Ontario 175/D2
James (bay), Québec 174/A2
James (isl.), S.C. 296/H6
James (riv.), S. Dak. 298/N5
James (riv.), Va. 307/O6
James A. Garfield Nat'l Hist. Site, Ohio 284/G2
James Bay, Ontario 177/E2
Jamesburg, N.J. (08831) 273/E3
James City, N.C. (28550) 281/R4
James City, Pa. (16734) 294/E2
James City (co.), Va. 307/P6
James Creek, Pa. (16657) 294/F5
Jameson, Mo. (64647) 261/E2
Jameson Park, S. Africa 118/J7
Jamesport, Mo. (64648) 261/E3
James Ross (isl.) 5/C16
James Ross (str.), N.W.T. 162/G1
James Ross (str.), N.W. Terrs. 187/J3
Jamestown, Ala. (†35973) 195/G2
Jamestown, Ark. (†72501) 202/G2
Jamestown, Calif. (95327) 204/E6
Jamestown, Colo. (80455) 208/J2
Jamestown, Ill. (†62238) 222/D5
Jamestown, Ind. (46147) 227/D5
Jamestown, Kansas (66948) 232/E2
Jamestown, Ky. (42629) 237/L7
Jamestown, La. (71045) 238/D2
Jamestown, Miss. (†39483) 256/E8
Jamestown, Mo. (65046) 261/G5
Jamestown, N.Y. 188/L2
Jamestown, N.Y. (14701) 276/B6
Jamestown, N.C. (27282) 281/K3
Jamestown, N. Dak. 188/G1
Jamestown, N. Dak. (58401) 282/N6
Jamestown (dam), N. Dak. 282/N6
Jamestown (res.), N. Dak. 282/N6
Jamestown, Ohio (45335) 284/C6
Jamestown, Pa. (16134) 294/A3
Jamestown, R.I. (02835) 249/J6
Jamestown○, R.I. (02835) 249/J6
Jamestown, S. Australia 88/F6
Jamestown, S. Australia 94/F5
Jamestown, S.C. (29453) 296/H5
Jamestown, Tenn. (38556) 237/M8
Jamestown, Va. (23081) 307/P6
Jamestown Nat'l Hist. Site, Va. 307/P6
Jamesville, N.Y. (13078) 276/H5
Jamesville, N.C. (27846) 281/R3
Jamesville, Va. (23398) 307/S5
Jamieson, Fla. (†32333) 212/B1
Jamieson, Oreg. (97909) 291/K3
Jamison, Nebr. (†68759) 264/E2
Jamison, S.C. (†29115) 296/F4
Jamma, Somalia 102/G4
Jammerbugt (bay), Denmark 21/C3
Jammu, India 54/J6
Jammu, India 68/D2
Jammu and Kashmir (state), India 68/D2
Jamnagar, India 68/B4
Jamnagar, India 54/H7
Jampur, Pakistan 59/K4
Jämsä, Finland 18/O6
Jamshedpur, India 54/K7
Jamshedpur, India 68/F4
Jämtland (co.), Sweden 18/J5
Jamursba (cape), Indonesia 85/J5
Janakpur, Nepal 68/F3
Jandaq, Iran 66/J3
Jandowae, Queensland 95/D5
Jane, Mo. (64846) 261/D9
Jane Lew, W. Va. (26378) 312/F4
Janesville, Calif. (96114) 204/E3
Janesville, Ill. (62435) 222/E4
Janesville, Iowa (50647) 229/J3
Janesville, Minn. (56048) 255/E6
Janesville (Smithmill), Pa. (16680) 294/F4
Janesville, Wis. 188/J2
Janesville, Wis. (53545) 317/H10
Janesville-Beloit, Wis. 317/80
Janetstown, Scotland 15/E2
Janeville, New Bruns. 170/E1
Jánico, Dom. Rep. 158/D5
Janikowo, Poland 47/C2
Janiuay, Philippines 82/D5
Jan Mayen (isl.), Norway 4/B10
Jan Mayen (isl.), Norway 7/D1
Janos, Mexico 150/F1
Jánoshalma, Hungary 41/E3
Jánosháza, Hungary 41/D3
Janów Lubelski, Poland 47/F3
Jansen, Colo. (†81082) 208/K8
Jansen, Nebr. (68377) 264/G4
Jansen, Sask. 181/G4
Jantetelco, Mexico 150/L2
Januária, Brazil 120/E4
Januária, Brazil 132/E6
Janvrin (isl.), Nova Scotia 168/G3
Jaora, India 68/D4
Japan 2/S4
Japan 54/R6
Japan (sea) 2/R4
Japan (sea) 54/P6
JAPAN 81
Japan (sea), Japan 81/G4
Japan (sea), N. Korea 81/G4
Japan (sea), S. Korea 81/G4
Japan (sea), U.S.S.R. 48/O6
Japurá, Brazil 132/G9
Japurá (riv.), Brazil 120/C3
Japurá (riv.), Brazil 132/G9
Jaquet (pt.), Dominica 161/E5
Jara, Cerrito (mt.), Bolivia 136/F6
Jara (hill), Paraguay 144/C1
Jarabacoa, Dom. Rep. 158/E5
Jarabub, Libya 102/E2
Jarabub, Libya 111/D2
Jaragua, Dom. Rep. 158/D6
Jaraíz de la Vera, Spain 33/D2
Jarales, N. Mex. (87023) 274/C4
Jarama (riv.), Spain 33/E2
Jaramillo, Argentina 143/C6
Jarandilla de la Vera, Spain 33/D2
Jarash, Jordan 65/D3
Jarbalo, Kansas (†66048) 232/G2
Jarbidge (riv.), Idaho 220/C7
Jarbidge, Nev. (89826) 266/F1
Jardim, Brazil 132/G4
Jardine, Mont. (†59030) 262/F5
Jardines de la Reina (arch.), Cuba 158/F3
Jardines de la Reina (arch.), Cuba 156/B2
Jargalant, Mongolia 77/J2
Jari (riv.), Brazil 120/D2
Jari (riv.), Brazil 132/C3
Järna, Sweden 18/G2
Jarnac, France 28/C5
Jaro, Philippines 82/E5
Jarocin, Poland 47/C3
Jaroměř, Czech. 41/C1
Jarosław, Poland 47/F4
Jaroso, Colo. (81138) 208/H8
Järpen, Sweden 18/H5
Jarrahdale, W. Australia 88/B3
Jarrahdale, W. Australia 92/B2
Jarratt, Va. (23867) 307/O7
Jarrettsville, Md. (21084) 245/M2
Jarrow, Alberta 182/E3
Jarrow, England 13/J3
Jarrow, England 10/F3
Jars (plain), Laos 72/D3
Jartai, China 77/G4
Jaruco, Cuba 158/C1
Jarud, China 77/K3
Järvenpää, Finland 18/O6
Jarvie, Alberta 182/D2
Jarvis (isl.), Pacific 87/K6
Jarvisburg, N.C. (27947) 281/T2
Jarvisville, W. Va. (†26462) 312/F4
Järvsö, Sweden 18/K6
Jask, Iran 59/G4

Jask, Iran 54/G7
Jask, Iran 66/K8
Jaslo, Poland 47/E4
Jasmin, Sask. 181/H4
Jasmine Estates, Fla. (†33568) 212/D3
Jason (isls.), 143/D7
Jason, Ky. (†41714) 237/06
Jason, N.C. (†28580) 281/04
Jasonville, Ind. (47438) 227/C6
Jasper, Ala. (35501) 195/D3
Jasper, Alberta 182/B3
Jasper, Alta. 162/E5
Jasper, Ark. (72641) 202/D1
Jasper, Fla. (32052) 212/D1
Jasper (co.), Georgia 217/E4
Jasper, Georgia (30143) 217/D2
Jasper (co.), Ill. 222/E4
Jasper (co.), Ind. 227/C2
Jasper, Ind. (47546) 227/D8
Jasper (co.), Iowa 229/G5
Jasper, Mich. (49248) 250/E7
Jasper, Minn. (56144) 255/B7
Jasper (co.), Miss. 256/F6
Jasper (co.), Mo. 261/D8
Jasper, Mo. (64755) 261/D8
Jasper, N.Y. (14855) 276/F6
Jasper, Ohio (45642) 284/D7
Jasper, Ontario 177/H3
Jasper, Oreg. (97401) 291/E3
Jasper (co.), S.C. 296/E6
Jasper, Tenn. (37347) 237/K10
Jasper (co.), Texas 303/K7
Jasper, Texas (75951) 303/L7
Jasper Nat'l Park, Alberta 182/A3
Jasper Nat'l Park, Alta. 162/E5
Jastrowie, Poland 47/C2
Jastrzębie Zdroj, Poland 47/D3
Jászapáti, Hungary 41/E3
Jászárokszállás, Hungary 41/E3
Jászberény, Hungary 41/E3
Jászfényszaru, Hungary 41/E3
Jászkarajenő, Hungary 41/E3
Jászkisér, Hungary 41/F3
Jászladány, Hungary 41/F3
Jataí, Brazil 120/D4
Jataí, Brazil 132/D7
Jatibonico, Cuba 158/F2
Jatibonico del Sur (riv.), Cuba 158/F3
Játiva, Spain 33/F3
Jaú, Brazil 132/D8
Jaú, Brazil 135/B3
Jauaperi (riv.), Brazil 132/A2
Jauari, Serra (mts.), Brazil 132/C3
Jauco, Cuba 158/K4
Jauf, Saudi Arabia 54/F7
Jauf, Saudi Arabia 59/C4
Jauja, Peru 128/E8
Jaumave, Mexico 150/K5
Jaun, Switzerland 39/D3
Jaunjelgava, U.S.S.R. 53/C2
Jaunpur, India 68/E3
Jauri, Iran 66/M6
Java (head), Indonesia 85/C7
Java (isl.), Indonesia 2/Q6
Java (isl.), Indonesia 54/M10
Java (isl.), Indonesia 85/J2
Java (sea), Indonesia 2/Q6
Java (sea), Indonesia 54/M10
Java (sea), Indonesia 85/D6
Java, S. Dak. (57452) 298/K3
Java, Va. (24565) 307/K7
Javari (riv.), Brazil 132/F9
Jávea, Spain 33/G3
Javier de Viana, Uruguay 145/C1
Jaworzno, Poland 47/B4
Jay, Fla. (32565) 212/B5
Jay (co.), Ind. 227/G4
Jay, Maine (04239) 243/C7
Jay○, Maine (04239) 243/C7
Jay, N.Y. (12941) 276/N2
Jay, Okla. (74346) 288/S2
Jay○, Vt. (†05859) 268/C2
Jay (peak), Vt. 268/B2
Jaya, Puncak (mt.), Indonesia 85/K6
Jayanca, Peru 128/B6
Jayapura, Indonesia 85/L6
Jayawijaya (range), Indonesia 85/K6
Jay Creek, North. Terr. 88/E4
Jay Em, Wyo. (82219) 319/H3
Jayess, Miss. (39641) 256/D8
Jayton, Texas (79528) 303/D4
Jayuya, P. Rico 156/G1
Jayuya, P. Rico 161/C2
Jaz Murian, Hamun-e (marsh), Iran 66/L7
Jaz Murian, Hamun-e (marsh), Iran 59/G4
Jean, Nev. (89019) 266/F7
Jean, Texas (†76374) 303/F4
Jean Côté, Alberta 182/B2
Jeanerette, La. (70544) 238/G7
Jeanette (bay), Newf. 166/C3
Jean Lafitte, La. (†70067) 238/K7
Jean Lafitte Nat'l Hist. Park, La. 238/P4
Jean-Marie River, N.W. Terrs. 187/F3
Jeanne Mance, New Bruns. 170/E1
Jeannette, Pa. (15644) 294/C5
Jean-Rabel, Haiti 158/B5
Jean-Rabel (pt.), Haiti 158/B5
Jebba, Nigeria 106/E7
Jebel Abyad (plat.), Sudan 111/E4
Jebel Aulia (dam), Sudan 102/F3
Jebel Aulia (dam), Sudan 111/F4
Jebel Dhanna, U.A.E. 59/F5
Jeberos, Peru 128/D5
Jeble, Syria 63/F5
Jedburg, S.C. (†29483) 296/G5
Jedburgh, Sask. 181/J4
Jedburgh, Scotland 10/E3
Jedburgh, Scotland 15/F5
Jeddah (Jidda), Saudi Arabia 59/C5
Jeddito, Ariz. (†86025) 198/E3
Jeddo, Mich. (48032) 250/G5
Jeddo, Pa. (†18224) 294/L3
Jeddore (cape), Nova Scotia 168/E4
Jeddore (harb.), Nova Scotia 168/F4
Jędrzejów, Poland 47/E3
Jefara (reg.), Libya 111/B1
Jefara (reg.), Tunisia 106/G2
Jeff, Ala. (†35804) 195/E1
Jeff, Ky. (41751) 237/P6
Jeff Davis (co.), Georgia 217/G7
Jeff Davis (co.), Texas 303/C11
Jeffers, Minn. (56145) 255/C6
Jeffers, Mont. (†59729) 262/E5
Jefferson (co.), Ala. 195/E3
Jefferson, Ala. (36745) 195/C6
Jefferson (co.), Ark. 202/G5
Jefferson, Ark. (72079) 202/F5
Jefferson (co.), Colo. 208/J3
Jefferson, Colo. (80456) 208/H4
Jefferson (co.), Fla. 212/C1
Jefferson (co.), Georgia 217/H4
Jefferson, Georgia (30549) 217/F2
Jefferson (co.), Idaho 220/F6
Jefferson (co.), Ill. 222/E5
Jefferson (co.), Ind. 227/G7
Jefferson, Ind. (†46041) 227/D4
Jefferson (co.), Iowa 229/K6
Jefferson, Iowa (50129) 229/E4
Jefferson (co.), Kansas 232/G2
Jefferson (co.), Ky. 237/K4
Jefferson (par.), La. 238/K7
Jefferson○, Maine (04348) 243/D7
Jefferson, Md. (21755) 245/J3
Jefferson, Mass. (01522) 249/G3
Jefferson (co.), Miss. 256/B7
Jefferson (co.), Mo. 261/L6
Jefferson (co.), Mont. 262/D4
Jefferson (riv.), Mont. 262/D5
Jefferson (co.), Nebr. 264/G4
Jefferson○, N.H. (03583) 268/D3
Jefferson (mt.), N.H. 268/E3
Jefferson (co.), N.Y. 276/J2
Jefferson, N.Y. (12093) 276/L6
Jefferson, N.C. (28640) 281/G2
Jefferson (co.), Ohio 284/J5
Jefferson (West Jefferson), Ohio (†43162) 284/D6
Jefferson, Ohio (44047) 284/J2
Jefferson (co.), Okla. 288/L6
Jefferson, Okla. (†73759) 288/L1
Jefferson (co.), Oreg. 291/F3
Jefferson, Oreg. (97352) 291/D3
Jefferson (mt.), Oreg. 291/F3
Jefferson (co.), Pa. 294/D3
Jefferson, Pa. (†15025) 294/B7
Jefferson, Pa. (15344) 294/B6
Jefferson (Codorus), Pa. (†17311) 294/J6
Jefferson, S.C. (29718) 296/G2
Jefferson, S. Dak. (57038) 298/S8
Jefferson (co.), Tenn. 237/P8
Jefferson (co.), Texas 303/K8
Jefferson, Texas (75657) 303/K5
Jefferson, Va. (†23139) 307/N5
Jefferson (co.), Wash. 310/B3
Jefferson (co.), W. Va. 312/L4
Jefferson (co.), Wis. 317/J9
Jefferson, Wis. (53549) 317/J10
Jefferson City (cap.), Mo. (65101) 261/H5
Jefferson City (cap.), Mo. 146/J6
Jefferson City (cap.), Mo. 188/H3
Jefferson City, Mont. (59638) 262/E4
Jefferson City, Tenn. (37760) 237/P8
Jefferson Davis (par.), La. 238/E6
Jefferson Davis (co.), Miss. 256/E7
Jefferson Heights, La. (70121) 238/O4
Jefferson Island, Mont. (†59721) 262/E5
Jefferson Manor, Va. (22303) 307/S3
Jefferson Nat'l Expansion Mem. Nat'l Hist. Site, Mo. 261/R3
Jefferson Proving Ground, Ind. 227/G7
Jeffersonton, Va. (22724) 307/N3
Jeffersontown, Ky. (40299) 237/L2
Jeffersonville, Georgia (31044) 217/F5
Jeffersonville, Ind. (47130) 227/F8
Jeffersonville, Ky. (40337) 237/O5
Jeffersonville, N.Y. (12748) 276/L7
Jeffersonville, Ohio (43128) 284/C6
Jeffersonville, Vt. (05464) 268/B2
Jeffrey (res.), Nebr. 264/D4
Jeffrey, W. Va. (25114) 312/C7
Jeffrey City, Wyo. (82310) 319/E3
Jeffrey's, Newf. 166/C4
Jef Jef es Seghin (plat.), Chad 111/D3
Jef Jef es Seghin (plat.), Libya 111/D3
Jega, Nigeria 106/E6
Jegenstorf, Switzerland 39/D2
Jeinemeni, Cerro (mt.), Chile 138/E6
Jeisyville, Ill. (†62568) 222/D4
Jejuí-Guazú (riv.), Paraguay 144/D4
Jēkabpils, U.S.S.R. 53/C2
Jēkabpils, U.S.S.R. 52/C3
Jekyll (isl.), Georgia 217/K8
Jeleń, Poland 47/B4
Jelenia Góra (prov.), Poland 47/B3
Jelenia Góra, Poland 47/B3
Jelgava, U.S.S.R. 7/G3
Jelgava, U.S.S.R. 53/B2
Jelgava, U.S.S.R. 52/B3
Jellico, Tenn. (37762) 237/N7
Jellico Creek, Ky. (†40769) 237/N7
Jellicoe, Ontario 177/H5
Jelling, Denmark 21/C6
Jelloway, Ohio (†43014) 284/F4
Jelm, Wyo. (82063) 319/G4
Jelšava, Czech. 41/F2
Jemaja (isl.), Indonesia 85/D5
Jemappes, Belgium 27/D8
Jember, Indonesia 85/K2
Jemez (riv.), N. Mex. 274/C3
Jemez Canyon (res.), N. Mex. 274/C3
Jemez Pueblo, N. Mex. (87024) 274/C3
Jemez Springs, N. Mex. (87025) 274/C3
Jeminay, China 77/C2
Jemison, Ala. (35085) 195/E5
Jemnice, Czech. 41/C2
Jemseg, New Bruns. 170/D3
Jemtland, Maine (†04783) 243/G1
Jena, E. Germany 22/D3
Jena, La. (71342) 238/F3
Jenaz, Switzerland 39/J3
Jenbach, Austria 41/A3
Jendouba, Tunisia 106/F1
Jeneponto, Indonesia 85/F7
Jenera, Ohio (45841) 284/C4
Jenifer, Ala. (†36268) 195/G3
Jenin, West Bank 65/C3
Jenison, Mich. (49428) 250/D6
Jenkinjones, W. Va. (24848) 312/D8
Jenkins (co.), Georgia 217/J5
Jenkins, Ky. (41537) 237/R6
Jenkins, Minn. (56456) 255/D4
Jenkins, Mo. (65677) 261/E9
Jenkinsburg, Georgia (30234) 217/E4
Jenkinsville, S.C. (29065) 296/E3
Jenkintown, Pa. (19046) 294/M5
Jenks, Okla. (74037) 288/P2
Jenner, Alberta 182/E4
Jennersdorf, Austria 41/C3
Jennerstown, Pa. (15547) 294/D5
Jennie, Ark. (71649) 202/H7
Jennie, Suriname 131/C3
Jennings, Ant. & Bar. 161/D11
Jennings, Fla. (32053) 212/C1
Jennings (co.), Ind. 227/F7
Jennings, Kansas (67643) 232/B2
Jennings, La. (70546) 238/E6
Jennings, Md. (†21536) 245/B2
Jennings, Mich. (†49651) 250/D4
Jennings, Mo. (63136) 261/R2
Jennings, N.S. Wales 97/F1
Jennings, Okla. (74038) 288/N2
Jennings Lodge, Oreg. (†97201) 291/B2
Jenny (creek), Oreg. 291/E5
Jenny Lake, Wyo. (†83012) 319/B2
Jenny Lind, Ark. (†72901) 202/B3
Jenny Lind, Calif. (†95252) 204/C9
Jenny Lind (isl.), N.W. Terrs. 187/H3
Jenolan Caves, N.S. Wales 97/E3
Jenpeg, Manitoba 179/J2
Jensen, Utah (84035) 304/E3
Jensen Beach, Fla. (33457) 212/F4
Jens Munk (isl.), N.W.T. 162/H2
Jens Munk (isl.), N.W. Terrs. 187/K3
Jepara, Indonesia 85/J2
Jequié, Brazil 120/E4
Jequié, Brazil 132/F6
Jequitinhonha, Brazil 132/F7
Jequitinhonha (riv.), Brazil 120/E4
Jequitinhonha (riv.), Brazil 132/F7
Jerablus, Syria 63/G4
Jerada, Morocco 106/D2
Jerauld (co.), S. Dak. 298/M5
Jérémie, Haiti 156/C3
Jérémie, Haiti 158/A6
Jeremoabo, Brazil 132/G5
Jeremy (riv.), Conn. 210/F2
Jerez, Spain 7/D5
Jerez de García Salinas, Mexico 150/H5
Jerez de la Frontera, Spain 33/C4
Jerez de los Caballeros, Spain 33/C3
Jericho, Ark. (†72327) 202/K3
Jericho, N.Y. (11753) 276/R6
Jericho, Queensland 95/C4
Jericho, Vt. (05465) 268/A2
Jericho○, Vt. (05465) 268/A2
Jericho, West Bank 65/C4
Jericho Center, Vt. (05465) 268/B3
Jerico Springs, Mo. (64756) 261/E7
Jeriel, Ky. (†41143) 237/R4
Jerilderie, N.S. Wales 97/C4
Jerimoth (hill), R.I. 249/G5
Jermyn, Pa. (18433) 294/L2
Jermyn, Texas (76057) 303/F4
Jerome, Ariz. (86331) 198/C4
Jerome, Ark. (71650) 202/G7
Jerome (co.), Idaho 220/D7
Jerome, Idaho (83338) 220/D7
Jerome, Ill. (†62701) 222/D4
Jerome, Mo. (65529) 261/J7
Jerome, Pa. (15937) 294/D5
Jeromesville, Ohio (44840) 284/F4
Jerry City, Ohio (43437) 284/C3
Jersey, Ark. (71651) 202/F7
Jersey (isl.), Chan. Is. 13/E8
Jersey (isl.), Chan. Is. 10/E6
Jersey, Georgia (30235) 217/E3
Jersey (co.), Ill. 222/C4
Jersey, Ohio (†43062) 284/E5
Jersey (bay), Virgin Is. (U.S.) 161/B4
Jersey City, N.J. 188/M2
Jersey City, N.J. (*07301) 273/B2
Jersey Mills, Pa. (17739) 294/H3
Jersey Shore, Pa. (17740) 294/H3
Jerseyside, Newf. 166/B3
Jerseytown, Pa. (†17815) 294/J3
Jersey Village, Texas (†77001) 303/J1
Jerseyville, Ill. (62052) 222/C4
Jerslev, Denmark 21/D3
Jerumenha, Brazil 132/F4
Jerusalem, Ark. (72080) 202/E3
Jerusalem (dist.), Israel 65/B4
Jerusalem (cap.), Israel 54/E6
Jerusalem (cap.), Israel 65/C4
Jerusalem (cap.), Israel 59/C3
Jerusalem, Ohio (43747) 284/H6
Jervis (inlet), Br. Col. 184/E5
Jervis (mt.), Chile 138/D8
Jervis Bay, Aust. Cap. Terr. 97/F4
Jervois Range, North. Terr. 88/F4
Jesenice, Yugoslavia 45/A2
Jeseník, Czech. 41/D1
Jeseníky (mts.), Czech. 41/D1
Jesenské, Czech. 41/F2
Jesi, Italy 34/D3
Jessamine (co.), Ky. 237/M5
Jesse, W. Va. (24849) 312/C7
Jessie, N. Dak. (58452) 282/O4
Jessieville, Ark. (71949) 202/D4
Jessnitz, E. Germany 22/E3
Jessore, Bangladesh 68/F4
Jessup, Pa. (18434) 294/F6
Jesterville, Md. (†21814) 245/P7
Jesuit Bend, La. (†70113) 238/K7
Jesup, Georgia (31545) 217/J7
Jesup, Iowa (50648) 229/J4
Jesús, Paraguay 144/E5
Jesús de Machaca, Bolivia 136/A5
Jesús de Otoro, Honduras 154/C3
Jesús María, Argentina 143/D3
Jesús María (reef), Mexico 150/L4
Jet, Okla. (73749) 288/K1
Jetersville, Va. (23083) 307/M6
Jetmore, Kansas (67854) 232/B3
Jett, Ky. (†40601) 237/M4
Jette, Belgium 27/B9
Jetts Creek, Ky. (†41382) 237/O6
Jever, W. Germany 22/B2
Jevíčko, Czech. 41/D2
Jewel Cave Nat'l Mon., S. Dak. 298/B6
Jewell, Georgia (31045) 217/G4
Jewell, Iowa (50130) 229/F4
Jewell (co.), Kansas 232/D2
Jewell, Kansas (66949) 232/D2
Jewell, Ohio (43530) 284/B3
Jewell, Oreg. (†97138) 291/D2
Jewell Ridge, Va. (24622) 307/E6
Jewett, Ill. (62436) 222/E4
Jewett, Ohio (43986) 284/H5
Jewett, Texas (75846) 303/H6
Jewett City, Conn. (06351) 210/H2
Jewish Aut. Obl., U.S.S.R. 48/O5
Jeypore, India 68/E5
Jhalawar, India 68/D4
Jhal Jhao, Pakistan 59/H4
Jhal Jhao, Pakistan 68/B3
Jhang Sadar, Pakistan 59/K3
Jhang Sadar, Pakistan 68/C2
Jhansi, India 68/D3
Jharsuguda, India 68/E4
Jhelum (riv.), India 68/C2
Jhelum, Pakistan 68/C2
Jhelum, Pakistan 59/K3
Jhelum (riv.), Pakistan 68/C2
Jhudo, Pakistan 68/B3
Jhunjhunu, India 68/D3
Jialing (riv.), China 54/M6
Jiamusi (Kiamusze), China 77/M2
Ji'an (Kian), China 77/J6
Jiande, China 77/J6
Jiangcheng, China 77/F7
Jiangmen (Kongmoon), China 77/H7
Jiangsu (Kiangsu) (prov.), China 77/K5
Jiangxi (Kiangsi) (prov.), China 77/J6
Jiangyou, China 77/G5
Jian'ou, China 77/J6
Jianshi, China 77/H5
Jianshui, China 77/F7
Jianyang, China 77/J6
Jiaohe, China 77/L3
Jiao Xian, China 77/K4
Jiaozuo (Tsiaotso), China 77/H4
Jiashan, China 77/J5
Jia Xian, China 77/H4
Jiaxing (Kashing), China 77/K5
Jiayin, China 77/M2
Jiayu, China 77/H6
Jiayuguan, China 77/E4
Jibaro, Cuba 158/F2
Jibhalanta (Uliastay), Mongolia 77/E2
Jibóia, Brazil 132/G8
Jibou, Romania 45/F2
Jibsh, Ras (cape), Oman 59/G5
Jicarilla, N. Mex. (†88313) 274/D5
Jicarilla Ind. Res., N. Mex. 274/B2
Jicarón (isl.), Panama 154/F7
Jičín, Czech. 41/C1
Jico, Mexico 150/P1
Jidda, Saudi Arabia 54/E7
Jidda, Saudi Arabia 59/C5
Jiexiu, China 77/H4
Jieyang, China 77/J7
Jifna, West Bank 65/C4
Jigalong Aboriginal Reserve, W. Australia 88/C4
Jigger, La. (71249) 238/G2
Jiggs, Nev. (†89801) 266/F2
Jiguaní, Cuba 158/H4
Jiguero (pt.), P. Rico 156/F1
Jiguero (pt.), P. Rico 161/A1
Jigüey (bay), Cuba 158/G2
Jigzhi, China 77/F5
Jihlava, Czech. 41/C2
Jihlava (riv.), Czech. 41/D2
Jihočeský (reg.), Czech. 41/C2
Jihomoravský (reg.), Czech. 41/D2
Jijel, Algeria 106/F1
Jijia (riv.), Romania 45/H2
Jijiga, Ethiopia 111/H6
Jijona, Spain 33/F3
Jilemnice, Czech. 41/C1
Jilib, Somalia 115/H3
Jilin (Kirin) (prov.), China 77/L3
Jilin (Kirin), China 77/L3
Jilin, China 54/O5
Jilotepec de Abasolo, Mexico 150/K7
Jim (lake), N. Dak. 282/N5
Jimaní, Dom. Rep. 158/C6
Jimbolia, Romania 45/E3
Jimena de la Frontera, Spain 33/D4
Jiménez, Chihuahua, Mexico 150/G3
Jiménez, Coahuila, Mexico 150/J2
Jim Falls, Wis. (54748) 317/D5
Jim Hogg (co.), Texas 303/F11
Jimma, Ethiopia 111/G6
Jimma, Ethiopia 102/F4
Jimsar, China 77/C3
Jim Thorpe, Pa. (18229) 294/L4
Jim Wells (co.), Texas 303/F10
Jim Woodruff (dam), Georgia 217/C9
Jinan (Tsinan), China 77/J4
Jinan, China 54/N6
Jincheng, China 77/H4
Jinchuan, China 77/F5
Jind, India 68/D3
Jindabyne, N.S. Wales 97/E5
Jindabyne (lake), N.S. Wales 97/E5
Jindalee, N.S. Wales 97/E4
Jindřichův Hradec, Czech. 41/C2
Jingbian, China 77/G4
Jingdezhen (Kingtehchen), China 77/J6
Jingellic, N.S. Wales 97/D4
Jinggu, China 77/F7
Jinghe, China 77/B3
Jinghong, China 77/F7
Jingtai, China 77/F4
Jingxi, China 77/G7
Jing Xian, Anhui, China 77/J5
Jing Xian, Hunan, China 77/H6
Jingyuan, China 77/F4
Jinhua (Kinhwa), China 77/J6
Jining (Tsining), Nei Monggol, China 77/H3
Jining (Tsining), Shandong, China 77/J4
Jinja, Uganda 115/F3
Jinja, Uganda 102/F4
Jinmen (Quemoy) (isl.), China 77/J7
Jinotega, Nicaragua 154/E4
Jinotepe, Nicaragua 154/D5
Jinping, China 77/G6
Jinsha Jiang (Yangtze) (riv.), China 77/E5
Jinshi (Tsingshih), China 77/H6
Jintotolo (chan.), Philippines 82/D5
Jinxi (Chinsi), China 77/K3
Jin Xian, China 77/K4
Jinzhou (Chinchow), China 77/K3
Jinzhou, China 54/N5
Jipijapa, Ecuador 128/B3
Jiran, Ethiopia 111/G6
Jirgalanta (Hovd), Mongolia 77/D2
Jiřkov, Czech. 41/B1
Jish, Israel 65/C1
Jishou, China 77/H6
Jisr esh Shughur, Syria 63/G5
Jiu (riv.), Romania 45/F3
Jiujiang (Kiukiang), China 77/J6
Jiulong, China 77/F6
Jiuquan (Kiuchüan), China 77/E4
Jixi (Kisi), China 77/M2
Jixi, China 54/P5
Ji Xian, China 77/H4
Jizan (Qizan), Saudi Arabia 59/D6
Jizera (riv.), Czech. 41/C1
Joaçaba, Brazil 132/D9
Joachimsthal, E. Germany 22/E2
Joachín, Mexico 150/Q2
Joana Peres, Brazil 132/D3
Joanico, Uruguay 145/B6
Joanna, S.C. (29351) 296/D3
João Monlevade, Brazil 135/E1
João Pessoa, Brazil 120/F3
João Pessoa, Brazil 132/H4
João Pinheiro, Brazil 132/E7
Joaquim Távora, Brazil 135/B3
Joaquin, Texas (75954) 303/L5
Joaquín Suárez, Canelones, Uruguay 145/B6
Joaquín Suárez, Colonia, Uruguay 145/B5
Joaquín V. González, Argentina 143/D2
Job (peak), Nev. 266/C3
Job, W. Va. (26274) 312/G5
Jobabo, Cuba 158/H3
Jobos, P. Rico 161/D3
Jobos (bay), P. Rico 161/D3
Job's Cove, Newf. 166/D2
Jobstown, N.J. (08041) 273/D3
Joch (pass), Switzerland 39/F3
Jódar, Spain 33/E4
Jo Daviess (co.), Ill. 222/C1
Jodhpur, India 54/J7
Jodhpur, India 68/C3
Jodie, W. Va. (26674) 312/D6
Jodoigne, Belgium 27/F7
Joe Batt's Arm, Newf. 166/D4
Joensuu, Finland 7/H2
Joensuu, Finland 18/R5
Joes, Colo. (80822) 208/O3
Joes (brook), Vt. 268/C3
Joetsu, Japan 81/H5
Joffre, Alberta 182/D3
Jofra (oasis), Libya 111/C2
Jõgeva, U.S.S.R. 53/D1
Joggins, Nova Scotia 168/D3
Joghatay, Kuh-e (mts.), Iran 66/K2
Jogjakarta (Yogyakarta), Indonesia 85/J2
Jogues, Ontario 175/D3
Johannesburg, Calif. (93528) 204/H8
Johannesburg, Mich. (49751) 250/E4
Johannesburg, S. Africa 102/E7
Johannesburg, S. Africa 2/L7
Johannesburg, S. Africa 118/H6
Johanngeorgenstadt, E. Germany 22/E3
John (riv.), Alaska 196/H1
John (cape), Nova Scotia 168/E3
John D. Rockefeller, Jr., Mem. Pkwy., Wyo. 319/B1
John Day, Oreg. (97845) 291/J3
John Day (dam), Oreg. 291/G2
John Day (riv.), Oreg. 291/G2
John Day (dam), Wash. 310/E5
John Day Fossil Beds Nat'l Mon., Oreg. 291/G3
John d'Or Prairie, Alberta 182/B5
Johnetta, Ky. (40439) 237/N6
John F. Kennedy Space Center, Fla. 212/F3
John F. Kennedy Nat'l Hist. Site, Mass. 249/C7
John H. Kerr (dam), Va. 307/M7
John Jay (mt.), Br. Col. 184/B2
John Martin (res.), Colo. 208/N6
John Muir Nat'l Hist. Site, Calif. 204/K1
John O'Groats, Scotland 15/E2
John Redmond (res.), Kansas 232/G3
Johns, Miss. (†39042) 256/E6
Johns, N.C. (†28352) 281/K5
Johns (isl.), S.C. 296/G6
Johnsburg, Minn. (†55909) 255/F7
Johnsburg, N.Y. (12843) 276/M3
Johnshaven, Scotland 15/F4
Johns Island, S.C. (29455) 296/G6
Johnson (co.), Ark. 202/C2
Johnson, Ark. (72741) 202/B1
Johnson (isl.), Chile 138/D5
Johnson (co.), Georgia 217/G5
Johnson (creek), Idaho 220/C5
Johnson (co.), Ill. 222/E6
Johnson (co.), Ind. 227/E6
Johnson, Ind. (†47565) 227/B8
Johnson (co.), Iowa 229/K5
Johnson (co.), Kansas 232/H3
Johnson, Kansas (67855) 232/A4
Johnson (co.), Ky. 237/R5
Johnson, Minn. (56250) 255/B5
Johnson (co.), Mo. 261/E5
Johnson (co.), Nebr. 264/H4
Johnson, Nebr. (68378) 264/J4
Johnson (lake), Nebr. 264/E4
Johnson (co.), Tenn. 237/T7
Johnson (co.), Texas 303/G5
Johnson, Vt. (05656) 268/B2
Johnson○, Vt. (05656) 268/B2
Johnson (co.), Wyo. 319/F1
Johnsonburg, N.J. (07846) 273/D2
Johnsonburg, Pa. (15845) 294/E3
Johnson City, N.Y. (13790) 276/J6
Johnson City, Oreg. (†97027) 291/B2
Johnson City, Tenn. (37601) 237/S8
Johnson City, Texas (78636) 303/F7
Johnson Creek, Wis. (53038) 317/J9
Johnsondale, Calif. (93238) 204/G8
Johnson Draw (dry riv.), Texas 303/C7
Johnsons (creek), Utah 304/B6
Johnsons Bayou, La. (†70631) 238/C7
Johnson's Crossing, Yukon 187/E3
Johnsons Landing, Br. Col. 184/J5
Johnsons Point, Ant. & Bar. 161/D11
Johnsonville, Ill. (62850) 222/E5
Johnsonville, N.Y. (12094) 276/O5
Johnsonville, S.C. (29555) 296/J4
Johnston (key), Fla. 212/E7
Johnston, Iowa (50131) 229/F5
Johnston (co.), N.C. 281/N4
Johnston (co.), Okla. 288/N6
Johnston (atoll), Pacific 87/K4
Johnston, S.C. (29832) 296/D4
Johnston (lakes), W. Australia 88/C6
Johnston, The (lakes), W. Australia 92/C6
Johnston City, Ill. (62951) 222/E6
Johnstone (str.), Br. Col. 184/D5
Johnstone, Scotland 10/A1
Johnstone, Scotland 15/B2
Johnstons Station, Miss. (†39666) 256/D8
Johnstown, Colo. (80534) 208/K2
Johnstown, Ireland 17/G6
Johnstown, Nebr. (69214) 264/D2
Johnstown, N.Y. (12095) 276/M4
Johnstown, N. Dak. (58245) 282/R3
Johnstown, Ohio (43031) 284/E5
Johnstown, Ontario 177/J3
Johnstown, Pa. 188/L2
Johnstown, Pa. (*15901) 294/D5
Johnsville, Ark. (†71648) 202/F7
Johnsville, Md. (†21791) 245/K2
Johor (Johore) (state), Malaysia 72/D7
Johor, Sungai (riv.), Malaysia 72/F5
Johor Baharu (Johore Bharu), Malaysia 72/F5
Johore (str.), Malaysia 72/E6
Johore Baharu, Malaysia 54/M9
Joice, Iowa (50446) 229/G2
Joigny, France 28/E3
Joiner, Ark. (72350) 202/K3
Joinville, Brazil 120/E5
Joinville, Brazil 132/D9
Joinville (isl.) 5/C16
Jojutla de Juárez, Mexico 150/L2
Jokkmokk, Sweden 18/L3
Jökulsá (riv.), Iceland 21/C1
Joli (pt.), Nova Scotia 168/D5
Jolicure, New Bruns. 170/F3
Joliet, Ill. 188/J2
Joliet, Ill. (*60431) 222/E2
Joliet, Mont. (59041) 262/G5
Joliette, N. Dak. (58246) 282/R2
Joliette (county), Québec 174/B3
Joliette (co.), Québec 172/C3
Joliette, Québec 172/D3
Jolietville, Ind. (†46074) 227/E4
Jolley, Iowa (50551) 229/D4
Jollytown, Pa. (†15352) 294/B6
Jolo, Philippines 82/C8
Jolo (isl.), Philippines 85/G4
Jolo (isl.), Philippines 82/C7
Jolon, Calif. (93928) 204/D7
Jomalig (isl.), Philippines 82/D3
Jo-Mary (lkes), Maine 243/E4
Jombang, Indonesia 85/K2
Jomda, China 77/E5
Jona, Switzerland 39/G2
Jonacatepec, Mexico 150/M2
Jonava, U.S.S.R. 53/C3
Jones (plain), Cambodia 72/E5
Jones (plain), Vietnam 72/E5
Jones, Ala. (36749) 195/E5
Jones (isls.), Alaska 196/J1
Jones (co.), Georgia 217/E5
Jones (co.), Iowa 229/L4
Jones, La. (71250) 238/G1
Jones, Mich. (49061) 250/D7
Jones (co.), Miss. 256/F7
Jones (beach), N.Y. 276/R7
Jones (co.), N.C. 281/P4
Jones (sound), N.W. Terrs. 187/K2
Jones (sound), N.W.T. 146/K2
Jones (sound), N.W.T. 162/M3
Jones, Okla. (73049) 288/M3
Jones (co.), S. Dak. 298/H6
Jones, Tenn. (†38006) 237/C9
Jones (co.), Texas 303/E5
Jonesboro, Ark. 188/H3
Jonesboro, Ark. (72401) 202/J2
Jonesboro, Georgia (30236) 217/D4
Jonesboro, Ill. (62952) 222/D6
Jonesboro, Ind. (46938) 227/F4
Jonesboro, La. (71251) 238/E2
Jonesboro○, Maine (04648) 243/J6
Jonesboro, Tenn. (37659) 237/R8
Jonesborough, N. Ireland 17/J3
Jonesburg, Mo. (63351) 261/K5

Jones Creek, Texas (†77541) 303/J9
Jonesdale, Wis. (†53565) 317/F10
Jones Mills, Ark. (72105) 202/E5
Jones Mills, Pa. (15646) 294/D5
Jonesport, Maine (04649) 243/H6
Jonesport○, Maine (04649) 243/H6
Jones Springs, W. Va. (25427) 312/K4
Jonestown, Miss. (38639) 256/D2
Jonestown, Pa. (17038) 294/K5
Jonesville, Alaska (99674) 196/B1
Jonesville, Ind. (47247) 227/F6
Jonesville, Ky. (41052) 237/M3
Jonesville, La. (71343) 238/D2
Jonesville, Mich. (49250) 250/E6
Jonesville, N.C. (28642) 281/H2
Jonesville, S.C. (29353) 296/D2
Jonesville, Va. (24263) 307/B7
Jonglei, Sudan 111/F6
Joniškis, U.S.S.R. 53/B2
Jönköping (co.), Sweden 18/H8
Jönköping, Sweden 7/F3
Jonquière, Que. 162/J6
Jonquière, Québec 172/F1
Jonquière, Québec 174/C3
Jonuta, Mexico 150/N7
Jonzac, France 28/C5
Joplin, Mo. (64801) 261/C8
Joplin, Mo. 146/J6
Joplin, Mo. 188/H3
Joplin, Mont. (59531) 262/F2
Joppa, Ala. (35087) 195/E2
Joppa, Ill. (62953) 222/E6
Joppa, Tenn. (37861) 237/O8
Joppatowne, Md. (†21085) 245/N3
Jorat (mt.), Switzerland 39/C3
Jordan 2/L4
Jordan 54/E6
JORDAN 59/C3
Jordan (dam), Ala. 195/F5
Jordan (lake), Ala. 195/F5
Jordan, (creek) Idaho 220/A7
Jordan, Iowa (50036) 229/F4
Jordan (riv.), Israel 65/D3
Jordan (riv.), Jordan 65/D3
Jordan, Minn. (55352) 255/E6
Jordan, Mont. (59337) 262/J3
Jordan, N.Y. (13080) 276/H4
Jordan, B. Everett (lake), N.C. 281/M3
Jordan (bay), Nova Scotia 168/C5
Jordan (lake), Nova Scotia 168/C5
Jordan (riv.), Nova Scotia 168/C5
Jordan (creek), Oreg. 291/K5
Jordan, S.C. (†29102) 296/G4
Jordan, Utah 304/C3
Jordan Falls, Nova Scotia 168/C5
Jordan River, Sask. 181/H2
Jordan Valley, Oreg. (97910) 291/K5
Jorge Montt (isl.), Chile 138/D9
Jorm, Afghanistan 68/C1
Jorm, Afghanistan 59/K2
Jörn, Sweden 18/M4
Jornada del Muerto (valley), N. Mex. 274/C5
Jorquera (riv.), Chile 138/B6
Jõrva-Jaani, U.S.S.R. 53/D1
Jos, Nigeria 106/F7
Jos, Nigeria 102/C4
Jos (plat.), Nigeria 106/F7
José Abad Santos, Philippines 82/E8
José Agustín Palacios, Bolivia 136/B3
José Batlle y Ordóñez, Uruguay 145/D4
José Cardel, Mexico 150/Q1
José de San Martín, Argentina 143/B5
José Enrique Rodó, Uruguay 145/D4
José Ignacio (lag.), Uruguay 145/E5
José M. Micheo, Argentina 143/D3
José Panganiban, Philippines 82/G3
José Pedro Varela, Uruguay 145/E4
Joseph (lake), Newf. 166/B3
Joseph (lake), Ontario 177/E2
Joseph, Oreg. (97846) 291/K2
Joseph (creek), Oreg. 291/K2
Joseph, Utah (84739) 304/B5
Joseph Bonaparte (gulf) 88/D2
Joseph Bonaparte (gulf), Australia 87/C7
Joseph Bonaparte (gulf), North. Terr. 93/A3
Joseph Bonaparte (gulf), W. Australia 92/E1
Joseph City, Ariz. (86032) 198/E4
Joseph, Ala. (†36530) 195/C10
Josephine (co.), Oreg. 291/D5
Josephine (creek), Oreg. 291/D5
Joshinetsu-Kogen National Park, Japan 81/J5
Joshua (pt.), Conn. 210/E4
Joshua Tree, Calif. (92252) 204/J9
Joshua Tree Nat'l Mon., Calif. 204/J10
Jostedal, Norway 18/E6
Jostedalsbreen (glac.), Norway 18/E6
Jost Van Dyke (isl.), Virgin Is. (Br.) 161/C3
Jost Van Dyke (isl.), Virgin Is. (Br.) 156/G1
Joubert, S. Dak. (†57344) 298/M7
Jourdanton, Texas (78026) 303/F9
Joure, Netherlands 27/H4
Joussard, Alberta 182/B2
Joux (lake), Switzerland 39/B3
Jovellanos, Cuba 156/B2
Jovellanos, Cuba 158/D1
Joveyn (riv.), Iran 66/K2
Joy, Ill. (61260) 222/C2
Joy, Ky. (†42047) 237/E6
Joyce, La. (71440) 238/D3
Joyce, Wash. (98343) 310/B2
Joyce's Country (dist.), Ireland 17/B4
Joyo, Japan 81/J7
Juab (co.), Utah 304/A4
Juana Díaz, P. Rico 161/C2
Juan Aldama, Mexico 150/H4

Juan D. Jackson, Uruguay 145/C4
Juan de Fuca (str.) 146/F5
Juan de Fuca (str.), Br. Col. 162/D6
Juan de Fuca (str.), Br. Col. 184/J4
Juan de Fuca (str.), Wash. 188/A1
Juan de Fuca (str.), Wash. 310/A2
Juan de Mena, Paraguay 144/D4
Juan de Nova (isl.), Réunion 102/G6
Juan de Nova (isl.), Réunion 118/G3
Juan Fernández (isls.), Chile 2/E7
Juan Fernández (isls.), Chile 120/B6
Juangriego, Venezuela 124/G2
Juani (isl.), Tanzania 115/G5
Juanita, N. Dak. (58453) 282/N4
Juanita, Wash. (98033) 310/B1
Juanjuí, Peru 128/D6
Juan L. Lacaze, Uruguay 145/B5
Juan Stuven (isl.), Chile 138/D7
Juárez, Argentina 143/D4
Juárez, Mexico 150/J3
Juazeiro, Brazil 132/G5
Juàzeiro, Brazil 120/E3
Juazeiro do Norte, Brazil 132/G4
Juàzeiro do Norte, Brazil 120/F3
Juba, Sudan 111/F7
Juba, Sudan 102/F4
Jubail, Saudi Arabia 59/F4
Jubba, Saudi Arabia 59/D4
Jubbada Hoose (prov.), Somalia 115/H3
Jubbulpore (Jabalpur), India 68/D4
Jubilee (lake), W. Australia 88/D5
Juby (cape), Morocco 106/B3
Júcar (riv.), Spain 7/D5
Júcar (riv.), Spain 33/F3
Júcaro, Cuba 158/F2
Juchipila, Mexico 150/H6
Juchique de Ferrer, Mexico 150/Q1
Juchitán de Zaragoza, Mexico 150/M8
Jucuarán, El Salvador 154/C4
Jucuapa, El Salvador 154/C4
Jud, N. Dak. (58454) 282/N6
Juda, Wis. (53550) 317/H10
Judaea (reg.), Israel 65/B5
Judaea (reg.), Jordan 65/C4
Judas (pt.), C. Rica 154/E6
Juquiá, Brazil 135/C4
Jur (riv.), Sudan 111/E6
Jupiter, Fla. (33458) 212/F5
Jupiter, N.C. (28787) 281/D4
Jupiter Island, Fla. (†33455) 212/F4
Juquiá, Brazil 135/C4
Jur (riv.), Sudan 111/E6
Judenburg, Austria 41/C3
Judibana, Venezuela 124/C2
Judique, Nova Scotia 168/G3
Judith (riv.), Mont. 262/G3
Judith (isl.), Scotland 10/D3
Judith, R. I. 249/J7
Judith Basin (co.), Mont. 262/F4
Judith Gap, Mont. (59453) 262/G4
Judson, Ind. (47856) 227/C5
Judson, Minn. (56055) 255/D6
Judson, N. Dak. (†58563) 282/H6
Judsonia, Ark. (72081) 202/G3
Judyville, Ind. (†47993) 227/C4
Juelsminde, Denmark 21/D6
Juhu, India 68/B7
Juichin (Ruijin), China 77/J6
Juigalpa, Nicaragua 154/E4
Juist (isl.), W. Germany 22/B2
Juiz de Fora, Brazil 120/E5
Juiz de Fora, Brazil 135/E2
Juiz de Fora, Brazil 132/F8
Jujuy (prov.), Argentina 143/C1
Jujuy, Argentina 143/C1
Jujuy, Argentina 120/C5
Jukskei (riv.), S. Africa 118/H6
Julesburg, Colo. (80737) 208/P1
Juli, Peru 128/H11
Juliaca, Peru 120/B4
Juliaca, Peru 128/G10
Julia Creek, Queensland 88/G4
Julia Creek, Queensland 95/B4
Julietta, Idaho (83535) 220/B3
Julian, Calif. (92036) 204/J10
Julian, Nebr. (68379) 264/J4
Julian, N.C. (27283) 281/K3
Julian, Pa. (16844) 294/G4
Julian Alps (range), Italy 34/D1
Julianatop (mt.), Suriname 131/C4
Julianehåb, Greenland. 4/D12
Julianehåb, Greenland 2/G2
Julianehåb, Greenland 146/P3
Jülich, W. Germany 22/B3
Juliette, Georgia (31046) 217/E4
Juliff, Texas (77583) 303/J3
Julio María Sanz, Uruguay 145/E4
Juliustown, N.J. (08042) 273/D3
Jullundur, India 68/D2
Jumbilla, Peru 128/D6
Jumbo, Okla. (†74523) 288/P6
Jumilla, Spain 33/F3
Jumla, Nepal 68/E2
Jumna (riv.), India 68/E3
Jump (riv.), Wis. 317/E5
Jumpertown, Miss. (†38829) 256/G1
Jumping Branch, W. Va. (25969) 312/E7
Jump River, Wis. (54434) 317/E5
Junagadh, India 68/B4
Junaina, Saudi Arabia 59/D5
Juncal, Argentina 143/F6
Juncos, P. Rico 161/E2
Juncos, P. Rico 156/G1
Junction, Ill. (62954) 222/E6
Junction, Texas (76849) 303/E7
Junction, Utah (84740) 304/B5
Junction, W. Va. (26824) 312/J4
Junction City, Ark. (71749) 202/E7
Junction City, Georgia (31812) 217/C5
Junction City, Ill. (61601) 222/D5
Junction City, Kansas (66441) 232/E2
Junction City, Ky. (40440) 237/M5
Junction City, La. (71749) 238/E1
Junction City, Mo. (†63645) 261/M7
Junction City, Ohio (43748) 284/F4
Junction City, Oreg. (97448) 291/D3
Junction City, Wis. (54443) 317/G6
Jundah, Queensland 88/G5
Jundah, Queensland 95/B5
Jundial, Brazil 135/C3
Jundial, Brazil 132/E8
Juneau, Alaska 146/E4
Juneau (cap.), Alaska 188/E6
Juneau (cap.), Alaska (99801) 196/N1
Juneau, U.S. 4/D16
Juneau, U.S. 2/C3
Juneau (co.), Wis. 317/F8

Juneau, Wis. (53039) 317/J9
Juneda, Spain 33/G2
Junee, N.S. Wales 97/D4
June in Winter (lake), Fla. 212/E4
June Lake, Calif. (93529) 204/G6
June Park, Fla. (†32901) 212/F3
Jungar, China 77/H4
Jungfrau (mt.), Switzerland 39/E3
Jungfraujoch, Switzerland 39/E3
Junggar Pendi (desert basin), China 77/C2
Junglei (prov.), Sudan 111/F6
Juniata, Nebr. (68955) 264/F4
Juniata (co.), Pa. 294/H4
Juniata (riv.), Pa. 294/G5
Juniata Terrace, Pa. (†17044) 294/G4
Junín, Argentina 143/F7
Junín, Argentina 120/C6
Junín (dept.), Peru 128/E8
Junín, Peru 128/E8
Junín (lake), Peru 128/E8
Junín de los Andes, Argentina 143/B4
Junior, W. Va. (26275) 312/G5
Juniper (mts.), Ariz. 198/C3
Juniper (mt.), Colo. 208/C1
Juniper, Georgia (31801) 217/C6
Juniper, New Bruns. 170/C2
Juniper (creek), S.C. 296/H2
Junius, S. Dak. (†57042) 298/P6
Juniye, Lebanon 63/F5
Junlian, China 77/F6
Juno, Georgia (30534) 217/D2
Juno, Tenn. (†38351) 237/E9
Juno, Texas (76943) 303/C7
Juno Beach, Fla. (†33404) 212/F5
Junuosuando, Sweden 18/M3
Juntura, Oreg. (97911) 291/K4
Jun Xian, China 77/H5
Juojärvi (lake), Finland 18/Q5
Jupiter, Fla. (33458) 212/F5
Jupiter, N.C. (28787) 281/D4
Jupiter Island, Fla. (†33455) 212/F4
Juquiá, Brazil 135/C4
Jur (riv.), Sudan 111/E6
Jura (dept.), France 28/F4
Jura (mts.), France 28/F4
Jura (isl.), Scotland 10/D3
Jura (isl.), Scotland 15/C5
Jura (riv.), Scotland 15/C5
Jura (sound), Scotland 15/C5
Jura (sound), Scotland 10/D3
Jura (canton), Switzerland /D2
Jura (mts.), Switzerland 39/B3
Jurado, Colombia 126/B4
Jurbarkas, U.S.S.R. 53/B3
Jurmala, U.S.S.R. 53/B2
Jurmala, U.S.S.R. 52/B3
Jurong, Singapore 72/E6
Juruá (riv.), Brazil 120/C3
Juruá (riv.), Brazil 132/G10
Juruá (riv.), Peru 128/F7
Juruena, Brazil 132/B6
Juruena (riv.), Brazil 120/D4
Juruena (riv.), Brazil 132/B5
Juruti, Brazil 132/B3
Jusepín, Venezuela 124/G2
Juskatla, Br. Col. 184/A3
Jussy, Switzerland 39/B4
Justice, Ill. (†60458) 222/B6
Justice, Manitoba 179/C4
Justice, W. Va. (24851) 312/C7
Justiceburg, Texas (79330) 303/C5
Justin, Texas (76247) 303/F1
Justus, Ohio (†44662) 284/F4
Jutai (riv.), Brazil 132/G9
Jüterbog, E. Germany 22/E3
Jutiapa, Guatemala 154/B3
Jutiapa, Honduras 154/D3
Juticalpa, Honduras 154/D3
Jutland (pen.), Denmark 21/C5
Jutland (pen.), Denmark 18/F9
Jutland, N.J. (08809) 273/D2
Juuka, Finland 18/Q5
Juventud (municipio especial), Cuba 158/C2
Juventud (isl.), Cuba 146/K7
Juventud, Isla de la (Pines), Cuba 158/B3
Juventud (Pines), Cuba 156/A2
Juwara, Oman 59/G6
Ju Xian, China 77/J4
Juye, China 77/J4
Jyderup, Denmark 21/E6
Jylland (Jutland) (pen.), Denmark 21/C5
Jyske Ås (hills), Denmark 21/D3
Jyväskylä, Finland 7/G2
Jyväskylä, Finland 18/O5

K

K2 (mt.) 54/J6
K2 (mt.), Pakistan 68/D1
Kaawawa, Hawaii (96730) 218/F1
Kaabong, Uganda 115/F3
Kaala (mt.), Hawaii 218/E1
Kaanapali, Hawaii (96761) 218/H2
Kaba (Habahe), China 77/C2
Kaba, Hungary 41/F2
Kabacan, Philippines 82/E7
Kabaena (isl.), Indonesia 85/G7
Kabala, S. Leone 106/B7
Kabale, Uganda 115/E4
Kabalo, Zaire 115/E5
Kabambare, Zaire 115/E4
Kabardin-Balkar A.S.S.R., U.S.S.R. 48/E5
Kabardin-Balkar A.S.S.R., U.S.S.R. 52/F6
Kabare, Zaire 115/E4
Kabarega Nat'l Park, Uganda 115/F3
Kabasalan, Philippines 82/D7
Kabba, Nigeria 106/F7
Kabetogama, Minn. (†56669) 255/F2
Kabetogama (lake), Minn. 255/E2

Kabinakagami (riv.), Ontario 177/J5
Kabin Buri, Thailand 72/D4
Kabinda, Zaire 115/D5
Kabir Kuh (mts.), Iran 66/E4
Kabompo, Zambia 115/D6
Kabompo (riv.), Zambia 115/D6
Kabong, Malaysia 85/E5
Kabongo, Zaire 115/E5
Kabud Gonbad, Iran 66/L2
Kabul (cap.), Afghanistan 59/J3
Kabul (cap.), Afghanistan 68/B2
Kabul (cap.), Afghanistan 54/J4
Kabul (cap.), Afghanistan 2/N4
Kabul (riv.), Afghanistan 59/K3
Kabul (riv.), Afghanistan 68/B2
Kabunda, Zaire 115/E6
Kabwe (harb.), N. Zealand 100/D2
Kabwe, Zambia 102/E6
Kabwe, Zambia 115/E6
Kabylia (reg.), Algeria 106/E1
Kachemak, Alaska (†99663) 196/B2
Kachemak (bay), Alaska 196/B2
Kachess (lake), Wash. 310/D3
Kachin (state), Burma 72/C1
Kachug, U.S.S.R. 48/L4
Kaçkar Daği (mt.), Turkey 63/J2
Kackley, Kansas (†66948) 232/E2
Kadañ, Czech. 41/B1
Kadan Kyun (isl.), Burma 72/C4
Kadavu (Kandavu) (isl.), Fiji 87/H7
Kadayanallur, India 68/D7
Kadei (riv.), Cameroon 115/C3
Kadei (riv.), Cent. Afr. Rep. 115/C3
Kadei (riv.), Congo 115/C3
Kadıköy, Turkey 63/D6
Kadina, S. Australia 88/F6
Kadina, S. Australia 94/F5
Kadınhanı, Turkey 63/E4
Kadiolo, Mali 106/C6
Kadiri, India 68/D6
Kadirli, Turkey 63/F4
Kadiyevka (Stakhanov), U.S.S.R. 52/E5
Kadmat (isl.), India 68/C6
Kadoka, S. Dak. (57543) 298/F6
Kadoma, Japan 81/J7
Kadoma (Gatooma), Zimbabwe 118/D3
Kadugli, Sudan 111/E5
Kadugli, Sudan 102/E3
Kaduna (state), Nigeria 106/F6
Kaduna, Nigeria 102/C3
Kaduna, Nigeria 106/F6
Kaduna (riv.), Nigeria 106/F7
Kadzherom, U.S.S.R. 52/J2
Kaech'ón, N. Korea 81/B4
Kaédi, Mauritania 106/B6
Kaédi, Mauritania 102/A5
Kaélé, Cameroon 115/B1
Kaena (pt.), Hawaii 218/D1
Kaeo, N. Zealand 100/D1
Kaesŏng, N. Korea 81/B5
Kaf, Saudi Arabia 59/C3
Kafan, U.S.S.R. 52/G7
Kafar Kanna, Israel 65/C2
Kaffa (prov.), Ethiopia 111/G6
Kaffrine, Senegal 106/A6
Kafia Kingi, Sudan 111/D6
Kafirévs (cape), Greece 45/G6
Kafr Yasif, Israel 65/C2
Kafue, Zambia 115/E7
Kafue (riv.), Zambia 115/E7
Kafue Nat'l Park, Zambia 115/E6
Kaga, Japan 81/H5
Kaga Bandoro, Cent. Afr. Rep. 115/C2
Kagalaska (isl.), Alaska 196/L4
Kagan, U.S.S.R. 48/G6
Kagawong, Ontario 177/B2
Kagawong (lake), Ontario 177/B2
Kagera Nat'l Park, Rwanda 115/F4
Kağithane, Turkey 63/D6
Kağizman, Turkey 63/K2
Kagoshima (pref.), Japan 81/E8
Kagoshima, Japan 81/E8
Kagoshima, Japan 54/O6
Kagoshima (bay), Japan 81/E8
Kagul, U.S.S.R. 52/C5
Kaguyak, Alaska (†99608) 196/H3
Kahala, Hawaii (†96801) 218/D5
Kahala (pt.), Hawaii 218/D1
Kahaluu, Hawaii (96744) 218/E2
Kahama, Tanzania 115/F4
Kahana, Hawaii (96717) 218/F1
Kahana (bay), Hawaii 218/F1
Kahayan (riv.), Indonesia 85/E6
Kahemba, Zaire 115/C5
Kahiltna (riv.), Alaska 196/B1
Kahlotus, Wash. (99335) 310/G4
Kah-Nee-Ta, Oreg. (†97761) 291/F3
Kahoka, Mo. (63445) 261/J2
Kahoolawe (isl.), Hawaii 188/F5
Kahoolawe (isl.), Hawaii 87/L4
Kahoolawe (isl.), Hawaii 218/H3
Kahouanne (isl.), Guadeloupe 161/A6
Kahramanmaraş (prov.), Turkey 63/G4
Kâhta, Turkey 63/H4
Kahuku, Hawaii (96731) 218/F1
Kahuku, Hawaii 188/F5
Kahuku (pt.), Hawaii (96732) 218/E1
Kahului, Hawaii (96732) 218/J2
Kahului, Hawaii 188/F5
Kahului (harb.), Hawaii 218/J1
Kai (isls.), Indonesia 85/J7
Kaiama, Nigeria 106/F7
Kaiapit, Papua N. G. 85/B7
Kaiapoi, N. Zealand 100/D5
Kaibab (plat.), Ariz. 198/C2
Kaibab Ind. Res., Ariz. 198/C2
Kaibito, Ariz. (86053) 198/D2
Kaibito (plat.), Ariz. 198/D2
Kaieteur (fall), Guyana 131/B3
Kaifeng, China 77/H5
Kaifeng, China 54/N6
Kaikohe, N. Zealand 100/D1
Kaikoura, N. Zealand 100/E5
Kaikoura (pen.), N. Zealand 100/E5
Kaikoura (range), N. Zealand 100/D5
Kaili, China 77/G6

Kailu, China 77/K3
Kailua (Kailua Kona), Hawaii, Hawaii (96740) 218/H5
Kailua, Hawaii (96734) 218/F2
Kailua (bay), Hawaii 218/F5
Kailua (bay), Hawaii 218/F5
Kailua Kona, Hawaii (96740) 218/F5
Kaimana, Indonesia 85/J6
Kaimanawa (range), N. Zealand 100/E3
Kaimu, Hawaii (†96778) 218/J6
Kaimuki, Hawaii (96816) 218/D4
Kainaliu, Hawaii (96750) 218/G5
Kainaliu, Hawaii 188/F6
Kainan, Japan 5/B10
Kainantu, N. Zealand 100/E7
Kainji (res.), Nigeria 106/E6
Kaipara (harb.), N. Zealand 100/D2
Kaipara (riv.), N. Zealand 100/A1
Kaiparowits (plat.), Utah 304/C6
Kaipokok (bay), Newf. 166/B2
Kaipokok (riv.), Newf. 166/B3
Kairouan, Tunisia 106/F1
Kairuku, Papua N. G. 85/B7
Kaiser, Mo. (65047) 261/K6
Kaiseregg (mt.), Switzerland 39/D3
Kaiserslautern, W. Germany 22/B4
Kaiserstuhl (mt.), W. Germany 22/B4
Kaitaia, N. Zealand 100/D1
Kaitangata, N. Zealand 100/C7
Kaitumälv (riv.), Sweden 18/M3
Kaiwi (chan.), Hawaii 218/E6
Kaiyuan, Liaoning, China 77/K3
Kaiyuan, Yunnan, China 77/F7
Kaiyuh (mts.), Alaska 196/G2
Kaizuka, Japan 81/H8
Kajaani, Finland 7/G2
Kajaani, Finland 18/P4
Kajabbi, Queensland 88/G3
Kajabbi, Queensland 95/A4
Kajiado, Kenya 115/G5
Kajok, Sudan 111/E6
Kaka, Cent. Afr. Rep. 115/E2
Kaka, Sudan 111/F5
Kakabeka Falls, Ontario 177/G5
Kakabeka Falls, Ontario 175/B3
Kakamega, Kenya 115/F3
Kake, Alaska (99830) 196/M1
Kakhk, Iran 66/K3
Kakhonak, Alaska (†99647) 196/H3
Kakhovka, U.S.S.R. 52/D5
Kakhovka (res.), U.S.S.R. 48/D5
Kakhovka (res.), U.S.S.R. 52/D5
Kakinada, India 54/K8
Kakinada, India 68/E5
Kakisa, N.W. Terrs. 187/G3
Kakogawa, Japan 81/G6
Kakkiviak (cape), Newf. 166/B1
Kaktovik, Alaska (99747) 196/K1
Kakwa (riv.), Alberta 182/A2
Kalaa-Kebira, Tunisia 106/F1
Kalabahi, Indonesia 85/G7
Kalabo, Zambia 115/D6
Kalach, U.S.S.R. 52/F4
Kalachinsk, U.S.S.R. 48/H4
Kalach-na-Donu, U.S.S.R. 52/F5
Kaladan (riv.), Burma 72/B2
Kaladar, Ontario 177/H3
Kalae, Hawaii (†96757) 218/G1
Ka La (cape), Hawaii 218/G7
Kalahari (des.) 102/E7
Kalahari (des.), Botswana 118/C4
Kalahari (des.), Namibia 118/C4
Kalahari Gemsbok Nat'l Park, S. Africa 118/C5
Kalaheo, Hawaii (96741) 218/C2
Kalajoki, Finland 18/N4
Kalajoki (riv.), Finland 18/O4
Kalakan, U.S.S.R. 48/M4
Kalaloch, Wash. (†98331) 310/A3
Kalam, Pakistan 68/C1
Kalama, Wash. (98625) 310/C4
Kalama (riv.), Wash. 310/C4
Kálamai, Greece 7/G5
Kálamai, Greece 45/F7
Kalamazoo, Mich. 188/J2
Kalamazoo (co.), Mich. 250/D6
Kalamazoo, Mich. (*49001) 250/D6
Kalamazoo, Mich. 250/D6
Kalamata (falls), Tanzania 115/F5
Kalambo (falls), Zambia 115/F5
Kalamo, Mich. (†49096) 250/D6
Kalampáka, Greece 45/E6
Kalamunda, W. Australia 88/B2
Kalan, Turkey 63/H3
Kalao (isl.), Indonesia 85/G7
Kalaoa, Hawaii (†96740) 218/G5
Kalaotoa (isl.), Indonesia 85/G5
Kalapana, Hawaii (†96778) 218/J6
Kalasin, Thailand 72/D3
Kalat (Qalat), Afghanistan 68/B2
Kalat (Qalat), Afghanistan 59/J3
Kalat, Pakistan 54/H7
Kalat, Pakistan 59/J4
Kalat, Pakistan 68/B3
Kalâtdlit-Nunât (Greenland) 4/B12
Kalâtdlit-Nunât (Greenland) 146/P2
Kalaupapa, Hawaii (96742) 218/G1
Kalaupapa (pen.), Hawaii 218/H1
Kalaupapa Nat'l Hist. Park, Hawaii 218/H1
Kalávrita, Greece 45/F6
Kalawao, Hawaii 218/G1
Kalbarri, W. Australia 92/A4
Kale, Turkey 63/D4
Kaledin, Br. Col. 184/H5
Kalegauk (isl.), Burma 72/C4
Kalehe, Zaire 115/E4
Kaleida, Manitoba 179/D5
Kalemie, Zaire 115/E5
Kalemyo, Burma 72/B2
Kaleva, Mich. (49645) 250/C4
Kalevala, U.S.S.R. 52/D1
Kalewa, Burma 72/B2
Kalgan (Zhangjiakou), China 77/J3
Kalgin (isl.), Alaska 196/B1
Kalgoorlie, Australia 2/R7

Kalgoorlie, W. Australia 88/C6
Kalgoorlie, W. Australia 92/C5
Kalgoorlie-Boulder, W. Australia 92/C5
Kaliakra (cape), Bulgaria 45/J4
Kalianda, Indonesia 85/D7
Kalibo, Philippines 82/D5
Kalida, Ohio (45853) 284/B4
Kalihi, Hawaii (†96801) 218/C4
Kalihi (stream), Hawaii 218/C3
Kalihi Entrance (str.), Hawaii 218/B4
Kalihiwai, Hawaii (†96754) 218/C1
Kalima, Zaire 115/E4
Kalimantan (reg.), Indonesia 85/E5
Kálimnos, Greece 45/H7
Kálimnos (isl.), Greece 45/H7
Kalinga, Queensland 88/K2
Kalinga-Apayao (prov.), Philippines 82/C1
Kalinin, U.S.S.R. 7/H3
Kalinin, U.S.S.R. 48/D4
Kalinin, U.S.S.R. 52/E3
Kaliningrad, U.S.S.R. 7/G3
Kaliningrad, U.S.S.R. 52/A3
Kaliningrad, Kaliningrad, U.S.S.R. 52/B4
Kaliningrad, Moscow Oblast, U.S.S.R. 52/E3
Kalininsk, U.S.S.R. 52/F4
Kalinkovichi, U.S.S.R. 52/C4
Kalispel Ind. Res., Wash. 310/H2
Kalispell, Mont. 188/C1
Kalispell, Mont. (59901) 262/D2
Kalisz (prov.), Poland 47/D3
Kalisz, Poland 7/F3
Kalisz, Poland 47/D3
Kaliua, Tanzania 115/F5
Kalix, Sweden 18/N4
Kalixälv (riv.), Sweden 18/N3
Kalkaska (co.), Mich. 250/D4
Kalkaska, Mich. (49646) 250/D4
Kalkfeld, Namibia 118/B3
Kalkfontein, Botswana 118/C4
Kallaste, U.S.S.R. 53/D1
Kallavesi (lake), Finland 18/P5
Kallsjö (lake), Sweden 18/H5
Kalmalo, Nigeria 106/F6
Kalmar (co.), Sweden 18/K8
Kalmar, Sweden 7/F3
Kalmar, Sweden 18/K8
Kalmarsund (sound), Sweden 18/K8
Kalmthout, Belgium 27/F6
Kalmuck A.S.S.R., U.S.S.R. 52/F5
Kalmuck A.S.S.R., U.S.S.R. 48/E5
Kalmunai, Sri Lanka 68/E7
Kalmykovo, U.S.S.R. 48/F5
Kalo, Iowa (†50569) 229/E4
Kalocsa, Hungary 41/E3
Kalohi (chan.), Hawaii 218/G1
Kaloko-Honokohau Nat'l. Hist. Park, Hawaii 218/F6
Kaloli (pt.), Hawaii 218/K5
Kalomo, Zambia 115/E6
Kalona, Iowa (52247) 229/K6
Kalpeni (isl.), India 68/C7
Kalpin, China 77/A3
Kalskag, Alaska (99607) 196/F2
Kaltag, Alaska (99748) 196/G2
Kaltbrunn, Switzerland 39/H2
Kaluaaha, Hawaii (†96748) 218/H1
Kaluga, U.S.S.R. 7/H3
Kaluga, U.S.S.R. 48/D4
Kaluga, U.S.S.R. 52/E4
Kalumburu Mission, W. Australia 88/D2
Kalumburu Mission, W. Australia 92/D1
Kalundborg, Denmark 21/E6
Kalundborg, Denmark 18/G9
Kalush, U.S.S.R. 52/B5
Kalutara, Sri Lanka 68/D7
Kalvarija, U.S.S.R. 53/B3
Kalvesta, Kansas (67856) 232/B3
Kalyan, India 68/B5
Kama, Burma 72/B3
Kama (res.), U.S.S.R. 52/J3
Kama (riv.), U.S.S.R. 7/K3
Kama (riv.), U.S.S.R. 52/H2
Kama, Zaire 115/E4
Kamaiki (pt.), Hawaii 218/H2
Kamaing, Burma 72/C1
Kamaishi, Japan 81/L4
Kamakou (peak), Hawaii 218/H1
Kamakura, Japan 81/O3
Kamakusa, Guyana 131/A3
Kamalino, Hawaii (96769) 218/A2
Kamalo, Hawaii (†96748) 218/H1
Kaman, Turkey 63/E3
Kamaniskeg (lake), Ontario 177/G2
Kamanjab, Namibia 118/A3
Kamaran (isl.), P.D.R. Yemen 59/D4
Kamarang, Guyana 131/A3
Kamaran (isl.), P.D.R. Yemen 59/D4
Kamaria (falls), Guyana 131/B2
Kamas, Utah (84036) 304/C3
Kamay, Texas (76369) 303/F4
Kambalda, W. Australia 88/C6
Kambalda, W. Australia 92/C5
Kambia, S. Leone 106/B7
Kambove, Zaire 115/E6
Kamchatka (pen.), U.S.S.R. 54/S4
Kamchatka (pen.), U.S.S.R. 2/T3
Kamchatka (pen.), U.S.S.R. 48/Q4
Kamela, Oreg. (†97859) 291/J2
Kamenets-Podol'skiy, U.S.S.R. 52/C5
Kamenice, Czech. 41/C2
Kamenjak (cape), Yugoslavia 45/A3
Kamenka, Archangel, U.S.S.R. 52/F1
Kamenka, Penza, U.S.S.R. 52/F4
Kamen'-na-Obi, U.S.S.R. 48/J4
Kamenskoye, U.S.S.R. 48/R3
Kamensk-Shakhtinskiy, U.S.S.R. 52/F5
Kamensk-Ural'skiy, U.S.S.R. 48/G4
Kamenz, E. Germany 22/F3
Kameoka, Japan 81/J7
Kames, Scotland 15/C5
Kamet (mt.), India 68/D2
Kamiah, Idaho (83536) 220/B3
Kamienna Góra, Poland 47/B3

Kamień Pomorski, Poland 47/B2
Kamiisco, Japan 81/K3
Kamil, Oman 59/G5
Kamilo (pt.), Hawaii 218/H7
Kamilukuak (lake), N.W. Terrs. 187/H3
Kamina, Zaire 102/E5
Kamina, Zaire 115/D5
Kaminak (lake), N.W. Terrs. 187/J3
Kaminoyama, Japan 81/J4
Kaminuriak (lake), N.W. Terrs. 187/J3
Kamishak (bay), Alaska 196/H3
Kamiyaku, Japan 81/E8
Kamloops, Br. Col. 162/D5
Kamloops, Br. Col. 146/G4
Kamloops, Br. Col. 184/G5
Kamo, Japan 81/J7
Kamoa (riv.), Guyana 131/B5
Kamouraska (co.), Québec 172/H2
Kamouraska, Québec 172/H2
Kamp (riv.), Austria 41/C2
Kampala (cap.), Uganda 2/L5
Kampala (cap.), Uganda 102/F4
Kampala (cap.), Uganda 115/F3
Kampar (riv.), Indonesia 85/C5
Kampar, Malaysia 72/D6
Kampen, Netherlands 27/H3
Kampen, W. Germany 22/C1
Kampene, Zaire 115/E4
Kampeska (lake), S. Dak. 298/P4
Kamphaeng Phet, Thailand 72/C3
Kampong Cham, Cambodia 54/M8
Kampong Cham, Cambodia 72/E4
Kampong Chhnang, Cambodia 72/D4
Kampong Khleang, Cambodia 72/E4
Kampong Kuala Besut, Malaysia (3) 72/D6
Kampong Saom, Cambodia 72/D5
Kampong Sedenak, Malaysia 72/E5
Kampong Sibuti, Malaysia 85/E5
Kampong Spoe, Cambodia 72/E5
Kampong Thum, Cambodia 72/E4
Kampong Trabek, Cambodia 72/E5
Kampot, Cambodia 72/E5
Kampsville, Ill. (62053) 222/C4
Kamptee, India 68/D4
KAMPUCHEA (CAMBODIA) 72
Kampung Baru (Tolitoli), Indonesia 85/G5
Kamrar, Iowa (50132) 229/F4
Kamsack, Sask. 162/F5
Kamsack, Sask. 181/K4
Kamsack Beach, Sask. 181/K4
Kamsar, Guinea 106/B6
Kamuela, Hawaii (96743) 218/G3
Kamui (cape), Japan 81/K2
Kamyshin, U.S.S.R. 7/J3
Kamyshin, U.S.S.R. 52/F4
Kamyshin, U.S.S.R. 48/E4
Kanaaupscow (riv.), Québec 174/B2
Kanab (creek), Ariz. 198/C2
Kanab (plat.), Ariz. 198/C2
Kanab, Utah (84741) 304/B6
Kanab (creek), Utah 304/B7
Kanabec (co.), Minn. 255/E5
Kanaga (isl.), Alaska 196/L4
Kanagawa (pref.), Japan 81/O2
Kanaio, Hawaii (†96790) 218/J3
Kanairiktok (riv.), Newf. 166/B3
Kanakanak, Alaska (†99576) 196/G3
Kananaskis, Alberta 182/C4
Kananga, Zaire 115/D5
Kananga, Zaire 102/E5
Kanapou (bay), Hawaii 218/J3
Kanaranzi, Minn. (56146) 255/B7
Kanaranzi (creek), Minn. 255/C7
Kanarraville, Utah (84742) 304/A6
Kanash, U.S.S.R. 52/G3
Kanata, Ontario 177/J2
Kanauga, Ohio (†45631) 284/F8
Kanawha, Iowa (50447) 229/F3
Kanawha (co.), W. Va. 312/C6
Kanawha (riv.), W. Va. 312/C5
Kanawha Falls, W. Va. (25115) 312/D6
Kanawha Head, W. Va. (26228) 312/F5
Kanazawa, Japan 81/H5
Kanazawa, Japan 54/P6
Kanbalu, Burma 72/B2
Kanchanaburi, Thailand 72/C4
Kanchenjunga (mt.), India 68/F3
Kanchenjunga (mt.), Nepal 68/F3
Kanchipuram, India 68/E6
Kanchow (Ganzhou), China 77/H6
Kanchrapara, India 68/F1
Kandahar (Qandahar), Afghanistan 59/J3
Kandahar (Qandahar), Afghanistan 68/B2
Kandahar, Sask. 181/G4
Kanda-Kanda, Zaire 115/D5
Kandalaksha, U.S.S.R. 7/H2
Kandalaksha, U.S.S.R. 52/D1
Kandalaksha, U.S.S.R. 48/C3
Kandalaksha (gulf), U.S.S.R. 52/D1
Kandangan, Indonesia 85/F6
Kándanos, Greece 45/F8
Kandava, U.S.S.R. 53/B2
Kandavu (Kadavu) (isl.), Fiji 87/H7
Kandavu (isl.), Fiji 86/Q11
Kandavu (passage), Fiji 86/Q11
Kander (riv.), Switzerland 39/E4
Kandersteg, Switzerland 39/E4
Kandi, Benin 106/E6
Kandıra, Turkey 63/D2
Kandiyohi (co.), Minn. 255/C5
Kandiyohi, Minn. (56251) 255/D5
Kandla, India 68/C4
Kandos, N.S. Wales 97/F3
Kandrach, Pakistan 68/A3
Kandrach, Pakistan 59/H4
Kandukur, India 68/E5
Kandy, Sri Lanka 54/K9
Kandy, Sri Lanka 68/E7
Kane (basin) 4/B13
Kane (co.), Ill. 222/E2
Kane, Ill. (62054) 222/C4
Kane, Manitoba 179/E5
Kane (basin), N.W.T. 162/N3
Kane (basin), N.W. Terrs. 187/L2
Kane, Pa. (16735) 294/E2
Kane (co.), Utah 304/B6
Kanem (reg.), Chad 111/C5
Kaneohe, Hawaii (96744) 218/F2
Kaneohe (bay), Hawaii 218/F2
Kaneohe Bay U.S.M.C. Air Station, Hawaii 218/F2
Kaneville, Ill. (60144) 222/E2
Kang, Botswana 118/C4
Kanga, Tanzania 115/G5
Kangaba, Mali 106/C6
Kangal, Turkey 63/G3
Kangan, Iran 59/F4
Kangan, Iran 66/G7
Kangar, Malaysia 72/D6
Kangarilla, S. Australia 94/B8
Kangaroo (isl.), Australia 87/D9
Kangaroo (isl.), S. Australia 88/F7
Kangaroo (isl.), S. Australia 94/E7
Kangaroo Ground, Victoria 97/J4
Kangaruma, Guyana 131/B3
Kangavar, Iran 59/E3
Kangavar, Iran 66/F3
Kangding, China 77/F5
Kangean (isl.), Indonesia 85/F7
Kangean (isls.), Indonesia 85/F7
Kanggye, N. Korea 81/C3
Kanghwa (bay), N. Korea 81/B5
Kanghwa (bay), S. Korea 81/B5
Kangley, Ill. (†61364) 222/E2
Kangnŭng, S. Korea 81/D5
Kango, Gabon 115/B3
Kangrinboqê Feng (mt.), China 77/B5
Kani, Burma 72/B2
Kaniama, Zaire 115/D5
Kanin (pen.), U.S.S.R. 7/J2
Kanin (pen.), U.S.S.R. 4/C7
Kanin (pen.), U.S.S.R. 48/E3
Kanin (pen.), U.S.S.R. 52/G1
Kaningo, Kenya 115/G4
Kanin Nos (cape), U.S.S.R. 48/E3
Kanin Nos (cape), U.S.S.R. 52/F1
Kaniva, Victoria 97/A5
Kanjiža, Yugoslavia 45/D2
Kankaanpää, Finland 18/M6
Kankakee, Ill. 188/J2
Kankakee (co.), Ill. 222/F2
Kankakee, Ill. (60901) 222/F2
Kankakee (riv.), Ill. 222/F2
Kankakee (riv.), Ind. 227/C2
Kankan, Guinea 106/C6
Kankan, Guinea 102/B3
Kanker, India 68/E4
Kankossa, Mauritania 102/A3
Kankossa, Mauritania 106/B5
Kannapolis, N.C. (28081) 281/H4
Kannata Valley, Sask. 181/G5
Kannauj, India 68/D3
Kano (state), Nigeria 106/F6
Kano, Nigeria 106/F6
Kano, Nigeria 102/C3
Kanon (pt.), Neth. Ant. 161/G9
Kanona, N.Y. (14856) 276/F6
Kanonji, Japan 81/F6
Kanopolis, Kansas (67454) 232/D3
Kanopolis (lake), Kansas 232/D3
Kanorado, Kansas (67741) 232/A2
Kanosh, Utah (84637) 304/B5
Kanosh Ind. Res., Utah 304/B5
Kanoya, Japan 81/E8
Kanpur, India 54/K7
Kanpur, India 68/E3
Kanrach, Pakistan 59/H4
Kansas 188/G3
KANSAS 232
Kansas, Ala. (35573) 195/C3
Kansas, Ill. (61933) 222/F4
Kansas (riv.), Kans. 188/G3
Kansas (riv.), Kansas 232/F2
Kansas, Ohio (44841) 284/D3
Kansas, Okla. (74347) 288/S2
Kansas (state), U.S. 146/J6
Kansas City, Kans. 188/G3
Kansas City, Kansas (*66101) 232/H2
Kansas City, Mo. (*64101) 261/P5
Kansas City, Mo. 188/H3
Kansas City, Mo. 146/J6
Kansasville, Wis. (53139) 317/L3
Kansk, U.S.S.R. 54/L4
Kansk, U.S.S.R. 48/K4
Kansu (Gansu) (prov.), China 77/E3
Kantishna (riv.), Alaska 196/H2
Kanton (Canton) (isl.), Kiribati 87/J6
Kantunilkin, Mexico 150/Q6
Kanturk, Ireland 10/B4
Kanturk, Ireland 17/D7
Kanuku (mts.), Guyana 131/B4
Kanuma, Japan 81/J5
Kanye, Botswana 102/E7
Kanye, Botswana 118/C5
Kanzi (cape), Tanzania 115/G5
Kaohsiung, China 77/J7
Kaohsiung, Taiwan 54/N7
Kaokoveld (reg.), Namibia 118/A3
Kaolack, Senegal 106/A6
Kaolack, Senegal 102/A3
Kaoma, Zambia 115/D6
Kao Prawa (mt.), Thailand 72/C3
Kapaa, Hawaii (96746) 218/D1
Kapaa, Hawaii 188/E5
Kapaahu, Hawaii (†96778) 218/J6
Kapaau, Hawaii (96755) 218/G3
Kapalama, Hawaii (96817) 218/C4
Kapalong, Philippines 82/E7
Kapanga, Zaire 115/D5
Kapchagay, U.S.S.R. 48/H5
Kapellen, Belgium 27/E6
Kapenguria, Kenya 115/G3
Kapfenberg, Austria 41/C3
Kapingamarangi (atoll), Micronesia 87/F5
Kapiri Mposhi, Zambia 115/E6
Kapiskau (riv.), Ontario 175/D2
Kapit, Malaysia 85/E5
Kapiti (isl.), N. Zealand 100/E4
Kaplan, La. (70548) 238/F6
Kaplice, Czech. 41/C2
Kapoeta, Sudan 111/F7
Kapoho, Hawaii (†96778) 218/K5
Kapos (riv.), Hungary 41/D3
Kaposvár, Hungary 41/D3
Kapowsin, Wash. (98344) 310/C4
Kappa, Ill. (†61738) 222/D3
Kappl, Austria 41/A3
Kaprun, Austria 41/B3
Kapsan, N. Korea 81/C3
Kapsukas, U.S.S.R. 53/B3
Kapsukas, U.S.S.R. 52/B4
Kapuas (riv.), Indonesia 85/D6
Kapulena, Hawaii (†96758) 218/H4
Kapunda, S. Australia 94/F6
Kapuskasing, Ont. 162/H6
Kapuskasing, Ontario 175/D3
Kapuskasing, Ontario 177/J5
Kapuskasing (riv.), Ontario 177/J5
Kapuskasing (riv.), Ontario 175/D3
Kapuvár, Hungary 41/D3
Kapydzhik (mt.), U.S.S.R. 52/G7
Kara, U.S.S.R. 48/G3
Kara, U.S.S.R. 52/L1
Kara (sea), U.S.S.R. 4/B6
Kara (sea), U.S.S.R. 54/H2
Kara (sea), U.S.S.R. 52/K1
Kara (sea), U.S.S.R. 48/G2
Kara-Bogaz-Gol (gulf), U.S.S.R. 48/F5
Karabük, Turkey 63/E2
Karaburun, Turkey 63/B3
Karacabey, Turkey 63/C2
Karaca Dağ (mt.), Turkey 63/H4
Karachay-Cherkess Aut. Obl., U.S.S.R. 48/E5
Karachay-Cherkess Aut. Obl., U.S.S.R. 52/F6
Karachayevsk, U.S.S.R. 52/F6
Karachev, U.S.S.R. 52/E4
Karachi, Pakistan 59/J5
Karachi, Pakistan 2/N4
Karachi, Pakistan 59/J5
Karachi, Pakistan 68/B4
Karachi, Pakistan 54/H7
Karád, Hungary 41/D3
Karad, India 68/C5
Karadağ (mt.), Turkey 59/B2
Karadağ (mt.), Turkey 63/E4
Karadeniz Boğazı (Bosporus) (str.), Turkey 63/C2
Karadeniz Boğazı (Bosporus) (str.), Turkey 59/A1
Karaganda, U.S.S.R. 2/N3
Karaganda, U.S.S.R. 54/J5
Karaganda, U.S.S.R. 48/H5
Karaginskiy (isl.), U.S.S.R. 54/T4
Karaginskiy (isl.), U.S.S.R. 48/R4
Karahallı, Turkey 63/C3
Karaikudi, India 68/D7
Karaisalı, Turkey 63/F4
Karaj, Iran 66/G3
Karakalpak A.S.S.R., U.S.S.R. 48/G5
Karakax (Kara Kash) (Moyu), China 77/A4
Karakax He (riv.), China 77/A4
Karakelong (isl.), Indonesia 85/H5
Karakhoto (ruins), China 77/F3
Karakoçan, Turkey 63/H3
Karakoram (mts.), India 68/D1
Karakoram (mts.), Pakistan 68/D1
Karakorum (ruins), Mongolia 54/M5
Karakorum (ruins), Mongolia 77/F2
Karaköse, Turkey 59/D2
Karaköse (Ağrı), Turkey 63/K3
Kara-Kum (canal), U.S.S.R. 48/F6
Kara-Kum (des.), U.S.S.R. 48/F5
Karakuwisa, Namibia 118/B3
Karaman, Turkey 59/B2
Karaman, Turkey 63/E4
Karamanlı, Turkey 63/C4
Karamay, China 77/B2
Karamay, China 54/K5
Karamea, N. Zealand 100/C4
Karamea (bight), N. Zealand 100/C4
Karamiran Shankou (pass), China 77/C4
Karangasem, Indonesia 85/F7
Karanja, India 68/D4
Karapelit, Bulgaria 45/H4
Karapınar, Turkey 63/E4
Karas, Namibia 118/B5
Karasabai, Guyana 131/B4
Karasavey (cape), U.S.S.R. 4/B6
Karasburg, Namibia 118/B5
Karasjok, Norway 18/O2
Karasu, Turkey 63/D2
Karasu (riv.), Turkey 63/J3
Karasu-Aras (mts.), Turkey 63/J3
Karasuk, U.S.S.R. 48/H4
Karat, Iran 66/M3
Karataş, Turkey 63/F4
Karataş (cape), Turkey 63/F4
Karatau, U.S.S.R. 48/H5
Karathuri, Burma 72/C5
Karatsu, Japan 81/D7
Karawanken (range), Austria 41/C3
Karayaka, Turkey 63/G2
Karayazı, Turkey 63/J3
Karazhal, U.S.S.R. 48/H5
Karbala (gov.), Iraq 66/B4
Karbal'a, Iraq 59/D3
Karbal'a, Iraq 54/F6
Karbal'a, Iraq 66/C4
Karbers Ridge, Ill. (62955) 222/E6
Karby, Denmark 21/B4
Karcag, Hungary 41/F3
Kardhítsa, Greece 45/E6
Kärdla, U.S.S.R. 53/B1
Karelian A.S.S.R., U.S.S.R. 48/D3
Karelian A.S.S.R., U.S.S.R. 52/D2
Karema, Tanzania 115/F5
Karen (state), Burma 72/C3
Karesuando, Sweden 18/M2
Kargasok, U.S.S.R. 48/J4
Karghalik (Yecheng), China 77/A4
Kargı, Turkey 63/F2
Kargil, India 68/D2
Kargopol', U.S.S.R. 52/E2
Karhula, Finland 18/P6
Kariá, Greece 45/E6
Kariaí, Greece 45/G5
Kariba (lake) 102/E6
Kariba (mt.), Japan 81/K2
Kariba (dam), Zambia 115/E7
Kariba (lake), Zambia 115/E7
Kariba, Zimbabwe 118/D3
Kariba (dam), Zimbabwe 118/D3
Kariba (lake), Zimbabwe 118/D3
Karibib, Namibia 118/B4
Karikal, India 68/E6
Karikari (cape), N. Zealand 100/D1
Karima, Sudan 59/B6
Karima, Sudan 111/F4
Karimata (arch.), Indonesia 85/D6
Karimata (isl.), Indonesia 85/D6
Karimata (str.), Indonesia 85/D6
Karimunjawa (isls.), Indonesia 85/J1
Karin, Somalia 115/J1
Karis, Finland 18/N6
Karise, Denmark 21/F7
Karisimbi (mt.), Rwanda 115/E4
Karisimbi (mt.), Zaire 115/E4
Káristos, Greece 45/G6
Kariz, Iran 66/M3
Karjaa (Karis), Finland 18/N6
Karkabat, Ethiopia 111/G4
Karkal, India 68/C6
Karkar (isl.), Papua N.G. 85/B6
Karkas, Kuh-e (mt.), Iran 66/G4
Karkheh (riv.), Iran 66/E4
Karkkila, Finland 18/N6
Karkur-Pardes Hanna, Israel 65/C3
Karlıova, Turkey 63/J3
Karl-Marx-Stadt, E. Germany 7/F3
Karl-Marx-Stadt (dist.), E. Germany 22/E3
Karl-Marx-Stadt, E. Germany 22/E3
Karlö (Hailuoto) (isl.), Finland 18/O4
Karlovac, Yugoslavia 45/B3
Karlovo, Bulgaria 45/G4
Karlovy Vary, Czech. 41/B1
Karlshamn, Sweden 18/J8
Karlskoga, Sweden 18/J7
Karlskrona, Sweden 18/J8
Karlsruhe, N. Dak. (58744) 282/J3
Karlsruhe, W. Germany 7/E4
Karlsruhe, W. Germany 22/C4
Karlstad, Minn. (56732) 255/B2
Karlstad, Sweden 7/F3
Karlstad, Sweden 18/H7
Karlstadt, W. Germany 22/C4
Karluk, Alaska (99608) 196/H3
Karnak (El Karnak), Egypt 111/F2
Karnak, Ill. (62956) 222/E6
Karnak, N. Dak. (†58448) 282/O5
Karnal, India 68/D3
Karnataka (state), India 68/D6
Karnes (co.), Texas 303/G9
Karnes City, Texas (78118) 303/G9
Karnobat, Bulgaria 45/H4
Karns, Tenn. (37921) 237/N9
Karns City, Pa. (16041) 294/C4
Karonga, Malawi 115/F5
Karora, Sudan 111/G4
Karoro, N. Zealand 100/C5
Karosa, Indonesia 85/F6
Karpakora, N.S. Wales 97/B3
Kárpathos, Greece 45/H8
Kárpathos (isl.), Greece 45/H8
Karpenísion, Greece 45/E6
Karpinsk, U.S.S.R. 48/F4
Karratha, W. Australia 88/B4
Karratha, W. Australia 92/B3
Kars, Ontario 177/J2
Kars (prov.), Turkey 63/K2
Kars, Turkey 59/D1
Kars, Turkey 63/K2
Kārsava, U.S.S.R. 53/D2
Karshi, U.S.S.R. 48/G6
Karşıyaka, Turkey 63/B3
Karskiye Vorota (str.), U.S.S.R. 4/B7
Karskiye Vorota (str.), U.S.S.R. 52/J1
Karskiye Vorota (str.), U.S.S.R. 48/F2
Kartal, Turkey 63/D6
Kartaly, U.S.S.R. 48/G4
Karthaus, Pa. (16845) 294/F3
Kartuzy, Poland 47/C1
Karumba, Queensland 88/G3
Karumba, Queensland 95/B3
Karun (riv.), Iran 59/E3
Karun (riv.), Iran 66/F5
Karunjie, W. Australia 92/D2
Karunki, Finland 18/O4
Karup, Denmark 21/C5
Karval, Colo. (80823) 208/N5
Karviná, Czech. 41/E2
Karwar, India 68/C6
Kaş, Turkey 63/C4
Kasaan, Alaska (99924) 196/N2
Kasabonika, Ontario 175/C2
Kasai (riv.) 102/E5
Kasai (riv.), Angola 115/D5
Kasai (riv.), Zaire 115/C4
Kasai-Occidental (prov.), Zaire 115/D4
Kasai-Oriental (prov.), Zaire 115/D5
Kasaji, Zaire 115/D6
Kasama, Zambia 102/F6
Kasama, Zambia 115/F6
Kasane, Botswana 118/D3
Kasanga, Tanzania 115/F5
Kasanga, Tanzania 102/F5
Kasangulu, Zaire 115/C4
Kasar, Ras (cape), Ethiopia 111/G4
Kasar, Ras (cape), Sudan 111/G4
Kasar, Ras (cape), Sudan 59/C6
Kasaragod, India 68/C6
Kasba (lake), N.W.T. 162/F3
Kasba (lake), N.W. Terrs. 187/H3
Kasbeer, Ill. (61328) 222/D2
Kaseda, Japan 81/D8
Kasempa, Zambia 115/E6
Kasenga, Zaire 115/E6
Kasenyi, Zaire 115/E3
Kasese, Uganda 115/F3
Kasese, Zaire 115/E4
Kasganj, India 68/D3
Kashabowie, Ontario 177/G5
Kashaf Rud (riv.), Iran 66/M2
Kashan, Iran 59/F3
Kashan, Iran 66/G3
Kashechewan, Ontario 175/D2
Kashegelok, Alaska (†99668) 196/G2
Kashi (Kashgar), China 77/A4
Kashi, China 54/J6
Kashihara, Japan 81/J8
Kashin, U.S.S.R. 52/E3
Kashing (Jiaxing), China 77/K5
Kashiwa, Japan 81/P2
Kashiwara, Japan 81/J8
Kashiwazaki, Japan 81/J5
Kashmar, Iran 66/L3
Kashmar, Iran 59/G2
Kashmor, Pakistan 68/C3
Kashunuk (riv.), Alaska 196/F2
Kasigluk, Alaska (99609) 196/F2
Kasilof, Alaska (99610) 196/B1
Kasimov, U.S.S.R. 52/F4
Kaskaskia, Ill. (†63673) 222/C6
Kaskaskia (riv.), Ill. 222/E4
Kaskinen (Kaskö), Finland 18/M5
Kaskö, Finland 18/M5
Kaslo, Br. Col. 162/E6
Kaslo, Br. Col. 184/J5
Kasongo, Zaire 115/E4
Kasongo-Lunda, Zaire 115/C5
Kásos (isl.), Greece 45/H8
Kasota, Minn. (56050) 255/D6
Kasper Creek, Sask. 181/F6
Kaspiysk, U.S.S.R. 52/G6
Kaspiyskiy, U.S.S.R. 52/G5
Kassala (prov.), Sudan 111/F4
Kassala, Sudan 111/G4
Kassala, Sudan 59/C6
Kassala, Sudan 102/F3
Kassándra (pen.), Greece 45/F6
Kassel, W. Germany 7/E3
Kassel, W. Germany 22/C3
Kasserine, Tunisia 106/F1
Kassinga, Angola 115/C7
Kasson, Minn. (55944) 255/F6
Kasson, W. Va. (26380) 312/G4
Kastamonu (prov.), Turkey 63/E2
Kastamonu, Turkey 59/B1
Kastamonu, Turkey 63/F2
Kastéllion, Greece 45/G8
Kastéllion (Kíssamos), Greece 45/F8
Kasterlee, Belgium 27/F6
Kastoría, Greece 45/E5
Kastrup, Denmark 21/F6
Kastrup, Denmark 18/H9
Kasugai, Japan 81/H6
Kasukabe, Japan 81/O2
Kasulu, Tanzania 115/F4
Kasumiga (lag.), Japan 81/K5
Kasungu, Malawi 115/F6
Kasur, Pakistan 68/C2
Kasur, Pakistan 59/K3
Kataba, Zambia 115/D7
Katahdin (mt.), Maine 243/F4
Katako-Kombe, Zaire 115/D4
Katákolon, Greece 45/E7
Katanga (reg.), Zaire 102/E5
Katangli, U.S.S.R. 48/P4
Katanning, W. Australia 88/B6
Katanning, W. Australia 92/B6
Katarnian Ghat, India 68/E3
Katchall (isl.), India 68/G7
Katemcy, Texas (76850) 303/E7
Katenga, Zaire 115/E5
Katepwa Beach, Sask. 181/H5
Katepwa Prov. Park, Sask. 181/H5
Katerini, Greece 45/F5
Kates Needle (mt.), Alaska 196/N1
Kates Needle (mt.), Br. Col. 184/A1
Katha, Burma 72/C1
Katherina, Jebel (mt.), Egypt 111/F2
Katherina, Jebel (mt.), Egypt 59/B4
Katherine, Ariz. (†86430) 198/A3
Katherine, Australia 87/D7
Katherine, North. Terr. 88/E2
Katherine, North. Terr. 93/B3
Katherine (riv.), North. Terr. 93/C3
Kathiawar (pen.), India 68/B4
Kathleen, Alberta 182/B2
Kathleen, Fla. (33849) 212/D3
Kathleen, Georgia (31047) 217/E6
Kathmandu (cap.), Nepal 54/K7
Kathmandu (cap.), Nepal 68/E3
Kathmandu, Nepal 68/E3
Kathryn, Alberta 182/D4
Kathryn, N. Dak. (58049) 282/P6
Kati, Mali 106/C6
Katihar, India 68/F3
Katima Mulilo, Namibia 118/C3
Katimik (lake), Manitoba 179/C2
Katiola, Ivory Coast 106/C7
Katipunan, Philippines 82/D6
Katmai (vol.), Alaska 196/H3
Katmai Nat'l Park, Alaska 196/H3
Katmai Nat'l Preserve, Alaska 196/H3
Katni (Murwara), India 68/E4
Katonah, N.Y. (10536) 276/N8
Katoomba-Wentworth Falls, N.S. Wales 97/F3
Katowice (prov.), Poland 47/D3
Katowice, Poland 7/F3
Katowice, Poland 47/B4
Katrime, Manitoba 179/D4
Katrine, Ontario 177/E2
Katrine, Loch (lake), Scotland 15/D4
Katrineholm, Sweden 18/K7
Katsina, Nigeria 102/C3
Katsina, Nigeria 106/F6
Katsina Ala, Nigeria 106/F7
Katsuta, Japan 81/K5
Katsuura, Japan 81/K6
Kattakurgan, U.S.S.R. 48/G5
Kattegat (str.) 7/F3
Kattegat (str.), Denmark 21/E4
Kattegat (str.), Denmark 18/G8
Kattegat (str.), Sweden 18/G8
Katwe, Uganda 115/F4
Katwijk aan Zee, Netherlands 27/E4
Katy, Texas (77450) 303/J8
Kau (des.), Hawaii 218/J6
Kau, Indonesia 85/H5
Kauai (co.), Hawaii 218/A1
Kauai (isl.), Hawaii 87/L3
Kauai (isl.), Hawaii 188/E5
Kauai (chan.), Hawaii 218/E6
Kauai (isl.), Hawaii 218/C1
Kaufbeuren, W. Germany 22/D5
Kaufman (co.), Texas 303/H5
Kaufman, Texas (75142) 303/H5
Kauhola (pt.), Hawaii 218/G3
Kauiki (head), Hawaii 218/K2
Kaukauna, Wis. (54130) 317/K7
Kaukauveld (mts.), Botswana 118/C3
Kaukauveld (mts.), Namibia 118/C3
Kaukonahua (stream), Hawaii 218/E1
Kaula (isl.), Hawaii 188/F6
Kaula (isl.), Hawaii 218/D6
Kaulakahi (chan.), Hawaii 218/B2
Kauliranta, Finland 18/O3
Kaumajet (mts.), Newf. 166/B2
Kaumakani, Hawaii (96747) 218/C2
Kaumalapau (harb.), Hawaii 218/G2
Kaumalapau Harbor, Hawaii (†96763) 218/G2
Kauna (pt.), Hawaii 218/G7
Kaunakakai, Hawaii (96748) 218/G1
Kaunakakai (harb.), Hawaii 218/G1
Kaunas, U.S.S.R. 7/G3
Kaunas, U.S.S.R. 53/C3
Kaunas, U.S.S.R. 48/C4
Kaunas, U.S.S.R. 52/B4
Kauniainen, Finland 18/O6
Kaunuopou (pt.), Hawaii 218/B2
Kaupakulua, Hawaii (†96708) 218/K2
Kaupo, Hawaii (†96713) 218/K2
Kaura Namoda, Nigeria 106/F6
Kauswagan, Philippines 82/E6
Kautokeino, Norway 18/N2
Kauttua, Finland 18/M6
Kavadarci, Yugoslavia 45/E5
Kavajë, Albania 45/D5
Kavak, Çanakkale, Turkey 63/C5
Kavak, Samsun, Turkey 63/F2
Kavalerovo, U.S.S.R. 48/O5
Kavalga (isl.), Alaska 196/K4
Kavali, India 68/E6
Kavála, Greece 7/G4
Kavála, Greece 45/G5
Kavanagh, Alberta 182/D3
Kavanayen, Venezuela 124/H5
Kavaratti, India 68/C6
Kavarna, Bulgaria 45/J4
Kaveri (riv.), India 68/D6
Kavieng, Papua N.G. 87/E6
Kavieng, Papua N.G. 86/B1
Kavir, Dasht-e (salt des.), Iran 59/G3
Kavir, Dasht-e (salt des.), Iran 66/J3
Kavir-e Namak (salt des.), Iran 59/G3
Kavirondo (gulf), Kenya 115/F4
Kaw, Fr. Guiana 131/E3
Kaw (lake), Okla. 288/N1
Kawachinagano, Japan 81/J8
Kawagama (lake), Ontario 177/F2
Kawagoe, Japan 81/O2
Kawaguchi, Japan 81/J6
Kawaihae, Hawaii (96743) 218/G4
Kawaihae (bay), Hawaii 218/G4
Kawaihoa (cape), Hawaii 218/A2
Kawaikini (peak), Hawaii 218/C1
Kawailoa, Hawaii (†96712) 218/E1
Kawakawa, N. Zealand 100/E1
Kawambwa, Zambia 115/E5
Kawanishi, Japan 81/H7
Kawardha, India 68/E4
Kawasaki, Japan 81/O2
Kawau (isl.), N. Zealand 100/E2
Kaw City, Okla. (74641) 288/N1
Kawerau, N. Zealand 100/F3
Kawhia, N. Zealand 100/E3
Kawhia (harb.), N. Zealand 100/E3
Kawi (mt.), Indonesia 85/K2
Kawich (peak), Nev. 266/E5
Kawich (range), Nev. 266/E5
Kawinaw (lake), Manitoba 179/C2
Kawio (isls.), Indonesia 85/G5
Kawkawlin, Mich. (48631) 250/F5
Kawlin, Burma 72/B2
Kawludo, Burma 72/C3
Kawthaung, Burma 72/C5
Kaxgar (Kashi), China 77/A4
Kay (co.), Okla. 288/M1
Kaya, Upper Volta 106/D6
Kayah (state), Burma 72/C3
Kayak (isl.), Alaska 196/K3
Kayan (riv.), Indonesia 54/N9
Kayan (riv.), Indonesia 85/F5
Kaycee, Wyo. (82639) 319/F2
Kayenta, Ariz. (86033) 198/E2
Kayes, Mali 106/B6
Kayes, Mali 102/A3
Kayjay, Ky. (†40906) 237/O7
Kaylor, Pa. (†16025) 294/C4
Kaylor, S. Dak. (57354) 298/O7
Kayser (mts.), Suriname 131/C4
Kayseri (prov.), Turkey 63/F3
Kayseri, Turkey 63/F3
Kayseri, Turkey 59/C2
Kayseri, Turkey 54/E6
Kaysville, Utah (84037) 304/B2
Kayuagung, Indonesia 85/D6
Kayville, Sask. 181/F6
Kazabazua, Québec 172/A4
Kazakh S.S.R., U.S.S.R. 54/H5
Kazakh S.S.R., U.S.S.R. 48/G5
Kazan (riv.), N.W.T. 162/F3
Kazan (riv.), N.W. Terrs. 187/H3
Kazan', U.S.S.R. 7/J3
Kazan', U.S.S.R. 48/F4
Kazan', U.S.S.R. 52/G3
Kazanlı, Turkey 63/F4
Kazanlŭk, Bulgaria 45/G4

Kazan-retto (Volcano) (isls.), Japan 81/M4
Kazatin, U.S.S.R. 52/C5
Kazbek (mt.), U.S.S.R. 52/F6
Kazerun, Iran 66/G6
Kazerun, Iran 59/F4
Kazhim, U.S.S.R. 52/H2
Kazimierza Wielka, Poland 47/E3
Kazımkarabekir, Turkey 63/E4
Kazincbarcika, Hungary 41/F2
Kazlu-Rūda, U.S.S.R. 53/B3
Kazumba, Zaire 115/D5
Kazvin (Qazvin), Iran 66/F2
Kbenhaven (co.), Denmark 21/F6
Kbenhavn (Copenhagen) (commune), Denmark 21/F6
Kdyně, Czech. 41/B2
Kéa, Greece 45/G7
Kéa (isl.), Greece 45/G7
Keaau, Hawaii 188/G6
Keaau, Hawaii (96749) 218/J5
Keady, N. Ireland 17/H3
Keahi (pt.), Hawaii 218/A4
Keahole (pt.), Hawaii 218/F5
Kealaikahiki (chan.), Hawaii 218/H3
Kealaikahiki (pt.), Hawaii 218/H3
Kealakekua, Hawaii (96750) 218/G5
Kealakekua (bay), Hawaii 218/F6
Kealia, Hawaii (†96704) 218/G6
Kealia, Kauai, Hawaii (96751) 218/D1
Keams Canyon, Ariz. (86034) 198/E3
Keanae, Hawaii (†96708) 218/K2
Keanapapa (pt.), Hawaii 218/G2
Keansburg, N.J. (07734) 273/E3
Kearney, Mo. (64060) 261/D4
Kearney, Nebr. 188/G2
Kearney (co.), Nebr. 264/F4
Kearney, Nebr. (68847) 264/E4
Kearney, Ontario 177/E2
Kearneysville, W. Va. (25430) 312/L4
Kearns, Utah (84118) 304/B3
Kearny, Ariz. (85237) 198/E5
Kearny (co.), Kansas 232/A3
Kearny, N.J. (07032) 273/B2
Kearny, Wyo. (†82832) 319/F1
Kearsarge, N.H. (03847) 268/E3
Kearsarge (mt.), N.H. 268/D5
Kearsarge, Pa. (†16501) 294/B1
Keasbey, N.J. (08832) 273/E2
Keatchie, La. (71046) 238/C2
Keating, Oreg. (†97814) 291/K3
Keating Summit, Pa. (16737) 294/F2
Keatley, Sask. 181/D3
Keaton, Ky. (41226) 237/P5
Keats, Kansas (†66502) 232/F2
Keats (mt.), W. Australia 92/A2
Keauhou, Hawaii (†96725) 218/F5
Keavy, Ky. (40737) 237/N6
Keawekaheka (pt.), Hawaii 218/F5
Keban, Turkey 63/H3
Kebang (mt.), S. Korea 81/D5
Ke Bao, Vietnam 72/E2
Kebbi (riv.), Nigeria 106/E6
Kebnekaise (mt.), Sweden 7/F2
Kebnekaise (mt.), Sweden 18/L3
Kebock (head), Scotland 15/B2
Kebumen, Indonesia 85/J2
Kecel, Hungary 41/E3
Kechi, Kansas (67067) 232/E4
Kechika (riv.), Br. Col. 184/L2
Keçiborlu, Turkey 63/D4
Kecskemét, Hungary 7/F4
Kecskemét, Hungary 41/E3
Kedah (state), Malaysia 72/D6
Kedainiai, U.S.S.R. 53/C3
Keddie, Calif. (95952) 204/E3
Kedges (strs), Md. 245/O8
Kedgwick, New Bruns. 170/C1
Kedgwick (riv.), New Bruns. 170/C1
Kedgwick Ouest, New Bruns. 170/C1
Kedgwick River, New Bruns. 170/C1
Kediri, Indonesia 85/K2
Kédougou, Senegal 106/B6
Kedron (brook), Queensland 95/D2
Kedzierzyn-Koźle, Poland 47/C3
Keechelus (lake), Wash. 310/D3
Keedysville, Md. (21756) 245/H3
Keefers, Br. Col. 184/G5
Keefton, Okla. (†74401) 288/R3
Keegan, Maine (†04785) 243/G1
Keego Harbor, Mich. (48030) 250/F6
Keehi (lag.), Hawaii 218/B4
Keel-Dooagh, Ireland 17/A4
Keele (riv.), N.W. Terrs. 187/F3
Keele (peak), Yukon 187/E3
Keeler, Calif. (93530) 204/H7
Keeler, Sask. 181/F5
Keeline, Wyo. (82220) 319/H3
Keeling (Cocos) (isls.), Australia 2/P6
Keeling, Va. (24566) 307/K7
Keels, Newf. 166/D1
Keelung, China 77/K6
Keenan, W. Va. (†24983) 312/F7
Keenan Siding, New Bruns. 170/E2
Keene, Calif. (93531) 204/G8
Keene, Ky. (40339) 237/M5
Keene, N.H. (03431) 268/C6
Keene, N.Y. (12942) 276/N2
Keene, N. Dak. (58847) 282/E4
Keene, Ohio (43828) 284/G5
Keene, Ontario 177/F3
Keene, Texas (76059) 303/G5
Keener, Ala. (†35954) 195/G2
Keenes, Ill. (62851) 222/E5
Keenesburg, Colo. (80643) 208/L2
Keene Valley, N.Y. (12943) 276/N2
Keensburg, Ill. (62852) 222/F5
Keeny (creek), Oreg. 291/K4
Keeper (hill), Ireland 17/E6
Keerweer (cape), Queensland 88/G2
Keerweer (cape), Queensland 95/B2
Keeseville, N.Y. (12944) 276/O2
Keesler A.F.B., Miss. 256/G10
Keetley, Utah (†84060) 304/C3
Keetmanshoop, Namibia 118/B5
Keetmanshoop, Namibia 102/D7
Keewatin, Minn. (55753) 255/E3
Keewatin (dist.), N.W.T. 162/G3
Keewatin (dist.), N.W. Terrs. 187/J3
Keewatin, Ontario 177/F5
Keewatin, Ontario 175/A3
Keewong, N.S. Wales 97/C3
Keezletown, Va. (22832) 307/L4
Kefallinía (isl.), Greece 45/E6
Kefar Blum, Israel 65/D1
Kefar Gil'adi, Israel 65/C1
Kefar Ruppin, Israel 65/D3
Kefar Sava, Israel 65/B3
Kefar Vitkin, Israel 65/B3
Kefar Zekhariya, Israel 65/B4
Keffi, Nigeria 106/F7
Keflavík, Iceland 21/B1
Kégashka, Québec 174/E2
Kegley, W. Va. (24731) 312/D8
Kegonsa (lake), Wis. 317/H10
Keg River, Alberta 182/A5
Kehl, W. Germany 22/B4
Kehoe, Ky. (†41144) 237/P4
Kehra, U.S.S.R. 53/C1
Keila, U.S.S.R. 53/C1
Keilor, Victoria 88/K7
Keilor, Victoria 97/H5
Keimoes, S. Africa 118/C5
Keirn, Miss. (†38924) 256/D4
Keiser, Ark. (72351) 202/K2
Keiss, Scotland 15/E2
Keitele (lake), Finland 18/O5
Keith (co.), Nebr. 264/C3
Keith, Scotland 10/E2
Keith, Scotland 15/F3
Keith, S. Australia 94/G7
Keith, W. Va. (†25148) 312/C6
Keith Arm (inlet), N.W. Terrs. 187/F3
Keithley Creek, Br. Col. 184/G4
Keithsburg, Ill. (61442) 222/B2
Keithville, La. (71047) 238/C2
Keizer, Oreg. (97303) 291/A3
Kejimkujik (lake), Nova Scotia 168/C4
Kejimkujik Nat'l Park, Nova Scotia 168/C4
Kekaa (pt.), Hawaii 218/H2
Kekaha, Hawaii (96752) 218/C2
Kekaha, Hawaii 188/E5
Kekertaluk (isl.), N.W. Terrs. 187/M3
Kékes (mt.), Hungary 41/E3
Kekoskee, Wis. (†53050) 317/J8
Kelang, Malaysia 72/D7
Kelantan (state), Malaysia 72/D6
Kelantan, Sungai (riv.), Malaysia 72/D6
Kelasa (str.), Indonesia 85/D6
Keldron, S. Dak. (57634) 298/F2
Keles, Turkey 63/C3
Kelfield, Sask. 181/C4
Kelford, N.C. (27847) 281/P2
Kelheim, W. Germany 22/D4
Kelkit, Turkey 63/H2
Kelkit (riv.), Turkey 59/C1
Kelkit (riv.), Turkey 63/G2
Kell, Ill. (62853) 222/E5
Kellé, Congo 115/B4
Keller (lake), N.W. Terrs. 187/F3
Keller, Texas (76248) 303/F2
Keller, Va. (23401) 307/S5
Keller, Wash. (99140) 310/G2
Kellerberrin, W. Australia 88/B6
Kellerberrin, W. Australia 92/B5
Kellerman, Ala. (35468) 195/D4
Kellerton, Iowa (50133) 229/E7
Kellerville, Texas (79057) 303/D2
Kellett (cape), N.W.T. 162/D1
Kellett (cape), N.W. Terrs. 187/F2
Kellett (str.), N.W. Terrs. 187/G2
Kellettville, Pa. (†16353) 294/D2
Kelley, Iowa (50134) 229/F5
Kelley (creek), Nev. 266/D1
Kelleys (isl.), Ohio 284/E2
Kelleys Island, Ohio (43438) 284/E2
Kelligrews, Newf. 166/D2
Kelliher, Minn. (56650) 255/D3
Kelliher, Sask. 181/H4
Kellnersville, Wis. (54215) 317/L7
Kellogg (mt.), Ariz. 198/E6
Kellogg, Idaho (83837) 220/B2
Kellogg, Iowa (50135) 229/H5
Kellogg, Minn. (55945) 255/G6
Kelloggsville, Ohio (†44048) 284/J2
Kelloselkä, Finland 18/Q3
Kells (Ceanannus Mór), Ireland 17/G4
Kells, Ireland 17/G6
Kells, N. Ireland 17/J2
Kelly, Georgia (31048) 217/E4
Kelly (creek), Idaho 220/C3
Kelly, Kansas (66446) 232/G2
Kelly, Ky. (†42240) 237/G7
Kelly, La. (71441) 238/F3
Kelly, N.C. (28448) 281/N6
Kelly, Wyo. (83011) 319/B2
Kelly A.F.B., Texas 303/J11
Kelly Lake, Br. Col. 184/G4
Kelly Lake, Minn. (55754) 255/F3
Kellys, N. Dak. (†58201) 282/R4
Kellysville, W. Va. (24732) 312/E8
Kellyton, Ala. (35089) 195/F5
Kellyville, Okla. (74039) 288/O3
Kelme, U.S.S.R. 53/B3
Kélo, Chad 111/C6
Kélo, Chad 102/D4
Kelowna, Br. Col. 146/G4
Kelowna, Br. Col. 162/E6
Kelowna, Br. Col. 184/H5
Kelsey, Alberta 182/D3
Kelsey, Minn. (55755) 255/F3
Kelsey Bay, Br. Col. 184/D5
Kelseyville, Calif. (95451) 204/C5
Kelso, Ark. (†71674) 202/H6
Kelso, Calif. (92351) 204/K8
Kelso, Mo. (63758) 261/O8
Kelso, Sask. 181/K6
Kelso, Scotland 10/E3
Kelso, Scotland 15/F5
Kelso, Tenn. (37348) 237/J10
Kelso, Wash. (98626) 310/C4
Kelstern, Sask. 181/E5
Kelston West, N. Zealand 100/B1
Keltie (cape) 5/C7
Keltner, Ky. (†42761) 237/K6
Kelton, S.C. (†29353) 296/D2
Kelty, Scotland 10/C1
Kelty, Scotland 15/D1
Keluang, Malaysia 72/D7
Kelvington, Sask. 181/H3
Kelwood, Manitoba 179/C4
Kem', U.S.S.R. 7/H2
Kem', U.S.S.R. 4/C8
Kem', U.S.S.R. 52/D2
Kem', U.S.S.R. 48/D3
Ké-Macina, Mali 106/C6
Kemah, Texas (77565) 303/K2
Kemah, Turkey 63/H3
Kemaliye, Turkey 63/H3
Kemalpaşa, Turkey 63/J2
Kemano, Br. Col. 184/D3
Kemasik, Malaysia 72/D6
Kembe, Cent. Afr. Rep. 115/D3
Kemble, Ontario 177/D3
Kemboma, Gabon 115/B3
Kemecse, Hungary 41/F2
Kemer, Turkey 63/D4
Kemerburgaz, Turkey 63/D5
Kemerovo, U.S.S.R. 54/K4
Kemerovo, U.S.S.R. 48/J4
Kemi, Finland 7/G2
Kemi, Finland 18/O4
Kemi (riv.), Finland 7/G2
Kemijärvi, Finland 18/P3
Kemijärvi (lake), Finland 18/Q3
Kemijoki (riv.), Finland 18/O3
Kemikli, Büyük (cape), Turkey 63/B6
Kemirhisar, Turkey 63/F4
Kemmerer, Wyo. (83101) 319/B4
Kemnay, Manitoba 179/B5
Kemnay, Scotland 15/F3
Kemp, Ill. (†61910) 222/E4
Kemp, Okla. (74747) 288/O7
Kemp, Texas (75143) 303/H5
Kemp (lake), Texas 303/E4
Kemp City (Hendrix), Okla. (†74741) 288/O7
Kemp Coast (reg.) 5/C3
Kemper (co.), Miss. 256/G5
Kemp Mill, Md. (†20901) 245/F3
Kempsey, N. S. Wales 88/J6
Kempsey, N.S. Wales 97/G2
Kempster, Wis. (54444) 317/H5
Kempston, England 13/G5
Kempt, Nova Scotia 168/C4
Kempt (lake), Québec 172/C2
Kempten, W. Germany 22/D5
Kempton, Ill. (60946) 222/E3
Kempton, Ind. (46049) 227/E4
Kempton, Md. (†26292) 245/A3
Kempton, N. Dak. (†58267) 282/P4
Kempton, Pa. (19529) 294/L4
Kempton, Tasmania 99/D4
Kempton Park, S. Africa 118/J6
Kemptown, Md. (†21770) 245/J3
Kemptown, Nova Scotia 168/E3
Kemptville, Nova Scotia 168/C4
Kemptville, Ontario 177/J2
Ken, Afghanistan 68/A2
Ken, Afghanistan 59/H3
Kenadsa, Algeria 106/D2
Kenai, Alaska (99611) 196/B1
Kenai (lake), Alaska 196/C1
Kenai (mt.), Alaska 196/C2
Kenai (pen.), Alaska 196/C2
Kenai Fjords Nat'l Park, Alaska 196/C3
Kenamu (riv.), Newf. 166/B3
Kenansville, Fla. (32739) 212/F4
Kenansville, N.C. (28349) 281/O5
Kenaston, N. Dak. (†58746) 282/F2
Kenaston, Sask. 181/E4
Kenbridge, Va. (23944) 307/M7
Kendal, Barbados 161/B8
Kendal, England 13/E3
Kendal, England 10/E3
Kendal, Indonesia 85/J2
Kendal, Sask. 181/H5
Kendall, Fla. (33156) 212/B5
Kendall (co.), Ill. 222/E2
Kendall, Kansas (67857) 232/A4
Kendall, N.S. Wales 97/G2
Kendall, N.Y. (14476) 276/E4
Kendall (cape), N.W. Terrs. 187/K3
Kendall (co.), Texas 303/F8
Kendall, Wash. (†98244) 310/C2
Kendall, Wis. (54638) 317/F8
Kendall Park, N.J. (08824) 273/D3
Kendallville, Ind. (46755) 227/G2
Kendallville, Iowa (†52136) 229/K2
Kendari, Indonesia 85/G6
Kendawangan, Indonesia 85/D6
Kendrapara, India 68/F4
Kendrick (peak), Ariz. 198/D3
Kendrick, Fla. (†32670) 212/D2
Kendrick, Idaho (83537) 220/B3
Kendrick, Okla. (74040) 288/N3
Kenduskeag○, Maine (04450) 243/E6
Kenedy (co.), Texas 303/G11
Kenedy, Texas (78119) 303/G9
Kenefic, Okla. (74748) 288/O6
Kenel, S. Dak. (†57642) 298/H2
Kenema, S. Leone 102/A4
Kenema, S. Leone 106/B7
Kenesaw, Nebr. (68956) 264/F4
Kengah (isls.), Indonesia 85/F7
Kenge, Zaire 115/C4
Keng Hkam, Burma 72/C2
Keng Tung, Burma 72/C2
Kenhardt, S. Africa 118/C5
Kenhorst, Pa. (†19607) 294/L5
Kéniéba, Mali 106/B6
Kenilworth, England 13/F5
Kenilworth, Ill. (60043) 222/B5
Kenilworth, N.J. (07033) 273/E2
Kenilworth Junction, Ontario 177/D4
Kenilworth, Utah (84529) 304/D4
Keningau, Malaysia 85/F4
Kenitra, Morocco 102/B1
Kenitra, Morocco 106/C2
Kenli, China 77/J4
Kenly, N.C. (27542) 281/N3
Kenmare, Ireland 10/B5
Kenmare, Ireland 17/B8
Kenmare (riv.), Ireland 17/A8
Kenmare, N. Dak. (58746) 282/G2
Kenmore, N.Y. (14271) 276/C5
Kenmore, Queensland 88/J3
Kenmore, Scotland 15/E4
Kenmore, Wash. (98028) 310/B1
Kenna, N. Mex. (88122) 274/F5
Kenna, W. Va. (25248) 312/C5
Kennan, Wis. (54537) 317/F5
Kennard, Ind. (47351) 227/G5
Kennard, Nebr. (68034) 264/H3
Kennard, Pa. (†16125) 294/B3
Kennard, Texas (75847) 303/J6
Kennebago Lake, Maine (†04970) 243/B5
Kennebec (co.), Maine 243/D7
Kennebec (riv.), Maine 243/D7
Kennebec, S. Dak. (57544) 298/K6
Kennebecasis (bay), New Bruns. 170/E3
Kennebecasis (riv.), New Bruns. 170/E3
Kennebunk, Maine (04043) 243/B9
Kennebunk○, Maine (04043) 243/B9
Kennebunk Beach, Maine (†04043) 243/C9
Kennebunkport, Maine (04046) 243/C9
Kennebunkport○, Maine (04046) 243/C9
Kennedale, Texas (76060) 303/F3
Kennedy, Ala. (35574) 195/B3
Kennedy (Canaveral) (cape), Fla. 212/F3
Kennedy, Minn. (56733) 255/B2
Kennedy, N.Y. (14747) 276/B6
Kennedy (chan.), N.W.T. 162/N3
Kennedy (chan.), N.W. Terrs. 187/M1
Kennedy, Sask. 181/J5
Kennedy Center, D.C. 245/A5
Kennedy Entrance (str.), Alaska 196/H3
Kennedyville, Md. (21645) 245/P3
Kenner, La. (70062) 238/N4
Kennesaw, Georgia (30144) 217/C2
Kennesaw Mtn. Nat'l Battlefield Park, Georgia 217/J1
Kennet (riv.), England 13/F6
Kennetcook, Nova Scotia 168/E3
Kennetcook (riv.), Nova Scotia 168/E3
Kenneth, Ind. (†46947) 227/E3
Kenneth, Minn. (56147) 255/B7
Kenneth City, Fla. (33709) 212/B3
Kennett, Mo. (63857) 261/M10
Kennett Square, Pa. (19348) 294/L6
Kennewick, Wash. (99336) 310/F4
Kenney (dam), Br. Col. 184/E3
Kenney, Ill. (61749) 222/D3
Kennisis (lake), Ontario 177/F2
Keno, Oreg. (97627) 291/F5
Kenogami (riv.), Ont. 162/H6
Kenogami (riv.), Ontario 177/H4
Kenogami (riv.), Ontario 175/C2
Kénogami (lake), Québec 172/F1
Keno Hill, Yukon 187/E3
Kenoma, Mo. (†64759) 261/D8
Kenora (terr. dist.), Ont. 177/G5
Kenora (terr. dist.), Ont. 175/C2
Kenora, Ont. 146/J4
Kenora, Ont. 162/G5
Kenora, Ontario 175/B3
Kenora, Ontario 177/F4
Kenosee Park, Sask. 181/J6
Kenosha (co.), Wis. 317/K10
Kenosha, Wis. (*53140) 317/M3
Kenova, W. Va. (25530) 312/A6
Kensal, N. Dak. (58455) 282/N5
Kenscoff, Haiti 158/C6
Kensett, Ark. (72082) 202/G3
Kensett, Iowa (50448) 229/G2
Kensington, Calif. (†94701) 204/J2
Kensington, Conn. (06037) 210/D2
Kensington, Kansas (66951) 232/C2
Kensington, Md. (20795) 245/E4
Kensington, Minn. (56343) 255/C5
Kensington, N.S. Wales 97/J4
Kensington, Ohio (44427) 284/J4
Kensington, Pr. Edward I. 168/E2
Kensington and Chelsea, England 13/G8
Kensington and Chelsea, England 10/B5
Kensington and Norwood, S. Australia 88/E8
Kensington and Norwood, S. Australia 94/B8
Kent, Ala. (36045) 195/G5
Kent, Br. Col. 184/M3
Kent○, Conn. (06757) 210/B2
Kent (co.), Del. 245/R4
Kent (co.), England 13/H6
Kent, Ill. (61044) 222/D1
Kent, Ind. (†47250) 227/F7
Kent, Iowa (50850) 229/E7
Kent (co.), Md. 245/O3
Kent (isl.), Md. 245/N5
Kent (pt.), Md. 245/N5
Kent (co.), Mich. 250/D5
Kent, Minn. (56553) 255/B4
Kent (co.), New Bruns. 170/E2
Kent (pen.), N.W. Terrs. 187/H3
Kent, Ohio (44240) 284/H3
Kent (county), Ontario 177/B5
Kent, Oreg. (97033) 291/G2
Kent, Pa. (15752) 294/D4
Kent (co.), R.I. 249/H6
Kent (co.), Texas 303/D4
Kent, Texas (79855) 303/C11
Kent, Wash. (98031) 310/C3
Kentau, U.S.S.R. 48/G5
Kent Bridge, Ontario 177/B5
Kent City, Mich. (49330) 250/D5
Kent Furnace, Conn. (†06757) 210/B2
Kent Group (isls.), Tasmania 99/D1
Kent Junction, New Bruns. 170/E2
Kent Lake, New Bruns. 170/E2
Kentland, Ind. (47951) 227/C3
Kenton, Del. (19955) 245/R4
Kenton (co.), Ky. 237/M3
Kenton, Ky. (41053) 237/N3
Kenton, Manitoba 179/B5
Kenton, Mich. (49943) 250/G2
Kenton, Ohio (43326) 284/C4
Kenton, Okla. (73946) 288/A1
Kenton, Tenn. (38233) 237/C8
Kenton Vale, Ky. (†41011) 237/S2
Kents Hill, Maine (04349) 243/D7
Kents Store, Va. (23084) 307/M5
Kentuck, W. Va. (25249) 312/C5
Kentucky 188/J3
Kentucky (lake) 188/J3
KENTUCKY 237
Kentucky (dam), Ky. 237/E7
Kentucky (lake), Ky. 237/E8
Kentucky (riv.), Ky. 237/M3
Kentucky (lake), Tenn. 237/E8
Kentucky (state), U.S. 146/K6
Kentville, Nova Scotia 168/D3
Kentwood, La. (70444) 238/J5
Kentwood, Mich. (49508) 250/D6
Kenville, Manitoba 179/A3
Kenvir, Ky. (40847) 237/P7
Kenwood, Georgia (†30214) 217/D3
Kenwood, Okla. (†74365) 288/S2
Kenya 2/L5
Kenya 102/F4
KENYA 115/G3
Kenya (mt.), Kenya 102/F4
Kenya (mt.), Kenya 115/G4
Kenyon, Minn. (55946) 255/E6
Kenyon, R.I. (02836) 249/H7
Kenyonville, Conn. (†06281) 210/G1
Keo, Ark. (72083) 202/G4
Keokea, Hawaii, Hawaii (†96704) 218/G6
Keokea, Maui, Hawaii (†96790) 218/J2
Keokee, Va. (24265) 307/C7
Keokuk (co.), Iowa 229/J6
Keokuk, Iowa 188/H2
Keokuk, Iowa (52632) 229/L8
Keoma, Alberta 182/D4
Keomah, Iowa (†52577) 229/J6
Keomuku, Hawaii (†96763) 218/H2
Keonjhar, India 68/F4
Keosauqua, Iowa (52565) 229/J7
Keota, Colo. (†80729) 208/L1
Keota, Iowa (52248) 229/K6
Keota, Okla. (74941) 288/S4
Keowee (lake), S.C. 296/B2
Keowee (riv.), S.C. 296/B2
Kepez, Turkey 63/B6
Kepi, Indonesia 85/K7
Kępno, Poland 47/C3
Keppel (harb.), Singapore 72/F6
Kepsut, Turkey 63/C3
Kerala (state), India 68/D6
Kerama (isls.), Japan 81/M6
Kerang, Victoria 97/B4
Kerava, Finland 18/O6
Kerby, Oreg. (97531) 291/D5
Kerch', U.S.S.R. 7/H4
Kerch', U.S.S.R. 52/E5
Kerchoual, Mali 106/E5
Kerema, Papua N.G. 85/B7
Keremeos, Br. Col. 184/G5
Kerempe (cape), Turkey 63/E1
Keren, Ethiopia 59/C6
Keren, Ethiopia 111/G4
Kerens, Texas (75144) 303/H5
Kerens, W. Va. (26276) 312/G4
Keret', U.S.S.R. 52/D1
Kerguélen (isl.), 2/N8
Kerhonkson, N.Y. (12446) 276/M7
Kericho, Kenya 115/F4
Kerinci (mt.), Indonesia 85/C6
Keriya (Yutian), China 77/B4
Keriya He (riv.), China 77/B4
Keriya Shankou (pass), China 77/B4
Kerkdriel, Netherlands 27/G5
Kerkennah (isls.), Tunisia 106/G2
Kerkhoven, Minn. (56252) 255/C5
Kerki, U.S.S.R. 48/G6
Kérkira, Greece 45/D6
Kérkira (isl.), Greece 7/F5
Kérkira (isl.), Greece 45/D6
Kerkrade, Netherlands 27/J7
Kerlin, Ark. (†71753) 202/D7
Kerma, Sudan 111/F4
Kerma, Sudan 59/B6
Kermadec (isls.), N. Zealand 2/T7
Kermadec (isls.), N. Zealand 87/J9
Kerman, Calif. (93600) 204/E7
Kerman (prov.), Iran 66/K6
Kerman, Iran 54/G6
Kerman, Iran 59/G3
Kerman, Iran 66/K5
Kermanshah, Iran 54/F6
Kermanshah, Iran 59/E3
Kermanshah, Iran 66/E3
Kermanshahan (prov.), Iran 66/E3
Kerme (gulf), Turkey 63/B4
Kermit, Texas (79745) 303/B6
Kermit, W. Va. (25674) 312/B7
Kern (co.), Calif. 204/G8
Kern (riv.), Calif. 204/G8
Kernan, Ill. (†61364) 222/E2
Kernersville, N.C. (27284) 281/J2
Kerns, Switzerland 39/F3
Kernville, Calif. (93238) 204/G8
Kernville, Oreg. (†97367) 291/D3
Kérouané, Guinea 106/C7
Kerr (lake), Fla. 212/E2
Kerr, N.C. (†28444) 281/N5
Kerr, W. Scott (res.), N.C. 281/G2
Kerr, Robert S. (res.), Okla. 288/S4
Kerr (co.), Texas 303/E7
Kerrera (isl.), Scotland 15/C4
Kerrick, Minn. (55756) 255/F4
Kerrick, Texas (79051) 303/B1
Kerrobert, Sask. 181/C4
Kerrville, Tenn. (†38053) 237/B10
Kerrville, Texas (78028) 303/E7
Kerry (co.), Ireland 17/B7
Kerry (head), Ireland 17/A7
Kerry, Wales 13/D5
Kersey, Colo. (80644) 208/L2
Kersey, Ind. (†46310) 227/C2
Kersey, Pa. (15846) 294/E3
Kershaw (co.), S.C. 296/F3
Kershaw, S.C. (29067) 296/G2
Kersley, Br. Col. 184/F4
Kerteminde, Denmark 21/D7
Kerulen (riv.) 54/N5
Kerulen (riv.), Mongolia 77/H2
Kerwood, Ontario 177/C5
Kerzaz, Algeria 106/D3
Kerzers, Switzerland 39/D3
Kesagami (lake), Ontario 175/E2
Keşan, Turkey 63/B2
Keşap, Turkey 63/H2
Kesch (peak), Switzerland 39/J3
Kesennuma, Japan 81/K4
Kesgrave, England 13/J5
Kesh, N. Ireland 17/F3
Keshena, Wis. (54135) 317/J6
Keşiş Tepesi (mt.), Turkey 63/H3
Keskin, Turkey 63/E3
Keski-Suomi (prov.), Finland 18/O5
Kesley, Iowa (50649) 229/H3
Kessel, W. Va. (†26818) 312/H4
Kesten'ga, U.S.S.R. 52/D1
Kesteren, Netherlands 27/G5
Keswick, England 13/D3
Keswick, England 10/E3
Keswick, Iowa (50136) 229/J6
Keswick, New Bruns. 170/D3
Keswick (riv.), New Bruns. 170/C2
Keswick, Ontario 177/E3
Keswick, Va. (22947) 307/M4
Keswick Grove, N.J. (†08759) 273/E4
Keszthely, Hungary 41/D3
Keta, Ghana 106/E7
Ketapang, Indonesia 85/E6
Ketchen, Sask. 181/J3
Ketch Harbour, Nova Scotia 168/E4
Ketchikan, Alaska 146/E4
Ketchikan, Alaska 188/E6
Ketchikan, Alaska (99901) 196/N2
Ketchum, Idaho (83340) 220/D6
Ketchum, Okla. (74349) 288/R1
Kétegyháza, Hungary 41/F3
Kete Krachi, Ghana 106/E7
Kętrzyn, Poland 47/E1
Kettering, England 13/G5
Kettering, England 10/F4
Kettering, Ohio (45429) 284/B6
Kettering, Tasmania 99/D5
Kettle (riv.), Br. Col. 184/H5
Kettle (riv.), Minn. 255/F4
Kettle (pt.), Ontario 177/B4
Kettle (riv.), Wash. 310/G2
Kettle Falls, Wash. (99141) 310/H2
Kettleman City, Calif. (93239) 204/E7
Kettle River, Minn. (55757) 255/E4
Kettle River (range), Wash. 310/G2
Kettlersville, Ohio (45336) 284/B5
Kettle Valley, Br. Col. 184/H5
Keuka (lake), N.Y. 276/F5
Keuka Park, N.Y. (14478) 276/F5
Keuterville, Idaho (83538) 220/B3
Kevelaer, W. Germany 22/B3
Kevil, Ky. (42053) 237/D6
Kevin, Mont. (59454) 262/D2
Kevisville, Alberta 182/C4
Kew, Victoria 88/L7
Kew, Victoria 97/J5
Kewa, Wash. (†99138) 310/G2
Kewanee, Ill. (61443) 222/C2
Kewanee, Miss. (†39364) 256/H6
Kewanee, Mo. (63860) 261/N9
Kewanna, Ind. (46939) 227/E2
Kewaskum, Wis. (53040) 317/K8
Kewaunee (co.), Wis. 317/L6
Kewaunee, Wis. (54216) 317/M7
Keweenaw (co.), Mich. 250/A1
Keweenaw (bay), Mich. 250/A1
Keweenaw (pt.), Mich. 250/B1
Keweenaw Bay, Mich. (49944) 250/G1
Key, Ala. (†35960) 195/G2
Key (lake), Ireland 17/E3
Keya Paha (co.), Nebr. 264/E2
Keya Paha (riv.), Nebr. 264/D1
Keyapaha, S. Dak. (57545) 298/J7
Keya Paha (riv.), S. Dak. 298/K7
Key Biscayne, Fla. (33149) 212/B5
Key Colony Beach, Fla. (33051) 212/F7
Keyes, Calif. (95328) 204/E6
Keyes, Manitoba 179/C4
Keyes, Okla. (73947) 288/B1
Keyesport, Ill. (62253) 222/D5
Keyhole (res.), Wyo. 319/H1
Key Largo, Fla. (33037) 212/F6
Key Largo (key), Fla. 212/F6
Keymar, Md. (21757) 245/K2
Keynsham, England 13/E6
Keyport, N.J. (07735) 273/E3
Keyport, Wash. (98345) 310/A2
Keysbrook, W. Australia 88/B3
Keyser, W. Va. (26726) 312/J4
Keystone, Ind. (46759) 227/G3
Keystone, Iowa (52249) 229/J5
Keystone, Nebr. (69144) 264/C3
Keystone (res.), Ohio 284/K2
Keystone (lake), Okla. 288/O2
Keystone, S. Dak. (57751) 298/C6
Keystone, W. Va. (24852) 312/D8
Keystone Heights, Fla. (32656) 212/E2
Keystown, Sask. 181/F5
Keysville, Georgia (30816) 217/H4
Keysville, Va. (23947) 307/M6
Keytesville, Mo. (65261) 261/G4
Key Vaca (key), Fla. 212/E7
Key West, Fla. 146/K7
Key West, Fla. 188/K6
Key West, Fla. (33040) 212/E7
Key West Naval Air Sta., Fla. 212/E7
Kezar (lake), Maine 243/B7
Kezar (pond), Maine 243/B7
Kezar Falls, Maine (04047) 243/B8
Kežmarok, Czech. 41/F2
Khabake (Habahe), China 77/C2
Khabarovsk, U.S.S.R. 54/P5
Khabarovsk, U.S.S.R. 2/R3
Khabarovsk, U.S.S.R. 48/O5
Khabur (riv.), Syria 63/J5
Khabur (riv.), Syria 59/D2
Khachmas, U.S.S.R. 52/G6

Khadyzhensk, U.S.S.R. 52/E6
Khaf, Iran 66/L3
Khaibar, 'Asir, Saudi Arabia 59/D5
Khaibar, Hejaz, Saudi Arabia 59/C4
Khairpur, Pakistan 68/B3
Khairpur, Pakistan 59/J4
Khakass Aut. Obl., U.S.S.R. 48/J4
Khálki (isl.), Greece 45/H7
Khalkís, Greece 45/F6
Khal'mer-Yu, U.S.S.R. 52/K1
Khaluf, Oman 59/G5
Khamgaon, India 68/D4
Khamis Mushait, Saudi Arabia 59/D6
Khamkeut, Laos 72/E3
Khamman, India 68/D5
Khanabad, Afghanistan 59/J2
Khanabad, Afghanistan 68/B1
Khanaqin, Iraq 59/D3
Khanaqin, Iraq 66/D3
Khancoban, N.S. Wales 97/E5
Khandwa, India 68/D4
Khandyga, U.S.S.R. 48/O3
Khan esh Shamat, Syria 63/G6
Khanewal, Pakistan 68/C2
Khanh Hoa, Vietnam 72/F4
Khanh Hung, Vietnam 72/E5
Khaniá, Greece 45/G8
Khaniá, Greece 7/G5
Khaniá (gulf), Greece 45/G8
Khanka (lake) 54/P5
Khanka (lake), China 77/M3
Khanka (lake), U.S.S.R. 48/O5
Khanpur, Pakistan 68/C3
Khanpur, Pakistan 59/K4
Khan Sheikhun, Syria 63/G5
Khanty-Mansi Aut. Okr., U.S.S.R. 48/H3
Khanty-Mansiysk, U.S.S.R. 54/J3
Khanty-Mansiysk, U.S.S.R. 48/H3
Khanu, Thailand 72/C3
Khan Yunis, Gaza Strip 65/A5
Khao Luang (mt.), Burma 72/C5
Khao Luang (mt.), Thailand 72/C5
Khapcheranga, U.S.S.R. 48/M5
Kharagpur, India 68/F4
Kharan, Pakistan 59/J4
Kharan Kalat, Pakistan 68/A3
Kharas, West Bank 65/C4
Kharasavey (cape), U.S.S.R. 48/G2
Khardah, India 68/F1
Khârga (oasis), Egypt 111/F2
Khârga (oasis), Egypt 59/B4
Khark (Kharg) (isl.), Iran 66/G6
Khar'kov, U.S.S.R. 7/H4
Khar'kov, U.S.S.R. 2/L3
Khar'kov, U.S.S.R. 52/E5
Khar'kov, U.S.S.R. 48/D4
Kharmanli, Bulgaria 45/H5
Kharovsk, U.S.S.R. 48/D3
Kharovsk, U.S.S.R. 52/F3
Khartoum (prov.), Sudan 111/F4
Khartoum (cap.), Sudan 2/L5
Khartoum (cap.), Sudan 59/B6
Khartoum (cap.), Sudan 111/F4
Khartoum (cap.), Sudan 102/F3
Khartoum North, Sudan 102/F3
Khartoum North, Sudan 59/B6
Khartoum North, Sudan 111/F4
Khasab, Oman 59/G4
Khasavyurt, U.S.S.R. 52/G6
Khash, Afghanistan 68/A2
Khash, Afghanistan 59/H3
Khash, Iran 59/H4
Khash, Iran 66/M6
Khashm el Girba, Sudan 111/G5
Khashuri, U.S.S.R. 52/F6
Khasi (hills), India 68/G3
Khaskovo, Bulgaria 45/G5
Khatanga, U.S.S.R. 4/B4
Khatanga, U.S.S.R. 54/M2
Khatanga, U.S.S.R. 48/L2
Khatuniye, Syria 63/J4
Khay, Saudi Arabia 59/D6
Khedive, Sask. 181/G6
Khemis Miliana, Algeria 106/E1
Khemmarat, Thailand 72/E4
Khenifra, Morocco 106/C2
Kherson, U.S.S.R. 7/H4
Kherson, U.S.S.R. 48/D5
Kherson, U.S.S.R. 52/D5
Khe Sanh, Vietnam 72/E3
Kheta (riv.), U.S.S.R. 48/K2
Khilok, U.S.S.R. 48/M4
Khíos, Greece 45/G6
Khíos (isl.), Greece 45/G6
Khirbet Qumran (site), Jordan 65/D4
Khiva, 48/F5
Khiyav, Iran 66/E1
Khmel'nitskiy, U.S.S.R. 7/G4
Khmel'nitskiy, U.S.S.R. 52/C5
Khoai, Hon (isl.), Vietnam 72/E5
Khodzheyli, U.S.S.R. 48/F5
Kholm, Afghanistan 68/B1
Kholm, Afghanistan 59/J2
Kholm, U.S.S.R. 52/D3
Kholmsk, U.S.S.R. 48/P5
Khoman, Iran 66/F2
Khon Kaen, Thailand 72/D3
Khoper (riv.), U.S.S.R. 52/F4
Khorasan (prov.), Iran 66/K3
Khóra Sfakíon, Greece 45/G8
Khorat (Nakhon Ratchasima), Thailand 72/D4
Khoreyver, U.S.S.R. 52/J1
Khorixas, Namibia 118/A4
Khorog, U.S.S.R. 48/H6
Khorol, U.S.S.R. 52/D5
Khorramabad, Iran 59/E3
Khorramabad, Iran 66/F4
Khorramshahr, Iran 59/E3
Khorramshahr, Iran 66/F5
Khotan (Hotan), China 77/B4
Khotin, U.S.S.R. 52/C5
Khouribga, Morocco 106/C2
Khowst, Afghanistan 68/B2
Khromtau, U.S.S.R. 48/F4
Khuaf, Iran 59/H3
Khugiani, Afghanistan 68/B2
Khugiani, Afghanistan 59/J3
Khuis, Botswana 118/C5
Khu Khan, Thailand 72/E4
Khulna, Bangladesh 68/F4
Khurda, India 68/F4
Khurma, Saudi Arabia 59/D5
Khusf Rud (riv.), Iran 66/L4
Khushab, Pakistan 68/C2
Khust, U.S.S.R. 52/B5
Khuzdar, Pakistan 59/J4
Khuzestan (prov.), Iran 66/F5
Khvaf, Iran 66/L3
Khvalynsk, U.S.S.R. 52/G4
Khvojeh Lak, Kuh-e (mt.), Iran 66/E3
Khvonsar, Iran 66/F4
Khvor, Iran 59/G3
Khvor, Iran 66/J4
Khvoy, Iran 59/E2
Khvoy (Khoi), Iran 66/D1
Khwae Noi, Mae Nam (riv.), Thailand 72/C4
Khyber (pass) 54/J6
Khyber (pass), Pakistan 59/K3
Khyber (pass), Pakistan 68/C2
Kia, Solomon Is. 86/D2
Kiahsville, W. Va. (25534) 312/B6
Kiama, N.S. Wales 97/F4
Kiamba, Philippines 82/E8
Kiambi, Zaire 115/E5
Kiambu, Kenya 115/G4
Kiamichi, Okla. (†74574) 288/R5
Kiamichi (mts.), Okla. 288/R5
Kiamichi (riv.), Okla. 288/R6
Kiamika, Québec 172/B3
Kiamika (lake), Québec 172/B3
Kiamika (riv.), Québec 172/B3
Kiamusze (Jiamusi), China 77/M2
Kian (Ji'an), China 77/J6
Kiana, Alaska (99749) 196/F1
Kiangsi (Jiangxi) (prov.), China 77/J6
Kiangsu (Jiangsu) (prov.), China 77/K5
Kiantajärvi (lake), Finland 18/Q4
Kiáton, Greece 45/F6
Kiawah (isl.), S.C. 296/G6
Kibaek, Denmark 21/B5
Kibangou, Congo 115/B4
Kibara, Tanzania 115/F4
Kibaya, Tanzania 115/G5
Kibbee, Georgia (†30474) 217/H6
Kibler, Ark. (†72956) 202/B3
Kibombo, Zaire 115/E4
Kibondo, Tanzania 115/F4
Kibre Mengist, Ethiopia 111/G6
Kibwezi, Kenya 115/G4
Kičevo, Yugoslavia 45/E5
Kickapoo (riv.), Wis. 317/E9
Kickapoo Ind. Res., Kansas 232/G2
Kicking Horse (pass), Alberta 182/B4
Kicking Horse (pass), Br. Col. 184/J4
Kidal, Mali 102/C3
Kidal, Mali 106/E5
Kidapawan, Philippines 82/E7
Kidder, Mo. (64649) 261/D3
Kidder (co.), N. Dak. 282/L6
Kidder, S. Dak. (†57430) 298/O2
Kidderminster, England 10/G3
Kidderminster, England 13/E5
Kidepo Nat'l Park, Uganda 115/F3
Kidnappers (cape), N. Zealand 100/F3
Kidron, Ohio (44636) 284/G4
Kidsgrove, England 13/E4
Kidwelly, Wales 13/C6
Kidwelly, Wales 10/D5
Kief, N. Dak. (58747) 282/J4
Kiefer, Okla. (74041) 288/O3
Kieffer, W. Va. (24950) 312/E7
Kiel, W. Germany 7/E3
Kiel, W. Germany 22/D1
Kiel (bay), W. Germany 22/D1
Kiel (Nord-Ostsee) (canal), W. Germany 22/C1
Kiel, Wis. (53042) 317/L8
Kielce (prov.), Poland 47/E3
Kielce, Poland 47/E3
Kielce, Poland 7/G3
Kieler, Wis. (53812) 317/E10
Kien Hung, Vietnam 72/E5
Kienyang (Qianyang), China 77/H6
Kiester, Minn. (56051) 255/E7
Kieta, Papua N.G. 86/C2
Kieta, Papua N.G. 87/F6
Kiev, U.S.S.R. 2/L3
Kiev, U.S.S.R. 7/H3
Kiev, U.S.S.R. 48/D4
Kiev, U.S.S.R. 52/D4
Kiev (res.), U.S.S.R. 52/C4
Kiffa, Mauritania 106/B5
Kifri, Iraq 66/D3
Kigali (cap.), Rwanda 115/F4
Kigali (cap.), Rwanda 102/F5
Kiger (creek), Oreg. 291/J5
Kiği, Turkey 63/J3
Kiglapait (cape), Newf. 166/B2
Kiglapait (mts.), Newf. 166/B2
Kigoma (reg.), Tanzania 115/F4
Kigoma-Ujiji, Tanzania 115/E4
Kigoma-Ujiji, Tanzania 102/F5
Kihei, Hawaii (96753) 218/J2
Kihnu (isl.), U.S.S.R. 53/B1
Kiholo, Hawaii (†96740) 218/G4
Kiholo (bay), Hawaii 218/F4
Kii (chan.), Japan 81/G7
Kiiminki, Finland 18/P4
Kikai (isl.), Japan 81/O5
Kikiktaksoak (isl.), Newf. 166/B2
Kikinda, Yugoslavia 45/E3
Kikino, Alberta 182/D2
Kikkertavak (isl.), Newf. 166/B2
Kikoira, N.S. Wales 97/D3
Kikonai, Japan 81/K3
Kikori, Papua N.G. 85/B7
Kikwit, Zaire 115/C5
Kikwit, Zaire 102/D5
Kila, Mont. (59920) 262/B2
Kilafors, Sweden 18/K6
Kilauea, Hawaii 188/E5
Kilauea, Hawaii (96754) 218/C1
Kilauea (crater), Hawaii 218/H6
Kilauea (pt.), Hawaii 218/C1
Kilbaha, Ireland 17/B6
Kilbarchan, Scotland 15/A2
Kilbeggan, Ireland 17/G5
Kilbirnie, Scotland 15/A2
Kilbourne, Ill. (62655) 222/D3
Kilbourne, La. (71253) 238/H1
Kilbrannan (sound), Scotland 15/C5
Kilbride, Newf. 166/D2
Kilbuck (mts.), Alaska 196/G2
Kilburn, New Bruns. 170/C2
Kilcar, Ireland 17/D2
Kilchoan, Scotland 15/B4
Kilchu, N. Korea 81/D3
Kilcock, Ireland 17/H5
Kilconnell, Ireland 17/E5
Kilcoole, Ireland 17/K5
Kilcormac, Ireland 17/F5
Kilcoy, Queensland 95/E5
Kilcullen, Ireland 17/H5
Kildare, Georgia (†30449) 217/K5
Kildare (co.), Ireland 17/H5
Kildare, Ireland 10/C4
Kildare, Ireland 17/H5
Kildare, Okla. (74601) 288/M1
Kildare (cape), Pr. Edward I. 168/E2
Kildare, Texas (75562) 303/K5
Kildeer, Ill. (†60069) 222/A5
Kil'din (isl.), U.S.S.R. 52/D1
Kildonan, Br. Col. 184/E5
Kildonan, Scotland 15/E2
Kildonan, Zimbabwe 118/E3
Kildurk, North. Terr. 93/A4
Kildysart, Ireland 17/C6
Kilembe, Uganda 115/F3
Kilembe, Zaire 115/C5
Kilfenora, Ireland 17/C6
Kilfinane, Ireland 17/D7
Kilgarvan, Ireland 17/B8
Kilgore, Idaho (†83423) 220/G5
Kilgore, Nebr. (69216) 264/D2
Kilgore, Ohio (†44615) 284/H5
Kilgore, Texas (75662) 303/K5
Kili (isl.), Marshall Is. 87/G5
Kilifi, Kenya 115/G4
Kilimanjaro (reg.), Tanzania 115/G4
Kilimanjaro (mt.), Tanzania 102/F5
Kilimanjaro (mt.), Tanzania 115/G4
Kilimli, Turkey 63/D2
Kilinailau (isls.), Papua N.G. 86/C2
Kilingi-Nõmme, U.S.S.R. 53/C1
Kilis, Turkey 63/G4
Kilitbahir, Turkey 63/B6
Kiliya, U.S.S.R. 52/C5
Kilkee, Ireland 10/B4
Kilkee, Ireland 17/B6
Kilkeel, N. Ireland 17/K3
Kilkelly, Ireland 17/D4
Kilkenny (co.), Ireland 17/G6
Kilkenny, Ireland 10/C4
Kilkenny, Ireland 17/G6
Kilkenny, Minn. (56052) 255/E6
Kilkieran (bay), Ireland 17/B5
Kilkís, Greece 45/F5
Killala, Ireland 17/C3
Killala (bay), Ireland 17/C3
Killaloe, Ireland 10/B4
Killaloe, Ireland 17/D6
Killaloe Station, Ontario 177/G2
Killaly, Sask. 181/J5
Killam, Alberta 182/E3
Killam, New Bruns. 170/E2
Killarney, Ireland 10/B4
Killarney, Ireland 17/C7
Killarney (lakes), Ireland 10/B4
Killarney, Man. 162/G6
Killarney, Manitoba 179/C5
Killarney, North. Terr. 93/B4
Killarney, Ontario 177/C2
Killarney Prov. Park, Ontario 177/C1
Killary (harb.), Ireland 17/A4
Killavullen, Ireland 17/D7
Killbear Point Prov. Park, Ontario 177/D2
Kill Buck, N.Y. (14748) 276/C6
Killbuck, Ohio (44637) 284/G5
Killbuck (creek), Ohio 284/G4
Killdeer, N. Dak. (58640) 282/E5
Killdeer, Sask. 181/E6
Kill Devil Hills, N.C. (27948) 281/T3
Killduff, Iowa (50137) 229/H5
Killearn, Scotland 15/B1
Killeen, Texas (76541) 303/G6
Killen, Ala. (35645) 195/D1
Killenaule, Ireland 17/F6
Killeshandra, Ireland 17/F3
Killian, La. (†70462) 238/M2
Killimor, Ireland 17/E5
Killin, Scotland 15/D4
Killinaboy, Ireland 17/C6
Killingly○, Conn. (†06241) 210/H1
Killington, Vt. (05751) 268/B4
Killington (peak), Vt. 268/B4
Killingworth○, Conn. (†06413) 210/E3
Killona, La. (70066) 238/M3
Killorglin, Ireland 17/B7
Killough, N. Ireland 17/K3
Killucan-Rathwire, Ireland 17/G4
Kill Van Kull (str.), N.J. 273/B2
Killybegs, Ireland 17/E2
Killyclogher, N. Ireland 17/G2
Killyleagh, N. Ireland 17/K3
Kilmacolm, Scotland 15/A2
Kilmacrennan, Ireland 17/F1
Kilmacthomas, Ireland 17/G7
Kilmallock, Ireland 17/D7
Kilmarnock, Scotland 15/D5
Kilmarnock, Scotland 10/D3
Kilmarnock, Va. (22482) 307/R5
Kilmaurs, Scotland 15/D5
Kilmeadan, Ireland 17/G7
Kilmichael, Miss. (39747) 256/E4
Kilmihill, Ireland 17/C6
Kilmoganny, Ireland 17/G7
Kilmore, Victoria 97/C5
Kilmore Quay, Ireland 17/H7
Kilmurry, Ireland 17/C6
Kiln, Miss. (39556) 256/F10
Kilnaleck, Ireland 17/G4
Kilninver, Scotland 15/C4
Kilo, Zaire 115/E3
Kilombero (riv.), Tanzania 115/G5
Kilosa, Tanzania 115/G5
Kilpisjärvi (lake), Finland 18/M2
Kilpisjärvi (lake), Sweden 18/M2
Kilrea, N. Ireland 17/H2
Kilrenny and Anstruther, Scotland 15/F4
Kilrenny and Anstruther, Scotland 10/E2
Kilronan, Ireland 17/B5
Kilrush, Ireland 10/B4
Kilrush, Ireland 17/C6
Kilsheelan, Ireland 17/F7
Kilsyth, Scotland 15/B1
Kilsyth, Scotland 10/B1
Kilsyth, W. Va. (25859) 312/D7
Kiltan (isl.), India 68/C6
Kiltimagh, Ireland 17/C4
Kilwa, Zaire 115/E5
Kilwa Kivinje, Tanzania 115/G5
Kilwa Masoko, Tanzania 115/G5
Kilwinning, Sask. 181/E2
Kilwinning, Scotland 15/D5
Kilworth, Ireland 17/E7
Kilyos, Turkey 63/D5
Kim, Colo. (81049) 208/N8
Kimba, S. Australia 88/F6
Kimba, S. Australia 94/E5
Kimball (mt.), Alaska 196/K2
Kimball, Alberta 182/D5
Kimball, Kansas (†66733) 232/G4
Kimball, Minn. (55353) 255/D5
Kimball (co.), Nebr. 264/A3
Kimball, Nebr. (69145) 264/A3
Kimball, S. Dak. (57355) 298/M6
Kimball, Tenn. (†37347) 237/K10
Kimball, W. Va. (24853) 312/C8
Kimballton, Iowa (51543) 229/D5
Kimballton, Va. (†24150) 307/G6
Kimbe, Papua N.G. 86/B2
Kimbe, Papua N.G. 87/F6
Kimberley, Br. Col. 184/K5
Kimberley, S. Africa 102/E7
Kimberley, S. Africa 118/C5
Kimberley (plat.), W. Australia 88/D3
Kimberley (plat.), W. Australia 92/D2
Kimberley Research Station, W. Australia 88/D3
Kimberling City, Mo. (65686) 261/F9
Kimberlin Heights, Tenn. (37920) 237/O9
Kimberly, Ala. (35091) 195/E3
Kimberly, Idaho (83341) 220/D7
Kimberly, Minn. (†56431) 255/E4
Kimberly, Oreg. (97848) 291/H3
Kimberly, Wis. (54136) 317/K7
Kimble (co.), Texas 303/E7
Kimbolton, Ohio (43749) 284/G5
Kimbrough, Ala. (36746) 195/C6
Kimch'aek, N. Korea 81/D3
Kimch'ŏn, S. Korea 81/D5
Kimesville, N.C. (†27298) 281/L3
Kimhae, Pakistan 81/D6
Kími, Greece 45/F6
Kimitsu, Japan 81/O3
Kimiwan (lake), Alberta 182/B2
Kimje, S. Korea 81/C6
Kimmell, Ind. (46760) 227/F2
Kimmins, Tenn. (38462) 237/F9
Kimmswick, Mo. (63053) 261/M6
Kímolos (isl.), Greece 45/G7
Kimovsk, U.S.S.R. 52/E4
Kimry, U.S.S.R. 52/E3
Kimsquit, Br. Col. 184/D4
Kinabalu (mt.), Malaysia 85/F4
Kinalı (isl.), Turkey 63/D6
Kinalung, N.S. Wales 97/B3
Kinango, Kenya 115/G4
Kinard, Fla. (32449) 212/D6
Kinards, S.C. (29355) 296/D3
Kinbrace, Scotland 15/E2
Kinbrae, Minn. (†56126) 255/C7
Kinburn, Ontario 177/H2
Kincaid, Ill. (62540) 222/D4
Kincaid, Kansas (66039) 232/G3
Kincaid, Sask. 181/D6
Kincardine, Ontario 177/C3
Kincardine, Scotland 10/B1
Kincardine, Scotland 15/C1
Kincardine, (trad. co.), Scotland, 15/A5
Kincheloe (pt.), Oreg. 291/C2
Kincheloe, W. Va. (†26378) 312/E4
Kincolith, Br. Col. 184/B2
Kincraig, Scotland 15/E3
Kinda, Zaire 115/D5
Kindama, Congo 115/C4
Kindberg, Austria 41/C3
Kinde, Mich. (48445) 250/G5
Kinder, La. (70648) 238/E6
Kinderhook, Ill. (62345) 222/B4
Kinderhook, N.Y. (12106) 276/N6
Kindersley, Sask. 162/F5
Kindersley, Sask. 181/B4
Kindia, Guinea 106/B6
Kindia, Guinea 102/A3
Kindred, N. Dak. (58051) 282/R6
Kinel', U.S.S.R. 52/H4
Kinel' (riv.), U.S.S.R. 52/H4
Kineshma, U.S.S.R. 7/J3
Kineshma, U.S.S.R. 52/F3
King (isl.), Alaska 196/E1
King (isl.), Australia 87/E10
King (isl.), Br. Col. 184/D4
King, Ky. (†40906) 237/O7
King (pt.), N.S. Wales 97/J2
King (cays), Nicaragua 154/F4
King, N.C. (27021) 281/J2
King (isl.), Tasmania 88/G7
King (isl.), Tasmania 99/A1
King (riv.), Tasmania 99/B4
King (co.), Texas 303/D4
King (co.), Wash. 310/D3
King (sound), W. Australia 88/C3
King (sound), W. Australia 92/C2
King, Wis. (54946) 317/H7
King and Queen (co.), Va. 307/P5
King and Queen Court House, Va. (23085) 307/P5
Kingaroy, Queensland 95/D5
Kingaroy, Queensland 88/J5
King Christian (isl.), N.W. Terrs. 187/H2
King Christian IX Land (reg.), Greenl. 4/C11
King Christian IX Land (reg.), Greenland 146/Q3
King Christian X Land (reg.), Greenl. 4/B11
King Christian X Land (reg.), Greenland 146/R2
King City, Calif. (93930) 204/D7
King City, Mo. (64463) 261/D2
King City, Ontario 177/J3
King City, Oreg. (97223) 291/A2
Kingcome Inlet, Br. Col. 184/D4
King Cove, Alaska (99612) 196/F4
Kingdom City, Mo. (65262) 261/J5
King Ferry, N.Y. (13081) 276/G5
Kingfield○, Maine (04947) 243/C6
Kingfisher (co.), Okla. 288/L3
Kingfisher, Okla. (73750) 288/L3
King Frederik VI Coast (reg.), Greenland 146/P3
King Frederik VIII Land (reg.), Greenl. 4/B11
King Frederik VIII Land (reg.), Greenland 146/R2
King George (isl.) 5/C16
King George (isls.), N.W. Terrs. 187/L4
King George (co.), Va. 307/O4
King George, Va. (22485) 307/O4
King George's (falls), S. Africa 118/B5
King Hill, Idaho (83633) 220/C6
Kinghorn, Scotland 15/D1
Kingisepp (Kuressaare), U.S.S.R. 53/B1
King Leopold (range), W. Australia 88/D3
King Leopold (range), W. Australia 92/D2
Kingman, Alberta 182/D3
Kingman, Ariz. (86401) 198/A3
Kingman, Ind. (47952) 227/C5
Kingman (co.), Kansas 232/D4
Kingman, Kansas (67068) 232/D4
Kingman, Maine (04451) 243/G4
Kingman (reef), Pacific 87/K5
Kingoonya, S. Australia 88/E6
Kingoonya, S. Australia 94/D4
Kings (co.), Calif. 204/G8
Kings (riv.), Calif. 204/F7
Kings (riv.), Nev. 266/C1
King's (co.), New Bruns. 170/E3
Kings (co.), N.Y. 276/N9
Kings (co.), Nova Scotia 168/D4
Kings (co.), Pr. Edward I. 168/F2
Kings (peak), Utah 304/D3
King Salmon, Alaska (99613) 196/G3
Kings Beach, Calif. (95719) 204/F4
Kingsbridge, England 13/D7
Kingsburg, Calif. (93631) 204/F7
Kingsburg, S.C. (†29555) 296/H4
Kingsburg, S. Dak. (†57062) 298/O8
Kingsbury, Ind. (46345) 227/D1
Kingsbury○, Maine (†04990) 243/D5
Kingsbury (pond), Maine 243/D5
Kingsbury, Québec 172/E4
Kingsbury (co.), S. Dak. 298/O5
Kingsbury, Texas (78638) 303/G8
Kings Canyon Nat'l Park, Calif. 204/G7
Kingsclear, New Bruns. 170/D3
Kingscliff, N.S. Wales 97/G1
Kingscote, S. Australia 88/F7
Kingscote, S. Australia 94/E6
Kingscourt, Ireland 17/H4
Kingscourt, Ireland 10/C4
King's Cove, Newf. 166/D1
Kings Creek, N.C. (†28645) 281/G3
Kings Creek, Ohio (†43078) 284/C5
Kings Creek, S.C. (29719) 296/E1
Kingsdale, Minn. (†55015) 255/F4
Kingsdown, Kansas (67858) 232/C4
Kingsey Falls, Québec 172/E4
Kingsford, Mich. (49801) 250/A3
Kingsford Heights, Ind. (46346) 227/D2
Kingsford-Smith Airport, N.S. Wales 88/L4
Kingsford-Smith Airport, N.S. Wales 97/J4
Kingsgate, Br. Col. 184/K5
Kingshill, Virgin Is. (U.S.) 161/F4
Kingsland, Ark. (71652) 202/D6
Kingsland, Georgia (31548) 217/J9
Kingsland, Texas (78639) 303/F7
Kings Landing, Ala. (†36775) 195/D6
King's Landing, New Bruns. 170/C3
Kingsley (lake), Fla. 212/E2
Kingsley, Iowa (51028) 229/A3
Kingsley, Ky. (†40201) 237/K2
Kingsley, Mich. (49649) 250/D4
Kingsley (dam), Nebr. 264/C3
Kingsley, New Bruns. 170/D2
King's Lynn, England 13/H5
King's Lynn, England 10/G4
Kingsmere (lake), Sask. 181/E1
Kingsmill, Texas (†79065) 303/D2
Kings Mills, Ohio (45034) 284/B7
Kings Mountain, Ky. (40442) 237/M6
Kings Mountain, N.C. (28086) 281/G4
Kings Mountain Nat'l Mil. Park, S.C. 296/E1
Kings Park, N.Y. (11754) 276/O9
King's Point, Newf. 166/C4
Kings Point, N.Y. (11024) 276/P6
Kingsport, Nova Scotia 168/D3
Kingsport, Tenn. (*37660) 237/R7
Kingston, Ark. (72742) 202/C1
Kingston, Australia 87/G8
Kingston, Georgia (30145) 217/C2
Kingston, Ill. (60145) 222/E1
Kingston, Ind. (†47240) 227/G6
Kingston, Iowa (†52637) 229/L7
Kingston (cap.), Jamaica 146/L8
Kingston (cap.), Jamaica 156/C3
Kingston (cap.), Jamaica 158/K6
Kingston, La. (†71032) 238/C2
Kingston, Md. (21834) 245/R8
Kingston, Mass. (02364) 249/M5
Kingston○, Mass. (02364) 249/M5
Kingston, Mich. (48741) 250/F5
Kingston, Minn. (55326) 255/D5
Kingston, Mo. (64650) 261/E3
Kingston○, N.H. (03848) 268/E6
Kingston, N.J. (08528) 273/D3
Kingston, N. Mex. (†88042) 274/B6
Kingston, N.Y. (12401) 276/M7
Kingston, N. Zealand 100/B6
Kingston, Norfolk I. 88/L6
Kingston, Nova Scotia 168/D4
Kingston, Ohio (45644) 284/E7
Kingston, Okla. (73439) 288/N7
Kingston, Ont. 162/J7
Kingston, Ont. 146/L5
Kingston, Ontario 177/H3
Kingston, Pa. (18704) 294/F7
Kingston, R.I. (02881) 249/J7
Kingston, S. Australia 94/G7
Kingston, Tasmania 99/D4
Kingston, Tenn. (37763) 237/N9
Kingston, Utah (84743) 304/B5
Kingston, Wash. (98346) 310/C3
Kingston, W. Va. (25120) 312/D7
Kingston, Wis. (53939) 317/H8
Kingston Mines, Ill. (61539) 222/D3
Kingston Springs, Tenn. (37082) 237/G8
Kingston upon Thames, England 10/B6
Kingston upon Thames, England 13/H8
Kingstown (Dún Laoghaire), Ireland 17/K5
Kingstown, N.S. Wales 97/F2
Kingstown (cap.), St. Vin. & Grens. 161/A9
Kingstown (cap.), St. Vin. & Grens. 156/G4
Kingstown (bay), St. Vin. & Grens. 161/A9
Kingstree, S.C. (29556) 296/H4
Kings Valley, Oreg. (†97361) 291/D3
Kingsville, Md. (21087) 245/N3
Kingsville, Mo. (64061) 261/D5
Kingsville, Ohio (44048) 284/J2
Kingsville, Ontario 177/B6
Kingsville, Texas (78363) 303/G10
Kingsville N.A.S., Texas 303/G10
Kingswood, England 13/E6
Kingswood, Ky. (†40144) 237/J5
Kingtehchen (Jingdezhen), China 77/J6
Kington, England 13/E5
Kington, England 10/E4
Kingurutik (mesa), Newf. 166/B2
Kingussie, Scotland 15/D3
Kingussie, Scotland 10/D2
Kingville, S.C. (†29052) 296/F4
King William (isl.), N.W.T. 162/G2
King William (isl.), N.W. Terrs. 187/J3
King William (lake), Tasmania 99/C4
King William (co.), Va. 307/O5
King William, Va. (23086) 307/O5
King William's Town, S. Africa 102/E8
King William's Town, S. Africa 118/D6
Kingwood, W. Va. (26537) 312/G4
Kinhwa (Jinhua), China 77/J6
Kiniama, Zaire 115/E6
Kınık, Turkey 63/B3
Kinistino, Sask. 181/F3
Kinkala, Congo 115/B4
Kinkora, Pr. Edward I. 168/E2
Kinley, Sask. 181/D3
Kinloch, Mo. (63140) 261/P2
Kinlochbervie, Scotland 15/D2
Kinlochewe, Scotland 15/C3
Kinlochleven, Scotland 15/D4
Kinloch Rannoch, Scotland 15/D4
Kinloss, Scotland 15/E3
Kinlough, Ireland 17/E3
Kinmount, Minn. (†55771) 255/F2
Kinmount, Ontario 177/F3
Kinmundy, Ill. (62854) 222/E5
Kinna, Sweden 18/H8
Kinnairds (head), Scotland 10/F2
Kinnairds (head), Scotland 15/G3
Kinnear, Wyo. (82516) 319/D2
Kinnegad, Ireland 17/G5
Kinnelon, N.J. (07405) 273/E2
Kinneret, Israel 65/D2
Kinney, Minn. (55758) 255/F3
Kinney (co.), Texas 303/D8
Kinnitty, Ireland 17/F5
Kino (riv.), Japan 81/G6
Kinoosao, Sask. 181/N3
Kinosota, Manitoba 179/D4
Kinrooi, Belgium 27/H6
Kinross, Iowa (52250) 229/J6
Kinross, Scotland 15/E4
Kinross, Scotland 10/E2
Kinross (trad. co.), Scotland, 15/A5
Kinsale, Ireland 10/B5
Kinsale, Ireland 17/D8
Kinsale (harb.), Ireland 17/E8
Kinsale, Old Head of (head), Ireland 17/E8
Kinsale, Va. (22488) 307/P4
Kinsella, Alberta 182/E3
Kinsey, Ala. (†36301) 195/H8
Kinsey, Mont. (59338) 262/L4
Kinshasa (prov.), Zaire 115/C4
Kinshasa (cap.), Zaire 2/K6
Kinshasa (cap.), Zaire 115/C4
Kinshasa (cap.), Zaire 102/D5
Kinsley, Kansas (67547) 232/C4
Kinsman, Ill. (60437) 222/E2
Kinsman (mt.), N.H. 268/D3
Kinsman, Ohio (44428) 284/J3
Kinsman Notch (pass), N.H. 268/D3
Kinston, Ala. (36453) 195/F8
Kinston, N.C. (28501) 281/O4
Kinta, Okla. (74552) 288/R4

Kintampo, Ghana 106/D7
Kintnersville, Pa. (18930) 294/M4
Kintore, Scotland 15/F3
Kintyre, N. Dak. (58549) 282/L6
Kintyre (pen.), Scotland 15/C5
Kintyre, Mull of (prom.), Scotland 15/C5
Kinuso, Alberta 182/C2
Kinvara, Ireland 17/D5
Kinwow (bay), Manitoba 179/E2
Kinyangiri, Tanzania 115/G4
Kinyeti (mt.), Sudan 111/F7
Kinzel Springs, Tenn. (†37882) 237/O9
Kinzua, Oreg. (97849) 291/H3
Kioa (isl.), Fiji 86/R10
Kioga (lake), Uganda 102/F4
Kioga (lake), Uganda 115/F3
Kiona, Wash. (†99320) 310/F4
Kiosk, Ontario 177/F1
Kiowa (co.), Colo. 208/O6
Kiowa, Colo. (80117) 208/L4
Kiowa (creek), Colo. 208/L3
Kiowa (co.), Kansas 232/C4
Kiowa, Kansas (67070) 232/D4
Kiowa (co.), Okla. 288/J5
Kiowa, Okla. (74553) 288/P5
Kiowa (creek), Okla. 288/F1
Kiowa (creek), Texas 303/D1
Kipabiskau, Sask. 181/G3
Kipahulu, Hawaii (†96713) 218/K2
Kiparissia, Greece 45/E7
Kiparissia (gulf), Greece 45/E7
Kipawa, Québec 174/B3
Kipili, Tanzania 115/F5
Kipini, Kenya 115/H4
Kipisa, N.W. Terrs. 187/M3
Kipling, N.C. (27543) 281/M4
Kipling, Sask. 181/J5
Kipnuk, Alaska (99614) 196/F2
Kipp, Alberta 182/D5
Kipp, Kansas (†67401) 232/E3
Kippel, Switzerland 39/E4
Kippen, Scotland 15/B1
Kippens, Newf. 166/C4
Kippure (mt.), Ireland 17/J5
Kipton, Ohio (44049) 284/F3
Kipushi, Zaire 115/E6
Kipushi, Zaire 102/E6
Kiput, Philippines 82/C8
Kira Kira, Solomon Is. 86/E3
Kiraz, Turkey 63/C3
Kirazlı, Turkey 63/C6
Kirby, Ark. (71950) 202/C5
Kirby, Mont. (†59016) 262/J5
Kirby, Ohio (43330) 284/D4
Kirby, Texas (†78109) 303/K11
Kirby, W. Va. (26729) 312/J4
Kirby, Wyo. (82430) 319/D2
Kirbyville, Texas (75956) 303/K7
Kirchberg, Bern, Switzerland 39/E2
Kirchberg, St. Gallen, Switzerland 39/G2
Kirchdorf an der Krems, Austria 41/C3
Kirchheim unter Teck, W. Germany 22/C4
Kircubbin, N. Ireland 17/K3
Kirensk, U.S.S.R. 48/L4
Kirgiz S.S.R., U.S.S.R. 54/J5
Kirgiz S.S.R., U.S.S.R. 48/H5
Kiri, Zaire 115/C4
Kiribati 2/A6
Kiribati 87/J6
Kiribati 2/T6
Kirigalpota (mt.), Sri Lanka 68/E7
Kırıkhan, Turkey 63/G4
Kırıkkale, Turkey 63/E3
Kirillov, U.S.S.R. 52/E2
Kirin (Jilin) (prov.), China 77/L3
Kirin (Jilin), China 77/L3
Kirishi, U.S.S.R. 52/D3
Kirishima-Yaku National Park, Japan 81/E7
Kiriwina (isl.), Papua N.G. 85/C7
Kirk, Colo. (80824) 208/P3
Kirk, Ky. (†40143) 237/H5
Kirk, W. Va. (†25671) 312/B7
Kırkağaç, Turkey 63/B3
Kirkburton, England 13/J2
Kirkby, England 13/G2
Kirkby, England 10/F2
Kirkby Lonsdale, England 13/E3
Kirkbymoorside, England 13/F3
Kirkby Stephen, England 13/E3
Kirkcaldy, Alberta 182/D4
Kirkcaldy, Scotland 15/D1
Kirkcaldy, Scotland 10/C1
Kirkcolm, Scotland 15/C6
Kirkconnel, Scotland 15/E5
Kirkcowan, Scotland 15/D6
Kirkcudbright, Scotland 15/E6
Kirkcudbright, Scotland 10/E3
Kirkcudbright (trad. co.), Scotland 15/A5
Kirkee, India 68/C5
Kirkella, Manitoba 179/A4
Kirkenes, Norway 18/Q2
Kirkersville, Ohio (43033) 284/E6
Kirkfield, Ontario 177/E3
Kirkham, England 13/G1B
Kirkham, England 10/F1
Kirkhill, Scotland 15/D3
Kirkinner, Scotland 15/D6
Kirkintilloch, Scotland 10/B1
Kirkintilloch, Scotland 15/B2
Kirkland, Ariz. (86332) 198/C4
Kirkland, Georgia (†31642) 217/G8
Kirkland, Ill. (60146) 222/E1
Kirkland, New Bruns. 170/C3
Kirkland, Québec 172/H4
Kirkland, Tenn. (†37046) 237/H9
Kirkland, Texas (79238) 303/D3
Kirkland, Wash. (98033) 310/B2
Kirkland Lake, Ont. 162/H6
Kirkland Lake, Ont. 146/K5
Kirkland Lake, Ontario 177/K5
Kirkland Lake, Ontario 175/D3
Kirklareli (prov.), Turkey 63/B2
Kirklareli, Turkey 63/B2
Kirklin, Ind. (46050) 227/E4
Kirkman, Iowa (51447) 229/C5
Kirkmansville, Ky. (†42216) 237/G6
Kirkmuirhill, Scotland 15/E5
Kirkpatrick (lake), Alberta 182/E4
Kirkpatrick (mt.) 5/A8
Kirkpatrick, Ind. (†47955) 227/D4
Kirkpatrick, Ohio (†43302) 284/D4
Kirksey, Ky. (42054) 237/E7
Kirksville, Ind. (†47401) 227/D6
Kirksville, Ky. (†40475) 237/N5
Kirksville, Mo. (63501) 261/H2
Kirkton of Glenisla, Scotland 15/E4
Kirkuk, Iraq 54/F6
Kirkuk, Iraq 59/D2
Kirkuk, Iraq 66/D3
Kirkville, Iowa (52566) 229/H6
Kirkville, Miss. (†38856) 256/H2
Kirkwall, Scotland 10/E1
Kirkwall, Scotland 15/E2
Kirkwood, Del. (19708) 245/R2
Kirkwood, Ill. (61447) 222/C3
Kirkwood, Mo. (63122) 261/O3
Kirkwood, N.J. (08043) 273/B4
Kirkwood, N.Y. (13795) 276/J6
Kirkwood, Pa. (17536) 294/K6
Kirkwood, S. Africa 118/D6
Kirmastı (riv.), Turkey 63/C3
Kirn, W. Germany 22/B4
Kiron, Iowa (51448) 229/C4
Kirov, U.S.S.R. 7/J3
Kirov, U.S.S.R. 48/E4
Kirov, Kirov, U.S.S.R. 52/G3
Kirov, Kaluga, U.S.S.R. 52/D4
Kirovabad, U.S.S.R. 7/J4
Kirovabad, U.S.S.R. 52/G6
Kirovabad, U.S.S.R. 48/E5
Kirovakan, U.S.S.R. 52/G6
Kirovo-Chepetsk, U.S.S.R. 52/H3
Kirovograd, U.S.S.R. 7/H4
Kirovograd, U.S.S.R. 48/D5
Kirovograd, U.S.S.R. 52/D5
Kirovsk, U.S.S.R. 52/D1
Kirovskiy, U.S.S.R. 48/H5
Kirriemuir, Alberta 182/E4
Kirriemuir, Scotland 10/E2
Kirriemuir, Scotland 15/E4
Kirs, U.S.S.R. 52/H3
Kirsanov, U.S.S.R. 52/F4
Kırşehir (prov.), Turkey 63/F3
Kırşehir, Turkey 63/F3
Kırşehir, Turkey 59/B2
Kirte, Turkey 63/B6
Kirtland, N. Mex. (87417) 274/A2
Kirtland, Ohio (†44094) 284/H2
Kirtland A.F.B., N. Mex. 274/C3
Kirtland Hills, Ohio (†44094) 284/H2
Kirton, England 13/H5
Kiruna, Sweden 4/C8
Kiruna, Sweden 7/G2
Kiruna, Sweden 18/L3
Kirundu, Zaire 115/E4
Kirwin, Kansas (67644) 232/C2
Kirwin (res.), Kansas 232/C2
Kiryu, Japan 81/J5
Kisa, Sweden 18/J7
Kisangani, Zaire 115/E3
Kisangani, Zaire 102/E4
Kisar (isl.), Indonesia 85/H7
Kisarazu, Japan 81/P3
Kisatchie, La. (†71468) 238/D4
K.I. Sawyer A.F.B., Mich. 250/B2
Kisbér, Hungary 41/D3
Kisbey, Sask. 181/J6
Kiselevsk, U.S.S.R. 48/J4
Kishangarh, India 68/D3
Kishinev, U.S.S.R. 7/G4
Kishinev, U.S.S.R. 48/C5
Kishinev, U.S.S.R. 52/C5
Kishiwada, Japan 81/J8
Kishorganj, Bangladesh 68/G4
Kishtwar, India 68/D2
Kisi (Jixi), China 77/M2
Kisii, Kenya 115/F4
Kisiju, Tanzania 115/G5
Kiska (isl.), Alaska 188/D6
Kiska (isl.), Alaska 196/J4
Kiska (vol.), Alaska 196/J4
Kiskatinaw (riv.), Br. Col. 184/G2
Kiskissink, Québec 172/E2
Kiskissink (lake), Québec 172/E2
Kiskőrös, Hungary 41/E3
Kiskunfélegyháza, Hungary 41/E3
Kiskunhalas, Hungary 41/E3
Kiskunmajsa, Hungary 41/E3
Kislovodsk, U.S.S.R. 7/J4
Kislovodsk, U.S.S.R. 52/F6
Kismayu (Chisimayu), Somalia 115/H4
Kismet, Kansas (67859) 232/B4
Kispest, Hungary 41/E3
Kissamos, Greece 45/G8
Kissee Mills, Mo. (65680) 261/G9
Kissidougou, Guinea 106/B7
Kissimmee, Fla. (32741) 212/E3
Kissimmee (lake), Fla. 212/E4
Kissimmee (riv.), Fla. 212/E4
Kississing (lake), Manitoba 179/H2
Kistelek, Hungary 41/E3
Kisten (pass), Switzerland 39/H3
Kisterenye, Hungary 41/E2
Kistler, Pa. (†17066) 294/G5
Kistler, W. Va. (25628) 312/C7
Kistna (riv.), India 54/J8
Kistna (Krishna) (riv.), India 68/D5
Kistrand, Norway 18/O1
Kisújszállás, Hungary 41/F3
Kisumu, Kenya 115/F3
Kisumu, Kenya 102/F5
Kisvárda, Hungary 41/G2
Kita, Mali 102/B3
Kita, Mali 106/C6
Kitaibaraki, Japan 81/K5
Kita Iwo (isl.), Japan 87/D3
Kita Iwo (isl.), Japan 81/M4
Kitakami, Japan 81/K4
Kitakami (riv.), Japan 81/K4
Kitakata, Japan 81/J5
Kitakyushu, Japan 81/E6
Kitakyushu, Japan 54/P6
Kitakyushu, Japan 2/R4
Kitale, Kenya 102/F4
Kitale, Kenya 115/G3
Kitami, Japan 81/L2
Kit Carson (co.), Colo. 208/O4
Kit Carson, Colo. (80825) 208/O5
Kit Carson (mt.), Colo. 208/H7
Kitchel, Ind. (†47353) 227/H5
Kitchener (mt.), Alberta 182/B3
Kitchener, Ontario 177/D4
Kite, Georgia (31049) 217/G5
Kite, Ky. (41828) 237/R6
Kitgum, Uganda 115/F3
Kithira, Greece 45/F7
Kithira (isl.), Greece 45/F7
Kithnos (isl.), Greece 45/G7
Kitim, Jordan 65/D3
Kitimat, Br. Col. 162/D5
Kitimat, Br. Col. 146/F4
Kitimat, Br. Col. 184/C3
Kitinen (riv.), Finland 18/P3
Kitsap (co.), Wash. 310/C3
Kitsault, Br. Col. 184/C2
Kitscoty, Alberta 182/E3
Kitt (peak), Ariz. 198/D7
Kittanning, Pa. (16201) 294/D4
Kittatinny (mts.), N.J. 273/D1
Kittery, Maine (03904) 243/B9
Kittery○, Maine (03904) 243/B9
Kittery Point, Maine (03905) 243/B9
Kittilä, Finland 18/O3
Kittitas (co.), Wash. 310/E3
Kittitas, Wash. (98934) 310/E4
Kittrell, N.C. (27544) 281/M2
Kitts, Ky. (40848) 237/P7
Kitts Hill, Ohio (45645) 284/E8
Kittson (co.), Minn. 255/B2
Kitty Hawk, N.C. (27949) 281/T2
Kitui, Kenya 115/G4
Kitunda, Tanzania 115/F5
Kitwanga, Br. Col. 184/D2
Kitwe, Zambia 115/E6
Kitwe, Zambia 102/E6
Kitzbühel, Austria 41/B3
Kitzingen, W. Germany 22/C4
Kitzmiller, Md. (21538) 245/B3
Kiuchüan (Jiuquan), China 77/E4
Kiukiang (Jiujiang), China 77/J6
Kiunga, Papua N.G. 85/B7
Kivalina, Alaska (99750) 196/E1
Kivijärvi (lake), Finland 18/O5
Kiviõli, U.S.S.R. 53/D1
Kivu (lake) 102/E5
Kivu (lake), Rwanda 115/E4
Kivu (prov.), Zaire 115/E4
Kivu (lake), Zaire 115/E4
Kiyiu (lake), Sask. 181/C4
Kizel', U.S.S.R. 7/K3
Kizel, U.S.S.R. 48/F4
Kizel, U.S.S.R. 52/J3
Kızılcahamam, Turkey 63/E2
Kızılhisar, Turkey 63/C4
Kızılırmak (riv.), Turkey 63/F2
Kızılırmak (riv.), Turkey 59/B1
Kızıltepe, Turkey 63/J4
Kızıltoprak, Turkey 63/D6
Kızılviran, Turkey 63/E4
Kizimkazi, Tanzania 115/G5
Kizlyar, U.S.S.R. 52/G6
Kizu, Japan 81/J7
Kizyl-Arvat, U.S.S.R. 48/F6
Kjeller, Norway 18/E3
Kjellerup, Denmark 21/C5
Kjölen (mts.) 7/F2
Kjölen (mts.), Norway 18/K3
Kladanj, Yugoslavia 45/D3
Kladno, Czech. 41/B1
Kladovo, Yugoslavia 45/F3
Klagenfurt, Austria 41/C3
Klagetoh, Ariz. (†86505) 198/F3
Klaipeda, U.S.S.R. 7/F3
Klaipeda, U.S.S.R. 53/A3
Klaipeda, U.S.S.R. 52/B3
Klaipeda, U.S.S.R. 48/B4
Klaksvík, Denmark 21/B2
Klamath, Calif. (95548) 204/B2
Klamath (riv.), Calif. 188/B2
Klamath (riv.), Calif. 204/B2
Klamath (co.), Oreg. 291/F5
Klamath (mts.), Oreg. 291/C5
Klamath (riv.), Oreg. 291/E5
Klamath Agency, Oreg. (97624) 291/F5
Klamath Falls, Oreg. 146/F5
Klamath Falls, Oreg. 188/B2
Klamath Falls, Oreg. (97601) 291/F5
Klappuis, S. Africa 118/F6
Klar (riv.), Sweden 7/F2
Klarälv (riv.), Sweden 18/H6
Klaten, Indonesia 85/J2
Klatovy, Czech. 41/B2
Klausen (pass), Switzerland 39/G3
Klawock, Alaska (99925) 196/M2
Klazienaveen, Netherlands 27/L3
Kleberg (co.), Texas 303/G10
Kleefeld, Manitoba 179/F5
Kleena Kleene, Br. Col. 184/E4
Klein, Mont. (†59072) 262/H4
Klein Bonaire (isl.), Neth. Ant. 161/E8
Kleine Emme (riv.), Switzerland 39/F3
Klein Karas, Namibia 118/B5
Kleinlützel, Switzerland 39/D2
Kleinmachnow, E. Germany 22/E4
Kleinmond, S. Africa 118/F7
Klemme, Iowa (50449) 229/F3
Klemtu, Br. Col. 184/C4
Klerksdorp, S. Africa 118/D5
Kleve, W. Germany 22/B3
Klickitat (co.), Wash. 310/E5
Klickitat, Wash. (98628) 310/D5
Klickitat (riv.), Wash. 310/D4
Klides (isls.), Cyprus 63/F5
Kliniklini (riv.), Br. Col. 184/E4
Kline, S.C. (29814) 296/E5
Kline, W. Va. (†26866) 312/H5
Kling, Philippines 82/E8
Klingenthal, E. Germany 22/E3
Klingnau, Switzerland 39/F1
Klintehamn, Sweden 18/K8
Klintsy, U.S.S.R. 52/D4
Klip (riv.), S. Africa 118/H6
Kliprivier, S. Africa 118/H7
Klitmøller, Denmark 21/B3
Ključ, Yugoslavia 45/C3
Kłobuck, Poland 47/D3
Kloch (lake), Br. Col. 184/E2
Kłodawa, Poland 47/D2
Kłodnica (riv.), Poland 47/A4
Kłodzko, Poland 47/C3
Klondike (riv.), Yukon 187/E3
Klondike Gold Rush Nat'l Hist. Park, Alaska 196/N1
Klondyke, Ariz. (85643) 198/E6
Klossner, Minn. (56053) 255/D6
Klosterneuburg, Austria 41/D2
Klosters Dorf, Switzerland 39/J3
Kloten, N. Dak. (58248) 282/O4
Kloten, Switzerland 39/G2
Klotzville, La. (†70341) 238/K3
Kluane (lake), Yukon 162/C3
Kluane (lake), Yukon 187/E3
Kluane Nat'l Park, Yukon 187/E3
Kluane Nat'l Park, Yukon 162/C3
Kluczbork, Poland 47/D3
Klukwan, Alaska (†99827) 196/M1
Klutina (lake), Alaska 196/D1
Klyuchevskaya Sopka (vol.), U.S.S.R. 48/Q4
Knapdale (dist.), Scotland 15/C5
Knapp, Minn. (†55321) 255/D5
Knapp, Wis. (54749) 317/B6
Knappa, Oreg. (†97103) 291/D1
Knapp Creek, N.Y. (14749) 276/C6
Knaresborough, England 13/F4
Knaresborough, England 10/F3
Knee (lake), Manitoba 179/J3
Knierim, Iowa (50552) 229/D4
Knife (lake), Minn. 255/G2
Knife (riv.), N. Dak. 282/G5
Knife R. Indian Villages Nat'l Hist. Site, N. Dak. 282/H5
Knife River, Minn. (55609) 255/G4
Knifley, Ky. (42753) 237/L6
Knight (isl.), Alaska 196/D1
Knight (inlet), Br. Col. 184/E5
Knightdale, N.C. (27545) 281/N3
Knight Inlet, Br. Col. 184/E5
Knighton, Wales 10/E4
Knighton, Wales 13/D5
Knightsen, Calif. (94548) 204/L1
Knights Landing, Calif. (95645) 204/B8
Knightstown, Ind. (46148) 227/F5
Knightstown, Ireland 17/A8
Knightsville, Ind. (47857) 227/C5
Knightville (res.), Mass. 249/C3
Knik Arm (inlet), Alaska 196/B1
Kniman, Ind. (†46392) 227/C2
Knin, Yugoslavia 45/C3
Knippa, Texas (78870) 303/E8
Knittelfeld, Austria 41/C3
Knjaževac, Yugoslavia 45/F4
Knobel, Ark. (72435) 202/J1
Knob Fork, W. Va. (26579) 312/E3
Knob Lick, Ky. (42154) 237/K6
Knob Lick, Mo. (63651) 261/M7
Knob Noster, Mo. (65336) 261/E5
Knock, Ireland 17/D4
Knockadoon (head), Ireland 17/F8
Knockanefune (mt.), Ireland 17/C7
Knockboy (mt.), Ireland 17/B8
Knocklayd (mt.), N. Ireland 17/J1
Knocklong, Ireland 17/D7
Knockmealdown (mts.), Ireland 17/F7
Knocknagashel, Ireland 17/C7
Knoke, Iowa (50553) 229/D3
Knokke-Heist, Belgium 27/C6
Knøsen (mt.), Denmark 21/D3
Knott (co.), Ky. 237/R6
Knott, Texas (79748) 303/C5
Knotts Island, N.C. (27950) 281/T2
Knottsville, Ky. (†42366) 237/H5
Knowles, Okla. (73847) 288/F1
Knowlesville, New Bruns. 170/C2
Knowlesville, N.Y. (14479) 276/D4
Knowlton, Mont. (†59336) 262/L4
Knowlton, Québec 172/E4
Knowlton, Wis. (†54455) 317/G6
Knox (cape), Br. Col. 162/C5
Knox (cape), Br. Col. 184/A3
Knox (co.), Ill. 222/C3
Knox (co.), Ind. 227/C7
Knox, Ind. (46534) 227/D2
Knox (co.), Ky. 237/O7
Knox (co.), Maine 243/E7
Knox○, Maine (†04986) 243/E6
Knox (co.), Mo. 261/H2
Knox (co.), Nebr. 264/G2
Knox (lake), Newf. 166/A3
Knox, N. Dak. (58343) 282/L3
Knox (co.), Ohio 284/F5
Knox, Pa. (16232) 294/C3
Knox (co.), Tenn. 237/O9
Knox (co.), Texas 303/E4
Knox, Victoria 97/K5
Knox, Victoria 88/M7
Knoxboro, N.Y. (13362) 276/J5
Knox Center, Maine (†04986) 243/E6
Knox City, Mo. (63446) 261/H2
Knox City, Texas (79529) 303/E4
Knox Coast (reg.) 5/C6
Knoxville, Ala. (35469) 195/C4
Knoxville, Ark. (72845) 202/D3
Knoxville, Georgia (31050) 217/E5
Knoxville, Ill. (61448) 222/C3
Knoxville, Iowa (50138) 229/G6
Knoxville, Md. (21758) 245/H3
Knoxville, Miss. (†39661) 256/B8
Knoxville, Mo. (†64084) 261/E4
Knoxville, Pa. (16928) 294/H2
Knoxville, Tenn. 146/K6
Knoxville, Tenn. 188/K3
Knoxville, Tenn. (*37901) 237/O9
Knud Rasmussen Land (reg.), Greenl. 4/B12
Knudshoved (pt.), Denmark 21/D7
Knurów, Poland 47/A4
Knutsford, Br. Col. 184/G5
Knutsford, England 13/H2
Knutsford, England 10/G2
Knutsford, Pr. Edward I. 168/D2
Knysna, S. Africa 118/C6
Koah Kong (isl.), Cambodia 72/D5
Koah Nhek, Cambodia 72/E4
Koah Rung (isl.), Cambodia 72/D5
Koah Tang (isl.), Cambodia 72/D5
Koali, Hawaii (†96713) 218/K2
Koa Mill, Hawaii (†96704) 218/G6
Koani, Tanzania 115/G5
Koartac, Québec 174/F1
Kobayashi, Japan 81/E8
Kobbfjorden (fjord), Norway 18/O1
Kobdo (Hovd), Mongolia 77/D2
Kobe, Japan 81/H7
Kobe, Japan 54/P6
København (Copenhagen) (cap.), Denmark 21/F6
Koblenz, Switzerland 39/F1
Koblenz, W. Germany 22/B3
Kobrin, U.S.S.R. 52/B4
Kobroor (isl.), Indonesia 85/K7
Kobuk, Alaska (99751) 196/G1
Kobuk (riv.), Alaska 188/C5
Kobuk (riv.), Alaska 196/G1
Kobuk Valley Nat'l Park, Alaska 196/F1
Kobuleti, U.S.S.R. 52/F6
Koca (riv.), Turkey 63/C3
Koca (riv.), Turkey 63/C6
Koca (riv.), Turkey 63/E2
Kocaeli (prov.), Turkey 63/C2
Kocaeli (İzmit), Turkey 63/D2
Kočani, Yugoslavia 45/F5
Koçarlı, Turkey 63/B4
Kočevje, Yugoslavia 45/B3
Koch (isl.), N.W. Terrs. 187/L3
Koch'ang, S. Korea 81/C6
Kochevo, U.S.S.R. 52/J3
Kochi (pref.), Japan 81/F7
Kochi, Japan 81/F7
Kodaira, Japan 81/O2
Kodak, Tenn. (37764) 237/O9
Kodiak, Alaska 188/D6
Kodiak, Alaska (99615) 196/H3
Kodiak (isl.), Alaska 146/C4
Kodiak (isl.), Alaska 196/H3
Kodiak, U.S. 4/D17
Kodiak (isl.), U.S. 4/D17
Kodok, Sudan 111/F6
Kodok, Sudan 102/F4
Koekelare, Belgium 27/B6
Koekelberg, Belgium 27/B9
Koenig, Mo. (†65013) 261/J6
Koenton, Ala. (†36558) 195/B7
Koes, Namibia 118/B5
Kofa (mts.), Ariz. 198/B5
Kofçaz, Turkey 63/B2
Koffiefontein, S. Africa 118/D5
Köflach, Austria 41/C3
Koforidua, Ghana 106/D7
Kofu, Japan 81/J6
Koga, Japan 81/J5
Kogaluc (riv.), Québec 174/E1
Kogaluk (riv.), Newf. 166/B2
Koganei, Japan 81/O2
Kogarah, N. S. Wales 88/K4
Kogarah, N.S. Wales 97/J4
Køge, Denmark 21/F7
Køge (bay), Denmark 21/F7
Koggiung, Alaska (†99633) 196/G3
Kohala (Kapaau), Hawaii (†96755) 218/G3
Kohala (mts.), Hawaii 218/G4
Kohala (peak), Hawaii 218/G4
Kohat, Pakistan 59/K3
Kohat, Pakistan 68/C2
Kohila, U.S.S.R. 53/C1
Kohima, India 68/G3
Kohler, Wis. (53044) 317/L8
Kohls Ranch, Ariz. (85538) 198/D4
Kohtla-Järve, U.S.S.R. 52/C3
Kohtla-Järve, U.S.S.R. 53/D1
Kohŭng, S. Korea 81/C6
Koidern, Yukon 187/D3
Koitere (lake), Finland 18/R5
Kŏje (isl.), S. Korea 81/D6
Kojetín, Czech. 41/D2
Kojonup, W. Australia 88/B6
Kojonup, W. Australia 92/B6
Kokadjo, Maine (†04411) 243/E4
Kokand, U.S.S.R. 48/H5
Kokanee Glacier Prov. Park, Br. Col. 184/J5
Kokava nad Rimavicou, Czech. 41/E2
Kokchetav, U.S.S.R. 54/J4
Kokchetav, U.S.S.R. 48/H4
Kokemäki, Finland 18/N6
Kokish, Br. Col. 184/D5
Kokiu (Gejiu), China 77/F7
Kokkola, Finland 18/N5
Koko (head), Hawaii 218/F2
Koko, Nigeria 106/F7
Kokoda, Papua N.G. 85/C7
Kokole (pt.), Hawaii 218/B2
Kokolik (riv.), Alaska 196/F1
Kokomo, Hawaii (†96708) 218/K2
Kokomo, Ind. 188/J2
Kokomo, Ind. (46901) 227/E4
Kokomo, Miss. (39643) 256/E8
Kokonau, Indonesia 85/K6
Kokopo, Papua N.G. 86/B2
Kokosing (riv.), Ohio 284/E5
Kokrines, Alaska (†99768) 196/G1
Kokrines (hills), Alaska 196/H1
Koksan, N. Korea 81/C4
Koksijde, Belgium 27/B6
Koksilah, Br. Col. 184/J3
Koksoak (riv.), Que. 162/K4
Koksoak (riv.), Québec 174/D1
Kokstad, S. Africa 118/D6
Kokubu, Japan 81/E8
Kola, Manitoba 179/A5
Kola (pen.), U.S.S.R. 7/H2
Kola (pen.), U.S.S.R. 4/C8
Kola (pen.), U.S.S.R. 52/E1
Kola (pen.), U.S.S.R. 48/D3
Kolahun, Liberia 106/C7
Kolaka, Indonesia 85/G6
Kolar, India 68/D6
Kolar Gold Fields, India 68/D6
Kolari, Finland 18/O3
Kolárovo, Czech. 41/D3
Kolašin, Yugoslavia 45/D4
Kolberg (Kołobrzeg), Poland 47/B1
Kolbio, Kenya 115/H4
Kolbuszowa, Poland 47/E3
Kolda, Senegal 106/B6
Kolding, Denmark 18/F9
Kolding, Denmark 21/C7
Kole, Haut-Zaïre, Zaire 115/E3
Kole, Kasai-Oriental, Zaire 115/D4
Koleen, Ind. (47439) 227/D7
Kolekole (stream), Hawaii 218/J4
Kölen (mts.), Sweden 18/K3
Kolepom (isl.), Indonesia 85/K7
Kolguyev (isl.), U.S.S.R. 7/J2
Kolguyev (isl.), U.S.S.R. 4/B7
Kolguyev (isl.), U.S.S.R. 52/G1
Kolguyev (isl.), U.S.S.R. 48/E3
Kolhapur, India 68/C5
Kolhapur, India 54/J8
Koliganek, Alaska (99576) 196/G3
Kolín, Czech. 41/C1
Kolin, Mont. (†59462) 262/G3
Kolind, Denmark 21/D5
Kölliken, Switzerland 39/F2
Kollum, Netherlands 27/J2
Kolmanskop, Namibia 118/B5
Köln (Cologne), W. Germany 22/B3
Kolno, Poland 47/F2
Koło, Poland 47/D2
Koloa, Hawaii 188/E5
Koloa, Hawaii (96756) 218/C2
Koloa Landing, Hawaii (†96756) 218/C2
Kołobrzeg, Poland 47/B1
Kologriv, U.S.S.R. 52/F3
Kolokani, Mali 106/C6
Kolola Springs, Miss. (†39740) 256/H3
Kolombangara (isl.), Solomon Is. 86/D2
Kolomiya, U.S.S.R. 52/B5
Kolomna, U.S.S.R. 7/H3
Kolomna, U.S.S.R. 48/D4
Kolomna, U.S.S.R. 52/E4
Kolondiéba, Mali 106/C6
Kolonia (cap.), Micronesia 87/F5
Kolonodale, Indonesia 85/G6
Kolovrat (mt.), Solomon Is. 86/E3
Kolpashevo, U.S.S.R. 54/K4
Kolpashevo, U.S.S.R. 48/J4
Kolpino, U.S.S.R. 52/D3
Kolva (riv.), U.S.S.R. 52/J1
Kolwezi, Zaire 102/E6
Kolwezi, Zaire 115/E6
Kolyma (range), U.S.S.R. 54/S3
Kolyma (range), U.S.S.R. 48/Q3
Kolyma (range), U.S.S.R. 4/C1
Kolyma (riv.), U.S.S.R. 48/Q3
Kolyma (riv.), U.S.S.R. 54/S3
Kolyma (riv.), U.S.S.R. 4/C2
Koma, Burma 72/C4
Komádi, Hungary 41/F3
Komadugu Yobe (riv.), Niger 106/G6
Komadugu Yobe (riv.), Nigeria 106/G6
Komaga (mt.), Japan 81/K2
Komagane, Japan 81/H6
Komandorskiye (isls.), U.S.S.R. 54/T4
Komandorskiye (isls.), U.S.S.R. 2/T3
Komandorskiye (isls.), U.S.S.R. 48/R4
Komárno, Czech. 41/D3
Komarno, Manitoba 179/E4
Komárom (co.), Hungary 41/E3
Komárom, Hungary 41/E3
Komatke, Ariz. (†85339) 198/C5
Komatsu, Japan 81/H5
Komba, Zaire 115/D3
Kŏmdŏk (mt.), N. Korea 81/D3
Komi A.S.S.R., U.S.S.R. 48/F3
Komi A.S.S.R., U.S.S.R. 52/H2
Komi-Permyak Aut. Okr., U.S.S.R. 52/H3
Komi-Permyak Aut. Okr., U.S.S.R. 48/F4
Komló, Hungary 41/E3
Kommetjie, S. Africa 118/E7
Kommunarsk, U.S.S.R. 52/E5
Komodo (isl.), Indonesia 85/F7
Komoka, Ontario 177/C5
Kôm Ombo, Egypt 111/F3
Kôm Ombo, Egypt 59/B5
Komono, Congo 115/B4
Komoran (isl.), Indonesia 85/K7
Komotiní, Greece 45/G5
Komrat, U.S.S.R. 52/C5
Komsomolets (isl.), U.S.S.R. 4/A5
Komsomolets (isl.), U.S.S.R. 54/M1
Komsomolets (isl.), U.S.S.R. 48/L1
Komsomol'sk, U.S.S.R. 54/P4
Komsomol'sk, U.S.S.R. 48/G4
Komsomol'skiy, U.S.S.R. 52/K1
Komsomol'sk-na-Amure, U.S.S.R. 48/O4
Kona, Ky. (41829) 237/R6
Konahuanui (peaks), Hawaii 218/C3
Konar (riv.), Afghanistan 68/C1
Konar (riv.), Afghanistan 59/K2
Konar (riv.), Pakistan 68/C1
Konawa, Okla. (74849) 288/N5
Kondoa, Tanzania 115/G4
Kondopoga, U.S.S.R. 52/D2
Kondopoga, U.S.S.R. 48/D3
Kondoros, Hungary 41/F3
Koné, New Caled. 86/G4
Kong, Koh (isl.), Cambodia 72/D5
Kong, Ivory Coast 106/D7
Kongiganak, Alaska (99559) 196/F3
Kongju, S. Korea 81/C5
Kong Karls Land (isl.), Norway 18/E1
Kongmoon (Jiangmen), China 77/H7
Kongolo, Zaire 115/E5
Kongolo, Zaire 102/E5
Kongor, Sudan 111/F6
Kongsberg, N. Dak. (†58792) 282/J4

Kongsberg, Norway 18/F7
Kongsfjorden (fjord), Norway 18/B2
Kongsvinger, Norway 18/H6
Kongwa, Tanzania 115/G5
Koni (pen.), U.S.S.R. 48/Q4
Koniecpol, Poland 47/D3
Königsberg (Kaliningrad), U.S.S.R. 52/B4
Königssee (lake), W. Germany 22/E5
Königswiesen, Austria 41/C2
Königswinter, W. Germany 22/B3
Königs Wusterhausen, E. Germany 22/E2
Konin, Poland 47/D2
Kónitsa, Greece 45/E5
Koniuji (isls.), Alaska 196/G3
Köniz, Switzerland 39/E2
Konjic, Yugoslavia 45/D4
Konkiep, Namibia 118/B5
Konnagar, India 68/F1
Konolfingen, Switzerland 39/E3
Konomoc (lake), Conn. 210/G3
Konotop, U.S.S.R. 52/F2
Konqi He (riv.), China 77/C3
Konosha, U.S.S.R. 52/F2
Końskie, Poland 47/E3
Konstantinovka, U.S.S.R. 52/E5
Konstantynów Łódzki, Poland 47/D3
Konstanz, W. Germany 22/C5
Kontagora, Nigeria 106/F6
Kontcha, Cameroon 115/B2
Kontich, Belgium 27/E6
Kontiomäki, Finland 18/Q4
Kontum, Vietnam 72/E4
Kontum (plat.), Vietnam 72/E4
Konya (prov.), Turkey 63/E4
Konya, Turkey 63/E4
Konya, Turkey 59/B2
Konya, Turkey 54/E6
Konza, Kenya 115/G5
Koocanusa (lake), Br. Col. 184/K6
Koocanusa (lake), Mont. 262/A2
Koochiching (co.), Minn. 255/E2
Koog aan de Zaan, Netherlands 27/A4
Koolan (isl.), W. Australia 88/C3
Koolan (isl.), W. Australia 92/C1
Koolau (range), Hawaii 218/G2
Kooline Station, W. Australia 92/B3
Koolpinyah, North. Terr. 93/B2
Koolyanobbing, W. Australia 88/B6
Koolyanobbing, W. Australia 92/B5
Koondrook, Victoria 97/K4
Koonibba, S. Australia 88/E6
Koonibba, S. Australia 94/C4
Koontz Lake, Ind. (†46574) 227/D2
Koorawatha, N.S. Wales 97/F2
Koosharem, Utah (84744) 304/C5
Koosharem Ind. Res., Utah 304/C5
Kooskia, Idaho 220/B2
Koostatak, Manitoba 179/E3
Kootenai (co.), Idaho 220/B2
Kootenai, Idaho (83840) 220/B1
Kootenai (riv.), Idaho 220/C1
Kootenai (riv.), Mont. 262/A2
Kootenay (lake), Br. Col. 162/E5
Kootenay (lake), Br. Col. 184/J5
Kootenay (riv.), Br. Col. 184/K5
Kootenay Nat'l Park, Br. Col. 184/J4
Kootenay Nat'l Pk., Br. Col. 162/E5
Kootingal, N.S. Wales 97/F2
Kópavogur, Iceland 21/B1
Köpenick, E. Germany 22/F4
Koper, Yugoslavia 45/A3
Kopervik, Norway 18/D7
Kopeysk, U.S.S.R. 48/G4
Köping, Sweden 18/J7
Koppal, India 68/D5
Koppang, Norway 18/G6
Kopparberg (co.), Sweden 18/J6
Kopparberg, Sweden 18/J7
Koppel, Pa. (†6136) 294/B4
Kopperston, W. Va. (24854) 312/C7
Koprivnica, Yugoslavia 45/C2
Köprü (riv.), Turkey 63/D4
Kor (riv.), Iran 66/H6
Korab (mt.), Albania 45/E5
Korab (mt.), Yugoslavia 45/E5
Koraka (cape), Fiji 86/S3
Koran, La. (†71037) 238/D4
Koraput, India 68/E5
Korba, India 68/E4
Korbach, W. Germany 22/C3
Korbel, Calif. (95550) 204/B3
Korçë, Albania 45/E5
Korčula (isl.), Yugoslavia 45/C4
Kordestan (Kurdistan) (prov.), Iran 66/E3
Kord Kuy, Iran 66/J2
Kordofan, Southern (prov.), Sudan 111/E5
Kordofan, Northern (prov.), Sudan 111/E5.
Korea (North) 2/R4
KOREA (NORTH) 81
Korea (South) 2/R4
KOREA (SOUTH) 81
Korea (bay), N. Korea 81/B4
Korea (str.), S. Korea 81/D6
Korenovsk, U.S.S.R. 52/E5
Korf, U.S.S.R. 54/T3
Korf, U.S.S.R. 48/R3
Korhogo, Ivory Coast 106/C7
Korhogo, Ivory Coast 102/B4
Körishegy (mt.), Hungary 41/D3
Koriyama, Japan 81/K5
Korkuteli, Turkey 63/D4
Korla, China 77/C3
Kormakiti (cape), Cyprus 59/B2
Kormakiti (cape), Cyprus 63/E5
Körmend, Hungary 41/D2
Kornat (isl.), Yugoslavia 45/B4
Korneuburg, Austria 41/D2
Kornsjö, Norway 18/G7
Kornwestheim, W. Germany 22/C4
Koro (isl.), Fiji 86/Q10
Koro (sea), Fiji 86/Q10

Köroğlu (mts.), Turkey 63/E2
Köroğlu Daği, Turkey 63/E2
Korogwe, Tanzania 115/G5
Koroit, Victoria 97/B6
Korona, Fla. (†32010) 212/E2
Koronadal, Philippines 82/E7
Koronowo, Poland 47/C2
Koropí, Greece 45/G7
Koror (cap.), Belau 87/D5
Kororoit (creek), Victoria 97/H5
Kororoit (creek), Victoria 88/K7
Korosten', U.S.S.R. 52/C4
Korostyshev, U.S.S.R. 52/C4
Koro Toro, Chad 111/C4
Korpilombolo, Sweden 18/N3
Korsakov, U.S.S.R. 48/P5
Korsnäs, Finland 18/M5
Korsør, Denmark 21/E7
Korsør, Denmark 18/G9
Kortemark, Belgium 27/C6
Korti, Sudan 59/B7
Korti, Sudan 111/F4
Kortrijk, Belgium 27/C7
Korumburra, Victoria 97/D6
Koryak (range), U.S.S.R. 54/U3
Koryak (range), U.S.S.R. 48/S3
Koryak Aut. Okr., U.S.S.R. 48/R3
Koryazhma, U.S.S.R. 52/G2
Kos, Greece 45/H7
Kos (isl.), Greece 45/H7
Kościan, Poland 47/C2
Kościerzyna, Poland 47/C1
Kosciusko (mt.), Australia 87/F9
Kosciusko (co.), Ind. 227/F2
Kosciusko, Miss. (39090) 256/E4
Kosciusko (mt.), N.S. Wales 88/H7
Kosciusko (mt.), N.S. Wales 97/E5
Koshigaya, Japan 81/P2
Koshiki (isls.), Japan 81/D8
Koshke-e Kohneh, Afghanistan 68/A2
Koshkonong, Mo. (65692) 261/J9
Koshkonong (lake), Wis. 317/H10
Koshima, Japan 81/E8
Kosi (riv.), India 68/D3
Košice, Czech. 7/G4
Košice, Czech. 41/F2
Koslan, U.S.S.R. 48/E3
Koslan, U.S.S.R. 52/G2
Köslin (Koszalin), Poland 47/C1
Kosoma, Okla. (†74557) 288/P6
Košöng, N. Korea 81/D4
Kosovo (aut. reg.), Yugoslavia 45/E4
Kosovska Mitrovica, Yugoslavia 45/E4
Kosrae (isl.), Micronesia 87/G5
Kosse, Texas (76653) 303/H6
Kössen, Austria 41/B3
Kossou, Lac de (lake), Ivory Coast 106/C7
Kossuth, La. (†47167) 227/E7
Kossuth (co.), Iowa 229/E3
Kossuth, Miss. (38834) 256/G1
Kostajnica, Yugoslavia 45/C3
Kostelec nad Černými Lesy, Czech. 41/C2
Kostelec nad Orlicí, Czech. 41/D1
Kosti, Sudan 59/B7
Kosti, Sudan 111/F5
Kostopol', U.S.S.R. 52/C4
Kostroma, U.S.S.R. 7/J3
Kostroma, U.S.S.R. 52/F3
Kostroma, U.S.S.R. 48/E3
Kostrzyń, Poland 47/B2
Koszalin (prov.), Poland 47/C1
Koszalin, Poland 7/F3
Koszalin, Poland 47/C1
Kőszeg, Hungary 41/D2
Koszta, Iowa (†52208) 229/J5
Kota, India 68/D3
Kota, India 68/D3
Kotaagung, Indonesia 85/C7
Kotabaharu, Indonesia 85/E6
Kota Baharu, Malaysia 54/M9
Kota Baharu, Malaysia 72/D6
Kotabaru, Indonesia 85/F6
Kotabumi, Indonesia 85/C7
Kota Kinabalu, Malaysia 54/N9
Kota Kinabalu, Malaysia 85/F4
Kotamobagu, Indonesia 85/G6
Kota Tinggi, Malaysia 72/F5
Kotawaringin, Indonesia 85/E6
Kotcho (lake), Br. Col. 184/M2
Kotcho (riv.), Br. Col. 184/M2
Kotel, Bulgaria 45/H4
Kotel'nich, U.S.S.R. 52/G3
Kotel'nikovo, U.S.S.R. 52/F5
Kotel'nyy (isl.), U.S.S.R. 4/B2
Kotel'nyy (isl.), U.S.S.R. 48/O2
Köthen, E. Germany 22/D3
Kotido, Uganda 115/F3
Kotka, Finland 7/G2
Kotka, Finland 18/P6
Kotlas, U.S.S.R. 7/J2
Kotlas, U.S.S.R. 48/E3
Kotlas, U.S.S.R. 52/G2
Kotlik, Alaska (99620) 196/F2
Kotor, Yugoslavia 45/D4
Kotovo, U.S.S.R. 52/G4
Kotovsk, Odessa, U.S.S.R. 52/C5
Kotovsk, Tambov, U.S.S.R. 52/F4
Kotri, Pakistan 68/B3
Kötschach-Mauthen, Austria 41/B3
Kottagudem, India 68/E5
Kottayam, India 68/D7
Kotto (riv.), Cent. Afr. Rep. 115/D2
Kotturu, India 68/D6
Kotuy (riv.), U.S.S.R. 4/B4
Kotuy (riv.), U.S.S.R. 54/M2
Kotuy (riv.), U.S.S.R. 48/L3
Kotzebue, Alaska (99752) 196/F1
Kotzebue, Alaska 146/B3
Kotzebue, Alaska 188/A5
Kotzebue (sound), Alaska 196/F1
Kotzebue, U.S. 4/C18
Kouango, Cent. Afr. Rep. 115/D2
Kouchibouguac, New Bruns. 170/F2
Kouchibouguac (bay), New Bruns. 170/F2

Kouchibouguacis (riv.), New Bruns. 170/F2
Kouchobouguac Nat'l Park, New Bruns. 170/F2
Koudougou, Upper Volta 106/D6
Kouilou (riv.), Congo 115/B4
Koukdjuak (riv.), N.W. Terrs. 187/L3
Kouki, Cent. Afr. Rep. 115/C2
Koula-Moutou, Gabon 115/B4
Koula-Moutou, Gabon 102/D5
Koulikoro, Mali 106/C6
Koulikoro, Mali 102/B3
Koumala, Queensland 95/D4
Koumbi Saleh (ruins), Mauritania 106/C5
Koumra, Chad 102/D4
Koumra, Chad 111/C6
Koundara, Guinea 106/B6
Kounde, Cent. Afr. Rep. 115/B2
Kouno, Chad 111/C6
Kounradskiy, U.S.S.R. 48/H5
Kountze, Texas (77625) 303/K7
Koupela, Upper Volta 106/D6
Kourou, Fr. Guiana 131/E3
Kourouba, Mali 106/B6
Kouroussa, Guinea 106/C6
Kousseri, Cameroon 115/B1
Koutiala, Mali 106/D6
Koutiala, Mali 102/B3
Kouts, Ind. (46347) 227/C2
Kouvola, Finland 18/P6
Kovdor, U.S.S.R. 52/D1
Kovel', U.S.S.R. 52/C4
Kovel', U.S.S.R. 48/C4
Kovrov, U.S.S.R. 52/F3
Kovrov, U.S.S.R. 48/E4
Kovur, India 68/E6
Kovylkino, U.S.S.R. 52/F4
Kowary, Poland 47/C3
Kowst, Afghanistan 68/B2
Kowt-e `Ashrow, Afghanistan 68/B2
Koyama, Japan 81/E8
Kōyceğiz, Turkey 63/C4
Kōyceğiz, Turkey 63/C4
Kōyceğiz (lake), Turkey 63/C4
Koyuk, Alaska (99753) 196/F1
Koyukuk, Alaska (99754) 196/G1
Koyukuk (riv.), Alaska 188/C5
Koyukuk (riv.), Alaska 196/G1
Koyulhisar, Turkey 63/G2
Kozakli, Turkey 63/F3
Kozan, Turkey 63/F4
Kozáni, Greece 45/F5
Kozhevnikovo, U.S.S.R. 48/L2
Kozhikode, India 68/D6
Kozhikode, India 54/J8
Kozhva, U.S.S.R. 48/G3
Kozienice, Poland 47/E3
Kozlu, Turkey 63/D2
Kozluk, Turkey 63/J3
Kozmin, Poland 47/D3
Kozuchów, Poland 47/B3
Kpalimé, Togo 106/E7
Kpandu, Ghana 106/D7
Kpémé, Togo 106/E7
Kra (isth.), Thailand 72/C5
Kraaifontein, S. Africa 118/F6
Kraainem, Belgium 27/C9
Krabi, Thailand 72/C5
Kra Buri, Thailand 72/C4
Kracheh, Cambodia 72/E4
Kraemer, La. (70371) 238/M4
Kragan, Indonesia 85/K2
Kragerø, Norway 18/F7
Kragujevac, Yugoslavia 45/E3
Kragujevac, Yugoslavia 7/G4
Krakatau (Rakata) (isl.), Indonesia 85/C7
Kraków, Mo. (†63090) 261/K6
Kraków (Cracow), Poland 47/E4
Krakow, Wis. (53147) 317/K6
Kralendijk (cap.), Bonaire, Neth. Ant. 161/E8
Kralendijk, Neth. Ant. 156/E4
Králíky, Czech. 41/D1
Kraljevo, Yugoslavia 45/E4
Kralovice, Czech. 41/B2
Král'ovský Chlmec, Czech. 41/G2
Kralupy nad Vltavou, Czech. 41/C1
Kramatorsk, U.S.S.R. 52/E5
Kramer, Ind. (†47918) 227/C4
Kramer, Nebr. (†68333) 264/H4
Kramer, N. Dak. (58748) 282/J2
Kramfors, Sweden 18/L5
Kranídhion, Greece 45/F7
Kranj, Yugoslavia 45/B2
Kranzburg, S. Dak. (57245) 298/R4
Krapkowice, Poland 47/D3
Krasino, U.S.S.R. 52/H1
Krasino, U.S.S.R. 48/F2
Kráslava, U.S.S.R. 53/D3
Kraslice, Czech. 41/B1
Krásná Lípa, Czech. 41/C1
Kraśnik Fabryczny, Poland 47/F3
Krasnoarmeysk, U.S.S.R. 52/G4
Krasnoborsk, U.S.S.R. 52/G2
Krasnodar, U.S.S.R. 7/H4
Krasnodar, U.S.S.R. 52/E5
Krasnodar, U.S.S.R. 48/E5
Krasnograd, U.S.S.R. 52/E5
Krasnokamensk, U.S.S.R. 48/M4
Krasnokamsk, U.S.S.R. 52/H3
Krasnokamsk, U.S.S.R. 48/F4
Krasnoperekopsk, U.S.S.R. 52/D5
Krasnoslobodsk, U.S.S.R. 52/F4
Krasnotur'insk, U.S.S.R. 48/G3
Krasnoural'sk, U.S.S.R. 52/J3
Krasnovishersk, U.S.S.R. 52/H2
Krasnovodsk, U.S.S.R. 48/F5
Krasnoyarsk, U.S.S.R. 54/K4
Krasnoyarsk, U.S.S.R. 2/P3
Krasnoyarsk, U.S.S.R. 48/K4
Krasnystaw, Poland 47/F3
Krasnyy Kut, U.S.S.R. 52/G4
Krasnyy Luch, U.S.S.R. 52/E5
Krasnyy Sulin, U.S.S.R. 52/F5
Krasnyy Yar, U.S.S.R. 52/G5
Kraulshavn, Greenl. 4/B13
Krause Lagoon (chan.), Virgin Is. (U.S.) 161/F4

Krawang, Indonesia 85/H2
Krebs, Okla. (74554) 288/P5
Krefeld, W. Germany 22/B3
Kremenchug, U.S.S.R. 7/H4
Kremenchug, U.S.S.R. 48/D5
Kremenchug, U.S.S.R. 52/D5
Kremlin, Mont. (59532) 262/F2
Kremlin, Okla. (73753) 288/L1
Kremmling, Colo. (80459) 208/G2
Krems an der Donau, Austria 41/C2
Krenitzin (isls.), Alaska 196/E4
Kresgeville, Pa. (18333) 294/L4
Kress, Texas (79052) 303/C3
Kretinga, U.S.S.R. 53/A3
Kreutzal, W. Germany 22/C3
Kreuzlingen, Switzerland 39/H1
Kribi, Cameroon 115/B3
Krichev, U.S.S.R. 52/D4
Krimml, Austria 41/B3
Krimpen aan den IJssel, Netherlands 27/C5
Kríos (cape), Greece 45/F8
Krishna (Kistna) (riv.), India 68/D5
Krishnanagar, India 68/F4
Kristiansand, Norway 18/F8
Kristiansand, Norway 7/E4
Kristiansand (C.), Sweden 18/J8
Kristianstad, Sweden 18/J9
Kristiansund, Norway 7/E2
Kristiansund, Norway 18/E5
Kristiinankaupunki (Kristinestad), Finland 18/N5
Kristinehamn, Sweden 18/H7
Kristinestad, Finland 18/N5
Kríti (Crete) (isl.), Greece 45/G8
Krivoy Rog, U.S.S.R. 7/H4
Krivoy Rog, U.S.S.R. 48/D5
Krivoy Rog, U.S.S.R. 52/D5
Križevci, Yugoslavia 45/C2
Krk, Yugoslavia 45/B3
Krk (isl.), Yugoslavia 45/B3
Krnov, Czech. 41/D1
Krolevets, U.S.S.R. 52/D4
Kroměříž, Czech. 41/D2
Krompachy, Czech. 41/F2
Kronach, W. Germany 22/D3
Kronau, Sask. 181/G5
Kronoberg (co.), Sweden 18/J8
Kronshtadt, U.S.S.R. 52/C3
Kroonstad, S. Africa 118/D5
Kropotkin, U.S.S.R. 52/F5
Kroschel, Minn. (†55037) 255/E4
Krosno (riv.), Poland 47/F4
Krosno, Poland 47/F4
Krosno Odrzanskie, Poland 47/B2
Krotoszyn, Poland 47/D3
Krotz Springs, La. (70750) 238/G5
Krško, Yugoslavia 45/B3
Kru Coast (reg.), Liberia 106/C8
Kruger Nat'l Park, S. Africa 118/E4
Krugersdorp, S. Africa 118/H6
Kruglói (pt.), Alaska 196/J3
Kruis (riv.), S. Africa 118/D5
Krujë, Albania 45/D5
Krum, Texas (76249) 303/G4
Krumbach, W. Germany 22/D4
Krummenau, Switzerland 39/H2
Krumovgrad, Bulgaria 45/G5
Krung Thep (Bangkok) (cap.), Thailand 72/D4
Krupina, Czech. 41/E2
Krupka, Czech. 41/B1
Krupp (Marlin), Wash. (†98832) 310/F3
Krusenstern (cape), Alaska 196/F1
Krusenstern (cape), N.W. Terrs. 187/G3
Kruševac, Yugoslavia 45/E4
Krušné Hory (Erzgebirge) (mts.), Czech. 41/B1
Kruszwica, Poland 47/D2
Kruzof (isl.), Alaska 196/M1
Krydor, Sask. 181/D3
Krymsk, U.S.S.R. 52/E5
Krynica, Poland 47/E4
Krypton, Ky. (41754) 237/P6
Krzyz, Poland 47/C2
Ksar el Boukhari, Algeria 106/E1
Ksar el Kebir, Morocco 106/C2
Ksar es Souk, Morocco 106/D2
Ktima, Cyprus 63/E5
Kuala Dungun, Malaysia 72/D6
Kualakapuas, Indonesia 85/E6
Kuala Kerai, Malaysia 72/D6
Kualakurun, Indonesia 85/E6
Kuala Lipis, Malaysia 72/D6
Kuala Lumpur (cap.), Malaysia 72/D7
Kuala Lumpur (cap.), Malaysia 54/M9
Kuala Lumpur (cap.), Malaysia 2/P5
Kuala Pilah, Malaysia 72/D7
Kualapuu, Hawaii (96757) 218/G1
Kuala Rompin, Malaysia 72/D7
Kuala Selangor, Malaysia 72/D7
Kuala Terengganu, Malaysia 72/D6
Kuancheng, China 77/J3
Kuantan, Malaysia 72/D6
Kuba, U.S.S.R. 52/G6
Kubachi, U.S.S.R. 52/G6
Kubaisa, Iraq 66/C4
Kuban' (riv.), U.S.S.R. 7/J4
Kuban' (riv.), U.S.S.R. 52/E5
Kubbum, Sudan 111/D5
Kubeno (lake), U.S.S.R. 52/F2
Kublis, Switzerland 39/J3
Kubokawa, Japan 81/F7
Kubrat, Bulgaria 45/H4
Kuching, Malaysia 54/N9
Kuching, Malaysia 85/E5
Kuchino (isl.), Japan 81/O4
Kuçovë (Stalin), Albania 45/D5
Küçükköy, Turkey 63/C3
Kudarebe (pt.), Neth. Ant. 161/D9
Kudat, Malaysia 85/F4
Kudowa Zdrój, Poland 47/C3
Kudus, Indonesia 85/J2

Kudymkar, U.S.S.R. 48/F4
Kudymkar, U.S.S.R. 52/H3
Kufra (oasis), Libya 102/E2
Kufra (oasis), Libya 111/D3
Kufstein, Austria 41/A3
Kuh (cape), Iran 66/K8
Kuhak, Iran 66/N7
Kuh (cap.), Gabon (80459) 208/G2
Kuhestan, Afghanistan 59/H3
Kuhestan, Afghanistan 68/A2
Kühlungsborn, E. Germany 22/D1
Kuhmo, Finland 18/Q4
Kuhpayeh, Iran 66/H4
Kuilsrivier, S. Africa 118/F6
Kuiseb (riv.), Namibia 118/B4
Kuiu (isl.), Alaska 196/M2
Kuivaniemi, Finland 18/O4
Kuji, Japan 81/K3
Kuju (mt.), Japan 81/E7
Kuk (riv.), Alaska 196/F1
Kukaiau, Hawaii (†96775) 218/H4
Kukaklek (lake), Alaska 196/G3
Kukalar, Kuh-e (mt.), Iran 59/F3
Kukalar, Kuh-e (mt.), Iran 66/G5
Kukalaya (riv.), Nicaragua 154/F4
Kukawa, Nigeria 106/G6
Kukës, Albania 45/E4
Kuki, Japan 81/O2
Kukpowruk (riv.), Alaska 196/F1
Kukui (riv.), Guyana 131/A3
Kukuihaele, Hawaii (96727) 218/H3
Kula, Bulgaria 45/F4
Kula, Hawaii (96790) 218/J2
Kula, Turkey 63/C3
Kulai, Malaysia 72/F5
Kula Kangri (mt.), Bhutan 68/G3
Kuldīga, U.S.S.R. 53/A2
Kuldja (Yining), China 77/B3
Kulebaki, U.S.S.R. 52/F3
Kulen, Cambodia 72/E4
Kulen Vakuf, Yugoslavia 45/B3
Kulga, Bulgaria 45/F4
Kulgera, North. Terr. 88/E5
Kulgera, North. Terr. 93/B6
Kulkyne (creek), N.S. Wales 97/C1
Kulm, N. Dak. (58456) 282/N7
Kulmbach, W. Germany 22/D3
Kuloy, U.S.S.R. 52/F2
Kulp, Turkey 63/J3
Kulpmont, Pa. (17834) 294/J4
Kulpsville, Pa. (19443) 294/M5
Kul'sary, U.S.S.R. 48/F5
Kulu, India 68/D2
Kulu, Turkey 63/E3
Kulunda, U.S.S.R. 48/H4
Kulyab, U.S.S.R. 48/H6
Küm (riv.), S. Korea 81/C5
Kuma (riv.), U.S.S.R. 7/J4
Kuma (riv.), U.S.S.R. 48/E5
Kuma (riv.), U.S.S.R. 52/G5
Kumai, U.S.S.R. 48/R5
Kumaka, Guyana 131/B4
Kumamoto (pref.), Japan 81/E7
Kumamoto, Japan 54/P6
Kumamoto, Japan 81/E7
Kumano, Japan 81/G7
Kumanovo, Yugoslavia 45/E4
Kumara, N. Zealand 100/C5
Kumasi, Ghana 100/C7
Kumasi, Ghana 102/B4
Kumba, Cameroon 115/A3
Kumbakonam, India 68/D6
Kumbo, Cameroon 115/B2
Kum-Dag, U.S.S.R. 48/F6
Kume (isl.), Japan 81/M6
Kumertau, U.S.S.R. 52/J4
Kumeu, N. Zealand 100/B1
Kümgang (mt.), N. Korea 81/D4
Kumiyama, Japan 81/J6
Kumkale, Turkey 63/B6
Kumköy, Turkey 63/B6
Kumla, Sweden 18/J7
Kumluca, Turkey 63/D4
Kummerowersee (lake), E. Germany 22/E2
Kumo, Finland 7/G2
Kumo, Nigeria 106/G7
Kumphawapi, Thailand 72/D3
Kumta, India 68/C6
Kumukahi (cape), Hawaii 218/K5
Kumul (Hami), China 77/D3
Kuna, Idaho (83634) 220/B6
Kunágota, Hungary 41/F3
Kunashir (isl.), U.S.S.R. 54/R5
Kunda, U.S.S.R. 52/C3
Kunda, U.S.S.R. 53/D1
Kundiawa, Papua N.G. 85/B7
Kundl, Austria 41/A3
Kunduz, Afghanistan 68/B1
Kundunga, Indonesia 85/H2
Künes (Xinyuan), China 77/B3
Künes He (riv.), China 77/B3
Kungälv, Sweden 18/G8
Kunghit (isl.), Br. Col. 184/B4
Kungsbacka, Sweden 18/G8
Kungu, Zaire 102/D4
Kungu, Zaire 115/C3
Kungur, U.S.S.R. 7/K3
Kungur, U.S.S.R. 48/G4
Kungur, U.S.S.R. 52/J3
Kunhegyes, Hungary 41/F3
Kunia, Hawaii (96759) 218/E2
Kuningan, Indonesia 85/H2
Kunkle, Ohio (43531) 284/C2
Kunkletown, Pa. (18058) 294/M4
Kunlong, Burma 72/C2
Kunlun (range), China 54/K6
Kunlun (range), China 77/B4
Kunlun Shan (range), China 77/B4
Kunmadaras, Hungary 41/F3
Kunming, China 77/F6
Kunming, China 2/Q4
Kunming, China 54/M7
Kunsan, S. Korea 81/C6
Kunszentmárton, Hungary 41/F3
Kunszentmiklós, Hungary 41/E3
Kununurra, W. Australia 92/E2
Kuolayarvi, U.S.S.R. 52/D1
Kuopio (prov.), Finland 18/P5

Kuopio, Finland 7/G2
Kuopio, Finland 18/Q5
Kupa (riv.), Yugoslavia 45/B3
Kupang, Indonesia 54/O11
Kupang, Indonesia 85/G8
Kuparuk (riv.), Alaska 196/H1
Kupino, U.S.S.R. 48/H4
Kupiškis, U.S.S.R. 53/C3
Kupreanof (isl.), Alaska 196/N1
Kupyansk, U.S.S.R. 52/E5
Kuqa, China 77/B3
Kur (isl.), Indonesia 85/J7
Kura (riv.), U.S.S.R. 48/E6
Kura (riv.), U.S.S.R. 52/G6
Kuraiyima, Jordan 65/D3
Kurang (riv.), Iran 66/G4
Kurashiki, Japan 81/F6
Kurayoshi, Japan 81/F6
Kurdistan (Kordestan) (prov.), Iran 66/E3
Kurdistan (reg.), Iran 59/D2
Kurdistan (reg.), Iran 66/D2
Kurdistan (reg.), Iran 66/E2
Kurdistan (reg.), Iraq 66/D2
Kurdistan (reg.), Turkey 59/D2
Kürdzhali, Bulgaria 45/G5
Kure (atoll), Hawaii 87/J3
Kure (atoll), Hawaii 218/A5
Kure (isl.), Hawaii 218/A5
Kure, Japan 81/F6
Küre, Turkey 63/E2
Küre (mts.), Turkey 63/E2
Kure Beach, N.C. (28449) 281/O7
Kuressaare, U.S.S.R. 53/B1
Kuressaare, U.S.S.R. 52/B3
Kurgan, U.S.S.R. 54/H4
Kurgan, U.S.S.R. 48/G4
Kurgan-Tyube, U.S.S.R. 48/G6
Kuria Muria (isls.), Oman 54/G8
Kuria Muria (isls.), Oman 59/G6
Kurikka, Finland 18/M5
Kuril (isls.), U.S.S.R. 2/S3
Kuril (isls.), U.S.S.R. 54/R5
Kuril (isls.), U.S.S.R. 48/P5
Kuril'sk, U.S.S.R. 54/R5
Ku-ring-gai, N.S. Wales 88/K4
Kuring Kuru, Namibia 118/B3
Kurla, India 68/B7
Kurmuk, Sudan 111/F5
Kurnell (pen.), N.S. Wales 97/J4
Kurnool, India 54/J8
Kurnool, India 68/D5
Kuroiso, Japan 81/K5
Kuroki, Sask. 181/H4
Kurow, N. Zealand 100/C6
Kurri Kurri-Weston, N.S. Wales 97/F3
Kuršėnai, U.S.S.R. 53/B2
Kursk, U.S.S.R. 7/H3
Kursk, U.S.S.R. 48/D4
Kursk, U.S.S.R. 52/E4
Kurşunlu, Turkey 63/E2
Kurtalan, Turkey 63/J3
Kurthwood, La. (71443) 238/D4
Kurtistown, Hawaii (96760) 218/J5
Kurtz, Ind. (†47249) 227/E7
Kurucaşile, Turkey 63/E2
Kuruçay (riv.), Turkey 63/K2
Kuruktag Shan (range), China 77/C3
Kuruman, S. Africa 118/C5
Kurume, Japan 81/E7
Kurunegala, Sri Lanka 68/E7
Kurungiku (mts.), Guyana 131/B3
Kurupukari, Guyana 131/B3
Kuş (lake), Turkey 63/B2
Kuşada (gulf), Turkey 63/B4
Kuşadasi, Turkey 63/B4
Kushchevskaya, U.S.S.R. 52/E5
Kushequa, Pa. (†16735) 294/E2
Kushikino, Japan 81/E8
Kushima, Japan 81/E8
Kushimoto, Japan 81/G7
Kushiro, Japan 54/R5
Kushiro, Japan 81/L3
Kushka, U.S.S.R. 48/G6
Kushog (lake), Ontario 177/F2
Kuskokwim (bay), Alaska 196/F3
Kuskokwim (mts.), Alaska 196/G2
Kuskokwim (riv.), Alaska 188/C6
Kuskokwim (riv.), Alaska 196/G2
Kuskokwim (riv.), Alaska 146/C3
Kuskokwim, North Fork (riv.), Alaska 196/H2
Kuskokwim, South Fork (riv.), Alaska 196/H2
Kuskokwim (riv.), U.S. 4/C17
Küsnacht, Switzerland 39/G2
Kusong, N. Korea 81/B4
Küssnacht am Rigi, Switzerland 39/F2
Kustanay, U.S.S.R. 54/H4
Kustanay, U.S.S.R. 48/G4
Kustatan, Alaska (†99682) 196/B1
Küstrin, Poland 47/B2
Kut, Iraq 66/D4
Kut, Ko (isl.), Thailand 72/D5
Kuta, Nigeria 106/F7
Kütahya (prov.), Turkey 63/C3
Kütahya, Turkey 59/B2
Kütahya, Turkey 63/C3
Kutaisi, U.S.S.R. 7/J4
Kutaisi, U.S.S.R. 48/E5
Kutaraja (Banda Aceh), Indonesia 85/A4
Kutari (riv.), Guyana 131/C4
Kutari (riv.), Suriname 131/C4
Kutch, Colo. (†80832) 208/M5
Kutch (gulf), India 68/B4
Kutch, Rann of (salt marsh), 54/H7
Kutch, Rann of (salt marsh), India 68/B4
Kutch, Rann of (salt marsh), Pakistan 68/B4
Kutch, Rann of (salt lake), Pakistan 59/K5
Kutcharo (lake), Japan 81/M2

Kutina, Yugoslavia 45/C3
Kutná Hora, Czech. 41/C2
Kutno, Poland 47/D2
Kutoarjo, Indonesia 85/J2
Kuttawa, Ky. (42055) 237/E6
Küttigen, Switzerland 39/F2
Kutu, Zaire 115/C4
Kutum, Sudan 111/D5
Kúty, Czech. 41/D2
Kutztown, Pa. (19530) 294/L4
Kuusamo, Finland 18/Q4
Kuusamojärvi (lake), Finland 18/Q4
Kuusankoski, Finland 18/P6
Kuvandyk, U.S.S.R. 52/J4
Kuwait 2/M4
Kuwait 54/F7
KUWAIT 59/E4
Kuybyshev, U.S.S.R. 2/M3
Kuybyshev, U.S.S.R. 7/K3
Kuybyshev, U.S.S.R. 48/F4
Kuybyshev, U.S.S.R. 52/H4
Kuybyshev, U.S.S.R. 48/H4
Kuybyshev (res.), U.S.S.R. 7/K3
Kuybyshev (res.), U.S.S.R. 48/F4
Kuybyshev (res.), U.S.S.R. 52/G4
Kuyto (lake), U.S.S.R. 52/D2
Kuytun, China 77/C3
Kuyucak, Turkey 63/C4
Kuyuwini (riv.), Guyana 131/B4
Kuznetsk, U.S.S.R. 52/G4
Kuzomen', U.S.S.R. 52/E1
Kvaenangen (fjord), Norway 18/N2
Kvaerndrup, Denmark 21/D7
Kvaløy (isl.), Norway 18/K2
Kvaløya (isl.), Norway 18/O1
Kvarner (gulf), Yugoslavia 45/B3
Kvichak, Alaska (†99625) 196/G3
Kvichak (bay), Alaska 196/G3
Kvikkjokk, Sweden 18/K3
Kvinnherad, Norway 18/E6
Kvissleby, Sweden 18/K5
Kviteseid, Norway 18/F7
Kwa (riv.), Zaire 115/C4
Kwai (Mae Nam Khwae Noi) (riv.), Thailand 72/C4
Kwajalein (atoll), Marshall Is. 87/G5
Kwakoegron, Suriname 131/D3
Kwakwani, Guyana 131/C3
Kwale, Kenya 102/F5
Kwale, Kenya 115/G4
Kwamouth, Zaire 115/C4
Kwangchow (Canton), China 77/H7
Kwangju, S. Korea 54/O6
Kwangju, S. Korea 81/C6
Kwango (riv.), Zaire 115/C5
Kwangsi Chuang Aut. Reg. (Guangxi Zhuangzu), China 77/G7
Kwangtung (Guangdong) (prov.), China 77/H7
Kwanmo (mt.), N. Korea 81/D3
Kwara (state), Nigeria 106/E7
Kweichow (Guizhou) (prov.), China 77/G6
Kweilin (Guilin), China 77/G6
Kweisui (Hohhot), China 77/H3
Kweiyang (Guiyang), China 77/G6
Kwekwe (Que Que), Zimbabwe 118/D3
Kwethluk, Alaska (99621) 196/F2
Kwidzin, Poland 47/D2
Kwigillingok, Alaska (99622) 196/F3
Kwilu (riv.), Angola 115/C5
Kwilu (riv.), Zaire 115/C5
Kwinana New Town, W. Australia 88/B2
Kwinana New Town, W. Australia 92/A1
Kwinitsa, Br. Col. 184/C3
Kwitaro (riv.), Guyana 131/B4
Kyabé, Chad 111/C6
Kyabram, Victoria 97/C5
Kyaikto, Burma 72/C3
Kya-in Seikkyi, Burma 72/C3
Kyakhta, U.S.S.R. 54/M4
Kyakhta, U.S.S.R. 48/L4
Kyalite, N.S. Wales 97/B4
Kyana, Ind. (47549) 227/D8
Kyancutta, S. Australia 94/D5
Kyangin, Burma 72/B3
Kyaukme, Burma 72/C2
Kyaukpadaung, Burma 72/B2
Kyaukpyu, Burma 72/B3
Kyaukse, Burma 72/C2
Kybartai, U.S.S.R. 53/B3
Kyeburn, N. Zealand 100/C6
Kyger, Ohio (†45620) 284/F8
Kyger, W. Va. (†25270) 312/D5
Kyjov, Czech. 41/D2
Kyle, Sask. 181/C5
Kyle, S. Dak. (57752) 298/E7
Kyle, Texas (78640) 303/G8
Kyleakin, Scotland 15/C3
Kylemore, Sask. 181/H4
Kyle of Lochalsh, Scotland 15/C3
Kyle of Tongue (inlet), Scotland 15/D2
Kyles Ford, Tenn. (37765) 237/R7
Kylestrome, Scotland 15/D2
Kymi (prov.), Finland 18/Q6
Kyneton, Victoria 97/C5
Kynšperk, Czech. 41/B1
Kynuna, Queensland 95/B4
Kyogle, N.S. Wales 97/G1
Kyonan, Japan 81/O3
Kyŏnghŭng, N. Korea 81/E2
Kyŏngju, S. Korea 81/D6
Kyoto (pref.), Japan 81/J7
Kyoto, Japan 54/P6
Kyoto, Japan 81/J7
Kyrenia, Cyprus 63/E5
Kyritz, E. Germany 22/E2
Kysucké Nové Mesto, Czech. 41/E2
Kythrea, Cyprus 63/E5
Kyuquot, Br. Col. 184/D5
Kyuquot (sound), Br. Col. 184/D5
Kyushu (isl.), Japan 2/R4
Kyushu (isl.), Japan 54/P6
Kyushu (isl.), Japan 81/E7
Kyustendil, Bulgaria 45/F4
Kyusyur, U.S.S.R. 48/N2
Kywebwe, Burma 72/C3
Kyzyl, U.S.S.R. 48/K4
Kyzyl (riv.), U.S.S.R. 54/L4
Kyzyl-Kum (des.), U.S.S.R. 48/G5
Kzyl-Orda, U.S.S.R. 54/H5
Kzyl-Orda, U.S.S.R. 48/G5

L

Laa an der Thaya, Austria 41/D2
La Aduana, Venezuela 124/D3
Laager, Tenn. (37349) 237/K10
La Aguja (cape), Colombia 126/C2
Laakirchen, Austria 41/B3
La Almunia de Doña Godina, Spain 33/F2
La Altagracia (prov.), Dom. Rep. 158/F6
La Anna, Pa. (†18326) 294/M3
La Antigua Veracruz, Mexico 150/Q1
La Araucanía (reg.), Chile 138/E2
La Asunción, Venezuela 124/G2
Laau (pt.), Hawaii 218/G1
Laax, Switzerland 39/H3
Laayoune, W. Sahara 102/A2
Laayoune, Western Sahara 106/B3
Labadie, Mo. (63055) 261/L5
Labadieville, La. (70372) 238/K4
La Baie, Québec 172/G1
La Baie-de-Shawinigan, Québec 172/E3
La Banda, Argentina 143/D2
La Bandera (pt.), P. Rico 161/F1
La Bañeza, Spain 33/C1
La Barca, Mexico 150/H6
La Barge, Wyo. (83123) 319/B3
La Barge (creek), Wyo. 319/B3
La Barra de Navidad, Mexico 150/G7
Labasheeda, Ireland 17/C6
La Baule-Escoublac, France 28/B4
L'Abbaye, Switzerland 39/B3
Labe (riv.), Czech. 41/C1
Labé, Guinea 106/B6
La Bella (lag.), Paraguay 144/B4
La Belle, Fla. (33935) 212/E5
La Belle, Mo. (63447) 261/J2
Labelle (co.), Québec 172/B3
Labelle, Québec 172/C3
Labelle (lake), Québec 172/C3
La Belle (lake), Wis. 317/H1
Laberge (lake), Yukon 162/C3
Laberinto de las Doce Leguas (cays), Cuba 158/F3
La Berra (mt.), Switzerland 39/D3
Labette (co.), Kansas 232/G4
Labette, Kansas (67350) 232/G4
Labinsk, U.S.S.R. 52/F6
La Bisbal, Spain 33/H1
La Blanquilla (isl.), Venezuela 124/F2
Labo, Philippines 82/D3
Labo (mt.), Philippines 82/D3
La Bolsa, Uruguay 145/C1
La Bolt, S. Dak. (57246) 298/R3
La Bonita, Ecuador 128/D2
La Boquilla (res.), Mexico 150/G3
Laborec (riv.), Czech. 41/F2
Laborie, St. Lucia 161/G7
La Bostonnais, Québec 172/E2
Labougle, Argentina 143/G5
Laboulaye, Argentina 143/D3
Labrador (sea) 146/N4
Labrador (sea) 162/L4
Labrador (reg.), Canada 2/G3
Labrador (sea), Newf. 166/C2
Labrador (reg.), Newf. 166/B2
Labrador (reg.), Newf. 146/M4
Labrador (reg.), Newf. 162/K4
Labrador City, Newf. 166/A3
La Branche, Mich. (†49873) 250/B3
Lábrea, Brazil 132/G10
La Brea, Trin. & Tob. 161/A11
Labrieville, Québec 174/C3
La Broquerie, Manitoba 179/F5
Labuan, Malaysia 85/E4
Labuan (isl.), Malaysia 85/E4
Labuha, Indonesia 85/H6
Labuhan, Indonesia 85/G2
Labuk (bay), Malaysia 85/F4
Labutta, Burma 72/B3
Labyrinth (canyon), Utah 304/D5
Labytnangi, U.S.S.R. 48/G3
Lac (bay), Neth. Ant. 161/D9
Lac-à-Beauce, Québec 172/E2
Lacadena, Sask. 181/C5
Lacadie, Québec 172/J4
La Cahouane, Haiti 158/A6
Lac-à-la-Croix, Québec 172/F1
La Calera, Chile 138/F2
Lac-Alouette, Québec 172/D4
Lacamp, La. (71444) 238/E4
La Canada, Calif. (91011) 204/C10
Lacanau (lake), France 28/C5
La Canoa, Venezuela 124/G3
Lacantum (riv.), Mexico 150/O8
La Capilla, 136/C8
La Carlota, Argentina 143/D3
La Carlota, Philippines 82/D5
La Carlota, Spain 33/D4
La Carolina, Spain 33/E3
Lacassine, La. (70650) 238/E6
Lac-au-Saumon, Québec 172/B2
Lac-aux-Sables, Québec 172/E3
Lac Baker, New Bruns. 170/B1
Lac-Beauport, Québec 172/F3
Lac-Bouchette, Québec 172/E1
Laccadive (isls.), India 54/H8
Laccadive (isls.), India 68/C6
Lac-Carré, Québec 172/C3
Lac-Cayamant, Québec 172/A3
Lac-Chat, Québec 172/E2
Lac Court Oreilles Ind. Res., Wis. 317/D4
Lac de Gras (lake), N.W. Terrs. 187/G3
Lac-Delage, Québec 172/H3
Lac-des-Aigles, Québec 172/J2
Lac des Arcs, Alberta 182/C4
Lac-des-Écorces, Québec 172/B3
Lac-des-Îles, Québec 172/B3
Lac-Drolet, Québec 172/G4
Lac du Bonnet, Manitoba 179/G4
Lac-du-Cerf, Québec 172/B3
Lac du Flambeau, Wis. (54538) 317/G4
Lac du Flambeau Ind. Res., Wis. 317/G3
Lac-Édouard, Québec 172/E2
La Ceiba, Hond. 146/K8
La Ceiba, Honduras 154/D3
La Ceiba, Apure, Venezuela 124/C4
La Ceiba, Trujillo, Venezuela 124/C3
La Center, Ky. (42056) 237/C6
La Center, Wash. (98629) 310/C5
Lacepede (bay), S. Australia 88/F7
Lacepede (bay), S. Australia 94/F7
Lacepede (isls.), W. Australia 92/C2
Lacepede (range), W. Australia 88/D3
La Cerbatana, Serranía de (mts.), Venezuela 124/E4
Lac-Etchemin, Québec 172/G3
Lacey, Ark. (†71655) 202/G7
Lacey, Wash. (98503) 310/C3
Lacey Spring, Va. (22833) 307/L3
Laceys Spring, Ala. (35754) 195/E1
Laceyville, Pa. (18623) 294/K2
Lac-Frontière, Québec 172/H3
Lac Giao (Ban Me Thuot), Vietnam 72/E4
Lacha (lake), U.S.S.R. 52/E2
La Chapelle, Haiti 158/C5
La Charité-sur-Loire, France 28/E4
La Châtre, France 28/D4
La Chaux-de-Fonds, Switzerland 39/C2
Lachay (pt.), Peru 128/D8
Lachen, Switzerland 39/G2
Lachenaie, Québec 172/D4
Lachine, Mich. (49753) 250/F3
Lachine, Québec 172/H4
Lachlan (range), N.S. Wales 97/C3
Lachlan (riv.), N. W. Wales 88/G6
Lachlan (riv.), N.S. Wales 97/C3
La Chorrera, Colombia 126/D8
La Chorrera, Panama 154/H6
Lac-Humqui, Québec 172/B2
Lachute, Québec 172/C4
La Ciénaga, Dom. Rep. 158/D6
La Ciotat, France 28/F6
Lackawanna, N.Y. (14218) 276/B5
Lackawanna (co.), Pa. 294/L3
Lackawaxen, Pa. (18435) 294/N3
Lackey, Ky. (41643) 237/R6
Lackland A.F.B., Texas 303/J11
Lac La Belle, Wis. (†53066) 317/H1
Lac La Biche, Alberta 182/E2
Lac La Biche, Alta. 162/E5
Lac La Hache, Br. Col. 184/G4
Lac la Martre, N.W. Terrs. 187/G3
La Clarita, Argentina 143/G5
Lac La Ronge Prov. Park, Sask. 181/M3
Laclede, Idaho (83841) 220/B1
La Clede, Ill. (62437) 222/E5
Laclede (co.), Mo. 261/G7
Laclede, Mo. (64651) 261/F3
Lac-Mégantic, Québec 172/G4
Lacolle, Québec 172/D4
La Colmena, Paraguay 144/B5
La Coloma, Cuba 158/B2
La Colonia, Chile 138/D7
Lacomb, Oreg. (†97355) 291/E3
Lacombe, Alberta 182/D3
Lacombe, Alta. 162/E5
Lacombe, La. (70445) 238/L6
Lacon, Ill. (61540) 222/D2
Lacona, Iowa (50139) 229/G6
Lacona, N.Y. (13083) 276/J3
La Concepción, Honduras 154/E3
La Concepción, Panama 154/F6
La Concepción, Venezuela 124/C2
La Concepción, Venezuela 124/B2
La Conception, Québec 172/C3
La Concordia, Mexico 150/N9
Laconia, Ind. (47135) 227/E8
Laconia, N.H. (03246) 268/E4
Laconia, Tenn. (38045) 237/C10
La Conner, Wash. (98257) 310/C2
La Conquista, Nicaragua 154/D5
Lacoochee, Fla. (33537) 212/D3
La Corey, Alberta 182/E2
La Coronilla, Uruguay 145/F4
La Coruña (prov.), Spain 33/B1
La Coruña, Spain 33/B1
La Coruña, Spain 7/D4
La Coste, Texas (78039) 303/J11
La Courneuve, France 28/B1
Lacovia, Jamaica 158/H6
Lac Pelletier, Sask. 181/C6
Lac-Poulin, Québec 172/G3
Lac qui Parle (co.), Minn. 255/B6
Lac qui Parle, Minn. (†56265) 255/B5
Lac qui Parle (lake), Minn. 255/C5
Lac qui Parle (riv.), Minn. 255/B6
Lacre (pt.), Neth. Ant. 161/E9
La Crescent, Minn. (55947) 255/G7
La Crescenta-Montrose, Calif. (91214) 204/C10
La Crete, Alberta 182/B5
La Croche, Québec 172/E2
La Croix (lake), Minn. 255/F2
La Crosse, Fla. (32658) 212/D2
La Crosse, Georgia (†31806) 217/D6
La Crosse, Ind. (46348) 227/D2
La Crosse, Kansas (67548) 232/C3
La Crosse, Va. (23950) 307/M7
Lacrosse, Wash. (99143) 310/H4
La Crosse (co.), Wis. 317/D8
La Crosse, Wis. 146/J5
La Crosse, Wis. 188/H2
La Crosse, Wis. (54601) 317/D8
La Cruz, Argentina 143/E2
La Cruz, Chile 138/F2
La Cruz, Colombia 126/B7
La Cruz, Chihuahua, Mexico 150/G3
La Cruz, Sinaloa, Mexico 150/F5
La Cruz, Nicaragua 154/E4
La Cruz, Uruguay 145/C4
Lac-Saguay, Québec 172/B3
Lac-Saint-Charles, Québec 172/H3
Lac-Sainte-Marie, Québec 172/A4
Lac-Saint-Jean-Est (co.), Québec 172/F1
Lac-Saint-Jean-Est (county), Québec 174/C3
Lac-Saint-Jean-Ouest (co.), Québec 172/E1
Lac-Saint-Jean-Ouest (county), Québec 174/C2
Lac-Saint-Joseph, Québec 172/F3
Lac-Saint-Paul, Québec 172/B3
Lac-Sergent, Québec 172/F3
Lac Seul, Ontario 175/B2
La Cuchilla, Uruguay 145/F3
La Cueva, N. Mex. (†87712) 274/D3
La Cumbre, Argentina 143/C3
La Cure, Switzerland 39/B4
Lacuy (pen.), Chile 138/D4
Lac Vert, Sask. 181/G3
La Cygne, Kansas (66040) 232/H3
Ladakh (reg.), India 68/D2
Ladd, Ill. (61329) 222/D2
Ladder (creek), Kansas 232/A3
Ladder (hills), Scotland 15/E3
Laddonia, Mo. (63352) 261/J4
La Decharge, Québec 172/F1
Ladelle, Ark. (†71655) 202/G7
Ladgasht (Qila Ladgasht), Pakistan 68/A3
Ladhar Bheinn (mt.), Scotland 15/C3
Ladiesburg, Md. (21759) 245/J2
La Digue (isl.), Seychelles 118/J5
Lâdik, Turkey 63/G2
Ladispoli, Italy 34/E6
Ladiz, Iran 66/M6
Ladner, S. Dak. (†57720) 298/B2
Lado, Sudan 111/F6
Ladoga, Ind. (47954) 227/D5
Ladoga (lake), U.S.S.R. 7/H2
Ladoga (lake), U.S.S.R. 48/D3
Ladoga (lake), U.S.S.R. 52/D2
La Dôle (mt.), Switzerland 39/B4
Ladonia, Texas (75449) 303/J4
Ladora, Iowa (52251) 229/J5
La Dorada, Colombia 126/C5
Ladrillero (gulf), Chile 138/C8
Ladrillero (mt.), Chile 138/E10
Ladrillo (pt.), Cuba 158/E3
Ladron (mts.), N. Mex. 274/B4
Ladrones (isls.), Panama 154/F7
Ladson, S.C. (29456) 296/G6
La Due, Mo. (†64735) 261/E6
Ladue, Mo. (†63124) 261/P3
La Durantaye, Québec 172/G3
Lady (pond), Newf. 166/D2
Lady Ann (str.), N.W. Terrs. 187/K2
Ladybank, Scotland 15/E4
Lady Barron, Tasmania 99/E2
Ladybrand, S. Africa 118/D5
Lady Franklin (bay), N.W. Terrs. 187/M1
Lady Franklin (isl.), N.W. Terrs. 187/M3
Lady Lake, Fla. (32659) 212/E3
Lady Lake, Sask. 181/J3
Lady's Island Lake (inlet), Ireland 17/J7
Ladysmith, Br. Col. 184/J3
Ladysmith, S. Africa 102/F7
Ladysmith, S. Africa 118/D5
Ladysmith, Va. (22501) 307/N4
Ladysmith, Wis. (54848) 317/D5
Ladywood, Manitoba 179/F4
Lae, Papua N.G. 85/B7
Lae, Papua N.G. 86/A2
Lae, Papua N.G. 87/E6
Lae, Thailand 72/D3
Laem Chong Phra (cape), Thailand 72/C5
Laem Pho (cape), Thailand 72/D6
Laem Talumphuk (cape), Thailand 72/D5
Laerdal, Norway 18/E6
La Esmeralda, Argentina 143/G5
La Esmeralda, Bolivia 136/D8
La Esmeralda, Venezuela 124/F6
Laesø (isl.), Denmark 18/G8
Laesø (isl.), Denmark 21/D3
La Esperanza, Argentina 143/B7
La Esperanza, Bolivia 136/D4
La Esperanza, Honduras 154/C3
La Esperanza, Venezuela 124/H3
La Estrada, Spain 33/B1
La Estrella, Chile 138/F5
La Estrelleta (prov.), Dom. Rep. 158/C5
La Falda, Argentina 143/D3
La Farge, Wis. (54639) 317/E8
La Fargeville, N.Y. (13656) 276/J2
Lafayette, Ala. (36862) 195/H5
Lafayette (co.), Ark. 202/C7
Lafayette, Calif. (94549) 204/K2
Lafayette, Colo. (80026) 208/K3
Lafayette (co.), Fla. 212/C2
La Fayette, Georgia (30728) 217/B1
La Fayette, Ill. (61449) 222/C2
Lafayette, Ind. 188/J2
Lafayette, Ind. (*47901) 227/D4
La Fayette, Ky. (42254) 237/F7
Lafayette (par.), La. 238/F6
Lafayette, La. (*70501) 238/F6
Lafayette, Minn. (56054) 255/D6
Lafayette (co.), Miss. 256/E2
Lafayette (co.), Mo. 261/E4
Lafayette (mt.), N.H. 268/D3
Lafayette, N.J. (07848) 273/D1
La Fayette, N.Y. (13084) 276/H5
Lafayette, Ohio (45854) 284/C4
Lafayette, Oreg. (97127) 291/A2
La Fayette, R.I. (†02852) 249/H6
Lafayette, Tenn. (37083) 237/J7
Lafayette (co.), Wis. 317/F10
Lafayette-Elliston, Va. (24108) 307/H6
Lafayette Springs, Miss. (38640) 256/F2
Lafe, Ark. (72436) 202/J1
La Fe, Cuba 158/A2
La Feria, Texas (78559) 303/G11
La Ferté-Macé, France 28/C3
Lafferty, Ohio (43951) 284/H5
Lafia, Nigeria 106/F7
Lafiagi, Nigeria 106/F7
Lafitte, La. (70067) 238/K7
La Flèche, France 28/C4
Lafleche, Sask. 181/E6
La Floresta, Uruguay 145/C7
Lafnitz (riv.), Austria 41/D3
La Follette, Tenn. (37766) 237/N8
La Fontaine, Ind. (46940) 227/F3
Lafontaine, Kansas (66750) 232/G4
Lafontaine, Québec 172/C4
Lafourche (par.), La. 238/K7
Lafourche, La. (†70301) 238/J7
Lafourche (bayou), La. 238/K8
La France, S.C. (29656) 296/B2
La Fría, Venezuela 124/B3
Laful, India 68/G7
Lagacéville, New Bruns. 170/E1
La Gallareta, Argentina 143/F5
Lagan (riv.), N. Ireland 17/K2
Lagarfljót (stream), Iceland 21/C1
La Garita, Colo. (†81132) 208/G7
La Garita (mts.), Colo. 208/F7
Lagawe, Philippines 82/C2
Lagayan, Philippines 82/C2
Lage, W. Germany 22/C3
Lågen (riv.), Norway 18/G6
Lages, Brazil 132/D9
Lages, Brazil 120/D5
Laggan, Scotland 15/D3
Laggan, Scotland 10/D2
Laggan (bay), Scotland 15/B5
Laggan, Loch (lake), Scotland 15/D4
Laghouat, Algeria 106/E2
Laghouat, Algeria 102/C1
Laghy, Ireland 17/E2
La Gineta, Spain 33/E3
Lagkadia, Greece 45/E7
La Glace, Alberta 182/A2
La Gloria, Colombia 126/D3
La Gloria, Cuba 158/G2
Lago, Neth. Ant. 161/E10
Lagoa, Portugal 33/B4
Lagoa da Prata, Brazil 135/D2
Lagoa Dourada, Brazil 135/D2
Lago Blanco, Argentina 143/B6
Lagoen, Neth. Ant. 161/F8
La Goleta, Colombia 126/B3
La Gomera, Guatemala 154/B3
Lagonegro, Italy 34/E4
Lagonoy (gulf), Philippines 82/E4
Lagoon (lake), S. Australia 88/F6
Lago Ranco, Chile 138/E3
Lagos (state), Nigeria 106/E7
Lagos (cap.), Nigeria 2/K5
Lagos (cap.), Nigeria 102/C4
Lagos (cap.), Nigeria 106/E7
Lagos, Portugal 33/B4
Lagos de Moreno, Mexico 150/J6
Lagosta (Lastovo) (isl.), Yugoslavia 45/C4
La Goulette (Halq el Oued), Tunisia 106/G1
Lago Verde, Chile 138/E5
La Grand-Combe, France 28/E5
La Grande, Oreg. 188/C1
La Grande, Oreg. (97850) 291/J2
La Grande, Wash. (98348) 310/C4
La Grand Rivière (riv.), Que. 162/J5
La Grange, Ark. (72352) 202/J4
La Grange, Australia 87/C7
La Grange, Calif. (95329) 204/E6
La Grange, Ga. 188/K4
La Grange, Georgia (30240) 217/B4
La Grange, Ill. (60525) 222/B6
Lagrange (co.), Ind. 227/G1
Lagrange, Ind. (46761) 227/F1
La Grange, Ky. (40031) 237/L4
La Grange, Maine (04453) 243/F5
La Grange○, Maine (04453) 243/F5
La Grange, Mo. (63448) 261/K2
La Grange, N.C. (28551) 281/O4
Lagrange, Ohio (44050) 284/F3
La Grange, Tenn. (38046) 237/C10
La Grange, Texas (78945) 303/G8
La Grange, W. Australia 92/C2
Lagrange, Wyo. (82221) 319/H4
La Grange Park, Ill. (60525) 222/B5
La Granja (San Ildefonso), Spain 33/E2
La Gran Sabana (plain), Venezuela 124/G5
La Grita, Venezuela 124/C3
Lagro, Ind. (46941) 227/F3
La Grue (bayou), Ark. 202/H5
La Guadeloupe, Québec 172/F4
La Guaira, Venezuela 124/E2
La Guaira, Venezuela 120/C1
La Guajira (dept.), Colombia 126/D2
La Guardia, Bolivia 136/D5
La Guardia, Spain 33/C2
Laguardia, Spain 33/E1
La Guata, Honduras 154/D3
Laguna (creek), Ariz. 198/E2
Laguna (dam), Ariz. 198/A6
Laguna (res.), Ariz. 198/A6
Laguna, Brazil 132/D10
Laguna (res.), Calif. 204/L11
Laguna, N. Mex. (87026) 274/B3
Laguna (prov.), Philippines 82/C3
Laguna Beach, Calif. (*92651) 204/G10
Laguna de Perlas, Nicaragua 154/F4
Laguna Hills, Calif. (92653) 204/D11
Laguna Niguel, Calif. (92677) 204/H10
Laguna Paiva, Argentina 143/F5
Lagunas, Chile 138/B3
Lagunas, Peru 128/E5
Laguna Yema, Argentina 143/D1
Lagunetas, Venezuela 124/C3
Lagunillas, Bolivia 136/D6
Lagunillas, Chile 138/F3
Lagunillas, Venezuela 124/C2
Lagunitas-Forest Knolls, Calif. (94938) 204/H1
La Habra, Calif. (90631) 204/D11
Lahad Datu, Malaysia 85/F5
Lahaina, Hawaii (96761) 218/H2
Lahaina, Hawaii 188/F5
Laham, Indonesia 85/F5
Lahan, Nong (lake), Thailand 72/D3
La Harpe, Ill. (61450) 222/C3
La Harpe, Kansas (66751) 232/G4
Lahat, Indonesia 85/C6
La Have, Nova Scotia 168/D4
La Have (isl.), Nova Scotia 168/D4
La Have (riv.), Nova Scotia 168/D4
Lahej, P.D.R. Yemen 59/E7
La Higuera, Chile 138/A7
Lahijan, Iran 66/G2
Lahinch, Ireland 17/C6
Lahmansville, W. Va. (26731) 312/H4
Lahn, W. Germany 22/C3
Lahn (riv.), W. Germany 22/C3
Lahnstein, W. Germany 22/B3
Laholm, Sweden 18/H8
Lahoma, Okla. (73754) 288/K2
La Honda, Calif. (94020) 204/J3
Lahontan (res.), Nev. 266/B3
Lahore, Pakistan 59/K3
Lahore, Pakistan 68/C2
Lahore, Pakistan 54/J6
Lahore, Va. (22502) 307/N4
La Horqueta, Venezuela 124/G3
Lahr, W. Germany 22/B4
Lahri, Pakistan 68/B3
Lahti, Finland 7/G2
Lahti, Finland 18/O6
La Huaca, Peru 128/B5
La Huerta, Mexico 150/G7
Laï, Chad 111/C6
Laï, Chad 102/D4
Lai Chau, Vietnam 72/D2
Laidlaw, Br. Col. 184/M3
Laidon, Loch (lake), Scotland 15/D4
Laie, Hawaii (96762) 218/E1
Laie (pt.), Hawaii 218/E1
L'Aigle, France 28/D3
Lai-hka, Burma 72/C2
Laila, Saudi Arabia 59/E5
Lailan, Iraq 66/D3
La Inglesa, Venezuela 124/G3
Laings, Ohio (43752) 284/J6
Laingsburg, Mich. (48848) 250/E6
Lainioälv (riv.), Sweden 18/N3
Lair, Ky. (†41031) 237/N4
Laird, Colo. (80739) 208/P2
Laird, Sask. 181/E3
Lairdsville, Pa. (17742) 294/J3
Lairg, Scotland 10/D1
Lairg, Scotland 15/D2
Lais, Philippines 82/E7
Laisamis, Kenya 115/G3
La Isla, Texas (†79838) 303/A10
Laiwui, Indonesia 85/H6
Laiyang, China 77/K4
Laja (riv.), Chile 138/E1
La Jalca, Peru 128/D6
La Jara, Colo. (81140) 208/H8
La Jara, N. Mex. (87027) 274/B2
Lajas, P. Rico 161/A2
Lajes do Pico, Portugal 33/B1
Lajinha, Brazil 135/F2
La Jolla, Calif. (92037) 204/H11
La Jolla Ind. Res., Calif. 204/J10
Lajord, Sask. 181/G5
La Jose, Pa. (15753) 294/E4
Lajosmizse, Hungary 41/E3
La Joya, Bolivia 136/B5
Lajoya, N. Mex. (87028) 274/C4
La Joya, Peru 128/G11
La Joya, Texas (78560) 303/F11
La Junta, Colo. 188/F3
La Junta, Colo. (81050) 208/M7
Lak Dera (dry riv.), Kenya 115/H3
Lak Dera (dry riv.), Somalia 115/H3
Lake (co.), Calif. 204/C4
Lake (co.), Colo. 208/G4
Lake (co.), Fla. 212/E3
Lake (co.), Ill. 222/E1
Lake (co.), Ind. 227/C2
Lake, Ky. (†40741) 237/O6
Lake (co.), Mich. 250/D5
Lake, Mich. (48632) 250/E5
Lake (co.), Minn. 255/G3
Lake, Miss. (39092) 256/F6
Lake (co.), Mont. 262/B3
Lake (co.), Ohio 284/H2
Lake (co.), Oreg. 291/G5
Lake (creek), Oreg. 291/J3
Lake (co.), S. Dak. 298/P5
Lake (riv.), Tasmania 99/D3
Lake (co.), Tenn. 237/B8
Lake (creek), Utah 304/A5
Lake (creek), Wash. 310/G3
Lake Alfred, Fla. (33850) 212/E3
Lake Alma, Sask. 181/G6
Lake Alpine, Calif. (†95223) 204/F5
Lake Andes, S. Dak. (57356) 298/M7
Lake Ann, Mich. (49650) 250/D4
Lake Ariel, Pa. (18436) 294/M3
Lake Arrowhead, Calif. (92352) 204/H9
Lake Arthur, La. (70549) 238/E6
Lake Arthur, N. Mex. (88253) 274/E5
Lake Arthur, N.S. Wales 97/D4
Lake Barcroft, Va. (†22041) 307/S3
Lake Barrington, Ill. (†60010) 222/A5
Lake Benton, Minn. (56149) 255/B6
Lake Beulah, Wis. (†53120) 317/J2
Lake Bluff, Ill. (60044) 222/B4
Lake Boga, Victoria 97/B4
Lake Bolac, Victoria 97/B5
Lake Bronson, Minn. (56734) 255/B2
Lake Bruce, Ind. (†46939) 227/E2
Lake Buena Vista, Fla. (†32830) 212/E3
Lake Butler, Fla. (32054) 212/D1
Lake Butte Des Morts (Butte Des Morts), Wis. (†54901) 317/J7
Lake Cargelligo, N.S. Wales 97/D3
Lake Carmel, N.Y. (10512) 276/N8
Lake Carroll, Fla. (†33601) 212/C2
Lake Catherine, Ill. (†60002) 222/E1
Lake Charles, La. 188/H4
Lake Charles, La. (*70601) 238/D6
Lake Chelan Nat'l Rec. Area, Wash. 310/E2

Lake Church, Wis. (†53004) 317/L9
Lake Cicott, Ind. (46942) 227/D3
Lake City, Ark. (72437) 202/K2
Lake City, Calif. (96115) 204/E2
Lake City, Colo. (81235) 208/E6
Lake City, Fla. (32055) 212/D1
Lake City, Georgia (†30260) 217/K2
Lake City, Ill. (†61937) 222/E4
Lake City, Iowa (51449) 229/D4
Lake City, Kansas (67071) 232/D4
Lake City, Mich. (49651) 250/D4
Lake City, Minn. (55041) 255/F6
Lake City, Pa. (16423) 294/B1
Lake City, S.C. (29560) 296/H4
Lake City, S. Dak. (57247) 298/O2
Lake City, Tenn. (37769) 237/N8
Lake City Arsenal, Mo. 261/R5
Lake Clark Nat'l Park, Alaska 196/H2
Lake Clark Nat'l Preserve, Alaska 196/H2
Lake Clear, N.Y. (12945) 276/M2
Lake Como, Fla. (32057) 212/E2
Lake Como, Miss. (†39422) 256/F7
Lake Como, Pa. (18437) 294/M2
Lake Cormorant, Miss. (38641) 256/D1
Lake Cowichan, Br. Col. 184/J3
Lakecreek, Oreg. (†97524) 291/E5
Lake Crystal, Minn. (56055) 255/D6
Lake Dallas, Texas (75065) 303/G1
Lake Delton, Wis. (53940) 317/G8
Lake District National Park, England 13/D3
Lake Dunmore, Vt. (†05769) 268/A4
Lake Elmo, Minn. (55042) 255/F6
Lake Elmore, Vt. (05657) 268/B2
Lake Elsinore, Calif. (92330) 204/F11
Lake End, La. (†71019) 238/D3
Lake Erie Beach, N.Y. (†14006) 276/B5
Lakefield, Minn. (56150) 255/C7
Lakefield, Ontario 177/F3
Lake-Fishing Bridge—Bridge Bay, Wyo. (†82190) 319/B1
Lake Forest, Fla. (32208) 212/B4
Lake Forest, Ill. (60045) 222/B4
Lake Forest Park, Wash. (†98101) 310/B1
Lake Fork, Gunnison (riv.), Colo. 208/E6
Lake Fork, Idaho (83635) 220/B5
Lake Fork, Ill. (62541) 222/D4
Lake Frances, Ark. (†72761) 202/B1
Lake Francis, Manitoba 179/E4
Lake Fremont (Zimmerman), Minn. (†55398) 255/E5
Lake Geneva, Wis. (53147) 317/K10
Lake George, Colo. (80827) 208/J5
Lake George, Mich. (48633) 250/E5
Lake George, Minn. (56458) 255/D3
Lake George, New Bruns. 170/C3
Lake George, N.Y. (12845) 276/N4
Lake Grace, W. Australia 88/B6
Lake Grace, W. Australia 92/B6
Lake Harbor, Fla. (33459) 212/F5
Lake Harbour, N.W.T. 162/J3
Lake Harbour, N.W. Terrs. 187/L3
Lake Hart, Ind. (†46157) /E5
Lake Havasu City, Ariz. 188/D4
Lake Havasu City, Ariz. (86403) 198/A4
Lakehead, Calif. (96051) 204/C3
Lake Helen, Fla. (32744) 212/E3
Lake Henry, Minn. (†56362) 255/D5
Lake Hiawatha, N.J. (07034) 273/E2
Lake Hopatcong, N.J. (07849) 273/D2
Lake Hubert, Minn. (56459) 255/D4
Lake Hughes, Calif. (93532) 204/G9
Lake Huntington, N.Y. (12752) 276/L7
Lakehurst, N.J. (08733) 273/E3
Lakehurst Naval Air Engineering Center, N.J. 273/E3
Lake in the Hills, Ill. (†60102) 222/E1
Lake Isabella, Calif. (93240) 204/G8
Lake Itasca, Minn. (56460) 255/C3
Lake Jackson, Texas (77566) 303/J8
Lake James, Ind. (†46703) 227/H1
Lake Jem, Fla. (32745) 212/E3
Lake Katrine, N.Y. (12449) 276/M7
Lake King, W. Australia 88/C6
Lakeland, Fla. 188/K5
Lakeland, Fla. (*33801) 212/D3
Lakeland, Georgia (31635) 217/F8
Lakeland, La. (70752) 238/H5
Lakeland, Manitoba 179/D4
Lakeland, Mich. (48143) 250/F6
Lakeland, Minn. (55043) 255/F6
Lakeland, Tenn. (†38134) 237/B10
Lake Landing, N.C. (†27824) 281/T4
Lakeland Village, Calif. (†92330) 204/E11
Lake Leelanau, Mich. (49653) 250/D4
Lake Lenore, Sask. 181/G3
Lake Lillian, Minn. (56253) 255/C6
Lake Linden, Mich. (49945) 250/A1
Lakeline, Ohio (†44094) 284/J8
Lake Lotawana, Mo. (64063) 261/R6
Lake Louise, Alberta 182/B4
Lake Louise, Alta. 162/E5
Lakelse Lake, Br. Col. 184/C3
Lake Lure, N.C. (28746) 281/E4
Lake Luzerne-Hadley, N.Y. (12846) 276/N4
Lake Mackay Aboriginal Reserve, North. Terr. 88/D4
Lake Mackay Aboriginal Res., North. Terr. 93/A6
Lake Macleod, W. Australia 92/A4
Lake Magdalene, Fla. (†33612) 212/D3
Lake Mary, Fla. (32746) 212/E3
Lake McDonald, Mont. (59921) 262/B2
Lake Mead Nat'l Rec. Area, Ariz. 198/A2
Lake Mead Nat'l Rec. Area, Nev. 266/G6
Lake Meredith Nat'l Rec. Area, Texas 303/C2
Lake Michigan Beach, Mich. (†49039) 250/C6
Lake Mills, Iowa (50450) 229/F2
Lake Mills, Wis. (53551) 317/H9
Lake Minchumina, Alaska (99623) 196/H2
Lake Minnewanka, Alberta 182/C4
Lake Mohawk, N.J. (†07871) 273/D1
Lake Monroe, Fla. (32747) 212/E3
Lakemont, Georgia (30552) 217/F1
Lakemont, Pa. (16602) 294/F5
Lake Montezuma, Ariz. (86342) 198/D4
Lakemoor, Ill. (†60050) 222/A4
Lakemore, Ohio (44250) 284/H3
Lake Moxie, Maine (†04985) 243/D5
Lake Nash, North. Terr. 88/F4
Lake Nash, North. Terr. 93/E6
Lake Nebagamon, Wis. (54849) 317/C3
Lakenheath, Sask. 181/E6
Lake Norden, S. Dak. (57248) 298/P4
Lake Odessa, Mich. (48849) 250/D6
Lake of the Woods (lake), Manitoba 179/H5
Lake of the Woods (co.), Minn. 255/D2
Lake of the Woods (lake), Minn. 255/D1
Lake of the Woods (lake), Ontario 177/F5
Lake of the Woods (lake), Ontario 175/B3
Lake Orion, Mich. (48035) 250/F6
Lake Oswego, Oreg. (97034) 291/B2
Lake Ozark, Mo. (65049) 261/G6
Lake Park, Fla. (33403) 212/F5
Lake Park, Georgia (31636) 217/F9
Lake Park, Iowa (51347) 229/C2
Lake Park, Minn. (56554) 255/B4
Lake Placid, Fla. (33852) 212/E4
Lake Placid, N.Y. (12946) 276/N2
Lake Pleasant, N.Y. (12108) 276/M4
Lake Pocotopaug, Conn. (†06424) 210/F2
Lakeport, Calif. (95453) 204/C4
Lakeport, Fla. (†33471) 212/E5
Lakeport, Ontario 177/G4
Lake Preston, S. Dak. (57249) 298/P5
Lake Providence, La. (71254) 238/H1
Lakes, England 13/E3
Lakes, England 10/E3
Lake Saint Croix Beach, Minn. (†55043) 255/F6
Lake Saint Louis, Mo. (†63336) 261/N2
Lake Saint Peter, Ontario 177/F2
Lakes Entrance, Victoria 97/E5
Lakeshire, Mo. (†63101) 261/P4
Lake Shore, Minn. (†56401) 255/D4
Lakeshore, Miss. (39558) 256/F10
Lakeside, Ariz. (85929) 198/E4
Lakeside, Conn. (06758) 210/B2
Lakeside, Iowa (50588) 229/C3
Lakeside, Mont. (59922) 262/B2
Lakeside, Nebr. (69351) 264/B2
Lakeside, Nova Scotia 168/E4
Lakeside, Ohio (43440) 284/E2
Lakeside, Oreg. (97449) 291/C4
Lakeside, Texas (†76108) 303/E2
Lakeside, Utah (†84401) 304/B2
Lakeside, Va. (23228) 307/N5
Lakeside Park, Ky. (†41017) 237/R2
Lakesite, Tenn. (†37379) 237/L10
Lake Spring, Mo. (65532) 261/J7
Lake Station, Ind. (46405) 227/C1
Lake Stevens, Wash. (98258) 310/D3
Lake Success, N.Y. (11040) 276/P7
Lake Superior Prov. Park, Ontario 175/D3
Lake Superior Prov. Park, Ontario 177/J5
Lakesville, Md. (†21622) 245/O7
Lake Tapawingo, Mo. (†64015) 261/R6
Lake Tomahawk, Wis. (54539) 317/H4
Laketon, Ind. (46943) 227/F3
Laketon, New Bruns. 170/E2
Laketown, Utah (84038) 304/C2
Lake Toxaway, N.C. (28747) 281/D4
Lakevale, Nova Scotia 168/G3
Lake Valley, Sask. 181/E5
Lakeview, Ala. (†35986) 195/G2
Lake View, Ark. (†72389) 202/J5
Lakeview, Ark. (72642) 202/E1
Lake View, Iowa (51450) 229/C4
Lake View○, Maine (†04463) 243/F5
Lakeview, Mich. (48850) 250/D5
Lake View, Miss. (†38680) 256/D1
Lakeview, Mont. (†59744) 262/E6
Lake View, N.Y. (14085) 276/B5
Lakeview, N.C. (28350) 281/L4
Lakeview, Ohio (43331) 284/C4
Lakeview, Oreg. (97630) 291/G5
Lake View, S.C. (29563) 296/J3
Lakeview, Texas (79239) 303/D3
Lakeview Heights, Mo. (†65338) 261/F6
Lake Villa, Ill. (60046) 222/A4
Lake Village, Ark. (71653) 202/H7
Lake Village, Ind. (46349) 227/C2
Lakeville, Conn. (06039) 210/B1
Lakeville, Ind. (46536) 227/E1
Lakeville, Mass. (02346) 249/L5
Lakeville○, Mass. (02346) 249/L5
Lakeville, Minn. (55044) 255/E6
Lakeville, New Bruns. 170/C2
Lakeville, N.Y. (14480) 276/E5
Lakeville, Ohio (44638) 284/F4
Lake Waccamaw, N.C. (28450) 281/M6
Lake Wales, Fla. (33853) 212/E4
Lake Waukomis, Mo. (†64152) 261/P5
Lake Way, W. Australia 88/C5
Lake Way, W. Australia 92/C4
Lake Wazeecha, Wis. (†54494) 317/G7
Lake Williams, N. Dak. (58478) 282/L5
Lake Wilson, Minn. (56151) 255/B7
Lake Winnebago, Mo. (64034) 261/R6
Lake Wissota, Wis. (†54729) 317/D6
Lakewood, Calif. (*90712) 204/C11
Lakewood, Colo. 188/E3
Lakewood, Colo. (80215) 208/J3
Lakewood, Fla. (†32536) 212/C5
Lakewood, Ill. (62438) 222/E4
Lakewood, N.J. (08701) 273/E3
Lakewood, N. Mex. (88254) 274/E6
Lakewood, N.Y. (14750) 276/B6
Lakewood, Ohio (44107) 284/G9
Lakewood, Pa. (18439) 294/M2
Lakewood, Tenn. (†37138) 237/H8
Lakewood, Wash. (98259) 310/C2
Lakewood, Wis. (54138) 317/K5
Lakewood Club, Mich. (†49440) 250/C5
Lake Worth, Fla. (*33460) 212/G5
Lake Worth, Texas (76135) 303/E2
Lake Zurich, Ill. (60047) 222/A5
Lakhdenpokh'ya, U.S.S.R. 52/C2
Lakhish (dry riv.), Israel 65/B4
Lakin, Kansas (67860) 232/A4
Lakonia (gulf), Greece 45/F7
Lakor (isl.), Indonesia 85/H7
Lakota, Iowa (50451) 229/E2
Lakota, N. Dak. (58344) 282/O3
Laksefjorden (fjord), Norway 18/P1
Lakselv, Norway 18/O2
Lakshadweep (terr.), India 68/C6
Lakshadweep (sea), India 68/C6
La Laguna, Chile 138/F2
La Laguna, Spain 106/A3
La Laja (lag.), Chile 138/E1
La Lande, N. Mex. (†88119) 274/E4
Lalapaşa, Turkey 63/B2
Lalara, Gabon 115/B3
La Lata, Uruguay 145/E2
Laleh Zar, Kuh-e (mt.), Iran 59/G4
Laleh Zar, Kuh-e (mt.), Iran 66/K6
La Leona, Venezuela 124/G3
Lalibela, Ethiopia 111/G5
La Libertad, Ecuador 128/B4
La Libertad, El Salvador 154/C4
La Libertad, Guatemala 154/B2
La Libertad, Nicaragua 154/E4
La Libertad (dept.), Peru 128/C6
La Ligua, Chile 138/A9
La Ligua (riv.), Chile 138/A9
Lalín, Spain 33/C1
La Línea de la Concepción, Spain 33/D4
Lalitpur, Nepal 68/E3
Lalitpur, Nepal 68/E3
La Loche, Sask. 181/L3
La Louvière, Belgium 27/E8
La Lune, Trin. & Tob. 161/B11
La Luz, N. Mex. (88337) 274/C6
La Luz, Venezuela 124/D3
La Macarena, Serranía de (mts.), Colombia 126/D6
La Macaza, Québec 172/C3
La Maddalena, Italy 34/B4
La Madera, N. Mex. (87539) 274/C2
Lamag, Malaysia 85/F4
Lama-Kara, Togo 106/E7
La Malbaie, Québec 172/G2
Lamaline, Newf. 166/C4
La Manche Valley Prov. Park, Newf. 166/D2
Lamar (co.), Ala. 195/B3
Lamar, Ark. (72846) 202/D3
Lamar, Colo. (81052) 208/O6
Lamar (co.), Georgia 217/D4
Lamar, Ind. (47550) 227/D8
Lamar, La. (†71232) 238/G2
Lamar (co.), Miss. 256/E8
Lamar, Miss. (38642) 256/F1
Lamar, Mo. (64759) 261/D8
Lamar, Nebr. (69035) 264/C4
Lamar, Okla. (74850) 288/O4
Lamar, Pa. (16848) 294/H4
Lamar, S.C. (29069) 296/G3
Lamar (co.), Texas 303/J4
Lamar (riv.), Wyo. 319/B1
La Margarita, Venezuela 124/H3
Lamar Heights, Mo. (†64759) 261/D8
La Marque, Texas (77568) 303/K3
Lamartine, Pa. (16375) 294/C3
Lamartine, Québec 172/G2
Lamartine, Wis. (†53065) 317/J8
La Martre (lake), N.W. Terrs. 187/G3
Lamas, Peru 128/D6
Lamasco, Ky. (†42038) 237/F7
La Maurice Nat'l Park, Québec 172/D3
La Maya, Cuba 158/J4
Lamb, Ind. (†47043) 227/G7
Lamb (co.), Texas 303/B3
Lambach, Austria 41/C2
Lamballe, France 28/B3
Lambaré, Paraguay 144/A4
Lambaréné, Gabon 115/B4
Lambaréné, Gabon 102/D5
Lambari, Brazil 135/D2
Lambasa, Fiji 86/Q10
Lambay (isl.), Ireland 17/K4
Lambayeque (dept.), Peru 128/B6
Lambayeque, Peru 128/B6
Lambeg, N. Ireland 17/J2
Lambert, Ark. (71929) 202/D5
Lambert, Miss. (38643) 256/D2
Lambert, Mo. (†63736) 261/O8
Lambert, Mont. (59243) 262/M3
Lambert, Okla. (†73728) 288/J1
Lambert Lake, Maine (04454) 243/H4
Lamberton, Minn. (56152) 255/C6
Lambert's Bay, S. Africa 118/B6
Lambertville, Mich. (48144) 250/F7
Lambertville, New Bruns. 170/C3
Lambertville, N.J. (08530) 273/D3
Lambeth, England 13/H8
Lambeth, England 10/B5
Lambeth, Ontario 177/C5
Lambric, Ky. (41340) 237/P5
Lamb's (head), Ireland 17/A8
Lambsburg, Va. (24351) 307/G7
Lambs Grove, Iowa (†50208) 229/G5
Lambton (cape), N.W. Terrs. 187/F2
Lambton (county), Ontario 177/B5
Lambton, Québec 172/F4
Lame Deer, Mont. (59043) 262/K5
Lamego, Portugal 33/C2
Lame Johnny (creek), S. Dak. 298/C6
Lamentin, Guadeloupe 161/A6
Lamèque, New Bruns. 170/F1
Lamèque (isl.), New Bruns. 170/F1
La Merced, Argentina 143/C2
La Merced, Bolivia 136/C8
Lameroo, S. Australia 94/G6
La Mesa, Calif. (92041) 204/H11
La Mesa, N. Mex. (88044) 274/C6
Lamesa, Texas (79331) 303/C5
Lamía, Greece 45/F6
La Minerve, Québec 172/C3
Lamington, N.J. (†08876) 273/D2
Lamington (riv.), N.J. 273/D2
La Mirada, Calif. (90638) 204/D11
Lamison, Ala. (36747) 195/C6
Lamitan, Philippines 82/D7
Lamkin, Texas (†76455) 303/F6
Lamlam (mt.), Guam 86/K7
Lamlash, Scotland 15/C5
Lammermuir (hills), Scotland 15/F5
Lamming Mills, Br. Col. 184/G3
La Moille, Ill. (61330) 222/D2
La Moille, Iowa (†50158) 229/G4
Lamoille, Minn. (†55987) 255/G6
Lamoille, Nev. (89828) 266/F2
Lamoille (co.), Vt. 268/B2
Lamoille (riv.), Vt. 268/A2
La Moine (riv.), Ill. 222/C3
Lamoine○, Maine (†04605) 243/G7
Lamon (bay), Philippines 82/C3
Lamona, Wash. (99144) 310/G3
Lamongan, Indonesia 85/K2
Lamoni, Iowa (50140) 229/E7
Lamont, Alberta 182/D3
Lamont, Calif. (93241) 204/G8
Lamont, Fla. (32336) 212/C1
Lamont, Idaho (†83420) 220/G6
Lamont, Iowa (50650) 229/K3
Lamont, Kansas (66855) 232/F3
Lamont, Miss. (38755) 256/B3
Lamont, Okla. (74643) 288/L1
Lamont, Wash. (99017) 310/H3
Lamont, Wis. (†53530) 317/G10
Lamont, Wyo. (†82301) 319/E3
La Monte, Mo. (65337) 261/F5
Lamotrek (atoll), Micronesia 87/E5
La Motte, Iowa (52054) 229/M4
Lamotte (peak), Utah 304/D3
LaMoure (co.), N. Dak. 282/N7
LaMoure, N. Dak. (58458) 282/O7
Lampa, Chile 138/G3
Lampa, Peru 128/G10
Lampang, Thailand 72/C3
Lampard, Sask. 181/G4
Lampasas (co.), Texas 303/F6
Lampasas, Texas (76550) 303/F6
Lampasas (riv.), Texas 303/G6
Lampe, Mo. (65681) 261/F9
Lampedusa (isl.), Italy 34/D7
Lampertheim, W. Germany 22/C4
Lampeter, Pa. (17537) 294/K6
Lampeter, Wales 13/D5
Lampeter, Wales 10/D4
Lamphun, Thailand 72/C3
Lampman, Sask. 181/J6
Lamprey (riv.), N.H. 268/E5
Lampson, Wis. (†54888) 317/C3
Lamu, Burma 72/B3
Lamu, Kenya 115/H4
Lamu, Kenya 102/G5
Lamud, Peru 128/C6
Lamy, N. Mex. (87540) 274/D3
Lamy, Québec 172/H2
Lanagan, Mo. (64847) 261/C9
Lanai (isl.), Hawaii 87/L3
Lanai (isl.), Hawaii 188/F5
Lanai (isl.), Hawaii 218/H2
Lanai City, Hawaii 188/F5
Lanai City, Hawaii (96763) 218/H2
Lanaihale (mt.), Hawaii 218/H2
Lanaken, Belgium 27/H7
Lanao (lake), Philippines 82/E7
Lanao del Norte (prov.), Philippines 82/E6
Lanao del Sur (prov.), Philippines 82/E7
Lanark, Ill. (61046) 222/D1
Lanark (county), Ontario 177/H3
Lanark, Ontario 177/H2
Lanark, Scotland 15/E5
Lanark, Scotland, 10/E3
Lanark (trad. co.), Scotland 15/A5
Lanark, W. Va. (25860) 312/D7
Lanbi Kyun (isl.), Burma 72/C5
Lancang, China 77/E7
Lancang Jiang (riv.), China 77/F7
Lancashire (co.), England 13/E4
Lancaster, Calif. (93534) 204/G9
Lancaster (sound), Canada 4/B14
Lancaster, England 13/E3
Lancaster, England 10/E3
Lancaster, Ill. (62855) 222/F5
Lancaster, Ind. (†47250) 227/F7
Lancaster, Kansas (66041) 232/G2
Lancaster, Ky. (40444) 237/M5
Lancaster○, Mass. (01523) 249/H3
Lancaster, Minn. (56735) 255/B2
Lancaster, Mo. (63548) 261/H1
Lancaster (co.), Nebr. 264/H4
Lancaster, N.H. (03584) 268/D3
Lancaster○, N.H. (03584) 268/D3
Lancaster, N.Y. (14086) 276/C5
Lancaster (sound), N.W.T. 162/H1
Lancaster (sound), N.W.T. 146/K2
Lancaster (sound), N.W. Terrs. 187/K2
Lancaster, Ohio (43130) 284/E6
Lancaster, Ontario 177/K2
Lancaster, Pa. 188/L2
Lancaster (co.), Pa. 294/K5
Lancaster, Pa. (*17601) 294/K5
Lancaster (co.), S.C. 296/F2
Lancaster, S.C. (29720) 296/F2
Lancaster, Texas (*75146) 303/G3
Lancaster (co.), Va. 307/R5
Lancaster, Va. (22503) 307/R5
Lancaster, Wis. (53813) 317/E10
Lancaster Mills, S.C. (†29720) 296/F2
Lance (creek), Wyo. 319/H2
Lance Creek, Wyo. (82222) 319/H2
Lancer, Sask. 181/C5
Lanchester, England 13/H3
Lanchow (Lanzhou), China 77/F4
Lanciano, Italy 34/E3
Lancing, Tenn. (37770) 237/M8
Lanco, Chile 138/D2
Lancret, Cuba 158/D1
Łańcut, Poland 47/F3
Lancy, Switzerland 39/B4
Land, Ala. (†36904) 195/B6
Landa, N. Dak. (58749) 282/J2
Landaff○, N.H. (†03585) 268/D3
Landaff Center, N.H. (†03585) 268/D3
Landau an der Isar, W. Germany 22/E4
Landau in der Pfalz, W. Germany 22/C4
Landay, Afghanistan 68/A2
Landay, Afghanistan 59/H3
Land Between The Lakes Rec. Area, Ky. 237/E7
Land Between The Lakes Rec. Area, Tenn. 237/E7
Landeck, Austria 41/A3
Landeck, Ohio (†45833) 284/B4
Landen, Belgium 27/G7
Landenberg, Pa. (19350) 294/L6
Lander (co.), Nev. 266/D3
Lander (riv.), North. Terr. 88/E4
Lander (riv.), North. Terr. 93/C6
Lander, Pa. (†16350) 294/D2
Lander, Wyo. 188/E2
Lander, Wyo. (82520) 319/D3
Landerneau, France 28/B3
Landersville, Ala. (†35650) 195/D2
Landes (dept.), France 28/C5
Landes, W. Va. (†26847) 312/H5
Landess, Ind. (46944) 227/F3
Landfall (isl.), India 68/G6
Landhi, Pakistan 59/J4
Landing, N.J. (07850) 273/D2
Landing (creek), N.J. 273/D4
Landis, N.C. (28088) 281/H3
Landis, Sask. 181/C3
Landisburg, Pa. (17040) 294/H5
Landisburg, W. Va. (†25831) 312/E7
Landisville, N.J. (08326) 273/D4
Landisville, Pa. (17538) 294/K5
Landmark, Manitoba 179/F5
Lando, S.C. (29724) 296/E2
Land O'Lakes, Fla. (33539) 212/D3
Land O'Lakes, Wis. (54540) 317/H3
Landover, Md. (20785) 245/G4
Landover Hills, Md. (20784) 245/G4
Landquart (riv.), Switzerland 39/J3
Landrum, S.C. (29356) 296/C1
Landry, New Bruns. 170/E1
Landsberg (Gorzów Wielkopolski), Poland 47/B2
Landsberg am Lech, W. Germany 22/D4
Land's End (prom.), England 7/D3
Land's End (prom.), England 13/B7
Land's End (prom.), England 10/D5
Lands End (cape), N.W. Terrs. 187/F2
Landshut, W. Germany 22/E4
Landskrona, Sweden 18/H9
Landsman (creek), Colo. 208/P4
Landsmeer, Netherlands 27/C4
Landstuhl, W. Germany 22/B4
Landusky, Mont. (59533) 262/H3
Landville, W. Va. (25629) 312/C7
Lane, Ill. (61750) 222/E3
Lane (co.), Kansas 232/B3
Lane, Kansas (66042) 232/G3
Lane, Okla. (74555) 288/O6
Lane (co.), Oreg. 291/E4
Lane, S.C. (29564) 296/H5
Lane, S. Dak. (57358) 298/N5
Lane, Tenn. (†38240) 237/C8
Laneburg, Ark. (71844) 202/D6
Lane Cove, N. S. Wales 88/L4
Lane Cove, N.S. Wales 97/J3
Lane Cove (riv.), N. W. Wales 88/K4
Lane Cove (riv.), N.S. Wales 97/J3
Lanes (creek), N.C. 281/J5
Lanesboro, Iowa (51451) 229/D4
Lanesboro○, Mass. (01237) 249/A2
Lanesboro, Minn. (55949) 255/G7
Lanesboro, Pa. (18827) 294/L2
Lanesborough-Ballyleague, Ireland 17/E4
Lanes Prairie, Mo. (†65013) 261/J6
Lanesville, Ind. (47136) 227/E8
Lanesville, Va. (†23086) 307/P5
Lanett, Ala. (36863) 195/H5
La Neuveville, Switzerland 39/D2
Lanfine, Alberta 182/E4
Lanford, S.C. (†29335) 296/C2
Lang, Ontario 177/F3
Lang, Sask. 181/G6
Langå, Denmark 21/C5
Langadhás, Greece 45/F5
Langádhia, Greece 45/F7
Langara (isl.), Br. Col. 184/A3
Langavat (lake), Scotland 15/B2
Langbank, Sask. 181/J5
Lang Bay, Br. Col. 184/E5
Lang Bian, Nui (mts.), Vietnam 72/E4
Langdale, Ala. (36864) 195/H5
Langdon, Alberta 182/D4
Langdon, Iowa (†51301) 229/C2
Langdon, Kansas (67549) 232/D4
Langdon, Mo. (†64446) 261/A2
Langdon○, N.H. (†03602) 268/C5
Langdon, N. Dak. (58249) 282/O2
Langdondale, Pa. (†16650) 294/F5
Langeac, France 28/E5
L'Ange-Gardien, Québec 172/F3
Langeland (isl.), Denmark 21/D8
Langelands Baelt (chan.), Denmark 21/D8
Längelmävesi (lake), Finland 18/O6
Langeloth, Pa. (15054) 294/A5
Langemark-Poelkapelle, Belgium 27/B7
Langen, W. Germany 22/C4
Langenburg, Sask. 181/K5
Längenfeld, Austria 41/A3
Langenhagen, W. Germany 22/C2
Langenlois, Austria 41/C2
Langenthal, Switzerland 39/E2
Langenwang, Austria 41/C3
Langeoog (isl.), W. Germany 22/B2
Langford, Miss. (†39042) 256/E6
Langford, S. Dak. (57454) 298/O2
Langham, Sask. 181/E3
Langholm, Scotland 15/E5
Langholm, Scotland 10/E3
Langhorne, Pa. (19047) 294/N5
Langjökull (glac.), Iceland 21/B1
Langkawi, Pulau (isl.), Malaysia 72/C6
Langlade (co.), Wis. 317/H5
Langlais, Québec 172/F1
Langley, Ark. (71952) 202/C5
Langley, Br. Col. 184/L3
Langley, Okla. (74350) 288/R2
Langley, S.C. (29834) 296/D4
Langley, Wash. (98260) 310/C2
Langley A.F.B., Va. 307/R6
Langley Park, Md. (20787) 245/F4
Langleyville, Ill. (62542) 222/D4
Langlois, Oreg. (97450) 291/C5
Langnau am Albis, Switzerland 39/G2
Langnau in Emmental, Switzerland 39/E3
Langness (prom.), I. of Man 13/C3
Langogne, France 28/E5
Langon, France 28/C5
Langøy (isl.), Norway 18/J2
Langres, France 28/F4
Langres (plat.), France 28/F4
Langruth, Manitoba 179/D4
Langsa, Indonesia 85/B5
Långsele, Sweden 18/K5
Långshyttan, Sweden 18/K6
Lang Son, Vietnam 72/E2
Langston, Ala. (35755) 195/G1
Langston, Okla. (73050) 288/M3
Lang Suan, Thailand 72/C5
Langsville, Ohio (45741) 284/F7
Langton, Ontario 177/D5
Langtry, Texas (78871) 303/C8
Languedoc (trad. prov.), France 29
L'Anguille (riv.), Ark. 202/J3
Langwies, Switzerland 39/J3
Langworthy, Iowa (52252) 229/L4
Lanham, Nebr. (†68415) 264/H4
Lanham-Seabrook, Md. (20801) 245/G4
Lanier (co.), Georgia 217/F8
Lanigan, Sask. 181/F4
Lanigan (creek), Sask. 181/F4
Lanín (vol.), Argentina 143/B4
Lanín (vol.), Chile 138/E2
Lanín Nat'l Park, Argentina 143/B4
Lankin, N. Dak. (58250) 282/P3
Lanlacuni Bajo, Peru 128/G9
Lannion, France 28/B3
Lannon, Wis. (53046) 317/K1
L'Annonciation, Québec 172/C3
Lanoka Harbor, N.J. (08734) 273/E4
Lanoraie, Québec 172/D4
Lansdale, Pa. (19446) 294/M5
Lansdowne, India 68/D3
Lansdowne, Ontario 177/H3
Lansdowne, Pa. (19050) 294/M7
Lansdowne-Baltimore Highlands, Md. (21227) 245/M3
Lansdowne House, Ontario 175/C2
L'Anse, Mich. (49946) 250/G1
Lanse, Pa. (16849) 294/F4
L'Anse-Amour, Newf. 166/C3
L'Anse-au-Clair, Newf. 166/C3
L'Anse-au-Loup, Newf. 166/C3
L'Anse au Meadow, Newf. 166/C3
L'Anse Ind. Res., Mich. 250/A2
Lansford, N. Dak. (58750) 282/H2
Lansford, Pa. (18232) 294/L4
Lansing, Ill. (60438) 222/C6
Lansing, Iowa (52151) 229/L2
Lansing, Kansas (66043) 232/H2
Lansing (cap.), Mich. 146/K5
Lansing (cap.), Mich. 188/K2
Lansing (cap.), Mich. (*48901) 250/E6
Lansing, Minn. (55950) 255/F7
Lansing, N.Y. (14882) 276/H5
Lansing, N.C. (28643) 281/F1
Lansing, Ohio (43934) 284/J5
Lanškroun, Czech. 41/D2
Lanta, Ko (isl.), Thailand 72/C6
Lantana, Fla. (33462) 212/F5
Lanton, Mo. (†65791) 261/J9
Lantry, S. Dak. (57636) 298/G3
Lantsch-Lenz, Switzerland 39/J3
Lantz, Md. (21760) 245/J2
Lantz, Nova Scotia 168/E4
Lantzville, Br. Col. 184/J3
Lanús, Argentina 143/H7
Lanús, Argentina 120/D6
Lanusei, Italy 34/B5
Lanuvio, Italy 34/F7
Lanuza, Philippines 82/F6
Lanuza (bay), Philippines 82/F6
Lanyon, Iowa (†50544) 229/E4
Lanza, Bolivia 136/B5
Lanzarote (isl.), Spain 102/A2
Lanzarote (isl.), Spain 106/B3
Lanzarote (isl.), Spain 33/C4
Lanzhou (Lanchow), China 77/F4
Lanzhou, China 54/M6
Lanzhou, China 2/P4
Laoag, Philippines 82/C1
Laoag, Philippines 54/N8
Laoag, Philippines 85/F2
Laoang, Philippines 82/E4
Lao Cai, Vietnam 54/M7
Lao Cai, Vietnam 72/E2
Laoha He (riv.), China 77/J3
Laoighis (co.), Ireland 17/G6
Laon, France 28/E3
Laona, Wis. (54541) 317/J4
La Orchila (isl.), Venezuela 124/F2
La Orotava, Spain 33/B4
La Oroya, Peru 120/B4
La Oroya, Peru 128/D8
Laos 2/Q5
Laos 54/M8
LAOS 72
Laotto, Ind. (46763) 227/G2
Lapa, Brazil 132/D9
Lapa, Philippines 82/C8
Lapalisse, France 28/E4
La Pallice, France 28/C4
La Palma, Colombia 126/C5

La Palma, El Salvador 154/C3
La Palma, Panama 154/H6
La Palma (isl.), Spain 102/A2
La Palma (isl.), Spain 106/A3
La Palma (isl.), Spain 33/A4
La Palma del Condado, Spain 33/C4
La Paloma, Uruguay 145/C4
La Pampa (prov.), Argentina 143/C4
Laparan (isl.), Philippines 82/B8
Laparan (isls.), Philippines 82/B8
La Passe, Ontario 177/H2
La Patrie, Québec 172/F4
La Paz, Entre Ríos, Argentina 143/G5
La Paz, Mendoza, Argentina 143/C3
La Paz (dept.), Bolivia 136/A4
La Paz (cap.), Bolivia 136/B5
La Paz (cap.), Bolivia 2/F6
La Paz (cap.), Bolivia 120/C4
La Paz, Honduras 154/D3
La Paz, Ind. (46537) 227/E2
La Paz, Mexico 146/G7
La Paz, Baja California Sur, Mexico 150/D5
La Paz, San Luis Potosí, Mexico 150/J5
La Paz (bay), Mexico 150/D4
La Paz, Philippines 82/E6
La Paz, Canelones, Uruguay 145/B6
La Paz, Colonia, Uruguay 145/B5
La Paz Central, Nicaragua 154/D4
La Paz de Oriente, Nicaragua 154/E5
La Pêche, Québec 172/B4
La Pedrera, Colombia 126/F8
La Pedrera, Uruguay 145/D5
Lapeer (co.), Mich. 250/F5
Lapeer, Mich. (48446) 250/F5
Lapel, Ind. (46051) 227/F4
La Pelada, Argentina 143/F5
La Pérade, Québec 172/E3
La Pérouse (str.) 54/R5
La Pérouse, N. S. Wales 88/L4
La Pérouse, N. S. Wales 97/J4
La Pérouse (str.), U.S.S.R. 48/P5
La Pesca, Mexico 150/L4
Lapeyrère (lake), Québec 172/E2
La Piedad Cavadas, Mexico 150/H6
Lapine, Ala. (36046) 195/F7
La Pine, Oreg. (97739) 291/J5
Lapinin (isl.), Philippines 82/E5
La Pintada, Panama 154/G6
Lapithos, Cyprus 63/E5
La Place, La. (70068) 238/N3
La Plaine, Dominica 161/F6
Lapland (reg.) 7/G2
Lapland (reg.), Finland 18/O2
Lapland (reg.), Norway 18/K2
Lapland (reg.), Sweden 18/M2
Lapland (reg.), U.S.S.R. 52/D1
La Plant, S. Dak. (57637) 298/H3
Laplante, New Bruns. 170/E1
La Plata (est.) 120/D6
La Plata, Argentina 143/H7
La Plata, Argentina 120/D6
La Plata, Río de (est.), Argentina 143/E4
La Plata, Colombia 126/C6
La Plata (co.), Colo. 208/D8
La Plata (peak), Colo. 208/G4
La Plata (riv.), Colo. 208/C8
La Plata, Md. (20640) 245/L6
La Plata, Mo. (63549) 261/H2
La Plata, N. Mex. (87418) 274/A2
La Plata (riv.), N. Mex. 274/A1
La Plume, Pa. (18440) 294/L2
La Pobla de Lillet, Spain 33/G1
La Pocatière, Québec 172/F2
La Poile, Newf. 166/C4
La Poile (bay), Newf. 166/C4 ·
Lapoint, Utah (84039) 304/F3
La Pointe, Wis. (54850) 317/E2
La Porte, Calif. (95981) 204/D4
Laporte, Colo. (80535) 208/J1
LaPorte (co.), Ind. 227/D1
LaPorte, Ind. (46350) 227/D1
Laporte, Mich. (48623) 250/E5
Laporte, Minn. (56461) 255/D3
Laporte, Pa. (18626) 294/K3
Laporte, Sask. 181/B4
La Porte, Texas (77571) 303/K2
La Porte City, Iowa (50651) 229/J4
Lappajärvi, Finland 18/O5
Lappajärvi (lake), Finland 18/O5
Lappeenranta, Finland 7/G2
Lappeenranta, Finland 18/P6
Lappi (prov.), Finland 18/P3
La Prairie, Ill. (62346) 222/B3
La Prairie, Minn. (†55444) 255/E3
Laprairie (co.), Québec 172/H4
La Prairie, Québec 172/J4
Laprida, Argentina 143/F4
La Protección, Honduras 154/D4
La Providence, Québec 172/E4
La Pryor, Texas (78872) 303/E9
La Puebla, Ill. 222/B3
La Puebla de Montalbán, Spain 33/D3
La Puente, Calif. (*91744) 204/D10
La Puerto, Cuba 158/C3
La Puntilla (peak), Ecuador 128/B4
La Purísima, Mexico 150/D3
La Push, Wash. (98350) 310/A3
Lapwai, Idaho (83540) 220/B3
Lapy, Poland 47/F2
Laqiya 'Umran (well), Sudan 111/E3
Laquey, Mo. (65534) 261/H7
La Quiaca, Argentina 143/C1
L'Aquila, Argentina 143/D3
L'Aquila (prov.), Italy 34/D3
L'Aquila, Italy 34/D3
Lar, Iran 59/H4
Lar, Iran 66/J7
Lara (state), Venezuela 124/C2

Lara, Victoria 97/C6
Larabee, Pa. (†16731) 294/F2
Larache, Morocco 106/C1
Laracor, Ireland 17/H4
Larak (isl.), Iran 66/K7
La Rambla, Spain 33/D4
Laramie (mts.), Colo. 208/H1
Laramie (riv.), Colo. 208/H1
Laramie (riv.), Wyo. 319/H4
Laramie, Wyo. 146/H5
Laramie, Wyo. 188/E2
Laramie, Wyo. (82070) 319/G4
Laramie (mts.), Wyo. 319/G3
Laramie (peak), Wyo. 319/G3
Laramie (riv.), Wyo. 319/G4
Laranjeiras do Sul, Brazil 132/C9
Larantuka, Indonesia 85/G7
Larat (isl.), Indonesia 85/J7
Larbert, Scotland 10/B1
Larbert, Scotland 15/C1
Lärbro, Sweden 18/L8
Larchmont, N.Y. (10538) 276/P7
Larchwood, Iowa (51241) 229/A2
Lardeau, Br. Col. 184/J5
Larder Lake, Ontario 175/E3
Larder Lake, Ontario 177/K5
L'Ardoise West, Nova Scotia 168/H3
La Rédemption, Québec 172/B2
Laredo (sound), Br. Col. 184/C4
Laredo, Mo. (64652) 261/G2
Laredo, Mont. (†59501) 262/G2
Laredo, Spain 33/E1
Laredo, Texas (*78040) 303/E10
Laredo, Texas 188/G5
Laredo, Texas 146/J7
La Reine, Québec 174/B3
Laren, Netherlands 27/G4
Larena, Philippines 82/D6
La Réole, France 28/C5
Lares, P. Rico 161/B2
Lares, P. Rico 156/F1
La Retuca, Chile 138/F3
Larew, W. Va. (†26537) 312/D4
Largentière, France 28/F5
Largo (cay), Cuba 158/D2
Largo (cay), Cuba 156/D5
Largo, Fla. (*33540) 212/B3
Largo (key), Fla. 212/F6
Largo, Md. (†20870) 245/G5
Largo, Cañon (creek), N. Mex. 274/B2
Largs, Scotland 15/A2
Largs, Scotland 10/A1
Lariat, Texas (79335) 303/B3
Larimer (co.), Colo. 208/H1
Larimer, Pa. (15647) 294/C5
Larimore, N. Dak. (58251) 282/P4
Larino, Italy 34/E4
La Rioja (prov.), Argentina 143/C2
La Rioja, Argentina 120/C5
La Rioja, Argentina 143/C2
La Rioja, Cuba 158/H3
Lárisa, Greece 45/F4
Lárisa, Greece 7/G5
Laristan (reg.), Iran 66/J7
La Rivière, Manitoba 179/D5
Lark, N. Dak. (58550) 282/H7
Larkana, Pakistan 59/J4
Larkana, Pakistan 68/B3
Larkhall, Scotland 10/B1
Larkhall, Scotland 15/E5
Lark Harbour, Newf. 166/C4
Larkinburg, Kansas (†66436) 232/G2
Larkinsville, Ala. (†35768) 195/F1
Larkspur, Calif. (94939) 204/H1
Larkspur, Colo. (80118) 208/K4
Larksville, Pa. (†18704) 294/E7
Larnaca, Cyprus 59/B3
Larnaca, Cyprus 63/B3
Larnaca (bay), Cyprus 63/E5
Larne (dist.), N. Ireland 17/K2
Larne, N. Ireland 17/K2
Larne, N. Ireland 10/D3
Larne (inlet), N. Ireland 17/K2
Larned, Kansas (67550) 232/C3
La Robla, Spain 33/D1
La Roche, Switzerland 39/D3
La Roche-en-Ardenne, Belgium 27/G8
La Rochelle, France 7/D4
La Rochelle, France 28/C4
La Rochelle, Manitoba 179/F5
La Roche-sur-Yon, France 28/C4
La Roda, Spain 33/E3
La Romana (prov.), Dom. Rep. 158/F6
La Romana, Dom. Rep. 158/F6
La Romana, Dom. Rep. 156/E3
La Ronge, Sask. 181/J3
La Rose, Ill. (61541) 222/D3
Larose, La. (70373) 238/K7
Larrabee, Iowa (51029) 229/B3
Larrimah, North. Terr. 88/E3
Larrimah, North. Terr. 93/C3
Larroque, Argentina 143/F5
Larry's River, Nova Scotia 168/G3
Larsen (sound), N.W. Terrs. 187/J2
Larsen, Wis. (54947) 317/G2
Larsen Bay, Alaska (99624) 196/H3
Larsen Ice Shelf, Ant. 2/J2
Larsen Ice Shelf, Ant. 5/C16
Larslan, Mont. (59244) 262/K2
Larsmont, Minn. (†55616) 255/G4
Larson, N. Dak. (58751) 282/E2
Larto, La. (71344) 238/G4
Larue (co.), Ky. 237/K5
La Rue, Ohio (43332) 284/D4
Laruns, France 28/C6
La Russell, Mo. (64848) 261/D8
Larvik, Norway 18/F4
Larwill, Ind. (46764) 227/F2
La Sal, Utah (84530) 304/F6
La Salle, Colo. (80645) 208/K2
La Salle (co.), Ill. 222/E4
La Salle, Ill. (61301) 222/E4
La Salle (par.), La. 238/F3
La Salle, Manitoba 179/F5
La Salle (riv.), Manitoba 179/E5
La Salle, Mich. (48145) 250/F7
La Salle, Minn. (56056) 255/D6
La Salle, Québec 172/H4

Lasalle (lake), Québec 172/E2
La Salle (co.), Texas 303/E9
Las Animas (co.), Colo. 208/L8
Las Animas, Colo. (81054) 208/N6
Las Animas (creek), N. Mex. 274/B5
Las Anod, Somalia 115/J2
La Sarraz, Switzerland 39/C3
La Sarre, Que. 162/J6
La Sarre, Québec 174/B3
Lasauces, Colo. (†81151) 208/H8
Las Aves (isls.), Venezuela 124/E2
Las Bonitas, Venezuela 124/F4
Las Breas, Chile 138/B7
Lasca, Ala. (†36784) 195/C6
Las Cabras, Chile 138/B7
Lascahobas, Haiti 156/D3
Lascahobas, Haiti 158/C6
Lascano, Uruguay 145/E4
Las Carreras, Bolivia 136/C7
Las Choapas, Mexico 150/M7
La Scie, Newf. 166/D1
Las Cruces, Chile 138/F3
Las Cruces, N. Mex. 146/H6
Las Cruces, N. Mex. 188/E4
Las Cruces, N. Mex. (88001) 274/C6
Las Dureh, Somalia 115/J1
La Selva Beach, Calif. (95076) 204/K4
La Serena, Chile 120/B5
La Serena, Chile 138/A8
La Seyne-sur-Mer, France 28/F6
Las Flores, Argentina 143/E4
Las Flores, Uruguay 145/D5
Las Hadas, Mexico 150/G7
Lashburn, Sask. 181/B2
Las Juntas, Colombia 126/E6
Las Juntas, C. Rica 154/E5
Lask, Poland 47/D3
Lasker, N.C. (27848) 281/P2
La Skhirra, Tunisia 106/G2
Las Khoreh, Somalia 115/J1
Las Lajas, Argentina 143/B4
Las Lajitas, Venezuela 124/F4
Las Lomas, Argentina 143/D1
Las Lomitas, Argentina 143/D1
Las Marias, P. Rico 161/B2
Las Martinas, Cuba 158/A2
Las Matas de Farfán, Dom. Rep. 158/D6
Las Matas de Farfán, Dom. Rep. 156/D3
Las Mercedes, Venezuela 124/E3
Las Navas del Marqués, Spain 33/D2
Las Nieves, Mexico 150/G3
La Solana, Spain 33/E3
La Sorcière (mt.), St. Lucia 161/G6
La Souterraine, France 28/D4
Las Palmas, Argentina 143/E2
Las Palmas (cap.), Canary Is., 102/A2
Las Palmas, Panama 154/G6
Las Palmas (prov.), Spain 33/C4
Las Palmas de Gran Canaria, Spain 33/B4
Las Palmas de Gran Canaria, Spain 106/B3
Las Pampitas, Bolivia 136/C3
Las Parejas, Argentina 143/F6
Las Pedroñeras, Spain 33/E3
Las Petas, Bolivia 136/F5
La Spezia (prov.), Italy 34/B2
La Spezia, Italy 34/B2
La Spezia, Italy 7/E4
Las Piedras, Peru 128/H9
Las Piedras, P. Rico 161/E2
Las Piedras (riv.), Peru 128/H9
Las Piedras, Uruguay 145/B6
Las Piedras, Falcón, Venezuela 124/C2
Las Piedras, Zulia, Venezuela 124/B2
Las Plumas, Argentina 143/C5
Lasqueti Island, Br. Col. 184/J2
Lastarria (vol.), Chile 138/B5
La Station-du-Coteau, Québec 172/C4
Last Chance (creek), Utah 304/C6
Last Mountain (lake), Sask. 181/F4
Las Termas, Argentina 143/D2
Las Toscas, Uruguay 145/E3
Las Tunas (prov.), Cuba 158/H3
Las Trincheras, Venezuela 124/F4
Las Truchas, Mexico 150/H7
Lastrup, Minn. (56347) 255/D4
Las Tunas, Cuba 158/H3
Las Varillas, Argentina 143/D3
Las Vegas, Nev. 146/G6
Las Vegas, Nev. (*89101) 266/F6
Las Vegas (range), Nev. 266/F6
Las Vegas, N. Mex. 188/E3
Las Vegas, N. Mex. (87701) 274/D3
Las Vegas, Venezuela 124/D3
La Tabatière, Québec 174/F2
Latacunga, Ecuador 128/C3
La Tagua, Colombia 126/C8
Latah (co.), Idaho 220/B3

Latah, Wash. (99018) 310/H3
Latah (creek), Wash. 310/H3
Latakia (prov.), Syria 63/G5
Latakia, Syria 54/E5
Latakia, Syria 63/F5
Latakia, Syria 59/C2
La Taste, Grenada 161/D8
Latchford, Ontario 177/K5
Laterrière, Québec 172/F1
Latexo, Texas (75849) 303/J6
Latham, Ala. (†36579) 195/C8
Latham, Ill. (62543) 222/E4
Latham, Kansas (67072) 232/F4
Latham, Mo. (65050) 261/G5
Latham, N.Y. (12110) 276/N5
Latham, Ohio (45646) 284/D7
Latham, Tenn. (†38225) 237/D8
Lathrop, Calif. (95330) 204/D6
Lathrop, Mich. (†49880) 250/B7
Lathrop, Mo. (64465) 261/D3
La Tigra, Venezuela 124/H4
Latimer (co.), Okla. 288/R5
Latimer, Kansas (†67449) 232/F3
Latimers (brook), Conn. 210/G3
Latina (prov.), Italy 34/D4
Latina, Italy 34/D4
Latium (Lazio) (reg.), Italy 34/D3
La Tola, Ecuador 128/C2
La Toma, Argentina 143/C3
La Tortuga (isl.), Venezuela 124/F2
Latouche Treville (cape), W. Australia 88/C3
Latouche Treville (cape), W. Australia 92/C2
Latour, Mo. (64760) 261/D5
La Tour-de-Peilz, Switzerland 39/C4
La Tour-du-Pin, France 28/F4
Latourell Falls, Oreg. (†97060) 291/E2
La Trinidad, Nicaragua 154/D4
La Trinidad, Philippines 82/C4
La Trinidad, Venezuela 124/E3
La Trinidad de Arauca, Venezuela 124/F3
La Trinidad de Orichuna, Venezuela 124/D4
La Trinité, Martinique 161/D6
La Trinité-des-Monts, Québec 172/J1
Latrobe, Pa. (15650) 294/D5
Latrobe, Tasmania 99/C3
Latta, S.C. (29565) 296/J3
Lattimore, N.C. (28089) 281/F4
Lattingtown, N.Y. (†11560) 276/R6
Latty, Ohio (45855) 284/A3
La Tuque, Que. 162/F3
La Tuque, Québec 172/E3
La Tuque, Québec 174/C3
Latur, India 68/C5
LATVIA 53/B2
Latvian S.S.R., U.S.S.R. 7/G3
Latvian S.S.R., U.S.S.R. 52/B3
Latvian S.S.R., U.S.S.R. 48/C4
Lauca (riv.), Bolivia 136/A6
Lauca (riv.), Chile 138/B1
Lauchhammer, E. Germany 22/E3
Laud, Ind. (†46725) 227/G2
Laudat, Dominica 161/E6
Lauder, Manitoba 179/C5
Lauder, Scotland 10/D3
Lauder, Scotland 15/F5
Lauderdale (co.), Ala. 195/C1
Lauderdale, Minn. (†55101) 255/G5
Lauderdale (co.), Miss. 256/G6
Lauderdale, Miss. (39335) 256/G5
Lauderdale, Tasmania 99/D4
Lauderdale (co.), Tenn. 237/B9
Lauderdale-by-the-Sea, Fla. (33308) 212/C4
Lauderdale Lakes, Fla. (†33313) 212/B3
Lauderhill, Fla. (33313) 212/B3
Lauenburg an der Elbe, W. Germany 22/D2
Lauenen, Switzerland 39/D4
Lauf an der Pegnitz, W. Germany 22/D4
Läufelfingen, Switzerland 39/E2
Laufen, Switzerland 39/D2
Laufen, W. Germany 22/E5
Laufenburg, Switzerland 39/F1
Laughery (creek), Ind. 227/G6
Laughing Fish (pt.), Mich. 250/B2
Laughlin A.F.B., Texas 303/D9
Lau Group (isls.), Fiji 87/J7
Lauingen, W. Germany 22/D4
Launceston, England 13/C7
Launceston, England 10/D5
Launceston, Tasmania 99/C3
Launceston, Tasmania 88/H8
Laune (riv.), Ireland 17/B7
Launglon Bok (isls.), Burma 72/C4
La Unión, Chile 138/D3
La Unión, Colombia 126/B7
La Unión, El Salvador 154/D4
La Unión, Mexico 150/J8
La Unión, N. Mex. (†88021) 274/C7
La Unión, Peru 128/D7
La Unión (prov.), Philippines 82/C2
La Unión, Spain 33/F4
La Unión, Venezuela 124/A7
La Victoria, Colombia 126/C6
La Victoria, Apure, Venezuela 124/D4
La Victoria, Apure, Venezuela 124/E4
La Victoria, Aragua, Venezuela 124/E2
Lavieille (lake), Ontario 177/J2
La Vieja (riv.), Chile 138/A11
Lavik, Norway 18/D3
Lavillette, New Bruns. 170/E1
Lavina, Mont. (59046) 262/H4
Lavinia, Tenn. (38348) 237/D9
Lavinia, Tenn. (38348) 237/D9
Lavina, Manitoba 179/B4
Lavonia, Georgia (30553) 217/F2
Lavos, Portugal 33/B2
Lavoy, Alberta 182/E3
Lavras, Brazil 132/E8
Lavras, Brazil 135/D2

Lávrion, Greece 45/G7
Lawa (riv.), Fr. Guiana 131/D4
Lawa (riv.), Surinam 131/D4
Lawang, Indonesia 85/K2
Lawal, Hawaii (96765) 218/C2
Lawen, Oreg. (97740) 291/J4
Lawler, Iowa (52154) 229/J2
Lawler, Minn. (†55760) 255/E4
Lawlers, W. Australia 92/C5
Lawley, Ala. (36793) 195/C5
Lawn, Newf. 166/C4
Lawn, Pa. (17041) 294/J5
Lawn, Texas (79530) 303/E5
Lawndale, Ill. (61751) 222/D3
Lawndale, Minn. (†56579) 255/B4
Lawndale, N.C. (28090) 281/F4
Lawnhill, Br. Col. 184/A3
Lawn Hill, Queensland 95/A3
Lawra, Ghana 106/D6
Lawrence (co.), Ala. 195/D1
Lawrence (co.), Ark. 202/H1
Lawrence (co.), Ill. 222/F5
Lawrence, Ind. 227/E7
Lawrence, Ind. (46226) 227/E5
Lawrence, Kans. 188/G3
Lawrence, Kansas (66044) 232/G3
Lawrence (co.), Ky. 237/R4
Lawrence, Mass. 188/M2
Lawrence, Mass. (*01840) 249/K2
Lawrence, Mich. (49064) 250/C6
Lawrence (co.), Miss. 256/F7
Lawrence, Miss. (39336) 256/F6
Lawrence (co.), Mo. 261/E8
Lawrence, Nebr. (68957) 264/F4
Lawrence, N.Y. (11559) 276/P7
Lawrence (co.), N.C. (†27886) 281/O2
Lawrence (co.), Ohio 284/E8
Lawrence (co.), Pa. 294/B4
Lawrence (co.), S. Dak. 298/B5
Lawrence (co.), Tenn. 237/G10
Lawrenceburg, Ind. (47025) 227/H6
Lawrenceburg, Ky. (40342) 237/M4
Lawrenceburg, Tenn. (38464) 237/G10
Lawrence Park○, Pa. (†16501) 294/C1
Lawrenceport, Ind. (†47446) 227/D7
Lawrence Station, New Bruns. 170/C3
Lawrencetown, Nova Scotia 168/C4
Lawrenceville, Georgia (30245) 217/D3
Lawrenceville, Ill. (62439) 222/F5
Lawrenceville, Ind. (†47041) 227/H6
Lawrenceville○, N.J. (08648) 273/D3
Lawrenceville, N.Y. (12949) 276/L1
Lawrenceville, Ohio (†45501) 284/C4
Lawrenceville, Pa. (16929) 294/H2
Lawrenceville, Québec 172/E4
Lawrenceville, Va. (23868) 307/N7
Lawson, Ark. (71750) 202/F7
Lawson, Colo. (†80452) 208/H3
Lawson, Mo. (64062) 261/D4
Lawson, Sask. 181/E5
Lawsonville, N.C. (27022) 281/J2
Lawtell, La. (70550) 238/F5
Lawtey, Fla. (32058) 212/D1
Lawton, Ind. (†46996) 227/D2
Lawton, Iowa (51030) 229/A4
Lawton, Kansas (66752) 232/H4
Lawton, Ky. (41153) 237/P4
Lawton, Mich. (49065) 250/D6
Lawton, N. Dak. (58345) 282/O3
Lawton, Okla. 146/J6
Lawton, Okla. (73501) 288/K5
Lawton, Pa. (18828) 294/K2
Lawton, W. Va. (25863) 312/E7
Lawtonka (lake), Okla. 288/K5
Lawu (mt.), Indonesia 85/J2
Lax, Georgia (†31650) 217/F8
Lax, Switzerland 39/F4
Laxå, Sweden 18/J7
Laxey, Isle of Man 13/C3
Laxford, Loch (inlet), Scotland 15/C2
Lay (dam), Ala. 195/E5
Lay (lake), Ala. 195/F4
Lay (pt.), Alaska 196/F1
Lay, Colo. (†81625) 208/D2
Lay, Mui (cape), Vietnam 72/E3
La Vega (prov.), Dom. Rep. 158/D6
La Vega, Dom. Rep. 158/E5
La Vega, Dom. Rep. 156/D3
La Vega, Spain 33/C1
La Vela (cape), Colombia 126/D1
La Vela de Coro, Venezuela 124/D2
Lavelanet, France 28/E6
Lavello, Italy 34/E4
Lavenham, Manitoba 179/D5
L'Avenir, Québec 172/E4
La Vergne, Tenn. (37086) 237/H9
La Verkin, Utah (84745) 304/A6
La Verne, Okla. (73848) 288/G1
La Vernia, Texas (78121) 303/K11
La Veta, Colo. (81055) 208/J8
Lavey-Morcles, Switzerland 39/D4
Lavezares, Philippines 82/E5
La Vista, Nebr. (†68046) 264/J3
La Urbana, Venezuela 124/E4
Lavon (lake), Texas 303/H1
Lavongai (isl.), Papua N.G. 87/F6
La Vengia N.G. 86/B1
Layandé Station, Ant. 5/C1
Lazarev Station, Ant. 5/C1
Lazdijai, U.S.S.R. 53/B3
Lazear, Colo. (81420) 208/D5
Lazi, Philippines 82/D6
Łaziska Górne, Poland 47/A4
Lazy Lake, Fla. (†33301) 212/B3
Lea (riv.), England 13/G6
Lea (co.), N. Mex. 274/F6
Leaburg, Oreg. (97401) 291/E3
Leach, Okla. (74351) 288/S2
Leachville, Ark. (72438) 202/K2
Lecaoss, Sask. 181/H2
Lead, S. Dak. 188/F2
Lead, S. Dak. (57754) 298/B5
Leadbetter, Minn. (†66442) 255/D4
Leader, Minn. (†66442) 255/D4
Leader, Sask. 181/B5
Lead Hill, Ark. (72644) 202/D1
Leadington, Mo. (†63640) 261/M7
Leadmine (brook), Conn. 210/C1
Lead Mine, W. Va. (†26290) 312/G4

Leadmine, Wis. (†53807) 317/F10
Leadore, Idaho (83464) 220/E5
Leadpoint, Wash. (†99114) 310/H2
Leadville, Colo. 188/E3
Leadville, Colo. (80461) 208/G4
Leadwood, Mo. (63653) 261/L7
Leaf, Georgia (†30528) 217/E1
Leaf (riv.), Manitoba 179/F2
Leaf (riv.), Minn. 255/C4
Leaf, Miss. (39450) 256/G8
Leaf (riv.), Miss. 256/F8
Leaf (lake), Sask. 181/J2
Leaf River, Ill. (61047) 222/D1
Leaf Valley, Minn. (†56332) 255/C4
League City, Texas (77573) 303/K2
Leah, Georgia (†30802) 217/H3
Leake (co.), Miss. 256/E5
Leakesville, Miss. (39451) 256/G8
Leakey, Texas (78873) 303/E8
Leal, N. Dak. (58459) 282/O5
Lealui, Zambia 115/D6
Leamington, Ontario 177/B5
Leamington, Utah (84638) 304/B4
Leamington Spa, England 10/F4
Leander, La. (71445) 238/E4
Leane (lake), Ireland 17/G4
Leane (lake), Ireland 17/B7
Leapwood, Tenn. (†38310) 237/E10
Learmonth, W. Australia 88/A4
Learmonth, W. Australia 92/A3
Learned, Miss. (39093) 256/C6
Leary, Georgia (31762) 217/C8
Leasburg, Mo. (65535) 261/K6
Leasburg, N.C. (27291) 281/L2
Leask, Sask. 181/E2
Leatherhead, England 13/G8
Leatherhead, England 10/B6
Leathersville, Georgia (†30817) 217/G3
Leatherwood, Ky. (41756) 237/P6
Leavenworth, Ind. (47137) 227/E8
Leavenworth (co.), Kansas 232/G2
Leavenworth, Kans. 188/G3
Leavenworth, Kansas (66048) 232/H2
Leavenworth, Wash. (98826) 310/E3
Leavitt, Alberta 182/D5
Leavittsburg, Ohio (44430) 284/J3
Leawood, Kansas (66206) 232/H3
Łeba, Poland 47/C1
Lebak, Philippines 82/D7
Lebam, Wash. (98554) 310/B4
Lebanon 2/L4
LEBANON 59/C3
LEBANON 63/F6
Lebanon 54/E6
Lebanon, Colo. (†81323) 208/B8
Lebanon○, Conn. (06249) 210/G2
Lebanon, Georgia (30146) 217/D2
Lebanon, Ill. (62254) 222/D5
Lebanon, Ind. (46052) 227/D4
Lebanon, Kansas (66952) 232/D2
Lebanon, Ky. (40033) 237/L5
Lebanon (mts.), Lebanon 63/F6
Lebanon, Mo. (65536) 261/G7
Lebanon, Nebr. (69036) 264/D4
Lebanon, N.H. (03766) 268/C4
Lebanon, N.J. (08833) 273/D2
Lebanon, Ohio (45036) 284/B7
Lebanon, Okla. (73440) 288/N7
Lebanon, Oreg. (97355) 291/E3
Lebanon (co.), Pa. 294/K5
Lebanon, Pa. (17042) 294/K5
Lebanon, S. Dak. (57455) 298/K3
Lebanon, Tenn. (37087) 237/J8
Lebanon, Va. (24266) 307/D7
Lebanon, Wis. (53047) 317/H1
Lebanon Church, Va. (22641) 307/L2
Lebanon Junction, Ky. (40150) 237/K5
Lebanon Springs, N.Y. (12114) 276/O6
Lebeau, La. (71345) 238/F5
Lebec, Calif. (93243) 204/G9
Lebedin, U.S.S.R. 52/D4
Lebedinyy, U.S.S.R. 48/N4
Lebel-sur-Quévillon, Québec 174/B3
Lébénymiklós, Hungary 41/D3
Lebesby, Norway 18/P1
Le Blanc, France 28/D4
Le Blanc, La. (70651) 238/E5
Le Blanc-Mesnil, France 28/B1
Lebo, Kansas (66856) 232/G3
Le Borgne, Haiti 158/C5
Lębork, Poland 47/C1
Le Bourget, France 28/B1
Le Brassus, Switzerland 39/B3
Lebret, Sask. 181/H5
Lebrija (riv.), Colombia 126/D4
Lebrija, Spain 33/D4
Lebu, Chile 138/D1
Lecanto, Fla. (32661) 212/D3
Le Carbet, Martinique 161/C6
Le Cateau, France 28/E2
Lecce (prov.), Italy 34/G4
Lecce, Italy 34/G4
Lecce, Italy 7/F4
Lecco, Italy 34/B2
Le Center, Minn. (56057) 255/E6
Lech (riv.), W. Germany 22/D4
Le Châble, Switzerland 39/D4
Le Chasseral (mt.), Switzerland 39/D2
Leche (lag.), Cuba 158/F2
Le Chenit (Le Brassus), Switzerland 39/B3
Le Chesnay, France 28/A2
Lechiguanas (isls.), Argentina 143/G6
Lechuguilla (des.), Ariz. 198/A6
Le Claire, Iowa (52753) 229/N5
Leclercville, Québec 172/F3
Lecointre (lake), Québec 172/B2
Lecoma, Mo. (65540) 261/J7
Lecompte, La. (71346) 238/F4
Lecompton, Kansas (66050) 232/G2
Lecontes Mills, Pa. (16850) 294/F3
Le Creusot, France 28/F4
Le Croisic, France 28/B4
Lecta, Ky. (†42141) 237/K6
Łęczyca, Poland 47/D2
Ledang, Gunong (mt.), Malaysia 72/D7
Ledbury, England 10/E5
Ledbury, England 13/E5
Lede, Belgium 27/D7
Ledeč, Czech. 41/C2
Ledesma, Spain 33/C2
Ledford, Ill. (†62946) 222/E6
Ledge, Bermuda 156/G2
Ledger, Mont. (59456) 262/E2
Ledgewood, N.J. (07852) 273/D2
Le Diamant, Martinique 161/D7
Ledoux, N. Mex. (87725) 274/D3
Leduc, Alberta 182/D3
Leduc, Alta. 162/E5
Ledyard○, Conn. (06339) 210/G3
Ledyard, Iowa (50556) 229/E2
Lee (co.), Ala. 195/H5
Lee (co.), Ark. 202/J4
Lee (co.), Fla. 212/E5
Lee, Fla. (32059) 212/C1
Lee (co.), Georgia 217/D7
Lee (co.), Ill. 222/D2
Lee, Ill. (60530) 222/E2
Lee, Ind. (†47978) 227/D3
Lee (co.), Iowa 229/L7
Lee (riv.), Ireland 17/D8
Lee (riv.), Ireland 10/B5
Lee (co.), Ky. 237/O5
Lee○, Maine (04455) 243/G5
Lee, Mass. (01238) 249/B3
Lee○, Mass. (01238) 249/B3
Lee (co.), Miss. 256/G2
Lee, Nev. (89829) 266/F2
Lee○, N.H. (†03857) 268/F5
Lee (co.), N.C. 281/L4
Lee (co.), S.C. 296/G3
Lee (co.), Texas 303/H7
Lee (co.), Va. 307/B7
Lee Bayou, La. (†71326) 238/G3
Lee Center, Ill. (61331) 222/D2
Lee Center, N.Y. (13363) 276/K4
Leech (lake), Minn. 188/G1
Leech (lake), Minn. 255/D3
Leech, New Bruns. 170/E1
Leech (lake), Sask. 181/J4
Leechburg, Pa. (15656) 294/C4
Leech Lake Ind. Res., Minn. 255/D3
Leechville, N.C. (†27810) 281/R3
Lee City, Ky. (41342) 237/P5
Leeco, Ky. (41343) 237/O5
Leecreek, Ark. (†72934) 202/B2
Leedale, Alberta 182/C3
Leedey, Okla. (73654) 288/H3
Leeds, Ala. (35094) 195/E3
Leeds, England 13/J1
Leeds, England 7/D3
Leeds, England 10/F4
Leeds, Maine (†04263) 243/C7
Leeds○, Maine (†04263) 243/C7
Leeds, Md. (†21757) 245/P2
Leeds, N.Y. (12451) 276/N6
Leeds, N. Dak. (58346) 282/M3
Leeds (county), Ontario 177/H3
Leeds, Québec 172/F3
Leeds, S.C. (†29031) 296/E2
Leeds, Utah (84746) 304/A6
Leeds Junction, Maine (†04236) 243/C7
Leeds Point, N.J. (08220) 273/E4
Leek, England 13/H2
Leek, England 10/E4
Leek, Netherlands 27/J2
Leelanau (co.), Mich. 250/D4
Leelanau (lake), Mich. 250/D4
Leenane, Ireland 17/B4
Leeper, Mo. (†63957) 261/L8
Leeper, Pa. (16233) 294/D3
Leer, W. Germany 22/B2
Leerdam, Netherlands 27/F5
Leesburg, Ala. (35983) 195/G2
Leesburg, Fla. (32748) 212/E3
Leesburg, Georgia (31763) 217/D7
Leesburg, Ind. (46538) 227/F2
Leesburg, Miss. (†39117) 256/E6
Leesburg, N.J. (08327) 273/D5
Leesburg, Ohio (45135) 284/D7
Leesburg, Pa. (†16156) 294/B3
Leesburg, Va. (22075) 307/N2
Lees Creek, Ohio (45138) 284/C7
Leesdale, Miss. (†39661) 256/B7
Lees Ferry, Ariz. (†86036) 198/D2
Leesport, Pa. (19533) 294/K5
Lee's Summit, Mo. (64063) 261/R6
Leeston, N. Zealand 100/D5
Leesville, Conn. (†06424) 210/F2
Leesville, Ind. (†47421) 227/E7
Leesville, La. (71446) 238/D4
Leesville, Ohio (44639) 284/H5
Leesville (lake), Ohio 284/H5
Leesville, S.C. (29070) 296/E4
Leesville, Va. (†24571) 307/K6
Leesville (lake), Va. 307/K6
Leet, W. Va. (25536) 312/B6
Leetes Island, Conn. (†06437) 210/E3
Leeton, Mo. (64761) 261/E5
Leeton, N. S. Wales 88/H6
Leeton, N. S. Wales 97/D4
Leeton, Utah (†84066) 304/E3
Leetonia, Ohio (44431) 284/J4
Leetsdale, Pa. (15056) 294/B4
Leetsville, Mich. (†49659) 250/D4
Leeuwarden, Netherlands 27/H2
Leeuwin (cape), Australia 87/B9
Leeuwin (cape), Australia 2/Q7
Leeuwin (cape), W. Australia 88/A6
Leeuwin (cape), W. Australia 92/A6
Lee Vining, Calif. (93541) 204/F6
Leeward (passage), Virgin Is. (U.S.) 161/B4
Leeward (isls.), W. Indies 156/F3
Lefaivre, Ontario 177/K2
Lefka, Cyprus 63/E5
Lefkara, Cyprus 63/E5
Leflore (co.), Miss. 256/D3
Le Flore, Miss. (†38940) 256/D3
Le Flore (co.), Okla. 288/S5
Leflore, Okla. (74942) 288/S5
Lefor, N. Dak. (58641) 282/F6
Lefors, Texas (79054) 303/D2
Le François, Martinique 161/D6
Lefroy, Ontario 177/E3
Lefroy (lake), W. Australia 88/C6
Lefroy (lake), W. Australia 92/C5
Left Hand, W. Va. (25251) 312/D5
Legal, Alberta 182/D3
Legana, Tasmania 99/C3
Leganés, Spain 33/F4
Legazpi, Philippines 82/D4
Legazpi, Philippines 85/G3
Legend, Alberta 182/E5
Legend (lake), Alberta 182/D1
Léger Brook, New Bruns. 170/F2
Légerville, New Bruns. 170/F2
Legerwood, Tasmania 99/D3
Legges Tor (mt.), Tasmania 99/D3
Leggett, N.C. (†27886) 281/O3
Leghorn (prov.), Italy 34/C3
Leghorn, Italy 7/E4
Leghorn, Italy 34/C3
Legionowo, Poland 47/E2
Léglise, Belgium 27/H9
Legnago, Italy 34/C2
Legnica (prov.), Poland 47/C4
Legnica, Poland 47/C3
Le Gore, Md. (21761) 245/J2
Le Goulet, New Bruns. 170/F1
Le Grand, Calif. (95333) 204/E6
Le Grand, Iowa (50142) 229/H5
Le Grand (cape), W. Australia 88/C6
Le Grand (cape), W. Australia 92/C6
Le Gros Crêt (mt.), Switzerland 39/B3
Leguan (isl.), Guyana 131/B2
Legune, North. Terr. 93/A3
Leh, India 68/D2
Le Havre, France 7/E4
Le Havre, France 28/C3
Lehew, W. Va. (26843) 312/K4
Lehi, Utah (84043) 304/C3
Lehigh, Alberta 182/D4
Lehigh, Iowa (50557) 229/E4
Lehigh, Kansas (67073) 232/E3
Lehigh, N. Dak. (†58601) 282/E6
Lehigh, Okla. (74556) 288/O6
Lehigh (co.), Pa. 294/L4
Lehigh (riv.), Pa. 294/L3
Lehigh Acres, Fla. (33936) 212/E5
Lehighton, Pa. (18235) 294/L4
Lehman, Pa. (18627) 294/E7
Lehman Caves Nat'l Mon., Nev. 266/G4
Lehr, N. Dak. (58460) 282/M7
Lehrte, W. Germany 22/D2
Lehua (isl.), Hawaii 218/A2
Lehututu, Botswana 118/C4
Leiah, Pakistan 68/C2
Leibnitz, Austria 41/C3
Leicester, England 13/F5
Leicester, England 10/F4
Leicester○, Mass. (01524) 249/G4
Leicester, N.Y. (14481) 276/D5
Leicester○, Vt. (†05752) 268/A4
Leicester Junction, Vt. (05752) 268/A4
Leicestershire (co.), England 13/F5
Leichhardt, N. S. Wales 88/L4
Leichhardt, N.S. Wales 97/J3
Leichhardt (range), Queensland 95/C4
Leichhardt (riv.), Queensland 88/F3
Leichhardt (riv.), Queensland 95/A3
Leiden, Netherlands 27/E4
Leidy (mt.), Wyo. 319/B2
Leigh, England 10/G2
Leigh, England 13/H2
Leigh, Nebr. (68643) 264/G3
Leigh Creek, S. Australia 88/F6
Leigh Creek, S. Australia 94/F4
Leighlindridge, Ireland 17/H6
Leighton, Ala. (35646) 195/D1
Leighton, Iowa (50143) 229/H6
Leighton-Linslade, England 13/F7
Leijun, P.D.R. Yemen 59/E6
Leimebamba, Peru 128/D6
Leinan, Sask. 181/D5
Leine (riv.), W. Germany 22/C2
Leinster (prov.), Ireland 17/G5
Leinster (trad. co.), Ireland, 17
Leinster (mt.), Ireland 17/H6
Leipers Fork, Tenn. (†37064) 237/G9
Leipsic, Del. (†19901) 245/S4
Leipsic (riv.), Del. 245/R4
Leipsic, Ind. (†47452) 227/E7
Leipsic, Ohio (45856) 284/C3
Leipzig, E. Germany 7/F3
Leipzig (dist.), E. Germany 22/E3
Leipzig, E. Germany 22/E3
Leipzig, Sask. 181/C3
Leiria (dist.), Portugal 33/B3
Leiria, Portugal 33/B3
Leisler (mt.), North. Terr. 93/A7
Leiston-cum-Sizewell, England 13/J5
Leisure, Ind. (†46036) 227/F4
Leisure City, Fla. (33033) 212/F6
Leitchfield, Ky. (42754) 237/J6
Leiter, Wyo. (82837) 319/F1
Leitersburg, Md. (†21740) 245/H2
Leiters Ford, Ind. (46945) 227/E2
Leith, N. Dak. (58551) 282/G7
Leith, Ontario 177/D3
Leitrim (co.), Ireland 17/E3
Leitrim, Ireland 17/F3
Leivasy, W. Va. (26676) 312/E6
Leix (Laoighis) (co.), Ireland 17/G6
Leixlip, Ireland 17/H5
Leiyang, China 77/H6
Leizhau Bandao (pen.), China 77/G7
Lejunior, Ky. (40849) 237/P7
Lek (riv.), Netherlands 27/F5
Leka (isl.), Norway 18/G4
Le Kef (El Kef), Tunisia 106/F1
Lekitobi, Indonesia 85/G6
Lekoni, Gabon 102/D5
Lekoni, Gabon 115/B4
Leksand, Sweden 18/J6
Leksula, Indonesia 85/H6
Lela, Okla. (†74058) 288/N2
Lela, Texas (†79079) 303/D2
Le Lamentin, Martinique 161/D6
Leland, Georgia (†30059) 217/J1
Leland, Ill. (60531) 222/E2
Leland, Iowa (50453) 229/F2
Leland, Mich. (49654) 250/D3
Leland, Miss. (38756) 256/C4
Leland, N.C. (28451) 281/N6
Leland, Oreg. (†97478) 291/D5
Le Landeron, Switzerland 39/C2
Leleiwi (pt.), Hawaii 218/K5
Leleque, Argentina 143/B5
Lelia Lake, Texas (79240) 303/D3
Le Lieu, Switzerland 39/B3
Le Locle, Switzerland 39/C2
Le Lorrain, Martinique 161/D6
Le Loup, Kansas (†66092) 232/G3
Lely (mts.), Suriname 131/D3
Lelydorp, Suriname 131/D3
Lelystad, Netherlands 27/H3
Lem, Denmark 21/B5
Léman (Geneva) (lake), Switzerland 39/C4
Le Mans, France 28/C3
Le Mans, France 7/E4
Le Marin, Martinique 161/D7
Le Mars, Iowa (51031) 229/A3
Lemasters, Pa. (17231) 294/G6
Lemay, Mo. (63125) 261/R4
Lemberg, Sask. 181/H5
Leme, Brazil 135/C3
Lemelerberg (hill), Netherlands 27/J4
Lemery, Philippines 82/C4
Lemesurier (isl.), Alaska 196/M1
Lemgo, W. Germany 22/C2
Lemhi (co.), Idaho 220/D4
Lemhi, Idaho (83465) 220/E5
Lemhi (pass), Idaho 220/E5
Lemhi (range), Idaho 220/E5
Lemhi (riv.), Idaho 220/E5
Lemhi (pass), Mont. 262/C6
Lemieux (isls.), N.W. Terrs. 187/M3
Lemington○, Vt. (†03576) 268/D2
Lemitar, N. Mex. (87823) 274/B4
Lemmer, Netherlands 27/H3
Lemmon (mt.), Ariz. 198/E6
Lemmon, S. Dak. (57638) 298/E2
Lemon (lake), Ind. 227/E6
Lemon, Miss. (†39074) 256/E6
Lemoncove, Calif. (93244) 204/G7
Lemon Grove, Calif. (92045) 204/J11
Lemons, Mo. (†63565) 261/F2
Lemon Springs, N.C. (28355) 281/L4
Lemont, Ill. (60439) 222/B6
Lemont, Pa. (16851) 294/G4
Lemont Furnace, Pa. (15456) 294/C6
Le Mont-sur-Lausanne, Switzerland 39/C3
Lemoore, Calif. (93245) 204/F7
Lemoore N.A.S., Calif. 204/F7
Le Morne-Rouge, Martinique 161/C5
Le Morne-Vert, Martinique 161/C6
Lemoyne, Nebr. (69146) 264/C3
Lemoyne, Ohio (43441) 284/D3
Lemoyne, Pa. (17043) 294/J5
Le Moyne, Québec 172/J4
Le Moyne (lake), Québec 174/D1
Lempa (riv.), El Salvador 154/C4
Lempster○, N.H. (03606) 268/C5
Lemreway, Scotland 15/B2
Lemsford, Sask. 181/B5
Lemvig, Denmark 18/F8
Lemvig, Denmark 21/B4
Lena, Ill. (61048) 222/D1
Lena, La. (71447) 238/E4
Lena, Manitoba 179/C5
Lena, Miss. (39094) 256/E5
Lena, Ohio (†45317) 284/B5
Lena, S.C. (†29918) 296/E6
Lena (riv.), U.S.S.R. 4/C3
Lena (riv.), U.S.S.R. 2/R2
Lena (riv.), U.S.S.R. 54/O3
Lena (riv.), U.S.S.R. 48/N3
Lena, Wis. (54139) 317/K6
Lenapah, Okla. (74042) 288/P1
Lenawee (co.), Mich. 250/E7
Lençóis, Brazil 132/F6
Lendery, U.S.S.R. 52/D2
Lendinara, Italy 34/C2
Lenexa, Kansas (66215) 232/H2
Leney, Sask. 181/D3
Lengau, Switzerland 39/D2
Lengby, Minn. (56651) 255/C3
Lengerich, W. Germany 22/B2
Lenghu, China 77/D4
Lengshuijiang, China 77/H6
Lengua de Vaca (pt.), Chile 138/A8
Lengyeltóti, Hungary 41/D3
Lenhartsville, Pa. (19534) 294/L4
Leninabad, U.S.S.R. 48/G5
Leninakan, U.S.S.R. 7/J4
Leninakan, U.S.S.R. 48/E5
Leninakan, U.S.S.R. 52/F6
Leningrad, U.S.S.R. 7/H3
Leningrad, U.S.S.R. 2/L3
Leningrad, U.S.S.R. 48/D4
Leningrad, U.S.S.R. 52/C3
Leninogorsk, U.S.S.R. 48/J5
Leninogorsk, U.S.S.R. 52/H4
Leninsk, U.S.S.R. 54/H5
Leninsk, U.S.S.R. 48/G5
Leninsk-Kuznetskiy, U.S.S.R. 54/K4
Leninsk-Kuznetskiy, U.S.S.R. 48/J4
Leninskoye, U.S.S.R. 48/O5
Leninváros, Hungary 41/F3
Lenk, Switzerland 39/D4
Lenkoran', U.S.S.R. 48/E6
Lenkoran', U.S.S.R. 52/G7
Lennard, Manitoba 179/A4
Lennep, Mont. (†59053) 262/F4
Lenni, Pa. (19052) 294/L7
Lennig, Va. (24568) 307/L7
Lennon, Mich. (48449) 250/E5
Lennox (isl.), Argentina 143/C8
Lennox (isl.), Chile 138/F11
Lennox (isl.), Pr. Edward I. 168/E2
Lennox (hills), Scotland 15/B1
Lennox, S. Dak. (57039) 298/R7
Lennox and Addington (county), Ontario 177/G3
Lennoxtown, Scotland 15/B1
Lennoxville, Québec 172/F4
Lenoir (co.), N.C. 281/O4
Lenoir, N.C. (28645) 281/G3
Lenoir City, Tenn. (37771) 237/N9
Le Noirmont, Switzerland 39/C2
Lenora, Kansas (67645) 232/C2
Lenora, Minn. (†55922) 255/G7
Lenorah, Texas (79749) 303/B5
Lenore, Manitoba 179/A5
Lenore (lake), Sask. 181/G3
Lenore (lake), Wash. 310/F3
Lenore, W. Va. (25676) 312/B7
Lenox, Ala. (36454) 195/D8
Lenox, Georgia (31637) 217/F8
Lenox, Iowa (50851) 229/D7
Lenox, Mass. (01240) 249/A3
Lenox○, Mass. (01240) 249/A3
Lenox, Mo. (65541) 261/J7
Lenox, Tenn. (38047) 237/C8
Lenox Dale, Mass. (01242) 249/B3
Lens, Belgium 27/D7
Lens, France 28/E2
Lens, Switzerland 39/D4
Lensk, U.S.S.R. 54/N3
Lensk, U.S.S.R. 48/M3
Lenswood, Manitoba 179/B2
Lenswood, S. Australia 94/C8
Lent, Netherlands 27/H5
Lenti, Hungary 41/D3
Lentini, Italy 34/E6
Lentner, Mo. (63450) 261/H3
Lenvik, Norway 18/L2
Lenwood, Calif. (†92311) 204/H9
Lenya, Burma 72/C5
Lenzburg, Ill. (62255) 222/D5
Lenzburg, Switzerland 39/F2
Lenzing, Austria 41/B3
Leo, Ind. (46765) 227/G2
Leo, S.C. (†29560) 296/H4
Léo, Upper Volta 106/D6
Leoben, Austria 41/C3
Léogâne, Haiti 156/D3
Léogâne, Haiti 158/C6
Leola, Ark. (72084) 202/E5
Leola, S. Dak. (57456) 298/M2
Leoma, Tenn. (38468) 237/G10
Leominster, England 13/E5
Leominster, England 10/E4
Leominster, Mass. (01453) 249/G2
Leon (co.), Fla. 212/B1
Leon, Iowa (50144) 229/F7
Leon, Kansas (67074) 232/F4
León, Mexico 146/H7
León, Mexico 150/J6
Leon, N.Y. (14751) 276/C6
León, Nic. 146/K8
León, Nicaragua 154/D4
Leon, Okla. (73441) 288/M7
León (mt.), Paraguay 144/B2
León (prov.), Spain 33/C1
León, Spain 7/D4
León, Spain 33/D1
León (reg.), Spain 33/C1
Leon (co.), Texas 303/J6
Leon (riv.), Texas 303/F6
Leon, W. Va. (25123) 312/C5
Leon, Wis. (†54656) 317/E8
Leona, Kansas (66448) 232/G2
Leona, Texas (75850) 303/H6
Leonard, Mich. (48038) 250/F6
Leonard, Minn. (56652) 255/C3
Leonard, Mo. (63451) 261/H3
Leonard, N. Dak. (58052) 282/R6
Leonard, Okla. (74043) 288/P3
Leonard, Texas (75452) 303/H4
Leonardo, N.J. (07737) 273/E3
Leonardsburg, Ohio (†43015) 284/D5
Leonardsville, N.Y. (13364) 276/K5
Leonardtown, Md. (20650) 245/M7
Leonardville, Kansas (66449) 232/F2
Leonardville, New Bruns. 170/C4
Leone (mt.), Switzerland 39/F4
Leonforte, Italy 34/E6
Leongatha, Victoria 97/C6
Leonia, Fla. (†32464) 212/C5
Leonia, N.J. (07605) 273/C2
Leonidas, Mich. (49066) 250/D6
Leonidas, Minn. (†55734) 255/F3
Leonídhion, Greece 45/F7
Leonora, W. Australia 88/C5
Leonora, W. Australia 92/C5
Leonore, Ill. (61332) 222/E2
Leon Springs, Texas (†78006) 303/J10
Leontes (Litani) (riv.), Lebanon 63/F6
Leon Valley, Texas (†78201) 303/J10
Leonville, La. (70551) 238/G6
Leopold, Ind. (47551) 227/D8
Leopold, Mo. (63760) 261/N8
Leopold, W. Va. (†26411) 312/E4
Leopoldina, Brazil 135/E2
Leopoldsburg, Belgium 27/G6
Leopolis, Wis. (54948) 317/J6
Leora, Mo. (†63960) 261/M9
Leota, Minn. (56153) 255/C7
Leota, Utah (†84059) 304/E3
Leoti, Kansas (67861) 232/A3
Leoville, Kansas (†67757) 232/B2
Leoville, Sask. 181/D2
Lepanto, Ark. (72354) 202/K2
Lepel', U.S.S.R. 52/C4
Lephepe, Botswana 118/D4
Leping, China 77/J6
L'Épiphanie, Québec 172/D4
Lepontine Alps (range), Italy 34/B1
Lepontine Alps (range), Switzerland 39/G4
Le Port, Réunion 118/F5
Lepreau, New Bruns. 170/D3
Lepreau (pt.), New Bruns. 170/D3
Lepreau (riv.), New Bruns. 170/D3
Le Prêcheur, Martinique 161/C5
Leptis Magna (ruins), Libya 111/B1
Le Puy, France 28/F5
Lequille, Nova Scotia 168/C4
Lequire, Okla. (74943) 288/R4
Le Raimeux (mt.), Switzerland 39/D2
Le Raysville, Pa. (18829) 294/K2
Lerdal, Minn. (†56007) 255/E7
Lerdo de Tejada, Mexico 150/M8
Léré, Chad 111/B6
Lere, Nigeria 106/F7
Leribe, Lesotho 118/D5
Lerici, Italy 34/B2
Lérida, Colombia 126/E7
Lérida (prov.), Spain 33/G2
Lérida, Spain 7/D4
Lérida, Spain 33/G2
Lerín, Spain 33/F1
Lerma, Mexico 150/O7
Lerma, Spain 33/E2
Lerna, Ill. (62440) 222/E4
Le Robert, Martinique 161/D6
Le Roeulx, Belgium 27/E8
Lerona, W. Va. (25971) 312/D8
Léros (isl.), Greece 45/H7
Leross, Sask. 181/H4
Leroy, Ala. (36548) 195/B8
Le Roy, Ill. (61752) 222/E3
Leroy, Ind. (46355) 227/C2
Le Roy, Iowa (†50123) 229/F7
Le Roy, Kansas (66857) 232/G3
Leroy, La. (†70555) 238/F6
Le Roy, Mich. (49655) 250/D4
Le Roy, Minn. (55951) 255/F7
Le Roy, N.Y. (14482) 276/E5
Leroy, N. Dak. (58252) 282/P2
Le Roy, Pa. (17743) 294/J2
Leroy, Sask. 181/G4
Leroy, Texas (76654) 303/G6
Le Roy, W. Va. (25252) 312/C5
Leroy Anderson (res.), Calif. 204/L4
Lerwick, Scotland 10/G1
Lerwick, Scotland 15/G2
Léry, Québec 172/H4
Les Abymes, Guadeloupe 161/B6
Lesage, W. Va. (25537) 312/B5
Le Saint-Esprit, Martinique 161/D6
Les Andelys, France 28/D3
Les Anglais, Haiti 158/A6
Les Anse-d'Arlets, Martinique 161/C7
Les Becquets, Québec 172/E3
Les Bois, Switzerland 39/C2
Les Cayes, Haiti 158/B6
Les Cayes, Haiti 156/C3
Les Éboulements, Québec 172/G2
Les Épey, Switzerland 39/D4
Les Étroits, Québec 172/J2
Leshan (Loshan), China 77/F6
Leshan, China 54/M6
Leshara, Nebr. (68035) 264/H3
Les Haudères, Switzerland 39/E4
Les Hauteurs-de-Rimouski, Québec 172/J1
Leshukonskoye, U.S.S.R. 52/G2
Lesina (lake), Italy 34/E3
Les Irois, Haiti 158/A6
Lesja, Norway 18/F5
Leskovac, Yugoslavia 45/E4
Leskovik, Albania 45/E5
Leslie, Ark. (72645) 202/E2
Leslie, Georgia (31764) 217/D7
Leslie, Idaho (83255) 220/E6
Leslie (co.), Ky. 237/P6
Leslie, Mich. (49251) 250/E6
Leslie, Mo. (63056) 261/K6
Leslie, Sask. 181/H4
Leslie, Scotland 15/E4
Leslie, W. Va. (25972) 312/E6
Leslieville, Alberta 182/C3
Lesmahagow, Scotland 15/E5
Les Méchins, Québec 172/B1
Lesnoy, U.S.S.R. 52/H3
Lesosibirsk, U.S.S.R. 48/K4
Lesotho 2/L7
Lesotho 102/E7
LESOTHO 118/D5
Lesozavodsk, U.S.S.R. 48/O5
Lesparre-Médoc, France 28/C5
Les Ponts-de-Martel, Switzerland 39/C2
Les Sables-d'Olonne, France 28/B4
Lesse (riv.), Belgium 27/F8
Lessebo, Sweden 18/J8
Lessen (Lessines), Belgium 27/D7
Lesser Antilles (isls.) 146/M8
Lesser Slave (lake), Alberta 182/C2
Lesser Slave (lake), Alta. 162/E4
Lessines, Belgium 27/D7
Lessley, Miss. (†39669) 256/B8
Lesslie, S.C. (†29730) 296/E2
Les Tantes (isls.), Grenada 161/D7
Lester, Ala. (35647) 195/D1
Lester, Iowa (51242) 229/A2
Lester, Pa. (19113) 294/M7
Lester, Wash. (98035) 310/D3
Lester, W. Va. (25865) 312/D7
Lester Prairie, Minn. (55354) 255/D6
Lesterville, Mo. (63654) 261/L8
Lesterville, S. Dak. (57040) 298/O7
Lestijärvi (lake), Finland 18/O5
Lestock, Sask. 181/G4
Les Trois-Îlets, Martinique 161/D6
Le Sueur (co.), Minn. 255/E6
Le Sueur, Minn. (56058) 255/E6
Les Verrières, Switzerland 39/B3
Lésvos (isl.), Greece 7/G5
Lésvos (isl.), Greece 45/G6
Leswalt, Scotland 15/C6
Leszczyny, Poland 47/A4
Leszno (prov.), Poland 47/C3
Leszno, Poland 47/C3
L'Étape, Québec 172/F2
Letart, W. Va. (25253) 312/C5
Letart Falls, Ohio (†45771) 284/F8
Létavértes, Hungary 41/G3
Letcher (co.), Ky. 237/R6
Letcher, S. Dak. (57359) 298/N6
Letchworth, England 13/G6
Le Teil, France 28/F5
Letellier, Manitoba 179/E5
Letenye, Hungary 41/D3
Letha, Idaho (83636) 220/B6
Letham, Scotland 15/F4
Lethbridge, Alberta 182/D5
Lethbridge, Alta. 146/G4
Lethbridge, Alta. 162/E6
Lethbridge, Newf. 166/D2
Lethem, Guyana 120/D2
Lethem, Guyana 131/B4

Leti (isls.), Indonesia 85/H7
Leticia, Colombia 126/F10
Leticia, Colombia 120/B3
L'Etivaz, Switzerland 39/D4
Letka, U.S.S.R. 52/H3
Leto, Fla. (†33614) 212/C2
Letohatchee, Ala. (36047) 195/E6
Leton, La. (†71072) 238/D1
Letona, Ark. (72085) 202/G3
Letong, Indonesia 85/D5
Le Touquet-Paris-Plage, France 28/D2
Letpadan, Burma 72/C3
Le Tréport, France 28/D2
Letsôk-aw Kyun (isl.), Burma 72/C5
Lette, N.S. Wales 97/B4
Letterkenny, Ireland 10/B3
Letterkenny, Ireland 17/F2
Letterkenny Army Depot, Pa. 294/G6
Lettermullan (isl.), Ireland 17/B5
Letts, Ind. (†47240) 227/F6
Letts, Iowa (52754) 229/L6
Lettsworth, La. (70753) 238/G5
Leucadia, Calif. (92024) 204/H10
Leucate (mts.), France 28/E6
Leuchars, Scotland 15/F4
Leuk, Switzerland 39/E4
Leukerbad, Switzerland 39/E4
Leupp, Ariz. (86035) 198/E3
Leurbost, Scotland 15/B2
Leuser (mt.), Indonesia 85/B5
Leuven, Belgium 27/F7
Leuze-en-Hainaut, Belgium 27/D7
Levádhia, Greece 45/F6
Levallois-Perret, France 28/A1
Levan, Utah (84639) 304/C4
Levanger, Norway 18/G5
Levant, Kansas (67743) 232/A2
Levant○, Maine (04456) 243/F6
Levanzo (isl.), Italy 34/D5
Levasy, Mo. (64066) 261/S5
Le Vauclin, Martinique 161/D6
Levee, Ky. (†40337) 237/O5
Level, Md. (†21078) 245/O2
Level Green, Ky. (†40456) 237/N6
Level Land, S.C. (†29655) 296/C3
Levelland, Texas (79336) 303/B4
Levelock, Alaska (99625) 196/G3
Level Plains, Ala. (†36322) 195/G8
Levels, W. Va. (25431) 312/J4
Leven, Scotland 10/E2
Leven, Scotland 15/F4
Leven, Loch (inlet), Scotland 15/D4
Leven (lake), Scotland 15/E4
Leven (riv.), Tasmania 99/B3
Leveque (cape), Australia 87/C7
L'Eveque (cape), N. Zealand 100/D7
Lévêque (cape), W. Australia 88/C3
Lévêque (cape), W. Australia 92/C2
Leverburgh, Scotland 15/B3
Le Verdon-sur-Mer, France 28/C5
Leverett○, Mass. (01054) 249/E3
Levering, Mich. (49755) 250/E3
Leverkusen, W. Germany 22/B3
Levesque, New Bruns. 170/C1
Levice, Czech. 41/E2
Levick (mt.) 5/B8
Levie, France 28/B7
Le Vigan, France 28/E5
Levin, N. Zealand 100/E4
Lévis (co.), Québec 172/J3
Lévis, Québec 172/J3
Lévis, Québec 174/C3
Levisa Fork (riv.), Va. 307/C5
Levítha (isl.), Greece 45/H7
Levittown, N.Y. (11756) 276/R7
Levittown, Pa. (*19053) 294/N5
Levittown, P. Rico 161/D1
Levkás, Greece 45/E6
Levkás (isl.), Greece 45/E6
Levoča, Czech. 41/F2
Lévrier (bay), Mauritania 106/A4
Levuka, Fiji 87/H7
Levukaa, Fiji 86/Q10
Levy (co.), Fla. 212/D2
Levy (lake), Fla. 212/D2
Levy, N. Mex. (†87752) 274/E2
Lewe, Burma 72/B3
Lewellen, Nebr. (69147) 264/B3
Lewes, Del. (19958) 245/T5
Lewes, England 13/H7
Lewes, England 10/G5
Lewis, Colo. (81327) 208/B8
Lewis (isl.), Fla. 212/B3
Lewis (co.), Idaho 220/B3
Lewis, Ind. (47858) 227/C6
Lewis, Iowa (51544) 229/C6
Lewis, Kansas (67552) 232/C4
Lewis (co.), Ky. 237/P3
Lewis, Manitoba 179/F5
Lewis (lake), Manitoba 179/G2
Lewis (co.), Mo. 261/J2
Lewis, Mo. (†64735) 261/E6
Lewis (range), Mont. 262/C2
Lewis (co.), N.Y. 276/K3
Lewis, N.Y. (12950) 276/N2
Lewis (dist.), Scotland 15/B2
Lewis (dist.), Scotland 10/C1
Lewis, Butt of (prom.), Scotland 15/B2
Lewis, Butt of (prom.), Scotland 10/C1
Lewis, S.C. (†29706) 296/E2
Lewis (co.), Tenn. 237/F9
Lewis (creek), Vt. 268/A3
Lewis (co.), Wash. 310/C4
Lewis (riv.), Wash. 310/C5
Lewis (co.), W. Va. 312/E4
Lewis, Wis. (54851) 317/B4
Lewis (lake), Wyo. 319/B1
Lewis and Clark (co.), Mont. 262/D3
Lewis and Clark (lake), Nebr. 264/G2
Lewis and Clark (lake), S. Dak. 298/O8
Lewis and Clark Village, Mo. (†64484) 261/C3
Lewisberry, Pa. (17339) 294/J5
Lewisburg, Ky. (42256) 237/G6
Lewisburg, La. (†70525) 238/F6
Lewisburg, Ohio (45338) 284/A6
Lewisburg, Pa. (17837) 294/J4
Lewisburg, Tenn. (37091) 237/H10
Lewisburg, W. Va. (24901) 312/E7
Lewis Center, Ohio (43035) 284/D5
Lewis Creek, Ind. (†47234) 227/F6
Lewisetta, Va. (22505) 307/R4
Lewisham, England 10/B5
Lewisham, England 13/H8
Lewis Hill (mt.), Newf. 166/C4
Lewisport, Ky. (42351) 237/H5
Lewisporte, Newf. 166/C4
Lewis Run, Pa. (16738) 294/E2
Lewis Smith (dam), Ala. 195/D3
Lewis Smith (lake), Ala. 195/D2
Lewiston, Calif. (96052) 204/C3
Lewiston, Idaho 188/C1
Lewiston, Idaho (83501) 220/A3
Lewiston, Maine 188/N2
Lewiston, Maine (04240) 243/C7
Lewiston, Mich. (49756) 250/E4
Lewiston, Minn. (55952) 255/G7
Lewiston, Nebr. (68380) 264/H4
Lewiston, N.Y. (14092) 276/B4
Lewiston, N.C. (27849) 281/P2
Lewiston, Utah (84320) 304/C2
Lewiston, Vt. (†05055) 268/C4
Lewistown, Ill. (61542) 222/C3
Lewistown, Md. (21701) 245/J2
Lewistown, Mo. (63452) 261/J2
Lewistown, Mont. (59457) 262/G3
Lewistown, Ohio (43333) 284/C5
Lewistown, Pa. (17044) 294/G4
Lewisville, Ark. (71845) 202/C7
Lewisville, Idaho (83431) 220/F6
Lewisville, Ind. (47352) 227/G5
Lewisville, Minn. (56060) 255/D7
Lewisville, Ohio (43754) 284/H6
Lewisville (Ulysses), Pa. (16948) 294/G2
Lewisville, Pa. (19351) 294/L6
Lewisville, Texas (*75067) 303/G1
Lewisville (lake), Texas 303/G1
Lewvan, Sask. 181/H5
Lexa, Ark. (72355) 202/J4
Lexie, Miss. (†39667) 256/D8
Lexington, Ala. (35648) 195/D1
Lexington, Ark. (†72153) 202/F2
Lexington, Georgia (30648) 217/F3
Lexington, Ill. (61753) 222/E3
Lexington, Ind. (47138) 227/F7
Lexington, Ky. (*40501) 237/N4
Lexington, Ky. 146/K6
Lexington, Ky. 188/K3
Lexington○, Mass. (02173) 249/B6
Lexington, Mich. (48450) 250/G5
Lexington, Minn. (†55014) 255/G5
Lexington, Miss. (39095) 256/D4
Lexington, Mo. (64067) 261/E4
Lexington, Nebr. (68850) 264/E4
Lexington, N.Y. (12452) 276/M6
Lexington, N.C. (27292) 281/J3
Lexington, Ohio (44904) 284/E4
Lexington, Okla. (73051) 288/M4
Lexington, Oreg. (97839) 291/H2
Lexington (co.), S.C. 296/E4
Lexington, S.C. (29072) 296/E4
Lexington, Tenn. (38351) 237/E9
Lexington, Texas (78947) 303/G7
Lexington (I.C.), Va. (24450) 307/J5
Lexington Blue Grass Army Depot, Ky. 237/N5
Lexington Park, Md. (20653) 245/M7
Lexsy, Georgia (†30401) 217/H6
Leyba, N. Mex. (87542) 274/D3
Leyburn, England 13/F3
Leyden○, Mass. (†01301) 249/D2
Leye, China 77/G7
Leyland, England 13/G1
Leyland, England 10/F1
Leyond (riv.), Manitoba 179/F3
Leysin, Switzerland 39/C4
Leyte (prov.), Philippines 82/E5
Leyte (isl.), Philippines 54/O8
Leyte (gulf), Philippines 82/E5
Leyte (isl.), Philippines 85/H3
Leyte (isl.), Philippines 82/E5
Lezajsk, Poland 47/F3
Lezama, Argentina 143/H7
Lézarde (riv.), Martinique 161/D6
Lezhë, Albania 45/D5
Lézignan-Corbières, France 28/E6
Lezuza, Spain 33/E3
L'gov, U.S.S.R. 52/E4
Lhanbryde, Scotland 15/E3
Lhari, China 77/D5
Lhasa, China 77/D6
Lhasa, China 2/P4
Lhasa, China 54/L7
Lhazê (Lhatse), China 77/C6
Lhazhong, China 77/C5
Lhokseumawe, Indonesia 85/B4
Lhorong, China 77/E5
Lhozhag, China 77/D6
Lhünzê, China 77/D6
Lhünzhub, China 77/D5
Liancheng, China 77/J6
Lianga, Philippines 82/E6
Lianga (bay), Philippines 82/F6
Lianping, China 77/H7
Lian Xian, China 77/H7
Lianyungang (Lienyünkang), China 77/J5
Lianyunggang, China 54/N6
Liao (riv.), China 54/O5
Liaodong Bandao (pen.), China 77/K3
Liao He (riv.), China 77/K3
Liaoning (prov.), China 77/K3
Liaoyang, China 77/K3
Liaoyuan, China 77/K3
Liard (riv.) 162/D3
Liard (riv.), Br. Col. 184/L2
Liard (riv.), Canada 146/F3
Liard (riv.), N.W. Terrs. 187/F4
Liard (riv.), Yukon 187/E3
Liard River, Br. Col. 184/L2
Libáň, Czech. 41/C1
Líbano, Colombia 126/C5
Libau, Manitoba 179/F4
Libby, Minn. (†55760) 255/E4
Libby, Mont. (59923) 262/A2
Libenge, Zaire 115/C3
Liberal, Kansas (67901) 232/B4
Liberal, Mo. (64762) 261/D7
Liberal, Oreg. (†97042) 291/B3
Liberdade, Brazil 135/D3
Liberec, Czech. 41/C1
Liberia 2/J5
Liberia 102/B4
LIBERIA 106/C7
Liberia, C. Rica 154/E5
Liberta, Ant. & Bar. 161/E11
Libertad, Belize 154/C1
Libertad, Uruguay 145/C5
Libertad, Barinas, Venezuela 124/D3
Libertad, Cojedes, Venezuela 124/D3
Liberty, Ariz. (†85326) 198/C5
Liberty (co.), Fla. 212/B1
Liberty (co.), Georgia 217/J7
Liberty, Ill. (62347) 222/B4
Liberty, Ind. (47353) 227/H5
Liberty, Kansas (67351) 232/G4
Liberty, Ky. (42539) 237/M6
Liberty, Maine (04949) 243/E7
Liberty○, Maine (04949) 243/E7
Liberty (lake), Md. 245/L3
Liberty, Miss. (39645) 256/C8
Liberty, Mo. (64068) 261/R5
Liberty (co.), Mont. 262/E2
Liberty, Nebr. (68381) 264/H4
Liberty (mt.), N.H. 268/D3
Liberty, N.Y. (12754) 276/L7
Liberty, N.C. (27298) 281/K3
Liberty, Pa. (16930) 294/H2
Liberty, Pa. (†15100) 294/C7
Liberty, Sask. 181/F4
Liberty, S.C. (29657) 296/B2
Liberty, Tenn. (37095) 237/K8
Liberty (co.), Texas 303/K7
Liberty, Texas (77575) 303/K7
Liberty, Wash. (†98922) 310/E3
Liberty, W. Va. (25124) 312/C5
Liberty Center, Ind. (46766) 227/G3
Liberty Center, Iowa (50145) 229/F6
Liberty Center, Ohio (43532) 284/B3
Liberty Corner, N.J. (07938) 273/D2
Liberty Grove, Md. (†21918) 245/O2
Liberty Hill, Conn. (†06249) 210/G2
Liberty Hill, La. (†71008) 238/E2
Liberty Hill, S.C. (29074) 296/F3
Liberty Lake, Wash. (99019) 310/J3
Liberty Mills, Ind. (46946) 227/F2
Liberty Pole, Wis. (†54665) 317/D8
Libertytown, Md. (21762) 245/J3
Libertyville, Ala. (†36420) 195/F8
Libertyville, Ill. (60048) 222/B4
Libertyville, Iowa (52567) 229/K7
Libiaz, Poland 47/D3
Libin, Belgium 27/G9
Libochovice, Czech. 41/B1
Libon, Philippines 82/D4
Libong, Ko (isl.), Thailand 72/C6
Libourne, France 28/C5
Libramont-Chevigny, Belgium 27/G9
Library, Pa. (15129) 294/B7
Libres, Mexico 150/O1
Libreville (cap.), Gabon 2/K6
Libreville (cap.), Gabon 115/A3
Libreville (cap.), Gabon 102/C4
Libuse, La. (71348) 238/F4
Libya 2/K4
Libya 102/D2
LIBYA 111/B2
Libyan (des.) 102/E2
Libyan (des.), Egypt 111/E2
Libyan (plat.), Egypt 111/E1
Libyan (des.), Libya 111/D2
Libyan (plat.), Libya 111/D1
Libyan (des.), Sudan 111/E3
Licancábur, Cerro (mt.), Chile 138/B4
Licantén, Chile 138/A10
Licata, Italy 34/D6
Lice, Turkey 63/J3
Lichfield, England 13/F5
Lichfield, England 10/G2
Lichinga, Mozambique 102/F6
Lichinga, Mozambique 118/F2
Lichtenberg, E. Germany 22/F4
Lichtenfels, W. Germany 22/D3
Lichtenrade, W. Germany 22/F4
Lichterfelde, W. Germany 22/E4
Lichtervelde, Belgium 27/C6
Lick (creek), Tenn. 237/R8
Lick Creek, Ill. (†62912) 222/D6
Licking, North Fork (riv.), Ky. 237/O3
Licking, South Fork (riv.), Ky. 237/N3
Licking (riv.), Ky. 237/N3
Licking, Mo. (65542) 261/J8
Licking (co.), Ohio 284/F5
Licking (riv.), Ohio 284/F5
Licking (creek), Pa. 294/F6
Licosa (cape), Italy 34/E4
Lida, Ky. (†40741) 237/O6
Lida (lake), Minn. 255/C4
Lida, U.S.S.R. 52/C4
Lidcombe, N.S. Wales 97/J3
Liddel Water (riv.), Scotland 15/F5
Lidderdale, Iowa (51452) 229/D4
Liddes, Switzerland 39/D5
Liddon (gulf), N.W. Terrs. 187/G4
Lidgerwood, N. Dak. (58053) 282/R7
Lidice, Czech. 41/C1
Lidingö, Sweden 18/H1
Lidköping, Sweden 18/H7
Lido di Ostia, Italy 34/F7
Lido di Venezia, Italy 34/D2
Lidzbark, Poland 47/D2
Lidzbark Warmiński, Poland 47/E1
Liebenthal, Kansas (67553) 232/C3
Liebenthal, Sask. 181/B5
Liechtensiteig, Switzerland 39/H2
Liechtenstein 7/F4
LIECHTENSTEIN 39/J2
Liedekerke, Belgium 27/D7
Liége (riv.), Alberta 182/D1
Liège (prov.), Belgium 27/H7
Liège, Belgium 7/E3
Liège, Belgium 27/H7
Liegnitz (Legnica), Poland 47/C3
Lieksa, Finland 18/R5
Lienyünkang (Lianyungang), China 77/J5
Lienz, Austria 41/B3
Liepāja, U.S.S.R. 7/F3
Liepāja, U.S.S.R. 53/A2
Liepāja, U.S.S.R. 48/B4
Liepāja, U.S.S.R. 52/B3
Lier, Belgium 27/F6
Lierneux, Belgium 27/H8
Lierre (Lier), Belgium 27/F6
Liestal, Switzerland 39/E2
Liévin, France 28/E2
Lièvre (riv.), Québec 172/B4
Lièvres (isl.), Québec 172/H2
Liezen, Austria 41/C3
Liffey (riv.), Ireland 17/H5
Liffey (riv.), Ireland 10/C4
Lifford, Ireland 10/C3
Lifford, Ireland 17/F2
Lifu (isl.), New Caled. 87/G8
Lifu (isl.), New Caled. 86/H4
Ligao, Philippines 82/D4
Ligatne, U.S.S.R. 53/C2
Liggett, Ky. (†40831) 237/P7
Lightfoot, Va. (23090) 307/P6
Lighthouse (pt.), Fla. 212/B2
Light House (pt.), Mich. 250/D3
Lighthouse Point, Fla. (33064) 212/F5
Lightning (creek), Oreg. 291/L2
Lightning (creek), Wyo. 319/G2
Lightning Ridge, N.S. Wales 97/E1
Lightsville, Ohio (†45362) 284/A5
Lignite, N. Dak. (58752) 282/F2
Ligon, Ky. (41646) 237/R6
Ligonha (riv.), Mozambique 118/F3
Ligonier, Ind. (46767) 227/F2
Ligonier, Pa. (15658) 294/D5
Liguria (reg.), Italy 34/B2
Ligurian (sea), Italy 34/B3
Lihir Group (isls.), Papua N.G. 86/C1
Lihou (cays), Coral Sea Is. Terr. 88/J3
Lihue, Hawaii (96766) 218/C2
Lihue, Hawaii 188/E5
Lihula, U.S.S.R. 53/C1
Lijiang, China 77/F6
Likasi, Panda-, Zaire 115/E6
Likati, Zaire 115/D3
Likely, Br. Col. 184/G4
Likely, Calif. (96116) 204/E2
Likhoslavl', U.S.S.R. 52/E3
Likouala (riv.), Congo 115/C3
Lila (lake), N.Y. 276/L2
Lilac, Sask. 181/D3
Lilbourn, Mo. (63862) 261/N9
Lilburn, Georgia (30247) 217/D3
Lileah, Tasmania 99/B2
L'Île-Rousse, France 28/B6
Liles (pt.), Chile 138/F2
Lilesville, N.C. (28091) 281/K5
Lilienfeld, Austria 41/C3
Lille, France 7/E3
Lille, France 28/E2
Lille, Maine (04749) 243/G1
Lilleå (riv.), Denmark 21/B5
Lille Baelt (chan.), Denmark 21/C7
Lillehammer, Norway 18/F6
Lillesand, Norway 18/F7
Lillestrøm, Norway 18/E3
Lillian, Ala. (36549) 195/D10
Lillie, La. (71256) 238/E1
Lilliesleaf, Scotland 15/F5
Lillington, N.C. (27546) 281/M4
Lillinonah (lake), Conn. 210/B3
Lilliwaup, Wash. (98555) 310/B3
Lillo, Spain 33/E3
Lillooet, Br. Col. 162/D5
Lillooet, Br. Col. 184/G5
Lillooet (riv.), Br. Col. 184/F5
Lilly, Georgia (31051) 217/E6
Lilly, Ill. (†61755) 222/D3
Lilly, Pa. (15938) 294/E5
Lilly Chapel, Ohio (†43162) 284/D6
Lillydale, Victoria 97/J4
Lilongwe (cap.), Malawi 2/L6
Lilongwe (cap.), Malawi 102/F6
Lilongwe (cap.), Malawi 115/F6
Liloy, Philippines 82/D6
Lily, Ky. (40740) 237/N6
Lily, S. Dak. (57250) 298/O3
Lily, Wis. (54445) 317/J5
Lilydale, Minn. (†55050) 255/G5
Lily Dale, N.Y. (14752) 276/B6
Lilydale, Tasmania 99/D3
Lily Plain, Sask. 181/E2
Lim (fjord), Denmark 18/E8
Lim (riv.), Yugoslavia 45/D4
Lima, Ill. (62348) 222/B3
Lima (isls.), Indonesia 85/F7
Lima, Pulau (isl.), Malaysia 72/F6
Lima, Mont. (59739) 262/D6
Lima (res.), Mont. 262/D6
Lima, N.Y. (14485) 276/E5
Lima, Ohio (*45801) 284/B4
Lima, Ohio 188/K2
Lima (New Lima), Okla. (†74858) 288/D4
Lima, Paraguay 144/D3
Lima, Pa. (†19037) 294/L7
Lima (dept.), Peru 128/D8
Lima (cap.), Peru 2/F6
Lima (cap.), Peru 120/B4
Lima (cap.), Peru 128/D8
Lima (riv.), Portugal 33/B2
Lima (pt.), P. Rico 161/F2
Lima Center, Wis. (†53190) 317/J10
Limache, Chile 138/F2
Lima Duarte, Brazil 135/E2
Limal, Bolivia 136/C8
Limanowa, Poland 47/E4
Limarí (riv.), Chile 138/A8
Limasawa (isl.), Philippines 82/E6
Limassol, Cyprus 59/B3
Limassol, Cyprus 63/E5
Limavady (dist.), N. Ireland 17/H1
Limavady, N. Ireland 10/C3
Limavady, N. Ireland 17/H1
Limaville, Ohio (44640) 284/H4
Limay (riv.), Argentina 120/C6
Limay (riv.), Argentina 143/C4
Limbach-Oberfrohna, E. Germany 22/E3
Limbani, Peru 128/H10
Limbaži, U.S.S.R. 53/C2
Limbé, Haiti 158/C5
Limbourg, Belgium 27/J7
Limbunya, North. Terr. 93/B4
Limburg (prov.), Belgium 27/G7
Limburg (Limbourg), Belgium 27/J7
Limburg (prov.), Netherlands 27/H6
Limburg an der Lahn, W. Germany 22/C3
Lime, Oreg. (†97907) 291/K3
Limedsforsen, Sweden 18/H6
Limeira, Brazil 132/E8
Limeira, Brazil 135/C3
Lime Kiln, Md. (†21701) 245/J3
Limekilns, Scotland 15/D1
Limenária, Greece 45/G5
Limerick (co.), Ireland 17/D7
Limerick, Ireland 17/D6
Limerick, Ireland 10/B4
Limerick, Ireland 7/D3
Limerick○, Maine (04048) 243/B8
Limerick, Sask. 181/E6
Limeridge, Wis. (53942) 317/F9
Lime Rock, Conn. (†06039) 210/B1
Lime Springs, Iowa (52155) 229/J2
Limestone (co.), Ala. 195/E1
Limestone, Ark. (72628) 202/D2
Limestone, Fla. (†33865) 212/E4
Limestone, Maine (04750) 243/H2
Limestone○, Maine (04750) 243/H2
Limestone, Mont. (†59028) 262/F5
Limestone, N.Y. (14753) 276/C6
Limestone, Tenn. (37681) 237/R8
Limestone (co.), Texas 303/H6
Lime Village, Alaska (†99673) 196/G2
Limfjorden (fjord), Denmark 21/D4
Limfjorden (fjord), Denmark 21/A4
Limington, Maine (04049) 243/B8
Limington○, Maine (04049) 243/B8
Limmat (riv.), Switzerland 39/F2
Limmen (bight), North. Terr. 88/F2
Limmen (bight), North. Terr. 93/D3
Limmen Bight (riv.), North. Terr. 88/F3
Limmen Bight (riv.), North. Terr. 93/D4
Límni, Greece 45/F6
Límnos (isl.), Greece 45/G6
Limoeiro, Brazil 132/H4
Limoeiro do Norte, Brazil 132/G4
Limoges, France 7/E4
Limoges, France 28/D5
Limoges, Ontario 177/J2
Limon, Colo. (80828) 208/M4
Limón, C. Rica 146/K8
Limón, C. Rica 154/F6
Limón, Honduras 154/E3
Limonade, Haiti 158/C5
Limonar, Cuba 158/D1
Limoquije, Bolivia 136/C4
Limousin (trad. prov.), France, 29
Limousin (reg.), France 28/D5
Limoux, France 28/E6
Limpio, Paraguay 144/B4
Limpopo (riv.) 102/E7
Limpopo (riv.), Botswana 118/D4
Limpopo (riv.), Mozambique 118/E4
Limpopo (riv.), S. Africa 118/D4
Lim Rock, Ala. (†35776) 195/F1
Linapacan (isl.), Philippines 82/B5
Linapacan (str.), Philippines 82/B5
Linard (peak), Switzerland 39/K3
Linares, Chile 138/A11
Linares, Chile 120/B6
Linares, Mexico 150/K4
Linares, Spain 33/E3
Linares, Spain 7/D5
Linaria, Alberta 182/C2
Lincang, China 77/E7
Linch, Wyo. (82640) 319/F2
Lincklaen, N.Y. (†13052) 276/J5
Lincoln (sea) 146/M1
Lincoln (sea) 4/A12
Lincoln, Ala. (35096) 195/F3
Lincoln, Argentina 143/F7
Lincoln (co.), Ark. 202/G6
Lincoln, Ark. (72744) 202/B2
Lincoln, Calif. (95648) 204/B8
Lincoln (isl.), China 85/E2
Lincoln (co.), Colo. 208/M5
Lincoln (mt.), Colo. 208/G4
Lincoln, Del. (19960) 245/S5
Lincoln, England 13/G4
Lincoln, England 10/F4
Lincoln (co.), Georgia 217/H3
Lincoln (co.), Idaho 220/D6
Lincoln, Ill. (62656) 222/D3
Lincoln, Ind. (†46994) 227/E3
Lincoln, Iowa (50652) 229/H4
Lincoln (co.), Kansas 232/D2
Lincoln, Kansas (67455) 232/D2
Lincoln (co.), Ky. 237/M6
Lincoln (par.), La. 238/E1
Lincoln (co.), Maine 243/D7
Lincoln, Maine (04457) 243/G5
Lincoln○, Maine (04457) 243/G5
Lincoln○, Mass. (01773) 249/B6
Lincoln, Mich. (48742) 250/F4
Lincoln (co.), Minn. 255/B6
Lincoln (co.), Miss. 256/D8
Lincoln (co.), Mo. 261/L4
Lincoln, Mo. (65338) 261/F6
Lincoln (co.), Mont. 262/A2
Lincoln, Mont. (59639) 262/D4
Lincoln (co.), Nebr. 264/D4
Lincoln (cap.), Nebr. 146/J5
Lincoln (cap.), Nebr. 188/G2
Lincoln (cap.), Nebr. (*68501) 264/H4
Lincoln (co.), Nev. 266/F5
Lincoln○, N.H. (03251) 268/D3
Lincoln (mt.), N.H. 268/D3
Lincoln (co.), N. Mex. 274/D5
Lincoln, N. Mex. (88338) 274/D5
Lincoln (co.), N.C. 281/G3
Lincoln, N. Dak. (†58501) 282/J6
Lincoln (sea), N.W. Terrs. 187/M1
Lincoln (co.), Okla. 288/N3
Lincoln, Ontario 177/E4
Lincoln (co.), Oreg. 291/D3
Lincoln, Pa. (†15037) 294/C7
Lincoln (co.), S. Dak. 298/R7
Lincoln (co.), Tenn. 237/H10
Lincoln, Texas (78948) 303/H7
Lincoln (creek), Utah 30r/C2
Lincoln○, Vt. (†05443) 268/B3
Lincoln (co.), Wash. 310/G3
Lincoln, Wash. (99147) 310/G3
Lincoln (co.), W. Va. 312/B6
Lincoln (co.), Wis. 317/G5
Lincoln (co.), Wyo. 319/B3
Lincoln Beach, Oreg. (†97341) 291/C3
Lincoln Boyhood Nat'l Mem., Ind. 227/C8
Lincoln Center, Maine (04458) 243/G5
Lincoln Center, Mass. (01773) 249/B6
Lincoln City, Ind. (47552) 227/C8
Lincoln City, Oreg. (97367) 291/C3
Lincoln Gap (pass), Vt. 268/B3
Lincoln Heights, Ohio (†45201) 284/C9
Lincolnia, Va. (†22313) 307/S3
Lincoln Park, Colo. (†81212) 208/J6
Lincoln Park, Georgia (†30286) 217/D5
Lincoln Park, Mich. (48146) 250/B7
Lincoln Park, N.J. (07035) 273/A1
Lincolnshire (co.), England 13/G4
Lincolnshire, Ill. (†60015) 222/B5
Lincolnton, Georgia (30817) 217/G3
Lincolnton, N.C. (28092) 281/G4
Lincoln University, Pa. (19352) 294/L6
Lincolnville, Ind. (†46992) 227/F3
Lincolnville, Kansas (66858) 232/F3
Lincolnville, Maine (04849) 243/E7
Lincolnville○, Maine (04849) 243/E7
Lincolnville, Nova Scotia 168/G3
Lincolnville, S.C. (†29483) 296/G6
Lincolnville Center, Maine (04850) 243/E7
Lincoln Wolds (hills), England 13/G4
Lincolnwood, Ill. (†60645) 222/B5
Lincroft, N.J. (07738) 273/E3
L'Incudine (mt.), France 28/B7
Lind, Wash. (99341) 310/G4
Linda, Calif. (†95901) 204/D4
Lindale, Alberta 182/C3
Lindale, Georgia (30147) 217/B2
Lindale, Texas (75771) 303/J5
Lindau, W. Germany 22/C5
Lindberg, Alberta 182/E3
Linden, Ala. (36748) 195/C6
Linden, Alberta 182/D4
Linden, Ariz. (†85901) 198/E4
Linden, Calif. (95236) 204/D5
Linden, Guyana 131/L2
Linden, Ind. (47955) 227/D4
Linden, Iowa (50146) 229/E5
Linden, Mich. (48451) 250/F6
Linden, N.J. (07036) 273/A3
Linden, N.C. (28356) 281/M4
Linden (mts.), Switzerland 39/F2
Linden, Tenn. (37096) 237/F9
Linden, Texas (75563) 303/K4
Linden, Va. (22642) 307/M3
Linden, W. Va. (25256) 312/D5
Linden, Wis. (53553) 317/F10
Linden Beach, Ontario 177/B6
Lindenhurst, Ill. (†60046) 222/B4
Lindenhurst, N.Y. (11757) 276/O9
Lindenwold, N.J. (08021) 273/B4
Lindenwood, Ill. (61049) 222/D1
Lindesberg, Sweden 18/J7
Lindesnes (cape), Norway 7/E3
Lindesnes (cape), Norway 18/E8
Lindi (reg.), Tanzania 115/G5
Lindi, Tanzania 102/F5
Lindi, Tanzania 115/G5
Lindi (riv.), Zaire 115/E3
Lindisfarne (Holy) (dist.), England 13/F2
Lindisfarne (Holy) (isl.), England 10/F3
Lindley, N.Y. (14858) 276/F6
Lindon, Colo. (80740) 208/N3
Lindon, Utah (†84062) 304/C3
Líndos, Greece 45/J7
Lindrith, N. Mex. (87029) 274/C2
Lindsay, Calif. (93247) 204/F7
Lindsay, La. (†70748) 238/H5
Lindsay, Mont. (59339) 262/L3
Lindsay, Nebr. (68644) 264/G3
Lindsay, Okla. (73052) 288/L5
Lindsay, Ontario 177/F3
Lindsborg, Kansas (67456) 232/E3
Lindsey, Ohio (43442) 284/D3
Lindsey, Wis. (†54449) 317/F6
Lindside, W. Va. (24951) 312/E8
Lindstrom, Minn. (55045) 255/F5
Line (isls.) 2/B6
Line (isls.), Pacific 87/K5
Lineboro, Md. (21088) 245/L2
Linesville, Pa. (16424) 294/A2
Lineville, Ala. (36266) 195/G4
Lineville, Iowa (50147) 229/G7
Linfen, China 77/H4
Linfield, Pa. (19468) 294/L5
Linganore (creek), Md. 245/J3
Lingao, China 77/G8
Lingayen, Philippines 85/F2
Lingayen, Philippines 82/C2
Lingayen (gulf), Philippines 82/C2
Lingen, W. Germany 22/B2
Lingga (arch.), Indonesia 85/D5
Lingga (isl.), Indonesia 85/D6
Lingle, Wyo. (82223) 319/H3
Linglestown, Pa. (17112) 294/J5
Lingling, China 77/H6
Lingo, N. Mex. (88123) 274/F5
Lingqui, China 77/H4
Lingshan, China 77/G7

Lingshui, China 77/H8
Linguère, Senegal 106/B5
Lingwu, China 77/G4
Linhai, China 77/K6
Linhares, Brazil 132/F7
Linhe, China 77/G3
Linière, Québec 172/G3
Linkebeek, Belgium 27/C10
Linköping, Sweden 18/K7
Linköping, Sweden 7/F3
Linkou, China 77/M2
Linkwood, Md. (21835) 245/P6
Linlithgow, Scotland 10/B1
Linlithgow, Scotland 15/C1
Linn (co.), Iowa 229/K4
Linn (co.), Kansas 232/H3
Linn, Kansas (66953) 232/E2
Linn (co.), Mo. 261/F3
Linn, Mo. (65051) 261/J5
Linn (co.), Oreg. 291/E3
Linn, Texas (78563) 303/F11
Linn, W. Va. (26384) 312/E4
Linn Creek, Mo. (65052) 261/G6
Linndale, Ohio (†44101) 284/G9
Linneus○, Maine (†04730) 243/H3
Linneus, Mo. (64653) 261/F3
Linn Grove, Ind. (46769) 227/H3
Linn Grove, Iowa (51033) 229/C3
Linnhe, Loch (inlet), Scotland 10/D2
Linnhe, Loch (inlet), Scotland 15/C4
Linnsburg, Ind. (†47933) 227/D5
Linntown, Pa. (†17837) 294/J4
Lino Lakes, Minn. (†55038) 255/G5
Linosa (isl.), Italy 34/D7
Linqing (Lintsing), China 77/J4
Lins, Brazil 132/D8
Lins, Brazil 135/B2
Linstead, Jamaica 158/J6
Linter, Belgium 27/G7
Linth (riv.), Switzerland 39/G3
Linthal, Switzerland 39/H3
Linthicum Heights, Md. (21090) 245/M4
Lintlaw, Sask. 181/H3
Linton, Georgia (†31087) 217/F4
Linton, Ind. (47441) 227/C6
Linton, Ky. (†42211) 237/F7
Linton, N. Dak. (58552) 282/K7
Linville, La. (71257) 238/F1
Linville, N.C. (28646) 281/F2
Linville, Va. (22834) 307/L3
Linville Falls, N.C. (28647) 281/F3
Linwood, Georgia (†30728) 217/B1
Linwood, Ind. (†46001) 227/F4
Linwood, Kansas (66052) 232/G2
Linwood, Ky. (†42765) 237/K6
Linwood, Md. (21764) 245/K2
Linwood, Mass. (01525) 249/H4
Linwood, Mich. (48634) 250/F5
Linwood, Nebr. (68036) 264/H3
Linwood, N.J. (08221) 273/D5
Linwood, N.C. (27299) 281/J3
Linwood, Ontario 177/D4
Linwood, Pa. (19061) 294/L7
Linwood, Scotland 15/B2
Linxi, China 77/J3
Linxia (Linsia), China 77/F4
Linyi, China 77/J4
Linz, Austria 7/F4
Linz, Austria 41/C2
Linze, China 77/F4
Linzee (cape), Nova Scotia 168/G2
Lionel, Scotland 15/B2
Lionel Town, Jamaica 158/J7
Lions (gulf) 7/E4
Lions (gulf), France 28/F6
Lion's Bay, Br. Col. 184/K3
Lion's Head, Ontario 177/C2
Lipa, Philippines 82/C4
Lipan, Texas (76462) 303/F5
Lipari, Italy 34/E5
Lipari (isl.), Italy 34/E5
Lipari (isls.), Italy 34/E5
Lipetsk, U.S.S.R. 7/H3
Lipetsk, U.S.S.R. 48/E4
Lipetsk, U.S.S.R. 52/E4
Lípez, Cordillera de (range), Bolivia 136/B8
Liping, China 77/G6
Lipník nad Bečvou, Czech. 41/D2
Lipno (res.), Czech. 41/C2
Lipno, Poland 47/D2
Lipoa (pt.), Hawaii 218/H1
Lipova, Romania 45/E2
Lippe (riv.), W. Germany 22/C3
Lippstadt, W. Germany 22/C3
Lipscomb, Ala. (35020) 195/E4
Lipscomb (co.), Texas 303/D1
Lipscomb, Texas (79056) 303/D1
Lipton, Sask. 181/H5
Liptovský Hrádok, Czech. 41/E2
Liptovský Mikuláš, Czech. 41/E2
Lipu, China 77/H7
Lira, Uganda 115/F3
Lircay, Peru 128/E9
Liri (riv.), Italy 34/D4
Liria, Spain 33/F3
Lisala, Zaire 115/D3
Lisbellaw, N. Ireland 17/K2
Lisbon○, Conn. (06351) 210/G2
Lisbon, Ill. (†60541) 222/E2
Lisbon, Ind. (†46755) 227/G2
Lisbon, Iowa (52253) 229/L5
Lisbon, La. (71048) 238/E1
Lisbon○, Maine (04250) 243/C7
Lisbon, Md. (21765) 245/K3
Lisbon, Mo. (†65254) 261/G4
Lisbon, N.H. (03585) 268/D3
Lisbon○, N.H. (03585) 268/D3
Lisbon, N.Y. (13658) 276/K1
Lisbon, N. Dak. (58054) 282/P7
Lisbon, Ohio (44432) 284/J4
Lisbon (dist.), Portugal 33/A1
Lisbon (cap.), Portugal 2/J4
Lisbon (cap.), Portugal 7/D5
Lisbon (Lisboa) (cap.), Portugal 33/A1
Lisbon Falls, Maine (04252) 243/C7
Lisbon-Lisbon Center, Maine (04250) 243/C7
Lisburn, Alberta 182/C3
Lisburn (dist.), N. Ireland 17/J2
Lisburn, N. Ireland 10/D3
Lisburn, N. Ireland 17/J2
Lisburne (cape), Alaska 196/E1
Lisburne (pen.), Alaska 196/E1
Liscannor (bay), Ireland 17/B6
Liscarroll, Ireland 17/D7
Lisco, Nebr. (69148) 264/B3
Liscomb, Iowa (50148) 229/H4
Liscomb (isl.), Nova Scotia 168/G4
Lisdoonvarna, Ireland 17/C5
Lishi, China 77/H4
Lishui, China 77/K6
Lisianski (isl.), Hawaii 188/E6
Lisianski (isl.), Hawaii 87/J3
Lisianski (isl.), Hawaii 218/B5
Lisichansk, U.S.S.R. 52/E5
Lisieux, France 28/D3
Lisieux, Sask. 181/E6
Liskeard, England 13/C7
Liskeard, England 10/D5
Lisle, Ill. (60532) 222/A6
Lisle, N.Y. (13797) 276/H6
Lisle, Ontario 177/E3
L'Islet (co.), Québec 172/G2
L'Islet, Québec 172/G2
L'Islet-sur-Mer, Québec 172/G2
L'Isle-Verte, Québec 172/G1
Lisman, Ala. (36912) 195/B6
Lisman, Ky. (†42404) 237/F6
Lismore, Australia 87/F8
Lismore, Ireland 10/B4
Lismore, Ireland 17/F7
Lismore, La. (†71343) 238/G3
Lismore, Minn. (56155) 255/B7
Lismore, N. S. Wales 88/J5
Lismore, N.S. Wales 97/G1
Lismore (isl.), Scotland 15/C4
Lisnaskea, N. Ireland 17/G3
Lišov, Czech. 41/C2
Lispeszentadorján, Hungary 41/D3
Lisse, Netherlands 27/F4
Lista (pen.), Norway 18/E7
Lister (mt.) 5/B8
Listie, Pa. (15549) 294/D5
Listowel, Ireland 17/C7
Listowel, Ireland 10/B4
Listowel, Ontario 177/D4
Litang, China 77/F6
Litani (riv.), Fr. Guiana 131/D4
Litani (riv.), Lebanon 63/F6
Litani (riv.), Suriname 131/D4
Litchfield (co.), Conn. 210/B1
Litchfield, Conn. (06759) 210/C2
Litchfield○, Conn. (06759) 210/C2
Litchfield, Ill. (62056) 222/D4
Litchfield○, Maine (04350) 243/D7
Litchfield, Mich. (49252) 250/E6
Litchfield, Minn. (55355) 255/D5
Litchfield, Nebr. (68852) 264/E3
Litchfield○, N.H. (†03051) 268/E6
Litchfield, North. Terr. 93/B2
Litchfield, Ohio (44253) 284/F3
Litchfield Park, Ariz. (85340) 198/C5
Litchville, N. Dak. (58461) 282/O6
Lith, Netherlands 27/G5
Litherland, England 13/G2
Litherland, England 10/F2
Lithgow, Australia 87/F9
Lithgow, N. S. Wales 88/J6
Lithgow, N.S. Wales 97/F3
Lithia, Va. (24066) 307/J6
Lithia Springs, Georgia (30057) 217/C3
Lithium, Mo. (†63775) 261/N7
Lithonia, Georgia (30058) 217/D3
Lithopolis, Ohio (43136) 284/E6
LITHUANIA 53/B3
Lithuanian S.S.R., U.S.S.R. 7/G3
Lithuanian S.S.R., U.S.S.R. 48/C4
Lithuanian S.S.R., U.S.S.R. 52/B3
Lititz, Pa. (17543) 294/K5
Litókhoron, Greece 45/F5
Litoměřice, Czech. 41/C1
Litomyšl, Czech. 41/D2
Litovel, Czech. 41/D2
Littau, Switzerland 39/F2
Littcarr, Ky. (41834) 237/R6
Little (riv.), Ala. 195/G2
Little (riv.), Ala. 195/C8
Little (riv.), Ark. 202/B6
Little (riv.), Conn. 210/G2
Little (riv.), Conn. 210/H1
Little (riv.), Ind. 227/G3
Little (riv.), La. 238/F3
Little (riv.), Mass. 249/C4
Little (riv.), New Bruns. 170/D2
Little (riv.), N.C. 281/N3
Little (riv.), N.C. 281/L4
Little (riv.), Okla. 288/R6
Little (riv.), Oreg. 291/E4
Little (riv.), S.C. 296/D3
Little (riv.), S.C. 296/C3
Little (riv.), Vt. 268/B3
Little (inlet), Va. 307/S6
Little (riv.), Va. 307/N5
Little (riv.), Va. 307/H7
Little Alföld (plain), Hungary 41/D3
Little America, Ant. 2/B10
Little America 5/B10
Little America, Wyo. (82929) 319/C4
Little Andaman (isl.), India 68/G6
Little Arkansas (riv.), Kansas 232/E3
Little Barrier (isl.), N. Zealand 100/E2
Little Bay de Noc (bay), Mich. 250/B3
Little Bay Islands, Newf. 166/C4
Little Beaver (creek), Kansas 232/A2
Little Beaver (creek), Ohio 284/J4
Little Bighorn (riv.), Mont. 262/J5
Little Birch (lake), Manitoba 179/E3
Little Birch, W. Va. (26629) 312/E5
Little Bitterroot (lake), Mont. 262/B2
Little Black (riv.), Maine 243/E1
Little Black, Wis. (†54451) 317/F5
Little Blue (riv.), Kansas 232/E1
Little Blue (riv.), Nebr. 264/H5
Little Boars Head, N.H. (†03871) 268/F6
Little Bow (riv.), Alberta 182/D4
Little Britain, Ontario 177/F3
Little Brook, Nova Scotia 168/B4
Little Brosna (riv.), Ireland 17/E5
Little Buffalo Lake, Alberta 182/B1
Little Bullhead, Manitoba 179/F3
Little Butter (creek), Oreg. 291/H2
Little Cadotte (riv.), Alberta 182/B1
Little Cape, New Bruns. 170/F2
Little Catalina, Newf. 166/D2
Little Cayman (isl.), Cayman Is. 156/B3
Little Cedar, Iowa (50454) 229/H2
Little Chief, Okla. (†74637) 288/N1
Little Choptank (riv.), Md. 245/N6
Little Chute, Wis. (54140) 317/K7
Little Coco (isl.), Burma 72/B4
Little Colorado (riv.), Ariz. 188/D3
Little Colorado (riv.), Ariz. 198/D3
Little Compton○, R.I. (02837) 249/K6
Little Corn (isl.), Nicaragua 154/F4
Little Creek, Del. (19961) 245/S4
Little Creek, La. (†71371) 238/F3
Little Creek (peak), Utah 304/B6
Little Current, Ontario 177/B2
Little Current (riv.), Ontario 175/C2
Little Deep (creek), N. Dak. 282/G2
Little Deer Isle, Maine (†04627) 243/F7
Little Diomede (isl.), Alaska 196/E1
Little Dover, Nova Scotia 168/G3
Little Dry (creek), Mont. 262/K3
Little Eagle, S. Dak. (57639) 298/H2
Little Egg (harb.), N.J. 273/E4
Little Egg (inlet), N.J. 273/E5
Little Elkhart (riv.), Ind. 227/F1
Little Falls, Minn. (56345) 255/D5
Little Falls○, N.J. (07424) 273/B2
Little Falls, N.Y. (13365) 276/L4
Little Falls-South Windham, Maine (04082) 243/C8
Little Farms, La. (†70123) 238/N4
Little Ferry, N.J. (07643) 273/B2
Littlefield, Ariz. (86432) 198/B2
Littlefield, Texas (79339) 303/B4
Little Flock, Ark. (†72712) 202/B1
Littlefork, Minn. (56653) 255/E2
Little Fork (riv.), Minn. 255/E2
Little Genesee, N.Y. (14754) 276/D6
Little Girl (pt.), Mich. 250/E1
Little Goose (dam), Wash. 310/G4
Little Grand Rapids, Manitoba 179/G2
Little Gunpowder Falls (creek), Md. 245/M2
Littlehampton, England 13/G7
Littlehampton, England 10/F5
Little Harbour, Nova Scotia 168/D5
Little Heart's Ease, Newf. 166/D2
Little Hocking, Ohio (45742) 284/G7
Little Humboldt (riv.), Nev. 266/D1
Little Inagua (isl.), Bahamas 156/D2
Little Kai (isl.), Indonesia 85/J7
Little Kanawha (riv.), W. Va. 312/D5
Little Knife (riv.), N. Dak. 282/F3
Little Lake, Calif. (93542) 204/H8
Little Lake, Mich. (49833) 250/B2
Little Laramie (riv.), Wyo. 319/G4
Little London, Jamaica 158/G6
Little Lorraine, Nova Scotia 168/J3
Little Lost (riv.), Idaho 220/E5
Littlelot, Tenn. (†38454) 237/G9
Little Lynches (riv.), S.C. 296/G3
Little Madawaska (riv.), Maine 243/G2
Little Makin (atoll), Kiribati 87/H5
Little Manitou (lake), Sask. 181/F4
Little Marais, Minn. (55611) 255/G3
Little Marsh, Pa. (16931) 294/H2
Little Meadows, Pa. (18830) 294/K2
Little Mecatina (riv.), Newf. 166/B3
Little Medicine Bow (riv.), Wyo. 319/F3
Little Miami (riv.), Ohio 284/B6
Little Minch (sound), Scotland 10/C2
Little Minch (sound), Scotland 15/B3
Little Missouri (riv.) 188/F1
Little Missouri (riv.), Ark. 202/D6
Little Missouri (riv.), Mont. 262/M5
Little Missouri (riv.), N. Dak. 282/D4
Little Missouri (riv.), S. Dak. 298/B1
Little Missouri (riv.), Wyo. 319/H1
Littlemore, England 13/F6
Little Moreau (riv.), S. Dak. 298/G3
Little Mountain, S.C. (29075) 296/E3
Little Muddy (riv.), N. Dak. 282/C3
Little Muddy (creek), Wyo. 319/B4
Little Muskingum (riv.), Ohio 284/H6
Little Narrows, Nova Scotia 168/H3
Little Nicobar (isl.), India 68/G7
Little Orleans, Md. (21766) 245/E2
Little Owyhee (riv.), Idaho 220/B7
Little Paint Branch (riv.), Md. 245/F4
Little Para (riv.), S. Australia 88/D7
Little Para (riv.), S. Australia 94/B7
Little Patuxent (riv.), Md. 245/L4
Little Pee Dee (riv.), N.C. 281/L6
Little Pee Dee (riv.), S.C. 296/J4
Little Pigeon (creek), Ind. 227/C9
Little Plymouth, Va. (23091) 307/P5
Little Popo Agie (riv.), Wyo. 319/D3
Littleport, England 13/H5
Littleport, Iowa (52055) 229/L3
Little Powder (riv.), Wyo. 319/G1
Little Prairie, Wis. (†53119) 317/H2
Little Red (riv.), Ark. 202/G3
Little River, Ala. (36550) 195/C8
Little River (co.), Ark. 202/B6
Littleriver, Calif. (95456) 204/B4
Little River, Kansas (67457) 232/E3
Little River, N. Zealand 100/D5
Little River, Nova Scotia 168/B4
Little River (harb.), Nova Scotia 168/B5
Little River, S.C. (29566) 296/K4
Little River (inlet), S.C. 296/L4
Little Rock (cap.), Ark. 188/H4
Little Rock (cap.), Ark. 146/J6
Little Rock (cap.), Ark. (*72201) 202/F4
Little Rock, Iowa (51243) 229/B2
Little Rock, Minn. (†56373) 255/D5
Little Rock (creek), Minn. 255/C7
Little Rock, Miss. (39337) 256/F5
Little Rock, S.C. (29567) 296/J3
Littlerock, Wash. (98556) 310/B4
Little Rock A.F.B., Ark. 202/F4
Little Sable (pt.), Mich. 250/C5
Little Saint Bernard (pass), France 28/G5
Little Saint George (isl.), Fla. 212/B2
Little Salmon (riv.), Idaho 220/B4
Little Salt (lake), Utah 304/A6
Little Sandy (creek), Wyo. 319/C3
Little Sauk, Minn. (56346) 255/D5
Little Sevogle (riv.), New Bruns. 170/D1
Little Shawmut, Ala. (†36876) 195/H5
Little Sheep (creek), Oreg. 291/K2
Little Shippegan, New Bruns. 170/F1
Little Silver, N.J. (07739) 273/F3
Little Sioux, Iowa (51545) 229/B5
Little Sioux (riv.), Iowa 229/B3
Little Sitkin (isl.), Alaska 196/K4
Little Smoky, Alberta 182/B2
Little Smoky (riv.), Alberta 182/B2
Little Smoky (valley), Nev. 266/E4
Little Southwest Miramichi (riv.), New Bruns. 170/D2
Little Spokane (riv.), Wash. 310/H3
Littlestown, Pa. (17340) 294/H6
Little Suamico, Wis. (54141) 317/L6
Little Summer (isl.), Mich. 250/C3
Little Tallahatchie (riv.), Miss. 256/D2
Little Tennessee (riv.), N.C. 281/B4
Little Tennessee (riv.), Tenn. 237/N10
Little Thunder (creek), Wyo. 319/G2
Little Tinicum (isl.), Pa. 294/M7
Little Tobago (isl.), Virgin Is. (Br.) 161/B3
Little Tobique (riv.), New Bruns. 170/C1
Littleton, Colo. (*80120) 208/K3
Littleton, Ill. (61452) 222/C3
Littleton, Iowa (†50648) 229/K3
Littleton, Ireland 17/F6
Littleton○, Maine (†04760) 243/H3
Littleton○, Mass. (01460) 249/H2
Littleton, N.H. (03561) 268/D3
Littleton○, N.H. (03561) 268/D3
Littleton, N.C. (27850) 281/O2
Littleton, Va. (†23890) 307/O7
Littleton, W. Va. (26581) 312/F3
Littleton Common, Mass. (†01460) 249/J2
Little Traverse (bay), Mich. 250/D3
Little Trout River (pond), Newf. 166/C4
Little Tupper (lake), N.Y. 276/L2
Little Valley, N.Y. (14755) 276/C6
Little Vermilion (riv.), Ind. 227/B5
Littleville, Ala. (†35653) 195/C1
Little Wabash (riv.), Ill. 222/E5
Little Weiser (riv.), Idaho 220/B5
Little White (riv.), S. Dak. 298/H7
Little Wood (riv.), Idaho 220/D6
Little Yenisey (riv.), U.S.S.R. 48/K4
Little York, Ill. (61453) 222/C2
Little York, Ind. (47139) 227/F7
Little York, N.J. (08834) 273/C2
Little Zab (riv.), Iraq 66/C3
Lituya (bay), Alaska 196/L1
Litvínov, Czech. 41/B1
Liuba, China 77/G5
Liukang Tenggaja (isls.), Indonesia 85/F7
Liuli, Tanzania 115/F6
Liuzhou (Liuchow), China 77/G7
Liuzhou, China 54/M7
Līvāni, U.S.S.R. 53/D2
Livelong, Sask. 181/C2
Lively (isl.), 143/E7
Lively, Va. (22507) 307/P5
Livengood, Alaska (†99701) 196/J1
Live Oak, Calif. (†95073) 204/K4
Live Oak, Calif. (95953) 204/D4
Live Oak, Fla. (32060) 212/D1
Live Oak (co.), Texas 303/F9
Live Oak, Texas (†78201) 303/K10
Liveringa, W. Australia 88/C3
Liveringa, W. Australia 92/D2
Livermore, Calif. (94550) 204/L2
Livermore, Colo. (80536) 208/J1
Livermore, Iowa (50558) 229/E3
Livermore, Ky. (42352) 237/G5
Livermore, Maine (04253) 243/C7
Livermore○, Maine (04253) 243/C7
Livermore (mt.), Texas 303/C11
Livermore Falls, Maine (04254) 243/C7
Livermore Falls○, Maine (04254) 243/C7
Livermore Falls, N.H. (†03264) 268/D4
Liverpool (swamp), Bolivia 136/D4
Liverpool, England 10/F2
Liverpool, England 13/G2
Liverpool, England 7/D3
Liverpool (bay), England 13/D4
Liverpool, Ill. (61543) 222/C3
Liverpool, N. S. Wales 88/K4
Liverpool, N.S. Wales 97/H4
Liverpool (range), N.S. Wales 97/F2
Liverpool, N.Y. (13088) 276/H4
Liverpool (bay), N.W. Terrs. 187/E2
Liverpool (cape), N.W. Terrs. 187/L2
Liverpool, Nova Scotia 168/D4
Liverpool (bay), Nova Scotia 168/D5
Liverpool, Pa. (17045) 294/H4
Liverpool, Texas (77577) 303/J3
Liverpool, W. Va. (25257) 312/C5
Livia, Ky. (†42376) 237/G5
Livigno, Italy 34/C1
Livingston, Ala. (35470) 195/B5
Livingston, Calif. (95334) 204/E6
Livingston, Guatemala 154/C3
Livingston (co.), Ill. 222/E3
Livingston, Ill. (62058) 222/D5
Livingston (co.), Ky. 237/E6
Livingston, Ky. (40445) 237/N6
Livingston (par.), La. 238/L2
Livingston, La. (70754) 238/L1
Livingston (co.), Mich. 250/F6
Livingston (co.), Mo. 261/E3
Livingston, Mont. 188/D1
Livingston, Mont. (59047) 262/F5
Livingston○, N.J. (07039) 273/E2
Livingston (co.), N.Y. 276/E5
Livingston, Scotland 15/C2
Livingston, Scotland 10/C1
Livingston, S.C. (29076) 296/E4
Livingston, Tenn. (38570) 237/L8
Livingston, Texas (77351) 303/K7
Livingston (lake), Texas 303/K7
Livingston, Wis. (53554) 317/E10
Livingstone (range), Alberta 182/C4
Livingstone (falls), Zaire 115/B5
Livingstone, Zambia 115/E7
Livingstone, Zambia 102/E6
Livingstonia, Malawi 115/F6
Livingston Manor, N.Y. (12758) 276/L7
Livingstonville, N.Y. (†12122) 276/M6
Livno, Yugoslavia 45/C4
Livny, U.S.S.R. 52/E4
Livona, N. Dak. (58501) 282/K6
Livonia, Ind. (†47108) 227/E7
Livonia, La. (70755) 238/G5
Livonia, Mich. (*48150) 250/F6
Livonia, Mo. (63551) 261/G1
Livonia, N.Y. (14487) 276/E5
Livonia, Pa. (†16872) 294/H4
Livorno (Leghorn), Italy 34/C3
Livry-Gargan, France 28/C1
Liwale, Tanzania 115/G5
Li Xian, China 77/H6
Lixoúrion, Greece 45/E6
Lizard, The (pen.), England 13/B8
Lizard (pt.), England 10/D6
Lizard (pt.), England 13/B8
Lizella, Georgia (31052) 217/E5
Lizemores, W. Va. (25125) 312/D6
Lizton, Ind. (46149) 227/D5
Ljubinje, Yugoslavia 45/D4
Ljubljana, Yugoslavia 45/B3
Ljubljana, Yugoslavia 7/F4
Ljubuški, Yugoslavia 45/C4
Ljugarn, Sweden 18/L8
Ljungan (riv.), Sweden 18/K5
Ljungby, Sweden 18/J8
Ljusdal, Sweden 18/J6
Ljusna (riv.), Sweden 7/F2
Ljusnan (riv.), Sweden 18/H5
Ljusne, Sweden 18/K6
Llagostera, Spain 33/H2
Llaima (vol.), Chile 138/E2
Llallagua, Bolivia 136/B6
Llallagua, Bolivia 120/C4
Llamara, Salar de (salt dep.), Chile 138/B3
Llanarth, Wales 13/C5
Llancanelo (lag.), Argentina 143/C4
Llancanelo, Salina y Laguna (salt dep.), Argentina 143/C4
Llandeilo, Wales 13/C6
Llandovery, Wales 13/D5
Llandovery, Wales 10/D5
Llandrindod Wells, Wales 10/E4
Llandrindod Wells, Wales 13/D5
Llandudno, Wales 13/D4
Llandudno, Wales 10/E4
Llandybie, Wales 13/C6
Llandyssul, Wales 13/C5
Llanelli, Wales 13/C6
Llanelli, Wales 10/D5
Llanes, Spain 33/D1
Llanfair Caereinion, Wales 13/D5
Llanfairfechan, Wales 13/D4
Llanfyllin, Wales 10/E4
Llanfyllin, Wales 13/D5
Llangefni, Wales 13/C4
Llangollen, Wales 13/D5
Llangollen, Wales 10/E4
Llanguicke, Wales 13/D6
Llanidloes, Wales 13/D5
Llanidloes, Wales 10/E4
Llanllyfni, Wales 13/C4
Llannon, Wales 13/C6
Llano, N.Mex. (87543) 274/D2
Llano (co.), Texas 303/F7
Llano, Texas (78643) 303/F7
Llano (riv.), Texas 303/D7
Llano Estacado (Staked) (plain), N. Mex. 274/F5
Llano Estacado (plain), Texas 303/B4
Llanon, Wales 13/C5
Llanos (plain) 120/B2
Llanos (plains), Colombia 126/D5
Llanquera, Bolivia 136/A6
Llanquihue (lake), Chile 138/E3
Llanrhaeadr, Wales 13/D5
Llanrhystyd, Wales 13/C5
Llanrian, Wales 13/B6
Llanrwst, Wales 13/D4
Llanrwst, Wales 10/E4
Llantrisant, Wales 13/A7
Llantwit Major, Wales 13/A7
Llanwnog, Wales 13/D5
Llanwrtyd Wells, Wales 13/D5
Llata, Peru 128/D7
Llay-Llay, Chile 138/G2
Llera de Canales, Mexico 150/K5
Llerena (pt.), C. Rica 154/F6
Llerena, Spain 33/C3
Lleyn (pen.), Wales 13/C5
Llica, Bolivia 136/A6
Llico, Chile 138/A10
Llivia, Spain 33/G1
Llobregat (riv.), Spain 33/H2
Llodio, Spain 33/E1
Llolleo, Chile 138/F4
Llorente, Philippines 82/E5
Lloyd, Fla. (32337) 212/C1
Lloyd (res.), Ga. 188/K4
Lloyd, Ky. (41156) 237/R3
Lloyd, Mont. (59535) 262/G2
Lloyd Harbor, N.Y. (†11743) 276/R6
Lloydminster, Alberta 182/E3
Lloydminster, Alta.-Sask. 162/E5
Lloydminster, Sask. 181/A2
Lluchmayor, Spain 33/H3
Lluidas Vale, Jamaica 158/J6
Llullaillaco (mt.) 120/C5
Llullaillaco (vol.), Argentina 143/C1
Llullaillaco (vol.), Chile 138/B5
Lluta (riv.), Chile 138/B1
Llwchwr, Wales 13/D6
Loa (riv.), Chile 120/C5
Loa (riv.), Chile 138/B3
Loa, Utah (84747) 304/C5
Loachapoka, Ala. (36865) 195/G5
Loami, Ill. (62661) 222/D4
Loange (riv.), Angola 115/C5
Loange (riv.), Zaire 115/C5
Loanhead, Scotland 10/C1
Loanhead, Scotland 15/D2
Lobatse, Botswana 118/D5
Löbau, E. Germany 22/F3
Lobaye (riv.), Cent. Afr. Rep. 115/C2
Lobdell, La. (†70767) 238/J1
Lobeco, S.C. (29931) 296/F6
Lobelia, W. Va. (†24946) 312/F6
Lobelville, Tenn. (37097) 237/F9
Lobenstein, E. Germany 22/D3
Lobería, Argentina 143/E4
Lobethal, S. Australia 94/C7
Łobez, Poland 47/B2
Lobito, Angola 115/B6
Lobito, Angola 102/D6
Lobitos, Peru 128/B5
Lobo, Philippines 82/C4
Lobo (cay), P. Rico 161/G1
Lobos, Argentina 143/G7
Lobos (pt.), Chile 138/A3
Lobos (cape), Mexico 150/C2
Lobos (pt.), Mexico 150/D3
Lobos (isl.), Uruguay 145/E6
Lobos de Afuera (isls.), Peru 128/B6
Lobos de Tierra (isl.), Peru 128/B6
Lobster (lake), Maine 243/E4
Locarno, Switzerland 39/G4
Locate, Mont. (†59336) 262/L4
Lochaber, Nova Scotia 168/F3
Lochaber (dist.), Scotland 15/D4
Lochailort, Scotland 15/C4
Lochaline, Scotland 15/C4
Lochans, Scotland 15/D6
Locharbriggs, Scotland 15/E5
Lochawe, Scotland 15/C4
Lochboisdale, Scotland 15/A3
Lochbuie, Colo. (†80601) 208/K2
Lochcarron, Scotland 10/D2
Lochcarron, Scotland 15/C3
Lochearnhead, Scotland 15/D4
Lochem, Netherlands 27/J4
Lochend, Scotland 15/D3
Loches, France 28/D4
Lochgelly, Scotland 10/C1
Lochgelly, Scotland 15/D1
Lochgelly, W. Va. (25866) 312/D6
Lochgilphead, Scotland 10/D2
Lochgilphead, Scotland 15/C4
Lochgoilhead, Scotland 15/D4
Lochindorb (lake), Scotland 15/E3
Lochinver, Scotland 10/D1
Lochinver, Scotland 15/C2
Lochloosa, Fla. (32662) 212/E2
Lochloosa (lake), Fla. 212/D2
Loch Lynn Heights, Md. (†21550) 245/A3
Lochmaben, Scotland 10/E3
Lochmaben, Scotland 15/E5
Lochmaddy, Scotland 15/A3
Lochmere, N.H. (03252) 268/D5
Lochnagar (mt.), Scotland 15/E4
Lochore, Scotland 15/D1
Lochranza, Scotland 15/C5
Loch Raven (res.), Md. 245/M3
Lochristi, Belgium 27/D6
Lochsa (riv.), Idaho 220/C3
Lochwinnoch, Scotland 15/A2
Lochy, Loch (lake), Scotland 15/D3
Lochy, Loch (lake), Scotland 10/D2
Lock, S. Australia 94/D5
Lockatong (creek), N.J. 273/C3
Lockbourne, Ohio (43137) 284/E6
Locke, Calif. (†95690) 204/B9
Locke, N.Y. (13092) 276/H5
Locke (mt.), Texas 303/D11
Lockeford, Calif. (95237) 204/C9
Locke Mills, Maine (04255) 243/B7
Lockeport, Nova Scotia 168/C5
Lockerbie, Scotland 10/E3
Lockerbie, Scotland 15/E5
Lockesburg, Ark. (71846) 202/B6
Lockhart, Ala. (36455) 195/F8
Lockhart, Minn. (†56510) 255/B3
Lockhart (mt.), Mont. 262/D3
Lockhart, N.S. Wales 97/D4
Lockhart (riv.), N.W. Terrs. 187/H3
Lockhart, S.C. (29364) 296/E2
Lockhart, Texas (78644) 303/G8
Lock Haven, Pa. (17745) 294/H3
Lockington, Ohio (†45356) 284/B5
Lockland, Ohio (45215) 284/C9
Lockney, Texas (79241) 303/C3
Lockney, W. Va. (25258) 312/E5
Lockport, Ill. (60441) 222/B6
Lockport, Ky. (40036) 237/M4
Lockport, La. (70374) 238/K7
Lockport, Manitoba 179/F4
Lockport, N.Y. (14094) 276/C4
Lockridge, Iowa (52635) 229/K7
Lock Springs, Mo. (64654) 261/E3
Lockwood, Mo. (65682) 261/E8
Lockwood, Sask. 181/G4
Lockwood, W. Va. (†26651) 312/D6
Loc Ninh, Vietnam 72/E5
Loco, Okla. (73442) 288/L6
Loco Hills, N. Mex. (88255) 274/F6

Locumba, Peru 128/G11
Locumba (riv.), Peru 128/G11
Locust, N.C. (28097) 281/J4
Locust Bayou, Ark. (†71701) 202/E6
Locust Fork, Ala. (35097) 195/E3
Locust Fork (riv.), Ala. 195/E3
Locust Grove, Ark. (72550) 202/G2
Locust Grove, Georgia (30248) 217/D4
Locust Grove, N.Y. (†11791) 276/R6
Locust Grove, Ohio (†45660) 284/D8
Locust Grove, Okla. (74352) 288/R2
Locust Hill, Ky. (40151) 237/J5
Locustville, Va. (23404) 307/S5
Lod (Lydda), Israel 65/B4
Loda, Ill. (60948) 222/E3
Lodar, P.D.R. Yemen 59/E7
Loddon, England 13/J5
Loddon (riv.), Victoria 97/B5
Loddon (riv.), Victoria 88/G7
Lodève, France 28/E6
Lodeynoye Pole, U.S.S.R. 52/D2
Lodge (creek), Mont. 262/G1
Lodge (creek), Sask. 181/B6
Lodge, S.C. (29082) 296/F5
Lodge Bay, Newf. 166/C3
Lodge Grass, Mont. (59050) 262/J5
Lodge Hill, Barbados 161/B8
Lodgepole, Alberta 182/C3
Lodge Pole, Mont. (†59524) 262/H2
Lodgepole, Nebr. (69149) 264/B3
Lodgepole (creek), Nebr. 264/A3
Lodgepole, S. Dak. (57640) 298/D2
Lodgepole (creek), Wyo. 319/H2
Lodgepole (creek), Wyo. 319/H4
Lodi, Calif. 188/B3
Lodi, Calif. (95240) 204/C9
Lodi, Italy 34/B2
Lodi, Miss. (†39767) 256/E3
Lodi, Mo. (63950) 261/M8
Lodi, N.J. (07644) 273/B2
Lodi, N.Y. (14860) 276/G5
Lodi, Ohio (44254) 284/F3
Lodi (cape), Tasmania 99/E3
Lodi, Texas (75564) 303/K5
Lodi, Wis. (53555) 317/G9
Lødingen, Norway 18/J2
Lodja, Zaire 102/E5
Lodja, Zaire 115/D4
Lodosa, Spain 33/E1
Lodrino, Switzerland 39/G4
Lodwar, Kenya 115/G3
Łódź (prov.), Poland 47/D3
Łódź (city), Poland 47/D3
Łódź, Poland 7/F3
Łódź, Poland 47/D3
Loei, Thailand 72/D3
Loen, Norway 18/E6
Lofer, Austria 41/B3
Lofoten (isls.), Norway 4/C9
Lofoten (isls.), Norway 7/F2
Lofoten (isls.), Norway 18/H2
Loftus, England 13/G3
Loftus, England 10/F3
Lofty (mt.), S. Australia 88/E8
Lofty (mt.), S. Australia 94/B8
Lofty (range), Tasmania 99/B3
Logan, Ala. (35098) 195/E2
Logan (lake), Alberta 182/E2
Logan (co.), Ark. 202/C3
Logan (mt.), Canada 4/C17
Logan (co.), Colo. 208/N1
Logan (co.), Ill. 222/D3
Logan, Ill. (62856) 222/E6
Logan, Ind. (†45030) 227/H6
Logan, Iowa (51546) 229/B5
Logan (co.), Kansas 232/A3
Logan, Kansas (67646) 232/C2
Logan (co.), Ky. 237/H7
Logan, Mont. (†59741) 262/E5
Logan (co.), Nebr. 264/D3
Logan (creek), Nebr. 264/H2
Logan, N. Mex. (88426) 274/F3
Logan (co.), N. Dak. 282/L7
Logan, N. Dak. (†58701) 282/H3
Logan (co.), Ohio 284/C5
Logan, Ohio (43138) 284/F6
Logan (co.), Okla. 288/M3
Logan, Okla. (73849) 288/F1
Logan, Oreg. (†97405) 291/B2
Logan, Utah (84321) 304/C2
Logan, Utah 188/D2
Logan (mt.), Wash. 310/E2
Logan (co.), W. Va. 312/C7
Logan, W. Va. (25601) 312/B7
Logan (mt.), Yukon 162/B3
Logan (mt.), Yukon 187/D3
Logan (mts.), Yukon 187/F3
Logandale, Nev. (89021) 266/G6
Logan Internat'l Airport, Mass. 249/D7
Logan Lake, Br. Col. 184/G5
Logan Martin (lake), Ala. 195/F4
Logansport, Ind. (46947) 227/E3
Logansport, Ky. (42258) 237/H6
Logansport, La. (71049) 238/C3
Loganton, Pa. (17747) 294/H3
Loganville, Georgia (30249) 217/E3
Loganville, Pa. (17342) 294/J6
Loganville, Wis. (53943) 317/F9
Loge (riv.), Angola 115/B5
Loggieville, New Bruns. 170/E1
Log Lane Village, Colo. (†80701) 208/M2
Logone (riv.) 102/D3
Logone (riv.), Cameroon 115/C2
Logone (riv.), Chad 111/C5
Logroño (prov.), Spain 33/E1
Logroño, Spain 7/D4
Logroño, Spain 33/E1
Logrosán, Spain 33/D3
Logsden, Oreg. (97357) 291/D3
Løgstør, Denmark 21/C4
Løgstør, Denmark 18/F8
Løgstør Bredning (fjord), Denmark 21/C4
Løgumkloster, Denmark 21/B7
Lohals, Denmark 21/D7
Lohardaga, India 68/E4
Lohatlha, S. Africa 118/C5
Lohman, Mo. (65053) 261/H5
Lohman, Mont. (†59523) 262/G2
Lohn, Texas (76852) 303/E6
Löhne, W. Germany 22/C2
Loho (Luohe), China 77/H5
Lohr am Main, W. Germany 22/C4
Lohrville, Iowa (51453) 229/D4
Lohrville, Wis. (†54970) 317/H7
Loica, Chile 138/F4
Loi-kaw, Burma 72/C3
Loi Leng (mt.), Burma 72/C2
Loimaa, Finland 18/N6
Loir (riv.), France 28/D4
Loire (dept.), France 28/F5
Loire (riv.), France 7/E4
Loire (riv.), France 28/C4
Loire-Atlantique (dept.), France 28/C4
Loiret (dept.), France 28/E4
Loir-et-Cher (dept.), France 28/D4
Loiza, P. Rico 161/E1
Loiza Aldea, P. Rico 161/E1
Loja (prov.), Ecuador 128/C4
Loja, Ecuador 128/C4
Loja, Ecuador 120/B3
Loja, Spain 33/D4
Løjt Kirkeby, Denmark 21/C7
Loka, Sudan 111/F7
Lokeren, Belgium 27/D6
Lokitaung, Kenya 115/G3
Lokka (res.), Finland 18/Q3
Løkken, Denmark 21/C3
Løkken, Norway 18/F5
Lokoja, Nigeria 106/F7
Lokolama, Zaire 115/D4
Lokoro (riv.), Zaire 115/C4
Lökösháza, Hungary 41/F3
Lokossa, Benin 106/E7
Loksa, U.S.S.R. 53/C1
Loks Land (isl.), N.W. Terrs. 187/M3
Lol (dry riv.), Sudan 111/E6
Lola, Ky. (42059) 237/E6
Loleta, Calif. (95551) 204/A3
Lolgorien, Kenya 115/G4
Lolita, Texas (77971) 303/H9
Lolland (isl.), Denmark 18/G9
Lolland (isl.), Denmark 21/E8
Lollie, Georgia (30433) 217/G6
Lolo (creek), Idaho 220/C3
Lolo (pass), Idaho 220/D3
Lolo, Mont. (59847) 262/B4
Lolo (pass), Mont. 262/B4
Lolo Hot Springs, Mont. (†59847) 262/B4
Lom, Bulgaria 45/F4
Lom (riv.), Cameroon 115/B2
Lom, Norway 18/F6
Loma, Colo. (81524) 208/B4
Loma, Mont. (59460) 262/F3
Loma, N. Dak. (†58311) 282/O2
Loma, Mansa (lag.), S. Leone 106/B7
Loma Alta, Bolivia 136/B2
Loma Bonita, Mexico 150/M7
Loma Linda, Calif. (92354) 204/F10
Loma Mar, Calif. (94021) 204/J3
Lomami (riv.), Zaire 115/D4
Loman, Minn. (56654) 255/E2
Loma Plata, Paraguay 144/C3
Lomas, Peru 128/E10
Lomas de Zamora, Argentina 143/G7
Lomax, Ala. (†35045) 195/E5
Lomax, Ill. (61454) 222/B3
Lomax, Texas (†77571) 303/K2
Lombard, Ill. (60148) 222/B5
Lombarda, Serra (mts.), Brazil 132/D2
Lombardville, Ill. (†61421) 222/D2
Lombardy (reg.), Italy 34/B2
Lombardy, S. Africa 118/H6
Lombez, France 28/D6
Lomblen (isl.), Indonesia 85/G7
Lombok (isl.), Indonesia 54/N10
Lombok (isl.), Indonesia 85/F7
Lombok (str.), Indonesia 85/E7
Lomé (cap.), Togo 102/C4
Lomé (cap.), Togo 106/E7
Lomela, Zaire 115/D4
Lomela (riv.), Zaire 115/D4
Lometa, Texas (76853) 303/F6
Lomié, Cameroon 115/B3
Lomira, Wis. (53048) 317/J8
Lo Miranda, Chile 138/G5
Lomita, Calif. (90717) 204/C11
Lommel, Belgium 27/G6
Lomnice, Czech. 41/C2
Lomond, Alberta 182/D4
Lomond, Loch (lake), Nova Scotia 168/H3
Lomond, Loch (lake), Scotland 15/D4
Lomond, Loch (lake), Scotland 10/A1
Lompoc, Calif. (93436) 204/E9
Lom Sak, Thailand 72/D3
Łomża (prov.), Poland 47/F2
Łomża, Poland 47/F2
Lonaconing, Md. (21539) 245/C2
Loncoche, Chile 138/D2
Loncopué, Argentina 143/B4
London, Ark. (72847) 202/D3
London, Greater, England 13/H8
London (cap.), England 7/D3
London (cap.), England 10/B5
London (cap.), England 13/H8
London, Ind. (†46126) 227/F5
London, Ky. (40741) 237/N6
London, Minn. (56061) 255/E7
London, Ohio (43140) 284/C6
London, Ont. 146/K5
London, Ont. 162/H7
London, Ontario 177/C5
London, Texas (76854) 303/E7
London (cap.), U.K. 2/J3
London, Wis. (†53523) 317/H9
Londonderry○, N.H. (03053) 268/E6
Londonderry, N. Ireland 7/D3
Londonderry (dist.), N. Ireland 17/G2
Londonderry, N. Ireland 10/C3
Londonderry, N. Ireland 17/G2
Londonderry, Nova Scotia 168/E3
Londonderry, Ohio (45647) 284/E7
Londonderry○, Vt. (05148) 268/B5
Londonderry (cape), W. Australia 88/D2
Londonderry (cape), W. Australia 92/D1
Londonderry Station, Nova Scotia 168/E3
London Mills, Ill. (61544) 222/C3
Londontowne, Md. (†21035) 245/M4
Londrina, Brazil 132/D8
Londrina, Brazil 120/D5
Lone (mt.), Mont. 262/E5
Lone (mt.), Nev. 266/D4
Lone Butte, Br. Col. 184/G4
Lone Cedar, W. Va. (†26153) 312/C4
Lone Cone (mt.), Colo. 208/C7
Lone Elm, Kansas (†66039) 232/G3
Lone Grove, Okla. (73443) 288/M6
Lone Jack, Mo. (64070) 261/S6
Lonely (lake), Manitoba 179/C3
Lonely (isl.), Ontario 177/C2
Lone Mountain, Tenn. (†37879) 237/O8
Lone Oak, Georgia (†30230) 217/C4
Lone Oak, Ky. (42001) 237/D6
Lone Oak, Texas (75453) 303/H5
Lone Pine, Alberta 182/C2
Lone Pine, Calif. (93545) 204/H7
Lone Pine (peak), Idaho 220/D5
Lonepine, La. (†71367) 238/F5
Lonepine, Mont. (59848) 262/B3
Lone Prairie, Br. Col. 184/G2
Lone Rock, Iowa (50559) 229/E2
Lonerock, Oreg. (†97823) 291/H2
Lone Rock, Sask. 181/A2
Lone Rock, Wis. (53556) 317/F9
Lone Star, S.C. (29077) 296/F4
Lone Tree (creek), Colo. 208/K1
Lone Tree, Iowa (52755) 229/L6
Lonetree, N. Dak. (†58718) 282/G3
Lonetree, Wyo. (82936) 319/B4
Lone Wolf, Okla. (73655) 288/H5
Long, Alaska (†99768) 196/G2
Long (isl.), Alaska 196/M2
Long (isl.), Ant. & Bar. 161/E11
Long (isl.), Bahamas 146/L7
Long (cay), Bahamas 156/C2
Long (isl.), Bahamas 156/C2
Long (bay), Barbados 161/B9
Long (mt.), Conn. 210/B2
Long (pond), Conn. 210/H3
Long (key), Fla. 212/F7
Long (key), Fla. 212/B3
Long (pond), Fla. 212/D2
Long (co.), Georgia 217/J7
Long (bay), Jamaica 158/H7
Long (lake), Maine 243/E2
Long (lake), Maine 243/G1
Long (lake), Maine 243/B7
Long (pond), Maine 243/E5
Long (pond), Maine 243/D6
Long (pond), Maine 243/C4
Long (lake), Manitoba 179/G4
Long (pt.), Manitoba 179/D1
Long (pt.), Manitoba 179/D4
Long (isl.), Martinique 161/D6
Long (isl.), Mass. 249/E7
Long (pond), Mass. 249/L5
Long (pt.), Mass. 249/O4
Long (lake), Mich. 250/F3
Long (lake), Minn. 255/D4
Long (lake), Minn. 255/F3
Long, Miss. (†38828) 256/C4
Long (valley), Nev. 266/B1
Long (isl.), New Bruns. 170/D3
Long (lake), New Bruns. 170/D1
Long (isl.), Newf. 166/C2
Long (isl.), Newf. 166/B3
Long (lake), Newf. 166/A3
Long (lake), Newf. 166/B3
Long (pt.), Newf. 166/C4
Long (mt.), N.H. 268/E2
Long (beach), N.J. 273/E4
Long (isl.), N.Y. 188/M2
Long (isl.), N.Y. 276/P9
Long (lake), N.Y. 276/M2
Long (isl.), N. Zealand 100/A7
Long (lake), N.C. 281/P5
Long (lake), N. Dak. 282/K6
Long (lake), N. Dak. 282/L3
Long (lake), N. Dak. 282/L2
Long (isl.), Nova Scotia 168/B4
Long (lake), Ontario 177/H5
Long (lake), Ontario 175/C3
Long (pt.), Ontario 177/D5
Long (isl.), Papua N.G. 85/B7
Long (isl.), Papua N.G. 86/A2
Long (creek), Sask. 181/H6
Long, Loch (inlet), Scotland 10/A1
Long, Loch (inlet), Scotland 15/D4
Long (lake), S. Dak. 298/L2
Long (pt.), Tasmania 99/E3
Long (str.), U.S.S.R. 48/S2
Long (pt.), Virgin Is. (U.S.) 161/B4
Long (pt.), Virgin Is. (U.S.) 161/E4
Long (isl.), Wash. 310/A4
Long (lake), Wash. 310/H3
Long (reef), W. Australia 92/D1
Long (lake), Wis. 317/C4
Longa, Angola 115/C6
Longa (isl.), Scotland 15/C3
Longaví, Chile 138/A11
Long Bay, Jamaica 158/K6
Long Beach, Br. Col. 184/E5
Long Beach, Calif. 146/F6
Long Beach, Calif. 188/C4
Long Beach, Calif. (*90801) 204/C11
Long Beach (pen.), Conn. 210/C4
Long Beach, Ind. (†46360) 227/D1
Long Beach, Minn. (†56334) 255/C5
Long Beach, Miss. (39560) 256/F10
Long Beach (isl.), N.J. 273/E4
Long Beach, N.Y. (11561) 276/R7
Long Beach, N.C. (28461) 281/N7
Long Beach, Ontario 177/E5
Long Beach, Wash. (98631) 310/A4
Longbenton, England 13/J3
Longboat (key), Fla. 212/D4
Longboat Key, Fla. (33548) 212/D4
Long Bottom, Ohio (45743) 284/G7
Long Branch, N.J. (07740) 273/F3
Longbranch, Wash. (98351) 310/C3
Longchuan, China 77/J7
Long Cove, Maine (†04857) 243/E8
Long Creek, Oreg. (97856) 291/H3
Longcreek, S.C. (29658) 296/A2
Longdale, Okla. (73755) 288/K2
Longde, China 77/G4
Long Eaton, England 13/F5
Long Eddy, N.Y. (12760) 276/K7
Long Falls (dam), Maine 243/C5
Longfellow (mts.), Maine 243/B6
Longford (co.), Ireland 17/F4
Longford, Ireland 10/C4
Longford, Ireland 17/F4
Longford, Kansas (67458) 232/E2
Longford, Ontario 177/E3
Longford, Tasmania 99/C3
Longford, Virgin Is. (U.S.) 161/F4
Long Green, Md. (21092) 245/M3
Long Grove, Ill. (60047) 222/B5
Long Grove, Iowa (52756) 229/M5
Long Harbour, Newf. 166/D2
Long Hill, Conn. (†06611) 210/C3
Longhua, China 77/J3
Longhui, China 77/H6
Longido, Tanzania 115/G4
Longiram, Indonesia 85/F5
Long Island (sound), Conn. 210/C4,
Long Island (bay), Ireland 17/B9
Long Island, Kansas (67647) 232/C2
Long Island (sound), N.Y. 276/P9
Longisland, N.C. (28648) 281/H3
Long Island, Tenn. (†37662) 237/S7
Long Island, Va. (24569) 307/K6
Longjiang, China 77/K2
Long Key, Fla. (33001) 212/F7
Longkou, China 77/J4
Longlac, Ontario 177/H5
Longlac, Ontario 175/C3
Long Lake, Mich. (48743) 250/F4
Long Lake, Minn. (55356) 255/F5
Long Lake, N.Y. (12847) 276/L3
Longlake, S. Dak. (57457) 298/L2
Long Lake, Wis. (54542) 317/J4
Long Lane, Mo. (65590) 261/G7
Longleaf, La. (71448) 238/E4
Long Meadow (pond), Conn. 210/C2
Longmeadow○, Mass. (01106) 249/D4
Longmire, Wash. (98397) 310/D4
Longmont, Colo. 188/E2
Longmont, Colo. (80501) 208/J2
Longnan, China 77/J7
Longnawan, Indonesia 85/F5
Long Neck (pt.), Conn. 210/B4
Long Pine, Nebr. (69217) 264/E2
Long Point, Ill. (61333) 222/E3
Long Point, Nova Scotia 168/G3
Long Point (bay), Ontario 177/D5
Long Point Beach, Ontario 177/D5
Long Pond, Maine (†04945) 243/C4
Long Pond, Pa. (18334) 294/L3
Longport, N.J. (08403) 273/D5
Long Prairie, Minn. (56347) 255/D5
Long Prairie (riv.), Minn. 255/D4
Long Range (mts.), Newf. 166/C4
Long Rapids, Mich. (†49753) 250/F3
Longreach, Australia 87/E8
Long Reach (inlet), New Bruns. 170/D3
Longreach, Queensland 88/G4
Longreach, Queensland 95/B4
Long Reef (pt.), N.S. Wales 97/K3
Longridge, England 13/H1
Longridge, England 10/G1
Longs (peak), Colo. 208/H2
Longs, S.C. (29568) 296/K4
Long Sault, Ontario 177/K2
Longshan, China 77/G6
Long Siding, Minn. (†55371) 255/E5
Long Society, Conn. (†06360) 210/G2
Long Spruce, Manitoba 179/K2
Longstreet, La. (71050) 238/B2
Longton, Kansas (67352) 232/F4
Longtown, Miss. (†38665) 256/D1
Longtown, Mo. (†63775) 261/N7
Longtown, S.C. (†29130) 296/F3
Longueuil, Québec 172/J4
Long Valley, N.J. (07853) 273/D2
Longvalley, S. Dak. (57547) 298/F7
Longview, Ala. (†35137) 195/E4
Longview, Alberta 182/C4
Longview, Colo. (†80135) 208/J4
Longview, Ill. (61852) 222/E4
Longview, Miss. (†39759) 256/G4
Longview, N.C. (28601) 281/F3
Longview, Texas 188/G4
Longview, Texas (*75601) 303/K5
Longview, Wash. 188/B1
Longview, Wash. (98632) 310/B4
Longville, La. (70652) 238/D5
Longville, Minn. (56655) 255/D4
Longwood, Fla. (32750) 212/E3
Longwood, Mo. (†65301) 261/F5
Longwood, N.C. (28452) 281/M7
Longwood Park, N.C. (†28345) 281/K5
Longworth, Br. Col. 184/G3
Longworth, Texas (†79604) 303/D5
Longwy, France 28/F3
Long Xian, China 77/G5
Long Xuyen, Vietnam 72/E5
Longyan, China 77/J6
Longyearbyen, Norway 18/D2
Longyearbyen, Norway 4/B8
Longzhen (Lungchen), China 77/L2
Loni Beach, Manitoba 179/F4
Lonigo, Italy 34/C2
Lonneker, Netherlands 27/K4
Lonoke (co.), Ark. 202/G4
Lonoke, Ark. (72086) 202/G4
Lonquimay, Chile 138/E2
Lonsdale, Ark. (72087) 202/E4
Lonsdale, Minn. (55046) 255/E6
Lonsdale, R.I. (†02864) 249/J5
Lons-le-Saunier, France 28/F4
Lonton, Burma 72/B1
Lontzen, Belgium 27/H9
Looe, England 13/C7
Loogootee, Ind. (47553) 227/D7
Lookeba, Okla. (73053) 288/K4
Lookingglass (riv.), Mich. 250/E6
Lookout (mt.), Ala. 195/G2
Lookout (ridge), Alaska 196/G1
Lookout, Calif. (96054) 204/D2
Lookout (mt.), Idaho 220/F5
Lookout (mt.), Idaho 220/D5
Lookout, Ky. (41542) 237/S6
Lookout (pt.), Md. 245/N8
Lookout (cape), N.C. 188/L4
Lookout (cape), N.C. 281/S5
Lookout, Okla. (†73842) 288/H1
Lookout (cape), Oreg. 188/B1
Lookout (cape), Oreg. 291/C2
Lookout, Pa. (†18417) 294/M2
Lookout, W. Va. (25868) 312/E6
Lookout, Wyo. (†82051) 319/G4
Lookout Mountain, Georgia (†30741) 217/B1
Lookout Mountain, Tenn. (37350) 237/L11
Lookout Point (lake), Oreg. 291/E4
Looma, Alberta 182/D3
Loomis, Calif. (95650) 204/C8
Loomis, Nebr. (68958) 264/E4
Loomis, Sask. 181/C6
Loomis, S. Dak. (57260) 298/N6
Loomis, Wash. (98827) 310/F2
Loomis, Wis. (†54157) 317/K5
Loon (lake), Alberta 182/C1
Loon (riv.), Alberta 182/C1
Loon (lake), Maine 243/D3
Loon (lake), Ontario 177/F3
Loon (creek), Sask. 181/E3
Loon (lake), Wash. 310/H2
Looneyville, W. Va. (25259) 312/D5
Loon Lake, Alberta 182/C1
Loon Lake, Maine (†04970) 243/B5
Loon Lake, N.Y. (†12968) 276/M1
Loon Lake, Sask. 181/B1
Loon Lake, Wash. (99148) 310/H2
Loon op Zand, Netherlands 27/G5
Loon Strait, Manitoba 179/F3
Loop (head), Ireland 17/A6
Loop (head), Ireland 10/A4
Loop, Texas (79342) 303/B5
Loos, Br. Col. 184/G3
Loosahatchie (riv.), Tenn. 237/B10
Loose Creek, Mo. (65054) 261/J5
Lo Ovalle, Chile 138/F3
Looxahoma, Miss. (†38668) 256/E1
Looz (Borgloon), Belgium 27/G7
Lopatka (cape), U.S.S.R. 54/S4
Lopatka (cape), U.S.S.R. 48/Q4
Lop Buri, Thailand 72/D4
Lopeno, Texas (78564) 303/E11
Lopez (pt.), Calif. 204/D7
Lopez (cape), Gabon 102/C5
Lopez (cape), Gabon 115/A4
Lopez, Pa. (18628) 294/K3
Lopez, Wash. (98261) 310/C2
Lopez (isl.), Wash. 310/C2
Lopi, Congo 115/C3
Lop Nor (Lop Nur) (lake), China 77/D3
Lopnur (Yuli), China 77/C3
Lop Nur (lake), China 54/L5
Lopphavet (bay), Norway 18/M1
Lora, Hamun-i- (swamp), Pakistan 68/B3
Lora, Hamun-i- (swamp), Pakistan 59/J4
Lora del Río, Spain 33/D4
Lorado, W. Va. (25630) 312/C7
Lorain (co.), Ohio 284/F3
Lorain, Ohio (*44052) 284/F3
Loraine, Ill. (62349) 222/B3
Loraine, N. Dak. (58761) 282/G2
Loraine, Texas (79532) 303/D5
Loraine, Wis. (†54825) 317/B4
Loralai, Pakistan 68/B2
Loralai, Pakistan 59/J3
Loramie (lake), Ohio 284/B5
Loranger, La. (70446) 238/N1
Lorca, Spain 33/F4
Lord Howe (isl.), Australia 87/G9
Lord Howe (isl.), Australia 2/T7
Lord Howe (isl.), N.S. Wales 97/J2
Lord Howe (Ontong Java) (isl.), Solomon Is. 87/G6
Lord Howe (Ontong Java) (isl.), Solomon Is. 86/D2
Lord Mayor (bay), N.W. Terrs. 187/J3
Lordsburg, N. Mex. (88045) 274/A6
Lords Point, Conn. (†06378) 210/H3
Lordstown, Ohio (†44481) 284/J3
Lords Valley, Pa. (†18428) 294/M3
Loreauville, La. (70552) 238/G6
Loreburn, Sask. 181/E4
Lore City, Ohio (43755) 284/H6
Lorena, Brazil 135/D3
Lorena, Miss. (†39153) 256/F6
Lorengau, Papua N.G. 86/A1
Lorengau, Papua N.G. 87/E6
Lo-Reninge, Belgium 27/B7
Lorentz, W. Va. (†26201) 312/F4
Lorenzo, Idaho (†83442) 220/G6
Lorenzo, Texas (79343) 303/C4
Lorenzo Geyres, Uruguay 145/B3
Lorestan (Luristan) (governorate), Iran 66/F4
Loreto, Bolivia 136/C4
Loreto, Colombia 126/E9
Loreto, Ecuador 128/D3
Loreto, Baja California, Mexico 150/D4
Loreto, Zacatecas, Mexico 150/J5
Loreto, Paraguay 144/D3
Loreto (dept.), Peru 128/E5
Loreto, Agusan del Sur, Philippines 82/E6
Loreto, Surigao del Norte, Philippines 82/E5
Loretta, Kansas (†67520) 232/C3
Loretta, Wis. (54852) 317/E4
Lorette, Manitoba 179/F5
Loretteville, Québec 172/H3
Loretto, Ky. (40037) 237/L5
Loretto, Mich. (49852) 250/B3
Loretto, Minn. (55357) 255/F5
Loretto, Nebr. (68646) 264/F3
Loretto, Pa. (15940) 294/E4
Loretto, Tenn. (38469) 237/G10
Loretto, Va. (22509) 307/O4
Lorian (swamp), Kenya 115/G3
Lorica, Colombia 126/C3
Lorida, Fla. (33857) 212/E4
Lorient, France 7/D4
Lorient, France 28/B4
L'Orignal, Ontario 177/K2
Lorimor, Iowa (50149) 229/E6
Lőrinci, Hungary 41/E3
Loring, Mont. (59537) 262/J2
Loring, Ontario 177/D2
Loring A.F.B., Maine 243/H2
Loris, S.C. (29569) 296/K3
Lorlie, Sask. 181/H5
Lorman, Miss. (39096) 256/B7
Lorne, New Bruns. 170/D1
Lorne, Nova Scotia 168/F3
Lorne (dist.), Scotland 15/C4
Lorne (firth), Scotland 10/D2
Lorne (firth), Scotland 15/C4
Loros (pt.), Chile 138/E3
Lörrach, W. Germany 22/B5
Lorrain (riv.), Martinique 161/D5
Lorraine (trad. prov.), France 29
Lorraine, Kansas (67459) 232/D3
Lorraine, N.Y. (13659) 276/J3
Lorraine, Québec 172/H4
Lorrainville, Québec 174/B3
Lorrha, Ireland 17/E5
Lort (riv.), W. Australia 88/C6
Lorton, Nebr. (68382) 264/H4
Lorton, Va. (22079) 307/O3
Lorze (riv.), Switzerland 39/F2
Los (isls.), Guinea 106/B7
Losada (riv.), Colombia 126/C6
Los Alamitos, Calif. (90720) 204/D11
Los Alamos, Calif. (93440) 204/E9
Los Alamos, N. Mex. 188/E3
Los Alamos (co.), N. Mex. 274/C3
Los Alamos, N. Mex. (87544) 274/C3
Los Alerces Nat'l Park, Argentina 143/C5
Los Algodones, Mexico 150/B1
Los Altos, Calif. (94022) 204/K3
Los Altos Hills, Calif. (94022) 204/J3
Los Amates, Guatemala 154/C3
Los Andes, Chile 138/B9
Los Andes, Colombia 126/B7
Los Angeles, Calif. 146/G6
Los Angeles, Calif. 188/C4
Los Angeles (co.), Calif. 204/G9
Los Angeles, Calif. (*90001) 204/C10
Los Ángeles, Chile 138/D1
Los Angeles, Chile 120/B6
Los Angeles, Texas (78051) 303/F9
Los Angeles, U.S. 2/D4
Los Angeles Aqueduct, Calif. 204/G8
Los Antiguos, Argentina 143/B6
Losantville, Ind. (47354) 227/G4
Los Arabos, Cuba 158/E1
Los Arroyos, Cuba 158/A2
Los Banos, Calif. (93635) 204/E6
Los Barcos (pt.), Cuba 158/B2
Los Canarreos (arch.), Cuba 158/C2
Los Castillos, Venezuela 124/G3
Los Choros (riv.), Chile 138/A7
Los Colorados (arch.), Cuba 158/A1
Los Conquistadores, Argentina 143/G5
Los Coyotes Ind. Res., Calif. 204/J10
Los Cusis, Bolivia 136/D4
Los Estados (isl.), Argentina 143/D7
Los Frailes (isl.), Dom. Rep. 158/C7
Los Fresnos, Texas (78566) 303/G11
Los Gatos, Calif. (95030) 204/K4
Los Glaciares Nat'l Park, Argentina 143/B6
Loshan (Leshan), China 77/F6
Los Hermanos (isls.), Venezuela 124/F2
Łosice, Poland 47/F2
Los Indios, Cuba 158/B2
Lošinj (isl.), Yugoslavia 45/B3
Los Lagos (reg.), Chile 138/D3
Los Lagos, Chile 138/D3
Los Llanos, Dom. Rep. 158/F6
Los Loros, Chile 138/B4
Los Lunas, N. Mex. (87031) 274/C4
Los Menucos, Argentina 143/C5
Los Mochis, Mexico 150/E4
Los Molinos, Calif. (96055) 204/D3
Los Monjes (isls.), Venezuela 124/C1
Los Muermos, Chile 138/D3
Los Navalmorales, Spain 33/D3
Los Navalucillos, Spain 33/D3
Los Negros, Cuba 158/F2
Løsning, Denmark 21/C6
Los Novillos, Uruguay 145/D2
Los Ojos, N. Mex. (87551) 274/C2
Los Olivos, Calif. (93441) 204/E9
Los Olmos (creek), Texas 303/F10
Los Olmos (creek), Texas 303/F11
Los Osos-Baywood Park, Calif. (†93402) 204/E8
Los Palacios, Cuba 158/B1
Los Palacios, Cuba 156/A2
Los Perales de Tapihue, Chile 138/F3
Los Pinos (riv.), Colo. 208/G8
Los Ranchos De Albuquerque, N. Mex. (†87101) 274/C3
Los Reyes de Salgado, Mexico 150/H7
Los Ríos (prov.), Ecuador 128/C3
Los Roques (isls.), Venezuela 124/E2
Los Santos, Panama 154/G7
Los Santos de Maimona, Spain 33/C3
Los Sauces, Chile 138/D2
Losser, Netherlands 27/L4
Lossiemouth and Branderburgh, Scotland 15/E3
Lossiemouth and Branderburgh, Scotland 10/E2
Lost (riv.), Calif. 204/D1

Lost (riv.), Ind. 227/D7
Lost (riv.), Minn. 255/C3
Lost (riv.), Oreg. 291/F5
Lost (creek), Utah 304/C5
Lostallo, Switzerland 39/H4
Lostant, Ill. (61334) 222/D2
Los Taques, Venezuela 124/C2
Lost Cabin, Wyo. (†82642) 319/E2
Lost City, W. Va. (26810) 312/J5
Lost Creek, Ky. (41348) 237/P6
Lost Creek, Wash. (†99180) 310/H2
Lost Creek, W. Va. (26385) 312/F4
Los Teques, Venezuela 120/C2
Los Teques, Venezuela 124/E2
Los Testigos (isls.), Venezuela 124/G2
Lost Hills, Calif. (93249) 204/F8
Lostine, Oreg. (97857) 291/K2
Lost Island (lake), Iowa 229/D2
Lost Nation, Iowa (52254) 229/M5
Lost River, Idaho (†83255) 220/E6
Lost River (range), Idaho 220/E5
Lost River, W. Va. (26811) 312/J5
Lost Springs, Kansas (66859) 232/E3
Lost Springs, Wyo. (82224) 319/G3
Lost Trail (pass), Idaho 220/E4
Lost Trail (pass), Mont. 262/B5
Lostwood, N. Dak. (†58784) 282/F3
Los Vilos, Chile 138/A9
Los Yébenes, Spain 33/E3
Lot (dept.), France 28/D5
Lot (riv.), France 28/D5
Lota, Chile 138/D1
Lotagipi Swamp (plain), Sudan 111/F6
Lotbinière (co.), Québec 172/F3
Lotbinière, Québec 172/F3
Lot-et-Garonne (dept.), France 28/D5
Lothair, Ky. (†41701) 237/P6
Lothair, Mont. (59461) 262/E2
Lothian, Md. (20820) 245/M5
Lothian (reg.), Scotland 15/E5
Lothain (trad. co.), Scotland 15/A5
Loto, Zaire 115/D4
Lötschberg (tunnel), Switzerland 39/E4
Lotsee, Okla. (†74063) 288/O2
Lott, Texas (76656) 303/H6
Lottie, Ala. (†36552) 195/C8
Lottie, La. (70756) 238/G5
Lottsville, Pa. (†16402) 294/D2
Lotus, Calif. (95651) 204/C8
Lotzwil, Switzerland 39/E2
Louang Namtha, Laos 72/D2
Louangphrabang, Laos 72/D3
Louangphrabang, Laos 54/M7
Louann, Ark. (71751) 202/E7
Loubomo, Congo 115/B4
Loubomo, Congo 102/D5
Loudéac, France 28/B3
Loudima, Congo 115/B4
Loudon◯, N.H. (03301) 268/E5
Loudon (co.), Tenn. 237/N9
Loudon, Tenn. (37774) 237/N9
Loudonville, Ohio (44842) 284/F4
Loudoun (co.), Va. 307/N2
Loudun, France 28/D4
Louellen, Ky. (40853) 237/P7
Louga, Senegal 106/A5
Loughborough, England 13/F5
Loughborough, England 10/F4
Loughbrickland, N. Ireland 17/J3
Lougheed, Alberta 182/E3
Lougheed (isl.), N.W. Terrs. 187/H2
Loughman, Fla. (33858) 212/E3
Loughrea, Ireland 17/E5
Loughrea, Ireland 10/B4
Loughros More (bay), Ireland 17/D2
Louin, Miss. (39338) 256/F6
Louisa (co.), Iowa 229/L6
Louisa, Ky. (41230) 237/R4
Louisa, La. (†70538) 238/G7
Louisa (lake), Ontario 177/F2
Louisa (co.), Va. 307/N5
Louisa, Va. (23093) 307/M4
Louisbourg, Nova Scotia 168/J3
Louisbourg Nat'l Hist. Park, Nova Scotia 168/J3
Louisburg, Kansas (66053) 232/H3
Louisburg, Minn. (56254) 255/B5
Louisburg, Mo. (65685) 261/F7
Louisburg, N.C. (27549) 281/N2
Louisburgh, Ireland 17/B4
Louis Creek, Br. Col. 184/H4
Louisdale, Nova Scotia 168/G3
Louise (lake), Alaska 196/C1
Louise (isl.), Br. Col. 184/B4
Louise, Miss. (39097) 256/C5
Louise (lake), Québec 172/C4
Louise, Texas (77455) 303/H8
Louiseville, Québec 172/E3
Louisiade (arch.), Papua N.G. 87/F7
Louisiade (arch.), Papua N.G. 85/D8
Louisiana 188/H4
LOUISIANA 238
Louisiana (pt.), La. 238/C7
Louisiana, Mo. (63353) 261/K4
Louisiana (state), U.S. 146/J6
Louis Trichardt, S. Africa 118/E4
Louisville, Ala. (36048) 195/G7
Louisville, Colo. (80027) 208/J3
Louisville, Georgia (30434) 217/H4
Louisville, Ill. (62858) 222/E5
Louisville, Kansas (66450) 232/F2
Louisville, Ky. (*40201) 237/J2
Louisville, Ky. 146/K6
Louisville, Miss. (39339) 256/G4
Louisville, Nebr. (68037) 264/H3
Louisville, Ohio (44641) 284/H4
Louisville, Tenn. (37777) 237/N9
Louisvillle, Ky. 188/J3
Louis XIV (pt.), Que. 162/H5
Louis XIV (pt.), Québec 174/B2
Loukhi, U.S.S.R. 52/D1
Loulé, Portugal 33/B4
Louny, Czech. 41/B1
Loup (co.), Nebr. 264/E3
Loup (riv.), Nebr. 264/F3
Loup (riv.), Québec 172/H2
Loup City, Nebr. (68853) 264/E3
Lourdes, France 28/C6
Lourdes, Newf. 166/C4
Lourdes, N. Mex. (†87701) 274/D3
Lourdes, Québec 172/F3
Louriçal, Portugal 33/B3
Lourinhã, Portugal 33/B3
Lousã, Portugal 33/B2
Lousana, Alberta 182/D3
Louth, England 13/H4
Louth, England 10/F4
Louth (co.), Ireland 17/J4
Louth, Ireland 17/J4
Louth, N.S. Wales 97/C2
Loutrá Aidhipsoú, Greece 45/F6
Louvain (Leuven), Belgium 27/F7
Louvale, Georgia (31814) 217/C6
Louviers, Colo. (80131) 208/K4
Louviers, France 28/D3
Lövånger, Sweden 18/M4
Lovango (cay), Virgin Is. (U.S.) 161/C4
Lovat' (riv.), U.S.S.R. 52/D3
Love, Miss. (†38632) 256/D1
Love (co.), Okla. 288/M7
Love, Sask. 181/G2
Lovech, Bulgaria 45/G4
Lovejoy, Georgia (30250) 217/D4
Lovejoy, Ill. (62059) 222/A2
Lovelaceville, Ky. (42060) 237/D7
Lovelady, Texas (75851) 303/J6
Loveland, Colo. 188/E2
Loveland, Colo. (80537) 208/J2
Loveland, Iowa (†51555) 229/B6
Loveland, Ohio (45140) 284/D9
Loveland, Okla. (73553) 288/J6
Lovell, Maine (04051) 243/B7
Lovell◯, Maine (04051) 243/B7
Lovell, Okla. (†73028) 288/L2
Lovell, Wyo. (82431) 319/D1
Lovells, Mich. (†49738) 250/E4
Lovelock, Nev. (89419) 266/C2
Lovely, Ky. (41231) 237/S5
Lovenia (mt.), Utah 304/D3
Love Point, Md. (†21617) 245/N4
Loverna, Sask. 181/B4
Loves Park, Ill. (61111) 222/E1
Lovett, Georgia (†31021) 217/G5
Lovett, Ind. (†47265) 227/F7
Lovettsville, Va. (22080) 307/N2
Love Valley, N.C. (28677) 281/H3
Loveville, Md. (20656) 245/M7
Lovewell, Kansas (†66942) 232/D2
Lovewell (res.), Kansas 232/D2
Lovilia, Iowa (50150) 229/H6
Loving, N. Mex. (88256) 274/E6
Loving (co.), Texas 303/A6
Lovingston, Va. (22949) 307/L5
Lovington, Ill. (61937) 222/E4
Lovington, N. Mex. (88260) 274/F6
Lovisa, Finland 18/P6
Lövö, Hungary 41/D3
Lovosice, Czech. 41/C1
Lóvua, Angola 115/D5
Low (cape), N.W.T. 162/H3
Low (cape), N.W. Terrs. 187/K3
Low, Québec 172/B4
Lowa (riv.), Zaire 115/E4
Low Bush River, Ontario 177/K5
Low Bush River, Ontario 175/E3
Lowden, Iowa (52255) 229/L5
Lowder, Ill. (62662) 222/D4
Lowe Farm, Manitoba 179/E5
Lowell, Ark. (72745) 202/B1
Lowell, Fla. (32663) 212/D2
Lowell, Idaho (†83539) 220/C3
Lowell (lake), Idaho 220/B6
Lowell, Ind. (46356) 227/C2
Lowell, Iowa (†52645) 229/L7
Lowell, Maine (†04433) 243/F5
Lowell◯, Maine (†04433) 243/F5
Lowell, Mass. 188/M2
Lowell, Mass. (*01850) 249/J2
Lowell, Mich. (49331) 250/D6
Lowell, N.C. (28098) 281/G4
Lowell, Ohio (45744) 284/H6
Lowell, Oreg. (97452) 291/E4
Lowell◯, Vt. (05847) 268/C2
Lowell, W. Va. (†24910) 312/E7
Lowell, Wis. (53557) 317/J9
Lowell Nat'l Hist. Park, Mass. 249/J2
Lowellville, Ohio (44436) 284/J3
Lower Alkali (lake), Calif. 204/E2
Lower Argyle, Nova Scotia 168/C5
Lower Arrow (lake), Br. Col. 184/H5
Lower Austria (prov.), Austria 41/C2
Lower Bank, N.J. (†08215) 273/E4
Lower Barneys River, Nova Scotia 168/F3
Lower Brule, S. Dak. (57548) 298/K5
Lower Brule Ind. Res., S. Dak. 298/K5
Lower Burrell, Pa. (15068) 294/C4
Lower Cabot, Vt. (†05658) 268/C3
Lower California (pen.), Mexico 2/D4
Lower California (pen.), Mexico 146/G7
Lower California (pen.), Mexico 150/C3
Lower Cloverdale, New Bruns. 170/F2
Lower Crab (creek), Wash. 310/F4
Lower Derby, New Bruns. 170/E2
Lower Durham, New Bruns. 170/D2
Lower East Pubnico, Nova Scotia 168/C5
Lower Elwah Ind. Res., Wash. 310/B2
Lower Engadine (valley), Switzerland 39/K3
Lower Goose Creek (res.), Idaho 220/D7
Lower Granite (lake), Idaho 220/A3
Lower Granite (dam), Wash. 310/H4
Lower Granite (lake), Wash. 310/H4
Lower Hainesville, New Bruns. 170/C2
Lower Hutt, N. Zealand 100/B2
Lower Island Cove, Newf. 166/D2
Lower Kalskag, Alaska (99626) 196/F2
Lower Kars, New Bruns. 170/E3
Lower Klamath (lake), Calif. 204/D2
Lower Lake, Calif. (95457) 204/C5
Lower L'Ardoise, Nova Scotia 168/H3
Lower Marlboro, Md. (†20836) 245/M6
Lower Matecumbe (key), Fla. 212/F7
Lower Millstream, New Bruns. 170/E3
Lower Montague, Pr. Edward I. 168/F2
Lower Monumental (dam), Wash. 310/G4
Lower Monumental (lake), Wash. 310/G4
Lower New York (bay), N.J. 273/E2
Lower Nicola, Br. Col. 184/G5
Lower Ohio, Nova Scotia 168/C5
Lower Paia, Hawaii (†96779) 218/J1
Lower Peach Tree, Ala. (36751) 195/C7
Lower Post, Br. Col. 184/K1
Lower Red (lake), Minn. 255/C3
Lower Red Rock (lake), Mont. 262/E6
Lower Rhine (riv.), Netherlands 27/H5
Lower Roach (pond), Maine 243/E4
Lower Saint Mary (lake), Mont. 262/C2
Lower Salem, Ohio (45745) 284/H6
Lower Sapin, New Bruns. 170/F2
Lower Saranac (lake), N.Y. 276/M2
Lower Saxony (state), W. Germany 22/C2
Lower Southampton, New Bruns. 170/C2
Lower South River, Nova Scotia 168/F3
Lower Sysladobsis (lake), Maine 243/G5
Lower Tonsina, Alaska (†99566) 196/J2
Lower Tunguska (riv.), U.S.S.R. 54/L3
Lower Tunguska (riv.), U.S.S.R. 48/K3
Lower Waterford, Vt. (05848) 268/D3
Lower Wedgeport, Nova Scotia 168/C5
Lower West Pubnico, Nova Scotia 168/C5
Lower Woods Harbour, Nova Scotia 168/C5
Lowery, Ala. (†36453) 195/F8
Lowery (lake), Fla. 212/E3
Lowes, Ky. (42061) 237/D7
Lowestoft, England 13/J5
Lowestoft, England 10/G4
Lowesville, Va. (22951) 307/K5
Lowgap, N.C. (27024) 281/H1
Łowicz, Poland 47/D2
Lowland, N.C. (28552) 281/S4
Lowman, Idaho (83637) 220/C5
Lowmansville, Ky. (41232) 237/R5
Low Moor, Iowa (52757) 229/N5
Lowmoor, Va. (24457) 307/J5
Lowndes (co.), Ala. 195/E6
Lowndes (co.), Georgia 217/F9
Lowndes (co.), Miss. 256/H4
Lowndes, Mo. (63951) 261/M8
Lowndesboro, Ala. (36752) 195/E6
Lowndesville, S.C. (29659) 296/B3
Lowpoint, Ill. (61545) 222/D3
Low Rocky (pt.), Tasmania 99/B4
Lowry, Minn. (56349) 255/C5
Lowry, S. Dak. (†57472) 298/K3
Lowry, Va. (24570) 307/K6
Lowry A.F.B., Colo. 208/K3
Lowry City, Mo. (64763) 261/E6
Lowrys, S.C. (†29706) 296/E2
Lowther (isl.), N.W. Terrs. 187/J2
Lowville, N.Y. (13367) 276/J3
Low Wassie, Mo. (†65588) 261/K9
Loxahatchee, Fla. (33470) 212/F5
Loxley, Ala. (36551) 195/C9
Loxton, S. Australia 94/G6
Loxton North, S. Australia 94/G6
Loyal, Okla. (73756) 288/K3
Loyal, Loch (lake), Scotland 15/D2
Loyal, Wis. (54446) 317/E6
Loyalhanna, Pa. (15661) 294/D5
Loyalist, Alberta 182/E4
Loyall, Ky. (40854) 237/P7
Loyalton, Calif. (96118) 204/E4
Loyalton, Pa. (†17048) 294/J4
Loyalton, S. Dak. (†57471) 298/L3
Loyalty (isls.), New Caled. 87/G8
Loyalty (isls.), New Caled. 86/H4
Loyang (Luoyang), China 77/H5
Loyd, Wis. (†53924) 317/F9
Loyne, Loch (lake), Scotland 15/C3
Loysburg, Pa. (16659) 294/F5
Loysville, Pa. (17047) 294/H5
Lozeau (lake), Newf. 166/B3
Lozère (dept.), France 28/E5
Loznica, Yugoslavia 45/D3
Lozovaya, U.S.S.R. 52/E5
Lua (riv.), Zaire 115/C3
Luacano, Angola 115/D6
Luachimo, Angola 115/D5
Lualaba (riv.), Zaire 115/E5
Lualaba (riv.), Zaire 115/E4
Lua Makika (mt.), Hawaii 218/J3
Lu'an, China 77/J5
Luana, Iowa (52156) 229/K2
Luana (pt.), Jamaica 158/G6
Luanchuan, China 77/H5
Luanda (dist.), Angola 115/B5
Luanda (cap.), Angola 2/K6
Luanda (cap.), Angola 115/B5
Luanda (cap.), Angola 102/D5
Luang, Thale (lag.), Thailand 72/D6
Luang (mt.), Thailand 72/C5
Luang Prabang (Loungphrabang), Laos 72/D3
Luangwa (Feira), Zambia 115/E7
Luangwa (riv.), Zambia 115/F6
Luanshya, Zambia 115/E6
Luanshya, Zambia 102/E6
Luao, Angola 115/D6
Luapula (riv.), Zaire 115/E6
Luapula (riv.), Zambia 115/E6
Luarca, Spain 33/C1
Luashi, Zaire 115/D6
Luba, Equat. Guinea 115/A3
Lubaczów, Poland 47/F3
Lubań, Poland 47/B3
Lubāna (lake), U.S.S.R. 53/D2
Lubang, Philippines 82/C4
Lubang (isls.), Philippines 85/F3
Lubang (isls.), Philippines 82/B4
Lubango, Angola 115/B6
Lubango, Angola 102/D6
Lubartów, Poland 47/F3
Lubawa, Poland 47/D2
Lübben, E. Germany 22/E3
Lübbenau, E. Germany 22/F3
Lubbock (co.), Texas 303/C4
Lubbock, Texas 146/H6
Lubbock, Texas (*79401) 303/C4
Lubbock, Texas 188/F4
Lubec, Maine (04652) 243/K6
Lubec◯, Maine (04652) 243/K6
Lübeck, W. Germany 7/E3
Lübeck, W. Germany 22/D2
Lubeck, W. Va. (†26101) 312/C4
Lubefu, Zaire 115/D4
Lubero, Zaire 115/E4
L'ubica, Czech. 41/F2
Lubicon (lake), Alberta 182/C1
Lubien Kujawski, Poland 47/D2
Lubilash (riv.), Zaire 115/D5
Lubin, Poland 47/C3
Lublin (prov.), Poland 47/F3
Lublin, Poland 47/F3
Lublin, Poland 7/G3
Lublin, Wis. (54447) 317/E5
Lubliniec, Poland 47/D3
Lubny, U.S.S.R. 52/D4
Luboń, Poland 47/C2
Lubrín, Spain 33/F4
Lubsko, Poland 47/B3
Lubuagan, Philippines 82/C2
Lubudi, Zaire 115/E5
Lubuklinggau, Indonesia 85/C6
Lubuksikaping, Indonesia 85/B5
Lubumbashi, Zaire 2/L6
Lubumbashi, Zaire 115/E6
Lubumbashi, Zaire 102/E6
Lubutu, Zaire 115/E4
Lübz, E. Germany 22/D2
Lucama, N.C. (27851) 281/N3
Lucan, Minn. (56255) 255/C6
Lucan, Ontario 177/C4
Luc An Chau, Vietnam 72/E2
Lucan-Doddsborough, Ireland 17/J5
Lucas (co.), Iowa 229/G6
Lucas, Iowa (50151) 229/G6
Lucas, Kansas (67648) 232/D2
Lucas, Ky. (42156) 237/K7
Lucas, Mich. (†49637) 250/D4
Lucas (co.), Ohio 284/C2
Lucas, Ohio (44843) 284/F4
Lucas, S. Dak. (57549) 298/L7
Lucas, Texas (†75069) 303/H1
Lucas E. de Peña, Dom. Rep. 158/D5
Lucas González, Argentina 143/G6
Lucasville, Ohio (45648) 284/E8
Lucban, Philippines 82/C3
Lucca (prov.), Italy 34/C3
Lucca, Italy 34/C3
Lucca, N. Dak. (†58027) 282/P6
Luce (co.), Mich. 250/D2
Luce, Minn. (†56573) 255/C4
Luce (bay), Scotland 10/D3
Luce (bay), Scotland 15/D6
Lucea, Jamaica 158/G5
Lucedale, Miss. (39452) 256/G9
Lucena, Philippines 85/G3
Lucena, Philippines 82/C4
Lucena, Spain 33/D4
Lucena del Cid, Spain 33/F2
Lučenec, Czech. 41/E2
Lucens, Switzerland 39/C3
Lucera, Italy 34/E4
Lucerna, Peru 128/H9
Lucerne, Calif. (95458) 204/C4
Lucerne, Colo. (80646) 208/K2
Lucerne, Ind. (46950) 227/E3
Lucerne, Mo. (64655) 261/F2
Lucerne, Québec 172/B4
Lucerne (Luzern) (canton), Switzerland 39/F2
Lucerne, Switzerland 39/F2
Lucerne (lake), Switzerland 39/F3
Lucerne, Wash. (†98816) 310/E2
Lucerne, Wyo. (†82443) 319/D2
Lucernemines, Pa. (15754) 294/D4
Lucerne Valley, Calif. (92356) 204/J9
Lucero (lake), N. Mex. 274/C6
Luceville, Québec 172/J1
Luchow (Luzhou), China 77/G6
Lüchow, W. Germany 22/D2
Lucia, Calif. (†93920) 204/D7
Lucie (riv.), Suriname 131/C4
Lucien, Miss. (39646) 256/C7
Lucien, Okla. (73757) 288/M2
Lucile, Georgia (†31723) 217/C8
Lucile, Idaho (83542) 220/B4
Lucile, Ky. (†41171) 237/P4
Lucinda, Pa. (16235) 294/D3
Lucira, Angola 115/B6
Luck, Wis. (54853) 317/B4
Luckau, E. Germany 22/E3
Luckenwalde, E. Germany 22/E2
Lucketts, Va. (†22075) 307/N2
Luckey, Ohio (43443) 284/D3
Lucknow, India 68/E3
Lucknow, India 54/K7
Lucknow, Ontario 177/C4
Lucky, La. (†71008) 238/E2
Lucky Lake, Sask. 181/D5
Lucky Peak (lake), Idaho 220/B6
Luçon, France 28/C4
Lucrecia (cape), Cuba 158/J3
Lucy, La. (†70049) 238/M3
Lucy, Tenn. (†38053) 237/B10
Lucy Creek, North. Terr. 93/E7
Lüda (Lüta), China 77/K4
Lüda, China 2/R4
Lüda, China 54/O6
Ludden, N. Dak. (58462) 282/O7
Ludell, Kansas (67744) 232/B2
Lüdenscheid, W. Germany 22/B3
Lüderitz, Namibia 118/A5
Lüderitz, Namibia 102/D7
Lüderitz (bay), Namibia 118/A5
Ludhiana, India 54/J6
Ludhiana, India 68/D2
Ludington, Mich. (49431) 250/C5
Ludlow, Calif. (†92365) 204/J9
Ludlow, England 10/E4
Ludlow, England 13/E5
Ludlow, Ill. (60949) 222/E3
Ludlow, Ky. (41016) 237/S2
Ludlow◯, Maine (†04730) 243/G3
Ludlow◯, Mass. (01056) 249/E4
Ludlow, Miss. (39098) 256/E5
Ludlow, Mo. (64656) 261/E3
Ludlow, New Bruns. 170/D2
Ludlow, Pa. (16333) 294/E2
Ludlow, S. Dak. (57755) 298/C2
Ludlow, Vt. (05149) 268/B5
Ludlow◯, Vt. (05149) 268/B5
Ludlow (mt.), Vt. 268/B5
Ludlow Center, Mass. (†01056) 249/E4
Ludlow Falls, Ohio (45339) 284/B6
Ludowici, Georgia (31316) 217/J7
Luduş, Romania 45/G2
Ludvika, Sweden 18/J6
Ludville, Georgia (†30175) 217/C2
Ludwigsburg, W. Germany 22/C4
Ludwigshafen am Rhein, W. Germany 22/C4
Ludwigslust, E. Germany 22/D2
Ludza, U.S.S.R. 53/D2
Lue, N.S. Wales 97/E3
Luebbering, Mo. (63061) 261/L6
Luebo, Zaire 115/D5
Lueders, Texas (79533) 303/E5
Luella, Georgia (†30248) 217/D4
Luena, Angola 115/C6
Luepa, Venezuela 124/H5
Lüeyang, China 77/G5
Lufeng, China 77/J7
Lufira (riv.), Zaire 115/E5
Lufkin, Texas (75901) 303/K6
Luga, U.S.S.R. 52/C3
Luga, U.S.S.R. 48/D4
Lugano, Switzerland 39/G4
Lugano (lake), Switzerland 39/H5
Luganville, Vanuatu 87/G7
Lugareño, Cuba 158/G3
Lugenda (riv.), Mozambique 118/F2
Lugerville, Wis. (†54555) 317/E4
Lugnaquillia (mt.), Ireland 17/J5
Lugo, Italy 34/D2
Lugo (prov.), Spain 33/C1
Lugo, Spain 33/C1
Lugoff, S.C. (29078) 296/F3
Lugoj, Romania 45/F3
Luhaiya, Yemen Arab Rep. 59/D6
Luiana, Angola 115/D7
Luiana, Angola 102/E6
Luik (Liège), Belgium 27/H7
Luilaka (riv.), Zaire 115/C4
Luimneach (Limerick), Ireland 10/B4
Luimneach (Limerick), Ireland 17/D6
Luina, Tasmania 99/B3
Luing (isl.), Scotland 15/C4
Luís Correia, Brazil 132/F3
Luis de Saboya, Cerro (mt.), Chile 138/F11
Luishia, Zaire 115/E6
Luitpold Coast (reg.) 5/B17
Luiza, Zaire 115/D5
Luján, Argentina 143/G7
Lujiang, China 77/J5
Lukachukai, Ariz. (86507) 198/F2
Lukachukai (mts.), Ariz. 198/F2
Lukapa, Angola 115/D5
Luke, Md. (21540) 245/B3
Luke A.F.B., Ariz. 198/C5
Lukenie (riv.), Zaire 115/D4
Lukeville, Ariz. (85341) 198/C7
Lukolela, Equateur, Zaire 115/C4
Lukolela, Kasai-Oriental, Zaire 115/D5
Lukovit, Bulgaria 45/G4
Łuków, Poland 47/F3
Lukuga (riv.), Zaire 115/E5
Lukula, Zaire 115/B5
Lukulu, Zambia 115/D6
Lula, Georgia (30554) 217/E2
Lula, Miss. (38644) 256/C2
Lula, Okla. (†74825) 288/O5
Lule (riv.), Sweden 7/G2
Luleå, Sweden 7/G2
Luleå, Sweden 18/N4
Luleälv (riv.), Sweden 18/M4
Lüleburgaz, Turkey 63/B2
Lules, Argentina 143/C2
Luling, La. (70070) 238/N4
Luling, Texas (78648) 303/G8
Lulu, Fla. (32061) 212/D1
Lulua (riv.), Zaire 115/D5
Lum, Mich. (48452) 250/F5
Lumajang, Indonesia 85/K2
Lumajangdong Co (lake), China 77/B5
Lumbala, Angola 115/D6
Lumber (riv.), N.C. 281/L6
Lumber (riv.), S.C. 296/J3
Lumber Bridge, N.C. (28357) 281/L5
Lumber City, Georgia (31549) 217/G7
Lumber City, Pa. (†16833) 294/E4
Lumberport, W. Va. (26386) 312/F4
Lumberton, Miss. (39455) 256/E8
Lumberton, N.J. (08048) 273/D4
Lumberton, N. Mex. (87547) 274/C2
Lumberton, N.C. (28358) 281/L5
Lumberton, Texas (†77656) 303/K7
Lumberville, Pa. (18933) 294/N5
Lumbo, Mozambique 118/G3
Lumbrales, Spain 33/C2
Lumbrein, Switzerland 39/H3
Lumby, Br. Col. 184/H5
Lumding, India 68/G3
Lummen, Belgium 27/G7
Lummi (isl.), Wash. 310/C2
Lummi Ind. Res., Wash. 310/C2
Lummi Island, Wash. (98262) 310/C2
Lumphat, Cambodia 72/E4
Lumpkin (co.), Georgia 217/D1
Lumpkin, Georgia (31815) 217/C6
Lumsden, Newf. 166/D4
Lumsden, N. Zealand 100/B6
Lumsden, Sask. 181/G5
Lumsden, Scotland 15/F3
Lumsden Beach, Sask. 181/F5
Lumut, Malaysia 72/D6
Luna, Ark. (†71653) 202/H7
Luna (co.), N. Mex. 274/B6
Luna, N. Mex. (87824) 274/A5
Luna Pier, Mich. (48157) 250/F7
Luncarty, Scotland 15/E4
Lund, Br. Col. 184/E5
Lund, Idaho (†83241) 220/G7
Lund, Nev. (89317) 266/F4
Lund, Sweden 18/H9
Lund, Utah (†84720) 304/A5
Lundale, W. Va. (25631) 312/C7
Lunda Norte (dist.), Angola 115/C5
Lundar, Manitoba 179/D4
Lunda Sul (dist.), Angola 115/D5
Lundazi, Zambia 115/F6
Lundbreck, Alberta 182/C5
Lundby, Denmark 21/E7
Lundell, Ark. (†72367) 202/H5
Lunderskov, Denmark 21/C7
Lundi (riv.), Zimbabwe 118/E4
Lunds Corner, Mass. (02745) 249/L6
Lundsvalley, N. Dak. (†58724) 282/E3
Lundy (isl.), England 13/C6
Lundy (isl.), England 10/D5
Lune (riv.), England 13/E3
Lüneburg, W. Germany 22/D2
Lüneburger Heide (dist.), W. Germany 22/C2
Lunel, France 28/E6
Lünen, W. Germany 22/B3
Lunenburg, Mass. (01462) 249/H2
Lunenburg◯, Mass. (01462) 249/H2
Lunenburg, N.S. 162/K7
Lunenburg (co.), Nova Scotia 168/D4
Lunenburg, Nova Scotia 168/D4
Lunenburg (bay), Nova Scotia 168/D4
Lunenburg◯, Vt. (05906) 268/D3
Lunenburg (co.), Va. 307/M7
Lunenburg, Va. (23952) 307/M7
Lunéville, France 28/G3
Lung (riv.), Ireland 17/D4
Lungchen (Longzhen), China 77/L2
Lungdo, China 77/B5
Lungern, Switzerland 39/F3
Lungi, S. Leone 106/B7
Lungleh, India 68/G4
Lungwebungu (riv.), Angola 115/D6
Lungwebungu (riv.), Zambia 115/D6
Luni (riv.), India 68/C3
Luninets, U.S.S.R. 52/C4
Luning, Nev. (89420) 266/C4
Lunita, La. (†70661) 238/C6
Lunsford, Ark. (†72437) 202/K2
Luocheng, China 77/G7
Luodian, China 77/G6
Luoding, China 77/H7
Luohe, China 77/H5
Luoyang (Loyang), China 77/H5
Luoyang, China 54/N6
Luozi, Zaire 115/B5
Lupeni, Romania 45/F3
Luperón, Dom. Rep. 158/D5
Lupon, Philippines 82/E7
Lupton, Ariz. (86508) 198/F3
Lupton, Mich. (48635) 250/F4
Lupus, Mo. (†65046) 261/H5
Luputa, Zaire 115/D5
Luqu, China 77/F5
Luque, Paraguay 144/B4
Luquillo, P. Rico 161/F1
Luquillo, Sierra de (mts.), P. Rico 161/E2
Lurah (riv.), Afghanistan 68/B2
Lurah (riv.), Afghanistan 59/J3
Luray, Kansas (67649) 232/D2
Luray, Mo. (63453) 261/J2
Luray, S.C. (29932) 296/E6
Luray, Tenn. (38352) 237/D9
Luray, Va. (22835) 307/M3
Lure, France 28/G4
Lurgan, N. Ireland 17/J3
Luribay, Bolivia 136/B5
Lurín, Peru 128/D9
Lúrio, Mozambique 118/G2
Lúrio (riv.), Mozambique 118/F2
Luristan (Lorestan) (gov.), Iran 66/F4
Lurton, Ark. (†72856) 202/D2
Lusaka (cap.), Zambia 115/E7
Lusaka (cap.), Zambia 102/E6
Lusaka (cap.), Zambia 2/L6
Lusambo, Zaire 102/E5
Lusambo, Zaire 115/D4
Lusatia (reg.), E. Germany 22/F3
Lusby, Md. (20657) 245/N7
Luseland, Sask. 181/B3
Lushi, China 77/H5
Lushnje, Albania 45/D5
Lushoto, Tanzania 115/G4
Lushton, Nebr. (†68371) 264/G4
Lushui, China 77/E6
Lüshun, China 77/K4
Lusk, Ireland 17/J4
Lusk, Wyo. (82225) 319/H3
Luso, Angola 102/E6
Luss, Scotland 15/D4
Lustenau, Austria 41/A3
Lustre, Mont. (59225) 262/K2
Lut, Dasht-e (des.), Iran 59/G3
Lut, Dasht-e (des.), Iran 66/L5
Lüta (Lüda), China 77/K4
Lutcher, La. (70071) 238/L3
Lutesville, Mo. (63762) 261/M8
Luther, Iowa (50152) 229/F5
Luther, Mich. (49656) 250/D4
Luther, Mont. (59051) 262/G5
Luther, Okla. (73054) 288/M3
Luther, Tenn. (†37869) 237/P8
Luthern, Switzerland 39/E2
Luthersburg, Pa. (15848) 294/E3
Luthersville, Georgia (30251) 217/C4
Lutherville-Timonium, Md. (21093) 245/M3
Lutie, Okla. (†74578) 288/R5
Luton, England 13/G6
Luton, England 10/F5
Luton, Iowa (†51052) 229/A4
Lutry, Switzerland 39/C3

Lutsen, Minn. (55612) 255/F2
Lutsk, U.S.S.R. 7/G3
Lutsk, U.S.S.R. 52/B4
Lutsk, U.S.S.R. 48/C4
Luttrell, Tenn. (37779) 237/O8
Lutts, Tenn. (38471) 237/F10
Lutz, Fla. (33549) 212/D3
Lützelflüh, Switzerland 39/E3
Lützow-Holm (bay) 5/C3
Luuq, Somalia 115/H3
Luverne, Ala. (36049) 195/F7
Lu Verne, Iowa (50560) 229/E3
Luverne Minn. (56156) 255/B7
Luverne, N. Dak. (58056) 282/P5
Luvua (riv.), Zaire 115/E5
Luwingu, Zambia 115/E6
Luwuk, Indonesia 85/G6
Lux, Miss. (†39401) 256/F8
Luxembourg 7/E4
Luxembourg (prov.), Belgium 27/G9
LUXEMBOURG 27/J9
Luxembourg (cap.), Luxembourg 27/J9
Luxemburg, Iowa (52056) 229/L3
Luxemburg, Minn. (†56301) 255/D5
Luxemburg, Wis. (54217) 317/L6
Luxeuil-les-Bains, France 28/G4
Luxi, China 77/E7
Luxi, China 77/F7
Luxor, Egypt 102/F2
Luxor, Egypt 59/B4
Luxor, Egypt 111/F2
Luxora, Ark. (72358) 202/K2
Luz, Brazil 135/D1
Luz (isl.), Chile 138/D6
Luza, U.S.S.R. 52/G2
Luzein, Switzerland 39/J3
Luzern (canton), Switzerland 39/F2
Luzern (Lucerne), Switzerland 39/F2
Luzerne, Iowa (52257) 229/J5
Luzerne, Mich. (48636) 250/E4
Luzerne (co.), Pa. 294/L3
Luzerne, Pa. (18709) 294/E7
Luzhai, China 77/G7
Luzhi, China 77/G6
Luzhou (Luchow), China 77/G6
Luziânia, Brazil 132/E7
Luzilândia, Brazil 132/F3
Lužnice (riv.), Czech. 41/C2
Luzon (isl.), Philippines 2/R5
Luzon (isl.), Philippines 54/O8
Luzon (isl.), Philippines 82/C3
Luzon (isl.), Philippines 85/G2
Luzon (sea), Philippines 82/B4
Luzon (str.), Philippines 82/A2
Luz-Saint-Sauveur, France 28/C6
L'vov, U.S.S.R. 7/G4
L'vov, U.S.S.R. 48/C4
L'vov (Lwów), U.S.S.R. 52/B5
Lyakhov (isls.), U.S.S.R. 4/B3
Lyal (isl.), Ontario 177/C3
Lyallpur (Faisalabad), Pakistan 68/C2
Lyallpur (Faisalabad), Pakistan 59/K3
Lyalta, Alberta 182/D4
Lyatkhovskiye (isls.), U.S.S.R. 48/O2
Lybster, Scotland 10/E1
Lybster, Scotland 15/E2
Lycan, Colo. (†81054) 208/P7
Lycksele, Sweden 18/L4
Lycoming, N.Y. (13093) 276/H3
Lycoming (co.), Pa. 294/H3
Lycoming (creek), Pa. 294/H3
Lydallville, Conn. (†06040) 210/F1
Lydd, England 13/H7
Lydda, Israel 65/B4
Lydenburg, S. Africa 118/E4
Lydia, Minn. (†55352) 255/E6
Lydia, S.C. (29079) 296/G3
Lydia Mills, S.C. (29325) 296/D3
Lydick, Ind. (†46601) 227/E1
Lyell (mt.), Alberta 182/B4
Lyell (isl.), Br. Col. 184/B4
Lyell (mt.), Br. Col. 184/J4
Lyell (mt.), Tasmania 99/B4
Lyerly, Georgia (30730) 217/B2
Lyford, Ind. (†47874) 227/C5
Lyford, Texas (78569) 303/G11
Lykens, Pa. (17048) 294/J4
Lyle, Minn. (55953) 255/F7
Lyle, Wash. (98635) 310/D5
Lyles, Tenn. (37098) 237/G9
Lyleton, Manitoba 179/A5
Lyman, Miss. (†39501) 256/F10
Lyman, Nebr. (69352) 264/A3
Lyman○, N.H. (†03585) 268/D3
Lyman, S.C. (29365) 296/C2
Lyman (co.), S. Dak. 298/J6
Lyman, Utah (84749) 304/C5
Lyman, Wash. (98263) 310/D2
Lyman, Wyo. (82937) 319/B4
Lyme (bay), England 13/D7
Lyme (bay), England 10/E5
Lyme○, N.H. (03768) 268/C4
Lyme Center, N.H. (03769) 268/C4
Lyme Regis, England 13/E7
Lyme Regis, England 10/E5
Lymington, England 10/F5
Lymington, England 13/F7
Lymm, England 13/H2
Lymm, England 10/G2
Lyn, Ontario 177/J3
Lyna (riv.), Poland 47/E1
Lynbrook, N.Y. (11563) 276/P7
Lynch, Ky. (40855) 237/R7
Lynch, Md. (21646) 245/O3
Lynch, Nebr. (68746) 264/F2
Lynchburg, Mo. (65543) 261/H7
Lynchburg, N. Dak. (†58023) 282/R6
Lynchburg, Ohio (45142) 284/C7
Lynchburg, S.C. (29080) 296/G3
Lynchburg, Tenn. (37352) 237/J10
Lynchburg, Va. 188/L3
Lynchburg, Va. 146/K6
Lynchburg (I.C.), Va. (*24501) 307/K6
Lynches (riv.), S.C. 296/H3
Lynch Station, Va. (24571) 307/K6
Lynd, Minn. (56157) 255/C6
Lynd, Queensland 95/C3
Lyndeborough○, N.H. (†03082) 268/D6
Lynden, Ontario 177/D4
Lynden, Wash. (98264) 310/C2
Lyndhurst○, N.J. (07071) 273/B2
Lyndhurst, N.S. Wales 97/E3
Lyndhurst, Ohio (44124) 284/J9
Lyndhurst, Ontario 177/H3
Lyndhurst, S. Australia 88/F6
Lyndhurst, S. Australia 94/E4
Lyndoch, S. Australia 94/C6
Lyndon, Ill. (61261) 222/D2
Lyndon, Kansas (66451) 232/G3
Lyndon, Ohio (45649) 284/D7
Lyndon○, Vt. (05849) 268/C2
Lyndon, W. Australia 92/A3
Lyndon B. Johnson Nat'l Hist. Site, Texas 303/F7
Lyndon B. Johnson Space Ctr., Texas 303/K2
Lyndon Center, Vt. (05850) 268/C2
Lyndon Station, Wis. (53944) 317/F8
Lyndonville, N.Y. (14098) 276/D4
Lyndonville, Vt. (05851) 268/D2
Lyndora, Pa. (16045) 294/B4
Lynedoch, Ontario 177/D5
Lyness, Scotland 15/E2
Lyngby, Denmark 21/F6
Lynhurst, Ontario 177/C5
Lynn, Ala. (35575) 195/C2
Lynn, Ark. (72440) 202/H2
Lynn, Ind. (47355) 227/H4
Lynn, Mass. (*01901) 249/D6
Lynn, N.C. (28750) 281/E4
Lynn (co.), Texas 303/C4
Lynn, Wis. (†54436) 317/F6
Lynn Canal (inlet), Alaska 196/M1
Lynn Center, Ill. (61262) 222/C2
Lynn Creek, Miss. (†39739) 256/G4
Lynndyl, Utah (84640) 304/B4
Lynnfield○, Mass. (01940) 249/D5
Lynnfield Center (Lynnfield P.O.), Mass. (†01940) 249/C5
Lynn Grove, Ky. (42062) 237/E7
Lynn Haven, Fla. (32444) 212/C6
Lynn Lake, Man. 162/G4
Lynn Lake, Man. 146/H4
Lynn Lake, Manitoba 179/H2
Lynnview, Ky. (†40201) 237/K4
Lynnville, Ill. (†62650) 222/C4
Lynnville, Ind. (47619) 227/C8
Lynnville, Iowa (50153) 229/H5
Lynnville, Ky. (42063) 237/D7
Lynnville, Tenn. (38472) 237/G10
Lynnwood, Wash. (98036) 310/C3
Lynton, England 10/E5
Lynton, England 13/D6
Lynwood, Calif. (90262) 204/C11
Lynwood, Ill. (†60411) 222/C6
Lynx (lake), N.W. Terrs. 187/H3
Lynxville, Wis. (54640) 317/D9
Lyon, France 7/E4
Lyon, France 28/F5
Lyon (co.), Iowa 229/A2
Lyon (co.), Kansas 232/F3
Lyon (co.), Ky. 237/E6
Lyon (co.), Minn. 255/C6
Lyon, Miss. (38645) 256/D2
Lyon (co.), Nev. 266/B3
Lyon (inlet), N.W. Terrs. 187/K3
Lyon, Loch (lake), Scotland 15/D4
Lyon (riv.), Scotland 15/D4
Lyon Mountain, N.Y. (12952) 276/N1
Lyonnais (trad. prov.) France 29
Lyons, Colo. (80540) 208/J2
Lyons, Georgia (30436) 217/H6
Lyons, Ill. (60534) 222/B6
Lyons, Ind. (47443) 227/C7
Lyons, Kansas (67554) 232/D3
Lyons, Ky. (†40051) 237/K5
Lyons, Mich. (48851) 250/E6
Lyons, Nebr. (68038) 264/H3
Lyons, N.J. (07939) 273/D2
Lyons, N.Y. (14489) 276/F4
Lyons, Ohio (43533) 284/B2
Lyons, Oreg. (97358) 291/E3
Lyons (Lyon Station), Pa. (19536) 294/L5
Lyons, S. Dak. (57041) 298/R6
Lyons (riv.), W. Australia 88/B4
Lyons (riv.), W. Australia 92/A4
Lyons, Wis. (53148) 317/K10
Lyons Brook, Nova Scotia 168/F3
Lyons Falls, N.Y. (13368) 276/K3
Lyons Plain, Conn. (†06880) 210/B4
Lyon Station, Pa. (19536) 294/L5
Lyra (reef), Papua N.G. 86/C1
Lys (riv.), Belgium 27/B7
Lys (riv.), France 28/E2
Lysaker, Norway 18/D3
Lysá nad Labem, Czech. 41/C1
Lysander, N.Y. (13094) 276/H4
Lysekil, Sweden 18/G7
Lysite, Wyo. (82642) 319/E2
Lyss, Switzerland 39/D2
Lyster, Québec 172/F3
Lys'va, U.S.S.R. 7/K3
Lys'va, U.S.S.R. 48/F4
Lys'va, U.S.S.R. 52/J3
Lytham Saint Anne's, England 13/G1
Lytham Saint Anne's, England 10/F1
Lytle, Texas (78052) 303/J11
Lyttelton, N. Zealand 100/D5
Lytton, Br. Col. 184/G5
Lytton, Iowa (50561) 229/D4
Lyubertsy, U.S.S.R. 52/E3
Lyubotin, U.S.S.R. 52/E4
Lyudinovo, U.S.S.R. 52/D4

M

Ma'ad, Jordan 65/D2
Maalaea, Hawaii (†96753) 218/J2
Maalaea (bay), Hawaii 218/J2
Ma'alot-Tarshiha, Israel 65/C1
Ma'an (dist.), Jordan 65/D5
Ma'an, Jordan 65/E5
Ma'an, Jordan 59/C3
Ma'anshan, China 77/J5
Maarianhamina (Mariehamn), Finland 18/M7
Maarssen, Netherlands 27/F4
Maas (riv.), Netherlands 27/G5
Maasbree, Netherlands 27/H6
Maaseik, Belgium 27/H6
Maasin, Philippines 82/E5
Maasmechelen, Belgium 27/H7
Maassluis, Netherlands 27/E5
Maastricht, Netherlands 27/H7
Maatsuyker (isls.), Tasmania 99/C5
Mababe (depr.), Botswana 118/C3
Mabalane, Mozambique 118/E4
Mabank, Texas (75147) 303/H5
Mabaruma, Guyana 131/B1
Mabay, Cuba 158/H4
Mabel (lake), Br. Col. 184/H5
Mabel, Minn. (55954) 255/G7
Mabelvale, Ark. (72103) 202/F4
Maben, Miss. (39750) 256/F3
Maben, W. Va. (25870) 312/D7
Maberly, Ontario 177/H3
Mabie, W. Va. (26278) 312/F5
Mabille (lake), Newf. 166/B3
Mabini, Philippines 82/E6
Mablethorpe and Sutton, England 13/H4
Mablethorpe and Sutton, England 10/G4
Mableton, Georgia (30059) 217/J1
Mabote, Mozambique 118/E4
Mabou, Nova Scotia 168/G2
Mabou (harb.), Nova Scotia 168/G2
Mabou Highlands (hills), Nova Scotia 168/G2
Mabrouk, Mali 106/D5
Mabscott, W. Va. (25871) 312/D7
Mabton, Wash. (98935) 310/E4
Macá (mt.), Chile 138/D5
Macachín, Argentina 143/D4
Macaé, Brazil 135/F3
Macaé, Brazil 132/F8
Macaíba, Brazil 132/H4
Macajalar (bay), Philippines 82/E6
Macalister, Br. Col. 184/F4
Macaloge, Mozambique 118/F2
MacAlpine (lake), N.W. Terrs. 187/H3
Macamic, Québec 174/B3
Macan (isls.), Indonesia 85/G7
Macanao (pen.), Venezuela 124/F2
Mação, Portugal 33/B3
Macao (Macau) 77
Macapá, Brazil 120/D2
Macapá, Brazil 132/D2
Macará, Ecuador 128/C5
Macaranaima, Colombia 126/E7
Macarena, Serranía de La (mts.), Colombia 126/D6
Macareo Santo Niño, Venezuela 124/H3
Macarthur, Victoria 97/A6
Macas, Ecuador 128/D4
Macassar, S. Africa 118/F6
Macau 54/N7
Macau, Brazil 120/F3
Macau, Brazil 132/G4
MACAU (MACAO) 77
Macau (Macao) (cap.), Macau 77/H7
Macau 2/Q4
Macaúbas, Brazil 132/F6
Macaya (mt.), Haiti 158/A6
Macbeth, S.C. (†29431) 296/H5
Maccan, Nova Scotia 168/D3
Maccarese, Italy 34/F6
Macclenny, Fla. (32063) 212/D1
Maccles (lake), Newf. 166/C1
Macclesfield, England 10/G2
Macclesfield, England 13/H2
Macclesfield, N.C. (27852) 281/O3
Macdiarmid, Ontario 177/H5
MacDill A.F.B., Fla. 212/C3
Macdoel, Calif. (96058) 204/D2
Macdona, Texas (78054) 303/J11
Macdonald, Manitoba 179/D4
Macdonald (lake), North. Terr. 93/B7
Macdonald (lake), W. Australia 88/D4
Macdonald (lake), W. Australia 92/E3
Macdonaldton, Pa. (†15530) 294/E6
Macdonnell (ranges), Australia 87/D8
Macdonnell (ranges), North. Terr. 88/E4
Macdonnell (ranges), North. Terr. 93/C7
Macdowall, Sask. 181/E2
Macduff, Scotland 10/E2
Macduff, Scotland 15/F3
Mace, Ind. (†47933) 227/D4
Mace, W. Va. (†26281) 312/F6
Macedon, N.Y. (14502) 276/F4
Macedonia, Ark. (†71753) 202/D7
Macedonia, Conn. (†06757) 210/A2
Macedonia (reg.), Greece 45/E5
Macedonia, Ill. (62860) 222/E5
Macedonia, Iowa (51549) 229/C6
Macedonia, Ohio (44056) 284/J10
Macedonia (rep.), Yugoslavia 45/E5
Maceió, Brazil 120/F3
Maceió, Brazil 132/H5
Macel, Miss. (†38950) 256/D3
Macenta, Guinea 102/B4
Macenta, Guinea 106/C7
Maceo, Cuba 158/H3
Maceo, Ky. (42355) 237/H5
Macerata (prov.), Italy 34/D3
Macerata, Italy 34/D3
Maces (bay), New Bruns. 170/D3
Maces Bay, New Bruns. 170/D3
Macfarlan, W. Va. (26148) 312/D4
Macfarlane (lake), S. Australia 88/F6
Macfarlane (lake), S. Australia 94/E5
Macgillicuddy's Reeks (mts.), Ireland 17/B7
MacGregor, Manitoba 179/D5
MacGregor's Bay, Ontario 177/G2
Mach, Pakistan 59/J4
Mach, Pakistan 68/B3
Macha, Bolivia 136/B6
Machacamarca, Bolivia 136/B5
Machachi, Ecuador 128/C3
Machado, Brazil 135/C2
Machakos, Kenya 115/G4
Machala, Ecuador 120/B3
Machala, Ecuador 128/B4
Machalí, Chile 138/G5
Machalilla, Ecuador 128/B3
Machaneng, Botswana 118/D4
Machanga, Mozambique 118/F4
Machareti, Bolivia 136/D7
Machattie (lake), Queensland 88/G5
Machattie (lake), Queensland 95/B5
Machaze, Mozambique 118/E4
Machelen, Belgium 27/C9
Macheng, China 77/J5
Machers, The (pen.), Scotland 15/D6
Machias, Maine (04654) 243/J6
Machias○, Maine (04654) 243/J6
Machias (bay), Maine 243/J6
Machias (riv.), Maine 243/H6
Machias (riv.), Maine 243/F2
Machias-Lime Lake, N.Y. (14101) 276/D6
Machiasport, Maine (04655) 243/H6
Machiasport○, Maine (04655) 243/H6
Machias Seal (isl.), Maine 243/J7
Machico, Portugal 33/A2
Machida, Japan 81/O2
Machilipatnam, India 68/E5
Machipongo, Va. (23405) 307/S6
Machiques, Venezuela 124/B3
Macho, Arroyo del (creek), N. Mex. 274/D5
Machrihanish, Scotland 15/C5
Machupicchu, Peru 128/F9
Machupo (riv.), Bolivia 136/C3
Machynlleth, Wales 13/D5
Machynlleth, Wales 10/D4
Macia, Mozambique 118/E4
Maciel, Argentina 143/F6
Maciel, Paraguay 144/D5
Maciel, Arroyo (riv.), Uruguay 145/C4
Macina (depr.), Mali 106/D6
Macintyre (riv.), N. S. Wales 88/J5
Macintyre (riv.), N.S. Wales 97/E1
Macintyre (riv.), Queensland 95/D6
Mack, Colo. (81525) 208/B4
Mack, Ohio (†45202) 284/B9
MacKay, Alberta 182/C3
MacKay (riv.), Alberta 182/D1
Mackay, Australia 87/F8
Mackay (lake), Australia 87/D8
Mackay, Idaho (83251) 220/E6
Mackay (res.), Idaho 220/E6
Mackay (lake), North. Terr. 88/D4
Mackay (lake), North. Terr. 93/A7
MacKay (lake), N.W. Terrs. 187/G3
Mackay, Queensland 88/H4
Mackay, Queensland 95/D4
Mackay (lake), W. Australia 92/E3
Mackenzie (bay) 5/C4
Mackenzie, Br. Col. 184/F2
Mackenzie, Br. Col. 184/F2
Mackenzie (bay), Canada 4/B16
Mackenzie (mts.), Canada 4/C16
Mackenzie (mts.), Canada 146/E3
Mackenzie (riv.), Canada 2/C2
Mackenzie (riv.), Canada 4/C16
Mackenzie, Mo. (†63101) 261/P3
Mackenzie (dist.), N.W.T. 162/E3
Mackenzie (bay), N.W. Terrs. 187/E3
Mackenzie (mts.), N.W. Terrs. 187/E3
Mackenzie (riv.), N.W.T. 162/C2
Mackenzie (riv.), N.W.T. 146/F3
Mackenzie (riv.), N.W. Terrs. 187/F3
Mackenzie (bay), Yukon 162/C2
Mackenzie (bay), Yukon 187/E3
Mackenzie (mts.), Yukon 187/E3
Mackenzie King (isl.), Canada 4/B15
Mackenzie King (isl.), N.W.T. 162/M3
Mackenzie King (isl.), N.W. Terrs. 187/G2
Mackey, Ind. (47654) 227/C8
Mackeys, N.C. (†27970) 281/R3
Mackeyville, Pa. (17750) 294/H3
Mackinac (co.), Mich. 250/D2
Mackinac (isl.), Mich. 250/E3
Mackinac (str.), Mich. 250/E3
Mackinac Island, Mich. (49757) 250/E3
Mackinaw, Ill. (61755) 222/D3
Mackinaw (riv.), Ill. 222/E3
Mackinaw City, Mich. (49701) 250/E3
Macklin, Sask. 181/A3
Macks, Ark. (72113) 202/H2
Macksburg, Iowa (50155) 229/E6
Macksburg, Ohio (45746) 284/G6
Macks Creek, Mo. (65786) 261/G7
Macks Inn, Idaho (83433) 220/G5
Macksville, Kansas (67557) 232/D4
Macksville, N.S. Wales 97/G2
Mackville, Ky. (40040) 237/L5
Maclean, N.S. Wales 97/G1
Maclean (str.), N.W. Terrs. 187/H2
Maclear, S. Africa 118/D6
Maclear (cape), S. Africa 118/F7
Macmillan (pass), N.W. Terrs. 187/F3
Macmillan (pass), Yukon 187/F3
Macmillan (riv.), Yukon 187/E3
Macnean (lake), Ireland 17/F3
Macnean (lake), N. Ireland 17/F3
MacNutt, Sask. 181/K4
Macomb, Ill. (61455) 222/C3
Macomb (co.), Mich. 250/G6
Macomb, Okla. (74852) 288/M4
Macomer, Italy 34/B4
Macomia, Mozambique 118/F2
Macon (co.), Ala. 195/G6
Mâcon, France 28/F4
Macon, Ga. 188/K4
Macon, Ga. 146/K6
Macon (co.), Georgia 217/D6
Macon, Georgia (*31201) 217/E5
Macon (co.), Ill. 222/E4
Macon, Ill. (62544) 222/E4
Macon (bayou), La. 238/H1
Macon, Miss. (39341) 256/G4
Macon (co.), Mo. 261/G3
Macon, Mo. (63552) 261/H3
Macon, Nebr. (†68939) 264/E4
Macon (co.), N.C. 281/B4
Macon, N.C. (27551) 281/N2
Macon, Ohio (†45171) 284/C8
Macon (co.), Tenn. 237/J7
Macon, Tenn. (38048) 237/B10
Macondo, Angola 115/D6
Macoris (cape), Dom. Rep. 158/E5
Macosquin, N. Ireland 17/H1
Macotera, Spain 33/D2
Macouba, Martinique 161/C5
Macoun, Sask. 181/H6
Macoupin (co.), Ill. 222/D4
Macoupin (riv.), Ill. 222/C4
Macouria, Fr. Guiana 131/E3
Macquarie (riv.), N. S. Wales 88/H6
Macquarie (lake), N. S. Wales 97/F3
Macquarie (riv.), N.S. Wales 97/D2
Macquarie (harb.), Tasmania 88/G8
Macquarie (harb.), Tasmania 99/B4
Macquarie (riv.), Tasmania 99/D3
Mac-Robertson Land (reg.) 5/B4
Macroom, Ireland 10/B5
Macroom, Ireland 17/C8
Macrorie, Sask. 181/E4
Mactan (isl.), Philippines 82/E5
Mactaquac (lake), New Bruns. 170/C3
MacTier, Ontario 177/E2
Macumba (riv.), S. Australia 88/F5
Macumba, The (riv.), S. Australia 94/E2
Macungie, Pa. (18062) 294/L4
Macurijes (pt.), Cuba 158/F3
Macuro, Venezuela 124/H2
Macusani, Peru 128/G10
Macuspana, Mexico 150/N8
Macuto, Venezuela 124/E2
Macwahoc○, Maine (†04451) 243/G4
Macy, Ind. (46951) 227/E3
Macy, Nebr. (68039) 264/H2
Mad (riv.), Calif. 204/B3
Mad (riv.), Conn. 210/C2
Mad (riv.), Conn. 210/C1
Mad (riv.), N.H. 268/D4
Mad (riv.), Ohio 284/C6
Mad (riv.), Vt. 268/B3
Ma'daba, Jordan 65/D4
Madadi, Chad 111/D4
Madagascal (pond), Maine 243/G5
Madagascar 2/M6
Madagascar 102/G7
MADAGASCAR 118/H3
Madaket, Mass. (†02554) 249/O7
Madama, Niger 106/G4
Madame (isl.), Nova Scotia 168/H3
Madang, Papua N.G. 85/B7
Madang, Papua N.G. 87/E6
Madaoua, Niger 106/F6
Madaras, Hungary 41/E3
Madaripur, Bangladesh 68/G4
Madauk, Burma 72/C3
Madawaska, Maine (04756) 243/G1
Madawaska○, Maine (04756) 243/G1
Madawaska (co.), New Bruns. 170/B1
Madawaska (riv.), New Bruns. 170/B1
Madawaska, Ontario 177/F2
Madawaska (riv.), Ontario 177/G2
Madawaska (riv.), Québec 172/J2
Madbury○, N.H. (†03820) 268/F5
Maddela, Philippines 82/C2
Madden, Alberta 182/C4
Madden, Miss. (39109) 256/F5
Maddock, N. Dak. (58348) 282/L4
Maddy, Loch (inlet), Scotland 15/A3
Madeira (riv.), Brazil 2/F6
Madeira (riv.), Brazil 120/C3
Madeira (riv.), Brazil 132/A4
Madeira, Ohio (45243) 284/C9
Madeira (isl.), Portugal 33/A2
Madeira (isl.), Portugal 106/A2
Madeira (isls.), Portugal 2/H4
Madeira (isls.), Portugal 102/A1
Madeira (isls.), Portugal 33/A2
Madeira (isls.) Portugal 106/A2
Madeira Beach, Fla. (33738) 212/B3
Madeira Park, Br. Col. 184/J2
Madeleine (cape), Québec 172/D1
Madelia, Minn. (56062) 255/D6
Madeline, Calif. (96119) 204/E2
Madeline (isl.), Wis. 317/E2
Maden, Turkey 63/H3
Madera (co.), Calif. 204/F6
Madera, Calif. (93637) 204/E7
Madera, Mexico 150/F2
Madera, Pa. (16661) 294/F4
Madera Canyon, Ariz. (†85637) 198/E7
Madh, India 68/B7
Madhubani, India 68/F3
Madhya Pradesh (state), India 68/D4
Madidi (riv.), Bolivia 136/A3
Madill, Okla. (73446) 288/N6
Madinat ash Sha'b, P.D.R. Yemen 59/E7
Madinat el-Thawra, Syria 63/H5
Madingo-Kayes, Congo 115/B4
Madingou, Congo 115/B4
Madirovalo, Madagascar 118/H3
Madison (co.), Ala. 195/E1
Madison, Ala. (35758) 195/E1
Madison (co.), Ark. 202/C1
Madison, Ark. (72359) 202/J4
Madison, Calif. (95653) 204/D5
Madison, Conn. (06443) 210/E3
Madison○, Conn. (06443) 210/E3
Madison (co.), Fla. 212/C1
Madison, Fla. (32340) 212/C1
Madison (co.), Georgia 217/F2
Madison, Georgia (30650) 217/F3
Madison (co.), Idaho 220/G6
Madison (co.), Ill. 222/D5
Madison, Ill. (62060) 222/A2
Madison (co.), Ind. 227/F4
Madison, Ind. (47250) 227/G7
Madison (co.), Iowa 229/E6
Madison, Kansas (66860) 232/F3
Madison (co.), Ky. 237/N5
Madison (par.), La. 238/H2
Madison, Maine (04950) 243/D6
Madison○, Maine (04950) 243/D6
Madison, Md. (21648) 245/O6
Madison, Minn. (56256) 255/B5
Madison (co.), Miss. 256/D5
Madison, Miss. (39110) 256/D6
Madison (co.), Mo. 261/M8
Madison, Mo. (65263) 261/H4
Madison (co.), Mont. 262/D5
Madison (riv.), Mont. 262/E5
Madison (co.), Nebr. 264/G3
Madison, Nebr. (68748) 264/G3
Madison○, N.H. (03849) 268/E4
Madison (mt.), N.H. 268/E3
Madison, N.J. (07940) 273/E2
Madison (co.), N.Y. 276/J5
Madison, N.Y. (13402) 276/J5
Madison (co.), N.C. 281/D3
Madison, N.C. (27025) 281/J2
Madison (co.), Ohio 284/D6
Madison, Ohio (44057) 284/H2
Madison, Sask. 181/B4
Madison, S.C. (29693) 296/A2
Madison, S.C. (†29829) 296/D4
Madison, S. Dak. (57042) 298/P6
Madison (lake), S. Dak. 298/P6
Madison (co.), Tenn. 237/D9
Madison (co.), Texas 303/J6
Madison (co.), Va. 307/M4
Madison, Va. (22727) 307/M4
Madison, W. Va. (25130) 312/C6
Madison (cap.), Wis. 146/K5
Madison (cap.), Wis. 188/J2
Madison (cap.), Wis. (*53701) 317/H9
Madison, Wyo. (†82190) 319/B1
Madison (plat.), Wyo. 319/B1
Madisonburg, Ohio (†44691) 284/G4
Madison Heights, Mich. (48071) 250/B6
Madison Heights, Va. (24572) 307/K6
Madison Lake, Minn. (56063) 255/E6
Madisonville, Ky. (42431) 237/F6
Madisonville, La. (70447) 238/K6
Madisonville, Tenn. (37354) 237/N9
Madisonville, Texas (77864) 303/J7
Madiun, Indonesia 85/K2
Madley (mt.), W. Australia 92/D4
Madoc, Mont. (†59222) 262/L2
Madoc, Ontario 177/G3
Mado Gashi, Kenya 115/G3
Madoi, China 77/E4
Madona, U.S.S.R. 53/C2
Madona, U.S.S.R. 52/C3
Madonna, Md. (†21161) 245/M2
Madraka, Ras (cape), Oman 59/G6
Madran, New Bruns. 170/E1
Madras, India 68/E6
Madras, India 54/K8
Madras, India 2/P5
Madras, Oreg. (97741) 291/F3
Madre (lag.), Mexico 150/L4
Madre (lag.), Texas 188/G5
Madre (lag.), Texas 303/G11
Madre de Dios (riv.) 120/C4
Madre de Diós (riv.), Bolivia 136/A3
Madre de Dios (isl.), Chile 120/B8
Madre de Dios (isl.), Chile 138/D8
Madre de Dios (dept.), Peru 128/G8
Madre de Dios, Peru 128/G9
Madre de Dios (riv.), Peru 128/G9
Madre del Sur, Sierra (mts.), Mexico 150/K8
Madre Occidental, Sierra (mts.), Mexico 150/F3
Madre Oriental, Sierra (mts.), Mexico 150/J4
Madrid, Ala. (36348) 195/H8
Madrid, Iowa (50156) 229/F5
Madrid○, Maine (†04966) 243/B6
Madrid, Nebr. (69150) 264/C4
Madrid, N. Mex. (†87010) 274/C3
Madrid, N.Y. (13660) 276/K1
Madrid (prov.), Spain 33/E2
Madrid (cap.), Spain 7/D4
Madrid (cap.), Spain 33/F4
Madrid (cap.), Spain 2/J4
Madridejos, Spain 33/E3
Madrigal de las Altas Torres, Spain 33/D2
Madrigalejo, Spain 33/D3
Madrisahorn (mt.), Switzerland 39/J3
Madroñera, Spain 33/D3
Madsen, Ontario 175/B2
Madugula, India 68/E5
Madura (isl.), Indonesia 54/N10
Madura (isl.), Indonesia 85/K2
Madura (str.), Indonesia 85/K2
Madura, W. Australia 92/D5
Madurai, India 54/J9
Madurai, India 68/D7
Madvar, Kuh-e (mt.), Iran 59/F3
Madvar, Kuh-e (mt.), Iran 66/F3
Maebashi, Japan 81/J5
Mae Hong Son, Thailand 72/C3
Mae Klong, Mae Nam (riv.), Thailand 72/C4
Mael, Norway 18/F6
Maella, Spain 33/G2
Maeser, Utah (†84078) 304/E3
Maesteg, Wales 13/D6
Maestra, Sierra (mts.), Cuba 158/H4
Maevatanana, Madagascar 118/H3
Maeystown, Ill. (62256) 222/C5
Mafeking, Manitoba 179/B2
Mafeking (Mafikeng), S. Africa 118/C5
Mafeteng, Lesotho 118/D5
Maffin (bay), Indonesia 85/K6
Maffra, Victoria 97/D5
Mafia (isl.), Tanzania 102/G5
Mafia (isl.), Tanzania 115/H5
Mafikeng (Mafeking), S. Africa 118/C5
Mafra, Brazil 132/D9
Mafra, Portugal 33/B3
Magadan, U.S.S.R. 2/S3
Magadan, U.S.S.R. 54/R4
Magadan, U.S.S.R. 48/P4
Magadi, Kenya 115/G4
Magadino, Switzerland 39/G4
Magaguadavic, New Bruns. 170/C3
Magaguadavic (lake), New Bruns. 170/C3
Magaguadavic (riv.), New Bruns. 170/C3

Magalia, Calif. (95954) 204/D4
Magaliesburg, S. Africa 118/G6
Magallanes (reg.), Chile 138/E10
Magallanes (Magellan) (str.), Chile 138/D10
Magallanes, Philippines 82/D4
Magallenes (Magellan) (str.), Argentina 143/C7
Magangué, Colombia 126/C3
Maganoy, Philippines 82/E7
Mağara, Turkey 63/G3
Magarabomba, Cuba 158/G2
Magaria, Niger 106/F6
Magazine, Ark. (72943) 202/C3
Magazine (mt.), Ark. 202/C3
Magdagachi, U.S.S.R. 48/N4
Magdala, Ethiopia 111/G5
Magdalen (isls.), Que. 162/K6
Magdalena, Argentina 143/H7
Magdalena, Bolivia 136/C3
Magdalena (isl.), Chile 138/D5
Magdalena (dept.), Colombia 126/C3
Magdalena (riv.), Colombia 120/B2
Magdalena (riv.), Colombia 126/C3
Magdalena (bay), Mexico 150/C4
Magdalena, N. Mex. (87825) 274/B4
Magdalena (mts.), N. Mex. 274/B4
Magdalena de Kino, Mexico 150/D1
Magdeburg, E. Germany 7/F3
Magdeburg (dist.), E. Germany 22/D2
Magdeburg, E. Germany 22/D2
Magdelaine (cays), Coral Sea Is. Terr. 88/H3
Magé, Brazil 135/E3
Magee, Miss. (39111) 256/E7
Magee, Island (pen.), N. Ireland 17/K2
Magelang, Indonesia 85/J2
Magellan (str.) 120/C8
Magellan (str.) 2/F8
Magellan (str.), Argentina 143/C7
Magellan (str.), Chile 138/D10
Magen, Israel 65/A5
Magens (bay), Virgin Is. (U.S.) 161/B4
Magerøya (isl.), Norway 18/P1
Magerrain (mt.), Switzerland 39/H2
Magetan, Indonesia 85/K2
Maggia, Switzerland 39/G4
Maggia (riv.), Switzerland 39/G4
Maggie Valley, N.C. (28751) 281/C3
Maggiore (lake), Fla. 212/B3
Maggiore (lake), Italy 34/B1
Maggiore (lake), Switzerland 39/G5
Maggotty, Jamaica 158/H6
Maghâgha, Egypt 59/B4
Maghâgha, Egypt 111/J4
Maghama, Mauritania 102/A3
Maghama, Mauritania 106/B5
Maghera, N. Ireland 17/H2
Magherafelt (dist.), N. Ireland 17/H2
Magherafelt, N. Ireland 10/C3
Magherafelt, N. Ireland 17/H2
Magic (res.), Idaho 220/D6
Magilligan (pt.), N. Ireland 17/H1
Maglaj, Yugoslavia 45/D3
Magley, Ind. (†46733) 227/G3
Maglie, Italy 34/G4
Magna, Utah (84044) 304/B3
Magna Bay, Br. Col. 184/H4
Magness, Ark. (72553) 202/H2
Magnet, Ark. (†72104) 202/E5
Magnet, Ind. (47555) 227/D8
Magnet, Manitoba 179/C3
Magnet, Nebr. (68749) 264/G2
Magnetawan, Ontario 177/E2
Magnetawan (riv.), Ontario 177/D2
Magnetic Springs, Ohio (43036) 284/D5
Magnitogorsk, U.S.S.R. 54/H4
Magnitogorsk, U.S.S.R. 48/G4
Magnolia, Ala. (36754) 195/C6
Magnolia, Ark. (71753) 202/D7
Magnolia, Del. (19962) 245/R4
Magnolia, Ill. (61336) 222/D2
Magnolia, Iowa (51550) 229/B5
Magnolia, Minn. (56158) 255/B7
Magnolia, Miss. (39652) 256/D8
Magnolia, N.J. (08049) 273/B3
Magnolia, N.C. (28453) 281/O5
Magnolia, Ohio (44643) 284/H4
Magnolia, Texas (77355) 303/J7
Magnolia, W. Va. (†25422) 312/K3
Magnolia Springs, Ala. (36555) 195/C10
Màgoé, Mozambique 118/E3
Magoffin (co.), Ky. 237/P5
Magog, Québec 172/E4
Magpie, Québec 174/E2
Magpie (lake), Québec 174/E2
Magrath, Alberta 182/D5
Magude, Mozambique 118/E5
Maguindanao (prov.), Philippines 82/E7
Maguse (lake), N.W. Terrs. 187/J3
Magwe (div.), Burma 72/B2
Magwe, Burma 72/B2
Mahabad, Iran 59/E2
Mahabad, Iran 66/D2
Mahabaleshwar, India 68/C5
Mahabo, Madagascar 118/G4
Mahaena, Fr. Poly. 86/T13
Mahaffey, Pa. (15757) 294/E4
Mahagi, Zaire 115/F3
Mahaica, Guyana 131/C2
Mahaicony Village, Guyana 131/C2
Mahajamba (bay), Madagascar 118/H2
Mahajanga (prov.), Madagascar 118/H3
Mahajanga, Madagascar 102/G6
Mahakam (riv.), Indonesia 85/F6
Mahalapye, Botswana 118/D4
Mahalapye, Botswana 102/E7
Mahalasville, Ind. (†46151) 227/E6
Mahallat, Iran 66/G4
Mahan, Iran 66/K5
Mahanadi (riv.), India 68/E4
Mahanoro, Madagascar 118/H3
Mahanoy City, Pa. (17948) 294/K4
Maharashtra (state), India 68/C5
Maha Sarakham, Thailand 72/D3
Mahaska (co.), Iowa 229/H6
Mahaska, Kansas (66955) 232/E2
Mahaxai, Laos 72/E3
Mahbubnagar, India 68/D5
Mahdia, Guyana 131/B3
Mahdia, Tunisia 106/G1
Mahe, India 68/D6
Mahé (isl.), Seychelles 118/H5
Mahébourg, Mauritius 118/G5
Mahenge, Tanzania 115/G5
Maheno, N. Zealand 100/C6
Maher, Colo. (81421) 208/D5
Maheshkhali, Bangladesh 68/G4
Mahia (pen.), N. Zealand 100/G3
Mahim, India 68/C5
Mahim (bay), India 68/B7
Mahkonce, Minn. (†56557) 255/C3
Mahlaing, Burma 72/B2
Mahmudiye, Turkey 63/D3
Mahnomen (co.), Minn. 255/C3
Mahnomen, Minn. (56557) 255/C3
Maho (bay), Virgin Is. (U.S.) 161/C4
Mahoba, India 68/D3
Mahomet, Ill. (61853) 222/E3
Mahón, Spain 33/J3
Mahone (bay), Nova Scotia 168/D4
Mahone Bay, Nova Scotia 168/D4
Mahoning (co.), Ohio 284/J4
Mahood (lake), Br. Col. 184/G4
Mahopac, N.Y. (10541) 276/N8
Mahout, Dominica 161/E6
Mahto, S. Dak. (57643) 298/H2
Mahtomedi, Minn. (55115) 255/F5
Mahtowa, Minn. (55762) 255/F4
Mahukona, Hawaii (†96719) 218/G3
Mahuva, India 68/C4
Mahwah○, N.J. (07430) 273/E1
Maia, Portugal 33/B2
Maicao, Colombia 126/D2
Maicuru (riv.), Brazil 132/C2
Maida, N. Dak. (58255) 282/O2
Maida, Yemen Arab Rep. 59/D6
Maidan, Iraq 66/D3
Maidan, Iraq 59/E3
Maidani, Ras (cape), Iran 59/G4
Maiden, N.C. (28650) 281/G3
Maidenhead, England 10/F5
Maidenhead, England 13/G8
Maiden Rock, Wis. (54750) 317/B6
Maidens, The (isls.), N. Ireland 17/K2
Maidens, Scotland 15/D5
Maidens, Va. (23102) 307/N5
Maidstone, England 13/J8
Maidstone, England 10/G5
Maidstone, Ontario 177/B5
Maidstone, Sask. 181/B2
Maidstone○, Vt. (†05905) 268/D2
Maidstone (lake), Vt. 268/D2
Maidsville, W. Va. (26541) 312/F3
Maiduguri, Nigeria 106/G6
Maiduguri, Nigeria 102/D3
Maienfeld, Switzerland 39/J2
Maigatari, Nigeria 106/F6
Maigualida, Sierra (range), Venezuela 124/F4
Maigue (riv.), Ireland 17/D6
Maihara, Japan 81/G6
Maili, Hawaii (†96792) 218/D2
Maillard, Québec 172/G2
Ma'in, Jordan 65/D4
Main (passage), La. 238/M8
Main (riv.), N. Ireland 17/J2
Main (chan.), Ontario 177/C2
Main (str.), Singapore 72/F6
Main (riv.), W. Germany 22/C4
Main-à-Dieu, Nova Scotia 168/J2
Main Barrier (range), N. S. Wales 88/G6
Main Barrier (range), N.S. Wales 97/A2
Main Brook, Newf. 166/C3
Main Centre, Sask. 181/D5
Mai-Ndombe (lake), Zaire 115/C4
Maine 188/N1
MAINE 243
Maine (gulf) 188/N2
Maine (gulf) 162/K7
Maine (trad. prov.), France, 29
Maine (riv.), Ireland 17/C7
Maine (gulf), Mass. 249/M2
Maine, N.Y. (13802) 276/H6
Maine (state), U.S. 146/M5
Maine-et-Loire (dept.), France 28/C4
Mainesburg, Pa. (16932) 294/J2
Maïné-Soroa, Niger 106/G6
Maineville, Ohio (45039) 284/C9
Maingard (lake), Québec 172/G1
Maingkwan, Burma 72/C1
Mainit, Philippines 82/E6
Mainit (lake), Philippines 82/E6
Mainland, Orkney Is. (isl.), Scotland 10/E1
Mainland, Shetland Is. (isl.), Scotland 10/G1
Mainland (isl.), Scotland 15/G2
Mainland (isl.), Scotland 15/E1
Mainling, China 77/D6
Mainoru, North. Terr. 93/C3
Mainstream, Maine (†04942) 243/D6
Maintirano, Madagascar 118/G3
Main Topsail (mt.), Newf. 166/C4
Mainz, W. Germany 22/C4
Maio (isl.), C. Verde 106/B8
Maipo (vol.), Argentina 143/C3
Maipo (riv.), Chile 138/F4
Maipú, Argentina 143/E4
Maipú, Chile 138/G3
Maipú (vol.), Chile 138/B10
Maipures, Colombia 126/F5
Maiquetía, Venezuela 120/C1
Maiquetía, Venezuela 124/E2
Mairana, Bolivia 136/D6
Maisí, Cuba 158/K4
Maisí (cape), Cuba 158/K4
Maisí (cape), Cuba 156/D2
Maison de Pierre (lake), Québec 172/C3
Maisonnette, New Bruns. 170/E1
Maisons-Alfort, France 28/B2
Maisons-Laffitte, France 28/A1
Maïssade, Haiti 158/C5
Maitencillo, Chile 138/A8
Maitland, Australia 87/F9
Maitland, Fla. (32751) 212/E3
Maitland, Mo. (64466) 261/B2
Maitland, N. S. Wales 88/J6
Maitland, N.S. Wales 97/F3
Maitland, Annapolis, Nova Scotia 168/C4
Maitland, Hants, Nova Scotia 168/E3
Maitland, Ontario 177/J3
Maitland, S. Australia 94/E6
Maitum, Philippines 82/E7
Maize, Kansas (67101) 232/E4
Maiz Grande (Great Corn) (isl.), Nicaragua 154/F4
Maizhokunggar, China 77/D6
Maiz Pequeña (Little Corn) (isl.), Nicaragua 154/F4
Maizuru, Japan 81/G6
Majagua, Cuba 158/F2
Majagual, Colombia 126/C3
Majalengka, Indonesia 85/H2
Majene, Indonesia 85/F6
Majenica, Ind. (†46750) 227/F3
Majes (riv.), Peru 128/F11
Majestic, Ky. (41547) 237/S5
Maji, Ethiopia 111/G6
Majma'a, Saudi Arabia 59/D4
Majoli, Suriname 131/D4
Major (co.), Okla. 288/K2
Major, Sask. 181/B4
Majorca (isl.), Spain 7/E5
Majorca (isl.), Spain 33/H3
Majorsville, W. Va. (†26036) 312/F3
Majunga, Madagascar 118/H3
Majuro (atoll) (cap.), Marshall Is. 87/H5
Makaha, Hawaii (†96792) 218/D2
Makaha (pt.), Hawaii 218/B1
Makah Ind. Res., Wash. 310/A2
Makahuena (pt.), Hawaii 218/C2
Makaiwa, Hawaii (†96763) 218/H2
Makakilo, Hawaii (96706) 218/E2
Makale, Ethiopia 102/F3
Makale, Ethiopia 111/G5
Makallé, Argentina 143/E2
Makanda, Ill. (62958) 222/D6
Makanza, Zaire 115/C3
Makapala, Hawaii (†96711) 218/G3
Makapuu (pt.), Hawaii 218/F2
Makara Beach, N. Zealand 100/A2
Makara-Ohariu, N. Zealand 100/A3
Makari, Cameroon 115/B1
Makaroff, Manitoba 179/A3
Makarov, U.S.S.R. 48/P5
Makarska, Yugoslavia 45/C4
Makar'yev, U.S.S.R. 52/F3
Makassar (Ujung Pandang), Indonesia 85/F7
Makassar (str.), Indonesia 54/N10
Makassar (str.), Indonesia 85/F6
Makatea (isl.), Fr. Poly. 87/L7
Makawao, Hawaii (96768) 218/K2
Makaweli, Hawaii (96769) 218/B2
Makena, Hawaii (†96790) 218/J2
Makeni, S. Leone 102/A4
Makeni, S. Leone 106/B7
Makepeace, Alberta 182/D4
Makeyevka, U.S.S.R. 7/H4
Makeyevka, U.S.S.R. 52/E5
Makgadikgadi (salt pan), Botswana 102/E7
Makgadikgadi (salt pan), Botswana 118/D3
Makhachkala, U.S.S.R. 7/J4
Makhachkala, U.S.S.R. 48/E5
Makhachkala, U.S.S.R. 52/G6
Makharadze, U.S.S.R. 52/F6
Makhmur, Iraq 66/C3
Makiki, Hawaii (96822) 218/C4
Makin (Butaritari) (atoll), Kiribati 87/H5
Makinak, Manitoba 179/C4
Makinen, Minn. (55763) 255/F3
Makinsk, U.S.S.R. 48/H4
Makinson (inlet), N.W. Terrs. 187/L2
Makkovik, Newf. 166/C2
Makkovik (cape), Newf. 166/C2
Makkum, Netherlands 27/G2
Makó, Hungary 41/F3
Makokou, Gabon 115/B3
Makoti, N. Dak. (58756) 282/G4
Makoua, Congo 115/C3
Maków Mazowiecki, Poland 47/E2
Makran (reg.), Iran 66/M8
Makteïr (des.), Mauritania 106/B4
Maku, Iran 66/D1
Makubetsu, Japan 81/L2
Makumbako, Tanzania 115/G5
Makurazaki, Japan 81/O3
Makurdi, Nigeria 102/C4
Makurdi, Nigeria 106/F7
Makushin (vol.), Alaska 196/E4
Makwa, Sask. 181/C2
Makwa (lake), Sask. 181/B1
Makwa (riv.), Sask. 181/B1
Mal, Mauritania 106/B5
Mala, Punta (cape), Panama 154/H7
Malå, Sweden 18/L4
Malabang, Philippines 82/D7
Malabar, Fla. (32950) 212/F3
Malabar (hill), India 68/B7
Malabar (pt.), India 68/B7
Malabar Coast (reg.), India 68/C6
Malabo (cap.), Equat. Guinea 102/C4
Malabo (cap.), Equat. Guinea 115/A3
Malabrigo, Argentina 143/F4
Mal Abrigo, Uruguay 145/C5
Malabungan, Philippines 82/A6
Malacca (str.) 54/M9
Malacca (str.), Indonesia 85/C5
Malacca (Melaka), Malaysia 72/D7
Malacca (str.), Malaysia 72/D7
Malacky, Czech. 41/D2
Malad (riv.), Idaho 220/F7
Malad, India 68/B6
Malad (creek), India 68/B7
Malad (riv.), Utah 304/B1
Malad City, Idaho (83252) 220/F7
Maladers, Switzerland 39/J3
Málaga, Colombia 126/D4
Malaga, N.J. (08328) 273/C4
Malaga, N. Mex. (88263) 274/E6
Malaga, Ohio (43757) 284/H6
Málaga (prov.), Spain 33/D4
Málaga, Spain 7/D5
Málaga, Spain 33/D4
Malaga, Wash. (98828) 310/E3
Malagash, Nova Scotia 168/E3
Malagash (pt.), Nova Scotia 168/E3
Malagón, Spain 33/E3
Malagueta (bay), Cuba 158/H3
Malahide, Ireland 17/J5
Malaita (isl.), 86/E3
Malaita (isl.), Solomon Is. 87/G6
Malakal, Sudan 111/F6
Malakal, Sudan 102/F4
Malakanagiri, India 68/E5
Malakand, Pakistan 68/C2
Malakand, Pakistan 59/K3
Malakoff, France 28/A2
Malakoff, Texas (75148) 303/H5
Malakwa, Br. Col. 184/H5
Malalag, Philippines 82/E7
Malamir (Izeh), Iran 66/F5
Malang, Indonesia 54/N10
Malang, Indonesia 85/K2
Malange (dist.), Angola 115/C6
Malange, Angola 102/D5
Malange, Angola 115/C5
Malangka (cape), Indonesia 85/G5
Malans, Switzerland 39/J3
Malanville, Benin 106/E6
Mälaren (lake), Sweden 18/G1
Malargüe, Argentina 120/C6
Malargüe, Argentina 143/C4
Malartic, Québec 174/B3
Malaspina (glac.), Alaska 196/K3
Malaspina (str.), Br. Col. 184/J2
Malatya (prov.), Turkey 63/H3
Malatya, Turkey 59/C2
Malatya, Turkey 54/E6
Malatya, Turkey 63/H3
Malawi 2/L6
Malawi 102/F6
MALAWI 115/F6
Malawi (Nyasa) (lake), Malawi 115/F6
Malay (pen.), Malaysia 72/D6
Malay (pen.), Malaysia 85/B4
Malay (pen.), Thailand 72/D6
MALAYA, MALAYSIA 72
Malaya (reg.), Malaysia 54/M9
Malaya (reg.), Malaysia 72/E6
Malaya Vishera, U.S.S.R. 52/D3
Malaybalay, Philippines 82/E6
Malayer, Iran 66/F3
Malaysia 2/Q5
Malaysia 54/M9
MALAYSIA 85/D4
Malazgirt, Turkey 63/K3
Malbaie (riv.), Québec 172/G2
Malbon, Queensland 95/B4
Malbork (Marienburg), Poland 47/D1
Malchinersee (lake), E. Germany 22/E2
Malchow, E. Germany 22/E2
Malcolm, Ala. (36556) 195/B8
Malcolm, Nebr. (68402) 264/H4
Malcom, Iowa (50157) 229/H5
Maldegem, Belgium 27/C6
Malden, Ill. (61337) 222/D2
Malden, Ind. (†46383) 227/C2
Malden (isl.), Kiribati 87/L6
Malden, Mass. (02148) 249/D6
Malden, Mo. (63863) 261/M9
Malden, New Bruns. 170/G2
Malden, Wash. (99149) 310/H3
Malden, W. Va. (25306) 312/C6
Maldives 2/N5
MALDIVES 68
Maldives 54/J9
Maldives (isls.) 68/C7
Maldon, England 10/G5
Maldon, England 13/H6
Maldon, Victoria 97/C5
Maldonado (pt.), Mexico 150/K8
Maldonado (dept.), Uruguay 145/E5
Maldonado, Uruguay 145/D6
Male (cap.), Maldives 54/J9
Male (cap.), Maldives 2/N5
Maléa (cape), Greece 45/F7
Malebo (Stanley Pool) (lake), Zaire 115/C4
Malegaon, India 54/J7
Malegaon, India 68/C4
Malekula (isl.), Vanuatu 87/G7
Malema, Mozambique 118/F2
Malemba-Nkulu, Zaire 115/E5
Malente, W. Germany 22/D1
Maler Kotla, India 68/D2
Malesus, Tenn. (†38301) 237/D9
Malgobek, U.S.S.R. 52/F6
Malhão da Estrela (mt.), Portugal 33/C2
Malheur (lake), Oreg. 188/C2
Malheur (co.), Oreg. 291/K4
Malheur (lake), Oreg. 291/J4
Malheur (riv.), Oreg. 291/J4
Mali 2/J5
Mali 102/B2
Mali (riv.), Burma 72/C1
Mali, Guinea 106/B6
MALI 106/D5
Malibu, Calif. (90265) 204/B10
Malignant Cove, Nova Scotia 168/F3
Maligne (lake), Alberta 182/B3
Mali Kyun (isl.), Burma 72/C4
Malili, Indonesia 85/G6
Malin, Ireland 17/G1
Malin (head), Ireland 17/F1
Malin (head), Ireland 10/C3
Malin, Oreg. (97632) 291/F5
Malin, U.S.S.R. 52/C4
Malinau, Indonesia 85/F5
Malindang (mt.), Philippines 82/D6
Malindi, Kenya 102/B2
Malindi, Kenya 115/H4
Malines (Mechelen), Belgium 27/F6
Malinta, Ohio (43535) 284/B3
Malita, Philippines 82/E7
Maliwun, Burma 72/C5
Malkapur, India 68/D4
Malkara, Turkey 63/B2
Malkiya, Israel 65/D1
Malko Tŭrnovo, Bulgaria 45/H4
Mallacoota, Victoria 97/E5
Mallaig, Alberta 182/E2
Mallaig, Scotland 10/D2
Mallaig, Scotland 15/C4
Mallanganee, N.S. Wales 97/G1
Mallard, Iowa (50562) 229/D3
Mallawi, Egypt 111/J4
Mallawi, Egypt 59/B4
Mallén, Spain 33/F2
Malleray, Switzerland 39/D2
Mallet Creek, Ohio (†44256) 284/G3
Malling, Denmark 21/D5
Mallnitz, Austria 41/B3
Malloa, Chile 138/G5
Malloch (cape), N.W. Terrs. 187/H2
Mallorca (Majorca) (isl.), Spain 33/H3
Mallory, N.Y. (13103) 276/H4
Mallory, W. Va. (25634) 312/C7
Mallorytown, Ontario 177/J3
Mallow, Ireland 17/D7
Mallow, Ireland 10/B4
Malmanoury, Fr. Guiana 131/E3
Malmberget, Sweden 18/M3
Malmédy, Belgium 27/J8
Malmesbury, England 13/E6
Malmesbury, S. Africa 118/B6
Malmköping, Sweden 18/F1
Malmo, Minn. (†56431) 255/E4
Malmo, Nebr. (68040) 264/H3
Malmö, Sweden 7/F3
Malmö, Sweden 18/H9
Malmöhus (co.), Sweden 18/H9
Malmok (mt.), Neth. Ant. 161/E8
Malmstrom A.F.B., Mont. 262/E3
Malo, Wash. (99150) 310/G2
Maloca, Brazil 132/C4
Maloca, Brazil 120/D3
Maloelap (atoll), Marshall Is. 87/H5
Malolos, Philippines 82/C3
Malone, Fla. (32445) 212/A1
Malone, Ky. (41451) 237/P5
Malone, N.Y. (12953) 276/M1
Malone, Texas (76660) 303/H6
Malone, Wash. (98559) 310/B4
Maloneton, Ky. (41158) 237/R3
Maloney (res.), Nebr. 264/D3
Malonton, Manitoba 179/E4
Malott, Wash. (98829) 310/F2
Maloy, Iowa (50852) 229/E7
Malpartida de Cáceres, Spain 33/C3
Malpartida de Plasencia, Spain 33/C2
Malpelo (isl.), Colombia 120/A2
Malpeque (bay), Pr. Edward I. 168/E2
Malta 2/K4
Malta 7/F5
Malta, Colo. (†80461) 208/G4
Malta, Idaho (83342) 220/E7
Malta, Ill. (60150) 222/E2
Malta (chan.), Italy 34/E6
MALTA 34
Malta (isl.), Malta 34/E7
Malta, Mont. (59538) 262/J2
Malta, Ohio (43758) 284/G6
Malta Bend, Mo. (65339) 261/F4
Maltahöhe, Namibia 118/B4
Maltepe, Turkey 63/D6
Malters, Switzerland 39/F2
Malton, England 13/G3
Malton, England 10/F3
Maltrata, Mexico 150/O2
Malung, Sweden 18/H6
Malvaglia, Switzerland 39/H4
Malvan, India 68/C5
Malvern, Ala. (36349) 195/G8
Malvern, Ark. (72104) 202/E5
Malvern, England 13/E5
Malvern, England 10/E4
Malvern, Iowa (51551) 229/B7
Malvern, Jamaica 158/H6
Malvern, Ohio (44644) 284/H4
Malvern, Pa. (19355) 294/L5
Malvern, Victoria 88/L7
Malvern, Victoria 97/J5
Malverne, N.Y. (11565) 276/R7
Malvina, Miss. (†38769) 256/C3
Malvinas (Falkland) (isls.), 143/D7
Malyye Karmakuly, U.S.S.R. 52/H1
Mama, U.S.S.R. 48/M4
Mamala (bay), Hawaii 218/B4
Mamalu (bay), Hawaii 218/K3
Mamanguape, Brazil 132/H4
Mamaroneck, N.Y. (10543) 276/P7
Mambahenauhan (isl.), Philippines 82/B7
Mambajao, Philippines 82/E6
Mambasa, Zaire 115/E3
Mamberamo (riv.), Indonesia 85/K6
Mambrui, Kenya 115/H4
Mamburao, Philippines 82/C4
Ma-Me-O Beach, Alberta 182/D3
Mamer, Luxembourg 27/H9
Mamers, France 28/D3
Mamers, N.C. (27552) 281/L4
Mamfé, Cameroon 115/A2
Mamie, N.C. (27952) 281/T2
Mamiña, Chile 138/B2
Mammoth, Ariz. (85618) 198/E6
Mammoth, Utah (†84628) 304/B4
Mammoth (creek), Utah 304/B6
Mammoth, W. Va. (25132) 312/D6
Mammoth Cave Nat'l Park, Ky. 237/J6
Mammoth Hot Springs (Yellowstone Nat'l Park), Wyo. (†82190) 319/B1
Mammoth Lakes, Calif. (93546) 204/G6
Mammoth Spring, Ark. (72554) 202/G1
Mamoré (riv.), Bolivia 120/C4
Mamoré (riv.), Bolivia 136/C2
Mamou, Guinea 106/B6
Mamou, La. (70554) 238/F5
Mampong, Ghana 106/D7
Mamry, Jezioro (lake), Poland 47/E1
Mamuju, Indonesia 85/F6
Man (isl.), I. of Man 13/C3
Man, Ivory Coast 106/C7
Man, Ivory Coast 102/B4
Man, W. Va. (25635) 312/C7
Mana, Fr. Guiana 131/E3
Mana (riv.), Fr. Guiana 131/E3
Mana (isl.), N. Zealand 100/B2
Manabí (prov.), Ecuador 128/B3
Manacacías (riv.), Colombia 126/D6
Manacapuru, Brazil 120/C3
Manacapuru, Brazil 132/H9
Manacas, Cuba 158/E1
Manacle (pt.), England 13/C7
Manacor, Spain 33/H3
Manado, Indonesia 54/O9
Manado, Indonesia 85/G5
Manage, Belgium 27/E7
Managua (cap.), Nic. 146/K8
Managua (cap.), Nicaragua 154/D4
Managua (lake), Nicaragua 154/E4
Manah, Oman 59/G5
Manahawkin, N.J. (08050) 273/E4
Manaia, N. Zealand 100/E3
Manakara, Madagascar 118/H4
Manakara, Madagascar 102/G7
Manakha, Yemen Arab Rep. 59/D6
Manakin-Sabot, Va. (23103) 307/N5
Manalapan, N.J. (†07746) 273/E3
Manama (cap.), Bahrain 59/F4
Manana (isl.), Hawaii 218/F2
Mananara, Madagascar 118/J3
Mananara (riv.), Madagascar 118/H4
Mananbao (riv.), Madagascar 118/G3
Mananjary, Madagascar 118/H4
Mananjary, Madagascar 102/G7
Manannah, Minn. (†56243) 255/D5
Manapire (riv.), Venezuela 124/E3
Manapouri (lake), N. Zealand 100/A6
Manaqil, Sudan 59/B7
Manar, Jebel (mt.), Yemen Arab Rep. 59/D7
Manare, Colombia 126/E4
Manas, China 77/C3
Manas He (riv.), China 77/C3
Manas Hu (lake), China 77/C2
Manasquan, N.J. (08736) 273/E3
Manasquan (riv.), N.J. 273/E3
Manassa, Colo. (81141) 208/H8
Manassas, Georgia (30438) 217/H6
Manassas (I.C.), Va. (22110) 307/O3
Manassas Nat'l Battlefield Park, Va. 307/K3
Manassas Park (I.C.), Va. (22110) 307/O3
Manatee (co.), Fla. 212/D4
Manatee (riv.), Fla. 212/D4
Manatí, Cuba 158/H3
Manatí, P. Rico 156/G1
Manatí, P. Rico 161/C1
Manaus, Brazil 2/F6
Manaus, Brazil 120/D3
Manaus, Brazil 132/H9
Manavgat, Turkey 63/D4
Manawa, Wis. (54949) 317/J7
Manay, Philippines 82/F7
Mancelona, Mich. (49659) 250/E4
Mancha, La (reg.), Spain 33/E3
Manchac (passage), La. 238/N2
Manchaca, Texas (78652) 303/G7
Mancha Real, Spain 33/E4
Manchaug, Mass. (01526) 249/G4
Manche (dept.), France 28/C3
Manche, La (English) (chan.), France 28/B3
Manchester, Ala. (†35501) 195/D3
Manchester, Calif. (95459) 204/B5
Manchester, Conn. (06040) 210/E1
Manchester○, Conn. (06040) 210/E1
Manchester, Greater (co.), England 13/H2
Manchester, England 7/D3
Manchester, England 10/G2
Manchester, England 13/H2
Manchester, Georgia (31816) 217/C5
Manchester, Ill. (62663) 222/C4
Manchester, Ind. (†47001) 227/H6
Manchester, Iowa (52057) 229/L3
Manchester, Kansas (67463) 232/E2
Manchester, Ky. (40962) 237/O6
Manchester○, Maine (04351) 243/D7
Manchester, Md. (21102) 245/L2
Manchester○, Mass. (01944) 249/F5
Manchester, Mich. (48158) 250/E6
Manchester, Minn. (56064) 255/E7
Manchester, Mo. (63011) 261/O3
Manchester, N.H. 188/M2
Manchester, N.H. (*03101) 268/E6
Manchester, N.Y. (14504) 276/F5
Manchester, Ohio (45144) 284/C8
Manchester, Okla. (73758) 288/L1
Manchester, Pa. (17345) 294/J5
Manchester, S. Dak. (†57353) 298/O5
Manchester, Tenn. (37355) 237/J10
Manchester, Vt. (05254) 268/A5
Manchester○, Vt. (05254) 268/A5
Manchester, Wash. (98353) 310/A2
Manchester, Wis. (†53946) 317/J8
Manchester Center, Vt. (05255) 268/A5
Manchester Depot, Vt. (†05254) 268/B5
Manchioneal, Jamaica 158/K6
Manchouli (Manzhouli), China 77/J2
Máncora, Peru 128/B5
Mancos, Colo. (81328) 208/C8
Mancos (riv.), Colo. 208/B8
Manda, Tanzania 115/F6
Mandabe, Madagascar 118/G4
Mandah, Mongolia 77/G3
Mandal, Norway 18/E7
Mandalay (div.), Burma 72/B2
Mandalay, Burma 54/L7
Mandalay, Burma 72/C2

Mandalgovĭ, Mongolia 77/G2
Mandali, Iraq 66/D4
Mandal-Ovoo, Mongolia 77/F3
Mandalya (gulf), Turkey 63/B4
Mandan, N. Dak. 188/F1
Mandan, N. Dak. (58554) 282/J6
Mandaon, Philippines 82/D4
Mandar (cape), Indonesia 85/F6
Mandaree, N. Dak. (58757) 282/E4
Mandaue, Philippines 82/E5
Mandeb, Bab el (str.), Saudi Arabia 59/D7
Mandeb, Bab el (str.), Yemen Arab Rep. 59/D7
Mandera, Kenya 115/H3
Manderson, S. Dak. (57756) 298/D7
Manderson, Wyo. (82432) 319/E1
Mandeville, Ark. (†75501) 202/C7
Mandeville, Jamaica 158/H6
Mandeville, La. (70448) 238/L6
Mandi, India 68/D2
Mandié, Mozambique 118/E3
Mandimba, Mozambique 118/F2
Mandinga, Panama 154/H6
Mandioré (lag.), Bolivia 136/F6
Mandla, India 68/E4
Mándok, Hungary 41/G2
Mandritsara, Madagascar 118/H3
Mand Rud (riv.), Iran 59/F4
Mand Rud (riv.), Iran 66/G6
Mandsaur, India 68/C4
Mandurah, W. Australia 88/B3
Mandurah, W. Australia 92/A2
Manduria, Italy 34/F4
Mandvi, India 68/B4
Manele (bay), Hawaii 218/H2
Manele Bay, Hawaii (†96763) 218/H2
Manendragarh, India 68/E4
Manes, Mo. (†65711) 261/H8
Manfalût, Egypt 111/J4
Manfalût, Egypt 59/B4
Manfred, N. Dak. (58465) 282/L4
Manfredonia, Italy 34/F4
Manfredonia (gulf), Italy 34/F4
Manga, Brazil 132/E6
Manga, Uruguay 145/B7
Mangai, Zaire 115/C4
Mangaia (isl.), Cook Is. 87/L8
Mangakino, N. Zealand 100/E3
Mangalia, Romania 45/J4
Mangalore, India 54/J8
Mangalore, India 68/C6
Mangareva (isl.), Fr. Poly. 87/N8
Mangaweka, N. Zealand 100/E3
Mangere (isl.), N. Zealand 100/E7
Mangerton (mt.), Ireland 17/C8
Mangham, La. (71259) 238/G2
Mangkalihat (cape), Indonesia 85/F5
Manglaralto, Ecuador 128/B3
Mangle (pt.), Cuba 158/J3
Manglillo (pt.), P. Rico 161/B3
Mangnai, China 77/D4
Mango, Fla. (33550) 212/D4
Mango, Togo 106/E6
Mangochi, Malawi 115/G6
Mangoky (riv.), Madagascar 102/G7
Mangoky (riv.), Madagascar 118/G4
Mangole (isl.), Indonesia 85/H6
Mangonui, N. Zealand 100/D1
Mangoro (riv.), Madagascar 118/H3
Mangotsfield, England 10/E5
Mangotsfield, England 13/E6
Mangrol, India 68/B4
Mangsee (isls.), Philippines 82/A7
Mangualde, Portugal 33/C2
Mangueigne, Chad 111/D5
Mangueira (lag.), Brazil 132/D11
Manguera Azul, Uruguay 145/D4
Mangui, China 77/K1
Manguito, Cuba 158/D1
Mangum, Okla. (73554) 288/G5
Mangyshlak (pen.), U.S.S.R. 48/F5
Manhan (riv.), Mass. 249/D4
Manhasset, N.Y. (11030) 276/P7
Manhattan, Ill. (60442) 222/F2
Manhattan, Ind. (†46171) 227/D5
Manhattan, Kansas (66502) 232/F2
Manhattan, Mont. (59741) 262/E5
Manhattan, Nev. (89022) 266/E4
Manhattan (borough), N.Y. (*10001) 276/M9
Manhattan (isl.), N.Y. 276/M9
Manhattan Beach, Calif. (90266) 204/B11
Manhattan Beach, Minn. (56463) 255/E4
Manhay, Belgium 27/H8
Manheim, Pa. (17545) 294/K5
Manheim, W. Va. (26403) 312/G4
Manhiça, Mozambique 118/E5
Man Hpang, Burma 72/C2
Manhuaçu, Brazil 132/F8
Manhuaçu, Brazil 135/E2
Manhumirim, Brazil 135/E2
Maní, Colombia 126/D5
Maniamba, Mozambique 102/F6
Maniamba, Mozambique 118/F2
Manibridge, Manitoba 179/J2
Manica (prov.), Mozambique 118/E4
Manica, Mozambique 118/E3
Manicani (isl.), Philippines 82/E5
Manicaragua, Cuba 158/E2
Manicoré, Brazil 120/C3
Manicoré, Brazil 132/H9
Manicouagan, Québec 174/D2
Manicouagan (pt.), Québec 172/B1
Manicouagan (res.), Québec 174/D2
Manicouagan (riv.), Que. 162/K5
Manicouagan (riv.), Québec 174/D2
Manifest, La. (†71340) 238/G3
Manifold (cape), Queensland 88/J4
Manifold (cape), Queensland 95/D4
Manigotagan, Manitoba 179/F3
Manigotagan (lake), Manitoba 179/G4
Manigotagan (riv.), Manitoba 179/G3
Manigouche, Québec 174/C3
Manihiki (atoll), Cook Is. 87/K7
Manila, Ala. (†36586) 195/C7
Manila, Ark. (72442) 202/K2
Manila (prov.), Philippines 82/C3
Manila (cap.), Philippines 2/R5
Manila (cap.), Philippines 85/G3
Manila (cap.), Philippines 82/C3
Manila (cap.), Philippines 54/N8
Manila (bay), Philippines 82/C3
Manila, Utah (84046) 304/E3
Manildra, N.S. Wales 97/E3
Manilla, Ind. (46150) 227/F5
Manilla, Iowa (51454) 229/C5
Manilla, N.S. Wales 97/F2
Maningrida, North. Terr. 93/C2
Manipa (str.), Indonesia 85/H6
Manipur (riv.), Burma 72/B2
Manipur (state), India 68/G4
Manisa (prov.), Turkey 63/B3
Manisa, Turkey 63/B3
Manisa, Turkey 59/A2
Manistee, Mich. 188/J2
Manistee (co.), Mich. 250/C4
Manistee, Mich. (49660) 250/C4
Manistee (riv.), Mich. 250/C4
Manistique, Mich. (49854) 250/C3
Manistique (lake), Mich. 250/D2
Manistique (riv.), Mich. 250/C2
Manito, Ill. (61546) 222/D3
Manito (lake), Sask. 181/B3
Manitoba (prov.) 162/G5
Manitoba (prov.), Canada 146/J4
MANITOBA 179
Manitoba (lake), Man. 146/H4
Manitoba (lake), Man. 162/G5
Manitoba (lake), Manitoba 179/D4
Manitou, Ky. (42436) 237/F6
Manitou, Manitoba 179/D5
Manitou (isl.), Mich. 250/B1
Manitou, N. Dak. (†58776) 282/E3
Manitou, Okla. (73555) 288/J5
Manitou (lake), Ontario 177/C2
Manitou (lake), Québec 172/C3
Manitou Beach, Sask. 181/F4
Manitoulin (terr. dist.), Ontario 175/D3
Manitoulin (terr. dist.), Ontario 177/B2
Manitoulin (isl.), Ont. 162/H6
Manitoulin (isl.), Ontario 175/D3
Manitoulin (isl.), Ontario 177/B2
Manitou Springs, Colo. (80829) 208/J5
Manitouwadge, Ontario 177/H5
Manitouwadge, Ontario 175/C3
Manitowaning, Ontario 177/C2
Manitowish, Wis. (†54547) 317/F3
Manitowoc (co.), Wis. 317/L7
Manitowoc, Wis. (54220) 317/L7
Maniwaki, Québec 174/B3
Maniwaki, Québec 172/B3
Manizales, Colombia 126/C5
Manizales, Colombia 120/B2
Manja, Jordan 65/D4
Manja, Madagascar 118/G4
Manjacaze, Mozambique 118/E5
Manjimup, W. Australia 88/B6
Manjimup, W. Australia 92/B6
Mankato, Kansas (66956) 232/D2
Mankato, Minn. 188/H2
Mankato, Minn. (56001) 255/E6
Mankono, Ivory Coast 106/C7
Mankota, Sask. 181/D6
Manley, Nebr. (68403) 264/H4
Manley Hot Springs, Alaska (99756) 196/H2
Manlius, Ill. (61338) 222/D2
Manlius, N.Y. (13104) 276/J5
Manlleu, Spain 33/H1
Manly, Iowa (50456) 229/G2
Manly, N. S. Wales 88/L4
Manly, N. S. Wales 97/K3
Manly, N.C. (†28387) 281/L4
Manly, Queensland 88/L2
Manmad, India 68/C4
Manmanoc (mt.), Philippines 82/C2
Mann (riv.), North. Terr. 93/D2
Manna, Indonesia 85/C6
Mannahill, S. Australia 94/F5
Mannar (gulf) 54/J9
Mannar (gulf), India 68/D7
Mannar, Sri Lanka 68/E7
Mannar (gulf), Sri Lanka 68/D7
Mannargudi, India 68/E6
Mannboro, Va. (23105) 307/N6
Männedorf, Switzerland 39/G2
Mannersdorf am Leithagebirge, Austria 41/D3
Manners Sutton, New Bruns. 170/D3
Mannford, Okla. (74044) 288/O2
Mannheim, W. Germany 7/E4
Mannheim, W. Germany 22/C4
Manning, Alberta 182/B1
Manning, Ark. (71757) 202/E5
Manning, Iowa (51455) 229/C5
Manning, Kansas (†67871) 232/B3
Manning (riv.), N.S. Wales 97/F2
Manning, N. Dak. (58642) 282/E5
Manning (cape), N.W. Terrs. 187/F2
Manning (str.), Solomon Is. 86/D2
Manning, S.C. (29102) 296/G4
Manning Prov. Park, Br. Col. 184/G5
Mannington, Ky. (42217) 237/G6
Mannington, W. Va. (26582) 312/F3
Männlifluh (mt.), Switzerland 39/E3
Manns Choice, Pa. (15550) 294/E6
Manns Harbor, N.C. (27953) 281/T3
Mannsville, Ky. (42758) 237/L6
Mannsville, N.Y. (13661) 276/H3
Mannsville, Okla. (73447) 288/N6
Mannu (riv.), Italy 34/B5
Mannum, S. Australia 94/F6
Mannville, Alberta 182/E3
Mano (riv.), Liberia 106/B7
Mano (riv.), S. Leone 106/B7
Manoa, Bolivia 136/C1
Manokin, Md. (21836) 245/P8
Manokin (riv.), Md. 245/P8
Manokotak, Alaska (99628) 196/G3
Manokwari, Indonesia 85/J6
Manola, Alberta 182/D2
Manombo, Madagascar 118/G4
Manomet, Mass. (02345) 249/M5
Manomet (pt.), Mass. 249/N5
Manono, Zaire 115/E5
Manono, Zaire 102/E5
Manor, Georgia (31550) 217/G8
Manor, Pa. (15665) 294/C5
Manor, Sask. 181/K6
Manor, Texas (78653) 303/G7
Manorhamilton, Ireland 17/E3
Manori, India 68/B6
Manori (creek), India 68/B7
Manorville, N.Y. (11949) 276/P9
Manorville, Pa. (16238) 294/C4
Manosque, France 28/G6
Manotick, Ontario 177/J2
Manouane, Québec 172/C2
Manouane (lake), Québec 174/C2
Manp'o, N. Korea 81/B3
Manquin, Va. (23106) 307/O5
Manra (Sydney) (isl.), Kiribati 87/K6
Manresa, Spain 33/G2
Mansa, Zambia 115/E6
Mansa, Zambia 102/E6
Mansalay, Philippines 82/C4
Mansavillagra, Uruguay 145/D4
Manseau, Québec 172/E3
Mansel (isl.), N.W.T. 162/H3
Mansel (isl.), N.W.T. 146/K3
Mansel (isl.), N.W. Terrs. 187/K3
Mansel'ka (mts.), U.S.S.R. 52/C1
Mansfield, Ark. (72944) 202/B3
Mansfield○, Conn. (†06250) 210/F1
Mansfield, England 13/K2
Mansfield, England 10/F4
Mansfield, Georgia (30255) 217/E4
Mansfield, Ill. (61854) 222/E3
Mansfield, Ind. (†47872) 227/C5
Mansfield, La. (71052) 238/C2
Mansfield, Mass. (02048) 249/J4
Mansfield○, Mass. (02048) 249/J4
Mansfield, Minn. (†56009) 255/E7
Mansfield, Mo. (65704) 261/G8
Mansfield, Ohio 188/K2
Mansfield, Ohio (*44901) 284/F4
Mansfield, Pa. (16933) 294/H2
Mansfield, S. Dak. (57460) 298/N3
Mansfield, Tenn. (38236) 237/E8
Mansfield, Texas (76063) 303/F3
Mansfield (mt.), Vt. 268/B2
Mansfield, Victoria 97/D5
Mansfield, Wash. (98830) 310/F3
Mansfield Center, Conn. (06250) 210/G1
Mansfield Depot, Conn. (06251) 210/F1
Mansfield Woodhouse, England 13/F4
Mansilla de las Mulas, Spain 33/D1
Manso (riv.), Brazil 132/C6
Manso (riv.), Chile 138/E4
Manson, Ind. (†46041) 227/D4
Manson, Iowa (50563) 229/D3
Manson, Manitoba 179/A4
Manson, N.C. (27553) 281/N2
Manson, Wash. (98831) 310/E3
Manson Creek, Br. Col. 184/E2
Mansonville, Québec 172/E4
Mansura, La. (71350) 238/G4
Manta, Ecuador 128/B3
Manta, Ecuador 120/A3
Manta (bay), Ecuador 128/B3
Mantachie, Miss. (38855) 256/H2
Mantador, N. Dak. (58058) 282/R7
Mantagao (lake), Manitoba 179/E3
Mantagao (riv.), Manitoba 179/E3
Mantalingajan (mt.), Philippines 82/A6
Mantario, Sask. 181/B4
Mantaro (riv.), Peru 128/E8
Mantas (well), Niger 106/E5
Manteca, Calif. (95336) 204/D6
Mantecal, Apure, Venezuela 124/D4
Mantecal, Bolívar, Venezuela 124/F4
Mantee, Miss. (39751) 256/F3
Manteigas, Portugal 33/C2
Manteno, Ill. (60950) 222/F2
Manteo, N.C. (27954) 281/T3
Manter, Kansas (67862) 232/A4
Mantes-la-Jolie, France 28/D3
Manti, Utah (84642) 304/C4
Mantiqueira (range), Brazil 135/D3
Manto, Honduras 154/D3
Mantoloking, N.J. (08738) 273/E3
Manton, Calif. (96059) 204/D3
Manton, Mich. (49663) 250/D4
Manton, R.I. (†02904) 249/J5
Mantorville, Minn. (55955) 255/F6
Mänttä, Finland 18/O6
Mantua, Ala. (35472) 195/C4
Mantua, Cuba 158/A2
Mantua (prov.), Italy 34/C2
Mantua, Italy 34/C2
Mantua○, N.J. (08051) 273/C4
Mantua, Ohio (44255) 284/H3
Mantua, Utah (†84302) 304/C2
Mantua, Va. (†22030) 307/S3
Manturovo, U.S.S.R. 52/F3
Manú, Peru 128/G9
Manú (riv.), Peru 128/G8
Manua (isls.), Amer. Samoa 87/K7
Manuae (atoll), Cook Is. 87/K7
Manuel Benavides, Mexico 150/H2
Manuelito, N. Mex. (†86506) 274/A3
Manuel Rodríguez (isl.), Chile 138/D10
Manuels, New Bruns. 170/F1
Manuels, Newf. 166/D2
Manui (isl.), Indonesia 85/G6
Manukan, Philippines 82/D6
Manukau, N. Zealand 100/C1
Manukau (harb.), N. Zealand 100/B1
Manulla, Ireland 17/C4
Manumuskin (riv.), N.J. 273/D5
Manunui, N. Zealand 100/E3
Manuripi (riv.), Bolivia 136/B2
Manus (isl.), Papua N.G. 87/E6
Manus (isl.), Papua N.G. 86/A1
Manutuke, N. Zealand 100/F3
Manvel, N. Dak. (58256) 282/R3
Manvel, Texas (77578) 303/J3
Manville, N.J. (08835) 273/D2
Manville, R.I. (02838) 249/H5
Manville, Wyo. (82227) 319/H3
Many, La. (71449) 238/C3
Manyara (lake), Tanzania 115/G4
Manyas, Turkey 63/B3
Manyberries, Alberta 182/E5
Manych-Gudilo (lake), U.S.S.R. 52/F5
Many Farms, Ariz. (86538) 198/F2
Manyoni, Tanzania 115/G5
Manzai, Pakistan 59/K3
Manzanar, Chile 138/F2
Manzanares, Spain 33/E3
Manzanares (riv.), Spain 33/F4
Manzanillo, Cuba 158/H4
Manzanillo, Cuba 156/C2
Manzanillo (bay), Dom. Rep. 158/C5
Manzanillo (bay), Haiti 158/C5
Manzanillo, Mexico 150/G7
Manzanillo, Mexico 146/H8
Manzanillo (pt.), Panama 154/H6
Manzanita, Oreg. (97130) 291/C2
Manzanita Ind. Res., Calif. 204/J11
Manzano, N. Mex. (†87016) 274/C4
Manzano (mts.), N. Mex. 274/C4
Manzano (peak), N. Mex. 274/C4
Manzanola, Colo. (81058) 208/M6
Manzhouli (Manchouli), China 77/J2
Manzini, Swaziland 118/E5
Mao, Chad 111/C5
Mao, Dom. Rep. 158/D5
Maoke (mts.), Indonesia 85/K6
Maoming (Mowming), China 77/H7
Mapai, Mozambique 118/E4
Mapararí, Venezuela 124/D2
Mapastepec, Mexico 150/N9
Mapes, N. Dak. (58349) 282/O3
Mapia (isls.), Indonesia 85/J5
Mapimí, Mexico 150/G4
Mapimí (depr.), Mexico 150/G3
Mapire, Venezuela 124/F4
Mapiri, Bolivia 136/B4
Mapiripán, Laguna (lake), Colombia 126/E6
Maple (peak), Ariz. 198/F5
Maple (riv.), Mich. 250/E5
Maple (lake), Minn. 255/B3
Maple (riv.), Minn. 255/E7
Maple (riv.), N. Dak. 282/O8
Maple (riv.), N. Dak. 282/R6
Maple (creek), Sask. 181/B5
Maple (riv.), S. Dak. 298/M1
Maple, Wis. (54854) 317/C2
Maple Bay, Br. Col. 184/K3
Maple Bay, Minn. (†56736) 255/B3
Maple City, Kansas (67102) 232/F4
Maple City, Mich. 250/D4
Maple Creek, Sask. 162/F6
Maple Creek, Sask. 181/B6
Maple Falls, Wash. (98266) 310/D2
Maple Grove, Minn. (†55369) 255/G5
Maple Grove, Ontario 177/F4
Maple Grove, Québec 172/H4
Maple Heights, Ohio (44137) 284/H9
Maple Hill, Iowa (50564) 229/D2
Maple Hill, Kansas (66507) 232/F2
Maple Hill, N.C. (28454) 281/O5
Maple Island, Minn. (†55082) 255/E7
Maple Lake, Minn. (55358) 255/D5
Maple Park, Ill. (60151) 222/E2
Maple Plain, Minn. (55359) 255/F5
Maple Rapids, Mich. (48853) 250/E5
Maple Ridge, Br. Col. 184/L3
Maple River, Iowa (†51401) 229/D4
Maples, Ind. (†46802) 227/H2
Maples, Mo. (†65542) 261/J7
Maple Shade○, N.J. (08052) 273/B3
Maplesville, Ala. (36750) 195/E5
Mapleton, Iowa (51034) 229/B4
Mapleton, Kansas (66754) 232/H3
Mapleton○, Maine (04757) 243/G2
Mapleton, Mich. (†49684) 250/D4
Mapleton, Minn. (56065) 255/E7
Mapleton, N.C. (†27855) 281/P2
Mapleton, N. Dak. (58059) 282/R6
Mapleton, Oreg. (97453) 291/C3
Mapleton (Mapleton Depot), Pa. (17052) 294/F5
Mapleton, Utah (†84663) 304/C3
Mapleton, Wis. (†53066) 317/J1
Mapleton Depot, Pa. (17052) 294/F5
Maple Valley, Wash. (98038) 310/C3
Mapleview, Minn. (†55912) 255/E7
Mapleview, New Bruns. 170/C2
Mapleville, Md. (†21713) 245/H2
Mapleville, R.I. (02839) 249/H5
Maplewood, La. (†70663) 238/D6
Maplewood, Minn. (55109) 255/G5
Maplewood, Mo. (63143) 261/P3
Maplewood, N.H. (†03574) 268/D3
Maplewood○, N.J. (07040) 273/E2
Maplewood, Ohio (45340) 284/B5
Maplewood, Wis. (54226) 317/M6
Mapocho (riv.), Chile 138/G3
Mapoon Mission Station, Queensland 88/G2
Mapoon Mission Station, Queensland 95/B1
Maporal, Venezuela 124/C4
Mapos (Amazones), Cuba 158/F2
Mappsville, Va. (23407) 307/T5
Mapuera (riv.), Brazil 132/B3
Maputo (city) (prov.), Mozambique 118/E5
Maputo (prov.), Mozambique 118/E5
Maputo (cap.), Mozambique 2/L7
Maputo (cap.), Mozambique 118/E5
Maputo (cap.), Mozambique 102/F7
Maqatin (ruins), P.D.R. Yemen 59/E7
Maqên, China 77/F5
Maqna, Saudi Arabia 59/C4
Ma Qu (Huang He) (riv.), China 77/F5
Maquapit (lake), New Bruns. 170/D3
Maqueda (chan.), Philippines 82/D3
Maquela do Zombo, Angola 102/D5
Maquela do Zombo, Angola 115/C5
Maquereau (pt.), Québec 172/D2
Maquinchao, Argentina 143/C5
Maquoketa, Iowa (52060) 229/M4
Maquon, Ill. (61458) 222/C3
Mar (mts.), Brazil 120/E5
Mar (range), Brazil 135/C4
Mar, Serra do (range), Brazil 132/E9
Mar (dist.), Scotland 15/F3
Mara, Guyana 131/C3
Mara (reg.), Tanzania 115/F4
Marabá, Brazil 132/D4
Marabá, Brazil 120/E3
Marabahan, Indonesia 85/E6
Marabella, Trin. & Tob. 161/A11
Maracá (isl.), Brazil 120/E2
Maracá (isl.), Brazil 132/D2
Maracaibo, Venezuela 124/C2
Maracaibo, Venezuela 120/B1
Maracaibo (lake), Venezuela 120/B2
Maracaibo (lake), Venezuela 124/C3
Maracaju, Brazil 132/C8
Maracas (bay), Trin. & Tob. 161/C10
Maracay, Venezuela 124/E2
Maracay, Venezuela 120/C2
Marada, Libya 111/C2
Maradi, Niger 106/F6
Maradi, Niger 102/C3
Maragheh, Iran 59/E2
Maragheh, Iran 66/E2
Maragogipe, Brazil 132/G6
Maraira (pt.), Philippines 82/C1
Marajó (est.), Brazil 120/E2
Marajó (isl.), Brazil 120/E3
Marajó (bay), Brazil 132/E2
Marajó (isl.), Brazil 132/D3
Maralal, Kenya 115/G3
Maralinga, S. Australia 88/E6
Maralwexi (Bachu), China 77/A4
Maramag, Philippines 82/E7
Maramec, Okla. (74045) 288/N2
Marampa, S. Leone 106/B7
Marana, Ariz. (85238) 198/D6
Marand, Iran 59/E2
Marand, Iran 66/D1
Marandellas, Zimbabwe 118/E3
Marang, Malaysia 72/D6
Maranguape, Brazil 132/G3
Maranhão (state), Brazil 132/E4
Maranoa (riv.), Queensland 95/C5
Marañón (riv.), Peru 120/B3
Marañón (riv.), Peru 128/E5
Marapanim, Brazil 132/E3
Maras (mt.), Indonesia 85/D6
Maraş, Turkey 59/C2
Maraş (Kahramanmaraş), Turkey 63/G4
Marathon, Fla. (33050) 212/E7
Marathón, Greece 45/G6
Marathon, Iowa (50565) 229/C3
Marathon, N.Y. (13803) 276/J6
Marathon, Ohio (45145) 284/C7
Marathon, Ont. 162/H6
Marathon, Ontario 177/H5
Marathon, Ontario 175/C3
Marathon, Texas (79842) 303/A7
Marathon (co.), Wis. 317/G6
Marathon, Wis. (54448) 317/G6
Maratua (isl.), Indonesia 85/F5
Maravillas, Bolivia 136/B2
Maravillas (creek), Texas 303/A7
Marawi, Philippines 85/G4
Marawi, Philippines 82/E6
Marbach, Switzerland 39/E3
Marbach am Neckar, W. Germany 22/C4
Marbella, Spain 33/D4
Marble, Ark. (72746) 202/C1
Marble, Colo. (†81623) 208/E4
Marble, Minn. (55764) 255/E3
Marble, N.C. (28905) 281/B4
Marble (isl.), N.W.T. 162/G3
Marble (isl.), N.W. Terrs. 187/J3
Marble Bar, Australia 87/C8
Marble Bar, W. Australia 88/B4
Marble Bar, W. Australia 92/C3
Marble Canyon, Ariz. (86036) 198/D2
Marble Canyon Nat'l Mon., Ariz. 198/D2
Marble City, Okla. (74945) 288/S3
Marble Dale, Conn. (06777) 210/B2
Marble Falls, Texas (78654) 303/F7
Marblehead, Ill. (†62301) 222/B4
Marblehead○, Mass. (01945) 249/E7
Marblehead (neck), Mass. 249/F6
Marblehead, Ohio (†43440) 284/E2
Marble Hill, Georgia (30148) 217/D2
Marble Hill, Mo. (63764) 261/N8
Marblemount, Wash. (98267) 310/D2
Marble Rock, Iowa (50653) 229/H3
Marbleton, Québec 172/F4
Marbleton, Wyo. (†83113) 319/B3
Marble Valley, Ala. (†35150) 195/F4
Marburg an der Lahn, W. Germany 22/C3
Marbury, Ala. (36051) 195/E5
Marbury, Md. (20658) 245/K6
Marcala, Honduras 154/C3
Marcali, Hungary 41/D3
Marcapata, Peru 128/G9
Marcelin, Sask. 181/E3
Marceline, Mo. (64658) 261/F3
Marcell, Minn. (56657) 255/E3
Marcella, Ark. (72555) 202/G2
Marcella, N.J. (†07866) 273/E2
Marcelline, Ill. (†62376) 222/B3
Marcellus, Mich. (49067) 250/D6
Marcellus, N.Y. (13108) 276/H5
Marcelville, New Bruns. 170/E2
March (riv.), Austria 41/D2
March, England 10/G4
March, England 13/H5
March A.F.B., Calif. 204/E11
Marchand, Manitoba 179/F5
Marche, Ark. (†72114) 202/F4
Marche (trad. prov.) France 29
Marche (reg.), Italy 34/D3
Marche-en-Famenne, Belgium 27/G8
Marchegg, Austria 41/D2
Marchena (isl.), Ecuador 128/B9
Marchena, Spain 33/D4
Marchfield, Barbados 161/B9
Marchigüe, Chile 138/F5
Marchin, Belgium 27/G8
Mar Chiquita (lake), Argentina 143/D3
Marchwell, Sask. 181/K5
Marco (Marco Island), Fla. (33937) 212/E6
Marco (isl.), Fla. 212/E6
Marco, Ind. (†47443) 227/C7
Marco, La. (†71447) 238/E3
Marcola, Oreg. (97454) 291/E3
Marcona, Peru 128/E10
Marcos Juárez, Argentina 143/D3
Marcus, Iowa (51035) 229/B3
Marcus (isl.), Japan 87/F3
Marcus, S. Dak. (57757) 298/E4
Marcus, Wash. (99151) 310/H2
Marcus Baker (mt.), Alaska 196/C1
Marcus Hook, Pa. (19061) 294/L7
Marcy, N.Y. (13403) 276/K4
Marcy (mt.), N.Y. 276/N2
Mardan, Pakistan 68/C2
Mardan, Pakistan 59/K3
Mardela Springs, Md. (21837) 245/P7
Mar del Plata, Argentina 143/E4
Mar del Plata, Argentina 120/D6
Mardin (prov.), Turkey 63/J4
Mardin, Turkey 63/J4
Mardin, Turkey 59/D2
Maré (isl.), New Caled. 87/G8
Mare (isl.), New Caled. 86/J4
Mareb (riv.), Ethiopia 59/C7
Marechal Deodoro, Brazil 132/H5
Maree, Loch (lake), Scotland 10/D2
Maree, Loch (lake), Scotland 15/C3
Mareeba, Queensland 95/C3
Mareeba, Queensland 88/G3
Mare Island Navy Yard, Calif. 204/J1
Marengo (co.), Ala. 195/C6
Marengo, Ala. (†36736) 195/C6
Marengo, Ill. (60152) 222/E1
Marengo, Ind. (47140) 227/E8
Marengo, Iowa (52301) 229/J5
Marengo, Ohio (43334) 284/E5
Marengo, Sask. 181/B4
Marengo, Wash. (†99004) 310/G3
Marengo, Wis. (54855) 317/E3
Marenisco, Mich. (49947) 250/F2
Marennes, France 28/C5
Mareth, Tunisia 106/F2
Marettimo (isl.), Italy 34/C6
Marfa, Texas (79843) 303/C12
Marfield, N.S. Wales 97/C3
Marfrance, W. Va. (25975) 312/E6
Margai Caka (lake), China 77/C4
Marganets, U.S.S.R. 52/E5
Margao, India 68/C5
Margaree, Nova Scotia 168/G2
Margaree (isl.), Nova Scotia 168/F4
Margaree Centre, Nova Scotia 168/H2
Margaree Forks, Nova Scotia 168/G2
Margaree Harbour, Nova Scotia 168/G2
Margaree Valley, Nova Scotia 168/H2
Margaret, Ala. (35112) 195/F3
Margaret (lake), Alberta 182/B5
Margaret, Manitoba 179/C5
Margaret, Texas (†79227) 303/E3
Margaret (riv.), W. Australia 88/D3
Margaret River, W. Australia 88/A6
Margaret River, W. Australia 92/A6
Margaret River Station, W. Australia 92/D2
Margaretsville, Nova Scotia 168/C3
Margaretville, N.Y. (12455) 276/L6
Margarita, Argentina 143/F5
Margarita (isl.), Venezuela 120/C1
Margarita (isl.), Venezuela 124/F2
Margate, England 13/J6
Margate, England 10/G5
Margate, Fla. (33063) 212/F5
Margate, S. Africa 118/E6
Margate, Tasmania 99/D4
Margate City, N.J. (08402) 273/E5
Margento, Colombia 126/C4
Margerum, Ala. (†35616) 195/B1
Margherita (Jamama), Somalia 115/H3
Margherita (mt.), Uganda 115/E3
Margherita (mt.), Zaire 102/E4
Margherita (mt.), Zaire 115/E3
Margie, Minn. (56658) 255/E2
Margo, Sask. 181/H4
Margos, Peru 128/D8
Margosatubig, Philippines 82/D7
Margow, Dasht-e (des.), Afghanistan 59/H3
Margow, Dasht-e (des.), Afghanistan 68/A2
Margraten, Netherlands 27/H7
Margret, Georgia (†30536) 217/D1
Margrethe (lake), Mich. 250/E4
Marguerite (bay) 5/C15
Maria (isl.), Fr. Poly. 87/L8
Maria (creek), Ind. 227/C7
Maria, Québec 172/C2
Maria (isls.), St. Lucia 161/G7
Maria (isl.), Tasmania 99/E4
Mari A.S.S.R., U.S.S.R. 52/G3
Mari A.S.S.R., U.S.S.R. 48/E4
María Albina, Uruguay 145/E4
María Cleófas (isl.), Mexico 150/F6
María Elena, Chile 138/B3
Mariager, Denmark 21/D4
Mariager, Denmark 18/G8
Mariager (fjord), Denmark 21/D4
Mariah Hill, Ind. (47556) 227/D8
María Madre (isl.), Mexico 150/F6
María Magdalena (isl.), Mexico 150/F6
Marian (lake), Fla. 212/E4
Marian (lake), N.W. Terrs. 187/G3
Marian, Queensland 88/H4
Marian, Queensland 95/D4
Mariana, Brazil 135/E2
Mariana Lake, Alberta 182/D2
Marianao, Cuba 158/C1
Marianao, Cuba 156/A2
Marianas, Northern 87/E4
Mariana Trench 87/E4
Marianna, Ark. (72360) 202/J4
Marianna, Fla. (32446) 212/A1
Marianna, Pa. (15345) 294/B5
Mariano I. Loza, Argentina 143/G4
Mariano Roque Alonso, Paraguay 144/A4

Mariánské Lázně, Czech. 41/B2
María Pinto, Chile 138/G3
Mariapolis, Manitoba 179/C5
Marías, Islas (isls.), Mexico 150/F6
Marias (riv.), Mont. 188/D1
Marias (riv.), Mont. 262/D2
Maria Stein, Ohio (45860) 284/A5
Mariato, Punta (cape), Panama 154/G7
María Trinidad Sánchez (prov.), Dom. Rep. 158/E5
Maria van Diemen (cape), N. Zealand 100/D1
Mariazell, Austria 41/C3
Marib, Saudi Arabia 59/E6
Marib, Yemen Arab Rep. 59/D6
Mariba, Ky. (40345) 237/O5
Maribel, Wis. (54227) 317/L7
Maribo, Denmark 21/E8
Maribor, Yugoslavia 7/F4
Maribor, Yugoslavia 45/B2
Maribyrnong (riv.), Victoria 97/H5
Maribyrnong (riv.), Victoria 88/K7
Maricao, P. Rico 161/B2
Maricopa (co.), Ariz. 198/C5
Maricopa, Ariz. (85239) 198/C5
Maricopa (mts.), Ariz. 198/C5
Maricopa, Calif. (93252) 204/F8
Maricopa Ind. Res., Ariz. 198/C6
Maricourt, Que. 162/J3
Maricourt, Québec 174/F1
Maricunga, Salar de (salt dep.), Chile 138/B6
Maridi, Sudan 111/E7
Marie (lake), Alberta 182/E2
Marie, Ark. (†72395) 202/K2
Marie, W. Va. (†24910) 312/E7
Marie Byrd Land (reg.), Ant. 2/D10
Marie Byrd Land (reg.) 5/B13
Mariefred, Sweden 18/F1
Marie-Galante (isl.), Guadeloupe 161/B7
Marie-Galante (isl.), Guadeloupe 156/G4
Mariehamn, Finland 18/M7
Mariel, Cuba 158/B1
Mariemont, Ohio (45227) 284/C9
Marienberg, Papua N.G. 85/B6
Marienburg (Malbork), Poland 47/D1
Mariënburg, Suriname 131/D2
Mariental, Namibia 118/B4
Mariental, Namibia 102/D7
Marienthal, Kansas (67863) 232/A3
Marienville, Pa. (16239) 294/D3
Marie-Reine, Alberta 182/B1
Maries (co.), Mo. 261/J6
Mariestad, Sweden 18/H7
Marietta, Georgia (*30060) 217/J1
Marietta, Ill. (61459) 222/C3
Marietta, Ind. (†46176) 227/F6
Marietta, Minn. (56257) 255/B5
Marietta, Miss. (38856) 256/H2
Marietta, N.C. (28362) 281/L6
Marietta, Ohio (45750) 284/G7
Marietta, Okla. (73448) 288/M7
Marietta, Pa. (17547) 294/J5
Marietta-Alderwood, Wash. (98268) 310/C2
Marietta-Slater, S.C. (29661) 296/C1
Marieval, Sask. 181/J5
Marieville, Québec 172/D4
Marigot, Dominica 161/F6
Marigot, Haiti 158/C6
Marigot, Martinique 161/D5
Marigot, St. Lucia 161/G6
Marigüitar, Venezuela 124/G2
Marihatag, Philippines 82/F6
Marília, Brazil 120/D5
Marília, Brazil 135/A3
Marília, Brazil 132/D8
Marilla, N.Y. (14102) 276/C5
Marin (co.), Calif. 204/C5
Marín, Spain 33/B1
Marina, Calif. (93933) 204/D7
Marinduque (prov.), Philippines 82/C4
Marinduque (isl.), Philippines 82/C4
Marine, Ill. (62061) 222/D5
Marine City, Mich. (48039) 250/G6
Marineland, Fla. (†32084) 212/E2
Marine on Saint Croix, Minn. (55047) 255/F5
Marinette (co.), Wis. 317/K5
Marinette, Wis. (54143) 317/L5
Maringá, Brazil 120/D5
Maringá, Brazil 132/D8
Maringouin, La. (70757) 238/G6
Marinha Grande, Portugal 33/B3
Marinhas, Portugal 33/B2
Marino, Italy 34/F7
Marion (reef), 95/E3
Marion (co.), Ala. 195/C2
Marion, Ala. (36756) 195/D5
Marion (lake), Alberta 182/D3
Marion (co.), Ark. 202/E1
Marion, Ark. (72364) 202/K3
Marion, Conn. (06444) 210/D2
Marion (co.), Fla. 212/D2
Marion (co.), Georgia 217/C6
Marion (co.), Ill. 222/E5
Marion, Ill. (62959) 222/E6
Marion (co.), Ind. 227/E5
Marion, Ind. 188/J2
Marion, Ind. (46952) 227/F3
Marion (co.), Iowa 229/G6
Marion, Iowa (52302) 229/K4
Marion (co.), Kansas 232/E3
Marion, Kansas (66861) 232/F3
Marion (lake), Kansas 232/E3
Marion (co.), Ky. 237/L5
Marion, Ky. (42064) 237/E6
Marion, La. (71260) 238/F1
Marion, Mass. (02738) 249/L6
Marion○, Mass. (02738) 249/L6
Marion, Mich. (49665) 250/D4
Marion (co.), Miss. 256/E8
Marion, Miss. (39342) 256/G6
Marion (co.), Mo. 261/J3
Marion, Mont. (59925) 262/B2
Marion, N.Y. (14505) 276/F4
Marion, N.C. (28752) 281/E3
Marion, N. Dak. (58466) 282/O6
Marion (co.), Ohio 284/D4
Marion, Ohio 188/K2
Marion, Ohio (43302) 284/D4
Marion (co.), Oreg. 291/E3
Marion, Oreg. (97359) 291/D3
Marion, Pa. (17235) 294/G6
Marion (res.), Queensland 88/J3
Marion, S. Australia 88/D8
Marion, S. Australia 94/A8
Marion (lake), S.C. 188/K4
Marion (co.), S.C. 296/J3
Marion, S.C. (29571) 296/J3
Marion (lake), S.C. 296/G5
Marion, S. Dak. (57043) 298/P7
Marion (bay), Tasmania 99/E4
Marion (co.), Tenn. 237/K10
Marion (co.), Texas 303/K5
Marion, Va. (24354) 307/E7
Marion (co.), W. Va. 312/F4
Marion, Wis. (54950) 317/J6
Marion Bridge, Nova Scotia 168/H3
Marion Center, Pa. (15759) 294/D4
Marion Junction, Ala. (36759) 195/D6
Marion Station, Md. (21838) 245/R8
Marionville, Mo. (65705) 261/E8
Maripa, Fr. Guiana 131/E4
Maripa, Venezuela 124/F4
Maripasoula, Fr. Guiana 131/D4
Mariposa (co.), Calif. 204/E6
Mariposa, Calif. (95338) 204/F6
Mariscala, Uruguay 145/E5
Mariscal Estigarribia, Paraguay 120/C5
Mariscal Estigarribia, Paraguay 144/B3
Marismas, Las (marsh), Spain 33/C4
Marissa, Ill. (62257) 222/D5
Maritime Alps (range), France 28/G5
Maritime Alps (range), Italy 34/A2
Maritsa, Bulgaria 45/H4
Maritsa (riv.), Bulgaria 45/G4
Mariupol' (Zhdanov), U.S.S.R. 52/E5
Marivan (Dezh Shahpur), Iran 66/E3
Mariveles, Philippines 82/C3
Märjamaa, U.S.S.R. 53/C1
Mark (riv.), Belgium 27/F6
Mark, Ill. (61340) 222/D2
Mark (riv.), Netherlands 27/F6
Marka (Merka), Somalia 115/H3
Marka, Somalia 102/G4
Markam, China 77/E6
Markaryd, Sweden 18/H8
Mark Center, Ohio (43536) 284/A3
Markdale, Ontario 177/D3
Marked Tree, Ark. (72365) 202/K2
Marken (isl.), Netherlands 27/G4
Markerwaard Polder, Netherlands 27/G3
Markesan, Wis. (53946) 317/J8
Market Drayton, England 10/E4
Market Drayton, England 13/E5
Market Harborough, England 13/G5
Markethill, N. Ireland 17/H3
Market Rasen, England 13/G4
Market Weighton, England 13/G4
Markha (riv.), U.S.S.R. 48/M3
Markham (mt.) 5/A8
Markham, Ill. (60426) 222/B6
Markham (bay), N.W. Terrs. 187/L3
Markham (inlet), N.W. Terrs. 187/L1
Markham, Ontario 177/K4
Markham, Va. (22643) 307/N3
Markham, Wash. (†98520) 310/B4
Markinch, Sask. 181/G5
Markinch, Scotland 15/E4
Markit, China 77/A4
Markkleeberg, E. Germany 22/E3
Markland, Ind. (†47020) 227/G7
Markland, Newf. 166/D2
Markle, Ind. (46770) 227/G3
Markleeville, Calif. (96120) 204/F5
Marklesburg (James Creek), Pa. (16657) 294/F5
Markleton, Pa. (15551) 294/D6
Markleville, Ind. (46056) 227/F5
Markleysburg, Pa. (15459) 294/C6
Markounda, Cent. Afr. Rep. 115/C2
Markovo, U.S.S.R. 4/C1
Markovo, U.S.S.R. 48/S3
Marks, Miss. (38646) 256/D2
Marks, U.S.S.R. 52/G4
Markstay, Ontario 177/D1
Marksville, La. (71351) 238/G4
Marktredwitz, W. Germany 22/E4
Markville, Minn. (55048) 255/F4
Marl, W. Germany 22/B3
Marland, Okla. (74644) 288/M1
Marlbank, Ontario 177/G3
Marlboro, Alberta 182/B3
Marlboro○, N.J. (07746) 273/E3
Marlboro, N.Y. (12542) 276/M7
Marlboro (co.), S.C. 296/H2
Marlboro○, Vt. (05344) 268/B6
Marlborough, Conn. (06447) 210/F2
Marlborough○, Conn. (06447) 210/F2
Marlborough, England 13/F6
Marlborough, England 10/F5
Marlborough, Mass. (01752) 249/H3
Marlborough, Mo. (†63101) 261/P3
Marlborough, N.H. (03455) 268/C6
Marlborough○, N.H. (03455) 268/C6
Marlborough, Queensland 95/D4
Marlette, Mich. (48453) 250/G5
Marlin, Texas (76661) 303/H6
Marlin, Wash. (98832) 310/F3
Marlinton, W. Va. (24954) 312/F6
Marlow, Ala. (†36580) 195/C10
Marlow, England 13/G8
Marlow, Georgia (†31312) 217/K6
Marlow○, N.H. (03456) 268/C5
Marlow, Okla. (73055) 288/K5
Marlton, N.J. (08053) 273/D4
Marmaduke, Ark. (72443) 202/K1
Marmagao, India 68/C5
Marmande, France 28/C5
Marmara (sea), Turkey 7/G4
Marmara (isl.), Turkey 63/B2
Marmara (sea), Turkey 63/C2
Marmara (sea), Turkey 59/A1
Marmaris, Turkey 63/C4
Marmarth, N. Dak. (58643) 282/B7
Mar Menor (lag.), Spain 33/F4
Marmet, W. Va. (25315) 312/C6
Marmolada (mt.), Italy 34/C1
Marmontana (mt.), Switzerland 39/H4
Marmora, N.J. (08223) 273/D5
Marmora, Ontario 177/G3
Marmot (bay), Alaska 196/H3
Marmot (isl.), Alaska 196/H3
Mar Muerto (lag.), Mexico 150/N9
Marne (dept.), France 28/F3
Marne (riv.), France 28/C2
Marne, Iowa (51552) 229/C6
Marne, Mich. (49435) 250/D5
Marne, W. Germany 22/C2
Maro (dry riv.), Chad 111/C4
Maro (reef), Hawaii 188/F6
Maro (reef), Hawaii 218/C6
Maroa, Ill. (61756) 222/E3
Maroa, Venezuela 124/E6
Maroantsetra, Madagascar 118/J3
Marolambo, Madagascar 118/H4
Maromokotro (mt.), Madagascar 102/G6
Maromokotro (mt.), Madagascar 118/H2
Maroni (riv.) 120/D2
Maroni (riv.), Fr. Guiana 131/D3
Maroochydore-Mooloolaba, Queensland 88/J5
Maroochydore-Mooloolaba, Queensland 95/E5
Maroon (peak), Colo. 208/F4
Maroon Town, Jamaica 158/H6
Maros (riv.), Hungary 41/F3
Maros, Indonesia 85/F6
Maroua, Cameroon 102/D3
Maroua, Cameroon 115/B1
Maroubra, N.S. Wales 97/K3
Marouini (riv.), Fr. Guiana 131/D4
Marovoay, Madagascar 102/G6
Marovoay, Madagascar 118/H3
Marowijne (dist.), Suriname 131/D4
Marowijne (riv.), Suriname 131/D3
Marquam, Oreg. (97362) 291/B3
Marquand, Mo. (63655) 261/M8
Marquesas (keys), Fla. 212/D7
Marquesas (isls.), Fr. Polynesia 2/B6
Marquesas (isls.), Fr. Poly. 87/N6
Marquette, Iowa (52158) 229/L2
Marquette, Kansas (67464) 232/E3
Marquette, Manitoba 179/E4
Marquette, Mich. 146/K5
Marquette, Mich. 188/J1
Marquette (co.), Mich. 250/B2
Marquette, Mich. (49855) 250/B2
Marquette (isl.), Mich. 250/E3
Marquette, Nebr. (68854) 264/G4
Marquette (co.), Wis. 317/H8
Marquette, Wis. (53947) 317/H8
Marquette Heights, Ill. (†61554) 222/D3
Marquez, Texas (77865) 303/H6
Marquis, Grenada 161/D8
Marquis, St. Lucia 161/G6
Marquis, Sask. 181/F5
Marra, Jebel (mt.), Sudan 102/E3
Marra (creek), N.S. Wales 97/D2
Marra, Jebel (mt.), Sudan 111/D5
Marracuene, Mozambique 118/E5
Marrakech, Morocco 102/B1
Marrakech, Morocco 106/C2
Marrawah, Tasmania 99/A2
Marree, S. Australia 88/F5
Marree, S. Australia 94/E3
Marrero, La. (70072) 238/O4
Marrickville, N. S. Wales 88/L4
Marrickville, N.S. Wales 97/J3
Marriott, Sask. 181/D4
Marromeu, Mozambique 118/F3
Marrowbone, Ky. (42759) 237/K7
Marrowie (creek), N.S. Wales 97/C3
Marrupa, Mozambique 118/F2
Mars, Pa. (16046) 294/C4
Mars (riv.), Québec 172/G1
Marsabit, Kenya 115/G3
Marsabit, Kenya 102/F4
Marsa el Brega, Libya 111/D1
Marsa el Hariga, Libya 111/D1
Marsala, Italy 34/D6
Marsa Oseif, Sudan 111/G3
Mars Bluff, S.C. (†29501) 296/H3
Marsciano, Italy 34/D3
Marsden, N.S. Wales 97/D3
Marsden, Sask. 181/B3
Marsden Park, N.S. Wales 97/H3
Marsdiep (chan.), Netherlands 27/F3
Marseille, France 7/E4
Marseille, France 28/F6
Marseilles, Ill. (61341) 222/E2
Marseilles, Ohio (†43351) 284/D4
Marsh (creek), Idaho 220/F7
Marsh (isl.), La. 238/G7
Marsh (lake), Minn. 255/B5
Marsh, Mont. (†59326) 262/M4
Marsh (peak), Utah 304/E3
Marshall (co.), Ala. 195/F2
Marshall, Ark. (72650) 202/E2
Marshall (co.), Ill. 222/D2
Marshall, Ill. (62441) 222/F4
Marshall (co.), Ind. 227/E2
Marshall, Ind. (47859) 227/C5
Marshall (co.), Iowa 229/G4
Marshall (co.), Kansas 232/F2
Marshall (co.), Ky. 237/E7
Marshall, Liberia 106/B7
Marshall (isl.), Maine 243/G7
Marshall, Mich. (49068) 250/E6
Marshall (co.), Minn. 255/B2
Marshall, Minn. (56258) 255/C6
Marshall (co.), Miss. 256/E1
Marshall, Mo. (65340) 261/F4
Marshall, N.C. (28753) 281/D3
Marshall, N. Dak. (58644) 282/F5
Marshall (riv.), North. Terr. 88/F4
Marshall (riv.), North. Terr. 93/D7
Marshall, Ohio (†45133) 284/C7
Marshall (co.), Okla. 288/N6
Marshall, Okla. (73056) 288/L2
Marshall (isls.), Pacific Is. Terr. 2/T5
Marshall, Sask. 181/B2
Marshall (co.), S. Dak. 298/O2
Marshall (co.), Tenn. 237/H10
Marshall, Texas 188/H4
Marshall, Texas (75670) 303/K5
Marshall, Va. (22115) 307/N3
Marshall (co.), W. Va. 312/E3
Marshall, Wis. (53559) 317/H9
Marshallberg, N.C. (28553) 281/S5
Marshall Hall, Md. (†20616) 245/K6
Marshall Islands 87/G4
Marshalls Creek, Pa. (18335) 294/M3
Marshallton, Del. (19808) 245/R2
Marshalltown, Iowa (50158) 229/G4
Marshalltown, Iowa 188/H2
Marshallville, Georgia (31057) 217/D6
Marshallville, Ohio (44645) 284/G4
Marshes Siding, Ky. (42631) 237/M7
Marshfield, Ind. (47956) 227/C4
Marshfield, Mass. (02050) 249/M4
Marshfield○, Mass. (02050) 249/M4
Marshfield, Mo. (65706) 261/G8
Marshfield, Vt. (05658) 268/C3
Marshfield○, Vt. (05658) 268/C3
Marshfield, Wis. (54449) 317/F6
Marshfield Hills, Mass. (02051) 249/M4
Marsh Hill, Pa. (†17722) 294/H3
Mars Hill○, Maine (04758) 243/H2
Mars Hill, N.C. (28754) 281/D3
Mars Hill-Blaine, Maine (04758) 243/H2
Marshland, Oreg. (†97016) 291/D1
Marshville, N.C. (28103) 281/J4
Marshy (lake), Manitoba 179/B5
Marshyhope (creek), Md. 245/P6
Marsing, Idaho (83639) 220/B6
Marsland, Nebr. (69354) 264/A2
Marsoui, Québec 172/C1
Märsta, Sweden 18/K7
Marstal, Denmark 21/D8
Marston, Mo. (63866) 261/N9
Marston, N.C. (28363) 281/K5
Marstons Mills, Mass. (02648) 249/N6
Marstrand, Sweden 18/G8
Mart, Texas (76664) 303/H6
Martaban, Burma 72/C3
Martaban (gulf), Burma 54/L8
Martaban (gulf), Burma 72/C4
Martapura, Indonesia 85/F6
Martel, Ohio (43335) 284/E4
Martel, Tenn. (†37771) 237/N9
Martelange, Belgium 27/H9
Martell, Calif. (95654) 204/C9
Martell, Wis. (†54767) 317/B6
Martelle, Iowa (52305) 229/L4
Marten (mt.), Alberta 182/C2
Martensdale, Iowa (50160) 229/F6
Martensville, Sask. 181/E3
Martha, Ky. (41159) 237/R4
Martha, Okla. (73556) 288/H5
Martha, Tenn. (†37087) 237/J8
Marthaguy (creek), N.S. Wales 97/D2
Marthasville, Mo. (63357) 261/L5
Martha's Vineyard (isl.), Mass. 188/N2
Martha's Vineyard (isl.), Mass. 249/M7
Marthaville, La. (71450) 238/D3
Martí, Camagüey, Cuba 158/G3
Martí, Matanzas, Cuba 158/D1
Martí, Cuba 156/C2
Martigny, Switzerland 39/C4
Martigues, France 28/F6
Martin (dam), Ala. 195/G5
Martin (lake), Ala. 195/G5
Martin, Czech. 41/E2
Martin (co.), Fla. 212/F4
Martin, Georgia (30557) 217/F2
Martin (co.), Ind. 227/D7
Martin (co.), Ky. 237/R5
Martin, Ky. (41649) 237/R5
Martin, La. (†71019) 238/D2
Martin, Mich. (49070) 250/D6
Martin (co.), Minn. 255/D7
Martin, New Bruns. 170/C1
Martin (head), New Bruns. 170/E3
Martin (co.), N.C. 281/P3
Martin, N. Dak. (58758) 282/K4
Martin, Ohio (43445) 284/D2
Martin, S.C. (29836) 296/D5
Martin, S. Dak. (57551) 298/F7
Martin, Tenn. (38237) 237/D8
Martin (co.), Texas 303/C5
Martin, W. Va. (†26702) 312/H4
Martina Franca, Italy 34/F4
Martinborough, N. Zealand 100/E4
Martín Chico, Uruguay 145/A5
Martinez, Calif. (94553) 204/K1
Martinez, Georgia (30907) 217/H3
Martínez de la Torre, Mexico 150/L6
Martín García (isl.), Argentina 143/H6
Martinique (isl.) 146/M8
Martinique (passage), Dominica 161/E7
MARTINIQUE 161/D5
MARTINIQUE 156/G4
Martinique (passage), Martinique 161/C5
Martin Luther King, Jr., Nat'l Hist. Site, Georgia 217/K1
Martinsburg, Ind. (†47165) 227/E8
Martinsburg, Iowa (52568) 229/J6
Martinsburg, Mo. (65264) 261/J4
Martinsburg, Nebr. (†68770) 264/H2
Martinsburg, N.Y. (13404) 276/J3
Martinsburg, Ohio (43037) 284/F5
Martinsburg, Pa. (16662) 294/F5
Martinsburg, W. Va. (25401) 312/K4
Martins Creek, Pa. (18063) 294/M4
Martinsdale, Mont. (59053) 262/F4
Martinsdale (res.), Mont. 262/F4
Martins Ferry, Ohio (43935) 284/J5
Martins Mills, Tenn. (†38471) 237/F10
Martins River, Nova Scotia 168/D4
Martinsville, Ill. (62442) 222/F4
Martinsville, Ind. (46151) 227/D6
Martinsville, Miss. (†39083) 256/D7
Martinsville, Mo. (64467) 261/D2
Martinsville, N.J. (08836) 273/D2
Martinsville, Ohio (45146) 284/C7
Martinsville (I.C.), Va. (24112) 307/J7
Martinton, Ill. (60951) 222/F3
Martintown, Ontario 177/K2
Martin Van Buren Nat'l Hist. Site, N.Y. 276/N6
Martinville, Ark. (†72039) 202/F3
Martinville, Québec 172/F4
Martock, England 13/E7
Martofte, Denmark 21/D6
Marton, N. Zealand 100/E4
Martos, Spain 33/E4
Martre, Lac la (LAKE(, N.W.T. 162/E3
Martwick, Ky. (†42330) 237/H6
Marty, S. Dak. (57361) 298/N8
Marudi (mts.), Guyana 131/B5
Marudi, Malaysia 85/E5
Ma'ruf, Afghanistan 68/B2
Ma'ruf, Afghanistan 59/J3
Maruim, Brazil 132/G5
Marulan, N.S. Wales 97/E4
Marungu (mts.), Zaire 115/E5
Marutea (atoll), Fr. Poly. 87/N8
Marv Dasht, Iran 66/H6
Marvejols, France 28/E5
Marvel, Ala. (†35115) 195/D4
Marvel, Colo. (81329) 208/C8
Marvell, Ark. (72366) 202/J4
Marvin, S. Dak. (57251) 298/R3
Marvindale, Pa. (†16749) 294/E2
Marvine (mt.), Utah 304/C5
Marvyn, Ala. (†36801) 195/H6
Marwayne, Alberta 182/E3
Marwood, Pa. (16047) 294/C4
Mary, Ky. (41350) 237/O5
Mary (lake), Minn. 255/C5
Mary (riv.), Queensland 95/E5
Mary, U.S.S.R. 54/H6
Mary (Merv), U.S.S.R. 48/G6
Maryborough, Australia 87/F8
Maryborough (Portlaoighise), Ireland 17/G5
Maryborough (Portlaoighise), Ireland 10/C4
Maryborough, Queensland 95/E5
Maryborough, Queensland 88/J5
Maryborough, Victoria 97/B5
Maryborough, Victoria 88/G7
Marydel, Md. (21649) 245/P4
Marydell, Miss. (†39051) 256/F5
Mary Esther, Fla. (32569) 212/B6
Maryfield, Sask. 181/K6
Maryhill, Wash. (†98620) 310/E5
Mary Kathleen, Queensland 88/G4
Mary Kathleen, Queensland 95/A4
Marykirk, Scotland 15/F4
Maryland 188/L3
MARYLAND 245
Maryland, N.Y. (12116) 276/L5
Maryland (state), U.S. 146/L6
Maryland City, Md. (†21113) 245/L4
Maryland Heights, Mo. (63043) 261/O2
Maryland Line, Md. (21105) 245/M2
Maryneal, Texas (79535) 303/D5
Maryport, England 10/E3
Maryport, England 13/D3
Mary Ronan (lake), Mont. 262/B3
Marys (creek), Idaho 220/C7
Marys (riv.), Nev. 266/F1
Mary's Harbour, Newf. 166/C3
Marystown, Newf. 166/D4
Marysvale, Utah (84750) 304/B5
Marysvale (peak), Utah 304/B5
Marysville, Calif. 188/B3
Marysville, Calif. (95901) 204/D4
Marysville, Ind. (47141) 227/F7
Marysville, Iowa (†50116) 229/G6
Marysville, Kansas (66508) 232/F2
Marysville, Mich. (48040) 250/G6
Marysville, Mont. (59640) 262/D4
Marysville, Ohio (43040) 284/D5
Marysville, Pa. (17053) 294/H5
Marysville, Wash. (98270) 310/C2
Marytown, W. Va. (†24889) 312/C8
Maryvale, Queensland 95/C3
Maryvale, Queensland 88/H3
Maryville, Ill. (62062) 222/B2
Maryville, Mo. (64468) 261/C2
Maryville, Tenn. (37801) 237/O9
Marzo (pt.), Colombia 126/B4
Masagua, Guatemala 154/B3
Masahim, Kuh-e (mt.), Iran 66/J5
Masai (steppe), Tanzania 115/G4
Masaka, Uganda 115/F4
Masamba, Indonesia 85/G6
Masan, S. Korea 81/D6
Masardis○, Maine (04759) 243/G3
Masaryktown, Fla. (33512) 212/D3
Masasi, Tanzania 115/G6
Masatepe, Nicaragua 154/D5
Masaya, Nicaragua 154/D5
Masbate (prov.), Philippines 82/D4
Masbate, Philippines 82/D4
Masbate (isl.), Philippines 82/D4
Masbate (isl.), Philippines 85/G3
Mascara, Algeria 106/D1
Mascarene (isls.), Mauritius 118/F5
Mascarene (isls.), Réunion 118/F5
Mascoma, N.H. (†03748) 268/C4
Mascoma (lake), N.H. 268/C4
Mascot, Tenn. (37806) 237/O8
Mascota, Mexico 150/G6
Mascotte, Fla. (32753) 212/E3
Mascouche, Québec 172/H4
Mascoutah, Ill. (62258) 222/D5
Masefield, Sask. 181/D6
Masela (isl.), Indonesia 85/H7
Maseru (cap.), Lesotho 102/E8
Maseru (cap.), Lesotho 118/D5
Mash 'Abbe Sade, Israel 65/B6
Mashabi (isl.), Saudi Arabia 59/C4
Masham, England 13/F3
Mashapaug, Conn. (†06076) 210/G1
Mashapaug (lake), Conn. 210/G1
Mashash, Wadi (dry riv.), Jordan 65/C4
Mashhad (Meshed), Iran 66/L2
Mashike, Japan 81/K2
Mashkel, Hamun-i- (swamp), Pakistan 68/A3
Mashkel, Hamun-i- (swamp), Pakistan 59/H4
Mashkid (riv.), Iran 66/N7
Mashkid (riv.), Iran 59/H4
Mashkid (riv.), Pakistan 59/H4
Mashkid (riv.), Pakistan 68/A3
Mashonaland (reg.), Zimbabwe 118/E3
Mashpee○, Mass. (02649) 249/M6
Mashulaville, Miss. (†39341) 256/G4
Masi-Manimba, Zaire 115/C4
Masindi, Uganda 115/F3
Masinloc, Philippines 82/B3
Masío (cay), Cuba 158/C2
Masira (isl.), Oman 54/G7
Masira (gulf), Oman 59/G6
Masira (isl.), Oman 59/G5
Masisea, Peru 128/E7
Masisi, Zaire 115/E4
Masjed Soleyman, Iran 66/F5
Mask (lake), Ireland 17/C4
Mask, Lough (lake), Ireland 10/B4
Maskell, Nebr. (68751) 264/H2
Maskinongé (co.), Québec 172/D3
Maskinongé (county), Québec 174/C3
Maskinongé, Québec 172/E3
Maskinongé (riv.), Québec 172/D3
Masoala (pen.), Madagascar 118/J3
Masoller, Uruguay 145/C2
Mason (isl.), Conn. 210/H3
Mason (co.), Ill. 222/D3
Mason, Ill. (62443) 222/E5
Mason (co.), Ky. 237/O3
Mason, Ky. (†41097) 237/M3
Mason (co.), Mich. 250/C4
Mason, Mich. (48854) 250/E6
Mason, Nev. (†89447) 266/B4
Mason (peak), Nev. 266/F1
Mason (bay), N. Zealand 100/A7
Mason, Ohio (45040) 284/B7
Mason, Okla. (74859) 288/O3
Mason, Tenn. (38049) 237/B10
Mason (co.), Texas 303/E7
Mason, Texas (76856) 303/E7
Mason (co.), Wash. 310/B3
Mason (co.), W. Va. 312/B5
Mason, W. Va. (25260) 312/B4
Mason, Wis. (54856) 317/D3
Mason City, Ill. (62664) 222/D3
Mason City, Iowa (50401) 229/G2
Mason City, Iowa 188/H2
Mason City, Nebr. (68855) 264/E3
Mason Hall, Tenn. (†38233) 237/C8
Mason Springs, Md. (†20640) 245/K6
Masontown, Pa. (15461) 294/C6
Masontown, W. Va. (26542) 312/G3
Masonville, Colo. (80541) 208/J2
Masonville, Iowa (50654) 229/K4
Masonville, N.Y. (13804) 276/K6
Masqat (Muscat) (cap.), Oman 59/G5
Massa, Italy 34/C2
Massabesic (lake), N.H. 268/E6
Massac (co.), Ill. 222/E6
Massa-Carrara (prov.), Italy 34/C2
Massachusetts 188/M2
MASSACHUSETTS 249
Massachusetts (bay), Mass. 249/M4
Massachusetts (state), U.S. 146/L5
Massacre (bay), Amer. Samoa 86/N9
Massacre (lake), Nev. 266/B1
Massafra, Italy 34/F4
Massakory, Chad 111/C5
Massa Marittima, Italy 34/C3
Massangena, Mozambique 118/E4
Massango (Forte República), Angola 115/C5
Massanutten (mt.), Va. 307/L3
Massapê, Brazil 132/G3
Massapeag, Conn. (†06382) 210/G3
Massapequa, N.Y. (11758) 276/R7
Massapequa Park, N.Y. (11762) 276/R7
Massaponax, Va. (†22553) 307/O4
Massawa, Ethiopia 111/G4
Massawa, Ethiopia 59/C6
Massawa, Ethiopia 102/F3
Massbach, Ill. (†61028) 222/C1
Mass City, Mich. (49948) 250/G1
Massena, Iowa (50853) 229/D6
Massena, N.Y. (13662) 276/L1
Massénya, Chad 111/C5
Masset, Br. Col. 184/B3
Masset (inlet), Br. Col. 184/A3
Massey, Md. (21650) 245/P3
Massey, N. Zealand 100/B1
Massey, Ontario 177/C1
Massies Mill, Va. (22954) 307/K5
Massillon, Ala. (†36759) 195/D6
Massillon, Iowa (†52255) 229/L5
Massillon, Ohio (44646) 284/H4
Massinga, Mozambique 118/F4
Massingir, Mozambique 118/E4
Massive (mt.), Colo. 208/F4
Masson, Québec 172/B4
Massueville, Québec 172/E4
Mastaba, Saudi Arabia 59/C5
Mastens Corner, Del. (†19943) 245/R5
Masters, Colo. (†80649) 208/L2
Masterton, N. Zealand 100/E4
Mastic Beach, N.Y. (11950) 276/P9
Mastuj, Pakistan 59/K2
Mastung, Pakistan 59/J4
Mastung, Pakistan 68/B3
Mastura, Saudi Arabia 59/C5
Masuda, Japan 81/E6
Masurian (lakes), Poland 47/E2
Masury, Ohio (44438) 284/J3
Masyaf, Syria 63/G5
Matabeleland (reg.), Zimbabwe 118/D3
Matachewan, Ontario 177/J5
Matachewan, Ontario 175/D3
Mata de São João, Brazil 132/G6
Matadi, Zaire 115/B5

Matadi, Zaire 102/D5
Matador, Sask. 181/D5
Matador, Texas (79244) 303/D3
Matagalpa, Nicaragua 154/E4
Matagami, Québec 174/B3
Matagami (lake), Québec 174/B3
Matagorda (co.), Texas 303/H9
Matagorda, Texas (77457) 303/J9
Matagorda (bay), Texas 188/G5
Matagorda (bay), Texas 303/H9
Matagorda (isl.), Texas 303/H9
Matagorda (pen.), Texas 303/J9
Matagorda Isl. Bombing and Gunnery Range, Texas 303/H9
Matakana (isl.), N. Zealand 100/F2
Matala (dam), Angola 115/B6
Matam, Senegal 106/B5
Matamoras, Pa. (18336) 294/N3
Matamoros, Mexico 146/J7
Matamoros, Coahuila, Mexico 150/H4
Matamoros, Tamaulipas, Mexico 150/L4
Matane (co.), Québec 172/B1
Matane (county), Québec 174/D3
Matane, Québec 174/D3
Matane, Québec 172/B1
Matane (riv.), Québec 172/B1
Matane Prov. Park, Québec 172/B1
Matanuska (riv.), Alaska 196/C1
Matanza, Colombia 126/D4
Matanzas (prov.), Cuba 158/D1
Matanzas, Cuba 146/K7
Matanzas, Cuba 158/C1
Matanzas, Cuba 156/B2
Matanzas (bay), Cuba 158/D1
Matanzas (inlet), Fla. 212/E2
Mata Palacio, Dom. Rep. 158/F6
Matapalo (cape), C. Rica 154/F6
Matapan (Taínaron) (cape), Greece 45/F7
Matapédia (county), Québec 174/D3
Matapédia (co.), Québec 172/B2
Matapédia, Québec 172/B2
Matapédia (lake), Québec 172/B1
Matapédia (riv.), Québec 172/B2
Mataquito (riv.), Chile 138/A10
Matara, Sri Lanka 68/E7
Mataram, Indonesia 85/F7
Matarani, Peru 120/B4
Matarani, Peru 128/F11
Mataranka, North. Terr. 93/C3
Matarinao (bay), Philippines 82/E5
Mataró, Spain 33/H2
Matatiele, S. Africa 118/D6
Matatindoc (pt.), Philippines 82/D6
Mataura, N. Zealand 100/B7
Mataura (riv.), N. Zealand 100/B6
Mata Utu (cap.), Wallis and Futuna 87/J7
Matawai, N. Zealand 100/F3
Matawan, Minn. (†56072) 255/E7
Matawan, N.J. (07747) 273/E3
Matawin (lake), Québec 172/C3
Matawin (riv.), Québec 172/D3
Mateare, Nicaragua 154/D4
Mateguá, Bolivia 136/D3
Matehuala, Mexico 150/J5
Matelot, Trin. & Tob. 161/B10
Matera (prov.), Italy 34/F4
Matera, Italy 34/F4
Maternillos (pt.), Cuba 158/H2
Mátészalka, Hungary 41/G3
Matetsi, Zimbabwe 118/D3
Mateur, Tunisia 106/F1
Matewan, W. Va. (25678) 312/B7
Matfield Green, Kansas (66862) 232/F3
Mather, Manitoba 179/C5
Mather, Wis. (54641) 317/F7
Mather A.F.B., Calif. 204/C8
Matherville, Ill. (61263) 222/C2
Matherville, Miss. (†39360) 256/G7
Matheson, Colo. (80830) 208/M4
Matheson, Ontario 177/K5
Matheson Island, Manitoba 179/E3
Mathews, Ala. (36052) 195/F6
Mathews (lake), Calif. 204/E11
Mathews, La. (70375) 238/J7
Mathews (co.), Va. 307/R6
Mathews, Va. (23109) 307/R6
Mathias, W. Va. (26812) 312/J5
Mathinna, Tasmania 99/E3
Mathis, Texas (78368) 303/G9
Mathiston, Miss. (39752) 256/F3
Mathoura, N.S. Wales 97/C4
Mathura, India 68/D3
Mati, Philippines 85/H4
Mati, Philippines 82/F7
Matías Romero, Mexico 150/M8
Matinenda (lake), Ontario 177/B1
Matinicus, Maine (04851) 243/F8
Matinicus Rock (isl.), Maine 243/F8
Matlock, England 10/F4
Matlock, England 13/J2
Matlock, Iowa (51244) 229/A2
Matlock, Wash. (98560) 310/B3
Matoaca, Va. (23803) 307/N6
Matoaka, W. Va. (24736) 312/D8
Matochkin Shar (str.), U.S.S.R. 48/F2
Mato Grosso (state), Brazil 132/B6
Mato Grosso, Brazil 120/D4
Mato Grosso, Brazil 132/B6
Mato Grosso (plat.), Brazil 120/D4
Mato Grosso, Planalto de (plat.), Brazil 132/B6
Mato Grosso do Sul (state), Brazil 132/C7
Matopos, Zimbabwe 118/D4
Matosinhos, Portugal 33/B2
Matoury, Fr. Guiana 131/E3
Mátra (mts.), Hungary 41/E3
Matrah, Oman 54/G7
Matrah, Oman 59/G5
Matrei in Osttirol, Austria 41/B3
Matruh, Egypt 59/A3
Matsqui, Br. Col. 184/L3
Matsu (Mazu) (isl.), China 77/K6
Matsubara, Japan 81/H8
Matsue, Japan 81/F6
Matsumae, Japan 81/J3
Matsumoto, Japan 81/H5
Matsusaka, Japan 81/H6
Matsuto, Japan 81/H5
Matsuyama, Japan 81/F7
Matsuyama, Japan 54/P6
Matt, Switzerland 39/H3
Mattabesset (riv.), Conn. 210/E2
Mattagami (riv.), Ontario 175/D3
Mattagami (riv.), Ontario 177/J5
Mattamiscontis (lake), Maine 243/F4
Mattamuskeet (lake), N.C. 281/S3
Mattapan, Mass. (02126) 249/C7
Mattapoisett, Mass. (02739) 249/L6
Mattapoisett○, Mass. (02739) 249/L6
Mattaponi (co.), Va. (23110) 307/P5
Mattaponi (riv.), Va. 307/O5
Mattaponi Ind. Res., Va. 307/P5
Mattawa, Ont. 162/J6
Mattawa, Ontario 177/F1
Mattawa, Ontario 175/E3
Mattawa, Wash. (99344) 310/F4
Mattawamkeag○, Maine (04459) 243/G5
Mattawamkeag (lake), Maine 243/G4
Mattawamkeag (riv.), Maine 243/G4
Mattawan, Mich. (49071) 250/D6
Mattawana, Pa. (17054) 294/G5
Mattawoman (creek), Md. 245/K6
Matterhorn (mt.), Switzerland 39/E4
Mattersburg, Austria 41/D3
Matteson, Ill. (60443) 222/B6
Matthew, Ky. (41454) 237/P5
Matthews, Georgia (30818) 217/H4
Matthews, Ind. (46957) 227/F4
Matthews, Mo. (63857) 261/N9
Matthews, N.C. (28105) 281/H4
Matthews Ridge, Guyana 131/B2
Mattice, Ontario 175/D3
Mattice, Ontario 177/J5
Mattighofen, Austria 41/B2
Mattituck, N.Y. (11952) 276/P9
Mattoon, Ill. (61938) 222/E4
Mattoon, Wis. (54450) 317/J5
Mattson, Miss. (38758) 256/C2
Matu, Venezuela 124/F4
Matucana, Peru 128/D8
Matuku (isl.), Fiji 86/Q11
Matún, Cuba 158/D2
Matura, Trin. & Tob. 161/B10
Matura (bay), Trin. & Tob. 161/B10
Maturín, Venezuela 120/C2
Maturín, Venezuela 124/G3
Matutum (mt.), Philippines 82/E7
Matutum (mt.), Philippines 85/G4
Matveyev (isl.), U.S.S.R. 52/J1
Mau, India 68/E3
Mauá, Brazil 135/C3
Maúa, Mozambique 118/F2
Maubeuge, France 28/F2
Ma-ubin, Burma 72/B3
Mauch Chunk (Jim Thorpe), Pa. (18229) 294/L4
Mauchline, Scotland 15/D5
Mauckport, Ind. (47142) 227/E8
Maud, Ala. (†35616) 195/B1
Maud, Ky. (40042) 237/L5
Maud, Miss. (†38626) 256/D1
Maud, Ohio (†45069) 284/B7
Maud, Okla. (74854) 288/N4
Maud, Scotland 15/F3
Maud, Texas (75567) 303/K4
Maude, N.S. Wales 97/C4
Maudlow, Mont. (59714) 262/E4
Mauerkirchen, Austria 41/B2
Maués, Brazil 132/B3
Maués, Brazil 120/D3
Maués-Açu (riv.), Brazil 132/B4
Maugansville, Md. (21767) 245/H2
Mauger (cay), Belize 154/D2
Maugerville, New Bruns. 170/D3
Maui (co.), Hawaii 218/J1
Maui (isl.), Hawaii 87/L3
Maui (isl.), Hawaii 188/F5
Maui (isl.), Hawaii 218/J2
Mauk, Georgia (31058) 217/D6
Mauke (isl.), Cook Is. 87/L8
Mauldin, S.C. (29662) 296/C2
Maule (reg.), Chile 138/A11
Maule (riv.), Chile 138/A11
Mauléon-Licharre, France 28/C6
Maullín, Chile 138/D4
Maullín (riv.), Chile 138/D3
Maumakeogh (mt.), Ireland 17/C3
Maumee (riv.), Ind. 227/H2
Maumee (bay), Mich. 250/F7
Maumee, Ohio (43537) 284/C2
Maumee (bay), Ohio 284/D2
Maumee (riv.), Ohio 284/A3
Maumelle (lake), Ark. 202/E4
Maumere, Indonesia 85/G7
Maumturk (mts.), Ireland 17/B5
Maun, Botswana 118/C4
Maunabo, P. Rico 161/E3
Mauna Kea (mt.), Hawaii 87/L4
Mauna Kea (mt.), Hawaii 188/G6
Mauna Kea (mt.), Hawaii 218/H4
Maunaloa, Hawaii (96770) 218/G1
Mauna Loa (mt.), Hawaii 188/G6
Mauna Loa (mt.), Hawaii 218/G6
Maunalua (bay), Hawaii 218/F2
Maunawili, Hawaii (†96744) 218/F2
Maungaturoto, N. Zealand 100/E1
Maungdaw, Burma 72/B2
Maunie, Ill. (62861) 222/E5
Maupin, Oreg. (97037) 291/F2
Maurepas, La. (70449) 238/M2
Maurepas (lake), La. 238/M2
Mauertown, Va. (22644) 307/L3
Mauriac, France 28/E5
Maurice, Iowa (51036) 229/A3
Maurice, La. (70555) 238/F6
Maurice (riv.), N.J. 273/C4
Maurice (lake), S. Australia 88/E5
Maurice (lake), S. Australia 94/B3
Mauricetown, N.J. (08325) 273/D5
Mauricio Hirsch, Argentina 143/F7
Maurine, S. Dak. (†57626) 298/E3
Mauritania 2/J4
Mauritania 102/A3
MAURITANIA 106/B5
Mauritius 2/M6
MAURITIUS 118/G5
Maury, N.C. (28554) 281/O4
Maury (co.), Tenn. 237/G9
Maury (riv.), Va. 307/K5
Maury City, Tenn. (38050) 237/C9
Mauston, Wis. (53948) 317/F8
Mautern in Steiermark, Austria 41/C3
Mauthausen, Austria 41/C2
Mauthen-Kötschach, Austria 41/B3
Mauvoisin (dam), Switzerland 39/D4
Mavaca (riv.), Venezuela 124/F6
Maverick, Ariz. (†85920) 198/F5
Maverick (co.), Texas 303/D9
Mavila, Peru 128/H8
Mavillette, Nova Scotia 168/B4
Mavinga, Angola 115/D7
Mavora (mt.), N. Zealand 100/B6
Mavqi'im, Israel 65/B4
Mawai, Malaysia 72/F5
Mawbanna, Tasmania 99/B2
Mawer, Sask. 181/E5
Mawkmai, Burma 72/C2
Mawlaik, Burma 72/B2
Mawlu, Burma 72/C1
Mawson 5/C4
Max, Minn. (56659) 255/D3
Max, Nebr. (69037) 264/C4
Max, N. Dak. (58759) 282/H4
Maxbass, N. Dak. (58760) 282/H2
Maxcanú, Mexico 150/O6
Maxeys, Georgia (30671) 217/F3
Maxie, La. (†70526) 238/F6
Maxie, Miss. (†39458) 256/F9
Máximo Gómez, Ciego de Ávila, Cuba 158/F2
Máximo Gómez, Matanzas, Cuba 158/D1
Máximo Paz, Argentina 143/F6
Maxinkuckee, Ind. (†46511) 227/E2
Maxinkuckee (lake), Ind. 227/E2
Maxixe, Mozambique 118/F4
Max Meadows, Va. (24360) 307/G6
Maxstone, Sask. 181/F6
Maxton, N.C. (28364) 281/L5
Maxville, Mont. (59858) 262/C4
Maxville, Ontario 177/K2
Maxwell, Calif. (95955) 204/C4
Maxwell, Ind. (46154) 227/F5
Maxwell, Iowa (50161) 229/G5
Maxwell, Nebr. (69151) 264/D3
Maxwell, New Bruns. 170/C3
Maxwell, N. Mex. (87728) 274/E2
Maxwell (bay), N.W. Terrs. 187/K2
Maxwell, Tenn. (†37306) 237/J10
Maxwell Air Force Base, Ala. 195/F6
Maxwelton, Queensland 88/G4
May, Idaho (83253) 220/E5
May (cape), N.J. 188/M3
May (cape), N.J. 273/C6
May, Okla. (73851) 288/G1
May, Isle of (isl.), Scotland 15/F4
May, Texas (76857) 303/F5
Maya (mts.), Belize 154/C2
Maya (riv.), U.S.S.R. 54/P4
Maya (riv.), U.S.S.R. 48/O4
Maya Beach, Belize 154/C2
Mayaguana (isl.), Bahamas 146/L7
Mayaguana (isl.), Bahamas 156/D2
Mayaguana (passage), Bahamas 156/D2
Mayagüez (dist.), P. Rico 161/B2
Mayagüez, P. Rico 161/A2
Mayagüez, P. Rico 156/F1
Mayagüez (bay), P. Rico 161/A2
Mayajigua, Cuba 158/F2
Mayáls, Spain 33/G2
Mayarí, Cuba 158/J3
Mayarí Arriba, Cuba 158/J4
Mayaro, Trin. & Tob. 161/B11
Mayaro (bay), Trin. & Tob. 161/B11
Maybee, Mich. (48159) 250/F6
Maybell, Colo. (81640) 208/C2
Mayberry, Md. (†21157) 245/K2
Maybeury, W. Va. (24861) 312/D8
Maybole, Scotland 10/D3
Maybole, Scotland 15/D5
Maybrook, N.Y. (12543) 276/M8
Mayburg, Pa. (†16347) 294/D2
Maydena, Tasmania 99/C4
Mayen, W. Germany 22/B3
Mayenne (dept.), France 28/C3
Mayenne, France 28/C3
Mayenne (riv.), France 28/C4
Mayer, Ariz. (86333) 198/C4
Mayer, Chile 138/E7
Mayer, Minn. (55360) 255/E6
Mayersville, Miss. (39113) 256/B5
Mayerthorpe, Alberta 182/C3
Mayes (co.), Okla. 288/R2
Mayesville, S.C. (29104) 296/G4
Mayetta, Kansas (66509) 232/G2
Mayetta, N.J. (†08092) 273/E4
Mayfair, Sask. 181/D2
Mayfield, Georgia (31087) 217/G4
Mayfield, Kansas (67103) 232/E4
Mayfield, Ky. (42066) 237/D7
Mayfield (creek), Ky. 237/C7
Mayfield, N.Y. (12117) 276/M4
Mayfield, Ohio (44124) 284/J9
Mayfield, Okla. (73656) 288/G4
Mayfield, Pa. (18433) 294/L2
Mayfield, Scotland 15/D2
Mayfield, Utah (84643) 304/C4
Mayfield (lake), Wash. 310/C4
Mayfield Heights, Ohio (44124) 284/J9
Mayflower, Ark. (72106) 202/F4
Mayger, Oreg. (†97016) 291/D1
Mayhew, Miss. (39753) 256/G4
Mayhill, N. Mex. (88339) 274/D6
Maykop, U.S.S.R. 7/H4
Maykop, U.S.S.R. 48/D5
Maykop, U.S.S.R. 52/F6
Mayland, Tenn. (†38555) 237/L8
Maylene, Ala. (35114) 195/E4
Maymont, Sask. 181/D3
Maymyo, Burma 72/C2
Mayna, La. (†71343) 238/G4
Maynard, Ark. (72444) 202/J1
Maynard, Iowa (50655) 229/K3
Maynard○, Mass. (01754) 249/J3
Maynard, Minn. (56260) 255/C6
Maynardville, Tenn. (37807) 237/O8
Mayne, Br. Col. 184/K3
Maynooth, Ireland 17/H5
Maynooth, Ontario 177/G2
Mayo, Canada 4/C16
Mayo, Fla. (32066) 212/C1
Mayo (co.), Ireland 17/C4
Mayo, Md. (21106) 245/M5
Mayo (riv.), Peru 128/D6
Mayo (bay), Philippines 82/F7
Mayo, S.C. (29368) 296/D1
Mayo, Yukon 187/E3
Mayo, Yukon 162/C3
Mayo (lake), Yukon 187/E3
Mayodan, N.C. (27027) 281/K2
Mayon (vol.), Philippines 82/D4
Mayor (isl.), N. Zealand 100/F2
Mayor (cape), Spain 33/E1
Mayor Martínez, Paraguay 144/C5
Mayor Pablo Lagerenza, Paraguay 144/B1
MAYOTTE 118/G2
Mayotte (isl), France 102/G6
Mayoworth, Wyo. (†82639) 319/F2
May Park, Oreg. (†97850) 291/J2
May Pen, Jamaica 158/J6
Mayport Naval Air Sta., Fla. 212/E1
Mayrhofen, Austria 41/A3
Mays, Ind. (46155) 227/G5
Maysan (gov.), Iraq 66/E5
Maysel, W. Va. (25133) 312/D5
Mays Landing, N.J. (08330) 273/D5
Mays Lick, Ky. (41055) 237/O3
Maysville, Ark. (72747) 202/A1
Maysville, Georgia (30558) 217/E2
Maysville, Iowa (†52773) 229/M5
Maysville, Ky. (41056) 237/O3
Maysville, Mo. (64469) 261/D3
Maysville, N.C. (28555) 281/P5
Maysville, Okla. (73057) 288/M5
Maysville, W. Va. (26833) 312/H4
Maytiguid (isl.), Philippines 82/B5
Maytown, Ky. (41455) 237/O5
Mayumba, Gabon 115/A4
Mayuram, India 68/D6
Mayview, Mo. (64071) 261/E4
Mayville, Mich. (48744) 250/F5
Mayville, N.Y. (14757) 276/A6
Mayville, N. Dak. (58257) 282/R4
Mayville, Oreg. (97830) 291/G2
Mayville, Wis. (53050) 317/K9
Maywood, Calif. (90201) 204/C10
Maywood, Ill. (60153) 222/B5
Maywood, Mo. (63454) 261/J3
Maywood, Nebr. (69038) 264/D4
Maywood, N.J. (07607) 273/B2
Maywood Park, Oreg. (97220) 291/B2
Maza, N. Dak. (†58324) 282/M3
Mazabuka, Zambia 115/E7
Mazabuka, Zambia 102/E6
Mazagan (El Jadida), Morocco 106/C2
Mazagão, Brazil 132/D3
Mazama, Wash. (98833) 310/E2
Mazamet, France 28/E6
Mazán, Peru 128/F4
Mazana (lake), Québec 172/C2
Mazana (riv.), Québec 172/C2
Mazandaran (prov.), Iran 66/H2
Mazangano, Uruguay 145/E3
Mazapil, Mexico 150/J4
Mazara del Vallo, Italy 34/D6
Mazar-e Sharif, Afghanistan 59/J2
Mazar-e Sharif, Afghanistan 68/B1
Mazarrón, Spain 33/F4
Mazaruni (riv.), Guyana 131/A2
Mazaruni-Potaro (dist.), Guyana 131/A2
Mazatán, Mexico 150/E2
Mazatenango, Guatemala 154/B3
Mazatlán, Mexico 146/H7
Mazatlán, Mexico 150/F5
Mazatzal (peak), Ariz. 198/D4
Mažeikiai, U.S.S.R. 53/A2
Mazenod, Sask. 181/E6
Mazeppa, Alberta 182/D4
Mazeppa, Minn. (55956) 255/F6
Mazgirt, Turkey 63/H3
Mazıdağı, Turkey 63/J4
Mazie, Okla. (74353) 288/R2
Mazinaw (lake), Ontario 177/G3
Mazirbe, U.S.S.R. 53/B2
Mazocruz, Peru 128/H11
Mazoe (riv.), Mozambique 118/E3
Mazoe, Zimbabwe 118/E3
Mazoe (riv.), Zimbabwe 118/E3
Mazomanie, Wis. (53560) 317/G9
Mazon, Ill. (60444) 222/E2
Mazra', Israel 65/C5
Mazu (Matsu) (isl.), China 77/K6
Mazzarino, Italy 34/E6
Mbabane (cap.), Swaziland 118/E5
Mbabane (cap.), Swaziland 102/F7
Mbaïki, Cent. Afr. Rep. 115/C3
Mbakou (res.), Cameroon 115/B2
Mbala, Zambia 102/F5
Mbala, Zambia 115/F5
Mbale, Uganda 115/F3
Mbale, Uganda 102/F4
Mbalmayo, Cameroon 115/B3
Mbamba Bay, Tanzania 115/G6
Mbandaka, Zaire 115/C3
Mbandaka, Zaire 102/D5
Mbanza Congo, Angola 115/B5
Mbanza-Ngungu, Zaire 102/D5
Mbanza-Ngungu, Zaire 115/C5
Mbaracayú, Cordillera de (mts.), Paraguay 144/E3
Mbarangandu (riv.), Tanzania 115/G5
Mbarara, Uganda 115/F4
Mbemkuru (riv.), Tanzania 115/G5
Mbengga (isl.), Fiji 86/Q11
Mbéré (riv.), Cameroon 115/B2
Mbéré (riv.), Cent. Afr. Rep. 115/B2
Mbéré (riv.), Chad 111/C6
Mbeya (reg.), Tanzania 115/F5
Mbeya, Tanzania 102/F5
Mbeya, Tanzania 115/F5
M'Bigou, Gabon 115/B4
Mbinda, Congo 115/B4
Mbini, Equat. Guinea 115/A3
Mbocayaty, Paraguay 144/C5
M'Bour, Senegal 106/A6
M'Bout, Mauritania 106/B5
Mbres, Cent. Afr. Rep. 115/D2
M'Bridge (riv.), Angola 115/B5
Mbuji-Mayi, Zaire 102/E5
Mbuji-Mayi, Zaire 115/D5
Mbulu, Tanzania 115/G4
Mburucuyá, Argentina 143/E2
Mbuyapey, Paraguay 144/D5
McAdam, New Bruns. 170/C3
McAdams, Miss. (39107) 256/E4
McAdoo, Pa. (18237) 294/L4
McAdoo, Texas (79243) 303/D4
McAfee, N.J. (07428) 273/D1
McAlester, Okla. 188/G4
McAlester, Okla. (74501) 288/P5
McAlester (lake), Okla. 288/P4
McAlister, N. Mex. (88427) 274/F4
McAlisterville, Pa. (17049) 294/H4
McAllaster, Kansas (†67755) 232/A3
McAllen, Texas 188/G5
McAllen, Texas (78501) 303/F11
McAllister, Mont. (59740) 262/E5
McAllister, Wis. (†54177) 317/L5
McAlpin, Fla. (32062) 212/D1
McAndrews, Ky. (41543) 237/S5
McArthur, Calif. (96056) 204/D2
McArthur, Ohio (45651) 284/F7
McArthur River, North. Terr. 88/F3
McAuley, Manitoba 179/A4
McBain, Mich. (49657) 250/D4
McBaine, Mo. (†65201) 261/H5
McBean, Georgia (†30908) 217/J4
McBee, S.C. (29101) 296/G3
McBride, Br. Col. 184/G3
McBride, Miss. (†39144) 256/C7
McBride, Mo. (63776) 261/N7
McBride, Okla. (†74441) 288/N7
McBride Lake, Sask. 181/J3
McBrides, Mich. (48852) 250/D5
McCabe, Mont. (59245) 262/M2
McCain, N.C. (28361) 281/L4
McCall, Idaho (83638) 220/C5
McCall, La. (†70346) 238/K3
McCalla, Ala. (35111) 195/E4
McCall Creek, Miss. (39647) 256/C7
McCallsburg, Iowa (50154) 229/G4
McCallum, Newf. 166/C4
McCamey, Texas (79752) 303/B6
McCammon, Idaho (83250) 220/F7
McCanna, N. Dak. (58253) 282/P3
McCarley, Miss. (38943) 256/E3
McCarr, Ky. (41544) 237/S5
McCarthy, Alaska (†99566) 196/K2
McCaskill, Ark. (71847) 202/C6
McCauley (isl.), Br. Col. 184/B3
McCaulley, Texas (79534) 303/E5
McCausland, Iowa (52758) 229/M5
McCaysville, Georgia (30555) 217/D1
McChord A.F.B., Wash. 310/C3
McClain (co.), Okla. 288/L5
McClave, Colo. (81057) 208/O6
McCleary, Wash. (98557) 310/B3
McClellan A.F.B., Calif. 204/B8
McClelland (lake), Alberta 182/E1
McClelland, Ark. (†72006) 202/H3
McClelland, Iowa (51548) 229/B6
McClellanville, S.C. (29458) 296/H5
McCloud, Calif. (96057) 204/C2
McCloud, Tenn. (†37857) 237/R8
McClure (lake), Calif. 204/E6
McClure, Ill. (62957) 222/D6
McClure, Ohio (43534) 284/C3
McClure, Pa. (17841) 294/H4
McClure, Va. (24269) 307/D6
McClusky, N. Dak. (58463) 282/K4
McColl, S.C. (29570) 296/H2
McComb, Miss. (39648) 256/D8
McComb, Ohio (45858) 284/C3
McConaughy, C. W. (lake), Nebr. 264/C3
McCondy, Miss. (38854) 256/G3
McCone (co.), Mont. 262/L3
McConnell, Ill. (61050) 222/D1
McConnell, Manitoba 179/B4
McConnell, Tenn. (†38237) 237/D8
McConnell A.F.B., Kansas 232/E4
McConnells, S.C. (29726) 296/E2
McConnellsburg, Pa. (17233) 294/F6
McConnellsville, N.Y. (13401) 276/J4
McConnelsville, Ohio (43756) 284/G6
McCook, Nebr. (69001) 264/D4
McCook (co.), S. Dak. 298/P6
McCool, Miss. (39108) 256/F4
McCool Junction, Nebr. (68401) 264/G4
McCord, Sask. 181/E6
McCordsville, Ind. (46055) 227/F5
McCorkle, W. Va. (†25564) 312/C6
McCormick (co.), S.C. 296/C4
McCormick, S.C. (29835) 296/C4
McCoy, Colo. (80463) 208/F3
McCoy (head), New Bruns. 170/E3
McCoy, Oreg. (†97338) 291/D2
McCoy (creek), Oreg. 291/J5
McCoy, Va. (24111) 307/G6
McCoy A.F.B., Fla. 212/E3
McCoysburg, Ind. (†47978) 227/C3
McCracken, Kansas (67556) 232/C3
McCracken (co.), Ky. 237/D6
McCrea, Pa. (†17241) 294/H5
McCreary (co.), Ky. 237/N7
McCreary, Manitoba 179/C4
McCrory, Ark. (72101) 202/H3
McCulloch (co.), Texas 303/E6
McCullom Lake, Ill. (†60050) 222/E1
McCullough, Ala. (36552) 195/D8
McCune, Kansas (66753) 232/G4
McCurtain (co.), Okla. 288/S6
McCurtain, Okla. (74944) 288/R4
McCutchenville, Ohio (44844) 284/D4
McDade, Texas (78650) 303/G7
McDaniel, Md. (21647) 245/N5
McDaniels, Ky. (40152) 237/J5
McDavid, Fla. (32568) 212/B5
McDermitt, Nev. (89421) 266/D1
McDermott, Ohio (45652) 284/D8
McDonald (isls.), Australia 2/N8
McDonald, Kansas (67745) 232/A2
McDonald (co.), Mo. 261/D9
McDonald (lake), Mont. 262/B2
McDonald, N. Mex. (88262) 274/F5
McDonald, N.C. (28340) 281/L5
McDonald, Ohio (44437) 284/J3
McDonald, Pa. (15057) 294/B5
McDonald, Tenn. (37353) 237/M10
McDonalds Corners, Ontario 177/H3
McDonnell, Queensland 95/B1
McDonough (lake), Conn. 210/D1
McDonough, Del. (†19709) 245/R3
McDonough, Georgia (30253) 217/D4
McDonough (co.), Ill. 222/C3
McDonough, N.Y. (13801) 276/J5
McDougal, Ark. (72441) 202/K1
McDougall (lake), New Bruns. 170/D3
McDowell, Ala. (†35450) 195/C5
McDowell, Ky. (41647) 237/R6
McDowell (co.), N.C. 281/E3
McDowell, Va. (24458) 307/J4
McDowell (co.), W. Va. 312/C8
McDowell, W. Va. (24858) 312/D8
McDuffie (co.), Georgia 217/H4
McElhattan, Pa. (17748) 294/H3
McElmo (creek), Colo. 208/B8
McElmo (creek), Utah 304/E6
McEwen, Tenn. (37101) 237/F8
McFadden, Wyo. (82080) 319/F4
McFall, Mo. (64657) 261/D2
McFarlan, N.C. (28102) 281/J5
McFarland, Calif. (93250) 204/F8
McFarland, Kansas (66501) 232/F2
McFarland, Mich. (†49880) 250/B2
McFarland, Wis. (53558) 317/H10
McFarlane (riv.), Sask. 181/L2
McGaffey, N. Mex. (†87316) 274/A3
McGaheysville, Va. (22840) 307/L4
McGee, Sask. 181/C4
McGees Mills, Pa. (15755) 294/E4
McGehee, Ark. (71654) 202/H6
McGill, Nev. (89318) 266/G3
McGivney, New Bruns. 170/D2
McGloughlin (peak), Mont. 262/C4
McGrath, Alaska 188/C5
McGrath, Alaska (99627) 196/H2
McGrath, Minn. (56350) 255/E4
McGraw, N.Y. (13101) 276/H5
McGraw Brook, New Bruns. 170/D2
McGrawsville, Ind. (†46911) 227/E3
McGregor (lake), Alberta 182/D4
McGregor, Br. Col. 184/G3
McGregor (riv.), Br. Col. 184/G3
McGregor, Iowa (52157) 229/L2
McGregor, Minn. (55760) 255/E4
McGregor (lake), Mont. 262/B3
McGregor, N. Dak. (58755) 282/D2
McGregor, Ontario 177/B5
McGregor, Texas (76657) 303/G6
McGrew, Nebr. (69353) 264/A3
McGuffey, Ohio (45859) 284/C4
McGuire (mt.), Idaho 220/D4
McGuire A.F.B., N.J. 273/D3
McHenry (co.), Ill. 222/E1
McHenry, Ill. (60050) 222/E1
McHenry, Ky. (42354) 237/H6
McHenry, Miss. (39561) 256/F9
McHenry (co.), N. Dak. 282/J3
McHenry, N. Dak. (58464) 282/N4
McHenry Shores, Ill. (†60050) 222/E1
Mchinga, Tanzania 115/H5
Mchinji, Malawi 115/F6
McIlwraith (range), Queensland 95/B2
McIndoe Falls, Vt. (05050) 268/C3
McIntire, Iowa (50455) 229/H2
McIntosh, Ala. (36553) 195/B8
McIntosh, Fla. (32664) 212/D2
McIntosh (co.), Georgia 217/K7
McIntosh, Georgia (†31320) 217/K7
McIntosh, Minn. (56556) 255/C3
McIntosh, N. Mex. (87032) 274/D4
McIntosh (co.), N. Dak. 282/L7
McIntosh (co.), Okla. 288/P4
McIntosh, Ontario 177/F4
McIntosh, Ontario 175/B3
McIntosh, S. Dak. (57641) 298/G2
McIntyre, Georgia (31054) 217/F5
McIvor, Mich. (†48748) 250/F4
McKague, Sask. 181/G3
McKamie, Ark. (†71860) 202/C7
McKay (lake), Manitoba 179/C2
McKay (res.), Oreg. 291/J2
McKean (co.), Pa. 294/E2
McKean, Pa. (16426) 294/B2
McKeand (riv.), N.W. Terrs. 187/M3
McKee (creek), Ill. 222/C4
McKee, Ky. (40447) 237/O6
McKee City, N.J. (†08232) 273/D5
McKeesport, Pa. 188/L2
McKeesport, Pa. (*15130) 294/C7
McKees Rocks, Pa. (15136) 294/B7
McKellar, Ontario 177/E2
McKendrick, New Bruns. 170/D1
McKenna, Wash. (98558) 310/C4
McKenney, Va. (23872) 307/N7
McKenzie, Ala. (36456) 195/E7
McKenzie (co.), N. Dak. 282/D4
McKenzie, N. Dak. (58553) 282/K6
McKenzie, South Fork (riv.), Oreg. 291/E3
McKenzie, Tenn. (38201) 237/E8
McKenzie Bridge, Oreg. (97401) 291/E3
McKerrow, Ontario 177/C1
McKinlay, Queensland 95/B4
McKinlay, Queensland 88/G4
McKinley, Ala. (†36743) 195/C6
McKinley (mt.), Alaska 146/C3
McKinley (mt.), Alaska 188/D5
McKinley (mt.), Alaska 196/H2
McKinley, Cuba 158/B2
McKinley, Minn. (55761) 255/F3
McKinley (co.), N. Mex. 274/A3
McKinley (mt.), U.S. 4/C17

McKinley, Wyo. (†82633) 319/G3
McKinley Park, Alaska (99755) 196/J2
McKinleyville, Calif. (95521) 204/A3
McKinney (lake), Kansas 232/A3
McKinney, Ky. (40448) 237/M6
McKinney, Texas (75069) 303/H4
McKinnon, Georgia (†31545) 217/J8
McKinnon, Tenn. (†37175) 237/F8
McKinnon, Wyo. (82938) 319/C4
McKittrick, Calif. (93251) 204/F8
McKittrick, Mo. (65056) 261/J5
McLain, Miss. (39456) 256/G8
McLane, Pa. (†16426) 294/B2
McLaughlin, Alberta 182/E3
McLaughlin, S. Dak. (57642) 298/H2
McLaurin, Miss. (†39401) 256/F8
McLean (co.), Ill. 222/E3
McLean, Ill. (61754) 222/D3
McLean (co.), Ky. 237/G5
McLean, Nebr. (68747) 264/G2
McLean, N.Y. (13102) 276/H5
McLean (co.), N. Dak. 282/G4
McLean, Sask. 181/G5
McLean, Texas (79057) 303/D2
McLean, Va. (*22101) 307/S2
McLeansboro, Ill. (62859) 222/E5
McLelan (str.), Newf. 166/B1
McLemoresville, Tenn. (38235) 237/D9
McLennan, Alberta 182/B2
McLennan (co.), Texas 303/G6
McLeod (riv.), Alberta 182/B3
McLeod (co.), Minn. 255/D6
McLeod, Mont. (59052) 262/G5
McLeod, N. Dak. (58057) 282/R7
McLeod (bay), N.W. Terrs. 187/G3
McLeod (lake), W. Australia 88/A4
McLeod (lake), W. Australia 92/A4
McLeod Lake, Br. Col. 184/F2
McLeod River, Alberta 182/B3
M'Clintock, Manitoba 179/K2
M'Clintock (chan.), N.W.T. 146/H2
M'Clintock (chan.), N.W.T. 162/F1
M'Clintock (bay), N.W. Terrs. 187/K1
M'Clintock (chan.), N.W. Terrs. 187/H2
McLoud, Okla. (74851) 288/M4
McLoughlin (mt.), Oreg. 291/E5
McLoughlin House Nat'l Hist. Site, Oreg. 291/B2
McLouth, Kansas (66054) 232/G2
McLure, Br. Col. 184/H4
M'Clure (str.), Canada 4/B15
M'Clure (cape), N.W.T. 162/D1
M'Clure (str.), N.W.T. 146/F2
M'Clure (str.), N.W.T. 162/E1
M'Clure (cape), N.W. Terrs. 187/F2
M'Clure (str.), N.W. Terrs. 187/G2
McMahon, Sask. 181/D5
McMechen, W. Va. (26040) 312/E3
McMillan, Mich. (49853) 250/D2
McMillan (lake), N.Mex. 188/F4
McMillan (lake), N. Mex. 274/E6
McMillan, Okla. (73445) 288/M6
McMinn (co.), Tenn. 237/M10
McMinnville, Oreg. (97128) 291/D2
McMinnville, Tenn. (37110) 237/K9
McMorran, Sask. 181/C4
McMullen (co.), Texas 303/F9
McMunn, Manitoba 179/G5
McMurdo (sound), Ant. 2/A10
McMurdo (sound) 5/B9
McMurdo, Br. Col. 184/J4
McMurray, Wash. (†98273) 310/C2
McNab, Alberta 182/D5
McNab, Ark. (†71838) 202/C6
McNabb, Ill. (61335) 222/D2
McNair, Texas (†77520) 303/K1
McNairy (co.), Tenn. 237/D10
McNairy, Tenn. (†38315) 237/D10
McNamee, New Bruns. 170/D2
McNary, Ariz. (85930) 198/F4
McNary, La. (†71433) 238/E5
McNary, Oreg. (97858) 291/H2
McNary (dam), Oreg. 291/H2
McNary, Texas (79839) 303/B11
McNary (dam), Wash. 310/F5
McNaughton, Wis. (54543) 317/H4
McNeal, Ariz. (85617) 198/F7
McNeil, Ark. (71752) 202/D7
McNeill, Miss. (39457) 256/E9
McNulty, Oreg. (†97053) 291/E2
McNutt (isl.), Nova Scotia 168/C5
McPhadyen (riv.), Newf. 166/A3
McPhail (riv.), Manitoba 179/F2
McPherson (co.), Kansas 232/E3
McPherson, Kansas (67460) 232/E3
McPherson (co.), Nebr. 264/C3
McPherson (range), N.S. Wales 97/G1
McPherson (co.), S. Dak. 298/L2
McQuady, Ky. (40153) 237/H5
McRae, Alberta 182/E2
McRae, Ark. (72102) 202/G3
McRae, Georgia (31055) 217/G6
McRoberts, Ky. (41835) 237/R6
McShan, Ala. (35471) 195/B4
McSherrystown, Pa. (17344) 294/H6
McTaggart, Sask. 181/H6
McTavish, Manitoba 179/E5
McTavish Arm (inlet), N.W. Terrs. 187/G3
McVeigh, Ky. (41546) 237/S5
McVeytown, Pa. (17051) 294/G4
McVicar Arm (inlet), N.W. Terrs. 187/F3
McVille, N. Dak. (58254) 282/O4
McWhorter, W. Va. (26401) 312/F4
McWilliams, Ala. (36753) 195/D7
Meacham (lake), N.Y. 276/M1
Meacham, Oreg. (97859) 291/J2
Meacham, Sask. 181/F3
Mead (lake) 188/D3
Mead (lake), Ariz. 198/A2
Mead, Colo. (80542) 208/K2
Mead, Nebr. (68041) 264/H3
Mead (lake), Nev. 266/G6
Mead, Okla. (73449) 288/O7
Mead (lake), U.S. 146/G6
Mead, Wash. (99021) 310/H3
Meade (riv.), Alaska 196/G1
Meade (peak), Idaho 220/G7
Meade (co.), Kansas 232/B4
Meade, Kansas (67864) 232/B4
Meade (co.), Ky. 237/J5
Meade (co.), S. Dak. 298/D5
Meador, W. Va. (†25678) 312/B7
Meadow (creek), Idaho 220/C4
Meadow (mt.), Md. 245/B2
Meadow (lake), Sask. 181/C1
Meadow, S. Dak. (57644) 298/E2
Meadow, Texas (79345) 303/B4
Meadow, Utah (84644) 304/B5
Meadow (riv.), W. Va. 312/E6
Meadow Bluff, W. Va. (24958) 312/E7
Meadow Bridge, W. Va. (25976) 312/E7
Meadowbrook, Ill. (†62010) 222/B2
Meadowbrook, W. Va. (26404) 312/F4
Meadow Creek, W. Va. (25977) 312/E7
Meadow Grove, Nebr. (68752) 264/G2
Meadow Lake, Sask. 181/C1
Meadow Lake Prov. Park, Sask. 181/K4
Meadowlands, Minn. (55765) 255/F3
Meadow Lands, Pa. (15347) 294/B5
Meadow Portage, Manitoba 179/C3
Meadows, Idaho (†83654) 220/B5
Meadows, Ill. (†61726) 222/E3
Meadows, Md. (†20870) 245/G5
Meadows, N.H. (03587) 268/E3
Meadows, S. Australia 94/B8
Meadows of Dan, Va. (24120) 307/H7
Meadow Vale, Ky. (†40201) 237/L1
Meadow Valley, Calif. (95956) 204/D4
Meadow Valley Wash (riv.), Nev. 266/G5
Meadowview-Emory, Va. (24361) 307/D7
Meadow Vista, N. Mex. (†79901) 274/C7
Meadville, Miss. (39653) 256/C8
Meadville, Mo. (64659) 261/F3
Meadville, Pa. 188/L2
Meadville, Pa. (16335) 294/B2
Meaford, Ontario 177/D3
Meagher (co.), Mont. 262/F4
Meaghers Grant, Nova Scotia 168/E4
Meakan (mt.), Japan 81/L2
Mealhada, Portugal 33/B2
Meally, Ky. (41234) 237/R5
Mealy (lake), Newf. 166/C3
Meander, Tasmania 99/C3
Meander River, Alberta 182/A5
Meanook, Alberta 182/D2
Means, Ky. (40346) 237/O5
Meansville, Georgia (30256) 217/D4
Meares (cape), Oreg. 291/C2
Mearim (riv.), Brazil 132/E4
Mearns, Alberta 182/D3
Mears, Mich. (49436) 250/C5
Meath (co.), Ireland 17/H4
Meathas Truim, Ireland 17/G4
Meath Park, Sask. 181/F2
Meaux, France 28/E3
Mebane, N.C. (27302) 281/L2
Mecca, Calif. (92254) 204/K10
Mecca, Ind. (47860) 227/C5
Mecca, Ohio (44410) 284/J3
Mecca (cap.), Saudi Arabia 2/M4
Mecca (cap.), Saudi Arabia 59/C5
Mecca (cap.), Saudi Arabia 54/F7
Mechanic Falls, Maine (04256) 243/C7
Mechanic Falls○, Maine (04256) 243/C7
Mechanicsburg, Ill. (62545) 222/D4
Mechanicsburg, Ind. (†47356) 227/G5
Mechanicsburg, Ohio (43044) 284/D5
Mechanicsburg, Pa. (17055) 294/H5
Mechanicsburg, Va. (†24315) 307/G6
Mechanicstown, Ohio (44651) 284/H4
Mechanicsville, Conn. (06252) 210/H1
Mechanicsville, Georgia (30040) 217/L1
Mechanicsville, Iowa (52306) 229/L5
Mechanicsville, Md. (20659) 245/M7
Mechanicsville, Va. (23111) 307/O5
Mechanicville, N.Y. (12118) 276/N5
Mechelen, Belgium 27/F6
Mecheria, Algeria 106/D2
Mechernich, W. Germany 22/B3
Mecidiye, Turkey 63/B5
Mecitözü, Turkey 63/F2
Mecklenburg (bay), E. Germany 22/D1
Mecklenburg (reg.), E. Germany 22/E2
Mecklenburg, N.Y. (14863) 276/G6
Mecklenburg (co.), N.C. 281/H4
Mecklenburg (co.), Va. 307/M7
Mecklenburg (bay), W. Germany 22/D1
Meckling, S. Dak. (57044) 298/R8
Meconta, Mozambique 118/F3
Mecosta (co.), Mich. 250/D5
Mecosta, Mich. (49332) 250/D5
Mecoya, Bolivia 136/C8
Mecsek (mts.), Hungary 41/D3
Mecúfi, Mozambique 118/G2
Mecula, Mozambique 118/F2
Medain Salih, Saudi Arabia 59/C4
Medan, Indonesia 54/L9
Medan, Indonesia 85/B5
Médanos, Buenos Aires, Argentina 143/D4
Médanos, Entre Ríos, Argentina 143/G6
Médanos (isth.), Venezuela 124/D2
Medanosa (pt.), Argentina 143/D6
Medaryville, Ind. (47957) 227/D2
Meddybemps○, Maine (04657) 243/J5
Meddybemps (lake), Maine 243/J5
Médéa, Algeria 106/E1
Medel (mt.), Switzerland 39/D2
Medellín, Colombia 120/B2
Medellín, Colombia 126/C4
Medellín de Bravo, Mexico 150/Q2
Medemblik, Netherlands 27/G3
Médenine, Tunisia 106/F2
Méderdra, Mauritania 106/A5
Mederville, Iowa (†52043) 229/K3
Medetsiz Tepe (mt.), Turkey 63/F4
Medfield, Mass. (02052) 249/B8
Medfield○, Mass. (02052) 249/B8
Medford, Maine (†04453) 243/F5
Medford○, Maine (†04453) 243/F5
Medford, Mass. (02155) 249/C6
Medford, Minn. (55049) 255/E6
Medford, N.J. (08055) 273/D4
Medford, Okla. (73759) 288/L1
Medford, Oreg. 188/B2
Medford, Oreg. 146/F5
Medford, Oreg. (97501) 291/E5
Medford, Wis. (54451) 317/F5
Medford Center, Maine (†04453) 243/F5
Medford Lakes, N.J. (08055) 273/D4
Medgidia, Romania 45/J3
Medgun (creek), N.S. Wales 97/E1
Media, Ill. (61460) 222/C3
Media, Pa. (*19063) 294/L7
Media Agua, Argentina 143/C3
Media Luna, Cuba 158/G4
Mediapolis, Iowa (52637) 229/L6
Mediaş, Romania 45/G2
Medical Lake, Wash. (99022) 310/H3
Medical Springs, Oreg. (97860) 291/K2
Medicine (lake), Mont. 262/M2
Medicine (creek), Nebr. 264/D4
Medicine (creek), S. Dak. 298/J6
Medicine Bow (range), Colo. 208/G1
Medicine Bow, Wyo. (82329) 319/F4
Medicine Bow (range), Wyo. 319/F4
Medicine Bow (riv.), Wyo. 319/F3
Medicine Creek (dam), Nebr. 264/D4
Medicine Hat, Alberta 182/E4
Medicine Hat, Alta. 146/H4
Medicine Hat, Alta. 162/E5
Medicine Knoll (creek), S. Dak. 298/J5
Medicine Lake, Minn. (55441) 255/G5
Medicine Lake, Mont. (59247) 262/M2
Medicine Lodge (creek), Idaho 220/F5
Medicine Lodge, Kansas (67104) 232/D4
Medicine Lodge (riv.), Kansas 232/D4
Medicine Mound, Texas (†79252) 303/E3
Medicine Park, Okla. (73557) 288/J5
Medill, Mo. (†63445) 261/J2
Medina, Colombia 126/D5
Medina (Hamel), Minn. (†55340) 255/F5
Medina, N.Y. (14103) 276/D4
Medina, N. Dak. (58467) 282/M6
Medina (co.), Ohio 284/G3
Medina, Ohio (44256) 284/G3
Medina, Saudi Arabia 54/F7
Medina, Saudi Arabia 59/D5
Medina, Tenn. (38355) 237/D9
Medina (co.), Texas 303/E8
Medina, Texas (78055) 303/E8
Medina (lake), Texas 303/E8
Medina (riv.), Texas 303/J11
Medina, Wash. (98039) 310/B2
Medinaceli, Spain 33/E2
Medina del Campo, Spain 33/D2
Medina de Rioseco, Spain 33/D2
Medina-Sidonia, Spain 33/D4
Mediodía, Colombia 126/D8
Mediterranean (sea) 2/K4
Mediterranean (sea) 7/E5
Mediterranean (sea), Algeria 106/E1
Mediterranean (sea), Egypt 111/E1
Mediterranean (sea), France 28/E7
Mediterranean (sea), Italy 34/B6
Mediterranean (sea), Libya 111/C1
Mediterranean (sea), Morocco 106/D1
Mediterranean (sea), Tunisia 106/F1
Medix Run, Pa. (†15868) 294/F3
Medjerda (riv.), Algeria 106/F1
Medjerda (riv.), Tunisia 106/F1
Medley, Fla. (†33101) 212/B4
Medley, W. Va. (26734) 312/H4
Mednogorsk, U.S.S.R. 52/J4
Mednogorsk, U.S.S.R. 48/F4
Médoc (reg.), France 28/C5
Mêdog, China 77/E6
Medon, Tenn. (38356) 237/D10
Medora, Ill. (62063) 222/C4
Medora, Ind. (47260) 227/E7
Medora, Kansas (67558) 232/E3
Medora, Manitoba 179/B5
Medora, N. Dak. (58645) 282/C6
Médouneu, Gabon 115/B3
Medstead, Sask. 181/C2
Meductic, New Bruns. 170/C3
Medveditsa (riv.), U.S.S.R. 52/F4
Medvezh'yegorsk, U.S.S.R. 48/D3
Medvezh'yegorsk, U.S.S.R. 52/D2
Medway (riv.), England 13/H6
Medway○, Maine (04460) 243/G4
Medway○, Mass. (02053) 249/J4
Medway (harb.), Nova Scotia 168/D4
Medway (riv.), Nova Scotia 168/C4
Medzilaborce, Czech. 41/F2
Meeandah, Queensland 88/K2
Meehan, Miss. (†39301) 256/G6
Meekatharra, Australia 87/B8
Meekatharra, W. Australia 88/B5
Meekatharra, W. Australia 92/B4
Meeker, Colo. (81641) 208/D2
Meeker, La. (71346) 238/F4
Meeker (co.), Minn. 255/D5
Meeker, Ohio (†43302) 284/D4
Meeker, Okla. (74855) 288/N4
Meeks, Georgia (†31049) 217/G5
Meeks Bay, Calif. (†95730) 204/E5
Meelpaeg (lake), Newf. 166/C4
Meerane, E. Germany 22/E3
Meerhout, Belgium 27/G6
Meers, Okla. (73558) 288/J5
Meersburg, W. Germany 22/C5
Meerssen, Netherlands 27/H7
Meerut, India 54/J7
Meerut, India 68/D3
Meeteetse, Wyo. (82433) 319/D1
Meeting (lake), Sask. 181/D2
Meeting Creek, Alberta 182/D3
Mega, Ethiopia 111/G7
Mega (isl.), Indonesia 85/C6
Megalópolis, Greece 45/E7
Mégantic (co.), Québec 172/F3
Mégantic (lake), Québec 172/G4
Mégara, Greece 45/F6
Megargel, Ala. (36457) 195/D8
Megargel, Texas (76370) 303/F4
Meggett, S.C. (29460) 296/G6
Meghalaya (state), India 68/G3
Megiddo, Israel 65/C2
Mehama, Oreg. (97384) 291/E3
Mehan, Okla. (†74074) 288/M2
Meherrin (riv.), N.C. 281/P1
Meherrin, Va. (23954) 307/M6
Meherrin (riv.), Va. 307/M7
Mehetia (isl.), Fr. Poly. 87/M7
Mehlville, Mo. (†63129) 261/P4
Mehoopany, Pa. (18629) 294/K2
Mehran, Iran 66/E4
Mehran (riv.), Iran 59/F4
Mehran (riv.), Iran 66/J7
Mehsana, India 68/C4
Mehun-sur-Yèvre, France 28/E4
Meifa, P.D.R. Yemen 59/E7
Meiganga, Cameroon 115/B2
Meighen (isl.), N.W.T. 162/M3
Meighen (isl.), N.W. Terrs. 187/H1
Meigle, Scotland 15/E4
Meigs, Georgia (31765) 217/D8
Meigs (co.), Ohio 284/F7
Meigs (co.), Tenn. 237/M9
Meikle (riv.), Alberta 182/A1
Meiktila, Burma 72/B2
Meilen, Switzerland 39/G2
Meiners Oaks-Mira Monte, Calif. (93023) 204/F9
Meiningen, E. Germany 22/D3
Meire Grove, Minn. (†56352) 255/C5
Meiringen, Switzerland 39/F3
Meiron (mt.), Israel 65/C1
Meise, Belgium 27/E7
Meissen, E. Germany 22/E3
Mei Xian, China 77/J7
Mejillones, Chile 120/B5
Mejillones, Chile 138/A4
Mejillones del Sur (bay), Chile 138/A4
Mekambo, Gabon 115/B3
Mekerrhane, Sebkha (salt lake), Algeria 106/E3
Mekili, Libya 111/D1
Mékinac (lake), Québec 172/E2
Mekinock, N. Dak. (58258) 282/R4
Meknès, Morocco 102/B1
Meknès, Morocco 106/C2
Mekong (riv.) 54/M8
Mekong (riv.) 2/Q4
Mekong (riv.), Burma 72/D2
Mekong (riv.), Cambodia 72/E4
Mekong (Lancang Jiang) (riv.), China 77/F7
Mekong (riv.), Laos 72/D3
Mekong (riv.), Thailand 72/E3
Mekong, Mouths of the (delta), Vietnam 72/E5
Mekoryuk, Alaska (99630) 196/E2
Melaka (state), Malaysia 72/D7
Melaka, Malaysia 54/M9
Melaka, Malaysia 72/D7
Melanesia (reg.), Pacific 87/E5
Melaval, Sask. 181/E6
Melba, Idaho (83641) 220/B6
Melber, Ky. (42069) 237/D7
Melbern, Ohio (†43506) 284/A3
Melbeta, Nebr. (69355) 264/A3
Melbourne, Ark. (72556) 202/G1
Melbourne, Australia 2/R7
Melbourne, Australia 87/E9
Melbourne, Fla. 188/K5
Melbourne, Fla. (*32901) 212/F3
Melbourne, Iowa (50162) 229/G5
Melbourne, Ky. (41059) 237/T2
Melbourne, Mo. (†64642) 261/E2
Melbourne (isl.), N.W. Terrs. 187/H3
Melbourne, Ontario 177/C5
Melbourne, Québec 172/E4
Melbourne (cap.), Victoria 88/H7
Melbourne (cap.), Victoria 97/H5
Melbourne, Wash. (†98563) 310/B4
Melbourne Airport, Victoria 88/K7
Melbourne Beach, Fla. (32951) 212/F3
Melby, Minn. (56351) 255/C4
Melcher, Honduras 154/D3
Melcher, Iowa (50163) 229/G6
Melchor (isl.), Chile 138/D6
Melchor Múzquiz, Mexico 150/H3
Melchor Ocampo, Mexico 150/H4
Melchor Ocampo del Balsas, Mexico 150/H8
Melder, La. (71451) 238/E4
Meldorf, W. Germany 22/C1
Meldrim, Georgia (31318) 217/K6
Meldrum Bay, Ontario 177/A2
Meldrum Creek, Br. Col. 184/F4
Meleb, Manitoba 179/E4
Melenki, U.S.S.R. 52/F3
Meleuz, U.S.S.R. 52/J4
Mélèzes (riv.), Québec 174/C1
Melfa, Va. (23410) 307/S5
Melfi, Chad 111/C5
Melfi, Italy 34/E4
Melfort, Sask. 162/F5
Melfort, Sask. 181/G3
Melfort, Loch (inlet), Scotland 15/C4
Melgaço, Portugal 33/B1
Melgar de Fernamental, Spain 33/D1
Melide, Switzerland 39/G5
Meligalá, Greece 45/F7
Melilla, Spain 106/D1
Melilla, Spain 102/B1
Melilla, Spain 7/D5
Melimoyu (mt.), Chile 138/D5
Melinca, Chile 138/D5
Melincué, Argentina /F6
Melipilla, Chile 138/F4
Melita, Manitoba 179/A5
Melitopol', U.S.S.R. 7/H4
Melitopol', U.S.S.R. 52/D5
Melitota, Md. (†21620) 245/O4
Melk, Austria 41/C2
Melkbosstrand, S. Africa 118/E6
Melksham, England 13/E6
Melksham, England 10/E5
Mellansel, Sweden 18/L5
Melle, W. Germany 22/C2
Mellen, Wis. (54546) 317/E3
Mellerud, Sweden 18/H7
Mellette (co.), S. Dak. 298/H6
Mellette, S. Dak. (57461) 298/N3
Mellingen, Switzerland 39/F2
Mellott, Ind. (47958) 227/C4
Mellwood, Ark. (72367) 202/H5
Melmore, Ohio (44845) 284/D3
Mělník, Czech. 41/C1
Melo, Uruguay 120/D6
Melo, Uruguay 145/E3
Melocheville, Québec 172/C4
Melozitna (riv.), Alaska 196/H1
Melrhir, Chott (salt lake), Algeria 106/F2
Melrose, Conn. (06049) 210/E1
Melrose, Fla. (32666) 212/D2
Melrose, Iowa (52569) 229/G7
Melrose, La. (71452) 238/E3
Melrose, Md. (†21102) 245/L2
Melrose, Mass. (02176) 249/D6
Melrose, Minn. (56352) 255/D5
Melrose, Mont. (59743) 262/D5
Melrose, New Bruns. 170/F2
Melrose, Newf. 166/D2
Melrose, N. Mex. (88124) 274/F4
Melrose, N.S. Wales 97/D3
Melrose, Ohio (45861) 284/B3
Melrose, Oreg. (†97470) 291/D4
Melrose, Scotland 15/F5
Melrose, Scotland 10/E3
Melrose, Wis. (54642) 317/E7
Melrose Park, Fla. (†33301) 212/B4
Melrose Park, Ill. (*60160) 222/B5
Melrose Park, N.Y. (†13021) 276/G5
Melrude, Minn. (55766) 255/F3
Mels, Switzerland 39/H2
Melsetter, Zimbabwe 118/E3
Melstone, Mont. (59054) 262/H4
Melsungen, W. Germany 22/C3
Melton, Victoria 97/C5
Melton Hill (lake), Tenn. 237/N9
Melton Mowbray, England 10/F4
Melton Mowbray, England 13/G5
Meltonville, Iowa (†50472) 229/G2
Melun, France 28/E3
Melut, Sudan 111/F5
Melvaig, Scotland 15/C3
Melvern, Kansas (66510) 232/G3
Melvern (lake), Kansas 232/G3
Melvern Square, Nova Scotia 168/C3
Melvich, Scotland 15/E2
Melville (isl.), Australia 87/D7
Melville (isl.), Canada 4/B15
Melville (pen.), Canada 4/C14
Melville (bay), Greenl. 4/B13
Melville, La. (71353) 238/G5
Melville, Mont. (59055) 262/F4
Melville (pen.), N.W.T. 146/K3
Melville (pen.), N.W.T. 162/H2
Melville (isl.), N.W.T. 146/G2
Melville (isl.), N.W.T. 162/E1
Melville (lake), N.W.T. 162/L5
Melville (lake), Newf. 166/C3
Melville, N.Y. (11746) 276/O9
Melville, N. Dak. (†58421) 282/M5
Melville (isl.), N.W. Terrs. 187/G2
Melville (pen.), N.W. Terrs. 187/K3
Melville (isl.), North. Terr. 88/E2
Melville (bay), North. Terr. 88/F2
Melville (bay), North. Terr. 93/E2
Melville (isl.), North. Terr. 93/B1
Melville (cape), Queensland 88/G2
Melville (cape), Queensland 95/C2
Melville, Sask. 162/F5
Melville, Sask. 181/J5
Melville, W. Australia 92/A1
Melvin, Ala. (36913) 195/B7
Melvin, Ill. (60952) 222/E3
Melvin, Iowa (51350) 229/B2
Melvin, Lough (lake), Ireland 10/B3
Melvin (lake), Ireland 17/E3
Melvin, Mich. (48454) 250/G5
Melvin, Minn. (†56540) 255/B3
Melvin, Lough (lake), N. Ireland 10/B3
Melvin, Texas (76858) 303/E6
Melvina, Wis. (54619) 317/E8
Melvindale, Mich. (48122) 250/B7
Melvin Mills, N.H. (†03278) 268/D5
Melvin Village, N.H. (03850) 268/E4
Mélykút, Hungary 41/E3
Memaliaj, Albania 45/D5
Memba, Mozambique 118/G2
Membij, Syria 63/G4
Memel (Klaipeda), U.S.S.R. 52/B3
Memel (Klaipeda), U.S.S.R. 53/A3
Memmingen, W. Germany 22/D5
Mempawah, Indonesia 85/D5
Memphis, Ala. (†35442) 195/B4
Memphis (ruins), Egypt 111/J3
Memphis, Fla. (†33561) 212/D4
Memphis, Ind. (47143) 227/F8
Memphis, Mich. (48041) 250/G6
Memphis, Miss. (†38680) 256/D1
Memphis, Mo. (63555) 261/H2
Memphis, Nebr. (68042) 264/H3
Memphis, Tenn. 146/K6
Memphis, Tenn. 188/J3
Memphis, Tenn. (*38101) 237/B10
Memphis, Texas (79245) 303/D3
Memphis Naval Air Sta., Tenn. 237/B10
Memphremagog (lake), Québec 172/E4
Memphremagog (lake), Vt. 268/C1
Memramcook, New Bruns. 170/F2
Mena, Ark. (71953) 202/B4
Menafra, Uruguay 145/B3
Menahga, Minn. (56464) 255/C4
Menai (str.), Wales 13/C4
Menai Bridge, Wales 13/C4
Menaik, Alberta 182/D3
Ménaka, Mali 106/E5
Menan, Idaho (83434) 220/F6
Menands, N.Y. (†12201) 276/N5
Menarandra (riv.), Madagascar 118/H4
Menard (co.), Ill. 222/D3
Menard (co.), Texas 303/E7
Menard, Texas (76859) 303/E7
Menasalbas, Spain 33/D3
Menasha, Wis. (54952) 317/J7
Mencué, Argentina 143/C5
Mendak, Saudi Arabia 59/D5
Mende, France 28/E5
Mendenhall (cape), Alaska 196/E3
Mendenhall, Miss. (39114) 256/E7
Menderes, Büyük (riv.), Turkey 59/A2
Menderes, Büyük (riv.), Turkey 63/C4
Mendes, Georgia (†30427) 217/H7
Méndez, Ecuador 128/C4
Mendham, N.J. (07945) 273/D2
Mendham, Sask. 181/B5
Mendi, Ethiopia 111/G6
Mendi, Papua N.G. 85/B7
Mendip (hills), England 13/E6
Mendocino (cape), Calif. 146/F5
Mendocino (cape), Calif. 188/A2
Mendocino (co.), Calif. 204/B4
Mendocino, Calif. (95460) 204/B4
Mendocino (cape), Calif. 204/A3
Mendon, Ill. (62351) 222/B3
Mendon○, Mass. (01756) 249/H4
Mendon, Mich. (49072) 250/D7
Mendon, Mo. (64660) 261/F3
Mendon, N.Y. (14506) 276/E4
Mendon, Ohio (45862) 284/A4
Mendon, Utah (84325) 304/B2
Mendon○, Vt. (†05701) 268/B4
Mendooran, N.S. Wales 97/E2
Mendota, Calif. (93640) 204/E7
Mendota, Ill. (61342) 222/D2
Mendota, Minn. (55050) 255/G5
Mendota, Va. (24270) 307/D7
Mendota (lake), Wis. 317/H9
Mendota Heights, Minn. (†55050) 255/G6
Mendoza (prov.), Argentina 143/C4
Mendoza, Argentina 120/C6
Mendoza, Argentina 143/C3
Mendoza (riv.), Argentina 143/C3
Mendoza, Cuba 158/A2
Mendoza, Peru 128/D6
Mendoza, Uruguay 145/C5
Mendrisio, Switzerland 39/G5
Mene de Mauroa, Venezuela 124/C2
Mene Grande, Venezuela 124/C3
Menemen, Turkey 63/B3
Menemsha, Mass. (02552) 249/L7
Menen, Belgium 27/C7
Meneses, Cuba 158/F2
Menfi, Italy 34/D6
Menfro, Mo. (63765) 261/N7
Mengcheng, China 77/J5
Mengen, Turkey 63/D2
Menggala, Indonesia 85/D6
Menghai, China 77/E7
Mengla, China 77/F7
Mengshan, China 77/H7
Mengzi, China 77/F7
Menifee, Ark. (72107) 202/E3
Menifee (co.), Ky. 237/O5
Menihek, Newf. 166/A3
Menihek (lakes), Newf. 166/A3
Menin (Menen), Belgium 27/C7
Menindee, N.S. Wales 97/B3
Menindee (lake), N.S. Wales 97/B3
Meningie, S. Australia 94/F6
Menisino, Manitoba 179/F5
Menistouc (lake), Newf. 166/A3
Menlo, Georgia (30731) 217/B2
Menlo, Iowa (50164) 229/E5
Menlo, Kansas (67746) 232/B2
Menlo, Wash. (98561) 310/B4
Menlo Park, Calif. (94025) 204/J3
Menlo Park, N.J. (08837) 273/E2
Menneval, New Bruns. 170/C1
Menno, S. Dak. (57045) 298/P7
Meno, Okla. (73760) 288/K2
Menoken, N. Dak. (58558) 282/J6
Menominee, Ill. (†61025) 222/C1
Menominee (co.), Mich. 250/B3
Menominee, Mich. (49858) 250/B3
Menominee (riv.), Mich. 250/B3
Menominee (co.), Wis. 317/J5
Menominee (riv.), Wis. 317/L5
Menominee Ind. Res., Wis. 317/J5
Menomonee Falls, Wis. (53051) 317/K1
Menomonie, Wis. (54751) 317/C6
Menongue, Angola 115/C6
Menorca (Minorca) (isl.), Spain 33/J2
Mentasta (pass), Alaska 196/K2
Mentasta Lake, Alaska (†99586) 196/K2
Mentawai (isls.), Indonesia 54/L10
Mentawai (isls.), Indonesia 85/B6
Mentmore, N. Mex. (87319) 274/A3
Menton, France 28/G6
Mentone, Ala. (35984) 195/G1
Mentone, Calif. (92359) 204/H9
Mentone, Ind. (46539) 227/E2
Mentone, Texas (79754) 303/D10
Mentor, Kansas (67465) 232/E3
Mentor, Ky. (†41060) 237/N3
Mentor, Minn. (56736) 255/B3
Mentor, Ohio (44060) 284/H2
Mentor-on-the-Lake, Ohio (44060) 284/G2
Menunketesuck (riv.), Conn. 210/E3
Menye, Turkey 63/C3
Menyuan, China 77/F4
Menzel Bourguiba, Tunisia 106/F1
Menzel Temime, Tunisia 106/G1
Menzie, Manitoba 179/B4
Menzies, W. Australia 88/C5
Menzies, W. Australia 92/C5
Menzingen, Switzerland 39/G2
Menznau, Switzerland 39/E2
Meoqui, Mexico 150/G2
Meota, Sask. 181/C2
Meppel, Netherlands 27/J3
Meppen, W. Germany 22/B2
Mequineza (res.), Spain 33/F2
Mequon, Wis. (53092) 317/L1
Mera, Ecuador 128/C3
Mera (riv.), Switzerland 39/E2
Merabéllou (gulf), Greece 45/H8
Merâia (reg.), Mauritania 106/C5
Meråker, Norway 18/G5
Meramangye (lake), S. Australia 94/C3
Meramec (riv.), Mo. 261/N3
Merano, Italy 34/C1
Merasheen (isl.), Newf. 166/C2

Merauke, Indonesia 85/K7
Merbein, Victoria 97/A4
Mercaderes, Colombia 126/B7
Mercara, India 68/D6
Merced (co.), Calif. 204/E6
Merced, Calif. (95340) 204/E6
Merced (riv.), Calif. 204/E6
Mercedario, Cerro (mt.), Argentina 143/B3
Mercedes, Buenos Aires, Argentina 143/G7
Mercedes, San Luis, Argentina 143/C3
Mercedes, Argentina 120/C6
Mercedes, Corrientes, Argentina 143/G4
Mercedes, Texas (78570) 303/F12
Mercedes, Uruguay 120/D6
Mercedes, Uruguay 145/B4
Merceditas, Chile 138/B7
Mercer (co.), Ill. 222/C2
Mercer (co.), Ky. 237/M5
Mercer○, Maine (04957) 243/D6
Mercer (co.), Mo. 261/E2
Mercer, Mo. (64661) 261/F2
Mercer (co.), N.J. 273/D3
Mercer, N. Zealand 100/E2
Mercer (co.), N. Dak. 282/G5
Mercer, N. Dak. (58559) 282/J5
Mercer (co.), Ohio 284/A4
Mercer, Ohio (†45862) 284/A4
Mercer (co.), Pa. 294/B3
Mercer, Pa. (16137) 294/B3
Mercer, Tenn. (38392) 237/D10
Mercer (co.), W. Va. 312/D8
Mercer, Wis. (54547) 317/F3
Mercer Island (city), Wash. (98040) 310/B2
Mercersburg, Pa. (17236) 294/G6
Mercerville-Hamilton Square, N.J. (08619) 273/D3
Merchantville, N.J. (08109) 273/B3
Merchtem, Belgium 27/E7
Mercier, Bolivia 136/B2
Mercier, Kansas (†66439) 232/G2
Mercier, Québec 172/H4
Mercier (dam), Québec 172/A3
Mercoal, Alberta 182/B3
Mercury, Nev. (89023) 266/E6
Mercury (bay), N. Zealand 100/F2
Mercury (isls.), N. Zealand 100/F2
Mercury, Texas (†76872) 303/E6
Mercy (bay), N.W. Terrs. 187/G2
Mercy (cape), N.W. Terrs. 187/M3
Mere, England 13/E6
Meredith, Colo. (81642) 208/F4
Meredith (lake), Colo. 208/M6
Meredith, N.H. (03253) 268/D4
Meredith○, N.H. (03253) 268/D4
Meredith Center, N.H. (†03253) 268/D4
Meredosia, Ill. (62665) 222/C4
Merefa, U.S.S.R. 52/E5
Meregh, Somalia 115/J3
Merelbeke, Belgium 27/D7
Merevari (riv.), Venezuela 124/F5
Mergui, Burma 54/L8
Mergui, Burma 72/C4
Mergui (arch.), Burma 72/C5
Meriç, Turkey 63/B2
Meriç (riv.), Turkey 63/B2
Merid, Sask. 181/B4
Mérida, Mexico 146/J7
Mérida, Mexico 150/P6
Mérida, Spain 7/D5
Mérida, Spain 33/C3
Mérida (state), Venezuela 124/C3
Mérida, Venezuela 120/B2
Mérida, Venezuela 124/C3
Mérida, Cordillera de (range), Venezuela 124/C3
Meriden, Conn. (06450) 210/D2
Meriden, Iowa (51037) 229/B3
Meriden, Kansas (66512) 232/G2
Meriden, Minn. (56067) 255/E6
Meriden, N.H. (03770) 268/C4
Meriden, Wyo. (82081) 319/H4
Meridian, Georgia (31319) 217/K8
Meridian, Idaho (83642) 220/B6
Meridian, Miss. 146/K6
Meridian, Miss. 188/J4
Meridian, Miss. (39301) 256/G6
Meridian, N.Y. (13113) 276/G4
Meridian, Okla. (73058) 288/M3
Meridian, Texas (76665) 303/G6
Meridian Naval Air Sta., Miss. 256/G5
Meridianville, Ala. (35759) 195/F1
Merigold, Miss. (38759) 256/C3
Merigomish, Nova Scotia 168/F3
Merigomish (harb.), Nova Scotia 168/F3
Merimbula, N.S. Wales 97/F5
Merín (lag.), Uruguay 145/F4
Merino, Colo. (80741) 208/N2
Merino, Victoria 97/A5
Merino Jarpa (isl.), Chile 138/D7
Merinos, Uruguay 145/C3
Merino Village, Mass. (†01570) 249/G4
Merion Station, Pa. (19066) 294/M6
Merir (isl.), Belau 87/D5
Meriwether (co.), Georgia 217/C4
Meriwether Lewis Park, Natchez Trace Pkwy., Tenn. 237/G10
Merj 'Uyun, Lebanon 63/F6
Mérk, Hungary 41/G3
Merkel, Texas (79536) 303/E5
Merksem, Belgium 27/E6
Merksplas, Belgium 27/F6
Merlin, Ontario 177/B5
Merlin, Oreg. (97532) 291/D5
Merlo, Argentina 143/G7
Mermentau, La. (70556) 238/E6
Mermentau (riv.), La. 238/E7
Merna, Nebr. (68856) 264/E3
Merna, Wyo. (†83115) 319/B3
Meroe (ruins), Sudan 111/F4
Merom, Ind. (47861) 227/B6
Merowe, Sudan 111/F4
Merowe, Sudan 59/B6
Merredin, W. Australia 88/B6
Merredin, W. Australia 92/B5
Merri (riv.), Victoria 88/L7
Merriam, Ind. (†46701) 227/G2
Merriam, Kansas (66203) 232/H3
Merrick (co.), Nebr. 264/F3
Merrick, N.Y. (11566) 276/R7
Merrick (mt.), Scotland 15/D5
Merrickville, Ontario 177/J3
Merricourt, N. Dak. (58469) 282/N7
Merrifield, Minn. (56465) 255/D4
Merrifield (bay), Newf. 166/B2
Merrifield, N. Dak. (†58201) 282/R4
Merrifield, Va. (22116) 307/S3
Merrill (pass), Alaska 196/H2
Merrill, Iowa (51038) 229/A3
Merrill, Mich. (48637) 250/E5
Merrill, Miss. (†39452) 256/G9
Merrill, N.Y. (12955) 276/N1
Merrill, Oreg. (97633) 291/F5
Merrill, Wis. (54452) 317/G5
Merrillan, Wis. (54754) 317/E7
Merrillville, Georgia (†31792) 217/E9
Merrillville, Ind. (46410) 227/C2
Merrimac, Ky. (†40009) 237/L6
Merrimac○, Mass. (01860) 249/L1
Merrimac, W. Va. (†25661) 312/B7
Merrimac, Wis. (53561) 317/G9
Merrimack (riv.), Mass. 249/K1
Merrimack (co.), N.H. 268/D5
Merrimack○, N.H. (03054) 268/D6
Merrimack (riv.), N.H. 268/D5
Merrimacport, Mass. (†01860) 249/L1
Merriman, Nebr. (69218) 264/C2
Merrimon, N.C. (†28516) 281/R5
Merrionette Park, Ill. (†60601) 222/B6
Merritt, Br. Col. 162/D5
Merritt, Br. Col. 184/G5
Merritt (isl.), Fla. 212/F3
Merritt, Ill. (†62650) 222/C4
Merritt, Mich. (49667) 250/D4
Merritt (res.), Nebr. 264/D2
Merritt, Wash. (†98826) 310/E3
Merritt Island, Fla. (32952) 212/F3
Merriwa, N.S. Wales 97/F3
Merriwagga, N.S. Wales 97/C3
Merriweather, Mich. (49947) 250/F1
Mer Rouge, La. (71261) 238/G1
Merrow, Conn. (06251) 210/F1
Merry Hill, N.C. (27957) 281/R2
Merrymeeting (lake), N.H. 268/E5
Merry Oaks, N.C. (†27559) 281/L3
Merryville, La. (70653) 238/D5
Mersa Fatma, Ethiopia 111/H5
Mersá Matrûh, Egypt 111/E1
Mersa Matrûh, Egypt 102/E1
Mersch, Luxembourg 27/J9
Mersea (dist.), England 13/J6
Merseburg, E. Germany 22/D3
Mersey (riv.), England 10/F2
Mersey (riv.), England 13/G2
Mersey (riv.), Nova Scotia 168/C4
Mersey (riv.), Tasmania 99/C3
Merseyside (co.), England 13/G2
Mershon, Georgia (31551) 217/H8
Mersin, Turkey 63/F4
Mersin, Turkey 59/B2
Mersin, Turkey 54/E6
Mersing, Malaysia 72/E7
Mērsrags, U.S.S.R. 53/B2
Mertert, Luxembourg 27/J9
Merthyr Tydfil, Wales 13/A6
Merthyr Tydfil, Wales 10/E5
Mértola, Portugal 33/C4
Merton, England 10/B5
Merton, England 13/H8
Merton, Wis. (53056) 317/K1
Mertz Glacier Tongue 5/C8
Mertzon, Texas (76941) 303/C6
Mertztown, Pa. (19539) 294/L4
Meru, Kenya 115/G3
Meru (mt.), Tanzania 115/G4
Merv (Mary), U.S.S.R. 48/F8
Merville, Br. Col. 184/E5
Mervin, Sask. 181/C2
Merwin, Mo. (†64723) 261/C6
Merwin (lake), Wash. 310/C5
Merzifon, Turkey 63/F2
Merzig, W. Germany 22/B4
Mesa, Ariz. 146/G6
Mesa, Ariz. 188/D4
Mesa, Ariz. (*85201) 198/D5
Mesa (co.), Colo. 208/B5
Mesa, Colo. (81005) 208/C4
Mesa, Idaho (83643) 220/B5
Mesa, Miss. (†39667) 256/D8
Mesa, Wash. (99343) 310/G4
Mesabi (range), Minn. 255/E3
Mesa Bolívar, Venezuela 124/C3
Mesachie Lake, Br. Col. 184/J3
Mesa del Seri, Mexico 150/D2
Mesagne, Italy 34/G4
Mesai (riv.), Colombia 126/D7
Mesará (gulf), Greece 45/G8
Mesa Verde National Park, Colo. (81330) 208/C8
Mesa Verde Nat'l Park, Colo. 208/C8
Mescalero, N. Mex. (88340) 274/D5
Mescalero (ridge), N. Mex. 274/F6
Mescalero (valley), N. Mex. 274/F5
Mescalero Apache Ind. Res., N. Mex. 274/D5
Meschede, W. Germany 22/C3
Mesena, Georgia (30819) 217/G4
Meservey, Iowa (50457) 229/G3
Meshed, Iran 54/G6
Meshed, Iran 66/L2
Meshed, Iran 59/H2
Meshed-i-Sar (Babol Sar), Iran 66/H2
Meshik, Alaska (†99579) 196/G3
Meshoppen, Pa. (18630) 294/K2
Meshra er Req, Sudan 111/E6
Mesic, N.C. (†28515) 281/R4
Mesick, Mich. (49668) 250/D4
Mesilla, N. Mex. (88046) 274/C6
Mesilla Park, N. Mex. (88047) 274/C6
Mesita, Colo. (81142) 208/H8
Meskanaw, Sask. 181/F3
Meskene, Syria 63/H5
Meskene, Syria 59/C2
Mesocco, Switzerland 39/H4
Mesolóngion, Greece 45/E6
Mesopotamia (reg.), Iraq 66/B3
Mesopotamia (reg.), Iraq 59/D3
Mesopotamia, Ohio (44439) 284/J3
Mesquite, Nev. (89024) 266/G6
Mesquite, N. Mex. (88048) 274/C6
Mesquite, Texas (*75149) 303/H2
Messancy, Belgium 27/H9
Messina (prov.), Italy 34/E5
Messina, Italy 34/F5
Messina, Italy 7/F5
Messina (str.), Italy 34/E6
Messina, S. Africa 118/D4
Messines, Québec 172/B3
Messíni, Greece 45/E7
Messíni (gulf), Greece 45/E7
Mesta (riv.), Bulgaria 45/F5
Mestre, Italy 34/D2
Mesudiye, Turkey 63/F4
Meta (riv.) 120/B2
Meta (dept.), Colombia 126/D6
Meta (riv.), Colombia 126/E5
Meta, Ky. (41501) 237/S5
Meta, Mo. (65058) 261/H6
Meta (riv.), Venezuela 124/E4
Metabetchouan, Québec 172/F1
Métabetchouane (riv.), Québec 172/F1
Metairie, La. (*70001) 238/O4
Metaline, Wash. (99152) 310/H2
Metaline Falls, Wash. (99153) 310/H2
Metamma, Ethiopia 111/G5
Metamora, Ill. (61548) 222/D3
Metamora, Ind. (47030) 227/G6
Metamora, Mich. (48455) 250/F6
Metamora, Ohio (43540) 284/C2
Metán, Argentina 143/D2
Metangula, Mozambique 118/F2
Metapán, El Salvador 154/C3
Métascouac (lake), Québec 172/F2
Metasville, Georgia (†30673) 217/G3
Metauro (riv.), Italy 34/D3
Metcalf, Georgia (†31792) 217/E9
Metcalf, Ill. (61940) 222/F4
Metcalfe (co.), Ky. 237/K7
Metcalfe, Miss. (38760) 256/B4
Metcalfe, Ontario 177/J2
Metchin (riv.), Newf. 166/B3
Metchosin, Br. Col. 184/K4
Metea, Ind. (†46950) 227/E3
Metedeconk (riv.), N.J. 273/E3
Meteghan, Nova Scotia 168/B4
Meteghan Centre, Nova Scotia 168/B4
Meteghan River, Nova Scotia 168/B4
Meteor (crater), Ariz. 198/E3
Metepec, Mexico 150/M2
Methlick, Scotland 15/F3
Methow, Wash. (98834) 310/E2
Methow (riv.), Wash. 310/E2
Methuen○, Mass. (01844) 249/K2
Methven, N. Zealand 100/C5
Methven, Scotland 15/E4
Metica (riv.), Colombia 126/D6
Metigoshe (lake), N. Dak. 282/K2
Metinic (isl.), Maine 243/E8
Metinota, Sask. 181/C2
Metiskow, Alberta 182/E3
Métis-sur-Mer, Québec 172/A1
Metlakatla, Alaska (99926) 196/N2
Metlakatla, Br. Col. 184/B3
Metlatonoc, Mexico 150/K8
Metlili Chaamba, Algeria 106/E2
Meto (bayou), Ark. 202/H5
Metolius, Oreg. (†97741) 291/F3
Metolius (riv.), Oreg. 291/F3
Metompkin (inlet), Va. 307/T5
Metompkin (isl.), Va. 307/T5
Metonga (lake), Wis. 317/J4
Metropolis, Ill. (62960) 222/E6
Metropolitan, Mich. (†49381) 250/A3
Métsovon, Greece 45/E6
Mettawa, Ill. (†60048) 222/B4
Mettawee (riv.), Vt. 268/A5
Metter, Georgia (30439) 217/H6
Mettet, Belgium 27/F8
Mettler, Calif. (93307) 204/G8
Metuchen, N.J. (08840) 273/E2
Metula, Israel 65/D1
Metz, France 7/E4
Metz, France 28/G3
Metz, Ind. (†46703) 227/H1
Metz, Mich. (†49776) 250/F3
Metz, Mo. (64765) 261/C6
Metz, W. Va. (26585) 312/F3
Metzger, Oreg. (†97223) 291/A2
Metzingen, W. Germany 22/C4
Meudon, France 28/A2
Meulaboh, Indonesia 85/B5
Meulebeke, Belgium 27/C7
Meung-sur-Loire, France 28/D4
Meurthe-et-Moselle (dept.), France 28/G3
Meuse (riv.), Belgium 27/F8
Meuse (dept.), France 28/F3
Meuse (riv.), France 28/F3
Meuselwitz, E. Germany 22/E3
Mexia, Ala. (36458) 195/D8
Mexia, Texas (76667) 303/H6
Mexiana (isl.), Brazil 132/D2
Mexicali, Mexico 150/B1
Mexicali, Mexico 146/G6
Mexican Hat, Utah (84531) 304/E6
Mexican Springs, N. Mex. (87320) 274/A3
Mexico 2/D4
MEXICO 150
Mexico 146/H7
Mexico (gulf) 188/J5
Mexico (gulf) 2/E4
Mexico (gulf) 146/K7
Mexico (gulf), Ala. 195/E10
Mexico (gulf), Cuba 158/A1
Mexico (gulf), Fla. 212/C4
Mexico, Ind. (46958) 227/E3
Mexico, Ky. (†41431) 237/E6
Mexico (gulf), La. 238/F8
Mexico, Maine (04257) 243/B6
Mexico○, Maine (04257) 243/B6
México (state), Mexico 150/K7
Mexico (gulf), Mexico 150/N4
Mexico, Mo. (65265) 261/J4
Mexico, N.Y. (13114) 276/H4
Mexico, Pa. (17056) 294/H4
Mexico (gulf), Texas 303/K9
Mexico Beach, Fla. (32410) 212/D6
Mexico City (cap.), Mexico 150/L1
Mexico City (cap.), Mexico 146/J7
Mexico City (cap.), Mexico 2/E5
Meyadin, Syria 59/C3
Meyadin, Syria 63/J5
Meybod, Iran 66/J4
Meydani, Ras-e (cape), Iran 59/G4
Meydani, Ras-e (cape), Iran 66/L8
Meyer, Iowa (†50455) 229/H2
Meyers Chuck, Alaska (99903) 196/N2
Meyersdale, Pa. (15552) 294/E6
Meyers Lake, Ohio (†44701) 284/H4
Meyerton, S. Africa 118/H7
Meymaneh, Afghanistan 68/A1
Meymaneh, Afghanistan 54/H6
Meymaneh, Afghanistan 59/H2
Meyrin, Switzerland 39/B4
Meyronne, Sask. 181/E6
Mezcala (riv.), Mexico 150/J8
Mezen', U.S.S.R. 4/C7
Mezen', U.S.S.R. 7/J2
Mezen' (riv.), U.S.S.R. 7/J2
Mezen', U.S.S.R. 48/E3
Mezen', U.S.S.R. 52/F1
Mezen' (bay), U.S.S.R. 52/F1
Mezen' (riv.), U.S.S.R. 52/G1
Mezen' (riv.), U.S.S.R. 48/E3
Mézenc (mt.), France 28/E5
Mezhdurechenskiy, U.S.S.R. 48/G4
Mezhdusharskiy (isl.), U.S.S.R. 52/G1
Meziadin (lake), Br. Col. 184/C2
Mézin, France 28/D5
Mezőberény, Hungary 41/F3
Mezőcsát, Hungary 41/F3
Mezőfalva, Hungary 41/E3
Mezőhegyes, Hungary 41/F3
Mezőkovácsháza, Hungary 41/F3
Mezőkövesd, Hungary 41/F3
Mezőszilas, Hungary 41/E3
Mezőtúr, Hungary 41/F3
Mezquital, Mexico 150/G5
Mezquital (riv.), Mexico 150/G5
Mhor, Loch (lake), Scotland 15/D3
Mhow, India 68/D4
Miacatlán, Mexico 150/K2
Miahuatlán de Porfirio Díaz, Mexico 150/L8
Miajadas, Spain 33/D3
Miami, Ariz. (85539) 198/E5
Miami, Fla. 188/K5
Miami, Fla. 146/K7
Miami, Fla. (*33101) 212/B5
Miami (canal), Fla. 212/F5
Miami (riv.), Fla. 212/B5
Miami (riv.), Ind. 227/E3
Miami (co.), Ind. 227/E3
Miami, Ind. (46959) 227/E3
Miami (co.), Kansas 232/H3
Miami, Manitoba 179/D5
Miami, Mo. (65344) 261/F4
Miami, N. Mex. (87729) 274/E2
Miami (co.), Ohio 284/B5
Miami, Okla. (74354) 288/S1
Miami, Texas (79059) 303/D2
Miami, U.S. 2/F4
Miami Beach, Fla. 188/L5
Miami Beach, Fla. (33139) 212/C5
Miami Lakes, Fla. (†33101) 212/B4
Miamisburg, Ohio (45342) 284/B6
Miami Shores, Fla. (33153) 212/B4
Miami Springs, Fla. (33166) 212/B5
Miamitown, Ohio (45041) 284/A9
Miamiville, Ohio (45147) 284/D9
Miandowab, Iran 66/E2
Miandrivazo, Madagascar 118/G3
Mianeh, Iran 59/E2
Mianeh, Iran 66/E2
Mianus, Conn. (†06830) 210/A4
Mianus (riv.), Conn. 210/A4
Mianwali, Pakistan 68/C2
Mianwali, Pakistan 59/K3
Mianyang, Hubei, China 77/H5
Mianyang, Sichuan, China 77/G5
Mianzhu, China 77/F5
Miass, U.S.S.R. 48/G4
Miastko, Poland 47/C2
Miazal, Ecuador 128/D4
Mica, Wash. (99023) 310/H3
Mica Creek, Br. Col. 184/H4
Micanopy, Fla. (32667) 212/D2
Micawber, Okla. (†74882) 288/N3
Micay, Colombia 126/B6
Micco, Fla. (†32960) 212/F4
Miccosukee, Fla. (32309) 212/B1
Miccosukee (lake), Fla. 212/B1
Michael, I. of Man 13/C3
Michael (lake), Newf. 166/C3
Michalovce, Czech. 41/G2
Michaud (pt.), Nova Scotia 168/H3
Michelago, N.S. Wales 97/E4
Michelson (mt.), Alaska 196/K1
Michelstadt, W. Germany 22/C4
Miches, Dom. Rep. 158/F6
Michiana, Mich. (†49117) 250/C7
Michiana Shores, Ind. (†49117) 227/D1
Michichi, Alberta 182/D4
Michie, Tenn. (38357) 237/E10
Michigamme, Mich. (49861) 250/B2
Michigamme (lake), Mich. 250/A2
Michigamme (res.), Mich. 250/B2
Michigamme (riv.), Mich. 250/A2
Michigan 188/J1
Michigan (lake) 188/J2
Michigan (lake), Ill. 222/F1
Michigan (lake), Ind. 227/C1
MICHIGAN 250/80
Michigan (lake), Mich. 250/B5
Michigan, N. Dak. (58259) 282/O3
Michigan (state), U.S. 146/K5
Michigan (lake), U.S. 146/K5
Michigan (isl.), Wis. 317/F2
Michigan (lake), Wis. 317/M9
Michigan Bar, Calif. (†95683) 204/C8
Michigan Center, Mich. (49254) 250/E6
Michigan City, Ind. (46360) 227/C1
Michigan City, Miss. (38647) 256/F1
Michigantown, Ind. (46057) 227/E4
Michigan Valley, Kansas (†66528) 232/G3
Michipicoten (isl.), Ontario 177/H5
Michipicoten (isl.), Ontario 175/C3
Michipicoten River, Ontario 177/H5
Michipicoten River, Ontario 175/C3
Michoacán (state), Mexico 150/H7
Michurin, Bulgaria 45/H4
Michurinsk, U.S.S.R. 52/F4
Michurinsk, U.S.S.R. 48/E4
Mickleton, N.J. (08056) 273/C4
Micotrin (mt.), Dominica 161/F6
Micoua, Québec 174/D3
Micoud, St. Lucia 161/G6
Micro, N.C. (27555) 281/N3
Micronesia, Federated States of 87/F5
Micronesia (reg.), Pacific 87/E4
Midale, Sask. 181/H6
Midas, Nev. (†89414) 266/E1
Middelburg, Netherlands 27/C6
Middelburg, C. of Good Hope, S. Africa 118/D6
Middelburg, Transvaal, S. Africa 118/D5
Middelfart, Denmark 21/C7
Middelfart, Denmark 18/G9
Middelharnis, Netherlands 27/E5
Middelkerke, Belgium 27/B6
Middelvlei, S. Africa 118/G7
Middenmeer, Netherlands 27/F3
Middle (riv.), Conn. 210/F1
Middle (pt.), Fla. 212/E6
Middle, Iowa (52307) 229/K5
Middle (riv.), Minn. 255/B2
Middle Alkali (lake), Calif. 204/E2
Middle Andaman (isl.), India 68/G6
Middle Arm, Newf. 166/C4
Middle Atlas (ranges), Morocco 106/C2
Middle Bass, Ohio (43446) 284/E2
Middle Bass (isl.), Ohio 284/E2
Middle Beaver (creek), Colo. 208/P4
Middleboro, Mass. (02346) 249/L5
Middleboro○, Mass. (02346) 249/L5
Middleboro (McKean), Pa. (16426) 294/B2
Middlebourne, W. Va. (26149) 312/E3
Middlebranch, Ohio (44652) 284/H4
Middlebro, Manitoba 179/G5
Middlebrook, Va. (24459) 307/K4
Middleburg, Fla. (32068) 212/E1
Middleburg, Ky. (42541) 237/M6
Middleburg, Md. (21768) 245/K2
Middleburg, N.C. (27556) 281/N2
Middleburg, Ohio (43336) 284/C5
Middleburg, Pa. (17842) 294/H4
Middleburg, Va. (22117) 307/N3
Middleburgh, N.Y. (12122) 276/M5
Middleburg Heights, Ohio (†44017) 284/G10
Middlebury○, Conn. (06762) 210/C2
Middlebury, Ind. (46540) 227/F1
Middlebury, Vt. (05753) 268/A3
Middlebury○, Vt. (05753) 268/A3
Middlebury Center, Pa. (16935) 294/H2
Middlebury Gap (pass), Vt. 268/B4
Middlebush, N.J. (08874) 273/D3
Middlechurch, Manitoba 179/E4
Middle Concho (riv.), Texas 303/C6
Middledam, Maine (†04216) 243/B6
Middle Falls, N.Y. (12848) 276/O4
Middlefield○, Conn. (06455) 210/E2
Middlefield○, Mass. (01243) 249/B3
Middlefield, Ohio (44062) 284/H3
Middle Fork (peak), Idaho 220/D5
Middlefork, Ind. (†46039) 227/E4
Middle Fork, Powder (riv.), Wyo. 319/F2
Middlegate, Norfolk Is. 88/L6
Middle Granville, N.Y. (12849) 276/O4
Middlegrove, Ill. (61549) 222/C3
Middle Grove, Mo. (†65263) 261/H4
Middle Haddam, Conn. (06456) 210/E2
Middle Harbour (creek), N. S. Wales 88/K3
Middle Harbour (creek), N.S. Wales 97/J3
Middle Hope, N.Y. (12550) 276/M7
Middle Inlet, Wis. (54148) 317/K5
Middle Lake, Sask. 181/F3
Middle Loch (inlet), Hawaii 218/A3
Middle Loup (riv.), Nebr. 264/D3
Middlemarch, N. Zealand 100/B6
Middle Musquodoboit, Nova Scotia 168/E3
Middle Patuxent (riv.), Md. 245/L3
Middle Piney (creek), Wyo. 319/B3
Middle Point, Ohio (45863) 284/B4
Middleport, N.Y. (14105) 276/C4
Middleport, Ohio (45760) 284/F7
Middleport, Pa. (17953) 294/K4
Middle River, Md. (21220) 245/N3
Middle River, Minn. (56737) 255/B2
Middle River, Nova Scotia 168/G2
Middle Saranac (lake), N.Y. 276/M2
Middlesboro, Ky. (40965) 237/O7
Middlesboro, Ky. 188/K3
Middlesbrough, England 7/D3
Middlesbrough, England 13/F3
Middlesbrough, England 10/F3
Middlesex (co.), Conn. 210/E3
Middlesex (co.), Mass. 249/J3
Middlesex (co.), N.J. 273/E3
Middlesex, N.J. (08846) 273/E2
Middlesex, N.Y. (14507) 276/F5
Middlesex, N.C. (27557) 281/N3
Middlesex (county), Ontario 177/C4
Middlesex○, Vt. (†05602) 268/B3
Middlesex (co.), Va. 307/R5
Middle Stewiacke, Nova Scotia 168/E3
Middleton (isl.), Alaska 196/J3
Middleton, England 13/H2
Middleton, England 10/G2
Middleton, Georgia (†30635) 217/G2
Middleton, Idaho (83644) 220/B6
Middleton○, Mass. (01949) 249/K2
Middleton, Mich. (48856) 250/E5
Middleton○, N.H. (†03887) 268/E5
Middleton, Nova Scotia 168/C4
Middleton, Tenn. (38052) 237/D10
Middleton, Wis. (53562) 317/G9
Middletown, Calif. (95461) 204/C5
Middletown, Conn. (06457) 210/E2
Middletown, Del. (19709) 245/R3
Middletown, Ill. (62666) 222/D3
Middletown, Ind. (47356) 227/F4
Middletown, Iowa (52638) 229/L7
Middletown, Ky. (40243) 237/L2
Middletown, Md. (21769) 245/J3
Middletown, Mo. (63359) 261/J4
Middletown○, N.J. (07748) 273/E3
Middletown, N.Y. (10940) 276/L8
Middletown, N.C. (†27824) 281/T4
Middletown, N. Ireland 17/H3
Middletown, Ohio (45042) 284/A6
Middletown, Pa. (17057) 294/J5
Middletown○, R.I. (02840) 249/J6
Middletown, Va. (22645) 307/M2
Middletown Springs○, Vt. (05757) 268/A5
Middle Valley, N.J. (†07853) 273/D2
Middleville, Mich. (49333) 250/D6
Middleville, N.J. (07855) 273/D1
Middleville, N.Y. (13406) 276/K4
Middleville, Ontario 177/H2
Middle Water, Texas (†79022) 303/B2
Middleway, W. Va. (†25430) 312/K4
Middlewich, England 13/H2
Middlewich, England 10/G2
Middlewood, Nova Scotia 168/D4
Midfield, Ala. (35228) 195/E4
Midgic Station, New Bruns. 170/F3
Mid Glamorgan, Wales 13/D6
Midhurst, Ontario 177/E3
Midian (dist.), Saudi Arabia 59/C4
Midkiff, W. Va. (25540) 312/B6
Midland, Ark. (72945) 202/B3
Midland, Ind. (47445) 227/C6
Midland, La. (70557) 238/F6
Midland, Md. (21542) 245/C2
Midland (co.), Mich. 250/E5
Midland, Mich. (48640) 250/E5
Midland, N.C. (28107) 281/J4
Midland, Ohio (45148) 284/C7
Midland, Ontario 177/D3
Midland, Oreg. (97634) 291/F5
Midland, Pa. (15059) 294/A4
Midland, S. Dak. (57552) 298/G5
Midland, Tex. 188/F4
Midland (co.), Texas 303/B6
Midland, Texas (*79701) 303/C6
Midland, Va. (22728) 307/N3
Midland City, Ala. (36350) 195/H8
Midland Park, N.J. (07432) 273/B1
Midlandvale, Alberta 182/D4
Midleton, Ireland 17/E8
Midleton, Ireland 10/B5
Midlothian, Ill. (60445) 222/B6
Midlothian (trad. co.), Scotland 15/B5
Midlothian, Texas (76065) 303/G5
Midlothian, Va. (23113) 307/N6
Midnapore, India 68/F4
Midnight, Miss. (39115) 256/C4
Midnight (lake), Sask. 181/C2
Midongy Atsimo, Madagascar 118/H4
Midvale, Idaho (83645) 220/B5
Midvale, Ohio (44653) 284/H5
Midvale, Utah (84047) 304/B3
Midville, Georgia (30441) 217/H5
Midway (isls.) 188/E6
Midway, Ala. (36053) 195/H6
Midway, Br. Col. 184/H6
Midway, Del. (†19971) 245/T6
Midway, Fla. (32343) 212/B1
Midway, Georgia (31320) 217/K7
Midway, Ind. (†47635) 227/C8
Midway, Ky. (40347) 237/M4
Midway (Sedalia), Ohio (†43151) 284/D6
Midway, Pa. (15060) 294/B5
Midway, Tenn. (37809) 237/P8
Midway (isls.), U.S. 87/J3
Midway, Utah (84049) 304/C3
Midway City, Calif. (92655) 204/D11
Midway Park, N.C. (28544) 281/O5
Midwest, Wyo. (82643) 319/F2
Midwest City, Okla. (73110) 288/M4
Midyat, Turkey 63/J4
Midye, Turkey 63/C2
Mid Yell, Scotland 15/G2
Midzhur (mt.), Bulgaria 45/F4
Midzhur (mt.), Yugoslavia 45/F4
Mie (pref.), Japan 81/H6
Miechów, Poland 47/E3
Międzychód, Poland 47/B2
Międzylesie, Poland 47/C3
Międzyrzec Podlaski, Poland 47/F3
Międzyrzecz, Poland 47/B2
Mielec, Poland 47/E3
Mier, Ind. (†46919) 227/F3
Mier, Mexico 150/K3
Miercurea Ciuc, Romania 45/G2
Mieres, Spain 33/D1
Miesso, Ethiopia 111/H6
Miesso, Ethiopia 102/G4
Miesville, Minn. (†55033) 255/F6
Miette, Alberta 182/B3
Mifflin, Ohio (†44805) 284/F4
Mifflin (co.), Pa. 294/G4
Mifflin, Pa. (17058) 294/H4
Mifflin, Wis. (†53580) 317/F10
Mifflinburg, Pa. (17844) 294/H4
Mifflintown, Pa. (17059) 294/H4
Miflin, Ala. (†36530) 195/C10
Migdal, Israel 65/C2
Migdal Ha 'Emeq, Israel 65/C2
Mignon, Ala. (†35150) 195/F4
Migori, Kenya 115/F4
Miguel Alves, Brazil 132/F4
Miguel Auza, Mexico 150/H4
Miguel de la Borda, Panama 154/G6
Miguelete, Uruguay 145/B5

Miguel Riglos, Argentina 143/D4
Migues, Uruguay 145/C6
Mihalıçcık, Turkey 63/D3
Mihara, Japan 81/F6
Mikado, Mich. (48745) 250/F4
Mikado, Sask. 181/J4
Mikana, Wis. (54857) 317/C4
Mikhaylovgrad, Bulgaria 45/F4
Mikhaylovka, U.S.S.R. 52/F4
Mikhmoret, Israel 65/B3
Miki, Japan 81/H7
Mikínai, Greece 45/F7
Mikkalo, Oreg. (97861) 291/G2
Mikkeli, Finland 18/P6
Mikkeli (prov.), Finland 18/P6
Mikkwa (riv.), Alberta 182/B5
Mikłów, Poland 47/B4
Mikulov, Czech. 41/C2
Mikumi Nat'l Park, Tanzania 115/G5
Mikun', U.S.S.R. 52/H2
Mikuni, Japan 81/H5
Milaca, Minn. (56353) 255/E5
Milagro, Ecuador 120/A3
Milagro, Ecuador 128/C4
Milagros, Philippines 82/D4
Milam (co.), Texas 303/H7
Milam, Texas 75959) 303/L6
Milam, W. Va. (26838) 312/H5
Milan, Georgia (31060) 217/G6
Milan, Ill. (61264) 222/C2
Milan, Ind. (47031) 227/G6
Milan (prov.), Italy 34/B2
Milan, Italy 7/E4
Milan, Italy 2/K3
Milan, Italy 34/B2
Milan, Kansas (67105) 232/E4
Milan, Mich. (48160) 250/F6
Milan, Minn. (56262) 255/C5
Milan, Mo. (63556) 261/F2
Milan○, N. H. (03588) 268/E2
Milan, N. Mex. (87021) 274/B3
Milan, Ohio (44846) 284/E3
Milan, Québec 172/F4
Milan, Tenn. (38358) 237/D9
Milan, Wash. (99003) 310/H3
Milan, Wis. (54453) 317/F6
Milanje, Mozambique 118/F3
Milano, Italy (76556) 303/H7
Milanville, Pa. (18443) 294/M2
Milâs, Turkey 63/B4
Milazzo, Italy 34/E5
Milbank, S. Dak. (57252) 298/K3
Milbanke (sound), Br. Col. 184/C4
Milberger, Kansas (†67665) 232/D3
Milbridge○, Maine (04658) 243/H6
Milburn, Ky. (42070) 237/D7
Milburn, Nebr. (68857) 264/E3
Milburn, Okla. (73450) 288/O6
Milden, Sask. 181/D4
Mildenhall, England 13/H5
Mildmay, Ontario 177/C3
Mildred, Kansas (66055) 232/G3
Mildred, Mont. (59341) 262/M4
Mildred, Pa. (18632) 294/K3
Mildred, Sask. 181/D2
Mildred Lake, Alberta 182/E1
Mildura, Victoria 88/G6
Mildura, Victoria 97/A4
Mile, China 77/F7
Miles, Iowa (52064) 229/N4
Miles, Queensland 88/H5
Miles, Texas (76861) 303/D6
Milesburg, Pa. (16853) 294/G4
Miles City, Mont. 188/E1
Miles City, Mont. (59301) 262/L4
Milestone, Sask. 181/G5
Milesville, S. Dak. (57553) 298/F5
Milevsko, Czech. 41/C2
Miley, S.C. (29934) 296/E6
Milfay, Okla. (74046) 288/N3
Milford, Calif. (96121) 204/E3
Milford, Conn. (06460) 210/C4
Milford (pt.), Conn. 210/C4
Milford, Del. (19963) 245/S5
Milford, Georgia (†31762) 217/C8
Milford, Ill. (60953) 222/F2
Milford, Ind. (46542) 227/F1
Milford, Ind. (†47240) 227/F6
Milford, Iowa (51351) 229/C2
Milford, Ireland 17/F1
Milford, Kansas (66514) 232/F2
Milford (lake), Kansas 232/E2
Milford, Ky. (41061) 237/N3
Milford○, Maine (04461) 243/F6
Milford○, Maine (04461) 243/F6
Milford, Mass. (01757) 249/H4
Milford○, Mass. (01757) 249/H4
Milford, Mich. (48042) 250/F6
Milford, Mo. (64766) 261/D7
Milford, Nebr. (68405) 264/H4
Milford, N. H. (03055) 268/D6
Milford○, N. H. (03055) 268/D6
Milford, N.J. (08848) 273/C2
Milford, N.Y. (13807) 276/K5
Milford (sound), N. Zealand 100/A6
Milford, Ohio (45150) 284/D9
Milford, Pa. (18337) 294/N3
Milford, Texas (76670) 303/H5
Milford, Utah (84751) 304/A5
Milford, Va. (22514) 307/O4
Milford, Wis. (†53038) 317/J9
Milford Bay, Ontario 177/E2
Milford Center, Ohio (43045) 284/D5
Milford Haven, Wales 13/B6
Milford Haven, Wales 10/D5
Milford Haven (inlet), Wales 13/B6
Milford Haven (inlet), Wales 10/D5
Milford Station, Nova Scotia 168/E3
Milh, Bahr al (lake), Iraq 66/C4
Mili (atoll), Marshall Is. 87/H5
Miliana, Algeria 106/E1
Milicz, Poland 47/C3
Milieu (riv.), Québec 172/C3
Mililani Town, Hawaii (96789) 218/E2
Milingimbi, North. Terr. 93/D2
Milk (riv.), Alberta 182/D5
Milk (riv.), Alberta 182/D5

Milk (riv.), Mont. 188/D1
Milk (riv.), Mont. 262/J2
Milk, Wadi el (dry riv.), Sudan 59/A6
Milk, Wadi el (dry riv.), Sudan 111/E4
Milk River, Alberta 182/D5
Mill (creek), Calif. 204/D3
Mill (riv.), Conn. 210/D3
Mill (riv.), Conn. 210/B4
Mill (creek), Ind. 227/D5
Mill (creek), Mass. 249/D3
Mill (riv.), Mass. 249/C3
Mill (creek), Mich. 250/G5
Mill (creek), N.J. 273/E4
Mill (isl.), N.W. Terrs. 187/L3
Mill (riv.), Vt. 268/B4
Mill (creek), W. Va. 312/C5
Milladore, Wis. (54454) 317/G6
Millard, Ky. (†41501) 237/S6
Millard, Mo. (†63501) 261/G2
Millard (co.), Utah 304/A4
Millarton, N. Dak. (58470) 282/N6
Millarville, Alberta 182/C4
Millau, France 28/E5
Millbank, Ontario 177/D4
Mill Bay, Br. Col. 184/K3
Millboro, S. Dak. (57554) 298/K7
Millboro, Va. (24460) 307/J5
Millboro Springs, Va. (24460) 307/J4
Millbourne, Pa. (†19082) 294/M6
Millbrae, Calif. (94030) 204/J2
Millbridge, Ontario 177/G3
Millbrook, Ala. (36054) 195/F6
Millbrook, Ill. (60536) 222/E2
Millbrook, Mich. (48860) 250/D5
Millbrook, N.Y. (12545) 276/N7
Millbrook, N.C. (27558) 281/M3
Millbrook, Ontario 177/F3
Millburn○, N.J. (07041) 273/E2
Millburne, Wyo. (82933) 319/B4
Millbury○, Mass. (01527) 249/H4
Millbury, Ohio (43447) 284/D2
Mill City, Nev. (†89418) 266/D2
Mill City, Oreg. (97360) 291/E3
Mill Cove, New Bruns. 170/D3
Mill Cove, Nova Scotia 168/D4
Millcreek, Ill. (62961) 222/D6
Millcreek, Mo. (†63645) 261/M7
Mill Creek, Okla. (74856) 288/N6
Mill Creek, Pa. (17060) 294/G5
Mill Creek, W. Va. (26280) 312/G5
Milldale, Conn. (06467) 210/C3
Millecoquins (lake), Mich. 250/D2
Milledgeville, Georgia (31061) 217/F4
Milledgeville, Ill. (61051) 222/D1
Milledgeville, Ohio (43142) 284/E4
Milledgeville, Tenn. (38359) 237/E10
Mille Iles (riv.), Québec 172/H4
Mille Lac Ind. Res., Minn. 255/E5
Mille Lacs (co.), Minn. 255/E5
Mille Lacs (lake), Minn. 188/H1
Mille Lacs (lake), Minn. 255/E4
Mille Lacs (lake), Ontario 177/G5
Mille Lacs (lake), Ontario 175/B3
Millen, Georgia (30442) 217/J5
Miller (peak), Ariz. 198/E7
Miller (co.), Ark. 202/C7
Miller (co.), Georgia 217/C8
Miller, Mo. 261/H6
Miller, Iowa (50438) 229/F2
Miller, Kansas (†66868) 232/F3
Miller, Miss. (†38654) 256/E1
Miller (co.), Mo. 261/H6
Miller, Mo. (65707) 261/E4
Miller, Nebr. (68858) 264/E4
Miller, Ohio (†45626) 284/F8
Miller (creek), Oreg. 291/F5
Miller, S. Dak. (57362) 298/L4
Miller City, Ohio (45864) 284/B3
Miller Dam Flowage (res.), Wis. 317/E5
Miller House, Alaska (99730) 196/J1
Millerovo, U.S.S.R. 48/G5
Millerovo, U.S.S.R. 52/F5
Millers, Md. (21107) 245/L2
Millers (riv.), Mass. 249/E2
Millersburg, Ind. (46543) 227/F1
Millersburg, Iowa (52308) 229/J5
Millersburg, Ky. (40348) 237/N4
Millersburg, Mich. (49759) 250/F3
Millersburg, Ohio (44654) 284/F4
Millersburg, Oreg. (†97321) 291/E3
Millersburg, Pa. (17061) 294/J4
Millers Falls, Mass. (01349) 249/E2
Millers Ferry, Ala. (36760) 195/D6
Millers Ferry, Ala. (†32437) 212/C6
Millersport, Ohio (43046) 284/E6
Millerstown, Ky. (†42754) 237/J6
Millerstown, Pa. (17062) 294/H4
Millersview, Texas (76862) 303/E6
Millersville, Md. (21108) 245/M4
Millersville, Ohio (43448) 284/D3
Millersville, Pa. (†17551) 294/K6
Millerton (lake), Calif. 204/F6
Millerton, Iowa (50165) 229/G7
Millerton, New Bruns. 170/E2
Millerton, N.Y. (12546) 276/O7
Millerton, Okla. (74750) 288/S7
Millerton, Pa. (16936) 294/J2
Millertown, Newf. 166/C4
Millerville, Ala. (36267) 195/G4
Millerville, Minn. (56315) 255/C4
Millet, Alberta 182/D3
Millett, S.C. (†29836) 296/D5
Millett, Texas (†78014) 303/E9
Millgrove, Ind. (†47348) 227/G4
Mill Grove, Mo. (†64673) 261/F2
Mill Hall, Pa. (17751) 294/G3
Millhaven, Georgia (†30467) 217/J5
Millheim, Pa. (16854) 294/H4
Millhousen, Ind. (47261) 227/E6
Millican, Oreg. (†97701) 291/F4
Millicent, Alberta 182/E4
Millicent, S. Australia 88/F7
Millicent, S. Australia 94/F7
Milligan, Fla. (32537) 212/C6
Milligan, Ind. (†47856) 227/C5
Milligan, Nebr. (68406) 264/G4

Milligan College, Tenn. (37682) 237/S8
Milliken, Colo. (80543) 208/F2
Millikin, La. (†71254) 238/H1
Millingen aan den Rijn, Netherlands 27/J5
Millington, Conn. (†06729) 210/C1
Millington, Conn. (†06423) 210/F3
Millington, Ill. (60537) 222/E2
Millington, Md. (21651) 245/P3
Millington, Mich. (48746) 250/F5
Millington, N.J. (07946) 273/D1
Millington, Tenn. (38053) 237/B10
Millinocket○, Maine (04462) 243/F4
Millinocket (lake), Maine 243/F4
Millinocket (lake), Maine 243/F3
Mill Iron, Mont. (59342) 262/M5
Millis○, Mass. (02054) 249/A8
Millis-Clicquot, Mass. (02054) 249/A8
Millisle, N. Ireland 17/K2
Millmerran, Queensland 95/D5
Millmont, Pa. (17845) 294/H4
Mill Neck, N.Y. (11765) 276/R6
Millom, England 13/D3
Mill Plain, Conn. (†06810) 210/A3
Millport, Ala. (35576) 195/B3
Millport, N.Y. (14864) 276/G6
Millport, Pa. (16739) 294/F2
Millport, Scotland 15/B2
Millport, Scotland 10/A1
Millrift, Pa. (18340) 294/N3
Mill River, Mass. (01244) 249/A4
Mill Road, Nova Scotia 168/D4
Millrose, W. Australia 92/D4
Mill Run, Pa. (15464) 294/C6
Millry, Ala. (36558) 195/B7
Mills (co.), Iowa 229/B6
Mills, Nebr. (68753) 264/E2
Mills, N. Mex. (87730) 274/E2
Mills (lake), N.W. Terrs. 187/G3
Mills, Pa. (16937) 294/G2
Mills (co.), Texas 303/F6
Mills, Utah (†84639) 304/B4
Mills, Wyo. (82644) 319/F3
Millsap, Texas (76066) 303/G5
Millsboro, Del. (19966) 245/S6
Millsboro, Pa. (15348) 294/B5
Mill Shoals, Ill. (62862) 222/E5
Mill Spring, Mo. (63952) 261/L8
Mill Springs, Ky. (42632) 237/M7
Millstadt, Ill. (62260) 222/B3
Millstone, Ky. (41838) 237/R6
Millstone, N.J. (08876) 273/D2
Millstone (riv.), N.J. 273/D3
Millstone, W. Va. (25261) 312/D5
Millstream Station, W. Australia 92/B3
Millstreet, Ireland 17/D7
Millthorpe, N.S. Wales 97/E5
Milltown, Ala. (†36855) 195/H4
Milltown, Ind. (47145) 227/E8
Milltown, Ireland 17/A7
Milltown, Ky. (42761) 237/L6
Milltown, Mont. (59851) 262/C4
Milltown, Newf. 166/C4
Milltown, N.J. (08850) 273/E3
Milltown, S. Dak. (†57366) 298/O7
Milltown, Wis. (54858) 317/B4
Millungera, Queensland 95/B3
Millvale, Pa. (15209) 294/H7
Mill Valley, Calif. (94941) 204/H2
Mill Village, Nova Scotia 168/D4
Mill Village, Pa. (16427) 294/C2
Millville, Del. (19967) 245/T6
Millville, Ind. (†47362) 227/G5
Millville, Iowa (52052) 229/L3
Millville○, Mass. (01529) 249/H4
Millville, Minn. (55957) 255/F6
Millville, New Bruns. 170/C2
Millville, N.J. (08332) 273/C5
Millville, Ohio (†45013) 284/A7
Millville, Pa. (17846) 294/J3
Millville, Utah (84326) 304/C2
Millwood○, Ark. 202/C6
Millwood, Georgia (31552) 217/G8
Millwood, Ky. (42762) 237/J6
Millwood, Manitoba 179/A4
Millwood, Ohio (†43014) 284/F5
Millwood, Va. (22646) 307/N2
Millwood, Wash. (†99210) 310/H3
Millwood, W. Va. (25262) 312/C5
Milly Milly, W. Australia 92/B4
Milmay, N.J. (08340) 273/D5
Milmine, Ill. (61855) 222/E4
Milmont Park, Pa. (19033) 294/M7
Milnathort, Scotland 15/D4
Milne (inlet), N.W. Terrs. 187/K2
Milne (riv.), N.W. Terrs. 187/K2
Milne (bay), Papua N.G. 85/D4
Milner, Colo. (80477) 208/F2
Milner, Georgia (30257) 217/D5
Milner Ridge, Manitoba 179/F4
Milnerton, S. Africa 118/F6
Milnesand, N. Mex. (88125) 274/F5
Milnes Landing, Br. Col. 184/J4
Milngavie, Scotland 10/B1
Milngavie, Scotland 15/B1
Milnor, N. Dak. (58060) 282/R7
Milo, Alberta 182/D4
Milo (riv.), Guinea 106/C7
Milo, Iowa (50166) 229/G6
Milo, Ky. (41055) 238/D1
Milo, Maine (04463) 243/F5
Milo○, Maine (04463) 243/F5
Milo, Mo. (64767) 261/D7
Milo, Okla. (73451) 288/M6
Milo, Oreg. (†97417) 291/E5
Milo, Tenn. (†37381) 237/L9
Milolii, Hawaii (†96704) 218/G6
Milos, Greece 45/G7
Milos (isl.), Greece 45/G7
Milpa, N.S. Wales 97/B2
Milparinka, N.S. Wales 97/A1
Milperra, N.S. Wales 97/H4
Milpitas, Calif. (95035) 204/L3
Milroy, Ind. (46156) 227/G6
Milroy, Minn. (56263) 255/C6

Milroy, Pa. (17063) 294/G4
Milstead, Ala. (†36075) 195/G6
Milstead, Georgia (30207) 217/D3
Milton (res.), Colo. 208/K2
Milton, Conn. (†06759) 210/C1
Milton, Del. (19968) 245/S5
Milton, Fla. (32570) 212/B6
Milton, Ill. (62352) 222/C4
Milton, Ind. (46542) 227/G5
Milton, Iowa (52570) 229/J7
Milton, Kansas (67106) 232/E4
Milton, Iowa (51554) 229/B6
Milton, Ky. (40045) 237/L3
Milton, La. (70558) 238/F4
Milton○, Mass. (02186) 249/D7
Milton, Newf. 166/C2
Milton, N.H. (03851) 268/F5
Milton, N.J. (07438) 273/D1
Milton, N.S. Wales 97/F4
Milton, N.Y. (12547) 276/M7
Milton, N.Y. (12020) 276/N4
Milton, N. Zealand 100/B7
Milton, N.C. (27305) 281/L1
Milton, N. Dak. (58260) 282/O2
Milton, Nova Scotia 168/D4
Milton, Okla. (†74944) 288/S4
Milton, Ontario 177/E4
Milton, Pa. (17847) 294/J3
Milton, Tenn. (37118) 237/J9
Milton, Vt. (05468) 268/A2
Milton○, Vt. (05468) 268/A2
Milton, Wash. (98354) 310/C3
Milton, W. Va. (25541) 312/B6
Milton, Wis. (53563) 317/J10
Miltona, Minn. (56354) 255/C4
Miltona (lake), Minn. 255/C4
Milton Center, Ohio (43541) 284/C3
Milton-Freewater, Oreg. (97862) 291/J2
Milton Keynes, England 13/F5
Milton Mills, N.H. (03852) 268/F4
Miltonsburg, Ohio (†43793) 284/H6
Miltonvale, Kansas (67466) 232/E2
Milton-Malbay, Ireland 17/C6
Milverton, Ontario 177/D4
Milwaukee, N.C. (27854) 281/P2
Milwaukee, Wis. 146/K5
Milwaukee (co.), Wis. 317/L9
Milwaukee, Wis. 188/J2
Milwaukee, Wis. (*53201) 317/M1
Milwaukie, Oreg. (97222) 291/B2
Mima, Ky. (41456) 237/P5
Mimbres, N. Mex. (88049) 274/B6
Mimbres (mts.), N. Mex. 274/B6
Mimbres (riv.), N. Mex. 274/B6
Miminegash, Pr. Edward I. 168/D2
Mimizan, France 28/C5
Mimoň, Czech. 41/C1
Mimongo, Gabon 115/B4
Mimosa Park, La. (†70070) 238/N4
Mimoso do Sul, Brazil 135/F2
Mims, Fla. (32754) 212/F3
Min (riv.), China 77/B4
Minford, Ohio (45653) 284/E8
Mingan, Que. 162/K5
Mingan (riv.), Que. 174/E2
Mingan (Jacques-Cartier) (pass), Québec 174/E3
Mingechaur, U.S.S.R. 52/G6
Mingenew, W. Australia 88/B5
Mingenew, W. Australia 92/A5
Mina, Nev. (89422) 266/C4
Mina, S. Dak. (57462) 298/M3
Mina al Ahmadi, Kuwait 59/E4
Mina al Fahal, Oman 59/G5
Minab, Iran 59/G2
Minaki, Ontario 177/F4
Minaki, U.S.S.R. 175/A3
Minam, Oreg. (†97827) 291/K2
Minamata, Japan 81/E7
Minami iwo (isl.), Japan 87/D3
Minami iwo (isl.), Japan 81/M5
Minapasuk, Philippines 82/D5
Minard, Ireland 17/A7
Minas (mts.), Guatemala 154/C3
Minas (basin), Nova Scotia 168/D3
Minas (chan.), Nova Scotia 168/D3
Minas, Uruguay 145/D2
Minas de Corrales, Uruguay 145/D2
Minas de Matahambre, Cuba 158/A1
Minas de Ríotinto, Spain 33/C4
Minas Gerais (state), Brazil 135/D2
Minas Gerais (state), Brazil 132/E7
Minas Novas, Brazil 132/F7
Minatare, Nebr. (69356) 264/A3
Minatare (lake), Nebr. 264/A3
Minatitlán, Mexico 150/M8
Minbu, Burma 72/B2
Minburn, Alberta 182/E3
Minburn, Iowa (50167) 229/E5
Minbya, Burma 72/B2
Mincha, Chile 138/A8
Minchinmávida (vol.), Chile 138/E4
Mincio (riv.), Italy 34/C2
Minco, Okla. (73059) 288/L4
Mindanao (isl.), Philippines 54/O9
Mindanao (isl.), Philippines 2/R5
Mindanao (isl.), Philippines 85/H4
Mindanao (isl.), Philippines 82/D7
Mindanao (riv.), Philippines 82/E7
Mindanao (sea), Philippines 85/G4
Mindanao (sea), Philippines 82/D7
Mindelheim, W. Germany 22/D4
Mindelo, C. Verde 106/A7
Mindemoya, Ontario 177/B2
Mindemoya (lake), Ontario 177/B2
Minden, Iowa (51553) 229/C6
Minden, Kansas (67865) 232/C4
Minden, La. (71055) 238/D1
Minden, Nebr. (68959) 264/F4
Minden, Nev. (89423) 266/B4
Minden, Ontario 177/F3
Minden, W. Germany 22/C2
Minden City, Mich. (48456) 250/G5
Mindenmines, Mo. (64769) 261/C8
Mindiptana, Indonesia 85/L7
Mindoro (isl.), Philippines 54/N8
Mindoro (isl.), Philippines 85/G3
Mindoro (isl.), Philippines 82/G4
Mindoro (str.), Philippines 85/F3
Mindoro (str.), Philippines 82/G4
Mindoro, Wis. (54644) 317/D7
Mindouli, Congo 115/B4

Mindszent, Hungary 41/F3
Mine (head), Ireland 17/F8
Mine Centre, Ontario 177/G5
Mine Centre, Ontario 175/B3
Minehead, England 13/D6
Minehead, England 10/E5
Mine Hill○, N. J. (†07801) 273/D2
Mineiros, Brazil 132/C7
Mineiros, Brazil 120/D4
Mine La Motte, Mo. (63659) 261/M7
Mineola, Iowa (51554) 229/B6
Mineola, Mo. (†63361) 261/J5
Mineola, N.Y. (11501) 276/R7
Mineola, Texas 75773) 303/J5
Miner, Mo. (163801) 261/N9
Miner (co.), S. Dak. 298/O5
Miner, Mont. (†59027) 262/E5
Miner, Calif. (46063) 204/D3
Mineral, Ill. (61344) 222/D2
Mineral (co.), Mont. 262/B3
Mineral (co.), Nev. 266/C4
Mineral, Ohio (†45766) 284/F7
Mineral, Texas (78125) 303/G9
Mineral (mts.), Utah 304/A5
Mineral, Va. (23117) 307/N4
Mineral, Wash. (98355) 310/C4
Mineral (co.), W. Va. 312/J4
Mineral Bluff, Georgia (30559) 217/D1
Mineral Center, Minn. (†55605) 255/G2
Mineral City, Ohio (44656) 284/H4
Mineral del Monte, Mexico 150/K6
Mineral Hills, Mich. (†49935) 250/G2
Mineral'nye Vody, U.S.S.R. 52/F6
Mineral Point, Mo. (63660) 261/L7
Mineral Point, Wis. (53565) 317/F10
Mineral Springs, Ark. (71851) 202/C6
Mineral Springs, N.C. (28108) 281/H5
Mineral Wells, Miss. (38648) 256/E1
Mineral Wells, Texas (76067) 303/F5
Mineralwells, W. Va. (26150) 312/C4
Minersville, Ohio (45769) 284/F7
Minersville, Pa. (17954) 294/K4
Minersville, Utah (84752) 304/A5
Mine Run, Va. (22568) 307/N4
Minerva, Ky. (41062) 237/O3
Minerva, N.Y. (12851) 276/N3
Minerva, Ohio (44657) 284/H4
Minerva (reefs), Tonga 87/H8
Minerva Park, Ohio (†43201) 284/E5
Minetto, N.Y. (13115) 276/H4
Mineville-Witherbee, N.Y. (12956) 276/O2
Minfeng (Niya), China 77/B4
Minford, Ohio (45653) 284/E8
Mingan, Que. 162/K5
Mingan (riv.), Que. 174/E2
Mingan (Jacques-Cartier) (pass), Québec 174/E3
Mingechaur, U.S.S.R. 52/G6
Mingenew, W. Australia 88/B5
Mingenew, W. Australia 92/A5
Minginish (dist.), Scotland 15/B3
Minglanilla, Spain 33/F3
Mingo, Iowa (50168) 229/G5
Mingo, Kansas (†67701) 232/B2
Mingo, Ohio (43047) 284/C5
Mingo (co.), W. Va. 312/B7
Mingo, W. Va. (26281) 312/E3
Mingo Junction, Ohio (43938) 284/J5
Mingshui, Gansu, China 77/E3
Mingshui, Heilongjiang, China 77/L2
Mingulay (isl.), Scotland 15/A4
Mingus, Texas (76453) 303/F5
Mingwal (lake), W. Australia 88/C5
Minho (riv.), Jamaica 158/J6
Minho (riv.), Portugal 33/B2
Minho (riv.), Portugal 33/B4
Minicoy (isl.), India 68/C7
Minidoka (co.), Idaho 220/E7
Minidoka, Idaho (83343) 220/F7
Minier, Ill. (61759) 222/D3
Minigwal (lake), W. Australia 92/C5
Minilya, W. Australia 92/A4
Miniota, Manitoba 179/B4
Minipi (lake), Newf. 166/B3
Minipi (riv.), Newf. 166/B3
Ministikwan (lake), Sask. 181/B1
Ministre (pt.), St. Lucia 161/G7
Minitonas, Manitoba 179/B2
Min Jiang (riv.), China 77/J6
Minlaton, S. Australia 94/E6
Minle, China 77/F4
Minna, Nigeria 106/F7
Minneapolis, Kansas (67467) 232/E2
Minneapolis, Minn. 146/J5
Minneapolis, Minn. (*55401) 255/G5
Minneapolis, U.S. 2/E3
Minneapolis-Saint Paul Airport, Minn. 255/G5
Minnechaduza (creek), S. Dak. 298/H7
Minnedosa, Manitoba 179/B4
Minnedosa (riv.), Manitoba 179/B4
Minnehaha (co.), S. Dak. 298/R6
Minnehaha Springs, W. Va. (24960) 312/G6
Minneiska, Minn. (55958) 255/G5
Minneola, Fla. (32755) 212/E3
Minneola, Kansas (67865) 232/C4
Minneota, Minn. (56264) 255/C6
Minnesota 188/H1
MINNESOTA 255
Minnesota (riv.), Minn. 188/G2
Minnesota (riv.), Minn. 255/C6
Minnesota (riv.), S. Dak. 298/S3
Minnesota (state), U.S. 146/H3
Minnesota City, Minn. (55959) 255/G6
Minnesota Lake, Minn. (56068) 255/E7
Minnesott Beach, N.C. (128510) 281/R5
Minnetonka, Minn. (55343) 255/G5
Minnetonka (lake), Minn. 255/F5
Minnetrista, Minn. (†55364) 255/F5
Minnewanka (lake), Minn. 255/F5
Minnewaukan, N. Dak. (58351) 282/M3
Minnigaff, Scotland 15/D6
Minnipa, S. Australia 88/F6

Minnipa, S. Australia 94/D5
Minnitaki (lake), Ontario 177/G4
Minnith, Mo. (†63673) 261/M7
Minnora, W. Va. (25263) 312/D5
Miño (riv.), Spain 7/D4
Miño (riv.), Spain 33/B1
Minoa, N.Y. (13116) 276/H4
Minobu, Japan 81/J6
Minocqua, Wis. (54548) 317/G4
Minong, Wis. (54859) 317/C3
Minonk, Ill. (61760) 222/D3
Minoo, Japan 81/J7
Minooka, Ill. (60447) 222/E2
Minorca (isl.), Spain 7/E4
Minorca (isl.), Spain 33/J2
Minor Hill, Tenn. (38473) 237/G10
Minor Lane Heights, Ky. (†40201) 237/K4
Minortown, Conn. (†06798) 210/C2
Minot, Maine (04258) 243/C7
Minot○, Maine (04258) 243/C7
Minot, Mass. (02055) 249/F8
Minot, N. Dak. 188/F1
Minot, N. Dak. 146/H5
Minot, N. Dak. (58701) 282/H3
Minot A.F.B., N. Dak. 282/H3
Minotola, Ohio (†08341) 273/D4
Minqin, China 77/F5
Minsk, U.S.S.R. 7/G3
Minsk, U.S.S.R. 2/L3
Minsk, U.S.S.R. 48/C4
Minsk, U.S.S.R. 52/C4
Mińsk Mazowiecki, Poland 47/E2
Minster, Ohio (45865) 284/B5
Minstrel Island, Br. Col. 184/D5
Minter City, Miss. (38944) 256/D3
Mint Hill, N.C. (28212) 281/H4
Mintlaw, Scotland 15/F3
Minto, Alaska (99758) 196/J2
Minto, Manitoba 179/B5
Minto, New Bruns. 170/D2
Minto, N. Dak. (58261) 282/R3
Minto (inlet), N.W. Terrs. 187/G2
Minto (lake), Que. 162/J4
Minto (lake), Québec 174/E2
Minto, Yukon 187/E3
Minton, Sask. 181/G6
Minturn, Ark. (72445) 202/H2
Minturn, Colo. (81645) 208/G3
Minturn, Maine (04659) 243/G7
Minturn, S.C. (29573) 296/J2
Minturno, Italy 34/D4
Minût, Egypt 111/J3
Minusinsk, U.S.S.R. 48/K4
Minusio, Switzerland 39/G4
Minute Man Nat'l Hist. Park, Mass. 249/B6
Minvoul, Gabon 115/B3
Min Xian, China 77/F5
Minyip, Victoria 97/B5
Mio, Mich. (48647) 250/E4
Miocene, Br. Col. 184/G4
Miquan, China 77/C3
Miquelon (isl.), 166/C4
Miquihuana, Mexico 150/J5
Miquillo (pt.), P. Rico 161/F1
Mira (riv.), Colombia 126/A7
Mira (riv.), Ecuador 128/C2
Mira, La. (71059) 238/C1
Mira (bay), Nova Scotia 168/J2
Mira (riv.), Nova Scotia 168/H3
Mira, Portugal 33/B2
Mira (riv.), Portugal 33/B4
Mirabad, Afghanistan 68/A2
Mirabel, Québec 172/H4
Mirabile, Mo. (†64671) 261/D3
Miracema, Brazil 135/E2
Miracema, Brazil 132/F8
Mirador, Brazil 132/E4
Mirador Nacional (mt.), Uruguay 145/D5
Miraflores, Boyacá, Colombia 126/D5
Miraflores, Vaupés, Colombia 126/D7
Miragoâne, Haiti 158/B6
Miragoâne, Haiti 156/C6
Miraj, India 68/D5
Mira Loma, Calif. (91752) 204/E10
Miramar, Argentina 143/E4
Miramar, C. Rica 154/E5
Miramar, Fla. (33023) 212/B4
Miramar, Panama 154/H6
Miramichi (bay), New Bruns. 170/E1
Miram Shah, Pakistan 68/C2
Miramichi (riv.), Brazil 132/C8
Miranda, Brazil 132/B8
Miranda, Colombia 126/B6
Miranda (riv.), Brazil 132/B8
Miranda, S. Dak. (57463) 298/M4
Miranda (state), Venezuela 124/E2
Miranda de Ebro, Spain 33/E1
Miranda do Corvo, Portugal 33/B2
Miranda do Douro, Portugal 33/C2
Mirande, France 28/D6
Mirandela, Portugal 33/C2
Mirando City, Texas (78369) 303/E10
Mirandola, Italy 34/C2
Mira Por Vos (cays), Bahamas 156/C2
Mira Road, Nova Scotia 168/H2
Mirassol, Brazil 135/B2
Mira Taglio, Italy 34/D2
Mirebalais, Haiti 158/C6
Mirecourt, France 28/G3
Mirgorod, U.S.S.R. 52/D5
Miri (hills), India 68/G3
Miri, Malaysia 85/E5
Mirik (Timiris) (cape), Mauritania 106/A5
Mirim (lake) 120/D6
Mirim (lag.), Brazil 132/C11
Mirimire, Venezuela 124/D2
Mírina, Greece 45/G6
Miritiparaná (riv.), Colombia 126/E8
Mirjaveh, Iran 59/H4
Mirjaveh, Iran 66/M6

Mirjaveh, Pakistan 59/H4
Mirnyy 5/C5
Mirnyy, U.S.S.R. 54/N3
Mirnyy, U.S.S.R. 48/M3
Mirpur, Pakistan 68/C2
Mirpur Khas, Pakistan 68/B3
Mirror, Alberta 182/D3
Mirror Lake, N.H. (03853) 268/E4
Mirtóön (sea), Greece 45/F7
Miryang, S. Korea 81/D6
Mirzapur-cum-Vindhyachal, India 68/E4
Misamis Occidental (prov.), Philippines 82/D6
Misamis Oriental (prov.), Philippines 82/E6
Misantla, Mexico 150/P1
Misawa, Japan 81/K3
Miscou (isl.), New Bruns. 170/F1
Miscou (pt.), New Bruns. 170/F1
Miscou Centre, New Bruns. 170/F1
Miscouche, Pr. Edward I. 168/D2
Miscou Harbour, New Bruns. 170/F1
Misenheimer, N.C. (28109) 281/J4
Misery (bay), Mich. 250/G1
Misery (riv.), Mich. 250/G1
Misery (mt.), St. Chris.-Nevis 161/C10
Misgar, Pakistan 68/C1
Misha'ab, Ras (cape), Saudi Arabia 59/E4
Mishagua, Peru 128/F8
Mishan, China 77/M2
Mishaum (pt.), Mass. 249/L6
Mishawaka, Ind. (46544) 227/E1
Misheguk (mt.), Alaska 196/F1
Mishicot, Wis. (54228) 317/L7
Mishmar Hanegev, Israel 65/B5
Mishmar Hayarden, Israel 65/D1
Mishmi (hills), India 68/H3
Misima (isl.), Papua N.G. 85/C8
Misiones (prov.), Argentina 143/F2
Misiones (dept.), Paraguay 144/D5
Miskitos (cays), Nicaragua 154/F3
Miskolc, Hungary 41/F2
Miskolc, Hungary 7/G4
Misool (isl.), Indonesia 85/J6
Mispec, New Bruns. 170/E3
Mispillion (riv.), Del. 245/S5
Misquah (hills), Minn. 255/F2
Missanabie, Ontario 177/J5
Missanabie, Ontario 175/D3
Missaukee (co.), Mich. 250/D4
Missi Falls, Manitoba 179/J2
Missinaibi (riv.), Ont. 162/H6
Missinaibi (lake), Ontario 175/D3
Missinaibi (riv.), Ontario 177/J5
Missinaibi (riv.), Ontario 175/D2
Mission, Br. Col. 184/L3
Mission, Kansas (66205) 232/H2
Mission (range), Mont. 262/C3
Mission, S. Dak. (57555) 298/H7
Mission, Texas (78572) 303/F11
Mission Beach, Alberta 182/C3
Mission City, Br. Col. 184/L3
Mission Hill, S. Dak. (57046) 298/P8
Mission Ridge, S. Dak. (57557) 298/H4
Mission Viejo, Calif. (92691) 204/D11
Missisa (lake), Ontario 175/D2
Missisquoi (co.), Québec 172/D4
Missisquoi (riv.), Vt. 268/B2
Mississagi (riv.), Ontario 177/A1
Mississagi (str.), Ontario 177/A2
Mississauga, Ontario 177/J4
Mississinewa (lake), Ind. 227/F3
Mississinewa (riv.), Ind. 227/F3
Mississippi 188/J4
MISSISSIPPI 256
Mississippi (riv.) 188/H4
Mississippi (sound), Ala. 195/B10
Mississippi (co.), Ark. 202/K2
Mississippi (riv.), Ark. 202/H7
Mississippi (riv.), Ill. 222/C5
Mississippi (riv.), Iowa 229/L7
Mississippi (riv.), Ky. 237/A10
Mississippi (delta), La. 146/K7
Mississippi (delta), La. 188/J5
Mississippi (delta), La. 238/M8
Mississippi (riv.), La. 238/H3
Mississippi (sound), La. 238/M6
Mississippi (riv.), Minn. 255/D4
Mississippi (riv.), Miss. 256/A8
Mississippi (sound), Miss. 256/G10
Mississippi (co.), Mo. 261/O9
Mississippi (riv.), Mo. 261/L4
Mississippi (lake), Ontario 177/H2
Mississippi (riv.), Tenn. 237/A10
Mississippi (state), U.S. 146/K6
Mississippi (riv.), U.S. 2/E4
Mississippi (riv.), U.S. 146/J6
Mississippi (riv.), Wis. 317/D10
Mississippi River Gulf Outlet (canal), La. 238/L7
Mississippi State, Miss. (39762) 256/G4
Missoula, Mont. 146/G5
Missoula, Mont. 188/D1
Missoula (co.), Mont. 262/C3
Missoula, Mont. (*59801) 262/C4
Missouri 188/H3
MISSOURI 261
Missouri (riv.) 188/H3
Missouri (riv.), Iowa 229/A4
Missouri (riv.), Kansas 232/G1
Missouri (riv.), Mo. 261/H5
Missouri (riv.), Mont. 262/L3
Missouri (riv.), Nebr. 264/H3
Missouri (riv.), N. Dak. 282/H5
Missouri (riv.), S. Dak. 298/P8
Missouri (state), U.S. 146/J6
Missouri (riv.), U.S. 2/D3
Missouri (riv.), U.S. 146/J5
Missouri Branch, W. Va. (†25511) 312/A7
Missouri City, Mo. (64072) 261/R5
Missouri City, Texas (77459) 303/J2
Missouri Coteau (hills), Sask. 181/F5
Missouri Valley, Iowa (51555) 229/B5
Mist, Ark. (†71646) 202/G7
Mist, Oreg. (97016) 291/D1
Mistake (bay), N.W. Terrs. 187/J3
Mistake Creek, North. Terr. 93/A4
Mistaken (pt.), Newf. 166/D2
Mistassibi (riv.), Que. 162/J5
Mistassibi (riv.), Québec 174/C3
Mistassini (lake), Que. 162/J5
Mistassini (lake), Que. 146/L4
Mistassini (terr.), Québec 174/B2
Mistassini, Québec 172/E1
Mistassini, Québec 174/C3
Mistassini (Baie-du-Poste), Québec 174/C2
Mistassini (lake), Québec 174/C2
Mistastin (lake), Newf. 166/B2
Mistastin (riv.), Newf. 166/B2
Mistatim, Sask. 181/H3
Mistehae (lake), Alberta 182/C2
Mistelbach an der Zaya, Austria 41/D2
Misteriosa (bank), Cayman Is. 156/A3
Misti, El (mt.), Peru 120/B4
Misti, El (mt.), Peru 128/G11
Mistinippi (lake), Newf. 166/B3
Miston, Tenn. (38056) 237/B8
Mistretta, Italy 34/E6
Misty Fjords Nat'l Mon., Alaska 196/N2
Misurata, Libya 102/D1
Misurata, Libya 111/C1
Mita (pt.), Mexico 150/G6
Mitaka, Japan 81/O2
Mitcham, S. Australia 88/D8
Mitcham, S. Australia 94/B8
Mitchell (lake), Ala. 188/J4
Mitchell (dam), Ala. 195/E5
Mitchell (lake), Ala. 195/F5
Mitchell, Ark. (†72583) 202/G1
Mitchell (co.), Georgia 217/D8
Mitchell, Georgia (30820) 217/G4
Mitchell, Ind. (47446) 227/E7
Mitchell (co.), Iowa 229/H2
Mitchell, Iowa (†50461) 229/H2
Mitchell (co.), Kansas 232/D2
Mitchell, La. (71453) 238/C3
Mitchell, Nebr. (69357) 264/A3
Mitchell (co.), N.C. 281/E2
Mitchell (mt.), N.C. 188/K3
Mitchell (mt.), N.C. 281/E3
Mitchell, Ontario 177/C4
Mitchell, Oreg. (97750) 291/G3
Mitchell, Queensland 88/H5
Mitchell, Queensland 95/C5
Mitchell (riv.), Queensland 88/G3
Mitchell (riv.), Queensland 95/B2
Mitchell, S. Dak. 188/G2
Mitchell, S. Dak. (57301) 298/N6
Mitchell (creek), S. Dak. 298/G5
Mitchell (co.), Texas 303/D5
Mitchell (riv.), Victoria 97/D5
Mitchell Bay, Ontario 177/B5
Mitchell Heights, W. Va. (†25601) 312/B7
Mitchells, Va. (22729) 307/N4
Mitchellsburg, Ky. (40452) 237/M5
Mitchellsville, Ill. (†62946) 222/E6
Mitchellton, Sask. 181/F6
Mitchellville, Ark. (†71639) 202/H6
Mitchellville, Iowa (50169) 229/G5
Mitchellville, Tenn. (37119) 237/J7
Mitchelstown, Ireland 10/B4
Mitchelstown, Ireland 17/E7
Mitchelton, Queensland 88/J2
Mitchelton, Queensland 95/D2
Mitchinamécus (res.), Québec 172/C2
Mithi, Pakistan 68/C4
Mithimna, Greece 45/G6
Mitiaro, Cook Is. 87/L7
Mitilíni, Greece 45/H6
Mitkof (isl.), Alaska 196/N2
Mitla (ruin), Mexico 150/M8
Mito, Japan 81/K5
Mitrofania (isl.), Alaska 196/G3
Mitsamiouli, Comoros 118/G2
Mitsinjo, Madagascar 118/H3
Mitsue, Alberta 182/C2
Mitsukaido, Japan 81/P2
Mittagong, N.S. Wales 97/F4
Mitta Mitta (riv.), Victoria 97/D5
Mittenwald, W. Germany 22/D5
Mittersill, Austria 41/B3
Mittie, La. (70654) 238/E5
Mittweida, E. Germany 22/E3
Mitú, Colombia 126/E7
Mitú, Colombia 120/B2
Mituas, Colombia 126/F6
Mitwaba, Zaire 115/E5
Mitzic, Gabon 115/B3
Miura, Japan 81/O3
Miura (pen.), Japan 81/O3
Mivtahim, Israel 65/A5
Mix, La. (†70760) 238/G5
Miyagi (pref.), Japan 81/K4
Miyako, Japan 81/L4
Miyako (isl.), Japan 81/L7
Miyako (isls.), Japan 81/L7
Miyakonojo, Japan 81/E8
Miyazaki (pref.), Japan 81/E8
Miyazaki, Japan 81/E8
Miyazu, Japan 81/G6
Miyoshi, Japan 81/F6
Mizan Teferi, Ethiopia 111/G6
Mizda, Libya 111/B1
Mize, Georgia (†30577) 217/F2
Mize, Miss. (39116) 256/E7
Mizen (head), Ireland 10/A5
Mizen (head), Ireland 17/B9
Mizen (head), Ireland 17/K6
Mizhi, China 77/H4
Mizil, Romania 45/H3
Mizo (hill), India 68/G4
Mizoram (terr.), India 68/G4
Mizpah, Minn. (56660) 255/D3
Mizpah, N.J. (08342) 273/D5
Mizpe Ramon, Israel 65/D5
Mizque, Bolivia 136/C5
Mizque (riv.), Bolivia 136/C6
Mizusawa, Japan 81/K4
Mjölby, Sweden 18/J7
Mkokotoni, Tanzania 115/G5
Mkushi, Zambia 115/E6
Mladá Boleslav, Czech. 41/C1
Mladá Vožice, Czech. 41/C2
Mława, Poland 47/E2
Mljet (isl.), Yugoslavia 45/C4
Mmabatho (cap.), Bophuthatswana, S. Africa 102/E7
Mmabatho, S. Africa 118/D5
Mnichovo Hradiště, Czech. 41/C1
Mo, Norway 7/F2
Mo, Norway 18/J3
Moa, Cuba 158/K3
Moa (riv.), Guinea 106/B7
Moa (isl.), Indonesia 85/H7
Moa (riv.), S. Leone 106/B7
Moab, Utah (84532) 304/E5
Moak Lake, Manitoba 179/J2
Moala (isl.), Fiji 86/Q11
Moama, N.S. Wales 97/C5
Moamba, Mozambique 118/E5
Moanalua (stream), Hawaii 218/B3
Moanda, Gabon 115/B4
Moanda, Zaire 115/B5
Moapa, Nev. (89025) 266/G6
Moapa River Ind. Res., Nev. 266/G6
Moar (lake), Manitoba 179/G2
Moark, Ark. (†72422) 202/J1
Moate, Ireland 17/F5
Moatsville, W. Va. (26405) 312/G4
Mobara, Japan 81/K6
Mobaye, Cent. Afr. Rep. 115/D3
Mobayi-Mbongo, Zaire 115/D3
Mobayi-Mbongo, Zaire 102/E4
Mobeetie, Texas (79061) 303/D2
Moberly, Br. Col. 184/J4
Moberly (lake), Br. Col. 184/F2
Moberly, Mo. (65270) 261/G4
Moberly, Mo. 188/H3
Moberly Lake, Br. Col. 184/G2
Mobile, Ala. 146/K6
Mobile, Ala. 188/J4
Mobile (bay), Ala. 188/J5
Mobile (co.), Ala. 195/B9
Mobile, Ala. (*36601) 195/B9
Mobile (bay), Ala. 195/B10
Mobile (pt.), Ala. 195/B10
Mobile (riv.), Ala. 195/C9
Mobile, Ariz. (†85239) 198/C5
Mobile, Newf. 166/D2
Mobile Big (pond), Newf. 166/D2
Mobjack, Va. (23118) 307/R6
Mobjack (bay), Va. 307/R6
Mobridge, S. Dak. (57601) 298/J2
Mobuto Sese Seko (lake) 102/F4
Mobuto Sese Seko (lake), Uganda 115/F3
Mobuto Sese Seko (lake), Zaire 115/F3
Moca, Dom. Rep. 156/D3
Moca, Dom. Rep. 158/D5
Moca, P. Rico 161/A1
Mocajuba, Brazil 132/D3
Moçambique, Mozambique 118/G3
Moçambique, Mozambique 102/F6
Moçâmedes (dist.), Angola 115/B7
Moçâmedes, Angola 102/D6
Moçâmedes, Angola 115/B7
Mocanaqua, Pa. (18655) 294/K3
Moccasin, Ariz. (†86022) 198/C2
Moccasin, Mont. (59462) 262/F3
Mocha (isl.), Chile 138/B2
Mocha, Yemen Arab Rep. 59/D7
Moc Hoa, Vietnam 72/E5
Mochudi, Botswana 118/D4
Mochudi, Botswana 102/E7
Mocímboa da Praia, Mozambique 118/G2
Mociu, Romania 45/G2
Mocksville, N.C. (27028) 281/H3
Moclips, Wash. (98562) 310/A3
Moco (mt.), Angola 115/C6
Mocoa, Colombia 126/B7
Mococa, Brazil 135/C2
Mocodome (cape), Nova Scotia 168/G3
Mocomoco, Bolivia 136/A4
Mocoretá, Argentina 143/G5
Mocorito, Mexico 150/F4
Moctezuma, San Luis Potosí, Mexico 150/J5
Moctezuma, Sonora, Mexico 150/E2
Moctezuma (riv.), Mexico 150/K6
Mocuba, Mozambique 118/F3
Modale, Iowa (51556) 229/B5
Modane, France 28/G5
Modasa, India 68/C4
Modderfontein, S. Africa 118/H6
Mode, Ill. (62444) 222/E4
Model, Colo. (81059) 208/L8
Modena (prov.), Italy 34/C2
Modena, Italy 7/F4
Modena, Italy 34/C2
Modena, Utah (84753) 304/A6
Modena, Wis. (†54755) 317/C7
Modeste, La. (70376) 238/K3
Modesto, Calif. 188/B3
Modesto, Calif. (*95350) 204/D6
Modesto, Ill. (62667) 222/D4
Modest Town, Va. (23412) 307/T5
Modica, Italy 34/E6
Mödling, Austria 41/D2
Modoc (co.), Calif. 204/E2
Modoc, Georgia (†30401) 217/H5
Modoc, Ill. (62261) 222/C5
Modoc, Ind. (47358) 227/G4
Modoc, Kansas (67866) 232/A3
Modoc, S.C. (29838) 296/C4
Modoc Point, Oreg. (†97624) 291/F5
Modra, Czech. 41/D2
Modriča, Yugoslavia 45/D3
Modrý Kameň, Czech. 41/E2
Mo Duc, Vietnam 72/F4
Moe, Victoria 88/H7
Moe, Victoria 97/D6
Moen (isl.), Micronesia 87/F5
Moencopi (plat.), Ariz. 198/D3
Moengo, Suriname 131/D3
Moenkopi, Ariz. (†86045) 198/D2
Moenkopi Wash (dry riv.), Ariz. 198/D2
Moerai, Fr. Poly. 87/L8
Moerdijk, Netherlands 27/F5
Moerewa, N. Zealand 100/E1
Moësa (riv.), Switzerland 39/H4
Moeskroen (Mouscron), Belgium 27/C7
Moffat (co.), Colo. 208/C1
Moffat, Colo. (81143) 208/H6
Moffat, Scotland 15/E5
Moffat, Scotland 10/E3
Moffet (peak), N. Zealand 100/B6
Moffett, Okla. (74946) 288/S4
Moffett Nav. Air Sta., Calif. 204/K3
Moffit, N. Dak. (58560) 282/K6
Mogadiscio (prov.), Somalia 115/J3
Mogadishu (cap.), Somalia 2/M5
Mogadishu (cap.), Somalia 102/G4
Mogadishu (cap.), Somalia 115/J3
Mogador (Essaouira), Morocco 106/B2
Mogadore, Ohio (44260) 284/H3
Mogadouro, Portugal 33/C2
Mogami (riv.), Japan 81/K4
Mogaung, Burma 72/C1
Møgeltønder, Denmark 21/B8
Mogi das Cruzes, Brazil 132/E9
Mogi das Cruzes, Brazil 135/C3
Mogi Guaçu (riv.), Brazil 135/C2
Mogi-Guaçu, Brazil 135/C3
Mogilev, U.S.S.R. 7/G3
Mogilev, U.S.S.R. 52/C4
Mogilev, U.S.S.R. 48/D4
Mogilev-Podol'skiy, U.S.S.R. 52/C5
Mogil Mogil, N.S. Wales 97/E1
Mogilno, Poland 47/C2
Mogi-Mirim, Brazil 135/C3
Mogincual, Mozambique 118/G3
Mogocha, U.S.S.R. 48/N4
Mogok, Burma 72/C2
Mogollon (plat.), Ariz. 198/D4
Mogollon, N. Mex. (†88039) 274/A5
Mogollon (mts.), N. Mex. 274/A5
Mogollon Baldy (peak), N. Mex. 274/A5
Mogollon Rim (cliffs), Ariz. 198/D4
Mogororo, Chad 111/D5
Mogotes (pt.), Argentina 143/E4
Moguer, Spain 33/C4
Mohács, Hungary 41/E4
Mohaka (riv.), N. Zealand 100/F3
Mohaleshoek, Lesotho 118/D6
Mohall, N. Dak. (58761) 282/G2
Mohammadia, Algeria 106/D1
Mohammedia, Morocco 106/C2
Mohave (co.), Ariz. 198/A3
Mohave (lake), Ariz. 198/A3
Mohave (mts.), Ariz. 198/A4
Mohave (lake), Nev. 266/G7
Mohawk (mts.), Ariz. 198/B6
Mohawk (mt.), Conn. 210/B1
Mohawk, Ind. (†46140) 227/F5
Mohawk, Mich. (49950) 250/A1
Mohawk (riv.), N.H. 268/E2
Mohawk (lake), N.J. 273/D1
Mohawk, N.Y. (13407) 276/L4
Mohawk (riv.), N.Y. 276/L5
Mohawk, Oreg. (†97477) 291/E3
Mohawk, Tenn. (37810) 237/P8
Mohawk, W. Va. (24862) 312/C7
Mohe, China 77/K1
Mohegan, Conn. (†06382) 210/G3
Mohéli (isl.), Comoros 102/G6
Mohéli (isl.), Comoros 118/G2
Mohelnice, Czech. 41/D2
Mohenjo Daro (ruins), Pakistan 68/B3
Moher (cliffs), Ireland 17/B6
Mohican (cape), Alaska 196/E2
Mohican (riv.), Ohio 284/F4
Mohill, Ireland 17/F4
Mohler, Wash. (99154) 310/G3
Möhlin, Switzerland 39/E1
Mohn, Kapp (cape), Norway 18/E1
Mohnton, Pa. (19540) 294/L5
Mohnyin, Burma 72/C1
Moho, Peru 128/H10
Mohoro, Tanzania 115/G5
Mohrsville, Pa. (19541) 294/K5
Moi, Norway 18/E7
Moidart (dist.), Scotland 15/C4
Moiese, Mont. (59824) 262/B3
Moiliili, Hawaii (96828) 218/C4
Moinești, Romania 45/H2
Moingona, Iowa (†50036) 229/F4
Moira, N.Y. (12957) 276/M1
Moîrai, Greece 45/G8
Moirones, Uruguay 145/E2
Mõisaküla, U.S.S.R. 53/C1
Moise (riv.), Que. 146/M4
Moisés Ville, Argentina 143/E5
Moisie (riv.), Que. 162/K5
Moisie, Québec 174/D2
Moisie (riv.), Québec 174/D2
Moissac, France 28/D5
Moïssala, Chad 111/C6
Moitaco, Venezuela 124/F4
Mojácar, Spain 33/F4
Mojave, Calif. (93501) 204/G8
Mojave (des.), Calif. 204/H9
Mojave (riv.), Calif. 204/J9
Mojo, Bolivia 136/C7
Mojocoya, Bolivia 136/C6
Mojokerto, Indonesia 85/K2
Mokane, Mo. (65059) 261/J5
Mokapu, Hawaii (†96734) 218/F2
Mokapu (pen.), Hawaii 218/F2
Mokau (riv.), N. Zealand 100/E3
Mokelumne (riv.), Calif. 204/C9
Mokelumne Hill, Calif. (95245) 204/E5
Mokena, Ill. (60448) 222/B6
Mokil (atoll), Micronesia 87/G5
Moknine, Tunisia 106/G1
Mokohinau (isl.), N. Zealand 100/E1
Mokokchung, India 68/G3
Mokolo, Cameroon 115/B1
Mokp'o, S. Korea 81/C6
Moksha (riv.), U.S.S.R. 52/F4
Mokuauia (isl.), Hawaii 218/E1
Mokuaweoweo (crater), Hawaii 218/H6
Mokuhooniki (isl.), Hawaii 218/J1
Mokuleia, Hawaii (†96791) 218/D1
Mol, Belgium 27/G6
Mola di Bari, Italy 34/F4
Molalla, Oreg. (97038) 291/B3
Molalla (riv.), Oreg. 291/B3
Moland, Minn. (†55946) 255/E6
Molanosa, Sask. 181/M4
Moláoi, Greece 45/F7
Molare (peak), Switzerland 39/G3
Mold, Wales 13/G2
Moldau (Vltava) (riv.), Czech. 41/C2
Moldava nad Bodvou, Czech. 41/F2
Moldavian S.S.R., U.S.S.R. 7/G4
Moldavian S.S.R., U.S.S.R. 52/C5
Moldavian S.S.R., U.S.S.R. 48/C5
Molde, Norway 18/E5
Moldova Nouă, Romania 45/E3
Moldoveanul (mt.), Romania 45/G3
Mole (riv.), England 13/H8
Môle (cape), Haiti 158/B5
Mole Creek, Tasmania 99/C3
Molega (lake), Nova Scotia 168/D4
Molena, Georgia (30258) 217/D4
Molenbeek-Saint-Jean, Belgium 27/B9
Molepolole, Botswana 118/D4
Molepolole, Botswana 102/E7
Môle Saint Nicolas, Haiti 158/B5
Molfetta, Italy 34/F4
Molina, Chile 138/A10
Molina, Colo. (81646) 208/D4
Molina, Spain 33/F2
Molinas, Argentina 143/C2
Moline, Ill. 188/J2
Moline, Ill. (61265) 222/C2
Moline, Kansas (67353) 232/F4
Moline, Manitoba 179/B4
Moline, Mich. (49335) 250/D6
Moline Acres, Mo. (†63101) 261/R2
Molinicos, Spain 33/E3
Molinière (pt.), Grenada 161/C8
Molino, Fla. (32577) 212/B6
Molinos (pt.), P. Rico 161/G1
Moliro, Zaire 115/E5
Molise (reg.), Italy 34/E4
Mollebjerg (mt.), Denmark 21/C6
Mollendo, Peru 120/B4
Mollendo, Peru 128/F11
Mollerusa, Spain 33/G2
Molles (pt.), Chile 138/A9
Mollis, Switzerland 39/H2
Mölln, W. Germany 22/D2
Mollusk, Va. (22517) 307/P5
Mollys Falls (pond), Vt. 268/C3
Mölndal, Sweden 18/H8
Moloaa, Hawaii (†96703) 218/D1
Molodechno, U.S.S.R. 48/C4
Molodechno, U.S.S.R. 52/C4
Molokai (isl.), Hawaii 87/L3
Molokai (isl.), Hawaii 188/F5
Molokai (isl.), Hawaii 218/G1
Molokini (isl.), Hawaii 218/J2
Molong, N.S. Wales 97/E3
Molopo (riv.), Botswana 118/C5
Molopo (riv.), S. Africa 118/C5
Molotov (Perm'), U.S.S.R. 52/J3
Moloundou, Cameroon 115/C3
Molson (lake), Manitoba 179/J3
Molson, Wash. (†98844) 310/F2
Molt, Mont. (59057) 262/H5
Molteno, S. Africa 118/D6
Molucca (isls.), Indonesia 54/O10
Molucca (sea), Indonesia 54/O10
Molucca (sea), Indonesia 85/H6
Moluccas (isls.), Indonesia 85/H6
Molunkus (lake), Maine 243/G4
Moma, Mozambique 118/F3
Mombasa, Kenya 115/G4
Mombasa, Kenya 102/G5
Mombetsu, Japan 81/L1
Mombo, Tanzania 115/G4
Momchilgrad, Bulgaria 45/G5
Momence, Ill. (60954) 222/F2
Momeyer, N.C. (†27856) 281/N3
Momignies, Belgium 27/E8
Momostenango, Guatemala 154/B3
Mompog (passage), Philippines 82/D4
Mompós, Colombia 126/C3
Mon (state), Burma 72/C3
Mon (riv.), Burma 72/B2
Møn (isl.), Denmark 21/F8
Møn (isl.), Denmark 18/H9
Mona (passg.) 146/M8
Mona, Cyprus 63/E5
Mona (passage), Dom. Rep. 156/E3
Mona (passage), Dom. Rep. 158/F6
Mona (isl.), P. Rico 156/E3
Mona (passage), P. Rico 156/E3
Mona (passage), P. Rico 161/A2
Mona, Utah (84645) 304/C4
Mona (res.), Utah 304/C4
Monaca, Pa. (15061) 294/B4
Monach (isls.), Scotland 15/A3
Monach (sound), Scotland 15/A3
Monaco 7/E4
MONACO 28/G6
Monadhliath (mts.), Scotland 15/D3
Monadnock (mt.), N.H. 268/C6
Monagas (state), Venezuela 124/G3
Monaghan (co.), Ireland 17/H3
Monaghan, Ireland 10/C3
Monaghan, Ireland 17/G3
Monahans, Texas (79756) 303/B6
Monango, N. Dak. (58471) 282/N7
Monapo, Mozambique 118/G2
Monar, Loch (lake), Scotland 15/C3
Monarch, Alberta 182/D5
Monarch, Mont. (59463) 262/F3
Monarch Mills, S.C. (†29379) 296/D2
Monarda, Maine (†04776) 243/G4
Monaro (range), N.S. Wales 97/E5
Monashee (mts.), Br. Col. 184/H4
Monasterevan, Ireland 17/H5
Monastery, Nova Scotia 168/G3
Monastir, Tunisia 106/G1
Monatélé, Cameroon 115/B3
Mona Vale, N. S. Wales 88/L3
Mona Vale, N.S. Wales 97/K2
Monaville, W. Va. (25636) 312/B7
Monavullagh (mts.), Ireland 17/F7
Monbetsu, Japan 81/L2
Moncalieri, Italy 34/A2
Monçao, Portugal 33/B1
Moncayo (mt.), Spain 33/F2
Moncayo, Sierra de (range), Spain 33/F2
Monchegorsk, U.S.S.R. 7/H2
Monchegorsk, U.S.S.R. 48/C3
Monchegorsk, U.S.S.R. 52/D1
Mönchengladbach, W. Germany 22/B3
Monches, Wis. (†53029) 317/J1
Monchique, Portugal 33/B4
Monchique, Serra de (mts.), Portugal 33/B4
Monción, Dom. Rep. 158/D5
Moncks Corner, S.C. (29461) 296/G5
Monclo, W. Va. (†25183) 312/C7
Monclova, Mexico 146/H7
Monclova, Mexico 150/J3
Monclova, Ohio (43542) 284/C2
Moncouche (lake), Québec 172/G1
Moncton, N. Br. 146/M5
Moncton, N. Br. 162/K6
Moncton, New Bruns. 170/F2
Moncure, N.C. (27559) 281/L3
Mondamin, Iowa (51557) 229/B5
Monday (riv.), Paraguay 144/E4
Mondego (cape), Portugal 33/B2
Mondego (riv.), Portugal 33/B2
Mondéjar, Spain 33/E2
Mondonac (lake), Québec 172/D2
Mondoñedo, Spain 33/C1
Mondovi, Wis. (54755) 317/C6
Mondovì Breo, Italy 34/A2
Mondragon, Philippines 85/H3
Mondragon, Philippines 82/E4
Mondsee, Austria 41/B3
Moneague, Jamaica 158/J6
Monee, Ill. (60449) 222/F2
Monero, N. Mex. (†87547) 274/C2
Monessen, Pa. (15062) 294/C5
Monesterio, Spain 33/C3
Moneta, Iowa (51352) 229/C2
Moneta, Va. (24121) 307/J6
Moneta, Wyo. (†82601) 319/E2
Monett, Mo. (65708) 261/E9
Monetta, S.C. (29105) 296/D4
Monette, Ark. (72447) 202/K2
Money (isl.), China 85/E2
Money, Miss. (38945) 256/D3
Moneygall, Ireland 17/F6
Moneymore, N. Ireland 17/H2
Monfalcone, Italy 34/D2
Monforte, Portugal 33/C3
Monforte, Spain 33/C1
Monga, Zaire 115/D3
Mongalla, Sudan 111/F6
Mong Cai, Vietnam 72/E2
Möng Hsat, Burma 72/C2
Monghyr, India 68/F3
Möng Maü, Burma 72/C3
Möng Mit, Burma 72/C2
Mongo, Chad 111/C5
Mongo, Chad 102/D3
Mongo, Ind. (46771) 227/G1
Mongolia 2/P3
Mongolia 54/M5
Mongoumba, Cent. Afr. Rep. 115/C3
Möng Pawn, Burma 72/C2
Möng Si, Burma 72/C2
Möng Tön, Burma 72/C2
Möng Tung, Burma 72/C2
Mongu, Zambia 102/E6
Mongu, Zambia 115/D7
Monhegan○, Maine (04852) 243/E8
Monhegan (isl.), Maine 243/E8
Mönhhaan, Mongolia 77/H2
Moniac, Georgia (†31646) 217/H9
Moniaive, Scotland 10/D3
Moniaive, Scotland 15/E5
Monica, Ill. (†61559) 222/D3
Monico, Wis. (54549) 317/H4
Monida, Mont. (†59739) 262/D6
Monie, Md. (†21853) 245/P8
Monifieth, Scotland 15/F4
Moniquirá, Colombia 126/D5
Moniteau (co.), Mo. 261/G5
Monitor, Alberta 182/E4
Monitor, Ind. (†47901) 227/D4
Monitor (range), Nev. 266/E4
Monitor, Oreg. (†97072) 291/B3
Monitor, Wash. (98836) 310/E3
Monivea, Ireland 17/D5
Monkayo, Philippines 82/E7
Monkey (pt.), Nicaragua 154/F5
Monkey (hill), St. Chris.-Nevis 161/C10
Monkey River Town, Belize 154/C2
Mońki, Poland 47/F2
Monkoto, Zaire 115/D4
Monkton, Md. (21111) 245/M2
Monkton, Ontario 177/C4
Monkton○, Vt. (05469) 268/A3
Monkton Ridge, Vt. (†05473) 268/A3
Monmouth, Ill. (61462) 222/C3
Monmouth, Ind. (†46733) 227/H3
Monmouth, Iowa (52309) 229/M4
Monmouth, Maine (04259) 243/D7
Monmouth○, Maine (04259) 243/D7
Monmouth (co.), N.J. 273/E3
Monmouth, Oreg. (97361) 291/D3
Monmouth, Wales 13/E6
Monmouth, Wales 10/E5
Monmouth Beach, N.J. (07750) 273/F3
Monmouth Junction, N.J. (08852) 273/D3
Monnickendam, Netherlands 27/C4
Mono (riv.), Benin 106/E7
Mono (lake), Calif. 188/C3
Mono (co.), Calif. 204/F5
Mono (lake), Calif. 204/G5
Mono (riv.), Togo 106/E7
Monocacy (riv.), Md. 245/J3
Monocacy Nat'l Battlefield, Md. 245/J3
Mono Lake, Calif. (†93541) 204/F5
Monolith, Calif. (†93561) 204/G8
Monólithos, Greece 45/H7
Monomonac (lake), Mass. 249/G2
Monomoy (isl.), Mass. 249/O6

Monomoy (pt.), Mass. 249/O6
Monon, Ind. (47959) 227/D3
Monona (co.), Iowa 229/B4
Monona, Iowa (52159) 229/L2
Monona, Wis. (53716) 317/H9
Monongah, W. Va. (26554) 312/F4
Monongahela, Pa. (15063) 294/B5
Monongahela (riv.), Pa. 294/C5
Monongahela (riv.), W. Va. 312/G3
Monongalia (co.), W. Va. 312/F3
Monopoli, Italy 34/F4
Monor, Hungary 41/E3
Monóvar, Spain 33/F3
Monos (isl.), Trin. & Tob. 161/A10
Monreal del Campo, Spain 33/F2
Monreale, Italy 34/D5
Monroe (co.), Ala. 195/D7
Monroe (co.), Ark. 202/H4
Monroe, Ark. (72108) 202/H4
Monroe (co.), Conn. (06468) 210/C3
Monroe (co.), Fla. 212/E7
Monroe (lake), Fla. 212/E3
Monroe (co.), Georgia 217/E4
Monroe, Georgia (30655) 217/E3
Monroe (co.), Ill. 222/C5
Monroe (co.), Ind. 227/D6
Monroe, Ind. (46772) 227/H3
Monroe (lake), Ind. 227/E6
Monroe (co.), Iowa 229/H7
Monroe, Iowa (50170) 229/G5
Monroe (co.), Ky. 237/K7
Monroe, La. 188/H4
Monroe, La. 146/J6
Monroe, La. (*71201) 238/F1
Monroe○, Maine (04951) 243/E6
Monroe (co.), Mich. 250/F7
Monroe, Mich. (48161) 250/F7
Monroe (co.), Miss. 256/H3
Monroe (co.), Mo. 261/H3
Monroe, Nebr. (68647) 264/G3
Monroe○, N.H. (03771) 268/C3
Monroe (mt.), N.H. 268/E3
Monroe○, N.J. (07434) 273/E3
Monroe (co.), N.Y. 276/E4
Monroe, N.Y. (10950) 276/M8
Monroe, N.C. (28110) 281/J5
Monroe (co.), Ohio 284/H6
Monroe, Ohio (45050) 284/B7
Monroe, Okla. (74947) 288/S4
Monroe, Oreg. (97456) 291/B4
Monroe (co.), Pa. 294/M3
Monroe (Monroeton), Pa. (18832) 294/J2
Monroe, S. Dak. (57047) 298/P7
Monroe (co.), Tenn. 237/N10
Monroe, Tenn. (38573) 237/L8
Monroe, Utah (84754) 304/B5
Monroe (peak), Utah 304/B5
Monroe, Va. (24574) 307/K6
Monroe, Wash. (98272) 310/D3
Monroe (co.), W. Va. 312/E7
Monroe (co.), Wis. 317/F8
Monroe, Wis. (53566) 317/G10
Monroe Bridge, Mass. (01350) 249/C2
Monroe Center, Ill. (61052) 222/E1
Monroe City, Ind. (47557) 227/C7
Monroe City, Mo. (63456) 261/J3
Monroe P.O. (Stepney), Conn. (06468) 210/B3
Monroeton, Pa. (18832) 294/J2
Monroeville, Ala. (36460) 195/D7
Monroeville, Ind. (46773) 227/H3
Monroeville, N.J. (08343) 273/C4
Monroeville, Ohio (44847) 284/E3
Monroeville, Pa. (15146) 294/C7
Monrovia, Ala. (†35804) 195/E1
Monrovia, Calif. (91016) 204/D10
Monrovia, Ind. (46157) 227/E5
Monrovia (cap.), Liberia 106/B7
Monrovia (cap.), Liberia 2/J5
Monrovia (cap.), Liberia 102/A4
Monrovia, Md. (21770) 245/J3
Mons, Belgium 27/E8
Monsanto, Portugal 33/C2
Monschau, W. Germany 22/B3
Monse, Wash. (†98812) 310/F2
Monsefú, Peru 128/C6
Monselice, Italy 34/C2
Monserrate (isl.), Mexico 150/D4
Monsey, N.Y. (10952) 276/J8
Mons Klint (cliff), Denmark 21/F8
Monson○, Maine (04464) 243/E5
Monson, Mass. (01057) 249/E4
Monson○, Mass. (01057) 249/E4
Mönsterås, Sweden 18/K8
Montagu, S. Africa 116/B3
Montague (isl.), Alaska 196/D1
Montague (str.), Alaska 196/D1
Montague, Calif. (96064) 204/C2
Montague○, Mass. (01351) 249/E2
Montague (isl.), Mexico 150/B1
Montague, Mich. (49437) 250/C5
Montague, Mont. (†59442) 262/F3
Montague, N.J. (†07851) 273/D1
Montague (co.), Texas (†28435) 281/N6
Montague (co.), Texas 303/G4
Montague, Texas (76251) 303/G4
Montague (sound), W. Australia 88/C2
Montague (sound), W. Australia 92/D1
Montague City, Mass. (†01351) 249/E2
Montalba, Texas (75853) 303/J6
Montalbán, Spain 33/F2
Montalcino, Italy 34/C3
Mont Alto, Pa. (17237) 294/G6
Montalto Uffugo, Italy 34/E5
Montalvão, Portugal 33/C3
Montalvo, Calif. (93003) 204/F9
Montana 188/E1
MONTANA 262
Montana, Alaska (†99676) 196/B1
Montaña, La (reg.), Peru 128/F8
Montana, Switzerland 39/D4
Montana (state), U.S. 146/H5
Montánchez, Spain 33/D3

Montanja di Reij, Neth. Ant. 161/G9
Montara, Calif. (94037) 204/H3
Montargil, Portugal 33/B3
Montargis, France 28/E3
Montauban, France 28/D5
Montauban, France 7/E4
Montauban, Québec 172/E3
Montauk, N.Y. (11954) 276/S8
Montauk (pt.), N.Y. 276/S8
Montbard, France 28/F4
Montbéliard, France 28/G4
Mont Belvieu, Texas (77580) 303/L1
Montblanch, Spain 33/G2
Montbrison, France 28/E5
Montbrook, Fla. (†32696) 212/D2
Montcalm (co.), Mich. 250/D5
Montcalm (co.), Québec 172/C3
Montcalm (county), Québec 174/B3
Mont-Carmel, Québec 172/H2
Montceau-les-Mines, France 28/F4
Mont Cenis (tunnel), France 28/G5
Mont Cenis (tunnel), Italy 34/A2
Montcerf, Québec 172/A3
Montclair, Calif. (91763) 204/D10
Montclair, N.J. (*07042) 273/B2
Montclare, S.C. (†29532) 296/H3
Montcoal, W. Va. (25135) 312/D6
Mont-de-Marsan, France 28/C6
Montdidier, France 28/E3
Mont-Dore, France 28/E5
Monteagle, Tenn. (37356) 237/K10
Monteagudo, Bolivia 136/D6
Monte Alegre, Brazil 132/C3
Montealegre del Castillo, Spain 33/F3
Monte Alegre de Minas, Brazil 132/D7
Monte Aprazível, Brazil 135/A2
Monte Azul, Brazil 132/F6
Monte Bello (isls.), Australia 87/B8
Montebello, Calif. (90640) 204/C10
Montebello, Québec 172/B4
Monte Bello (isls.), W. Australia 88/A1
Monte Bello (isls.), W. Australia 92/A3
Montebelluna, Italy 34/D2
Monte Carlo, Monaco 28/G6
Monte Caseros, Argentina 143/G5
Montecito, Calif. (93103) 204/F9
Monte Comán, Argentina 143/C3
Monte Creek, Br. Col. 184/G3
Montecristi (prov.), Dom. Rep. 158/D5
Monte Cristi, Dom. Rep. 156/D2
Montecristi, Dom. Rep. 158/C5
Montecristi, Ecuador 128/B3
Monte Cristo, Bolivia 136/E4
Montecristo (isl.), Italy 34/C3
Monte Cristo (range), Nev. 266/D4
Monte Dourado, Brazil 132/C3
Montebasscone, Italy 34/D3
Montefrío, Spain 33/D4
Montego (bay), Jamaica 158/G5
Montego Bay, Jamaica 158/H5
Montego Bay, Jamaica 156/B3
Montego Bay (pt.), Jamaica 158/G5
Montegut, La. (70377) 238/J8
Montehermoso, Spain 33/C2
Monteiro, Brazil 132/G4
Monteith, Iowa (†50115) 229/D5
Montejinnie, North. Terr. 93/C4
Monte Lake, Br. Col. 184/G5
Montélimar, France 28/F5
Montelindo (riv.), Paraguay 144/C3
Montellano, Spain 33/D4
Montello, Nev. (89830) 266/G1
Montello, Wis. (53949) 317/H8
Montemayor (plat.), Argentina 143/C5
Montemorelos, Mexico 150/K3
Montemor-o-Novo, Portugal 33/B3
Montemor-o-Velho, Portugal 33/B2
Monte Ne, Ark. (†72756) 202/B1
Montenegro, Brazil 132/D10
Montenegro, Chile 138/G2
Montenegro (rep.), Yugoslavia 45/D4
Monte Patria, Chile 138/A8
Monte Plata, Dom. Rep. 158/E6
Montepuez, Mozambique 118/F2
Montepulciano, Italy 34/C3
Monte Quemado, Argentina 143/D2
Monte Real, Brazil 132/A5
Monterey, Ala. (†36030) 195/E7
Monterey, Calif. 188/B3
Monterey (bay), Calif. 188/B3
Monterey (co.), Calif. 204/D7
Monterey, Calif. (93940) 204/D7
Monterey (bay), Calif. 204/K4
Monterey, Ind. (46960) 227/D2
Monterey, La. (71354) 238/G4
Monterey○, Mass. (†40359) 237/M4
Monterey, La. (71354) 238/G4
Monterey○, Mass. (01245) 249/B4
Monterey, Tenn. (38574) 237/L8
Monterey, Va. (24465) 307/J4
Monterey, Wis. (†53066) 317/J1
Monterey Park, Calif. (91754) 204/C10
Montería, Colombia 120/B2
Montería, Colombia 126/B3
Monte Rio, Calif. (95462) 204/B5
Montero, Bolivia 136/D5
Monteros, Argentina 143/C2
Monterotondo, Italy 34/F6
Monterrey, Mexico 2/D4
Monterrey, Mexico 146/J7
Monterrey, Mexico 150/J4
Montes, Uruguay 145/B3
Montesano, Wash. 188/B4
Monte Sant'Angelo, Italy 34/F4
Montes Claros, Brazil 120/E4
Montes Claros, Brazil 132/E7
Monte Sereno, Calif. (95030) 204/K4
Montevallo, Ala. (35115) 195/E4
Montevarchi, Italy 34/C3
Montevideo, Minn. (56265) 255/C6
Montevideo, Uruguay 145/B7
Montevideo, Uruguay 145/B7
Montevideo (cap.), Uruguay 145/B7
Montevideo (cap.), Uruguay 2/G7
Monteview, Idaho (83435) 220/F6

Monte Vista, Colo. (81144) 208/G7
Montezuma, Colo. (†80435) 208/H3
Montezuma (peak), Colo. 208/F8
Montezuma, Georgia (31063) 217/E6
Montezuma, Ind. (47862) 227/C5
Montezuma, Iowa (50171) 229/H5
Montezuma, Kansas (67867) 232/B4
Montezuma (co.), N.Mex. (†87731) 274/D3
Montezuma, Ohio (45866) 284/A4
Montezuma, Tenn. (†38340) 237/D10
Montezuma (creek), Utah 304/E6
Montezuma Castle Nat'l Mon., Ariz. 198/D4
Montezuma Creek, Utah (84534) 304/E6
Montfoort, Netherlands 161/G5
Montfort, France 28/C3
Montfort, Wis. (53569) 317/E10
Montgomery (cap.), Ala. 188/J4
Montgomery (cap.), Ala. 146/K6
Montgomery (co.), Ala. 195/F6
Montgomery (co.), Ala. (*36101) 195/F6
Montgomery (co.), Ark. 202/C4
Montgomery (co.), Georgia 217/G6
Montgomery (co.), Ill. 222/D4
Montgomery, Ill. (60538) 222/E2
Montgomery (co.), Ind. 227/D4
Montgomery, Ind. (47558) 227/C7
Montgomery (co.), Iowa 229/C7
Montgomery, Iowa (51353) 229/C2
Montgomery (co.), Kansas 232/G4
Montgomery (co.), Ky. 237/O4
Montgomery, La. (71454) 238/E3
Montgomery (co.), Md. 245/J4
Montgomery, Mich. (49255) 250/E7
Montgomery, Minn. (56069) 255/E6
Montgomery (co.), Miss. 256/E4
Montgomery (co.), Mo. 261/K5
Montgomery (co.), N.Y. 276/M5
Montgomery, N.Y. (12549) 276/M7
Montgomery (co.), N.C. 281/H4
Montgomery (co.), Ohio 284/C9
Montgomery, Ohio (45242) 284/C9
Montgomery (co.), Pa. 294/M5
Montgomery, Pa. (17752) 294/H3
Montgomery (co.), Tenn. 237/G8
Montgomery (co.), Texas 303/J7
Montgomery, Texas (77356) 303/J7
Montgomery○, Vt. (05470) 268/B2
Montgomery (co.), Va. 307/H5
Montgomery, Wales 13/D5
Montgomery, Wales 10/E4
Montgomery, W. Va. (25136) 312/D6
Montgomery Center, Vt. (05471) 268/B2
Montgomery Creek, Calif. (91214) 204/C10
Montgomery City, Mo. (63361) 261/K5
Monthey, Switzerland 39/C4
Monticello, Ark. (71655) 202/G6
Monticello, Fla. (32344) 212/C1
Monticello, Georgia (31064) 217/E4
Monticello, Ill. (61856) 222/E3
Monticello, Ind. (47960) 227/D3
Monticello, Iowa (52310) 229/L4
Monticello, Ky. (42633) 237/M7
Monticello○, Maine (04760) 243/H3
Monticello, Minn. (55362) 255/E5
Monticello, Miss. (39654) 256/D7
Monticello, Mo. (63457) 261/J2
Monticello, N. Mex. (87939) 274/B5
Monticello, N.Y. (12701) 276/L7
Monticello○, Ohio (†45887) 284/B4
Monticello, S.C. (29106) 296/E4
Monticello, Utah (84535) 304/E6
Monticello, Wis. (53570) 317/G10
Mont Ida, Kansas (†66091) 232/G3
Montier, Mo. (65546) 261/J8
Montigny-les-Metz, France 28/G3
Montigny-le-Tilleul, Belgium 27/E8
Montijo, Panama 154/G6
Montijo (gulf), Panama 154/G7
Montijo, Portugal 33/B3
Montijo, Spain 33/C3
Montilla, Spain 33/D4
Montjoie (lake), Québec 172/B3
Mont-Joli, Que. 162/K6
Mont-Joli, Québec 174/D3
Mont-Joli, Québec 172/J1
Mont-Laurier, Que. 162/J6
Mont-Laurier, Québec 172/B3
Mont-Laurier, Québec 174/B3
Mont-Louis, Québec 172/C1
Montluçon, France 28/E4
Montmagny (co.), Québec 172/G3
Montmagny, Québec 174/C3
Montmartre, Sask. 181/H5
Montmédy, France 28/F3
Montmorenci, Ind. (47962) 227/D4
Montmorenci, S.C. (29839) 296/D4
Montmorency (co.), Mich. 250/E3
Montmorency, Québec 172/J3
Montmorency (riv.), Québec 172/F2
Montmorency, Victoria 97/J4
Montmorency No. 1 (co.), Québec 172/F2
Montmorency No. 2 (co.), Québec 172/G3
Montmorency No. 1 (county), Québec 174/C3
Montmorillon, France 28/D4
Mont Nebo, Sask. 181/E2
Montney, Br. Col. 184/G2
Monto, Queensland 95/D5
Monto, Queensland 88/J4
Montoire-sur-le-Loir, France 28/D4
Montoro, Spain 33/D3
Montosa (mesa), N. Mex. 274/E3
Montour, Iowa (50173) 229/H5
Montour (co.), Pa. 294/J3
Montour Falls, N.Y. (14865) 276/G6
Montoursville, Pa. (17754) 294/H3
Montowese, Conn. (†06473) 210/D3
Montoya, N. Mex. (†88401) 274/F3
Montoz (mt.), Switzerland 39/D2
Montpelier, Idaho 188/D2
Montpelier, Idaho (83254) 220/G7
Montpelier, Ind. (47359) 227/G3

Montpelier, Iowa (52759) 229/M6
Montpelier, Jamaica 158/H6
Montpelier, Miss. (39754) 256/G3
Montpelier, N. Dak. (58472) 282/N6
Montpelier, Ohio (43543) 284/A2
Montpelier (co.), Vt. (*05602) 268/B3
Montpelier (cap.), Vt. 146/L5
Montpelier, Vt. 188/M2
Montpelier (cap.), Vt. 146/L5
Montpellier, France 7/E4
Montpellier, France 28/E6
Montpellier, Québec 172/B4
Montpellier, Québec 172/H4
Montpon, France 28/D5
Montréal (riv.), Mich. 250/F1
Montreal, Mo. (65591) 261/G7
Montréal, Que. 146/L5
Montréal, Que. 162/J7
Montréal, Québec 172/J4
Montreal (lake), Sask. 181/F1
Montréal-Est, Québec 172/J4
Montreal (lake), Sask. 181/H6
Montreal, Wis. (54550) 317/F3
Montreal (riv.), Wis. 317/F2
Montreal, Wis. (54550) 317/F3
Montreal (riv.), Wis. 317/F2
Montreal River Harbor, Ontario 177/J5
Montréal-Nord, Québec 172/H4
Montreat, N.C. (28757) 281/E3
Montreuil, Pas-de-Calais, France 28/D2
Montreuil, Seine-Saint-Denis, France 28/B2
Montreux, Switzerland 39/C4
Montrichter, Switzerland 39/B3
Mont-Rolland, Québec 172/G4
Montrose, Ala. (36559) 195/C9
Montrose, Ark. (71658) 202/H7
Montrose, Br. Col. 184/G3
Montrose (co.), Colo. 208/C6
Montrose, Colo. (81401) 208/D6
Montrose (co.), Colo. 208/D6
Montrose, Georgia (31065) 217/F5
Montrose, Ill. (62445) 222/E4
Montrose, Iowa (52639) 229/L7
Montrose, Kansas (†66956) 232/D2
Montrose, La. (†71457) 238/D3
Montrose, Md. (†20850) 245/K4
Montrose, Mich. (48457) 250/F5
Montrose, Minn. (55363) 255/E5
Montrose, Miss. (†39338) 256/F6
Montrose, Mo. (64770) 261/E6
Montrose, Pa. (18801) 294/L2
Montrose, Scotland 10/E2
Montrose, Scotland 15/F4
Montrose, S. Dak. (57048) 298/P6
Montrose, Victoria 97/K5
Montrose, W. Va. (26283) 312/G4
Montrose, Wis. (†52250) 307/P4
Montrouge, France 28/B2
Mont-Royal, Québec 172/H4
Monts (pt.), Québec 172/B1
Mont-Saint-Hilaire, Québec 172/D3
Mont-Saint-Michel, France 28/C3
Mont-Saint-Michel, Québec 172/B3
Mont-Saint-Pierre, Québec 172/C1
MONTSERRAT 156/G3
Montserrat (mt.), Spain 33/G2
Montsinéry, Fr. Guiana 131/E3
Mont-Tremblant, Québec 172/C3
Mont-Tremblant Prov. Park, Québec 172/C3
Mont-Tremblant Prov. Park, Québec 174/C3
Montvale, N.J. (07645) 273/B1
Montvale, Va. (24122) 307/J6
Montverde, Fla. (32756) 212/E3
Mont Vernon○, N.H. (03057) 268/D6
Montville, Conn. (06353) 210/G3
Montville○, Conn. (06353) 210/G3
Montville, Maine (†04941) 243/E7
Montville○, Maine (†04941) 243/E7
Montville, Mass. (†01255) 249/B4
Montville○, N.J. (07045) 273/E2
Montville, Ohio (44064) 284/H2
Montz, La. (†70468) 238/M3
Monument, Colo. (80132) 208/K4
Monument (peak), Idaho 220/B4
Monument, Kansas (67747) 232/A2
Monument, N. Mex. (88265) 274/F6
Monument, Oreg. (97864) 291/H3
Monument (valley), Utah 304/D6
Monument Beach, Mass. (02553) 249/M6
Monument Valley, Utah (84536) 304/D6
Monywa, Burma 72/B2
Monza, Italy 34/B2
Monze, Zambia 115/E7
Monzón, Spain 33/G2
Mooar, Iowa (†52632) 229/L8
Moodie (is.), N.W. Terrs. 187/M3
Moodus, Conn. (06469) 210/F2
Moodus, Conn. 210/F2
Moody, Ala. (†35125) 195/F3
Moody, Mo. (64054) 243/B9
Moody, Mo. (65770) 261/J9
Moody (co.), S. Dak. 298/R5
Moody, Texas (76557) 303/G6
Moodys, Okla. (74444) 288/S2
Moodyville, Tenn. (†38549) 237/L7
Mooers, N.Y. (12958) 276/N1
Mooka, Japan 81/K5
Mooleyville, Ky. (40154) 237/H4
Mooloo Downs, W. Australia 92/B4
Moomin (creek), N.S. Wales 97/E1
Moon (lake), Calif. 204/E2
Moon (lake), Nebr. 264/E2
Moon, Okla. (†71821) 288/S7
Moonachie, N.J. (†07070) 273/B2
Moonah (township), Queensland 95/A4
Moonbeam, Ontario 177/J5
Mooncoin, Ireland 17/G7
Moonie, N.S. Wales 97/E1
Moonie, Queensland 95/D5
Moon Run (pt.), (15244) 294/B6
Moonta, S. Australia 94/E5
Moora, W. Australia 88/B6
Moora, W. Australia 92/B5
Moorabbin, Victoria 88/L7
Moorabbin, Victoria 97/J5

Moorcroft, Wyo. (82721) 319/H1
Moore (co.), Idaho (83265) 220/E6
Moore, Mont. (59464) 262/G4
Moore (dam), N.H. 268/D3
Moore (co.), N.C. 281/K4
Moore, Okla. (73160) 288/M4
Moore, S.C. (29369) 296/D2
Moore (co.), Tenn. 237/J10
Moore, Texas (78057) 303/E9
Moore (co.), Texas 303/C2
Moore, Utah (84523) 304/C5
Moore (res.), Vt. 268/D3
Moore (co.), W. Australia 88/B5
Moore (lake), W. Australia 92/B5
Moorefield (co.), Iowa (†72501) 202/G2
Moorefield, Ind. (†47043) 227/G7
Moorefield, Ky. (40350) 237/O4
Moorefield, Nebr. (69039) 264/D4
Moorefield, Ontario 177/D4
Moorefield, W. Va. (26836) 312/J4
Moore Haven, Fla. (33471) 212/E5
Mooreland, Ind. (47360) 227/G5
Mooreland, Okla. (73852) 288/H2
Moore Park, Manitoba 179/C4
Mooresboro, N.C. (28114) 281/F4
Moores Bridge, Ala. (†35458) 195/C4
Mooresburg, Tenn. (37811) 237/P8
Moores Creek, Ky. (40453) 237/O6
Moores Creek Nat'l Battlefield, N.C. 281/N6
Moores Hill, Ind. (47032) 227/G6
Moores Mills, New Bruns. 170/C3
Moorestown, Mich. (49651) 250/D4
Moorestown, N.J. (08057) 273/B3
Mooresville, Ala. (35649) 195/E1
Mooresville, Ind. (46158) 227/E5
Mooresville, Mo. (64664) 261/E3
Mooresville, N.C. (28115) 281/H3
Moore Town, Jamaica 158/K6
Mooreton, N. Dak. (58061) 282/S7
Mooreville, Miss. (38857) 256/G2
Moorfoot (hills), Scotland 15/E5
Moorhead, Iowa (51558) 229/B5
Moorhead, Minn. (56560) 255/B4
Moorhead, Miss. (38761) 256/D5
Mooringsport, La. (71060) 238/B1
Moorland (par.), La. 238/G1
Moorland, Iowa (50566) 229/E4
Moorland, Ky. (†40223) 237/L2
Moorman, Ky. (42357) 237/G6
Moorooka, Queensland 88/K3
Moorooka, Queensland 95/B5
Mooroopna, Victoria 97/C5
Moorpark, Calif. (93021) 204/G9
Mooresburg, Louisiana 118/B6
Moorslede, Belgium 27/B7
Moosburg an der Isar, W. Germany 22/D4
Moose (creek), Idaho 220/D3
Moose (pond), Maine 243/B7
Moose (riv.), Maine 243/E4
Moose (isl.), Manitoba 179/E3
Moose (co.), Minn. 255/C2
Moose (riv.), N.Y. 276/K3
Moose (riv.), Sask. 181/J6
Moose (riv.), Vt. 268/D2
Moose (lake), Wis. 317/F3
Moose (lake), Wis. 317/F3
Moose, Wyo. (83012) 319/B2
Moose Creek, Ontario 177/K2
Moose Factory, Ontario 175/D2
Moosehead, Maine (†04478) 243/E4
Moosehead (lake), Maine 243/E4
Mooseheart, Ill. (60539) 222/E2
Moose Heights, Br. Col. 184/F3
Moosehorn, Manitoba 179/D3
Moose Jaw, Sask. 146/H4
Moose Jaw, Sask. 162/F6
Moose Jaw, Sask. 181/F5
Moose Jaw (riv.), Sask. 181/G5
Moose Lake, Manitoba 179/H3
Moose Lake, Minn. (55767) 255/F4
Mooseland, Nova Scotia 168/F4
Mooseleuk (stream), Maine 243/F2
Mooselookmeguntic (lake), Maine 243/B6
Moose Mountain (creek), Sask. 181/J6
Moose Mountain Prov. Park, Sask. 181/J6
Moose Pass, Alaska (99631) 196/C1
Moose Range, Sask. 181/H2
Moose River○, Maine (†04945) 243/C4
Moose River, Ontario 175/D2
Moosic, Pa. (18507) 294/F7
Moosilauke (mt.), N.H. 268/D3
Moosomin, Sask. 162/F6
Moosomin, Sask. 181/K5
Moosonee, Ont. 162/H6
Moosonee, Ont. 146/K4
Moosonee, Ontario 175/D2
Moosup, Conn. (06354) 210/H2
Moosup (riv.), Conn. 210/H2
Mopang (lake), Maine 243/H6
Mopeia, Mozambique 118/F3
Mopti, Mali 102/B3
Mopti, Mali 106/D6
Moqatta, Sudan 59/C7
Moqor, Afghanistan 68/B2
Moqor, Afghanistan 59/J3
Moquah, Wis. (†54806) 317/D2
Moquegua (dept.), Peru 128/G11
Moquegua, Peru 128/B8
Moquegua, Peru 128/G11
Mór, Hungary 41/E3
Mora, Cameroon 115/B1
Mora, India 68/B7
Mora, La. (71455) 238/E4
Mora, Minn. (55051) 255/E5
Mora, Mo. (65345) 261/F5
Mora, N. Mex. (87732) 274/D3
Mora (co.), N. Mex. 274/D3
Mora (riv.), N. Mex. 274/E3

Mora, Portugal 33/B3
Mora, Spain 33/E3
Mora, Sweden 18/J6
Moradabad, India 54/J7
Moradabad, India 68/D3
Mora de Rubielos, Spain 33/F2
Morado, Quebrado (riv.), Chile 138/A6
Morafenobe, Madagascar 118/G3
Morafenobe, Madagascar 118/G3
Moraga, Poland 47/E2
Moraga, Calif. (94556) 204/K2
Moraine, Ohio (†45439) 284/B6
Moraleda (chan.), Chile 138/A6
Morales, Guatemala 154/C3
Morales, Peru 128/D6
Moramanga, Madagascar 118/H3
Moramanga, Madagascar 102/G6
Moran, Ind. (†46041) 227/D4
Moran, Kansas (66755) 232/G4
Moran, Mich. (49760) 250/E2
Moran, Texas (76464) 303/E5
Moran, Wyo. (83013) 319/B2
Moranbah, Queensland 95/C9
Morane (isl.), Fr. Poly. 87/N8
Morane (isl.), Fr. Poly. 87/N8
Morant (pt.), Jamaica 156/C3
Morant Bay, Jamaica 158/K7
Morar, Scotland 15/C4
Morar, Loch (lake), Scotland 15/C4
Morat (lake), Switzerland 39/D3
Morata de Tajuña, Spain 33/G4
Moratalla, Spain 33/E3
Morattico, Va. (22523) 307/P5
Moratuwa, Sri Lanka 68/D7
Morava (riv.), Czech. 41/D2
Morava (riv.), Yugoslavia 45/E3
Moravia, Iowa (52571) 229/H7
Moravia, N.Y. (13118) 276/H5
Moravian Falls, N.C. (28654) 281/G4
Moravská Třebová, Czech. 41/D2
Moravské Budějovice, Czech. 41/D2
Morawa, W. Australia 88/B5
Morawa, W. Australia 92/B5
Morawhanna, Guyana 120/D2
Morawhanna, Guyana 120/D2
Moray (firth), Scotland 7/D3
Moray (firth), Scotland 15/D3
Moray (firth), Scotland 10/E2
Moray (trad. co.) – Scotland 15/A5
Morazán, Honduras 154/C2
Morbihan (dept.), France 28/B4
Mörbylånga, Sweden 18/K8
Morden, Man. 162/G6
Morden, Manitoba 179/D5
Mordialloc, Victoria 97/J6
Mordialloc, Victoria 88/L7
Mordvinian A.S.S.R., U.S.S.R. 52/G4
Mordvinian A.S.S.R., U.S.S.R. 48/E4
More, Loch (lake), Scotland 15/E2
More, Loch (lake), Scotland 15/D2
Morea, Victoria 97/C5
Moreau (riv.), S. Dak. 298/G3
Moreauville, La. (71355) 238/G4
Morebattle, Scotland 15/F5
Morecambe, Alberta 182/E3
Morecambe (bay), England 10/oe3
Morecambe (bay), England 13/D3
Moree, N.S. Wales 88/H5
Moree, N.S. Wales 97/E1
Morehead, Kansas (†66776) 232/G4
Morehead, Ky. (40351) 237/P4
Morehead City, N.C. (28557) 281/R5
Morehouse (par.), La. 238/G1
Morehouse, Mo. (63868) 261/N9
Moreland, Ark. (72849) 202/E3
Moreland, Georgia (30259) 217/C4
Moreland, Idaho (83256) 220/F6
Moreland Hills, Ohio (†44022) 284/J9
Morelia, Mexico 150/J7
Morelia, Mexico 146/H8
Morelia, Queensland 95/B4
Morell, Pr. Edward I. 168/F2
Morella, Queensland 88/G4
Morella, Spain 33/F2
Morelos (state), Mexico 150/K7
Morelos, Mexico 150/J4
Morelos Cañada, Mexico 150/O2
Morena, India 68/D3
Morena, Sierra (mts.), Spain 7/D5
Morena, Sierra (range), Spain 33/C3
Morenci, Ariz. (85540) 198/F5
Morenci, Mich. (49256) 250/E7
Moreni, Romania 45/G3
Moreno, Bolivia 136/B2
Moreno, Calif. (92360) 204/H10
Moreno (bay), Chile 138/A4
Møre og Romsdal (co.), Norway 18/E5
Mores (creek), Idaho 220/D5
Moresby, Br. Col. 184/B3
Moresby (isl.), Br. Col. 184/B4
Moreton (isl.), Queensland 88/J5
Moreton (bay), Queensland 88/K2
Moreton (bay), Queensland 95/E5
Moreton (isl.), Queensland 95/E5
Moretonhampstead, England 13/C7
Moreton-in-Marsh, England 13/F6
Moretown○, Vt. (05660) 268/B3
Morewood, Ontario 177/J2
Morgan (co.), Ala. 195/E2
Morgan (co.), Colo. 208/M2
Morgan (pt.), Conn. 210/D4
Morgan (co.), Georgia 217/C7
Morgan, Georgia (31766) 217/C7
Morgan (co.), Ill. 222/C4
Morgan (co.), Ind. 227/E6
Morgan (co.), Ky. 237/P5
Morgan (co.), Mo. 261/G6
Morgan, Mo. (†65706) 261/G3
Morgan (co.), Ohio 284/G6
Morgan (co.), Tenn. 237/M8
Morgan, Texas (76671) 303/G6
Morgan (co.), Utah 304/C2
Morgan, Utah (84050) 304/C2
Morgan○, Vt. (05853) 268/C2
Morgan (co.), W. Va. 312/K3
Morgan Center, Vt. (05854) 268/D2
Morgan City, La. (70380) 238/H7
Morgan City, Miss. (38946) 256/D4

Morgan Falls (dam), Georgia 217/K1
Morganfield, Ky. (42437) 237/E5
Morgan Hill, Calif. (95037) 204/L4
Morganito, Venezuela 124/E5
Morgans Point, Texas (†77571) 303/K2
Morgansville, W. Va. (†26456) 312/E4
Morganton, Ark. (72109) 202/F3
Morganton, Georgia (30560) 217/D1
Morganton, N.C. (28655) 281/F3
Morgantown, Ind. (46160) 227/E6
Morgantown, Ky. (42261) 237/H6
Morgantown, Miss. (39484) 256/E8
Morgantown, Miss. (†39120) 256/B7
Morgantown, Ohio (†45612) 284/D7
Morgantown, Pa. (19543) 294/L5
Morgantown, W. Va. (26505) 312/G3
Morganville, Kansas (67468) 232/E2
Morganville, N.J. (07751) 273/E3
Morganza, La. (70759) 238/G5
Morges, Switzerland 39/B3
Morguilla (pt.), Chile 138/D1
Mori, China 77/D3
Mori, Japan 81/K2
Moriah, N.Y. (12960) 276/N2
Moriah Center, N.Y. (12961) 276/N2
Moriarty, N. Mex. (87035) 274/D4
Morice (lake), Br. Col. 184/D3
Morice (riv.), Br. Col. 184/D3
Morichal, Colombia 126/E6
Morichal Largo (riv.), Venezuela 124/G3
Morien (cape), Nova Scotia 168/J2
Moriguchi, Japan 81/J7
Morin Creek, Sask. 181/C1
Morin Dawa Daurzu, China 77/K2
Morin Heights, Québec 172/C4
Morinville, Alberta 182/D3
Morioka, Japan 81/K4
Morisset, N.S. Wales 97/F3
Morisset, Québec 172/G3
Moriston (riv.), Scotland 15/D3
Morjärv, Sweden 18/N3
Morlaix, France 28/B3
Morland, Kansas (67650) 232/B2
Morley, Alberta 182/C4
Morley, Iowa (52312) 229/L4
Morley, Mich. (49336) 250/D5
Morley, Mo. (63767) 261/N8
Morley, N.Y. (13617) 276/K1
Morley, Tenn. (37812) 237/O7
Mormon (lake), Ariz. 198/D4
Mormon (mt.), Idaho 220/D4
Mormon (mts.), Nev. 266/G5
Mormon Lake, Ariz. (86038) 198/D4
Morne-à-l'Eau, Guadeloupe 161/A6
Morne Seychellois (mt.), Seychelles 118/H5
Morningside, Alberta 182/D3
Morningside, Md. (†20028) 245/G5
Morningside, Queensland 88/K2
Morningside Park, Conn. (†06385) 210/G3
Morning Sun, Iowa (52640) 229/L6
Mornington (isl.), Chile 138/D8
Mornington (isl.), Queensland 88/F3
Mornington (isl.), Queensland 95/A3
Mornington, Victoria 97/C6
Mornington (pen.), Victoria 97/C6
Morning View, Ky. (41063) 237/N3
Moro, Ark. (72368) 202/H4
Moro (creek), Ark. 202/F7
Moro, Oreg. (97039) 291/G2
Moro (gulf), Philippines 82/D7
Moro (gulf), Philippines 85/G4
Moro (mt.), Switzerland 39/E5
Moro Bay, Ark. (†71651) 202/F7
Morobe, Papua N.G. 85/C7
Morocco 2/J4
Morocco 102/B1
Morocco, Ind. (47963) 227/C3
MOROCCO 106/C2
Morocelí, Honduras 154/D3
Morochata, Bolivia 136/B5
Morococha, Peru 128/D8
Morogoro (reg.), Tanzania 115/G5
Morogoro, Tanzania 115/G5
Morogoro, Tanzania 102/F5
Moroleón, Mexico 150/J6
Morombe, Madagascar 118/G4
Moromoro, Bolivia 136/C6
Morón, Argentina 143/G7
Morón, Cuba 158/F2
Morón, Cuba 156/B2
Morón, Haiti 158/A6
Mörön (Muren), Mongolia 77/F2
Moron (mt.), Switzerland 39/D2
Morón, Venezuela 124/D2
Morona, Ecuador 128/C4
Morona (riv.), Peru 128/D5
Morona-Santiago (prov.), Ecuador 128/C4
Morondava, Madagascar 118/G3
Morondava, Madagascar 102/G7
Morón de la Frontera, Spain 33/D4
Morongo Ind. Res., Calif. 204/J10
Moroni (cap.), Comoros 118/G2
Moroni (cap.), Comoros 102/G6
Moroni, Utah (84646) 304/C4
Moron Us He (riv.), China 77/D5
Morotai (isl.), Indonesia 54/O9
Morotai (isl.), Indonesia 85/H5
Moroto, Uganda 115/F3
Morovis, P. Rico 161/D1
Morpeth, England 13/F2
Morpeth, England 10/F3
Morpeth, Ontario 177/C5
Morphou, Cyprus 63/E5
Morphou (bay), Cyprus 63/E5
Morral, Ohio (43337) 284/D4
Morrice, Mich. (48857) 250/E6
Morrill, Kansas (66515) 232/G2
Morrill○, Maine (04952) 243/E7
Morrill, Minn. (†56329) 255/E5
Morrill (co.), Nebr. 264/A3
Morrill, Nebr. (69358) 264/A3
Morrilton, Ark. (72110) 202/E3
Morrin, Alberta 182/D4
Morrinhos, Brazil 132/D7
Morrinsville, N. Zealand 100/E2
Morris, Ala. (35116) 195/E3
Morris (res.), Calif. 204/D10
Morris○, Conn. (06763) 210/C2
Morris, Georgia (31767) 217/C7
Morris, Ill. (60450) 222/E2
Morris, Ind. (47033) 227/G6
Morris (co.), Kansas 232/F3
Morris, Manitoba 179/E5
Morris, Minn. (56267) 255/C5
Morris (co.), N.J. 273/D2
Morris, N.Y. (13808) 276/K5
Morris, Okla. (74445) 288/P3
Morris, Pa. (16938) 294/H2
Morris (mt.), S. Australia 94/B2
Morris (isl.), S.C. 296/H6
Morris (co.), Texas 303/K4
Morris, W. Va. (†26639) 312/E5
Morrisburg, Ontario 177/J3
Morris Chapel, Tenn. (38361) 237/E10
Morrisdale, New Bruns. 170/D3
Morrisdale, Pa. (16858) 294/F4
Morrisey, Wyo. (†82701) 319/H2
Morris Fork, Ky. (41353) 237/O6
Morris Jesup (cape), Greenl. 4/A11
Morrison, Colo. (80465) 208/J3
Morrison, Ill. (61270) 222/C2
Morrison, Iowa (50657) 229/H4
Morrison (lake), Manitoba 179/C1
Morrison (co.), Minn. 255/D4
Morrison, Mo. (65061) 261/J5
Morrison, Okla. (73061) 288/M2
Morrison, Tenn. (37357) 237/K9
Morrison Bluff, Ark. (†72863) 202/D3
Morrison City, Tenn. (†37660) 237/R7
Morrisonville, Ill. (62546) 222/D4
Morrisonville, N.Y. (12962) 276/N1
Morrisonville, Wis. (53571) 317/G9
Morris Plains, N.J. (07950) 273/D2
Morris Run, Pa. (16939) 294/J2
Morriston, Ark. (72559) 202/D3
Morriston, Fla. (32668) 212/D2
Morriston, Ontario 177/D4
Morristown, Ariz. (85342) 198/C5
Morristown, Ill. (†61101) 222/D1
Morristown, Ind. (46161) 227/F5
Morristown, Minn. (55052) 255/E6
Morristown, N.J. (07960) 273/D2
Morristown, N.Y. (13664) 276/J1
Morristown, Ohio (43759) 284/H5
Morristown, S. Dak. (57645) 298/F2
Morristown, Tenn. (37814) 237/P8
Morristown○, Vt. (†05661) 268/B2
Morristown Nat'l Hist. Park, N.J. 273/D2
Morrisvale, W. Va. (25542) 312/C6
Morrisville, Mo. (65710) 261/F8
Morrisville, N.Y. (13408) 276/J5
Morrisville, N.C. (27560) 281/M3
Morrisville, Pa. (19067) 294/N5
Morrisville, Vt. (05661) 268/B2
Morrito, Nicaragua 154/E5
Morro (pt.), Chile 138/A6
Morro Bay, Calif. (93442) 204/D8
Morro do Chapéu, Brazil 132/F5
Morropón, Peru 128/C5
Morros, Brazil 132/F3
Morrosquillo (gulf), Colombia 126/C3
Morrow, Ark. (72749) 202/B2
Morrow, Georgia (30260) 217/K2
Morrow, La. (71356) 238/F5
Morrow (co.), Ohio 284/E4
Morrow, Ohio (45152) 284/B7
Morrow (co.), Oreg. 291/H2
Morrow Point (res.), Colo. 208/E6
Morrowville, Kansas (66958) 232/E2
Morrumbala, Mozambique 118/F3
Morrumbene, Mozambique 118/F4
Mors (isl.), Denmark 21/B4
Morse (res.), Ind. 227/E4
Morse, La. (70559) 238/F6
Morse, Sask. 181/D5
Morse, Texas (79062) 303/C1
Morse, Wis. (†54527) 317/E3
Morse Bluff, Nebr. (68648) 264/H3
Morse Mill, Mo. (63066) 261/L6
Morses Line, Vt. (†05459) 268/A2
Morshansk, U.S.S.R. 52/F4
Mortagne-au-Perche, France 28/D3
Mortara, Italy 34/B2
Morte (pt.), England 13/C6
Morteau, France 28/G4
Morteros, Argentina 143/D3
Mortes (Manso) (riv.), Brazil 132/D6
Mortlach, Sask. 181/E5
Mortlake, Victoria 97/B6
Morton, Ill. (61550) 222/D3
Morton (co.), Kansas 232/A4
Morton, Minn. (56270) 255/C6
Morton, Miss. (39117) 256/E6
Morton (co.), N. Dak. 282/H6
Morton, Ontario 177/H3
Morton, Pa. (19070) 294/M7
Morton, Texas (79346) 303/B4
Morton, Wash. (98356) 310/C4
Morton, Wyo. (82522) 319/D2
Morton Grove, Ill. (60053) 222/B5
Morton Mills, Iowa (†50864) 229/C6
Mortons Gap, Ky. (42440) 237/F6
Mortsel, Belgium 27/E6
Moruga, Trin. & Tob. 161/B11
Moruka (riv.), Guyana 131/B2
Morundah, N.S. Wales 97/D4
Moruya, N.S. Wales 97/F4
Morvan (plat.), France 28/F4
Morven, Georgia (31638) 217/E9
Morven, N. Zealand 100/C6
Morven, N.C. (28119) 281/J5
Morven, Queensland 95/C5
Morven (dist.), Scotland 15/C4
Morven (mt.), Scotland 15/E2
Morvi, India 68/C4
Morvin, Ala. (36762) 195/C7
Morwell, Victoria 97/D5
Morwell, Victoria 88/H7
Mosbach, W. Germany 22/C4
Mosby, Mo. (64073) 261/R4
Mosby, Mont. (59058) 262/J4
Mosca, Colo. (81146) 208/H7
Moscavide, Portugal 33/A1
Moscow, Ark. (71659) 202/G5
Moscow, Idaho (83843) 220/B3
Moscow, Idaho 188/C1
Moscow, Ind. (†46156) 227/F6
Moscow, Iowa (52760) 229/L5
Moscow, Kansas (67952) 232/A4
Moscow, Ky. (†42031) 237/D7
Moscow, Miss. (†39328) 256/G5
Moscow, Ohio (45153) 284/B8
Moscow, Pa. (18444) 294/F7
Moscow, Tenn. (38057) 237/C10
Moscow (cap.), U.S.S.R. 2/L3
Moscow (cap.), U.S.S.R. 7/H3
Moscow (cap.), U.S.S.R. 48/D4
Moscow (Moskva) (cap.), U.S.S.R. 52/E3
Moscow, Vt. (05662) 268/B3
Moscow Mills, Md. (†21521) 245/B2
Moscow Mills, Mo. (63362) 261/K5
Mosel (riv.), Luxembourg 27/J9
Mosel (riv.), W. Germany 22/B3
Moseley, Sask. 181/G3
Moseley, Va. (23120) 307/N6
Moselle (dept.), France 28/G3
Moselle (riv.), France 28/G3
Moselle, Miss. (39459) 256/F8
Moselle, Mo. (†63084) 261/L6
Moser River, Nova Scotia 168/F4
Moses (lake), Wash. 310/F3
Moses Coulee (canyon), Wash. 310/F3
Moses Lake, Wash. (98837) 310/F3
Moses Point, Alaska (†99762) 196/F2
Mosetenes, Cordillera de (range), Bolivia 136/B5
Mosgiel, N. Zealand 100/C6
Mosgrove, Pa. (†16259) 294/D4
Moshannon, Pa. (16859) 294/F3
Mosheim, Tenn. (37818) 237/R8
Mosher, S. Dak. (57558) 298/J7
Moshi, Tanzania 115/G4
Moshi, Tanzania 102/F5
Mosier, Oreg. (97040) 291/F2
Mosina, Poland 47/C2
Mosinee, Wis. (54455) 317/G6
Mosi-Oa-Tunya (falls) 102/E6
Mosi-Oa-Tunya (Victoria) (falls), Zambia 115/E7
Mosi-Oa-Tunya (Victoria) (falls), Zimbabwe 118/C3
Mosjøen, Norway 18/H4
Moskenesøya (isl.), Norway 18/H3
Moskva (Moscow) (cap.), U.S.S.R. 52/E3
Moskva (riv.), U.S.S.R. 52/E3
Mosman, N. S. Wales 88/L4
Mosman, N.S. Wales 97/J3
Mosonmagyaróvár, Hungary 41/D3
Mosquera, Colombia 126/A6
Mosquero, N. Mex. (87733) 274/F3
Mosquic (lake), Québec 172/C3
Mosquito (lag.), Fla. 212/F3
Mosquito, Riacho (riv.), Paraguay 144/C3
Mosquito Creek (lake), Ohio 284/J3
Mosquitos, Costa de (reg.), Nicaragua 154/E4
Mosquitos, Golfo de los (gulf), Panama 154/G6
Moss, Miss. (39460) 256/F7
Moss, Norway 18/D4
Moss, Tenn. (38575) 237/K7
Mossaka, Congo 115/C4
Mossbank, Sask. 181/E6
Moss Beach, Calif. (94038) 204/H3
Mossel Bay, S. Africa 102/E8
Mossel Bay, S. Africa 118/C6
Mossendjo, Congo 115/B4
Mossgiel, N.S. Wales 97/C4
Moss Landing, Calif. (95039) 204/C7
Mossleigh, Alberta 182/D4
Mossman, Queensland 88/G3
Mossman, Queensland 95/C3
Mossoró, Brazil 120/F3
Mossoró, Brazil 132/G4
Moss Point, Miss. (39563) 256/G10
Mossuril, Mozambique 118/G2
Moss Vale, N.S. Wales 97/F4
Mossville, Ill. (61552) 222/D3
Mossy (riv.), Manitoba 179/C3
Mossy (riv.), Sask. 181/H1
Mossy Head, Fla. (32434) 212/C6
Mossyrock, Wash. (98564) 310/C4
Most, Czech. 41/B1
Mostaganem, Algeria 102/C1
Mostaganem, Algeria 106/D1
Mostar, Yugoslavia 7/F4
Mostar, Yugoslavia 45/D4
Mosty, U.S.S.R. 52/B4
Mosul, Iraq 66/C2
Mosul, Iraq 59/D2
Mosul, Iraq 54/F6
Motacucito, Bolivia 136/E5
Mota del Cuervo, Spain 33/E3
Motagua (riv.), Guatemala 154/C3
Motala, Sweden 18/J7
Motherwell and Wishaw, Scotland 10/B1
Motherwell and Wishaw, Scotland 15/C2
Motilla del Palancar, Spain 33/E3
Motiti (isl.), N. Zealand 100/F2
Motley, Minn. (56466) 255/D4
Motley (co.), Texas 303/D3
Motobu, Japan 81/N6
Motozintla de Mendoza, Mexico 150/N9
Motril, Spain 33/E4
Motsuta (cape), Japan 81/J2
Mott, N. Dak. (58646) 282/F7
Motu (riv.), N. Zealand 100/F3
Motueka, N. Zealand 100/D4
Motuhora (isl.), N. Zealand 100/F2
Motuihe (isl.), N. Zealand 100/C1
Motul de Felipe Carillo Puerto, Mexico 150/P6
Motupe, Peru 128/C6
Motutapu (isl.), N. Zealand 100/C1
Motygino, U.S.S.R. 48/K4
Mouchoir (passage), Turks & Caicos 156/D2
Moúdhros, Greece 45/G6
Moudjéria, Mauritania 106/B5
Moudon, Switzerland 39/C3
Mouila, Gabon 115/B4
Mouka, Cent. Afr. Rep. 115/D2
Moulamein, N.S. Wales 97/C4
Moulamein (creek), N.S. Wales 97/C4
Mould Bay, Canada 4/B16
Mould Bay, N.W. Terrs. 187/F2
Moule, Guadeloupe 161/B6
Moule à Chique (cape), St. Lucia 161/G7
Moulin-Morneault, New Bruns. 170/B1
Moulins, France 28/E4
Moulmein, Burma 72/C3
Moulmein, Burma 54/L8
Moulouya (riv.), Morocco 106/D2
Moulton, Ala. (35650) 195/D2
Moulton, Iowa (52572) 229/H7
Moulton, Mont. (†59423) 262/G3
Moulton, Texas (77975) 303/H8
Moultonboro○, N.H. (03254) 268/E4
Moultrie, Fla. (†32084) 212/E2
Moultrie, Georgia (31768) 217/E8
Moultrie (co.), Ill. 222/E4
Moultrie (lake), S.C. 188/K4
Moultrie (lake), S.C. 296/G5
Mounana, Gabon 115/B4
Mound, La. (71262) 238/H2
Mound, Minn. (55364) 255/E6
Mound Bayou, Miss. (38762) 256/C3
Mound City, Ill. (62963) 222/D6
Mound City, Kansas (66056) 232/H3
Mound City, Mo. (64470) 261/B2
Mound City, S. Dak. (57646) 298/K2
Mound City Group Nat'l Mon., Ohio 284/E7
Moundou, Chad 111/C6
Moundou, Chad 102/D4
Moundridge, Kansas (67107) 232/E3
Mounds, Ill. (62964) 222/D6
Mounds, Okla. (74047) 288/O3
Mound Station (Timewell), Ill. (†62375) 222/C3
Mounds View, Minn. (†55112) 255/G5
Moundsville, W. Va. (26041) 312/E3
Mound Valley, Kansas (67354) 232/G4
Moundville, Ala. (35474) 195/C5
Moundville, Mo. (64771) 261/C7
Moung Roessei, Cambodia 72/D4
Mounlapamôk, Laos 72/E4
Mount (cape), Liberia 106/B7
Mountain, N. Dak. (58262) 282/P2
Mountain (riv.), N.W. Terrs. 187/F3
Mountain, Ontario 177/J2
Mountain (prov.), Philippines 82/C2
Mountain, W. Va. (26407) 312/E4
Mountain, Wis. (54149) 317/K5
Mountainair, N. Mex. (87036) 274/C4
Mountain Ash, Ky. (†40769) 237/N7
Mountain Ash, Wales 13/A6
Mountain Ash, Wales 10/E5
Mountainboro, Ala. (†35957) 195/F2
Mountain Brook, Ala. (35223) 195/E4
Mountainburg, Ark. (72946) 202/B2
Mountain City, Georgia (30562) 217/F1
Mountain City, Nev. (89831) 266/F1
Mountain City, Tenn. (37683) 237/T8
Mountain Creek, Ala. (†36051) 195/E5
Mountain Creek (lake), Texas 303/G2
Mountaindale, Md. (†21701) 245/J2
Mountain Dale, N.Y. (12763) 276/L7
Mountaindale, Oreg. (†97106) 291/A1
Mountain Fork (riv.), Ark. 202/A5
Mountain Fork (riv.), Okla. 288/S6
Mountain Grove, Mo. (65711) 261/H8
Mountain Grove, Ontario 177/H3
Mountain Home, Ark. (72653) 202/F1
Mountain Home, Idaho (83647) 220/C6
Mountain Home (res.), Idaho 220/C6
Mountainhome, Pa. (18342) 294/M3
Mountain Home, Utah (84051) 304/D3
Mountain Home A.F.B., Idaho 220/C6
Mountain Iron, Minn. (55768) 255/F3
Mountain Lake, Minn. (56159) 255/D7
Mountain Lake Park, Md. (21550) 245/A3
Mountain Lakes, N.J. (07046) 273/E2
Mountain Meadows (res.), Calif. 204/E3
Mountain Park, Georgia (†30075) 217/D2
Mountain Park, Okla. (73559) 288/J5
Mountain Pine, Ark. (71956) 202/D4
Mountain Point, Alaska (†99901) 196/N2
Mountain Rest, S.C. (29664) 296/A2
Mountain Road, Manitoba 179/C4
Mountainside, N.J. (07092) 273/E2
Mountain Valley, Ark. (†71901) 202/D4
Mountain View, Alberta 182/D5
Mountain View, Ark. (72560) 202/F2
Mountain View, Calif. (*94042) 204/K3
Mountain View, Georgia (30070) 217/K2
Mountainview, Hawaii (96771) 218/J5
Mountain View, Mo. (65548) 261/J8
Mountain View, N.J. (†07470) 273/B2
Mountain View, Okla. (73062) 288/J4
Mountain View, Wyo. (82939) 319/B4
Mountain View, Wyo. (†82601) 319/F3
Mountain Village, Alaska (99632) 196/E2
Mountain Zebra Nat'l Park, S. Africa 118/C6
Mount Airy, Georgia (30563) 217/F1
Mount Airy, La. (70076) 238/M3
Mount Airy, Md. (21771) 245/K3
Mount Airy, N.C. (27030) 281/H1
Mount Airy, Tenn. (†37327) 237/L10
Mount Albert, N. Zealand 100/B1
Mount Albert, Ontario 177/E3
Mount Alto, W. Va. (25264) 312/C5
Mount Alton, Pa. (†16738) 294/E2
Mount Andrew, Ala. (†36053) 195/H7
Mount Angel, Oreg. (97362) 291/B3
Mount Apo National Park, Philippines 82/E7
Mount Arlington, N.J. (07856) 273/D2
Mount Arrowsmith, N.S. Wales 97/A2
Mount Assiniboine Prov. Park, Br. Col. 184/K5
Mount Auburn, Ill. (62547) 222/D4
Mount Auburn, Ind. (†47327) 227/G5
Mount Auburn, Iowa (52313) 229/J4
Mount Aukum, Calif. (95656) 204/E5
Mount Ayr, Ind. (47964) 227/C3
Mount Ayr, Iowa (50854) 229/E7
Mount Barker, S. Australia 94/C8
Mount Barker, W. Australia 88/B6
Mount Barker, W. Australia 92/B6
Mount Beauty, Victoria 97/D5
Mount Bellew, Ireland 17/D5
Mount Berry, Georgia (30149) 217/B2
Mount Bethel, Georgia (†30060) 217/K1
Mount Blanchard, Ohio (45867) 284/D4
Mount Bold (res.), S. Australia 94/B8
Mount Brydges, Ontario 177/C5
Mount Calm, Texas (76673) 303/H6
Mount Calvary, Wis. (53057) 317/K8
Mount Carbon, W. Va. (25139) 312/D7
Mount Carleton Prov. Park, New Bruns. 170/D1
Mount Carmel, Ala. (†36047) 195/F6
Mount Carmel, Ill. (62863) 222/F5
Mount Carmel, Ind. (†47012) 227/H6
Mount Carmel, Miss. (†39474) 256/E7
Mount Carmel, Newf. 166/C2
Mount Carmel, N. Dak. (†58249) 282/O2
Mount Carmel, Ohio (45244) 284/C10
Mount Carmel, Pa. (17851) 294/K4
Mount Carmel, Pr. Edward I. 168/D2
Mount Carmel, S.C. (29840) 296/C3
Mount Carmel, Tenn. (37642) 237/R8
Mount Carmel, Utah (84755) 304/B6
Mount Carroll, Ill. (61053) 222/D1
Mount Cavenagh, North. Terr. 93/C8
Mountcharles, Ireland 17/E2
Mount Clare, W. Va. (26408) 312/F4
Mount Clemens, Mich. (48043) 250/G6
Mount Cory, Ohio (45868) 284/C4
Mount Crawford, Va. (22841) 307/L4
Mount Croghan, S.C. (29727) 296/G2
Mount Currie, Br. Col. 184/F5
Mount Darwin, Zimbabwe 118/E3
Mount Desert, Maine (04660) 243/G7
Mount Desert○, Maine (04660) 243/G7
Mount Desert (isl.), Maine 243/G7
Mount Desert Rock (isl.), Maine 243/G8
Mount Dora, Fla. (32757) 212/E3
Mount Dora, N. Mex. (88429) 274/F2
Mount Doreen, North. Terr. 93/B7
Mount Douglas, Queensland 95/C4
Mount Drysdale, N.S. Wales 97/C2
Mount Eaton, Ohio (44659) 284/G4
Mount Eba, S. Australia 94/D4
Mount Eden, Ky. (40046) 237/L4
Mount Eden, N. Zealand 100/B1
Mount Edziza Prov. Park and Rec. Area, Br. Col. 184/B1
Mount Elgin, Ontario 177/D5
Mount Emu (creek), Victoria 97/B5
Mount Enterprise, Texas (75681) 303/K6
Mount Ephraim, N.J. (08059) 273/B3
Mount Erie, Ill. (62446) 222/E5
Mount Etna, Ind. (†46750) 227/F3
Mount Etna, Iowa (50855) 229/D6
Mount Everard, Guyana 131/B2
Mount Forest, Mich. (†48650) 250/F5
Mount Forest, Ontario 177/D4
Mount Freedom, N.J. (07970) 273/D2
Mount Gambier, Australia 87/D9
Mount Gambier, S. Australia 88/G7
Mount Gambier, S. Australia 94/G7
Mount Gay, W. Va. (25637) 312/C7
Mount Gilead, N.C. (27306) 281/K4
Mount Gilead, Ohio (43338) 284/E4
Mount Gravatt, Queensland 88/K3
Mount Hagen, Papua N.G. 85/B7
Mount Hamill, Iowa (†52625) 229/K7
Mount Healthy, Ohio (45231) 284/B9
Mount Hermon, Calif. (95041) 204/K4
Mount Hermon, La. (70450) 238/K5
Mount Hermon, Mass. (01354) 249/D2
Mount Holly, Ark. (71758) 202/E7
Mount Holly○, N.J. (08060) 273/D4
Mount Holly, N.C. (28120) 281/H4
Mount Holly, S.C. (29463) 296/H5
Mount Holly○, Vt. (05758) 268/B5
Mount Holly, Va. (22524) 307/P4
Mount Holly Springs, Pa. (17065) 294/H5
Mount Hood, Oreg. (97041) 291/F2
Mount Hope, Ala. (35651) 195/D2
Mount Hope (riv.), Conn. 210/G1
Mount Hope, Kansas (67108) 232/E4
Mount Hope (bay), Mass. 249/K6
Mount Hope, N.J. (†07885) 273/D2
Mount Hope, N.S. Wales 97/C3
Mount Hope, Ohio (44660) 284/G4
Mount Hope, Ontario 177/E4
Mount Hope (bay), R.I. 249/K6
Mount Hope, W. Va. (25880) 312/D7
Mount Hope, Wis. (53816) 317/D10
Mount Horeb, Wis. (53572) 317/G10
Mount Ida, Ark. (71957) 202/C4
Mount Ida, Wis. (†53809) 317/E10
Mount Isa, Queensland 95/A4
Mount Isa, Queensland 88/F4
Mount Jackson, Va. (22842) 307/L3
Mount Jewett, Pa. (16740) 294/E2
Mount Joy, Pa. (17552) 294/K5
Mount Judea, Ark. (72655) 202/D2
Mount Juliet, Tenn. (37122) 237/H8
Mount Kisco, N.Y. (10549) 276/N8
Mount Kuring-gai, N.S. Wales 97/J3
Mountlake Terrace, Wash. (98043) 310/B1
Mount Laurel○, N.J. (†08054) 273/D4
Mount Lebanon, La. (71028) 238/D2
Mount Lebanon○, Pa. (15228) 294/B7
Mount Lemmon, Ariz. (85619) 198/E6
Mount Leonard, Mo. (†65339) 261/F4
Mount Liberty, Ohio (43048) 284/E5
Mount Lofty (range), S. Australia 88/F6
Mount Lookout, W. Va. (26678) 312/E6
Mount Magnet, W. Australia 88/B5
Mount Magnet, W. Australia 92/B5
Mount Margaret, Queensland 95/A3
Mount Margaret, W. Australia 88/C5
Mount Margaret, W. Australia 92/C5
Mount Maunganui, N. Zealand 100/E2
Mount Meigs, Ala. (36057) 195/F6
Mountmellick, Ireland 10/C4
Mountmellick, Ireland 17/G5
Mount Meridian, Ind. (†46135) 227/D5
Mount Molloy, Queensland 95/C3
Mount Montgomery, Nev. (†89422) 266/C5
Mount Morgan, Queensland 88/J4
Mount Morgan, Queensland 95/D4
Mount Moriah, Mo. (64665) 261/E2
Mount Morris, Ill. (61054) 222/D1
Mount Morris, Mich. (48458) 250/F5
Mount Morris, N.Y. (14510) 276/E5
Mount Morris, Pa. (15349) 294/B6
Mount Mourne, N.C. (28123) 281/H3
Mount Nebo, W. Va. (26679) 312/E6
Mount Olga Nat'l Park, North. Terr. 93/B8
Mount Olive, Ill. (62069) 222/D4
Mount Olive, Miss. (39119) 256/E7
Mount Olive○, N.J. (†07828) 273/D2
Mount Olive, N.C. (28365) 281/O4
Mount Oliver, Pa. (15210) 294/B7
Mount Olivet, Ky. (41064) 237/N3
Mount Orab, Ohio (45154) 284/C7
Mount Pearl, Newf. 166/D2
Mount Penn, Pa. (19606) 294/L5
Mount Pleasant, Ark. (72561) 202/G2
Mount Pleasant, Del. (†19709) 245/R2
Mount Pleasant, Fla. (32352) 212/B1
Mount Pleasant, Ind. (47559) 227/D8
Mount Pleasant, Iowa (52641) 229/L7
Mount Pleasant, Md. (†21701) 245/J3
Mount Pleasant, Mich. (48858) 250/E5
Mount Pleasant, Miss. (38649) 256/E1
Mount Pleasant, N.C. (28124) 281/J4
Mount Pleasant, Nova Scotia 168/C4
Mount Pleasant, Ohio (43939) 284/J5
Mount Pleasant, Ontario 177/D4
Mount Pleasant, Pa. (15666) 294/D5
Mount Pleasant, S.C. (29464) 296/H6
Mount Pleasant, Tenn. (38474) 237/G9
Mount Pleasant, Texas (75455) 303/K4
Mount Pleasant, Utah (84647) 304/C4
Mount Pocono, Pa. (18344) 294/M3
Mount Prospect, Ill. (60056) 222/B5
Mount Pulaski, Ill. (62548) 222/D3
Mountrail (co.), N. Dak. 282/E3
Mount Rainier, Md. (20822) 245/F4
Mount Rainier Nat'l Park, Wash. 310/D4
Mountrath, Ireland 17/F5
Mount Revelstoke Nat'l Park, Br. Col. 184/H4
Mount Robson, Br. Col. 184/H3
Mount Robson Prov. Park, Br. Col. 184/H3
Mount Rogers Nat'l Rec. Area, Va. 307/F7
Mount Roskill, N. Zealand 100/B1
Mount Royal, N.J. (08061) 273/C4
Mount Royal (range), N.S. Wales 97/F2
Mount Rushmore Nat'l Mem., S. Dak. 298/B6
Mounts (bay), England 10/D6
Mounts (bay), England 13/B7
Mount Salem, Ky. (†40437) 237/M6
Mount Savage, Md. (21545) 245/C2
Mount Shasta, Calif. (96067) 204/C5
Mount Sherman, Ark. (†72641) 202/D1
Mount Sherman, Ky. (42764) 237/K6
Mount Sidney, Va. (24467) 307/K4
Mount Solon, Va. (22843) 307/K4
Mount Standfast, Barbados 161/B8
Mount Sterling, Ala. (†36904) 195/B6
Mount Sterling, Ill. (62353) 222/C4
Mount Sterling, Ind. (†47043) 227/G7
Mount Sterling, Iowa (52573) 229/J7
Mount Sterling, Ky. (40353) 237/N4
Mount Sterling, Ohio (43143) 284/D6
Mount Sterling, Wis. (54645) 317/D9
Mount Stewart, Pr. Edward I. 168/F2
Mount Storm, W. Va. (26739) 312/H4
Mount Storm (lake), W. Va. 312/H4
Mount Summit, Ind. (47361) 227/G4
Mount Sunapee, N.H. (03772) 268/C5
Mount Surprise, Queensland 95/C3
Mount Tabor○, Vt. (†05739) 268/B5
Mount Tabor, Wis. (†54638) 317/F8
Mount Tivoli, Grenada 161/D8
Mount Tom, Mass. (01058) 249/D3
Mount Trumbull, Ariz. (†84770) 198/B2
Mount Uniacke, Nova Scotia 168/E4
Mount Union, Iowa (52644) 229/L6
Mount Union, Pa. (17066) 294/G5
Mount Upton, N.Y. (13809) 276/K6
Mount Vernon, Ala. (36560) 195/B8
Mount Vernon, Ark. (72111) 202/F3
Mount Vernon, Georgia (30445) 217/G6
Mount Vernon, Ill. (62864) 222/E5
Mount Vernon, Ind. (47620) 227/B9
Mount Vernon, Iowa (52314) 229/K5
Mount Vernon, Ky. (40456) 237/N6
Mount Vernon○, Maine (04352) 243/D7
Mount Vernon, Md. (†21853) 245/P8
Mount Vernon, Mo. (65712) 261/E8
Mount Vernon, N.Y. (*10550) 276/O7
Mount Vernon, Ohio (43050) 284/E5
Mount Vernon, Oreg. (97865) 291/H3
Mount Vernon, S. Dak. (57363) 298/N6
Mount Vernon, Tenn. (37358) 237/N10
Mount Vernon, Texas (75457) 303/J4
Mount Vernon, Va. (22121) 307/O3
Mount Vernon, Wash. (98273) 310/C2
Mount Vernon, W. Australia 92/B4
Mount Vernon, Wis. (†53572) 317/G10
Mount Vernon Springs, N.C. (27345) 281/L3
Mount Victory, Ohio (43340) 284/D4
Mountview, W. Va. (†25825) 312/D7
Mountville, Georgia (30261) 217/C4
Mountville, Pa. (17554) 294/K5
Mountville, S.C. (29370) 296/C3
Mount Washington, Ky. (40047) 237/K4

Mount Washington○, Mass. (†12517) 249/A4
Mount Wellington, N. Zealand 100/C1
Mount Willing, Ala. (†36012) 195/E6
Mount Wolf, Pa. (17347) 294/J5
Mount Zion, Georgia (30150) 217/B3
Mount Zion, Ill. (62549) 222/E4
Mount Zion, Ind. (†46792) 227/G3
Mount Zion, Iowa (†52565) 229/K7
Mount Zion, W. Va. (26151) 312/D5
Moura, Portugal 33/C3
Moura, Queensland 95/D5
Moura, Queensland 88/J4
Mourão, Portugal 33/C3
Mourdi (depr.), Chad 111/D4
Mourne (Newry and Mourne) (dist.), N. Ireland17/J3
Mourne (mts.), N. Ireland 17/J3
Mourne (riv.), N. Ireland 17/G2
Mouscron, Belgium 27/C7
Moussoro, Chad 111/C5
Mouthcard, Ky. (41548) 237/S6
Mouth of Wilson, Va. (24363) 307/F7
Moutier, Switzerland 39/D2
Moûtiers, France 28/G5
Mouton (isl.), Nova Scotia 168/D5
Mouydir (mts.), Algeria 106/E3
Moville, Iowa (51039) 229/A4
Moville, Ireland 17/G1
Mowbray, Manitoba 179/D5
Mowdok Mual (mt.), Bangladesh 68/G4
Moweaqua, Ill. (62550) 222/E4
Mower (co.), Minn. 255/F7
Mowming (Maoming), China 77/H7
Mowrystown, Ohio (45155) 284/C7
Moxahala, Ohio (43761) 284/F6
Moxee City, Wash. (98936) 310/E4
Moxico (dist.), Angola 115/D6
Moxie (lake), Maine 243/D5
Moxley, Georgia (†30477) 217/H5
Moy (riv.), Ireland 17/C3
Moy, N. Ireland 17/H3
Moyale, Ethiopia 111/G7
Moyale, Kenya 115/G3
Moyamba, S. Leone 106/B7
Moycullen, Ireland 17/C5
Moyers, Okla. (74557) 288/P6
Moyers, W. Va. (26813) 312/H6
Moyeuvre-Grande, France 28/G3
Moygashel, N. Ireland 17/H3
Moyie, Br. Col. 184/K5
Moyie (riv.), Idaho 220/B1
Moyie Springs, Idaho (83845) 220/B1
Moyle (dist.), N. Ireland 17/J1
Moynalty, Ireland 17/H4
Moyo, Uganda 115/F3
Moyobamba, Peru 120/B3
Moyobamba, Peru 128/D6
Moyock, N.C. (27958) 281/S1
Moyogalpa, Nicaragua 154/E5
Moyu (Karakax), China 77/A4
Moza Illit, Israel 65/C4
Mozambique 2/L6
Mozambique 102/F6
Mozambique (chan.) 2/L7
Mozambique (chan.) 102/G6
Mozambique (pt.), La. 238/M7
Mozambique (chan.), Madagascar 118/G3
MOZAMBIQUE 118/E4
Mozambique (chan.), Mozambique 118/G3
Mozart, Sask. 181/G4
Mozer, W. Va. (†26866) 312/H5
Mozhaysk, U.S.S.R. 52/E3
Mozhga, U.S.S.R. 52/H3
Mozier, Ill. (62070) 222/C4
Mozyr', U.S.S.R. 48/C4
Mozyr', U.S.S.R. 52/C4
Mpanda, Tanzania 115/F5
Mpika, Zambia 115/F6
Mporokoso, Zambia 115/F5
M'Pouya, Congo 115/C4
Mpraeso, Ghana 106/D7
Mpulungu, Zambia 115/F5
Mpwapwa, Tanzania 115/G5
Mrągowo, Poland 47/E2
Msaken, Tunisia 106/G1
M'Sila, Algeria 106/E1
Msta (riv.), U.S.S.R. 52/D3
Mtakuja, Tanzania 115/F5
Mtsensk, U.S.S.R. 52/E4
Mtwara (reg.), Tanzania 115/G5
Mtwara-Mikindani, Tanzania 102/G6
Mtwara-Mikindani, Tanzania 115/H6
Mu (riv.), Burma 72/B2
Mualama, Mozambique 118/F3
Muang Hinboun, Laos 72/E3
Muang Kènthao, Laos 72/D3
Muang Khammouan, Laos 72/E3
Muang Không, Laos 72/E4
Muang Khôngxédôn, Laos 72/E4
Muang Khoua, Laos 72/D2
Muang May, Laos 72/E4
Muang Ou Tai, Laos 72/D2
Muang Pak-Lay, Laos 72/D3
Muang Paktha, Laos 72/D2
Muang Pakxan, Laos 72/D2
Muang Phin, Laos 72/E3
Muang Sing, Laos 72/D2
Muang Tahoi, Laos 72/E3
Muang Vangviang, Laos 72/D3
Muang Vapi, Laos 72/E4
Muang Xaignabouri (Sayaboury), Laos 72/D3
Muang Xay, Laos 72/D3
Muang Xépôn, Laos 72/E3
Muang Xon, Laos 72/D2
Muar, Malaysia 72/D7
Muarabungo, Indonesia 85/C6
Muarasiberut, Indonesia 85/B6
Muaratewe, Indonesia 85/F6
Muari, Ras (cape), Pakistan 68/B4
Muari, Ras (cape), Pakistan 59/J5
Mubarraz, Saudi Arabia 59/E4
Mubende, Uganda 115/F3
Mubi, Nigeria 106/G6
Muchanes, Bolivia 136/B4
Mücheln, E. Germany 22/D3
Muck (isl.), Scotland 10/C2
Muck (isl.), Scotland 15/B4
Muckamore, N. Ireland 17/J2
Muckle Flugga (isl.), Scotland 15/G2
Muckleshoot Ind. Res., Wash. 310/C3
Muckno (lake), Ireland 17/H3
Muco (riv.), Colombia 126/E5
Mucojo, Mozambique 118/G2
Mucope, Angola 115/B7
Mucuchachí, Venezuela 124/C3
Mucuchíes, Venezuela 124/C3
Mucugê, Brazil 132/F6
Mucur, Turkey 63/F3
Mucurapo, Trin. & Tob. 161/A10
Mucuri, Brazil 132/G7
Mucuripe (pt.), Brazil 132/G3
Mucusso, Angola 115/D7
Mud (lake), Idaho 220/F6
Mud (riv.), Ky. 237/H7
Mud (lake), La. 238/D7
Mud (lake), Minn. 255/C2
Mud (lake), Minn. 255/B5
Mud (riv.), Minn. 255/C2
Mud (isl.), Nova Scotia 168/B5
Mud (creek), Okla. 288/L6
Mud (creek), Oreg. 291/K2
Mud (creek), S. Dak. 298/N3
Mud (lake), S. Dak. 298/R2
Mud, W. Va. (†25565) 312/C6
Mud (riv.), W. Va. 312/B6
Mudanjiang (Mutankiang), China 77/M3
Mudanjiang, China54/O5
Mudan Jiang (riv.), China 77/L3
Mudanya, Turkey 63/C2
Mudauwara, Jordan 59/C4
Mud Bay, Br. Col. 184/H2
Mud Butte, S. Dak. (57758) 298/D4
Muddy (creek), Colo. 208/E4
Muddy (brook), Conn. 210/H1
Muddy (pond), Conn. 210/G1
Muddy (riv.), Conn. 210/D3
Muddy, Ill. (62965) 222/E6
Muddy (mts.), Nev. 266/G6
Muddy (creek), N. Dak. 282/G6
Muddy (pt.), St. Chris.-Nevis 161/C10
Muddy (lake), Sask. 181/B3
Muddy (creek), Utah 304/A2
Muddy (creek), Utah 304/C4
Muddy (creek), Wyo. 319/F3
Muddy (creek), Wyo. 319/D2
Muddy (mt.), Wyo. 319/F3
Muddy Boggy (creek), Okla. 288/O5
Mudge (pond), Conn. 210/B1
Mudgee, N. S. Wales 88/J6
Mudgee, N.S. Wales 97/E3
Mudhnib, Saudi Arabia 59/D4
Mudjatik (riv.), Sask. 181/L3
Mud Lake, Idaho (†83450) 220/F6
Mud Lake, Newf. 166/B3
Mud Lake (res.), S. Dak. 298/N2
Mud Mountain (lake), Wash. 310/D3
Mudon, Burma 72/C3
Mudug (prov.), Somalia 115/J2
Mudurnu, Turkey 63/D2
Muecate, Mozambique 118/F2
Mueda, Mozambique 118/F2
Muenster, Sask. 181/F3
Muenster, Texas (76252) 303/G4
Muerto, Mar (lag.), Mexico 150/N9
Muezerskiy, U.S.S.R. 52/D2
Muff, Ireland 17/G1
Mufulira, Zambia 115/E6
Mufulira, Zambia 102/E6
Muge, Portugal 33/B3
Mugford (cape), Newf. 166/B2
Mughar, Israel 65/C2
Muğla (prov.), Turkey 63/C4
Muğla, Turkey 59/A2
Muğla, Turkey 63/C4
Muglad, Sudan 111/E5
Muhammad, Ras (cape), Egypt 59/B4
Muhammad, Ras (cape), Egypt 111/F2
Muhammad Qol, Sudan 59/C5
Muhammad Qol, Sudan 111/G3
Muharraq, Bahrain 59/F4
Mühldorf am Inn, W. Germany 22/E4
Muhlenberg (co.), Ky. 237/G6
Mühlhausen (Thomas-Müntzer-Stadt), E. Germany 22/D3
Mühlviertel (reg.), Austria 41/C2
Muhu (isl.), U.S.S.R. 53/B1
Mui Bai Bung (pt.), Vietnam 54/M9
Muiden, Netherlands 27/G4
Muinebeag, Ireland 17/H6
Muir (glac.), Alaska 196/M1
Muir, Mich. (48860) 250/D5
Muir, Pa. (17957) 294/J4
Muirkirk, Md. (†20705) 245/L4
Muirkirk, Scotland 15/E5
Muir of Ord, Scotland 15/D3
Muiron (isls.), W. Australia 88/A4
Muiron (isls.), W. Australia 92/A3
Muir Woods Nat'l Mon., Calif. 204/H2
Muizenberg, S. Africa 118/E7
Muju, S. Korea 81/C6
Mukachevo, U.S.S.R. 52/B5
Mukah, Malaysia 85/E5
Mukalla, P.D.R. Yemen 54/F8
Mukalla, P.D.R. Yemen 59/E7
Mukdahan, Thailand 72/E3
Mukden, Bolivia 136/A2
Mukden (Shenyang), China 77/K3
Mukilteo, Wash. (98275) 310/C3
Mukinbudin, W. Australia 92/B5
Muko, Japan 81/J7
Muko (isl.), Japan 81/M3
Muko (riv.), Japan 81/H7
Mukutawa (lake), Manitoba 179/G2
Mukutawa (riv.), Manitoba 179/E1
Mukwonago, Wis. (53149) 317/J2
Mula, Spain 33/F3
Mulaló, Ecuador 128/C3
Mulanje (mt.), Malawi 102/F6
Mulanje (mts.), Malawi 115/G7
Mulatos, Colombia 126/B3
Mulberry (creek), Ala. 195/E5
Mulberry, Ark. (72947) 202/B2
Mulberry (riv.), Ark. 202/C2
Mulberry, Calif. (†95926) 204/D4
Mulberry, Fla. (33860) 212/E4
Mulberry, Ind. (46058) 227/D4
Mulberry, Kansas (66756) 232/H4
Mulberry, Ohio (†45150) 284/B7
Mulberry, Tenn. (37359) 237/H10
Mulberry Fork (riv.), Ala. 195/E3
Mulberry Grove, Ill. (62262) 222/D5
Mulchatna (riv.), Alaska 196/G2
Mulchén, Chile 138/E1
Mulde (riv.), E. Germany 22/E3
Muldon, Miss. (†39730) 256/G3
Muldoon, Texas (78949) 303/G8
Muldraugh, Ky. (40155) 237/J5
Muldrow, Okla. (74948) 288/S4
Mule (mts.), Ariz. 198/E7
Mule (creek), Kansas 232/C4
Mule Creek, N. Mex. (88051) 274/A5
Mule Creek, Wyo. (†57735) 319/H2
Muleculus, Nicaragua 154/E4
Mulegé, Mexico 150/C3
Mulegns, Switzerland 39/J3
Muleshoe, Texas (79347) 303/B3
Mulgrave, Nova Scotia 168/G3
Mulgrave (lake), Nova Scotia 168/F3
Mulhacén (mt.), Spain 33/E4
Mulhall, Okla. (73063) 288/M2
Mülheim an der Ruhr, W. Germany 22/B3
Mulhouse, France 7/E4
Mulhouse, France 28/G4
Mulhurst, Alberta 182/D3
Muli, China 77/F6
Muli (str.), Indonesia 85/K7
Mulino, Oreg. (97042) 291/B2
Mulinu'u (cape), W. Samoa 86/L8
Mulkear (riv.), Ireland 17/E6
Mull (head), Scotland 15/F2
Mull (head), Scotland 15/F1
Mull (isl.), Scotland 10/D2
Mull (isl.), Scotland 15/C4
Mull (sound), Scotland 15/C4
Mullagh, Ireland 17/H4
Mullaghareirk (mts.), Ireland 17/C7
Mullaghearn (mt.), N. Ireland 17/G2
Mullaghmore, Ireland 17/D3
Mullaittivu, Sri Lanka 68/E7
Mullaley, N.S. Wales 97/E2
Mullan, Idaho (83846) 220/C2
Mullardoch, Loch (lake), Scotland 15/C3
Mullen, Nebr. (69152) 264/C2
Mullens, W. Va. (25882) 312/D7
Müller (mts.), Indonesia 85/E5
Mullet (key), Fla. 212/D4
Mullett (lake), Mich. 250/E3
Mullett Lake, Mich. (49761) 250/E3
Mullewa, W. Australia 88/B5
Mullewa, W. Australia 92/A5
Müllheim, Switzerland 39/G1
Müllheim, W. Germany 22/B5
Mullica (riv.), N.J. 273/D4
Mullica Hill, N.J. (08062) 273/C4
Mulliken, Mich. (48861) 250/E6
Mullin, Texas (76864) 303/F6
Mullinahone, Ireland 17/F7
Mullinavat, Ireland 17/G7
Mullingar, Ireland 10/C4
Mullingar, Ireland 17/G4
Mullingar, Sask. 181/D2
Mullins, S.C. (29574) 296/J3
Mullinville, Kansas (67109) 232/C4
Mullion, England 13/B7
Mull of Galloway (prom.), Scotland 15/D6
Mull of Kintyre (prom.), Scotland 15/C5
Mull of Oa (prom.), Scotland 15/B5
Mullumbimby, N.S. Wales 97/G1
Mulobezi, Zambia 115/E7
Mulongo, Zaire 115/E5
Mulroy (bay), Ireland 17/F1
Multan, Pakistan 54/J6
Multan, Pakistan 68/C2
Multan, Pakistan 59/K3
Multnomah (co.), Oreg. 291/E2
Mulund, India 68/B6
Mulungushi (dam), Zambia 115/E6
Mulvane, Kansas (67110) 232/E4
Mulvihill, Manitoba 179/D4
Mulwala, N.S. Wales 97/D4
Mumbwa, Zambia 115/E6
Mumford, N.Y. (14511) 276/E4
Mümliswil-Ramiswil, Switzerland 39/E2
Mumra, U.S.S.R. 52/G5
Mun, Mae Nam (riv.), Thailand 72/D4
Muna (isl.), Indonesia 85/G7
Muna, Mexico 150/P6
Munbura, Queensland 95/C2
Munbura, Queensland 88/G2
Müncheberg, E. Germany 22/F2
München (Munich), W. Germany 22/D4
Münchenbuchsee, Switzerland 39/E2
Muncho Lake, Br. Col. 184/L2
Muncho Lake Prov. Park, Br. Col. 184/L2
Muncie, Ill. (61857) 222/F3
Muncie, Ind. 188/J2
Muncie, Ind. (*47302) 227/G4
Muncy, Pa. (17756) 294/J3
Muncy Valley, Pa. (17758) 294/J3
Munda, Solomon Is. 86/D3
Mundabullangana, W. Australia 92/B3
Mundare, Alberta 182/D3
Mundaring (res.), W. Australia 88/B2
Munday, Texas (76371) 303/E4
Mundelein, Ill. (60060) 222/A4
Munden, Kansas (66959) 232/E2
Münden, W. Germany 22/C3
Mundesley, England 13/J5
Mundijong, W. Australia 88/B3
Mundijong, W. Australia 92/A2
Mundiwindi, W. Australia 92/B3
Mundo Novo, Brazil 132/F5
Mundrabilla, W. Australia 92/E5
Mundubbera, Queensland 88/J5
Munera, Spain 33/E3
Munford, Ala. (36268) 195/F3
Munford, Tenn. (38058) 237/B10
Munfordville, Ky. (42765) 237/J6
Mungbere, Zaire 115/E3
Munger, Mich. (48747) 250/F5
Mungindi, N.S. Wales 97/E1
Mungindi, Queensland 95/D6
Munguba, Brazil 132/C2
Munhall, Pa. (15120) 294/C7
Munhango, Angola 115/C6
Munich, N. Dak. (58352) 282/N2
Munich, W. Germany 7/F4
Munich, W. Germany 22/D4
Munising, Mich. (49862) 250/C2
Munith, Mich. (49259) 250/E6
Munjor, Kansas (†67601) 232/C3
Munku-Sardyk (mt.), Mongolia 77/F1
Munnerlyn, Georgia (†30830) 217/H5
Munning (pt.), N. Zealand 100/E7
Munnsville, N.Y. (13409) 276/J4
Muñoz Gamero (pen.), Chile 138/D10
Munro (mt.), Tasmania 99/E2
Munroe Falls, Ohio (44262) 284/H3
Münsingen, Switzerland 39/E3
Munson, Alberta 182/D4
Munson, Fla. (†32570) 212/B5
Munson, Pa. (16860) 294/F4
Munsonville, N.H. (03457) 268/C5
Munster, Ind. (46321) 227/B1
Munster (prov.), Ireland 17/D7
Munster (trad. prov.), Ireland 17
Munster, Ontario 177/J2
Münster, Switzerland 39/F4
Münster, W. Germany 22/B3
Munsungan (lake), Maine 243/E3
Muntendam, Netherlands 27/K2
Muntok, Indonesia 85/D6
Munuscong (lake), Mich. 250/E2
Muojärvi (lake), Finland 18/R4
Muong Khuong, Vietnam 72/E2
Muonio (riv.) 7/G2
Muonio, Finland 18/O3
Muonio (riv.), Finland 18/M2
Muonioälv (riv.), Sweden 18/M2
Muota (riv.), Switzerland 39/G3
Muotathal, Switzerland 39/G3
Muqaddam, Wadi (dry riv.), Sudan 111/F4
Muqdadiyah, Iraq 66/D4
Muqdisho (Mogadishu) (cap.), Somalia 115/J3
Muqdisho (Mogadishu) (cap.), Somalia 102/G4
Muqeible, Israel 65/C2
Muqui, Brazil 132/F8
Mur (riv.), Austria 41/C3
Mur (riv.), Yugoslavia 45/B2
Mura (riv.), Hungary 41/D3
Muradiye, Turkey 63/K3
Murakami, Japan 81/J4
Murallón, Cerro (mt.), Argentina 143/B6
Murallón, Cerro (mt.), Chile 138/D8
Murarrie, Queensland 88/K2
Murashi, U.S.S.R. 52/G3
Murat, France 28/E5
Murat (riv.), Turkey 59/C2
Murat (riv.), Turkey 63/H3
Murat Dağı (mt.), Turkey 63/C3
Murau, Austria 41/C3
Murbat, Oman 59/G6
Murchison, N. Zealand 100/D4
Murchison (range), North. Terr. 93/D6
Murchison (falls), Uganda 115/F3
Murchison (riv.), W. Australia 88/B5
Murchison (mt.), W. Australia 92/B4
Murchison (riv.), W. Australia 92/B4
Murchison Downs, W. Australia 88/B5
Murcia (prov.), Spain 33/F4
Murcia, Spain 7/D5
Murcia, Spain 33/F4
Murcia (reg.), Spain 33/F3
Murderers (creek), Oreg. 291/H3
Murderkill (riv.), Del. 245/R5
Murdo, S. Dak. (57559) 298/H6
Murdochville, Québec 172/C1
Murdock, Fla. (33938) 212/D4
Murdock, Ill. (61941) 222/E4
Murdock, Kansas (67111) 232/E4
Murdock, Minn. (56271) 255/C5
Murdock, Nebr. (68407) 264/H4
Muren (Mörön), Mongolia 77/F2
Mureş (riv.), Romania 45/E2
Muret, France 28/D6
Muretto (pass), Switzerland 39/J4
Murfreesboro, Ark. (71958) 202/C5
Murfreesboro, N.C. (27855) 281/R2
Murfreesboro, Tenn. (37130) 237/J9
Murg (riv.), Switzerland 39/G1
Murgab, U.S.S.R. 48/H6
Murgab (riv.), U.S.S.R. 48/G6
Murghab (riv.), Afghanistan 59/H2
Murgon, Queensland 88/J5
Murgon, Queensland 95/D5
Murgoo, W. Australia 92/B4
Muri, Switzerland 39/F2
Muriaé, Brazil 135/E2
Muriaé, Brazil 132/F8
Murias de Paredes, Spain 33/C1
Muri bei Bern, Switzerland 39/E3
Muriel (lake), Alberta 182/E2
Murindó, Colombia 126/B4
Müritzsee (lake), E. Germany 22/E2
Murjek, Sweden 18/M3
Murl, Ky. (†42633) 237/M7
Murle, Ethiopia 111/G6
Murmansk, U.S.S.R. 7/H2
Murmansk, U.S.S.R. 2/L2
Murmansk, U.S.S.R. 4/C8
Murmansk, U.S.S.R. 48/D3
Murmansk, U.S.S.R. 52/D1
Murnau, W. Germany 22/D5
Murnpeowie, S. Australia 94/F3
Murom, U.S.S.R. 52/F3
Murongo, Tanzania 115/F4
Muroran, Japan 81/K2
Muros, Spain 33/B1
Muroto, Japan 81/G7
Muroto (pt.), Japan 81/G7
Murphy, Idaho (83650) 220/B6
Murphy, Miss. (†38748) 256/C4
Murphy, Mo. (†63088) 261/O4
Murphy, N.C. (28906) 281/B4
Murphy, Oreg. (97533) 291/D5
Murphy (isl.), S.C. 296/J5
Murphy, Texas (†75074) 303/H1
Murphys, Calif. (95247) 204/E5
Murphysboro, Ill. (62966) 222/D6
Murphytown, W. Va. (†26142) 312/D4
Murra Murra, Queensland 95/C6
Murray (riv.) 88/G6
Murray (river), Australia 87/E9
Murray (riv.), Br. Col. 184/G3
Murray (co.), Georgia 217/C1
Murray, Idaho (83874) 220/C2
Murray, Ind. (†46714) 227/G3
Murray, Iowa (50174) 229/F6
Murray, Ky. (42071) 237/E7
Murray (co.), Minn. 255/C6
Murray, Nebr. (68409) 264/J4
Murray (riv.), N.S. Wales 97/A4
Murray (co.), Okla. 288/M6
Murray (lake), Okla. 288/M6
Murray (lake), Papua N.G. 85/B7
Murray (riv.), S. Australia 94/F6
Murray (lake), S.C. 188/K4
Murray (lake), S.C. 296/D4
Murray, Utah (84107) 304/C3
Murray, Utah 188/D2
Murray (riv.), Victoria 97/A4
Murray (bay), Victoria 88/H7
Murray (riv.), W. Australia 92/A2
Murray Bridge, S. Australia 88/F7
Murray Bridge, S. Australia 94/F6
Murray City, Ohio (43144) 284/F6
Murray Corner, New Bruns. 170/G2
Murray Downs, North. Terr. 93/D6
Murray Harbour, Pr. Edward I. 168/F2
Murray Lake Hills, Tenn. (37416) 237/L10
Murray River, Pr. Edward I. 168/F2
Murraysville, W. Va. (26153) 312/C4
Murrayville, Georgia (30564) 217/E2
Murrayville, Ill. (62668) 222/C4
Murrayville, Victoria 97/A4
Murree, Pakistan 68/C2
Murrells Inlet, S.C. (29576) 296/K4
Mürren, Switzerland 39/E3
Murrieta, Calif. (92362) 204/H10
Murringo, N.S. Wales 97/E4
Murrumbidgee (riv.), N. S. Wales 88/H6
Murrumbidgee (riv.), N.S. Wales 97/C4
Murrumburrah, N.S. Wales 97/E4
Murrupula, Mozambique 118/F3
Murrurundi, N.S. Wales 97/F2
Murrysville, Pa. (15668) 294/C5
Murska Sobota, Yugoslavia 45/C2
Murtaröl (peak), Switzerland 39/K3
Murtaugh, Idaho (83344) 220/D7
Murten, Switzerland 39/D3
Murtle (lake), Br. Col. 184/H4
Murtoa, Victoria 97/B5
Murud, India 68/C5
Murupara, N. Zealand 100/F3
Mururoa (isl.), Fr. Poly. 87/M8
Murwara, India 68/E4
Murwillumbah, N. S. Wales 88/J5
Murwillumbah, N.S. Wales 97/G1
Muryo (mt.), Indonesia 85/J2
Mürz (riv.), Austria 41/C3
Murzuk, Libya 102/D2
Murzuk, Libya 111/B2
Mürzzuschlag, Austria 41/C3
Muş (prov.), Turkey 63/J3
Muş, Turkey 63/J3
Muş, Turkey 59/D2
Musa Khel Bazar, Pakistan 59/K3
Musa Khel Bazar, Pakistan 68/B2
Musala (mt.), Bulgaria 45/F4
Musan, N. Korea 81/D2
Musandam, Ras (cape), Oman 59/G4
Musashino, Japan 81/O2
Muscadine, Ala. (36269) 195/H3
Muscat (cap.), Oman 54/G7
Muscat (cap.), Oman 2/M4
Muscat (cap.), Oman 59/G5
Muscatatuck (riv.), Ind. 227/E7
Muscatine (co.), Iowa 229/L5
Muscatine, Iowa 188/H2
Muscatine, Iowa (52761) 229/L6
Muscle Shoals, Ala. (35660) 195/C1
Muscoda, Wis. (53573) 317/F9
Muscogee (co.), Georgia 217/C6
Musconetcong (beach), N.J. 273/C2
Muscongus (bay), Maine 243/E8
Muscotah, Kansas (66058) 232/G2
Muscoy, Calif. (92405) 204/E10
Muse, Okla. (74949) 288/S5
Musella, Georgia (31066) 217/E5
Musgrave (ranges), Australia 87/D8
Musgrave, Queensland 95/B2
Musgrave (range), S. Australia 88/E5
Musgrave (ranges), S. Australia 94/B2
Musgrave Harbour, Newf. 166/D4
Musgravetown, Newf. 166/C2
Mushaboom, Nova Scotia 168/F4
Mushandike Nat'l Park, Zimbabwe 118/D4
Mushie, Zaire 102/D5
Mushie, Zaire 115/C4
Musi (riv.), Indonesia 85/C6
Musidora, Alberta 182/E3
Muskeg (bay), Manitoba 179/G6
Muskeg (bay), Minn. 255/C2
Muskeget (chan.), Mass. 249/N7
Muskeget (isl.), Mass. 249/N7
Muskego, Wis. (53150) 317/K2
Muskegon (co.), Mich. 250/C5
Muskegon, Mich. 188/J2
Muskegon, Mich. (*49440) 250/C5
Muskegon (riv.), Mich. 250/C5
Muskegon Heights, Mich. (49444) 250/C5
Muskeg River, Alberta 182/A3
Muskingum (co.), Ohio 284/G5
Muskingum (riv.), Ohio 284/G6
Muskogee, Okla. 188/H3
Muskogee (co.), Okla. 288/R3
Muskogee, Okla. (74401) 288/R3
Muskoka (dist. munic.), Ontario 177/E3
Muskoka (lake), Ontario 177/E2
Muskrat (creek), Wyo. 319/E2
Muskwa (lake), Alberta 182/C1
Muskwa (riv.), Alberta 182/C1
Muskwa (riv.), Br. Col. 184/M2
Muslimiya, Syria 63/G4
Musmar, Sudan 111/G4
Musoma, Tanzania 115/F4
Musoma, Tanzania 102/F5
Musquacook (lakes), Maine 243/E2
Musquash (harb.), New Bruns. 170/D3
Musquodoboit (riv.), Nova Scotia 168/E4
Musquodoboit Harbour, Nova Scotia 168/E4
Mussau (isl.), Papua N.G. 86/B1
Musselburgh, Scotland 15/D2
Musselburgh, Scotland 10/C1
Musselshell (riv.) 188/E1
Musselshell (co.), Mont. 262/H4
Musselshell, Mont. (59059) 262/H4
Musselshell (riv.), Mont. 262/J3
Mustafakemalpaşa, Turkey 63/C3
Mustahil, Ethiopia 111/H6
Müstair, Switzerland 39/K3
Mustang, Nepal 68/E3
Mustang, Okla. (73064) 288/L4
Mustang (creek), Texas 303/A1
Mustang (isl.), Texas 303/G10
Mustang Draw (dry riv.), Texas 303/B5
Musters (lake), Argentina 143/C6
Mustinka (riv.), Minn. 255/B5
Mustoe, Va. (24468) 307/J4
Mustvee, U.S.S.R. 53/D1
Muswellbrook, N. S. Wales 88/J6
Muswellbrook, N.S. Wales 97/F3
Mût, Egypt 111/E2
Mût, Egypt 59/A4
Mût, Egypt 102/E2
Mut, Turkey 63/E4
Mutankiang (Mudanjiang), China 77/M3
Mutarara (Dona Ana), Mozambique 118/F3
Mutare (Umtali), Zimbabwe 118/E3
Muthanna (gov.), Iraq 66/D5
Muthill, Scotland 15/E4
Muting, Indonesia 85/K7
Mutki, Turkey 63/J3
Mutrie, Sask. 181/H5
Mutsamudu, Comoros 118/G2
Mutshatsha, Zaire 115/D6
Mutsu, Japan 81/K3
Mutsu (bay), Japan 81/K3
Muttaburra, Queensland 95/C4
Muttalip, Turkey 63/D3
Muttenz, Switzerland 39/E1
Muttler (mt.), Switzerland 39/K3
Mutton (isl.), Ireland 17/B6
Mutton Bird (isl.), N.S. Wales 97/J2
Muttonville, Mich. (†48062) 250/G6
Mutual, Ohio (†43078) 284/C5
Mutual, Okla. (73853) 288/H2
Mutum, Brazil 135/F1
Mu Us Shamo (des.), China 77/G4
Muwailih, Saudi Arabia 59/C4
Muwale, Tanzania 115/F5
Muxima, Angola 115/B5
Muy Muy, Nicaragua 154/E4
Muy Muy Viejo, Nicaragua 154/E4
Muynak, U.S.S.R. 48/F5
Muyumba, Zaire 115/E5
Muzaffarabad, Pakistan 68/C2
Muzaffarnagar, India 68/D3
Muzaffarpur, India 68/E3
Muzambinho, Brazil 135/C2
Muzo, Colombia 126/D5
Muzon (cape), Alaska 196/M2
Muztag (mt.), China 77/B4
Muztagata (mt.), China 77/A4
Mvadhi-Ousyé, Gabon 115/B3
M'Vouti, Congo 115/B4
Mwadingusha, Zaire 115/E6
Mwadui, Tanzania 115/F4
Mwanza, Malawi 115/F7
Mwanza (reg.), Tanzania 115/F4
Mwanza, Tanzania 115/F4
Mwanza, Tanzania 102/F5
Mwanza, Zaire 115/E5
Mwaya, Tanzania 115/F5
Mweelrea (mt.), Ireland 17/B4
Mweenish (isl.), Ireland 17/B5
Mweka, Zaire 115/D4
Mwene-Ditu, Zaire 115/D5
Mwenga, Zaire 115/E4
Mweru (lake) 102/E5
Mweru (lake), Zaire 115/E5
Mweru (lake), Zambia 115/E5
Mwesi, Tanzania 115/F5
Mwinilunga, Zambia 115/D6
Mya, Wadi (dry riv.), Algeria 106/E2
Myakka (riv.), Fla. 212/D4
Myakka City, Fla. (33551) 212/D4
Myall (lake), N.S. Wales 97/G3
Myanaung, Burma 72/B3
Myaungmya, Burma 72/B3
Myebon, Burma 72/B2
Myers, Ky. (†40311) 237/O4
Myers, Mont. (†59038) 262/J4
Myerstown, Pa. (17067) 294/K5
Myersville, Md. (21773) 245/H3
Myingyan, Burma 72/B2
Myitkyina, Burma 54/L7
Myitkyina, Burma 72/C1
Myitnge, Burma 72/C2
Myitnge (riv.), Burma 72/C2
Myjava, Czech. 41/D2
Mylo, N. Dak. (†58317) 282/L2
Mymensingh (Nasirabad), Bangladesh 68/G4
Mynyddislwyn, Wales 13/B6
Myohaung, Burma 72/B2
Myohyang (mt.), N. Korea 81/C3
Myŏngch'ŏn, N. Korea 81/D3

Myra, Texas (76253) 303/G4
Myra, W. Va. (25544) 312/B6
Myricks, Mass. (†02780) 249/K5
Myrnam, Alberta 182/E3
Myrtle, Idaho (†83540) 220/B3
Myrtle, Manitoba 179/E5
Myrtle, Minn. (56070) 255/E7
Myrtle, Miss. (38650) 256/F1
Myrtle, Mo. (65778) 261/K9
Myrtle (lake), N. Dak. 282/L5
Myrtle Beach, S.C. (29577) 296/K4
Myrtle Beach A.F.B., S.C. 296/K4
Myrtle Creek, Oreg. (97457) 291/D4
Myrtleford, Victoria 97/D5
Myrtle Grove, Fla. (32506) 212/B6
Myrtle Grove, La. (†70083) 238/K7
Myrtle Point, Oreg. (97458) 291/C4
Myrtlewood, Ala. (36763) 195/C6
Mysen, Norway 18/G7
Myślenice, Poland 47/E4
Myślibórz, Poland 47/B2
Mysłowice, Poland 47/C4
Mysore, India 68/D6
Mysore, India 54/J8
Mys Shmidta, U.S.S.R. 4/C1
Mys Shmidta, U.S.S.R. 48/T3
Mystery Lake, Manitoba 179/J2
Mystic, Conn. (06355) 210/H3
Mystic (riv.), Conn. 210/H3
Mystic, Georgia (31769) 217/F7
Mystic, Iowa (52574) 229/H7
Mystic (lake), Mass. 249/C6
Mystic (riv.), Mass. 249/C6
Mystic, S. Dak. (†57778) 298/B5
Mystic Islands, N.J. (08087) 273/E4
Myszków, Poland 47/D3
My Tho, Vietnam 72/E5
Mytishchi, U.S.S.R. 52/E3
Myton, Utah (84052) 304/D3
M'zab (oasis), Algeria 106/E2
Mže (riv.), Czech. 41/B2
Mzimba, Malawi 115/F6
Mzimba, Malawi 102/F6

N

Naab (riv.), W. Germany 22/E4
Naafkopf (mt.), Switzerland 39/J2
Naaldwijk, Netherlands 27/E4
Naalehu, Hawaii (96772) 218/H7
Naalehu, Hawaii 188/G6
Naantali, Finland 18/M6
Naarden, Netherlands 27/G4
Naas, Ireland 10/C4
Naas, Ireland 17/H5
Naba, Burma 72/B1
Nababeep, S. Africa 118/B5
Nabari, Kiribati 87/J6
Nabb, Ind. (47147) 227/F7
Nabburg, W. Germany 22/E4
Naberezhnye Chelny, U.S.S.R. 52/H3
Nabesna, Alaska (†99764) 196/K2
Nabeul, Tunisia 106/G1
Nabiac, N.S. Wales 97/G3
Nabire, Indonesia 85/K6
Nablus (Nabulus), West Bank 65/C3
Nabnasset, Mass. (01861) 249/J2
Nabua, Philippines 82/D4
Nacala, Mozambique 118/G2
Nacala, Mozambique 102/G6
Nacaome, Honduras 154/D4
Naches, Wash. (98937) 310/E4
Naches (pass), Wash. 310/D3
Naches (riv.), Wash. 310/E4
Nachikatsuura, Japan 81/H7
Nachingwea, Tanzania 115/G6
Náchod, Czech. 41/D1
Nachusa, Ill. (61057) 222/D2
Nachvak (fjord), Newf. 166/B2
Nacimiento (riv.), Calif. 204/D8
Nacimiento, Chile 138/D1
Nacimiento (mts.), N. Mex. 274/C3
Nacimiento (peak), N. Mex. 274/C2
Nacka, Sweden 18/H1
Nackawic, New Bruns. 170/C2
Nacmine, Alberta 182/D4
Naco, Ariz. (85620) 198/E7
Naco, Mexico 150/D1
Nacogdoches (co.), Texas 303/K6
Nacogdoches, Texas (75961) 303/J6
Nacozari, Mexico 150/E1
Ñacunday, Paraguay 144/E5
Nadadores, Mexico 150/H3
Nadawah, Ala. (†36726) 195/D7
Nadeau, Mich. (49863) 250/B3
Nadi, Fiji 86/P10
Nadi, Fiji 87/H7
Nadiad, India 68/C4
Nădlac, Romania 45/E2
Nador, Morocco 106/D1
Nádudvar, Hungary 41/F3
Nadvoitsy, U.S.S.R. 52/D2
Nadym, U.S.S.R. 48/H3
Nadym (riv.), U.S.S.R. 48/H3
Naestved, Denmark 21/E7
Naestved, Denmark 18/G9
Naf, Idaho (83342) 220/E7
Näfels, Switzerland 39/H2
Nafenen, Switzerland 39/H3
Naft-e Shah, Iran 66/D4
Naft Kaneh, Iraq 66/D3
Naga, Philippines 85/G3
Naga, Philippines 54/O8
Naga, Philippines 82/D4
Nagahama, Ehime, Japan 81/F7
Nagahama, Shiga, Japan 81/H6
Nagai (isl.), Alaska 196/F4
Nagaland (state), India 68/G3
Nagambie, Victoria 97/C5
Nagano (pref.), Japan 81/J5
Nagano, Japan 81/J5
Nagaoka, Kyoto, Japan 81/J7
Nagaoka, Niigata, Japan 81/J5
Nagaokakyo, Japan 81/J7
Nagapattinam, India 68/E6
Nagar, Pakistan 68/D1
Nagarote, Nicaragua 154/D4
Nagar Parkar, Pakistan 68/C4
Nagarzê, China 77/C6
Nagasaki (pref.), Japan 81/D7
Nagasaki, Japan 54/O6
Nagasaki, Japan 81/D7
Nagato, Japan 81/E6
Nagaur, India 68/C3
Nagawicka (lake), Wis. 317/J1
Nagele, Netherlands 27/H3
Nagercoil, India 68/D7
Nagina, India 68/D3
Nagishot, Sudan 111/F7
Nagles (mts.), Ireland 17/E7
Nago, Japan 81/N6
Nagold, W. Germany 22/C4
Nagorno-Karabakh Aut. Obl., U.S.S.R. 48/E5
Nagorno-Karabakh Aut. Obl., U.S.S.R. 52/G7
Nagornyy, U.S.S.R. 48/N4
Nagoya, Japan 81/H6
Nagoya, Japan 2/R4
Nagoya, Japan 54/P6
Nagpur, India 54/J7
Nagpur, India 68/D4
Nagqu, China 77/D5
Nags Head, N.C. (27959) 281/T3
Nagua, Dom. Rep. 158/E5
Naguabo, P. Rico 161/F2
Naguabo, P. Rico 156/G1
Nagyatád, Hungary 41/D3
Nagybajom, Hungary 41/D3
Nagyecsed, Hungary 41/G3
Nagyhalász, Hungary 41/F2
Nagykálló, Hungary 41/F3
Nagykanizsa, Hungary 41/D3
Nagykáta, Hungary 41/E3
Nagykőrös, Hungary 41/E3
Nagyszénás, Hungary 41/F3
Naha, Japan 54/O7
Naha, Japan 81/N6
Nahan, India 68/D2
Nahang (riv.), Iran 66/N7
Nahanni Butte, N.W. Terrs. 187/F3
Nahanni Nat'l Park, N.W.T. 162/D3
Nahanni Nat'l Park, N.W. Terrs. 187/F3
Nahant○, Mass. (01908) 249/E6
Nahant (bay), Mass. 249/E6
Nahariyya, Israel 65/C1
Nahavand, Iran 59/E3
Nahavand, Iran 66/F3
Nahcotta, Wash. (98537) 310/A4
Nahhalin, West Bank 65/C4
Nahiku, Hawaii (†96713) 218/K2
Nahma, Mich. (49864) 250/C3
Nahmakanta (lake), Maine 243/E4
Nahuel Huapi (lake), Argentina 120/B7
Nahuel Huapi (lake), Argentina 143/B5
Nahuel Huapi Nat'l Park, Argentina 143/B5
Nahunta, Georgia (31553) 217/H8
Naica, Mexico 150/G2
Naicam, Sask. 181/G3
Naihati, India 68/F1
Nailsworth, England 13/E6
Naiman, China 77/K3
Na'in, Iran 66/H4
Na'in, Iran 59/F3
Nain, Jamaica 158/H6
Nain, Newf. 166/B2
Nain, Newf. 162/K4
Naini Tal, India 68/D3
Nainpur, India 68/E4
Naipo (isl.), Colombia 126/F6
Nairn, La. (†70082) 238/L8
Nairn, Ontario 177/C1
Nairn, Scotland 15/E3
Nairn, Scotland 10/E2
Nairn (trad. co.), Scotland 15/B5
Nairn (riv.), Scotland 15/D3
Nairne, S. Australia 94/C8
Nairobi, Kenya 115/G4
Nairobi (cap.), Kenya 2/L6
Nairobi (cap.), Kenya 115/G4
Nairobi (cap.), Kenya 102/F5
Naivasha, Kenya 115/G4
Najafabad, Iran 59/F3
Najafabad, Iran 66/G4
Najayo Abajo, Dom. Rep. 158/E6
Najin, N. Korea 81/E2
Najran (Aba as Sa'ud), Saudi Arabia 59/D6
Naka (riv.), Japan 81/K5
Nakalele (pt.), Hawaii 218/J1
Nakaminato, Japan 81/K5
Nakamti, Ethiopia 102/F4
Nakamti, Ethiopia 111/G6
Nakamura, Japan 81/F7
Nakasato, Japan 81/K3
Nakatane, Japan 81/E8
Nakatsu, Japan 81/E7
Na Keal, Loch (inlet), Scotland 15/B4
Naked (isl.), Alaska 196/D1
Nakfa, Ethiopia 111/G4
Nakhichevan', U.S.S.R. 7/J5
Nakhichevan', U.S.S.R. 48/E6
Nakhichevan', U.S.S.R. 52/F7
Nakhichevan' A.S.S.R., U.S.S.R. 52/F7
Nakhichevan' A.S.S.R., U.S.S.R. 48/E6
Nakhodka, U.S.S.R. 54/P5
Nakhodka, U.S.S.R. 48/O5
Nakhon Nayok, Thailand 72/D4
Nakhon Pathom, Thailand 72/C4
Nakhon Phanom, Thailand 72/D3
Nakhon Ratchasima, Thailand 72/D4
Nakhon Ratchasima, Thailand 54/M8
Nakhon Sawan, Thailand 72/D4
Nakhon Si Thammarat, Thailand 54/M9
Nakhon Si Thammarat, Thailand 72/D5
Nakina, N.C. (28455) 281/M6
Nakina, Ont. 162/H5
Nakina, Ontario 177/H4
Nakina, Ontario 175/C2
Nakło nad Notecia, Poland 47/C2
Naknek, Alaska (99633) 196/G3
Naknek (lake), Alaska 196/G3
Nakonde, Zambia 115/F5
Nakop, Namibia 118/B5
Nakskov, Denmark 21/E8
Nakskov, Denmark 18/G9
Naktong (riv.), S. Korea 81/D6
Nakuru, Kenya 102/F5
Nakuru, Kenya 115/G4
Nakusp, Br. Col. 184/J5
Nal, Pakistan 59/J4
Nal, Pakistan 68/B3
Nal (riv.), Pakistan 59/J4
Nal (riv.), Pakistan 68/B3
Nalate, Turkey 63/G4
Nalayh (Nalaikha), Mongolia 77/G2
Nalcayec (isl.), Chile 138/D6
Nal'chik, U.S.S.R. 7/J4
Nal'chik, U.S.S.R. 48/E5
Nal'chik, U.S.S.R. 52/F6
Nalgonda, India 68/D5
Nallen, W. Va. (26680) 312/E6
Nallıhan, Turkey 63/D2
Nalut, Libya 111/B1
Namacurra, Mozambique 118/F3
Namak, Daryacheh-ye (salt lake), Iran 59/F3
Namak, Daryacheh-ye (salt lake), Iran 66/G3
Namaka, Alberta 182/D4
Namaksar (salt lake), Afghanistan 59/H3
Namaksar (salt lake), Afghanistan 68/A2
Namaksar (lake), Iran 66/M4
Namaksar (salt lake), Iran 59/H3
Namakzar-e Shahdad (salt lake), Iran 59/G3
Namakzar-e Shahdad (salt lake), Iran 66/L5
Namanga, Kenya 115/G4
Namangan, U.S.S.R. 48/H5
N'amaniya, Iraq 66/D4
Namapa, Mozambique 118/F2
Namaqualand (reg.), S. Africa 118/B5
Namarrói, Mozambique 118/F3
Namasagali, Uganda 115/F3
Namasigüe, Honduras 154/D4
Namatanai, Papua N.G. 87/F6
Namatanai, Papua N.G. 86/C1
Nambe, N. Mex. (†87501) 274/D3
Nambour, Queensland 88/J5
Nambour, Queensland 95/E5
Nambucca Heads, N.S. Wales 97/G2
Nam Co (lake), China 77/D5
Nam Dinh, Vietnam 72/E2
Namekagon (lake), Wis. 317/D3
Namekagon (riv.), Wis. 317/C3
Namen (Namur), Belgium 27/F8
Námestovo, Czech. 41/E2
Nametil, Mozambique 118/F3
Namhkam, Burma 72/C2
Namib (des.), Namibia 118/A3
Namibia 2/K7
Namibia 102/D7
Namibia (des.) 102/D6
NAMIBIA (SOUTH-WEST AFRICA) 118/B3
Naminga, U.S.S.R. 48/M4
Namiquipa, Mexico 150/F2
Namlan, Burma 72/C2
Namlea, Indonesia 85/H6
Namoi (riv.), N. S. Wales 88/H6
Namoi (riv.), N.S. Wales 97/E2
Namonuito (atoll), Micronesia 87/E5
Namorik (atoll), Marshall Is. 87/G5
Nampa, Alberta 182/B1
Nampa, Idaho 146/G5
Nampa, Idaho (83651) 220/B6
Nampa, Idaho 188/C2
Nampala, Mali 106/C5
Namp'o, N. Korea 81/B4
Nampo-Shoto (isls.), Japan 81/M3
Nampula (prov.), Mozambique 118/F2
Nampula, Mozambique 118/F3
Nampula, Mozambique 102/F6
Namsen (riv.), Norway 18/H4
Namsos, Norway 7/F2
Namsos, Norway 18/G4
Nam Tram, Mui (cape), Vietnam 72/F4
Namtu, Burma 72/C2
Namu, Br. Col. 184/D4
Namuac, Philippines 82/C1
Namuli, Serra (mt.), Mozambique 118/F3
Namuno, Mozambique 118/F2
Namur (lake), Alberta 182/D1
Namur (prov.), Belgium 27/F8
Namur, Belgium 27/F8
Namur, Québec 172/C4
Namutoni, Namibia 118/B3
Namwala, Zambia 115/E7
Namwŏn, S. Korea 81/C6
Namysłów, Poland 47/C3
Namzha Parwa (mt.), China 77/E6
Nan, Thailand 72/D3
Nan, Mae Nam (riv.), Thailand 72/D3
Nanacamilpa, Mexico 150/M1
Nana Candundo, Angola 115/D6
Nanafalia, Ala. (36764) 195/B6
Nanaimo, Br. Col. 146/F5
Nanaimo, Br. Col. 162/D6
Nanaimo, Br. Col. 184/J3
Nanakuli, Hawaii (96792) 218/D2
Nanao, Japan 81/H5
Nanay (riv.), Peru 128/E4
Nancagua, Chile 138/F6
Nance (co.), Nebr. 264/F3
Nanchang, China 77/J6
Nanchang, China 54/N7
Nancheng, China 77/J6
Nanchong (Nanchung), China 77/G5
Nanchong, China 54/M6
nan Clar, Loch (lake), Scotland 15/D2
Nancowry (isl.), India 68/G7
Nancy, France 28/G3
Nancy, France 7/E4
Nancy, Ky. (42544) 237/M6
Nanda Devi (mt.), India 68/D2
Nandaime, Nicaragua 154/E5
Nander, India 68/D5
Nandi (Nadi), Fiji 87/H7
Nando, Uruguay 145/F3
Nandurbar, India 68/C4
Nandyal, India 68/D5
Nanga-Eboko, Cameroon 115/B3
Nanga Parbat (mt.), Pakistan 68/D1
Nangapinoh, Indonesia 85/E6
Nangatayap, Indonesia 85/E6
Nangnim-sanmaek (range), N. Korea 81/C3
Nangong, China 77/H4
Nangqên, China 77/E5
Nang Rong, Thailand 72/D4
Nangwarry, S. Australia 94/G7
Nang Xian, China 77/D6
Nanika (dam), Br. Col. 184/D3
Nanika (lake), Br. Col. 184/D3
Nanisivik, N.W. Terrs. /K2
Nanjemoy, Md. (20662) 245/K7
Nanjing (Nanking), China 77/J5
Nanjing, China 2/Q4
Nanjing, China 54/N6
Nanking (Nanjing), China 77/J5
Nankoku, Japan 81/F7
Nan Ling (mts.), China 77/H6
Nannine, W. Australia 92/B4
Nanning, China 77/G7
Nanning, China 54/M7
Nannup, W. Australia 92/B6
Nanoose Bay, Br. Col. 184/J3
Nanortalik, Greenl. 4/D12
Nanpan Jiang (riv.), China 77/F7
Nanping, China 77/J6
Nansei Shoto (Ryukyu) (isls.), Japan 81/M6
Nansen (sound), N.W. Terrs. 187/J1
Nanson, N. Dak. (58354) 282/L2
Nantahala, N.C. (†28702) 281/B4
Nantahala (lake), N.C. 281/B4
Nantai (mt.), Japan 81/J5
Nantasket Beach, Mass. (†02045) 249/E7
Nanterre, France 28/A1
Nantes, France 28/C4
Nantes, France 7/D4
Nantes, Québec 172/F4
Nanticoke (riv.), Del. 245/R6
Nanticoke, Md. (21840) 245/P7
Nanticoke (riv.), Md. 245/P7
Nanticoke, Ontario 177/E5
Nanticoke, Pa. (18634) 294/E7
Nanton, Alberta 182/D4
Nantong, China 77/K5
Nantua, France 28/F4
Nantucket (co.), Mass. 249/O7
Nantucket, Mass. (02554) 249/O7
Nantucket○, Mass. (02554) 249/O7
Nantucket (isl.), Mass. 188/N2
Nantucket (isl.), Mass. 249/O8
Nantucket (sound), Mass. 249/N6
Nanty Glo, Pa. (15943) 294/E5
Nantyglo and Blaina, Wales 13/B6
Nanuet, N.Y. (10954) 276/K8
Nanuktok (isls.), Newf. 166/C2
Nanuku (passage), Fiji 86/R10
Nanumea (atoll), Tuvalu 87/H6
Nanuque, Brazil 132/F7
Nanuque, Brazil 120/E4
Nanxiong, China 77/H6
Nanyang, China 77/H5
Nanyuki, Kenya 115/G3
Nanzhang, China 77/H5
Nanzhao, China 77/H5
Nao (cape), Spain 33/G3
Naococane (lake), Québec 174/C2
Naolinco de Victoria, Mexico 150/P1
Naomi, Ky. (†42544) 237/M6
Náousa, Greece 45/F5
Napa, Calif. 188/B3
Napa (co.), Calif. 204/C5
Napa, Calif. (94558) 204/C5
Napadogan, New Bruns. 170/D2
Napa Junction, Calif. (†94590) 204/J1
Napakiak, Alaska (99634) 196/F2
Napaktok (bay), Newf. 166/B2
Napanee, Ontario 177/G3
Napanoch, N.Y. (12458) 276/M7
Napaskiak, Alaska (99559) 196/F2
Napata (ruins), Sudan 111/F4
Napavine, Wash. (98565) 310/C4
Napè, Laos 72/E3
Naper, Nebr. (68755) 264/E2
Naperville, Ill. (60540) 222/A6
Napf (mt.), Switzerland 39/E3
Napier, Ky. (†40851) 237/P7
Napier, N. Zealand 100/F3
Napier, N. Zealand 87/H9
Napier (mt.), North Terr. 93/A4
Napier, W. Va. (26631) 312/E5
Napier Field, Ala. (36303) 195/H8
Napierville (co.), Québec 172/D4
Napierville, Québec 172/D4
Napili-Honokowai, Hawaii (†96761) 218/H1
Napinka, Manitoba 179/B5
Naplate, Ill. (†61350) 222/E2
Naples, Fla. (*33940) 212/E5
Naples, Idaho (83847) 220/B1
Naples, Ill. (62669) 222/C4
Naples (prov.), Italy 34/E4
Naples, Italy 7/F4
Naples, Italy 34/E4
Naples○, Maine (04055) 243/B8
Naples, N.Y. (14512) 276/F5
Naples, S. Dak. (†57271) 298/O4
Naples, Texas (75568) 303/K4
Naples Park, Fla. (†33940) 212/E5
Napo (riv.) 120/B3
Napo, China 77/G7
Napo (prov.), Ecuador 128/D3
Napo (riv.), Ecuador 128/D3
Napo (riv.), Peru 128/F4
Napoleon, Ind. (47034) 227/G6
Napoleon, Mich. (49261) 250/E6
Napoleon, Mo. (64074) 261/E4
Napoleon, N. Dak. (58561) 282/L6
Napoleon, Ohio (43545) 284/B3
Naponee, Nebr. (68960) 264/E4
Nappa Merri, Queensland 95/B5
Nappan, Nova Scotia 168/D3
Nappanee, Ind. (46550) 227/F2
Napperby, North. Terr. 93/C7
Napton, Mo. (65346) 261/F4
Naqa (ruins), Sudan 111/F4
Nara (pref.), Japan 81/J8
Nara, Japan 81/J8
Nara, Mali 106/C5
Naracoopa, Tasmania 99/B1
Naracoorte, S. Australia 88/F7
Naracoorte, S. Australia 94/G7
Naradhan, N.S. Wales 97/D3
Naramata, Br. Col. 184/H5
Naranja, Fla. (33032) 212/F6
Naranjal (riv.), Ecuador 128/C4
Naranjito, Honduras 154/C3
Naranjito, P. Rico 161/D1
Naranjos, Mexico 150/L6
Naraq, Iran 66/G3
Narashino, Japan 81/P2
Narathiwat, Thailand 72/D6
Nara Visa, N. Mex. (88430) 274/F3
Narayanganj, Bangladesh 68/G4
Narayanpet, India 68/D5
Narberth, Pa. (19072) 294/M6
Narberth, Wales 13/C6
Narbonne, France 28/E6
Narcissa, Okla. (†74354) 288/S1
Narcisse, Manitoba 179/E4
Narcondam (isl.), India 68/G6
Narcoossee, Fla. (†32769) 212/E3
Nardin, Okla. (74646) 288/M1
Nardò, Italy 34/F4
Naré, Argentina 143/F5
Nare, Colombia 126/C4
Narellan, N.S. Wales 97/F3
Nares (str.) 146/L2
Nares (str.), N.W.T. 162/N3
Nares (str.), N.W. Terrs. 187/L2
Narew (riv.), Poland 47/E2
Naricual, Venezuela 124/F2
Narinda, Madagascar 118/H3
Nariño (dept.), Colombia 126/B7
Nariva (swamp), Trin. & Tob. 161/B10
Narka, Kansas (66960) 232/E2
Narmada (riv.), India 54/J7
Narmada (riv.), India 68/D4
Narman, Turkey 63/J2
Narnaul, India 68/D3
Narni, Italy 34/D3
Naro, Italy 34/D6
Narodnaya (mt.), U.S.S.R. 7/K2
Narodnaya (mt.), U.S.S.R. 48/G3
Narodnaya (mt.), U.S.S.R. 52/J1
Narok, Kenya 115/G4
Narooma, N.S. Wales 97/F5
Narrabeen, N.S. Wales 88/L3
Narrabeen, N. S. Wales 97/K3
Narrabri, N. S. Wales 88/J6
Narrabri, N.S. Wales 97/E2
Narragansett, R.I. (02882) 249/J7
Narragansett○, R.I. (02882) 249/J7
Narragansett (bay), R.I. 249/J6
Narran (lake), N.S. Wales 97/D1
Narran (riv.), N.S. Wales 97/D1
Narrandera, N. S. Wales 88/H6
Narrandera, N.S. Wales 97/D4
Narre Warren North, Victoria 97/K5
Narrogin, W. Australia 88/B6
Narrogin, W. Australia 92/B2
Narromine, N. S. Wales 88/H6
Narromine, N.S. Wales 97/E3
Narrows, Ky. (42358) 237/H5
Narrows, Oreg. (†97721) 291/H4
Narrows, The (str.), St. Christ.-Nevis 161/D11
Narrows, The (str.), Virgin Is. (Br.) 161/C4
Narrows, The (str.), Virgin Is. (U.S.) 161/C4
Narrows, Va. (24124) 307/G6
Narrowsburg, N.Y. (12764) 276/L7
Narrows Park-La Vale, Md. (†21502) 245/C2
Narsimhapur, India 68/D4
Narsinghgarh, India 68/D4
Narssaq, Greenl. 4/C12
Naruna, Va. (24576) 307/L6
Narva, U.S.S.R. 52/C3
Narva, U.S.S.R. 53/E1
Narva (res.), U.S.S.R. 53/D1
Narvik, Norway 4/C9
Narvik, Norway 7/F2
Narvik, Norway 18/K2
Nary, Minn. (†56601) 255/D3
Nar'yan-Mar, U.S.S.R. 4/C7
Nar'yan-Mar, U.S.S.R. 7/K2
Nar'yan-Mar, U.S.S.R. 48/F3
Nar'yan-Mar, U.S.S.R. 52/H1
Naryn, U.S.S.R. 48/H5
Nasarawa, Nigeria 106/F7
Năsăud, Romania 45/G2
Naseby, N. Zealand 100/C6
Naseby, Sask. 181/C3
Naselle, Wash. (98638) 310/B4
Naselle (riv.), Wash. 310/B4
Nash (stream), N.H. 268/E2
Nash (co.), N.C. 281/O2
Nash, N. Dak. (58264) 282/P3
Nash, Okla. (73761) 288/K1
Nash, Texas (75569) 303/K4
Nashawena (isl.), Mass. 249/L7
Nash Creek, New Bruns. 170/D1
Nashoba, Okla. (74558) 288/R6
Nashotah, Wis. (53058) 317/J1
Nashport, Ohio (43830) 284/F5
Nashua, Iowa (50658) 229/J3
Nashua (riv.), Mass. 249/H3
Nashua, Minn. (56565) 255/B4
Nashua, Mont. (59248) 262/K2
Nashua, N.H. 188/M2
Nashua, N.H. (03060) 268/D6
Nashville, Ark. (71852) 202/C6
Nashville, Georgia (31639) 217/F8
Nashville, Ill. (62263) 222/D5
Nashville, Ind. (47448) 227/E6
Nashville, Kansas (67112) 232/D4
Nashville, Mich. (49073) 250/D6
Nashville, Mo. (†64855) 261/D8
Nashville, N.C. (27856) 281/O3
Nashville, Ohio (44661) 284/F4
Nashville, Oreg. (†97370) 291/D3
Nashville (cap.), Tenn. 146/K6
Nashville (cap.), Tenn. 188/J3
Nashville (cap.), Tenn. (*37201) 237/H8
Nashwaak (riv.), New Bruns. 170/D2
Nashwaak Bridge, New Bruns. 170/D2
Nashwaak Village, New Bruns. 170/D2
Nashwauk, Minn. (55769) 255/E3
Našice, Yugoslavia 45/C3
Nasielsk, Poland 47/E2
Näsijärvi (lake), Finland 18/O6
Nasik, India 68/C5
Nasik, India 54/J8
Nasir, Sudan 111/F6
Nasirabad, Bangladesh 68/G4
Nasirabad, India 68/C3
Naskaupi (riv.), Newf. 166/B3
Naso (pt.), Philippines 82/C5
Nason, Ill. (†62816) 222/D5
Nasonville, R.I. (†02830) 249/H5
Nasratabad (Zabol), Iran 59/H3
Nasratabad (Zabol), Iran 66/M5
Nass (riv.), Br. Col. 184/C2
Nassau (cap.), Bahamas 146/L7
Nassau (cap.), Bahamas 156/C1
Nassau (bay), Chile 120/C8
Nassau (bay), Chile 138/F11
Nassau (isl.), Cook Is. 87/K7
Nassau, Del. (19969) 245/T6
Nassau (co.), Fla. 212/E1
Nassau (riv.), Fla. 212/E1
Nassau (sound), Fla. 212/E1
Nassau, Minn. (56272) 255/B5
Nassau (co.), N.Y. 276/N9
Nassau, N.Y. (12123) 276/N5
Nassau Bay, Texas (†77598) 303/K2
Nassawadox, Va. (23413) 307/S6
Nassawango (creek), Md. 245/S8
Nasser (lake), Egypt 102/F2
Nasser (lake), Egypt 111/F3
Nasser (lake), Egypt 59/F3
Nassereith, Austria 41/A3
Nässjö, Sweden 18/J8
Nassogne, Belgium 27/G8
Nasty (creek), S. Dak. 298/C2
Nasu (mt.), Japan 81/J5
Nata, Botswana 118/D4
Natá, Panama 154/G6
Natagaima, Colombia 126/C6
Natal, Brazil 2/H6
Natal, Brazil 132/H4
Natal, Brazil 120/F3
Natal, Br. Col. 184/K5
Natal (prov.), S. Africa 102/F7
Natal (prov.), S. Africa 118/E5
Natalbany, La. (70451) 238/N1
Natalia, Texas (78059) 303/J11
Natalicio Talavera, Paraguay 144/D4
Natanz, Iran 59/F3
Natanz, Iran 66/H4
Natashquan (riv.) 162/K5
Natashquan (riv.), Newf. 166/B3
Natashquan, Québec 174/E3
Natashquan (riv.), Québec 174/E2
Natashquan-Est (riv.), Newf. 166/B3
Natchaug (riv.), Conn. 210/G1
Natchez, Ala. (†36425) 195/D7
Natchez, La. (71456) 238/D3
Natchez, Miss. 188/H4
Natchez, Miss. (39120) 256/B7
Natchitoches (par.), La. 238/D3
Natchitoches, La. (71457) 238/D3
Naters, Switzerland 39/E4
Natewa (bay), Fiji 86/Q10
Nathalia, Victoria 97/C5
Nathalie, Va. (24577) 307/L7
Nathan, Mich. (†49821) 250/B3
Nathrop, Colo. (81236) 208/H5
Natick○, Mass. (01760) 249/A7
Natick, R.I. (†02887) 249/H6
Natimuk, Victoria 97/A5
Nation (riv.), Br. Col. 184/F2
National Agricultural Research Center, Md. 245/G3
National Capital Region (Manila) (prov.), Philippines 82/C3
National City, Calif. (92050) 204/J11
National City, Mich. (48748) 250/F4
National Gardens, Fla. (†32074) 212/E2
National Mills, Manitoba 179/A2
National Mine, Mich. (49865) 250/B2
National Park, N.J. (08063) 273/B3
National Park, Switzerland 39/K3
National Reactor Testing Sta. (U.S.A.E.C), Idaho 220/F6
National Stock Yards, Ill. (62071) 222/A2
Natitingou, Benin 106/E6
Natividade, Brazil 132/E5
Natmauk, Burma 72/B2
Natoma, Kansas (67651) 232/D2
Natron (lake), Kenya 115/G4
Natron (lake), Tanzania 115/G4
Natrona (co.), Wyo. 319/F3
Natrona, Wyo. (82646) 319/F2
Natrona Heights, Pa. (15065) 294/C4
Nattavaara, Sweden 18/M3
Natuna (isls.), Indonesia 54/M9
Natuna (isls.), Indonesia 85/D5
Natural Bridge, Ala. (35577) 195/C2
Natural Bridge, N.Y. (13665) 276/K2
Natural Bridge, Va. (24578) 307/J5
Natural Bridge Nat'l Mon., Utah 304/E6
Natural Bridge Station, Va. (24579) 307/K5
Natural Dam, N.Y. (†13642) 276/J2
Naturaliste (cape), Tasmania 99/E2
Naturaliste (cape), W. Australia 88/A6
Naturaliste (chan.), W. Australia 88/A5
Naturaliste (cape), W. Australia 92/A6

Naturaliste (chan.), W. Australia 92/A4
Natural Steps, Ark. (†72135) 202/F4
Naturita, Colo. (81422) 208/B6
Naubinway, Mich. (49762) 250/D2
Naucalpan de Juárez, Mexico 150/L1
Nauders, Austria 41/A3
Nauen, E. Germany 22/E2
Naugatuck, Conn. (06770) 210/C3
Naugatuck (riv.), Conn. 210/C3
Naugatuck, W. Va. (25685) 312/B7
Nauhcampatépetl (mt.), Mexico 150/O1
Naujan (lake), Philippines 82/C4
Naujoji-Akmene, U.S.S.R. 53/B2
Naumburg, E. Germany 22/D3
Na'ur, Jordan 65/D4
Nauru 2/T6
Nauru 87/G6
Naushon (isl.), Mass. 249/L7
Naustdal, Norway 18/E6
Nauta, Peru 128/F5
Nautla, Mexico 150/L6
Nauvoo, Ala. (35578) 195/D3
Nauvoo, Ill. (62354) 222/B3
Nauwigewauk, New Bruns. 170/E3
Nava, Mexico 150/J2
Nava del Rey, Spain 33/D2
Navajo (co.), Ariz. 198/E3
Navajo, Ariz. (86509) 198/F3
Navajo (creek), Ariz. 198/D2
Navajo (peak), Colo. 208/F8
Navajo (res.), Colo. 208/E8
Navajo, Mont. (†59222) 262/M2
Navajo, N. Mex. (87328) 274/A3
Navajo (dam), N. Mex. 274/B2
Navajo (res.), N. Mex. 274/B2
Navajo (mt.), Utah 304/D6
Navajo Ind. Res., Ariz. 198/D2
Navajo Ind. Res., N. Mex. 274/A2
Navajo Ind. Res., Utah 304/D7
Navajo Nat'l Mon., Ariz. 198/E2
Navajo Ord. Depot, Ariz. 198/D3
Naval Academy, U.S., Md. 245/N5
Naval Air Sta., La. 238/O4
Naval Air Station, Calif. 204/J2
Naval Air Station, Va. 307/R7
Naval Base, S.C. 296/H6
Navalcarnero, Spain 33/F4
Naval Medical Center, Md. 245/E4
Navalmoral de la Mata, Spain 33/D3
Naval Submarine Base, Conn. 210/G3
Naval Support Ctr., Wash. 310/B1
Naval Weapons Center, Md. 245/F3
Naval Yard, D.C. 245/F5
Navan (An Uaimh), Ireland 17/H4
Navan, Ontario 177/J2
Navarin (cape), U.S.S.R. 4/C18
Navarin (cape), U.S.S.R. 48/T3
Navarino (isl.), Chile 138/F11
Navarino, Wis. (54108) 317/J6
Navarra (prov.), Spain 33/F1
Navarre, Kansas (67469) 232/E3
Navarre, Ohio (44662) 284/H4
Navarro, Argentina 143/G7
Navarro, Calif. (95463) 204/B4
Navarro (riv.), Calif. 204/B4
Navarro (co.), Texas 303/H5
Navasota, Texas (77868) 303/J7
Navasota (riv.), Texas 303/H7
Navassa, N.C. (†28404) 281/O6
Navassa (isl.), Virgin Is. (U.S.) 156/C3
Naver, Loch (lake), Scotland 15/D2
Naver (riv.), Scotland 15/D2
Navesink, N.J. (07752) 273/E3
Navesink (riv.), N.J. 273/E3
Navia (riv.), Spain 33/C1
Navidad, Chile 138/A10
Navidad (riv.), Texas 303/H8
Navin, Manitoba 179/F5
Navoi, U.S.S.R. 48/G6
Navojoa, Mexico 150/E3
Navolato, Mexico 150/E4
Návpaktos, Greece 45/F6
Návplion, Greece 45/F7
Navrongo, Ghana 106/D6
Navsari, India 68/C4
Navy Board (inlet), N.W. Terrs. 187/K2
Navy Yard City, Wash. (†98310) 310/A2
Nawabganj, Bangladesh 68/F4
Nawabshah, Pakistan 68/B3
Nawabshah, Pakistan 59/J4
Nawiliwili (bay), Hawaii 218/D2
Naxera, Va. (23122) 307/R6
Náxos, Greece 45/G7
Náxos (isl.), Greece 45/G7
Naya, Colombia 126/B6
Nayarit (state), Mexico 150/G6
Nayarit, Sierra (mts.), Mexico 150/G5
Nay Band, Iran 59/F4
Nay Band, Iran 59/G3
Nay Band, Bushehr, Iran 66/H7
Nay Band, Khorasan, Iran 66/K4
Naylor, Georgia (31641) 217/F9
Naylor, Mo. (63953) 261/L9
Nayoro, Japan 81/L1
Naytahwaush, Minn. (56566) 255/C3
Nazaré, Brazil 132/G6
Nazaré, Portugal 33/B3
Nazareth, Belgium 27/D7
Nazareth, Israel 65/C2
Nazareth, Pa. (18064) 294/M4
Nazareth, Texas (79063) 303/B3
Nazarovo, U.S.S.R. 48/K4
Nazas, Mexico 150/G4
Nazas (riv.), Mexico 150/G4
Nazca, Peru 128/E10
Naze, The (prom.), England 13/J6
Naze, Japan 81/O5
Nazerat 'Illit, Israel 65/C2
Nazilli, Turkey 63/C4
Nazko, Br. Col. 184/F3
Nazret, Ethiopia 111/G6
Nazret, Ethiopia 102/F4
Nazyvayevsk, U.S.S.R. 48/H4
Ncheu (Ntcheu), Malawi 115/F6
Ndalatando, Angola 115/B5
Ndele, Cent. Afr. Rep. 115/D2
N'Dendé, Gabon 115/B4
Ndeni (isl.), Solomon Is. 87/G7
N'Djamena (cap.), Chad 111/C5
N'Djamena (cap.), Chad 2/K5
N'Djamena (cap.), Chad 102/D3
N'Djolé, Gabon 115/B4
N'Dogo (lag.), Gabon 115/B4
Ndola, Zambia 115/E6
Ndola, Zambia 102/E6
Nead, Ind. (†46970) 227/E3
Neagh (lake), N. Ireland 17/J2
Neagh, Lough (lake), N. Ireland 10/C3
Neah Bay, Wash. (98357) 310/A2
Neal, Kansas (66863) 232/F4
Neale (lake), North. Terr. 88/D4
Neale (lake), North. Terr. 93/A8
Neales, The (riv.), S. Australia 94/E3
Neales, The (riv.), S. Australia 88/F5
Neápolis, Greece 45/F7
Neapolis, Ohio (43547) 284/C3
Near (isls.), Alaska 196/H3
Neath, Wales 10/E5
Neath, Wales 13/D6
Neavitt, Md. (21652) 245/N6
Nebikon, Switzerland 39/F2
Nebish, Minn. (†56667) 255/D3
Nebit-Dag, U.S.S.R. 48/F6
Neblina, Pico da (peak), Brazil 132/G8
Neblina (Phelps) (peak), Venezuela 124/E7
Nebo (mt.), Ark. 202/D3
Nebo, Ill. (62355) 222/C4
Nebo (mt.), Jordan 65/D4
Nebo, Ky. (42441) 237/F6
Nebo, La. (†71342) 238/F3
Nebo, Mo. (65471) 261/H7
Nebo (mt.), Utah 304/C4
Nebo, W. Va. (25141) 312/D5
Nebraska 188/F2
NEBRASKA 264
Nebraska, Ind. (47262) 227/F6
Nebraska (state), U.S. 146/J5
Nebraska City, Nebr. (68410) 264/J4
Necedah, Wis. (54646) 317/F7
Nechako (riv.), Br. Col. 184/E3
Neche, N. Dak. (58265) 282/P2
Neches, Texas (75779) 303/J6
Neches (riv.), Texas 303/K6
Nechí (riv.), Colombia 126/C4
Neckar (riv.), W. Germany 22/C4
Neckarsulm, W. Germany 22/C4
Neck City, Mo. (†64755) 261/C8
Necker (isl.), Hawaii 87/K3
Necker (isl.), Hawaii 188/F6
Necker (isl.), Hawaii 218/D6
Necochea, Argentina 143/E4
Necochea, Argentina 120/D6
Nectar, Ala. (†35049) 195/E3
Necum Teuch (harb.), Nova Scotia 168/F4
Ned, Ky. (41355) 237/P6
Neded, Czech. 41/D2
Nederland, Colo. (80466) 208/H3
Nederland, Texas (77627) 303/K8
Nedgera (creek), N.S. Wales 97/E2
Nedlands, W. Australia 88/B2
Nedlands, W. Australia 92/A1
Neeb, Sask. 181/C1
Neebish (isl.), Mich. 250/E2
Neede, Netherlands 27/K4
Needham, Ala. (36915) 195/B7
Needham, Ind. (46162) 227/E5
Needham○, Mass. (02192) 249/B7
Needham Heights, Mass. (02194) 249/B7
Needle (mt.), Wyo. 319/C1
Needles, Calif. 188/C4
Needles, Calif. (92363) 204/L9
Needles (pt.), N. Zealand 100/E2
Needmore, Ind. (†47421) 227/E7
Needmore, N.C. (†28713) 281/B4
Needmore, Pa. (17238) 294/F6
Needmore, W. Va. (†26801) 312/J4
Needville, Texas (77461) 303/J8
Neelin, Manitoba 179/C5
Neely, Miss. (39461) 256/G8
Neely Henry (lake), Ala. 195/F3
Neelys Landing, Mo. (†63755) 261/O7
Neelyton, Pa. (17239) 294/G5
Neelyville, Mo. (63954) 261/M9
Ñeembucú (dept.), Paraguay 144/C-D5
Neenah, Wis. (54956) 317/J7
Neepawa, Manitoba 179/C4
Neerlandia, Alberta 182/C2
Neerpelt, Belgium 27/G6
Neeses, S.C. (29107) 296/E4
Nee Soon, Singapore 72/F6
Nee so Pah (res.), Colo. 208/D6
Neffs, Ohio (43940) 284/J5
Neffs Mills, Pa. (†16669) 294/G4
Nefta, Tunisia 106/F2
Neftekamsk, U.S.S.R. 52/J3
Nefteyugansk, U.S.S.R. 48/H3
Nefud (des.), Saudi Arabia 54/F7
Nefud (des.), Saudi Arabia 59/D4
Nefud Dahi (des.), Saudi Arabia 59/D5
Nefusa, Jebel (mts.), Libya 111/B1
Nefyn, Wales 13/C5
Negara, Indonesia 85/E7
Negaunee, Mich. (49866) 250/B2
Negba, Israel 65/B4
Negelli, Ethiopia 111/G6
Negeri Sembilan (state), Malaysia 72/D7
Negev (reg.), Israel 65/D5
Negley, Ohio (44441) 284/J4
Negomane, Mozambique 118/F2
Negombo, Sri Lanka 68/D7
Negotin, Yugoslavia 45/F3
Negra, Cordillera (mts.), Peru 128/D7
Negra (pt.), Peru 128/B6
Negra (pt.), P. Rico 161/G2
Negra (lag.), Uruguay 145/F5
Negra (range), Uruguay 145/D2
Negrais (cape), Burma 72/B3
Negreet, La. (71460) 238/C4
Negreiros, Chile 138/B2
Negreşti, Romania 45/H2
Negril, Jamaica 158/G6
Negrillos, Bolivia 136/A6
Negritos, Peru 128/B5
Negro (riv.) 2/F5
Negro (cape), Angola 115/B7
Negro (riv.), Argentina 120/C6
Negro (riv.), Argentina 143/D4
Negro (riv.), Bolivia 136/D4
Negro (riv.), Brazil 120/C3
Negro (riv.), Brazil 132/H9
Negro (riv.), Colombia 126/G7
Negro (riv.), Paraguay 144/C4
Negro (bay), Somalia 115/J2
Negro (riv.), Uruguay 120/D6
Negro, Arroyo (riv.), Uruguay 145/B3
Negro (riv.), Uruguay 145/B4
Negro (riv.), Venezuela 124/E7
Negro Bay, Virgin Is. (U.S.) 161/E4
Negros (isl.), Philippines 54/O9
Negros (isl.), Philippines 85/G4
Negros (isl.), Philippines 82/D6
Negros Occidental (prov.), Philippines 82/D6
Negros Oriental (prov.), Philippines 82/D6
Neguac, New Bruns. 170/E1
Nehalem, Oreg. (97131) 291/D2
Nehalem (riv.), Oreg. 291/D2
Nehawka, Nebr. (68413) 264/H4
Nehbandan, Iran 59/G3
Nehbandan, Iran 66/L5
Nehe, China 77/L2
Neheim-Hüsten, W. Germany 22/C3
Neiafu, Tonga 87/J7
Neiba, Dom. Rep. 158/D6
Neiba, Dom. Rep. 156/D3
Neiba (bay), Dom. Rep. 158/D6
Neiba, Sierra de (mts.), Dom. Rep. 158/D6
Neiber, Wyo. (†82401) 319/D2
Neidpath, Sask. 181/D5
Neiges (lake), Québec 172/F2
Neigette, Québec 172/J1
Neihart, Mont. (59465) 262/F4
Neijiang (Neikiang), China 77/G6
Neilburg, Sask. 181/B3
Neillsville, Wis. (54456) 317/E6
Neil's Harbour, Nova Scotia 168/H2
Neilston, Scotland 15/B2
Neilton, Wash. (98566) 310/B3
Nei Monggol (Inner Mongolian Aut. Reg.), China 77/H3
Neis Beach, Sask. 181/E2
Neisse (riv.), E. Germany 22/F3
Neisse (Nysa), Poland 47/C3
Neisse (riv.), Poland 47/B3
Neiva, Colombia 120/B2
Neiva, Colombia 126/C6
Nejanilini (lake), Manitoba 179/J1
Nejd (reg.), Saudi Arabia 59/D4
Nejdek, Czech. 41/B1
Nejo, Ethiopia 111/G6
Nekoma, Kansas (67559) 232/C3
Nekoma, N. Dak. (58355) 282/O2
Nekoosa, Wis. (54457) 317/G7
Neksø, Denmark 18/J9
Neksø, Denmark 21/F9
Nelagoney, Okla. (†74056) 288/O1
Nelas, Portugal 33/C2
Nelchina, Alaska (†99588) 196/C1
Nelidovo, U.S.S.R. 52/D3
Neligh, Nebr. (68756) 264/G2
Nel'kan, U.S.S.R. 48/O4
Nellie, Ohio (†43844) 284/F5
Nellis, W. Va. (25142) 312/C6
Nellis A.F.B., Nev. 266/F6
Nellis Air Force Range and AEC Nuclear Testing Sit, Nev. 266/E5
Nelliston, N.Y. (13410) 276/L5
Nellore, India 54/K8
Nellore, India 68/E6
Nellysford, Va. (22958) 307/L5
Nelma, Wis. (†49935) 317/J3
Nelse, Ky. (41550) 237/R6
Nelson (isl.), Alaska 196/E2
Nelson, Argentina 143/F5
Nelson, Ariz. (†86434) 198/B3
Nelson, Br. Col. 162/E6
Nelson, Br. Col. 184/J5
Nelson (str.), Chile 138/D9
Nelson, England 13/H1
Nelson, England 10/G1
Nelson, Georgia (30151) 217/D2
Nelson, Ill. (61058) 222/D2
Nelson (co.), Ky. 237/K5
Nelson, Ky. (†42330) 237/G6
Nelson (riv.), Man. 146/J4
Nelson (riv.), Man. 162/G4
Nelson (riv.), Manitoba 179/J2
Nelson, Minn. (56355) 255/C5
Nelson, Mo. (65347) 261/F4
Nelson (res.), Mont. 262/J2
Nelson, Nebr. (68961) 264/F4
Nelson, Nev. (†89046) 266/G7
Nelson (creek), Nev. 266/G2
Nelson○, N.H. (†03457) 268/C5
Nelson, N. Zealand 87/H10
Nelson, N. Zealand 100/D4
Nelson (co.), N. Dak. 282/O4
Nelson, Pa. (16940) 294/H2
Nelson (cape), Victoria 97/A6
Nelson (co.), Va. 307/L5
Nelson, Wis. (54756) 317/C7
Nelson Forks, Br. Col. 184/M2
Nelson Head (prom.), N.W. Terrs. 187/F2
Nelson House, Manitoba 179/J2
Nelson Lagoon, Alaska (†99571) 196/F3
Nelson-Miramichi, New Bruns. 170/E2
Nelsonville, Ark. (†72569) 202/H1
Nelsonville, Ky. (†40051) 237/K5
Nelsonville, N.Y. (10516) 276/N8
Nelsonville, Ohio (45764) 284/F7
Nelsonville, Wis. (54458) 317/H7
Nelspruit, S. Africa 118/E5
Néma, Mauritania 106/C5
Néma, Mauritania 102/B3
Nemacolin, Pa. (15351) 294/B6
Nemadji (riv.), Minn. 255/F4
Nemaha, Iowa (50567) 229/C3
Nemaha (co.), Kansas 232/F2
Nemaha (riv.), Kansas 232/G1
Nemaha (co.), Nebr. 264/J4
Nemaha, Nebr. (68414) 264/J4
Neméa, Greece 45/F7
Nemi, Italy 34/F7
Nemiskam, Alberta 182/E5
Nemo, S. Dak. (57759) 298/B5
Nemours, France 28/E3
Nemrut Dağı (mt.), Turkey 63/J3
Nemunas (Niemen) (riv.), U.S.S.R. 53/A3
Nemuro, Japan 81/M2
Nemuro (str.), Japan 81/M1
Nen (riv.), China 54/O5
Nenagh, Ireland 10/B4
Nenagh, Ireland 17/E6
Nenagh (riv.), Ireland 17/E6
Nenana, Alaska (99760) 196/J2
Nendaz, Switzerland 39/D4
Nene (riv.), England 10/F4
Nene (riv.), England 13/H5
Nenets Aut. Okr., U.S.S.R. 48/F3
Nenets Aut. Okr., U.S.S.R. 52/H1
Nenjiang, China 77/L2
Nen Jiang (riv.), China 77/K2
Nenzel, Nebr. (69219) 264/C2
Neodesha, Kansas (66757) 232/G4
Neoga, Ill. (62447) 222/E4
Neola, Iowa (51559) 229/B6
Neola, Utah (84053) 304/D3
Neola, W. Va. (24961) 312/F7
Neon-Fleming, Ky. (41840) 237/R6
Néon Karlóvasi, Greece 45/H7
Neopit, Wis. (54150) 317/J6
Neópolis, Brazil 132/G5
Neosho (riv.) 188/G3
Neosho (co.), Kansas 232/G4
Neosho (riv.), Kansas 232/G4
Neosho, Mo. (64850) 261/D9
Neosho (riv.), Okla. 288/R1
Neosho, Wis. (53059) 317/J9
Neosho Falls, Kansas (66758) 232/G3
Neosho Rapids, Kansas (66864) 232/F3
Neotsu, Oreg. (97364) 291/C2
Nepa, U.S.S.R. 48/L4
Nepal 2/P4
Nepal 54/K7
NEPAL 68/E3
Nepalganj, Nepal 68/E3
Nepaug (res.), Conn. 210/D1
Nepaug (riv.), Conn. 210/C1
Nepean (isl.), Norfolk I. 88/L6
Nephi, Utah (84648) 304/C4
Nephin (mt.), Ireland 17/C3
Nephin Beg (mt.), Ireland 17/B3
Nephton, Ontario 177/G3
Nepisiguit (bay), New Bruns. 170/E1
Nepisiguit (lakes), New Bruns. 170/D1
Nepisiguit (riv.), New Bruns. 170/D1
Nepomuk, Czech. 41/B2
Neponset, Ill. (61345) 222/D2
Neponset, Mass. (†02122) 249/D7
Neponset (riv.), Mass. 249/C8
Nepton, Ky. (†41039) 237/O4
Neptune○, N.J. (07753) 273/E3
Neptune, Ohio (†45822) 284/A4
Neptune (isls.), S. Australia 94/D6
Neptune Beach, Fla. (32233) 212/E1
Neptune City, N.J. (07753) 273/E3
Nera (riv.), Italy 34/D3
Nérac, France 28/D5
Nerekhta, U.S.S.R. 52/F3
Nerepis (riv.), New Bruns. 170/D3
Neresheim, W. Germany 22/D4
Nereta, U.S.S.R. 53/C2
Neretva (riv.), Yugoslavia 45/D4
Neriquinha, Angola 115/D7
Nerja, Spain 33/E4
Nerka (lake), Alaska 196/G3
Ñermete (pt.), Peru 128/B5
Nerpio, Spain 33/E3
Nerstrand, Minn. (55053) 255/E6
Nerva, Spain 33/C4
Neryungri, U.S.S.R. 48/N4
Nes (Neskaupstadhur), Iceland 21/D1
Nes, Netherlands 27/H2
Nesbit, Miss. (38651) 256/D1
Nesbitt, Manitoba 179/C5
Nescopeck, Pa. (18635) 294/K3
Nesebūr, Bulgaria 45/H4
Neshanic Station, N.J. (†08853) 273/D3
Nesher, Israel 65/C2
Neshkoro, Wis. (54960) 317/H8
Neshoba (co.), Miss. 256/F5
Neshoba, Miss. (39365) 256/F5
Neskaupstadhur, Iceland 7/C2
Neskaupstadhur, Iceland 21/D1
Neskowin, Oreg. (97149) 291/D2
Nesmith, S.C. (29580) 296/H4
Nespelem, Wash. (99155) 310/G2
Nesquehoning, Pa. (18240) 294/L4
Ness (co.), Kansas 232/C3
Ness, Loch (lake), Scotland 15/D3
Ness, Loch (lake), Scotland 10/D2
Ness (riv.), Scotland 15/D3
Ness City, Kansas (67560) 232/C3
Nesselrode (mt.), Alaska 196/N1
Nesselwang, W. Germany 22/D5
Nesslau, Switzerland 39/H2
Neston, England 13/G2
Neston, England 10/F2
Nestor, Trin. & Tob. 161/B10
Nestor Falls, Ontario 177/F5
Nestor Falls, Ontario 175/B3
Nestoria, Mich. (†49861) 250/A2
Nestórion, Greece 45/E5
Nestorville, W. Va. (†26380) 312/G4
Néstos (riv.), Greece 45/G5
Nestow, Alberta 182/D2
Nesttun, Norway 18/D6
Nestucca (riv.), Oreg. 291/D2
Nesvady, Czech. 41/E3
Nes Ziyyona, Israel 65/B4
Net (riv.), Mich. 250/G2
Netanya, Israel 65/B3
Netarts, Oreg. (97143) 291/C2
Netawaka, Kansas (66516) 232/G2
Netcong, N.J. (07857) 273/D2
Nethe (riv.), Belgium 27/F6
Netherhill, Sask. 181/C4
Netherlands 2/K3
Netherlands 7/E3
NETHERLANDS 27/G4
Netherlands Antilles 120/C1
Netherlands Antilles 146/M8
NETHERLANDS ANTILLES 161
NETHERLANDS ANTILLES 156/E4
Nethy Bridge, Scotland 15/E3
Netivot, Israel 65/B5
Netolice, Czech. 41/C2
Netstal, Switzerland 39/H2
Nett (lake), Minn. 255/E2
Nettie, W. Va. (26681) 312/E6
Nettilling (lake), Canada 4/C13
Nettilling (fjord), N.W. Terrs. 187/M3
Nettilling (lake), N.W. Terrs. 187/L3
Nett Lake, Minn. (55772) 255/E2
Nett Lake Ind. Res., Minn. 255/E2
Nettleham, England 13/G4
Nettleton, Miss. (38858) 256/G2
Nettuno, Italy 34/D4
Netzahualcóyotl, Mexico 150/L1
Neuanlage, Sask. 181/E3
Neuberg an der Mürz, Austria 41/C3
Neubert, Tenn. (†37901) 237/O9
Neubois, Québec 172/F3
Neubrandenburg (dist.), E. Germany 22/E2
Neubrandenburg, E. Germany 22/E2
Neuburg an der Donau, W. Germany 22/D4
Neuchâtel (canton), Switzerland 39/C3
Neuchâtel, Switzerland 39/C3
Neuchâtel (lake), Switzerland 39/C3
Neudorf, Sask. 181/J5
Neuenegg, Switzerland 39/D3
Neuenhagen bei Berlin, E. Germany 22/F4
Neufchâteau, Belgium 27/G9
Neufchâteau, France 28/F3
Neufchâtel-en-Bray, France 28/D3
Neugersdorf, E. Germany 22/F3
Neuhausen am Rheinfall, Switzerland 39/G1
Neuhorst, Sask. 181/E3
Neuilly-sur-Seine, France 28/A1
Neu-Isenburg, W. Germany 22/C3
Neumarkt am Wallersee, Austria 41/B3
Neumarkt in der Oberpfalz, W. Germany 22/D4
Neumarkt in Steiermark, Austria 41/C3
Neumünster, W. Germany 22/C1
Neunkirch, Switzerland 39/F1
Neunkirchen, Austria 41/C3
Neunkirchen, W. Germany 22/B4
Neuquén (prov.), Argentina 143/C4
Neuquén, Argentina 143/C4
Neuquén, Argentina 120/C6
Neuquén (riv.), Argentina 143/C4
Neuruppin, E. Germany 22/E2
Neuse, N.C. (27561) 281/M3
Neuse (riv.), N.C. 281/O5
Neusiedl am See, Austria 41/D3
Neusiedler See (lake), Austria 41/D3
Neusiedler See (lake), Hungary 41/D3
Neuss, W. Germany 22/B3
Neustadt, E. Germany 22/D3
Neustadt, Ontario 177/D3
Neustadt (Titisee-Neustadt), W. Germany 22/C5
Neustadt an der Aisch, W. Germany 22/D4
Neustadt an der Weinstrasse, W. Germany 22/B4
Neustadt bei Coburg, W. Germany 22/D3
Neustadt-Glewe, E. Germany 22/D2
Neustadt in Holstein, W. Germany 22/D1
Neustift im Stubaital, Austria 41/A3
Neustrelitz, E. Germany 22/E2
NEUTRAL ZONE 59/E4
Neutral Zone 54/F7
Neu-Ulm, W. Germany 22/D4
Neuville, Québec 172/F3
Neuwerk (isl.), W. Germany 22/C2
Neuwied, W. Germany 22/B3
Neva, Tenn. (†37689) 237/T8
Nevada 188/C3
NEVADA 266
Nevada (co.), Ark. 202/D6
Nevada (co.), Calif. 204/E4
Nevada, Sierra (mts.), Calif. 204/E4
Nevada, Iowa (50201) 229/G5
Nevada, Mo. (64772) 261/D7
Nevada, Ohio (44849) 284/D4
Nevada, Sierra (mts.), Spain 33/E4
Nevada (state), U.S. 146/G6
Nevada City, Calif. (95959) 204/D4
Nevatim, Israel 65/B5
Nevel', U.S.S.R. 52/D3
Nevele, Belgium 27/D6
Nevel'sk, U.S.S.R. 48/P5
Nevers, France 28/E4
Neversink (res.), N.Y. 276/L7
Nevertire, N.S. Wales 97/D2
Nevesinje, Yugoslavia 45/D4
Neville, Ohio (45156) 284/B8
Neville, Sask. 181/D6
Nevils, Georgia (†30458) 217/J6
Nevinnomyssk, U.S.S.R. 52/F6
Nevinville, Iowa (50856) 229/D6
Nevis, Alberta 182/D3
Nevis, Minn. (56467) 255/D4
Nevis (isl.), St. Chris.-Nevis 161/D11
Nevis (isl.), St. Chris.-Nevis 156/F3
Nevis (peak), St. Chris.-Nevis 161/D11
Nevis, Loch (inlet), Scotland 15/C3
Nevisdale, Ky. (40754) 237/N7
Nevşehir (prov.), Turkey 63/F3
Nevşehir, Turkey 59/B2
Nevşehir, Turkey 63/F3
New (riv.), Belize 154/C2
New (riv.), Calif. 204/K11
New (for.), England 13/F6
New (riv.), Fla. 212/D1
New (riv.), Fla. 212/B1
New (riv.), Guyana 131/C4
New, South Fork (riv.), N.C. 281/G2
New (riv.), N.C. 281/O5
New (riv.), S.C. 296/E6
New (inlet), Va. 307/S6
New (riv.), Va. 307/F8
New (riv.), W. Va. 312/E7
New Abbey, Scotland 15/E6
New Agat, Guam 86/K7
Newagen, Maine (04552) 243/D8
Newala, Tanzania 115/G6
New Albany, Ind. 188/J3
New Albany, Ind. (47150) 227/F8
New Albany, Kansas (66759) 232/G4
New Albany, Miss. (38652) 256/G2
New Albany, Ohio (43054) 284/E5
New Albany, Pa. (18833) 294/J2
New Albin, Iowa (52160) 229/L2
Newald, Wis. (54551) 317/J4
New Alexandria, Ohio (†43938) 284/J5
New Alexandria, Pa. (15670) 294/C5
Newalla, Okla. (74857) 288/M4
New Alluwe, Okla. (74049) 288/R1
New Almaden, Calif. (95042) 204/L4
New Almelo, Kansas (67652) 232/B2
New Amsterdam, Guyana 120/D2
New Amsterdam, Guyana 131/B2
New Amsterdam, Ind. (†47110) 227/E8
New Amsterdam, Wis. (†54636) 317/C8
New Angledool, N.S. Wales 97/E1
Newark, Ark. (72562) 202/H2
Newark, Calif. (94560) 204/K3
Newark, Del. (19711) 245/P2
Newark, England 13/G4
Newark, England 10/F4
Newark, Ill. (60541) 222/E2
Newark, Md. (21841) 245/S7
Newark, Mo. (63458) 261/H2
Newark, N.J. 188/L2
Newark, N.J. (*07101) 273/B2
Newark (bay), N.J. 273/B2
Newark, N.Y. (14513) 276/G4
Newark, Ohio (43055) 284/F5
Newark, S. Dak. (†57453) 298/O2
Newark, Texas (76071) 303/E1
Newark○, Vt. (†05871) 268/D2
Newark, W. Va. (†26143) 312/D4
Newark Int'l Airport, N.J. 273/B2
Newark Valley, N.Y. (13811) 276/H6
Newarthill, Scotland 15/C2
New Athens, Ill. (62264) 222/D5
New Athens, Ohio (43981) 284/H5
New Auburn, Minn. (55366) 255/D6
New Auburn, Wis. (54757) 317/D5
New Augusta, Miss. (39462) 256/F8
Newaygo (co.), Mich. 250/D5
Newaygo, Mich. (49337) 250/D5
New Baden, Ill. (62265) 222/D5
New Baltimore, Mich. (48047) 250/G6
New Baltimore, N.Y. (12124) 276/N6
New Baltimore, Pa. (15553) 294/E6
New Baltimore, Va. (†22186) 307/N3
New Bavaria, Ohio (43548) 284/B3
New Beaver, Pa. (†16141) 294/B4
New Bedford, Ill. (61346) 222/D2
New Bedford, Mass. 188/N2
New Bedford, Mass. (*02740) 249/K6
New Bedford, Ohio (†43824) 284/G5
New Bedford, Pa. (16140) 294/A3
New Bellsville, Ind. (†47448) 227/E6
Newberg, Oreg. (97132) 291/A2
New Berlin, Ill. (62670) 222/D4
New Berlin, N.Y. (13411) 276/K5
New Berlin, Pa. (17855) 294/J4
New Berlin, Wis. (53151) 317/K2
Newbern, Ala. (36765) 195/C5
Newbern, Ind. (†47201) 227/F6
New Bern, N.C. 188/L4
New Bern, N.C. (28560) 281/P4
Newbern, Tenn. (38059) 237/C8
Newberne, W. Va. (26409) 312/E4
Newberry, Fla. (32669) 212/D2
Newberry, Ind. (47449) 227/C7
Newberry, Mich. (49868) 250/D2
Newberry (co.), S.C. 296/D3
Newberry, S.C. (29108) 296/D3
Newberry Springs, Calif. (92365) 204/J9
New Bethlehem, Pa. (16242) 294/D3
Newbiggin-by-the-Sea, England 13/F2
New Blaine, Ark. (72851) 202/D3
Newbliss, Ireland 17/G3
New Bloomfield, Mo. (65063) 261/J5
New Bloomfield, Pa. (17068) 294/H5
New Bloomington, Ohio (43341) 284/D4
New Bonaventure, Newf. 166/D2
Newborn, Georgia (30262) 217/E3
Newboro, Ontario 177/H3
New Boston, Ill. (61272) 222/B2
New Boston, Mich. (48164) 250/F6
New Boston, Mo. (63557) 261/G3
New Boston○, N.H. (03070) 268/D6
New Boston, Ohio (45662) 284/E8
New Boston, Texas (75570) 303/K4
New Bothwell, Manitoba 179/F5
New Braintree○, Mass. (01531) 249/F3
New Braunfels, Texas (78130) 303/K10
New Bremen, N.Y. (13412) 276/K3
New Bremen, Ohio (45869) 284/B5
Newbridge (Droichead Nua), Ireland 17/H5
New Bridge, Oreg. (†97870) 291/K3
New Brigden, Alberta 182/E4
New Brighton, Minn. (55112) 255/G5
New Brighton, Pa. (15066) 294/B4
New Britain, Conn. (*06050) 210/E2
New Britain (isl.), Papua N.G. 87/F6
New Britain (isl.), Papua N.G. 85/C7
New Britain (isl.), Papua N.G. 86/B2

New Britain, Pa. (18901) 294/M5
New Brockton, Ala. (36351) 195/G8
Newbrook, Alberta 182/D2
New Brunswick (prov.) 162/K6
New Brunswick (prov.), Canada 146/M5
NEW BRUNSWICK 170
New Brunswick, N.J. (*08901) 273/E3
New Buena Vista, Pa. (†15550) 294/E5
New Buffalo, Mich. (49117) 250/C7
New Buffalo, Pa. (17069) 294/H5
Newburg, Ark. (72556) 202/G1
Newburg, Iowa (†50135) 229/H5
Newburg, Md. (20664) 245/L7
Newburg, Mo. (65550) 261/J7
Newburg, N. Dak. (58762) 282/J2
Newburg, Pa. (17240) 294/G5
Newburg (La Jose), Pa. (†15753) 294/E4
Newburg, W. Va. (26410) 312/G4
Newburg, Wis. (53060) 317/K9
Newburgh, Ind. (47630) 227/C9
Newburgh○, Maine (†04445) 243/F6
Newburgh, N.Y. 188/M2
Newburgh, N.Y. (12550) 276/M7
Newburgh, Ontario 177/H3
Newburgh, Scotland 10/E2
Newburgh, Grampian, Scotland 15/G3
Newburgh, Fife, Scotland 15/E4
Newburgh Heights, Ohio (†44101) 284/H9
New Burlington, Ind. (†47302) 227/G4
New Burlington, Ohio (†45201) 284/B9
New Burnside, Ill. (62967) 222/E6
Newbury, England 13/F6
Newbury, England 10/F5
Newbury○, Mass. (01950) 249/L1
Newbury○, N.H. (03255) 268/C5
Newbury, Ohio (44065) 284/H3
Newbury, Ontario 177/C5
Newbury, Vt. (05051) 268/C3
Newbury○, Vt. (05051) 268/C3
Newburyport, Mass. (01950) 249/L1
New Bussa, Nigeria 106/E6
New Caledonia (isl.) 2/T7
NEW CALEDONIA 86
New Caledonia 87/G8
New Caledonia (isl.), New Caled. 87/G8
New Caledonia (isl.), New Caled. 86/G4
New Cambria, Kansas (67470) 232/E3
New Cambria, Mo (63558) 261/G3
New Canaan○, Conn. (06840) 210/B4
New Canton, Ill. (62356) 222/B4
New Canton, Va. (23123) 307/M5
New Carlisle, Ind. (46552) 227/E1
New Carlisle, Ohio (45344) 284/C6
New Carlisle, Québec 172/D2
New Carlisle, Québec 174/E3
New Carrollton, Md. (20784) 245/G4
New Castile (reg.), Spain 33/E3
Newcastle, Australia 2/S7
Newcastle, Calif. (95658) 204/C8
New Castle, Colo. (81647) 208/E3
New Castle (co.), Del. 245/R2
New Castle, Del. (19720) 245/R2
New Castle, Ind. (47362) 227/G5
Newcastle, Ireland 10/B4
Newcastle, Ireland 17/D7
New Castle, Ky. (40050) 237/L4
Newcastle○, Maine (04553) 243/D7
Newcastle, N. Br. 162/K6
Newcastle, N. S. Wales 88/J6
Newcastle, Nebr. (68757) 264/H2
Newcastle, New Bruns. 170/E2
New Castle○, N.H. (03854) 268/F5
Newcastle, N.S. Wales 97/F3
Newcastle, N. Ireland 10/D3
Newcastle, N. Ireland 17/J3
Newcastle (creek), North. Terr. 93/C4
New Castle, Ohio (†43843) 284/F5
Newcastle, Okla. (73065) 288/L4
Newcastle, Ontario 177/F4
New Castle, Pa. 188/K2
New Castle, Pa. (*16101) 294/B3
Newcastle, St. Chris.-Nevis 161/D11
Newcastle, S. Africa 118/E5
Newcastle, Texas (76372) 303/F4
Newcastle, Utah (84756) 304/A6
New Castle, Va. (24127) 307/H5
Newcastle, Wyo. (82701) 319/H2
Newcastle Creek, New Bruns. 170/D2
Newcastle-Damariscotta, Maine (04553) 243/E7
Newcastle Emlyn, Wales 13/C5
Newcastleton, Scotland 15/F5
Newcastle-under-Lyme, England 13/E4
Newcastle-under-Lyme, England 10/E4
Newcastle upon Tyne, England 7/D3
Newcastle upon Tyne, England 10/E3
Newcastle upon Tyne, England 13/H3
Newcastle Waters, North. Terr. 93/C4
New Centerville, Pa. (†15557) 294/D6
New Chelsea, Newf. 166/D2
New Chicago, Ind. (†46342) 227/C1
New Church, Va. (23415) 307/S5
New Cinema, Br. Col. 184/F3
New City, N.Y. (10956) 276/K8
New Columbia, Pa. (17856) 294/H3
New Columbus, Pa. (†17878) 294/K3
Newcomb, N. Mex. (†87325) 274/A2
Newcomb, N.Y. (12852) 276/M3
Newcomb, Tenn. (37819) 237/N7
Newcomerstown, Ohio (43832) 284/G5
New Concord, Ky. (42076) 237/E7
New Concord, Ohio (43762) 284/G6
New Cordell (Cordell), Okla. (†73632) 288/H4
New Corydon, Ind. (†47326) 227/H3
New Court, Mo. (†63452) 261/J2
New Creek, W. Va. (26743) 312/J4
New Cumberland, Pa. (17070) 294/J5
New Cumberland, W. Va. (26047) 312/E2
New Cumnock, Scotland 15/D5
Newdale, Idaho (83436) 220/G6
Newdale, Manitoba 179/B4
New Dayton, Alberta 182/D5
New Deal, Texas (79350) 303/C4
New Deer, Scotland 15/F3
Newdegate, W. Australia 92/B6
New Delhi (cap.), India 2/N4
New Delhi, India 54/J7
New Delhi (cap.), India 68/D3
New Denmark, New Bruns. 170/C1
New Denver, Br. Col. 184/J5
New Diggings, Wis. (†61075) 317/F10
New Douglas, Ill. (62074) 222/D5
New Dover, Ohio (†43040) 284/D5
New Durham○, N.H. (03855) 268/E5
New Eagle, Pa. (15067) 294/B5
New Edinburg, Ark. (71660) 202/F6
New Effington, S. Dak. (57255) 298/R2
New Egypt, N.J. (08533) 273/E3
Newell, Ala. (36270) 195/H4
Newell (lake), Alberta 182/E4
Newell, Iowa (50568) 229/D3
Newell, S. Dak. (57760) 298/C4
Newell, W. Va. (26050) 312/E1
New Ellenton, S.C. (29809) 296/D5
Newellton, La. (71357) 238/H2
Newellton, Nova Scotia 168/C5
New England (range), N.S. Wales 97/F1
New England, N. Dak. (58647) 282/E6
New England, W. Va. (26154) 312/C4
Newenham (cape), Alaska 196/F3
New Enterprise, Pa. (16664) 294/F5
New Era, La. (†71354) 238/G4
New Era, Mich. (49446) 250/C5
New Era, Oreg. (†97013) 291/B2
Newe Yam, Israel 65/B2
Newe Zohar, Israel 65/C5
New Fairfield○, Conn. (06810) 210/B3
Newfane, N.Y. (14108) 276/C4
Newfane, Vt. (05345) 268/B6
Newfane○, Vt. (05345) 268/B6
Newfield, Maine (04056) 243/B8
Newfield○, Maine (04056) 243/B8
Newfield, N.J. (08344) 273/D4
Newfield, N.Y. (14867) 276/G6
Newfields○, N.H. (03856) 268/F5
New Fish Creek, Alberta 182/B2
New Florence, Mo. (63363) 261/K5
New Florence, Pa. (15944) 294/D5
Newfolden, Minn. (56738) 255/B2
New Fork (lakes), Wyo. 319/C2
Newfound (lake), N.H. 268/D4
Newfoundland (prov.) 162/L5
Newfoundland (isl.) 162/L6
Newfoundland (prov.), Canada 146/M4
Newfoundland (isl.), Canada 2/G3
Newfoundland, Ky. (41162) 237/P4
NEWFOUNDLAND 166
Newfoundland (isl.), Newf. 166/C4
Newfoundland (isl.), Newf. 146/N5
Newfoundland, N.J. (07435) 273/D1
Newfoundland, Pa. (18445) 294/M3
Newfoundland (mts.), Utah 304/A2
New Franken, Wis. (54229) 317/L6
New Frankfort, Mo. (†65349) 261/F4
New Franklin, Mo. (65274) 261/G4
New Freedom, Pa. (17349) 294/J6
New Freeport, Pa. (15352) 294/B6
New Galilee, Pa. (16141) 294/A4
New Galloway, Scotland 10/D3
New Galloway, Scotland 15/D5
Newgate, Br. Col. 184/K5
New Georgia (isl.), Solomon Is. 87/F6
New Georgia (isl.), Solomon Is. 86/D3
New Germantown, Pa. (17071) 294/G5
New Germany, Minn. (55367) 255/E6
New Germany, Nova Scotia 168/D4
New Glarus, Wis. (53574) 317/G10
New Glasgow, Nova Scotia 168/F3
New Glasgow, Québec 172/D4
New Gloucester, Maine (04260) 243/C8
New Gloucester○, Maine (04260) 243/C8
New Goshen, Ind. (47863) 227/B5
New Gretna, N.J. (08224) 273/E4
New Guinea (isl.) 2/S6
New Guinea (isl.) 54/P10
New Guinea (isl.) 87/E6
New Guinea (isl.), Papua N.G. 86/B2
Newgulf, Texas (77462) 303/J8
Newhalem, Wash. (†98283) 310/D2
Newhalen, Alaska (†99606) 196/H3
Newhall, Calif. (91321) 204/G9
Newhall, Iowa (52315) 229/K5
Newhall, W. Va. (24866) 312/C8
Newham, England 13/H8
Newham, England 10/B5
New Hamburg, Mo. (†63736) 261/O8
New Hamburg, Ontario 177/D4
New Hampshire 188/M2
NEW HAMPSHIRE 268
New Hampshire, Ohio (45870) 284/C4
New Hampshire (state), U.S. 146/L5
New Hampton, Iowa (50659) 229/J2
New Hampton, Mo. (64471) 261/D2
New Hampton○, N.H. (03256) 268/D4
New Hampton, N.J. (†08827) 273/D2
New Hanover (co.), N.C. 281/O6
New Hanover (Lavongai) (isl.), Papua N.G. 87/F6
New Hanover (isl.), Papua N.G. 86/B1
New Harbor, Maine (04554) 243/E8
New Harbour, Newf. 166/C4
New Harbour, Newf. 166/D2
New Harbour, Nova Scotia 168/G3
New Harmony, Ind. (47631) 227/B8
New Harmony, Utah (84757) 304/A6
New Hartford, Conn. (06057) 210/C1
New Hartford○, Conn. (06057) 210/C1
New Hartford, Iowa (50660) 229/H3
New Hartford, Mo. (63364) 261/K4
New Hartford, N.Y. (13413) 276/K4
New Haven, Conn. 188/M2
New Haven (co.), Conn. 210/D3
New Haven, Conn. (*06501) 210/D3
New Haven (harb.), Conn. 210/D3
Newhaven, England 10/F5
Newhaven, England 13/H7
New Haven, Ill. (62857) 222/E6
New Haven, Ind. (46774) 227/H2
New Haven, Ky. (40051) 237/K5
New Haven, Mich. (48048) 250/G6
New Haven, Mo. (63068) 261/K5
New Haven, N.Y. (13121) 276/H4
New Haven, Nova Scotia 168/H2
New Haven, Ohio (44850) 284/E3
New Haven○, Vt. (05472) 268/A3
New Haven, W. Va. (25265) 312/C5
New Haven, Wyo. (†82720) 319/H1
New Hazelton, Br. Col. 184/D2
New Hebrides (Vanuatu) 87/G7
Newhebron, Miss. (39140) 256/D7
New Hill, N.C. (27562) 281/M3
New Holland, Georgia (†30501) 217/E2
New Holland, Ill. (62671) 222/D3
New Holland, N.C. (27885) 281/S4
New Holland, Ohio (43145) 284/D6
New Holland, Pa. (17557) 294/K5
New Holland, S. Dak. (57364) 298/M7
New Holstein, Wis. (53061) 317/K8
New Home, Texas (79383) 303/C4
New Hope, Ala. (35760) 195/F1
Newhope, Ark. (71959) 202/C5
New Hope, Ky. (40052) 237/L5
New Hope, Minn. (†55428) 255/G5
New Hope, Ohio (†45320) 284/A6
New Hope, Pa. (18938) 294/N5
New Hope, Tenn. (†37380) 237/K11
New Hope, Va. (24469) 307/L4
New Horse Springs, N. Mex. (†87821) 274/A5
New Houlka (Houlka), Miss. (38850) 256/G2
New Hradec, N. Dak. (58648) 282/E5
New Hyde Park, N.Y. (11040) 276/P7
New Iberia, La. (70560) 238/G6
Newington○, Conn. (06111) 210/E2
Newington, Georgia (30446) 217/J5
Newington○, N.H. (†03801) 268/F5
Newington, Ontario 177/K2
Newington, Va. (22122) 307/S3
New Ipswich○, N.H. (03071) 268/D6
New Ireland (isl.), Papua N.G. 87/F6
New Ireland (isl.), Papua N.G. 86/B1
New Jersey 188/M3
NEW JERSEY 273
New Jersey, New Bruns. 170/E1
New Jersey (state), U.S. 146/L5
New Johnsonville, Tenn. (37134) 237/E8
New Kensington, Pa. (15068) 294/C4
New Kent (co.), Va. 307/P5
New Kent, Va. (23124) 307/P5
Newkirk, N. Mex. (88431) 274/E3
Newkirk, Okla. (74647) 288/N1
New Knoxville, Ohio (45871) 284/B5
New Laguna, N. Mex. (87038) 274/B4
New Lancaster, Kansas (†66040) 232/H3
Newland, Ind. (†47978) 227/C2
Newland, N.C. (28657) 281/F2
New Lebanon, Ind. (47864) 227/C6
New Lebanon, N.Y. (12125) 276/O6
New Lebanon, Ohio (45345) 284/B6
New Lebanon, Pa. (†16145) 294/B3
New Leipzig, N. Dak. (58562) 282/G7
New Lenox, Ill. (60451) 222/B6
New Lexington, Ohio (43764) 284/F6
New Liberty, Iowa (52765) 229/M5
New Liberty, Ky. (40355) 237/M3
New Lima, Okla. (74884) 288/O4
New Limerick○, Maine (04761) 243/G3
Newlin, Texas (†79245) 303/D3
New Lisbon, Ind. (47366) 227/G5
New Lisbon, N.J. (08064) 273/D4
New Lisbon, Wis. (53950) 317/F8
New Liskeard, Ont. 162/H6
New Liskeard, Ontario 177/K5
New Liskeard, Ontario 175/E3
Newllano, La. (71461) 238/D4
New London, Ark. (†71765) 202/F7
New London, Conn. 188/M2
New London (co.), Conn. 210/G2
New London, Conn. (06320) 210/G3
New London, Ind. (†46979) 227/E4
New London, Iowa (52645) 229/L7
New London, Minn. (56273) 255/C5
New London, Mo. (63459) 261/K3
New London, N.H. (03257) 268/D5
New London○, N.H. (03257) 268/D5
New London, N.C. (28127) 281/J4
New London, Ohio (44851) 284/F3
New London (bay), Pr. Edward I. 168/E2
New London, Texas (75682) 303/K5
New London, Wis. (54961) 317/J7
New Lothrop, Mich. (48460) 250/F5
New Lowell, Ontario 177/E3
New Lyme, Ohio (44066) 284/J2
New Lynn, N. Zealand 100/B1
New Madison, Ohio (45346) 284/A6
New Madrid (co.), Mo. 261/N9
New Madrid, Mo. (63869) 261/O9
Newmains, Scotland 15/C2
Newman, Calif. (95360) 204/D6
Newman, Ill. (61942) 222/B6
Newman, Ky. (†42301) 237/G5
Newman (sound), Newf. 166/D2
Newman, W. Australia 88/C4
Newman (lake), Wash. 310/H3
Newman, W. Australia 92/B3
New Manchester, W. Va. (26056) 312/E1
Newman Grove, Nebr. (68758) 264/G3
Newman Lake, Wash. (99025) 310/J3
Newmans Cove, Newf. 166/D2
New Marion, Ind. (†47023) 227/G6
New Market, Ala. (35761) 195/F1
Newmarket, England 13/H5
Newmarket, England 10/G4
New Market, Ind. (47965) 227/D5
New Market, Iowa (51646) 229/D7
Newmarket, Ireland 10/B4
Newmarket, Ireland 17/C7
Newmarket, Jamaica 158/H6
New Market, Md. (21774) 245/J3
New Market, Minn. (55054) 255/E6
New Market, Mo. (†64439) 261/C4
New Market, New Bruns. 170/D3
Newmarket, N.H. (03857) 268/F5
Newmarket○, N.H. (03857) 268/F5
New Market, Ohio (†45133) 284/C7
Newmarket, Ontario 177/E3
Newmarket, Queensland 88/K2
Newmarket, Queensland 95/D2
Newmarket, Scotland 15/B2
New Market, Tenn. (37820) 237/O8
New Market, Va. (22844) 307/L3
Newmarket-on-Fergus, Ireland 17/D6
New Marlborough○, Mass. (†01230) 249/B4
New Martinsburg, Ohio (†43160) 284/D7
New Martinsville, W. Va. (26155) 312/E3
New Maryland, New Bruns. 170/D3
New Matamoras, Ohio (45767) 284/J6
New Meadows, Idaho (83654) 220/B4
New Melle, Mo. (63365) 261/L5
New Memphis, Ill. (62266) 222/D5
Newmerella, Victoria 97/E5
New Mexico 188/E4
NEW MEXICO 274
New Mexico (state), U.S. 146/H6
New Miami, Ohio (45011) 284/A7
New Middleton, Tenn. (†38563) 237/J8
New Middletown, Ind. (47160) 227/E8
New Middletown, Ohio (44442) 284/J4
New Milford, Conn. (06776) 210/B2
New Milford○, Conn. (06776) 210/B2
New Milford, N.J. (07646) 273/B1
New Milford, Ohio (†44272) 284/H3
New Milford, Pa. (18834) 294/L2
Newmill, Scotland 15/E3
New Mills, England 13/J2
New Mills, England 10/G2
Newmilns and Greenholm, Scotland 15/D5
New Milton, W. Va. (26411) 312/E4
New Minas, Nova Scotia 168/D3
New Minden, Ill. (†62263) 222/D5
New Mount Pleasant, Ind. (†47371) 227/G4
New Munich, Minn. (56356) 255/D5
Newnan, Georgia (30263) 217/C4
Newnans (lake), Fla. 212/D2
New Norcia, W. Australia 92/A5
New Norfolk, Tasmania 88/H8
New Norfolk, Tasmania 99/C4
New Norway, Alberta 182/D3
New Offenburg, Mo. (63661) 261/M7
New Orleans, La. 188/H5
New Orleans, La. 146/K7
New Orleans, La. (*70101) 238/O4
New Orleans, U.S. 2/E4
New Osgoode, Sask. 181/H3
New Oxford, Pa. (17350) 294/H6
New Palestine, Ind. (46163) 227/F5
New Pallas, Ireland 17/E6
New Paltz, N.Y. (12561) 276/M7
New Paris, Ind. (46553) 227/F2
New Paris, Ohio (45347) 284/A6
New Paris, Pa. (15554) 294/E5
New Pass (range), Nev. 266/D3
New Pekin, Ind. (†47165) 227/F7
New Perlican, Newf. 166/D2
New Petersburg, Ohio (†45123) 284/D7
New Philadelphia, Ill. (†61459) 222/C3
New Philadelphia, Ind. (†47167) 227/F7
New Philadelphia, Ohio (44663) 284/G5
New Philadelphia, Pa. (17959) 294/K4
New Pine Creek, Oreg. (97635) 291/G5
New Pitsligo, Scotland 15/F3
New Pittsburg, Ohio (†44691) 284/F4
New Plymouth, Idaho (83655) 220/B6
New Plymouth, N. Zealand 100/D3
New Plymouth, Ohio (45654) 284/F7
New Point, Ind. (47263) 227/G6
New Point, Mo. (64473) 261/B2
Newport, Ark. (72112) 202/H2
Newport, Del. (19804) 245/R2
Newport, England 13/F7
Newport, England 13/E5
Newport, England 10/F5
Newport, Ind. (47966) 227/C5
Newport, Mayo, Ireland 17/C4
Newport, Tipperary, Ireland 17/E6
Newport, Ky. (*41071) 237/S2
Newport, Ky. 188/K3
Newport, Maine (04953) 243/E6
Newport○, Maine (04953) 243/E6
Newport, Md. (†20622) 245/L7
Newport, Minn. (55055) 255/F6
Newport, Miss. (†38641) 256/D1
Newport, Nebr. (68759) 264/E2
New Port, Neth. Ant. 161/G9
Newport, N.H. (03773) 268/C5
Newport○, N.H. (03773) 268/C5
Newport, N.J. (08345) 273/C5
Newport, N.Y. (13416) 276/K4
Newport, N.C. (28570) 281/R5
Newport, Nova Scotia 168/E3
Newport, Ohio (†43140) 284/C6
Newport, Ohio (45768) 284/H7
Newport, Oreg. (97365) 291/C3
Newport, Pa. (17074) 294/H5
Newport, Québec 172/D2
Newport, R.I. 188/M2
Newport (co.), R.I. 249/K6
Newport, R.I. (02840) 249/J7
Newport, Tenn. (37821) 237/P9
Newport, Texas (76254) 303/F4
Newport, Vt. (05855) 268/C2
Newport○, Vt. (05855) 268/C2
Newport, Va. (24128) 307/H6
Newport, Wales 10/E5
Newport, Dyfed, Wales 13/C5
Newport, Gwent, Wales 13/B6
Newport, Wash. (99156) 310/H2
Newport Beach, Calif. (*92660) 204/D11
Newport Center, Vt. (05857) 268/C2
New Portland, Maine (04954) 243/C6
New Portland○, Maine (04954) 243/C6
Newport News, Va. 188/L3
Newport News (I.C.), Va. (*23601) 307/P6
Newport-on-Tay, Scotland 15/F4
Newport Pagnell, England 13/G5
New Port Richey, Fla. (*33552) 212/D3
New Prague, Minn. (56071) 255/E6
New Preston, Conn. (06777) 210/B2
New Providence (isl.), Bahamas 156/C1
New Providence (Borden), Ind. (†47106) 227/F8
New Providence, Iowa (50206) 229/G4
New Providence, N.J. (07974) 273/E2
New Providence, Pa. (17560) 294/K6
New Prue (Prue), Okla. (†74060) 288/O2
Newquay, England 10/D5
Newquay, England 13/B7
New Quay, Wales 10/D4
New Quay, Wales 13/C5
New Raymer, Colo. (80742) 208/M1
New Richland, Minn. (56072) 255/E7
New Richmond, Ind. (47967) 227/D4
New Richmond, Ohio (45157) 284/B8
New Richmond, Québec 172/C2
New Richmond, Wis. (54017) 317/A5
New Riegel, Ohio (44853) 284/D3
New River, N.C. (28540) 281/O5
New River (inlet), N.C. 281/P6
New River, Tenn. (†37755) 237/M8
New River, Va. (24129) 307/G6
New River Beach, New Bruns. 170/D3
New Road, Nova Scotia 168/E4
New Roads, La. (70760) 238/G5
New Rochelle, N.Y. (*10801) 276/P7
New Rockford, N. Dak. (58356) 282/N4
New Romney, England 13/J7
New Ross, Ind. (47968) 227/D5
New Ross, Ireland 10/C4
New Ross, Ireland 17/H7
New Ross, Nova Scotia 168/D4
Newry, Maine (04261) 243/B6
Newry○, Maine (04261) 243/B6
Newry, N. Ireland 17/J3
Newry, N. Ireland 10/C3
Newry, North. Terr. 93/A3
Newry, Pa. (16665) 294/F5
Newry, S.C. (29665) 296/B2
New Salem, Ill. (62357) 222/C4
New Salem, Ind. (†46173) 227/G5
New Salem, Kansas (†67156) 232/F4
New Salem○, Mass. (01355) 249/E2
New Salem, N. Dak. (58563) 282/G6
New Salem, Nova Scotia 168/D3
New Salem, Ohio (†43148) 284/E6
New Salem, Pa. (15468) 294/C6
New Salem (Delmont), Pa. (†15626) 294/D5
New Salisbury, Ind. (47161) 227/E8
New Sarepta, Alberta 182/D3
New Sarpy, La. (70078) 238/N4
New Schwabenland (reg.) 5/B1
New Scone, Scotland 15/E4
New Sharon, Iowa (50207) 229/H6
New Sharon○, Maine (04955) 243/C6
New Sharon, N.J. (†08691) 273/D3
New Shoreham (Block Island)○, R.I. (†02807) 249/H8
New Siberian (isls.), U.S.S.R. 54/R2
New Siberian (isls.), U.S.S.R. 4/B2
New Siberian (isls.), U.S.S.R. 2/S2
New Siberian (isls.), U.S.S.R. 48/P2
New Site, Ala. (†35010) 195/G4
New Site, Miss. (38859) 256/H1
New Smyrna Beach, Fla. (32069) 212/F2
Newsoms, Va. (23874) 307/O7
New South Wales, /H6
New South Wales (state), Australia 87/E9
NEW SOUTH WALES 97
New Spadra, Ark. (†72830) 202/C3
New Square, N.Y. (†10901) 276/K8
New Stanton, Pa. (15672) 294/C5
New Straitsville, Ohio (43766) 284/F6
New Strawn (Strawn), Kansas (66839) 232/G3
New Stuyahok, Alaska (99636) 196/G3
New Sweden, Maine (04762) 243/G2
New Sweden○, Maine (04762) 243/G2
New Tazewell, Tenn. (37825) 237/O8
Newtok, Alaska (99681) 196/F2
Newton, Ala. (36352) 195/G8
Newton (co.), Ark. 202/D2
Newton (co.), Georgia 217/E3
Newton, Georgia (31770) 217/D8
Newton, Ill. (62448) 222/E5
Newton (co.), Ind. 227/C3
Newton, Iowa 188/H2
Newton, Iowa (50208) 229/H5
Newton, Kansas (67114) 232/E3
Newton, Mass. (†02158) 249/C7
Newton (co.), Miss. 256/F6
Newton, Miss. (39345) 256/F6
Newton (co.), Mo. 261/D9
Newton○, N.H. (03858) 268/E6
Newton, N.J. (07860) 273/D1
Newton, N.C. (28658) 281/G3
Newton, Québec 172/C4
Newton, Scotland 15/E5
Newton (co.), Texas 303/L7
Newton, Texas (75966) 303/L7
Newton, Utah (84327) 304/C2
Newton, W. Va. (25266) 312/D5
Newton Abbot, England 13/D7
Newton Abbot, England 10/E5
Newton Center, Mass. (02159) 249/C7
Newton Falls, N.Y. (13666) 276/K2
Newton Falls, Ohio (44444) 284/J3
Newtongrange, Scotland 15/D2
Newton Grove, N.C. (28366) 281/N4
Newton Hamilton, Pa. (17075) 294/G5
Newton Highlands, Mass. (02161) 249/C7
Newtonia, Mo. (64853) 261/D9
Newton Junction, N.H. (03859) 268/E6
Newton-le-Willows, England 13/H2
Newton Lower Falls, Mass. (†02162) 249/B7
Newton Mearns, Scotland 15/B2
Newton Mills, Nova Scotia 168/F3
Newtonmore, Scotland 15/D3
Newton Siding, Manitoba 179/D5
Newton Stewart, Scotland 10/D3
Newton Stewart, Scotland 15/D6
Newtonsville, Ohio (45158) 284/B7
Newton Upper Falls, Mass. (†02164) 249/C7
Newtonville, Ind. (47632) 227/D8
Newtonville, Mass. (02160) 249/C7
Newtonville, N.J. (08346) 273/D4
Newtown, Conn. (06470) 210/B3
Newtown○, Conn. (06470) 210/B3
Newtown, Ind. (47969) 227/C4
Newtown, Ky. (†40324) 237/N4
Newtown, Mo. (64667) 261/F2
Newtown, New Bruns. 170/E3
Newtown, Newf. 166/D4
Newtown, N.S. Wales 97/C6
New Town, N. Dak. (58763) 282/F4
Newtown, Ohio (45244) 284/C10
Newtown, Pa. (18940) 294/N5
New Town, S.C. (†29536) 296/J3
Newtown, Victoria 97/C6
Newtown, Wales 13/D5
Newtown, Wales 10/E4
Newtownabbey (dist.), N. Ireland 17/J2
Newtownabbey, N. Ireland 17/K2
Newtownards, N. Ireland 17/K2
Newtownbutler, N. Ireland 17/G3
Newtown Forbes, Ireland 17/F4
Newtownhamilton, N. Ireland 17/H3
Newtownmountkennedy, Ireland 17/J5
Newtown Saint Boswells, Scotland 15/F5
Newtownsandes, Ireland 17/C6
Newtown Square○, Pa. (19073) 294/L6
Newtownstewart, N. Ireland 17/G2
New Trenton, Ind. (47035) 227/H6
New Trier, Minn. (†55031) 255/F6
New Tripoli, Pa. (18066) 294/L4
New Troy, Mich. (49119) 250/C7
New Tulsa, Okla. (†74080) 288/P2
Newtyle, Scotland 15/E4
New Ulm, Minn. (56073) 255/D6
New Ulm, Texas (78950) 303/H8
New Underwood, S. Dak. (57761) 298/D5
New Vernon, N.J. (07976) 273/D2
New Victoria, Nova Scotia 168/H2
New Vienna, Iowa (52065) 229/L3
New Vienna, Ohio (45159) 284/C7
Newville, Ala. (36353) 195/H8
Newville, Ind. (†46721) 227/H2
Newville, Pa. (17241) 294/H5
Newville, W. Va. (26632) 312/E5
New Vineyard○, Maine (04956) 243/C6
New Virginia, Iowa (50210) 229/F6
New Washington, Ind. (47162) 227/F7
New Washington, Ohio (44854) 284/E4
New Washington, Philippines 82/D5
New Waterford, Nova Scotia 168/J2
New Waterford, Ohio (44445) 284/J4
New Waverly, Ind. (46961) 227/E3
New Waverly, Texas (77358) 303/J7
New Westminster, Br. Col. 162/D6
New Westminster, Br. Col. 184/K3
New Weston, Ohio (45348) 284/A5
New Whiteland, Ind. (46184) 227/E5
New Wilmington, Pa. (16142) 294/B3
New Winchester, Ind. (†46122) 227/D5
New Winchester, Ohio (†44820) 284/D4
New Windsor, England 13/G8
New Windsor, England 10/F5
New Windsor, Ill. (61465) 222/C2
New Windsor, Md. (21776) 245/K2
New Windsor, N.Y. (12550) 276/N8
New Witten, S. Dak. (†57584) 298/K7
New Woodstock, N.Y. (13122) 276/J5
New World (isl.), Newf. 166/C4
New York 188/L2
NEW YORK 276
New York, N.Y. 146/L5
New York, N.Y. 188/M2
New York (co.), N.Y. 276/M9
New York, N.Y. (*10001) 276/M9
New York (state), U.S. 146/L5
New York, U.S. 2/F3
New York Mills, Minn. (56567) 255/C4
New York Mills, N.Y. (13417) 276/K4
New York State Barge (canal), N.Y. 276/C4
New Zealand 2/T8
New Zealand 87/G9
NEW ZEALAND 100
New Zion, New Bruns. 170/D2
New Zion, S.C. (29111) 296/H4
Ney, Ohio (43549) 284/B3
Neyagawa, Japan 81/J7
Neyland, Wales 13/B6
Neyriz, Iran 66/J6
Neyshabur, Iran 59/G2
Neyshabur, Iran 66/L2
Nezhin, U.S.S.R. 52/D4
Nez Perce (co.), Idaho 220/B3
Nezperce, Idaho (83543) 220/B3
Nez Perce Nat'l Hist. Park, Idaho 220/B-C3
Nezwar (mt.), Iran 66/H3
Ngabang, Indonesia 85/D5
N'gage, Angola 115/C5
Ngage, Angola 102/D5
Ngahere, N. Zealand 100/C5
Ngami (lake), Botswana 118/C4
Ngamiland (reg.), Botswana 118/C3
Ngamring, China 77/C6
Ngangla Ringco (lake), China 77/B5
Ngangzê Co (lake), China 77/C5
Ngao, Thailand 72/D3
Ngaoundéré, Cameroon 115/B2
Ngaoundéré, Cameroon 102/D4
Ngapara, N. Zealand 100/C6
Ngara, Tanzania 115/F4
Ngaruawahia, N. Zealand 100/E2
Ngatapa, N. Zealand 100/F3
Ngatik (atoll), Micronesia 87/F5
Ngau (isl.), Fiji 86/Q10
Ngauruhoe (mt.), N. Zealand 100/E3
Ngawi, Indonesia 85/K2
Nghia Lo, Vietnam 72/D2
Ngiva, Angola 102/D6
Ngiva, Angola 115/C7
Ngoc Linh (mt.), Vietnam 72/E4
Ngom Qu (riv.), China 77/E5
Ngong, Kenya 115/G4
Ngoring Hu (lake), China 77/E4
Ngorongoro (crater), Tanzania 115/F4
N'Gounié (riv.), Congo 115/B4

N'Gounié (riv.), Gabon 115/B4
Ngourou, Cent. Afr. Rep. 115/D2
N'Guigmi, Niger 106/G6
Ngulu (atoll), Micronesia 87/D5
Ngunju (cape), Indonesia 85/F8
Ngunza, Angola 102/D6
Ngunza, Angola 115/B6
Nguru, Nigeria 102/D3
Nguru, Nigeria 106/G6
Nhâmundá (riv.), Brazil 120/D3
Nhamundá (riv.), Brazil 132/B3
Nharêa, Angola 115/C6
Nharêa, Angola 102/D6
Nha Trang, Vietnam 72/F4
Nha Trang, Vietnam 54/M8
Nhava-Sheva, India 68/B7
Nhill, Victoria 88/G7
Nhill, Victoria 97/A5
Nhulunbuy, North. Terr. 88/F2
Nhulunbuy, North. Terr. 93/E2
Ni (riv.), Va. 307/N4
Niafunké, Mali 106/D5
Niagara (co.), N.Y. 276/C4
Niagara (riv.), N.Y. 276/B4
Niagara, N. Dak. (58266) 282/P4
Niagara (reg. munic.), Ontario 177/E4
Niagara (riv.), Ontario 177/E4
Niagara, Wis. (54151) 317/K4
Niagara Falls, N.Y. 188/K2
Niagara Falls, N.Y. (*14301) 276/C4
Niagara Falls, Ont. 162/J7
Niagara Falls, Ontario 177/E4
Niagara-on-the-Lake, Ontario 177/E4
Niamey (cap.), Niger 2/K5
Niamey (cap.), Niger 102/C3
Niamey (cap.), Niger 106/E6
Niangara, Zaire 115/E3
Niangua, Mo. (65713) 261/G8
Niantic, Conn. (06357) 210/G3
Niantic (riv.), Conn. 210/G3
Niantic, Ill. (62551) 222/D4
Niarada, Mont. (59852) 262/B3
Niari (riv.), Congo 115/B4
Nias (isl.), Indonesia 54/L9
Nias (isl.), Indonesia 85/B5
Niassa (prov.), Mozambique 118/F2
Nibbe, Mont. (†59088) 262/H4
Nibe, Denmark 21/C4
Nibe, Denmark 18/F8
Nibley, Utah (†84321) 304/C2
Nicaragua 2/E5
Nicaragua 146/K8
Nicaragua (lake), Nic. 146/K8
NICARAGUA 154/E4
Nicaragua (lake), Nicaragua 154/E5
Nicaro, Cuba 158/J3
Nicasio, Calif. (94946) 204/H1
Nicastro, Italy 34/F5
Nicatous (lake), Maine 243/G5
Nice, France 7/E4
Nice, France 28/G6
Niceville, Fla. (32578) 212/C6
Nichinan, Japan 81/E8
Nichol (isl.), Nova Scotia 168/F4
Nicholas (chan.), Cuba 156/B2
Nicholas (chan.), Cuba 158/E1
Nicholas (co.), Ky. 237/N4
Nicholas (co.), W. Va. 312/E6
Nicholas Denys, New Bruns. 170/D1
Nicholasville, Ky. (40356) 237/N5
Nicholls, Georgia (31554) 217/G7
Nichols, Conn. (†06611) 210/C4
Nichols, Fla. (33863) 212/E4
Nichols, Iowa (52766) 229/L6
Nichols, Minn. (†56431) 255/E4
Nichols, N.Y. (13812) 276/H6
Nichols, S.C. (29581) 296/J3
Nichols, Wis. (54152) 317/K6
Nichols Hills, Okla. (†73116) 288/L3
Nicholson (riv.) 88/F3
Nicholson, Br. Col. 184/J4
Nicholson, Georgia (30565) 217/F2
Nicholson, Miss. (39463) 256/E10
Nicholson, Port (inlet), N. Zealand 100/B3
Nicholson (riv.), North. Terr. 93/E5
Nicholson, Pa. (18446) 294/L2
Nicholson (riv.), Queensland 95/A3
Nicholson, W. Australia 92/E2
Nicholsville, Ala. (†36784) 195/C6
Nicholville, N.Y. (12965) 276/L1
Nickel Centre, Ontario 175/D3
Nickel Centre, Ontario 177/D1
Nickelsville, Va. (24271) 307/D7
Nickerie (dist.), Suriname 131/C3
Nickerie (riv.), Suriname 131/C3
Nickerson, Kansas (67561) 232/D3
Nickerson, Minn. (†55797) 255/F4
Nickerson, Nebr. (68044) 264/H3
Nicobar (isls.), India 54/L9
Nicobar (isls.), India 68/G7
Nicodemus, Kansas (†67625) 232/C2
Nicola, Br. Col. 184/G5
Nicolaus, Calif. (95659) 204/B8
Nicolet (co.), Québec 172/E3
Nicolet, Québec 172/E3
Nicolet (lake), Québec 172/F4
Nicolet (riv.), Québec 172/E3
Nicollet (co.), Minn. 255/D6
Nicollet, Minn. (56074) 255/D6
Nicoma Park, Okla. (73066) 288/M4
Nicomen Island, Br. Col. 184/L3
Nico Pérez, Uruguay 145/D4
Nicosia (cap.), Cyprus 63/E5
Nicosia (cap.), Cyprus 59/B2
Nicosia (cap.), Cyprus 54/E6
Nicosia, Italy 34/E6
Nicoya, C. Rica 154/E5
Nicoya (gulf), C. Rica 154/E6
Nicoya (pen.), C. Rica 154/E6
Nictau, New Bruns. 170/C1
Nictaux, Nova Scotia 168/D4
Nidau, Switzerland 39/D2
Nidd (riv.), England 10/F3
Nidwalden (canton), Switzerland 39/F3
Nidzica, Poland 47/E2
Niebüll, W. Germany 22/C1
Niederbipp, Switzerland 39/E2
Niedere Tauern (range), Austria 41/B3
Niederurnen, Switzerland 39/G2
Nielsville, Minn. (56568) 255/B3
Niemba, Zaire 115/E5
Niemen (riv.), U.S.S.R. 7/G3
Niemen (riv.), U.S.S.R. 52/B4
Niemen (riv.), U.S.S.R. 53/A3
Nienburg, W. Germany 22/C2
Nieuport (Nieuwpoort), Belgium 27/B6
Nieuw-Amsterdam, Suriname 131/D2
Nieuw-Buinen, Netherlands 27/K3
Nieuwegein, Netherlands 27/G4
Nieuwendam, Netherlands 27/C4
Nieuwe-Pekela, Netherlands 27/L2
Nieuweschans, Netherlands 27/L2
Nieuwkoop, Netherlands 27/F4
Nieuw-Nickerie, Suriname 120/D2
Nieuw-Nickerie, Suriname 131/C2
Nieuwpoort, Belgium 27/B6
Nieuw-Schoonebeek, Netherlands 27/L3
Nieuwveld (range), S. Africa 118/C6
Nieves, Mexico 150/H5
Nièvre (dept.), France 28/E4
Nigadoo, New Bruns. 170/E1
Niğde (prov.), Turkey 63/F4
Niğde, Turkey 59/B2
Niğde, Turkey 63/F4
Nigel, S. Africa 118/J7
Niger 2/K5
Niger 102/C3
Niger (riv.) 2/K5
Niger (riv.) 102/C4
Niger (riv.), Benin 106/E6
Niger (riv.), Guinea 106/C6
Niger (riv.), Mali 106/D5
NIGER 106/F5
Niger (riv.), Niger 106/E6
Niger (state), Nigeria 106/F7
Niger (delta), Nigeria 106/F8
Niger (riv.), Nigeria 106/F7
Nigeria 2/K5
Nigeria 102/C4
NIGERIA 106/F6
Nightcaps, N. Zealand 100/B6
Nighthawk, Wash. (†98855) 310/F2
Nightingale, Alberta 182/D4
Nightingale (mts.), Nev. 266/B2
Nightingale (Bach Long Vi) (isl.), Vietnam 72/F2
Nightmute, Alaska (99690) 196/F2
Nigrita, Greece 45/F5
Nigua (riv.), P. Rico 161/D2
Nihoa (isl.), Hawaii 87/K3
Nihoa (isl.), Hawaii 188/F6
Nihoa (isl.), Hawaii 218/D6
Nii (isl.), Japan 81/J6
Niigata (pref.), Japan 81/J5
Niigata, Japan 54/P6
Niigata, Japan 81/J5
Niihama, Japan 81/F6
Niihau (isl.), Hawaii 87/K3
Niihau (isl.), Hawaii 188/E5
Niihau (isl.), Hawaii 218/A2
Niimi, Japan 81/F6
Niitsu, Japan 81/J5
Níjar, Spain 33/E4
Nijkerk, Netherlands 27/H4
Nijmegen, Netherlands 27/H5
Nijvel (Nivelles), Belgium 27/E7
Nijverdal, Netherlands 27/J4
Nikel', U.S.S.R. 52/C1
Nikep, Md. (21546) 245/C2
Nikki, Benin 106/E7
Nikko National Park, Japan 81/J5
Nikolai, Alaska (99691) 196/H2
Nikolayev, U.S.S.R. 7/H4
Nikolayev, U.S.S.R. 48/D5
Nikolayev, U.S.S.R. 52/D5
Nikolayevsk, U.S.S.R. 2/S3
Nikolayevsk, U.S.S.R. 4/D2
Nikolayevsk, U.S.S.R. 54/P4
Nikolayevsk, U.S.S.R. 52/G4
Nikolayevsk-na-Amure, U.S.S.R. 48/P4
Nikol'sk, U.S.S.R. 52/G3
Nikol'sk, U.S.S.R. 52/G4
Nikolski, Alaska (99638) 196/E4
Nikol'skoye, U.S.S.R. 48/R4
Nikopol, Bulgaria 45/G4
Nikopol', U.S.S.R. 52/D5
Niksar, Turkey 63/G2
Nikshahr, Iran 59/H4
Nikshahr, Iran 66/L7
Nikšić, Yugoslavia 45/D4
Nikumaroro (Gardner) (isl.), Kiribati 87/J6
Nila (isl.), Indonesia 85/H7
Nilahue, Chile 138/E6
Niland, Calif. (92257) 204/K10
Nilaveli, Sri Lanka 68/E7
Nile (riv.) 2/L5
Nile (riv.) 102/F2
Nile (riv.), Egypt 111/F2
Nile (riv.), Egypt 59/B4
Nile (prov.), Sudan 111/F4
Nile (riv.), Sudan 59/B6
Nile (riv.), Sudan 111/F4
Niles, Ill. (60648) 222/B5
Niles, Kansas (†67480) 232/E2
Niles, Mich. (49120) 250/C7
Niles, Ohio (44446) 284/J3
Ni'lin, West Bank 65/C4
Nilópolis, Brazil 135/E3
Nilwood, Ill. (62672) 222/D4
Nimach, India 68/D4
Nimba (lag.), Guinea 106/C7
Nimba (lag.), Ivory Coast 106/C7
Nimba (lag.), Liberia 106/C7
Nîmes, France 28/F6
Nîmes, France 7/E4
Nimmitabel, N.S. Wales 97/E5
Nimmons, Ark. (72461) 202/K1
Nimnyrskiy, U.S.S.R. 48/N4
Nimrod, Ark. (†72126) 202/D4
Nimrod (lake), Ark. 202/D4
Nimrod, Minn. (56478) 255/D4
Nimule, Sudan 111/F7
Nin (bay), Philippines 82/D4
Nin, Yugoslavia 45/B3
Ninaview, Colo. (†81054) 208/N7
Ninawa (gov.), Iraq 66/B3
Nine Degree (chan.), India 68/C7
Ninemile (pt.), Mich. 250/E3
Nine Mile (creek), Utah 304/D4
Nine Mile Falls, Wash. (99026) 310/H3
Nine Mile River, Nova Scotia 168/E3
Ninepipe (res.), Mont. 262/C3
Nine Times, S.C. (†29685) 296/B2
Ninette, Manitoba 179/C5
Ninety Mile (beach), N. Zealand 100/D1
Ninety Mile (beach), Victoria 97/D6
Ninety Six, S.C. (29666) 296/C3
Ninety Six Nat'l Hist. Site, S.C. 296/C3
Nineveh (ruins), Iraq 66/C2
Nineveh, N.Y. (13813) 276/J6
Nineveh, Pa. (15353) 294/B6
Ninfas (pt.), Argentina 143/D5
Ninga, Manitoba 179/C5
Ning'an, China 77/L3
Ningbo (Ningpo), China 77/K6
Ningbo, China 54/O7
Ningde, China 77/K6
Ningdu, China 77/J6
Ninghua, China 77/J6
Ningpo (Ningbo), China 77/K6
Ningsia (Yinchuan, Yinchwan), China 77/G4
Ningsia Hui Aut. Reg. (Ningxia Huizu), China 77/F3
Ningwu, China 77/H4
Ningxia Huizu (Ningsia Hui Aut. Reg.), China 77/F3
Ning Xian, China 77/G4
Ninh Binh, Vietnam 72/E3
Ninigo Group (isls.), Papua N.G. 87/E6
Ninilchik, Alaska (99639) 196/B1
Ninini (pt.), Hawaii 218/D2
Ninnekah, Okla. (73067) 288/L5
Ninnescah (riv.), Kansas 232/E4
Ninnis Glacier Tongue, Ant. 5/C8
Ninole, Hawaii (96773) 218/J4
Ninove, Belgium 27/D7
Nioaque, Brazil 132/C8
Niobe, N.Y. (14758) 276/B6
Niobe, N. Dak. (†58746) 282/F2
Niobrara (riv.), Nebr. 188/F2
Niobrara, Nebr. (68760) 264/G2
Niobrara (riv.), Nebr. 264/E2
Niobrara (co.), Wyo. 319/H2
Niobrara (riv.), Wyo. 319/J3
Niono, Mali 106/C6
Nioro, Mali 106/C5
Nioro, Mali 102/B3
Nioro-du-Rip, Senegal 106/A6
Niort, France 28/C4
Niota, Ill. (62358) 222/B3
Niota, Tenn. (37826) 237/M9
Niotaze, Kansas (67355) 232/F4
Nipani, India 68/C5
Nipawin, Sask. 181/H2
Nipawin Prov. Park, Sask. 181/G1
Nipe (bay), Cuba 158/J3
Nipigon, Ont. 162/H6
Nipigon (lake), Ont. 146/K5
Nipigon (lake), Ont. 162/H6
Nipigon, Ontario 177/H5
Nipigon, Ontario 175/C3
Nipigon (lake), Ontario 175/C3
Nipigon (lake), Ontario 177/H5
Nipinnawasee, Calif. (†93601) 204/F6
Nipishish (lake), Newf. 166/B3
Nipissing (terr. dist.), Ontario 177/F2
Nipissing (terr. dist.), Ontario 175/E3
Nipissing, Ontario 177/E1
Nipissing (lake), Ontario 177/E1
Nipissing (lake), Ontario 175/E3
Nipomo, Calif. (93444) 204/E8
Nippers Harbour, Newf. 166/C4
Nipton, Calif. (92364) 204/K8
Niquelândia, Brazil 132/D6
Niquén, Chile 138/E1
Niquero, Cuba 158/G4
Niquero, Cuba 156/C2
Niquivil, Argentina 143/C3
Nirgua, Venezuela 124/D2
Nirmal, India 68/D5
Nirvana, Mich. (†49642) 250/D5
Nir Yitzhaq, Israel 65/A5
Niš, Yugoslavia 7/G4
Niš, Yugoslavia 45/F4
Nisa, Portugal 33/C3
Nisab, P.D.R. Yemen 59/E7
Nisab, Saudi Arabia 59/D4
Nisbet, Pa. (17759) 294/H3
Niscemi, Italy 34/E6
Nishapur (Neyshabur), Iran 66/L2
Nishino (isl.), Japan 81/M3
Nishinomiya, Japan 81/H8
Nishinoomote, Japan 81/E8
Nísiros (isl.), Greece 45/H7
Niskayuna, N.Y. (†12301) 276/N5
Nisko, Poland 47/E3
Nisku, Alberta 182/D3
Nisland, S. Dak. (57762) 298/C4
Nisqually, Wash. (†98501) 310/C3
Nisqually (riv.), Wash. 310/C4
Nisqually Ind. Res., Wash. 310/C4
Nissan (isl.), Papua N.G. 86/C2
Nissum (fjord), Denmark 21/A5
Nisswa, Minn. (56468) 255/D4
Niterói, Brazil 132/F8
Niterói, Brazil 120/E5
Niterói, Brazil 135/E3
Nith (riv.), Scotland 15/E5
Nith (riv.), Scotland 10/E3
Nitil, Jordan 65/D4
Nitinat, Br. Col. 184/H3
Nitinat (lake), Br. Col. 184/H3
Niton Junction, Alberta 182/C3
Nitra, Czech. 41/E2
Nitra (riv.), Czech. 41/E2
Nitro, W. Va. (25143) 312/C6
Nitta Yuma, Miss. (38763) 256/C4
Nittedal, Norway 18/D3
Niuafo'ou (isl.), Tonga 87/J7
Niuatoputapu (isl.), Tonga 87/J7
Niue (isl.) 87/K7
Niue (isl.), N. Zealand 2/A6
Niutao (atoll), Tuvalu 87/H6
Nivala, Finland 18/O5
Nive (riv.), Tasmania 99/C4
Nivelles, Belgium 27/E7
Nivernais (trad. prov.), France 29
Niverville, Manitoba 179/F5
Niwot, Colo. (80544) 208/J2
Nixa, Mo. (65714) 261/F8
Nixburg, Ala. (36058) 195/F5
Nixon, Nev. (89424) 266/B3
Nixon, N.J. (08817) 273/E2
Nixon, Texas (78140) 303/G8
Nixonville, S.C. (†29526) 296/K4
Niya (Minfeng), China 77/B4
Nizamabad, India 68/D5
Nizao, Dom. Rep. 158/E6
Nizhnekamsk, U.S.S.R. 7/K3
Nizhnekamsk, U.S.S.R. 52/H3
Nizhneudinsk, U.S.S.R. 48/K4
Nizhnevartovsk, U.S.S.R. 48/H3
Nizhneyansk, U.S.S.R. 48/O3
Nizhniy Lomov, U.S.S.R. 52/F4
Nizhniy Novgorod (Gor'kiy), U.S.S.R. 52/F3
Nizhniy Tagil, U.S.S.R. 54/H4
Nizhniy Tagil, U.S.S.R. 48/G4
Nizhnyaya Pesha, U.S.S.R. 52/G1
Nizina, Alaska (†99566) 196/K2
Nizip, Turkey 63/G4
Nizwa, Oman 59/G5
Nizza Monferrato, Italy 34/B2
Nizzanim, Israel 65/B4
Njombe, Tanzania 115/F5
Njombe (riv.), Tanzania 115/F5
Nkambe, Cameroon 115/B2
Nkayi, Congo 115/B4
Nkhata Bay, Malawi 115/F6
Nkhotakota, Malawi 115/F6
N'Komi (lag.), Gabon 115/A4
Nkongsamba, Cameroon 115/B3
Nkongsamba, Cameroon 102/D4
Nmai (riv.), Burma 72/C1
Nnewi, Nigeria 106/F7
Noah, Tenn. (†37355) 237/J9
Noakhali, Bangladesh 68/G4
Noank, Conn. (06340) 210/G3
Noatak, Alaska (99761) 196/F1
Noatak (riv.), Alaska 196/F1
Noatak Nat'l Preserve, Alaska 196/F1
Nobel, Ontario 177/D2
Nobeoka, Japan 81/E7
Noble, Georgia (†30728) 217/B1
Noble, Ill. (62868) 222/E5
Noble (co.), Ind. 227/G2
Noble, Iowa (†52641) 229/K6
Noble, La. (71462) 238/C3
Noble, Mo. (65715) 261/G9
Noble (co.), Ohio 284/G6
Noble (co.), Okla. 288/M2
Noble, Okla. (73068) 288/M4
Nobleboro◯, Maine (04555) 243/D7
Nobleford, Alberta 182/D5
Noble Lake, Ark. (†71601) 202/G5
Nobles (co.), Minn. 255/C7
Noblesville, Ind. (46060) 227/F4
Nobleton, Fla. (33554) 212/D3
Nobleton, Ontario 177/J3
Noboribetsu, Japan 81/K2
Nocatee, Fla. (33864) 212/E4
Noccundra, Queensland 95/B5
Nocera Inferiore, Italy 34/E4
Nochistlán, Mexico 150/H6
Nocona, Texas (76255) 303/G4
Noctor, Ky. (41357) 237/P5
Noda, Japan 81/P2
Nodaway, Iowa (50857) 229/D7
Nodaway (riv.), Iowa 229/D7
Nodaway (co.), Mo. 261/C2
Nodaway, Mo. (†64421) 261/C3
Node, Wyo. (82228) 319/H3
Nodine, Minn. (†55925) 255/G7
Noel, Mo. (64854) 261/D9
Noel, Nova Scotia 168/E3
Noel Road, Nova Scotia 168/E3
Noelville, Ontario 177/D1
Nogal, N. Mex. (88341) 274/D5
Nogal (reg.), Somalia 115/J2
Nogales, Ariz. 188/D4
Nogales, Ariz. (85621) 198/E7
Nogales, Chile 138/F2
Nogales, Mexico 150/P2
Nogamut, Alaska (†99668) 196/G2
Nogata, Japan 81/E7
Nogent-le-Rotrou, France 28/D3
Nogent-sur-Seine, France 28/E3
Nogoa (riv.), Queensland 88/H4
Nogoa (riv.), Queensland 95/C5
Nogoyá, Argentina 143/F6
Nógrád (co.), Hungary 41/E3
Nohili (pt.), Hawaii 218/B1
Nohkú (pt.), Mexico 150/Q7
Noinville, New Bruns. 170/E2
Noir (isl.), Chile 138/E11
Noires (mts.), Dom. Rep. 158/C5
Noires (mts.), Haiti 158/C5
Noirmont (mt.), Switzerland 39/B4
Noirmoutier (isl.), France 28/B4
Noisy-le-Sec, France 28/B1
Nojima (cape), Japan 81/K6
Nokesville, Va. (22123) 307/N3
Nokhowch, Kuh-e (mt.), Iran 66/M7
Nokia, Finland 18/N6
Nok Kundi, Pakistan 68/A3
Nok Kundi, Pakistan 59/H4
Nokomis, Ala. (†36502) 195/D8
Nokomis, Fla. (33555) 212/D4
Nokomis, Ill. (62075) 222/D4
Nokomis, Sask. 181/F4
Nokou, Chad 111/B5
Nola, Ark. (†72838) 202/C4
Nola, Cent. Afr. Rep. 115/C3
Nola, Miss. (†39665) 256/D7
Nolan (co.), Texas 303/D5
Nolan, W. Va. (25687) 312/B7
Nolichucky (riv.), N.C. 281/E2
Nolichucky (riv.), Tenn. 237/R8
Nolin, Ky. (†42776) 237/K5
Nolin (lake), Ky. 237/K6
Nolin (riv.), Ky. 237/J6
Nolinsk, U.S.S.R. 52/H3
Nollesemic (lake), Maine 243/F4
Noma, Fla. (32452) 212/C5
Nomans Land (isl.), Mass. 249/L7
Nombre de Dios, Mexico 150/G5
Nome, Alaska 146/B3
Nome, Alaska (99762) 196/E2
Nome, Alaska 188/C5
Nome, Alaska 2/A2
Nome, N. Dak. (58062) 282/P6
Nome, U.S. 4/C18
Nomgon, Mongolia 77/G3
Nominingue, Québec 172/B3
Nominingue (lake), Québec 172/B3
Nomoi (isls.), Micronesia 87/F5
Nonacho (lake), N.W.T. 162/F3
Nonacho (lake), N.W. Terrs. 187/H3
Nonamesset (isl.), Mass. 249/M6
Nondalton, Alaska (99640) 196/G2
Nong Het, Laos 72/E3
Nong Khai, Thailand 72/D3
Nong Lahan (lake), Thailand 72/D3
Nonoava, Mexico 150/F3
Nonouti (atoll), Kiribati 87/H6
Nonquitt, Mass. (02748) 249/L6
Nonsan, S. Korea 81/C5
Nontron, France 28/D5
Nooksack, Wash. (98276) 310/C2
Nooksack (riv.), Wash. 310/C2
Noonan, N. Dak. (58765) 282/D2
Noord (pt.), Neth. Ant. 161/F8
Noord (pt.), Neth. Ant. 161/D8
Noord di Salinja, Neth. Ant. 161/E8
Noordwijk, Netherlands 27/E4
Noorvik, Alaska (99763) 196/F1
Nootka, Br. Col. 184/D5
Nootka (isl.), Br. Col. 184/D5
Nootka (sound), Br. Col. 184/D5
Nopalucan de la Granja, Mexico 150/O1
Nopeming, Minn. (†55810) 255/F4
Nopiming Prov. Park, Manitoba 179/G4
Nóqui, Angola 115/B5
Noquochoke P.O. (Westport), Mass. (02790) 249/K6
Nora, Ill. (61059) 222/D1
Nora, Nebr. (68962) 264/G4
Nora, Sask. 181/H3
Nora, S. Dak. (†57001) 298/R8
Nora, Sweden 18/J7
Nora, Va. (24272) 307/D6
Noranda, Que. 162/J6
Noranda, Québec 174/B3
Noranside, Queensland 95/A4
Nora Springs, Iowa (50458) 229/H2
Norbeck, S. Dak. (†57480) 298/L3
Norberg, Sweden 18/K6
Norberto de la Riestra, Argentina 143/G7
Norbertville, Québec 172/F3
Norborne, Mo. (64668) 261/E4
Norcatur, Kansas (67653) 232/B2
Norco, Calif. (91760) 204/E11
Norco, La. (70079) 238/N3
Norcross, Georgia (*30071) 217/D3
Norcross, Maine (†04462) 243/F4
Norcross, Minn. (56274) 255/B5
Nord (dept.), France 28/E2
Nord, Greenl. 4/A10
Nord (pt.), Guadeloupe 161/B7
Nord (dept.), Haiti 158/C5
Nord (riv.), Québec 172/C4
Nordaustlandet (isl.), Norway 18/D1
Nordborg, Denmark 21/C7
Nordby, Århus, Denmark 21/D5
Nordby, Ribe, Denmark 21/B7
Norddeich, W. Germany 22/B2
Nordegg, Alberta 182/B3
Nordegg (riv.), Alberta 182/C3
Norden, Nebr. (†68778) 264/D2
Norden, W. Germany 22/B2
Nordenham, W. Germany 22/C2
Norderney, W. Germany 22/B2
Norderney (isl.), W. Germany 22/B2
Norderstedt, W. Germany 22/D2
Nord-Est (bay), Guadeloupe 161/B6
Nordfjord (fjord), Norway 18/E6
Nordhausen, E. Germany 22/D3
Nordheim, Texas (78141) 303/G9
Nordhorn, W. Germany 22/B2
Nordin, New Bruns. 170/E1
Nordjylland (co.), Denmark 21/D4
Nordkapp (cape), Norway 7/G1
Nordkapp (pt.), Norway 18/C1
Nordkinn (headland), Norway 18/Q1
Nordkinn (pen.), Norway 18/P1
Nordland (co.), Norway 18/J3
Nordland, Wash. (98358) 310/C2
Nordli, Norway 18/H4
Nördlingen, W. Germany 22/D4
Nordmaling, Sweden 18/L5
Nordman, Idaho (83848) 220/B1
Nord-Ostsee (canal), W. Germany 22/C1
Nord-Ouest (dept.), Haiti 158/B5
Nordstrand (isl.), W. Germany 22/C1
Nord-Trøndelag (co.), Norway 18/H4
Nordvik-Ugol'naya, U.S.S.R. 4/B4
Nordvik-Ugol'naya, U.S.S.R. 48/M2
Nore (riv.), Ireland 10/C4
Nore (riv.), Ireland 17/G7
Norene, Tenn. (37136) 237/J8
Norfield, Miss. (†39629) 256/C8
Norfolk (isl.), Australia 2/T7
Norfolk◯, Conn. (06058) 210/C1
Norfolk (co.), England 13/H5
Norfolk (co.), Mass. 249/K4
Norfolk◯, Mass. (02056) 249/J4
Norfolk, Nebr. 188/G2
Norfolk, Nebr. (68701) 264/G2
Norfolk, N.Y. (13667) 276/K1
Norfolk (bay), Tasmania 99/D4
Norfolk, Va. 188/L3
Norfolk, Va. 146/L6
Norfolk (I.C.), Va. (*23501) 307/R7
Norfolk Island, /L5
Norfolk Island (terr.), Australia 87/G8
Norfork, Ark. (72658) 202/F1
Norfork (lake), Ark. 202/F1
Norfork (lake), Mo. 261/H10
Norg, Netherlands 27/J2
Norge, Okla. (†73018) 288/K4
Norge, Va. (23127) 307/P6
Norglenwold, Alberta 182/C3
Noril'sk, U.S.S.R. 2/P2
Noril'sk, U.S.S.R. 54/L3
Noril'sk, U.S.S.R. 4/B5
Noril'sk, U.S.S.R. 48/J3
Norland, Fla. (†33169) 212/B4
Norland, Ontario 177/F3
Norlina, N.C. (27563) 281/N2
Norma, N.J. (08347) 273/C4
Norma, N. Dak. (58766) 282/G2
Norma, Tenn. (†37827) 237/N8
Normal, Ill. (61761) 222/E3
Normalville, Pa. (15469) 294/D5
Norman, Ark. (71960) 202/C5
Norman, Ind. (47264) 227/E7
Norman (co.), Minn. 255/B3
Norman, Nebr. (68963) 264/F4
Norman (cape), Newf. 166/C3
Norman, N.C. (28367) 281/K4
Norman (lake), N.C. 281/H3
Norman, Okla. (*73069) 288/M4
Norman (riv.), Queensland 88/G3
Norman (creek), Queensland 95/D3
Norman (riv.), Queensland 95/B3
Norman (isl.), Virgin Is. (Br.) 161/D4
Normanby, Queensland 88/K2
Normand (lake), Québec 172/D2
Normandale, Ontario 177/D5
Normandin, Québec 172/E1
Normandy (trad. prov.), France 29
Normandy, Mo. (63121) 261/R2
Normandy (riv.), Queensland 95/C2
Normandy, Tenn. (37360) 237/J10
Normandy, Texas (78852) 303/D9
Normandy Beach, N.J. (08739) 273/E3
Normandy Park, Wash. (†98100) 310/A2
Normangee, Texas (77871) 303/H6
Norman Park, Georgia (31771) 217/E8
Norman's Cove, Newf. 166/D2
Normanton, Australia 87/E7
Normanton, Queensland 95/B3
Normanton, Queensland 88/G3
Normantown, Georgia (†30474) 217/H6
Normantown, W. Va. (25267) 312/E5
Norman Wells, Canada 4/C16
Norman Wells, N.W.T. 146/F3
Norman Wells, N.W.T. 162/D2
Norman Wells, N.W. Terrs. 187/F3
Normétal, Québec 174/B3
Noroton, Conn. (†06820) 210/B4
Noroton Heights, Conn. (†06820) 210/B4
Norphlet, Ark. (71759) 202/E7
Norquay, Sask. 181/J4
Ñorquincó, Argentina 143/B5
Norrbotten (co.), Sweden 18/L3
Nørre Åby, Denmark 21/C7
Nørre Alslev, Denmark 21/E8
Nørre Broby, Denmark 21/D7
Nørre Nebel, Denmark 21/B6
Nørre Snede, Denmark 21/C6
Nørre Vorupør, Denmark 21/B4
Norridge, Ill. (†60656) 222/B5
Norridgewock, Maine (04957) 243/D6
Norridgewock◯, Maine (04957) 243/D6
Norrie, Wis. (†54414) 317/H6
Norris, Ill. (61553) 222/C3
Norris, Miss. (†39074) 256/F6
Norris, Mont. (59745) 262/E5
Norris, S.C. (29667) 296/B2
Norris, S. Dak. (57560) 298/G7
Norris (lake), Tenn. 188/K3
Norris, Tenn. (37828) 237/N8
Norris (dam), Tenn. 237/N8
Norris (lake), Tenn. 237/O8
Norris, Wyo. (†82190) 319/B1
Norris Arm, Newf. 166/C4
Norris City, Ill. (62869) 222/E6
Norris Point, Newf. 166/C4
Norristown, Ark. (†72801) 202/D3
Norristown, Georgia (30447) 217/H5
Norristown, Ind. (†47234) 227/F6
Norristown, Pa. (*19401) 294/M5
Norrisville, Md. (†21161) 245/N2
Norrköping, Sweden 7/F3
Norrköping, Sweden 18/K7
Norrsundet, Sweden 18/K6
Norrtälje, Sweden 18/L7
Norseland, Minn. (†56082) 255/D6
Norseman, W. Australia 88/C6
Norseman, W. Australia 92/C6
Norsjö, Sweden 18/L4
Norte (pt.), Argentina 143/D5
Norte (chan.), Brazil 120/E2
Norte (chan.), Brazil 132/D2
Norte, Serra do (range), Brazil 132/B5
Norte del Cabo San Antonio (pt.), Argentina 143/E4
Norte de Santander (dept.), Colombia 126/D3
North (sea) 2/K3
North (sea) 7/E3
North (cape), Alaska 196/L4
North (pt.), Barbados 161/B8
North (sea), Belgium 27/D4
North (rocks), Bermuda 156/H2
North (sea), Denmark 21/B9
North (sea), England 13/J4
North (sea), France 28/E1
North (cape), Ice. 4/C11
North (Horn) (cape), Iceland 21/B1
North (sound), Ireland 17/B5
North (isls.), La. 238/M7
North (pass), La. 238/N8
North (pt.), La. 238/M7
North (pt.), Md. 245/N4

North (riv.), Mass. 249/D2
North (riv.), Mass. 249/L4
North (chan.), Mich. 250/F2
North (pt.), Mich. 250/F3
North (lake), Minn. 255/F1
North (sea), Netherlands 27/E3
North (lake), New Bruns. 170/C3
North (riv.), Newf. 166/C3
North (riv.), Newf. 166/B2
North (isl.), N. Zealand 87/H9
North (cape), N. Zealand 87/H9
North (cape), N. Zealand 100/D1
North (isl.), N. Zealand 100/F1
North (lake), N. Dak. 282/J3
North (chan.), N. Ireland 10/D3
North (chan.), N. Ireland 17/K1
North (Nordkapp) (cape), Norway 7/G1
North (cape), Norway 4/B8
North (cape), N.S. 162/K6
North (cape), Nova Scotia 168/H1
North (mt.), Nova Scotia 168/D3
North (chan.), Ontario 177/A1
North (chan.), Ontario 175/D3
North (mt.), Pa. 294/K3
North (pt.), Pr. Edward I. 168/E1
North (chan.), Scotland 10/D3
North (chan.), Scotland 15/C5
North (sound), Scotland 15/G4
North (sound), Scotland 15/F1
North (isl.), Seychelles 118/H5
North, S.C. (29112) 296/E4
North (inlet), S.C. 296/J5
North (isl.), S.C. 296/J5
North (pt.), Tasmania 99/E1
North (creek), Utah 304/C6
North (lake), Utah 304/B2
North (riv.), Wash. 310/B4
North (sea), W. Germany 22/B2
North (riv.), W. Va. 312/J4
North (lake), Wis. 317/J1
North Abington, Mass. (02351) 249/L4
North Acton, Mass. (†01720) 249/J2
North Adams, Mass. (01247) 249/B2
North Adams, Mich. (49262) 250/E7
Northallerton, England 10/F3
Northallerton, England 13/F3
Northam, England 13/C6
Northam, W. Australia 88/B6
Northam, W. Australia 92/B1
NORTH AMERICA 146
North America 2/C4
North Amherst, Mass. (01059) 249/E3
North Amity, Maine (04465) 243/H4
Northampton, England 13/F5
Northampton, England 10/F4
Northampton, Mass. (01060) 249/D3
Northampton (co.), N.C. 281/P2
Northampton (co.), Pa. 294/M4
Northampton, Pa. (18067) 294/M4
Northampton (co.), Va. 307/S6
Northampton, W. Australia 88/A5
Northampton, W. Australia 92/A5
Northamptonshire (co.), England 13/G5
North Andaman (isl.), India 68/G6
North Andover○, Mass. (01845) 249/K2
North Anna (riv.), Va. 307/M4
North Anson, Maine (04958) 243/D6
North Apollo, Pa. (15673) 294/D4
North Arlington, N.J. (07032) 273/B2
North Arm (inlet), N.W. Terrs. 187/G3
North Asheboro, N.C. (†27203) 281/K3
North Ashford, Conn. (†06282) 210/G1
North Aspy (riv.), Nova Scotia 168/H2
North Atlantic Ocean 2/H3
North Attleboro○, Mass. (*02760) 249/J5
North Augusta, Ontario 177/J3
North Augusta, S.C. (29841) 296/C5
North Aulatsivik (isl.), Newf. 166/B2
North Aurora, Ill. (60542) 222/E2
North Avondale, Colo. (†81022) 208/L6
North Ballachulish, Scotland 15/C4
North Baltimore, Ohio (45872) 284/C3
North Bangor, N.Y. (12966) 276/M1
North Barrington, Ill. (†60010) 222/A5
North Bass (isl.), Ohio 284/E2
North Battleford, Sask. 146/H4
North Battleford, Sask. 162/F5
North Battleford, Sask. 181/C3
North Bay, N.Y. (13123) 276/J4
North Bay, Ont. 146/L5
North Bay, Ont. 162/J6
North Bay, Ontario 177/E1
North Bay, Ontario 175/E3
North Bay, Wis. (†53401) 317/M3
North Bay Ingonish (bay), Nova Scotia 168/H2
North Bay Village, Fla. (33141) 212/B4
North Beach, Md. (20831) 245/N6
North Belgrade, Maine (†04963) 243/D7
North Bellingham, Mass. (†02019) 249/J4
North Bend, Br. Col. 184/G5
North Bend, Nebr. (68649) 264/H3
North Bend, Ohio (45052) 284/B9
North Bend, Oreg. (97459) 291/C4
North Bend, Pa. (17760) 294/G3
North Bend, Wash. (98045) 310/D3
North Bend, Wis. (†54642) 317/D7
North Bennington, Vt. (05257) 268/A6
North Bergen○, N.J. (07047) 273/B2
North Berwick, Maine (03906) 243/B9
North Berwick○, Maine (03906) 243/B9
North Berwick, Scotland 15/F4
North Berwick, Scotland 10/E2
North Beveland (isl.), Netherlands 27/D5
North Billerica, Mass. (01862) 249/J2
North Bloomfield, Conn. (†06002) 210/E1
North Bloomfield, Ohio (44450) 284/J3
North Bonneville, Wash. (98639) 310/C5
North Borneo (Sabah) (state), Malaysia 85/F3
Northboro, Iowa (51647) 229/C7
Northborough, Mass. (01532) 249/H3
Northborough○, Mass. (01532) 249/H3
North Boston, N.Y. (14110) 276/C5
North Bourke, N.S. Wales 97/C2
North Brabant (prov.), Netherlands 27/F5
North Braddock, Pa. (15104) 294/C7
North Bradford, Maine (†04410) 243/F5
North Bradley, Mich. (†48618) 250/E5
Northbranch, Kansas (†66936) 232/D2
North Branch, Md. (†21502) 245/D2
North Branch, Mich. (48461) 250/F5
North Branch, Minn. (55056) 255/F5
North Branch, N.H. (†03440) 268/D5
North Branch, N.J. (08876) 273/D2
North Branch Oromocto (riv.), New Bruns. 170/D3
North Branford○, Conn. (06471) 210/E3
North Brentwood, Md. (†20722) 245/F4
Northbridge○, Mass. (01534) 249/H4
North Bridgton, Maine (04057) 243/B7
Northbrook, Ill. (60062) 222/B5
North Brook, Ontario 177/G3
North Brookfield, Mass. (01535) 249/F3
North Brookfield○, Mass. (01535) 249/F3
North Brooksville, Maine (†04617) 243/F7
North Brunswick○, N.J. (08902) 273/D3
North Bruny (isl.), Tasmania 99/D5
North Buena Vista, Iowa (52066) 229/L3
North Calais, Vt. (†05648) 268/C3
North Caldwell, N.J. (†07006) 273/B2
North Calling Lake, Alberta 182/D2
North Canadian (riv.) 188/G3
North Canadian (riv.), Okla. 288/K3
North Canton, Conn. (06059) 210/D1
North Canton, Georgia (†30114) 217/C2
North Canton, Ohio (44720) 284/H4
North Cape (Nordkapp) (pt.), Norway 18/P1
North Cape May, N.J. (08204) 273/C6
North Caribou (lake), Ontario 175/B2
North Carolina 188/L3
NORTH CAROLINA 281
North Carolina (state), U.S. 146/K6
North Carrizo (creek), Colo. 208/N8
North Carrizo (riv.), Okla. 288/A1
North Carrollton, Miss. (38947) 256/E3
North Carter (mt.), N.H. 268/E3
North Carver, Mass. (02355) 249/L5
North Cascades Nat'l Park, Wash. 310/D2
North Catasauqua, Pa. (†18032) 294/L4
North Charleston, S.C. (29406) 296/G6
North Charlestown, N.H. (†03603) 268/C5
North Chatham, Mass. (02650) 249/O6
North Chatham, N.H. (†04058) 268/E3
North Chelmsford, Mass. (01863) 249/J2
North Chesterville, Maine (†04938) 243/C6
North Chicago, Ill. (60064) 222/B4
North Chichester, N.H. (†03263) 268/E5
North Chili, N.Y. (14514) 276/E4
North City (Coello), Ill. (†62825) 222/E5
North Clarendon, Vt. (05759) 268/B4
Northcliffe, W. Australia 92/B6
North Cohasset, Mass. (†02025) 249/F7
North Colebrook, Conn. (†06021) 210/C1
North College Hill, Ohio (45239) 284/B9
North Collins, N.Y. (14111) 276/C5
North Concho (riv.), Texas 303/C6
North Concord, Vt. (05858) 268/D3
North Conway, N.H. (03860) 268/E3
North Cooking Lake, Alberta 182/D3
North Cotabato (prov.), Philippines 82/E7
Northcote, Minn. (†56728) 255/A2
Northcote, N. Zealand 100/B1
Northcote, Victoria 88/L7
Northcote, Victoria 97/J5
North Cove, N.C. (†28752) 281/F3
North Cove, Wash. (†98590) 310/A4
North Cowichan, Br. Col. 184/J3
North Creek, N.Y. (15853) 276/M3
North Crossett, Ark. (71635) 202/G7
North Cutler, Maine (†04626) 243/J6
North Dakota 188/F1
NORTH DAKOTA 282
North Dakota (state), U.S. 146/H5
North Dandalup, W. Australia 88/B3
North Danger (reef), Philippines 85/E3
North Danville, Vt. (†05819) 268/C3
North Dartmouth, Mass. (02747) 249/K6
North Dexter, Maine (†04930) 243/E5
North Dighton, Mass. (02764) 249/K5
North Dixmont, Maine (†04932) 243/E6
North Down (dist.), N. Ireland 17/K2
North Downs (hills), England 13/G6
North Eagle Butte, S. Dak. (†57625) 298/G3
Northeast (cape), Alaska 196/E2
North East (pt.), Jamaica 158/K6
Northeast (pass), La. 238/M8
North East, Md. (21901) 245/P2
North East, Pa. (16428) 294/C1
North East Breakers, Bermuda 156/H2
North East Cape Fear (riv.), N.C. 281/O4
North East Carry, Maine (†04441) 243/D4
North-Eastern (prov.), Kenya 115/G3
Northeast Foreland (pen.), Greenl. 4/A10
North Eastham, Mass. (02651) 249/O5
Northeast Harbor, Maine (04662) 243/G7
Northeast Land (isl.), Norway 4/B8
North East Margaree (riv.), Nova Scotia 168/H2
North Easton, Mass. (02356) 249/K4
North East Polder, Netherlands 27/H3
North East Providence (chan.), Bahamas 156/C1
North Edisto (riv.), S.C. 296/G6
North Edwards, Calif. (93523) 204/H8
North Egremont, Mass. (01252) 249/A4
Northeim, W. Germany 22/C3
North English, Iowa (52316) 229/J5
North Enid, Okla. (†73701) 288/L2
Northern (dist.), Israel 65/C2
Northern (head), New Bruns. 170/D4
Northern (prov.), Sudan 111/E3
Northern Cheyenne Ind. Res., Mont. 262/K5
Northern Dvina (riv.), U.S.S.R. 52/F2
Northern Dvina (riv.), U.S.S.R. 48/E3
Northern Indian (lake), Manitoba 179/J2
NORTHERN IRELAND 17
NORTHERN IRELAND 10/C3
Nottoway (riv.), Va. 307/O7
Notukeu (creek), Sask. 181/D6
Notus, Idaho (83656) 220/B6
Nouadhibou, Mauritania 106/A4
Nouadhibou, Mauritania 102/A2
Nouakchott (cap.), Mauritania 106/A5
Nouakchott (cap.), Mauritania 102/A3
Nouakchott (cap.), Mauritania 2/J5
Nouméa (cap.), New Caled. 87/G8
Nouméa (cap.), New Caledonia 2/T7
Nouméa (cap.), New Caled. 86/H5
Nounan, Idaho (†83254) 220/G7
Noup (head), Scotland 15/E1
Noupoort, S. Africa 118/C6
Nouveau-Comptoir, Québec 174/B2
Northern Ireland, U.K. 7/D3
Northern Marianas 87/E4
Northern Marianas, U.S. 2/S5
Northern Peninsula Aboriginal Reserve, Queensland 88/G2
Northern Peninsula Aboriginal Res., Queensland 95/B1
Northern Samar (prov.), Philippines 82/E4
Northern Sporades (isls.), Greece 45/F6
Northern Territory, 88/E3
NORTHERN TERRITORY 93
Northern Territory (terr.), Australia 87/D7
North Esk (riv.), Scotland 15/F4
North Esk (riv.), Tasmania 99/D3
North Fairfield, Ohio (44855) 284/E3
North Falmouth, Mass. (02556) 249/M6
North Ferrisburg, Vt. (05473) 268/A3
Northfield, Conn. (06778) 210/C2
Northfield, Ill. (60093) 222/B5
Northfield, Ky. (†40201) 237/K1
Northfield○, Maine (†04654) 243/H6
Northfield, Mass. (01360) 249/E2
Northfield○, Mass. (01360) 249/E2
Northfield, Minn. (55057) 255/E6
Northfield○, N.H. (†03276) 268/D5
Northfield, N.J. (08225) 273/D5
Northfield, Ohio (44067) 284/J10
Northfield, Texas (79246) 303/D3
Northfield, Vt. (05663) 268/B3
Northfield○, Vt. (05663) 268/B3
Northfield, Wis. (†54635) 317/D7
Northfield Falls, Vt. (05664) 268/B3
Northfield Farms, Mass. (†01360) 249/E2
Northfield-Tilton, N.H. (†03276) 268/D5
Northfleet, England 10/C5
Northfleet, England 13/J8
North Fond du Lac, Wis. (†54935) 317/J8
Northford, Conn. (06472) 210/D3
North Foreland (prom.), England 10/G5
North Foreland (prom.), England 13/J6
North Fork, Calif. (93643) 204/F6
North Fork, Frenchman (creek), Colo. 208/O1
North Fork, Gunnison (riv.), Colo. 208/D5
North Fork, Smoky Hill (riv.), Colo. 208/P4
North Fork, Idaho (83466) 220/D4
North Fork (riv.), Idaho 220/B7
North Fork, Flathead (riv.), Mont. 262/B2
North Fork, Little Humboldt (riv.), Nev. 266/D1
North Fork, Grand (riv.), N. Dak. 282/E8
Northfork, W. Va. (24868) 312/D8
North Fork, Powder (riv.), Wyo. 319/F2
North Fork, Shoshone (riv.), Wyo. 319/C1
North Fork, Wind (riv.), Wyo. 319/C2
North Fort Myers, Fla. (33903) 212/E5
North Foster, R.I. (†02857) 249/H5
North Fourchu, Nova Scotia 168/H3
North Fox (isl.), Mich. 250/D3
North Franklin, Conn. (06254) 210/G2
North Freedom, Wis. (53951) 317/G9
North Friars (bay), St. Chris.-Nevis 161/D10
North Friesland (reg.), W. Germany 22/C1
North Frisian (isls.), Denmark 21/B7
North Frisian (isls.), W. Germany 22/B1
North Fryeburg, Maine (04058) 243/B7
North Galiano, Br. Col. 184/K3
North Garden, Va. (22959) 307/L5
Northgate, N. Dak. (58767) 282/F2
Northgate, Sask. 181/J6
Northglenn, Colo. (80233) 208/K3
North Gorham, Maine (†04075) 243/B8
North Gosforth, England 13/J3
North Gower, Ontario 177/J2
North Grafton, Mass. (01536) 249/H4
North Granby, Conn. (06060) 210/D1
North Grant, Nova Scotia 168/G3
North Grosvenor Dale, Conn. (06255) 210/H1
North Groton, N.H. (†03266) 268/D4
North Grove, Ind. (†46911) 227/F3
North Guilford, Conn. (†06437) 210/E3
North Hadley, Mass. (01035) 249/D3
North Haledon, N.J. (07508) 273/B1
North Hampton○, N.H. (03862) 268/F6
North Hampton, Ohio (45349) 284/C5
North Hanover, Mass. (†02339) 249/L4
North Hansel (mts.), Utah 304/B2
North Harbour, Newf. 166/D2
North Harlowe, N.C. (†28532) 281/R5
North Hartland, Vt. (05052) 268/C4
North Hartsville, S.C. (†29550) 296/G3
North Harwich, Mass. (†02645) 249/O6
North Hatfield, Mass. (01066) 249/D3
North Hatley, Québec 172/F4
North Haven○, Conn. (06473) 210/D3
North Haven, Maine (04853) 243/F7
North Haven○, Maine (04853) 243/F7
North Haverhill, N.H. (03774) 268/D3
North Havre, Mont. (†59501) 262/G2
North Hayden, Ind. (†46356) 227/B2
North Head, New Bruns. 170/D4
North Henderson, Ill. (61466) 222/C2
North Hero, Vt. (05474) 268/A2
North Highlands, Calif. (95660) 204/B8
North High Shoals, Georgia (†30645) 217/F3
North Hills, W. Va. (†26101) 312/D4
North Hodge, La. (†71247) 238/E2
North Holland (prov.), Netherlands 27/F3
North Holland (canal), Netherlands 27/C4
North Hollywood, Calif. (*91601) 204/B10
North Hornell, N.Y. (†14843) 276/E6
North Horr, Kenya 115/G3
North Hudson, N.Y. (12855) 276/N3
North Hudson, Wis. (†54016) 317/A5
North Hyde Park, Vt. (05665) 268/B2
North Hykeham, England 13/G4
North Industry, Ohio (44707) 284/H4
North Inishkea (isl.), Ireland 17/A3
North Java, N.Y. (14113) 276/D5
North Jay, Maine (04262) 243/C6
North Johns, Ala. (35006) 195/D4
North Judson, Ind. (46366) 227/D2
North Kansas City, Mo. (64116) 261/P5
North Kedgwick (riv.), New Bruns. 170/C1
North Kent, Conn. (†06757) 210/B1
North Kent (isl.), N.W. Terrs. 187/J2
North Kingstown○, R.I. (02852) 249/J6
North Kingsville, Ohio (44068) 284/J2
North Knife (lake), Manitoba 179/J2
North Knife Lake, Manitoba 179/J2
North Korea 54/O5
North La Junta, Colo. (†81050) 208/N7
Northlake, Ill. (60164) 222/B5
Northlake, Texas (75238) 303/F1
North Lake, Wis. (53064) 317/J1
North Lakhimpur, India 68/G3
Northland, Mich. (49869) 250/B2
Northland, Wis. (†54945) 317/H6
North Landgrove, Vt. (†05148) 268/B5
North Laramie (riv.), Wyo. 319/G3
North Las Vegas, Nev. (89030) 266/F6
North Lauderdale, Fla. (†33063) 212/B3
North Lawrence, N.Y. (12967) 276/L1
North Lawrence, Ohio (44666) 284/G4
North Leeds, Maine (04263) 243/C7
North Lewisburg, Ohio (43060) 284/C5
North Liberty, Ind. (46554) 227/E1
North Liberty, Iowa (52317) 229/K5
North Lima, Ohio (44452) 284/J4
North Limington, Maine (†04049) 243/B8
North Little Rock, Ark. (*72114) 202/F4
North Livermore, Maine (†04254) 243/C7
North Loup, Nebr. (68859) 264/F3
North Loup (riv.), Nebr. 264/E3
North Lovell, Maine (†04231) 243/B7
North Lubec, Maine (†04652) 243/J6
North Luconia (shoals), Philippines 85/E4
North Madison, Conn. (†06443) 210/E3
North Madison, Ohio (†44057) 284/H2
North Magnetic Pole (dist.) 162/F1
North Magnetic Pole, Canada 4/B15
North Magnetic Pole, N.W. Terrs. 187/H2
North Manchester, Ind. (46962) 227/F3
North Manitou, Mich. (†49654) 250/C3
North Manitou (isl.), Mich. 250/C3
North Mankato, Minn. (56001) 255/D6
North Marshfield, Mass. (02059) 249/M4
North Merritt (isl.), Fla. 212/F3
North Miami, Fla. (33161) 212/B4
North Miami, Okla. (74358) 288/R1
North Miami Beach, Fla. (33161) 212/C4
North Middleboro, Mass. (02346) 249/L5
North Middletown, Ky. (40357) 237/N4
North Minch (sound), Scotland 10/D1
North Minch (sound), Scotland 15/B3
North Montpelier, Vt. (05666) 268/C3
Northmoor, Mo. (†64152) 261/P5
North Motton, Tasmania 99/C3
North Mountain, W. Va. (†25427) 312/K3
North Muskegon, Mich. (49445) 250/C5
North Myrtle Beach, S.C. (29582) 296/K4
North Naples, Fla. (33940) 212/E5
North Natuna (isl.), Indonesia 85/D4
North Negril (pt.), Jamaica 158/G6
North Newport, N.H. (†03773) 268/C5
North New Portland, Maine (04961) 243/C6
North New River (canal), Fla. 212/F5
North Newry, Maine (†04261) 243/B6
North Newton, Kansas (67117) 232/E3
North Oaks, Minn. (†55101) 255/G5
North Ogden, Utah (†84404) 304/C2
North Olmsted, Ohio (44070) 284/G9
Northome, Minn. (56661) 255/D3
North Ossetian A.S.S.R., U.S.S.R. 48/E5
North Ossetian A.S.S.R., U.S.S.R. 52/F6
North Oxford, Mass. (01537) 249/G4
North Pacific (ocean) 87/F4
North Pacific Ocean 2/B5
North Pacific Ocean 2/T4
North Pagai (isl.), Indonesia 85/C6
North Palm Beach, Fla. (33403) 212/F5
North Park, Ill. (†61111) 222/D1
North Parsonfield, Maine (†04047) 243/A8
North Pease (riv.), Texas 303/D3
North Pekin, Ill. (†61554) 222/D3
North Pembroke, Mass. (02358) 249/M4
North Pender Island, Br. Col. 184/K3
North Penobscot, Maine (†04476) 243/F7
North Perry, Maine (†04667) 243/J5
North Perry, Ohio (†44081) 284/H2
North Petherton, England 13/D6
North Pine, Br. Col. 184/G2
North Plain, Conn. (†06371) 210/F3
North Plainfield, N.J. (07060) 273/E2
North Plains, Oreg. (97133) 291/A2
North Platte (riv.) 188/F2
North Platte (riv.), Colo. 208/G1
North Platte, Nebr. 188/F2
North Platte, Nebr. (69101) 264/D3
North Platte (riv.), Nebr. 264/A3
North Platte (riv.), U.S. 146/H5
North Platte (riv.), Wyo. 319/H3
North Plymouth, Mass. (02360) 249/L5
North Pole 4/A1
North Pole 2/F1
North Pole, Alaska (99705) 196/J2
North Pole (brook), New Bruns. 170/D1
North Pomfret, Vt. (05053) 268/B4
Northport, Ala. (35476) 195/C4
North Port, Fla. (33595) 212/D4
Northport○, Maine (†04849) 243/E7
Northport, Mich. (49670) 250/D3
Northport, Nebr. (†69336) 264/B3
Northport, N.Y. (11768) 276/O9
Northport, Nova Scotia 168/E3
Northport, Wash. (99157) 310/H2
North Portal, Sask. 181/J6
North Potomac, Md. (†20857) 245/K4
North Powder, Oreg. (97867) 291/K2
North Pownal, Vt. (05260) 268/A6
North Prairie, Wis. (53153) 317/J2
North Providence○, R.I. (02908) 249/J5
North Pulaski, Va. (†24301) 307/G6
North Randall, Ohio (†44101) 284/H9
North Randolph, Vt. (†05061) 268/B4
North Raymond, Maine (†04274) 243/C8
North Reading○, Mass. (01864) 249/C5
North Redington Beach, Fla. (†33708) 212/B3
North Redwood, Minn. (56275) 255/D6
North Renous (riv.), New Bruns. 170/D2
North Rhine-Westphalia (state), W. Germany 22/B3
North Richland Hills, Texas (76118) 303/F2
Northridge, Ohio (45414) 284/B6
North Ridgeville, Ohio (44039) 284/F3
North Rim, Ariz. (86052) 198/C2
North River, Newf. 166/D2
North River, N.Y. (12856) 276/M3
North River, N. Dak. (†58102) 282/S6
North River, Nova Scotia 168/D4
North Riverside, Ill. (60546) 222/B5
North Robinson, Ohio (44856) 284/E4
North Ronaldsay (firth), Scotland 15/F1
North Ronaldsay (isl.), Scotland 15/F1
North Ronaldsay (isl.), Scotland 10/E1
Northrop, Minn. (56075) 255/D7
North Rose, N.Y. (14516) 276/G4
North Roxboro, N.C. (†27573) 281/L2
North Royalton, Ohio (44133) 284/H10
North Rustico, Pr. Edward I. 168/E2
North Saanich, Br. Col. 184/K3
North Saint Paul, Minn. (55109) 255/G5
North Salem, Ind. (46165) 227/D5
North Salem, N.H. (03073) 268/E6
North Salt Lake, Utah (†84010) 304/C3
North Sandwich, N.H. (03259) 268/E4
North San Juan, Calif. (95960) 204/E4
North Santiam (riv.), Oreg. 291/E3
North Saskatchewan (riv.) (dist.) 162/E5
North Saskatchewan (riv.), Alberta 182/E3
North Saskatchewan (riv.), Canada 146/G4
North Saskatchewan (riv.), Sask. 181/D3
North Scituate, Mass. (02060) 249/F8
North Scituate, R.I. (02857) 249/H5
North Sea (canal), Netherlands 27/F4
North Seal (riv.), Manitoba 179/H2
North Searsmont, Maine (†04973) 243/E7
North Sentinel (isl.), India 68/G6
North Sevogle (riv.), New Bruns. 170/D1
North Shapleigh, Maine (04060) 243/B8
North Shoal (lake), Manitoba 179/E4
North Shore, Wis. 317/M1
Northside, N.C. (27564) 281/M2
Northside, Sask. 181/F2
North Sioux City, S. Dak. (57049) 298/R8
North Skunk (riv.), Iowa 229/H5
North Somercotes, England 10/G4
North Somercotes, England 13/H4
North Somers, Conn. (†06071) 210/F1
North Spectacle (lake), Conn. 210/B2
North Spirit Lake, Ontario 175/B2
North Springfield, Pa. (16430) 294/A1
North Springfield, Vt. (05150) 268/B5
North Springfield, Va. (22151) 307/S3
North Star, Alberta 182/B1
North Star, Mich. (48862) 250/E5
North Star, Ohio (45350) 284/A5
North Stonington○, Conn. (06359) 210/H3
North Stratford, N.H. (03590) 268/D2
North Sunderland, England 13/F2
North Sutton, N.H. (03260) 268/D5
North Swansea, Mass. (†02777) 249/K5
North Sydney, N. S. Wales 88/L4
North Sydney, N.S. Wales 97/J3
North Sydney, Nova Scotia 168/H2
North Syracuse, N.Y. (13212) 276/H4
North Taranaki (bight), N. Zealand 100/D3
North Tarrytown, N.Y. (10591) 276/O6
North Terre Haute, Ind. (47805) 227/C5
North Thetford, Vt. (05054) 268/C4
North Thompson (riv.), Br. Col. 184/G4
North Tidworth, England 13/F6
North Tiverton, R.I. (†02722) 249/K6
North Tolsta, Scotland 15/B2
Northton, Scotland 15/B3
North Tonawanda, N.Y. (14120) 276/C4
North Trap (isl.), N. Zealand 100/B7
North Troy, Vt. (05859) 268/C2
North Truchas (peak), N. Mex. 274/D3
North Truro, Mass. (02652) 249/O4
North Tunbridge, Vt. (†05077) 268/C4
North Turner, Maine (04266) 243/C7
North Twin (mt.), N.H. 268/D3
North Tyne (riv.), England 13/E2
North Uist (isl.), Scotland 15/A3
North Uist (isl.), Scotland 10/C2
Northumberland (co.), England 13/E2
Northumberland (co.), New Bruns. 170/D2
Northumberland (str.), New Bruns. 170/F2
Northumberland○, N.H. (†03582) 268/D2
Northumberland (str.), Nova Scotia 168/E2
Northumberland (county), Ontario 177/G3
Northumberland (co.), Pa. 294/J4
Northumberland, Pa. (17857) 294/J4
Northumberland (str.), Pr. Edward I. 168/D2
Northumberland (isls.), Queensland 95/D4
Northumberland (cape), S. Australia 94/F8
Northumberland (co.), Va. 307/R5
Northumberland National Park, England 13/E2
North Umpqua (riv.), Oreg. 291/E4
North Ural (mts.), U.S.S.R. 52/K1
North Utica (Utica), Ill. (†61373) 222/E2
North Uxbridge, Mass. (01538) 249/H4
Northvale, N.J. (07647) 273/F1
North Vancouver, Br. Col. 162/D6
North Vancouver, Br. Col. 184/K3
North Vassalboro, Maine (04962) 243/D7
North Vernon, Ind. (47265) 227/F6
Northview, Mo. (†65706) 261/G8
Northville, Conn. (†06776) 210/B2
Northville, Mich. (48167) 250/F6
Northville, N.Y. (12134) 276/M4
Northville, S. Dak. (57465) 298/M3
North Wabasca (lake), Alberta 182/D1
North Wakefield, N.H. (†03872) 268/E4
North Waldoboro, Maine (†04572) 243/E7
North Wales, Pa. (19454) 294/M5
North Walpole, N.H. (†03608) 268/C5
North Walsham, England 13/J5
North Walsham, England 10/G4
North Waltham, Mass. (02154) 249/B6
North Warren, Pa. (†16365) 294/D2
North Washington, Iowa (50661) 229/J2
North Waterboro, Maine (04061) 243/B8
North Waterford, Maine (04267) 243/B7
Northway, Alaska (99764) 196/K2
North Wayne, Maine (†04284) 243/C7
North Weare, N.H. (†03281) 268/D5
North Webster, Ind. (46555) 227/F2
Northwest (pt.), Fla. 212/E6
North West (dist.), Guyana 131/A2
North West (pt.), Jamaica 158/G5
North West (cape), Australia 87/B8
North West (cape), W. Australia 88/A4
North West (cape), W. Australia 92/A3
North-West Aboriginal Reserve, S. Australia 88/E5
North-West Aboriginal Res., W. Australia 92/E4
North West Arm (inlet), Newf. 166/D2
North West Brook, Newf. 166/C2
North West Brook (riv.), Newf. 166/D2
North Westchester, Conn. (06474) 210/F2
Northwestern (sen. dist.), Alaska 196/E2
North-West Frontier (prov.), Pakistan 68/C2
North West Gander (riv.), Newf. 166/C4
North Westminster, Vt. (†05101) 268/B5
Northwest Miramichi (riv.), New Bruns. 170/D1
Northwest Oromocto (riv.), New Bruns. 170/D3
North Westport, Mass. (02790) 249/K6
North West Providence (chan.), Bahamas 156/B1
North West River, Newf. 166/B3
Northwest Territories 162/E2
Northwest Territories (prov.), Canada 146/G3
NORTHWEST TERRITORIES 187
Northwest Upsalquitch (riv.), New Bruns. 170/D1
North Weymouth, Mass. (02191) 249/D8
North Whitefield, Maine (04353) 243/D7
Northwich, England 13/H2
Northwich, England 10/G2
North Wilbraham, Mass. (†01095) 249/E4
North Wildwood, N.J. (08260) 273/D6
North Wilkesboro, N.C. (28659) 281/G2
North Williston, Vt. (†05495) 268/A3
North Wilton, Conn. (†06897) 210/B4
North Windham, Conn. (06256) 210/G1
North Windham, Maine (04062) 243/C8
North Wolcott, Vt. (†05680) 268/C2
Northwood, Iowa (50459) 229/G2
Northwood○, N.H. (03261) 268/E5
Northwood, N. Dak. (58267) 282/P4

Northwood, Ohio (†43619) 284/D2
North Woodbury, Conn. (†06798) 210/C2
Northwood Center, N.H. (†03261) 268/E5
Northwood Narrows (Northwood P.O.), N.H. (03261) 268/E5
Northwoods, Mo. (†63101) 261/R2
North Woodstock, Conn. (†06281) 210/G1
North Woodstock, Maine (†04219) 243/B7
North Woodstock, N.H. (03262) 268/D3
Northwye, Mo. (†65401) 261/J7
North Yarmouth, Maine (†04096) 243/C8
North Yarmouth○, Maine (†04096) 243/C8
North York, Ontario 177/J4
North York Moors National Park, England 13/G3
North Yorkshire (co.), England 13/F3
North Zanesville, Ohio (†43701) 284/G6
Norton (sound), Alaska 146/B3
Norton (sound), Alaska 188/C5
Norton (bay), Alaska 196/F2
Norton (sound), Alaska 196/E2
Norton, England 13/G3
Norton (peak), Idaho 220/D6
Norton (co.), Kansas 232/C2
Norton, Kansas (67654) 232/C2
Norton (res.), Kansas 232/C2
Norton, Mass. (02766) 249/K5
Norton○, Mass. (02766) 249/K5
Norton, New Bruns. 170/E3
Norton, Ohio (44203) 284/G3
Norton, Texas (76865) 303/E6
Norton (sound), U.S. 4/C18
Norton○, Vt. (05907) 268/D2
Norton (pond), Vt. 268/D2
Norton (I.C.), Va. (24273) 307/C7
Norton, W. Va. (26285) 312/G5
Norton A.F.B., Calif. 204/F10
Norton-Radstock, England 13/E6
Norton Shores, Mich. (†49441) 250/C5
Nortonville, Kansas (66060) 232/G2
Nortonville, Ky. (42442) 237/G6
Nortonville, N. Dak. (58473) 282/N6
Norumbega, Argentina 143/F7
Norvegia (cape) 5/B18
Norvell, Arkm (†72386) 202/K3
Norvelt, Pa. (15674) 294/D5
Norwalk, Calif. (90650) 204/C11
Norwalk, Conn. (*06850) 210/B4
Norwalk (isls.), Conn. 210/B4
Norwalk (riv.), Conn. 210/B4
Norwalk, Iowa (50211) 229/F6
Norwalk, Mich. (†49660) 250/C4
Norwalk, Ohio (44857) 284/E3
Norwalk, Wis. (54648) 317/E8
Norway 2/K2
Norway 4/C9
NORWAY 18
Norway 7/E2
Norway, Ind. (†47960) 227/D3
Norway, Iowa (52318) 229/K5
Norway, Kansas (66961) 232/E2
Norway, Maine (04268) 243/B7
Norway○, Maine (04268) 243/B7
Norway, Mich. (49870) 250/B3
Norway (bay), N.W. Terrs. 187/H2
Norway, Oreg. (97460) 291/C4
Norway, S.C. (29113) 296/E5
Norway House, Man. 162/G5
Norway House, Manitoba 179/J3
Norway Lake, Maine (†04268) 243/B7
Norwegian (sea) 4/C10
Norwegian (sea) 7/D2
Norwegian (bay), N.W. Terrs. 187/J2
Norwegian (sea), Norway 18/F3
Norwell○, Mass. (02061) 249/F8
Norwich, Conn. (06360) 210/G2
Norwich, England 13/J5
Norwich, England 10/G4
Norwich, Kansas (67118) 232/E4
Norwich, N.Y. (13815) 276/J5
Norwich, N. Dak. (58768) 282/J3
Norwich, Ohio (43767) 284/G6
Norwich, Ontario 177/D5
Norwich○, Vt. (05055) 268/C4
Norwichtown, Conn. (†06360) 210/G2
Norwood, Colo. (81423) 208/C6
Norwood, Georgia (30821) 217/G4
Norwood, La. (70761) 238/H5
Norwood○, Mass. (02062) 249/B8
Norwood, Minn. (55368) 255/E6
Norwood, Mo. (65717) 261/H8
Norwood, N.J. (07648) 273/C1
Norwood, N.Y. (13668) 276/L1
Norwood, N.C. (28128) 281/J4
Norwood, Ohio (45212) 284/C9
Norwood, Ontario 177/F3
Norwood, Pa. (19074) 294/M7
Norwood, R.I. (†02887) 249/J6
Norwood, Va. (24581) 307/L5
Nosappu (pt.), Japan 81/N2
Nosbonsing (lake), Ontario 177/E1
Nose, Japan 81/J7
Noshiro, Japan 81/J3
Nosovka, U.S.S.R. 52/D4
Nosratabad, Iran 66/L6
Noss (head), Scotland 15/F2
Nossa Senhora do Livramento, Brazil 132/B6
Nossob (riv.), Botswana 118/B4
Nossob (riv.), Namibia 118/B4
Nosy Be (isl.), Madagascar 118/H2
Nosy Boraha (isl.), Madagascar 118/J3
Nosy-Varika, Madagascar 118/H4
Notakwanon (riv.), Newf. 166/B2
Notasulga, Ala. (36866) 195/G5
Noteć (riv.), Poland 47/B2
Notikewin, Alberta 182/B1
Notikewin (riv.), Alberta 182/A1
Noto, Italy 34/E6
Noto, Japan 81/H5
Noto (pen.), Japan 81/H5
Notodden, Norway 18/F7
Notre Dame, Ind. (46556) 227/E1
Notre-Dame, New Bruns. 170/F2
Notre Dame (bay), Newf. 166/C4
Notre-Dame-de-Ham, Québec 172/F4
Notre-Dame-de-la-Doré, Québec 172/E1
Notre-Dame-de-la-Paix, Québec 172/C4
Notre-Dame-de-la-Salette, Québec 172/B4
Notre Dame de Lourdes, Manitoba 179/D5
Notre-Dame-de-Pierreville, Québec 172/D3
Notre-Dame-des-Anges, Québec 172/E3
Notre-Dame-des-Bois, Québec 172/G4
Notre-Dame-des-Laurentides, Québec 172/H3
Notre-Dame-des-Monts, Québec 172/G2
Notre-Dame-des-Prairies, Québec 172/D3
Notre-Dame-de-Stanbridge, Québec 172/D4
Notre-Dame-du-Bon-Conseil, Québec 172/E4
Notre-Dame-du-Lac, Québec 172/J2
Notre-Dame-du-Laus, Québec 172/B3
Notre-Dame-du-Portage, Québec 172/H2
Notre-Dame-du-Rosaire, Québec 172/F1
Nottawa, Ontario 177/D3
Nottawasaga (bay), Ontario 177/D3
Nottawasaga (riv.), Ontario 177/E3
Nottaway (riv.), Que. 162/J5
Nottaway (riv.), Québec 174/B2
Nottely (lake), Georgia 217/D1
Nøtterøy, Norway 18/D4
Nottingham, Ala. (†35014) 195/F4
Nottingham, England 13/F5
Nottingham, England 10/F4
Nottingham○, N.H. (03290) 268/E5
Nottingham (isl.), N.W.T. 162/H3
Nottingham (isl.), N.W. Terrs. 187/L3
Nottinghamshire (co.), England 13/F4
Nottinghill, Mo. (65718) 261/G9
Nottoway (co.), Va. 307/M6
Nouveau-Québec (terr.), Québec 174/E1
Nouveau-Québec (crater), Québec 174/F1
Nouvelle, Québec 172/C2
Nouvelle (riv.), Québec 172/C2
Nouvelle-France (cape), Québec 174/F1
Nouvelle-Ouest, Québec 172/C2
Nova, Hungary 41/D3
Nova, Ohio (44859) 284/F3
Nová Baňa, Czech. 41/E2
Nová Bystrica, Czech. 41/E2
Nová Bystřice, Czech. 41/C2
Nova Chaves, Angola 115/D6
Nova Cruz, Brazil 132/H4
Nova Era, Brazil 135/E1
Nova Friburgo, Brazil 132/F8
Nova Friburgo, Brazil 135/E3
Nova Gaia, Angola 115/C5
Nova Goa (Panaji), India 68/C5
Nova Gorizia, Yugoslavia 45/A2
Nova Gradiška, Yugoslavia 45/C3
Nova Granada, Brazil 135/B2
Nova Iguaçu, Brazil 120/E5
Nova Iguaçu, Brazil 135/E3
Nova Iguaçu, Brazil 132/F8
Nova Iorque, Brazil 132/E4
Nova Lima, Brazil 135/E2
Nova Lusitânia, Mozambique 118/E3
Nova Mambone, Mozambique 118/F4
Novar, Ontario 177/E2
Novara (prov.), Italy 34/B2
Novara, Italy 34/B2
Nova Russas, Brazil 132/F4
Nova Scotia (prov.) 162/K7
Nova Scotia (prov.), Canada 146/M5
NOVA SCOTIA 168
Nova Sofala, Mozambique 118/F4
Novato, Calif. (94947) 204/H1
Novaya Kakhovka, U.S.S.R. 52/D5
Novaya Kazanka, U.S.S.R. 48/F5
Novaya Sibir' (isl.), U.S.S.R. 4/B2
Novaya Sibir' (isl.), U.S.S.R. 48/Q2
Novaya Zemlya (isl.), U.S.S.R. 2/L2
Novaya Zemlya (isls.), U.S.S.R. 4/B7
Novaya Zemlya (isls.), U.S.S.R. 48/F2
Novaya Zemlya (isls.), U.S.S.R. 52/H1
Nova Zagora, Bulgaria 45/H4
Nové Hrady, Czech. 41/C2
Novelda, Spain 33/F3
Novelty, Mo. (63460) 261/H2
Nové Město nad Váhom, Czech. 41/D2
Nové Město na Moravě, Czech. 41/D2
Nové Strašeci, Czech. 41/B1
Nové Zámky, Czech. 41/D3
Novgorod, U.S.S.R. 7/H3
Novgorod, U.S.S.R. 52/D3
Novgorod, U.S.S.R. 48/D4
Novgorod-Severskiy, U.S.S.R. 52/D4
Novi, Mich. (48050) 250/F6
Novi, Yugoslavia 45/B3
Novice, Texas (79538) 303/E5
Novi Ligure, Italy 34/B2
Novinger, Mo. (63559) 261/G2
Novi Pazar, Bulgaria 45/H4
Novi Pazar, Yugoslavia 45/E4
Novi Sad, Yugoslavia 45/D3
Novi Sad, Yugoslavia 7/F4
Nóvita, Colombia 126/B5
Novoannenskiy, U.S.S.R. 52/F4
Novo Aripuanã, Brazil 120/D3
Novocherkassk, U.S.S.R. 7/J4
Novocherkassk, U.S.S.R. 52/F5
Novodvinsk, U.S.S.R. 52/F2
Novograd-Volynskiy, U.S.S.R. 52/C4
Novogrudok, U.S.S.R. 52/C4
Novo Hamburgo, Brazil 132/D10
Novo Horizonte, Brazil 135/B2
Novokazalinsk, U.S.S.R. 48/G5
Novokuybyshevsk, U.S.S.R. 7/K3
Novokuybyshevsk, U.S.S.R. 52/G4
Novokuznetsk, U.S.S.R. 54/K4
Novokuznetsk, U.S.S.R. 48/J4
Novo Mesto, Yugoslavia 45/B3
Novomoskovsk, U.S.S.R. 7/H3
Novomoskovsk, U.S.S.R. 52/E4
Novomoskovsk, U.S.S.R. 48/E4
Novopolotsk, U.S.S.R. 52/C3
Novorossiysk, U.S.S.R. 7/H4
Novorossiysk, U.S.S.R. 48/D5
Novorossiysk, U.S.S.R. 52/E6
Novosergiyevka, U.S.S.R. 52/H4
Novoshakhtinsk, U.S.S.R. 52/E5
Novosibirsk, U.S.S.R. 54/J4
Novosibirsk, U.S.S.R. 2/P3
Novosibirsk, U.S.S.R. 48/J4
Novotroitsk, U.S.S.R. 52/J4
Novoukrainka, U.S.S.R. 52/D5
Novouzensk, U.S.S.R. 52/G4
Novovolynsk, U.S.S.R. 52/B4
Novovyatsk, U.S.S.R. 52/G3
Novozybkov, U.S.S.R. 48/D4
Novozybkov, U.S.S.R. 52/D4
Novra, Manitoba 179/B2
Novska, Yugoslavia 45/C3
Nový Bohumín, Czech. 41/E2
Nový Bor, Czech. 41/C1
Nový Bydžov, Czech. 41/C1
Nový Hrozenkov, Czech. 41/E2
Nový Jičín, Czech. 41/E2
Novyy Bug, U.S.S.R. 52/D5
Novyy Port, U.S.S.R. 4/C6
Novyy Port, U.S.S.R. 48/G3
Novyy Urengoy, U.S.S.R. 48/H3
Novyy Uzen', U.S.S.R. 48/F5
Nowa Dęba, Poland 47/E3
Nowa Nowa, Victoria 97/E5
Nowa Ruda, Poland 47/C3
Nowa Sól, Poland 47/B3
Nowata (co.), Okla. 288/P1
Nowata, Okla. (74048) 288/P1
Nowater (creek), Wyo. 319/E2
Nowe, Poland 47/D2
Nowe Miasto Lubawskie, Poland 47/D2
Nowendoc, N.S. Wales 97/F2
Nowgong, Assam, India 68/G3
Nowgong, Madhya Pradesh, India 68/D3
Nowitna (riv.), Alaska 196/H2
Nowogard, Poland 47/B2
Nowood (riv.), Wyo. 319/E1
Nowra-Bomaderry, N.S. Wales 88/J6
Nowra-Bomaderry, N.S. Wales 97/F4
Now Shahr, Iran 66/G2
Nowshera, Pakistan 59/K3
Nowshera, Pakistan 68/C2
Nowy Dwór Gdański, Poland 47/D1
Nowy Dwór Mazowiecki, Poland 47/E2
Nowy Sącz (prov.), Poland 47/E4
Nowy Sącz, Poland 47/E4
Nowy Staw, Poland 47/D1
Nowy Targ, Poland 47/E4
Nowy Tomyśl, Poland 47/C2
Now Zad, Afghanistan 68/A2
Now Zad, Afghanistan 59/H3
Noxapater, Miss. (39346) 256/F5
Noxen, Pa. (18636) 294/E7
Noxon, Mont. (59853) 262/A3
Noxubee (co.), Miss. 256/G4
Noxubee (riv.), Miss. 256/G4
Noya, Spain 33/B1
Noyes (isl.), Alaska 196/M2
Noyes, Minn. (56740) 255/A2
Noyes (pt.), R.I. 249/H7
Noyo (riv.), Calif. 204/B4
Noyon, France 28/E3
Noyon, Mongolia 77/F3
Nsanje, Malawi 115/G7
Nsawam, Ghana 106/D7
Nsukka, Nigeria 106/F7
Nsuta, Ghana 106/D7
Nuanetsi, Zimbabwe 118/E4
Nuba (mts.), Sudan 111/E5
Nubanusit (lake), N.H. 268/C5
Nubeena, Tasmania 99/D5
Nuberg, Georgia (†30634) 217/G2
Nubia (lake), Sudan 111/F3
Nubian (des.), Sudan 102/F2
Nubian (des.), Sudan 111/F3
Nubian (des.), Sudan 59/B5
Nubieber, Calif. (96068) 204/D2
Nuckolls (co.), Nebr. 264/F4
Nuckols, Ky. (†42352) 237/G5
Nucla, Colo. (81424) 208/B6
Nudgee, Queensland 88/K2
Nueces (co.), Texas 303/G10
Nueces (riv.), Texas 188/G5
Nueces (riv.), Texas 303/F9
Nueltin (lake) 162/G3
Nueltin (lake), Manitoba 179/H1
Nueltin (lake), N.W. Terrs. 187/H3
Nuestra Señora (bay), Chile 138/A5
Nueva (isl.), Argentina 143/C8
Nueva (isl.), Chile 138/F11
Nueva Alejandría, Peru 128/F5
Nueva Antioquia, Colombia 126/F5
Nueva Armenia, Honduras 154/D4
Nueva Asunción (dept.), Paraguay 144/B2
Nueva Casas Grandes, Mexico 150/F1
Nueva Ciudad Guerrero, Mexico 150/K3
Nueva Colombia, Paraguay 144/B4
Nueva Ecija (prov.), Philippines 82/C3
Nueva Esparta (state), Venezuela 124/G2
Nueva Germania, Paraguay 144/D3
Nueva Gerona, Cuba 156/A2
Nueva Gerona, Cuba 158/B2
Nueva Helvecia, Uruguay 145/B5
Nueva Imperial, Chile 138/D2
Nueva Italia, Paraguay 144/B5
Nueva Italia de Ruiz, Mexico 150/J7
Nueva Ocotepeque, Honduras 154/C3
Nueva Palmira, Uruguay 145/A4
Nueva Rosita, Mexico 150/J2
Nueva San Salvador, El Salvador 154/C4
Nueva Vizcaya (prov.), Philippines 82/C2
Nueve de Julio, Argentina 143/F7
Nuevitas, Cuba 156/C2
Nuevitas, Cuba 158/G2
Nuevitas (bay), Cuba 158/H2
Nuevo (gulf), Argentina 143/D5
Nuevo, Bajo (reef), Mexico 150/O6
Nuevo Berlín, Uruguay 145/B3
Nuevo Chagres, Panama 154/G6
Nuevo Ideal, Mexico 150/G4
Nuevo Juncal, Chile 138/B5
Nuevo Laredo, Mexico 146/H7
Nuevo Laredo, Mexico 150/J3
Nuevo León (state), Mexico 150/K4
Nuevo Mamo, Venezuela 124/G3
Nuevo Rocafuerte, Ecuador 128/E3
Nufenen, Switzerland 39/H3
Nugaal (prov.), Somalia 115/J2
Nugget (pt.), N. Zealand 100/C7
Nugrus, Jebel (mt.), Egypt 59/B4
Nuguria (isls.), Papua N.G. 86/C1
Nui (atoll), Tuvalu 87/H6
Nuiqsut, Alaska (99723) 196/H1
Nu Jiang (riv.), China 77/E6
Nûk (cap.), Greenl. 4/C12
Nûk (cap.), Greenland 146/N3
Nûk (cap.), Greenland 2/G2
Nuka (bay), Alaska 196/C2
Nuka (isl.), Alaska 196/C2
Nukey Bluff (mt.), S. Australia 94/D5
Nukheila (oasis), Sudan 111/E4
Nuku'alofa (cap.), Tonga 87/J8
Nukuhiva (isl.), Fr. Poly. 87/M6
Nukulaelae (atoll), Tuvalu 87/H6
Nukumanu (atoll), Papua N.G. 87/F6
Nukumanu (isls.), Papua N.G. 86/D2
Nukunonu (atoll), Tokelau Is. 87/J6
Nukuoro (atoll), Micronesia 87/F5
Nukus, U.S.S.R. 54/H5
Nukus, U.S.S.R. 48/G5
Nulato, Alaska (99765) 196/G2
Nules, Spain 33/F3
Nulhegan (riv.), Vt. 268/D2
Nulhegan, East Branch (riv.), Vt. 268/D2
Nullagine, W. Australia 88/C4
Nullagine, W. Australia 92/C3
Nullarbor (plain) 88/D6
Nullarbor (plain), Australia 87/C9
Nullarbor, S. Australia 94/B4
Nullarbor (plain), S. Australia 94/A4
Nullarbor (plain), W. Australia 92/D5
Nulliberg (mt.), Virgin Is. (U.S.) 161/B4
Nulltown, Ind. (†47331) 227/G5
Numa, Iowa (52575) 229/G7
Numan, Nigeria 106/G7
Numancia, Philippines 82/D5
Numansdorp, Netherlands 27/E5
Numata, Japan 81/J5
Numazu, Japan 81/J6
Numbulwar, North. Terr. 88/F2
Numbulwar, North. Terr. 93/D3
Numfoor (isl.), Indonesia 85/J6
Ñumí, Paraguay 144/D4
Nu Mine, Pa. (16244) 294/D4
Numurkah, Victoria 97/C5
Nunaksaluk (isl.), Newf. 166/B2
Nunapitchuk, Alaska (99641) 196/F2
Nunawading, Victoria 88/L7
Nunawading, Victoria 97/J5
Nunchia, Colombia 126/D5
Nunda, N.Y. (14517) 276/E5
Nunda, S. Dak. (57050) 298/P5
Nundah, Queensland 88/K2
Nundah, Queensland 95/E2
Nundle, N.S. Wales 97/F2
Nuneaton, England 13/F5
Nuneaton, England 10/F4
Núñez (isl.), Chile 138/D10
Nunez, Georgia (30448) 217/H5
Nungarin, W. Australia 92/B5
Nungesser (lake), Ontario 175/B2
Nunivak (isl.), Alaska 146/B4
Nunivak (isl.), Alaska 188/C6
Nunivak (isl.), Alaska 196/E3
Nunivak (isl.), U.S. 4/D18
Nunley, Ark. (†71953) 202/B4
Nunn, Colo. (80648) 208/K1
Nunnelly, Tenn. (37137) 237/G9
Nunningen, Switzerland 39/E2
Nuñoa, Peru 128/G10
Nunspeet, Netherlands 27/H4
Nuoro (prov.), Italy 34/B4
Nuoro, Italy 34/B4
Nuqub, P.D.R. Yemen 59/E6
Nuquí, Colombia 126/B5
Nuremberg, Pa. (18241) 294/K4
Nuremberg, W. Germany 7/F4
Nuremberg, W. Germany 22/D4
Nurestan (reg.), Afghanistan 59/K2
Nurhak, Turkey 63/G4
Nuri, Mexico 150/E3
Nuri (ruins), Sudan 111/F4
Nuria, Sierra de (mts.), Venezuela 124/H4
Nuriootpa, S. Australia 94/F6
Nuristan (reg.), Afghanistan 68/C1
Nurlat, U.S.S.R. 52/H4
Nurmes, Finland 18/Q5
Nürnberg (Nuremberg), W. Germany 22/D4
Nurrari (lakes), S. Australia 94/B3
Nursery, Texas (77976) 303/H9
Nürtingen, W. Germany 22/C4
Nuruhak Dağı (mt.), Turkey 63/G3
Nus, Ras (cape), Oman 59/G6
Nusa Barung (isl.), Indonesia 85/K3
Nusaybin, Turkey 63/J4
Nushagak (bay), Alaska 196/G3
Nushagak (riv.), Alaska 196/G2
Nushki, Pakistan 68/B3
Nushki, Pakistan 59/J4
Nutimik, Manitoba 179/G5
Nutley, N.J. (07110) 273/B2
Nut Mountain, Sask. 181/H3
Nutrioso, Ariz. (85932) 198/F5
Nuttby (mt.), Nova Scotia 168/E3
Nutter Fort, W. Va. (26301) 312/F4
Nutting Lake, Mass. (01865) 249/B5
Nutwood Downs, North. Terr. 93/D3
Nuuanu (stream), Hawaii 218/C4
Nuupere (pt.), Fr. Poly. 86/S13
Nuwara Eliya, Sri Lanka 68/E7
Nuweiba, Egypt 111/F2
Nuyakuk (lake), Alaska 196/F3
Nuyts (cape), S. Australia 88/E6
Nuyts (arch.), S. Australia 94/C5
Nuyts (cape), S. Australia 94/C5
Nyabing, W. Australia 88/B6
Nyabisindu, Rwanda 115/F4
Nyack, Mont. (†59936) 262/C2
Nyack, N.Y. (10960) 276/K8
Nyah, Victoria 97/B4
Nyah West, Victoria 97/B4
Nyainqêntanglha Shan (range), China 77/D5
Nyainrong, China 77/D5
Nyala, Sudan 111/E5
Nyala, Sudan 102/E3
Ny-Ålesund, Norway 18/C2
Nyamlell, Sudan 111/E6
Nyandoma, U.S.S.R. 7/J2
Nyandoma, U.S.S.R. 48/E3
Nyandoma, U.S.S.R. 52/F2
Nyanga, Gabon 115/A4
Nyanga, S. Africa 118/F6
Nyanza (prov.), Kenya 115/F4
Nyasa (lake) 102/F6
Nyasa (lake) 2/L6
Nyasa (lake), Malawi 115/F6
Nyasa (lake), Mozambique 118/E2
Nyasa (lake), Tanzania 115/F6
Nyborg, Denmark 21/D7
Nyborg, Denmark 18/G9
Nybro, Sweden 18/J8
Nye, Mont. (59061) 262/G5
Nye (co.), Nev. 266/E4
Nyeri, Kenya 115/G4
Nyerol, Sudan 111/F6
Nyima, China 77/C5
Nyingchi, China 77/D6
Nyírábrány, Hungary 41/G3
Nyíradony, Hungary 41/G3
Nyírbátor, Hungary 41/G3
Nyíregyháza, Hungary 41/F3
Nyírmada, Hungary 41/F2
Nyiru (mt.), Kenya 115/G3
Nykarleby, Finland 18/N5
Nykøbing, Denmark 18/H9
Nykøbing, Denmark 18/G9
Nykøbing, Storstrøm, Denmark 21/F8
Nykøbing, Vestsjaelland, Denmark 21/E6
Nykøbing, Viborg, Denmark 21/B4
Nykøbing, Denmark 18/F8
Nyköping, Sweden 18/K7
Nylstroom, S. Africa 118/D4
Nymagee, N.S. Wales 97/D3
Nymboida, N.S. Wales 97/G1
Nymboida (riv.), N.S. Wales 97/G1
Nymburk, Czech. 41/C1
Nynäshamn, Sweden 18/L7
Nyngan, N. S. Wales 88/H6
Nyngan, N.S. Wales 97/D2
Nyon, Switzerland 39/B4
Nyons, France 28/F5
Nýřany, Czech. 41/B2
Nyrob, U.S.S.R. 52/J2
Nýrsko, Czech. 41/B2
Nysa, Poland 47/C3
Nysa Kłodzka (riv.), Poland 47/C3
Nysa Łuzycka (Neisse) (riv.), Poland 47/B3
Nyssa, Oreg. (97913) 291/K4
Nysted, Denmark 21/E8
Nytva, U.S.S.R. 52/H3
Nyudo (cape), Japan 81/J4
Nyukhcha, U.S.S.R. 52/G2
Nyunzu, Zaire 115/E5
Nyurba, U.S.S.R. 48/M3
Nyuvchim, U.S.S.R. 52/H2
Nzega, Tanzania 115/F4
N'Zérékoré, Guinea 106/C7
Nzeto, Angola 115/B5

O

Oa, Mull of (prom.), Scotland 10/C3
Oa, Mull of (prom.), Scotland 15/B5
Oacoma, S. Dak. (57365) 298/L6
Oadby, England 13/F5
Oahe (lake), N. Dak. 282/J7
Oahe (lake), S. Dak. 188/F1
Oahe (dam), S. Dak. 298/J5
Oahe (lake), S. Dak. 298/J1
Oahe (lake), U.S. 146/J5
Oahu (isl.), Hawaii 87/L3
Oahu (isl.), Hawaii 188/F5
Oahu (isl.), Hawaii 218/E2
Oak (lake), Manitoba 179/B5
Oak (pt.), Mich. 250/F5
Oak, Nebr. (68964) 264/G4
Oak (creek), N. Dak. 282/J8
Oak (isl.), Nova Scotia 168/E3
Oak (creek), S. Dak. 298/J6
Oak (creek), S. Dak. 298/H2
Oak (isl.), Wis. 317/E2
Oakbank, Manitoba 179/F5
Oak Bay, Br. Col. 184/K4
Oak Bay, New Bruns. 170/C3
Oak Bluffs, Mass. (02557) 249/M7
Oak Bluffs○, Mass. (02557) 249/M7
Oak Bluff Station, Manitoba 179/E5
Oakboro, N.C. (28129) 281/J4
Oak Brae, Manitoba 179/E3
Oak Brook, Ill. (60521) 222/B6
Oakbrook Terrace, Ill. (†60181) 222/B5
Oakburn, Manitoba 179/B4
Oak Center, Minn. (†55041) 255/F6
Oak City, N.C. (27857) 281/P3
Oak City, Utah (84649) 304/B4
Oak Creek, Colo. (80467) 208/F2
Oak Creek, Wis. (53154) 317/M2
Oakdale, Calif. (95361) 204/E6
Oakdale, Conn. (06370) 210/G3
Oakdale, Ill. (62268) 222/D5
Oakdale, Iowa (52319) 229/K5
Oakdale, La. (71463) 238/E5
Oakdale, Mass. (01539) 249/G3
Oakdale, Minn. (†55109) 255/E5
Oakdale, Nebr. (68761) 264/F2
Oakdale, Pa. (15071) 294/B5
Oakdale, Tenn. (37829) 237/M9
Oakdale, Wis. (54649) 317/F8
Oakes, N. Dak. (58474) 282/O7
Oakesdale, Wash. (99158) 310/H3
Oakfield, Georgia (31772) 217/E7
Oakfield○, Maine (04763) 243/G3
Oakfield, N.Y. (14125) 276/D4
Oakfield, Tenn. (38362) 237/D9
Oakfield, Wis. (53065) 317/J8
Oakford, Ill. (62673) 222/D3
Oakford, Ind. (46965) 227/E4
Oak Forest, Ill. (60452) 222/B6
Oak Grove, Ala. (†36612) 195/B9
Oak Grove, Ala. (35006) 195/F4
Oak Grove, Ark. (72660) 202/C1
Oak Grove, Del. (†19973) 245/R6
Oak Grove, Ill. (†61264) 222/C2
Oak Grove, Ky. (42262) 237/G7
Oak Grove, La. (71263) 238/H1
Oak Grove, Mich. (48863) 250/F6
Oak Grove, Mo. (64075) 261/S6
Oak Grove, Mo. (†63080) 261/K6
Oak Grove, Oreg. (97268) 291/B2
Oak Grove, Va. (22443) 307/O4
Oak Grove Fork, Clackamas (riv.), Oreg. 291/F2
Oak Hall, Pa. (†16827) 294/G4
Oak Hall, Va. (23416) 307/S5
Oakham, England 13/G5
Oakham, England 10/F4
Oakham○, Mass. (01068) 249/F3
Oak Harbor, Ohio (43449) 284/D2
Oak Harbor, Wash. (98277) 310/C2
Oak Harbor Naval Air Sta., Wash. 310/C2
Oakhaven, Ark. (†71801) 202/C6
Oak Hill, Ala. (36766) 195/D7
Oak Hill, Fla. (32759) 212/F3
Oak Hill, Ill. (61518) 222/D3
Oakhill, Kansas (67472) 232/E2
Oak Hill, Mo. (†65453) 261/K6
Oak Hill, Ohio (45656) 284/E8
Oak Hill, Tenn. (†37201) 237/H8
Oak Hill, W. Va. (25901) 312/D6
Oakhurst, Calif. (93644) 204/F6
Oakhurst, N.J. (07755) 273/E3
Oakhurst, Okla. (74050) 288/P2
Oakhurst, Texas (77359) 303/J7
Oak Island, Minn. (56741) 255/D1
Oak Lake, Manitoba 179/B5
Oakland, Calif. 146/F6
Oakland, Calif. 188/B3
Oakland, Calif. (*94601) 204/J2
Oakland, Conn. (†06040) 210/E1
Oakland, Fla. (32760) 212/E3
Oakland, Ill. (61943) 222/F4
Oakland, Iowa (51560) 229/C6
Oakland, Ky. (42159) 237/J6
Oakland, La. (†71260) 238/F1
Oakland, Maine (04963) 243/D6
Oakland○, Maine (04963) 243/D6
Oakland, Md. (†21784) 245/L3
Oakland, Md. (21550) 245/A3
Oakland (co.), Mich. 250/F6
Oakland, Miss. (38948) 256/E2
Oakland, Mo. (†63101) 261/P3
Oakland, Mo. (†65536) 261/G7
Oakland, Nebr. (68045) 264/H3
Oakland, N.J. (07436) 273/B1
Oakland, Okla. (73452) 288/N6
Oakland, Ontario 177/D4
Oakland, Oreg. (97462) 291/D4
Oakland, Pa. (†18847) 294/L2
Oakland, R.I. (02858) 249/H5
Oakland, Tenn. (38060) 237/B10
Oakland Acres, Iowa (†50112) 229/H5
Oakland Army Base, Calif. 204/J2
Oakland Beach, R.I. (†02887) 249/J6
Oakland City, Ind. (47660) 227/C8
Oakland Mills, Iowa (†52641) 229/K7
Oakland Park, Fla. (33334) 212/B3
Oaklands, N.S. Wales 97/D4
Oak Lawn, Ill. (*60453) 222/B6
Oakleigh, Victoria 88/L7
Oakleigh, Victoria 97/J5
Oakley, Calif. (94561) 204/L1
Oakley, Idaho (83346) 220/D7
Oakley, Ill. (62552) 222/E4
Oakley, Iowa (†50049) 229/G6
Oakley, Kansas (67748) 232/B2
Oakley, Mich. (48649) 250/E5
Oakley, Miss. (†39154) 256/D6
Oakley, N.C. (†27871) 281/P3
Oakley, Scotland 15/C1
Oakley, S.C. (29466) 296/G5
Oakley, Tenn. (†38541) 237/L8
Oakley, Utah (84055) 304/C3
Oaklyn, N.J. (08107) 273/B3
Oakman, Ala. (35579) 195/D3
Oakman, Georgia (30732) 217/C1
Oakmont, Pa. (15139) 294/C6
Oakmulgee (creek), Ala. 195/D5
Oakner, Manitoba 179/B4
Oak Orchard, Del. (†19966) 245/T6
Oakover (riv.), W. Australia 88/C4
Oakover (riv.), W. Australia 92/C3
Oak Park, Georgia (31903) 217/H6
Oak Park, Ill. (*60303) 222/B5
Oak Park, Mich. (48237) 250/B6
Oak Park, Minn. (56357) 255/E5
Oakpark, Va. (22730) 307/M4
Oak Point, Manitoba 179/B5
Oak Point, New Bruns. 170/D3
Oak Ridge, La. (71264) 238/G1
Oak Ridge, Miss. (†39180) 256/C6
Oak Ridge, Mo. (63769) 261/N7
Oak Ridge, N.J. (07438) 273/E1
Oak Ridge (res.), N.J. 273/D1
Oak Ridge, N.C. (27310) 281/K2
Oakridge, Oreg. (97463) 291/E4
Oak Ridge, Pa. (16245) 294/D3
Oak Ridge, Tenn. 188/J3
Oak Ridge, Tenn. (37830) 237/N8

Oak River, Manitoba 179/D4
Oaks, Mo. (†64116) 261/P5
Oaks, Okla. (74359) 288/S2
Oakshela, Sask. 181/J5
Oakton, Ky. (42077) 237/C7
Oakton, Va. (22124) 307/R3
Oaktown, Ind. (47561) 227/C7
Oak Vale, Miss. (39656) 256/E8
Oakvale, W. Va. (24739) 312/D8
Oak Valley, Kansas (†67352) 232/G4
Oak View, Calif. (93022) 204/F9
Oakview, Manitoba 179/D3
Oakview, Mo. (†64116) 261/P5
Oakville, Conn. (06779) 210/C2
Oakville, Ind. (47367) 227/G4
Oakville, Iowa (52646) 229/L6
Oakville, Ky. (42263) 237/H7
Oakville, Manitoba 179/D5
Oakville, Ontario 177/E4
Oakville, Pa. (†17257) 294/H5
Oakville, Texas (78060) 303/G9
Oakville, Wash. (98568) 310/B4
Oakway, S.C. (†29694) 296/A2
Oakwood, Georgia (30566) 217/E2
Oakwood, Ill. (61858) 222/F3
Oakwood, Mo. (63401) 261/P5
Oakwood, N. Dak. (†58237) 282/R3
Oakwood, Ohio (†44146) 284/H9
Oakwood, Ohio (†45419) 284/B6
Oakwood, Ohio (45873) 284/B3
Oakwood, Okla. (73658) 288/J3
Oakwood, Ontario 177/F3
Oakwood, Texas (75855) 303/J6
Oakwood, Va. (24631) 307/E6
Oakwood Heights, Ill. (†62095) 222/B2
Oakwood Manor, Mo. (†64101) 261/P5
Oakwood Park, Mo. (†64116) 261/P5
Oamaru, N. Zealand 100/C6
Oani (riv.), Japan 81/K3
Oasis, Nev. (†89830) 266/G1
Oasis, Utah (84650) 304/B4
Oates Coast (reg.) 5/B8
Oatlands, Tasmania 99/D4
Oatman, Ariz. (86433) 198/A3
Oatsville, Ind. (†47567) 227/C8
Oaxaca (state), Mexico 150/L8
Oaxaca, Mexico 146/J8
Oaxaca de Juárez, Mexico 150/L8
Ob' (riv.), U.S.S.R. 2/N2
Ob' (riv.), U.S.S.R. 54/H3
Ob' (gulf), U.S.S.R. 54/J3
Ob' (gulf), U.S.S.R. 4/B6
Ob' (riv.), U.S.S.R. 4/C6
Ob' (gulf), U.S.S.R. 48/H3
Ob' (riv.), U.S.S.R. 48/G3
Oba, Ont. 162/H6
Oba, Ontario 177/J5
Oba, Ontario 175/D3
Obama, Japan 81/G6
Oban (Half Moon Bay), N. Zealand 100/B7
Oban, Sask. 181/C3
Oban, Scotland 15/C4
Oban, Scotland 10/D2
Obbia, Somalia 115/J2
Obed, Alberta 182/B3
Obed (riv.), Tenn. 237/M8
Ober, Ind. (†46534) 227/D2
Oberá, Argentina 143/F2
Oberägeri, Switzerland 39/G2
Oberalp (pass), Switzerland 39/G3
Oberalpstock (mt.), Switzerland 39/G3
Oberammergau, W. Germany 22/D5
Oberburg, Switzerland 39/E2
Oberdiessbach, Switzerland 39/E3
Oberdorf, Switzerland 39/E2
Ober Grafendorf, Austria 41/C2
Oberhausen, W. Germany 22/B3
Oberhof, E. Germany 22/D3
Oberlin, Kansas (67749) 232/B2
Oberlin, La. (70655) 238/E5
Oberlin, Mich. (†48624) 250/E4
Oberlin, Ohio (44074) 284/F3
Oberndorf bei Salzburg, Austria 41/B3
Oberon, Manitoba 179/C4
Oberon, N.S. Wales 97/E3
Oberon, N. Dak. (58357) 282/M4
Oberpfälzer Wald (for.), W. Germany 22/E4
Oberriet, Switzerland 39/J2
Obersaxen, Switzerland 39/H3
Obersiggenthal, Switzerland 39/F1
Oberstammheim, Switzerland 39/G1
Oberstdorf, W. Germany 22/D5
Obert, Nebr. (68762) 264/G2
Oberursel, W. Germany 22/C3
Obervellach, Austria 41/B3
Oberwald, Switzerland 39/F3
Oberwart, Austria 41/D3
Oberwil, Switzerland 39/D3
Oberwölz, Austria 41/C3
Oberzwil, Switzerland 39/H2
Obetz, Ohio (†43201) 284/E6
Obi (isl.), Indonesia 85/H6
Obi (isls.), Indonesia 85/H6
Óbidos, Brazil 120/D3
Óbidos, Brazil 132/C3
Óbidos, Portugal 33/B3
Obihiro, Japan 81/L2
Obion (creek), Ky. 237/C7
Obion (co.), Tenn. 237/C8
Obion, Tenn. (38240) 237/C8
Obion, Middle Fork (riv.), Tenn. 237/D8
Obion, North Fork (riv.), Tenn. 237/D8
Obion, South Fork (riv.), Tenn. 237/D8
Obion (riv.), Tenn. 237/C8
Obispos, Venezuela 124/D3
Obitsu (riv.), Japan 81/P3
Oblong, Ill. (62449) 222/F5
Obluch'ye, U.S.S.R. 48/N5
Obninsk, U.S.S.R. 52/E3
Obo, Cent. Afr. Rep. 115/E2
Obock, Djibouti 111/H5
Oborniki, Poland 47/C2
Oboyan', U.S.S.R. 52/E4
Obozerskiy, U.S.S.R. 52/E2
O'Brien, Fla. (32071) 212/D1
O'Brien (co.), Iowa 229/B2
O'Brien, Oreg. (97534) 291/D5
O'Brien, Texas (79539) 303/E4
O'Briensbridge-Montpelier, Ireland 17/D6
Observatory (inlet), Br. Col. 184/C2
Obsidian, Idaho (†83278) 220/D6
Obuasi, Ghana 106/D7
Obukowin (lake), Manitoba 179/G3
Obwalden (canton), Switzerland 39/F3
Ocala, Fla. (*32670) 212/D2
Ocamo (riv.), Venezuela 124/F6
Ocampo, Chihuahua, Mexico 150/E2
Ocampo, Coahuila, Mexico 150/H3
Ocampo, Tamaulipas, Mexico 150/K5
Ocaña, Colombia 126/D3
Ocaña, Spain 33/E3
Ocate, N. Mex. (87734) 274/E2
Ocate (creek), N. Mex. 274/E2
Occidental, Cordillera (range), Bolivia 136/A6
Occidental, Cordillera (range), Colombia 126/B5
Occidental, Cordillera (range), Peru 128/F10
Occidental Mindoro (prov.), Philippines 82/C4
Occoquan, Va. (22125) 307/O3
Occum, Conn. (†06360) 210/G2
Ocean (cape), Alaska 196/K3
Ocean (pond), Fla. 212/D1
Ocean (Banaba) (isl.), Kiribati 87/G6
Ocean (co.), N.J. 273/E4
Ocean (lake), Nova Scotia 168/G3
Ocean (lake), Wyo. 319/D2
Oceana (co.), Mich. 250/C5
Oceana, W. Va. (24870) 312/C7
Oceana N.A.S., Va. 307/S7
Ocean Beach, N.Y. (11770) 276/O9
Ocean Bluff-Brant Rock, Mass. (02065) 249/M4
Ocean Breeze Park, Fla. (†33457) 212/F4
Ocean City, Md. (21842) 245/T7
Ocean City, N.J. (08226) 273/D5
Ocean City, Wash. (98569) 310/A3
Ocean Falls, Br. Col. 184/D4
Ocean Gate, N.J. (08740) 273/E4
Ocean Grove, Mass. (02777) 249/K6
Ocean Grove, N.J. (07756) 273/F3
Ocean Isle Beach, N.C. (28459) 281/N7
Oceano, Calif. (93445) 204/E8
Oceanographic Office, Md. 245/F5
Ocean Park, Maine (04063) 243/C9
Ocean Park, Wash. (98640) 310/A4
Oceanport, N.J. (07757) 273/F3
Ocean Ridge, Fla. (33444) 212/F5
Ocean Shores, Wash. (98551) 310/A3
Oceanside, Calif. (92054) 204/H10
Oceanside, N.Y. (11572) 276/R7
Oceanside, Oreg. (97134) 291/C2
Ocean Springs, Miss. (39564) 256/G10
Ocean View, Del. (19970) 245/T6
Ocean View, N.J. (08230) 273/D5
Oceanville, N.J. (08231) 273/D5
Oceola, Ohio (44860) 284/D4
Ochamchira, U.S.S.R. 52/F6
Ochelata, Okla. (74051) 288/P1
Ocheyedan, Iowa (51354) 229/B2
Ochil (hills), Scotland 15/E4
Ochiltree (co.), Texas 303/D1
Ochlocknee, Georgia (31773) 217/E9
Ochlockonee (riv.), Fla. 212/B1
Ochlockonee (riv.), Georgia 217/C10
Ochoco (creek), Oreg. 291/G3
Ochopee, Fla. (33943) 212/E6
Ocho Rios, Jamaica 158/J6
Ochre River, Manitoba 179/C3
Ochsen (mt.), Switzerland 39/D3
Ochsenfurt, W. Germany 22/D4
Ochsenkopf (mt.), Liecht. 39/J2
Ocie, Mo. (65719) 261/G9
Ocilla, Georgia (31774) 217/F7
Ockelbo, Sweden 18/K6
Ocklawaha (lake), Fla. 212/E2
Ockley, Ind. (†46923) 227/D4
Ocmulgee (riv.), Georgia 217/E5
Ocmulgee Nat'l Mon., Georgia 217/F5
Ocna Mureș, Romania 45/G2
Ocoa, Chile 138/G2
Ocoa (bay), Dom. Rep. 158/D6
Ocoee, Fla. (32761) 212/E3
Ocoee (riv.), Tenn. 237/M10
Ocoña, Peru 128/F11
Ocoña (riv.), Peru 128/F11
Oconee (co.), Georgia 217/F3
Oconee, Georgia (31067) 217/G5
Oconee (riv.), Georgia 217/F5
Oconee, Ill. (62553) 222/D4
Oconee (co.), S.C. 296/A2
Oconomowoc, Wis. (53066) 317/H1
Oconomowoc (lake), Wis. 317/H1
Oconomowoc Lake, Wis. (†53066) 317/H1
Oconto, Nebr. (68860) 264/E3
Oconto (co.), Wis. 317/K6
Oconto, Wis. (54153) 317/L6
Oconto (riv.), Wis. 317/K5
Oconto Falls, Wis. (54154) 317/K6
Ocós, Guatemala 154/A3
Ocosingo, Mexico 150/O8
Ocosta, Wash. (†98520) 310/B4
Ocotal, Segovia, Nicaragua 154/D4
Ocotal, Zelaya, Nicaragua 154/E4
Ocotlán, Mexico 150/H6
Ocotlán de Morelos, Mexico 150/L8
Ocqueoc, Mich. (†49759) 250/F3
Ocracoke, N.C. (27960) 281/T4
Ocracoke (inlet), N.C. 281/T5
Ocracoke (isl.), N.C. 281/T4
Ocre, Ala. (†36274) 195/H4
Ocros, Peru 128/D8
Octa, Mo. (†63876) 261/M10
Octa, Ohio (†43160) 284/C6
Octagon, Ala. (†36748) 195/C6
Octavia, Nebr. (68650) 264/G3
Octavia, Okla. (74958) 288/S5
October Revolution (isl.), U.S.S.R. 54/K2
October Revolution (isl.), U.S.S.R. 4/B5
October Revolution (isl.), U.S.S.R. 48/L2
Ocú, Panama 154/G7
Ocumare de la Costa, Venezuela 124/E2
Ocumare del Tuy, Venezuela 124/E2
Ocuri, Bolivia 136/C6
Oda, Ghana 106/D7
Oda, Japan 81/F6
Oda, Jebel (mt.), Sudan 59/C5
Oda, Jebel (mt.), Sudan 111/G3
Odanah, Wis. (54861) 317/E2
Odate, Japan 81/K3
Odawara, Japan 81/J6
Odd, W. Va. (25902) 312/D7
Odda, Norway 18/E6
Odder, Denmark 21/D6
Oddur, Somalia 115/H3
Odebolt, Iowa (51458) 229/C4
Odell, Ill. (60460) 222/E2
Odell, Ind. (†47992) 227/C4
Odell, Nebr. (68415) 264/H4
Odell, Oreg. (97044) 291/F2
Odell (lake), Oreg. 291/E4
Odell, Texas (79247) 303/E3
Odell River, New Bruns. 170/C2
Odemira, Portugal 33/B4
Ödemiș, Turkey 63/C3
Oden, Ark. (71961) 202/C4
Odenburg, La. (†71369) 238/G5
Odendaalsrus, S. Africa 118/D5
Odense, Denmark 7/F3
Odense, Denmark 18/G9
Odense, Denmark 21/D7
Odense (fjord), Denmark 21/D7
Odenton, Md. (21113) 245/M4
Odenville, Ala. (35120) 195/F3
Odenwald (for.), W. Germany 22/C4
Oder (riv.) 7/F3
Oder (Odra) (riv.), Czech. 41/D2
Oder (riv.), E. Germany 22/F2
Oder (riv.), Poland 47/B2
Oder-Haff (mts.), E. Germany 22/F2
Oder-Haff (lag.), Poland 47/B2
Odessa, Del. (19730) 245/R3
Odessa, Fla. (33556) 212/D3
Odessa, Minn. (56276) 255/B5
Odessa, Mo. (64076) 261/E5
Odessa, Nebr. (68861) 264/E4
Odessa, N.Y. (14869) 276/G6
Odessa, Ontario 177/H3
Odessa, Sask. 181/H5
Odessa, Tex. 188/F4
Odessa, Texas 146/H6
Odessa, Texas (79760) 303/B6
Odessa, U.S.S.R. 2/L3
Odessa, U.S.S.R. 7/H4
Odessa, U.S.S.R. 48/D5
Odessa, U.S.S.R. 52/D5
Odessa, Wash. (99159) 310/G3
Odessadale, Georgia (†30222) 217/C5
Odgen, Utah 188/D2
Odgensburg, N.Y. 188/M2
Odiel (riv.), Spain 33/C4
Odienné, Ivory Coast 106/C7
Odin, Ill. (62870) 222/D5
Odin, Kansas (67562) 232/D3
Odin, Minn. (56160) 255/D7
Odiongan, Philippines 82/C4
Odivelas, Portugal 33/A1
Odobești, Romania 45/H3
Odon, Ind. (47562) 227/C7
Odongk, Cambodia 72/E5
O'Donnell, Texas (79351) 303/C5
O'Donnells, Newf. 166/D2
Odoorn, Netherlands 27/K3
Odorheiu Secuiesc, Romania 45/G2
Odra (Oder) (riv.), Poland 47/B2
Odry, Czech. 41/D2
Odum, Georgia (31555) 217/H7
Odweina, Somalia 115/J2
Oebisfelde, E. Germany 22/D2
Oeiras, Brazil 132/F4
Oeiras, Portugal 33/B3
Oelemari (riv.), Suriname 131/D4
Oella, Md. (†21228) 245/L3
Oelrichs, S. Dak. (57763) 298/C7
Oelsnitz, E. Germany 22/E3
Oelsnitz im Erzgebirge, E. Germany 22/E3
Oelwein, Iowa (50662) 229/K3
Oeno (isl.), Pitcairn Is. 87/O8
Oenpelli, North. Terr. 93/C2
Oensingen, Switzerland 39/E2
Of, Turkey 63/J2
Ofahoma, Miss. (39141) 256/E5
O'Fallon, Ill. (62269) 222/B2
O'Fallon, Mo. (63366) 261/L5
O'Fallon (creek), Mont. 262/L4
Ofanto (riv.), Italy 34/E4
Ofaqim, Israel 65/B5
Ofen (pass), Switzerland 39/K3
Ofenhorn (mt.), Switzerland 39/F4
Offa, Nigeria 106/E7
Offaly (co.), Ireland 17/F5
Offenbach am Main, W. Germany 22/C3
Offenburg, W. Germany 22/B4
Offerle, Kansas (67563) 232/C4
Offerman, Georgia (31556) 217/H8
Offutt, Ky. (41237) 237/R5
Offutt A.F.B., Nebr. 264/J3
Ofotfjorden (fjord), Norway 18/K2
Ofqui (isth.), Chile 138/D6
Oftringen, Switzerland 39/E2
Ofunato, Japan 81/K4
Oga, Japan 81/J4
Oga (pen.), Japan 81/J4
Ogaden (reg.), Ethiopia 102/G4
Ogaden (reg.), Ethiopia 111/H6
Ogaki, Japan 81/H6
Ogallah, Kansas (67656) 232/C3
Ogallala, Nebr. (69153) 264/C3
Ogasawara-gunto (Bonin) (isls.), Japan 81/M3
Ogbomosho, Nigeria 102/C4
Ogbomosho, Nigeria 106/E7
Ogden, Ark. (71853) 202/B6
Ogden, Ill. (61859) 222/F3
Ogden, Iowa (50212) 229/E4
Ogden, Kansas (66517) 232/F2
Ogden (bay), N.W. Terrs. 187/H3
Ogden, Utah 146/G6
Ogden, Utah (*84401) 304/C2
Ogden Dunes, Ind. (†46401) 227/C1
Ogdensburg, N.J. (07439) 273/D1
Ogdensburg, N.Y. (13669) 276/K1
Ogdensburg, Wis. (54962) 317/J7
Ogeechee (riv.), Georgia 217/J5
Ogema, Minn. (56569) 255/C3
Ogema, Sask. 181/G6
Ogema, Wis. (54459) 317/F5
Ogemaw, Ark. (†71764) 202/E7
Ogemaw (co.), Mich. 250/E4
Ogi, Japan 81/J5
Ogidaki (mt.), Ontario 175/D3
Ogidaki (mt.), Ontario 177/J5
Ogilvie, Manitoba 179/D4
Ogilvie, Minn. (56358) 255/E5
Ogilvie (mts.), Yukon 187/E3
Ogilvie (riv.), Yukon 187/E3
Oglala, S. Dak. (57764) 298/D7
Ogle (co.), Ill. 222/D1
Ogle, Ky. (40971) 237/O6
Oglesby, Ill. (61348) 222/D2
Oglesby, Texas (76561) 303/G6
Oglethorpe (co.), Georgia 217/F3
Oglethorpe, Georgia (31406) 217/D6
Oglio (riv.), Italy 34/C2
Ogmore, Queensland 88/J4
Ogmore and Garw, Wales 13/A6
Ogoja, Nigeria 106/F7
Ogoki (riv.), Ont. 162/H5
Ogoki (riv.), Ontario 175/C2
Ogooué (riv.), Congo 115/A4
Ogooué (riv.), Gabon 115/A4
Ogre, U.S.S.R. 53/C2
Ogulin, Yugoslavia 45/B3
Ogun (state), Nigeria 106/E7
Ogunquit, Maine (03907) 243/B9
Oğuzeli, Turkey 63/G4
Ohai, N. Zealand 100/A6
Ohakune, N. Zealand 100/E3
O'Hare Field-Chicago International Airport, Ill. 222/B5
Ohariu (stream), N. Zealand 100/B2
Ohata, Japan 81/K3
Ohatchee, Ala. (36271) 195/G3
Ohaton, Alberta 182/D3
Ohau (lake), N. Zealand 100/B6
Ohaupo, N. Zealand 100/E2
Ohey, Belgium 27/G8
O'Higgins (lake), Chile 138/D7
Ohio 188/K2
OHIO 284
Ohio (riv.) 188/J3
Ohio, Colo. (81237) 208/F5
Ohio, Ill. (61349) 222/D2
Ohio (co.), Ill. 222/E6
Ohio (co.), Ind. 227/H7
Ohio (riv.), Ind. 227/B9
Ohio (co.), Ky. 237/H6
Ohio (riv.), Ky. 237/F5
Ohio, Nova Scotia 168/F3
Ohio (riv.), Nova Scotia 168/D4
Ohio (riv.), Ohio 284/B8
Ohio (riv.), Pa. 294/A4
Ohio (state), U.S. 146/K5
Ohio (riv.), U.S. 146/K6
Ohio (riv.), U.S. 2/E4
Ohio (co.), W. Va. 312/E2
Ohio (riv.), W. Va. 312/B5
Ohio Brush (creek), Ohio 284/D8
Ohio City, Ohio (45874) 284/A4
Ohiopyle, Pa. (15470) 294/D6
Ohioville, Pa. (†15059) 294/B4
Ohiowa, Nebr. (68416) 264/G4
Ohley, W. Va. (25147) 312/D6
Ohlman, Ill. (62076) 222/D4
Ohoopee, Georgia (†30436) 217/H6
Ohopoho, Namibia 118/A3
Ohře (riv.), Czech. 41/B1
Ohrid (lake), Albania 45/E5
Ohrid, Yugoslavia 45/E5
Ohrid (lake), Yugoslavia 45/E5
Ohura, N. Zealand 100/E3
Oiapoque (Oyapock) (riv.), Brazil 132/C2
Oich, Loch (lake), Scotland 15/D3
Oich (riv.), Scotland 15/D3
Oies (isl.), Québec 172/G2
Oil (creek), Pa. 294/C2
Oil Center, N. Mex. (88266) 274/F6
Oil City, La. (71061) 238/C1
Oil City, Ontario 177/B5
Oil City, Pa. 188/L2
Oil City, Pa. (16301) 294/F8
Oildale, Calif. (93308) 204/F8
Oilmont, Mont. (59466) 262/E2
Oil Spring Ind. Res., N.Y. 276/D6
Oil Springs, Ky. (41238) 237/P5
Oil Springs, Ontario 177/B5
Oilton, Okla. (74052) 288/N2
Oilton, Texas (78371) 303/F10
Oil Trough, Ark. (72564) 202/G2
Oinói, Greece 45/F6
Oise (dept.), France 28/E3
Oise (riv.), France 28/E3
Oiseau (lake), Manitoba 179/G4
Oiseau (riv.), Manitoba 179/G4
Oisterwijk, Netherlands 27/G5
Oistins, Barbados 161/B9
Oistins (bay), Barbados 161/B9
Oita (pref.), Japan 81/E7
Oita, Japan 81/E7
Ojai, Calif. (93023) 204/F9
Ojibwa, Wis. (54862) 317/D4
Ojinaga, Mexico 150/G2
Ojiya, Japan 81/J5
Ojocaliente, Mexico 150/H5
Ojo Caliente, N. Mex. (87549) 274/D2
Ojo del Toro (mt.), Cuba 158/G4
Ojo Feliz, N. Mex. (87735) 274/E2
Ojo Sarco, N. Mex. (87550) 274/D2
Ojos del Salado (mt.) 120/C5
Ojos del Salado, Cerro (mt.), Argentina 143/C2
Ojos del Salado, Nevado (mt.), Chile 138/B6
Ojos Negros, Spain 33/F2
Ojus, Fla. (33163) 212/B4
Oka, Québec 172/C4
Oka (riv.), U.S.S.R. 7/J3
Oka (riv.), U.S.S.R. 48/L4
Oka (riv.), U.S.S.R. 52/F4
Okaba, Indonesia 85/K7
Okabena, Minn. (56161) 255/C7
Okahandja, Namibia 102/D7
Okahandja, Namibia 118/B4
Okahumpka, Fla. (32762) 212/D3
Okak (bay), Newf. 166/B2
Okak (isls.), Newf. 166/B2
Okaloacoochee Slough (swamp), Fla. 212/E5
Okaloosa (co.), Fla. 212/C6
Okamanpeedan (lake), Iowa 229/D2
Okanagan (lake), Br. Col. 162/D6
Okanagan (lake), Br. Col. 184/H5
Okanagan Centre, Br. Col. 184/H5
Okanagan Falls, Br. Col. 184/H5
Okanagan Landing, Br. Col. 184/H5
Okanagan Mission, Br. Col. 184/H5
Okanagan Mtn. Prov. Park, Br. Col. 184/G5
Okanogan (riv.), Br. Col. 184/H6
Okanogan (co.), Wash. 310/F2
Okanogan, Wash. (98840) 310/F2
Okanogan (riv.), Wash. 310/F2
Okarche, Okla. (73762) 288/L3
Okatibbee (creek), Miss. 256/G5
Okatibbee (lake), Miss. 256/G5
Okato, N. Zealand 100/D3
Okaton, S. Dak. (57562) 298/H6
Okauchee, Wis. (53069) 317/J1
Okauchee (lake), Wis. 317/J1
Okaukuejo, Namibia 118/B3
Okawa, Japan 81/E7
Okawville, Ill. (62271) 222/D5
Okay, Okla. (74446) 288/R3
Okaya, Japan 81/H5
Okayama (pref.), Japan 81/F6
Okayama, Japan 81/F6
Okazaki, Japan 81/H6
O'Kean, Ark. (72449) 202/J1
Okeana, Ohio (45053) 284/A7
Okee, Wis. (†53555) 317/H9
Okeechobee (lake), Fla. 188/K5
Okeechobee (co.), Fla. 212/F4
Okeechobee, Fla. (33472) 212/F4
Okeechobee (lake), Fla. 212/F5
O'Keeffe Nat'l Hist. Site, N. Mex. 274/C2
Okeene, Okla. (73763) 288/K2
Okefenokee (swamp), Fla. 212/D1
Okefenokee (swamp), Georgia 217/H9
Okehampton, England 13/D7
Okehampton, England 10/D5
Okemah, Okla. (74859) 288/O4
Okemo (Ludlow) (mt.), Vt. 268/B5
Okemos, Mich. (48864) 250/E6
Okene, Nigeria 106/F7
Oker (riv.), W. Germany 22/D2
Okesa, Okla. (†74003) 288/O1
Oketo, Kansas (66518) 232/F2
Okfuskee (co.), Okla. 288/O3
Okha, U.S.S.R. 54/R4
Okha, U.S.S.R. 48/P4
Okha Port, India 68/B4
Okhotsk (sea) 2/S3
Okhotsk (sea), Japan 81/M1
Okhotsk, U.S.S.R. 54/R4
Okhotsk (sea), U.S.S.R. 54/R4
Okhotsk, U.S.S.R. 48/P4
Okhotsk (sea), U.S.S.R. 48/P4
Oki (isls.), Japan 81/F5
Okiep, S. Africa 118/B5
Okinawa (pref.), Japan 81/N6
Okinawa (isls.), Japan 54/O7
Okinawa (isl.), Japan 81/N6
Okinawa (isls.), Japan 81/N6
Okinoerabu (isl.), Japan 81/N5
Okkan, Burma 72/B3
Okla, Sask. 181/H3
Oklahoma 188/G3
Oklahoma (co.), Okla. 288/M3
OKLAHOMA 288
Oklahoma (state), U.S. 146/J6
Oklahoma City (cap.), Okla. 146/J6
Oklahoma City (cap.), Okla. 188/G3
Oklahoma City (cap.), Okla. (*73101) 288/L4
Oklaunion, Texas (76373) 303/F3
Oklawaha, Fla. (32679) 212/E2
Oklawaha (riv.), Fla. 212/E2
Oklee, Minn. (56742) 255/C3
Okmulgee, Okla. 188/G3
Okmulgee (co.), Okla. 288/P3
Okmulgee, Okla. (74447) 288/O3
Okoboji, Iowa (51355) 229/C2
Okobojo (creek), S. Dak. 298/J4
Okolona, Ark. (71962) 202/D5
Okolona, Ky. (40219) 237/K4
Okolona, Miss. (38860) 256/G2
Okolona, Ohio (43550) 284/B3
Okondja, Gabon 115/B4
Okotoks, Alberta 182/C4
Okovango (riv.) 102/D6
Okovango (riv.), Botswana 118/C3
Okovango (swamps), Botswana 118/C3
Okovango (riv.), Namibia 118/C3
Okoyo, Congo 115/C4
Okpo, Burma 72/C3
Okreek, S. Dak. (57563) 298/J7
Oksino, U.S.S.R. 52/H1
Oktaha, Okla. (74450) 288/R3
Oktibbeha (co.), Miss. 256/G4
Oktyabr'sk, U.S.S.R. 52/G4
Oktyabr'skiy, U.S.S.R. 52/H4
Okulovka, U.S.S.R. 52/D3
Okushiri (isl.), Japan 81/J2
Ola, Ark. (72853) 202/D3
Ola, Georgia (†30253) 217/E4
Ola, Idaho (83657) 220/B5
Olá, Panama 154/G6
Ólafsfjördhur, Iceland 21/C1
Ola Grande (pt.), P. Rico 161/D3
Olalla, Br. Col. 184/H5
Olalla, Wash. (98359) 310/A2
Olamon, Maine (04467) 243/F5
Olancha, Calif. (93549) 204/H7
Olanchito, Honduras 154/D3
Öland (isl.), Sweden 7/F3
Öland (isl.), Sweden 18/K8
Olanta, Pa. (16863) 294/F4
Olanta, S.C. (29114) 296/H4
Olar, S.C. (29843) 296/E5
Olary, S. Australia 94/G5
Olathe, Colo. (81425) 208/D5
Olathe, Kansas (66061) 232/H3
Olathe Nav. Air Sta., Kansas 232/H3
Olavarría, Argentina 143/D4
Olavarría, Argentina 120/C6
Oława, Poland 47/C3
Olberg, Ariz. (†85247) 198/D5
Olbernhau, E. Germany 22/E3
Olbia, Italy 34/B4
Olcott, N.Y. (14126) 276/C4
Old (riv.), Calif. 204/L1
Old (riv.), La. 238/G5
Old (stream), Maine 243/H6
Oldany (isl.), Scotland 15/C2
Old Appleton, Mo. (63770) 261/N7
Old Bahama (chan.), Bahamas 156/B2
Old Bahama (chan.), Cuba 158/G1
Old Bahama (chan.), Cuba 156/B2
Old Bar, N.S. Wales 97/G2
Old Barkerville, Br. Col. 184/G3
Old Bennington, Vt. (†05201) 268/A6
Old Bonaventure, Newf. 166/D2
Old Bridge, N.J. (08857) 273/E3
Old Castile (reg.), Spain 33/D2
Oldcastle, Ireland 10/C4
Oldcastle, Ireland 17/G4
Old Crow, Yukon 187/E3
Oldemarkt, Netherlands 27/J3
Olden, Mo. (†65789) 261/J9
Olden, Norway 18/E6
Olden, Texas (76466) 303/F5
Oldenburg, Ind. (47036) 227/G6
Oldenburg, Miss. (†39661) 256/C7
Oldenburg, W. Germany 22/C2
Oldenburg in Holstein, W. Germany 22/D1
Old England, Jamaica 158/H6
Old Entrance, Alberta 182/B3
Oldenzaal, Netherlands 27/K4
Old Faithful, Wyo. (82190) 319/B1
Oldfield, La. (†70754) 238/L1
Old Fields, W. Va. (26845) 312/J4
Old Forge, N.Y. (13420) 276/L3
Old Forge, Pa. (18518) 294/F7
Old Fort, N.C. (28762) 281/E3
Old Fort, Ohio (44861) 284/D3
Oldfort, Tenn. (37362) 237/M10
Old Glory, Texas (79540) 303/D4
Old Greenwich, Conn. (06870) 210/A4
Oldham, England 13/H2
Oldham, England 10/G2
Oldham (co.), Ky. 237/L4
Oldham, S. Dak. (57051) 298/P5
Oldham (co.), Texas 303/B2
Old Harbor, Alaska (99643) 196/H3
Old Harbour, Jamaica 158/J6
Old Harbour (bay), Jamaica 158/J6
Old Harbour Bay, Jamaica 158/J6
Old Hickory (dam), Tenn. 237/H8
Old Hickory (lake), Tenn. 237/J8
Old Kilpatrick, Scotland 15/B2
Old Landing, Ky. (41358) 237/O5
Old Leighlin, Ireland 17/G6
Old Lodge (creek), S. Dak. 298/K6
Old Lyme○, Conn. (06371) 210/F3
Old Main Centre, Sask. 181/D5
Oldman (riv.), Alberta 182/D5
Oldman (riv.), Sask. 181/L2
Oldmans (creek), N.J. 273/C4
Old Marsh Bed, North. Terr. 93/B6
Oldmeldrum, Scotland 15/F3
Oldmeldrum, Scotland 10/E2
Old Mill Creek, Ill. (†60083) 222/B4
Old Mission, Mich. (49673) 250/D4
Old Monroe, Mo. (63369) 261/L5
Old Mystic, Conn. (06372) 210/H3
Old Orchard Beach, Maine (04064) 243/C9
Old Orchard Beach○, Maine (04064) 243/C9
Old Perlican, Newf. 166/D2
Old Rhine (riv.), Netherlands 27/E4
Old Rhodes (key), Fla. 212/F6
Old Ripley, Ill. (†62275) 222/D5
Old Road, Ant. & Bar. 161/D11
Old Road Town, St. Chris.-Nevis 161/C10
Olds, Alberta 182/D4
Olds, Iowa (52647) 229/K6
Old Saybrook, Conn. (06475) 210/F3
Old Saybrook○, Conn. (06475) 210/F3
Old Shawneetown, Ill. (†62984) 222/E6
Oldsmar, Fla. (33557) 212/B2
Old Spring Hill, Ala. (†36742) 195/C6
Old Sturbridge Village, Mass. (†01566) 249/F4
Old Tampa (bay), Fla. 212/B3
Old Tappan, N.J. (07675) 273/C1
Old Town, Fla. (32680) 212/D2
Oldtown, Idaho (†99156) 220/A1
Oldtown, Ky. (41163) 237/R4
Old Town, Maine (04468) 243/F6
Oldtown, Md. (21555) 245/D2
Old Trap, N.C. (†27974) 281/T2
Olduvai Gorge (canyon), Tanzania 115/G4

Old Washington, Ohio (43768) 284/H5
Oldwick, N.J. (08858) 273/D2
Old Wives, Sask. 181/E5
Old Wives (lake), Sask. 181/E5
Old Woman (creek), Wyo. 319/H3
Olean, Mo. (65064) 261/G6
Olean, N.Y. (14760) 276/D6
O'Leary (peak), Ariz. 198/D5
O'Leary, Pr. Edward I. 168/D2
Olecko, Poland 47/F1
Oleiros, Portugal 33/B3
Olëkma (riv.), U.S.S.R. 48/N4
Olëkminsk, U.S.S.R. 48/N3
Olema, Calif. (94950) 204/H1
Olenegorsk, U.S.S.R. 52/D1
Olenek, U.S.S.R. 4/C4
Olenëk (riv.), U.S.S.R. 54/N3
Olënek, U.S.S.R. 48/M3
Olënek (bay), U.S.S.R. 48/N2
Olënek (riv.), U.S.S.R. 48/M3
Olentangy (riv.), Ohio 284/D4
Oléron (isl.), France 28/C5
Oleśnica, Poland 47/C3
Olesno, Poland 47/D3
Oleta, Okla. (74751) 288/R6
Olex, Oreg. (†97812) 291/G2
Oley, Pa. (19547) 294/L5
Olga, N. Dak. (†58221) 282/O2
Olga (mt.), North. Terr. 93/B8
Olga, Wash. (98279) 310/C2
Ölgiy (Ulegei), Mongolia 77/C2
Ølgod, Denmark 21/B6
Olha, Manitoba 179/B4
Olhão, Portugal 33/C4
Oliena, Italy 34/B4
Olifants (riv.), Mozambique 118/D4
Olifants (riv.), S. Africa 118/D4
Olimar, Uruguay 145/E3
Olimar Grande (riv.), Uruguay 145/E4
Olímpia, Brazil 135/B2
Olin, Iowa (52320) 229/L5
Olin, Ky. (†40447) 237/N6
Olin, N.C. (28660) 281/H3
Olinda, Brazil 120/F3
Olinda, Brazil 132/H4
Olinda, Calif. (96007) 204/C3
Olinda, Victoria 97/K5
Oliva, Argentina 143/D3
Oliva, Spain 33/F3
Oliva de la Frontera, Spain 33/C3
Olivais, Portugal 33/A1
Olivar Alto, Chile 138/G5
Olivares, Cerro de (mt.), Argentina 143/B3
Olivares, Cerro de (mt.), Chile 138/B8
Olive, Mont. (59343) 262/L5
Olive, Okla. (†74030) 288/O2
Olive Branch, Ill. (62969) 222/D6
Olive Branch, Miss. (38654) 256/E1
Olive Branch, Ohio (†45103) 284/D10
Olive Hill, Ky. (41164) 237/P4
Olivehill, Tenn. (38475) 237/E10
Oliveira, Brazil 135/D2
Olivenza, Spain 33/C3
Oliver (dam), Ala. 195/J5
Oliver, Br. Col. 184/H5
Oliver, Georgia (30449) 217/J5
Oliver (dam), Georgia 217/B6
Oliver (lake), Georgia 217/B5
Oliver, Ind. (†47620) 227/B8
Oliver (co.), N. Dak. 282/H5
Oliver, Pa. (15472) 294/C6
Oliver, Wis. (†54880) 317/B2
Oliver Springs, Tenn. (37840) 237/N8
Olivet, Ill. (†61846) 222/F4
Olivet, Kansas (†66856) 232/G3
Olivet, Md. (†20657) 245/N7
Olivet, Mich. (49076) 250/E6
Olivet, S. Dak. (57052) 298/O7
Olivet, Wis. (†54767) 317/B6
Olivette, Mo. (63124) 261/O2
Olivia, Minn. (56277) 255/C6
Olivia, N.C. (28368) 281/L4
Olivier, La. (†70560) 238/G7
Olivone, Switzerland 39/G3
Olkusz, Poland 47/D3
Olla, La. (71465) 238/F3
Ollachea, Peru 128/G9
Ollagüe (vol.), Bolivia 136/B7
Ollagüe, Chile 120/C5
Ollagüe, Chile 138/B3
Ollantaytambo, Peru 128/F9
Ollie, Iowa (52576) 229/J6
Ollon, Switzerland 39/D4
Olmedo, Spain 33/D2
Olmitz, Kansas (67564) 232/D3
Olmos, Peru 128/C5
Olmos Park, Texas (78212) 303/F8
Olmstead, Ky. (42265) 237/H7
Olmsted, Ill. (62970) 222/D6
Olmsted (co.), Minn. 255/F7
Olmsted Falls, Ohio (44138) 284/G9
Olmstedville, N.Y. (12857) 276/N3
Olmué, Chile 138/F2
Olney, Ill. (62450) 222/E5
Olney, Md. (20832) 245/K4
Olney, Mo. (63370) 261/K4
Olney, Mont. (59927) 262/B2
Olney, Okla. (†74538) 288/O6
Olney, Oreg. (†97103) 291/D1
Olney, Texas (76374) 303/F4
Olney Springs, Colo. (81062) 208/M6
Olofström, Sweden 18/J8
Oloh, Miss. (†39482) 256/E8
Olomouc, Czech. 7/F4
Olomouc, Czech. 41/D2
Olonets, U.S.S.R. 52/D2
Olongapo, Philippines 82/C3
Oloron-Sainte-Marie, France 28/C6
Olot, Spain 33/H1
Olowalu, Hawaii (†96761) 218/H2
Oloy (range), U.S.S.R. 48/R3
Olpe, Kansas (66865) 232/F3
Olsa (riv.), Austria 41/C3
Olsburg, Kansas (66520) 232/F2
Olst, Netherlands 27/J4
Olsztyn (prov.), Poland 47/E2
Olsztyn, Poland 7/G3
Olsztyn, Poland 47/E2
Olsztynek, Poland 47/E2
Olt (riv.), Romania 7/G4
Olt (riv.), Romania 45/G3
Olta, Argentina 143/C3
Olten, Switzerland 39/E2
Oltenița, Romania 45/H3
Olton, Texas (79064) 303/B3
Oltu, Turkey 63/J2
Olur, Turkey 63/K2
Olustee, Fla. (32072) 212/D1
Olustee (riv.), Fla. 212/D1
Olustee, Okla. (73560) 288/H5
Olutanga (isl.), Philippines 82/D7
Olutanga (isl.), Philippines 85/G4
Olvera, Spain 33/D4
Olvey, Ark. (†72601) 202/E1
Olwampi (cape), China 77/K7
Olympia (isls.), Greece 45/E7
Olympia, Ky. (40358) 237/O4
Olympia (cap.), Wash. 146/E5
Olympia (cap.), Wash. 188/B1
Olympia (cap.), Wash. (*98501) 310/C3
Olympia Fields, Ill. (60461) 222/B6
Olympian Village, Mo. (†63050) 261/M6
Olympic (mts.), Wash. 310/B3
Olympic Nat'l Park, Wash. 188/A1
Olympic Nat'l Park, Wash. 310/B3
Olympic Valley, Calif. (95730) 204/E4
Olympus (mt.), Greece 45/F5
Olympus (mt.), Wash. 310/B3
Olyphant, Ark. (72020) 202/H3
Olyphant, Pa. (18447) 294/F7
Olyphic, N.C. (†28463) 281/M7
Olyutorskiy (cape), U.S.S.R. 54/U4
Olyutorskiy (cape), U.S.S.R. 48/S4
Oma (cape), Japan 81/K3
Oma, Miss. (†39654) 256/D7
Omagari, Japan 81/K4
Omagh (dist.), N. Ireland 17/G2
Omagh, N. Ireland 10/C3
Omagh, N. Ireland 17/G2
Omaguas, Peru 128/F5
Omaha, Ala. (†36274) 195/H4
Omaha, Ark. (72662) 202/D1
Omaha (beach), France 28/C3
Omaha, Georgia (31821) 217/C6
Omaha, Ill. (62871) 222/E6
Omaha, Nebr. 188/G2
Omaha, Nebr. 146/J5
Omaha, Nebr. (*68101) 264/J3
Omaha Ind. Res., Nebr. 264/H2
Omak, Wash. (98841) 310/F2
Omak (lake), Wash. 310/F2
Oman 2/M5
Oman 54/G8
Oman (gulf) 54/G7
Oman (gulf), Iran 59/G5
Oman (gulf), Iran 66/M8
OMAN 59/G6
Oman (gulf), Oman 59/G5
Oman (reg.), Oman 59/G5
Oman (gulf), U.A.E. 59/G5
Omar, W. Va. (25638) 312/C7
Omaruru, Namibia 118/B4
Omas, Peru 128/D9
Omatako (riv.), Namibia 118/B3
Omate, Peru 128/G11
Ombai (str.), Indonesia 85/H7
Omboué, Gabon 115/A4
Ombrone (riv.), Italy 34/C3
Ombúes de Lavalle, Uruguay 145/B4
Ombúes de Oribe, Uruguay 145/C4
Omdurman, Sudan 102/F3
Omdurman, Sudan 59/B6
Omdurman, Sudan 111/F4
Omega, Georgia (31775) 217/E8
Omega, Ind. (†46030) 227/F4
Omega, Ohio (†45690) 284/E7
Omega, Okla. (73764) 288/K3
Omemee, N. Dak. (†58739) 282/K2
Omemee, Ontario 177/F3
Omena, Mich. (49674) 250/D3
Omeo, Victoria 97/D5
'Omer, Israel 65/B5
Omer, Mich. (48749) 250/F4
Ömerli, Turkey 63/J4
Omerville, Québec 172/E4
Ometepe (isl.), Nicaragua 154/E5
Ometepec, Mexico 150/K8
Omey (isl.), Ireland 17/A5
Omineca (mts.), Br. Col. 184/E2
Omineca (riv.), Br. Col. 184/E2
Omiš, Yugoslavia 45/C4
Omiya, Japan 81/O2
Ommaney (cape), Alaska 196/M2
Ommanney (bay), N.W. Terrs. 187/H2
Omme (riv.), Denmark 21/B6
Ommen, Netherlands 27/J3
Ömnögovĭ, Mongolia 77/F3
Omø (isl.), Denmark 21/E7
Omo (riv.), Ethiopia 111/G6
Omoa, Honduras 154/C3
Omolon (riv.), U.S.S.R. 54/S3
Omolon (riv.), U.S.S.R. 4/C1
Omolon (riv.), U.S.S.R. 48/Q3
Omoloy (riv.), U.S.S.R. 48/O3
Omono (riv.), Japan 81/J4
Ompah, Ontario 177/H2
Omps, W. Va. (†25411) 312/K4
Omro, Wis. (54963) 317/J7
Omsk, U.S.S.R. 54/J4
Omsk, U.S.S.R. 2/N3
Omsk, U.S.S.R. 48/H4
Omsukchan, U.S.S.R. 48/Q3
Omu, Japan 81/L1
Omura, Bonin Is., Japan 81/M3
Omura, Nagasaki, Japan 81/E7
Omurtag, Bulgaria 45/H4
Omuta, Japan 81/E7
Omutninsk, U.S.S.R. 48/F4
Omutninsk, U.S.S.R. 52/H3
Ona, Fla. (33865) 212/E4
Ona, W. Va. (25545) 312/B6
Onaga, Kansas (66521) 232/F2
Onagawa, Japan 81/K4
Onaka, S. Dak. (57466) 298/L3
Onalaska, Texas (77360) 303/J7
Onalaska, Wash. (98570) 310/C4
Onalaska, Wis. (54650) 317/D8
Onaman (lake), Ontario 177/H4
Onamia, Minn. (56359) 255/E4
Onancock, Va. (23417) 307/S5
Onangué (lake), Gabon 115/A4
Onanole, Manitoba 179/C4
Onaping Falls, Ontario 177/J5
Onaping Falls, Ontario 175/D3
Onaqui, Utah (†84080) 304/B3
Onarga, Ill. (60955) 222/F3
Onawa, Iowa (51040) 229/A4
Onawa, Maine (†04464) 243/E5
Onawa (lake), Maine 243/E5
Onaway, Idaho (†83855) 220/B3
Onaway, Mich. (49765) 250/E3
Onchan, I. of Man 13/C3
Onchiota, N.Y. (12968) 276/M2
Oncócua, Angola 115/B7
Onda, Spain 33/F3
Ondangua, Namibia 118/B3
Ondava (riv.), Czech. 41/F2
Ondo (state), Nigeria 106/F7
Ondo, Nigeria 106/F7
Öndörhaan (Undur Khan), Mongolia 77/G2
Öndörhaan, Mongolia 54/N5
Öndverdharnes (mt.), Iceland 21/A1
O'Neals, Calif. (93645) 204/F6
Oneco, Conn. (06373) 210/H2
Oneco, Fla. (33558) 212/D4
Onefour, Alberta 182/E5
Onega, U.S.S.R. 7/H2
Onega (lake), U.S.S.R. 7/H2
Onega (riv.), U.S.S.R. 7/H2
Onega, U.S.S.R. 52/E2
Onega, U.S.S.R. 48/D3
Onega (bay), U.S.S.R. 52/E2
Onega (lake), U.S.S.R. 48/D3
Onega (lake), U.S.S.R. 52/E2
Onega (riv.), U.S.S.R. 48/D3
Onega (riv.), U.S.S.R. 52/E2
Onego, W. Va. (26886) 312/H5
One Hundred and Fifty Mile House, Br. Col. 184/G4
One Hundred Mile House, Br. Col. 184/G4
Onehunga, N. Zealand 100/B1
Oneida, Ark. (72369) 202/J5
Oneida (co.), Idaho 220/F7
Oneida, Ill. (61467) 222/C2
Oneida, Iowa (†52057) 229/L3
Oneida, Kansas (66522) 232/G2
Oneida, Ky. (40972) 237/O6
Oneida (co.), N.Y. 276/J4
Oneida, N.Y. (13421) 276/J4
Oneida (lake), N.Y. 276/J4
Oneida, Pa. (18242) 294/K4
Oneida, Tenn. (37841) 237/N7
Oneida (co.), Wis. 317/G4
Oneida, Wis. (54155) 317/K7
O'Neill, Nebr. (68763) 264/F2
Onekama, Mich. (49675) 250/C4
Oneonta, Ala. (35121) 195/E3
Oneonta, N.Y. (13820) 276/K6
One Tree Hill, N. Zealand 100/B1
Ong, Nebr. (68452) 264/G4
Ongjin, N. Korea 81/B5
Ongniud, China 77/J3
Ongole, India 68/E5
Ongwediva, Namibia 118/B3
Onhaye, Belgium 27/F8
Oni, U.S.S.R. 52/F6
Onida, S. Dak. (57564) 298/K4
Onilahy (riv.), Madagascar 118/G4
Onima, Neth. Ant. 161/E8
Onion Lake, Sask. 181/B2
Onitsha, Nigeria 106/F7
Onitsha, Nigeria 102/C4
Onkaparinga (riv.), S. Australia 88/D8
Onkaparinga (riv.), S. Australia 94/B8
Onkivesi (lake), Finland 18/P5
Onley, Va. (23418) 307/S5
Only, Tenn. (37140) 237/F9
Ono, Calif. (†96001) 204/C3
Ono, Japan 81/H6
Ono (riv.), Japan 81/E7
Ono, Pa. (17077) 294/J5
Onoda, Japan 81/E6
Onomea, Hawaii (†96781) 218/J4
Onomichi, Japan 81/F6
Onon, Mongolia 77/H2
Onondaga, Mich. (49264) 250/E6
Onondaga (co.), N.Y. 276/H5
Onondaga Ind. Res., N.Y. 276/H5
Onota (lake), Mass. 249/A3
Onoto, Venezuela 124/F3
Onotoa (atoll), Kiribati 87/H6
Onoway, Alberta 182/C3
Onrusrivier, S. Africa 118/G7
Onset, Mass. (02558) 249/M6
Onslow, Australia 87/B8
Onslow, Iowa (52321) 229/M4
Onslow (co.), N.C. 281/P5
Onslow (bay), N.C. 281/P6
Onslow, W. Australia 88/B4
Onslow, W. Australia 92/A3
Onsŏng, N. Korea 81/E2
Onsted, Mich. (49265) 250/E6
Onstwedde, Netherlands 27/K2
Ontake (mt.), Japan 81/H6
Ontario (prov.) 162/H5
Ontario (lake) 146/L5
Ontario (lake) 162/J7
Ontario, Calif. (*91761) 204/D10
Ontario (prov.), Canada 146/K4
Ontario, Ind. (†46746) 227/G1
Ontario, Iowa (†50010) 229/F4
Ontario (lake), N.Y. 188/L2
Ontario (co.), N.Y. 276/F5
Ontario, N.Y. (14519) 276/F4
Ontario (lake), N.Y. 276/F3
Ontario, Ohio (44862) 284/E4
ONTARIO 177
Ontario (lake), Ontario 177/G4
Ontario, Oreg. (97914) 291/K3
Ontario, Wis. (54651) 317/E8
Onteniente, Spain 33/F3
Onton, Ky. (†42455) 237/G5
Ontonagon (co.), Mich. 250/F1
Ontonagon, Mich. (49953) 250/F1
Ontonagon (riv.), Mich. 250/G1
Ontonagon Ind. Res., Mich. 250/F1
Ontong Java (isl.), Solomon Is. 87/G6
Ontong Java (isls.), Solomon Is. 86/D2
Onverwacht, Suriname 131/D3
Onward, Ind. (46967) 227/E3
Onward, Miss. (†39159) 256/C5
Onycha, Ala. (†36467) 195/F8
Onyx, Ark. (†72860) 202/D4
Onyx, Calif. (93255) 204/G8
Oobagooma, W. Australia 92/D2
Oodnadatta, Australia 87/D8
Oodnadatta, S. Australia 88/E5
Oodnadatta, S. Australia 94/D2
Ookala, Hawaii (96774) 218/J4
Oola, Ireland 17/E6
Ooldea, S. Australia 94/B4
Oolitic, Ind. (47451) 227/E7
Oologah, Okla. (74053) 288/P2
Oologah (lake), Okla. 288/P1
Ooltewah, Tenn. (37363) 237/M10
Oona River, Br. Col. 184/C3
Oosterwolde, Netherlands 27/J2
Oostanaula, Georgia (†30701) 217/B1
Oostanaula (riv.), Georgia 217/B2
Oostburg, Netherlands 27/C6
Oostburg, Wis. (53070) 317/L8
Oostende (Ostend), Belgium 27/B6
Oosterend, Netherlands 27/G2
Oosterhout, Netherlands 27/F5
Oostkamp, Belgium 27/C6
Oostmahorn, Netherlands 27/J2
Oost-Vlieland, Netherlands 27/F2
Oostzaan, Netherlands 27/C4
Oostzaan Polder, Netherlands 27/B4
Ootacamund, India 68/D6
Ootmarsum, Netherlands 27/K4
Ootsa (lake), Br. Col. 184/D3
Ootsa Lake, Br. Col. 184/E3
Oozewekwun, Manitoba 179/B5
Opal, Alberta 182/D3
Opal, S. Dak. (57765) 298/D4
Opal, Wyo. (83124) 319/B4
Opala, Zaire 115/D4
Opal Cliffs, Calif. (†95060) 204/K4
Opa Locka, Fla. (33054) 212/B4
Opalton, Queensland 95/B4
Opari, Sudan 111/F7
Oparino, U.S.S.R. 52/G3
Opasatika, Ontario 177/J5
Opasatika, Ontario 175/D3
Opatija, Yugoslavia 45/A3
Opatów, Poland 47/E3
Opava, Czech. 41/E2
Opazatika (riv.), Ontario 175/D3
Opdyke, Ill. (62872) 222/E5
Opelika, Ala. (36801) 195/H5
Opelousas, La. (70570) 238/G5
Opeongo (lake), Ontario 177/F2
Opfikon, Switzerland 39/G2
Opheim, Ill. (61468) 222/C2
Opheim, Mont. (59250) 262/K2
Ophir, Alaska 196/G2
Ophir, Colo. (81426) 208/D7
Ophir, Oreg. (97464) 291/C5
Ophir, Utah (†84074) 304/B3
Opihikao, Hawaii (†96778) 218/K6
Opinaca (riv.), Que. 162/J5
Opinaca (riv.), Québec 174/B2
Opine, Ala. (†36784) 195/C7
Opinnagau (riv.), Ontario 175/D2
Opiscotéo (lake), Québec 174/D2
Opladen, W. Germany 22/B3
Opochka, U.S.S.R. 52/C3
Opoco, Bolivia 136/B6
Opoczno, Poland 47/E3
Opole, Minn. (†56340) 255/D5
Opole (prov.), Poland 47/C3
Opole, Poland 47/C3
Opolis, Kansas (66760) 232/H4
Oporto (Porto) (dist.), Portugal 33/B2
Oporto (Porto), Portugal 33/B2
Opotiki, N. Zealand 100/F3
Opp, Ala. (36467) 195/F8
Oppdal, Norway 18/F5
Oppeln, Poland 47/C3
Oppelo, Ark. (†72110) 202/E3
Oppenheim, W. Germany 22/C4
Oppland (co.), Norway 18/F6
Opportunity, Wash. (99214) 310/H3
Oppy, Ky. (†25685) 237/S5
Optima, Okla. (73948) 288/D1
Optima (lake), Okla. 288/D1
Opua, N. Zealand 100/D1
Opunake, N. Zealand 100/D3
Opuntia (lake), Sask. 181/C4
Opwijk, Belgium 27/F7
'Oqair, Saudi Arabia 59/E4
Oquawka, Ill. (61469) 222/C3
Oquossoc, Maine (04964) 243/B6
Ora, Ind. (46968) 227/D2
Ora, Miss. (†39428) 256/E7
Ora, S.C. (29371) 296/D2
Oracabessa, Jamaica 158/J5
Oracle, Ariz. (85623) 198/E6
Oradea, Romania 7/G4
Oradea, Romania 45/E2
Oradell, N.J. (07649) 273/B1
Oradell (res.), N.J. 273/B1
Orai, India 68/D3
Oraibi, Ariz. (86039) 198/E3
Oraibi Wash (dry riv.), Ariz. 198/E3
Oral, S. Dak. (57766) 298/C7
Oran, Algeria 106/D1
Oran, Algeria 102/B1
Oran, Iowa (50664) 229/J3
Oran, Mo. (63771) 261/N8
Orange (riv.) 2/K7
Orange (riv.) 102/D7
Orange, Australia 87/E9
Orange (riv.), Botswana 118/B5
Orange (cape), Brazil 132/D1
Orange (co.), Calif. 204/H10
Orange, Calif. (*92666) 204/D11
Orange○, Conn. (06477) 210/C3
Orange (co.), Fla. 212/E3
Orange, Fla. (†32321) 212/B1
Orange (lake), Fla. 212/D2
Orange, France 28/F5
Orange, Georgia (†30114) 217/D2
Orange (co.), Ind. 227/E7
Orange, Ind. (†47343) 227/G5
Orange, Mass. (01364) 249/E2
Orange○, Mass. (01364) 249/E2
Orange (canal), Netherlands 27/K3
Orange○, N.H. (†03741) 268/D4
Orange, N.J. (*07050) 273/B2
Orange, N. S. Wales 88/H6
Orange, N.S. Wales 97/E3
Orange (co.), N.Y. 276/M8
Orange (co.), N.C. 281/L2
Orange, Ohio (†44101) 284/J9
Orange (riv.), S. Africa 118/B5
Orange (mts.), Suriname 131/D4
Orange (co.), Texas 303/L7
Orange, Texas (77630) 303/L7
Orange (butte), Utah 304/D5
Orange (co.), Vt. 268/C3
Orange○, Vt. (†05649) 268/C3
Orange (co.), Va. 307/M4
Orange, Va. (22960) 307/M4
Orange Beach, Ala. (36561) 195/C10
Orangeburg, N.Y. (10962) 276/K8
Orangeburg (co.), S.C. 296/F5
Orangeburg, S.C. (29115) 296/F4
Orange City, Fla. (32763) 212/E3
Orange City, Iowa (51041) 229/A2
Orange Cove, Calif. (93646) 204/F7
Orangedale, Nova Scotia 168/G3
Orange Free State (prov.), S. Africa 102/E7
Orange Free State (prov.), S. Africa 118/D5
Orange Grove, Miss. (†39501) 256/H10
Orange Grove, Texas (78372) 303/F10
Orange Hill, St. Vin. & Grens. 161/A8
Orange Lake, Fla. (32681) 212/D2
Orange Park, Fla. (32073) 212/E1
Orange Springs, Fla. (32682) 212/E2
Orangeville, Ill. (61060) 222/D1
Orangeville, Ind. (†47452) 227/D7
Orangeville, Mich. (†49344) 250/D6
Orangeville, Ohio (44453) 284/J3
Orangeville, Ontario 177/D4
Orangeville, Pa. (17859) 294/K3
Orangeville, Utah (84537) 304/C4
Orange Walk Town, Belize 154/C1
Oranienburg, E. Germany 22/E2
Oranjemund, Namibia 102/D7
Oranjemund, Namibia 118/B5
Oranjestad (cap.), Aruba, Neth. Ant. 161/D10
Oranjestad, Neth. Ant. 156/D4
Oranmore, Ireland 17/D5
Orapa, Botswana 118/D4
Oras, Philippines 82/E4
Orăștie, Romania 45/F3
Orava (res.), Czech. 41/E2
Orava (riv.), Czech. 41/E2
Orava (res.), Poland 47/D4
Oraville, Ill. (62971) 222/D6
Oravița, Romania 45/E3
Orb (riv.), France 28/E6
Orbe, Switzerland 39/C3
Orbe (riv.), Switzerland 39/C3
Orbetello, Italy 34/C3
Órbigo (riv.), Spain 33/D1
Orbisonia, Pa. (17243) 294/G5
Orbost, Victoria 88/H7
Orbost, Victoria 97/E5
Örbyhus, Sweden 18/K6
Orca, Alaska (†99574) 196/J2
Orcadia, Sask. 181/J4
Orcas (isl.), Wash. 310/C2
Orcera, Spain 33/E3
Orchard, Colo. (80649) 208/L2
Orchard, Iowa (50460) 229/H2
Orchard, Nebr. (68764) 264/F2
Orchard Beach, Md. (†21122) 245/M4
Orchard Hill, Georgia (30266) 217/D4
Orchard Lake, Mich. (48033) 250/F6
Orchard Mesa, Colo. (†81501) 208/C4
Orchard Park, N.Y. (14127) 276/C5
Orchards, Wash. (98662) 310/C5
Orchard Valley, Wyo. (†82001) 319/H4
Orchid, Fla. (†32970) 212/F4
Orchid, Va. (†23117) 307/N5
Orchy (riv.), Scotland 15/D4
Orcotuna, Peru 128/E8
Orcutt, Calif. (93948) 204/E9
Orcuttville, Conn. (†06076) 210/F1
Ord (mt.), Ariz. 198/D5
Ord (riv.), Australia 87/C7
Ord, Nebr. (68862) 264/F3
Ord (riv.), W. Australia 88/D3
Ord (mt.), W. Australia 92/D2
Ord (riv.), W. Australia 92/E2
Órdenes, Spain 33/B1
Orderville, Utah (84758) 304/B6
Ordoqui, Argentina 143/F7
Ordos (reg.), China 77/G4
Ord River, W. Australia 92/E2
Ordu (prov.), Turkey 63/G2
Ordu, Turkey 59/C1
Ordu, Turkey 63/G2
Ordway, Colo. (81063) 208/M6
Ordway, S. Dak. (†57433) 298/N2
Ordzhonikidze, U.S.S.R. 7/J4
Ordzhonikidze, U.S.S.R. 48/E5
Ordzhonikidze, U.S.S.R. 52/F6
Oreälla, Guyana 131/C3
Oreälv (riv.), Sweden 18/L4
Oreana, Idaho (83659) 220/B6
Oreana, Ill. (62554) 222/E4
Oreana, Nev. (†89419) 266/C2
Orebank, Tenn. (†37660) 237/R7
Örebro (co.), Sweden 18/J7
Örebro, Sweden 7/F3
Örebro, Sweden 18/J7
Ore City, Texas (75683) 303/K5
Oregon 188/B2
OREGON 291
Oregon, Ill. (61061) 222/D1
Oregon (co.), Mo. 261/K9
Oregon, Mo. (64473) 261/B2
Oregon (inlet), N.C. 281/U3
Oregon, Ohio (43616) 284/D2
Oregon (creek), Oreg. 291/K5
Oregon (state), U.S. 146/F5
Oregon, Wis. (53575) 317/H10
Oregon Caves Nat'l Mon., Oreg. 291/D5
Oregon City, Oreg. 188/B1
Oregon City, Oreg. (97045) 291/B2
Oregon Dunes Nat'l Rec. Area, Oreg. 291/C4
Oregonia, Ohio (45054) 284/B7
Öregrund, Sweden 18/L6
Orel, U.S.S.R. 7/H3
Orel, U.S.S.R. 52/E4
Orel, U.S.S.R. 48/D4
Orellana, Peru 128/E6
Orellana la Vieja, Spain 33/D3
Orem, Utah (84057) 304/C3
Orenburg, U.S.S.R. 7/K3
Orenburg, U.S.S.R. 48/F4
Orenburg, U.S.S.R. 52/J4
Orenco, Oreg. (†97123) 291/A2
Orense (prov.), Spain 33/C1
Orense, Spain 33/C1
Orense, Spain 7/D4
Orestes, Ind. (46063) 227/F4
Orestías, Greece 45/H5
Øresund (sound), Denmark 21/F6
Öresund (sound), Denmark 18/H9
Öresund (sound), Sweden 18/H9
Oreti (riv.), N. Zealand 100/B6
Oretta, La. (†70660) 238/D5
Orewa, N. Zealand 100/E2
Orford○, N.H. (03777) 268/C4
Orford, Tasmania 99/D4
Orford Ness (prom.), England 13/J5
Orfordville, N.H. (†03777) 268/C4
Orfordville, Wis. (53576) 317/H10
Organ, N. Mex. (88052) 274/C6
Organabo, Fr. Guiana 131/E3
Organ Pipe Cactus Nat'l Mon., Ariz. 198/C6
Órgãos (range), Brazil 135/E3
Orgas, W. Va. (25148) 312/C6
Orgaz, Spain 33/E3
Orgeyev, U.S.S.R. 52/C5
Orhaneli, Turkey 63/C3
Orhangazi, Turkey 63/C2
Orhon Gol (riv.), Mongolia 77/F2
Oria, Spain 33/E4
Orick, Calif. (95555) 204/A2
Orient, Ill. (62874) 222/E6
Orient, Iowa (50858) 229/E6
Orient○, Maine (04471) 243/H4
Orient, N.Y. (11957) 276/R8
Orient (pt.), N.Y. 276/R8
Orient, Ohio (43146) 284/D6
Orient, S. Dak. (57467) 298/L4
Orient, Wash. (99160) 310/G2
Orienta, Okla. (73765) 288/J2
Oriental, Cordillera (range), Bolivia 136/C5
Oriental, Cordillera (range), Colombia 126/D5
Oriental, Cordillera (range), Dom. Rep. 158/F6
Oriental, Mexico 150/O1
Oriental, N.C. (28571) 281/R4
Oriental, Cordillera (range), Peru 128/H10
Oriental Mindoro (prov.), Philippines 82/C4
Orihuela, Spain 33/F3
Orihvesi (lake), Finland 18/Q5
Orillia, Ontario 177/E3
Orin, Wash. (†99114) 310/H2
Orin, Wyo. (†82633) 319/G3
Orinda, Calif. (94563) 204/J2
Orinoca, Bolivia 136/B6
Orinoco (riv.) 120/C2
Orinoco (riv.) 2/F5
Orinoco (riv.), Colombia 126/G5
Orinoco (delta), Venezuela 120/C2
Orinoco (delta), Venezuela 124/H3
Orinoco (riv.), Venezuela 124/G3
Oriole, Md. (21848) 245/P8
Orion, Ala. (†36081) 195/F7
Orion, Alberta 182/E5
Orion, Ill. (61273) 222/C2
Oriska, N. Dak. (58063) 282/P6
Oriskany, N.Y. (13424) 276/K4
Oriskany, Va. (24130) 307/J5
Oriskany Falls, N.Y. (13425) 276/J5
Orissa (state), India 68/E5
Oristano, Italy 34/B5
Oristano (gulf), Italy 34/B5
Orituco (riv.), Venezuela 124/E3
Oriximiná, Brazil 132/C3
Orizaba, Mexico 146/J8
Orizaba, Mexico 150/P2
Orizaba (Citlaltépetl) (mt.), Mexico 150/O2
Órjiva, Spain 33/E4
Orkanger, Norway 18/F5
Örkény, Hungary 41/E3
Orkney, Sask. 181/D6
Orkney (islands area), Scotland 15/E1
Orkney (trad. co.), Scotland 15/B4
Orkney (isls.), Scotland 7/D3
Orkney (isls.), Scotland 15/F1
Orkney (isls.), Scotland 10/E1
Orla, Texas (79770) 303/D10
Orland, Calif. (95963) 204/C4
Orland, Ind. (46776) 227/G1
Orland, Maine (04472) 243/F6
Orland○, Maine (04472) 243/F6
Orlândia, Brazil 135/C2
Orlando, Fla. 146/K7
Orlando, Fla. 188/K5
Orlando, Fla. (*32801) 212/E3
Orlando, Ky. (40460) 237/N6

Orlando, Okla. (73073) 288/M2
Orlando, W. Va. (26412) 312/E5
Orland Park, Ill. (60462) 222/B6
Orleães, Brazil 132/D10
Orléanais (trad. prov.) France 29
Orléans, Calif. (95556) 204/B2
Orleans, France 7/E4
Orléans, France28/D3
Orleans, Ind. (47452) 227/D7
Orleans, Iowa (†51360) 229/C2
Orleans, (par.), La. 238/L6
Orleans, Minn. (56743) 255/B2
Orleans, Nebr.(68966) 264/E4
Orleans (co.), N.Y. 276/D4
Orleans Ontario 177/J2
Orléans (isl.), Québec 172/F3
Orleans (co.), Vt. 268/C2
Orleans, Vt. (05860) 268/C2
Orleans Cross Roads, W. Va. (†25422)
  312/K3
Orléansville (El Asnam), Algeria
  106/E1
Orlice (riv.), Czech. 41/D1
Orlická (res.), Czech. 41/C2
Orlinda, Tenn. (37141) 237/H7
Orlová, Czech. 41/E2
Orly, France 28/B2
Orma, W. Va. (25268) 312/D5
Ormara, Pakistan 59/J4
Ormara, Pakistan 68/A3
Orme, Tenn. (35740) 237/K10
Ormea, Italy 34/A2
Ormiston, Sask. 181/F6
Ormoc, Philippines 82/E5
Ormoc (bay), Philippines 82/E5
Ormond Beach, Fla. (32074) 212/E2
Ormond-by-the-Sea, Fla. (32074)
  212/E2
Ormont-Dessus, Switzerland 39/D4
Ormsby, Minn. (56162) 255/D7
Ormsby, Pa. (16741) 294/E2
Ormskirk, England 10/E2
Ormskirk, England 13/G2
Ormstown, Québec 177/D4
Orne (dept.), France 28/C3
Orne (riv.), France 28/C3
Orneta, Poland 47/E1
Ornö (isl.), Sweden 18/J2
Örnsköldsvik, Sweden 18/L5
Orobayaya, Bolivia 136/D3
Orocovis, P. Rico 161/G1
Orocué, Colombia 126/E5
Orofino, Idaho (83564) 220/B3
Orofino (creek), Idaho 220/C3
Oro Grande, Calif. (92368) 204/H9
Orogrande, N. Mex. (88342) 274/D6
Orohena (mt.), Fr. Poly. 86/†13
Oro Ingenio, Bolivia 136/C7
Oroluk (atoll), Micronesia 87/F5
Oromocto, New Bruns. 170/D3
Oromocto (lake), New Bruns. 170/C3
Oromocto (riv.), New Bruns. 170/D3
Oron, Israel 65/C6
Oron, Nigeria 106/F8
Orona (Hull) (isl.), Kiribati 87/J6
Orongorongo (riv.), N. Zealand 100/B3
Orondo, Wash. (98843) 310/E3
Orono◯, Maine (04473) 243/F6
Orono, Minn. (†55323) 255/F5
Oronoco, Minn. (55960) 255/F6
Oronogo, Mo. (64855) 261/D8
Oronsay (isl.), Scotland 15/B4
Orontes (riv.), Syria 59/D2
Orontes (riv.), Syria 63/G5
Oropesa, Spain 33/D3
Oropuche (riv.), Trin. & Tob. 161/B10
Oroqen, China 77/K1
Oroquieta, Philippines 85/G4
Oroquieta, Philippines 82/D6
Orosei (gulf), Italy 34/B4
Orosháza, Hungary 41/F3
Orosi, Calif. (93647) 204/F7
Orotina, C. Rica 154/E6
Orotukan, U.S.S.R. 48/Q3
Orovada, Nev. (89425) 266/D1
Oro Valley, Ariz. (†85704) 198/E6
Oroville, Calif. (95965) 204/D4
Oroville (lake), Calif. 204/D4
Oroville, Wash. (98844) 310/F2
Orozco, Cuba 158/B1
Orpha, Wyo. (†82633) 319/G3
Orr, Minn. (55771) 255/F2
Orr, N. Dak. (58244) 282/P3
Orr, Okla. (†73456) 288/M6
Orrefors, Sweden 18/J8
Orrick, Mo. (64077) 261/D4
Orrin, N. Dak. (58359) 282/K3
Orrin, (riv.), Scotland 15/D3
Orrington, Maine (04474) 243/F6
Orrington◯, Maine (04474) 243/F6
Orrs Island, Maine (04066) 243/D8
Orrstown, Pa. (17244) 294/H5
Orrtanna, Pa. (17353) 294/J6
Orrum, N.C. (28369) 281/L6
Orrville, Ala. (36767) 195/D6
Orrville, Ohio (44667) 284/G4
Orrville, Ontario 177/E2
Orsa, U.S.S.R. 18/J6
Orsainville, Québec 172/H3
Orsha, U.S.S.R. 7/G3
Orsha, U.S.S.R. 52/C4
Orsières, Switzerland 39/D4
Orsk, U.S.S.R. 7/K3
Orsk, U.S.S.R. 48/F4
Orsk, U.S.S.R. 52/J4
Orson, Pa. (18449) 294/M2
Orsonnens, Switzerland 39/D3
Orşova, Romania 45/F3
Orta, Turkey 63/E2

Ortaca, Turkey 63/C4
Ortakaraviran, Turkey 63/E4
Ortaköy, Çorum, Turkey 63/F2
Ortaköy, Niğde, Turkey 63/F3
Ortega, Colombia 126/C6
Orteguaza (riv.), Colombia 126/C7
Orthez, France 28/C4
Ortigueira, Spain 33/C1
Orting, Wash. (98360) 310/C3
Ortiz, Colo. (†81120) 208/H8
Ortiz, Mexico 150/D2
Ortiz, Venezuela 124/E3
Ortles (range), Italy 34/C1
Ortley, S. Dak. (57256) 298/P3
Ortoire (riv.), Trin. & Tob. 161/B11
Ortona, Italy 34/F3
Ortonville, Mich. (48462) 250/F6
Ortonville, Minn. (56278) 255/B5
Oruro (dept.), Bolivia 136/A6
Oruro, Bolivia 136/A6
Oruro, Bolivia 7/D6
Oruzgan (Hazar Qadam), Afghanistan
  68/B2
Orvieto, Italy 34/D3
Orville, Ky. (40057) 237/M4
Orviston (mt.), Tennessee 88/H8
Orwell, N.Y. (13426) 276/J3
Orwell, Ohio (44076) 284/J2
Orwell◯, Vt. (05760) 268/A4
Orwigsburg, Pa. (17961) 294/K4
Oryakhovo, Bulgaria 45/G4
Or Yehuda, Israel 65/B4
Orzesze, Poland 47/A4
Orzysz, Poland 47/E2
Osa, U.S.S.R. 52/J3
Osage (riv.) 188/H3
Osage, Ark. (†72638) 202/D1
Osage, Iowa (50461) 229/H2
Osage (co.), Kansas 232/G3
Osage, Minn. (56570) 255/C4
Osage (co.), Mo. 261/J6
Osage (riv.), Mo. 261/E6
Osage (co.), Okla. 288/O1
Osage, Okla. (74054) 288/O2
Osage, Sask. 181/H6
Osage, W. Va. (26543) 312/F3
Osage, Wyo. (82723) 319/H2
Osage Beach, Mo. (65065) 261/G6
Osage City, Kansas (66523) 232/G4
Osage Ind. Res., Okla. 288/O1
Osaka (pref.), Japan 81/J8
Osaka, Japan 2/R4
Osaka, Japan 54/P6
Osaka, Japan 81/J8
Osaka (bay), Japan 81/H8
Osakis, Minn. (56360) 255/C5
Osawatomie, Kansas (66064) 232/H3
Osborn, Miss. (†39759) 256/G3
Osborn, Mo. (64474) 261/D3
Osborn, S.C. (†29426) 296/G6
Osborne (co.), Kansas 232/D2
Osborne, Kansas (67473) 232/D2
Osborne, Pa. (†15143) 294/B4
Osbornsville, N.J. (08723) 273/E3
Osburn, Idaho (83849) 220/B2
Oscar, Fr. Guiana 131/E4
Oscar, La. (70762) 238/H5
Oscar, Okla. (73561) 288/L7
Oscarville, Alaska (†99559) 196/F2
Osceola, Ark. (72370) 202/K2
Osceola ◯, Fla. 212/E3
Osceola, Ind. (46561) 227/E1
Osceola (co.), Iowa 229/B2
Osceola, Iowa (50213) 229/F6
Osceola (co.), Mich. 250/D5
Osceola, Mo. (64776) 261/E6
Osceola, Nebr. (68651) 264/G3
Osceola (mt.), N.H. 268/E3
Osceola, N.Y. (†13316) 276/J3
Osceola, Pa. (16942) 294/H2
Osceola, S. Dak. (†57353) 298/O5
Osceola, Wis. (54020) 317/A5
Osceola Mills, Pa. (16666) 294/F4
Oschatz, E. Germany 22/E3
Oschersleben, E. Germany 22/D2
Oscoda (co.), Mich. 250/E4
Oscoda, Mich. (48750) 250/F4
Oscura, (mts.), N. Mex. 274/C4
Oscuro, N. Mex. (†88301) 274/C5
Ösel (Saaremaa) (isl.), U.S.S.R.
  52/B3
Osgood, Ind. (47037) 227/G6
Osgood, Mo. (63650) 261/F2
Osgood, Ohio (45351) 284/A5
Osgoode, Ontario 177/J2
Osh, U.S.S.R. 54/J5
Osh, U.S.S.R. 48/H5
Osha (peak), N. Mex. 274/C4
Oshawa, Ontario 177/H4
Oshikango, Namibia 118/A3
O-Shima (isl.), Japan 81/J6
Oshkosh, Nebr. (69154) 264/B3
Oshkosh, Wis. 188/J2
Oshkosh, Wis. (54901) 317/J8
Oshnoviyeh, Iran 66/D2
Oshogbo, Nigeria 102/G4
Oshogbo, Nigeria 106/F7
Oshoto, Wyo. (82724) 319/G1
Oshtoran Kuh (mt.), Iran 66/F4
Oshwe, Zaire 115/C4
Osierfield, Georgia (†31798) 217/F7
Osijek, Yugoslavia 7/F4
Osijek, Yugoslavia 45/D3
Osimo, Italy 34/E3
Osipenko (Berdyansk), U.S.S.R. 52/E5
Osipovichi, U.S.S.R. 52/C4
Oskaloosa, Iowa (52577) 229/H6
Oskaloosa, Iowa 188/H4
Oskaloosa, Kansas (66066) 232/G2
Oskaloosa, Iowa (†66711) 261/D7
Oskarshamn, Sweden 18/K8
Oskélaneo◯, Québec 174/C3
Oslavany, Czech. 41/D2
O.T. Downs, North Terr. 93/D4
Oteen, Tenn. (28805) 281/E3
Otego, N.Y. (13825) 276/K6
Otematata, N. Zealand 100/B6
Otero (co.), Colo. 208/M7
Otero (co.), N. Mex. 274/D6
Othello, Wash. (99344) 310/F4
Otho, Iowa (50569) 229/E4
Oti (riv.), Ghana 106/E7
Oti (riv.), Togo 106/E7
Oti (riv.), Upper Volta 106/E7
Otira, N. Zealand 100/C5
Otis, Colo. (80743) 208/O2
Otis, Ind. (46367) 227/D1
Osler, Sask. 181/E3
Oslo, Minn. (56744) 255/A2

Oslo (city), Norway 18/D3
Oslo (cap.), Norway 2/K2
Oslo (cap.), Norway 7/F2
Oslo (cap.), Norway 18/D3
Oslofjord (fjord), Norway 18/D4
Osmanabad, India 68/D5
Osmancık, Turkey 63/F2
Osmaneli, Turkey 63/D2
Osmaniye, Turkey 63/G4
Osmond, Nebr. (68765) 264/G2
Osnabrock, N. Dak. (58269) 282/O2
Osnabrück, W. Germany 22/C2
Osnaburgh House, Ontario 175/B2
Oso, Wash. (98223) 310/D2
Osogna, Switzerland 39/H4
Osorno, Chile 120/B7
Osorno, Chile 138/D3
Osorno, Spain 33/D1
Osoyoos, Br. Col. 184/H5
Osoyoos (lake), Wash. 310/F1
Ospino, Venezuela 124/D3
Osprey (reef), 95/C2
Osprey, Fla. (33559) 212/D4
Osprey (reef), Queensland 88/H2
Oss, Netherlands 27/H5
Ossa, Serra da (mts.), Portugal 33/C3
Ossa (mt.), Tasmania 88/H8
Ossa (mt.), Tasmania 99/C3
Ossabaw (isl.), Georgia 217/K7
Ossabaw (sound), Georgia 217/K7
Osse (riv.), Nigeria 106/F7
Osseo, Mich. (49266) 250/F7
Osseo, Minn. (55369) 255/G5
Osseo, Wis. (54758) 317/D6
Ossian, Ind. (46777) 227/G3
Ossian, Iowa (52161) 229/K2
Ossineke, Mich. (49078) 250/F4
Ossining, N.Y. (10562) 276/N8
Ossipee◯, N.H. (03864) 268/E4
Ossipee (lake), N.H. 268/E4
Ossipee (mts.), N.H. 268/E4
Ossipee (riv.), N.H. 268/F4
Ossokmanuan (res.), Newf. 166/B3
Ostashkov, U.S.S.R. 52/D3
Oste (riv.), W. Germany 22/C2
Osteen, Fla. (32764) 212/E3
Ostend, Belgium 27/B6
Osterburg, Pa. (16667) 294/E5
Österdalälven, Sweden 18/H6
Osterdock, Iowa (†52035) 229/L3
Östergötland (co.), Sweden 18/J7
Osterholz-Scharmbeck, W. Germany
  22/C2
Osterode am Harz, W. Germany 22/D3
Östersund, Sweden 18/J5
Östersund, Sweden 18/J5
Osterville, Mass. (02655) 249/N6
Osterwick, Manitoba 179/D5
Østfold (co.), Norway 18/G7
Osthammar, Sweden 18/L6
Ostia Antica, Italy 34/F7
Ostrander, Minn. (55961) 255/F7
Ostrander, Ohio (43061) 284/D5
Ostrava, Czech. 7/F4
Ostrava, Czech. 41/E2
Ostróda, Poland 47/D2
Ostrogozhsk, U.S.S.R. 48/D4
Ostrogozhsk, U.S.S.R. 52/E4
Ostrołęka (prov.), Poland 47/E2
Ostrołęka, Poland 47/E2
Ostrov, Czech. 41/B1
Ostrov, U.S.S.R. 52/C3
Ostrowiec Świętokrzyski, Poland 47/E3
Ostrów Mazowiecka, Poland 47/E2
Ostrów Wielkopolski, Poland 47/C3
Ostrzeszów, Poland 47/C3
Ostuni, Italy 34/G4
O'Sullivan (dam), Wash. 310/F4
Osüm (riv.), Bulgaria 45/H4
Osumi (isls.), Japan 81/E8
Osumi (pen.), Japan 81/E8
Osumi (str.), Japan 81/E8
Osuna, Spain 33/D4
Oswaldtwistle, England 13/H1
Oswayo, Pa. (16915) 294/G2
Oswegatchie, N.Y. (13670) 276/K2
Oswegatchie (riv.), N.Y. 276/K2
Oswego, Ill. (60543) 222/E2
Oswego, Ill. (†46538) 227/F4
Oswego, Kansas (67356) 232/G4
Oswego, Mont. (59251) 262/L2
Oswego (riv.), N.J. 273/E4
Oswego, N.Y. 188/L2
Oswego (co.), N.Y. 276/H4
Oswego, N.Y. (13126) 276/H4
Oswego (riv.), N.Y. 276/H4
Oswego, S.C. (29121) 296/G3
Oswestry, England 10/E4
Oswestry, England 13/E5
Oświęcim, Poland 47/D3
Osyka, Miss. (39657) 256/D8
Ota, Japan 81/J5
Otago (harb.), N. Zealand 100/C6
Otago (pen.), N. Zealand 100/C6
Otahuhu, N. Zealand 100/C1
Otaki, N. Zealand 100/B4
Otakine (mt.), Japan 81/K5
Otaru, Japan 81/H2
Otautau, N. Zealand 100/A7
Otava (riv.), Czech. 41/B2
Otavalo, Ecuador 128/C2
Otavi, Namibia 118/B3
Otawara, Japan 81/K5
Otdia (atoll), Marshall Is. 87/G4
Otego, N.Y. (13825) 276/K6

Otis, Kansas (67565) 232/C3
Otis, La. (71466) 238/E4
Otis◯, Mass. (†71763) 249/B4
Otis, Mass. 249/B4
Otis, N. Mex. (88220) 274/E6
Otis, Oreg. (97368) 291/C3
Otis, Québec 172/G1
Otisco, Ind. (47163) 227/F7
Otisco, Minn. (56077) 255/E7
Otisco (lake), N.Y. 276/H5
Otisfield, Maine (†04270) 243/B7
Otisfield◯, Maine (†04270) 243/B7
Otish (mts.), Québec 174/C2
Otis Orchards-East Farms, Wash. (99027)
  310/H3
Otisville, Mich. (48463) 250/F5
Otisville, N.Y. (10963) 276/L8
Otley, Iowa (50214) 229/G6
Oto, Iowa (51044) 229/B4
Otoe, Nebr. 264/H4
Otoe (co.), Nebr. 264/H4
Otofuke, Japan 81/L2
Otog, China 77/G3
Otorohanga, N. Zealand 100/E3
Otoskwin (riv.), Ontario 175/B2
Otra (riv.), Norway 18/E7
Otrabanda, Neth. Ant. 161/F9
Otranto (str.), Albania 45/D5
Otranto, Iowa (†50472) 229/H2
Otranto, Italy 34/G4
Otranto (str.), Italy 34/G5
Otsego (co.), Mich. 250/E3
Otsego, Mich. (49078) 250/D6
Otsego (lake), Mich. 250/E4
Otsego (co.), N.Y. 276/K5
Otsego (lake), N.Y. 276/K5
Otsego, Ohio (†43762) 284/G5
Otsego Lake, Mich. (†49735) 250/E4
Otselic (riv.), N.Y. 276/J5
Otsu, Japan 81/J7
Otta, Norway 18/F6
Ottauquechee (riv.), Vt. 268/B4
Ottawa (riv.) 162/J6
Ottawa◯, Canada 2/F3
Ottawa (cap.), Canada 146/L5
Ottawa (cap.), Canada 162/J6
Ottawa (riv.), Canada 146/L5
Ottawa, Ill. (61350) 222/E2
Ottawa (co.), Kansas 232/G3
Ottawa, Kansas (66067) 232/G3
Ottawa (co.), Mich. 250/C6
Ottawa, Minn. (†56058) 255/E6
Ottawa (isls.), N.W.T. 146/J4
Ottawa (isls.), N.W.T. 162/H4
Ottawa (isls.), N. Terrs. 187/K4
Ottawa (co.), Ohio 284/D2
Ottawa, Ohio (45875) 284/B3
Ottawa (co.), Okla. 288/S1
Ottawa (cap.), Canada, Ontario 177/J2
Ottawa (riv.), Ontario 177/H2
Ottawa (riv.), Ontario 177/H2
Ottawa (prov.), Québec 174/B3
Ottawa (riv.), Québec 174/B3
Ottawa Beach, Mich. (†49423) 250/C6
Ottawa-Carleton (reg. munic.), Ontario
  177/J2
Ottawa Hills, Ohio (†43601) 284/C2
Ottawa Lake, Mich. (49267) 250/F7
Otter (lakes), Alberta 182/B1
Otter, Mont. (59062) 262/K5
Otter (creek), Utah 304/C2
Otter (creek), Vt. 268/A3
Otterbein, Ind. (47970) 227/C4
Otterburn, Manitoba 179/E5
Otterburn Park, Québec 172/D4
Otter Creek, Fla. (32683) 212/D2
Otter Creek, Maine (04665) 243/G7
Otter Creek (res.), Utah 304/C5
Otter Lake, Mich. (48464) 250/F5
Otterlo, Netherlands 27/H4
Otterøya (isl.), Norway 18/E5
Otter River, Mass. (†01440) 249/F2
Otter Rock, Oreg. (97369) 291/C3
Ottertail (co.), Minn. 255/C4
Otter Tail (co.), Minn. 255/C4
Otter Tail (lake), Minn. 255/C4
Otter Tail (riv.), Minn. 255/B4
Otterup, Denmark 21/D7
Otterville, Ill. (†62052) 222/C4
Otterville, Iowa (†50644) 229/K3
Otterville, Mo. (65348) 261/G5
Otterville, Ontario 177/F5
Ottery Saint Mary, England 13/D7
Ottery Saint Mary, England 10/E5
Otthon, Sask. 181/J4
Ottleys (creek), N.S. Wales 97/F1
Otto, Ind. (†47162) 227/G7
Otto, Mo. (63052) 261/M6
Otto, N.Y. (14766) 276/D6
Otto, N.C. (28763) 281/C4
Otto (riv.), N.W. Terrs. 187/K1
Otto, Wyo. (82434) 319/D1
Ottosen, Iowa (50570) 229/E3
Ottoville, Ohio (45876) 284/B4
Ottsville, Pa. (18942) 294/M5
Ottumwa, Iowa (52501) 229/J6
Ottumwa, Iowa 188/H2
Ottweiler, W. Germany 22/B4
Otukpo, Nigeria 106/F7
Otumba de Gómez Farías, Mexico 150/M1
Oturkpo, Nigeria 106/F7
Otuquis, Bolivia 136/F6
Oturzco, Peru 128/C5
Otway (bay), Chile 138/E10
Otway (sound), Chile 138/E10
Otway, Ohio (45657) 284/D8
Otway (cape), Victoria 97/B6
Otway (cape), Victoria 88/G7
Otwell, Ark. (†72401) 202/J2
Otwell, Ind. (47564) 227/C8
Otwock, Poland 47/E2
Ötztal Alps (mts.), Austria 41/A3
Ötztal Alps (range), Italy 34/C1
Otis, Colo. (80743) 208/O2
Otis, Ind. (46367) 227/D1

Ouachita (riv.) 188/H4
Ouachita (co.), Ark. 202/E6
Ouachita, Ark. (†71763) 202/D5
Ouachita (lake), Ark. 202/C4
Ouachita (mts.), Ark. 202/B4
Ouachita (par.), La. 238/F1
Ouachita (riv.), La. 238/F2
Ouadane, Mauritania 106/B4
Ouadda, Cent. Afr. Rep. 115/D2
Ouagadougou (cap.), Upper Volta
  106/D6
Ouagadougou (cap.), Upper Volta
  102/D3
Ouahigouya, Upper Volta 106/D6
Ouahigouya, Upper Volta 102/B3
Oualata, Mauritania 106/C5
Ouallene, Algeria 106/E4
Ouanaminthe, Haiti 158/E4
Ouanary, Fr. Guiana 131/F3
Ouanda Djalle, Cent. Afr. Rep. 115/D2
Ouango, Cent. Afr. Rep. 115/D3
Ouaqui, Fr. Guiana 131/D4
Ouarane (reg.), Mauritania 106/B4
Ouareau (lake), Québec 172/D3
Ouareau (riv.), Québec 172/D3
Ouargla, Algeria 106/F2
Ouargla, Algeria 102/C1
Ouarzazate, Morocco 106/C2
Ouchy, Switzerland 39/C4
Oud-Beijerland, Netherlands 27/E5
Ouddorp, Netherlands 27/D5
Oudenaarde, Belgium 27/D7
Oudenbosch, Netherlands 27/F5
Oude (str.), Italy 34/G5
Oude-Pekela, Netherlands 27/K2
Ouderkerk, Netherlands 27/F2
Oude-Tonge, Netherlands 27/E5
Oudewater, Netherlands 27/F4
Oudtshoorn, S. Africa 102/E8
Oudtshoorn, S. Africa 118/C6
Oued Zem, Morocco 106/C2
Ouelle (riv.), Québec 172/H2
Ouellette, Maine (†04743) 243/G1
Ouémé (riv.), Benin 106/E7
Ouessant (isl.), France 28/A3
Ouesso, Congo 115/C3
Ouest (dept.), Haiti 158/C6
Ouest (pt.), Haiti 158/B6
Ouest (pt.), Haiti 158/B6
Ouezzane, Morocco 106/C2
Oughter (lake), Ireland 17/G3
Oughterard, Ireland 17/G5
Ouham (riv.), Cent. Afr. Rep. 115/C2
Ouham (riv.), Chad 111/C6
Ouidah, Benin 106/E7
Oujaf, Mauritania 106/C5
Oujda, Morocco 106/D2
Oujda, Morocco 102/B1
Oujeft, Mauritania 106/B4
Oulainen, Finland 18/O4
Ouled Djellal, Algeria 106/F2
Oullins, France 28/F5
Oulu, Finland 18/P4
Oulu, Finland 7/G2
Oulu (lake), Finland 7/G2
Oulujärvi (lake), Finland 18/P4
Oulujoki (riv.), Finland 18/O4
Oum Chalouba, Chad 111/D4
Oum el Asel (well), Mali 106/D3
Oum Hadjer, Chad 102/E5
Oum Hadjer, Chad 111/D5
Ounas (riv.), Finland 7/G2
Ounasjoki (riv.), Finland 18/O3
Oundle, England 13/G5
Ounianga-Kébir, Chad 111/D4
Oupeye, Belgium 27/H7
Oupu, China 77/L1
Our (riv.), Luxembourg 27/J9
Our (riv.), W. Germany 22/B3
Ouray, Colo. (81427) 208/D6
Ouray (co.), Colo. 208/D6
Ouray, Utah (†84026) 304/E3
Ourinhos, Brazil 132/B8
Ourinhos, Brazil 135/B3
Ourique, Portugal 33/B4
Ouro Fino, Brazil 132/E8
Ouro Fino, Brazil 135/E2
Ouro Preto, Brazil 132/E4
Ouro Preto, Brazil 135/E2
Ourthe (riv.), Belgium 27/G8
Ouse (riv.), England 13/G6
Ouse (riv.), England 13/G4
Ouse, Tasmania 99/C4
Ouse, Tasmania 99/C4
Ousley, Georgia (†31601) 217/F9
Outagamie (co.), Wis. 317/K7
Outardes (riv.), Québec 174/D2
Outer (isl.), Wis. 317/F1
Outer Harbor, S. Australia 94/A7
Outer Hebrides (isls.), Scotland
  15/A3
Outer Santa Barbara (passage), Calif.
  204/G10
Outing, Minn. (56662) 255/E4
Outjo, Namibia 118/B4
Outjo, Namibia 102/D7
Outlook, Mont. (59252) 262/M2
Outlook, Sask. 181/H4
Outlook, Wash. (98938) 310/F4
Outram, Sask. 181/H6
Ouvéa (isl.), New Caledonia 89/M6
Ouyen, Victoria 88/F2
Ouyen, Victoria 97/B4
Ouzinkie, Alaska (99644) 196/H3
Ovacık, Çankırı, Turkey 63/E2
Ovacık, İçel, Turkey 63/F4
Ovacık, Tunceli, Turkey 63/H3
Ovalau (isl.), Fiji 86/Q10
Ovalle, Chile 120/A3
Ovalle, Chile 138/A8
Ovamboland (reg.), Namibia 118/B3

Ovando, Mont. (59854) 262/C3
Ovar, Portugal 33/B2
Ovens (riv.), Victoria 97/D5
Overall, Tenn. (†37130) 237/J9
Overbrook, Kansas (66524) 232/G3
Overbrook, Okla. (73453) 288/M6
Overflakkee (isl.), Netherlands 27/E5
Overflow (bay), Manitoba 179/A1
Overflowing (riv.), Sask. 181/K2
Overflowing River, Manitoba 179/A1
Overgaard, Ariz. (85933) 198/E4
Overhills, N.C. (28370) 281/L4
Overijse, Belgium 27/F7
Overijssel (prov.), Netherlands 27/J4
Overisel, Mich. (†49423) 250/C6
Överkalix, Sweden 18/N3
Overland, Mo. (63114) 261/O2
Overland Park, Kansas (66204) 232/H3
Overlea, Md. (21206) 245/K3
Overloon, Netherlands 27/H5
Overly, N. Dak. (58360) 282/K2
Overpelt, Belgium 27/G6
Overstreet, Fla. (32453) 212/D6
Overton, Nebr. (68863) 264/E4
Overton, Nev. (89040) 266/G6
Overton, Pa. (†18833) 294/K2
Overton (co.), Tenn. 237/L8
Overton, Texas (75684) 303/K5
Övertorneå, Sweden 18/N3
Överum, Sweden 18/K7
Ovett, Miss. (39464) 256/F6
Ovid, Colo. (80744) 208/P1
Ovid, Idaho (83260) 220/G7
Ovid, Mich. (48866) 250/E5
Ovid, N.Y. (14521) 276/G5
Oviedo, Dom. Rep. 158/D7
Oviedo, Fla. (32765) 212/E3
Oviedo (prov.), Spain 33/C1
Oviedo, Spain 33/C1
Oviedo, Spain 7/D4
Ovilla, Texas (†76065) 303/G2
Ovoca (riv.), Ireland 17/J6
Övörhangay, Mongolia 77/F2
Øvre-Sirdal, Norway 18/E7
Ovruch, U.S.S.R. 52/C4
Ovtrup, Denmark 21/B6
Owaka, N. Zealand 100/B7
Owando, Congo 115/C4
Owando, Congo 102/D5
Owaneco, Ill. (62555) 222/D4
Owasa, Iowa (†50473) 229/G4
Owasco, N.Y. (13130) 276/G5
Owasco (lake), N.Y. 276/G5
Owase, Japan 81/H6
Owassa, Ala. (†36401) 195/E8
Owassa (lake), N.J. 273/C1
Owasso, Okla. (74055) 288/P2
Owatonna, Minn. (55060) 255/E6
Owbeh, Afghanistan 68/H4
Owbeh, Afghanistan 59/H3
Owego, N.Y. (13827) 276/H6
Owell (lake), Ireland 17/G4
Owen (co.), Ind. 227/D6
Owen (co.), Ky. 237/M3
Owen (mt.), N. Zealand 100/D4
Owen (chan.), Ontario 177/C2
Owen (sound), Ontario 177/D2
Owen, Wis. (54460) 317/F6
Owen (lake), Wis. 317/D3
Owendale, Alberta 182/D5
Owendale, Mich. (48754) 250/F5
Owendo, Gabon 115/A3
Owen Falls (dam), Uganda 115/F3
Owenga, N. Zealand 100/F4
Owenkillew (riv.), N. Ireland 17/G2
Owenmore (riv.), Ireland 17/B3
Owenmore (riv.), Ireland 17/B3
Owens (lake), Calif. 188/C3
Owens (lake), Calif. 204/H7
Owens (peak), Calif. 204/H8
Owens (riv.), Calif. 204/G6
Owensboro, Ky. (42301) 237/G5
Owensboro, Ky. 188/J3
Owensburg, Ind. (47453) 227/D7
Owens Cross Roads, Ala. (35763) 195/E1
Owen Sound, Ont. 162/H7
Owen Sound, Ontario 177/D3
Owensville, Ark. (†72087) 202/E4
Owensville, Ind. (47665) 227/B8
Owensville, Mo. (65066) 261/K6
Owensville, Ohio (45160) 284/B7
Owenton, Ky. (40359) 237/M3
Owenton, Va. (†23077) 307/O5
Owerri, Nigeria 106/F7
Owia (bay), St. Vin. & Grens. 161/A8
Owikeno (lake), Br. Col. 184/D4
Owings, Md. (20836) 245/M6
Owings, S.C. (†29645) 296/C2
Owings Mills, Md. (21117) 245/L3
Owingsville, Ky. (40360) 237/O4
Owl (creek), Colo. 208/K1
Owl (riv.), Manitoba 179/K2
Owl (creek), S. Dak. 298/B4
Owl, North Fork (creek), Wyo. 319/D2
Owl Creek (mts.), Wyo. 319/C2
Owl River, Alberta 182/E2
Owlshead◯, Maine (04854) 243/F7
Owo, Nigeria 106/F7
Owosso, Mich. (48867) 250/E5
Owraman, Iran 66/E4
Owsley (co.), Ky. 237/O6
Owyhee (co.), Idaho 220/B7
Owyhee (mts.), Idaho 220/B8
Owyhee, East Fork (riv.), Idaho
  220/B7
Owyhee, Nev. (89832) 266/F1
Owyhee (riv.), Nev. 266/E1
Owyhee (dam), Oreg. 291/K4
Owyhee (lake), Oreg. 291/K4
Owyhee (mts.), Oreg. 291/K4
Owyhee, North Fork (riv.), Oreg.
  291/K5
Owyhee (riv.), Oreg. 291/K5

Ox (Slieve Gamph) (mts.), Ireland 17/D3
Oxapampa, Peru 128/E8
Oxbow (dam), Idaho 220/B5
Oxbow○, Maine (04764) 243/G3
Oxbow, Oreg. (97840) 291/L2
Oxbow (dam), Oreg. 291/L3
Oxbow, Sask. 181/J6
Oxelösund, Sweden 18/K7
Oxford, Ala. (36203) 195/G3
Oxford, Ark. (72565) 202/G1
Oxford○, Conn. (06483) 210/C3
Oxford, England 13/F6
Oxford, England 10/F5
Oxford, Fla. (32684) 212/D3
Oxford, Georgia (30267) 217/E3
Oxford, Idaho (†83263) 220/F7
Oxford, Ind. (47971) 227/C3
Oxford, Iowa (52322) 229/K5
Oxford, Kansas (67119) 232/E4
Oxford, La. (†71052) 238/C3
Oxford (co.), Maine 243/B7
Oxford, Maine (04270) 243/B7
Oxford○, Maine (04270) 243/B7
Oxford (lake), Manitoba 179/J3
Oxford, Md. (21654) 245/O6
Oxford, Mass. (01540) 249/G4
Oxford○, Mass. (01540) 249/G4
Oxford, Mich. (48051) 250/F6
Oxford, Miss. (38655) 256/F2
Oxford, Nebr. (68967) 264/E4
Oxford, N.J. (07863) 273/C2
Oxford, N.Y. (13830) 276/J6
Oxford, N. Zealand 100/D5
Oxford, N.C. (27565) 281/M2
Oxford, Nova Scotia 168/E3
Oxford, Ohio (45056) 284/A6
Oxford, Pa. (19363) 294/K6
Oxford, W. Va. (†26456) 312/E4
Oxford, Wis. (53952) 317/H8
Oxford House, Manitoba 179/J3
Oxford Junction, Iowa (52323) 229/M4
Oxford Junction, Nova Scotia 168/E3
Oxford Mills, Iowa (†52323) 229/L5
Oxford Mills, Ontario 177/J3
Oxfordshire (co.), England 13/F6
Oxkutzcab, Mexico 150/P6
Oxley, N.S. Wales 97/C4
Oxley (creek), Queensland 95/D3
Oxly, Mo. (63955) 261/L9
Oxnard, Calif. (93030) 204/F9
Oxnard A.F.B., Calif. 204/F9
Oxon Hill, Md. (20745) 245/F6
Oxon Run (riv.), Md. 245/F5
Oxton, Scotland 15/F5
Oxtongue Lake, Ontario 177/E2
Oyabe, Japan 81/H5
Oyahue (vol.), Chile 138/C3
Oyama, Br. Col. 184/H5
Oyama, Japan 81/J5
Oyapock (riv.) 120/D2
Oyapock (riv.), Brazil 132/C2
Oyapock (riv.), Fr. Guiana 131/E4
Oyem, Gabon 102/D4
Oyem, Gabon 115/B3
Oyen, Alberta 182/E4
Oyens, Iowa (51045) 229/A3
Oykel (riv.), Scotland 15/D3
Oykel Bridge, Scotland 15/D3
Oylen, Minn. (†56481) 255/D4
Oymyakon, U.S.S.R. 4/C2
Oymyakon, U.S.S.R. 48/O3
Oyo, Congo 115/C4
Oyo (state), Nigeria 106/E7
Oyo, Nigeria 106/E7
Oyón, Peru 128/D8
Oyonnax, France 28/F4
Oyster (bay), Tasmania 88/H8
Oyster (bay), Tasmania 99/E4
Oyster, Va. (23419) 307/S6
Oyster Bay, N.Y. (11771) 276/R6
Oyster River (pt.), Conn. 210/D4
Oysterville, Wash. (98641) 310/A4
Ozalp, Turkey 63/K3
Ozamiz, Philippines 82/D6
Ozan, Ark. (71855) 202/C6
Ozark (mts.) 188/H3
Ozark, Ala. (36360) 195/G8
Ozark, Ark. (72949) 202/C3
Ozark (lake), Ark. 202/C3
Ozark (plat.), Ark. 202/C1
Ozark (res.), Ark. 202/C3
Ozark, Ill. (62972) 222/E6
Ozark (co.), Mo. 261/H9
Ozark, Mo. (65721) 261/F8
Ozark (plat.), Mo. 261Œ9
Ozark Nat'l Scenic Riverways, Mo. 261/K8
Ozarks, Lake of the (lake), Mo. 261/F6
Ozaukee (co.), Wis. 317/L9
Ozawkie, Kansas (66070) 232/G2
Özd, Hungary 41/F2
Ozernovskiy, U.S.S.R. 48/Q4
Ozernoy (cape), U.S.S.R. 48/R4
Ozette, Wash. (†98326) 310/A2
Ozette (lake), Wash. 310/A2
Ozette Ind. Res., Wash. 310/A2
Ozieri, Italy 34/B4
Ozona, Fla. (33560) 212/D3
Ozona, Texas (76943) 303/C7
Ozone, Ark. (72854) 202/D2
Ozone, Tenn. (37842) 237/M9
Ozorków, Poland 47/D3
Ozu, Japan 81/F7
Ozuluama, Mexico 150/L6
Ozumba de Alzate, Mexico 150/M1

P

Pa-an, Burma 72/C3
Paarden (bay), Neth. Ant. 161/D10
Paarl, S. Africa 102/D8
Paarl, S. Africa 118/F6
Paauhau, Hawaii (96775) 218/H4
Paauilo, Hawaii (96776) 218/H4
Paavola, Finland 18/O4
Pabbay (isl.), Scotland 15/A4
Pabbay (isl.), Scotland 15/A3
Pabianice, Poland 47/D3
Pablo, Mont. (59855) 262/B3
Pabna, Bangladesh 68/F4
Pabos, Québec 172/D2
Pabos-Mills, Québec 172/D2
Pabrade, U.S.S.R. 53/C3
Pacajá Grande (riv.), Brazil 132/D4
Pacaraimã, Serra da (mts.), Brazil 132/H8
Pacaraima, Sierra (mts.), Venezuela 124/G5
Pacasmayo, Peru 128/C6
Pace, Fla. (32570) 212/B6
Pace, Miss. (38764) 256/C3
Pachaug, Conn. (†06351) 210/H2
Pachaug (pond), Conn. 210/H2
Pachaug (riv.), Conn. 210/H2
Pacheco, Calif. (94553) 204/K1
Pachino, Italy 34/E6
Pachitea (riv.), Peru 128/E7
Pachiza, Peru 128/D6
Pachmarhi, India 68/D4
Pacho, Colombia 126/C5
Pachuca de Soto, Mexico 150/K6
Pachuta, Miss. (39347) 256/G6
Pacific (ocean) 54/T5
Pacific (ocean) 146/E6
Pacific, Mo. (63069) 261/L5
Pacific (co.), Wash. 310/B4
Pacific, Wash. (98047) 310/C3
Pacifica, Calif. (94044) 204/H2
Pacific Beach, Calif. (92109) 204/H11
Pacific Beach, Wash. (98571) 310/A3
Pacific City, Oreg. (97135) 291/C2
Pacific Grove, Calif. (93950) 204/C7
Pacific Heights, Hawaii (†96801) 218/C4
Pacific Islands, Terr. of the 87/F5
Pacific Islands, Territory of the 2/S5
Pacific Junction, Iowa (51561) 229/B6
Pacific Palisades, Hawaii (†96782) 218/E2
Pacific Rim Nat'l Park, Br. Col. 184/E6
Pacitan, Indonesia 85/J2
Pack (riv.), Idaho 220/B1
Pack (creek), Utah 304/E5
Packanack Lake, N.J. (07470) 273/B1
Packertown, Ind. (†46510) 227/F2
Packerville, Conn. (†06331) 210/H2
Packington, Québec 172/J2
Packsville, W. Va. (25151) 312/C7
Packwaukee, Wis. (53953) 317/G8
Packwood, Iowa (52580) 229/J6
Packwood, Wash. (98361) 310/D4
Paco, Philippines 82/C2
Pacoa, Colombia 126/F7
Paço de Arcos, Portugal 33/A1
Pacoima, Calif. (91331) 204/B10
Pacolet, S.C. (29372) 296/D2
Pacolet (riv.), S.C. 296/D1
Pacolet Mills, S.C. (29373) 296/D2
Pacov, Czech. 41/C2
Pacsa, Hungary 41/D3
Pacsan (mt.), Philippines 82/C2
Pactolus, N.C. (†27834) 281/P3
Padada, Philippines 82/E7
Padang, Indonesia 54/L10
Padang, Indonesia 85/B6
Padangpanjang, Indonesia 85/B6
Padangsidempuan, Indonesia 85/B5
Padany, U.S.S.R. 52/D2
Padborg, Denmark 21/C8
Padcaya, Bolivia 136/C7
Paddle Prairie, Alberta 182/A5
Paddock Lake, Wis. (†53168) 317/K10
Paddockwood, Sask. 181/F2
Paden, Miss. (38861) 256/H1
Paden, Okla. (74860) 288/N3
Paden City, W. Va. (26159) 312/D3
Paderborn, W. Germany 22/C3
Padgett, S.C. (†29481) 296/F5
Padiham, England 13/H1
Padilla, Bolivia 136/C6
Padilla, Mexico 150/K5
Padilla (creek), N. Mex. 274/D5
Padilla (bay), Wash. 310/C2
Padloping (isl.), N.W.T. 162/K2
Padloping (isl.), N.W. Terrs. 187/M3
Padre (isl.), Texas 188/G5
Padre (isl.), Texas 303/G10
Padre Island Nat'l Seashore, Texas 303/G11
Padre Las Casas, Dom. Rep. 158/D6
Padrón, Spain 33/B1
Padroni, Colo. (80745) 208/N1
Padstow, England 10/D5
Padstow, England 13/B7
Padua (prov.), Italy 34/C2
Padua, Italy 7/F4
Padua, Italy 34/C2
Padua, Minn. (†56378) 255/C5
Paducah, Ky. (42001) 237/D6
Paducah, Ky. 146/K6
Paducah, Ky. 188/J3
Paducah, Texas (79248) 303/D4
Padul, Spain 33/E4
Paekam, N. Korea 81/D3
Paektu (mt.), N. Korea 81/C3
Paeroa, N. Zealand 100/E2
Páez, Colombia 126/C6
Pafúri, Mozambique 118/E4
Pag, Yugoslavia 45/B3
Pag (isl.), Yugoslavia 45/B3
Pagadian, Philippines 82/D7
Pagalungan, Philippines 82/E7
Pagan, Burma 72/B2
Pagan (isl.), No. Marianas 87/E4
Page, Ariz. (86040) 198/D2
Page (co.), Iowa 229/C7
Page, Nebr. (68766) 264/F2
Page, N. Dak. (58064) 282/P5
Page, Okla. (†74939) 288/S5
Page (co.), Va. 307/M3
Page, W. Va. (25152) 312/D6
Page City, Kansas (†67764) 232/A2
Pagedale, Mo. (†63101) 261/P2
Pageland, S.C. (29728) 296/G2
Pago (bay), Guam 86/K7
Pagoda (peak), Colo. 208/E2
Pago Pago (Cap.), Amer. Samoa 86/N9
Pago Pago (cap.), Amer. Samoa 87/J7
Pagosa Junction, Colo. (†81147) 208/E8
Pagosa Springs, Colo. (81147) 208/E8
Pagoua (bay), Dominica 161/F6
Paguate, N. Mex. (87040) 274/B3
Pagwa River, Ontario 177/J5
Pagwa River, Ontario 175/D3
Pahala, Hawaii 188/G6
Pahala, Hawaii (96777) 218/H6
Pahang (state), Malaysia 72/D7
Pahang, Sungai (riv.), Malaysia 72/D7
Pahaska, Wyo. (82414) 319/C1
Pahiatua, N. Zealand 100/F4
Pahlevi (Enzeli), Iran 59/F2
Pahlevi (Enzeli), Iran 66/F2
Pahoa, Hawaii (96778) 218/J5
Pahokee, Fla. (33476) 212/F5
Pahranagat (range), Nev. 266/F5
Pahrock (range), Nev. 266/F5
Pah-rum (peak), Nev. 266/B2
Pahrump, Nev. (89041) 266/E6
Pahrump (valley), Nev. 266/F6
Pahsimeroi (riv.), Idaho 220/E5
Pahute (mesa), Nev. 266/E5
Paia, Hawaii (96779) 218/J2
Paicheng (Baicheng), China 77/K2
Paicines, Calif. (95043) 204/D7
Paide, U.S.S.R. 53/C1
Paige, Texas (78659) 303/G7
Paihia, N. Zealand 100/D1
Paihuano, Chile 138/B8
Paiján, Peru 128/C6
Päijänne (lake), Finland 18/O6
Pailin, Cambodia 72/D4
Paillaco, Chile 138/D3
Pailolo (chan.), Hawaii 218/H1
Paimboeuf, France 28/C4
Paimpol, France 28/B3
Painan, Indonesia 85/C6
Paincourt, Ontario 177/B5
Paincourtville, La. (70391) 238/K3
Paine, Chile 138/G4
Paine, Cerro (mt.), Chile 138/D9
Painesdale, Mich. (49955) 250/G1
Painesville, Ohio (44077) 284/H2
Painswick, Ontario 177/E3
Paint (lake), Manitoba 179/J2
Paint (riv.), Mich. 250/A2
Paint (creek), Ohio 284/D7
Paint, Pa. (†15963) 294/E5
Paint Bank, Va. (24131) 307/H5
Paint Branch (riv.), Md. 245/F4
Painted (des.), Ariz. 198/D2
Painted Desert Section (Petrified Forest), Ariz. 198/F3
Painted Post, N.Y. (14870) 276/F6
Painted Rock (dam), Ariz. 198/C5
Painter, Ala. (†35962) 195/F2
Painter, Va. (23420) 307/S5
Painter Ridge (hills), Conn. 210/B2
Painters Hill, Fla. (†32036) 212/E2
Paintersville, Ohio (†45335) 284/C6
Paint Lick, Ky. (40461) 237/N5
Paint Lick (riv.), Ky. 237/M5
Paint Rock, Ala. (35764) 195/F1
Paint Rock (riv.), Ala. 195/F1
Paint Rock, Texas (76866) 303/E6
Paintsville, Ky. (41240) 237/R5
Paipa, Colombia 126/D5
Paipote, Chile 138/B6
Paipote, Quebrada de (riv.), Chile 138/B6
Paisley, Ontario 177/C3
Paisley, Oreg. (97636) 291/G5
Paisley, Scotland 10/A1
Paisley, Scotland 15/B2
Paita, Peru 128/B5
Paita (bay), Peru 128/B5
Paiute Ind. Res., Calif. 204/G6
Pajala, Sweden 18/N3
Paján, Ecuador 128/B3
Pajarito (creek), N. Mex. 274/A2
Pajaro, Calif. (†95076) 204/D7
Pájaros (isls.), Chile 138/A7
Pakanbaru, Indonesia 54/M9
Pakanbaru, Indonesia 85/C5
Pakaraima (mts.), Guyana 131/A3
Pakawau, N. Zealand 100/D4
Pakchan (riv.), Burma 72/C5
Pakchan (riv.), Thailand 72/C5
Pakch'ŏn, N. Korea 81/B4
Pakenham, Ontario 177/H2
Pakhoi (Beihai), China 77/G7
Pakistan 2/N4
Pakistan 54/H7
PAKISTAN 59/J4
PAKISTAN 68/B3
Pakokku, Burma 72/B2
Pakowki (lake), Alberta 182/E5
Paks, Hungary 41/E3
Pakwach, Uganda 115/F3
Pakxé, Laos 72/E4
Pala, Chad 111/B6
Palacios, Texas (77465) 303/H9
Palafrugell, Spain 33/H2
Palagruža (Pelagosa) (isl.), Yugoslavia 45/C4
Pala Ind. Res., Calif. 204/H10
Palamós, Spain 33/H2
Palana, U.S.S.R. 54/S4
Palana, U.S.S.R. 48/R4
Palanan, Philippines 82/D2
Palanan, Philippines 85/G2
Palanan (bay), Philippines 82/D2
Palanda, Ecuador 128/C5
Palanga, U.S.S.R. 53/A3
Palangkaraya, Indonesia 85/E6
Palanpur, India 68/C4
Palaoa (pt.), Hawaii 218/G2
Palapag, Philippines 82/E4
Palapye, Botswana 118/D4
Palas de Rey, Spain 33/C1
Palatine, Ill. (60067) 222/B5
Palatka, Ark. (†72422) 202/J1
Palatka, Fla. (32077) 212/E2
Palau (Belau) 87/D5
Palauli (bay), W. Samoa 86/L8
Palaumerak, Indonesia 85/G1
Palaw, Burma 72/C4
Palawan (prov.), Philippines 82/B6
Palawan (isl.), Philippines 2/Q5
Palawan (isl.), Philippines 54/N8
Palawan (isl.), Philippines 85/F4
Palawan (isl.), Philippines 82/B6
Palawan (passage), Philippines 85/F4
Palawan (passage), Philippines 82/A6
Palaya, Bolivia 136/A6
Palayan, Philippines 82/C3
Palayankottai, India 68/D7
Palazzolo Acreide, Italy 34/E6
Palca, Bolivia 136/A5
Palco, Kansas (67657) 232/C2
Paldiski, U.S.S.R. 53/B1
Paldiski, U.S.S.R. 52/B3
Paleleh, Indonesia 85/G5
Palembang, Indonesia 54/M10
Palembang, Indonesia 85/D6
Palena, Chile 138/E5
Palena (lake), Chile 138/E5
Palena (riv.), Chile 138/D5
Palencia (prov.), Spain 33/D1
Palencia, Spain 33/D2
Palenque (pt.), Dom. Rep. 158/E6
Palenque, Mexico 150/O8
Palenque (ruin), Mexico 150/O8
Palenville, N.Y. (12463) 276/M6
Palermo, Calif. (95968) 204/D4
Palermo (prov.), Italy 34/D5
Palermo, Italy 7/F5
Palermo, Italy 34/D5
Palermo○, Maine (04354) 243/E7
Palermo, N.J. (†08226) 273/D5
Palermo, N. Dak. (58769) 282/F3
Palermo, Uruguay 145/D4
Palestina, Chile 138/B4
Palestine, Ala. (†36252) 195/H3
Palestine, Ark. (72372) 202/J4
Palestine, Ill. (62451) 222/F4
Palestine, Ind. (†46508) 227/F2
Palestine, Ohio (45352) 284/A5
Palestine, Texas 188/H4
Palestine, Texas (75801) 303/J6
Palestine, W. Va. (26160) 312/D4
Palestrina, Italy 34/F7
Paletwa, Burma 72/B2
Palghat, India 68/D6
Palha, Mar da (bay), Portugal 33/A1
Pali, India 68/C3
Palidoro, Italy 34/F6
Paliocabe (Payocabe), Chile 138/F4
Palisade, Colo. (81526) 208/C4
Palisade, Minn. (56469) 255/E4
Palisade, Nebr. (69040) 264/C4
Palisade, Nev. (†89822) 266/E2
Palisades, Idaho (83437) 220/G6
Palisades (res.), Idaho 220/G6
Palisades, N.J. 273/C1
Palisades, N.Y. (10964) 276/K8
Palisades, Wash. (98845) 310/E3
Palisades (res.), Wyo. 319/A2
Palisades Park, N.J. (07650) 273/C2
Paliseul, Belgium 27/G9
Palizada, Mexico 150/O7
Palk (str.), India 68/D7
Palk (str.), Sri Lanka 68/D7
Pallamallawa, N.S. Wales 97/F1
Palling, Br. Col. 184/D3
Palliser (bay), N. Zealand 100/C3
Palliser (cape), N. Zealand 100/E4
Pall Mall, Tenn. (38577) 237/M7
Palm (beach), Neth. Ant. 161/D10
Palma (bay), Alaska 196/L1
Palma, Mozambique 118/G2
Palma, Spain 7/E5
Palma, Spain 33/H3
Palma (bay), Spain 33/H3
Palma del Río, Spain 33/D4
Palma di Montechiaro, Italy 34/D6
Palmarejo, Venezuela 124/C2
Palmares, Brazil 132/H5
Palmares, C. Rica 154/F6
Palmarito, Apure, Venezuela 124/D4
Palmarito, Guárico, Venezuela 124/F3
Palmarito, Mérida, Venezuela 124/C3
Palmarola (isl.), Italy 34/D4
Palmas (cape) 102/B4
Palmas, Brazil 132/C9
Palmas (cape), Liberia 106/C8
Palmas Altas (pt.), P. Rico 161/C1
Palma Soriano, Cuba 158/J4
Palm Bay, Fla. (32905) 212/F3
Palm Beach, Fla. 188/L5
Palm Beach (co.), Fla. 212/F5
Palm Beach, Fla. (33480) 212/G4
Palm Beach Gardens, Fla. (†33403) 212/F5
Palm Beach Shores, Fla. (†33404) 212/G5
Palm City, Fla. (33490) 212/F4
Palm Coast, Fla. (32037) 212/E2
Palmdale, Calif. (93550) 204/G9
Palmdale, Fla. (33944) 212/E5
Palm Desert, Calif. (92260) 204/J10
Palmeira, Brazil 132/D9
Palmeira, Brazil 135/B4
Palmeira das Missões, Brazil 132/C9
Palmeiras, Brazil 132/F6
Palmeirinhas (pt.), Angola 115/B5
Palmer, Alaska (99645) 196/C1
Palmer, Alaska 188/D5
Palmer (arch.) 5/C15
Palmer, Ill. (62556) 222/D4
Palmer, Ind. (†46307) 227/C2
Palmer, Iowa (50571) 229/D3
Palmer, Kansas (66962) 232/E2
Palmer, Mass. (01069) 249/E4
Palmer○, Mass. (01069) 249/E4
Palmer, Mich. (49871) 250/B2
Palmer, Nebr. (68864) 264/F3
Palmer (head), N. Zealand 100/B3
Palmer, P. Rico 161/F1
Palmer (riv.), Queensland 95/B2
Palmer, Sask. 181/E6
Palmer, Tenn. (37365) 237/K10
Palmer, Wash. (98048) 310/D3
Palmer (lake), Wash. 310/F2
Palmer Lake, Colo. (80133) 208/J4
Palmer Land (reg.), Ant. 2/F9
Palmer Land (reg.), Ant. 5/B15
Palmer Rapids, Ontario 177/G2
Palmers, Minn. (†55801) 255/G4
Palmers Crossing, Miss. (†39401) 256/F8
Palmer Station, Ant. 5/C15
Palmerston (atoll), Cook Is. 87/K7
Palmerston, N. Zealand 100/C6
Palmerston, Ontario 177/D4
Palmerston North, N. Zealand 87/H10
Palmerston North, N. Zealand 100/E4
Palmersville, Tenn. (38241) 237/D8
Palmerton, Pa. (18071) 294/L4
Palmerville, Queensland 95/B3
Palmetto, Fla. (33561) 212/D4
Palmetto, Georgia (30268) 217/C3
Palmetto, La. (71358) 238/G5
Palmetto (pt.), St. Chris.-Nevis 161/C10
Palm Harbor, Fla. (33563) 212/D3
Palmi, Italy 34/E5
Palmiet (riv.), S. Africa 118/F7
Palmilla, Chile 138/F6
Palmillas (pt.), Dom. Rep. 158/F6
Palmillas, Mexico 150/K5
Palmira, Colombia 120/B2
Palmira, Colombia 126/B6
Palmira, Cuba 158/E2
Palmitas, Uruguay 145/B4
Palmito de la Virgen (isl.), Mexico 150/F5
Palmito del Verde (isl.), Mexico 150/F5
Palm River-Clair Mel, Fla. (33619) 212/C3
Palms, Mich. (48465) 250/G5
Palms, Isle of (isl.), S.C. 296/H6
Palm Shores, Fla. (†32901) 212/F3
Palm Springs, Calif. 188/C4
Palm Springs, Calif. (92262) 204/J10
Palm Springs, Fla. (33460) 212/F5
Palmyra, Ill. (62674) 222/C4
Palmyra, Ind. (47164) 227/E8
Palmyra○, Maine (04965) 243/E6
Palmyra, Mich. (49268) 250/E7
Palmyra, Mo. (63461) 261/J3
Palmyra, Nebr. (68418) 264/H4
Palmyra, N.J. (08065) 273/B3
Palmyra, N.Y. (14522) 276/F4
Palmyra (atoll), Pacific 87/K5
Palmyra, Pa. (17078) 294/J5
Palmyra (ruin), Syria 59/C3
Palmyra (Tadmor) (ruins), Syria 63/H5
Palmyra, Tenn. (37142) 237/G8
Palmyra (isl.), U.S. 2/A5
Palmyra, Va. (22963) 307/M5
Palmyra, Wis. (53156) 317/H2
Palmyras (pt.), India 68/F4
Palnackie, Scotland 15/E6
Palni, India 68/D6
Palo, Iowa (52324) 229/K4
Palo, Mich. (48870) 250/E5
Palo, Minn. (†55705) 255/F3
Palo, Philippines 82/E5
Palo Alto, Calif. 188/B3
Palo Alto, Calif. (*94301) 204/K3
Palo Alto, Cuba 158/F3
Palo Alto (co.), Iowa 229/D2
Palo Alto (lake), Iowa 229/D2
Palo Bola, Mexico 150/D4
Palo Duro (creek), Texas 303/B2
Palo Duro (creek), Texas 303/C1
Paloemeu (riv.), Suriname 131/D4
Palolo (stream), Hawaii 218/D4
Paloma, Ill. (62359) 222/B3
Palomar (mt.), Calif. 204/J10
Palomas, Mexico 150/F1
Palomas, Uruguay 145/B2
Palombara Sabina, Italy 34/F6
Palometas, Bolivia 136/D5
Palompon, Philippines 82/E5
Palo Pinto (co.), Texas 303/F5
Palo Pinto, Texas (76072) 303/F5
Palopo, Indonesia 85/F6
Palos (cape), Spain 33/F4
Palo Santo, Argentina 143/E2
Palo Seco, P. Rico 161/D1
Palo Seco, Trin. & Tob. 161/A11
Palos Heights, Ill. (60463) 222/B6
Palos Hills, Ill. (60465) 222/B6
Palos Park, Ill. (60464) 222/B6
Palos Verdes Estates, Calif. (90274) 204/B11
Palotás, Hungary 41/E3
Palourde (lake), La. 238/H7
Palouse (riv.), Idaho 220/B3
Palouse, Wash. (99161) 310/H4
Palouse (riv.), Wash. 310/G4
Palo Verde, Ariz. (85343) 198/C5
Palo Verde, Calif. (92266) 204/L10
Palpa, Nepal 68/E3
Palpa, Peru 128/E10
Palsagua, Nicaragua 154/E4
Palsen (riv.), Manitoba 179/G2
Palu, Indonesia 85/F6
Palu, Turkey 63/H3
Paluan, Philippines 82/C4
Pama, Upper Volta 106/E6
Pamangkat, Indonesia 85/D5
Pamar, Colombia 126/E8
Pambrun, Sask. 181/D6
Pambula, N.S. Wales 97/E5
Pamekasan, Indonesia 85/L2
Pameungpeuk, Indonesia 85/H2
Pamiers, France 28/D6
Pamir (plat.) 54/J6
Pamlico (sound), N.C. 188/L3
Pamlico (co.), N.C. 281/R4
Pamlico (riv.), N.C. 281/R4
Pamlico (sound), N.C. 281/S4
Pampa, Texas 188/F3
Pampa, Texas (79065) 303/D2
Pampa Aullagas, Bolivia 136/B6
Pampachiri, Peru 128/F10
Pampacolca, Peru 128/F10
Pampa de la Salina (salt dep.), Argentina 143/C3
Pampa de las Salinas, Argentina 143/C3
Pampa de la Tres Hermanas (plain), Argentina143/C6
Pampa del Infierno, Argentina 143/D2
Pampa Grande, Bolivia 136/D5
Pampanga (prov.), Philippines 82/C3
Pampas (plain), Argentina 120/C6
Pampas (plain), Argentina 143/D4
Pampas, Peru 128/E9
Pampas (riv.), Peru 128/E9
Pampilhosa da Serra, Portugal 33/C3
Pamplico, S.C. (29583) 296/H4
Pamplin, Va. (23958) 307/L6
Pamplona, Colombia 126/D4
Pamplona, Spain 33/F1
Pamplona, Spain 7/D4
Pamunkey (riv.), Va. 307/O5
Pamunkey Ind. Res., Va. 307/P5
Pana, Ill. (62557) 222/D4
Panabá, Mexico 150/P6
Panabo, Philippines 82/E7
Panaca, Nev. (89042) 266/G5
Panacachi, Bolivia 136/B6
Panacea, Fla. (32346) 212/B1
Panache (lake), Ontario 177/C1
Panagyurishte, Bulgaria 45/F4
Panaitan (isl.), Indonesia 85/C7
Panaji, India 68/C5
Panama 2/E5
Panama 146/K9
Panama (canal) 2/E5
Panama, Ill. (62077) 222/D4
Panama, Iowa (51562) 229/B5
Panama, Nebr. (68419) 264/H4
Panama, N.Y. (14767) 276/A6
Panama, Okla. (74951) 288/S4
Panamá (cap.), Pan. 146/L9
Panama (canal), Pan. 146/L8
Panamá (gulf), Pan. 146/L9
PANAMA 154/G6
Panamá (cap.), Panama 154/H6
Panamá (gulf), Panama 154/H7
Panama City, Fla. 188/K4
Panama City, Fla. (*32401) 212/C6
Panama City Beach, Fla. (32407) 212/C6
Panamint (range), Calif. 204/H7
Panamint (valley), Calif. 204/H7
Panao, Peru 128/E7
Panaon (isl.), Philippines 82/E5
Panarea (isl.), Italy 34/E5
Panaro (riv.), Italy 34/C2
Panarukan, Indonesia 85/K2
Panay (isl.), Philippines 54/O8
Panay (isl.), Philippines 85/G3
Panay (isl.), Philippines 82/D5
Pancake (range), Nev. 266/F4
Pančevo, Yugoslavia 45/E3
Panchor, Malaysia 72/F5
Panchur, India 68/F2
Panciu, Romania 45/H3
Panda, Mozambique 118/E4
Pandale, Texas (76944) 303/C7
Panda-Likasi, Zaire 115/E6
Panda-Likasi, Zaire 102/E6
Pandan, Antique, Philippines 82/C5
Pandan, Catanduanes, Philippines 82/E3
Pan de Azúcar, Quebrado (riv.), Chile 138/B5
Pan de Azúcar, Uruguay 145/D5
Pandeglang, Indonesia 85/G1
Pandharpur, India 68/D5
Pandi Pandi, S. Australia 94/F2
Pando (dept.), Bolivia 136/B2
Pando, Cerro (mt.), Panama 154/F6
Pando, Uruguay 145/B6
Pando (riv.), Uruguay 145/B6
Pandora, Ohio (45877) 284/C4
Pandrup, Denmark 21/C3
Panevežys, U.S.S.R. 52/B3
Panevežys, U.S.S.R. 53/C3
Panfilov, U.S.S.R. 48/H5
Pangai, Tonga 87/J7
Pangala, Congo 115/B4
Pangalanes (canal), Madagascar 118/H4
Pangani, Tanzania 115/G5
Pangani (riv.), Tanzania 115/G4
Panganiban, Philippines 82/E4
Pangasinan (prov.), Philippines 82/C3
Pangburn, Ark. (72121) 202/G3
Pangi, Zaire 115/E4
Pangkalanberandan, Indonesia 85/B5
Pangkalanbuun, Indonesia 85/E6
Pangkalpinang, Indonesia 85/D6
Pangkor, Pulau (isl.), Malaysia 72/D6
Panglao (isl.), Philippines 82/D6
Pangman, Sask. 181/G6
Pangnirtung, Canada 4/C13
Pangnirtung, N.W.T. 162/K2
Pangnirtung, N.W. Terrs. 187/M3
Pangong Tso (lake), India 68/D2
Pangsau (pass), Burma 72/C1
Panguipulli, Chile 138/E2
Panguitch, Utah (84759) 304/B6
Panguitch (creek), Utah 304/B6
Pangutaran, Philippines 82/C7
Pangutaran (isl.), Philippines 82/C7
Pangutaran Group (isls.), Philippines 82/C7
Pangutaran Group (isls.), Philippines 85/G4
Panhandle, Texas (79068) 303/C2
Paniau (peak), Hawaii 218/A2
Panié (mt.), New Caled. 86/G4
Panihati, India 68/F1
Panipat, India 68/D3
Paniqui, Philippines 82/C3
Panj (riv.), Afghanistan 68/C1

Panjab, Afghanistan 68/B2
Panjab, Afghanistan 59/J3
Panjang, Hon (Hon Tho Chau) (isl.), Vietnam 72/D5
Panjgur, Pakistan 68/A3
Panjgur, Pakistan 59/H4
Panjim, India 54/J8
Panjim (Panaji), India 68/C5
Pankow, E. Germany 22/F3
Pankshin, Nigeria 106/F7
P'anmunjŏm, N. Korea 81/C5
P'anmunjŏm, S. Korea 81/C5
Panmure (isl.), Pr. Edward I. 168/F2
Panna, India 68/E4
Pannawonica, W. Australia 92/B3
Pannonhalma, Hungary 41/D3
Panny (riv.), Alberta 182/C1
Panola, Ala. (35477) 195/B5
Panola, Ill. (†61738) 222/E3
Panola (co.), Miss. 256/E2
Panola, Okla. (74559) 288/R5
Panola (co.), Texas 303/K5
Panora, Iowa (50216) 229/E5
Panorama Park, Iowa (†52722) 229/N5
Panquehue, Chile 138/G2
Panruti, India 68/D6
Pansey, Ala. (36370) 195/H8
Pantanal (reg.), Brazil 120/D4
Pantar (isl.), Indonesia 85/G7
Pantego, N.C. (27860) 281/R3
Pantego, Texas (76013) 303/F2
Pantelleria, Italy 34/C6
Pantelleria (isl.), Italy 7/F5
Pantelleria (isl.), Italy 34/D6
Pantha, Burma 72/B2
Panther (creek), Idaho 220/D4
Panther (creek), Ky. 237/G5
Panther, W. Va. (24872) 312/C8
Panther Burn, Miss. (38765) 256/C4
Panthersville, Georgia (†30032) 217/L1
Pantin, France 28/B1
Pantoja, Peru 128/E3
Panton○, Vt. (†05491) 268/A3
Pánuco, Mexico 150/K6
Pánuco (riv.), Mexico 150/K5
Panuke (lake), Nova Scotia 168/D4
Pan Xian, China 77/G6
Panyam, Nigeria 106/F7
Panzós, Guatemala 154/C3
Pao (riv.), Venezuela 124/D3
Pao (riv.), Venezuela 124/F3
Paoki (Baoji), China 77/G5
Paola, Italy 34/E5
Paola, Kansas (66071) 232/H3
Paoli, Colo. (80746) 208/P1
Paoli, Ind. (47454) 227/E7
Paoli, Okla. (73074) 288/M5
Paoli, Pa. (19301) 294/M5
Paoli, Wis. (†53508) 317/G10
Paonia, Colo. (81428) 208/D5
Paopao (bay), Fr. Poly. 86/S12
Paoting (Baoding), China 77/J4
Paotow (Baotou), China 77/G3
Paoua, Cent. Afr. Rep. 115/C2
Paoy Pet, Cambodia 72/D4
Papa, Hawaii (†96704) 218/G6
Pápa, Hungary 41/D3
Papaaloa, Hawaii (96780) 218/J4
Papagaio (riv.), Brazil 132/B6
Papagayo (gulf), C. Rica 154/E5
Papago Ind. Res., Ariz. 198/C6
Papaikou, Hawaii 188/G6
Papaikou, Hawaii (96781) 218/J5
Papakura, N. Zealand 100/E2
Papallacta, Ecuador 128/D3
Papanoa, Mexico 150/J8
Papantla de Olarte, Mexico 150/L6
Papar, Malaysia 85/F4
Papara, Fr. Poly. 86/S13
Papa Stour (isl.), Scotland 10/G1
Papa Stour (isl.), Scotland 15/F2
Papatoetoe, N. Zealand 100/C1
Papa Westray (isl.), Scotland 15/F1
Papa Westray (isl.), Scotland 10/E1
Papeete (cap.), Fr. Polynesia 2/B6
Papeete (cap.), Fr. Poly. 86/S13
Papeete (cap.), Fr. Poly. 87/M7
Papelón, Venezuela 124/D3
Papenburg, W. Germany 22/B2
Papenoo, Fr. Poly. 86/T12
Papetoai, Fr. Poly. 86/S12
Paphos, Cyprus 63/E5
Papillion, Nebr. (68046) 264/J3
Papineau, Ill. (60956) 222/F3
Papineau (lake), Ontario 177/G2
Papineau (co.), Québec 172/B4
Papineau (lake), Québec 172/C4
Papineauville, Québec 172/C4
Paposo, Chile 138/A5
Papradno, Czech. 41/E2
Paps, The (mt.), Ireland 17/C7
Paps of Jura (mt.), Scotland 15/C5
Papua (gulf), Papua N.G. 87/E6
Papua New Guinea 2/S6
PAPUA NEW GUINEA 86/B1
PAPUA NEW GUINEA 85/B7
Papua New Guinea 87/E6
Papudo, Chile 138/A9
Papun, Burma 72/C3
Papunáua (riv.), Colombia 126/E6
Papunya, North. Terr. 93/B7
Papurí (riv.), Colombia 126/F7
Paquera, C. Rica 154/E6
Paquette, Québec 172/F4
Paquetville, New Bruns. 170/E1
Pará (state), Brazil 132/C4
Pará (Belém), Brazil 132/E3
Pará (est.), Brazil 120/E3
Pará (riv.), Brazil 132/D3
Para (dist.), Suriname 131/D3
Paraburdoo, W. Australia 88/B4
Paraburdoo, W. Australia 92/B3
Paracale, Philippines 82/D3
Paracas (pen.), Peru 128/D9
Paracatu, Brazil 132/E7
Paracatu (riv.), Brazil 132/E7
Paracel (isls.), China 85/E2
Parachilna, S. Australia 88/F6
Parachilna, S. Australia 94/F4
Parachute, Colo. 208/C4
Paraćin, Yugoslavia 45/E4
Parada Esperanza, Uruguay 145/B3
Parada Liebigs, Uruguay 145/A4
Parada Rivas, Uruguay 145/B2
Parade, S. Dak. (57647) 298/G3
Pará de Minas, Brazil 132/E7
Pará de Minas, Brazil 135/D1
Paradip, India 68/F4
Paradis, La. (70080) 238/M4
Paradise, Ariz. (†85632) 198/F7
Paradise, Calif. (95969) 204/D4
Paradise, Guyana 131/C3
Paradise, Kansas (67658) 232/D2
Paradise, Mich. (49768) 250/D2
Paradise (lake), Mich. 250/E3
Paradise, Mo. (†64089) 261/D4
Paradise, Mont. (59856) 262/B3
Paradise, Newf. 166/D2
Paradise (riv.), Newf. 166/C3
Paradise, Nova Scotia 168/C4
Paradise (lake), Nova Scotia 168/C4
Paradise, Pa. (17562) 294/K5
Paradise, Texas (76073) 303/G5
Paradise, Utah (84328) 304/C2
Paradise, W. Va. (†25124) 312/C5
Paradise Hill, Okla. (†74435) 288/R3
Paradise Hill, Sask. 181/B2
Paradise Inn, Wash. (98398) 310/D4
Paradise River, Newf. 166/C3
Paradise Valley, Alberta 182/E3
Paradise Valley, Ariz. (85253) 198/D5
Paradise Valley, Nev. (89119) 266/F6
Paradise Valley, Nev. (89426) 266/D1
Paradise Valley, Wyo. (†82601) 319/F3
Paradisino (peak), Switzerland 39/K4
Paradiso, Switzerland 39/G5
Paradox, Colo. (81429) 208/B6
Paragon, Ind. (46166) 227/D6
Paragonah, Utah (84760) 304/B6
Paragould, Ark. (72450) 202/J1
Paraguá (riv.), Bolivia 136/E4
Paragua (riv.), Venezuela 124/G4
Paraguaçu (riv.), Brazil 120/F4
Paraguaçu (riv.), Brazil 132/F6
Paraguaçu Paulista, Brazil 132/D8
Paraguai (riv.) 120/D4
Paraguai (riv.), Brazil 132/B8
Paraguaipoa, Venezuela 124/C2
Paraguaná (pen.), Venezuela 124/C1
Paraguarí (dept.), Paraguay 144/D4-5
Paraguay 2/F7
Paraguay 120/D5
PARAGUAY 144
Paraguay (riv.), Argentina 143/E1
Paraguay (riv.), Bolivia 136/F7
Paraguay (riv.), Paraguay 144/D4
Paraíba (state), Brazil 132/G4
Paraíba (riv.), Brazil 120/E5
Paraíba (riv.), Brazil 135/E2
Paraíba do Sul, Brazil 135/E3
Parainen, Finland 18/M6
Paraíso, C. Rica 154/F6
Paraíso, Dom. Rep. 158/D7
Paraíso, Mexico 150/N7
Paraíso de Chabasquén, Venezuela 124/D3
Parakou, Benin 106/E7
Parallel, Kansas (†66933) 232/F2
Paraloma, Ark. (†71846) 202/B6
Paramaribo (dist.), Suriname 131/D2
Paramaribo (cap.), Suriname 131/D2
Paramaribo (cap.), Suriname 2/G5
Paramaribo (cap.), Suriname 120/D2
Paramithiá, Greece 45/E6
Paramonga, Peru 128/C8
Paramount, Calif. (90723) 204/C11
Paramus, N.J. (07652) 273/B1
Paramushir (isl.) 54/S5
Paramushir (isl.), U.S.S.R. 48/Q4
Paran (dry riv.), Israel 65/D5
Paraná (riv.) 2/G7
Paraná (riv.) 120/D5
Paraná, Argentina 143/F5
Paraná, Argentina 120/D6
Paraná (riv.), Argentina 143/E2
Paraná (state), Brazil 132/D9
Paraná (state), Brazil 135/B4
Paranã, Brazil 132/E6
Paranã (riv.), Brazil 132/C8
Paranã (riv.), Brazil 132/E6
Paranaguá, Brazil 120/E5
Paranaguá, Brazil 132/E9
Paranaguá, Brazil 135/B4
Paranaíba, Brazil 132/D7
Paranam, Suriname 131/D3
Paranapanema (riv.), Brazil 132/C8
Paranapanema (riv.), Brazil 135/B3
Paranapiacaba (range), Brazil 135/B4
Paranatinga (riv.), Brazil 120/D4
Paranatinga (riv.), Brazil 132/C6
Parang, Maguindanao, Philippines 82/E7
Parang, Sulu, Philippines 82/C8
Parao (riv.), Uruguay 145/E3
Paraparaumu, N. Zealand 100/E4
Parapetí (riv.), Bolivia 136/D6
Para Station, N. S. Wales 97/B3
Parati, Brazil 135/D3
Paratinga, Brazil 132/F6
Paray-le-Monial, France 28/F4
Parbhani, India 68/D5
Parchim, E. Germany 22/D2
Parchman, Miss. (38738) 256/D3
Parchment, Mich. (49004) 250/D6
Parczew, Poland 47/F3
Pardee (res.), Calif. 204/C9
Pardee, Va. (†24285) 307/C6
Pardeeville, Wis. (53954) 317/H8
Pardes Hanna-Karkur, Israel 65/B2
Parding, China 77/C5
Pardo (riv.), Brazil 132/D8
Pardo (riv.), Brazil 132/F6
Pardo (riv.), Brazil 135/B2
Pardo (riv.), Brazil 132/C8
Pardoe, Pa. (†16137) 294/B3
Pardoo, W. Australia 92/B3
Pardubice, Czech. 41/C1
Pare, Indonesia 85/K2
Parece Vela (isl.), Japan 54/P7
Parece Vela (isl.), Japan 87/D3
Parecis (mts.), Brazil 120/C4
Parecis, Serra dos (range), Brazil 132/B6
Paredes de Nava, Spain 33/D1
Paredones, Chile 138/A10
Parent, Québec 174/C3
Pareora, N. Zealand 100/C6
Parepare, Indonesia 85/F6
Parguera, P. Rico 161/A3
Parham, Ant. & Bar. 161/E11
Parham, Ontario 177/H3
Parhams, La. (†71343) 238/G4
Paria (gulf) 120/C1
Paria (plat.), Ariz. 198/D2
Paria (riv.), Ariz. 198/D1
Paria, Bolivia 136/B5
Paria (gulf), Trin. & Tob. 156/G5
Paria (gulf), Trin. & Tob. 161/A11
Paria (riv.), Utah 304/B6
Paria (gulf), Venezuela 124/H2
Paria (pen.), Venezuela 124/G2
Pariaguán, Venezuela 124/F3
Pariaman, Indonesia 85/B6
Paricutín (vol.), Mexico 150/H7
Parida (isl.), Panama 154/F6
Parika, Guyana 131/B2
Parikkala, Finland 18/Q6
Parima, Sierra (mts.), Venezuela 124/F6
Parinacochas (lake), Peru 128/F10
Parinacota, Cerro (mt.), Chile 138/B1
Parinari, Peru 128/E5
Pariñas (pt.), Peru 128/B5
Parintins, Brazil 120/D3
Parintins, Brazil 132/B3
Paris, Ark. (72855) 202/C3
Paris (city) (dept.), France 28/B2
Paris (cap.), France 2/J3
Paris (cap.), France 7/E4
Paris (cap.), France 28/B2
Paris, Idaho (83261) 220/G7
Paris, Ill. (61944) 222/F4
Paris, Iowa (†52214) 229/K4
Paris, Ky. (40361) 237/N4
Paris○, Maine (04271) 243/B7
Paris, Mich. (49338) 250/D5
Paris, Miss. (38949) 256/F2
Paris, Mo. (65275) 261/J4
Paris, Ohio (44669) 284/H4
Paris, Ontario 177/D4
Paris, Tenn. (38242) 237/E8
Paris, Texas (75460) 303/J4
Paris, Texas 188/G4
Paris, Va. (22130) 307/N3
Paris Crossing, Ind. (47270) 227/F7
Parish, N.Y. (13131) 276/H4
Parish, Uruguay 145/C3
Parishville, N.Y. (13672) 276/L1
Parisville, Mich. (†48470) 250/G5
Parisville, Québec 172/F3
Parita, Panama 154/G6
Parita (bay), Panama 154/G6
Park (co.), Colo. 208/H4
Park (range), Colo. 208/F1
Park (riv.), Conn. 210/E2
Park, Kansas (67751) 232/B2
Park (co.), Mont. 262/F5
Park (riv.), N. Dak. 282/R3
Park (dist.), Scotland 15/B2
Park (co.), Wyo. 319/C1
Parkano, Finland 18/N6
Parkbeg, Sask. 181/E5
Park City, Ill. (†60085) 222/B4
Park City, Kansas (†67201) 232/E4
Park City, Ky. (42160) 237/J6
Park City, Mont. (59063) 262/H5
Park City, Utah (84060) 304/C3
Parkdale, Ark. (71661) 202/H7
Parkdale, Colo. (†81212) 208/H6
Parkdale, Oreg. (97041) 291/F2
Parkdale, Pr. Edward I. 168/E2
Parke (co.), Ind. 227/C5
Parker, Ariz. (85344) 198/A4
Parker (dam), Ariz. 198/A4
Parker, Colo. (80134) 208/K4
Parker, Fla. (32401) 212/C6
Parker, Idaho (83438) 220/G6
Parker, Kansas (66072) 232/H3
Parker, Pa. (16049) 294/C3
Parker, S. Dak. (57053) 298/P7
Parker (lake), S. Dak. 298/P3
Parker (co.), Texas 303/G5
Parker, Texas (†75069) 303/H1
Parker, Wash. (98939) 310/E4
Parker City, Ind. (47368) 227/G4
Parker Dam, Calif. (92267) 204/L9
Parkersburg, Ill. (62452) 222/F5
Parkersburg, Ind. (†47954) 227/D5
Parkersburg, Iowa (50665) 229/H3
Parkersburg, W. Va. 188/K3
Parkersburg, W. Va. (26101) 312/D4
Parkers Cove, Newf. 166/D4
Parkers Cove, Nova Scotia 168/C4
Parkers Lake, Ky. (42634) 237/M7
Parkers Prairie, Minn. (56361) 255/C4
Parkertown, N.J. (†08087) 273/E4
Parkerview, Sask. 181/H4
Parkerville, Kansas (†66846) 232/F3
Parkes, N. S. Wales 88/H6
Parkes, N.S. Wales 97/E3
Parkesburg, Pa. (19365) 294/L6
Park Falls, Wis. (54552) 317/F4
Park Forest, Ill. (60466) 222/B6
Park Forest South, Ill. (60466) 222/F2
Park Hall, Md. (20667) 245/N8
Park Hill, Okla. (74451) 288/R3
Parkhill, Ontario 177/C4
Park Hills, Ky. (†41011) 237/S2
Parkin, Ark. (72373) 202/J3
Parkland, Alberta 182/D4
Parkland, Fla. (†33441) 212/F5
Parkland, Okla. (†74824) 288/N3
Parkland, Wash. (98444) 310/C3
Parkman○, Maine (†04443) 243/D5
Parkman, Ohio (44080) 284/H3
Parkman, Sask. 181/K6
Parkman, Wyo. (82838) 319/E1
Park Place, Oreg. (†97045) 291/B2
Park Rapids, Minn. (56470) 255/D4
Park Rapids, Wash. (†99114) 310/H2
Park Ridge, Ill. (60068) 222/B5
Park Ridge, N.J. (07656) 273/B1
Park Ridge, Wis. (†54481) 317/H6
Park River, N. Dak. (58270) 282/P3
Parks, Ariz. (86018) 198/C3
Parks, Ark. (72950) 202/B4
Parks, La. (70582) 238/G6
Parks, Nebr. (69041) 264/C4
Parkside, Pa. (†19013) 294/M7
Parkside, Sask. 181/E2
Parksley, Va. (23421) 307/S5
Parkston, S. Dak. (57366) 298/O7
Parksville, Br. Col. 184/J3
Parksville, Ky. (40464) 237/M5
Parksville, N.Y. (12768) 276/L7
Parksville, S.C. (29844) 296/C4
Parkton, Md. (21120) 245/M2
Parkton, N.C. (28371) 281/M5
Park Valley, Utah (84329) 304/A2
Parkview (mt.), Colo. 208/G2
Parkville, Md. (21234) 245/M3
Parkville, Mo. (64152) 261/O5
Parkville, Pa. (†17331) 294/J6
Parkville, Victoria 97/H5
Parkway, Mo. (64130) 261/L6
Parkway Village, Ky. (†40201) 237/J2
Parkwood, N.C. (27707) 281/M3
Parlakhemundi, India 68/E5
Parlier, Calif. (93648) 204/F7
Parlin, Colo. (81239) 208/F6
Parlin (pond), Maine 243/C4
Parma, Idaho (83660) 220/B6
Parma (prov.), Italy 34/C2
Parma, Italy 7/E4
Parma, Italy 34/C2
Parma (riv.), Italy 34/C2
Parma, Mich. (49269) 250/E6
Parma, Mo. (63870) 261/N9
Parma, Ohio (44129) 284/H9
Parmachenee (lake), Maine 243/B5
Parma Heights, Ohio (†44130) 284/G9
Parmana, Venezuela 124/F4
Parmele, N.C. (27861) 281/P3
Parmelee, S. Dak. (57566) 298/G7
Parmer (co.), Texas 303/B3
Parnaguá, Brazil 132/E5
Parnaíba, Brazil 132/F3
Parnaíba, Brazil 120/E3
Parnaíba (riv.), Brazil 120/E3
Parnaíba (riv.), Brazil 132/F3
Parnamirim, Brazil 132/F4
Parnassus (mt.), Greece 45/F6
Parnassus, N. Zealand 100/D5
Parndana, S. Australia 94/E8
Parnell, Iowa (52325) 229/J5
Parnell, Mo. (64475) 261/C2
Pärnu, U.S.S.R. 7/G3
Pärnu, U.S.S.R. 53/C1
Pärnu, U.S.S.R. 52/C3
Pärnu, U.S.S.R. 48/C4
Paro, Bhutan 68/F3
Paron, Ark. (72132) 202/E4
Paroo (riv.), N. S. Wales 88/G5
Paroo (chan.), N.S. Wales 97/B2
Paroo (riv.), N.S. Wales 97/C1
Paroo (riv.), Queensland 95/C6
Paropamisus (mts.), Afghanistan 59/H3
Paropamisus (range), Afghanistan 68/A2
Páros (isl.), Greece 45/G7
Parow, S. Africa 118/F6
Parowan, Utah (84761) 304/B6
Parpan, Switzerland 39/J3
Parr, Ind. (†47978) 227/C2
Parr, S.C. (29066) 296/E3
Parral, Chile 138/A11
Parral, Mexico 150/G3
Parral, Ohio (†44622) 284/G4
Parramatta, N. S. Wales 88/K4
Parramatta (riv.), N. S. Wales 88/K4
Parramatta, N.S. Wales 97/J3
Parramatta (riv.), N.S. Wales 97/J3
Parramore (isl.), Va. 307/S5
Parran, Md. (†20639) 245/M6
Parras de la Fuente, Mexico 150/H4
Parratah, Tasmania 99/D4
Parrett (riv.), England 13/E6
Parrish, Ala. (35580) 195/D3
Parrish, Fla. (33564) 212/D4
Parrish, Wis. (†54435) 317/H5
Parris Island Marine Base, S.C. 296/F7
Parrott, Georgia (31777) 217/D7
Parrott, Va. (24132) 307/G6
Parrottsville, Tenn. (37843) 237/P8
Parrsboro, Nova Scotia 168/D3
Parry (isls.), N.W.T. 146/G2
Parry (chan.), N.W.T. 146/G2
Parry (chan.), N.W.T. 162/E-H1
Parry (bay), N.W. Terrs. 187/K3
Parry (cape), N.W. Terrs. 187/F2
Parry (chan.), N.W. Terrs. 187/G2
Parry (isls.), N.W. Terrs. 187/G2
Parry (pen.), N.W. Terrs. 187/F2
Parry (isl.), Ontario 177/D2
Parry (sound), Ontario 177/D2
Parry, Sask. 181/G6
Parry Sound, Ont. 162/J6
Parry Sound (terr. dist.), Ontario 175/E3
Parry Sound, Ontario 175/D3
Parry Sound, Ontario 177/E2
Parseierspitze (mt.), Austria 41/A3
Parshall, Colo. (80468) 208/G2
Parshall, N. Dak. (58770) 282/F4
Parsippany-Troy Hills○, N.J. (07054) 273/E2
Parsnip (riv.), Br. Col. 184/F3
Parson, Br. Col. 184/J4
Parsons, Kans. 188/G3
Parsons, Kansas (67357) 232/G4
Parsons, Tenn. (38363) 237/E9
Parsons, W. Va. (26287) 312/G4
Parsonsburg, Md. (21849) 245/R7
Parson's Pond, Newf. 166/C3
Partanna, Italy 34/D6
Partapgarh, India 68/C4
Parthenay, France 28/C4
Partinico, Italy 34/D6
Partizansk, U.S.S.R. 48/O5
Partizánske, Czech. 41/E2
Partlow, Va. (22534) 307/N4
Partridge, Kansas (67566) 232/D4
Partridge (riv.), Minn. 255/G3
Partridge (bay), Newf. 166/C3
Partridge (pt.), Newf. 166/C3
Partry (mts.), Ireland 17/C4
Paru (riv.), Brazil 132/C3
Paru de Oeste (riv.), Brazil 120/D3
Paru de Oeste (riv.), Brazil 132/B3
Paruro, Peru 128/F9
Parvatipuram, India 68/E5
Parys, S. Africa 118/D5
Pas, De (riv.), Québec 174/D1
Pasadena, Calif. 188/C4
Pasadena, Calif. (*91101) 204/C10
Pasadena, Md. (21122) 245/M4
Pasadena, Newf. 166/C4
Pasadena, Texas (*77501) 303/J2
Pasado (cape), Ecuador 128/B3
Pasaje, Ecuador 128/C4
Pa Sak, Mae Nam (riv.), Thailand 72/D4
Pasangkayu, Indonesia 85/F6
Pasargadae (ruins), Iran 66/H5
Pasatiempo, Calif. (†95060) 204/K4
Pasawng, Burma 72/C3
Pasayten (riv.), Wash. 310/E2
Pascagoula, Miss. (39567) 256/G10
Pascagoula (riv.), Miss. 256/G9
Pascalis, Québec 174/B3
Paşcani, Romania 45/H2
Paschall, N.C. (†27589) 281/N1
Pasco (co.), Fla. 212/D3
Pasco (dept.), Peru 128/E8
Pasco, Wash. (99301) 310/F4
Pascoag, R.I. (02859) 249/H5
Pascola, Mo. (63871) 261/N10
Pascua (riv.), Chile 138/D7
Pas-de-Calais (dept.), France 28/E2
Pasewalk, E. Germany 22/F2
Pasighat, India 68/G3
Pasinler, Turkey 63/J3
Pasión (riv.), Guatemala 154/B2
Paskenta, Calif. (96074) 204/C4
Pasłęk, Poland 47/D1
Pasley (bay), N.W. Terrs. 187/J2
Pasni, Pakistan 68/A3
Pasni, Pakistan 54/H7
Pasni, Pakistan 59/H4
Paso Ataques, Uruguay 145/D2
Paso Barreto, Paraguay 144/D3
Paso de Andrés Pérez, Uruguay 145/B3
Paso de Indios, Argentina 143/C5
Paso de la Laguna, Salto, Uruguay 145/B2
Paso de la Laguna, Tacuarembó, Uruguay 145/D3
Paso de las Piedras, Uruguay 145/C2
Paso del Borracho, Uruguay 145/D2
Paso del Cerro, Uruguay 145/C2
Paso de León, Uruguay 145/B1
Paso del Horno, Uruguay 145/C2
Paso de Los Libres, Argentina 143/E2
Paso de los Toros, Uruguay 145/C3
Paso del Parque, Uruguay 145/B2
Paso de Ovejas, Mexico 150/Q2
Paso de Patria, Paraguay 144/C5
Paso de Ramos, Uruguay 145/C1
Paso de Uleste, Uruguay 145/B3
Paso Flores, Argentina 143/C5
Paso Hondo, Uruguay 145/B4
Paso Potrero, Uruguay 145/C2
Pasorapa, Bolivia 136/C6
Paso Real, Honduras 154/E3
Paso Robles, Calif. (93446) 204/E8
Paspébiac, Québec 172/D2
Pasqua, Sask. 181/F5
Pasque (isl.), Mass. 249/L7
Pasquia (hills), Sask. 181/J2
Pasquia (riv.), Sask. 181/K2
Pasquotank (co.), N.C. 281/S2
Pass (creek), Wyo. 319/F4
Passaconaway (mt.), N.H. 268/E4
Passadumkeag○, Maine (04475) 243/F5
Passage (isl.), Mich. 250/E1
Passage East, Ireland 17/G7
Passagem Franca, Brazil 132/E4
Passage West, Ireland 17/E8
Passage West, Ireland 10/B5
Passaic, Mo. (64777) 261/D6
Passaic (co.), N.J. 273/E1
Passaic, N.J. (07055) 273/E2
Passaic (riv.), N.J. 273/E2
Passamaquoddy (bay), Maine 243/J5
Passamaquoddy (bay), New Bruns. 170/C3
Passamaquoddy Ind. Res., Maine 243/J6
Passau, W. Germany 22/E4
Pass Christian, Miss. (39571) 256/F10
Passero (cape), Italy 7/F5
Passero (cape), Italy 34/E6
Passes (lake), Québec 172/F2
Passi, Philippines 82/D5
Passo Fundo, Brazil 120/D5
Passo Fundo, Brazil 132/D10
Passos, Brazil 132/E8
Passos, Brazil 135/C2
Passumpsic, Vt. (05861) 268/D3
Passumpsic (riv.), Vt. 268/D2
Pastaza (riv.) 120/B3
Pastaza (prov.), Ecuador 128/D3
Pastaza (riv.), Ecuador 128/D4
Pastaza (riv.), Peru 128/D5
Pasto, Colombia 120/B2
Pasto, Colombia 126/B7
Pastol (bay), Alaska 196/F2
Pastora (peak), Ariz. 198/F2
Pastos Bons, Brazil 132/E4
Pastrana, Spain 33/E2
Pastura, N. Mex. (88435) 274/E4
Pasuquin, Philippines 82/C1
Pasuruan, Indonesia 85/K2
Pasvalys, U.S.S.R. 53/C2
Pasvikelv (riv.), Norway 18/Q2
Paswegin, Sask. 181/H4
Pásztó, Hungary 41/E3
Pata, Bolivia 136/A4
Patacamaya, Bolivia 136/B5
Patagonia (reg.), Argentina 120/C7
Patagonia (reg.), Argentina 143/C5
Patagonia, Ariz. (85624) 198/E7
Pataguanset (lake), Conn. 210/G3
Pataha, Wash. (†99347) 310/H4
Pataha (creek), Wash. 310/H4
Patan, India 68/C4
Patapédia (riv.), New Bruns. 170/C1
Patapédia (riv.), Québec 172/B2
Patapsco, Md. (†21048) 245/L2
Patapsco (riv.), Md. 245/M4
Pataskala, Ohio (43062) 284/E5
Pataz, Peru 128/D6
Patchewollock, Victoria 97/A4
Patch Grove, Wis. (53817) 317/D10
Patchogue, N.Y. (11772) 276/P9
Patea, N. Zealand 100/E3
Paternion, Austria 41/B3
Paterno, Italy 34/E6
Pateros, Wash. (98846) 310/E2
Pateros (lake), Wash. 310/F2
Paterson, N.J. 188/M2
Paterson, N.J. (*07501) 273/B2
Paterson, Wash. (99345) 310/F5
Patesville, Ky. (†42364) 237/H5
Pathankot, India 68/D2
Pathfinder (res.), Wyo. 188/E2
Pathfinder (res.), Wyo. 319/F3
Pathiu, Thailand 72/C5
Pathlow, Sask. 181/G3
Pati (pt.), Guam 86/K6
Pati, Indonesia 85/J2
Patía, Colombia 126/B6
Patía (riv.), Colombia 126/B6
Patiala, India 68/D2
Patillas, P. Rico 161/E2
Patillas (lake), P. Rico 161/E2
Pativilca (riv.), Peru 128/D8
Patmos, Ark. (†71801) 202/C7
Pátmos (isl.), Greece 45/H7
Patna, India 54/K7
Patna, India 68/F3
Patna, Scotland 15/D5
Patnanongan (isl.), Philippines 82/D3
Patnos, Turkey 63/K3
Patoka, Ill. (62875) 222/D5
Patoka, Ind. (47666) 227/B8
Patoka (riv.), Ind. 227/C8
Paton, Iowa (50217) 229/E4
Patos, Brazil 120/F3
Patos, Brazil 132/G4
Patos (lake), Brazil 120/D6
Patos (lag.), Brazil 132/D10
Patos de Minas, Brazil 120/E4
Patos de Minas, Brazil 132/E7
Patoutville, La. (†70544) 238/G7
Patquía, Argentina 143/C3
Pátrai, Greece 7/G5
Pátrai, Greece 45/E6
Patricia, Alberta 182/E4
Patricia, S. Dak. (†57551) 298/G7
Patricia, Texas (79352) 303/B5
Patricio Lynch (isl.), Chile 138/D7
Patrick, Neth. Ant. 161/F8
Patrick, S.C. (29584) 296/G2
Patrick (co.), Va. 307/H7
Patrick A.F.B., Fla. 212/F3
Patricksburg, Ind. (47455) 227/D6
Patrick's Cove, Newf. 166/C2
Patrick Springs, Va. (24133) 307/H7
Patrickswell, Ireland 17/D6
Patriot, Ind. (47038) 227/H7
Patriot, Ohio (45658) 284/F8
Patrocínio, Brazil 132/E7
Patronville, Ind. (†47635) 227/C9
Patroon, Texas (75967) 303/L6
Patsaliga (creek), Ala. 195/F7
Patsburg, Ala. (†36049) 195/F7
Patta (isl.), Kenya 115/H4
Pattani, Thailand 72/D6
Patten, Maine (04765) 243/F4
Patten○, Maine (04765) 243/F4
Pattenburg, N.J. (†08802) 273/C2
Patterson, Ark. (72123) 202/H3
Patterson, Calif. (95363) 204/D6
Patterson, Georgia (31557) 217/H8
Patterson, Idaho (†83253) 220/E5
Patterson, Ill. (62078) 222/C4
Patterson, Iowa (50218) 229/F6
Patterson, La. (70392) 238/H7
Patterson (pt.), Mich. 250/D3
Patterson, Mo. (63956) 261/L8
Patterson, N.Y. (12563) 276/N7
Patterson, N.C. (28661) 281/F3
Patterson, Edward A. (lake), N. Dak. 282/E6
Patterson, Ohio (45843) 284/C4
Patterson, Va. (24633) 307/D6
Patterson (creek), W. Va. 312/J4
Patterson Creek, W. Va. (26746) 312/J3
Pattersonville, N.Y. (12137) 276/M5
Patti, Italy 34/E5
Pattison, Miss. (39144) 256/C7
Patton, Mo. (63662) 261/M8
Patton, Pa. (16668) 294/E4
Pattonsburg, Mo. (64670) 261/D2
Patuanak, Sask. 181/L3
Patuca, Honduras 154/E3
Patuca (pt.), Honduras 154/E3
Patuca (riv.), Honduras 154/E3
Patuha (mt.), Indonesia 85/H2
Pătulele, Romania 45/F3
Patutahi, N. Zealand 100/F3
Patuxent (riv.), Md. 245/M7
Patuxent River Nav. Air Test Ctr., Md. 245/N7
Patzau, Wis. (†54836) 317/B3
Pátzcuaro, Mexico 150/J7

Pátzcuaro (lake), Mexico 150/J7
Pau, France 28/C6
Pau, France 7/D4
Paucarbamba, Peru 128/E9
Paucartambo, Cusco, Peru 128/G9
Paucartambo, Pasco, Peru 128/E8
Paudash (lake), Ontario 177/F3
Pau dos Ferros, Brazil 132/G4
Pauillac, France 28/C5
Paukaa, Hawaii (†96781) 218/J5
Paul, Ala. (36469) 195/E8
Paul, Idaho (83347) 220/E7
Paul (isl.), Newf. 166/B2
Paul (stream), Vt. 268/D2
Paulatuk, N.W.T. 162/D2
Paulatuk, N.W. Terrs. 187/F3
Paulaya (riv.), Honduras 154/E3
Paulden, Ariz. (86334) 198/C4
Paulding (co.), Georgia 217/C3
Paulding, Miss. (39348) 256/F6
Paulding (co.), Ohio 284/A3
Paulding, Ohio (45879) 284/A3
Paulette, Miss. (39349) 256/H4
Paulina, La. (70763) 238/L3
Paulina, Oreg. (97751) 291/G3
Paulina (lake), Oreg. 291/F4
Pauline, Kansas (66619) 232/G3
Pauline, Nebr. (†68941) 264/F4
Pauline, S.C. (29374) 296/D2
Paulins Kill (riv.), N.J. 273/D1
Paul Isnard, Fr. Guiana 131/D3
Paulistana, Brazil 132/F4
Paullina, Iowa (51046) 229/B3
Paulo Afonso, Brazil 120/F3
Paulo Afonso, Brazil 132/G5
Paulo Afonso (falls), Brazil 120/F3
Paulo de Faria, Brazil 135/B2
Paulsboro, N.J. (08066) 273/C4
Paul Smiths, N.Y. (12970) 276/M2
Paul Spur, Ariz. (†85607) 198/F7
Pauls Valley, Okla. (73075) 288/M5
Paungassi, Manitoba 179/G2
Paungde, Burma 72/B3
Pauni, India 68/E4
Paunsaugunt (plat.), Utah 304/B6
Paupack, Pa. (18451) 294/M3
Paute, Ecuador 128/C4
Pauto (riv.), Colombia 126/E5
Pauwalu (pt.), Hawaii 218/K2
Pauwela, Hawaii (†96708) 218/K2
Pavant (mts.), Utah 304/B5
Pavia (prov.), Italy 34/B2
Pavia, Italy 34/B2
Pavilion (key), Fla. 212/E6
Pavilion, N.Y. (14525) 276/D5
Pavillion, Wyo. (82523) 319/D2
Pāvilosta, U.S.S.R. 53/A2
Pāvilosta, U.S.S.R. 52/B3
Pavlodar, U.S.S.R. 54/J4
Pavlodar, U.S.S.R. 48/H4
Pavlof (bay), Alaska 196/F3
Pavlof (vol.), Alaska 196/F3
Pavlograd, U.S.S.R. 52/E5
Pavlovo, U.S.S.R. 52/F3
Pavo, Georgia (31778) 217/E9
Pavón, Colombia 126/D6
Pavullo nel Frignano, Italy 34/C2
Pawcatuck, Conn. (06379) 210/H3
Pawcatuck (riv.), Conn. 210/H3
Pawcatuck (riv.), R.I. 249/G7
Pawhuska, Okla. (74056) 288/O1
Pawlet○, Vt. (05761) 268/A5
Pawleys Island, S.C. (29585) 296/J5
Pawling, N.Y. (12564) 276/N7
Pawn, Nam (riv.), Burma 72/C2
Pawnee (creek), Colo. 208/M1
Pawnee, Ill. (62558) 222/D4
Pawnee (co.), Kansas 232/C3
Pawnee (riv.), Kansas 232/B3
Pawnee (co.), Nebr. 264/H4
Pawnee (co.), Okla. 288/N2
Pawnee, Okla. (74058) 288/N2
Pawnee City, Nebr. (68420) 264/H4
Pawnee Rock, Kansas (67567) 232/D3
Pawpaw, Ill. (61353) 222/E2
Paw Paw, Mich. (49079) 250/D6
Paw Paw (riv.), Mich. 250/C6
Paw Paw, W. Va. (25434) 312/K3
Paw Paw Lake, Mich. (†49038) 250/C6
Pawtuckaway (pond), N.H. 268/E5
Pawtucket, R.I. (*02860) 249/J5
Pax, W. Va. (25904) 312/D7
Paxico, Kansas (66526) 232/F2
Paxoí (isl.), Greece 45/D6
Paxson, Alaska (99737) 196/J2
Paxton, Fla. (32538) 212/C5
Paxton, Ill. (60957) 222/E3
Paxton, Ind. (47865) 227/C6
Paxton○, Mass. (01612) 249/G3
Paxton, Nebr. (69155) 264/C3
Paxville, S.C. (29102) 296/G4
Payakumbuh, Indonesia 85/C6
Paya Lebar, Singapore 72/F6
Payerne, Switzerland 39/C3
Payette (co.), Idaho 220/B5
Payette, Idaho (83661) 220/B5
Payette (lake), Idaho 220/C4
Payette (mts.), Idaho 220/B5
Payette (riv.), Idaho 220/B6
Payne, Georgia (†31201) 217/E5
Payne, Minn. (†55765) 255/F3
Payne, Ohio (45880) 284/A3
Payne (co.), Okla. 288/N2
Payne (lake), Que. 162/J4
Payne (lake), Québec 174/E1
Payneham, S. Australia 88/E8
Payneham, S. Australia 94/B7
Paynes, Miss. (†38920) 256/D3
Paynes Creek, Calif. (96080) 204/D3
Payne's Find, W. Australia 92/B5
Paynesville, Ind. (†47250) 227/F7
Paynesville, Mich. (†49912) 250/G2
Paynesville, Minn. (56362) 255/D5
Paynesville, Mo. (63371) 261/L4
Payneville, Ky. (40157) 237/J5
Paynton, Sask. 181/B2
Paysandú (dept.), Uruguay 145/B3
Paysandú, Uruguay 145/A3
Paysandú, Uruguay 120/D6
Payson, Ariz. (85541) 198/D4
Payson, Ill. (62360) 222/B4
Payson, Utah (84651) 304/C3
Pay-Yer (mt.), U.S.S.R. 52/K1
Pazanan, Iran 66/F5
Pazar, Rize, Turkey 63/J2
Pazar, Tokat, Turkey 63/G2
Pazarcık, Turkey 63/G4
Pazardzhik, Bulgaria 45/G4
Pazaryeri, Turkey 63/C3
Paz de Ariporo, Colombia 126/E5
Paz de Río, Colombia 126/D4
Pazña, Bolivia 136/B6
Pea (riv.), Ala. 195/F8
Peabody, Kansas (66866) 232/E3
Peabody, Ky. (40974) 237/O6
Peabody, Mass. (01960) 249/E5
Peace (riv.), Alberta 182/B1
Peace (riv.), Alta. 162/E4
Peace (riv.), Br. Col. 162/D4
Peace (riv.), Br. Col. 184/G2
Peace (riv.), Canada 146/G4
Peace (riv.), Fla. 212/E4
Peace Dale-Wakefield, R.I. (02883) 249/J7
Peace Garden, Manitoba 179/C5
Peace River, Alberta 182/B1
Peace River, Alta. 146/G4
Peace River, Alta. 162/E4
Peace Valley, Mo. (65788) 261/J9
Peach (co.), Georgia 217/E5
Peacham○, Vt. (05862) 268/C3
Peach Bottom, Pa. (17563) 294/K6
Peachland, Br. Col. 184/G5
Peachland, N.C. (28133) 281/J5
Peach Orchard, Ark. (72453) 202/J1
Peach Springs, Ariz. (86434) 198/B3
Peachtree (creek), Georgia 217/K1
Peachtree, North Fork (creek), Georgia 217/L1
Peachtree City, Georgia (30269) 217/C4
Peacock, Alberta 182/D4
Peacock, Mich. (†49938) 250/D4
Peacock, Texas (79542) 303/D4
Peahi, Hawaii (†96708) 218/K2
Peak, The (mt.), England 13/J2
Peak (range), Queensland 95/C4
Peak, S.C. (29122) 296/E3
Peak District National Park, England 13/F4
Peak Hill, N.S. Wales 97/E3
Peak Hill, W. Australia 92/B4
Peaks, Va. (†23069) 307/O5
Peale (mts.), Idaho 220/G7
Peale (mt.), Utah 304/E5
Peapack-Gladstone, N.J. (07977) 273/D2
Pear, W. Va. (†25955) 312/E7
Pearblossom, Calif. (93553) 204/H9
Pearce, Alberta 182/D5
Pearce, Ariz. (85625) 198/F7
Pearcy, Ark. (71964) 202/D5
Peard (bay), Alaska 196/G1
Pea Ridge, Ark. (72751) 202/B1
Pea Ridge Nat'l Mil. Park, Ark. 202/B1
Pearisburg, Va. (24134) 307/G6
Pearl (riv.) 188/J4
Pearl (harb.), Hawaii 188/F5
Pearl (harb.), Hawaii 218/A3
Pearl, Idaho (†83616) 220/B6
Pearl, Ill. (62361) 222/C4
Pearl (riv.), La. 238/L5
Pearl, Miss. (39208) 256/D6
Pearl (riv.), Miss. 256/D8
Pearl (cays), Nicaragua 154/F4
Pearl (creek), S. Dak. 298/N5
Pearl, Texas (†76528) 303/F6
Pearland, Texas (77581) 303/J2
Pearl and Hermes (reef), Hawaii 188/E5
Pearl and Hermes (atoll), Hawaii 218/B5
Pearl Beach, Mich. (48052) 250/G6
Pearl City, Hawaii (96782) 218/B3
Pearl City, Ill. (61062) 222/D1
Pearl Harbor Naval Station, Hawaii 218/B3
Pearlington, Miss. (39572) 256/E10
Pearl Lake, Québec 172/E2
Pearl River, La. (70452) 238/L6
Pearl River (co.), Miss. 256/E9
Pearl River, N.Y. (10965) 276/K8
Pearsall, Texas (78061) 303/E9
Pearse (canal), Alaska 196/N2
Pearson, Ark. (†72131) 202/F3
Pearson, Georgia (31642) 217/G8
Pearson, Okla. (†74826) 288/N4
Pearson, Wis. (54462) 317/H5
Peary (chan.), N.W.T. 162/M3
Peary (chan.), N.W. Terrs. 187/H2
Peary Land (reg.), Greenl. 4/A11
Peary Land (reg.), Greenland 146/Q1
Pease, Minn. (56363) 255/E5
Pease (riv.), Texas 303/D3
Pease A.F.B., N.H. 268/F5
Peasleeville, N.Y. (†12972) 276/N1
Peason, La. (†71429) 238/D4
Pebane, Mozambique 118/F3
Pebble (isl.), 143/E7
Pebble Beach, Calif. (93953) 204/C7
Pebworth, Ky. (41359) 237/O5
Peć, Yugoslavia 45/E4
Peçanha, Brazil 132/F7
Pecan Island, La. (†70548) 238/F7
Pecan Point, Ark. (†72350) 202/L3
Pecatonica, Ill. (61063) 222/D1
Pecatonica (riv.), Wis. 317/H11
Peccia, Switzerland 39/G4
Pechea, Romania 45/H3
Pechenga, U.S.S.R. 4/C8
Pechenga, U.S.S.R. 48/D2
Pechenga, U.S.S.R. 52/D1
Pechora, U.S.S.R. 7/K2
Pechora (riv.), U.S.S.R. 4/C7
Pechora (riv.), U.S.S.R. 7/K2
Pechora, U.S.S.R. 48/F3
Pechora, U.S.S.R. 52/J1
Pechora (bay), U.S.S.R. 52/H1
Pechora (riv.), U.S.S.R. 52/H1
Pechora (riv.), U.S.S.R. 48/F3
Pechora (sea), U.S.S.R. 52/H1
Pecica, Romania 45/E2
Peck, Idaho (83545) 220/B3
Peck, Kansas (67120) 232/E4
Peck, La. (†71368) 238/G3
Peck, Mich. (48466) 250/G5
Peckerwood (lake), Ark. 202/G4
Peckham, Okla. (74647) 288/M1
Pecks Mill, W. Va. (25547) 312/B7
Peconic, N.Y. (11958) 276/P8
Peconic (bay), N.Y. 276/R9
Pecos (riv.) 188/F4
Pecos, N. Mex. (87552) 274/D3
Pecos (riv.), N. Mex. 274/E5
Pecos (co.), Texas 303/B7
Pecos, Texas (79772) 303/D10
Pecos (riv.), Texas 303/C7
Pecos (riv.), U.S. 146/H6
Pecos Nat'l Mon., N. Mex. 274/D3
Pécs, Hungary 7/F4
Pécs, Hungary 41/E3
Pécsvárad, Hungary 41/E3
Peculiar, Mo. (64078) 261/D5
Pedasí, Panama 154/G7
Pedder (lake), Tasmania 99/B4
Pedee, Oreg. (†97361) 291/D3
Pedernal, Argentina 143/G5
Pedernales, Salar de (salt dep.), Chile 138/B5
Pedernales (prov.), Dom. Rep. 158/D7
Pedernales, Dom. Rep. 158/C7
Pedernales, Ecuador 128/B2
Pedernales (riv.), Texas 303/F7
Pedernales, Venezuela 124/G3
Pederneiras, Brazil 135/B3
Pedlar Mills, Va. (†24574) 307/K5
Pedley, Alberta 182/B3
Pedley, Calif. (†92509) 204/E10
Pedra Azul, Brazil 132/F6
Pedraza, Colombia 126/C2
Pedregal, Venezuela 124/C2
Pedreiras, Brazil 132/E4
Pedrera, Uruguay 145/C6
Pedricktown, N.J. (08067) 273/C4
Pedrika, S. Australia 94/D2
Pedro (bank), Jamaica 156/B3
Pedro (cays), Jamaica 156/C3
Pedro, Ohio (45659) 284/E8
Pedro, S. Dak. (†57729) 298/E5
Pedro (pt.), Sri Lanka 68/E6
Pedro Afonso, Brazil 132/E5
Pedro Antonio de los Santos, Mexico 150/Q7
Pedro Bay, Alaska (99647) 196/H3
Pedro Betancourt, Cuba 158/D1
Pedro Chico, Colombia 126/E7
Pedro de Valdivia, Chile 138/B4
Pedro Díaz Colodrero, Argentina 143/G5
Pedro Juan Caballero, Argentina 120/D5
Pedro Juan Cabellero, Paraguay 144/E3
Pedro Luro, Argentina 143/D4
Pedro Montoya, Mexico 150/K6
Pedro Segundo, Brazil 132/F4
Peduyim, Israel 65/B5
Peebles, Ohio (45660) 284/D8
Peebles, Sask. 181/J5
Peebles, Scotland 10/E3
Peebles (trad. co.), Scotland 15/B5
Peebles, Scotland 15/E5
Pee Dee, N.C. (†28091) 281/K5
Pee Dee (riv.), N.C. 281/J4
Peedee, S.C. (29586) 296/H3
Pee Dee (riv.), S.C. 296/H2
Peekskill, N.Y. (10566) 276/N8
Peeksville, Wis. (†54514) 317/E3
Peel (riv.) 162/C2
Peel, I. of Man 10/D3
Peel, I. of Man 13/C3
Peel (sound), N.W.T. 162/G1
Peel, New Bruns. 170/C2
Peel (sound), N.W. Terrs. 187/J2
Peel (reg. munic.), Ontario 177/E4
Peel (inlet), W. Australia 92/A2
Peel (riv.), Yukon 187/E3
Peel Fell (mt.), England 13/E2
Peel Fell (mt.), Scotland 15/F5
Pe Ell, Wash. (98572) 310/B4
Peene (riv.), E. Germany 22/E2
Peenemünde, E. Germany 22/E1
Peer, Belgium 27/G6
Peera Peera Poolanna (lake), S. Australia 94/F2
Peerless (lake), Alberta 182/C1
Peerless, Mont. (59253) 262/L2
Peerless, Sask. 181/L4
Peerless Lake, Alberta 182/C1
Peerless Park, Mo. (†63088) 261/N4
Peers, Alberta 182/B3
Peers, Mo. (†63357) 261/K5
Peery (lake), N.S. Wales 97/B2
Peesane, Sask. 181/H3
Peetz, Colo. (80747) 208/N1
Peever, S. Dak. (57257) 298/R2
Pefferlaw, Ontario 177/E3
Pegarah, Tasmania 99/B1
Pegasus (bay), N. Zealand 100/D5
Pegasus, Port (inlet), N. Zealand 100/B7
Peggs, Okla. (74452) 288/R2
Pegnitz, W. Germany 22/D4
Pego, Spain 33/F3
Pegram, Tenn. (37143) 237/H8
Pegu (div.), Burma 72/C3
Pegu, Burma 54/L8
Pegu, Burma 72/C3
Pegun (isl.), Indonesia 85/J5
Pegu Yoma (mts.), Burma 72/B3
Pehan (Bei'an), China 77/L2
Pehuajó, Argentina 143/D4
Pehuajó, Argentina 120/C6
Peine, W. Germany 22/D2
Pelplin, Poland 47/D2
Peipus (lake), U.S.S.R. 7/G3
Peipus (lake), U.S.S.R. 53/D1
Peipus (lake), U.S.S.R. 48/C4
Peipus (lake), U.S.S.R. 52/C3
Peixe, Brazil 132/D6
Pei Xian, China 77/J5
Pejepscot, Maine (04067) 243/D8
Pejivalle, C. Rica 154/F6
Pekalongan, Indonesia 85/J2
Pekan, Malaysia 72/D7
Pekan Nanas, Malaysia 72/E5
Pekelmeer (lake), Neth. Ant. 161/E9
Pekin, Ill. (61554) 222/D3
Pekin, Ind. (47165) 227/E7
Pekin, N. Dak. (58361) 282/O4
Peking (cap.), People's Rep. of China 54/N5
Peking (cap.), People's Rep. of China 2/Q3
Peking (Beijing) (cap.), People's Rep. of China 77/J3
Pekisko, Alberta 182/C4
Pelabuhan Ratu (bay), Indonesia 85/G2
Pelagie (isls.), Italy 34/D7
Pelagosa (Palagruža) (isl.), Yugoslavia 45/C4
Pelahatchie, Miss. (39145) 256/E6
Peleaga (mt.), Romania 45/F3
Pelechuco, Bolivia 136/A4
Peleduy, U.S.S.R. 48/M4
Pelée (vol.), Martinique 161/C5
Pelée (vol.), Martinique 156/G4
Pelee (pt.), Ontario 177/B6
Peleihari, Indonesia 85/E6
Peleliu (isl.), Belau 87/D5
Peleng (isl.), Indonesia 85/G6
Pelequén, Chile 138/G5
Pelham, Ala. (35124) 195/E4
Pelham, Georgia (31779) 217/D8
Pelham○, Mass. (†01002) 249/E3
Pelham○, N.H. (03076) 268/E6
Pelham, N.Y. (10803) 276/O7
Pelham, N.Y. (†10803) 276/O7
Pelham, N.C. (27311) 281/L1
Pelham, Ontario 177/E4
Pelham, Queensland 95/B3
Pelham, Tenn. (37366) 237/K10
Pelhřimov, Czech. 41/C2
Pelican, Alaska (99832) 196/M1
Pelican (lake), Alberta 182/D2
Pelican (mt.), Alberta 182/D2
Pelican (mts.), Alberta 182/D2
Pelican (isl.), Barbados 161/B9
Pelican, La. (71063) 238/C3
Pelican (bay), Manitoba 179/B2
Pelican (lake), Manitoba 179/C5
Pelican (lake), Manitoba 179/B2
Pelican (lake), Minn. 255/C4
Pelican (lake), Minn. 255/D4
Pelican (lake), Minn. 255/F2
Pelican (riv.), Minn. 255/B4
Pelican (riv.), Minn. 255/F2
Pelican (lake), Nebr. 264/D2
Pelican (lake), Sask. 181/E5
Pelican (lake), Wis. 317/H4
Pelican Lake, Wis. (54463) 317/H4
Pelican Lakes (Breezy Point), Minn. (†56472)255/D4
Pelican Narrows, Sask. 181/N3
Pelican Portage, Alberta 182/D2
Pelican Rapids, Manitoba 179/B2
Pelican Rapids, Minn. (56572) 255/B4
Pelileo, Ecuador 128/C3
Pelion, S.C. (29123) 296/E4
Pelkie, Mich. (49958) 250/G1
Pelkosenniemi, Finland 18/P3
Pella, Iowa (50219) 229/H6
Pella, Wis. (†54950) 317/J6
Pell City, Ala. (35125) 195/F3
Pellegrini, Argentina 143/D4
Pelletier Mills, New Bruns. 170/B1
Pell Lake, Wis. (53157) 317/K10
Pello, Finland 18/O3
Pellston, Mich. (49769) 250/E3
Pellville, Ky. (42364) 237/H5
Pellworm (isl.), W. Germany 22/C1
Pelly (bay), N.W. Terrs. 187/J3
Pelly (lake), N.W. Terrs. 187/H3
Pelly, Sask. 181/K4
Pelly (riv.), Yukon 162/C3
Pelly (mts.), Yukon 187/E3
Pelly (riv.), Yukon 187/E3
Pelly Bay, N.W.T. 162/G2
Pelly Bay, N.W. Terrs. 187/K3
Pelly Crossing, Yukon 187/E3
Peloncillo (mts.), Ariz. 198/F6
Peloncillo (mts.), N. Mex. 274/A6
Pelopónnisos (reg.), Greece 45/F7
Pelotas, Brazil 120/D6
Pelotas, Brazil 132/C10
Pelsor, Ark. (72856) 202/D2
Pelzer, Ind. (†47601) 227/C8
Pelzer, S.C. (29669) 296/B2
Pemadumcook (lake), Maine 243/E4
Pemalang, Indonesia 85/J2
Pemaquid, Maine (04558) 243/E8
Pematangsiantar, Indonesia 54/L9
Pematangsiantar, Indonesia 85/B5
Pemba, Mozambique 118/G2
Pemba, Mozambique 102/F6
Pemba (reg.), Tanzania 115/H5
Pemba (isl.), Tanzania 102/G5
Pemba (isl.), Tanzania 115/H5
Pemberton, Br. Col. 184/F5
Pemberton, Minn. (56078) 255/E7
Pemberton, N.J. (08068) 273/D4
Pemberton, Ohio (45353) 284/B5
Pemberton, W. Australia 92/A6
Pemberton, W. Va. (25905) 312/D7
Pemberville, Ohio (43450) 284/C3
Pembina (riv.), Alberta 182/C3
Pembina (hills), Manitoba 179/D5
Pembina (riv.), Manitoba 179/C5
Pembina (co.), N. Dak. 282/P2
Pembina, N. Dak. (58271) 282/R2
Pembina (riv.), N. Dak. 282/P2
Pembine, Wis. (54156) 317/L4
Pembrey, Wales 13/C6
Pembroke, Georgia (31321) 217/J6
Pembroke, Ill. (†60944) 222/F2
Pembroke, Ky. (42266) 237/G7
Pembroke, Maine (04666) 243/J6
Pembroke○, Maine (04666) 243/J6
Pembroke○, Mass. (02359) 249/L4
Pembroke○, N.H. (†03275) 268/E5
Pembroke, N.C. (28372) 281/L5
Pembroke, Ont. 162/J6
Pembroke, Ontario 177/G2
Pembroke, Ontario 175/E3
Pembroke, Va. (24136) 307/G6
Pembroke, Wales 13/C6
Pembroke, Wales 10/D5
Pembroke Park, Fla. (†33023) 212/B4
Pembroke Pines, Fla. (33024) 212/B4
Pembrokeshire Coast National Park, Wales 13/C6
Pembuang (riv.), Indonesia 85/E6
Pemigewasset, East Branch (riv.), N.H. 268/D3
Pemigewasset (riv.), N.H. 268/D4
Pemiscot (co.), Mo. 261/N10
Pemuco, Chile 138/E1
Peñablanca, Chile 138/F2
Pena Blanca, N. Mex. (87041) 274/C3
Penafiel, Portugal 33/B2
Peñafiel, Spain 33/E2
Peñaflor, Chile 138/G4
Peñal, Trin. & Tob. 161/B11
Peñalara (mt.), Spain 33/E2
Penalosa, Kansas (67121) 232/D4
Penalva, Brazil 132/E3
Penamacor, Portugal 33/C2
Penápolis, Brazil 135/A2
Peñaranda de Bracamonte, Spain 33/D2
Pen Argyl, Pa. (18072) 294/M4
Peñarroya (peak), Spain 33/F2
Peñarroya-Pueblonuevo, Spain 33/D3
Penarth, Wales 13/B7
Penas (gulf), Chile 120/B7
Penas (gulf), Chile 138/D7
Peñas (cape), Spain 33/D1
Penasco, N. Mex. (87553) 274/D2
Peña Vieja (mt.), Spain 33/D1
Penbrook, Pa. (17103) 294/J5
Pencarrow (head), N. Zealand 100/B3
Pence, Ind. (47973) 227/C4
Pence, Wis. (54553) 317/F3
Pencer, Minn. (56746) 255/C2
Pence Springs, W. Va. (24962) 312/E7
Pencil Bluff, Ark. (71965) 202/C4
Penco, Chile 138/D1
Pendant d'Oreille, Alberta 182/E5
Pendé (riv.), Cent. Afr. Rep. 115/C2
Pendé (riv.), Chad 111/C6
Pendembu, S. Leone 106/B7
Pender, Nebr. (68047) 264/H2
Pender (co.), N.C. 281/O5
Pendergrass, Georgia (30567) 217/E2
Pendleton, Camp, Calif. 204/H10
Pendleton, Ind. (46064) 227/F5
Pendleton (co.), Ky. 237/N3
Pendleton, N.Y. (†14094) 276/C4
Pendleton, N.C. (27862) 281/P2
Pendleton, Oreg. 188/C1
Pendleton, Oreg. (97801) 291/J2
Pendleton, S.C. (29670) 296/B2
Pendleton (co.), W. Va. 312/H5
Pendleton Bay, Br. Col. 184/E3
Pendletons, Va. (23117) 307/N5
Pend Oreille (riv.), Br. Col. 184/J6
Pend Oreille (lake), Idaho 188/C1
Pend Oreille (lake), Idaho 220/B1
Pend Oreille (mt.), Idaho 220/B1
Pend Oreille (riv.), Idaho 220/A1
Pend Oreille (co.), Wash. 310/H2
Pend Oreille (riv.), Wash. 310/H2
Pendroy, Mont. (59467) 262/D2
Pendryl, Alberta 182/C3
Penedo, Brazil 132/G5
Penetanguishene, Ontario 177/D3
Penfield, Georgia (30658) 217/F3
Penfield, Ill. (61862) 222/F3
Penfield, N.Y. (14526) 276/F4
Penfield, Pa. (15849) 294/E3
Penganga (riv.), India 68/D5
Penge, Zaire 115/D5
Penghu (Pescadores) (isls.), China 77/J7
Pengilly, Minn. (55775) 255/E3
Penglai, China 77/K4
Pengpu (Bengbu), China 77/J5
Pengshui, China 77/G6
Penguin, Tasmania 99/C3
Penhold, Alberta 182/D3
Penhook, Va. (24137) 307/J7
Penibética, Sistema (range), Spain 33/E4
Peniche, Portugal 33/B3
Penicuik, Scotland 10/C1
Penicuik, Scotland 15/D2
Peninsula (pt.), Mich. 250/C3
Peninsula (pt.), N.Y. 276/H3
Peninsula, Ohio (44264) 284/G3
Peñíscola, Spain 33/G2
Penitente, Serra do (range), Brazil 132/E5
Pénjamo, Mexico 150/J6
Penk (riv.), England 10/G2
Penki (Benxi), China 77/K3
Penmaenmawr, Wales 13/D4
Penmarch (pt.), France 28/A4
Penn, N. Dak. (58362) 282/M3
Penn, Pa. (15675) 294/C5
Pennant (pt.), Nova Scotia 168/E4
Pennant, Sask. 181/C5
Penndel, Pa. (19047) 294/N5
Penne, Italy 34/D3
Pennell (mt.), Utah 304/D6
Penner (riv.), India 68/D6
Penney Farms, Fla. (32079) 212/E2
Pennfield, New Bruns. 170/D3
Penn Hills○, Pa. (15235) 294/C7
Penniac, New Bruns. 170/D2
Pennine Alps (range), Italy 34/A2
Pennine Alps (range), Switzerland 39/D5
Pennine Chain (range), England 13/E3
Pennington, Ala. (36916) 195/B6
Pennington (co.), Minn. 255/B2
Pennington, Minn. (56663) 255/D3
Pennington, N.J. (08534) 273/D3
Pennington (co.), S. Dak. 298/C6
Pennington, Texas (75856) 303/J6
Pennington Gap, Va. (24277) 307/C7
Pennline, Pa. (†16424) 294/A2
Pennock, Minn. (56279) 255/C5
Penn Run, Pa. (15765) 294/D4
Pennsauken○, N.J. (08110) 273/B3
Pennsauken (creek), N.J. 273/B3
Pennsboro, Mo. (†65752) 261/E8
Pennsboro, W. Va. (26415) 312/E4
Pennsburg, Pa. (18073) 294/M5
Pennsdale, Pa. (17761) 294/J3
Penns Grove, N.J. (08069) 273/C4
Penns Park, Pa. (18943) 294/N5
Pennsuco, Fla. (†33010) 212/B4
Pennsville, N.J. (08070) 273/C4
Pennsville, Ohio (43770) 284/G6
Pennsylvania 188/L2
PENNSYLVANIA 294
Pennsylvania (state), U.S. 146/L5
Pennsylvania Furnace, Pa. (16865) 294/G4
Pennville, Ind. (47369) 227/G4
Pennville, Pa. (†17331) 294/J6
Penn Wynne, Pa. (†19151) 294/M6
Penny, Br. Col. 184/G3
Penny (str.), N.W. Terrs. 187/J2
Penn Yan, N.Y. (14527) 276/F5
Penobscot (co.), Maine 243/F5
Penobscot, Maine (04476) 243/F7
Penobscot○, Maine (04476) 243/F7
Penobscot (bay), Maine 243/F7
Penobscot (lake), Maine 243/C4
Penobscot (riv.), Maine 243/F5
Penobscot Ind. Res., Maine 243/F6
Penobsquis, New Bruns. 170/E3
Penokee, Kansas (67659) 232/C2
Penola, S. Australia 88/G7
Penola, S. Australia 94/G7
Peñón (pt.), P. Rico 161/B1
Peñón Blanco, Mexico 150/H4
Penong, S. Australia 88/E6
Penong, S. Australia 94/C4
Penonomé, Panama 154/G6
Penpont, Scotland 15/E5
Penrhyn (Tongareva) (atoll), Cook Is. 87/L6
Penrith, England 10/E3
Penrith, England 13/E3
Penrith, N. S. Wales 88/J6
Penrith, N.S. Wales 97/F3
Penrod, Ky. (42365) 237/G6
Penrose, Colo. (81240) 208/K6
Penrose, N.C. (28766) 281/D4
Penryn, Calif. (95663) 204/C8
Penryn, England 13/B7
Pensacola, Fla. 146/K6
Pensacola, Fla. 188/J4
Pensacola, Fla. (*32501) 212/B6
Pensacola (bay), Fla. 212/B6
Pensacola, Okla. (†74301) 288/R2
Pensacola N.A.S., Fla. 212/B6
Pensamiento, Bolivia 136/E4
Pensaukee, Wis. (†54153) 317/L6
Pense, Sask. 181/G5
Penshurst, Victoria 97/B5
Pentagon, Va. 307/T3
Penthalaz, Switzerland 39/C3
Penticton, Br. Col. 184/H5
Pentland (firth), Scotland 10/E1
Pentland (firth), Scotland 15/E2
Pentland (hills), Scotland 15/D2
Penton, Miss. (†38664) 256/D1
Pentress, W. Va. (26544) 312/F3
Pentwater, Mich. (49449) 250/C5
Peñuelas, Chile 138/F3
Peñuelas (lake), Chile 138/F2
Peñuelas, P. Rico 161/B2
Penwell, Texas (79776) 303/B6
Penyu (isls.), Indonesia 85/H7
Penza, U.S.S.R. 7/J3
Penza, U.S.S.R. 52/G4
Penza, U.S.S.R. 48/E4
Penzance, England 10/D5
Penzance, England 13/B7
Penzance, Sask. 181/F4
Penzberg, W. Germany 22/D5
Penzhina (bay), U.S.S.R. 48/R3
Peoa, Utah (84061) 304/C3
Peonan (pt.), Manitoba 179/D3
Peoples, Ky. (40467) 237/N6
Peoria, Alberta 182/A2
Peoria, Ariz. (85345) 198/C5
Peoria, Ill. 188/J2
Peoria, Ill. 146/K5
Peoria (co.), Ill. 222/D3
Peoria, Ill. (*61601) 222/D3
Peoria, Iowa (†50219) 229/H6
Peoria, Kansas (†66067) 232/G3
Peoria, Miss. (†39645) 256/C8
Peoria, Ohio (†43067) 284/D5
Peoria, Okla. (†66713) 288/S1
Peoria Heights, Ill. (61614) 222/D3
Peosta, Iowa (52068) 229/M4
Peotone, Ill. (60468) 222/F2
Pep, N. Mex. (†88115) 274/F5
Pepacton (res.), N.Y. 276/L6
Pepe (cape), Cuba 158/B3
Pepeekeo, Hawaii (96783) 218/J4
Pepeekeo (pt.), Hawaii 218/J4
Pepel, S. Leone 106/B7
Pepin (lake), Minn. 255/F6
Pepin (co.), Wis. 317/C6
Pepin, Wis. (54759) 317/B7
Pepin (lake), Wis. 317/B7
Pepperell, Mass. (01463) 249/H2
Pepperell○, Mass. (01463) 249/H2
Pepper Pike, Ohio (†44124) 284/J9
Pepperwood, Calif. (†95569) 204/A3
Peqin, Albania 45/D5
Pequabuck, Conn. (06781) 210/C2
Pequabuck (riv.), Conn. 210/D2
Pequaming, Mich. (†49946) 250/A2
Pequannock○, N.J. (07440) 273/B1

Pequea, Pa. (17565) 294/K6
Pequest (riv.), N.J. 273/D2
Pequonnock (riv.), Conn. 210/C3
Pequop (mts.), Nev. 266/G2
Pequot Lakes, Minn. (56472) 255/D4
Pera (head), Queensland 88/G2
Pera (head), Queensland 95/B2
Pera (Beyoğlu), Turkey 63/D6
Perabumulih, Indonesia 85/C6
Peraitepuí, Venezuela 124/H5
Perak (state), Malaysia 72/D6
Perak, Gunong (mt.), Malaysia 72/D6
Perales (riv.), Spain 33/F4
Peralta, Dom. Rep. 158/D6
Peralta, N. Mex. (87042) 274/C4
Peralta, Spain 33/F1
Peralta, Uruguay 145/C3
Peravia (prov.), Dom. Rep. 158/E6
Percé (cape), Nova Scotia 168/J2
Percé, Québec 172/D1
Percé, Québec 174/E3
Perch (lake), Mich. 250/G2
Perch (riv.), Mich. 250/G2
Perche (reg.), France 28/D3
Percival, Iowa (51648) 229/B7
Percival, Sask. 181/J5
Percival (lakes), W. Australia 88/C4
Percival (lakes), W. Australia 92/D3
Percy, Ill. (62272) 222/D5
Percy, Miss. (†38748) 256/C4
Percy, N.H. (†03582) 268/E2
Perdido, Ala. (36562) 195/C8
Perdido (bay), Ala. 195/D10
Perdido (riv.), Ala. 195/C9
Perdido (riv.), Fla. 212/B6
Perdido (mt.), Spain 33/G1
Perdido Beach, Ala. (36530) 195/C10
Pérdika, Greece 45/E6
Perdue, Sask. 181/D3
Perdue Hill, Ala. (36470) 195/C8
Pereira, Colombia 126/C5
Pereira, Colombia 120/B2
Perelló, Spain 33/G2
Pere Marquette (riv.), Mich. 250/D5
Perené (riv.), Peru 128/E8
Perenjori, W. Australia 92/B5
Pérez, Argentina 143/F6
Pérez (isl.), Mexico 150/P5
Perg, Austria 41/C2
Pergamino, Argentina 143/F6
Pergamino, Argentina 120/C6
Pergine Valsugana, Italy 34/C1
Pergola, Italy 34/D3
Perham○, Maine (04766) 243/G2
Perham, Minn. (56573) 255/C4
Perhentian, Kepulauan (isls.), Malaysia 72/D6
Periam, Romania 45/E2
Péribonca (riv.), Que. 162/J5
Péribonca (riv.), Québec 174/C3
Péribonca (riv.), Québec 172/F1
Péribonka, Québec 172/E1
Perico, Cuba 158/D1
Perico, Texas (†79087) 303/B1
Pericos, Mexico 150/F4
Peridot, Ariz. (85542) 198/E5
Perigord, Sask. 181/H3
Périgueux, France 28/D5
Perijá, Serranía de (mts.), Colombia 126/D2
Perijá, Sierra de (mts.), Venezuela 124/B2
Perim (isl.), P.D.R. Yemen 59/D7
Perintown, Ohio (†45150) 284/B7
Perito F.P. Moreno Nat'l Park, Argentina 143/B6
Perito Moreno, Argentina 143/B6
Periyar (lake), India 68/D7
Perkam (cape), Indonesia 85/K6
Perkasie, Pa. (18944) 294/M5
Perkatkin, U.S.S.R. 48/T2
Perkins (mt.), Conn. 210/F1
Perkins, Georgia (30822) 217/J5
Perkins, Iowa (†51239) 229/A2
Perkins, Mich. (49872) 250/B3
Perkins, Mo. (63774) 261/N8
Perkins (co.), Nebr. 264/C4
Perkins, Okla. (74059) 288/M3
Perkins, Québec 172/B4
Perkins (co.), S. Dak. 298/D3
Perkins, S. Dak. (†57062) 298/O8
Perkins, W. Va. (26634) 312/E5
Perkinsfield, Ontario 177/E3
Perkinston, Miss. (39573) 256/F9
Perkinstown, Wis. (†54451) 317/E5
Perkinsville, Ind. (†46011) 227/F4
Perkinsville, N.Y. (14529) 276/E5
Perkinsville, Vt. (05151) 268/B5
Perks, Ill. (62973) 222/D6
Perla, Ark. (†72104) 202/E5
Perlas (lag.), Nicaragua 154/F4
Perlas (pt.), Nicaragua 154/F4
Perlas (arch.), Panama 154/H6
Perleberg, E. Germany 22/D2
Perley, Minn. (56574) 255/B3
Perlis (state), Malaysia 72/D6
Perm', U.S.S.R. 2/M3
Perm', U.S.S.R. 7/K3
Perm', U.S.S.R. 52/J3
Perm', U.S.S.R. 48/F4
Perma, Mont. (59857) 262/B3
Përmet, Albania 45/E5
Pernambuco (state), Brazil 132/G5
Pernambuco (Recife), Brazil 132/H5
Pernell, Okla. (73076) 288/M5
Pernik, Bulgaria 45/F4
Peron (isls.), North. Terr. 88/D2
Peron (isls.), North. Terr. 93/A2
Peron (cape), Tasmania 99/E4
Peron (cape), W. Australia 88/A2
Peron (pen.), W. Australia 92/A4
Péronne, France 28/E3
Perote, Ala. (36061) 195/G7
Perote, Mexico 150/O1
Perow, Br. Col. 184/D3
Perpetua (cape), Oreg. 291/C3
Perpignan, France 7/E4
Perpignan, France 28/E6
Perquilauquén (riv.), Chile 138/A11
Perquimans (co.), N.C. 281/S2
Perrin, Mo. (†64477) 261/D3
Perrin, Texas (76075) 303/G5
Perrin, Va. (†23072) 307/R6
Perrine, Fla. (33157) 212/F6
Perrineville, N.J. (08535) 273/E3
Perrinton, Mich. (48871) 250/E5
Perris, Calif. (92370) 204/F11
Perro (mts.), N. Mex. 274/D4
Perronville, Mich. (49873) 250/B3
Perros (bay), Cuba 158/G2
Perry (co.), Ala. 195/D5
Perry (isl.), Alaska 196/C1
Perry (co.), Ark. 202/E4
Perry, Ark. (72125) 202/E3
Perry, Fla. (32347) 212/C1
Perry, Georgia (31069) 217/E6
Perry (co.) Ill. 222/D5
Perry, Ill. (62362) 222/C4
Perry (co.), Ind. 227/D8
Perry, Iowa (50220) 229/E5
Perry, Kansas (66073) 232/G2
Perry (lake), Kansas 232/G2
Perry (co.), Ky. 237/P6
Perry, La. (70575) 238/F7
Perry○, Maine (04667) 243/J6
Perry, Mich. (48872) 250/E6
Perry (co.), Miss. 256/G8
Perry (co.), Mo. 261/N7
Perry, Mo. (63462) 261/J4
Perry (stream), N.H. 268/E1
Perry, N.Y. (14530) 276/D5
Perry (co.), Ohio 284/F6
Perry, Ohio (44081) 284/H2
Perry, Okla. (73077) 288/M2
Perry, Oreg. (†97850) 291/J2
Perry (co.), Pa. 294/H5
Perry, S.C. (29124) 296/E4
Perry (co.), Tenn. 237/F9
Perry, Utah (†84302) 304/C2
Perry, W. Va. (†26851) 312/J5
Perrydale, Oreg. (†97101) 291/D2
Perrygo Place, Wis. (†53511) 317/J10
Perry Hall, Md. (21128) 245/N3
Perryman, Md. (21130) 245/O3
Perryopolis, Pa. (15473) 294/C5
Perrysburg, Ind. (†46951) 227/E3
Perrysburg, N.Y. (14129) 276/B6
Perrysburg, Ohio (43551) 284/C2
Perry's Cove, Newf. 166/D2
Perry's Victory and Int'l Peace Mem., Ohio 284/E2
Perrysville, Ind. (47974) 227/C4
Perrysville, Ohio (44864) 284/F4
Perrysville, Pa. (15237) 294/B6
Perryton, Ohio (†43822) 284/F5
Perryton, Texas (79070) 303/D1
Perrytown, Ark. (71801) 202/C6
Perryvale, Alberta 182/D2
Perryville, Alaska (99648) 196/G3
Perryville, Ark. (72126) 202/E3
Perryville, Ky. (40468) 237/M5
Perryville, La. (†71220) 238/G1
Perryville, Md. (21903) 245/O2
Perryville, Mo. (63775) 261/N7
Perryville, Tenn. (38364) 237/F9
Perşembe, Turkey 63/G2
Persepolis (ruins), Iran 66/H6
Perseverance (bay), Virgin Is. (U.S.) 161/A4
Perseverancia, Bolivia 136/D4
Pershing, Ind. (†46975) 227/E2
Pershing, Ind. (47370) 227/G5
Pershing, Iowa (50221) 229/G6
Pershing (co.), Nev. 266/C2
Pershing, Okla. (†74002) 288/O1
Pershore, England 13/E5
Persia, Iowa (51563) 229/B5
Persia, Tenn. (†37857) 237/P8
Persian (gulf) 54/G7
Persian (gulf), Bahrain 59/F4
Persian (gulf), Iran 66/F6
Persian (gulf), Iran 59/F4
Persian (gulf), Iraq 59/F4
Persian (gulf), Kuwait 59/F4
Persian (gulf), Qatar 59/F4
Persian (gulf), Saudi Arabia 59/F4
Persinger, W. Va. (†26651) 312/E6
Person (co.), N.C. 281/M2
Pertek, Turkey 63/H3
Perth, Australia 2/Q7
Perth, Kansas (†67152) 232/E4
Perth, N. Dak. (58363) 282/M2
Perth (county), Ontario 177/C4
Perth, Ontario 177/H3
Perth, Scotland 10/E2
Perth, Scotland 15/E4
Perth (trad. co.), Scotland 15/A5
Perth, Tasmania 99/D3
Perth (cap.), W. Australia 88/B2
Perth (cap.), W. Australia 92/A1
Perth Airport, W. Australia 88/B2
Perth Amboy, N.J. (*08861) 273/E2
Perth-Andover, New Bruns. 170/C2
Perthshire, Miss. (†38746) 256/C3
Perthville, N.S. Wales 97/E3
Pertominsk, U.S.S.R. 52/E2
Peru 2/F6
Peru 120/B4
PERU 128
Peru, Ill. (61354) 222/D2
Peru, Ind. (46970) 227/E3
Peru, Iowa (50222) 229/F6
Peru, Kansas (67360) 232/F4
Peru, Maine (04272) 243/C6
Peru○, Maine (04272) 243/C6
Peru, Nebr. (68421) 264/J4
Peru, N.Y. (12972) 276/N1
Peru○, Vt. (05152) 268/B5
Perugia (prov.), Italy 34/D3
Perugia, Italy 34/D3
Perugia, Italy 7/F4
Perugorría, Argentina 143/G4
Péruwelz, Belgium 27/D8
Pervari, Turkey 63/K4
Pervomaysk, U.S.S.R. 52/F3
Pervomaysk, U.S.S.R. 52/D5
Pervoural'sk, U.S.S.R. 48/F4
Perwez, Belgium 27/F7
Péry, Switzerland 39/D2
Pesaro, Italy 34/D3
Pesaro e Urbino (prov.), Italy 34/D3
Pescadero, Calif. (94060) 204/J4
Pescadero (creek), Calif. 204/J3
Pescadero (pt.), Calif. 204/J3
Pescadero, Mexico 150/D5
Pescadores (Penghu) (isls.), China 77/J7
Pescara (prov.), Italy 34/E3
Pescara, Italy 34/E3
Pescara, Italy 7/F4
Pescara (riv.), Italy 34/D3
Pescia, Italy 34/C3
Peseux, Switzerland 39/C3
Peshastin, Wash. (98847) 310/E3
Peshawar, Pakistan 54/H6
Peshawar, Pakistan 59/K3
Peshawar, Pakistan 68/C2
Peshkopi, Albania 45/E5
Peshtera, Bulgaria 45/G4
Peshtigo, Wis. (54157) 317/L5
Peshtigo (riv.), Wis. 317/K5
Peskovka, U.S.S.R. 52/H3
Peskowesk (lake), Nova Scotia 168/C4
Peso da Régua, Portugal 33/C2
Pesotum, Ill. (61863) 222/E4
Pespire, Honduras 154/D4
Pessac, France 28/C5
Pest (co.), Hungary 41/E3
Pestel, Haiti 158/A6
Pestovo, U.S.S.R. 52/D3
Péta, Greece 45/E6
Petaca, N. Mex. (87554) 274/C2
Petacalco (bay), Mexico 150/H8
Petah Tiqwa, Israel 65/B3
Petal, Miss. (39465) 256/F8
Petaluma, Calif. (94952) 204/H1
Pétange, Luxembourg 27/H9
Petas, Las (riv.), Bolivia 136/F5
Petatlán, Mexico 150/J8
Petauke, Zambia 115/F6
Petawaga (lake), Québec 172/A2
Petawawa, Ontario 177/G2
Petawawa (riv.), Ontario 177/G2
Petén-Itzá (lake), Guatemala 154/B2
Petenwell (lake), Wis. 317/G7
Peter (isl.), Virgin Is. (Br.) 156/H1
Peter (isl.), Virgin Is. (Br.) 161/D4
Peterborough, England 13/G5
Peterborough, England 10/F4
Peterborough, N.H. (03458) 268/D6
Peterborough○, N.H. (03458) 268/D6
Peterborough, Ont. 162/J7
Peterborough (county), Ontario 177/F3
Peterborough, Ontario 177/F3
Peterborough, S. Australia 88/F6
Peterborough, S. Australia 94/F5
Peterculter, Scotland 15/F3
Peterhead, Scotland 15/G3
Peterhead, Scotland 10/E2
Peter I (isl.) 5/B14
Peter I (isl.), Norway 2/E9
Peterlee, England 13/J3
Peterman, Ala. (36471) 195/D7
Petermann (ranges), North. Terr. 93/A8
Petermann (ranges), W. Australia 92/E4
Petermann Ranges Aboriginal Reserve, North. Terr. 88/D4
Petermann Ranges Aboriginal Res., North. Terr. 93/A8
Peteroa (vol.), Argentina 143/B4
Peteroa (vol.), Chile 138/B10
Peter Pond (lake), Sask. 181/L3
Peter's (riv.), Newf. 166/D2
Petersburg, Alaska 188/E6
Petersburg, Alaska (99833) 196/N2
Petersburg, Ill. (62675) 222/D4
Petersburg, Ind. (47567) 227/C7
Petersburg, Ky. (41080) 237/M2
Petersburg, Mich. (49270) 250/F7
Petersburg, Minn. (†56143) 255/C7
Petersburg, Nebr. (68652) 264/G3
Petersburg, N.J. (†08270) 273/D5
Petersburg, N.Y. (12138) 276/O5
Petersburg, N. Dak. (58272) 282/P3
Petersburg, Pa. (16669) 294/G4
Petersburg, Tenn. (37144) 237/H10
Petersburg, Texas (79250) 303/C4
Petersburg, Va. 188/L3
Petersburg (I.C.), Va. (23803) 307/N6
Petersburg, W. Va. (26847) 312/H5
Petersburg Nat'l Battlefield, Va. 307/O6
Petersfield, England 13/F6
Petersfield, Jamaica 158/G6
Petersfield, Manitoba 179/E4
Petersham○, Mass. (01366) 249/F3
Peterson, Ala. (35478) 195/D4
Peterson, Iowa (51047) 229/C3
Peterson, Minn. (55962) 255/G7
Peterson, Sask. 181/F3
Peterson Air Force Base, Colo. 208/K5
Peterstown, W. Va. (24963) 312/E8
Petersville, Ind. (†47201) 227/F6
Persville, Ky. (†41179) 237/P4
Petersville, Md. (†21758) 245/H3
Pétervására, Hungary 41/F3
Peterview, Newf. 166/C4
Pétionville, Haiti 158/C6
Petit Bois (isl.), Miss. 256/H10
Petit-Bourg, Guadeloupe 161/A6
Petit-Canal, Guadeloupe 161/A6
Petit Cap, Québec 172/D1
Petitcodiac, New Bruns. 170/E3
Petitcodiac (riv.), New Bruns. 170/F3
Petit Cul-de-Sac Marin (bay), Guadeloupe 161/A6
Petit-de-Grat, Nova Scotia 168/H3
Petit-de-Grat (isl.), Nova Scotia 168/H3
Petite Cascapédia (riv.), Québec 172/C1
Petite-Matane, Québec 172/B1
Petite Nation (riv.), Québec 172/B4
Petite Riviere Bridge, Nova Scotia 168/D4
Petite Rivière de l'Artibonite, Haiti 158/B5
Petite-Rivière-de-l'Île, New Bruns. 170/F1
Petite-Rivière-Ouest, Québec 172/D2
Petites, Newf. 166/C4
Petit-Étang, Nova Scotia 168/G2
Petite-Terre (isls.), Guadeloupe 161/B6
Petite-Vallée, Québec 172/D1
Petit-Goâve, Haiti 156/D3
Petit Goâve, Haiti 158/B6
Petit Jean (mt.), Ark. 202/C3
Petit Jean (riv.), Ark. 202/D3
Petitjean (Sidi-Kacem), Morocco 106/C2
Petit Mécatina (isl.), Québec 174/F2
Petit Mécatina (riv.), Québec 174/E2
Petitot (riv.), Br. Col. 184/M2
Petitot (riv.), N.W. Terrs. 187/F4
Petit Piton (mt.), St. Lucia 161/G6
Petit Rocher, New Bruns. 170/E1
Petit Rocher Sud, New Bruns. 170/E1
Petit-Saguenay (Saint-François-d'Assise), Québec 172/G1
Petitsikapau (lake), Newf. 166/A3
Petit Soufrière, Dominica 161/F6
Petley, Newf. 166/D2
Peto, Mexico 150/P6
Petone, N. Zealand 100/B2
Petorca, Chile 138/A9
Petoskey, Mich. (49770) 250/E3
Petpeswick (head), Nova Scotia 168/E4
Petra (ruins), Jordan 65/D5
Petre (bay), N. Zealand 100/D7
Petre (pt.), Ontario 177/G4
Petrey, Ala. (36062) 195/F7
Petrich, Bulgaria 45/F5
Petrified Forest, Ariz. (86028) 198/F3
Petrified Forest Nat'l Park, Ariz. 198/F4
Petrila, Romania 45/F3
Petrinja, Yugoslavia 45/B3
Petrohué, Chile 138/E3
Petrokrepost', U.S.S.R. 52/D3
Petrólea, Colombia 126/D3
Petroleum, Ind. (46778) 227/G3
Petroleum, Ky. (†42120) 237/J7
Petroleum (co.), Mont. 262/H3
Petroleum, W. Va. (26161) 312/D4
Petrolia, Kansas (†66720) 232/G4
Petrolia, Ontario 177/B5
Petrolia, Pa. (16050) 294/C3
Petrolia, Texas (76377) 303/F4
Petrolina, Brazil 120/E3
Petrolina, Brazil 132/G5
Petrona (pt.), P. Rico 161/D3
Petropavlovsk, U.S.S.R. 54/J4
Petropavlovsk, U.S.S.R. 48/G4
Petropavlovsk-Kamchatskiy, U.S.S.R. 2/S3
Petropavlovsk-Kamchatskiy, U.S.S.R. 54/T4
Petropavlovsk-Kamchatskiy, U.S.S.R. 48/R4
Petrópolis, Brazil 132/F8
Petrópolis, Brazil 135/E3
Petros, Tenn. (37845) 237/M8
Petroşeni, Romania 45/F3
Petrovsk, U.S.S.R. 52/G4
Petrovsk-Zabaykal'skiy, U.S.S.R. 48/L4
Petrozavodsk, U.S.S.R. 7/H2
Petrozavodsk, U.S.S.R. 48/D3
Petrozavodsk, U.S.S.R. 52/D2
Petsamo (Pechenga), U.S.S.R. 52/D1
Pettibone, N. Dak. (58475) 282/L5
Pettigo, Ireland 17/F2
Pettigo, N. Ireland 17/F2
Pettigrew, Ark. (72752) 202/C2
Pettis (co.), Mo. 261/F5
Pettisville, Ohio (43553) 284/B2
Pettus, Texas (78146) 303/G9
Petty Harbour, Newf. 166/D2
Petworth, D.C. (20011) 245/F4
Peu, Solomon Is. 87/G7
Peuco, Chile 138/G4
Peuerbach, Austria 41/B2
Peumo, Chile 138/F5
Pevas, Peru 128/G4
Pevek, U.S.S.R. 54/U3
Pevek, U.S.S.R. 48/S3
Pevely, Mo. (63070) 261/M6
Pewamo, Mich. (48873) 250/E5
Pewaukee, Wis. (53072) 317/K1
Pewaukee (lake), Wis. 317/K1
Pewee Valley, Ky. (40056) 237/L4
Peyrano, Argentina 143/F6
Peyton, Colo. (80831) 208/K4
Peytona, W. Va. (25154) 312/C6
Pézenas, France 28/E6
Pezinok, Czech. 41/D2
Pfaffenhofen an der Ilm, W. Germany 22/D4
Pfäffnau, Switzerland 39/E2
Pfarrkirchen, W. Germany 22/E4
Pfeifer, Kansas (67660) 232/C3
Pflugerville, Texas (78660) 303/G7
Pforzheim, W. Germany 22/C4
Pfronten, W. Germany 22/D5
Pfullingen, W. Germany 22/C4
Pfunds, Austria 41/A3
Pha Hom Pok, Doi (mt.), Thailand 72/C2
Phalaborwa, S. Africa 118/E4
Phalodi, India 68/C3
Phanat Nikhom, Thailand 72/D4
Phangan, Ko (isl.), Thailand 72/D5
Phangnga, Thailand 72/C5
Phan Rang, Vietnam 72/F5
Phan Thiet, Vietnam 72/F5
Phan Thiet, Vietnam 54/M8
Phantom (lake), Wis. 317/J2
Pharoah, Okla. (74862) 288/O4
Pharr, Texas (78577) 303/F11
Phatthalung, Thailand 72/D6
Phayao, Thailand 72/C3
Pheasant (hills), Sask. 181/J5
Pheba, Miss. (39755) 256/G3
Phelps (mt.) 120/C2
Phelps, Ky. (41553) 237/S6
Phelps (co.), Mo. 261/J7
Phelps (co.), Nebr. 264/E4
Phelps, N.Y. (14532) 276/F5
Phelps (lake), N.C. 281/S3
Phelps (peak), Venezuela 124/E7
Phelps, Wis. (54554) 317/H3
Phelps City, Mo. (†64482) 261/A2
Phenix, Va. (23959) 307/L6
Phenix City, Ala. 188/J4
Phenix City, Ala. (36867) 195/H6
Phet Buri, Thailand 72/C4
Phetchabun, Thailand 72/D3
Phiafai, Laos 72/E4
Phichai, Thailand 72/D3
Phichit, Thailand 72/D3
Phil, Ky. (†42539) 237/M6
Philadelphia, Ill. (†62612) 222/C4
Philadelphia, Miss. (39350) 256/F5
Philadelphia, Mo. (63463) 261/J3
Philadelphia, N.Y. (13673) 276/J2
Philadelphia, Pa. 188/M2
Philadelphia, Pa. 146/L6
Philadelphia (city county), Pa. 294/M6
Philadelphia, Pa. (*19101) 294/N6
Philadelphia, Tenn. (37846) 237/M9
Philadelphia, U.S. 2/F4
Philbrook, Minn. (†56466) 255/D4
Phil Campbell, Ala. (35581) 195/C2
Philip (riv.), Nova Scotia 168/E3
Philip, S. Dak. (57567) 298/F5
Philipp, Miss. (38950) 256/D3
Philippeville (Skikda), Algeria 106/F1
Philippeville, Belgium 27/E8
Philippi, W. Va. (26416) 312/G4
Philippine (sea) 54/O8
Philippine (sea), Guam 86/K6
Philippine (sea), Philippines 85/G2
Philippine (sea), Philippines 82/D3
Philippines 2/R5
Philippines 54/O8
PHILIPPINES 85/H4
PHILIPPINES 82
Philipsburg, Mont. (59858) 262/C4
Philipsburg, Pa. (16866) 294/F4
Philipsburg, Québec 172/D4
Philip Smith (mts.), Alaska 196/J1
Phillip (pt.), N.S. Wales 97/J2
Phillip (isl.), Victoria 97/C6
Philippy, Tenn. (†38079) 237/C8
Phillips (co.), Ark. 202/J5
Phillips (co.), Colo. 208/P1
Phillips (co.), Kansas 232/C2
Phillips○, Maine (04966) 243/C6
Phillips (co.), Mont. 262/J2
Phillips, Nebr. (68865) 264/F4
Phillips (bay), N.W. Terrs. 187/J1
Phillips, Okla. (†74538) 288/O6
Phillips, Texas (†79007) 303/C2
Phillips, Wis. (54555) 317/E4
Phillipsburg, Georgia (†31794) 217/E8
Phillipsburg, Kansas (67661) 232/C2
Phillipsburg, Mo. (65722) 261/G7
Phillipsburg, N.J. (08865) 273/C2
Phillipsburg, Ohio (45354) 284/B6
Phillipston○, Mass. (†01331) 249/F2
Phillipstown, Ill. (†62827) 222/F5
Phillipsville, N.C. (†28716) 281/D3
Philmont, N.Y. (12565) 276/N6
Philo, Calif. (95466) 204/B4
Philo, Ill. (61864) 222/E3
Philo, Ohio (43771) 284/G6
Philomath, Georgia (30659) 217/G3
Philomath, Oreg. (97370) 291/D3
Philomont, Va. (22131) 307/N2
Philpot, Ky. (42366) 237/H5
Philpots (isl.), N.W. Terrs. 187/L2
Philpott (lake), Va. 307/H7
Phippen, Sask. 181/C3
Phippsburg, Colo. (80469) 208/F2
Phippsburg, Maine (04562) 243/D8
Phippsburg○, Maine (04562) 243/D8
Phitsanulok, Thailand 72/D3
Phlox, Wis. (54464) 317/J5
Phnom Penh (cap.), Cambodia 54/M8
Phnom Penh (cap.), Cambodia 72/E5
Phnum Tbeng Meanchey, Cambodia 72/E4
Phoenix (cap.), Ariz. 146/G6
Phoenix (cap.), Ariz. 188/D4
Phoenix (cap.), Ariz. (*85001) 198/C5
Phoenix, Ill. (†60426) 222/C6
Phoenix (isls.), Kiribati 87/J6
Phoenix, La. (†70042) 238/L7
Phoenix, Md. (21131) 245/M2
Phoenix, N.Y. (13135) 276/H4
Phoenix, Oreg. (97535) 291/E5
Phoenixville, Conn. (†06235) 210/G1
Phoenixville, Pa. (19460) 294/L5
Phôngsali, Laos 72/D2
Phon Phisai, Thailand 72/D3
Phoques (bay), Tasmania 99/A1
Phou Bia (mt.), Laos 72/D3
Phou Cô Pi (mt.), Laos 72/E3
Phou Loi (mt.), Laos 72/D2
Phou San (mt.), Laos 72/D3
Phrae, Thailand 72/D3
Phra Nakhon Si Ayutthaya, Thailand 72/D4
Phsar Ream, Cambodia 72/D5
Phuc Loi, Vietnam 72/E3
Phu Cuong, Vietnam 72/E5
Phu Dien, Vietnam 72/E3
Phuket, Thailand 54/L9
Phuke, Thailand 72/C6
Phuket, Ko (isl.), Thailand 72/C5
Phu Lang Thuong (Bac Giang), Vietnam 72/E2
Phulbani, India 68/E4
Phu Ly, Vietnam 72/E2
Phumi Banam, Cambodia 72/E5
Phumi Phsar, Cambodia 72/E4
Phumi Prek Kak, Cambodia 72/E4
Phumi Samraong, Cambodia 72/D4
Phu My, Vietnam 72/F4
Phu Qui, Vietnam 72/E3
Phu Quoc, Dao (isl.), Vietnam 72/D5
Phu Rieng, Vietnam 72/E5
Phu Tho, Vietnam 72/E2
Phutthaisong, Thailand 72/D4
Phu Vinh, Vietnam 72/E5
Piaçabuçu, Brazil 132/H5
Piacenza (prov.), Italy 34/B2
Piacenza, Italy 34/B2
Piacoa, Venezuela 124/H3
Pia Fai, Doi (mt.), Thailand 72/D4
Piai, Tanjong (pt.), Malaysia 72/E6
Pian (creek), N.S. Wales 97/E1
Piankatank (riv.), Va. 307/R5
Pianosa (isl.), Italy 34/C3
Pianosa (isl.), Italy 34/E3
Piapoco, Colombia 126/F6
Piapot, Sask. 181/B6
Piarco, Trin. & Tob. 161/B10
Piat, Philippines 82/C2
Piatra Neamț, Romania 45/G2
Piatt (co.), Ill. 222/E4
Piauí (state), Brazil 132/F4
Piauí, Serra do (range), Brazil 132/F5
Piauí (riv.), Brazil 132/F5
Piave (riv.), Italy 34/D2
Piave, Miss. (†39476) 256/G8
Piazza Armerina, Italy 34/E6
Piazzi (isl.), Chile 138/D9
Pibor (riv.), Sudan 111/F6
Pibor Post, Sudan 111/F6
Pibrac, Québec 172/F1
Pibroch, Alberta 182/D2
Pica, Chile 138/B2
Picabo, Idaho (83348) 220/D6
Picacho, Ariz. (85241) 198/D6
Picacho, N. Mex. (88343) 274/D5
Picadilly, Newf. 166/C4
Picara (pt.), Virgin Is. (U.S.) 161/B4
Picard (lake), Québec 172/D2
Picardville, Alberta 182/D2
Picardy (trad. prov.), France 29
Picatinny Arsenal, N.J. 273/D2
Picayune, Miss. (39466) 256/E9
Piceance (creek), Colo. 208/C3
Picher, Okla. (74360) 288/S1
Pichidegua, Chile 138/F5
Pichilemu, Chile 138/A10
Pichincha (prov.), Ecuador 128/C3
Pichis (riv.), Peru 128/E8
Pichones (cays), Honduras 154/F3
Pichucalco, Mexico 150/N8
Pickard, Ind. (†46069) 227/E4
Pickaway (co.), Ohio 284/D6
Pickaway, W. Va. (24964) 312/E7
Pick City, N. Dak. (†58545) 282/G5
Pickens (co.), Ala. 195/B4
Pickens, Ark. (71662) 202/H6
Pickens (co.), Georgia 217/D2
Pickens, Miss. (39146) 256/E5
Pickens, Okla. (74752) 288/S6
Pickens (co.), S.C. 296/B2
Pickens, S.C. (29671) 296/B2
Pickens, W. Va. (26230) 312/F5
Pickensville, Ala. (†35447) 195/B4
Pickerel (lake), Conn. 210/F2
Pickerel (lake), Manitoba 179/C2
Pickerel, Wis. (54465) 317/J5
Pickering, England 13/G3
Pickering, Mo. (64476) 261/C2
Pickering, Ontario 177/K4
Pickerington, Ohio (43147) 284/E6
Pickersgill, Guyana 131/B2
Pickert, N. Dak. (†58230) 282/P5
Pickett (co.), Tenn. 237/M7
Pickett, Wis. (54964) 317/J8
Pickford, Mich. (49774) 250/E2
Pickle Lake, Ontario 175/C2
Pickrell, Nebr. (68422) 264/H4
Pickstown, S. Dak. (57367) 298/M7
Pickton, Texas (75471) 303/J5
Pickwick (lake), Ala. 195/B1
Pickwick, Minn. (†55948) 255/G7
Pickwick (lake), Miss. 256/H1
Pickwick (lake), Tenn. 237/E11
Pickwick Dam, Tenn. (38365) 237/E10
Pico (isl.), Portugal 33/C1
Pico (peak), Vt. 268/B4
Pico Rivera, Calif. (90660) 204/C10
Picos, Brazil 132/F4
Picos, Brazil 120/E3
Picota, Peru 128/D6
Pico Truncado, Argentina 143/C6
Pictograph (rocks), Ariz. 198/B5
Picton (isl.), Argentina 143/C8
Picton (isl.), Chile 138/F11
Picton, N.S. Wales 97/F4
Picton, N. Zealand 100/D4
Picton, Ontario 177/G3
Picton (mt.), Tasmania 99/C5
Pictou (co.), Nova Scotia 168/F3
Pictou, Nova Scotia 168/F3
Pictou (harb.), Nova Scotia 168/F3
Pictou (isl.), Nova Scotia 168/F3
Pictou Landing, Nova Scotia 168/F3
Picture Butte, Alberta 182/D5
Pictured Rocks (cliff), Mich. 250/C2
Pictured Rocks Nat'l Lakeshore, Mich. 250/C2
Picture Rocks, Pa. (17762) 294/J3
Pidurutalagala (mt.), Sri Lanka 68/E7
Pie, W. Va. (25689) 312/B7
Piedade, Brazil 135/C3
Piedade do Rio Grande, Brazil 135/D2
Piedecuesta, Colombia 126/D4
Piedmont, Ala. (36272) 195/G3
Piedmont, Calif. (94611) 204/J2
Piedmont, Georgia (†30204) 217/D4
Piedmont (reg.), Italy 34/A2
Piedmont, Kansas (67122) 232/F4
Piedmont, Mo. (63957) 261/L8
Piedmont, Ohio (43983) 284/H5

Piedmont (lake), Ohio 284/H5
Piedmont, Okla. (73078) 288/L3
Piedmont, S.C. (29673) 296/C2
Piedmont, S. Dak. (57769) 298/C5
Piedmont, W. Va. (26750) 312/H4
Piedmont, Wyo. (†82933) 319/B4
Piedra, Calif. (93649) 204/F7
Piedra (riv.), Colo. 208/E8
Piedra Blanca, Dom. Rep. 158/E6
Piedrabuena, Spain 33/D3
Piedrahita, Spain 33/D2
Piedras, Las (riv.), Peru 128/G8
Piedras Blancas (pt.), Calif. 204/D8
Piedras Coloradas, Uruguay 145/B3
Piedras Negras, Mexico 146/H7
Piedras Negras, Coahuila, Mexico 150/J2
Piedras Negras, Veracruz, Mexico 150/Q2
Piedra Sola, Uruguay 145/C3
Piekary Śląskie, Poland 47/B4
Pieksämäki, Finland 18/P5
Pielinen (lake), Finland 18/Q5
Pieman (riv.), Tasmania 99/B3
Piendamó, Colombia 126/B6
Pierce, Colo. (80650) 208/K1
Pierce, Fla. (†33860) 212/E4
Pierce (co.), Georgia 217/H8
Pierce, Idaho (83546) 220/C3
Pierce, Ky. (†42743) 237/K6
Pierce (pond), Maine 243/C5
Pierce (co.), Nebr. 264/G2
Pierce, Nebr. (68767) 264/G2
Pierce (co.), N. Dak. 282/K3
Pierce (co.), Wash. 310/C3
Pierce, W. Va. (†26292) 312/H4
Pierce (co.), Wis. 317/B6
Pierce City, Mo. (65723) 261/E8
Piercefield, N.Y. (12973) 276/L2
Pierceland, Sask. 181/K4
Pierceton, Ind. (46562) 227/F2
Pierceville, Ind. (†47031) 227/G6
Pierceville, Kansas (67868) 232/B4
Piermont○, N.H. (03779) 268/C4
Piermont, N.Y. (10968) 276/K8
Pierowall, Scotland 15/E1
Pierpont, Ohio (44082) 284/J2
Pierpont, S. Dak. (57468) 298/O3
Pierre (bayou), Miss. 256/C7
Pierre (cap.), S. Dak. 146/J5
Pierre (cap.), S. Dak. 188/F2
Pierre (cap.), S. Dak. (57501) 298/J5
Pierrefonds, Québec 172/H4
Pierrepont Manor, N.Y. (13674) 276/J3
Pierreville, Québec 172/E3
Pierron, Ill. (62273) 222/D5
Pierson, Fla. (32080) 212/E2
Pierson, Iowa (51048) 229/B3
Pierson, Manitoba 179/A5
Pierson, Mich. (49339) 250/D5
Pierson Station, Ill. (†61929) 222/E4
Pierz, Minn. (†56364) 255/D5
Piešťany, Czech. 41/D2
Pietarsaari (Jakobstad), Finland 18/N5
Pieterlen, Switzerland 39/D2
Pietermaritzburg (cap.), Natal, S. Africa 102/F7
Pietermaritzburg, S. Africa 118/E5
Pietersburg, S. Africa 102/F7
Pietersburg, S. Africa 118/D4
Pie Town, N. Mex. (87827) 274/A4
Pietrasanta, Italy 34/B3
Piet Retief, S. Africa 118/D5
Pietrosul (mt.), Romania 45/G2
Pigeon (creek), Ala. 195/E7
Pigeon (lake), Alberta 182/D3
Pigeon, Guadeloupe 161/A6
Pigeon (creek), Ind. 227/C8
Pigeon (riv.), Ind. 227/F1
Pigeon (riv.), Manitoba 179/F2
Pigeon, Mich. (48755) 250/F5
Pigeon (riv.), Mich. 250/D7
Pigeon (riv.), Mich. 250/E3
Pigeon (riv.), Minn. 255/G2
Pigeon (riv.), N.C. 281/C3
Pigeon (isl.), St. Lucia 161/G5
Pigeon, W. Va. (25155) 312/D5
Pigeon (creek), W. Va. 312/B7
Pigeon Cove, Mass. (01966) 249/M2
Pigeon Creek, Ala. (†36037) 195/E7
Pigeon Falls, Wis. (54760) 317/D7
Pigeon Forge, Tenn. (37863) 237/O9
Pigeon Hill, New Bruns. 170/F1
Pigeonroost, Ky. (†40962) 237/O6
Pigg (riv.), Va. 307/J7
Piggott, Ark. (72454) 202/K1
Pignon, Haiti 158/C5
Pigs (Cochinos) (bay), Cuba 158/D2
Pigüé, Argentina 143/D4
Piippola, Finland 18/P4
Pija, Sierra de (mts.), Honduras 154/D3
Pijijiapan, Mexico 150/N9
Pik, Iran 66/G3
Pike (co.), Ala. 195/G7
Pike (co.), Ark. 202/C5
Pike (co.), Georgia 217/D4
Pike (co.), Ill. 222/C4
Pike (co.), Ind. 227/C8
Pike (co.), Ky. 237/S6
Pike (riv.), Minn. 255/F3
Pike (co.), Miss. 256/D8
Pike (co.), Mo. 261/K4
Pike, N.H. (03780) 268/C3
Pike, N.Y. (14130) 276/D5
Pike (co.), Ohio 284/D7
Pike (co.), Pa. 294/M3
Pike, W. Va. (†26346) 312/D4
Pike Bay, Ontario 177/C3
Pike City, Ark. (†71940) 202/C5
Pike Lake, Sask. 181/E4
Pike Road, Ala. (36064) 195/F6
Pike Road, N.C. (27860) 281/R3
Pikes (peak), Colo. 188/E3
Pikes (peak), Colo. 208/J5
Pikes Peak, Ind. (†47201) 227/E6
Pikesville, Md. (21208) 245/M3
Piketberg, S. Africa 118/B6
Piketon, Ohio (45661) 284/E7
Pike View, Ky. (42770) 237/K6
Pikeville, Ind. (†47590) 227/C8
Pikeville, Ky. (41501) 237/S6
Pikeville, N.C. (27863) 281/N4
Pikeville, Tenn. (37367) 237/L9
Pikol'skiy, U.S.S.R. 48/G5
Pikwitonei, Manitoba 179/J3
Pila, Argentina 143/H7
Piła (prov.), Poland 47/C2
Piła, Poland 47/C2
Pilão Arcado, Brazil 132/F5
Pilar, Argentina 143/F5
Pilar, Brazil 132/H5
Pilar, Paraguay 120/D5
Pilar, Paraguay 144/C5
Pilas (isl.), Philippines 82/C7
Pilate, Haiti 158/C5
Pilatus (mt.), Switzerland 39/F2
Pilaya (riv.), Bolivia 136/C7
Pilcomayo (riv.) 120/C5
Pilcomayo (riv.), Argentina 143/E1
Pilcomayo (riv.), Bolivia 136/D7
Pilcomayo (riv.), Paraguay 144/C4
Pilger, Nebr. (68768) 264/G2
Pilger, Sask. 181/F3
Pilgrim, Ky. (41250) 237/S5
Pili, Philippines 82/D4
Pilibhit, India 68/D3
Pilica (riv.), Poland 47/D3
Pilis, Hungary 41/E3
Pilisvörösvár, Hungary 41/E3
Pillager, Minn. (56473) 255/D4
Pillar (pt.), Calif. 204/H3
Pillar (cape), Tasmania 88/H8
Pillar (cape), Tasmania 99/E5
Pillar (pt.), Wash. 310/A2
Píllaro, Ecuador 128/C3
Pilliga, N.S. Wales 97/E2
Pillow, Pa. (17080) 294/J4
Pillsbury (lake), Calif. 204/C4
Pillsbury, N. Dak. (58065) 282/P5
Pillsbury (sound), Virgin Is. (U.S.) 161/B4
Pilmaiquén (riv.), Chile 138/D3
Pilón, Cuba 158/H4
Pílos, Greece 45/E7
Pilot (peak), Idaho 220/C6
Pilot (peak), Idaho 220/C4
Pilot, Ky. (†40380) 237/O5
Pilot (peak), Nev. 266/C4
Pilot, Va. (24138) 307/H6
Pilot Butte, Sask. 181/G5
Pilot Butte (res.), Wyo. 319/D2
Pilote (riv.), Martinique 161/D7
Pilot Grove, Iowa (52648) 229/L7
Pilot Grove, Minn. (†56027) 255/D7
Pilot Grove, Mo. (65276) 261/G5
Pilot Knob (mt.), Idaho 220/C4
Pilot Knob, Ind. (†47118) 227/E8
Pilot Knob, Mo. (63663) 261/L7
Pilot Mound, Iowa (50223) 229/F4
Pilot Mound, Manitoba 179/D5
Pilot Mound, Minn. (†55923) 255/F7
Pilot Mountain, N.C. (27041) 281/J2
Pilotos, Cuba 158/B1
Pilot Point, Alaska (99649) 196/G3
Pilot Point, Texas (76258) 303/H4
Pilot Rock, Oreg. (97868) 291/J2
Pilot Station, Alaska (99650) 196/F2
Pilottown, La. (70081) 238/M8
Pilsen, Wis. (†54217) 317/L7
Piltene, U.S.S.R. 53/A2
Piltown, Ireland 17/G7
Pima (co.), Ariz. 198/D6
Pima, Ariz. (85543) 198/F6
Pimenta, Brazil 135/D2
Pimentel, Dom. Rep. 158/E5
Pimentel, Peru 128/B6
Pimento, Ind. (47866) 227/C6
Pimichín, Venezuela 124/E6
Pimmit, Va. (22043) 307/S2
Pina, Spain 33/F2
Pinal (co.), Ariz. 198/D6
Pinal (peak), Ariz. 198/E5
Pinaleno (mts.), Ariz. 198/F6
Pinamalayan, Philippines 82/C4
Pinang (Penang) (state), Malaysia 72/D6
Pinang (George Town), Malaysia 72/C6
Pinang, Pulau (isl.), Malaysia 72/C6
Pınarbaşı, Turkey 63/G3
Pinar del Río (prov.), Cuba 158/A2
Pinar del Río, Cuba 146/K7
Pinar del Río, Cuba 156/A2
Pinar del Río, Cuba 158/B2
Pınarhisar, Turkey 63/B2
Piñas, Ecuador 128/C4
Piñas (pt.), Panama 154/H7
Pinatubo (mt.), Philippines 82/C3
Pinawa, Manitoba 179/G4
Pinch, W. Va. (25156) 312/D6
Pinchbeck, England 13/G5
Pincher Creek, Alberta 182/D5
Pincher Creek, Br. Col. 162/E6
Pincher Station, Alberta 182/D5
Pinchi (lake), Br. Col. 184/E3
Pinckard, Ala. (36371) 195/G8
Pinckney, Mich. (48169) 250/F6
Pinckneyville, Ill. (62274) 222/D5
Pinckneyville, Miss. (†39669) 256/B8
Pinconning, Mich. (48650) 250/F5
Pincota, Romania 45/E2
Pincourt, Québec 172/D4
Pińczów, Poland 47/E3
Pindall, Ark. (72669) 202/E1
Pindamonhangaba, Brazil 135/D3
Pindi Gheb, Pakistan 68/C2
Pindo (riv.), Ecuador 128/D3
Píndus (mts.), Greece 45/E6
Pine, Ariz. (85544) 198/D4
Pine (riv.), Br. Col. 184/G2
Pine (creek), Calif. 204/D3
Pine, Colo. (80470) 208/J4
Pine (brook), Conn. 210/E2
Pine (isl.), Fla. 212/D5
Pine (pt.), Fla. 212/C2
Pine, Idaho (†83647) 220/C6
Pine (mt.), Ky. 237/O7
Pine (lake), Mich. 250/F4
Pine (riv.), Mich. 250/D4
Pine (riv.), Mich. 250/E5
Pine (co.), Minn. 255/F4
Pine, Mo. (†63935) 261/K9
Pine (creek), Nev. 266/E2
Pine (cape), Newf. 166/D2
Pine (riv.), N.H. 268/E4
Pine (creek), Oreg. 291/L3
Pine (creek), Oreg. 291/J4
Pine (creek), Pa. 294/H2
Pine (creek), Utah 304/C6
Pine (creek), Wash. 310/H3
Pine (lake), Wis. 317/J1
Pine Alcove (creek), Utah 304/D6
Pine Apple, Ala. (36768) 195/E7
Pine Bank, Pa. (15354) 294/B6
Pine Beach, N.J. (08741) 273/E4
Pine Bluff, Ark. 188/H4
Pine Bluff, Ark. (*71601) 202/F5
Pinebluff, N.C. (28373) 281/K4
Pine Bluff Arsenal, Ark. 202/F5
Pine Bluffs, Wyo. (82082) 319/H4
Pine Brook, N.J. (07058) 273/E2
Pine Bush, N.Y. (12566) 276/M7
Pine City, Ark. (†72069) 202/H4
Pine City, Minn. (55063) 255/F5
Pine City, Wash. (†99170) 310/H3
Pinecliffe, Colo. (80471) 208/J3
Pine Creek (pt.), Conn. 210/C4
Pinecreek, Minn. (†56753) 255/C2
Pine Creek, North. Terr. 88/E2
Pine Creek, North. Terr. 93/C2
Pine Creek (lake), Okla. 288/R6
Pinecrest, Calif. (95364) 204/F5
Pinedale, Ariz. (85934) 198/E4
Pinedale, Calif. (93650) 204/F7
Pinedale, Wyo. (82941) 319/C3
Pine Dock, Manitoba 179/F3
Pine Falls, Manitoba 179/F4
Pine Flat (lake), Calif. 204/F7
Pine Forest (range), Nev. 266/C1
Pinega, U.S.S.R. 52/F2
Pinega (riv.), U.S.S.R. 52/G2
Pine Grove, Ark. (†71763) 202/E6
Pine Grove, Georgia (†31513) 217/H7
Pine Grove, Ky. (40470) 237/N5
Pine Grove, La. (70453) 238/J5
Pine Grove, Pa. (17963) 294/K4
Pine Grove (res.), Pa. 294/K6
Pine Grove, W. Va. (26419) 312/E3
Pine Grove Furnace, Pa. (†17324) 294/H5
Pine Grove Mills, Pa. (16868) 294/G4
Pine Hall, N.C. (27042) 281/K2
Pinehaven (Heretaunga-Pinehaven), N. Zealand 100/C2
Pine Hill, Ala. (36769) 195/C7
Pine Hill, Ky. (40364) 237/N6
Pine Hill, N.J. (08021) 273/D4
Pine Hill, N.Y. (12465) 276/M6
Pine Hills, Fla. (32808) 212/E3
Pine House, Sask. 181/M3
Pinehurst (lake), Alberta 182/E2
Pinehurst, Georgia (31070) 217/E6
Pinehurst, Idaho (83850) 220/B2
Pinehurst, Mass. (01866) 249/B5
Pinehurst, N.C. (28374) 281/K4
Pine Island (sound), Fla. 212/D5
Pine Island, Minn. (55963) 255/F6
Pine Island, N.Y. (10969) 276/L8
Pine Knoll Shores, N.C. (†28557) 281/R5
Pine Knot, Ky. (42635) 237/M7
Pine Lake, Alberta 182/B4
Pine Lake, Georgia (30072) 217/D3
Pine Lake, Ind. (†46350) 227/D1
Pine Lake Prov. Park, Sask. 181/E4
Pineland, Fla. (33945) 212/D5
Pineland, S.C. (29934) 296/E6
Pineland, Texas (75968) 303/L6
Pinelands, S. Africa 118/F6
Pine Lawn, Mo. (†63120) 261/R2
Pine Level, Ala. (36065) 195/F6
Pine Level, N.C. (27568) 281/N4
Pinellas (co.), Fla. 212/D4
Pinellas (pt.), Fla. 212/C3
Pinellas Park, Fla. (33565) 212/B3
Pine Log (creek), Fla. 212/C6
Pine Log, Georgia (30152) 217/C2
Pine Meadow, Conn. (06061) 210/D1
Pine Mountain, Georgia (31822) 217/C5
Pineola, N.C. (28662) 281/F2
Pineora, Georgia (†31312) 217/K6
Pine Orchard, Conn. (†06405) 210/D3
Pine Park, Georgia (†31728) 217/D9
Pine Plains, N.Y. (12567) 276/N7
Pine Point, Maine (†04074) 243/C8
Pine Point, N.W. Terrs. 187/G3
Pine Prairie, La. (70576) 238/E5
Piñera, Uruguay 145/C3
Pine Ridge, Ark. (71961) 202/C4
Pine Ridge, Miss. (†39120) 256/B7
Pineridge, S.C. (†29169) 296/E4
Pine Ridge, S. Dak. (57770) 298/E7
Pine Ridge Ind. Res., S. Dak. 298/D7
Pine River, Manitoba 179/B3
Pine River, Minn. (56474) 255/D4
Pine River, Wis. (54965) 317/H7
Pinerolo, Italy 34/A2
Pines (Isla de la Juventud) (isl.), Cuba 158/B3
Pines (isl.), Cuba 156/A2
Pines (isl.), New Caled. 87/G8
Pines, New Caled. 86/H5
Pines (lake), N.J. 273/B1
Pine Springs, Texas (†88220) 303/C10
Pinetop, Ariz. (85935) 198/F4
Pinetops, N.C. (27864) 281/O3
Pinetown, N.C. (27865) 281/R3
Pinetown, S. Africa 118/E6
Pinetta, Fla. (32350) 212/C1
Pine Valley, Br. Col. 184/F2
Pine Valley, Calif. (92062) 204/J11
Pine Valley, N.J. (†08021) 273/C4
Pine Valley, N.Y. (14872) 276/G6
Pine Valley, Utah (†84722) 304/A6
Pineview, Georgia (31071) 217/F6
Pineview, N.C. (†27330) 281/L4
Pine Village, Ind. (47975) 227/C4
Pineville, Ark. (72566) 202/F1
Pineville, Ky. (40977) 237/O7
Pineville, La. (71360) 238/F4
Pineville, Miss. (†39074) 256/F6
Pineville, Mo. (64856) 261/D9
Pineville, N.C. (28134) 281/H4
Pineville, S.C. (29468) 296/H5
Pineville, W. Va. (24874) 312/C7
Pinewood, Minn. (56664) 255/C3
Pinewood, S.C. (29125) 296/G4
Piney, Ark. (†72847) 202/D3
Piney (isl.), Fla. 212/B1
Piney (pt.), Fla. 212/C2
Piney, Manitoba 179/F5
Piney Flats, Tenn. (37686) 237/S8
Piney Fork, Ohio (43941) 284/J5
Piney Park, Mo. (†63077) 261/L6
Piney Point, Md. (20674) 245/M8
Piney Point Village, Texas (†77001) 303/J1
Piney River, Va. (22964) 307/L5
Piney Woods, Miss. (39148) 256/D6
Ping, Mae Nam (riv.), Thailand 72/C3
Pingdingshan, China 77/H5
Pingelap (atoll), Micronesia 87/G5
Pingelly, W. Australia 88/B6
Pingelly, W. Australia 92/B2
Pingelly West, W. Australia 92/B2
Pinger (pt.), N.W. Terrs. 187/K3
Pingguo, China 77/G7
Pingjiang, China 77/H6
Pingle, China 77/H7
Pingliang, China 77/G4
Pingluo, China 77/G4
Pingquan, China 77/J3
Pingree, Idaho (83262) 220/F6
Pingree, N. Dak. (58476) 282/N5
Pingtan (isl.), China 77/K6
Pingtung, China 77/K7
Pingwu, China 77/F5
Pingxiang, Guangxi Zhuangzu, China 77/G7
Pingxiang, Jiangxi, China 77/H6
Pingyang, China 77/K6
Pinhal, Brazil 135/C3
Pinheiro, Brazil 132/E3
Pinheiro, Brazil 120/E3
Pinhel, Portugal 33/C2
Piniós (riv.), Greece 45/E6
Pinjarra, W. Australia 88/B6
Pinjarra, W. Australia 92/A2
Pink (cliffs), Ariz. 198/E4
Pink, Okla. (†74873) 288/M4
Pinkafeld, Austria 41/C3
Pinke Gat (chan.), Netherlands 27/H2
Pinkham, Sask. 181/B4
Pinkham Notch (pass), N.H. 268/E3
Pink Hill, N.C. (28572) 281/O4
Pink Mountain, Br. Col. 184/F1
Pinkstaff, Ill. (†62439) 222/F5
Pinnacle, N.C. (27043) 281/J2
Pinnacles Nat'l Mon., Calif. 204/D7
Pinnaroo, S. Australia 88/G7
Pinnaroo, S. Australia 94/G6
Pinneberg, W. Germany 22/C2
Pinnebog, Mich. (†48445) 250/F5
Pinney's (beach), St. Chris.-Nevis 161/D11
Pinola, Miss. (39149) 256/E7
Pinole, Calif. (94564) 204/J1
Pinon, Ariz. (86510) 198/E2
Pinon, Colo. (†81001) 208/K6
Pinon, N. Mex. (88344) 274/D6
Pinopolis, S.C. (29469) 296/G5
Pinopolis (dam), S.C. 296/G5
Pinos (riv.), Colo. 208/D8
Pinos, Rio de los (riv.), N. Mex. 274/B2
Pinos Altos, N. Mex. (88053) 274/A6
Pinos-Puente, Spain 33/E4
Pinquén, Peru 128/G9
Pinrang, Indonesia 85/F6
Pins (pt.), Ontario 177/C5
Pinsk, U.S.S.R. 48/C4
Pinsk, U.S.S.R. 52/C4
Pinson, Ala. (35126) 195/E3
Pinson, Tenn. (38366) 237/D10
Pinta (isl.), Ecuador 128/B9
Pintada, N. Mex. (†88435) 274/D4
Pintada Arroyo (creek), N. Mex. 274/E4
Pintado, Artigas, Uruguay 145/C1
Pintado, Florida, Uruguay 145/C4
Pintados, Chile 138/B2
Pintados, Salar de (salt dep.), Chile 138/B2
Pintendre, Québec 172/J3
Pinto, Chile 138/A11
Pinto, Md. (21556) 245/C2
Pinto (creek), Sask. 181/D6
Pintura, Utah (84720) 304/A6
Pintuyan, Philippines 82/E6
Pintwater (range), Nev. 266/F6
Pinware (riv.), Newf. 166/C3
Pinware River, Newf. 166/C3
Pinyon (peak), Idaho 220/C5
Pinzón (isl.), Ecuador 128/B9
Pioche, Nev. (89043) 266/G5
Piombino, Italy 34/C3
Pioneer (mts.), Idaho 220/D6
Pioneer, Iowa (†50541) 229/E3
Pioneer, La. (71266) 238/H1
Pioneer, Mo. (†65734) 261/E9
Pioneer, Ohio (43554) 284/A2
Pioneer, Tenn. (37847) 237/N8
Pioneer (isl.), U.S.S.R. 48/J2
Pionerskiy, U.S.S.R. 48/G3
Pionki, Poland 47/E3
Piopio, N. Zealand 100/E3
Piopolis, Québec 172/F4
Piotrków (prov.), Poland 47/D3
Piotrków Trybunalski, Poland 47/D3
Piove di Sacco, Italy 34/C2
Pipe (creek), Ind. 227/F4
Piper (peak), Nev. 266/D5
Piper City, Ill. (60959) 222/E3
Piper City, Kansas (66109) 232/H2
Pipersville, Pa. (18947) 294/M5
Piperton, Tenn. (†38017) 237/B10
Pipe Spring Nat'l Mon., Ariz. 198/C2
Pipestem (riv.), N. Dak. 282/M5
Pipestem, W. Va. (25979) 312/E7
Pipestone, Manitoba 179/B5
Pipestone (creek), Manitoba 179/A5
Pipestone (co.), Minn. 255/B6
Pipestone, Minn. (56164) 255/B7
Pipestone (riv.), Ontario 175/B2
Pipestone (creek), Sask. 181/K6
Pipestone (riv.), Sask. 181/L2
Pipestone Nat'l Mon., Minn. 255/B6
Pipinas, Argentina 143/H7
Pipinui (pt.), N. Zealand 100/B2
Pipmuacan (res.), Québec 174/D3
Piqan (Shanshan), China 77/D3
Piqua, Kansas (66761) 232/G4
Piqua, Ohio (45356) 284/B5
Piquete, Brazil 135/D3
Piquiri (riv.), Brazil 132/C7
Piquiri (riv.), Brazil 132/C9
Piracanjuba, Brazil 132/D7
Piracicaba, Brazil 120/E5
Piracicaba, Brazil 135/C3
Piracicaba, Brazil 132/E8
Piracuruca, Brazil 132/F3
Piraí do Sul, Brazil 132/D9
Piraí do Sul, Brazil 135/B4
Piraiévs, Greece 7/G5
Piraiévs (Piraeus), Greece 45/F7
Piraju, Brazil 135/B3
Pirajuba, Brazil 135/B1
Pirajuí, Brazil 135/B2
Pirámide, Cerro (mt.), Chile 138/D8
Piran, Yugoslavia 45/A3
Pirané, Argentina 143/E2
Pirapora, Brazil 120/E4
Pirapora, Brazil 132/E7
Piraraja, Uruguay 145/E4
Pirassununga, Brazil 135/C2
Pirata (mt.), P. Rico 161/F2
Piraúba, Brazil 135/E2
Piray (riv.), Bolivia 136/D5
Pirayú, Paraguay 144/B5
Pirdop, Bulgaria 45/G4
Pirenópolis, Brazil 132/D6
Pires do Rio, Brazil 132/D7
Pírgos, Greece 45/E7
Piriápolis, Uruguay 145/D5
Piribebuy, Paraguay 144/B5
Piribebuy (riv.), Paraguay 144/B4
Piripiri, Brazil 132/F4
Píritu, Anzoátegui, Venezuela 124/F2
Píritu, Falcón, Venezuela 124/D2
Píritu, Portuguesa, Venezuela 124/D3
Pirmasens, W. Germany 22/B4
Pirna, E. Germany 22/E3
Pirongia (mt.), N. Zealand 100/E3
Pirot, Yugoslavia 45/F4
Piru, Calif. (93040) 204/G9
Piru, Indonesia 85/H6
Piryatin, U.S.S.R. 52/D4
Piryí, Greece 45/G6
Pisa (prov.), Italy 34/C3
Pisa, Italy 34/C3
Pisac, Peru 128/G9
Pisagua, Chile 138/A2
Piscadera (bay), Neth. Ant. 161/F9
Piscataqua (riv.), Maine 243/B9
Piscataqua (riv.), N.H. 268/F5
Piscataquis (co.), Maine 243/E4
Piscataquis (riv.), Maine 243/E5
Piscataquog (riv.), N.H. 268/D5
Piscataway, Md. (†20735) 245/L6
Piscataway (creek), Md. 245/G6
Piscataway○, N.J. (08854) 273/D2
Piscataway Park, Md. 245/K6
Piscatosine (lake), Québec 172/B3
Pisco, Peru 128/D9
Pisco, Peru 120/B4
Pisco (bay), Peru 128/D9
Pisco (riv.), Peru 128/D9
Piseco, N.Y. (12139) 276/L4
Piseco (lake), N.Y. 276/M4
Písek, Czech. 41/C2
Pisek, N. Dak. (58273) 282/P3
Pisgah, Ala. (35765) 195/G1
Pisgah, Iowa (51564) 229/B5
Pisgah, Md. (20640) 245/K6
Pisgah Forest, N.C. (28768) 281/D4
Pishan (Guma), China 77/A4
Pishin, Iran 66/M7
Pishin, Pakistan 68/B2
Pishin, Pakistan 59/J3
Pishkun (res.), Mont. 262/D3
Pisinimo, Ariz. (85634) 198/C6
Pismo Beach, Calif. (93449) 204/E8
Piso Firme, Bolivia 136/D3
Pisoniano, Italy 34/F6
Pissis (mt.), Argentina 143/C2
Pistakee (lake), Ill. 222/A4
Pistapaug (pond), Conn. 210/E3
Pisticci, Italy 34/F4
Pistoia (prov.), Italy 34/C2
Pistoia, Italy 34/C2
Pistolet (bay), Newf. 166/C3
Pistol River, Oreg. (97444) 291/C5
Pisz, Poland 47/E2
Pit (riv.), Calif. 204/D2
Pitalito, Colombia 126/B7
Pitangui, Brazil 135/D1
Pitarpunga (lake), N.S. Wales 97/B4
Pitcairn (isl.) 87/O8
Pitcairn (isl.), (Br.) 2/C7
Pitcairn, Pa. (15140) 294/C5
Pitch (lake), Trin. & Tob. 161/A11
Piteå, Sweden 18/M4
Piteälv (riv.), Sweden 18/M4
Pitești, Romania 45/G3
Pitești, Romania 7/G4
Pithion, Greece 45/H5
Pithiviers, France 28/E3
Pitiquito, Mexico 150/D1
Pitkas Point, Alaska (†99658) 196/F2
Pitkin (co.), Colo. 208/F4
Pitkin, Colo. (81241) 208/F5
Pitkin, La. (70656) 238/E5
Pitlochry, Scotland 15/E4
Pitlochry, Scotland 10/E2
Pitman, N.J. (08071) 273/C4
Pitman, Sask. 181/G5
Pitmedden, Scotland 15/F3
Pitogo, Philippines 82/C4
Piton des Neiges (mt.), Réunion 118/G5
Pitrufquén, Chile 138/D2
Pitsburg, Ohio (45358) 284/A6
Pitt (isl.), Br. Col. 184/C3
Pitt (lake), Br. Col. 184/L2
Pitt, Minn. (56665) 255/D2
Pitt (isl.), N. Zealand 100/E7
Pitt (str.), N. Zealand 100/E7
Pitt (co.), N.C. 281/P3
Pittenweem, Scotland 15/F4
Pitti (isl.), India 68/C6
Pittman Center, Tenn. (†37738) 237/P9
Pitt Meadows, Br. Col. 184/L3
Pittock, Pa. (†15136) 294/B7
Pitts, Ark. (†72421) 202/J2
Pitts, Georgia (31072) 217/E7
Pittsboro, Ind. (46167) 227/D5
Pittsboro, Miss. (38951) 256/F3
Pittsboro, N.C. (27312) 281/L3
Pittsburg, Calif. (94565) 204/L1
Pittsburg, Georgia (†30084) 217/L1
Pittsburg, Ill. (62974) 222/E6
Pittsburg, Ind. (†46923) 227/D3
Pittsburg, Kansas (66762) 232/H4
Pittsburg, Ky. (40755) 237/N6
Pittsburg, Mo. (65724) 261/F7
Pittsburg○, N.H. (03592) 268/E1
Pittsburg (co.), Okla. 288/P5
Pittsburg, Okla. (74560) 288/P5
Pittsburg, Texas (75686) 303/J4
Pittsburgh, Pa. 146/K5
Pittsburgh, Pa. 188/L2
Pittsburgh, Pa. (*15201) 294/B7
Pittsfield, Ill. (62363) 222/C4
Pittsfield, Maine (04967) 243/E6
Pittsfield○, Maine (04967) 243/E6
Pittsfield, Mass. 188/M2
Pittsfield, Mass. (01201) 249/A3
Pittsfield, N.H. (03263) 268/E5
Pittsfield○, N.H. (03263) 268/E5
Pittsfield, Pa. (16340) 294/D2
Pittsfield○, Vt. (05762) 268/B4
Pittsford, Mich. (49271) 250/E7
Pittsford, N.Y. (14534) 276/E4
Pittsford, Vt. (05763) 268/A4
Pittsford○, Vt. (05763) 268/A4
Pittston○, Maine (†04345) 243/D7
Pittston, Pa. (*18640) 294/F7
Pittstown, N.J. (08867) 273/C2
Pittsview, Ala. (36871) 195/H6
Pittsville, Md. (21850) 245/S7
Pittsville, Mo. (†64040) 261/E5
Pittsville, Va. (24139) 307/K7
Pittsville, Wis. (54466) 317/F7
Pittsylvania (co.), Va. 307/K7
Pittville, Calif. (†96056) 204/D2
Pittwood, Ill. (†60970) 222/F3
Piúí, Brazil 132/E8
Piúí, Brazil 135/D2
Piura (dept.), Peru 128/B5
Piura, Peru 120/A3
Piura, Peru 128/B5
Piura (riv.), Peru 128/B5
Piute (co.), Utah 304/B5
Piute (res.), Utah 304/B5
Pivijay, Colombia 126/C2
Pixley, Calif. (93256) 204/F8
Piyas (lake), S. Dak. 298/P2
Pizacoma, Peru 128/H11
Pizarro, Colombia 126/B5
Pizol (peak), Switzerland 39/H3
Place, Ky. (†40734) 237/N7
Placentia, Calif. (92670) 204/D11
Placentia, Newf. 166/C2
Placentia (bay), Newf. 166/C2
Placentia (sound), Newf. 166/C2
Placer (co.), Calif. 204/E4
Placer, Philippines 82/E6
Placerville, Calif. (95667) 204/C8
Placerville, Colo. (81430) 208/D6
Placerville, Idaho (83666) 220/C6
Placetas, Cuba 158/E2
Placid (lake), Fla. 212/E4
Placid (lake), N.Y. 276/N2
Placida, Fla. (33946) 212/D5
Placilla, Chile 138/F6
Placilla de Caracoles, Chile 138/B4
Placilla de Peñuelas, Chile 138/F2
Placitas, N. Mex. (87043) 274/C3
Plad, Mo. (†65764) 261/G7
Pladda (isl.), Scotland 15/C5
Plaffeien, Switzerland 39/D3
Plahn, Liberia 106/C7
Plain, Wash. (†98826) 310/E3
Plain, Wis. (53577) 317/F9
Plain City, Ohio (43064) 284/D5
Plain City, Utah (†84401) 304/B2
Plain Dealing, La. (71064) 238/C1
Plainfield, Ark. (†71740) 202/D7
Plainfield, Conn. (06374) 210/H2
Plainfield○, Conn. (06374) 210/H2
Plainfield, Georgia (31073) 217/F6
Plainfield, Ill. (60544) 222/A6
Plainfield, Ind. (46168) 227/E5
Plainfield, Iowa (50666) 229/J3
Plainfield○, Mass. (01070) 249/C2
Plainfield○, N.H. (03781) 268/C4
Plainfield, N.J. (*07060) 273/E2
Plainfield, Ohio (43836) 284/G5
Plainfield, Vt. (05667) 268/C3
Plainfield○, Vt. (05667) 268/C3
Plainfield, Wis. (54966) 317/G7
Plains, Georgia (31780) 217/D6
Plains, Kansas (67869) 232/B4
Plains, Mont. (59859) 262/B3
Plains, Pa. (†18701) 294/F7
Plains, Texas (79355) 303/B4
Plainsboro, N.J. (08536) 273/D3

Plainview, Ark. (72857) 202/D4
Plain View, Iowa (†52773) 229/M5
Plainview, Minn. (55964) 255/F6
Plainview, Nebr. (68769) 264/G2
Plainview, N.Y. (11803) 276/R
Plainview, S. Dak. (57771) 298/E4
Plainview, Texas (79072) 303/C3
Plainville○, Conn. (06062) 210/D2
Plainville, Georgia (30733) 217/C2
Plainville, Ill. (62365) 222/B4
Plainville, Ind. (47568) 227/C7
Plainville, Kansas (67663) 232/C2
Plainville○, Mass. (02762) 249/J4
Plainwell, Mich. (49080) 250/D6
Plaisance, Haiti 158/C5
Plaisance, Québec 172/B4
Plaisted, Maine (04767) 243/F1
Plaistow○, N.H. (03865) 268/E6
Plaju, Indonesia 85/D6
Plamondon, Alberta 182/D2
Plana (cays), Bahamas 156/D2
Planá, Czech. 41/B2
Planada, Calif. (95365) 204/E6
Planeta Rica, Colombia 126/C3
Plánice, Czech. 41/B2
Plankinton, S. Dak. (57368) 298/N6
Plano, Ill. (60545) 222/E2
Plano, Iowa (52581) 229/G7
Plano, Texas (75074) 303/G1
Plant, Tenn. (†37054) 237/F9
Plantagenet, Ontario 177/K2
Plantation, Fla. (33317) 212/B4
Plantation (key), Fla. 212/F7
Plantation, Ky. (†40201) 237/K1
Plant City, Fla. (33566) 212/D3
Plantersville, Ala. (36758) 195/E5
Plantersville, Miss. (38862) 256/G2
Plantersville, S.C. (29441) 296/J4
Plantsite, Ariz. (†85540) 198/F5
Plantsville, Conn. (06479) 210/D2
Plaquemine, La. (70764) 238/J2
Plaquemines (par.), La. 238/L8
Plasencia, Spain 33/C2
Plaster City, Calif. (92269) 204/K11
Plaster Rock, New Bruns. 170/C2
Plastun, U.S.S.R. 48/O5
Plasy, Czech. 41/B2
Plat, Wis. (†53017) 317/K1
Plata (riv.) 2/G7
Plata, Río de la (est.), Argentina 143/E4
Plata (riv.), P. Rico 161/D2
Plata, La (riv.), Uruguay 145/B5
Platanal, Venezuela 124/F6
Platanilla, C. Rica 154/F6
Platea, Pa. (†16417) 294/B2
Plateau (creek), Colo. 208/C4
Plateau (state), Nigeria 106/F7
Plateau, Nova Scotia 168/H2
Plateau City, Colo. (†81624) 208/D4
Plate Cove, Newf. 166/D2
Platen, Kapp (pt.), Norway 18/D1
Platina, Calif. (96076) 204/B3
Platinum, Alaska (99651) 196/F3
Platner, Colo. (†80743) 208/N2
Plato, Colombia 126/C3
Plato, Minn. (55370) 255/D6
Plato, Mo. (65552) 261/H8
Plato, Sask. 181/C4
Platoro (res.), Colo. 208/F8
Platte (riv.), Iowa 229/D8
Platte (lake), Mich. 250/C4
Platte (co.), Mo. 261/C4
Platte (riv.), Mo. 261/C3
Platte (riv.), Nebr. 146/J5
Platte (riv.), Nebr. 188/G2
Platte (co.), Nebr. 264/G3
Platte (riv.), Nebr. 264/E4
Platte, S. Dak. (57369) 298/M7
Platte (lake), S. Dak. 298/M6
Platte (co.), Wyo. 319/H4
Platte Center, Nebr. (68653) 264/G3
Platte City, Mo. (64079) 261/C4
Plattenville, La. (70393) 238/K4
Platter, Okla. (74753) 288/O7
Platteville, Colo. (80651) 208/K2
Platteville, Wis. (53818) 317/F10
Platte Woods, Mo. (†64152) 261/O5
Platt Nat'l Park, Okla. 288/N6
Plattsburg, Mo. (64477) 261/D3
Plattsburgh, N.Y. (12901) 276/O1
Plattsburgh A.F.B., N.Y. 276/N1
Plattsmouth, Nebr. (68048) 264/J3
Plattsville, Ontario 177/D4
Plau, E. Germany 22/E2
Plaucheville, La. (71362) 238/G5
Plauen, E. Germany 22/E3
Plauersee (lake), E. Germany 22/E2
Plav, Yugoslavia 45/D4
Plavinas, U.S.S.R. 53/C2
Playa (pt.), Guyana 131/B1
Playa Azul, Mexico 150/H7
Playa Bonita, C. Rica 154/E6
Playa de Fajardo, P. Rico 161/F1
Playa de Humacao, P. Rico 161/F2
Playa Grande, Nicaragua 154/D4
Playas, Ecuador 128/B4
Playas (lake), N. Mex. 274/A7
Playón Chico, Panama 154/H6
Playón Grande, Panama 154/H6
Plaza, N. Dak. (58771) 282/G3
Plaza, Wash. (99028) 310/H3
Plaza Huincul, Argentina 143/B4
Pleasant (isl.), Alaska 196/M1
Pleasant (lake), Ariz. 198/C5
Pleasant, Ind. (†47043) 227/G7
Pleasant (lake), Maine 243/H5
Pleasant (lake), Maine 243/E3
Pleasant (lake), Maine 243/G3
Pleasant (riv.), Maine 243/H6
Pleasant (mt.), New Bruns. 170/D3
Pleasant (lake), N.Y. 276/M4
Pleasant (bay), Nova Scotia 168/H2
Pleasant Bay, Nova Scotia 168/H2
Pleasant City, Ohio (43772) 284/G6
Pleasant Dale, Nebr. (68423) 264/G4
Pleasantdale, Sask. 181/G3
Pleasant Gap, Ala. (†36272) 195/H3
Pleasant Gap, Pa. (16823) 294/G4
Pleasant Green, Mo. (†65276) 261/F5
Pleasant Grove, Ala. (35127) 195/D4
Pleasant Grove, Calif. (95668) 204/B8
Pleasant Grove, Miss. (38657) 256/D2
Pleasant Grove, N.J. (†07865) 273/D2
Pleasant Grove, Utah (84062) 304/C3
Pleasant Hill, Ala. (†36701) 195/E6
Pleasant Hill, Calif. (94523) 204/K2
Pleasant Hill, Ill. (62366) 222/C4
Pleasant Hill, La. (71065) 238/C3
Pleasant Hill, Miss. (†38651) 256/E1
Pleasant Hill, Mo. (64080) 261/D5
Pleasant Hill, N.C. (27866) 281/O1
Pleasant Hill, Ohio (45359) 284/B5
Pleasant Hill, S.C. (†29058) 296/F2
Pleasant Hill, Tenn. (38578) 237/L9
Pleasant Hills, Md. (†21087) 245/N3
Pleasant Hills, Pa. (15236) 294/B7
Pleasant Hope, Mo. (65725) 261/F8
Pleasant Island, Maine (†04964) 243/B5
Pleasant Lake, Ind. (46779) 227/H1
Pleasant Lake, Mass. (†02645) 249/O6
Pleasant Lake, Minn. (†56301) 255/D5
Pleasant Lake, N. Dak. (58364) 282/L3
Pleasant Lane, S.C. (†29824) 296/D4
Pleasant Mills, Ind. (46780) 227/H3
Pleasant Mound, Ill. (†62284) 222/D5
Pleasant Mount, Pa. (18453) 294/M2
Pleasanton, Calif. (94566) 204/L2
Pleasanton, Iowa (50224) 229/F7
Pleasanton, Kansas (66075) 232/H3
Pleasanton, Nebr. (68866) 264/E4
Pleasanton, N. Mex. (†88039) 274/A5
Pleasanton, Texas (78064) 303/F9
Pleasant Plain, Iowa (†52540) 229/K6
Pleasant Plain, Ohio (45162) 284/B7
Pleasant Plains, Ark. (72568) 202/G2
Pleasant Plains, Ill. (62677) 222/D4
Pleasant Point, N. Zealand 100/C6
Pleasant Pond, Maine (†04925) 243/D5
Pleasant Prairie, Wis. (53158) 317/L10
Pleasant Ridge, Mich. (48069) 250/B6
Pleasants (co.), W. Va. 312/D4
Pleasant Shade, Tenn. (37145) 237/K8
Pleasant Valley, Conn. (06063) 210/C1
Pleasant Valley, Md. (†21157) 245/L2
Pleasant Valley, Mo. (†64836) 261/R5
Pleasant Valley, Oreg. (†97814) 291/K3
Pleasant Valley (creek), S. Dak. 298/B6
Pleasant Valley, Va. (22848) 307/L4
Pleasant View, Colo. (81331) 208/B7
Pleasant View, Ill. (†62681) 222/C3
Pleasant View, Ky. (40769) 237/N7
Pleasant View, Tenn. (37146) 237/G8
Pleasant View, Utah (†84401) 304/B2
Pleasantville, Ind. (†47838) 227/C7
Pleasantville, Iowa (50225) 229/G6
Pleasantville, N.J. (08232) 273/D5
Pleasantville, N.Y. (10570) 276/N8
Pleasantville, Ohio (43148) 284/F6
Pleasantville (Alum Bank), Pa. (†15521) 294/E5
Pleasantville, Pa. (16341) 294/C2
Pleasantville, Tenn. (37147) 237/F9
Pleasure Beach, Conn. (†06385) 210/G3
Pleasure Ridge Park, Ky. (40258) 237/J4
Pleasureville, Ky. (40057) 237/L4
Pleiku, Vietnam 72/E4
Plenița, Romania 45/F3
Plenty (bay), N. Zealand 100/F2
Plenty, Sask. 181/C4
Plenty (riv.), Victoria 97/J4
Plenty (riv.), Victoria 88/L6
Plenty River Mine, North. Terr. 93/D7
Plentywood, Mont. (59254) 262/M2
Plesetsk, U.S.S.R. 52/F2
Plessis, N.Y. (13675) 276/J2
Plessisville, Québec 172/F3
Plessur (riv.), Switzerland 39/J3
Pleszew, Poland 47/C3
Pletcher, Ala. (†36750) 195/E5
Plétipi (lake), Que. 162/J5
Plétipi (lake), Québec 174/C2
Plettenberg (bay), S. Africa 118/C6
Plettenberg, W. Germany 22/C3
Pleven, Bulgaria 7/G4
Pleven, Bulgaria 45/G4
Plevna, Ala. (†35761) 195/F1
Plevna, Ind. (†46901) 227/E3
Plevna, Kansas (67568) 232/D4
Plevna, Mo. (63464) 261/H3
Plevna, Mont. (59344) 262/M4
Plevna, Ontario 177/H3
Pliny, W. Va. (25158) 312/B5
Pljevlja, Yugoslavia 45/D4
Ploče, Yugoslavia 45/C4
Płock (prov.), Poland 47/D2
Płock, Poland 47/D2
Plockton, Scotland 15/C3
Ploërmel, France 28/B4
Ploiești, Romania 45/H3
Ploiești, Romania 7/G4
Plomárion, Greece 45/H6
Plomb du Cantal (mt.), France 28/E5
Plombières, Belgium 27/F7
Plomer (pt.), N.S. Wales 97/G2
Plomosa (mts.), Ariz. 198/A5
Plön, W. Germany 22/D1
Płonia (riv.), Poland 47/B2
Płońsk, Poland 47/E2
Plovdiv, Bulgaria 7/G4
Plovdiv, Bulgaria 45/G4
Plover, Iowa (50573) 229/D3
Plover, Wis. (54467) 317/G7
Pluckemin, N.J. (07978) 273/D2
Plum (riv.), Ill. 222/C1
Plum (creek), Manitoba 179/B5
Plum (lake), Manitoba 179/B5
Plum (isl.), Mass. 249/L2
Plum (isl.), N.Y. 276/R8
Plum, Pa. (15239) 294/C5
Plumas (co.), Calif. 204/E4
Plumas, Manitoba 179/D4
Plumber (creek), Utah 304/C2
Plum Branch, S.C. (29845) 296/C4
Plum City, Wis. (54761) 317/B6
Plum Coulee, Manitoba 179/E5
Plumerville, Ark. (72127) 202/E3
Plummer, Idaho (83851) 220/B2
Plummer, Ind. (†47424) 227/C7
Plummer, Minn. (56748) 255/B3
Plummers Landing, Ky. (41081) 237/P4
Plum Point, Md. (†20639) 245/N6
Plum Springs, Ky. (†42101) 237/J7
Plumsteadville, Pa. (18949) 294/M5
Plum Tree, Ind. (†46792) 227/G3
Plumtree, Zimbabwe 118/D4
Plumville, Pa. (16246) 294/D4
Plumwood, Ohio (†43140) 284/D6
Plunge, U.S.S.R. 53/B3
Plunkett, Sask. 181/F4
Plunkettville, Okla. (†74963) 288/S6
Plush, Oreg. (97637) 291/H5
Plymouth (cap.), 156/F3
Plymouth, Calif. (95669) 204/C8
Plymouth○, Conn. (06782) 210/C2
Plymouth, England 7/D3
Plymouth, England 10/E5
Plymouth, England 13/C7
Plymouth (sound), England 13/C7
Plymouth, Fla. (32768) 212/E3
Plymouth, Ill. (62367) 222/C3
Plymouth, Ind. (46563) 227/E2
Plymouth (co.), Iowa 229/A3
Plymouth, Iowa (50464) 229/G2
Plymouth○, Maine (04969) 243/E6
Plymouth (co.), Mass. 249/L5
Plymouth, Mass. (02360) 249/M5
Plymouth○, Mass. (02360) 249/M5
Plymouth (bay), Mass. 249/M5
Plymouth, Mich. (*48170) 250/F6
Plymouth, Minn. (*55441) 255/G5
Plymouth, Nebr. (68424) 264/G4
Plymouth, N.H. (03264) 268/D4
Plymouth○, N.H. (03264) 268/D4
Plymouth, N.C. (27962) 281/R3
Plymouth, Ohio (44865) 284/E4
Plymouth, Pa. (18651) 294/E7
Plymouth, Utah (84330) 304/B2
Plymouth○, Vt. (05056) 268/B4
Plymouth, Wash. (99346) 310/F5
Plymouth, W. Va. (†25011) 312/C5
Plymouth, Wis. (53073) 317/L8
Plymouth Union, Vt. (†05056) 268/B4
Plympton○, Mass. (02367) 249/L5
Plympton, Nova Scotia 168/C4
Plynlimon (mt.), Wales 13/D5
Plzeň, Czech. 7/F4
Plzeň, Czech. 41/B2
Pniel, S. Africa 118/F6
Pniewy, Poland 47/C2
Po (riv.), Italy 7/E4
Po (riv.), Italy 34/C2
Po, Upper Volta 106/D6
Po (riv.), Va. 307/N4
Poá, Brazil 135/C3
Poatina, Tasmania 99/C3
Pobeda (peak), China 77/A3
Pobeda (peak), U.S.S.R. 48/J5
Población, Chile 138/F5
Pobla de Segur, Spain 33/G1
Poca, W. Va. (25159) 312/C6
Pocahontas, Alberta 182/B3
Pocahontas, Ark. (72455) 202/H1
Pocahontas, Ill. (62275) 222/D5
Pocahontas (co.), Iowa 229/D3
Pocahontas, Iowa (50574) 229/D3
Pocahontas, Miss. (39072) 256/D6
Pocahontas, Mo. (63779) 261/N8
Pocahontas, Tenn. (38061) 237/D10
Pocahontas, Va. (24635) 307/F6
Pocahontas (co.), W. Va. 312/F6
Pocasset, Mass. (02559) 249/M6
Pocasset, Okla. (73079) 288/L4
Pocatalico (riv.), W. Va. 312/C5
Pocatello, Idaho 146/G5
Pocatello, Idaho (*83201) 220/F7
Pocatello, Idaho 188/D2
Počátky, Czech. 41/C2
Pochep, U.S.S.R. 52/D4
Pöchlarn, Austria 41/C2
Pocklington, England 13/G4
Pocoata, Bolivia 136/B6
Poções, Brazil 132/F6
Pocola, Okla. (74902) 288/T4
Pocologan, New Bruns. 170/D3
Pocomoke (riv.), Md. 245/S8
Pocomoke (sound), Md. 245/P9
Pocomoke (sound), Va. 307/S5
Pocomoke City, Md. (21851) 245/R8
Pocomoonshine (lake), Maine 243/H5
Pocona, Bolivia 136/C5
Poconé, Brazil 132/B7
Pocono (mts.), Pa. 294/M3
Pocono Lake, Pa. (18347) 294/L3
Pocono Pines, Pa. (18350) 294/M3
Poços de Caldas, Brazil 120/E5
Poços de Caldas, Brazil 135/C2
Poços de Caldas, Brazil 132/E8
Pocotalago, Georgia (†30633) 217/F2
Pocotalico, W. Va. (†25301) 312/C6
Pocotaligo (riv.), S.C. 296/G4
Pocotopaug (lake), Conn. 210/E2
Pocpo, Bolivia 136/C6
Podbořany, Czech. 41/B1
Poděbrady, Czech. 41/C1
Podol'sk, U.S.S.R. 7/H3
Podol'sk, U.S.S.R. 52/E3
Podol'sk, U.S.S.R. 48/D4
Podor, Senegal 106/B5
Podporozh'ye, U.S.S.R. 52/D2
Podunk (riv.), Conn. 210/E1
Poe, Alberta 182/D3
Poe, Ind. (†46802) 227/G3
Poel (isl.), E. Germany 22/D1
Poenari Burchi, Romania 45/G3
Poge (cape), Mass. 249/N7
Poggibonsi, Italy 34/C3
Pogradec, Albania 45/E5
Pohakuloa (pt.), Hawaii 218/H2
P'ohang, S. Korea 81/D5
Pohatcong (creek), N.J. 273/C2
Pohénégamook, Québec 172/H2
Pohjois-Karjala (prov.), Finland 18/Q5
Pohořelice, Czech. 41/D2
Pohsien (Bo Xian), China 77/J5
Poiana Mare, Romania 45/F4
Poigan (lake), Québec 172/A2
Poinsett (co.), Ark. 202/J2
Poinsett (lake), Fla. 212/F3
Poinsett (lake), S. Dak. 298/P4
Point, La. (†71234) 238/F1
Point (lake), N.W. Terrs. 187/G3
Point, Texas (75472) 303/J5
Point Alexander, Ontario 177/G1
Point Arena, Calif. (95468) 204/B5
Point au Fer (isl.), La. 238/H8
Point au Fer (pt.), La. 238/H8
Point Baker, Alaska (99927) 196/M2
Point Cedar, Ark. (†71921) 202/D5
Point Clear, Ala. (36564) 195/C10
Point Comfort, Texas (77978) 303/H9
Point Cross, Nova Scotia 168/G2
Point de Bute, New Bruns. 170/F3
Point du Bois, Manitoba 179/G4
Pointe-à-la-Croix, Québec 172/C2
Pointe-à-la-Frégate, Québec 172/D1
Pointe-à-la-Garde, Québec 172/B2
Pointe a la Hache, La. (70082) 238/L7
Pointe-à-Pitre, Guadeloupe 161/B6
Pointe-à-Pitre, Guadeloupe 156/G3
Pointe à Raquette, Haiti 158/B6
Pointe au Baril Station, Ontario 177/D2
Pointe-au-Chêne, Québec 172/C4
Pointe-au-Père, Québec 172/J1
Pointe-au-Pic, Québec 172/G2
Pointe Aux Barques, Mich. (48467) 250/G4
Pointe-aux-Outardes, Québec 172/A1
Pointe Aux Pins, Mich. (49775) 250/E3
Pointe-aux-Trembles, Québec 172/J4
Pointe-Bleue, Québec 172/E1
Pointe-Calumet, Québec 172/G4
Pointe-Claire, Québec 172/H4
Pointe Coupee (par.), La. 238/G5
Pointe du Bout, Martinique 161/C6
Pointe-du-Chêne, New Bruns. 170/F2
Pointe-du-Lac, Québec 172/E3
Pointe-du-Moulin, Québec 172/H4
Point Edward, Ontario 177/B4
Pointe-Gatineau, Québec 172/B4
Pointe-Lebel, Québec 172/A1
Pointe-Noire, Congo 102/D5
Pointe-Noire, Congo 115/B4
Pointe-Noire, Guadeloupe 161/A6
Pointe-Sapin, New Bruns. 170/F2
Pointe-Verte, New Bruns. 170/E1
Point Fortin, Trin. & Tob. 161/A11
Point Harbor, N.C. (27964) 281/T2
Point Hope, Alaska 188/C5
Point Hope, Alaska (99766) 196/E1
Point Isabel, Ind. (†46928) 227/F4
Point La Haye, Newf. 166/D2
Point Lance, Newf. 166/C2
Point Lay, Alaska (†99723) 196/F1
Point Leamington, Newf. 166/C4
Point Marion, Pa. (15474) 294/C6
Point Mugu Pacific Missile Test Center, Calif. 204/F9
Point of Rocks, Md. (21777) 245/J3
Point of Rocks, Wyo. (82942) 319/D4
Point Pelee, Ontario 177/B6
Point Pelee Nat'l Park, Ontario 177/B5
Point Pleasant, Mo. (63873) 261/O10
Point Pleasant, N.J. (08742) 273/E3
Point Pleasant, Ohio (45163) 284/B8
Point Pleasant, Pa. (18950) 294/N5
Point Pleasant, W. Va. (25550) 312/B5
Point Pleasant Beach, N.J. (08742) 273/E3
Point Reyes Nat'l Seashore, Calif. 204/H1
Point Reyes Station, Calif. (94956) 204/H1
Point Roberts, Wash. (98281) 310/B2
Points, W. Va. (25437) 312/J4
Point Salvation Aboriginal Reserve, W. Australia 88/C5
Point Salvation Aboriginal Res., W. Australia 92/D5
Point Tupper, Nova Scotia 168/G3
Point Verde, Newf. 166/C2
Point Washington, Fla. (32454) 212/C6
Poipu, Hawaii (†96756) 218/C2
Poison (creek), Wyo. 319/E2
Poison Spider (creek), Wyo. 319/F3
Poisson Blanc (lake), Québec 172/B4
Poissons (riv.), Newf. 166/A3
Poitiers, France 7/E4
Poitiers, France 28/D4
Poitou (trad. prov.) France 29
Pojo, Bolivia 136/C5
Pojoaque, N. Mex. (†87501) 274/C3
Pokaran, India 68/C3
Pokataroo, N.S. Wales 97/E1
Pokegama (lake), Minn. 255/E3
Pokemouche (riv.), New Bruns. 170/E1
Pokesudie (isl.), New Bruns. 170/F1
Pokhara, Nepal 68/E3
Pokhvistnevo, U.S.S.R. 52/H4
Poko, Zaire 115/E3
Pokrovsk, U.S.S.R. 48/N3
Pola, Philippines 82/C4
Pola (Pula), Yugoslavia 45/A3
Polacca, Ariz. (86042) 198/E3
Polacca Wash (dry riv.), Ariz. 198/E3
Pola de Lena, Spain 33/C1
Pola de Siero, Spain 33/D1
Polanco del Yi, Uruguay 145/D4
Poland 2/K3
Poland 7/F3
POLAND 47
Poland (riv.), Conn. 210/C2
Poland, Ind. (47868) 227/C6
Poland, Maine (04273) 243/C7
Poland○, Maine (04273) 243/C7
Poland, N.Y. (13431) 276/L4
Poland, Ohio (44514) 284/J3
Poland Spring, Maine (04274) 243/C7
Polar, Wis. (54418) 317/H5
Polar Bear Prov. Park, Ontario 175/D2
Polaris, Mont. (59746) 262/C5
Polatlı, Turkey 63/E3
Polatlı, Turkey 59/B2
Połczyn-Zdrój, Poland 47/C2
Polebridge, Mont. (59928) 262/B2
Pol-e Khomri, Afghanistan 68/B1
Pol-e Khomri, Afghanistan 59/J2
Polgár, Hungary 41/F3
Polgárdi, Hungary 41/E3
Poli, Cameroon 115/B2
Policastro (gulf), Italy 34/E5
Police, Poland 47/B2
Polička, Czech. 41/D2
Poligny, France 28/F4
Poligus, U.S.S.R. 48/K3
Polikastron, Greece 45/F5
Polikhnitos, Greece 45/G6
Polillo, Philippines 82/D3
Polillo (isl.), Philippines 85/G3
Polillo (isl.), Philippines 82/C3
Polillo (str.), Philippines 82/C3
Polis, Cyprus 63/E5
Poliyiros, Greece 45/F5
Polk (co.), Ark. 202/B5
Polk (co.), Fla. 212/E4
Polk (co.), Georgia 217/B3
Polk (co.), Iowa 229/F5
Polk (co.), Minn. 255/B3
Polk (co.), Mo. 261/F7
Polk, Mo. (65727) 261/F7
Polk (co.), Nebr. 264/G3
Polk, Nebr. (68654) 264/G3
Polk (co.), N.C. 281/E4
Polk, Ohio (44866) 284/F4
Polk (co.), Oreg. 291/D3
Polk, Pa. (16342) 294/C3
Polk (co.), Tenn. 237/N10
Polk (co.), Texas 303/K7
Polk (co.), Wis. 317/B5
Polk City, Fla. (33868) 212/E3
Polk City, Iowa (50226) 229/F5
Polkowice, Poland 47/C3
Polkton, N.C. (28135) 281/J4
Polkville, Miss. (39118) 256/E6
Polkville, N.C. (28136) 281/F4
Pollaphuca (res.), Ireland 17/J5
Pollard, Ala. (†36441) 195/D8
Pollard, Ark. (72456) 202/K1
Pollards Point, Newf. 166/C4
Pöllau, Austria 41/C3
Pollensa, Spain 33/H3
Pollett (riv.), New Bruns. 170/E3
Pollett River, New Bruns. 170/E3
Pollock, Idaho (83547) 220/B4
Pollock, La. (71467) 238/F3
Pollock, Mo. (63560) 261/F2
Pollock, S. Dak. (57648) 298/J2
Pollock Pines, Calif. (95726) 204/E5
Pollocksville, N.C. (28573) 281/P5
Pollockville, Alberta 182/E4
Polmak, Norway 18/Q2
Polná, Czech. 41/C2
Polo, Dom. Rep. 158/D6
Polo, Ill. (61074) 222/D1
Polo, Mo. (64671) 261/D3
Polomka, Czech. 41/E2
Polonia, Chile 138/G6
Polonia, Wis. (†54423) 317/H6
Polonio, Uruguay 145/F5
Polonio (cape), Uruguay 145/F5
Polonnaruwa, Sri Lanka 68/E7
Polonnoye, U.S.S.R. 52/C4
Polotsk, U.S.S.R. 52/C3
Polperro, England 13/C7
Polson, Mont. (59860) 262/B3
Poltava, U.S.S.R. 7/H4
Poltava, U.S.S.R. 52/D5
Poltava, U.S.S.R. 48/D5
Poltimore, Québec 172/B4
Põltsamaa, U.S.S.R. 53/D1
Polvadera, N. Mex. (87828) 274/B4
Polyarnyy, U.S.S.R. 52/D1
Polyarnyy, U.S.S.R. 48/D3
Polynesia (reg.), Pacific 87/K7
Pomabamba, Peru 128/D7
Pomaire, Chile 138/F4
Pomán, Argentina 143/C2
Pomaria, S.C. (29126) 296/E3
Pombal, Brazil 132/G4
Pombal, Portugal 33/B3
Pomerania (reg.), E. Germany 22/E2
Pomeranian (bay), E. Germany 22/F1
Pomeranian (bay), Poland 47/B2
Pomerene, Ariz. (85627) 198/E6
Pomeroon (riv.), Guyana 131/B2
Pomeroy, Iowa (50575) 229/D3
Pomeroy, N. Ireland 17/H2
Pomeroy, Ohio (45769) 284/G7
Pomeroy, Wash. (99347) 310/H4
Pomezia, Italy 34/F7
Pomfret○, Conn. (06258) 210/H1
Pomfret, Md. (20675) 245/L6
Pomfret○, Vt. (†05067) 268/B4
Pomfret Center, Conn. (06259) 210/H1
Pomme de Terre (riv.), Minn. 255/C5
Pomme de Terre (lake), Mo. 261/E7
Pomona, Calif. 188/C4
Pomona, Calif. (*91766) 204/D10
Pomona, Kansas (66076) 232/G3
Pomona (lake), Kansas 232/G3
Pomona, Md. (†21620) 245/O4
Pomona, Mo. (65789) 261/J9
Pomona, N.J. (08240) 273/D5
Pomona Park, Fla. (32081) 212/E2
Pomonkey, Md. (†20640) 245/K6
Pomorie, Bulgaria 45/H4
Pomos (pt.), Cyprus 63/E5
Pompano Beach, Fla. (*33060) 212/F5
Pompanoosuc, Vt. (†05078) 268/C4
Pompéia, Brazil 135/A3
Pompeii (ruins), Italy 34/E4
Pompeii, Mich. (48874) 250/E5
Pomperaug, Conn. (†06798) 210/C2
Pomperaug (riv.), Conn. 210/C3
Pompey, N.Y. (13138) 276/J5
Pompeys Pillar, Mont. (59064) 262/J5
Pompton (lake), N.J. 273/B1
Pompton Lakes, N.J. (07442) 273/A1
Pompton Plains, N.J. (07444) 273/B1
Pomquet, Nova Scotia 168/G3
Pomy, Switzerland 39/C3
Ponape (isl.), Micronesia 87/F5
Ponass (lakes), Sask. 181/H3
Ponca, Ark. (72670) 202/D1
Ponca, Nebr. (68770) 264/H2
Ponca (creek), S. Dak. 298/L7
Ponca City, Okla. 188/G3
Ponca City, Okla. (74601) 288/M1
Ponce (dist.), P. Rico 161/C2
Ponce, P. Rico 161/C3
Ponce, P. Rico 156/F1
Ponce de Leon, Fla. (32455) 212/C6
Ponce de Leon (bay), Fla. 212/E6
Ponce Inlet, Fla. (†32019) 212/F2
Poncha Springs, Colo. (81242) 208/G6
Ponchatoula, La. (70454) 238/N2
Pond (pt.), Conn. 210/C4
Pond (riv.), Ky. 237/G6
Pond, Miss. (†39669) 256/B8
Pond, Mo. (†63038) 261/M3
Pond (inlet), N.W. Terrs. 187/L2
Pond Creek, Okla. (73766) 288/L1
Pond Eddy, Pa. (†12770) 294/N3
Pondera (co.), Mont. 262/D2
Ponderay, Idaho (83852) 220/B1
Ponderosa, N. Mex. (87044) 274/C3
Pond Fork (riv.), W. Va. 312/C6
Pondicherry (terr.), India 68/E6
Pondicherry, India 68/E6
Pond Inlet, Canada 4/B13
Pond Inlet, N.W.T. 162/J1
Pond Inlet, N.W. Terrs. 187/L2
Pondoland (reg.), S. Africa 118/D6
Pondosa, Calif. (96007) 204/D2
Ponds (isl.), Newf. 166/C3
Poneloya, Nicaragua 154/D4
Ponemah, Minn. (56666) 255/D2
Ponemah, N.H. (†03055) 268/D6
Poneto, Ind. (46781) 227/G3
Ponferrada, Spain 33/C1
Pongara (pt.), Gabon 115/A3
Ponhook (lake), Nova Scotia 168/D4
Poniatowa, Poland 47/E3
Ponnani, India 68/D6
Ponoka, Alberta 182/D3
Ponomarevka, U.S.S.R. 52/H4
Ponorogo, Indonesia 85/J2
Ponoy, U.S.S.R. 48/E3
Ponoy, U.S.S.R. 52/F1
Ponoy (riv.), U.S.S.R. 52/E1
Pons, France 28/C5
Ponset, Conn. (†06441) 210/E3
Ponsford, Minn. (56575) 255/C4
Pont-à-Celles, Belgium 27/E8
Ponta Delgada (dist.), Portugal 33/D2
Ponta Delgada, Portugal 33/C2
Ponta de Pedras, Brazil 132/D3
Ponta do Sol, Portugal 33/A2
Ponta Grossa, Brazil 120/D5
Ponta Grossa, Brazil 132/D9
Ponta Grossa, Brazil 135/B4
Pont-à-Mousson, France 28/G3
Ponta Porã, Brazil 132/C8
Ponta Porã, Paraguay 144/E3
Pontarlier, France 28/G4
Pontbriand, Québec 172/F3
Pont Canavese, Italy 34/A2
Pontchartrain (lake), La. 188/J5
Pontchartrain (lake), La. 238/O3
Pontchartrain Causeway, La. 238/O3
Pontecorvo, Italy 34/D4
Ponte de Sor, Portugal 33/C3
Ponte do Lima, Portugal 33/B2
Ponteix, Sask. 181/D6
Ponteland, England 13/H3
Ponte Nova, Brazil 132/F8
Ponte Nova, Brazil 135/E2
Pontevedra, Philippines 82/D5
Pontevedra (prov.), Spain 33/B1
Pontevedra, Spain 33/B1
Ponte Vedra Beach, Fla. (32082) 212/E1
Pontgrave, New Bruns. 170/F1
Pontiac, Ill. (61764) 222/E3
Pontiac, Mich. 188/K2
Pontiac, Mich. (*48053) 250/F6
Pontiac, Mo. (65729) 261/G9
Pontiac (co.), Québec 172/A3
Pontiac (county), Québec 174/B3
Pontiac, R.I. (†02887) 249/J6
Pontiac, S.C. (†29045) 296/F3
Pontianak, Indonesia 85/D6
Pontianak, Indonesia 54/N10
Pontian Kechil, Malaysia 72/E5
Pontic (mts.), Turkey 59/C1
Pontic (mts.), Turkey 63/H2
Pontine (isls.), Italy 34/D4
Pontinia, Italy 34/D4
Pontivy, France 28/B3
Pont-l'Abbe, France 28/A4
Pont-Lafrance, New Bruns. 170/E1
Pont-Landry, New Bruns. 170/F1
Pont-l'Évêque, France 28/D3
Ponto da Divisão, Brazil 132/B5
Pontoise, France 28/E3
Pontoon Beach, Ill. (†62040) 222/A2
Pontoosuc, Ill. (†62330) 222/B3
Pontoosuc (lake), Mass. 249/A3
Pontorson, France 28/C3
Pontotoc (co.), Miss. 256/F2
Pontotoc, Miss. (38863) 256/G2
Pontotoc (co.), Okla. 288/N5
Pontotoc, Okla. (74863) 288/N6
Pontotoc, Texas (76869) 303/E7
Pontremoli, Italy 34/B2
Pontresina, Switzerland 39/J3
Pontrilas, Sask. 181/H2
Pont-Rouge, Québec 172/F3
Pontypool, Ontario 177/F3
Pontypool, Wales 13/B6
Pontypridd, Wales 13/A6
Pontypridd, Wales 10/E5
Pony, Mont. (59747) 262/E5
Pony (creek), Okla. 288/C1

Ponza (isl.), Italy 34/D4
Ponzer, N.C. (†27810) 281/S3
Poole, England 13/E7
Poole, England 10/E5
Poole (bay), England 13/F7
Poole, Ky. (42444) 237/F5
Poole, Nebr. (68867) 264/F4
Pooler, Georgia (31322) 217/K6
Pooles (isl.), Md. 245/O3
Poolesville, Md. (20837) 245/J4
Pooleville, Okla. (73451) 288/M6
Poolewe, Scotland 15/C3
Poolville, N.Y. (13432) 276/K5
Poona, India 54/J8
Pooncarie, N.S. Wales 97/B3
Poopeloe (lake), N.S. Wales 97/C2
Poopó (lake), Bolivia 120/C4
Poopó, Bolivia 136/B6
Poopó (lake), Bolivia 136/B6
Poor Knights (isls.), N. Zealand 100/E1
Poorman, Alaska (†99768) 196/G2
Poosepatuck Ind. Res., N.Y. 276/P9
Popa Hill (mt.), Burma 72/B2
Popayán, Colombia 120/B2
Popayán, Colombia 126/B6
Pope (co.), Ark. 202/D3
Pope (co.), Ill. 222/E6
Pope (creek), Ill. 222/C2
Pope, Manitoba 179/B4
Pope (co.), Minn. 255/C5
Pope, Miss. (38568) 256/E2
Pope A.F.B., N.C. 281/L4
Popejoy, Iowa (50227) 229/G3
Poperinge, Belgium 27/B7
Popes Creek, Md. (†20632) 245/L7
Popham Beach, Maine (04562) 243/D8
Popilta (lake), N.S. Wales 97/A3
Popiltah, N.S. Wales 97/A3
Popina, Bulgaria 45/H3
Popkum, Br. Col. 184/M3
Poplar (pt.), Manitoba 179/E2
Poplar (riv.), Manitoba 179/E2
Poplar (isl.), Md. 245/N5
Poplar (riv.), Minn. 255/C3
Poplar, Mont. (59255) 262/L2
Poplar (riv.), Mont. 262/L2
Poplar (riv.), Sask. 181/E6
Poplar, Wis. (54864) 317/C2
Poplar Bluff, Mo. (63901) 261/L9
Poplar Branch, N.C. (27965) 281/T2
Poplar-Cotton Center, Calif. (93257) 204/F7
Poplar Creek, Br. Col. 184/J5
Poplar Creek, Miss. (†39747) 256/E4
Poplarfield, Manitoba 179/E4
Poplar Grove, Ark. (72374) 202/J4
Poplar Grove, Ill. (61065) 222/E1
Poplar Hill, Ontario 175/B2
Poplar Point, Manitoba 179/D4
Poplarville, Ky. (42548) 237/N6
Poplarville, Miss. (39470) 256/E9
Popo Agie (riv.), Wyo. 319/D3
Popocatépetl (mt.), Mexico 150/M1
Popokabaka, Zaire 115/C5
Popoli, Italy 34/E3
Popomanatseu (mt.), Solomon Is. 86/E3
Popondetta, Papua N.G. 85/C7
Popondetta, Papua N.G. 87/E6
Popovo, Bulgaria 45/H4
Poprad, Czech. 41/F2
Poprad (riv.), Czech. 41/F2
Poquetanuck, Conn. (†06360) 210/G3
Poquis, Nevados de (mt.), Chile 138/C4
Poquonock, Conn. (06064) 210/E1
Poquonock Bridge, Conn. (†06340) 210/G3
Poquoson (I.C.), Va. (23662) 307/R6
Porangaatu, Brazil 132/D6
Porangaatu, Brazil 120/E4
Porangahau, N. Zealand 100/F4
Porbandar, India 68/B4
Porcher (isl.), Br. Col. 184/B3
Porciúncula, Brazil 135/E2
Porco, Bolivia 136/B6
Porcuna, Spain 33/D4
Porcupine (riv.) 146/D3
Porcupine (riv.), Alaska 196/K1
Porcupine (hills), Alberta 182/C4
Porcupine (hills), Manitoba 179/A2
Porcupine (mts.), Mich. 250/F1
Porcupine (creek), Mont. 262/K2
Porcupine (cape), Newf. 166/C3
Porcupine (creek), N. Dak. 282/J7
Porcupine, Ontario 175/D3
Porcupine (hills), Sask. 181/K3
Porcupine, S. Dak. (57772) 298/E7
Porcupine (creek), Wyo. 319/G2
Porcupine (riv.), Yukon 187/E3
Porcupine Plain, Sask. 181/H3
Pordenone (prov.), Italy 34/D2
Pordenone, Italy 34/D2
Pore, Colombia 126/D5
Poreč, Yugoslavia 45/A3
Porepunkah, Victoria 97/D5
Pores Knob, N.C. (†28654) 281/G2
Pori, Finland 18/M6
Pori, Finland 7/G2
Porirua, N. Zealand 100/B2
Porirua (harb.), N. Zealand 100/B2
Porjus, Sweden 18/M3
Porkhov, U.S.S.R. 52/C3
Porkkala (pen.), Finland 18/O7
Porlamar, Venezuela 124/G2
Porlock, England 13/D6
Pornic, France 28/C4
Poroma, Bolivia 136/C6
Poronaysk, U.S.S.R. 54/R5
Poronaysk, U.S.S.R. 48/P5
Póros, Greece 45/F7
Poroto (pt.), Chile 138/A7
Porrentruy, Switzerland 39/C2
Porsangen (fjord), Norway 18/O1
Porsgrunn, Norway 18/G7
Porsuk (riv.), Turkey 63/D3
Portachuelo, Bolivia 136/D5
Port Adelaide, S. Australia 88/D7
Port Adelaide, S. Australia 94/A7
Portadown, N. Ireland 17/H3
Portaferry, N. Ireland 17/K3
Portage, Ind. (46368) 227/C1
Portage○, Maine (04768) 243/G2
Portage (lake), Maine 243/F2
Portage (bay), Manitoba 179/D3
Portage, Mich. (49081) 250/D6
Portage (isl.), New Bruns. 170/F1
Portage (co.), Ohio 284/H3
Portage, Ohio (43451) 284/C3
Portage (riv.), Ohio 284/D3
Portage, Pa. (15946) 294/E5
Portage, Utah (84331) 304/B2
Portage (co.), Wis. 317/G6
Portage, Wis. (53901) 317/G8
Portage-des-Roches, Québec 172/F1
Portage Des Sioux, Mo. (63373) 261/M5
Portage la Prairie, Man. 162/G5
Portage la Prairie, Manitoba 179/D4
Portageville, Mo. (63873) 261/N10
Portageville, N.Y. (14536) 276/D5
Portal, Ariz. (85632) 198/F7
Portal, Georgia (30450) 217/J5
Portal, N. Dak. (58772) 282/E2
Portalban, Switzerland 39/C3
Port Alberni, Br. Col. 184/H3
Port Albert, Ontario 177/C4
Port Albert, Victoria 97/D6
Port Albion, Br. Col. 184/E6
Portalegre (dist.), Portugal 33/C3
Portalegre, Portugal 33/C3
Portales, N. Mex. 188/F4
Portales, N. Mex. (88130) 274/F4
Port Alexander, Alaska (99836) 196/M2
Port-Alfred, Québec 172/G1
Port Alfred, S. Africa 118/D6
Port Alice, Br. Col. 184/D5
Port Allegany, Pa. (16743) 294/F2
Port Allen, La. (70767) 238/J2
Port Alma, Ontario 177/B5
Port Angeles, Wash. 188/B1
Port Angeles, Wash. (98362) 310/B2
Port Angeles Ind. Res., Wash. 310/B2
Port Antonio, Jamaica 158/K6
Port Antonio, Jamaica 156/C3
Port-à-Piment, Haiti 158/A6
Port Appin, Scotland 15/C4
Port Aransas, Texas (78373) 303/H10
Portarlington, Ireland 17/G5
Portarlington, Ireland 10/C4
Port Arthur, Texas 146/J7
Port Arthur, Texas 188/H5
Port Arthur, Texas (77640) 303/K8
Port Askaig, Scotland 15/B5
Port au Bras, Newf. 166/D4
Port au Choix, Newf. 166/C3
Port Augusta, Austrlia 87/D9
Port Augusta, S. Australia 88/F6
Port Augusta, S. Australia 94/E5
Port-au-Persil, Québec 172/G2
Port au Port, Newf. 166/C4
Port au Port (bay), Newf. 166/C4
Port au Port (pen.), Newf. 166/C4
Port-au-Prince (cap.), Haiti 156/D3
Port-au-Prince (cap.), Haiti 146/L8
Port-au-Prince (cap.), Haiti 158/C6
Port Austin, Mich. (48467) 250/F4
Portavogie, N. Ireland 17/K3
Port Bannatyne, Scotland 15/A2
Port Barre, La. (70577) 238/G5
Port-Bergé, Madagascar 118/H3
Port Blair, India 54/L8
Port Blair, India 68/G6
Port Blakely, Wash. (†98101) 310/A2
Port Blandford, Newf. 166/C2
Port Bolivar, Texas (77650) 303/L3
Port-Bou, Spain 33/H1
Port-Bouet, Ivory Coast 106/D7
Port Broughton, S. Australia 94/F5
Port Bruce, Ontario 177/D5
Port Burwell, N.W. Terrs. 187/M3
Port Burwell, Ontario 177/D5
Port Byron, Ill. (61275) 222/C2
Port Byron, N.Y. (13140) 276/G4
Port Carbon, Pa. (17965) 294/K4
Port Carling, Ontario 177/E2
Port-Cartier, Que. 162/K5
Port-Cartier, Québec 174/D2
Port-Cartier-Ouest, Québec 174/D3
Port Castries (harb.), St. Lucia 161/G6
Port Chalmers, N. Zealand 100/C6
Port Charlotte, Fla. (33952) 212/D5
Port Charlotte, Scotland 15/B5
Port Chester, N.Y. (10573) 276/P7
Port Clarence (inlet), Alaska 196/E1
Port Clements, Br. Col. 184/B3
Port Clinton, Ohio (43452) 284/E2
Port Clinton, Pa. (19549) 294/K4
Port Clyde, Maine (04855) 243/E8
Port Clyde, Nova Scotia 168/C5
Port Colborne, Ontario 177/E5
Port Coquitlam, Br. Col. 184/L3
Port Costa, Calif. (94569) 204/J1
Port Darwin (inlet), North. Terr. 93/B2
Port Davey (inlet), Tasmania 88/G8
Port Davey (inlet), Tasmania 99/B5
Port-de-Bouc, France 28/F6
Port-de-Paix, Haiti 156/D3
Port-de-Paix, Haiti 158/B5
Port Deposit, Md. (21904) 245/O2
Port Dickinson, N.Y. (†13901) 276/J6
Port Dickson, Malaysia 72/D7
Port Dufferin, Nova Scotia 168/F4
Porte des Morts (str.), Wis. 317/N5
Port Edward, Br. Col. 184/B3
Port Edwards, Wis. (54469) 317/G7
Portel, Brazil 132/D3
Port Elgin, New Bruns. 170/F2
Port Elgin, Ontario 177/C3
Port Elizabeth, N.J. (08348) 273/D5
Port Elizabeth, S. Africa 102/E8
Port Elizabeth, S. Africa 118/D6
Port Ellen, Scotland 15/B5
Porter (co.), Ind. 227/C2
Porter, Ind. (46304) 227/C1
Porter, Maine (04068) 243/B8
Porter○, Maine (04068) 243/B8
Porter, Minn. (56280) 255/B6
Porter, Ohio (†45614) 284/F8
Porter, Okla. (74454) 288/R3
Porter (pt.), St. Vin. & Grens. 161/A8
Porter, Texas (77365) 303/J7
Porter, Wash. (†98541) 310/B4
Porterdale, Georgia (30270) 217/E3
Porterfield, Wis. (54159) 317/L5
Port Erin, I. of Man 13/C3
Porters (lake), Nova Scotia 168/E4
Porters Lake, Nova Scotia 168/E4
Portersville, Ala. (†35961) 195/G2
Portersville, Ind. (†47546) 227/C8
Portersville, Pa. (16051) 294/B4
Porterville, Calif. (93257) 204/G7
Porterville, Miss. (39352) 256/G5
Porterwood, W. Va. (†26283) 312/G4
Port Everglades (harb.), Fla. 212/C4
Port Ewen, N.Y. (12466) 276/N7
Port Fairy, Victoria 97/B6
Port Felix, Nova Scotia 168/G3
Port Franks, Ontario 177/C4
Port Fuad, Egypt 111/K3
Port Gamble, Wash. (98364) 310/C3
Port Gamble Ind. Res., Wash. 310/C3
Port-Gentil, Gabon 115/A4
Port-Gentil, Gabon 102/C5
Port Gibson, Miss. (39150) 256/B7
Port Glasgow, Scotland 10/A1
Port Glasgow, Scotland 15/A2
Portglenone, N. Ireland 17/H2
Portgordon, Scotland 15/F3
Port Graham, Alaska (99603) 196/B2
Port Greville, Nova Scotia 168/D3
Port Hamilton (So) (isl.), S. Korea 81/C6
Port Hammond, Br. Col. 184/L3
Port Harcourt, Nigeria 106/F8
Port Harcourt, Nigeria 102/C4
Port Hardy, Br. Col. 184/D5
Port Hastings, Nova Scotia 168/G3
Port Hawkesbury, Nova Scotia 168/G3
Porthcawl, Wales 13/D6
Port Hebert (harb.), Nova Scotia 168/D5
Port Hedland, Australia 2/R7
Port Hedland, Australia 87/B7
Port Hedland, W. Australia 88/B3
Port Hedland, W. Australia 92/B3
Port Heiden, Alaska (†99579) 196/G3
Port Heiden (inlet), Alaska 196/G3
Port Henry, N.Y. (12974) 276/O2
Porthill, Idaho (83853) 220/B1
Porthmadog, Wales 10/D4
Porthmadog, Wales 13/C5
Port Hood, Nova Scotia 168/G2
Port Hood (isl.), Nova Scotia 168/G2
Port Hope, Mich. (48468) 250/G5
Port Hope, Ontario 177/F4
Port Hope Simpson, Newf. 166/C3
Port Houghton (inlet), Alaska 196/N1
Port Howe, Nova Scotia 168/E3
Port Hudson, La. (†70791) 238/J1
Port Hueneme, Calif. (93041) 204/F9
Port Huron, Mich. 188/K2
Port Huron, Mich. (48060) 250/G6
Portia, Ark. (72457) 202/H1
Portimão, Portugal 33/B4
Portis, Kansas (67474) 232/D2
Port Isabel, Texas (78578) 303/G11
Portishead, England 13/E6
Port Jackson (inlet), N. S. Wales 88/L4
Port Jackson (inlet), N.S. Wales 97/J3
Port Jefferson, N.Y. (11777) 276/P9
Port Jefferson, Ohio (45360) 284/C5
Port Jervis, N.Y. (12771) 276/L8
Port Joli, Nova Scotia 168/D5
Port Joli (harb.), Nova Scotia 168/D5
Port Kaiser, Jamaica 158/H7
Port Kaituma, Guyana 131/A2
Port Keats, North. Terr. 93/A3
Port Keats Mission, North. Terr. 88/D2
Port Kelang, Malaysia 72/D7
Port Kembla, N.S. Wales 97/F4
Port Kenny, S. Australia 94/D5
Port Kent, N.Y. (12975) 276/O1
Port Kindu, Zaire 115/E4
Port Kindu, Zaire 102/E5
Port Kirwan, Newf. 166/D2
Portknockie, Scotland 15/F3
Port Lambton, Ontario 177/B5
Portland, Ark. (71663) 202/H7
Portland, Barbados 161/B8
Portland (canal), Br. Col. 162/C4
Portland (canal), Br. Col. 184/B2
Portland (inlet), Br. Col. 184/C3
Portland, Colo. (†81226) 208/K6
Portland, Colo. (†81427) 208/D6
Portland, Conn. (06480) 210/E2
Portland○, Conn. (06480) 210/E2
Portland, England 13/E7
Portland, Bill of (prom.), England 10/E5
Portland, Bill of (prom.), England 13/E7
Portland, Fla. (†32439) 212/C6
Portland, Ind. (47371) 227/H4
Portland (pt.), Jamaica 156/C3
Portland (pt.), Jamaica 158/J7
Portland, Maine 188/N2
Portland, Maine 146/M5
Portland, Maine (*04101) 243/C8
Portland, Mich. (48875) 250/E6
Portland, Mo. (65067) 261/J5
Portland, N.S. Wales 97/E3
Portland, N.Y. (14769) 276/B6
Portland, N. Zealand 100/E1
Portland (isl.), N. Zealand 100/G3
Portland, N. Dak. (58274) 282/R5
Portland, Ohio (45770) 284/G7
Portland, Ontario 177/H3
Portland, Oreg. 146/F5
Portland, Oreg. 188/B1
Portland, Oreg. (*97201) 291/B2
Portland, Pa. (18351) 294/M4
Portland (cape), Tasmania 99/D2
Portland, Tenn. (37148) 237/H7
Portland, Texas (78374) 303/G10
Portland, U.S. 2/G3
Portland, Victoria 88/G7
Portland, Victoria 97/A6
Portland (bay), Victoria 97/A6
Portland (bay), Victoria 88/G7
Portland Canal (inlet), Alaska 196/N2
Portland Creek (pond), Newf. 166/C3
Portland Int'l Airport, Oreg. 291/B2
Portlaoighise, Ireland 17/G5
Portlaoighise, Ireland 10/C4
Port Latta, Tasmania 99/B2
Port Lavaca, Texas (77979) 303/H9
Portlaw, Ireland 17/G7
Port Leyden, N.Y. (13433) 276/K3
Port Lincoln, Australia 87/D9
Port Lincoln, S. Australia 88/E6
Port Lincoln, S. Australia 94/E6
Port Lions, Alaska (99550) 196/H3
Portlock, Alaska (†99603) 196/C2
Port Loko, S. Leone 106/B7
Port Loring, Ontario 177/E2
Port Lorne, Nova Scotia 168/C4
Port-Louis, France 28/B4
Port-Louis, Guadeloupe 161/B5
Port-Louis, Guadeloupe 156/G3
Port Louis (cap.), Mauritius 118/G5
Port Ludlow, Wash. (98365) 310/C3
Port-Lyautey (Kénitra), Morocco 106/C2
Port Macquarie, N. S. Wales 88/J6
Port Macquarie, N. S. Wales 97/G2
Port Madison Ind. Res., Wash. 310/A1
Portmahomack, Scotland 15/E3
Port Maitland, Nova Scotia 168/B5
Port Manvers (harb.), Newf. 166/B2
Port Margot, Haiti 158/C5
Port Maria, Jamaica 158/J6
Port Maria, Jamaica 156/C3
Portmarnock, Ireland 17/J5
Port Matilda, Pa. (16870) 294/F4
Port Mayaca, Fla. (†33438) 212/F5
Port McNeill, Br. Col. 184/D5
Port McNicoll, Ontario 177/E3
Port Medway, Nova Scotia 168/D4
Port Melbourne, Victoria 88/K7
Port Melbourne, Victoria 97/H5
Port Mellon, Br. Col. 184/K2
Port-Menier, Que. 162/K6
Port-Menier, Québec 174/E3
Port Moller (inlet), Alaska 196/F3
Port Monmouth, N.J. (07758) 273/E3
Port Moody, Br. Col. 184/L3
Port Morant, Jamaica 158/K6
Port Moresby (cap.), Papua N.G. 87/E6
Port Moresby (cap.), Papua N.G. 85/B7
Port Morien, Nova Scotia 168/J2
Port Morris, N.J. (†07850) 273/D2
Port Mouton, Nova Scotia 168/D5
Port Mouton (harb.), Nova Scotia 168/D5
Port Murray, N.J. (07865) 273/D2
Port Neches, Texas (77651) 303/K7
Port Nellie Juan, Alaska (†99501) 196/C1
Portneuf (res.), Idaho 220/F7
Portneuf (riv.), Idaho 220/F7
Portneuf (county), Québec 174/C3
Portneuf (co.), Québec 172/E3
Portneuf, Québec 172/F3
Portneuf (riv.), Québec 172/H1
Portneuf-sur-Mer, Québec 172/H1
Port Neville, Br. Col. 184/E5
Port Nicholson (inlet), N. Zealand 100/B3
Port Nolloth, S. Africa 102/D7
Port Nolloth, S. Africa 118/B5
Port Norris, N.J. (08349) 273/C5
Port-Nouveau-Québec, Que. 162/K4
Port-Nouveau-Québec, Québec 174/F2
Porto, Brazil 132/F3
Porto (dist.), Portugal 33/B2
Porto, Portugal 7/D4
Porto, Portugal 33/B2
Porto Alegre, Brazil 132/D10
Porto Alegre, Brazil 120/D6
Porto Alegre, Brazil 2/G7
Porto Alexandre, Angola 115/B7
Porto Amboim, Angola 115/B6
Porto Amboim, Angola 102/D6
Portobelo, Panama 154/H6
Portocivitanova, Italy 34/D3
Porto Colômbia (res.), Brazil 135/B2
Port O'Connor, Texas (77982) 303/H9
Porto de Mós, Portugal 33/B3
Porto de Moz, Brazil 132/D3
Porto Empedocle, Italy 34/D6
Porto Esperança, Brazil 132/B7
Porto Feliz, Brazil 135/C3
Portoferraio, Italy 34/C3
Portofino, Italy 34/B2
Porto Franco, Brazil 132/E4
Port-of-Spain (cap.), Trin. & Tob. 161/A10
Port-of-Spain (cap.), Trin. & Tob. 156/G5
Portogruaro, Italy 34/D2
Portola, Calif. (96122) 204/E4
Pôrto La Cruz, Venezuela 120/C1
Portola Valley, Calif. (94025) 204/J3
Portomaggiore, Italy 34/C2
Porto Moniz, Portugal 33/A2
Porto Moroco, Brazil 132/C6
Porto Murtinho do Sul, Brazil 132/B8
Porto Nacional, Brazil 120/E4
Porto Nacional, Brazil 132/E5
Porto-Novo (cap.), Benin 106/E7
Porto-Novo (cap.), Benin 102/C4
Porto Novo, India 68/E6
Porto Poet, Brazil 132/C2
Port Orange, Fla. (32019) 212/F2
Port Orchard, Wash. (98366) 310/A2
Porto Recanati, Italy 34/D3
Port Orford, Oreg. (97465) 291/C5
Porto Santo, Portugal 33/A2
Porto Santo (isl.), Portugal 102/A1
Pôrto Santo (isl.), Portugal 106/A2
Porto Santo (isl.), Portugal 33/B2
Porto Seguro, Brazil 132/G7
Porto Tolle, Italy 34/D2
Porto Torres, Italy 34/B4
Porto União, Brazil 132/D9
Porto-Vecchio, France 28/B7
Porto Velho, Brazil 132/H10
Porto Velho, Brazil 120/C3
Portoviejo, Ecuador 120/A3
Portoviejo, Ecuador 128/B3
Portpatrick, Scotland 15/C6
Portpatrick, Scotland 10/D3
Port Pegasus (inlet), N. Zealand 100/B7
Port Penn, Del. (19731) 245/R2
Port Perry, Ontario 177/E3
Port Phaeton (bay), Fr. Poly. 86/T13
Port Phillip (bay), Victoria 97/C6
Port Phillip (bay), Victoria 88/K7
Port Pirie, Australia 87/D9
Port Pirie, S. Australia 88/F6
Port Pirie, S. Australia 94/E5
Port Praslin (bay), St. Lucia 161/G6
Port Radium, Canada 4/C15
Port Radium, N.W.T. 162/E2
Port Radium, N.W.T. 146/G3
Port Radium, N.W. Terrs. 187/G3
Port Reading, N.J. (07064) 273/E2
Portree, Scotland 10/C2
Portree, Scotland 15/B3
Portreeve, Sask. 181/B5
Port Renfrew, Br. Col. 184/J3
Port Republic, Md. (20676) 245/N6
Port Republic, N.J. (08241) 273/D4
Port Republic, Va. (24471) 307/L4
Port Rexton, Newf. 166/D2
Port Rhoades, Jamaica 158/H5
Port Richey, Fla. (33568) 212/D3
Port Rowan, Ontario 177/D5
Port Royal, Jamaica 158/J6
Port Royal, Ky. (40058) 237/L3
Port Royal, Nova Scotia 168/H3
Port Royal, Pa. (17082) 294/H4
Port Royal, S.C. (29935) 296/F7
Port Royal (sound), S.C. 296/F7
Port Royal, Va. (22535) 307/O4
Portrush, N. Ireland 10/C3
Portrush, N. Ireland 17/H1
Port Safâga, Egypt 59/B4
Port Safâga, Egypt 111/F2
Port Said, Egypt 102/F1
Port Said, Egypt 59/B3
Port Said, Egypt 111/K2
Port Saint Joe, Fla. (32456) 212/D6
Port Saint Johns (Umzimbuvu), S. Africa 118/D6
Port-Saint-Louis-du-Rhône, France 28/F6
Port Saint Lucie, Fla. (33452) 212/F4
Port Saint Mary, I. of Man 13/C3
Port Salerno, Fla. (33492) 212/F4
Port-Salut, Haiti 158/A6
Port Sanilac, Mich. (48469) 250/G5
Port Saunders, Newf. 166/C3
Pörtschach am Wörthersee, Austria 41/C3
Port Severn, Ontario 177/E3
Port Severn, Ontario 175/C1
Port Shepstone, S. Africa 102/F8
Port Shepstone, S. Africa 118/E6
Port Simpson, Br. Col. 184/B3
Portslade-by-Sea, England 13/G7
Portsmouth, Dominica 156/G4
Portsmouth, Dominica 161/E5
Portsmouth, England 10/F5
Portsmouth, England 7/D3
Portsmouth, England 13/F7
Portsmouth, Iowa (51565) 229/C5
Portsmouth, N.H. 188/N2
Portsmouth, N.H. (03801) 268/F5
Portsmouth, N.C. (†27960) 281/S4
Portsmouth (isl.), N.C. 281/T5
Portsmouth, Ohio 188/K3
Portsmouth, Ohio (45662) 284/D8
Portsmouth○, R.I. (02871) 249/J6
Portsmouth, Va. 188/L3
Portsmouth (I.C.), Va. (*23701) 307/R7
Port Sorell, Tasmania 99/C3
Portsoy, Scotland 15/F3
Port Stanley, Ontario 177/C5
Port Stephens (inlet), N.S. Wales 97/G3
Portstewart, N. Ireland 17/H1
Port Sudan, Sudan 111/G4
Port Sudan, Sudan 102/F3
Port Sudan, Sudan 59/C6
Port Sulphur, La. (70083) 238/L8
Port Talbot, Wales 10/E5
Port Talbot, Wales 13/D6
Port Tampa (harb.), Fla. 212/B3
Port Taufiq, Egypt 111/K3
Port Tobacco, Md. (20677) 245/K6
Port Townsend, Wash. (98368) 310/C2
Port Trevorton, Pa. (17864) 294/J4
Portugal 2/J3
Portugal 7/D5
PORTUGAL 33/B3
Portugal Cove, Newf. 166/D2
Portugal Cove South, Newf. 166/D2
Portugalete, Bolivia 136/B7
Portugalete, Spain 33/E1
Portuguesa (state), Venezuela 124/D3
Portuguesa (riv.), Venezuela 124/D3
Portuguese Cove, Nova Scotia 168/E4
Portumna, Ireland 17/E5
Port Union, Newf. 166/D2
Port-Valais, Switzerland 39/C4
Port-Vendres, France 28/E6
Portville, N.Y. (14770) 276/D6
Port Vincent, La. (†70726) 238/L2
Port Vue, Pa. (†15133) 294/C7
Port Washington, N.Y. (11050) 276/R6
Port Washington, Ohio (43837) 284/G5
Port Washington, Wis. (53074) 317/L9
Port Weld, Malaysia 72/D6
Port Wells (inlet), Alaska 196/C1
Port Wentworth, Georgia (31407) 217/K6
Port William, Ohio (45164) 284/C6
Port William, Scotland 15/D6
Port Williams, Nova Scotia 168/D3
Port Wing, Wis. (54865) 317/D2
Porum, Okla. (74455) 288/R4
Porus, Jamaica 158/H6
Porvenir, Pando, Bolivia 136/A2
Porvenir, Santa Cruz, Bolivia 136/E4
Porvenir, Chile 138/E10
Porvenir, Peru 128/E5
Porvenir, Uruguay 145/B3
Porz am Rhein, W. Germany 22/B3
Posadas, Argentina 143/E2
Posadas, Argentina 120/D5
Posadas, Paraguay 144/E5
Posadas, Spain 33/D4
Poschiavo, Switzerland 39/J4
Poschiavo (riv.), Switzerland 39/K4
Poschiavo (valley), Switzerland 39/K4
Posen, Ill. (60469) 222/B6
Posen, Mich. (49776) 250/F3
Posey (co.), Ind. 227/B8
Poseyville, Ind. (47633) 227/B8
Posio, Finland 18/Q3
Poskin, Wis. (54866) 317/C5
Poso, Indonesia 85/G6
Posof, Turkey 63/K2
Posŏng, S. Korea 81/C6
Posorja, Ecuador 128/B4
Posse, Brazil 132/E6
Possel, Cent. Afr. Rep. 115/C2
Pössneck, E. Germany 22/D3
Possum Kingdom (lake), Texas 303/F5
Post, Oreg. (97752) 291/G3
Post, Texas (79356) 303/C4
Postavy, U.S.S.R. 52/C3
Poste (riv.), Québec 172/D3
Poste-de-la-Baleine, Que. 162/J4
Poste-de-la-Baleine, Québec 174/B1
Postelle, Ark. (†72366) 202/J4
Postelle, Tenn. (†37317) 237/N10
Poste Maurice Cortier, Algeria 106/E4
Poste Weygand, Algeria 106/D4
Post Falls, Idaho (83854) 220/A2
Postmasburg, S. Africa 118/C5
Post Mills, Vt. (05058) 268/C4
Postoak, Mo. (†64761) 261/E5
Postoak, Texas (†76230) 303/F4
Postojna, Yugoslavia 45/B3
Poston, Ariz. (85371) 198/A4
Poston, S.C. (29588) 296/J4
Postrervalle, Bolivia 136/D6
Postville, Iola (52162) 229/K2
Postville, Newf. 166/B3
Pot (creek), Colo. 208/A1
Pot (mt.), Idaho 220/C3
Pot (creek), Utah 304/E3
Potagannissing (bay), Mich. 250/F2
Potam, Mexico 150/D3
Potaro (riv.), Guyana 131/B3
Potato Creek, S. Dak. (†57750) 298/F6
Potawatomi Ind. Res., Kansas 232/G2
Potchefstroom, S. Africa 118/D5
Poteau (mt.), Ark. 202/B4
Poteau, Okla. (74953) 288/S4
Poteau (riv.), Okla. 288/S5
Poteca, Nicaragua 154/E4
Poteet, Texas (78065) 303/F8
Potenza (prov.), Italy 34/E4
Potenza, Italy 34/E4
Potes, Spain 33/D1
Potgietersrus, S. Africa 118/D4
Poth, Texas (78147) 303/F8
Potholes (res.), Wash. 310/F3
Poti, U.S.S.R. 52/F6
Potlatch, Idaho (83855) 220/A3
Potlatch (riv.), Idaho 220/B3
Potlatch, Wash. (†98584) 310/B3
Poto, Peru 128/H10
Potomac (riv.) 188/L3
Potomac, Ill. (61865) 222/F3
Potomac (riv.), Md. 245/M8
Potomac, Mont. (†59823) 262/C4
Potomac (riv.), Va. 307/O4
Potomac (riv.), W. Va. 312/L3
Potomac, North Branch (riv.), W. Va. 312/J4
Potomac, South Branch (riv.), W. Va. 312/J4
Potomac Beach, Va. (†22443) 307/P4
Potomac Heights, Md. (20640) 245/K6
Potomac Park-Bowling Green, Md. (†21502) 245/C2
Potomac Valley, Md. (†20768) 245/E4
Potosí (dept.), Bolivia 136/B7
Potosí, Bolivia 136/C6
Potosí, Bolivia 120/C4
Potosí, Colombia 126/C7
Potosi, Mo. (63664) 261/L7
Potosi (mt.), Nev. 266/F7
Potosi, Wis. (53820) 317/E10
Potrerillo (peak), Cuba 158/E2
Potrerillos, Chile 138/B6
Potrero, Calif. (92063) 204/J11
Potrillo (mts.), N. Mex. 274/B7
Potro, Cerro del (mt.), Argentina 143/C2
Potro, Cerro del (mt.), Chile 138/B7
Potsdam (dist.), E. Germany 22/E2
Potsdam, E. Germany 22/E2
Potsdam, Minn. (†55932) 255/F6
Potsdam, N.Y. (13676) 276/K1
Potsdam, Ohio (45361) 284/B6
Pottageville, Ontario 177/J3
Pottawatomie (co.), Kansas 232/F2
Pottawatomie (co.), Okla. 288/N4
Pottawattamie (co.), Iowa 229/B6
Pottawattamie Park, Ind. (†46360) 227/C1
Potter, Kansas (66077) 232/G2
Potter, Nebr. (69156) 264/A3
Potter (co.), Pa. 294/G2
Potter (co.), S. Dak. 298/J3
Potter (c.), Texas 303/C2
Potter, Wis. (54160) 317/K7

Potter Hill, R.I. (†02891) 249/H7
Potters Bar, England 13/H7
Potters Bar, England 10/B5
Pottersdale, Pa. (16871) 294/F3
Pottersville, Mo. (65790) 261/H9
Pottersville, N.J. (07979) 273/D2
Pottersville, N.Y. (12860) 276/N3
Potterville, Mich. (48876) 250/E6
Potts (creek), W. Va. 312/F7
Pottsboro, Texas (75076) 303/H4
Potts Camp, Miss. (38659) 256/F1
Pottstown, Pa. (19464) 294/L5
Pottsville, Ark. (72858) 202/D3
Pottsville, Pa. (17901) 294/K4
Pottsville, Texas (76565) 303/F6
Potwin, Kansas (67123) 232/F4
Pouce-Coupé, Br. Col. 184/G2
Pouch Cove, Newf. 166/D2
Poudre d'Or, Mauritius 118/G5
Poughkeepsie, Ark. (72569) 202/H1
Poughkeepsie, N.Y. (*12601) 276/N7
Pouillon, France 28/C6
Poulan, Georgia (31781) 217/E8
Poulet Cove (bay), Nova Scotia 168/H2
Poulin-de-Courval (lake), Québec 172/G1
Poulo Wai (isls.), Cambodia 72/D5
Poulsbo, Wash. (98370) 310/A1
Poultney, Vt. (05764) 268/A4
Poultney○, Vt. (05764) 268/A4
Poultney (riv.), Vt. 268/A4
Poulton le Fylde, England 10/F1
Poulton-le-Fylde, England 13/G1
Pound, Va. (24279) 307/C6
Pound, Wis. (54161) 317/L5
Pounding Mill, Va. (24637) 307/E6
Pouso Alegre, Brazil 135/D3
Pouthisat, Cambodia 72/D4
Považská Bystrica, Czech. 41/E2
Povenets, U.S.S.R. 52/E2
Poverty (isl.), Mich. 250/C3
Poverty (bay), N. Zealand 100/G3
Póvoa de Varzim, Portugal 33/B2
Povorino, U.S.S.R. 52/F4
Povungnituk, Que. 162/J3
Povungnituk, Québec 174/E1
Powassan, Ontario 177/E1
Poway, Calif. (92064) 204/J11
Powder (riv.) 188/E2
Powder (riv.), Mont. 262/L4
Powder (riv.), Oreg. 291/K3
Powder (riv.), Wyo. 319/F2
Powderhorn, Colo. (81243) 208/E6
Powderly, Ky. (42367) 237/G6
Powderly, Texas (75473) 303/J4
Powder River (co.), Mont. 262/L5
Powder River, Wyo. (82648) 319/F2
Powder Springs, Georgia (30073) 217/C3
Powder Springs, Tenn. (37848) 237/O8
Powderville, Mont. (59345) 262/L5
Powe, Mo. (†63822) 261/M9
Powell (lake) 188/D3
Powell (lake), Ariz. 198/E1
Powell (co.), Ky. 237/O5
Powell, Manitoba 179/A2
Powell (co.), Mont. 262/D4
Powell, Nebr. (†68352) 264/G4
Powell, Ohio (43065) 284/D5
Powell, Pa. (†18832) 294/J2
Powell, Tenn. (37849) 237/N8
Powell (riv.), Tenn. 237/P8
Powell (lake), U.S. 146/G6
Powell (lake), Utah 304/D6
Powell (riv.), Va. 307/B7
Powell, Wyo. (82435) 319/D1
Powell Butte, Oreg. (97753) 291/G3
Powell Creek, North. Terr. 93/C5
Powell River, Br. Col. 184/E5
Powell's Crossroads, Ala. (†35986) 195/G1
Powells Crossroads, Tenn. (†37397) 237/L10
Powells Point, N.C. (27966) 281/T2
Powellsville, N.C. (27967) 281/R2
Powellton, W. Va. (25161) 312/D6
Powellville, Md. (21852) 245/S7
Powelton, Georgia (†31059) 217/G4
Power (co.), Idaho 220/F7
Power, Mont. (59468) 262/E3
Powers (lake), Conn. 210/G3
Powers, Mich. (49874) 250/B3
Powers, Oreg. (97466) 291/D5
Powers Lake, N. Dak. (58773) 282/E2
Powersville, Georgia (31074) 217/E5
Powersville, Iowa (†50636) 229/H3
Powersville, Mo. (64672) 261/F1
Powerview, Manitoba 179/F4
Poweshiek (co.), Iowa 229/H5
Powhatan, Ark. (72458) 202/H1
Powhatan, La. (71066) 238/D3
Powhatan (co.), Va. 307/N5
Powhatan, Va. (23139) 307/N5
Powhatan, W. Va. (24877) 312/D8
Powhatan Point, Ohio (43942) 284/J6
Powhattan, Kansas (66527) 232/G2
Pownal○, Maine (04069) 243/C8
Pownal○, Vt. (05261) 268/A6
Pownal Center, Vt. (†05261) 268/A6
Powys (co.), Wales 13/D5
Poxoréo, Brazil 132/C6
Poyang (lake), China 54/N7
Poyang Hu (lake), China 77/J6
Poyen, Ark. (72128) 202/E5
Poygan (lake), Wis. 317/J7
Poynette, Wis. (53955) 317/G9
Poynor, Mo. (63959) 261/L9
Poynor, Texas (75782) 303/J5
Poysdorf, Austria 41/D2
Poy Sippi, Wis. (54967) 317/J7
Pozanti, Turkey 63/F4
Požarevac, Yugoslavia 45/E3
Poza Rica de Hidalgo, Mexico 150/L6
Poznań (prov.), Poland 47/C2
Poznań, Poland 47/C2
Poznań, Poland 7/F3
Pozo Almonte, Chile 138/B2
Pozoblanco, Spain 33/D3
Pozo Colorado, Paraguay 144/C3
Pozo Hondo, Argentina 143/D2
Pozohondo, Spain 33/F3
Pozuelo de Alarcón, Spain 33/D2
Pozuelos, Venezuela 124/F2
Pozuzo, Peru 128/E8
Pozzallo, Italy 34/E6
Pozzuoli, Italy 34/D4
Prabuty, Poland 47/D2
Prachatice, Czech. 41/B2
Prachin Buri, Thailand 72/D4
Prachuap Khiri Khan, Thailand 72/D5
Pradera, Colombia 126/B6
Prades, France 28/E6
Prado (dam), Calif. 204/E11
Praestø, Denmark 21/F7
Pragel (pass), Switzerland 39/G2
Prague (cap.), Czech. 7/F3
Prague (Praha) (cap.), Czech. 41/C1
Prague, Nebr. (68050) 264/H3
Prague, Okla. (74864) 288/N4
Praha (city), Czech. 41/C1
Prahan, Victoria 88/L7
Prahran, Victoria 97/J5
Praia (cap.), C. Verde 106/B8
Prainha, Amazonas, Brazil 132/A4
Prainha, Pará, Brazil 132/C3
Prairie, Ala. (36771) 195/D6
Prairie (co.), Ark. 202/G4
Prairie (creek), Ind. 227/C7
Prairie (riv.), Mich. 250/D7
Prairie (riv.), Minn. 255/E3
Prairie, Miss. (39756) 256/G3
Prairie (co.), Mont. 262/L4
Prairie, Queensland 95/C4
Prairie (lake), S. Dak. 298/P3
Prairieburg, Iowa (52219) 229/L4
Prairie City, Ill. (61470) 222/C3
Prairie City, Iowa (50228) 229/G5
Prairie City, Oreg. (97869) 291/J3
Prairie City, S. Dak. (57649) 298/D2
Prairie Creek, Ind. (47869) 227/C6
Prairie Dog Town Fork, Red (riv.), Okla. 288/F5
Prairie Dog Town Fork, Red (riv.), Texas 303/C3
Prairie du Chien, Wis. (53821) 317/D9
Prairie du Rocher, Ill. (62277) 222/C5
Prairie du Sac, Wis. (53578) 317/G9
Prairie Farm, Wis. (54762) 317/C5
Prairie Grove, Ark. (72753) 202/B2
Prairie Grove, Ill. (†60050) 222/E1
Prairie Home, Mo. (65068) 261/G5
Prairie Point, Miss. (39353) 256/H4
Prairie River, Sask. 181/J3
Prairies (riv.), Québec 172/H4
Prairieton, Ind. (47870) 227/B6
Prairietown, Ill. (†62097) 222/B2
Prairie View, Ark. (72859) 202/C3
Prairie View, Kansas (67664) 232/C2
Prairie View, Texas (77445) 303/J7
Prairie Village, Kansas (66208) 232/H2
Prairieville, La. (70769) 238/K2
Pran Buri, Thailand 72/D4
Prangins, Switzerland 39/B4
Prapat, Indonesia 85/B5
Praslin, St. Lucia 161/G6
Praslin (isl.), Seychelles 118/H5
Prat (isl.), Chile 138/D7
Pratas (Dongsha) (isl.), China 77/J7
Prathersville, Mo. (†64024) 261/R4
Prato, Italy 34/C3
Prato-Sornico, Switzerland 39/G4
Pratt (co.), Kansas 232/D4
Pratt, Kansas (67124) 232/D4
Pratt, Manitoba 179/D5
Pratt, Minn. (†55060) 255/E6
Pratt, W. Va. (25162) 312/D6
Pratteln, Switzerland 39/E1
Prattsburg, Georgia (†31039) 217/D5
Prattsburg○, N.Y. (14873) 276/F5
Prattsville, Ark. (72129) 202/F5
Prattsville, N.Y. (12468) 276/M6
Prattville, Ala. (36067) 195/E6
Pratum, Oreg. (†97301) 291/A3
Pravia, Spain 33/C1
Prawda, Manitoba 179/G5
Prawle (pt.), England 13/D7
Prawle (pt.), England 10/E5
Praxedis G. Guerrero, Mexico 150/G1
Pray, Mont. (59065) 262/F5
Praya, Indonesia 85/F7
Preble, Ind. (46782) 227/H3
Preble, N.Y. (13141) 276/H5
Preble (co.), Ohio 284/A6
Preeceville, Sask. 181/J4
Preemption, Ill. (61276) 222/C2
Preesall, England 13/E4
Preetz, W. Germany 22/D1
Pregarten, Austria 41/C2
Pregnall, S.C. (†29437) 296/G5
Pregonero, Venezuela 124/C3
Preiji, U.S.S.R. 53/D2
Prek Pouthi, Cambodia 72/E5
Prelate, Sask. 181/B5
Přelouč, Czech. 41/C1
Premier, W. Va. (24878) 312/C8
Premium, Ky. (41845) 237/R6
Premont, Texas (78375) 303/F10
Prentice, Ill. (†62612) 222/C4
Prentice, Wis. (54556) 317/F4
Prentiss○, Maine (†04487) 243/G5
Prentiss (co.), Miss. 256/G1
Prentiss, Miss. (39474) 256/E7
Prentiss, N.C. (†28734) 281/C4
Prenzlau, E. Germany 22/E2
Preparis (isl.), Burma 72/B4
Preparis North (chan.), Burma 72/B4
Preparis South (chan.), Burma 72/B4
Přerov, Czech. 41/D2
Presanella (mt.), Italy 34/C1
Prescott, Ariz. 188/D4
Prescott, Ariz. (86301) 198/C4
Prescott, Ark. (71857) 202/D6
Prescott, Ind. (†46176) 227/F6
Prescott, Iowa (50859) 229/D6
Prescott, Kansas (66767) 232/H3
Prescott, Mich. (48756) 250/F4
Prescott (county), Ontario 177/K2
Prescott, Ontario 177/J3
Prescott, Oreg. (†97048) 291/D1
Prescott, Wash. (99348) 310/G4
Prescott, Wis. (54021) 317/A6
Prescott Valley, Ariz. (†86301) 198/C4
Preseli (mts.), Wales 13/C6
Preševo, Yugoslavia 45/E4
Presho, S. Dak. (57568) 298/J6
Presidencia de la Plaza, Argentina 143/D2
Presidencia R. Sáenz Peña, Argentina 120/C5
Presidencia Roque Sáenz Peña, Argentina 143/D2
President, Pa. (†16353) 294/C3
Presidente Dutra, Brazil 132/E4
Presidente Hayes, Paraguay 144/C3
Presidente Prudente, Brazil 132/D8
Presidente Prudente, Brazil 120/D5
Presidente Ríos (lake), Chile 138/D6
Presidente Venceslau, Brazil 132/D8
Presidential (range), N.H. 268/E3
Presidio, Calif. 204/J2
Presidio (co.), Texas 303/C12
Presidio, Texas (79845) 303/C12
Presidio Modelo, Cuba 158/C2
Prešov, Czech. 41/F2
Prespa (lake), Albania 45/E5
Prespa (lake), Greece 45/E5
Prespa (lake), Yugoslavia 45/E5
Presque Isle, Maine (04769) 243/H2
Presque Isle (co.), Mich. 250/F3
Presque Isle, Mich. (49777) 250/F3
Presque Isle (riv.), Mich. 250/F1
Presque Isle, Wis. (54557) 317/G3
Presque Isle A.F.B., Maine 243/G2
Presqu'île Prov. Park, Ontario 177/G4
Prestatyn, Wales 13/D4
Prestea, Ghana 106/D7
Presteigne, Wales 10/E4
Presteigne, Wales 13/D5
Přeštice, Czech. 41/B2
Presto, Bolivia 136/C6
Preston○, Conn. (†06360) 210/H2
Preston, England 10/F1
Preston, England 13/G1
Preston, Georgia (31824) 217/C6
Preston, Idaho (83263) 220/G7
Preston, Ill. (†62242) 222/D5
Preston, Iowa (52069) 229/N4
Preston, Kansas (67569) 232/D4
Preston, Ky. (40366) 237/O4
Preston, Md. (21655) 245/P6
Preston, Minn. (55965) 255/F7
Preston, Miss. (39354) 256/G5
Preston, Mo. (65732) 261/F7
Preston, Nebr. (†68355) 264/J4
Preston, Nev. (†89301) 266/G4
Preston, Okla. (74456) 288/P3
Preston, Scotland 15/F5
Preston, Victoria 88/L7
Preston, Victoria 97/J4
Preston, Wash. (98050) 310/D3
Preston (co.), W. Va. 312/G4
Preston City, Conn. (†06360) 210/H2
Preston Hollow, N.Y. (12469) 276/M6
Prestonpans, Scotland 15/D1
Prestonsburg, Ky. (41653) 237/R5
Prestonville, Ky. (†41008) 237/L3
Prestwich, England 13/H2
Prestwick, Scotland 10/D3
Prestwick, Scotland 15/D5
Prêto (riv.), Brazil 132/E5
Prêto (riv.), Brazil 135/E3
Prêto (riv.), Brazil 132/A4
Pretoria (cap.), S. Africa 2/L7
Pretoria (cap.), S. Africa 102/E7
Pretoria (cap.), S. Africa 118/D5
Prettyboy (res.), Md. 245/M2
Pretty Prairie, Kansas (67570) 232/D4
Préveza, Greece 45/E6
Prévost, Québec 172/C4
Prevost, Wash. (†98250) 310/B2
Prewitt (res.), Colo. 208/N2
Prewitt, N. Mex. (87045) 274/B3
Prey Veng, Cambodia 72/E5
Pribilof (isls.), Alaska 188/C6
Pribilof (isls.), Alaska 196/D3
Pribilof (isls.), U.S. 4/D18
Priboj, Yugoslavia 45/D4
Příbor, Czech. 41/E2
Příbram, Czech. 41/B2
Price (isl.), Br. Col. 184/C4
Price, Md. (21656) 245/P4
Price, N. Dak. (†58547) 282/H5
Price, Québec 172/A1
Price, Utah (84501) 304/D4
Price, Utah 188/D3
Price (riv.), Utah 304/D4
Price (co.), Wis. 317/F4
Price, Wis. (†54741) 317/E6
Pricedale, Miss. (†39666) 256/D8
Price Hill, W. Va. (†25880) 312/D7
Priceville, Ala. (†35601) 195/E1
Priceville, Ky. (†42713) 237/K6
Priceville, Ontario 177/D3
Prichard, Ala. (36610) 195/B9
Prichard, Miss. (†38676) 256/D1
Prichard, W. Va. (25555) 312/A6
Prickly (pt.), Grenada 161/C9
Priddis, Alberta 182/C4
Priddy, Texas (76870) 303/F6
Pride, Ky. (†42404) 237/F5
Pride, La. (70770) 238/K1
Pridgen, Georgia (†31519) 217/G7
Priego, Spain 33/E2
Priego de Córdoba, Spain 33/D4
Priekule, U.S.S.R. 53/A2
Priekule, U.S.S.R. 53/A3
Prien am Chiemsee, W. Germany 22/E5
Prieska, S. Africa 118/C5
Priest (lake), Idaho 220/B1
Priest (riv.), Idaho 220/B1
Priest, J. Percy (lake), Tenn. 237/J8
Priestly (lake), Maine 243/E2
Priest Rapids (dam), Wash. 310/F4
Priest Rapids (lake), Wash. 310/E4
Priest River, Idaho (83856) 220/A1
Prievidza, Czech. 41/E2
Prijedor, Yugoslavia 45/C3
Prijepolje, Yugoslavia 45/D4
Prikumsk, U.S.S.R. 52/F6
Prikumsk, U.S.S.R. 48/E5
Prilep, Yugoslavia 45/E5
Priluki, U.S.S.R. 52/D4
Prim (pt.), Nova Scotia 168/C4
Prim (pt.), Pr. Edward I. 168/E2
Prima Porta, Italy 34/F6
Primate, Sask. 181/B3
Prime, New Bruns. 170/B1
Primero de Marzo, Paraguay 144/B4
Primero Enero, Cuba 158/F2
Primghar, Iowa (51245) 229/B2
Primm Springs, Tenn. (38476) 237/G9
Primorsk, U.S.S.R. 52/C2
Primorsko-Akhtarsk, U.S.S.R. 52/E5
Primos, Pa. (19018) 294/M7
Primrose (lake), Alberta 182/E2
Primrose, Georgia (†30222) 217/C4
Primrose, Iowa (†52625) 229/K7
Primrose, Nebr. (68655) 264/F3
Primrose (lake), Sask. 181/L3
Primrose Lake Air Weapons Range, Sask. 181/L3
Prince (co.), Pr. Edward I. 168/D2
Prince, Sask. 181/C3
Prince Albert (sound), N.W.T. 162/E1
Prince Albert (pen.), N.W.T. 162/E1
Prince Albert (pen.), N.W. Terrs. 187/G2
Prince Albert (sound), N.W. Terrs. 187/G2
Prince Albert, Sask. 162/F5
Prince Albert, Sask. 146/H4
Prince Albert, Sask. 181/F2
Prince Albert, S. Africa 118/C6
Prince Albert Nat'l Park, Sask. 162/F5
Prince Albert Nat'l Park, Sask. 181/E1
Prince Alfred (cape), N.W. Terrs. 187/F2
Prince Charles (isl.), Canada 4/C13
Prince Charles (isl.), N.W.T. 162/J2
Prince Charles (isl.), N.W. Terrs. 187/L3
Prince Edward (isls.) 5/E2
Prince Edward (county), Ontario 177/G3
Prince Edward (isls.), S. Africa 2/L8
Prince Edward (co.), Va. 307/M6
Prince Edward Island (prov.) 162/K6
Prince Edward Island (prov.), Canada 146/M5
PRINCE EDWARD ISLAND 168
Prince Edward Island Nat'l Park, Pr. Edward I. 168/E2
Prince Frederick, Md. (20678) 245/M6
Prince George, Br. Col. 146/F4
Prince George, Br. Col. 162/D5
Prince George, Br. Col. 184/F3
Prince George (co.), Va. 307/O6
Prince George, Va. (23875) 307/O6
Prince Georges (co.), Md. 245/L5
Prince Gustav Adolf (sea), N.W. Terrs. 187/H2
Prince of Wales (cape), Alaska 196/E1
Prince of Wales (isl.), Alaska 196/N2
Prince of Wales (isl.), Canada 4/B14
Prince of Wales, New Bruns. 170/D3
Prince of Wales (str.), N.W.T. 162/D1
Prince of Wales (isl.), N.W.T. 146/J2
Prince of Wales (isl.), N.W.T. 162/F1
Prince of Wales (isl.), N.W. Terrs. 187/J2
Prince of Wales (str.), N.W. Terrs. 187/G2
Prince of Wales (isl.), Queensland 88/G2
Prince of Wales (isl.), Queensland 95/B1
Prince of Wales (cape), U.S. 4/C18
Prince Olav Coast (reg.) 5/C3
Prince Patrick (isl.), Canada 4/B16
Prince Patrick (isl.), N.W.T. 146/F2
Prince Patrick (isl.), N.W.T. 162/M3
Prince Patrick (isl.), N.W. Terrs. 187/F2
Prince Regent (inlet), N.W.T. 162/G1
Prince Regent (inlet), N.W. Terrs. 187/J2
Prince Rupert, Br. Col. 162/C5
Prince Rupert, Br. Col. 146/F4
Prince Rupert, Br. Col. 184/B3
Prince Rupert (bay), Dominica 161/E5
Princes Lakes, Ind. (†46164) 227/E6
Princess Anne, Md. (21853) 245/P8
Princess Astrid Coast (reg.) 5/B1
Princess Charlotte (bay), Queensland 88/G2
Princess Charlotte (bay), Queensland 95/C2
Princess Harbour, Manitoba 179/F3
Princess Martha Coast (reg.) 5/B18
Princess Ragnhild Coast (reg.) 5/B2
Princess Royal (isl.), Br. Col. 184/C3
Princes Town, Trin. & Tob. 161/B11
Princeton, Ala. (35766) 195/F1
Princeton, Ark. (†71725) 202/E6
Princeton, Br. Col. 184/G5
Princeton, Calif. (95970) 204/C4
Princeton (mt.), Colo. 208/G5
Princeton, Fla. (33032) 212/F6
Princeton, Idaho (83857) 220/B3
Princeton, Ill. (61356) 222/D2
Princeton, Ind. (47670) 227/B8
Princeton, Iowa (52768) 229/N5
Princeton, Kansas (66078) 232/G3
Princeton, Ky. (42445) 237/F6
Princeton, La. (71067) 238/C1
Princeton○, Maine (04668) 243/H5
Princeton○, Mass. (01541) 249/G3
Princeton, Mich. (49875) 250/B2
Princeton, Minn. (55371) 255/E5
Princeton, Mo. (64673) 261/E2
Princeton, Newf. 166/D2
Princeton, N.J. (08540) 273/D3
Princeton, N.C. (27569) 281/N4
Princeton, Ontario 177/D4
Princeton, Oreg. (97721) 291/J4
Princeton, S.C. (29674) 296/C2
Princeton, W. Va. (24740) 312/D8
Princeton, Wis. (54968) 317/H8
Princeton Junction, N.J. (08550) 273/D3
Princeville, Hawaii (†96714) 218/C1
Princeville, Ill. (61559) 222/D3
Princeville, N.C. (†27886) 281/P3
Princeville, Québec 172/F3
Prince William (sound), Alaska 196/D1
Prince William, New Bruns. 170/C3
Prince William (co.), Va. 307/O3
Principe (chan.), Br. Col. 184/C3
Principe (isl.), Sao Tomé e Príncipe 106/F8
Principio Furnace, Md. (†21903) 245/P2
Prineville, Oreg. (97754) 291/G3
Prineville (res.), Oreg. 291/G3
Pringle, S. Dak. (57773) 298/B6
Prinkipo (Adalar) (isl.), Turkey 63/D6
Prinsburg, Minn. (56281) 255/C6
Prins Karls Forland (isl.), Norway 18/B2
Prinzapolca (riv.), Nicaragua 154/F4
Prinzapolka, Nicaragua 154/F4
Prior (cape), Spain 33/B1
Prior Lake, Minn. (55372) 255/E6
Priozersk, U.S.S.R. 52/D2
Pripet (marshes), U.S.S.R. 52/C4
Pripyat' (riv.), U.S.S.R. 7/G3
Pripyat' (riv.), U.S.S.R. 52/C4
Priština, Yugoslavia 45/E4
Pritchards (isl.), S.C. 296/G7
Pritchardville, S.C. (†29927) 296/E7
Pritchett, Colo. (81064) 208/O8
Pritzwalk, E. Germany 22/E2
Privas, France 28/F5
Privateer (pt.), Virgin Is. (U.S.) 161/D4
Priverno, Italy 34/D4
Privolzhskiy, U.S.S.R. 52/G4
Priyutnoye, U.S.S.R. 52/F5
Priyutovo, U.S.S.R. 52/H3
Prizren, Yugoslavia 45/E4
Probolinggo, Indonesia 85/K2
Procious, W. Va. (25164) 312/D5
Procter, Br. Col. 184/J5
Proctor, Ark. (72376) 202/K3
Proctor, Colo. (†80736) 208/N1
Proctor, Minn. (55810) 255/F4
Proctor, Mont. (59929) 262/B3
Proctor, Okla. (74457) 288/S3
Proctor, Pa. (†17701) 294/J3
Proctor, Texas (76468) 303/F5
Proctor○, Vt. (05765) 268/A4
Proctor, W. Va. (26055) 312/E3
Proctorsville, Vt. (05153) 268/B5
Proctorville, N.C. (28375) 281/M6
Proctorville, Ohio (45669) 284/F9
Proddatur, India 68/D6
Proença-a-Nova, Portugal 33/B3
Profesor Rafael Ramírez, Mexico 150/O1
Profondeville, Belgium 27/F8
Progreso, Mexico 150/P6
Progreso, Uruguay 145/B6
Progress, Br. Col. 184/G2
Progress, Oreg. (†97233) 291/A2
Progress, U.S.S.R. 48/O5
Progress Village, Fla. (†33619) 212/C3
Project City, Calif. (96079) 204/C3
Prokhladnyy, U.S.S.R. 52/F6
Prokop'yevsk, U.S.S.R. 54/K4
Prokop'yevsk, U.S.S.R. 48/J4
Prokuplje, Yugoslavia 45/E4
Prole, Iowa (50229) 229/F6
Prome (Pye), Burma 72/B3
Promise City, Iowa (52583) 229/G7
Promissão, Brazil 135/B2
Promontory, Utah (†84307) 304/B2
Prompton, Pa. (18456) 294/M2
Prongua, Sask. 181/C3
Prony (bay), New Caled. 86/H5
Prophet (riv.), Br. Col. 184/M2
Prophetstown, Ill. (61277) 222/D2
Propriá, Brazil 132/G5
Proserpine, Queensland 88/H4
Proserpine, Queensland 95/D4
Prosit, Minn. (†55702) 255/F4
Prosna (riv.), Poland 47/C3
Prospect, Ala. (†35578) 195/D3
Prospect○, Conn. (06712) 210/D2
Prospect, Ky. (40059) 237/K4
Prospect○, Maine (†04981) 243/F6
Prospect (res.), N. S. Wales 88/K4
Prospect (res.), N. S. Wales 97/H3
Prospect, N.Y. (13435) 276/K4
Prospect, Nova Scotia 168/E4
Prospect, Ohio (43342) 284/D5
Prospect, Oreg. (97536) 291/E5
Prospect, Pa. (16052) 294/B4
Prospect, S. Australia 88/D8
Prospect, S. Australia 94/B7
Prospect, Tenn. (38477) 237/G10
Prospect, Va. (23960) 307/L6
Prospect Harbor, Maine (04669) 243/H7
Prospect Heights, Ill. (60070) 222/B5
Prospect Hill, N.C. (27314) 281/L2
Prospect Park, N.J. (†07885) 273/B1
Prospect Park, Pa. (19076) 294/M7
Prosper, Mich. (†49632) 250/E4
Prosper, Minn. (55966) 255/G7
Prosper, N. Dak. (†58042) 282/R6
Prosper, Oreg. (†97411) 291/C4
Prosperidad, Philippines 82/F6
Prosperity, Pa. (15329) 294/B5
Prosperity, S.C. (29127) 296/D3
Prosser, Nebr. (68868) 264/F4
Prosser, Wash. (99350) 310/F4
Prostějov, Czech. 41/D2
Protection, Kansas (67127) 232/C4
Protem, Mo. (65733) 261/G9
Protivín, Czech. 41/C2
Protivin, Iowa (52163) 229/J2
Proulxville, Québec 172/E3
Prouts Neck, Maine (04074) 243/C8
Provadiya, Bulgaria 45/H4
Provencal, La. (71468) 238/D3
Provence (trad. prov.), France 29
Providence, Ala. (†36748) 195/C6
Providence (mts.), Calif. 204/K8
Providence, Fla. (†32061) 212/D2
Providence, Grenada 161/D9
Providence, Ind. (†46106) 227/E6
Providence, Ky. (42450) 237/F6
Providence (cape), N. Zealand 100/A7
Providence (cap.), R.I. 188/M2
Providence (cap.), R.I. 146/L5
Providence (co.), R.I. 249/H5
Providence (cap.), R.I. (*02901) 249/H5
Providence, Utah (84332) 304/C2
Providence Bay, Ontario 177/B2
Providence Forge, Va. (23140) 307/P6
Providencia (isl.), Colombia 126/B9
Providenciales (isl.), Turks & Caicos 156/D2
Provideniya, U.S.S.R. 4/C18
Provideniya, U.S.S.R. 48/T3
Province Lake, N.H. (†03888) 268/E4
Provincetown, Mass. (02657) 249/O4
Provincetown○, Mass. (02657) 249/O4
Provins, France 28/E3
Provo, Ark. (†71846) 202/B5
Provo, S. Dak. (57774) 298/B7
Provo, Utah 146/G6
Provo, Utah (84601) 304/C3
Provo, Utah 188/D2
Provo (peak), Utah 304/C3
Provo (riv.), Utah 304/C3
Provost, Alberta 182/E3
Prowers (co.), Colo. 208/P7
Prozor, Yugoslavia 45/C4
Pruden, Tenn. (37851) 237/O7
Prudence (isl.), R.I. 249/J6
Prudence Island, R.I. (02872) 249/J6
Prudentópolis, Brazil 132/D9
Prudenville, Mich. (48651) 250/E4
Prudhoe (bay), Alaska 146/D2
Prudhoe (bay), Alaska 196/J1
Prudhoe, England 13/H3
Prudhoe Bay, Alaska (†99723) 196/J1
Prud'homme, Sask. 181/F3
Prudnik, Poland 47/C3
Prue, Okla. (74060) 288/O2
Pruitt, Ark. (72671) 202/D1
Prüm, W. Germany 22/B3
Pruntytown, W. Va. (†26354) 312/F4
Pruszcz Gdanski, Poland 47/D1
Pruszków, Poland 47/E2
Prut (riv.) 7/G4
Prut (riv.), Romania 45/J2
Prut (riv.), U.S.S.R. 52/C5
Prydz (bay) 5/C4
Pryor, Colo. (81065) 208/K8
Pryor, Mont. (59066) 262/H5
Pryor, Okla. (74361) 288/R2
Pryorsburg, Ky. (†42066) 237/D7
Pryse, Ky. (40471) 237/O5
Przasnysz, Poland 47/E2
Przemkow, Poland 47/B2
Przemsza (riv.), Poland 47/B4
Przemyśl (prov.), Poland 47/F4
Przemyśl, Poland 7/G4
Przmyśl, Poland 47/F4
Przeworsk, Poland 47/F3
Przheval'sk, U.S.S.R. 48/H5
Psakhná, Greece 45/F6
Psará (isl.), Greece 45/G6
Psári, Greece 45/E7
Psel (riv.), U.S.S.R. 52/D4
Psevdhókavos (cape), Greece 45/G6
Pskov, U.S.S.R. 7/G3
Pskov, U.S.S.R. 48/C4
Pskov U.S.S.R. 52/C3
Pskov (lake), U.S.S.R. 53/D1
Ptolemaís, Greece 45/E5
Ptuj, Yugoslavia 45/C2
Puako, Hawaii (†96743) 218/G4
Puán, Argentina 143/D4
Puangue, Chile 138/F4
Puangue, Estero de (riv.), Chile 138/F3
Pubnico, Nova Scotia 168/C5
Pubnico (harb.), Nova Scotia 168/C5
Puca Barranca, Peru 128/E4
Pucallpa, Peru 128/E7
Pucallpa, Peru 120/B3
Pucara, Bolivia 136/C6
Pucará, Peru 128/G10
Pucarani, Bolivia 136/A5
Pucatrihue, Chile 138/D3
Pucaurco, Peru 128/G4
Puce, Ontario 177/B5
Pucheng, China 77/J6
Púchov, Czech. 41/E2
Puchuncaví, Chile 138/F2
Pucio (pt.), Philippines 82/C5
Pucioasa, Romania 45/G3
Puck, Poland 47/D1
Puckaway (lake), Wis. 317/H8
Puckett, Miss. (39151) 256/E6
Pucón, Chile 138/E2
Pucusana, Peru 128/D9
Pudahuel, Chile 138/G3
Pudasjärvi, Finland 18/P4
Pudding (riv.), Oreg. 291/A3
Pudozh, U.S.S.R. 52/E2
Puduchcheri (Pondicherry), India 68/E6
Pudukkottai, India 68/D6
Puebla (state), Mexico 150/L7
Puebla, Mexico 146/J8
Puebla de Alcocer, Spain 33/D3
Puebla de Don Fadrique, Spain 33/E4
Puebla del Caramiñal, Spain 33/B1
Puebla de Sanabria, Spain 33/C1
Puebla de Trives, Spain 33/C1
Puebla de Zaragoza, Mexico 150/N2

Pueblo, Colo. 146/H6
Pueblo, Colo. 188/F3
Pueblo (co.), Colo. 208/K6
Pueblo, Colo. (*81001) 208/K6
Pueblo (res.), Colo. 208/K6
Pueblo (mts.), Oreg. 291/J5
Pueblo Army Depot, Colo. 208/L6
Pueblo Colorado Wash (dry riv.), Ariz. 198/F3
Pueblo del Sauce, Uruguay 145/E4
Pueblo Hondo, Venezuela 124/B3
Pueblo Hundido, Chile 138/B6
Pueblo Ind. Res., N. Mex. 274/B4
Pueblo Ind. Res., N. Mex. 274/D2
Pueblo Ind. Res., N. Mex. 274/C4
Pueblo Ind. Res., N. Mex. 274/D3
Pueblo La Paloma, Uruguay 145/D3
Pueblo Nuevo, Uruguay 145/B2
Pueblo Nuevo, Venezuela 124/D1
Pueblo of Acoma, N. Mex. (†87034) 274/B3
Puebloviejo, Colombia 126/D5
Puelches, Argentina 143/C4
Puelén, Argentina 143/C4
Puelo (riv.), Chile 138/E4
Puente Alto, Chile 138/B10
Puenteareas, Spain 33/B1
Puente de Ixtla, Mexico 150/K2
Puente del Inca, Argentina 143/B3
Puentedeume, Spain 33/C1
Puente-Genil, Spain 33/D4
Puente Nacional, Colombia 126/D5
Pueo (pt.), Hawaii 218/A2
Pu'er, China 77/F7
Puerca (pt.), P. Rico 161/F2
Puerco (riv.), Ariz. 198/F3
Puerco (riv.), N. Mex. 274/A3
Puercos, Morro de (head), Panama 154/H7
Puerto Acosta, Bolivia 136/A4
Puerto Adela, Paraguay 144/E4
Puerto Aisén, Chile 138/E6
Puerto Aisén, Chile 120/B7
Puerto Alegre, Bolivia 136/E3
Puerto Alianza, Peru 128/D5
Puerto Almacen, Bolivia 136/C4
Puerto Amaro, Uruguay 145/F3
Puerto América, Peru 128/D5
Puerto Ángel, Mexico 150/L9
Puerto Antioquia, Colombia 126/C4
Puerto Arazatí, Uruguay 145/C5
Puerto Argentina, Colombia 126/C7
Puerto Armuelles, Panama 154/F6
Puerto Arturo, Peru 128/F3
Puerto Asís, Colombia 126/B7
Puerto Ayacucho, Venezuela 120/C2
Puerto Ayacucho, Venezuela 124/E5
Puerto Ayora, Ecuador 128/B9
Puerto Bahía Negra, Paraguay 144/C2
Puerto Ballivián, Bolivia 136/C4
Puerto Barrios, Guatemala 154/C3
Puerto Bermúdez, Peru 128/E8
Puerto Berrío, Colombia 126/C4
Puerto Bertrand, Chile 138/E7
Puerto Boy, Colombia 126/C7
Puerto Caballas, Peru 128/E10
Puerto Caballo, Paraguay 144/C2
Puerto Cabello, Venezuela 124/E2
Puerto Cabello, Venezuela 120/C1
Puerto Cabezas, Nicaragua 154/F3
Puerto Calvimonte, Bolivia 136/C4
Puerto Carlos Pfannl, Paraguay 144/D4
Puerto Carranza, Colombia 126/F9
Puerto Carreño, Colombia 120/C2
Puerto Carreño, Colombia 126/G4
Puerto Casado, Paraguay 144/C3
Puerto Castilla, Honduras 154/D3
Puerto Chacabuco, Chile 138/D6
Puerto Chicama, Peru 128/C6
Puerto Cisnes, Chile 138/E5
Puerto Coig, Argentina 143/C7
Puerto Colombia, Colombia 126/C2
Puerto Colón, Paraguay 144/C3
Puerto Cooper, Paraguay 144/C3
Puerto Córdoba, Colombia 126/E8
Puerto Cortés, C. Rica 154/F6
Puerto Cortés, Honduras 154/D2
Puerto Cortés, Mexico 150/D4
Puerto Crevaux, Colombia 126/C6
Puerto Cristal, Chile 138/E6
Puerto Cumarebo, Venezuela 124/D2
Puerto de Cayo, Ecuador 128/B3
Puerto de la Concordia, El Salvador 154/C4
Puerto de Luna, N. Mex. (88432) 274/E4
Puerto de Nutrias, Venezuela 124/D3
Puerto Deseado, Argentina 143/D6
Puerto Deseado, Argentina 120/C7
Puerto El Carmen, Ecuador 128/E3
Puerto Escondido, Colombia 126/B3
Puerto Escondido, Mexico 150/L9
Puerto Esperanza, Cuba 158/A1
Puerto Esperanza, Paraguay 144/C2
Puerto Estrella, Colombia 126/E1
Puerto Eten, Peru 128/B6
Puerto Fonciere, Paraguay 144/D3
Puerto Frey, Bolivia 136/E4
Puerto Galileo, Paraguay 144/C4
Puerto General Busch, Bolivia 136/G7
Puerto General Ovando, Bolivia 136/50
Puerto Grether, Bolivia 136/C5
Puerto Guachalla, Bolivia 136/F6
Puerto Guaraní, Paraguay 144/C2
Puerto Harberton, Argentina 143/C7
Puerto Heath, Bolivia 136/A3
Puerto Hierro, Venezuela 124/H2
Puerto Huitoto, Colombia 126/D7
Puerto Iguazú, Argentina 143/F2
Puerto Inca, Peru 128/E7
Puerto Indio, Paraguay 144/E4
Puerto Ingeniero Ibáñez, Chile 138/E6
Puerto Iradier, Equat. Guinea 115/A3
Puerto Irigoyen, Argentina 143/D1
Puerto Isabel, Bolivia 136/F6
Puerto Izozog, Bolivia 136/D6
Puerto José Pardo, Peru 128/D4
Puerto Juárez, Mexico 150/Q6
Puerto La Concordia, Colombia 126/D6
Puerto La Cruz, Venezuela 124/F2
Puerto Leguía, Loreto, Peru 128/D4
Puerto Leguía, Puno, Peru 128/G9
Puerto Leguízamo, Colombia 126/C8
Puerto Lempira, Honduras 154/F3
Puerto Liberador General San Martín, Argentina 143/E2
Puerto Limón, Colombia 126/B7
Puertollano, Spain 33/D3
Puerto López, La Guajira, Colombia 126/E2
Puerto López, Meta, Colombia 126/D5
Puerto Madero, Mexico 150/N9
Puerto Madryn, Argentina 143/C5
Puerto Madryn, Argentina 120/C7
Puerto Maldonado, Peru 128/H9
Puerto Maldonado, Peru 120/C4
Puerto Mamoré, Bolivia 136/C5
Puerto Manglares, Colombia 126/B5
Puerto María, Paraguay 144/C2
Puerto Max, Paraguay 144/D3
Puerto Mayor Otaño, Paraguay 144/E5
Puerto Medio Mundo (bay), P. Rico 161/F2
Puerto Mercedes, Colombia 126/D7
Puerto Mihanovich, Paraguay 144/C2
Puerto Miranda, Venezuela 124/E4
Puerto Montt, Chile 120/B7
Puerto Montt, Chile 138/E4
Puerto Morelos, Mexico 150/Q6
Puerto Morín, Peru 128/C7
Puerto Mosquito, Colombia 126/C3
Puerto Murillo, Colombia 126/G4
Puerto Mutis, Colombia 126/B4
Puerto Napo, Ecuador 128/D3
Puerto Nare, Colombia 126/D7
Puerto Nariño, Colombia 126/F5
Puerto Natales, Chile 120/B8
Puerto Natales, Chile 138/E9
Puerto Niño, Colombia 126/C5
Puerto Nuevo, Colombia 126/F5
PUERTO NUEVO, Paraguay 144/C2
Puerto Nuevo, P. Rico 161/D1
Puerto Nuevo (pt.), P. Rico 161/C1
Puerto Obaldía, Panama 154/J6
Puerto Ocopa, Peru 128/E8
Puerto Olaya, Colombia 126/C4
Puerto Ospina, Colombia 126/C7
Puerto Padre, Cuba 158/H3
Puerto Padre, Cuba 156/C2
Puerto Páez, Venezuela 124/E4
Puerto Palena, Chile 138/D5
Puerto Palma, Paraguay 144/E4
Puerto Pando, Bolivia 136/B4
Puerto Paraíso, Paraguay 144/A5
Puerto Pardo, Peru 128/F7
Puerto Patiño, Bolivia 136/C5
Puerto Paulina, Colombia 126/D7
Puerto Peñasco, Mexico 150/C1
Puerto Pinasco, Paraguay 144/C3
Puerto Pirámides, Argentina 143/D5
Puerto Píritu, Venezuela 124/F2
Puerto Pizarro, Colombia 126/D8
Puerto Pizarro, Peru 128/B4
Puerto Plata (prov.), Dom. Rep. 158/D5
Puerto Plata, Dom. Rep. 156/D3
Puerto Plata, Dom. Rep. 158/D5
Puerto Portillo, Peru 128/F7
Puerto Prado, Peru 128/E8
Puerto Presidente Franco, Paraguay 144/E4
Puerto Presidente Stroessner, Paraguay 144/E4
Puerto Princesa, Philippines 85/F4
Puerto Princesa, Philippines 82/B6
Puerto Quellón, Chile 138/D4
Puerto Quijarro, Bolivia 136/G5
Puerto Ramírez, Chile 138/E5
Puerto Real, P. Rico 161/A2
Puerto Real (Playa de Fajardo), P. Rico 161/F1
Puerto Real, Spain 33/D4
Puerto Reyes, Colombia 126/B5
Puerto Rico 2/F5
Puerto Rico 146/M8
PUERTO RICO 161
PUERTO RICO 156/G1
Puerto Rico, Argentina 143/D1
Puerto Rico, Bolivia 136/B2
Puerto Rico, Caquetá, Colombia 126/C7
Puerto Rico, Meta, Colombia 126/C6
Puerto Rondón, Colombia 126/E4
Puerto Rosario, Paraguay 144/D4
Puerto Ruiz, Argentina 143/G6
Puerto Saavedra, Chile 138/D2
Puerto Salgar, Colombia 126/C5
Puerto Samanco, Peru 128/C7
Puerto San Francisco, Bolivia 136/C5
Puerto San José, Paraguay 144/E5
Puerto San Rafael, Paraguay 144/E5
Puerto Sastre, Paraguay 144/C3
Puerto Saucedo, Bolivia 136/D3
Puerto Siles, Bolivia 136/C3
Puerto Suárez, Bolivia 136/F6
Puerto Tacurú Pytá, Paraguay 144/D3
Puerto Tahuantinsuyo, Peru 128/G9
Puerto Tarafa, Cuba 158/H3
Puerto Tejada, Colombia 126/B6
Puerto Toledo, Colombia 126/C8
Puerto Torno, Bolivia 136/C5
Puerto Tres Palmas, Paraguay 144/C2
Puerto Vallarta, Mexico 150/G6
Puerto Varas, Chile 138/E3
Puerto Velarde, Bolivia 136/D5
Puerto Victoria, Peru 128/E7
Puerto Villarroel, Bolivia 136/C5
Puerto Villazón, Bolivia 136/D3
Puerto Wilches, Colombia 126/D4
Puerto Williams, Chile 138/F11
Puerto Yartou, Chile 138/E10
Puerto Ybapobó, Paraguay 144/D3
Pueyrredón (lake), Argentina 143/B6
Puffin (isl.), Ireland 17/A8
Pugachev, U.S.S.R. 52/G4
Puget (isl.), Wash. 310/B4
Puget (sound), Wash. 310/C3
Puget Sound Navy Yard, Wash. 310/A2
Pugwash, Nova Scotia 168/E3
Pugwash (harb.), Nova Scotia 168/E3
Puha, N. Zealand 100/F3
Puhi, Hawaii (96766) 218/C2
Puigcerdá, Spain 33/G1
Puina, Bolivia 136/A4
Puinagua, Canal de (riv.), Peru 128/E5
Pujada (bay), Philippines 82/F7
Pujehun, S. Leone 106/B7
Pujili, Ecuador 128/C3
Pujut (pt.), Indonesia 85/G1
Pukaki (lake), N. Zealand 100/B6
Pukalani, Hawaii (96788) 218/J2
Pukapuka (atoll), Cook Is. 87/K7
Puka-Puka (atoll), Fr. Poly. 87/N7
Pukaskwa Nat'l Park, Ont. 162/H6
Pukaskwa Prov. Park, Ontario 175/C3
Pukaskwa Prov. Park, Ontario 177/H5
Pukch'ŏng, N. Korea 81/C3
Pukë, Albania 45/E4
Pukekohe, N. Zealand 100/E2
Pukoo, Hawaii (†96748) 218/H1
Puksubaek (mt.), N. Korea 81/C3
Pukwana, S. Dak. (57370) 298/L6
Pula, Yugoslavia 45/A3
Pulacayo, Bolivia 120/C5
Pulacayo, Bolivia 136/B7
Pulai, Sungai (riv.), Malaysia 72/E5
Pulanduta (pt.), Philippines 82/D5
Pulangi (riv.), Philippines 82/E7
Pulap (atoll), Micronesia 87/E5
Púlar, Cerro (mt.), Chile 138/B4
Pulaski (co.), Ark. 202/F4
Pulaski (co.), Georgia 217/E6
Pulaski, Georgia (30451) 217/J6
Pulaski (co.), Ill. 222/D6
Pulaski, Ill. (62976) 222/D6
Pulaski (co.), Ind. 227/D2
Pulaski, Ind. (†46996) 227/D3
Pulaski, Iowa (52584) 229/J7
Pulaski (co.), Ky. 237/M6
Pulaski, Ky. (42550) 237/M6
Pulaski, Miss. (39152) 256/E6
Pulaski (co.), Mo. 261/H7
Pulaski, N.Y. (13142) 276/H3
Pulaski, Ohio (†43506) 284/A2
Pulaski, Pa. (16143) 294/B3
Pulaski, Tenn. (38478) 237/G10
Pulaski (co.), Va. 307/G6
Pulaski, Va. (24301) 307/G6
Pulaski, Wis. (54162) 317/K6
Puławy, Poland 47/F3
Pulcifer, Wis. (54164) 317/K6
Pulehu, Hawaii (†96788) 218/J2
Pulicat (lake), India 68/E6
Pull (pt.), Virgin Is. (U.S.) 161/F3
Pullman, Mich. (49450) 250/C6
Pullman, Wash. (99163) 310/H4
Pullman, W. Va. (26421) 312/D4
Pully, Switzerland 39/C4
Pulo Anna (isl.), Belau 87/D5
Pulog (mt.), Philippines 82/C2
Pulpit Harbor, Maine (†04853) 243/F7
Pulteney, Alberta 182/D4
Pulteney, N.Y. (14874) 276/F5
Pulteneyville, N.Y. (14538) 276/F4
Pułtusk, Poland 47/E2
Pülümür, Turkey 63/H3
Pulusuk (atoll), Micronesia 87/F5
Puluwat (atoll), Micronesia 87/E5
Pumanque, Chile 138/F6
Pumphrey, Md. (†21090) 245/M4
Pumpkin (creek), Nebr. 264/A3
Pumpville, Texas (†78851) 303/C8
Puna, Bolivia 136/C6
Puná (isl.), Ecuador 128/B4
Punaauia, Fr. Poly. 86/S13
Puna de Atacama (reg.), Argentina 143/C2
Punakha, Bhutan 68/G3
Punaluu, Hawaii (†96777) 218/H7
Punaluu (harb.), Hawaii 218/H7
Punata, Bolivia 136/C5
Punchaw, Br. Col. 184/F3
Punchbowl (hill), Hawaii 218/C4
Punchestown, Ireland 17/H5
Pune, India 68/C5
Pungo, N.C. (†27860) 281/R3
Pungo (lake), N.C. 281/S3
Pungo (riv.), N.C. 281/R4
Pungoteague, Va. (23422) 307/S5
P'ungsan, N. Korea 81/D3
Punia, Zaire 115/E4
Punitaqui, Chile 138/A8
Punjab (state), India 68/D2
Punjab (prov.), Pakistan 68/C2
Punk (isl.), Manitoba 179/F3
Punnichy, Sask. 181/G4
Puno (dept.), Peru 128/G10
Puno, Peru 128/G10
Puno, Peru 120/B4
Punta, Cerro de (mt.), P. Rico 161/C2
Punta Abreojos, Mexico 150/C3
Punta Alta, Argentina 143/D4
Punta Alta, Argentina 120/C6
Punta Arenas, Chile 120/B8
Punta Arenas, Chile 138/E10
Punta Cardón, Venezuela 124/C2
Punta de Bombón, Peru 128/F11
Punta de Díaz, Chile 138/B7
Punta del Este, Uruguay 145/E6
Punta de Mata, Venezuela 124/G3
Punta de Piedras, Venezuela 124/F2
Punta Gorda, Belize 154/C2
Punta Gorda (pt.), Calif. 204/A3
Punta Gorda, Fla. (*33950) 212/E5
Punta Medanosa, Argentina 143/C6
Punta Moreno, Peru 128/C6
Punta Negra, Salar de (salt dep.), Chile 138/B5
Puntarenas, C. Rica 154/E6
Punta Santiago (Playa de Humacao), P. Rico 161/F2
Puntas de Maciel, Uruguay 145/C4
Punto Fijo, Venezuela 120/C1
Punto Fijo, Venezuela 124/D2
Punxsutawney, Pa. (15767) 294/E4
Puolanka, Finland 18/P4
Puolo (pt.), Hawaii 218/C2
Pupiales, Colombia 126/B7
Puposky, Minn. (56667) 255/C3
Pupuke (lake), N. Zealand 100/B1
Pupuya, Nevada (mt.), Bolivia 136/A4
Puqi, China 77/H6
Puquina, Peru 128/G11
Puquintica, Nevado (mt.), Bolivia 136/A6
Puquintica, Cerro (mt.), Chile 138/B1
Puquio, Peru 128/F10
Puquios, Chile 138/B1
Pur (riv.), U.S.S.R. 48/H3
Puracé (vol.), Colombia 126/B6
Purbeck, Isle of (pen.), England 13/F7
Purbolinggo, Indonesia 85/J2
Purcell (mts.), Br. Col. 184/J5
Purcell (mts.), Idaho 220/B1
Purcell, Kansas (†66038) 232/G2
Purcell, Mo. (64857) 261/D8
Purcell (mts.), Mont. 262/A2
Purcell, Okla. (73080) 288/M4
Purcellville, Va. (22132) 307/N2
Purchase, N.Y. (10577) 276/P6
Purdin, Mo. (64674) 261/F3
Purdum, Nebr. (69157) 264/D2
Purdy, Mo. (65734) 261/E9
Purdy, Va. (†23847) 307/N7
Pure Air, Mo. (†63559) 261/G2
Purén, Chile 138/D2
Purgatoire (riv.), Colo. 208/M8
Purgitsville, W. Va. (26852) 312/J4
Puri, India 68/F5
Purial, Sierra de (mts.), Cuba 158/K4
Purificación, Colombia 126/C6
Purificación, Mexico 150/G7
Purli, India 68/D5
Purmerend, Netherlands 27/F4
Purnea, India 68/F3
Purple Springs, Alberta 182/E5
Purple Valley, Ontario 177/C3
Purranque, Chile 138/D3
Pursat (Pouthisat), Cambodia 72/D4
Puruándiro, Mexico 150/J7
Puruey, Venezuela 124/F4
Purukcahu, Indonesia 85/E6
Purulia, India 68/F4
Puruname, Venezuela 124/E6
Puruni (riv.), Guyana 131/B2
Purus (riv.), Brazil 120/C3
Purus (riv.), Brazil 132/H9
Purús (riv.), Peru 128/G8
Puruvesi (lake), Finland 18/Q6
Purves, Manitoba 179/D5
Purvis, Miss. (39475) 256/F8
Purwakarta, Indonesia 85/H2
Purwodadi, Indonesia 85/J2
Purwokerto, Indonesia 85/H2
Purworejo, Indonesia 85/J2
Puryear, Tenn. (38251) 237/E8
Pusan, S. Korea 54/06
Pusan, S. Korea 81/D6
Pushaw (lake), Maine 243/F6
Pushkin, U.S.S.R. 48/C4
Pushkin, U.S.S.R. 52/C3
Pushmataha, Ala. (36917) 195/B6
Pushmataha (co.), Okla. 288/R6
Püspökladány, Hungary 41/F3
Pustunich, Mexico 150/O7
Pusztaszabolcs, Hungary 41/E3
Putaendo, Chile 138/A9
Putao, Burma 72/C1
Putaruru, N. Zealand 100/E3
Putbus, E. Germany 22/E1
Puteau, France 28/A2
Putian, China 77/J6
Putignano, Italy 34/F4
Putina, Peru 128/H10
Put-in-Bay, Ohio (43456) 284/E2
Puting, Borneo (cape), Indonesia 85/E6
Puting, Sumatra (cape), Indonesia 85/C7
Putla de Guerrero, Mexico 150/L8
Putnam, Ala. (36772) 195/B6
Putnam, Conn. (06260) 210/H1
Putnam○, Conn. (06260) 210/H1
Putnam (co.), Fla. 212/E2
Putnam (co.), Georgia 217/F4
Putnam, Georgia (†31803) 217/D6
Putnam (co.), Ill. 222/D2
Putnam, Ill. (61560) 222/D2
Putnam (co.), Ind. 227/D5
Putnam (co.), Mo. 261/F2
Putnam (co.), N.Y. 276/N8
Putnam (co.), Ohio 284/B3
Putnam, Okla. (73659) 288/J3
Putnam (co.), Tenn. 237/K8
Putnam, Texas (76469) 303/E5
Putnam (co.), W. Va. 312/C6
Putnam Heights, Conn. (†06260) 210/H1
Putnam Valley○, N.Y. (10579) 276/N8
Putnamville, Ind. (46170) 227/D5
Putney, Georgia (31782) 217/D8
Putney, S. Dak. (†57445) 298/N2
Putney○, Vt. (05346) 268/B6
Putnok, Hungary 41/F2
Putre, Chile 138/B1
Puttalam, Sri Lanka 68/D7
Putte, Belgium 27/F6
Putten, Netherlands 27/H4
Puttgarden, W. Germany 22/D1
Puttur, India 68/D6
Putumayo (riv.) 120/B3
Putumayo (inten.), Colombia 126/C7
Putumayo (riv.), Colombia 126/E9
Putumayo, Ecuador 128/E3
Putumayo (riv.), Ecuador 128/E2
Putumayo (riv.), Peru 128/G4
Pütürge, Turkey 63/H3
Putussibau, Indonesia 85/E5
Puuanahulu, Hawaii (†96740) 218/G4
Puuhonua O Honaunau Nat'l Hist. Park, Hawaii 218/F6
Puuiki, Hawaii (†96713) 218/K2
Puu Keahiakahoe (mt.), Hawaii 218/D3
Puukohola Heiau Nat'l Hist. Site, Hawaii 218/G4
Puu Kukui (mt.), Hawaii 218/J2
Puu Lanihuli (mt.), Hawaii 218/D3
Puulavesi (lake), Finland 18/P5
Puunene, Hawaii (96784) 218/J2
Puunui, Hawaii (†96801) 218/C4
Puuwaawaa, Hawaii (†96740) 218/G5
Puuwai, Hawaii (†96769) 218/A2
Pu Xian, China 77/H4
Puxico, Mo. (63960) 261/M9
Puyallup, Wash. 188/B1
Puyallup, Wash. (98371) 310/C3
Puyallup (riv.), Wash. 310/C4
Puy-de-Dôme (dept.), France 28/E5
Puy-de-Dôme (mt.), France 28/E5
Puyehue, Chile 138/E3
Puyehue (lake), Chile 138/E3
Puyo, Ecuador 128/D3
Puysegur (pt.), N. Zealand 100/A7
Pwani (Coast) (reg.), Tanzania 115/G5
Pweto, Zaire 115/E5
Pwllheli, Wales 10/D4
Pwllheli, Wales 13/C5
Pyapon, Burma 72/B3
Pyasina (riv.), U.S.S.R. 48/J2
Pyatigorsk, U.S.S.R. 52/F6
Pyatt, Ark. (72672) 202/E1
Pye (isls.), Alaska 196/C2
Pye, Burma 72/B3
Pye, Burma 54/L8
Pyengana, Tasmania 99/E3
Pygmalion (pt.), India 68/G7
Pyhäjärvi (lake), Finland 18/M6
Pyhäjärvi (lake), Finland 18/O5
Pyinmana, Burma 72/C3
Pyland, Miss. (†38851) 256/F3
Pymatuning (res.), Ohio 284/J2
Pymatuning (res.), Pa. 294/A2
P'yŏnggang, N. Korea 81/C4
P'yŏngsan, N. Korea 81/C4
P'yongyang (cap.), N. Korea 54/O6
P'yŏngyang (cap.), N. Korea 81/C4
Pyote, Texas (79777) 303/A6
Pyramid (peak), Idaho 220/E4
Pyramid, Ky. (41656) 237/R5
Pyramid (lake), Nev. 266/B2
Pyramid (isl.), N. Zealand 100/E7
Pyramid Lake Ind. Res., Nev. 266/B2
Pyramids (ruins), Egypt 111/J3
Pyrenees (mts.) 7/D4
Pyrenees (range), France 28/C6
Pyrenees (range), Spain 33/F1
Pyrénées-Atlantiques (dept.), France 28/C6
Pyrénées-Orientales (dept.), France 28/E6
Pyrites, N.Y. (13677) 276/K1
Pyriton, Ala. (†36266) 195/G4
Pyrzyce, Poland 47/B2
Pytalovo, U.S.S.R. 52/C3
Pyu, Burma 72/C4
Pyuthan, Nepal 68/E3

Q

Qabalan, West Bank 65/C3
Qabatiya, West Bank 65/C3
Qabr Hud, P.D.R. Yemen 59/E6
Qadhima, Saudi Arabia 59/C5
Qadima, Israel 65/B3
Qadisiya (gov.), Iraq 66/D4
Qafar, Saudi Arabia 59/D4
Qaffin, West Bank 65/C3
Qagan Nur, China 77/H3
Qâhira, El (Cairo), Egypt 111/J3
Qaidam (basin), China 54/L6
Qaidam Pendi (basin) (swamp), China 77/D4
Qala'en Nahl, Sudan 59/C7
Qala'en Nahl, Sudan 111/F5
Qalansuwa, Israel 65/B3
Qal'a Sharqat, Iraq 59/D2
Qal'a Sharqat, Iraq 66/C3
Qalat, Afghanistan 68/B2
Qalat, Afghanistan 59/J3
Qa'lat Diza, Iraq 66/D2
Qal'at es Salihiye, Syria 63/J5
Qal'eh Mureh (riv.), Iran 66/J3
Qal'eh-ye Now, Afghanistan 68/A1
Qal'eh-ye Now, Afghanistan 59/H3
Qal'eh-ye Panjeh, Afghanistan 59/K2
Qal'eh-ye Panjeh, Afghanistan 68/C1
Qalqiliya, West Bank 65/C3
Qalyub, Egypt 111/J3
Qamdo, China 77/E5
Qamdo, China 54/L6
Qamishliye, Syria 63/J4
Qamr (bay), P.D.R. Yemen 59/F6
Qandahar, Afghanistan 54/H6
Qandahar, Afghanistan 68/B2
Qandahar, Afghanistan 59/J3
Qantara, El, Egypt 111/K3
Qaranqu (riv.), Iran 66/E2
Qareh Dagh (mts.), Iran 66/E1
Qareh Su (riv.), Iran 66/E1
Qareh Su (riv.), Iran 66/G3
Qarkilik (Ruoqiang), China 77/C4
Qarn (riv.), Israel 65/C1
Qarqan (Qiemo), China 77/C4
Qarqan He (riv.), China 77/C4
Qasr al Haiyanya, Saudi Arabia 59/D4
Qasr al Khubbaz, Iraq 66/B4
Qasr-e Qand, Iran 59/H4
Qasr-e Qand, Iran 66/M7
Qasr-e Shirin, Iran 66/E3
Qasr Farâfra, Egypt 111/E2
Qasr Farâfra, Egypt 102/E2
Qasr Farâfra, Egypt 59/A4
Qatar 2/M4
Qatar 54/G7
QATAR 59/F4
Qatif, Saudi Arabia 59/E4
Qattara (depr.), Egypt 59/A4
Qattara (depr.), Egypt 111/E2
Qayen, Iran 66/L4
Qayen, Iran 59/G3
Qazvin, Iran 54/F6
Qazvin, Iran 59/E2
Qazvin, Iran 66/F2
Qedma, Israel 65/B4
Qena, Egypt 111/F2
Qena, Egypt 102/F2
Qena, Egypt 59/B4
Qeshm (isl.), Iran 59/G4
Qeshm (isl.), Iran 66/J7
Qeys (isl.), Iran 59/F4
Qeys (isl.), Iran 66/J7
Qezel Owzam (riv.), Iran 66/F2
Qezel Owzan (riv.), Iran 59/E2
Qianyang (Kienyang), China 77/H6
Qiaowan, China 77/E3
Qibya, West Bank 65/C4
Qiebei, China 77/F7
Qiema, China 54/K6
Qiemo (Qarqan), China 77/C4
Qijiang, China 77/G6
Qila Ladgasht, Pakistan 68/A3
Qila Ladgasht, Pakistan 59/H4
Qilian, China 77/E4
Qilian Shan (range), China 54/L6
Qilian Shan (range), China 77/E4
Qimen, China 77/J5
Qina (Qena), Egypt 59/B4
Qingdao (Tsingtao), China 77/K4
Qingdao, China 54/O6
Qinghai (Tsinghai) (prov.), China 77/E4
Qinghai (lake), China 54/L6
Qinghai Hu (lake), China 77/E4
Qinghe (Qinggil), China 77/D2
Qingjiang, Jiangxi, China 77/J6
Qingjiang, Anhui, China 77/J5
Qingtongxia, China 77/F4
Qingyuan, China 77/H7
Qinhuangdao (Chinwangtao), China 77/K4
Qinzhou, China 77/G7
Qionghai, China 77/H8
Qiongshan, China 77/H8
Qiongzhou Haixia (str.), China 77/G7
Qiqihar (Tsitsihar), China 77/K2
Qiqihar, China 54/O5
Qira, China 77/B4
Qiryat Atta, Israel 65/C2
Qiryat Bialik, Israel 65/C2
Qiryat Gat, Israel 65/B4
Qiryat Mal'akhi, Israel 65/B4
Qiryat Motzkin, Israel 65/C2
Qiryat Shemona, Israel 65/C1
Qiryat Tiv'on, Israel 65/C2
Qiryat Yam, Israel 65/C2
Qishn, P.D.R. Yemen 59/F6
Qishon (riv.), Israel 65/C2
Qitai, China 77/C3
Qitaihe, China 77/M2
Qizan, Saudi Arabia 59/D6
Qog, China 77/G3
Qom (Qum), Iran 54/G6
Qom, Iran 59/F3
Qom, Iran 66/G3
Qonduz, Afghanistan 68/B1
Qonduz, Afghanistan 59/J2
Qonduz (riv.), Afghanistan 68/B1
Qonduz (riv.), Afghanistan 59/J2
Qoqek (Tacheng), China 77/B2
Qorveh, Iran 66/E3
Qotur, Iran 66/D1
Quabbin (res.), Mass. 249/E3
Quaboag (riv.), Mass. 249/F4
Quaco (head), New Bruns. 170/E3
Quaddick, Conn. (†06277) 210/H1
Quaddick (res.), Conn. 210/H1
Quadeville, Ontario 177/G2
Quail, Texas (79251) 303/D3
Quairading, W. Australia 92/B1
Quajote Wash (dry riv.), Ariz. 198/D6
Quakenbrück, W. Germany 22/C2
Quaker City, Ohio (43773) 284/H6
Quaker Farms, Conn. (†06483) 210/C3
Quaker Hill, Conn. (06375) 210/G3
Quakertown, N.J. (08868) 273/D2
Quakertown, Pa. (18951) 294/M5
Qualicum Beach, Br. Col. 184/J3
Quality, Ky. (42268) 237/H6
Quallam, Niger 106/E6
Quamba, Minn. (55064) 255/E5
Quambatook, Victoria 97/B4
Quambone, N.S. Wales 97/E2
Quanah, Texas (79252) 303/E3
Quan Dao Nam Duj (isls.), Vietnam 72/D5
Quandary (peak), Colo. 208/G4
Quandialla, N.S. Wales 97/D4
Quanduck (brook), Conn. 210/H1
Quang Nam, Vietnam 72/E4
Quang Ngai, Vietnam 72/F4
Quang Tri, Vietnam 72/E3
Quang Yen, Vietnam 72/E2
Quan Long, Vietnam 72/E5
Quantico, Md. (21856) 245/R7
Quantico, Va. (22134) 307/O3
Quantico Marine Corps Air Sta., Va. 307/O4
Quanzhou, China 77/H6
Quanzhou (Chüanchow), China 77/J7
Quapaw, Okla. (74363) 288/S1
Qu'Appelle, Sask. 181/H5
Qu'Appelle (riv.), Sask. 181/J5
Quaraí, Brazil 132/C10
Quaregnon, Belgium 27/D8
Quarryville, New Bruns. 170/E2
Quarryville, Pa. (17566) 294/K6
Quartu Sant'Elena, Italy 34/B5
Quartz (peak), Calif. 204/L11
Quartz Hill, Calif. (93534) 204/G9
Quartz Mountain, Oreg. (†97630) 291/G5
Quartzsite, Ariz. (85346) 198/A5
Quasqueton, Iowa (52326) 229/K4
Quassapaug (pond), Conn. 210/C2
Quatervals (peak), Switzerland 39/K3
Quathiaski Cove, Br. Col. 184/E5
Quatre Bornes, Mauritius 118/G5
Quatsino (sound), Br. Col. 184/C5

Quay (co.), N. Mex. 274/F3
Quay, N. Mex. (88433) 274/F4
Quay, Okla. (†74085) 288/N2
Quchan, Iran 66/L2
Quchan, Iran 59/G2
Quealy, Wyo. (†82901) 319/C4
Queanbeyan, N. S. Wales 88/H7
Queanbeyan, N.S. Wales 97/E4
Québec (prov.) 162/J5
QUÉBEC 172
Québec, Canada 2/F3
Québec (prov.), Que. 146/L4
Québec (cap.), Que. 146/L5
Québec (cap.) Que. 162/J6
Québec (co.), Québec 172/F3
Québec (county), Québec 174/C3
Québec (cap.), Québec 172/H3
Québec (cap.), Québec 174/C3
Quebeck, Tenn. (38579) 237/K9
Quebracho, Uruguay 145/B2
Quebracho Coto, Argentina 143/D2
Quebrada de Alvarado, Chile 138/F2
Quebradillas, P. Rico 161/B1
Quechee, Vt. (05059) 268/C4
Quechisla, Bolivia 136/C7
Quecholac, Mexico 150/02
Quecreek, Pa. (15555) 294/D5
Quedlinburg, E. Germany 22/D3
Queen (cape), N.W. Terrs. 187/L3
Queen, Pa. (16670) 294/E5
Queen Anne, Md. (21657) 245/05
Queen Annes (co.), Md. 245/P4
Queenborough, England 13/H6
Queenborough, England 10/G5
Queen Charlotte (isls.), Br. Col. 146/E4
Queen Charlotte (isls.), Br. Col. 162/C5
Queen Charlotte, Br. Col. 184/A3
Queen Charlotte (isls.), Br. Col. 184/B3
Queen Charlotte (sound), Br. Col. 184/C4
Queen Charlotte (str.), Br. Col. 184/D5
Queen Charlotte (sound), N.W.T. 162/D5
Queen City, Mo. (63561) 261/H2
Queen City, Texas (75572) 303/L4
Queen Creek, Ariz. (85242) 198/D5
Queen Elizabeth (isls.), Canada 2/C2
Queen Elizabeth (isls.), Canada 4/B15
Queen Elizabeth (isls.), N.W.T. 146/G2
Queen Elizabeth (isls.), N.W.T. 162/M3
Queen Elizabeth (isls.), N.W. Terrs. 187/H1
Queen Mary Coast (reg.) 5/C5
Queen Maud (mts.) 5/A12
Queen Maud (gulf), N.W.T. 162/F2
Queen Maud (gulf), N.W. Terrs. 187/H3
Queen Maud Land (reg.), Ant. 2/K10
Queen Maud Land (reg.) 5/B1
Queens (sound), Br. Col. 184/C4
Queen's (co.), New Bruns. 170/D3
Queens (co.), N.Y. 276/N9
Queens (borough), N.Y. (*11101) 276/N9
Queens (chan.), N.W. Terrs. 187/J2
Queens (co.), Nova Scotia 168/C4
Queens (co.), Pr. Edward I. 168/E2
Queens, W. Va. (†26237) 312/F5
Queensberry (mt.), Scotland 15/E5
Queenscliff, Victoria 97/C6
Queensferry, Scotland 10/C1
Queensferry, Scotland 15/D1
Queen Shoals, W. Va. (†25045) 312/D6
Queensland, 88/G4
QUEENSLAND 95
Queensland (state), Australia 87/E8
Queensport, Nova Scotia 168/G3
Queenstown, Alberta 182/D4
Queenstown, Guyana 131/B2
Queenstown (Cóbh), Ireland 10/B5
Queenstown (Cóbh), Ireland 17/E8
Queenstown, Md. (21658) 245/05
Queenstown, New Bruns. 170/D3
Queenstown, N. Zealand 100/B6
Queenstown, S. Africa 102/E8
Queenstown, S. Africa 118/D6
Queenstown, Tasmania 99/B4
Queenstown, Tasmania 88/H8
Queensville, Ind. (†47265) 227/F6
Queets, Wash. (†98331) 310/A3
Queets (riv.), Wash. 310/A3
Queilén, Chile 138/D4
Queimadas, Brazil 132/F5
Quela, Angola 115/C5
Quelimane, Mozambique 118/F3
Quelimane, Mozambique 102/F6
Quelpart (Cheju) (isl.), S. Korea 81/C7
Queluz, Portugal 33/A1
Quemado (pt.), Cuba 158/K4
Quemado, N. Mex. (87829) 274/A4
Quemado, Texas (78877) 303/D9
Quemado de Güines, Cuba 158/E1
Quemchi, Chile 138/D4
Quemoy (Jinmen) (isl.), China 77/J7
Quemú-Quemú, Argentina 143/D4
Quenemo, Kansas (66528) 232/G3
Quentin, Miss. (39647) 256/C8
Quepos, C. Rica 154/E6
Quequay Chico (riv.), Uruguay 145/B3
Quequay Grande (riv.), Uruguay 145/B3
Que Que, Zimbabwe 118/D3
Que Que, Zimbabwe 102/E6
Quequén, Argentina 143/E4
Querecotillo, Peru 128/B5
Querétaro (state), Mexico 150/J6
Querétaro, Mexico 146/J7
Querétaro, Mexico 150/J6
Quesada, Spain 33/E4
Queshan, China 77/H5
Quesnel, Br. Col. 162/D5
Quesnel (lake), Br. Col. 162/D5
Quesnel, Br. Col. 184/F4
Quesnel (lake), Br. Col. 184/G4
Quesnel (riv.), Br. Col. 184/F4
Quesnel (lake), Manitoba 179/G4
Questa, N. Mex. (87556) 274/D2
Queteña, Bolivia 136/B8
Quetico Prov. Park, Ontario 175/B3
Quetico Prov. Park, Ontario 177/G5
Quetta, Pakistan 54/H6
Quetta, Pakistan 59/J3
Quetta, Pakistan 68/B2
Queule, Chile 138/D2
Quevedo, Ecuador 128/C3
Quévy, Belgium 27/D8
Quezaltenango, Guatemala 154/B3
Quezaltepeque, Guatemala 154/C3
Quezon (prov.), Philippines 82/C3
Quibala, Angola 115/C6
Quibaxe, Angola 115/B5
Quibdó, Colombia 126/B5
Quiberon, France 28/B4
Quibor, Venezuela 124/D3
Quicacha, Peru 128/F10
Quick, Br. Col. 184/D3
Quick, W. Va. (25045) 312/D6
Quicksand, Ky. (41363) 237/P5
Quicksburg, Va. (22847) 307/L3
Quiebra Hacha, Cuba 158/B1
Quiévrain, Belgium 27/D8
Quigley, Alberta 182/E1
Quiindy, Paraguay 144/B5
Quijotoa, Ariz. (†85634) 198/C6
Quilali, Nicaragua 154/E4
Quilán (cape), Chile 138/D4
Quilán (isl.), Chile 138/D5
Quilca, Peru 128/F11
Quilcene, Wash. (98376) 310/B3
Quilchena, Br. Col. 184/G5
Quilengues, Angola 115/B6
Quilicura, Chile 138/G3
Quill (lakes), Sask. 181/G4
Quillabamba, Peru 128/F9
Quillacas, Bolivia 136/B6
Quillacollo, Bolivia 136/B5
Quillacollo, Bolivia 120/C4
Quillagua, Chile 138/B3
Quillaicillo, Chile 138/A8
Quillan, France 28/E6
Quillayute Ind. Res., Wash. 310/A3
Quilleco, Chile 138/E1
Quill Lake, Sask. 181/G3
Quillota, Chile 138/F2
Quilon, India 68/D7
Quilpie, Queensland 88/G5
Quilpie, Queensland 95/C5
Quilpué, Chile 138/F2
Quimby, Iowa (51049) 229/B3
Quimby, Maine (04770) 243/F2
Quime, Bolivia 136/B5
Quimili, Argentina 143/D2
Quimper, France 28/A4
Quimperlé, France 28/B4
Quinault, Wash. (98575) 310/B3
Quinault (lake), Wash. 310/B3
Quinault (riv.), Wash. 310/A3
Quinault Ind. Res., Wash. 310/A3
Quinby, S.C. (†29501) 296/H3
Quinby, Va. (23423) 307/S5
Quinby (inlet), Va. 307/S6
Quince Mil, Peru 128/G9
Quincy, Calif. (95971) 204/E4
Quincy, Fla. (32351) 212/B1
Quincy, Ill. 188/H3
Quincy, Ill. (62301) 222/B4
Quincy, Ind. (47456) 227/D6
Quincy, Kansas (†66870) 232/F4
Quincy, Ky. (41166) 237/P3
Quincy, Mass. (02169) 249/D7
Quincy (bay), Mass. 249/D7
Quincy, Mich. (49082) 250/E7
Quincy, Miss. (†38848) 256/H3
Quincy, Mo. (65735) 261/F6
Quincy, N.H. (†03266) 268/D4
Quincy, Ohio (43343) 284/C5
Quincy, Wash. (98848) 310/F3
Quincy, W. Va. (†25015) 312/C6
Quindío (dept.), Colombia 126/C5
Quinebaug, Conn. (06262) 210/H1
Quinebaug (riv.), Conn. 210/H2
Quinebaug (riv.), Mass. 249/F4
Quines, Argentina 143/C3
Quinhagak, Alaska (99655) 196/F3
Qui Nhon, Vietnam 72/F4
Qui Nhon, Vietnam 54/M8
Quiniluban (isls.), Philippines 82/C5
Quinlan, Okla. (†73852) 288/J2
Quinlan, Texas (75474) 303/H5
Quinn (riv.), Nev. 266/D1
Quinn, S. Dak. (57775) 298/E5
Quinn Canyon (range), Nev. 266/F4
Quinnesec, Mich. (49876) 250/A3
Quinnimont, W. Va. (25910) 312/D7
Quinnipiac, Conn. (†06492) 210/D3
Quinnipiac (riv.), Conn. 210/D3
Quinta de Tilcoco, Chile 138/G5
Quintana de la Serena, Spain 33/D3
Quintanar de la Orden, Spain 33/E3
Quintana Roo (state), Mexico 150/P7
Quintay, Chile 138/F3
Quinter, Kansas (67752) 232/B2
Quintero, Chile 138/F2
Quinto (riv.), Argentina 143/D3
Quinto, Spain 33/F2
Quinto, Switzerland 39/G3
Quinton, Ky. (†42518) 237/M7
Quinton, N.J. (08072) 273/C4
Quinton, Okla. (74561) 288/R4
Quinton, Sask. 181/G4
Quinton, Va. (23141) 307/O5
Quinwood, W. Va. (25981) 312/E6
Quinzau, Angola 115/B5
Quionga, Mozambique 118/G2
Quipapá, Brazil 132/G5
Quirey, Colombia 126/F5
Quirihue, Chile 138/E1
Quirindi, N.S. Wales 97/F2
Quirino (prov.), Philippines 82/C2
Quirino, Philippines 82/C2
Quiriquire, Venezuela 124/G3
Quirke (lake), Ontario 177/B1
Quiroga, Argentina 143/F7
Quiroga, Bolivia 136/C6
Quiroga, Spain 33/C1
Quirusillas, Bolivia 136/D6
Quisiro, Venezuela 124/C2
Quispamsis, New Bruns. 170/E3
Quissanga, Mozambique 118/G2
Quissett, Mass. (†02540) 249/M6
Quissico, Mozambique 118/F4
Quitaque, Texas (79255) 303/C3
Quitasueño (bank), Colombia 126/A8
Quitilipi, Argentina 143/D2
Quitman, Ark. (72131) 202/F3
Quitman (co.), Georgia 217/B7
Quitman, Georgia (31643) 217/E9
Quitman, La. (71268) 238/E2
Quitman (co.), Miss. 256/D2
Quitman, Miss. (39355) 256/G6
Quitman, Mo. (†64428) 261/C2
Quitman, Texas (75783) 303/J5
Quitman (mts.), Texas 303/B11
Quito (cap.), Ecuador 2/F6
Quito (cap.), Ecuador 128/C3
Quito (cap.), Ecuador 120/B2
Quixadá, Brazil 120/F3
Quixadá, Brazil 132/G4
Quixeramobim, Brazil 132/F4
Qujing, China 77/F6
Qulin, Mo. (63961) 261/M9
Qum, Iran 54/G6
Qum (Qom), Iran 59/F3
Qum (Qom), Iran 66/G3
Qumar He (riv.), China 77/D4
Qumarlêb, China 77/D5
Qumeim, Jordan 65/D2
Qunfidha, Saudi Arabia 59/D6
Quogue, N.Y. (11959) 276/P9
Quoich (riv.), N.W. Terrs. 187/J3
Quoich, Loch (lake), Scotland 15/C3
Quonnipaug (lake), Conn. 210/E3
Quorn, S. Australia 88/F6
Quorn, S. Australia 94/F5
Quryat, Oman 59/G5
Qusaiba, Saudi Arabia 59/D4
Quteife, Syria 63/G6
Qu Xian, Sichuan, China 77/G5
Qu Xian, Zhejiang, China 77/J6
Qüxü, China 77/D6
Quyon, Québec 172/A4
Quyquyó, Paraguay 144/D5

R

Raab (riv.), Austria 41/C3
Raabs an der Thaya, Austria 41/C2
Raahe, Finland 18/O4
Raalte, Netherlands 27/J4
Ra'anana, Israel 65/B3
Raanes (pen.), N.W. Terrs. 187/K2
Raasay (isl.), Scotland 15/C3
Raasay (sound), Scotland 15/B3
Rab, Yugoslavia 45/B3
Rab (isl.), Yugoslavia 45/B3
Rába (riv.), Hungary 41/D3
Raba, Indonesia 85/F7
Rabat (cap.), Morocco 2/J4
Rabat (cap.), Morocco 106/C2
Rabat (cap.), Morocco 102/B1
Rabaul, Papua N.G. 87/F6
Rabaul, Papua N.G. 86/B2
Rabbit (riv.), Mich. 250/D6
Rabbit (isl.), N.S. Wales 97/J2
Rabbit (creek), S. Dak. 298/E3
Rabbit Ears (peak), Colo. 208/G2
Rabbit Ears (range), Colo. 208/F2
Rabbithash, Ky. (†41091) 237/M3
Rabbit Lake, Sask. 181/D2
Rabbit Lake, Sask. 181/M2
Rabigh, Saudi Arabia 59/C5
Rabinal, Guatemala 154/B3
Rabka, Poland 47/D4
Rabocheostrovsk, U.S.S.R. 52/D1
Rabun, Ala. (†36507) 195/C8
Rabun (co.), Georgia 217/F1
Rabun (lake), Georgia 217/F1
Rabun Gap, Georgia (30568) 217/F1
Raccoon (pt.), Fla. 212/D3
Raccoon, Ind. (†46172) 227/D5
Raccoon (riv.), Iowa 229/D4
Raccoon (pt.), La. 238/H8
Raccoon (creek), N.J. 273/C4
Raccoon (creek), Ohio 284/F8
Race (pt.), Mass. 249/N4
Race (cape), Newf. 166/D2
Race (cape.), Newf. 146/N5
Race (cape), Newf. 162/L6
Raceland, Ky. (41169) 237/R3
Raceland, La. (70394) 238/J7
Racepond, Georgia (†31537) 217/H8
Rachel, W. Va. (26587) 312/F3
Rach Gia, Vietnam 72/E5
Racibórz, Poland 47/C3
Racine, Minn. (55967) 255/F7
Racine, Mo. (64858) 261/C9
Racine, Ohio (45771) 284/G8
Racine, Québec 172/E4
Racine, W. Va. (25165) 312/C6
Racine, Wis. 188/J2
Racine (co.), Wis. 317/K10
Racine, Wis. (*53401) 317/M3
Ráckeve, Hungary 41/E3
Rackham, Manitoba 179/B4
Raco, Mich. (49778) 250/E2
Racola, Mo. (†63630) 261/L6
Radama (isls.), Madagascar 118/H2
Rădăuţi, Romania 45/G2
Radbuza (riv.), Czech. 41/B2
Radcliff, Ky. (40160) 237/K5
Radcliff, Ohio (45670) 284/F7
Radcliffe, England 13/H2
Radcliffe, Iowa (50230) 229/G4
Radeberg, E. Germany 22/E3
Radebeul, E. Germany 22/E3
Radenthein, Austria 41/B3
Rader, Tenn. (†37743) 237/R8
Radersburg, Mont. (59641) 262/E4
Radford (I.C.), Va. (24141) 307/G6
Radhanpur, India 68/C4
Radiant, Va. (22732) 307/M4
Radisson, Québec 174/B2
Radisson, Sask. 181/D3
Radisson, Wis. (54867) 317/D4
Radium, Colo. (80472) 208/G3
Radium, Kansas (67571) 232/D3
Radium, Minn. (56749) 255/B2
Radium Hill, S. Australia 88/G6
Radium Hill, S. Australia 94/G5
Radium Hot Springs, Br. Col. 184/J5
Radium Springs, N. Mex. (88054) 274/B6
Radkersburg, Austria 41/C3
Radley, Ind. (†46938) 227/F4
Radnice, Czech. 41/B2
Radnor, Ind. (†46923) 227/D3
Radnor, Ohio (43066) 284/D5
Radnor (for.), Wales 13/D5
Radnor, W. Va. (25556) 312/A6
Radolfzell, W. Germany 22/C5
Radom, Ill. (62876) 222/D5
Radom (prov.), Poland 47/E3
Radom, Poland 7/G3
Radom, Poland 47/E3
Radomir, Bulgaria 45/F4
Radomsko, Poland 47/D3
Radoviš, Yugoslavia 45/F5
Radstadt, Austria 41/B3
Radviliškis, U.S.S.R. 53/B3
Radville, Sask. 162/F6
Radville, Sask. 181/G6
Radway, Alberta 182/D2
Radziejów, Poland 47/D2
Radzyń Podlaski, Poland 47/F3
Rae (isth.), N.W.T. 162/H2
Rae (isth.), N.W. Terrs. 187/K3
Rae (riv.), N.W. Terrs. 187/G3
Rae (str.), N.W. Terrs. 187/J3
Rae-Edzo, N.W.T. 162/E3
Rae-Edzo, N.W. Terrs. 187/G3
Raeford, N.C. (28376) 281/L5
Rae Lake, N.W. Terrs. 187/G3
Raeren, Belgium 27/J7
Raeside (lake), W. Australia 88/C5
Raeside (lake), W. Australia 92/C5
Raetihi, N. Zealand 100/E3
Raeville, Nebr. (68656) 264/F3
Rafaela, Argentina 143/F5
Rafaela, Argentina 120/C6
Rafah, Gaza Strip 65/A5
Rafai, Cent. Afr. Rep. 115/D2
Rafidiya, West Bank 65/C3
Rafsanjan, Iran 59/G3
Rafsanjan, Iran 66/K5
Raft (riv.), Idaho 220/E7
Raft (riv.), Utah 304/A1
Raft River (mts.), Utah 304/A2
Rafz, Switzerland 39/G1
Raga, Sudan 111/E6
Ragan, Nebr. (68969) 264/E4
Ragang (vol.), Philippines 82/E7
Ragay (gulf), Philippines 82/D4
Ragged (isl.), Bahamas 156/C2
Ragged (pt.), Barbados 161/C8
Ragged (isl.), Maine 243/F8
Ragged (lake), Maine 243/E4
Ragged (isls.), Newf. 166/C2
Raglan, N. Zealand 100/E2
Raglan (harb.), N. Zealand 100/E2
Ragland, Ala. (35131) 195/F3
Ragley, La. (70657) 238/D5
Rago, Kansas (67128) 232/D4
Ragsdale, Ind. (47573) 227/C7
Ragusa (prov.), Italy 34/E6
Ragusa, Italy 34/E6
Ragusa (Dubrovnik), Yugoslavia 45/C4
Raha, Indonesia 85/G6
Raheng (Tak), Thailand 72/C3
Rahan, Ireland 17/F5
Rahimyar Khan, Pakistan 68/C3
Rahotu, N. Zealand 100/D3
Rahue (riv.), Chile 138/D3
Rahway, N.J. (*07065) 273/E2
Raiatea (isl.), Fr. Poly. 87/L7
Raices, Argentina 143/G6
Raichur, India 68/D5
Raiford, Fla. (32083) 212/D1
Raigarh, India 68/E4
Railley (mt.), Mont. 262/C3
Railroad (valley), Nev. 266/F4
Railroad, Pa. (17355) 294/J6
Railroad Canyon (res.), Calif. 204/E11
Railton, Tasmania 99/C3
Rainbow (lake), Alberta 182/A5
Rainbow (plat.), Ariz. 198/D2
Rainbow, Conn. (†06095) 210/E1
Rainbow (mt.), Idaho 220/C4
Rainbow (lake), Maine 243/E4
Rainbow, Victoria 97/A4
Rainbow Bridge Nat'l Mon., Utah 304/C6
Rainbow City, Ala. (35901) 195/F3
Rainbow Lake, Alberta 182/A5
Rainelle, W. Va. (25962) 312/E7
Rainier, Alberta 182/D4
Rainier, Oreg. (97048) 291/E1
Rainier (mt.), Wash. 188/B1
Rainier, Wash. (98576) 310/C4
Rainier (mt.), Wash. 310/D4
Rains, S.C. (29589) 296/J3
Rains (co.), Texas 303/J5
Rainsboro, Ohio (45165) 284/C7
Rainsburg, Pa. (†15522) 294/F6
Rainsville, Ala. (35986) 195/G2
Rainsville, Ind. (†47918) 227/C4
Rainsville, N. Mex. (87736) 274/D2
Rainy (riv.), Minn. 188/H1
Rainy (lake), Minn. 188/H1
Rainy (lake), Minn. 255/E2
Rainy (riv.), Minn. 255/D2
Rainy (lake), Ont. 162/G6
Rainy (lake), Ontario 177/G5
Rainy (lake), Ontario 175/B3
Rainy River, Ont. 162/G6
Rainy River (terr. dist.), Ontario 177/G5
Rainy River (terr. dist.), Ontario 175/B3
Rainy River, Ontario 175/A3
Rainy River, Ontario 177/F5
Raipur, India 54/K7
Raipur, India 68/E4
Raisin, Calif. (93652) 204/E7
Raisin (riv.), Mich. 250/F7
Raisio, Finland 18/M6
Raith, Ontario 175/C3
Raith, Ontario 177/G5
Raivavae (isl.), Fr. Poly. 87/M8
Raja Ampat Group (isls.), Indonesia 85/H6
Rajahmundry, India 68/E5
Rajang (riv.), Malaysia 85/E5
Rajapalaiyam, India 68/D7
Rajapur, India 68/C5
Rajasthan (state), India 68/C3
Rajec, Czech. 41/E2
Rajgarh, India 68/D4
Rajka, Hungary 41/D3
Rajkot, India 54/H7
Rajkot, India 68/C4
Rajnandgaon, India 68/E4
Rajpipla, India 68/C4
Rajpur, India 68/F2
Rajpura, India 68/D2
Rajshahi, Bangladesh 68/F4
Rakahanga (atoll), Cook Is. 87/K7
Rakaia, N. Zealand 100/C5
Rakaia (riv.), N. Zealand 100/C5
Rakamaz, Hungary 41/F2
Rakan, Ras (cape), Qatar 59/F4
Rakaposhi (mt.), Pakistan 68/C1
Rakaposhi (mt.), Pakistan 59/K2
Rakata (isl.), Indonesia 85/C7
Rake, Iowa (50465) 229/F2
Rakhov, U.S.S.R. 52/B5
Rakino (isl.), N. Zealand 100/C1
Rakitu (isl.), N. Zealand 100/E2
Rakkestad, Norway 18/G7
Rakof (isls.), Alaska 196/M1
Rákospalota, Hungary 41/E3
Rakovnik, Czech. 41/B1
Rakvere, U.S.S.R. 53/D1
Rakvere, U.S.S.R. 52/C3
Raleigh, Fla. (†32696) 212/D2
Raleigh, Georgia (†30293) 217/C5
Raleigh, Ill. (62977) 222/E6
Raleigh, Ind. (†46173) 227/G5
Raleigh, Miss. (39153) 256/F6
Raleigh, Newf. 166/C3
Raleigh (cap.), N.C. 146/L6
Raleigh (cap.), N.C. 188/L3
Raleigh (cap.), N.C. (*27601) 281/M3
Raleigh (bay), N.C. 281/S5
Raleigh, N. Dak. (58564) 282/H7
Raleigh, Tenn. (38128) 237/B10
Raleigh (co.), W. Va. 312/D7
Raleigh, W. Va. (25911) 312/D7
Ralik Chain (isls.), Marshall Is. 87/G5
Ralls (co.), Mo. 261/J3
Ralls, Texas (79357) 303/C4
Ralph, Ala. (35480) 195/C4
Ralph, Mich. (49877) 250/B2
Ralph, Sask. 181/H6
Ralph, S. Dak. (57650) 298/C2
Ralphton, Pa. (†15522) 294/D5
Ralston, Alberta 182/E4
Ralston, Iowa (51459) 229/D4
Ralston, Nebr. (68127) 264/J3
Ralston, N.J. (†07945) 273/D2
Ralston, Okla. (74650) 288/N2
Ralston, Pa. (17763) 294/H2
Ralston, Tenn. (†38237) 237/D8
Ralston, Wash. (†99169) 310/G4
Ralston, Wyo. (82440) 319/D1
Ram (head), Virgin Is. (U.S.) 161/C5
Rama, Nicaragua 154/E4
Rama, Sask. 181/H4
Ramadi, Iraq 59/D3
Ramadi, Iraq 66/C4
Ramage, W. Va. (25166) 312/C7
Ramah, Colo. (80832) 208/L4
Ramah (bay), Newf. 166/B2
Ramah, N. Mex. (87321) 274/A3
Ramallah, West Bank 65/C4
Ramallo, Argentina 143/F6
Ramapo (riv.), N.J. 273/E1
Ramat Gan, Israel 65/B3
Ramat Hasharon, Israel 65/B3
Rambi (isl.), Fiji 86/R10
Ramblewood, N.J. (†08054) 273/D4
Rambouillet, France 28/D3
Rame, Israel 65/C2
Ramea, Newf. 166/C4
Ramea (isls.), Newf. 166/C4
Ramechhap, Nepal 68/F3
Ramelton, Ireland 17/F1
Ramer, Ala. (36069) 195/F6
Ramer, Tenn. (38367) 237/D10
Rameswarem, India 68/D7
Ramey, Minn. (†56329) 255/E5
Ramey, Pa. (16671) 294/F4
Ramey A.F.B., P. Rico 161/A1
Ramhormoz, Iran 66/F5
Ramhurst, Georgia (†30705) 217/C1
Ramièrs (isl.), Martinique 161/C6
Ramla, Israel 65/B4
Ramm, Jebel (mt.), Jordan 65/D5
Ramme, Denmark 21/B4
Rammun, West Bank 65/C4
Ramnäs, Sweden 18/J7
Ramon (mt.), Israel 65/D5
Ramon, N. Mex. (†88136) 274/D4
Ramona, Calif. (92065) 204/J10
Ramona, Kansas (67475) 232/E3
Ramona, Okla. (74061) 288/P1
Ramona, S. Dak. (57054) 298/P5
Ramón Castilla, Peru 128/G5
Ramón de las Yaguas, Cuba 158/J4
Ramón Santana, Dom. Rep. 158/F6
Ramón Trigo, Uruguay 145/E3
Ramor (lake), Ireland 17/G4
Ramore, Ontario 177/K5
Ramos (riv.), Mexico 150/G4
Ramos Arizpe, Mexico 150/J4
Ramosch, Switzerland 39/K3
Ramotswa, Botswana 118/C4
Rampart, Alaska (99767) 196/H1
Ramparts (riv.), N.W. Terrs. 187/E3
Rampur, Him. Pradesh, India 68/D2
Rampur, Uttar Pradesh, India 68/D3
Ramree (isl.), Burma 72/B3
Ramsar, Iran 66/G2
Ramsay, Mich. (49959) 250/F2
Ramsay, Mont. (59748) 262/D4
Ramsay, Ontario 177/J5
Ramsbottom, England 13/H2
Ramsele, Sweden 18/K5
Ramsen, Switzerland 39/G1
Ramseur, N.C. (27316) 281/K3
Ramsey, England 10/F4
Ramsey, England 13/G5
Ramsey, Ill. (62080) 222/D4
Ramsey, Ind. (47166) 227/E8
Ramsey, I. of Man 13/C3
Ramsey, I. of Man 10/D3
Ramsey (bay), I. of Man 13/C3
Ramsey (co.), Minn. 255/E5
Ramsey, Minn. (†55303) 255/E5
Ramsey, N.J. (07446) 273/B1
Ramsey (co.), N. Dak. 282/N3
Ramsey (mt.), Tasmania 99/B3
Ramsey (isl.), Wales 13/B6
Ramsgate, England 10/G5
Ramsgate, England 13/J6
Ramsjö, Sweden 18/J5
Ramu (riv.), Papua N.G. 85/B7
Ramunia, Tanjong (pt.), Malaysia 72/F6
Ramville (isl.), Martinique 161/D6
Rana (fjord), Norway 18/H3
Rana (riv.), Norway 18/J3
Ranau, Malaysia 85/F4
Ranburne, Ala. (36273) 195/H3
Rancagua, Chile 138/G5
Rancagua, Chile 120/B6
Ranches of Taos, N. Mex. (87557) 274/D2
Ranchester, Wyo. (82839) 319/E1
Ranchi, India 68/F4
Rancho Cordova, Calif. (95670) 204/C8
Rancho Cucamonga, Calif. (91730) 204/E10
Rancho Mirage, Calif. (92270) 204/J10
Rancho Palos Verdes, Calif. (90274) 204/B11
Rancho Santa Clarita, Calif. (†91321) 204/G9
Rancho Santa Fe, Calif. (92067) 204/H10
Rancho Veloz, Cuba 158/D1
Ranchuelo, Cuba 158/E2
Ranchwood Manor, Okla. (†73160) 288/L4
Ranco (lake), Chile 138/E3
Rancocas, N.J. (08073) 273/D3
Rancocas (creek), N.J. 273/D3
Rand, Colo. (80473) 208/G2
Randalia, Iowa (52164) 229/K3
Randall, Iowa (50231) 229/F4
Randall, Kansas (66963) 232/D2
Randall, Minn. (56475) 255/D4
Randall (co.), Texas 303/C2
Randall (mt.), W. Australia 88/B3
Randallstown, Md. (21133) 245/L3
Randalstown, N. Ireland 17/J2
Randburg, S. Africa 118/H6
Randers, Denmark 18/G8
Randers, Denmark 21/C5
Randfontein, S. Africa 118/G6
Randle, Wash. (98377) 310/D4
Randleman, N.C. (27317) 281/K3
Randles, Mo. (†63740) 261/N8
Randlett, Okla. (73562) 288/K6
Randlett, Utah (84063) 304/E3
Randolph (co.), Ala. 195/H4
Randolph, Ala. (36792) 195/E5
Randolph, Ariz. (85243) 198/D6
Randolph (co.), Ark. 202/H1
Randolph (co.), Georgia 217/C7
Randolph (co.), Ill. 222/D5
Randolph (co.), Ind. 227/G4
Randolph, Iowa (51649) 229/B7
Randolph, Kansas (66554) 232/F2
Randolph○, Maine (†04345) 243/D7
Randolph, Md. (†20853) 245/K4
Randolph○, Mass. (02368) 249/D8
Randolph, Minn. (55065) 255/E6
Randolph, Miss. (38864) 256/F2
Randolph (co.), Mo. 261/G3
Randolph, Mo. (†64101) 261/P5
Randolph, Nebr. (68771) 264/G2
Randolph○, N.H. (03593) 268/E3
Randolph○, N.J. (†07801) 273/D2
Randolph, N.Y. (14772) 276/C6
Randolph (co.), N.C. 281/K3
Randolph, Ohio (44265) 284/H3
Randolph, Utah (84064) 304/C2
Randolph, Vt. (05060) 268/B4
Randolph○, Vt. (05060) 268/B4
Randolph, Va. (23962) 307/L7
Randolph (co.), W. Va. 312/G5
Randolph, Wis. (53956) 317/H8
Randolph A.F.B., Texas 303/K10
Randolph Center, Vt. (05061) 268/B4
Random (isl.), Newf. 166/D2
Random (sound), Newf. 166/D2
Random Lake, Wis. (53075) 317/K8
Randsburg, Calif. (93554) 204/H8
Randwick, N. S. Wales 88/L4
Randwick, N.S. Wales 97/J3
Ranelagh, Tasmania 99/C4
Ranfurly, Alberta 182/E3
Ranfurly, N. Zealand 100/B6
Rangamati, Bangladesh 68/G4
Rangasa (cape), Indonesia 85/F6
Rangatira (isl.), N. Zealand 100/E7
Range, Ala. (36473) 195/D8
Range (creek), Utah 304/D4
Rangeley, Maine (04970) 243/B6

Rangeley○, Maine (04970) 243/B6
Rangeley (lake), Maine 243/B6
Rangely, Colo. (81648) 208/B2
Ranger, Georgia (30734) 217/C2
Ranger (peak), Idaho 220/D3
Ranger, N.C. (†28906) 281/A4
Ranger, Sask. 181/D2
Ranger, Texas (76470) 303/F5
Ranger, W. Va. (25557) 312/B6
Rangiauria (Pitt) (isl.), N. Zealand 100/E7
Rangiora, N. Zealand 100/D5
Rangiroa (atoll), Fr. Poly. 87/M7
Rangitaiki (riv.), N. Zealand 100/F3
Rangitata (riv.), N. Zealand 100/C5
Rangitikei (riv.), N. Zealand 100/E3
Rangitoto (isl.), N. Zealand 100/C1
Rangkasbitung, Indonesia 85/G2
Rangoon (div.), Burma 72/C3
Rangoon (cap.), Burma 2/P5
Rangoon (cap.), Burma 54/L8
Rangoon (cap.), Burma 72/C3
Rangoon, W. Va. (†26238) 312/F4
Rangpur, Bangladesh 68/F3
Rania, Iraq 66/D2
Ranier, Minn. (56668) 255/E2
Ranken (riv.), North. Terr. 93/E6
Rankin, Ill. (60960) 222/F3
Rankin (co.), Miss. 256/E6
Rankin, Pa. (†15104) 294/C7
Rankin, Texas (79778) 303/B6
Rankine Store, North. Terr. 93/E5
Rankin Inlet, N.W. Terrs. 187/J3
Rankins Springs, N.S. Wales 97/D3
Rankweil, Austria 41/A3
Ranlo, N.C. (28052) 281/G4
Rannoch (dist.), Scotland 15/D4
Rannoch, Loch (lake), Scotland 10/D2
Rannoch, Loch (lake), Scotland 15/D4
Ranong, Thailand 72/C5
Ransiki, Indonesia 85/J6
Ransom, Ill. (60470) 222/E2
Ransom, Kansas (67572) 232/C3
Ransom (co.), N. Dak. 282/P7
Ransom, Pa. (18653) 294/F7
Ransomville, N.Y. (14131) 276/C4
Ranson, W. Va. (25438) 312/L4
Rantauprapat, Indonesia 85/C5
Rantekombola (mt.), Indonesia 85/F6
Rantis, West Bank 65/C3
Rantoul, Ill. (61866) 222/E3
Rantoul, Kansas (66079) 232/G3
Ranua, Finland 18/P4
Ranui, N. Zealand 100/B1
Ranum, Denmark 21/C4
Ranya, Wadi (dry riv.), Saudi Arabia 59/D5
Rao Co (mt.), Laos 72/E3
Rao Co (mt.), Vietnam 72/E3
Raohe, China 77/M2
Raoui, Erg er (des.), Algeria 106/D3
Raoul (isl.), N. Zealand 87/J8
Raoul (cape), Tasmania 99/D5
Rapa (isl.), Fr. Poly. 87/M8
Rapallo, Italy 34/B2
Rapa Nui (Easter) (isl.), Chile 87/Q8
Rapch (riv.), Iran 66/L8
Rapel, Chile 138/F4
Rapel (riv.), Chile 138/F4
Rapelje, Mont. (59067) 262/G5
Raper (cape), N.W.T. 162/K2
Raper (cape), N.W. Terrs. 187/M3
Raphine, Va. (24472) 307/K5
Raphoe, Ireland 17/F2
Rapid (riv.), Mich. 250/B2
Rapid (riv.), Minn. 255/D2
Rapidan, Va. (22733) 307/M4
Rapidan (riv.), Va. 307/M4
Rapid City, Manitoba 179/B4
Rapid City, Mich. (49676) 250/D4
Rapid City, S. Dak. 146/H5
Rapid City, S. Dak. 188/F2
Rapid City, S. Dak. (57701) 298/C5
Rapide-Blanc, Québec 174/C3
Rapides (par.), La. 238/E4
Rapide Taureau (dam), Québec 172/D3
Rapid River, Mich. (49878) 250/C3
Rapids City, Ill. (61278) 222/C2
Rapid View, Sask. 181/C1
Räpina, U.S.S.R. 53/D1
Raposos, Brazil 135/E2
Rappahannock (co.), Va. 307/M3
Rappahannock (riv.), Va. 307/P4
Rapperswil, Switzerland 39/G2
Rápulo (riv.), Bolivia 136/C4
Rapu-Rapu (isl.), Philippines 82/E4
Raqqa (El Rashid), Syria 63/H5
Raquette (lake), N.Y. 276/L3
Raquette (riv.), N.Y. 276/L1
Raquette Lake, N.Y. (13436) 276/L3
Raraka (atoll), Fr. Poly. 87/M7
Rarden, Ohio (45671) 284/D8
Rardin, Ill. (61948) 222/E4
Raritan, Ill. (61471) 222/C3
Raritan, N.J. (08869) 273/D2
Raritan (bay), N.J. 273/E3
Raritan (riv.), N.J. 273/D2
Raroia (atoll), Fr. Poly. 87/M7
Raron, Switzerland 39/E4
Rarotonga (isl.), Cook Is. 87/K8
Rasa (isl.), Philippines 82/B6
Ra's al Khafji, Saudi Arabia 59/E4
Ras al Khaimah, U.A.E. 59/G4
Rasar, Tenn. (†37878) 237/O9
Ras Asèr (cape), Somalia 102/H3
Ras Dashan (mt.), Ethiopia 102/F3
Ras Dashan (mt.), Ethiopia 111/G5
Raseiniai, U.S.S.R. 53/B3
Ra's en Naqb, Jordan 65/E5
Ras Ghârib, Egypt 111/F2
Rashad, Sudan 111/F5
Ras Hafun (cape), Somalia 102/H3
Rasharkin, N. Ireland 17/J2
Rasheiya, Lebanon 63/F6
Rashid (Rosetta), Egypt 111/J2
Rashid (Rosetta), Egypt 59/B3
Rashid (prov.), Syria 63/H5
Rasht, Iran 59/E2
Rasht, Iran 54/G6
Rasht, Iran 66/F2
Rask, Iran 66/M7
Raška, Yugoslavia 45/E4
Ras Lanuf, Libya 111/C1
Rason (lake), W. Australia 88/C5
Rason (lake), W. Australia 92/D5
Rasskazovo, U.S.S.R. 52/F4
Ras Tanura, Saudi Arabia 59/F4
Rastatt, W. Germany 22/C4
Rastede, W. Germany 22/C2
Rat (isls.), Alaska 196/K4
Rat (riv.), Manitoba 179/F5
Ratak Chain (isls.), Marshall Is. 87/G5
Ratangarh, India 68/C3
Rat Buri, Thailand 72/C4
Ratcliff, Ark. (72951) 202/C3
Ratcliff, Texas (75858) 303/J6
Ratcliffe, Sask. 181/G6
Rathangan, Ireland 17/G5
Rathbun, Iowa (52545) 229/H7
Rathbun (lake), Iowa 229/G7
Rathcoole, Ireland 17/J5
Rathcormac, Ireland 17/E7
Rathdowney, Ireland 17/F6
Rathdrum, Idaho (83858) 220/A2
Rathdrum, Ireland 17/J6
Rathedaung, Burma 72/B2
Rathenow, E. Germany 22/E2
Rathfriland, N. Ireland 17/J3
Rathgormuck, Ireland 17/F7
Rathkeale, Ireland 17/D7
Rathkeale, Ireland 10/B4
Rathlin (isl.), N. Ireland 10/C3
Rathlin (isl.), N. Ireland 17/J1
Rathlin (sound), N. Ireland 17/J1
Rathlin O'Birne (isl.), Ireland 17/C2
Rathluirc, Ireland 10/B4
Rathluirc, Ireland 17/D7
Rathmore, Ireland 17/J5
Rathmullen, Ireland 17/F1
Rathnew-Merrymeeting, Ireland 17/J6
Rathowen, Ireland 17/F4
Rathvilly, Ireland 17/H6
Rathwell, Manitoba 179/D5
Ratibor (Racibórz), Poland 47/C3
Ratingen, W. Germany 22/B3
Ratio, Ark. (†72333) 202/J5
Ratlam, India 68/C4
Ratliff City, Okla. (73081) 288/M6
Ratnagiri, India 68/C5
Ratnapura, Sri Lanka 68/D7
Ratoath, Ireland 17/J5
Raton, N. Mex. 188/F3
Raton, N. Mex. (87740) 274/E2
Rats (riv.), Québec 172/D2
Rattan, Okla. (74562) 288/R6
Ratten, Austria 41/C3
Rattlesnake (creek), Kansas 232/D4
Rattlesnake (creek), Ohio 284/C7
Rattlesnake (creek), Oreg. 291/K5
Rattlesnake (hills), Wyo. 319/E3
Rattlesnake (range), Wyo. 319/E3
Rattray (head), Scotland 15/G3
Rättvik, Sweden 18/J6
Ratzeburg, W. Germany 22/D2
Raub, Ind. (47976) 227/C3
Raub, Malaysia 72/D7
Raub, N. Dak. (58774) 282/F4
Rauch, Argentina 143/E4
Rauch, Minn. (†55740) 255/E3
Raukumara (range), N. Zealand 100/F3
Raul Leoni (dam), Venezuela 124/G4
Raul Soares, Brazil 135/E2
Rauma, Finland 18/M6
Rauma (riv.), Norway 18/F5
Raunds, England 13/G5
Raung (mt.), Indonesia 85/L2
Raurkela, India 68/F4
Rausu, Japan 81/M1
Rauville, S. Dak. (†57201) 298/P3
Ravalli (co.), Mont. 262/B4
Ravalli, Mont. (59863) 262/B3
Ravanna, Ark. (†75556) 202/C7
Ravanna, Mo. (†64673) 261/E2
Ravar, Iran 59/G3
Ravar, Iran 66/K5
Ravelo, Bolivia 136/C6
Ravels, Belgium 27/G6
Raven, Va. (24639) 307/E6
Ravena, N.Y. (12143) 276/N6
Ravencliff, W. Va. (25913) 312/C7
Ravendale, Calif. (96123) 204/E3
Ravenden, Ark. (72459) 202/H1
Ravenden Springs, Ark. (72460) 202/H1
Ravenel, S.C. (29470) 296/G6
Ravenna (prov.), Italy 34/D2
Ravenna, Italy 34/D2
Ravenna, Ky. (40472) 237/O5
Ravenna, Mich. (49451) 250/D5
Ravenna, Nebr. (68869) 264/F4
Ravenna, Ohio (44266) 284/H3
Ravenna, Texas (75476) 303/H4
Raven Rock, W. Va. (†26170) 312/D4
Ravensburg, W. Germany 22/C5
Ravenscrag, Sask. 181/C6
Ravensdale, Wash. (98051) 310/D3
Ravenshoe, Queensland 88/H3
Ravenshoe, Queensland 95/C3
Ravensthorpe, W. Australia 88/C6
Ravensthorpe, W. Australia 92/B6
Ravenswood, W. Va. (26164) 312/C5
Ravenwood, Mo. (64479) 261/C2
Ravi (riv.), Pakistan 68/C2
Ravia, Okla. (73455) 288/N6
Ravine, Pa. (17966) 294/K4
Ravinia, S. Dak. (57357) 298/N7
Ravne na Koroškem, Yugoslavia 45/B2
Rawalpindi, Pakistan 54/J6
Rawalpindi, Pakistan 68/C2
Rawalpindi, Pakistan 59/K3
Rawa Mazowiecka, Poland 47/E3
Rawdon, Québec 172/D3
Rawene, N. Zealand 100/D2
Rawhide (creek), Wyo. 319/G1
Rawhide (creek), Wyo. 319/H3
Rawi, Ko (isl.), Thailand 72/C6
Rawicz, Poland 47/C3
Rawlings, Md. (21557) 245/C2
Rawlings, Va. (23876) 307/N7
Rawlinna, W. Australia 88/C6
Rawlinna, W. Australia 92/E5
Rawlins (co.), Kansas 232/A2
Rawlins, Wyo. 188/E2
Rawlins, Wyo. (82301) 319/E4
Rawson, Argentina 120/C7
Rawson, Buenos Aires, Argentina 143/F7
Rawson, Chubut, Argentina 143/D5
Rawson, N. Dak. (†58831) 282/C4
Rawson, Ohio (45881) 284/C4
Rawtenstall, England 13/H1
Rawtenstall, England 10/G2
Raxaul, India 68/E3
Ray (mts.), Alaska 196/H1
Ray, Ill. (†62681) 222/C3
Ray, Ind. (46737) 227/H1
Ray, Minn. (56669) 255/E2
Ray (co.), Mo. 261/E4
Ray (cape), Newf. 166/C4
Ray (cape), Newf. 162/K6
Ray, N. Dak. (58849) 282/D3
Ray, Ohio (45672) 284/E7
Raya (mt.), Indonesia 85/E6
Rayagada, India 68/E5
Rayak, Lebanon 63/G6
Raybon, Georgia (†31553) 217/H8
Rayborn, Mo. (†65703) 261/H8
Raychikhinsk, U.S.S.R. 48/N5
Ray City, Georgia (31645) 217/F8
Ray Hubbard (lake), Texas 303/H2
Rayland, Ohio (43943) 284/J5
Rayle, Georgia (30660) 217/G3
Rayleigh, Br. Col. 184/G5
Rayleigh, England 13/J8
Raymer (New Raymer), Colo. (80742) 208/M1
Raymond, Alberta 182/D5
Raymond, Alta. 162/E6
Raymond, Calif. (93653) 204/F6
Raymond, Idaho (†83114) 220/G7
Raymond, Ill. (62560) 222/D4
Raymond, Ind. (†45056) 227/H6
Raymond, Iowa (50667) 229/J4
Raymond, Kansas (67573) 232/D3
Raymond, Maine (04071) 243/B8
Raymond○, Maine (04071) 243/B8
Raymond, Minn. (56282) 255/C5
Raymond, Miss. (39154) 256/D6
Raymond, Mont. (59256) 262/M2
Raymond, Nebr. (68428) 264/H4
Raymond, N.H. (03077) 268/E5
Raymond○, N.H. (03077) 268/E5
Raymond, Ohio (43067) 284/C5
Raymond, S. Dak. (57258) 298/O4
Raymond, Wash. (98577) 310/B4
Raymond, Wis. (†53126) 317/L2
Raymond City, W. Va. (†25159) 312/C6
Raymond Terrace, N.S. Wales 97/F3
Raymondville, Mo. (65555) 261/J8
Raymondville, N.Y. (13678) 276/L1
Raymondville, Texas (78580) 303/G11
Raymore, Mo. (64083) 261/D5
Raymore, Sask. 181/G4
Rayne, La. (70578) 238/F6
Raynesford, Mont. (59469) 262/F3
Raynham○, Mass. (02767) 249/K5
Raynham, N.C. (†28340) 281/L5
Raynham Center, Mass. (02768) 249/K5
Rayón, San Luis Potosí, Mexico 150/K6
Rayón, Sonora, Mexico 150/D2
Rayong, Thailand 72/D4
Rays (lake), Idaho 220/F6
Rays Crossing, Ind. (†46176) 227/F5
Rayside-Balfour, Ontario 177/K5
Raystown (lake), Pa. 294/F5
Raystown Branch, Juniata (riv.), Pa. 294/F5
Raysut (Risut), Oman 59/F6
Raytown, Mo. (64133) 261/P6
Rayville, La. (71269) 238/G2
Rayville, Mo. (64084) 261/E4
Raywick, Ky. (40060) 237/L5
Razan, Iran 66/F3
Razaza (res.), Iraq 66/C4
Razdan, U.S.S.R. 52/G6
Razgrad, Bulgaria 45/H4
Razlog, Bulgaria 45/F5
Ré (isl.), France 28/C4
Rea, Mo. (64480) 261/C2
Reader, Ark. (71726) 202/D6
Reader, W. Va. (26167) 312/E3
Readfield, Maine (04355) 243/D7
Readfield○, Maine (04355) 243/D7
Readfield, Wis. (54969) 317/J7
Reading, England 13/G8
Reading, England 10/F5
Reading, Kansas (66868) 232/F3
Reading○, Mass. (01867) 249/C5
Reading, Mich. (49274) 250/E7
Reading, Minn. (56165) 255/C7
Reading, Ohio (45215) 284/C9
Reading, Pa. 188/L2
Reading, Pa. (*19601) 294/L5
Reading○, Vt. (05062) 268/B5
Readington, N.J. (08870) 273/D2
Readland, Ark. (71664) 202/H7
Readlyn, Iowa (50668) 229/J3
Readlyn, Sask. 181/F6
Readsboro, Vt. (05350) 268/B6
Readsboro○, Vt. (05350) 268/B6
Reads Landing, Minn. (55968) 255/F6
Reads Mill, Ala. (†36279) 195/G3
Readstown, Wis. (54652) 317/E9
Readsville, Mo. (†65067) 261/J5
Readville, Mass. (02137) 249/C8
Ready, Ky. (†42721) 237/J6
Readyville, Tenn. (37149) 237/J9
Reagan, Okla. (†73460) 288/N6
Reagan, Tenn. (38368) 237/E9
Reagan (co.), Texas 303/C6
Reagan, Texas (76680) 303/H6
Real, Cordillera (range), Bolivia 136/A5
Real (co.), Texas 303/E8
Real de San Carlos, Uruguay 145/A5
Realitos, Texas (78376) 303/F10
Realp, Switzerland 39/F3
Reamstown, Pa. (17567) 294/K5
Reao (atoll), Fr. Poly. 87/N7
Reardan, Wash. (99029) 310/H3
Reasnor, Iowa (50232) 229/G5
Reaville, N.J. (†08822) 273/D3
Reay, Scotland 15/E2
Rebecca, Georgia (31783) 217/E7
Rebecca (lake), W. Australia 88/C6
Rebecca (lake), W. Australia 92/C5
Rebecq, Belgium 27/E7
Rebersburg, Pa. (16872) 294/H4
Rebiana (oasis), Libya 111/D3
Rebiana Sand Sea (des.), Libya 111/D3
Reboledo, Uruguay 145/D4
Rebun (isl.), Japan 81/K1
Recanati, Italy 34/D3
Recherche (arch.), Australia 87/C9
Recherche (arch.), W. Australia 88/C6
Recherche (arch.), W. Australia 92/C6
Rechitsa, U.S.S.R. 52/C4
Rechnitz, Austria 41/D3
Rechthalten, Switzerland 39/D3
Recife, Brazil 132/H5
Recife, Brazil 120/F3
Recife, Brazil 2/H6
Reckingen, Switzerland 39/F4
Recklinghausen, W. Germany 22/B3
Recluse, Wyo. (82725) 319/G1
Reconquista, Argentina 143/F4
Recreo, Argentina 143/C2
Rector, Ark. (72461) 202/K1
Rectortown, Va. (22140) 307/N3
Recuay, Peru 128/D7
Red (riv.) 146/J5
Red (riv.) 188/H4
Red (sea) 2/L4
Red (sea) 54/E7
Red (riv.) 54/M7
Red (sea) 102/F2
Red (riv.), Ark. 202/C6
Red (mt.), Conn. 210/B1
Red (sea), Egypt 111/G3
Red (sea), Ethiopia 111/H4
Red (riv.), Idaho 220/C4
Red (riv.), Ky. 237/G7
Red (riv.), Ky. 237/O5
Red (riv.), La. 238/G4
Red (riv.), Manitoba 179/F4
Red (lake), Minn. 188/H1
Red (isl.), Newf. 166/C2
Red (bay), N. Ireland 17/K1
Red, North Fork (riv.), Okla. 288/H4
Red (riv.), Okla. 288/R7
Red (lake), Ontario 175/B2
Red (sea), Saudi Arabia 59/C5
Red (lake), S. Dak. 298/L6
Red (sea), Sudan 111/G3
Red (riv.), Tenn. 237/G7
Red (riv.), Texas 303/F3
Red (riv.), U.S. 146/J6
Red (riv.), Utah 304/D3
Red (riv.), Vietnam 72/E2
Red (pt.), Virgin Is. (U.S.) 161/D4
Red (sea), Yemen Arab Rep. 59/C5
Reda, Poland 47/D1
Redang, Pulau (isl.), Malaysia 72/D6
Redange, Luxembourg 27/H9
Red Ash, Va. (24640) 307/E6
Red Bank, New Bruns. 170/E2
Red Bank, N.J. (07701) 273/E3
Redbank (creek), Pa. 294/E3
Red Bank, Tenn. (37415) 237/L10
Red Banks, Miss. (38661) 256/F1
Red Bay, Ala. (35582) 195/B2
Red Bay, Fla. (†32455) 212/C6
Red Bay, Newf. 166/C3
Red Bay, Ontario 177/C3
Red Beach, Maine (04670) 243/J5
Red Bird, Mo. (†65014) 261/J6
Redbird, Okla. (74458) 288/P3
Red Bluff, Calif. (96080) 204/C3
Red Bluff (lake), N. Mex. 274/E7
Red Bluff (lake), Texas 188/F4
Red Bluff (lake), Texas 303/A6
Red Boiling Springs, Tenn. (37150) 237/K7
Redbridge, England 13/H8
Redbridge, England 10/C5
Red Bud, Ill. (62278) 222/D5
Redbush, Ky. (41251) 237/P5
Redby, Minn. (56670) 255/D3
Redcar, England 10/F3
Redcar, England 13/F3
Red Cedar (riv.), Wis. 317/C5
Red Chute (bayou), La. 238/C1
Redcliff, Alberta 182/E4
Red Cliff, Colo. (81649) 208/G4
Red Cliff, Wis. (†54814) 317/E2
Redcliffe, Queensland 95/E5
Red Cliffe, Queensland 88/J5
Red Cliff Ind. Res., Wis. 317/E2
Red Cliffs, Victoria 97/B4
Redcloud (peak), Colo. 208/E6
Red Cloud, Nebr. (68970) 264/F4
Red Creek, N.Y. (13143) 276/G4
Red Creek, W. Va. (26289) 312/H4
Redcrest, Calif. (95569) 204/A3
Red Deer, Alberta 182/D3
Red Deer (lake), Alberta 182/D3
Red Deer (riv.), Alberta 182/D4
Red Deer, Alta. 146/G4
Red Deer, Alta. 162/E5
Red Deer (lake), Manitoba 179/A2
Red Deer (riv.), Manitoba 179/A2
Red Deer (riv.), Sask. 181/K3
Red Deer (riv.), Sask. 181/A5
Red Deer Hill, Sask. 181/F2
Reddell, La. (70580) 238/F5
Redden, Del. (†19947) 245/S5
Red Devil, Alaska (99656) 196/G2
Reddick, Fla. (32686) 212/D2
Reddick, Ill. (60961) 222/E2
Redding, Calif. 146/F5
Redding, Calif. 188/B2
Redding, Calif. (96001) 204/C3
Redding○, Conn. (06875) 210/B3
Redding, Iowa (50860) 229/E7
Redding Ridge, Conn. (06876) 210/B3
Reddington, Ind. (†47274) 227/F6
Redditch, England 13/E5
Redditch, England 10/G3
Red Earth Creek, Alberta 182/C1
Red Elm, S. Dak. (†57623) 298/F3
Redeye (riv.), Minn. 255/D4
Red Feather Lakes, Colo. (80545) 208/H1
Redfield, Ark. (72132) 202/F5
Redfield, Iowa (50233) 229/E5
Redfield, Kansas (66769) 232/H4
Redfield, N.Y. (13437) 276/J3
Redfield, Sask. 181/D2
Redfield, S. Dak. (57469) 298/N4
Redfish (lake), Idaho 220/D5
Redford, Mo. (63665) 261/L8
Redford, Texas (79846) 303/C12
Red Fork, Powder (riv.), Wyo. 319/F2
Redgranite, Wis. (54970) 317/J7
Redhead, Trin. & Tob. 161/B10
Red Head Cove, Newf. 166/D2
Red Hill (mt.), Hawaii 218/K2
Red Hill, Pa. (18076) 294/L5
Red Hook, N.Y. (12571) 276/N7
Redhouse, Ky. (†40475) 237/N5
Red House, Nev. (†89414) 266/D2
Red House, Va. (23963) 307/L6
Red House, W. Va. (25168) 312/C5
Redig, S. Dak. (57776) 298/C3
Red Indian (lake), Newf. 166/C4
Redington, Nebr. (†69336) 264/A3
Redington Beach, Fla. (33708) 212/B3
Redington Shores, Fla. (†33708) 212/B3
Red Jacket, W. Va. (25692) 312/B7
Redkey, Ind. (47373) 227/G4
Red Lake (co.), Minn. 255/B3
Redlake, Minn. (56671) 255/C3
Red Lake (riv.), Minn. 255/B2
Red Lake, Ont. 162/G6
Red Lake, Ontario 175/B2
Red Lake Falls, Minn. (56750) 255/B3
Red Lake Ind. Res., Minn. 255/C2
Red Lake Road, Ontario 177/G5
Red Lake Road, Ontario 175/B2
Redland, Alberta 182/D4
Redland, Oreg. (†97045) 291/B2
Redlands, Calif. (92373) 204/H9
Red Level, Ala. (36474) 195/E8
Red Lick, Miss. (†39096) 256/B7
Red Lion, Del. (†19701) 245/R2
Red Lion, N.J. (08088) 273/D4
Red Lion, Ohio (†45005) 284/B7
Red Lion, Pa. (17356) 294/J6
Red Lodge, Mont. (59068) 262/G5
Redman, Mich. (†48468) 250/G5
Red Mesa, Colo. (†81326) 208/C8
Redmon, Ill. (61949) 222/F4
Redmond, Oreg. (97756) 291/F3
Redmond, Utah (84652) 304/C4
Redmond, Wash. (98052) 310/B1
Red Mountain, Calif. (93558) 204/H8
Red Oak, Georgia (30272) 217/J2
Red Oak, Iowa (51566) 229/C6
Red Oak, Mich. (†49756) 250/E4
Red Oak, N.C. (27868) 281/N2
Red Oak, Okla. (74563) 288/R5
Red Oak, Texas (75154) 303/H5
Red Oak, Va. (23964) 307/L7
Red Oaks Mill, N.Y. (†12601) 276/N7
Redon, France 28/C4
Redonda (isl.), Ant. & Bar. 156/F3
Redondela, Spain 33/B1
Redondo, Portugal 33/C3
Redondo, Wash. (98054) 310/C3
Redondo Beach, Calif. (*90277) 204/B11
Redoubt (vol.), Alaska 196/H2
Redowl, S. Dak. (57777) 298/D4
Red Owl (creek), S. Dak. 298/E4
Redpa, Tasmania 99/A2
Red Pass, Br. Col. 184/H4
Redridge, Mich. (†49931) 250/G1
Red River (par.), La. 238/D2
Red River, N. Mex. (87558) 274/D2
Red River, Nova Scotia 168/H2
Red River, S.C. (†29730) 296/F2
Red River (co.), Texas 303/J4
Red River Hot Springs, Idaho (†83525) 220/C4
Red River of the North (riv.) 188/G1
Red River of the North (riv.), Minn. 255/A2
Red River of the North (riv.), N. Dak. 282/S4
Red Rock, Ariz. (85245) 198/D6
Red Rock, Br. Col. 184/F3
Red Rock (lake), Iowa 229/G6
Red Rock (lakes), Mont. 262/E6
Red Rock (riv.), Mont. 262/D6
Redrock, N. Mex. (88055) 274/A6
Red Rock, Okla. (74651) 288/M2
Red Rock, Ontario 177/H5
Red Rock, Ontario 175/C3
Red Rock, Texas (78662) 303/G8
Red Scaffold (creek), S. Dak. 298/F4
Red Sea (prov.), Sudan 111/G4
Red Sea Hills (mts.), Sudan 59/C5
Red Springs, N.C. (28377) 281/L5
Red Springs, Texas (76378) 303/E4
Redstar, W. Va. (25914) 312/D7
Redstone, Br. Col. 184/F4
Redstone, Colo. (†81623) 208/E4
Redstone, Mont. (59257) 262/M2
Redstone, N.H. (†03813) 268/E3
Redstone (riv.), N.W. Terrs. 187/F3
Redstone (lake), Ontario 177/F2
Redstone (creek), S. Dak. 298/O5
Redstone Arsenal, Ala. 195/E1
Red Sucker Lake, Manitoba 179/K3
Red Sulphur Springs, W. Va. (†24963) 312/E7
Redtop, Minn. (†56342) 255/E4
Redvale, Colo. (81431) 208/B6
Redvers, Sask. 181/K6
Red Volta (riv.), Ghana 106/D6
Red Volta (riv.), Upper Volta 106/D6
Redwater, Alberta 182/D3
Redwater (riv.), Mont. 262/L3
Redwater (creek), S. Dak. 298/A4
Redway, Calif. (95560) 204/B3
Red Willow, Alberta 182/D3
Red Willow (co.), Nebr. 264/D4
Redwine, Ky. (†41477) 237/P4
Red Wine (riv.), Newf. 166/B3
Red Wing, Colo. (81066) 208/J7
Redwing, Kansas (†67544) 232/D3
Red Wing, Minn. (55066) 255/F6
Redwood (co.), Minn. 255/C6
Redwood (riv.), Minn. 255/C6
Redwood, Miss. (39156) 256/C6
Redwood, N.Y. (13679) 276/J2
Redwood City, Calif. (*94061) 204/J3
Redwood Estates-Chemeketa Park, Calif. (95044) 204/K4
Redwood Falls, Minn. (56283) 255/C6
Redwood Nat'l Park, Calif. 204/A2
Redwood Valley, Calif. (95470) 204/B4
Ree, Lough (lake), Ireland 10/B4
Ree (lake), Ireland 17/F5
Reece, Kansas (†67045) 232/F4
Reece City, Ala. (†35954) 195/G2
Reed, Ark. (71670) 202/H6
Reed, Ky. (42451) 237/G5
Reed, Okla. (73563) 288/G5
Reed (mt.), Québec 174/D2
Reed City, Mich. (49677) 250/D5
Reeder, Manitoba 179/A4
Reeder, N. Dak. (58649) 282/E7
Reedley, Calif. (93654) 204/F7
Reedpoint, Mont. (59069) 262/G5
Reeds, Mo. (64859) 261/D8
Reedsburg (res.), Mich. 250/E4
Reedsburg, Ohio (†44691) 284/F4
Reedsburg, Wis. (53959) 317/G8
Reeds Ferry, N.H. (†03054) 268/D6
Reedsport, Oreg. (97467) 291/C4
Reeds Spring, Mo. (65737) 261/F9
Reedsville, Ohio (45772) 284/G7
Reedsville, Pa. (17084) 294/G4
Reedsville, W. Va. (26547) 312/G3
Reedsville, Wis. (54230) 317/L7
Reedville, Oreg. (†97005) 291/A2
Reedville, Va. (22539) 307/R5
Reedy (lake), Fla. 212/E4
Reedy (riv.), S.C. 296/C2
Reedy, W. Va. (25270) 312/D5
Reedy (creek), W. Va. 312/D5
Reedyville, Ky. (†42275) 237/H6
Reef (bay), Virgin Is. (U.S.) 161/C4
Reefton, N. Zealand 100/C5
Ree Heights, S. Dak. (57371) 298/L4
Reelfoot (lake), Tenn. 237/C8
Reelsville, Ind. (46171) 227/D5
Reeman, Mich. (†49412) 250/D5
Reese, Mich. (48757) 250/F5
Reese (riv.), Nev. 266/D3
Reese A.F.B., Texas 303/B4
Reeseville, Wis. (53579) 317/J9
Reesville, Ohio (45166) 284/C7
Reeves, La. (70658) 238/D5
Reeves (co.), Texas 303/D11
Reeves Knob (mt.), Ark. 202/E2
Reevesville, Ill. (†62943) 222/E6
Reevesville, S.C. (29471) 296/F5
Refahiye, Turkey 63/H3
Refa'i, Iraq 66/E5
Reform, Ala. (35481) 195/C4
Reform, Miss. (39757) 256/F4
Refton, Pa. (17568) 294/K6
Refuge Cove, Br. Col. 184/E5
Refugio (isl.), Chile 138/D5
Refugio (co.), Texas 303/G9
Refugio, Texas (78377) 303/G9
Rega (riv.), Poland 47/B2
Regal, Minn. (†56312) 255/D5
Regan, N. Dak. (58477) 282/K5
Regen, W. Germany 22/E4
Regen (riv.), W. Germany 22/E4
Regeneração, Brazil 132/F4
Regensburg, W. Germany 7/F4
Regensburg, W. Germany 22/E4
Regensdorf, Switzerland 39/F2
Regent, Manitoba 179/B5
Regent, N. Dak. (58650) 282/E7
Reger, Mo. (65556) 261/F2
Reggane, Algeria 106/D3
Regge (riv.), Netherlands 27/K4
Reggio, La. (†70085) 238/L7
Reggio di Calabria (prov.), Italy 34/E5
Reggio di Calabria, Italy 7/F5
Reggio di Calabria, Italy 34/E5
Reggio nell'Emilia (prov.), Italy 34/C2
Reggio nell'Emilia, Italy 34/C2
Reghin, Romania 45/G2
Régina, Fr. Guiana 131/E3
Regina, Mont. (59539) 262/J3
Regina, N. Mex. (87046) 274/B2
Regina (cap.), Sask. 162/F5
Regina (cap.), Sask. 146/H4
Regina (cap.), Sask. 181/G5
Regina Beach, Sask. 181/F5
Register, Georgia (30452) 217/J6
Registro, Brazil 135/C4
Regla, Cuba 158/C1
Regnitz (riv.), W. Germany 22/D4
Reguengos de Monsaraz, Portugal 33/C3
Regway, Sask. 181/G6
Rehau, W. Germany 22/D3
Rehoboth, Ala. (†36720) 195/D6
Rehoboth○, Mass. (02769) 249/K5
Rehoboth, Namibia 118/B4
Rehoboth, Namibia 102/D7
Rehoboth, N. Mex. (87322) 274/A3
Rehoboth, Va. (†23974) 307/M7
Rehoboth Beach, Del. (19971) 245/T6
Rehovot, Israel 65/B4
Rehrersburg, Pa. (19550) 294/K5
Reichenau an der Rax, Austria 41/C3
Reichenbach, E. Germany 22/E3
Reichenbach im Kandertal, Switzerland 39/E3

Reid, Md. (†21740) 245/H2
Reid (lake), S. Dak. 298/O3
Reid (rocks), Tasmania 99/B1
Reid, W. Australia 92/E5
Reiden, Switzerland 39/F2
Reids Grove, Md. (†21869) 245/P6
Reidsville, Georgia (30453) 217/H6
Reidsville, N.C. (27320) 281/K2
Reidville, S.C. (29375) 296/C2
Reigate, England 13/H8
Reigate, England 10/F5
Reile's Acres, N. Dak. (†58078) 282/S6
Reily, Ohio (45060) 284/A7
Re'im, Israel 65/A5
Reims, France 7/E4
Reims, France 28/E3
Reina Adelaida (arch.), Chile 120/B8
Reina Adelaida (arch.), Chile 138/D9
Reinach in Aargau, Switzerland 39/F2
Reinach in Baselland, Switzerland 39/E2
Reinbeck, Iowa (50669) 229/H4
Reindeer (lake) 162/F4
Reindeer (lake), Canada 146/H4
Reindeer (isl.), Manitoba 179/E2
Reindeer (lake), Manitoba 179/H2
Reindeer (lake), Sask. 181/N3
Reindeer (riv.), Sask. 181/M3
Reinersville, Ohio (43756) 284/G6
Reinfeld, W. Germany 22/D2
Reinga (cape), N. Zealand 100/D1
Reinland, Manitoba 179/E5
Reinosa, Spain 33/D1
Reisaelv (riv.), Norway 18/M2
Reisduoddarhal'di (Haltiatunturi), Norway 18/M2
Reiss, Scotland 15/E2
Reisterstown, Md. (21136) 245/L3
Reitz, S. Africa 118/D5
Rejaf, Sudan 111/F7
Reliance, Md. (†19973) 245/P6
Reliance, N.W. Terrs. 187/H3
Reliance, S. Dak. (57569) 298/K6
Reliance, Tenn. (37369) 237/N10
Reliance, Va. (22649) 307/M3
Reliance, Wyo. (82943) 319/C4
Relief, Ky. (41463) 237/P5
Relizane, Algeria 106/E1
Reloncaví (bay), Chile 138/D4
Remada, Tunisia 106/F2
Remagen, W. Germany 22/B3
Remanso, Brazil 132/F5
Remates, Cuba 158/A2
Rembang, Indonesia 85/K2
Rembert, S.C. (29128) 296/G3
Rembrandt, Iowa (50576) 229/C3
Rembrandt, Manitoba 179/E4
Remedios, Colombia 126/C4
Remedios, Cuba 156/B2
Remedios, Cuba 158/E2
Remedios (pt.), El Salvador 154/B4
Remer, Minn. (56672) 255/E3
Remerton, Georgia (31601) 217/F9
Remich, Luxembourg 27/J9
Reminderville, Ohio (†44202) 284/J10
Remington, Ind. (47977) 227/C3
Remington, Ohio (†45202) 284/C9
Remington, Va. (22734) 307/N3
Rémire, Fr. Guiana 131/E3
Rémire (isls.), Fr. Guiana 131/F3
Remiremont, France 28/G3
Remlap, Ala. (35133) 195/E3
Remmel (mt.), Wash. 310/E2
Remo, Br. Col. 184/C3
Remolino, Colombia 126/C2
Remote, Oreg. (97468) 291/D5
Remscheid, W. Germany 22/B3
Remsen, Iowa (51050) 229/B3
Remsen, N.Y. (13438) 276/K4
Remus, Mich. (49340) 250/D5
Remus, Okla. (†74801) 288/N4
Remy, La. (†70763) 238/L3
Rena, Ark. (†72956) 202/B3
Rena, Norway 18/G6
Reñaca, Chile 138/F2
Renaix (Ronse), Belgium 27/D7
Rena Lara, Miss. (38767) 256/C2
Renan, Switzerland 39/C2
Renault, Ill. (62279) 222/C5
Renca, Chile 138/G3
Rencona, N. Mex. (†87562) 274/D3
Rencontre East, Newf. 166/C4
Rend (lake), Ill. 222/E5
Rendeux, Belgium 27/H8
Rendova (isl.), Solomon Is. 86/D3
Rendsburg, W. Germany 22/C1
Rendville, Ohio (†43730) 284/F6
Renens, Switzerland 39/C3
Renews, Newf. 166/D2
Renforth, New Bruns. 170/E3
Renfrew, Ont. 162/J6
Renfrew (county), Ontario 177/G2
Renfrew (county), Ontario 175/E3
Renfrew, Ontario 177/H2
Renfrew, Ontario 175/E3
Renfrew, Pa. (16053) 294/C4
Renfrew, Scotland 10/A1
Renfrew, Scotland 15/B2
Renfrew (trad. co.), Scotland 15/A5
Renfroe, Ala. (†35160) 195/F4
Renfroe, Georgia (†31805) 217/C6
Renfroe, Miss. (†39051) 256/F5
Renfrow, Okla. (†73759) 288/L1
Rengam, Malaysia 72/E5
Rengat, Indonesia 85/C6
Rengo, Chile 138/G5
Reni, U.S.S.R. 52/C5
Renick, Mo. (65278) 261/H4
Renick, W. Va. (24966) 312/F6
Renigunta, India 68/E6
Renish (pt.), Scotland 15/B3
Renk, Sudan 111/F5
Renkum, Netherlands 27/H5
Renmark, S. Australia 88/G6
Renmark, S. Australia 94/G5
Rennell (isl.), Solomon Is. 87/F7
Rennell (isl.), Solomon Is. 86/E3
Renner, S. Dak. (57055) 298/R6
Rennert, N.C. (†28386) 281/L5
Rennes, France 7/D4
Rennes, France 28/C3
Rennie, Manitoba 179/G5
Rennie (lake), N.W. Terrs. 187/H3
Renno, S.C. (†29325) 296/D2
Reno, Alberta 182/B2
Reno, Georgia (†31728) 217/D9
Reno, Ill. (†62086) 222/D5
Reno (co.), Kansas 232/D4
Reno (lake), Minn. 255/C5
Reno, Nev. 146/G6
Reno, Nev. 188/C3
Reno, Nev. (*89501) 266/B3
Reno, Ohio (45773) 284/H7
Reno, Texas (†76020) 303/E2
Reno Beach, Ohio (†43412) 284/D2
Renous, New Bruns. 170/E2
Renous (riv.), New Bruns. 170/D2
Renova, Miss. (†38732) 256/C3
Renovo, Pa. (17764) 294/G3
Renown, Sask. 181/F4
Rensburg, S. Africa 118/J7
Rensselaer, Ind. (47978) 227/C3
Rensselaer, Mo. (†63401) 261/J3
Rensselaer (co.), N.Y. 276/O5
Rensselaer, N.Y. (12144) 276/N5
Rensselaer Falls, N.Y. (13680) 276/K1
Rentchler, Ill. (†62220) 222/B3
Rentiesville, Okla. (74459) 288/R4
Renton, Scotland 15/A1
Renton, Wash. (98055) 310/B2
Rentz, Georgia (31075) 217/G6
Renville (co.), Minn. 255/C6
Renville, Minn. (56284) 255/C6
Renville (co.), N. Dak. 282/G2
Renwer, Manitoba 179/B2
Renwick, Iowa (50577) 229/E3
Répcelak, Hungary 41/D3
Repentigny, Québec 172/J4
Replete, W. Va. (†26222) 312/F5
Repos (lake), Québec 172/C2
Repton, Ala. (36475) 195/D8
Republic, Ala. (†35203) 195/E3
Republic (co.), Kansas 232/E2
Republic, Kansas (66964) 232/E2
Republic, Mich. (49879) 250/B2
Republic, Mo. (65738) 261/E8
Republic, Ohio (44867) 284/D3
Republic, Wash. (99166) 310/G2
República Dominicana, Cuba 158/F2
Republican (riv.) 188/F2
Republican (riv.), Colo. 208/P3
Republican (riv.), Kansas 232/E2
Republican (riv.), Nebr. 264/G5
Republican City, Nebr. (68971) 264/E4
Republican Grove, Va. (24585) 307/K7
Repulse (bay), Queensland 88/H4
Repulse Bay, Canada 4/C14
Repulse Bay, N.W.T. 162/H2
Repulse Bay, N.W. Terrs. 187/K3
Requa, Calif. (†95548) 204/A2
Requegua, Chile 138/G5
Requena, Peru 128/F5
Requena, Spain 33/F3
Requinoa, Chile 138/G5
Rera, Brazil 132/A1
Resaca, Georgia (30735) 217/C1
Reşadiye, Turkey 63/G2
Research, Victoria 97/J4
Reseda, Calif. (91335) 204/B10
Resende, Brazil 135/D3
Resende, Portugal 33/B2
Reserve, Kansas (66529) 232/G2
Reserve, La. (70084) 238/M3
Reserve, Mont. (59258) 262/M2
Reserve, N. Mex. (87830) 274/A5
Reserve, Sask. 181/J3
Reserve, Wis. (†54876) 317/D4
Reserve Mines, Nova Scotia 168/H2
Resht (Rasht), Iran 66/F2
Reshui, China 77/E4
Resistencia, Argentina 143/E2
Resistencia, Argentina 120/D5
Reşiţa, Romania 45/E3
Resolute, Canada 4/B14
Resolute Bay, N.W.T. 162/G1
Resolute Bay, N.W. Terrs. 187/J2
Resolution (isl.), N.W.T. 146/M3
Resolution (isl.), N.W.T. 162/K3
Resolution (isl.), N. Zealand 100/A6
Resolution (isl.), N.W. Terrs. 187/M3
Resolution Island, N.W. Terrs. 187/M3
Resort, Loch (inlet), Scotland 15/A2
Resource, Sask. 181/G3
Respenda de la Peña, Spain 33/D1
Restauración, Dom. Rep. 158/D5
Rest Haven, Georgia (†30518) 217/E2
Restigouche (co.), New Bruns. 170/C1
Restigouche (riv.), New Bruns. 170/C1
Restigouche, Québec 172/C2
Reston, Manitoba 179/A5
Reston, Va. (22090) 307/R2
Restoule, Ontario 177/E1
Restoule (lake), Ontario 177/E1
Restrepo, Colombia 126/D5
Reszel, Poland 47/E1
Retalhuleu, Guatemala 154/B3
Retamosa, Uruguay 145/E4
Rethel, France 28/F3
Réthimnon, Greece 45/G8
Retie, Belgium 27/G6
Retiro, Chile 138/A11
Retlaw, Alberta 182/D4
Rétság, Hungary 41/E3
Retsil, Wash. (98378) 310/A2
Retsof, N.Y. (14539) 276/E5
Retz, Austria 41/D2
Reubens, Idaho (83548) 220/B3
Réunion (isl.), (Fr.) 2/M7
RÉUNION 118/F5
Reus, Spain 33/G2
Reusel, Netherlands 27/G6
Reuss (riv.), Switzerland 39/F2
Reutlingen, W. Germany 22/C4
Reutte, Austria 41/A3
Reva, S. Dak. (57651) 298/C2
Revadim, Israel 65/B4
Reveille (peak), Nev. 266/E5
Reveille (range), Nev. 266/E4
Revel, France 28/E6
Revel (Tallinn), U.S.S.R. 52/B3
Revelo, Ky. (42638) 237/N7
Revelstoke, Br. Col. 162/E5
Revelstoke, Br. Col. 184/J5
Reventazón, Peru 128/B6
Revenue, Sask. 181/B3
Revere, Mass. (02151) 249/D6
Revere, Minn. (56166) 255/C6
Revere, Mo. (63465) 261/J2
Revere, N. Dak. (†58484) 282/O5
Revere, W. Va. (†26158) 312/E5
Reverie, Tenn. (38062) 237/A9
Revillagigedo (chan.), Alaska 196/N2
Revillagigedo (isl.), Alaska 196/N2
Revillagigedo (isls.), Mexico 146/G8
Revillagigedo (isls.), Mexico 2/D5
Revillagigedo (isls.), Mexico 150/C7
Revillo, S. Dak. (57259) 298/R3
Révin, France 28/F3
Revivim, Israel 65/D5
Revúca, Czech. 41/F2
Revuelto (creek), N. Mex. 274/F3
Rew, Pa. (16744) 294/F2
Rewa, India 68/E4
Reward, Sask. 181/B3
Rewataya (reef), Indonesia 85/F7
Rewey, Wis. (53580) 317/F10
Rex, N.C. (28378) 281/M5
Rex, Oreg. (†97132) 291/A2
Rexburg, Idaho (83440) 220/G6
Rexford, Kansas (67753) 232/B2
Rexford, Mont. (59930) 262/A2
Rexton, Mich. (†49734) 250/D2
Rexton, New Bruns. 170/F2
Rexville, Ind. (†47250) 227/G7
Rexville, N.Y. (14877) 276/E6
Rey, Iran 59/F2
Rey, Iran 66/G3
Rey (isl.), Panama 154/H6
Rey Bouba, Cameroon 115/B2
Reydell, Ark. (72133) 202/G5
Reydon, Okla. (73660) 288/G3
Reyes, Bolivia 136/B4
Reyes (pt.), Calif. 204/B6
Reyhanlı, Turkey 63/G4
Reykjanestá (cape), Iceland 7/B2
Reykjanestá (cape), Iceland 21/A2
Reykjavík (cap.), Iceland 4/C11
Reykjavík (cap.), Iceland 2/J2
Reykjavík (cap.), Iceland 21/B1
Reykjavík (cap.), Iceland 7/B2
Reynaud, Sask. 181/F3
Reyno, Ark. (72462) 202/J1
Reynolds, Georgia (31076) 217/D5
Reynolds (creek), Idaho 220/B6
Reynolds, Ill. (61279) 222/C2
Reynolds, Ind. (47980) 227/D3
Reynolds (co.), Mo. 261/L8
Reynolds, Mo. (63666) 261/K8
Reynolds, Nebr. (68429) 264/G4
Reynolds, N. Dak. (58275) 282/R4
Reynolds Bridge, Conn. (†06787) 210/C2
Reynoldsburg, Ohio (43068) 284/E6
Reynolds Station, Ky. (42368) 237/H5
Reynoldsville, Pa. (15851) 294/D3
Reynosa, Mexico 150/K3
Rezaiyeh (Urmia), Iran 66/D2
Reza'iyeh (Urmia), Iran 59/D2
Rezé, France 28/C4
Rēzekne, U.S.S.R. 52/C3
Rēzekne, U.S.S.R. 53/D2
Rhaetian Alps (range), Switzerland 39/J3
Rhame, N. Dak. (58651) 282/C7
Rhätikon (mts.), Liecht. 39/J2
Rhätikon (mts.), Switzerland 39/J2
Rhayader, Wales 13/D5
Rhea (creek), Oreg. 291/H2
Rhea (co.), Tenn. 237/M9
Rheatown, Tenn. (†37641) 237/R8
Rheda-Wiedenbrück, W. Germany 22/C3
Rheden, Netherlands 27/J4
Rheims (Reims), France 28/E4
Rhein, Sask. 181/J4
Rheinau, Switzerland 39/G1
Rheine, W. Germany 22/B2
Rheineck, Switzerland 39/J2
Rheinfeld, Sask. 181/D5
Rheinfelden, Switzerland 39/E1
Rheinfelden, W. Germany 22/B5
Rheinsberg, E. Germany 22/E2
Rheinwaldhorn (mt.), Switzerland 39/G4
Rhems, S.C. (†29440) 296/H4
Rhenen, Netherlands 27/H5
Rhéris, Wadi (dry riv.), Morocco 106/D2
Rheydt, W. Germany 22/B3
Rhine (riv.) 7/E4
Rhine (riv.), Austria 41/A3
Rhine (riv.), France 28/G3
Rhine, Georgia (31077) 217/F7
Rhine (riv.), Liecht. 39/J2
Rhine (riv.), Netherlands 27/J5
Rhine (riv.), Switzerland 39/J2
Rhine (riv.), W. Germany 22/B3
Rhinebeck, N.Y. (12572) 276/N7
Rhinecliff, N.Y. (12574) 276/N7
Rhineland, Mo. (65069) 261/J5
Rhineland, Sask. 181/D5
Rhinelander, Wis. (54501) 317/H4
Rhineland-Palatinate (state), W. Germany 22/B4
Rhinns, The (pen.), Scotland 15/C6
Rhinns (pt.), Scotland 15/B5
Rhino Camp, Uganda 115/F3
Rhir, Wadi (dry riv.), Algeria 106/F2
Rhir (cape), Morocco 106/B2
Rho, Italy 34/B2
Rhode Island 188/M2
RHODE ISLAND 249
Rhode Island (isl.), R.I. 249/J6
Rhode Island (sound), R.I. 249/J7
Rhode Island (state), U.S. 146/M5
Rhodell, W. Va. (25915) 312/D7
Rhodes (Ródhos), Greece 45/J7
Rhodes (isl.), Greece 7/G5
Rhodes (isl.), Greece 45/H7
Rhodes (peak), Idaho 220/D3
Rhodes, Iowa (50234) 229/G5
Rhodes, Mich. (48652) 250/E5
Rhodes Inyanga Nat'l Park, Zimbabwe 118/E3
Rhodes Point, Md. (21858) 245/O9
Rhodhiss, N.C. (28667) 281/F3
Rhododendron, Oreg. (97073) 291/F2
Rhodope (mts.), Bulgaria 45/G5
Rhodope (mts.), Greece 45/G5
Rhome, Texas (76078) 303/E1
Rhön (mts.), E. Germany 22/D3
Rhön (mts.), W. Germany 22/D3
Rhondda, Wales 13/A6
Rhondda, Wales 10/E5
Rhône (dept.), France 28/F5
Rhône (riv.), France 7/E4
Rhône (riv.), France 28/F5
Rhône (riv.), Switzerland 39/D4
Rhosllanerchrugog, Wales 13/D4
Rhu, Scotland 15/A1
Rhu Coigeach (cape), Scotland 15/C2
Rhyl, Wales 13/D4
Rhymney, Wales 13/A6
Rhymney (riv.), Wales 13/B6
Rhynie, Scotland 15/F3
Rhyolite (Ghost Town), Nev. (†89003) 266/E6
Riachão, Brazil 132/E4
Riachuelo, Uruguay 145/B5
Rialto, Calif. (92376) 204/E10
Riana, Tasmania 99/B3
Riaño, Spain 33/D1
Riau (arch.), Indonesia 85/C5
Riaza, Spain 33/E2
Rib (mt.), Wis. 317/G6
Ribadavia, Spain 33/B1
Ribadeo, Spain 33/C1
Ribamar, Brazil 132/F3
Ribas do Rio Pardo, Brazil 132/C8
Ribat Qila, Pakistan 68/A3
Ribat Qila, Pakistan 59/H4
Ribáuè, Mozambique 118/F3
Ribble (riv.), England 10/E4
Ribble (riv.), England 13/E4
Ribe (co.), Denmark 21/B7
Ribe, Denmark 21/B7
Ribe, Denmark 18/F9
Ribeira, Brazil 135/B4
Ribeira (riv.), Brazil 135/B4
Ribeira Brava, Portugal 33/A2
Ribeira de Iguape, Brazil 135/C4
Ribeira de Pena, Portugal 33/C2
Ribeira Grande, C. Verde 106/B7
Ribeirão Preto, Brazil 120/E5
Ribeirão Preto, Brazil 135/C2
Ribeirão Preto, Brazil 132/E8
Ribera, N. Mex. (87560) 274/D3
Ribérac, France 28/D5
Riberach an der Riss, W. Germany 22/C4
Riberalta, Bolivia 136/C2
Riberalta, Bolivia 120/C4
Rib Falls, Wis. (†54426) 317/G6
Ribla, Kuh-e (riv.), Iran 66/J6
Rib Lake, Wis. (54470) 317/F5
Ribnitz-Damgarten, E. Germany 22/E1
Ribstone, Alberta 182/E3
Říčany u Prahy, Czech. 41/C2
Ricaurte, Colombia 126/A7
Riccarton, N. Zealand 100/D5
Rice, Calif. (†92280) 204/L9
Rice (co.), Kansas 232/D3
Rice, Kansas (66965) 232/E2
Rice (co.), Minn. 255/E6
Rice, Minn. (56367) 255/D5
Rice (lake), Minn. 255/E4
Rice (mt.), N.H. 268/E2
Rice (co.), Tenn. 237/M9
Rice (lake), Ontario 177/F3
Rice, Texas (75155) 303/H5
Rice, Va. (23966) 307/M6
Rice, Wash. (99167) 310/G2
Riceboro, Georgia (31323) 217/K7
Rice Lake, Wis. (54868) 317/C5
Rices Landing, Pa. (15357) 294/C6
Riceton, Sask. 181/G5
Ricetown, Ky. (41364) 237/O6
Riceville, Iowa (50466) 229/H2
Riceville, Pa. (16432) 294/C2
Riceville, Tenn. (37370) 237/M10
Rich, Miss. (38662) 256/D2
Rich (cape), Ontario 177/D3
Rich (co.), Utah 304/C2
Richard, Sask. 181/D3
Richard City, Tenn. (†37380) 237/K11
Richard Collinson (inlet), N.W. Terrs. 187/G2
Richards, Iowa (†50579) 229/D4
Richards, Mo. (64778) 261/D7
Richards (isl.), N.W. Terrs. 187/E3
Richards Bay, S. Africa 118/E5
Richards Gebaur A.F.B., Mo. 261/P6
Richards Landing, Ontario 177/J5
Richardson (riv.), Alberta 182/C5
Richardson, Ky. (41253) 237/R5
Richardson (lakes), Maine 243/B6
Richardson (co.), Nebr. 264/J4
Richardson (isls.), N.W. Terrs. 187/G3
Richardson (mts.), N.W. Terrs. 187/E3
Richardson, Sask. 181/G5
Richardson, Texas (75080) 303/G2
Richardson, W. Va. (†26151) 312/D5
Richardson (mts.), Yukon 187/E3
Richardsons Landing, Tenn. (†38023) 237/B10
Richardsville, Ky. (42270) 237/J6
Richardsville, New Bruns. 170/D1
Richard Toll, Senegal 106/A5
Richardton, N. Dak. (58652) 282/F6
Richburg, N.Y. (14774) 276/D6
Richburg, S.C. (29729) 296/E2
Rich Creek, Va. (24147) 307/G6
Richdale, Alberta 182/E4
Riche (pt.), Newf. 166/C3
Richelieu, Ky. (42271) 237/H7
Richelieu (co.), Québec 172/D4
Richelieu, Québec 172/D4
Richer, Manitoba 179/F5
Richey, Mont. (59259) 262/L3
Richfield, Idaho (83349) 220/D6
Richfield, Kansas (67953) 232/A4
Richfield, Minn. (55423) 255/G6
Richfield, N.C. (28137) 281/J4
Richfield, Nova Scotia 168/C4
Richfield, Ohio (44286) 284/G3
Richfield, Pa. (17086) 294/H4
Richfield, Utah (84701) 304/B5
Richfield, Wis. (53076) 317/K1
Richfield Springs, N.Y. (13439) 276/K5
Richford, N.Y. (13835) 276/H6
Richford, Vt. (05476) 268/B2
Richford○, Vt. (05476) 268/B2
Richford, Wis. (†54930) 317/H7
Rich Fountain, Mo. (65070) 261/J6
Richgrove, Calif. (93261) 204/F8
Rich Hill, Mo. (64779) 261/D6
Richhill, N. Ireland 17/H3
Richibucto, New Bruns. 170/E2
Richibucto (harb.), New Bruns. 170/F2
Richibucto (riv.), New Bruns. 170/E2
Richibucto Village, New Bruns. 170/F2
Rich Lake, Alberta 182/E2
Richland, Fla. (†33599) 212/D3
Richland, Georgia (31825) 217/C6
Richland (co.), Ill. 222/E5
Richland, Ind. (47634) 227/C9
Richland (creek), Ind. 227/D6
Richland, Iowa (52585) 229/K6
Richland, Kansas (†66409) 232/G3
Richland (par.), La. 238/G2
Richland, Mich. (49083) 250/D6
Richland, Miss. (†39218) 256/D6
Richland, Mo. (65556) 261/H7
Richland (co.), Mont. 262/M3
Richland, Mont. (59260) 262/K2
Richland, Nebr. (68657) 264/G3
Richland, N.J. (08350) 273/D5
Richland, N.Y. (13144) 276/H3
Richland (co.), N. Dak. 282/R7
Richland (co.), Ohio 284/E4
Richland, Oreg. (97870) 291/K3
Richland, Pa. (17087) 294/K5
Richland (co.), S.C. 296/F4
Richland, S.C. (29675) 296/A2
Richland, S. Dak. (†57025) 298/R8
Richland (creek), Tenn. 237/G10
Richland, Texas (76681) 303/H6
Richland, Wash. 188/B1
Richland, Wash. (99352) 310/F4
Richland (co.), Wis. 317/F9
Richland Balsam (mt.), N.C. 281/D4
Richland Center, Wis. (53581) 317/F9
Richland Hills, Texas (76118) 303/F2
Richland-Kennewick, Wash. 310/B0
Richlands, N.C. (28574) 281/O5
Richlands, Va. (24641) 307/E6
Richland Springs, Texas (76871) 303/F6
Richlandtown, Pa. (18955) 294/M5
Richlea, Sask. 181/C4
Richmond, Ala. (†36761) 195/D6
Richmond, Ark. (†71822) 202/B6
Richmond, Br. Col. 184/K3
Richmond, Calif. (*94801) 204/J1
Richmond, England 13/F3
Richmond, England 10/E3
Richmond (co.), Georgia 217/H4
Richmond, Ill. (60071) 222/E1
Richmond, Ind. (47374) 227/H5
Richmond, Iowa (52247) 229/K6
Richmond, Jamaica 158/J6
Richmond, Kansas (66080) 232/G3
Richmond, Ky. (40475) 237/N5
Richmond, La. (†71282) 238/H2
Richmond, Maine (04357) 243/D7
Richmond○, Maine (04357) 243/D7
Richmond○, Mass. (01254) 249/A3
Richmond, Mich. (48062) 250/G6
Richmond, Minn. (56368) 255/D5
Richmond, Mo. (64085) 261/D4
Richmond○, N.H. (†03470) 268/C6
Richmond (range), N.S. Wales 97/G1
Richmond (riv.), N.S. Wales 97/G1
Richmond (co.), N.Y. 276/M9
Richmond (Staten Island) (borough), N.Y. 276/M9
Richmond, N. Zealand 100/D4
Richmond (range), N. Zealand 100/D4
Richmond (co.), N.C. 281/K4
Richmond (co.), Nova Scotia 168/H3
Richmond (Grand River), Ohio (†44045) 284/H2
Richmond, Ohio (43944) 284/J5
Richmond, Ontario 177/J2
Richmond (co.), Québec 172/E4
Richmond, Québec 172/E4
Richmond, Queensland 88/G4
Richmond, Queensland 95/B4
Richmond (peak), St. Vin. & Grens. 161/A8
Richmond, S. Africa 118/C6
Richmond, Tasmania 99/D4
Richmond, Texas (77469) 303/J8
Richmond, Utah (84333) 304/C2
Richmond, Vt. (05477) 268/A3
Richmond○, Vt. (05477) 268/A3
Richmond, Victoria 88/L7
Richmond, Victoria 97/J5
Richmond (cap.), Va. 188/L3
Richmond (cap.), Va. 146/L6
Richmond (co.), Va. 307/P5
Richmond (cap.) (I.C.), Va. (*23201) 307/O5
Richmond Beach-Innis Arden, Wash. (98160) 310/A1
Richmond Corner, Maine (†04357) 243/D7
Richmond Corner, New Bruns. 170/C2
Richmond Dale, Ohio (45673) 284/E7
Richmond Furnace, Mass. (†01254) 249/A3
Richmond Heights, Fla. (†33158) 212/F6
Richmond Heights, Mo. (63117) 261/P3
Richmond Heights, Ohio (44143) 284/H9
Richmond Highlands, Wash. (†98133) 310/A1
Richmond Hill, Georgia (31324) 217/K7
Richmond Hill, Ontario 177/J4
Richmond Nat'l Battlefield Park, Va. 307/O6
Richmond upon Thames, England 10/B5
Richmond upon Thames, England 13/H8
Richmondville, N.Y. (12149) 276/M5
Richmond-Windsor, N.S. Wales 97/F3
Richmound, Sask. 181/B5
Rich Mountain, Ark. (†71953) 202/B4
Rich Square, N.C. (27869) 281/P2
Richterswil, Switzerland 39/G2
Richthofen (mt.), Colo. 208/G2
Richton, Miss. (39476) 256/G8
Richton Park, Ill. (60471) 222/B6
Richvale, Calif. (95974) 204/D4
Richvalley, Ind. (†46992) 227/F3
Richview, Ill. (62877) 222/D5
Richville, Mich. (48758) 250/F5
Richville, Minn. (56576) 255/C4
Richville, N.Y. (13681) 276/K2
Richwood, La. (†71201) 238/F2
Richwood, Minn. (56577) 255/C4
Richwood, N.J. (08074) 273/C4
Richwood, Ohio (43344) 284/D5
Richwood, W. Va. (26261) 312/F6
Richwood, Wis. (†53094) 317/J9
Richwoods, Mo. (63071) 261/L6
Rickardsville, Iowa (†52039) 229/M3
Rickenbacker Air Force Base, Ohio 284/E6
Ricketts, Iowa (51460) 229/B4
Ricketts (pt.), Victoria 97/J6
Ricketts (pt.), Victoria 88/L8
Rickman, Tenn. (38580) 237/L8
Rickmansworth, England 13/G8
Rickmansworth, England 10/A5
Rickreall, Oreg. (97371) 291/D3
Ricla, Spain 33/F2
Rico, Colo. (81332) 208/C7
Ricobayo (res.), Spain 33/D2
Ricse, Hungary 41/G2
Ridderkerk, Netherlands 27/F5
Riddle, Idaho (†89832) 220/B7
Riddle, Oreg. (97469) 291/D5
Riddlesburg, Pa. (16672) 294/F5
Riddleton, Tenn. (37151) 237/J8
Riddleville, Georgia (†31018) 217/G5
Riddon, Loch (inlet), Scotland 15/C5
Rideau (lake), Ontario 177/H3
Riderwood, Ala. (†36904) 195/B6
Ridge, Md. (20680) 245/N8
Ridge, Mont. (†59314) 262/M5
Ridgecrest, Calif. (93555) 204/H8
Ridgecrest, La. (†71334) 238/G3
Ridgedale, Mo. (65739) 261/F9
Ridgedale, Sask. 181/H2
Ridge Farm, Ill. (61870) 222/F4
Ridgefield, Conn. (06877) 210/B3
Ridgefield○, Conn. (06877) 210/B3
Ridgefield, N.J. (07657) 273/B2
Ridgefield, Wash. (98642) 310/C5
Ridgefield Park, N.J. (07660) 273/B2
Ridgeland, Miss. (39157) 256/D6
Ridgeland, S.C. (29936) 296/E7
Ridgeland, Wis. (54763) 317/B5
Ridgeley, W. Va. (26753) 312/J3
Ridgely, Md. (21660) 245/P5
Ridgely, Mo. (†64444) 261/C4
Ridgely, Tenn. (38080) 237/B8
Ridgeside, Tenn. (†37401) 237/L10
Ridge Spring, S.C. (29129) 296/D4
Ridgetop, Tenn. (37152) 237/H8
Ridgetown, Ontario 177/C5
Ridgeview, S. Dak. (57652) 298/H3
Ridgeville, Georgia (†31331) 217/K8
Ridgeville, Ind. (47380) 227/G4
Ridgeville, Manitoba 179/E5
Ridgeville, S.C. (29472) 296/G5
Ridgeville Corners, Ohio (43555) 284/B3
Ridgeway, Iowa (52165) 229/K2
Ridgeway, Minn. (†55943) 255/G7
Ridgeway, Mo. (64481) 261/D2
Ridgeway, N.C. (27570) 281/N2
Ridgeway, Ohio (43345) 284/C4
Ridgeway, S.C. (29130) 296/F3
Ridgeway, Va. (24148) 307/J7
Ridgeway, W. Va. (25440) 312/K4
Ridgeway, Wis. (53582) 317/F10
Ridgeway Branch, Toms (riv.), N.J. 273/E3
Ridgewood, N.J. (*07450) 273/B1
Ridgley, Tasmania 99/B3
Ridgway, Colo. (81432) 208/D6
Ridgway, Ill. (62979) 222/E6
Ridgway, Pa. (15853) 294/E3
Ridi, Nepal 68/E3
Riding (mt.), Manitoba 179/B4
Riding Mountain, Manitoba 179/C4
Riding Mountain Nat'l Park, Man. 162/F5
Riding Mountain Nat'l Park, Manitoba 179/B4
Ridley, Tenn. (†38474) 237/G9
Ridley Park, Pa. (19078) 294/M7
Ridott, Ill. (61067) 222/D1
Ridotto, Iowa (†50546) 229/D3
Ried im Innkreis, Austria 41/B2
Riegelsville, N.J. (†08865) 273/C2
Riegelsville, Pa. (18077) 294/M4
Riegelwood, N.C. (28456) 281/N6
Riehen, Switzerland 39/E1
Rienzi, Miss. (38865) 256/G1
Riesa, E. Germany 22/E3
Riesco (isl.), Chile 138/E10
Riesel, Texas (76682) 303/H6
Riesi, Italy 34/E6
Rietavas, U.S.S.R. 53/A3
Rietberg, W. Germany 22/C3
Rietfontein, Namibia 118/C4
Rieth, Oreg. (†97801) 291/J2
Rieti (prov.), Italy 34/D3
Rieti, Italy 34/D3
Rif, Er (range), Morocco 106/D2

Riffelalp, Switzerland 39/E5
Rifle, Colo. (81650) 208/D3
Rifle (creek), Colo. 208/D3
Rifle (riv.), Mich. 250/E4
Rifle (lake), Wash. 310/C4
Rifstangi (cape), Iceland 21/C1
Rift Valley (prov.), Kenya 115/G3
Riga (lake), Conn. 210/B1
Riga, U.S.S.R. 2/L3
Riga, U.S.S.R. 7/G3
Riga (gulf), U.S.S.R. 7/G3
Riga (cap.), U.S.S.R. 53/C2
Riga, U.S.S.R. 48/C4
Riga, U.S.S.R. 52/B3
Riga (gulf), U.S.S.R. 52/B3
Riga (gulf), U.S.S.R. 53/B2
Riga (gulf), U.S.S.R. 48/C4
Rigan, Iran 66/L6
Rigaud, Québec 172/C4
Rigby, Idaho (83442) 220/F6
Rigdon, Ind. (†46928) 227/F4
Rigestan (reg.), Afghanistan 59/H3
Riggins, Idaho (83549) 220/B4
Riggisberg, Switzerland 39/E3
Rigi (mt.), Switzerland 39/F2
Rigo, Papua N.G. 85/C7
Rigolet, Newf. 166/C3
Rigolet, Newf. 162/L5
Rig Rig, Chad 111/B5
Rigside, Scotland 15/E5
Riihimäki, Finland 18/O6
Riiser-Larsen (pen.), Ant. 2/L9
Riiser-Larsen (pen.) 5/C2
Rijeka, Yugoslavia 45/B3
Rijeka, Yugoslavia 7/F4
Rijen, Netherlands 27/F5
Rijnsburg, Netherlands 27/F4
Rijssen, Netherlands 27/J4
Rijswijk, Netherlands 27/E4
Rikitea, Fr. Poly. 87/N8
Rikuchu-Kaigan National Park, Japan 81/L4
Rikuzentakata, Japan 81/K4
Riley, Ind. (47871) 227/C6
Riley (co.), Kansas 232/F2
Riley, Kansas (66531) 232/F2
Riley, Ky. (†40328) 237/L5
Riley, Maine (†04262) 243/C6
Riley, Oreg. (97758) 291/H4
Riley Brook, New Bruns. 170/C1
Rileysburg, Ind. (†47932) 227/B4
Rillito, Ariz. (85246) 198/D6
Rillton, Pa. (15678) 294/C5
Rima (riv.), Niger 106/F6
Rima (riv.), Nigeria 106/F6
Rima, Wadi (dry riv.), Saudi Arabia 59/D4
Rimac (riv.), Peru 128/D9
Rimal, Ar (des.), Saudi Arabia 59/F5
Rimatara (isl.), Fr. Poly. 87/L8
Rimavská Sobota, Czech. 41/E2
Rimbey, Alberta 182/C3
Rimbo, Sweden 18/L7
Rimersburg, Pa. (16248) 294/D3
Rimini, Italy 34/D2
Rimini, S.C. (29131) 296/G4
Rîmnicu Sărat, Romania 45/H3
Rîmnicu Vîlcea, Romania 45/G3
Rimouski, Que. 162/K6
Rimouski (co.), Québec 172/J1
Rimouski (county), Québec 174/D3
Rimouski, Québec 172/J1
Rimouski, Québec 174/D3
Rimouski (riv.), Québec 172/J1
Rimouski-Est, Québec 172/J1
Rimpfischhorn (mt.), Switzerland 39/E4
Rimrock, Ariz. (86335) 198/D4
Rimrock (lake), Wash. 310/D4
Rimutaka (range), N. Zealand 100/B3
Rinard, Ill. (62878) 222/E5
Rinard, Iowa (50587) 229/D4
Rincón, Cerro (mt.), Argentina 143/C1
Rincon (peak), Ariz. 198/E6
Rincón, Cerro (mt.), Chile 138/C4
Rincón, Dom. Rep. 158/E5
Rincón (bay), Dom. Rep. 158/F5
Rincon, Georgia (31326) 217/K6
Rincon, Neth. Ant. 161/E8
Rincon, N. Mex. (87940) 274/C6
Rincón (pt.), Panama 154/G6
Rincón, P. Rico 161/A1
Rincón (bay), P. Rico 161/D3
Rinconada, Argentina 143/C1
Rinconada San Martín, Chile 138/G2
Rincón de Romos, Mexico 150/H5
Rindge○, N.H. (03461) 268/C6
Riner, Va. (24149) 307/H6
Rineyville, Ky. (40162) 237/K5
Ringarooma, Tasmania 99/D3
Ringarooma (bay), Tasmania 99/D2
Ringe, Denmark 21/D7
Ringebu, Norway 18/G6
Ringelspitz (mt.), Switzerland 39/H3
Ringerike, Norway 18/C3
Ringgold, Georgia (30736) 217/B1
Ringgold (co.), Iowa 229/E7
Ringgold, La. (71068) 238/D2
Ringgold, Md. (†21783) 245/H2
Ringgold, Nebr. (†69167) 264/D3
Ringgold, Texas (76261) 303/G4
Ringgold, Va. (24586) 307/K7
Ringim, Nigeria 106/F6
Ringkøbing (co.), Denmark 21/B5
Ringkøbing, Denmark 18/E8
Ringkøbing, Denmark 21/A5
Ringkøbing (fjord), Denmark 21/B6
Ringling, Mont. (59642) 262/F4
Ringling, Okla. (73456) 288/L6
Ringmer, England 13/H7
Ringoes, N.J. (08551) 273/D3
Ringold, Okla. (74754) 288/R6
Ringsted, Denmark 21/E7
Ringsted, Iowa (50578) 229/D2
Ringtown, Pa. (17967) 294/K4
Ringvassøy (isl.), Norway 18/L2
Ringwood, England 13/F7
Ringwood, Ill. (60072) 222/E1
Ringwood, N.J. (07456) 273/E1
Ringwood, N.C. (†27823) 281/O2
Ringwood, North. Terr. 93/D7
Ringwood, Okla. (73768) 288/K2
Ringwood, Victoria 88/M7
Ringwood, Victoria 97/K5
Rinn (lake), Ireland 17/F4
Rinteln, W. Germany 22/C2
Rio, Ill. (61472) 222/C2
Rio, La. (†70427) 238/L5
Rio, W. Va. (26755) 312/J4
Rio, Wis. (53960) 317/H9
Rio Arriba (co.), N. Mex. 274/B2
Riobamba, Ecuador 128/C3
Riobamba, Ecuador 120/B3
Río Blanco, Chile 138/B9
Rio Blanco (co.), Colo. 208/C3
Rio Blanco, Colo. (†81650) 208/C3
Río Blanco, P. Rico 161/F2
Rio Bonito, Brazil 135/E3
Rio Branco, Brazil 120/C3
Rio Branco, Brazil 132/G10
Rio Branco, Uruguay 145/F3
Rio Brazos (riv.), N. Mex. 274/C2
Rio Brilhante, Brazil 132/C8
Río Bueno, Chile 138/D3
Rio Bueno, Jamaica 158/H5
Río Caribe, Venezuela 124/G2
Río Cauto, Cuba 158/H4
Rio Chama (riv.), N. Mex. 274/C2
Río Chico, Venezuela 124/F2
Río Cisnes, Chile 138/E5
Rio Claro, Brazil 132/E8
Rio Claro, Brazil 135/C3
Rio Claro, Trin. & Tob. 161/B11
Río Claro, Venezuela 124/D3
Río Colorado, Argentina 120/C6
Río Colorado, La Pampa, Argentina 143/D4
Río Colorado, Río Negro, Argentina 143/D4
Rio Creek, Wis. (54231) 317/L6
Río Cuarto, Argentina 143/D3
Río Cuarto, Argentina 120/C6
Rio de Janeiro (state), Brazil 135/E3
Rio de Janeiro (state), Brazil 132/F8
Rio de Janeiro, Brazil 2/G7
Rio de Janeiro, Brazil 120/E5
Rio de Janeiro, Brazil 135/E3
Rio de Janeiro, Brazil 132/F8
Rio Dell, Calif. (95562) 204/A3
Río de Oro, Colombia 126/D3
Rio do Sul, Brazil 132/D9
Rio Felix (riv.), N. Mex. 274/E5
Río Gallegos, Argentina 120/C8
Río Gallegos, Argentina 143/C7
Rio Grande (riv.) 2/D4
Rio Grande (riv.) 146/H7
Rio Grande (riv.) 188/F5
Río Grande, Argentina 143/C7
Río Grande, Bolivia 136/B7
Rio Grande, Brazil 120/D6
Rio Grande, Brazil 132/D11
Rio Grande (co.), Colo. 208/G7
Rio Grande (res.), Colo. 208/E7
Rio Grande (riv.), Colo. 208/H8
Rio Grande, N.J. (08242) 273/D5
Rio Grande (riv.), N. Mex. 274/C5
Rio Grande, Ohio (45674) 284/F8
Río Grande, P. Rico 161/E1
Rio Grande (riv.), Texas 303/D9
Rio Grande City, Texas (78582) 303/F11
Rio Grande do Norte (state), Brazil 132/G4
Rio Grande do Sul (state), Brazil 132/C10
Rio Grande Pyramid (mt.), Colo. 208/E7
Rio Grande Wild and Scenic River, Texas 303/B8
Riohacha, Colombia 120/B1
Riohacha, Colombia 126/D2
Río Hondo, Guatemala 154/C3
Rio Hondo (riv.), N. Mex. 274/E5
Rio Hondo, Texas (78583) 303/G11
Rioja, Peru 128/D6
Río Lagartos, Mexico 150/P6
Rio Linda, Calif. (95673) 204/B8
Rio Maior, Portugal 33/B3
Río Mulato, Bolivia 136/B6
Río Muni (terr.), Equat. Guinea 115/B3
Rion, S.C. (29132) 296/E3
Riondel, Br. Col. 184/J5
Río Negro (prov.), Argentina 143/C5
Rio Negro, Brazil 132/D9
Río Negro, Chile 138/D3
Rionegro, Antioquia, Colombia 126/C4
Rionegro, Santander, Colombia 126/D4
Río Negro (dept.), Uruguay 145/B3
Río Negro (res.), Uruguay 145/D3
Rionero in Vulture, Italy 34/E4
Rio Pardo, Brazil 132/C10
Rio Pardo de Minas, Brazil 132/F6
Rio Penasco (riv.), N. Mex. 274/E6
Río Piedras, P. Rico 161/E1
Rio Pomba, Brazil 135/E2
Rio Puerco (riv.), N. Mex. 274/C4
Rio Rancho, N. Mex. (87124) 274/C3
Rio Real, Brazil 132/G5
Rio Rico, Ariz. (85621) 198/E7
Rio Salado (riv.), N. Mex. 274/B4
Río San Juan, Dom. Rep. 158/E5
Río Seco, Cuba 158/A2
Río Segundo, Argentina 143/D3
Riosucio, Caldas, Colombia 126/C5
Riosucio, Chocó, Colombia 126/B4
Río Tercero, Argentina 143/D3
Río Tigre, Ecuador 128/D4
Rio Tinto, Brazil 132/H4
Río Tocuyo, Venezuela 124/C2
Riou (lake), Sask. 181/M2
Rio Verde, Brazil 132/D7
Rio Verde, Brazil 120/D4
Río Verde, Chile 138/E10
Ríoverde, Mexico 150/J6
Rio Verde de Mato Grosso, Brazil 132/C7
Rio Vista, Calif. (94571) 204/L1
Riparia, Wash. (†99359) 310/G4
Riparius, N.Y. (12862) 276/M3
Ripley, Calif. (92272) 204/L10
Ripley, England 13/F4
Ripley, Ill. (†62353) 222/C3
Ripley (co.), Ind. 227/G6
Ripley○, Maine (†04930) 243/E5
Ripley, Miss. (38663) 256/G1
Ripley (co.), Mo. 261/L9
Ripley, N.Y. (14775) 276/A6
Ripley, Ohio (45167) 284/C8
Ripley, Okla. (74062) 288/N2
Ripley, Ontario 177/C3
Ripley, Tenn. (38063) 237/B9
Ripley, W. Va. (25271) 312/C5
Riplinger, Wis. (†54479) 317/E6
Ripoll, Spain 33/H1
Ripon, Calif. (95366) 204/D6
Ripon, England 10/F3
Ripon, England 13/F3
Ripon, Québec 172/B4
Ripon, Wis. (54971) 317/J8
Rippey, Iowa (50235) 229/E5
Ripplemead, Va. (24150) 307/G6
Ripples, New Bruns. 170/D3
Rippon, W. Va. (25441) 312/L4
Rippowam (riv.), Conn. 210/A4
Ripton○, Vt. (05766) 268/A4
Ririe, Idaho (83443) 220/G6
Risafe, Syria 63/H5
Risalpur Cantonment, Pakistan 68/C2
Risaralda (dept.), Colombia 126/B5
Risca, Wales 13/B6
Risco, Mo. (63874) 261/N9
Rishiri (isl.), Japan 81/K1
Rishon Le Ziyyon, Israel 65/B4
Rishra, India 68/F1
Rising City, Nebr. (68658) 264/G3
Rising Fawn, Georgia (30738) 217/A1
Rising Star, Texas (76471) 303/F5
Rising Sun, Ind. (47040) 227/H7
Rising Sun, Md. (21911) 245/O2
Risingsun, Ohio (43457) 284/C3
Rising Sun, Wis. (†54628) 317/D9
Risle (riv.), France 28/D3
Rison, Ark. (71665) 202/F6
Risør, Norway 18/F7
Risoux (mt.), Switzerland 39/B3
Ristigouche (riv.), Québec 172/B2
Ristijärvi, Finland 18/Q4
Rita Blanca (creek), Texas 303/B2
Ritchey, Mo. (†64844) 261/D9
Ritchie, Md. (†20027) 245/G5
Ritchie (co.), W. Va. 312/D4
Ritchies (arch.), India 68/G6
Ritidian (pt.), Guam 86/K6
Ritner, Ky. (42639) 237/M7
Ritter, Oreg. (97872) 291/H3
Ritter, S.C. (29488) 296/F6
Rittman, Ohio (44270) 284/G4
Ritzville, Wash. (99169) 310/G3
Rivadavia, Mendoza, Argentina 143/C3
Rivadavia, Salta, Argentina 143/D1
Rivadavia, San Juan, Argentina 143/C3
Rivadavia, Chile 138/A7
Riva del Garda, Italy 34/C2
Rivanna (riv.), Va. 307/M5
Rivas, Nicaragua 154/E5
Riva San Vitale, Switzerland 39/G5
Rive-de-Gier, France 28/F5
Rivera, Switzerland 39/G4
Rivera (dept.), Uruguay 145/D2
Rivera, Uruguay 145/D1
Rivera, Uruguay 120/D6
Riverbank, Calif. (95367) 204/E6
River Bourgeois, Nova Scotia 168/H3
River Cess, Liberia 106/C7
Rivercourse, Alberta 182/E3
Riverdale, Calif. (93656) 204/E7
Riverdale, Georgia (*30274) 217/K2
Riverdale, Ill. (60627) 222/C6
Riverdale, Iowa (†52722) 229/N5
Riverdale, Kansas (†67152) 232/E4
Riverdale, Md. (20840) 245/F4
Riverdale, Mich. (48877) 250/E5
Riverdale, Nebr. (68870) 264/E4
Riverdale, N.H. (†03045) 268/D5
Riverdale, N.J. (07457) 273/A1
Riverdale, N. Dak. (58565) 282/H4
Riverdale Heights, Md. (†20840) 245/G4
River de Chute, New Bruns. 170/C2
River Denys, Nova Scotia 168/G3
River Edge, N.J. (07661) 273/B1
River Falls, Ala. (36476) 195/E8
River Falls, Wis. (54022) 317/A6
River Forest, Ill. (60305) 222/B5
River Forest, Ind. (†46011) 227/F4
River Glade, New Bruns. 170/E3
River Grove, Ill. (60171) 222/B5
River Grove, Oreg. (†97223) 291/B2
Riverhead, Newf. 166/D2
Riverhead, N.Y. (11901) 276/P9
Riverhead, N. Zealand 100/B1
River Hébert, Nova Scotia 168/D3
River Heights, Utah (†84321) 304/C2
River Hills, Manitoba 179/G4
River Hills, Wis. (†53201) 317/M1
Riverhurst, Sask. 181/E5
Riverina (reg.), N. S. Wales 88/H7
Riverina (reg.), N.S. Wales 97/C4
River John, Nova Scotia 168/E3
River Jordan, Br. Col. 184/J3
Riverland, Fla. (†33301) 212/B4
Riverlea, Ohio (†43085) 284/D5
Rivermines, Mo. (†63601) 261/L7
Rivero (isl.), Chile 138/D6
River Oaks, Texas (77019) 303/E2
River of Ponds, Newf. 166/C3
River of Ponds (lake), Newf. 166/C3
Riverport, Nova Scotia 168/D4
River Rouge, Mich. (48218) 250/B7
Rivers (inlet), Br. Col. 184/D4
Rivers, Manitoba 179/B4
Rivers (state), Nigeria 106/F8
Rivers (lake), Sask. 181/F6
Riversdale, Jamaica 158/J6
Riversdale, S. Africa 118/H7
Riversdale, S. Africa 118/C6
Riverside, Ala. (35135) 195/F3
Riverside, Calif. 188/C4
Riverside (co.), Calif. 204/J10
Riverside, Calif. (*92501) 204/E11
Riverside (res.), Colo. 208/L2
Riverside, Conn. (06878) 210/A4
Riverside, Georgia (†30759) 217/B2
Riverside, Georgia (†31768) 217/E8
Riverside, Ill. (60546) 222/B6
Riverside, Ind. (†47918) 227/C4
Riverside, Iowa (52327) 229/K6
Riverside, Ky. (42272) 237/J6
Riverside, Md. (†20662) 245/K7
Riverside, Mass. (†01376) 249/D2
Riverside, Mich. (49084) 250/C6
Riverside, Mo. (64168) 261/O5
Riverside○, N.J. (08075) 273/B3
Riverside, N.Y. (†14830) 276/F6
Riverside, N. Dak. (†58078) 282/S6
Riverside, Oreg. (97917) 291/J4
Riverside, Pa. (17868) 294/J4
Riverside, R.I. (02915) 249/J5
Riverside, Sask. 181/G6
Riverside, Texas (77367) 303/J7
Riverside, Utah (84334) 304/B2
Riverside, Wash. (98849) 310/F2
Riverside, Wyo. (†82325) 319/F4
Riverside-Albert, New Bruns. 170/F3
Riverside Stage Stop, Ariz. (85237) 198/D5
Rivers Inlet, Br. Col. 184/D4
River Sioux, Iowa (†51545) 229/B5
Riverstown, Ireland 17/E3
Riverton, Conn. (06065) 210/D1
Riverton, Ill. (62561) 222/D4
Riverton, Ind. (†47861) 227/B6
Riverton, Iowa (51650) 229/B7
Riverton, Kansas (66770) 232/H4
Riverton, La. (†71418) 238/F2
Riverton, Man. 162/G5
Riverton, Manitoba 179/E3
Riverton, Minn. (†56455) 255/D4
Riverton, Nebr. (68972) 264/F4
Riverton, N.J. (08077) 273/B3
Riverton, N. Zealand 100/B7
Riverton, Nova Scotia 168/F3
Riverton, Oreg. (†97423) 291/C4
Riverton, Utah (84065) 304/B3
Riverton, Vt. (05668) 268/B3
Riverton, Va. (22651) 307/M3
Riverton, Wash. (†98188) 310/B2
Riverton, W. Va. (26814) 312/H5
Riverton, Wyo. (82501) 319/D2
Riverton Heights, Wash. (98188) 310/B2
Rivervale, Ark. (72377) 202/K2
River Vale○, N.J. (07675) 273/B1
River Valley, Ontario 177/D1
Riverview, Ala. (†36426) 195/D8
River View, Ala. (36872) 195/H5
Riverview, Fla. (33569) 212/D4
Riverview, Mich. (48192) 250/B7
Riverview, Mo. (†63101) 261/R2
Riverview, New Bruns. 170/F2
Riverville, Va. (†24553) 307/L5
Riverwood, Ky. (†40222) 237/K1
Riverwoods, Ill. (†60015) 222/B5
Rives, Mo. (63875) 261/M10
Rives, Tenn. (38253) 237/C8
Rives Junction, Mich. (49277) 250/E6
Rivesville, W. Va. (26588) 312/F3
Riviera (reg.), France 28/G6
Riviera, Texas (78379) 303/G10
Riviera Beach, Fla. (33404) 212/G5
Riviera Beach, Md. (†21061) 245/N4
Riviera-Bullhead, Ariz. (86440) 198/A3
Rivière-à-Claude, Québec 172/C1
Rivière-à-Pierre, Québec 172/E3
Rivière-au-Renard, Québec 172/D1
Rivière-au-Tonnerre, Québec 174/D2
Rivière-Bleue, Québec 172/J2
Rivière-Bois-Clair, Québec 172/F3
Rivière-du-Loup, Que. 162/K6
Rivière-du-Loup (co.), Québec 172/H2
Rivière-du-Loup, Québec 174/D3
Rivière-du-Loup, Québec 172/H2
Rivière-du-Moulin, Québec 172/G1
Rivière-du-Portage, New Bruns. 170/F1
Rivière-Éternité, Québec 172/G1
Rivière-la-Madeleine, Québec 172/C1
Rivière-Matawin, Québec 172/E3
Rivière-Ouelle, Québec 172/G2
Rivière-Pentecôte, Québec 174/D3
Rivière-Pilote, Martinique 161/D7
Rivière-Port-Daniel, Québec 172/D2
Rivière-Portneuf, Québec 172/H1
Rivière-Saint-Paul, Québec 174/F2
Rivière-Salée, Martinique 161/D7
Rivière-Trois-Pistoles, Québec 172/J1
Rivière Verte, New Bruns. 170/B1
Rivière-Verte, Québec 172/H2
Riwaka, N. Zealand 100/D4
Riwoqê, China 77/E5
Rixeyville, Va. (22737) 307/M3
Rixford, Pa. (16745) 294/F2
Riyadh (cap.), Saudi Arabia 2/M4
Riyadh (cap.), Saudi Arabia 54/F7
Riyadh (cap.), Saudi Arabia 59/E5
Riyan, P.D.R. Yemen 59/E7
Rizal (prov.), Philippines 82/C3
Rize (prov.), Turkey 63/J2
Rize, Turkey 59/D1
Rize, Turkey 63/J2
Rizokarpaso, Cyprus 63/F5
Rjukan, Norway 18/F7
Roa, Norway 18/G6
Roa, Spain 33/E2
Roachdale, Ind. (46172) 227/D5
Road (bay), Virgin Is. (Br.) 161/D3
Roadside, Scotland 15/E4
Roadstown, N.J. (†08302) 273/C5
Road Town (cap.), Virgin Is. (Br.) 161/D3
Road Town (cap.), Virgin Is. (Br.) 156/H1
Roag, Loch (inlet), Scotland 15/B2
Roan (creek), Colo. 208/C4
Roan (plat.), Colo. 208/B3
Roan, Norway 18/G4
Roan (isl.), Scotland 15/D2
Roan (cliffs), Utah 304/E4
Roane (co.), Tenn. 237/M9
Roane (co.), W. Va. 312/D5
Roan Mountain, Tenn. (37687) 237/S8
Roann, Ind. (46974) 227/F3
Roanne, France 28/E4
Roanoke (riv.) 188/L3
Roanoke, Ala. (36274) 195/H4
Roanoke, Ill. (61561) 222/D3
Roanoke, Ind. (46783) 227/G3
Roanoke, La. (70581) 238/E6
Roanoke, Mo. (†65230) 261/G4
Roanoke (isl.), N.C. 281/T3
Roanoke (riv.), N.C. 281/P2
Roanoke, Texas (76262) 303/F1
Roanoke, Va. 146/L6
Roanoke, Va. 188/K3
Roanoke (co.), Va. 307/H6
Roanoke (I.C.), Va. (*24001) 307/H6
Roanoke (riv.), Va. 307/N8
Roanoke, W. Va. (26423) 312/F5
Roanoke Rapids, N.C. (27870) 281/O2
Roaring (brook), Conn. 210/F1
Roaring (brook), Conn. 210/E2
Roaring Branch, Pa. (17765) 294/J2
Roaring Fork, Colorado (riv.), Colo. 208/E4
Roaring Gap, N.C. (28668) 281/H2
Roaring River, N.C. (28669) 281/G2
Roaring Spring, Pa. (16673) 294/F5
Roaring Springs, Texas (79256) 303/D4
Roaringwater (bay), Ireland 17/B9
Roark, Ky. (40979) 237/P6
Roatán, Honduras 154/D2
Roatán (isl.), Honduras 154/D2
Roba, Ala. (†36089) 195/G6
Robards, Ky. (42452) 237/F5
Robat Karim, Iran 66/G3
Robb, Alberta 182/B3
Robben (isl.), S. Africa 118/E6
Robbins, Calif. (95676) 204/B8
Robbins, Ill. (60472) 222/B6
Robbins, N.C. (27325) 281/K4
Robbins (isl.), Tasmania 99/B2
Robbins, Tenn. (37852) 237/M8
Robbinsdale, Minn. (55422) 255/G5
Robbinston, Maine (04671) 243/J5
Robbinston○, Maine (04671) 243/J5
Robbinsville, N.J. (08691) 273/D3
Robbinsville, N.C. (28771) 281/B4
Robe (mt.), N.S. Wales 97/A2
Robe, S. Australia 94/F7
Robe, Wash. (†98252) 310/D2
Robeline, La. (71469) 238/D3
Roberdel, N.C. (†28379) 281/K5
Robersonville, N.C. (27871) 281/P3
Robert (isl.), China 85/E2
Robert, La. (70455) 238/N1
Robert (harb.), Martinique 161/D6
Roberta, Georgia (31078) 217/D5
Roberta, Okla. (†74701) /K7
Robert Lee, Texas (76945) 303/D6
Roberto Payán, Colombia 126/A7
Roberts, Idaho (83444) 220/F6
Roberts, Ill. (60962) 222/E3
Roberts, Mont. (59070) 262/G5
Roberts (co.), S. Dak. 298/P2
Roberts (co.), Texas 303/D2
Roberts, Wis. (54023) 317/A6
Robert's Arm, Newf. 166/C4
Robertsburg, W. Va. (25172) 312/C5
Roberts Creek, Br. Col. 184/J3
Robertsdale, Ala. (36567) 195/C9
Robertsdale, Pa. (16674) 294/F5
Roberts Field Int'l Airport, Liberia 106/C7
Robertsfors, Sweden 18/M4
Robertsganj, India 68/E4
Robertson (co.), Ky. 237/N3
Robertson, S. Africa 118/C6
Robertson (co.), Tenn. 237/H7
Robertson (co.), Texas 303/H6
Robertson, Wyo. (82944) 319/B4
Robertsonville, Québec 172/F3
Robertsport, Liberia 102/A4
Robertsport, Liberia 106/B7
Robertstown, Georgia (†30545) 217/E1
Robertsville, Conn. (†06098) 210/C1
Robertsville, Ohio (44670) 284/H4
Robertville, New Bruns. 170/E1
Roberval, Que. 162/J6
Roberval, Québec 174/C3
Roberval, Québec 172/E1
Robeson (co.), N.C. 281/L5
Robeson (chan.), N.W. Terrs. 187/M1
Robesonia, Pa. (19551) 294/K5
Robichaud, New Bruns. 170/F2
Robinhood, Sask. 181/C2
Robins, Iowa (52328) 229/K4
Robins, Ohio (†43723) 284/H6
Robins A.F.B., Georgia 217/F5
Robinson, Ill. (62454) 222/F5
Robinson, Iowa (†52330) 229/K4
Robinson, Kansas (66532) 232/G2
Robinson, Ky. (41082) 237/N4
Robinson, N. Dak. (58478) 282/L5
Robinson (riv.), North. Terr. 93/E4
Robinson, Pa. (15949) 294/D5
Robinson (lake), S.C. 296/G3
Robinson (ranges), W. Australia 92/B4
Robinson Creek, Ky. (41560) 237/S6
Robinson Crusoe (isl.), Chile 120/B6
Robinson River, North. Terr. 93/E4
Robinsons, Maine (†04734) 243/H3
Robinsonville, Miss. (38664) 256/D1
Robinsonville, New Bruns. 170/C1
Robinvale, Victoria 97/B4
Robles, Colombia 126/D2
Roblin, Manitoba 179/A3
Roblin, Ontario 177/G3
Roboré, Bolivia 136/F6
Roboré, Bolivia 120/D4
Rob Roy, Ind. (†47918) 227/C4
Robsart, Sask. 181/B6
Robson (mt.), Br. Col. 162/D5
Robson, Br. Col. 184/J5
Robson (mt.), Br. Col. 184/H3
Robstown, Texas (78380) 303/G10
Roby, Mo. (65557) 261/H7
Roby, Texas (79543) 303/D5
Roca, Nebr. (68430) 264/H4
Roca (cape), Portugal 33/B3
Rocafuerte, Ecuador 128/B3
Rocanville, Sask. 181/K5
Roca Partida (isl.), Mexico 150/C7
Roca que Vela (cay), Colombia 126/B8
Rocas (isl.), Brazil 120/F3
Rocas de Santo Domingo, Chile 138/F4
Roccastrada, Italy 34/C3
Rocha (dept.), Uruguay 145/E4
Rocha, Uruguay 145/E5
Rocha (lag.), Uruguay 145/E5
Rochdale, England 13/H2
Rochdale, England 10/G2
Rochdale, Mass. (01542) 249/G4
Roche, Switzerland 39/C4
Rochechouart, France 28/D5
Rochefort, Belgium 27/G8
Rochefort, France 28/C4
Roche Harbor, Wash. (98250) 310/B2
Rochelle, Georgia (31079) 217/F7
Rochelle, Ill. (61068) 222/D2
Rochelle, Texas (76872) 303/E6
Rochelle, Wyo. (†82701) 319/H2
Rochelle Park○, N.J. (07662) 273/B2
Roche Percé, Sask. 181/J6
Rocheport, Mo. (65279) 261/H5
Rocher River, N.W.T. 162/E3
Rocher River, N.W. Terrs. 187/G3
Rochert, Minn. (56578) 255/C4
Rochester, Alberta 182/D2
Rochester, England 13/J8
Rochester, England 10/G5
Rochester, Ill. (62563) 222/D4
Rochester, Ind. (46975) 227/E2
Rochester, Iowa (†52772) 229/L5
Rochester, Ky. (42273) 237/H6
Rochester○, Mass. (02770) 249/L6
Rochester, Mich. (48063) 250/F6
Rochester, Minn. 188/H2
Rochester, Minn. (55901) 255/F6
Rochester, N.H. (03867) 268/E5
Rochester, N.Y. 188/L2
Rochester, N.Y. 146/L5
Rochester, N.Y. (*14601) 276/E4
Rochester, Ohio (†44090) 284/F3
Rochester, Pa. (15074) 294/B4
Rochester, Texas (79544) 303/E4
Rochester○, Vt. (05767) 268/B4
Rochester, Victoria 97/C5
Rochester, Wash. (98579) 310/C4
Rochester, Wis. (53167) 317/K3
Rochester Mills, Pa. (15771) 294/D4
Rochford, S. Dak. (57778) 298/B5
Rochfort Bridge, Alberta 182/C3
Rochon Sands, Alberta 182/D3
Rociada, N. Mex. (87742) 274/D3
Rock (creek), Idaho 220/F7
Rock (creek), Ill. 222/D2
Rock (riv.), Ill. 222/C2
Rock (riv.), Iowa 229/A2
Rock, Kansas (67131) 232/F4
Rock (lake), Manitoba 179/C5
Rock (creek), Md. 245/K4
Rock, Mass. (†02346) 249/L5
Rock, Mich. (49880) 250/B2
Rock (co.), Minn. 255/B7
Rock (riv.), Minn. 255/B7
Rock (creek), Mont. 262/C4
Rock (co.), Nebr. 264/E2
Rock (creek), Nev. 266/E2
Rock (creek), Oreg. 291/E4
Rock (creek), Oreg. 291/G2
Rock (creek), Oreg. 291/H3
Rock (creek), S. Dak. 298/O6
Rock (creek), Wash. 310/H3
Rock (lake), Wash. 310/H3
Rock (co.), Wis. 317/H10
Rock (riv.), Wis. 317/J9
Rockall (isl.), Scotland 7/C3
Rockaway, N.J. (07866) 273/D2
Rockaway, Oreg. (97136) 291/C2
Rockaway Beach, Mo. (65740) 261/F9
Rock Bluff, Fla. (†32321) 212/B1
Rockbridge, Ill. (62081) 222/C4
Rockbridge, Mo. (65741) 261/H9
Rockbridge, Ohio (43149) 284/E6
Rockbridge (co.), Va. 307/K5
Rockbridge, Wis. (†53581) 317/F9
Rockcastle (co.), Ky. 237/N6
Rockcastle (riv.), Ky. 237/N6
Rock Castle, W. Va. (25272) 312/C5
Rock Cave, W. Va. (26234) 312/F5
Rock City, Ill. (61070) 222/D1
Rockcliffe Park, Ontario 177/J2
Rockcorry, Ireland 17/H3
Rock Creek, Br. Col. 184/H6
Rock Creek, Kansas (†66512) 232/G2
Rock Creek, Minn. (55067) 255/F5
Rock Creek, Ohio (44084) 284/J2
Rock Creek, Yukon 187/E3
Rockdale (co.), Georgia 217/D3
Rockdale, Ill. (60436) 222/E2
Rockdale, N. S. Wales 88/K4
Rockdale, N.S. Wales 97/J4
Rockdale, Texas (76567) 303/G7
Rockdale, Wis. (†53523) 317/J10
Rock Dell, Minn. (†55920) 255/F7
Rockerville, S. Dak. (†57701) 298/C6
Rockfall, Conn. (06481) 210/E2
Rock Falls, Ill. (61071) 222/D2
Rock Falls, Iowa (50467) 229/G2
Rock Falls, Wis. (54764) 317/C6
Rockfield, Ind. (46977) 227/D3
Rockfield, Ky. (42274) 237/J7
Rockfield, Wis. (53077) 317/L1
Rockfish, N.C. (†28302) 281/L5
Rockford, Ala. (35136) 195/F5
Rockford, Idaho (†83221) 220/F6
Rockford, Ill. 146/K5
Rockford, Ill. 188/J2
Rockford, Ill. (*61101) 222/D1
Rockford, Iowa (50468) 229/H2
Rockford, Mich. (49341) 250/D5

Rockford, Minn. (55373) 255/F5
Rockford, N.C. (27044) 281/H2
Rockford, Ohio (45882) 284/A4
Rockford, Sask. 181/J3
Rockford, Tenn. (37853) 237/O9
Rockford, Wash. (99030) 310/H3
Rock Forest, Québec 172/F4
Rock Glen, Pa. (18246) 294/K4
Rockglen, Sask. 181/F6
Rock Grove, Ill. (†61070) 222/D1
Rock Hall, Md. (21661) 245/O4
Rockham, S. Dak. (57470) 298/M4
Rockhampton, Australia 2/S7
Rockhampton, Australia 87/F8
Rockhampton, Queensland 88/H4
Rockhampton, Queensland 95/D4
Rockhampton Downs, North. Terr. 93/D5
Rockhaven, Sask. 181/B3
Rock Hill, Mo. (†63119) 261/P3
Rock Hill, S.C. 188/K4
Rock Hill, S.C. (29730) 296/E2
Rockholds, Ky. (40759) 237/N7
Rockingham, Georgia (†31510) 217/H7
Rockingham (co.), N.H. 268/E5
Rockingham (co.), N.C. 281/K2
Rockingham, N.C. (28379) 281/K5
Rockingham○, Vt. (†05101) 268/B5
Rockingham (co.), Va. 307/L4
Rockingham, W. Australia 88/B2
Rockingham, W. Australia 92/A2
Rock Island, Ill. 188/J2
Rock Island (co.), Ill. 222/C2
Rock Island, Ill. (61201) 222/C2
Rock Island, Okla. (†74932) 288/T4
Rock Island, Québec 172/E4
Rock Island, Tenn. (38581) 237/K9
Rock Island, Texas (77470) 303/H8
Rock Island, Wash. (†98801) 310/E3
Rock Island (dam), Wash. 310/E3
Rock Island Arsenal, Ill. 222/C2
Rocklake, N. Dak. (58365) 282/M2
Rockland, Conn. (†06443) 210/E3
Rockland, Del. (19732) 245/R1
Rockland, Idaho (83271) 220/F7
Rockland, Maine (04841) 243/E7
Rockland○, Mass. (02370) 249/L4
Rockland, Mich. (49960) 250/G1
Rockland (co.), N.Y. 276/M8
Rockland, Ontario 177/J2
Rockland, Texas (75970) 303/K6
Rockland, Wis. (54653) 317/D8
Rocklands (res.), Victoria 97/B5
Rockledge, Fla. (32955) 212/F3
Rockledge, Georgia (30454) 217/G6
Rockledge, Pa. (†19101) 294/M5
Rockleigh, N.J. (07647) 273/C1
Rocklin, Calif. (95677) 204/B8
Rockmart, Georgia (30153) 217/B2
Rock Mills, Ala. (36274) 195/H4
Rock Oak, W. Va. (†26756) 312/J4
Rock Point, Md. (20682) 245/L7
Rockport, Ark. (†72104) 202/E5
Rockport, Calif. (†95488) 204/B4
Rockport, Ill. (62370) 222/B4
Rockport, Ind. 227/C9
Rockport, Ky. (42369) 237/H6
Rockport, Maine (04856) 243/F7
Rockport○, Maine (04856) 243/F7
Rockport○, Mass. (01966) 249/M2
Rockport, Miss. (†39083) 256/D7
Rock Port, Mo. (64482) 261/B2
Rockport, Texas (78382) 303/H9
Rockport (lake), Utah 304/C3
Rockport, Wash. (98283) 310/D2
Rockport, W. Va. (26169) 312/C4
Rock Rapids, Iowa (51246) 229/A2
Rock River, Wyo. (82083) 319/G4
Rock Run, Ala. (†36272) 195/G2
Rocks, Md. (†21084) 245/N2
Rocks (pt.), N. Zealand 100/C4
Rock Springs, Mont. (59312) 262/K4
Rocksprings, Texas (78880) 303/D8
Rock Springs, Wis. (53961) 317/F8
Rock Springs, Wyo. 146/H5
Rock Springs, Wyo. 188/E2
Rock Springs, Wyo. (82901) 319/C4
Rockstone, Guyana 131/B2
Rockton, Ill. (61072) 222/E1
Rockvale, Colo. (81244) 208/J6
Rockvale, Mont. (†59080) 262/H5
Rockvale, Tenn. (37153) 237/J9
Rock Valley, Iowa (51247) 229/A2
Rockville, Conn. (†06066) 210/F1
Rockville, Ind. (47872) 227/C5
Rockville, Maine (†04841) 243/E7
Rockville, Md. (*20850) 245/K4
Rockville, Mass. (†02054) 249/A8
Rockville, Minn. (56369) 255/D5
Rockville, Mo. (64780) 261/D6
Rockville, Nebr. (68871) 264/F3
Rockville, Nova Scotia 168/B5
Rockville, R.I. (02873) 249/G6
Rockville, S.C. (†29487) 296/G6
Rockville, Utah (84763) 304/A6
Rockville, Va. (23146) 307/N5
Rockville, Wis. (†53820) 317/E10
Rockville Centre, N.Y. (*11570) 276/R7
Rockwall (co.), Texas 303/H5
Rockwall, Texas (75087) 303/H5
Rockwell, Iowa (50469) 229/G3
Rockwell, N.C. (28138) 281/J3
Rockwell City, Iowa (50579) 229/D4
Rockwood, Ala. (†35653) 195/C2
Rockwood, Ill. (62280) 222/D6
Rockwood, Maine (04478) 243/D4
Rockwood, Mich. (48173) 250/F6
Rockwood, Ontario 177/D4
Rockwood, Pa. (15557) 294/D6
Rockwood, Tenn. (37854) 237/M9
Rockwood, Texas (76873) 303/E6
Rocky (mts.) 162/D4
Rocky (mts.) 146/F4
Rocky (mts.) 188/E3
Rocky (mts.), Alberta 182/BC4
Rocky (mts.), Br. Col. 184/F2
Rocky (mts.), Canada 4/D16
Rocky (mts.), Colo. 208/F1
Rocky (mts.), Idaho 220/D1
Rocky (lake), Maine 243/J6
Rocky (mts.), Mont. 262/D4
Rocky (bay), Newf. 166/C3
Rocky (riv.), Newf. 166/D2
Rocky (mts.), N. Mex. 274/C1
Rocky (pt.), Norfolk I. 88/K6
Rocky (riv.), N.C. 281/H4
Rocky (riv.), Ohio 284/G9
Rocky, Okla. (73661) 288/J4
Rocky (riv.), S.C. 296/B3
Rocky (cape), Tasmania 99/B2
Rocky (mts.), Wash. 310/H2
Rocky (mts.), Wyo. 319/C1
Rocky (mts.), Yukon 187/F4
Rocky Bottom, S.C. (†29685) 296/B1
Rocky Boy, Mont. (†59521) 262/G2
Rocky Boy's Ind. Res., Mont. 262/G2
Rocky Comfort, Mo. (64861) 261/D9
Rocky Face, Georgia (30740) 217/C1
Rockyford, Alberta 182/D4
Rocky Ford, Colo. (81067) 208/M6
Rocky Ford, Georgia (30455) 217/J5
Rocky Fork (lake), Ohio 284/D7
Rocky Gap, Va. (24366) 307/F6
Rocky Gorge (res.), Md. 245/L4
Rocky Harbour, Newf. 166/C4
Rocky Hill○, Conn. (06067) 210/E2
Rocky Hill, Ky. (42163) 237/J6
Rocky Hill, N.J. (08553) 273/D3
Rocky Lane, Alberta 182/B5
Rocky Mount, Georgia (†30251) 217/C4
Rocky Mount, La. (†71064) 238/C1
Rocky Mount, Mo. (65072) 261/G6
Rocky Mount, N.C. 188/L3
Rocky Mount, N.C. (27801) 281/O3
Rocky Mount, Va. (24151) 307/J7
Rocky Mountain Arsenal, Colo. 208/K3
Rocky Mountain House, Alberta 182/C3
Rocky Mountain House, Alta. 162/E5
Rocky Mountain Nat'l Park, Colo. 208/H2
Rocky Point, N.C. (28457) 281/O6
Rocky Point, Wash. (†98626) 310/A2
Rockypoint, Wyo. (†82721) 319/G1
Rocky Rapids, Alberta 182/C3
Rocky Reach (dam), Wash. 310/E3
Rocky Ridge (mt.), Idaho 220/C3
Rocky Ridge, Ohio (43458) 284/D2
Rocky River, Ohio (44116) 284/G9
Rodanthe, N.C. (27968) 281/U3
Rodarte, N. Mex. (87561) 274/D2
Rodas, Cuba 158/E2
Rødby, Denmark 21/E8
Rødby, Denmark 18/G9
Roddickton, Newf. 166/C3
Rødding, Denmark 21/B7
Roddy, Tenn. (†37381) 237/M9
Rødekro, Denmark 21/C7
Roden, Netherlands 27/J2
Rodeo, Calif. (94572) 204/J1
Rodeo, Mexico 150/G4
Rodeo, N. Mex. (88056) 274/A7
Roderfield, W. Va. (24881) 312/C8
Roderick (isl.), Br. Col. 184/C4
Rodessa, La. (71069) 238/B1
Rodez, France 28/E5
Ródhos, Greece 45/J7
Roding (riv.), England 13/J7
Rodinga, North. Terr. 93/D8
Rødkaersbro, Denmark 21/C5
Rodman, Iowa (50580) 229/D2
Rodman, N.Y. (13682) 276/J3
Rodman, S.C. (†29706) 296/E2
Rodney, Mich. (49342) 250/D5
Rodney, Miss. (†39096) 256/B7
Rodney, Ontario 177/C5
Rodney Village, Del. (19901) 245/R4
Rodrigues, Brazil 132/F10
Rodríguez, Uruguay 145/C5
Rødvig, Denmark 21/F7
Roe, Ark. (72134) 202/H4
Roe (riv.), N. Ireland 17/H1
Roebling-Florence, N.J. (08554) 273/D3
Roebourne, W. Australia 88/B4
Roebourne, W. Australia 92/B3
Roebuck (bay), W. Australia 88/C3
Roebuck (bay), W. Australia 92/C2
Roebuck Plains, W. Australia 92/C2
Roeland Park, Kansas (†66205) 232/H2
Roer (riv.), Netherlands 27/J6
Roermond, Netherlands 27/J6
Roeselare, Belgium 27/C7
Roes Welcome (sound), N.W.T. 162/H2
Roes Welcome (sound), N.W. Terrs. 187/K3
Roff, Okla. (74865) 288/N5
Rogachev, U.S.S.R. 52/D4
Rogagua (lake), Bolivia 136/B3
Rogaguado (lake), Bolivia 136/C3
Rogaland (co.), Norway 18/E7
Rogatica, Yugoslavia 45/D4
Roger Mills (co.), Okla. 288/G3
Rogers, Ark. (72756) 202/B1
Rogers, Br. Col. 184/J4
Rogers (lake), Calif. 204/H9
Rogers, Conn. (06263) 210/H1
Rogers (lake), Conn. 210/E3
Rogers, La. (†71342) 238/F3
Rogers, Minn. (55374) 255/E5
Rogers, Nebr. (68659) 264/H3
Rogers, N. Mex. (88132) 274/F5
Rogers, N. Dak. (58479) 282/O5
Rogers, Ohio (44455) 284/J4
Rogers (co.), Okla. 288/P2
Rogers, Texas (76569) 303/G7
Rogers (mt.), Va. 307/E7
Rogers City, Mich. (49779) 250/F3
Rogerson, Idaho (83302) 220/D7
Rogers Springs, Tenn. (†38052) 237/D10
Rogersville, Ala. (35652) 195/D1
Rogersville, Mo. (65742) 261/G8
Rogersville, New Bruns. 170/E2
Rogersville, Pa. (15359) 294/B6
Rogersville, Tenn. (37857) 237/P8
Roger Williams Nat'l Mem., R.I. 249/J5
Roggen, Colo. (80652) 208/L2
Roggwil, Switzerland 39/E2
Rogliano, France 28/B6
Rogozno, Poland 47/C2
Rogue (riv.), Oreg. 291/C5
Rogue River, Oreg. (97537) 291/D5
Roha, India 68/C5
Rohnert Park, Calif. (94928) 204/C5
Rohnerville, Calif. (†95540) 204/B3
Rohrbach in Oberösterreich, Austria 41/B2
Rohrersville, Md. (21779) 245/H3
Rohri, Pakistan 68/B3
Rohtak, India 68/C5
Rohwer, Ark. (71666) 202/H6
Roi Et, Thailand 72/D4
Roja, U.S.S.R. 53/B2
Rojas, Argentina 143/F7
Rojo (cape), Mexico 150/L6
Rojo (cape), P. Rico 161/A3
Rojo (cape), P. Rico 156/F1
Rokan (riv.), Indonesia 85/C5
Rokeby, Sask. 181/J4
Rokiškis, U.S.S.R. 53/C2
Rokycany, Czech. 41/B2
Rokytnice nad Jizerou, Czech. 41/C1
Rola Co (lake), China 77/C4
Roland, Ark. (72135) 202/E4
Roland, Iowa (50236) 229/F4
Roland, Manitoba 179/D5
Roland, Okla. (74954) 288/S4
Røldal, Norway 18/E7
Roldán, Argentina 143/F6
Rolecha, Chile 138/D4
Rolesville, N.C. (27571) 281/N3
Rolette (co.), N. Dak. 282/L2
Rolette, N. Dak. (58366) 282/L2
Roleystone, W. Australia 88/B2
Rolfe, Iowa (50581) 229/D3
Roll, Ariz. (85347) 198/A6
Rolla, Ark. (†72104) 202/E5
Rolla, Br. Col. 184/G2
Rolla, Kansas (67954) 232/A4
Rolla, Mo. (65401) 261/J7
Rolla, N. Dak. (58367) 282/L2
Rollag, Minn. (†56549) 255/B4
Rolle, Switzerland 39/B4
Rollingbay, Wash. (98061) 310/A2
Rollingdam, New Bruns. 170/C3
Rolling Fields, Ky. (†40201) 237/K2
Rolling Fork (riv.), Ky. 237/L5
Rolling Fork, Miss. (39159) 256/C5
Rolling Hills, Alberta 182/E4
Rolling Hills, Calif. (90274) 204/B11
Rolling Hills, Ky. (†40201) 237/L1
Rolling Hills Estates, Calif. (90274) 204/B11
Rolling Meadows, Ill. (60008) 222/A5
Rolling Prairie, Ind. (46371) 227/D1
Rollingstone, Minn. (55969) 255/G6
Rollins, Mont. (59931) 262/B3
Rollo (bay), Pr. Edward I. 168/F2
Rolphton, Ontario 177/G1
Roma, Australia 87/E8
Roma (Rome) (cap.), Italy 34/F6
Roma, Queensland 88/H5
Roma, Queensland 95/D5
Roma, Sweden 18/L8
Romain (cape), S.C. 296/J6
Romaine (riv.), Newf. 166/B3
Romaine (riv.), Que. 162/K5
Romaine, Québec 174/E2
Romaine (riv.), Québec 174/E2
Roma-Los Saenz, Texas (78584) 303/E11
Roman, Romania 45/H2
Romance, Ark. (72136) 202/F3
Romance, Sask. 181/G3
Romance, W. Va. (25175) 312/C5
Romang, Argentina 143/F4
Romang (isl.), Indonesia 85/H7
Romania 2/L3
Romania 7/G4
ROMANIA 45/F3
Romano (cay), Cuba 158/G2
Romano (cay), Cuba 156/C2
Romano (cape), Fla. 212/E6
Romanshorn, Switzerland 39/H1
Romans-sur-Isère, France 28/F5
Romanzof (cape), Alaska 196/E2
Rombauer, Mo. (63962) 261/M9
Romblon (prov.), Philippines 82/D4
Romblon, Philippines 82/D4
Romblon (isl.), Philippines 82/D4
Rome, Ga. 188/K4
Rome, Georgia (30161) 217/B2
Rome, Ill. (61562) 222/D3
Rome, Ind. (47574) 227/D9
Rome, Iowa (52642) 229/K7
Rome (prov.), Italy 34/F6
Rome (cap.), Italy 7/F4
Rome (cap.), Italy 34/F6
Rome (cap.), Italy 2/K3
Rome○, Maine (†04957) 243/D6
Rome, Miss. (38768) 256/C3
Rome, N.Y. 188/M2
Rome, N.Y. (13440) 276/J4
Rome (Stout), Ohio (†45684) 284/D8
Rome, Ohio (44085) 284/J2
Rome, Oreg. (†97910) 291/K5
Rome, Pa. (18837) 294/K2
Rome, Wis. (†53178) 317/H1
Rome City, Ind. (46784) 227/G1
Romeo, Colo. (81148) 208/G8
Romeo, Mich. (48065) 250/F6
Romeoville, Ill. (60441) 222/B6
Romeroville, N. Mex. (†87701) 274/D3
Romeville, La. (†70723) 238/L3
Romilly-sur-Seine, France 28/F3
Romney, Ind. (47981) 227/D4
Romney, W. Va. (26757) 312/J4
Romny, U.S.S.R. 52/D4
Rømø, Denmark 21/B7
Rømø (isl.), Denmark 21/B7
Rømø (isl.), Denmark 18/F9
Romont, Switzerland 39/C3
Romorantin-Lanthenay, France 28/D4
Romsdalsfjorden (fjord), Norway 18/E5
Romsey, England 10/F5
Romsey, England 13/F6
Romulus, Mich. (48174) 250/F6
Romulus, N.Y. (14541) 276/G5
Ron, Vietnam 72/E3
Ron, Mui (cape), Vietnam 72/E3
Rona (isl.), Scotland 15/B3
Ronald, Wash. (98940) 310/E3
Ronan, Mont. (59864) 262/C3
Ronay (isl.), Scotland 15/A3
Roncador, Serra do (range), Brazil 132/D5
Roncador (cays), Colombia 126/B9
Ronceverte, W. Va. (24970) 312/F7
Ronciglione, Italy 34/C3
Ronda, N.C. (28670) 281/H2
Ronda, Spain 33/D4
Rønde, Denmark 21/D5
Ronde (isl.), Grenada 161/D7
Rondeau Prov. Park, Ontario 177/C5
Rondo, Ark. (†72355) 202/J4
Rondônia (terr.), Brazil 132/H10
Rondônia, Brazil 132/H10
Rondonópolis, Brazil 120/D4
Rondout (res.), N.Y. 276/M7
Rondu, Pakistan 68/D1
Rong, Koh (isl.), Cambodia 72/D5
Rong'an, China 77/G6
Ronge, Lac La (lake), Sask. 162/F4
Ronge, La (lake), Sask. 181/M3
Rongelap (atoll), Marshall Is. 87/G4
Rongjiang, China 77/G6
Rong Kwang, Thailand 72/D3
Rong Xian, China 77/H7
Ronju (mt.), Fr. Poly. 86/T13
Ronkonkoma, N.Y. (11779) 276/O9
Rønne, Denmark 21/F9
Rønne, Denmark 18/J9
Ronneby, Minn. (†56324) 255/E5
Ronneby, Sweden 18/J8
Ronne Entrance (inlet) 5/B15
Ronne Ice Shelf, Ant. 2/F10
Ronne Ice Shelf 5/B15
Ronse, Belgium 27/D7
Ronuro (riv.), Brazil 132/C6
Roodeport, S. Africa 118/H6
Roodhouse, Ill. (62082) 222/C4
Roof Butte (mt.), Ariz. 198/F2
Rooi, Neth. Ant. 161/E8
Rooks (co.), Kansas 232/C2
Roopville, Georgia (30170) 217/B4
Roosendaal, Netherlands 27/F5
Roosevelt (isl.), Ant. 2/A10
Roosevelt (isl.) 5/A10
Roosevelt (res.), Ariz. 188/D4
Roosevelt, Ariz. (85545) 198/D5
Roosevelt (riv.), Brazil 120/C3
Roosevelt (riv.), Brazil 132/A5
Roosevelt, La. (†71276) 238/H1
Roosevelt, Minn. (56673) 255/C2
Roosevelt (co.), Mont. 262/L2
Roosevelt, N.J. (08555) 273/E3
Roosevelt (co.), N. Mex. 274/F4
Roosevelt, N.Y. (11575) 276/R7
Roosevelt, Okla. (73564) 288/J5
Roosevelt, Texas (76874) 303/D7
Roosevelt, Utah (84066) 304/D3
Roosevelt, Wash. (99356) 310/E5
Roosevelt Campobello Int'l Park, New Bruns. 170/D4
Roosevelt City, Ala. (35020) 195/E4
Roosevelt Park, Mich. (49444) 250/C5
Roosevelt Road Naval Res., P. Rico 161/F2
Roosville, Br. Col. 184/K5
Root (riv.), Minn. 255/G7
Rootstown, Ohio (44272) 284/H3
Roper, N.C. (27970) 281/R3
Roper (riv.), North. Terr. 88/E3
Roper (riv.), North. Terr. 93/C3
Roper River, North. Terr. 93/D3
Roper River Mission, North. Terr. 88/E2
Roper Valley, North. Terr. 93/D3
Ropesville, Texas (79358) 303/B4
Roque Bluffs○, Maine (†04654) 243/H6
Roque González de Santa Cruz, Paraguay 144/B5
Roque Pérez, Argentina 143/G7
Roquetas, Spain 33/G2
Rora (head), Scotland 15/E2
Roraima (mt.) 120/C2
Roraima (terr.), Brazil 132/H8
Roraima (mt.), Guyana 131/A3
Roraima (mt.), Venezuela 124/H5
Rørby, Denmark 21/E6
Rorketon, Manitoba 179/C3
Røros, Norway 18/G5
Rorschach, Switzerland 39/H2
Rosa, Ala. (†35049) 195/E3
Rosa (cape), Ecuador 128/B10
Rosa (mt.), Italy 34/A1
Rosa, La. (71364) 238/G5
Rosa, Manitoba 179/F5
Rosa (mt.), Switzerland 39/E5
Rosaire, Québec 172/G3
Rosaireville, New Bruns. 170/E2
Rosalia, Kansas (67132) 232/F4
Rosalia, Wash. (99170) 310/H3
Rosalie, Dominica 161/F6
Rosalie, Nebr. (68055) 264/H2
Rosalina, Paraguay 144/D3
Rosalind, Alberta 182/D3
Rosamond, Calif. (93560) 204/G9
Rosamond (lake), Calif. 204/G9
Rosamond, Ill. (62083) 222/D4
Rosamorada, Mexico 150/G5
Rosapenna, Ireland 17/F1
Rosario, Argentina 143/F6
Rosario, Argentina 120/C6
Rosário, Brazil 132/F3
Rosario, Chile 138/F5
Rosario (cay), Cuba 158/C2
Rosario, Sinaloa, Mexico 150/G5
Rosario, Sonora, Mexico 150/E3
Rosario, Paraguay 144/D4
Rosario, P. Rico 161/A2
Rosario, Uruguay 145/B5
Rosario, Venezuela 124/B2
Rosario (str.), Wash. 310/C2
Rosario de la Frontera, Argentina 143/D2
Rosario de Lerma, Argentina 143/C1
Rosario del Tala, Argentina 143/G6
Rosáriodo Sul, Brazil 132/C10
Rosário Oeste, Brazil 132/C6
Rosas, Spain 33/H1
Rosas (gulf), Spain 33/H1
Rosati, Mo. (†65559) 261/J6
Rosa Zárate, Ecuador 128/C2
Rosburg, Wash. (98643) 310/B4
Rosbys Rock, W. Va. (†26041) 312/E3
Roscoe, Ill. (61073) 222/D1
Roscoe, Minn. (56371) 255/D5
Roscoe, Mo. (64781) 261/E7
Roscoe, Mont. (59071) 262/G5
Roscoe, Nebr. (†69153) 264/C3
Roscoe, N.Y. (12776) 276/L7
Roscoe, Pa. (15477) 294/C5
Roscoe, S. Dak. (57471) 298/L3
Roscoe, Texas (79545) 303/D5
Roscoff, France 28/A3
Roscommon (co.), Ireland 17/E4
Roscommon, Ireland 10/B4
Roscommon, Ireland 17/E4
Roscommon (co.), Mich. 250/E4
Roscommon, Mich. (48653) 250/E4
Roscrea, Ireland 10/B4
Roscrea, Ireland 17/F6
Rose (peak), Ariz. 198/F5
Rose (pt.), Br. Col. 184/B3
Rose (pt.), Martinique 161/D6
Rose, Nebr. (68772) 264/E2
Rose, N.Y. (14542) 276/G4
Rose (riv.), North. Terr. 93/D2
Rose, Okla. (74364) 288/R2
Roseau (cap.), Dominica 156/G4
Roseau (cap.), Dominica 161/E7
Roseau (riv.), Dominica 161/E7
Roseau (co.), Minn. 255/C2
Roseau, Minn. (56751) 255/C2
Roseau (riv.), Minn. 255/B2
Roseau (riv.), St. Lucia 161/G6
Roseau River, Manitoba 179/F5
Roseaux, Haiti 158/A6
Rosebank, Manitoba 179/D5
Roseberry, Br. Col. 184/J5
Rosebery, Tasmania 99/B3
Rose Blanche, Newf. 166/C4
Roseboom, N.Y. (13450) 276/L5
Roseboro, N.C. (28382) 281/N5
Rosebud, Alberta 182/D4
Rosebud (riv.), Alberta 182/D4
Rose Bud, Ark. (72137) 202/F3
Rosebud, Mo. (63091) 261/K6
Rosebud (co.), Mont. 262/K4
Rosebud, Mont. (59347) 262/K4
Rosebud (creek), Mont. 262/K4
Rosebud, S. Dak. (57570) 298/H7
Rosebud, Texas (76570) 303/G7
Rosebud (creek), Utah 304/A2
Rosebud Ind. Res., S. Dak. 298/H7
Roseburg, Oreg. (97470) 291/D4
Rosebush Mich. (48878) 250/E5
Rose City, Mich. (48654) 250/E4
Rose Creek, Minn. (55970) 255/F7
Rosedale, Ind. (47874) 227/C5
Rosedale, La. (70772) 238/G6
Rosedale, Md. (21237) 245/M3
Rosedale, Miss. (38769) 256/B3
Rosedale, Okla. (†74831) 288/M5
Rosedale, Tenn. (†37728) 237/N8
Rosedale, Va. (24280) 307/E7
Rosedale, W. Va. (26636) 312/E5
Rosefield, La. (†71435) 238/G3
Roseglen, N. Dak. (58775) 282/G4
Rosehearty, Scotland 15/F3
Rosehill, Ala. (†36028) 195/F8
Rose Hill, Barbados 161/B8
Rose Hill, Ill. (†62448) 222/E4
Rose Hill, Iowa (52586) 229/J6
Rose Hill, Kansas (67133) 232/E4
Rose Hill, Ky. (†40330) 237/M5
Rose Hill, Miss. (39356) 256/F6
Rose Hill, N.C. (28458) 281/N5
Rose Hill, Va. (24281) 307/B7
Roseisle, Manitoba 179/D5
Rose Lake, Br. Col. 184/E3
Roseland, Ark. (72463) 202/K2
Roseland, Fla. (32957) 212/F4
Roseland, Ind. (†46601) 227/E1
Roseland, Kansas (†66773) 232/H4
Roseland, La. (70456) 238/J5
Roseland, Minn. (†56216) 255/C6
Roseland, Nebr. (68973) 264/F4
Roseland, N.J. (07068) 273/A2
Roseland, Va. (22967) 307/K5
Roselawn, Ind. (†46310) 227/C2
Roselle, Ill. (60172) 222/A5
Roselle, N.J. (07203) 273/B2
Roselle Park, N.J. (07204) 273/A2
Rose Lodge, Oreg. (97372) 291/D3
Rose Lynn, Alberta 182/E4
Rosemark, Tenn. (38053) 237/B10
Rosemary, Alberta 182/E4
Rosemead, Calif. (91770) 204/C10
Rosemère, Québec 172/H4
Rosemont, Ill. (60018) 222/B5
Rosemont, Md. (†21758) 245/H3
Rosemont, N.J. (08556) 273/D3
Rosemont, Pa. (19010) 294/M5
Rosemount, Minn. (55068) 255/E6
Rosemount, Ohio (†45662) 284/D8
Rosen, Minn. (†56212) 255/B5
Rosenberg, Texas (77471) 303/J8
Rosendale, Minn. (†56243) 255/D5
Rosendale, Mo. (64483) 261/C2
Rosendale, N.Y. (12472) 276/M7
Rosendale, Wis. (54974) 317/J8
Rosenfeld, Manitoba 179/E5
Rosengart, Manitoba 179/E5
Rosenhayn, N.J. (08352) 273/C5
Rosenheim, W. Germany 22/D5
Rosenhof, Sask. 181/D5
Rosenlaui, Switzerland 39/F3
Rosenort, Manitoba 179/E5
Rosepine, La. (70659) 238/D5
Rose Prairie, Br. Col. 184/G2
Roseray, Sask. 181/C5
Roseto, Pa. (18013) 294/M4
Rosetown, Sask. 162/F5
Rosetown, Sask. 181/D4
Rosetta, Egypt 111/J2
Rosetta, Egypt 59/B3
Rosetta, Miss. (†39633) 256/B8
Rosette, Utah (†84329) 304/A2
Rose Valley, Pa. (†19065) 294/L7
Rose Valley, Sask. 181/H3
Roseville, Calif. (95678) 204/B8
Roseville, Ill. (61473) 222/C3
Roseville, Mich. (48066) 250/B6
Roseville, Minn. (55113) 255/G5
Roseville, Ohio (43777) 284/F6
Roseville, Pa. (†16933) 294/H2
Roseway (riv.), Nova Scotia 168/C4
Rosewood, North. Terr. 93/A4
Rosewood, Ohio (43070) 284/C5
Rosewood Heights, Ill. (†62024) 222/B2
Roseworthy, S. Australia 94/B6
Roshage (cape), Denmark 18/F8
Rosharon, Texas (77583) 303/J3
Rosh Ha 'Ayin, Israel 65/B3
Rosholt, S. Dak. (57260) 298/R2
Rosholt, Wis. (54473) 317/H6
Rosh Pinna, Israel 65/D2
Rosice, Czech. 41/D2
Rosiclare, Ill. (62982) 222/E6
Rosie, Ark. (72571) 202/G2
Rosier, Georgia (†30434) 217/H5
Rosignano Marittimo, Italy 34/C3
Rosignol, Guyana 131/C2
Rosine, Ky. (42370) 237/H6
Roşiori de Vede, Romania 45/G3
Rositsa, Bulgaria 45/H4
Roskilde (co.), Denmark 21/E6
Roskilde, Denmark 21/E6
Roskilde, Denmark 18/G9
Roslavl', U.S.S.R. 52/D4
Roslev, Denmark 21/B4
Roslin, Ontario 177/G3
Roslin, Tenn. (†38556) 237/M8
Roslyn, N.Y. (11576) 276/R6
Roslyn, S. Dak. (57261) 298/P2
Roslyn, Wash. (98941) 310/E3
Rosman, N.C. (28772) 281/D4
Rosmaninhal, Portugal 33/C3
Røsnaes (pen.), Denmark 21/D6
Rosneath, Scotland 15/A1
Rosneath, Scotland 10/A1
Ross (isl.), Ant. 2/T10
Ross (sea), Ant. 2/A10
Ross (isl.) 5/B9
Ross (sea) 5/B10
Ross, Calif. (94957) 204/H1
Ross, Iowa (†50025) 229/D5
Ross, Manitoba 179/F5
Ross (isl.), Manitoba 179/J3
Ross, Minn. (56753) 255/C2
Ross, New Bruns. 170/D4
Ross, N. Zealand 100/C5
Ross (pt.), Norfolk I. 88/L6
Ross, N. Dak. (58776) 282/E3
Ross (co.), Ohio 284/D7
Ross, Ohio (45061) 284/B9
Ross, Tasmania 99/D4
Ross (dam), Wash. 310/D2
Ross (lake), Wash. 310/D2
Rossa, Switzerland 39/H4
Rossall (pt.), England 13/D4
Rossan (pt.), Ireland 10/B3
Ross and Cromarty (trad. co.), Scotland 15/A5
Rossano, Italy 34/F5
Rossarden, Tasmania 99/D3
Ross Barnett (res.), Miss. 256/D6
Ross Bay Junction, Newf. 166/A3
Rossbear (lake), Alberta 182/C1
Rossburg, Ohio (45362) 284/A5
Rossburn, Manitoba 179/B4
Rosscarbery, Ireland 17/C8
Rosscarbery (bay), Ireland 10/B5
Rosscarbery (bay), Ireland 17/D9
Rosseau, Ontario 177/E2
Rosseau (lake), Ontario 177/E2
Rossel (isl.), Papua N.G. 85/D8
Rossendale, Manitoba 179/D5
Rosser, Manitoba 179/E4
Rosses (bay), Ireland 17/D1
Rosses Point, Ireland 17/D3
Rossford, Ohio (43460) 284/C2
Ross Fork, Mont. (†59457) 262/G3
Ross Ice Shelf, Ant. 2/A11
Ross Ice Shelf 5/A10
Rossie, Iowa (51356) 229/C2
Rossie, N.Y. (†13646) 276/J2
Rossignol (lake), Nova Scotia 168/C4
Rossing, Namibia 118/B4
Rossiter, Pa. (15772) 294/E4
Rosskeeragh (pt.), Ireland 17/D3
Ross Lake Nat'l Rec. Area, Wash. 310/E2
Rossland, Br. Col. 162/E6
Rossland, Br. Col. 184/H6
Rosslare, Ireland 10/C4
Rosslare, Ireland 17/J7
Rosslare (bay), Ireland 17/J7
Rosslare Harbour (Ballygeary), Ireland 17/J7
Rosslau, E. Germany 22/E3
Rosslyn Farms, Pa. (†15106) 294/B7
Rosslyn Village, Ontario 177/G5
Rosslyn Village, Ontario 175/C3
Rossmore, W. Va. (25643) 312/C7
Rossmoyne, Ohio (45236) 284/C9
Rosso, Mauritania 106/A5
Rosso, Mauritania 102/A3
Ross of Mull (pen.), Scotland 15/B4
Ross-on-Wye, England 10/E5
Ross-on-Wye, England 13/E6
Rossosh', U.S.S.R. 52/E4
Rossport, Ontario 177/H5
Ross River, Yukon 187/E3
Rosstock (mt.), Switzerland 39/G3
Rosston, Ark. (71858) 202/D6
Rosston, Ind. (†46077) 227/E4
Rosston, Okla. (73855) 288/G1

Rossville, Georgia (30741) 217/B1
Rossville, Ill. (60963) 222/F3
Rossville, Ind. (46065) 227/D4
Rossville, Iowa (†52172) 229/L2
Rossville, Kansas (66533) 232/G2
Rossville, Tenn. (38066) 237/B10
Rosswein, E. Germany 22/E3
Rostaq, Afghanistan 68/B1
Rostaq, Afghanistan 59/J2
Rosthern, Sask. 162/F5
Rosthern, Sask. 181/E3
Rostock, E. Germany 7/F3
Rostock (dist.), E. Germany 22/E1
Rostock, E. Germany 22/E1
Rostov, U.S.S.R. 2/M3
Rostov, U.S.S.R. 7/J4
Rostov, U.S.S.R. 52/E3
Rostov-na-Donu, U.S.S.R. 48/E5
Rostov-na-Donu, U.S.S.R. 52/F5
Rostrevor, N. Ireland 17/J3
Røsvatn (lake), Norway 18/H4
Roswell, Georgia (*30075) 217/D2
Roswell, Idaho (†83660) 220/A6
Roswell, N. Mex. 146/H6
Roswell, N. Mex. 188/E4
Roswell, N. Mex. (88201) 274/E5
Roswell, Ohio (†44663) 284/H5
Roswell, S. Dak. (57372) 298/O6
Rosyth, Alberta 182/E3
Rota (isl.), No. Marianas 87/E4
Rota, Spain 33/C4
Rotan, Texas (79546) 303/D5
Rotenburg, W. Germany 22/C2
Rotenburg an der Fulda, W. Germany 22/C3
Roth, N. Dak. (†58783) 282/J2
Roth bei Nürnberg, W. Germany 22/D4
Rothbury, England 10/E3
Rothbury, England 13/E2
Rothbury, Mich. (49452) 250/C5
Rothenburg ob der Tauber, W. Germany 22/D4
Rotherham, England 13/K2
Rotherham, England 10/F4
Rothes, Scotland 10/E2
Rothes, Scotland 15/E3
Rothesay, New Bruns. 170/E3
Rothesay, Scotland 10/D3
Rothesay, Scotland 15/A2
Rothorn (mt.), Switzerland 39/D4
Rothrist, Switzerland 39/E2
Rothsay, Minn. (56579) 255/B4
Rothschild, Wis. (54474) 317/G6
Rothville, Mo. (64676) 261/F3
Roti (isl.), Indonesia 85/G8
Roto, N.S. Wales 97/C3
Rotoroa (lake), N. Zealand 100/D4
Rotorua, N. Zealand 100/F3
Rotorua (lake), N. Zealand 100/F3
Rottenburg am Neckar, W. Germany 22/C4
Rottenmann, Austria 41/C3
Rotterdam, Netherlands 27/E5
Rotterdam, Netherlands 7/E3
Rotterdam Junction, N.Y. (12150) 276/N5
Rottershausen, W. Germany 22/D3
Rottnest (isl.), W. Australia 92/A1
Rottumeplaat (isl.), Netherlands 27/J1
Rottumeroog (isl.), Netherlands 27/J1
Rottweil, W. Germany 22/C4
Rotuma (isl.), Fiji 87/H7
Roubaix, France 28/E2
Roudnice nad Labem, Czech. 41/C1
Rouen, Barbados 161/B9
Rouen, France 7/E4
Rouen, France 28/D3
Rouge (riv.), Québec 172/C3
Rougemont, N.C. (27572) 281/L2
Rougemont, Québec 172/D4
Rougemont, Switzerland 39/D4
Rough (riv.), Ky. 237/H5
Rough River (lake), Ky. 237/J5
Rouka, Papua N.G. 85/B7
Rouleau, Sask. 181/G5
Roulers (Roeselare), Belgium 27/C7
Roulette, Pa. (16746) 294/F2
Roumania (Romania) 45
Round (pond), Maine 243/E2
Round (isl.), Miss. 256/G10
Round (pond), Newf. 166/C4
Round, The (mt.), N.S. Wales 97/G2
Round (lake), N.Y. 276/L2
Round (lake), N. Dak. 282/K3
Round (lak), Ontario 177/G2
Round (lake), Wis. 317/F4
Round (lake), Wis. 317/D3
Roundaway, Miss. (†38614) 256/C2
Round Grove, Ill. (†61270) 222/D2
Round Grove, Mo. (†65707) 261/E8
Roundhead, Ohio (43346) 284/C4
Round Hill, Alberta 182/D3
Round Hill, Conn. (†06830) 210/A4
Round Hill, Nova Scotia 168/C4
Round Hill, Va. (22141) 307/N2
Round Lake, Fla. (†32446) 212/D6
Round Lake, Ill. (60673) 222/A4
Round Lake, Minn. (56167) 255/C7
Roundlake, Miss. (†38740) 256/C2
Round Lake, N.Y. (12151) 276/N5
Round Lake Beach, Ill. (†60673) 222/A4
Round Lake Centre, Ontario 177/G2
Round Lake Heights, Ill. (†60673) 222/E1
Round Lake Park, Ill. (†60673) 222/A4
Round Mountain, Ala. (†35959) 195/G2
Round Mountain, Calif. (96084) 204/D3
Round Mountain, Nev. (89045) 266/E4
Round O, S.C. (29474) 296/F6
Round Oak, Georgia (†31038) 217/E4
Round Pond, Ark. (72378) 202/J3
Round Pond, Maine (04564) 243/E8
Round Prairie, Minn. (†56347) 255/D5
Round Rock, Texas (78664) 303/G7
Round Spring, Mo. (65467) 261/J8
Roundstone, Ireland 17/A5
Roundtop (mt.), Hawaii 218/C4
Round Top, Texas (78954) 303/H7
Roundup, Mont. (59072) 262/H4
Round Valley (res.), N.J. 273/D2
Round Valley Ind. Res., Calif. 204/B4
Roundwood, Ireland 17/J5
Rounthwaite, Manitoba 179/C5
Roura, Fr. Guiana 131/E3
Rousay (isl.), Scotland 10/E1
Rousay (isl.), Scotland 15/E1
Rouses Point, N.Y. (12979) 276/O1
Rouseville, Pa. (16344) 294/C3
Roussillon (trad. prov.) France 29
Routhierville, Québec 172/B2
Routt (co.), Colo. 208/E1
Rouville (co.), Québec 172/D4
Rouvroy, Belgium 27/G9
Rouyn, Que. 162/J6
Rouyn, Québec 174/B3
Rouzerville, Pa. (17250) 294/G6
Rovaniemi, Finland 18/O3
Rover, Ark. (72860) 202/D4
Rover, Mo. (†65775) 261/J9
Roveredo, Switzerland 39/H4
Rovereto, Italy 34/C2
Roviano, Italy 34/F6
Rovieng Tbong, Cambodia 72/E4
Rovigo (prov.), Italy 34/C2
Rovigo, Italy 34/C2
Rovinj, Yugoslavia 45/A3
Rovira, Colombia 126/C5
Rovno, U.S.S.R. 7/G3
Rovno, U.S.S.R. 48/C4
Rovno, U.S.S.R. 52/C4
Rovuma (riv.) 102/F6
Rovuma (riv.), Mozambique 118/F2
Rovuma (riv.), Tanzania 115/G6
Rowan, Iowa (50470) 229/F3
Rowan (co.), Ky. 237/P4
Rowan (co.), N.C. 281/H3
Rowans Ravine Prov. Park, Sask. 181/F4
Rowatt, Sask. 181/G5
Rowayton, Conn. (06853) 210/B4
Rowe (lake), Maine 243/F2
Rowe○, Mass. (01367) 249/C2
Rowe, N. Mex. (87562) 274/D3
Rowe, Va. (24646) 307/D6
Rowell, Ark. (†71665) 202/F6
Rowena, Georgia (†31713) 217/C8
Rowena, Ky. (†42629) 237/L7
Rowena, New Bruns. 170/C2
Rowena, N.S. Wales 97/E1
Rowena, S. Dak. (57056) 298/R6
Rowena, Texas (76875) 303/D6
Rowesville, S.C. (29133) 296/F5
Rowland, Ky. (†40484) 237/M5
Rowland, Nev. (†89831) 266/F1
Rowland, N.C. (28383) 281/L5
Rowlesburg, W. Va. (26425) 312/G4
Rowlett, Texas (75088) 303/H2
Rowletta, Sask. 181/F5
Rowletts, Ky. (42772) 237/K6
Rowley, Alberta 182/D4
Rowley, Iowa (52329) 229/K4
Rowley, Mass. (01969) 249/L2
Rowley○, Mass. (01969) 249/L2
Rowley (isl.), N.W. Terrs. 187/K3
Rowley (shoals), W. Australia 92/B2
Roxabell, Ohio (†45628) 284/D7
Roxana, Del. (†19945) 245/T6
Roxana, Ill. (62084) 222/B2
Roxas, Philippines 85/G3
Roxas, Capiz, Philippines 82/D5
Roxas, Isabela, Philippines 82/C2
Roxas, Or. Mindoro, Philippines 82/C4
Roxas, Palawan, Philippines 82/B5
Roxboro, N.C. (27573) 281/M2
Roxboro, Québec 172/H4
Roxburgh, N. Zealand 100/B6
Roxburgh (trad. co.), Scotland 15/B5
Roxbury○, Conn. (06783) 210/B2
Roxbury, Kansas (67476) 232/E3
Roxbury, Maine (04275) 243/B6
Roxbury○, Maine (04275) 243/B6
Roxbury, Mass. (†02119) 249/C7
Roxbury○, N.H. (†03431) 268/C6
Roxbury○, N.J. (†07876) 273/D2
Roxbury, N.Y. (12474) 276/L6
Roxbury, Pa. (17251) 294/G5
Roxbury○, Vt. (05669) 268/B3
Roxbury, Wis. (†53583) 317/G9
Roxbury Center, N.H. (†03431) 268/C6
Roxbury Falls, Conn. (†06783) 210/B2
Roxbury Station, Conn. (†06783) 210/B2
Roxie, Miss. (39661) 256/B8
Roxobel, N.C. (27872) 281/P2
Roxton, Texas (75477) 303/J4
Roxton Falls, Québec 172/E4
Roxton Pond, Québec 172/E4
Roy, Georgia (†30540) 217/D1
Roy, Idaho (†83271) 220/F7
Roy, Mont. (59471) 262/H3
Roy, New Bruns. 170/F2
Roy, N. Mex. (87743) 274/E3
Roy, Oreg. (†97106) 291/A2
Roy, Utah (84067) 304/C2
Roy, Wash. (98580) 310/C4
Royal, Ill. (61871) 222/E3
Royal, Iowa (51357) 229/C2
Royal (canal), Ireland 17/G4
Royal, Nebr. (68773) 264/F2
Royal, N.C. (†27806) 281/R4
Royal Center, Ind. (46978) 227/E3
Royal City, Wash. (99357) 310/F4
Royal Cotton Mills, N.C. (†27587) 281/M2
Royale (isl.), Mich. 188/J1
Royale (isl.), Mich. 250/E1
Royal Geographical Society (isls.), N.W. Terrs. 187/J3
Royal Gorge (canyon), Colo. 208/J6
Royal Lakes, Ill. (†62685) 222/C4
Royal Leamington Spa, England 13/F5
Royal Natal Nat'l Park, S. Africa 118/D5
Royal Oak, Md. (21662) 245/O6
Royal Oak, Mich. (*48067) 250/B6
Royal Park, Alberta 182/D3
Royal Pines, N.C. (†28803) 281/D4
Royal Road, New Bruns. 170/D2
Royalston○, Mass. (01368) 249/F2
Royalties, Alberta 182/C4
Royalton, Ill. (62983) 222/D6
Royalton, Ind. (†46077) 227/E5
Royalton, Ky. (41464) 237/R5
Royalton, Minn. (56373) 255/D5
Royalton, Ohio (†43130) 284/E6
Royalton, Pa. (†17057) 294/J5
Royalton○, Vt. (†05068) 268/B4
Royalton, Wis. (54975) 317/J7
Royal Tunbridge Wells, England 13/H6
Royal Tunbridge Wells, England 10/G5
Royalty, Texas (79779) 303/B6
Royan, France 28/C5
Royersford, Pa. (19468) 294/L5
Royerton, Ind. (†47302) 227/G4
Roy Hill, W. Australia 92/B3
Royse City, Texas (75089) 303/H4
Royston, Br. Col. 184/H2
Royston, England 13/G5
Royston, England 10/F4
Royston, Georgia (30662) 217/F2
Rožaj, Yugoslavia 45/E4
Rozel, Kansas (67574) 232/C3
Rozellville, Wis. (†54484) 317/G6
Rozet, Wyo. (82727) 319/G1
Rožňava, Czech. 41/F2
Rožnov pod Radhoštěm, Czech. 41/E2
Rtishchevo, U.S.S.R. 52/F4
Ruacana Falls (dam), Angola 115/B7
Ruacana Falls (falls), Namibia 118/A3
Ruaha Nat'l Park, Tanzania 115/F5
Ruahine (range), N. Zealand 100/F4
Ruapehu (mt.), N. Zealand 100/E3
Ruapuke (isl.), N. Zealand 100/B7
Ruatapu, N. Zealand 100/C5
Ruawai, N. Zealand 100/D2
Rub al Khali (des.), Saudi Arabia 54/G7
Rub al Khali (des.), Saudi Arabia 59/F5
Rubezhnoye, U.S.S.R. 52/E5
Rubicon, Wis. (53078) 317/K9
Rubidoux, Calif. (92509) 204/E10
Rubin (dry riv.), Israel 65/B4
Rubio, Iowa (52587) 229/K6
Rubio, Venezuela 124/B4
Rubottom, Okla. (73457) 288/M7
Rubtsovsk, U.S.S.R. 54/K4
Rubtsovsk, U.S.S.R. 48/J4
Ruby, Alaska (99768) 196/G2
Ruby, La. (71365) 238/F4
Ruby (riv.), Mont. 262/D5
Ruby (lake), Nev. 266/F2
Ruby (mts.), Nev. 266/F2
Ruby, S.C. (29741) 296/G2
Ruby, Va. (22545) 307/N3
Ruby, Wash. (†99119) 310/H2
Ruby Creek, Br. Col. 184/M3
Ruby Lake, Sask. 181/J3
Ruby River (res.), Mont. 262/D5
Rubys Inn, Utah (84764) 304/B6
Ruby Valley, Nev. (89833) 266/F2
Rucava, U.S.S.R. 53/A2
Ruch, Oreg. (†97530) 291/E5
Ruch'i, U.S.S.R. 48/E3
Ruch'i, U.S.S.R. 52/F1
Rucia (pt.), Dom. Rep. 158/D5
Ruckersville, Georgia (†30635) 217/G2
Ruda Śląska, Poland 47/B4
Rudbar, Afghanistan 59/H3
Rudbar, Afghanistan 68/A2
Rudd, Iowa (50471) 229/H2
Ruddell, Sask. 181/D3
Ruddle, W. Va. (†26807) 312/H5
Rudeis, Egypt 111/F2
Rüdersdorf bei Berlin, E. Germany 22/E2
Rüdesheim am Rhein, W. Germany 22/B3
Rudha Hunish (cape), Scotland 15/B3
Rudh Re (cape), Scotland 15/C3
Rudkøbing, Denmark 18/G9
Rudkøbing, Denmark 21/D8
Rudnik, Poland 47/F3
Rudnyy, U.S.S.R. 48/G4
Rudolf (Turkana) (lake), Ethiopia 111/G7
Rudolf (Turkana) (lake), Kenya 115/G3
Rudolf (isl.), U.S.S.R. 4/A7
Rudolph, Ohio (43462) 284/C3
Rudolph, Wis. (54475) 317/G7
Rudolstadt, E. Germany 22/D3
Rud Sar, Iran 66/G2
Ruds Vedby, Denmark 21/E7
Rudy, Ark. (72952) 202/B2
Rudyard, Mich. (49780) 250/E2
Rudyard, Mont. (59540) 262/F2
Rue, Switzerland 39/C3
Rueda, Spain 33/D2
Ruedi (res.), Colo. 208/F4
Rüeggisberg, Switzerland 39/E3
Rueil-Malmaison, France 28/A2
Rueter, Mo. (65744) 261/G9
Rueun, Switzerland 39/H3
Rufa'a, Sudan 102/F3
Rufa'a, Sudan 59/B7
Rufa'a, Sudan 111/F5
Rufe, Okla. (74755) 288/R6
Ruff, Wash. (†98832) 310/F3
Ruffec, France 28/D5
Ruffin, N.C. (27326) 281/K2
Ruffin, S.C. (29475) 296/F6
Rufiji (riv.), Tanzania 102/F5
Rufiji (riv.), Tanzania 115/G5
Rufino, Argentina 143/D3
Rufisque, Senegal 106/A6
Rufus, Oreg. (97050) 291/G2
Rufus Woods (lake), Wash. 310/F2
Rugby, England 13/F5
Rugby, England 10/F4
Rugby, N. Dak. (58368) 282/L3
Rugby, Tenn. (37733) 237/M8
Rugeley, England 13/E5
Rugeley, England 10/G2
Rügen (isl.), E. Germany 7/F3
Rügen (isl.), E. Germany 22/E1
Rugless, Ky. (†41127) 237/P3
Ruhama, Israel 65/B4
Ruhnu (isl.), U.S.S.R. 53/B2
Ruhr (riv.), W. Germany 22/B3
Ruhtuk (riv.), Iran 66/M7
Ruidosa, Texas (†79843) 303/C12
Ruidoso, N. Mex. (88345) 274/D5
Ruidoso Downs, N. Mex. (88346) 274/D5
Ruijin (Juichin), China 77/J6
Ruili, China 77/E7
Ruislede, Belgium 27/C6
Ruiz, Mexico 150/G6
Rujen (mt.), Bulgaria 45/F4
Rujen (mt.), Yugoslavia 45/F4
Rūjiena, U.S.S.R. 53/C2
Rukwa (reg.), Tanzania 115/F5
Rukwa (lake), Tanzania 102/F5
Rukwa (lake), Tanzania 115/F5
Rule (creek), Colo. 208/N7
Rule, Texas (79547) 303/E4
Ruleton, Kansas (†67735) 232/A2
Ruleville, Miss. (38771) 256/D3
Rulhieres (cape), W. Australia 88/D2
Rulhieres (cape), W. Australia 92/D1
Rulo, Nebr. (68431) 264/J4
Rum (cay), Bahamas 156/C2
Rum (riv.), Minn. 255/E5
Rum (isl.), Scotland 10/C2
Rum (isl.), Scotland 15/B3
Rum (sound), Scotland 15/B4
Ruma, Ill. (†62278) 222/C5
Ruma, Yugoslavia 45/D3
Rumah, Saudi Arabia 59/E4
Rumaitha, Iraq 66/D5
Ruman, Venezuela 124/H5
Rumania (Romania) 45
Rumbalara, North. Terr. 93/D8
Rumbek, Sudan 111/E6
Rumbley, Md. (†21867) 245/P8
Rumburk, Czech. 41/C1
Rumelifeneri, Turkey 63/D5
Rumford, Maine 188/M2
Rumford, Maine (04276) 243/B6
Rumford○, Maine (04276) 243/B6
Rumford, R.I. (02916) 249/J5
Rumford, S. Dak. (†57774) 298/B7
Rumford Center, Maine (04278) 243/B7
Rumford Point, Maine (04279) 243/B6
Rumia, Poland 47/D1
Rum Jungle, North. Terr. 88/E2
Rum Jungle, North. Terr. 93/B2
Rumlang, Switzerland 39/G2
Rummerfield, Pa. (†18853) 294/K2
Rumney○, N.H. (03266) 268/D4
Rumney Depot, N.H. (†03266) 268/D4
Rumoi, Japan 81/K2
Rumphi, Malawi 115/F6
Rumsey, Alberta 182/D4
Rumsey, Ky. (42371) 237/G5
Rumson, N.J. (07760) 273/F3
Rumuruti, Kenya 115/G3
Runa, W. Va. (26688) 312/E6
Runanga, N. Zealand 100/C5
Runaway (cape), N. Zealand 100/G2
Runaway Bay, Jamaica 158/J5
Runcorn, England 10/G2
Runcorn, England 13/G2
Runge, Texas (78151) 303/G9
Rungue, Chile 138/G2
Rungwa, Tanzania 115/F5
Rungwa (riv.), Tanzania 115/F5
Rungwe (mt.), Tanzania 115/F5
Runnells, Iowa (50237) 229/G5
Runnels (co.), Texas 303/E6
Runnelstown, Miss. (†39401) 256/F8
Runnemede, N.J. (08078) 273/B3
Running Water, S. Dak. (†57062) 298/N8
Runnymede, Sask. 181/K4
Runtu, Namibia 118/B3
Ruoqiang (Qarkilik), China 77/C4
Rupanco (lake), Chile 138/D3
Rupat (isl.), Indonesia 85/C5
Rupel (riv.), Belgium 27/F7
Rupert, Georgia (31081) 217/D6
Rupert, Idaho (83350) 220/E7
Rupert (riv.), Québec 174/B2
Rupert○, Vt. (05768) 268/A5
Rupert, W. Va. (25984) 312/E7
Rupert House, Que. 162/J5
Rupert House, Québec 174/B2
Rupununi (dist.), Guyana 131/B4
Rupununi (riv.), Guyana 131/B4
Rural Hall, N.C. (27045) 281/J2
Rural Hill, Miss. (†39108) 256/F4
Rural Retreat, Va. (24368) 307/F7
Rural Valley, Pa. (16249) 294/D4
Rurrenabaque, Bolivia 136/B4
Rurutu (isl.), Fr. Poly. 87/L8
Rusagonis, New Bruns. 170/D3
Rusanovo, U.S.S.R. 52/J1
Rusape, Zimbabwe 118/E3
Rüschegg, Switzerland 39/D3
Ruse, Bulgaria 7/G4
Ruse, Bulgaria 45/H4
Rush, Colo. (80833) 208/L5
Rush (creek), Colo. 208/N5
Rush (co.), Ind. 227/G5
Rush, Ireland 10/C4
Rush, Ireland 17/J4
Rush (co.), Kansas 232/C3
Rush, Ky. (41168) 237/R4
Rush, N.Y. (14543) 276/E5
Rush (lake), N. Dak. 282/N2
Rush (riv.), N. Dak. 282/R5
Rush Center, Kansas (67575) 232/C3
Rush City, Minn. (55069) 255/F5
Rushden, England 13/G5
Rushden, England 10/F4
Rushford, Minn. (55971) 255/G7
Rushford, N.Y. (14777) 276/D6
Rush Hill, Mo. (65280) 261/J4
Rush Lake, Sask. 181/D5
Rush Lake, Wis. (†54971) 317/J8
Rushmere, Va. (†23430) 307/P6
Rushmore, Minn. (56168) 255/C7
Rushoon, Newf. 166/D4
Rush River, Minn. (†56058) 255/D6
Rush Run, Ohio (†43943) 284/J5
Rush Springs, Okla. (73082) 288/L5
Rushsylvania, Ohio (43347) 284/C5
Rushville, Ill. (62681) 222/C3
Rushville, Ind. (46173) 227/G5
Rushville, Mo. (64484) 261/B3
Rushville, Nebr. (69360) 264/B2
Rushville, N.Y. (14544) 276/F5
Rushville, Ohio (43150) 284/F6
Rushworth, Victoria 97/C5
Rusk, Ind. (†47581) 227/D7
Rusk (co.), Texas 303/K5
Rusk, Texas (75785) 303/J6
Rusk (co.), Wis. 317/D5
Rusk, Wis. (†54751) 317/C6
Ruskin, Fla. (33570) 212/C3
Ruskin, Nebr. (68974) 264/G4
Ruskington, England 13/G4
Ruso, N. Dak. (58778) 282/J4
Russas, Brazil 132/G4
Russell (co.), Ala. 195/H6
Russell (lake), Alberta 182/C1
Russell, Ark. (72139) 202/G3
Russell, Georgia (†30680) 217/E3
Russell, Iowa (50238) 229/G7
Russell (co.), Kansas 232/D3
Russell, Kansas (67665) 232/D3
Russell (co.), Ky. 237/L7
Russell, Ky. (41169) 237/R3
Russell, Manitoba 179/A4
Russell○, Mass. (01071) 249/C4
Russell, Minn. (56169) 255/C6
Russell, Miss. (†39301) 256/G6
Russell, N.Y. (13684) 276/K2
Russell, N. Zealand 100/E1
Russell, N. Dak. (†58762) 282/J2
Russell (cape), N.W. Terrs. 187/G2
Russell (isl.), N.W. Terrs. 187/J2
Russell (pt.), N.W. Terrs. 187/G2
Russell (county), Ontario 177/J2
Russell, Ontario 177/J2
Russell, Pa. (16345) 294/D2
Russell (isls.), Solomon Is. 86/D3
Russell (co.), Va. 307/D7
Russell Cave Nat'l Mon., Ala. 195/G1
Russell Fork (riv.), Va. 307/C5
Russells Point, Ohio (43348) 284/C5
Russell Springs, Kansas (67755) 232/A3
Russell Springs, Ky. (42642) 237/L6
Russellton, Pa. (15076) 294/C4
Russellville, Ala. (35653) 195/C2
Russellville, Ark. (72801) 202/D3
Russellville, Ill. (†47591) 222/F5
Russellville, Ind. (46175) 227/D5
Russellville, Ky. (42276) 237/H7
Russellville, Mo. (65074) 261/H6
Russellville, Ohio (45168) 284/C8
Russellville, S.C. (29476) 296/H5
Russellville, Tenn. (37860) 237/P8
Russellville, W. Va. (26689) 312/E6
Rüsselsheim, W. Germany 22/C4
Russia, Ohio (45363) 284/B5
Russian (riv.), Calif. 204/B4
Russian Mission, Alaska (99657) 196/F2
Russian S.F.S.R., U.S.S.R. 7/H3
Russian S.F.S.R., U.S.S.R. 52/F3
Russian S.F.S.R., U.S.S.R. 48/D4
Russian Soviet Federated Socialist Republic, U.S.S 54/L3
Russiaville, Ind. (46975) 227/E4
Russkiy Zavorot (cape), U.S.S.R. 52/H1
Russum, Miss. (†39150) 256/B7
Rust, Austria 41/D3
Rustad, Minn. (†56560) 255/B4
Rustavi, U.S.S.R. 52/G6
Rustburg, Va. (24588) 307/K6
Rustenburg, S. Africa 118/D5
Ruston, La. (71270) 238/E1
Ruston, Wash. (†98401) 310/C3
Ruswil, Switzerland 39/F2
Rutan, Sask. 181/F3
Rutba, Iraq 59/D3
Rutba, Iraq 66/B4
Rute, Spain 33/D4
Ruteng, Indonesia 85/G7
Ruth, Mich. (48470) 250/G5
Ruth, Miss. (39662) 256/D8
Ruth, Nev. (89319) 266/F3
Ruth, N.C. (†28139) 281/E4
Rutherford, Ala. (†36860) 195/H6
Rutherford, N.J. (*07070) 273/B2
Rutherford (co.), N.C. 281/E4
Rutherford (co.), Tenn. 237/J9
Rutherford, Tenn. (38369) 237/C8
Rutherford College, N.C. (28671) 281/F3
Rutherford Fork, Obion (riv.), Tenn. 237/D8
Rutherfordton, N.C. (28139) 281/E4
Rutherglen, Ontario 177/F1
Rutherglen, Scotland 15/B2
Rutherglen, Scotland 10/B1
Rutherglen, Victoria 97/D5
Ruther Glen, Va. (22546) 307/O5
Rutheron, N. Mex. (87563) 274/C2
Rüthi, Switzerland 39/J2
Ruthilda, Sask. 181/C4
Ruthin, Wales 10/E4
Ruthin, Wales 13/D4
Ruthsburg, Md. (†21617) 245/P4
Ruthton, Minn. (56170) 255/B6
Ruthven, Iowa (51358) 229/D2
Ruthven, Ontario 177/B6
Ruthville, Va. (23147) 307/P6
Rüti, Glarus, Switzerland 39/H3
Rüti, Zürich, Switzerland 39/G2
Rutland, Br. Col. 184/H5
Rutland, Ill. (61358) 222/D3
Rutland (isl.), India 68/G6
Rutland, Iowa (50582) 229/E3
Rutland, Mass. (01543) 249/G3
Rutland○, Mass. (01543) 249/G3
Rutland, N. Dak. (58067) 282/P7
Rutland, Ohio (45775) 284/F7
Rutland, Sask. 181/B3
Rutland, S. Dak. (57057) 298/P5
Rutland (co.), Vt. 268/A4
Rutland, Vt. (05701) 268/B4
Rutland○, Vt. (05701) 268/B4
Rutland, Vt. 188/M2
Rutland Plains, Queensland 95/B2
Rutledge, Ala. (35228) 195/F7
Rutledge, Georgia (30663) 217/E3
Rutledge, Minn. (55778) 255/F4
Rutledge, Mo. (63563) 261/H2
Rutledge, Pa. (19070) 294/M7
Rutledge, Tenn. (37861) 237/P8
Rutog, China 77/A5
Rutshuru, Zaire 115/E4
Rutten, Netherlands 27/H3
Ruurlo, Netherlands 27/J4
Ruus al Jibal (dist.), Oman 59/G4
Ruvo di Puglia, Italy 34/F4
Ruvuma (reg.), Tanzania 115/G6
Ruwais, U.A.E. 59/F5
Ruwandiz, Iraq 66/D2
Ruwaq, Jebel er (mts.), Syria 63/G5
Ruwenzori (range), Uganda 115/E3
Ruwenzori (range), Zaire 115/E3
Ruzayevka, U.S.S.R. 52/F4
Ruzizi (riv.), Burundi 115/E4
Ruzizi (riv.), Rwanda 115/E4
Ruzizi (riv.), Zaire 115/E4
Ružomberok, Czech. 41/E2
Rwanda 2/L6
Rwanda 102/F5
RWANDA 115/E4
Ry, Denmark 21/C5
Ryan (peak), Idaho 220/D6
Ryan, Iowa (52330) 229/K4
Ryan, Okla. (73565) 288/L6
Ryan, Loch (inlet), Scotland 15/C5
Ryan Park, Wyo. (82331) 319/F4
Ryans (bay), Newf. 166/B2
Ryazan', U.S.S.R. 7/H3
Ryazan', U.S.S.R. 48/E4
Ryazan', U.S.S.R. 52/E4
Ryazhsk, U.S.S.R. 52/F4
Rybachiy (pen.), U.S.S.R. 48/D2
Rybachiy (pen.), U.S.S.R. 52/D1
Rybach'ye, U.S.S.R. 48/H5
Rybinsk, U.S.S.R. 7/H3
Rybinsk (res.), U.S.S.R. 7/H3
Rybinsk, U.S.S.R. 52/E3
Rybinsk, U.S.S.R. 48/D4
Rybinsk (res.), U.S.S.R. 52/E3
Rybinsk (res.), U.S.S.R. 48/D4
Rybnik, Poland 47/D3
Rybnitsa, U.S.S.R. 52/C5
Rychnov nad Kněžnou, Czech. 41/D1
Rycroft, Alberta 182/A2
Rydal, Georgia (30171) 217/C2
Ryde, Calif. (95680) 204/B9
Ryde, England 10/F5
Ryde, England 13/F7
Ryde, N.S. Wales 88/K4
Ryde, N.S. Wales 97/J3
Ryder, N. Dak. (58779) 282/G4
Ryderwood, Wash. (98581) 310/B4
Rye, Ark. (†71671) 202/F6
Rye, Colo. (81069) 208/K7
Rye, England 13/H7
Rye, England 10/G5
Rye (bay), England 13/H7
Rye○, N.H. (03870) 268/F5
Rye, N.Y. (10580) 276/P6
Rye, Texas (77369) 303/K7
Rye Beach, N.H. (03871) 268/F6
Ryegate, Mont. (59074) 262/G4
Ryegate○, Vt. (05042) 268/C3
Rye North Beach, N.H. (†03870) 268/F5
Rye Patch (res.), Nev. 266/C2
Ryerson, Sask. 181/K6
Rye Valley, Oreg. (†97907) 291/K3
Ryggebyen, Norway 18/D4
Ryki, Poland 47/F3
Ryland, Ala. (35767) 195/F1
Ryland, N.C. (†27980) 281/R2
Ryland Heights, Ky. (†41015) 237/M3
Ryley, Alberta 182/D3
Rylstone, N.S. Wales 97/E3
Rýmařov, Czech. 41/D2
Ryomgård, Denmark 21/D5
Ryotsu, Japan 81/J4
Rypin, Poland 47/D2
Rysy (mt.), Poland 47/D4
Ryton, England 13/H3
Ryugasaki, Japan 81/P2
Ryukyu (isls.), Japan 54/O7
Ryukyu (isls.), Japan 2/R4
Ryukyu (isls.), Japan 81/L7
Rzepin, Poland 47/B2
Rzeszów (prov.), Poland 47/F4
Rzeszów, Poland 47/F4
Rzhev, U.S.S.R. 7/H3
Rzhev, U.S.S.R. 48/D4
Rzhev, U.S.S.R. 52/D3

S

Sa'ad, Israel 65/B5
Sa'ada, Yemen Arab Rep. 59/D6
Saale (riv.), E. Germany 22/D3
Saalfeld, E. Germany 22/D3
Saalfelden am Steinernen Meer, Austria 41/B3
Saane (Sarine) (riv.), Switzerland 39/D3
Saanen, Switzerland 39/D4
Saanich, Br. Col. 184/K3
Saar (riv.), France 28/G3
Saar (riv.), W. Germany 22/B4
Saarbrücken, W. Germany 7/E4
Saarbrücken, W. Germany 22/B4
Saarburg, W. Germany 22/B4
Saaremaa (isl.), U.S.S.R. 7/G3
Saaremaa (isl.), U.S.S.R. 53/B1
Saaremaa (isl.), U.S.S.R. 52/B3
Saaremaa (isl.), U.S.S.R. 48/B4
Saarijärvi, Finland 18/O5
Saarland (state), W. Germany 22/B4
Saarlouis, W. Germany 22/B4

Saas, Switzerland 39/J3
Saas Fee, Switzerland 39/E4
Saba (isl.), Neth. Ant. 156/F3
Saba (isl.), Virgin Is. (U.S.) 161/A4
Šabac, Yugoslavia 45/D3
Sabadell, Spain 7/E4
Sabadell, Spain 33/H2
Sabae, Japan 81/H5
Sabah (state), Malaysia 2/Q5
Sabah (state), Malaysia 85/F4
Sabah (reg.), Malaysia 54/N9
Sábalo, Cuba 158/A2
Sabana, Cuba 158/K4
Sabana (arch.), Cuba 158/E1
Sabana de la Mar, Dom. Rep. 156/E3
Sabana de la Mar, Dom. Rep. 158/F5
Sabana Grande, Dom. Rep. 158/E6
Sabanagrande, Honduras 154/D4
Sabana Grande, P. Rico 161/B2
Sabanalarga, Colombia 126/C2
Sabana Seca, P. Rico 161/D1
Sabancuy, Mexico 150/O7
Sabaneta, Dom. Rep. 158/D5
Sabaneta, Barinas, Venezuela 124/D3
Sabaneta, Falcón, Venezuela 124/D2
Sabang, Celebes, Indonesia 85/F5
Sabang, Weh, Indonesia 85/B4
Şabanözü, Turkey 63/E2
Sabará, Brazil 135/E1
Sabattus, Maine (04280) 243/C7
Sabattus○, Maine (04280) 243/C7
Sabaudia, Italy 34/D4
Sabaya, Bolivia 136/A6
Saberi, Hamun-e (lake), Iran 66/M5
Sabetha, Kansas (66534) 232/G2
Sabi (riv.), Zimbabwe 118/E3
Sabile, U.S.S.R. 53/B2
Sabillasville, Md. (21780) 245/J2
Sabin, Minn. (56580) 255/B4
Sabina, Ohio (45169) 284/C7
Sabinal (cay), Cuba 158/H2
Sabinal, Texas (78881) 303/E8
Sabinas, Mexico 150/J3
Sabinas (riv.), Mexico 150/J3
Sabinas Hidalgo, Mexico 150/J3
Sabine (riv.) 188/H4
Sabine (mt.) 5/B9
Sabine (par.), La. 238/C3
Sabine (lake), La. 238/C7
Sabine (passage), La. 238/C7
Sabine (riv.), La. 238/C5
Sabine (pen.), N.W. Terrs. 187/H2
Sabine (co.), Texas 303/L6
Sabine, Texas (†77640) 303/L8
Sabine (lake), Texas 303/L8
Sabine (riv.), Texas 303/L7
Sabine Pass, Texas (77655) 303/L8
Sabinópolis, Brazil 132/F7
Sabinoso, N. Mex. (†87746) 274/E3
Sabinov, Czech. 41/F2
Sabinsville, Pa. (16943) 294/G2
Sabir, Jebel (mt.), Yemen Arab Rep. 59/D7
Sabirabad, U.S.S.R. 52/G6
Sabkha, Syria 63/H5
Sablayan, Philippines 82/C4
Sable (cape), Fla. 188/K5
Sable (cape), Fla. 212/E6
Sable (cape), N.S. 146/M5
Sable (isl.), N.S. 146/N5
Sable (cape), N.S. 162/K7
Sable (isl.), N.S. 162/L7
Sable (cape), Nova Scotia 168/C5
Sable (isl.), Nova Scotia 168/J5
Sable (riv.), Ontario 177/B1
Sable (riv.), Québec 174/D1
Sable River, Nova Scotia 168/C5
Sables (lake), Québec 172/B3
Sables (lake), Québec 172/H1
Sablé-sur-Sarthe, France 28/C4
Sabougla, Miss. (†38955) 256/F3
Sabra (cape), Indonesia 85/J6
Sabrathaa, Libya 111/B1
Sabrina Coast (reg.) 5/C6
Sabtang, Philippines 82/B2
Sabtang (isl.), Philippines 82/B2
Sabugal, Portugal 33/C2
Sabula, Iowa (52070) 229/N4
Sabula, Mo. (†63620) 261/L8
Sabula, Pa. (†15801) 294/E3
Sabya, Saudi Arabia 59/D6
Sabzevar, Iran 54/G6
Sabzevar, Iran 59/G2
Sabzevar, Iran 66/K2
Sabzvaran, Iran 66/K6
Sabzvaran, Iran 59/G4
Sac (co.), Iowa 229/C4
Sac (riv.), Mo. 261/E7
Sacaba, Bolivia 136/C5
Sacaca, Bolivia 136/B6
Sacajawea (peak), Oreg. 291/K2
Sacajawea (lake), Wash. 310/G4
Sácama, Colombia 126/D4
Sacandaga (lake), N.Y. 276/L3
Sac and Fox Ind. Res., Iowa 229/H5
Sacapulas, Guatemala 154/B3
Sacaton, Ariz. (85247) 198/D5
Sacavém, Portugal 33/A1
Sac City, Iowa (50583) 229/C4
Sacedón, Spain 33/E2
Săcele, Romania 45/G3
Sac-Fox-Iowa Ind. Res., Kansas 232/G2
Sacheen (lake), Wash. 310/H2
Sachem (head), Conn. 210/E4
Sachem Head, Conn. (†06437) 210/E3
Sachigo (riv.), Ont. 162/G5
Sachigo (riv.), Ontario 175/B2
Sachojere, Bolivia 136/C4
Sachse, Texas (†75040) 303/H2
Sachseln, Switzerland 39/F3
Sachs Harbour, Canada 4/B16
Sachs Harbour, N.W.T. 162/D1
Sachs Harbour, N.W. Terrs. 187/F2
Sackets (harb.), N.Y. 276/H3
Sackets Harbor, N.Y. (13685) 276/H3
Säckingen, W. Germany 22/C5
Sackville, New Bruns. 170/F3
Sackville, Nova Scotia 168/E4
Saco, Ala. (†36081) 195/G7
Saco, Maine (04072) 243/C8
Saco (riv.), Maine 243/B8
Saco, Mo. (†63645) 261/M8
Saco, Mont. (59261) 262/J2
Saco (riv.), N.H. 268/E3
Sacol (isl.), Philippines 82/D7
Sacramento, Brazil 132/D7
Sacramento, Brazil 135/C1
Sacramento (cap.), Calif. 146/F6
Sacramento (cap.), Calif. 188/B3
Sacramento (riv.), Calif. 188/B3
Sacramento (co.), Calif. 204/D5
Sacramento (cap.), Calif. (*95801) 204/B8
Sacramento (riv.), Calif. 204/D5
Sacramento, Ky. (42372) 237/G6
Sacramento, N. Mex. (88347) 274/D6
Sacramento (mts.), N. Mex. 274/D6
Sacramento Army Depot, Calif. 204/B8
Sacramento Wash (dry riv.), Ariz. 198/A4
Sacratif (cape), Spain 33/E4
Sacré-Coeur-de-Saguenay, Québec 172/H1
Sacred Heart, Minn. (56285) 255/C6
Sacul, Texas (75788) 303/K6
Sádaba, Spain 33/F1
Sadani, Tanzania 115/G5
Saddle (hills), Alberta 182/A2
Saddle, Ark. (†72554) 202/G1
Saddle (mt.), Idaho 220/F6
Saddle (mt.), Idaho 220/D3
Saddle (riv.), N.J. 273/B1
Saddle (mts.), Wash. 310/E4
Saddle Brook○, N.J. (07662) 273/B1
Saddle Mountain, Okla. (†73023) 288/J5
Saddle River, N.J. (07458) 273/B1
Saddlestring, Wyo. (82840) 319/F1
Saddleworth, England 13/J2
Saddleworth, England 10/G2
Sa Dec, Vietnam 72/E5
Sadhoowa, Trin. & Tob. 161/B11
Sadieville, Ky. (40370) 237/M4
Sadij (riv.), Iran 66/L8
Sadiya, India 68/H3
Sa'diya, Iraq 66/D3
Sa'diya, Hor (lake), Iraq 66/E4
Sadlers Village, St. Chris.-Nevis 161/C10
Sadlersville, Tenn. (37154) 237/G7
Sado (isl.), Japan 81/J4
Sado (riv.), Portugal 33/B3
Sadon, Burma 72/C1
Sadorus, Ill. (61872) 222/E4
Saeby, Denmark 18/G8
Saeby, Denmark 21/D3
Saegertown, Pa. (16433) 294/B2
Saetermoen, Norway 18/L2
Saetermoen, Norway 18/L2
Safad (Zefat), Israel 65/C2
Safaniya, Ras (cape), Saudi Arabia 59/E4
Šafárikovo, Czech. 41/F2
Safata (bay), W. Samoa 86/M9
Safe, Mo. (†65559) 261/J6
Safety Harbor, Fla. (33572) 212/B2
Säffle, Sweden 18/H7
Safford, Ala. (36773) 195/D6
Safford, Ariz. (85546) 198/F6
Saffordville, Kansas (†66869) 232/F3
Saffron Walden, England 10/G4
Saffron Walden, England 13/H5
Safi, Jordan 65/E5
Safi, Morocco 102/B1
Safi, Morocco 106/C2
Safidar, Kuh-e (mt.), Iran 59/F4
Safidar, Kuh-e (mt.), Iran 66/H6
Safid Rud (riv.), Iran 66/F2
Safien, Switzerland 39/H3
Safita, Syria 63/G5
Safonovo, U.S.S.R. 52/D3
Safranbolu, Turkey 63/E2
Safut, Jordan 65/D3
Saga, China 77/B6
Saga (pref.), Japan 81/E7
Saga, Japan 81/E7
Sagadahoc (co.), Maine 243/D7
Sagaing (div.), Burma 72/B1
Sagaing, Burma 72/B2
Sagami (bay), Japan 81/O3
Sagami (riv.), Japan 81/O2
Sagami (sea), Japan 81/J6
Sagamihara, Japan 81/O2
Sagamore, Mass. (02561) 249/M5
Sagamore, Pa. (16250) 294/D4
Sagamore Hill Nat'l Hist. Site, N.Y. 276/R6
Sagamore Hills, Ohio (†44067) 284/J10
Saganaga (lake), Minn. 255/H2
Saganaga (lake), Ontario 175/B3
Sagar, India 68/D4
Sagavanirktok (riv.), Alaska 196/J1
Sagay, Camiguin, Philippines 82/E6
Sagay, Negros Occ., Philippines 82/D5
Sage, Ark. (72573) 202/G1
Sage (creek), Mont. 262/F2
Sage (mt.), Virgin Is. (Br.) 161/D4
Sage, Wyo. (†83101) 319/B4
Sagemace (bay), Manitoba 179/B3
Sagerton, Texas (79548) 303/E4
Sageville, Iowa (†52001) 229/M3
Sag Harbor, N.Y. (11963) 276/R8
Saginaw, Ala. (35137) 195/E4
Saginaw, Mich. 188/K2
Saginaw (bay), Mich. 188/K2
Saginaw (co.), Mich. 250/E5
Saginaw, Mich. (*48601) 250/F5
Saginaw (bay), Mich. 250/F5
Saginaw (riv.), Mich. 250/F5
Saginaw, Minn. (55779) 255/F4
Saginaw, Mo. (64864) 261/C8
Saginaw, Oreg. (97472) 291/E4
Saginaw, Texas (76179) 303/E2
Sagle, Idaho (83860) 220/B1
Saglek (bay), Newf. 166/B2
Saglek (fiord), Newf. 166/B2
Saglouc, Que. 162/J3
Saglouc, Québec 174/E1
Sagnay, Philippines 82/D4
Sagola, Mich. (49881) 250/B2
Saguache (co.), Colo. 208/G6
Saguache, Colo. (81149) 208/G6
Saguache (creek), Colo. 208/F6
Sagua de Tánamo, Cuba 158/K3
Sagua la Grande, Cuba 156/B2
Sagua la Grande, Cuba 158/E1
Sagua la Grande (riv.), Cuba 158/E1
Saguaro (lake), Ariz. 198/D5
Saguaro Nat'l Mon., Ariz. 198/E6
Saguenay (county), Québec 174/D2
Saguenay (co.), Québec 172/H1
Saguenay (riv.), Québec 174/C3
Saguenay (riv.), Québec 172/G1
Saguia el Hamra (dry riv.), Western Sahara 106/B3
Sagunto, Spain 33/F3
Sa'gya, China 77/C6
Sahagún, Colombia 126/C3
Sahagún, Spain 33/D1
Sahand, Kuh-e (mt.), Iran 66/E2
Sahara (desert) 2/J4
Sahara (des.) 102/C2
Sahara (des.), Algeria 106/E4
Sahara (des.), Chad 111/C3
Sahara (des.), Egypt 111/E3
Sahara (des.), Libya 111/C3
Sahara (des.), Mali 106/D4
Sahara (des.), Mauritania 106/C4
Sahara (des.), Niger 106/F4
Sahara (des.), Sudan 111/E3
Saharan Atlas (ranges), Algeria 106/E2
Saharanpur, India 68/D3
Saharsa, India 68/F3
Şahinli, Turkey 63/C6
Sahiwal, Pakistan 68/C2
Sahiwal, Pakistan 59/K3
Sahuaripa, Mexico 150/E2
Sahuarita, Ariz. (85629) 198/E7
Sahuayo de Díaz, Mexico 150/H7
Šahy, Czech. 41/E2
Saïda, Algeria 106/E2
Saida, Lebanon 63/F6
Sa'idabad, Iran 66/J6
Sa'idabad, Iran 59/G4
Saïdia, Morocco 106/D2
Saidor, Papua N.G. 85/B7
Saidu, Pakistan 68/C2
Saignelégier, Switzerland 39/D2
Saigo, Japan 81/F5
Saigon (Ho Chi Minh City), Vietnam 54/M8
Saihut, P.D.R. Yemen 54/G8
Saihut, P.D.R. Yemen 59/F6
Saikai National Park, Japan 81/D7
Saiki, Japan 81/E7
Sailes, La. (†71028) 238/D2
Sailor (creek), Idaho 220/C7
Sailor Springs, Ill. (62879) 222/E5
Saimaa (lake), Finland 18/Q6
Saimbeyli, Turkey 63/G4
Sain Alto, Mexico 150/H5
Sain-ni, N. Korea 81/B4
Saint Abbs, Scotland 15/F5
Saint Abbs (head), Scotland 15/F5
Saint-Adalbert, Québec 172/H3
Saint-Adelme, Québec 172/B1
Saint-Adelphe, Québec 172/E3
Saint Adolphe, Manitoba 179/E5
Saint-Adolphe, Québec 172/F2
Saint-Adolphe-d'Howard, Québec 172/C4
Saint-Adrien, Québec 172/F4
Saint-Affrique, France 28/E6
Saint-Agapitville, Québec 172/F3
Saint Agatha○, Maine (04772) 243/G1
Saint Agnes, England 13/B7
Saint-Aimé-des-Lacs, Québec 172/G2
Saint-Alban, Québec 172/E3
Saint Albans, England 13/H7
Saint Albans, England 10/F5
Saint Alban's (head), England 13/F7
Saint Albans○, Maine (04971) 243/E6
Saint Albans, Mo. (63073) 261/L5
Saint Alban's, Newf. 166/C4
Saint Albans, Vt. (05478) 268/A2
Saint Albans○, Vt. (05478) 268/A2
Saint Albans, W. Va. (25177) 312/C6
Saint Albans Bay, Vt. (05481) 268/A2
Saint Albert, Alberta 182/D3
Saint Albert, Ontario 177/J2
Saint-Albert, Québec 172/E3
Saint-Alexandre, Québec 172/D4
Saint-Alexandre-de-Kamouraska, Québec 172/H2
Saint-Alexis, Québec 172/D4
Saint-Alexis-de-Matapédia, Québec 172/B2
Saint-Alexis-des-Monts, Québec 172/D3
Saint Almo, New Bruns. 170/C2
Saint Alphonse, Manitoba 179/C5
Saint-Alphonse, Québec 172/D3
Saint Alphonse de Clare, Nova Scotia 168/B4
Saint-Alphonse-de-Caplan, Québec 172/C2
Saint-Amable, Québec 172/J4
Saint-Amand-Mont-Rond, France 28/E4
Saint Amant, La. (70774) 238/L2
Saint Ambroise, Manitoba 179/E4
Saint-Ambroise, Québec 172/F1
Saint-Anaclet, Québec 172/J1
Saint-André (cape), Madagascar 118/G3
Saint-André, New Bruns. 170/C1
Saint-André, Québec 172/B2
Saint-André, Réunion 118/G5
Saint-André-Avellin, Québec 172/B4
Saint-André-de-Kamouraska, Québec 172/H2
Saint-André-du-Lac-Saint-Jean, Québec 172/E1
Saint-André-Est, Québec 172/C4
Saint Andrew (pt.), Fla. 212/D6
Saint Andrew (sound), Georgia 217/K9
Saint Andrew (lake), Manitoba 179/E3
Saint Andrew (mt.), St. Vin. & Grens. 161/A9
Saint Andrews, New Bruns. 170/C3
Saint Andrew's, Newf. 166/C4
Saint Andrews, Nova Scotia 168/F3
Saint Andrews (chan.), Nova Scotia 168/H2
Saint Andrews, Scotland 15/F4
Saint Andrews, Scotland 10/E2
Saint Andrews (bay), Scotland 15/F4
Saint Andrews, S.C. (29407) 296/G6
Saint Andrews, Tenn. (37372) 237/K10
Saint-Anicet, Québec 172/C4
Saint Ann, Mo. (63074) 261/O2
Saint Anne, Chan. Is. 13/E8
Saint Anne, Ill. (60964) 222/F2
Saint Anns (bay), Nova Scotia 168/H2
Saint Anns, Ontario 177/E4
Saint Ann's Bay, Jamaica 156/C3
Saint Ann's Bay, Jamaica 158/J5
Saint-Anselme, Québec 172/F3
Saint Ansgar, Iowa (50472) 229/H2
Saint Anthony, Idaho (83445) 220/G6
Saint Anthony, Ind. (47575) 227/D8
Saint Anthony, Iowa (50239) 229/G4
Saint Anthony, Minn. (†56307) 255/D5
Saint Anthony, Minn. (55414) 255/G5
Saint Anthony, Newf. 166/C3
Saint Anthony, N. Dak. (58566) 282/H6
Saint-Antoine, New Bruns. 170/F2
Saint-Antoine, Québec 172/H4
Saint-Antoine-Abbé, Québec 172/D4
Saint-Antoine-sur-Richelieu, Québec 172/D4
Saint-Antonin, Québec 172/H2
Saint-Antonin-Noble-Val, France 28/D5
Saint Arnaud, Victoria 97/B5
Saint-Arsène, Québec 172/H2
Saint Arthur, New Bruns. 170/D1
Saint-Astier, France 28/D5
Saint-Athanase, Québec 172/H2
Saint-Aubert, Québec 172/G2
Saint Aubin, Chan. Is. 13/E8
Saint-Aubin-Sauges, Switzerland 39/C3
Saint Augustin (riv.), Newf. 166/C3
Saint-Augustin, Québec 172/G4
Saint-Augustin, Québec 174/F2
Saint-Augustin (riv.), Québec 174/F2
Saint-Augustin-de-Québec, Québec 172/E3
Saint Augustine, Fla. 188/K5
Saint Augustine, Fla. 146/K7
Saint Augustine, Fla. (32084) 212/E2
Saint Augustine, Ill. (61474) 222/C3
Saint Augustine, Md. (†21915) 245/P3
Saint Augustine Beach, Fla. (32084) 212/E2
Saint Austell (bay), England 13/C7
Saint Austell-with-Fowey, England 13/C7
Saint Austell with Fowey, England 10/D5
Saint-Barnabé, Québec 172/D4
Saint-Barthélemy (isl.), Guadeloupe 156/F3
Saint-Barthélemy, Québec 172/D3
Saint Basile, New Bruns. 170/B1
Saint-Basile-le-Grand, Québec 172/J4
Saint-Basile-Sud, Québec 172/F3
Saint Bees (head), England 13/D3
Saint Benedict, Kansas (†66538) 232/F2
Saint Benedict, La. (70457) 238/K5
Saint Benedict, Oreg. (97373) 291/B3
Saint Benedict, Pa. (15773) 294/E4
Saint Benedict, Sask. 181/F3
Saint-Benjamin, Québec 172/G3
Saint-Benoît, Réunion 118/G5
Saint-Benoît-Labre, Québec 172/G3
Saint Bernard, Ala. (35138) 195/E2
Saint Bernard (par.), La. 238/L7
Saint Bernard, La. (70085) 238/L7
Saint Bernard, Nova Scotia 168/B4
Saint Bernard, Ohio (45217) 284/B9
Saint-Bernard, Québec 172/F3
Saint-Bernard, Great (pass), Switzerland 39/D5
Saint-Bernard-sur-Mer, Québec 172/G2
Saint Bernice, Ind. (47875) 227/C5
Saint Bethlehem, Tenn. (37155) 237/G7
Saint Blaise, Québec 172/E4
Saint-Blaise, Switzerland 39/D2
Saint-Bonaventure-de-Yamaska, Québec 172/E4
Saint-Boniface-de-Shawinigan, Québec 172/D3
Saint Bonifacius, Minn. (55375) 255/F6
Saint Brendan's, Newf. 166/D4
Saint Brides, Alberta 182/E2
Saint Bride's, Newf. 166/C2
Saint Brides (bay), Wales 10/D5
Saint Brides (bay), Wales 13/B6
Saint-Brieuc, France 28/B3
Saint Brieux, Sask. 181/G3
Saint-Bruno, Québec 172/F1
Saint-Bruno-de-Montarville, Québec 172/J4
Saint-Calais, France 28/D4
Saint-Calixte-de-Kilkenny, Québec 172/D4
Saint-Camille, Québec 172/F4
Saint-Camille-de-Bellechasse, Québec 172/G3
Saint-Casimir, Québec 172/E3
Saint Catharine, Mo. (64677) 261/G3
Saint Catharines, Ontario 177/E4
Saint Catherine, Fla. (†33513) 212/D3
Saint Catherine (mt.), Grenada 161/D8
Saint Catherine (lake), Vt. 268/A5
Saint Catherines (isl.), Georgia 217/K7
Saint Catherines (sound), Georgia 217/K7
Saint-Céré, France 28/D5
Saint-Cergue, Switzerland 39/B4
Saint-Césaire, Québec 172/D4
Saint-Chamond, France 28/F5
Saint Charles, Ark. (72140) 202/H5
Saint Charles, Idaho (83272) 220/G7
Saint Charles, Ill. (60174) 222/E2
Saint Charles, Iowa (50240) 229/F6
Saint Charles, Ky. (42453) 237/F6
Saint Charles (par.), La. 238/K7
Saint Charles, Mich. (48655) 250/E5
Saint Charles, Minn. (55972) 255/F7
Saint Charles (co.), Mo. 261/M2
Saint Charles, Mo. (63301) 261/N1
Saint-Charles, New Bruns. 170/F2
Saint Charles, Ontario 177/D1
Saint-Charles, Bellechasse, Québec 172/G3
Saint Charles, S.C. (29134) 296/G3
Saint Charles, S. Dak. (57571) 298/L7
Saint Charles, Va. (24282) 307/B7
Saint-Charles-de-Mandeville, Québec 172/D3
Saint-Charles-Garnier, Québec 172/J1
Saint-Charles-sur-Richelieu, Québec 172/D4
SAINT CHRISTOPHER-NEVIS 156/F3
SAINT CHRISTOPHER (SAINT KITTS)-NEVIS 161/D11
Saint Christopher (isl.), St. Chris.-Nevis 156/F3
Saint Christopher (isl.), St. Chris.-Nevis 161/D10
Saint Chrysostom, Pr. Edward I. 168/E2
Saint-Chrysostome, Québec 172/D4
Saint Clair (co.), Ala. 195/F3
Saint Clair, Ala. (36774) 195/E6
Saint Clair, Georgia (†30816) 217/H4
Saint Clair (co.), Ill. 222/D5
Saint Clair (lake), Mich. 188/K2
Saint Clair (co.), Mich. 250/G6
Saint Clair, Mich. (48079) 250/G6
Saint Clair (lake), Mich. 250/G6
Saint Clair (riv.), Mich. 250/G6
Saint Clair, Minn. (56080) 255/E6
Saint Clair (co.), Mo. 261/E6
Saint Clair, Mo. (63077) 261/K6
Saint Clair (lake), Ontario 177/B5
Saint Clair (riv.), Ontario 177/B5
Saint Clair, Pa. (17970) 294/K4
Saint Clair (lake), Tasmania 99/C4
Saint Clair Beach, Ontario 177/B5
Saint Clair Shores, Mich. (*48080) 250/B6
Saint Clair Springs, Ala. (†35146) 195/F3
Saint Clairsville, Ohio (43950) 284/J5
Saint Clairsville, Pa. (16676) 294/F5
Saint-Claude, France 28/F4
Saint-Claude, Guadeloupe 161/A7
Saint Claude, Manitoba 179/D5
Saint-Claude, Québec 172/F4
Saint Clears, Wales 13/C6
Saint-Clément, Québec 172/H2
Saint Clements, Ontario 177/D4
Saint-Cléophas, Québec 172/D3
Saint-Clet, Québec 172/C4
Saint Cloud, Fla. (32769) 212/E3
Saint-Cloud, France 28/A2
Saint Cloud, Minn. 188/H1
Saint Cloud, Minn. (56301) 255/D5
Saint Cloud, Wis. (53079) 317/K8
Saint Columb Major, England 13/B7
Saint Combs, Scotland 15/G3
Saint-Côme, Québec 172/D3
Saint-Constant, Québec 172/H4
Saint Croix (riv.) 188/H1
Saint Croix, Ind. (47576) 227/D8
Saint Croix (riv.), Maine 243/J5
Saint Croix (riv.), Minn. 255/F5
Saint Croix, New Bruns. 170/C3
Saint Croix (riv.), New Bruns. 170/C3
Saint Croix, Nova Scotia 168/E4
Saint Croix (isl.), Virgin Is. (U.S.) 156/H2
Saint Croix (isl.), Virgin Is. (U.S.) 161/C4
Saint Croix (co.), Wis. 317/B5
Saint Croix (lake), Wis. 317/A6
Saint Croix (riv.), Wis. 317/A4
Saint Croix Falls, Wis. (54024) 317/A5
Saint Croix Flowage (res.), Wis. 317/C3
Saint Croix Isl. Nat'l Mon., Maine 243/J5
Saint-Cuthbert, Québec 172/D3
Saint-Cyprien, Québec 172/J2
Saint-Cyrille, Québec 172/E4
Saint-Cyrille-de-L'Islet, Québec 172/G3
Saint Cyrus, Scotland 15/F4
Saint-Damase, Québec 172/B1
Saint-Damase-des-Aulnaies, Québec 172/C2
Saint-Damien-de-Brandon, Québec 172/D3
Saint-Damien-de-Buckland, Québec 172/G3
Saint David, Ariz. (85630) 198/E7
Saint David, Ill. (61563) 222/C3
Saint David, Maine (04773) 243/G1
Saint-David, Québec 172/J3
Saint-David-de-Falardeau, Québec 172/F1
Saint-David-d'Yamaska, Québec 172/E4
Saint Davids (isl.), Bermuda 156/H2
Saint David's, Wales 13/B6
Saint David's (head), Wales 10/D5
Saint David's (head), Wales 13/B6
Saint-Denis, France 28/B1
Saint-Denis, Québec 172/D4
Saint-Denis (cap.), Réunion 118/F5
Saint Denis, Sask. 181/F3
Saint-Denis-de-la-Bouteillerie, Québec 172/G2
Saint-Didace, Québec 172/D3
Saint-Dié, France 28/G3
Saint-Dizier, France 28/F3
Saint-Dominique, Québec 172/E4
Saint-Donat-de-Montcalm, Québec 172/C3
Saint-Donat-de-Rimouski, Québec 172/J1
Saint Donatus, Iowa (52071) 229/M4
Sainte-Adèle, Québec 172/C4
Sainte Agathe, Manitoba 179/E5
Sainte-Agathe, Québec 172/F3
Sainte-Agathe-des-Monts, Québec 172/C3
Sainte-Agnes-de-Charlevoix, Québec 172/G2
Sainte Amélie, Manitoba 179/C4
Sainte-Anastasie, Québec 172/F3
Sainte-Angèle-de-Mérici, Québec 172/J1
Sainte Anne (lake), Alberta 182/C3
Sainte-Anne, Guadeloupe 161/B6
Sainte Anne, Manitoba 179/F5
Sainte-Anne, Martinique 161/D7
Sainte-Anne, New Bruns. 170/E1
Sainte-Anne (lake), Québec 172/H2
Sainte-Anne (riv.), Québec 172/C1
Sainte-Anne (riv.), Québec 172/G2
Sainte-Anne (riv.), Québec 172/F3
Sainte Anne (isl.), Seychelles 118/H5
Sainte-Anne-de-Beaupré, Québec 172/F2
Sainte-Anne-de-Bellevue, Québec 172/H4
Sainte-Anne-de-Kent, New Bruns. 170/F2
Sainte-Anne-de-Madawaska, New Bruns. 170/B1
Sainte-Anne-des-Monts, Québec 172/C1
Sainte-Anne-des-Plaines, Québec 172/H4
Sainte-Anne-du-Lac, Québec 172/B3
Sainte-Apolline, Québec 172/G3
Sainte-Aurélie, Québec 172/G3
Sainte-Béatrix, Québec 172/D3
Sainte-Bernadette, Québec 172/G1
Sainte-Blandine, Québec 172/J1
Sainte-Brigide, Québec 172/D4
Sainte-Catherine, Québec 172/F3
Sainte-Cécile-de-Frontenac, Québec 172/G4
Sainte-Cécile-de-Masham, Québec 172/A4
Sainte-Claire, Québec 172/G3
Sainte-Clothilde-de-Horton, Québec 172/E4
Sainte-Croix, Québec 172/F3
Sainte-Croix, Switzerland 39/B3
Saint-Édouard-de-Kent, New Bruns. 170/F2
Saint-Édouard-de-Maskinongé, Québec 172/D3
Saint-Édouard-de-Napierville, Québec 172/D4
Saint Edward, Nebr. (68660) 264/G3
Saint Edward, Pr. Edward I. 168/D2
Sainte-Edwidge, Québec 172/F4
Sainte-Élisabeth, Québec 172/D3
Sainte-Émélie-de-l'Énergie, Québec 172/D3
Sainte-Eulalie, Québec 172/E3
Sainte-Euphémie, Québec 172/G3
Sainte-Famille-d'Aumond, Québec 172/B3
Sainte-Famille-d'Orléans, Québec 172/G3
Sainte-Félicité, Québec 172/B1
Sainte-Flavie, Québec 172/J1
Sainte-Florence, Québec 172/B2
Sainte-Foy, Québec 172/H3
Sainte-Françoise, Québec 172/H1
Sainte-Geneviève, Manitoba 179/F5
Sainte Genevieve (co.), Mo. 261/M7
Sainte Genevieve, Mo. (63670) 261/M6
Sainte-Geneviève, Québec 172/H4
Sainte-Geneviève-de-Batiscan, Québec 172/E3
Sainte-Hedwidge-de-Roberval, Québec 172/E1
Sainte-Hélène-de-Bagot, Québec 172/E4
Sainte-Hélène-de-Kamouraska, Québec 172/H2
Sainte-Hénédine, Québec 172/F3
Sainte-Julie-de-Verchères, Québec 172/J4
Sainte-Julienne, Québec 172/D4
Sainte-Julie-Station, Québec 172/F3
Sainte-Justine, Québec 172/G3
Sainte-Justine-de-Newton, Québec 172/C4
Saint Eleanors, Pr. Edward I. 168/E2
Sainte-Éleuthère, Québec 172/H2
Saint Elias (mt.), Alaska 188/D5
Saint Elias (cape), Alaska 196/K3
Saint Elias (mt.), Alaska 196/K2
Saint Elias (mts.), Alaska 196/L2
Saint Elias (mts.), Yukon 162/B3
Saint Elias (mt.), Yukon 187/D3
Saint Elias (mts.), Yukon 187/E3
Saint-Élie, Fr. Guiana 131/E3
Saint-Élie, Québec 172/E3
Saint Elizabeth, Mo. (65075) 261/H6
Saint Elmo, Ala. (36568) 195/B10
Saint Elmo, Colo. (†81236) 208/G5
Saint Elmo, Ill. (62458) 222/E4
Saint-Éloi, Québec 172/H1
Sainte-Louise, Québec 172/G2
Sainte-Luce, Martinique 161/D7
Sainte-Luce, Québec 172/J1
Sainte-Lucie-de-Beauregard, Québec 172/H3
Sainte-Lucie-de-Doncaster, Québec 172/C3
Saint-Elzéar, Québec 172/F3
Saint-Elzéar-de-Bonaventure, Québec 172/C2
Sainte-Marguerite, Guadeloupe 161/B6
Sainte-Marguerite-de-Dorchester, Québec 172/G3
Sainte-Marguerite-Marie, Québec 172/B2
Sainte-Marguerite (riv.), Québec 172/G1
Sainte-Marguerite Nord-Est (riv.), Québec 172/H1
Sainte-Marguerite (riv.), Québec 174/D2

Sainte-Marie, Guadeloupe 161/A6
Sainte Marie, Ill. (62459) 222/E5
Sainte-Marie (cape), Madagascar 2/M4
Sainte-Marie (Vohimena) (cape), Madagascar 102/G7
Sainte-Marie (Vohimena) (cape), Madagascar 118/G5
Sainte-Marie (Nosy Boraha) (isl.), Madagascar 118/J3
Sainte-Marie, Martinique 161/D5
Sainte-Marie, Beauce, Québec 172/G3
Sainte-Marie, Nicolet, Québec 172/E3
Sainte-Marie (lake), Québec 172/B4
Sainte-Marie-de-Kent, New Bruns. 170/F2
Sainte-Marie-sur-Mer, New Bruns. 170/F1
Sainte-Marthe-de-Gaspé, Québec 172/C1
Sainte-Martine, Québec 172/D4
Sainte-Menehould, France 28/F3
Sainte-Mère-Église, France 28/C3
Saint-Émile, Québec 172/H3
Saint-Émile-de-Suffolk, Québec 172/B4
Sainte-Monique, Nicolet, Québec 172/E3
Sainte-Monique, Lac-St-Jean-E., Québec 172/F1
Sainte-Ode, Belgium 27/H8
Sainte-Perpétue, Québec 172/E3
Sainte-Perpétue-de-L'Islet, Québec 172/H2
Saint-Éphrem-de-Tring, Québec 172/G3
Saint-Épiphane, Québec 172/H2
Sainte-Pudentienne, Québec 172/E4
Sainte Rita, Manitoba 179/F5
Sainte-Rosalie, Québec 172/E4
Sainte-Rose, Guadeloupe 161/A6
Sainte-Rose-de-Watford, Québec 172/G3
Sainte Rose du Lac, Manitoba 179/C3
Sainte-Rose-du-Nord, Québec 172/G1
Sainte-Rose-Gloucester, New Bruns. 170/F1
Saintes, France 28/C5
Saintes (chan.), Guadeloupe 161/A7
Saintes (isls.), Guadeloupe 161/A7
Sainte-Sabine-de-Bellechasse, Québec 172/G3
Sainte-Savine, France 28/E3
Sainte-Sophie-de-Lévrard, Québec 172/E3
Sainte-Sophie-de-Mégantic, Québec 172/F3
Saint-Esprit, Québec 172/D4
Sainte-Thècle, Québec 172/E3
Sainte-Thérèse, Québec 172/H4
Sainte-Thérèse (isl.), Québec 172/J4
Sainte-Thérèse-de-L'Enfant-Jésus, Québec 172/D2
Sainte-Thérèse-Ouest, Québec 172/H4
Saint-Étienne, France 7/E4
Saint-Étienne, France 28/F5
Saint-Étienne-de-Grès, Québec 172/E3
Saint-Étienne-de-Lauzon, Québec 172/J3
Saint-Eugène, Ontario 177/K2
Saint-Eugène-de-Grantham, Québec 172/E4
Sainte-Ursule, Québec 172/D3
Saint-Eusèbe, Québec 172/J2
Saint Eustache, Manitoba 179/E5
Saint-Eustache, Québec 172/H4
Saint Eustatius (isl.), Neth. Ant. 156/F3
Saint-Évariste-de-Forsyth, Québec 172/F4
Sainte-Véronique, Québec 172/C3
Sainte-Victoire, Québec 172/D4
Saint-Fabien, Québec 172/J1
Saint-Fabien-de-Panet, Québec 172/G3
Saint-Félicien, Québec 172/E1
Saint-Félicien, Québec 174/C3
Saint-Félix-de-Valois, Québec 172/D3
Saint Fergus, Scotland 15/G3
Saint-Fidèle, Québec 172/H2
Saintfield, N. Ireland 17/K3
Saint Finan's (bay), Ireland 17/A8
Saint-Flavien, Québec 172/F3
Saint-Florent (gulf), France 28/B6
Saint-Florent-sur-Cher, France 28/E4
Saint Florian, Ala. (†35630) 195/C1
Saint-Flour, France 28/E5
Saint-Fortunat, Québec 172/F4
Saint Francis (co.), Ark. 202/J3
Saint Francis, Ark. (72464) 202/K1
Saint Francis (riv.), Ark. 202/J4
Saint Francis, Kansas (67756) 232/A2
Saint Francis, Ky. (40062) 237/L5
Saint Francis○, Maine (04774) 243/E1
Saint Francis (riv.), Maine 243/E1
Saint Francis, Minn. (55070) 255/E5
Saint Francis (riv.), Mo. 261/M9
Saint Francis (riv.), New Bruns. 170/A1
Saint Francis (cape), Newf. 166/D2
Saint Francis (bay), S. Africa 118/D6
Saint Francis, S. Dak. (57572) 298/H7
Saint Francis, Wis. (†53207) 317/M2
Saint Francis Harbour, Nova Scotia 168/G3
Saint Francisville, Ill. (62460) 222/F5
Saint Francisville, La. (70775) 238/H5
Saint Francisville, Mo. (†63465) 261/J2
Saint-François, Guadeloupe 161/B6
Saint Francois (co.), Mo. 261/M7
Saint François (mts.), Mo. 261/L7
Saint-François (lake), Québec 172/F4
Saint-François (riv.), Québec 172/E4
Saint-François-d'Assise, Québec 172/G1
Saint François de Madawaska, New Bruns. 170/B1
Saint-François-de-Montmagny, Québec 172/G3
Saint-François-de-Sales, Québec 172/E1
Saint-François-du-Lac, Québec 172/E3
Saint-Frédéric, Québec 172/G3
Saint Froid (lake), Maine 243/F2
Saint Front, Sask. 181/G3
Saint-Fulgence, Québec 172/G1
Saint Gabriel, La. (70776) 238/K2
Saint-Gabriel, Québec 172/D3
Saint-Gabriel-de-Rimouski, Québec 172/J1
Saint Gall (Sankt Gallen), Switzerland 39/H2
Saint-Gaudens, France 28/D6
Saint-Gaudens Nat'l Hist. Site, N.H. 268/B4
Saint-Gédéon, Frontenac, Québec 172/G4
Saint-Gédéon, Lac-St-Jean-E., Québec 172/F1
Saint George, Alaska (†99660) 196/E3
Saint George (isl.), Alaska 196/D3
Saint George (head), Aust. Cap. Terr. 97/F4
Saint George, Bermuda 156/H2
Saint George (pt.), Calif. 204/A2
Saint George (cape), Fla. 212/A2
Saint George (isl.), Fla. 212/B2
Saint George (sound), Fla. 212/B2
Saint George, Georgia (31646) 217/H9
Saint George, Kansas (66535) 232/F2
Saint George, Maine (04857) 243/E7
Saint George○, Maine (04857) 243/E7
Saint George, Manitoba 179/F4
Saint George (lake), Manitoba 179/E3
Saint George (isl.), Md. 245/M8
Saint George, Minn. (†56073) 255/D6
Saint George, Mo. (†63101) 261/P4
Saint George, New Bruns. 170/D3
Saint George (cape), Newf. 166/C4
Saint George, N.Y. (10301) 276/M9
Saint George, Ontario 177/D4
Saint George (cape), Papua N.G. 86/C2
Saint George, Queensland 95/D5
Saint George, Queensland 88/H5
Saint George, S.C. (29477) 296/F5
Saint George, Utah (84770) 304/A6
Saint George, Utah 188/D3
Saint George○, Vt. (†05401) 268/A2
Saint George (ranges), W. Australia 92/D2
Saint George, W. Va. (26290) 312/G4
Saint George's (chan.) 7/D3
Saint Georges (cay), Belize 154/D2
Saint George's (isl.), Bermuda 156/H2
Saint Georges, Del. (19733) 245/R2
Saint-Georges, Fr. Guiana 131/F4
Saint George's (cap.), Grenada 156/F5
Saint George's (cap.), Grenada 161/C9
Saint George's (chan.), Ireland 10/D4
Saint George's (chan.), Ireland 17/K7
Saint George's, Newf. 166/C4
Saint George's (bay), Newf. 166/C4
Saint Georges (bay), Nova Scotia 168/G3
Saint George's (chan.), Papua N.G. 86/C2
Saint-Georges, Beauce, Québec 172/G3
Saint-Georges, Champlain, Québec 172/E3
Saint George's (chan.), Wales 13/B5
Saint George's (chan.), Wales 10/D4
Saint-Georges-de-Malbaie, Québec 172/D1
Saint-Georges-de-Windsor, Québec 172/F4
Saint-Georges-Ouest, Québec 172/G3
Saint-Georges-sur-Meuse, Belgium 27/G7
Saint-Gérard, Québec 172/F4
Saint-Germain-de-Grantham, Québec 172/E4
Saint-Germain-de-Kamouraska, Québec 172/H2
Saint-Germain-en-Laye, France 28/D3
Saint-Gervais, Québec 172/G3
Saint-Gilles, Belgium 27/B9
Saint-Gilles, Québec 172/F3
Saint-Gilles-Croix-de-Vie, France 28/B4
Saint-Gingolph, Switzerland 39/C4
Saint-Girons, France 28/D6
Saint-Godefroi, Québec 172/D2
Saint Gotthard (pass), Switzerland 39/G3
Saint Gotthard (tunnel), Switzerland 39/G3
Saint Gowans (head), Wales 13/C6
Saint-Grégoire, Québec 172/D4
Saint-Grégoire-de-Greenlay, Québec 172/E4
Saint Gregor, Sask. 181/G3
Saint Gregory (cape), Newf. 166/C4
Saint-Guillaume-Nord, Québec 172/D3
Saint Hedwig, Texas (78152) 303/K11
Saint Helen, Mich. (48656) 250/E4
Saint Helena (isl.), (Br.) 2/J6
Saint Helena, Calif. (94574) 204/C5
Saint Helena (par.), La. 238/J5
Saint Helena, Nebr. (68774) 264/G2
Saint Helena (bay), S. Africa 118/B6
Saint Helena (isl.), S.C. 296/F7
Saint Helena (sound), S.C. 296/G7
Saint Helena (isl.), U.K. 102/B6
Saint Helens, England 10/F2
Saint Helens, England 13/G2
Saint Helens, Ky. (41368) 237/O5
Saint Helens, Oreg. (97051) 291/E2
Saint Helens, Tasmania 99/E3
Saint Helens (pt.), Tasmania 99/E3
Saint Helens (mt.), Wash. 188/B1
Saint Helens (mt.), Wash. 310/C4
Saint Helier (cap.), Jersey, Chan. Is. 13/E8
Saint Helier (cap.), Jersey, Chan. Is. 10/E6
Saint-Henri, Québec 172/J3
Saint Henry, Ind. (†47532) 227/D8
Saint Henry, Ohio (45883) 284/A5
Saint-Herménégilde, Québec 172/F4
Saint Hilaire, Minn. (56554) 255/B2
Saint Hilaire, New Bruns. 170/B1
Saint-Hilarion, Québec 172/G2
Saint Hippolyte, Sask. 181/C2
Saint-Honoré, Beauce, Québec 172/G4
Saint-Honoré, Chicoutimi, Québec 172/F1
Saint-Honoré-de-Témiscouata, Québec 172/H2
Saint-Hubert, Belgium 27/G8
Saint-Hubert, Québec 172/J4
Saint-Hubert-de-Témiscouata, Québec 172/J2
Saint Hubert Mission, Sask. 181/J5
Saint-Hugues, Québec 172/E4
Saint-Hyacinthe (co.), Québec 172/D4
Saint-Hyacinthe, Québec 172/D4
Saint Ignace, Mich. (49781) 250/E3
Saint-Ignace, New Bruns. 170/F2
Saint Ignace (isl.), Ontario 177/H5
Saint Ignace (isl.), Ontario 175/C3
Saint Ignatius, Mont. (59865) 262/C3
Saint-Imier, Switzerland 39/D2
Saint Inigoes, Md. (20684) 245/N8
Saint-Irénée, Québec 172/G2
Saint Isidore, Alberta 182/B1
Saint-Isidore, New Bruns. 170/E1
Saint-Isidore, Québec 172/F3
Saint-Isidore-d'Auckland, Québec 172/F4
Saint-Isidore-de-Gaspé, Québec 172/D1
Saint-Isidore-de-Laprairie, Québec 172/D4
Saint Isidore de Prescott, Ontario 177/K2
Saint Issells, Wales 13/C6
Saint Ives, Cornwall, England 13/B7
Saint Ives, England 10/G4
Saint Ives, England 10/D5
Saint Ives, Cambs., England 13/G5
Saint Ives (bay), England 13/B7
Saint Jacob, Ill. (62281) 222/D5
Saint Jacobs, Ontario 177/D4
Saint-Jacques, New Bruns. 170/B1
Saint-Jacques, Québec 172/D4
Saint-Jacques-le-Mineur, Québec 172/H4
Saint James, Ark. (†72560) 202/F2
Saint James (cape), Br. Col. 184/B4
Saint James, Ill. (†62857) 222/E5
Saint James (par.), La. 238/L3
Saint James, La. (70086) 238/L3
Saint James, Md. (21781) 245/G2
Saint James, Mich. (49782) 250/D3
Saint James, Minn. (56081) 255/D7
Saint James, Mo. (65559) 261/J6
Saint James, N.Y. (11780) 276/O9
Saint James (isls.), Virgin Is. (U.S.) 161/B4
Saint James City, Fla. (33956) 212/D5
Saint-Jean, Fr. Guiana 131/D3
Saint-Jean (lake), Que. 162/J6
Saint-Jean (co.), Québec 172/D4
Saint-Jean, Québec 172/D4
Saint-Jean (lake), Québec 172/E1
Saint-Jean (lake), Québec 174/C3
Saint-Jean (riv.), Québec 172/D1
Saint Jean Baptiste, Manitoba 179/E5
Saint-Jean-Baptiste-de-Restigouche, New Bruns. 170/C1
Saint-Jean-Chrysostome, Québec 172/J3
Saint-Jean-d'Angély, France 28/C4
Saint-Jean-de-Dieu, Québec 172/J1
Saint-Jean-de-Luz, France 28/C6
Saint-Jean-de-Matha, Québec 172/D3
Saint-Jean-de-Maurienne, France 28/G5
Saint-Jean-de-Monts, France 28/B4
Saint-Jean-des-Piles, Québec 172/E3
Saint-Jean-d'Orléans, Québec 172/G3
Saint-Jean du Sud, Haiti 158/B6
Saint-Jean-Pied-de-Port, France 28/C6
Saint-Jean-Port-Joli, Québec 172/G2
Saint-Jean-sur-le-Lac, Québec 172/B3
Saint-Jérôme, Terrebonne, Québec 172/H4
Saint Jo, Texas (76265) 303/G4
Saint-Joachim, Québec 172/G2
Saint Joe, Ark. (72675) 202/E1
Saint Joe, Idaho (†83861) 220/B2
Saint Joe (riv.), Idaho 220/B2
Saint Joe, Ind. (46785) 227/H2
Saint John, Ind. (46383) 227/C2
Saint John, Kansas (67576) 232/D3
Saint John○, Maine (†04743) 243/F1
Saint John (riv.), Maine 188/N1
Saint John (pond), Maine 243/D3
Saint John (riv.), Maine 243/G1
Saint John, Mo. (63114) 261/P2
Saint John (cap.), N. Br. 146/M5
Saint John, N. Br. 162/K6
Saint John (co.), New Bruns. 170/E3
Saint John, New Bruns. 170/E3
Saint John (harb.), New Bruns. 170/E3
Saint John (riv.), New Bruns. 170/C2
Saint John (bay), Newf. 166/C3
Saint John (cape), Newf. 166/C3
Saint John (isl.), Newf. 166/C3
Saint John, N. Dak. (58369) 282/L2
Saint John, Utah (84069) 304/B3
Saint John (isl.), Virgin Is. (U.S.) 161/C4
Saint John (isl.), Virgin Is. (U.S.) 156/H1
Saint John, Wash. (99171) 310/H3
Saint John's (cap.), Ant. & Bar. (21814) 161/E11
Saint John's, Ant. & Bar. 156/G3
Saint John's (harb.), Ant. & Bar. 161/C11
Saint Johns, Ariz. (85936) 198/F4
Saint John's, Canada 2/G3
Saint Johns (co.), Fla. 212/E2
Saint Johns (riv.), Fla. 212/E2
Saint Johns, Ill. (†62832) 222/D5
Saint John's (pt.), Ireland 17/D2
Saint Johns, Mich. (48879) 250/E6
Saint John's (cap.), Newf. 166/D2
Saint John's (cap.), Newf. 146/N5
Saint John's (cap.), Newf. 162/L6
Saint John's (pt.), N. Ireland 17/K3
Saint Johns, Ohio (45884) 284/B4
Saint John's (head), Scotland 15/E2
Saint Johns Branch, Nanticoke (riv.), Del. 245/R6
Saint Johnsbury, Vt. (05819) 268/D3
Saint Johnsbury○, Vt. (05819) 268/D3
Saint Johnsbury Center, Vt. (05863) 268/D3
Saint Johnston, Ireland 17/F2
Saint Johnsville, N.Y. (13452) 276/L5
Saint John the Baptist (par.), La. 238/M3
Saint Jones (riv.), Del. 245/R4
Saint Joseph, Dominica 161/E6
Saint Joseph (bay), Fla. 212/D6
Saint Joseph (pen.), Fla. 212/D7
Saint Joseph (pt.), Fla. 212/D6
Saint Joseph, Ill. (61873) 222/E3
Saint Joseph (co.), Ind. 227/E1
Saint Joseph (riv.), Ind. 227/H2
Saint Joseph (riv.), Ind. 227/E1
Saint Joseph, Kansas (†66938) 232/E2
Saint Joseph, Ky. (42373) 237/G5
Saint Joseph, La. (71366) 238/H3
Saint Joseph, Manitoba 179/E5
Saint-Joseph, Martinique 161/D6
Saint Joseph (co.), Mich. 250/D7
Saint Joseph, Mich. (49085) 250/C6
Saint Joseph (riv.), Mich. 250/C7
Saint Joseph, Minn. (56374) 255/D5
Saint Joseph, Mo. (*64501) 261/C3
Saint Joseph, Mo. 188/H3
Saint Joseph, Mo. 146/J6
Saint-Joseph, New Bruns. 170/F3
Saint Joseph (riv.), Ohio 284/A3
Saint Joseph (lake), Ont. 162/G5
Saint Joseph (lake), Ontario 175/B2
Saint-Joseph, Réunion 118/G6
Saint Joseph, Tenn. (38481) 237/G10
Saint Joseph, Trin. & Tob. 161/B10
Saint Joseph, Trin. & Tob. 161/B11
Saint-Joseph-de-Beauce, Québec 172/G3
Saint-Joseph-de-Kamouraska, Québec 172/H2
Saint-Joseph-de-la-Rive, Québec 172/G2
Saint-Joseph-de-Madawaska, New Bruns. 170/B1
Saint-Joseph-de-Mékinac, Québec 172/E3
Saint-Joseph-de-Sorel, Québec 172/D3
Saint-Joseph-du-Lac, Québec 172/C4
Saint-Joseph-du-Moine, Nova Scotia 168/G2
Saint Joseph Ridge, Wis. (†54601) 317/D8
Saint Joseph's, Newf. 166/D2
Saint-Josse-ten-Noode, Belgium 27/C9
Saint-Jovite, Québec 172/C3
Saint-Jude, Québec 172/E4
Saint-Junien, France 28/D5
Saint Just, England 13/B7
Saint-Justin, Québec 172/D3
Saint Kilda, N. Zealand 100/C7
Saint Kilda (isl.), Scotland 15/A2
Saint Kilda, Victoria 88/L7
Saint Kilda, Victoria 97/J5
Saint Kilian, Minn. (†56185) 255/C7
Saint Kitts (Saint Christopher) (isl.), St. Chris.-Nevis 156/F3
Saint Kitts (Saint Christopher) (isl.), St. Chris.-Nevis 161/D10
Saint-Lambert, Québec 172/J4
Saint Landry (par.), La. 238/F5
Saint Landry, La. (71367) 238/F5
Saint Laurent, Manitoba 179/D4
Saint-Laurent, Québec 172/H4
Saint-Laurent-de-la-Salanque, France 28/E6
Saint-Laurent-d'Orléans, Québec 172/G3
Saint-Laurent-du-Maroni, Fr. Guiana 120/D2
Saint-Laurent du Maroni (dist.), Fr. Guiana 131/E4
Saint-Laurent du Maroni, Fr. Guiana 131/E3
Saint Lawrence (riv.) 162/K6
Saint Lawrence (gulf) 146/M5
Saint Lawrence (riv.) 146/L5
Saint Lawrence (isl.), Alaska 146/A3
Saint Lawrence (isl.), Alaska 196/D2
Saint Lawrence (isl.), Alaska 196/D2
Saint Lawrence, Barbados 161/B9
Saint Lawrence (gulf), Canada 2/G3
Saint Lawrence (gulf), New Bruns. 170/F1
Saint Lawrence, Newf. 166/C4
Saint Lawrence (gulf), Newf. 166/B4
Saint Lawrence (riv.), N.Y. 188/N1
Saint Lawrence (co.), N.Y. 276/K2
Saint Lawrence (lake), N.Y. 276/K1
Saint Lawrence (riv.), N.Y. 276/J2
Saint Lawrence (bay), Nova Scotia 168/H1
Saint Lawrence (cape), Nova Scotia 168/H1
Saint Lawrence (lake), Ontario 177/K3
Saint Lawrence (riv.), Ontario 177/J3
Saint Lawrence (gulf), Pr. Edward I. 168/F2
Saint Lawrence (gulf), Que. 162/K6
Saint Lawrence (gulf), Québec 172/D2
Saint Lawrence (gulf), Québec 174/E3
Saint Lawrence (riv.), Québec 174/D3
Saint Lawrence (riv.), Québec 172/H1
Saint Lawrence, Queensland 95/D4
Saint Lawrence, S. Dak. (57373) 298/M4
Saint Lawrence (isl.), U.S. 4/C18
Saint Lawrence Is. Nat'l Park, Ontario 177/J3
Saint Lazare, Manitoba 179/A4
Saint-Lazare, Québec 172/G3
Saint-Léandre, Québec 172/B1
Saint-Léger, Belgium 27/H9
Saint-Légier-La Chiésaz, Switzerland 39/C4
Saint Leo, Fla. (33574) 212/D3
Saint Leo, Minn. (56286) 255/C6
Saint-Léolin, New Bruns. 170/E1
Saint Leon, Ind. (46373) 227/H6
Saint Leon, Manitoba 179/D5
Saint-Léon, Québec 172/D3
Saint Leonard, Md. (20685) 245/N7
Saint Leonard, New Bruns. 170/C1
Saint-Léonard, Québec 172/H4
Saint-Léonard-d'Aston, Québec 172/E3
Saint-Léonard-de-Noblat, France 28/D5
Saint-Léonard-de-Portneuf, Québec 172/E3
Saint-Léon-de-Chicoutimi, Québec 172/F1
Saint-Léon-de-Standon, Québec 172/G3
Saint-Léon-le-Grand, Québec 172/B2
Saint Lewis (cape), Newf. 166/C3
Saint Lewis (riv.), Newf. 166/C3
Saint-Liboire, Québec 172/E4
Saint Libory, Ill. (62282) 222/D5
Saint Libory, Nebr. (68872) 264/F3
Saint Lina, Alberta 182/E2
Saint-Lô, France 28/C3
Saint-Louis, Guadeloupe 161/B7
Saint Louis, Mich. (48880) 250/E5
Saint Louis (co.), Minn. 255/F3
Saint Louis (riv.), Minn. 255/F4
Saint Louis (bay), Miss. 256/F10
Saint Louis (co.), Mo. 261/O3
Saint Louis (city county), Mo. 261/P3
Saint Louis, Mo. (*63101) 261/R3
Saint Louis, Mo. 146/K6
Saint Louis, Mo. 188/H3
Saint Louis, Okla. (74866) 288/N4
Saint Louis, Oreg. (†97026) 291/A3
Saint Louis, Pr. Edward I. 168/D2
Saint-Louis (lake), Québec 172/H4
Saint-Louis, Réunion 118/F5
Saint Louis, Sask. 181/F3
Saint-Louis, Senegal 102/A3
Saint-Louis, Senegal 106/A5
Saint Louis, U.S. 2/E4
Saint Louis (riv.), Wis. 317/A2
Saint Louis Crossing, Ind. (†47201) 227/F6
Saint-Louis-de-Gonzague, Québec 172/D4
Saint-Louis-de-Kent, New Bruns. 170/F2
Saint-Louis-de-Terrebonne, Québec 172/H4
Saint-Louis-du-Ha! Ha!, Québec 172/H2
Saint-Louis du Nord, Haiti 158/B5
Saint-Louis du Sud, Haiti 158/B6
Saint Louis Park, Minn. (55426) 255/G5
Saint Louisville, Ohio (43071) 284/F5
Saint-Luc, Québec 172/D4
Saint Lucas, Iowa (52166) 229/K2
Saint-Luc-de-Matane, Québec 172/B1
Saint Lucia 2/G5
Saint Lucia 146/M8
SAINT LUCIA 161/G5
SAINT LUCIA 156/G4
Saint Lucia, Queensland 88/K3
Saint Lucia, Queensland 95/D3
Saint Lucia (chan.), St. Lucia 161/G5
Saint Lucia (cape), S. Africa 118/E5
Saint Lucia (lake), S. Africa 118/E5
Saint Lucie (co.), Fla. 212/F4
Saint Lucie, Fla. (33452) 212/F4
Saint Lucie (canal), Fla. 212/F4
Saint Lucie (inlet), Fla. 212/F4
Saint-Ludger, Québec 172/G4
Saint Lunaire-Griquet, Newf. 166/C3
Saint-Magloire, Québec 172/G3
Saint Magnus (bay), Scotland 10/G1
Saint Magnus (bay), Scotland 15/F2
Saint-Malachie, Québec 172/G3
Saint-Malo, France 28/B3
Saint-Malo (gulf), France 28/B3
Saint Malo, Manitoba 179/F5
Saint-Malo, Québec 172/F4
Saint-Mandé, France 28/B2
Saint-Marc, Haiti 158/B5
Saint-Marc, Haiti 156/D3
Saint-Marc (chan.), Haiti 158/B6
Saint-Marc (pt.), Haiti 158/B5
Saint-Marc, Québec 172/D4
Saint-Marc-des-Carrières, Québec 172/E3
Saint-Marcel (mt.), Fr. Guiana 131/E4
Saint-Marcel-de-L'Islet, Québec 172/G3
Saint-Marcellin, France 28/F5
Saint Margarets, New Bruns. 170/E2
Saint Margarets (bay), Nova Scotia 168/E4
Saint Margaret's Bay, Jamaica 158/K6
Saint Margaret's Hope, Scotland 15/F2
Saint Margaret Village, Nova Scotia 168/H2
Saint Maries, Idaho (83861) 220/B2
Saint Maries (riv.), Idaho 220/B2
Saint Marks, Fla. (32355) 212/B1
Saint Marks, Georgia (†30230) 217/C4
Saint Marks, Manitoba 179/E4
Saint Martin (isl.), Guadeloupe 156/F3
Saint Martin (par.), La. 238/G6
Saint Martin (lake), Manitoba 179/D3
Saint-Martin (cape), Martinique 161/C5
Saint Martin, Md. (†21811) 245/T7
Saint Martin (bay), Mich. 250/E3
Saint Martin (isl.), Mich. 250/C3
Saint Martin, Minn. (56376) 255/D5
Saint Martin (Sint Maarten) (isl.), Neth. Ant. 156/F3
Saint Martin, Ohio (45118) 284/C7
Saint-Martin, Switzerland 39/E4
Saint Martin de Restigouche, New Bruns. 170/C1
Saint Martins, Barbados 161/C9
Saint Martin's (isl.), England 13/A8
Saint Martins, Mo. (†65101) 261/H5
Saint Martins, New Bruns. 170/E3
Saint Martin Station, Manitoba 179/D3
Saint Martinville, La. (70582) 238/G6
Saint Mary (res.), Alberta 182/D5
Saint Mary (riv.), Alberta 182/D5
Saint Mary, Ky. (40063) 237/L5
Saint Mary (par.), La. 238/H7
Saint Mary (lake), Mont. 262/C2
Saint Mary (riv.), Mont. 262/C1
Saint Mary, Nebr. (68432) 264/H4
Saint Mary (cape), Nova Scotia 168/B4
Saint Mary (peak), S. Australia 94/F4
Saint Mary-of-the-Woods, Ind. (47876) 227/B6
Saint Marys, Alaska (99658) 196/F2
Saint Mary's (isl.), England 13/A8
Saint Marys (riv.), Fla. 212/D1
Saint Marys, Georgia (31558) 217/J9
Saint Marys (riv.), Georgia 217/J9
Saint Marys, Ind. (†46556) 227/E1
Saint Marys (lake), Ind. 227/H3
Saint Marys (riv.), Ind. 227/H3
Saint Marys, Iowa (50241) 229/F6
Saint Marys, Kansas (66536) 232/G2
Saint Marys (co.), Md. 245/M7
Saint Marys (riv.), Md. 245/N8
Saint Marys (riv.), Mich. 250/E2
Saint Marys, Mo. (63673) 261/M7
Saint Mary's, Newf. 166/D2
Saint Mary's (bay), Newf. 166/C2
Saint Mary's (cape), Newf. 166/C2
Saint Marys (bay), Nova Scotia 168/B4
Saint Mary's (riv.), Nova Scotia 168/F3
Saint Marys, Ohio (45885) 284/B4
Saint Marys (lake), Ohio 284/A4
Saint Marys (riv.), Ohio 284/A4
Saint Mary's, Ontario 177/C4
Saint Marys, Pa. (15857) 294/E3
Saint Mary's, Scotland 15/F2
Saint Mary's (lake), Scotland 15/E5
Saint Marys, Tasmania 99/E3
Saint Marys, W. Va. (26170) 312/D4
Saint Marys (peak), Wyo. 319/D3
Saint Marys City, Md. (20686) 245/N8
Saint Marys Entrance (inlet), Fla. 212/E1
Saint-Mathieu, Québec 172/J1
Saint-Mathieu (lake), Québec 172/J1
Saint Matthew (isl.), Alaska 188/C5
Saint Matthew (isl.), Alaska 146/A4
Saint Matthew (isl.), Alaska 196/D2
Saint Matthew (isl.), U.S. 4/C18
Saint Matthews, Ky. (40207) 237/K2
Saint Matthews, S.C. (29135) 296/F4
Saint Matthias Group (isls.), Papua N.G. 86/B1
Saint-Maur-des-Fossés, France 28/B2
Saint Maurice, Ind. (†47240) 227/G6
Saint Maurice, La. (71471) 238/E3
Saint-Maurice (co.), Québec 172/D3
Saint-Maurice (county), Québec 174/C3
Saint-Maurice (riv.), Québec 172/E2
Saint-Maurice, Switzerland 39/C4
Saint-Médard, Québec 172/J1
Saint Meinrad, Ind. (47577) 227/D8
Saint-Méthode, Québec 172/E1
Saint-Méthode-de-Frontenac, Québec 172/F3
Saint Michae, N. Dak. (58370) 282/N4
Saint Michael, Alaska (99659) 196/F2
Saint Michael, Alberta 182/D3
Saint Michael, Minn. (55376) 255/E5
Saint Michael, Pa. (15951) 294/E5
Saint Michaels, Ariz. (86511) 198/F3
Saint Michaels, Md. (21663) 245/N5
Saint Michaels (bay), Newf. 166/C3
Saint-Michel-de-Bellechasse, Québec 172/G3
Saint-Michel de l'Atalaye, Haiti 158/C5
Saint-Michel-des-Saints, Québec 172/D3
Saint-Michel du Sud, Haiti 158/B6
Saint-Mihiel, France 28/F3
Saint-Modeste, Québec 172/H2
Saint-Moïse, Québec 172/B1
Saint Monance, Scotland 15/F4
Saint Moritz, Switzerland 39/J3
Saint-Narcisse-de-Rimouski, Québec 172/J1
Saint-Nazaire, France 7/D4
Saint-Nazaire, France 28/B4
Saint-Nazaire, Fr. Guiana 131/E3
Saint-Nazaire, Québec 172/E4
Saint-Nazaire-de-Buckland, Québec 172/G3
Saint-Nazaire-de-Chicoutimi, Québec 172/F1
Saint Nazianz, Wis. (54232) 317/L7
Saint Neots, England 13/G5
Saint Neots, England 10/F4
Saint-Nérée, Québec 172/G3
Saint-Nicolas, Belgium 27/G7
Saint Niklaus, Switzerland 39/E4
Saint-Noël, Québec 172/B1
Saint-Octave, Québec 172/B1
Saint-Odilon, Québec 172/G3
Saint Olaf, Iowa (52072) 229/L3
Saint-Omer, France 28/E2
Saint Omer, Ind. (†47272) 227/F6
Saint-Omer, Québec 172/C2
Saint-Onésime, Québec 172/H2
Saintonge (trad. prov.) France 29
Saint Onge, S. Dak. (57779) 298/B4
Saint-Ouen, France 28/B1
Saint-Ours, Québec 172/D4
Saint-Pacôme, Québec 172/G2
Saint-Pamphile, Québec 172/H3
Saint Paris, Ohio (43072) 284/C5
Saint-Pascal, Québec 174/D3
Saint-Pascal, Québec 172/H2
Saint-Patrice-de-Beaurivage, Québec 172/F3
Saint Patrick (lake), Manitoba 179/E3
Saint Patrick, Mo. (63466) 261/J2
Saint Patrick (chan.), Nova Scotia 168/G3
Saint Paul (isl.), (Fr.) 2/P7
Saint Paul (isl.), Alaska 196/D3

Saint Paul, Alberta 182/E3
Saint Paul, Alta. 162/E5
Saint Paul, Ark. (72760) 202/C2
Saint Paul (cape), Ghana 106/E7
Saint Paul, Ind. (47272) 227/F6
Saint Paul, Iowa (52657) 229/L7
Saint Paul, Kansas (66771) 232/G4
Saint Paul (cap.), Minn. 188/H1
Saint Paul (cap.), Minn. 146/J5
Saint Paul (cap.), Minn. (*55101) 255/G6
Saint Paul, Mo. (63366) 261/L5
Saint Paul, Nebr. (68873) 264/F3
Saint-Paul, New Bruns. 170/E2
Saint Paul (riv.), Newf. 166/C3
Saint Paul (isl.), Nova Scotia 168/H1
Saint Paul, Oreg. (97137) 291/A3
Saint-Paul, Québec 172/E4
Saint Paul (riv.), Québec 174/F2
Saint Paul, S.C. (†29148) 296/G4
Saint Paul, Texas (†75098) 303/H1
Saint Paul, Va. (24283) 307/D7
Saint-Paul-de-Montminy, Québec 172/G3
Saint-Paul-du-Nord, Québec 172/H1
Saint-Paulin, Québec 172/D3
Saint Paul Island, Alaska (99660) 196/D3
Saint-Paul-l'Ermite, Québec 172/J4
Saint Paul Park, Minn. (55071) 255/G6
Saint Paul's, Newf. 166/C4
Saint Pauls, N.C. (28384) 281/M5
Saint Peter, Ill. (62880) 222/E5
Saint Peter, Ind. (†47012) 227/H6
Saint Peter, Kansas (†67650) 232/C2
Saint Peter, Minn. (56082) 255/E6
Saint Peter Port (cap.), Guernsey, Chan. Is. 13/E8
Saint Peter Port (cap.), Guernsey, Chan. Is. 10/E6
Saint Peters, Mo. (63376) 261/M1
Saint Peters, Nova Scotia 168/H3
Saint Peters (bay), Nova Scotia 168/H3
Saint Peters, Pr. Edward I. 168/F2
Saint Peters (bay), Pr. Edward I. 168/F2
Saint Peters (isl.), Pr. Edward I. 168/E2
Saint Peters, S. Australia 88/E8
Saint Petersburg, Fla. 188/K5
Saint Petersburg, Fla. 146/K7
Saint Petersburg, Fla. (*33701) 212/B3
Saint Petersburg, Pa. (16054) 294/C3
Saint Petersburg Beach, Fla. (33736) 212/B3
Saint-Petronille, Québec 172/J3
Saint-Philémon, Québec 172/G3
Saint Philip, Ind. (†47620) 227/B9
Saint-Philippe-de-Laprairie, Québec 172/J4
Saint-Philippe-de-Néri, Québec 172/H2
Saint Phillips, Sask. 181/K4
Saint Phillips, Mont. (†59353) 262/M4
Saint Phillips, Newf. 166/D2
Saint-Pie, Québec 172/E4
Saint-Pierre, Martinique 161/C6
Saint-Pierre, Martinique 156/G4
Saint-Pierre (bay), Martinique 161/C6
Saint-Pierre, Ile-de-Mont., Québec 172/H4
Saint-Pierre, Joliette, Québec 172/D3
Saint-Pierre (lake), Québec 172/E3
Saint-Pierre (pt.), Québec 172/D1
Saint-Pierre, Réunion 118/F6
Saint Pierre & Miquelon (cap.), 166/C4
Saint Pierre & Miquelon (isl.), 166/C4
Saint Pierre & Miquelon (isls.) 162/L6
SAINT PIERRE AND MIQUELON 166/C4
Saint Pierre and Miquelon (isls.) 146/N5
Saint-Pierre-Baptiste, Québec 172/F3
Saint-Pierre-de-Broughton, Québec 172/F3
Saint-Pierre-d'Orléans, Québec 172/G3
Saint Pierre-Jolys, Manitoba 179/F5
Saint-Pierre-Montmagny, Québec 172/G3
Saint-Placide, Québec 172/C4
Saint-Pol-de-Léon, France 28/A3
Saint-Pol-sur-Ternoise, France 28/E2
Saint-Polycarpe, Québec 172/C4
Saint-Pons, France 28/E6
Saint-Prex, Switzerland 39/B4
Saint-Prime, Québec 172/E1
Saint-Prosper-de-Dorchester, Québec 172/G3
Saint-Quentin, France 28/E3
Saint Quentin, New Bruns. 170/C1
Saint-Raphaël, France 28/G6
Saint-Raphaël, Haiti 158/C5
Saint-Raphaël, Québec 172/G3
Saint-Raphaël-sur-Mer, New Bruns. 170/F1
Saint-Raymond, Québec 172/F3
Saint-Rédempteur, Québec 172/J3
Saint Regis, Mont. (59866) 262/A3
Saint Regis (riv.), N.Y. 276/L1
Saint-Régis, Québec 172/C4
Saint Regis Falls, N.Y. (12980) 276/M1
Saint Regis Ind. Res., N.Y. 276/M1
Saint Regis Park, Ky. (†40201) 237/K2
Saint-Rémi, Québec 172/D4
Saint-Rémi-d'Amherst, Québec 172/C3
Saint-Rémi-de-Tingwick, Québec 172/F4
Saint-René-de-Matane, Québec 172/B1
Saint Robert, Mo. (65583) 261/H7
Saint-Roch-de-l'Achigan, Québec 172/D4
Saint-Roch-de-Mékinac, Québec 172/E3
Saint-Roch-de-Richelieu, Québec 172/D4
Saint-Romain, Québec 172/F4
Saint-Romuald-d'Etchemin, Québec 172/J3
Saint Rosa, Minn. (†56331) 255/D5
Saint Rose, Ill. (†62230) 222/D5
Saint Rose, La. (70087) 238/N4
Saint Sampson's, Chan. Is. 13/E8
Saints-Anges, Québec 172/G3
Saint Sauveur, New Bruns. 170/E1
Saint-Sauveur-des-Monts, Québec 172/C4
Saint-Sébastien (cape), Madagascar 118/H2
Saint-Sébastien, Frontenac, Québec 172/G4
Saint-Sébastien, Iberville, Québec 172/D4
Saint-Sever, France 28/C6
Saint-Séverin-de-Beaurivage, Québec 172/G3
Saint Shotts, Newf. 166/D2
Saint-Siméon, Québec 172/G2
Saint-Siméon-de-Bonaventure, Québec 172/C2
Saint-Simon, Québec 172/H1
Saint-Simon-de-Bagot, Québec 172/E4
Saint Simons, Georgia 217/K8
Saint Simons Island, Georgia (31522) 217/K8
Saint-Stanislas, Québec 172/E3
Saint Stephan, Switzerland 39/D3
Saint Stephen, Minn. (56375) 255/D5
Saint Stephen, N. Br. 162/K6
Saint Stephen, New Bruns. 170/C3
Saint Stephen, S.C. (29479) 296/H5
Saint Stephen-in-Brannel, England 13/B7
Saint Stephens, Ala. (36569) 195/B7
Saint Stephens, Wyo. (82524) 319/D3
Saint Stephens Church, Va. (23148) 307/O5
Saint-Sylvère, Québec 172/E3
Saint Tammany (par.), La. 238/L6
Saint Tammany, La. (†70445) 238/L6
Saint Teresa, Fla. (†32327) 212/B2
Saint-Théodore, Québec 172/D3
Saint-Théodore-d'Acton, Québec 172/E4
Saint-Théophile, Québec 172/G4
Saint Theresa Point, Manitoba 179/J3
Saint Thomas, Mo. (65076) 261/H6
Saint Thomas, N. Dak. (58276) 282/R2
Saint Thomas, Ontario 177/C5
Saint Thomas, Pa. (17252) 294/G6
Saint Thomas (harb.), Virgin Is. (U.S.) 161/B4
Saint Thomas (isl.), Virgin Is. (U.S.) 156/G1
Saint Thomas (isl.), Virgin Is. (U.S.) 161/A4
Saint-Thomas-de-Joliette, Québec 172/D3
Saint-Thuribe, Québec 172/E3
Saint-Timothée, Québec 172/D4
Saint-Tite, Québec 172/E3
Saint-Tite-des-Caps, Québec 172/G2
Saint-Trond (Sint-Truiden), Belgium 27/G7
Saint-Tropez, France 28/G6
Saint-Ubald, Québec 172/E3
Saint-Ulric, Québec 172/B1
Saint-Urbain-de-Charlevoix, Québec 172/G2
Saint-Ursanne, Switzerland 39/D2
Saint-Valentin, Québec 172/D4
Saint-Valère-de-Bulstrode, Québec 172/E3
Saint-Valérien, Québec 172/E4
Saint-Valérien-de-Rimouski, Québec 172/J1
Saint-Valéry-sur-Somme, France 28/D2
Saint-Vallier, France 28/F5
Saint-Vallier, Québec 172/G3
Saint-Victor, Québec 172/G3
Saint Victor, Sask. 181/F6
Saint Victor Petroglyphs Hist. Park, Sask. 181/E6
Saint Vincent, Alberta 182/E2
Saint Vincent (isl.), Fla. 212/D7
Saint Vincent, Italy 34/A2
Saint Vincent, Minn. (56755) 255/A2
Saint Vincent (São Vicente)(cape) Portugal 33/B4
Saint Vincent (gulf), S. Australia 88/D8
Saint Vincent (chan.), St. Lucia 161/G7
SAINT VINCENT & THE GRENADINES 161/A8
Saint Vincent (isl.), St. Vin. & Grens. 156/G4
Saint Vincent (passage), St. Vin. & Grens. 161/A8
Saint Vincent (gulf), S. Australia 94/F6
Saint Vincent (cape), Tasmania 99/B5
Saint Vincent and The Grenadines 2/F5
Saint Vincent and the Grenadines 146/M8
SAINT VINCENT & THE GRENADINES 156/G4
Saint Vincent's, Newf. 166/D2
Saint-Vith (Sankt Vith), Belgium 27/J8
Saint Vrain, N. Mex. (88133) 274/F4
Saint Walburg, Sask. 162/F5
Saint Walburg, Sask. 181/B2
Saint-Wenceslas, Québec 172/E3
Saint Wendel, Ind. (†47638) 227/B8
Saint Wilfred, New Bruns. 170/E1
Saint Williams, Ontario 177/D5
Saint Xavier, Mont. (59075) 262/J5
Saint-Yrieix-la-Perche, France 28/D5
Saint-Yvon, Québec 172/D1
Saint-Zacharie, Québec 172/G3
Saint-Zénon, Québec 172/D3
Saint-Zéphirin, Québec 172/E3
Saint-Zotique, Québec 172/C4
Saio (Dembidollo), Ethiopia 111/F6
Saipan (isl.), No. Marianas 87/E4
Saipina, Bolivia 136/C6
Saitama (pref.), Japan 81/O2
Saito, Japan 81/E7
Sajama, Bolivia 136/A6
Sajama, Nevada (mt.), Bolivia 136/A6
Sajó (riv.), Hungary 41/F2
Sajószentpéter, Hungary 41/F2
Sak (riv.), S. Africa 118/C6
Sakado, Japan 81/O2
Sakai, Ibaraki, Japan 81/P1
Sakai, Osaka, Japan 81/J8
Sakaide, Japan 81/G6
Sakaiminato, Japan 81/F6
Sakaka, Saudi Arabia 59/D4
Sakakawea (lake), N. Dak. 146/H5
Sakakawea (lake), N. Dak. 188/F1
Sakakawea (lake), N. Dak. 282/G5
Sakami (lake), Québec 174/B2
Sakami (riv.), Québec 174/B2
Sakania, Zaire 115/E6
Sakarya (prov.), Turkey 63/D2
Sakarya (Adapazarı), Turkey 63/D2
Sakarya (riv.), Turkey 59/B1
Sakarya (riv.), Turkey 63/D2
Sakata, Japan 81/J4
Sakhalin (isl.), U.S.S.R. 2/S3
Sakhalin (isl.), U.S.S.R. 54/R4
Sakhalin (gulf), U.S.S.R. 48/P4
Sakhalin (isl.), U.S.S.R. 48/P4
Sakhar, Afghanistan 68/B2
Sakhar, Afghanistan 59/J3
Sakhnin, Israel 65/C2
Saki, U.S.S.R. 52/D5
Šakiai, U.S.S.R. 53/B3
Sakishima (isls.) Japan 54/O7
Sakishima (isls.), Japan 81/K7
Sakon Nakhon, Thailand 72/E3
Sakonnet (pt.), R.I. 249/K7
Sakonnet (riv.), R.I. 249/K7
Sakrivier, S. Africa 118/C6
Saksaul'skiy, U.S.S.R. 48/F5
Sakskøbing, Denmark 21/E8
Saku, Japan 81/J5
Sakurai, Japan 81/J8
Sakwatamau (riv.), Alberta 182/C2
Sal (isl.), C. Verde 106/B7
Šal'a, Czech. 41/D2
Sala, Sweden 18/K7
Salabangka (isls.), Indonesia 85/G6
Salacgrīva, U.S.S.R. 53/C2
Sala Consilina, Italy 34/E4
Saladas, Argentina 143/E2
Saladillo, Argentina 143/G7
Saladillo (riv.), Argentina 143/D2
Saladillo, Bolivia 136/D7
Salado (riv.), Argentina 120/C6
Salado (riv.), Argentina 143/H7
Salado (riv.), Argentina 143/C4
Salado, Ark. (72575) 202/G2
Salado, Chile 138/A6
Salado, Quebrado del (riv.), Chile 138/B6
Salado, Cuba 158/H3
Salado, Honduras 154/D3
Salado del Norte (riv.), Argentina 120/C5
Salado del Norte (riv.), Argentina 143/D2
Salaga, Ghana 106/D7
Salahuddin (gov.), Iraq 66/C3
Sala'ilua, W. Samoa 86/L8
Salala, Oman 59/F6
Salala, Oman 54/G8
Salamá, Guatemala 154/B3
Salamanca, Chile 138/A9
Salamanca, Mexico 150/J6
Salamanca, N.Y. (14779) 276/C6
Salamanca (prov.), Spain 33/C2
Salamanca, Spain 33/D2
Salamanca, Spain 7/D4
Salamat, Bahr (riv.), Chad 111/C6
Salamina, Colombia 126/C5
Salamís, Greece 45/F6
Salamonia, Ind. (47381) 227/H4
Salamonie (lake), Ind. 227/F3
Salamonie (riv.), Ind. 227/G4
Salas, Spain 33/C1
Salas de los Infantes, Spain 33/E2
Salatiga, Indonesia 85/J2
Salavat, U.S.S.R. 7/K3
Salavat, U.S.S.R. 52/H4
Salaverry, Peru 128/C7
Salawati (isl.), Indonesia 85/J6
Salay, Philippines 82/E6
Sala y Gómez (isl.), Chile 2/D7
Sala y Gómez (isl.), Chile 87/P8
Salazar, Colombia 126/D4
Salcantay (mt.), Peru 128/F9
Salcedo (prov.), Dom. Rep. 158/E5
Salcedo, Dom. Rep. 158/E5
Salcombe, England 13/D7
Salcombe, England 10/D5
Saldaña (riv.), Colombia 126/C6
Saldaña, Spain 33/D1
Saldanha, S. Africa 118/B6
Saldee, Ky. (41369) 237/P6
Saldus, U.S.S.R. 53/B2
Sale, England 13/H2
Sale (riv.), Manitoba 179/E5
Salé, Morocco 106/C2
Sale, Victoria 88/H7
Sale, Victoria 97/D6
Sale City, Georgia (31784) 217/D8
Sale Creek, Tenn. (37373) 237/L10
Salée (riv.), Guadeloupe 161/A6
Salekhard, U.S.S.R. 54/H3
Salekhard, U.S.S.R. 4/C6
Salekhard, U.S.S.R. 2/N2
Salekhard, U.S.S.R. 48/G3
Salem, Ala. (36874) 195/H5
Salem, Ark. (72576) 202/G1
Salem○, Conn. (†06415) 210/F3
Salem, Fla. (32356) 212/C2
Salem, Ill. (62881) 222/E5
Salem, India 54/J8
Salem, India 68/D6
Salem, Ind. (47167) 227/E7
Salem, Ind. (†47362) 227/H3
Salem, Iowa (52649) 229/K7
Salem, Ky. (42078) 237/E6
Salem, Maine (†04983) 243/C6
Salem, Md. (†21869) 245/P7
Salem, Mass. (01970) 249/E5
Salem, Mich. (48175) 250/F6
Salem, Mo. (65560) 261/J7
Salem, Nebr. (68433) 264/J4
Salem○, N.H. (03079) 268/E6
Salem (co.), N.J. 273/C4
Salem, N.J. (08079) 273/C4
Salem (riv.), N.J. 273/C4
Salem, N. Mex. (87941) 274/B6
Salem, N.Y. (12865) 276/O4
Salem, Ohio (44460) 284/J4
Salem, Ontario 177/D4
Salem (cap.), Oreg. 146/F5
Salem (cap.), Oreg. 188/B1
Salem (cap.), Oreg. (*97301) 291/A3
Salem, S.C. (29676) 296/A2
Salem, S. Dak. (57058) 298/P6
Salem, Utah (84653) 304/C3
Salem (lake), Vt. 268/C2
Salem (I.C.), Va. (24153) 307/H6
Salem, W. Va. (26426) 312/E4
Salemburg, N.C. (28385) 281/N4
Salem Center, Ind. (†46747) 227/G1
Salem Depot, N.H. (†03079) 268/E6
Salemi, Italy 34/D6
Salem Maritime Nat'l Hist. Site, Mass. 249/E5
Salen, Scotland 15/C4
Saleratus Wash (creek), Utah 304/D4
Salernes, France 28/G6
Salerno (prov.), Italy 34/E4
Salerno, Italy 7/F4
Salerno, Italy 34/E4
Salerno (gulf), Italy 34/E4
Salesville, Ark. (†72658) 202/F1
Salesville, Ohio (43778) 284/H6
Salfit, West Bank 65/C3
Salford, England 13/H2
Salford, England 10/G2
Salgótarján, Hungary 41/E2
Salgueiro, Brazil 132/G5
Sal (riv.), Argentina 143/C2
Salida, Colo. (81201) 208/H6
Salies-de-Béarn, France 28/C6
Salihli, Turkey 63/C3
Salima, Malawi 115/F6
Salina (isl.), Italy 34/E5
Salina, Kans. 188/G3
Salina, Kansas (67401) 232/E3
Salina, Kansas 146/J6
Salina, Okla. (74365) 288/R2
Salina, Utah (84654) 304/C5
Salina (creek), Utah 304/C5
Salina Cruz, Mexico 150/M9
Salinas, Brazil 132/F7
Salinas, Calif. 188/B3
Salinas, Calif. (93901) 204/D7
Salinas (riv.), Calif. 204/D7
Salinas, Chile 138/B4
Salinas (bay), C. Rica 154/D5
Salinas (pt.), Dom. Rep. 158/E6
Salinas, Ecuador 128/B4
Salinas (riv.), Guatemala 154/B2
Salinas, Mexico 150/J5
Salinas (riv.), Mexico 150/O8
Salinas (bay), Nicaragua 154/D5
Salinas, P. Rico 156/G1
Salinas, P. Rico 161/D3
Salinas (pt.), P. Rico 161/D1
Salinas (cape), Spain 33/H3
Salinas de Garci Mendoza, Bolivia 136/B6
Salinas de Santiago, Bolivia 136/E6
Salinas Grandes (salt dep.), Argentina 143/D2
Salinas Nat'l Mon., N. Mex. 274/C4
Saline (co.), Ark. 202/E4
Saline (riv.), Ark. 202/E5
Saline (riv.), Ark. 202/B6
Saline (pt.), Grenada 161/C9
Saline (co.), Ill. 222/E6
Saline (riv.), Ill. 222/E6
Saline (co.), Kansas 232/E3
Saline (riv.), Kansas 232/D3
Saline, La. (71070) 238/E2
Saline (lake), La. 238/E3
Saline, Mich. (48176) 250/F6
Saline (co.), Mo. 261/F4
Saline, Mo. (†64632) 261/E1
Saline (co.), Nebr. 264/G4
Saline, Scotland 15/C1
Saline City, Ind. (†47840) 227/C6
Salineno, Texas (78585) 303/E11
Salines (pt.), Martinique 161/D7
Salineville, Ohio (43945) 284/J4
Salinópolis, Brazil 132/E3
Salins-les-Bains, France 28/F4
Salisbury (sound), Alaska 196/M1
Salisbury○, Conn. (06068) 210/B1
Salisbury, England 13/F6
Salisbury, England 10/F5
Salisbury, Md. (21801) 245/R7
Salisbury, Mass. (01950) 249/L1
Salisbury○, Mass. (01950) 249/L1
Salisbury, Mo. (65281) 261/G4
Salisbury, New Bruns. 170/E2
Salisbury (bay), New Bruns. 170/F3
Salisbury○, N.H. (03268) 268/D5
Salisbury, N.C. (28144) 281/H3
Salisbury (isl.), N.W.T. 162/J3
Salisbury (isl.), N.W. Terrs. 187/L3
Salisbury, Pa. (15558) 294/D6
Salisbury, S. Australia 88/E7
Salisbury, S. Australia 94/B7
Salisbury○, Vt. (05769) 268/A4
Salisbury (cap.), Zimbabwe 118/E3
Salisbury (Harare) (cap.), Zimbabwe 102/E6
Salisbury (cap.), Zimbabwe 2/L6
Salisbury Beach, Mass. (01950) 249/L1
Salisbury Center, N.Y. (13454) 276/L4
Salisbury Downs, N.S. Wales 97/B1
Salitpa, Ala. (36570) 195/C7
Salix, Iowa (51052) 229/A4
Salkehatchie (riv.), S.C. 296/E5
Salkum, Wash. (98582) 310/C4
Salladasburg, Pa. (17740) 294/H3
Sallent, Spain 33/H2
Salley, S.C. (29137) 296/E4
Salliqueló, Argentina 143/D4
Sallis, Miss. (39160) 256/E4
Sallisaw, Okla. (74955) 288/S4
Sallyan, Nepal 68/E3
Sally's Cove, Newf. 166/C4
Salma, Jebel (mts.), Saudi Arabia 59/D4
Salmo, Br. Col. 184/J5
Salmon (riv.), Br. Col. 184/F3
Salmon (riv.), Calif. 204/B2
Salmon (brook), Conn. 210/D1
Salmon (creek), Conn. 210/B1
Salmon (riv.), Conn. 210/F2
Salmon, Idaho (83467) 220/D4
Salmon, Idaho 188/D1
Salmon (riv.), Idaho 188/C1
Salmon (falls), Idaho 220/C7
Salmon (riv.), Idaho 220/B4
Salmon (riv.), New Bruns. 170/C1
Salmon (riv.), New Bruns. 170/E2
Salmon (res.), N.Y. 276/J3
Salmon (riv.), N.Y. 276/H3
Salmon (riv.), N.Y. 276/M1
Salmon (riv.), Nova Scotia 168/G3
Salmon (riv.), Nova Scotia 168/E3
Salmon Arm, Br. Col. 184/H5
Salmon Arm, Br. Col. 184/H5
Salmon Arm (inlet), Br. Col. 184/J2
Salmon Beach, New Bruns. 170/E1
Salmon Cove, Newf. 166/D2
Salmon Creek, New Bruns. 170/E2
Salmon Falls (creek), Idaho 220/D7
Salmon Falls (riv.), Maine 243/B9
Salmon Falls, N.H. (†03820) 268/F5
Salmon Falls (riv.), N.H. 268/F5
Salmon Falls Creek (res.), Idaho 220/D7
Salmon Gums, W. Australia 88/C6
Salmon Gums, W. Australia 92/C6
Salmonier (riv.), Newf. 166/D2
Salmon River (mts.), Idaho 220/C5
Salmon River, Colchester, Nova Scotia 168/E3
Salmon River, Digby, Nova Scotia 168/B4
Salmons, Ky. (†42134) 237/H7
Salmon Valley, Br. Col. 184/F3
Salo, Finland 18/N6
Salobreña, Spain 33/E4
Salol, Minn. (56756) 255/C2
Saloma, Ky. (†42718) 237/L6
Salome, Ariz. (85348) 198/B5
Salomon (pt.), Martinique 161/C7
Salona, Pa. (17767) 294/H3
Salon-de-Provence, France 28/F6
Salong, Philippines 82/D5
Salonga Nat'l Park, Zaire 115/D4
Salonika (Thessaloníki), Greece 45/F5
Salonika (Thermaic) (gulf), Greece 45/F6
Salonta, Romania 45/E2
Salop (co.), England 13/E5
Sal Rei, C. Verde 106/B8
Salsette (isl.), India 68/B7
Sal'sk, U.S.S.R. 52/F5
Sal'sk, U.S.S.R. 48/E5
Salso (riv.), Italy 34/D6
Salsomaggiore Terme, Italy 34/B2
Salt (riv.), Ariz. 188/D4
Salt (riv.), Ariz. 198/D5
Salt (creek), Calif. 204/J7
Salt (lake), Hawaii 218/B3
Salt (creek), Ill. 222/D3
Salt (creek), Ind. 227/E6
Salt (riv.), Ky. 237/K5
Salt (pt.), Mich. 250/E2
Salt (riv.), Mo. 261/J3
Salt (creek), N. Mex. 274/E5
Salt (lake), N. Mex. 274/F4
Salt, The (lake), N.S. Wales 97/B2
Salt (creek), Ohio 284/E7
Salt (creek), Oreg. 291/E4
Salt, Spain 33/H1
Salt (cay), Virgin Is. (U.S.) 161/A4
Salt (riv.), Virgin Is. (U.S.) 161/F4
Salt (lake), W. Australia 88/B5
Salt (lake), W. Australia 92/B5
Salt (riv.), Wyo. 319/B3
Salta (prov.), Argentina 143/D1
Salta, Argentina 120/C5
Salta, Argentina 143/C1
Saltair, Br. Col. 184/J3
Saltaire, N.Y. (11706) 276/O9
Saltash, England 10/D5
Saltash, England 13/C7
Saltburn and Marske-by-the-Sea, England 13/G3
Saltcoats, Sask. 181/J4
Saltcoats, Scotland 15/D5
Saltcoats, Scotland 10/D3
Saltee (isls.), Ireland 17/H7
Saltelv (riv.), Norway 18/J3
Salters, S.C. (29590) 296/H4
Saltese, Mont. (59867) 262/A3
Saltfjorden (fjord), Norway 18/J3
Salt Fork (creek), Ohio 284/H5
Salt Fork (lake), Ohio 284/H5
Salt Fork, Okla. (†74640) 288/L1
Salt Fork, Arkansas (riv.), Okla. 288/J1
Salt Fork, Red (riv.), Okla. 288/G5
Salt Fork, Red (riv.), Texas 303/D3
Saltillo, Ind. (†47108) 227/E7
Saltillo, Mexico 146/H7
Saltillo, Mexico 150/J4
Saltillo, Miss. (38866) 256/G2
Saltillo, Pa. (17253) 294/G5
Saltillo, Tenn. (38370) 237/E10
Salt Lake (co.), Utah 304/B3
Salt Lake City (cap.), Utah (*84101) 304/C3
Salt Lake City (cap.), Utah 188/D2
Salt Lake City (cap.), Utah 146/G6
Salt Lick, Ky. (40371) 237/O4
Salt Marsh (lake), Utah 304/A4
Salto, Argentina 143/F7
Salto (riv.), Argentina 143/F7
Salto, Brazil 135/C3
Salto (dept.), Uruguay 145/B2
Salto, Uruguay 120/D6
Salto, Uruguay 145/B2
Salto Angel (fall), Venezuela 124/G5
Salto del Guairá, Paraguay 144/E4
Salto Grande (falls), Colombia 120/B3
Salto Grande (falls), Colombia 126/D8
Salto Grande (falls), Uruguay 145/A2
Salton Sea (lake), Calif. 188/C4
Salton Sea (lake), Calif. 204/K10
Saltpetre, W. Va. (25558) 312/A6
Salt River, N.W. Terrs. 187/G3
Salt River (bay), Virgin Is. (U.S.) 161/F3
Salt River (range), Wyo. 319/B3
Salt River Ind. Res., Ariz. 198/D5
Salt Rock, W. Va. (25559) 312/B6
Saltsburg, Pa. (15681) 294/C4
Saltsjöbaden, Sweden 18/J1
Salt Sulphur Springs, W. Va. (†24945) 312/E7
Saltville, Va. (24370) 307/E7
Salt Wells (creek), Wyo. 319/D4
Saluda, N.C. (28773) 281/E4
Saluda (co.), S.C. 296/D3
Saluda, S.C. (29138) 296/D4
Saluda (dam), S.C. 296/E3
Saluda (riv.), S.C. 296/D3
Saluda, Va. (23149) 307/P5
Salûm, Egypt 102/E1
Salûm, Egypt 111/E1
Salus, Ark. (72854) 202/D2
Salut (isls.), Fr. Guiana 131/F3
Saluzzo, Italy 34/A2
Salvador, Brazil 132/G6
Salvador, Brazil 120/F4
Salvador, Brazil 2/H6
Salvador (lake), La. 238/K7
Salvador, Sask. 181/B3
Salvage (isls.), Portugal 106/A2
Salvaterra de Magos, Portugal 33/B3
Salvatierra, Mexico 150/J6
Salvation (creek), Utah 304/C5
Salvisa, Ky. (40372) 237/M5
Salvo, N.C. (27972) 281/U3
Salwa, Saudi Arabia 59/F5
Salween (riv.) 54/L8
Salween (riv.), Burma 72/C3
Salween (Nu Jiang) (riv.), China 77/E6
Sal'yany, U.S.S.R. 52/G7
Salybia, Dominica 161/F6
Salyer, Calif. (95563) 204/B3
Salyersville, Ky. (41465) 237/P5
Salzach (riv.), Austria 41/B2
Salzach (riv.), W. Germany 22/E5
Salzburg (prov.), Austria 41/B3
Salzburg, Austria 7/F4
Salzburg, Austria 41/B3
Salzgitter, W. Germany 22/D2
Salzkammergut (reg.), Austria 41/B3
Salzwedel, E. Germany 22/D2
Samá, Cuba 158/J3
Sama (riv.), Peru 128/G11
Sama, Spain 33/D1
Samagaltay, U.S.S.R. 48/K4
Samaipata, Bolivia 136/D6
Samal (isl.), Philippines 82/E7
Samales Group (isls.), Philippines 82/D7
Samalkot, India 68/E5
Samalût, Egypt 111/J4
Samalût, Egypt 59/B4
Samana (cay), Bahamas 156/D2
Samaná (prov.), Dom. Rep. 158/E5
Samaná, Dom. Rep. 158/F5
Samaná, Dom. Rep. 156/E3
Samaná (bay), Dom. Rep. 156/E3
Samaná (bay), Dom. Rep. 158/F5
Samaná (cape), Dom. Rep. 158/F5
Samandağı, Turkey 63/F4
Samaniego, Colombia 126/B7
Samantha, Ala. (35482) 195/C4
Samar, Jordan 65/D2
Samar (isls.), Philippines 54/O8
Samar (isl.), Philippines 85/H3
Samar (isl.), Philippines 82/E5
Samar (sea), Philippines 82/E4
Samara (Kuybyshev), U.S.S.R. 52/H4
Samara (riv.), U.S.S.R. 52/H4
Samarai, Papua N.G. 87/E7
Samarai, Papua N.G. 85/C8
Samaria, Idaho (83252) 220/F7
Samaria, Ind. (†46181) 227/E6
Samaria (reg.), Jordan 65/C3
Samariapo, Venezuela 124/E5
Samarinda, Indonesia 85/F6
Samarinda, Indonesia 54/N10
Samarkand, U.S.S.R. 54/H6
Samarkand, U.S.S.R. 48/G6
Samarra, Iraq 59/D3
Samarra, Iraq 66/C3
Samawa, Iraq 66/D5
Samawa, Iraq 59/D3
Samba, Zaire 115/E4
Sambalpur, India 68/E4
Sambas, Indonesia 85/D5
Sambava, Madagascar 118/J2
Sambhal, India 68/D3
Sambhar (lake), India 68/C3
Sambiase, Italy 34/F5
Samboja, Indonesia 85/F6
Sambor, Cambodia 72/E4
Sambor, U.S.S.R. 52/B5
Samborombón (bay), Argentina 143/E4
Sambre (riv.), Belgium 27/D8
Sambre (riv.), France 28/F2
Sambro, Nova Scotia 168/E4
Samburg, Tenn. (38254) 237/C8
Samch'ŏk, S. Korea 81/D5
Same, Tanzania 115/G4
Samedan, Switzerland 39/J3
Sámi, Greece 45/E6
Samish (lake), Wash. 310/C2
Samka, Burma 72/C2
Sam Lord's Castle, Barbados 161/C9

Sammamish (lake), Wash. 310/B2
Samnangjin, S. Korea 81/D6
Samnaun, Switzerland 39/K3
Samnorwood, Texas (79077) 303/D2
Samoa, Calif. (95564) 204/A3
Samoa (isls.), Pacific 87/J7
Samo Alto, Chile 138/A8
Samobor, Yugoslavia 45/B3
Samokov, Bulgaria 45/F4
Samorín, Czech. 41/D2
Sámos, Greece 45/H7
Sámos (isl.), Greece 45/H7
Samoset, Fla. (†33508) 212/D4
Samothráki, Greece 45/G5
Samothrace (isl.), Greece 45/G5
Sampang, Indonesia 85/K2
Samper de Calanda, Spain 33/F2
Sampit, Indonesia 85/E6
Sampson (co.), N.C. 281/N4
Sampwe, Zaire 115/E5
Sam Rayburn (res.), Tex. 188/G4
Sam Rayburn (res.), Texas 303/K6
Samsang, China 77/B5
Samsat, Turkey 63/H4
Samsø (isl.), Denmark 21/D6
Samsø Baelt (chan.), Denmark 21/D6
Samson, Ala. (36477) 195/F8
Samsula, Fla. (†32069) 212/E2
Samsun (prov.), Turkey 63/F2
Samsun, Turkey 59/C1
Samsun, Turkey 63/F2
Samsun, Turkey 54/E5
Sams Valley, Oreg. (†97525) 291/E5
Samtown, La. (71301) 238/F4
Samu, West Bank 65/C5
Samuels, Idaho (83862) 220/B1
Samuels, Ky. (40064) 237/L5
Samui, Ko (isl.), Thailand 72/D5
Samui (str.), Thailand 72/D5
Samut Prakan, Thailand 72/D4
Samut Sakhon, Thailand 72/D4
Samut Songkhram, Thailand 72/C4
San, Se (riv.), Cambodia 72/E4
San, Mali 106/D6
San, Mali 102/B3
San (riv.), Poland 47/F3
Sana, P.D.R. Yemen 59/F6
Saña, Peru 128/C6
San'a (cap.), Yemen Arab Rep. 54/F8
San'a (cap.), Yemen Arab Rep. 59/D6
Sanaag (prov.), Somalia 115/J1
San Acacia, N. Mex. (87831) 274/B4
San Acacio, Colo. (81150) 208/J8
Sanae Station, Ant. 5/B18
Sanaga (riv.), Cameroon 115/B3
San Agustín, Bolivia 136/B7
San Agustín, Colombia 126/B7
San Agustin (plains), N. Mex. 274/B5
San Agustin (cape), Philippines 85/H4
San Agustin (cape), Philippines 82/F7
Sanak (isl.), Alaska 196/F4
Sanam, Saudi Arabia 59/D5
San Ambrosio (isl.), Chile 120/B5
Sanana, Indonesia 85/H6
Sanandaj, Iran 59/E2
Sanandaj, Iran 66/E3
Sanandita, Bolivia 136/D7
San Andreas, Calif. (95249) 204/E5
San Andreas (lake), Calif. 204/H2
San Andrés, Bolivia 136/C4
San Andrés, Antioquia, Colombia 126/C4
San Andrés, San Andrés y Providencia, Colombia 126/A9
San Andrés (isl.), Colombia 126/A10
San Andrés, Cuba 158/H3
San Andrés, Guatemala 154/B2
San Andres (mts.), N. Mex. 274/C6
San Andrés de Machaca, Bolivia 136/A5
San Andrés Tuxtla, Mexico 150/M7
San Andrés y Providencia (inten.), Colombia 126/B10
San Angelo, Texas 146/H6
San Angelo, Texas (*76901) 303/D6
San Angelo, Texas 188/F4
San Anselmo, Calif. (94960) 204/H1
San Antero, Colombia 126/C3
San Antonio (cape), Argentina 120/D6
San Antonio (cape), Argentina 143/E4
San Antonio, El Beni, Bolivia 136/C4
San Antonio (lake), Calif. 204/E8
San Antonio, Chile 138/F3
San Antonio, Colombia 126/B6
San Antonio (cape), Cuba 156/A2
San Antonio (cape), Cuba 158/A2
San Antonio, Fla. (33576) 212/D3
San Antonio, Mexico 150/D5
San Antonio (reef), Mexico 150/L4
San Antonio, N. Mex. (87832) 274/B5
San Antonio (peak), N. Mex. 274/C2
San Antonio, Paraguay 144/A5
San Antonio, Philippines 82/B3
San Antonio (bay), Philippines 82/A6
San Antonio, P. Rico 161/A1
San Antonio, Texas 146/J7
San Antonio, Texas (*78201) 303/J11
San Antonio, Texas 188/G5
San Antonio (bay), Texas 303/H9
San Antonio (mt.), Texas 303/B10
San Antonio (riv.), Texas 303/K11
San Antonio, Canelones, Uruguay 145/B6
San Antonio, Salto, Uruguay 145/B2
San Antonio, Amazones, Venezuela 124/E6
San Antonio, Monagas, Venezuela 124/G2
San Antonio, Zulia, Venezuela 124/C3
San Antonio Abad, Spain 33/G3
San Antonio de Areco, Argentina 143/G7
San Antonio de Caparo, Venezuela 124/C4
San Antonio de Lípez, Bolivia 136/B7
San Antonio de los Baños, Cuba 158/C1
San Antonio de los Baños, Cuba 156/A2
San Antonio de los Cobres, Argentina 143/C1
San Antonio del Parapetí, Bolivia 136/D7
San Antonio del Táchira, Venezuela 124/B4
San Antonio de Tabasca, Venezuela 124/G3
San Antonio Missions Nat'l Hist. Park, Texas 303/J11
San Antonio Oeste, Argentina 120/C7
San Antonio Oeste, Argentina 143/C5
San Ardo, Calif. (93450) 204/E7
Sanare, Venezuela 124/D3
Sanator, S. Dak. (†57730) 298/B6
Sanatorium, Miss. (39112) 256/E7
San Augustine (co.), Texas 303/K6
San Augustine, Texas (75972) 303/K6
San Bartolomeo in Galdo, Italy 34/E4
San Bautista, Uruguay 145/B6
San Benedetto del Tronto, Italy 34/E3
San Benedicto (isl.), Mexico 150/D7
San Benito (co.), Calif. 204/D7
San Benito (riv.), Calif. 204/D7
San Benito (isl.), Mexico 150/B2
San Benito, Texas (78586) 303/G12
San Bernardino, Calif. 146/G6
San Bernardino, Calif. 188/C4
San Bernardino (co.), Calif. 204/J9
San Bernardino, Calif. (*92401) 204/E10
San Bernardino (mts.), Calif. 204/J10
San Bernardino, Paraguay 144/B4
San Bernardino (str.), Philippines 82/E4
San Bernardino, Switzerland 39/H4
San Bernardino (pass), Switzerland 39/H3
San Bernardo, Chile 138/G4
San Bernardo (isls.), Colombia 126/C3
San Blas (cape), Fla. 212/D7
San Blas, Nayarit, Mexico 150/G6
San Blas, Sinaloa, Mexico 150/E3
San Blas, Archipiélago de (arch.), Panama 154/J7
San Blas, Cordillero de (mts.), Panama 154/H6
San Blas, Golfo de (bay), Panama 154/H6
San Blas, Pta. de (pt.), Panama 154/H6
San Borja, Bolivia 136/B4
Sanborn, Iowa (51248) 229/B2
Sanborn, Minn. (56083) 255/C6
Sanborn, N.Y. (14132) 276/C4
Sanborn, N. Dak. (58480) 282/O6
Sanborn (co.), S. Dak. 298/N5
Sanborn, Wis. (†54806) 317/E3
Sanbornton◯, N.H. (03269) 268/D5
Sanbornville, N.H. (03872) 268/F4
San Bruno, Calif. (94066) 204/J2
San Bruno, Mexico 150/C3
San Buenaventura, Bolivia 136/A4
San Buenaventura, Mexico 150/J3
San Carlos, Mendoza, Argentina 143/C3
San Carlos, Santa Fe, Argentina 143/F6
San Carlos, Ariz. (85550) 198/E5
San Carlos (lake), Ariz. 188/E4
San Carlos (lake), Ariz. 198/E5
San Carlos (riv.), Ariz. 198/E5
San Carlos, Bolivia 136/D5
San Carlos, Calif. (94070) 204/J3
San Carlos, Chile 138/E1
San Carlos, Coahuila, Mexico 150/J2
San Carlos, Tamaulipas, Mexico 150/K4
San Carlos, Nicaragua 154/E5
San Carlos, Panama 154/H6
San Carlos, Paraguay 144/D3
San Carlos, Negros Occ., Philippines 82/D5
San Carlos, Pangasinan, Philippines 82/C3
San Carlos, Uruguay 145/E5
San Carlos, Cojedes, Venezuela 124/D3
San Carlos, Zulia, Venezuela 124/C2
San Carlos de Bariloche, Argentina 143/B5
San Carlos de Bariloche, Argentina 120/B7
San Carlos de la Rápita, Spain 33/G2
San Carlos del Zulia, Venezuela 124/C3
San Carlos de Río Negro, Venezuela 124/E7
San Carlos Ind. Res., Ariz. 198/E5
San Casimiro, Venezuela 124/E3
San Cataldo, Italy 34/D6
San Cayetano, Argentina 143/E4
Sancerre, France 28/E4
Sanchez (res.), Colo. 208/H8
Sánchez, Dom. Rep. 156/E3
Sánchez, Dom. Rep. 158/E5
Sánchez, Uruguay 145/B3
Sánchez Ramírez (prov.), Dom. Rep. 158/E5
San Clara, Manitoba 179/A3
San Clemente, Calif. (92672) 204/H10
San Clemente (isl.), Calif. 204/G11
San Clemente, Chile 138/A11
San Clemente, Spain 33/E3
Sancoins, France 28/E4
San Cosme y Damián, Paraguay 144/D5
San Cristóbal, Argentina 143/F5
San Cristóbal, Potosí, Bolivia 136/B7
San Cristóbal, Santa Cruz, Bolivia 136/E3
San Cristóbal, Cuba 158/B1
San Cristóbal (prov.), Dom. Rep. 158/E6
San Cristóbal, Dom. Rep. 158/E6
San Cristóbal (isl.), Ecuador 128/C9
San Cristobal, N. Mex. (87564) 274/D2
San Cristóbal, Panama 154/G6
San Cristobal (isl.), Solomon Is. 87/G7
San Cristobal (isl.), Solomon Is. 86/E3
San Cristóbal, Venezuela 124/B4
San Cristóbal, Venezuela 120/B2
San Cristóbal de las Casas, Mexico 150/N8
Sancti Spíritus (prov.), Cuba 158/F2
Sancti Spíritus, Cuba 158/E2
Sancti Spíritus, Cuba 156/B2
Sanctuary, Sask. 181/D4
Sand (mt.), Ala. 195/G1
Sand (key), Fla. 212/B3
Sand (isl.), Hawaii 218/B4
Sand (creek), Ind. 227/F6
Sand (pt.), Mich. 250/F5
Sand (creek), Minn. 255/F5
Sand, Norway 18/E7
Sand (riv.), S. Africa 118/D4
Sand (creek), S. Dak. 298/M5
Sand (creek), S. Dak. 298/C2
Sand (isl.), Wash. 310/A4
Sand (isl.), Wis. 317/D2
Sanda, Japan 81/H7
Sanda (isl.), Scotland 15/C5
Sandakan, Malaysia 54/N9
Sandakan, Malaysia 85/F4
Sandalwood (Sumba) (isl.), Indonesia 85/F7
Sandane, Norway 18/D6
Sandanski, Bulgaria 45/F5
Sand Arroyo (dry riv.), Colo. 208/O8
Sanday (isl.), Scotland 10/E1
Sanday (isl.), Scotland 15/B3
Sanday (isl.), Scotland 15/F1
Sanday (sound), Scotland 15/F1
Sandbach, England 10/G2
Sandbach, England 13/H2
Sandbank, Scotland 15/A1
Sandborn, Ind. (47578) 227/C7
Sand Brook, N.J. (†08559) 273/D3
Sand Coulee, Mont. (59472) 262/E3
Sand Creek, Wis. (54765) 317/C5
Sand Draw, Wyo. (†82501) 319/D3
Sandefjord, Norway 18/C4
San de Fuca, Wash. (†98239) 310/C2
Sandel, La. (†71429) 238/D4
Sanders, Ariz. (86512) 198/F3
Sanders, Idaho (†83870) 220/B2
Sanders, Ind. (†47401) 227/E6
Sanders, Ky. (41083) 237/M3
Sanders (co.), Mont. 262/A3
Sanders, Mont. (59076) 262/J4
Sanderson, Fla. (32087) 212/D1
Sanderson, Texas (79848) 303/B7
Sandersville, Georgia (31082) 217/G5
Sandersville, Miss. (39477) 256/F7
Sandford, Ind. (47877) 227/B5
Sand Fork, W. Va. (26430) 312/E5
Sandgap, Ky. (40481) 237/N6
Sandgate, Queensland 88/K2
Sandgate, Queensland 95/D2
Sandgate◯, Vt. (†05250) 268/A5
Sandhead, Scotland 15/D6
Sand Hill (riv.), Minn. 255/B3
Sandhill, Miss. (39161) 256/E5
Sand Hill (riv.), Newf. 166/C3
Sand Hills, Mass. (†02066) 249/M4
Sandhurst, England 13/G8
Sandia (peak), N. Mex. 274/C3
Sandia, Peru 128/H10
Sandia, Texas (78383) 303/F9
Sandia Park, N. Mex. (87047) 274/C3
San Diego (cape), Argentina 120/E8
San Diego (cape), Argentina 143/D7
San Diego, Bolivia 136/D7
San Diego (co.), Calif. 204/J10
San Diego, Calif. 146/G6
San Diego, Calif. 188/C4
San Diego, Calif. (*92101) 204/H11
San Diego (bay), Calif. 204/H11
San Diego, Texas (78384) 303/F10
San Diego de Cabrutica, Venezuela 124/F3
San Diego de los Baños, Cuba 158/B1
Sandıklı, Turkey 63/D3
Sandilands, Manitoba 179/F5
San Dimas, Calif. (91773) 204/D10
Sanding (isl.), Indonesia 85/C6
Sandisfield◯, Mass. (01255) 249/B4
Sand Lake, Mich. (49343) 250/D5
Sand Lake, N.Y. (12153) 276/O5
Sandnes, Norway 18/D7
Sandoa, Zaire 115/D5
Sandomierz, Poland 47/E3
Sandoná, Colombia 126/B7
Sándorfalva, Hungary 41/F3
Sandoval, Ill. (62882) 222/D5
Sandoval (co.), N. Mex. 274/C3
Sandover (riv.), North. Terr. 93/D6
Sandoway, Burma 72/B3
Sandown◯, N.H. (03873) 268/E6
Sandown (bay), S. Africa 118/F7
Sandown-Shanklin, England 13/F7
Sandoy (isl.), Denmark 21/B3
Sand Point, Alaska (99661) 196/G3
Sandpoint, Idaho (83864) 220/B1
Sand Point, Okla. (†73449) 288/N7
Sandray (isl.), Scotland 15/A4
Sandridge, Manitoba 179/E4
Sand Ridge, N.Y. (†13132) 276/H4
Sandringham, Victoria 88/L7
Sandringham, Victoria 97/J5
Sands (key), Fla. 212/F6
Sands, Mich. (†49841) 250/B2
Sand Shoal (inlet), Va. 307/S6
Sandspit, Br. Col. 184/B3
Sands Point, N.Y. (†11050) 276/P6
Sand Springs, Iowa (†52237) 229/L4
Sand Springs, Ky. (†40456) 237/N6
Sand Springs, Mont. (59077) 262/J3
Sand Springs (salt flat), Nev. 266/C3
Sand Springs, Okla. (74063) 288/O2
Sandston, Va. (23150) 307/O5
Sandstone, Minn. (55072) 255/F4
Sandstone, W. Australia 88/B5
Sandstone, W. Australia 92/B4
Sandstone, W. Va. (25985) 312/E7
Sand Tank (mt.), Ariz. 198/C6
Sandusky, Ill. (†62988) 222/D6
Sandusky, Ind. (†47240) 227/G6
Sandusky, Mich. (48471) 250/G5
Sandusky, N.Y. (14133) 276/D6
Sandusky (co.), Ohio 284/D3
Sandusky, Ohio (44870) 284/E3
Sandusky, Ohio 188/E3
Sandusky (bay), Ohio 284/E3
Sandusky (riv.), Ohio 284/D3
Sandvika, Norway 18/C3
Sandviken, Sweden 18/K6
Sandwich, England 13/J6
Sandwich, Ill. (60548) 222/E2
Sandwich, Mass. (02563) 249/N5
Sandwich◯, Mass. (02563) 249/N5
Sandwich (bay), Newf. 166/C3
Sandwich◯, N.H. (03270) 268/E4
Sandwich (mt.), N.H. 268/E4
Sandwich (range), N.H. 268/E4
Sandwick, Scotland 15/B2
Sandwith, Sask. 181/C2
Sandy (creek), Ala. 195/H7
Sandy (brook), Conn. 210/C1
Sandy, England 13/G5
Sandy (riv.), Maine 243/C6
Sandy (isls.), Manitoba 179/D2
Sandy (creek), Mont. 262/F2
Sandy (lake), Newf. 166/C4
Sandy (lake), Ont. 162/G5
Sandy (lake), Ontario 175/B2
Sandy, Oreg. (97055) 291/E2
Sandy, Pa. (†15801) 294/E3
Sandy (cape), Queensland 88/J4
Sandy (cape), Queensland 95/E5
Sandy (pt.), R.I. 249/H8
Sandy (pt.), S.C. 296/H6
Sandy (riv.), S.C. 296/E2
Sandy (cape), Tasmania 99/A3
Sandy, Utah (*84070) 304/C3
Sandy (pt.), Virgin Is. (U.S.) 161/D4
Sandy Bay, Jamaica 158/G5
Sandy Bay, Nicaragua 154/F3
Sandy Bay, Sask. 181/N3
Sandy Beach, Alberta 182/C3
Sandy Beach, Sask. 181/H5
Sandy Cove, Nova Scotia 168/B4
Sandy Creek, Maine (†04009) 243/B7
Sandy Creek, N.Y. (13145) 276/H3
Sandy Hook, Conn. (06482) 210/B3
Sandy Hook, Ky. (41171) 237/P4
Sandy Hook, Manitoba 179/E4
Sandy Hook, Miss. (39478) 256/E8
Sandy Hook (spit), N.J. 273/F3
Sandy Hook, Va. (23153) 307/M5
Sandy Lake, Alberta 182/D2
Sandy Lake, Manitoba 179/B4
Sandy Lake, Pa. (16145) 294/B3
Sandy Point, Maine (04972) 243/F7
Sandy Point, Nova Scotia 168/C5
Sandy Point, St. Chris.-Nevis 161/C10
Sandy Ridge, Ala. (†36047) 195/E6
Sandy Ridge, N.C. (27046) 281/J1
Sandy Ridge, Pa. (16677) 294/F4
Sandy Spring-Ashton, Md. (20860) 245/K4
Sandy Springs, Georgia (30358) 217/K1
Sandy Springs, S.C. (29677) 296/B2
Sandyville, Iowa (†50001) 229/G6
Sandyville, Ohio (44671) 284/H4
Sandyville, W. Va. (25275) 312/C5
San Elizario, Texas (79849) 303/A10
San Estanislao, Paraguay 144/D4
San Esteban (gulf), Chile 138/D7
San Esteban, Honduras 154/D3
San Esteban de Gormaz, Spain 33/E2
San Felipe, Chile 138/G2
San Felipe, Colombia 126/G7
San Felipe (cays), Cuba 156/A2
San Felipe (cays), Cuba 158/B2
San Felipe, Guatemala 154/B3
San Felipe, Baja California, Mexico 150/B1
San Felipe, Guanajuanto, Mexico 150/J6
San Felipe, Philippines 82/B3
San Felipe, Yaracuy, Venezuela 124/D2
San Felipe, Zulia, Venezuela 124/B3
San Felipe Pueblo, N. Mex. (†87001) 274/C3
San Feliu de Guixols, Spain 33/H2
San Félix, Chile 138/A7
San Félix (isl.), Chile 120/A5
San Félix, Panama 154/G6
San Félix, Venezuela 124/C2
San Fernando, Argentina 143/G7
San Fernando (riv.), Bolivia 136/F5
San Fernando, Calif. (*91340) 204/C10
San Fernando, Chile 138/G6
San Fernando, Tamaulipas, Mexico 150/L4
San Fernando, La Union, Philippines 82/C2
San Fernando, Masbate, Philippines 82/D4
San Fernando, Pampanga, Philippines 82/C3
San Fernando, Spain 33/C4
San Fernando, Trin. & Tob. 161/A11
San Fernando, Trin. & Tob. 156/G5
San Fernando, Venezuela 120/C2
San Fernando de Apure, Venezuela 124/E4
San Fernando de Atabapo, Venezuela 124/E5
San Fidel, N. Mex. (87049) 274/B3
Sanford, Ala. (†36420) 195/F8
Sanford (mt.), Alaska 196/K2
Sanford, Colo. (81151) 208/H8
Sanford, Fla. 188/K5
Sanford, Fla. (32771) 212/E3
Sanford, Maine (04073) 243/B9
Sanford◯, Maine (04073) 243/B9
Sanford, Manitoba 179/E5
Sanford, Mich. (48657) 250/E5
Sanford, Miss. (*39479) 256/F8
Sanford, N.C. (27330) 281/L4
Sanford, Nova Scotia 168/B5
Sanfrancique, Trin. & Tob. 161/A11
San Francisco, Argentina 120/C6
San Francisco, Córdoba, Argentina 143/D3
San Francisco, San Luis, Argentina 143/C3
San Francisco (riv.), Ariz. 198/F5
San Francisco, Bolivia 136/C4
San Francisco, Calif. 146/F6
San Francisco, Calif. 188/B3
San Francisco (city county), Calif. 204/J2
San Francisco, Calif. (*94101) 204/H2
San Francisco (bay), Calif. 204/J2
San Francisco, Colombia 126/B7
San Francisco (cape), Ecuador 128/B2
San Francisco, Honduras 154/D3
San Francisco (riv.), N. Mex. 274/A5
San Francisco, Nicaragua 154/E5
San Francisco, Panama 154/G6
San Francisco (creek), Texas 303/B8
San Francisco, U.S. 2/C4
San Francisco, Lara, Venezuela 124/C2
San Francisco de la Paz, Honduras 154/D3
San Francisco del Chañar, Argentina 143/C2
San Francisco del Oro, Mexico 150/F3
San Francisco del Rincón, Mexico 150/H6
San Francisco de Macorís, Dom. Rep. 158/E5
San Francisco de Macorís, Dom. Rep. 156/E3
San Francisco de Mostazal, Chile 138/G4
San Francisco Gotera, El Salvador 154/C4
Sanga (riv.), Cameroon 115/C3
Sanga (riv.), Cent. Afr. Rep. 115/C3
San Gabriel, Calif. (*91775) 204/C10
San Gabriel (res.), Calif. 204/D10
San Gabriel, Ecuador 128/D2
San Gabriel Chilac, Mexico 150/K7
San Gallán (isl.), Peru 128/D9
Sangamner, India 68/C5
Sangamon (co.), Ill. 222/D4
Sangamon (riv.), Ill. 222/C3
Sangan, Iran /M3
Sangar, U.S.S.R. 54/O3
Sangar, U.S.S.R. 48/N3
Sangaredyi, Guinea 106/B6
Sangay (mt.), Ecuador 128/C4
Sangeang (isl.), Indonesia 85/F7
San Genaro, Argentina 143/F6
Sanger, Calif. (93657) 204/F7
Sanger, N. Dak. (†58547) 282/H5
Sanger, Texas (76266) 303/G4
Sangerhausen, E. Germany 22/D3
San Germán, Cuba 158/J3
San Germán, P. Rico 161/A2
San Germán, P. Rico 156/F1
Sangerville◯, Maine (04479) 243/E5
Sang-e Sar, Iran 66/H3
Sangestan, Kuh-e (mt.), Iran 66/H5
Sanggabuwana (mt.), Indonesia 85/G2
Sanggau, Indonesia 85/E5
Sangha (riv.), Congo 115/C3
Sangihe (isls.), Indonesia 54/O9
Sangihe (isl.), Indonesia 85/H5
Sangihe (isls.), Indonesia 85/G5
San Gil, Colombia 126/D4
San Giovanni in Fiore, Italy 34/F5
San Giovanni in Persiceto, Italy 34/C2
San Giuliano Terme, Italy 34/C3
Sangju, S. Korea 81/D5
Sangkulirang, Indonesia 85/F5
Sangli, India 68/C5
Sangmélima, Cameroon 115/B3
Sangolquí, Ecuador 128/C3
Sangre de Cristo (mts.), Colo. 208/H6
Sangre de Cristo (mts.), N. Mex. 274/D3
San Gregorio, Calif. (94074) 204/J3
San Gregorio, San José, Uruguay 145/C4
San Gregorio, Tacuarembó, Uruguay 145/D3
Sangre Grande, Trin. & Tob. 161/B10
Sangre Grande, Trin. & Tob. 156/G5
Sangri, China 77/D6
Sangro (riv.), Italy 34/E4
Sangudo, Alberta 182/C3
Sangue (riv.), Brazil 132/B6
Sangüesa, Spain 33/F1
Sanhe, China 77/K1
Sanibel, Fla. (33957) 212/D5
Sanibel (isl.), Fla. 212/D5
San Ignacio, Argentina 143/E2
San Ignacio, Belize 154/C2
San Ignacio, El Beni, Bolivia 136/C4
San Ignacio, Santa Cruz, Bolivia 136/E5
San Ignacio, Chile 138/E1
San Ignacio, C. Rica 154/E6
San Ignacio, Baja California Sur, Mexico 150/C3
San Ignacio, Sinaloa, Mexico 150/F5
San Ignacio, Paraguay 144/D5
San Ignacio, Venezuela 124/B2
Sanilac (co.), Mich. 250/G5
San Ildefonso, N. Mex. (†87501) 274/C3
San Ildefonso (cape), Philippines 82/D2
San Ildefonso, Spain 33/E2
San'in Kaigan National Park, Japan 81/G6
San Isabel, Colo. (†81069) 208/K7
Sanish, N. Dak. (58780) 282/E4
San Isidro, Argentina 143/C2
San Isidro, Philippines 82/E5
Saniya, Hor (lake), Iraq 66/E5
San Jacinto, Calif. (92383) 204/H10
San Jacinto, Colombia 126/C3
San Jacinto, Nev. (†89825) 266/G1
San Jacinto, Philippines 82/D4
San Jacinto (co.), Texas 303/J7
San Jacinto, Uruguay 145/C6
San Jaime de la frontera, Argentina 143/G5
San Javier, Río Negro, Argentina 143/D5
San Javier, Santa Fe, Argentina 143/F5
San Javier, El Beni, Bolivia 136/C4
San Javier, Santa Cruz, Bolivia 136/D5
San Javier, Chile 138/A11
San Javier, Uruguay 145/A3
San Jerónimo de Juárez, Mexico 150/J8
Sanjo, Japan 81/J5
San Joaquín, Bolivia 136/C3
San Joaquin (riv.), Calif. 188/C3
San Joaquin (co.), Calif. 204/D6
San Joaquin, Calif. (93660) 204/E7
San Joaquin (riv.), Calif. 204/E6
San Joaquin (valley), Calif. 204/D6
San Joaquín, Paraguay 144/E4
San Jon, N. Mex. (88434) 274/F3
San Jorge (gulf), Argentina 120/C7
San Jorge (gulf), Argentina 143/C6
San Jorge (riv.), Colombia 126/C3
San Jorge (bay), Mexico 150/C1
San Jorge, Nicaragua 154/E5
San Jorge (gulf), Spain 33/G2
San José, Belize 154/C2
San Jose, Calif. 146/F6
San Jose, Calif. 188/B3
San Jose, Calif. (*95101) 204/L3
San José, Colombia 126/F6
San José (cap.), C. Rica 146/K9
San José (cap.), C. Rica 154/F5
San José, Guatemala 154/B4
San Jose, Ill. (62682) 222/D3
San José (isl.), Mexico 150/D4
San Jose, N. Mex. (87565) 274/D3
San Jose (riv.), N. Mex. 274/B3
San José (isl.), Panama 154/H6
San José, Paraguay 144/B5
San José, Peru 128/B6
San Jose, Philippines 85/G3
San Jose, Nueva Ecija, Philippines 82/C3
San Jose, Occ. Mindoro, Philippines 82/C4
San José (lag.), P. Rico 161/E1
San Jose (isl.), Texas 303/H9
San José (dept.), Uruguay 145/C5
San José (riv.), Uruguay 145/C4
San José, Amazonas, Venezuela 124/E5
San José, Zulia, Venezuela 124/B3
San José de Amacuro, Venezuela 124/H3
San José de Areocuar, Venezuela 124/G2
San Jose de Buenavista, Philippines 82/C5
San José de Chiquitos, Bolivia 136/E5
San José de Feliciano, Argentina 143/G5
San José de Guanipa, Venezuela 124/G3
San José de la Costa, Venezuela 124/D2
San José de la Mariquina, Chile 138/D2
San José de las Lajas, Cuba 158/C1
San José de las Matas, Dom. Rep. 158/D5
San José del Cabo, Mexico 150/D5
San José del Guaviare, Colombia 126/D6
San Jose del Monte, Philippines 82/C3
San José del Ocune, Colombia 126/E5
San José de los Ramos, Cuba 158/D1
San José de Maipo, Chile 138/B10
San José de Mayo, Uruguay 145/C5
San José de Ocoa, Dom. Rep. 158/E6
San José de Río Chico, Venezuela 124/F2
San José de Sisa, Peru 128/D6
San José de Tiznados, Venezuela 124/E3
San José de Uchupiamonas, Bolivia 136/A4
San Juan (riv.), 188/E3
San Juan (prov.), Argentina 143/C3
San Juan, Argentina 143/C3
San Juan, Argentina 120/C6
San Juan (riv.), Argentina 143/C3
San Juan, Potosí, Bolivia 136/B7
San Juan, Santa Cruz, Bolivia 136/F5
San Juan (riv.), Bolivia 136/C7
San Juan (riv.), Br. Col. 184/J3
San Juan (creek), Calif. 204/E8
San Juan (riv.), Colombia 126/B5
San Juan (co.), Colo. 208/D7
San Juan (mts.), Colo. 208/F7
San Juan (riv.), Colo. 208/E8
San Juan (riv.), C. Rica 154/E5
San Juan (prov.), Dom. Rep. 158/D6
San Juan, Dom. Rep. 158/D6
San Juan, Mexico 150/K6
San Juan (co.), N. Mex. 274/A2
San Juan (riv.), N. Mex. 274/B2
San Juan (riv.), Nicaragua 154/E5
San Juan, Peru 128/E10
San Juan, Philippines 82/E5
San Juan (cap.), P. Rico 146/M8
San Juan (dist.), P. Rico 161/D1
San Juan (cap.), P. Rico 161/E1
San Juan (cap.), P. Rico 156/G1
San Juan (cape), P. Rico 156/G1
San Juan, Cabezas de (prom.), P. Rico 161/F1
San Juan, Texas (78589) 303/F11
San Juan, Trin. & Tob. 161/A10
San Juan (co.), Utah 304/E6
San Juan (riv.), Utah 304/D6
San Juan (co.), Wash. 310/C2
San Juan (isl.), Wash. 310/B2
San Juan Bautista, Calif. (95045) 204/D7
San Juan Bautista, Paraguay 144/D5
San Juan Bautista de Neembucú, Paraguay 144/D5
San Juan Capistrano, Calif. (†92691) 204/H10
San Juan de Colón, Venezuela 124/B3
San Juan de Flores, Honduras 154/D3
San Juan de las Galdonas, Venezuela 124/G2
San Juan del César, Colombia 126/D2
San Juan del Norte, Nicaragua 154/F5

San Juan del Norte (bay), Nicaragua 154/F5
San Juan de los Cayos, Venezuela 124/D2
San Juan de los Lagos, Mexico 150/H6
San Juan de los Morros, Venezuela 124/E3
San Juan de los Planes, Mexico 150/D4
San Juan del Piray, Bolivia 136/C7
San Juan del Potrero, Bolivia 136/C5
San Juan del Sur, Nicaragua 154/D5
San Juan de Manapiare, Venezuela 124/E5
San Juan de Payara, Venezuela 124/E4
San Juan Island Nat'l Hist. Park, Wash. 310/B2
San Juan Nat'l Hist. Site, P. Rico 161/D1
San Juan Nepomuceno, Paraguay 144/E5
San Juan Pueblo, N. Mex. (87566) 274/C2
San Juan Xiutetelco, Mexico 150/O1
San Juan y Martínez, Cuba 158/B2
San Julián, Argentina 143/C6
San Julián, Argentina 120/C7
San Justo, Argentina 143/F5
Sankrail, India 68/E2
Sankt Aegyd am Neuwalde, Austria 41/C3
Sankt Anton am Arlberg, Austria 41/A3
Sankt Blasien, W. Germany 22/C5
Sankt Gallen (canton), Switzerland 39/H2
Sankt Gallen, Switzerland 39/H2
Sankt Goar, W. Germany 22/B3
Sankt Ingbert, W. Germany 22/B4
Sankt Johann in Tirol, Austria 41/B3
Sankt Margrethen, Switzerland 39/J2
Sankt Michael im Lungau, Austria 41/B3
Sankt Michael in Obersteiermark, Austria 41/C3
Sankt Paul im Lavanttal, Austria 41/C3
Sankt Peter-Ording, W. Germany 22/C1
Sankt Pölten, Austria 41/C2
Sankt Valentin, Austria 41/C2
Sankt Veit an der Glan, Austria 41/C3
Sankt Vith, Belgium 27/J8
Sankt Wendel, W. Germany 22/B4
Sankt Wolfgang im Salzkammergut, Austria 41/B3
Sankuru (riv.), Zaire 102/E5
Sankuru (riv.), Zaire 115/D4
San Lázaro (cape), Mexico 150/C4
San Lázaro, Paraguay 144/D3
San Leandro, Calif. (*94577) 204/J2
San Leon, Texas (77539) 303/L2
San Lorenzo, Argentina 143/F6
San Lorenzo, Cerro (mt.), Argentina 143/B6
San Lorenzo, El Beni, Bolivia 136/C4
San Lorenzo, Pando, Bolivia 136/B2
San Lorenzo, Tarija, Bolivia 136/C7
San Lorenzo, Serranía (mts.) Bolivia 136/E5
San Lorenzo, Calif. (94580) 204/K2
San Lorenzo (riv.), Calif. 204/K4
San Lorenzo, Cerro (Cochrane) (mt.), Chile 138/E7
San Lorenzo, Ecuador 128/C2
San Lorenzo (cape), Ecuador 128/B3
San Lorenzo, N. Mex. (88041) 274/B6
San Lorenzo, Paraguay 144/B4
San Lorenzo, Peru 128/H8
San Lorenzo (isl.), Peru 128/D9
San Lorenzo, P. Rico 161/E2
San Lorenzo, P. Rico 156/G1
San Lorenzo, Falcón, Venezuela 124/D2
San Lorenzo, Zulia, Venezuela 124/C3
San Lorenzo de El Escorial, Spain 33/E2
Sanlúcar de Barrameda, Spain 33/C4
Sanlúcar la Mayor, Spain 33/C4
San Lucas, Bolivia 136/C7
San Lucas, Calif. (93954) 204/E7
San Lucas (cape), Mexico 146/G7
San Lucas (cape), Mexico 2/D4
San Lucas (cape), Mexico 150/E5
San Luis (prov.), Argentina 143/C3
San Luis, Argentina 143/C3
San Luis, Argentina 120/C6
San Luis, Ariz. (85349) 198/A6
San Luis (lake), Bolivia 136/C3
San Luis (res.), Calif. 204/E7
San Luis, Colo. (81152) 208/J8
San Luis (creek), Colo. 208/H6
San Luis (lake), Colo. 208/H7
San Luis (peak), Colo. 208/F6
San Luis, Cuba 156/C2
San Luis, Pinar del Río, Cuba 158/B2
San Luis, Santiago de Cuba, Cuba 158/J4
San Luis, Guatemala 154/C2
San Luis, Honduras 154/C3
San Luis, Philippines 82/E6
San Luis (passage), Texas 303/K8
San Luis, Venezuela 124/D2
San Luis de la Paz, Mexico 150/J6
San Luis del Cordero, Mexico 150/H4
San Luis Jilotepeque, Guatemala 154/C3
San Luis Obispo, Calif. 188/B3
San Luis Obispo, Calif. 146/F6
San Luis Obispo (co.), Calif. 204/E8
San Luis Obispo, Calif. (93401) 204/E8
San Luis Potosí (state), Mexico 150/J5
San Luis Potosí, Mexico 150/J5
San Luis Potosí, Mexico 146/H7
San Luis Río Colorado, Mexico 150/B1
San Manuel, Ariz. (85631) 198/E6
San Marcelino, Philippines 82/B3
San Marco in Lamis, Italy 34/E4
San Marcos, Calif. (92069) 204/H10
San Marcos, Colombia 126/C3
San Marcos, C. Rica 154/E6
San Marcos, Guatemala 154/B3
San Marcos, Honduras 154/C3
San Marcos, Mexico 150/K8
San Marcos (isl.), Mexico 150/D3
San Marcos, Texas (78666) 303/F8
San Mariano, Philippines 82/D2
San Marino 7/F4
San Marino, Calif. (91108) 204/D10
SAN MARINO 34
San Marino (cap.), San Marino 34/D3
San Martín (lake) 120/B7
San Martín, Argentina 143/C3
San Martín (lake), Argentina 143/B6
San Martín (riv.), Bolivia 136/D3
San Martin, Calif. (95046) 204/L4
San Martín (cape), Calif. 204/D8
San Martín (lake), Chile 138/E7
San Martín, Colombia 126/D6
San Martín (dept.), Peru 128/D6
San Martín, Peru 128/E3
San Martín de las Pirámides, Mexico 150/M1
San Martín de los Andes, Argentina 143/C5
San Martín de Valdeiglesias, Spain 33/D2
San Martine Draw (dry riv.), Texas 303/C11
San Martín Jilotepeque, Guatemala 154/B3
San Martín Texmelucan, Mexico 150/M1
San Mateo (co.), Calif. 204/C3
San Mateo, Calif. (*94401) 204/J3
San Mateo, Fla. (32088) 212/E2
San Mateo, N. Mex. (87050) 274/B3
San Mateo (mts.), N. Mex. 274/B5
San Mateo, Spain 33/F2
San Mateo, Venezuela 124/F3
San Mateo Ixtatán, Guatemala 154/B3
San Matías (gulf), Argentina 120/C7
San Matías (gulf), Argentina 143/D5
San Matías, Bolivia 136/F5
Sanmaur, Québec 174/C3
San Mauricio, Venezuela 124/E3
Sanmenxia, China 77/H5
San Miguel, Argentina 143/E2
San Miguel, Bolivia 136/E5
San Miguel (riv.), Bolivia 136/D4
San Miguel, Calif. (93451) 204/E8
San Miguel (isl.), Calif. 204/E9
San Miguel (riv.), Colombia 126/B7
San Miguel (co.), Colo. 208/C6
San Miguel (mts.), Colo. 208/C7
San Miguel (riv.), Colo. 208/B6
San Miguel, Cuba 158/H3
San Miguel, Ecuador 128/C3
San Miguel (riv.), Ecuador 128/D2
San Miguel, El Salvador 154/D4
San Miguel (co.), N. Mex. 274/D3
San Miguel, N. Mex. (88058) 274/C6
San Miguel, Golfo de (bay), Panama 154/H6
San Miguel, Paraguay 144/D5
San Miguel, Ayacucho, Peru 128/F9
San Miguel, Cajamarca, Peru 128/C6
San Miguel (bay), Philippines 82/D3
San Miguel (isls.), Philippines 85/F4
San Miguel (isls.), Philippines 82/B7
San Miguel (swamp), Uruguay 145/F4
San Miguel de Allende, Mexico 150/J6
San Miguel de Huachi, Bolivia 136/B4
San Miguel del Monte, Argentina 143/G7
San Miguel de Salcedo, Ecuador 128/C3
San Miguel de Tucumán, Argentina 143/D2
San Miguel de Tucumán, Argentina 120/C5
San Miguelito, Bolivia 136/A2
San Miguelito, Nicaragua 154/E5
San Miguel Nuevo, Colombia 126/B7
Sanming, China 77/J6
San Miniato, Italy 34/C3
San Narciso, Philippines 82/D4
Sannicandro Garganico, Italy 34/E4
San Nicolás, Argentina 143/F6
San Nicolás, Argentina 120/D6
San Nicolas (isl.), Calif. 204/F10
San Nicolás, Cuba 158/C1
San Nicolás (bay), Peru 128/E10
San Nicolás de los Garza, Mexico 150/J3
Sannikova (str.), U.S.S.R. 48/O2
San Nua (Sam Neua), Laos 72/E2
Sano, Ky. (†42728) 237/L6
Sanok, Poland 47/F4
San Onofre, Colombia 126/C3
San Pablo, Potosí, Bolivia 136/B7
San Pablo, Santa Cruz, Bolivia 136/D4
San Pablo, Calif. (94806) 204/J1
San Pablo (bay), Calif. 204/J1
San Pablo, Chile 138/D3
San Pablo, Colombia 126/B7
San Pablo, Colo. (81153) 208/J8
San Pablo, Sierra (mts.), Honduras 154/E3
San Pablo, Laguna, Philippines 82/C3
San Pablo, Negros Occ., Philippines 82/D5
San Pascual, Philippines 82/D4
San Patricio, N. Mex. (88348) 274/D5
San Patricio, Paraguay 144/D5
San Patricio (co.), Texas 303/G10
San Pedro, Buenos Aires, Argentina 143/F6
San Pedro, Jujuy, Argentina 143/D1
San Pedro (riv.), Ariz. 188/D4
San Pedro (riv.), Ariz. 198/E6
San Pedro, Belize 154/D2
San Pedro, Chuquisaca, Bolivia 136/C6
San Pedro, El Beni, Bolivia 136/C4
San Pedro, Pando, Bolivia 136/B2
San Pedro, Santa Cruz, Bolivia 136/D5
San Pedro, Calif. (*90731) 204/C11
San Pedro (bay), Calif. 204/C11
San Pedro (chan.), Calif. 204/G10
San Pedro, Santiago, Chile 138/F4
San Pedro, Valparaíso, Chile 138/F2
San Pedro (pt.), Chile 138/A5
San Pedro, Colombia 126/E5
San Pedro, Cuba 158/B2
San Pedro (riv.), Cuba 158/G3
San Pedro (riv.), Guatemala 154/B2
San Pedro, Ivory Coast 106/C8
San Pedro, Nicaragua 154/E4
San Pedro (dept.), Paraguay 144/D4-5
San Pedro, Paraguay 144/D3
San Pedro (bay), Philippines 82/E5
San Pedro, Sierra de (range), Spain 33/C3
San Pedro Carchá, Guatemala 154/B3
San Pedro de Arimena, Colombia 126/E5
San Pedro de Atacama, Chile 138/C4
San Pedro de Buena Vista, Bolivia 136/C6
San Pedro de las Bocas, Venezuela 124/G4
San Pedro de las Colonias, Mexico 150/H4
San Pedro del Gallo, Mexico 150/G4
San Pedro de Lloc, Peru 128/C6
San Pedro del Paraná, Paraguay 144/D5
San Pedro de Macorís (prov.), Dom. Rep. 158/F6
San Pedro de Macorís, Dom. Rep. 156/E3
San Pedro de Macorís, Dom. Rep. 158/F6
San Pedro de Quemes, Bolivia 136/A7
San Pedro Pochutla, Mexico 150/L9
San Pedro Sula, Honduras 154/C3
San Pedro Zacapa, Honduras 154/D3
San Perlita, Texas (78590) 303/G11
Sanpete (co.), Utah 304/C4
San Pierre, Ind. (46374) 227/D2
San Pietro (isl.), Italy 34/B5
San Pitch (riv.), Utah 304/C4
Sanpoil (riv.), Wash. 310/G2
San Quentin, Calif. (94964) 204/H1
Sanquhar, Scotland 15/D5
San Quintín, Philippines 82/C3
San Rafael, Argentina 143/C3
San Rafael, Argentina 120/C6
San Rafael, Bolivia 136/E5
San Rafael, Calif. (*94901) 204/J1
San Rafael (cape), Dom. Rep. 158/F5
San Rafael, Mexico 150/M1
San Rafael (reef), Mexico 150/L4
San Rafael, N. Mex. (87051) 274/A3
San Rafael (riv.), Utah 304/D4
San Rafael, Venezuela 124/C2
San Rafael de Atamaica, Venezuela 124/E4
San Rafael del Norte, Nicaragua 154/E4
San Rafael del Sur, Nicaragua 154/D5
San Rafael del Yuma, Dom. Rep. 158/F6
San Rafael de Orituco, Venezuela 124/E3
San Rafael Swell (mts.), Utah 304/D5
San Ramón, El Beni, Bolivia 136/C3
San Ramón, Santa Cruz, Bolivia 136/D5
San Ramon, Calif. (94583) 204/K2
San Ramón, C. Rica 154/E5
San Ramón, Cuba 158/H4
San Ramón, Nicaragua 154/E4
San Ramón, Peru 128/E8
San Ramón, Uruguay 145/D5
San Ramón de la Nva. Orán, Argentina 143/D1
San Remo, Italy 34/A3
San Roque, Colombia 126/C4
San Roque, Spain 33/D4
San Rosendo, Chile 138/E1
San Saba (co.), Texas 303/F6
San Saba, Texas (76877) 303/F6
San Saba (riv.), Texas 303/D7
San Salvador, Argentina 143/G5
San Salvador (isl.), Bahamas 156/D1
San Salvador (isl.), Ecuador 128/B9
San Salvador (cap.), El Salvador 154/C4
San Salvador (cap.), El Salvador 146/J8
San Salvador, Paraguay 144/C5
San Salvador (riv.), Uruguay 145/B4
San Salvador el Seco, Mexico 150/O1
Sans Bois (mts.), Okla. 288/R4
San Sebastián, Argentina 143/C7
San Sebastián, Chile 138/F3
San Sebastián, P. Rico 161/B1
San Sebastián, Spain 7/D4
San Sebastián, Spain 33/E1
San Sebastián, Venezuela 124/E2
Sansepolcro, Italy 34/C3
San Servando, Uruguay 145/F3
San Severino Marche, Italy 34/D3
San Severo, Italy 34/E4
San Simeon, Calif. (93452) 204/D8
San Simon, Ariz. (85632) 198/F6
San Simon (riv.), Ariz. 198/F6
San Simón, Serranía (mts.), Bolivia 136/D4
San Simón del Cocuy, Venezuela 124/E3
Sanski Most, Yugoslavia 45/C3
Sansom Park Village, Texas (†76101) 303/E2
Sans Souci, Trin. & Tob. 161/B10
Sans Toucher (mt.), Guadeloupe 161/A6
Santa, Idaho (83866) 220/B2
Santa, Peru 128/C7
Santa (riv.), Peru 128/C7
Santa Ana, El Beni, Bolivia 136/C3
Santa Ana, La Paz, Bolivia 136/B4
Santa Ana, Santa Cruz, Bolivia 136/F6
Santa Ana, Santa Cruz, Bolivia 136/E5
Santa Ana, Calif. 188/C4
Santa Ana, Calif. (*92701) 204/D11
Santa Ana (riv.), Calif. 204/E11
Santa Ana, Colombia 126/F6
Santa Ana, Ecuador 128/B3
Santa Ana, El Salvador 154/C4
Santa Ana (mt.), El Salvador 154/C4
Santa Ana, Guatemala 154/C2
Santa Ana, Mexico 150/D1
Santa Ana (reef), Mexico 150/N7
Santa Ana, Uruguay 145/B1
Santa Ana (range), Uruguay 145/D2
Santa Ana, Anzoátegui, Venezuela 124/F3
Santa Ana, Táchira, Venezuela 124/B4
Santa Ana Chiautempan (Chiautempan), Mexico 150/N1
Santa Anna, Texas (76878) 303/E6
Santa Barbara, Calif. 146/F6
Santa Barbara, Calif. 188/C4
Santa Barbara (isls.), Calif. 146/F6
Santa Barbara (isls.), Calif. 188/C4
Santa Barbara (co.), Calif. 204/E9
Santa Barbara, Calif. (*93101) 204/F9
Santa Barbara (chan.), Calif. 204/E9
Santa Barbara (isl.), Calif. 204/G10
Santa Barbara (isls.), Calif. 204/F10
Santa Bárbara, Chile 138/E1
Santa Bárbara, Colombia 126/C5
Santa Bárbara, Cuba 158/B2
Santa Bárbara, Honduras 154/C3
Santa Bárbara, Mexico 150/F3
Santa Barbara, Neth. Ant. 161/G9
Santa Bárbara, Amazonas, Venezuela 124/E6
Santa Bárbara, Barinas, Venezuela 124/C4
Santa Bárbara, Monagas, Venezuela 124/G3
Santa Bárbara, Zulia, Venezuela 124/C3
Santa Catalina, Argentina 143/C1
Santa Catalina (mts.), Ariz. 198/E6
Santa Catalina (gulf), Calif. 204/G11
Santa Catalina (isl.), Calif. 204/G10
Santa Catalina (isl.), Colombia 126/A9
Santa Catalina (isl.), Mexico 150/D4
Santa Catalina, Philippines 82/D6
Santa Catalina, Uruguay 145/B4
Santa Catalina, Barinas, Venezuela 124/D4
Santa Catalina, Delta Amacuro, Venezuela 124/H3
Santa Catarina (state), Brazil 132/D9
Santa Catarina (isl.), Brazil 120/E5
Santa Catarina (isl.), Brazil 132/E9
Santa Catharina, Neth. Ant. 161/G9
Santa Clara (co.), Calif. 204/D6
Santa Clara, Calif. (*95050) 204/K3
Santa Clara, Colombia 126/F9
Santa Clara, Cuba 158/E2
Santa Clara, Cuba 146/K7
Santa Clara, Cuba 156/B2
Santa Clara (bay), Cuba 158/D1
Santa Clara, Mexico 150/H4
Santa Clara, Utah (84765) 304/A6
Santa Clara (riv.), Utah 304/A6
Santa Clara de Olimar, Uruguay 145/D3
Santa Claus, Georgia (†20436) 217/H6
Santa Claus, Ind. (47579) 227/D8
Santa Clotilde, Peru 128/E4
Santa Cruz (prov.), Argentina 143/C6
Santa Cruz, Argentina 120/C8
Santa Cruz, Argentina 143/C7
Santa Cruz (riv.), Argentina 120/B7
Santa Cruz (riv.), Argentina 143/B7
Santa Cruz (co.), Ariz. 198/E7
Santa Cruz (riv.), Ariz. 198/D6
Santa Cruz (dept.), Bolivia 136/E5
Santa Cruz, Bolivia 120/C4
Santa Cruz, Santa Cruz, Bolivia 136/D5
Santa Cruz, Brazil 132/G4
Santa Cruz, Calif. 188/B3
Santa Cruz (co.), Calif. 204/C6
Santa Cruz, Calif. (*95060) 204/K4
Santa Cruz (chan.), Calif. 204/F10
Santa Cruz (isl.), Calif. 204/F10
Santa Cruz (cap.), Canary Is., Spain 102/A2
Santa Cruz, Chile 138/F6
Santa Cruz, C. Rica 154/E5
Santa Cruz (isl.), Ecuador 128/C9
Santa Cruz, India 68/B7
Santa Cruz, Jamaica 158/H6
Santa Cruz, Mexico 150/D1
Santa Cruz (isl.), Mexico 150/D4
Santa Cruz, N. Mex. (87567) 274/D2
Santa Cruz, Nicaragua 154/E4
Santa Cruz, Cajamarca, Peru 128/C6
Santa Cruz, Loreto, Peru 128/E5
Santa Cruz, Davao del Sur, Philippines 82/E7
Santa Cruz, Laguna, Philippines 82/C3
Santa Cruz, Marinduque, Philippines 82/D4
Santa Cruz, Zambales, Philippines 82/B3
Santa Cruz, Portugal 33/A2
Santa Cruz (isls.), Solomon Is. 87/G6
Santa Cruz, Venezuela 124/C3
Santa Cruz das Flores, Portugal 33/A1
Santa Cruz de Bucaral, Venezuela 124/D2
Santa Cruz de la Palma, Spain 33/B4
Santa Cruz de la Palma, Spain 106/A3
Santa Cruz de la Zarza, Spain 33/E3
Santa Cruz del Norte, Cuba 158/C1
Santa Cruz de los Pinos, Cuba 158/B1
Santa Cruz del Quiché, Guatemala 154/B3
Santa Cruz del Sur, Cuba 158/G3
Santa Cruz del Sur, Cuba 156/B2
Santa Cruz del Valle Ameno, Bolivia 136/A4
Santa Cruz del Zulia, Venezuela 124/B3
Santa Cruz de Mara, Venezuela 124/C2
Santa Cruz de Mudela, Spain 33/E3
Santa Cruz de Orinoco, Venezuela 124/F3
Santa Cruz de Tenerife (prov.), Spain 33/B5
Santa Cruz de Tenerife, Spain 33/B4
Santa Cruz de Tenerife, Spain 106/A3
Santa Cruz de Yojoa, Honduras 154/D3
Santa Cruz do Rio Pardo, Brazil 135/B3
Santa Cruz do Sul, Brazil 132/C10
Santa Elena, Argentina 143/F5
Santa Elena, Bolivia 136/C7
Santa Elena (cape), C. Rica 154/D5
Santa Elena, Ecuador 128/B4
Santa Elena (bay), Ecuador 128/B3
Santa Elena, Paraguay 144/B5
Santa Elena, Peru 128/F5
Santa Elena, Venezuela 124/H5
Santa Eugenia (pt.), Mexico 146/G7
Santa Eugenia (pt.), Mexico 150/B3
Santa Eugenia, Spain 33/B1
Santa Fe (prov.), Argentina 143/D3
Santa Fe, Argentina 120/C6
Santa Fe, Argentina 143/F5
Santa Fe, Bolivia 136/D6
Santa Fe (peak), Colo. 208/H4
Santa Fe, Cuba 158/B2
Santa Fe, Cuba 156/A2
Santa Fe (isl.), Ecuador 128/C9
Santa Fe, Fla. (32616) 212/D2
Santa Fe (lake), Fla. 212/D2
Santa Fe (riv.), Fla. 212/D2
Santa Fe, Ind. (†46970) 227/E3
Santa Fe, Mo. (65282) 261/J4
Santa Fe (cap.), N. Mex. 146/H6
Santa Fe (cap.), N. Mex. 188/E3
Santa Fe (co.), N. Mex. 274/C3
Santa Fe (cap.), N. Mex. (87501) 274/C3
Santa Fe, Panama 154/G6
Santa Fe, Philippines 82/C2
Santa Fé, Spain 33/E4
Santa Fe, Tenn. (38482) 237/G9
Santa Fe, Texas 303/K3
Santa Fe Springs, Calif. (90670) 204/C11
Santa Filomena, Brazil 132/E5
Santa Helena de Goiás, Brazil 132/D7
Santai, China 77/G5
Santa Inés (isl.), Chile 120/B8
Santa Inés (isl.), Chile 138/D10
Santa Inés, Anzoátegui, Venezuela 124/F3
Santa Inés, Barinas, Venezuela 124/C3
Santa Isabel, Bolivia 136/B7
Santa Isabel, Brazil 132/B5
Santa Isabel, Colombia 126/B9
Santa Isabel, Ecuador 128/C4
Santa Isabel, P. Rico 161/C3
Santa Isabel (isl.), Solomon Is. 87/G6
Santa Isabel (isl.), Solomon Is. 86/D2
Santa Isabel (creek), Texas 303/E10
Santa Isabel, Venezuela 124/F7
Santa Isabel de las Lajas, Cuba 158/E2
Santa Isabel de Sihuas, Peru 128/F11
Santa Leopoldina, Brazil 132/G7
Santa Lucía, Buenos Aires, Argentina 143/F6
Santa Lucía, Corrientes, Argentina 143/E2
Santa Lucía, Camagüey, Cuba 158/H3
Santa Lucía, Holguín, Cuba 158/J3
Santa Lucía, Pinar del Río, Cuba 158/A1
Santa Lucía, Uruguay 145/B6
Santa Lucía (riv.), Uruguay 145/D5
Santa Lucía, Venezuela 124/D3
Santa Lucía Chico (riv.), Uruguay 145/D4
Santa Luzia, Brazil 135/E1
Santa Luzia (isl.), C. Verde 106/B8
Santa Margarita, Calif. (93453) 204/E8
Santa Margarita (isl.), Mexico 150/C4
Santa Maria (cape), Angola 115/B6
Santa María, Argentina 143/C2
Santa Maria (riv.), Ariz. 198/B4
Santa Maria, Brazil 132/C10
Santa Maria, Brazil 120/D5
Santa Maria, Calif. (93454) 204/E9
Santa Maria (riv.), Calif. 204/E9
Santa Maria, C. Verde 106/B8
Santa María, Chile 138/G2
Santa María (isl.), Chile 138/D1
Santa María (cay), Cuba 158/F1
Santa María (isl.), Ecuador 128/B10
Santa María (lake), Mexico 150/F1
Santa María (riv.), Mexico 150/F1
Santa María, Paraguay 144/D5
Santa Maria, Philippines 82/C7
Santa Maria (cape), Portugal 33/C4
Santa Maria (isl.), Portugal 33/D2
Santa Maria, Texas (78592) 303/F12
Santa María (cape), Uruguay 145/F5
Santa María, Bolívar, Venezuela 124/G3
Santa Maria Capua Vetere, Italy 34/E4
Santa Maria da Vitória, Brazil 132/F6
Santa María de Erebató, Venezuela 124/F5
Santa María de Ipire, Venezuela 124/F3
Santa María del Orinoco, Venezuela 124/E4
Santa María del Oro, Mexico 150/G3
Santa María del Río, Mexico 150/J6
Santa María del Tule, Mexico 150/L8
Santa María de Nanay, Peru 128/F4
Santa Maria di Leuca (cape), Italy 34/G5
Santa Maria im Münstertal, Switzerland 39/K3
Santa Marta, Colombia 126/D2
Santa Marta, Colombia 120/B1
Santa Marta, Sierra Nevada de (range), Colombia 126/D2
Santa Monica, Calif. (*90401) 204/B10
Santa Monica (bay), Calif. 204/B11
Santana, Brazil 132/E6
Santana, Portugal 33/A2
Santana do Ipanema, Brazil 132/G5
Santana do Jacaré, Brazil 135/D2
Santana do Livramento, Brazil 132/C10
Santana do Livramento, Brazil 120/D6
Santander (dept.), Colombia 126/D4
Santander, Colombia 126/B6
Santander, Philippines 82/D6
Santander (prov.), Spain 33/D1
Santander, Spain 33/D1
Santander, Spain 7/D4
Santander Jiménez, Mexico 150/K4
Sant'Antioco (pen.), Italy 34/B5
Santañy, Spain 33/H3
Santa Olalla del Cala, Spain 33/C4
Santa Paula, Calif. (93060) 204/F9
Santaquin, Utah (84655) 304/C4
Santarém, Brazil 132/C3
Santarém, Brazil 120/D3
Santarém (dist.), Portugal 33/B3
Santarém, Portugal 33/B3
Santarén (chan.), Bahamas 156/B1
Santarén (chan.), Cuba 156/B1
Santa Rita, Cuba 158/H4
Santa Rita, Guam 86/K7
Santa Rita, Honduras 154/D3
Santa Rita, Mont. (59473) 262/D2
Santa Rita, N. Mex. (†88041) 274/B6
Santa Rita, Philippines 82/E5
Santa Rita, Guárico, Venezuela 124/E3
Santa Rita, Zulia, Venezuela 124/C2
Santa Rita do Sapucaí, Brazil 135/D3
Santa Rosa, La Pampa, Argentina 143/C4
Santa Rosa, San Luis, Argentina 143/C3
Santa Rosa, Argentina 120/C6
Santa Rosa, Córdoba, Argentina 143/D3
Santa Rosa, Cochabamba, Bolivia 136/B5
Santa Rosa, Cochabamba, Bolivia 136/C5
Santa Rosa, El Beni, Bolivia 136/B4
Santa Rosa, Pando, Bolivia 136/B2
Santa Rosa, Santa Cruz, Bolivia 136/D5
Santa Rosa, Brazil 132/C4
Santa Rosa, Calif. 188/B3
Santa Rosa, Calif. (*95401) 204/C5
Santa Rosa (isl.), Calif. 204/E10
Santa Rosa, C. Rica 154/E5
Santa Rosa, Ecuador 128/C4
Santa Rosa (co.), Fla. 212/B6
Santa Rosa (isl.), Fla. 212/B6
Santa Rosa (sound), Fla. 212/B6
Santa Rosa, Mo. (†64670) 261/D3
Santa Rosa (range), Nev. 266/D1
Santa Rosa, N. Mex. (88435) 274/E4
Santa Rosa, Paraguay 144/D5
Santa Rosa, Uruguay 145/B6
Santa Rosa, Anzoátegui, Venezuela 124/F3
Santa Rosa, Apure, Venezuela 124/D4
Santa Rosa, Barinas, Venezuela 124/D3
Santa Rosa Beach, Fla. (32459) 212/C6
Santa Rosa de Aguán, Honduras 154/E2
Santa Rosa de Amanadona, Venezuela 124/E7
Santa Rosa de Cabal, Colombia 126/C5
Santa Rosa de Copán, Honduras 154/C3
Santa Rosa de la Mina, Bolivia 136/D5
Santa Rosa de la Roca, Bolivia 136/E5
Santa Rosa de Lima, El Salvador 154/D4
Santa Rosa de Lima, Guatemala 154/B3
Santa Rosa del Palmar, Bolivia 136/E5
Santa Rosa de Osos, Colombia 126/C4
Santa Rosa Ind. Res., Calif. 204/J10
Santa Rosalía, Mexico 150/C3
Santa Rosalía, Venezuela 124/F4
Santa Rosa Wash (dry riv.), Ariz. 198/D6
Santa Teresa, North. Terr. 93/D8
Santa Teresa, Venezuela 124/E2
Santa Venetia, Calif. (†94901) 204/J1
Santa Victoria, Argentina 143/D1
Santa Vitória do Palmar, Brazil 132/C11
Santa Ynez (riv.), Calif. 204/E9
Santa Ysabel Ind. Res., Calif. 204/J10
Santee, Calif. (92071) 204/J11
Santee, Nebr. (†68760) 264/G2
Santee (riv.), S.C. 188/L4
Santee, S.C. (29142) 296/F5
Santee (dam), S.C. 296/G4
Santee (riv.), S.C. 296/H5
Santee Ind. Res., Nebr. 264/G2
Santeetlah (lake), N.C. 281/B4
Sant'Elpidio a Mare, Italy 34/E3
Santeramo in Colle, Italy 34/F4
Sant'Eufemia (gulf), Italy 34/F5
Santiago, Serranía de (mts.), Bolivia 136/F6
Santiago, Potosí, Bolivia 136/A7
Santiago, Santa Cruz, Bolivia 136/F6
Santiago, Brazil 132/C10
Santiago, Región Metropolitana de (Santiago Metropolitan Region) (reg.) Chile 138/A9
Santiago (cap.), Chile 2/F7
Santiago (cap.), Chile 120/B6
Santiago (cap.), Chile 138/G3
Santiago (prov.), Dom. Rep. 158/D5
Santiago, Dom. Rep. 158/D5
Santiago, Dom. Rep. 156/D3
Santiago (San Salvador) (isl.), Ecuador 128/B9
Santiago, Mexico 150/E5
Santiago (riv.), Mexico 146/H7
Santiago, Minn. (55377) 255/E5
Santiago, Panama 154/G6
Santiago, Cerro (mt.), Panama 154/G6
Santiago, Paraguay 144/D5
Santiago, Peru 128/E10
Santiago (riv.), Peru 128/D4
Santiago, Philippines 82/C2
Santiago, Spain 33/B1
Santiago (mts.), Texas 303/A8
Santiago (peak), Texas 303/D12
Santiago de Cao, Peru 128/C6
Santiago de Chocorvos, Peru 128/E9
Santiago de Chuco, Peru 128/C7
Santiago de Cuba (prov.), Cuba 158/H4
Santiago de Cuba, Cuba 146/L8

Santiago de Cuba, Cuba 156/C3
Santiago de Cuba, Cuba 158/J4
Santiago de Huata, Bolivia 136/A5
Santiago de las Vegas, Cuba 158/C1
Santiago del Estero (prov.), Argentina 143/D2
Santiago del Estero, Argentina 120/C5
Santiago del Estero, Argentina 143/D2
Santiago de Machaca, Bolivia 136/A5
Santiago de Pacaguaras, Bolivia 136/A3
Santiago do Cacem, Portugal 33/B3
Santiago Ixcuintla, Mexico 150/G6
Santiago Jamiltepec, Mexico 150/K8
Santiago Juxtlahuaca, Mexico 150/K8
Santiago Miahuatlán, Mexico 150/O2
Santiago Papasquiaro, Mexico 150/F4
Santiago Pinotepa Nacional, Mexico 150/K8
Santiago Rodríguez (prov.), Dom. Rep. 158/D5
Santiago Tuxtla, Mexico 150/M7
Santiago Vázquez, Uruguay 145/A7
Santiaguillo (lake), Mexico 150/G4
San Timoteo, Venezuela 124/C3
Santipur, India 68/F4
Säntis (mt.), Switzerland 39/H2
Santo, Texas (76472) 303/F5
Santo Amaro, Brazil 132/G6
Santo André, Brazil 135/C3
Santo Ângelo, Brazil 132/C10
Santo Antão (isl.), C. Verde 106/A7
Santo António, São Tomé e Príncipe 106/F8
Santo Antônio da Platina, Brazil 132/D8
Santo Antônio da Platina, Brazil 135/A3
Santo Antônio do Leverger, Brazil 132/C6
Santo Corazón, Bolivia 136/F5
Santo Domingo, C. Rica 154/F6
Santo Domingo, Cuba 158/E1
Santo Domingo (cap.), Dom. Rep. 146/L8
Santo Domingo (cap.), Dom. Rep. 156/E3
Santo Domingo (cap.), Dom. Rep. 158/E6
Santo Domingo, Nicaragua 154/E4
Santo Domingo de la Calzada, Spain 33/E1
Santo Domingo de los Colorados, Ecuador 128/C3
Santo Domingo Pueblo, N. Mex. (87052) 274/C3
San Tomé, Venezuela 124/F3
Santoña, Spain 33/E1
Santos, Brazil 2/G7
Santos, Brazil 132/E9
Santos, Brazil 120/E5
Santos, Brazil 135/C3
Santos, Fla. (†32670) 212/D2
Santos Dumont, Brazil 132/F8
Santos Dumont, Brazil 135/E2
Santos Mercado, Bolivia 136/B1
Santo Tomás, Mexico 150/A1
Santo Tomás, Nicaragua 154/E5
Santo Tomás, Amazonas, Peru 128/C6
Santo Tomás, Cusco, Peru 128/G10
Santo Tomas, Davao, Philippines 82/E7
Santo Tomas, La Union, Philippines 82/C2
Santo Tomas (mt.), Philippines 82/C2
Santo Tomás de Andoas, Peru 128/D4
Santo Tomás de Castilla, Guatemala 154/C3
Santo Tomé, Corrientes, Argentina 143/E2
Santo Tomé, Santa Fe, Argentina 143/F5
Santuck, S.C. (†29031) 296/D2
Santuit, Mass. (†02635) 249/N6
Santurce, P. Rico 161/E1
San Urbano, Argentina 143/F6
San Valentín, Cerro (mt.), Chile 138/D6
San Vicente, Chile 138/F4
San Vicente (San Vicente de Tagua Tagua), Chile 138/F5
San Vicente, El Salvador 154/C4
San Vicente, Mexico 150/B1
San Vicente, Amazonas, Venezuela 124/E5
San Vicente, Apure, Venezuela 124/D4
San Vicente de Alcántara, Spain 33/C3
San Vicente de Cañete, Peru 128/D9
San Vicente del Caguán, Colombia 126/C6
San Vito, Italy 34/B5
San Vito (cape), Italy 34/D5
San Vito al Tagliamento, Italy 34/D2
San Vito dei Normanni, Italy 34/F4
San Vito Romano, Italy 34/F6
San Xavier Ind. Res., Ariz. 198/D6
Sanyati (riv.), Zimbabwe 118/D3
San Ygnacio, Texas (78067) 303/E10
San Ysidro, N. Mex. (87053) 274/C3
Sanyuan, China 77/G5
Sanza Pombo, Angola 115/C5
São Bento, Brazil 132/E3
São Bernardo do Campo, Brazil 135/C3
São Borja, Brazil 132/C10
São Brás de Alportel (Alportel), Portugal 33/C4
São Carlos, Brazil 135/C3
São Cristóvão, Brazil 132/G5
São Domingos, Brazil 132/E6
São Félix, Brazil 132/G6
São Fidélis, Brazil 132/F8
São Fidélis, Brazil 135/F2
São Francisco, Brazil 132/E6
São Francisco (riv.), Brazil 120/E4
São Francisco (riv.), Brazil 2/G6
São Francisco (riv.), Brazil 132/F5
São Francisco (riv.), Brazil 135/D2
São Francisco do Sul, Brazil 132/E9
São Gabriel, Brazil 132/C10
São Gonçalo, Brazil 132/F8
São Gonçalo, Brazil 135/E3
São Gonçalo do Sapucaí, Brazil 135/D2
São João da Boa Vista, Brazil 132/E8
São João da Boa Vista, Brazil 135/C2
São João da Madeira, Portugal 33/B2
São João da Pesqueira, Portugal 33/C2
São João del Rei, Brazil 132/E8
São João del Rei, Brazil 135/D2
São João de Meriti, Brazil 135/E3
São João do Piauí, Brazil 132/F5
São João dos Patos, Brazil 132/F4
São João Nepomuceno, Brazil 135/E2
São Joaquim da Barra, Brazil 135/C2
São Jorge (isl.), Portugal 33/B1
São José, Brazil 132/D9
São José da Laje, Brazil 132/H5
São José do Gurupi, Brazil 132/E3
São José do Rio Pardo, Brazil 135/C2
São José do Rio Preto, Brazil 120/E5
São José do Rio Preto, Brazil 132/D8
São José do Rio Preto, Brazil 135/B2
São José dos Campos, Brazil 135/D3
São José dos Pinhais, Brazil 132/D9
São Leopoldo, Brazil 132/D10
São Lourenço, Brazil 135/D3
São Lourenço (riv.), Brazil 132/C7
São Lourenço do Sul, Brazil 132/C10
São Luís, Brazil 132/F3
São Luís, Brazil 120/E3
São Luís Gonzaga, Brazil 132/C10
São Manuel, Brazil 135/B3
São Marcos (bay), Brazil 120/E3
São Marcos (bay), Brazil 132/F3
São Martinho do Porto, Portugal 33/B3
São Mateus, Brazil 132/G7
São Miguel (isl.), Portugal 33/B1
São Miguel Arcanjo, Brazil 135/C3
São Miguel do Guamá, Brazil 132/E3
São Miguel dos Campos, Brazil 132/G5
Saona (isl.), Dom. Rep. 156/E3
Saona (isl.), Dom. Rep. 158/F6
Saône (riv.), France 28/F4
Saône-et-Loire (dept.), France 28/F4
Saonek, Indonesia 85/J6
São Nicolau, Angola 115/B6
São Nicolau (isl.), C. Verde 106/B8
São Paulo (state), Brazil 135/B3
São Paulo (state), Brazil 132/D8
São Paulo, Brazil 120/E5
São Paulo, Brazil 2/G6
São Paulo, Brazil 132/E8
São Paulo, Brazil 135/C3
São Paulo de Olivença, Brazil 132/G9
São Pedro do Piauí, Brazil 132/F4
São Pedro do Sul, Portugal 33/B2
São Raimundo das Mangabeiras, Brazil 132/E4
São Raimundo Nonato, Brazil 132/F5
São Romão, Brazil 132/E7
São Roque, Brazil 135/C3
São Roque (cape), Brazil 2/H6
São Roque (cape), Brazil 120/F3
São Roque (cape), Brazil 132/H4
São Sebastião, Brazil 135/D3
São Sebastião (isl.), Brazil 120/E5
São Sebastião (isl.), Brazil 132/D3
São Sebastião (isl.), Brazil 132/E8
São Sebastião (pt.), Mozambique 118/F4
São Sebastião do Paraíso, Brazil 135/C2
São Sebastião do Paraíso, Brazil 132/E8
São Simão, Brazil 135/C2
São Teotónio, Portugal 33/B4
São Tiago (isl.), C. Verde 106/B8
São Tomé (cape), Brazil 120/F5
São Tomé (cape), Brazil 132/F8
São Tomé (cap.), São Tomé e Príncipe 106/F8
São Tomé (isl.), São T. & Pr. 106/F8
São Tomé e Príncipe 102/C4
SÃO TOMÉ E PRÍNCIPE 106/F8
Saoura, Wadi (dry riv.), Algeria 106/D3
São Vicente, Brazil 135/C4
São Vicente (isl.), C. Verde 106/B7
São Vicente, Portugal 33/A2
São Vicente (cape), Portugal 7/C5
São Vicente Ferrer, Brazil 132/E3
São Vincent (cape), Portugal 33/B4
Sapahaqui, Bolivia 136/B5
Sápai, Greece 45/G5
Sapanca, Turkey 63/D2
Saparua, Indonesia 85/H6
Sapawe, Ontario 177/G5
Sapawe, Ontario 175/B3
Sapele, Nigeria 106/F7
Sapello, N. Mex. (87745) 274/D3
Sapelo (isl.), Georgia 217/K8
Sapelo (sound), Georgia 217/K7
Sapelo Island, Georgia (31327) 217/K8
Şaphane, Turkey 63/C3
Saponac, Maine (†04417) 243/G5
Saposoa, Peru 128/D6
Sappa (creek), Kansas 232/B2
Sappemeer-Hoogezand, Netherlands 27/K2
Sapphire, N.C. (28774) 281/D4
Sappho, Wash. (†98305) 310/A2
Sappington, Mo. (63126) 261/O4
Sapporo, Japan 2/S3
Sapporo, Japan 54/P5
Sapporo, Japan 81/K2
Sapse, Bolivia 136/C6
Sapucaí (riv.), Brazil 135/D2
Sapucaí, Paraguay 144/B5
Sapulpa, Okla. (74066) 288/O3
Saqqez, Iran 59/E2
Saqqez, Iran 66/E2
Saquena, Peru 128/F5
Sara (riv.), Cent. Afr. Rep. 115/C2
Sara (riv.), Chad 111/C6
Sara, Philippines 82/D5
Sarab, Iran 66/E2
Sara Buri, Thailand 72/D4
Saragosa, Texas (79780) 303/D11
Saragossa, Ala. (†35578) 195/D3
Saragossa, Spain 7/D4
Saragossa, Spain 33/F2
Saraguro, Ecuador 128/C4
Sarah (lake), Minn. 255/F5
Sarah, Miss. (38665) 256/D1
Sarahsville, Ohio (43779) 284/H6
Sarajevo, Yugoslavia 7/F4
Sarajevo, Yugoslavia 45/D4
Sarakhs, Iran 66/M2
Saraland, Ala. (36571) 195/B9
Saramacca (dist.), Suriname 131/C3
Saramacca (riv.), Suriname 131/D3
Sarampiuni, Bolivia 136/A4
Saran', U.S.S.R. 48/H5
Saranac, Mich. (48881) 250/D6
Saranac, N.Y. (12981) 276/N1
Saranac (lakes), N.Y. 276/M2
Saranac (riv.), N.Y. 276/N1
Saranac Lake, N.Y. (12983) 276/M2
Sarandë, Albania 45/E6
Sarandí del Yi, Uruguay 145/D4
Sarandí de Navarro, Uruguay 145/C3
Sarandí Grande, Uruguay 145/C4
Sarangani (bay), Philippines 82/E8
Sarangani (isls.), Philippines 82/E8
Sarangani (isls.), Philippines 85/G4
Sarangani (str.), Philippines 82/E8
Saransk, U.S.S.R. 7/J3
Saransk, U.S.S.R. 48/E4
Saransk, U.S.S.R. 52/G4
Sarapul, U.S.S.R. 7/K3
Sarapul, U.S.S.R. 48/F4
Sarapul, U.S.S.R. 52/H3
Sarare, Venezuela 124/D3
Sarare (riv.), Venezuela 124/C4
Sarasota, Fla. 188/K5
Sarasota (co.), Fla. 212/D4
Sarasota, Fla. (*33577) 212/D4
Sarasota (pt.), Fla. 212/D4
Sarasota Springs, Fla. (†33577) 212/D4
Saraswati (riv.), India 68/F1
Saratoga, Ark. (71859) 202/C6
Saratoga, Calif. (95070) 204/K4
Saratoga, Ind. (47382) 227/H4
Saratoga, Iowa (52167) 229/J2
Saratoga, Miss. (†39111) 256/E7
Saratoga (co.), N.Y. 276/N4
Saratoga (lake), N.Y. 276/N4
Saratoga, Texas (77585) 303/K7
Saratoga, Wyo. (82331) 319/F4
Saratoga Nat'l Hist. Park, N.Y. 276/N4
Saratoga Springs, N.Y. (12866) 276/N4
Saratov, U.S.S.R. 7/J3
Saratov, U.S.S.R. 48/E4
Saratov, U.S.S.R. 52/G4
Saravan, Iran 59/H4
Saravan, Iran 66/N7
Saravan, Laos 72/E4
Sarawak (state), Malaysia 2/Q5
Sarawak (state), Malaysia 85/E5
Sarawak (reg.), Malaysia 54/N9
Sarayacu, Ecuador 128/D3
Sarayköy, Turkey 63/C4
Sarayönü, Turkey 63/E3
Sarbaz, Iran 66/M7
Sarbaz, Iran 59/H4
Sarben, Nebr. (†69155) 264/C3
Sárbogárd, Hungary 41/E3
Sarco (bay), Chile 138/A7
Sarcoxie, Mo. (64862) 261/D8
Sardarshahr, India 68/C3
Sar Dasht, Iran 66/D2
Sardina (pt.), P. Rico 161/A1
Sardinata, Colombia 126/D3
Sardinia, Ind. (†47283) 227/F6
Sardinia (reg.), Italy 34/B4
Sardinia (isl.), Italy 7/E4
Sardinia (isl.), Italy 34/B4
Sardinia, N.Y. (14134) 276/C5
Sardinia, Ohio (45171) 284/C7
Sardinia, S.C. (29143) 296/G4
Sardis, Ala. (36775) 195/E6
Sardis, Ala. (†35957) 195/F2
Sardis, Br. Col. 184/M3
Sardis, Georgia (30456) 217/J5
Sardis, Ky. (41056) 237/O3
Sardis (lake), Miss. 188/J4
Sardis, Miss. (38666) 256/E2
Sardis (dam), Miss. 256/E2
Sardis (lake), Miss. 256/E2
Sardis, Ohio (43946) 284/J6
Sardis, Okla. (74564) 288/R5
Sardis, Tenn. (38371) 237/E10
Sardis, W. Va. (†26461) 312/F4
Sar-e Pol, Afghanistan 68/B1
Sar-e Pol, Afghanistan 59/J2
Sarepta, La. (71071) 238/D1
Sarepta, Miss. (38867) 256/F2
Sar Eskand Khan, Iran 66/E2
Sargans, Switzerland 39/H2
Sargeant, Minn. (55973) 255/F7
Sargent, Georgia (30275) 217/C4
Sargent, Nebr. (68874) 264/E3
Sargent (co.), N. Dak. 282/P7
Sargents, Colo. (81248) 208/F6
Sargodha, Pakistan 59/K3
Sargodha, Pakistan 68/C2
Sarh, Chad 111/C6
Sarh, Chad 102/D4
Sarhro, Jebel (mts.), Morocco 106/C2
Sari, Iran 59/F2
Sari, Iran 66/H2
Sariá (isl.), Greece 45/H8
Sarigan (isl.), No. Marianas 87/E4
Sarıgöl, Turkey 63/C3
Sarih, Jordan 65/D2
Sarıkamış, Turkey 63/K2
Sarıkamış, Turkey 59/D1
Sarıkaya, Turkey 63/F3
Sarıköy, Turkey 63/B2
Sarina, Queensland 88/H4
Sarina, Queensland 95/D4
Sarine (Saane) (riv.), Switzerland 39/D3
Sariñena, Spain 33/F2
Sarıoğlan, Turkey 63/G3
Sarita, Texas (78385) 303/G10
Sariwŏn, N. Korea 81/B4
Sarıyer, Turkey 63/D5
Sarız, Turkey 63/G3
Sark (isl.), Chan. Is. 13/E8
Sark (isl.), Chan. Is. 10/E6
Sarkad, Hungary 41/F3
Sarkand, U.S.S.R. 48/J5
Şarkikaraağaç, Turkey 63/D3
Şarkışla, Turkey 63/G3
Şarköy, Turkey 63/B2
Sarlat-La-Canéda, France 28/D5
Sarles, N. Dak. (58372) 282/N2
Sarmi, Indonesia 85/K6
Sarmiento, Argentina 143/B6
Sarmiento, Argentina 120/C7
Sarmiento, Cerro (mt.), Chile 138/E11
Särna, Sweden 18/H6
Sarnath, India 68/E3
Sarnen, Switzerland 39/F3
Sarnen (lake), Switzerland 39/F3
Sarnia, Ont. 162/H7
Sarnia, Ontario 177/B5
Sarny, U.S.S.R. 52/C4
Sarona, Wis. (54870) 317/C4
Saronic (gulf), Greece 45/F7
Saronno, Italy 34/B2
Saronville, Nebr. (68975) 264/G4
Saros (gulf), Turkey 63/B2
Sárospatak, Hungary 41/F2
Sarpsborg, Norway 18/D4
Sarpy (co.), Nebr. 264/H3
Sarra (well), Libya 111/D3
Sarralbe, France 28/G3
Sarrebourg, France 28/G3
Sarreguemines, France 28/G3
Sarria, Spain 33/C1
Sarroch, Italy 34/B5
Sarsati (riv.), India 68/F1
Sarstún (riv.), Belize 154/C3
Sarstún (riv.), Guatemala 154/C3
Sartell, Minn. (56377) 255/D5
Sartène, France 28/B7
Sarthe (dept.), France 28/D3
Sarthe (riv.), France 28/D4
Sartrouville, France 28/A1
Sarufutsu, Japan 81/L1
Sarur, Oman 59/G5
Sárvár, Hungary 41/D3
Sarver, Pa. (16055) 294/C4
Sárviz csatorna (canal), Hungary 41/E3
Saryshagan, U.S.S.R. 48/H5
Sary Su (riv.), U.S.S.R. 48/H5
Sasabe, Ariz. (85633) 198/D7
Sasabe, Mexico 150/C1
Sasaginnigak (lake), Manitoba 179/G3
Sasakwa, Okla. (74867) 288/N5
Sasaram, India 68/E4
Sasebo, Japan 81/D7
Saseenos, Br. Col. 184/J3
Saskatchewan (prov.) 162/F5
Saskatchewan (riv.) 162/F5
Saskatchewan (prov.), Canada 146/H4
Saskatchewan (riv.), Canada 2/D3
SASKATCHEWAN 181
Saskatchewan (riv.), Sask. 181/H2
Saskatchewan Beach, Sask. 181/G5
Saskatchewan Landing Prov. Park, Sask. 181/C5
Saskatoon, Sask. 162/F5
Saskatoon, Sask. 146/H4
Saskatoon, Sask. 181/E3
Saskeram (riv.), Sask. 181/K2
Sason, Turkey 63/J3
Sasovo, U.S.S.R. 52/F4
Saspamco, Texas (78153) 303/K11
Sassafras, Md. (†21637) 245/P3
Sassafras (riv.), Md. 245/P3
Sassafras (mt.), S.C. 296/B1
Sassafras, Tasmania 99/C3
Sassandra, Ivory Coast 106/C8
Sassandra (riv.), Ivory Coast 106/C7
Sassari (prov.), Italy 34/B4
Sassari, Italy 34/B4
Sassari, Italy 7/E4
Sasseneire (mt.), Switzerland 39/E4
Sasser, Georgia (31785) 217/D7
Sassnitz, E. Germany 22/E1
Sasstown, Liberia 106/C8
Sassuolo, Italy 34/C2
Sástago, Spain 33/F2
Sasu (mt.), N. Korea 81/C4
Sata (cape), Japan 81/E8
Satadougou, Mali 106/B6
Satanta, Kansas (67870) 232/B4
Satara, India 68/C5
Satartia, Miss. (39162) 256/C5
Satawal (isl.), Micronesia 87/E5
Satellite Beach, Fla. (32935) 212/F3
Säter, Sweden 18/J6
Saticoy, Calif. (93004) 204/F9
Satigny, Switzerland 39/A4
Satilla (riv.), Georgia 217/G8
Satipo, Peru 128/E8
Satluj (Sutlej) (riv.), 68/C3
Satna, India 68/E4
Sátoraljaújhely, Hungary 41/F2
Satpura (range), India 68/D4
Satsop, Wash. (98583) 310/B3
Satsuma, Ala. (36572) 195/B9
Satsuma, Fla. (32189) 212/E2
Satsuma, La. (†70754) 238/L1
Satte, Japan 81/O1
Satu Mare, Romania 45/F2
Satu Mare, Romania 7/G4
Satun, Thailand 72/C6
Satupaitea, W. Samoa 86/L8
Saturna Island, Br. Col. 184/K3
Saturnia, Italy 34/C3
Satus (creek), Wash. 310/E4
Sauble Beach, Ontario 177/C3
Sauce, Argentina 143/G5
Sauce, Peru 128/D6
Sauce, Canelones, Uruguay 145/B6
Sauce, Rocha, Uruguay 145/E5
Sauce (lag.), Uruguay 145/D5
Sauceda (mts.), Ariz. 198/C6
Sauce de Luna, Argentina 143/G5
Sauce del Yi, Uruguay 145/D4
Saucedo, Uruguay 145/B2
Sauchie, Scotland 15/C1
Saucier, Miss. (39574) 256/F9
Saucillo, Mexico 150/G2
Sauda, Qurnet es (mt.), Lebanon 63/G5
Saudhárkrókur, Iceland 21/B1
Saudi Arabia 2/M4
Saudi Arabia 54/F7
SAUDI ARABIA 59/D4
Sauer (riv.), Luxembourg 27/J9
Sauer (riv.), W. Germany 22/B4
Sauerland (reg.), W. Germany 22/B3
Saugatuck, Conn. (†06880) 210/B4
Saugatuck (res.), Conn. 210/B3
Saugatuck (riv.), Conn. 210/B3
Saugatuck, Mich. (49453) 250/C6
Saugeen (riv.), Ontario 177/C3
Saugerties, N.Y. (12477) 276/M6
Sauget, Ill. (62201) 222/A2
Saugus○, Mass. (01906) 249/D6
Saugus Iron Works Nat'l Hist. Site, Mass. 249/D6
Sauiá, Brazil 132/B3
Sauk (riv.), Wash. 310/D2
Sauk (co.), Wis. 317/G9
Sauk Centre, Minn. (56378) 255/C5
Sauk City, Wis. (53583) 317/G9
Sauk Rapids, Minn. (56379) 255/D5
Sauk Village, Ill. (60411) 222/C6
Saukville, Wis. (53080) 317/L9
Saül, Fr. Guiana 131/E4
Saulgau, W. Germany 22/C5
Saulkrasti, U.S.S.R. 53/C2
Saulnierville, Nova Scotia 168/B4
Saulsbury, Tenn. (38067) 237/C10
Saulsville, W. Va. (25876) 312/C7
Sault-au-Mouton, Québec 172/H1
Sault au Mouton (riv.), Québec 172/H1
Saulteaux (riv.), Alberta 182/C2
Sault Sainte Marie, Mich. 188/J1
Sault Sainte Marie, Mich. (49783) 250/E2
Sault Sainte Marie, Ont. 162/H6
Sault Sainte Marie, Ontario 177/J5
Sault Sainte Marie, Ontario 175/D3
Sault Ste. Marie, Ont. 146/K5
Saum, Minn. (56674) 255/D3
Saumarez (reef), 95/E4
Saumarez (reef), Coral Sea Is. Terr. 88/J4
Saumarez, New Bruns. 170/E1
Saumâtre (lake), Haiti 158/C6
Saumlaki, Indonesia 85/J7
Saumur, France 28/D4
Saunders (isl.), 143/E7
Saunders (co.), Nebr. 264/H3
Saundersville, Mass. (†01560) 249/G4
Saunemin, Ill. (61769) 222/E3
Sauqira (bay), Oman 59/G6
Sauqira, Ras (cape), Oman 59/G6
Sauquoit, N.Y. (13456) 276/K5
Saurimo, Angola 115/D5
Saurimo, Angola 102/E6
Sausalito, Calif. (94965) 204/H2
Sautatá, Colombia 126/B4
Sautee-Nacoochee, Georgia (30571) 217/E1
Sauteurs, Grenada 161/D8
Saut-Tigre, Fr. Guiana 131/E3
Sauzal, Chile 138/G5
Sava (riv.) 7/F4
Sava (riv.), Yugoslavia 45/D3
Savage, (riv.), Md. 245/B2
Savage, Minn. (55337) 255/G6
Savage, Miss. (38667) 256/D1
Savage, Mont. (59262) 262/M3
Savage (harb.), Pr. Edward I. 168/F2
Savage (riv.), Tasmania 99/B3
Savage-Guilford, Md. (20863) 245/L4
Savage River (lake), Md. 245/B2
Savage River, Tasmania 99/B3
Savageton, Wyo. (†82716) 319/G2
Savah, Ind. (†47620) 227/B8
Savai'i (isl.), W. Samoa 87/J7
Savai'i (isl.), W. Samoa 86/L8
Savalou, Benin 106/E7
Savana (isl.), Virgin Is. (U.S.) 161/A4
Savanat (Estahbanat), Iran 66/J6
Savaneta, Neth. Ant. 161/E10
Savanette, Haiti 158/C6
Savanna, Ill. (61074) 222/C1
Savanna, Okla. (74565) 288/P5
Savanna Army Depot, Ill. 222/C1
Savannah (riv.) 188/K4
Savannah, Ga. 146/K6
Savannah, Ga. 188/K4
Savannah, Georgia (*31401) 217/L6
Savannah (riv.), Georgia 217/K5
Savannah, Mo. (64485) 261/C3
Savannah○, N.Y. (13146) 276/G4
Savannah, Ohio (44874) 284/F4
Savannah (riv.), S.C. 296/E6
Savannah, Tenn. (38372) 237/E10
Savannah, U.S. 2/F4
Savannah (riv.), U.S. 146/K6
Savannah River Plant Atomic Energy Commission, S.C. 296/D5
Savannakhét, Laos 72/E3
Savanna-la-Mar, Jamaica 158/G6
Savanna-la-Mar, Jamaica 156/B3
Savannes (bay), St. Lucia 161/G7
Savant (lake), Ontario 177/G4
Savant (lake), Ontario 175/B2
Savant Lake, Ontario 177/G4
Savant Lake, Ontario 175/B2
Savantvadi, India 68/C5
Savanur, India 68/D6
Savaştepe, Turkey 63/B3
Save (riv.) 102/F7
Savé, Benin 106/E7
Save (riv.), Mozambique 118/E4
Saveh, Iran 59/F2
Saveh, Iran 66/G3
Săveni, Romania 45/H1
Saverne, France 28/G3
Saverton, Mo. (63467) 261/K3
Savery, Wyo. (82332) 319/E4
Savery (creek), Wyo. 319/E4
Savièse, Switzerland 39/D4
Savigliano, Italy 34/A2
Savo (isl.), Solomon Is. 86/D3
Savognin, Switzerland 39/J3
Savoie (Savoy) (dept.), France 28/G5
Savona, Br. Col. 184/G5
Savona (prov.), Italy 34/B2
Savona, Italy 34/B2
Savona, N.Y. (14879) 276/F6
Savonburg, Kansas (66772) 232/G4
Savonet, Neth. Ant. 161/F8
Savonlinna, Finland 18/Q6
Savoonga, Alaska (99769) 196/E2
Savoy, Ill. (61874) 222/E3
Savoy, Ky. (†40769) 237/N7
Savoy○, Mass. (01256) 249/B2
Savoy, Mont. (†59526) 262/H2
Savoy, S. Dak. (†57754) 298/B5
Şavşat, Turkey 63/K2
Sävsjö, Sweden 18/J8
Savu (sea), Indonesia 54/O10
Savukoski, Finland 18/Q3
Savur, Turkey 63/J4
Savusavu (bay), Fiji 86/Q10
Sawahlunto, Indonesia 85/C6
Sawankhalok, Thailand 72/C3
Sawara, Japan 81/K6
Sawatch (range), Colo. 208/G4
Sawbill, Newf. 166/A3
Sawbill Landing, Minn. (†55603) 255/G3
Sawbridgeworth, England 13/H7
Saweba (cape), Indonesia 85/J6
Sawi, India 68/G7
Sawmill Bay, N.W. Terrs. 187/G3
Sawpit, Colo. (†81430) 208/D7
Sawston, England 13/H5
Sawtell, N.S. Wales 97/G2
Sawtooth (range), Idaho 220/C6
Sawtooth (ridge), Wash. 310/E2
Sawtooth Nat'l Rec. Area, Idaho 220/D5
Sawu (isl.), Indonesia 85/G8
Sawu (isls.), Indonesia 85/G8
Sawu (sea), Indonesia 85/G7
Sawyer, Kansas (67134) 232/D4
Sawyer, Ky. (42643) 237/N7
Sawyer, Mich. (49125) 250/C7
Sawyer, Minn. (55780) 255/F4
Sawyer, N. Dak. (58781) 282/H3
Sawyer, Okla. (74756) 288/R7
Sawyer (co.), Wis. 317/D4
Sawyers Bar, Calif. (96027) 204/B2
Sawyerville, Ala. (36776) 195/C5
Sawyerville, Ill. (62085) 222/D4
Sawyerville, Québec 172/F4
Saxapahaw, N.C. (27340) 281/L3
Saxe, Va. (23967) 307/L7
Saxeville, Wis. (54976) 317/H7
Saxis, Va. (23427) 307/S5
Saxman, Alaska (†99901) 196/N2
Saxmundham, England 13/J5
Saxon, S.C. (†29301) 296/D2
Saxon, Switzerland 39/D4
Saxon, Wis. (54559) 317/F3
Saxonburg, Pa. (16056) 294/C4
Saxonville, Mass. (01701) 249/A7
Saxony (reg.), E. Germany 22/E3
Saxton, Ky. (†40769) 237/N7
Saxton, Pa. (16678) 294/F5
Saxtons River, Vt. (05154) 268/B5
Say, Niger 106/E6
Saya, Bolivia 136/B5
Sayabec, Québec 172/B2
Sayaboury (Muang Xaignabouri), Laos 72/D3
Sayama, Japan 81/O2
Sayán, Peru 128/D8
Sayan (mts.), U.S.S.R. 48/K4
Saybrook, Ill. (61770) 222/E3
Saybrook, Pa. (†16347) 294/D2
Saybrook Point, Conn. (†06475) 210/F3
Sayhan-Ovoo, Mongolia 77/F2
Saylesville, R.I. (†02865) 249/J5
Saylorsburg, Pa. (18353) 294/M4
Saylorville (lake), Iowa 229/F5
Sayner, Wis. (54560) 317/H4
Saynshand, Mongolia 77/H3
Saynshand, Mongolia 54/M5
Sayre, Ala. (35139) 195/E3
Sayre, Okla. (73662) 288/G4
Sayre, Pa. (18840) 294/K2
Sayreville, N.J. (08872) 273/E3
Sayula, Mexico 150/H7
Sayula de Alemán, Mexico 150/M8
Sayville, N.Y. (11782) 276/O9
Sayward, Br. Col. 184/D5
Sazan (isl.), Albania 45/D5
Sázava (riv.), Czech. 41/C2
Sbaa, Algeria 106/D3
Sbeitla, Tunisia 106/F1
Scafell Pike (mt.), England 13/D3
Scafell Pike (mt.), England 10/E3
Scalasaig, Scotland 15/B4
Scalby, England 13/G3
Scales Mound, Ill. (61075) 222/C1
Scaletta (pass), Switzerland 39/J3
Scalf, Ky. (40982) 237/O7
Scalloway, Scotland 10/G1
Scalloway, Scotland 15/G2
Scalpay (isl.), Scotland 15/B3
Scalpay (isl.), Scotland 15/C3
Scalp Level, Pa. (†15963) 294/E5
Scaly Mountain, N.C. (28775) 281/C4
Scammon, Kansas (66773) 232/H4
Scammon Bay, Alaska (99662) 196/E2
Scandia, Alberta 182/E4
Scandia, Kansas (66966) 232/E2
Scandia, Minn. (55073) 255/F5
Scandia, Wash. (†98370) 310/A1
Scandinavia 18
Scandinavia, Wis. (54977) 317/H7
Scanlon, Minn. (†55720) 255/F4
Scanterbury, Manitoba 179/F4
Scantic, Conn. (†06097) 210/E1
Scantic (riv.), Conn. 210/E1

Scapa, Alberta 182/D4
Scapa Flow (chan.), Scotland 15/E2
Scapa Flow (chan.), Scotland 10/E1
Scappoose, Oreg. (97056) 291/E2
Scarba (isl.), Scotland 15/C4
Scarboro, Barbados 161/B9
Scarboro, Georgia (†30442) 217/J5
Scarborough, England 10/F3
Scarborough, England 13/G3
Scarborough, Maine (04074) 243/C8
Scarborough○, Maine (04074) 243/C8
Scarborough, Ontario 177/K4
Scarborough, Trin. & Tob. 156/G5
Scarbro, W. Va. (25917) 312/D7
Scarinish, Scotland 15/B4
Scarp (isl.), Scotland 15/A2
Scarriff, Ireland 17/E6
Scarriff (isl.), Ireland 17/A8
Scarsdale, N.Y. (10583) 276/P6
Scarth, Manitoba 179/B5
Scarville, Iowa (50473) 229/F2
Scatarie (isl.), Nova Scotia 168/J2
Scavaig, Loch (inlet), Scotland 15/B3
Sceaux, France 28/A2
Scenic, S. Dak. (57780) 298/D6
Scenic, Wash. (†98288) 310/D3
Sceptre, Sask. 181/B5
Schaal, Ark. (†71851) 202/C6
Schaalsee (lake), E. Germany 22/D2
Schaalsee (lake), W. Germany 22/D2
Schaan, Liecht. 39/H2
Schaefferstown, Pa. (17088) 294/K5
Schaerbeek, Belgium 27/C9
Schaffer, Mich. (49882) 250/B3
Schaffhausen (canton), Switzerland 39/G1
Schaffhausen, Switzerland 39/G1
Schagen, Netherlands 27/F3
Schaghticoke, N.Y. (12154) 276/N5
Schaller, Iowa (51053) 229/C4
S-chanf, Switzerland 39/J3
Schangnau, Switzerland 39/E3
Schänis, Switzerland 39/H2
Scharans, Switzerland 39/J3
Schärding, Austria 41/B2
Scharhörn (isl.), W. Germany 22/C2
Schattdorf, Switzerland 39/G3
Schaumburg, Ill. (60194) 222/A5
Schawana, Wash. (†99321) 310/F4
Schefferville, Que. 146/L4
Schefferville, Que. 162/K5
Schefferville, Québec 174/D2
Scheibbs, Austria 41/C2
Scheinfeld, W. Germany 22/D4
Schelde (Scheldt) (riv.), Belgium 27/C7
Scheldt (riv.), Belgium 27/C7
Schell City, Mo. (64783) 261/D6
Schell Creek (range), Nev. 266/G3
Schellsburg, Pa. (15559) 294/E5
Schellville, Calif. (†95476) 204/J1
Schenectady, N.Y. 188/M2
Schenectady (co.), N.Y. 276/M5
Schenectady, N.Y. (*12301) 276/M5
Schenevus, N.Y. (12155) 276/L5
Schererville, Ind. (46375) 227/C2
Scherhorn (mt.), Switzerland 39/G3
Schertz, Texas (78154) 303/K10
Scherzingen, Switzerland 39/H1
Schesaplana (mt.), Switzerland 39/J2
Scheveningen, Netherlands 27/E4
Schichallion (mt.), Scotland 15/D4
Schiedam, Netherlands 27/E5
Schiermonnikoog, Netherlands 27/J1
Schiermonnikoog (isl.), Netherlands 27/J1
Schiers, Switzerland 39/J3
Schijndel, Netherlands 27/G5
Schiller Park, Ill. (60176) 222/B5
Schinznach-Dorf, Switzerland 39/F2
Schio, Italy 34/C2
Schiphol, Netherlands 27/B5
Schkeuditz, E. Germany 22/E3
Schladming, Austria 41/B3
Schlater, Miss. (38952) 256/D3
Schlei (inlet), W. Germany 22/C1
Schleicher (co.), Texas 303/D7
Schleitheim, Switzerland 39/G1
Schleswig, Iowa (51461) 229/B4
Schleswig, W. Germany 22/C1
Schleswig-Holstein (state), W. Germany 22/C1
Schleusingen, E. Germany 22/D3
Schley (co.), Georgia 217/D6
Schley, Minn. (†56633) 255/D3
Schlieren, Switzerland 39/F2
Schliersee, W. Germany 22/D5
Schlitz, W. Germany 22/C3
Schlüchtern, W. Germany 22/C3
Schmalkalden, E. Germany 22/D3
Schmölln, E. Germany 22/E3
Schnecksville, Pa. (18078) 294/L4
Schneeberg, E. Germany 22/E3
Schneeberg (mt.), W. Germany 22/D3
Schnee Eifel (plat.), Belgium 27/J8
Schneidemühl (Piła), Poland 47/C2
Schneider, Ind. (46376) 227/C2
Schnellville, Ind. (47580) 227/D8
Schoelcher, Martinique 161/C6
Schoenchen, Kansas (67667) 232/C3
Schoenfeld, Sask. 181/D5
Schoen Lake Prov. Park, Br. Col. 184/E5
Schofield, Wis. (54476) 317/H6
Schofield Barracks, Hawaii (96786) 218/E2
Schoharie (co.), N.Y. 276/M5
Schoharie, N.Y. (12157) 276/M5
Schoharie (creek), N.Y. 276/M6
Schoharie (res.), N.Y. 276/M6
Scholle, N. Mex. (†87036) 274/C4
Scholls, Oreg. (†97123) 291/A2
Schomberg, Ontario 177/J3
Schönberg, E. Germany 22/D2
Schönberg, W. Germany 22/D1
Schönebeck, E. Germany 22/D2
Schöneberg, W. Germany 22/E4
Schöneiche, E. Germany 22/E2
Schönenwerd, Switzerland 39/E2
Schongau, W. Germany 22/D5
Schöningen, W. Germany 22/D2
Schoodic (lake), Maine 243/F5
Schoolcraft (co.), Mich. 250/C2
Schoolcraft, Mich. (49087) 250/D6
Schoolcraft (riv.), Minn. 255/D3
Schooleys Mountain, N.J. (07870) 273/D2
School Hill, Wis. (†53042) 317/L8
Schoonhoven, Netherlands 27/F5
Schoten, Belgium 27/F6
Schottegat (bay), Neth. Ant. 161/G9
Schouten (isls.), Indonesia 85/K6
Schouten (isls.), Papua N.G. 85/B6
Schouten (isl.), Tasmania 99/E4
Schouwen (isl.), Netherlands 27/D5
Schramberg, W. Germany 22/C4
Schram City, Ill. (†62049) 222/D4
Schreckhorn (mt.), Switzerland 39/F3
Schreiber, Ontario 177/H5
Schreiber, Ontario 175/C3
Schrems, Austria 41/C2
Schriever, La. (70395) 238/J7
Schroeder, Minn. (55613) 255/G3
Schroon (lake), N.Y. 276/N3
Schroon (riv.), N.Y. 276/N3
Schroon Lake, N.Y. (12870) 276/N3
Schruns, Austria 41/A3
Schübelbach, Switzerland 39/G2
Schulenburg, Texas (78956) 303/H8
Schuler, Alberta 182/E4
Schull, Ireland 10/B5
Schull, Ireland 17/B8
Schulter, Okla. (74460) 288/P3
Schultz (lake), N.W. Terrs. 187/J3
Schumacher, Ontario 175/D3
Schüpfheim, Switzerland 39/F3
Schurz, Nev. (89427) 266/C4
Schussenried, W. Germany 22/C4
Schuyler (co.), Ill. 222/C3
Schuyler (co.), Mo. 261/G2
Schuyler, Nebr. (68661) 264/G3
Schuyler (co.), N.Y. 276/G6
Schuyler, Va. (22969) 307/L5
Schuyler Lake, N.Y. (13457) 276/L5
Schuylerville, N.Y. (12871) 276/N4
Schuylkill (co.), Pa. 294/K4
Schuylkill (riv.), Pa. 294/M5
Schuylkill Haven, Pa. (17972) 294/K4
Schwaan, E. Germany 22/E2
Schwabach, W. Germany 22/D4
Schwäbisch Gmünd, W. Germany 22/C4
Schwäbisch Hall, W. Germany 22/C4
Schwalmstadt, W. Germany 22/C3
Schwanden, Switzerland 39/H2
Schwandorf im Bayern, W. Germany 22/E4
Schwaner (mts.), Indonesia 85/E6
Schwarzach im Pongau, Austria 41/B3
Schwarzenburg, Switzerland 39/D3
Schwarzhorn (mt.), Switzerland 39/E4
Schwarzhorn (mt.), Switzerland 39/F3
Schwarzwald (Black) (for.), W. Germany 22/C4
Schwatka (mts.), Alaska 196/G1
Schwaz, Austria 41/A3
Schwechat, Austria 41/D2
Schwedt, E. Germany 22/F2
Schweidnitz (Świdnica), Poland 47/C3
Schweinfurt, W. Germany 22/D3
Schwelm, W. Germany 22/B3
Schwenksville, Pa. (19473) 294/L5
Schwerin (dist.), E. Germany 22/D2
Schwerin, E. Germany 22/D2
Schwerinersee (lake), E. Germany 22/D2
Schwertberg, Austria 41/C2
Schwetzingen, W. Germany 22/C4
Schwyz (canton), Switzerland 39/G2
Schwyz, Switzerland 39/G2
Sciacca, Italy 34/D6
Scicli, Italy 34/E6
Science Hill, Ky. (42553) 237/M6
Scilly (isls.), England 13/A7
Scilly (isls.), England 10/C6
Scio, N.Y. (14880) 276/E6
Scio, Ohio (43988) 284/H5
Scio, Oreg. (97374) 291/E3
Sciota, Ill. (61475) 222/C3
Sciota, Pa. (18354) 294/M4
Scioto (co.), Ohio 284/D8
Scioto (riv.), Ohio 284/D8
Sciotodale, Ohio (†45662) 284/E8
Scioto Furnace, Ohio (45677) 284/E8
Scipio, Ind. (47273) 227/F6
Scipio, Ind. (†45053) 227/H6
Scipio, Okla. (†74501) 288/P4
Scipio, Utah (84656) 304/B4
Scircleville, Ind. (46066) 227/E4
Scitico, Conn. (†06036) 210/E1
Scituate, Mass. (02066) 249/F8
Scituate○, Mass. (02066) 249/F8
Scituate (res.), R.I. 249/H5
Sclater, Manitoba 179/B3
Scobey, Miss. (38953) 256/E3
Scobey, Mont. (59263) 262/L2
Scofield, Utah (†84526) 304/C4
Scofield (res.), Utah 304/C4
Scollard, Alberta 182/D4
Scone, N.S. Wales 97/F3
Scooba, Miss. (39358) 256/G5
Scopi (mt.), Switzerland 39/G3
Scopus, Mo. (63762) 261/N8
Scoresby (sound), Greenl. 4/B10
Scoresbysund, Greenl. 4/B10
Scotch Grove, Iowa (52331) 229/L4
Scotch Plains○, N.J. (07076) 273/E2
Scotchtown, Nova Scotia 168/H2
Scotch Village, Nova Scotia 168/E3
Scotfield, Alberta 182/E4
Scotia (sea) 2/G8
Scotia (sea) 5/D16
Scotia, Calif. (95565) 204/A3
Scotia, Nebr. (68875) 264/F3
Scotia, N.Y. (12302) 276/N5
Scotia, S.C. (29939) 296/E6
SCOTLAND 15
SCOTLAND 10/D2
Scotland, Ark. (72141) 202/E2
Scotland○, Conn. (06264) 210/G2
Scotland, Georgia (31083) 217/G6
Scotland, Ind. (47457) 227/D7
Scotland, Md. (20687) 245/N8
Scotland (co.), Mo. 261/H2
Scotland (co.), N.C. 281/L5
Scotland, Ontario 177/D4
Scotland, Pa. (17254) 294/G6
Scotland, S. Dak. (57059) 298/O7
Scotland, Texas (76379) 303/F4
Scotland, U.K. 7/D3
Scotland Neck, N.C. (27874) 281/P2
Scotlandville, La. (70807) 238/J1
Scots (bay), Nova Scotia 168/D3
Scots Bay, Nova Scotia 168/D3
Scotsburn, Nova Scotia 168/F3
Scotsguard, Sask. 181/C6
Scotstown, Ireland 17/H3
Scotstown, Québec 172/F4
Scotsville, Nova Scotia 168/G2
Scott (isl.), Ant. 2/A9
Scott (isl.), Ant. 5/C10
Scott (co.), Ark. 202/B4
Scott, Ark. (72142) 202/F4
Scott (cape), Br. Col. 162/D5
Scott (cape), Br. Col. 184/C5
Scott (isls.), Br. Col. 184/C5
Scott (riv.), Calif. 204/B2
Scott, Georgia (31095) 217/G5
Scott (co.), Ill. 222/C4
Scott (co.), Ind. 227/F7
Scott, Ind. (†46746) 227/F1
Scott (co.), Iowa 229/M5
Scott (co.), Kansas 232/B3
Scott (co.), Ky. 237/M4
Scott, La. (70583) 238/F6
Scott (co.), Minn. 255/E6
Scott (co.), Miss. 256/E6
Scott, Miss. (38772) 256/B3
Scott (co.), Mo. 261/N8
Scott, Ohio (45886) 284/A4
Scott (mt.), Okla. 288/K5
Scott, Sask. 181/C3
Scott (lake), Sask. 181/M2
Scott (co.), Tenn. 237/M8
Scott (co.), Va. 307/C7
Scott A.F.B., Ill. 222/B3
Scott City, Kansas (67871) 232/B3
Scott City, Mo. (63780) 261/O8
Scottdale, Georgia (30079) 217/L1
Scottdale, Pa. (15683) 294/C5
Scott-Jonction, Québec 172/F3
Scottland, Ill. (†61924) 222/F4
Scotts (head), Dominica 161/E7
Scotts, N.C. (28699) 281/H3
Scottsbluff, Nebr. 188/F2
Scottsbluff, Nebr. 146/H5
Scotts Bluff (co.), Nebr. 264/A3
Scottsbluff, Nebr. (69361) 264/A3
Scotts Bluff Nat'l Mon., Nebr. 264/A3
Scottsboro, Ala. (35768) 195/F1
Scottsburg, Ind. (47170) 227/F7
Scottsburg, Ky. (†42445) 237/F6
Scottsburg, Oreg. (97473) 291/D4
Scottsburg, Va. (24589) 307/L7
Scottsdale, Ariz. (*85251) 198/D5
Scottsdale, Tasmania 99/D3
Scotts Hill, N.C. (†28401) 281/O6
Scotts Hill, Tenn. (38374) 237/E10
Scotts Mills, Oreg. (97375) 291/B3
Scottsmoor, Fla. (32775) 212/F3
Scotts Ridge (hills), Conn. 210/A3
Scott Station, Ant. 5/B9
Scotts Valley, Calif. (95060) 204/K4
Scottsville, Ark. (†72843) 202/D3
Scottsville, Kansas (67477) 232/D2
Scottsville, Ky. (42164) 237/J7
Scottsville, N.Y. (14546) 276/E4
Scottsville, Va. (24590) 307/L5
Scottville, Ill. (62683) 222/C4
Scottville, Mich. (49454) 250/C5
Scoudouc, New Bruns. 170/F2
Scourie, Scotland 15/C2
Scout Lake, Sask. 181/F6
Scrabster, Scotland 15/E2
Scraggly (lake), Maine 243/H5
Scraggly (lake), Maine 243/F3
Scranage, Ala. (†36552) 195/C8
Scranton, Ark. (72863) 202/C3
Scranton, Iowa (51462) 229/D4
Scranton, Kansas (66537) 232/G3
Scranton, Ky. (40373) 237/O5
Scranton, N.Y. (†14075) 276/C5
Scranton, N.C. (27875) 281/S4
Scranton, N. Dak. (58653) 282/D7
Scranton, Pa. 188/L2
Scranton, Pa. 146/L5
Scranton, Pa. (*18501) 294/F7
Scranton, S.C. (29591) 296/H4
Scraper, Okla. (†74359) 288/S2
Screven (co.), Georgia 217/J5
Screven, Georgia (31560) 217/H7
Scriba, N.Y. (†13827) 276/H4
Scribner, Nebr. (68057) 264/H3
Scridain, Loch (inlet), Scotland 15/B4
Scugog (lake), Ontario 177/F3
Scullin, Okla. (†73086) 288/N5
Scunthorpe, England 10/F4
Scunthorpe, England 13/G4
Scuol, Switzerland 39/K3
Scurdie Ness (prom.), Scotland 15/F4
Scurrival (pt.), Scotland 15/A3
Scurry (co.), Texas 303/D5
Scurry, Texas (75158) 303/H5
Scusciuban, Somalia 115/J1
Scutari (lake), Albania 45/D4
Scutari (lake), Yugoslavia 45/D4
Scyrene, Ala. (†36436) 195/C7
Sea (isls.), Georgia 217/K9
Sea (isls.), S.C. 296/G7
Seabeck, Wash. (98380) 310/C3
Seaboard, N.C. (27876) 281/O1
Seabold, Wash. (†98110) 310/A1
Sea Breeze, N.Y. (†14617) 276/F4
Sea Bright, N.J. (07760) 273/F3
Seabrook○, N.H. (03874) 268/F6
Seabrook, N.J. (08302) 273/C5
Seabrook, S.C. (29940) 296/F6
Seabrook (isl.), S.C. 296/G6
Seabrook, Texas (77586) 303/K2
Seabrook-Lanham, Md. (20801) 245/G4
Sea Cliff, N.Y. (11579) 276/R6
Seadrift, Texas (77983) 303/H9
Seaford, Del. (19973) 245/R6
Seaford, England 10/G5
Seaford, England 13/H7
Seaford, N.Y. (11783) 276/R7
Seaford, Va. (23696) 307/R6
Seaforth, Minn. (56287) 255/C6
Seaforth, Ontario 177/C4
Seaforth, Loch (inlet), Scotland 15/B3
Sea Girt, N.J. (08750) 273/E3
Seagoville, Texas (75159) 303/H3
Seagraves, Texas (79359) 303/B5
Seagrove, N.C. (27341) 281/K3
Seaham, England 10/F3
Seaham, England 13/J3
Seahorse (lake), Newf. 166/A3
Seahorse (pt.), N.W. Terrs. 187/L3
Seahurst, Wash. (98062) 310/A2
Sea Island, Georgia (31561) 217/K8
Sea Isle City, N.J. (08243) 273/D5
Seal (isl.), Maine 243/F8
Seal (riv.), Man. 162/G4
Seal (riv.), Manitoba 179/J2
Seal (lake), Newf. 166/B3
Seal (isl.), Nova Scotia 168/B5
Seal (isl.), S. Africa 118/F7
Sea Lake, Victoria 97/B4
Seal Beach, Calif. (90740) 204/C11
Seal Cove, Maine (04674) 243/G7
Seal Cove, New Bruns. 170/D4
Seal Cove, Newf. 166/C4
Seal Cove, Newf. 166/C3
Seale, Ala. (36875) 195/H6
Sealevel, N.C. (28577) 281/S5
Seal Harbor, Maine (04675) 243/G7
Seal Rock, Oreg. (97376) 291/C3
Sealston, Va. (22547) 307/O4
Sealy, Texas (77474) 303/H8
Seaman, Ohio (45679) 284/C8
Seamer, England 13/G3
Sea Pines, S.C. (†29928) 296/F7
Sea Ranch Lakes, Fla. (†33301) 212/C3
Searchlight, Nev. (89046) 266/F7
Searchmont, Ontario 177/J5
Searcy (co.), Ark. 202/E2
Searcy, Ark. (72143) 202/G3
Searight, Ala. (†36028) 195/F8
Searles, Ala. (†35468) 195/D4
Searles (lake), Calif. 204/H8
Searles, Minn. (56084) 255/D6
Sears, Mich. (49679) 250/D5
Searsboro, Iowa (50242) 229/H5
Searsburg○, Vt. (†05363) 268/A6
Searsmont, Maine (04973) 243/E7
Searsmont○, Maine (04973) 243/E7
Searsport, Maine (04974) 243/F7
Searsport○, Maine (04974) 243/F7
Seascale, England 10/E2
Seascale, England 13/D3
Seaside, Calif. (93955) 204/D7
Seaside, Oreg. (97138) 291/D2
Seaside Heights, N.J. (08751) 273/E4
Seaside Park, N.J. (08752) 273/E4
Seaton, England 13/D7
Seaton, Ill. (61476) 222/C2
Seaton Valley, England 13/J3
Seatonville, Ill. (61359) 222/D2
Seat Pleasant, Md. (20027) 245/G5
Seattle, U.S. 2/C3
Seattle, Wash. 146/F5
Seattle, Wash. 188/B1
Seattle, Wash. (*98101) 310/A2
Seaview, Wash. (98644) 310/A4
Seaward Kaikouras (range), N. Zealand 100/D5
Seawell, Barbados 161/B9
Seba, Indonesia 85/G8
Seba Beach, Alberta 182/C3
Sebago (lake), Maine 243/B8
Sebago Lake, Maine (04075) 243/B8
Sebastian (co.), Ark. 202/B3
Sebastian, Fla. (32958) 212/F4
Sebastian (cape), Oreg. 291/C5
Sebastián Vizcaíno (bay), Mexico 150/B2
Sebasticook (lake), Maine 243/E6
Sebastopol, Calif. (95472) 204/C5
Sebastopol, Miss. (39359) 256/F5
Sebastopol, Victoria 97/B5
Sebatik (isl.), Indonesia 85/F5
Sebatik (isl.), Malaysia 85/F5
Sebec, Maine (04481) 243/E5
Sebec○, Maine (04481) 243/E5
Sebec Lake, Maine (04482) 243/E5
Sebec Station, Maine (†04426) 243/E5
Sebeka, Minn. (56477) 255/C4
Seben, Turkey 63/D2
Sebeş, Romania 45/F3
Sebes Körös (riv.), Hungary 41/F3
Sebewaing, Mich. (48759) 250/F5
Sebha, Libya 102/D2
Sebha, Libya 111/B2
Şebinkarahisar, Turkey 63/H2
Şebiş, Romania (2) 45/F2
Sebnitz, E. Germany 22/F3
Seboeis○, Maine (04484) 243/F5
Seboeis (lake), Maine 243/F5
Seboeis (riv.), Maine 243/F3
Seboomook, Maine (†04478) 243/D4
Seboomook (lake), Maine 243/D4
Seboruco, Venezuela 124/B3
Sebou (riv.), Morocco 106/C2
Seboyeta, N. Mex. (87055) 274/B3
Sebree, Ky. (42455) 237/F5
Sebrell, Va. (†23837) 307/O7
Sebring, Fla. (33870) 212/E4
Sebring, Ohio (44672) 284/H4
Sebringville, Ontario 177/C4
Sebuku (bay), Indonesia 85/F5
Secane, Pa. (†19018) 294/M7
Secas (isls.), Panama 154/G7
Secaucus, N.J. (07094) 273/B2
Secesh (riv.), Idaho 220/C4
Sechelt, Br. Col. 184/J2
Sechura, Peru 128/B5
Sechura (bay), Peru 128/B5
Seco, Ky. (41849) 237/R6
Second (lake), N.H. 268/E1
Second Cataract, Sudan 59/B5
Second Cataract, Sudan 111/F3
Secondcreek, W. Va. (24974) 312/F7
Second Mesa, Ariz. (86043) 198/E3
Secor, Ill. (61771) 222/D3
Sečovce, Czech. 41/F2
Secretan, Sask. 181/E5
Secretary, Md. (21664) 245/P6
Secretary (isl.), N. Zealand 100/A6
Section, Ala. (35771) 195/G1
Secunderabad, India 68/D5
Sécure (riv.), Bolivia 136/C4
Security-Widefield, Colo. (80911) 208/K5
Sedalia, Alberta 182/E4
Sedalia, Colo. (80135) 208/K4
Sedalia, Ind. (46067) 227/E4
Sedalia, Ky. (42079) 237/D7
Sedalia, Mo. (65301) 261/F5
Sedalia, Mo. 188/H3
Sedalia, Ohio (43151) 284/D6
Sedalia, S.C. (†29472) 296/D2
Sedan, France 28/F3
Sedan, Ind. (†46793) 227/G2
Sedan, Kansas (67361) 232/F4
Sedan, Minn. (56380) 255/C5
Sedan, N. Mex. (88436) 274/F2
Sedano, Spain 33/E1
Sedbergh (co.), England 10/E3
Sedbergh, England 13/E3
Seddon, N. Zealand 100/E4
Seddonville, N. Zealand 100/C4
Seddülbahir, Turkey 63/B6
Sede Boqer, Israel 65/D5
Sederot, Israel 65/B4
Sedgewick, Alberta 182/E3
Sedgewickville, Mo. (63781) 261/N7
Sedgwick, Ark. (72465) 202/J2
Sedgwick (co.), Colo. 208/P1
Sedgwick, Colo. (80749) 208/O1
Sedgwick (co.), Kansas 232/E4
Sedgwick, Kansas (67135) 232/E4
Sedgwick○, Maine (04676) 243/F7
Sedhiou, Senegal 106/A6
Sedili Kechil, Tanjong (pt.), Malaysia 72/F5
Sedlčany, Czech. 41/C2
Sedley, Sask. 181/H5
Sedley, Va. (23878) 307/P7
Sedom, Israel 65/D5
Sedona, Ariz. (86336) 198/D4
Sedot Yam, Israel 65/B3
Sedro-Woolley, Wash. (98284) 310/C2
Šeduva, U.S.S.R. 53/B3
Seebe, Alberta 182/C4
Seebert, W. Va. (24975) 312/F6
Seechelt (inlet), Br. Col. 184/J2
Seechelt (pen.), Br. Col. 184/J2
Seeheim, Namibia 118/B5
Seeis, Namibia 118/B4
Seekonk○, Mass. (02771) 249/J5
Seeley, Calif. (92273) 204/K11
Seeley, Wis. (†54843) 317/D3
Seeley Lake, Mont. (59868) 262/C3
Seeleys Bay, Ontario 177/H3
Seelyville, Ind. (47878) 227/C6
Seelyville, Pa. (18431) 294/M2
Seesen, W. Germany 22/D3
Seewis im Prättigau, Switzerland 39/J2
Seez (riv.), Switzerland 39/H2
Şefaatli, Turkey 63/F3
Seferihisar, Turkey 63/B3
Seffner, Fla. (33584) 212/D4
Sefrou, Morocco 106/D2
Sefton, N. Zealand 100/D5
Seg (lake), U.S.S.R. 52/D2
Segamat, Malaysia 72/D7
Segarcea, Romania 45/F3
Segezha, U.S.S.R. 52/D2
Segezha, U.S.S.R. 48/D3
Segnes (pass), Switzerland 39/H3
Segni, Italy 34/F7
Segorbe, Spain 33/F3
Ségou, Mali 106/C6
Ségou, Mali 102/B3
Segovia, Colombia 126/C4
Segovia (Coco) (riv.), Honduras 154/E3
Segovia (Coco) (riv.), Nicaragua 154/E3
Segovia (prov.), Spain 33/D2
Segovia, Spain 33/D2
Segré, France 28/C4
Segre (riv.), Spain 33/G2
Segreganset, Mass. (02773) 249/K5
Seguam (isl.), Alaska 196/D4
Seguam (passage), Alaska 196/D4
Séguéla, Ivory Coast 106/C7
Segul, Argentina 143/F6
Seguin, Kansas (†67740) 232/B2
Séguin (lake), Québec 172/B2
Seguin, Texas (78155) 303/G8
Segula (isl.), Alaska 196/K4
Segundo, Colo. (81070) 208/K8
Segura (riv.), Spain 33/F3
Sehore, India 68/D4
Sehwan, Pakistan 59/J4
Sehwan, Pakistan 68/B3
Seiad Valley, Calif. (96086) 204/B2
Seibert (lake), Alberta 182/E2
Seibert, Colo. (80834) 208/O4
Seibo, Dom. Rep. 156/E3
Seil (isl.), Scotland 15/C4
Seiland (isl.), Norway 18/N1
Seiling, Okla. (73663) 288/J2
Sein (isl.), France 28/A3
Seinäjoki, Finland 18/N5
Seine (riv.), France 7/E4
Seine (bay), France 28/C3
Seine (riv.), France 28/D3
Seine (riv.), Ontario 175/B3
Seine-et-Marne (dept.), France 28/E3
Seine-Saint-Denis (dept.), France 28/C1
Seistan (reg.), Iran 66/M5
Seixal, Portugal 33/A1
Seiyun, P.D.R. Yemen 59/E6
Sejerø (isl.), Denmark 21/E6
Sejny, Poland 47/F1
Seke-Banza, Zaire 115/B5
Sekenke, Tanzania 115/F4
Se Khong (riv.), Cambodia 72/E4
Se Khong (riv.), Laos 72/E4
Sekiu, Wash. (98381) 310/A2
Sekkane, Erg (des.), Mali 106/D4
Sekondi, Ghana 106/D8
Selah, Wash. (98942) 310/E4
Selama, Malaysia 72/D6
Selangor (state), Malaysia 72/D7
Selaphum, Thailand 72/E3
Selaru (isl.), Indonesia 85/J7
Selatan (cape), Indonesia 85/E6
Selawik, Alaska (99770) 196/G1
Selawik (lake), Alaska 196/F1
Selayar (isl.), Indonesia 85/G7
Selb, W. Germany 22/E3
Selby, England 13/F4
Selby, S. Dak. (57472) 298/J3
Selby, Victoria 97/K5
Selby-on-the-Bay, Md. (†21037) 245/N5
Selbyville, Del. (19975) 245/S7
Selbyville, W. Va. (26236) 312/F5
Selçuk, Turkey 63/B3
Selden, Kansas (67757) 232/B2
Seldom, Newf. 166/D4
Seldovia, Alaska (99663) 196/B2
Sele (riv.), Italy 34/E4
Selebi-Pikwe, Botswana 118/D4
Selemdzha (riv.), U.S.S.R. 48/O4
Selemiya, Syria 63/G5
Selendi, Turkey 63/C3
Selenga (riv.) 54/M5
Selenge, Mongolia 77/G2
Selenge, Mongolia 77/F2
Selenge (Selenga) Mörön (riv.), Mongolia 77/G2
Sélestat, France 28/G3
Selfridge, N. Dak. (58568) 282/J7
Sélibaby, Mauritania 106/B5
Seligman, Ariz. (86337) 198/B3
Seligman, Mo. (65745) 261/D9
Selim, Turkey 63/K2
Selima (oasis), Sudan 111/E3
Selima (oasis), Sudan 59/A5
Selimiye, Turkey 63/B4
Selinsgrove, Pa. (17870) 294/J4
Selje, Norway 18/D5
Selkirk (mts.), Br. Col. 184/J4
Selkirk (mts.), Idaho 220/B1
Selkirk, Kansas (67873) 232/A3
Selkirk, Man. 162/G5
Selkirk, Manitoba 179/F4
Selkirk (isl.), Manitoba 179/C1
Selkirk, Mich. (†48661) 250/E4
Selkirk, Scotland 10/E3
Selkirk, Scotland 15/F5
Selkirk (trad. co.) Scotland 15/B5
Sella, Bolivia 136/C7
Selle (peak), Haiti 158/C6
Selleck, Wash. (†98051) 310/D3
Sellers, Ala. (†36046) 195/F6
Sellers, S.C. (29592) 296/H3
Sellersburg, Ind. (47172) 227/F8
Sellersville, Pa. (18960) 294/M5
Sells, Ariz. (85634) 198/D7
Sells, Georgia (†30548) 217/E2
Sellye, Hungary 41/D4
Selma, Ala. 188/J4
Selma, Ala. (36701) 195/E6
Selma, Ark. (†71670) 202/G6
Selma, Calif. (93662) 204/F7
Selma, Ind. (47383) 227/G4
Selma, Iowa (52588) 229/J7
Selma, Miss. (†39120) 256/B7
Selma, N.C. (27576) 281/N3
Selma, Ohio (45364) 284/C6
Selma, Oreg. (97538) 291/D5
Selma, Texas (†78201) 303/K10
Selma, Va. (24474) 307/J5
Selmah, Nova Scotia 168/E3
Selman, Okla. (73856) 288/H1
Selmer, Tenn. (38375) 237/D10
Selmont, Ala. (†36701) 195/E6
Selous (mt.), Yukon 187/E3
Selsey, England 13/G7
Selsey Bill (prom.), England 13/G7
Selukwe, Zimbabwe 118/E3
Selva, Argentina 143/D2
Selvas (for.), Brazil 120/C3
Selvin, Ind. (†47523) 227/C8
Selway (riv.), Idaho 220/C3
Selwyn (lake), N.W. Terrs. 187/H4
Selwyn, Queensland 95/B4
Selwyn (range), Queensland 95/B4
Selwyn (lake), Sask. 181/M2
Selwyn, W. Va. (†25674) 312/B7
Selwyn (mts.), Yukon 187/E3
Selz, N. Dak. (58373) 282/L4
Seman, Ala. (†36092) 195/F5
Semans, Sask. 181/G4
Semara, W. Sahara 102/A2
Semara, Western Sahara 106/B3
Semarang, Indonesia 54/N10
Semarang, Indonesia 85/J2
Sematan, Malaysia 85/D5
Sembé, Congo 115/B3
Sembrancher, Switzerland 39/D4
Şemdinli, Turkey 63/L4
Semenov, U.S.S.R. 52/F3
Semeru (mt.), Indonesia 85/K2
Semichi (isls.), Alaska 196/J3
Semidi (isls.), Alaska 196/G3
Semiluki, U.S.S.R. 52/E4
Semily, Czech. 41/C1
Seminary, Miss. (39479) 256/E7
Seminoe (res.), Wyo. 188/E2

Seminoe (mts.), Wyo. 319/E3
Seminoe (res.), Wyo. 319/F3
Seminoe Dam, Wyo. (†82334) 319/E3
Seminole, Ala. (36574) 195/D10
Seminole (co.), Fla. 212/E3
Seminole, Fla. (33542) 212/B3
Seminole (lake), Fla. 212/B1
Seminole (co.), Georgia 217/C9
Seminole (lake), Georgia 217/B9
Seminole (co.), Okla. 288/N4
Seminole, Okla. (74868) 288/N4
Seminole, Pa. (16253) 294/D4
Seminole, Texas (79360) 303/B5
Seminole Ind. Res., Fla. 212/F5
Seminole Ind. Res., Fla. 212/E4
Semipalatinsk, U.S.S.R. 54/K4
Semipalatinsk, U.S.S.R. 48/H4
Semirara (isls.), Philippines 82/C5
Semisopochnoi (isl.), Alaska 196/K4
Semitau, Indonesia 85/E5
Semmering (pass), Austria 41/C3
Semmes, Ala. (36575) 195/B9
Semnan (governorate), Iran 66/J3
Semnan, Iran 59/F2
Semnan, Iran 66/H3
Semois (riv.), Belgium 27/G9
Semora, N.C. (27343) 281/L2
Sempach, Switzerland 39/F2
Sempach (lake), Switzerland 39/F2
Semporna, Malaysia 85/F5
Semsales, Switzerland 39/C3
Semur-en-Auxois, France 28/F4
Sen, Stoeng (riv.), Cambodia 72/E4
Sena, Bolivia 136/B2
Sena, N. Mex. (87568) 274/D3
Senador Pompeu, Brazil 132/G4
Senai, Malaysia 72/F5
Sena Madureira, Brazil 132/G10
Senanga, Zambia 115/D7
Senate, Sask. 181/B6
Senath, Mo. (63876) 261/M10
Senatobia, Miss. (38668) 256/E1
Sendai, Japan 54/R6
Sendai, Kagoshima, Japan 81/E8
Sendai, Miyagi, Japan 81/K4
Senec, Czech. 41/D2
Seneca, Ill. (61360) 222/E2
Seneca, Kansas (66538) 232/F2
Seneca, Miss. (†39455) 256/F8
Seneca, Mo. (64865) 261/C9
Seneca, Nebr. (69161) 264/D2
Seneca, N. Mex. (88437) 274/F2
Seneca (co.), N.Y. 276/G5
Seneca (lake), N.Y. 276/G5
Seneca (riv.), N.Y. 276/G5
Seneca (co.), Ohio 284/D3
Seneca, Oreg. (97873) 291/J3
Seneca, S.C. (29678) 296/A2
Seneca (riv.), S.C. 296/B2
Seneca, S. Dak. (57473) 298/L3
Seneca, Wis. (54654) 317/E9
Seneca Falls, N.Y. (13148) 276/G5
Seneca Gardens, Ky. (†40201) 237/K2
Senécal (lake), Newf. 166/B3
Senecaville, Ohio (43780) 284/H6
Senecaville (lake), Ohio 284/H6
Senegal 2/J5
Senegal 102/A3
Seegal (riv.) 102/A3
Senegal (riv.), Mali 106/B5
Senegal (riv.), Mauritania 106/B5
SENEGAL 106/A5
Senegal (riv.), Senegal 106/B5
Senekal, S. Africa 118/D5
Seney, Iowa (†51031) 229/A3
Seney, Mich. (49883) 250/C2
Senftenberg, E. Germany 22/F3
Sengiley, U.S.S.R. 52/G4
Senguerr (riv.), Argentina 143/B6
Senhor do Bonfim, Brazil 120/F4
Senhor do Bonfim, Brazil 132/F5
Senica, Czech. 41/D2
Senigallia, Italy 34/D3
Senirkent, Turkey 63/D4
Senj, Yugoslavia 45/B3
Senja (isl.), Norway 7/F2
Senja (isl.), Norway 18/K2
Şenkaya, Turkey 63/K2
Senlac, Sask. 181/B3
Senlis, France 28/E3
Senmonoron, Cambodia 72/E4
Sennar, Sudan 59/B7
Sennar, Sudan 111/F5
Sennar, Sudan 102/F3
Sennar (dam), Sudan 59/B7
Sennar (dam), Sudan 111/F5
Senne (riv.), Belgium 27/E7
Sennestadt, W. Germany 22/C3
Senneterre, Québec 174/B3
Senneville, Québec 172/G4
Senoia, Georgia (30276) 217/C4
Sens, France 28/E3
Sense (riv.), Switzerland 39/D3
Sensuntepeque, El Salvador 154/C4
Sent, Switzerland 39/K3
Senta, Yugoslavia 45/D3
Sentery, Zaire 115/E5
Sentinel, Alberta 182/C5
Sentinel, Ariz. (†85333) 198/B6
Sentinel (butte), N. Dak. 282/C6
Sentinel, Okla. (73664) 288/H4
Sentinel Butte, N. Dak. (58654) 282/C6
Senyavin (isls.), Micronesia 87/F5
Seo de Urgel, Spain 33/G1
Seon, Switzerland 39/F2
Seoni, India 68/D4
Seoul (cap.), S. Korea 2/R4
Seoul (cap.), S. Korea 54/O6
Seoul (cap.), S. Korea 81/C5
Separ, N. Mex. (†88045) 274/A6
Sepetiba (bay), Brazil 135/D3
Sepik (riv.), Papua N.G. 85/B6
Sępólno Krajenskie, Poland 47/C2
Septentrional, Cordillera (range), Dom. Rep. 158/D5
Sept-Îles, Que. 146/M4
Sept-Îles (Seven Is.), Que. 162/K5
Sept-Îles, Québec 174/D2
Septimer (pass), Switzerland 39/J4
Sepulga (riv.), Ala. 195/E7
Sepulveda, Calif. (91343) 204/B10
Sequatchie (co.), Tenn. 237/L10
Sequatchie, Tenn. (37374) 237/K10
Sequatchie (riv.), Tenn. 237/L10
Sequeira, Uruguay 145/C1
Sequeros, Spain 33/C2
Sequim, Wash. (98382) 310/B2
Sequoia Nat'l Park, Calif. 204/G7
Sequoyah (co.), Okla. 288/S3
Sera (isl.), Indonesia 85/J7
Serafimovich, U.S.S.R. 52/F5
Serafina, N. Mex. (87569) 274/D3
Seraing, Belgium 27/G7
Serakhs, U.S.S.R. 48/G6
Serampore, India 68/F1
Serang, Indonesia 85/G1
Serangoon, Singapore 72/F6
Serasan (isl.), Indonesia 85/D5
Serbia (rep.), Yugoslavia 45/E3
Serçiler, Turkey 63/C6
Serdobol (Sortavala), U.S.S.R. 52/D2
Serdobsk, U.S.S.R. 52/F4
Sered', Czech. 41/D2
Şereflikoçhisar, Turkey 63/E3
Seremban, Malaysia 72/D7
Serena, Ill. (60549) 222/E2
Serengeti Nat'l Park, Tanzania 115/F4
Serenje, Zambia 115/F6
Sergach, U.S.S.R. 52/F3
Sergeant, Pa. (†16735) 294/E2
Sergeant Bluff, Iowa (51054) 229/A4
Sergeantsville, N.J. (08557) 273/D3
Sergeya Kirova (isls.), U.S.S.R. 48/J2
Sergipe (state), Brazil 132/G5
Seria, Brunei 85/E5
Serian, Malaysia 85/E5
Sérifos (isl.), Greece 45/G7
Sérigny (riv.), Québec 174/D1
Serik, Turkey 63/D4
Seringapatam, India 68/D6
Sermata (isl.), Indonesia 85/H7
Serón, Spain 33/E4
Serós, Spain 33/G2
Serov, U.S.S.R. 54/H4
Serov, U.S.S.R. 48/G4
Serowe, Botswana 118/D4
Serowe, Botswana 102/E7
Serpa, Portugal 33/C4
Serpa Pinto, Angola 102/D6
Serpentine (lake), New Bruns. 170/D1
Serpentine (lake), S. Australia 88/D5
Serpentine (lakes), S. Australia 94/A3
Serpentine (riv.), W. Australia 88/B3
Serpents Mouth (passage), Trin. & Tob. 156/G5
Serpents Mouth (passage), Trin. & Tob. 161/A11
Serpents Mouth (passage), Venezuela 124/H3
Serpukhov, U.S.S.R. 7/H3
Serpukhov, U.S.S.R. 48/D4
Serpukhov, U.S.S.R. 52/E4
Serra do Navio, Brazil 132/C2
Sérrai, Greece 7/G4
Sérrai, Greece 45/F5
Serrana (bank), Colombia 126/B9
Serra Namuli (mt.), Mozambique 102/F6
Serranilla (bank), Colombia 126/B8
Serra Talhada, Brazil 132/G4
Serres, France 28/F5
Serrinha, Brazil 120/F4
Serrinha, Brazil 132/G5
Sertã, Portugal 33/B3
Sertânia, Brazil 132/G5
Sertãozinho, Brazil 135/B2
Serua (isl.), Indonesia 85/H7
Serui, Indonesia 85/K6
Serule, Botswana 118/D4
Sérvia, Greece 45/F5
Servia, Ind. (46980) 227/F3
Servia, W. Va. (†26623) 312/E5
Service Creek, Oreg. (†97874) 291/G3
Serviceton, Victoria 97/A5
Sêrxü, China 77/E5
Se San (riv.), Vietnam 72/E4
Sese (isls.), Uganda 115/F4
Sesegenaga (lake), Ontario 177/G4
Sesfontein, S. Africa 118/J6
Sesheke, Zambia 115/D7
Sesimbra, Portugal 33/B3
Sesser, Ill. (62884) 222/D5
Sessums, Miss. (39758) 256/G4
Sesto Fiorentino, Italy 34/C3
Sestri Levante, Italy 34/B2
Sesvenna (peak), Switzerland 39/K3
Sète, France 28/E6
Sete Lagoas, Brazil 120/E4
Sete Lagoas, Brazil 132/E7
Sete Quedas (falls) 120/D5
Sete Quedas (falls), Brazil 132/C9
Sete Quedas, (Grande) (isl.), Brazil 132/C8
Seth, W. Va. (25181) 312/C6
Sétif, Algeria 106/F1
Sétif, Algeria 102/C1
Setit (riv.), Sdan 111/G5
Seto, Japan 81/H6
Setonaikai National Park, Japan 81/H7
Seton Portage, Br. Col. 184/F5
Setouchi, Japan 81/O5
Settat, Morocco 106/C2
Settebagni, Italy 34/F6
Setté-Cama, Gabon 115/A4
Settecamini, Italy 34/F6
Setting (lake), Manitoba 179/H3
Settle, England 13/E3
Settsu, Japan 81/J8
Setúbal (dist.), Portugal 33/B3
Setúbal, Portugal 7/D5
Setúbal, Portugal 33/B3
Setúbal (bay), Portugal 33/B3
Seul (lake), Ontario 177/G4
Seul (lake), Ontario 175/B2
Seul Choix (pt.), Mich. 250/D3
Seuzach, Switzerland 39/G1
Sevan (lake), U.S.S.R. 7/J4
Sevan (lake), U.S.S.R. 52/G6
Sevaruyo, Bolivia 136/B6
Sevastopol', U.S.S.R. 7/H4
Sevastopol', U.S.S.R. 48/D5
Sevastopol', U.S.S.R. 52/D6
Sevelen, Switzerland 39/H2
Seven (heads), Ireland 17/D8
Seven Corners, Va. (22044) 307/S3
Seven Devils (res.), Ark. 202/G6
Seven Devils (mts.), Idaho 220/B4
Seven Hills, Ohio (†44131) 284/H9
Seven Hogs, The (isls.), Ireland 17/A7
Seven Islands (bay), Newf. 166/B2
Seven Islands (Sept-Îles), Québec 174/D2
Seven Mile, Ohio (45062) 284/A7
Sevenmile (creek), Ohio 284/A6
Seven Mile Ford, Va. (24373) 307/E7
Sevenoaks, England 13/J8
Sevenoaks, England 10/C6
Seven Persons, Alberta 182/E5
Seven Rivers (riv.), N. Mex. 274/E6
Seven Sisters, Texas (†78357) 303/F9
Seven Sisters Falls, Manitoba 179/G4
Seven Springs, N.C. (28578) 281/O4
Seven Springs, Pa. (†15557) 294/D6
Seventy Mile House, Br. Col. 184/G4
Seven Valleys, Pa. (17360) 294/J6
Severance, Colo. (80546) 208/K1
Severance, Kansas (66081) 232/G2
Severn (riv.), England 13/E6
Severn (riv.), England 10/E5
Severn, Md. (21144) 245/M4
Severn (riv.), Md. 245/N4
Severn (riv.), N.S. Wales 97/F1
Severn, N.C. (27877) 281/P2
Severn (riv.), Ont. 146/J4
Severn (riv.), Ont. 162/G5
Severn (lake), Ontario 175/B2
Severn (riv.), Ontario 177/E3
Severn (riv.), Ontario 175/B2
Severn, Mouth of the (est.), Wales 13/B7
Severn (riv.), Wales 13/E5
Severna Park, Md. (21146) 245/M4
Severnaya Zemlya (isls.), U.S.S.R. 4/A4
Severnaya Zemlya (isls.), U.S.S.R. 54/M1
Severnaya Zemlya (isls.), U.S.S.R. 2/P1
Severnaya Zemlya (isls.), U.S.S.R. 48/L1
Severnyy, U.S.S.R. 52/K1
Severobaykal'sk, U.S.S.R. 48/M4
Severočeský (reg.), Czech. 41/C1
Severodonetsk, U.S.S.R. 52/E5
Severodvinsk, U.S.S.R. 7/H2
Severodvinsk, U.S.S.R. 48/E3
Severodvinsk, U.S.S.R. 52/E2
Severo-Kuril'sk, U.S.S.R. 48/Q4
Severomoravský (reg.), Czech. 41/D2
Severomorsk, U.S.S.R. 52/D1
Severoural'sk, U.S.S.R. 48/G3
Severo-Yeniseysk, U.S.S.R. 48/K3
Severy, Kansas (67137) 232/F4
Sevier (co.), Ark. 202/B6
Sevier, N.C. (†28752) 281/E3
Sevier (co.), Tenn. 237/O9
Sevier (co.), Utah 304/C5
Sevier, Utah (84766) 304/B5
Sevier (lake), Utah 188/D3
Sevier (riv.), Utah 188/D3
Sevier (des.), Utah 304/B4
Sevier (lake), Utah 304/A5
Sevier (riv.), Utah 304/B4
Sevier, East Fork (riv.), Utah 304/B6
Sevier Bridge (res.), Utah 304/C4
Sevierville, Tenn. (37862) 237/P9
Sevilla, Colombia 126/C5
Sevilla (prov.), Spain 33/D4
Seville, Fla. (32090) 212/E2
Seville, Georgia (31084) 217/E7
Seville, Ohio (44273) 284/G3
Seville, Spain 7/D5
Seville, Spain 33/D4
Sevlievo, Bulgaria 45/G4
Sèvres, France 28/A2
Sewal, Iowa (52589) 229/G7
Sewalls Point, Fla. (†33457) 212/F4
Sewanee, Tenn. (37375) 237/K10
Seward, Alaska 146/D3
Seward, Alaska 188/D6
Seward, Alaska (99664) 196/C1
Seward (pen.), Alaska 146/B3
Seward (pen.), Alaska 196/E1
Seward (co.), Kansas 232/B4
Seward, Kansas (67577) 232/D3
Seward (co.), Nebr. 264/G4
Seward, Nebr. (68434) 264/H4
Seward, Okla. (†73044) 288/M3
Seward, Pa. (15954) 294/E5
Seward, U.S. 4/D17
Seward (pen.), U.S. 4/C18
Sewaren, N.J. (07077) 273/E2
Sewart A.F.B., Tenn. 237/J8
Sewell, Br. Col. 184/A3
Sewell, Ky. (†41385) 237/P5
Sewell, N.J. (08080) 273/C4
Sewickley, Pa. (15143) 294/B4
Sexsmith, Alberta 182/A2
Sexton, Ind. (†46173) 227/G5
Sexton, Iowa (†50483) 229/E2
Sextons Creek, Ky. (40983) 237/O6
Sextonville, Wis. (53584) 317/F9
Seybaplaya, Mexico 150/O7
Seychelles 2/M6
SEYCHELLES 118/H5
Seydhisfjördhur, Iceland 21/D1
Seydişehir, Turkey 63/D4
Seyfe (lake), Turkey 63/F3
Seyhan (riv.), Turkey 63/F4
Seyhan (riv.), Turkey 59/C2
Seyitgazi, Turkey 63/D3
Seym (riv.), U.S.S.R. 52/D4
Seymour (canal), Alaska 196/N1
Seymour (inlet), Br. Col. 184/D4
Seymour○, Conn. (06483) 210/C3
Seymour, Ill. (61875) 222/E3
Seymour, Ind. (47274) 227/F7
Seymour, Iowa (52590) 229/G7
Seymour, Mo. (65746) 261/G8
Seymour, Tenn. (37865) 237/O9
Seymour, Texas (76380) 303/E4
Seymour (lake), Vt. 268/D2
Seymour, Victoria 88/H7
Seymour, Victoria 97/C5
Seymour, Wis. (54165) 317/K6
Seymour Johnson A.F.B., N.C. 281/O4
Seymourville, La. (†70764) 238/J2
Seymourville, Manitoba 179/F3
Seyppel, Ark. (†72348) 202/K4
Sézanne, France 28/E3
Sezze, Italy 34/D4
Sfax, Tunisia 106/G2
Sfax, Tunisia 102/D1
Sfîntu Gheorghe, Romania 45/G3
Sfîntu Gheorghe, Romania 45/J3
's-Gravenbrakel (Braine-le-Comte), Belgium 27/D7
's Gravendeel, Netherlands 27/E5
's Gravenhage (The Hague) (cap.), Netherlands 27/E4
's Gravenzande, Netherlands 27/E4
Sgurr a Choire Ghlais (mt.), Scotland 15/D3
Sgurr Alasdair (mt.), Scotland 15/B3
Sgurr Mor (mt.), Scotland 15/C3
Sgurr na Ciche (mt.), Scotland 15/C3
Sgurr na Lapaich (mt.), Scotland 15/C3
Shaanxi (Shensi) (prov.), China 77/G5
Shaba (prov.), Zaire 115/E5
Shabani, Zimbabwe 102/E7
Shabani, Zimbabwe 118/E4
Shabasha, Sudan 59/B7
Shabbona, Ill. (60550) 222/E2
Shabeellaha Dhexe (prov.), Somalia 115/J3
Shabeellaha Hoose (prov.), Somalia 115/H3
Shabla, Bulgaria 45/J4
Shabo, Newf. 166/A3
Shabogamo (lake), Newf. 166/A3
Shabunda, Zaire 115/E4
Shabwa, P.D.R. Yemen 59/E6
Shache (Yarkand), China 77/A4
Shache, China 54/J6
Shackelford (co.), Texas 303/E5
Shackleton, Sask. 181/D5
Shackleton Ice Shelf, Ant. 2/P9
Shackleton Ice Shelf, Ant. 5/C5
Shade, Ohio (45776) 284/G7
Shadegan, Iran 66/F5
Shade Gap, Pa. (17255) 294/G5
Shadehill, S. Dak. (57653) 298/E2
Shadehill (res.), S. Dak. 298/E2
Shadeland, Ind. (†47901) 227/C4
Shader, Scotland 15/B2
Shadrinsk, U.S.S.R. 48/G4
Shady Bend, Kansas (†67455) 232/D2
Shady Cove, Oreg. (97539) 291/E5
Shady Dale, Georgia (31085) 217/E4
Shady Grove, Ala. (†36036) 195/F7
Shady Grove, Fla. (32357) 212/C1
Shady Grove, Ky. (†42064) 237/F6
Shady Grove, Pa. (17256) 294/G6
Shady Point, Okla. (74956) 288/S4
Shady Side, Md. (20867) 245/M5
Shadyside, Ohio (43947) 284/J6
Shady Valley, Tenn. (37688) 237/T7
Shafer (lake), Ind. 227/D3
Shafer, Minn. (55074) 255/F5
Shafter, Calif. (93263) 204/F8
Shafter, Nev. (†89835) 266/G2
Shafter, Texas (79850) 303/C12
Shaftesbury, England 13/E7
Shaftesbury, England 10/E5
Shaftsbury○, Vt. (05262) 268/A6
Shageluk, Alaska (99665) 196/G2
Shag Harbour, Nova Scotia 168/C5
Shahabad, Iran 66/E3
Shahat, Libya 102/E1
Shahat, Libya 111/D1
Shahbandar, Pakistan 68/B4
Shahdad, Iran 59/G3
Shahdad, Iran 66/K5
Shahdol, India 68/E4
Shahi, Iran 66/H2
Shahin Dezh, Iran 66/E2
Shahistan (Saravan), Iran 66/N7
Shah Jahan, Kuh-e (mts.), Iran 66/L2
Shahjahanpur, India 68/E3
Shah Juy, Afghanistan 59/J3
Shahpur, Iran 66/D1
Shahrakht, Iran 66/M4
Shahreza, Iran 59/F3
Shahreza, Iran 66/H4
Shahr Kord, Iran 66/G4
Shahrud, Iran 59/G2
Shahrud, Iran 66/J2
Shahsavar, Iran 59/F2
Shahsavar, Iran 66/G2
Shaibara (isl.), Saudi Arabia 59/C4
Sha'ib Hisb, Wadi (dry riv.), Iraq 66/C5
Shaikh Sa'ad, Iraq 66/E4
Shaikh Shu'aib (isl.), Iran 66/H7
Shaikh Shu'aib (isl.), Iran 59/F4
Shailerville, Conn. (†06438) 210/E3
Shajapur, India 68/D4
Shakawe, Botswana 118/C3
Shaker Heights, Ohio (44120) 284/H9
Shakespeare, Ontario 177/D4
Shakhtinsk, U.S.S.R. 48/H5
Shakhty, U.S.S.R. 7/J4
Shakhty, U.S.S.R. 48/E5
Shakhty, U.S.S.R. 52/F5
Shakhun'ya, U.S.S.R. 52/G3
Shaki, Nigeria 106/E7
Shâkir (isl.), Egypt 59/B4
Shakopee, Minn. (55379) 255/F6
Shakopee (creek), Minn. 255/C5
Shaktoolik, Alaska (99771) 196/F2
Shalath, Br. Col. 184/F5
Shaler (mts.), N.W. Terrs. 187/G2
Shalimar, Fla. (32579) 212/C6
Shallala, Wadi esh (dry riv.), Jordan 65/D2
Shallotte, N.C. (28459) 281/N7
Shallow (lake), Maine 243/E3
Shallowater, Texas (79363) 303/B4
Shallow Lake, Ontario 177/C3
Shallow Water, Kansas (†67871) 232/B3
Sham, Jebel (mt.), Oman 59/G5
Shamattawa, Manitoba 179/K2
Shamattawa (riv.), Ontario 175/C2
Shambaugh, Iowa (51651) 229/D7
Shambe, Sudan 111/F6
Shamil, Iran 66/K7
Shammar, Jebel (plat.), Saudi Arabia 59/D4
Shamokin, Pa. (17872) 294/J4
Shamokin Dam, Pa. (17876) 294/J4
Shamrock, Fla. (†32628) 212/C2
Shamrock, Okla. (74068) 288/N3
Shamrock, Sask. 181/E5
Shamrock, Texas (79079) 303/D2
Shamrock Lakes, Ind. (†47348) 227/G4
Shamva, Zimbabwe 118/E3
Shan (state), Burma 72/C2
Shan (plat.), Burma 72/C2
Shanagolden, Ireland 17/C6
Shandan, China 77/F4
Shandon, Calif. (93461) 204/E8
Shandong (Shantung) (prov.), China 77/J4
Shangani (riv.), Zimbabwe 118/D3
Shangdu, China 77/H3
Shanghai, China 2/R4
Shanghai, China 54/O6
Shanghai, China 77/K5
Shanghai, Va. (23158) 307/P5
Shanghai, W. Va. (†25427) 312/K4
Shanghang, China 77/J6
Shangnan, China 77/H5
Shangqui (Shangkiu), China 77/J5
Shangrao (Shangjao), China 77/J6
Shangshui, China 77/J5
Shang Xian, China 77/H5
Shangzhi, China 77/L2
Shaniko, Oreg. (97057) 291/G3
Shanks, W. Va. (26761) 312/J4
Shanksville, Pa. (15560) 294/E5
Shannock, R.I. (02875) 249/H7
Shannon, Georgia (30172) 217/B2
Shannon (isl.), Greenl. 4/B10
Shannon, Ill. (61078) 222/D1
Shannon, Mouth of the (est.), Ireland 17/B6
Shannon (riv.), Ireland 10/B4
Shannon (riv.), Ireland 17/E6
Shannon, Miss. (38868) 256/G2
Shannon (co.), Mo. 261/K8
Shannon, New Bruns. 170/E3
Shannon, N. Zealand 100/E4
Shannon, N.C. (28386) 281/L5
Shannon, Québec 172/F3
Shannon (co.), S. Dak. 298/D7
Shannon (lake), Wash. 310/D2
Shannon Airport, Ireland 17/D6
Shannon Bridge, Ireland 17/F5
Shannon City, Iowa (50861) 229/E7
Shannondale, Ind. (†47933) 227/D4
Shannon Hills, Ark. (†72103) 202/F4
Shannontown, S.C. (†29150) 296/G4
Shannonville, Ontario 177/G3
Shanshan (Piqan), China 77/D3
Shansi (Shanxi) (prov.), China 77/H4
Shantar (isls.), U.S.S.R. 54/P4
Shantar (isls.), U.S.S.R. 48/O4
Shantou (Swatou), China 77/J7
Shantou, China 54/N7
Shantung (Shandong) (prov.), China 77/J4
Shanty Bay, Ontario 177/E3
Shanxi (Shansi) (prov.), China 77/H4
Shanyang, China 77/G5
Shanyin, China 77/H4
Shaoguan (Shiukwan), China 77/H7
Shaowu, China 77/J6
Shaoxing (Shaohing), China 77/K5
Shaoyang, China 77/H6
Shap, England 13/E3
Shapinsay (isl.), Scotland 15/F1
Shapio (lake), Newf. 166/B3
Shapleigh, Maine (04076) 243/B8
Shapleigh○, Maine (04076) 243/B8
Shaqlawa, Iraq 66/D2
Shaqra, Saudi Arabia 54/F7
Shaqra, Saudi Arabia 59/D4
Sharafkhaneh, Iran 66/D1
Sharbatat, Ras (cape), Oman 59/G6
Sharbot Lake, Ontario 177/H3
Shari (riv.) 102/D4
Shari (riv.), Cent. Afr. Rep. 115/C2
Shari (riv.), Chad 111/C5
Shari, Japan 81/M2
Sharifabad, Iran 66/L2
Sharjah, U.A.E. 59/F4
Shark (pt.), Fla. 212/E6
Shark (bay), W. Australia 88/A5
Shark (bay), W. Australia 92/A4
Shark Bay, W. Australia 88/A5
Sharkey (co.), Miss. 256/C5
Sharlyk, U.S.S.R. 52/H4
Sharon○, Conn. (06069) 210/B1
Sharon, Georgia (30664) 217/G3
Sharon, Kansas (67138) 232/D4
Sharon, Mass. (02067) 249/K4
Sharon○, Mass. (02067) 249/K4
Sharon, Miss. (39163) 256/E5
Sharon○, N.H. (†03458) 268/D6
Sharon, N. Dak. (58277) 282/P4
Sharon, Ohio (43781) 284/G6
Sharon, Okla. (73857) 288/H2
Sharon, Pa. (16146) 294/B3
Sharon, S.C. (29742) 296/E2
Sharon, Tenn. (38255) 237/D8
Sharon○, Vt. (05065) 268/C4
Sharon, W. Va. (25182) 312/D6
Sharon, Wis. (53585) 317/J11
Sharon Center, Ohio (44274) 284/G3
Sharon Grove, Ky. (42280) 237/G7
Sharon Hill, Pa. (19079) 294/N7
Sharon Springs, Kansas (67758) 232/A3
Sharon Springs, N.Y. (13459) 276/L5
Sharon Valley, Conn. (†06069) 210/B1
Sharonville, Ohio (45241) 284/C9
Sharp (co.), Ark. 202/G1
Sharpe, Kansas (†66871) 232/G3
Sharpe (lake), S. Dak. 298/J5
Sharpe Army Depot, Calif. 204/D6
Sharpes, Fla. (32959) 212/F3
Sharples, W. Va. (25183) 312/C7
Sharps (isl.), Md. 245/N6
Sharps, Va. (22548) 307/P5
Sharpsburg, Georgia (30277) 217/C4
Sharpsburg, Iowa (50862) 229/D7
Sharpsburg, Ky. (40374) 237/O4
Sharpsburg, Md. (21782) 245/G3
Sharpsburg, N.C. (27878) 281/O3
Sharpsburg, Pa. (15215) 294/B6
Sharps Chapel, Tenn. (37866) 237/O8
Sharpsville, Ind. (46068) 227/E4
Sharpsville, Pa. (16150) 294/A3
Sharptown, Md. (21861) 245/R6
Sharptown, N.J. (†08098) 273/C4
Shar'ya, U.S.S.R. 48/E4
Shar'ya, U.S.S.R. 52/G3
Shashe, Botswana 118/D4
Shashe (riv.), Botswana 118/D4
Shashe (riv.), Zimbabwe 118/D4
Shashi (Shasi), China 77/H5
Shasta (res.), Calif. 188/B2
Shasta (mt.), Calif. 188/B2
Shasta (co.), Calif. 204/C3
Shasta, Calif. (96087) 204/C3
Shasta (dam), Calif. 204/C3
Shasta (lake), Calif. 204/C3
Shasta (mt.), Calif. 204/C2
Shasta (riv.), Calif. 204/C2
Shati, Wadi esh (dry riv.), Libya 111/B2
Shatra, Iraq 66/E5
Shatt-al-'Arab (riv.), 66/E4
Shattuc, Ill. (62283) 222/D5
Shattuck, Okla. (73858) 288/G2
Shattuckville, Mass. (01369) 249/D2
Shauck, Ohio (43349) 284/E4
Shaughnessy, Alberta 182/D5
Shaunavon, Sask. 162/F6
Shaunavon, Sask. 181/C6
Shavano Park, Texas (†78201) 303/J10
Shaver Lake, Calif. (93664) 204/F6
Shavers Fork (riv.), W. Va. 312/G5
Shave Ziyyon, Israel 65/B2
Shaw, Kansas (†66733) 232/G4
Shaw, La. (†71373) 238/G4
Shaw, Minn. (†55717) 255/F3
Shaw, Miss. (38773) 256/C3
Shaw (mt.), N.H. 268/E4
Shaw, Oreg. (†97325) 291/A3
Shaw A.F.B., S.C. 296/F4
Shawan, China 77/B3
Shawanese, Pa. (18654) 294/E7
Shawano (co.), Wis. 317/J6
Shawano, Wis. (54166) 317/J6
Shawano (lake), Wis. 317/K6
Shawboro, N.C. (27973) 281/S2
Shawbost, Scotland 15/B2
Shawbridge, Québec 172/C4
Shawinigan, Que. 162/J6
Shawinigan, Québec 174/C3
Shawinigan, Québec 172/E3
Shawinigan (riv.), Québec 172/E3
Shawinigan-Sud, Québec 172/E3
Shaw Island, Wash. (98286) 310/B2
Shawmut, Ala. (36876) 195/H5
Shawmut, Maine (04975) 243/D6
Shawmut, Mont. (59078) 262/G4
Shawmut, Pa. (†15823) 294/E3
Shawnee, Colo. (80475) 208/H4
Shawnee (co.), Kansas 232/G2
Shawnee, Kansas (*66202) 232/H2
Shawnee, Ohio (43782) 284/F6
Shawnee, Okla. 188/G3
Shawnee, Okla. (74801) 288/N4
Shawnee, Wyo. (82229) 319/G3
Shawnee Hills, Ohio (43065) 284/D5
Shawnee on Delaware, Pa. (18356) 294/N3
Shawneetown, Ill. (62984) 222/E6
Shawnigan Lake, Br. Col. 184/J3
Shawomet, R.I. (†02886) 249/J6
Shawsheen Village, Mass. (01810) 249/K2
Shawshine (riv.), Mass. 249/K2
Shawsville, Md. (†21161) 245/M2
Shawsville, Va. (24162) 307/H6
Shawville, Québec 172/A4
Shay Gap, W. Australia 92/C3
Shayib, Jebel (mt.), Egypt 59/B4
Shay Juy, Afghanistan 68/B2
Shchekino, U.S.S.R. 52/E4
Shchel'yayur, U.S.S.R. 52/H1
Shchigry, U.S.S.R. 52/E4
Shchuchinsk, U.S.S.R. 48/H4
Sheakleyville, Pa. (16151) 294/B3
Sheaville, Oreg. (†97910) 291/K4
Shebandowan, Ontario 177/G5
Sheberghan, Afghanistan 54/H6
Sheberghan, Afghanistan 68/B1
Sheberghan, Afghanistan 59/H2
Sheboygan, Wis. 188/J2
Sheboygan (co.), Wis. 317/L8
Sheboygan, Wis. (53081) 317/L8
Sheboygan Falls, Wis. (53085) 317/L8
Shedd, Oreg. (97377) 291/D3
Shedden, Ontario 177/C5
Shediac, New Bruns. 170/F2
Shediac (isl.), New Bruns. 170/F2
Shediac Bridge, New Bruns. 170/F2
Sheeffry (hills), Ireland 17/B4
Sheelin (lake), Ireland 17/G4
Sheenjek (riv.), Alaska 196/K1
Sheep (mt.), Colo. 208/E6
Sheep (mt.), Mont. 262/C2
Sheep (range), Nev. 266/F6
Sheep (creek), Oreg. 291/L2

Sheep (creek), Utah 304/E3
Sheep Creek, Alberta 182/A2
Sheep Haven (harb.), Ireland 17/F1
Sheeps (head), Ireland 17/B8
Sheepscott, Maine (†04579) 243/D7
's Heerenberg, Netherlands 27/J5
Sheerness, Alberta 182/E4
Sheet (harb.), Nova Scotia 168/F4
Sheet Harbour, Nova Scotia 168/F4
Shefar'am, Israel 65/C2
Shefayim, Israel 65/B3
Sheffield, Ala. (35660) 195/C1
Sheffield, England 7/D3
Sheffield, England 10/F4
Sheffield, England 13/J2
Sheffield, Ill. (61361) 222/D2
Sheffield, Iowa (50475) 229/G3
Sheffield○, Mass. (01257) 249/A4
Sheffield, Mont. (†59347) 262/K4
Sheffield, New Bruns. 170/D3
Sheffield, Ohio (†44052) 284/F3
Sheffield, Pa. (16347) 294/D2
Sheffield, Tasmania 99/C3
Sheffield, Texas (79781) 303/B7
Sheffield○, Vt. (05866) 268/C2
Sheffield Lake, Ohio (44054) 284/F3
Shefford (co.), Québec 172/E4
Sheguiandah, Ontario 177/C2
Sheho, Sask. 181/H4
Shehy (mts.), Ireland 17/C8
Sheikh Sa'id, Yemen Arab Rep. 59/D7
Sheila, New Bruns. 170/F1
Sheki, U.S.S.R. 52/G6
Shelagh (riv.), Iran 66/M5
Shelagskiy (cape), U.S.S.R. 48/R2
Shelbiana, Ky. (41562) 237/R6
Shelbina, Mo. (63468) 261/H3
Shelburn, Ind. (47879) 227/C6
Shelburne, N.H. (†03581) 268/E3
Shelburne○, N.H. (†03581) 268/E3
Shelburne (co.), Nova Scotia 168/C5
Shelburne, Nova Scotia 168/C5
Shelburne, Ontario 177/D3
Shelburne○, Vt. (05482) 268/A3
Shelburne (pond), Vt. 268/A3
Shelburne Falls, Mass. (01370) 249/D2
Shelby (co.), Ala. 195/E4
Shelby, Ala. (35143) 195/E4
Shelby (co.), Ill. 222/E4
Shelby (co.), Ind. 227/F5
Shelby, Ind. (46377) 227/C2
Shelby (co.), Iowa 229/C5
Shelby, Iowa (51570) 229/C5
Shelby (co.), Ky. 237/L4
Shelby, Mich. (49455) 250/C5
Shelby, Miss. (38774) 256/C3
Shelby (co.), Mo. 261/H3
Shelby, Mont. (59474) 262/E2
Shelby, Nebr. (68662) 264/G3
Shelby, N.C. (28150) 281/G4
Shelby (co.), Ohio 284/B5
Shelby, Ohio (44875) 284/E4
Shelby (co.), Tenn. 237/B10
Shelby (co.), Texas 303/K6
Shelby Center, N.Y. (†14103) 276/D4
Shelbyville, Ill. (62565) 222/E4
Shelbyville (lake), Ill. 222/E4
Shelbyville, Ind. (46176) 227/F6
Shelbyville, Ky. (40065) 237/L4
Shelbyville, Mo. (63469) 261/H3
Shelbyville, Tenn. (37160) 237/H10
Shelbyville, Texas (75973) 303/L6
Sheldahl, Iowa (50243) 229/F5
Sheldon, Ill. (60966) 222/F3
Sheldon, Iowa (51201) 229/B2
Sheldon, Minn. (†55921) 255/G7
Sheldon, Mo. (64784) 261/D7
Sheldon, N. Dak. (58068) 282/P6
Sheldon, S.C. (29941) 296/F6
Sheldon, Texas (†77001) 303/K1
Sheldon○, Vt. (05483) 268/B2
Sheldon, Wis. (54766) 317/D5
Sheldon Junction, Vt. (†05483) 268/B2
Sheldon Point, Alaska (99666) 196/E2
Sheldon Springs, Vt. (05485) 268/A2
Sheldonville, Mass. (02070) 249/J4
Shelekhov (gulf), U.S.S.R. 54/S3
Shelekhov (gulf), U.S.S.R. 48/Q4
Shelikof (str.), Alaska 196/H3
Shell (pt.), Fla. 212/B1
Shell (riv.), Minn. 255/C4
Shell (creek), N. Dak. 282/F3
Shell, Loch (inlet), Scotland 15/B3
Shell (lake), Wis. 317/C4
Shell, Wyo. (82441) 319/E1
Shell (creek), Wyo. 319/E1
Shellbrook, Sask. 162/F5
Shellbrook, Sask. 181/E2
Shelley, Br. Col. 184/F3
Shelley, Idaho (83274) 220/F6
Shellharbour, N.S. Wales 97/F4
Shell Knob, Mo. (65747) 261/E9
Shell Lake, Sask. 181/D2
Shell Lake, Wis. (54871) 317/C4
Shellman, Georgia (31786) 217/C7
Shellmouth, Manitoba 179/A4
Shell Rock, Iowa (50670) 229/H3
Shellsburg, Iowa (52332) 229/K4
Shelltown, Md. (†21838) 245/R9
Shelly, Minn. (56581) 255/B3
Shelmerdine, N.C. (†27834) 281/P4
Shelocta, Pa. (15774) 294/D4
Shelter (isl.), N.Y. 276/R8
Shelton, Conn. (06484) 210/C3
Shelton, Nebr. (68876) 264/F4
Shelton, S.C. (†29015) 296/E3
Shelton, Wash. (98584) 310/B3
Shemakha, U.S.S.R. 52/G6
Shemogue, New Bruns. 170/F2
Shemya (isl.), Alaska 196/J3
Shemya Air Force Base, Alaska 196/J3
Shenandoah, Iowa (51601) 229/C7
Shenandoah, Pa. (17976) 294/K4
Shenandoah (co.), Va. 307/L3
Shenandoah, Va. (22849) 307/L4
Shenandoah (mt.), Va. 307/K3
Shenandoah (riv.), Va. 307/N2
Shenandoah (riv.), W. Va. 312/K4
Shenandoah Junction, W. Va. (25442) 312/L4
Shenandoah Nat'l Park, Va. 307/L3
Shenango, Pa. (†16125) 294/A3
Shenango River (lake), Pa. 294/B3
Shendam, Nigeria 106/F7
Shendi, Sudan 59/B6
Shendi, Sudan 102/F3
Shendi, Sudan 111/F4
Shëngjin, Albania 45/D5
Sheng Xian, China 77/K6
Shenipsit (lake), Conn. 210/F1
Shenkursk, U.S.S.R. 52/F2
Shenkursk, U.S.S.R. 48/E3
Shenmu, China 77/G4
Shennington, Wis. (†54618) 317/F7
Shennongjia, China 77/H5
Shensi (Shaanxi) (prov.), China 77/G5
Shenyang (Mukden), China 77/K3
Shenyang, China 54/O5
Shenyang, China 2/R3
Sheopur, India 68/D3
Shepard, Alberta 182/D4
Shepardsville, Ind. (47880) 227/B5
Shepaug (dam), Conn. 210/B3
Shepaug (riv.), Conn. 210/B2
Shepetovka, U.S.S.R. 52/C4
Shepherd, Mich. (48883) 250/E5
Shepherd, Mont. (59079) 262/H5
Shepherd (bay), N.W. Terrs. 187/J3
Shepherd, Texas (77371) 303/K7
Shepherdstown, W. Va. (25443) 312/L4
Shepherdsville, Ky. (40165) 237/K4
Shepody, New Bruns. 170/F3
Shepody (bay), New Bruns. 170/F3
Sheppard A.F.B., Texas 303/F3
Shepparton, Victoria 88/G7
Shepparton, Victoria 97/C5
Sheppey (isl.), England 13/J6
Sheppton, Pa. (18248) 294/K4
Shepshed, England 13/F5
Shepton Mallet, England 13/E6
Shepton Mallet, England 10/E5
Sheqi, China 77/H5
Sherack, Minn. (†56722) 255/B2
Sherard, Miss. (38669) 256/C2
Sherard (cape), N.W. Terrs. 187/L2
Sherborn○, Mass. (01770) 249/A8
Sherborne, England 10/E5
Sherborne, England 13/E7
Sherbro (isl.), S. Leone 106/B7
Sherbrooke, Nova Scotia 168/G3
Sherbrooke (lake), Nova Scotia 168/D4
Sherbrooke (riv.), Nova Scotia 168/D4
Sherbrooke, Que. 162/J7
Sherbrooke (co.), Québec 172/E4
Sherbrooke, Québec 172/E4
Sherburn, Minn. (56171) 255/D7
Sherburne (co.), Minn. 255/E5
Sherburne, N.Y. (13460) 276/K5
Shercock, Ireland 17/G4
Shereik, Sudan 111/F4
Sheridan, Ark. (72150) 202/F5
Sheridan, Calif. (95681) 204/D5
Sheridan, Colo. (†80110) 208/J3
Sheridan, Ill. (60551) 222/E2
Sheridan, Ind. (46069) 227/E4
Sheridan (co.), Kansas 232/B2
Sheridan, Maine (04775) 243/F2
Sheridan, Mich. (48884) 250/D5
Sheridan, Mo. (64486) 261/C1
Sheridan (co.), Mont. 262/M2
Sheridan, Mont. (59749) 262/D5
Sheridan (co.), Nebr. 264/B2
Sheridan, N.Y. (14135) 276/B5
Sheridan (co.), N. Dak. 282/K4
Sheridan, Oreg. (97378) 291/D2
Sheridan, W. Va. (†25506) 312/B6
Sheridan, Wis. (†54981) 317/H7
Sheridan, Wyo. 188/E2
Sheridan, Wyo. 146/H5
Sheridan (co.), Wyo. 319/F1
Sheridan, Wyo. (82801) 319/F1
Sheridan Lake, Colo. (81071) 208/P6
Sheringham, England 13/J5
Sheringham, England 10/G4
Sherkin (isl.), Ireland 17/C9
Sherman (mt.), Colo. 208/G4
Sherman○, Conn. (06784) 210/B2
Sherman, Ill. (62684) 222/D4
Sherman (co.), Kansas 232/A2
Sherman, Kansas (†67356) 232/H4
Sherman, Ky. (†41035) 237/M3
Sherman, Maine (†04777) 243/G4
Sherman○, Maine (†04777) 243/G4
Sherman, Mich. (†49668) 250/D4
Sherman, Miss. (38869) 256/G2
Sherman, Mo. (63078) 261/N3
Sherman (co.), Nebr. 264/F3
Sherman (res.), Nebr. 264/E3
Sherman, N. Mex. (†88057) 274/B6
Sherman, N.Y. (14781) 276/A6
Sherman (inlet), N.W. Terrs. 187/J3
Sherman (co.), Oreg. 291/G2
Sherman, S. Dak. (57060) 298/S6
Sherman (co.), Texas 303/C1
Sherman, Texas (75090) 303/H4
Sherman, Texas 188/G4
Sherman, W. Va. (26173) 312/C5
Sherman City, Mich. (†48632) 250/D5
Sherman Mills, Maine (04776) 243/G4
Shermans Dale, Pa. (17090) 294/H5
Sherman Station, Maine (04777) 243/F4
Sherrard, Ill. (61281) 222/C2
Sherrard, W. Va. (†26003) 312/E3
Sherridon, Man. 162/G4
Sherridon, Manitoba 179/H3
Sherrill, Ark. (72152) 202/F5
Sherrill, Iowa (52073) 229/M3
Sherrill, N.Y. (13461) 276/J4
Sherrington, Québec 172/D4
Sherrodsville, Ohio (44675) 284/H4
Sherry, Wis. (†54454) 317/G6
's Hertogenbosch, Netherlands 27/G5
Sherwood, Ark. (72116) 202/F4
Sherwood (pt.), Conn. 210/B4
Sherwood (for.), England 13/F4
Sherwood, Mich. (49089) 250/D6
Sherwood, N. Dak. (58782) 282/G2
Sherwood, Ohio (43556) 284/A3
Sherwood, Okla. (†74728) 288/S6
Sherwood, Oreg. (97140) 291/A2
Sherwood, Pr. Edward I. 168/E2
Sherwood, Tenn. (37376) 237/K10
Sherwood, Texas (†76941) 303/D6
Sherwood, Wis. (54169) 317/K7
Sherwood Park, Alberta 182/D3
Sheslay (riv.), Br. Col. 184/J2
Shetek (lake), Minn. 255/C6
Shetland (islands area), Scotland 15/F2
Shetland (isls.), Scotland 7/D2
Shetland (isls.), Scotland 10/G1
Shetland (isls.), Scotland 15/G2
Shetucket (riv.), Conn. 210/G2
Shevchenko, U.S.S.R. 54/G5
Shevchenko, U.S.S.R. 48/F5
Shevlin, Manitoba 179/A3
Shevlin, Minn. (56676) 255/C3
Sheyenne, N. Dak. (58374) 282/M4
Sheyenne (riv.), N. Dak. 188/G1
Sheyenne (riv.), N. Dak. 282/O6
Sheykh Sho'eyb (isl.), Iran 66/H7
Shiant (isls.), Scotland 15/B3
Shiant (sound), Scotland 15/B3
Shiawassee (co.), Mich. 250/E6
Shiawassee (riv.), Mich. 250/E5
Shibam, P.D.R. Yemen 59/E6
Shibata, Japan 81/J5
Shibetsu, Japan 81/M2
Shibin el Kom, Egypt 111/J3
Shibogama (lake), Ontario 175/C2
Shickley, Nebr. (68436) 264/G4
Shickshinny, Pa. (18655) 294/K3
Shideler, Ind. (†47338) 227/G4
Shidler, Okla. (74652) 288/N1
Shiel, Loch (lake), Scotland 10/D2
Shiel, Loch (lake), Scotland 15/C4
Shieldaig, Scotland 15/C3
Shields, Kansas (67874) 232/B3
Shields (riv.), Mont. 262/F4
Shields, N. Dak. (58569) 282/H7
Shieldsville, Minn. (†55021) 255/E6
Shifnal, England 13/E5
Shiga (pref.), Japan 81/J7
Shigatse (Xigazê), China 77/C6
Shigawake, Québec 172/D2
Shihezi (Shihhotzu), China 77/C3
Shihr, P.D.R. Yemen 59/E7
Shijak, Albania 45/D5
Shijiazhuang (Shihkiachwang), China 77/J4
Shijiazhuang, China 54/N6
Shikarpur, Pakistan 68/B3
Shikarpur, Pakistan 59/J4
Shikoku, Japan 54/P6
Shikoku (isl.), Japan 2/R4
Shikoku (isl.), Japan 81/F7
Shikotan (isl.), Japan 81/N2
Shikotsu (lake), Japan 81/K2
Shikotsu-Toya National Park, Japan 81/K2
Shilbottle, England 13/F2
Shildon, England 13/F3
Shilka (riv.), U.S.S.R. 54/N4
Shilka, U.S.S.R. 48/M4
Shillelagh, Ireland 17/J6
Shillelagh, Ireland 10/C4
Shillington, Pa. (19607) 294/K5
Shillong, India 68/G3
Shilo, Manitoba 179/C5
Shiloh, Ala. (†35979) 195/G2
Shiloh, Ala. (†36754) 195/C6
Shiloh, Georgia (31826) 217/C5
Shiloh, Ill. (†62220) 222/B3
Shiloh, N.J. (08353) 273/C5
Shiloh, Ohio (44878) 284/E4
Shiloh, S.C. (†29080) 296/G4
Shiloh, Tenn. (38376) 237/E10
Shiloh, Va. (22549) 307/O4
Shiloh Nat'l Mil. Park, Tenn. 237/E10
Shilovo, U.S.S.R. 52/F4
Shimabara, Japan 81/E7
Shimamoto, Japan 81/J7
Shimane (pref.), Japan 81/F6
Shimane (pen.), Japan 81/F6
Shimanovsk, U.S.S.R. 48/N4
Shimbir Berris (mt.), Somalia 115/J1
Shimizu, Japan 81/J6
Shimoda, Japan 81/J6
Shimoga, India 68/D6
Shimokita (pen.), Japan 81/K3
Shimonoseki, Japan 81/E6
Shin (falls), Scotland 15/D3
Shin, Loch (lake), Scotland 15/D2
Shin, Loch (lake), Scotland 10/D1
Shin (riv.), Scotland 15/D3
Shinano (riv.), Japan 81/J5
Shinas, Oman 59/G5
Shindand, Afghanistan 59/H3
Shindand, Afghanistan 68/A2
Shindler, S. Dak. (†57101) 298/R7
Shiner, Texas (77984) 303/G8
Shingbwiyang, Burma 72/B1
Shinglehouse, Pa. (16748) 294/F2
Shingler, Georgia (†31781) 217/E7
Shingle Springs, Calif. (95682) 204/C8
Shingleton, Mich. (49884) 250/C2
Shingu, Japan 81/H7
Shining Tree, Ontario 177/J5
Shinjo, Japan 81/K4
Shinko (riv.), Cent. Afr. Rep. 115/D2
Shinnecock Ind. Res., N.Y. 276/R9
Shinnston, W. Va. (26431) 312/F4
Shin Pond, Maine (†04765) 243/F3
Shinrone, Ireland 17/F5
Shinyanga (reg.), Tanzania 115/F4
Shinyanga, Tanzania 115/F4
Shinyanga, Tanzania 102/F5
Shiocton, Wis. (54170) 317/K7
Shiogama, Japan 81/K4
Shiono (cape), Japan 81/H7
Ship (isl.), Miss. 256/G10
Ship Bottom, N.J. (08008) 273/E4
Ship Harbour, Newf. 166/D2
Ship Harbour, Nova Scotia 168/F4
Shiping, China 77/F7
Shipki (pass), India 68/D2
Shipman, Ill. (62685) 222/C4
Shipman, Sask. 181/F2
Shipman, Va. (22971) 307/L5
Shippan (pt.), Conn. 210/A4
Shippegan, New Bruns. 170/F1
Shippegan (gully), New Bruns. 170/F1
Shippegan (bay), New Bruns. 170/E1
Shippensburg, Pa. (17257) 294/H5
Shippenville, Pa. (16254) 294/D3
Shiprock, N. Mex. (87420) 274/A2
Ship Rock (peak), N. Mex. 274/A2
Shipshaw (riv.), Québec 172/F1
Shipshewana, Ind. (46565) 227/F1
Ship Shoal (isl.), Va. 307/S6
Shipston on Stour, England 13/F5
Shiqian, China 77/G6
Shiqma (riv.), Israel 65/B4
Shiquan, China 77/G5
Shiquanhe, China 77/A5
Shiragami (cape), Japan 81/J3
Shirakawa, Japan 81/K5
Shirane (mt.), Japan 81/H6
Shirane (mt.), Japan 81/J5
Shiranuka, Japan 81/M2
Shiraz, Iran 54/G7
Shiraz, Iran 66/H6
Shiraz, Iran 59/F4
Shire (riv.), Malawi 115/G7
Shire (riv.), Mozambique 118/E3
Shiretoko (cape), Japan 81/M1
Shiriya (cape), Japan 81/K3
Shir Kuh (mt.), Iran 59/F3
Shir Kuh (mt.), Iran 66/J5
Shirland, Ill. (61079) 222/D1
Shirley, Ark. (72153) 202/F2
Shirley, Ill. (61772) 222/E3
Shirley, Ind. (47384) 227/F5
Shirley, Mass. (01464) 249/H2
Shirley○, Mass. (01464) 249/H2
Shirley, Mo. (†63664) 261/L7
Shirley, W. Va. (26434) 312/E4
Shirley (basin), Wyo. 319/F3
Shirley Basin, Wyo. (82615) 319/F3
Shirley Center, Mass. (01465) 249/H2
Shirley City (Woodburn), Ind. (†46797) 227/H2
Shirley Mills, Maine (04485) 243/D5
Shirley Mills○, Maine (04485) 243/D5
Shirleysburg, Pa. (17260) 294/G5
Shiro, Texas (77876) 303/J7
Shiroishi, Japan 81/K4
Shirvan, Iran 59/G2
Shirvan, Iran 66/K2
Shirvan (riv.), Iran 66/E3
Shishaldin (vol.), Alaska 196/E4
Shishmaref, Alaska (99772) 196/E1
Shithatha, Iraq 59/D3
Shithatha, Iraq 66/C4
Shitike (creek), Oreg. 291/F3
Shiukwan (Shaoyuan), China 77/H7
Shively, Calif. (†95565) 204/B3
Shively, Ky. (40216) 237/K4
Shivers, Miss. (39164) 256/E7
Shivpuri, India 68/D3
Shivwits (plat.), Ariz. 198/B2
Shivwits Ind. Res., Utah 304/A6
Shiyan, China 77/H5
Shizuishan (Shihsuishan), China 77/G4
Shizunai, Japan 81/L2
Shizuoka (pref.), Japan 81/H6
Shizuoka, Japan 54/P6
Shizuoka, Japan 81/H6
Shkodër, Albania 7/G4
Shkodër, Albania 45/D5
Shoa (prov.), Ethiopia 111/G6
Shoal (riv.), Fla. 212/C6
Shoal (creek), Ill. 222/D5
Shoal (lake), Manitoba 179/B4
Shoal (lake), Manitoba 179/G5
Shoal (riv.), Manitoba 179/B2
Shoal (bay), Newf. 166/D2
Shoal (bay), Nova Scotia 168/F4
Shoal (creek), Tenn. 237/F10
Shoal (creek), Utah 304/A6
Shoal Branch, Wading (riv.), N.J. 273/D4
Shoal Cove, Newf. 166/C3
Shoal Harbour, Newf. 166/C2
Shoalhaven (riv.), N.S. Wales 97/E4
Shoal Lake, Manitoba 179/B4
Shoals, Ind. (47581) 227/D7
Shoals (isls.), N.H. 268/F6
Shoals, N.C. (†27043) 281/J2
Shoals, W. Va. (25562) 312/B6
Shoals Junction, S.C. (29638) 296/C3
Shoalwater (bay), Queensland 88/J4
Shoalwater (cape), Wash. 310/A4
Shoalwater Ind. Res., Wash. 310/B4
Shobara, Japan 81/F6
Shobonier, Ill. (62885) 222/D5
Shock, W. Va. (26638) 312/D5
Shoemakersville, Pa. (19555) 294/K4
Shoffner, Ark. (†72112) 202/H2
Shohola, Pa. (18458) 294/N3
Sholapur, India 54/J8
Sholapur, India 68/D5
Sholes, Nebr. (†68771) 264/G2
Shona (isl.), Scotland 15/C4
Shongaloo, La. (71072) 238/D1
Shonkin, Mont. (59476) 262/F3
Shonto, Ariz. (86054) 198/E2
Shonto (plat.), Ariz. 198/E2
Shook, Mo. (63963) 261/M8
Shooting Creek, N.C. (†28904) 281/B4
Shopiere, Wis. (†53525) 317/H10
Shop Springs, Tenn. (†37184) 237/J8
Shorapur, India 68/D5
Shoreacres, Br. Col. 184/J5
Shore Acres, N.J. (†08723) 273/E3
Shore Acres, Texas (†77571) 303/K2
Shoreham, Mich. (†49085) 250/C6
Shoreham, Minn. (†56501) 255/C4
Shoreham, N.Y. (11786) 276/P8
Shoreham○, Vt. (05770) 268/A4
Shoreham-by-Sea, England 13/G7
Shoreham-by-Sea, England 10/F5
Shoreview, Minn. (†55112) 255/G5
Shorewood, Ill. (60435) 222/E2
Shorewood, Minn. (†55331) 255/F5
Shorewood, Wis. (53211) 317/M1
Shorewood Hills, Mich. (†49125) 250/C7
Shorewood Hills, Wis. (†53701) 317/G9
Short, Okla. (†72955) 288/S3
Short Beach, Conn. (†06405) 210/D3
Short Creek, Ohio (43989) 284/J5
Shortdale, Manitoba 179/A3
Shorter, Ala. (36075) 195/G6
Shorterville, Ala. (36373) 195/H7
Short Falls, N.H. (†03234) 268/E5
Short Hills, N.J. (07078) 273/E2
Shortland (isls.), Solomon Is. 86/D2
Shortleaf, Ala. (†36732) 195/C6
Shortsville, N.Y. (14548) 276/F5
Shoshone, Calif. (92384) 204/J8
Shoshone (co.), Idaho 220/B2
Shoshone, Idaho (83352) 220/D7
Shoshone (falls), Idaho 220/D7
Shoshone (mt.), Nev. 266/E6
Shoshone (mts.), Nev. 266/D3
Shoshone (range), Nev. 266/E2
Shoshone (lake), Wyo. 319/B1
Shoshone (riv.), Wyo. 319/D1
Shoshong, Botswana 118/D4
Shoshoni, Wyo. (82649) 319/D2
Shostka, U.S.S.R. 52/D4
Shotley, England 13/J6
Shotts, Scotland 15/C2
Shouldice, Alberta 182/D4
Shoultes, Wash. (†98270) 310/C2
Shouns, Tenn. (†37683) 237/T8
Shoup, Idaho (83469) 220/D4
Shoval, Israel 65/B5
Shovel Lake, Minn. (†55785) 255/E4
Showak, Sudan 111/G5
Showell, Md. (21862) 245/T7
Show Low, Ariz. (85901) 198/F4
Shoyna, U.S.S.R. 52/F1
Shpola, U.S.S.R. 52/D5
Shreve, Ohio (44676) 284/F4
Shreveport, La. 146/J6
Shreveport, La. 188/H4
Shreveport, La. (*71101) 238/C2
Shrewsbury, England 13/E5
Shrewsbury, England 10/E4
Shrewsbury○, Mass. (01545) 249/H3
Shrewsbury, Mo. (†63101) 261/P3
Shrewsbury, N.J. (07701) 273/E3
Shrewsbury, Pa. (17361) 294/J6
Shrewsbury○, Vt. (†05738) 268/B4
Shrule, Ireland 17/C4
Shuangcheng, China 77/L2
Shuangliao, China 77/K3
Shuangyashan, China 77/M2
Shubenacadie, Nova Scotia 168/E3
Shubenacadie (lake), Nova Scotia 168/E4
Shubenacadie (riv.), Nova Scotia 168/E3
Shubert, Nebr. (68437) 264/J4
Shubuta, Miss. (39360) 256/G7
Shue (creek), S. Dak. 298/N5
Shu'eib, Wadi (dry riv.), Jordan 65/D4
Shueyville, Iowa (†52401) 229/K5
Shu'fat, West Bank 65/C4
Shuicheng, China 77/G6
Shuksan (mt.), Wash. 310/D2
Shulan, China 77/L3
Shulerville, S.C. (29480) 296/H5
Shullsburg, Wis. (53586) 317/F10
Shumagin (isls.), Alaska 196/G4
Shumen, Bulgaria 45/H4
Shumerlya, U.S.S.R. 52/G3
Shumway, Ill. (62461) 222/E4
Shunat Nimrin, Jordan 65/D4
Shunchang, China 77/J6
Shungnak, Alaska (99773) 196/G1
Shungopavy (Shongopovi), Ariz. (†86043) 198/E3
Shunk, Pa. (17768) 294/J2
Shunock (riv.), Conn. 210/H3
Shuo Xian, China 77/H4
Shuqaiq, Saudi Arabia 59/D6
Shuqra, P.D.R. Yemen 59/E7
Shuqualak, Miss. (39361) 256/G5
Shur (riv.), Iran 66/J7
Shusf, Iran 66/L5
Shush, Iran 66/F4
Shushan, N.Y. (12873) 276/O4
Shushenskoye, U.S.S.R. 48/K4
Shushtar, Iran 66/F4
Shushtar, Iran 59/E3
Shuswap (lake), Br. Col. 162/E5
Shuswap (lake), Br. Col. 184/H4
Shutesbury○, Mass. (01072) 249/E3
Shuttle Meadow (res.), Conn. 210/D2
Shutty Bench, Br. Col. 184/J5
Shuweika, West Bank 65/C3
Shuya, U.S.S.R. 52/F3
Shuyak (isl.), Alaska 196/H3
Shwebo, Burma 72/B2
Shwegyin, Burma 72/C3
Shwell (riv.), Burma 72/C2
Shwenyaung, Burma 72/C2
Shyok, India 68/D2
Si (riv.), China 54/N7
Siahan (mts.), Pakistan 59/H4
Siahan (range), Pakistan 68/A3
Siah Kuh (mt.), Iran 66/L3
Siak (riv.), Indonesia 85/C5
Siaksriinderapura, Indonesia 85/C5
Siakwan (Xiaguan), China 77/E6
Sialkot, Pakistan 68/C2
Sialkot, Pakistan 59/K3
Siam (Thailand) (gulf), Thailand 72/D5
Sian (Xi'an), China 77/G5
Siangfan (Xiangfan), China 77/H5
Siangtan (Xiangtan), China 77/H6
Siapa (riv.), Venezuela 124/E7
Siargao (isl.), Philippines 85/H4
Siargao (isl.), Philippines 82/F6
Siasconset, Mass. (02564) 249/P7
Siasi, Philippines 82/C8
Siasi (isl.), Philippines 82/C8
Siátista, Greece 45/E5
Siaton, Philippines 82/D6
Siaton (pt.), Philippines 82/D6
Siau (isl.), Indonesia 85/H5
Šiauliai, U.S.S.R. 7/G3
Šiauliai, U.S.S.R. 53/B3
Šiauliai, U.S.S.R. 52/B3
Šiauliai, U.S.S.R. 48/C4
Sib, Iran 66/N7
Sibalom, Philippines 82/C5
Sibanicú, Cuba 158/G3
Sibay (isl.), Philippines 82/C5
Sibay, U.S.S.R. 52/J4
Sibbald, Alberta 182/E4
Šibenik, Yugoslavia 45/C4
Siberia, Ind. (47582) 227/D8
Siberia (reg.), U.S.S.R. 4/C2
Siberia (reg.), U.S.S.R. 2/P2
Siberia (reg.), U.S.S.R. 54/M4
Siberia (reg.), U.S.S.R. 48/M3
Sibert, Ky. (†40962) 237/O6
Siberut (isl.), Indonesia 54/L10
Siberut (isl.), Indonesia 85/B6
Siberut (str.), Indonesia 85/B6
Sibi, Pakistan 68/B3
Sibi, Pakistan 59/J4
Sibiti, Congo 115/B4
Sibiu, Romania 7/G4
Sibiu, Romania 45/G3
Sibley, Ill. (61773) 222/E3
Sibley, Iowa (51249) 229/B2
Sibley, La. (71073) 238/D1
Sibley (co.), Minn. 255/D6
Sibley, Miss. (39165) 256/B8
Sibley, Mo. (64088) 261/S5
Sibley, N. Dak. (†58429) 282/P5
Sibley Prov. Park, Ontario 175/C3
Sibley Prov. Park, Ontario 177/H5
Sibolga, Indonesia 85/B5
Siboney, Cuba 158/J4
Sibsagar, India 68/H3
Sibu, Malaysia 85/E5
Sibu, Malaysia 54/N9
Sibube, C. Rica 154/F6
Sibuco, Philippines 82/C7
Sibuguey (bay), Philippines 82/D7
Sibundoy, Colombia 126/B7
Sibut, Cent. Afr. Rep. 115/C2
Sibutu (passage), Philippines 85/F4
Sibutu (passage), Philippines 82/B8
Sibutu Group (isls.), Philippines 82/B8
Sibuyan (isl.), Philippines 85/G3
Sibuyan (isl.), Philippines 82/D4
Sibuyan (sea), Philippines 82/D4
Sibuyan (sea), Philippines 85/G3
Sicamous, Br. Col. 184/H5
Sicasica, Bolivia 136/B5
Siccus (riv.), S. Australia 88/F6
Sichuan (Szechwan) (prov.), China 77/F5
Sicily (reg.), Italy 34/D6
Sicily (isl.), Italy 7/F5
Sicily (isl.), Italy 34/E6
Sicily (str.), Italy 34/D6
Sicily Island, La. (71368) 238/G3
Sicklerville, N.J. (08081) 273/D4
Sico (riv.), Honduras 154/E3
Sicuani, Peru 128/G10
Šid, Yugoslavia 45/D3
Sidamo (prov.), Ethiopia 111/G7
Siddipet, India 68/D5
Sideby, Finland 18/M5
Side Lake, Minn. (55781) 255/E3
Sidell, Ill. (61876) 222/F4
Siderno, Italy 34/F5
Sidewood, Sask. 181/C5
Sídheros (cape), Greece 45/H8
Sidhi, India 68/E4
Sidhirókastron, Greece 45/F5
Sidhpur, India 68/C4
Sidi Barrani, Egypt 111/E1
Sidi Barrani, Egypt 59/A3
Sidi Bel-Abbes, Algeria 106/D1
Sidi Bel Abbes, Algeria 102/C1
Sidi Kacem, Morocco 106/C2
Siding Springs, N.S. Wales 97/E2
Sidlaw (hills), Scotland 15/E4
Sidley (mt.) 5/B12
Sidmouth, England 13/D7
Sidmouth, England 10/E5
Sidmouth (cape), Queensland 95/C2
Sidnaw, Mich. (49961) 250/G2
Sidney, Ark. (72577) 202/G1
Sidney, Br. Col. 184/K3
Sidney, Ill. (61877) 222/E3
Sidney, Ind. (46566) 227/F2
Sidney, Iowa (51652) 229/B7
Sidney○, Maine (†04330) 243/D7
Sidney, Manitoba 179/C5
Sidney, Mich. (48885) 250/D5
Sidney, Mont. (59270) 262/M3
Sidney, Nebr. (69162) 264/B3
Sidney, N.Y. (13838) 276/K6
Sidney, Ohio (45365) 284/B5
Sidney Center, N.Y. (13839) 276/K6
Sidney Lanier (lake), Georgia 217/D2
Sidoharjo, Indonesia 85/K2
Sidon, Ark. (†72137) 202/G3
Sidon (Saida), Lebanon 63/F6
Sidon, Miss. (38954) 256/D4
Sidonia, Tenn. (†38255) 237/D8
Sidra (gulf) 102/D1
Sidra (gulf), Libya 111/C1
Siedlce (prov.), Poland 47/F2
Siedlce, Poland 47/F2
Siegas, New Bruns. 170/C1
Siegburg, W. Germany 22/B3
Siegen, W. Germany 22/C3
Siemianowice Śląskie, Poland 47/B4
Siemiatycze, Poland 47/F2
Siempang, Cambodia 72/E4
Siemreab, Cambodia 72/D4
Siena (prov.), Italy 34/C3
Siena, Italy 34/C3
Sienyang (Xianyang), China 77/G5
Sieper, La. (71472) 238/E4
Sieradz (prov.), Poland 47/D3

Sieradz, Poland 47/D3
Sierning, Austria 41/C2
Sierpc, Poland 47/D2
Sierra (co.), Calif. 204/E4
Sierra (co.), N. Mex. 274/B5
Sierra Ancha (mts.), Ariz. 198/E5
Sierra Apache (mts.), Ariz. 198/E5
Sierra Army Depot, Calif. 204/E3
Sierra Blanca, Texas 303/B11
Sierra Blanca (peak), N. Mex. 274/C5
Sierra City, Calif. (96125) 204/E4
Sierra Colorada, Argentina 143/C5
Sierra Diablo (mts.), Texas 303/C10
Sierra Gorda, Chile 138/B4
Sierra Grande, Argentina 143/C5
Sierra Leone 2/J5
Sierra Leone 102/A4
SIERRA LEONE 106/B7
Sierra Madre, Calif. (91024) 204/D10
Sierra Madre (mts.), Wyo. 319/E4
Sierra Madre (mts.), Philippines 82/D2
Sierra Madre Occidental (mts.), Mexico 146/H7
Sierra Madre Oriental (mts.), Mexico 146/H7
Sierra Mojada, Mexico 150/H3
Sierra Morena (mts.), Spain 7/D5
Sierra Nevada (mts.) 188/B3
Sierra Nevada (range), Calif. 146/F6
Sierra Nevada (mts.), Calif. 204/E4
Sierra Vieja (mts.), Texas 303/C11
Sierraville, Calif. (96126) 204/E4
Sierra Vista, Ariz. (85635) 198/E7
Sierre, Switzerland 39/G4
Siesta Key, Fla. (†33578) 212/D4
Siete Cerros, Uruguay 145/E5
Sífnos (isl.), Greece 45/G7
Sifton, Manitoba 179/B3
Sigean, France 28/E6
Sigel, Ill. (62462) 222/E4
Sigel, Pa. (15860) 294/D3
Sighetu Marmaţiei, Romania 45/F2
Sighişoara, Romania 45/G2
Sigiriya, Sri Lanka 68/E7
Sigli, Indonesia 85/B4
Siglufjördhur, Iceland 21/C1
Sigmaringen, W. Germany 22/C4
Sigmundsherberg, Austria 41/C2
Signal Hill, Calif. (†90806) 204/C11
Signal Mountain, Tenn. (37377) 237/L10
Signau, Switzerland 39/E3
Signy, France 28/F6
Sigourney, Iowa (52591) 229/J6
Sigriswil, Switzerland 39/E3
Sigsig, Ecuador 128/C4
Sigtuna, Sweden 18/H1
Siguanea (bay), Cuba 158/B2
Siguatepeque, Honduras 154/D3
Sigüe, Ecuador 128/D2
Sigüenza, Spain 33/E2
Siguiri, Guinea 106/C6
Sigulda, U.S.S.R. 53/C2
Sigurd, Utah (84657) 304/B5
Sihlsee (lake), Switzerland 39/G2
Sihuas, Peru 128/D7
Siikajoki (riv.), Finland 18/O4
Siipyy (Sideby), Finland 18/M5
Siirt (prov.), Turkey 63/J4
Siirt, Turkey 63/J4
Siirt, Turkey 59/D2
Sikanni Chief (riv.), Br. Col. 184/F1
Sikar, India 68/D3
Sikasso, Mali 106/C6
Sikasso, Mali 102/B3
Sikes, La. (71473) 238/F2
Sikeston, Mo. (63801) 261/N9
Sikhote-Alin' (range), U.S.S.R. 54/P5
Sikhote-Alin' (range), U.S.S.R. 48/O5
Sikkim (state), India 68/G3
Siklós, Hungary 41/E4
Siktyakh, U.S.S.R. 48/N3
Sil (riv.), Spain 33/C1
Silao, Mexico 150/J6
Silas, Ala. (36919) 195/B7
Silat Dharr, West Bank 65/C3
Silay, Philippines 82/D5
Silchar, India 68/G4
Şile, Turkey 63/C2
Silenen, Switzerland 39/G3
Siler City, N.C. (27344) 281/L3
Silerton, Tenn. (38377) 237/D10
Silesia, Mont. (59080) 262/H5
Silet, Algeria 106/E4
Siletz, Oreg. (97380) 291/D3
Silex, Mo. (63377) 261/K4
Silhouette (isl.), Seychelles 118/H5
Silica, Kansas (†67526) 232/D3
Silifke, Turkey 59/B2
Silifke, Turkey 63/E4
Siliguri, India 68/F3
Siling Co (lake), China 77/C5
Silinhot (Abnagar), China 77/J3
Silisili (mt.), W. Samoa 86/L8
Silistra, Bulgaria 45/H3
Silivri, Turkey 63/C2
Siljan (lake), Sweden 18/J6
Silkeborg, Denmark 21/C5
Silkeborg, Denmark 18/F8
Sillajguay, Cordillera (mt.), Chile 138/D2
Sillajhuay, Cordillera (mt.), Bolivia 136/A6
Sillamäe, U.S.S.R. 53/D1
Silleda, Spain 33/B1
Sillery, Québec 172/J3
Sillian, Austria 41/B3
Sillikers, New Bruns. 170/E2
Silloth, England 13/D3
Silo, Okla. (†47701) 288/N6
Siloam, Georgia (30665) 217/F3
Siloam, Ky. (†41175) 237/R2
Siloam, N.C. (27047) 281/H2
Siloam Springs, Ark. (72761) 202/B1
Siloam Springs, Mo. (†65775) 261/H9
Silopi, Turkey 63/K4
Silsbee, Texas (77656) 303/K7
Sils im Domleschg, Switzerland 39/H3
Silt, Colo. (81652) 208/D4

Siltcoos, Oreg. (†97441) 291/C4
Silton, Sask. 181/G5
Silup (riv.), Iran 66/M8
Šilute, U.S.S.R. 53/A3
Silva, Mo. (63964) 261/M8
Silva, N. Dak. (58375) 282/L3
Silvan, Turkey 63/J3
Silvana, Wash. (98287) 310/C2
Silvânia, Brazil 132/D7
Silvaplana, Switzerland 39/J4
Silvassa, India 68/C4
Silver (creek), Ariz. 198/E4
Silver (creek), Ill. 222/B2
Silver (creek), Ind. 227/F8
Silver, Manitoba 179/E4
Silver, Newf. 166/A3
Silver (lake), Mass. 249/L4
Silver, Newf. 166/A3
Silver (lake), N.Y. 276/N1
Silver (creek), Oreg. 291/F4
Silver (creek), Oreg. 291/H4
Silver (lake), Oreg. 291/G4
Silver (lake), Oreg. 291/G4
Silver (lake), Oreg. 291/H7
Silver (bank), Turks & Caicos 156/E2
Silver (lake), Wash. 310/C4
Silverado, Calif. (92676) 204/E11
Silver Bank (passage), Turks & Caicos 156/E2
Silver Bay, Minn. (55614) 255/G3
Silver Beach, Alberta 182/D2
Silver Bell, Ariz. (85270) 198/D6
Silver Bow (riv.), Mont. 262/D5
Silver City, Georgia (†30501) 217/D2
Silver City, Idaho (†83650) 220/B6
Silver City, Iowa (51571) 229/B6
Silver City, Miss. (39166) 256/C4
Silver City, N. Mex. 188/E4
Silver City, N. Mex. (88061) 274/A6
Silver City, S. Dak. (57781) 298/B5
Silver Cliff, Colo. (81249) 208/J6
Silver Creek, Georgia (30173) 217/B2
Silver Creek, Minn. (55380) 255/D5
Silver Creek, Miss. (39663) 256/C7
Silver Creek, Nebr. (68663) 264/G3
Silver Creek, N.Y. (14136) 276/B5
Silver Creek, Wash. (98585) 310/C4
Silverdale, Kansas (†67005) 232/F4
Silverdale, N.C. (†28539) 281/P5
Silverdale, Wash. (98383) 310/A2
Silver Dollar City, Mo. (65616) 261/F9
Silver Grove, Ky. (41085) 237/T2
Silverhill, Ala. (36576) 195/C9
Silverhill, Ky. (41467) 237/R5
Silver Hill, W. Va. (†26155) 312/E3
Silver Hill-Suitland, Md. (20023) 245/F5
Silver Island (mts.), Utah 304/A3
Silver Lake, Newf. (46982) 217/F4
Silver Lake, Kansas (66539) 232/G2
Silver Lake, Mass. (†02364) 249/C5
Silver Lake, Minn. (55381) 255/D6
Silver Lake, Mo. (†63775) 261/M7
Silver Lake, N.H. (03875) 268/E4
Silver Lake, Ohio (†44221) 284/G3
Silver Lake, Oreg. (97638) 291/F4
Silverlake, Wash. (98645) 310/C4
Silver Lake, Wis. (53170) 317/K10
Silverleaf, N. Dak. (†58436) 282/D7
Silvermine (riv.), Conn. 210/B4
Silvermine (mts.), Ireland 17/E6
Silver Park, Sask. 181/G3
Silverpeak, Nev. (89047) 266/D5
Silver Peak (range), Nev. 266/D5
Silver Plume, Colo. (80476) 208/H3
Silver Point, Tenn. (38582) 237/K8
Silver Run, Ala. (†36268) 195/G3
Silver Run, Md. (†21157) 245/K2
Silvers Mills, Maine (†04930) 243/E5
Silver Spring, Md. (*20901) 245/F4
Silver Spring, Md. (*20901) 245/F4
Silver Springs, Fla. (32688) 212/D2
Silver Springs, N.Y. (14550) 276/C5
Silver Star, Mont. (59751) 262/D5
Silver Star Prov. Park, Br. Col. 184/H5
Silverstreet, S.C. (29145) 296/D3
Silverthorne, Colo. (80498) 208/G3
Silverton, Br. Col. 184/J5
Silverton, Colo. (81433) 208/J7
Silverton, N.S. Wales 97/A2
Silverton, Ohio (†45236) 284/C9
Silverton, Oreg. (97381) 291/B3
Silverton, Texas (79257) 303/C3
Silverton, Wash. (98252) 310/D2
Silverton, W. Va. (†26155) 312/C5
Silverton Station, Manitoba 179/A4
Silver Valley, Alberta 182/A1
Silverville, Ind. (†47470) 227/D7
Silverwood, Ind. (†47952) 227/C5
Silverwood, Mich. (48760) 250/F5
Silver Water, Ontario 177/B2
Silves, Portugal 33/B4
Silvia, Colombia 126/B6
Silvies, Oreg. (†97720) 291/H3
Silvies (riv.), Oreg. 291/H4
Silvis, Ill. (61282) 222/C2
Silvretta (mts.), Switzerland 39/K3
Sim (cape), Morocco 106/B3
Simanggang, Malaysia 85/E5
Simao (Fusingchen), China 77/F7
Simara (isl.), Philippines 82/C3
Simaraña, Venezuela 124/G5
Simav, Turkey 63/C3
Simav (riv.), Turkey 63/C3
Simcoe, N. Dak. (†58741) 282/J3
Simcoe (county), Ontario 177/E3
Simcoe, Ontario 177/D5
Simcoe (lake), Ontario 177/E3
Simen (mts.), Ethiopia 111/G5
Simeonovgrad (Maritsa), Bulgaria 45/H4
Simeto (riv.), Italy 34/E6
Simeulue, Indonesia 54/L9
Simeulue (isl.), Indonesia 85/A5
Simferopol', U.S.S.R. 57/H4
Simferopol', U.S.S.R. 52/D6
Simferopol', U.S.S.R. 48/D5
Sími, Greece 45/H7
Sími (isl.), Greece 45/H7

Simikot, Nepal 68/E3
Similkameen (riv.), Wash. 310/F1
Simiti, Colombia 126/C4
Simi Valley, Calif. (†93065) 204/G9
Singora (Songkhla), Thailand 72/C6
Simla, Colo. (80835) 208/M4
Simla, India 68/D2
Şimleu Silvaniei, Romania 45/F2
Simme (riv.), Switzerland 39/D3
Simmern, W. Germany 22/B4
Simmesport, La. (†13960) 238/G5
Simmie, Sask. 181/C6
Simmons, Mo. (†65689) 261/H8
Simms, Mont. (59477) 262/E3
Simnasho, Oreg. (†97761) 291/F3
Simojärvi (lake), Finland 18/P3
Simojoki (riv.), Finland 18/O4
Simojovel de Allende, Mexico 150/N8
Simon (lake), Québec 172/F3
Simonds, New Bruns. 170/C2
Simonette (riv.), Alberta 182/A2
Simonstown, S. Africa 118/E7
Simonsville, Vt. (†05143) 268/B5
Simontornya, Hungary 41/E3
Simoom Sound, Br. Col. 184/D5
Simpelveld, Netherlands 27/H7
Simplício Mendes, Brazil 132/F4
Simplon, Switzerland 39/F4
Simplon (pass), Switzerland 39/F4
Simplon (tunnel), Switzerland 39/F4
Simpson (riv.), Chile 138/C6
Simpson, Ill. (62985) 222/E6
Simpson (co.), Ky. 237/H7
Simpson, Ky. (†41301) 237/P5
Simpson (prov.), Turkey 63/F2
Simpson (co.), Miss. 256/E7
Simpson, Mont. (†59501) 262/H2
Simpson (des.), North. Terr. 88/F4
Simpson (des.), North. Terr. 93/E8
Simpson (des.), N.W. Terrs. 187/K3
Simpson, Pa. (18407) 294/L2
Simpson (des.), Queensland 95/A5
Simpson, Sask. 181/F4
Simpson (des.), S. Australia 94/E1
Simpson, W. Va. (26435) 312/F4
Simpson Park (mts.), Nev. 266/E3
Simpsons, Va. (24072) 307/H6
Simpsonville, Ky. (40067) 237/L4
Simpsonville, S.C. (29681) 296/C2
Simrishamn, Sweden 18/J9
Sims, Ark. (71969) 202/C4
Sims, Ill. (62886) 222/E5
Sims (lake), Newf. 166/A3
Sims, N.C. (27880) 281/N3
Sims Chapel, Ala. (†36553) 195/B8
Simunul (isl.), Philippines 82/B8
Sinabang, Indonesia 85/B5
Sinai (mt.), Egypt 111/F2
Sinai (pen.), Egypt 111/F2
Sinai (pen.), Egypt 59/B4
Sinai (mt.), Grenada 161/G7
Sinai, S. Dak. (57061) 298/P5
Sinaia, Romania 45/G3
Sinait, Philippines 82/C2
Sinajana, Guam 86/K7
Sinaloa (state), Mexico 150/F4
Sinaloa de Leyva, Mexico 150/E4
Sinamaica, Venezuela 124/B2
Sinanju, Libya 111/B1
Sincanli, Turkey 63/D3
Sincé, Colombia 126/C3
Sincelejo, Colombia 126/C3
Sinch'ŏn, N. Korea 81/B4
Sinchu, China 77/K7
Sinclair (lake), Georgia 217/F4
Sinclair, Maine (04779) 243/G1
Sinclair (head), N. Zealand 100/A3
Sinclair, Wyo. (82334) 319/E4
Sinclair Mills, Br. Col. 184/G3
Sinclair Station, Manitoba 179/A5
Sinclairville, N.Y. (14782) 276/B6
Sind (prov.), Pakistan 68/B3
Sind (reg.), Pakistan 59/J4
Sindal, Denmark 21/D3
Sindangan, Philippines 82/D6
Sindangan (bay), Philippines 82/D6
Sindangbarang, Indonesia 85/D7
Sindelfingen, W. Germany 22/C4
Sindi, U.S.S.R. 53/C1
Sındırgı, Turkey 63/C3
Sines, Portugal 33/B4
Sines (cape), Portugal 33/B4
Sinewit (mt.), Papua N.G. 86/B2
Sinfra, Ivory Coast 106/C7
Singa, Sudan 111/F5
Singa, Sudan 102/F3
Singa, Sudan 59/B7
Singac, N.J. (†07424) 273/B2
Singapore 2/Q5
Singapore 54/M9
Singapore (str.), Malaysia 72/F6
SINGAPORE 85/C5
Singapore (cap.), Philippines 85/C5
SINGAPORE 72
Singapore (str.), Singapore 72/F6
Singapore (cap.), Singapore 72/F6
Singaraja, Indonesia 85/F7
Sing Buri, Thailand 72/C4
Singen, W. Germany 22/C4
Singer, La. (70660) 238/D5
Singers Glen, Va. (22850) 307/K3
Şiran, Turkey 63/H2
Singhampton, Ontario 177/D3
Singida (reg.), Tanzania 115/F5
Singida, Tanzania 115/F4
Singida, Tanzania 102/F4
Singkaling Hkamti, Burma 72/B1
Singkang, Indonesia 85/G6
Singkawang, Indonesia 85/D5

Singkep (isl.), Indonesia 85/D6
Singleton, N.S. Wales 97/F3
Singleton (mt.), North. Terr. 93/B6
Singora (Songkhla), Thailand 72/C6
Singsås, Norway 18/G5
Singtai (Xingtai), China 77/H4
Singur, Burma 72/C1
Singu, Burma 72/B2
Sinhalien (Lianyungang), China 77/J5
Sining (Xining), China 77/F4
Siniscola, Italy 34/B4
Sinj, Yugoslavia 45/C4
Sinjai, Indonesia 85/G7
Sinjar, Iraq 66/B2
Sinjar, Jebel (mts.), Iraq 66/B2
Sinjar, Jebel (mts.), Syria 63/J4
Sinjil, West Bank 65/C3
Sinkat, Sudan 59/C6
Sinkat, Sudan 111/G4
Sinkiang (reg.), China 54/K5
Sinkiang (reg.), China 2/P3
Sinkiang-Uigur Aut. Reg. (Xinjiang Uygur), China 77/B3
Sinking Spring, Ohio (45172) 284/D7
Sinking Spring, Pa. (19608) 294/K5
Sinks Grove, W. Va. (24976) 312/F7
Sinlumkaba, Burma 72/C1
Sinnai, Italy 34/B5
Sinnamahoning, Pa. (15861) 294/G3
Sinnamary, Fr. Guiana 131/E3
Sinnamary (riv.), Fr. Guiana 131/E3
Sinneh (Sanandaj), Iran 66/F3
Sinnemahoning (creek), Pa. 294/F3
Sinnett, Sask. 181/G4
Sinnicolaui Mare, Romania 45/E2
Sinnúris, Egypt 111/J3
Sinoia, Zimbabwe 118/D3
Sinop (prov.), Turkey 63/F2
Sinop, Turkey 59/C1
Sinop, Turkey 63/F2
Sinop (cape), Turkey 63/F1
Sinp'o, N. Korea 81/D4
Sins, Switzerland 39/F2
Sinsiang (Xinxiang), China 77/H4
Sintaluta, Sask. 181/H5
Sintang, Indonesia 85/E5
Sint Anna, Neth. Ant. 161/D10
Sint Anna (bay), Neth. Ant. 161/F9
Sint Annaland, Netherlands 27/E5
Sint Christoffel (mt.), Neth. Ant. 161/F8
Sint Jacobiparochie, Netherlands 27/H2
Sint Jan, Neth. Ant. 161/D8
Sint Joris (bay), Neth. Ant. 161/G9
Sint Kruis, Neth. Ant. 161/F8
Sint-Laureins, Belgium 27/D6
Sint Maarten (Saint Martin) (isl.), Neth. Ant. 156/F3
Sint Martha, Neth. Ant. 161/F8
Sint Michiel, Neth. Ant. 161/F9
Sint Nicolaas, Neth. Ant. 161/E10
Sint-Niklaas, Belgium 27/E6
Sint-Pieters-Leeuw, Belgium 27/B9
Sint-Truiden, Belgium 27/G7
Sint Willebrordus, Neth. Ant. 161/F8
Sinú (riv.), Colombia 126/B3
Sinuapa, Honduras 154/D2
Sinŭiju, N. Korea 81/B3
Sinyang (Xinyang), China 77/H5
Siocon, Philippines 82/D7
Siocon (Siokun), Philippines 85/G4
Sió csatorna (canal), Hungary 41/E3
Siófok, Hungary 41/E3
Sion, Switzerland 39/D4
Sion Mills, N. Ireland 17/G2
Siparia, Trin. & Tob. 156/G5
Siparia, Trin. & Tob. 161/B11
Sipacate, Guatemala 154/B4
Sipalay, Philippines 82/D6
Sipaliwini (riv.), Suriname 131/C4
Sipapo (riv.), Venezuela 124/E5
Siping (Szeping), China 77/K3
Sipiwesk (lake), Manitoba 179/J3
Siple (mt.), Ant. 5/B12
Sipocot, Philippines 82/D4
Sippola, Finland 18/P6
Sipsey, Ala. (35584) 195/D3
Sipsey (riv.), Ala. 195/B4
Sipsey Fork (riv.), Ala. 195/D2
Sip Song Chau Thai (mts.), Vietnam 72/D7
Sipura (isl.), Indonesia 85/B6
Siquijor (prov.), Philippines 82/D6
Siquijor, Philippines 82/D6
Siquijor (isl.), Philippines 82/D6
Siquirres, C. Rica 154/F5
Siracusa (Syracuse), Italy 34/E6
Sirajganj, Bangladesh 68/F4
Siran, Turkey 63/H2
Singer, La. (70660) 238/D5
Sirdar, Br. Col. 184/J5
Sir Edward Pellew Group (isls.), North. Terr. 88/F3
Sir Edward Pellew Group (isls.), North. Terr. 93/F3
Siren, Wis. (54872) 317/B4
Siret (riv.) 7/G4
Siret, Romania 45/G1
Siret (riv.), Romania 45/H2

Siret (riv.), Romania 45/H2
Sir Francis Drake (chan.), Virgin Is. (Br.) 161/F4
Sirhan, Wadi (dry riv.), Saudi Arabia 59/C3
Sirik, Iran 66/K7
Sirik (cape), Malaysia 85/E5
Siris, West Bank 65/C3
Sirius (pt.), Alaska 196/J4
Sir James MacBrien (mt.), N.W. Terrs. 187/F3
Sirjan (Sa'idabad), Iran 66/J6
Sirjan (Sa'idabad), Iran 66/J6
Sir John's (peak), Jamaica 158/K6
Sir Johns Run, W. Va. (†25411) 312/K3
Sirnach, Switzerland 39/G2
Sırnak, Turkey 63/K4
Sirohi, India 68/C3
Sironj, India 68/D4
Siros, Greece 45/G7
Sironj, India 68/D4
Sirri (isl.), Iran 66/J8
Sirsa, India 68/D3
Sir Sandford (mt.), Br. Col. 184/H4
Sirsi, India 68/D6
Siruma, Philippines 82/D3
Şırvan, Turkey 63/J3
Sirvintos, U.S.S.R. 53/C3
Sisak, Yugoslavia 45/C3
Sisaket, Thailand 72/E4
Sisal, Mexico 150/O6
Sishen, S. Africa 118/C5
Sisi (cape), Indonesia 85/F6
Sisib (lake), Manitoba 179/C2
Sisikon, Switzerland 39/G3
Siskiwit (bay), Mich. 250/E1
Siskiyou (co.), Calif. 204/C2
Siskiyou (mts.), Calif. 204/C2
Siskiyou (mts.), Oreg. 291/C5
Sisophon, Cambodia 72/D4
Sissach, Switzerland 39/F2
Sisseton, S. Dak. (57262) 298/R2
Sissiboo (riv.), Nova Scotia 168/B3
Sisson Branch, Tobique (riv.), New Bruns. 170/C1
Sisson Ridge, New Bruns. 170/C2
Sissonville, W. Va. (25320) 312/C5
Sistan and Baluchestan (prov.), Iran 66/M6
Sister Bay, Wis. (54234) 317/M5
Sister Lakes, Mich. (49047) 250/C6
Sisteron, France 28/F5
Sisters, The (isls.), N. Zealand 100/D6
Sisters, Oreg. (97759) 291/F3
Sistersville, W. Va. (26175) 312/D3
Siteki, Swaziland 118/E5
Sites, Calif. (†95979) 204/C4
Sitges, Spain 33/G2
Sithoniá (pen.), Greece 45/F5
Sitía, Greece 45/H8
Sítio d'Abadia, Brazil 132/E6
Sitionuevo, Colombia 126/C2
Sitka, Alaska 146/K4
Sitka, Alaska 188/D6
Sitka, Alaska (99835) 196/M1
Sitka (sound), Alaska 196/M1
Sitka, Alaska (†72482) 202/H1
Sitka, Kansas (†67831) 232/C4
Sitka, U.S. 4/D16
Sitkalidak (isl.), Alaska 196/H3
Sitka Nat'l Hist. Park, Alaska 196/M1
Sitkinak (isl.), Alaska 196/H3
Sitkinak (str.), Alaska 196/H3
Sitkum, Oreg. (†97458) 291/D4
Sittang (riv.), Burma 72/C3
Sittard, Netherlands 27/H6
Sittingbourne and Milton, England 13/H6
Sittwe, Burma 54/L7
Sittwe, Burma 72/B2
Situbondo, Indonesia 85/L2
Siuna, Nicaragua 154/E4
Siuslaw (riv.), Oreg. 291/D4
Sivand, Iran 66/H5
Sivas (prov.), Turkey 63/G3
Sivas, Turkey 54/E5
Sivas, Turkey 63/G3
Sivas, Turkey 59/C1
Sivasli, Turkey 63/C3
Siverek, Turkey 63/H4
Sivirlez, Switzerland 39/C3
Sivry-Rance, Belgium 27/E8
Sivrihisar, Turkey 63/D3
Siwa, Egypt 111/E2
Siwa, Egypt 102/E2
Siwa (oasis), Egypt 111/E2
Sixes, Oreg. (97476) 291/C5
Sixes (riv.), Oreg. 291/C5
Six Flags Over Georgia, Georgia (†30336) 217/L3
Six Flags Over Mid America, Mo. (†63025)261/M4
Six Lakes, Mich. (48886) 250/D5
Six Mile, S.C. (29682) 296/B2
Six Mile, U.S. 195/B4
Six Mile Run, N.J. (†08536) 273/D3
Sixmilebridge, Ireland 17/D6
Sixmilecross, N. Ireland 17/C8
Six Mile Run (Pa. 16679) 294/F5
Six Roads, New Bruns. 170/F1
Six Run (creek), N.C. 281/N4
Sixteen, Mont. (†59642) 262/F4
Sixteen Island Lake, Québec 172/G4
Sixth Cataract, Sudan 111/F4
Sixth Cataract, Sudan 59/B6
Sixth Cataract (dam), Sudan 102/F3
Siyah Kuh (mt.), Iran 66/D2
Siyeh (mt.), Mont. 262/C2
Sizerock, Ky. (41762) 237/P6
Siziwang, China 77/H3
Sjaelland (isl.), Denmark 21/E6
Sjaelland (isl.), Denmark 18/H9
Sjaellands Odde (pen.), Denmark 21/E5
Sjenica, Yugoslavia 45/E4
Skadovsk, U.S.S.R. 52/D5
Skaelskör, Denmark 21/E7

Skærbæk, Denmark 21/B7
Skagen, Denmark 21/D2
Skagen, Denmark 18/G8
Skagens Odde (cape), Denmark 21/D2
Skagens Odde (cape), Denmark 18/G8
Skagerrak (str.) 7/E3
Skagerrak (str.), Denmark 21/C2
Skagerrak (str.), Denmark 18/F8
Skagerrak (str.), Norway 18/F8
Skagerrak (str.), Sweden 18/F8
Skagit (riv.), Br. Col. 184/G6
Skagit (riv.), Wash. 310/C2
Skagway, Alaska 146/K4
Skagway, Alaska (99840) 196/M1
Skalica, Czech. 41/D2
Skals, Denmark 21/C4
Skamania (co.), Wash. 310/D5
Skamania, Wash. (†98648) 310/C5
Skanderborg, Denmark 21/D5
Skanderborg, Denmark 18/F8
Skaneateles, N.Y. (13152) 276/H5
Skaneateles (lake), N.Y. 276/H5
Skanee, Mich. (49962) 250/A2
Skåneviksjöen, Norway 18/E7
Skanör med Falsterbo, Sweden 18/H9
Skara, Sweden 18/H7
Skaraborg (co.), Sweden 18/H7
Skardu, Pakistan 68/D1
Skårup, Denmark 21/D7
Skateraw, Scotland 15/F3
Skaudvile, U.S.S.R. 53/B3
Skaw, The (Skagens Odde) (cape), Denmark 18/G8
Skaw, The (Skagens Odde) (cape), Denmark 21/D2
Skawina, Poland 47/D4
Skedee, Okla. (74069) 288/N2
Skeena (riv.), Br. Col. 162/D5
Skeena (riv.), Br. Col. 184/C2
Skeena (riv.), Br. Col. 184/C3
Skegness, England 13/H4
Skegness, England 10/G4
Skeldon (lake), Ontario 177/E2
Skeleton Coast (reg.), Namibia 118/A3
Skellefteå, Sweden 7/F2
Skellefteå, Sweden 18/M4
Skellefteälv (riv.), Sweden 18/L4
Skellytown, Texas (79080) 303/C2
Skelmersdale, England 10/F2
Skelmersdale, England 13/G2
Skelmorlie, Scotland 15/F2
Skelton, Ind. (†47570) 227/B8
Skelton and Brotton, England 13/G3
Skene, Miss. (38775) 256/C3
Skerries, Ireland 17/J4
Skerries, Ireland 10/H4
Skerries, The (isls.), Wales 13/C4
Ski, Norway 18/D4
Skiathos, Greece 45/F6
Skiatook, Okla. (74070) 288/O2
Skibbereen, Ireland 10/B5
Skibbereen, Ireland 17/C8
Skibby, Denmark 21/E6
Skidaway (isl.), Georgia 217/L7
Skiddaw (mt.), England 13/D3
Skiddy, Kansas (†66872) 232/F3
Skidegate, Br. Col. 184/B3
Skidegate (inlet), Br. Col. 184/B3
Skidmore, Mo. (64487) 261/B2
Skidmore, Texas (78389) 303/G9
Skien, Norway 7/E3
Skien, Norway 18/F7
Skierniewice (prov.), Poland 47/E3
Skierniewice, Poland 47/E2
Skiff, Alberta 182/E5
Skiff (lake), New Bruns. 170/C2
Skikda, Algeria 102/C1
Skikda, Algeria 106/G1
Skilak (lake), Alaska 196/C1
Skillet Fork (riv.), Ill. 222/E5
Skillman, N.J. (08558) 273/D3
Skinners Eddy, Pa. (†18623) 294/K2
Skipness, Scotland 15/F3
Skippack, Pa. (19474) 294/M5
Skippers, Va. (23879) 307/O7
Skipperville, Ala. (36374) 195/G7
Skipsea, England 13/G4
Skipton, England 13/H1
Skipton, England 10/E4
Skipwith, Va. (23968) 307/L7
Skir Dhu, Nova Scotia 168/H2
Skíros, Greece 45/G6
Skíros (Skyros) (isl.), Greece 45/G6
Skive, Denmark 18/F8
Skive, Denmark 21/C4
Skive (riv.), Denmark 21/C5
Skjåk, Norway 18/F6
Skjálfandafljót (stream), Iceland 21/C1
Skjern, Denmark 21/B6
Skodborg, Denmark 21/C7
Škofja Loka, Yugoslavia 45/A2
Skokie, Ill. (*60076) 222/B5
Skokomish (riv.), Wash. 310/B3
Skokomish Ind. Res., Wash. 310/B3
Skomer (isl.), Wales 13/B6
Skookumchuck, Br. Col. 184/K5
Skootamatala (lake), Ontario 177/G3
Skópelos, Greece 45/F6
Skopin, U.S.S.R. 52/F4
Skopje, Yugoslavia 7/G4
Skopje, Yugoslavia 45/E5
Skørping, Denmark 21/C4
Skövde, Sweden 18/H7
Skovorodino, U.S.S.R. 54/O4
Skovorodino, U.S.S.R. 48/N4
Skowhegan, Maine (04976) 243/D6
Skowhegan○, Maine (04976) 243/D6
Skownan, Manitoba 179/C3
Skradin, Yugoslavia 45/B4
Skreia, Norway 18/G6
Skudeneshavn, Norway 18/D7
Skull Valley, Ariz. (86338) 198/C4

Skull Valley Ind. Res., Utah 304/B3
Skuna (riv.), Miss. 256/F2
Skungamaug (riv.), Conn. 210/F1
Skunk (riv.), Iowa 229/K6
Skuteč, Czech. 41/D2
Skutskär, Sweden 18/K6
Skwentna, Alaska (99667) 196/B1
Skwentna (riv.), Alaska 196/A1
Skwierzyna, Poland 47/B2
Skye, Isle of (isl.), Scotland 15/B3
Skye (isl.), Scotland 10/C2
Skykomish, Wash. (98288) 310/D3
Skykomish (riv.), Wash. 310/D3
Skyland, N.C. (28776) 281/D4
Skylight (mt.), N.Y. 276/M2
Skyline, Minn. (†56001) 255/D6
Skyring (bay), Chile 138/E10
Skytop, Pa. (18357) 294/M3
Sky Valley, Georgia (30525) 217/F1
Slab Fork, W. Va. (25920) 312/D7
Slade, Ky. (40376) 237/O5
Sládečkovce, Czech. 41/D2
Slag (bay), Neth. Ant. 161/D8
Slagelse, Denmark 21/E7
Slagelse, Denmark 18/G9
Slagle, La. (71475) 238/D4
Slakow, Poland 47/B4
Slamannan, Scotland 15/C2
Slamet (mt.), Indonesia 85/J2
Slana, Alaska 196/K2
Slaná (riv.), Czech. 41/F2
Slane, Ireland 17/H4
Slanesville, W. Va. (25444) 312/K4
Slaney (riv.), Ireland 17/H7
Slangerup, Denmark 21/E6
Slangkop (pt.), S. Africa 118/E7
Slănic, Romania 45/G3
Slantsy, U.S.S.R. 52/C3
Slaný, Czech. 41/C1
Slate (mt.), Ariz. 198/D3
Slate (riv.), Colo. 208/E5
Slate (creek), Idaho 220/B4
Slate (isls.), Ontario 175/C3
Slate (riv.), Va. 307/L5
Slate, W. Va. (†26143) 312/D4
Slate (creek), Wyo. 319/C3
Slatedale, Pa. (18079) 294/L4
Slater, Colo. (81653) 208/E1
Slater, Iowa (50244) 229/F5
Slater, Mo. (65349) 261/G4
Slater, Wyo. (82201) 319/H4
Slater-Marietta, S.C. (29683) 296/C1
Slatersville, R.I. (02876) 249/H4
Slate Run, Pa. (17769) 294/H3
Slaterville Springs, N.Y. (14881) 276/H6
Slate Spring, Miss. (38955) 256/F3
Slatina, Romania 45/G3
Slatington, Pa. (18080) 294/L4
Slaton, Texas (79364) 303/C4
Slaughter, La. (70777) 238/H5
Slaughter Beach, Del. (†19963) 245/S5
Slaughters, Ky. (42456) 237/F6
Slaughterville, Okla. (†73051) 288/M4
Slave (riv.) 162/E3
Slave (riv.), Alberta 182/C5
Slave (riv.), Canada 146/G3
Slave (riv.), N.W. Terrs. 187/G3
Slave Coast (reg.), Benin 106/E7
Slave Coast (reg.), Nigeria 106/E7
Slave Coast (reg.), Togo 106/E7
Slave Lake, Alberta 182/C2
Slavgorod, U.S.S.R. 48/H4
Slavkov, Czech. 41/D2
Slavonia (reg.), Yugoslavia 45/C3
Slavonska Požega, Yugoslavia 45/C3
Slavonski Brod, Yugoslavia 45/D3
Slavuta, U.S.S.R. 52/C4
Slavyansk, U.S.S.R. 52/E5
Slavyansk-na-Kubani, U.S.S.R. 52/E5
Slawno, Poland 47/C1
Slayden, Miss. (†38642) 256/F1
Slayden, Tenn. (37165) 237/G8
Slayton, Minn. (56172) 255/C7
Sleaford, England 13/G5
Sleaford, England 10/F4
Sleat (dist.), Scotland 15/C3
Sleat (pt.), Scotland 15/B4
Sleat (sound), Scotland 15/C3
Sledge, Miss. (38670) 256/D2
Sleeper, Mo. (†65536) 261/G7
Sleeping Bear Dunes Nat'l Lakeshore, Mich. 250/C4
Sleeping Deer (mt.), Idaho 220/D5
Sleepy Creek, W. Va. (†25411) 312/K3
Sleepy Eye, Minn. (56085) 255/D6
Sleepy Eye (creek), Minn. 255/C6
Sleepy Hollow, Ill. (†60118) 222/E1
Sleetmute, Alaska (99668) 196/G2
Sleeve (lake), Manitoba 179/E3
Slemish (mt.), N. Ireland 17/J2
Slemon (lake), Manitoba 179/G1
Slemp, Ky. (41763) 237/P6
Slick, Okla. (74071) 288/O3
Slickford, Ky. (†42633) 237/M7
Slickville, Pa. (15684) 294/C5
Slide (mt.), N.Y. 276/L6
Slidell, La. (70458) 238/L6
Sliedrecht, Netherlands 27/F5
Sliema, Malta 34/E7
Slieve Anierin (mt.), Ireland 17/F3
Slieve Aughty (mts.), Ireland 17/D5
Slieve Beagh (mt.), N. Ireland 17/G3
Slieve Bernagh (mt.), Ireland 17/D6
Slieve Bloom (mts.), Ireland 17/F5
Slieve Callan (mt.), Ireland 17/C6
Slieve Car (mt.), Ireland 17/B3
Slieve Donard (mt.), N. Ireland 10/D3
Slieve Donard (mt.), N. Ireland 17/K3
Slieve Elva (mt.), Ireland 17/C5
Slievefelim (mts.), Ireland 17/E6
Slieve Gamph (mts.), Ireland 17/D3
Slieve Gullion (mt.), N. Ireland 17/J3
Slieve League (mt.), Ireland 17/D2
Slieve Mishkish (mts.), Ireland 17/B8
Slievenamon (mt.), Ireland 17/F7
Sligo (co.), Ireland 17/D3
Sligo, Ireland 17/E3
Sligo, Ireland 10/B3
Sligo (bay), Ireland 10/B3
Sligo (bay), Ireland 17/D3
Sligo, La. (†71037) 238/C2
Sligo, Pa. (16255) 294/C3
Slinger, Wis. (53086) 317/K9
Slipper (isl.), N. Zealand 100/F2
Slippery Rock, Pa. (16057) 294/B3
Slite, Sweden 18/L8
Sliven, Bulgaria 45/H4
Sliven, Bulgaria 7/G4
Sloan, Iowa (51055) 229/A4
Sloan, Nev. (†89114) 266/F7
Sloan, N.Y. (†14201) 276/C5
Sloans Valley, Ky. (42555) 237/N7
Sloat, Calif. (†96103) 204/E4
Sloatsburg, N.Y. (10974) 276/M8
Slobodskoy, U.S.S.R. 48/E4
Slobodskoy, U.S.S.R. 52/H3
Slobozia, Romania 45/H3
Slocan, Br. Col. 184/J5
Slocan (lake), Br. Col. 184/J5
Slocan Park, Br. Col. 184/J5
Slochteren, Netherlands 27/K2
Slocomb, Ala. (36375) 195/G8
Slocum, R.I. (02877) 249/H6
Slocum, Texas (†75839) 303/J6
Slonim, U.S.S.R. 52/B4
Slope (co.), N. Dak. 282/C7
Slot, The (chan.), Solomon Is. 86/D3
Sloten, Friesland, Netherlands 27/H3
Sloten, North Holland, Netherlands 27/B5
Sloterdijk, Netherlands 27/B4
Slotermeer (lake), Netherlands 27/H3
Slough, England 13/G8
Sloughhouse, Calif. (95683) 204/C8
Slovak Socialist Rep., Czech. 41/E2
Slovenia (rep.), Yugoslavia 45/B2
Slovenské Rudohorie (mts.), Czech. 41/E2
Słubice, Poland 47/B2
Sluis, Netherlands 27/C6
Słupca, Poland 47/D2
Słupia (riv.), Poland 47/C1
Słupsk (prov.), Poland 47/C1
Słupsk, Poland 47/C1
Słupsk, Poland 7/F3
Slutsk, U.S.S.R. 52/C4
Slyne (head), Ireland 10/A4
Slyne (head), Ireland 17/A5
Slyudyanka, U.S.S.R. 48/L4
Smackover, Ark. (71762) 202/E7
Smale, Ark. (†72021) 202/H4
Small, Idaho (†83423) 220/F5
Small (cape), Maine 243/D8
Small Isles (isls.), Scotland 15/B4
Small Point, Maine (04567) 243/D8
Smallwood (res.), Newf. 166/B3
Smallwood (res.), Newf. 166/M4
Smallwood (res.), Newf. 162/K5
Smarr, Georgia (31086) 217/E5
Smarts (mt.), N.H. 268/C4
Smartt, Tenn. (37378) 237/K9
Smartville, Calif. (95977) 204/D4
Smeaton, Sask. 181/G2
Smederevo, Yugoslavia 45/E3
Smederevska Palanka, Yugoslavia 45/E3
Smedjebacken, Sweden 18/J6
Smela, U.S.S.R. 52/D5
Smelterville, Idaho (83868) 220/B2
Smerwick (harb.), Ireland 17/A7
Smethport, Pa. (16749) 294/F2
Smicksburg, Pa. (16256) 294/D4
Smilax, Ky. (41764) 237/P6
Smilde, Netherlands 27/K3
Smiley, Sask. 181/B4
Smiley, Texas (78159) 303/G8
Smiltene, U.S.S.R. 53/C2
Smith (bay), Alaska 196/H1
Smith, Alberta 182/D2
Smith (sound), Br. Col. 184/C4
Smith (riv.), Calif. 204/A2
Smith (creek), Idaho 220/B1
Smith (co.), Kansas 232/D2
Smith (isl.), Md. 245/O8
Smith (co.), Miss. 256/E6
Smith (riv.), Mont. 262/E3
Smith, Nev. (89430) 266/B4
Smith (sound), Newf. 166/D2
Smith (isl.), N.C. 281/O7
Smith (basin), N.W.T. 162/N3
Smith (cape), N.W.T. 162/H3
Smith (bay), N.W. Terrs. 187/L2
Smith (cape), N.W. Terrs. 187/L3
Smith (sound), N.W. Terrs. 187/L2
Smith (cape), Ontario 177/C2
Smith (riv.), Oreg. 291/D4
Smith (creek), S. Dak. 298/L6
Smith (co.), Tenn. 237/J8
Smith (co.), Texas 303/J5
Smith (isl.), Va. 307/S6
Smith (riv.), Va. 307/J7
Smith Arm (inlet), N.W. Terrs. 187/F3
Smithboro, Ill. (62284) 222/D5
Smithburg, N.J. (†07728) 273/E3
Smithburg, W. Va. (26436) 312/E4
Smith Center, Kansas (66967) 232/D2
Smith Creek (valley), Nev. 266/D3
Smith Creek, W. Va. (†26807) 312/H5
Smithdale, Miss. (39664) 256/C8
Smithers, Br. Col. 146/F4
Smithers, Br. Col. 162/D5
Smithers, Br. Col. 184/D3
Smithers, W. Va. (25186) 312/D6
Smithfield, Ill. (61477) 222/C3
Smithfield, Ky. (40068) 237/L4
Smithfield◯, Maine (04978) 243/D6
Smithfield, Nebr. (68976) 264/E4
Smithfield, N.C. (27577) 281/N3
Smithfield, Ohio (43948) 284/J5
Smithfield, Ontario 177/G3
Smithfield, Pa. (15478) 294/C6
Smithfield, Texas (†76180) 303/F2
Smithfield, Utah (84335) 304/C2
Smithfield, Va. (23430) 307/P7
Smithfield, W. Va. (26437) 312/E4
Smith Hill, Manitoba 179/C5
Smithland, Iowa (51056) 229/B4
Smithland, Ky. (42081) 237/E6
Smithmill, Pa. (16680) 294/F4
Smith Mills, Ky. (42457) 237/F5
Smith Mountain (lake), Va. 307/J6
Smithonia, Georgia (†30628) 217/F2
Smithport (lake), La. 238/C2
Smith River, Calif. (95567) 204/A2
Smiths, Ala. (36877) 195/H5
Smithsburg, Md. (21783) 245/H2
Smiths Cove, Nova Scotia 168/C4
Smiths Creek, Mich. (48074) 250/G6
Smiths Creek, New Bruns. 170/E3
Smiths Falls, Ontario 177/H3
Smiths Ferry, Idaho (†83611) 220/C5
Smiths Fork (riv.), Wyo. 319/B3
Smiths Grove, Ky. (42171) 237/J6
Smithshire, Ill. (61478) 222/C3
Smiths Station, Miss. (†39066) 256/C6
Smithton, Ark. (†71743) 202/D6
Smithton, Ill. (62285) 222/C5
Smithton, Mo. (65350) 261/F5
Smithton, Pa. (15479) 294/C5
Smithton, Tasmania 99/A2
Smithton, Tasmania 88/H8
Smith Town, Ky. (†42647) 237/M7
Smithtown, N.H. (†03874) 268/F6
Smithtown, N.Y. (11787) 276/O9
Smithtown-Gladstone, N.S. Wales 97/G2
Smith Valley, Ind. (†46142) 227/E5
Smithville, Ark. (72466) 202/H1
Smithville, Georgia (31787) 217/D7
Smithville, Ind. (47458) 227/D6
Smithville, Miss. (38870) 256/H2
Smithville, Mo. (64089) 261/D4
Smithville, N.J. (†08060) 273/D4
Smithville, N.J. (08201) 273/E5
Smithville, Ohio (44677) 284/G4
Smithville, Okla. (74957) 288/S6
Smithville, Ontario 177/E4
Smithville, Tenn. (37166) 237/K9
Smithville, Texas (78957) 303/G7
Smithville, W. Va. (26178) 312/D4
Smithville Flats, N.Y. (13841) 276/J6
Smithwick, S. Dak. (57782) 298/C7
Smoaks, S.C. (29481) 296/F5
Smoke Bend, La. (†70346) 238/K3
Smoke Creek (des.), Nev. 266/B2
Smoke Hole, W. Va. (†26866) 312/H5
Smokey Burn, Sask. 181/H2
Smoky (riv.), Alberta 182/A2
Smoky (riv.), Alta. 162/E5
Smoky (mts.), Idaho 220/D6
Smoky (cape), N.S. Wales 97/G2
Smoky (lake), N. Dak. 282/K3
Smoky (cape), Nova Scotia 168/H2
Smoky Bay, S. Australia 88/E6
Smoky Bay, S. Australia 94/D5
Smoky Hill (riv.) 188/G3
Smoky Hill (riv.), Colo. 208/P5
Smoky Hill, North Fork (riv.), Kansas 232/A2
Smoky Hill (riv.), Kansas 232/C3
Smoky Junction, Tenn. (†37827) 237/N8
Smoky Lake, Alberta 182/D2
Smøla (isl.), Norway 18/E5
Smolan, Kansas (67479) 232/E3
Smolensk, U.S.S.R. 7/H3
Smolensk, U.S.S.R. 48/D4
Smolensk, U.S.S.R. 52/D4
Smolyan, Bulgaria 45/G5
Smoot, W. Va. (24977) 312/E7
Smoot, Wyo. (83126) 319/B3
Smooth Rock Falls, Ontario 177/J5
Smooth Rock Falls, Ontario 175/D3
Smugglers Notch (pass), Vt. 268/B2
Smuts, Sask. 181/F3
Smyadovo, Bulgaria 45/H4
Smyer, Ala. (†36727) 195/B7
Smyrna, Del. (19977) 245/R3
Smyrna (res.), Del. 245/R3
Smyrna, Georgia (30080) 217/K1
Smyrna, Mich. (48887) 250/D5
Smyrna, N.Y. (13464) 276/J5
Smyrna, N.C. (28579) 281/R5
Smyrna, S.C. (29743) 296/E1
Smyrna, Tenn. (37167) 237/H9
Smyrna (Izmir), Turkey 63/B3
Smyrna, Wash. (†99357) 310/F4
Smyrna Mills, Maine (04780) 243/G3
Smyrna Mills◯, Maine (04780) 243/G3
Smyth (co.), Va. 307/E7
Snaefell (mt.), I. of Man 13/C3
Snaefell (mt.), I. of Man 10/D3
Snake (riv.) 188/C1
Snake (riv.), Idaho 220/A3
Snake (riv.), Minn. 255/A2
Snake (riv.), Minn. 255/E4
Snake (riv.), Nebr. 264/C2
Snake (mts.), Nev. 266/F1
Snake (range), Nev. 266/G3
Snake (riv.), Oreg. 291/K3
Snake (creek), S. Dak. 298/F4
Snake (creek), S. Dak. 298/F5
Snake (creek), S. Dak. 298/M3
Snake (riv.), U.S. 146/G5
Snake (isl.), Victoria 97/D6
Snake (riv.), Wash. 310/G4
Snake (riv.), Wyo. 319/B2
Snake Creek (canal), Fla. 212/B4
Snake Indian (riv.), Alberta 182/A3
Snake River (plain), Idaho 220/D7
Snake River (range), Idaho 220/G6
Snake River, Wash. (†99301) 310/G4
Snare (riv.), N.W. Terrs. 187/G3
Snare Lake, N.W. Terrs. 187/G3
Snares, The (isls.), N. Zealand 100/A7
Snåsa, Norway 18/H4
Snåsavatn (lake), Norway 18/H4
Snead, Ala. (35952) 195/F2
Sneads, Fla. (32460) 212/B1
Sneads Ferry, N.C. (28460) 281/P5
Snedsted, Denmark 21/B4
Sneedville, Tenn. (37869) 237/P7
Sneek, Netherlands 27/H2
Sneekermeer (lake), Netherlands 27/H2
Sneem, Ireland 17/B8
Sneeuwkop (mt.), S. Africa 118/F6
Sneffels (mt.), Colo. 208/D7
Snegamook (lake), Newf. 166/B3
Snell, Va. (22553) 307/N4
Snelling, Calif. (95369) 204/E6
Snelling, S.C. (†29812) 296/E5
Snellville, Georgia (30278) 217/D3
Snezhnogorsk, U.S.S.R. 48/J3
Śniardwy, Jezioro (lake), Poland 47/E2
Sniečkus, U.S.S.R. 53/D3
Snina, Czech. 41/G2
Snipe (lake), Alberta 182/B2
Snipe Lake, Sask. 181/B4
Snizort, Loch (inlet), Scotland 15/B3
Snohomish (co.), Wash. 310/D2
Snohomish, Wash. (98290) 310/D3
Snohomish (riv.), Wash. 310/C3
Snoqualmie, Wash. (†98065) 310/D3
Snoqualmie (pass), Wash. 310/D3
Snoqualmie (riv.), Wash. 310/D3
Snoqualmie Falls, Wash. (†98065) 310/D3
Snover, Mich. (48472) 250/G5
Snow, Okla. (74567) 288/R6
Snow (mt.), Vt. 268/B6
Snow (peak), Wash. 310/G2
Snowball, Ark. (†72650) 202/E2
Snowbird (lake), N.W. Terrs. 187/H3
Snow Camp, N.C. (27349) 281/L3
Snowden, N.C. (†27929) 281/S2
Snowden, Sask. 181/G2
Snowdon (mt.), Wales 13/D4
Snowdon (mt.), Wales 10/D4
Snowdonia Nat'l Park, Wales 13/D4
Snowdoun, Ala. (†36104) 195/F6
Snowdrift, N.W.T. 162/E3
Snowdrift, N.W. Terrs. 187/G3
Snowfield (peak), Wash. 310/D2
Snowflake, Ariz. (85937) 198/E4
Snowflake, Manitoba 179/D5
Snow Hill, Ala. (36778) 195/E7
Snow Hill, Ark. (†71751) 202/E7
Snow Hill, Md. (21863) 245/S8
Snow Hill, N.C. (28580) 281/O4
Snow Lake, Ark. (72379) 202/H5
Snow Lake, Man. 162/G5
Snow Lake, Manitoba 179/H3
Snowmass, Colo. (81654) 208/E4
Snowshoe (lake), Manitoba 179/G4
Snow Shoe, Pa. (16874) 294/G3
Snowtown, S. Australia 94/E5
Snowville, N.H. (†03849) 268/E4
Snowville, Utah (84336) 304/B2
Snow Water (lake), Nev. 266/G2
Snowy (mts.), N.S. Wales 97/E5
Snowy (riv.), N.S. Wales 97/E5
Snowy (riv.), Victoria 88/H7
Snug, Tasmania 99/D5
Snyder, Ark. (†71658) 202/G7
Snyder, Colo. (80750) 208/M2
Snyder, Mo. (†65286) 261/F3
Snyder, Nebr. (68664) 264/H3
Snyder, Okla. (73566) 288/J5
Snyder (co.), Pa. 294/H4
Snyder, Texas (79549) 303/D5
Snydertown, Pa. (17877) 294/J4
So (isl.), S. Korea 81/C6
Soalala, Madagascar 118/H3
Soalara, Madagascar 118/G4
Soanierana-Ivongo, Madagascar 118/H3
Soap (lake), Wash. 310/F3
Soap Lake, Wash. (98851) 310/F3
Soasiu, Indonesia 85/H5
Soatá, Colombia 126/D4
Soay (isl.), Scotland 15/A2
Soay (isl.), Scotland 15/B3
Sobat (riv.), Sudan 111/F6
Sober (isl.), Nova Scotia 168/F4
Soběslav, Czech. 41/C2
Sobieski, Minn. (†56345) 255/D5
Sobieski, Wis. (54171) 317/L6
Sobotka, Czech. 41/C1
Sobral, Brazil 120/E3
Sobral, Brazil 132/G3
Sobrance, Czech. 41/G2
Soca, Uruguay 145/C6
Sochaczew, Poland 47/E2
Soche (Shache), China 77/A4
Sochi, U.S.S.R. 7/H4
Sochi, U.S.S.R. 48/D5
Sochi, U.S.S.R. 52/E6
Social Circle, Georgia (30279) 217/E3
Society (isls.), Fr. Poly. 87/L7
Society Hill, Ala. (†36801) 195/H6
Society Hill, S.C. (29593) 296/H2
Socompa (vol.), Chile 138/B4
Socorro, Brazil 135/C3
Socorro, Colombia 126/D4
Socorro (isl.), Mexico 150/D7
Socorro, N. Mex. 188/E4
Socorro (co.), N. Mex. 274/C5
Socorro, N. Mex. (87801) 274/C4
Socotra (isl.), P.D.R. Yemen 54/G8
Socotra (isl.), P.D.R. Yemen 2/M5
Socotra (isl.), P.D.R. Yemen 59/F7
Socuéllamos, Spain 33/E3
Soda (lake), Calif. 204/K8
Soda (plains), India 68/D1
Soda, Jebel es (mts.), Libya 111/C2
Soda Creek, Br. Col. 184/F4
Sodankylä, Finland 18/P3
Soda Plains, Pakistan 68/D1
Soda Springs, Calif. (95728) 204/E4
Soda Springs, Idaho (83276) 220/G7
Sodaville, Oreg. (†97355) 291/E3
Soddu, Ethiopia 111/G6
Soddy-Daisy, Tenn. (37319) 237/L10
Söderhamn, Sweden 18/K6
Söderköping, Sweden 18/K7
Södermanland (co.), Sweden 18/K7
Södertälje, Sweden 18/G1
Sodiri, Sudan 111/E5
Sodus, Mich. (49126) 250/C6
Sodus, N.Y. (14551) 276/G4
Sodus Point, N.Y. (14555) 276/G4
Soe, Indonesia 85/G7
Soest, Netherlands 27/G4
Soest, W. Germany 22/C3
Soesterberg, Netherlands 27/G4
Soeurs (isl.), Québec 172/H4
Sofala (prov.), Mozambique 118/E3
Sofia (cap.), Bulgaria 7/G4
Sofia (cap.), Bulgaria 45/F4
Sofia (riv.), Madagascar 118/H3
Sofkee, Georgia (†31201) 217/E5
Soft Shell, Ky. (41853) 237/P6
Sogamoso, Colombia 126/D5
Sogamoso (riv.), Colombia 126/D4
Soğanlı (mts.), Turkey 63/H2
Soğanlı (riv.), Turkey 63/E2
Sognafjorden (fjord), Norway 18/D6
Sognefjorden (fjord), Norway 7/E2
Sogn og Fjordane (co.), Norway 18/E6
Sogod, Philippines 82/E5
Sogod (bay), Philippines 82/E5
Söğüt, Turkey 63/D3
Söğüt (lake), Turkey 63/D4
Sog Xian, China 77/D5
Soh, Iran 66/G4
Sohâg, Egypt 111/F2
Sohâg, Egypt 59/B4
Sohâg, Egypt 102/F2
Soham, N. Mex. (†87565) 274/D3
Sohar, Oman 59/G5
Sŏhŭng, N. Korea 81/C4
Soignies, Belgium 27/D7
Sointula, Br. Col. 184/D5
Soissons, France 28/E3
Soka, Japan 81/O2
Sokch'o, S. Korea 81/D4
Söke, Turkey 63/B4
Söke, Turkey 59/A2
Sokna, Libya 111/C2
Sokodé, Togo 106/E7
Sokol, U.S.S.R. 52/F3
Sokol, U.S.S.R. 48/E4
Sokółka, Poland 47/F2
Sokolo, Mali 106/C6
Sokolov, Czech. 41/B1
Sokołów Podlaski, Poland 47/F2
Sokota, Ethiopia 111/G5
Sokoto (state), Nigeria 106/F6
Sokoto, Nigeria 102/C3
Sokoto, Nigeria 106/F6
Sokoto (riv.), Nigeria 106/F6
Sola, Cuba 158/G2
Solana Beach, Calif. (92075) 204/H11
Solander (isl.), N. Zealand 100/A7
Solano (co.), Calif. 204/D5
Solano (pt.), Colombia 126/B4
Solano, N. Mex. (87746) 274/E3
Solano, Philippines 82/C2
Solano, Venezuela 124/E6
Solbad Hall in Tirol, Austria 41/A3
Solca, Romania 45/G2
Soldado (pt.), P. Rico 161/G2
Soldier, Iowa (51572) 229/B5
Soldier, Kansas (66540) 232/G2
Soldier, Ky. (41173) 237/P4
Soldier Pond, Maine (04781) 243/F1
Soldiers Cove, Nova Scotia 168/H3
Soldiers Grove, Wis. (54655) 317/E9
Soldier Summit, Utah (†84601) 304/C4
Soldotna, Alaska (99669) 196/B1
Soledad, Argentina 143/F5
Soledad, Calif. (93960) 204/D7
Soledad, Colombia 126/C2
Soledad, Venezuela 124/G3
Soledad de Doblado, Mexico 150/Q2
Soledad Díez Gutiérrez, Mexico 150/J5
Soleduck (riv.), Wash. 310/A3
Solen, N. Dak. (58570) 282/J7
Solent (chan.), England 13/F7
Solentiname (isls.), Nicaragua 154/E5
Soleure (Solothurn) (canton), Switzerland 39/E2
Solgohachia, Ark. (72156) 202/E3
Solhan, Turkey 63/J3
Soligalich, U.S.S.R. 52/F3
Soligorsk, U.S.S.R. 52/C4
Solihull, England 13/F5
Solihull, England 10/G3
Solikamsk, U.S.S.R. 7/K3
Solikamsk, U.S.S.R. 48/F3
Solikamsk, U.S.S.R. 52/J3
Sol'-Iletsk, U.S.S.R. 52/J4
Solingen, W. Germany 22/B3
Solis, Uruguay 145/D5
Solís de Mataojo, Uruguay 145/D5
Solitary (isl.), N.S. Wales 97/G1
Sollefteå, Sweden 18/K5
Sollentuna, Sweden 18/H1
Sóller, Spain 33/H3
Søllested, Denmark 21/E8
Solna, Sweden 18/H1
Solo (Surakarta), Indonesia 85/J2
Solo, Mo. (65564) 261/J8
Sologne (reg.), France 28/E4
Solok, Indonesia 85/C6
Sololá, Guatemala 154/B3
Solomon (sea) 87/F6
Solomon, Alaska (†99762) 196/F2
Solomon, Ariz. (85551) 198/F6
Solomon, Kansas (67480) 232/E3
Solomon (riv.), Kansas 232/E2
Solomon (isls.), Pacific 87/F6
Solomon (isls.), Papua N.G. 86/C3
Solomon (sea), Papua N.G. 85/C7
Solomon (isls.), Solomon Is. 86/C3
Solomon (sea), Solomon Is. 86/D3
Solomon Islands 2/T6
SOLOMON ISLANDS 86/D2
Solomon Islands 87/G6
Solomons, Md. (20688) 245/N7
Solon, China 77/K2
Solon, Ind. (†47111) 227/F7
Solon, Iowa (52333) 229/L5
Solon◯, Maine (04979) 243/D6
Solon, Ohio (44139) 284/J9
Solon Springs, Wis. (54873) 317/C3
Solor (isl.), Indonesia 85/G7
Solothurn (elec. div.), Switzerland 39/E2
Solothurn (Soleure), Switzerland 39/E2
Solovetskiye (isls.), U.S.S.R. 52/E1
Solsberry, Ind. (47459) 227/D6
Solsgirth, Manitoba 179/B4
Solsona, Philippines 82/C1
Solsona, Spain 33/G2
Solt, Hungary 41/E3
Šolta (isl.), Yugoslavia 45/C4
Soltau, W. Germany 22/C2
Soltvadkert, Hungary 41/E3
Soluk, Libya 111/D1
Solund, Norway 18/D6
Solvang, Calif. (93463) 204/E9
Solvay, N.Y. (13209) 276/H4
Sölvesborg, Sweden 18/J9
Solway (firth), England 13/D3
Solway, Minn. (56678) 255/C3
Solway (firth), Scotland 10/D3
Solway (firth), Scotland 15/E6
Solwezi, Zambia 115/E6
Soma, Japan 81/K5
Soma, Turkey 63/B3
Somabula, Zimbabwe 118/D3
Somalia 2/M5
Somalia 102/G4
SOMALIA 115/J2
Sombor, Yugoslavia 45/D3
Sombra, Ontario 177/B5
Sombrerete, Mexico 150/H5
Sombrero (chan.), India 68/G7
Sombrero (isl.), St. Chris.-Nevis 156/F3
Somerdale, N.J. (08083) 273/B4
Somers, Conn. (06071) 210/F1
Somers◯, Conn. (06071) 210/F1
Somers, Iowa (50586) 229/E4
Somers, Mont. (59932) 262/B2
Somers, Wis. (53171) 317/M3
Somerset, (22972) 307/M4
Somerset (isl.), Bermuda 156/G3
Somerset (isl.), Canada 4/B14
Somerset, Colo. (81434) 208/E5
Somerset (co.), England 13/E6
Somerset, Ind. (46984) 227/F3
Somerset, Ky. (42501) 237/M6
Somerset, La. (†71357) 238/H2
Somerset (co.), Maine 243/C4
Somerset, Manitoba 179/D5
Somerset (co.), Md. 245/R8
Somerset, Md. (†20015) 245/E4
Somerset◯, Mass. (02725) 249/K5
Somerset (co.), N.J. 273/D2
Somerset, N.Y. (†14012) 276/C4
Somerset (isl.), N.W.T. 146/J2
Somerset (isl.), N.W.T. 162/G1
Somerset (isl.), N.W. Terrs. 187/J2
Somerset, Nova Scotia 168/D3
Somerset, Ohio (43784) 284/F6
Somerset (co.), Pa. 294/D6
Somerset, Pa. (15501) 294/D6
Somerset, Texas (78069) 303/J11
Somerset (res.), Vt. 268/A5
Somerset, Wis. (54025) 317/A5
Somerset East, S. Africa 118/D6
Somerset West, S. Africa 118/F6
Somers Point, N.J. (08244) 273/D5
Somersville, Conn. (06072) 210/F1
Somersworth, N.H. (03878) 268/F5
Somerton, Ariz. (85350) 198/A6
Somerton, England 13/E6
Somerton, Ohio (43784) 284/H6
Somervell (co.), Texas 303/G5
Somerville, Ala. (35670) 195/E2
Somerville, Ind. (47683) 227/C8
Somerville◯, Maine (†04341) 243/D7
Somerville, Mass. (02143) 249/C6
Somerville, New Bruns. 170/C2
Somerville, N.J. (08876) 273/D2
Somerville, Ohio (45064) 284/A6
Somerville, Tenn. (38068) 237/C10
Somerville, Texas (77879) 303/H7
Somes (isl.), N. Zealand 100/B2
Someş (riv.), Romania 45/F2
Somesbar, Calif. (95568) 204/B2
Somesville (Mount Desert), Maine (†04660) 243/G7
Somme (dept.), France 28/E3
Somme (riv.), France 28/D2
Somme, Sask. 181/J3
Somme-Leuze, Belgium 27/G8
Sommen (lake), Sweden 18/J8
Sömmerda, E. Germany 22/D3
Somogy (co.), Hungary 41/D3
Somonauk, Ill. (60552) 222/E2
Somotillo, Nicaragua 154/D4
Somoto, Nicaragua 154/D4
Somvix, Switzerland 39/G3
Son (riv.), India 68/E3
Son, Norway 18/D4
Son, Con (isls.), Vietnam 72/E5
Soná, Panama 154/G6
Sonaguera, Honduras 154/D3
Sŏnch'ŏn, N. Korea 81/B4
Sønderborg, Denmark 21/C8
Sønderborg, Denmark 18/G9
Sønderho, Denmark 21/B7
Sønderjylland (co.), Denmark 21/C7
Sønder Nissum, Denmark 21/A5
Sønder Omme, Denmark 21/B6
Sondershausen, E. Germany 22/D3
Søndersø, Denmark 21/D7
Sondheimer, La. (71276) 238/H1
Søndre Strømfjord, Greenl. 4/C12
Sondrio (prov.), Italy 34/B1
Sondrio, Italy 34/B1
Sonepur, India 68/E4
Sonestown, Pa. (17770) 294/K3
Song Ba (riv.), Vietnam 72/F4
Song Ca (riv.), Vietnam 72/E3
Song Cai (riv.), Vietnam 72/E4
Song Cau, Vietnam 72/F4
Song Da (Black) (riv.), Vietnam 72/E2
Songea, Tanzania 115/G6
Songea, Tanzania 102/F6
Song Hong (Red) (riv.), Vietnam 72/E2
Songhua (riv.), China 54/P5
Songhua Hu (lake), China 77/L3

Songhua Jiang (Sungari) (riv.), China 77/M2
Songkhla, Thailand 54/L9
Songkhla, Thailand 72/D6
Songling, China 77/K2
Songnim, N. Korea 81/B4
Songo, Angola 115/C5
Songo, Angola 102/D5
Songo, Mozambique 118/E3
Songololo, Zaire 115/B5
Songololo, Zaire 102/D5
Songpan, China 77/F5
Songxi, China 77/J6
Son Ha, Vietnam 72/F4
Sonid Youqi, China 77/H3
Sonid Zuoqi, China 77/H3
Son La, Vietnam 72/D2
Sonmiani, Pakistan 68/B3
Sonmiani, Pakistan 59/J4
Sonneberg, E. Germany 22/D3
Sonnenhorn (mt.), Switzerland 39/F4
Sonnette, Mont. (59348) 262/L5
Sonningdale, Sask. 181/D3
Sono (riv.), Brazil 132/E5
Sonobe, Japan 81/J7
Sonoita, Ariz. (85637) 198/E7
Sonoma (co.), Calif. 204/C5
Sonoma, Calif. (95476) 204/C5
Sonoma (range), Nev. 266/D2
Sonora, Calif. (95370) 204/E6
Sonora, Ky. (42776) 237/K5
Sonora (state), Mexico 150/D2
Sonora (riv.), Mexico 150/D2
Sonora, Nova Scotia 168/G3
Sonora, Texas (76950) 303/D7
Sonoyta, Mexico 150/C1
Sonqor, Iran 66/E3
Sonseca, Spain 33/D3
Sonsón, Colombia 126/C5
Sonsonate, El Salvador 154/C4
Sonsorol (isl.), Belau 87/D5
Sontag, Miss. (39665) 256/D7
Son Tay, Vietnam 72/E2
Sonthofen, W. Germany 22/D5
Sonvico, Switzerland 39/G4
Soochow (Suzhou), China 77/K5
Sooke, Br. Col. 184/J4
Sopachuy, Bolivia 136/C6
Sopas, Arroyo (riv.), Uruguay 145/C2
Sopchoppy, Fla. (32358) 212/B1
Soper, Okla. (74759) 288/P6
Soperton, Georgia (30457) 217/G6
Sopetrán, Colombia 126/C4
Sophia, N.C. (27350) 281/K3
Sophia, W. Va. (25921) 312/D7
Sophie, Fr. Guiana 131/E4
Sopi (cape), Indonesia 85/H5
Sopot, Poland 47/D1
Sopron, Hungary 41/D3
Soquel, Calif. (95073) 204/K4
Sora, Italy 34/D4
Sorah, Pakistan 68/B3
Sorata, Bolivia 136/A4
Sorbas, Spain 33/E4
Sorcière, La (mt.), St. Lucia 161/G6
Sorel, Québec 172/D4
Sorell (cape), Tasmania 99/B4
Sorell (lake), Tasmania 99/D4
Sorell-Midway Point, Tasmania 99/D4
Sorento, Ill. (62086) 222/D5
Soresina, Italy 34/C2
Sorgun, Turkey 63/F3
Soria (prov.), Spain 33/E2
Soria, Spain 33/E2
Soriano (dept.), Uruguay 145/B4
Soriano, Uruguay 145/A4
Sorikmerapi (mt.), Indonesia 85/B5
Sørkapp (pt.), Norway 18/C2
Sorø, Denmark 21/E7
Sorocaba, Brazil 132/D8
Sorocaba, Brazil 120/E5
Sorocaba, Brazil 135/C3
Sorochinsk, U.S.S.R. 52/H4
Soroki, U.S.S.R. 52/C5
Sorol (atoll), Micronesia 87/D5
Sorong, Indonesia 54/P10
Sorong, Indonesia 85/J6
Sororieng (mt.), Guyana 131/B2
Soroti, Uganda 115/F3
Söröya (isl.), Norway 7/G1
Soröya (isl.), Norway 18/N1
Sorrento, Br. Col. 184/H5
Sorrento, Fla. (32776) 212/E3
Sorrento, Italy 34/E4
Sorrento, La. (70778) 238/L3
Sorrento○, Maine (04677) 243/G7
Sorsele, Sweden 18/K4
Sorso, Italy 34/B4
Sorsogon (prov.), Philippines 82/E4
Sorsogon, Philippines 82/E4
Sorsogon, Philippines 85/G3
Sort, Spain 33/G1
Sortavala, U.S.S.R. 48/C3
Sortavala, U.S.S.R. 52/D2
Sør-Trøndelag (co.), Norway 18/G5
Sorum, S. Dak. (57654) 298/D3
Sŏsan, S. Korea 81/C5
Sos del Rey Católico, Spain 33/F1
Sosnogorsk, U.S.S.R. 7/K2
Sosnogorsk, U.S.S.R. 48/F3
Sosnogorsk, U.S.S.R. 52/H2
Sosnovka, U.S.S.R. 52/F4
Sosnovo-Ozerskoye, U.S.S.R. 48/M4
Sosnowiec, Poland 47/B4
Soso, Miss. (39480) 256/F7
Sosúa, Dom. Rep. 158/E5
Sosumav, Madagascar 118/H2
Sotkamo, Finland 18/Q4
Soto la Marina, Mexico 150/L4
Sotomayor, Bolivia 136/C6
Sotrondio, Spain 33/D1
Sotteville-les-Rouen, France 28/D3
Sotuta, Mexico 150/P6
Souanké, Congo 115/B3
Soubey, Switzerland 39/D2
Soudan, Ark. (†72360) 202/J4
Soudan, Minn. (55782) 255/F3
Soudan, North. Terr. 93/E6
Souderton, Pa. (18964) 294/M5
Souf (oasis), Algeria 106/F2
Souflion, Greece 45/H5
Soufrière, Dominica 161/E7
Soufrière (bay), Dominica 161/E7
Soufrière (mt.), Guadeloupe 161/A7
Soufrière, St. Lucia 161/F6
Soufrière, St. Lucia 156/G4
Soufrière (bay), St. Lucia 161/F6
Soufrière (mt.), St. Vin. & Grens. 161/A8
Souhegan (riv.), N.H. 268/D6
Souillac, Mauritius 118/G5
Souk Ahras, Algeria 106/F1
Soul (lake), Manitoba 179/C2
Soulanges (co.), Québec 172/C4
Soul City, N.C. (†27553) 281/N2
Sounding (creek), Alberta 182/E4
Sound View, Conn. (†06371) 210/F3
Sourdnahunk (lake), Maine 243/F3
Sourdough, Alaska (†99586) 196/J2
Soure, Brazil 132/D3
Soure, Portugal 33/B2
Souris, Man. 162/F6
Souris, Manitoba 179/B5
Souris (riv.), Manitoba 179/B5
Souris (riv.), N. Dak. 188/F1
Souris, N. Dak. (58783) 282/J2
Souris (riv.), N. Dak. 282/J2
Souris, P.E.I. 162/K6
Souris, Prince Edward I. 168/F2
Souris (riv.), Sask. 181/H6
Sour Lake, Texas (77659) 303/K7
Sousa, Brazil 120/F3
Sousel, Portugal 33/C3
Sousse, Tunisia 102/D1
Sousse, Tunisia 106/G1
Soustons, France 28/C6
South, Ala. (†36474) 195/E8
South (pt.), Barbados 161/B9
South (riv.), Georgia 217/K2
South (Ka Lae) (cape), Hawaii 218/G7
South (sound), Ireland 17/C5
South, Ky. (42777) 237/J6
South (pass), La. 238/M8
South (pt.), La. 238/G8
South (riv.), Mass. 249/D2
South (bay), Mich. 250/C2
South (chan.), Mich. 250/E3
South (pt.), Mich. 250/F4
South (isl.), N. Zealand 87/G10
South (cape), N. Zealand 100/A7
South (isl.), N. Zealand 100/B5
South (riv.), N.C. 281/M5
South (bay), N.W. Terrs. 187/K3
South (bay), Ontario 177/C2
South (mt.), Pa. 294/H6
South (isl.), S.C. 296/J5
South (cape), Tasmania 99/C5
South Acton, Maine (†04027) 243/B8
South Acton, Mass. (†01720) 249/J3
South Acworth, N.H. (03607) 268/C5
South Addison, Maine (†04606) 243/H6
South Africa 2/L7
South Africa 102/E7
SOUTH AFRICA 118/C5
South Alexandria, N.H. (†03222) 268/D4
South Allan, Sask. 181/E4
South Alligator (riv.), North. Terr. 88/E2
South Alligator (riv.), North. Terr. 93/C2
Southam, N. Dak. (†58327) 282/N3
South Amana, Iowa (52334) 229/J5
South Amboy, N.J. (08879) 273/E3
South America 2/D6
South Amherst, Mass. (†01002) 249/E3
South Amherst, Ohio (†44001) 284/F3
Southampton, England 7/D3
Southampton, England 10/F5
Southampton, England 13/F7
Southampton○, Mass. (01073) 249/C4
Southampton, N.Y. (11968) 276/R9
Southampton (isl.), N.W.T. 162/H2
Southampton (cape), N.W. Terrs. 187/K3
Southampton (isl.), N.W. Terrs. 187/K3
Southampton, Nova Scotia 168/D3
Southampton, Ontario 177/C3
Southampton (co.), Va. 307/O7
South Andaman (isl.), India 68/G6
South Anna (riv.), Va. 307/N5
Southard, Okla. (73770) 288/K2
South Ashburnham, Mass. (01466) 249/G2
South Athol, Mass. (01331) 249/F2
South Atlantic Ocean 2/J6
South Aulatsivik (isl.), Newf. 166/B2
South Australia 88/F6
South Australia (state), Australia 87/D8
Southaven, Miss. (38671) 256/E1
South Bancroft, Maine (†04424) 243/G4
Southbank, Br. Col. 184/E3
South Barre, Mass. (01074) 249/F3
South Barre, Vt. (05670) 268/B3
South Barrington, Ill. (†60010) 222/A5
South Barwon, Victoria 97/C6
South Bass (isl.), Ohio 284/E2
South Bay, Fla. (33493) 212/F5
South Bay Aqueduct, Calif. 204/L2
South Baymouth, Ontario 177/B2
Southbeach, Oreg. (97366) 291/C3
South Belmar, N.J. (†07719) 273/E3
South Beloit, Ill. (61080) 222/E1
South Bend, Ind. 188/J2
South Bend, Ind. (*46601) 227/E1
South Bend, Nebr. (68058) 264/H4
South Bend, Texas (76081) 303/F5
South Bend, Wash. (98586) 310/B4
South Bennettsville, S.C. (†29512) 296/H2
South Bentinck Arm (inlet), Br. Col. 184/D4
South Berlin, Mass. (01549) 249/H3
South Berwick, Maine (03908) 243/B9
South Berwick○, Maine (03908) 243/B9
South Bethany, Del. (†19930) 245/T6
South Bethlehem, N.Y. (12161) 276/N5
South Bethlehem, Pa. (†16242) 294/D4
South Beveland (isl.), Netherlands 27/D6
South Bloomfield, Ohio (†43103) 284/D6
South Bloomingville, Ohio (43152) 284/E7
South Boardman, Mich. (49680) 250/D4
South Bolton, Québec 172/E4
Southborough, England 13/J8
Southborough○, Mass. (01772) 249/H3
South Boston (I.C.), (24592) 307/L7
South Boston, Ind. (†47167) 227/F7
South Bound Brook, N.J. (08880) 273/E2
South Braintree, Mass. (†02185) 249/D8
South Branch, Mich. (48761) 250/E4
South Branch, Minn. (†56081) 255/D7
South Branch, New Bruns. 170/F2
South Branch, Newf. 166/C4
South Branch, N.J. (†08876) 273/D2
South Branch Oromocto (riv.), New Bruns. 170/D3
Southbridge, Mass. (01550) 249/G4
Southbridge○, Mass. (01550) 249/G4
Southbridge, N. Zealand 100/D5
South Bridgton, Maine (†04009) 243/B8
South Bristol○, Maine (04568) 243/D8
South Britain, Conn. (06487) 210/B3
South Broadway, Wash. (†98901) 310/E4
South Brook, Green Bay Dist., Newf. 166/C4
South Brook, Humber Dist., Newf. 166/C4
South Brookfield, Nova Scotia 168/D4
South Brooksville, Maine (†04617) 243/F7
South Brunswick○, N.J. (†08852) 273/E3
South Bruny (isl.), Tasmania 99/D5
South Burlington, Vt. (05401) 268/A3
South Burro (mt.), Utah 304/D3
Southbury○, Conn. (06488) 210/C3
South Calling Lake, Alberta 182/D2
South Canaan, Conn. (†06031) 210/B1
South Carolina 188/K4
SOUTH CAROLINA 296
South Carolina (state), U.S. 146/K6
South Carrollton, Ky. (42374) 237/G6
South Carthage, Tenn. (†37030) 237/K8
South Carver, Mass. (02366) 249/M5
South Casco, Maine (04077) 243/B8
South Central (sen. dist.), Alaska 196/G3
South Charleston, Ohio (45368) 284/C6
South Charleston, W. Va. (25303) 312/C6
South Chatham, Mass. (02659) 249/O6
South Chatham, N.H. (†04037) 268/E3
South Cheyenne (riv.), Wyo. 319/H2
South Chicago Heights, Ill. (60411) 222/C6
South China (sea) 54/N8
South China (sea) 2/Q5
South China (sea), China 77/J7
South China (sea), Indonesia 85/D4
South China, Maine (04358) 243/D7
South China (sea), Malaysia 85/D4
South China (sea), Philippines 85/D4
South China (sea), Philippines 82/B3
South China (sea), Vietnam 72/F4
South Cle Elum, Wash. (98943) 310/D3
South Cleveland, Tenn. (†37311) 237/M10
South Clinton, Tenn. (†37716) 237/N8
South Coffeyville, Okla. (74072) 288/P1
South Colby, Wash. (98384) 310/A2
South Colton, N.Y. (13687) 276/L1
South Congaree, S.C. (†29169) 296/E4
South Connellsville, Pa. (15425) 294/C6
South Corning, N.Y. (14830) 276/F6
South Cotabato (prov.), Philippines 82/E7
South Coventry (Coventry), Conn. (†06238) 210/F1
South Cow (creek), Calif. 204/C3
South Dakota 188/F2
SOUTH DAKOTA 298
South Dakota (state), U.S. 146/H5
South Danbury, N.H. (†03230) 268/D5
South Danville, N.H. (03881) 268/E6
South Dartmouth, Mass. (02748) 249/L6
South Dayton, N.Y. (14138) 276/C6
South Daytona, Fla. (32021) 212/F2
South Deerfield, Mass. (01373) 249/D3
South Deerfield, N.H. (†03037) 268/E5
South Dennis, Mass. (02660) 249/O6
South Dennis, N.J. (08245) 273/D5
South Dorset, Vt. (05263) 268/A5
South Dos Palos, Calif. (93665) 204/E7
South Downs (hills), England 13/G7
South Dum Dum, India 68/F2
South Duxbury, Mass. (†02332) 249/M4
Southeast (cape), Alaska 196/E2
South East (cape), Australia 87/E10
South East (pt.), Jamaica 158/K6
Southeast (pass), La. 238/M8
South East (cape), Tasmania 88/H8
South East (cape), Tasmania 99/C5
South East (pt.), Victoria 97/D6
SOUTHEAST ASIA 85
Southeastern (sen. dist.), Alaska 196/L3
Southeast Loch (inlet), Hawaii 218/B3
South Easton, Mass. (02375) 249/K4
Southeast Upsalquitch (riv.), New Bruns. 170/D1
South Effingham, N.H. (03882) 268/E4
South Egremont, Mass. (01258) 249/A4
South Elgin, Ill. (60177) 222/E2
South Eliot, Maine (†03903) 243/B9
South El Monte, Calif. (91733) 204/C10
Southend, Scotland 15/C5
Southend-on-Sea, England 10/G5
Southend-on-Sea, England 13/H6
South English, Iowa (52335) 229/J6
Southern (dist.), Israel 65/B5
Southern Alps (range), N. Zealand 100/C5
Southern Cross, Mont. (†59711) 262/C4
Southern Cross, W. Australia 88/E6
Southern Cross, W. Australia 92/B5
Southern Harbour, Newf. 166/C2
Southern Indian (lake), Man. 162/G4
Southern Indian (lake), Man. 146/J4
Southern Indian (lake), Manitoba 179/H2
Southern Leyte (prov.), Philippines 82/E5
Southern Pines, N.C. (28387) 281/L4
Southern Ute Ind. Res., Colo. 208/D8
South Esk (riv.), Scotland 15/F4
South Esk (riv.), Tasmania 99/D3
Southesk Tablelands, W. Australia 92/D3
South Euclid, Ohio (44121) 284/H9
South Exeter, Maine (†04928) 243/E6
Southey, Sask. 181/G5
South Fallsburg, N.Y. (12779) 276/L7
Southfield, Mass. (†01259) 249/B4
Southfield, Mich. (*48034) 250/F6
Southfields, N.Y. (10975) 276/M8
South Flomaton, Fla. (†36441) 212/B5
Southford, Conn. (†06488) 210/C3
South Foreland (prom.), England 13/J6
South Fork, Colo. (81154) 208/F7
South Fork, Frenchman, Colo. 208/O1
South Fork, Mo. (65776) 261/J9
South Fork, Flathead (riv.), Mont. 262/C3
South Fork, Humboldt (riv.), Nev. 266/F2
South Fork, Owyhee (riv.), Nev. 266/E1
South Fork, Pa. (15956) 294/E5
South Fork, Sask. 181/C6
South Fork, Powder (riv.), Wyo. 319/F2
South Fork, Shoshone (riv.), Wyo. 319/C1
South Foster, R.I. (†02857) 249/H5
South Fowl (lake), Minn. 255/G1
South Fox (isl.), Mich. 250/D3
South Friars (bay), St. Chris.-Nevis 161/C10
South Fulton, Tenn. (†42041) 237/D8
Southgate, Calif. (90280) 204/C11
Southgate, Ky. (41071) 237/T2
South Gate, Md. (†21113) 245/M4
Southgate, Mich. (48195) 250/F6
South Georgia (isl.) 2/H8
South Georgia (isl.), Ant. 5/D17
South Gifford, Mo. (†63549) 261/G2
South Glamorgan, Wales 13/B7
South Glastonbury, Conn. (06073) 210/E2
South Glens Falls, N.Y. (†12801) 276/N4
South Goldsboro, N.C. (†27530) 281/N4
South Grafton, Mass. (01560) 249/H4
South Greenfield, Mo. (65752) 261/E8
South Groveland, Mass. (†01830) 249/L2
South Hadley○, Mass. (†01075) 249/D4
South Hadley Falls, Mass. (01075) 249/D4
Southhampton (isl.), N.W.T. 146/K3
South Hampton○, N.H. (†01913) 268/F6
South Hanover, Mass. (†02339) 249/L4
South Harbour, Nova Scotia 168/H2
South Harpswell, Maine (04079) 243/C8
South Harwich, Mass. (02661) 249/O6
South Haven, Kansas (67140) 232/E4
South Haven, Mich. (49090) 250/C6
South Haven, Minn. (55382) 255/D5
South Haven, Nova Scotia 168/H2
South Hazelton, Br. Col. 184/D2
South Heart, N. Dak. (58655) 282/D6
South Heights, Pa. (15081) 294/B4
South Hero○, Vt. (05486) 268/A2
South Hill, Va. (23970) 307/M7
South Hiram, Maine (04080) 243/B8
South Holland, Ill. (60473) 222/C6
South Holland (prov.), Netherlands 27/E5
South Holston (lake), Tenn. 237/S7
South Holston (lake), Va. 307/E7
South Hope, Maine (†04862) 243/E7
South Horr, Kenya 115/G3
South Houston, Texas (77587) 303/J2
South Hutchinson, Kansas (†67501) 232/D3
South Indian Lake, Manitoba 179/H2
Southington○, Conn. (06489) 210/D2
South International Falls, Minn. (56679) 255/E2
South Irvine, Ky. (†40336) 237/N5
South Jacksonville, Ill. (†62650) 222/C4
South Jordan, Utah (†84065) 304/B3
South Junction, Manitoba 179/G5
South Junction, Oreg. (†97037) 291/F3
South Kedgwick (riv.), New Bruns. 170/B1
South Kensington, Md. (†20795) 245/E4
South Kent, Conn. (06785) 210/B2
South Killingly, Conn. (†06239) 210/H1
South Knife (riv.), Manitoba 179/J2
South Knife Lake, Manitoba 179/J2
South Korea 54/O6
South La Grange, Maine (†04453) 243/F5
Southlake, Texas (76051) 303/F2
South Lake Tahoe, Calif. (95705) 204/F5
South Lancaster, Mass. (01561) 249/H3
Southland, Texas (79368) 303/C4
South Laurel, Md. (†20810) 245/L4
South Lead Hill, Ark. (†72644) 202/D1
South Lebanon, Maine (†03901) 243/A9
South Lebanon, Ohio (45065) 284/B7
South Lee, Mass. (01260) 249/A3
South Lee, N.H. (†03042) 268/E5
South Liberty, Maine (†04949) 243/E7
South Lincoln, Maine (†04457) 243/F5
South Lincoln, Vt. (†05443) 268/B3
South Lineville, Mo. (†50147) 261/E1
South Londonderry, Vt. (05155) 268/B5
South Loup (riv.), Nebr. 264/E3
South Luconia (shoal), Philippines 85/E4
South Lunenburg, Vt. (05908) 268/D3
South Lyme, Conn. (06376) 210/F3
South Lyndeboro, N.H. (03082) 268/D6
South Lynnfield, Mass. (01940) 249/D5
South Lyon, Mich. (48178) 250/F6
South Magnetic Pole, Ant. 2/R9
South Magnetic Pole, Ant. 5/C8
South Maitland, Nova Scotia 168/E3
South Manitou, Mich. (†49654) 250/C3
South Manitou (isl.), Mich. 250/C3
South Mansfield, La. (†71052) 238/C3
South Marsh (isl.), Md. 245/O8
South Mayo (riv.), Va. 307/H7
South Medford, Oreg. (†97501) 291/E5
South Melbourne, Victoria 97/J5
South Melbourne, Victoria 88/K7
South Merrimack, N.H. (03083) 268/D6
South Miami, Fla. (33143) 212/B5
South Miami Heights, Fla. (†33157) 212/F6
South Middleboro, Mass. (†02346) 249/L5
South Milford, Ind. (46786) 227/G1
South Mills, N.C. (27976) 281/S2
South Milwaukee, Wis. (53172) 317/M2
South Molton, England 10/E5
South Molton, England 13/D6
South Monmouth, Maine (†04259) 243/D7
South Monroe, Mich. (†48161) 250/F7
Southmont, N.C. (27351) 281/J3
South Mound, Kansas (†67357) 232/G4
South Mountain, Ontario 177/J3
South Mountain, Pa. (17261) 294/H6
South Nahanni (riv.), N.W. Terrs. 187/F3
South Naknek, Alaska (99670) 196/G3
South Natick, Mass. (†01760) 249/A7
South Natuna (isls.), Indonesia 85/D5
South Negril (pt.), Jamaica 158/G6
South Negril (pt.), Jamaica 156/C3
South New Berlin, N.Y. (13843) 276/K5
South Newbury, N.H. (03272) 268/D5
South Newbury, Vt. (05066) 268/C3
South Newfane, Vt. (05351) 268/B6
South Newport, Georgia (†31323) 217/K7
South New River (canal), Fla. 212/F5
South Norfolk, Conn. (†06058) 210/C1
South Norwalk, Conn. (†06850) 210/B4
South Nyack, N.Y. (†10960) 276/K8
South Ogden, Utah (†84403) 304/C2
South Ohio, Nova Scotia 168/B5
Southold, N.Y. (11971) 276/P8
South Olive, Ohio (43724) 284/G6
South Orange○, N.J. (07079) 273/A2
South Orkney (isls.) 2/G9
South Orkney (isls.), Ant. 5/C16
South Orleans, Mass. (02662) 249/O5
South Oromocto (lake), New Bruns. 170/D3
South Oroville, Calif. (†95965) 204/D4
South Orrington, Maine (†04474) 243/F6
South Ossetian Aut. Obl., U.S.S.R. 48/E5
South Ossetian Aut. Obl., U.S.S.R. 52/F6
South Otselic, N.Y. (13155) 276/J5
South Pacific (ocean) 87/H8
South Pacific Ocean 2/C8
South Padre Island, Texas (78597) 303/G11
South Pagai (isl.), Indonesia 85/C6
South Para (riv.), S. Australia 94/C7
South Paris, Maine (04281) 243/C7
South Pasadena, Calif. (91030) 204/C10
South Pasadena, Fla. (33707) 212/B3
South Pass City, Wyo. (82520) 319/D3
South Patrick Shores, Fla. (†32901) 212/F3
South Pekin, Ill. (61564) 222/D3
South Pender Island, Br. Col. 184/K3
South Penobscot, Maine (†04476) 243/F7
South Perry, Ohio (†43135) 284/E6
South Perth, W. Australia 88/B2
South Perth, W. Australia 92/A1
South Philipsburg, Pa. (†16866) 294/F4
South Piney (creek), Wyo. 319/B3
South Pittsburg, Tenn. (37380) 237/K10
South Pittsfield, N.H. (†03263) 268/E5
South Plainfield, N.J. (07080) 273/E2
South Plains, Texas (79258) 303/C3
South Platte (riv.) 188/F2
South Platte (riv.), Colo. 208/N1
South Platte (riv.), Nebr. 264/C3
South Platte (riv.), U.S. 146/H6
South Point, Ohio (45680) 284/E9
South Polar (plat.), Ant. 5/A1
South Pole 2/E11
South Pole, Ant. 5/A4
South Pomfret, Vt. (05067) 268/B4
South Porcupine, Ontario 175/D3
Southport, Conn. (06490) 210/B4
Southport, England 10/F2
Southport, England 13/G1
Southport, Fla. (†32401) 212/C6
Southport, Ind. (†46201) 227/E5
Southport, Maine (04569) 243/D8
Southport○, Maine (†04538) 243/D8
Southport, N.Y. (†14901) 276/G6
Southport, N.C. (28461) 281/N7
South Portland, Maine (04106) 243/C8
South Portsmouth, Ky. (41174) 237/P3
South Prairie, Wash. (98385) 310/D3
South Pugwash, Nova Scotia 168/E3
South Range, Mich. (49963) 250/G1
South Range, Wis. (54874) 317/B2
South Renous (riv.), New Bruns. 170/D2
South Renovo, Pa. (†17764) 294/G3
South River (peak), Colo. 208/F7
South River, Newf. 166/D2
South River, N.J. (08882) 273/E3
South River, Ontario 177/E2
South Rockwood, Mich. (48179) 250/F7
South Ronaldsay (isl.), Scotland 10/E1
South Ronaldsay (isl.), Scotland 15/F2
South Roxana, Ill. (62087) 222/B2
South Royalston, Mass. (†01331) 249/F2
South Royalton, Vt. (05068) 268/C4
South Russell, Ohio (†44022) 284/H3
South Ryegate, Vt. (05069) 268/C3
South Sacramento, Calif. (†95823) 204/B8
South Saint Paul, Minn. (55075) 255/G6
South Salem, Ohio (45681) 284/D7
South Salt Lake, Utah (84115) 304/C3
South Sandisfield, Mass. (†01255) 249/B4
South Sandwich (isls.) 2/H8
South Sandwich (isls.), Ant. 5/D17
South Sanford, Maine (†04073) 243/B9
South San Francisco, Calif. (94080) 204/J2
South Santiam (riv.), Oreg. 291/E3
South Saskatchewan (riv.), Alberta 182/E4
South Saskatchewan (riv.), Canada 146/G4
South Saskatchewan (riv.), Sask. 181/C5
South Seabrook, N.H. (†03874) 268/F6
South Seal (riv.), Manitoba 179/J2
South Seaville, N.J. (08246) 273/D5
South Sevogle (riv.), New Bruns. 170/D1
South Shaftsbury, Vt. (†05262) 268/A6
South Shetland (isls.) 2/F9
South Shetland (isls.), Ant. 5/C15
South Shields, England 13/J3
South Shields, England 10/F3
South Shore, Ky. (41175) 237/R3
South Shore, S. Dak. (57263) 298/P3
Southside, Ala. (†35901) 195/F3
Southside, Tenn. (37171) 237/G8
Southside Place, Texas (†77001) 303/J2
South Sioux City, Nebr. (68776) 264/H2
South Skunk (riv.), Iowa 229/H6
South Slocan, Br. Col. 184/J5
South Solon, Ohio (43153) 284/C6
South Spectacle (lake), Conn. 210/B2
South Stoddard, N.H. (†03464) 268/C5
South Strafford, Vt. (05070) 268/C4
South Suburban, India 68/F2
South Sudbury, Mass. (†01776) 249/J3
South Superior, Wyo. (†82945) 319/D4
South Sutton, N.H. (03273) 268/D5
South Sydney, N. S. Wales 88/L4
South Sydney, N.S. Wales 97/J3
South Taft, Calif. (†93268) 204/F8
South Tamworth, N.H. (03883) 268/E4
South Taranaki (bight), N. Zealand 100/D3
South Thomaston○, Maine (04858) 243/
South Toms River, N.J. (08753) 273/E4
South Trap (isl.), N. Zealand 100/B7
South Tucson, Ariz. (85713) 198/D6
South Tunnel, Tenn. (†37066) 237/H7
South Twin (mt.), N.H. 268/D3
South Tyne (riv.), England 13/E3
South Uist (isl.), Scotland 10/C2
South Uist (isl.), Scotland 15/A3
South Umpqua (riv.), Oreg. 291/E4
South Union, Ky. (42283) 237/H7
South Union, Maine (†04864) 243/E7
South Ural (mts.), U.S.S.R. 52/J4
South Venice, Fla. (33595) 212/D4
South Vienna, Ohio (45369) 284/C6
Southville, Mass. (†01772) 249/H3
South Wabasca (lake), Alberta 182/D2
South Wadesboro, N.C. (†28170) 281/J5
South Waldoboro, Maine (†04572) 243/E
South Wallingford, Vt. (05773) 268/A5
South Walpole, Mass. (02071) 249/K4
South Wanatah, Ind. (†46390) 227/D2
Southwark, England 13/H8
Southwark, England 10/B5
South Warren, Maine (†04864) 243/E7
South Waterford, Maine (04081) 243/B7
South Waverly, Pa. (†18840) 294/J2
South Wayne, Wis. (53587) 317/G10
South Weare, N.H. (†03281) 268/D5
South Webster, Ohio (45682) 284/E8
South Weldon, N.C. (†27890) 281/O2
South Wellfleet, Mass. (02663) 249/P5
South Wellington, Br. Col. 184/J3
Southwest (pass), La. 238/L8
Southwest (head), New Bruns. 170/D4
South West (brook), Newf. 166/C2
South West (cape), Tasmania 88/G7
South West (cape), Tasmania 99/B5
Southwest (cape), Virgin Is. (U.S.) 161/E4
South-West Africa (Namibia) 2/K6
South-West Africa (Namibia) 102/D7
South West Arm (inlet), Newf. 166/D2
South West City, Mo. (64863) 261/D9
South West Gander (riv.), Newf. 166/C4
Southwest Harbor, Maine (04679) 243/G7
Southwest Harbor○, Maine (04679) 243/G7
South West Margaree (riv.), Nova Scotia 168/G2
Southwest Miramichi (riv.), New Bruns. 170/D2
South Westport, Mass. (02790) 249/K6
South West Port Mouton, Nova Scotia 168/D5
South West Rocks, N.S. Wales 97/G2
South Weymouth, Mass. (†02190) 249/E8
South Whitley, Ind. (46787) 227/F2
Southwick, England 13/G7
Southwick, Idaho (†83537) 220/B3
Southwick○, Mass. (01077) 249/C4
South Williamson, Ky. (25661) 237/S5
South Williamsport, Pa. (17701) 294/J3

South Willington, Conn. (06265) 210/F1
South Wilmington, Ill. (60474) 222/E2
South Wilton, Conn. (†06897) 210/B4
South Windham, Conn. (06266) 210/G2
South Windham (Little Falls-South Windham), Maine (04082) 243/C8
South Windham, Vt. (†05359) 268/B5
South Windsor○, Conn. (06074) 210/E1
Southwold, England 13/J5
Southwold, England 10/G4
South Wolf (isl.), Newf. 166/C3
South Wolfeboro, N.H. (†03894) 268/E4
South Woodbury, Vt. (†05681) 268/C3
South Woodstock, Conn. (06267) 210/G1
South Woodstock, Vt. (05071) 268/B4
Southworth, Wash. (98386) 310/A2
South Worthington, Mass. (†01098) 249/C3
South Yadkin (riv.), N.C. 281/H3
South Yarmouth, Mass. (02664) 249/O6
South Yorkshire (co.), England 13/F4
South Zanesville, Ohio (43701) 284/F6
Sovata, Romania 45/G2
Sovereign, Sask. 181/D4
Sovetsk, U.S.S.R. 7/J3
Sovetsk, U.S.S.R. 52/G3
Sovetsk (Tilsit), U.S.S.R. 52/B4
Sovetskaya Gavan', U.S.S.R. 54/R5
Sovetskaya Gavan', U.S.S.R. 48/P5
SOVIET UNION (U.S.S.R.) 48
Sowerby Bridge, England 13/H1
Sowerby Bridge, England 10/G2
Soweto, S. Africa 118/H6
Soya (pt.), Japan 81/L1
Soyhières, Switzerland 39/D2
Soyo, Angola 115/B5
Soyo, Angola 102/D5
Sozopol, Bulgaria 45/H4
Spa, Belgium 27/H8
Spades, Ind. (†47041) 227/G6
Spain 2/J3
Spain 7/D4
SPAIN 33
Spalding, England 13/G5
Spalding, England 10/F4
Spalding (co.), Georgia 217/D4
Spalding, Mich. (49886) 250/B3
Spalding, Mo. (†63401) 261/J3
Spalding, Nebr. (68665) 264/F3
Spalding, Sask. 181/G3
Spaldings, Jamaica 158/H6
Spallumcheen, Br. Col. 184/H5
Spanaway, Wash. (98387) 310/C3
Spandau, W. Germany 22/E3
Spangle, Wash. (99031) 310/H3
Spangler, Pa. (15775) 294/E4
Spaniard's Bay, Newf. 166/D2
Spanish (head), I. of Man 13/C3
Spanish, Ontario 177/J5
Spanish (riv.), Ontario 177/C1
Spanishburg, W. Va. (25922) 312/D8
Spanish Fork, Utah (84660) 304/C3
Spanish Fork (riv.), Utah 304/C3
Spanish Fort, Ala. (36527) 195/C9
Spanish Fort, Texas (†76255) 303/G4
Spanish Lake, Mo. (†63138) 261/R1
Spanish Ship Bay, Nova Scotia 168/G4
Spanish Town, Jamaica 158/J6
Spanish Town, Jamaica 156/C3
Sparkill, N.Y. (10976) 276/K8
Sparkman, Ark. (71763) 202/E6
Sparks, Georgia (31647) 217/F8
Sparks, Kansas (†66035) 232/G2
Sparks, Nebr. (69220) 264/D2
Sparks, Nev. 188/C3
Sparks, Nev. (89431) 266/B3
Sparks, Okla. (74869) 288/N3
Sparks (lake), Oreg. 291/F3
Sparksville, Ky. (42778) 237/L6
Sparland, Ill. (61565) 222/D2
Sparlingville, Mich. (†48060) 250/G6
Sparr, Fla. (32690) 212/D2
Sparrow Bush, N.Y. (12780) 276/L8
Sparrows Point, Md. (21219) 245/N4
Sparta, Georgia (31087) 217/F4
Sparta, Greece 45/F7
Sparta, Ill. (62286) 222/D5
Sparta, Ky. (41086) 237/M3
Sparta, Mich. (49345) 250/D5
Sparta, Mo. (65753) 261/F9
Sparta○, N.J. (07871) 273/D1
Sparta, N.C. (28675) 281/G1
Sparta, Ohio (43350) 284/E5
Sparta, Ontario 177/C5
Sparta, Oreg. (†97870) 291/K3
Sparta, Tenn. (38583) 237/K9
Sparta, Va. (22552) 307/O4
Sparta, Wis. (54656) 317/E8
Spartanburg, Ind. (†47355) 227/H4
Spartanburg, S.C. 188/K4
Spartanburg (co.), S.C. 296/D2
Spartanburg, S.C. (*29301) 296/C1
Spartansburg, Pa. (16434) 294/C2
Spartivento (cape), Italy 34/B5
Spartivento (cape), Italy 34/F6
Sparwood, Br. Col. 184/K5
Spassk-Dal'niy, U.S.S.R. 48/O5
Spátha (cape), Greece 45/F8
Spaulding, Ill. (†62561) 222/D4
Spavinaw, Okla. (74366) 288/R2
Spavinaw (lake), Okla. 288/S2
Spean (riv.), Scotland 15/D4
Spean Bridge, Scotland 15/D4
Spear (cape), New Bruns. 170/G2
Spear (cape), Newf. 166/D2
Spearfish, S. Dak. (57783) 298/B5
Spearman, Texas (79081) 303/C1
Spearsville, Ind. (†46181) 227/E6
Spearsville, La. (71277) 238/E1
Spearville, Kansas (67876) 232/C4
Spectacle (lakes), Conn. 210/B2
Specter (range), Nev. 266/C1
Speculator, N.Y. (12164) 276/M3
Spedden, Alberta 182/E2
Spednik (lake), New Bruns. 170/C3
Speed, Ind. (47172) 227/F8
Speed, Kansas (†67639) 232/C2
Speed, N.C. (27881) 281/P3
Speedway, Ind. (46224) 227/E5
Speedwell, Tenn. (37870) 237/O8
Speedwell, Va. (24374) 307/F8
Speer (mt.), Switzerland 39/H2
Speers, Sask. 181/D3
Speightstown, Barbados 161/B8
Speightstown, Barbados 156/G4
Speigner, Ala. (†36025) 195/F5
Spelterville, Ind. (†47808) 227/C5
Spelve, Loch (inlet), Scotland 15/C4
Spenard, Alaska (99503) 196/C1
Spenborough, England 13/J1
Spence Bay, N.W. Terrs. 187/J3
Spencer (cape), Alaska 196/L1
Spencer (pt.), Alaska 196/E1
Spencer (lake), Alberta 182/E2
Spencer (gulf), Australia 87/D9
Spencer, Idaho (†83423) 220/F5
Spencer (co.), Ind. 227/C9
Spencer, Ind. (47460) 227/D6
Spencer, Iowa (51301) 229/C2
Spencer (co.), Ky. 237/L4
Spencer, La. (71278) 238/F1
Spencer (pond), Maine 243/D4
Spencer (stream), Maine 243/C5
Spencer, Mass. (01562) 249/F3
Spencer○, Mass. (01562) 249/F3
Spencer, Nebr. (68777) 264/F2
Spencer (cape), New Bruns. 170/E3
Spencer, N.Y. (14883) 276/H6
Spencer, N.C. (28159) 281/H3
Spencer, Ohio (44275) 284/F3
Spencer, Okla. (73084) 288/M3
Spencer (creek), Oreg. 291/E5
Spencer (gulf), S. Australia 88/F6
Spencer (cape), S. Australia 88/F7
Spencer (cape), S. Australia 94/E6
Spencer (gulf), S. Australia 94/E6
Spencer, S. Dak. (57374) 298/O6
Spencer, Tenn. (38585) 237/L9
Spencer, Va. (24165) 307/J7
Spencer, W. Va. (25276) 312/D5
Spencer, Wis. (54479) 317/F6
Spencerburg, Mo. (†63441) 261/K4
Spencerport, N.Y. (14559) 276/E4
Spencers Island, Nova Scotia 168/D3
Spencerville, Ind. (46788) 227/G2
Spencerville, Ohio (45887) 284/B4
Spencerville, Okla. (74760) 288/R6
Spencerville, Ontario 177/J3
Spences Bridge, Br. Col. 184/G5
Spennymoor, England 13/F3
Spennymoor, England 10/F3
Spenser (mts.), N. Zealand 100/D5
Sperling, Manitoba 179/E5
Sperrin (mts.), N. Ireland 17/G2
Sperry, Iowa (52650) 229/L7
Sperry, Okla. (74073) 288/P2
Sperryville, Va. (22740) 307/M3
Spessart (range), W. Germany 22/C4
Spétsai, Greece 45/F7
Spey (riv.), Scotland 10/E2
Spey (riv.), Scotland 15/E3
Speyer, W. Germany 22/C4
Sphinx (mt.), Mont. 262/E5
Spiceland, Ind. (47385) 227/F5
Spicer, Minn. (56288) 255/C5
Spicer (isls.), N.W. Terrs. 187/L3
Spicewood, Texas (78669) 303/F7
Spickard, Mo. (64679) 261/F2
Spiddal, Ireland 17/C5
Spider (lake), Maine 243/E3
Spider (lake), Wis. 317/D3
Spiekeroog (isl.), W. Germany 22/B2
Spies, N.C. (†27325) 281/K4
Spiez, Switzerland 39/E3
Spili, Greece 45/G8
Spillimacheen, Br. Col. 184/J5
Spillville, Iowa (52168) 229/J2
Spilsby, England 13/H4
Spin Buldak, Afghanistan 68/B2
Spin Buldak, Afghanistan 59/J3
Spindale, N.C. (28160) 281/F4
Spink (co.), S. Dak. 298/N4
Spink, S. Dak. (†57010) 298/R8
Spinnerstown, Pa. (18968) 294/M5
Spirit (lake), Idaho 220/B2
Spirit (lake), Iowa 229/C2
Spirit (lake), S. Dak. 298/O4
Spirit (lake), Wash. 310/C4
Spirit, Wis. (†54513) 317/F5
Spirit Lake, Idaho (83869) 220/A2
Spirit Lake, Iowa (51360) 229/C2
Spirit River, Alberta 182/A2
Spirit River, Alta. 162/E4
Spiritwood, N. Dak. (58481) 282/N6
Spiritwood, Sask. 181/D2
Spiro, Okla. (74959) 288/S4
Spišská Belá, Czech. 41/F2
Spišská Nová Ves, Czech. 41/F2
Spital am Pyhrn, Austria 41/C3
Spithead (chan.), England 13/F7
Spitsbergen (isl.), Norway 4/B9
Spitsbergen (isl.), Norway 18/C2
Spittal an der Drau, Austria 41/B3
Spitz, Austria 41/C2
Spivey, Kansas (67142) 232/D4
Splendora, Texas (77372) 303/J7
Split (lake), Manitoba 179/J2
Split (cape), Nova Scotia 168/D3
Split, Yugoslavia 7/F4
Split, Yugoslavia 45/C4
Split Lake, Manitoba 179/J2
Split Rock, Wis. (†54486) 317/H6
Splügen (pass), Italy 34/B1
Splügen, Switzerland 39/H3
Splügen (pass), Switzerland 39/H3
Spofford, N.H. (03462) 268/C6
Spofford, Texas (78877) 303/D8
Spokane, Mo. (65754) 261/F9
Spokane, Wash. 146/G5
Spokane, Wash. 188/C1
Spokane (co.), Wash. 310/H3
Spokane, Wash. (*99201) 310/H3
Spokane (mt.), Wash. 310/H3
Spokane (riv.), Wash. 310/H3
Spokane Ind. Res., Wash. 310/G3
Spöl (riv.), Switzerland 39/F2
Spoleto, Italy 34/D3
Spoon (riv.), Ill. 222/C3
Spooner, Wis. (54801) 317/B4
Spot (pond), Mass. 249/C6
Spotswood, N.J. (08884) 273/E3
Spotsylvania (co.), Va. 307/N4
Spotsylvania, Va. (22553) 307/N4
Spotted (range), Nev. 266/F6
Spotted Horse, Wyo. (†82831) 319/G1
Spottsville, Ky. (42458) 237/G5
Spottswood, Va. (24475) 307/K5
Spotville, Ark. (†71753) 202/D7
Spragge, Ontario 177/J5
Sprague, Ala. (36076) 195/F6
Sprague, Manitoba 179/G5
Sprague, Nebr. (68438) 264/H4
Sprague (riv.), Oreg. 291/F5
Sprague, Wash. (99032) 310/G3
Sprague (lake), Wash. 310/G3
Sprague River, Oreg. (97639) 291/F5
Spragueville, Iowa (52074) 229/N4
Spratly (isl.), Philippines 85/E4
Spratt, Mich. (†49753) 250/F3
Spray (mts.), Alberta 182/C4
Spray, Oreg. (97874) 291/H3
Spray Lakes, Alberta 182/C4
Spraytown, Ind. (†47228) 227/E6
Spread Eagle, Wis. (†54121) 317/K4
Spreckelsville, Hawaii (96779) 218/J1
Spree (riv.), E. Germany 22/F3
Spreewald (for.), E. Germany 22/F3
Spremberg, E. Germany 22/F3
Sprent, Tasmania 99/C3
Sprigg, W. Va. (25693) 312/B7
Sprimont, Belgium 27/H8
Spring (riv.), Ark. 202/H1
Spring (creek), Nev. 266/D2
Spring (mts.), Nev. 266/F6
Spring (valley), Nev. 266/G3
Spring (creek), N. Dak. 282/E5
Spring (creek), S. Dak. 298/J2
Spring (creek), S. Dak. 298/C6
Spring, Texas (*77373) 303/J7
Spring Arbor, Mich. (49283) 250/E6
Spring Bay, Ill. (†61601) 222/D3
Spring Bay, Ontario 177/B2
Springbok, S. Africa 118/B5
Springboro, Ohio (45066) 284/B6
Springboro, Pa. (16435) 294/B2
Springbrook, Iowa (52075) 229/N4
Spring Brook, N. Dak. (58850) 282/D3
Springbrook, Ontario 177/G3
Springbrook, Oreg. (†97132) 291/A2
Springbrook, Wis. (54875) 317/C4
Spring City, Mo. (†64801) 261/C9
Spring City, Pa. (19475) 294/L5
Spring City, Tenn. (37381) 237/M9
Spring City, Utah (84662) 304/C4
Spring Coulée, Alberta 182/D5
Spring Creek, Pa. (16436) 294/D2
Spring Creek, Tenn. (38378) 237/D9
Spring Creek, W. Va. (†24966) 312/F7
Springdale, Ark. (72764) 202/B1
Springdale, Iowa (†52776) 229/L5
Springdale, Mont. (59082) 262/F5
Springdale, Newf. 166/C4
Springdale, Ohio (45246) 284/B9
Springdale, Pa. (15144) 294/C6
Springdale, S.C. (†29720) 296/F2
Springdale, S.C. (29169) 296/E4
Springdale, Utah (84767) 304/B6
Springdale, Wash. (99173) 310/H2
Springe, W. Germany 22/C2
Springer (mt.), Georgia 217/D1
Springer (lake), Ill. 222/E4
Springer, N. Mex. (87747) 274/E2
Springer, Okla. (73458) 288/M6
Springerton, Ill. (62887) 222/E5
Springerville, Ariz. (85938) 198/F4
Springfield, Ark. (72157) 202/E3
Springfield, Colo. (81073) 208/O8
Springfield, Fla. (32401) 212/D6
Springfield, Georgia (31329) 217/K6
Springfield, Idaho (83277) 220/F6
Springfield (cap.), Ill. 146/J6
Springfield (cap.), Ill. 188/H3
Springfield (cap.), Ill. (*62701) 222/D4
Springfield (lake), Ill. 222/D4
Springfield, Ind. (†47638) 227/B8
Springfield, Ky. (40069) 237/L5
Springfield, La. (70462) 238/M2
Springfield○, Maine (04487) 243/G5
Springfield, Mass. 188/M2
Springfield, Mass. (*01101) 249/D4
Springfield, Mich. (49015) 250/D6
Springfield, Minn. (56087) 255/C6
Springfield, Mo. (*65801) 261/F8
Springfield, Mo. 188/H3
Springfield, Mo. 146/J6
Springfield, Nebr. (68059) 264/H3
Springfield, King's, New Bruns. 170/E3
Springfield, York, New Bruns. 170/C2
Springfield○, N.H. (†03284) 268/C4
Springfield○, N.J. (07081) 273/E2
Springfield, Nova Scotia 168/D4
Springfield, Ohio 188/K2
Springfield, Ohio (*45501) 284/C6
Springfield, Ontario 177/C5
Springfield, Oreg. (97477) 291/E3
Springfield○, Pa. (19064) 294/M7
Springfield, Queensland 88/G5
Springfield, Queensland 95/B5
Springfield, S.C. (29146) 296/E4
Springfield, S. Dak. (57062) 298/N8
Springfield, Tenn. (37172) 237/H8
Springfield, Vt. (05156) 268/B5
Springfield○, Vt. (05156) 268/B5
Springfield, Va. (*22150) 307/S3
Springfield, W. Va. (26763) 312/J4
Springfield Armory Nat'l Hist. Site, Mass. 249/D4
Springford, Ontario 177/D5
Spring Garden, Ala. (36275) 195/G3
Spring Garden, Calif. (95971) 204/D4
Spring Garden, Ill. (†62846) 222/E5
Spring Green, Wis. (53588) 317/G9
Spring Grove, Ill. (60081) 222/E1
Spring Grove, Ind. (†47374) 227/H5
Spring Grove, Minn. (55974) 255/G7
Spring Grove, Pa. (17362) 294/J6
Spring Grove, Va. (23881) 307/P6
Spring Hall, Barbados 161/B8
Springhaven, Nova Scotia 168/C5
Spring Hill, Ark. (†71801) 202/C6
Spring Hill, Iowa (†50125) 229/F6
Spring Hill, Kansas (66083) 232/H3
Springhill, La. (71075) 238/D1
Spring Hill, Minn. (†56352) 255/D5
Springhill, Nova Scotia 168/E3
Spring Hill, Tenn. (37174) 237/H9
Springhill Junction, Nova Scotia 168/D3
Springhills, Ohio (†43357) 284/C5
Springholm, Scotland 15/E5
Spring Hope, N.C. (27882) 281/N3
Springhouse, Br. Col. 184/G4
Spring Lake, Ind. (†46140) 227/F5
Spring Lake, Mich. (49456) 250/C5
Spring Lake, Minn. (†55056) 255/E5
Spring Lake, Minn. (56680) 255/E3
Spring Lake, N.J. (07762) 273/F3
Spring Lake, N.C. (28390) 281/M4
Springlake, Texas (79082) 303/B3
Spring Lake, Wis. (†54960) 317/H8
Spring Lake Heights, N.J. (†07762) 273/E3
Spring Lake Park, Minn. (†55432) 255/E5
Springlee, Ky. (†40201) 237/K2
Spring Lick, Ky. (42779) 237/H6
Spring Mills, Pa. (16875) 294/G4
Spring Mills, S.C. (†29067) 296/F2
Spring Park, Minn. (55384) 255/F5
Spring Place, Georgia (†30705) 217/C1
Springport, Ind. (47386) 227/G4
Springport, Mich. (49284) 250/E6
Spring Ridge, La. (†71047) 238/B2
Springs, S. Africa 118/J6
Springside, Sask. 181/J4
Springstein, Manitoba 179/E5
Springsure, Queensland 95/D5
Springton (res.), Pa. 294/L6
Springtown, Ark. (72767) 202/B1
Springtown, Texas (76082) 303/G5
Springvale, Georgia (†31788) 217/C7
Springvale, Maine (04083) 243/B9
Springvale, Victoria 88/L7
Springvale, Victoria 97/J5
Spring Valley, Ala. (†35674) 195/C1
Spring Valley, Ill. (61362) 222/D2
Spring Valley, Minn. (55975) 255/F7
Spring Valley, N.Y. (10977) 276/K8
Spring Valley, Ohio (45370) 284/C6
Spring Valley, Sask. 181/F6
Spring Valley, Texas (†77001) 303/J1
Spring Valley, Wis. (54767) 317/B6
Springview, Nebr. (68778) 264/E2
Springville, Ala. (35146) 195/E3
Springville, Calif. (93265) 204/G7
Springville, Ind. (47462) 227/D7
Springville, Iowa (52336) 229/L4
Springville, La. (†70754) 238/L2
Springville, Miss. (†38863) 256/F2
Springville, N.Y. (14141) 276/C5
Springville, Pa. (18844) 294/L2
Springville, Tenn. (38256) 237/E8
Springville, Utah (84663) 304/C3
Springwater, N.Y. (14560) 276/E5
Springwater, Sask. 181/C4
Springwood, Va. (†24066) 307/J5
Sproat Lake, Br. Col. 184/H3
Sprott, Ala. (36779) 195/D5
Sprowston, England 13/J5
Spruce (isl.), Manitoba 179/B1
Spruce, Mich. (48762) 250/F4
Spruce (mt.), Vt. 268/C3
Spruce Creek, Pa. (16683) 294/F4
Sprucedale, Ontario 177/E2
Spruce Grove, Alberta 182/D3
Spruce Home, Sask. 181/F2
Spruce Knob (mt.), W. Va. 312/G5
Spruce Knob-Seneca Rocks Nat'l Rec. Area, W. Va. 312/H5
Spruce Lake, Sask. 181/B2
Spruce Pine, Ala. (35585) 195/C2
Spruce Pine, N.C. (28777) 281/E3
Spruce Run (res.), N.J. 273/D2
Spruce View, Alberta 182/C3
Spruce Woods, Manitoba 179/C5
Spruce Woods Prov. Park, Manitoba 179/C5
Sprule, Ky. (40986) 237/O7
Spry (harb.), Nova Scotia 168/F4
Spry Harbour, Nova Scotia 168/F4
Spur, Texas (79370) 303/D4
Spurgeon, Ind. (47584) 227/C8
Spurlockville, W. Va. (25565) 312/B6
Spurn (head), England 13/H4
Spurn (head), England 10/G4
Spurr (mt.), Alaska 196/B1
Spur Tree, Jamaica 158/H6
Spuzzum, Br. Col. 184/G5
Spy (pond), Mass. 249/C6
Spy Hill, Sask. 181/K5
Squam (lake), N.H. 268/E4
Squamish, Br. Col. 184/F5
Squa Pan, Maine (†04732) 243/G2
Squa Pan (lake), Maine 243/G2
Square (lake), Maine 243/G1
Square Butte, Mont. (†59442) 262/F3
Square Islands, Newf. 166/C3
Squatec, Québec 172/J2
Squatec (lake), Québec 172/J2
Squaw (creek), Idaho 220/B5
Squaw (peak), Idaho 220/D4
Squaw (creek), Oreg. 291/F3
Squaw (creek), S. Dak. 298/B3
Squaw Harbor, Alaska (†99661) 196/F3
Squaw Lake, Minn. (56681) 255/D3
Squaw Rapids, Sask. 181/H2
Squibnocket (pt.), Mass. 249/M7
Squillace (gulf), Italy 34/F5
Squinzano, Italy 34/G4
Squire, W. Va. (24884) 312/C8
Squires, Mo. (65755) 261/G9
Squires Mem. Park, Newf. 166/C4
Squirrel, Idaho (83447) 220/G5
Sragen, Indonesia 85/J2
Sre Ambel, Cambodia 72/D5
Srebrenica, Yugoslavia 45/D3
Srednekolymsk, U.S.S.R. 4/C2
Srednekolymsk, U.S.S.R. 48/Q3
Sre Khtum, Cambodia 72/E4
Srem, Poland 47/C2
Sremska Mitrovica, Yugoslavia 45/D3
Srepok (riv.), Cambodia 72/E4
Sretensk, U.S.S.R. 54/N4
Sretensk, U.S.S.R. 48/M4
Srikakulam, India 68/E5
Sri Lanka 54/K9
SRI LANKA (CEYLON) 68/E7
Srinagar, India 68/D2
Srinagar, India 54/J6
Srivardhan, India 68/C5
Środa Śląska, Poland 47/C3
Środa Wielkopolska, Poland 47/C2
Staaten (riv.), Queensland 88/G3
Staaten (riv.), Queensland 95/B3
Staatsburg, N.Y. (12580) 276/N7
Stab, Ky. (42557) 237/N6
Stacks (mts.), Ireland 17/B7
Stacy, Minn. (55079) 255/E5
Stacy, N.C. (28581) 281/S5
Stacy, Va. (24616) 307/E6
Stacyville, Iowa (50476) 229/H2
Stacyville, Maine (04782) 243/F4
Stacyville○, Maine (04782) 243/F4
Stade, W. Germany 22/C2
Staden, Belgium 27/B7
Stadskanaal, Netherlands 27/L3
Stadthagen, W. Germany 22/C2
Stäfa, Switzerland 39/G2
Staffa (isl.), Scotland 15/B4
Staffelstein, W. Germany 22/D3
Staffhorst, W. Germany 22/C2
Stafford○, Conn. (06075) 210/F1
Stafford, England 10/G2
Stafford, England 13/E5
Stafford (lake), Fla. 212/D2
Stafford (co.), Kansas 232/D3
Stafford, Kansas (67578) 232/D4
Stafford, N.Y. (14143) 276/D5
Stafford, Ohio (43786) 284/H6
Stafford, Okla. (†73601) 288/H3
Stafford, Queensland 88/K2
Stafford, Queensland 95/D2
Stafford, Texas (77477) 303/J2
Stafford (co.), Va. 307/O4
Stafford, Va. (22554) 307/O4
Staffordshire (co.), England 13/E5
Stafford Springs, Conn. (06076) 210/F1
Staffordsville, Ky. (41256) 237/R5
Staffordville, Conn. (06077) 210/G1
Staffordville, N.J. (†08092) 273/E4
Staines (pen.), Chile 138/D9
Staines, England 10/B5
Staines, England 13/G8
Stainville, Tenn. (†37710) 237/N8
Staked (Llano Estacado) (plain), N. Mex. 274/F5
Staked (Llano Estacado) (plain), Texas 303/B4
Stakhanov, U.S.S.R. 52/E5
Stalden, Switzerland 39/E4
Staley, N.C. (27355) 281/K3
Stalham, England 13/J5
Stalheim, Norway 18/E6
Stalin, Albania 45/D5
Stalingrad (Volgograd), U.S.S.R. 7/J4
Stalingrad (Volgograd), U.S.S.R. 48/E5
Stalingrad (Volgograd), U.S.S.R. 52/F5
Stallings, N.C. (†28079) 281/H4
Stallo, Miss. (†39350) 256/F5
Stallworthy (cape), N.W. Terrs. 187/J1
Stalowa Wola, Poland 47/F3
Stalwart, Mich. (49789) 250/E2
Stalwart, Sask. 181/F4
Stambaugh, Mich. (49964) 250/G2
Stamford, Conn. (*06901) 210/A4
Stamford, England 13/G5
Stamford, England 10/F4
Stamford, Nebr. (68977) 264/E4
Stamford, N.Y. (12167) 276/L6
Stamford, Queensland 88/G4
Stamford, Texas (79553) 303/E5
Stamford (lake), Texas 303/E4
Stamford○, Vt. (05352) 268/A6
Stampa, Switzerland 39/J4
Stamping Ground, Ky. (40379) 237/M4
Stampriet, Namibia 118/B4
Stamps, Ark. (71860) 202/D7
Stanardsville, Va. (22973) 307/L4
Stanberry, Mo. (64489) 261/C2
Stanbridge-Est, Québec 172/D4
Stanchfield, Minn. (55080) 255/E5
Standard, Alberta 182/D4
Standard, Calif. (95373) 204/E6
Standard, Ill. (61363) 222/D2
Standard, La. (†71465) 238/F3
Standard City, Ill. (62686) 222/D4
Standerton, S. Africa 118/D5
Standfast (pt.), Ant. & Bar. 161/E11
Standing Rock, Ala. (36878) 195/H4
Standing Rock Ind. Res., N. Dak. 282/J7
Standish, Calif. (96128) 204/E3
Standish, Maine (04084) 243/B8
Standish○, Maine (04084) 243/B8
Standish, Mich. (48658) 250/F5
Standish-with-Langtree, England 13/G2
Stand Off, Alberta 182/D5
Stanfield, Ariz. (85272) 198/C6
Stanfield, N.C. (28163) 281/J4
Stanfield, Oreg. (97875) 291/H2
Stanford, Calif. (94305) 204/J3
Stanford, Ill. (61774) 222/D3
Stanford, Ind. (47463) 227/D6
Stanford, Ky. (40484) 237/M5
Stanford, Mont. (59479) 262/F3
Stanfordville, N.Y. (12581) 276/N7
Stangelville, Wis. (†54208) 317/L7
Stanger, S. Africa 118/E5
Stanhope, England 13/E3
Stanhope, Iowa (50246) 229/F4
Stanhope, N.J. (07874) 273/D2
Stanhope, Pr. Edward I. 168/E2
Stanhope, Québec 172/F4
Stanislaus (co.), Calif. 204/D6
Stanke Dimitrov, Bulgaria 45/F4
Stanley, England 13/H3
Stanley (cap.), Falk. Is. 120/D8
Stanley (cap.), Falk. Is. 143/E7
Stanley, Idaho (83278) 220/D5
Stanley, Iowa (50671) 229/K3
Stanley, Kansas (†66223) 232/H3
Stanley, Ky. (42375) 237/G5
Stanley, La. (†71049) 238/C3
Stanley, New Bruns. 170/D2
Stanley, N. Mex. (87056) 274/D3
Stanley, N.Y. (14561) 276/F5
Stanley, N.C. (28164) 281/G4
Stanley, N. Dak. (58784) 282/F3
Stanley (mt.), North. Terr. 93/B7
Stanley, Nova Scotia 168/E3
Stanley, Okla. (†74536) 288/R5
Stanley, Scotland 15/E4
Stanley (co.), S. Dak. 298/H5
Stanley, Tasmania 99/B2
Stanley (mt.), Tasmania 99/A1
Stanley, Va. (22851) 307/L3
Stanley, Wis. (54768) 317/E6
Stanley (falls), Zaire 102/E5
Stanley (falls), Zaire 115/D3
Stanley Pool (lake), Zaire 115/C4
Stanleytown, Va. (24168) 307/H7
Stanleyville, N.C. (†27045) 281/J2
Stanly (co.), N.C. 281/J4
Stanmore, Alberta 182/E4
Stannards, N.Y. (†14895) 276/E6
Stann Creek Town, Belize 154/C2
Stanovoy (range), U.S.S.R. 54/O4
Stanovoy (range), U.S.S.R. 48/N4
Stans, Switzerland 39/F3
Stanstead (co.), Québec 172/F4
Stanstead Plain, Québec 172/F4
Stanthorpe, Queensland 88/J5
Stanthorpe, Queensland 95/D6
Stanton, Ala. (36790) 195/E5
Stanton, Calif. (90680) 204/D11
Stanton, England 13/H5
Stanton, Iowa (51573) 229/C7
Stanton (co.), Kansas 232/A4
Stanton, Ky. (40380) 237/O5
Stanton, Mich. (48888) 250/D5
Stanton, Miss. (†39120) 256/B7
Stanton, Mo. (63079) 261/K6
Stanton (co.), Nebr. 264/G3
Stanton, Nebr. (68779) 264/G3
Stanton, N.J. (08885) 273/D2
Stanton, N. Dak. (58571) 282/H5
Stanton, Tenn. (38069) 237/C10
Stanton, Texas (79782) 303/C5
Stantonsburg, N.C. (27883) 281/O3
Stantonville, Tenn. (38379) 237/E10
Stanwood, Iowa (52337) 229/L5
Stanwood, Mich. (49346) 250/D5
Stanwood, Wash. (98292) 310/C2
Stanzel, Iowa (†50849) 229/E6
Staphorst, Netherlands 27/J3
Staplehurst, Nebr. (68439) 264/G4
Staples, Minn. (56479) 255/D4
Staples, Ontario 177/B5
Stapleton, Ala. (36578) 195/C9
Stapleton, Georgia (30823) 217/H4
Stapleton, Nebr. (69163) 264/D3
Stapylton (bay), N.W. Terrs. 187/G3
Star, Alberta 182/D3
Star, Idaho (83669) 220/B6
Star (lake), Minn. 255/C4
Star, Miss. (39167) 256/D6
Star, N.C. (27356) 281/K4
Star, Texas (76880) 303/F6
Starachowice, Poland 47/E3
Stará L'ubovňa, Czech. 41/F2
Staraya Russa, U.S.S.R. 52/D3
Stara Zagora, Bulgaria 7/G4
Stara Zagora, Bulgaria 45/G4
Starbuck (isl.), Kiribati 87/L6
Starbuck, Manitoba 179/E5
Starbuck, Minn. (56381) 255/C5
Starbuck, Wash. (99359) 310/G4
Star City, Ark. (71667) 202/G6
Star City, Ind. (46985) 227/D3
Star City, Sask. 181/G3
Star City, W. Va. (26505) 312/F3
Staré Město, Czech. 41/D2
Stargard Szczeciński, Poland 47/B2
Stargo, Ariz. (†85540) 198/F5
Starhill, La. (†70748) 238/H5
Stark (co.), Ill. 222/D2
Stark, Kansas (66775) 232/G4
Stark, Minn. (†55032) 255/E5
Stark, Mont. (†59846) 262/B3
Stark○, N.H. (†03582) 268/E2
Stark (co.), N. Dak. 282/E6
Stark (co.), Ohio 284/H4
Stark City, Mo. (64866) 261/D9
Starke, Fla. (32091) 212/D2
Starke (co.), Ind. 227/D2
Starkey, Oreg. (†97850) 291/J2
Star Keys (isls.), N. Zealand 100/E7
Starks, La. (70661) 238/C6
Starks○, Maine (†04980) 243/D6
Starks, Wis. (†54501) 317/H4
Starksboro○, Vt. (05487) 268/A3
Starkville, Colo. (81074) 208/L8
Starkville, Miss. (39759) 256/G4
Starkweather, N. Dak. (58377) 282/N3
Star Lake, N.Y. (13690) 276/K2
Star Lake, Wis. (54561) 317/G3
Starlight, Ind. (†47119) 227/F8
Starnberg, W. Germany 22/D4
Starnberger See (lake), W. Germany 22/D5
Starodub, U.S.S.R. 52/D4
Starogard Gdański, Poland 47/D2

Star Prairie, Wis. (54026) 317/A5
Starr, S.C. (29684) 296/B3
Starr (co.), Texas 303/F11
Starr King (mt.), N.H. 268/E3
Starrucca, Pa. (18462) 294/M2
Start, La. (71279) 238/G2
Start (pt.), Scotland 15/F1
Star Tannery, Va. (22654) 307/L2
Startex, S.C. (29377) 296/C2
Startup, Wash. (98293) 310/D3
Starvation (res.), Utah 304/D3
Stary Sacz, Poland 47/E4
Staryy Oskol, U.S.S.R. 52/E4
Staser, Ind. (†47539) 227/B8
Stassfurt, E. Germany 22/D3
Staszów, Poland 47/E3
State Center, Iowa (50247) 229/G5
State College, Pa. (16801) 294/G4
State Line, Idaho (†83854) 220/A2
State Line, Ind. (47982) 227/C4
State Line, Mass. (01261) 249/A3
State Line, Miss. (39362) 256/G8
State Line, Pa. (17263) 294/G6
Staten (Los Estados) (isl.), Argentina 143/D7
Staten (isl.), N.Y. 276/M9
Staten Island (borough), N.Y. (*10301) 276/M9
Statenville, Georgia (31648) 217/G9
State Road, N.C. (28676) 281/H2
Statesboro, Georgia (30458) 217/J6
Statesville, N.C. (28677) 281/H3
Statesville, Tenn. (†37184) 237/J8
Statham, Georgia (30666) 217/E3
Static, Tenn. (†38549) 237/L7
Station Camp, Ky. (†40336) 237/N5
Statts Mills, W. Va. (25279) 312/C5
Statue of Liberty Nat'l Mon., N.J. 273/B2
Statue of Liberty Nat'l Mon., N.Y. 276/M9
Staunton, Ill. (62088) 222/D5
Staunton, Ind. (47881) 227/C6
Staunton (I.C.), Va. (24401) 307/K4
Staunton (Roanoke) (riv.), Va. 307/K6
Stavanger, Norway 18/D7
Stavanger, Norway 7/E3
Stave (lake), Br. Col. 184/L3
Staveley, England 13/K2
Stavelot, Belgium 27/H8
Stavely, Alberta 182/D4
Staveren, Netherlands 27/G3
Stavern, Norway 18/D4
Stavropol', U.S.S.R. 7/J4
Stavropol', U.S.S.R. 52/F5
Stavropol', U.S.S.R. 48/E5
Stavrós, Greece 45/F5
Stawell, Victoria 97/B5
Stayner, Ontario 177/E3
Stayton, Oreg. (97383) 291/E3
Stead, Manitoba 179/F4
Stead, N. Mex. (88438) 274/F2
Steamboat, Ariz. (†86505) 198/F3
Steamboat (mt.), Idaho 220/C4
Steamboat Rock, Iowa (50672) 229/G4
Steamboat Springs, Colo. (80477) 208/F2
Steamburg, N.Y. (14783) 276/C6
Stearns, Ky. (42647) 237/N7
Stearns (co.), Minn. 255/D5
Stebbins, Alaska (99671) 196/F2
Steckborn, Switzerland 39/G1
Stecoah, N.C. (†28771) 281/B4
Stedman, N.C. (28391) 281/M4
Steedman, S.C. (†29070) 296/E4
Steeds, N.C. (†27341) 281/K4
Steel (mt.), Idaho 220/C6
Steele, Ala. (35987) 195/F3
Steele (co.), Minn. 255/E7
Steele, Mo. (63877) 261/N10
Steele (co.), N. Dak. 282/P4
Steele, N. Dak. (58482) 282/L6
Steele City, Nebr. (68440) 264/G4
Steele Narrows Hist. Park, Sask. 181/B2
Steele's (pt.), Norfolk I. 88/L5
Steeles Tavern, Va. (24476) 307/K5
Steeleville, Ill. (62288) 222/D6
Steelman, Sask. 181/J6
Steelmanville, N.J. (†08270) 273/D5
Steelton, Pa. (17113) 294/J5
Steelville, Mo. (65565) 261/K7
Steen, Minn. (56173) 255/B7
Steen, Sask. 181/H3
Steenbergen, Netherlands 27/E5
Steenkool, Indonesia 85/J6
Steenokkerzeel, Belgium 27/C9
Steen River, Alberta 182/B4
Steens, Miss. (39766) 256/H3
Steens (mt.), Oreg. 291/J5
Steensby (inlet), N.W. Terrs. 187/L2
Steenwijk, Netherlands 27/J3
Steep (pt.), W. Australia 88/A5
Steep (pt.), W. Australia 92/A4
Steep Falls, Maine (04085) 243/B8
Steep Rock, Manitoba 179/D3
Ştefăneşti, Romania 45/H2
Stefanie (lake), Ethiopia 111/G7
Stefansson (isl.), N.W.T. 162/F1
Stefansson (isl.), N.W. Terrs. 187/H2
Steffenville, Mo. (63470) 261/J3
Steffisburg, Switzerland 39/E3
Stege, Denmark 21/F8
Stege, Denmark 18/H9
Steger, Ill. (60475) 222/F2
Stehekin, Wash. (98852) 310/E2
Steigerwald (for.), W. Germany 22/D4
Steilacoom, Wash. (98388) 310/C3
Stein, Switzerland 39/E1
Steinach, Austria 41/A3
Steinach, E. Germany 22/D3
Stein am Rhein, Switzerland 39/G1
Steinauer, Nebr. (68441) 264/H4
Steinbach, Manitoba 179/F5
Steinhatchee, Fla. (32359) 212/C2
Steinhausen, Namibia 118/B4
Steinhuder Meer (lake), W. Germany 22/C2
Steinkjer, Norway 18/G4
Steinneset (cape), Norway 18/E2
Stekene, Belgium 27/E6
Stella, Mo. (64867) 261/D9
Stella, Nebr. (68442) 264/J4
Stella, N.C. (28582) 281/P5
Stella, Wash. (†98632) 310/B4
Stellarton, Nova Scotia 168/F3
Stellenbosch, S. Africa 118/F6
Steller (mt.), Alaska 196/K2
Stelvio (pass), Switzerland 39/K3
Stem, N.C. (27581) 281/M2
Stendal, E. Germany 22/D2
Stendal, Ind. (47585) 227/C8
Stenen, Sask. 181/J4
Stenhousemuir, Scotland 15/C1
Stenlille, Denmark 21/E6
Stennett, Iowa (†51566) 229/C6
Stenstrup, Denmark 21/D7
Stenungsund, Sweden 18/G7
Stepanakert, U.S.S.R. 48/E6
Stepanakert, U.S.S.R. 52/G7
Stepaside, Ireland 17/J5
Stephan, S. Dak. (57346) 298/K5
Stephen, Minn. (56757) 255/A2
Stephenfield, Manitoba 179/D5
Stephens (passage), Alaska 196/N1
Stephens, Ark. (71764) 202/E7
Stephens (isl.), Br. Col. 184/B3
Stephens (co.), Georgia 217/F1
Stephens, Georgia (30667) 217/F3
Stephens (isl.), N. Zealand 100/D4
Stephens (co.), Okla. 288/L6
Stephens (co.), Texas 303/F5
Stephensburg, Ky. (42781) 237/J5
Stephensburg, N.J. (†07865) 273/D2
Stephens City, Va. (25279) 307/M2
Stephen's Creek, N.S. Wales 97/A2
Stephenson (co.), Ill. 222/D1
Stephenson, Mich. (49887) 250/B3
Stephensport, Ky. (40170) 237/H5
Stephentown, N.Y. (12168) 276/O5
Stephenville, Newf. 166/C4
Stephenville, Newf. 162/L6
Stephenville, Texas (76401) 303/F5
Stephenville Crossing, Newf. 166/C4
Stepney, Conn. (†06468) 210/B3
Stepovak (bay), Alaska 196/G3
Steprock, Ark. (72159) 202/G3
Steptoe, Wash. (99174) 310/H3
Sterling, Alaska (99672) 196/B1
Sterling, Colo. 188/F2
Sterling, Colo. (80751) 208/N1
Sterling (res.), Colo. 208/N1
Sterling○, Conn. (06377) 210/H2
Sterling, Georgia (†31520) 217/K8
Sterling, Idaho (83279) 220/F6
Sterling, Ill. (61081) 222/D2
Sterling, Kansas (67579) 232/D3
Sterling○, Mass. (01564) 249/G3
Sterling, Mich. (48659) 250/E4
Sterling, Nebr. (68443) 264/H4
Sterling, N. Dak. (58572) 282/K6
Sterling, Ohio (44276) 284/G4
Sterling, Okla. (73567) 288/K5
Sterling, Pa. (18463) 294/M3
Sterling (co.), Texas 303/C6
Sterling, Utah (84665) 304/C4
Sterling, Va. (22170) 307/O2
Sterling City, Texas (76951) 303/D6
Sterling Heights, Mich. (48077) 250/B6
Sterling Run, Pa. (†15832) 294/F3
Sterlington, La. (71280) 238/F1
Sterlitamak, U.S.S.R. 7/K3
Sterlitamak, U.S.S.R. 48/F4
Sterlitamak, U.S.S.R. 52/J4
Šternberg, E. Germany 22/D2
Šternberk, Czech. 41/D2
Sterrett, Ala. (35147) 195/F4
Stet, Mo. (64680) 261/E4
Stetson○, Maine (04488) 243/E6
Stetsonville, Wis. (54480) 317/F5
Stettin (Szczecin), Poland 47/B2
Stettler, Alberta 182/D3
Stettler, Alta. 162/E5
Steytlerville (mt.), S. Africa 118/G6
Steuben (co.), Ind. 227/G1
Steuben, Maine (04680) 243/H6
Steuben○, Maine (04680) 243/H6
Steuben, Mich. (†49854) 250/C2
Steuben (co.), N.Y. 276/F6
Steuben, Wis. (54657) 317/E9
Steubenville, Ohio 188/K2
Steubenville, Ohio (43952) 284/J5
Steve, Ark. (†72857) 202/D4
Stevenage, England 13/G6
Stevenage, England 10/F5
Stevens (co.), Kansas 232/A4
Stevens (co.), Minn. 255/B5
Stevens (creek), S.C. 296/C4
Stevens (co.), Wash. 310/H2
Stevens (pass), Wash. 310/D3
Stevensburg, Va. (22741) 307/N4
Stevenson, Ala. (35772) 195/G1
Stevenson, Conn. (06491) 210/C3
Stevenson (lake), Manitoba 179/J3
Stevenson, The (riv.), S. Australia 94/D2
Stevenson, Wash. (98648) 310/C5
Stevenson Entrance (str.), Alaska 196/H3
Stevens Point, Wis. (54481) 317/G7
Stevens Pottery, Georgia (31088) 217/F5
Stevenston, Scotland 15/D5
Stevenstown, Wis. (†54636) 317/D7
Stevens Village, Alaska (99774) 196/J1
Stevensville, Md. (21666) 245/N5
Stevensville, Mich. (49127) 250/C6
Stevensville, Mont. (59870) 262/C4
Stevns Klint (cliff), Denmark 21/F7
Steward, Ill. (60553) 222/D2
Stewardson, Ill. (62463) 222/E4
Stewart, Ala. (35484) 195/C5
Stewart, Br. Col. 162/D4
Stewart, Br. Col. 184/C2
Stewart (isl.), Chile 138/E11
Stewart (co.), Georgia 217/C6
Stewart, Minn. (55385) 255/D6
Stewart, Miss. (39767) 256/F4
Stewart (isl.), N. Zealand 87/G10
Stewart (isl.), N. Zealand 100/A7
Stewart (cape), North. Terr. 93/D1
Stewart, Ohio (45778) 284/G7
Stewart (co.), Tenn. 237/F7
Stewart, Tenn. (37175) 237/F8
Stewart (riv.), Yukon 162/C3
Stewart (riv.), Yukon 187/E3
Stewart Crossing, Yukon 187/E3
Stewarton, Scotland 15/D5
Stewart River, Yukon 187/D3
Stewart River, Yukon 162/B3
Stewarts Point, Calif. (95480) 204/B5
Stewartstown○, N.H. (†03576) 268/E2
Stewartstown, N. Ireland 17/H2
Stewartstown, Pa. (17363) 294/K6
Stewartsville, Ind. (47636) 227/B8
Stewartsville, Mo. (64490) 261/C3
Stewartsville, N.J. (08886) 273/C2
Stewartsville, Ohio (43960) 284/J6
Stewart Town, Jamaica 158/H6
Stewart Valley, Sask. 181/D5
Stewartville, Ala. (†35150) 195/F4
Stewartville, Minn. (55976) 255/F7
Stewiacke, Nova Scotia 168/E3
Stewiacke (riv.), Nova Scotia 168/E3
Steyer, Md. (†21550) 245/A3
Steyr, Austria 41/C2
Stia, Italy 34/C3
Stickney, Ill. (60402) 222/B6
Stickney, New Bruns. 170/C2
Stickney, S. Dak. (57375) 298/M6
Stickney, W. Va. (25188) 312/D7
Stidham, Okla. (74461) 288/P4
Stiens, Netherlands 27/H2
Stigler, Okla. (74462) 288/R4
Stikine (riv.), Alaska 196/N2
Stikine (str.), Alaska 196/N2
Stikine (riv.), Br. Col. 162/C4
Stikine (riv.), Br. Col. 184/B1
Stiles, Iowa (†52537) 229/J7
Stiles, Wis. (54172) 317/L6
Stilesville, Ind. (46180) 227/D5
Stilis, Greece 45/F6
Still (riv.), Conn. 210/B3
Still (riv.), Conn. 210/C1
Still (riv.), Conn. 210/G1
Stillman Valley, Ill. (61084) 222/D1
Stillmore, Georgia (30464) 217/H6
Still Pond, Md. (21667) 245/O3
Still River, Mass. (01467) 249/H3
Stillwater, Ky. (†41301) 237/O5
Stillwater, Maine (04489) 243/F6
Stillwater (riv.), Mass. 249/G3
Stillwater, Minn. (55082) 255/F5
Stillwater (co.), Mont. 262/G5
Stillwater (riv.), Mont. 262/G5
Stillwater, Nev. (†89406) 266/C3
Stillwater (range), Nev. 266/C3
Stillwater, N.J. (07875) 273/D1
Stillwater, N.Y. (12170) 276/N5
Stillwater, Ohio (44679) 284/H5
Stillwater (riv.), Ohio 284/B5
Stillwater, Okla. (74074) 288/N2
Stillwater, Pa. (17878) 294/K3
Stillwater (res.), R.I. 249/C2
Stillwell, Ill. (†62380) 222/B3
Stillwell, Ind. (46351) 227/D1
Stilson, Georgia (30415) 217/J6
Stilwell, Kansas (66085) 232/H3
Stilwell, Okla. (74960) 288/S3
Stimson (mt.), Mont. 262/C2
Stinchar (riv.), Scotland 15/D5
Stinesville, Ind. (47464) 227/D6
Stinnett, Texas (79083) 303/C2
Stinson Beach, Calif. (94970) 204/H2
Stinson Lake, N.H. (03274) 268/D4
Stintino, Italy 34/B4
Štip, Yugoslavia 45/F5
Stiring-Wendel, France 28/G3
Stirling, Alberta 182/D5
Stirling, N.J. (07980) 273/E2
Stirling (creek), North. Terr. 88/D3
Stirling (creek), North. Terr. 93/A4
Stirling, Ontario 177/G3
Stirling, Scotland 10/B1
Stirling, Scotland 15/C1
Stirling (trad. co.), Scotland 15/A5
Stirling, W. Australia 88/B2
Stirling, W. Australia 92/A1
Stirling City, Calif. (95978) 204/D4
Stirling North, S. Australia 94/E5
Stirling Station, North. Terr. 93/C6
Stirrat, W. Va. (25645) 312/C7
Stirum, N. Dak. (58069) 282/P7
Stites, Idaho (83552) 220/C3
Stittsville, Ontario 177/J2
Stittville, N.Y. (13469) 276/K4
Stitzer, Wis. (53825) 317/E10
Stockbridge, Georgia (30281) 217/D3
Stockbridge, Mass. (01262) 249/A3
Stockbridge○, Mass. (01262) 249/A3
Stockbridge, Mich. (49285) 250/E6
Stockbridge○, Vt. (05772) 268/B4
Stockbridge, Wis. (53088) 317/K7
Stockbridge Ind. Res., Wis. 317/J6
Stockdale, Ohio (45683) 284/E8
Stockdale, Texas (78160) 303/G8
Stockerau, Austria 41/D2
Stockertown, Pa. (18083) 294/M4
Stockett, Mont. (59480) 262/E3
Stockham, Nebr. (†68818) 264/F4
Stockholm○, Maine (04783) 243/G1
Stockholm, N.J. (07460) 273/D1
Stockholm, Sask. 181/J5
Stockholm, S. Dak. (57264) 298/R3
Stockholm (co.), Sweden 18/L7
Stockholm (cap.), Sweden 2/K3
Stockholm (cap.), Sweden 7/F3
Stockholm (cap.), Sweden 18/G1
Stockholm, Wis. (54769) 317/B7
Stockhorn (mt.), Switzerland 39/E3
Stockland, Ill. (60967) 222/F3
Stockley, Del. (†19947) 245/S6
Stockport, England 13/H2
Stockport, England 10/G2
Stockport, Iowa (52651) 229/K7
Stockport, Ohio (43787) 284/G6
Stocksbridge, England 13/J2
Stockton, Ala. (36579) 195/C9
Stockton, Calif. 188/B3
Stockton, Calif. (*95201) 204/D6
Stockton, Georgia (31649) 217/G9
Stockton, Ill. (61085) 222/C1
Stockton, Iowa (52769) 229/M5
Stockton, Kansas (67669) 232/C2
Stockton, Manitoba 179/C5
Stockton, Md. (21864) 245/S8
Stockton, Minn. (55988) 255/G6
Stockton, Mo. (65785) 261/E7
Stockton (lake), Mo. 261/E7
Stockton, N.J. (08559) 273/D3
Stockton, N.Y. (14784) 276/B6
Stockton (plat.), Texas 303/B7
Stockton, Utah (84071) 304/B3
Stockton (isl.), Wis. 317/F2
Stockton-on-Tees, England 13/F3
Stockton-on-Tees, England 10/F3
Stockton Springs, Maine (04981) 243/F7
Stockton Springs○, Maine (04981) 243/F7
Stockville, Nebr. (69042) 264/D4
Stockwell, Ind. (47983) 227/D4
Stoco, Ontario 177/G3
Stod, Czech. 41/B2
Stoddard (co.), Mo. 261/N9
Stoddard○, N.H. (03464) 268/C5
Stoddard, Wis. (54658) 317/D8
Stoeng Treng, Cambodia 72/E4
Stoer (pt.), Scotland 15/C2
Stoholm, Denmark 21/C5
Stoke-on-Trent, England 13/E4
Stoke-on-Trent, England 10/E4
Stokes (bay), Chile 138/D10
Stokes (co.), N.C. 281/J2
Stokes, N.C. (27884) 281/P3
Stokes (pt.), Tasmania 99/A1
Stokes Bay, Ontario 177/C2
Stokesdale, N.C. (27357) 281/K2
Stolac, Yugoslavia 45/D4
Stolberg, W. Germany 22/B3
Stolp (Słupsk), Poland 47/C1
Ston, Yugoslavia 45/C4
Stone (co.), Ark. 202/F2
Stone, England 10/E4
Stone, England 13/E5
Stone, Idaho (83280) 220/F7
Stone, Ky. (41567) 237/S5
Stone (co.), Miss. 256/F9
Stone (co.), Mo. 261/F9
Stone (mts.), N.C. 281/F2
Stone (mts.), Tenn. 237/T8
Stone Bank, Wis. (†53066) 317/J1
Stonebluff, Ind. (†47987) 227/C4
Stonebluff, Okla. (†74436) 288/P3
Stoneboro, Pa. (16153) 294/B3
Stoneboro, S.C. (†29058) 296/F2
Stone City, Iowa (†52205) 229/L4
Stonecliffe, Ontario 177/G1
Stone Creek, Ohio (43840) 284/G5
Stonefort, Ill. (62987) 222/E6
Stonega, Va. (24285) 307/C7
Stoneham, Colo. (80754) 208/M1
Stoneham○, Mass. (02180) 249/C6
Stoneham, Québec 172/F2
Stone Harbor, N.J. (08247) 273/D5
Stonehaven, Scotland 10/E2
Stonehaven, Scotland 15/F4
Stonehenge (ruins), England 13/F6
Stonehenge, Queensland 95/B5
Stonehenge, Queensland 88/G4
Stonehenge, Sask. 181/F6
Stonehouse, Scotland 15/D5
Stone Lake, Wis. (54876) 317/C4
Stone Mountain, Georgia (*30083) 217/D3
Stone Mountain Prov. Park, Br. Col. 184/L2
Stone Park, Ill. (†60160) 222/B5
Stoner, Br. Col. 184/F3
Stones (riv.), Tenn. 237/H9
Stones River Nat'l Battlefield, Tenn. 237/H9
Stoneville, Miss. (38776) 256/C4
Stoneville, N.C. (27048) 281/K2
Stoneville, S. Dak. (57784) 298/D4
Stonewall, Ark. (72468) 202/J1
Stonewall, Georgia (†30291) 217/J2
Stonewall, La. (71078) 238/C2
Stonewall, Manitoba 179/E4
Stonewall, Miss. (39363) 256/G6
Stonewall, N.C. (28583) 281/R4
Stonewall, Okla. (74871) 288/O5
Stonewall (co.), Texas 303/D4
Stonewall, Texas (78671) 303/F7
Stonewood, W. Va. (26301) 312/F4
Stoney Creek, Ontario 177/E4
Stoney Point, Ontario 177/B5
Stonington, Colo. (81075) 208/P8
Stonington, Conn. (06378) 210/H3
Stonington○, Conn. (06378) 210/H3
Stonington, Ill. (62567) 222/D4
Stonington○, Maine (04681) 243/F7
Stonington, Mich. (†49878) 250/C3
Stono (inlet), S.C. 296/H6
Stony (riv.), Alaska 196/G2
Stony (isl.), Newf. 166/C3
Stony (brook), N.J. 273/D3
Stony (ranges), N.S. Wales 97/B2
Stony (isl.), N.Y. 276/H3
Stony (pt.), N.Y. 276/H3
Stony (lake), Ontario 177/E2
Stony (lake), Ontario 177/G3
Stony (head), Tasmania 99/C2
Stony (creek), Va. 307/N6
Stony (riv.), W. Va. 312/H4
Stony Beach, Sask. 181/F5
Stony Bottom, W. Va. (24979) 312/F6
Stony Brook, N.Y. (11790) 276/O9
Stony Creek, Conn. (†06405) 210/E3
Stony Creek, N.Y. (12878) 276/M4
Stony Creek, Va. (23882) 307/N7
Stonyford, Calif. (95979) 204/C4
Stony Gorge (res.), Calif. 204/C4
Stony Island, Nova Scotia 168/C5
Stony Lake, Mich. (49455) 250/C5
Stony Mountain, Manitoba 179/E4
Stony Plain, Alberta 182/C3
Stony Point, N.Y. (10980) 276/M8
Stony Point, N.C. (28678) 281/G3
Stony Rapids, Sask. 181/M2
Stony River, Alaska (†99557) 196/G2
Stony Tunguska (riv.), U.S.S.R. 54/L3
Stony Tunguska (riv.), U.S.S.R. 48/K3
Stony Wold, N.Y. (†12968) 276/M1
Storå (riv.), Denmark 21/B5
Stora Lulevatten (lake), Sweden 18/L3
Storden, Minn. (56174) 255/C6
Store Baelt (chan.), Denmark 18/G9
Store Baelt (chan.), Denmark 21/D6
Store Heddinge, Denmark 21/F7
Store-Heddinge, Denmark 18/H9
Stor-Elvdal, Norway 18/G6
Støren, Norway 18/F5
Storey (co.), Nev. 266/B3
Storeys Creek, Tasmania 99/D3
Storeytown, New Bruns. 170/D2
Storfjorden (fjord), Norway 18/D2
Storjorm (lake), Sweden 18/J4
Storkerson (bay), N.W. Terrs. 187/F2
Storla, S. Dak. (†57359) 298/M6
Storm (lake), Iowa 229/C3
Storm (bay), Tasmania 99/D5
Storm Lake, Iowa (50588) 229/C3
Stormont (county), Ontario 177/K2
Stornoway, Québec 172/F4
Stornoway, Sask. 181/K4
Stornoway, Scotland 15/B2
Stornoway, Scotland 10/C1
Stornoway (harb.), Scotland 15/B2
Storøya (isl.), Norway 18/E1
Storozhevsk, U.S.S.R. 52/H2
Storr, The (mt.), Scotland 15/B3
Storrs, Conn. (06268) 210/F1
Storsjön (lake), Sweden 18/J5
Storstrøm (co.), Denmark 21/E7
Stort (riv.), England 13/H7
Storthoaks, Sask. 181/K6
Storuman, Sweden 18/K4
Storuman (lake), Sweden 18/K4
Storvik, Sweden 18/K6
Story, Ind. (†47448) 227/E6
Story (co.), Iowa 229/G4
Story, Wyo. (82842) 319/F1
Story City, Iowa (50248) 229/F4
Stosch (isl.), Chile 138/C8
Stotesbury, Mo. (64786) 261/C7
Stotesbury, W. Va. (†25921) 312/D7
Stotts City, Mo. (65756) 261/E8
Stottville, N.Y. (12172) 276/N6
Stoughton○, Mass. (02072) 249/K4
Stoughton, Sask. 181/J6
Stoughton, Wis. (53589) 317/H10
Stoumont, Belgium 27/H8
Stour (riv.), England 13/J6
Stour (riv.), England 13/H6
Stour (riv.), England 13/E7
Stour (riv.), England 10/G4
Stourbridge, England 13/E5
Stourbridge, England 10/G3
Stourport-on-Severn, England 10/G3
Stourport-on-Severn, England 13/E5
Stout, Iowa (50673) 229/H3
Stout, Ohio (45684) 284/D8
Stout (lake), Ontario 175/B2
Stoutland, Mo. (65567) 261/G7
Stoutsville, Mo. (65283) 261/J3
Stoutsville, Ohio (43154) 284/E6
Stovall, Miss. (38672) 256/C2
Stovall, N.C. (27582) 281/M2
Stover, Mo. (65078) 261/G6
Støvring, Denmark 21/C4
Stow○, Maine (†04058) 243/A7
Stow○, Mass. (01775) 249/H3
Stow (creek), N.J. 273/C5
Stow, Ohio (44224) 284/H3
Stow, Scotland 15/F5
Stowe, Pa. (19464) 294/L5
Stowe, Vt. (05672) 268/B3
Stowe○, Vt. (05672) 268/B3
Stowmarket, England 13/J5
Stowmarket, England 10/G4
Stowport, Tasmania 99/B3
Stoy, Ill. (62464) 222/F5
Stoystown, Pa. (15563) 294/E5
Strabane (dist.), N. Ireland 17/G2
Strabane, N. Ireland 17/G2
Strabane, N. Ireland 10/C3
Strachan, Scotland 15/F3
Strachur Bay, Scotland 15/C4
Stradbally, Laoighis, Ireland 17/G5
Stradbally, Waterford, Ireland 17/F7
Strafford, Mo. (65757) 261/F8
Strafford (co.), N.H. 268/E5
Strafford○, N.H. (03884) 268/E5
Strafford○, Vt. (05072) 268/C4
Straffordville, Ontario 177/D5
Strahan, Iowa (†51540) 229/B7
Strahan, Tasmania 99/B4
Strait (butte), Utah 304/C6
Straits, N.C. (†28579) 281/R5
Straits Pond, Mass. (†02045) 249/F7
Straitsville, Conn. (†06770) 210/C3
Strakonice, Czech. 41/B2
Stralsund, E. Germany 22/E1
Strand, S. Africa 118/F7
Strandburg, S. Dak. (57265) 298/R3
Strandby, Denmark 21/D3
Strandquist, Minn. (56758) 255/B2
Strang, Nebr. (68444) 264/G4
Strang, Okla. (74367) 288/R2
Strange Creek, W. Va. (26639) 312/E5
Strangford, N. Ireland 17/K3
Strangford (inlet), N. Ireland 17/K3
Strängnäs, Sweden 18/F1
Stranraer, Sask. 181/C4
Stranraer, Scotland 10/D3
Stranraer, Scotland 15/C6
Strasbourg, France 28/H3
Strasbourg, France 7/E4
Strasbourg, Sask. 181/G4
Strasburg, Colo. (80136) 208/L3
Strasburg, Ill. (62465) 222/E4
Strasburg, Mo. (64090) 261/D5
Strasburg, N. Dak. (58573) 282/K7
Strasburg, Ohio (44680) 284/G4
Strasburg, Pa. (17579) 294/K6
Strasburg, Va. (22657) 307/M3
Strassburg, Austria 41/C3
Strafford, Calif. (93266) 204/F7
Stratford○, Conn. (06497) 210/C4
Stratford (pt.), Conn. 210/C4
Stratford, Iowa (50249) 229/F4
Stratford○, N.H. (†03590) 268/D2
Stratford, N.J. (08084) 273/B4
Stratford, N.Y. (13470) 276/L4
Stratford, N. Zealand 100/E3
Stratford, Okla. (74872) 288/M5
Stratford, Ontario 177/C4
Stratford, S. Dak. (57474) 298/N3
Stratford, Texas (79084) 303/C1
Stratford, Va. (22558) 307/P4
Stratford, Wash. (98853) 310/F3
Stratford, Wis. (54484) 317/F6
Stratford-Centre, Québec 172/F4
Stratford-upon-Avon, England 13/F5
Stratford-upon-Avon, England 10/F4
Strathalbyn, S. Australia 94/F6
Stratham○, N.H. (03885) 268/F5
Strathaven, Scotland 15/D5
Strathbogie (dist.), Scotland 15/F3
Strathclair, Manitoba 179/B4
Strathclyde (reg.), Scotland 15/C4
Strathcona, Minn. (56759) 255/B2
Strathcona Prov. Park, Br. Col. 184/E5
Strathfield, N. S. Wales 88/K4
Strathfield, N.S. Wales 97/J3
Strathfoyle, N. Ireland 17/G1
Strathgordon, Tasmania 99/C4
Strathlorne, Nova Scotia 168/G2
Strathmere, N.J. (08248) 273/D5
Strathmoor Village, Ky. (†40201) 237/J2
Strathmore, Alberta 182/D4
Strathmore, Calif. (93267) 204/F7
Strathmore, N.J. (†07747) 273/E3
Strathmore, Scotland 15/E4
Strathnaver, Br. Col. 184/F3
Strathpeffer, Scotland 15/D3
Strathroy, Ontario 177/C5
Strathspey (dist.), Scotland 15/E3
Strathy (pt.), Scotland 15/E2
Strathy (pt.), Scotland 10/D1
Strathyre, Scotland 15/D4
Strattanville, Pa. (16258) 294/D3
Stratton, Colo. (80836) 208/O4
Stratton, Maine (04982) 243/B5
Stratton, Nebr. (69043) 264/C4
Stratton, Ohio (43961) 284/J4
Stratton, Ontario 177/F5
Stratton, Ontario 175/B3
Stratton○, Vt. (†05360) 268/B5
Stratton (mt.), Vt. 268/B5
Straubing, W. Germany 22/E4
Straubville, N. Dak. (58070) 282/O7
Straughn, Ind. (47387) 227/G5
Strausberg, E. Germany 22/F2
Strausstown, Pa. (19559) 294/K5
Straw, Mont. (†59418) 262/G4
Strawberry, Ark. (72469) 202/H2
Strawberry (lake), N. Dak. 282/J4
Strawberry (res.), Utah 304/C3
Strawberry (riv.), Utah 304/D3
Strawberry Plains, Tenn. (37871) 237/O8
Strawberry Point, Iowa (52076) 229/K3
Strawn, Ill. (61775) 222/E3
Strawn, Kansas (66839) 232/G3
Strawn, Texas (76475) 303/F5
Strayhorn, Miss. (†38665) 256/D1
Strážnice, Czech. 41/D2
Streaky (bay), S. Australia 88/E6
Streaky (bay), S. Australia 94/C5
Streaky Bay, S. Australia 88/E6
Streaky Bay, S. Australia 94/D5
Streamstown, Alberta 182/E3
Streamwood, Ill. (60103) 222/A5
Streator, Ill. (61364) 222/E2
Středočeský (reg.), Czech. 41/C2
Středoslovenský (reg.), Czech. 41/E2
Street, England 10/E5
Street, England 13/E6
Street, Md. (21154) 245/N2
Streeter, N. Dak. (58483) 282/M6
Streeter, Texas (†76856) 303/E7
Streetman, Texas (75859) 303/H6
Streetsboro, Ohio (44240) 284/H3
Strehaia, Romania 45/F3
Stresa, Italy 34/B2
Stretford, England 13/H2
Stretford, England 10/G2
Streymoy (isl.), Denmark 21/B3
Strezhevoy, U.S.S.R. 48/H3
Stříbro, Czech. 41/B2
Strichen, Scotland 15/F3
Strickler, Ark. (†72774) 202/B2
Strike, C.J. (res.), Idaho 220/C7
Strimón (gulf), Greece 45/G5
Stringer, Miss. (39481) 256/F7
Stringtown, Miss. (38777) 256/C3
Stringtown, Okla. (74569) 288/P6
Stripe (lake), Sask. 181/C4
Striven, Loch (inlet), Scotland 15/A2
Stroeder, Argentina 143/D5
Strofádhes (isls.), Greece 45/E7
Stroh, Ind. (46789) 227/G1
Strokestown, Ireland 17/E4
Stroma (isl.), Scotland 15/E2
Stromboli (isl.), Italy 34/E5
Strome, Alberta 182/E3
Stromeferry, Scotland 15/C3
Stromness, Scotland 10/E1
Stromness, Scotland 15/E2
Stromsburg, Nebr. (68666) 264/G3
Strömstad, Sweden 18/G7
Strömsund, Sweden 18/K5
Stronach, Mich. (49681) 250/C4
Strong, Ark. (71765) 202/F7
Strong○, Maine (04983) 243/C6

Strong (riv.), Miss. 256/D7
Strong City, Kansas (66869) 232/F3
Strong City, Okla. (73665) 288/G3
Strongfield, Sask. 181/E4
Stronghurst, Ill. (61480) 222/C3
Strongs, Mich. (49790) 250/E2
Strongs, Miss. (†39730) 256/G3
Strongsville, Ohio (44136) 284/G10
Stronsay (firth), Scotland 15/F1
Stronsay (isl.), Scotland 10/E1
Stronsay (isl.), Scotland 15/F1
Strontian, Scotland 15/C4
Stropkov, Czech. 41/F2
Stroud, Ala. (†36855) 195/H4
Stroud, England 13/E6
Stroud, England 10/E5
Stroud, N.S. Wales 97/G3
Stroud, Okla. (74079) 288/N3
Stroud, Ontario 177/E3
Stroudsburg, Pa. (18360) 294/M4
Struan, Sask. 181/D3
Struan, Scotland 15/B3
Struble, Iowa (51057) 229/A3
Struer, Denmark 21/B5
Struer, Denmark 18/F8
Struga, Yugoslavia 45/E5
Strum, Wis. (54770) 317/D6
Struma (riv.), Bulgaria 45/F5
Strumble (head), Wales 13/B5
Strumica, Yugoslavia 45/F5
Strunk, Ky. (42649) 237/N7
Struthers, Ohio (44471) 284/J3
Stryker, Mont. (59933) 262/B2
Stryker, Ohio (43557) 284/B3
Strykersville, N.Y. (14145) 276/C5
Strzegom, Poland 47/C3
Strzelce Krajeńskie, Poland 47/B2
Strzelce Opolskie, Poland 47/D3
Strzelecki (creek), S. Australia 88/G5
Strzelecki (creek), S. Australia 94/G3
Strzelecki (mt.), Tasmania 99/D2
Strzelin, Poland 47/C3
Strzelno, Poland 47/D2
Strzyżów, Poland 47/E4
Stuart (isl.), Alaska 196/F2
Stuart (lake), Br. Col. 184/E3
Stuart, Fla. (33494) 212/F4
Stuart, Iowa (50250) 229/E6
Stuart, Nebr. (68780) 264/E2
Stuart, Okla. (74570) 288/O5
Stuart (range), S. Australia 94/D3
Stuart, Va. (24171) 307/H7
Stuart (mt.), Wash. 310/E3
Stuartburn, Manitoba 179/F5
Stuart Island, Br. Col. 184/E5
Stuarts Draft, Va. (24477) 307/L4
Stuart Town, N.S. Wales 97/E3
Stubbekøbing, Denmark 21/F8
Stubbenkammer (pt.), E. Germany 22/E1
Stub Hill (mt.), N.H. 268/E1
Stuckey, S.C. (29554) 296/H4
Studénka, Czech. 41/D2
Studley, Kansas (67759) 232/B2
Studley, Va. (23162) 307/O5
Stukely-Sud, Québec 172/E4
Stump (lake), N. Dak. 282/O4
Stumptown, W. Va. (25280) 312/E5
Stumpy Point, N.C. (27978) 281/T3
Stupino, U.S.S.R. 52/E4
Stura (riv.), Italy 34/A2
Sturbridge, Mass. (01566) 249/F4
Sturbridge○, Mass. (01566) 249/F4
Sturdivant, Mo. (63782) 261/M8
Sturgeon (lake), Alberta 182/B2
Sturgeon (bay), Manitoba 179/E3
Sturgeon (riv.), Mich. 250/C2
Sturgeon (riv.), Minn. 255/F3
Sturgeon, Mo. (65284) 261/H4
Sturgeon (lake), Ontario 177/G5
Sturgeon, Pa. (15082) 294/B5
Sturgeon, Pr. Edward I. 168/F2
Sturgeon (riv.), Sask. 181/E2
Sturgeon Bay, Wis. (54235) 317/M6
Sturgeon Falls, Ont. 162/H6
Sturgeon Falls, Ontario 175/E3
Sturgeon Falls, Ontario 177/E1
Sturgeon Heights, Alberta 182/B2
Sturgeon Lake, Minn. (55783) 255/F4
Sturgeon Point, Ontario 177/F3
Sturgeon Weir, Sask. 181/N4
Sturgis, Ky. (42459) 237/F5
Sturgis, Mich. (49091) 250/D7
Sturgis, Miss. (39769) 256/G4
Sturgis, Sask. 181/J4
Sturgis, S. Dak. (57785) 298/B5
Štúrovo, Czech. 41/E3
Sturt (mt.), N.S. Wales 97/A1
Sturt (plain), North. Terr. 93/C4
Sturt (des.), Queensland 88/G5
Sturt (des.), Queensland 95/B3
Sturt (riv.), S. Australia 88/D8
Sturt (des.), S. Australia 94/G3
Sturt (riv.), S. Australia 94/B8
Sturt (creek), W. Australia 88/D3
Sturt (creek), W. Australia 92/D2
Sturtevant, Wis. (53177) 317/M3
Stutsman (co.), N. Dak. 282/M5
Stutterheim, S. Africa 118/D6
Stuttgart, Ark. (72160) 202/H4
Stuttgart, Kansas (67670) 232/C2
Stuttgart, W. Germany 7/E4
Stuttgart, W. Germany 22/C4
Styria (prov.), Austria 41/C3
Suai, Malaysia 85/E5
Suakin, Sudan 102/F3
Suakin, Sudan 59/C6
Suakin, Sudan 111/G4
Suakin (arch.), Sudan 111/G4
Suamico, Wis. (54173) 317/K6
Suao, China 77/K7
Suapi, Bolivia 136/B4
Suapure (riv.), Venezuela 124/E4
Suaqui, Mexico 150/E2
Suárez (riv.), Colombia 126/D4
Subang, Indonesia 85/H2
Subata, U.S.S.R. 53/D2
Subei, China 77/E4
Subeihi, Jordan 65/D3
Subh, Jebel (mt.), Saudi Arabia 59/C5
Subiaco, Ark. (72865) 202/C3
Subiaco, W. Australia 88/B2
Subi Besar (isl.), Indonesia 85/D5
Subic (bay), Philippines 82/C3
Sublett (mts.), Idaho 220/E7
Sublett, Ky. (41470) 237/P5
Sublette, Ill. (61367) 222/D2
Sublette, Kansas (67877) 232/B4
Sublette, Mo. (†63546) 261/G2
Sublette (co.), Wyo. 319/C3
Subligna, Georgia (†30747) 217/B1
Sublimity, Oreg. (97385) 291/E3
Subotica, Yugoslavia 7/F4
Subotica, Yugoslavia 45/D2
Subtle, Ky. (42129) 237/L7
Sucarnochee, Miss. (†39352) 256/H5
Sucarnoochee (creek), Miss. 256/G5
Succasunna, N.J. (07876) 273/D2
Success, Ark. (72470) 202/J1
Success, Mo. (65570) 261/H8
Success, Sask. 181/D5
Succor (creek), Oreg. 291/K4
Suceava, Romania 45/G2
Suchedniów, Poland 47/E3
Suches, Bolivia 136/A4
Suches (riv.), Bolivia 136/A4
Suches, Georgia (30572) 217/E1
Suchitoto, El Salvador 154/C4
Süchow (Xuzhou), China 77/J5
Sucia (bay), P. Rico 161/A3
Sucia (isl.), Wash. 310/C2
Sucio (riv.), Colombia 126/B4
Suck (riv.), Ireland 17/E5
Sucre (cap.), Bolivia 2/F6
Sucre (cap.), Bolivia 120/C4
Sucre (cap.), Bolivia 136/C6
Sucre (dept.), Colombia 126/C3
Sucre, Bolívar, Colombia 126/C3
Sucre, Caquetá, Colombia 126/C7
Sucre, Ecuador 128/B3
Sucre (state), Venezuela 124/G2
Sucre, Venezuela 124/D3
Sucúa, Ecuador 128/C4
Sucuriju, Brazil 132/D2
Sud (dept.), Haiti 158/A6
Sud (chan.), Haiti 158/B6
Suda (riv.), U.S.S.R. 52/E3
Sudak, U.S.S.R. 52/E6
Sudan 2/L5
Sudan 102/E3
SUDAN 111/E4
SUDAN 59/B6
Sudan (reg.) 102/D3
Sudan (reg.), Benin 106/E6
Sudan (reg.), Chad 111/C5
Sudan (reg.), Mali 106/D6
Sudan (reg.), Niger 106/F6
Sudan (reg.), Nigeria 106/F6
Sudan (reg.), Sudan 111/E5
Sudan, Texas (79371) 303/B3
Sudan (reg.), Upper Volta 106/D6
Sudbury, England 10/G4
Sudbury, England 13/H5
Sudbury○, Mass. (01776) 249/A6
Sudbury (res.), Mass. 249/H3
Sudbury (riv.), Mass. 249/A6
Sudbury, Ont. 162/H6
Sudbury, Ont. 146/K5
Sudbury (reg. munic.), Ontario 175/D3
Sudbury (terr. dist.), Ontario 175/D3
Sudbury (terr. dist.), Ontario 177/J5
Sudbury (reg. munic.), Ontario 177/K6
Sudbury, Ontario 177/K5
Sudbury, Ontario 175/D3
Sudbury○, Vt. (†05733) 268/A4
Sudd (swamp), Sudan 102/F4
Sudd (swamp), Sudan 111/F6
Suddie, Guyana 131/B2
Sudeten (mts.), Czech. 41/C1
Sudeten (range), Poland 47/B3
Sudhuroy (isl.), Denmark 21/B3
Sudirman (range), Indonesia 85/K6
Sudith, Ky. (40381) 237/O4
Sudlersville, Md. (21668) 245/P4
Sue (riv.), Sudan 111/E6
Sueca, Spain 33/F3
Suemez (isl.), Alaska 196/M2
Suez (canal) 2/L4
Suez, Egypt 111/K3
Suez, Egypt 59/B4
Suez, Egypt 102/F2
Suez (canal), Egypt 102/F1
Suez (canal), Egypt 111/K3
Suez (canal), Egypt 59/B3
Suez (gulf), Egypt 111/F2
Suez (gulf), Egypt 59/B4
Suf, Jordan 65/D3
Sufeina, Saudi Arabia 59/D5
Sufers, Switzerland 39/H3
Suffern, N.Y. (10901) 276/J8
Suffield, Alberta 182/E4
Suffield, Conn. (06078) 210/E1
Suffield○, Conn. (06078) 210/E1
Suffield, Ohio (†44260) 284/H3
Suffolk (co.), England 13/H5
Suffolk (co.), Mass. 249/K3
Suffolk, Mont. (†59451) 262/G3
Suffolk (co.), N.Y. 276/P9
Suffolk (I.C.), Va. (*23432) 307/P7
Sufian, Iran 66/E1
Sugar (creek), Ind. 227/B3
Sugar (creek), Ind. 227/F5
Sugar (creek), Ind. 227/C5
Sugar (isl.), Mich. 250/E2
Sugar (riv.), N.H. 268/C5
Sugar (creek), Pa. 294/J2
Sugar (riv.), Wis. 317/H10
Sugar Bush, Wis. (†54961) 317/J7
Sugarbush Hill (mt.), Wis. 317/J4
Sugar City, Colo. (81076) 208/M6
Sugar City, Idaho (83448) 220/G6
Sugarcreek, Ohio (44681) 284/G5
Sugar Creek, Pa. (†16323) 294/C3
Sugar Grove, Ark. (†72927) 202/C3
Sugar Grove, Ill. (60554) 222/E2
Sugar Grove, Ohio (43155) 284/E6
Sugargrove, Pa. (16350) 294/D1
Sugar Grove, Va. (24375) 307/E7
Sugar Grove, W. Va. (26815) 312/H5
Sugar Hill, Georgia (†30518) 217/E2
Sugar Hill○, N.H. (03585) 268/D3
Sugar Island, Mich. (†49783) 250/E2
Sugar Land, Texas (77478) 303/J8
Sugarloaf (key), Fla. 212/E7
Sugarloaf (hill), Hawaii 218/C4
Sugarloaf (mt.), Ireland 17/B8
Sugarloaf (pt.), N. S. Wales 88/J6
Sugarloaf (passage), N.S. Wales 97/J1
Sugarloaf (pt.), N.S. Wales 97/G3
Sugarloaf P.O. (Big Bear City), Calif. (92314) 204/J9
Sugar Notch, Pa. (18706) 294/E7
Sugartown, La. (70662) 238/D5
Sugar Tree, Tenn. (38380) 237/E9
Sugar Tree Ridge, Ohio (45133) 284/C7
Sugar Valley, Georgia (30746) 217/C1
Sugbai (passage), Philippines 82/C8
Sugden, Okla. (†73565) 288/L6
Suggsville, Ala. (†36482) 195/C7
Sühbaatar, Mongolia 77/H2
Sühbaatar (Sukhe Bator), Mongolia 77/G1
Sühbaatar, Mongolia 54/M5
Suheli Par (atoll), India 68/C6
Suhl (dist.), E. Germany 22/D3
Suhl, E. Germany 22/D3
Suhr, Switzerland 39/F2
Şuhut, Turkey 63/D3
Sui, Pakistan 68/B3
Suiattle (riv.), Wash. 310/D2
Suichang, China 77/J6
Suide, China 77/G4
Suifenhe, China 77/M3
Suihua, China 77/L2
Suijiang, China 77/F6
Suileng, China 77/L2
Suining, China 77/G5
Suipacha, Argentina 143/G7
Suipacha, Bolivia 136/C7
Suir (riv.), Ireland 10/C4
Suir (riv.), Ireland 17/G7
Suisun (bay), Calif. 204/K1
Suisun City, Calif. (94585) 204/K1
Suit, N.C. (28906) 281/A4
Suita, Japan 81/J7
Suitland-Silver Hill, Md. (†20746) 245/F5
Sui Xian, China 77/H5
Suizhong, China 77/K3
Sukabumi, Indonesia 85/H2
Sukadana, Indonesia 85/E6
Sukagawa, Japan 81/K5
Sukhana, U.S.S.R. 48/M3
Sukhinichi, U.S.S.R. 52/E4
Sukhona (riv.), U.S.S.R. 52/F2
Sukhothai, Thailand 72/D3
Sukhumi, U.S.S.R. 7/H4
Sukhumi, U.S.S.R. 52/F6
Sukhumi, U.S.S.R. 48/D5
Suk, Sudan 111/F5
Sukkertoppen, Greenl. 4/C12
Sukkur, Pakistan 54/H7
Sukkur, Pakistan 59/J4
Sukkur, Pakistan 68/B3
Süköśd, Hungary 41/E3
Sukumo, Japan 81/F7
Sul (chan.), Brazil 120/E2
Sul (chan.), Brazil 132/D2
Sula (isls.), Indonesia 54/O10
Sula (isls.), Indonesia 85/H6
Sula, Mont. (59871) 262/B5
Sulaco, Honduras 154/D3
Sulaco (riv.), Honduras 154/D3
Sulaiman (range), Pakistan 68/C3
Sulaimaniya (gov.), Iraq 66/D3
Sulaimaniya, Iraq 59/E2
Sulaimaniya, Iraq 66/D3
Sulaiyil, Saudi Arabia 59/E5
Sulakyurt, Turkey 63/E2
Sulanheer, Mongolia 77/G3
Sulawesi (isl.), Indonesia 85/G6
Sulechów, Poland 47/B2
Sulęcin, Poland 47/B2
Sulgen, Switzerland 39/H1
Sulina, Romania 45/J3
Sulitelma (mt.), Sweden 18/K3
Sulitjelma, Norway 18/K3
Sulitjelma (mt.), Norway 18/J3
Sullana, Peru 128/B5
Sullana, Peru 120/A3
Sulligent, Ala. (35586) 195/B3
Sullivan (lake), Alberta 182/D3
Sullivan, Ill. (61951) 222/E4
Sullivan (co.), Ind. 227/C6
Sullivan, Ind. (47882) 227/C6
Sullivan, Ky. (42460) 237/E6
Sullivan, Maine (†04689) 243/G6
Sullivan○, Maine (†04689) 243/G6
Sullivan (co.), Mo. 261/F2
Sullivan, Mo. (63080) 261/K6
Sullivan (co.), N.H. 268/C5
Sullivan○, N.H. (†03445) 268/C5
Sullivan (co.), N.Y. 276/L7
Sullivan, Ohio (44880) 284/F3
Sullivan (co.), Pa. 294/J3
Sullivan (co.), Tenn. 237/S7
Sullivan (lake), Wash. 310/H2
Sullivan, W. Va. (†25847) 312/D7
Sullivan, Wis. (53178) 317/H1
Sullivan Gardens, Tenn. (†37660) 237/R8
Sullivan Mines, Québec 174/B3
Sullivans Island, S.C. (29482) 296/H6
Sully, Iowa (50251) 229/H5
Sully, Québec 172/H2
Sully (co.), S. Dak. 298/J4
Sulmona, Italy 34/D3
Sulphide, Ontario 177/G3
Sulphur (riv.), Ark. 202/B7
Sulphur, Ind. (47174) 227/E8
Sulphur, Ky. (40070) 237/L4
Sulphur, La. (70663) 238/D6
Sulphur, Nev. (†89445) 266/C2
Sulphur, Okla. (73086) 288/N5
Sulphur (creek), S. Dak. 298/D4
Sulphur (riv.), Texas 303/J4
Sulphur City, Ark. (†72701) 202/B2
Sulphur Creek, Tasmania 99/C3
Sulphur Draw (dry riv.), Texas 303/B4
Sulphur Fork, Red (riv.), Tenn. 237/H8
Sulphur Rock, Ark. (72579) 202/H2
Sulphur Spring (valley), Ariz. 198/F6
Sulphur Spring (range), Nev. 266/E3
Sulphur Springs, Ala. (†30738) 195/G1
Sulphur Springs, Ark. (72768) 202/B1
Sulphur Springs, Ind. (47388) 227/G4
Sulphur Springs, Iowa (†50588) 229/C3
Sulphur Springs, Mo. (63083) 261/M6
Sulphur Springs, Ohio (44881) 284/E4
Sulphur Springs, Texas (75482) 303/J4
Sulphur Springs (creek), Texas 303/B4
Sulphur Well, Ky. (42129) 237/K6
Sultan, Ontario 175/D3
Sultan, Ontario 177/J5
Sultan (mts.), Turkey 63/D3
Sultan, Wash. (98294) 310/D3
Sultan (riv.), Wash. 310/D3
Sultanabad (Kashmar), Iran 66/L3
Sultandağı, Turkey 63/D3
Sultanhanı, Turkey 63/E3
Sultan Kudarat (prov.), Philippines 82/E7
Sulu (prov.), Philippines 82/C7
Sulu (arch.), Philippines 54/O9
Sulu (sea), Philippines 54/N9
Sulu (arch.), Philippines 82/B8
Sulu (arch.), Philippines 85/G4
Sulu (sea), Philippines 85/G4
Sulu (sea), Philippines 82/B6
Suluan (isl.), Philippines 82/F5
Suluova, Turkey 63/F2
Sulz, Switzerland 39/F1
Sulzbach, W. Germany 22/B4
Sulzbach-Rosenberg, W. Germany 22/D4
Sulzberger (bay) 5/B11
Sulzflüh (mt.), Switzerland 39/J2
Sumami Auma, Brazil 132/B4
Sumampa, Argentina 143/D2
Sumas, Wash. (98295) 310/C2
Sumatra, Fla. (32335) 212/B1
Sumatra (isl.), Indonesia 2/P6
Sumatra (isl.), Indonesia 54/L9
Sumatra (isl.), Indonesia 85/B5
Sumatra, Mont. (59083) 262/J4
Sumava Resorts, Ind. (46379) 227/C2
Sumba (isl.), Indonesia 54/N11
Sumba (isl.), Indonesia 85/F7
Sumba (str.), Indonesia 85/F7
Sumbawa (isl.), Indonesia 54/N11
Sumbawa (isl.), Indonesia 85/F7
Sumbawa Besar, Indonesia 85/F7
Sumbawanga, Tanzania 115/F5
Sumbay, Peru 128/G10
Sumbilca, Peru 128/D8
Sumbing (mt.), Indonesia 85/J2
Sumburgh (head), Scotland 15/G2
Sumedang, Indonesia 85/H2
Sümeg, Hungary 41/D3
Sumenep, Indonesia 85/L2
Sumgait, U.S.S.R. 7/J4
Sumgait, U.S.S.R. 52/G6
Sumidero, Cuba 158/A2
Sumiswald, Switzerland 39/E2
Sumiton, Ala. (35148) 195/D3
Summan (plat.), Saudi Arabia 59/E4
Summer (isl.), Mich. 250/C3
Summer (lake), Oreg. 188/C2
Summer (lake), Oreg. 291/G5
Summerberry, Sask. 181/J5
Summerdale, Ala. (36580) 195/C10
Summerfield, Ala. (†36701) 195/E5
Summerfield, Fla. (32691) 212/D2
Summerfield, Ill. (62289) 222/D5
Summerfield, Kansas (66541) 232/F2
Summerfield, La. (71079) 238/E1
Summerfield, Mo. (†65013) 261/J6
Summerfield, N.C. (27358) 281/K2
Summerfield, Ohio (43788) 284/H6
Summerfield, Okla. (74966) 288/S5
Summerfield, Texas (79085) 303/B3
Summerford, Newf. 166/C4
Summerford, Ohio (†43140) 284/D6
Summer Hill, Ill. (62372) 222/C4
Summerhill, Pa. (15958) 294/E5
Summer Isles (isls.), Scotland 15/C2
Summer Lake, Oreg. (97640) 291/G5
Summerland, Br. Col. 184/G5
Summerland, Calif. (93067) 204/F9
Summerland, Miss. (†39168) 256/F7
Summerland Key, Fla. (33042) 212/E7
Summers, Ark. (72769) 202/A2
Summers (co.), W. Va. 312/E7
Summerset, Iowa (†50125) 229/F6
Summer Shade, Ky. (42166) 237/K7
Summerside, Pr. Edward I. 168/E2
Summersville, Ky. (42782) 237/K6
Summersville, Mo. (65571) 261/J8
Summersville, Ohio (†43344) 284/D5
Summersville, W. Va. (26651) 312/E6
Summersville (lake), W. Va. 312/E6
Summerton, S.C. (29148) 296/G4
Summertown, Georgia (30466) 217/H5
Summertown, Tenn. (38483) 237/G10
Summerville, Georgia (30747) 217/B2
Summerville, La. (†71465) 238/F3
Summerville, Newf. 166/D2
Summerville, Oreg. (97876) 291/K2
Summerville, Pa. (15864) 294/D3
Summerville, S.C. (29483) 296/G5
Summerville Centre, Nova Scotia 168/D5
Summit, Ala. (†35031) 195/F2
Summit, Alaska (†99729) 196/J2
Summit, Ark. (72677) 202/E1
Summit (co.), Colo. 208/G3
Summit (peak), Colo. 208/F8
Summit (lake), Iowa 229/E6
Summit, Miss. (39666) 256/D8
Summit (lake), Nev. 266/C1
Summit, N.J. (07901) 273/E2
Summit (co.), Ohio 284/G3
Summit (creek), Oreg. 291/J3
Summit, Québec 172/E2
Summit, R.I. (†02827) 249/H6
Summit, S.C. (†29054) 296/E4
Summit, S. Dak. (57266) 298/P3
Summit (co.), Utah 304/D3
Summit, Utah (84772) 304/B6
Summit-Argo, Ill. (60501) 222/B6
Summit Bridge, Del. (†19709) 245/R2
Summit City, Calif. (96089) 204/C3
Summit City, Mich. (†49649) 250/D4
Summit Hill, Pa. (18250) 294/L4
Summit Lake, Wis. (54485) 317/H5
Summit Lake Ind. Res., Nev. 266/B1
Summit Point, W. Va. (25446) 312/K4
Summitville, Ind. (46070) 227/F4
Summitville, Iowa (†52632) 229/K8
Summitville, N.Y. (12781) 276/L7
Summitville, Ohio (43962) 284/J4
Summitville, Tenn. (37382) 237/K9
Summum, Ill. (†61501) 222/C3
Sumner (str.), Alaska 196/M2
Sumner, Georgia (31789) 217/E7
Sumner, Ill. (62466) 222/F5
Sumner, Iowa (50674) 229/J3
Sumner (co.), Kansas 232/E4
Sumner○, Maine (†04292) 243/C7
Sumner, Mich. (48889) 250/E5
Sumner, Miss. (38957) 256/D3
Sumner, Mo. (64681) 261/F3
Sumner, Nebr. (68878) 264/E4
Sumner (dam), N. Mex. 274/E4
Sumner (lake), N. Mex. 274/E4
Sumner (lake), N. Zealand 100/D5
Sumner, Oreg. (†97420) 291/C4
Sumner (co.), Tenn. 237/J8
Sumner, Wash. (98390) 310/C3
Sumner-East Sumner, Maine (04232) 243/C7
Sumoto, Japan 81/G6
Šumperk, Czech. 41/D1
Sumprabum, Burma 72/C1
Sumpter, Ark. (†71647) 202/F7
Sumpter, Oreg. (97877) 291/J3
Sumrall, Miss. (39482) 256/E8
Sumter (co.), Ala. 195/B5
Sumter, Ala. (†35086) 195/D4
Sumter (co.), Fla. 212/D3
Sumter (co.), Georgia 217/D6
Sumter (co.), S.C. 296/G4
Sumter, S.C. (29150) 296/G4
Sumterville, Ala. (†35460) 195/B5
Sumy, U.S.S.R. 7/H3
Sumy, U.S.S.R. 52/E4
Sumy, U.S.S.R. 48/D4
Sun, La. (70463) 238/L5
Sun (riv.), Mont. 262/D3
Sunagawa, Japan 81/K2
Sunapee○, N.H. (03782) 268/C5
Sunapee (lake), N.H. 268/C5
Sunart, Loch (inlet), Scotland 15/C4
Sunbeam, Colo. (†81640) 208/C1
Sunbeam, Idaho (†83278) 220/D5
Sunbright, Tenn. (37872) 237/M8
Sunburg, Minn. (56289) 255/C5
Sunburst, Mont. (59482) 262/E2
Sunbury, Iowa (†52778) 229/M5
Sunbury (co.), New Bruns. 170/D3
Sunbury, N.C. (27979) 281/R2
Sunbury, Ohio (43074) 284/E5
Sunbury, Pa. (17801) 294/J4
Sunbury, Victoria 97/C5
Sunbury-on-Thames, England 13/G8
Sunbury-on-Thames, England 10/B6
Sunchales, Argentina 143/F5
Suncho Corral, Argentina 143/D2
Sunch'ŏn, N. Korea 81/B4
Sunch'ŏn, S. Korea 81/C6
Sun City, Ariz. (*85351) 198/C5
Sun City, Calif. (92381) 204/F11
Sun City, Fla. (33586) 212/D4
Sun City, Kansas (67143) 232/D4
Sun City Center, Fla. (†33570) 212/C3
Suncook, N.H. (03275) 268/D5
Suncook (lakes), N.H. 268/E5
Suncook (riv.), N.H. 268/E5
Sunda (isls.), Indonesia 54/L10
Sunda (str.), Indonesia 54/M10
Sunda (str.), Indonesia 85/C7
Sundahl, Minn. (†56545) 255/B3
Sundance, Wyo. (82729) 319/H1
Sundarbans (reg.), Bangladesh 68/F4
Sundarbans (reg.), India 68/F4
Sundargarh, India 68/E4
Sunday (riv.), Maine 243/B6
Sundbyberg, Sweden 18/G1
Sunderland, England 13/J3
Sunderland, England 10/F3
Sunderland○, Mass. (01375) 249/D3
Sunderland, Ontario 177/E3
Sunderland○, Vt. (†05250) 268/A5
Sundown, Manitoba 179/F5
Sundown, Texas (79372) 303/B4
Sundra, S. Africa 118/J6
Sundre, Alberta 182/C4
Sundridge, Ontario 177/E2
Sundsvall, Sweden 18/K5
Sundsvall, Sweden 7/F2
Sunfield, Ill. (†62832) 222/D5
Sunfield, Mich. (48890) 250/D6
Sunfish Lake, Minn. (†55075) 255/E6
Sunflower, Ala. (36581) 195/B8
Sunflower (mt.), Kansas 232/A2
Sunflower (co.), Miss. 256/C3
Sunflower, Miss. (38778) 256/C3
Sunflower (riv.), Miss. 256/C5
Sungaipenuh, Indonesia 85/C6
Sungai Petani, Malaysia 72/C6
Sungurlu, Turkey 63/F2
Sunland, Calif. (91040) 204/C10
Sunland Gardens, Fla. (†33450) 212/F4
Sunman, Ind. (47041) 227/G6
Sunndalsøra, Norway 18/F5
Sunne, Sweden 18/H7
Sunnybrae, Nova Scotia 168/F3
Sunnybrook, Alberta 182/C3
Sunny Corner, New Bruns. 170/E2
Sunnydale, Wash. (†98101) 310/B2
Sunny Isles, Fla. (33160) 212/C4
Sunnymead, Calif. (92388) 204/F11
Sunnynook, Alberta 182/E4
Sunny Point Mil. Ocean Term., N.C. 281/O6
Sunnyside, Fla. (32461) 212/C6
Sunny Side, Georgia (30284) 217/D4
Sunnyside, Ill. (†60050) 222/A4
Sunnyside, New Bruns. 170/D1
Sunnyside, Newf. 166/D2
Sunnyside, Utah (84539) 304/D4
Sunnyside, Wash. (98944) 310/F4
Sunnyslope, Alberta 182/D4
Sunny South, Ala. (36780) 195/C7
Sunnyvale, Calif. (*94086) 204/K3
Sunnyvale, Texas (†75149) 303/H2
Sunny Valley, Oreg. (97478) 291/D5
Sunol, Calif. (94586) 204/L2
Sunol, Nebr. (†69149) 264/B3
Sun Prairie, Wis. (53590) 317/H9
Sunray, Texas (79086) 303/C1
Sunrise, Fla. (33313) 212/B4
Sunrise, Wyo. (†82215) 319/H3
Sunrise Beach, Mo. (65079) 261/G6
Sunrise Manor, Nev. (†89110) 266/F6
Sunrise Ridge, Ill. (†60097) 222/E1
Sunrise Valley, Br. Col. 184/G2
Sun River, Mont. (59483) 262/E3
Sunsas, Serranía de (mts.), Bolivia 136/F5
Sunset, Ark. (†72364) 202/K3
Sunset (peak), Idaho 220/E6
Sunset, La. (70584) 238/F6
Sunset, Maine (04683) 243/F7
Sunset, S.C. (29685) 296/B2
Sunset, Texas (79407) 303/G4
Sunset, Utah (†84015) 304/B2
Sunset Beach, Calif. (90742) 204/C11
Sunset Beach, Hawaii (†96712) 218/E1
Sunset Beach, N.C. (28459) 281/N7
Sunset Crater Nat'l Mon., Ariz. 198/D3
Sunset Hills, Mo. (†63101) 261/O4
Sunset Hills, Va. (22090) 307/R2
Sunset House, Alberta 182/B2
Sunset Prairie, Br. Co. 184/G2
Sunshine, La. (†70776) 238/K2
Sunshine, Maine (†04627) 243/G7
Sunshine, Victoria 88/K7
Sunshine, Victoria 97/H5
Sunspot, N. Mex. (88349) 274/D6
Suntar, U.S.S.R. 48/M3
Suntrana, Alaska (†99743) 196/J2
Sun Valley, Idaho (83353) 220/D6
Sun Valley, Nev. (†89431) 266/B3
Sun Valley, Sask. 181/F5
Sunyani, Ghana 106/D7
Sunzu (mt.), Zambia 115/F5
Suo (sea), Japan 81/E7
Suolahti, Finland 18/O5
Suomussalmi, Finland 18/Q4
Suonenjoki, Finland 18/P5
Suong, Cambodia 72/E5
Suoyarvi, U.S.S.R. 52/D2
Supai, Ariz. (86435) 198/C2
Supe, Peru 128/D8
Superb, Sask. 181/B4
Superior (lake) 162/G-H6
Superior (lake) 146/K5
Superior (lake) 188/J1
Superior, Ariz. (85273) 198/D5
Superior, Colo. (†80027) 208/J3
Superior, Iowa (51363) 229/D2
Superior (lag.), Mexico 150/M9
Superior (lake), Mich. 250/C2
Superior (lake), Minn. 255/G3
Superior, Mont. (59872) 262/B3
Superior, Nebr. (68978) 264/F4
Superior (lake), Ontario 177/H5
Superior (lake), Ontario 175/C3
Superior, Wis. 188/H1
Superior, Wis. (54880) 317/C2
Superior (lake), Wis. 317/F1
Superior, Wyo. (82945) 319/D4
Superior Village, Wis. (†54880) 317/B2
Superstition (mts.), Ariz. 198/D5
Suphan Buri, Thailand 72/C4
Süphan Dağı (mt.), Turkey 59/D2
Süphan Dağı (mt.), Turkey 63/K3
Supiori (isl.), Indonesia 85/K6
Supply, N.C. (28462) 281/N6
Supreme, La. (†70372) 238/K4
Supung (res.), N. Korea 81/B3
Suqian, China 77/J5
Suquamish, Wash. (98392) 310/A1
Sur (pt.), Calif. 204/D7
Sur, Lebanon 63/F6
Sur, Oman 59/G5
Sura, Ras (cape), Somalia 115/J1
Sura (riv.), U.S.S.R. 52/G4
Surab, Pakistan 68/B3
Surab, Pakistan 59/J4
Surabaya, Indonesia 54/N10
Surabaya, Indonesia 85/K2
Surada, India 68/E5
Surahammar, Sweden 18/J7
Surakarta, Indonesia 54/N10
Surakarta, Indonesia 85/J2
Šurany, Czech. 41/E2
Surat, India 54/J7
Surat, India 68/C4
Surat, Queensland 88/H5
Suratgarh, India 68/C3
Surat Thani, Thailand 72/C5
Sur del Cabo San Antonio (pt.), Argentina 143/E4
Surdulica, Yugoslavia 45/F4
Surendranagar, India 68/C4
Suresnes, France 28/A2
Suretka, C. Rica 154/F6
Surette Island, Nova Scotia 168/B5
Surf, Calif. (†93436) 204/E9
Surf City, N.J. (08008) 273/E4
Surf City, N.C. (28445) 281/O6
Surfside, Fla. (33154) 212/B4
Surfside Beach, S.C. (29577) 296/K4

Surgidero de Batabanó, Cuba 156/A2
Surgidero de Batabanó, Cuba 158/C1
Surgoinsville, Tenn. (37873) 237/R8
Surgut, U.S.S.R. 48/H3
Surigao, Philippines 82/E6
Surigao, Philippines 85/H4
Surigao (str.), Philippines 82/E6
Surigao del Norte (prov.), Philippines 82/F5
Surigao del Sur (prov.), Philippines 82/F6
Surimena, Colombia 126/D6
Surin, Thailand 72/D4
Suriname 2/G5
Suriname 120/D2
SURINAME 131/C3
Suriname (dist.), Suriname 131/D3
Suriname (riv.), Suriname 131/D3
Suring, Wis. (54174) 317/K5
Suripa, Venezuela 124/D4
Suripá (riv.), Venezuela 124/C4
Surire, Salar de (salt dep.), Chile 138/B2
Sürmene, Turkey 63/J2
Surprise, Ariz. (85345) 198/C5
Surprise, Ind. (†47228) 227/E7
Surprise, Nebr. (68667) 264/G3
Surrency, Georgia (31563) 217/H7
Surrey, Br. Col. 184/K3
Surrey (co.), England 13/G6
Surrey, N. Dak. (58785) 282/H3
Surry○, Maine (04684) 243/F7
Surry○, N.H. (03431) 268/C5
Surry (co.), N.C. 281/H2
Surry (co.), Va. 307/P6
Surry, Va. (23883) 307/P6
Surry Mountain (lake), N.H. 268/C5
Sursee, Switzerland 39/F2
Surtsey (isl.), Iceland 21/B2
Sürüç, Turkey 63/H4
Surud Ad (mt.), Somalia 115/J1
Suruga (bay), Japan 81/J6
Surup, Philippines 82/F7
Surveyor, W. Va. (25932) 312/D7
Surwakwima (fall), Guyana 131/A2
Susa (riv.), Denmark 21/E7
Susa (ruins), Iran 66/F4
Susa, Italy 34/A2
Susa, Libya 111/D1
Susa Creek, Alberta 182/A3
Susaki, Japan 81/F7
Susan (riv.), Calif. 204/D3
Susan, Va. (23163) 307/R6
Susangerd, Iran 59/E3
Susangerd, Iran 66/F5
Susank, Kansas (67580) 232/D3
Susanville, Calif. (96130) 204/E3
Susch, Switzerland 39/K3
Suşehri, Turkey 63/H2
Sušice, Czech. 41/B2
Susie, Ky. (†42633) 237/M7
Susitna, Alaska (†99501) 196/B1
Susitna (riv.), Alaska 196/B1
Susquehanna (riv.), Md. 245/N1
Susquehanna (riv.), N.Y. 276/H6
Susquehanna (co.), Pa. 294/L2
Susquehanna, Pa. (18847) 294/L2
Susquehanna (riv.), Pa. 294/K6
Susquehanna, West Branch (riv.), Pa. 294/G3
Susques, Argentina 143/C1
Sussex (co.), Del. 245/S6
Sussex, East (co.), England 13/H7
Sussex, West (co.), England 13/G7
Sussex, New Bruns. 170/E3
Sussex (co.), N.J. 273/D1
Sussex, N.J. (07461) 273/D1
Sussex (co.), Va. 307/O7
Sussex, Va. (23884) 307/O7
Sussex, Wis. (53089) 317/K1
Sussex, Wyo. (†82639) 319/F2
Sussex Corner, New Bruns. 170/E3
Sussex Inlet, N.S. Wales 97/F4
Susten (pass), Switzerland 39/G3
Sustenhorn (mt.), Switzerland 39/G3
Sustut (riv.), Br. Col. 184/D2
Susuman, U.S.S.R. 4/C2
Susuman, U.S.S.R. 54/R3
Susuman, U.S.S.R. 48/P3
Susurluk, Turkey 63/C3
Susuz, Turkey 63/K2
Sütçüler, Turkey 63/D4
Sutherland, Iowa (51058) 229/B3
Sutherland, Nebr. (69165) 264/C3
Sutherland (res.), Nebr. 264/C3
Sutherland, N. S. Wales 88/K5
Sutherland, N.S. Wales 97/J4
Sutherland (trad. co.), Scotland 15/A4
Sutherland Springs, Texas (78161) 303/K11
Sutherlin, Oreg. (97479) 291/D4
Sutherlin, Va. (24594) 307/O7
Sutlej (riv.) 54/J6
Sutlej (riv.), India 68/C3
Sutlej (riv.), Pakistan 68/C3
Sutlej (riv.), Pakistan 59/K4
Sutter (co.), Calif. 204/D4
Sutter, Calif. (95982) 204/D4
Sutter Creek, Calif. (95685) 204/C9
Sutton, Ala. (†36701) 195/D5
Sutton, Alaska (99674) 196/C1
Sutton, England 13/H8
Sutton, England 10/B6
Sutton, Mass. (†01527) 249/G4
Sutton, Nebr. (68979) 264/G4
Sutton○, N.H. (†03260) 268/D5
Sutton, N. Dak. (58484) 282/O5
Sutton, Ontario 177/E3
Sutton (lake), Ontario 175/D2
Sutton (riv.), Ontario 175/D2
Sutton, Québec 172/E4
Sutton (co.), Texas 303/D7
Sutton○, Vt. (05867) 268/C2
Sutton, W. Va. (26601) 312/E5
Sutton (lake), W. Va. 312/F5
Sutton Bridge, England 13/H5
Sutton in Ashfield, England 10/F4
Sutton in Ashfield, England 13/K2
Suttons Bay, Mich. (49682) 250/D3
Suttor (riv.), Queensland 88/H4
Suttor (riv.), Queensland 95/C4
Suttsu, Japan 81/J2
Sutwik (isl.), Alaska 196/G3
Suure-Jaani, U.S.S.R. 53/C1
Suva (cap.), Fiji 87/H7
Suva (cap.), Fiji 2/T6
Suva (Cap.), Fiji 86/Q11
Suvarlı, Turkey 63/G4
Suwa, Japan 81/H6
Suwaiq, Oman 59/G5
Suwaiqiya, Hor as (lake), Iraq 66/D4
Suwałki (prov.), Poland 47/F1
Suwałki, Poland 47/F1
Suwanee, Georgia (30174) 217/D2
Suwannaphum, Thailand 72/D4
Suwannee (co.), Fla. 212/C1
Suwannee, Fla. (32692) 212/C2
Suwannee (riv.), Fla. 212/C2
Suwannee (sound), Fla. 212/C2
Suwannee (riv.), Georgia 217/G10
Suwannee River (Fanning Springs), Fla. (†32693) 212/D2
Suwanose (isl.), Japan 81/O4
Suwar, Syria 63/J5
Suwarrow (atoll), Cook Is. 87/K7
Suweilih, Jordan 65/D3
Suweima, Jordan 65/D4
Suwŏn, S. Korea 81/C5
Su Xian, China 77/J5
Suzhou (Soochow), China 77/K5
Suzhou, China 54/O6
Suzu, Japan 81/H5
Suzu (pt.), Japan 81/H5
Suzuka, Japan 81/H6
Suzzara, Italy 34/C2
Svalbard (isls.), Norway 2/J2
Svalbard (isls.), Norway 4/B9
Svalbard (isls.), Norway 18/C3
Svaneke, Denmark 21/F8
Svanstein, Sweden 18/N3
Svanvik, Norway 18/R2
Svärov, Czech. 41/C1
Svay Rieng, Cambodia 72/E5
Svea, Minn. (56290) 255/C6
Sveagruva, Norway 18/D2
Sveg, Sweden 18/J5
Svelvik, Norway 18/D4
Švenčionis, U.S.S.R. 53/D3
Svendborg, Denmark 18/G9
Svendborg, Denmark 21/D7
Svenljunga, Sweden 18/H8
Svensen, Oreg. (†97103) 291/D1
Svenstavik, Sweden 18/H5
Sverdlovsk, U.S.S.R. 2/N3
Sverdlovsk, U.S.S.R. 54/H4
Sverdlovsk, U.S.S.R. 48/F4
Sverdrup (isls.), N.W.T. 146/J2
Sverdrup (chan.), N.W. Terrs. 187/J1
Sverdrup (isls.), N.W. Terrs. 187/J2
Svetlogorsk, U.S.S.R. 52/C4
Svetlograd, U.S.S.R. 52/F5
Svetlyy, U.S.S.R. 52/A4
Svetozarevo, Yugoslavia 45/E4
Svidník, Czech. 41/F2
Svilajnac, Yugoslavia 45/E3
Svilengrad, Bulgaria 45/G5
Svinninge, Denmark 21/E6
Svir' (riv.), U.S.S.R. 52/D2
Svishtov, Bulgaria 45/G4
Svitava (riv.), Czech. 41/D2
Svitavy, Czech. 41/D2
Svobodnyy, U.S.S.R. 54/O4
Svobodnyy, U.S.S.R. 48/N4
Svolvaer, Norway 18/J2
Svratka (riv.), Czech. 41/D2
Swabian Jura (range), W. Germany 22/C4
Swadlincote, England 10/G2
Swadlincote, England 13/F5
Swaffham, England 13/H5
Swaffham, England 10/G4
Swaim, Ala. (†35764) 195/F1
Swain (co.), N.C. 281/B3
Swain (reefs), Queensland 88/J4
Swain (reefs), Queensland 95/E4
Swains (isl.), Amer. Samoa 87/K7
Swainsboro, Georgia (30401) 217/H5
Swakop (riv.), Namibia 118/B4
Swakopmund, Namibia 118/A4
Swakopmund, Namibia 102/D7
Swale (riv.), England 13/F3
Swale (riv.), England 10/F3
Swale (isl.), Newf. 166/D1
Swaledale, Iowa (50477) 229/G3
Swalwell, Alberta 182/D4
Swampscott○, Mass. (01907) 249/E6
Swan (hills), Alberta 182/C2
Swan (riv.), Alberta 182/C2
Swan (Cisne) (isls.), Honduras 154/F2
Swan, Iowa (50252) 229/G6
Swan (lake), Manitoba 179/B2
Swan (lake), Manitoba 179/D5
Swan (riv.), Manitoba 179/A3
Swan (lake), Minn. 255/D6
Swan (lake), Mont. 262/C3
Swan (lake), Nebr. 264/B3
Swan (riv.), Sask. 181/J3
Swan (creek), S. Dak. 298/J3
Swan (lake), S. Dak. 298/K3
Swan (lake), S. Dak. 298/O3
Swan (isl.), Tasmania 99/E2
Swan (lake), Utah 304/B4
Swan, W. Australia 88/B2
Swan (riv.), W. Australia 92/A1
Swanage, England 10/F5
Swanage, England 13/E7
Swan Creek, Ill. (†61473) 222/C3
Swandale, W. Va. (†25043) 312/E5
Swan Hill, Victoria 88/G7
Swan Hill, Victoria 97/B4
Swan Hills, Alberta 182/C2
Swanington, Ind. (†47944) 227/C3
Swanlake, Idaho (83281) 220/F7
Swan Lake, Manitoba 179/D5
Swan Lake, Miss. (38958) 256/D3
Swan Lake, Mont. (59911) 262/C3
Swanlinbar, Ireland 17/F3
Swannanoa, N.C. (28778) 281/E3
Swan Plain, Sask. 181/K3
Swanquarter, N.C. (27885) 281/S4
Swan River, Man. 162/F5
Swan River, Manitoba 179/A2
Swan River, Minn. (55784) 255/E3
Swansboro, N.C. (28584) 281/P5
Swanscombe, England 10/C5
Swanscombe, England 13/J8
Swansea, Ill. (62221) 222/B3
Swansea○, Mass. (02777) 249/K5
Swansea, S.C. (29160) 296/E4
Swansea, Tasmania 99/D4
Swansea, Wales 7/D3
Swansea, Wales 13/C6
Swansea, Wales 10/D5
Swansea (bay), Wales 13/D6
Swansea Center, Mass. (†02777) 249/K5
Swans Island, Maine (04685) 243/G7
Swans Island○, Maine (04685) 243/G7
Swanson (lake), Nebr. 264/D4
Swanson, Sask. 181/D4
Swanton, Md. (21561) 245/A3
Swanton, Nebr. (68445) 264/H4
Swanton, Ohio (43558) 284/C2
Swanton, Vt. (05488) 268/A2
Swanton○, Vt. (05488) 268/A2
Swan Valley, Idaho (83449) 220/G6
Swan View, W. Australia 88/B2
Swanville○, Maine (†04915) 243/E6
Swanville, Minn. (56382) 255/D5
Swanwick, Ill. (62290) 222/D5
Swanzey○, N.H. (†03431) 268/C6
Swarthmore, Pa. (19081) 294/M7
Swartswood, N.J. (07877) 273/D1
Swartswood (lake), N.J. 273/D1
Swartz, La. (71281) 238/G1
Swartz Creek, Mich. (48473) 250/F6
Swarzędz, Poland 47/C2
Swatara, Minn. (55785) 255/E4
Swatara, Pa. (17111) 294/J5
Swatow (Shantou), China 77/J7
Swayzee, Ind. (46986) 227/F4
Swaziland 2/L7
Swaziland 102/F7
SWAZILAND 118/E5
Swea City, Iowa (50590) 229/E2
Sweatman, Miss. (†38925) 256/E3
Swedeborg, Mo. (65572) 261/H7
Swedeburg, Nebr. (†68066) 264/H3
Sweden 2/K2
Sweden 4/D9
Sweden 7/F2
SWEDEN 18
Sweden, Maine (†04040) 243/B7
Sweden○, Maine (†04040) 243/B7
Swedesboro, N.J. (08085) 273/C4
Swedesburg, Iowa (52652) 229/L6
Sweeden, Ky. (42285) 237/J6
Sween, Loch (inlet), Scotland 15/C5
Sweeny, Texas (77480) 303/J8
Sweet, Idaho (83670) 220/B6
Sweet Bay, Newf. 166/D2
Sweet Briar, Va. (24595) 307/K5
Sweet Chalybeate, Va. (†24426) 307/H5
Sweet Grass (co.), Mont. 262/G5
Sweetgrass, Mont. (59484) 262/E2
Sweet Home, Ark. (72164) 202/F4
Sweet Home, Oreg. (97386) 291/E3
Sweet Home, Texas (77987) 303/H8
Sweet Lake, La. (†70601) 238/D7
Sweetser, Ind. (46987) 227/F3
Sweet Springs, Mo. (65351) 261/F5
Sweet Springs, W. Va. (24980) 312/F7
Sweet Valley, Pa. (18656) 294/K3
Sweet Water, Ala. (36782) 195/C6
Sweetwater, Fla. (†33144) 212/B5
Sweetwater, Nebr. (†68844) 264/E3
Sweetwater (lake), N. Dak. 282/N3
Sweetwater, Okla. (73666) 288/G4
Sweetwater, Tenn. (37874) 237/N9
Sweetwater, Texas 188/F4
Sweetwater, Texas (79556) 303/D5
Sweetwater (co.), Wyo. 319/D4
Sweetwater (riv.), Wyo. 319/D3
Sweetwater Creek, Fla. (†33614) 212/B2
Swellendam, S. Africa 118/C6
Swenson, Texas (†79502) 303/D4
Swett, S. Dak. (†57551) 298/E7
Świdnica, Poland 47/C3
Świdnik, Poland 47/F3
Świdwin, Poland 47/B2
Świebodzice, Poland 47/C3
Świebodzin, Poland 47/B2
Świecie, Poland 47/D2
Świętochłowice, Poland 47/A4
Swift (riv.), Maine 243/B6
Swift (riv.), Mass. 249/E4
Swift (co.), Minn. 255/C5
Swift, Minn. (56682) 255/C2
Swift (creek), Va. 307/O6
Swift Creek (res.), Wash. 310/C4
Swift Current, Newf. 166/C2
Swift Current, Sask. 146/H4
Swift Current, Sask. 162/F5
Swift Current, Sask. 181/D5
Swift Current (creek), Sask. 181/D5
Swifterbant, Netherlands 27/H3
Swift Falls, Minn. (†56215) 255/C5
Swifton, Ark. (72471) 202/H2
Swiftown, Miss. (38959) 256/D4
Swift River, Yukon 187/E3
Swifts Creek, Victoria 97/D5
Swiftwater, N.H. (†03785) 268/D3
Swilly (inlet), Ireland 17/F1
Swilly, Lough (inlet), Ireland 10/C3
Swilly (riv.), Ireland 17/F2
Swinburne (cape), N.W. Terrs. 187/J2
Swindon, England 10/E5
Swindon, England 13/F6
Swinemünde (Świnoujście), Poland 47/A2
Swinford, Ireland 17/C4
Swink, Colo. (81077) 208/M7
Swink, Okla. (74761) 288/R6
Swinomish Ind. Res., Wash. 310/C2
Świnoujście (Swinemünde), Poland 47/B1
Swinton, Scotland 15/F5
Swisher, Iowa (52338) 229/K5
Swisher (co.), Texas 303/C3
Swiss, W. Va. (26690) 312/D6
Swisshome, Oreg. (97480) 291/D3
Swissvale, Pa. (15218) 294/C7
Switz City, Ind. (47465) 227/C6
Switzer, Ky. (†40379) 237/M4
Switzer, W. Va. (25647) 312/B7
Switzerland 2/K3
Switzerland 7/E4
Switzerland, Fla. (†32043) 212/E1
Switzerland (co.), Ind. 227/G7
Switzerland, S.C. (†29936) 296/E7
SWITZERLAND 39
Swona (isl.), Scotland 15/E2
Swords, Ireland 17/J5
Swords Creek, Va. (24649) 307/E6
Sybille (creek), Wyo. 319/G4
Sybouts, Sask. 181/G6
Sycamore, Ala. (35149) 195/F4
Sycamore, Georgia (31790) 217/E7
Sycamore, Ill. (60178) 222/E2
Sycamore, Ind. (†46936) 227/F4
Sycamore, Kansas (67363) 232/G4
Sycamore, Mo. (65758) 261/H9
Sycamore, Ohio (44882) 284/D4
Sycamore, Pa. (15364) 294/B6
Sycamore, S.C. (29846) 296/E5
Sycan (riv.), Oreg. 291/F5
Syców, Poland 47/C3
Sydenham, Ontario 177/H3
Sydenham (riv.), Ontario 177/B5
Sydney, Australia 2/S7
Sydney, Australia 87/F9
Sydney (isl.), Kiribati 87/K6
Sydney (cap.), N. S. Wales 88/L4
Sydney (bay), Norfolk I. 88/L6
Sydney, N. Dak. (†58401) 282/N6
Sydney, Nova Scotia 168/H2
Sydney, N.S. 162/K6
Sydney, N.S. 146/N5
Sydney (harb.), Nova Scotia 168/H2
Sydney Mines, Nova Scotia 168/H2
Sydney River, Nova Scotia 168/H2
Sykeston, N. Dak. (58486) 282/M5
Sykesville, Md. (21784) 245/K3
Sykesville, Pa. (15865) 294/E3
Syktyvkar, U.S.S.R. 7/K2
Syktyvkar, U.S.S.R. 48/F3
Syktyvkar, U.S.S.R. 52/H2
Sylacauga, Ala. (35150) 195/F4
Sylhet, Bangladesh 68/G4
Sylt (isl.), W. Germany 22/C1
Sylva, N.C. (28779) 281/C4
Sylvan, Oreg. (†97201) 291/B2
Sylvan, Pa. (†17236) 294/G6
Sylvan (lake), Wash. 310/G3
Sylvan Beach, N.Y. (13157) 276/J4
Sylvan Grove, Kansas (67481) 232/D2
Sylvania, Ala. (35988) 195/G1
Sylvania, Georgia (30467) 217/J5
Sylvania, Ind. (†47985) 227/C5
Sylvania, Ohio (43560) 284/C2
Sylvania, Pa. (16945) 294/J2
Sylvania, Sask. 181/G3
Sylvan Lake, Alberta 182/C3
Sylvan Lake, Mich. (†48053) 250/F6
Sylvarena, Miss. (39153) 256/F6
Sylvatus, Va. (†24343) 307/G7
Sylvester, Georgia (31791) 217/E7
Sylvester (lake), North. Terr. 93/D5
Sylvester, Texas (79560) 303/D5
Sylvester, W. Va. (25193) 312/C6
Sylvia, Kansas (67581) 232/D4
Sylvia, Tenn. (†37055) 237/G8
Symco, Wis. (†54949) 317/J6
Symerton, Ill. (†60481) 222/F2
Symmes (creek), Ohio 284/F8
Symonds, Miss. (†38769) 256/C3
Syosset, N.Y. (11791) 276/R6
Syracuse, Ind. (46567) 227/F2
Syracuse (prov.), Italy 34/E6
Syracuse, Italy 7/F5
Syracuse, Italy 34/E6
Syracuse, Kansas (67878) 232/A3
Syracuse, Mo. (65354) 261/G5
Syracuse, Nebr. (68446) 264/H4
Syracuse, N.Y. 188/L2
Syracuse, N.Y. (*13201) 276/H4
Syracuse, Ohio (45779) 284/G7
Syracuse, Utah (†84041) 304/B2
Syrdar'ya (riv.), U.S.S.R. 2/N3
Syrdar'ya (riv.), U.S.S.R. 54/H5
Syrdar'ya (riv.), U.S.S.R. 48/G5
Syre, Minn. (†56584) 255/B3
Syria 2/L4
Syria 54/E6
SYRIA 63/G5
SYRIA 59/C3
Syria, Va. (22743) 307/M4
Syriam, Burma 72/C3
Syrian (des.), Iraq 66/B4
Syrian (El Hamad) (des.), Iraq 59/D3
Syrte, Libya 102/D1
Syrte, Libya 111/C1
Sysladobsis, Lower (lake), Maine 243/G5
Sysola (riv.), U.S.S.R. 52/H2
Syssleback, Sweden 18/H6
Syzran', U.S.S.R. 7/J3
Syzran', U.S.S.R. 48/E4
Syzran', U.S.S.R. 52/G4
Szabadszállás, Hungary 41/E3
Szabolcs-Szatmár (co.), Hungary 41/G3
Szamotuły, Poland 47/C2
Szarvas, Hungary 41/F3
Százhalombatta, Hungary 41/E3
Szczebrzeszyn, Poland 47/F3
Szczecin (prov.), Poland 47/B2
Szczecin, Poland 7/F3
Szczecin, Poland 47/B2
Szczecinek, Poland 47/C2
Szczytno, Poland 47/E2
Szechwan (Sichuan) (prov.), China 77/F5
Szécsény, Hungary 41/E2
Szeged, Hungary 41/E3
Szeged, Hungary 7/F4
Szeghalom, Hungary 41/F3
Szegvár, Hungary 41/F3
Székesfehérvár, Hungary 41/E3
Szekszárd, Hungary 41/E3
Szendrő, Hungary 41/F2
Szentendre, Hungary 41/E3
Szentendreisziget (isl.), Hungary 41/E3
Szentes, Hungary 41/F3
Szentgotthárd, Hungary 41/D3
Szentlőrinc, Hungary 41/E3
Szeping (Siping), China 77/K3
Szerencs, Hungary 41/F2
Szigetvár, Hungary 41/D3
Szikszó, Hungary 41/F2
Szil, Hungary 41/D3
Szirák, Hungary 41/E3
Szolnok (co.), Hungary 41/F3
Szolnok, Hungary 41/F3
Szombathely, Hungary 41/D3
Szprotawa, Poland 47/B3
Sztum, Poland 47/D2
Szubin, Poland 47/C2
Szydłowiec, Poland 47/E3

T

Taal (lake), Philippines 82/C4
Tab, Hungary 41/E3
Tab, Ind. (†47917) 227/C4
Taba, Bir, Egypt 59/B4
Tabacal (pt.), Cuba 158/H4
Tabacundo, Ecuador 128/C2
Tabaquite, Trin. & Tob. 161/B11
Tabar (isls.), Papua N.G. 86/C1
Tabarka, Tunisia 106/F1
Tabas, Iran 59/G3
Tabas, Iran 66/L4
Tabas, Iran 66/K4
Tabasará (mts.), Panama 154/G6
Tabasco (state), Mexico 150/N7
Tabasco, Mexico 150/H6
Tabask, Kuh-e (mt.), Iran 66/G6
Tabas-Masina (Tabas), Iran 59/H3
Tabb, Va. (23602) 307/R6
Tabelbala, Algeria 106/D3
Taber, Alberta 182/E5
Taberg, N.Y. (13471) 276/J4
Tabernacle, St. Chris.-Nevis 161/C10
Tabernash, Colo. (80478) 208/H3
Tabernes de Valldigna, Spain 33/G3
Taberville, Mo. (64787) 261/E6
Tabiang, Kiribati 87/G6
Tabiona, Utah (84072) 304/D3
Tabiteuea (atoll), Kiribati 87/H6
Tablas (cape), Chile 138/A9
Tablas (isl.), Philippines 82/D4
Tablas (str.), Philippines 82/C4
Table (mt.), Nev. 266/C3
Table (bay), Newf. 166/C3
Table (bay), S. Africa 118/E6
Table (mt.), S. Africa 118/E6
Table (peak), Wyo. 319/C2
Table Grove, Ill. (61482) 222/C3
Tableland, Trin. & Tob. 161/B11
Tableland Station, W. Australia 92/D2
Tabler, Okla. (†73018) 288/L4
Table Rock (riv.), Ark. 202/D1
Table Rock (res.), Mo. 261/E9
Table Rock, Nebr. (68447) 264/H4
Tablers Station, W. Va. (†25428) 312/K4
Taboada, Spain 33/C1
Taboga (isl.), Panama 154/H6
Tábor, Czech. 41/C2
Tabor, Iowa (51653) 229/B7
Tabor (mt.), Israel 65/C2
Tabor, Minn. (†56712) 255/B2
Tabor, N.J. (07878) 273/E2
Tabor, S. Dak. (57063) 298/O8
Tabor (mt.), Vt. 268/B5
Tabora (reg.), Tanzania 115/F5
Tabora, Tanzania 115/F5
Tabora, Tanzania 102/F5
Tabor City, N.C. (28463) 281/M6
Tabou, Ivory Coast 106/C8
Tabriz, Iran 54/F6
Tabriz, Iran 66/D2
Tabriz, Iran 59/E2
Tabuk, Philippines 82/C2
Tabusintac, New Bruns. 170/E1
Tabusintac (gully), New Bruns. 170/F1
Tabusintac (riv.), New Bruns. 170/E1
Täby, Sweden 18/H1
Tacajó, Cuba 158/J3
Tacámbaro de Codallos, Mexico 150/J7
Tacaná, Guatemala 154/A3
Tacaná (vol.), Guatemala 154/A3
Tacarigua, Trin. & Tob. 161/B10
Taché (lake), Québec 172/J1
Tachen (Taizhou) (isls.), China 77/K6
Tacheng (Qoqek), China 77/B2
Tachikawa, Japan 81/O2
Tachina, Ecuador 128/C2
Táchira (state), Venezuela 124/C4
Tachov, Czech. 41/B2
Tacloban, Philippines 82/E5
Tacloban, Philippines 85/H3
Tacna, Ariz. (85352) 198/B6
Tacna (dept.), Peru 128/G11
Tacna, Peru 128/G11
Tacna, Peru 120/B4
Tacobamba, Bolivia 136/C6
Tacoma, Va. (†24230) 307/C7
Tacoma, Wash. 146/F5
Tacoma, Wash. 188/B1
Tacoma, Wash. (*98401) 310/C3
Tacoma Park, S. Dak. (†57433) 298/N2
Taconic, Conn. (06079) 210/B1
Taconic (mts.), Mass. 249/A2
Taconite, Minn. (55786) 255/E3
Taconite Harbor, Minn. (†55613) 255/H3
Tacopaya, Bolivia 136/B5
Tacora (vol.), Chile 138/B1
Tacotalpa, Mexico 150/N8
Tacuaras, Paraguay 144/C5
Tacuarembó (dept.), Uruguay 145/D3
Tacuarembó, Uruguay 145/D2
Tacuarembó (riv.), Uruguay 145/D2
Tacuarí (riv.), Uruguay 145/E3
Tacuatí, Paraguay 144/D3
Tacutu (riv.), Brazil 132/B2
Tadcaster, England 13/K1
Tademaït (plat.), Algeria 102/C2
Tademaït, Plateau du (plat.), Algeria 106/E3
Tadine, New Caled. 86/H4
Tadjnout Hagguerete (well), Mali 106/D4
Tadjoura, Djibouti 111/H5
Tadley, England 13/F6
Tadmor (Palmyra) (ruin), Syria 59/C3
Tadmore, Sask. 181/J4
Tadmur, Syria 59/C3
Tadmur, Syria 63/H5
Tadó, Colombia 126/B5
Tadoule (lake), Manitoba 179/J2
Tadoussac, Que. 162/J6
Tadoussac, Québec 174/C3
Tadoussac, Québec 172/H1
Tadzhik S.S.R., U.S.S.R. 54/H6
Tadzhik S.S.R., U.S.S.R. 48/H6
Taebaek (mts.), S. Korea 81/D5
Taedong (riv.), N. Korea 81/C4
Taegu, S. Korea 54/O6
Taegu, S. Korea 81/D6
Taejŏn, S. Korea 54/O6
Taejŏn, S. Korea 81/C5
Tafalla, Spain 33/F1
Tafassasset, Wadi (dry riv.), Algeria 106/F4
Tafassasset, Wadi (dry riv.), Niger 106/F4
Tafers, Switzerland 39/D3
Taff (riv.), Wales 13/B7
Tafí Viejo, Argentina 120/C5
Tafí Viejo, Argentina 143/C2
Taft, Calif. (93268) 204/F8
Taft, Fla. (32809) 212/E3
Taft, Iran 66/J5
Taft, La. (†70057) 238/N4
Taft, Okla. (74463) 288/R3
Taft, Philippines 82/E5
Taft, Tenn. (38488) 237/H10
Taft, Texas (78390) 303/G9
Taftan, Kuh-e (mt.), Iran 66/M6
Taftan, Kuh-e (mt.), Iran 59/H4
Taftsville, Vt. (05073) 268/C4
Taftville, Conn. (06380) 210/G2
Tagab, Afghanistan 59/J3
Tagab, Afghanistan 68/B2
Taga Dzong, Bhutan 68/G3
Taganrog, U.S.S.R. 7/H4
Taganrog, U.S.S.R. 52/E5
Taganrog, U.S.S.R. 48/D5
Tagant (reg.), Mauritania 106/B5
Tagapula (isl.), Philippines 82/E4
Tagawa, Japan 81/E7
Tagaytay, Philippines 82/C3
Tagbilaran, Philippines 82/E6
Taghit, Algeria 106/D2
Taghmon, Ireland 17/H7
Tagish (lake), Br. Col. 184/J1
Tagish, Yukon 187/E3
Tagliamento (riv.), Italy 34/D1
Tagolo (pt.), Philippines 82/D6
Tagolo (pt.), Philippines 85/G4
Tagoloan, Philippines 82/E6
Tagomago (isl.), Spain 33/G3
Tagounite, Morocco 106/C3
Tagua, Bolivia 136/B6
Taguatinga, Brazil 120/E4
Taguatinga, Fed. Dist., Brazil 132/D6
Taguatinga, Goiás, Brazil 132/E6
Tague (bay), Virgin Is. (U.S.) 161/G4
Tagula (isl.), Papua N.G. 85/C8
Tagum, Philippines 82/E7
Tagus (riv.) 7/D5
Tagus, N. Dak. (†58720) 282/G3
Tagus (riv.), Portugal 33/B3
Tagus (riv.), Spain 33/D3
Tahaa (isl.), Fr. Poly. 87/L7
Tahakopa, N. Zealand 100/B7
Tahan, Gunong (mt.), Malaysia 72/D6
Tahat (mt.), Algeria 102/C2
Tahat (mt.), Algeria 106/F4
Tahawus, N.Y. (12879) 276/M2
Tahiryuak (lake), N.W. Terrs. 187/G2
Tahiti (isl.), Fr. Polynesia 2/B6
TAHITI, Fr. Poly. 86/S13
Tahiti (isl.), Fr. Poly. 87/L7
Tahiti (isl.), Fr. Poly. 86/S13
Tahlequah, Okla. (74464) 288/R3
Tahma (riv.), Turkey 63/G3
Tahoe (lake) 188/C3
Tahoe (lake), Calif. 204/F4
Tahoe (lake), Nev. 266/B3
Tahoe City, Calif. (95730) 204/E4
Tahoka, Texas (79373) 303/C4
Taholah, Wash. (98587) 310/A3
Tahoma, Calif. (95733) 204/E4
Tahoua, Niger 102/C3
Tahoua, Niger 106/F6
Tahquamenon (falls), Mich. 250/D2
Tahquamenon (riv.), Mich. 250/D2
Tahsis, Br. Col. 184/D5
Tahta, Egypt 59/B4
Tahta, Egypt 111/F2
Tahtsa (lake), Br. Col. 184/D3
Tahua, Bolivia 136/B3
Tahuamanu (riv.), Bolivia 136/A2
Tahuamanu, Peru 128/H8
Tahuamanu (riv.), Peru 128/H8
Tahulandang (isl.), Indonesia 85/H5
Tahuna, Indonesia 85/G5
Tahuya, Wash. (98588) 310/B3
Tai'an, China 77/J4
Taiarapu (pen.), Fr. Poly. 86/T13
Taiban, N. Mex. (88134) 274/F4
Taibus, China 77/J3
Taichow (Taizhou), China 77/K5

Taichung, China 77/K7
Taichung, Taiwan 54/O7
Taieri (riv.), N. Zealand 100/C7
Taif, Saudi Arabia 54/F7
Taif, Saudi Arabia 59/D5
Taigu, China 77/H4
Taihape, N. Zealand 100/E3
Taihe, China 77/J6
Tai Hu (lake), China 77/J5
Tailem Bend, S. Australia 88/F7
Tailem Bend, S. Australia 94/F6
Tailfingen, W. Germany 22/C4
Taima, Saudi Arabia 59/C4
Tain, Scotland 15/D3
Tain, Scotland 10/D2
Tainan, China 77/J7
Tainaron (cape), Greece 7/G5
Tainaron (cape), Greece 45/F7
Taintor, Iowa (50253) 229/H6
Taipei, China 77/K7
Taipei (cap.), Rep. of China 54/O7
Taipei (cap.), Rep. of China 2/R4
Taiping, Malaysia 72/D6
Taitao (pen.), Chile 120/B7
Taitao (cape), Chile 138/D6
Taitao (pen.), Chile 138/D6
Taits Gap, Ala. (†35121) 195/F3
Taitung, China 77/K7
Taivalkoski, Finland 18/P4
Taiwan 54/N7
Taiwan 2/R4
Taiwan (str.) 54/N7
Taiwan (isl.) 54/N7
Taiwan, China 77/K7
Taiwan (Formosa) (isl.), China 77/K7
Taiwan (Formosa) (str.), China 77/J7
Taiyuan, China 77/H4
Taiyuan, China 54/N6
Taizhou (Tachen) (isls.), China 77/K6
Taizhou (Taichow), China 77/K5
Ta'izz, Yemen Arab Rep. 54/F8
Ta'izz, Yemen Arab Rep. 59/D7
Tajimi, Japan 81/H6
Tajique, N. Mex. (87057) 274/C4
Tajo (Tagus) (riv.), Spain 33/D3
Tajrish, Iran 66/G3
Tajumulco (vol.), Guatemala 154/B3
Tak, Thailand 72/C3
Takaishi, Japan 81/H8
Takaka, N. Zealand 100/D4
Takalar, Indonesia 85/F7
Takama, Guyana 131/C3
Takamatsu, Japan 81/F6
Takaoka, Japan 81/H5
Takapau, N. Zealand 100/F4
Takapuna, N. Zealand 100/B1
Takarazuka, Japan 81/H7
Takaroa (atoll), Fr. Poly. 87/M7
Takasaki, Japan 81/J5
Takatsuki, Japan 81/J7
Takayama, Japan 81/H5
Takefu, Japan 81/G6
Takeshima (isls.), Japan 81/F5
Takestan, Iran 66/F2
Takev, Cambodia 72/E5
Takhiatash, U.S.S.R. 48/F5
Takhta-Bazar, U.S.S.R. 48/G6
Takikawa, Japan 81/K2
Takingeun, Indonesia 85/B5
Takitimu (mts.), N. Zealand 100/A6
Takkaze (riv.), Ethiopia 111/G5
Takla (lake), Br. Col. 184/D2
Takla Makan (des.), China 54/K6
Takla Makan (Taklimakan Shamo) (des.), China 77/B4
Taklimakan Shamo (des.), China 77/B4
Tako, Sask. 181/B3
Takoma Park, Md. (20912) 245/F4
Takoradi, Ghana 106/D8
Takoradi-Sekondi, Ghana 102/B4
Taksimo, U.S.S.R. 48/M4
Taku (glac.), Alaska 196/N1
Taku (riv.), Alaska 196/N1
Taku (riv.), Br. Col. 184/J2
Takua Pa, Thailand 72/C5
Takutu (riv.), Guyana 131/B4
Tala, Mexico 150/H6
Tala, Uruguay 145/D5
Talab (riv.), Iran 59/H4
Talab (riv.), Iran 66/N6
Talab (riv.), Pakistan 68/A3
Talagante, Chile 138/G4
Talai (Da'an, Dalai), China 77/K2
Talak (reg.), Niger 106/E5
Talala, Okla. (74080) 288/P1
Talamanca (range), C. Rica 154/F6
Talangbetutu, Indonesia 85/C6
Talara, Peru 128/B5
Talara, Peru 120/A3
Talaud (isls.), Indonesia 54/O9
Talaud (isls.), Indonesia 85/H5
Talavera de la Reina, Spain 33/D2
Talawe (mt.), Papua N.G. 86/B2
Talbert, Ky. (41377) 237/P6
Talbingo, N.S. Wales 97/E4
Talbot, Alberta 182/E3
Talbot (isl.), Fla. 212/E1
Talbot (co.), Georgia 217/C5
Talbot, Ind. (47984) 227/C3
Talbot (co.), Md. 245/O5
Talbot (inlet), N.W. Terrs. 187/L2
Talbot (cape), W. Australia 88/D2
Talbot (cape), W. Australia 92/D1
Talbott, Tenn. (37877) 237/P8
Talbotton, Georgia (31827) 217/C5
Talca, Chile 138/A11
Talca, Chile 120/B6
Talca (pt.), Chile 138/E3
Talcahuano, Chile 138/D1
Talcahuano, Chile 120/B6
Talcán (isl.), Chile 138/D4
Talco, Texas (75487) 303/K4
Talcott (range), Conn. 210/D1
Talcott, W. Va. (24981) 312/E7
Talcottville, Conn. (†06066) 210/F1
Taldy-Kurgan, U.S.S.R. 54/J5
Taldy-Kurgan, U.S.S.R. 48/H5
Taleh, Somalia 115/J2
Talent, Oreg. (97540) 291/E5
Talgar, U.S.S.R. 48/H5
Talgarth, Wales 13/D5
Tali (Dali), China 77/E6
Taliabu (isl.), Indonesia 85/G6
Taliaferro (co.), Georgia 217/G3
Talibon, Philippines 82/E5
Talihina, Okla. (74571) 288/S5
Talina, Bolivia 136/B7
Tali Post, Sudan 111/F6
Talisayan, Philippines 82/E6
Talisheek, La. (70464) 238/L5
Talita, Uruguay 145/D4
Tal Kaif, Iraq 66/C2
Talkeetna, Alaska (99676) 196/B1
Talkeetna (mts.), Alaska 196/J2
Talkheh (riv.), Iran 66/E1
Talking Rock, Georgia (30175) 217/D1
Talaboa, P. Rico 161/B3
Talladega (co.), Ala. 195/F4
Talladega, Ala. (35160) 195/F4
Talladega Springs, Ala. (†35150) 195/F4
Tallaght, Ireland 17/J5
Tallahaga (creek), Miss. 256/F4
Tallahala (creek), Miss. 256/F7
Tallahassee (cap.), Fla. 146/K6
Tallahassee (cap.), Fla. 188/K4
Tallahassee (cap.), Fla. (*32301) 212/B1
Tallahatchie (co.), Miss. 256/D3
Tallahatchie (riv.), Miss. 256/D3
Tallahatta Springs, Ala. (†36784) 195/C7
Tallangatta, Victoria 97/D5
Tallant, Okla. (†74002) 288/O1
Tallapoosa (co.), Ala. 195/G5
Tallapoosa (riv.), Ala. 195/G5
Tallapoosa, Georgia (30176) 217/B3
Tallapoosa, Mo. (63878) 261/N9
Tallassee, Ala. (36078) 195/G5
Tallinn, U.S.S.R. 7/G3
Tallinn (cap.), U.S.S.R. 53/C1
Tallinn, U.S.S.R. 48/C4
Tallinn, U.S.S.R. 52/B3
Tallmadge, Ohio (44278) 284/H3
Tallman, N.Y. (10982) 276/J8
Tallman, Sask. 181/E3
Tallmansville, W. Va. (26237) 312/F5
Tallow, Ireland 17/F7
Tallula, Ill. (62688) 222/D4
Tallulah, La. (71282) 238/H2
Tallulah Falls, Georgia (30573) 217/F1
Talma, Ind. (†46975) 227/E2
Talmage, Kansas (67482) 232/E2
Talmage, Nebr. (68448) 264/H4
Talmage, Sask. 181/H6
Talmage, Utah (84073) 304/D3
Talmo, Georgia (30575) 217/E2
Talmo, Kansas (†66935) 232/E2
Talmoon, Minn. (56637) 255/E3
Talodi, Sudan 111/F5
Talofofo (bay), Guam 86/K7
Taloga, Okla. (73667) 288/J2
Talon (lake), Ontario 177/E1
Taloqan, Afghanistan 68/B1
Taloqan, Afghanistan 59/J2
Talpa, Texas (76882) 303/E6
Talpa de Allende, Mexico 150/G6
Talparo, Trin. & Tob. 161/B10
Talquin (lake), Fla. 212/B1
Talsi, U.S.S.R. 53/B2
Taltal, Chile 138/A5
Taltal, Chile 120/B5
Taltal, Quebrada de (riv.), Chile 138/B5
Taltson (riv.), N.W. Terrs. 187/G3
Talvik, Norway 18/N2
Talyawalka (creek), N.S. Wales 97/B2
Talyawalka Ana Branch, Darling (riv.), N.S. Wales 97/B3
Tama (co.), Iowa 229/H4
Tama, Iowa (52339) 229/H5
Tama (riv.), Japan 81/O2
Tamaha, Okla. (†74462) 288/S4
Tamaki (str.), N. Zealand 100/C1
Tamale, Ghana 102/B4
Tamale, Ghana 106/D7
Tamalpais (mt.), Calif. 204/H1
Tamana (mt.), Trin. & Tob. 161/B10
Tamanrasset, Algeria 106/F4
Tamanrasset, Algeria 102/C2
Tamanrasset, Wadi (dry riv.), Algeria 106/E4
Tamaqua, Pa. (18252) 294/L4
Tamar (riv.), England 13/C7
Tamar (riv.), England 10/D5
Tamar (riv.), Tasmania 99/D3
Támara, Colombia 126/D5
Tamarac, Fla. (†33321) 212/B3
Tamarac (riv.), Minn. 255/A2
Tamarack, Idaho (†83654) 220/B5
Tamarack (isl.), Manitoba 179/F3
Tamarack, Minn. (55787) 255/E4
Tamarack (riv.), Minn. 255/D2
Tamarack, Pa. (†17729) 294/G3
Tamarite de Litera, Spain 33/G2
Tamaro (mt.), Switzerland 39/G4
Tamaroa, Ill. (62888) 222/D5
Tamarugal, Pampa del (plain), Chile 138/B3
Tamási, Hungary 41/E3
Tamassee, S.C. (29686) 296/A2
Tamatama, Venezuela 124/F6
Tamatave (Toamasina), Madagascar 118/H3
Tamaulipas (state), Mexico 150/K4
Tamaya, Chile 138/A8
Tamayo, Dom. Rep. 158/D6
Tamazula, Mexico 150/F4
Tamazulapan del Progreso, Mexico 150/L8
Tamazunchale, Mexico 150/K6
Tambacounda, Senegal 106/B6
Tambar Springs, N.S. Wales 97/E2
Tambelan (isls.), Indonesia 85/D5
Tamberías, Argentina 143/C3
Tambey, U.S.S.R. 48/G2
Tambo (riv.), Peru 128/G11
Tambo, Queensland 88/H4
Tambo, Queensland 95/C5
Tambo de Mora, Peru 128/D9
Tambo Grande, Peru 128/B5
Tambohorano, Madagascar 118/G3
Tambopata (riv.), Peru 128/H9
Tambores, Uruguay 145/C2
Tamboril, Dom. Rep. 158/D5
Tamboritha (mt.), Victoria 97/D5
Tambov, U.S.S.R. 7/J3
Tambov, U.S.S.R. 52/F4
Tambov, U.S.S.R. 48/E4
Tambura, Sudan 111/E6
Tamchakett, Mauritania 106/B5
Tame, Colombia 126/E4
Tame (riv.), England 10/G3
Tâmega (riv.), Portugal 33/C2
Tamentit, Algeria 106/D3
Tamiahua, Mexico 150/L6
Tamiami (canal), Fla. 212/E6
Tamil Nadu (state), India 68/D6
Tamin (gov.), Iraq 66/D3
Tamina (riv.), Switzerland 39/H3
Tamins, Switzerland 39/H3
Tamise (Temse), Belgium 27/E6
Tam Ky, Vietnam 72/F4
Tammisaari (Ekenäs), Finland 18/N6
Tamms, Ill. (62988) 222/D6
Tammun, West Bank 65/C3
Tamo, Ark. (71644) 202/G5
Tamora, Nebr. (†68434) 264/G4
Tampa, Fla. 188/K5
Tampa, Fla. 146/K7
Tampa (bay), Fla. 188/K5
Tampa, Fla. (*33601) 212/C2
Tampa (bay), Fla. 212/D4
Tampa, Kansas (67483) 232/E3
Tampere, Finland 7/G2
Tampere, Finland 18/N6
Tampico, Ill. (61283) 222/D2
Tampico, Ind. (†47220) 227/F7
Tampico, Mexico 146/J7
Tampico, Mexico 150/L5
Tampico, Mont. (†59230) 262/K2
Tampico, Wash. (†98901) 310/E4
Tampoc (riv.), Fr. Guiana 131/E4
Tam Quan, Vietnam 72/F4
Tamra, Saudi Arabia 59/E5
Tams, W. Va. (25933) 312/D7
Tamsagbulag, Mongolia 77/J2
Tamsagout, Mauritania 106/C4
Tamshiyacu, Peru 128/F5
Tamsweg, Austria 41/B3
Tamuín, Mexico 150/K6
Tamuning, Guam 86/K7
Tamworth, Australia 87/E9
Tamworth, England 13/F5
Tamworth, England 10/G3
Tamworth○, N.H. (03886) 268/E4
Tamworth, N.S. Wales 88/J6
Tamworth, N.S. Wales 97/F2
Tamworth, Ontario 177/H3
Tamyang, S. Korea 81/C6
Tana (lake), Ethiopia 102/F3
Tana (lake), Ethiopia 111/G5
Tana (riv.), Finland 18/P2
Tana (riv.), Kenya 102/G5
Tana (riv.), Kenya 115/G4
Tana, Norway 18/Q1
Tana (riv.), Norway 18/P1
Tanabe, Kyoto, Japan 81/J7
Tanabe, Wakayama, Japan 81/G7
Tanacross, Alaska (99776) 196/K2
Tanafjord (fjord), Norway 18/Q1
Tanaga (isl.), Alaska 196/K4
Tanaga (vol.), Alaska 196/K4
Tanahgrogot, Indonesia 85/F6
Tanahmerah, Indonesia 85/K7
Tanah Merah, Malaysia 72/D6
Tanamá (riv.), P. Rico 161/B1
Tanami (des.), North. Terr. 88/E3
Tanami, North. Terr. 93/A5
Tanami (des.), North. Terr. 93/C5
Tánamo, Cuba 158/J3
Tan An, Vietnam 72/E5
Tanana, Alaska 188/D5
Tanana, Alaska (99777) 196/H1
Tanana (riv.), Alaska 146/D3
Tanana (riv.), Alaska 188/D5
Tanana (riv.), Alaska 196/J2
Tananarive (Antananarivo) (cap.), Madagascar 118/H3
Tanaro (riv.), Italy 34/B2
Tanch'ŏn, N. Korea 81/D3
Tancook Island, Nova Scotia 168/D4
Tanda, India 68/E3
Tandag, Philippines 82/F6
Tandil, Argentina 143/E4
Tandil, Argentina 120/D6
Tando Adam, Pakistan 68/B3
Tando Allahyar, Pakistan 68/B3
Tandou (lake), N.S. Wales 97/A3
Tandragee, N. Ireland 17/J3
Tanega (isl.), Japan 81/E8
Taney (co.), Mo. 261/F9
Taneycomo (lake), Mo. 261/F9
Taneytown, Md. (21787) 245/K2
Taneyville, Mo. (65759) 261/F9
Tanezrouft (des.), Algeria 102/C2
Tanezrouft (des.), Algeria 106/E4
Tang, Kas (isl.), Cambodia 72/D5
Tanga (isls.), Papua N.G. 86/C1
Tanga (reg.), Tanzania 115/G5
Tanga, Tanzania 115/G4
Tanga, Tanzania 102/F5
Tangainony, Madagascar 118/H4
Tangalla, Sri Lanka 68/E7
Tanganyika (lake) 102/F5
Tanganyika (lake), Burundi 115/E5
Tanganyika (lake), Tanzania 115/E5
Tanganyika (lake), Zaire 115/E5
Tanganyika (lake), Zambia 115/E5
Tangent (pt.), Alaska 196/H1
Tangent, Alberta 182/B2
Tangent, Oreg. (97389) 291/D3
Tangerang, Indonesia 85/G1
Tangermünde, E. Germany 22/D2
Tanggula Shan (range), China 77/D5
Tangier, Ind. (47985) 227/C5
Tangier (sound), Md. 245/P8
Tangier (Tanger), Morocco 106/C1
Tangier, Morocco 102/B1
Tangier, Nova Scotia 168/F4
Tangier (riv.), Nova Scotia 168/F4
Tangier, Okla. (†73801) 288/G2
Tangier, Va. (23440) 307/R5
Tangier (isl.), Va. 307/R5
Tangier (sound), Va. 307/S5
Tangipahoa (par.), La. 238/K5
Tangipahoa, La. (70465) 238/J5
Tangipahoa (riv.), La. 238/N1
Tangra Yumco (lake), China 77/C5
Tangshan, China 77/J4
Tangshan, China 54/N5
Tangub, Philippines 82/D6
Tangyanika (lake) 2/L6
Tangyuan, China 77/L2
Tanimbar (isls.), Indonesia 54/P10
Tanimbar (isls.), Indonesia 85/J7
Tanjay, Philippines 82/D6
Tanjore (Thanjavur), India 68/D6
Tanjungbalai, Indonesia 85/C5
Tanjungkarang, Indonesia 54/M10
Tanjungkarang, Indonesia 85/C7
Tanjungpandan, Indonesia 85/D6
Tanjungpinang, Indonesia 85/C5
Tanjungpriok, Indonesia 85/H1
Tanjungpura, Indonesia 85/B5
Tanjungredeb, Indonesia 85/F5
Tanjungselor, Indonesia 85/F5
Tanna (isl.), Vanuatu 87/H7
Tanner, Ala. (35671) 195/E1
Tanner, W. Va. (26179) 312/E5
Tannersville, N.Y. (12485) 276/M6
Tannersville, Pa. (18372) 294/M3
Tannis (bay), Denmark 21/D2
Tannu-Ola (range), Mongolia 77/D1
Tannu-Ola (range), U.S.S.R. 48/K5
Tanon (str.), Philippines 82/D6
Tanout, Niger 106/F6
Tanque Verde, Ariz. (†85701) 198/E6
Tanta, Egypt 111/J3
Tanta, Egypt 59/B3
Tantallon, Sask. 181/K5
Tantalus (mt.), Hawaii 218/D4
Tan-Tan, Morocco 106/B3
Tantoyuca, Mexico 150/L6
Tantung (Dandong), China 77/K3
Tanumshede, Sweden 18/G7
Tanunda, S. Australia 94/C6
Tanzania 2/L6
Tanzania 102/F5
TANZANIA 115/F5
Tao, Ko (isl.), Thailand 72/C5
Tao'an, China 77/K2
Taole, China 77/G4
Taongi (atoll), Marshall Is. 87/G4
Taopi, Minn. (55977) 255/F7
Taormina, Italy 34/E6
Taos, Mo. (†65101) 261/H5
Taos (co.), N. Mex. 274/D2
Taos, N. Mex. (87571) 274/D2
Taos Pueblo, N. Mex. (†87571) 274/D2
Taoudenni, Mali 106/D4
Taoudenni, Mali 102/B2
Taourirt, Algeria 106/E3
Taourirt, Morocco 106/D2
Taouz, Morocco 106/D2
Taoyuan, China 77/K6
Tapa, U.S.S.R. 53/C1
Tapacarí, Bolivia 136/B5
Tapachula, Mexico 150/N9
Tapajós (riv.), Brazil 2/G6
Tapajós (riv.), Brazil 120/D3
Tapajós (riv.), Brazil 132/B4
Tapaktuan, Indonesia 85/B5
Tapalquén, Argentina 143/E4
Tapanahoni (riv.), Suriname 131/D4
Tapani (lake), Québec 172/B3
Tapanui, N. Zealand 100/B6
Tapaz, Philippines 82/D5
Tapera do Jeronimo, Brazil 132/C2
Tapeta, Liberia 106/C7
Tapi, Mae Nam (riv.), Thailand 72/C5
Tapiantana Group (isls.), Philippines 82/D7
Tapiche (riv.), Peru 128/E6
Taping (riv.), Burma 72/C1
Tápiószele, Hungary 41/E3
Tapirapecó, Sierra (mts.), Venezuela 124/F7
Tapiutan (isl.), Philippines 82/B5
Tapoco, N.C. (28780) 281/A4
Tapolca, Hungary 41/D3
Tappahannock, Va. (22560) 307/O5
Tappan (lake), N.J. 273/C1
Tappan, N.Y. (10983) 276/K8
Tappan (lake), Ohio 284/H5
Tappen, N. Dak. (58487) 282/L6
Tappi (cape), Japan 81/K3
Tapti (riv.), India 68/D4
Tapul (isl.), Philippines 82/C8
Tapul Group (isls.), Philippines 85/G4
Tapul Group (isls.), Philippines 82/C8
Taputapu (cape), Amer. Samoa 86/N9
Taquari (riv.), Brazil 132/C7
Taquaritinga, Brazil 132/D8
Taquaritinga, Brazil 135/B2
Tar (riv.), N.C. 281/O3
Tara (hill), Ireland 17/H4
Tara, Ontario 177/C3
Tara (isl.), Philippines 82/C4
Tara, Queensland 95/D5
Tara, Queensland 88/J5
Tara, U.S.S.R. 48/H4
Tara (riv.), Yugoslavia 45/D4
Tarabuco, Bolivia 136/C6
Tarabulus, Lebanon 63/F5
Tarabulus, Lebanon 59/C3
Taradale, N. Zealand 100/F3
Taraira (riv.), Colombia 126/F8
Tarairí, Bolivia 136/D7
Tarakan, Indonesia 54/N9
Tarakan, Indonesia 85/F5
Taralga, N.S. Wales 97/E4
Tarama (isl.), Japan 81/L7
Tarancón, Spain 33/E3
Tarangire Nat'l Park, Tanzania 115/G4
Taranna, Tasmania 99/D5
Taransay (isl.), Scotland 15/A3
Taranto (prov.), Italy 34/F4
Taranto, Italy 34/F4
Taranto, Italy 7/F4
Taranto (gulf), Italy 7/F5
Taranto (gulf), Italy 34/F5
Tarapacá (reg.), Chile 138/B2
Tarapacá, Chile 138/B2
Tarapacá, Colombia 126/F9
Tarapaya, Bolivia 136/B6
Tarapoto, Peru 120/B3
Tarapoto, Peru 128/D6
Tarare, France 28/F5
Tararua (range), N. Zealand 100/E4
Tarascon, France 28/F6
Tarasp, Switzerland 39/K3
Tarata, Bolivia 136/C5
Tarata, Peru 128/H11
Tarauacá, Brazil 132/G10
Tarauaca, Brazil 120/C3
Taravao (bay), Fr. Poly. 86/T13
Taravao (isth.), Fr. Poly. 86/T13
Tarawa (atoll), Kiribati 87/H5
Tarazona, Spain 33/E2
Tarazona de la Mancha, Spain 33/F3
Tarbat Ness (prom.), Scotland 15/E3
Tarbert, Ireland 17/C6
Tarbert, Strathclyde, Scotland 15/C5
Tarbert, W. Isles, Scotland 15/B3
Tarbert, East Loch (inlet), Scotland 15/B3
Tarbert, Loch (inlet), Scotland 15/B5
Tarbert, West Loch (inlet), Scotland 15/C5
Tarbert, West Loch (inlet), Scotland 15/A3
Tarbes, France 7/E4
Tarbes, France 28/D6
Tarbolton, Scotland 15/D5
Tarboro, Georgia (†31568) 217/J8
Tarboro, N.C. (27886) 281/O3
Tarbot, Nova Scotia 168/H2
Tarcoola, S. Australia 88/E6
Tarcoola, S. Australia 94/D4
Tarcutta, N.S. Wales 97/D4
Tardienta, Spain 33/F2
Tardošked, Czech. 41/E2
Taree, N.S. Wales 88/J6
Taree, N.S. Wales 97/G2
Tärendö, Sweden 18/N3
Tarentum, Pa. (15084) 294/C4
Tarfaya, Morocco 106/B3
Tarfaya, Morocco 102/A2
Tar Heel, N.C. (28392) 281/M5
Tarhuna, Libya 102/D1
Tarhuna, Libya 111/B1
Tariana, Colombia 126/F7
Táriba, Venezuela 124/B4
Tarifa, Spain 33/D4
Tariff, W. Va. (25281) 312/D5
Tariffville, Conn. (06081) 210/E1
Tarija (riv.), Argentina 143/D1
Tarija (dept.), Bolivia 136/D7
Tarija, Bolivia 120/C5
Tarija, Bolivia 136/C7
Tarija, Rio Grande de (riv.), Bolivia 136/C8
Tariku (riv.), Indonesia 85/K6
Tarim (riv.), China 54/K5
Tarim, P.D.R. Yemen 59/E6
Tarim He (riv.), China 77/B3
Tarim Pendi (basin), China 77/B4
Tar Island, Alberta 182/E1
Taritatu (riv.), Indonesia 85/K6
Tarkio, Mo. (64491) 261/B2
Tarkio, Mont. (†59872) 262/B4
Tarko-Sale, U.S.S.R. 48/H3
Tarkwa, Ghana 106/D7
Tarlac (prov.), Philippines 82/C3
Tarlac, Philippines 82/C3
Tarlac, Philippines 85/G2
Tarland, Scotland 15/F3
Tarleton (lake), N.H. 268/D4
Tarlton, Ohio (43156) 284/E6
Tarlton, Tenn. (†37301) 237/K9
Tarlton Downs, North. Terr. 93/E7
Tarm, Denmark 21/B6
Tarma, Peru 128/E8
Tarn (dept.), France 28/E6
Tarn (riv.), France 28/E6
Tarna (riv.), Hungary 41/F3
Tärnaby, Sweden 18/J4
Tarnak (riv.), Afghanistan 68/B2
Tårnby, Denmark 21/F6
Tarn-et-Garonne (dept.), France 28/D5
Tarnobrzeg (prov.), Poland 47/E3
Tarnobrzeg, Poland 47/E3
Tarnopol, Sask. 181/E3
Tarnov, Nebr. (68642) 264/G3
Tarnów (prov.), Poland 47/E4
Tarnów, Poland 7/G3
Tarnów, Poland 47/E4
Tarnowskie Góry, Poland 47/A3
Tarom, Iran 66/J6
Tarom, Iran 59/G4
Taroom, Queensland 95/D5
Tarouca, Portugal 33/C2
Taroudannt, Morocco 106/C2
Taroudant, Morocco 102/A2
Tarpa, Hungary 41/G2
Tarpon Springs, Fla. (*33589) 212/D3
Tarqui, Peru 128/E3
Tarquinia, Italy 34/C3
Tarqumiya, West Bank 65/C4
Tarragona (prov.), Spain 33/G2
Tarragona, Spain 33/G2
Tarragona, Spain 7/E4
Tarraleah, Tasmania 99/C4
Tarrant (co.), Texas 303/G5
Tarrant City, Ala. (35217) 195/E3
Tarrants, Mo. (†63334) 261/K4
Tarrasa, Spain 33/G2
Tárrega, Spain 33/G2
Tarryall (creek), Colo. 208/H4
Tarrytown, Georgia (30470) 217/H6
Tarrytown, N.Y. (10591) 276/O6
Tarsney Lakes, Mo. (†64063) 261/R6
Tarsus, Turkey 59/C2
Tarsus, Turkey 63/F4
Tart, China 77/D4
Tartagal, Argentina 143/D1
Tartagal, Argentina 120/C5
Tartas, France 28/C6
Tartu, U.S.S.R. 7/G3
Tartu, U.S.S.R. 53/D1
Tartu, U.S.S.R. 48/C4
Tartu, U.S.S.R. 52/C3
Tartus (prov.), Syria 63/G5
Tartus, Syria 63/F5
Tarutung, Indonesia 85/B5
Tarver, Georgia (†31648) 217/G9
Tarzan, Texas (79783) 303/B5
Tarzana, Calif. (91356) 204/B10
Täsch, Switzerland 39/E4
Tasco, Kansas (†67740) 232/B2
Tashauz, U.S.S.R. 48/F5
Tashk (lake), Iran 59/F4
Tashk (lake), Iran 66/J6
Tashkent, U.S.S.R. 54/H5
Tashkent, U.S.S.R. 2/N3
Tashkent, U.S.S.R. 48/G5
Tasikmalaya, Indonesia 85/H2
Tasisuak (lake), Newf. 166/B2
Taşkent, Turkey 63/E4
Taşköprü, Turkey 63/F2
Taşlıçay, Turkey 63/K3
Tasman (sea) 2/S7
Tasman (sea) 87/G9
Tasman (sea) 88/J7
Tasman (sea), N.S. Wales 97/F5
Tasman (bay), N. Zealand 100/D4
Tasman (mt.), N. Zealand 100/C5
Tasman (mts.), N. Zealand 100/D4
Tasman (sea), N. Zealand 100/B4
Tasman (pen.), Tasmania 88/H8
Tasman (head), Tasmania 99/D5
Tasman (pen.), Tasmania 99/E5
Tasman (sea), Tasmania 99/E4
Tasman (sea), Victoria 97/F5
Tasmania, 87/H8
Tasmania (state), Australia 87/E10
TASMANIA 99
Tasmania (isl.), Australia 2/S8
Tăşnad, Romania 45/F2
Taşova, Turkey 63/F4
Tassili N'Ahagger (plat.), Algeria 106/E4
Tassili N'Ajjer (plat.), Algeria 106/F3
Tåstrup, Denmark 21/F6
Tasu, Br. Col. 184/A4
Taşucu (gulf), Turkey 63/E4
Taswell, Ind. (47175) 227/D8
Tata, Hungary 41/E3
Tataa (pt.), Fr. Poly. 86/S13
Tatabánya, Hungary 41/E3
Tatahouine, Tunisia 106/G2
Tatalrose, Br. Col. 184/D3
Tatamagouche, Nova Scotia 168/E3
Tatamba, Solomon Is. 86/D3
Tatamy, Pa. (18085) 294/M4
Tatar (str.), U.S.S.R. 54/R5
Tatar (str.), U.S.S.R. 48/P4
Tatar A.S.S.R., U.S.S.R. 52/G3
Tatar A.S.S.R., U.S.S.R. 48/F4
Tatarsk, U.S.S.R. 48/H4
Tate, Georgia (30177) 217/D2
Tate (co.), Miss. 256/E1
Tate, Sask. 181/G4
Tateville, Ky. (42558) 237/M7
Tateyama, Japan 81/K6
Tathlina (lake), N.W. Terrs. 187/G3
Tathlith, Saudi Arabia 59/D6
Tathra, N.S. Wales 97/F5
Tati (riv.), Botswana 118/D4
Tatitlek, Alaska (99677) 196/D1
Tatla Lake, Br. Col. 184/E4
Tatlatui (lake), Br. Col. 184/D2
Tatlayoko (lake), Br. Col. 184/E4
Tatnam (cape), Manitoba 179/K2
Tatnum (cape), Man. 162/G4
Tatoosh (isl.), Wash. 310/A2
Tatra, High (mts.), Czech. 41/E2
Tatra, High (range), Poland 47/D4
Tatta, Pakistan 59/J5
Tatta, Pakistan 68/B4
Tattnall (co.), Georgia 217/J6
Tatuí, Brazil 135/C3
Tatum, N. Mex. (88267) 274/F5
Tatum, S.C. (29594) 296/H2
Tatum, Texas (75691) 303/K5
Tatums, Okla. (73087) 288/M6
Tatung (Datong), China 77/H3
Tatura, Victoria 97/C5
Tatvan, Turkey 63/K3
Taubaté, Brazil 132/E8
Taubaté, Brazil 135/D3
Tauber (riv.), W. Germany 22/C4
Täuffelen, Switzerland 39/D2
Taumarunui, N. Zealand 100/E3
Taum Sauk (mt.), Mo. 261/L7
Taung, S. Africa 118/C5
Taungdwingyi, Burma 72/C2
Taunggyi, Burma 72/C2
Taungthonton (mt.), Burma 72/B1
Taungup, Burma 72/B3
Taunton, England 13/D6
Taunton, England 10/E5
Taunton, Mass. (02780) 249/K5
Taunton (riv.), Mass. 249/K5
Taunton, Minn. (56291) 255/B6
Taunus (range), W. Germany 22/C3
Taupo, N. Zealand 100/F3
Taupo (lake), N. Zealand 100/F3
Tauq, Iraq 66/D3
Taurage, U.S.S.R. 53/B3
Taurage, U.S.S.R. 52/B3
Tauranga, N. Zealand 100/F2
Taureau (res.), Québec 172/D3
Taurianova, Italy 34/E5

Tauroa (pt.), N. Zealand 100/D1
Taurus (mts.), Turkey 59/B2
Taurus (mts.), Turkey 63/D4
Tauste, Spain 33/F2
Tautira, Fr. Poly. 86/T13
Tautira (pt.), Fr. Poly. 86/T13
Tauu (isls.), Papua N.G. 86/D2
Tavai, Paraguay 144/E5
Tavan Bogd Uul (mt.), Mongolia 77/C2
Tavannes, Switzerland 39/D2
Tavaputs (plat.), Utah 304/D4
Tavares, Fla. (32778) 212/E3
Tavas, Turkey 63/C4
Tavda, U.S.S.R. 48/G4
Tavernier, Fla. (33070) 212/F6
Taveta, Kenya 115/G4
Tavetsch, Switzerland 39/G3
Taveuni (isl.), Fiji 87/H7
Taveuni (isl.), Fiji 86/R10
Tavignano (riv.), France 28/B6
Tavira, Portugal 33/C4
Tavistock, England 10/D5
Tavistock, England 13/C7
Tavistock, N.J. (†08033) 273/B3
Tavistock, Ontario 177/D4
Tavoy, Burma 54/L8
Tavoy, Burma 72/C4
Tavoy (pt.), Burma 72/C4
Tavrichanka, U.S.S.R. 48/O5
Tavşanlı, Turkey 63/C3
Taw (riv.), England 13/D7
Taw (riv.), England 10/D5
Tawa, N. Zealand 100/B2
Tawas (lake), Mich. 250/F4
Tawas (pt.), Mich. 250/F4
Tawas City, Mich. (48763) 250/F4
Tawatinaw, Alberta 182/D2
Tawau, Malaysia 85/F5
Tawin (isl.), Ireland 17/C5
Tawi-Tawi (prov.), Philippines 82/B8
Tawi-Tawi (isl.), Philippines 82/B8
Tawitawi Group (isls.), Philippines 85/G4
Taxco de Alarcón, Mexico 150/K7
Taxila (ruins), Pakistan 68/C2
Taxis River, New Bruns. 170/D2
Taxkorgan, China 77/A4
Tay (firth), Scotland 15/F4
Tay (firth), Scotland 10/E2
Tay, Loch (lake), Scotland 10/D2
Tay, Loch (lake), Scotland 15/D4
Tay (riv.), Scotland 10/E2
Tay (riv.), Scotland 15/E4
Tay (lake), W. Australia 88/C6
Tayabamba, Peru 128/D7
Tayabas (bay), Philippines 82/C4
Tayasan, Philippines 82/D6
Taycheedah, Wis. (53090) 317/K8
Tay Creek, New Bruns. 170/D2
Tayibe, Israel 65/C3
Tayinloan, Scotland 15/C5
Taylor, Ala. (†36301) 195/H8
Taylor (mts.), Alaska 196/G2
Taylor, Ariz. (85939) 198/E4
Taylor, Ark. (71861) 202/D7
Taylor, Br. Col. 184/G2
Taylor (peak), Colo. 208/F5
Taylor (riv.), Colo. 208/F5
Taylor (co.), Fla. 212/C1
Taylor (co.), Georgia 217/D5
Taylor (mt.), Idaho 220/D5
Taylor (co.), Iowa 229/D7
Taylor (co.), Ky. 237/L6
Taylor, La. (71080) 238/D1
Taylor, Mich. (48180) 250/B7
Taylor, Miss. (38673) 256/E2
Taylor, Mo. (63471) 261/J3
Taylor, Nebr. (68879) 264/E3
Taylor (mt.), N. Mex. 274/B3
Taylor, N. Dak. (58656) 282/F6
Taylor (head), Nova Scotia 168/F4
Taylor, Pa. (18517) 294/F7
Taylor (co.), Texas 303/E5
Taylor, Texas (76574) 303/G7
Taylor, Texas 188/G4
Taylor (co.), W. Va. 312/F4
Taylor (co.), Wis. 317/E5
Taylor, Wis. (54659) 317/E7
Taylor Lake Village, Texas (†77586) 303/K2
Taylor Mill, Ky. (†41011) 237/S2
Taylor Park (res.), Colo. 208/F5
Taylors, S.C. (29687) 296/C2
Taylor's Arm, N.S. Wales 97/G2
Taylors Falls, Minn. (55084) 255/F5
Taylors Island, Md. (21669) 245/N7
Taylorsport, Ky. (†41048) 237/R2
Taylor Springs, Ill. (62089) 222/D4
Taylorstown, Pa. (15365) 294/A5
Taylorsville, Calif. (95983) 204/E3
Taylorsville, Georgia (30178) 217/C2
Taylorsville, Ind. (47280) 227/F6
Taylorsville, Ky. (40071) 237/L4
Taylorsville, Md. (†21157) 245/K3
Taylorsville, Miss. (39168) 256/F7
Taylorsville, N.C. (28681) 281/G3
Taylorsville (Philo), Ohio (†43771) 284/G6
Taylorsville, Utah (†84101) 304/B3
Taylortown, La. (†71010) 238/C2
Taylorville, Ill. (62568) 222/D4
Taymouth, New Bruns. 170/D2
Taymyr (lake), U.S.S.R. 4/B5
Taymyr (pen.), U.S.S.R. 4/B4
Taymyr (pen.), U.S.S.R. 54/L2
Taymyr (lake), U.S.S.R. 54/M2
Taymyr (lake), U.S.S.R. 48/K2
Taymyr (pen.), U.S.S.R. 48/L2
Taymyr (riv.), U.S.S.R. 48/K2
Taymyr Aut. Okr., U.S.S.R. 48/K2
Tay Ninh, Vietnam 72/E5
Tayoltita, Mexico 150/G4
Tayport, Scotland 10/E2
Tayport, Scotland 15/F4
Tayshet, U.S.S.R. 48/K4
Tayside (reg.), Scotland 15/E4
Taytay, Philippines 85/G3
Taytay, Philippines 82/B5
Taytay (bay), Philippines 82/B5
Tayyebat, Iran 66/M3
Taz (riv.), U.S.S.R. 54/K3
Taz (river), U.S.S.R. 4/C5
Taz (riv.), U.S.S.R. 48/J3
Taza, Morocco 106/D2
Tazadit, Mauritania 106/B4
Taza Khurmatu, Iraq 66/D3
Tazawa (lake), Japan 81/K4
Tazerbo (oasis), Libya 111/D2
Tazewell, Georgia (31828) 217/D6
Tazewell (co.), Ill. 222/D3
Tazewell, Tenn. (37879) 237/O8
Tazewell (co.), Va. 307/E6
Tazewell, Va. (24651) 307/E6
Tazin (lake), Sask. 181/L2
Tazlina (lake), Alaska 196/D1
Tazlina (riv.), Alaska 196/D1
Tazlina Glacier Lodge, Alaska (†99588) 196/C1
Tazovskiy, U.S.S.R. 48/J3
Tbilisi, U.S.S.R. 7/J4
Tbilisi, U.S.S.R. 2/M3
Tbilisi, U.S.S.R. 52/F6
Tbilisi, U.S.S.R. 48/E5
Tchentlo (lake), Br. Col. 184/E2
Tchibanga, Gabon 115/B4
Tchien, Liberia 106/C7
Tcholliré, Cameroon 115/B2
Tchula, Miss. (39169) 256/D4
Tchula (lake), Miss. 256/D4
Tczew, Poland 47/D1
Tea, S. Dak. (57064) 298/R7
Teacapán (inlet), Mexico 150/F5
Teachey, N.C. (28464) 281/N5
Teague, Texas (75860) 303/H6
Te Anau, N. Zealand 100/A6
Te Anau (lake), N. Zealand 100/A6
Teaneck○, N.J. (07666) 273/B2
Teapa, Mexico 150/N8
Teapot Dome (mt.), Wyo. 319/F2
Te Araroa, N. Zealand 100/G2
Te Aroha, N. Zealand 100/E2
Teasdale, Utah (84773) 304/C5
Te Atatu, N. Zealand 100/B1
Teaticket, Mass. (02536) 249/M6
Tea Tree Gully, S. Australia 88/E7
Tea Tree Gully, S. Australia 94/B7
Tea Tree Well, North. Terr. 93/C7
Te Awamutu, N. Zealand 100/E3
Teays, W. Va. (25569) 312/B6
Tebenkof (bay), Alaska 196/M2
Tébessa, Algeria 102/C1
Tébessa, Algeria 106/F1
Tebicuary (riv.), Paraguay 144/C5
Tebicuary Mí, Paraguay 144/B5
Tebicuary Mí (riv.), Paraguay 144/C5
Tebingtinggi, Indonesia 85/B5
Tebuk (Tabuk), Saudi Arabia 59/C4
Tecamachalco, Mexico 150/O2
Tecate, Mexico 150/A1
Tecer (mts.), Turkey 63/G3
Techirghiol, Romania 45/J3
Tecka, Argentina 143/B5
Tecolote (creek), N. Mex. 274/D3
Tecomán, Mexico 150/H7
Tecopa, Calif. (92389) 204/J8
Tecpan de Galeana, Mexico 150/J8
Tecuala, Mexico 150/G5
Tecuci, Romania 45/H3
Tecumseh, Kansas (66542) 232/G2
Tecumseh, Mich. (49286) 250/E7
Tecumseh, Mo. (65760) 261/H9
Tecumseh, Nebr. (68450) 264/H4
Tecumseh (mt.), N.H. 268/D4
Tecumseh, Okla. (74873) 288/N4
Tecumseh, Ontario 177/B5
Tedrow, Ohio (†43567) 284/B2
Teduzara, Bolivia 136/B2
Tedzhen, U.S.S.R. 48/F6
Teec Nos Pos, Ariz. (86514) 198/F2
Teeds Grove, Iowa (52771) 229/N4
Teegarden, Ind. (†46574) 227/E2
Teepee Creek, Alberta 182/A2
Tees, Alberta 182/D3
Tees (riv.), England 10/F3
Tees (riv.), England 13/F3
Teeswater, Ontario 177/C3
Teeterville, Ontario 177/D5
Tefé, Brazil 132/G9
Tefé, Brazil 120/C3
Tefé (riv.), Brazil 132/G9
Tefenni, Turkey 63/C4
Tefft, Ind. (46380) 227/D2
Tegal, Indonesia 85/J2
Tegel, W. Germany 22/E3
Tegelen, Netherlands 27/J6
Tegernsee (lake), W. Germany 22/D5
Tegucigalpa (cap.), Hond. 146/K8
Tegucigalpa (cap.), Honduras 154/D3
Tehachapi, Calif. (93561) 204/G8
Tehachapi (mts.), Calif. 204/G9
Tehama (co.), Calif. 204/C3
Tehama, Calif. (96090) 204/C3
Tehchow (Dezhou), China 77/J4
Tehek (lake), N.W. Terrs. 187/J3
Tehkummah, Ontario 177/B2
Tehran (cap.), Iran 2/M4
Tehran (cap.), Iran 66/G3
Tehran (cap.), Iran 59/F2
Tehran (cap.), Iran 54/G6
Tehri, India 68/D2
Tehuacán, Mexico 150/L7
Tehuantepec, Mexico 146/J8
Tehuantepec, Mexico 150/M8
Tehuantepec (gulf), Mexico 150/M9
Tehuantepec (isth.), Mexico 150/M8
Tehuipango, Mexico 150/P2
Teide, Pico de (peak), Spain 33/B5
Teifi (riv.), Wales 13/C5
Teifi (riv.), Wales 10/D4
Teigen, Mont. (59084) 262/H3
Teignmouth, England 10/C5
Teignmouth, England 13/D7
Teith (riv.), Scotland 15/D4
Tejerri, Libya 111/B3
Tejerri, Libya 102/D2
Tejo (Tagus) (riv.), Portugal 33/B3
Tejutla, Guatemala 154/B3
Tekamah, Nebr. (68061) 264/H3
Te Kao, N. Zealand 100/D1
Tekapo (lake), N. Zealand 100/C5
Te Karaka, N. Zealand 100/G3
Te Kauwhata, N. Zealand 100/E2
Tekax de Álaro Obregón, Mexico 150/P6
Tekeli, U.S.S.R. 48/H5
Tekes, China 77/B3
Tekirdağ (prov.), Turkey 63/B2
Tekirdağ, Turkey 59/A1
Tekirdağ, Turkey 63/B2
Tekman, Turkey 63/J3
Teknaf, Bangladesh 68/G4
Tekoa, Wash. (99033) 310/H3
Tekong Besar, Pulau (isl.), Singapore 72/F6
Tekonsha, Mich. (49092) 250/E6
Te Kopuru, N. Zealand 100/D2
Te Kuiti, N. Zealand 100/E3
Tel (riv.), India 68/E4
Tela, Honduras 154/D3
Telanaipura, Indonesia 54/M10
Telavi, U.S.S.R. 52/G6
Tel Aviv (dist.), Israel 65/B3
Tel Aviv-Jaffa, Israel 59/B3
Tel Aviv-Jaffa, Israel 65/B3
Tel Aviv-Jaffa, Israel 54/E6
Telč, Czech. 41/C2
Telde, Spain 33/B5
Telefomin, Papua N.G. 85/B7
Telegraph, Texas (76883) 303/E7
Telegraph Creek, Br. Col. 184/K2
Telemark (co.), Norway 18/F7
Telephone, Texas (75488) 303/J4
Telescope (peak), Calif. 204/H7
Telescope (pt.), Grenada 161/D8
Teles Pires (riv.), Brazil 120/D3
Teles Pires (riv.), Brazil 132/B5
Telfair (co.), Georgia 217/G7
Telford, England 13/E5
Telford, Pa. (18969) 294/M5
Telford, Tenn. (37690) 237/S8
Telfordville, Alberta 182/C3
Telfs, Austria 41/A3
Telgte, W. Germany 22/B3
Télimélé, Guinea 106/B6
Telkalakh, Syria 63/F5
Tel Kotchek, Syria 59/D2
Tel Kotchek, Syria 63/K4
Telkwa, Br. Col. 184/D3
Tell, Georgia (†30304) 217/J2
Tell, Texas (79259) 303/D3
Tell 'Asur (mt.), Jordan 65/C4
Tell City, Ind. (47586) 227/D9
Teller, Alaska (99778) 196/E1
Teller (co.), Colo. 208/J5
Tellicherry, India 68/C6
Tellico (riv.), Tenn. 237/N10
Tellico Plains, Tenn. (37385) 237/N10
Tellin, Belgium 27/G8
Telluride, Colo. (81435) 208/D7
Telma, Wash. (†98826) 310/E3
Telocaset, Oreg. (†97883) 291/K2
Telogia, Fla. (32360) 212/B1
Telok Anson, Malaysia 72/D6
Teloloapan, Mexico 150/J7
Telpaneca, Nicaragua 154/D4
Tel'pos-Iz (mt.), U.S.S.R. 52/K2
Telsen, Argentina 143/C5
Telšiai, U.S.S.R. 53/B2
Telšiai, U.S.S.R. 52/B3
Teltow, E. Germany 22/E4
Teltown, Ireland 17/H4
Telukbayur, Indonesia 85/C6
Tema, Ghana 106/E7
Tema, Ghana 102/C4
Temacine, Algeria 106/F2
Temae (lake), Fr. Poly. 86/S12
Temanggung, Indonesia 85/J2
Temascalapa, Mexico 150/M1
Tematangi (isl.), Fr. Poly. 87/M8
Temax, Mexico 150/P6
Tembisa, S. Africa 118/H6
Temblador, Venezuela 124/G3
Tembuè, Mozambique 118/E2
Temecula, Calif. (92390) 204/H10
Temerloh, Malaysia 72/D7
Temiang, Bukit (mt.), Malaysia 72/D6
Temirtau, U.S.S.R. 54/J4
Temirtau, U.S.S.R. 48/H4
Témiscamingue (county), Québec 174/B3
Témiscaming, Québec 174/B3
Témiscouata (co.), Québec 172/J2
Témiscouata (lake), Québec 172/H2
Temma, Tasmania 99/A3
Temora, N. S. Wales 88/H6
Temora, N.S. Wales 97/D4
Temoris, Mexico 150/E3
Temósachic, Mexico 150/E2
Tempe, Ariz. (*85282) 198/D5
Tempe Downs, North. Terr. 93/C8
Tempelhof, W. Germany 22/F4
Temperance, Mich. (48182) 250/F7
Temperance Hall, Tenn. (†37095) 237/K8
Temperance Vale, New Bruns. 170/C2
Temperanceville, Va. (23442) 307/T5
Tempio Pausania, Italy 34/B4
Temple (mt.), Alberta 182/B4
Temple, Georgia (30179) 217/B3
Temple, La. (71476) 238/E4
Temple○, Maine (04984) 243/C6
Temple, Mich. (†48625) 250/E4
Temple○, N.H. (03084) 268/D6
Temple, Okla. (73568) 288/K6
Temple, Pa. (19560) 294/L5
Temple, Texas (76501) 303/G6
Temple, Texas 188/G4
Temple Bar, Ariz. (86443) 198/A2
Temple City, Calif. (91780) 204/D10
Templemore, Ireland 10/C4
Templemore, Ireland 17/F6
Templestowe and Doncaster, Victoria 97/J5
Temple Terrace, Fla. (33617) 212/C2
Templeton, Calif. (93465) 204/E8
Templeton, Ind. (47986) 227/C3
Templeton, Iowa (51463) 229/D5
Templeton○, Mass. (01468) 249/F2
Templeton, Pa. (16259) 294/C4
Templeton, Québec 172/B4
Templetuohy, Ireland 17/F6
Templeville, Md. (21670) 245/P4
Templin, E. Germany 22/E2
Tempo, N. Ireland 17/G3
Temryuk, U.S.S.R. 52/E5
Temse, Belgium 27/E6
Temuco, Chile 120/B6
Temuco, Chile 138/E2
Temuka, N. Zealand 100/C6
Temvik, N. Dak. (†58552) 282/K7
Ten, Colombia 126/D5
Tena, Ecuador 128/D3
Tenabo, Mexico 150/P6
Tenafly, N.J. (07670) 273/C1
Tenaha, Texas (75974) 303/K6
Tenakee Springs, Alaska (99841) 196/M1
Tenali, India 68/E5
Tenancingo de Degollado, Mexico 150/K7
Tenango de Río Blanco, Mexico 150/O2
Tenants Harbor, Maine (04860) 243/E8
Tenares, Dom. Rep. 158/E5
Tenasserim (div.), Burma 72/C4
Tenasserim, Burma 72/C5
Tenasserim (isl.), Burma 72/C4
Tenbury, England 13/E5
Tenby, Manitoba 179/C4
Tenby, Wales 10/D5
Tenby, Wales 13/C6
Tendal, La. (†71290) 238/H2
Ten Degree (chan.), India 68/G7
Tendelti, Sudan 111/F5
Tendelti, Sudan 102/F3
Tendo, Japan 81/K4
Tendoy, Idaho (83468) 220/E5
Tendrara, Morocco 106/D2
Tendre (peak), Switzerland 39/B3
Tenecape, Nova Scotia 168/E3
Ténenkou, Mali 106/C6
Ténéré (des.), Niger 106/G5
Tenerife (isl.), Spain 102/A2
Tenerife (isl.), Spain 106/A3
Tenerife (isl.), Spain 33/B5
Ténès, Algeria 106/E1
Teng, Nam (riv.), Burma 72/C2
Tengchong, China 77/E6
Tenggarong, Indonesia 85/F6
Tenggol, Pulau (isl.), Malaysia 72/D6
Tengiz (lake), U.S.S.R. 48/G4
Teng Xian, China 77/H7
Tenigerbad, Switzerland 39/G3
Tenino, Wash. (98589) 310/C4
Tenke, Zaire 115/E6
Tenkiller Ferry (lake), Okla. 288/S3
Tenkodogo, Upper Volta 106/E6
Tenleytown, D.C. (†20008) 245/E4
Ten Mile (lake), Newf. 166/C3
Tenmile, Oreg. (97481) 291/D4
Tenmile (creek), Oreg. 291/K5
Ten Mile, Tenn. (37880) 237/M9
Tenmile (creek), Texas 303/G3
Tennant, Calif. (†96066) 204/C2
Tennant, Iowa (51574) 229/C5
Tennant Creek, Australia 87/D7
Tennant Creek, North. Terr. 88/E3
Tennant Creek, North. Terr. 93/C5
Tennent, N.J. (07763) 273/E3
Tennessee 188/J3
Tennessee (riv.) 188/J3
Tennessee (riv.), Ala. 195/C1
Tennessee, Ill. (62374) 222/C3
Tennessee (riv.), Ky. 237/D6
TENNESSEE 237
Tennessee (riv.), Tenn. 237/E10
Tennessee (state), U.S. 146/K6
Tennessee City, Tenn. (†37055) 237/F8
Tennessee Pass, Colo. (†80461) 208/G4
Tennessee Ridge, Tenn. (37178) 237/F8
Tennessee-Tombigbee Waterway, Ala. 195/B4
Tennessee-Tombigbee Waterway, Miss. 256/H2
Tenneville, Belgium 27/H8
Tenney, Minn. (56582) 255/B4
Tennga, Georgia (30751) 217/C1
Tennille, Ala. (†36010) 195/G7
Tennille, Georgia (31089) 217/G5
Tennyson, Ind. (47637) 227/C8
Tennyson, Texas (76953) 303/D6
Tennyson, Wis. (†53820) 317/E10
Teno, Chile 138/A10
Tenosique de Pino Suárez, Mexico 150/O8
Tenquehuen (isl.), Chile 138/D6
Tenri, Japan 81/J8
Tensas (par.), La. 238/H2
Tensas (riv.), La. 238/G3
Tensaw, Ala. (†36507) 195/C8
Tensaw (riv.), Ala. 195/C9
Tensed, Idaho (83870) 220/B2
Ten Sleep, Wyo. (82442) 319/E1
Tenstrike, Minn. (56683) 255/D3
Tenterden, England 13/H6
Tenterfield, N. S. Wales 88/J5
Tenterfield, N.S. Wales 97/G1
Ten Thousand (isls.), Fla. 188/K5
Ten Thousand (isls.), Fla. 212/E6
Ten Thousand Smokes (valley), Alaska 196/G3
Teocaltiche, Mexico 150/H6
Teocelo, Mexico 150/P1
Teófilo Otoni, Brazil 120/E4
Teófilo Otoni, Brazil 132/F7
Te One, N. Zealand 100/D7
Teotihuacán (ruin), Mexico 150/M1
Teotihuacán de Arista, Mexico 150/L1
Teotitlán del Camino, Mexico 150/L8
Tepa, Indonesia 85/H7
Tepache, Mexico 150/E2
Tepalcingo, Mexico 150/M2
Tepatitlán de Morelos, Mexico 150/H6
Tepeaca, Mexico 150/N2
Tepeapulco, Mexico 150/M1
Tepehuanes, Mexico 150/G4
Tepeji del Río, Mexico 150/L1
Tepelenë, Albania 45/D5
Tepetlán, Mexico 150/P1
Tepexi de Rodríguez, Mexico 150/N2
Tepeyahualco, Mexico 150/O1
Tepic, Mexico 150/G6
Teplá u Toužimě, Czech. 41/B1
Teplice, Czech. 41/B1
Tepoztlán, Mexico 150/L1
Te Puke, N. Zealand 100/F2
Tequeje (riv.), Bolivia 136/B3
Tequendama (falls), Colombia 126/C5
Tequesquite (creek), N. Mex. 274/E2
Tequesta, Fla. (33458) 212/F5
Tequixquitla, Mexico 150/O1
Ter (riv.), Spain 33/H1
Téra, Niger 106/E6
Teramo (prov.), Italy 34/D3
Teramo, Italy 34/D3
Terán, Mexico 150/N8
Terang, Victoria 97/B6
Ter Apel, Netherlands 27/L3
Terawhiti (cape), N. Zealand 100/A3
Tercan, Turkey 63/J3
Terceira (isl.), Portugal 33/C1
Tercero (riv.), Argentina 143/D3
Terchová, Czech. 41/E2
Terempa, Indonesia 85/D5
Terence Bay, Nova Scotia 168/E4
Terengganu (state), Malaysia 72/D6
Teresina, Brazil 120/E3
Teresina, Brazil 132/F4
Teresita, Mo. (65573) 261/J9
Teresópolis, Brazil 135/E3
Terespol, Poland 47/F2
Teressa (isl.), India 68/G7
Terevinto, Bolivia 136/D5
Terhazza (ruins), Mali 106/C4
Terhune, Ind. (†46069) 227/E4
Teriberka, U.S.S.R. 52/E1
Terlingua, Texas (79852) 303/D12
Terlingua (creek), Texas 303/D12
Terlton, Okla. (74081) 288/O2
Termas de Cauquenes, Chile 138/B10
Terme, Turkey 63/G2
Termez, U.S.S.R. 48/G6
Termini Imerese, Italy 34/D6
Términos (lag.), Mexico 150/O7
Termo, Calif. (96132) 204/E3
Termoli, Italy 34/E3
Termonde (Dendermonde), Belgium 27/E6
Termonfeckin, Ireland 17/J4
Termunten, Netherlands 27/K2
Ternate, Indonesia 54/O9
Ternate, Indonesia 85/H5
Terneuzen, Netherlands 27/D6
Terni (prov.), Italy 34/D3
Terni, Italy 34/D3
Terni, Italy 7/F4
Ternitz, Austria 41/D3
Ternopol', U.S.S.R. 48/C5
Ternopol', U.S.S.R. 52/C5
Te Roto, N. Zealand 100/D7
Terowie, S. Australia 94/F5
Terpeniye (cape), U.S.S.R. 48/P5
Terra Alta, W. Va. (26764) 312/H4
Terra Bella, Calif. (93270) 204/G8
Terrabona, Nicaragua 154/E4
Terrace, Br. Col. 162/D5
Terrace, Br. Col. 184/C3
Terrace, Br. Col. 184/C3
Terrace, Minn. (†56380) 255/C5
Terrace Bay, Ontario 177/H5
Terrace Bay, Ontario 175/C3
Terrace Heights, Wash. (98901) 310/E4
Terra Ceia, Fla. (33591) 212/D4
Terrace Park, Ohio (45174) 284/D9
Terracina, Italy 34/D4
Terra Corra, Neth. Ant. 161/E8
Terråk, Norway 18/H4
Terral, Okla. (73569) 288/L7
Terralba, Italy 34/B5
Terra Nova, Newf. 166/C2
Terra Nova (riv.), Newf. 166/C2
Terra Nova Nat'l Park, Newf. 166/D2
Terrebonne (par.), La. 238/J8
Terrebonne (bay), La. 238/J8
Terrebonne, Minn. (†56750) 255/B3
Terrebonne, Oreg. (97760) 291/F3
Terrebonne (co.), Québec 172/H4
Terrebonne, Québec 172/H4
Terre-de-Bas (isl.), Guadeloupe 161/A7
Terre-de-Haut (isl.), Guadeloupe 161/A7
Terre Haute, Ill. (†61454) 222/C3
Terre Haute, Ind. 188/J3
Terre Haute, Ind. (*47801) 227/C6
Terre Hill, Pa. (17581) 294/L5
Terrell (co.), Georgia 217/D7
Terrell, N.C. (28682) 281/G3
Terrell (co.), Texas 303/B7
Terrell, Texas 188/G4
Terrell, Texas (75160) 303/H5
Terrell Hills, Texas (†78201) 303/K11
Terrenate, Mexico 150/N1
Terrenceville, Newf. 166/D4
Terre Neuve, Haiti 158/B5
Terreton, Idaho (83450) 220/F6
Terrey Hills, N. S. Wales 88/L3
Terrey Hills, N.S. Wales 97/J3
Terri (mt.), Switzerland 39/D2
Terri (peak), Switzerland 39/H3
Terrigal-Wamberal, N.S. Wales 97/F3
Terril, Iowa (51364) 229/C2
Terrill (mt.), Utah 304/C5
Territok (cape), Newf. 166/B2
Terry, La. (71285) 238/H1
Terry, Miss. (39170) 256/D6
Terry, Mont. (59349) 262/L4
Terry (co.), Texas 303/B4
Terry Town, La. (†70053) 238/O4
Terrytown, Nebr. (†69341) 264/A3
Terryville, Conn. (06786) 210/C2
Terschelling (isl.), Netherlands 27/G2
Teruel (prov.), Spain 33/F2
Teruel, Spain 33/F2
Terutao, Ko (isl.), Thailand 72/C6
Tervola, Finland 18/O3
Tesawa, Libya 111/B2
Tescott, Kansas (67484) 232/E2
Teshekpuk (lake), Alaska 196/H1
Teshio, Japan 81/K1
Teshio (mt.), Japan 81/L1
Teshio (riv.), Japan 81/L1
Tesla, W. Va. (26640) 312/E5
Teslić, Yugoslavia 45/C3
Teslin (lake) 162/C3
Teslin (lake), Br. Col. 184/K1
Teslin, Yukon 187/E3
Teslin (lake), Yukon 187/E4
Teslin (riv.), Yukon 187/E3
Tessalit, Mali 106/E4
Tessaoua, Niger 106/F6
Tessenderlo, Belgium 27/G6
Tessenei, Ethiopia 59/C6
Tessenei, Ethiopia 111/G4
Tessier, Sask. 181/D4
Test (riv.), England 13/F6
Testa (cape), Italy 34/B4
Testa del Gargano (cape), Italy 34/F4
Tesuque, N. Mex. (87574) 274/C3
Tét, Hungary 41/D3
Tetachuck (lake), Br. Col. 184/E3
Tetagouche (riv.), New Bruns. 170/D1
Tetas (pt.), Chile 138/A4
Tete (prov.), Mozambique 118/E3
Tete, Mozambique 102/F6
Tete, Mozambique 118/E3
Tête-à-la-Baleine, Québec 174/E2
Tete Jaune Cache, Br. Col. 184/H4
Te Teko, N. Zealand 100/F3
Teterboro, N.J. (07608) 273/B2
Teterow, E. Germany 22/E2
Teterton, W. Va. (†26886) 312/H5
Teteven, Bulgaria 45/G4
Tetiaroa (atoll), Fr. Poly. 87/M7
Tetlin, Alaska (99779) 196/K2
Teton (co.), Idaho 220/G6
Teton, Idaho (83451) 220/G6
Teton (riv.), Idaho 220/G6
Teton (co.), Mont. 262/D3
Teton (riv.), Mont. 262/E3
Teton (co.), Wyo. 319/B2
Teton (range), Wyo. 319/B2
Tetonia, Idaho (83452) 220/G6
Teton Village, Wyo. (83025) 319/B2
Tetotum, Va. (†22485) 307/P4
Tétouan, Morocco 106/C1
Tétouan, Morocco 102/B1
Tetovo, Yugoslavia 45/E5
Teuco (riv.), Argentina 143/D1
Teufen, Switzerland 39/H2
Teulada (cape), Italy 7/E5
Teulada (cape), Italy 34/B5
Teulon, Manitoba 179/E4
Teupasenti, Honduras 154/D3
Teustepe, Nicaragua 154/E4
Teutoburger Wald (for.), W. Germany 22/C2
Teutopolis, Ill. (62467) 222/E4
Teuva, Finland 18/M5
Teviot (riv.), Scotland 15/F5
Teviot (riv.), Scotland 10/E3
Te Waewae (bay), N. Zealand 100/A7
Tewantin-Noosa, Queensland 95/E5
Te Whanga (mts.), N. Zealand 100/E7
Tewkesbury, England 10/F5
Tewkesbury, England 13/E6
Tewksbury, Mass. (01876) 249/K2
Texa (isl.), Scotland 15/B5
Texada (isl.), Br. Col. 184/J2
Texarkana, Ark. 188/H4
Texarkana, Ark. (75502) 202/C7
Texarkana, Texas 188/H4
Texarkana, Texas 146/J6
Texarkana, Texas (*75501) 303/L4
Texas 188/G4
TEXAS 303
Texas, Georgia (†30217) 217/B4
Texas, Ky. (†40069) 237/L5
Texas, Md. (†21030) 245/M3
Texas (co.), Mo. 261/J8
Texas (co.), Okla. 288/C1
Texas (state), U.S. 146/J6
Texas City, Texas (77590) 303/K3
Texas Creek, Colo. (81250) 208/H6
Texcoco de Mora, Mexico 150/M1
Texel (isl.), Netherlands 27/F2
Texhoma, Okla. (73949) 288/C1
Texhoma, Texas (73949) 303/C1
Texico, N. Mex. (88135) 274/F4
Texistepeque, El Salvador 154/C3
Texline, Texas (79087) 303/B1
Texola, Okla. (73668) 288/G4
Texoma (lake) 188/G4
Texoma (lake), Okla. 288/N7
Texoma (lake), Texas 303/H3
Texon, Texas (76954) 303/C6
Teykovo, U.S.S.R. 52/E3
Teyvareh, Afghanistan 68/A2
Teyvareh, Afghanistan 59/H3
Teziutlán, Mexico 150/O1
Tezonapa, Mexico 150/P2
Tezontepec, Mexico 150/M1
Tezpur, India 68/G3
Tezu, India 68/H3
Tezzeron (lake), Br. Col. 184/E3
Tha, Nam (riv.), Laos 72/D2
Tha-anne (riv.), N.W. Terrs. 187/J3
Thabazimbi, S. Africa 118/D4
Thacher (isl.), Mass. 249/M2
Tha Chin, Mae Nam (riv.), Thailand 72/C4
Thacker, W. Va. (25694) 312/B7
Thackeray, Ill. (†62859) 222/E5
Thackeray, Ohio (†43078) 284/C5
Thackerville, Okla. (73459) 288/M7
Thai Binh, Vietnam 72/E2
Thailand 2/Q5
Thailand 54/M8
Thailand (gulf) 2/Q5
Thailand (gulf) 54/M9

Thailand (gulf), Cambodia 72/D5
THAILAND (SIAM) 72
Thailand (gulf), Thailand 72/D5
Thai Nguyen, Vietnam 72/E2
Thakhek (Muang Khammouan), Laos 72/K3
Thal, Pakistan 59/K3
Thal, Switzerland 39/J2
Thalberg, Manitoba 179/F4
Thale, E. Germany 22/D3
Thale Luang (lag.), Thailand 72/D6
Thalia, Texas 79227/ 303/E4
Thalmann, Georgia (†31520) 217/J8
Thalu, Ko (isls.), Thailand 72/C4
Thalwil, Switzerland 39/G2
Thame, England 13/G6
Thame (riv.), England 13/G7
Thames, Conn. 210/G3
Thames (riv.), England 11/F5
Thames (riv.), England 13/H6
Thames, N. Zealand 100/E2
Thames (firth), N. Zealand 100/E2
Thames (riv.), Ontario 177/B5
Thamesford, Ontario 177/C4
Thamesville, Conn. (†06360) 210/G2
Thamesville, Ontario 177/C5
Thana, India 68/B6
Thana (creek), India 68/B7
Thane, Alaska (†99801) 196/N1
Thangool, Queensland 95/D5
Thanh Hoa, Vietnam 72/E3
Thanh Tri, Vietnam 72/E5
Thanjavur, India 68/D6
Thann, France 28/G4
Thar (des.), Pakistan 68/C3
Thargomindah, Queensland 88/G5
Thargomindah, Queensland 95/C5
Tharrawaddy, Burma 72/C3
Tharthar, Wadi (dry riv.), Iraq 66/C3
Tharthar (res.), Iraq 66/C3
Thásos, Greece 45/G5
Thásos (isl.), Greece 45/G5
Thatch (cay), Virgin Is. (U.S.) 161/B4
Thatcham, England 13/F6
Thatcher, Ariz. (85552) 198/F6
Thatcher, Colo. (†81082) 208/L7
Thatcher, Idaho (83283) 220/H7
That Khe, Vietnam 72/E2
Thaton, Burma 72/C3
Thau (mts.), France 28/F6
Thaungdut, Burma 72/B1
Thawville, Ill. (60968) 222/E3
Thaxton, Miss. (38871) 256/F2
Thaxton, Va. (24174) 307/K6
Thaya (riv.), Austria 41/C2
Thayawthadangyi Kyun (isl.), Burma 72/C4
Thayer, Ill. (62689) 222/D4
Thayer, Ind. (46381) 227/C2
Thayer, Iowa (50254) 229/E6
Thayer, Kansas (66776) 232/G4
Thayer, Mo. (65791) 261/J9
Thayer (co.), Nebr. 264/G4
Thayer, Nebr. (†68467) 264/G4
Thayer, W. Va. (†25936) 312/E7
Thayer Junction, Wyo. (†82901) 319/D4
Thayetmyo, Burma 72/B3
Thayne, Wyo. (83127) 319/A3
Thayngen, Switzerland 39/G1
Thazi, Burma 72/C2
The Alberga (riv.), S. Australia 94/F2
Thealka, Ky. (41259) 237/R5
Theano (pt.), Ontario 177/J5
Thebarton, S. Australia 88/D8
Thebarton, S. Australia 94/A7
Thebes, Ill. (62990) 222/D6
The Colony, Texas 303/G1
The Coorong (lag.), S. Australia 94/F6
The Dalles, Oreg. 188/B1
The Dalles, Oreg. (97058) 291/F2
The Dalles (dam), Oreg. 291/F2
The Dalles (dam), Wash. 310/D5
Thedford, Nebr. (69166) 264/D3
Thedford, Ontario 177/C4
The Entrance, N. S. Wales 88/J6
The Entrance, N.S. Wales 97/F3
The Gap, N.S. Wales 97/A2
The Gap, Queensland 88/J2
The Glen, N.Y. (†12885) 276/N3
The Granites, North. Terr. 93/B6
The Hamilton (riv.), S. Australia 94/D2
The Hawk, Nova Scotia 168/C5
The Heads (prom.), Oreg. 291/C5
The Hermitage, N. Zealand 100/C5
Theilman, Minn. (55978) 255/F6
Thelon (riv.), N.W.T. 146/H3
Thelon (riv.), N.W.T. 162/F3
Thelon (riv.), N.W. Terrs. 187/H3
Them, Denmark 21/C5
The Macumba (riv.), S. Australia 94/E2
The Narrows (str.), N.J. 273/E2
Thendara, N.Y. (13472) 276/K3
The Neales (riv.), S. Australia 94/E3
Theodore, Ala. (36582) 195/B9
Theodore, Queensland 95/D5
Theodore, Sask. 181/J4
Theodore Roosevelt (dam), Ariz. 198/D5
Theodore Roosevelt (lake), Ariz. 198/D5
Theodore Roosevelt Nat'l Park, N. Dak. 282/D4
Theodore Roosevelt Nat'l Park, N. Dak. 282/C5
Theodore Roosevelt Nat'l Park, N. Dak. 282/D6
Theodosia, Mo. (65761) 261/G9
The Pas, Man. 162/F5
The Pas, Manitoba 179/F3
The Plains, Ohio (45780) 284/F7
The Plains, Va. (22171) 307/N3

The Range, New Bruns. 170/E2
Theresa, N.Y. (13691) 276/J2
Theresa, Wis. (53091) 317/K8
Therien, Alberta 182/E2
Theriot, La. (70397) 238/J8
Thermaic (gulf), Greece 45/F5
Thermal, Calif. (92274) 204/J10
Thermalito, Calif. (†95965) 204/D4
Thermopolis, Wyo. (82443) 319/D2
The Rock, Georgia (30285) 217/D5
The Rock, N.S. Wales 97/D4
The Round (mt.), N.S. Wales 97/G2
Therwil, Switzerland 39/E1
The Salt (lake), N.S. Wales 97/B2
Thesiger (bay), N.W. Terrs. 187/F2
The Skaw (Skagens Odde) (cape), Denmark 21/D2
Thessalon, Ont. 162/H6
Thessalon, Ontario 175/D3
Thessalon, Ontario 177/J5
Thessaloníki, Greece 7/G4
Thessaloníki, Greece 45/F5
Thessaly (reg.), Greece 45/F6
The Stevenson (riv.), S. Australia 94/D2
Thetford, England 13/H5
Thetford, England 10/G4
Thetford◯, Vt. (05074) 268/C4
Thetford Center, Vt. (05075) 268/C4
Thetford Mines, Québec 172/F3
Thetis Island, Br. Col. 184/J3
The Twins (mt.), Alberta 182/B3
Theux, Belgium 27/H8
The Village, Okla. (73120) 288/L3
The Warburton (riv.), S. Australia 94/F2
Thibault, New Bruns. 170/C1
Thibodaux, La. (70301) 238/J7
Thicket Portage, Manitoba 179/J3
Thickwood (hills), Alberta 182/D1
Thickwood (hills), Sask. 181/D2
Thida, Ark. (72165) 202/H2
Thief (lake), Minn. 255/C2
Thief (riv.), Minn. 255/B2
Thief River Falls, Minn. (56701) 255/B2
Thielsen (mt.), Oreg. 291/F4
Thiensville, Wis. (53092) 317/L1
Thiers, France 28/E5
Thiès, Senegal 106/A5
Thiès, Senegal 102/A3
Thika, Kenya 102/F5
Thika, Kenya 115/G4
Thimphu (cap.), Bhutan 54/L7
Thimphu (cap.), Bhutan 68/G3
Thio, Ethiopia 111/H5
Thio, New Caled. 86/H4
Thionville, France 28/G3
Thíra, Greece 45/G7
Thíra (isl.), Greece 45/G7
Third (lake), Maine 243/H5
Third (lake), N.H. 268/E1
Third Cataract, Sudan 111/E4
Third Cataract (dam), Sudan 102/E3
Third Cataract, Sudan 59/B6
Third Lake, Ill. (†60046) 222/B4
Thirsk, England 13/F3
Thirty Mile (lake), N. Dak. 282/F6
Thirtymile (creek), Oreg. 291/G2
Thisted, Denmark 21/B4
Thisted, Denmark 18/F8
Thistle (isl.), S. Australia 94/E6
Thistle, Utah (†84629) 304/C4
Thithia (isl.), Fiji 86/R10
Thívai, Greece 45/F6
Thiviers, France 28/D5
Thjórsá (riv.), Iceland 21/C1
Thlewiaza (riv.), N.W. Terrs. 187/J3
Thoa (riv.), N.W. Terrs. 187/H3
Tho Chau, Hon (isl.), Vietnam 72/D5
Thoen, Thailand 72/G1
Thohoyandou, S. Africa 118/E4
Thohoyandou (cap.), Venda, S. Africa 102/F7
Tholen, Netherlands 27/E5
Thomas (co.), Georgia 217/E9
Thomas (co.), Kansas 232/A2
Thomas, Md. (†21613) 245/N6
Thomas (co.), Nebr. 264/D3
Thomas, Okla. (73669) 288/J3
Thomas (co.), Oreg. 291/C5
Thomas, S. Dak. (†57242) 298/P4
Thomas (lake), Texas 303/C5
Thomas (range), Utah 304/A4
Thomas, W. Va. (26292) 312/H4
Thomasboro, Ill. (61878) 222/E3
Thomas-Müntzer-Stadt, E. Germany 22/D3
Thomas Stone Nat'l Hist. Site, Md. 245/K6
Thomaston, Ala. (36783) 195/C6
Thomaston◯, Conn. (06787) 210/C2
Thomaston (bay), Conn. 210/C2
Thomaston, Georgia (30286) 217/D5
Thomaston, Maine (04861) 243/E7
Thomaston◯, Maine (04861) 243/E7
Thomaston, N.Y. (†11020) 276/P7
Thomastown, Ireland 10/C4
Thomastown, Ireland 17/G7
Thomastown, La. (†71262) 238/H2
Thomastown, Miss. (39171) 256/E5
Thomastown, Victoria 97/J4
Thomasville, Ala. (36784) 195/C7
Thomasville, Ga. 188/K4
Thomasville, Georgia (31792) 217/E9
Thomasville, Miss. (†39073) 256/F6
Thomasville, Mo. (†65438) 261/J9
Thomasville, N.C. (†27360) 281/J3
Thomasville, Pa. (17364) 294/J6
Thomonde, Haiti 158/C6
Thompson, Ala. (†36089) 195/G6
Thompson (riv.), Br. Col. 184/G5
Thompson◯, Conn. (06277) 210/H1
Thompson (lake), Conn. 210/H1
Thompson (peak), Colo. 208/C5
Thompson, Iowa (50478) 229/F2
Thompson (riv.), Iowa 229/E5
Thompson, Man. 162/G4
Thompson, Man. 146/J4

Thompson, Manitoba 179/J2
Thompson (isl.), Mass. 249/D7
Thompson, Mich. (49889) 250/C3
Thompson (creek), Miss. 256/G8
Thompson, Mo. (65285) 261/J4
Thompson (peak), N. Mex. 274/D3
Thompson, N. Dak. (58278) 282/R4
Thompson, Ohio (44086) 284/H2
Thompson, Pa. (18465) 294/L2
Thompson (riv.), Queensland 95/B5
Thompson (lake), Sask. 88/O5
Thompson, Utah (84540) 304/E5
Thompson (mt.), Wyo. 319/B3
Thompson Falls, Mont. (59873) 262/A3
Thompsons (creek), S.C. 296/G2
Thompsons Station, Tenn. (37179) 237/H9
Thompsontown, Pa. (17094) 294/H4
Thompson Valley (res.), Oreg. 291/F5
Thompsonville, Conn. (†06082) 210/E1
Thompsonville, Ill. (62890) 222/E6
Thompsonville, Mich. (49683) 250/C4
Thomsen (riv.), N.W. Terrs. 187/F2
Thomson, Georgia (30824) 217/H4
Thomson, Ill. (61285) 222/C2
Thomson, Minn. (†56319) 255/F4
Thomson (riv.), Queensland 88/G4
Thomson's Falls, Kenya 115/G3
Thon Buri, Thailand 72/D4
Thongwa, Burma 72/C3
Thonon-les-Bains, France 28/G4
Thonotosassa, Fla. (33592) 212/D3
Thor, Iowa (50591) 229/E3
Thorburn, Nova Scotia 168/F3
Thoreau, N. Mex. (87323) 274/A3
Thoresby (riv.), Newf. 166/B2
Thorhild, Alberta 182/D2
Thorn, Miss. (†38851) 256/F3
Thornaby-on-Tees, England 10/F3
Thornaby-on-Tees, England 13/F3
Thornburg, Iowa (50255) 229/J6
Thornburg, Va. (22565) 307/N4
Thornbury, England 13/E6
Thornbury, Ontario 177/D3
Thorndale, Ontario 177/C4
Thorndale, Texas (76577) 303/G7
Thorndike◯, Maine (04986) 243/E6
Thorndike, Mass. (01079) 249/E4
Thorne, England 13/F4
Thorne, N. Dak. (†58366) 282/L2
Thorne, Ontario 175/C3
Thorne Bay, Alaska (†99901) 196/M2
Thornfield, Mo. (65762) 261/G9
Thornhill, Br. Col. 184/C3
Thornhill, Ky. (†40222) 237/K1
Thornhill, Manitoba 179/D5
Thornhill, Central, Scotland 15/D4
Thornhill, Dumf. & Gall., Scotland 15/E5
Thornley, England 13/J4
Thornhurst, Pa. (†18424) 294/L3
Thornloe, Ontario 175/D3
Thornloe, Ontario 177/K5
Thornton, Ark. (71766) 202/F6
Thornton, Calif. (95686) 204/B9
Thornton, Colo. (80229) 208/K3
Thornton, Idaho (†83453) 220/G6
Thornton, Ill. (60476) 222/C6
Thornton, Iowa (50479) 229/G3
Thornton, Miss. (39172) 256/D4
Thornton◯, N.H. (†03285) 268/D4
Thornton, Ontario 177/E3
Thornton, Texas (76687) 303/H6
Thornton, Wash. (99176) 310/H3
Thornton, W. Va. (26440) 312/G4
Thornton Cleveleys, England 13/G1
Thornton Cleveleys, England 10/F1
Thorntown, Ind. (46071) 227/D4
Thornville, Ohio (43076) 284/F6
Thornwell, La. (†70549) 238/E6
Thornwood, W. Va. (†24920) 312/G5
Thorofare, N.J. (08086) 273/B4
Thorold, Ontario 177/E4
Thorp, Wash. (98946) 310/E3
Thorp, Wis. (54771) 317/E6
Thorpe (lake), N.C. 281/C4
Thorpe, W. Va. (24888) 312/D8
Thorp Spring, Texas (†76048) 303/F5
Thorsby, Ala. (35171) 195/E5
Thorsby, Alberta 182/C3
Thouars, France 28/C4
Thouin (pt.), W. Australia 88/B4
Thouin (pt.), W. Australia 92/B3
Thousand (isls.), N.Y. 276/H2
Thousand (isls.), Ontario 177/H2
Thousand Island Park, N.Y. (13692) 276/J2
Thousand Lake (mt.), Utah 304/C5
Thousand Oaks, Calif. (*91360) 204/G9
Thousand Palms, Calif. (92276) 204/J10
Thousand Spring (creek), Nev. 266/G1
Thousand Springs, Nev. (†89835) 266/G1
Thrace (reg.), Greece 45/G5
Thrall, Kansas (†66853) 232/F3
Thrasher, Miss. (†38829) 256/G1
Three (isls.), New Bruns. 170/D4
Three Bridges, N.J. (08887) 273/D2
Three Churches, W. Va. (26765) 312/J4
Three Creek, Idaho (†83302) 220/C7
Three Creeks, Alberta 182/B1
Three Creeks, Ark. (71749) 202/E7
Three Forks, Mont. (59752) 262/E5
Three Guardsmen (mt.), Br. Col. 184/H1
Three Hills, Alberta 182/D4
Three Hummock (isl.), Tasmania 99/B2
Three Kings (isls.), N. Zealand 100/D1
Three Lakes, Wis. (54562) 317/H4
Three Mile Bay, N.Y. (13693) 276/H2
Three Mile Plains, Nova Scotia 168/D4
Three Notch, Ala. (†36053) 195/G6
Three Oaks, Mich. (49128) 250/C7
Three Pagodas (pass), Burma 72/C4

Thompson, Manitoba 179/J2
Three Pagodas (pass), Thailand 72/C4
Three Points (cape), Ghana 106/D8
Three Rivers, Mass. (01080) 249/E4
Three Rivers, Mich. (49093) 250/D7
Three Rivers, N. Mex. (†88352) 274/C5
Three Rivers, Texas (78071) 303/F9
Three Rivers (Trois-Rivières), Que. 172/E3
Three Sisters (mt.), Oreg. 291/F3
Three Springs, Pa. (17264) 294/G5
Three Springs, W. Australia 88/B5
Three Springs, W. Australia 92/A5
Throckmorton (co.), Texas 303/E4
Throckmorton, Texas (76083) 303/F4
Throne, Alberta 182/E3
Throop, Pa. (18512) 294/F7
Thrums, Br. Col. 184/J5
Thrumster (bay), Scotland 15/E2
Thuin, Belgium 27/E8
Thule, Greenl. 4/B13
Thule, Greenland 2/F2
Thule, Greenland 146/M2
Thule A.F.B. (Dundas), Greenl. 4/B13
Thule A.F.B. (Dundas), Greenland 146/M2
Thumail, Iraq 66/C4
Thun, Switzerland 39/E3
Thunder (bay), Mich. 250/F4
Thunder (hills), Sask. 181/L4
Thunder (creek), S. Dak. 298/N4
Thunder (lake), Wis. 317/H4
Thunder Bay (riv.), Mich. 250/F3
Thunder Bay, Ont. 162/H6
Thunder Bay, Ont. 146/K5
Thunder Bay (terr. dist.), Ontario 175/C3
Thunder Bay (terr. dist.), Ontario 177/H5
Thunder Bay, Ontario 175/C3
Thunder Bay, Ontario 177/H5
Thunderbird (lake), Okla. 288/M4
Thunderbolt, Georgia (†31404) 217/K6
Thunder Butte (creek), S. Dak. 298/E3
Thunder Hawk, S. Dak. (†57638) 298/F2
Thunder Lake, Alberta 182/C2
Thunersee (lake), Switzerland 39/E3
Thunstetten, Switzerland 39/E2
Thur (riv.), Switzerland 39/G1
Thurgau (canton), Switzerland 39/H1
Thüringer Wald (for.), E. Germany 22/D3
Thuringia (reg.), E. Germany 22/D3
Thurles, Ireland 10/B4
Thurles, Ireland 17/F6
Thurloo Downs, N.S. Wales 97/B1
Thurlow (dam), Ala. 195/G5
Thurlow, Mont. (†59347) 262/K4
Thurman, Iowa (51654) 229/B7
Thurman, N.Y. (†12885) 276/M3
Thurman, Ohio (45685) 284/F8
Thurmond, W. Va. (25936) 312/D7
Thurmont, Md. (21788) 245/J2
Thurrock, England 13/J8
Thurrock, England 10/C5
Thursday Island, Queensland 88/G2
Thursday Island, Queensland 95/B1
Thurso, Québec 172/B4
Thurso, Scotland 10/E1
Thurso, Scotland 15/E2
Thurso (riv.), Scotland 15/E2
Thurston (isl.) 5/C14
Thurston (co.), Nebr. 264/H2
Thurston, Nebr. (68062) 264/H2
Thurston, Ohio (43157) 284/E6
Thurston (co.), Wash. 310/C4
Thusis, Switzerland 39/H3
Thutade (lake), Br. Col. 184/D2
Thyatira, Miss. (†38668) 256/E1
Thyborön, Denmark 21/A4
Thyolo, Malawi 115/F3
Thyregod, Denmark 21/C6
Tia, N.S. Wales 97/G2
Tiahuanacu, Bolivia 136/A5
Tia Juana, Venezuela 124/C2
Tiandong, China 77/G7
Tianjin, China 77/J4
Tianjin, China 2/Q4
Tianjin (Tientsin), China 77/J4
Tianjun, China 77/E4
Tianlin, China 77/G7
Tian Shan (range), China 77/C3
Tianshui, China 77/F5
Tianzhu, China 77/F4
Tiaret, Algeria 106/E1
Tiatucurá, Uruguay 145/C3
Tiavea, W. Samoa 86/M8
Tiawah, Okla. (†74017) 288/P2
Tib, Ras el (Bon) (cape), Tunisia 106/G1
Tibagi, Brazil 135/A4
Tibagi (riv.), Brazil 135/A4
Tibaná, Colombia 126/D5
Tibati, Cameroon 115/B2
Tibbie, Ala. (36583) 195/B8
Tibbita, N.S. Wales 97/C4
Tiber (riv.), Italy 7/F4
Tiber (riv.), Italy 34/D3
Tiberias, Israel 65/C2
Tiberias (lake), Israel 65/D3
Tibesti (mts.) 102/D2
Tibesti (mts.), Chad 111/C3
Tibesti (des.), Chad 111/C3
Tibesti, Serir (des.), Libya 111/C3
Tibet (reg.), China 2/P6
Tibet (reg.), China 54/K6
Tibet (reg.), China 77/B5
Tibet Aut. Reg. (Xizang), China 77/B5
Tibooburra, N.S. Wales 88/G5
Tibooburra, N.S. Wales 97/B1
Tibro, Sweden 18/J7
Tibugá (gulf), Colombia 126/B5
Tiburon, Calif. (94920) 204/J2
Tiburon, Haiti 158/A6
Tiburón (cape), Haiti 158/A6
Tiburón (isl.), Mexico 150/C2
Tiburón (pt.), Panama 154/J6
Ticaco, Peru 128/H11
Ticao (isl.), Philippines 82/D4
Tice, Fla. (33905) 212/E5
Ticehurst, England 13/H6
Tichfield, Sask. 181/D4
Tichigan, Wis. (†53185) 317/K2
Tichigan (lake), Wis. 317/K2
Tichitt, Mauritania 106/C5
Tichla (well), Western Sahara 106/A4
Tichnor, Ark. (72166) 202/H5
Ticino (canton), Switzerland 39/G4
Ticino (riv.), Switzerland 39/G4
Tickfaw, La. (70466) 238/M1
Tickfaw (riv.), La. 238/M1
Tickle (bay), Newf. 166/D2
Ticonderoga, N.Y. (12883) 276/N3
Ticonic, Iowa (†51010) 229/B4
Ticul, Mexico 150/P6
Tidaholm, Sweden 18/J7
Tide Head, New Bruns. 170/D1
Tidewater, Oreg. (97390) 291/D3
Tidikelt (oasis), Algeria 106/E3
Tidioute, Pa. (16351) 294/D2
Tidjikja, Mauritania 102/A3
Tidjikja, Mauritania 106/B5
Tidnish, Nova Scotia 168/E3
Tidore (isl.), Indonesia 85/H5
Tidra (isl.), Mauritania 106/A5
Tiedemann (mt.), Br. Col. 184/E4
Tiefencastel, Switzerland 39/J3
Tiel, Netherlands 27/G5
Tiel, Belgium 27/C7
Tielt-Winge, Belgium 27/F7
Tienen, Belgium 27/F7
Tieling, China 77/K3
Tienshui (Tianshui), China 77/F5
Tientsin (Tianjin), China 77/J4
Tien Yen, China 72/E2
Tie Plant, Miss. (38960) 256/E3
Tiernan, Oreg. (†97453) 291/C3
Tierp, Sweden 18/K6
Tierra Amarilla, Chile 138/A6
Tierra Amarilla, N. Mex. (87575) 274/C2
Tierra Blanca, Mexico 150/L7
Tierra Blanca (creek), N. Mex. 274/B6
Tierra Blanca (creek), Texas 303/B3
Tierra del Fuego (isl.) 2/F8
Tierra del Fuego (isl.) 120/C8
Tierra del Fuego, Antártida, e Islas del Atlántico del sur (prov.) Argentina 143/C7
Tierra del Fuego, Grande de (isl.), Argentina 143/C7
Tierra del Fuego, Grande de (isl.), Chile 138/E11
Tierralta, Colombia 126/C3
Tie Siding, Wyo. (82084) 319/G4
Tietê, Brazil 135/C3
Tietê (riv.), Brazil 120/E5
Tietê (riv.), Brazil 132/D8
Tietê (riv.), Brazil 135/B2
Tieton, Wash. (98947) 310/E4
Tieton (riv.), Wash. 310/E4
Tieyon, S. Australia 94/D3
Tiff, Mo. (63674) 261/L6
Tiffany, Colo. (†81137) 208/D8
Tiffany (mt.), Wash. 310/F2
Tiffin, Iowa (52340) 229/K5
Tiffin, Mo. (64744) 261/E7
Tiffin, Ohio (44883) 284/D3
Tiffin (riv.), Ohio 284/B3
Tiff City, Mo. (64868) 261/C9
Tiflis (Tbilisi), U.S.S.R. 52/F6
Tift (co.), Georgia 217/E7
Tifton, Georgia (31794) 217/F8
Tiftona, Tenn. (†37401) 237/L11
Tigalda (isl.), Alaska 196/F4
Tigard, Oreg. (97223) 291/A2
Tiger, Georgia (30576) 217/F1
Tiger (falls), Guyana 131/C4
Tiger (Macan) (isls.), Indonesia 85/G7
Tiger (falls), Suriname 131/C4
Tiger, Wash. (†99180) 310/H2
Tiger Lily, Alberta 182/C2
Tigerton, Wis. (54486) 317/H6
Tigerville, S.C. (29688) 296/C1
Tighina (Bendery), U.S.S.R. 52/C5
Tigiegio, Somalia 111/H5
Tignall, Georgia (30668) 217/G3
Tignamar, Chile 138/B1
Tigné, Cameroon 115/B2
Tignish, Pr. Edward I. 168/D2
Tigre, Argentina 143/G7
Tigre (prov.), Ethiopia 111/H5
Tigre (riv.), Peru 128/E4
Tigre (riv.), Uruguay 145/A7
Tigre (riv.), Venezuela 124/G3
Tigrett, Tenn. (38070) 237/C9
Tigris (riv.) 54/F6
Tigris (riv.), Iraq 59/E3
Tigris (riv.), Iraq 66/E4
Tigris (riv.), Syria 59/E3
Tigris (riv.), Syria 63/K4
Tigris (riv.), Turkey 59/E3
Tigris (riv.), Turkey 59/E3
Tigris (Dicle) (riv.), Turkey 63/J4
Tiguentourine, Algeria 106/F3
Tihama (reg.), Saudi Arabia 59/C5
Tihama (reg.), Yemen Arab Rep. 59/C5
Tihany, Hungary 41/D3
Tihwa (Ürümqi), China 77/C3
Tijamuchi (riv.), Bolivia 136/C4
Tijeras, N. Mex. (87059) 274/C3
Tijuana, Mexico 150/A1
Tijuana, Mexico 166/G6
Tijucas, Brazil 132/D9
Tikal, Guatemala 154/C2
Tikamgarh, India 68/D4
Tikchik (lkes.), Alaska 196/G2
Tikhoretsk, U.S.S.R. 52/F5
Tikhvin, U.S.S.R. 52/D3
Tiki (riv.), Bulgaria 45/F3
Tikok (riv.), Yugoslavia 45/F3
Tikoleague, Ireland 17/D8
Timon, Brazil 132/F4
Tikrit, Iraq 59/D3
Tikrit, Iraq 66/C3

Tiksi, U.S.S.R. 4/B3
Tiksi, U.S.S.R. 54/P2
Tiksi, U.S.S.R. 48/N2
Tila, Mexico 150/N8
Tilburg, Netherlands 27/G5
Tilbury, Ontario 177/B5
Tilcara, Argentina 143/C1
Tilcha, S. Australia 94/G3
Tilden, Ill. (62292) 222/D5
Tilden, Ill. (†38843) 256/H2
Tilden, Ky. (†42409) 237/F5
Tilden, Nebr. (68781) 264/G2
Tilden, Texas (78072) 303/F9
Tilehurst, England 13/F6
Tilemsi (valley), Mali 106/E5
Tilford, S. Dak. (†57769) 298/C5
Tilghman, Md. (21671) 245/N6
Tilin, Burma 72/B2
Tiline, Ky. (42083) 237/E6
Till (riv.), England 13/E2
Tillabéry, Niger 106/E6
Tillamook (co.), Oreg. 291/D2
Tillamook, Oreg. (97141) 291/D2
Tillamook (head), Oreg. 291/C2
Tillanchong (isl.), India 68/G7
Tillar, Ark. (71670) 202/H6
Tillatoba, Miss. (38961) 256/E3
Tilleda, Wis. (54978) 317/J6
Tiller, Oreg. (97484) 291/E5
Tillery, N.C. (27887) 281/O2
Tillery (lake), N.C. 281/J4
Tilley, Alberta 182/E4
Tilley, New Bruns. 170/C2
Tillicoultry, Scotland 10/B1
Tillicoultry, Scotland 15/C1
Tillicum, Wash. (98492) 310/C3
Tillman (co.), Okla. 288/J6
Tillman, Miss. (†39150) 256/C7
Tillman (co.), Okla. 288/J6
Tillman, S.C. (29943) 296/E7
Tillson, N.Y. (12486) 276/M7
Tillsonburg, Ontario 177/D5
Tilney, Sask. 181/F5
Tilomonte, Chile 138/B4
Tílos (isl.), Greece 45/H7
Tilpa, N.S. Wales 97/C2
Tilsit (Sovetsk), U.S.S.R. 52/B4
Tilston, Manitoba 179/A5
Tilt (riv.), Scotland 15/E4
Tiltagara, N.S. Wales 97/C2
Tiltil, Chile 138/G2
Tilting, Newf. 166/D2
Tilton, Ark. (†72347) 202/J3
Tilton, Georgia (†30720) 217/B1
Tilton, Ill. (†61832) 222/F3
Tilton◯, N.H. (03276) 268/D5
Tilton-Northfield, N.H. (03276) 268/D5
Tiltonsville, Ohio (43963) 284/J5
Tim, Denmark 21/B5
Timagami, Ontario 177/K5
Timagami, Ontario 175/E3
Timagami (lake), Ontario 177/K5
Timagami (lake), Ontario 175/D3
Timan (ridge), U.S.S.R. 52/G1
Timaná, Colombia 126/C7
Timane (riv.), Paraguay 144/B2
Timaru, N. Zealand 100/C6
Timashevsk, U.S.S.R. 52/E5
Timbákion, Greece 45/G8
Timbalier (bay), La. 238/K8
Timbalier (isl.), La. 238/K8
Timbarra, N.S. Wales 97/G1
Timbédra, Mauritania 106/C5
Timber (mt.), Nev. 266/E5
Timber (mt.), Nev. 266/F4
Timber, Oreg. (97144) 291/D2
Timber Bay, Sask. 181/F1
Timber Creek, North. Terr. 88/E3
Timber Lake, S. Dak. (†57583) 281/M2
Timberlake, Ohio (†44094) 284/H2
Timber Lake, S. Dak. (57656) 298/H3
Timberlea, Nova Scotia 168/E4
Timberville, Va. (22853) 307/L3
Timbio, Colombia 126/B6
Timbiqui, Colombia 126/B6
Timblin, Pa. (15778) 294/D4
Timbo, Ark. (72680) 202/F2
Timboulaga (well), Niger 106/F5
Timbuktu (Tombouctou), Mali 106/D5
Timbuktu, Mali 102/B3
Time, Ill. (62375) 222/C3
Times Beach, Mo. (†63025) 261/N4
Timewell, Ill. (62375) 222/C3
Timgad, Algeria 106/F1
Timimoun, Algeria 102/C2
Timimoun, Algeria 106/E3
Timiris (cape), Mauritania 106/A5
Timiş (riv.), Romania 45/E3
Timiskaming (lake) 162/J6
Timiskaming (terr. dist.), Ontario 177/K5
Timiskaming (terr. dist.), Ontario 175/D3
Timiskaming (lake), Ontario 175/E3
Timişoara, Romania 7/G4
Timişoara, Romania 45/E3
Timken, Kansas (67582) 232/C3
Timmendorfer Strand, W. Germany 22/D1
Timmins, Ont. 146/K5
Timmins, Ont. 162/H6
Timmins, Ontario 177/J5
Timmins, Ontario 175/D3
Timmissao (well), Algeria 106/E4
Timmonsville, S.C. (29161) 296/H3
Timms Hill (mt.), Wis. 317/F5
Timnath, Colo. (80547) 208/J2
Timok (riv.), Bulgaria 45/F3
Timok (riv.), Yugoslavia 45/F3
Timoleague, Ireland 17/D8
Timon, Brazil 132/F4
Timonium-Lutherville, Md. (21093) 245/M3
Timor (sea) 54/O11
Timor (sea) 88/D2
Timor (isl.), Indonesia 54/O10
Timor (isl.), Indonesia 2/R6

Timor (reg.), Indonesia 85/H7
Timor (sea), Indonesia 85/H7
Timor (sea), North. Terr. 93/A2
Timor (sea), W. Australia 92/D1
Timote, Uruguay 145/D4
Timotes, Venezuela 124/C3
Timothy, Tenn. (†38568) 237/L8
Timpahute (range), Nev. 266/F5
Timpanogos Cave Nat'l Mon., Utah 304/C3
Timpas, Colo. (†81034) 208/M7
Timpas (creek), Colo. 208/M7
Timpson, Texas (75975) 303/K6
Timrå, Sweden 18/K5
Tims Ford (lake), Tenn. 237/J10
Tina (mt.), Dom. Rep. 158/D6
Tina, Mo. (64682) 261/F3
Tinaca (pt.), Philippines 54/O9
Tinaca (pt.), Philippines 85/H4
Tinaca (pt.), Philippines 82/E8
Tinaco, Venezuela 124/D3
Tinahely, Ireland 17/H6
Tinaquillo, Venezuela 124/D3
Tinca, Romania 45/F2
Tin City, Alaska (†99778) 196/E1
Tincup, Colo. (†81210) 208/F5
Tindall, Mo. (†64683) 261/E2
Tindouf, Algeria 102/B2
Tindouf, Algeria 106/C3
Tindouf, Sebkha de (salt lake), Algeria 106/C3
Tinela, Ala. (†36452) 195/D7
Tineo, Spain 33/C1
Tinggi, Pulau (isl.), Malaysia 72/E7
Tingha, N.S. Wales 97/F1
Tinghert Hamada (Tinrhert) (des.), Libya 111/B2
Tinglev, Denmark 21/C8
Tingley, Iowa (50863) 229/E7
Tingmerkpuk (mt.), Alaska 196/F1
Tingmiarmiut, Greenl. 4/C12
Tingo María, Peru 128/D7
Tingri, China 77/C6
Tinguipaya, Bolivia 136/C6
Tinguiririca, Chile 138/G6
Tinguiririca (riv.), Chile 138/F5
Tingwick, Québec 172/F4
Tinian (isl.), No. Marianas 87/E4
Tinjoub, Algeria 106/C3
Tinker A.F.B., Okla. 288/M4
Tinkers (creek), Md. 245/F6
Tinley Park, Ill. (60477) 222/B6
Tinmouth○, Vt. (†05773) 268/A5
Tinnie, N. Mex. (88351) 274/D5
Tinogasta, Argentina 143/C2
Tínos, Greece 45/G7
Tínos (isl.), Greece 45/G7
Tinrhert, Hamada de (des.), Algeria 106/F3
Tinrhert Hamada (Tinghert) (des.), Libya 111/B2
Tinsley, Ky. (40993) 237/O7
Tinsman, Ark. (71767) 202/F6
Tinsukia, India 68/H3
Tintagel, Br. Col. 184/E3
Tintagel (head), England 13/C7
Tintah, Minn. (56583) 255/B5
Tintigny, Belgium 27/G9
Tintina, Argentina 143/D2
Tinto (riv.), Spain 33/C4
Tinton Falls, N.J. (07724) 273/E3
Tinui, N. Zealand 100/F4
Tiny, Sask. 181/J4
Tin-Zaouatene, Algeria 106/E5
Tioga, Ill. (†62351) 222/B3
Tioga, La. (71477) 238/F4
Tioga (co.), N.Y. 276/H6
Tioga, N. Dak. (58852) 282/E3
Tioga (co.), Pa. 294/H2
Tioga, Pa. (16946) 294/H2
Tioga (riv.), Pa. 294/H1
Tioga, W. Va. (26691) 312/E6
Tioman, Pulau (isl.), Malaysia 72/E7
Tiona, Pa. (16352) 294/D2
Tionesta, Pa. (16353) 294/C2
Tionesta Creek (lake), Pa. 294/D3
Tiosa, Ind. (†46975) 227/E2
Tioughnioga (riv.), N.Y. 276/H6
Tipitapa, Nicaragua 154/E4
Tipler, Wis. (†49935) 317/J4
Tiplersville, Miss. (38674) 256/G1
Tippah (co.), Miss. 256/G1
Tipp City, Ohio (45371) 284/B6
Tippecanoe (co.), Ind. 227/D4
Tippecanoe, Ind. (46570) 227/E2
Tippecanoe (riv.), Ind. 227/E2
Tippecanoe, Ohio (44699) 284/H5
Tipperary (co.), Ireland 17/F6
Tipperary, Ireland 10/B4
Tipperary, Ireland 17/E7
Tippettville, Georgia (†31092) 217/E6
Tippo, Miss. (38962) 256/D3
Tipton, Calif. (93272) 204/F7
Tipton (co.), Ind. 227/E4
Tipton, Ind. (46072) 227/E4
Tipton, Iowa (52772) 229/L5
Tipton, Kansas (67485) 232/D2
Tipton, Mo. (65081) 261/G5
Tipton, Okla. (73570) 288/H6
Tipton, Pa. (16684) 294/F4
Tipton (co.), Tenn. 237/B9
Tipton, Tenn. (38071) 237/B10
Tiptonville, Tenn. (38079) 237/B8
Tiptop, Ky. (†41409) 237/P5
Tiptop, Va. (24655) 307/F6
Tiptree, England 13/H6
Tipuani, Bolivia 136/B4
Tiracambu, Serra (range), Brazil 132/E3
Tiran (str.), Egypt 111/F2
Tiran (str.), Egypt 59/C4
Tiran (isl.), Saudi Arabia 59/C4
Tiran (str.), Saudi Arabia 59/C4
Tirana (Tiranë) (cap.), Albania 45/E5
Tiranë (Tirana) (cap.), Albania 45/E5
Tiranë (cap.), Albania 7/F4
Tirano, Italy 34/C1
Tiraque, Bolivia 136/C5
Tiraspol', U.S.S.R. 7/G4
Tiraspol', U.S.S.R. 52/D5
Tirat Hakarmel, Israel 65/B2
Tirat Zevi, Israel 65/D3
Tire, Turkey 59/A2
Tire, Turkey 63/B3
Tirebolu, Turkey 63/H2
Tiree (isl.), Scotland 15/B4
Tiree (isl.), Scotland 10/C2
Tîrgovişte, Romania 45/G3
Tîrgu Cărbuneşti, Romania 45/F3
Tîrgu Frumos, Romania 45/H2
Tîrgu Jiu, Romania 45/F3
Tîrgu Mures, Romania 7/G4
Tîrgu Mureş, Romania 45/G2
Tîrgu Neamţ, Romania 45/G2
Tîrgu Ocna, Romania 45/H2
Tîrgu Secuiesc, Romania 45/H2
Tirich Mir (mt.), Pakistan 68/C1
Tirich Mir (mt.), Pakistan 59/K2
Tirlemont (Tienen), Belgium 27/F7
Tîrnava Mare (riv.), Romania 45/G2
Tîrnăveni, Romania 45/F2
Tírnavos, Greece 45/F6
Tiro, Ohio (44887) 284/E4
Tirol (prov.), Austria 41/A3
Tirschenreuth, W. Germany 22/E4
Tirso (riv.), Italy 34/B4
Tiruchchirappalli, India 54/J8
Tiruchchirappalli, India 68/D6
Tiruchendur, India 68/D7
Tirunelveli, India 68/D7
Tiruntán, Peru 128/E6
Tirupati, India 68/D6
Tiruppattur, India 68/D6
Tiruppur, India 68/D6
Tiruvannamalai, India 68/D6
Tirzah, S.C. (†29745) 296/E2
Tisa (riv.), Yugoslavia 45/E3
Tisbury, England 13/E6
Tisch Mills, Wis. (54240) 317/L7
Tisdale, Sask. 162/F5
Tisdale, Sask. 181/H3
Tishomingo (co.), Miss. 256/H1
Tishomingo, Miss. (38873) 256/H1
Tishomingo, Okla. (73460) 288/N6
Tisisat (fall), Ethiopia 111/G5
Tiskilwa, Ill. (61368) 222/D2
Tišnov, Czech. 41/D2
Tisovec, Czech. 41/E2
Tistrup, Denmark 21/B6
Tisza (riv.), Hungary 41/F3
Tiszacsege, Hungary 41/F3
Tiszaföldvár, Hungary 41/F3
Tiszafüred, Hungary 41/F3
Tiszakécske, Hungary 41/E3
Tiszalök, Hungary 41/F2
Tiszavasvári, Hungary 41/F3
Titagarh, India 68/F1
Titicaca (lake) 2/F6
Titicaca (lake) 120/C4
Titicaca (lake), Bolivia 136/A4
Titicaca (lake), Peru 128/H10
Titicus, Conn. (†06877) 210/A3
Titicus (riv.), Conn. 210/A3
Titicus (res.), Conn. 210/A3
Titihira (head), N. Zealand 100/B5
Titirangi, N. Zealand 100/B1
Titisee-Neustadt, W. Germany 22/C5
Titlagarh, India 68/E4
Titlis (mt.), Switzerland 39/F3
Titograd, Yugoslavia 7/F4
Titograd, Yugoslavia 45/D4
Titonka, Iowa (50480) 229/E2
Titov Užice, Yugoslavia 45/D4
Titov Veles, Yugoslavia 45/E5
Tittabawassee (riv.), Mich. 250/E5
Titule, Zaire 115/E3
Titus, Ala. (36080) 195/F5
Titus (mt.), Conn. 210/B1
Titus, Georgia (†30546) 217/E1
Titus (lake), N.Y. 276/M1
Titus (co.), Texas 303/K4
Titusville, Fla. 188/K5
Titusville, Fla. (32780) 212/F3
Titusville, N.J. (08560) 273/D3
Titusville, Pa. (16354) 294/C2
Tiumpan (head), Scotland 15/B2
Tivaouane, Senegal 106/A5
Tiverton, England 10/E5
Tiverton, England 13/D7
Tiverton, Nova Scotia 168/B4
Tiverton, Ontario 177/C3
Tiverton, R.I. (02878) 249/K6
Tiverton○, R.I. (02878) 249/K6
Tiverton Four Corners, R.I. (†02878) 249/K6
Tivoli, Italy 34/F6
Tivoli, N.Y. (12583) 276/N6
Tivoli, Texas (77990) 303/G9
Tiwi, Philippines 82/D4
Tixtla de Guerrero, Mexico 150/K8
Tizard (bank), Philippines 85/E3
Tizayuca, Mexico 150/L1
Tizimín, Mexico 150/Q6
Tizi Ouzou, Algeria 106/E1
Tiznit, Morocco 106/B3
Tjeukemeer (lake), Netherlands 27/H3
Tjilatjap (Cilacap), Indonesia 85/H2
Tjirebon (Cirebon), Indonesia 85/H2
Tjonger (riv.), Netherlands 27/J3
Tjuvfjorden (fjord), Norway 18/D2
Tlachichuca, Mexico 150/O1
Tlacolula de Matamoros, Mexico 150/L8
Tlacotepec de Mejía, Mexico 150/P1
Tlahualilo de Zaragoza, Mexico 150/H3
Tlalancaneca, Mexico 150/M1
Tlalixcoyan, Mexico 150/Q2
Tlalmanalco de Velásquez, Mexico 150/L1
Tlalnepantla de Comonfort, Mexico 150/L1
Tlalpan, Mexico 150/L1
Tlaltenango de Sánchez Román, Mexico 150/H6
Tlaltizapan, Mexico 150/L2
Tlapacoyan, Mexico 150/P1
Tlapa de Comonfort, Mexico 150/K8
Tlaquepaque, Mexico 150/H6
Tlaquiltenango, Mexico 150/L2
Tlatlauquitepec, Mexico 150/O1
Tlaxcala (state), Mexico 150/N1
Tlaxcala de Xicotencatl, Mexico 150/M1
Tlaxco, Mexico 150/N1
Tlaxiaco, Mexico 150/L8
Tlayacapan, Mexico 150/L1
Tlell, Br. Col. 184/B3
Tlemcen, Algeria 106/D2
Tlemcen, Algeria 102/B1
Tmessa, Libya 111/C2
Tmessa, Libya 102/D2
Tni Haïa (well), Algeria 106/D4
Toa, Cuchillas de (mts.), Cuba 158/K4
Toa Alta, P. Rico 161/D1
Toa Baja, P. Rico 161/D1
Toad (riv.), Br. Col. 184/L2
Toadlena, N. Mex. (87324) 274/A2
Toamasina (prov.), Madagascar 118/H3
Toamasina, Madagascar 118/H3
Toamasina, Madagascar 102/G6
Toana (range), Nev. 266/G2
Toano, Va. (23168) 307/P6
Toast, N.C. (27049) 281/H2
Toay, Argentina 143/D4
Toba, Argentina 143/F4
Toba (inlet), Br. Col. 184/E5
Toba (lake), Indonesia 85/B5
Toba, Japan 81/H6
Tobago (isl.), Trin. & Tob. 156/G5
Tobarra, Spain 33/F3
Tobasco, Ohio (†45245) 284/C10
Tobatí, Paraguay 144/B4
Tobeatic (lake), Nova Scotia 168/C4
Tobelo, Indonesia 85/H5
Tobermory, Ontario 177/C2
Tobermory, Scotland 10/C2
Tobermory, Scotland 15/B4
Tobetsu, Japan 81/K2
Tobi (isl.), Belau 87/D5
Tobi (isl.), Japan 81/J4
Tobias, Nebr. (68453) 264/G4
Tobin (lake), Sask. 181/H2
Tobin (lake), W. Australia 88/D4
Tobin Lake, Sask. 181/H2
Tobinsport, Ind. (47587) 227/D9
Tobique (riv.), New Bruns. 170/C2
Tobique Narrows, New Bruns. 170/C2
Tobol (riv.), U.S.S.R. 54/H4
Tobol (riv.), U.S.S.R. 48/G4
Tobol'sk, U.S.S.R. 54/J4
Tobol'sk, U.S.S.R. 48/G4
Tobruk, Libya 102/E1
Tobruk, Libya 111/D1
Tobseda, U.S.S.R. 52/H1
Toby (mt.), Mass. 249/E3
Tobyhanna, Pa. (18466) 294/M3
Tocache, Peru 128/D7
Tocantínia, Brazil 132/D5
Tocantinópolis, Brazil 120/E3
Tocantinópolis, Brazil 132/D4
Tocantins (riv.), Brazil 2/G6
Tocantins (riv.), Brazil 120/E4
Tocantins (riv.), Brazil 132/D4
Toccoa, Georgia (30577) 217/F1
Toccopola, Miss. (38874) 256/F2
Tochcha (lake), Br. Col. 184/E3
Tochigi (pref.), Japan 81/K5
Tochimilco, Mexico 150/M2
Toco, Chile 138/B3
Toco, Trin. & Tob. 161/B10
Tocoa, Honduras 154/E3
Toco Hills, Georgia (†30329) 217/K1
Tocomechi, Bolivia 136/D5
Toconao, Chile 138/C4
Tocópero, Venezuela 124/D2
Tocopilla, Chile 138/A3
Tocopilla, Chile 120/B5
Tocorpuri, Cerros de (mt.), Bolivia 136/A8
Tocorpuri, Cerro de (mts.), Chile 138/B3
Tocsin, Ind. (46790) 227/G3
Tocuco (riv.), Venezuela 124/B3
Tocumen, Panama 154/H6
Tocumwal, N.S. Wales 97/C4
Tocuyo (riv.), Venezuela 124/D2
Tocuyo de la Costa, Venezuela 124/D2
Todd (co.), Ky. 237/G7
Todd (co.), Minn. 255/D4
Todd (mt.), New Bruns. 170/D2
Todd, N.C. (28684) 281/F2
Todd (riv.), North. Terr. 93/D8
Todd (co.), S. Dak. 298/H7
Toddville, Iowa (52341) 229/K4
Toddville, Md. (21672) 245/O7
Toddville, S.C. (†29526) 296/J4
Todenyang, Kenya 115/G3
Todi, Italy 34/D3
Tödi (mt.), Switzerland 39/G3
Todmorden, England 13/H1
Todmorden, England 10/G2
Todos Santos, Cochabamba, Bolivia 136/C5
Todos Santos, La Paz, Bolivia 136/B3
Todos Santos, Oruro, Bolivia 136/A6
Todos Santos, Mexico 150/D5
Toe (head), Ireland 17/C9
Toe (head), Scotland 15/A3
Toekomstig (res.), Suriname 131/C3
Toeterville, Iowa (50481) 229/H2
Tofield, Alberta 182/D3
Tofino, Br. Col. 184/E5
Tofte, Minn. (55615) 255/H3
Toftlund, Denmark 21/B7
Togane, Japan 81/K6
Togdheer (prov.), Somalia 115/J2
Toggenburg (dist.), Switzerland 39/H2
Togiak, Alaska (99678) 196/F3
Togiak (bay), Alaska 196/F3
Togian (isls.), Indonesia 85/G6
Togliatti, U.S.S.R. 7/J3
Togliatti (Tol'yatti), U.S.S.R. 52/G4
Togliatti (Tol'yatti), U.S.S.R. 48/F4
Togo 2/J5
Togo 102/C4
Togo, Minn. (55788) 255/E3
Togo, Sask. 181/K4
TOGO 106/E7
Togtoh, China 77/H3
Tohamiyam, Sudan 59/C6
Tohatchi, N. Mex. (87325) 274/A3
Tohivea (mt.), Fr. Poly. 86/S13
Toi, Japan 81/J6
Toibalawe, India 68/G6
Toijala, Finland 18/N6
Toimi, Minn. (†55602) 255/G3
Toivola, Minn. (55789) 255/F3
Toiyabe (range), Nev. 266/D3
Tojo, Japan 81/F6
Tok, Alaska (99780) 196/K2
Tokachi (mt.), Japan 81/L2
Tokachi (riv.), Japan 81/L2
Tokaj (Tokay), Hungary 41/F2
Tokamachi, Japan 81/J5
Tokanui, N. Zealand 100/B7
Tokar, Sudan 111/G4
Tokar, Sudan 102/F3
Tokar, Sudan 59/C6
Tokara (isls.), Japan 81/O5
Tokat (prov.), Turkey 63/G2
Tokat, Turkey 59/C1
Tokat, Turkey 63/G2
Tokeen, Alaska (†99901) 196/M2
Tokeland, Wash. (98590) 310/A4
Tokelau (isls.) 87/J6
Tokewamna (peak), Utah 304/D3
Tokio, N. Dak. (58379) 282/N4
Tokio, Texas (79376) 303/B4
Tokke (riv.), Norway 18/F7
Tokmak, U.S.S.R. 48/H5
Tokmak, U.S.S.R. 52/E5
Toko, N. Zealand 100/E3
Tokomaru Bay, N. Zealand 100/G3
Tokoroa, N. Zealand 100/F3
Tokorozawa, Japan 81/O2
Tokra, Libya 111/D1
Toksook Bay, Alaska (99637) 196/E2
Toksu (Xinhe), China 77/B3
Toksun, China 77/C3
Tokuno (isl.), Japan 81/O5
Tokunoshima, Japan 81/O5
Tokushima (pref.), Japan 81/G7
Tokushima, Japan 81/G7
Tokuyama, Japan 81/F6
Tokyo (pref.), Japan 81/O2
Tokyo (cap.), Japan 2/R4
Tokyo (cap.), Japan 54/R6
Tokyo (cap.), Japan 81/O2
Tokyo (bay), Japan 81/O2
Tolaga Bay, N. Zealand 100/G3
Tolar, Texas (76476) 303/G5
Tolbukhin, Bulgaria 45/H4
Tolchester Beach, Md. (†21620) 245/N4
Tolé, Panama 154/G6
Toledo, Bolivia 136/B6
Toledo, Colombia 126/D4
Toledo, Ill. (62468) 222/E4
Toledo, Iowa (52342) 229/H4
Toledo, Ohio 146/K5
Toledo, Ohio (*43601) 284/D2
Toledo, Ohio 188/K2
Toledo, Ontario 177/H3
Toledo, Oreg. (97391) 291/D3
Toledo, Philippines 82/D5
Toledo (prov.), Spain 33/D3
Toledo, Spain 7/D5
Toledo, Spain 33/D3
Toledo (mts.), Spain 33/E3
Toledo, Uruguay 145/B6
Toledo, Wash. (98591) 310/C4
Toledo Bend (res.) 188/H4
Toledo Bend (dam), La. 238/C4
Toledo Bend (res.), La. 238/C3
Toledo Bend (dam), Texas 303/L6
Toledo Bend (res.), Texas 303/L6
Tolentino, Italy 34/D3
Toli, China 77/B2
Toliary (prov.), Madagascar 118/G4
Toliary (Tuléar), Madagascar 118/G4
Toliary, Madagascar 102/G7
Tolima (dept.), Colombia 126/C5
Tolima, Colombia 126/D8
Tolima (mt.), Colombia 120/B2
Tolima, Nevada del (mt.), Colombia 126/C5
Tolimán, Mexico 150/K6
Tolitoli, Indonesia 85/G5
Tolland, Alberta 182/E3
Tolland (co.), Conn. 210/F1
Tolland○, Conn. (06084) 210/F1
Tolland○, Mass. (†01034) 249/B4
Tollensesee (lake), E. Germany 22/E2
Tollesboro, Ky. (41189) 237/O3
Tolleson, Ariz. (85353) 198/C5
Tollette, Ark. (71851) 202/C6
Tolley, N. Dak. (58787) 282/G2
Toll Gate, W. Va. (26442) 312/E4
Tølløse, Denmark 21/E6
Tollville, Ark. (72041) 202/G4
Tollygunge, India 68/F2
Tolly's Nullah (riv.), India 68/F2
Tolna (co.), Hungary 41/E3
Tolna, Hungary 41/E3
Tolna, N. Dak. (58380) 282/O4
Tolo (gulf), Indonesia 85/G6
Tolob, Scotland 15/G2
Tolona, Mo. (†63450) 261/J2
Tolong (bay), Philippines 82/D6
Tolono, Ill. (61880) 222/E3
Tolosa, Spain 33/F1
Tolovana Park, Oreg. (97145) 291/C2
Tolsta (head), Scotland 15/B2
Tolstoi, Manitoba 179/F5
Tolstoy, S. Dak. (57475) 298/K3
Toltén, Chile 138/D2
Toltén (riv.), Chile 138/D2
Tolt River (res.), Wash. 310/D3
Tolú, Colombia 126/C3
Tolu, Ky. (42084) 237/E6
Toluca, Ill. (61369) 222/D2
Toluca, Mexico 146/H8
Toluca de Lerdo, Mexico 150/K7
Tom (mt.), Conn. 210/B1
Tom (mt.), Mass. 249/D4
Tom (mt.), N.H. 268/E3
Tom, Okla. (†74740) 288/S7
Tomah, Wis. (54660) 317/F8
Tomahawk, Alberta 182/C3
Tomahawk, N.C. (28465) 281/N5
Tomahawk, Wis. (54487) 317/G5
Tomakomai, Japan 81/K2
Tomales, Calif. (94971) 204/C5
Tomales (pt.), Calif. 204/B5
Tomanao, Philippines 82/E8
Tomaniivi (mt.), Fiji 86/Q10
Tomar, Portugal 33/B3
Tomarza, Turkey 63/F3
Tomasaki (mt.), Utah 304/E5
Tomás Barrón, Bolivia 136/B5
Tomás Gomensoro, Uruguay 145/B1
Tomaszów Lubelski, Poland 47/F3
Tomaszów Mazowiecki, Poland 47/E3
Tomatin, Scotland 15/D3
Tomatlán, Mexico 150/G6
Tomave, Bolivia 136/B7
Tombador, Serra do (range), Brazil 132/B6
Tomball, Texas (77375) 303/J7
Tombe, Sudan 111/F6
Tombigbee (riv.) 188/J4
Tombigbee (riv.), Ala. 195/B7
Tombigbee (riv.), Miss. 256/H4
Tombstone, Ariz. (85638) 198/F7
Tomé, Chile 138/D1
Tome, N. Mex. (87060) 274/C4
Tomelilla, Sweden 18/J9
Tomelloso, Spain 33/E3
Tom Green (co.), Texas 303/D6
Tomhannock (res.), N.Y. 276/O5
Tomichi (creek), Colo. 208/F5
Tomifobia, Québec 172/E4
Tomina, Bolivia 136/C6
Tomingley, N.S. Wales 97/E3
Tomini (gulf), Indonesia 54/O10
Tomini (gulf), Indonesia 85/G6
Tomintoul, Scotland 15/E3
Tomiyama, Japan 81/O3
Tomkinson (ranges), W. Australia 92/E4
Tommerup, Denmark 21/D7
Tommot, U.S.S.R. 48/N4
Tomnolen, Miss. (39770) 256/F4
Tomo (riv.), Colombia 126/F5
To Mo, Thailand 72/D6
Tompa, Hungary 41/E3
Tompkins (co.), N.Y. 276/H6
Tompkins, Sask. 181/C5
Tompkinsville, Ky. (42167) 237/K7
Tompkinsville, Md. (†20664) 245/L7
Tom Price, W. Australia 88/B4
Tom Price, W. Australia 92/B3
Toms (riv.), N.J. 273/E3
Toms Brook, Va. (22660) 307/L3
Toms Creek, Va. (†24230) 307/D7
Tomsk, U.S.S.R. 54/K4
Tomsk, U.S.S.R. 48/J4
Tomslake, Br. Col. 184/H2
Toms River, N.J. (08753) 273/E4
Tom Steed (res.), Okla. 288/J5
Tömük, Turkey 63/F4
Tonalá, Mexico 150/N8
Tonalea, Ariz. (86044) 198/E2
Tonasket, Wash. (98855) 310/F2
Tonate, Fr. Guiana 131/E3
Tonawanda, N.Y. (14150) 276/B4
Tonawanda Ind. Res., N.Y. 276/D4
Tonbridge, England 13/H8
Tonckens (falls), Suriname 131/C3
Tondabayashi, Japan 81/J8
Tondano, Indonesia 85/H5
Tønder, Denmark 21/B8
Tønder, Denmark 18/F9
Tone (riv.), Japan 81/K6
Tonegrama, Peru 128/D4
Toney, Ala. (35773) 195/E1
Toney River, Nova Scotia 168/F3
Tonga 2/A6
Tonga 87/J8
Tonga, Sudan 111/F6
Tongala, Victoria 97/C5
Tonganoxie, Kansas (66086) 232/G2
Tongareva (atoll), Cook Is. 87/L6
Tongatapu (isls.), Tonga 87/J8
Tongcheng, China 77/J5
T'ongch'ŏn, N. Korea 81/D4
Tongchuan (Tungchwan), China 77/G5
Tongde, China 77/F4
Tongeren, Belgium 27/G7
Tonghai, China 77/F7
Tonghe, China 77/L2
Tonghua (Tunghwa), China 77/L3
Tongjiang (Tungkiang), China 77/M2
Tongliao, China 77/K3
Tongling, China 77/J5
Tongo, N.S. Wales 97/B2
Tongo (lake), N.S. Wales 97/B2
Tongoy, Chile 138/A8
Tongoy (bay), Chile 138/A8
Tongquil (isl.), Philippines 82/D8
Tongren, Guizhou, China 77/G6
Tongren, Qinghai, China 77/F4
Tongres (Tongeren), Belgium 27/G7
Tongs, Ky. (†41175) 237/R3
Tongsa Dzong, Bhutan 68/G3
Tongtian He (Zhi Qu) (riv.), China 77/E5
Tongue (riv.), Mont. 262/K5
Tongue (pt.), N. Zealand 100/A3
Tongue (riv.), N. Dak. 282/P2
Tongue, Scotland 15/D2
Tongue (riv.), Wyo. 319/E1
Tongue of the Ocean (chan.), Bahamas 156/C1
Tongxin, China 77/G4
Tongyu, China 77/K3
Tongzi, China 77/G6
Tonica, Ill. (61370) 222/E2
Tonj, Sudan 111/E6
Tonk, India 68/D3
Tonka Bay, Minn. (†55331) 255/F5
Tonkawa, Okla. (74653) 288/M1
Tonkin (gulf) 54/M8
Tonkin (gulf), China 77/G7
Tonkin, Sask. 181/J4
Tonkin (gulf), Vietnam 72/E3
Tonle Sap (lake), Cambodia 72/D4
Ton Mhor (pt.), Scotland 15/B5
Tonneins, France 28/D5
Tonnerre, France 28/E4
Tönning, W. Germany 22/C1
Tonopah, Ariz. (85354) 198/B5
Tonopah, Nev. 188/C3
Tonopah, Nev. (89049) 266/D4
Tonosí, Panama 154/G7
Tonota, Botswana 118/D4
Tonota, Botswana 102/E7
Tønsberg, Norway 18/D4
Tonsina, Alaska (†99686) 196/J2
Tontitown, Ark. (72770) 202/B1
Tonto (basin), Ariz. 198/D4
Tonto (creek), Ariz. 198/D4
Tonto Basin, Ariz. (85553) 198/D5
Tontogany, Ohio (43565) 284/C3
Tonto Nat'l Mon., Ariz. 198/D5
Tony, Wis. (54563) 317/E5
Tonya, Turkey 63/H2
Toodyay, W. Australia 88/B2
Toodyay, W. Australia 92/B1
Tooele (co.), Utah 304/A3
Tooele, Utah (84074) 304/B3
Tooele, Utah 188/D2
Tooele Army Depot, Utah 304/B3
Toole (co.), Mont. 262/E2
Tooleybuc, N.S. Wales 97/B4
Toombs (co.), Georgia 217/H6
Toomevara, Ireland 17/E6
Tooms (lake), Tasmania 99/D4
Toomsboro, Georgia (31090) 217/F5
Toomsuba, Miss. (39364) 256/G6
Toone, Tenn. (38381) 237/D10
Tooraweenah, N.S. Wales 97/E2
Toowoomba, Australia 87/F8
Toowoomba, Queensland 88/J5
Toowoomba, Queensland 95/D5
Top (lake), U.S.S.R. 52/D1
Topador, Uruguay 145/C1
Topanga, Calif. (90290) 204/B10
Topanga Beach, Calif. (†90290) 204/B10
Topará, Peru 128/D9
Topawa, Ariz. (85639) 198/D7
Topaz (lake), Nev. 266/B4
Topeka, Ill. (61567) 222/D3
Topeka, Ind. (46571) 227/F1
Topeka (cap.), Kans. 188/G3
Topeka (cap.), Kansas 146/J6
Topeka (cap.), Kansas (*66601) 232/G2
Topia, Mexico 150/F4
Topinabee, Mich. (49791) 250/E3
Topl'a (riv.), Czech. 41/F2
Topley, Br. Col. 184/D3
Toplița, Romania 45/G2
Topocalma (pt.), Chile 138/A10
Topock, Ariz. (86436) 198/A4
Top Of The World Prov. Park, Br. Col. 184/K5
Topographic Center, Md. 245/E4
Topol'čany, Czech. 41/D2
Topolobampo, Mexico 150/E4
Topolovgrad, Bulgaria 45/H4
Toponas, Colo. (80479) 208/F2
Toppenish, Wash. (98948) 310/E4
Toppenish (creek), Wash. 310/E4
Topsail Beach, N.C. (28445) 281/O6
Topsfield○, Maine (04490) 243/H5
Topsfield, Mass. (01983) 249/L2
Topsfield○, Mass. (01983) 249/L2
Topsham, Maine (04086) 243/D8
Topsham○, Maine (04086) 243/D8
Topsham○, Vt. (05076) 268/C3
Top Springs, North. Terr. 93/C4
Topton, N.C. (28781) 281/B4
Topton, Pa. (19562) 294/L5
Toquepala, Peru 128/G11
Toquerville, Utah (84774) 304/A6
Toquima (range), Nev. 266/E4
Tor (bay), Nova Scotia 168/G3
Torata, Peru 128/G11
Torawitan (cape), Indonesia 85/G5
Torbali, Turkey 63/B3
Torbat-e-Heydariyeh, Iran 66/L3
Torbat-e Heydariyeh, Iran 59/G2
Torbat-e Jam, Iran 66/M3
Torbat-e Jam, Iran 59/H2
Torbay, England 10/E5
Torbay, England 13/D7
Torbay, Newf. 166/D2
Torbay (pt.), Newf. 166/D2
Torbeck, Haiti 158/A6
Torch (key), Fla. 212/E7
Torch (lake), Mich. 250/D3
Torch, Ohio (45781) 284/G7
Torch (riv.), Sask. 181/H2
Torch River, Sask. 181/G2
Tordesillas, Spain 33/D2
Torgau, E. Germany 22/E3
Torgelow, E. Germany 22/F2
Torhout, Belgium 27/C6
Tori, Ethiopia 111/F6
Torino (Turin), Italy 34/A2
Torit, Sudan 111/F7
Torne (riv.) 7/G2
Torneälv (riv.), Sweden 18/M3
Tor Ness (prom.), Scotland 15/E2
Torneträsk (lake), Sweden 18/L2
Torngat (mts.), Newf. 166/B2
Tornillo, Texas (79853) 303/A10
Tornio, Finland 18/O4
Tornio (riv.), Finland 18/O3
Tornquist, Argentina 143/D4
Toro, Cerro del (mt.), Argentina 143/B2
Toro (lake), Chile 138/D9
Toro, Cerro del (mt.), Chile 138/B7
Toro (pt.), Chile 138/A10
Toro, La. (†71429) 238/C4
Toro, El (mt.), P. Rico 161/F2
Toro, Spain 33/D2
Törökszentmiklós, Hungary 41/F3
Toronaic (gulf), Greece 45/F5

Toronto, Canada 2/F3
Toronto, Iowa (52343) 229/M5
Toronto, Kansas (66777) 232/G4
Toronto (lake), Kansas 232/F4
Toronto (res.), N.Y. 276/L7
Toronto, Ohio (43964) 284/J5
Toronto (cap.), Ont. 146/K5
Toronto (cap.), Ont. 162/H7
Toronto (metro. munic.), Ontario 177/K4
Toronto (cap.), Ontario 177/K4
Toronto, S. Dak. (57268) 298/R4
Toropalca, Bolivia 136/B7
Toropets, U.S.S.R. 52/D3
Tororo, Uganda 115/F3
Torote (riv.), Spain 33/G4
Torotoro, Bolivia 136/C6
Torpedo, Pa. (†16340) 294/D2
Torphins, Scotland 15/F3
Torpoint, England 13/C7
Torquay (Torbay), England 13/D7
Torquay, Sask. 181/H6
Torquemada, Spain 33/D1
Torr (head), N. Ireland 17/K1
Torrance, Calif. 188/C4
Torrance, Calif. (*90501) 204/C11
Torrance (co.), N. Mex. 274/D4
Torrance, Ontario 177/E3
Torrance, Pa. (15779) 294/D5
Torre, Cerro de la (mt.), Chile 138/E4
Torre Annunziata, Italy 34/E4
Torreblanca, Spain 33/G2
Torrecilla (lag.), P. Rico 161/E1
Torre del Greco, Italy 34/E4
Torre de Moncorvo, Portugal 33/C2
Torredonjimeno, Spain 33/D4
Torre Gaia, Italy 34/F6
Torrejón (res.), Spain 33/D3
Torrejoncillo, Spain 33/C3
Torrejón de Ardoz, Spain 33/G4
Torrelaguna, Spain 33/E2
Torrelavega, Spain 33/D1
Torremaggiore, Italy 34/E4
Torremolinos, Spain 33/D4
Torrens (riv.) 88/E7
Torrens (lake), Australia 87/D9
Torrens (isl.), S. Australia 88/D7
Torrens (lake), S. Australia 88/F6
Torrens (lake), S. Australia 94/E4
Torrens (riv.), S. Australia 94/C7
Torrente, Spain 33/F3
Torreón, Mexico 146/H7
Torreón, Mexico 150/H4
Torreon, N. Mex. (87061) 274/C4
Torre-Pacheco, Spain 33/F4
Torres (strait) 87/E7
Torres (str.), Papua N.G. 85/A7
Torres (str.), Queensland 88/G2
Torres (str.), Queensland 95/B1
Torres (isls.), Vanuatu 87/G7
Torres Martinez Ind. Res., Calif. 204/J10
Torres Novas, Portugal 33/B3
Torres Vedras, Portugal 33/B3
Torrevieja, Spain 33/F4
Torrey, Utah (84775) 304/C5
Torridge (riv.), England 13/C7
Torridon, Loch (inlet), Scotland 15/C3
Torriente, Cuba 158/D1
Torrijos, Philippines 82/D4
Torrijos, Spain 33/D2
Tørring, Denmark 21/C6
Torringford, Conn. (†06790) 210/C1
Torrington, Alberta 182/D4
Torrington, Conn. (06790) 210/C1
Torrington, Wyo. (82240) 319/H3
Torroella de Montgrí, Spain 33/H1
Torrowangee, N.S. Wales 97/A2
Torrox, Spain 33/E4
Torsby, Sweden 18/H6
Tors Cove, Newf. 166/D2
Torshälla, Sweden 18/K7
Tórshavn, Denmark 7/D2
Tórshavn (cap.), Faerøe Is., Denmark 21/A3
Tortilla Flat, Ariz. (85290) 198/D5
Tortola (isl.), Virgin Is. (Br.) 161/D3
Tortola (isl.), Virgin Is. (Br.) 156/H1
Tórtolas, Cerro de las (mt.), Chile 138/B8
Tortona, Italy 34/B2
Tortorici, Italy 34/E6
Tortosa, Spain 33/G2
Tortosa (cape), Spain 33/G2
Tortue (chan.), Haiti 158/C5
Tortue (Tortuga) (isl.), Haiti 156/D2
Tortue (Tortuga) (isl.), Haiti 158/C4
Tortuga (isl.), Haiti 158/C4
Tortuga (isl.), Haiti 156/D2
Tortugas (gulf), Colombia 126/B6
Tortuguero (lag.), P. Rico 161/D1
Tortuguilla (pt.), Cuba 158/K4
Tortum, Turkey 63/J2
Torud, Iran 59/F2
Torud, Iran 66/J3
Torul, Turkey 63/H2
Toruń (prov.), Poland 47/D2
Toruń, Poland 7/F3
Toruń, Poland 47/D2
Torunos, Venezuela 124/C3
Tõrva, U.S.S.R. 53/C1
Tory (isl.), Ireland 17/E1
Tory (isl.), Ireland 10/B3
Tory (sound), Ireland 17/E1
Torysa (riv.), Czech. 41/F2
Torzhok, U.S.S.R. 52/D3
Tosa, Japan 81/F7
Tosa (bay), Japan 81/F7
Tosashimizu, Japan 81/F7
Toson Hu (lake), China 77/E4
Töss (riv.), Switzerland 39/G1
Tostado, Argentina 143/D2
Toston, Mont. (59643) 262/E4
Tosu, Japan 81/E7
Tosya, Turkey 63/F2
Tota, Laguna de (lake), Colombia 126/D5
Totana, Spain 33/F4
Tótkomlós, Hungary 41/F3
Tot'ma, U.S.S.R. 48/E4
Tot'ma, U.S.S.R. 52/F3
Totnes, England 13/D7
Totnes, England 10/E5
Totnes, Sask. 181/C4
Totness, Suriname 131/C3
Toto, Ind. (†46534) 227/D2
Totoket, Conn. (†06405) 210/D3
Totonicapán, Guatemala 154/B3
Totora, Cochabamba, Bolivia 136/C5
Totora, Oruro, Bolivia 136/A5
Totoral, Chile 138/A6
Totoral, Quebrada (riv.), Chile 138/A6
Totoral, Uruguay 145/C3
Totowa, N.J. (07512) 273/B1
Totoya (isl.), Fiji 86/R11
Tottenham, N.S. Wales 97/D3
Tottenham, Ontario 177/E3
Tottori (pref.), Japan 81/G6
Tottori, Japan 81/G6
Touat (oasis), Algeria 106/E3
Touba, Ivory Coast 106/C7
Touba, Senegal 106/A6
Toubkal, Jebel (mt.), Morocco 102/B1
Toubkal, Jebel (mt.), Morocco 106/C2
Touchet, Wash. (99360) 310/G4
Touchet (riv.), Wash. 310/G4
Touchwood (lake), Alberta 182/E2
Touchwood (hills), Sask. 181/G4
Toufourine (well), Mali 106/C4
Tougaloo, Miss. (39174) 256/D6
Tougan, Upper Volta 106/D6
Touggourt, Algeria 106/F2
Touggourt, Algeria 102/C1
Toughkenamon, Pa. (19374) 294/L6
Tougué, Guinea 106/B6
Touila (well), Algeria 106/C3
Touila (well), Mauritania 106/C3
Toukoto, Mali 106/C6
Toul, France 28/F3
Touladi, Grand Lac (lake), Québec 172/J1
Toulnustouc (riv.), Québec 174/D2
Toulon, France 7/E4
Toulon, France 28/F6
Toulon, Ill. (61483) 222/D2
Toulouse, France 7/E4
Toulouse, France 28/D6
Toumodi, Ivory Coast 106/D7
Toungo, Nigeria 106/G7
Toungoo, Burma 72/C3
Touraine (trad. prov.), France 29
Tourakom, Laos 72/D3
Tourbis (lake), Québec 172/C2
Tourcoing, France 28/E2
Tour d'Ai (mt.), Switzerland 39/C4
Tourelle, Québec 172/C1
Tournai, Belgium 27/C7
Tournavista, Peru 128/E7
Tournon, France 28/F5
Tournus, France 28/F4
Touros, Brazil 132/H4
Touro Synagogue Nat'l Hist. Site, R.I. 249/J7
Tours, France 28/D4
Tours, France 7/E4
Tourville, Québec 172/H2
Toutes Aides, Manitoba 179/C3
Toutle, Wash. (98649) 310/C4
Toutle, North Fork (riv.), Wash. 310/C4
Toutle, South Fork (riv.), Wash. 310/C4
Toužim, Czech. 41/B1
Töv, Mongolia 77/G2
Tovar, Venezuela 124/C3
Tovey, Ill. (62570) 222/D4
Towaco, N.J. (07082) 273/E2
Towada, Japan 81/K3
Towada (lake), Japan 81/K3
Towada-Hachimantai National Park, Japan 81/K3
Towakaima, Guyana 131/B2
Towanda, Ill. (61776) 222/E3
Towanda, Kansas (67144) 232/E4
Towanda, Pa. (18848) 294/J2
Towanda (creek), Pa. 294/J2
Towaoc, Colo. (81334) 208/B8
Towcester, England 13/F5
Tower, Mich. (49792) 250/E3
Tower, Minn. (55790) 255/F3
Tower, Wyo. (†82190) 319/B1
Tower City, N. Dak. (58071) 282/P6
Tower City, Pa. (17980) 294/J4
Tower Hamlets, England 13/H8
Tower Hill, Ill. (62571) 222/E4
Tower Lakes, Ill. (†60010) 222/A4
Towers of Silence, India 68/B7
Tow Law, England 13/H4
Town (creek), Ala. 195/C1
Town (creek), Md. 245/E2
Town and Country, Mo. (†63101) 261/O3
Town and Country, Wash. (†99218) 310/H3
Town Creek, Ala. (35672) 195/D1
Towner, Colo. (81080) 208/P6
Towner (co.), N. Dak. 282/M2
Towner, N. Dak. (58788) 282/K3
Townley, Ala. (35587) 195/D3
Town of Pines, Ind. (†46360) 227/D1
Town Point, Md. (†21915) 245/P3
Towns (co.), Georgia 217/E1
Towns, Georgia (†31055) 217/G7
Townsend, Del. (19734) 245/R3
Townsend, Georgia (31331) 217/J7
Townsend, Mass. (01469) 249/H2
Townsend○, Mass. (01469) 249/H2
Townsend, Mont. (59644) 262/E4
Townsend (inlet), N.J. 273/D5
Townsend, Tenn. (37882) 237/O9
Townsend, Va. (23443) 307/R6
Townsend, Wis. (54175) 317/K5
Townsend Harbor, Mass. (†01469) 249/G2
Townsends Inlet, N.J. (†08243) 273/D5
Townshend○, Vt. (05353) 268/B5
Townsville, Australia 2/S6
Townsville, Australia 87/E7
Townsville, N.C. (27584) 281/N1
Townsville, Queensland 88/H3
Townsville, Queensland 95/C3
Townville, Pa. (16360) 294/C2
Townville, S.C. (29689) 296/B2
Towot, Sudan 111/F6
Towraghondi, Afghanistan 68/A1
Towson, Md. (21204) 245/M3
Towuti (lake), Indonesia 85/G6
Towy (riv.), Wales 13/D6
Towy (riv.), Wales 10/E5
Toxey, Ala. (36921) 195/B7
Toya (lake), Japan 81/K2
Toyah, Texas (79785) 303/D11
Toyah (creek), Texas 303/D11
Toyah (lake), Texas 303/A6
Toyahvale, Texas (79786) 303/D11
Toyama (pref.), Japan 81/H5
Toyama, Japan 81/H5
Toyama (bay), Japan 81/H5
Toyohashi, Japan 81/H6
Toyonaka, Japan 81/J7
Toyooka, Japan 81/G6
Toyota, Japan 81/H6
Tozeur, Tunisia 106/F2
Trabzon (prov.), Turkey 63/H2
Trabzon, Turkey 54/E5
Trabzon, Turkey 63/H2
Trabzon, Turkey 59/C1
Tracadie, New Bruns. 170/F1
Tracadie, Nova Scotia 168/G3
Tracadie (bay), Pr. Edward I. 168/F2
Trachselwald, Switzerland 39/E2
Tracy, Calif. (95376) 204/D6
Tracy, Conn. (†06492) 210/D2
Tracy, Iowa (50256) 229/H6
Tracy, Ky. (†42123) 237/K7
Tracy, Minn. (56175) 255/C6
Tracy, Mo. (64079) 261/C4
Tracy, New Bruns. 170/D3
Tracy, Québec 172/D3
Tracy Arm (inlet), Alaska 196/N1
Tracy City, Tenn. (37387) 237/K10
Tracyton, Wash. (98393) 310/A2
Trade, Tenn. (37691) 237/T8
Trade Lake, Wis. (†54837) 317/A4
Tradespark, Scotland 15/E3
Tradesville, S.C. (†29720) 296/F2
Tradewater (riv.), Ky. 237/F6
Trading (bay), Alaska 196/B1
Trading Post, Kansas (†66075) 232/H3
Traer, Iowa (50675) 229/J4
Traer, Kansas (†67749) 232/B2
Trafalgar, Ind. (46181) 227/E6
Trafalgar, Nova Scotia 168/F3
Trafalgar (cape), Spain 33/C4
Trafaria, Portugal 33/A1
Trafford, Ala. (35172) 195/E3
Trafford, Pa. (15085) 294/C5
Traghen, Libya 111/B2
Traiguén, Chile 138/D2
Traiguén (isl.), Chile 138/D6
Trail, Br. Col. 162/E6
Trail, Br. Col. 146/G4
Trail, Br. Col. 184/J6
Trail, Minn. (56684) 255/C3
Trail, Oreg. (97541) 291/E5
Trail City, S. Dak. (57657) 298/H3
Trail Creek, Ind. (†46360) 227/D1
Traill (isl.), Greenl. 4/B10
Traill (co.), N. Dak. 282/R5
Traîne (lake), Québec 172/D2
Trainer, Pa. (†19013) 294/L7
Traiskirchen, Austria 41/D2
Trakai, U.S.S.R. 53/C3
Tralake, Miss. (38757) 256/C4
Tralee, Ireland 10/B4
Tralee, Ireland 17/B7
Tralee (bay), Ireland 17/B7
Tramán-tepuí (mt.), Venezuela 124/G5
Tramelan, Switzerland 39/D2
Trammel, Va. (24289) 307/D6
Tramore, Ireland 10/C4
Tramore, Ireland 17/G7
Tramore (bay), Ireland 17/G7
Trampas, N. Mex. (87576) 274/D2
Tramperos (creek), N. Mex. 274/F2
Tramping (lake), Sask. 181/C3
Tramping Lake, Sask. 181/B3
Tranås, Sweden 18/J7
Trancoso, Portugal 33/C2
Tranebjerg, Denmark 21/D6
Tranebjerg (mt.), Denmark 21/C6
Tranent, Scotland 15/F5
Trang, Thailand 72/C6
Trangan (isl.), Indonesia 85/J7
Trangie, N.S. Wales 97/D3
Trani, Italy 34/F4
Tranquebar, India 68/E6
Tranqueras, Uruguay 145/D2
Tranqui (isl.), Chile 138/D4
Tranquility, N.J. (07879) 273/D2
Tranquillity, Calif. (93668) 204/E7
Transantarctic (mts.) 5/B17
Trans-Carpathian Oblast, U.S.S.R. 52/B5
Transfer, Pa. (16154) 294/A3
Transkei (aut. rep.), S. Africa 102/E8
Transkei (aut. rep.), S. Africa 118/D6
Transquaking (riv.), Md. 245/P7
Transvaal (prov.), S. Africa 102/E7
Transvaal (prov.), S. Africa 118/D4
Transylvania, La. (71286) 238/H1
Transylvania (co.), N.C. 281/D4
Transylvanian Alps (mts.), Romania 45/G3
Trapani (prov.), Italy 34/D5
Trapani, Italy 7/F5
Trapani, Italy 34/D5
Trap Falls (res.), Conn. 210/C3
Traphill, N.C. (28685) 281/H2
Trappe, Md. (21673) 245/O6
Trappers (lake), Colo. 208/E3
Traralgon, Victoria 97/D6
Traralgon, Victoria 88/H7
Trarza (reg.), Mauritania 106/A5
Trasimeno (lake), Italy 34/D3
Traskwood, Ark. (72167) 202/E5
Trat, Thailand 72/D4
Traun, Austria 41/C2
Traun (riv.), Austria 41/C2
Traun See (lake), Austria 41/B3
Traunstein, W. Germany 22/E5
Travancore (reg.), India 68/D7
Travelers Rest, S.C. (29690) 296/C2
Travellers (lake), N.S. Wales 97/B3
Travellers Rest, Ky. (†41314) 237/O6
Travemünde, W. Germany 22/D2
Travers, Alberta 182/D4
Travers (res.), Alberta 182/D4
Traverse (bay), Manitoba 179/F4
Traverse (isl.), Mich. 250/A1
Traverse (pt.), Mich. 250/A1
Traverse (co.), Minn. 255/B5
Traverse, Minn. (†56082) 255/D6
Traverse (lake), Minn. 255/B5
Traverse (lake), S. Dak. 298/R2
Traverse City, Mich. 188/K2
Traverse City, Mich. (49684) 250/D4
Tra Vinh (Phu Vinh), Vietnam 72/E5
Travis (co.), Texas 303/G7
Travis (lake), Texas 303/G7
Travis A.F.B., Calif. 204/L1
Travnik, Yugoslavia 45/C3
Trawbreaga (bay), Ireland 17/F1
Traynor, Sask. 181/C3
Traytown, Newf. 166/D1
Trbovlje, Yugoslavia 45/B2
Treadway, Tenn. (37883) 237/P8
Treasure (isl.), Fla. 212/B3
Treasure (co.), Mont. 262/J4
Treasure Island, Fla. (33740) 212/B3
Treasury (isls.), Solomon Is. 86/C2
Treaty, Ind. (†46992) 227/F3
Trebbia (riv.), Italy 34/B2
Třebíč, Czech. 41/C2
Trebinje, Yugoslavia 45/D4
Trebišov, Czech. 41/F2
Trebizond (Trabzon), Turkey 63/H2
Trebloc, Miss. (38875) 256/G3
Třeboň, Czech. 41/C2
Trece Martires, Philippines 82/C3
Tredegar, Wales 13/B6
Treece, Kansas (66778) 232/H4
Treelon, Sask. 181/C6
Trees, La. (71081) 238/B1
Treesbank, Manitoba 179/C5
Tregaron, Wales 13/D5
Tregaron, Wales 10/E4
Tregarva, Sask. 181/G5
Trego (co.), Kansas 232/C3
Trego, Mont. (59934) 262/B2
Trego, Wis. (54888) 317/C4
Treherne, Manitoba 179/D5
Treig, Loch (lake), Scotland 15/D4
Treinta y Tres (dept.), Uruguay 145/E4
Treinta y Tres, Uruguay 145/E4
Trelew, Argentina 143/C5
Trelleborg, Sweden 18/H9
Tremadoc (bay), Wales 10/D4
Tremadoc (prom.), Wales 13/C5
Tremblant (lake), Québec 172/C3
Trembleur (lake), Br. Col. 184/E3
Trementina, N. Mex. (88439) 274/E3
Tremiti (isls.), Italy 34/E3
Tremont, Ill. (61568) 222/D3
Tremont, Maine (†04653) 243/G7
Tremont○, Maine (†04653) 243/G7
Tremont, Miss. (38876) 256/H2
Tremont, Pa. (17981) 294/K4
Tremont City, Ohio (45372) 284/C5
Tremonton, Utah (84337) 304/B2
Tremp, Spain 33/G1
Trempealeau (co.), Wis. 317/D7
Trempealeau, Wis. (54661) 317/C8
Trempealeau (riv.), Wis. 317/C7
Trenary, Mich. (49891) 250/C2
Trenčín, Czech. 41/E2
Trenel, Argentina 143/D4
Trenggalek, Indonesia 85/K2
Trenque Lauquen, Argentina 143/D4
Trent (riv.), England 13/G4
Trent (riv.), England 10/F4
Trent (riv.), N.C. 281/P4
Trent, Oreg. (†97431) 291/E4
Trent, S. Dak. (57065) 298/R6
Trent, Texas (79561) 303/D5
Trente et un Milles (lake), Québec 172/B3
Trentham, Manitoba 179/F5
Trentham Cliffs, N.S. Wales 97/B4
Trentino-Alto Adige (reg.), Italy 34/C1
Trento (prov.), Italy 34/C1
Trento, Italy 34/C1
Trenton, Ala. (35774) 195/F1
Trenton, Ark. (†72374) 202/J5
Trenton, Fla. (32693) 212/D2
Trenton, Georgia (30752) 217/A1
Trenton, Ill. (62293) 222/D5
Trenton, Iowa (†52641) 229/K6
Trenton, Ky. (42286) 237/G7
Trenton, Maine (†04605) 243/G7
Trenton○, Maine (†04605) 243/G7
Trenton, Md. (†21155) 245/L2
Trenton, Mich. (48183) 250/B7
Trenton, Miss. (†39153) 256/E6
Trenton, Mo. (64683) 261/E2
Trenton, Nebr. (69044) 264/D4
Trenton (cap.), N.J. 146/L5
Trenton (cap.), N.J. 188/M2
Trenton (cap.), N.J. (*08601) 273/D3
Trenton, N.C. (28585) 281/P4
Trenton, N. Dak. (58853) 282/C3
Trenton, Nova Scotia 168/F3
Trenton, Ohio (45067) 284/B7
Trenton, Ontario 177/G3
Trenton, S.C. (29847) 296/D4
Trenton, Tenn. (38382) 237/D9
Trenton, Texas (75490) 303/H4
Trenton, Utah (84338) 304/B2
Trent Woods, N.C. (†28560) 281/P4
Trepassey, Newf. 166/D2
Treptow, E. Germany 22/F4
Tres Árboles, Uruguay 145/C3
Tres Arroyos, Argentina 143/D4
Tres Arroyos, Argentina 120/C6
Tres Bocas, Uruguay 145/B2
Tresckow, Pa. (18254) 294/K4
Tresco (isl.), England 13/A8
Três Corações, Brazil 132/E8
Três Corações, Brazil 135/D2
Tres Cruces, Nevado (mt.), Chile 138/B6
Tres Esquinas, Colombia 126/C7
Treshnish (isls.), Scotland 15/B4
Tres Islas, Uruguay 145/E3
Três Lagoas, Brazil 120/D5
Três Lagoas, Brazil 132/C8
Três Marias (res.), Brazil 120/E4
Tres Montes (cape), Chile 120/B7
Tres Montes (cape), Chile 138/C7
Tres Montes (gulf), Chile 138/D6
Tres Montes (pen.), Chile 138/C6
Tres Palmas, Colombia 126/B3
Trespassey (bay), Newf. 166/D2
Tres Picos, Cerro (mt.), Argentina 143/B5
Tres Piedras, N. Mex. (87577) 274/D2
Tres Pinos, Calif. (95075) 204/D7
Três Pontas, Brazil 135/D2
Tres Puntas (cape), Argentina 120/C7
Tres Puntas (cape), Argentina 143/D6
Tres Puntas (cape), Guatemala 154/C3
Três Rios, Brazil 135/E3
Três Rios, Brazil 132/F8
Tres Ritos, N. Mex. (†87579) 274/D2
Třešť, Czech. 41/C2
Treuchtlingen, W. Germany 22/D4
Treungen, Norway 18/F7
Treutlen (co.), Georgia 217/G6
Trevelín, Argentina 143/B5
Trevett, Maine (04571) 243/D8
Treviglio, Italy 34/B2
Treviño, Spain 33/E1
Treviso (prov.), Italy 34/D2
Treviso, Italy 34/D2
Trevlac, Ind. (†47448) 227/E6
Trevorton, Pa. (17881) 294/J4
Trevose (head), England 13/B7
Trévoux, France 28/F5
Treynor, Iowa (51575) 229/B6
Treyvaux, Switzerland 39/D3
Trezevant, Tenn. (38258) 237/D8
Trhové Sviny, Czech. 41/C2
Triabunna, Tasmania 99/D4
Triadelphia (lake), Md. 245/L4
Triadelphia, W. Va. (26059) 312/E2
Triana, Ala. (†35758) 195/E1
Triangle, Alberta 182/B2
Triangle, Va. (22172) 307/O3
Triángulo Este (isl.), Mexico 150/N6
Triángulo Oeste (isl.), Mexico 150/N6
Tribbett, Miss. (38779) 256/C4
Tribbey, Okla. (†74852) 288/M4
Triberg im Schwarzwald, W. Germany 22/C4
Tribune, Kansas (67879) 232/A3
Tribune, Sask. 181/H6
Tricase, Italy 34/G5
Trichur, India 68/D6
Trida, N.S. Wales 97/C3
Tridell, Utah (84076) 304/E3
Trident, Mont. (59752) 262/E5
Trident (peak), Nev. 266/C1
Trieben, Austria 41/C3
Trier, W. Germany 22/B4
Triesen, Liecht. 39/H2
Trieste (prov.), Italy 34/E2
Trieste, Italy 34/E2
Trieste, Italy 7/F4
Trieste (gulf), Italy 34/D2
Trigal, Bolivia 136/C6
Trigg (co.), Ky. 237/F7
Triglav (mt.), Yugoslavia 45/A2
Trigueros, Spain 33/C4
Trikkala, Greece 45/E6
Tri Lakes, Ind. (†46725) 227/G2
Trilby, Fla. (33593) 212/D3
Trilla, Ill. (62469) 222/E4
Trillick, N. Ireland 17/G3
Trim, Ireland 17/H4
Trim, Ireland 10/C4
Trimble, Ill. (†62454) 222/F4
Trimble (co.), Ky. 237/L3
Trimble, Ky. (42559) 237/M6
Trimble, Mo. (64492) 261/D4
Trimble, Ohio (45782) 284/F7
Trimble, Tenn. (38259) 237/C8
Trim Cane (creek), Miss. 256/G4
Trimmis, Switzerland 39/J3
Trimont, Minn. (56176) 255/D7
Trin, Switzerland 39/H3
Trinchera, Colo. (81081) 208/M8
Trinchera (peak), Colo. 208/J8
Trinchera (riv.), Colo. 208/H8
Trincheras, Mexico 150/D1
Trincomalee, Sri Lanka 54/K9
Trincomalee, Sri Lanka 68/E7
Trindade, Brazil 132/D7
Třinec, Czech. 41/E2
Tring, England 13/G7
Tring, England 10/F5
Tring-Jonction, Québec 172/F3
Trinidad (isl.), Argentina 143/D4
Trinidad, Bolivia 120/C4
Trinidad, Bolivia 136/C4
Trinidad, Calif. (95570) 204/A2
Trinidad (head), Calif. 204/A2
Trinidad (gulf), Chile 138/D8
Trinidad, Colombia 126/E5
Trinidad, Colo. 146/H6
Trinidad, Colo. 188/F3
Trinidad, Colo. (81082) 208/L8
Trinidad, Cuba 158/E2
Trinidad, Cuba 156/B2
Trinidad, Honduras 154/C3
Trinidad, Paraguay 144/E5
Trinidad, Texas (75163) 303/J5
Trinidad (isl.), Trin. & Tob. 156/G5
Trinidad (isl.), Trin. & Tob. 161/A9
Trinidad, Uruguay 145/B4
Trinidad, Wash. (†98848) 310/F3
Trinidad and Tobago 2/G5
Trinidad and Tobago 146/N8
TRINIDAD and TOBAGO 161
TRINIDAD and TOBAGO 156/G5
Trinity, Ala. (35673) 195/D1
Trinity (isls.), Alaska 196/H3
Trinity (co.), Calif. 204/B3
Trinity (riv.), Calif. 204/B3
Trinity (mt.), Idaho 220/C6
Trinity, Ky. (†41179) 237/O3
Trinity (range), Nev. 266/C2
Trinity, Newf. 166/D4
Trinity, Newf. 166/D2
Trinity (bay), Newf. 166/D2
Trinity (bay), Queensland 88/H3
Trinity (bay), Queensland 95/C3
Trinity (co.), Texas 303/J6
Trinity, Texas (75862) 303/J7
Trinity (bay), Texas 303/L2
Trinity (riv.), Texas 188/G4
Trinity (riv.), Texas 303/H5
Trinity, West Fork (riv.), Texas 303/G2
Trinity Center, Calif. (96091) 204/C2
Trinity Springs, Ind. (†47581) 227/D7
Trinity Ville, Jamaica 158/K6
Trinkitat, Sudan 111/G4
Trinkitat, Sudan 59/C6
Trino, Italy 34/B2
Trinway, Ohio (43842) 284/F5
Trio, S.C. (29595) 296/H5
Trion, Georgia (30753) 217/B1
Triplet, Va. (23886) 307/N7
Triplett, Mo. (65286) 261/F4
Tripoli, Iowa (50676) 229/J3
Tripoli (Tarabulus), Lebanon 59/C3
Tripoli (Tarabulus), Lebanon 63/F5
Tripoli (cap.), Libya 2/K4
Tripoli (cap.), Libya 102/D1
Tripoli (cap.), Libya 111/B1
Tripoli, Wis. (54564) 317/G4
Trípolis, Greece 45/F7
Tripolitania (reg.), Libya 102/D1
Tripolitania (reg.), Libya 111/B1
Tripp (co.), S. Dak. 298/K7
Tripp, S. Dak. (57376) 298/N7
Tripura (state), India 68/G4
Trischen (isl.), W. Germany 22/C1
Tristan da Cunha (isl.), St. Helena 2/J7
Triste (gulf), Venezuela 124/D2
Triton (isl.), China 85/E2
Triumph, Ill. (61371) 222/E2
Triumph-Buras, La. (†70041) 238/L8
Triune, Tenn. (†37014) 237/H9
Trivandrum, India 54/J9
Trivandrum, India 68/D7
Trivoli, Ill. (61569) 222/D3
Trnava, Czech. 41/D2
Trobriand (isls.), Papua N.G. 87/F6
Trobriand (isls.), Papua N.G. 85/C7
Trochu, Alberta 182/D4
Troense, Denmark 21/D7
Trofaiach, Austria 41/C3
Trogir, Yugoslavia 45/C4
Troisdorf, W. Germany 22/B3
Trois-Pistoles, Québec 172/H1
Trois Pitons, Morne (mt.), Dominica 161/E6
Trois-Ponts, Belgium 27/H8
Trois-Rivières, Guadeloupe 161/A7
Trois-Rivières (riv.), Haiti 158/B5
Trois-Rivières, Que. 162/J6
Trois-Rivières, Que. 146/L5
Trois-Rivières, Que. 172/E3
Trois-Rivières-Ouest, Québec 172/E3
Trois-Saumons, Québec 172/G2
Troistorrents, Switzerland /C4
Troisvierges, Luxembourg 27/J9
Troitsa (lake), Br. Col. 184/D3
Troitsk, U.S.S.R. 48/G4
Troitsko-Pechorsk, U.S.S.R. 52/J2
Trojan, S. Dak. (†57754) 298/B5
Trollhättan, Sweden 18/H7
Trombay, India 68/B7
Trombetas (riv.), Brazil 132/B3
Tromie (riv.), Scotland 15/D4
Trommald, Minn. (†56455) 255/D4
Troms (co.), Norway 18/L2
Tromsø, Norway 4/B9
Tromsø, Norway 7/F2
Tromsø, Norway 18/L2
Trona, Calif. (93562) 204/H8
Tronador (mt.), Argentina 143/B5
Tronador, Cerro (mt.), Chile 138/E3
Trondheim, Norway 7/F2
Trondheim, Norway 18/F5
Trondheimsfjorden (fjord), Norway 7/F2
Trondheimsfjorden (fjord), Norway 18/G5
Troodos (mt.), Cyprus 63/E5
Troon, Scotland 10/D3
Troon, Scotland 15/D5
Tropic, Utah (84776) 304/B6
Trosa, Sweden 18/K7
Trosky, Minn. (56177) 255/B7
Trossachs, Sask. 181/G6
Trossachs, The, (valley), Scotland 15/D4
Trostan (mt.), N. Ireland 17/J1
Trotternish (dist.), Scotland 15/B3
Trotters, N. Dak. (58657) 282/C5
Trotwood, Ohio (45426) 284/B6
Trou Bonbon, Haiti 158/A6
Trou du Nord, Haiti 158/C5
Troup (co.), Georgia 217/B4
Troup (head), Scotland 15/F3
Troup, Texas (75789) 303/J5
Troupsburg, N.Y. (14885) 276/F6
Trousdale, Kansas (†67059) 232/C4
Trousdale, Okla. (†74878) 288/M4

Trousdale (co.), Tenn. 237/J8
Trousers (lake), New Bruns. 170/C1
Trout (mt.), Alberta 182/C1
Trout (riv.), Alberta 182/C1
Trout (creek), Ariz. 198/B3
Trout (creek), Colo. 208/E2
Trout (creek), Idaho 220/B1
Trout, La. (71371) 238/F3
Trout (lake), Minn. 255/F2
Trout (lake), N.W.T. 162/D3
Trout (lake), N.W. Terrs. 187/F3
Trout (lake), Ontario 175/B2
Trout (lake), Ontario 177/F1
Trout (lake), Ontario 177/E1
Trout (creek), Oreg. 291/F3
Trout (creek), Oreg. 291/J5
Trout (riv.), Sask. 181/L2
Trout (riv.), Vt. 268/B2
Trout, W. Va. (24982) 312/F6
Trout (lake), Wis. 317/G3
Trout Creek, Mich. (49967) 250/G2
Trout Creek, Mont. (59874) 262/A3
Trout Creek, Ontario 177/E2
Trout Creek, Utah (84077) 304/A4
Troutdale, Maine (†04985) 243/D5
Troutdale, Oreg. (97060) 291/E2
Trout Dale, Va. (24378) 307/F7
Trout Hall, Jamaica 158/J6
Trout Lake, Alberta 182/C1
Trout Lake, Br. Col. 184/J5
Trout Lake, Mich. (49793) 250/E2
Trout Lake, N.W. Terrs. 187/F3
Trout Lake, Wash. (98650) 310/D5
Troutman, Georgia (†31721) 217/C7
Troutman, N.C. (28166) 281/H3
Trout River, Newf. 166/C4
Trout Run, Pa. (17771) 294/H3
Troutville, Pa. (15866) 294/E3
Troutville, Va. (24175) 307/J6
Trowbridge, England 13/E6
Trowbridge, England 10/E5
Trowutta, Tasmania 99/B3
Troxelville, Pa. (17882) 294/H4
Troy, Ala. (36081) 195/G7
Troy, Idaho (83871) 220/B3
Troy, Ill. (62294) 222/B6
Troy, Ind. (47588) 227/D9
Troy, Iowa (52537) 229/J7
Troy, Kansas (66087) 232/G2
Troy○, Maine (04987) 243/E6
Troy, Mich. (*48084) 250/B6
Troy, Miss. (†38863) 256/G2
Troy, Mo. (63379) 261/L5
Troy, Mont. (59935) 262/A2
Troy, N.H. (03465) 268/C6
Troy○, N.H. (03465) 268/C6
Troy, N.Y. 188/M2
Troy, N.Y. (*12180) 276/N5
Troy, N.C. (27371) 281/K4
Troy, Nova Scotia 168/G3
Troy, Ohio (45373) 284/B5
Troy, Okla. (†74856) 288/N6
Troy, Oreg. (†97885) 291/K2
Troy, Pa. (16947) 294/J2
Troy, S.C. (29848) 296/C4
Troy, S. Dak. (†57265) 298/R3
Troy, Tenn. (38260) 237/C8
Troy (Ilium) (ruins), Turkey 63/B6
Troy○, Vt. (05868) 268/C2
Troy, Va. (22974) 307/M5
Troy, W. Va. (26443) 312/E4
Troyan, Bulgaria 45/G4
Troy Center, Wis. (53180) 317/J2
Troyes, France 7/E4
Troyes, France 28/F3
Troyes (lake), Québec 172/C2
Troy Grove, Ill. (61372) 222/E2
Troy Mills, Iowa (52344) 229/K4
Trstená, Czech. 41/E2
Trstenik, Yugoslavia 45/E4
Truandó (riv.), Colombia 126/B4
Truax, Sask. 181/G6
Trub, Switzerland 39/E3
Truba, Saudi Arabia 59/D4
Truc Giang, Vietnam 72/E5
Truchas, N. Mex. (87578) 274/D2
Truckee, Calif. (95734) 204/E4
Truckee (riv.), Calif. 204/F4
Truckee (riv.), Nev. 266/B3
Truckton, Colo. (†80864) 208/L5
Truesdail, Mo. (63383) 261/K5
Truesdale, Iowa (50592) 229/C3
Trufant, Mich. (49347) 250/D5
Truganina, Victoria 97/H5
Trujillo, Honduras 154/E3
Trujillo, N. Mex. (†87701) 274/E3
Trujillo, Peru 120/B3
Trujillo, Peru 128/C7
Trujillo, Spain 33/D3
Trujillo (creek), Texas 303/A2
Trujillo (state), Venezuela 124/C3
Trujillo, Venezuela 120/C2
Trujillo, Venezuela 124/C3
Trujillo Alto, P. Rico 161/E1
Truk (isls.), Micronesia 87/F5
Truman, Minn. (56088) 255/D7
Truman, Ark. (72472) 202/J3
Trumansburg, N.Y. (14886) 276/G5
Trumbauersville, Pa. (18970) 294/M5
Trumbull (mt.), Ariz. 198/B2
Trumbull○, Conn. (06611) 210/C4
Trumbull (lake), Iowa 229/C2
Trumbull, Nebr. (68980) 264/F4
Trumbull (co.), Ohio 284/J3
Trŭn, Bulgaria 45/F4
Trun, Switzerland 39/G3
Trundle, N.S. Wales 97/D3
Trung Khanh Phu, Vietnam 72/E2
Truro, England 13/B7
Truro, England 10/D5
Truro, Iowa (50257) 229/F6
Truro○, Mass. (02666) 249/O5
Truro, N.S. 162/K6
Truro, Nova Scotia 168/E3
Truscott, Texas (79260) 303/E4
Truskmore (mt.), Ireland 17/E3
Trussville, Ala. (35173) 195/E3
Trustrup, Denmark 21/D5
Trutch, Br. Col. 184/F1
Truth or Consequences, N. Mex. (87901) 274/B5
Trutnov, Czech. 41/D1
Truxno, La. (†71260) 238/F1
Truxton, Mo. (63381) 261/K4
Truxton, N.Y. (13158) 276/H5
Trwyn Cilan (prom.), Wales 13/C5
Tryon, Nebr. (69167) 264/C3
Tryon, N.C. (28782) 281/E4
Tryon, Okla. (74875) 288/N3
Tryonville, Pa. (†16404) 294/C2
Trysil, Norway 18/H6
Trysilelv (riv.), Norway 18/H6
Trzcianka, Poland 47/C2
Trzebiatów, Poland 47/B1
Trzebinia-Siersza, Poland 47/C4
Tržič, Yugoslavia 45/B2
Tsagaannuur, Mongolia 77/C2
Tsagaan-Ovoo, Mongolia 77/H2
Tsagaan-Uul, Mongolia 77/E2
Tsaile, Ariz. (†86503) 198/F2
Tsala Apopka (lake), Fla. 212/D3
Tsamkong (Zhanjiang), China 77/H7
Tsangchow (Cangzhou), China 77/J4
Tsau, Botswana 118/C4
Tsavo, Kenya 115/G4
Tsavo Nat'l Park, Kenya 115/G4
Tschida (lake), N. Dak. 282/G6
Tschierv, Switzerland 39/K3
Tschlin, Switzerland 39/K3
Tseelim (dry riv.), Israel 65/C5
Tselinograd, U.S.S.R. 54/J4
Tselinograd, U.S.S.R. 48/H4
Tsenhermandal, Mongolia 77/H2
Tses, Namibia 118/B5
Tsetseg, Mongolia 77/D2
Tsetserleg, Mongolia 77/F2
Tsetserleg, Mongolia 54/L5
Tshabong, Botswana 118/C5
Tshane, Botswana 118/C4
Tshela, Zaire 115/B4
Tshikapa, Zaire 115/D5
Tshofa, Zaire 115/D5
Tshuapa (riv.), Zaire 115/D4
Tsiafajavona (mt.), Madagascar 102/G6
Tsiafajavona (mt.), Madagascar 118/H3
Tsiaotso (Jiaozuo), China 77/H4
Tsihombe, Madagascar 118/H5
Tsil'ma (riv.), U.S.S.R. 52/H1
Tsimlyansk (res.), U.S.S.R. 7/J4
Tsimlyansk (res.), U.S.S.R. 52/F5
Tsimlyansk (res.), U.S.S.R. 48/E5
Tsin (dry riv.), Israel 65/D5
Tsinan (Jinan), China 77/J4
Tsinghai (Qinghai) (prov.), China 77/E4
Tsingkiang (Qingjiang), China 77/J5
Tsingshih (Jinshi), China 77/H6
Tsingtao (Qingdao), China 77/K4
Tsining (Jining), Nei Monggol, China 77/H3
Tsining (Jining), Shandong, China 77/J4
Tsiribihina (riv.), Madagascar 118/G3
Tsiroanomandidy, Madagascar 118/H3
Tsitsihar (Qiqihar), China 77/K2
Tsivory, Madagascar 118/H4
Tskhinvali, U.S.S.R. 48/E5
Tskhinvali, U.S.S.R. 52/F6
Tsodilo Hill (mt.), Botswana 118/C3
Tsu, Japan 81/H6
Tsu (isls.), Japan 81/D6
Tsubame, Japan 81/J5
Tsuchiura, Japan 81/J5
Tsugaru (str.), Japan 81/K3
Tsumeb, Namibia 118/B3
Tsumeb, Namibia 102/D6
Tsunyi (Zunyi), China 77/G6
Tsuruga, Japan 81/G6
Tsurugi (mt.), Japan 81/G7
Tsuruoka, Japan 81/J4
Tsushima (strait), Japan 81/D7
Tsuyama, Japan 81/F6
Tuadook (lake), New Bruns. 170/D2
Tuakau, N. Zealand 100/E2
Tual, Indonesia 85/J7
Tualatin, Oreg. (97062) 291/A2
Tualatin (riv.), Oreg. 291/A2
Tuam, Ireland 10/B4
Tuam, Ireland 17/D4
Tuamotu (arch.), Fr. Polynesia 2/C6
Tuamotu (arch.), Fr. Poly. 87/M7
Tuapse, U.S.S.R. 52/E6
Tuatapere, N. Zealand 100/A7
Tuath, Loch (inlet), Scotland 15/B4
Tubac, Ariz. (85640) 198/E7
Tuba City, Ariz. (86045) 198/D2
Tubal, Wadi al (dry riv.), Iraq 66/B4
Tuban, Indonesia 85/K2
Tubarão, Brazil 120/E5
Tubarão, Brazil 132/D10
Tubarão, Brazil 132/G8
Tubas, West Bank 65/C3
Tubbataha (reefs), Philippines 85/G4
Tubbataha (reefs), Philippines 82/B6
Tubbercurry, Ireland 17/D3
Tubeke (Tubize), Belgium 27/E7
Tuberose, Sask. 181/C5
Tubigon, Philippines 82/D6
Tübingen, W. Germany 22/C4
Tubize, Belgium 27/E7
Tubmanburg, Liberia 106/B7
Tubuai (Austral) (isls.), Fr. Poly. 87/M8
Tubuai (isl.), Fr. Poly. 87/M8
Tuburan, Philippines 82/D5
Tucacas, Venezuela 124/D2
Tucannon (riv.), Wash. 310/G4
Tucano, Brazil 132/G5
Tucavaca, Bolivia 136/F6
Tucavaca (riv.), Bolivia 136/F6
Tuchitua Lake, Yukon 187/E3
Tuchola, Poland 47/C2
Tuckahoe (creek), Md. 245/P5
Tuckahoe, N.J. (08250) 273/D5
Tuckahoe (riv.), N.J. 273/D5
Tuckahoe, N.Y. (10707) 276/O7
Tucker, Ark. (72168) 202/G5
Tucker (co.), W. Va. 312/F4
Tuckerman, Ark. (72473) 202/H2
Tuckernuck (isl.), Mass. 249/N7
Tuckerton, N.J. (08087) 273/E4
Tucson, Ariz. 188/D4
Tucson, Ariz. 146/G6
Tucson, Ariz. (*85701) 198/D6
Tucumán (prov.), Argentina 143/C2
Tucumán, San Miguel de, Arg. 143/D2
Tucumcari, N. Mex. 188/F3
Tucumcari, N. Mex. (88401) 274/F3
Tucupido, Venezuela 124/F3
Tucupita, Venezuela 124/H3
Tucuruí, Brazil 132/D3
Tudela, Spain 33/F1
Tudela de Duero, Spain 33/D2
Tudor (lake), 174/D1
Tudor, Alberta 182/D4
Tuena, N.S. Wales 97/E3
Tuffnell, Sask. 181/H4
Tufi, Papua N.G. 85/C7
Tuftonboro○, N.H. (†03816) 268/E4
Tugaloo (riv.), Georgia 217/F1
Tugaloo (riv.), S.C. 296/A2
Tugaske, Sask. 181/E5
Tug Fork (riv.), Ky. 237/S5
Tug Fork (riv.), Va. 307/D5
Tug Fork (riv.), W. Va. 312/B7
Tuggerah (lake), N.S. Wales 97/F3
Tugidak (isl.), Alaska 196/G3
Tugnug (pt.), Philippines 82/E5
Tuguegarao, Philippines 85/G2
Tuguegarao, Philippines 82/C2
Tuichi (riv.), Bolivia 136/A4
Tujunga, Calif. (91042) 204/C10
Tukangbesi (isls.), Indonesia 85/G7
Tuktoyaktuk, Canada 4/C16
Tuktoyaktuk, N.W.T. 162/C2
Tuktoyaktuk, N.W. Terrs. 187/E3
Tukuhnikivatz (mt.), Utah 304/E5
Tukums, U.S.S.R. 53/B2
Tukums, U.S.S.R. 52/B3
Tukuran, Philippines 82/D7
Tukuyu, Tanzania 115/F5
Tukwila, Wash. (98188) 310/B2
Tula, Tamaulipas, Mexico 150/K5
Tula (riv.), Mexico 150/L1
Tula, Miss. (38675) 256/F2
Tula, U.S.S.R. 7/H3
Tula, U.S.S.R. 48/D4
Tula, U.S.S.R. 52/E4
Tula de Allende, Mexico 150/K6
Tulagi, Solomon Is. 86/E3
Tulak, Afghanistan 68/A2
Tulak, Afghanistan 59/H3
Tulalip, Wash. (†98270) 310/C2
Tulalip Ind. Res., Wash. 310/C2
Tulameen, Br. Col. 184/G5
Tulancingo, Mexico 150/K7
Tulare (co.), Calif. 204/G7
Tulare, Calif. (93274) 204/F7
Tulare (lake), Calif. 188/B3
Tulare (lake), Calif. 204/F7
Tulare, S. Dak. (57476) 298/N4
Tularosa, N. Mex. (88352) 274/C5
Tularosa (valley), N. Mex. 274/C6
Tulcán, Ecuador 128/D2
Tulcea, Romania 45/J3
Tul'chin, U.S.S.R. 52/C5
Tulcingo del Valle, Mexico 150/M2
Tule (lake), Calif. 204/D2
Tule (des.), Nev. 266/G5
Tule (lake), Wash. 310/G3
Tuléar (Toliary), Madagascar 118/G4
Tulelake, Calif. (96134) 204/D2
Tule River Ind. Res., Calif. 204/G7
Tuli, Zimbabwe 118/D4
Tulia, Texas (79088) 303/C3
Tulip, Ark. (†71725) 202/E5
Tulip, Ind. (†47424) 227/D6
Tulkarm, West Bank 65/C3
Tull, Ark. (†72015) 202/E5
Tulla, Ireland 17/D6
Tullah, Tasmania 99/B3
Tullahassee, Okla. (74466) 288/P3
Tullahoma, Tenn. (37388) 237/J10
Tullamore, Ireland 10/C4
Tullamore, Ireland 17/G5
Tullamore, N.S. Wales 97/D3
Tullaroan, Ireland 17/G6
Tulle, France 28/D5
Tullibigeal, N.S. Wales 97/D3
Tullibody, Scotland 15/C1
Tulliby Lake, Alberta 182/E3
Tullis, Sask. 181/D4
Tulln, Austria 41/D2
Tullos, La. (71479) 238/F3
Tullow, Ireland 10/C4
Tullow, Ireland 17/H6
Tully, Mass. (†01364) 249/E2
Tully, N.Y. (13159) 276/H5
Tully, Queensland 88/H3
Tully, Queensland 95/C3
Tully (falls), Queensland 95/C3
Tullytown, Pa. (19007) 294/N5
Tuloma (riv.), U.S.S.R. 52/D1
Tulot, Ark. (†72472) 202/K2
Tulsa, Okla. 188/G3
Tulsa, Okla. 146/J6
Tulsa (co.), Okla. 288/P2
Tulsa, Okla. (*74101) 288/O2
Tulsi (lake), India 68/B6
Tultepec, Mexico 150/L1
Tuluá, Colombia 126/B5
Tuluksak, Alaska (99679) 196/F2
Tulun, U.S.S.R. 54/M4
Tulun, U.S.S.R. 48/L4
Tulungagung, Indonesia 85/K2
Tuma (riv.), Nicaragua 154/E4
Tumacacori, Ariz. (85640) 198/D7
Tumacacori Nat'l Mon., Ariz. 198/E7
Tumaco, Colombia 120/B2
Tumaco, Colombia 126/A7
Tumaco, Rada de (bay), Colombia 126/A6
Tumalo, Oreg. (†97701) 291/F3
Tumalo (creek), Oreg. 291/F3
Tumatumari, Guyana 131/B3
Tumba (lake), Zaire 115/C4
Tumbarumba, N.S. Wales 97/D4
Tumbes (pen.), Chile 138/D1
Tumbes (riv.), Ecuador 128/B4
Tumbes (dept.), Peru 128/B4
Tumbes, Peru 120/A3
Tumbes, Peru 128/B4
Tumbes (riv.), Peru 128/B4
Tumblong, N.S. Wales 97/E4
Tumby Bay, S. Australia 88/F6
Tumby Bay, S. Australia 94/E6
Tumd Youqi, China 77/H3
Tumd Zuoqi, China 77/G3
Tumegl-Tomils, Switzerland 39/H3
Tumeka (lake), Br. Col. 184/C1
Tumen, China 77/M3
Tumen, China 54/O5
Tumen (riv.), China 77/L3
Tumen (riv.), N. Korea 81/D2
Tumeremo, Venezuela 124/H4
Tumereng, Guyana 131/B2
Tumindao (isl.), Philippines 82/B8
Tumkur, India 68/D6
Tummel (lake), Scotland 15/E4
Tummo (El War) (well), Niger 106/G4
Tumon (bay), Guam 86/K6
Tump, Pakistan 68/A3
Tump, Pakistan 59/H4
Tumpat, Malaysia 72/D6
Tumtum (lake), Oreg. 291/J5
Tumtum, Wash. (99034) 310/H3
Tumu, Ghana 106/D6
Tumuc-Humac (mts.), Fr. Guiana 131/D3
Tumuc-Humac (mts.), Suriname 131/D4
Tumucumaque, Serra de (range), Brazil 132/C2
Tumupasa, Bolivia 136/B4
Tumusla, Bolivia 136/C7
Tumut, N.S. Wales 88/H7
Tumut, N.S. Wales 97/E4
Tumwater, Wash. (98501) 310/B3
Tun (Ferdows), Iran 66/K3
Tuna (pt.), P. Rico 161/E3
Tunahí, Sierra (mts.), Colombia 126/E7
Tunapuna, Trin. & Tob. 161/B10
Tunas, Mo. (65764) 261/F7
Tunas de Zaza, Cuba 158/E2
Tunas de Zaza, Cuba 156/B2
Tunbridge, N. Dak. (†58368) 282/K3
Tunbridge○, Vt. (05077) 268/C4
Tunceli (prov.), Turkey 63/H3
Tunceli (Kalan), Turkey 63/H3
Tunduru, Tanzania 115/G6
Tundzha (riv.), Bulgaria 45/G4
Tungabhadra (riv.), India 68/D5
Tungchwan (Tongchuan), China 77/G5
Tunghwa (Tonghua), China 77/L3
Tungkiang (Tongjiang), China 77/M2
Tungliao (Tongliao), China 77/K3
Tungsten, N.W. Terrs. 187/F3
Tungurahua (prov.), Ecuador 128/C3
Tunhwa (Dunhua), China 77/L3
Tunhwang (Dunhuang), China 77/E3
Tuni, India 68/E5
Tunica, La. (70782) 238/G5
Tunica (co.), Miss. 256/D1
Tunica, Miss. (38676) 256/D1
Tunis, N.C. (†27986) 281/R2
Tunis (cap.), Tunisia 102/D1
Tunis (cap.), Tunisia 106/G1
Tunis (cap.), Tunisia 2/K4
Tunis (gulf), Tunisia 106/G1
Tunisia 2/K4
Tunisia 102/D1
TUNISIA 106/F1
Tunis Mills, Md. (†21601) 245/O5
Tunja, Colombia 120/B2
Tunja, Colombia 126/D5
Tunker, Ind. (†46724) 227/F2
Tunkhannock, Pa. (18657) 294/L2
Tunki, Nicaragua 154/E4
Tunnack, Tasmania 99/D4
Tunnel, Alaska (†99587) 196/C1
Tunnel City, Wis. (54662) 317/E7
Tunnel Hill, Georgia (30755) 217/C1
Tunnel Springs, Ala. (†36471) 195/D7
Tunnelton, Ind. (47467) 227/E7
Tunnelton, W. Va. (26444) 312/G4
Tunnsjøen (lake), Norway 18/H4
Tunstall, Va. (†23124) 307/O5
Tuntutuliak, Alaska (99680) 196/F2
Tununak, Alaska (99681) 196/E2
Tunungayualok (isl.), Newf. 166/B2
Tunuyán, Argentina 143/C3
Tunuyán (riv.), Argentina 143/C3
Tunxi (Tunki), China 77/J6
Tuolumne (co.), Calif. 204/F5
Tuolumne, Calif. (95379) 204/E6
Tuotuo He (riv.), China 77/D5
Tupã, Brazil 132/D8
Tupã, Brazil 135/A2
Tupambaé, Uruguay 145/E3
Tupanciretã, Brazil 132/C10
Tupelo, Ark. (72169) 202/H3
Tupelo, Miss. (38801) 256/G2
Tupelo, Okla. (74572) 288/O5
Tupelo Nat'l Battlefield, Miss. 256/G2
Tupí, Venezuela 124/D2
Tupiza, Bolivia 120/C5
Tupiza, Bolivia 136/C7
Tupman, Calif. (93276) 204/F8
Tupper, Br. Col. 184/G2
Tupper (lake), N.Y. 276/M2
Tupper (lake), Nova Scotia 168/D4
Tupper Lake, N.Y. (12986) 276/M2
Tuppers Plains, Ohio (45783) 284/G7
Tupungato, Cerro (mt.), Argentina 143/B3
Tupungato, Cerro (mt.), Chile 138/B9
Tuquan, China 77/K2
Túquerres, Colombia 126/B7
Tur, West Bank 65/C4
Tura, Hungary 41/E3
Tura, India 68/G3
Tura, U.S.S.R. 54/M3
Tura, U.S.S.R. 48/L3
Turaba, Saudi Arabia 59/D5
Turagua, Serranía (mts.), Venezuela 124/F4
Turakina, N. Zealand 100/E4
Turakirae (head), N. Zealand 100/B3
Turan, Iran 59/G2
Turan, Iran 66/K3
Tur'an, Israel 65/C2
Turan, U.S.S.R. 48/K4
Turangi, N. Zealand 100/E3
Turbaco, Colombia 126/C2
Turbat, Pakistan 59/H4
Turbat, Pakistan 68/A3
Turbat-i-Shaikh Jam, Iran 59/H2
Turbat-i-Shaikh Jam, Iran 66/M3
Turbenthal, Switzerland 39/G2
Turbeville, S.C. (29162) 296/G4
Turbeville, Va. (24596) 307/K7
Turbo, Colombia 126/B3
Turbotville, Pa. (17772) 294/J3
Turčianske Teplice, Czech. 41/E2
Turco, Bolivia 136/A6
Turda, Romania 45/F2
Tureia (atoll), Fr. Poly. 87/N8
Turek, Poland 47/D2
Turen, Indonesia 85/K2
Turén, Venezuela 124/D3
Turfan (Turpan), China 77/C3
Turgay, U.S.S.R. 48/G5
Türgovishte, Bulgaria 45/H4
Turgutlu, Turkey 63/B3
Turhal, Turkey 63/F2
Türi, U.S.S.R. 53/C1
Turia (riv.), Spain 33/F3
Turiaçu, Brazil 132/E3
Turiaçu (riv.), Brazil 132/E3
Turiamo, Venezuela 124/E2
Turin, Alberta 182/D5
Turin, Georgia (30289) 217/C4
Turin, Iowa (51059) 229/B4
Turin (prov.), Italy 34/A2
Turin, Italy 34/A2
Turin, Italy 7/E4
Turin, N.Y. (13473) 276/K3
Turkana (lake), Ethiopia 111/G7
Turkana (lake), Kenya 102/F4
Turkana (lake), Kenya 115/G3
Türkeli, Turkey 63/F2
Turkestan, U.S.S.R. 48/G5
Túrkeve, Hungary 41/F3
Turkey 2/L4
Turkey 7/H5
Turkey 54/E6
TURKEY 59/B2
TURKEY 63/D3
Turkey (riv.), Iowa 229/K2
Turkey, Ky. (41382) 237/P6
Turkey, N.C. (28393) 281/N4
Turkey (creek), Okla. 288/L2
Turkey (creek), S.C. 296/E2
Turkey, Texas (79261) 303/D3
Turkey Creek, La. (70585) 238/F5
Turkey Creek (lake), La. 238/G3
Turkey Creek, W. Australia 92/E2
Turkey Point, Ontario 177/D5
Türkmen Dağı (mt.), Turkey 63/D3
Turkmen S.S.R., U.S.S.R. 54/G6
Turkmen S.S.R., U.S.S.R. 48/F6
Türkoğlu, Turkey 63/G4
Turks (isls.), Turks & Caicos 156/D2
Turks and Caicos (isls.) 146/L7
TURKS and CAICOS ISLANDS 156/D2
Turks Island (passage), Turks & Caicos 156/D2
Turku, Finland 7/G2
Turku, Finland 18/N6
Turku ja Pori (prov.), Finland 18/N6
Turlock, Calif. (95380) 204/E6
Turmero, Venezuela 124/E2
Turnagain (riv.), Br. Col. 184/K2
Turnagain (cape), N. Zealand 100/F4
Turnagain Arm (inlet), Alaska 196/B1
Turnavik (isls.), Newf. 166/C2
Turnberry, Scotland 15/D5
Turneffe (isls.), Belize 154/D2
Turnen (mt.), Switzerland 39/D3
Turner, Ark. (72383) 202/H5
Turner (co.), Georgia 217/E7
Turner, Maine (04282) 243/C7
Turner○, Maine (04282) 243/C7
Turner, Mich. (48765) 250/F4
Turner, Mont. (59542) 262/H2
Turner, Oreg. (97392) 291/E3
Turner (co.), S. Dak. 298/P7
Turner, Wash. (†99328) 310/H4
Turner Center, Maine (04283) 243/C7
Turner Hole (bay), Virgin Is. (U.S.) 161/G4
Turners, Mo. (65765) 261/F8
Turnersburg, N.C. (28688) 281/H3
Turners Falls, Mass. (01376) 249/D2
Turners Station, Ky. (40075) 237/L3
Turnersville, N.J. (†08012) 273/C4
Turnersville, Texas (76580) 303/G6
Turner Valley, Alberta 182/C4
Turnerville, Wyo. (†83112) 319/A3
Turney, Mo. (64493) 261/D3
Turnhout, Belgium 27/F6
Turnor Lake, Sask. 181/L3
Turnov, Czech. 41/C1
Turnu Măgurele, Romania 45/G4
Turon, Kansas (67583) 232/D4
Turpan (Turfan), China 77/C3
Turpan, China 54/L5
Turpin, Okla. (73950) 288/E1
Turquino (peak), Cuba 158/H4
Turrell, Ark. (72384) 202/K3
Turrialba, C. Rica 154/F6
Turriff, Scotland 10/E2
Turriff, Scotland 15/F3
Turtle (Penju) (isls.), Indonesia 85/H7
Turtle (mts.), Manitoba 179/B5
Turtle (riv.), Manitoba 179/C3
Turtle (lake), Mich. 250/F4
Turtle (lake), N. Dak. 282/H4
Turtle (mts.), N. Dak. 282/K2
Turtle (isls.), Philippines 82/B7
Turtle (lake), Sask. 181/C2
Turtle (creek), S. Dak. 298/M4
Turtle Creek, New Bruns. 170/F3
Turtle Creek, Pa. (15145) 294/C7
Turtle Creek, W. Va. (25203) 312/C6
Turtleford, Sask. 181/B2
Turtle Lake, N. Dak. (58575) 282/J4
Turtle Lake, Wis. (54889) 317/B5
Turtle Mountain Ind. Res., N. Dak. 282/L2
Turtle Mountain Prov. Park, Manitoba 179/B5
Turtlepoint, Pa. (16750) 294/F2
Turtle River, Minn. (†56601) 255/D3
Turtletown, Tenn. (37391) 237/N10
Turtola, Finland 18/O3
Turton, England 13/H2
Turton, S. Dak. (57477) 298/N3
Turukhansk, U.S.S.R. 48/J3
Turvo (riv.), Brazil 135/B2
Turzovka, Czech. 41/E2
Tuscaloosa, Ala. 188/J4
Tuscaloosa (co.), Ala. 195/C4
Tuscaloosa, Ala. (*35401) 195/C4
Tuscaloosa (lake), Ala. 195/D4
Tuscan (arch.), Italy 34/B3
Tuscany (reg.), Italy 34/C3
Tuscarawas (co.), Ohio 284/H5
Tuscarawas, Ohio (44682) 284/H5
Tuscarawas (riv.), Ohio 284/H5
Tuscarora, Nev. (89834) 266/E1
Tuscarora (mts.), Nev. 266/E1
Tuscarora (mt.), Pa. 294/G5
Tuscarora Ind. Res., N.Y. 276/B4
Tuscola, Ill. (61953) 222/E4
Tuscola (co.), Mich. 250/F5
Tuscola, Mich. (48769) 250/F5
Tuscola, Texas (79562) 303/E5
Tuscor, Mont. (†59874) 262/A3
Tusculum, Georgia (†31329) 217/K6
Tusculum, Tenn. (37743) 237/R8
Tuscumbia, Ala. (35674) 195/C1
Tuscumbia, Mo. (65082) 261/H6
Tushar (mts.), Utah 304/B5
Tushka, Okla. (74573) 288/O6
Tuskahoma, Okla. (74574) 288/R5
Tuskegee, Ala. (36083) 195/G6
Tuskegee Institute, Ala. (36088) 195/G6
Tuskegee Institute Nat'l Hist. Park, Ala. 195/G6
Tusket, Nova Scotia 168/C5
Tusket (isl.), Nova Scotia 168/B5
Tusket (riv.), Nova Scotia 168/C4
Tussy, Okla. (73088) 288/L6
Tustin, Calif. (92680) 204/D11
Tustin, Mich. (49688) 250/D4
Tustin, Wis. (†54940) 317/J7
Tustumena (lake), Alaska 196/C1
Tutak, Turkey 63/K3
Tutamoe (range), N. Zealand 100/D1
Tutayev, U.S.S.R. 52/E3
Tuthill, S. Dak. (57574) 298/G7
Tuticorin, India 68/D7
Tutóia, Brazil 132/F3
Tutrakan, Bulgaria 45/H4
Tuttle, Idaho (†83332) 220/D7
Tuttle, N. Dak. (58488) 282/L5
Tuttle, Okla. (73089) 288/L4
Tuttle Creek (lake), Kansas 232/F2
Tuttlingen, W. Germany 22/C5
Tutuila (isl.), Amer. Samoa 87/J7
Tutuila (isl.), Amer. Samoa 86/N9
Tutwiler, Miss. (38963) 256/D2
Tuun (mt.), N. Korea 81/C3
Tuvalu 2/T6
Tuvalu 87/H6
Tuvinian A.S.S.R., U.S.S.R. 48/K4
Tuwaiq, Jebel (range), Saudi Arabia 59/E5
Tuxedo, Md. (†20780) 245/G5
Tuxedo Park, N.Y. (10987) 276/M8
Tuxford, Sask. 181/F5
Tuxpan, Jalisco, Mexico 150/H7
Tuxpan, Nayarit, Mexico 150/G6
Tuxpan de Rodríguez Cano, Mexico 150/L6
Tuxtepec, Mexico 150/L7
Tuxtla Gutiérrez, Mexico 146/J8
Tuxtla Gutiérrez, Mexico 150/N8
Túy, Spain 33/B1
Tuy (riv.), Venezuela 124/E2
Tuya (riv.), Br. Col. 184/K2
Tuyen Quang, Vietnam 72/E2
Tuy Hoa, Vietnam 72/F4
Tuymazy, U.S.S.R. 52/H4
Tuysarkan, Iran 66/F3
Tuyün (Duyun), China 77/G6
Tuz (lake), Turkey 63/E3
Tuz (lake), Turkey 59/B2
Tuzigoot Nat'l Mon., Ariz. 198/D4
Tuz Khurmatu, Iraq 66/D3
Tuzla, Yugoslavia 7/F4
Tuzla, Yugoslavia 45/D3
Tuzluca, Turkey 63/K3
Tuzlukçu, Turkey 63/D3
Tvedestrand, Norway 18/F7
Tver (Kalinin), U.S.S.R. 52/E3
Tversted, Denmark 21/D2
Twain, Calif. (95984) 204/D4
Twain Harte, Calif. (95383) 204/E6
Tway, Sask. 181/F3
Tweed (riv.), England 13/E2
Tweed, Ontario 177/G3
Tweed (riv.), Scotland 15/F5
Tweed (riv.), Scotland 10/E3
Tweed Heads, N.S. Wales 97/G1
Tweedie, Alberta 182/E2
Tweedside, New Bruns. 170/C3
Tweedsmuir, Sask. 181/F2
Tweedsmuir, Scotland 15/E5
Tweedsmuir Prov. Park, Br. Col. 184/D3
Twello, Netherlands 27/J4
Twelve Mile, Ind. (46988) 227/E3
Twelvemile (lake), Sask. 181/E6
Twelve Mile (creek), Utah 304/C2
Twelve Pins (mt.), Ireland 17/B4

Twelvepole (creek), W. Va. 312/A6
Twentynine Palms, Calif. (92277) 204/K9
Twentynine Palms Marine Base, Calif. 204/J9
Twig, Minn. (55791) 255/F4
Twiggs (co.), Georgia 217/F5
Twila, Ky. (†40873) 237/P7
Twillingate, Newf. 166/C4
Twin (lakes), Conn. 210/B1
Twin (falls), Idaho 220/D7
Twin (lakes), Maine 243/F4
Twin (lakes), Wash. 310/G2
Twin Bridges, Mont. (59754) 262/D5
Twin Brooks, S. Dak. (57269) 298/R3
Twin City, Georgia (30471) 217/H5
Twin Falls (co.), Idaho 220/D7
Twin Falls, Idaho (83301) 220/D7
Twin Falls, Idaho 188/C2
Twin Falls, Idaho 146/G5
Twin Falls, Newf. 166/B3
Twin Hills, Alaska (†99576) 196/F3
Twining, Mich. (48766) 250/F4
Twin Lake, Mich. (49457) 250/C5
Twin Lakes, Calif. (†95060) 204/K4
Twin Lakes, Colo. (81251) 208/G4
Twin Lakes (res.), Colo. 208/G4
Twin Lakes, Minn. (56089) 255/E7
Twin Lakes, Wis. (53181) 317/K11
Twin Mountain, N.H. (03595) 268/D3
Twin Oaks, Mo. (†63088) 261/N3
Twin Peaks, Calif. (92391) 204/H9
Twin Peaks (mt.), Idaho 220/D5
Twin Rocks, Oreg. (†97136) 291/C2
Twin Rocks, Pa. (15960) 294/E4
Twinsburg, Ohio (44087) 284/J10
Twin Sisters (mt.), Wash. 310/D2
Twin Valley, Minn. (56584) 255/B3
Twisp, Wash. (98856) 310/E2
Twisp (pass), Wash. 310/E2
Twisp (riv.), Wash. 310/E2
Twitchell (res.), Calif. 204/E9
Two Arm (bay), Alaska 196/C2
Two Butte (creek), Colo. 208/N7
Two Buttes, Colo. (81084) 208/P7
Two Buttes (res.), Colo. 208/O7
Twodot, Mont. (59085) 262/F4
Twofold (bay), N.S. Wales 97/F5
Two Harbors, Minn. (55616) 255/G3
Two Hearted (riv.), Mich. 250/D2
Two Hills, Alberta 182/E3
Two Rivers (riv.), Minn. 255/A1
Two Rivers (res.), N. Mex. 274/E5
Two Rivers, Wis. (54241) 317/M7
Two Water (creek), Utah 304/E4
Twynholm, Scotland 15/D6
Tyaskin, Md. (21865) 245/P7
Tybee Island, Georgia (31328) 217/L6
Tybee Roads (chan.), S.C. 296/F7
Tychy, Poland 47/B4
Tye, Texas (79563) 303/E5
Tye River, Va. (22975) 307/L5
Tygart (lake), W. Va. 312/G4
Tygart Valley (riv.), W. Va. 312/F5
Tyger (riv.), S.C. 296/D2
Tygh Valley, Oreg. (97063) 291/F2
Tyler, Ala. (36785) 195/E6
Tyler (lake), Conn. 210/B1
Tyler, Minn. (56178) 255/B6
Tyler, Mo. (†63877) 261/N10
Tyler, N. Dak. (†58075) 282/S7
Tyler, Pa. (†15849) 294/F3
Tyler (co.), Texas 303/K7
Tyler, Texas 188/H4
Tyler, Texas (*75701) 303/J5
Tyler, Wash. (†99004) 310/H3
Tyler (co.), W. Va. 312/E4
Tylersburg, Pa. (16361) 294/D3
Tylersville, Pa. (17773) 294/G4
Tylertown, Miss. (39667) 256/D8
Tylerville, Conn. (†06438) 210/F3
Tym (riv.), U.S.S.R. 48/J3
Tymovskoye, U.S.S.R. 48/P4
Týn, Czech. 41/C2
Tynagh, Ireland 17/E5
Tynan, Texas (78391) 303/G9
Tynda, U.S.S.R. 48/N4
Tyndall, Manitoba 179/F4
Tyndall, S. Dak. (57066) 298/O8
Tyndall A.F.B., Fla. 212/C6
Tyndrum, Scotland 15/D4
Tyne (riv.), England 13/F3
Tyne (riv.), England 10/F3
Tyne (riv.), Scotland 15/F5
Tyne and Wear (co.), England 13/H3
Tynemouth, England 13/J3
Tynemouth, England 10/F3
Tyner, Ind. (46572) 227/E2
Tyner, Ky. (40486) 237/O6
Tyner, N.C. (27980) 281/R2
Tyner, Sask. 181/C4
Tyne Valley, Pr. Edward I. 168/E2
Tyngsboro○, Mass. (01879) 249/J2
Tynset, Norway 18/G5
Tyntynder South, Victoria 97/B4
Tyonek, Alaska (99682) 196/B1
Tyra (cays), Nicaragua 154/F4
Tyre (Sur), Lebanon 63/F6
Tyrifjord (lake), Norway 18/C3
Tyringham○, Mass. (01264) 249/A4
Tyrnyauz, U.S.S.R. 52/F6
Tyro, Kansas (67364) 232/G4
Tyro, Miss. (†38668) 256/E1
Tyro, Va. (22976) 307/K5
Tyrol (Tirol) (prov.), Austria 41/A3
Tyrone, Colo. (†81059) 208/L8
Tyrone, Georgia (30290) 217/C4
Tyrone, Ky. (†40342) 237/M4
Tyrone, Mo. (†65564) 261/J8
Tyrone, N. Mex. (88065) 274/A6
Tyrone, Okla. (73951) 288/D1
Tyrone, Pa. (16686) 294/F4
Tyronza, Ark. (72386) 202/K3
Tyronza (riv.), Ark. 202/K2
Tyrrell (co.), N.C. 281/S3
Tyrrell (lake), Victoria 97/B4
Tyrrellspass, Ireland 17/G5
Tyrrhenian (sea) 7/F4
Tyrrhenian (sea), Italy 34/C4
Tysnes, Norway 18/D6
Tyson, Vt. (†05149) 268/B5
Tyson Wash (dry riv.), Ariz. 198/A5
Ty Ty, Georgia (31795) 217/E8
Tyumen, U.S.S.R. 54/H4
Tyumen', U.S.S.R. 48/G4
Tyung (riv.), U.S.S.R. 48/M3
Tyvan, Sask. 181/H5
Tywyn, Wales 13/C5
Tywyn, Wales 10/D4
Tzaneen, S. Africa 118/E4
Tzekung (Zigong), China 77/F6
Tzepo (Zibo), China 77/J4
Tzucabab, Mexico 150/P7

U

Uahuka (isl.), Fr. Poly. 87/N6
Uanda, Queensland 95/C4
Uanle Uen, Somalia 115/H3
Uanle Uen, Somalia 102/G4
Uapou (isl.), Fr. Poly. 87/M6
Uatumã (riv.), Brazil 132/B3
Uaupés (riv.), Brazil 132/G9
Ub, Yugoslavia 45/E3
Ubá, Brazil 135/E2
Ubá, Brazil 132/F8
Übach-Palenberg, W. Germany 22/B3
Ubaíra, Brazil 132/G6
Ubaitaba, Brazil 132/G6
Ubaiyidh, Wadi (dry riv.), Iraq 66/B5
Ubangi (riv.) 102/D4
Ubangi (riv.), Cent. Afr. Rep. 115/C3
Ubangi (riv.), Congo 115/C3
Ubangi (riv.), Zaire 115/C3
Ubari, Libya 102/D2
Ubari, Libya 111/B2
Ubaté, Colombia 126/D5
Ubatuba, Brazil 135/D3
Ubay, Philippines 82/E5
Ube, Japan 81/E6
Úbeda, Spain 33/E3
Uberaba (lag.), Bolivia 136/G5
Uberaba, Brazil 120/E4
Uberaba, Brazil 132/D7
Uberaba, Brazil 135/C1
Uberlândia, Brazil 120/E4
Uberlândia, Brazil 132/E7
Überlingen, W. Germany 22/C5
Ubina, Bolivia 136/B7
Ubinas, Peru 128/G11
Ubly, Mich. (48475) 250/G5
Ubombo, S. Africa 118/E5
Ubon, Thailand 54/M8
Ubon, Thailand 72/E4
Ubrique, Spain 33/D4
Ubundu, Zaire 115/E4
Ucayali (dept.), Peru 128/E6
Ucayali (riv.), Peru 2/F6
Ucayali (riv.), Peru 120/B3
Ucayali (riv.), Peru 128/F5
Uccle, Belgium 27/B9
Uch, Pakistan 68/B3
Uchaly, U.S.S.R. 7/K3
Uchaly, U.S.S.R. 52/J4
Ucharonidge, North. Terr. 93/D4
Uchee, Ala. (†36858) 195/H6
Uchiura (bay), Japan 81/K2
Uchiza, Peru 128/D7
Uch Turfan (Wushi), China 77/A3
Uckange, France 28/G3
Ücker (riv.), E. Germany 22/E2
Uckfield, England 10/G5
Uckfield, England 13/H7
Ucluelet, Br. Col. 184/E6
Ucon, Idaho (83454) 220/F6
Ucross, Wyo. (†82835) 319/F1
Ucumasi, Bolivia 136/B6
Uda (riv.), U.S.S.R. 48/O4
Udaipur, India 68/C4
Udall, Kansas (67146) 232/E4
Udaypur, India 54/J7
Uddevalla, Sweden 18/G7
Uddingston, Scotland 15/B2
Uddjaur (lake), Sweden 18/L4
Udell, Iowa (52593) 229/H7
Uden, Netherlands 27/H5
Udhampur, India 68/D2
Udine (prov.), Italy 34/D1
Udine, Italy 34/D2
Udine, Italy 7/F4
Udipi, India 68/C6
Udmurt A.S.S.R., U.S.S.R. 48/F4
Udmurt A.S.S.R., U.S.S.R. 52/H3
Udon Thani, Thailand 54/M8
Udon Thani, Thailand 72/D3
Udora, Ontario 177/E3
Ueckermünde, E. Germany 22/F2
Ueda, Japan 81/J5
Uehling, Nebr. (68063) 264/H3
Uele (riv.) 102/E4
Uele (riv.), Zaire 115/E3
Uelen, U.S.S.R. 54/V3
Uelen, U.S.S.R. 4/C18
Uelen, U.S.S.R. 48/T3
Uelzen, W. Germany 22/D2
Uen (isl.), New Caled. 86/H5
Uetendorf, Switzerland 39/E3
Uetersen, W. Germany 22/C2
Ufa, U.S.S.R. 2/M3
Ufa, U.S.S.R. 7/K3
Ufa, U.S.S.R. 48/F4
Ufa, U.S.S.R. 52/J4
Ufa (riv.), U.S.S.R. 52/J3
Ugab (riv.), Namibia 118/A4
Uganda 2/L5
Uganda 102/F4
UGANDA 115/F3
Ugashik, Alaska (†99649) 196/G3
Ugashik (lakes), Alaska 196/G3
Ugie (riv.), Scotland 15/G3
Ugijar, Spain 33/E4
Ugjoktok (bay), Newf. 166/B2
Uglegorsk, U.S.S.R. 48/P5
Uglich, U.S.S.R. 52/E3
Ugo, Japan 81/K4
Ugod, Hungary 41/D3
Uherské Hradiště, Czech. 41/D2
Uherský Brod, Czech. 41/D2
Úhlava (riv.), Czech. 41/B2
Uhlířské Janovice, Czech. 41/C2
Uhrichsville, Ohio (44683) 284/H5
Uig, Scotland 10/C2
Uig, Highland, Scotland 15/B3
Uig, W. Isles, Scotland 15/A2
Uige (dist.), Angola 115/B5
Uige, Angola 115/C5
Uiju, N. Korea 81/B3
Uinkaret (plat.), Ariz. 198/B2
Uinta (mts.), Utah 304/D3
Uinta (riv.), Utah 304/D3
Uinta (co.), Wyo. 319/B4
Uintah (co.), Utah 304/E3
Uintah, Utah (†84401) 304/C2
Uintah and Ouray Ind. Res., Utah 304/D3
Ŭisŏng, S. Korea 81/D5
Uitenhage, S. Africa 102/E8
Uitenhage, S. Africa 118/C6
Uithoorn, Netherlands 27/F4
Uithuizen, Netherlands 27/K2
Uitkijk, Suriname 131/D3
Uivak (cape), Newf. 166/B2
Ujelang (atoll), Marshall Is. 87/F5
Újfehértó, Hungary 41/F3
Uji, Honduras 154/F3
Uji, Japan 81/J7
Ujiji (Kigoma-Ujiji), Tanz. 115/E4
Ujjain, India 68/D4
Újpest, Hungary 41/E3
Újszász, Hungary 41/F2
Ujung Pandang, Indonesia 54/N10
Ujung Pandang, Indonesia 85/F7
Ukasiksalik (isl.), Newf. 166/B2
Ukhta, U.S.S.R. 48/F3
Ukiah, Calif. (95482) 204/B4
Ukiah, Oreg. (97880) 291/J2
Ukkel (Uccle), Belgium 27/B9
Ukmerge, U.S.S.R. 53/C3
Ukmerge, U.S.S.R. 52/C3
Ukrainian S.S.R., U.S.S.R. 7/G4
Ukrainian S.S.R., U.S.S.R. 48/C5
Ukrainian S.S.R., U.S.S.R. 52/D5
Ula, Turkey 63/C4
Ulaanbaatar (Ulan Bator) (cap.), Mongolia 77/G2
Ulaanbaatar (cap.), Mongolia 54/M5
Ulaanbaatar (cap.), Mongolia 2/Q3
Ulaangom (Ulangom), Mongolia 77/D2
Ulaangom, Mongolia 54/L5
Ulah, N.C. (†27203) 281/K3
Ulak (isl.), Alaska 196/K4
Ulan, China 77/E4
Ulanhot (Horquin Youyi Qianqi), China 77/K2
Ulan-Ude, U.S.S.R. 54/M4
Ulan-Ude, U.S.S.R. 2/Q3
Ulan-Ude, U.S.S.R. 48/L4
Ulapes, Argentina 143/C3
Ulaş, Turkey 63/G3
Ulchin, S. Korea 81/D5
Ulcinj, Yugoslavia 45/D5
Uldum, Denmark 21/C6
Ulegei (Ölgiy), Mongolia 77/C2
Ulen, Ind. (†46052) 227/E4
Ulen, Minn. (56585) 255/B3
Uler, W. Va. (25282) 312/D5
Ulfborg, Denmark 21/B5
Ulhasnagar, India 68/C5
Uliastay (Jibhalanta), Mongolia 77/E2
Uliastay, Mongolia 54/L5
Ulindi (riv.), Zaire 115/E4
Ulithi (atoll), Micronesia 87/D4
Ulla (riv.), Spain 33/B1
Ulladulla, N.S. Wales 97/F4
Ullapool, Scotland 10/D2
Ullapool, Scotland 15/C3
Ulla Ulla, Bolivia 136/A4
Ulldecona, Spain 33/G2
Ullensvang, Norway 18/E6
Ullin, Ill. (62992) 222/D6
Ulloma, Bolivia 136/A5
Ullŭng (isl.), S. Korea 81/E5
Ulm, Ark. (72170) 202/H4
Ulm, Mont. (59485) 262/E3
Ulm, W. Germany 22/C4
Ulm, Wyo. (†82835) 319/F1
Ulman, Mo. (65083) 261/H6
Ulmarra, N.S. Wales 97/G1
Ulmer, Iowa (51464) 229/D4
Ulmer, S.C. (29849) 296/E5
Ulongue, Mozambique 118/E2
Ulricehamn, Sweden 18/H8
Ulrichen, Switzerland 39/F3
Ulriksfors, Sweden 18/K5
Ulrum, Netherlands 27/J2
Ulsan, S. Korea 81/D6
Ulster (part) (prov.), Ireland 17/G2
Ulster (trad. prov.), Ireland 17/K8
Ulster (co.), N.Y. 276/M7
Ulster (part) (prov.), N. Ireland 17/G2
Ulster, Pa. (18850) 294/J2
Ulster Spring, Jamaica 158/H6
Última Esperanza (sound), Chile 138/E9
Ulúa (riv.), Honduras 154/D3
Ulubat (lake), Turkey 63/C2
Ulubey, Turkey 63/C3
Uluborlu, Turkey 63/D3
Uludağ (mt.), Turkey 63/C3
Uludere, Turkey 63/K4
Ulugan (bay), Philippines 82/B5
Ulughchat (Wuqia), China 77/A4
Ulukışla, Turkey 63/F4
Ulumalu, Hawaii (†96708) 218/K2
Ulu Muztag (mt.), China 77/C4
Ulundi, S. Africa 118/E5
Ulungur He (riv.), China 77/C2
Ulungur Hu (lake), China 77/C2
Ulupalakua, Hawaii (†96790) 218/J2
Ulus, Turkey 63/E2
Ulutau (mts.), U.S.S.R. /G5
Ulu Tiram, Malaysia 72/F5
Ulva (isl.), Scotland 15/B4
Ulverston, England 13/D3
Ulverston, England 10/E3
Ulverstone, Tasmania 99/C3
Ulvik, Norway 18/E6
Ulvila, Finland 18/N6
Ul'yanovsk, U.S.S.R. 7/J3
Ul'yanovsk, U.S.S.R. 48/E4
Ul'yanovsk, U.S.S.R. 52/G4
Ulysses, Kansas (67880) 232/A4
Ulysses, Ky. (†41232) 237/R5
Ulysses, Nebr. (68669) 264/G3
Ulysses, Pa. (16948) 294/G2
Umag, Yugoslavia 45/A3
Umala, Bolivia 136/B5
Umán, Mexico 150/P6
Uman', U.S.S.R. 52/D5
Umanun (pt.), Philippines 82/F6
Umapine, Oreg. (97881) 291/J2
Umarkot, Pakistan 59/J4
Umatilla, Fla. (32784) 212/E3
Umatilla (co.), Oreg. 291/J2
Umatilla, Oreg. (97882) 291/H2
Umatilla (lake), Oreg. 291/G2
Umatilla (riv.), Oreg. 291/H2
Umatilla (lake), Wash. 310/E5
Umatilla Army Depot, Oreg. 291/H2
Umatilla Ind. Res., Oreg. 291/J2
Umba, U.S.S.R. 52/D1
Umbagog (lake), Maine 243/A6
Umbagog (lake), N.H. 268/E2
Umbakumba, North. Terr. 93/E3
Umbarger, Texas (79091) 303/B3
Umbeara, North. Terr. 93/C8
Umbertide, Italy 34/D3
Umboi (isl.), Papua N.G. 86/A2
Umbrail (peak), Switzerland 39/K3
Umbria, Colombia 126/B7
Umbria (reg.), Italy 34/D3
Umcalcus (lake), Maine 243/G3
Ume (riv.), Sweden 7/F2
Umeå, Sweden 7/F2
Umeå, Sweden 18/M5
Umeälv (riv.), Sweden 18/L4
Umiakovik (lake), Newf. 166/B2
Umiat, Alaska (†99701) 196/H1
Umikoa, Hawaii (†96776) 218/H4
Um Jauza, Jordan 65/D3
Umm al Qaiwain, U.A.E. 59/G4
Umm el Abid, Libya 111/C2
Umm el Fahm, Israel 65/C2
Umm Hajar, Ethiopia 111/G5
Umm Keddada, Sudan 111/E5
Umm Lajj, Saudi Arabia 59/C4
Umm Qasr, Iraq 66/E5
Umm Ruwaba, Sudan 111/F5
Umm Ruwaba, Sudan 102/F3
Umm Ruwaba, Sudan 59/B7
Umm Sa'id, Qatar 59/F5
Umnak (isl.), Alaska 196/E4
Umnak (passage), Alaska 196/E4
Umnak (isl.), U.S. 4/D18
Umpire, Ark. (71971) 202/B5
Umpqua, Oreg. (97486) 291/D4
Umpqua (riv.), Oreg. 291/D4
Umrer, India 68/D4
Umsaskis (lake), Maine 243/E2
Umtali, Zimbabwe 102/F6
Umtali, Zimbabwe 118/E3
Umtata, S. Africa 118/D6
Umtata (cap.), Transkei, S. Africa 102/E8
Umurbey, Turkey 63/C6
Umvukwe (range), Zimbabwe 118/E3
Umvuma, Zimbabwe 118/D3
Umzimbuvu, S. Africa 118/D6
Umzinto, S. Africa 118/E6
Una (mt.), N. Zealand 100/D5
Una (riv.), Yugoslavia 45/C3
Unadilla, Georgia (31091) 217/E6
Unadilla, Nebr. (68454) 264/H4
Unadilla, N.Y. (13849) 276/K6
Unadilla (riv.), N.Y. 276/K5
Unaí, Brazil 132/E7
Unaka, N.C. (28908) 281/A4
Unaka (mts.), N.C. 281/E2
Unaka (mts.), Tenn. 237/S8
Unalakleet, Alaska (99684) 196/G2
Unalakleet, Alaska 188/C5
Unalaska, Alaska 188/C6
Unalaska, Alaska (99685) 196/E4
Unalaska (isl.), Alaska 196/E4
Unalaska (isl.), U.S. 4/D18
Unare (riv.), Venezuela 124/F3
Uncas, Okla. (†74601) 288/M1
Uncastillo, Spain 33/F1
Uncasville, Conn. (06382) 210/G3
Uncertain, Texas (†75661) 303/K5
Uncía, Bolivia 136/B6
Uncompahgre (peak), Colo. 208/E6
Uncompahgre (plat.), Colo. 208/B5
Uncompahgre (riv.), Colo. 208/D5
Underbool, Victoria 97/A4
Underhill, Manitoba 179/B5
Underhill○, Vt. (05489) 268/B2
Underhill, Wis. (54176) 317/K6
Underhill Center, Vt. (05490) 268/B2
Underwood, Ind. (47177) 227/F7
Underwood, Iowa (51576) 229/B6
Underwood, Minn. (56586) 255/C4
Underwood, N. Dak. (58576) 282/H5
Underwood, Ontario 177/C3
Underwood, Wash. (98651) 310/D5
Undu (pt.), Fiji 86/R10
Undzha (riv.), U.S.S.R. 52/F3
Unecha, U.S.S.R. 52/D4
Uneeda, W. Va. (25205) 312/C6
Unga (isl.), Alaska 196/F3
Ungalik, Alaska (†99684) 196/F2
Ungarie, N.S. Wales 97/D3
Ungava (bay), Canada 146/M4
Ungava (bay), N.W.T. 162/K4
Ungava (bay), N.W. Terrs. 187/M4
Ungava (bay), Québec 174/F1
Ungava (pen.), Que. 146/L3
Ungava (pen.), Que. 162/J3
Ungava (pen.), Québec 174/E1
Ungeny, U.S.S.R. 52/C5
Unger, W. Va. (25447) 312/K4
Unggi, N. Korea 81/E2
União, Brazil 132/F4
União da Vitória, Brazil 132/D9
União dos Palmares, Brazil 132/H5
Unicoi (mts.), N.C. 281/A4
Unicoi (co.), Tenn. 237/S8
Unicoi, Tenn. (37692) 237/S8
Unicoi (mts.), Tenn. 237/N10
Uničov, Czech. 41/D2
Unimak (isl.), Alaska 188/C6
Unimak (bight), Alaska 196/F4
Unimak (isl.), Alaska 196/E4
Unimak (passage), Alaska 196/F4
Unimak (isl.), U.S. 4/D18
Unini, Peru 128/F8
Union, Ala. (†35462) 195/C5
Union (mt.), Ariz. 198/C4
Union (co.), Ark. 202/E7
Union, Ark. (†72576) 202/G1
Union○, Conn. (†06076) 210/G1
Union (co.), Fla. 212/D1
Union (co.), Georgia 217/E1
Union, Grenada 161/D8
Union (co.), Ill. 222/D6
Union, Ill. (60180) 222/E1
Union (co.), Ind. 227/H5
Union, Ind. (†47540) 227/C8
Union (co.), Iowa 229/E7
Union, Iowa (50258) 229/G4
Union (co.), Ky. 237/F5
Union, Ky. (41091) 237/M3
Union (par.), La. 238/F1
Union, La. (†70723) 238/L3
Union, Maine (04862) 243/E7
Union○, Maine (04862) 243/E7
Union, West Branch (riv.), Maine 243/G6
Union (co.), Miss. 256/F2
Union, Miss. (39365) 256/F5
Union, Mo. (63084) 261/L6
Union, Nebr. (68455) 264/J4
Union, N.H. (03887) 268/E5
Union (co.), N.J. 273/E2
Union○, N.J. (07083) 273/A2
Union (lake), N.J. 273/C5
Union (co.), N. Mex. 274/F2
Union (co.), N.C. 281/H4
Union, N. Dak. (58279) 282/O2
Union (co.), Ohio 284/D5
Union, Ohio (45322) 284/B6
Union, Ontario 177/C5
Union (co.), Oreg. 291/J2
Union, Oreg. (97883) 291/K2
Unión, Paraguay 144/D4
Union (co.), Pa. 294/H4
Union (isl.), St. Vin. & Grens. 156/G4
Union (co.), S.C. 296/D2
Union, S.C. (29379) 296/D2
Union (co.), S. Dak. 298/R8
Union (co.), Tenn. 237/O8
Unión, Uruguay 145/B7
Union, Wash. (98592) 310/B3
Union (lake), Wash. 310/B2
Union, W. Va. (24983) 312/E7
Union Bay, Br. Col. 184/H2
Union Beach, N.J. (07735) 273/E3
Union Bridge, Md. (21791) 245/K2
Union Center, S. Dak. (57787) 298/D4
Union Center, Wis. (53962) 317/F8
Union Church, Miss. (39668) 256/C7
Union City, Calif. (94587) 204/K2
Union City, Conn. (†06770) 210/C2
Union City, Georgia (30291) 217/J2
Union City, Ind. (47390) 227/H4
Union City, Mich. (49094) 250/D6
Union City, N.J. (07087) 273/C2
Union City, Ohio (†47390) 284/A5
Union City, Okla. (73090) 288/L4
Union City, Pa. (16438) 294/C2
Union City, Tenn. (38261) 237/C8
Union Creek, Oreg. (†97536) 291/E5
Uniondale, Ind. (46791) 227/G3
Uniondale, N.Y. (11553) 276/R7
Union Dale, Pa. (18470) 294/M2
Unión de Reyes, Cuba 158/C1
Unión de Tula, Mexico 150/G7
Union Furnace, Ohio (43158) 284/F7
Union Gap, Wash. (98903) 310/E4
Union Grove, Ala. (35175) 195/E2
Union Grove, N.C. (28689) 281/H2
Union Grove, Wis. (53182) 317/L3
Union Hall, Va. (24176) 307/J6
Unión Hidalgo, Mexico 150/M8
Union Hill, Ill. (60969) 222/E2
Union Hill, N.Y. (14563) 276/F4
Union Level, Va. (23973) 307/M7
Union Mills, Ind. (46382) 227/D2
Union Mills, Md. (†21157) 245/K2
Union Mills, N.C. (28167) 281/F3
Union of Soviet Socialist Republics 2/L2
Union of Soviet Socialist Republics 4/C2
Union of Soviet Socialist Republics 54/L3
Union of Soviet Socialist Republics 7/H2
UNION OF SOVIET SOCIALIST REPUBLICS 48
UNION OF SOVIET SOCIALIST REPUBLICS, EUROPEAN 52
Union Pier, Mich. (49129) 250/C7
Union Point, Georgia (30669) 217/F3
Unionport, Ind. (†47340) 227/G4
Unionport, Ohio (43966) 284/J5
Union Springs, Ala. (36089) 195/G6
Union Springs, N.Y. (13160) 276/G5
Union Star, Ky. (40171) 237/H5
Union Star, Mo. (64494) 261/C2
Uniontown, Ala. (36786) 195/D6
Uniontown, Ark. (72955) 202/B2
Uniontown, Ind. (†47515) 227/D
Uniontown, Kansas (66779) 232/G4
Uniontown, Ky. (42461) 237/F5
Uniontown, Md. (21157) 245/K2
Uniontown, Mo. (63783) 261/N7
Uniontown, Ohio (44685) 284/H4
Uniontown, Pa. (15401) 294/C6
Uniontown, Wash. (99179) 310/H4
Union Village, Vt. (†05075) 268/C4
Unionville, Conn. (06085) 210/D1
Unionville, Georgia (†31794) 217/F8
Unionville, Ill. (†61270) 222/E6
Unionville, Ind. (47468) 227/E6
Unionville, Iowa (52594) 229/H7
Unionville, Maine (†04622) 243/H6
Unionville, Md. (21792) 245/K3
Unionville, Mich. (48767) 250/F5
Unionville, Mo. (63565) 261/G2
Unionville, Nev. (†89418) 266/C2
Unionville, N.Y. (10988) 276/L8
Unionville, N.C. (†28110) 281/J4
Unionville, Ohio (44088) 284/J2
Unionville (Fleming), Pa. (19375) 294/G4
Unionville, Tenn. (37180) 237/H9
Unionville, Va. (22567) 307/N4
Unionville Center, Ohio (43077) 284/D5
Uniopolis, Ohio (45888) 284/B4
United, Pa. (15689) 294/D5
United Arab Emirates 2/M4
United Arab Emirates 54/G7
UNITED ARAB EMIRATES 59/F5
UNITED KINGDOM 10
United Kingdom 2/J3
United Kingdom 7/D3
United States 2/D4
United States 146/H5
United States 4/C17
UNITED STATES 188
U.S. Capitol, D.C. 245/F5
U.S.S. Arizona Memorial, Hawaii 218/B3
U.S. Nav. Air Sta., Virgin Is. (U.S.) 161/A4
U.S. Naval Base, Va. 307/R7
Unity○, Maine (04988) 243/E6
Unity, Md. (†20729) 245/K4
Unity, Mo. (64063) 261/R6
Unity○, N.H. (†03743) 268/C5
Unity, Ohio (†44413) 284/J4
Unity, Oreg. (97884) 291/J3
Unity, Sask. 181/B3
Unity, Wis. (54488) 317/F6
Unityville, Pa. (17774) 294/K3
Unityville, S. Dak. (†57058) 298/P6
Universal, Ind. (47884) 227/C5
Universal City, Texas (78148) 303/K10
University, Fla. (33620) 212/C2
University, N.C. (†27701) 281/L2
University City, Mo. (63130) 261/P3
University City, Mo. 188/H3
University Heights, Iowa (†52240) 229/K5
University Heights, Ohio (44118) 284/H9
University Park, Iowa (52595) 229/H6
University Park, Md. (†20740) 245/F4
University Park, N. Mex. (88003) 274/C6
University Park, Texas (†75205) 303/G2
Unley, S. Australia 88/D8
Unley, S. Australia 94/B8
Unnao, India 68/E3
Uno, Manitoba 179/B4
Unsan, N. Korea 81/C4
Unst (isl.), Scotland 15/G2
Unst (isl.), Scotland 10/H1
Unstrut (riv.), E. Germany 22/D3
Unterägeri, Switzerland 39/G2
Unteriberg, Switzerland 39/G2
Unterkulm, Switzerland 39/F2
Untermann (mt.), Utah 304/E3
Untersee (lake), Switzerland 39/H1
Unterseen, Switzerland 39/E3
Untervaz, Switzerland 39/H3
Unterwalden (reg.), Switzerland 39/F3
Unuk (riv.), Alaska 196/N2
Unuk (riv.), Br. Col. 184/B2
Ünye, Turkey 59/C1
Ünye, Turkey 63/G2
Unzen (mt.), Japan 81/D7
Unzen-Amakusa National Park, Japan 81/D7
Uozu, Japan 81/H5
Upalco, Utah (†84007) 304/D3
Upata, Venezuela 124/G3
Upemba (lake), Zaire 115/E5
Upemba Nat'l Park, Zaire 115/E5
Upernavik, Greenl. 4/B12
Uphall, Scotland 15/C1
Upham, New Bruns. 170/E3
Upham, N. Dak. (58789) 282/J2
Upía (riv.), Colombia 126/D5
Úpice, Czech. 41/C1
Upington, S. Africa 102/E7
Upington, S. Africa 118/C5
Upland, Calif. (91786) 204/E10
Upland, Ind. (46989) 227/F4
Upland, Kansas (†67431) 232/E2
Upland, Nebr. (68981) 264/F4
Upland, Pa. (†19013) 294/L7
Upolu (pt.), Hawaii 218/G3
Upolu (isl.), W. Samoa 87/J7
Upolu (isl.), W. Samoa 86/M8
Upolu (isl.), W. Samoa 86/M8
Upper Alkali (lake), Calif. 204/E2
Upper Amherst Cove, Newf. 166/D2
Upper Ammonoosuc (riv.), N.H. 268/E2
Upper Arlington, Ohio (43221) 284/D6
Upper Arrow (lake), Br. Col. 184/H5
Upper Austria (prov.), Austria 41/B2
Upper Black Eddy, Pa. (18972) 294/N4
Upper Blackville, New Bruns. 170/E2
Upper Buctouche, New Bruns. 170/F2
Upper Chateaugay (lake), N.Y. 276/M1
Upperco, Md. (21155) 245/L2
Upper Dam, Maine (†04293) 243/B6
Upper Darby○, Pa. (*19082) 294/M6
Upper Des Lacs (lake), N. Dak. 282/F2
Upper Engadine (valley), Switzerland 39/J4
Upper Fairmount, Md. (21867) 245/P8

Upper Falls, Md. (21156) 245/N3
Upper Fraser, Br. Col. 184/G3
Upper Frenchville, Maine (04784) 243/G1
Upper Gagetown, New Bruns. 170/D3
Upperglade, W. Va. (26266) 312/F6
Upper Greenwood Lake, N.J. (†07421) 273/E1
Upper Hainesville, New Bruns. 170/C2
Upper Horton, N.S. Wales 97/F2
Upper Hutt, N. Zealand 100/B2
Upper Iowa (riv.), Iowa 229/K2
Upper Island Cove, Newf. 166/D2
Upper Jay, N.Y. (12987) 276/N2
Upper Kennetcook, Nova Scotia 168/E3
Upper Kent, New Bruns. 170/C2
Upper Klamath (lake), Oreg. 188/B2
Upper Klamath (lake), Oreg. 291/E5
Upper Lake, Calif. (95485) 204/C4
Upper Liard, Yukon 187/F3
Upper Lough Erne (lake), N. Ireland 17/F3
Upper Lough Erne (lake), N. Ireland 10/C3
Upper Macopin, N.J. (†07435) 273/E1
Upper Manzanilla, Trin. & Tob. 161/B10
Upper Marlboro, Md. (20870) 245/M5
Upper Matecumbe (key), Fla. 212/F7
Upper Maugerville, New Bruns. 170/D3
Upper Mills, New Bruns. 170/C3
Upper Musquodoboit, Nova Scotia 168/F3
Upper New York (bay), N.J. 273/B2
Upper Nile (prov.), Sudan 111/F6
Upper Rawdon, Nova Scotia 168/E3
Upper Red (lake), 255/D2
Upper Red Rock (lake), Mont. 262/E6
Upper Rexton, New Bruns. 170/F3
Upper Saddle River, N.J. (†07458) 273/B1
Upper Saint Claire○, Pa. (15241) 294/B7
Upper Sandusky, Ohio (43351) 284/D4
Upper Saranac (lake), N.Y. 276/M2
Upper Seven Sisters, Manitoba 179/G4
Upper Sheikh, Somalia 115/J2
Upper Sheila, New Bruns. 170/E1
Upper South River, Nova Scotia 168/G3
Upper Stepney, Conn. (†06468) 210/B3
Upper Stewiacke, Nova Scotia 168/F3
Upper Strasburg, Pa. (17265) 294/G5
Upper Tract, W. Va. (26866) 312/H5
Upper Tygart, Ky. (41178) 237/P4
Upper Vaughan, Nova Scotia 168/D4
Upperville, Va. (22176) 307/N2
Upper Volta 2/J5
Upper Volta 102/B3
UPPER VOLTA 106/D6
Upper Woodstock, New Bruns. 170/C2
Uppingham, England 13/G5
Uppsala (co.), Sweden 18/K7
Uppsala, Sweden 7/F3
Uppsala, Sweden 18/K7
Upright (cape), Alaska 196/D2
Upsala, Minn. (56384) 255/D5
Upsala, Ontario 177/G5
Upsala, Ontario 175/B3
Upsalquitch, New Bruns. 170/D1
Upsalquitch (riv.), New Bruns. 170/D1
Upshur (co.), Texas 303/K5
Upshur (co.), W. Va. 312/F5
Upson (co.), Georgia 217/D5
Upson, Wis. (54565) 317/F3
Upton, Ky. (42784) 237/K6
Upton○, Maine (04261) 243/B6
Upton○, Mass. (01568) 249/H4
Upton, Québec 172/E4
Upton (co.), Texas 303/B6
Upton, Wyo. (82730) 319/H1
Upton upon Severn, England 13/E5
Upton-West Upton, Mass. (01568) 249/H4
Ur (ruins), Iraq 66/E5
Urabá (gulf), Colombia 126/B3
Urachiche, Venezuela 124/D2
Uracoa, Venezuela 124/G3
Urad Qianqi, China 77/G3
Urad Zhonghou Lianheqi, China 77/G3
Uraidla, S. Australia 94/B8
Urakawa, Japan 81/L2
Ural (mts.), U.S.S.R. 4/C6
Ural (mts.), U.S.S.R. 54/G4
Ural (mts.), U.S.S.R. 7/L2
Ural (mts.), U.S.S.R. 52/J2
Ural (mts.), U.S.S.R. 48/F4
Ural (riv.), U.S.S.R. 54/G5
Ural (riv.), U.S.S.R. 48/F5
Ural (riv.), U.S.S.R. 52/J4
Uralla, N.S. Wales 97/F2
Ural'sk, U.S.S.R. 54/G4
Ural'sk, U.S.S.R. 48/F4
Urambo, Tanzania 115/F4
Uran, India 68/B7
Urana, N.S. Wales 97/D4
Urana (lake), N.S. Wales 97/D4
Urania, La. (71480) 238/F3
Uranium City, Sask. 162/F4
Uranium City, Sask. 146/H4
Uranium City, Sask. 181/L2
Urapunga, North. Terr. 93/D3
Uraricoera, Brazil 132/H8
Uraricoera (riv.), Brazil 132/H8
Uravan, Colo. (81436) 208/B6
Urawa, Japan 81/O2
Uray, U.S.S.R. 48/G3
Urban, Ky. (40765) 237/O6
Urban, Pa. (†17830) 294/J4
Urban, Wash. (†98221) 310/C2
Urbana, Ark. (71768) 202/E7
Urbana, Ill. (61801) 222/E3
Urbana, Ind. (46990) 227/F3
Urbana, Iowa (52345) 229/K4
Urbana, Md. (†21701) 245/J3
Urbana, Mo. (65767) 261/F7
Urbana, Ohio (43078) 284/C5
Urbancrest, Ohio (†43123) 284/D6
Urbandale, Iowa (50322) 229/F5
Urbanette, Ark. (†72616) 202/D1
Urbankn Minn. (†56361) 255/C4
Urbanna, Va. (23175) 307/P5
Urbenville, N.S. Wales 97/G1
Urbino, Italy 34/D3
Urcos, Peru 128/G9
Urda, Spain 33/E3
Urdinarrain, Argentina 143/G6
Ure (riv.), England 13/F3
Uren, Sask. 181/E5
Urenui, N. Zealand 100/E3
Ures, Mexico 150/D2
Urfa (prov.), Turkey 63/H4
Urfa, Turkey 59/C2
Urfa, Turkey 63/H4
Urga (Ulaanbaatar) (cap.), Mongolia 77/G2
Urgel, Llanos de (plain), Spain 33/G2
Urgench, U.S.S.R. 54/G5
Urgench, U.S.S.R. 48/G5
Ürgüp, Turkey 63/F3
Uri (canton), Switzerland 39/G3
Uriah, Ala. (36480) 195/D8
Uribe, Colombia 126/C6
Uribia, Colombia 126/D2
Urica, Venezuela 124/F3
Urich, Mo. (64788) 261/E6
Urim, Israel 65/B5
Urimán, Venezuela 124/G5
Uriondo, Bolivia 136/C7
Urique (riv.), Mexico 150/F3
Urirotstock (mt.), Switzerland 39/G3
Urituyacu (riv.), Peru 128/D5
Urk, Netherlands 27/H3
Urla, Turkey 63/B3
Urlaţa, Romania 45/H3
Urlingford, Ireland 17/F6
Urmia, Iran 59/D2
Urmia, Iran 66/D2
Urmia (lake), Iran 54/F6
Urmia (lake), Iran 59/E2
Urmia (lake), Iran 66/D2
Urmston, England 13/H2
Urnäsch, Switzerland 39/H2
Uromi, Nigeria 106/F7
Uroševac, Yugoslavia 45/E4
Urrao, Colombia 126/B4
Ursa, Ill. (62376) 222/B3
Ursina, Pa. (15485) 294/D6
Ursine, Nev. (†89043) 266/G5
Úrsulo Galván, Mexico 150/Q1
Uruáchic, Mexico 150/E3
Uruaçu, Brazil 132/D6
Uruaçu, Brazil 120/E4
Uruana, Brazil 132/D6
Uruapan del Progreso, Mexico 150/H7
Urubamba, Peru 128/F9
Urubamba (riv.), Peru 128/F8
Urubichá, Bolivia 136/D4
Urubu (riv.), Brazil 132/A3
Urubupungá (dam), Brazil 132/C8
Urubupungá (dam), Brazil 120/D5
Urucará, Brazil 132/B3
Uruçuí, Brazil 132/E4
Urucún, Morro do (mt.), Brazil 132/B7
Urucurituba, Brazil 132/B3
Uruguai (riv.), Brazil 132/C9
Uruguaiana, Brazil 120/D5
Uruguaiana, Brazil 132/B10
Uruguay 2/G7
Uruguay 120/D6
URUGUAY 145
Uruguay (riv.) 120/D5
Uruguay (riv.), Argentina 143/E3
Uruguay (riv.), Uruguay 145/A3
Urumaco, Venezuela 124/C2
Ürümqi (Urumchi), China 77/C3
Ürümqi, China 2/P3
Ürümqi, China 54/K5
Urunga, N.S. Wales 97/G2
Urup (isl.), U.S.S.R. 54/S5
Urup (isl.), U.S.S.R. 48/Q5
Uruyén, Venezuela 124/G5
Uryupinsk, U.S.S.R. 52/F4
Urzhum, U.S.S.R. 52/G3
Urziceni, Romania 45/H3
Usa (riv.), U.S.S.R. 7/K2
Usa (riv.), U.S.S.R. 52/K1
Uşak (prov.), Turkey 63/C3
Uşak, Turkey 59/A2
Uşak, Turkey 63/C3
Usakos, Namibia 118/B4
Usakos, Namibia 102/D7
Usedom (isl.), E. Germany 22/F1
Usedom (Uznam) (isl.), Poland 47/B1
Ushakov (isl.), U.S.S.R. 4/B5
Ushant (isl.), France 7/D4
Ushant (Ouessant) (isl.), France 28/A3
Usherville, Sask. 181/J3
Ushibuka, Japan 81/D7
Ushtobe, U.S.S.R. 48/H5
Ushuaia, Argentina 143/C7
Ushuaia, Argentina 120/C8
Usibelli, Alaska (99787) 196/J2
Usinsk, U.S.S.R. 52/J1
Usk, Br. Col. 184/C3
Usk, Wales 13/B6
Usk (riv.), Wales 10/E5
Usk (riv.), Wales 13/B6
Usk, Wash. (99180) 310/H2
Üsküdar, Turkey 63/D6
Uslar, W. Germany 22/C3
Usman', U.S.S.R. 52/E4
Usol'ye-Sibirskoye, U.S.S.R. 48/L4
Usquepaug, R.I. (†02892) 249/H6
Ussel, France 28/E5
Ussuri (riv.) 54/P5
Ussuri (Wusuli Jiang) (riv.), China 77/M2
Ussuri (riv.), U.S.S.R. 48/O5
Ussuriysk, U.S.S.R. 54/P5
Ussuriysk, U.S.S.R. 48/O5
Úštěk, Czech. 41/C1
Uster, Switzerland 39/G2
Ustica (isl.), Italy 34/D5
Ust'-Ilimsk, U.S.S.R. 48/L4
Ústí nad Labem, Czech. 41/C1
Ústí nad Orlicí, Czech. 41/D2
Ustka, Poland 47/C1
Ust'-Kamchatsk, U.S.S.R. 54/T4
Ust'-Kamchatsk, U.S.S.R. 48/R4
Ust'-Kamenogorsk, U.S.S.R. 54/J5
Ust'-Kamenogorsk, U.S.S.R. 48/J5
Ust'-Kulom, U.S.S.R. 52/H2
Ust'-Kut, U.S.S.R. 54/M4
Ust'-Kut, U.S.S.R. 48/L4
Ust'-Kuyga, U.S.S.R. 4/C3
Ust'-Kuyga, U.S.S.R. 48/O3
Ust'-Maya, U.S.S.R. 48/O3
Ust'-Nera, U.S.S.R. 54/R3
Ust'-Nera, U.S.S.R. 48/P3
Ust'-Olenëk, U.S.S.R. 48/M2
Ust'-Omchug, U.S.S.R. 48/P3
Ust'-Ordynskiy, U.S.S.R. 48/L4
Ust'-Ordynskiy Buryat Aut. Okr., U.S.S.R. 48/L4
Ust'-Pinega, U.S.S.R. 52/F2
Ust'-Port, U.S.S.R. 48/J2
Ustrzyki Dolne, Poland 47/F4
Ust'-Tsil'ma, U.S.S.R. 52/H1
Ust'-Urt (plat.), U.S.S.R. 54/G5
Ust'-Urt (plat.), U.S.S.R. 48/F5
Ustyuzhna, U.S.S.R. 52/E3
Usu, China 77/B3
Usuki, Japan 81/E7
Usulután, El Salvador 154/C4
Usumacinta (riv.), Guatemala 154/B2
Usumacinta (riv.), Mexico 150/O8
Utah 188/D3
UTAH 304
Utah (beach), France 28/C3
Utah (state), U.S. 146/G6
Utah (co.), Utah 304/C3
Utah (lake), Utah 188/D2
Utah (lake), Utah 304/C3
Utajärvi, Finland 18/P4
Ute, Iowa (51060) 229/B4
Ute (creek), N. Mex. 274/F3
Ute (peak), N. Mex. 274/D2
Ute (res.), N. Mex. 274/F3
Ute Mountain Ind. Res., Colo. 208/B8
Ute Mountain Ind. Res., N. Mex. 274/A1
Utena, U.S.S.R. 53/C3
Utena, U.S.S.R. 52/C3
Ute Park, N. Mex. (87749) 274/D2
Utete, Tanzania 115/G5
Uthai Thani, Thailand 72/C4
Uthal, Pakistan 68/B3
Uthal, Pakistan 59/J4
Utica, Ill. (61373) 222/E2
Utica, Ind. (†47130) 227/F8
Utica, Kansas (67584) 232/B3
Utica, Ky. (42376) 237/G5
Utica, Mich. (*48087) 250/F6
Utica, Minn. (55979) 255/G7
Utica, Miss. (39175) 256/C6
Utica, Mo. (64686) 261/E3
Utica, Mont. (59452) 262/F4
Utica, Nebr. (68456) 264/G4
Utica, N.Y. 188/M2
Utica, N.Y. (*13501) 276/K4
Utica, Ohio (43080) 284/F5
Utica, Okla. (74763) 288/O7
Utica, Pa. (16362) 294/C3
Utica, S.C. (†29678) 296/B2
Utica, S. Dak. (57067) 298/P8
Utica, Wis. (†53589) 317/H10
Utica Junior College, Miss. (39175) 256/C6
Utiel, Spain 33/F3
Utikuma (lake), Alberta 182/C2
Utikuma (riv.), Alberta 182/C1
Utikumasis (lake), Alberta 182/C2
Utila, Honduras 154/D2
Utila (isl.), Honduras 154/D2
Utleyville, Colo. (†81064) 208/O8
Utopia, Alaska (†99745) 196/H1
Utopia (lake), New Bruns. 170/D3
Utopia, North. Terr. 93/D7
Utopia, Texas (78884) 303/E8
Utrecht (prov.), Netherlands 27/G4
Utrecht, Netherlands 27/G4
Utrecht, Netherlands 7/E3
Utrera, Spain 33/D4
Utsjoki, Finland 18/P2
Utsunomiya, Japan 81/K5
Uttaradit, Thailand 72/D3
Uttarpara-Kotrung, India 68/F1
Uttar Pradesh (state), India 68/D3
Uttoxeter, England 13/E5
Uttoxeter, England 10/F4
Utuado, P. Rico 156/F1
Utuado, P. Rico 161/B2
Utukok (riv.), Alaska 196/F1
Uturoa, Fr. Poly. 87/L7
Utzenstorf, Switzerland 39/E2
Uusikaarlepyy (Nykarleby), Finland 18/N5
Uusikaupunki, Finland 18/M6
Uusimaa (prov.), Finland 18/O6
Uva, Laguna (lake), Colombia 126/E6
Uva (riv.), Colombia 126/E6
Uvalda, Georgia (30473) 217/H6
Uvalde (co.), Texas 303/E8
Uvalde, Texas (78801) 303/E8
Uvarovo, U.S.S.R. 52/F4
Uvéa (isl.), New Caled. 87/G7
Uvéa (isl.), New Caled. 86/H4
Uvéa (bay), New Caled. 86/H4
Uverito, Venezuela 124/F3
Uvinza, Tanzania 115/F5
Uvira, Zaire 115/E4
Uvs, Mongolia 77/D2
Uvs Nuur (lake), Mongolia 54/L4
Uvs Nuur (lake), Mongolia 77/D1
Uwajima, Japan 81/F7
Uwchland, Pa. (19480) 294/L5
'Uweinat, Jebel (mt.), Egypt 111/E3
'Uweinat, Jebel (mt.), Libya 111/E3
'Uweinat, Jebel (mt.), Sudan 111/E3
Uxbridge○, Mass. (01569) 249/H4
Uxbridge, Ontario 177/E3
Uxin, China 77/G4
Uxmal (ruins), Mexico 150/P6
Uyak, Alaska (†99624) 196/H3
Uyuni, Bolivia 120/C5
Uyuni, Bolivia 136/B7
Uyuni (salt dep.), Bolivia 136/B7
Už (riv.), Czech. 41/G2
Uzbek, S.S.R., U.S.S.R. 54/H5
Uzbek, S.S.R., U.S.S.R. 48/G5
Uzès, France 28/F5
Uzhgorod, U.S.S.R. 7/G4
Uzhgorod, U.S.S.R. 52/B5
Uzlovaya, U.S.S.R. 52/E4
Uznach, Switzerland 39/H2
Uznam (Usedom) (isl.), Poland 47/B1
Üzümlü, Turkey 63/D4
Uzunköprü, Turkey 63/B2
Uzwil, Switzerland 39/H2
Uzza, Israel 65/B4

V

Vaaifetu (mt.), W. Samoa 86/M8
Vaal (riv.), S. Africa 102/E7
Vaal (riv.), S. Africa 118/D5
Vaala, Finland 18/P4
Vaals, Netherlands 27/H7
Vaalserberg (mt.), Belgium 27/J7
Vaalserberg (mt.), Netherlands 27/J7
Vaalspan, S. Africa 118/C5
Vaasa (prov.), Finland 18/N5
Vaasa, Finland 7/G2
Vaasa, Finland 18/M5
Vaassen, Netherlands 27/H4
Vác, Hungary 41/E3
Vaca (key), Fla. 212/E7
Vacaria, Brazil 132/D10
Vacaville, Calif. (95688) 204/D5
Vaccarès (lag.), France 28/F6
Vache (isl.), Haiti 158/B6
Vache (isl.), Haiti 156/D3
Vacherie, La. (70090) 238/L3
Vacía Talega (pt.), P. Rico 161/E1
Vada, Ky. (41383) 237/O5
Vaden, Ark. (†71923) 202/E6
Vader, Wash. (98593) 310/B4
Vadis, W. Va. (26445) 312/E4
Vadito, N. Mex. (87579) 274/D2
Vadnais Heights, Minn. (†55101) 255/G5
Vado, N. Mex. (88072) 274/C6
Vadodara, India 68/C4
Vadret (peak), Switzerland 39/J3
Vadsø, Norway 7/H1
Vadsø, Norway 18/Q1
Vadstena, Sweden 18/J7
Vaduz (cap.), Liecht. 39/H2
Vaea (mt.), W. Samoa 86/M8
Vaerøy (isl.), Norway 18/H3
Vaga (riv.), U.S.S.R. 52/F2
Vagåvatn (lake), Norway 18/F6
Vaggeryd, Sweden 18/J8
Vagos, Portugal 33/B2
Vagthus (pt.), Virgin Is. (U.S.) 161/F4
Váh (riv.), Czech. 41/D2
Vahitahi (atoll), Fr. Poly. 87/N7
Vaiden, Miss. (39176) 256/E4
Vaihiria (lake), Fr. Poly. 86/T13
Vail, Ariz. (85641) 198/E6
Vail, Colo. (81657) 208/G3
Vail, Iowa (51465) 229/C4
Vails, N.J. (†07832) 273/D2
Vaitupu (atoll), Tuvalu 87/H6
Vakfıkebir, Turkey 63/H2
Vakh (riv.), U.S.S.R. 48/J3
Vakhan (reg.), Afghanistan 59/K2
Vakhtan, U.S.S.R. 52/G3
Vál, Hungary 41/E3
Valais (canton), Switzerland 39/D4
Val-Alain, Québec 172/F3
Valašské Meziříčí, Czech. 41/D2
Valatie, N.Y. (12184) 276/N6
Val-Barrette, Québec 172/B3
Val-Brillant, Québec 172/B1
Valcheta, Argentina 143/C5
Val-Comeau, New Bruns. 170/F1
Valcour (isl.), N.Y. 276/N1
Valcourt, Québec 172/E4
Valdagno, Italy 34/C2
Val d'Amour, New Bruns. 170/D1
Val-David, Québec 172/C3
Valday, U.S.S.R. 52/D3
Valday (Valdai) (hills), U.S.S.R. 52/D3
Valdecañas (res.), Spain 33/D3
Val-de-Marne (dept.), France 28/C1
Valdemarsvik, Sweden 18/K7
Valdemoro, Spain 33/F4
Valdepeñas, Spain 33/E3
Valderas, Spain 33/D1
Valderrobres, Spain 33/F2
Valders, Wis. (54245) 317/L7
Valdés, Argentina 143/C3
Valdés (pen.), Argentina 120/C7
Valdés (pen.), Argentina 143/D5
Valdes (isl.), Br. Col. 184/K3
Val-des-Bois, Québec 172/B4
Valdese, N.C. (28690) 281/F3
Valdeverdeja, Spain 33/D3
Valdez, Alaska (99686) 196/D1
Valdez, Colo. (†81082) 208/K8
Valdez, Ecuador 128/C2
Valdez, N. Mex. (87580) 274/D2
Valdivia, Chile 2/F7
Valdivia, Chile 138/D3
Valdivia, Chile 120/B6
Valdivia, Colombia 126/C4
Val-d'Oise (dept.), France 28/E3
Val-d'Or, Que. 162/J6
Val d'Or, Québec 174/B3
Valdosta, Ga. 188/K4
Valdosta, Georgia (31601) 217/F9
Val Doucet, New Bruns. 170/E1
Vale, N.C. (28168) 281/G3
Vale, Oreg. (97918) 291/K4
Vale, S. Dak. (57788) 298/C4
Vale, Tenn. (†38317) 237/E8
Valeda, Kansas (†67337) 232/G4
Valeene, Ind. (†47125) 227/E8
Valemount, Br. Col. 184/H4
Valença, Brazil 120/F4
Valença, Brazil 135/E3
Valença, Brazil 132/G6
Valença, Portugal 33/B2
Valença do Piauí, Brazil 132/F4
Valence, France 7/E4
Valence, France 28/F5
Valencia, Calif. (91355) 204/G9
Valencia (Valentia) (isl.), Ireland 10/A5
Valencia (isl.), Ireland 17/A8
Valencia (co.), N. Mex. 274/C4
Valencia, N. Mex. (†87031) 274/C4
Valencia, Pa. (16059) 294/C4
Valencia, Philippines 82/E7
Valencia (prov.), Spain 33/F3
Valencia, Spain 7/D5
Valencia, Spain 33/F3
Valencia (gulf), Spain 33/G3
Valencia, Albufera de (lag.), Spain 33/G3
Valencia (reg.), Spain 33/F3
Valencia, Trin. & Tob. 161/B10
Valencia, Venezuela 120/C2
Valencia, Venezuela 124/E2
Valencia (lake), Venezuela 124/E2
Valencia de Alcántara, Spain 33/C3
Valencia de Don Juan, Spain 33/D1
Valenciennes, France 28/E2
Valendas, Switzerland 39/H3
Valentia (isl.), Ireland 10/A5
Valentine, Ariz. (86437) 198/B3
Valentine, Ind. (†46761) 227/G1
Valentine, Nebr. (69201) 264/D2
Valentine, Texas (79854) 303/C11
Valentines, Uruguay 145/E4
Valentines, Va. (23887) 307/N7
Valenza, Italy 34/B2
Valenzuela, Paraguay 144/B5
Valera, Texas (76884) 303/E6
Valera, Venezuela 124/C3
Valera, Venezuela 120/B2
Valeria, Iowa (†50054) 229/G5
Vale Summit, Md. (†21532) 245/C2
Valga, U.S.S.R. 52/C3
Valga, U.S.S.R. 53/D2
Valhalla, Alberta 182/A2
Valhalla, N.Y. (10595) 276/P6
Valhalla Centre, Alberta 182/A2
Valhermoso Springs, Ala. (35775) 195/E2
Valiente (pen.), Panama 154/G6
Valier, Ill. (62891) 222/D5
Valier, Mont. (59486) 262/D2
Valier, Pa. (15780) 294/D4
Valjean, Sask. 181/E5
Valjevo, Yugoslavia 45/D3
Valka, U.S.S.R. 53/C2
Valkeakoski, Finland 18/N6
Valkenswaard, Netherlands 27/H6
Valladolid, Mexico 146/K7
Valladolid, Mexico 150/P6
Valladolid (prov.), Spain 33/D2
Valladolid, Spain 33/D2
Valladolid, Spain 7/E4
Vallay (isl.), Scotland 15/A3
Vall de Uxó, Spain 33/F3
Valle, Norway 18/E7
Valle Alegre, Chile 138/F2
Vallecas, Spain 33/G4
Vallecito (res.), Colo. 208/D8
Vallecitos, N. Mex. (87581) 274/C2
Valle de Allende, Mexico 150/G3
Valle de Bravo, Mexico 150/J7
Valle de Guanape, Venezuela 124/F3
Valle de la Pascua, Venezuela 124/F3
Valle del Cauca (dept.), Colombia 126/B6
Valledupar, Colombia 120/B1
Valledupar, Colombia 126/D2
Vallée-Jonction, Québec 172/G3
Vallegrande, Bolivia 120/C4
Vallegrande, Bolivia 136/C6
Valle Hermoso, Mexico 150/L4
Vallehermoso, Spain 33/A5
Vallejo, Calif. 188/B3
Vallejo, Calif. (94590) 204/J1
Valle Mí, Paraguay 144/D3
Vallenar, Chile 138/A7
Vallentuna, Sweden 18/H1
Valle San Telmo, Mexico 150/A1
Valles Mines, Mo. (63087) 261/L6
Valletta (cap.), Malta 7/F5
Valletta (cap.), Malta 34/E7
Valley (co.), Idaho 220/C5
Valley (riv.), Manitoba 179/B3
Valley (co.), Mont. 262/K2
Valley (co.), Nebr. 264/E3
Valley, Nebr. (68064) 264/H3
Valley, Nova Scotia 168/E3
Valley, Wash. (99181) 310/H2
Valley, Wis. (†54639) 317/F8
Valley, Wyo. (†82414) 319/C1
Valley Bend, W. Va. (26293) 312/F5
Valley Brook, Okla. (†73101) 288/M4
Valley Center, Kansas (67147) 232/E4
Valley Centre, Sask. 181/D4
Valley City, Ill. (†62340) 222/C4
Valley City, N. Dak. (58072) 282/P6
Valley City, Ohio (44280) 284/G3
Valley Cottage, N.Y. (10989) 276/K8
Valley East, Ontario 177/J5
Valley East, Ontario 175/D3
Valley Falls, Kansas (66088) 232/G2
Valley Falls, N.Y. (12185) 276/N5
Valley Falls, Oreg. (†97630) 291/G5
Valley Falls, R.I. (†02864) 249/J5
Valley Farms, Ariz. (85291) 198/D6
Valleyfield, Québec 172/C4
Valleyford, Wash. (99036) 310/H3
Valley Forge, Pa. (19481) 294/L5
Valley Grove, W. Va. (26060) 312/E2
Valley Head, Ala. (35989) 195/G1
Valley Head, W. Va. (26294) 312/G5
Valley Hi, Ohio (†43360) 284/C5
Valley Lee, Md. (20692) 245/M8
Valley Mills, Texas (76689) 303/G6
Valley Park, Miss. (39177) 256/C5
Valley Park, Mo. (63088) 261/O3
Valley Point, W. Va. (†26519) 312/G3
Valley River, Manitoba 179/B3
Valley Spring, Texas (76885) 303/F7
Valley Springs, Ark. (72682) 202/D1
Valley Springs, Calif. (95252) 204/C9
Valley Springs, S. Dak. (57068) 298/S6
Valley Station, Ky. (40272) 237/K4
Valley Stream, N.Y. (*11580) 276/P7
Valleyview, Alberta 182/B2
Valley View, Ill. (†60120) 222/E2
Valley View, Ky. (†40475) 237/N5
Valley View, Ohio (†43201) 284/D6
Valley View, Ohio (†44101) 284/H9
Valley View, Pa. (17983) 294/J4
Valley View, Texas (76272) 303/H4
Vallgrund (isl.), Finland 18/M5
Valliant, Okla. (74764) 288/R6
Vallières, Haiti 158/C5
Vallimanca (riv.), Argentina 143/F7
Vallonia, Ind. (47281) 227/E7
Vallon-Pont-d'Arc, France 28/F5
Vallorbe, Switzerland 39/B3
Valls, Spain 33/G2
Val Marie, Sask. 181/D6
Valmeyer, Ill. (62295) 222/C5
Valmiera, U.S.S.R. 53/C2
Valmiera, U.S.S.R. 52/C3
Valmont, Québec 172/E3
Valmontone, Italy 34/F7
Valmora, N. Mex. (87750) 274/D3
Valmy, Nev. (89438) 266/D2
Valmy, Wis. (†54235) 317/M6
Valognes, France 28/C3
Valona, Georgia (31332) 217/K8
Valor, Sask. 181/E6
Valpaços, Portugal 33/C2
Valparaíso (reg.), Chile 138/A9
Valparaíso, Chile 2/F7
Valparaíso, Chile 120/B6
Valparaíso, Chile 138/E2
Valparaiso, Fla. (32580) 212/C6
Valparaiso, Ind. (46383) 227/C2
Valparaiso, Nebr. (68065) 264/H3
Valparaiso, Sask. 181/G3
Val-Racine, Québec 172/G4
Vals (cape), Indonesia 85/K7
Vals, Switzerland 39/H3
Valsad, India 68/C4
Valsequillo (res.), Mexico 150/N2
Valserrhein (riv.), Switzerland 39/H3
Valsetz, Oreg. (97393) 291/D3
Value, Miss. (39178) 256/D6
Valuyki, U.S.S.R. 52/E4
Valverda, La. (†70757) 238/G5
Valverde (prov.), Dom. Rep. 158/D5
Valverde, Dom. Rep. 158/D5
Val Verde (co.), Texas 303/C8
Valverde del Camino, Spain 33/C4
Vamdrup, Denmark 21/C7
Vammala, Finland 18/N6
Vámos, Greece 45/F8
Vámospércs, Hungary 41/F3
Van, Ky. (41857) 237/R6
Van (lake), N. Dak. 282/L5
Van, Oreg. (†97904) 291/J4
Van, Pa. (†16319) 294/C3
Van, Texas (75790) 303/J5
Van (prov.), Turkey 63/K3
Van, Turkey 59/D2
Van, Turkey 63/K3
Van (lake), Turkey 54/F6
Van (lake), Turkey 59/D2
Van (lake), Turkey 63/K3
Van, W. Va. (25206) 312/C7
Vanadium, N. Mex. (88073) 274/A6
Van Alstyne, Texas (75095) 303/H4
Vananda, Br. Col. 184/E5
Vananda, Mont. (†59237) 262/K4
Vanatta, Ohio (†43055) 284/E5
Vanavara, U.S.S.R. 48/L3
Van Blommenstein (lake), Suriname 120/D2
Van Blommestein (lake), Suriname 131/D3
Van Bruyssel, Québec 172/E2
Van Buren (co.), Ark. 202/E2
Van Buren, Ark. (72956) 202/B3
Van Buren, Ind. (46991) 227/F3
Van Buren (co.), Iowa 229/K7
Van Buren, Maine (04785) 243/G1
Van Buren○, Maine (04785) 243/G1
Van Buren (co.), Mich. 250/C6
Van Buren, Mo. (63965) 261/L8
Van Buren, Ohio (45889) 284/C3
Van Buren (co.), Tenn. 237/L9
Vance, Ala. (35490) 195/D4
Vance, Miss. (38964) 256/D2
Vance (co.), N.C. 281/N2
Vance, S.C. (29163) 296/G5
Vance A.F.B., Okla. 288/K2
Vanceboro○, Maine (04491) 243/J4
Vanceboro, N.C. (28586) 281/P4
Vanceburg, Ky. (41179) 237/P3
Vancleave, Miss. (†39564) 256/G9
Van Cleve, Iowa (†50162) 229/G5
Vancourt, Texas (76955) 303/D6
Vancouver (mt.), Alaska 196/L2
Vancouver, Br. Col. 146/F4
Vancouver, Br. Col. 162/D6
Vancouver, Br. Col. 184/K3
Vancouver (Greater), Br. Col. 184/K3
Vancouver (isl.), Br. Col. 146/F5
Vancouver (isl.), Br. Col. 162/D6
Vancouver (isl.), Br. Col. 184/D5
Vancouver, Canada 2/C3
Vancouver (isl.), Canada 2/C3
Vancouver, Wash. 188/B1
Vancouver, Wash. (*98660) 310/C5
Vancouver (lake), Wash. 310/C5
Vandalia, Ill. (62471) 222/D5
Vandalia, Mich. (49095) 250/D7
Vandalia, Mo. (63382) 261/J4
Vandalia, Mont. (59273) 262/J2
Vandalia, Ohio (45377) 284/B6
Vandalia, W. Va. (†26423) 312/F5
Vandemere, N.C. (28587) 281/R4

Vandenberg A.F.B., Calif. 204/E9
Vanderbijl Park, S. Africa 118/D5
Vanderbilt, Mich. (49795) 250/E3
Vanderbilt, Pa. (15486) 294/C5
Vanderbilt, Texas (77991) 303/H9
Vanderburgh (co.), Ind. 227/B8
Vandergrift, Pa. (15690) 294/D4
Vanderhoof, Br. Col. 162/D5
Vanderhoof, Br. Col. 184/E3
Vanderlin (isl.), North. Terr. 88/F3
Vanderlin (isl.), North. Terr. 93/E3
Vanderpool, Texas (78885) 303/E8
Vanderpool, Va. (†24465) 307/J4
Vandervoort, Ark. (71972) 202/B5
Van Diemen (cape), North. Terr. 88/D2
Van Diemen (cape), North. Terr. 93/A1
Van Diemen (gulf), North. Terr. 88/E2
Van Diemen (gulf), North. Terr. 93/B1
Vandiver, Ala. (35176) 195/F4
Vandiver, Mo. (†65265) 261/J4
Vandling, Pa. (18421) 294/M2
Vändra, U.S.S.R. 53/C1
Vandura, Sask. 181/K5
Vanduser, Mo. (63784) 261/N9
Vanegas, Mexico 150/J5
Vänern (lake), Sweden 7/F3
Vänern (lake), Sweden 18/H7
Vänersborg, Sweden 18/G7
Van Etten, N.Y. (14889) 276/G6
Vanga, Kenya 115/G4
Vangaindrano, Madagascar 118/H4
Vanguard, Sask. 181/D6
Vangunu (isl.), Solomon Is. 86/D3
Van Hoa, Vietnam 72/E2
Van Horn, Texas (79855) 303/C11
Van Horne, Iowa (52346) 229/J4
Van Hornesville, N.Y. (13475) 276/L5
Vanier, Ontario 177/J2
Vanier, Québec 172/J3
Vanikoro (isl.), Solomon Is. 87/G7
Vanil Noir (mt.), Switzerland 39/D3
Vanimo, Papua N.G. 87/E6
Vanimo, Papua N.G. 85/B6
Vanino, U.S.S.R. 48/P5
Vaniyambadi, India 68/D6
Vankleek Hill, Ontario 177/K2
Van Lear, Ky. (41265) 237/R5
Vanleer, Tenn. (37181) 237/G8
Vanlue, Ohio (45890) 284/D4
Van Meter, Iowa (50261) 229/E5
Vanna, Georgia (30672) 217/F2
Vännäs, Sweden 18/L5
Vanndale, Ark. (72387) 202/J3
Vannes, France 28/B4
Van Ninh, Vietnam 72/F4
Vannøy (isl.), Norway 18/L1
Van Nuys, Calif. (*91401) 204/B10
Van Orin, Ill. (61374) 222/D2
Vanoss, Okla. (†74820) 288/N5
Vanrhynsdorp, S. Africa 118/B6
Van Rook, Queensland 95/B3
Vansant, Va. (24656) 307/D6
Vansbro, Sweden 18/H6
Vanscoy, Sask. 181/D4
Vansittart (isl.), N.W. Terrs. 187/K3
Vansittart (isl.), Tasmania 99/E2
Vantage, Sask. 181/F6
Vantage, Wash. (98950) 310/E4
Van Tassell, Wyo. (82242) 319/H3
Vanua Levu (isl.), Fiji 87/H7
Vanua Levu (isl.), Fiji 86/Q10
Vanuatu 2/T6
Vanuatu 87/G7
Van Vleet, Miss. (†38851) 256/G3
Vanvoorhis, W. Va. (†26505) 312/G3
Van Wert, Georgia (†30153) 217/B3
Van Wert, Iowa (50262) 229/F7
Van Wert (co.), Ohio 284/A4
Van Wert, Ohio (45891) 284/A4
Van Wyck, S.C. (29744) 296/F2
Van Yen, Vietnam 72/E2
Vanylven, Norway 18/E5
Van Zandt (co.), Texas 303/J5
Van Zandt, Wash. (†98244) 310/C2
Vanzant, Mo. (65768) 261/H9
Var (dept.), France 28/G6
Vara, Sweden 18/H7
Vara de María, Venezuela 124/C4
Varadero, Cuba 158/D1
Varakļāni, U.S.S.R. 53/D2
Varallo Pombia, Italy 34/B2
Varamin, Iran 66/G3
Varanasi, India 54/K7
Varanasi, India 68/E3
Varangerfjorden (fjord), Norway 18/Q2
Varangerfjorden (fjord), Norway 7/H1
Varangerhalvøya (pen.), Norway 18/Q1
Varano (lake), Italy 34/F3
Varaždin, Yugoslavia 45/B2
Varazze, Italy 34/B2
Varberg, Sweden 18/G8
Vardaman, Miss. (38878) 256/F3
Vardar (riv.), Greece 45/E5
Vardar (riv.), Yugoslavia 45/E5
Varde, Denmark 18/F9
Varde, Denmark 21/B6
Varde (riv.), Denmark 21/B6
Vardø, Norway 18/R1
Varel, W. Germany 22/C2
Varella, Mui (cape), Vietnam 72/F4
Varena, U.S.S.R. 53/C3
Varennes, Québec 172/J4
Vareš, Yugoslavia 45/D3
Varese (prov.), Italy 34/B2
Varese, Italy 34/B2
Vargem Bonita, Brazil 135/E3
Varginha, Brazil 135/D2
Varginha, Brazil 132/E8
Varina, Iowa (50593) 229/D3
Varkaus, Finland 18/Q5
Värmland (co.), Sweden 18/H7
Varna, Bulgaria 7/G4
Varna, Bulgaria 45/J4
Varna, Ill. (61375) 222/D2
Varnado, La. (70467) 238/L5
Värnamo, Sweden 18/J8
Varnek, U.S.S.R. 52/J1
Varnell, Georgia (30756) 217/C1
Varner, Kansas (†67068) 232/D4
Varney, Ontario 177/D3
Varney, W. Va. (25696) 312/B7
Varnsdorf, Czech. 41/C1
Varnville, S.C. (29944) 296/E6
Várpalota, Hungary 41/E3
Vars, Ontario 177/J2
Vartholomión, Greece 45/E7
Varto, Turkey 63/J3
Varysburg, N.Y. (14167) 276/D5
Varzarin, Kuh-e (mt.), Iran 59/E3
Varzarin, Kuh-e (mt.), Iran 66/E4
Vas (co.), Hungary 41/D3
Vasa (Vaasa), Finland 18/M5
Vasa, Minn. (†55089) 255/F6
Vasa Barris (riv.), Brazil 132/G5
Vásárosnamény, Hungary 41/G2
Vascongadas (reg.), Spain 33/E1
Vashi, India 68/B7
Vashka (riv.), U.S.S.R. 52/G2
Vashon, Wash. (98070) 310/A2
Vasile Roaită, Romania 45/J3
Vasil'kov, U.S.S.R. 52/D4
Vaslui, Romania 45/H2
Vass, N.C. (28394) 281/L4
Vassalboro, Maine (04989) 243/D7
Vassalboro○, Maine (04989) 243/D7
Vassar, Kansas (66543) 232/G3
Vassar, Manitoba 179/G5
Vassar, Mich. (48768) 250/F5
Vassouras, Brazil 135/E3
Vastenjaure (lake), Sweden 18/K3
Västerås, Sweden 7/F3
Västerås, Sweden 18/K7
Västerbotten (co.), Sweden 18/K4
Västerdalälven (riv.), Sweden 18/H6
Västerhaninge, Sweden 18/H1
Västernorrland (co.), Sweden 18/K5
Västervik, Sweden 18/K8
Västmanland (co.), Sweden 18/K7
Vasto, Italy 34/E3
Vasvár, Hungary 41/D3
Vaternish (dist.), Scotland 15/B3
Vaternish (pt.), Scotland 15/B3
Vatersay (isl.), Scotland 15/A4
Vathí, Greece 45/H7
Vatican City 7/F4
VATICAN CITY 34
Vatican City, Vatican City 34/B6
Vaticano (cape), Italy 34/E5
Vatnajökull (glac.), Iceland 21/C1
Vatomandry, Madagascar 118/H3
Vatra Dornei, Romania 45/G2
Vättern (lake), Sweden 7/F3
Vättern (lake), Sweden 18/J7
Vatukoula, Fiji 86/P10
Vatulele (isl.), Fiji 86/P11
Vauclin (mt.), Martinique 161/D6
Vaucluse (dept.), France 28/F6
Vaucluse, S.C. (29850) 296/D7
Vaud (canton), Switzerland 39/B3
Vaudreuil (co.), Québec 172/C4
Vaudreuil, Québec 172/C4
Vaughan, Miss. (39179) 256/D5
Vaughan, N.C. (27586) 281/N2
Vaughan, Ontario 177/E4
Vaughan, W. Va. (†26656) 312/D6
Vaughn, Mont. (59487) 262/E3
Vaughn, N. Mex. (88353) 274/D4
Vaughn, Wash. (98394) 310/C3
Vaughnsville, Ohio (45893) 284/B4
Vaupés (comm.), Colombia 126/E7
Vaupés (riv.), Colombia 120/B2
Vaupés (riv.), Colombia 126/E7
Vauxhall, Alberta 182/D4
Vauxhall, N.J. (07088) 273/A2
Vaux-sur-Sûre, Belgium 27/H9
Vava'u Group (isls.), Tonga 87/J7
Vavenby, Br. Col. 184/H4
Vavuniya, Sri Lanka 68/E7
Vawn, Sask. 181/C2
Vaxholm, Sweden 18/J1
Växjö, Sweden 7/F3
Växjö, Sweden 18/J8
Vaygach (isl.), U.S.S.R. 4/C6
Vaygach (isl.), U.S.S.R. 52/K1
Vayland, S. Dak. (†57381) 298/M5
Važec, Czech. 41/E2
Vazhgort, U.S.S.R. 52/G2
Vaz-Obervaz, Switzerland 39/J3
Vázquez, Cuba 158/H3
Veagh (lake), Ireland 17/F1
Vealmoor, Texas (79720) 303/C5
Veazie○, Maine (04401) 243/F6
Veblen, S. Dak. (57270) 298/P2
Vechigen, Switzerland 39/E3
Vecht (riv.), Netherlands 27/F4
Vechta, W. Germany 22/C2
Vechte (riv.), Netherlands 27/J3
Vechte (riv.), W. Germany 22/B2
Vecsés, Hungary 41/E3
Vedaranniyam, India 68/E6
Vedia, Argentina 143/F7
Veedersburg, Ind. (47987) 227/C4
Veendam, Netherlands 27/K2
Veenendaal, Netherlands 27/G4
Veenhuizen, Netherlands 27/J2
Veere, Netherlands 27/D5
Veersche Meer (lake), Netherlands 27/D5
Vega (pt.), Alaska 196/J4
Vega, Alberta 182/C2
Vega (isl.), Norway 18/G4
Vega, Texas (79092) 303/B2
Vega Alta, P. Rico 161/D1
Vega Baja, P. Rico 161/D1
Vegafjorden (fjord), Norway 18/G4
Vegas Creek, Nev. (89121) 266/G6
Veghel, Netherlands 27/H5
Vegreville, Alberta 182/E3
Vegreville, Alta. 162/E5
Veguita, N. Mex. (87062) 274/C4
Vehar (lake), India 68/B7
Veinticinco (25) de Agosto, Uruguay 145/A6
Veinticinco (25) de Diciembre, Paraguay 144/D4
Veinticinco de Mayo, Argentina 143/F7
Veinticinco de Mayo, Ecuador 128/C4
Veinticinco (25) de Mayo, Uruguay 145/C5
Veintiocho de Noviembre, Argentina 143/B7
Vejen, Denmark 21/C7
Vejer de la Frontera, Spain 33/C4
Vejle (co.), Denmark 21/C6
Vejle, Denmark 21/C6
Vejle, Denmark 18/F9
Vejle (fjord), Denmark 21/C6
Vejprty, Czech. 41/B1
Vela, La (cape), Colombia 126/D1
Vela, Roca que (cay), Colombia 126/B8
Vélan (mt.), Switzerland 39/D5
Velarde, N. Mex. (87582) 274/C2
Velas (cape), C. Rica 154/D5
Velasco, Ciego de Ávila, Cuba 158/G2
Velasco, Holguín, Cuba 158/H3
Velázquez, Uruguay 145/E5
Velda, Mo. (†63101) 261/P2
Velden am Wörthersee, Austria 41/C3
Veldhoven, Netherlands 27/G6
Veldrif, S. Africa 118/B6
Velence, Hungary 41/E3
Velenje, Yugoslavia 45/B2
Vélez, Colombia 126/D4
Vélez-Blanco, Spain 33/E4
Vélez-Málaga, Spain 33/E4
Vélez-Rubio, Spain 33/E4
Velhas (riv.), Brazil 132/E7
Velika Plana, Yugoslavia 45/E3
Velikaya (riv.), U.S.S.R. 48/S3
Velikaya (riv.), U.S.S.R. 52/C3
Veliki Bečkerek (Zrenjanin), Yugoslavia 45/E3
Velikiye Luki, U.S.S.R. 7/H3
Velikiye Luki, U.S.S.R. 52/D3
Velikiye Luki, U.S.S.R. 48/D4
Velikiy Ustyug, U.S.S.R. 7/J2
Velikiy Ustyug, U.S.S.R. 48/E3
Velikiy Ustyug, U.S.S.R. 52/F2
Veliko Tŭrnovo, Bulgaria 45/G4
Velikovisochnoye, U.S.S.R. 52/H1
Velizh, U.S.S.R. 52/D3
Velká Bíteš, Czech. 41/D2
Velká Bystřice, Czech. 41/D2
Veľké Kapušany, Czech. 41/G2
Velké Meziříčí, Czech. 41/D2
Veľké Rovné, Czech. 41/E2
Vella Lavella (isl.), Solomon Is. 86/D2
Velletri, Italy 34/F7
Vellore, India 68/D6
Velluda, Sierra (mt.), Chile 138/E1
Velma, Okla. (73091) 288/L6
Velp, Netherlands 27/J5
Velpen, Ind. (47590) 227/C8
Velsen, Netherlands 27/F4
Vel'sk, U.S.S.R. 52/F2
Vel'sk, U.S.S.R. 48/E3
Velten, E. Germany 22/E2
Veluwe (reg.), Netherlands 27/H4
Velva, N. Dak. (58790) 282/J3
Velvendós, Greece 45/F5
Vemb, Denmark 21/B5
Véménd, Hungary 41/E3
Venadillo, Colombia 126/C5
Venado, Mexico 150/J5
Venado Tuerto, Argentina 143/D3
Venafro, Italy 34/E4
Venaissin (trad. prov.), France 29
Venamo (mt.), Guyana 131/A3
Venamo, Cerro (mt.), Venezuela 124/H4
Venamo (riv.), Venezuela 124/H4
Venango, Nebr. (69168) 264/C4
Venango (co.), Pa. 294/C3
Venango, Pa. (16440) 294/B2
Vena Park, Queensland 95/B3
Vence, France 28/G6
Venda (aut. rep.), S. Africa 102/F7
Venda (aut. rep.), S. Africa 118/E4
Vendas Novas, Portugal 33/B3
Vendée (dept.), France 28/C4
Vendôme, France 28/D4
Vendrell, Spain 33/G2
Venedocia, Ohio (45894) 284/B4
Venedy, Ill. (62296) 222/D5
Veneta, Oreg. (97487) 291/D3
Venetie, Alaska (99781) 196/J1
Veneto (reg.), Italy 34/D3
Venezia (Venice), Italy 34/D2
Venezuela 2/F5
Venezuela 120/C2
Venezuela, Cuba 158/F2
VENEZUELA 124
Venezuela (gulf), Venezuela 120/B1
Venezuela (gulf), Venezuela 124/C2
Vengurla, India 68/C5
Veniaminof (crater), Alaska 196/F3
Venice, Alberta 182/E2
Venice, Calif. (90291) 204/B11
Venice, Fla. (*33595) 212/D4
Venice, Ill. (62090) 222/A2
Venice (prov.), Italy 34/D2
Venice, Italy 7/F4
Venice, Italy 34/D2
Venice (gulf), Italy 34/D2
Venice, La. (70091) 238/M8
Venice, Utah (†84701) 304/C5
Vénissieux, France 28/F5
Venkatagiri, India 68/D6
Venlo, Netherlands 27/J6
Venn, Sask. 181/F4
Venosa, Italy 34/E4
Venraij, Netherlands 27/H6
Venta (riv.), U.S.S.R. 53/B2
Venterspos, S. Africa 118/G6
Ventimiglia, Italy 34/A3
Ventnor, England 10/F5
Ventnor, England 13/F7
Ventnor City, N.J. (08406) 273/E5
Ventotene (isl.), Italy 34/D4
Ventspils, U.S.S.R. 48/B4
Ventspils, U.S.S.R. 53/A2
Ventspils, U.S.S.R. 52/B3
Ventuari (riv.), Venezuela 124/E5
Ventura (co.), Calif. 204/F9
Ventura, Calif. (*93001) 204/F9
Ventura, Iowa (50482) 229/F2
Venturia, N. Dak. (58489) 282/L7
Venus, Fla. (33960) 212/E4
Venus (pt.), Fr. Poly. 86/T12
Venus, Pa. (16364) 294/C3
Venus (bay), Victoria 97/C6
Venustiano Carranza, Mexico 150/N8
Venustiano Carranza (res.), Mexico 150/J3
Ver (riv.), England 13/H7
Vera, Argentina 143/F5
Vera, Ill. (†62080) 222/D4
Vera, Okla. (74082) 288/P2
Vera (lag.), Paraguay 144/D5
Vera, Spain 33/F4
Vera, Texas (76383) 303/E4
Vera, Va. (†24522) 307/L6
Vera Cruz, Brazil 135/B3
Vera Cruz, Ind. (†46714) 227/G3
Veracruz (state), Mexico 150/L7
Veracruz, Mexico 150/Q1
Veracruz, Mexico 2/E5
Veracruz, Mexico 146/J8
Veradale, Wash. (99037) 310/H3
Veragua Abajo, Dom. Rep. 158/E5
Veras, Uruguay 145/C2
Veraval, India 68/C4
Verbania, Italy 34/B2
Verbena, Ala. (36091) 195/E5
Verboort, Oreg. (†97116) 291/A2
Vercelli (prov.), Italy 34/B2
Vercelli, Italy 34/B2
Verchères (co.), Québec 172/J4
Verchères, Québec 172/J4
Verçinin Tepesi (mt.), Turkey 63/J2
Verda, Ky. (†40828) 237/P7
Verda, La. (71481) 238/E3
Verde (riv.), Ariz. 188/D4
Verde (riv.), Ariz. 198/D5
Verde (cay), Bahamas 156/C2
Verde (riv.), Brazil 132/C7
Verde (riv.), Mexico 150/F3
Verde (riv.), Mexico 150/L8
Verde (riv.), Paraguay 144/C3
Verde (cape), Senegal 102/A3
Verde (cape), Senegal 106/A6
Verde Island (passage), Philippines 82/C4
Verdel, Nebr. (68782) 264/F2
Verden, Okla. (73092) 288/K4
Verden, W. Germany 22/C2
Verdery, S.C. (†29819) 296/C3
Verdi, Minn. (56179) 255/B6
Verdi, Nev. (89439) 266/B3
Verdigre, Nebr. (68783) 264/F2
Verdigris (riv.), Kansas 232/G5
Verdigris, Okla. (†74017) 288/P2
Verdigris (riv.), Okla. 288/P2
Verdinho (riv.), Brazil 132/D7
Verdon, Nebr. (68457) 264/J4
Verdon, S. Dak. (57478) 298/N3
Verdun, Québec 172/H4
Verdún, Uruguay 145/D5
Verdun-sur-Meuse, France 28/F3
Verdunville, W. Va. (25649) 312/B7
Vereeniging, S. Africa 102/E7
Vereeniging, S. Africa 118/D5
Veregin, Sask. 181/K4
Verendrye, N. Dak. (†58717) 282/J3
Vereshchagino, U.S.S.R. 52/H3
Vergara (cape), Guinea 106/B6
Vergara, Argentina 143/H7
Vergara, Spain 33/E1
Vergara, Uruguay 145/E3
Vergas, Minn. (56587) 255/C4
Vergeletto, Switzerland 39/G4
Vergennes, Ill. (62994) 222/D6
Vergennes, Vt. (05491) 268/A3
Veribest, Texas (76886) 303/D6
Verín, Spain 33/C2
Veríssimo, Brazil 135/B1
Verkhnevilyuysk, U.S.S.R. 48/N3
Verkhnyaya Toyma, U.S.S.R. 52/G2
Verkhoyansk, U.S.S.R. 2/R2
Verkhoyansk, U.S.S.R. 4/C3
Verkhoyansk, U.S.S.R. 54/P3
Verkhoyansk, U.S.S.R. 48/N3
Verkhoyansk (range), U.S.S.R. 48/N3
Verkhoyansk (range), U.S.S.R. 4/C3
Verkhoyansk (range), U.S.S.R. 54/O3
Verkniy At-Uryakh, U.S.S.R. 48/Q3
Verlo, Sask. 181/C5
Vermejo (riv.), N. Mex. 274/E2
Vermejo Park, N. Mex. (†81091) 274/D2
Vermilion, Alberta 182/E3
Vermilion (riv.), Alberta 182/E3
Vermilion (cliffs), Ariz. 198/D2
Vermilion (co.), Ill. 222/F3
Vermilion, Ill. (61955) 222/F4
Vermilion (riv.), Ind. 227/B4
Vermilion (par.), La. 238/F7
Vermilion (bay), La. 238/F7
Vermilion (lake), Minn. 188/H1
Vermilion (lake), Minn. 255/F3
Vermilion (range), Minn. 255/F3
Vermilion (riv.), Minn. 255/F2
Vermilion, Ohio (44089) 284/F3
Vermilion (riv.), Ohio 284/F3
Vermilion (hills), Sask. 181/E5
Vermilion (cliffs), Utah 304/B6
Vermilion Bay, Ontario 177/G4
Vermilion Bay, Ontario 175/B3
Vermilion Grove, Ill. (†61870) 222/F4
Vermillion (co.), Ind. 227/C5
Vermillion, Kansas (66544) 232/F2
Vermillion, Minn. (55085) 255/F6
Vermillion, S. Dak. (57069) 298/R8
Vermillion (riv.), S. Dak. 298/P6
Vermillion (riv.), Québec 172/D2
Vermont 188/M2
Vermont, Ill. (61484) 222/C3
Vermont (state), U.S. 146/L5
VERMONT 268
Vermontville, Mich. (49096) 250/E6
Vernal, Utah (84078) 304/E3
Vernayaz, Switzerland 39/D4
Verndale, Minn. (56481) 255/C4
Verndon, Minn. (†55752) 255/E4
Verner, Ontario 177/D1
Verneuil-sur-Avre, France 28/D3
Vernon, Ala. (35592) 195/B3
Vernon, Ariz. (85940) 198/F4
Vernon, Br. Col. 162/E5
Vernon, Br. Col. 184/H5
Vernon, Colo. (80755) 208/P3
Vernon○, Conn. (06066) 210/F1
Vernon, Fla. (32462) 212/C6
Vernon, France 28/D3
Vernon, Ill. (62892) 222/D5
Vernon, Ind. (47282) 227/F7
Vernon, Ky. (†42151) 237/L7
Vernon (par.), La. 238/D4
Vernon, La. (†71228) 238/E2
Vernon (lake), La. 238/D4
Vernon, Mich. (48476) 250/F6
Vernon (co.), Mo. 261/D7
Vernon, N.J. (07462) 273/E1
Vernon, Okla. (74877) 288/P4
Vernon, Ontario 177/J2
Vernon (lake), Ontario 177/E2
Vernon, Pr. Edward I. 168/E2
Vernon, Texas (76384) 303/E3
Vernon, Utah (84080) 304/B3
Vernon○, Vt. (05354) 268/B6
Vernon (co.), Wis. 317/E8
Vernonburg, Georgia (†31401) 217/K7
Vernon Center, Conn. (†06066) 210/F1
Vernon Center, Minn. (56090) 255/D7
Vernon Fork (creek), Ind. 227/F7
Vernon Hill, Va. (24597) 307/K7
Vernon Hills, Ill. (60061) 222/B4
Vernonia, Oreg. (97064) 291/D2
Vero Beach, Fla. (32960) 212/F4
Veroli, Italy 34/D4
Verona, Ill. (60479) 222/E2
Verona (prov.), Italy 34/C2
Verona, Italy 34/C2
Verona, Italy 7/F4
Verona, Ky. (41092) 237/M3
Verona, Miss. (38879) 256/G2
Verona, Mo. (65769) 261/E9
Verona, N.J. (07044) 273/B2
Verona, N. Dak. (58490) 282/O7
Verona, Ohio (45378) 284/A6
Verona, Ontario 177/H3
Verona, Pa. (15147) 294/C6
Verona, Va. (24482) 307/K4
Verona, Wis. (53593) 317/G9
Verónica, Argentina 143/H7
Verpelét, Hungary 41/F2
Verret (lake), La. 238/H7
Verret, New Bruns. 170/B1
Verrettes, Haiti 158/C5
Vérroia, Greece 45/F5
Versailles, Conn. (06383) 210/G2
Versailles, France 28/A2
Versailles, France 7/E4
Versailles, Ill. (62378) 222/C4
Versailles, Ind. (47042) 227/G6
Versailles, Ky. (40383) 237/M4
Versailles, Mo. (65084) 261/G6
Versailles, N.Y. (14168) 276/B6
Versailles, Ohio (45380) 284/A5
Versailles, Pa. (15132) 294/C7
Versalles, Bolivia 136/D3
Vershire○, Vt. (05079) 268/C4
Versoix, Switzerland 39/B4
Verte (bay), New Bruns. 170/G2
Verte (bay), Nova Scotia 168/D2
Verte (isl.), Québec 172/H1
Vertientes, Cuba 158/G3
Vert-Pré, Martinique 161/D6
Vertus, France 28/E3
Verviers, Belgium 27/H7
Verwoerd, Hendrik (dam), S. Africa 118/D6
Verwood, Sask. 181/F6
Vesava, India 68/B7
Vesdre (riv.), Belgium 27/H7
Veseleyville, N. Dak. (†58237) 282/R3
Veseli, Minn. (55086) 255/E6
Veselí nad Lužnicí, Czech. 41/C2
Veselí nad Moravou, Czech. 41/D2
Vesoul, France 28/F4
Vesper, Kansas (†67455) 232/D2
Vesper, Sask. 181/D5
Vesper, Wis. (54489) 317/F7
Vesta, C. Rica 154/F6
Vesta, Minn. (56292) 255/C6
Vesta, Va. (24177) 307/H7
Vestaburg, Mich. (48891) 250/E5
Vest-Agder (co.), Norway 18/E7
Vestal○, N.Y. (13850) 276/H6
Vestavia Hills, Ala. (35216) 195/E4
Vesterålen (isls.), Norway 7/F2
Vesterålen (isls.), Norway 18/J2
Vester Skerninge, Denmark 21/D7
Vestervig, Denmark 21/B4
Vestfjord (fjord), Norway 18/H3
Vestfjorden (fjord), Norway 7/F2
Vestfold (co.), Norway 18/G7
Vestmannaeyjar, Iceland 7/C2
Vestmannaeyjar, Iceland 21/B2
Vestsjaelland (co.), Denmark 21/E6
Vestvågøya (isl.), Norway 18/H3
Vesuvius (vol.), Italy 34/E4
Vesuvius, Va. (24483) 307/K5
Veszprém (co.), Hungary 41/D3
Veszprém, Hungary 41/D3
Vésztő, Hungary 41/F3
Vetal, S. Dak. (57575) 298/G7
Veteran, Alberta 182/E3
Veteran, Wyo. (82243) 319/H4
Vetlanda, Sweden 18/J8
Vetluga (riv.), U.S.S.R. 52/G3
Veurne, Belgium 27/B6
Vevay, Ind. (47043) 227/G7
Vevey, Switzerland 39/C4
Vex, Switzerland 39/D4
Veyo, Utah (†84722) 304/A6
Veys, Iran 66/F5
Veytaux, Switzerland 39/C4
Vezirköprü, Turkey 63/F2
Viacha, Bolivia 120/C4
Viacha, Bolivia 136/A5
Viadana, Italy 34/C2
Viale, Argentina 143/F5
Vian, Okla. (74962) 288/S4
Viana, Brazil 132/E3
Viana del Bollo, Spain 33/C1
Viana do Alentejo, Spain 33/C3
Viana do Castelo, Portugal 33/B2
Vianden, Luxembourg 27/J9
Vianen, Netherlands 27/G5
Viangchan (Vientiane), Laos 72/D3
Viangphoukha, Laos 72/D2
Viano do Castelo (dist.), Portugal 33/B2
Viareggio, Italy 34/C3
Vibank, Sask. 181/H5
Vibbard, Mo. (†64062) 261/D4
Viborg (co.), Denmark 21/C4
Viborg, Denmark 18/F8
Viborg, Denmark 21/C5
Viborg, S. Dak. (57070) 298/P7
Vibo Valentia, Italy 34/F5
Viburnum, Mo. (65566) 261/K7
Viby, Denmark 21/F6
Vicálvaro, Spain 33/G4
Vicam, Mexico 150/D3
Vicco, Ky. (41773) 237/P6
Vicente Guerrero, Baja California, Mexico 150/A1
Vicente Guerrero, Durango, Mexico 150/G5
Vicente López, Argentina 143/G7
Vicenza (prov.), Italy 34/C2
Vicenza, Italy 34/C2
Viceroy, Sask. 181/F6
Vich, Spain 33/H2
Vichacla, Bolivia 136/C7
Vichada (comm.), Colombia 126/F5
Vichada (riv.), Colombia 126/F5
Vichadero, Uruguay 145/E2
Vichaya, Bolivia 136/A5
Viche, Ecuador 128/C2
Vichuga, U.S.S.R. 52/F3
Vichy, France 28/E4
Vichy, Mo. (65580) 261/J6
Vici, Okla. (73859) 288/H2
Vick, Ark. (†71648) 202/F7
Vick, La. (71372) 238/F4
Vickers (lake), Manitoba 179/F3
Vickery, Ohio (43464) 284/D3
Vicksburg, Ariz. (†85348) 198/B5
Vicksburg, Ind. (†47441) 227/C6
Vicksburg, Mich. (49097) 250/D6
Vicksburg, Miss. 188/H4
Vicksburg, Miss. (39180) 256/C6
Vicksburg Nat'l Mil. Park, Miss. 256/C6
Viçosa, Brazil 135/E2
Viçosa, Brazil 132/G5
Vicosoprano, Switzerland 39/J4
Vicovaro, Italy 34/F6
Victoire, Sask. 181/E3
Victor, Calif. (95253) 204/C9
Victor, Colo. (80860) 208/J5
Victor, Idaho (83455) 220/G6
Victor, Iowa (52347) 229/J5
Victor, Mont. (59875) 262/B4
Victor, N.Y. (14564) 276/F5
Victor, S. Dak. (†57260) 298/R2
Victor, W. Va. (25938) 312/D6
Victor Harbor, S. Australia 88/F7
Victor Harbor, S. Australia 94/F6
Victoria 88/G7
Victoria (lake) 2/L6
Victoria (lake) 102/F5
Victoria (Mosi-Oa-Tunya) (falls) 102/E6
Victoria, Argentina 143/F6
Victoria, Ark. (72388) 202/K2
Victoria (state), Australia 87/E9
Victoria (cap.), Br. Col. 146/F5
Victoria (cap.), Br. Col. 162/D6
Victoria (cap.), Br. Col. 184/K4
Victoria (mt.), Burma 72/B2
Victoria (Limbe), Cameroon 115/A3
Victoria (isl.), Canada 2/D2
Victoria (isl.), Canada 4/B15
Victoria, Malleco, Chile 138/D2
Victoria, Tarapacá, Chile 138/A3
Victoria, Grenada 161/D8
Victoria, Guinea 106/B6
Victoria, Ill. (61485) 222/C2
Victoria, Kansas (67671) 232/C3
Victoria (lake), Kenya 115/F4
Victoria, Malta 34/E6
Victoria, Minn. (55386) 255/F6
Victoria, Miss. (38679) 256/E1
Victoria, Mo. (†63020) 261/M6
Victoria (co.), New Bruns. 170/C1
Victoria, Newf. 166/D2
Victoria (lake), Newf. 166/C4
Victoria (lake), N.S. Wales 97/A3
Victoria (riv.), North. Terr. 88/E3
Victoria (riv.), North. Terr. 93/B3
Victoria (isl.), N.W.T. 146/G2
Victoria (isl.), N.W.T. 162/E1
Victoria (isl.), N.W. Terrs. 187/G2
Victoria (str.), N.W.T. 162/F2
Victoria (str.), N.W. Terrs. 187/H3
Victoria (co.), Nova Scotia 168/H2
Victoria (county), Ontario 177/F3
Victoria (lake), Ontario 177/F2
Victoria (peaks), Philippines 82/B6
Victoria, Pr. Edward I. 168/E2
Victoria (cap.), Seychelles 118/H5
Victoria (lake), Tanzania 115/F4
Victoria, Tenn. (†37397) 237/K10
Victoria (co.), Texas 303/H9
Victoria, Texas 188/G5
Victoria, Texas (77901) 303/H9
Victoria (lake), Uganda 115/F4
VICTORIA 97
Victoria, Va. (23974) 307/M6
Victoria (falls), Zambia 115/E7
Victoria (falls), Zimbabwe 118/C3
Victoria Beach, Manitoba 179/F4
Victoria Beach, Nova Scotia 168/C4
Victoria de las Tunas, Cuba 158/H3
Victoria Harbour, Ontario 177/E3

Victoria Land (reg.), Ant. 2/T10
Victoria Land (reg.) 5/B8
Victoria River Downs, North. Terr. 88/E3
Victoria River Downs, North. Terr. 93/B4
Victoria Road, Ontario 177/F3
Victorias, Philippines 82/D5
Victoriaville, Québec 172/F3
Victoria West, S. Africa 118/C6
Victorica, Argentina 143/C4
Victorino, Colombia 126/F6
Victorino, Venezuela 124/E6
Victor Rosales, Mexico 150/H5
Victorville, Calif. (92392) 204/H9
Victory, Ky. (40767) 237/N6
Victory, Wis. (54663) 317/D9
Victory Mills (Victory), N.Y. (12884) 276/N4
Vicuña, Chile 138/A8
Vicuña Mackenna, Argentina 143/D3
Vida, Mo. (†65401) 261/J7
Vida, Mont. (59274) 262/L3
Vida, Oreg. (97488) 291/E3
Vidal, Calif. (92280) 204/L9
Vidal Gormaz (isl.), Chile 138/D9
Vidalia, Georgia (30474) 217/H6
Vidalia, La. (71373) 238/G3
Videbaek, Denmark 21/B5
Videle, Romania 45/G3
Vidette, Georgia (†30830) 217/H4
Vidigueira, Portugal 33/C3
Vidin, Bulgaria 45/F4
Vidisha, India 68/D4
Vidor, Texas (77662) 303/L7
Vidora, Sask. 181/B6
Vidrine, La. (†70586) 238/F5
Viechtach, W. Germany 22/E4
Viedma, Argentina 143/D5
Viedma, Argentina 120/C7
Viedma (lake), Argentina 120/B7
Viedma (lake), Argentina 143/B6
Vieille Case, Dominica 161/E5
Vieira de Leiria, Portugal 33/B3
Vieja, Sierra (mts.), Texas 303/C11
Viella, Spain 33/G1
Vielsalm, Belgium 27/H8
Vienna (city), Austria 41/D2
Vienna (cap.), Austria 41/D2
Vienna (cap.), Austria 7/F4
Vienna, Georgia (31092) 217/E6
Vienna, Ill. (62995) 222/E6
Vienna, Ind. (†47170) 227/F7
Vienna, La. (†71270) 238/E1
Vienna○, Maine (04360) 243/D6
Vienna, Md. (21869) 245/P7
Vienna, Mo. (65582) 261/H6
Vienna, N.J. (07880) 273/D2
Vienna, Ohio (44473) 284/J3
Vienna, Ontario 177/D5
Vienna, S. Dak. (57271) 298/O4
Vienna, Va. (22180) 307/R2
Vienna, W. Va. (26105) 312/D4
Vienne (dept.), France 28/D4
Vienne, France 28/F5
Vienne (riv.), France 28/D4
Vientiane (Viangchan) (cap.), Laos 72/D3
Vientiane (cap.), Laos 54/M8
Viento (pt.), P. Rico 161/E3
Vieques, P. Rico 156/G1
Vieques (Isabel Segunda), P. Rico 161/G2
Vieques (isl.), P. Rico 161/G2
Vieques (isl.), P. Rico 156/G1
Vieques (passage), P. Rico 161/F2
Vieques (sound), P. Rico 161/G2
Vierge (pt.), St. Lucia 161/G6
Vierkant (pt.), Neth. Ant. 161/E8
Viersen, W. Germany 22/B3
Vierzon, France 28/D4
Viesca, Mexico 150/H4
Vietnam 2/Q5
Vietnam 54/M8
VIETNAM 72
Vieux Desert (lake), Mich. 250/G2
Vieux Desert (lake), Wis. 317/J3
Vieux-Fort, Guadeloupe 161/B7
Vieux-Fort (pt.), Guadeloupe 161/A7
Vieux Fort, St. Lucia 156/G4
Vieux Fort, St. Lucia 161/G7
Vieux Fort (riv.), St. Lucia 161/G6
Vieux-Habitants, Guadeloupe 161/A7
Vievis, U.S.S.R. 53/C3
Viewpark, Scotland 15/C2
Vieytes, Argentina 143/H7
Vig, Denmark 21/E6
Viga, Philippines 82/E4
Vigan, Philippines 85/F2
Vigan, Philippines 82/C2
Vigevano, Italy 34/B2
Vigia, Brazil 132/E3
Vigía (cay), Colombia 126/A10
Vignemale (mt.), France 28/C6
Vigo (co.), Ind. 227/C6
Vigo, Spain 7/D4
Vigo, Spain 33/B1
Vigrestad, Norway 18/D7
Vigsø (bay), Denmark 21/B3
Viipuri (Vyborg), U.S.S.R. 52/C2
Vijayawada, India 54/K8
Vijayawada, India 68/D5
Vijosë (riv.), Albania 45/D5
Vik, Norway 18/E6
Viking, Alberta 182/E3
Viking (lake), Manitoba 179/G3
Viking, Minn. (56760) 255/B2
Vikna (isls.), Norway 18/G4
Vila (cap.), Vanuatu 87/G7
Vila Arminda Monteiro, Indonesia 85/H7
Vilacaya, Bolivia 136/C6
Vila de Maganja, Mozambique 118/F3
Vila de Sena, Mozambique 118/E3
Vila do Bispo, Portugal 33/B4
Vila do Conde, Portugal 33/B2
Vila do Porto, Portugal 33/D2
Vila Fontes, Mozambique 118/E3
Vilafranca del Penedés, Spain 33/G2
Vila Franca de Xira, Portugal 33/B3
Vila Guilherme Capelo, Angola 115/B5
Vilaine (riv.), France 28/C4
Vilaka, U.S.S.R. 53/D2
Vilanculos, Mozambique 118/E4
Vila Nova de Foz Côa, Portugal 33/C2
Vila Nova de Gaia, Portugal 33/B2
Vila Nova de Milfontes, Portugal 33/B4
Vila Nova do Seles, Angola 115/B6
Vila Paiva de Andrada, Mozambique 118/E3
Vila Pouca de Aguiar, Portugal 33/C2
Vila Real (dist.), Portugal 33/C2
Vila Real, Portugal 33/C2
Vila Real de Santo António, Portugal 33/C4
Vilar Formoso, Portugal 33/C2
Vilas, Colo. (81087) 208/P8
Vilas, N.C. (28692) 281/F2
Vilas, S. Dak. (†57349) 298/O6
Vilas (co), Wis. 317/G3
Vila Salazar, Indonesia 85/H7
Vila Velha, Brazil 132/D2
Vila Velha Argolas, Brazil 132/F8
Vila Velha de Ródão Portugal 33/C3
Vila Verde, Portugal 33/B2
Vila Viçosa, Portugal 33/C3
Vilcabamba, Cordillera (mts.), Peru 128/F9
Vilcanota (mt.), Peru 128/G10
Vildbjerg, Denmark 21/B5
Vildo, Tenn. (†38075) 237/C10
Vileyka, U.S.S.R. 52/C4
Vilhelmina, Sweden 18/K4
Vilhena, Brazil 132/H10
Viliya (riv.), U.S.S.R. 53/C3
Viljandi, U.S.S.R. 53/C1
Vilkija, U.S.S.R. 53/B3
Vil'kitskogo (str.), U.S.S.R. 4/B4
Vil'kitskogo (str.), U.S.S.R. 48/L2
Villa, Switzerland 39/H3
Villa Abecia, Bolivia 136/C7
Villa Acuña, Mexico 150/J2
Villa Alemana, Chile 138/F2
Villa Alhué, Chile 138/G4
Villa Altagracia, Dom. Rep. 158/E6
Villa Amazónica, Colombia 126/B7
Villa Ana, Argentina 143/E2
Villa Ángela, Argentina 143/D2
Villa Atamisqui, Argentina 143/D2
Villa Atuel, Argentina 143/C3
Villa Bella, Bolivia 136/C2
Villablino, Spain 33/C1
Villa Bruzual, Venezuela 124/D3.
Villabruzzi (Johar), Somalia 115/J3
Villa Bustos, Argentina 143/C2
Villa Cañas, Argentina 143/F6
Villacañas, Spain 33/E3
Villacarriedo, Spain 33/D1
Villacarrillo, Spain 33/E3
Villach, Austria 41/B3
Villacidro, Italy 34/B5
Villa Cisneros (Dakhla), W. Sahara 102/A2
Villa Cisneros (Dakhla), Western Sahara 106/A4
Villa Clara, Argentina 143/G5
Villa Clara (prov.), Cuba 158/E1
Villa Constitución, Argentina 143/F6
Villa Cuauhtémoc, Mexico 150/L5
Villada, Spain 33/D1
Villa Darwin, Uruguay 145/B4
Villa de Cos, Mexico 150/H5
Villa de Cura, Venezuela 124/E2
Villa de Guadelupe Hidalgo, Mexico 150/L1
Villa del Cerro, Uruguay 145/A7
Villa del Rosario, Argentina 143/D3
Villa de María, Argentina 143/D2
Villa de San Antonio, Honduras 154/D3
Villa Diego, Argentina 143/F6
Villadiego, Spain 33/D1
Villa Dolores, Argentina 143/C3
Villa Elisa, Argentina 143/G6
Villa E. Viscarra, Bolivia 136/C6
Villafamés, Spain 33/G2
Villa Federal, Argentina 143/G5
Villa Florida, Paraguay 144/D5
Villa Franca, Paraguay 144/C5
Villafranca, Spain 33/F1
Villafranca del Bierzo, Spain 33/C1
Villafranca del Cid, Spain 33/F2
Villafranca de los Barros, Spain 33/C3
Villafranca di Verona, Italy 34/C2
Villa Frontado, Venezuela 124/G2
Villa Frontera, Mexico 150/J3
Villa García, Mexico 150/J5
Villagarcía, Spain 33/B1
Village, Ark. (71769) 202/D7
Village, N.S. Wales 97/J2
Village-Blier, Québec 172/H2
Village-des-Aulnaies, Québec 172/G2
Village Mills, Texas (77663) 303/K7
Villa General Pérez, Bolivia 136/A4
Villa General Ramírez, Argentina 143/F6
Villa General Roca, Argentina 143/C3
Village-Saint-Laurent, New Bruns. 170/E1
Villagran, Mexico 150/K4
Villa Grove, Colo. (81155) 208/G6
Villa Grove, Ill. (61956) 222/E4
Villaguay, Argentina 143/G5
Villa Guillermina, Argentina 143/D2
Villa Hayes, Paraguay 144/A4
Villahermosa, Mexico 146/J8
Villahermosa, Mexico 150/N8
Villahermosa, Spain 33/E3
Villa Hidalgo, Durango, Mexico 150/G3
Villa Hidalgo, Sonora, Mexico 150/E1
Villa Hills, Ky. (†41017) 237/R2
Villa Huidobro, Argentina 143/D3
Villa Industrial, Chile 138/B1
Villa Ingavi, Bolivia 136/D7
Villajoyosa, Spain 33/F3
Villa Krause, Argentina 143/C3
Villalba, P. Rico 161/C2
Villalba, Spain 33/C1
Villalbín, Paraguay 144/D5
Villaldama, Mexico 150/J3
Villalon, Philippines 82/E5
Villalón de Campos, Spain 33/D1
Villalpando, Spain 33/D2
Villa Mantero, Argentina 143/G6
Villa María, Argentina 143/D3
Villa María, Argentina 120/C6
Villa María Grande, Argentina 143/F5
Villa Martín, Bolivia 136/B7
Villa Matamoros, Mexico 150/G3
Villamil, Ecuador 128/B9
Villa Montes, Bolivia 120/C5
Villa Montes, Bolivia 136/D7
Villanova, Pa. (19085) 294/M6
Villanow, Georgia (†30728) 217/B1
Villa Nueva, Argentina 143/C3
Villanueva, Colombia 126/D2
Villanueva, Mexico 150/H5
Villanueva, N. Mex. (87583) 274/D3
Villanueva de Córdoba, Spain 33/D3
Villanueva del Arzobispo, Spain 33/E3
Villanueva de la Serena, Spain 33/D3
Villanueva de los Infantes, Spain 33/E3
Villanueva y Geltrú, Spain 33/G2
Villány, Hungary 41/E4
Villa Ocampo, Argentina 143/D2
Villa Oliva, Paraguay 144/C5
Villa Orías, Bolivia 136/C6
Villa Park, Calif. (92667) 204/D11
Villa Park, Ill. (60181) 222/B5
Villar, Bolivia 136/C6
Villa Ranchaero, S. Dak. (†57701) 298/C5
Villarcayo, Spain 33/E1
Villard, Minn. (56385) 255/C5
Villar del Arzobispo, Spain 33/F3
Villareal, Philippines 82/E5
Villa Regina, Argentina 143/C4
Villa Rica, Georgia (30180) 217/C3
Villa Ridge, Ill. (62996) 222/D6
Villa Riva, Dom. Rep. 158/E5
Villa Rosario, Colombia 126/D4
Villarreal de los Infantes, Spain 33/G3
Villarrica, Chile 138/E2
Villarrica (lake), Chile 138/E2
Villarrica, Paraguay 120/D5
Villarrica, Paraguay 144/C5
Villarrobledo, Spain 33/E3
Villarrubia de los Ojos, Spain 33/E3
Villas, N.J. (08251) 273/D5
Villasana de Mena, Spain 33/E1
Villa San Agustín, Argentina 143/C3
Villa San José, Argentina 143/G6
Villa San Martín, Argentina 143/D2
Villa Serrano, Bolivia 136/C6
Villa Tasso, Fla. (†32548) 212/C6
Villa Tunari, Bolivia 136/C5
Villa Unión, Argentina 143/C2
Villa Unión, Coahuila, Mexico 150/J2
Villa Unión, Durango, Mexico 150/H5
Villa Unión, Sinaloa, Mexico 150/F5
Villa Vaca Guzmán, Bolivia 136/D6
Villaverde, Spain 33/F4
Villavicencio, Colombia 120/B2
Villavicencio, Colombia 126/D5
Villa Vicente Guerrero, Mexico 150/N1
Villaviciosa, Spain 33/D1
Villawood, N.S. Wales 88/K4
Villawood, N.S. Wales 97/H3
Villa y Puerto del Son, Spain 33/B1
Villazón, Bolivia 120/C5
Villazón, Bolivia 136/C7
Ville Bonheur, Haiti 158/C6
Villefranche, France 28/G6
Villefranche-de-Lauragais, France 28/D6
Villefranche-de-Rouergue, France 28/E5
Villefranche-sur-Saône, France 28/F4
Villegreen, Colo. (81088) 208/M8
Villejuif, France 28/B2
Ville-Marie, Québec 174/B3
Villemomble, France 28/C1
Villena, Spain 33/F3
Villeneuve, Québec 172/J3
Villeneuve, Switzerland 39/C4
Villeneuve-Saint-Georges, France 28/E3
Villeneuve-sur-Lot, France 28/D5
Ville Platte, La. (70586) 238/F5
Villeroy, Québec 172/F3
Villeta, Colombia 126/C5
Villeta, Paraguay 144/A5
Villeurbanne, France 28/F5
Villiers (lake), Québec 172/C2
Villiersdorp, S. Africa 118/G6
Villingen-Schwenningen, W. Germany 22/C4
Villisca, Iowa (50864) 229/C7
Villupuram, India 68/D6
Vilna, Alberta 182/E2
Vilna, U.S.S.R. 7/G3
Vilna (Vilnius) (cap.), U.S.S.R. 53/C3
Vilna (Vilnius), U.S.S.R. 52/C4
Vilna (Vilnius), U.S.S.R. 48/C4
Vilnius (Vilna) (cap.), U.S.S.R. 53/C3
Vilonia, Ark. (72173) 202/F3
Vils, Austria 41/A3
Vilvoorde, Belgium 27/F7
Vilvorde (Vilvoorde), Belgium 27/F7
Vilyuy (range), U.S.S.R. 48/M3
Vilyuy (res.), U.S.S.R. 48/M3
Vilyuy (riv.), U.S.S.R. 48/L3
Vilyuy (riv.), U.S.S.R. 54/N3
Vilyuysk, U.S.S.R. 48/N3
Vimianzo, Spain 33/B1
Vimioso, Portugal 33/C2
Vimmerby, Sweden 18/J8
Vimperk, Czech. 41/B2
Vimy, Alberta 182/D2
Vimy Ridge, Ark. (†72002) 202/F4
Vina, Ala. (35593) 195/B2
Vina, Calif. (96092) 204/D4
Viña del Mar, Chile 120/B6
Viña del Mar, Chile 138/F2
Viñales, Cuba 156/A2
Viñales, Cuba 158/A1
Vinalhaven○, Maine (04863) 243/F7
Vinalhaven (isl.), Maine 243/F7
Vinaroz, Spain 33/G2
Vincennes (bay) 5/C6
Vincennes, France 28/B2
Vincennes, Ind. (47591) 227/C7
Vincennes, Iowa (†52619) 229/K7
Vincent, Ala. (35178) 195/F4
Vincent, Ark. (†72327) 202/K3
Vincent, Iowa (50594) 229/E3
Vincent (pt.), Norfolk I. 88/K5
Vincent, Ohio (45784) 284/G7
Vincentown, N.J. (08088) 273/D4
Vinces, Ecuador 128/C3
Vinchina, Argentina 143/C2
Vinchos, Peru 128/E9
Vindelälven (riv.), Sweden 18/L4
Vindeln, Sweden 18/L4
Vinderup, Denmark 21/B5
Vindhya (range), India 68/D4
Vinegar Bend, Ala. (36584) 195/B8
Vine Grove, Ky. (40175) 237/K5
Vineland, Colo. (†81001) 208/K6
Vineland, N.J. (08360) 273/C5
Vinemont, Ala. (35179) 195/E2
Vineyard (sound), Mass. 249/L7
Vineyard Haven, Mass. (02568) 249/M7
Vinh, Vietnam 72/E3
Vinh, Vietnam 54/M8
Vinhais, Portugal 33/C2
Vinh Long, Vietnam 72/E5
Vinh Yen, Vietnam 72/E2
Vining, Iowa (52348) 229/J5
Vining, Kansas (†66937) 232/E2
Vining, Minn. (56588) 255/C4
Vinings, Georgia (†30080) 217/K1
Vinita, Okla. (74301) 288/R1
Vinita Park, Mo. (†63101) 261/P2
Vinkovci, Yugoslavia 45/D3
Vinland, Kansas (†66006) 232/G3
Vinnitsa, U.S.S.R. 7/G4
Vinnitsa, U.S.S.R. 52/C5
Vinnitsa, U.S.S.R. 48/C5
Vinogradov, U.S.S.R. 52/B5
Vinson, Okla. (73571) 288/G5
Vinson Massif (mt.) 5/B14
Vinsulla, Br. Col. 184/G5
Vint Hill Farms Mil. Res., Va. 307/N3
Vinton, Iowa (52349) 229/J4
Vinton, La. (70668) 238/C6
Vinton (co.), Ohio 284/E7
Vinton, Ohio (45686) 284/F8
Vinton, Va. (24179) 307/J6
Vintondale, Pa. (15961) 294/E5
Vinukonda, India 68/D5
Viola, Ark. (72583) 202/G1
Viola, Del. (19979) 245/R4
Viola, Idaho (83872) 220/B3
Viola, Ill. (61486) 222/C2
Viola, Iowa (52350) 229/L4
Viola, Kansas (67149) 232/E4
Viola, Ky. (†42051) 237/D7
Viola, Minn. (55980) 255/F6
Viola, Mo. (†65747) 261/E9
Viola, N.Y. (†10901) 276/J8
Viola, Tenn. (37394) 237/K9
Viola, Wis. (54664) 317/E8
Violet, La. (70092) 238/P4
Violet Grove, Alberta 182/C3
Violet Hill, Ark. (72584) 202/G1
Violette Brook, New Bruns. 170/C1
Violet Valley Aboriginal Reserve, W. Australia 88/D3
Viqueque, Indonesia 85/H7
Virac, Philippines 82/E4
Virago (sound), Br. Col. 184/A3
Virajpet, India 68/D6
Viramgam, India 68/C4
Viranşehir, Turkey 63/H4
Virden, Ill. (62690) 222/D4
Virden, Manitoba 179/A5
Virden, N. Mex. (†88055) 274/A6
Vire, France 28/C3
Virgelle, Mont. (†59460) 262/F2
Virgie, Ky. (41572) 237/R6
Virgil, Kansas (66870) 232/F4
Virgil, S. Dak. (57379) 298/N5
Virgilina, Va. (24598) 307/L7
Virgin (mts.), Ariz. 198/B2
Virgin (riv.), Ariz. 198/B2
Virgin (mts.), Nev. 266/G6
Virgin (peak), Nev. 266/G6
Virgin (riv.), Nev. 266/G6
Virgin (creek), S. Dak. 298/H3
Virgin, Utah (84779) 304/A6
Virgin (riv.), Utah 304/A6
Virgin, East Fork (riv.), Utah 304/B6
Virgin (str.), Virgin Is. (U.S.) 161/A4
Virgin Gorda (isl.), Virgin Is. (Br.) 156/H1
Virginia 188/L3
VIRGINIA 307
Virginia (key), Fla. 212/B5
Virginia, Idaho (†83234) 220/F7
Virginia, Ill. (62691) 222/C4
Virginia, Ireland 17/G4
Virginia, Minn. 188/H1
Virginia, Minn. (55792) 255/F3
Virginia, Nebr. (68458) 264/H4
Virginia (range), Nev. 266/B3
Virginia, S. Australia 94/B7
Virginia (state), U.S. 146/L6
Virginia Beach, Va. 188/L3
Virginia Beach (I.C.), Va. (*23450) 307/S7
Virginia City, Mont. (59755) 262/E5
Virginia City, Nev. (89440) 266/B3
Virginia Dale, Colo. (80548) 208/J1
Virginia Gardens, Fla. (†33166) 212/B5
Virginiatown, Ontario 177/K5
Virginiatown, Ontario 175/E3
Virgin Islands (isls.) 146/M8
VIRGIN ISLANDS (U.S.) 161
VIRGIN ISLANDS (U.S.) 156/H1
VIRGIN ISLANDS (Br.) 161
VIRGIN ISLANDS (British) 156/H1
Virgin Isls. Nat'l Park, Virgin Is. (U.S.) 161/C4
Virginville, W. Va. (†26035) 312/F2
Virochey, Cambodia 72/E4
Viroinval, Belgium 27/F8
Viroqua, Wis. (54665) 317/D8
Virovitica, Yugoslavia 45/C3
Virserum, Sweden 18/J8
Virton, Belgium 27/H9
Virtsu, U.S.S.R. 53/B1
Virú, Peru 128/C7
Virunga (range), Rwanda 115/E4
Virunga (range), Uganda 115/E4
Virunga, Zaire 115/E5
Virunga (range), Zaire 115/E4
Virunga Nat'l Park, Uganda 115/F4
Virunga Nat'l Park, Zaire 115/E4
Vis (isl.), Yugoslavia 45/C4
Visakhapatnam, India 54/K8
Visakhapatnam, India 68/E5
Visalia, Calif. 188/C3
Visalia, Calif. (93277) 204/F7
Visalia, Ky. (†41063) 237/N3
Visayan (sea), Philippines 82/D5
Visayan (sea), Philippines 85/G3
Visby, Sweden 18/L8
Visby, Sweden 7/F3
Viscondé dos Rio Branco, Brazil 135/E2
Viscount, Sask. 181/F4
Viscount Melville (sound), Canada 4/B15
Viscount Melville (sound), N.W.T. 162/E1
Viscount Melville (sound), N.W.T. 146/H2
Viscount Melville (sound), N.W. Terrs. 187/G2
Visé, Belgium 27/H7
Višegrad, Yugoslavia 45/D4
Viseu, Brazil 132/E3
Viseu (dist.), Portugal 33/C2
Viseu, Portugal 33/C2
Vişeul de Sus, Romania 45/F2
Vishera (riv.), U.S.S.R. 52/J2
Vishoek, S. Africa 118/E7
Vislanda, Sweden 18/H8
Visnagar, India 68/C4
Viso (mt.), Italy 34/A2
Visoko, Yugoslavia 45/D4
Visp, Switzerland 39/E4
Visp (riv.), Switzerland 39/E4
Vissoie, Switzerland 39/E4
Vista, Calif. (92083) 204/H10
Vista, Manitoba 179/B4
Vista, Mo. (64789) 261/E7
Vista Hermosa, Cuba 158/G3
Vistula, Ind. (†46507) 227/F1
Vistula (riv.), Poland 7/F3
Vistula (riv.), Poland 47/D2
Vistula (spit), Poland 47/D1
Vit (riv.), Bulgaria 45/G4
Vita, Manitoba 179/F5
Vitali (isl.), Philippines 82/D7
Vitebsk, U.S.S.R. 7/H3
Vitebsk, U.S.S.R. 52/C3
Vitebsk, U.S.S.R. 48/D4
Viterbo (prov.), Italy 34/C3
Viterbo, Italy 34/C3
Vitiaz (str.), Papua N.G. 85/B7
Vitiaz (str.), Papua N.G. 86/A2
Vitichi, Bolivia 136/C7
Vitigudino, Spain 33/C2
Viti Levu (isl.), Fiji 87/H7
Viti Levu (isl.), Fiji 86/P11
Vitim (riv.), U.S.S.R. 54/N4
Vitim (riv.), U.S.S.R. 48/M4
Vitimskiy, U.S.S.R. 48/M4
Vítkov, Czech. 41/D2
Vitor, Quebrado (riv.), Chile 138/A1
Vitor, Peru 128/G11
Vitor (riv.), Peru 128/F11
Vitória, Brazil 120/F5
Vitória, Brazil 132/G8
Vitoria, Spain 33/E1
Vitoria, Spain 7/D4
Vitória da Conquista, Brazil 120/E4
Vitória da Conquista, Brazil 132/F6
Vitória de Santo Antão, Brazil 132/G4
Vitória de Sto. Antao, Brazil 120/F3
Vitré, France 28/C3
Vitry-le-François, France 28/F3
Vitry-sur-Seine, France 28/B2
Vittangi, Sweden 18/M3
Vittel, France 28/F3
Vittoria, Italy 34/E6
Vittoria, Ontario 177/D5
Vittorio Veneto, Italy 34/D1
Vitu (isls.), Papua N.G. 86/B2
Vivero, Spain 33/C1
Vivian, La. (71082) 238/B1
Vivian, S. Dak. (57576) 298/J6
Vivian, W. Va. (24891) 312/D8
Vivorillo (cays), Honduras 154/F3
Vixen, La. (†71418) 238/F2
Vizagapatam (Visakhapatnam), India 68/E5
Vizcaino (cape), Calif. 204/B4
Vizcaya (prov.), Spain 33/E1
Vize, Turkey 63/B2
Vizianagaram, India 68/E5
Vizille, France 28/F5
Vizinga, U.S.S.R. 52/G2
Viziru, Romania 45/H3
Vizovice, Czech. 41/D2
Vizzini, Italy 34/E6
Vlaardingen, Netherlands 27/E5
Vladimir, U.S.S.R. 7/J3
Vladimir, U.S.S.R. 48/D4
Vladimir, U.S.S.R. 52/F3
Vladimir-Volynskiy, U.S.S.R. 52/B4
Vladivostok, U.S.S.R. 54/P5
Vladivostok, U.S.S.R. 2/R3
Vladivostok, U.S.S.R. 48/O5
Vlagtwedde, Netherlands 27/L3
Vlasenica, Yugoslavia 45/D3
Vlašim, Czech. 41/C2
Vleteren, Belgium 27/B7
Vlieland (isl.), Netherlands 27/F2
Vliestroom (str.), Netherlands 27/G2
Vliets, Kansas (66545) 232/F2
Vlijmen, Netherlands 27/G5
Vlissingen (Flushing), Netherlands 27/C6
Vlorë, Albania 45/D5
Vltava (riv.), Czech. 41/C2
Voca, Texas (76887) 303/E7
Vöcklabruck, Austria 41/B2
Voda, Kansas (†67631) 232/C2
Vodl (lake), U.S.S.R. 52/E2
Vodňany, Czech. 41/C2
Vogar, Manitoba 179/D4
Vogel Center, Mich. (†49657) 250/E4
Vogelkop (Doberai) (pen.), Indonesia 85/J6
Vogelsberg (mts.), W. Germany 22/C3
Voghera, Italy 34/B2
Voglers Cove, Nova Scotia 168/D4
Voh, New Caled. 86/G4
Vohibinany, Madagascar 118/H3
Vohimarina (Vohémar), Madagascar 118/J2
Vohimena (cape), Madagascar 102/G7
Vohimena (cape), Madagascar 118/G5
Vohipeno, Madagascar 118/H4
Voi, Kenya 115/G4
Voi, Kenya 102/F5
Voil, Loch (lake), Scotland 15/D4
Voiron, France 28/F5
Voisey (bay), Newf. 166/B2
Voitsberg, Austria 41/C3
Voïvís (lake), Greece 45/F6
Vojens, Denmark 21/C7
Vojmsjön (lake), Sweden 18/J4
Vojnice, Czech. 41/E3
Vojvodina (aut. prov.), Yugoslavia 45/D3
Volador, Colombia 126/C3
Volant, Pa. (16156) 294/B3
Volary, Czech. 41/B2
Volborg, Mont. (59351) 262/L5
Volcano, Calif. (95689) 204/E5
Volcano, Hawaii (96785) 218/J6
Volcano (isls.), Japan 87/E3
Volcano (isls.), Japan 81/M4
Volda, Norway 18/E5
Volendam-Edam, Netherlands 27/G4
Volga, Iowa (52077) 229/L3
Volga, S. Dak. (57071) 298/R5
Volga (riv.), U.S.S.R. 7/J4
Volga (riv.), U.S.S.R. 2/M3
Volga (riv.), U.S.S.R. 48/E5
Volga (riv.), U.S.S.R. 52/G5
Volga, W. Va. (26238) 312/F4
Volga-Don (canal), U.S.S.R. 52/F5
Volgodonsk, U.S.S.R. 52/F5
Volgograd, U.S.S.R. 2/M3
Volgograd, U.S.S.R. 7/J4
Volgograd, U.S.S.R. 48/E5
Volgograd, U.S.S.R. 52/F5
Volgograd (res.), U.S.S.R. 52/G5
Volin, S. Dak. (57072) 298/P8
Völkermarkt, Austria 41/C3
Volkhov, U.S.S.R. 52/D3
Volkhov (riv.), U.S.S.R. 52/D3
Völklingen, W. Germany 22/B4
Volkovysk, U.S.S.R. 52/B4
Volksrust, S. Africa 118/D5
Volney, Mich. (†49309) 250/D5
Volney, Va. (24379) 307/F7
Volochanka, U.S.S.R. 48/K2
Vologda, U.S.S.R. 7/J3
Vologda, U.S.S.R. 48/E4
Vologda, U.S.S.R. 52/F3
Vólos, Greece 7/G5
Vólos, Greece 45/F6
Vol'sk, U.S.S.R. 7/J3
Vol'sk, U.S.S.R. 52/G4
Volta (lake), Ghana 102/B4
Volta (lake), Ghana 106/D7
Volta (riv.), Ghana 102/C4
Volta (riv.), Ghana 106/E7
Volta Grande (res.), Brazil 135/B1
Voltaire, N. Dak. (58792) 282/J3
Volta Redonda, Brazil 120/E5
Volta Redonda, Brazil 132/E8
Volta Redonda, Brazil 135/D3
Volterra, Italy 34/C3
Volturno (riv.), Italy 34/E4
Voluntown○, Conn. (06384) 210/H2
Volusia (co.), Fla. 212/E2
Vólvi (lake), Greece 45/F5
Volyně, Czech. 41/B2
Volyn Oblast, U.S.S.R. 52/C4
Volzhsk, U.S.S.R. 52/G3
Volzhskiy, U.S.S.R. 7/J4
Volzhskiy, U.S.S.R. 52/G5
Vom, Nigeria 106/F7
Vona, Colo. (80861) 208/O4
Vonda, Sask. 181/F3
Vónitsa, Greece 45/E6
Vonore, Tenn. (37885) 237/N9
Von Ormy, Texas (78073) 303/J11
Voorburg, Netherlands 27/E4
Voorhees○, N.J. (†08043) 273/B3
Voorheesville, N.Y. (12186) 276/M5
Voorhies, Iowa (†50643) 229/J4
Voorne (isl.), Netherlands 27/D5
Voorst, Netherlands 27/J4
Vopnafjördhur (fjord), Iceland 21/D1
Vorab (mt.), Switzerland 39/H3
Vorarlberg (prov.), Austria 41/A3
Vorbasse, Denmark 21/B6
Vorden, Netherlands 27/J4
Vordernberg, Austria 41/C3
Vorderrhein (riv.), Switzerland 39/G3

Vordingborg, Denmark 21/E7
Vordingborg, Denmark 18/G9
Vorgod (riv.), Denmark 21/B6
Vorkuta, U.S.S.R. 4/C6
Vorkuta, U.S.S.R. 7/L2
Vorkuta, U.S.S.R. 52/K1
Vorkuta, U.S.S.R. 48/G3
Vormsi (isl.), U.S.S.R. 53/B1
Vorona (riv.), U.S.S.R. 52/F4
Voronezh, U.S.S.R. 7/H3
Voronezh, U.S.S.R. 52/E4
Voronezh, U.S.S.R. 48/E4
Voroshilovgrad, U.S.S.R. 7/H4
Voroshilovgrad, U.S.S.R. 52/E5
Voroshilovgrad, U.S.S.R. 48/E5
Vorskla (riv.), U.S.S.R. 52/E4
Vorst (Forest), Belgium 27/B9
Võrtsjärv (lake), U.S.S.R. 53/D1
Võru, U.S.S.R. 52/C2
Võru, U.S.S.R. 53/D2
Vosges (dept.), France 28/G3
Vosges (mts.), France 28/G3
Voskresensk, U.S.S.R. 52/E3
Voss, N. Dak. (58280) 282/R3
Voss, Norway 18/E6
Vossburg, Miss. (39366) 256/F7
Vostochnyy, U.S.S.R. 48/O5
Vostok (isl.), Kiribati 2/B6
Vostok (isl.), Kiribati 87/L7
Votamo (riv.), Venezuela 124/F6
Votice, Czech. 41/C2
Votkinsk, U.S.S.R. 48/F4
Votkinsk, U.S.S.R. 52/H3
Votuporanga, Brazil 135/B2
Vouvry, Switzerland 39/C4
Voúxa (cape), Greece 45/F8
Vouziers, France 28/F3
Voyageurs Nat'l Park, Minn. 255/F2
Voy-Vozh, U.S.S.R. 48/F3
Voy-Vozh, U.S.S.R. 52/H2
Vozhe (lake), U.S.S.R. 52/F2
Vozhega, U.S.S.R. 52/F2
Vozhma, U.S.S.R. 52/G3
Voznesensk, U.S.S.R. 52/D5
Vrå, Denmark 21/C3
Vráble, Czech. 41/E2
Vracov, Czech. 41/D2
Vrangelya (isl.), U.S.S.R. 54/U2
Vranje, Yugoslavia 45/F4
Vranov nad Teplou, Czech. 41/F2
Vratsa, Bulgaria 45/F4
Vrbas, Yugoslavia 45/D3
Vrbas (riv.), Yugoslavia 45/C3
Vrbno pod Pradědem, Czech. 41/D1
Vrbovce, Czech. 41/D1
Vrbové, Czech. 41/D2
Vrchlabí, Czech. 41/C1
Vrede, S. Africa 118/D5
Vredenburg, S. Africa 118/B6
Vredenburgh, Ala. (36481) 195/D7
Vredendal, S. Africa 118/B6
Vreed-en-Hoop, Guyana 131/B2
Vresse-sur-Semois, Belgium 27/F9
Vriezenveen, Netherlands 27/K4
Vrondádhes, Greece 45/G6
Vršac, Yugoslavia 45/E3
Vrútky, Czech. 41/E2
Vryburg, S. Africa 118/C5
Vryheid, S. Africa 118/E5
Vsetín, Czech. 41/D2
Vsevidof (mt.), Alaska 196/E4
Vuadens, Switzerland 39/C3
Vučitrn, Yugoslavia 45/E4
Vught, Netherlands 27/G5
Vukovar, Yugoslavia 45/D3
Vulcan, Alberta 182/D4
Vulcan, Mich. (49892) 250/B3
Vulcan, Mo. (63675) 261/L8
Vulcan, W. Va. (25697) 312/B7
Vulcano (isl.), Italy 34/E5
Vu Liet, Vietnam 72/E3
Vung Tau, Vietnam 72/E5
Vuollerim, Sweden 18/M3
Vuolvojaure (lake), Sweden 18/L3
Vuotso, Finland 18/P2
Vya, Nev. (†96104) 266/B1
Vyatka (riv.), U.S.S.R. 52/H3
Vyatskiye Polyany, U.S.S.R. 52/H3
Vyazemskiy, U.S.S.R. 48/O5
Vyaz'ma, U.S.S.R. 52/D3
Vyborg, U.S.S.R. 7/G2
Vyborg, U.S.S.R. 52/C2
Vyborg, U.S.S.R. 48/C3
Vychegda (riv.), U.S.S.R. 52/G2
Východočeský (reg.), Czech. 41/C1
Východoslovenský (reg.), Czech. 41/F2
Vyg (lake), U.S.S.R. 52/E2
Vyksa, U.S.S.R. 52/F3
Vym' (riv.), U.S.S.R. 52/H2
Vyshniy Volochek, U.S.S.R. 7/H2
Vyshniy Volochek, U.S.S.R. 52/D3
Vyshniy Volochek, U.S.S.R. 48/D4
Vyškov, Czech. 41/D2
Vysoké Mýto, Czech. 41/D2
Vysoké Tatry, Czech. 41/F2
Vyšší Brod, Czech. 41/C2
Vytegra, U.S.S.R. 52/E2

W

Wa, Ghana 106/D6
Waal (riv.), Netherlands 27/G5
Waalre, Netherlands 27/G6
Waalwijk, Netherlands 27/F5
Waarschoot, Belgium 27/D6
Waas (mt.), Utah 304/E5
Waasis, New Bruns. 170/D3
Wabamun, Alberta 182/C3
Waban, Mass. (†02168) 249/B7
Wabana, Newf. 166/D2
Wabaningo, Mich. (49463) 250/C5
Wabasca, Alberta 182/D2
Wabasca (riv.), Alberta 182/C1
Wabasca (riv.), Alta. 162/E4
Wabash (riv.) 188/J3
Wabash, Ark. (72389) 202/J5
Wabash (co.), Ill. 222/F5
Wabash (riv.), Ill. 222/F5
Wabash (co.), Ind. 227/F3
Wabash, Ind. (46992) 227/F3
Wabash (riv.), Ind. 227/B7
Wabash, Ohio (†45822) 284/A4
Wabash (riv.), Ohio 284/A5
Wabasha (co.), Minn. 255/F6
Wabasha, Minn. (55981) 255/G6
Wabasso, Fla. (32970) 212/F4
Wabasso, Minn. (56293) 255/C6
Wabatawangang (lake), Minn. 255/D3
Wabaunsee (co.), Kansas 232/F3
Wabaunsee, Kansas (†66547) 232/F2
Wabbaseka, Ark. (72175) 202/G5
Wabeno, Wis. (54566) 317/J5
Wabi (riv.), Ethiopia 111/H6
Wabigoon, Ontario 175/B3
Wabigoon, Ontario 177/G5
Wabi Shebelle (riv.) 102/G4
Wabi Shebelle (riv.), Ethiopia 111/H6
Wabowden, Manitoba 179/J3
Wągbrzeźno, Poland 47/D2
Wabuk (pt.), Ontario 175/D1
Wabush, Newf. 166/A3
Wabush, Newf. 162/K5
Wabuska, Nev. (†89447) 266/B3
Waccamaw (lake), N.C. 281/N6
Waccamaw (riv.), N.C. 281/M7
Waccamaw (riv.), S.C. 296/J5
Waccasassa (bay), Fla. 212/D2
Waccasassa (riv.), Fla. 212/D2
Wachapreague, Va. (23480) 307/S5
Wachapreague (inlet), Va. 307/T6
Wachtebeke, Belgium 27/D6
Wachusett (mt.), Mass. 249/G3
Wachusett (res.), Mass. 249/G3
Wacissa, Fla. (32361) 212/B1
Waco, Georgia (30182) 217/B3
Waco, Ky. (40385) 237/N5
Waco, Mo. (63869) 261/C8
Waco, Nebr. (68460) 264/G4
Waco, N.C. (28169) 281/G4
Waco, Texas 188/G4
Waco, Texas 146/J6
Waco, Texas (*76701) 303/G6
Waconda (lake), Kansas 232/D2
Waconia, Minn. (55387) 255/E6
Wadai (reg.), Chad 111/D5
Waddamana, Tasmania 99/C4
Waddan, Libya 102/D2
Waddan, Libya 111/C2
Waddell, Ariz. (85355) 198/C5
Waddenzee (sound), Netherlands 27/G2
Waddington (mt.), Br. Col. 162/D5
Waddington (mt.), Br. Col. 184/E4
Waddington, N.Y. (13694) 276/K1
Waddy, Ky. (40076) 237/L4
Wade, Miss. (†39567) 256/G9
Wade (lake), Newf. 166/A3
Wade, N.C. (28395) 281/M4
Wade, Okla. (†74723) 288/O7
Wadebridge, England 13/C7
Wade-Hampton, S.C. (†29607) 296/C2
Wadena, Ind. (†47944) 227/C3
Wadena, Iowa (52169) 229/K3
Wadena (co.), Minn. 255/D4
Wadena, Minn. (56482) 255/C4
Wadena, Sask. 181/H4
Wädenswil, Switzerland 39/G2
Wadesboro, La. (†70454) 238/M2
Wadesboro, N.C. (28170) 281/J5
Wadestown, W. Va. (26589) 312/F3
Wadesville, Ind. (47638) 227/B8
Wadeville, N.C. (†27306) 281/J4
Wadhams, N.Y. (12990) 276/N2
Wadi Dra, Morocco 102/B2
Wadi es Sir, Jordan 65/D4
Wadi Halfa, Sudan 111/F3
Wadi Musa, Jordan 65/E5
Wading (riv.), N.J. 273/D4
Wading River, N.Y. (11792) 276/P9
Wadley, Ala. (36276) 195/G4
Wadley, Georgia (30477) 217/H5
Wadmalaw (isl.), S.C. 296/G6
Wad Medani, Sudan 111/F5
Wad Medani, Sudan 59/B7
Wad Medani, Sudan 102/F3
Wadowice, Poland 47/D4
Wadsworth, Ala. (†36022) 195/E5
Wadsworth, Ill. (60083) 222/B4
Wadsworth, Nev. (89442) 266/B3
Wadsworth, Ohio (44281) 284/G3
Wadsworth, Texas (77483) 303/J9
Waelder, Texas (78959) 303/G8
Wagait Aboriginal Res., North. Terr. 93/B2
Wagarville, Ala. (36585) 195/B8
Wagener, S.C. (29164) 296/E4
Wageningen, Netherlands 27/H5
Wageningen, Suriname 131/C3
Wager (bay), N.W.T. 146/J3
Wager (bay), N.W.T. 162/G2
Wager (bay), N.W. Terrs. 187/K3
Wagga Wagga, Australia 87/E9
Wagga Wagga, N. S. Wales 88/H7
Wagga Wagga, N.S. Wales 97/D4
Waggoner, Ill. (62572) 222/D4
Waggrakine, W. Australia 92/A5
Wagin, W. Australia 88/B6
Wagin, W. Australia 92/B2
Wagner, Alberta 182/C2
Wagner, Mont. (59543) 262/H2
Wagner, S. Dak. (57380) 298/N7
Wagoner (co.), Okla. 288/P3
Wagoner, Okla. (74467) 288/R3
Wagon Mound, N. Mex. (87752) 274/E2
Wagontire, Oreg. (†97720) 291/H4
Wagon Wheel Gap, Colo. (†81130) 208/F7
Wagram, N.C. (28396) 281/L5
Wągrowiec, Poland 47/C2
Wah, Pakistan 68/C2
Wahai, Indonesia 85/H6
Wahalak, Miss. (†39358) 256/G5
Wahiawa, Hawaii (96786) 218/E2
Wahiawa, Hawaii 188/F5
Wahkiacus, Wash. (98670) 310/D5
Wahkiakum (co.), Wash. 310/B4
Wahkon, Minn. (56386) 255/E4
Wahlern, Switzerland 39/D3
Wahoo, Nebr. (68066) 264/H3
Wahpeton, Iowa (†51360) 229/C2
Wahpeton, N. Dak. 188/G1
Wahpeton, N. Dak. (58075) 282/S7
Wahsatch, Utah (†82930) 304/C2
Wah Wah (mts.), Utah 304/A5
Wahwashkesh (lake), Ontario 177/D2
Wahweap (creek), Utah 304/C6
Wai, Poulo (isls.), Vietnam 72/E4
Waiakoa, Hawaii (†96788) 218/J2
Waialae, Hawaii (96816) 218/D4
Waialeale (mt.), Hawaii 218/C1
Waialee, Hawaii (†96731) 218/E1
Waialua, Hawaii 188/F5
Waialua, Molokai, Hawaii (†96748) 218/H1
Waialua, Oahu, Hawaii (96791) 218/E1
Waianae, Hawaii (96792) 218/D2
Waiau, N. Zealand 100/D5
Waiau (riv.), N. Zealand 100/A6
Waidhofen an der Thaya, Austria 41/C2
Waidhofen an der Ybbs, Austria 41/C3
Waidsboro, Va. (†24088) 307/J7
Waigama, Indonesia 85/H6
Waigeo (isl.), Indonesia 85/J5
Waihee, Hawaii (†96793) 218/J2
Waiheke (isl.), N. Zealand 100/E2
Waihi, N. Zealand 100/E2
Waikabubak, Indonesia 85/F7
Waikanae, N. Zealand 100/E4
Waikane, Hawaii (†96744) 218/F2
Waikapu, Hawaii (†96793) 218/J2
Waikaremoana (lake), N. Zealand 100/F3
Waikari, N. Zealand 100/D5
Waikato (riv.), N. Zealand 100/E2
Waikawa, N. Zealand 100/B7
Waikerie, S. Australia 94/F6
Waikiki, Hawaii (†96743) 218/H4
Waikiki (canton), Hawaii (96815) 218/C4
Waikiki (beach), Hawaii 218/C4
Waikouaiti, N. Zealand 100/C6
Wailau, Hawaii (†96748) 218/H1
Wailea, Hawaii (†96710) 218/J4
Wailea, Maui, Hawaii (†96790) 218/J2
Wailua, Hawaii (†96746) 218/D2
Wailuku, Hawaii (96793) 218/J2
Wailuku, Hawaii 188/F5
Wailuku (riv.), Hawaii 218/J5
Waimakariri (riv.), N. Zealand 100/D5
Waimalu, Hawaii (†96701) 218/B3
Waimanalo, Hawaii (96795) 218/F2
Waimanalo Bch., Hawaii (†96795) 218/F2
Waimangaroa, N. Zealand 100/C4
Waimate, N. Zealand 100/C6
Waimea (Kamuela), Hawaii, (†96743) 218/G3
Waimea, Kauai, Hawaii (96796) 218/B2
Waimea, Oahu, Hawaii (†96712) 218/E1
Waimea (bay), Hawaii 218/B2
Waimea (riv.), Hawaii 218/C2
Waimes, Belgium 27/J8
Wainaku, Hawaii (†96720) 218/J5
Wainfleet, Ontario 177/E4
Wainfleet All Saints, England 13/H4
Waingapu, Indonesia 85/G7
Waini (riv.), Guyana 131/B2
Wainiha, Hawaii (†96714) 218/C1
Wainiha (riv.), Hawaii 218/C1
Wainuiomata, N. Zealand 100/B3
Wainui-o-mata (riv.), N. Zealand 100/B3
Wainwright, Alaska (99782) 196/F1
Wainwright, Alberta 182/E3
Wainwright, Ohio (44686) 284/G5
Wainwright, Okla. (74468) 288/R3
Wainwright, U.S. 4/B18
Waiohinu, Hawaii (†96772) 218/G7
Waipa (riv.), N. Zealand 100/E2
Waipahu, Hawaii 188/F5
Waipahu, Hawaii (96797) 218/A3
Waipara, N. Zealand 100/D5
Waipawa, N. Zealand 100/F3
Waipio, Hawaii (†96758) 218/H3
Waipio (bay), Hawaii 218/H3
Waipio (pen.), Hawaii 218/A3
Waipio (pt.), Hawaii 218/A4
Waipio Acres, Hawaii (†96786) 218/E2
Waipiro Bay, N. Zealand 100/G3
Waipukurau, N. Zealand 100/F4
Wairau (riv.), N. Zealand 100/D4
Wairoa, N. Zealand 100/F3
Wairoa (riv.), N. Zealand 100/E1
Waitakere, N. Zealand 100/B1
Waitakere (range), N. Zealand 100/A1
Waitaki (riv.), N. Zealand 100/C6
Waitangi, N. Zealand 100/D7
Waitara, N. Zealand 100/E3
Waite○, Maine (04492) 243/H5
Waite Hill, Ohio (†44094) 284/H2
Waitemata (harb.), N. Zealand 100/B1
Waite Park, Minn. (56387) 255/D5
Waiteville, W. Va. (24984) 312/F8
Waitotara, N. Zealand 100/E3
Waits (riv.), Vt. 268/C3
Waitsburg, Wash. (99361) 310/G4
Waitsfield○, Vt. (05673) 268/B3
Waits River, Vt. (†05076) 268/C3
Waitville, Sask. 181/F3
Waiuku, N. Zealand 100/E2
Waiyevu, Fiji 86/R10
Wajabula, Indonesia 85/H5
Wajima, Japan 81/H5
Wajir, Kenya 115/H3
Wajir, Kenya 102/F4
Waka, Ethiopia 111/G6
Waka, Texas (79093) 303/D1
Waka, Zaire 115/D3
Wakarusa, Ind. (46573) 227/F1
Wakarusa, Kansas (66546) 232/G3
Wakasa, Japan 81/G6
Wakasa (bay), Japan 81/G6
Wakatipu (lake), N. Zealand 100/B6
Wakaw, Sask. 181/F3
Wakaw Lake, Sask. 181/F3
Wakayama (pref.), Japan 81/G6
Wakayama, Japan 54/P6
Wakayama, Japan 81/G6
Wakde (isl.), Indonesia 85/K6
Wake (co.), N.C. 281/M3
Wake (isl.), Pacific 87/G4
WaKeeney, Kansas (67672) 232/C2
Wakefield, England 10/F4
Wakefield, England 13/J2
Wakefield, Kansas (67487) 232/E2
Wakefield, La. (70784) 238/H5
Wakefield○, Mass. (01880) 249/C5
Wakefield, Mich. (49968) 250/F2
Wakefield, Nebr. (68784) 264/H2
Wakefield○, N.H. (†03872) 268/F4
Wakefield, Ohio (45687) 284/E8
Wakefield, Va. (23888) 307/O7
Wakefield-Peace Dale, R.I. (*02879) 249/J7
Wake Forest, N.C. (27587) 281/M3
Wakema, Burma 72/B3
Wakeman, Ohio (44889) 284/F3
Wakenda, Mo. (64687) 261/F4
Wake Village, Texas (75501) 303/K4
Wakita, Okla. (73771) 288/L1
Wakkanai, Japan 81/K1
Wakonda, S. Dak. (57073) 298/P7
Wakool, N.S. Wales 97/C4
Wakopa, Manitoba 179/C5
Wakpala, S. Dak. (57658) 298/H2
Wakulla (co.), Fla. 212/B1
Wakulla, Fla. (†32327) 212/B1
Wakwekobi (lake), Ontario 177/A1
Wala, Kuh-i- (mt.), Afghanistan 59/H3
Walbridge, Ohio (43465) 284/C2
Wałbrzych (prov.), Poland 47/C3
Wałbrzych, Poland 47/C3
Walcha, N.S. Wales 97/F2
Walchensee (lake), W. Germany 22/D5
Walcheren (isl.), Netherlands 27/C5
Walcott, Ark. (72474) 202/J1
Walcott, Br. Col. 184/D3
Walcott (lake), Idaho 220/E7
Walcott, Iowa (52773) 229/M5
Walcott, N. Dak. (58077) 282/R6
Walcott, Wyo. (82335) 319/F4
Walcourt, Belgium 27/F8
Walcz, Poland 47/C2
Wald, Switzerland 39/G2
Waldeck, Sask. 181/D5
Walden, Colo. (80480) 208/G1
Walden, Georgia (†31201) 217/E5
Walden, Ky. (40768) 237/N7
Walden (pond), Mass. 249/A6
Walden, N.Y. (12586) 276/M7
Walden, Ontario 175/D3
Walden, Ontario 177/J5
Walden, Tenn. (†37377) 237/L10
Walden○, Vt. (†05873) 268/C3
Waldenburg, Ark. (72475) 202/J2
Waldenburg (Wałbrzych), Poland 47/C3
Waldenburg, Switzerland 39/E2
Walden Heights, Vt. (†05873) 268/C3
Waldersee, Manitoba 179/D4
Waldheim, E. Germany 22/E3
Waldheim, La. (†70433) 238/L5
Waldheim, Sask. 181/E3
Waldia, Ethiopia 111/G5
Waldkirch, Switzerland 39/H2
Waldkirch, W. Germany 22/B4
Waldkraiburg, W. Germany 22/E4
Waldo, Ala. (†35150) 195/F4
Waldo, Ark. (71770) 202/D7
Waldo, Br. Col. 184/K5
Waldo, Fla. (32694) 212/D2
Waldo, Kansas (67673) 232/D2
Waldo (co.), Maine 243/E6
Waldo○, Maine (†04915) 243/E7
Waldo, Ohio (43356) 284/D5
Waldo (lake), Oreg. 291/E4
Waldo, Wis. (53093) 317/L8
Waldoboro, Maine (04572) 243/E7
Waldoboro○, Maine (04572) 243/E7
Waldorf, Md. (20601) 245/L6
Waldorf, Minn. (56091) 255/E7
Waldport, Oreg. (97394) 291/C3
Waldron, Ark. (72958) 202/B4
Waldron, Ind. (46182) 227/F6
Waldron, Kansas (67150) 232/D4
Waldron, Mich. (49288) 250/E7
Waldron, Mo. (64092) 261/O5
Waldron, Sask. 181/J5
Waldron, Wash. (98297) 310/B2
Waldrup, Miss. (†39422) 256/F7
Waldsassen, W. Germany 22/E3
Waldshut-Tiengen, W. Germany 22/C5
Waldwick, N.J. (07463) 273/B1
Waldwick, Wis. (†53565) 317/G10
Walensee (lake), Switzerland 39/H2
Walenstadt, Switzerland 39/H2
Wales, Alaska (99783) 196/E1
Wales, Alaska 188/C5
Wales○, Mass. (01081) 249/F4
Wales, Minn. (†55616) 255/G3
Wales, N. Dak. (58281) 282/N2
Wales (isl.), N.W. Terrs. 187/K3
Wales, Tenn. (†38478) 237/G10
Wales, U.K. 7/D3
Wales, Utah (84667) 304/C4
WALES 13
WALES, Wales 10/E4
Wales, Wis. (53183) 317/J1
Walesboro, Ind. (†47201) 227/F6
Waleska, Georgia (30183) 217/D2
Walford, Iowa (52351) 229/K5
Walford, Ontario 177/B1
Walgett, N. S. Wales 88/H6
Walgett, N.S. Wales 97/E2
Walgreen Coast (reg.) 5/B13
Walhachin, Br. Col. 184/G5
Walhalla, Mich. (49458) 250/C5
Walhalla, N. Dak. (58282) 282/P2
Walhalla, S.C. (29691) 296/A2
Walhonding, Ohio (43843) 284/F5
Walikale, Zaire 115/E4
Walker (co.), Ala. 195/D3
Walker (creek), Ariz. 198/F2
Walker (mt.), Ark. 202/E2
Walker (co.), Georgia 217/B1
Walker, Iowa (52352) 229/K4
Walker, Kansas (67674) 232/C3
Walker, Ky. (40997) 237/O7
Walker, La. (70785) 238/L1
Walker, Mich. (49504) 250/D6
Walker, Minn. (56484) 255/D3
Walker, Mo. (64790) 261/D7
Walker (lake), Nev. 188/C3
Walker (lake), Nev. 266/C4
Walker (riv.), Nev. 266/C3
Walker, N.Y. (†14468) 276/E4
Walker (bay), N.W. Terrs. 187/G2
Walker, Oreg. (†97426) 291/D4
Walker, S. Dak. (57659) 298/G2
Walker (isl.), Tasmania 99/B2
Walker (co.), Texas 303/J7
Walker (creek), Va. 307/F6
Walker, W. Va. (26180) 312/D4
Walkerburn, Scotland 15/F5
Walker Mill, Md. (†20023) 245/F5
Walker River Ind. Res., Nev. 266/C3
Walker Springs, Ala. (36586) 195/C7
Walkerston, Queensland 88/H4
Walkerston, Queensland 95/D4
Walkersville, Md. (21793) 245/J3
Walkersville, W. Va. (26447) 312/F5
Walkerton, Ind. (46574) 227/E2
Walkerton, Ontario 177/C3
Walkerton, Va. (23177) 307/O5
Walkertown, N.C. (27051) 281/J2
Walkerville, Mich. (49459) 250/C5
Walkerville, Mont. (59701) 262/D4
Walkerville, S. Australia 88/E8
Wall○, N.J. (07719) 273/E3
Wall, Pa. (†15148) 294/C5
Wall, S. Dak. (57790) 298/E6
Wall, Texas (76957) 303/D6
Wallace, Ala. (†36426) 195/D8
Wallace (mt.), Alberta 182/C2
Wallace, Calif. (95254) 204/C9
Wallace, Idaho (83873) 220/C2
Wallace, Ind. (47988) 227/C5
Wallace (co.), Kansas 232/A3
Wallace, Kansas (67761) 232/A3
Wallace, La. (†70049) 238/M3
Wallace (lake), La. 238/C2
Wallace, Mich. (49893) 250/B3
Wallace, Nebr. (69169) 264/C4
Wallace, N.Y. (14890) 276/E6
Wallace, N.C. (28466) 281/N5
Wallace, Nova Scotia 168/E3
Wallace (harb.), Nova Scotia 168/E3
Wallace, S.C. (29596) 296/H2
Wallace, S. Dak. (57272) 298/P3
Wallace, W. Va. (26448) 312/E4
Wallaceburg, Ontario 177/B5
Wallace Lake, Manitoba 179/G3
Wallaceton, Pa. (16876) 294/F4
Wallacetown, Ontario 177/C5
Wallaga (prov.), Ethiopia 111/G6
Wallal Station, W. Australia 92/C2
Walland, Tenn. (37886) 237/O9
Wallaroo, S. Australia 94/E5
Wallasey, England 13/G2
Wallasey, England 10/F2
Walla Walla, N.S. Wales 97/D4
Walla Walla (riv.), Oreg. 291/J1
Walla Walla, Wash. 188/C1
Walla Walla (co.), Wash. 310/G4
Walla Walla, Wash. (99362) 310/G4
Walla Walla (riv.), Wash. 310/G4
Wallback, W. Va. (25285) 312/D5
Wallburg, N.C. (27373) 281/J3
Walldürn, W. Germany 22/C4
Walled Lake, Mich. (48088) 250/F6
Wallen, Ind. (†46802) 227/G2
Wallendbeen, N.S. Wales 97/E4
Wallenpaupack (lake), Pa. 294/M3
Waller (co.), Texas 303/J8
Wallerawang, N.S. Wales 97/F3
Wallerville, Miss. (†38652) 256/G2
Wallibu, St. Vin. & Grens. 161/A8
Walling, Tenn. (38587) 237/K9
Wallingford, Conn. (06492) 210/D3
Wallingford○, Conn. (06492) 210/D3
Wallingford, England 13/F6
Wallingford, Iowa (51365) 229/D2
Wallingford, Ky. (41093) 237/O4
Wallingford, Pa. (19086) 294/L7
Wallingford, Vt. (05773) 268/B5
Wallingford○, Vt. (05773) 268/B5
Wallington, N.J. (07057) 273/B2
Wallins Creek, Ky. (40873) 237/O7
Wallis (lake), N.S. Wales 97/G3
Wallis, Texas (77485) 303/H8
Wallis (isls.), Wallis and Futuna 87/J7
Wallis and Futuna 87/J7
Wallisellen, Switzerland 39/G2
Wallisville, Texas (77597) 303/L1
Wallkill (riv.), N.J. 273/D1
Wallkill, N.Y. (12589) 276/M7
Wallkill (riv.), N.Y. 276/L8
Wall Lake, Iowa (51466) 229/C4
Wallo (prov.), Ethiopia 111/H5
Wallonia, Ky. (†42405) 237/F7
Walloon (lake), Mich. 250/E3
Walloon Lake, Mich. (49796) 250/E3
Wallops (isl.), Va. 307/T5
Wallowa (co.), Oreg. 291/K2
Wallowa, Oreg. (97885) 291/K2
Wallowa (mts.), Oreg. 291/K2
Wallowa (riv.), Oreg. 291/K2
Wallpack Center, N.J. (07881) 273/D1
Walls, Miss. (38680) 256/D1
Walls, Scotland 15/G2
Wallsburg, Utah (84082) 304/C3
Wallsend, England 13/J3
Wallula (lake), Oreg. 291/H1
Wallula, Wash. (99363) 310/G4
Wallula (lake), Wash. 310/F4
Walney, Isle of (isl.), England 13/D3
Walney, Isle of (isl.), England 10/E3
Walnut, Calif. (91789) 204/D10
Walnut (creek), Calif. 204/K1
Walnut, Ill. (61376) 222/D2
Walnut, Iowa (51577) 229/C6
Walnut, Kansas (66780) 232/G4
Walnut (creek), Kansas 232/B3
Walnut (riv.), Kansas 232/E4
Walnut, Miss. (38683) 256/G1
Walnut, N.C. (28753) 281/D3
Walnut, Pa. (†17082) 294/G4
Walnut (creek), Texas 303/F3
Walnut Bottom, Pa. (17266) 294/H5
Walnut Canyon Nat'l Mon., Ariz. 198/D3
Walnut Cove, N.C. (27052) 281/J2
Walnut Creek, Calif. (*94595) 204/K2
Walnut Creek, N.C. (†27530) 281/O4
Walnut Creek, Ohio (44687) 284/G4
Walnut Grove, Ala. (35990) 195/F2
Walnut Grove, Calif. (95690) 204/B9
Walnut Grove, Georgia (†30209) 217/E3
Walnut Grove, Ill. (†61470) 222/C3
Walnut Grove, Ky. (42563) 237/M6
Walnut Grove, Minn. (56180) 255/C6
Walnut Grove, Miss. (39189) 256/F5
Walnut Grove, Mo. (65770) 261/F8
Walnut Hill, Ark. (†71826) 202/C7
Walnut Hill, Fla. (32568) 212/B5
Walnut Hill, Ill. (62893) 222/E5
Walnut Hill, Maine (†04021) 243/C8
Walnutport, Pa. (18088) 294/L4
Walnut Ridge, Ark. (72476) 202/J1
Walnut Springs, Texas (76690) 303/G5
Walpole, Mass. (02081) 249/B8
Walpole○, Mass. (02081) 249/B8
Walpole○, N.H. (03608) 268/C5
Walpole (isl.), Ontario 177/B5
Walpole, Sask. 181/K6
Walpole, W. Australia 92/B6
Walrus (isl.), Alaska 196/E3
Walrus (isls.), Alaska 196/F3
Walsall, England 10/G3
Walsall, England 13/E5
Walsenburg, Colo. (81089) 208/K7
Walsh, Alberta 182/E5
Walsh, Colo. (81090) 208/P8
Walsh (co.), N. Dak. 282/P3
Walsh, Queensland 95/B3
Walshville, Ill. (62091) 222/D4
Walsingham, England 13/H5
Walsingham (cape), N.W.T. 162/K2
Walsingham (cape), N.W. Terrs. 187/M3
Walsrode, W. Germany 22/C2
Walston, Pa. (15781) 294/D4
Walstonburg, N.C. (27888) 281/O3
Walterboro, S.C. (29488) 296/F6
Walter F. George (dam), Ala. 195/H7
Walter F. George (res.), Ala. 195/H7
Walter F. George (dam), Georgia 217/B7
Walter F. George (res.), Georgia 217/B7
Walterhill, Tenn. (†37130) 237/J9
Walter Reed Army Med. Ctr., D.C. 245/E4
Walter Reed Army Med. Ctr. Annex, Md. 245/E4
Walters, La. (71374) 238/G3
Walters, Minn. (56092) 255/E7
Walters, Okla. (73572) 288/K6
Walters Falls, Ontario 177/D3
Waltershausen, E. Germany 22/D3
Waltersville, Ky. (†40312) 237/N5
Waltersville, Miss. (†39180) 256/C6
Walterville, Oreg. (97489) 291/E3
Walthall (co.), Miss. 256/D8
Walthall, Miss. (39771) 256/F3
Waltham○, Maine (†04605) 243/G6
Waltham, Mass. (02154) 249/B6
Waltham, Minn. (55982) 255/F7
Waltham○, Vt. (†05491) 268/A3
Waltham Forest, England 13/H8
Waltham Forest, England 10/B5
Waltham Holy Cross, England 13/H7
Waltham Holy Cross, England 10/B5
Walthill, Nebr. (68067) 264/H2
Walthourville, Georgia (31333) 217/J7
Waltman, Wyo. (†82648) 319/E2
Walton (co.), Fla. 212/C6
Walton, Fla. (†33457) 212/F4
Walton (co.), Georgia 217/E3
Walton, Ind. (46994) 227/E3
Walton, Kansas (67151) 232/E3
Walton, Ky. (41094) 237/M3
Walton, Nebr. (68461) 264/H4
Walton, N.Y. (13856) 276/K6
Walton, Nova Scotia 168/E3
Walton, Ontario 177/C4
Walton, Oreg. (97490) 291/D3
Walton, W. Va. (25286) 312/D5
Walton and Weybridge, England 13/G8
Walton and Weybridge, England 10/B6
Walton Hills, Ohio (†44146) 284/J10
Walton-le-Dale, England 13/G1
Walton-le-Dale, England 10/F1
Waltonville, Ill. (62894) 222/D5
Waltreak, Ark. (†72833) 202/C4
Waltz, Mich. (†48164) 250/F6
Walum, N. Dak. (†58448) 282/O5
Walupt (lake), Wash. 310/D4
Walvis (bay), S. Africa 118/A4
Walvis Bay, S. Africa 2/K7
Walvis Bay, S. Africa 102/D7
Walvis Bay, S. Africa 118/A4
Walworth, N.Y. (14568) 276/F4
Walworth (co.), S. Dak. 298/J3
Walworth (co.), Wis. 317/J10
Walworth, Wis. (53184) 317/J10
Walzenhausen, Switzerland 39/J2
Wamac, Ill. (†62801) 222/D5
Wamba, Kenya 115/G3
Wamba, Nigeria 106/F7
Wamba, Zaire 115/E3
Wamego, Kansas (66547) 232/F2
Wamel, Netherlands 27/H5
Wamena, Indonesia 85/K6
Wamgumbaug (lake), Conn. 210/F1
Wami (riv.), Tanzania 115/G5
Wamic, Oreg. (97063) 291/F2

Wampee, S.C. (†29582) 296/K4
Wampsville, N.Y. (13163) 276/J4
Wampum, Manitoba 179/G5
Wampum, Pa. (16157) 294/B4
Wamsutter, Wyo. (82336) 319/E4
Wana, Pakistan 68/C2
Wana, Pakistan 59/J3
Wana, W. Va. (26590) 312/F3
Wanaaring, N.S. Wales 97/B1
Wanaka, N. Zealand 100/B6
Wanaka (lake), N. Zealand 100/B6
Wanakah, N.Y. (†14075) 276/C5
Wanakena, N.Y. (13695) 276/K2
Wanamassa, N.J. (†07712) 273/E3
Wanamingo, Minn. (55983) 255/F6
Wan'an, China 77/H6
Wanapitei (riv.), Ontario 177/D1
Wanapum (dam), Wash. 310/E4
Wanapum (lake), Wash. 310/E3
Wanaque, N.J. (07465) 273/B1
Wanaque (res.), N.J. 273/E1
Wanatah, Ind. (46390) 227/D2
Wanblee, S. Dak. (57577) 298/F6
Wanchese, N.C. (27981) 281/T3
Wanda, Minn. (56294) 255/C6
Wandel (sea), Greenl. 4/A10
Wandering, W. Australia 92/B2
Wandering River, Alberta 182/D2
Wanderoos, Wis. (†54001) 317/B5
Wandfluhhorn (mt.), Switzerland 39/G4
Wando, S.C. (29492) 296/H6
Wando (riv.), S.C. 296/H6
Wandoan, Queensland 95/D5
Wandsworth, England 13/H8
Wandsworth, England 10/B5
Wanette, Okla. (74878) 288/M5
Wang, Mae Nam (riv.), Thailand 72/C3
Wanganella, N.S. Wales 97/C4
Wanganui, N. Zealand 87/H9
Wanganui, N. Zealand 100/E3
Wanganui (riv.), N. Zealand 100/E3
Wangaratta, Victoria 88/H7
Wangaratta, Victoria 97/D5
Wangen an der Aare, Switzerland 39/E2
Wangen im Allgäu, W. Germany 22/C5
Wangerooge (isl.), W. Germany 22/B2
Wängi, Switzerland 39/H1
Wangi-Rathmines, N.S. Wales 97/F3
Wangiwangi (isl.), Indonesia 85/G7
Wangqing, China 77/M3
Wangum (lake), Conn. 210/B1
Wanham, Alberta 182/A2
Wanhsien (Wanxian), China 77/G5
Wanilla, S. Australia 94/D6
Wanipigow, Manitoba 179/F3
Wanipigow (riv.), Manitoba 179/G3
Wankai, Sudan 111/E6
Wankie, Zimbabwe 118/D3
Wankie, Zimbabwe 102/E6
Wanks (Coco) (riv.), Honduras 154/E3
Wanks (Coco) (riv.), Nicaragua 154/E3
Wanless, Manitoba 179/H3
Wann, Okla. (74083) 288/P1
Wanna (lakes), W. Australia 92/E5
Wannaska, Minn. (56761) 255/C2
Wanne-Eickel, W. Germany 22/B3
Wanneroo, W. Australia 88/B2
Wanneroo, W. Australia 92/A1
Wanning, China 77/H8
Wanship, Utah (†84017) 304/C3
Wantage, England 13/F6
Wantage, England 10/F5
Wantagh, N.Y. (11793) 276/R7
Wanxian (Wanhsien), China 77/G5
Wanzai, China 77/H6
Wao, Philippines 82/E7
Wapakoneta, Ohio (45895) 284/B4
Wapanucka, Okla. (73461) 288/N6
Wapato, Wash. (98951) 310/E4
Wapawekka (hills), Sask. 181/M4
Wapella, Ill. (61777) 222/E3
Wapella, Sask. 181/K5
Wapello (co.), Iowa 229/J6
Wapello, Iowa (52653) 229/L6
Wapinitia, Oreg. (†97037) 291/F2
Wapiti (riv.), Alberta 182/A2
Wapiti (riv.), Br. Col. 184/H3
Wapiti, Wyo. (82450) 319/C1
Wappapello, Mo. (63966) 261/M9
Wappapello (lake), Mo. 261/L8
Wappau (lake), Alberta 182/E2
Wappingers Falls, N.Y. (12590) 276/N7
Wapsipinicon (riv.), Iowa 229/J3
Wapske, New Bruns. 170/C2
Wapwallopen, Pa. (18660) 294/K3
Waqqas, Jordan 65/D2
Waquoit, Mass. (02536) 249/M6
War, W. Va. (24892) 312/C8
Warabi, Japan 81/O2
Waramaug (lake), Conn. 210/B2
Waranga (res.), Victoria 97/C5
Warangal, India 54/J8
Warangal, India 68/D5
Waratah, Tasmania 99/B3
Waratah (bay), Victoria 97/C6
Warba, Minn. (55743) 255/E3
Warburg, Alberta 182/C3
Warburg, W. Germany 22/C3
Warburton, The (riv.), S. Australia 94/F2
Warburton, The (riv.), S. Australia 88/F5
Warburton, Victoria 97/C5
Warburton Aboriginal Reserve, W. Australia 88/D5
Warburton Aboriginal Res., W. Australia 92/D4
Ward, Ala. (36922) 195/B6
Ward, Ark. (72176) 202/F3
Ward, Colo. (80481) 208/H2
Ward (peak), Mont. 262/A3
Ward, N. Zealand 100/E4
Ward (co.), N. Dak. 282/G3
Ward, S.C. (29166) 296/D4
Ward, S. Dak. (57074) 298/R5
Ward (co.), Texas 303/A6
Ward, W. Va. (†25039) 312/D6
Ward Cove, Alaska (99928) 196/N2
Wardell, Mo. (63879) 261/N10
Warden, La. (71289) 238/H1
Warden, Québec 172/E4
Warden, Wash. (98857) 310/F4
Wardensville, W. Va. (26851) 312/J4
Wardere, Ethiopia 111/J6
Wardha, India 68/D4
Wardha (riv.), India 68/D4
Wardlow, Alberta 182/E4
Wardner, Br. Col. 184/K5
Wardner, Idaho (†83837) 220/B2
Ward Ridge, Fla. (†32456) 212/D6
Wardsboro○, Vt. (05355) 268/B5
Ward Springs, Minn. (†56336) 255/D5
Wardsville, Mo. (†65101) 261/H6
Wardsville, Ontario 177/C5
Wardville, La. (†71301) 238/F4
Wardville, Okla. (74576) 288/P5
Ware, Br. Col. 184/E1
Ware, England 13/H7
Ware, England 10/F5
Ware (co.), Georgia 217/H8
Ware, Mass. (01082) 249/E3
Ware○, Mass. (01082) 249/E3
Ware (riv.), Mass. 249/E3
War Eagle, Ark. (†72756) 202/C1
War Eagle, W. Va. (†24862) 312/C7
Waregem, Belgium 27/C7
Wareham, England 13/E7
Wareham, England 10/E5
Wareham, Mass. (02571) 249/L5
Wareham○, Mass. (02571) 249/L5
Wareham Center, Mass. (02571) 249/L5
Warehouse Point, Conn. (06088) 210/E1
Waremme, Belgium 27/G7
Waren, E. Germany 22/E2
Waren, Indonesia 85/K6
Warenda, Queensland 95/B4
Warendorf, W. Germany 22/B3
Ware Neck, Va. (23178) 307/R6
Waresboro, Georgia (31564) 217/H8
Ware Shoals, S.C. (29692) 296/C3
Waretown, N.J. (08758) 273/E4
Warfield, Br. Col. 184/J5
Warfield, Ky. (41267) 237/S5
Warfield, Va. (23889) 307/N7
Warfordsburg, Pa. (17267) 294/F6
Warialda, N.S. Wales 97/F1
Warin Chamrap, Thailand 72/E4
Waring (mts.), Alaska 196/G1
Waring, Texas (78074) 303/F8
Warka, Poland 47/E3
Warkworth, N. Zealand 100/E2
Warkworth, Ontario 177/G3
Warley, England 13/E5
Warley, England 10/G3
Warm (creek), Utah 304/C6
Warman, Sask. 181/E3
Warmbad, Namibia 118/B5
Warmbad, S. Africa 118/D5
Warm Beach, Wash. (†98292) 310/C2
Warmenhuizen, Netherlands 27/F3
Warmia (reg.), Poland 47/D1
Warminster, England 10/E5
Warminster, England 13/E6
Warm Lake, Idaho (83611) 220/C5
Warm River, Idaho (†83420) 220/G5
Warm Springs, Ark. (72478) 202/H1
Warm Springs, Georgia (31830) 217/C5
Warmsprings, Mont. (59756) 262/D4
Warm Springs, Oreg. (97761) 291/F3
Warm Springs (res.), Oreg. 291/J4
Warm Springs, Va. (24484) 307/J4
Warm Springs Ind. Res., Oreg. 291/F3
Warne, N.C. (28909) 281/B5
Warner, Alberta 182/D5
Warner○, N.H. (03278) 268/D5
Warner (riv.), N.H. 268/D5
Warner, Ohio (45785) 284/H6
Warner, Okla. (74469) 288/R4
Warner, S. Dak. (57479) 298/M3
Warner Robins, Georgia (31093) 217/E5
Warners, N.Y. (13164) 276/H4
Warnerton, La. (†70438) 238/K5
Warnes, Bolivia 136/D5
Warnow (riv.), E. Germany 22/D2
Waroona, W. Australia 92/A2
Warrabri, North. Terr. 93/D6
Warrabri Aboriginal Reserve, North. Terr. 88/E4
Warracknabeal, Victoria 97/B5
Warracknabeal, Victoria 88/G7
Warr Acres, Okla. (73132) 288/L3
Warragamba, N.S. Wales 97/F3
Warragul, Victoria 97/D6
Warrandyte, Victoria 97/J4
Warrandyte, Victoria 88/M6
Warrego (riv.), N.S. Wales 97/C1
Warrego, North. Terr. 93/C5
Warrego (range), Queensland 88/H5
Warrego (range), Queensland 95/C5
Warrego (riv.), Queensland 88/H5
Warrego (riv.), Queensland 95/C5
Warren, Ark. (71671) 202/F6
Warren○, Conn. (06754) 210/B2
Warren (co.), Georgia 217/G4
Warren, Idaho (83671) 220/C4
Warren (co.), Ill. 222/C3
Warren, Ill. (61087) 222/C1
Warren (co.), Ind. 227/C4
Warren, Ind. (46792) 227/G3
Warren (co.), Iowa 229/F6
Warren (co.), Ky. 237/H6
Warren, Maine (04864) 243/E7
Warren○, Maine (04864) 243/E7
Warren, Manitoba 179/E4
Warren, Mass. (01083) 249/F4
Warren○, Mass. (01083) 249/F4
Warren, Mich. (*48089) 250/B6
Warren, Minn. (56762) 255/B2
Warren (co.), Miss. 256/C6
Warren (co.), Mo. 261/K5
Warren, Mo. (†63456) 261/J3
Warren○, N.H. (03279) 268/D4
Warren (co.), N.J. 273/C2
Warren○, N.J. (†07060) 273/D2
Warren, N.S. Wales 88/H6
Warren, N.S. Wales 97/D2
Warren (co.), N.Y. 276/N3
Warren (co.), N.C. 281/N2
Warren, Nova Scotia 168/D3
Warren (co.), Ohio 284/B7
Warren, Ohio (*44481) 284/J3
Warren, Ontario 177/D1
Warren, Oreg. (97053) 291/E2
Warren (co.), Pa. 294/D2
Warren, Pa. (16365) 294/D2
Warren○, R.I. (02885) 249/J6
Warren (res.), S. Australia 94/C7
Warren (co.), Tenn. 237/K9
Warren○, Vt. (05674) 268/B3
Warren (co.), Va. 307/M3
Warren Center, Pa. (18851) 294/K2
Warrenpoint, N. Ireland 17/J3
Warrens, Wis. (54666) 317/E7
Warrensburg, Ill. (62573) 222/D4
Warrensburg, Mo. (64093) 261/E5
Warrensburg, N.Y. (12885) 276/N3
Warrensville, Alberta 182/B1
Warrensville, N.C. (28693) 281/F2
Warrensville, Pa. (†17701) 294/J3
Warrensville Heights, Ohio (44128) 284/H9
Warrenton, Georgia (30828) 217/G4
Warrenton, Ind. (†47539) 227/B8
Warrenton, Mo. (63383) 261/K5
Warrenton, N.C. (27589) 281/N2
Warrenton, Oreg. (97146) 291/C1
Warrenton, S. Africa 118/C5
Warrenton, Va. (22186) 307/N3
Warrenville, Conn. (†06278) 210/G1
Warrenville, Ill. (60555) 222/A6
Warrenville, S.C. (29851) 296/D4
Warri, Nigeria 106/F7
Warrick (co.), Ind. 227/C8
Warrick, Mont. (†59520) 262/G2
Warrina, S. Australia 94/D3
Warringah, N.S. Wales 97/K3
Warrington, England 13/G2
Warrington, England 10/F2
Warrington, Fla. (32507) 212/B6
Warrington, Ind. (†46186) 227/F5
Warrior, Ala. (35180) 195/E3
Warrior (dam), Ala. 195/C5
Warrior Run, Pa. (18706) 294/E7
Warriors Mark, Pa. (16877) 294/F4
Warrnambool, Australia 87/D9
Warrnambool, Victoria 97/B6
Warrnambool, Victoria 88/G7
Warroad, Minn. (56763) 255/C2
Warrumbungle (range), N. S. Wales 88/H6
Warsaw, Ill. (62379) 222/B3
Warsaw, Ind. (46580) 227/F2
Warsaw, Ky. (41095) 237/M3
Warsaw, Minn. (55087) 255/E6
Warsaw, Mo. (65355) 261/F6
Warsaw, N.Y. (14569) 276/D5
Warsaw, N.C. (28398) 281/N4
Warsaw, N. Dak. (†58261) 282/R3
Warsaw, Ohio (43844) 284/G5
Warsaw, Ontario 177/F3
Warsaw (city), Poland 47/E2
Warsaw (prov.), Poland 47/E2
Warsaw (cap.), Poland 7/G3
Warsaw (cap.), Poland 2/L3
Warsaw (Warszawa) (cap.), Poland 47/E2
Warsaw, Va. (22572) 307/P5
Warson Woods, Mo. (†63101) 261/O3
Warsop, England 13/F4
Warspite, Alberta 182/D2
Warta (riv.), Poland 7/F3
Warta (riv.), Poland 47/B2
Wartau, Switzerland 39/H2
Wartburg, Tenn. (37887) 237/M8
Warthen, Georgia (31094) 217/G4
Wartime, Sask. 181/C4
Wartrace, Tenn. (37183) 237/J9
Warwick, Alberta 182/D3
Warwick, England 10/F4
Warwick, England 13/F5
Warwick, Georgia (31796) 217/E7
Warwick, Md. (21912) 245/P3
Warwick○, Mass. (†01364) 249/E2
Warwick, N.Y. (10990) 276/M8
Warwick, N. Dak. (58381) 282/N4
Warwick (chan.), North. Terr. 93/E3
Warwick, Okla. (†74834) 288/M3
Warwick, Québec 172/F4
Warwick, Queensland 88/J5
Warwick, Queensland 95/D6
Warwick, R.I. (*02886) 249/J6
Warwickshire (co.), England 13/F5
Wasa, Br. Col. 184/K5
Wasaga Beach, Ontario 177/D3
Wasagaming, Manitoba 179/C4
Wasatch (range), Idaho 220/G7
Wasatch (co.), Utah 304/C3
Wasatch (plat.), Utah 304/C4
Wasatch (range), Utah 304/C3
Wasco, Calif. (93280) 204/F8
Wasco (co.), Oreg. 291/F2
Wasco, Oreg. (97065) 291/G2
Wascott, Wis. (54890) 317/C3
Waseca (co.), Minn. 255/E6
Waseca, Minn. (56093) 255/E6
Waseca, Sask. 181/B2
Wasen, Switzerland 39/E2
Wash, The (bay), England 13/H5
Wash, The (bay), England 10/G4
Washademoak (lake), New Bruns. 170/E3
Washago, Ontario 177/E3
Washakie (co.), Wyo. 319/E2
Washakie Ind. Res., Utah 304/B2
Washburn, Ark. (†72936) 202/B3
Washburn, Ill. (61570) 222/D3
Washburn, Iowa (50706) 229/J4
Washburn, Maine (04786) 243/G2
Washburn○, Maine (04786) 243/G2
Washburn, Mo. (65772) 261/E9
Washburn, N. Dak. (58577) 282/J5
Washburn (co.), Pakistan 317/C4
Washburn, Tenn. (37888) 237/O8
Washburn, W. Va. (†26362) 312/D4
Washburn, Wis. (54891) 317/D2
Washburn (mt.), Wyo. 319/B1
Washdyke, N. Zealand 100/C6
Washington 188/B1
WASHINGTON 310
Washington (co.), Ala. 195/B8
Washington (co.), Ark. 202/B2
Washington, Ark. (71862) 202/C6
Washington (co.), Colo. 208/N3
Washington○, Conn. (06793) 210/B2
Washington, England 13/J3
Washington (co.), Fla. 212/C6
Washington (lake), Fla. 212/F3
Washington (co.), Georgia 217/G4
Washington, Georgia (30673) 217/G3
Washington (co.), Idaho 220/B5
Washington (co.), Ill. 222/D5
Washington, Ill. (61571) 222/D3
Washington (co.), Ind. 227/E7
Washington, Ind. (47501) 227/C7
Washington (co.), Iowa 229/K6
Washington, Iowa (52353) 229/K6
Washington (co.), Kansas 232/E2
Washington, Kansas (66968) 232/F2
Washington (co.), Ky. 237/L5
Washington, Ky. (41096) 237/O3
Washington (isl.), Kiribati 87/L5
Washington (par.), La. 238/K5
Washington, La. (70589) 238/G5
Washington (co.), Maine 243/H6
Washington○, Maine (04574) 243/E7
Washington (co.), Md. 245/G2
Washington○, Mass. (†01223) 249/B3
Washington (co.), Minn. 255/F5
Washington (co.), Miss. 256/C4
Washington, Miss. (39190) 256/B7
Washington (co.), Mo. 261/L7
Washington, Mo. (63090) 261/K5
Washington (co.), Nebr. 264/H3
Washington, Nebr. (68068) 264/H3
Washington○, N.H. (03280) 268/C5
Washington (mt.), N.H. 268/E3
Washington, N.J. (07882) 273/D2
Washington (co.), N.Y. 276/O4
Washington (co.), N.C. 281/R3
Washington, N.C. (27889) 281/R3
Washington (co.), Ohio 284/H7
Washington (Old Washington), Ohio (†43768) 284/H5
Washington (co.), Okla. 288/P1
Washington, Okla. (73093) 288/L4
Washington (co.), Oreg. 291/D2
Washington (co.), Pa. 294/B5
Washington, Pa. (15301) 294/B5
Washington (co.), R.I. 249/H7
Washington (Coventry), R.I. (†02816) 249/H6
Washington (co.), Tenn. 237/R8
Washington (co.), Texas 303/H7
Washington, Texas (77880) 303/J7
Washington (state), U.S. 146/F5
Washington, D.C. (cap.), U.S. 146/L6
Washington, D.C. (cap.), U.S. 188/L3
Washington, D.C. (cap.), U.S., (*20001) 245/F5
Washington (cap.), U.S. 2/F4
Washington (co.), Utah 304/A6
Washington, Utah (84780) 304/A6
Washington (co.), Vt. 268/B3
Washington○, Vt. (05675) 268/C3
Washington (co.), Va. 307/D7
Washington, Va. (22747) 307/M3
Washington (lake), Wash. 310/B2
Washington, W. Va. (26181) 312/C4
Washington (co.), Wis. 317/K9
Washington (isl.), Wis. 317/M5
Washington Court House, Ohio (43160) 284/D6
Washington Crossing, N.J. (†08560) 273/D3
Washington Crossing, Pa. (18977) 294/N5
Washington Depot, Conn. (06794) 210/B2
Washington Grove, Md. (20880) 245/K4
Washington Island, Wis. (54246) 317/M5
Washington Lands, W. Va. (†26041) 312/E3
Washington Park, Ill. (62204) 222/B2
Washington Park, N.C. (†27889) 281/R3
Washington Terrace, Utah (†84403) 304/B2
Washingtonville, N.Y. (10992) 276/M8
Washingtonville, Ohio (44490) 284/J4
Washingtonville, Pa. (17884) 294/J3
Washita, Ark. (†71957) 202/C4
Washita (co.), Okla. 288/J4
Washita, Okla. (73094) 288/K
Washita (riv.), Okla. 288/M5
Washita (riv.), Texas 303/D2
Washoe, Mont. (†59007) 262/G5
Washoe (co.), Nev. 266/B2
Washoe (lake), Nev. 266/B3
Washougal, Wash. (98671) 310/C5
Washow (bay), Manitoba 179/F3
Washta, Iowa (51061) 229/B3
Washtenaw (co.), Mich. 250/F6
Washtucna, Wash. (99371) 310/G4
Washunga, Okla. (†74641) 288/N1
Wasilkow, Poland 47/F2
Wasilla, Alaska (99687) 196/B1
Wasior, Indonesia 85/K6
Wasit (gov.), Iraq 66/D4
Waskada, Manitoba 179/B5
Waskana (creek), Sask. 181/G5
Waskatenau, Alberta 182/D2
Waskesiu (lake), Sask. 181/E2
Waskesiu Lake, Sask. 181/E2
Waskigomog (lake), Ontario 177/F2
Waskish, Minn. (56685) 255/D2
Waskom, Texas (75692) 303/L5
Waspán, Nicaragua 154/E3
Waspuk (riv.), Nicaragua 154/E3
Wassataquoik (stream), Maine 243/F4
Wassaw (sound), Georgia 217/L7
Wassen, Switzerland 39/G3
Wasser, Namibia 118/B5
Wasserbillig, Luxembourg 27/J9
Wasserburg am Inn, W. Germany 22/E4
Wasserkuppe (mt.), W. Germany 22/C3
Wasson, Ill. (†62930) 222/E6
Wassuk (range), Nev. 266/C4
Wassy, France 28/F3
Wasta, S. Dak. (57791) 298/D5
Wataga, Ill. (61488) 222/C2
Watampone, Indonesia 85/G6
Watauga (co.), N.C. 281/F2
Watauga, S. Dak. (57660) 298/F2
Watauga, Tenn. (37694) 237/S8
Watauga (lake), Tenn. 237/T8
Watauga, Texas (76248) 303/F2
Watauga Valley, Tenn. (†37643) 237/S8
Watchet, England 13/D6
Watch Hill, R.I. (02891) 249/G7
Watch Hill (pt.), R.I. 249/G7
Watchman (isl.), Newf. 166/B2
Watchung, N.J. (07060) 273/E2
Watchusk (lake), Alberta 182/E1
Water (isl.), Virgin Is. (U.S.) 161/A4
Waterberg, Namibia 118/B4
Waterboro, Maine (04087) 243/B8
Waterboro○, Maine (04087) 243/B8
Waterbury, Conn. 188/M2
Waterbury, Conn. (*06701) 210/C2
Waterbury, Nebr. (68785) 264/H2
Waterbury, Vt. (05676) 268/B3
Waterbury○, Vt. (05676) 268/B3
Waterbury (res.), Vt. 268/B3
Waterbury Center, Vt. (05677) 268/B3
Waterdown, Ontario 177/D4
Wateree, S.C. (†29044) 296/F4
Wateree (lake), S.C. 296/F3
Wateree (riv.), S.C. 296/F3
Waterflow, N. Mex. (87421) 274/A2
Waterford, Calif. (95386) 204/E6
Waterford, Conn. (06385) 210/G3
Waterford○, Conn. (06385) 210/G3
Waterford (co.), Ireland 17/F7
Waterford, Ireland 17/G7
Waterford, Ireland 10/C4
Waterford (harb.), Ireland 10/C4
Waterford (harb.), Ireland 17/G7
Waterford, Maine (04088) 243/B7
Waterford○, Maine (04088) 243/B7
Waterford, Miss. (38685) 256/E1
Waterford, New Bruns. 170/E3
Waterford, N.Y. (12188) 276/N5
Waterford, Ohio (45786) 284/G6
Waterford, Pa. (16441) 294/B2
Waterford, Va. (22190) 307/N2
Waterford, Wis. (53185) 317/K3
Waterford Works, N.J. (08089) 273/D4
Watergap, Ky. (41665) 237/R5
Waterhen (lake), Manitoba 179/C2
Waterhouse (isl.), Tasmania 99/D2
Waterloo, Ala. (35677) 195/B1
Waterloo, Ark. (†71858) 202/D6
Waterloo, Belgium 27/E7
Waterloo, Ill. (62298) 222/C5
Waterloo, Ind. (46793) 227/G2
Waterloo, Iowa 188/H2
Waterloo, Iowa 146/J5
Waterloo, Iowa (*50701) 229/J4
Waterloo, Kansas (†67111) 232/E4
Waterloo, Mont. (†59759) 262/D5
Waterloo, Nebr. (68069) 264/H3
Waterloo, N.Y. (13165) 276/G5
Waterloo, North. Terr. 93/A4
Waterloo, Ohio (45688) 284/F8
Waterloo (reg. munic.), Ontario 177/D4
Waterloo, Ontario 177/D4
Waterloo, Oreg. (†97355) 291/E3
Waterloo, Québec 172/E4
Waterloo, S.C. (29384) 296/C3
Waterloo, Trin. & Tob. 161/A10
Waterloo, Wis. (53594) 317/J9
Watermaal-Bosvoorde (Watermael-Boitsfort), Belgium 27/C9
Watermael-Boitsfort, Belgium 27/C9
Waterman, Ill. (60556) 222/E2
Waterman, Ind. (†47952) 227/C5
Waterpocket Fold (butte), Utah 304/D6
Waterport, N.Y. (14571) 276/D4
Waterproof, La. (71375) 238/H3
Waters, Mich. (49797) 250/E4
Watersmeet, Mich. (49969) 250/G2
Waterton-Glacier Int'l Peace Park, Alberta 182/C5
Waterton-Glacier International Peace Park, Alta. 162/E6
Waterton-Glacier Int'l Peace Park, Mont. 262/C2
Waterton Lakes Nat'l Park, Alberta 182/C5
Waterton Park, Alberta 182/D5
Watertown○, Conn. (06795) 210/C2
Watertown, Fla. (†32055) 212/D1
Watertown○, Mass. (02172) 249/C6
Watertown, Minn. (55388) 255/E6
Watertown, N.Y. 188/M2
Watertown, N.Y. (13601) 276/J3
Watertown, Ohio (45787) 284/G7
Watertown, S. Dak. 188/G1
Watertown, S. Dak. (57201) 298/P4
Watertown, Tenn. (37184) 237/J8
Watertown, Wis. (53094) 317/J9
Waterval-Bo, S. Africa 118/D5
Water Valley, Ala. (†36908) 195/B7
Water Valley, Alberta 182/C4
Water Valley, Ky. (42085) 237/D7
Water Valley, Miss. (38965) 256/E2
Water Valley, Texas (76958) 303/C6
Waterview, Ky. (42786) 237/L7
Waterview, Md. (†21840) 245/P8
Water View, Va. (23180) 307/P5
Waterville, Iowa (52170) 229/L2
Waterville, Ireland 17/A8
Waterville, Kansas (66548) 232/F2
Waterville, Maine 188/N2
Waterville, Maine (04901) 243/D6
Waterville, Mass. (†01475) 249/F2
Waterville, Minn. (56096) 255/E6
Waterville, New Bruns. 170/C2
Waterville, N.Y. (13480) 276/K5
Waterville, Nova Scotia 168/D3
Waterville, Ohio (43566) 284/C3
Waterville, Québec 172/F4
Waterville○, Vt. (05492) 268/B2
Waterville, Wash. (98858) 310/E3
Waterville Valley○, N.H. (03223) 268/D4
Watervliet, Mich. (49098) 250/C6
Watervliet, N.Y. (12189) 276/N5
Waterways, Alberta 182/E1
Watford, England 10/B5
Watford, England 13/H7
Watford, Ontario 177/C5
Watford City, N. Dak. (58854) 282/D4
Watha, N.C. (28471) 281/O5
Wathaman (riv.), Sask. 181/M3
Wathena, Kansas (66090) 232/H2
Watheroo, W. Australia 92/A5
Watino, Alberta 182/B2
Watkins, Iowa (52354) 229/J5
Watkins, Minn. (55389) 255/D5
Watkins Glen, N.Y. (14891) 276/G6
Watkinsville, Georgia (30677) 217/E3
Watling (San Salvador) (isl.), Bahamas 156/C1
Watonga, Okla. (73772) 288/K3
Watonwan (co.), Minn. 255/D7
Watova, Okla. (†74048) 288/P1
Watrous, N. Mex. (87753) 274/D3
Watrous, Sask. 162/F5
Watrous, Sask. 181/F4
Watsa, Zaire 115/E3
Watseka, Ill. (60970) 222/F3
Watson, Ark. (71674) 202/H6
Watson, Ill. (62473) 222/E4
Watson, Ind. (†47130) 227/F8
Watson, La. (70786) 238/L1
Watson, Minn. (56295) 255/C5
Watson, Mo. (64496) 261/A1
Watson, Okla. (74963) 288/S6
Watson, Sask. 181/G3
Watson (mt.), Utah 304/C3
Watson Lake, Yukon 187/F3
Watson Lake, Yukon 162/D3
Watsontown, Pa. (17777) 294/J3
Watsonville, Calif. (95076) 204/D7
Watten, Scotland 15/E2
Watten, Loch (lake), Scotland 15/E2
Wattensaw (bayou), Ark. 202/G4
Watton, England 13/H5
Watton, Mich. (49970) 250/G2
Watts, Okla. (74964) 288/S2
Watts Bar (dam), Tenn. 237/M9
Watts Bar (lake), Tenn. 237/M9
Watts Bar Dam, Tenn. (37395) 237/M9
Wattsburg, Pa. (16442) 294/C1
Watt Section Sheet Harbour, Nova Scotia 168/F4
Watts Mills, S.C. (†29360) 296/D2
Wattsview, Manitoba 179/A4
Wattsville, Ala. (35182) 195/F3
Wattwil, Switzerland 39/H2
Watubela (isls.), Indonesia 85/J6
Watuppa (pond), Mass. 249/K6
Watzmann (mt.), W. Germany 22/E5
Wau, Papua N.G. 85/B7
Wau, Papua N.G. 87/E6
Wau, Sudan 111/E6
Wau, Sudan 102/E4
Waubamik, Ontario 177/E2
Waubaushene, Ontario 177/E3
Waubay, S. Dak. (57273) 298/P3
Waubay (lake), S. Dak. 298/O3
Waubeek, Iowa (†52214) 229/K4
Waubeka, Wis. (53021) 317/L9
Waubun, Minn. (56589) 255/C3
Waucedah, Mich. (†49892) 250/B3
Wauchope, N.S. Wales 97/G2
Wauchope, Sask. 181/K6
Wauchula, Fla. (33873) 212/E4
Waucoma, Iowa (52171) 229/J2
Wauconda, Ill. (60084) 222/A4
Wauconda, Wash. (98859) 310/F2
Wau el Kebir, Libya 111/C2
Waugh, Ala. (†36104) 195/F6
Waugh (mt.), Idaho 220/D4
Waugh, Manitoba 179/G5
Waukarlycarly (lake), W. Australia 88/C4
Waukee, Iowa (50263) 229/F5
Waukeenah, Fla. (†32344) 212/C1
Waukegan, Ill. (60085) 222/B4
Waukesha (co.), Wis. 317/K9
Waukesha, Wis. (53186) 317/K1
Waukomis, Okla. (73773) 288/K2
Waukon, Iowa (52172) 229/L2
Waukon, Wash. (†99008) 310/H3
Waukon Junction, Iowa (†52146) 229/L2
Waumandee, Wis. (†54622) 317/C7
Waumbek (mt.), N.H. 268/E3
Wauna, Oreg. (†97016) 291/D1
Wauna, Wash. (98395) 310/C3
Waunakee, Wis. (53597) 317/G9
Wauneta, Kansas (†67024) 232/F4
Wauneta, Nebr. (69045) 264/C4
Waupaca (co.), Wis. 317/J6
Waupaca, Wis. (54981) 317/H7
Waupun, Wis. (53963) 317/J8
Wauregan, Conn. (06387) 210/H2
Waurika, Okla. (73573) 288/L6
Waurika (lake), Okla. 288/K6
Wausa, Nebr. (68786) 264/G2
Wausau, Fla. (32463) 212/D6
Wausau, Wis. 188/J2
Wausau, Wis. (54401) 317/G6
Wausau○, Wis. 317/80
Wausaukee, Wis. (54177) 317/K5
Wauseon, Ohio (43567) 284/B2
Waushara (co.), Wis. 317/H7
Wautoma, Wis. (54982) 317/H7
Wauwatosa, Wis. (53226) 317/L1
Wauzeka, Wis. (53826) 317/E9
Wave Hill, North. Terr. 88/E3
Wave Hill, North. Terr. 93/B4
Waveland, Ark. (72867) 202/C3
Waveland, Ind. (47989) 227/D5
Waveland, Miss. (39576) 256/F10

Waver (Wavre), Belgium 27/F7
Waverley, Mass. (02179) 249/B6
Waverley, N. S. Wales 88/L4
Waverley, N.S. Wales 97/K3
Waverley, N. Zealand 100/E3
Waverley, Nova Scotia 168/E4
Waverley, Ontario 177/E3
Waverley, Victoria 97/J5
Waverley, Victoria 88/L7
Waverley Downs, N.S. Wales 97/B1
Waverly, Ala. (36879) 195/G5
Waverly, Fla. (33877) 212/E4
Waverly, Georgia (31565) 217/J8
Waverly, Ill. (62692) 222/D4
Waverly, Iowa (50677) 229/J3
Waverly, Kansas (66871) 232/G3
Waverly, Ky. (42462) 237/F5
Waverly, La. (71232) 238/H2
Waverly, Minn. (55390) 255/E5
Waverly, Mo. (64096) 261/E4
Waverly, Nebr. (68462) 264/H4
Waverly, N.Y. (14892) 276/G7
Waverly, Ohio (45690) 284/D7
Waverly, S. Dak. (57202) 298/R3
Waverly, Tenn. (37185) 237/F8
Waverly, Va. (23890) 307/O6
Waverly, Wash. (99039) 310/H3
Waverly, W. Va. (26184) 312/D4
Waverly Hall, Georgia (31831) 217/C5
Waves, N.C. (27982) 281/U3
Wavre, Belgium 27/F7
Wawa (riv.), Nicaragua 154/E3
Wawa, Ontario 175/C3
Wawa, Ontario 177/J5
Wawaka, Ind. (46794) 227/F2
Wawanesa, Manitoba 179/C5
Wawasee, Ind. (†46567) 227/F2
Wawasee (lake), Ind. 227/F2
Wawayanda (lake), N.J. 273/E1
Waweig, New Bruns. 170/C3
Wawina, Minn. (55794) 255/E3
Wawota, Sask. 181/J6
Wawpecong, Ind. (†46901) 227/F3
Wax, Ky. (42787) 237/J6
Waxahachie, Texas (75165) 303/H5
Waxhaw, N.C. (28173) 281/H5
Way, Miss. (†39046) 256/E5
Way (lake), W. Australia 88/C5
Way (lake), W. Australia 92/C4
Wayagamac (lake), Québec 172/E2
Wayan, Idaho (83285) 220/G7
Wayatinah, Tasmania 99/C4
Waycross, Ga. 188/K4
Waycross, Georgia (31501) 217/H8
Wayerton, New Bruns. 170/E1
Wayland, Iowa (52654) 229/K6
Wayland, Ky. (41666) 237/R6
Wayland○, Mass. (01778) 249/A7
Wayland, Mich. (49348) 250/D6
Wayland, Mo. (63472) 261/J2
Wayland, N.Y. (14572) 276/E5
Wayland, Ohio (44285) 284/H3
Waymansville, Ind. (†47201) 227/E6
Waymart, Pa. (18472) 294/M2
Wayne, Ala. (†36763) 195/C6
Wayne, Alberta 182/D4
Wayne (co.), Georgia 217/J7
Wayne (co.), Ill. 222/E5
Wayne, Ill. (60184) 222/E2
Wayne (co.), Ind. 227/G5
Wayne (co.), Iowa 229/G7
Wayne, Kansas (†66930) 232/E2
Wayne (co.), Ky. 237/M7
Wayne, Maine (04284) 243/D7
Wayne○, Maine (04284) 243/D7
Wayne (co.), Mich. 250/F6
Wayne, Mich. (48184) 250/F6
Wayne (co.), Miss. 256/G7
Wayne (co.), Mo. 261/L8
Wayne (co.), Nebr. 264/G2
Wayne, Nebr. (68787) 264/G2
Wayne○, N.J. (07470) 273/A1
Wayne (co.), N.Y. 276/F4
Wayne, N.Y. (14893) 276/F6
Wayne (co.), N.C. 281/N4
Wayne (co.), Ohio 284/G4
Wayne, Ohio (43466) 284/C3
Wayne, Okla. (73095) 288/M5
Wayne (co.), Pa. 294/M2
Wayne, Pa. (19087) 294/M6
Wayne (co.), Tenn. 237/F10
Wayne (co.), Utah 304/C5
Wayne (co.), W. Va. 312/B6
Wayne, W. Va. (25570) 312/B6
Wayne City, Ill. (62895) 222/E5
Waynesboro, Georgia (30830) 217/J4
Waynesboro, Miss. (39367) 256/G7
Waynesboro, Pa. (17268) 294/G6
Waynesboro, Tenn. (38485) 237/F10
Waynesboro (I.C.), Va. (22980) 307/K4
Waynesburg, Ky. (40489) 237/M6
Waynesburg, Ohio (44688) 284/H4
Waynesburg, Pa. (15370) 294/B6
Waynesfield, Ohio (45896) 284/C4
Waynesville, Georgia (31566) 217/J8
Waynesville, Ill. (61778) 222/D3
Waynesville, Ind. (†47201) 227/F6
Waynesville, Mo. (65583) 261/H7
Waynesville, N.C. (28786) 281/D4
Waynesville, Ohio (45068) 284/B6
Waynetown, Ind. (47990) 227/C4
Waynoka, Okla. (73860) 288/J1
Wayside, Georgia (†31032) 217/E4
Wayside, Kansas (†67301) 232/G4
Wayside, Miss. (38780) 256/C4
Wayside, Texas (79094) 303/C3
Wayside, Wis. (†54126) 317/L7
Wayzata, Minn. (55391) 255/G5
Wazirabad, Pakistan 59/K3
We (isl.), Indonesia 85/B4
Wé, New Caled. 86/H4
Weagamow Lake, Ontario 175/B2
Weakley (co.), Tenn. 237/D8
Weald, The (reg.), England 13/H6
Wear (riv.), England 13/F3
Wear (riv.), England 10/F3
Weare○, N.H. (03281) 268/D5
Weare P.O. (North Weare), N.H. (03281) 268/D5
Weatherby, Mo. (64497) 261/D3
Weatherby Lake, Mo. (†64152) 261/O5
Weatherford, Okla. (73096) 288/J4
Weatherford, Texas (76086) 303/G5
Weatherly, Pa. (18255) 294/L4
Weathers, Okla. (†74560) 288/P5
Weathersby, Miss. (†39114) 256/E7
Weatogue, Conn. (06089) 210/D1
Weaubleau, Mo. (65774) 261/F7
Weaver, Ala. (36277) 195/G3
Weaver (riv.), England 13/G2
Weaver (lake), Manitoba 179/F2
Weaver, Minn. (†55958) 255/G6
Weaver, New Bruns. 170/E2
Weaver, N. Dak. (†58352) 282/N2
Weaverville, Calif. (96093) 204/B3
Weaverville, N.C. (28787) 281/D3
Webb, Ala. (36376) 195/H8
Webb, Iowa (51366) 229/D3
Webb (lake), Maine 243/C6
Webb, Miss. (38966) 256/D3
Webb (bay), Newf. 166/B2
Webb, Sask. 181/C5
Webb (co.), Texas 303/E10
Webb, Texas (†76010) 303/F3
Webb City, Ark. (†72949) 202/C3
Webb City, Mo. (64870) 261/C8
Webb City, Okla. (74654) 288/N1
Webber, Kansas (66970) 232/D2
Webbers Falls, Okla. (74470) 288/R3
Webbers Falls (res.), Okla. 288/R3
Webberville, Mich. (48892) 250/E6
Webb Lake, Wis. (54892) 317/B3
Webbs Cross Roads, Ky. (42652) 237/L6
Webbville, Ky. (41180) 237/R4
Webbwood, Ontario 177/C1
Webequie, Ontario 175/C2
Weber (co.), Utah 304/B2
Weber (riv.), Utah 304/C3
Weber City, Va. (24251) 307/C7
Webi Shabelle (riv.), Somalia 115/H3
Webster, Fla. (33597) 212/D3
Webster (co.), Georgia 217/C6
Webster, Ind. (47392) 227/H5
Webster (co.), Iowa 229/E4
Webster, Iowa (52355) 229/J6
Webster (res.), Kansas 232/C2
Webster (co.), Ky. 237/F5
Webster, Ky. (40176) 237/J5
Webster (par.), La. 238/D1
Webster (brook), Maine 243/E3
Webster, Mass. (01570) 249/G4
Webster○, Mass. (01570) 249/G4
Webster (lake), Mass. 249/G4
Webster, Minn. (55088) 255/E6
Webster (co.), Miss. 256/F3
Webster (co.), Mo. 261/G8
Webster (co.), Nebr. 264/F4
Webster○, N.H. (†03301) 268/D5
Webster, N.Y. (14580) 276/F4
Webster, N.C. (28788) 281/C4
Webster, N. Dak. (58382) 282/N3
Webster, Pa. (15087) 294/C5
Webster, S. Dak. (57274) 298/P3
Webster, Texas (77598) 303/K2
Webster (co.), W. Va. 312/F6
Webster, Wis. (54893) 317/B4
Webster City, Iowa (50595) 229/F4
Webster Groves, Mo. (63119) 261/P3
Webster Mills, Pa. (†17233) 294/F6
Webster Springs, W. Va. (26288) 312/F6
Websterville, Vt. (05678) 268/B3
Wecota, S. Dak. (57480) 298/L3
Weda, Indonesia 85/H5
Wedau, Papua N.G. 85/C7
Weddel (isl.), 143/D7
Weddell (sea), Ant. 2/H10
Weddell (sea) 5/C16
Wedderburn, Oreg. (97491) 291/C5
Wedderburn, Victoria 97/B5
Weddington, Ark. (†72701) 202/B1
Wedel, W. Germany 22/C2
Wedgefield, S.C. (29168) 296/F4
Wedgeport, Nova Scotia 168/C5
Wedgeworth, Ala. (†36776) 195/C5
Wedowee, Ala. (36278) 195/H4
Weed, Calif. (96094) 204/C2
Weed, N. Mex. (88354) 274/D6
Weed (hills), Sask. 181/J5
Weed Heights, Nev. (89443) 266/B4
Weedon-Centre, Québec 172/F4
Weedsport, N.Y. (13166) 276/G4
Weedville, Pa. (15868) 294/F3
Weehawken○, N.J. (07087) 273/C2
Week (isls.), Chile 138/D10
Weekapaug, R.I. (02891) 249/G7
Weekes, Sask. 181/J3
Weeki Wachee, Fla. (†33512) 212/D3
Weeks, La. (†70569) 238/G7
Weeks, Nev. (†89447) 266/B3
Weeks (isl.), N. Zealand 100/B1
Weeksbury, Ky. (41667) 237/R6
Weeks Mills, Maine (04361) 243/E7
Weeksville, N.C. (27909) 281/S2
Weems, Va. (22576) 307/P5
Weeping Water, Nebr. (68463) 264/J4
Weert, Netherlands 27/H6
Weesatche, Texas (77993) 303/G9
Weesen, Switzerland 39/H2
Weesp, Netherlands 27/C5
Weethalle, N.S. Wales 97/D3
Wee Waa, N.S. Wales 97/E2
Wegdahl, Minn. (†56265) 255/C6
Weggis, Switzerland 39/F2
Węgorzewo, Poland 47/E1
Wegra-Flat Creek, Ala. (†35129) 195/D3
Węgrów, Poland 47/E2
Weichang, China 77/J3
Weida, E. Germany 22/D3
Weiden in der Oberpfalz, W. Germany 22/D4
Weidman, Mich. (48893) 250/D5
Weifang, China 77/J4
Weihai (Weihaiwei), China 77/K4
Wei He (riv.), China 77/G5
Weilburg, W. Germany 22/C3
Weilheim im Oberbayern, W. Germany 22/D5
Weimar, E. Germany 22/D3
Weimar, Texas (78962) 303/H8
Weinan, China 77/H5
Weiner, Ark. (72479) 202/J2
Weinert, Texas (76388) 303/E4
Weinfelden, Switzerland 39/H1
Weingarten, W. Germany 22/C5
Weinheim, W. Germany 22/C4
Weining, China 77/F6
Weinsberg, W. Germany 22/C4
Weipa, Queensland 88/G1
Weipa, Queensland 95/B2
Weippe, Idaho (83553) 220/C3
Weir (lake), Fla. 212/E2
Weir, Kansas (66781) 232/H4
Weir, Miss. (39772) 256/F4
Weirdale, Sask. 181/F2
Weirgor, Wis. (†54835) 317/D4
Weir River, Manitoba 179/J2
Weirsdale, Fla. (32695) 212/D3
Weirton, W. Va. (26062) 312/E2
Weirwood, Va. (23484) 307/S6
Weisburg, Ind. (†47041) 227/H6
Weiser, Idaho (83672) 220/B5
Weiser (riv.), Idaho 220/B5
Weishan, China 77/F6
Weismes (Waimes), Belgium 27/J8
Weiss (lake), Ala. 195/G2
Weiss (lake), Georgia 217/A2
Weissenburg im Bayern, W. Germany 22/D4
Weissenfels, E. Germany 22/D3
Weissensee, E. Germany 22/F3
Weissenstein (mts.), Switzerland 39/D2
Weisserstein (mt.), Belgium 27/J8
Weissert, Nebr. (68880) 264/E3
Weisshorn (mt.), Switzerland 39/J3
Weisshorn (mt.), Switzerland 39/E4
Weissmies (mt.), Switzerland 39/F4
Weisswasser, E. Germany 22/F3
Weitchpec, Calif. (†95546) 204/B2
Weitensfeld-Flattnitz, Austria 41/B3
Weitra, Austria 41/C2
Weixi, China 77/E6
Weixin, China 77/F6
Weiz, Austria 41/C3
Wejh, Saudi Arabia 59/C4
Wejh, Saudi Arabia 54/E7
Wejherowo, Poland 47/D1
Welaka, Fla. (32093) 212/E2
Welbekend, S. Africa 118/J6
Welch, Okla. (74369) 288/R1
Welch, Texas (79377) 303/B5
Welch, W. Va. (24801) 312/C8
Welches, Oreg. (†97067) 291/E2
Welchman Hall, Barbados 161/B8
Welchville, Maine (†04270) 243/C7
Welcome, La. (†70086) 238/L3
Welcome, Md. (20693) 245/K7
Welcome, Minn. (56181) 255/D7
Welcome, N.C. (27374) 281/J3
Welcome, Ontario 177/F4
Welcome All, Georgia (†30304) 217/J2
Weld (co.), Colo. 208/L1
Weld○, Maine (04285) 243/C6
Weld (range), W. Australia 92/B4
Welda, Kansas (66091) 232/G3
Weldon, Ark. (72177) 202/H3
Weldon, Calif. (93283) 204/G8
Weldon, Ill. (61882) 222/E3
Weldon, Iowa (50264) 229/F7
Weldon, New Bruns. 170/F3
Weldon, N.C. (27890) 281/O2
Weldon, Sask. 181/F2
Weldon, Texas (75863) 303/J6
Weldona, Colo. (80653) 208/M2
Weldon Spring Heights, Mo. (†63301) 261/M2
Weleetka, Okla. (74880) 288/O4
Welford, Queensland 95/C5
Welkom, S. Africa 102/E7
Welkom, S. Africa 118/D5
Welland (riv.), England 13/G5
Welland (riv.), England 10/F4
Welland, Ontario 177/E5
Welland (canal), Ontario 177/E5
Wellandport, Ontario 177/E4
Wellborn, Fla. (32094) 212/D1
Wellersburg, Pa. (15564) 294/E6
Wellesley (isls.), Australia 87/D7
Wellesley○, Mass. (02181) 249/B7
Wellesley, Ontario 177/D4
Wellesley (isls.), Queensland 88/F3
Wellesley (isls.), Queensland 95/A3
Wellesley Hills, Mass. (02181) 249/B7
Wellfleet○, Mass. (02667) 249/O5
Wellfleet (harb.), Mass. 249/O5
Wellfleet, Nebr. (69170) 264/D4
Wellford, S.C. (29385) 296/C2
Wellin, Belgium 27/G8
Welling, Alberta 182/D5
Welling, Okla. (74471) 288/S3
Wellingborough, England 13/G5
Wellingborough, England 10/F4
Wellington, Ala. (36279) 195/G3
Wellington (isl.), Chile 120/B7
Wellington (isl.), Chile 138/D8
Wellington, Colo. (80549) 208/K1
Wellington, England 13/D7
Wellington, England 10/E5
Wellington, Ill. (60973) 222/F3
Wellington, Kansas (67152) 232/E4
Wellington, Ky. (†40201) 237/K2
Wellington, Ky. (40387) 237/O5
Wellington○, Maine (04990) 243/D5
Wellington, Mo. (64097) 261/E4
Wellington, Nev. (89444) 266/B4
Wellington, N.S. Wales 97/E3
Wellington (cap.), N. Zealand 2/T8
Wellington (cap.), N. Zealand 87/H10
Wellington (cap.), N. Zealand 100/A3
Wellington (bay), N.W. Terrs. 187/H3
Wellington (chan.), N.W.T. 162/G1
Wellington (chan.), N.W. Terrs. 187/J2
Wellington, Nova Scotia 168/E4
Wellington, Ohio (44090) 284/F3
Wellington (county), Ontario 177/D4
Wellington, Ontario 177/G4
Wellington, Pr. Edward I. 168/D2
Wellington, S. Africa 118/B6
Wellington, Texas (79095) 303/D3
Wellington, Utah (84542) 304/D4
Wellington (lake), Victoria 97/D6
Wellington, Va. (†22308) 307/T3
Wellman, Iowa (52356) 229/K6
Wellman (lake), Manitoba 179/B3
Wellman, Texas (79378) 303/B5
Wellpinit, Wash. (99040) 310/G3
Wells, Br. Col. 184/G3
Wells, England 13/E6
Wells, England 10/E5
Wells (co.), Ind. 227/G3
Wells, Kansas (67488) 232/E2
Wells, Maine (04090) 243/B9
Wells○, Maine (04090) 243/B9
Wells, Mich. (49894) 250/B3
Wells, Minn. (56097) 255/E7
Wells, Nev. (89835) 266/G1
Wells, N.Y. (12190) 276/M4
Wells (co.), N. Dak. 282/L4
Wells, Texas (75976) 303/J6
Wells○, Vt. (05774) 268/A5
Wells (riv.), Vt. 268/C3
Wells (dam), Wash. 310/F3
Wells (lake), W. Australia 88/C5
Wells (lake), W. Australia 92/C4
Wells Beach, Maine (04090) 243/B9
Wellsboro, Ind. (†46382) 227/D1
Wellsboro, Pa. (16901) 294/H2
Wells Bridge, N.Y. (13859) 276/K6
Wellsburg, Iowa (50680) 229/H4
Wellsburg, N.Y. (14894) 276/G6
Wellsburg, N. Dak. (†58341) 282/L4
Wellsburg, W. Va. (26070) 312/E2
Wellsford, N. Zealand 100/E2
Wells Gray Prov. Park, Br. Col. 184/H4
Wells-next-the-Sea, England 13/H5
Wells-next-the-Sea, England 10/G4
Wells River, Vt. (05081) 268/C3
Wellston, Mich. (49689) 250/D4
Wellston, Mo. (63112) 261/R2
Wellston, Ohio (45692) 284/F7
Wellston, Okla. (74881) 288/M3
Wellsville, Kansas (66092) 232/G3
Wellsville, Mo. (63384) 261/K4
Wellsville, N.Y. (14895) 276/E6
Wellsville, Ohio (43968) 284/J4
Wellsville, Pa. (17365) 294/J5
Wellsville, Utah (84339) 304/C2
Wellton, Ariz. (85356) 198/A6
Wellwood, Manitoba 179/C4
Wels, Austria 41/C2
Welsford, New Bruns. 170/D3
Welsford, Nova Scotia 168/E3
Welsh, La. (70591) 238/E6
Welshfield, Ohio (†44021) 284/H3
Welshpool, New Bruns. 170/D4
Welshpool, Wales 10/E4
Welshpool, Wales 13/D5
Welton, Iowa (52774) 229/M5
Welty, Okla. (74882) 288/O3
Welwyn, England 13/H7
Welwyn, England 10/F5
Welwyn, Sask. 181/K5
Wem, England 13/E5
Wembere (riv.), Tanzania 115/F4
Wembley, Alberta 182/A2
Wemmel, Belgium 27/B9
Wemyss Bay, Scotland 15/A2
Wenamu (riv.), Guyana 131/A2
Wenas (creek), Wash. 310/E4
Wenasoga, Miss. (†38834) 256/G1
Wenatchee, Wash. 188/B1
Wenatchee, Wash. (98801) 310/E3
Wenatchee (lake), Wash. 310/E3
Wenatchee (mts.), Wash. 310/E3
Wenatchee (riv.), Wash. 310/E3
Wenchi, Ghana 106/D7
Wenchow (Wenzhou), China 77/J6
Wendel, Calif. (96136) 204/E3
Wendel, W. Va. (26450) 312/F4
Wendell, Idaho (83355) 220/D7
Wendell○, Mass. (01379) 249/E2
Wendell, Minn. (56590) 255/B4
Wendell, N.H. (03783) 268/C5
Wendell, N.C. (27591) 281/N3
Wendell Depot, Mass. (01380) 249/E2
Wenden, Ariz. (85357) 198/B5
Wendeng, China 77/K4
Wendover, England 13/G7
Wendover, Ontario 177/J2
Wendover, Utah (84083) 304/A3
Wendover, Wyo. (†82214) 319/H3
Wendron, England 13/B7
Wendte, S. Dak. (†57532) 298/H5
Wenham○, Mass. (01984) 249/L2
Wenling, China 77/K6
Wenlock (riv.), Queensland 88/G2
Wenman (isl.), Ecuador 128/B8
Wenona, Georgia (†31015) 217/E7
Wenona, Ill. (61377) 222/E2
Wenona, Md. (21870) 245/P8
Wenona, N.C. (†27860) 281/R3
Wenonah, Ill. (†62075) 222/D4
Wenonah, N.J. (08090) 273/C4
Wenquan, Qinghai, China 77/D5
Wenquan, Xinjiang Uygur, China 77/B3
Wenshan, China 77/F7
Wensum (riv.), England 13/J5
Wentworth, Mo. (64873) 261/D8
Wentworth○, N.H. (03282) 268/D4
Wentworth (lake), N.H. 268/E4
Wentworth, N.S. Wales 97/B4
Wentworth, N.C. (27375) 281/K2
Wentworth, Nova Scotia 168/E3
Wentworth, S. Dak. (57075) 298/R6
Wentworth, Wis. (54894) 317/C2
Wentworths Location○, N.H. (†03579) 268/E2
Wentzville, Mo. (63385) 261/L5
Wen Xian, China 77/G5
Wenzhou (Wenchow), China 77/J6
Wenzhou, China 54/N7
Weogufka, Ala. (35183) 195/F4
Weohyakapka (lake), Fla. 212/E4
Weott, Calif. (95571) 204/A3
Wepawaug (riv.), Conn. 210/C3
Wequetequock, Conn. (†02891) 210/H3
Werdau, E. Germany 22/E3
Werder, E. Germany 22/E2
Werner Lake, Ontario 175/A2
Wernersville, Pa. (19565) 294/K5
Wernigerode, E. Germany 22/D3
Werra (riv.), E. Germany 22/D3
Werra (riv.), W. Germany 22/D3
Werribee, Victoria 88/G7
Werrimull, Victoria 97/A4
Werris Creek, N.S. Wales 97/F2
Wertheim, W. Germany 22/C4
Wervik, Belgium 27/B7
Wesco, Mo. (65586) 261/K7
Wesel, W. Germany 22/B3
Weser (riv.), Germany 7/E3
Weser (riv.), W. Germany 22/C2
Weskan, Kansas (67762) 232/A3
Weslaco, Texas (78596) 303/F11
Weslemkoon (lake), Ontario 177/G2
Wesley, Ark. (72773) 202/C1
Wesley, Dominica 161/E5
Wesley, Georgia (†30401) 217/H6
Wesley, Iowa (50483) 229/E2
Wesley, Maine (04686) 243/H6
Wesley○, Maine (04686) 243/H6
Wesley Vale, Tasmania 99/C3
Wesleyville, Newf. 166/D4
Wesleyville, Pa. (16510) 294/C1
Wes-Rand, S. Africa 118/G6
Wessel (isls.), Australia 87/D7
Wessel (cape), North. Terr. 88/F2
Wessel (cape), North. Terr. 93/E1
Wessel (isls.), North. Terr. 88/F2
Wessel (isls.), North. Terr. 93/E1
Wessington, S. Dak. (57381) 298/M5
Wessington Springs, S. Dak. (57382) 298/M5
Wesson, Ark. (†71749) 202/E7
Wesson, Miss. (39191) 256/D7
West (riv.), Conn. 210/D3
West (riv.), Conn. 210/E3
West, Iowa (52357) 229/J5
West (bay), La. 238/M8
West (isl.), Mass. 249/L6
West (riv.), Mass. 249/H4
West, Miss. (39192) 256/E4
West (isls.), New Bruns. 170/D4
West (cape), N. Zealand 100/A6
West (bay), Nova Scotia 168/G3
West (pt.), Nova Scotia 168/H5
West (riv.), Nova Scotia 168/F3
West (pt.), Pr. Edward I. 168/D2
West (pt.), Tasmania 99/A2
West, Texas (76691) 303/G6
West (bay), Texas 303/K3
West (riv.), Vt. 268/B5
West Acton, Mass. (01720) 249/H3
West Alexander, Pa. (15376) 294/B5
West Alexandria, Ohio (45381) 284/A6
West Allis, Wis. (53214) 317/L1
West Alton, Mo. (63386) 261/M5
West Alton, N.H. (†03246) 268/E4
West Amboy, N.Y. (†13493) 276/J4
West Arichat, Nova Scotia 168/G3
West Ashford, Conn. (†06251) 210/G1
West Aspetuck (riv.), Conn. 210/B2
West Athens, Maine (†04912) 243/D6
West Augusta, Va. (24485) 307/K4
West Avon, Conn. (†06001) 210/D1
West Baden Springs, Ind. (47469) 227/D7
West Baines (riv.), North. Terr. 93/A4
West Baldwin, Maine (04091) 243/B8
Westbank, Br. Col. 184/H5
WEST BANK 59/C3
WEST BANK 65/C3
West Bank (reg.), 65/C3
West Baraboo, Wis. (†53913) 317/G9
West Barnet, Vt. (05870) 268/C3
West Barns, Scotland 15/F5
West Barnstable, Mass. (02668) 249/N6
West Barrington, R.I. (†02806) 249/J5
West Bath○, Maine (†04530) 243/D8
West Baton Rouge (par.), La. 238/H6
West Bay, Fla. (32407) 212/C6
West Bay, Nova Scotia 168/G3
West Bay Road, Nova Scotia 168/G3
West Bend, Iowa (50597) 229/D3
Westbend, Ky. (40388) 237/N5
West Bend, Sask. 181/H4
West Bend, Wis. (53095) 317/K9
West Bengal (state), India 68/F4
West Berkshire, Vt. (†05450) 268/B2
West Berlin, Mass. (†01503) 249/H3
West Berlin, N.J. (08091) 273/D4
West Bethel, Maine (04286) 243/B7
West Blocton, Ala. (35184) 195/D4
West Bloomfield, Wis. (†54983) 317/J7
Westboro, Mo. (64498) 261/B1
Westboro, Ohio (†45148) 284/C7
Westboro, Wis. (54490) 317/F5
Westborough, Mass. (01581) 249/H3
Westborough○, Mass. (01581) 249/H3
West Bountiful, Utah (†84087) 304/B3
Westbourne, Manitoba 179/D4
Westbourne, Tenn. (†37766) 237/O7
West Boxford, Mass. (01885) 249/K2
West Boylston○, Mass. (01583) 249/G3
West Braintree, Vt. (†05669) 268/B4
West Branch (res.), Conn. 210/C1
West Branch, Iowa (52358) 229/L5
West Branch, Farmington (riv.), Mass. 249/B4
West Branch, Mich. (48661) 250/E4
West Branch, Rocky (riv.), Ohio 284/G10
West Brattleboro, Vt. (05301) 268/B6
West Brentwood, N.H. (†03848) 268/E6
West Brewster, Mass. (†02631) 249/O5
Westbridge, Br. Col. 184/H5
West Bridgewater○, Mass. (02379) 249/K4
West Bridgewater, Vt. (†05034) 268/B4
West Bridgford, England 13/F5
West Bromwich, England 13/F5
West Bromwich, England 10/G3
Westbrook, Conn. (06498) 210/F3
Westbrook○, Conn. (06498) 210/F3
Westbrook, Maine (04092) 243/C8
Westbrook, Minn. (56183) 255/C6
West Brook, Nova Scotia 168/D3
Westbrook, Texas (79565) 303/C5
West Brookfield, Mass. (01585) 249/F4
West Brookfield○, Mass. (01585) 249/F4
West Brooklyn, Ill. (61378) 222/D2
West Brooksville, Maine (†04617) 243/F7
West Brownsville, Pa. (15417) 294/C5
West Buechel, Ky. (†40218) 237/K2
West Burke, Vt. (05871) 268/C2
West Burlington, Iowa (52655) 229/L7
West Burra (isl.), Scotland 15/G2
Westbury, England 10/E5
Westbury, England 13/E6
Westbury, N.Y. (11590) 276/R7
Westbury, Tasmania 99/C3
West Buxton, Maine (04093) 243/B8
Westby, Mont. (59275) 262/M2
Westby, Wis. (54667) 317/E8
West Calder, Scotland 15/C2
West Caldwell, N.J. (07006) 273/A2
West Campton, N.H. (03228) 268/D4
West Canaan, N.H. (03741) 268/C4
West Cape May, N.J. (†08204) 273/D6
West Carroll (par.), La. 238/H1
West Carrollton, Ohio (45449) 284/B6
West Carthage, N.Y. (†13619) 276/J3
West Charleston, Vt. (05872) 268/C2
West Chatham, Mass. (02669) 249/O6
West Chazy, N.Y. (12992) 276/N1
West Chelmsford, Mass. (†01824) 249/J2
Westchester, Conn. (†06474) 210/F2
Westchester, Ill. (60153) 222/B5
West Chester, Iowa (52359) 229/K6
Westchester (co.), N.Y. 276/N8
West Chester, Ohio (45069) 284/C9
West Chester, Pa. (19380) 294/L6
West Chesterfield, Mass. (01084) 249/C3
Westchester Station, Nova Scotia 168/E3
West Chicago, Ill. (60185) 222/A5
West Chop (pt.), Mass. 249/M7
West City, Ill. (†62812) 222/E5
Westcliffe, Colo. (81252) 208/H6
West College Corner, Ind. (†47353) 227/H5
West Columbia, S.C. (29169) 296/E4
West Columbia, Texas (77486) 303/J8
West Columbia, W. Va. (25287) 312/B5
West Concord, Mass. (†01742) 249/A6
West Concord, Minn. (55985) 255/F6
West Corinth, Vt. (†05039) 268/C3
West Cornwall, Conn. (06796) 210/B1
West Cornwall, Vt. (†05753) 268/A4
West Cote Blanche (bay), La. 238/G7
Westcott, Alberta 182/C4
Westcott Cove (bay), Conn. 210/A4
West Covina, Calif. (*91790) 204/D10
Westcreek, Colo. (†80135) 208/J4
West Creek, N.J. (08092) 273/E4
West Crossett, Ark. (†71635) 202/F7
West Cummington, Mass. (†01026) 249/B3
West Danville, Vt. (05873) 268/C3
West Dean, England 13/E6
West Demerara-Essequibo Coast (dist.), Guyana 131/B2
West Dennis, Mass. (02670) 249/O6
West Deptford○, N.J. (†08086) 273/B3
West Des Moines, Iowa (50318) 229/F5
West Dover, Nova Scotia 168/E4
West Dover, Vt. (05356) 268/B6
West Dublin, Nova Scotia 168/D4
West Dudley, Mass. (†01550) 249/F4
West Dummerston, Vt. (05357) 268/B6
West Dundee (Dundee), Ill. (†60118) 222/E1
West Eau Gallie, Fla. (32935) 212/F3
West Elizabeth, Pa. (15088) 294/C5
West Elkton, Ohio (45070) 284/A6
West Elmira, N.Y. (†14901) 276/G6
West Eminence, Mo. (†65466) 261/J8
Westend, Calif. (†93562) 204/H8
West End, N.C. (27376) 281/K4
West End, Sask. 181/J5
West End, Virgin Is. (Br.) 161/C4
West End-Cobb Town, Ala. (†36201) 195/G3
Westend Saltpond (lag.), Virgin Is. (U.S.) 161/E4
West Enfield, Maine (04493) 243/F5
West Epping, N.H. (†03042) 268/E5
Wester Eems (chan.), Netherlands 27/K1
Westerland, W. Germany 22/C1
Westerlo, Belgium 27/F6
Westerlo, N.Y. (12193) 276/M6
Westerly, R.I. (02891) 249/G7
Westerly○, R.I. (02891) 249/G7
Western (prov.), Kenya 115/G3
Western, Nebr. (68464) 264/G4
Western (head), Nova Scotia 168/D5
Western Australia 88/B5
WESTERN AUSTRALIA 92
Western Australia (state), Australia 87/C8
Western Bay, Newf. 166/D2
Western Channel (str.), Japan 81/D6
Western Dvina (riv.), U.S.S.R. 53/C2
Western Dvina (riv.), U.S.S.R. 52/C3
Western Dvina (riv.), U.S.S.R. 48/C4
Western Ghats (mts.), India 68/C5
Western Grove, Ark. (72685) 202/D1
Western Institute, Tenn. (38074) 237/C10
Western Isles (islands area), Scotland 15/A3
Westernport, Md. (21562) 245/B3

Western Port (inlet), Victoria 97/C6
Western Sahara 2/J4
Western Sahara 102/A2
WESTERN SAHARA 106/A4
Western Samar (prov.), Philippines 82/E5
Western Samoa 2/A6
WESTERN SAMOA 86/M8
Western Samoa 87/J7
Western Scheldt (De Honte) (bay), Netherlands 27/D6
Western Shore, Nova Scotia 168/D4
Western Shoshone Ind. Res., Idaho 220/B7
Western Shoshone Ind. Res., Nev. 266/E1
Western Springs, Ill. (60558) 222/B6
Westernville, N.Y. (13486) 276/K4
Westerose, Alberta 182/C3
Westerstede, W. Germany 22/B2
Westervelt, Ill. (62574) 222/E4
Westerville, Nebr. (68881) 264/E3
Westerville, Ohio (43081) 284/D5
Westerville, S. Dak. (†57069) 298/P8
Westerwald (for.), W. Germany 22/B3
West Fairlee○, Vt. (05083) 268/C4
West Falkland (isl.), 143/D7
West Falkland (isl.), Falk. Is. 120/C8
Westfall, Kansas (67489) 232/D3
Westfall, Oreg. (97920) 291/K3
West Falmouth, Mass. (02574) 249/M6
West Fargo, N. Dak. (58078) 282/S6
West Farmington, Maine (04992) 243/C6
West Farmington, Ohio (44491) 284/J3
West Feliciana (par.), La. 238/H5
Westfield, Conn. (†06457) 210/E2
Westfield, Ill. (62474) 222/F4
Westfield, Ind. (46074) 227/E4
Westfield, Iowa (51062) 229/A3
Westfield○, Maine (04787) 243/G2
Westfield, Mass. (01085) 249/D4
Westfield (riv.), Mass. 249/C3
Westfield, New Bruns. 170/D3
Westfield, N.J. (*07090) 273/E2
Westfield, N.Y. (14787) 276/A6
Westfield, N.C. (27053) 281/H2
Westfield, N. Dak. (†58542) 282/K7
Westfield, Nova Scotia 168/C4
Westfield, Pa. (16950) 294/H2
Westfield○, Vt. (05874) 268/C2
Westfield, Wis. (53964) 317/H8
Westfield Center, Ohio (44251) 284/G3
West Finley, Pa. (15377) 294/B5
Westfir, Oreg. (97492) 291/E4
West Flanders (prov.), Belgium 27/B7
Westford, Conn. (†06076) 210/G1
Westford○, Mass. (01886) 249/J2
Westford, Pa. (16134) 294/A2
Westford○, Vt. (05494) 268/A2
West Fork, Ark. (72774) 202/B2
West Fork, Ind. (47178) 227/D8
West Fork, Bruneau (riv.), Nev. 266/F1
West Fork (riv.), W. Va. 312/E5
West Forks○, Maine (04985) 243/D5
West Frankfort, Ill. (62896) 222/E6
West Franklin, Ind. (†47620) 227/B9
West Franklin, Maine (†04634) 243/G6
West Frisian (isls.), Netherlands 27/F2
Westgat (chan.), Netherlands 27/F3
Westgate, Iowa (50681) 229/K3
Westgate, Manitoba 179/A2
West Germany 7/E3
WEST GERMANY 22
West Glacier, Mont. (59936) 262/C2
West Glamorgan, Wales 13/D6
West Glens Falls, N.Y. (†12801) 276/N4
West Glocester, R.I. (†02814) 249/G5
West Glover, Vt. (05875) 268/C2
West Gorham, Maine (†04038) 243/C8
West Goshen, Conn. (†06756) 210/B1
West Gouldsboro, Maine (†04607) 243/G7
West Granby, Conn. (06090) 210/D1
West Grand (lake), Maine 243/H5
West Granville, Mass. (†01034) 249/C4
West Green, Georgia (31567) 217/G7
West Greene, Ala. (35491) 195/B5
West Green Harbour, Nova Scotia 168/C5
West Groton, Mass. (01472) 249/H2
West Grove, Iowa (52538) 229/J7
West Grove, Pa. (19390) 294/L6
West Gulfport (North Gulfport), Miss. (†39501) 256/F10
West Halifax, Vt. (05358) 268/B6
West Hamlin, W. Va. (25571) 312/B6
West Hampstead, N.H. (†03841) 268/E6
Westhampton○, Mass. (†01027) 249/C3
Westhampton, N.Y. (11977) 276/P9
Westhampton Beach, N.Y. (11978) 276/P9
West Hanover, Mass. (02339) 249/L4
West Harrison, Ind. (†45030) 227/H6
West Hartford○, Conn. (06107) 210/D1
West Hartford, Vt. (05084) 268/C4
West Hartland, Conn. (06091) 210/D1
West Harwich, Mass. (02671) 249/O6
West Haven, Conn. (06516) 210/D3
Westhaven, Ill. (†60462) 222/B6
West Haven○, Vt. (†05743) 268/A4
West Hawk (lake), Manitoba 179/G5
Westhawk Lake, Manitoba 179/G5
West Hawley, Mass. (†01339) 249/C2
West Hazleton, Pa. (18201) 294/K4
West Helena, Ark. (72390) 202/J4
West Henniker, N.H. (†03242) 268/D5
West Hickory, Pa. (16370) 294/C2
West Hill (pond), Conn. 210/C1
Westhoff, Texas (77994) 303/G8
West Hollywood, Calif. (†90069) 204/B10
Westholme, Br. Col. 184/J3
Westhope, N. Dak. (58793) 282/H2
West Hopkinton, N.H. (†03229) 268/D5
West Hurley, N.Y. (12491) 276/M6
West Ice Shelf, Ant. 5/C5
West Indies (isls.) 2/G5
West Indies (isls.) 146/M7
WEST INDIES 156
West Irvine, Ky. (40491) 237/N5
West Jefferson, Ala. (†35005) 195/D4
West Jefferson, N.C. (28694) 281/F2
West Jefferson, Ohio (43162) 284/D6
West Jersey, Ill. (†61483) 222/D2
West Jonesport, Maine (†04649) 243/H6
West Jordan, Utah (84084) 304/B3
Westkapelle, Netherlands 27/C5
West Kennebunk, Maine (04094) 243/B9
West Kilbride, Scotland 15/D5
West Kingston, R.I. (02892) 249/H7
West Kittanning, Pa. (†16201) 294/C4
West Lafayette, Ind. (47906) 227/D4
West Lafayette, Ohio (43845) 284/G5
Westlake, La. (70669) 238/D6
Westlake, Ohio (44145) 284/G9
Westlake, Oreg. (97493) 291/C4
Westlake, Texas (†76101) 303/F1
Westland, Mich. (48185) 250/F6
Westland, Pa. (15378) 294/B5
West Lanham Hills, Md. (†20784) 245/G4
West Laurel, Md. (†20810) 245/L4
West Lawn, Pa. (19609) 294/K5
West Lebanon, Ind. (47991) 227/C4
West Lebanon, Maine (04027) 243/B9
West Lebanon, N.H. (03784) 268/C4
West Ledge, Bermuda 156/G3
West Leechburg, Pa. (†15656) 294/C4
West Leipsic, Ohio (†45856) 284/B3
West Leyden, N.Y. (13489) 276/J4
West Liberty, Ill. (62475) 222/E5
West Liberty, Iowa (52776) 229/L5
West Liberty, Ky. (41472) 237/P5
West Liberty, Ohio (43357) 284/C5
West Liberty, Pa. (†16057) 294/B4
West Liberty, W. Va. (26074) 312/E2
West Lima, Wis. (†54639) 317/E8
West Line, Mo. (64791) 261/C5
Westline, Pa. (16751) 294/E2
West Linn, Oreg. (97068) 291/B2
West Linton, Scotland 15/D5
West Liscomb (riv.), Nova Scotia 168/F3
West Little Owyhee (riv.), Oreg. 291/K5
West Loch (inlet), Hawaii 218/A3
West Loch Tarbert (inlet), Scotland 15/A3
West Loch Tarbert (inlet), Scotland 15/C5
Westlock, Alberta 182/C2
West Logan, W. Va. (25601) 312/C7
West Long (lake), New Bruns. 170/D3
West Long Branch, N.J. (07764) 273/F3
West Lorne, Ontario 177/C5
West Los Angeles, Calif. (90025) 204/B10
West Lothian (trad. co.), Scotland 15/B5
West Louisville, Ky. (42377) 237/G5
West Lubec, Maine (†04652) 243/J6
Westmalle, Belgium 27/F6
West Manchester, Ohio (45382) 284/A6
West Mansfield, Mass. (†02048) 249/K5
West Mansfield, Ohio (43358) 284/C5
West Maui (mts.), Hawaii 218/H2
Westmeath (co.), Ireland 17/G5
Westmeath, Ontario 177/H2
West Medway, Mass. (†02053) 249/J4
West Melbourne, Fla. (†32901) 212/F3
West Memphis, Ark. (72301) 202/K3
West Mersea, England 13/H6
West Mersea, England 10/G5
West Miami, Fla. (†33101) 212/B5
West Middlesex, Pa. (16159) 294/B3
West Middleton, Ind. (46995) 227/E4
West Middletown, Pa. (15379) 294/A5
West Midlands (co.), England 13/F5
West Mifflin, Pa. (15122) 294/C7
West Milan, N.H. (†03588) 268/E2
West Milford, N.J. (07480) 273/E1
West Milford, W. Va. (26451) 312/F4
West Millbury, Mass. (01586) 249/G4
West Millgrove, Ohio (43467) 284/C3
West Mills, Maine (†04938) 243/C6
West Milton, Ohio (45383) 284/B6
West Milton, Pa. (17886) 294/J4
West Milwaukee, Wis. (†53201) 317/L1
West Mineral, Kansas (66782) 232/H4
West Minot, Maine (04288) 243/C7
Westminster, Calif. (92683) 204/D11
Westminster, Colo. (80030) 208/J3
Westminster, Conn. (†06331) 210/G2
Westminster, England 10/B5
Westminster, England 13/H8
Westminster, Md. (21157) 245/L2
Westminster○, Mass. (01473) 249/G2
Westminster, S.C. (29693) 296/A2
Westminster, Vt. (05158) 268/C5
Westminster○, Vt. (05158) 268/C5
Westminster Station, Vt. (05159) 268/B5
Westminster West, Vt. (†05158) 268/B5
West Monroe, La. (71291) 238/F1
Westmont, Calif. (†90047) 204/C11
Westmont, Ill. (60559) 222/B6
Westmont, N.J. (08108) 273/B3
Westmont, Pa. (†15905) 294/D5
West Monterey, Pa. (†16049) 294/C3
Westmore○, Vt. (†05860) 268/C2
Westmoreland, Kansas (66549) 232/F2
Westmoreland○, N.H. (03467) 268/C6
Westmoreland (co.), Pa. 294/D5
Westmoreland, Queensland 95/A3
Westmoreland, Tenn. (37186) 237/J7
Westmoreland (co.), Va. 307/P4
Westmorland, Calif. (92281) 204/K10
Westmorland (co.), New Bruns. 170/F2
Westmount, Nova Scotia 168/H2
Westmount, Québec 172/H4
Westmuir, Scotland 15/E4
West Musquash (lake), Maine 243/H5
West Mystic, Conn. (06388) 210/H3
West Newbury○, Mass. (01985) 249/L1
West Newbury, Vt. (05085) 268/C3
West Newfield, Maine (04095) 243/B8
West Newton, Mass. (†02165) 249/B7
West Newton, Pa. (15089) 294/C5
West New York, N.J. (07093) 273/C2
West Nicholson, Zimbabwe 118/D4
West Nishnabotna (riv.), Iowa 229/C6
West Norwalk, Conn. (†06856) 210/B4
West Nottingham, N.H. (03291) 268/E5
West Nyack, N.Y. (10994) 276/K8
West Okoboji, Iowa (†51351) 229/C2
West Olive, Mich. (49460) 250/C6
Weston, Ala. (†35570) 195/B2
Weston, Colo. (81091) 208/K8
Weston○, Conn. (06883) 210/B4
Weston, Georgia (31832) 217/C7
Weston, Idaho (83286) 220/F7
Weston, Ill. (†61726) 222/E3
Weston, Iowa (†51576) 229/B6
Weston○, Maine (04424) 243/H4
Weston, Malaysia 85/F4
Weston○, Mass. (02193) 249/B6
Weston, Mich. (49289) 250/E7
Weston, Mo. (64098) 261/C4
Weston, Nebr. (68070) 264/H3
Weston, Ohio (43569) 284/C3
Weston, Oreg. (97886) 291/J2
Weston, Pa. (18256) 294/K4
Weston○, Vt. (05161) 268/B5
Weston, W. Va. (26452) 312/F4
Weston, Wis. (†54751) 317/C6
Weston, Wis. (†54476) 317/G6
Weston (co.), Wyo. 319/H2
Weston, Wyo. (82731) 319/G1
Westonaria, S. Africa 118/H7
Westons Mills, N.Y. (14788) 276/D6
Weston-super-Mare, England 13/D6
Weston-super-Mare, England 10/E5
West Orange, N.J. (07052) 273/A2
West Orange, Texas (77630) 303/L7
West Ossipee, N.H. (03890) 268/E4
Westover, Ala. (35185) 195/E4
Westover, Md. (21871) 245/R8
Westover, Pa. (16692) 294/E4
Westover, S. Dak. (†57559) 298/H6
Westover, W. Va. (26505) 312/G3
Westover A.F.B., Mass. 249/D4
Westover Hills, Texas (†76101) 303/E2
West Paducah, Ky. (42086) 237/D6
West Palm Beach, Fla. 188/K5
West Palm Beach, Fla. 146/K7
West Palm Beach, Fla. (*33401) 212/F5
West Palm Beach (canal), Fla. 212/F5
West Paris○, Maine (04289) 243/B7
West Paterson, N.J. (07424) 273/B2
West Pawlet, Vt. (05775) 268/A5
West Pelzer, S.C. (29669) 296/B2
West Pembroke, Maine (†04666) 243/J6
West Pensacola, Fla. (†32502) 212/B6
West Peru, Maine (04290) 243/C7
West Peterborough, N.H. (03468) 268/C6
West Petersburg, Alaska (†99833) 196/M1
Westphalia, Ind. (47596) 227/C7
Westphalia, Iowa (51578) 229/C5
Westphalia, Kansas (66093) 232/G3
Westphalia, Mich. (48894) 250/E6
Westphalia, Mo. (65085) 261/J6
West Pittsburg, Calif. (†94565) 204/K1
West Pittsburg, Pa. (16160) 294/B4
West Pittston, Pa. (18643) 294/F7
West Plains (Plains), Kansas (67869) 232/B4
West Plains, Mo. (65775) 261/J9
West Point, Ala. (†35179) 195/D2
West Point (lake), Ala. 195/H4
West Point (mt.), Alaska 196/K2
West Point, Ark. (72178) 202/G3
West Point, Calif. (95255) 204/E5
West Point, Georgia (31833) 217/B5
West Point (lake), Georgia 217/B4
West Point, Ill. (62380) 222/B3
Westpoint, Ind. (47992) 227/C4
West Point, Iowa (52656) 229/K7
West Point, Ky. (40177) 237/J4
West Point, Miss. (39773) 256/G3
West Point, Nebr. (68788) 264/H3
West Point, N.Y. (10996) 276/M8
West Point, Ohio (44492) 284/J4
Westpoint, Tenn. (38486) 237/G10
West Point, Va. (23181) 307/P5
West Pointe a la Hache, La. (†70082) 238/L7
West Poland, Maine (04291) 243/C7
West Poplar, Sask. 181/E6
Westport, Calif. (95488) 204/B4
Westport○, Conn. (06880) 210/B4
Westport, Ind. (47283) 227/F6
Westport, Ireland 17/C4
Westport, Ireland 10/B4
Westport, Ky. (40077) 237/K4
Westport○, Mass. (02790) 249/K6
Westport, Minn. (†56385) 255/C5
Westport, N.H. (†03470) 268/C6
Westport, N.Y. (12993) 276/N2
Westport, N. Zealand 100/C4
Westport, Nova Scotia 168/B4
Westport, Okla. (†74020) 288/O2
Westport, Ontario 177/H3
Westport, Oreg. (97016) 291/D1
Westport, Pa. (17778) 294/G3
Westport, S. Dak. (57481) 298/M2
Westport, Tenn. (38387) 237/E9
Westport, Wash. (98595) 310/A4
West Portal, N.J. (†08802) 273/D2
Westport Point, Mass. (02791) 249/K6
West Portsmouth, Ohio (†45662) 284/D8
West Pubnico, Nova Scotia 168/C5
Westpunt, Aruba, Neth. Ant. 161/D10
Westpunt, Curaçao, Neth. Ant. 161/F8
West Quaco, New Bruns. 170/E3
West Quoddy (head), Maine 243/K6
Westray (firth), Scotland 15/E1
Westray (isl.), Scotland 15/E1
Westray (isl.), Scotland 10/E1
West Redding, Conn. (06896) 210/B3
West Richland, Wash. (†99352) 310/F4
West Ridge, Ark. (72391) 202/K2
West Rindge, N.H. (†03461) 268/C6
West River, Md. (20881) 245/M5
West Road (riv.), Br. Col. 184/E3
West Rockport, Maine (04865) 243/E7
West Rock Ridge (hills), Conn. 210/D3
West Rumney, N.H. (†03266) 268/D4
West Rupert, Vt. (05776) 268/A5
West Rushville, Ohio (43163) 284/E6
West Rutland, Vt. (05777) 268/A4
West Rutland○, Vt. (05777) 268/A4
West Rye, N.H. (†03870) 268/F6
West Sacramento, Calif. (95691) 204/B8
West Saint Mary's (riv.), Nova Scotia 168/F3
West Saint Modeste, Newf. 166/C3
West Saint Paul, Minn. (55118) 255/G5
West Salem, Ill. (62476) 222/F5
West Salem, Ohio (44287) 284/F4
West Salem, Wis. (54669) 317/D8
West Salisbury, Pa. (15565) 294/D6
West Salisbury, Vt. (†05769) 268/A4
West Sayville, N.Y. (11796) 276/O9
West Scarborough, Maine (04074) 243/C8
West Seboois, Maine (†04484) 243/F4
West Seneca, N.Y. (14224) 276/C5
West Shoal (lake), Manitoba 179/E4
Westside, Iowa (51467) 229/C4
West Side, Oreg. (†97630) 291/G5
West Siloam Springs, Okla. (†72761) 288/S2
West Simsbury, Conn. (06092) 210/D1
West Sister (isl.), Ohio 284/D2
West Sister (isl.), Tasmania 99/D1
West Somerset, Ky. (†42501) 237/M6
West Springfield○, Mass. (01089) 249/D4
West Springfield, N.H. (03284) 268/C5
West Springfield, Pa. (16443) 294/B2
West Springfield, Va. (22153) 307/S3
West Springs, S.C. (†29374) 296/D2
West Stafford, Conn. (†06076) 210/F1
West Statesville, N.C. (28677) 281/G3
West Stewartstown, N.H. (03597) 268/E2
West Stockbridge○, Mass. (01266) 249/A3
West Stockholm, N.Y. (13696) 276/K1
West Suffield, Conn. (06093) 210/E1
West Sullivan, Maine (04689) 243/G6
West Sumner, Maine (04292) 243/B7
West Sunbury, Pa. (16061) 294/C3
West Sussex (co.), England 13/G7
West Swan (riv.), Minn. 255/F3
West Swanzey, N.H. (03469) 268/C6
West Terre Haute, Ind. (47885) 227/B6
West-Terschelling, Netherlands 27/G2
West Thompson, Conn. (06255) 210/H1
West Thornton, N.H. (03285) 268/D4
West Thumb-Grant Village, Wyo. (†82190) 319/B1
West Tisbury○, Mass. (02575) 249/M7
West Torrens, S. Australia 88/D8
West Torrens, S. Australia 94/A8
West Torrington, Conn. (†06790) 210/C1
West Townsend, Mass. (01474) 249/H2
West Townshend, Vt. (05359) 268/B5
West Tremont, Maine (04690) 243/G7
West Trenton, N.J. (08628) 273/D3
West Union, Ill. (62477) 222/F4
West Union, Iowa (52175) 229/K3
West Union, Minn. (56389) 255/C5
West Union, Ohio (45693) 284/C8
West Union, S.C. (29696) 296/B2
West Union, W. Va. (26456) 312/E4
West Unity, Ohio (43570) 284/B2
West University Place, Texas (†77005) 303/J2
West Upton-Upton, Mass. (01587) 249/H4
West Valley, N.Y. (14171) 276/C6
West Vancouver, Br. Col. 184/K3
West Van Lear, Ky. (41268) 237/R5
West View, Pa. (15229) 294/B6
West View, Sask. 181/J5
Westview, S.C. (†29301) 296/C2
Westville, Fla. (32464) 212/C6
Westville, Ill. (61883) 222/F3
Westville, Ind. (46391) 227/D1
Westville, N.H. (03865) 268/E6
Westville, N.J. (08093) 273/B3
Westville, Nova Scotia 168/F3
Westville, Okla. (74965) 288/S2
Westville, Pa. (15869) 294/E3
Westville, S.C. (29175) 296/F3
West Virginia 188/K3
WEST VIRGINIA 313
West Virginia (state), U.S. 146/K6
Westward Ho, Alberta 182/C4
West Wardsboro, Vt. (05360) 268/B5
West Wareham, Mass. (02576) 249/L5
West Warren, Mass. (01092) 249/F4
West Warwick, R.I. (02893) 249/H6
West Weber, Utah (†84401) 304/B2
Westwego, La. (70094) 238/O4
West Wenatchee, Wash. (†98801) 310/E3
West Wildwood, N.J. (†08260) 273/D6
West Willington, Conn. (06279) 210/F1
West Windham, N.H. (†03087) 268/E6
West Winfield, N.Y. (13491) 276/K5
West Winterport, Maine (†04496) 243/E6
Westwold, Br. Col. 184/G5
Westwood, Lassen, Calif. (96151) 204/D3
Westwood, Ky. (41101) 237/R4
Westwood, Ky. (†40207) 237/L1
Westwood○, Mass. (02090) 249/B8
Westwood, Mo. (†63101) 261/O3
Westwood, N.J. (07675) 273/B1
West Woodburn, Oreg. (†97071) 291/A3
Westwood Lakes, Fla. (†33165) 212/B5
West Woodstock, Conn. (†06260) 210/G1
West Woodstock, Vt. (†05091) 268/B4
Westwood Village, Los Angeles, Calif. (90024) 204/B10
Westworth, Texas (†76101) 303/E2
West Wyalong, N.S. Wales 88/H6
West Wyalong, N.S. Wales 97/D3
West Wyoming, Pa. (18644) 294/E7
West Yarmouth, Mass. (02673) 249/N6
West Yellowstone, Mont. (59758) 262/E6
West York, Ill. (62478) 222/F4
West York, Pa. (†17401) 294/J6
West Yorkshire (co.), England 13/J1
West Yuma, Ariz. (†85364) 198/A6
Westzaan, Netherlands 27/A4
Wet (mts.), Colo. 208/J6
Wetar (isl.), Indonesia 54/O10
Wetar (isl.), Indonesia 85/H7
Wetaskiwin, Alberta 182/D3
Wetaskiwin, Alta. 162/E5
Wete, Tanzania 115/G4
Wetheral, England 13/E3
Wethersfield○, Conn. (06109) 210/E2
Wetipquin, Md. (†21856) 245/P7
Wetmore, Colo. (81253) 208/J6
Wetmore, Kansas (66550) 232/G2
Wetmore, Mich. (49895) 250/C2
Wetmore, Tenn. (†37325) 237/N10
Wetmore, Texas (78163) 303/K10
Wetonka, S. Dak. (57482) 298/M2
Wetteren, Belgium 27/D7
Wetterhorn (peak), Colo. 208/D6
Wetterhorn (mt.), Switzerland 39/F3
Wettingen, Switzerland 39/F2
Wetumka, Okla. (74883) 288/O4
Wetumpka, Ala. (36092) 195/F5
Wetuppa, N.S. Wales 97/B4
Wetzel (co.), W. Va. 312/E3
Wetzikon, Switzerland 39/G2
Wetzlar, W. Ger. 22/C3
Wever, Iowa (52658) 229/L7
Weverton, Md. (†21758) 245/H3
Wewahitchka, Fla. (32465) 212/D6
Wewak, Papua N.G. 87/E6
Wewak, Papua N.G. 85/B6
Weweantic (riv.), Mass. 249/L5
Wewela, S. Dak. (57578) 298/K7
Wewoka, Okla. (74884) 288/O4
Wexford (co.), Ireland 17/H7
Wexford, Ireland 17/H7
Wexford, Ireland 10/C4
Wexford (bay), Ireland 17/J7
Wexford (harb.), Ireland 17/J7
Wexford (harb.), Ireland 10/C4
Wexford (co.), Mich. 250/D4
Wey (riv.), England 13/G6
Weyanoke, La. (70787) 238/H5
Weyauwega, Wis. (54983) 317/H7
Weybridge○, Vt. (†05753) 268/A3
Weyburn, Sask. 162/F6
Weyburn, Sask. 181/H6
Weyerhaeuser, Wis. (54895) 317/D5
Weyer Markt, Austria 41/C3
Weyers Cave, Va. (24486) 307/L4
Weymouth (bay), England 13/E7
Weymouth, Mass. (02188) 249/D8
Weymouth, Nova Scotia 168/C4
Weymouth and Melcombe Regis, England 13/E7
Weymouth and Melcombe Regis, England 10/E5
Weymouth North, Nova Scotia 168/C4
Wezembeek-Oppem, Belgium 27/D9
Wezet (Visé), Belgium 27/H7
Whakatane, N. Zealand 100/F2
Whalan, Minn. (55986) 255/G7
Whalan (creek), N.S. Wales 97/E1
Whale (bay), Alaska 196/M1
Whaleback (mt.), W. Australia 92/B3
Whale Cove, N.W. Terrs. 187/J3
Whaletown, Br. Col. 184/E5
Whaley Bridge, England 13/J2
Whaley Bridge, England 10/G2
Whaleysville, Md. (21872) 245/S7
Whallonsburg, N.Y. (12994) 276/O2
Whalsay (isl.), Scotland 10/H1
Whalsay (isl.), Scotland 15/G2
Whangamata, N. Zealand 100/F2
Whangarei, N. Zealand 87/H9
Whangarei, N. Zealand 100/E1
Wharfe (riv.), England 13/E3
Wharfe (riv.), England 10/F3
Wharncliffe, W. Va. (25651) 312/C7
Wharton (pen.), Chile 138/D8
Wharton, N.J. (07885) 273/D2
Wharton (lake), N.W. Terrs. 187/H3
Wharton, Ohio (43359) 284/D4
Wharton, Pa. (†16720) 294/G2
Wharton (co.), Texas 303/H8
Wharton, Texas (77488) 303/J8
Wharton, W. Va. (25208) 312/C7
Whataroa, N. Zealand 100/C5
Whatatutu, N. Zealand 100/F3
What Cheer, Iowa (50268) 229/J6
Whatcom (co.), Wash. 310/D2
Whatcom (lake), Wash. 310/C2
Whately○, Mass. (01093) 249/D3
Whatley, Ala. (36482) 195/C7
Wheatcroft, Ky. (42463) 237/F5
Wheatfield, Ind. (46392) 227/C2
Wheatland, Calif. (95692) 204/D4
Wheatland, Ind. (47597) 227/C7
Wheatland, Iowa (52777) 229/M5
Wheatland, Manitoba 179/B4
Wheatland, Mo. (65779) 261/F7
Wheatland (co.), Mont. 262/G4
Wheatland, N. Mex. (†88120) 274/F4
Wheatland, N. Dak. (58079) 282/R6
Wheatland, Pa. (16161) 294/B3
Wheatland, Wyo. (82201) 319/H3
Wheatland (res.), Wyo. 319/G4
Wheatley, Ark. (72392) 202/H4
Wheatley, Ky. (40389) 237/M3
Wheatley, Ontario 177/B5
Wheaton, Ill. (60187) 222/A5
Wheaton, Kansas (66551) 232/F2
Wheaton, Minn. (56296) 255/B5
Wheaton, Mo. (64874) 261/E9
Wheaton-Glenmont, Md. (20902) 245/E3
Wheat Ridge, Colo. (80033) 208/J3
Wheeler, Ala. (†35618) 195/D1
Wheeler (dam), Ala. 195/D1
Wheeler (lake), Ala. 188/J4
Wheeler (lake), Ala. 195/D1
Wheeler (peak), Calif. 204/F5
Wheeler (co.), Georgia 217/G6
Wheeler, Ill. (62479) 222/E4
Wheeler, Ind. (46393) 227/C1
Wheeler, Kansas (67763) 232/A2
Wheeler, Ky. (†40906) 237/O7
Wheeler, Mich. (48662) 250/E5
Wheeler, Miss. (38880) 256/G1
Wheeler (co.), Nebr. 264/F3
Wheeler (peak), Nev. 266/G4
Wheeler (peak), N. Mex. 188/F3
Wheeler (peak), N. Mex. 274/D2
Wheeler (co.), Oreg. 291/G3
Wheeler, Oreg. (97147) 291/D2
Wheeler (riv.), Québec 174/D1
Wheeler (co.), Texas 303/D2
Wheeler, Texas (79096) 303/D2
Wheeler, Wash. (†98837) 310/F3
Wheeler, Wis. (54772) 317/C5
Wheeler A.F.B., Hawaii 218/E1
Wheelersburg, Ohio (45694) 284/E8
Wheeless, Okla. (†73933) 288/A1
Wheeling, Ill. (60090) 222/B5
Wheeling, Ind. (†47342) 227/G4
Wheeling, Ind. (†47534) 227/C8
Wheeling, Mo. (64688) 261/F3
Wheeling, W. Va. 188/K2
Wheeling, W. Va. (26003) 312/E2
Wheelock, N. Dak. (58855) 282/D3
Wheelock○, Vt. (†05851) 268/C2
Wheelwright, Ky. (41669) 237/R6
Wheelwright, Mass. (01094) 249/F3
Whelen Springs, Ark. (71772) 202/D6
Whetstone (buttes), N. Dak. 282/E7
Whetstone (creek), S. Dak. 298/R3
Whick, Ky. (41390) 237/P6
Whickham, England 13/J3
Whidbey (isls.), S. Australia 94/D6
Whidbey (isl.), Wash. 310/C2
Whigham, Georgia (31797) 217/D9
Whigville, Conn. (†06010) 210/D2
Whipholt, Minn. (56485) 255/D3
Whippany, N.J. (07981) 273/E2
Whipple (mts.), Calif. 204/L9
Whipple, Ohio (45788) 284/H6
Whiskeytown-Shasta-Trinity Nat'l Rec. Area, Calif. 204/C3
Whispering Pines, N.C. (†28389) 281/L4
Whistler, Br. Col. 184/F5
Whitaker, Ind. (†46166) 227/D6
Whitaker, Pa. (15120) 294/C7
Whitakers, N.C. (27891) 281/O2
Whitbourne, Newf. 166/D2
Whitburn, Scotland 10/C1
Whitburn, Scotland 15/C2
Whitby, England 13/G3
Whitby, England 10/F3
Whitby, Ontario 177/F4
Whitchurch, England 13/E5
Whitchurch-Stouffville, Ontario 177/J3
Whitcomb, Ind. (†47012) 227/H6
Whitcombe (mt.), N. Zealand 100/C5
White (mts.), Alaska 196/J1
White (pass), Alaska 196/N1
White (riv.), Alaska 196/K2
White (riv.), Ariz. 198/E5
White (co.), Ark. 202/G3
White, Ark. (†71635) 202/G7
White (riv.), Ark. 188/H3
White (riv.), Ark. 202/H5
White (riv.), Colo. 208/B2
White (co.), Georgia 217/E1
White, Georgia (30184) 217/C2
White (co.), Ill. 222/E5
White (co.), Ind. 227/D3
White (riv.), Ind. 227/B8
White, East Fork (riv.), Ind. 227/C7
White, West Fork (riv.), Ind. 227/C7
White (lake), La. 238/E7
White (riv.), Mich. 250/C5
White (riv.), Mo. 261/G10
White (riv.), Nebr. 264/A2
White (riv.), Nev. 266/F4
White (bay), Newf. 166/C3
White (bay), Newf. 162/L5
White (isl.), N.H. 268/F6
White (mts.), N.H. 268/E3
White (isl.), N. Zealand 100/F2
White (butte), N. Dak. 282/D7
White (lake), North. Terr. 88/D4
White (lake), North. Terr. 93/A6
White (isl.), N.W. Terrs. 187/K3
White (lake), Ontario 177/F2
White (lake), Ontario 177/H2
White (riv.), Oreg. 291/F2
White, S. Dak. (57276) 298/R5
White (lake), S. Dak. 298/M6
White (riv.), S. Dak. 298/D7
White (co.), Tenn. 237/L9
White (riv.), Texas 303/C3
White (sea), U.S.S.R. 4/C8
White (sea), U.S.S.R. 7/H2
White (sea), U.S.S.R. 48/D3
White (sea), U.S.S.R. 52/E1
White (riv.), Utah 304/E3
White (riv.), Vt. 268/C4
White, First Branch (riv.), Vt. 268/B4
White, Second Branch (riv.), Vt. 268/B4
White (pass), Wash. 310/D4
White (riv.), Wash. 310/D3
White (riv.), Yukon 187/D3
White Bear (isl.), Newf. 166/B2
White Bear (lake), Newf. 166/C4
White Bear (riv.), Newf. 166/C4
White Bear, Sask. 181/C5
White Bear Lake, Minn. (55110) 255/G5
Whitebeech, Sask. 181/K3
White Bird, Idaho (83554) 220/B4
White Bluff, Tenn. (37187) 237/G8
White Butte (mt.), N. Dak. 282/D7
White Butte, S. Dak. (†57638) 298/E2
White Carpathians (mts.), Czech. 41/E2
White Castle, La. (70788) 238/J3
White Center-Shorewood, Wash. (98146) 310/A2
White City, Fla. (†32465) 212/F4
White City, Ill. (†62069) 222/D4
White City, Kansas (66872) 232/F3
White City, Oreg. (97503) 291/E5

White City, Sask. 181/G5
Whiteclay, Nebr. (69365) 264/B2
White Cliffs, N.S. Wales 97/B2
White Cloud, Ind. (†47112) 227/E8
White Cloud, Kansas (66094) 232/G2
White Cloud, Mich. (49349) 250/D5
White Coomb (mt.), Scotland 15/E5
White Cottage, Ohio (43791) 284/F6
Whitecourt, Alberta 182/C2
White Deer, Pa. (17887) 294/J3
White Deer, Texas (79097) 303/C2
White Earth, Minn. (56591) 255/C3
White Earth, N. Dak. (58794) 282/E3
White Earth (riv.), N. Dak. 282/E3
White Earth Ind. Res., Minn. 255/C3
White Elster (riv.), E. Germany 22/E3
Whiteface (riv.), Minn. 255/F3
Whiteface, N.H. (†03259) 268/E4
Whiteface (mt.), N.H. 268/E4
Whiteface (mt.), N.Y. 276/N2
Whiteface, Texas (79379) 303/B4
White Face (mt.), Vt. 268/B2
Whitefield, Maine (04362) 243/D7
Whitefield○, Maine (04362) 243/D7
Whitefield, N.H. (03598) 268/D3
Whitefield○, N.H. (03598) 268/D3
Whitefield, Okla. (74472) 288/R4
Whitefish (bay), Mich. 250/E2
Whitefish (pt.), Mich. 250/E2
Whitefish (riv.), Mich. 250/C2
Whitefish (lake), Minn. 255/D4
Whitefish, Mont. (59937) 262/B2
Whitefish (lake), Mont. 262/B2
Whitefish Falls, Ontario 177/C1
Whitefish Point, Mich. (†49768) 250/E2
Whiteflat, Texas (†79234) 303/D3
Whiteford, Md. (21160) 245/N2
White Fox, Sask. 181/H2
White Fox (riv.), Sask. 181/G2
Whitegate, Ireland 17/E8
White Gull (creek), Sask. 181/G2
White Hall, Ala. (†36040) 195/E6
Whitehall, Ark. (†72432) 202/J3
White Hall, Ark. (71602) 202/F5
White Hall, Georgia (†30601) 217/F3
White Hall, Ill. (62092) 222/C4
Whitehall, Ind. (†47401) 227/D6
Whitehall, La. (†70462) 238/M2
White Hall, Md. (21161) 245/M2
Whitehall, Mich. (49461) 250/C5
Whitehall, Mont. (59759) 262/D5
Whitehall, N.Y. (12887) 276/O3
Whitehall, Ohio (43213) 284/E6
Whitehall, Pa. (†15234) 294/B7
Whitehall, Scotland 15/F1
White Hall, S.C. (†29945) 296/F6
White Hall, Va. (22987) 307/L4
Whitehall, Wis. (54773) 317/D7
White Handkerchief (cape), Newf. 166/B2
Whitehaven, England 13/D3
Whitehaven, England 10/E3
Whitehaven, Md. (21873) 245/P7
Whitehaven (harb.), Nova Scotia 168/G3
White Haven, Pa. (18661) 294/L3
White Head, New Bruns. 170/D4
White Head (isl.), New Bruns. 170/D4
Whitehead, N. Ireland 17/K2
Whitehead, Nova Scotia 168/G3
White Heath, Ill. (61884) 222/E3
Whitehills, Scotland 15/F3
White Horn, Tenn. (†37711) 237/R8
Whitehorse, Canada 4/C16
Whitehorse, Canada 2/C3
Whitehorse, S. Dak. (57661) 298/H3
Whitehorse (cap.), Yukon 187/E3
Whitehorse (cap.), Yukon 162/C3
Whitehorse (cap.), Yukon 146/E3
White Horse Lake, N. Mex. (87073) 274/B3
Whitehouse, Ky. (41269) 237/R5
Whitehouse, N.J. (08888) 273/D2
Whitehouse, Ohio (43571) 284/C2
White House, Tenn. (37188) 237/H8
White House Station, N.J. (08889) 273/D2
White Iron (lake), Minn. 255/G3
White Knob (mts.), Idaho 220/E6
White Lake, N.C. (28337) 281/M5
White Lake, Ontario 177/H2
White Lake, S. Dak. (57383) 298/M6
White Lake, Wis. (54491) 317/J5
Whiteland, Ind. (46184) 227/E5
Whitelaw, Alberta 182/A1
Whitelaw, Wis. (54247) 317/L7
Whiteman A.F.B., Mo. 261/E5
Whitemark, Tasmania 99/D2
White Marsh, Md. (21162) 245/N3
White Meadow Lake, N.J. (†07866) 273/D2
White Mills, Ky. (42788) 237/J5
White Mills, Pa. (18473) 294/M2
White Mountain, Alaska (99784) 196/F2
White Mountains Nat'l Rec. Area, Alaska 196/J1
Whitemouth, Manitoba 179/G5
Whitemouth (lake), Manitoba 179/G5
Whitemouth (riv.), Manitoba 179/G5
Whitemud (riv.), Alberta 182/A1
Whiten (head), Scotland 15/D2
White Nile (riv.) 2/L5
White Nile (riv.) 102/F4
White Nile (prov.), Sudan 111/F5
White Nile (riv.), Sudan 111/F5
White Nile (riv.), Sudan 59/B7
White Oak (lake), Ark. 202/D6
White Oak, Georgia (31568) 217/J8
White Oak, Md. (†20901) 245/F3
Whiteoak, Mo. (63880) 261/M10
White Oak, N.C. (28399) 281/M5
Whiteoak (swamp), N.C. 281/P5
Whiteoak (creek), Ohio 284/C7
White Oak, Okla. (†74301) 288/R1
White Oak, Pa. (15131) 294/C7
White Oak, S.C. (29176) 296/E3
Whiteoak (creek), Tenn. 237/F8
White Oak, Texas (75693) 303/K5
White Oaks, Conn. (†06488) 210/C2
White Oaks, N. Mex. (†88301) 274/D5
White Owl, S. Dak. (57792) 298/E4
White Partridge (lake), Ontario 177/G2
White Pass, Wash. (†98937) 310/D4
White Pigeon, Mich. (49099) 250/D7
White Pine, Mich. (49971) 250/F1
Whitepine, Mont. (†59874) 262/A3
White Pine (co.), Nev. 266/F3
White Pine (range), Nev. 266/F3
White Pine, Tenn. (37890) 237/P8
White Pines, Calif. (†95223) 204/E5
White Plains, Ala. (†36862) 195/G3
White Plains, Georgia (30678) 217/F4
White Plains, Ky. (42464) 237/G6
White Plains, Md. (20695) 245/L6
White Plains, N.Y. (*10601) 276/P6
White Plains, N.C. (27031) 281/H2
White Plains, Va. (23893) 307/N7
White Pond, S.C. (29854) 296/D5
White Post, Va. (22663) 307/M2
White Quartz Hill, North. Terr. 93/D7
White Rapids, New Bruns. 170/E2
Whiteriver, Ariz. (85941) 198/E5
White River, Ont. 162/H6
White River, Ontario 175/C3
White River, Ontario 177/J5
White River, S. Dak. (57579) 298/H6
White River (lake), Texas 303/C4
White River Junction, Vt. (05001) 268/C4
White Rock, Br. Col. 184/K3
White Rock (creek), Kansas 232/D2
White Rock, N. Mex. (87544) 274/C3
Whiterock, N.C. (†28753) 281/D3
White Rock, S.C. (29177) 296/E3
White Rock, S. Dak. (†57260) 298/R2
White Rock (creek), Texas 303/G2
Whiterocks, Utah (84085) 304/E3
White Russian S.S.R., U.S.S.R. 7/G3
White Russian S.S.R., U.S.S.R. 52/C4
White Russian S.S.R., U.S.S.R. 48/C4
Whites, Wash. (†98541) 310/B3
Whitesail (lake), Br. Col. 184/D3
White Salmon, Wash. (98672) 310/D5
White Salmon (riv.), Wash. 310/D4
White Sands (des.), N. Mex. 274/C5
White Sands Missile Range, N. Mex. (88002) 274/C6
White Sands Missile Range, N. Mex. 274/C5
White Sands Nat'l Mon., N. Mex. 274/C6
Whitesbog, N.J. (†08015) 273/E4
Whitesboro, N.J. (08252) 273/D5
Whitesboro, N.Y. (13492) 276/K4
Whitesboro, Okla. (74577) 288/S5
Whitesboro, Texas (76273) 303/H4
Whitesburg, Georgia (30185) 217/B4
Whitesburg, Ky. (41858) 237/R6
Whitesburg, Tenn. (37891) 237/P8
Whites Chapel, Ala. (†35094) 195/F3
Whites City, N. Mex. (88268) 274/E6
Whites Creek, W. Va. (†25530) 312/A6
White Settlement, Texas (76108) 303/E2
Whiteshell Prov. Park, Manitoba 179/G4
White Shield, N. Dak. (†58534) 282/G4
Whiteshore (lake), Sask. 181/C4
Whiteside (chan.), Chile 138/E10
Whiteside (co.), Ill. 222/D2
Whiteside, Mo. (63387) 261/K4
Whiteside, Tenn. (37396) 237/K10
Whites Lake, Nova Scotia 168/E4
Whiteson, Oreg. (†97128) 291/D2
White Springs, Fla. (32096) 212/D1
Whitestone, Georgia (30186) 217/C1
White Stone, Va. (22578) 307/R5
Whitestown, Ind. (46075) 227/E5
White Sulphur Springs, Georgia (†31822) 217/C5
White Sulphur Springs, La. (†71371) 238/F3
White Sulphur Springs, Mont. (59645) 262/E4
White Sulphur Springs, W. Va. (24986) 312/F7
Whitesville, Georgia (†31833) 217/C5
Whitesville, Ky. (42378) 237/H5
Whitesville, Mo. (†64480) 261/C2
Whitesville, N.J. (†08701) 273/E3
Whitesville, N.Y. (14897) 276/E6
Whitesville, W. Va. (25209) 312/C6
Whiteswan (lakes), Sask. 181/F1
White Swan, Wash. (98952) 310/E4
Whitetail, Mont. (59276) 262/L2
Whitetop, Va. (24292) 307/E7
Whiteville, La. (71376) 238/F5
Whiteville, N.C. (28472) 281/M6
Whiteville, Tenn. (38075) 237/C10
White Volta (riv.) 102/B4
White Volta (riv.), Ghana 106/D6
White Volta (riv.), Upper Volta 106/D6
Whitewater, Colo. (81527) 208/C5
Whitewater (bay), Fla. 212/F6
Whitewater, Ind. (†47374) 227/H5
Whitewater (riv.), Ind. 227/H6
Whitewater, Kansas (67154) 232/E4
Whitewater, Manitoba 179/B5
Whitewater (lake), Manitoba 179/B5
Whitewater, Mo. (63785) 261/N8
Whitewater, Mont. (59544) 262/J2
Whitewater, Wis. (53190) 317/J10
Whitewater Baldy (mt.), N. Mex. 274/A5
Whitewood, Sask. 181/J5
Whitewood, S. Dak. (57793) 298/B5
Whitewood (creek), S. Dak. 298/B4
Whitewood, Va. (24657) 307/E6
Whitewright, Texas (75491) 303/H4
Whitfield, Ala. (†36925) 195/B6
Whitfield (co.), Georgia 217/B1
Whitfield, Miss. (39193) 256/E6
Whitford, Alberta 182/D3
Whitharral, Texas (79380) 303/B4
Whithorn, Scotland 10/D3
Whithorn, Scotland 15/D6
Whitianga, N. Zealand 100/E2
Whiting, Ind. (46394) 227/C1
Whiting, Iowa (51063) 229/A4
Whiting, Kansas (66552) 232/G2
Whiting○, Maine (04691) 243/J6
Whiting, Mo. (†63845) 261/O9
Whiting, N.J. (08759) 273/E4
Whiting○, Vt. (05778) 268/A4
Whiting, Wis. (†54481) 317/H7
Whiting Bay, Scotland 15/C5
Whiting Field Naval Air Sta., Fla. 212/B6
Whitingham○, Vt. (05361) 268/B6
Whitinsville, Mass. (01588) 249/H4
Whitkow, Sask. 181/D3
Whitla, Alberta 182/E5
Whitlash, Mont. (59545) 262/E2
Whitley (co.), Ind. 227/F2
Whitley (co.), Ky. 237/N7
Whitley Bay, England 13/J3
Whitley City, Ky. (42653) 237/N7
Whitleyville, Tenn. (38588) 237/K8
Whitlock, Tenn. (†38242) 237/E8
Whitman○, Mass. (02382) 249/L4
Whitman (riv.), Mass. 249/G2
Whitman, Nebr. (69366) 264/C2
Whitman, N. Dak. (58283) 282/O3
Whitman (co.), Wash. 310/H4
Whitman Mission Nat'l Hist. Site, Wash. 310/G4
Whitmer, W. Va. (26296) 312/G5
Whitmire, S.C. (29178) 296/D3
Whitmore, Calif. (96096) 204/D3
Whitmore Lake, Mich. (48189) 250/F6
Whitmore Village, Hawaii (†96786) 218/E1
Whitnel, N.C. (28645) 281/F3
Whitney (mt.), Calif. 188/C3
Whitney (mt.), Calif. 204/G7
Whitney (lake), Conn. 210/D3
Whitney, Nebr. (69367) 264/A2
Whitney, New Bruns. 170/E2
Whitney, Ontario 177/F2
Whitney, Pa. (15693) 294/D5
Whitney, S.C. (29303) 296/D1
Whitney, Texas (76692) 303/G6
Whitney Point, N.Y. (13862) 276/J6
Whitney Point (lake), N.Y. 276/J6
Whitneyville, Conn. (06517) 210/D3
Whitneyville○, Maine (04692) 243/H6
Whitsett, Texas (78075) 303/F9
Whitsunday (isl.), Queensland 88/H4
Whitsunday (isl.), Queensland 95/D4
Whitt, Texas (76090) 303/G5
Whittaker, Mich. (48190) 250/F6
Whittemore, Iowa (50598) 229/E2
Whittemore, Mich. (48770) 250/F4
Whitten, Iowa (50269) 229/H4
Whittier, Alaska (99693) 196/C1
Whittier, Calif. (*90601) 204/D11
Whittier, Iowa (52360) 229/K4
Whittier, N.C. (28789) 281/C4
Whittle (cape), Québec 174/F2
Whittlesea, Victoria 97/C5
Whittlesey, England 13/G5
Whittlesey, Wis. (†54451) 317/F5
Whitton, N.S. Wales 97/C4
Whitwell, Tenn. (37397) 237/K10
Wholdaia (lake), N.W. Terrs. 187/H3
Why, Ariz. (85321) 198/C6
Whyalla, Australia 87/D9
Whyalla, S. Australia 94/E5
Whycocomagh, Nova Scotia 168/G3
Whyjonta, N.S. Wales 97/B1
Wiarton, Ontario 177/C3
Wiau (lake), Alberta 182/E2
Wiawso, Ghana 106/D7
Wiay (isl.), Scotland 15/A3
Wibaux (co.), Mont. 262/M4
Wibaux, Mont. (59353) 262/M3
Wichabai, Guyana 131/B4
Wichita, Kans. 188/G3
Wichita (co.), Kansas 232/A3
Wichita, Kansas 146/J6
Wichita, Kansas (*67201) 232/E4
Wichita (mts.), Okla. 288/J5
Wichita (co.), Texas 303/F3
Wichita (riv.), Texas 303/F4
Wichita Falls, Texas 146/H6
Wichita Falls, Texas (*76301) 303/F4
Wichita Falls, Texas 188/G4
Wick, Iowa (†50240) 229/F6
Wick, Scotland 10/E1
Wick, Scotland 15/E2
Wick (riv.), Scotland 15/E2
Wick, W. Va. (26185) 312/E4
Wickahoney (creek), Idaho 220/C7
Wickatunk, N.J. (07765) 273/E3
Wicked (pt.), Manitoba 179/D2
Wickenburg, Ariz. (85358) 198/C5
Wickepin, W. Australia 92/B2
Wickersham, Wash. (†98284) 310/C2
Wickes, Ark. (71973) 202/B5
Wickes, Mont. (†59638) 262/D4
Wickett, Texas (79788) 303/A6
Wickham, New Bruns. 170/D3
Wickham, Québec 172/E4
Wickham (cape), Tasmania 99/A1
Wickham, W. Australia 92/B3
Wickiup (res.), Oreg. 291/F4
Wickliffe, Ky. (42087) 237/C7
Wickliffe, Ohio (44092) 284/J9
Wicklow (co.), Ireland 17/J5
Wicklow, Ireland 10/C4
Wicklow, Ireland 17/K6
Wicklow (head), Ireland 17/K6
Wicklow (head), Ireland 10/D4
Wicklow (mts.), Ireland 17/J6
Wicklow, New Bruns. 170/C2
Wicksburg, Ala. (†36352) 195/G8
Wicomico (co.), Md. 245/R7
Wicomico, Md. (†20611) 245/L7
Wicomico (riv.), Md. 245/R7
Wicomico (riv.), Md. 245/L7
Wicomico Church, Va. (22579) 307/R5
Wiconisco, Pa. (17097) 294/J4
Wide (chan.), Chile 138/D8
Wide (bay), Papua N.G. 86/C2
Wide (bay), Queensland 95/E5
Wideman, Ark. (72585) 202/G1
Widemouth, W. Va. (†24736) 312/D8
Widen, W. Va. (25211) 312/E6
Widener, Ark. (72394) 202/J3
Widewater, Alberta 182/C2
Widgiemooltha, W. Australia 88/C6
Widgiemooltha, W. Australia 92/C5
Widnes, England 10/F2
Widnes, England 13/G2
Widnoon, Pa. (16261) 294/D4
Więcbork, Poland 47/C2
Wiederkehr Village, Ark. 202/C3
Wiehl, W. Germany 22/B3
Wiek, E. Germany 22/E1
Wieliczka, Poland 47/E3
Wieluń, Poland 47/D3
Wien (Vienna) (cap.), Austria 41/D2
Wiener Neustadt, Austria 41/D3
Wieprz (riv.), Poland 47/F3
Wierden, Netherlands 27/K4
Wieringermeer Polder, Netherlands 27/G3
Wierum, Netherlands 27/H2
Wieruszów, Poland 47/D3
Wiesbaden, W. Germany 7/E3
Wiesbaden, W. Germany 22/B3
Wiese (isl.), U.S.S.R. 4/B6
Wiese (isl.), U.S.S.R. 48/H2
Wiesmoor, W. Germany 22/B2
Wigan, England 13/G2
Wigan, England 10/G2
Wiggins, Colo. (80654) 208/L2
Wiggins, Miss. (39577) 256/F9
Wiggins, S.C. (†29446) 296/F6
Wight (isl.), England 13/F5
Wight (isl.), England 10/F5
Wigston, England 13/F5
Wigton, England 13/D3
Wigtown, Scotland 10/D3
Wigtown, Scotland 15/D6
Wigtown (trad. co.), Scot. 15/A5
Wigtown (bay), Scotland 10/D3
Wigtown (bay), Scotland 15/D6
Wijhe, Netherlands 27/J4
Wijk bij Duurstede, Netherlands 27/G5
Wijk en Aalburg, Netherlands 27/F5
Wikel, W. Va. (†24945) 312/E7
Wikieup, Ariz. (85360) 198/B4
Wikwemikong, Ontario 177/C2
Wil, Switzerland 39/H2
Wilawana, Pa. (†18840) 294/J2
Wilbarger (co.), Texas 303/E3
Wilber, Nebr. (68465) 264/G4
Wilberforce, Ontario 177/F3
Wilbert, Minn. (†56031) 255/D7
Wilbraham, Mass. (01095) 249/E4
Wilbraham○, Mass. (01095) 249/E4
Wilbur, Ind. (†46151) 227/D5
Wilbur, Ky. (†41124) 237/R5
Wilbur, Oreg. (97494) 291/D4
Wilbur, Wash. (99185) 310/G3
Wilbur, W. Va. (26459) 312/E4
Wilbur Park, Mo. (†63101) 261/P3
Wilburton, Kansas (†67950) 232/A4
Wilburton, Okla. (74578) 288/R5
Wilcannia, N. S. Wales 88/G6
Wilcannia, N.S. Wales 97/B2
Wilchingen, Switzerland 39/F1
Wilcox (co.), Ala. 195/D7
Wilcox, Fla. (†32693) 212/D2
Wilcox (co.), Georgia 217/F7
Wilcox, Mo. (†64468) 261/C2
Wilcox, Nebr. (68982) 264/E4
Wilcox, Pa. (15870) 294/E2
Wilcox, Sask. 181/G5
Wilczek Land (isl.), U.S.S.R. 4/B6
Wilczek Land (isl.), U.S.S.R. 48/G1
Wild Ammonoosuc (riv.), N.H. 268/D3
Wildbad im Schwarzwald, W. Germany 22/C4
Wildcat (creek), Ind. 227/E4
Wild Cat, Ky. (40998) 237/O6
Wild Cherry, Ark. (†72576) 202/F1
Wild Cove, Newf. 166/C3
Wilder, Idaho (83676) 220/A6
Wilder, Minn. (56184) 255/C7
Wilder (dam), N.H. 268/C4
Wilder, Tenn. (38589) 237/L8
Wilder, Vt. (05088) 268/C4
Wilder (dam), Vt. 268/C4
Wilderness, Va. (†22553) 307/N4
Wilders, Ky. (†41071) 237/S2
Wildersville, Tenn. (38388) 237/E9
Wilderswil, Switzerland 39/E3
Wildervank, Netherlands 27/K2
Wilderville, Oreg. (97543) 291/D5
Wildeshausen, W. Germany 22/C2
Wild Goose, Ontario 177/H5
Wild Goose, Ontario 175/C3
Wildhaus, Switzerland 39/H2
Wildhay (riv.), Alberta 182/B3
Wildhorn (mt.), Switzerland 39/D4
Wild Horse, Colo. (80862) 208/N5
Wild Horse (res.), Nev. 266/E1
Wildhorse (creek), Okla. 288/L5
Wildie, Ky. (40492) 237/N6
Wildomar, Calif. (92395) 204/H10
Wildon, Austria 41/C3
Wildorado, Texas (79098) 303/B2
Wild Rice (lake), Minn. 255/F4
Wild Rice (riv.), Minn. 255/B3
Wild Rice, N. Dak. (†58047) 282/S6
Wild Rice (riv.), N. Dak. 282/R7
Wildrose, N. Dak. (58795) 282/D2
Wild Rose, Wis. (54984) 317/H7
Wildspitze (mt.), Austria 41/A3
Wildstrubel (mt.), Switzerland 39/E4
Wildsville, La. (71377) 238/G3
Wildwood, Alberta 182/C3
Wildwood, Fla. (32785) 212/D3
Wildwood, Minn. (†56663) 255/E3
Wildwood, N.J. (08260) 273/D6
Wildwood Crest, N.J. (08260) 273/D6
Wileville, Nova Scotia 168/D4
Wiley, Colo. (81092) 208/O6
Wiley, Georgia (30581) 217/F1
Wiley (creek), Oreg. 291/E3
Wiley City, Wash. (98906) 310/E4
Wiley Ford, W. Va. (26767) 312/J3
Wileyville, W. Va. (26186) 312/E3
Wilfred, Ind. (†47879) 227/C6
Wilhelm (mt.), Papua N.G. 85/B7
Wilhelm II Coast (reg.) 5/C5
Wilhelmina (canal), Netherlands 27/G6
Wilhelmina (mts.), Suriname 131/C4
Wilhelm-Pieck-Stadt, E. Germany 22/F3
Wilhelmsburg, Austria 41/C2
Wilhelmshaven, W. Germany 22/B2
Wilkes (co.), Georgia 217/G3
Wilkes (co.), N.C. 281/G2
Wilkes (lake), Ontario 177/F1
Wilkes-Barre, Pa. 188/L2
Wilkes-Barre, Pa. (*18701) 294/F7
Wilkesboro, N.C. (28697) 281/G2
Wilkes Land (reg.), Ant. 2/R10
Wilkes Land (reg.) 5/B7
Wilkeson, Wash. (98396) 310/D3
Wilkesville, Ohio (45695) 284/F7
Wilke, Sask. 181/C3
Wilkin (co.), Minn. 255/B4
Wilkins, Nev. (†89835) 266/G1
Wilkinsburg, Pa. (15221) 294/C7
Wilkinson (co.), Georgia 217/F5
Wilkinson, Ind. (46186) 227/F5
Wilkinson, Minn. (†56633) 255/D3
Wilkinson (co.), Miss. 256/B8
Wilkinson, Miss. (†39669) 256/B8
Wilkinson (lakes), S. Australia 94/C3
Wilkinson, W. Va. (25653) 312/B7
Wilkinsonville, Mass. (01590) 249/G4
Will (co.), Ill. 222/F2
Willacoochee, Georgia (31650) 217/G8
Willacy (co.), Texas 303/G11
Willamette (riv.), Oreg. 291/A3
Willamette, Middle Fork (riv.), Oreg. 291/E4
Willamina, Oreg. (97396) 291/D2
Willandra Billabong (creek), N.S. Wales 97/C3
Willapa, Wash. (†98577) 310/B4
Willapa (bay), Wash. 310/A4
Willard, Kansas (†66601) 232/G2
Willard, Mich. (†48611) 250/E5
Willard, Mo. (65781) 261/F8
Willard, Mont. (59354) 262/M4
Willard, N. Mex. (87063) 274/D4
Willard, N.Y. (14588) 276/G5
Willard, N.C. (28478) 281/O5
Willard, Ohio (44890) 284/E3
Willard, Utah (84340) 304/C2
Willard, Wis. (54493) 317/E6
Willards, Md. (21874) 245/S7
Willaumez (pen.), Papua N.G. 86/B2
Willaura, Victoria 97/B5
Willcox, Ariz. (85643) 198/F6
Willebroek, Belgium 27/E6
Willems (canal), Netherlands 27/G5
Willemstad, Netherlands 27/F5
Willemstad (cap.), Neth. Ant. 161/F9
Willemstad (cap.), Neth. Ant. 156/E4
Willen, Manitoba 179/A4
Willernie, Minn. (55090) 255/G5
Willeroo, North. Terr. 93/B3
Willette, Tenn. (†37150) 237/K8
Willey, Iowa (†51401) 229/D5
Willey House, N.H. (†03812) 268/E3
William (riv.), Sask. 181/L2
William Creek, S. Australia 94/E3
William H. Taft Nat'l Hist. Site, Ohio 284/C10
William L. Springer (lake), Ill. 222/E4
Williams, Ariz. (86046) 198/C3
Williams, Calif. (95987) 204/C4
Williams, Ind. (47470) 227/D7
Williams, Iowa (50271) 229/F3
Williams, Minn. (56686) 255/D2
Williams (co.), N. Dak. 282/C3
Williams (co.), Ohio 284/A2
Williams, Okla. (†74932) 288/T4
Williams, Oreg. (97544) 291/D5
Williams, S.C. (29493) 296/F5
Williams, W. Australia 92/B2
Williams (riv.), W. Va. 312/F6
Williams A.F.B., Ariz. 198/D5
Williams Bay, Wis. (53191) 317/J10
Williamsboro, N.C. (†27536) 281/M2
Williamsburg, Colo. (†81226) 208/J6
Williamsburg, Iowa (52361) 229/J5
Williamsburg, Kansas (66095) 232/G3
Williamsburg, Ky. (40769) 237/N7
Williamsburg, Md. (21674) 245/P6
Williamsburg○, Mass. (01096) 249/C3
Williamsburg, Mich. (49690) 250/D4
Williamsburg, Mo. (63388) 261/J5
Williamsburg, New Bruns. 170/D2
Williamsburg, N. Mex. (87942) 274/B5
Williamsburg, Ohio (45176) 284/B7
Williamsburg, Ontario 177/J3
Williamsburg, Pa. (16693) 294/F5
Williamsburg (co.), S.C. 296/H4
Williamsburg (I.C.), Va. (23185) 307/P6
Williamsburg, W. Va. (24991) 312/F7
Williamsfield, Ill. (61489) 222/C3
Williamsfield, Jamaica 158/H6
Williamsfield, Ohio (44093) 284/J2
Williamsford, Ontario 177/D3
Williamsford, Tasmania 99/B3
Williams Fork, Colorado (riv.), Colo. 208/G3
Williams Fork, Yampa (riv.), Colo. 208/E2
Williams Harbour, Newf. 166/C3
Williams Lake, Br. Col. 162/D5
Williams Lake, Br. Col. 184/F4
Williamson, Georgia (30292) 217/D4
Williamson (co.), Ill. 222/E6
Williamson, Ill. (†62088) 222/D5
Williamson, Iowa (50272) 229/G6
Williamson, N.Y. (14589) 276/F4
Williamson (riv.), Oreg. 291/F5
Williamson (co.), Tenn. 237/H9
Williamson (co.), Texas 303/G7
Williamson, W. Va. (25661) 312/B7
Williamsport, Ind. (47993) 227/C4
Williamsport, Ky. (41271) 237/R5
Williamsport, Md. (21795) 245/G2
Williamsport, Newf. 166/C3
Williamsport, Ohio (43164) 284/D6
Williamsport, Pa. 188/L2
Williamsport, Pa. (17701) 294/H3
Williamsport, Tenn. (38487) 237/G9
Williamston, Mich. (48895) 250/E6
Williamston, N.C. (27892) 281/R3
Williamston, S.C. (29697) 296/B2
Williamstown, Kansas (†66073) 232/G2
Williamstown, Ky. (41097) 237/M3
Williamstown, Mass. (01267) 249/B2
Williamstown○, Mass. (01267) 249/B2
Williamstown, Mo. (63473) 261/J2
Williamstown, New Bruns. 170/C2
Williamstown, N.J. (08094) 273/D4
Williamstown, N.Y. (13493) 276/J4
Williamstown, Ontario 177/K2
Williamstown, Pa. (17098) 294/J4
Williamstown, S. Australia 94/C7
Williamstown○, Vt. (05679) 268/B3
Williamstown, Victoria 97/H5
Williamstown, Victoria 88/K7
Williamstown, W. Va. (26187) 312/C4
Williamsville, Ill. (62693) 222/D4
Williamsville, Miss. (†39090) 256/F4
Williamsville, Mo. (63977) 261/L9
Williamsville, N.Y. (14221) 276/C5
Williamsville, Vt. (05362) 268/B6
Williamsville, Va. (24487) 307/J4
Willies (range), Queensland 95/C6
Williford, Ark. (72482) 202/H1
Willimantic, Conn. (06226) 210/G2
Willimantic (riv.), Conn. 210/F1
Willimantic, Maine (†04443) 243/E5
Willimantic○, Maine (†04443) 243/E5
Willingboro○, N.J. (08046) 273/D3
Willingdon, Alberta 182/E3
Willington○, Conn. (†06279) 210/F1
Willington, S.C. (29852) 296/C4
Willis (islets), Australia 87/F7
Willis (isls.), Coral Sea Is. Terr. 88/J3
Willis, Kansas (66435) 232/G2
Willis, Mich. (48191) 250/F6
Willis, Okla. (†73439) 288/N7
Willis, Texas (77378) 303/J7
Willis, Va. (24380) 307/H7
Willis (riv.), Va. 307/M5
Willisau, Switzerland 39/F2
Willisburg, Ky. (40078) 237/L5
Williston (lake), Br. Col. 162/D4
Williston (lake), Br. Col. 184/F2
Williston, Fla. (32696) 212/D2
Williston, N. Dak. 188/F1
Williston, N. Dak. (58801) 282/C3
Williston, S.C. (29853) 296/E5
Williston, Tenn. (38076) 237/C10
Williston○, Vt. (05495) 268/A3
Williston Park, N.Y. (11596) 276/R7
Willisville, Ark. (71864) 202/D6
Willisville, Ill. (62997) 222/D6
Willisville, Ontario 177/C1
Willis Wharf, Va. (23486) 307/S5
Williton, England 13/D6
Willits, Calif. (95490) 204/B4
Willmar, Minn. (56201) 255/C5
Willmar, Sask. 181/J6
Willmathsville, Mo. (†63546) 261/G2
Willmore Wilderness Prov. Park, Alberta 182/A3
Willoughby (bay), Ant. & Bar. 161/E11
Willoughby, N. S. Wales 88/K3
Willoughby, N.S. Wales 97/J3
Willoughby, Ohio (44094) 284/J8
Willoughby, Vt. (†05822) 268/C2
Willoughby (lake), Vt. 268/D2
Willoughby Hills, Ohio (†44094) 284/J9
Willow, Alaska (99688) 196/B1
Willow, Ark. (†72084) 202/E5
Willow (creek), Calif. 204/E3
Willow (creek), Idaho 220/G6
Willow (riv.), Minn. 255/E4
Willow (creek), Mont. 262/E2
Willow, Okla. (73673) 288/G4
Willow (creek), Oreg. 291/H2
Willow (creek), Oreg. 291/K3
Willow (creek), S. Dak. 298/C4
Willow (creek), Utah 304/E4
Willow (res.), Wis. 317/F4
Willow (creek), Wyo. 319/F2
Willow (lake), Wyo. 319/C2
Willow Bend, W. Va. (24992) 312/F7
Willow Branch, Ind. (46187) 227/F5
Willowbrook, Ill. (†60521) 222/B6
Willowbrook, Kansas (†67501) 232/D3
Willowbrook, Sask. 181/J4
Willow Bunch, Sask. 181/F6
Willow Bunch (lake), Sask. 181/F6
Willow City, N. Dak. (58384) 282/K2
Willow City, Texas (78675) 303/F7
Willow Creek, Calif. (95573) 204/B3
Willow Creek, Mont. (59760) 262/E5
Willow Creek (res.), Mont. 262/D3
Willowcreek, Oreg. (†97918) 291/K3
Willow Creek, Sask. 181/B6
Willowdale, Oreg. (†97741) 291/G3
Willow Grove, Del. (†19934) 245/R4
Willow Grove, New Bruns. 170/E3
Willow Grove, Pa. (19090) 294/M5
Willow Hill, Ill. (62480) 222/E5
Willow Hill, Pa. (17271) 294/G5
Willowick, Ohio (44094) 284/J8
Willow Island, Nebr. (69171) 264/D4
Willowlake (riv.), N.W. Terrs. 187/F3
Willow Lake, S. Dak. (57278) 298/O4
Willowmore, S. Africa 118/C6
Willowra, North. Terr. 93/C6
Willow Ranch, Calif. (96108) 204/E2
Willow River, Br. Col. 184/F3
Willow River, Minn. (55795) 255/F4
Willows, Calif. (95988) 204/C4
Willows, Md. (†20732) 245/M6
Willows, Sask. 181/F6
Willow Springs, Ill. (60480) 222/B6

Willow Springs, Mo. (65793) 261/H9
Willowton, W. Va. (†24740) 312/E8
Willow Tree, N.S. Wales 97/F2
Wills (creek), Ohio 284/G5
Wills (lake), W. Australia 88/D4
Willsboro, N.Y. (12996) 276/N2
Wills Creek (lake), Ohio 284/G5
Willshire, Ohio (45898) 284/A4
Wills Point, La. (†70040) 238/L7
Wills Point, Texas (75169) 303/J5
Willunga, S. Australia 94/F6
Wilma, Fla. (†32351) 212/B1
Wilmar, Ark. (71675) 202/G6
Wilmer, Ala. (36587) 195/B9
Wilmer, Br. Col. 184/J5
Wilmer, La. (†70444) 238/K5
Wilmer, Texas (75172) 303/H3
Wilmerding, Pa. (15148) 294/C5
Wilmette, Ill. (60091) 222/B5
Wilmington, Calif. (*90744) 204/C11
Wilmington, Del. 188/M3
Wilmington, Del. (*19801) 245/R2
Wilmington, Ill. (60481) 222/E2
Wilmington (Patterson), Ill. (†62078) 222/C4
Wilmington, Ind. (†47001) 227/H6
Wilmington○, Mass. (01887) 249/C5
Wilmington, N.Y. (12997) 276/N2
Wilmington, N.C. 188/L4
Wilmington, N.C. 146/L6
Wilmington, N.C. (28401) 281/N6
Wilmington, Ohio (45177) 284/C7
Wilmington, S. Australia 94/F6
Wilmington○, Vt. (05363) 268/B6
Wilmington Island, Georgia (31410) 217/L7
Wilmont, Minn. (56185) 255/C7
Wilmore, Kansas (67155) 232/C4
Wilmore, Ky. (40390) 237/M5
Wilmore, Pa. (15962) 294/E5
Wilmot, Ark. (71676) 202/G7
Wilmot, Ind. (†46562) 227/F2
Wilmot, Kansas (†67131) 232/F4
Wilmot, New Bruns. 170/C3
Wilmot○, N.H. (†03287) 268/D5
Wilmot, Ohio (44689) 284/G4
Wilmot, Pr. Edward I. 168/E2
Wilmot, S. Dak. (57279) 298/R3
Wilmot, Tasmania 99/C3
Wilmot Flat, N.H. (03287) 268/D5
Wilmot Station, Nova Scotia 168/D4
Wilmslow, England 10/G2
Wilmslow, England 13/H2
Wilno, Ontario 177/G2
Wilpen, Pa. (15694) 294/D5
Wilrijk, Belgium 27/E6
Wilsall, Mont. (59086) 262/F5
Wilsey, Kansas (66873) 232/F3
Wilson (dam), Ala. 195/C1
Wilson, Ark. (72395) 202/K2
Wilson (mt.), Calif. 204/D10
Wilson (mt.), Colo. 188/E3
Wilson (mt.), Colo. 208/C7
Wilson, Conn. (†06095) 210/E1
Wilson (co.), Kansas 232/G4
Wilson, Kansas (67490) 232/D3
Wilson (lake), Kansas 232/D3
Wilson, La. (70789) 238/H5
Wilson (ponds), Maine 243/E5
Wilson, Mich. (49896) 250/B3
Wilson, Minn. (†55987) 255/G7
Wilson, N.Y. (14172) 276/C4
Wilson (co.), N.C. 281/O3
Wilson, N.C. (27893) 281/O3
Wilson (cape), N.W. Terrs. 187/K3
Wilson, Ohio (†43716) 284/H6
Wilson, Okla. (73463) 288/M6
Wilson (riv.), Oreg. 291/D2
Wilson, Pa. (15025) 294/M4
Wilson (riv.), Queensland 95/B5
Wilson, S.C. (29179) 296/G4
Wilson (co.), Tenn. 237/J8
Wilson (co.), Texas 303/F8
Wilson, Texas (79381) 303/C4
Wilson (creek), Wash. 310/F3
Wilson, W. Va. (26768) 312/H4
Wilson, Wis. (54027) 317/B6
Wilson, Wyo. (83014) 319/B2
Wilson Bluff (prom.), S. Australia 94/A4
Wilsonburg, W. Va. (26461) 312/F4
Wilson City, Mo. (†63834) 261/O9
Wilson Creek, Br. Col. 184/J2
Wilson Creek, Wash. (98860) 310/F3
Wilsondale, W. Va. (25699) 312/B7
Wilson Lake (res.), Idaho 220/D7
Wilson Landing, Br. Col. 184/H5
Wilson Point, New Bruns. 170/F1
Wilsons (prom.), Victoria 97/D6
Wilsons (prom.), Victoria 88/H7
Wilsons, Va. (23894) 307/N6
Wilsons Beach, New Bruns. 170/D4
Wilson's Creek Nat'l Battlefield, Mo. 261/F8
Wilsons Mills, Maine (04293) 243/B6
Wilsons Mills, N.C. (27593) 281/M3
Wilsonville, Ala. (35186) 195/E4
Wilsonville, Conn. (†06255) 210/H1
Wilsonville, Ill. (62093) 222/D4
Wilsonville, Nebr. (69046) 264/D4
Wilsonville, Oreg. (97070) 291/A2
Wilton, Ala. (35187) 195/E4
Wilton, Ark. (71865) 202/B6
Wilton, Calif. (95693) 204/C9
Wilton○, Conn. (06897) 210/B4
Wilton, England 13/F6
Wilton, Iowa (52778) 229/M5
Wilton, Maine (04294) 243/C6
Wilton○, Maine (04294) 243/C6
Wilton, Minn. (56687) 255/C3
Wilton, N.H. (03086) 268/D6
Wilton○, N.H. (03086) 268/D6
Wilton, N.Y. (†12866) 276/N4
Wilton, N. Dak. (58579) 282/J5
Wilton, Wis. (54670) 317/F8
Wilton Manors, Fla. (33334) 212/B3
Wiltshire (co.), England 13/E6
Wiltz, Luxembourg 27/H9
Wiluna, Australia 87/C8
Wiluna, W. Australia 88/C5
Wiluna, W. Australia 92/C4
Wimauma, Fla. (33598) 212/D4
Wimbledon, N. Dak. (58492) 282/O5
Wimborne, Alberta 182/D4
Wimborne Minster, England 13/E7
Wimico (lake), Fla. 212/A2
Wimmer, Sask. 181/G3
Wimmera (riv.), Victoria 97/A5
Wimmis, Switzerland 39/E3
Winagami (lake), Alberta 182/B2
Winam (bay), Kenya 115/F4
Winamac, Ind. (46996) 227/D2
Winborn, Miss. (†38659) 256/F1
Winburg, S. Africa 118/D5
Winburne, Pa. (16879) 294/F4
Wincanton, England 13/E6
Winchburgh, Scotland 15/D1
Winchelsea, Victoria 97/B6
Winchendon, Mass. (01475) 249/F2
Winchendon○, Mass. (01475) 249/F2
Winchendon Springs, Mass. (01477) 249/G2
Winchester, Ark. (71677) 202/G6
Winchester○, Conn. (06094) 210/C1
Winchester, England 13/F6
Winchester, England 10/F5
Winchester, Idaho (83555) 220/B3
Winchester, Ill. (62694) 222/C4
Winchester, Ind. (47394) 227/G4
Winchester, Kansas (66097) 232/G2
Winchester, Ky. (40391) 237/N5
Winchester○, Mass. (01890) 249/C6
Winchester, Mo. (†63435) 261/N3
Winchester, Nev. (†89109) 266/F6
Winchester, N.H. (03470) 268/C6
Winchester○, N.H. (03470) 268/C6
Winchester, N. Zealand 100/C6
Winchester, Ohio (45697) 284/C8
Winchester, Ontario 177/J2
Winchester, Oreg. (97495) 291/D4
Winchester (bay), Oreg. 291/C4
Winchester, Tenn. (37398) 237/J10
Winchester (I.C.), Va. (22601) 307/M2
Winchester, Wash. (†98848) 310/F3
Winchester, Wis. (54567) 317/G3
Winchester Bay, Oreg. (97467) 291/C4
Winchester Center, Conn. (06094) 210/C1
Wind (riv.), Wash. 310/D5
Wind (lake), Wis. 317/K2
Wind (riv.), Wyo. 319/C2
Windber, Pa. (15963) 294/E5
Wind Cave Nat'l Park, S. Dak. 298/B6
Windcrest, Texas (†78201) 303/K11
Windemere, Conn. (†06066) 210/F1
Winder (lake), Fla. 212/F3
Winder, Georgia (30680) 217/E3
Windermere, Br. Col. 184/K5
Windermere, England 13/E3
Windermere, England 10/E3
Windermere, Fla. (32786) 212/E3
Windermere, Ontario 177/E2
Windfall, Ind. (46076) 227/F4
Windgap, Pa. (18091) 294/M4
Windham (co.), Conn. 210/H1
Windham○, Conn. (06280) 210/G2
Windham, Mont. (†59479) 262/F3
Windham○, N.H. (03087) 268/E6
Windham, N.Y. (12496) 276/M6
Windham, Ohio (44288) 284/H3
Windham (co.), Vt. 268/B5
Windham○, Vt. (05359) 268/B5
Windham Depot, N.H. (†03087) 268/E6
Windhoek (cap.), Namibia 102/D7
Windhoek (cap.), Namibia 118/B4
Windhoek (cap.), Namibia 2/K7
Winding Gulf, W. Va. (25941) 312/D7
Windisch, Switzerland 39/F1
Windischgarsten, Austria 41/C3
Wind Lake, Wis. (53185) 317/K2
Windom (peak), Colo. 208/D7
Windom, Ind. (†47581) 227/D7
Windom, Kansas (67491) 232/E3
Windom, Minn. (56101) 255/C7
Windorah, Queensland 88/G5
Windorah, Queensland 95/B5
Window Rock, Ariz. (86515) 198/F3
Wind Point, Wis. (†53401) 317/M2
Wind River (canyon), Wyo. 319/D2
Wind River (range), Wyo. 319/C2
Wind River Ind. Res., Wyo. 319/C2
Windsor, Calif. (95492) 204/C5
Windsor, Colo. (80550) 208/J2
Windsor, Conn. (06095) 210/E1
Windsor○, Conn. (06095) 210/E1
Windsor, New, Eng. 13/G6
Windsor (New Windsor), Ill. (†61465) 222/C2
Windsor, Ill. (61957) 222/E4
Windsor, Ind. (†47368) 227/G4
Windsor○, Maine (04363) 243/D7
Windsor○, Mass. (01270) 249/B2
Windsor, Mo. (65360) 261/E5
Windsor, New Bruns. 170/C2
Windsor, Newf. 166/C4
Windsor, N.J. (08561) 273/D3
Windsor, N.Y. (13865) 276/J6
Windsor, N.C. (27983) 281/P2
Windsor, N. Dak. (†58424) 282/N6
Windsor, N.S. 162/K7
Windsor, Nova Scotia 168/D3
Windsor, Ohio (44099) 284/J2
Windsor, Ont. 162/H7
Windsor, Ontario 177/B5
Windsor, Pa. (17366) 294/J6
Windsor, Québec 172/F4
Windsor, Queensland 88/K2
Windsor, Queensland 95/D2
Windsor, S.C. (29856) 296/E5
Windsor (co.), Vt. 268/B4
Windsor, Vt. (05089) 268/C5
Windsor○, Vt. (05089) 268/C5
Windsor, Va. (23487) 307/P7
Windsor, Wis. (53598) 317/H9
Windsor Forest, Georgia (31406) 217/K7
Windsor Heights, Iowa (50311) 229/F5
Windsor Heights, W. Va. (26075) 312/E2
Windsor Locks○, Conn. (06096) 210/E1
Windsorville, Conn. (06097) 210/E1
Windthorst, Sask. 181/J5
Windthorst, Texas (76389) 303/F4
Windward (passg.) 146/L8
Windward (passage), Cuba 156/C3
Windward (passage), Grenada 156/C3
Windward (passage), Haiti 158/A5
Windward (isls.), W. Indies 156/G4
Windy, W. Va. (†26143) 312/C4
Windy Hill, S.C. (†29501) 296/H3
Windy Hills, Ky. (†40201) 237/K1
Windyville, Mo. (65783) 261/F7
Winefred (lake), Alberta 182/E2
Winefred (riv.), Alberta 182/E2
Winesburg, Ohio (44690) 284/G4
Winfall, N.C. (27985) 281/S2
Winfield, Ala. (35594) 195/C3
Winfield, Alberta 182/C3
Winfield, Ark. (†72958) 202/B4
Winfield, Br. Col. 184/H5
Winfield, Ill. (60190) 222/A5
Winfield, Iowa (52659) 229/L6
Winfield, Kansas (67156) 232/F4
Winfield, Md. (†21157) 245/K3
Winfield, Mo. (63389) 261/L5
Winfield○, N.J. (†07036) 273/B2
Winfield, Pa. (17889) 294/J4
Winfield, Tenn. (37892) 237/M7
Winfield, Texas (75493) 303/K4
Winfield, W. Va. (25213) 312/C5
Winfred, S. Dak. (57076) 298/P6
Wing, Ala. (36483) 195/E8
Wing, Ill. (†61741) 222/E3
Wing, N. Dak. (58494) 282/K5
Wingard, Sask. 181/E3
Wingate, England 13/J4
Wingate, Ind. (47994) 227/C4
Wingate, Md. (21675) 245/O7
Wingate, N.C. (28174) 281/J5
Wingate, Pa. (16880) 294/G4
Wingate, Texas (79566) 303/D5
Wingate Army Depot, N. Mex. 274/A3
Wingdale, N.Y. (12594) 276/N7
Wingene, Belgium 27/C6
Winger, Minn. (56592) 255/B3
Winger, Ontario 177/E5
Wingham, N.S. Wales 97/G2
Wingham, Ontario 177/C4
Wingo, Ky. (42088) 237/D7
Winifred, Kansas (66553) 232/F2
Winifred, Ky. (†41219) 237/R5
Winifred, Mont. (59489) 262/G3
Winifreda, Argentina 143/D4
Winifrede, W. Va. (25214) 312/C6
Winigan, Mo. (63566) 261/G2
Winisk, Ontario 175/C1
Winisk (lake), Ontario 175/C2
Winisk (riv.), Ont. 162/H5
Winisk (riv.), Ontario 175/C2
Wink, Texas (79789) 303/A6
Winkel, Netherlands 27/F3
Winkelman, Ariz. (85292) 198/E6
Winkle, Ill. (†62290) 222/D5
Winkler, Manitoba 179/E5
Winkler (co.), Texas 303/A6
Winlaw, Br. Col. 184/J5
Winlock, Oreg. (†97874) 291/H3
Winlock, Wash. (98596) 310/C4
Winn, Ala. (†36545) 195/C7
Winn (par.), La. 238/E3
Winn, Maine (04495) 243/G5
Winn○, Maine (04495) 243/G5
Winn, Mich. (48896) 250/E5
Winnabow, N.C. (28479) 281/N6
Winnaleah, Tasmania 99/D3
Winneba, Ghana 106/D7
Winnebago (co.), Ill. 222/D1
Winnebago, Ill. (61088) 222/D1
Winnebago (co.), Iowa 229/F2
Winnebago, Minn. (56098) 255/D7
Winnebago, Nebr. (68071) 264/H2
Winnebago (co.), Wis. 317/J8
Winnebago, Wis. (54985) 317/J8
Winnebago (lake), Wis. 317/K7
Winnebago Ind. Res., Nebr. 264/H2
Winnecke (creek), North. Terr. 93/B5
Winneconne, Wis. (54986) 317/J7
Winnecook (lake), Maine 243/E6
Winnemaug (lake), Conn. 210/C2
Winnemucca, Nev. 188/C2
Winnemucca, Nev. (89445) 266/D2
Winnemucca (lake), Nev. 188/C2
Winnemucca (lake), Nev. 266/B2
Winnemucca Ind. Res., Nev. 266/D2
Winner, S. Dak. (57580) 298/K7
Winneshiek (co.), Iowa 229/K2
Winnetka, Ill. (60093) 222/B5
Winnetoon, Nebr. (68789) 264/F2
Winnett, Mont. (59087) 262/H4
Winnfield, La. (71483) 238/E3
Winnibigoshish (lake), Minn. 255/D3
Winnie, Texas (77665) 303/K8
Winnifred, Alberta 182/E5
Winning Pool, W. Australia 92/A3
Winnipauk, Conn. (†06856) 210/B4
Winnipeg, Canada 2/E3
Winnipeg (cap.), Man. 146/J5
Winnipeg (cap.), Man. 162/G6
Winnipeg (cap.), Manitoba 179/E5
Winnipeg (lake), Man. 162/G5
Winnipeg (lake), Man. 146/J4
Winnipeg (lake), Manitoba 179/E2
Winnipeg (riv.), Manitoba 179/G4
Winnipeg (riv.), Ontario 175/A2
Winnipeg Beach, Manitoba 179/F4
Winnipegosis, Man. 162/F5
Winnipegosis, Manitoba 179/B3
Winnipegosis (lake), Man. 146/H4
Winnipegosis (lake), Man. 162/F5
Winnipesaukee, N.H. (†03254) 268/E4
Winnipesaukee (lake), N.H. 268/E4
Winnipesaukee (riv.), N.H. 268/D5
Winnisquam, N.H. (03289) 268/E5
Winnisquam (lake), N.H. 268/D4
Winnsboro, La. (71295) 238/G2
Winnsboro, S.C. (29180) 296/E3
Winnsboro, Texas (75494) 303/J5
Winnsboro Mills, S.C. (†29180) 296/E3
Winokapau (lake), Newf. 166/B3
Winokur, Georgia (†31537) 217/H8
Winona, Ariz. (†86001) 198/D3
Winona (lake), Ark. 202/E4
Winona, Kansas (67764) 232/A2
Winona, Mich. (49972) 250/G1
Winona, Minn. 188/H2
Winona (co.), Minn. 255/G6
Winona, Minn. (55987) 255/G6
Winona, Miss. (38967) 256/E4
Winona, Mo. (65588) 261/K8
Winona, Ohio (44493) 284/J4
Winona, Tenn. (37893) 237/N8
Winona, Texas (75792) 303/J5
Winona, Wash. (†99125) 310/H4
Winona, W. Va. (25942) 312/E6
Winona Lake, Ind. (46590) 227/F2
Winooski, Vt. (05404) 268/A2
Winooski (riv.), Vt. 268/B3
Winschoten, Netherlands 27/L2
Winscombe, England 13/D6
Winsen, W. Germany 22/D2
Winsford, England 13/G2
Winsford, England 10/G2
Winside, Nebr. (68790) 264/G2
Winslow, Ariz. (86047) 198/E3
Winslow, Ark. (72959) 202/B2
Winslow, Ill. (61089) 222/D1
Winslow, Ind. (47598) 227/C8
Winslow, Maine (†04901) 243/D6
Winslow○, Maine (†04901) 243/D6
Winslow, Nebr. (68072) 264/H3
Winslow, N.J. (08095) 273/D4
Winslow (Bainbridge Island-Winslow) Wash. (†98110) 310/A2
Winslows Mills, Maine (†04572) 243/E7
Winsted, Conn. (06098) 210/C1
Winsted, Minn. (55395) 255/D6
Winston (co.), Ala. 195/D2
Winston, Georgia (30187) 217/C3
Winston, Ky. (40495) 237/N5
Winston (co.), Miss. 256/F4
Winston, Mo. (64689) 261/D3
Winston, Mont. (59647) 262/E4
Winston, N. Mex. (87943) 274/B5
Winston, Oreg. (97496) 291/D4
Winston Park, Ky. (†41011) 237/S2
Winston-Salem, N.C. 188/K3
Winston-Salem, N.C. (*27101) 281/J2
Winstonville, Miss. (38781) 256/C3
Winsum, Netherlands 27/K2
Winter (harb.), N.W. Terrs. 187/H2
Winter (isl.), N.W. Terrs. 187/K3
Winter, Sask. 181/B3
Winter, Wis. (54896) 317/E4
Winter Beach, Fla. (32971) 212/F4
Winterdale, Pa. (†13783) 294/M2
Winter Garden, Fla. (32787) 212/E3
Winter Harbor○, Maine (04693) 243/G7
Winter Harbour, Br. Col. 184/C5
Winterhaven, Calif. (92283) 204/L11
Winter Haven, Fla. (33880) 212/E3
Winter I. Coast Guard Air Sta., Mass. 249/E5
Winter Park, Colo. (80482) 208/H3
Winter Park, Fla. (*32789) 212/E3
Winterpock, Va. (†23832) 307/N6
Winterport, Maine (04496) 243/F6
Winterport○, Maine (04496) 243/F6
Winters, Calif. (95694) 204/D5
Winters, Texas (79567) 303/E6
Wintersburg, Ariz. (†85322) 198/B5
Winterset, Iowa (50273) 229/E6
Winterset, Ohio (†43755) 284/H5
Winter Springs, Fla. (†32801) 212/E3
Winters Run (creek), Md. 245/N2
Wintersville, Ohio (43952) 284/J5
Winterswijk, Netherlands 27/K5
Winterthur, Switzerland 39/G1
Winterton, Newf. 166/D2
Winterville, Georgia (30683) 217/F3
Winterville, Maine (04788) 243/F2
Winterville○, Maine (04788) 243/F2
Winterville, Miss. (38782) 256/B4
Winterville, N.C. (28590) 281/P3
Winthrop, Ark. (71866) 202/B6
Winthrop, Conn. (†06417) 210/E3
Winthrop, Iowa (50682) 229/K4
Winthrop, Maine (04364) 243/C7
Winthrop○, Maine (04364) 243/C7
Winthrop○, Mass. (02152) 249/D6
Winthrop, Minn. (55396) 255/D6
Winthrop, Wash. (98862) 310/E2
Winthrop-Brasher Falls, N.Y. (13697) 276/L1
Winthrop Harbor, Ill. (60096) 222/F1
Winton, Calif. (95388) 204/E6
Winton, Minn. (55796) 255/G3
Winton, N. Zealand 100/B7
Winton, N.C. (27986) 281/P2
Winton, Queensland 88/G4
Winton, Queensland 95/B4
Winton, Wash. (†98826) 310/E3
Winyah (bay), S.C. 296/J5
Wiota, Iowa (50274) 229/D6
Wiota, Wis. (†53587) 317/G10
Wirksworth, England 13/F4
Wirral, England 13/G2
Wirral, England 10/F2
Wirral (pen.), England 13/G2
Wirral, New Bruns. 170/D3
Wirrulla, S. Australia 94/D5
Wirt, Ind. (†47250) 227/G7
Wirt, Minn. (56688) 255/E3
Wirt, Okla. (†73438) 288/L6
Wirt (co.), W. Va. 312/D4
Wirth, Ark. (†72554) 202/H1
Wirtz, Va. (24184) 307/J6
Wirulla, S. Australia 88/E6
Wisacky, S.C. (29183) 296/G3
Wisbech, England 10/G4
Wisbech, England 13/H5
Wiscasset, Maine (04578) 243/D7
Wiscasset○, Maine (04578) 243/D7
Wisconsin 188/J2
WISCONSIN 317
Wisconsin (state), U.S. 146/K5
Wisconsin (riv.), Wis. 188/H2
Wisconsin (riv.), Wis. 317/E9
Wisconsin Dells, Wis. (53965) 317/G8
Wisconsin Rapids, Wis. (54494) 317/G7
Wisdom, Ky. (†42129) 237/K7
Wisdom, Mont. (59761) 262/C5
Wise, N.C. (27594) 281/N2
Wise (co.), Texas 303/G4
Wise (co.), Va. 307/C6
Wise, Va. (24293) 307/C7
Wiseman, Alaska (†99726) 196/H1
Wiseman, Ark. (72587) 202/G1
Wise River, Mont. (59762) 262/C5
Wiseton, Sask. 181/D4
Wishart, Mo. (†65710) 261/F7
Wishart, Sask. 181/H4
Wishek, N. Dak. (58495) 282/L7
Wishram, Wash. (98673) 310/D5
Wisła, Poland 47/D4
Wisła (Vistula) (riv.), Poland 47/D2
Wismar, E. Germany 22/D2
Wisner, La. (71378) 238/G3
Wisner, Mich. (†48701) 250/F5
Wisner, Nebr. (68791) 264/H3
Wissembourg, France 28/G3
Wistaria, Br. Col. 184/D3
Wister, Okla. (74966) 288/S5
Wister (lake), Okla. 288/S5
Witbank, S. Africa 118/D5
Witchekan (lake), Sask. 181/D2
Witham, England 13/H6
Witham (riv.), England 10/F4
Witham (riv.), England 13/G4
Withamsville, Ohio (45245) 284/C10
Withee, Wis. (54498) 317/E6
Witherbee-Mineville, N.Y. (12998) 276/N2
Withernsea, England 13/H4
Withernsea, England 10/G4
Witherspoon (mt.), Alaska 196/C1
Withlacoochee (riv.), Fla. 212/C1
Withlacoochee (riv.), Fla. 212/D2
Withrow, Wash. (†98858) 310/F3
Witkowo, Poland 47/C2
Witless Bay, Newf. 166/D2
Witney, England 13/F6
Witnica, Poland 47/B2
Witoka, Minn. (†55987) 255/G7
Witt, Ill. (62094) 222/D4
Witten, S. Dak. (57584) 298/J7
Witten, W. Germany 22/B3
Wittenberg, E. Germany 22/E3
Wittenberg, Mo. (63786) 261/O7
Wittenberg, Wis. (54499) 317/H6
Wittenberge, E. Germany 22/D2
Wittenoom, W. Australia 88/B4
Wittenoom, W. Australia 92/B3
Witter, Ark. (72776) 202/C2
Wittingen, W. Germany 22/D2
Wittlich, W. Germany 22/B3
Wittman, Md. (21676) 245/N5
Wittmann, Ariz. (85361) 198/C5
Witts Springs, Ark. (72686) 202/E2
Wittstock, E. Germany 22/E2
Witu, Kenya 115/H4
Witvlei, Namibia 118/B4
Witwatersberg (range), S. Africa 118/G6
Witwatersrand (reg.), S. Africa 118/H7
Witzenhausen, W. Germany 22/C3
Wivenhoe, England 13/J6
Wivenhoe, Manitoba 179/K2
Wiville, Ark. (†72101) 202/H3
Wixom, Mich. (48096) 250/F6
Wkra (riv.), Poland 47/E2
Władysławowo, Poland 47/D1
Włocławek (prov.), Poland 47/D2
Włocławek, Poland 47/D2
Włocławskie (lake), Poland 47/D2
Włodawa, Poland 47/F3
Włoszczowa, Poland 47/D3
Woburn, Grenada 161/C9
Woburn, Mass. (01801) 249/C6
Woburn, Québec 172/G4
Woden, Iowa (50484) 229/F2
Wodonga, Victoria 88/H7
Wodonga, Victoria 97/D5
Wodzisław Śląski, Poland 47/D4
Woensdrecht, Netherlands 27/E6
Woerden, Netherlands 27/F4
Wohlen, Switzerland 39/F2
Wohlen bei Bern, Switzerland 39/D3
Wojkowice, Poland 47/B3
Wokam (isl.), Indonesia 85/K7
Woking, Alberta 182/A2
Woking, England 13/G8
Woking, England 10/B6
Wokingham, England 13/G8
Wokingham, England 10/F5
Wolbach, Nebr. (68882) 264/F3
Wolbrom, Poland 47/D3
Wolco, Okla. (†74002) 288/O1
Wolcott, Colo. (81655) 208/F3
Wolcott○, Conn. (06716) 210/D2
Wolcott, Ind. (47995) 227/C3
Wolcott, N.Y. (14590) 276/G4
Wolcott○, Vt. (05680) 268/C2
Wolcottsville, N.Y. (†14001) 276/C4
Wolcottville, Ind. (46795) 227/G1
Wolds, The (hills), England 13/G4
Woleai (atoll), Micronesia 87/E5
Wolf (lake), Alberta 182/E2
Wolf (Wenman) (isl.), Ecuador 128/B8
Wolf (lake), Ill. 222/C6
Wolf (riv.), Miss. 256/F9
Wolf, Okla. (†74868) 288/N4
Wolf (creek), Okla. 288/G2
Wolf (creek), S. Dak. 298/L4
Wolf (riv.), Tenn. 237/B10
Wolf (creek), Texas 303/D1
Wolf (creek), Va. 307/F6
Wolf (riv.), Wis. 317/J5
Wolf, Wyo. (82844) 319/E1
Wolf Creek, Alberta 182/B3
Wolf Creek, Ky. (†40104) 237/J4
Wolf Creek, Mont. (59648) 262/D3
Wolf Creek, Oreg. (97497) 291/D5
Wolf Creek, Wis. (†54024) 317/A4
Wolfdale, Pa. (†15301) 294/B5
Wolfe (co.), Ky. 237/O5
Wolfe (co.), Québec 172/F4
Wolfeboro, N.H. (03894) 268/E4
Wolfeboro○, N.H. (03894) 268/E4
Wolfeboro Falls, N.H. (03896) 268/E4
Wolfe City, Texas (75496) 303/J4
Wolfe Island, Ontario 177/H3
Wolfen, E. Germany 22/E3
Wolfenbüttel, W. Germany 22/D2
Wolfenschiessen, Switzerland 39/F3
Wolfforth, Texas (79382) 303/C4
Wolf Island, Mo. (63881) 261/O9
Wolf Lake, Ill. (62998) 222/D6
Wolflake, Ind. (46796) 227/F2
Wolf Lake, Mich. (†49440) 250/D5
Wolf Lake, Minn. (56593) 255/C4
Wolford, N. Dak. (58385) 282/L3
Wolf Pen, W. Va. (24896) 312/C7
Wolf Point, Mont. (59201) 262/L2
Wolfsberg, Austria 41/C3
Wolfsburg, W. Germany 22/D2
Wolf Summit, W. Va. (26462) 312/F4
Wolfsville, Md. (†21773) 245/H2
Wolfton, S.C. (†29115) 296/E4
Wolftown, Va. (22748) 307/M4
Wolf Trap Farm Park, Va. 307/S2
Wolfville, Nova Scotia 168/D3
Wolgast, E. Germany 22/E1
Wolhusen, Switzerland 39/F2
Wolin, Poland 47/B2
Wolin (Wollin) (isl.), Poland 47/B2
Wollaston (isl.), Chile 138/F11
Wollaston (pen.), N.W. Terrs. 187/G3
Wollaston (lake), Sask. 162/F4
Wollaston (lake), Sask. 146/H4
Wollaston (lake), Sask. 181/N2
Wollaston Lake, Sask. 181/N2
Wollogorang, North. Terr. 88/F3
Wollogorang, North. Terr. 93/F5
Wollomombi, N.S. Wales 97/G2
Wollondilly (riv.), N.S. Wales 97/F4
Wollongong, Australia 87/F9
Wollongong, N.S. Wales 88/J6
Wollongong, N.S. Wales 97/F4
Wolmaransstad, S. Africa 118/D5
Wołomin, Poland 47/E2
Wołów, Poland 47/C3
Wolseley, Sask. 181/H5
Wolseth, N. Dak. (†58740) 282/H3
Wolsey, S. Dak. (57384) 298/N5
Wolstenholme (cape), Québec 174/E1
Wolsztyn, Poland 47/C2
Wolta, Ethiopia 111/G6
Woluwe-Saint-Lambert, Belgium 27/C9
Woluwe-Saint-Pierre, Belgium 27/C9
Wolvega, Netherlands 27/J3
Wolverhampton, England 13/E5
Wolverhampton, England 10/G3
Wolverine (riv.), Alberta 182/B1
Wolverine, Ky. (41394) 237/P5
Wolverine, Mich. (49799) 250/E3
Wolverton, Minn. (56594) 255/B4
Wolves, The (isls.), New Bruns. 170/D4
Womack, Mo. (†63645) 261/M7
Womack Hill, Ala. (†36908) 195/B7
Womboota, N.S. Wales 97/C4
Wombwell, England 13/K2
Womelsdorf, Pa. (19567) 294/K5
Womelsdorf (Coalton), W. Va. (†26257) 312/G5
Womer, Kansas (†66934) 232/D2
Wonalancet, N.H. (03897) 268/E4
Wonder, Oreg. (†97526) 291/D5
Wonewoc, Wis. (53968) 317/F8
Wongalarra (lake), N.S. Wales 97/C2
Wongan Hills, W. Australia 92/B5
Wŏnju, S. Korea 81/D5
Wonogiri, Indonesia 85/J2
Wononpakook (lake), Conn. 210/B1
Wononskopomuc (lake), Conn. 210/B1
Wonosobo, Indonesia 85/J2
Wonowon, Br. Col. 184/G2
Wonreli, Indonesia 85/H7
Wŏnsan, N. Korea 54/O6
Wŏnsan, N. Korea 81/C4
Wonthaggi, Australia 87/E9
Wonthaggi, Victoria 97/C6
Wonthaggi, Victoria 88/G7
Wood (isls.), Chile 138/E11
Wood (isl.), Mich. 250/C2
Wood, N.C. (†27549) 281/N2
Wood (co.), Ohio 284/C3
Wood (isls.), Pr. Edward I. 168/F3
Wood (mt.), Sask. 181/E6
Wood (riv.), Sask. 181/E6
Wood, S. Dak. (57585) 298/J6
Wood (co.), Texas 303/J5
Wood (co.), W. Va. 312/D4
Wood (co.), Wis. 317/F7
Wood (riv.), Wyo. 319/C2
Woodall (mt.), Miss. 256/H1
Woodberry, Ark. (†71767) 202/E6
Woodberry Forest, Va. (22989) 307/M4
Woodbine, Georgia (31569) 217/J9
Woodbine, Ill. (†61085) 222/C1
Woodbine, Iowa (51579) 229/B5
Woodbine, Kansas (67492) 232/E3
Woodbine, Ky. (40771) 237/N7
Woodbine, Md. (21797) 245/K3
Woodbine, N.J. (08270) 273/D5
Woodbourne, N.Y. (12788) 276/M7
Woodbridge, Calif. (95258) 204/B9
Woodbridge○, Conn. (†06515) 210/D3
Woodbridge, England 13/J5
Woodbridge, England 10/G4
Woodbridge○, N.J. (07095) 273/E2
Woodbridge, Tasmania 99/D5
Woodbridge, Va. (*22191) 307/O3
Wood Buffalo Nat'l Park, Alberta 182/B5
Wood Buffalo Nat'l Park, Alta. 162/E4
Wood Buffalo Nat'l Park, N.W. Terrs. 187/G3
Woodburn, Ind. (46797) 227/H2

Woodburn, Iowa (50275) 229/F7
Woodburn, Ky. (42170) 237/J7
Woodburn, N.S. Wales 97/G1
Woodburn, Oreg. (97071) 291/A3
Woodbury, Conn. (06798) 210/C2
Woodbury○, Conn. (06798) 210/C2
Woodbury, Georgia (30293) 217/C5
Woodbury (co.), Iowa 229/B4
Woodbury, Ky. (42288) 237/H6
Woodbury, Minn. (†55798) 255/F6
Woodbury, N.J. (08096) 273/B4
Woodbury, Pa. (16695) 294/F5
Woodbury, Tenn. (37190) 237/J9
Woodbury○, Vt. (05681) 268/C3
Woodbury Heights, N.J. (08097) 273/B4
Woodbury P.O. (North Woodbury), Conn. (06798) 210/C2
Woodchopper, Alaska (†99733) 196/K1
Woodcliff, Georgia (†30467) 217/J5
Woodcliff Lake, N.J. (07675) 273/B1
Woodcock, Pa. (†16335) 294/B2
Woodcrest, Calif. (†92504) 204/E11
Wood Dale, Ill. (60191) 222/B5
Wooden Ball (isl.), Maine 243/F8
Woodenbong, N.S. Wales 97/G1
Woodenbridge, Ireland 17/J6
Woodend, Victoria 97/C5
Woodfibre, Br. Col. 184/K2
Woodfin, N.C. (†28804) 281/D3
Woodford, Grenada 161/C8
Woodford (co.), Ill. 222/D3
Woodford, Ireland 17/E5
Woodford (co.), Ky. 237/M4
Woodford, Okla. (†73451) 288/M6
Woodford, S.C. (†29112) 296/E4
Woodford○, Vt. (†05201) 268/A6
Woodford, Wis. (†53504) 317/G10
Woodgate, N.Y. (13494) 276/K3
Woodhall Spa, England 13/G4
Woodhaven, La. (†70466) 238/M1
Woodhaven, Mich. (†48183) 250/F6
Woodhouse, Alberta 182/D5
Woodhull, Ill. (61490) 222/C2
Woodhull, N.Y. (14898) 276/F6
Woodhull (lake), N.Y. 276/L3
Woodington, Ohio (†45331) 284/A5
Woodinville, Wash. (98072) 310/B1
Wood Islands, Pr. Edward I. 168/F2
Woodlake, Calif. (93286) 204/G7
Wood Lake, Minn. (56297) 255/C6
Wood Lake, Nebr. (69221) 264/D2
Woodland, Ala. (36280) 195/H4
Woodland, Calif. (95695) 204/B8
Woodland, Georgia (31836) 217/D5
Woodland, Ill. (60974) 222/F3
Woodland, Ind. (†46624) 227/E1
Woodland, La. (†70722) 238/J5
Woodland○, Maine (04694) 243/H5
Woodland, Mich. (48897) 250/D6
Woodland, Miss. (39776) 256/F3
Woodland, N.C. (27897) 281/P2
Woodland, Pa. (16881) 294/F4
Woodland, Utah (†84036) 304/C3
Woodland, Wash. (98674) 310/C5
Woodland Hills, Calif. (*91364) 204/B10
Woodland Hills, Ky. (†40201) 237/L2
Woodland Mills, Tenn. (38271) 237/C8
Woodland Park, Colo. (80863) 208/J4
Woodlands, Manitoba 179/E4
Woodlands, Singapore 72/F6
Woodlands, W. Va. (†26055) 312/E3
Woodlark (isl.), Papua N.G. 85/C7
Woodlawn, Hawaii (†96801) 218/D4
Woodlawn, Ill. (62898) 222/D5
Woodlawn, Ky. (†41071) 237/T2
Woodlawn, La. (†70647) 238/E6
Woodlawn, Md. (†21201) 245/M3
Woodlawn, Ohio (†45201) 284/C9
Woodlawn, Tenn. (37191) 237/G7
Woodlawn, Va. (24381) 307/G7
Woodlawn Heights, Ind. (†46011) 227/F4
Woodlawn-Oakdale, Ky. (†42001) 237/D6
Woodlawn Park, Ky. (†40201) 237/K2
Woodleaf, N.C. (27054) 281/H3
Woodley, Sask. 181/J6
Woodley and Sandford, England 13/G8
Woodlyn, Pa. (19094) 294/M7
Wood-Lynne, N.J. (†08107) 273/B3
Woodman, Wis. (53827) 317/E9
Woodmere, N.Y. (11598) 276/P7
Woodmere, Ohio (†44101) 284/J9
Woodmont, Conn. (†06460) 210/D4
Woodmoor, Md. (†21207) 245/L3
Wood Mountain, Sask. 181/E6
Wood Mountain Hist. Park, Sask. 181/E6
Woodnorth, Manitoba 179/A5
Woodport, N.J. (†07885) 273/D2
Woodridge, Ill. (†60517) 222/B6
Woodridge, Manitoba 179/G5
Wood-Ridge, N.J. (07075) 273/B2
Woodridge, N.Y. (12789) 276/L7
Wood River, Ill. (62095) 222/B2
Wood River, Nebr. (68883) 264/F4
Wood River Junction, R.I. (02894) 249/H7
Woodroffe (mt.), S. Australia 88/E5
Woodroffe (mt.), S. Australia 94/B2
Woodrow, Colo. (80757) 208/M3
Woodrow, Sask. 181/E6
Woodruff, Ariz. (85942) 198/E4
Woodruff (co.), Ark. 202/H3
Woodruff, Kansas (†67661) 232/C2
Woodruff, S.C. (29388) 296/D2
Woodruff, Utah (84086) 304/C2
Woodruff, Wis. (54568) 317/G4
Woods (lake) 146/J5
Woods (lake) 162/G6
Woods (lake), Ind. 227/E2
Woods (lake), Manitoba 179/H5
Woods (lake), Minn. 188/G1
Woods (lake), Minn. 255/D1
Woods (lake), Newf. 166/B3
Woods (lake), North. Terr. 88/E3
Woods (lake), North. Terr. 93/C4
Woods (co.), Okla. 288/J1
Woods (lake), Ontario 177/F5
Woods (lake), Ontario 175/B3
Woods (res.), Tenn. 237/J10
Woodsbend, Ky. (†41472) 237/P5
Woodsboro, Md. (21798) 245/J2
Woodsboro, Texas (78393) 303/G9
Woods Cross, Utah (84087) 304/B3
Woodsdale, N.C. (27595) 281/M2
Woodsfield, Ohio (43793) 284/H6
Woods Heights, Mo. (†64024) 261/S4
Woods Hole, Mass. (02543) 249/M6
Woodside, Calif. (94062) 204/J3
Woodside, Del. (19980) 245/R4
Woodside, Manitoba 179/D4
Woodside, Mont. (†59875) 262/B4
Woodside, S. Australia 94/C8
Woodside, Utah (†84501) 304/D4
Woodson, Ark. (72180) 202/F4
Woodson, Ill. (62695) 222/C4
Woodson (co.), Kansas 232/G4
Woodson, Texas (76091) 303/E5
Woodson Terrace, Mo. (†63101) 261/P2
Wood's Point, Victoria 97/D5
Woodstock, Ala. (35188) 195/D4
Woodstock○, Conn. (06281) 210/H1
Woodstock, England 13/F6
Woodstock, England 10/F5
Woodstock, Georgia (30188) 217/D2
Woodstock, Ill. (60098) 222/E1
Woodstock, Md. (21163) 245/L3
Woodstock, Minn. (56186) 255/B7
Woodstock, N. Br. 162/K6
Woodstock, New Bruns. 170/C2
Woodstock○, N.H. (03293) 268/D4
Woodstock, N.S. Wales 97/E3
Woodstock, N.Y. (12498) 276/M6
Woodstock, Ohio (43084) 284/C5
Woodstock, Ontario 177/D4
Woodstock, Vt. (05091) 268/B4
Woodstock○, Vt. (05091) 268/B4
Woodstock, Va. (22664) 307/L3
Woodstock Valley, Conn. (06282) 210/G1
Woodston, Kansas (67675) 232/C2
Woodstown, N.J. (08098) 273/C4
Woodsville, N.H. (03785) 268/C3
Wood Village, Oreg. (†97060) 291/B2
Woodville, Ala. (35776) 195/F1
Woodville, Conn. (†06777) 210/B2
Woodville, Fla. (32362) 212/B1
Woodville, Georgia (30670) 217/F3
Woodville, Mass. (01784) 249/H4
Woodville, Miss. (39669) 256/B8
Woodville, N.Y. (13698) 276/H3
Woodville, N. Zealand 100/F4
Woodville, N.C. (†27849) 281/P2
Woodville, Ohio (43469) 284/D3
Woodville, Okla. (73466) 288/N7
Woodville, Ontario 177/F3
Woodville, Pa. (†15106) 294/B7
Woodville, S. Australia 88/D7
Woodville, S. Australia 94/A7
Woodville, S.C. (†29669) 296/C2
Woodville, Texas (75979) 303/K7
Woodville, Va. (22749) 307/M3
Woodville, W. Va. (25572) 312/C6
Woodville, Wis. (54028) 317/B6
Woodward, Iowa (50276) 229/E5
Woodward (co.), Okla. 288/H2
Woodward, Okla. (73801) 288/H2
Woodward, S.C. (†29014) 296/E2
Woodwards Cove, New Bruns. 170/D4
Woodway, Va. (†24277) 307/C7
Woodway, Wash. (†98020) 310/C3
Woodworth, Ill. (†60953) 222/F3
Woodworth, La. (71485) 238/E4
Woodworth, N. Dak. (58496) 282/M5
Woody (mt.), Ariz. 198/D3
Woody, Calif. (93287) 204/G8
Woody (isl.), China 85/E2
Woody Creek, Colo. (81656) 208/F4
Woody Island, Alaska (†99615) 196/H3
Woody Island, Newf. 166/C2
Woody Point, Newf. 166/C4
Wool, England 13/E7
Wooldridge, Mo. (65287) 261/G5
Wooler, England 13/F2
Woolford, Alberta 182/D5
Woolford, Md. (21677) 245/O7
Woolgar, Queensland 95/B3
Woolgoolga, N.S. Wales 97/G2
Wooli, N.S. Wales 97/G1
Woollahra, N.S. Wales 88/L4
Woollahra, N.S. Wales 97/K3
Woollum, Ky. (40999) 237/O6
Woolrich, Pa. (17779) 294/H3
Woolsey, Georgia (30294) 217/D4
Woolsington, England 13/H3
Woolstock, Iowa (50599) 229/F3
Wooltana, S. Australia 88/F6
Wooltana, S. Australia 94/F4
Woolwich○, Maine (04579) 243/D8
Woolwine, Va. (24185) 307/H7
Woomera, Australia 87/D9
Woomera, S. Australia 88/F6
Woomera, S. Australia 94/E4
Woonsocket, R.I. (02895) 249/J4
Woonsocket, S. Dak. (57385) 298/N5
Wooramel, W. Australia 92/A4
Wooramel (riv.), W. Australia 88/A5
Wooramel (riv.), W. Australia 92/A4
Wooroloo, W. Australia 88/B2
Wooster, Ark. (72181) 202/F3
Wooster, Ohio (44691) 284/G4
Woosung, Ill. (61091) 222/D2
Wooton, Ky. (41776) 237/P6
Wootton Basset, England 13/E6
Woqooyi Galbeed (prov.), Somalia 115/H1
Worb, Switzerland 39/E3
Worcester, England 13/E5
Worcester, England 10/E4
Worcester (co.), Md. 245/S8
Worcester, Mass. 188/M2
Worcester (co.), Mass. 249/G3
Worcester, Mass. (*01601) 249/H3
Worcester, N.Y. (12197) 276/L5
Worcester, S. Africa 102/D8
Worcester, S. Africa 118/B6
Worcester○, Vt. (05682) 268/B3
Worden, Ark. (†72010) 202/H3
Worden, Ill. (62097) 222/B2
Worden, Kansas (†66006) 232/G3
Worden, Mont. (59088) 262/H5
Worden, Oreg. (†97601) 291/F5
Wordsworth, Sask. 181/J6
Wörgl, Austria 41/A3
Work (chan.), Br. Col. 184/C3
Workai (isl.), Indonesia 85/K7
Workington, England 13/D3
Workington, England 10/E3
Worksop, England 10/F4
Worksop, England 13/F4
Workum, Netherlands 27/G3
Worland, Mo. (†64752) 261/C6
Worland, Wyo. (82401) 319/E1
WORLD 2
Worley, Idaho (83876) 220/B2
Wormerveer, Netherlands 27/F4
Worms, W. Germany 22/C4
Woronoco, Mass. (01097) 249/C4
Woronora (riv.), N. S. Wales 88/K5
Woronora (riv.), N.S. Wales 97/J4
Worpswede, W. Germany 22/C2
Worsbrough, England 13/J2
Worsley, Alberta 182/A1
Worsley, England 13/H2
Worth (co.), Georgia 217/E8
Worth, Georgia (†31714) 217/E7
Worth, Ill. (60482) 222/B6
Worth (co.), Iowa 229/G2
Worth (co.), Mo. 261/D2
Worth, Mo. (†64456) 261/C2
Worth (lake), Texas 303/E2
Wortham, Texas (76693) 303/H6
Worthing, England 13/G7
Worthing, England 10/F5
Worthing, S. Dak. (57077) 298/R7
Worthington, Ind. (47471) 227/C6
Worthington, Iowa (52078) 229/L4
Worthington, Ky. (41183) 237/R3
Worthington○, Mass. (01098) 249/C3
Worthington, Minn. (56187) 255/C7
Worthington, Mo. (63567) 261/G2
Worthington, Ohio (43085) 284/E5
Worthington, Pa. (16262) 294/C4
Worthington, W. Va. (26591) 312/F4
Worthington Springs, Fla. (32697) 212/D2
Worthville, Ky. (41098) 237/L3
Worthville, N.C. (27378) 281/K3
Worthville, Pa. (15784) 294/D3
Worton, Md. (21678) 245/O3
Woss Lake, Br. Col. 184/D5
Wostok, Alberta 182/D3
Wotje (atoll), Marshall Is. 87/H5
Wottonville, Québec 172/F4
Wounded Knee, S. Dak. (57794) 298/D7
Wounded Knee (creek), S. Dak. 298/E7
Wour, Chad 111/C3
Wowoni (isl.), Indonesia 85/G6
Wragby, England 13/G4
Wrangel (isl.), U.S.S.R. 4/B18
Wrangel (isl.), U.S.S.R. 48/T2
Wrangell, Alaska (99929) 196/N2
Wrangell (cape), Alaska 196/H3
Wrangell (isl.), Alaska 196/N2
Wrangell (mts.), Alaska 196/K2
Wrangell-St. Elias Nat'l Preserve, Alaska 196/K2
Wrangell-St. Elias Nat'l Park, Alaska 196/K2
Wrangle, England 13/H4
Wrath (cape), Scotland 15/C2
Wrath (cape), Scotland 10/D1
Wray, Colo. (80758) 208/P2
Wray, Georgia (31798) 217/F7
Wreck Cove, Nova Scotia 168/H2
Wren, Ala. (†35650) 195/D2
Wren, Miss. (†39730) 256/G3
Wren, Ohio (45899) 284/A4
Wrens, Georgia (30833) 217/H4
Wrenshall, Minn. (55797) 255/F4
Wrentham, Alberta 182/D5
Wrentham○, Mass. (02093) 249/J4
Wrexham, Wales 13/E4
Wrexham, Wales 10/E4
Wright, Ala. (†35677) 195/C1
Wright, Ark. (72182) 202/F5
Wright (co.), Iowa 229/F3
Wright, Kansas (67882) 232/C4
Wright, La. (†70548) 238/F6
Wright (co.), Minn. 255/D5
Wright, Minn. (55798) 255/E4
Wright (co.), Mo. 261/H8
Wright (mt.), Québec 174/D2
Wright (lake), S. Australia 94/A2
Wright, Wyo. (82732) 319/G2
Wright Brothers Nat'l Mem., N.C. 281/T2
Wright City, Mo. (63390) 261/K5
Wright City, Okla. (74766) 288/R6
Wright Patman (lake), Texas 303/K4
Wright-Patterson Air Force Base, Ohio 284/B6
Wrights, Ill. (62098) 222/C4
Wrights, Pa. (†16743) 294/F2
Wrightstown, Minn. (†56453) 255/C4
Wrightstown, N.J. (08562) 273/D3
Wrightstown, Wis. (54180) 317/K7
Wrightsville, Ark. (72183) 202/F4
Wrightsville, Georgia (31096) 217/G5
Wrightsville, Pa. (17368) 294/J5
Wrightsville Beach, N.C. (28480) 281/O6
Wrightwood, Calif. (92397) 204/D10
Wrigley, Ky. (41477) 237/P4
Wrigley, N.W.T. 162/D3
Wrigley, N.W. Terrs. 187/F3
Wrigley, Tenn. (37098) 237/G9
Wrocław (prov.), Poland 47/C3
Wrocław, Poland 47/C3
Wrocław, Poland 7/F3
Wrong (lake), Manitoba 179/F2
Wroughton, England 13/F6
Wroxeter, Ontario 177/C4
Wroxton, Sask. 181/K4
Września, Poland 47/C2
Wschowa, Poland 47/C3
W. Scott Kerr (res.), N.C. 281/G2
Wuchang, China 77/L3
Wuchow (Wuzhou), China 77/H7
Wuchuan, Guizhou, China 77/G6
Wuchuan, Nei Monggol, China 77/H3
Wuchung (Wuzhong), China 77/G4
Wuda, China 77/G4
Wudaoliang, China 77/D5
Wuding, China 77/F6
Wudinna, S. Australia 88/E6
Wudinna, S. Australia 94/D5
Wudu, China 77/F5
Wugang, China 77/H6
Wuhai, China 77/G4
Wuhan, China 77/H5
Wuhan, China 54/N6
Wuhan, China 2/Q4
Wuhing (Wuxing), China 77/K5
Wuhu, China 77/J5
Wuhu, China 54/N6
Wu Jiang (riv.), China 77/G6
Wukari, Nigeria 106/F7
Wum, Cameroon 115/A2
Wun, India 68/D5
Wundowie, W. Australia 88/C2
Wundowie, W. Australia 92/B1
Wünnewil, Switzerland 39/D3
Wunnummin Lake, Ontario 175/C2
Wunsiedel, W. Germany 22/E3
Wunstorf, W. Germany 22/C2
Wupatki Nat'l Mon., Ariz. 198/D3
Wuppertal, W. Germany 22/B3
Wuqi, China 77/G4
Wuqia, China 77/A4
Würmsee (Starnbergersee) (lake), W. Germany 22/D5
Wurong, Queensland 95/B3
Wurtland, Ky. (41144) 237/R3
Wurtsboro, N.Y. (12790) 276/L7
Wurtsmith A.F.B., Mich. 250/F4
Würzburg, W. Germany 22/C4
Wurzen, E. Germany 22/E3
Wushi, China 77/A3
Wusih (Wuxi), China 77/J5
Wusuli Jiang (Ussuri) (riv.), China 77/M2
Wutai, China 77/H4
Wuwei, China 77/F4
Wuxi (Wusih), China 77/K5
Wuxi, China 54/O6
Wuxing (Wuhing), China 77/K5
Wuyang, China 77/H4
Wuyiling, China 77/L2
Wuyi Shan (range), China 77/J6
Wuyuan, China 77/G3
Wuzhong (Wuchung), China 77/G4
Wuzhou (Wuchow), China 77/H7
Wyaconda, Mo. (63474) 261/J2
Wyalkatchem, W. Australia 88/B6
Wyalkatchem, W. Australia 92/B5
Wyalla, S. Australia 88/F6
Wyalusing, Pa. (18853) 294/K2
Wyalusing, Wis. (†53801) 317/D10
Wyandanch, N.Y. (11798) 276/N9
Wyandot (co.), Ohio 284/D4
Wyandotte, Ind. (†47137) 227/E8
Wyandotte (co.), Kansas 232/H2
Wyandotte, Mich. (48192) 250/B7
Wyandotte, Okla. (74370) 288/S1
Wyandra, Queensland 95/C5
Wyanet, Ill. (61379) 222/D2
Wyangala (res.), N.S. Wales 97/E3
Wyarno, Wyo. (82845) 319/G2
Wyassup (lake), Conn. 210/H3
Wyatt, Ind. (46595) 227/E1
Wyatt, La. (†71251) 238/E2
Wyatt, Mo. (63882) 261/O9
Wycheproof, Victoria 97/B5
Wyckoff○, N.J. (07481) 273/B1
Wye (riv.), England 13/J2
Wye (riv.), England 13/D5
Wye (riv.), Wales 13/D5
Wye (riv.), Wales 10/E4
Wye Mills, Md. (21679) 245/O5
Wyeville, Wis. (54671) 317/F7
Wyk auf Föhr, W. Germany 22/C1
Wykoff, Minn. (55990) 255/F7
Wylie, Minn. (†56750) 255/B3
Wylie (lake), S.C. 296/E1
Wylie, Texas (75098) 303/H1
Wylliesburg, Va. (23976) 307/L7
Wyman, Iowa (†52621) 229/L6
Wyman (lake), Maine 243/C5
Wyman Dam, Maine (†04920) 243/D5
Wymark, Sask. 181/D5
Wymer, W. Va. (26297) 312/G5
Wymondham, England 13/J5
Wymore, Nebr. (68466) 264/H4
Wynberg, S. Africa 118/E6
Wynbring, S. Australia 94/C4
Wyndham, Australia 87/C7
Wyndham, N. Zealand 100/B7
Wyndham, W. Australia 88/D3
Wyndham, W. Australia 92/E1
Wyndmere, N. Dak. (58081) 282/R7
Wynigen, Switzerland 39/E2
Wynnburg, Tenn. (38077) 237/C8
Wynndel, Br. Col. 184/J5
Wynne, Ark. (72396) 202/J3
Wynne, Md. (†20680) 245/N8
Wynnewood, Okla. (73098) 288/M5
Wynnewood, Pa. (19096) 294/M6
Wynniatt (bay), N.W. Terrs. 187/G2
Wynnum, Queensland 95/E5
Wynnum, Queensland 88/L2
Wynona, Okla. (74084) 288/O1
Wynoochee (lake), Wash. 310/B3
Wynoochee (riv.), Wash. 310/B3
Wynot, Nebr. (68792) 264/G2
Wynyard, Sask. 162/F5
Wynyard, Sask. 181/G4
Wynyard, Tasmania 88/H8
Wynyard, Tasmania 99/B3
Wyocena, Wis. (53969) 317/H9
Wyola, Mont. (59089) 262/J5
Wyoming 188/E2
WYOMING 319
Wyoming, Del. (19934) 245/R4
Wyoming, Ill. (61491) 222/D2
Wyoming, Iowa (52362) 229/L4
Wyoming, Mich. (49509) 250/D6
Wyoming, Minn. (55092) 255/F5
Wyoming (co.), N.Y. 276/D5
Wyoming, N.Y. (14591) 276/D5
Wyoming, Ohio (45215) 284/C9
Wyoming, Ontario 177/B5
Wyoming (co.), Pa. 294/K2
Wyoming, Pa. (18644) 294/E7
Wyoming, R.I. (02898) 249/H6
Wyoming (state), U.S. 146/H5
Wyoming (co.), W. Va. 312/C7
Wyoming (peak), Wyo. 319/B3
Wyoming (range), Wyo. 319/B2
Wyomissing, Pa. (19610) 294/K5
Wyong, N.S. Wales 97/F3
Wyre (riv.), England 13/G1
Wyre (isl.), Scotland 15/F1
Wyrzysk, Poland 47/C2
Wysokie Mazowieckie, Poland 47/F2
Wysox, Pa. (18854) 294/K2
Wyszków, Poland 47/E2
Wythe (co.), Va. 307/F7
Wytheville, Va. (24382) 307/G7
Wytopitlock, Maine (04497) 243/G4
Wytopitlock (lake), Maine 243/G4

X

Xainza, China 77/C5
Xaitongmoin, China 77/C6
Xai-Xai, Mozambique 118/E5
Xai-Xai, Mozambique 102/F7
Xaltocan, Mexico 150/N1
Xangongo, Angola 115/C7
Xanten, W. Germany 22/B3
Xánthi, Greece 45/G5
Xapuri, Brazil 132/G10
Xar Moron He (riv.), China 77/J3
Xarrama (riv.), Portugal 33/B3
Xau (lake), Botswana 118/C4
Xavantina, Brazil 132/C6
Xcalak, Mexico 150/Q7
Xenia, Ill. (62899) 222/E5
Xenia, Ohio (45385) 284/C6
Xiadong, China 77/E3
Xiaguan (Siakwan), China 77/E6
Xiamen (Amoy), China 77/J7
Xiamen, China 54/N7
Xi'an (Sian), China 77/G5
Xi'an, China 54/M6
Xi'an, China 2/Q4
Xianfeng, China 77/G6
Xiangfan (Siangfan), China 77/H5
Xianghoang (plat.), Laos 72/D3
Xiang Jiang (riv.), China 77/H6
Xiangkhoang, Laos 72/D3
Xiangshan, China 77/K6
Xiangtan (Siangtan), China 77/H6
Xiangtan, China 54/N7
Xianyang (Sienyang), China 77/G5
Xiaogan, China 77/H5
Xiapu (Siapu), China 77/K6
Xichang (Sichang), China 77/F6
Xicoténcatl, Mexico 150/K5
Xicotepec de Juárez, Mexico 150/L6
Xicute, Colombia 126/E7
Xigazê (Shigatse), China 77/C6
Xigazê, China 54/K7
Xiji, China 77/G4
Xi Jiang (riv.), China 77/H7
Xilin, China 77/G7
Ximiao, China 77/F3
Xin Barag Zuoqi, China 77/J2
Xinghai, China 77/E4
Xingtai (Singtai), China 77/H4
Xingu (riv.), Brazil 120/D3
Xingu (riv.), Brazil 132/C3
Xingyi, China 77/G6
Xinhe (Toksu), China 77/B3
Xining (Sining), China 77/F4
Xining, China 54/M6
Xinjiang Uygur (Sinkiang-Uigur Aut. Reg.), China 77/B3
Xinjin, China 77/K4
Xintai, China 77/J4
Xin Xian, China 77/H4
Xinxiang (Sinsiang), China 77/H4
Xinyang (Sinyang), China 77/H5
Xinyi, China 77/J5
Xinyi He (riv.), China 77/J5
Xinyuan (Künes), China 77/B3
Xique-Xique, Brazil 132/F5
Xisha (isls.), China 85/E2
Xishui, China 77/G6
Xi Ujimqin, China 77/J3
Xiushui, China 77/H6
Xiuyan, China 77/K3
Xixia, China 77/H5
Xizang (Tibet Aut. Reg.), China 77/B5
Xochihuehuetlán, Mexico 150/K8
Xochimilco, Mexico 150/L1
Xochitlán, Mexico 150/N2
Xpujil, Mexico 150/P7
Xuanhan, China 77/G5
Xuan Loc, Vietnam 72/E5
Xuanwei, China 77/F6
Xuchang (Hsüchang), China 77/H5
Xuguit, China 77/K2
Xunke, China 77/L2
Xuwen, China 77/H7
Xuzhou (Süchow), China 77/J5
Xuzhou, China 54/N6

Y

Yaak, Mont. (†59935) 262/A2
Ya'an, China 77/F6
Yaapeet, Victoria 97/B4
Ya'bad, West Bank 65/C3
Yaballo, Ethiopia 111/G6
Yabassi, Cameroon 115/B3
Yabebyry, Paraguay 144/D5
Yabis, Wadi el (dry riv.), Jordan 65/D3
Yablis, Nicaragua 154/F4
Yablonovyy (range), U.S.S.R. 54/N4
Yablonovyy (range), U.S.S.R. 48/M4
Yabrud, West Bank 65/C4
Yabucoa, P. Rico 161/E2
Yachats, Oreg. (97498) 291/C3
Yacimientos de Río Turbio, Argentina 120/B8
Yaco, Bolivia 136/B5
Yacolt, Wash. (98675) 310/C5
Yacuiba, Bolivia 120/C5
Yacuiba, Bolivia 136/D7
Yacuma (riv.), Bolivia 136/B3
Yadgir, India 68/D5
Yadkin (co.), N.C. 281/H2
Yadkin (riv.), N.C. 281/J3
Yadkinville, N.C. (27055) 281/H2
Yad Mordekhai, Israel 65/A4
Yadong, China 77/D6
Yaeyama (isls.), Japan 81/K7
Yagoua, Cameroon 115/B1
Yagra, China 77/B5
Yagradagzê Shan (mt.), China 77/D4
Yaguachi Nuevo, Ecuador 128/B4
Yaguajay, Cuba 158/F2
Yaguaraparo, Venezuela 124/G2
Yaguarí, Uruguay 145/E2
Yaguarón, Paraguay 144/B5
Yaguarón (riv.), Uruguay 145/F3
Yaguarú, Bolivia 136/D4
Yaguas (riv.), Peru 128/G4
Yaguate, Dom. Rep. 158/E6
Yagüez (riv.), P. Rico 161/A2
Yagur, Israel 65/C2
Yahav, Israel 65/D5
Yahk, Br. Col. 184/J5
Yahuma, Zaire 115/D3
Yahyalı, Turkey 63/F3
Yaizu, Japan 81/J6
Yajalón, Mexico 150/N8
Yakima, Wash. 146/F5
Yakima, Wash. 188/B1
Yakima (co.), Wash. 310/E4
Yakima, Wash. (*98901) 310/E4
Yakima (ridge), Wash. 310/E4
Yakima (riv.), Wash. 310/F4
Yakima Ind. Res., Wash. 310/E4
Yako, Upper Volta 106/D6
Yakobi (isl.), Alaska 196/M1
Yakoma, Zaire 115/D3
Yaku (isl.), Japan 81/E8
Yakumo, Japan 81/J2
Yakut A.S.S.R., U.S.S.R. 48/N3
Yakutat, Alaska 188/D6
Yakutat, Alaska (99689) 196/L3
Yakutat (bay), Alaska 196/K3
Yakutsk, U.S.S.R. 54/O3
Yakutsk, U.S.S.R. 2/R2
Yakutsk, U.S.S.R. 48/N3
Yala, Thailand 72/D6
Yalaha, Fla. (32797) 212/E3
Yalata Aboriginal Reserve, S. Australia 88/E6
Yalata Aboriginal Res., S. Australia 94/B4
Yale, Br. Col. 184/M2
Yale (mt.), Colo. 208/G5
Yale (lake), Fla. 212/E3
Yale, Ill. (62481) 222/E4
Yale, Iowa (50277) 229/E5
Yale, Mich. (48097) 250/G5
Yale, Okla. (74085) 288/N2
Yale, S. Dak. (57386) 298/O5
Yale, Va. (23897) 307/O7
Yale (lake), Wash. 310/C4
Yalesville, Conn. (†06492) 210/D3
Yalgoo, W. Australia 92/B5
Yali, Estero (riv.), Chile 138/F4
Yalinga, Cent. Afr. Rep. 115/D2
Yallahs, Jamaica 158/K6
Yallock, N.S. Wales 97/C3
Yallourn, Victoria 97/D6
Yalmer, Mich. (†49885) 250/B2
Yalobusha (co.), Miss. 256/E2
Yalobusha (riv.), Miss. 256/E3
Yalong (riv.), China 54/M7
Yalong Jiang (riv.), China 77/F6
Yalova, Çanakkale, Turkey 63/B6
Yalova, Istanbul, Turkey 63/C2
Yalpunga, N.S. Wales 97/A1
Yalta, U.S.S.R. 7/H4
Yalta, U.S.S.R. 52/D6
Yalu (riv.) 54/O5
Yalu (riv.), China 77/L3
Yalu (riv.), N. Korea 81/C3
Yalutorsk, U.S.S.R. 48/G4
Yalvaç, Turkey 63/D3
Yalvaç, Turkey 59/B2
Yamachiche, Québec 172/E3
Yamagata (pref.), Japan 81/K4
Yamagata, Japan 81/K4
Yamaguchi (pref.), Japan 81/E6
Yamaguchi, Japan 81/E6
Yamal (pen.), U.S.S.R. 54/H2
Yamal (pen.), U.S.S.R. 4/B6
Yamal (pen.), U.S.S.R. 48/G2
Yamal-Nenets Aut. Okr., U.S.S.R. 48/H3
Yamama, Saudi Arabia 59/E5
Yamanashi (pref.), Japan 81/J6
Yamantau (mt.), U.S.S.R. 52/J4
Yamarna Aboriginal Reserve, W. Australia 88/C5
Yamarna Aboriginal Res., W. Australia 92/D4
Yamasá, Dom. Rep. 158/E6
Yamaska (co.), Québec 172/E3
Yamaska, Québec 172/E4
Yamaska (riv.), Québec 172/E4
Yamaska-Est, Québec 172/E4
Yamato, Japan 81/O2
Yamatokoriyama, Japan 81/J8

Yamatotakada, Japan 81/J8
Yamba, N.S. Wales 97/G1
Yambah, North. Terr. 93/C7
Yambio, Sudan 111/E7
Yambio, Sudan 102/E4
Yambol, Bulgaria 45/H4
Yambou (head), St. Vin. & Grens. 161/A9
Yambrasbamba, Peru 128/D5
Yamdena (isl.), Indonesia 85/J7
Yamethin, Burma 72/C2
Yamhill (co.), Oreg. 291/D2
Yamhill, Oreg. (97148) 291/D2
Y'Ami (isl.), Philippines 82/B2
Yamma Yamma (lake), Queensland 88/G5
Yamma Yamma (lake), Queensland 95/B5
Yampa, Colo. (80483) 208/F2
Yampa (riv.), Colo. 208/B2
Yamparaéz, Bolivia 136/C6
Yampi Sound, W. Australia 88/C3
Yampi Sound, W. Australia 92/C2
Yamsk, U.S.S.R. 48/Q4
Yamun, West Bank 65/C3
Yamuna (Jumna) (riv.), Pakistan 68/E3
Yamzho Yumco (lake), China 77/C6
Yan, Nigeria 106/G7
Yana, U.S.S.R. 54/P3
Yana (riv.), U.S.S.R. 4/C3
Yana (riv.), U.S.S.R. 48/O3
Yanac, Victoria 97/A5
Yanacachi, Bolivia 136/B5
Yanahuanca, Peru 128/D8
Yanam, India 68/E5
Yan'an (Yenan), China 77/G4
Yanaoca, Peru 128/G10
Yanaul, U.S.S.R. 52/J3
Yancannia, N.S. Wales 97/B2
Yancey, Ky. (†40831) 237/P7
Yancey (co.), N.C. 281/E3
Yanceyville, N.C. (27379) 281/L2
Yancheng, China 77/K5
Yanchi, China 77/G4
Yanco, N.S. Wales 97/D4
Yandé (isl.), New Caled. 86/F4
Yandeyarra Aboriginal Reserve, W. Australia 88/B4
Yandina, Solomon Is. 86/D3
Yandoon, Burma 72/B3
Yanfolila, Mali 106/C6
Yanga, Mexico 150/P2
Yangambi, Zaire 115/D3
Yangambi, Zaire 102/E4
Yangcheng, China 77/H4
Yangchow (Yangzhou), China 77/J5
Yangchüan (Yangquan), China 77/H4
Yangchun, China 77/H7
Yangdŏk, N. Korea 81/C4
Yanggao, China 77/H3
Yanggu, S. Korea 81/C4
Yangjiang, China 77/H7
Yangquan (Yangchüan), China 77/H4
Yangshan, China 77/H7
Yang Sin, Chu (mt.), Vietnam 72/F4
Yangtze (riv.), China 54/N6
Yangtze (riv.), China 2/Q4
Yangtze (Chang Jiang) (riv.), China 77/K5
Yangyang, S. Korea 81/D4
Yangzhou (Yangchow), China 77/J5
Yanhuqu, China 77/B5
Yanji (Yenki), China 77/L3
Yankee Fork, Salmon (riv.), Idaho 220/D5
Yankee Lake, Ohio (†44403) 284/J3
Yankeetown, Fla. (32698) 212/D2
Yankeetown, Ind. (†47630) 227/C9
Yanko (creek), N.S. Wales 97/C4
Yankton, S. Dak. 188/G2
Yankton (co.), S. Dak. 298/P7
Yankton, S. Dak. (57078) 298/P8
Yanqi, China 77/C3
Yanrey, W. Australia 92/A3
Yantabulla, N.S. Wales 97/C1
Yantai (Chefoo), China 77/K4
Yantai, China 54/O6
Yantara, N.S. Wales 97/B1
Yantara (lake), N.S. Wales 97/B1
Yantic, Conn. (06389) 210/G2
Yantic (riv.), Conn. 210/G2
Yantis, Texas (75497) 303/J5
Yantley, Ala. (36924) 195/B6
Yanush, Okla. (†74574) 288/R5
Yao, Japan 81/J8
Yaoundé (cap.), Cameroon 2/K5
Yaoundé (cap.), Cameroon 102/D4
Yaoundé (cap.), Cameroon 115/B3
Yap (isl.), Micronesia 87/D5
Yapacani (riv.), Bolivia 136/C5
Yapei, Ghana 106/D7
Yapen (isl.), Indonesia 85/K6
Yapen (str.), Indonesia 85/K6
Yapraklı, Turkey 63/E2
Yaque del Norte (riv.), Dom. Rep. 158/D5
Yaque del Sur (riv.), Dom. Rep. 158/D6
Yaqui, Mexico 150/D3
Yaqui (riv.), Mexico 146/H7
Yaqui (riv.), Mexico 150/E2
Yaquina, Oreg. (†97365) 291/C3
Yara, Cuba 158/H4
Yaracuy (state), Venezuela 124/D2
Yaraka, Queensland 95/C5
Yaraligöz Daği (mt.), Turkey 59/B1
Yaraligöz Daği (mt.), Turkey 63/F2
Yaransk, U.S.S.R. 52/G3
Yarbo, Ala. (†36558) 195/B7
Yarbo, Sask. 181/K5
Yarboutenda, Senegal 106/B6
Yarda, Chad 111/C4
Yardley, Pa. (19067) 294/N5
Yardville, N.J. (08620) 273/D3
Yare (riv.), England 13/J5
Yare (riv.), England 10/G4
Yarega, U.S.S.R. 52/H2
Yaretas de Vizcachas, Cerro (mt.), Chile 138/G3
Yarí, Colombia 126/D7
Yarí (riv.), Colombia 126/D8
Yarim, Yemen Arab Rep. 59/D7
Yaritagua, Venezuela 124/D2
Yarkand (Shache), China 77/A4
Yarkand (riv.), China 54/K6
Yarkant He (riv.), China 77/A4
Yarker, Ontario 177/H3
Yarle (lakes), S. Australia 94/B4
Yarmouth, Iowa (52660) 229/L6
Yarmouth, Maine (04096) 243/C8
Yarmouth○, Maine (04096) 243/C8
Yarmouth○, Mass. (02675) 249/O6
Yarmouth, N.S. 162/K7
Yarmouth (co.), Nova Scotia 168/C5
Yarmouth, Nova Scotia 168/B5
Yarmouth (sound), Nova Scotia 168/B5
Yarmouth Port, Mass. (02675) 249/N6
Yarmuk (riv.), Israel 65/D2
Yarnell, Ariz. (85362) 198/C4
Yaroslavl', U.S.S.R. 7/H3
Yaroslavl', U.S.S.R. 48/D4
Yaroslavl', U.S.S.R. 52/E3
Yarqon (riv.), Israel 65/B3
Yarra (riv.), Victoria 97/C5
Yarra (riv.), Victoria 88/L6
Yarram, Victoria 97/D6
Yarrawonga, Victoria 97/C5
Yarrow, Br. Col. 184/M3
Yarrow, Mo. (†63501) 261/G2
Yarrow (riv.), Scotland 15/E5
Yarrowitch, N.S. Wales 97/F2
Yarrow Point, Wash. (†98004) 310/B2
Yartsevo, U.S.S.R. 48/J4
Yartsevo, U.S.S.R. 52/D3
Yarumal, Colombia 126/C4
Yaruu, Mongolia 77/E2
Yas (isl.), U.A.E. 59/F5
Yasawa Group (isls.), Fiji 86/P10
Yásica Abajo, Dom. Rep. 158/E5
Yasin, Pakistan 59/K2
Yasin, Pakistan 68/C1
Yasnyy, U.S.S.R. 52/J4
Yasothon, Thailand 72/D4
Yass, N.S. Wales 97/E4
Yasuj, Iran 66/G5
Yasun (cape), Turkey 63/G2
Yata (riv.), Bolivia 136/C3
Yatabe, Japan 81/P2
Yatağan, Turkey 63/C4
Yataity, Paraguay 144/C5
Yateley, England 13/G8
Yates (dam), Ala. 195/G5
Yates (co.), N.Y. 276/F5
Yates Center, Kansas (66783) 232/G4
Yates City, Ill. (61572) 222/C3
Yatesville, Georgia (31097) 217/D5
Yathkyed (lake), N.W.T. 162/F3
Yathkyed (lake), N.W. Terrs. 187/J3
Yatina, Bolivia 136/C7
Yatsushiro, Japan 81/E7
Yatta, West Bank 65/C5
Yatton, England 13/E6
Yatua (riv.), Venezuela 124/E7
Yauca, Peru 128/E10
Yauco, P. Rico 161/B2
Yauco, P. Rico 156/F1
Yauco (lake), P. Rico 161/B2
Yauli, Peru 128/D8
Yaúna Moloca, Colombia 126/E8
Yaupi, Ecuador 128/D4
Yaupon Beach, N.C. (†28461) 281/N7
Yauri, Peru 128/G10
Yautepec, Mexico 150/L2
Yauyos, Peru 128/E9
Yava, Ariz. (†86301) 198/C4
Yavapai (co.), Ariz. 198/C4
Yavapai Ind. Res., Ariz. 198/C4
Yavaraté, Colombia 126/F7
Yavarí (riv.) 120/B3
Yavarí (riv.), Peru 128/G5
Yavaros, Mexico 150/E3
Yavero (riv.), Peru 128/F9
Yavita, Venezuela 124/E6
Yavne, Israel 65/B4
Yavne'el, Israel 65/C2
Yawata, Japan 81/J7
Yawatahama, Japan 81/F7
Yawkey, W. Va. (25573) 312/C6
Yawri (bay), S. Leone 106/B7
Ya Xian, China 77/G8
Yaxley, England 13/G5
Yayladağı, Turkey 63/F5
Yazd (governorate), Iran 66/J5
Yazd (Yezd), Iran 66/J5
Yazd, Iran 59/F3
Yazd, Iran 54/G6
Yazdan, Iran 66/M4
Yazdan, Iran 59/H3
Yazd-e Khvast, Iran 66/H5
Yazoo (co.), Miss. 265/D5
Yazoo (riv.), Miss. 188/H4
Yazoo (riv.), Miss. 256/C5
Yazoo City, Miss. (39194) 256/D5
Ybbs an der Donau, Austria 41/C2
Ybycuí, Paraguay 144/B5
Ybytymí, Paraguay 144/B5
Yding Skovhøj (mt.), Denmark 21/C6
Ye, Burma 72/C4
Yea, Victoria 97/C5
Yeaddiss, Ky. (41777) 237/P6
Yeadon, Pa. (19050) 294/N7
Yeager, Okla. (†74848) 288/O4
Yeagertown, Pa. (17099) 294/G4
Yebbi-Bou, Chad 111/C3
Yecheng, China 77/A4
Yecla, Spain 33/F3
Yécora, Mexico 150/E2
Yecuatla, Mexico 150/P1
Yeddo, Ind. (†47952) 227/C4
Yeeda River, W. Australia 92/C2
Yeelirrie, W. Australia 92/C4
Yefremov, U.S.S.R. 52/E4
Yegros, Paraguay 144/D5
Yeguas (pt.), P. Rico 161/F3
Yehualtepec, Mexico 150/O2
Yehud, Israel 65/B3
Yei, Sudan 111/F7
Yelabuga, U.S.S.R. 52/H3
Yelan', U.S.S.R. 52/F4
Yelcho (lake), Chile 138/E4
Yelets, U.S.S.R. 7/H3
Yelets, U.S.S.R. 48/D4
Yelets, U.S.S.R. 52/E4
Yélimané, Mali 106/B5
Yelizavety (cape), U.S.S.R. 54/R4
Yelizavety (cape), U.S.S.R. 48/P4
Yelizovo, U.S.S.R. 48/Q4
Yell (co.), Ark. 202/D3
Yell (isl.), Scotland 15/G2
Yell (isl.), Scotland 10/G1
Yell (sound), Scotland 15/G2
Yellamanchili, India 68/E5
Yelleq, Jebel (mt.), Egypt 59/B3
Yellow (sea) 54/O6
Yellow (Huang He) (riv.), China 77/J4
Yellow (sea), China 77/K4
Yellow (creek), Colo. 208/C3
Yellow (riv.), Fla. 212/B6
Yellow (riv.), Ind. 227/D2
Yellow (brook), N.J. 273/E3
Yellow (sea), N. Korea 81/B6
Yellow (creek), Ohio 284/J4
Yellow (sea), S. Korea 81/B6
Yellow (creek), Tenn. 237/F8
Yellow (creek), W. Va. 312/E1
Yellow (lake), Wis. 317/B4
Yellow (riv.), Wis. 317/F7
Yellow Bluff, Ala. (†36769) 195/C7
Yellowbud, Ohio (†45601) 284/D7
Yellowcreek, N.C. (†28771) 281/A4
Yellow Creek, Sask. 181/F3
Yellow Dog (riv.), Mich. 250/B2
Yellow Grass, Sask. 181/H6
Yellowhead (pass), Alberta 182/A3
Yellowhead (pass), Br. Col. 184/H4
Yellow Jacket, Colo. (81335) 208/B7
Yellowknife, Canada 4/C15
Yellowknife, Canada 2/D2
Yellowknife, N.W.T. 146/G3
Yellowknife (cap.), N.W.T. 162/E3
Yellowknife (cap.), N.W. Terrs. 187/G3
Yellowknife (riv.), N.W. Terrs. 187/G3
Yellow Medicine (co.), Minn. 255/B6
Yellow Pine, Ala. (36588) 195/B8
Yellow Pine, Idaho (83677) 220/C4
Yellow Pine, La. (†71039) 238/D2
Yellow Spring, W. Va. (26865) 312/J4
Yellow Springs, Md. (†21701) 245/H3
Yellow Springs, Ohio (45387) 284/C6
Yellowstone (riv.) 188/E1
Yellowstone (co.), Mont. 262/H4
Yellowstone (riv.), Mont. 262/M3
Yellowstone (riv.), N. Dak. 282/B4
Yellowstone (riv.), U.S. 146/H5
Yellowstone (lake), Wyo. 188/E2
Yellowstone (lake), Wyo. 319/B1
Yellowstone (riv.), Wyo. 319/B1
Yellowstone Nat'l Park, Idaho 262/F6
Yellowstone Nat'l Park, Mont. 262/F6
Yellowstone Nat'l Park, Wyo. (82190) 31/B1
Yellowstone Nat'l Park, Wyo. 188/E2
Yellowstone Nat'l Park, Wyo. 319/B1
Yellville, Ark. (72687) 202/E1
Yelm, Wash. (98597) 310/C4
Yelverton (bay), N.W. Terrs. 187/K1
Yelwa, Nigeria 106/F6
Yemassee, S.C. (29945) 296/F6
Yemen, People's Dem. Rep. of 2/M5
Yemen, People's Democratic Republic of 54/F8
YEMEN, PEOPLE'S DEM. REPUBLIC OF, 59/E7
YEMEN ARAB REP. 59/D7
Yemen Arab Republic 2/M5
Yemen Arab Republic 54/F8
Yemetsk, U.S.S.R. 52/F2
Yenakiyevo, U.S.S.R. 52/E5
Yenan (Yan'an), China 77/G4
Yenangyaung, Burma 72/B2
Yen Bai, Vietnam 72/E2
Yenbo, Saudi Arabia 54/E7
Yenbo, Saudi Arabia 59/C5
Yenda, N.S. Wales 97/D4
Yendi, Ghana 106/D7
Yengisar, China 77/A4
Yenice, Çanakkale, Turkey 63/B3
Yenice, İçel, Turkey 63/F4
Yenice, Zonguldak, Turkey 63/E2
Yeniceoba, Turkey 63/E3
Yeniköy, Çanakkale, Turkey 63/B6
Yeniköy, Çanakkale, Turkey 63/C5
Yeniköy, Istanbul, Turkey 63/D6
Yenimahalle, Turkey 63/E3
Yenişehir, Turkey 63/C2
Yenisey (riv.), U.S.S.R. 4/C5
Yenisey (riv.), U.S.S.R. 2/P2
Yenisey (riv.), U.S.S.R. 54/K3
Yenisey (riv.), U.S.S.R. 48/J3
Yeniseysk, U.S.S.R. 54/L4
Yeniseysk, U.S.S.R. 48/K4
Yenki (Yanji), China 77/L3
Yen Minh, Vietnam 72/E2
Yentai (Yantai), China 77/K4
Yentna (riv.), Alaska 196/A1
Yeo (lake), W. Australia 88/D5
Yeo (lake), W. Australia 92/D5
Yeola, India 68/C4
Yeoman, Ind. (47997) 227/D3
Yeotmal, India 68/D4
Yeoval, N.S. Wales 97/E3
Yeovil, England 10/E5
Yeovil, England 13/E7
Yeppoon, Queensland 95/D4
Yeppoon, Queensland 88/J4
Yerevan (Erivan), U.S.S.R. 52/F6
Yerichaña, Venezuela 124/F5
Yerington, Nev. (89447) 266/B4
Yerington Ind. Res., Nev. 266/B3
Yerkesik, Turkey 63/C4
Yerköy, Turkey 63/F3
Yerlisu, Turkey 63/C5
Yermak, U.S.S.R. 48/H4
Yermentau, 48/H4
Yermo, Calif. (92398) 204/J9
Yeroham, Israel 65/B6
Yerolimin, Greece 45/F7
Yeronga, Queensland 95/D3
Yeronga, Queensland 88/K3
Yerseke, Netherlands 27/E6
Yershov, U.S.S.R. 52/G4
Yesagyo, Burma 72/B2
Yeshbum, P.D.R. Yemen 59/E7
Yesil', 48/G4
Yeşilhisar, Turkey 63/F3
Yeşilırmak (riv.), Turkey 63/G2
Yeşilköy, Turkey 63/D6
Yeşilova, Burdur, Turkey 63/C4
Yeşilova, Niğde, Turkey 63/E3
Yeşilyurt, Turkey 63/H3
Yeso, N. Mex. (88136) 274/E4
Yeso (creek), N. Mex. 274/E4
Yesodot, Israel 65/B4
Yessentuki, U.S.S.R. 52/F6
Yessey, U.S.S.R. 48/L3
Yeste, Spain 33/E3
Yesud Hama'ala, Israel 65/D1
Yetholm, Scotland 15/F5
Yetman, N.S. Wales 97/F1
Yettem, Calif. (93670) 204/F7
Yetter, Iowa (51433) 229/D4
Ye-u, Burma 72/B2
Yeu (isl.), France 28/B4
Yevlakh, U.S.S.R. 52/G6
Yevpatoriya, U.S.S.R. 52/D5
Ye Xian, China 77/K4
Yeysk, U.S.S.R. 52/E5
Ygatimí, Paraguay 144/E4
Yhú, Paraguay 144/E4
Yí (riv.), Uruguay 145/B4
Yialousa, Cyprus 63/F5
Yiannitsá, Greece 45/F5
Yibin (Ipin), China 77/F6
Yibin, China 54/M7
Yibug Caka (lake), China 77/C5
Yichang (Ichang), China 77/H5
Yichun, Heilongjiang, China 77/L2
Yichun, Jiangxi, China 77/H6
Yidu, Hubei, China 77/H5
Yidu, Shandong, China 77/J4
Yiftah, Israel 65/D1
Yığılca, Turkey 63/D2
Yıldızeli, Turkey 63/G3
Yiliang, China 77/F7
Yinchuan (Ningsia, Yinchwan), China 77/G4
Yinchuan, China 54/M6
Yingjiang, China 77/E7
Yingkou (Yinkow), China 77/K3
Yingshan, Hubei, China 77/H5
Yingshan, Sichuan, China 77/G5
Yining, China 77/B3
Yining, China 54/K5
Yinjiang, China 77/G6
Yin Shan (mts.), China 77/G3
Yirga-Alam, Ethiopia 102/F4
Yirga Alam, Ethiopia 111/G6
Yirka, Israel 65/C2
Yirol, Sudan 111/F6
Yirrkala, North. Terr. 93/E2
Yishan, China 77/G7
Yíthion, Greece 45/F7
Yiwu (Aratürük), China 77/D3
Yiyang, China 77/H6
Ylikitka (lake), Finland 18/Q3
Ylitornio, Finland 18/O3
Ylivieska, Finland 18/O4
Ymir, Br. Col. 184/J5
Ynys Môn (Anglesey) (ridge), Wales 13/C4
Ynys Môn (Anglesey) (isl.), Wales 10/D4
Yoakum (co.), Texas 303/B4
Yoakum, Texas (77995) 303/G8
Yocalla, Bolivia 136/B6
Yocemento, Kansas (†67601) 232/C3
Yockanookany (riv.), Miss. 256/E5
Yoco, Venezuela 124/G2
Yocón, Honduras 154/D3
Yocum, Ky. (41478) 237/P5
Yoder, Colo. (80864) 208/L5
Yoder, Ind. (46798) 227/G3
Yoder, Kansas (67585) 232/E4
Yoder, Wyo. (82244) 319/H4
Yodo (riv.), Japan 81/J7
Yog (pt.), Philippines 82/E3
Yogyakarta, Indonesia 54/M10
Yogyakarta, Indonesia 85/J2
Yoho Nat'l Park, Br. Col. 162/E5
Yoho Nat'l Park, Br. Col. 184/J4
Yoichi, Japan 81/K2
Yojoa (lake), Honduras 154/D3
Yokadouma, Cameroon 115/B3
Yokawa, Japan 81/H7
Yokena, Miss. (†39180) 256/C6
Yokkaichi, Japan 81/H6
Yoko, Cameroon 115/B2
Yokohama, Japan 2/R4
Yokohama, Japan 81/O3
Yokohama, Japan 54/R6
Yokosuka, Japan 81/O3
Yokote, Japan 81/K4
Yola, Nigeria 106/G7
Yola, Nigeria 102/D4
Yolla, Tasmania 99/B3
Yolo (co.), Calif. 204/D5
Yolo, Calif. (95697) 204/B8
Yolyn, W. Va. (25654) 312/C7
Yona, Guam 86/K7
Yonago, Japan 81/F6
Yonaguni (isl.), Japan 81/K7
Yoncalla, Oreg. (97499) 291/D4
Yonezawa, Japan 81/K5
Yongamp'o, N. Korea 81/B4
Yongchang, China 77/F4
Yŏngch'ŏn, S. Korea 81/D6
Yongchuan, China 77/G6
Yongdeng, China 77/F4
Yŏngdŏk, S. Korea 81/D5
Yonges Island, S.C. (29494) 296/G6
Yonghe, China 77/H4
Yŏnghŭng, N. Korea 81/C4
Yŏngju, S. Korea 81/D5
Yongning, China 77/G4
Yongren, China 77/F6
Yongxin, China 77/H6
Yongxing, China 77/H6
Yonkers, Georgia (†31014) 217/F6
Yonkers, N.Y. (*10701) 276/O6
Yonne (dept.), France 28/E4
Yonne (riv.), France 28/E3
Yono, Japan 81/O2
Yopal, Colombia 126/D5
Yorba Linda, Calif. (92686) 204/D11
Yorito, Honduras 154/D3
York, Ala. (36925) 195/B6
York (cape), Australia 2/S6
York (cape), Australia 87/E7
York, England 13/F4
York, England 10/F4
York (cape), Greenl. 4/B13
York, Ky. (41184) 237/P3
York (co.), Maine 243/B9
York, Maine (03909) 243/B9
York○, Maine (03909) 243/B9
York (co.), Nebr. 264/G4
York, Nebr. (68467) 264/G4
York (co.), New Bruns. 170/C3
York, N.Y. (14592) 276/E5
York, N. Dak. (58386) 282/L3
York (reg. munic.), Ontario 177/E4
York, Ontario 177/J4
York, Pa. 188/L3
York (co.), Pa. 294/J6
York, Pa. (*17401) 294/J6
York (riv.), Québec 172/D1
York (cape), Queensland 88/G2
York (cape), Queensland 95/B1
York (co.), S.C. 296/E2
York, S.C. (29745) 296/E1
York (co.), Va. 307/P6
York (riv.), Va. 307/P6
York, W. Australia 88/B6
York, W. Australia 92/B1
York (sound), W. Australia 88/C2
York (sound), W. Australia 92/D1
York, Wis. (†54758) 317/D7
York Beach, Maine (03910) 243/B9
Yorke (pen.), S. Australia 88/F7
Yorke (pen.), S. Australia 94/E6
Yorketown, S. Australia 88/F6
Yorketown, S. Australia 94/E6
York Factory, Man. 162/G4
York Factory, Manitoba 179/K2
York Harbor, Maine (03911) 243/B9
York Haven, Pa. (17370) 294/J5
York Landing, Man. 146/J4
York Landing, Manitoba 179/J2
Yorklyn, Del. (19736) 245/R1
Yorkshire, North (co.), England 13/F3
Yorkshire, South (co.), England 13/F4
Yorkshire, West (co.), England 13/J1
Yorkshire, N.Y. (14173) 276/D5
Yorkshire, Ohio (45388) 284/B5
Yorkshire Dales National Park, England 13/E3
York Springs, Pa. (17372) 294/H6
Yorkton, Sask. 146/H4
Yorkton, Sask. 162/F5
Yorkton, Sask. 181/J4
Yorktown, Ark. (71678) 202/G5
Yorktown, Ind. (47396) 227/G4
Yorktown, Iowa (51656) 229/C7
Yorktown, N.J. (†08098) 273/C4
Yorktown, Texas (78164) 303/G9
Yorktown, Va. (23690) 307/R6
Yorktown Heights, N.Y. (10598) 276/N8
Yorkville, Ill. (60560) 222/E2
Yorkville, Ind. (†47022) 227/H6
Yorkville, N.Y. (13495) 276/K4
Yorkville, Ohio (43971) 284/J5
Yorkville, Tenn. (38389) 237/C8
Yoro, Honduras 154/D3
Yoron (isl.), Japan 81/N6
Yorosso, Mali 106/C6
Yosemite, Ky. (42566) 237/M6
Yosemite National Park, Calif. (95389) 204/F6
Yosemite Nat'l Park, Calif. 188/C3
Yosemite Nat'l Park, Calif. 204/F6
Yoshino (riv.), Japan 81/G6
Yoshino-Kumano National Park, Japan 81/H7
Yoshkar-Ola, U.S.S.R. 7/J3
Yoshkar-Ola, U.S.S.R. 52/G3
Yoshkar-Ola, U.S.S.R. 48/E4
Yost, Utah (†84329) 304/A2
Yŏsu, S. Korea 81/C6
Yotala, Bolivia 136/C6
Yotaú, Bolivia 136/D5
Yotvata, Israel 65/D5
Youanmi, W. Australia 92/B5
Youbou, Br. Col. 184/J3
Youghal, Ireland 10/B5
Youghal, Ireland 17/F8
Youghal (bay), Ireland 10/C5
Youghal (bay), Ireland 17/F8
Youghiogheny (riv.), Md. 245/A3
Youghiogheny (dam), Pa. 294/D6
Youghiogheny River (lake), Md. 245/A2
Youghiogheny River (lake), Pa. 294/D6
Young, Ariz. (85554) 198/D4
Young, N.S. Wales 88/H6
Young, N.S. Wales 97/E4
Young (cape), N. Zealand 100/D7
Young (mt.), North. Terr. 93/D3
Young, Sask. 181/F4
Young (co.), Texas 303/F4
Young, Uruguay 145/B3
Young America, Ind. (46998) 227/E3
Young America, Minn. (55397) 255/E6
Youngcane, Georgia (†30512) 217/D1
Young Cove, Nova Scotia 168/C4
Young Harris, Georgia (30582) 217/E1
Youngs Cove, New Bruns. 170/E3
Youngs Creek, Ind. (†47454) 227/D8
Youngs Creek, Ky. (†40759) 237/N7
Youngstown, Alberta 182/E4
Youngstown, Fla. (32466) 212/D6
Youngstown, Ind. (†47808) 227/C6
Youngstown, N.Y. (14174) 276/C4
Youngstown, Ohio (*44501) 284/J3
Youngstown, Ohio 188/K2
Youngsville, La. (70592) 238/G6
Youngsville, N. Mex. (87064) 274/C2
Youngsville, N.Y. (12791) 276/L7
Youngsville, N.C. (27596) 281/N2
Youngsville, Pa. (16371) 294/D2
Youngtown, Ariz. (85363) 198/C5
Youngwood, Pa. (15697) 294/D5
Yountville, Calif. (94599) 204/C5
Youshashan, China 77/D4
Youssoufia, Morocco 106/C2
Youyang, China 77/G6
Yozgat (prov.), Turkey 63/F3
Yozgat, Turkey 59/B2
Yozgat, Turkey 63/F3
Ypacaraí, Paraguay 144/B5
Ypané, Paraguay 144/B5
Ypané (riv.), Paraguay 144/D3
Ypé Jhú, Paraguay 144/E3
Ypoá (lake), Paraguay 144/B5
Ypres (Ieper), Belgium 27/B7
Ypsilanti, Georgia (†31827) 217/D5
Ypsilanti, Mich. (48197) 250/F6
Ypsilanti, N. Dak. (58497) 282/N6
Yreka, Calif. 188/B2
Yreka, Calif. (96097) 204/C2
Yser (riv.), Belgium 27/B7
Yssingeaux, France 28/F5
Ystad, Sweden 18/H9
Ystradgynlais, Wales 13/D6
Ythan (riv.), Scotland 15/F3
Yuan (riv.), China 54/M7
Yuan Jiang (riv.), China 77/H6
Yuanling, China 77/G6
Yuanmou, China 77/F6
Yuanping, China 77/H4
Yuba (co.), Calif. 204/D4
Yuba (riv.), Calif. 204/D4
Yuba, Okla. (†74721) 288/O7
Yuba, Wis. (54672) 317/F8
Yuba City, Calif. (95991) 204/D4
Yubari, Japan 81/L2
Yubetsu, Japan 81/L1
Yucaipa, Calif. (92399) 204/J9
Yucatán (chan.) 146/K7
Yucatán (state), Mexico 150/P6
Yucatán (pen.), Mexico 146/K7
Yucatán (pen.), Mexico 150/P7
Yucca, Ariz. (86438) 198/A4
Yucca Flat (basin), Nev. 266/E6
Yucca House Nat'l Mon., Colo. 208/B8
Yucca Valley, Calif. (92284) 204/J9
Yuci (Yütze), China 77/H4
Yudu, China 77/J6
Yuendumu, North. Terr. 93/B7
Yuexi, China 77/F6
Yueyang, China 77/H6
Yug (riv.), U.S.S.R. 52/G2
Yugorskiy (pen.), U.S.S.R. 52/K1
Yugoslavia 2/K3
Yugoslavia 7/F4
YUGOSLAVIA 45/C3
Yuhuan (isl.), China 77/K6
Yukon (riv.) 2/B2
Yukon (riv.) 146/C3
Yukon (riv.) 4/C17
Yukon (riv.), Alaska 188/C5
Yukon (riv.), Alaska 196/F2
Yukon, Mo. (65589) 261/J8
Yukon, Okla. (73099) 288/L3
Yukon, Pa. (15698) 294/C5
Yukon (riv.), Yukon 162/C3
Yukon (riv.), Yukon 187/E3
Yukon-Charley Rivers Nat'l Preserve, Alaska 196/K2
Yukon Territory 162/C3
Yukon Territory (terr.), Canada 146/E3
YUKON TERRITORY 187
Yüksekova, Turkey 63/L4
Yukuhashi, Japan 81/E7
Yule (riv.), W. Australia 92/B3
Yulee, Fla. (32097) 212/E1
Yuli (Lopnur), China 77/C3
Yulin, Guangxi Zhuangzu, China 77/G7
Yulin, Shanxi, China 77/G4
Yuma, Ariz. 188/D4
Yuma, Ariz. 146/G6
Yuma (co.), Ariz. 198/A5
Yuma, Ariz. (85364) 198/A6
Yuma (des.), Ariz. 198/A6
Yuma (co.), Colo. 208/P2
Yuma, Colo. (80759) 208/O2
Yuma (bay), Dom. Rep. 158/F6
Yuma, Tenn. (38390) 237/E9
Yuma Ind. Res., Calif. 204/L11
Yuma Marine Corps Air Sta., Ariz. 198/A6
Yuma Proving Ground, Ariz. 198/A6
Yumbel, Chile 138/E1
Yumbo, Colombia 126/B6
Yumen, China 77/E4
Yumen, China 54/L6
Yumenzhen, China 77/E3
Yumin, China 77/B2
Yumurtalık, Turkey 63/F4
Yuna (riv.), Dom. Rep. 158/E5
Yuna, W. Australia 92/A5
Yunak, Turkey 63/D3
Yunan, China 77/H7
Yunaska (isl.), Alaska 196/D4
Yuncheng, China 77/H4
Yungas, Las (reg.), Bolivia 136/B5
Yungay, Chile 138/E1
Yungkia (Wenzhou), China 77/J6
Yunguyo, Peru 128/H11
Yunnan (prov.), China 77/F7
Yunta, S. Australia 94/F5
Yunxi, China 77/G5
Yunxiao, China 77/J7
Yunyang, China 77/G5
Yupukari, Guyana 131/B4

Yura, Bolivia 136/B7
Yuraguanal, Cuba 158/G2
Yurga, U.S.S.R. 48/J4
Yurimaguas, Peru 128/E5
Yuruá (riv.), Peru 128/F7
Yuruari (riv.), Venezuela 124/H4
Yurungkax He (riv.), China 77/A4
Yur'yevets, U.S.S.R. 52/F3
Yuscarán, Honduras 154/D4
Yushan (isls.), China 77/K6
Yü Shan (mt.), China 77/K7
Yushu, Jilin, China 77/L3
Yushu, Qinghai, China 77/E5
Yusufeli, Turkey 63/J2
Yutan, Nebr. (68073) 264/H3
Yutian, Hebei, China 77/J4
Yutian, Xinjiang Uygur, China 77/B4
Yuty, Paraguay 144/D5
Yütze (Yuci), China 77/H4
Yuxi, China 77/F7
Yu Xian, China 77/H4
Yuzawa, Japan 81/K4
Yuzhno-Kuril'sk, U.S.S.R. 48/P5
Yuzhno-Sakhalinsk, U.S.S.R. 54/R5
Yuzhno-Sakhalinsk, U.S.S.R. 48/P5
Yvelines (dept.), France 28/D3
Yverdon, Switzerland 39/C3
Yvetot, France 28/D3
Yvoir, Belgium 27/F8
Yvonand, Switzerland 39/C3
Ywathit, Burma 72/C3

Z

Zaachila, Mexico 150/L8
Zaandam (Zaanstad), Netherlands 27/B4
Zaandijk, Netherlands 27/B4
Zabaykal'sk, U.S.S.R. 48/M5
Zabid, Yemen Arab Rep. 59/D7
Ząbki, Poland 47/E2
Ząbkowice, Poland 47/B3
Ząbkowice Śląskie, Poland 47/C3
Žabljak, Yugoslavia 45/D4
Zabol, Iran 59/H3
Zabol, Iran 66/M5
Zabré, Upper Volta 106/D6
Zábřeh, Czech. 41/D2
Zabrze, Poland 7/F3
Zabrze, Poland 47/A4
Zacapa, Guatemala 154/C3
Zacapoaxtla, Mexico 150/O1
Zacapu, Mexico 150/J7
Zacatecas (state), Mexico 150/H5
Zacatecas, Mexico 150/H5
Zacatecoluca, El Salvador 154/C4
Zacatelco, Mexico 150/N1
Zacatepec, Mexico 150/L2
Zacatlán, Mexico 150/N1
Zach, Tenn. (†38320) 237/E8
Zachariah, Ky. (41396) 237/O5
Zachary, La. (70791) 238/K1
Zachow, Wis. (54182) 317/K6
Zacoalco de Torres, Mexico 150/H6
Zadar, Yugoslavia 45/B3
Zadetkyi Kyun (isl.), Burma 72/C5
Zadi, Burma 72/C4
Zadoi, China 77/E5
Zafra, Spain 33/C3
Żagań, Poland 47/B3
Žagare, U.S.S.R. 53/B2
Zagarolo, Italy 34/F7
Zagazig, Egypt 59/B3
Zagazig, Egypt 111/K3
Zagheh, Iran 66/F4
Zagora, Morocco 106/C2
Zagorsk, U.S.S.R. 7/H3
Zagorsk, U.S.S.R. 52/E3
Zagreb, Yugoslavia 7/F4
Zagreb, Yugoslavia 45/C3
Zagros (mts.), Iran 59/E3
Zagros (mts.), Iran 66/E4
Zagyva (riv.), Hungary 41/F3
Zahedan, Iran 59/H4
Zahedan, Iran 66/M6
Zahedan, Iran 54/G7
Zahl, N. Dak. (58856) 282/C2
Zahle, Lebanon 63/F6
Záhony, Hungary 41/G2
Zahran, Saudi Arabia 59/D6
Zaidín, Spain 33/G2
Zaire 2/K6
Zaire 102/E5
ZAIRE 115/D4
Zaire (Congo) (riv.) 102/E4
Zaire (dist.), Angola 115/B5
Zaire (Congo) (riv.), Zaire 115/C4
Zaječar, Yugoslavia 45/E4
Zakamensk, U.S.S.R. 48/L4
Zakho, Iraq 66/C2
Zákinthos, Greece 45/E7
Zákinthos (Zante) (isl.), Greece 45/E7
Zako, Cent. Afr. Rep. 115/D2
Zakopane, Poland 47/D4
Zala (co.), Hungary 41/D3
Zala (riv.), Hungary 41/D3
Zalaegerszeg, Hungary 41/D3
Zalamea de la Serena, Spain 33/D3
Zalamea la Real, Spain 33/C4
Zalaszentgrót, Hungary 41/D3
Zalău, Romania 45/F2
Zaleski, Ohio (45698) 284/F7
Zalim, Saudi Arabia 59/D5
Zalingei, Sudan 111/D5
Zalma, Mo. (63787) 261/N8
Zaltbommel, Netherlands 27/G5
Zalun, Burma 72/B3
Zama (lake), Alberta 182/A5
Zama, Miss. (†39090) 256/F5
Zambales (prov.), Philippines 82/C3
Žamberk, Czech. 41/D1
Zambezi (riv.) 2/L6
Zambezi (riv.) 102/E6
Zambezi (riv.), Angola 115/D6
Zambezi (riv.), Mozambique 118/E3
Zambezi (riv.), Namibia 118/C3
Zambezi, Zambia 115/D6
Zambezi (riv.), Zambia 115/D7
Zambezi (riv.), Zimbabwe 118/E3
Zambézia (prov.), Mozambique 118/F3
Zambia 2/L6
Zambia 102/E6
ZAMBIA 115/E7
Zamboanga, Philippines 85/G4
Zamboanga, Philippines 82/C7
Zamboanga, Philippines 54/N9
Zamboanga del Norte (prov.), Philippines 82/D6
Zamboanga del Sur (prov.), Philippines 82/D7
Zambrów, Poland 47/E2
Zamora, Calif. (95698) 204/C5
Zamora, Ecuador 128/C5
Zamora (riv.), Ecuador 128/B4
Zamora (prov.), Spain 33/D2
Zamora, Spain 33/D2
Zamora-Chinchipe (prov.), Ecuador 128/C5
Zamora de Hidalgo, Mexico 150/H7
Zamość (prov.), Poland 47/F3
Zamość, Poland 47/F3
Zams, Austria 41/A3
Zamtang, China 77/F5
Zanaga, Congo 115/B4
Zanda, China 77/A5
Zanderij, Suriname 131/D3
Zanderij, Suriname 120/D2
Zandvoort, Netherlands 27/E4
Zanesfield, Ohio (43360) 284/C5
Zanesville, Ind. (46799) 227/G3
Zanesville, Ohio 188/K3
Zanesville, Ohio (43701) 284/G6
Zanja de Lira, Venezuela 124/E3
Zanjan (governorate), Iran 66/F2
Zanjan, Iran 59/E2
Zanjan, Iran 66/K5
Zanjan (riv.), Iran 66/F2
Zanoni, Mo. (65784) 261/H9
Zante (Zákinthos), Greece 45/E7
Zanthus, W. Australia 92/C5
Zanzibar, Tanzania 102/F5
Zanzibar, Tanzania 115/G5
Zanzibar (isl.), Tanzania 2/M6
Zanzibar (isl.), Tanzania 102/F5
Zanzibar (isl.), Tanzania 115/G5
Zanzibar Mjini (reg.), Tanzania 115/G5
Zanzibar Shambani North (reg.), Tanzania 115/G5
Zanzibar Shambani South (reg.), Tanzania 115/G5
Zao (mt.), Japan 81/K5
Zaouiet Kounta, Algeria 106/D3
Zaoyang, China 77/H5
Zaozernyy, U.S.S.R. 48/K4
Zaozhuang, China 77/J5
Zap, N. Dak. (58580) 282/G5
Západočeský (reg.), Austria 41/B2
Západoslovenský (reg.), Austria 41/D2
Zapala, Argentina 143/B4
Zapala, Argentina 120/B6
Zapaleri, Cerro (mt.), Argentina 143/C1
Zapaleri, Cerro (mt.), Bolivia 136/B8
Zapaleri, Cerro (mt.), Chile 138/C4
Zapallar, Chile 138/A9
Zapata (pen.), Cuba 158/C2
Zapata (co.), Texas 303/E11
Zapata, Texas (78076) 303/E11
Zapata Occidental (swamp), Cuba 158/D2
Zapata Oriental (swamp), Cuba 158/D2
Zapatera (isl.), Nicaragua 154/E5
Zapatoca, Colombia 126/D4
Zapatosa, Ciénaga de (swamp), Colombia 126/D3
Zapicán, Uruguay 145/E4
Zapiga, Chile 138/B2
Zapolyarnyy, U.S.S.R. 52/D1
Zaporozh'ye, U.S.S.R. 7/H4
Zaporozh'ye, U.S.S.R. 48/D5
Zaporozh'ye, U.S.S.R. 52/E5
Zapotillo, Ecuador 128/B5
Zapucay, Uruguay 145/D2
Zapug, China 77/B5
Za Qu (riv.), China 77/E5
Zara, Turkey 59/C1
Zara, Turkey 63/G3
Zara (Zadar), Yugoslavia 45/B3
Zarafshan, 48/G5
Zaragoza, Colombia 126/C4
Zaragoza, Chihuahua, Mexico 150/F1
Zaragoza, Coahuila, Mexico 150/J2
Zaragoza, Puebla, Mexico 150/O1
Zaragoza (prov.), Spain 33/F2
Zaragoza (Saragossa), Spain 33/F2
Zarand, Iran 59/G3
Zarand, Iran 66/H6
Zaranj, Afghanistan 68/A2
Zaranj, Afghanistan 59/H3
Zarasai, U.S.S.R. 53/C3
Zárate, Argentina 143/G6
Zaraza, Venezuela 124/F3
Zard Kuh (mt.), Iran 66/F4
Zarembo (isl.), Alaska 196/N2
Zarephath, N.J. (08890) 273/D2
Zaria, Nigeria 106/F6
Zaria, Nigeria 102/C3
Zarineh (riv.), Iran 66/E2
Zărnești, Romania 45/G3
Zarqa' (riv.), Jordan 65/D3
Zarqam, Iran 66/H6
Zaruma, Ecuador 128/C4
Zarumilla, Peru 128/B4
Żary, Poland 47/B3
Zarzal, Colombia 126/B5
Zarza la Mayor, Spain 33/C3
Zarzis, Tunisia 106/G2
Zarzis, Tunisia 102/D1
Zaskar (mts.), India 68/D2
Zastron, S. Africa 118/D6
Žatec, Czech. 41/B1
Zavala (co.), Texas 303/E9
Zavalla, Argentina 143/F6
Zavalla, Texas (75980) 303/K6
Zavdi'el, Israel 65/B4
Zaventem, Belgium 27/C9
Zavitinsk, U.S.S.R. 48/O4
Zawi, Zimbabwe 118/D3
Zawia, Libya 102/D1
Zawia, Libya 111/B1
Zawiercie, Poland 47/D3
Zayandeh (riv.), Iran 66/H4
Zayar, China 77/B3
Zaysan, U.S.S.R. 48/J5
Zaysan (lake), U.S.S.R. 54/K5
Zaysan (lake), U.S.S.R. 48/J5
Zayü, China 77/E6
Zaza del Medio, Cuba 158/F2
Zázrivá, Czech. 41/E2
Zbąszyń, Poland 47/B2
Zbiroh, Czech. 41/B2
Zborov, Czech. 41/F2
Žďár nad Sázavou, Czech. 41/C2
Zduńska Wola, Poland 47/D3
Zealand (Sjaelland) (isl.), Den. 21/E6
Zealand, New Bruns. 170/D2
Zealandia, Sask. 181/D4
Zearing, Iowa (50278) 229/G4
Zeballos, Br. Col. 184/D5
Zebdani, Syria 63/G6
Zebirget (isl.), Egypt 59/C5
Zebulon, Georgia (30295) 217/D4
Zebulon, Ky. (†41501) 237/S5
Zebulon, N.C. (27597) 281/N3
Zedelgem, Belgium 27/C6
Zeebrugge, Belgium 27/C6
Zeehan, Tasmania 99/B3
Zeeland, Mich. (49464) 250/D6
Zeeland (prov.), Netherlands 27/D6
Zeeland, N. Dak. (58581) 282/L8
Ze'elim, Israel 65/A5
Zeerust, S. Africa 118/D5
Zeewolde, Netherlands 27/G4
Zefat, Israel 65/C2
Zegharta, Lebanon 63/G5
Zegrzyńskie (lake), Poland 47/E2
Zehdenick, E. Germany 22/E2
Zehner, Sask. 181/G5
Zeigler, Ill. (62999) 222/D6
Zeila, Somalia 115/H1
Zeil am Main, W. Germany 22/D4
Zeist, Netherlands 27/G4
Zeitz, E. Germany 22/E3
Zekiah Swamp (riv.), Md. 245/L7
Zêkog, China 77/F5
Zele, Belgium 27/E6
Zelenoborskiy, U.S.S.R. 52/D1
Zelenodol'sk, U.S.S.R. 52/G3
Zelenokumsk, U.S.S.R. 52/F6
Zelienople, Pa. (16063) 294/B4
Želiezovce, Czech. 41/E2
Zell, S. Dak. (57483) 298/M4
Zell, Luzern, Switzerland 39/E2
Zell, Zürich, Switzerland 39/G2
Zell, W. Germany 22/B4
Zella, Libya 102/D2
Zella, Libya 111/C2
Zella-Mehlis, E. Germany 22/D3
Zell am See, Austria 41/B3
Zell am Ziller, Austria 41/A3
Zellersee (lake), Switzerland 39/G1
Zellwood, Fla. (32798) 212/E3
Zelma, Sask. 181/F4
Zelow, Poland 47/D3
Zelten, Jebel (mts.), Libya 111/D2
Zeltweg, Austria 41/C3
Zelzate, Belgium 27/D6
Zemio, Cent. Afr. Rep. 115/D2
Zemongo, Cent. Afr. Rep. 115/E2
Zemple, Minn. (†56636) 255/E3
Zempoala, Mexico 150/Q1
Zemst, Belgium 27/E7
Zenas, Ind. (†47223) 227/G6
Zenda, Kansas (67159) 232/D4
Zeneta, Sask. 181/J5
Zenia, Calif. (95495) 204/B3
Zenica, Yugoslavia 45/C3
Zenith, Ill. (†62899) 222/E5
Zenith, Kansas (†67578) 232/D4
Zenith, W. Va. (†24951) 312/F7
Zenith-Saltwater, Wash. (†98101) 310/C3
Zenjan (Zanjan), Iran 66/F2
Zenobia (peak), Colo. 208/B1
Zenon Park, Sask. 181/H2
Zenoria, La. (†71371) 238/F3
Zent, Ark. (†72021) 202/H4
Zenta (Senta), Yugo. 45/D3
Zeona, S. Dak. (57795) 298/D3
Žepče, Yugoslavia 45/D3
Zepernick, E. Germany 22/F3
Zephyr, Ontario 177/E3
Zephyr, Texas (76890) 303/F6
Zephyr Cove, Nev. (89448) 266/A3
Zephyrhills, Fla. (33599) 212/D3
Zepp, Va. (†22654) 307/L3
Zerbst, E. Germany 22/E3
Zereh, Gowd-e (depr.), Afghanistan 68/A3
Zermatt, Switzerland 39/E4
Zernez, Switzerland 39/K3
Zernograd, U.S.S.R. 52/F5
Zessfontein, Namibia 118/A3
Zêtang, China 77/D6
Zetland (trad. co.), Scot. 15/B4
Zeulenroda, E. Germany 22/D3
Zeven, W. Germany 22/C2
Zevenaar, Netherlands 27/J5
Zevenbergen, Netherlands 27/E5
Zeya, U.S.S.R. 48/N4
Zeya (riv.), U.S.S.R. 48/N4
Zeytinburnu, Turkey 63/D6
Zeytindağ, Turkey 63/B3
Zgierz, Poland 47/D3
Zgorzelec, Poland 47/B3
Zhanang, China 77/D6
Zhanghei, China 77/J3
Zhangjiakou (Kalgan), China 77/J3
Zhangjiakou, China 54/N5
Zhangping, China 77/J6
Zhangye (Changyeh), China 77/F4
Zhangye, China 54/M6
Zhangzhou (Changchow), China 77/J7
Zhanjiang (Chankiang), China 77/H7
Zhanjiang, China 54/N7
Zhanyi, China 77/F6
Zhaodong, China 77/K2
Zhaojue, China 77/F6
Zhaoqing, China 77/H7
Zhaosu, China 77/B3
Zhaotong (Chaotung), China 77/F6
Zhari Namco (lake), China 77/C5
Zhashui, China 77/G5
Zhatay, U.S.S.R. 48/O3
Zhaxi Co (lake), China 77/C5
Zhdanov, U.S.S.R. 7/H4
Zhdanov, U.S.S.R. 48/D5
Zhdanov, U.S.S.R. 52/E5
Zhejiang (Chekiang) (prov.), China 77/K6
Zhelaniye (cape), U.S.S.R. 48/H2
Zheleznodorozhnyy, U.S.S.R. 52/H2
Zheleznogorsk, U.S.S.R. 52/E4
Zheleznogorsk-Ilimskiy, U.S.S.R. 48/L4
Zhenba, China 77/G5
Zheng'an, China 77/G6
Zhenglan, China 77/J3
Zhengzhou (Chengchow), China 77/H5
Zhengzhou, China 54/N6
Zhenjiang (Chinkiang), China 77/J5
Zhenxiong, China 77/F6
Zhenyuan, China 77/G6
Zhido, China 77/E5
Zhigalovo, U.S.S.R. 48/L4
Zhigansk, U.S.S.R. 4/C3
Zhigansk, U.S.S.R. 54/N3
Zhigansk, U.S.S.R. 48/N3
Zhigulevsk, U.S.S.R. 52/G4
Zhi Qu (Tongtian He) (riv.), China 77/E5
Zhirnovsk, U.S.S.R. 52/G4
Zhitomir, U.S.S.R. 7/G3
Zhitomir, U.S.S.R. 48/C4
Zhitomir, U.S.S.R. 52/C4
Zhlobin, U.S.S.R. 52/D4
Zhmerinka, U.S.S.R. 52/C5
Zhob (riv.), Pakistan 59/J3
Zhob (riv.), Pakistan 68/B2
Zhodino, U.S.S.R. 52/C4
Zhongba, China 77/B6
Zhongdian, China 77/F6
Zhongning, China 77/G4
Zhongshan (Chungshan), China 77/H7
Zhongwei, China 77/F4
Zhoushan (arch.), China 77/K5
Zhovtnevoye, U.S.S.R. 52/D5
Zhuanghe, China 77/K4
Zhucheng, China 77/J4
Zhukovka, U.S.S.R. 52/D4
Zhumadian (Chumatien), China 77/H5
Zhushan, China 77/G5
Zhuzhou (Chuchow), China 77/H6
Zhuzhou, China 54/N7
Zia Pueblo, N. Mex. (†87053) 274/C3
Žiar nad Hronom, Czech. 41/E2
Zibak, Afghanistan 59/K2
Zibak, Afghanistan 68/C1
Zibo (Tzepo), China 77/J4
Zibo, China 54/N6
Zichang, China 77/G4
Židlochovice, Czech. 41/D2
Ziebach (co.), S. Dak. 298/F4
Ziębice, Poland 47/C3
Ziel (mt.), North. Terr. 88/E4
Ziel (mt.), North. Terr. 93/C7
Zielona Góra (prov.), Poland 47/B3
Zielona Góra, Poland 47/B3
Zierikzee, Netherlands 27/D5
Zifta, Egypt 111/J3
Zigong (Tzekung), China 77/F6
Ziguei, Chad 111/C5
Zigui, China 77/H5
Ziguinchor, Senegal 106/A6
Ziguinchor, Senegal 102/A3
Zihuatanejo, Mexico 150/J8
Zikhron Ya'aqov, Israel 65/B2
Zilbir (riv.), Iran 66/D1
Zile, Turkey 59/C1
Zile, Turkey 63/G2
Zilfi, Saudi Arabia 59/E4
Žilina, Czech. 41/E2
Zillah, Wash. (98953) 310/E4
Zillis-Reischen, Switzerland 39/H3
Zilupe, U.S.S.R. 53/D2
Zilwaukee, Mich. (†48601) 250/F5
Zim, Minn. (55799) 255/F3
Zima, U.S.S.R. 48/L4
Zimatlán de Álvarez, Mexico 150/L8
Zimbabwe 2/L6
Zimbabwe 102/E6
ZIMBABWE 118/D4
Zimbabwe Nat'l Park, Zimbabwe 118/E4
Zimmerdale, Kansas (†67117) 232/E3
Zimmerman, La. (†71409) 238/E4
Zimmerman, Minn. (55398) 255/E5
Zimnicea, Romania 45/G4
Zimnitsa, Bulgaria 45/H4
Zinal, Switzerland 39/E4
Zinc, Ark. (†72601) 202/E1
Zinder, Niger 106/F6
Zinder, Niger 102/C3
Zingren, China 77/G6
Zingst, E. Germany 22/E1
Zinhui, China 77/H7
Zinjibar, P.D.R. Yemen 59/E7
Zinnik (Soignies), Belgium 27/D7
Zion, Ark. (72589) 202/G1
Zion, Ill. (60099) 222/F1
Zion, Md. (†21901) 245/P2
Zion, Mo. (†63645) 261/M8
Zion, N.J. (†08853) 273/D3
Zion, S.C. (†29574) 296/J3
Zion Hill, St. Chris.-Nevis 161/D11
Zion National Park, Utah (84767) 304/B6
Zion Nat'l Park, Utah 304/A6
Zionsville, Ind. (46077) 227/E5
Zionville, N.C. (28698) 281/F2
Zipaquirá, Colombia 126/D5
Zippori, Israel 65/C2
Zirc, Hungary 41/D3
Žirje (isl.), Yugoslavia 45/B4
Zirkel (mt.), Colo. 208/F1
Zirko (isl.), U.A.E. 59/F5
Zirl, Austria 41/A3
Zirndorf, W. Germany 22/D4
Zistersdorf, Austria 41/D2
Zitácuaro, Mexico 150/J7
Zittau, E. Germany 22/F3
Zitterwald (plat.), Belgium 27/J8
Zivarık, Turkey 63/E3
Ziwa Magharibi (West Lake) (reg.), Tanzania 115/F4
Ziyang, China 77/F5
Ziz, Wadi (dry riv.), Morocco 106/D2
Zizers, Switzerland 39/J3
Zlaté Moravce, Czech. 41/E2
Zlatograd, Bulgaria 45/G5
Zlatoust, U.S.S.R. 54/G4
Zlatoust, U.S.S.R. 48/F4
Zlín (Gottwaldov), Czech. 41/D2
Zliten, Libya 111/C1
Złocieniec, Poland 47/C2
Złotoryja, Poland 47/C3
Złotów, Poland 47/C2
Žlutice, Czech. 41/B1
Znamenka, U.S.S.R. 52/D5
Znin, Poland 47/C2
Znojmo, Czech. 41/D2
Zoar (lake), Conn. 210/C3
Zoar, Ohio (44697) 284/H4
Zoarville, Ohio (44698) 284/H4
Zofingen, Switzerland 39/F2
Zogang, China 77/E5
Zohreh (riv.), Iran 66/F5
Zoigê, China 77/F5
Zolfo Springs, Fla. (33890) 212/E4
Zollikofen, Switzerland 39/E3
Zollikon, Switzerland 39/G2
Zolotonosha, U.S.S.R. 52/D5
Zomba, Malawi 115/G7
Zomba, Malawi 102/F6
Zonderend (riv.), S. Africa 118/G6
Zongo, Bolivia 136/B5
Zongo, Zaire 115/C3
Zongolica, Mexico 150/P2
Zonguldak (prov.), Turkey 63/D2
Zonguldak, Turkey 63/D2
Zonguldak, Turkey 59/B1
Zonhoven, Belgium 27/G6
Zoo Baba (well), Niger 106/G5
Zook, Kansas (†67550) 232/C3
Zorbatiya, Iraq 66/D4
Zorita, Spain 33/D3
Zorritos, Peru 128/B4
Zortman, Mont. (59546) 262/H3
Zottegem, Belgium 27/D7
Zouar, Chad 111/C3
Zouîrât, Mauritania 106/B4
Zoutkamp, Netherlands 27/J2
Zrenjanin, Yugoslavia 45/E3
Zuata, Venezuela 124/F3
Zuata (riv.), Venezuela 124/F3
Zububa, West Bank 65/C2
Zucchero (mt.), Switzerland 39/G4
Zudáñez, Bolivia 136/C6
Zug (canton), Switzerland 39/G2
Zug, Switzerland 39/G2
Zugdidi, U.S.S.R. 52/F6
Zugersee (lake), Switzerland 39/F2
Zugspitze (mt.), Austria 41/A3
Zugspitze (mt.), W. Germany 22/D5
Zuidelijke IJsselmeerpolders (prov.), Netherlands 27/H4
Zuienkerke, Belgium 27/C6
Zuila, Libya 111/C2
Zújar, Spain 33/E4
Zújar (res.), Spain 33/D3
Zula, Ethiopia 111/G4
Zula, Ethiopia 59/C6
Zula, Ky. (†42603) 237/M7
Zulia (state), Venezuela 124/B2
Zulia (riv.), Venezuela 124/B3
Zülpich, W. Germany 22/B3
Zulu, Ind. (†46773) 227/H2
Zulueta, Cuba 158/E2
Zululand (reg.), S. Africa 118/E5
Zumba, Ecuador 128/C5
Zumbo, Mozambique 118/E3
Zumbro (riv.), Minn. 255/F6
Zumbro Falls, Minn. (55991) 255/F6
Zumbrota, Minn. (55992) 255/F6
Zumpango del Río, Mexico 150/J8
Zumpango de Ocampo, Mexico 150/L1
Zundert, Netherlands 27/F6
Zungeru, Nigeria 106/F7
Zunhua, China 77/J3
Zuni (riv.), Ariz. 198/F4
Zuni, N. Mex. (87327) 274/A3
Zuni (mts.), N. Mex. 274/A3
Zuni (riv.), N. Mex. 274/A3
Zuni, Va. (23898) 307/P7
Zuni Ind. Res., N. Mex. 274/A3
Zunyi (Tsunyi), China 77/G6
Zunyi, China 54/M7
Zuoz, Switzerland 39/J3
Zuqar (isl.), Yemen Arab Rep. 59/D7
Zurabad, Iran 66/M3
Zurich, Kansas (67676) 232/C2
Zurich, Mont. (59547) 262/G2
Zurich, Ontario 177/C4
Zürich (canton), Switzerland 39/G2
Zürich, Switzerland 39/F2
Zürich, Switzerland 7/E4
Zürichsee (lake), Switzerland 39/G2
Żuromin, Poland 47/E2
Zurzach, Switzerland 39/F1
Zushi, Japan 81/O3
Zutphen, Netherlands 27/J4
Zuweiza, Jordan 65/D4
Zuyevka, U.S.S.R. 52/H3
Zvolen, Czech. 41/E2
Zvornik, Yugoslavia 45/D3
Zwai (lake), Ethiopia 111/G6
Zwanenburg, Netherlands 27/A4
Zwara, Libya 111/B1
Zwart (riv.), S. Africa 118/G7
Zwartsluis, Netherlands 27/H3
Zweibrücken, W. Germany 22/B4
Zweisimmen, Switzerland 39/D3
Zwelitsha, S. Africa 118/D6
Zwenkau, E. Germany 22/E3
Zwettl-Niederösterreich, Austria 41/C2
Zwickau, E. Germany 22/E3
Zwijndrecht, Netherlands 27/E5
Zwingle, Iowa (52079) 229/M4
Zwischenahn, W. Germany 22/B2
Zwoleń, Poland 47/E3
Zwolle, La. (71486) 238/C3
Zwolle, Netherlands 27/J3
Zychlin, Poland 47/D2
Żyrardów, Poland 47/E2
Zyryanka, U.S.S.R. 4/C2
Zyryanka, U.S.S.R. 54/S3
Zyryanka, U.S.S.R. 48/Q3
Zyryanovsk, U.S.S.R. 48/J5
Żywiec, Poland 47/D4
Zzyzx, Calif. (†92309) 204/J8

GEOGRAPHICAL TERMS

A. = Arabic Burm. = Burmese Camb. = Cambodian Ch. = Chinese Czech. = Czechoslovakian Dan. = Danish Du. = Dutch Finn. = Finnish Fr. = French Ger. = German Ice. = Icelandic It. = Italian Jap. = Japanese Mong. = Mongol Nor. = Norwegian Per. = Persian Port. = Portuguese Russ. = Russian Sp. = Spanish Sw. = Swedish Turk. = Turkish

Å	Nor., Sw.	Stream
Aas	Dan., Nor.	Hills
Abajo	Sp.	Lower
Ada, Adasi	Turk.	Island
Altipiano	It.	Plateau
Altiplano	Sp.	Plateau
Alv, Alf, Elf	Sw.	River
Arrecife	Sp.	Reef
Asa	Nor., Sw.	Hill
Asaga	Turk.	Lower
Austral	Sp.	Southern
Baai	Du.	Bay
Bab	Arabic	Gate or Strait
Bahia	Sp.	Bay
Bahr	Arabic	Marsh, Lake, Sea, River
Baia	Port.	Bay
Baie	Fr.	Bay, Gulf
Baizo	Port.	Low
Bakke	Dan.	Hill
Bana	Jap.	Cape
Bañados	Sp.	Marshes
Band	Per.	Mt. Range
Bandao	Ch.	Peninsula
Bandar	Per.	Harbor
Barra	Sp.	Reef
Bel	Turk.	Pass
Belt	Ger.	Strait
Ben	Gaelic	Mountain
Bera	Du.	Mountain
Berg	Ger., Du.	Mountain
Bir	Arabic	Well
Boca	Sp.	Gulf, Inlet
Boğhaz	Turk.	Strait
Bolshoi, Bolshaya	Russ.	Big
Bolson	Sp.	Depression
Bong	Korean	Mountain
Boreal	Sp.	Northern
Breen	Nor.	Glacier
Bro	Dan., Nor., Sw.	Bridge
Bucht	Ger.	Bay
Bugt	Dan.	Bay
Bukhta	Russ.	Bay
Bukit	Malay	Hill, Mountain
Bukt	Nor., Sw.	Bay, Gulf
Burnu, Burun	Turk.	Cape, Point
By	Dan., Nor., Sw.	Town
Cabo	Port., Sp.	Cape
Campos	Port.	Plains
Canal	Port., Sp.	Channel
Cap, Capo	Fr., It.	Cape
Cataratas	Sp.	Falls
Catena	It.	Mt. Range
Catingas	Port.	Open Woodlands
Cayos	Sp.	Islands
Central, Centrale	Fr., It.	Middle
Cerrito, Cerro	Sp.	Hill
Cerros	Sp.	Hills, Mountains
Chai	Turk.	River
Chott	Arabic	Salt Lake
Ciénaga	Sp.	Swamp
Ciudad	Sp.	City
Col	Fr.	Pass
Cordillera	Sp.	Mt. Range, Mts.
Côte	Fr.	Coast
Csatoria	Magyar	Canal
Cuchilla	Sp.	Mt. Range
Curiche	Sp.	Swamp
Dağ, Dağı	Turk.	Mountain, Peak
Dağlari	Turk.	Mt. Range
Dal	Nor., Sw.	Valley
Dar	Arabic	Land
Dar'ya	Russ.	River
Daryacheh	Per.	Marshy Lake
Dasht	Per.	Desert, Plain
Deniz, Denizi	Turk.	Sea, Lake
Desierto	Sp.	Desert
Détroit	Fr.	Strait
Djeziret	Arabic, Turk.	Island
Do	Korean	Island
Doi	Thai	Mountain
Eiland	Du.	Island
Elv	Dan., Nor.	River
Embalse	Sp.	Reservoir
Emi	Berber	Mountain
Erg	Arabic	Dune, Desert
Eski	Turk.	Old
Est, Este	Fr., Port., Sp.	East
Estero	Sp.	Estuary, Creek
Estrecho, Estreito	Sp., Port.	Strait
Etang	Fr.	Pond, Lagoon, Lake
Feng	Ch.	Mountain
Fiume	It.	River
Fjäll	Sw.	Mountain
Fjeld, Fjell	Nor.	Hills, Mountain
Fjord	Dan., Nor., Sw.	Fiord
Fleuve	Fr.	River
Fljót	Ice.	Stream
Fluss	Ger.	River
Fors	Sw.	Waterfall
Fos, Foss	Dan., Nor.	Waterfall
Gamla	Nor.	Old
Gamle	Dan.	Old
Gata	Jap.	Lake
Gawa	Jap.	River
Gebel	Arabic	Mountain
Gebergte	Du.	Mt. Range
Gebirge	Ger.	Mt. Range
Gobi	Mongol	Desert
Goe	Jap.	Pass
Gol	Mongol, Turk.	Lake, Stream
Golf	Ger., Du.	Gulf
Golfe	Fr.	Gulf
Golfo	Sp., It., Port.	Gulf
Gölü	Turk.	Lake
Gora	Russ.	Mountain
Grand, Grande	Fr., Sp.	Big
Groot	Du.	Big
Gross	Ger.	Big
Grosso	It., Port.	Big
Guba	Russ.	Bay, Gulf
Gunto	Jap.	Archipelago
Gunung	Malay	Mountain
Hai	Ch.	Sea
Haixia	Ch.	Strait
Halbinsel	Ger.	Peninsula
Hamáda, Hammada	Arabic	Rocky Plateau
Hamn	Sw.	Harbor
Hamún	Per.	Marsh
Hanto	Jap.	Peninsula
Has, Hassi	Arabic	Well
Hav	Dan., Nor., Sw.	Sea, Ocean
Havet	Nor.	Bay
Havn	Dan., Nor.	Harbor
Havre	Fr.	Harbor
He	Ch.	River, Stream
Higashi, Higasi	Jap.	East
Hochebene	Ger.	Plateau
Hoek	Du.	Cape
Hoku	Jap.	North
Holm	Dan., Nor., Sw.	Island
Hory	Czech	Mountains
Hoved	Dan., Nor.	Cape, Promontory
Hu	Ch.	Lake
Huang	Ch.	Yellow
Huk	Dan., Nor., Sw.	Point
Hus, Huus	Dan., Nor., Sw.	House
Idehan	Arabic	Desert
Ile	Fr.	Island
Ilet	Fr.	Islet
Ilot	Fr.	Islet
Indre	Dan., Nor.	Inner
Inferieur, Inferiore	Fr., It.	Lower
Inner, Inre	Sw.	Inner
Insel	Ger.	Island
Irmak	Turk.	River
Isla	Sp.	Island
Isola	It.	Island
Jabal, Jebel	Arabic	Mountains
Järvi	Finn.	Lake
Jaure	Sw.	Lake
Jiang	Ch.	River, Stream
Jima	Jap.	Island
Joki	Finn.	River
Kaap	Du.	Cape
Kabir, Kebir	Arabic	Big
Kai	Jap.	Sea
Kaikyo	Jap.	Strait
Kami	Turk.	Upper
Kanaal	Du.	Canal
Kanal	Russ., Ger.	Canal, Channel
Kao	Thai	Mountain
Kap, Kapp	Nor., Sw., Ice.	Cape
Kaupunki	Finn.	Town
Kawa	Jap.	River
Khao	Thai	Mountain
Khrebet	Russ.	Mt. Range
Kita	Jap.	North
Klein	Du., Ger.	Small
Klint	Dan.	Promontory
Kô	Jap.	Lake
Ko	Thai	Island
Koh	Camb., Khmer	Island
Kop	Du.	Peak, Head
Köping	Sw.	Market, Borough
Körfez, Körfezi	Turk.	Gulf
Kosa	Russ.	Spit
Kosui	Jap.	Lake
Kraal	Du.	Native Village
Kuchuk	Turk.	Small
Kuh, Kuhha	Per.	Mt. Range, Mts.
Kul	Sinkiang Turki	Lake
Kum	Turk.	Desert
Kuro	Jap.	Black
Laag	Du.	Low
Lac	Fr.	Lake
Lago	Port., Sp., It.	Lake
Lagoa	Port.	Lagoon
Laguna	Sp.	Lagoon
Lagune	Fr.	Lagoon
Lahti	Finn.	Bay, Bight
Län	Sw.	County
Liedao	Ch.	Islands, Archipelago
Lilla	Sw.	Small
Lille	Dan., Nor.	Small
Ling	Ch.	Mountain
Llanos	Sp.	Plains
Mae Nam	Thai	River
Mali, Malaya	Russ.	Small
Man	Korean	Bay
Mar	Sp., Port.	Sea
Mare	It.	Sea
Medio	Sp.	Middle
Meer	Du.	Lake
Meer	Ger.	Sea
Mer	Fr.	Sea
Meridionale	It.	Southern
Meseta	Sp.	Plateau
Middelst, Midden	Du.	Middle
Minami	Jap.	Southern
Mis	Russ.	Cape
Misaki	Jap.	Cape
Mittel	Ger.	Middle
Mont	Fr.	Mountain
Montagne	Fr.	Mountain
Montaña	Sp.	Mountains
Monte	Sp., It., Port.	Mountain
More	Russ.	Sea
Mörön	Mong.	Stream
Morro	Port., Sp.	Mountain, Promontory
Morue	Fr.	Hill
Moyen	Fr.	Middle
Muang	Siamese	Town
Mui	Vietnamese	Cape, Point
Mys	Russ.	Cape
Nada	Jap.	Sea
Naka	Jap.	Middle
Nam	Burm., Lao	River
Namakzar	Per.	Salt Waste
Nan	Jap.	South
Nes	Nor.	Cape, Point
Nevado	Sp.	Snow-covered Peak
Nieder	Ger.	Lower
Nishi, Nisi	Jap.	West
Nizhni, Nizhnyaya	Russ.	Lower
Njarga	Finn.	Peninsula, Promontory
Nong	Thai	Lake
Noord	Du.	North
Nord	Fr., Ger.	North
Norte	Sp., It., Port.	North
Nos	Russ.	Cape
Novi, Novaya	Russ.	New
Nur, Nuur	Ch., Mong.	Lake
Nuruu	Mong.	Mountains
Nusa	Malay	Island
Ny, Nya	Nor., Sw.	New
O	Jap.	Big
Ö	Nor., Sw.	Island
Ober	Ger.	Upper
Occidental, Occidentale	Sp., It.	Western
Odde	Dan.	Point
Oeste	Port.	West
Ooster	Du.	Eastern
Opper, Over	Du.	Upper
Oriental	Sp., Fr.	Eastern
Orientale	It.	Eastern
Orta	Turk.	Middle
Ost	Ger.	East
Ostrov	Russ.	Island
Ouest	Fr.	West
öy	Nor.	Island
Ozero	Russ.	Lake
Pampa	Sp.	Plain
Pas	Fr.	Channel, Strait
Paso	Sp.	Pass
Passo	It., Port.	Pass
Peña	Sp.	Rock, Mountain
Pendi	Ch.	Basin
Penisola	It.	Peninsula
Pequeño	Sp.	Small
Pereval	Russ.	Pass
Peski	Russ.	Desert
Petit, Petite	Fr.	Small
Phu	Lao, Annamese	Mtn.
Pic	Fr.	Mountain
Piccolo	It.	Small
Pico	Port., Sp.	Mountain, Peak
Pik	Russ.	Mountain, Peak
Piton	Fr.	Mountain, Peak
Planalto	Port.	Plateau
Plato	Russ.	Plateau
Pointe	Fr.	Point
Poluostrov	Russ.	Peninsula
Ponta	Port.	Point
Presa	Sp.	Reservoir
Presqu'île	Fr.	Peninsula
Proliv	Russ.	Strait
Pulou, Pulo	Malay	Island
Punt	Du.	Point
Punta	Sp., It., Port.	Point
Qiryat	Hebrew	City, Settlement
Qum	Turk.	Desert
Qundao	Ch.	Islands
Rada	Sp.	Inlet
Rade	Fr.	Bay, Inlet
Ras	Arabic	Cape
Reka	Russ.	River
Retto	Jap.	Archipelago
Ria	Sp.	Estuary
Río	Sp.	River
Rivier, Rivière	Du., Fr.	River
Rud	Per.	River
Sai	Jap.	West
Saki	Jap.	Cape
Salar, Salina	Sp.	Salt Deposit
Salto	Sp., Port.	Falls
San	Jap., Korean	Hill
Sanmaek	Korean	Mt. Range
Schiereiland	Du.	Peninsula
Se	Camb., Khmer	River
See	Ger.	Sea, Lake
Selvas	Sp., Port.	Woods, Forest
Seno	Sp.	Bay, Gulf
Serra	Port.	Mts.
Serranía	Sp.	Mts.
Seto	Jap.	Strait
Settentrionale	It.	Northern
Severni, Severnaya	Russ.	North
Shamo	Ch.	Desert
Shan	Ch., Jap.	Hill, Mts.
Shankou	Ch.	Pass
Shatt	Arabic	River
Shima	Jap.	Island
Shimo	Jap.	Lower
Shin	Jap.	Land
Shiro	Jap.	White
Shoto	Jap.	Islands
Si	Ch.	West
Sierra	Sp.	Mt. Range, Mts.
Sjö	Nor., Sw.	Lake, Sea
Sok, Suk, Souk	Arabic	Market
Song	Annamese	River
Sopka	Russ.	Volcano
Spitze	Ger.	Mt. Peak
Sredni, Srednyaya	Russ.	Middle
Stad	Dan., Nor., Sw.	City
Stari, Staraya	Russ.	Old
Step	Russ.	Treeless Plain
Straat	Du.	Strait
Strasse	Ger.	Strait
Stretto	It.	Strait
Ström	Dan., Nor., Sw.	Sound
Stung	Camb., Khmer	River
Su	Turk.	River
Sud, Süd	Sp., Fr., Ger.	South
Suido	Jap.	Strait, Channel
Sul	Port.	South
Sund	Dan., Nor., Sw.	Sound
Sungei	Malay	River
Supérieur	Fr.	Upper
Superior, Superiore	Sp., It.	Upper
Sur	Sp.	South
Suyu	Turk.	River
Ta	Ch.	Big
Tafelland	Du.	Plateau
Tagh	Turk.	Mt. Range
Take	Jap.	Peak, Ridge
Takht	Arabic	Lower
Tal	Ger.	Valley
Tanjung	Malay	Cape, Point
Tell	Arabic	Hill
Thale	Thai	Sea, Lake
Tind	Nor.	Peak
Tö	Jap.	East
To	Jap.	Island
Toge	Jap.	Pass
Trask	Finn.	Lake
Tugh	Somali	Dry River
Ujung	Malay	Point
Umi	Jap.	Bay
Unter	Ger.	Lower
Ura	Jap.	Inlet
Uul	Mong.	Mountain
Val	Fr.	Valley
Vatn	Nor.	Lake
Vecchio	It.	Old
Veld	Du.	Plain, Field
Velho	Port.	Old
Verkhni	Russ.	Upper
Vesi	Finn.	Lake
Viejo	Sp.	Old
Vik	Nor., Sw.	Bay
Vishni, Vishnyaya	Russ.	High
Vodokhranilishche	Russ.	Reservoir
Volcán	Sp.	Volcano
Vostochni, Vostochnaya	Russ.	East, Eastern
Wadi	Arabic	Dry River
Wald	Ger.	Forest
Wan	Jap.	Bay
Westersch	Du.	Western
Wüste	Ger.	Desert
Yama	Jap.	Mountain
Yug, Yuzhni, Yuzhnaya	Russ.	South, Southern
Zaki	Jap.	Cape
Zaliv	Russ.	Bay, Gulf
Zangbo	Tibetan	River, Stream
Zapadni, Zapadnaya	Russ.	Western
Zee	Du.	Sea
Zemlya	Russ.	Land
Zizhiqu	Ch.	Autonomous Region
Zuid	Du.	South

MAP PROJECTIONS

by Erwin Raisz

Our earth is rotating around its *axis* once a day. The two end points of its axis are the *poles*; the line circling the earth midway between the poles is the *equator*. The arc from either of the poles to the equator is divided into 90 *degrees*. The distance, expressed in degrees, from the equator to any point is its *latitude* and circles of equal latitude are the *parallels*. On maps it is customary to show parallels of evenly-spaced degrees such as every fifth or every tenth.

The equator is divided into 360 degrees. Lines circling from pole to pole through the degree points on the equator are called *meridians*. They are all equal in length but by international agreement the meridian passing through the Greenwich Observatory in London has been chosen as *prime meridian*. The distance, expressed in degrees, from the prime meridian to any point is its *longitude*. While meridians are all equal in length, parallels become shorter and shorter as they approach the poles. Whereas one degree of latitude represents everywhere approximately 69 miles, one degree of longitude varies from 69 miles at the equator to nothing at the poles.

Each degree is divided into 60 minutes and each minute into 60 seconds. One minute of latitude equals a nautical mile.

The map is flat but the earth is nearly spherical. Neither a rubber ball nor any part of a rubber ball may be flattened without stretching or tearing unless the part is very small. To present the curved surface of the earth on a flat map is not difficult as long as the areas under consideration are small, but the mapping of countries, continents, or the whole earth requires some kind of *projection*. Any regular set of parallels and meridians upon which a map can be drawn makes a map projection. Many systems are used.

In any projection only the parallels or the meridians or some other set of lines can be *true* (the same length as on the globe of corresponding scale); all other lines are too long or too short. Only on a globe is it possible to have both the parallels and the meridians true. The scale given on a flat map cannot be true everywhere. The construction of the various projections begins usually with laying out the parallels or meridians which have true lengths.

RECTANGULAR PROJECTION — This is a set of evenly-placed meridians and horizontal parallels. The central or *standard parallel* and all meridians are true. All other parallels are either too long or too short. The projection is used for simple maps of small areas, as city plans, etc.

MERCATOR PROJECTION — In this projection the meridians are evenly-spaced vertical lines. The parallels are horizontal, spaced so that their length has the same relation to the meridians as on a globe. As the meridians converge at higher latitudes on the globe, while on the map they do not, the parallels have to be drawn also farther and farther apart to maintain the correct relationship. When every very small area has the same shape as on a globe we call the projection *conformal*. The most interesting quality of this projection is that all *compass directions* appear as straight lines. For this reason it is generally used for marine charts. It is also frequently used for world maps in spite of the fact that the high latitudes are very much exaggerated in size. Only the equator is true to scale; all other parallels and meridians are too long. The Mercator projection did *not* derive from projecting a globe upon a cylinder.

SINUSOIDAL PROJECTION — The parallels are truly-spaced horizontal lines. They are divided truly and the connecting curves make the meridians. It does not make a good world map because the outer regions are distorted, but the

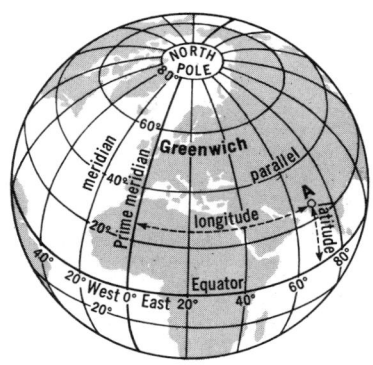

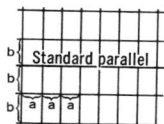

Rectangular Projection

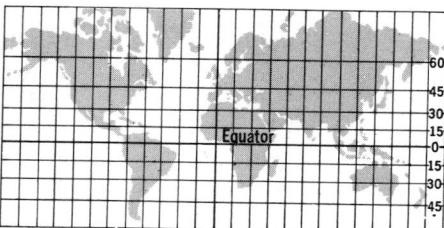

Mercator Projection

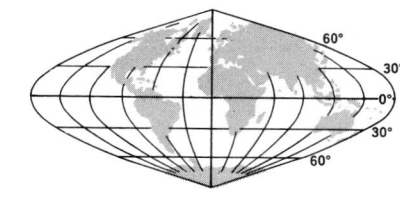

Sinusoidal Projection

central portion is good and this part is often used for maps of Africa and South America. Every part of the map has the same area as the corresponding area on the globe. It is an *equal-area* projection.

MOLLWEIDE PROJECTION — The meridians are equally-spaced ellipses; the parallels are horizontal lines spaced so that every belt of latitude should have the same area as on a globe. This projection is popular for world maps, especially in European atlases.

GOODE'S INTERRUPTED PROJECTIONS—Only the good central part of the Mollweide or sinusoidal (or both) projection is used and the oceans are cut. This makes an equal-area map with little distortion of shape. It is commonly used for world maps.

ECKERT PROJECTIONS — These are similar to the sinusoidal or the Mollweide projections, but the poles are shown as lines half the length of the equator. There are several variants; the meridians are either sine curves or ellipses; the parallels are horizontal and spaced either evenly or so as to make the projection equal area. Their use for world maps is increasing. The figure shows the elliptical equal-area variant.

CONIC PROJECTION — The original idea of the conic projection is that of capping the globe by a cone upon which both the parallels and meridians are projected from the center of the globe. The cone is then cut open and laid flat. A cone can be made tangent to any chosen *standard parallel*.

The actually-used conic projection is a modification of this idea. The radius of the standard parallel is obtained as above. The meridians are straight radiating lines spaced truly on the standard parallel. The parallels are concentric circles spaced at true distances. All parallels except the standard are too long. The projection is used for maps of countries in middle latitudes, as it presents good shapes with small scale error.

There are several variants: The use of *two standard parallels*, one near the top, the other near the bottom of the map, reduces the scale error. In the *Albers projection* the parallels are spaced unevenly, to make the projection equal-area. This is a good projection for the United States. In the *Lambert conformal conic projection* the parallels are spaced so that any small quadrangle of the grid should have the same shape as on the globe. This is the best projection for air-navigation charts as it has relatively straight azimuths.

An *azimuth* is a great-circle direction reckoned clockwise from north. A *great-circle direction* points to a place along the shortest line on the earth's surface. This is not the same as compass direction. The center of a great circle is the center of the globe.

BONNE PROJECTION — The parallels are laid out exactly as in the conic projection. All parallels are divided truly and the connecting curves make the meridians. It is an equal-area projection. It is used for maps of the northern continents, as Asia, Europe, and North America.

POLYCONIC PROJECTION — The central meridian is divided truly. The parallels are non-concentric circles, the radii of which are obtained by drawing tangents to the globe as though the globe were covered by several cones rather than by only one. Each parallel is divided truly and the connecting curves make the meridians. All meridians except the central one are too long. This projection is used for large-scale topographic sheets — less often for countries or continents.

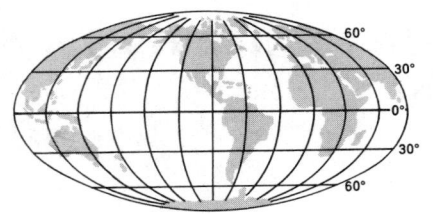

Mollweide Projection

Goode's Interrupted Projection

Eckert Projection

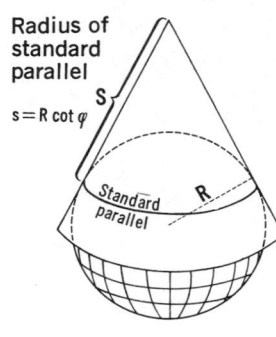

Radius of standard parallel

$s = R \cot \varphi$

Conic Projection

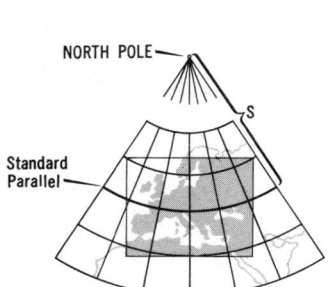

Albers Projection

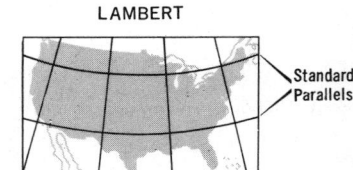

Lambert Conformal Conic Projection

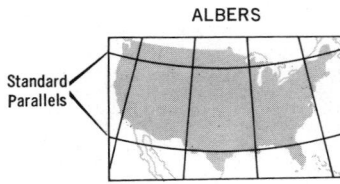

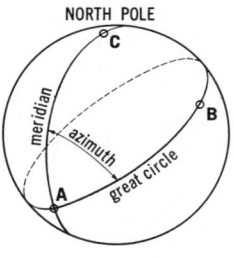

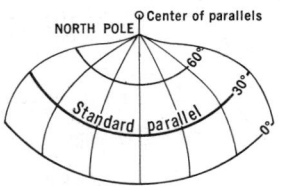

Bonne Projection

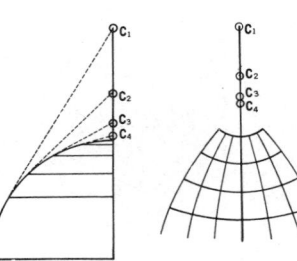

Polyconic Projection

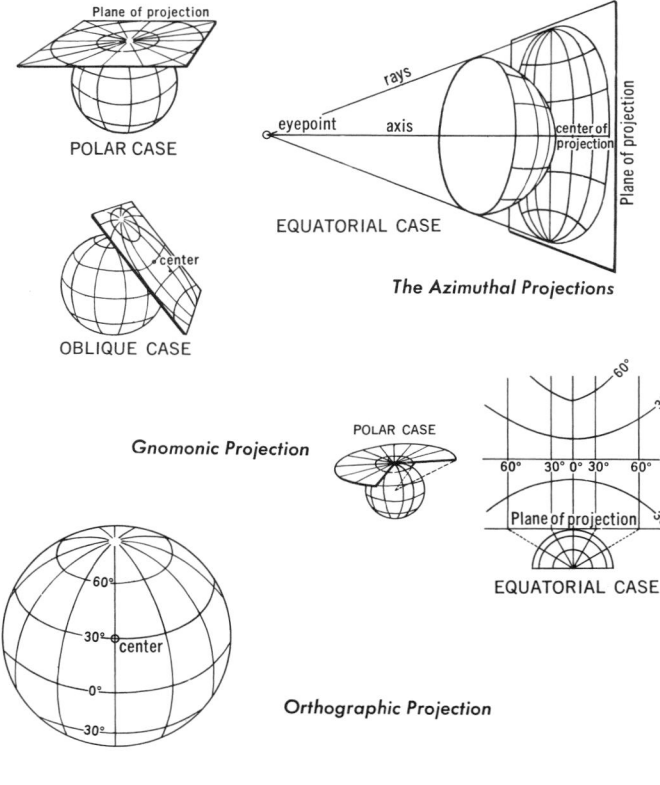

The Azimuthal Projections

Gnomonic Projection

Orthographic Projection

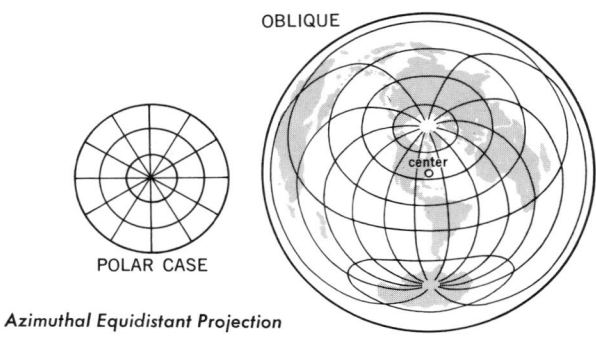

Azimuthal Equidistant Projection

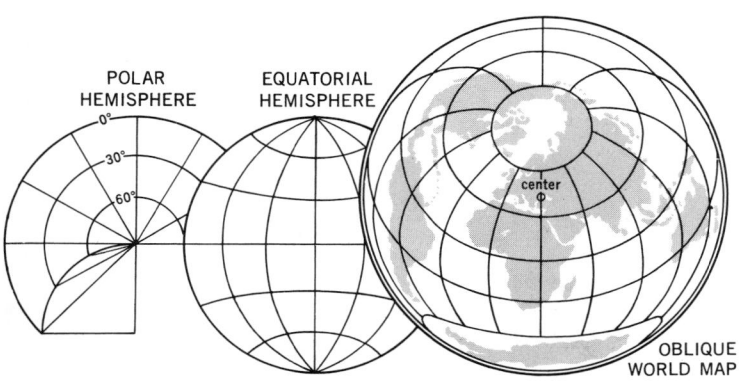

Lambert Azimuthal Equal-Area Projection

THE AZIMUTHAL PROJECTIONS — In this group a part of the globe is projected from an eyepoint onto a plane. The eyepoint can be at different distances, making different projections. The plane of projection can be tangent at the equator, at a pole, or at any other point on which we want to focus attention. The most important quality of all azimuthal projections is that they show every point at its true direction (azimuth) from the center point, and all points equally distant from the center point will be equally distant on the map also.

GNOMONIC PROJECTION — This projection has the eyepoint at the center of the globe Only the central part is good; the outer regions are badly distorted. Yet the projection has one important quality, all great circles being shown as straight lines. For this reason it is used for laying out the routes for long range flying or trans-oceanic navigation.

ORTHOGRAPHIC PROJECTION — This projection has the eyepoint at infinite distance and the projecting rays are parallel. The polar or equatorial varieties are rare but the oblique case became very popular on account of its visual quality. It looks like a picture of a globe. Although the distortion on the peripheries is extreme, we see it correctly because the eye perceives it not as a map but as a picture of a three-dimensional globe. Obviously only a hemisphere (half globe) can be shown.

Some azimuthal projections do not derive from the actual process of projecting from an eyepoint, but are arrived at by other means:

AZIMUTHAL EQUIDISTANT PROJECTION — This is the only projection in which every point is shown both at true great-circle direction and at true distance from the center point, but all other directions and distances are distorted. The principle of the projection can best be understood from the polar case. Most polar maps are in this projection. The oblique case is used for radio direction finding, for earthquake research, and in long-distance flying. A separate map has to be constructed for each central point selected.

LAMBERT AZIMUTHAL EQUAL-AREA PROJECTION—The construction of this projection can best be understood from the polar case. All three cases are widely used. It makes a good polar map and it is often extended to include the southern continents. It is the most common projection used for maps of the Eastern and Western Hemispheres, and it is a good projection for continents as it shows correct areas with relatively little distortion of shape. Most of the continent maps in this atlas are in this projection.

IN THIS ATLAS, on almost all maps, parallels and meridians have been marked because they are useful for the following:

(a) They show the north-south and east-west directions which appear on many maps at oblique angles especially near the margins.

(b) With the help of parallels and meridians every place can be exactly located; for instance, New York City is at 41° N and 74° W on any map.

(c) They help to measure distances even in the distorted parts of the map. The scale given on each map is true only along certain lines which are specified in the foregoing discussion for each projection. One degree of latitude equals nearly 69 statute miles or 60 nautical miles. The length of one degree of longitude varies (1° long. = 1° lat. × cos lat.).

WORLD STATISTICAL TABLES

Elements of the Solar System

	Mean Distance from Sun: in Miles	in Kilometers	Period of Revolution around Sun	Period of Rotation on Axis	Equatorial Diameter: in Miles	in Kilometers	Surface Gravity (Earth = 1)	Mass (Earth = 1)	Mean Density (Water = 1)	Number of Satellites
MERCURY	35,990,000	57,900,000	87.97 days	59 days	3,032	4,880	0.38	0.055	5.5	0
VENUS	67,240,000	108,200,000	224.70 days	243 days†	7,523	12,106	0.90	0.815	5.25	0
EARTH	93,000,000	149,700,000	365.26 days	23h 56m	7,926	12,755	1.00	1.00	5.5	1
MARS	141,730,000	228,100,000	687.00 days	24h 37m	4,220	6,790	0.38	0.107	4.0	2
JUPITER	483,880,000	778,700,000	11.86 years	9h 50m	88,750	142,800	2.87	317.9	1.3	16
SATURN	887,130,000	1,427,700,000	29.46 years	10h 14m	74,580	120,020	1.32	95.2	0.7	17
URANUS	1,783,700,000	2,870,500,000	84.01 years	10h 49m†	31,600	50,900	0.93	14.6	1.3	5
NEPTUNE	2,795,500,000	4,498,800,000	164.79 years	15h 48m	30,200	48,600	1.23	17.2	1.8	3
PLUTO	3,667,900,000	5,902,800,000	247.70 years	6.39 days (?)	1,500	2,400	0.03 (?)	0.01(?)	0.7(?)	1

†Retrograde motion

Facts About the Sun

Equatorial diameter	865,000 miles	1,392,000 kilometers
Period of rotation on axis	25-35 days*	
Orbit of galaxy	every 225 million years	
Surface gravity (Earth = 1)	27.8	
Mass (Earth = 1)	333,000	
Density (Water = 1)	1.4	
Mean distance from Earth	93,000,000 miles	149,700,000 kilometers

*Rotation of 25 days at Equator, decreasing to about 35 days at the poles.

Facts About the Moon

Equatorial diameter	2,160 miles	3,476 kilometers
Period of rotation on axis	27 days, 7 hours, 43 minutes	
Period of revolution around Earth (sidereal month)	27 days, 7 hours, 43 minutes	
Phase period between new moons (synodic month)	29 days, 12 hours, 44 minutes	
Surface gravity (Earth = 1)	0.16	
Mass (Earth = 1)	0.0123	
Density (Water = 1)	3.34	
Maximum distance from Earth	252,710 miles	406,690 kilometers
Minimum distance from Earth	221,460 miles	356,400 kilometers
Mean distance from Earth	238,860 miles	384,400 kilometers

Dimensions of the Earth

	Area in Sq. Miles	Sq. Kilometers
Superficial area	197,751,000	512,175,090
Land surface	57,970,000	150,142,300
Water surface	139,781,000	362,032,790

	Miles	Kilometers
Equatorial circumference	24,902	40,075
Polar circumference	24,860	40,007
Equatorial diameter	7,926.68	12,756.4
Polar diameter	7,899.99	12,713.4
Equatorial radius	3,963.34	6,378.2
Polar radius	3,949.99	6,356.7
Volume of the Earth	2.6×10^{11} cubic miles	10.84×10^{11} cubic kilometers
Mass or weight	6.6×10^{21} short tons	6.0×10^{21} metric tons
Maximum distance from Sun	94,600,000 miles	152,000,000 kilometers
Minimum distance from Sun	91,300,000 miles	147,000,000 kilometers

The Continents

	Area in: Sq. Miles	Sq. Km.	Percent of World's Land
Asia	17,128,500	44,362,815	29.5
Africa	11,707,000	30,321,130	20.2
North America	9,363,000	24,250,170	16.2
South America	6,875,000	17,806,250	11.8
Antarctica	5,500,000	14,245,000	9.5
Europe	4,057,000	10,507,630	7.0
Australia	2,966,136	7,682,300	5.1

Oceans and Major Seas

	Area in: Sq. Miles	Sq. Km.	Greatest Depth in: Feet	Meters
Pacific Ocean	64,186,000	166,241,700	36,198	11,033
Atlantic Ocean	31,862,000	82,522,600	28,374	8,648
Indian Ocean	28,350,000	73,426,500	25,344	7,725
Arctic Ocean	5,427,000	14,056,000	17,880	5,450
Caribbean Sea	970,000	2,512,300	24,720	7,535
Mediterranean Sea	969,000	2,509,700	16,896	5,150
Bering Sea	875,000	2,266,250	15,800	4,800
Gulf of Mexico	600,000	1,554,000	12,300	3,750
Sea of Okhotsk	590,000	1,528,100	11,070	3,370
East China Sea	482,000	1,248,400	9,500	2,900
Sea of Japan	389,000	1,007,500	12,280	3,740
Hudson Bay	317,500	822,300	846	258
North Sea	222,000	575,000	2,200	670
Black Sea	185,000	479,150	7,365	2,245
Red Sea	169,000	437,700	7,200	2,195
Baltic Sea	163,000	422,170	1,506	459

Major Ship Canals

	Length in: Miles	Kms.	Minimum Feet	Depth in: Meters
Volga-Baltic, U.S.S.R.	225	362	—	—
Baltic-White Sea, U.S.S.R.	140	225	16	5
Suez, Egypt	100.76	162	42	13
Albert, Belgium	80	129	16.5	5
Moscow-Volga, U.S.S.R.	80	129	18	6
Volga-Don, U.S.S.R.	62	100	—	—
Göta, Sweden	54	87	10	3
Kiel (Nord-Ostsee), W. Ger.	53.2	86	38	12
Panama Canal, Panama	50.72	82	41.6	13
Houston Ship, U.S.A.	50	81	36	11

Largest Islands

	Area in: Sq. Mi.	Sq. Km.		Area in: Sq. Mi.	Sq. Km.		Area in: Sq. Mi.	Sq. Km.
Greenland	840,000	2,175,600	South I., New Zealand	58,393	151,238	Hokkaido, Japan	28,983	75,066
New Guinea	305,000	789,950	Java, Indonesia	48,842	126,501	Banks, Canada	27,038	70,028
Borneo	290,000	751,100	North I., New Zealand	44,187	114,444	Ceylon, Sri Lanka	25,332	65,610
Madagascar	226,400	586,376	Newfoundland, Canada	42,031	108,860	Tasmania, Australia	24,600	63,710
Baffin, Canada	195,928	507,454	Cuba	40,533	104,981	Svalbard, Norway	23,957	62,049
Sumatra, Indonesia	164,000	424,760	Luzon, Philippines	40,420	104,688	Devon, Canada	21,331	55,247
Honshu, Japan	88,000	227,920	Iceland	39,768	103,000	Novaya Zemlya (north isl.), U.S.S.R.	18,600	48,200
Great Britain	84,400	218,896	Mindanao, Philippines	36,537	94,631	Marajó, Brazil	17,991	46,597
Victoria, Canada	83,896	217,290	Ireland	31,743	82,214	Tierra del Fuego, Chile & Argentina	17,900	46,360
Ellesmere, Canada	75,767	196,236	Sakhalin, U.S.S.R.	29,500	76,405	Alexander, Antarctica	16,700	43,250
Celebes, Indonesia	72,986	189,034	Hispaniola, Haiti & Dom. Rep.	29,399	76,143			

Principal Mountains of the World

	Feet	Meters		Feet	Meters		Feet	Meters
Everest, Nepal-China	29,028	8,848	Pissis, Argentina	22,241	6,779	Kazbek, U.S.S.R.	16,512	5,033
Godwin Austen (K2),			Mercedario, Argentina	22,211	6,770	Puncak Jaya, Indonesia	16,503	5,030
Pakistan-China	28,250	8,611	Huascarán, Peru	22,205	6,768	Tyree, Antarctica	16,289	4,965
Kanchenjunga, Nepal-India	28,208	8,598	Llullaillaco, Chile-Argentina	22,057	6,723	Blanc, France	15,771	4,807
Lhotse, Nepal-China	27,923	8,511	Nevada Ancohuma, Bolivia	21,489	6,550	Klyuchevskaya Sopka, U.S.S.R.	15,584	4,750
Makalu, Nepal-China	27,824	8,481	Illampu, Bolivia	21,276	6,485	Fairweather (Br. Col., Canada)	15,300	4,663
Dhaulagiri, Nepal	26,810	8,172	Chimborazo, Ecuador	20,561	6,267	Dufourspitze (Mte. Rosa), Italy-		
Nanga Parbat, Pakistan	26,660	8,126	McKinley, Alaska	20,320	6,194	Switzerland	15,203	4,634
Annapurna, Nepal	26,504	8,078	Logan, Canada (Yukon)	19,524	5,951	Ras Dashan, Ethiopia	15,157	4,620
Gasherbrum, Pakistan-China	26,740	8,068	Cotopaxi, Ecuador	19,347	5,897	Matterhorn, Switzerland	14,691	4,478
Nanda Devi, India	25,645	7,817	Kilimanjaro, Tanzania	19,340	5,895	Whitney, California, U.S.A.	14,494	4,418
Rakaposhi, Pakistan	25,550	7,788	El Misti, Peru	19,101	5,822	Elbert, Colorado, U.S.A.	14,433	4,399
Kamet, India	25,447	7,756	Pico Cristóbal Colón, Colombia	19,029	5,800	Rainier, Washington, U.S.A.	14,410	4,392
Gurla Mandhada, China	25,355	7,728	Huila, Colombia	18,865	5,750	Shasta, California, U.S.A.	14,162	4,350
Kongur Shan, China	25,325	7,719	Citlaltépetl (Orizaba), Mexico	18,855	5,747	Pikes Peak, Colorado, U.S.A.	14,110	4,301
Tirich Mir, Pakistan	25,230	7,690	El'brus, U.S.S.R.	18,510	5,642	Finsteraarhorn, Switzerland	14,022	4,274
Gongga Shan, China	24,790	7,556	Damavand, Iran	18,376	5,601	Mauna Kea, Hawaii, U.S.A.	13,796	4,205
Muztagata, China	24,757	7,546	St. Elias, Alaska-Canada			Mauna Loa, Hawaii, U.S.A.	13,677	4,169
Communism Peak, U.S.S.R.	24,599	7,498	(Yukon)	18,008	5,489	Jungfrau, Switzerland	13,642	4,158
Pobeda Peak, U.S.S.R.	24,406	7,439	Vilcanota, Peru	17,999	5,486	Cameroon, Cameroon	13,350	4,069
Chomo Lhari, Bhutan-China	23,997	7,314	Popocatépetl, Mexico	17,887	5,452	Grossglockner, Austria	12,457	3,797
Muztag, China	23,891	7,282	Dykhtau, U.S.S.R.	17,070	5,203	Fuji, Japan	12,389	3,776
Cerro Aconcagua, Argentina	22,831	6,959	Kenya, Kenya	17,058	5,199	Cook, New Zealand	12,349	3,764
Ojos del Salado, Chile-Argentina	22,572	6,880	Ararat, Turkey	16,946	5,165	Etna, Italy	11,053	3,369
Bonete, Chile-Argentina	22,541	6,870	Vinson Massif, Antarctica	16,864	5,140	Kosciusko, Australia	7,310	2,228
Tupungato, Chile-Argentina	22,310	6,800	Margherita (Ruwenzori), Africa	16,795	5,119	Mitchell, North Carolina, U.S.A.	6,684	2,037

Longest Rivers of the World

	Length in:			Length in:			Length in:	
	Miles	Kms.		Miles	Kms.		Miles	Kms.
Nile, Africa	4,145	6,671	São Francisco, Brazil	1,811	2,914	Ohio-Allegheny, U.S.A.	1,306	2,102
Amazon, S. Amer.	3,915	6,300	Indus, Asia	1,800	2,897	Kama, U.S.S.R.	1,262	2,031
Chang Jiang (Yangtze), China	3,900	6,276	Danube, Europe	1,775	2,857	Red, U.S.A.	1,222	1,966
Mississippi-Missouri-Red Rock, U.S.A.	3,741	6,019	Salween, Asia	1,770	2,849	Don, U.S.S.R.	1,222	1,967
Ob'Irtysh-Black Irtysh, U.S.S.R.	3,362	5,411	Brahmaputra, Asia	1,700	2,736	Columbia, U.S.A.-Canada	1,214	1,953
Yenisey-Angara, U.S.S.R.	3,100	4,989	Euphrates, Asia	1,700	2,736	Saskatchewan, Canada	1,205	1,939
Huang He (Yellow), China	2,877	4,630	Tocantins, Brazil	1,677	2,699	Peace-Finlay, Canada	1,195	1,923
Amur-Shilka-Onon, Asia	2,744	4,416	Xi (Si), China	1,650	2,655	Tigris, Asia	1,181	1,901
Lena, U.S.S.R.	2,734	4,400	Amudar'ya, Asia	1,616	2,601	Darling, Australia	1,160	1,867
Congo (Zaire), Africa	2,718	4,374	Nelson-Saskatchewan, Canada	1,600	2,575	Angara, U.S.S.R.	1,135	1,827
Mackenzie-Peace-Finlay, Canada	2,635	4,241	Orinoco, S. Amer.	1,600	2,575	Sungari, Asia	1,130	1,819
Mekong, Asia	2,610	4,200	Zambezi, Africa	1,600	2,575	Pechora, U.S.S.R.	1,124	1,809
Missouri-Red Rock, U.S.A.	2,564	4,125	Paraguay, S. Amer.	1,584	2,549	Snake, U.S.A.	1,000	1,609
Niger, Africa	2,548	4,101	Kolyma, U.S.S.R.	1,562	2,514	Churchill, Canada	1,000	1,609
Paraná-La Plata, S. Amer.	2,450	3,943	Ganges, Asia	1,550	2,494	Pilcomayo, S. Amer.	1,000	1,609
Mississippi, U.S.A.	2,348	3,778	Ural, U.S.S.R.	1,509	2,428	Magdalena, Colombia	1,000	1,609
Murray-Darling, Australia	2,310	3,718	Japurá, S. Amer.	1,500	2,414	Uruguay, S. Amer.	994	1,600
Volga, U.S.S.R.	2,194	3,531	Arkansas, U.S.A.	1,450	2,334	Platte-N. Platte, U.S.A.	990	1,593
Madeira, S. Amer.	2,013	3,240	Colorado, U.S.A.-Mexico	1,450	2,334	Ohio, U.S.A.	981	1,578
Purus, S. Amer.	1,995	3,211	Negro, S. Amer.	1,400	2,253	Pecos, U.S.A.	926	1,490
Yukon, Alaska-Canada	1,979	3,185	Dnieper, U.S.S.R.	1,368	2,202	Oka, U.S.S.R.	918	1,477
St. Lawrence, Canada-U.S.A.	1,900	3,058	Orange, Africa	1,350	2,173	Canadian, U.S.A.	906	1,458
Rio Grande, Mexico-U.S.A.	1,885	3,034	Irrawaddy, Burma	1,325	2,132	Colorado, Texas, U.S.A.	894	1,439
Syrdar'ya-Naryn, U.S.S.R.	1,859	2,992	Brazos, U.S.A.	1,309	2,107	Dniester, U.S.S.R.	876	1,410

Principal Natural Lakes

	Area in:		Max. Depth in:			Area in:		Max. Depth in:	
	Sq. Miles	Sq. Km.	Feet	Meters		Sq. Miles	Sq. Km.	Feet	Meters
Caspian Sea, U.S.S.R.-Iran	143,243	370,999	3,264	995	Lake Eyre, Australia	3,500-0	9,000-0	—	—
Lake Superior, U.S.A.-Canada	31,820	82,414	1,329	405	Lake Titicaca, Peru-Bolivia	3,200	8,288	1,000	305
Lake Victoria, Africa	26,724	69,215	270	82	Lake Nicaragua, Nicaragua	3,100	8,029	230	70
Aral Sea, U.S.S.R.	25,676	66,501	256	78	Lake Athabasca, Canada	3,064	7,936	400	122
Lake Huron, U.S.A.-Canada	23,010	59,596	748	228	Reindeer Lake, Canada	2,568	6,651	—	—
Lake Michigan, U.S.A.	22,400	58,016	923	281	Lake Turkana (Rudolf), Africa	2,463	6,379	240	73
Lake Tanganyika, Africa	12,650	32,764	4,700	1,433	Issyk-Kul', U.S.S.R.	2,425	6,281	2,303	702
Lake Baykal, U.S.S.R.	12,162	31,500	5,316	1,620	Lake Torrens, Australia	2,230	5,776	—	—
Great Bear Lake, Canada	12,096	31,328	1,356	413	Vänern, Sweden	2,156	5,584	328	100
Lake Nyasa (Malawi), Africa	11,555	29,928	2,320	707	Nettiling Lake, Canada	2,140	5,543	—	—
Great Slave Lake, Canada	11,031	28,570	2,015	614	Lake Winnipegosis, Canada	2,075	5,374	38	12
Lake Erie, U.S.A.-Canada	9,940	25,745	210	64	Lake Mobutu Sese Seko (Albert),				
Lake Winnipeg, Canada	9,417	24,390	60	18	Africa	2,075	5,374	160	49
Lake Ontario, U.S.A.-Canada	7,540	19,529	775	244	Kariba Lake, Zambia-Zimbabwe	2,050	5,310	295	90
Lake Ladoga, U.S.S.R.	7,104	18,399	738	225	Lake Nipigon, Canada	1,872	4,848	540	165
Lake Balkhash, U.S.S.R.	7,027	18,200	87	27	Lake Mweru, Zaire-Zambia	1,800	4,662	60	18
Lake Maracaibo, Venezuela	5,120	13,261	100	31	Lake Manitoba, Canada	1,799	4,659	12	4
Lake Chad, Africa	4,000-	10,360-			Lake Taymyr, U.S.S.R.	1,737	4,499	85	26
	10,000	25,900	25	8	Lake Khanka, China-U.S.S.R.	1,700	4,403	33	10
Lake Onega, U.S.S.R.	3,710	9,609	377	115	Lake Kioga, Uganda	1,700	4,403	25	8

Foreign City Weather

Two figures are given for each of the months, thus 88/73. The first figure is the average daily high temperature (°F) and the second is the average daily low temperature (°F) for the month. The boldface figures indicate the average number of days with rain for each month.

City	January	February	March	April	May	June	July	August	September	October	November	December
ABIDJAN, Ivory Coast	88/73 3	90/75 4	90/75 6	90/75 9	88/75 16	85/73 18	83/73 8	82/71 7	83/73 8	85/74 13	87/74 13	88/74 6
ACAPULCO, Mexico	85/70 0	87/70 0	87/70 4	87/71 1	89/74 4	89/76 15	89/75 14	89/75 14	88/75 18	88/74 12	88/72 4	87/70 1
ACCRA, Ghana	87/73 1	88/75 2	88/76 4	88/76 6	87/75 9	84/74 10	81/73 4	80/71 3	81/73 4	85/74 6	87/75 3	88/75 2
ADDIS ABABA, Ethiopia	75/43 2	76/47 5	77/49 8	77/50 10	77/50 10	74/49 20	69/50 28	69/50 27	72/49 21	75/45 3	73/43 3	73/41 2
ALGIERS, Algeria	59/49 11	61/49 9	63/52 9	68/55 5	73/59 5	78/65 2	83/70 1	85/71 1	81/69 4	74/63 7	66/56 11	60/51 12
AMSTERDAM, Netherlands	40/34 19	41/34 15	46/37 13	52/43 14	60/50 12	65/55 12	69/59 14	68/59 14	64/56 15	56/48 18	47/41 19	41/35 19
ANKARA, Turkey	39/24 8	42/26 8	51/31 7	63/40 7	73/49 7	78/53 5	86/59 9	87/59 1	78/52 3	69/44 5	57/37 6	43/29 9
APIA, Western Samoa	86/75 22	85/76 19	86/74 19	86/75 14	85/74 12	85/74 7	84/75 9	84/75 11	84/74 11	85/75 14	86/74 16	85/74 19
ATHENS, Greece	54/42 7	55/43 6	60/46 5	67/52 3	77/60 3	85/73 2	90/72 1	90/72 1	83/66 2	74/60 4	64/52 6	57/46 7
BAGHDAD, Iraq	60/39 4	64/42 4	71/48 4	85/57 7	97/67 0	105/73 0	110/76 0	110/76 0	104/70 0	92/61 1	77/51 3	64/42 5
BALI, Indonesia	88/74 19	88/74 14	88/74 13	88/74 7	88/73 5	87/71 3	87/70 1	87/70 1	89/71 1	90/73 2	90/75 6	88/74 14
BANGKOK, Thailand	89/68 1	91/72 1	93/75 3	95/77 3	93/77 9	91/76 10	90/76 13	90/76 13	89/76 15	88/75 14	87/72 7	87/68 1
BARCELONA, Spain	56/42 5	57/44 7	61/47 7	64/51 8	71/57 8	77/63 5	81/69 4	82/69 5	78/65 7	71/58 8	62/50 7	57/44 6
BEIRUT, Lebanon	62/51 15	63/51 12	66/54 9	72/58 5	78/64 2	84/74 11	83/73 11	84/74 11	84/74 9	85/74 9	85/73 12	85/73 15
BELFAST, Northern Ireland	45/34 22	47/34 18	49/35 20	53/39 18	59/43 17	64/49 10	66/51 18	65/51 20	62/48 17	55/42 19	52/39 21	46/35 25
BELGRADE, Yugoslavia	37/27 8	41/27 6	53/35 7	64/45 9	74/53 9	79/58 9	84/61 6	83/60 7	76/55 6	65/47 8	52/39 7	40/30 9
BERLIN, Germany	35/26 10	38/27 8	46/32 9	55/38 9	55/46 8	70/51 9	74/55 10	72/54 10	65/47 8	55/41 8	43/33 8	37/29 11
BIARRITZ, France	54/40 10	52/38 11	63/43 11	63/44 11	69/44 11	72/56 10	72/56 10	77/61 7	77/58 9	74/55 11	58/44 12	53/41 14
BOGOTA, Colombia	67/48 6	68/49 7	67/50 13	67/51 20	66/51 17	65/51 16	64/50 18	65/50 16	66/49 13	66/50 20	66/50 16	66/49 15
BOMBAY, India	83/67 1	83/67 1	86/72 1	89/76 1	91/80 1	89/79 14	85/77 21	85/76 19	85/76 13	89/76 3	89/73 1	87/69 1
BONN, West Germany	39/30 7	37/26 6	50/35 7	58/39 14	67/46 13	69/52 19	73/56 16	72/55 17	67/50 16	58/45 16	47/37 15	44/36 16
BRASILIA, Brazil	80/65 17	81/64 20	82/64 7	82/62 10	79/56 5	77/52 2	78/51 2	80/64 2	82/60 2	82/64 16	82/66 17	78/64 16
BRINDISI, Italy	55/43 10	57/43 5	60/45 5	65/50 5	73/57 5	80/64 2	84/68 4	84/69 3	80/65 4	70/58 8	64/52 10	58/46 8
BUCHAREST, Romania	33/20 6	38/24 5	51/33 6	63/41 6	74/51 8	81/58 9	86/61 7	86/60 5	76/53 5	65/44 5	49/35 6	37/26 6
BUDAPEST, Hungary	35/26 7	40/28 6	51/36 7	62/44 8	72/52 9	78/57 8	82/61 7	81/59 6	74/53 7	69/50 8	47/37 8	38/31 9
BUENOS AIRES, Argentina	85/63 7	83/63 6	79/60 7	72/53 8	64/47 7	57/41 7	57/42 8	60/43 9	64/46 8	69/50 7	76/56 9	82/61 8
CAIRO, Egypt	65/47 1	69/48 1	75/52 1	83/57 1	91/63 1	95/68 0	96/70 0	95/71 0	90/68 0	86/65 1	78/58 1	68/50 1
CALCUTTA, India	80/55 1	84/59 1	93/69 2	97/75 3	96/77 2	92/79 13	89/79 18	89/78 18	90/78 13	89/74 6	84/64 1	79/55 1
CAPE TOWN, South Africa	78/60 3	79/60 2	77/58 3	72/53 6	67/49 9	65/46 9	63/45 10	64/46 9	65/49 7	70/52 5	73/55 3	78/58 10
CARACAS, Venezuela	75/56 6	77/56 7	79/58 3	81/60 4	80/62 9	78/62 14	78/61 15	79/61 15	80/61 13	79/61 12	77/60 13	78/60 13
CHARLOTTE AMALIE, Virgin Islands	82/73 18	81/72 13	82/73 12	83/74 13	85/76 15	86/77 18	88/78 19	88/78 19	87/78 17	87/77 18	85/76 19	83/74 18
COLOMBO, Sri Lanka	86/72 7	87/72 6	88/74 8	88/76 14	87/78 19	85/77 11	85/77 12	85/77 11	85/77 13	85/75 19	85/73 16	85/72 11
COPENHAGEN, Denmark	36/29 9	36/28 7	41/31 8	50/37 7	61/44 8	67/51 8	72/55 9	69/54 12	63/49 8	53/42 9	43/35 10	38/32 11
DARWIN, Australia	90/77 20	90/77 18	91/77 17	92/76 6	91/73 1	88/69 1	87/67 0	89/70 0	91/74 2	93/77 5	94/78 10	92/78 15
DJAKARTA, Indonesia	84/74 18	84/74 17	86/74 15	87/75 11	87/75 9	87/74 7	87/73 5	87/73 4	88/73 4	87/74 12	86/74 12	86/74 12
DUBLIN, Ireland	47/35 13	47/35 11	51/36 10	54/38 11	59/42 11	65/48 11	67/51 13	67/51 13	63/47 12	57/43 12	51/38 12	47/36 13
EDINBURGH, Scotland	43/35 18	43/35 15	47/36 15	50/39 16	55/43 15	62/48 15	65/52 17	64/52 17	60/48 16	53/44 18	47/39 18	44/36 17
FLORENCE, Italy	49/35 9	53/36 9	60/40 7	68/46 7	75/53 9	84/58 5	89/63 4	88/62 4	81/58 6	69/51 9	58/42 10	50/37 9
GENEVA, Switzerland	39/29 10	43/30 9	51/35 10	58/41 11	66/48 12	73/55 11	77/58 9	76/57 10	69/52 10	58/44 11	47/37 11	40/31 10
GUAYAQUIL, Ecuador	88/70 20	88/70 20	88/72 24	89/71 14	88/68 9	87/68 4	84/67 2	86/65 2	87/66 2	86/68 3	88/68 4	88/70 10
HAMBURG, West Germany	35/28 12	37/30 10	42/33 10	51/39 11	60/47 9	67/53 10	69/56 12	67/55 13	63/51 10	53/44 11	44/36 11	38/31 12
HAMILTON, Bermuda	68/58 14	68/57 13	68/57 12	71/59 9	76/64 9	81/69 9	85/73 10	89/75 10	84/72 10	79/69 12	74/63 13	70/60 15
HAVANA, Cuba	79/65 6	79/65 8	81/67 4	84/69 4	86/72 7	88/74 10	89/75 9	88/75 11	88/75 11	85/73 11	81/69 7	79/67 6
HELSINKI, Finland	27/17 11	26/15 8	32/22 8	43/31 8	55/41 8	63/49 9	71/57 7	66/55 12	57/46 11	45/37 12	37/30 11	31/22 11
HONG KONG	64/56 4	63/55 5	67/60 7	75/67 8	82/74 13	85/78 18	87/78 17	87/78 15	85/77 12	81/73 7	74/65 2	68/59 3
JERUSALEM, Israel	55/41 9	56/42 11	65/46 3	73/50 3	81/57 1	85/60 1	87/63 0	87/64 0	85/62 1	81/59 1	70/53 4	59/45 7
JOHANNESBURG, South Africa	78/58 12	77/58 9	75/55 9	72/50 4	66/43 3	62/39 1	63/39 1	68/43 2	73/48 2	77/53 7	77/55 10	78/57 11
KARACHI, Pakistan	77/55 1	79/58 1	85/67 1	90/73 7	93/79 1	93/82 3	91/81 2	88/79 2	88/77 1	91/72 1	87/64 1	80/57 1
KINGSTON, Jamaica	86/67 3	86/67 3	86/68 2	87/70 3	87/72 4	89/74 5	90/73 4	90/73 7	89/73 6	88/73 9	87/71 5	87/69 4
LAGOS, Nigeria	88/74 2	89/77 3	89/78 7	89/77 10	87/76 16	85/74 20	83/74 16	82/73 10	83/74 14	85/74 16	88/75 7	88/75 2
LA PAZ, Bolivia	63/43 21	63/43 18	64/42 16	65/40 9	64/37 5	62/34 2	62/33 1	63/35 4	64/38 9	66/40 9	67/42 11	65/42 18

Foreign City Weather

	January	February	March	April	May	June	July	August	September	October	November	December
LAS PALMAS, Canary Is.	70/58 8	71/58 5	71/59 5	71/61 3	73/62 1	75/65 1	77/67 1	79/70 1	79/69 1	79/67 5	76/64 7	72/60 8
LENINGRAD, USSR	23/12 17	24/12 15	33/18 13	45/31 11	58/42 12	66/51 12	71/57 13	66/53 15	57/45 14	45/37 15	34/27 17	26/18 18
LIMA, Peru	82/66 1	83/67 1	83/66 1	80/63 1	74/60 1	68/58 1	67/57 1	66/56 1	68/57 1	71/58 1	74/60 1	78/62 1
LISBON, Portugal	56/46 9	58/47 8	61/49 10	64/52 7	69/56 6	75/60 2	79/63 1	80/64 1	76/62 4	69/57 7	62/52 10	57/47 10
LIVERPOOL, England	44/36 18	44/36 13	48/38 13	52/41 14	58/46 14	63/51 13	66/55 15	65/55 16	61/51 15	55/46 17	48/41 17	45/37 18
LONDON, England	44/35 17	45/35 13	51/47 11	56/40 14	63/45 13	69/51 11	73/55 13	72/54 13	67/51 13	58/44 14	49/39 16	45/36 16
MADRID, Spain	47/33 9	51/35 9	57/40 11	64/44 9	71/50 9	80/57 6	87/62 3	86/62 2	77/56 6	66/48 8	54/40 10	48/35 9
MANILA, Philippines	86/69 6	88/69 3	91/71 4	93/73 4	93/75 12	91/75 17	88/75 24	87/75 23	88/75 22	88/74 19	87/72 14	86/70 11
MARACAIBO, Venezuela	90/73 1	90/73 1	91/74 1	92/76 1	92/77 6	93/77 6	94/76 5	94/77 7	94/77 6	92/76 9	91/76 8	91/75 2
MARSEILLE, France	53/38 10	52/37 9	55/38 8	59/41 10	65/46 10	72/52 9	78/58 7	83/61 4	82/61 5	76/57 7	67/50 10	59/43 11
MELBOURNE, Australia	78/57 9	78/57 9	75/55 9	68/51 13	62/47 14	57/44 16	56/42 17	59/43 17	63/46 15	67/48 14	71/51 13	75/54 11
MEXICO CITY, Mexico	66/42 4	69/43 5	75/47 9	77/51 14	78/54 17	76/55 21	73/53 27	73/54 27	74/53 23	70/50 13	68/46 6	66/43 4
MILAN, Italy	40/29 7	47/33 6	56/38 6	66/46 6	72/54 9	80/61 6	84/64 6	82/63 6	76/58 6	64/49 7	51/39 7	42/33 7
MONTEVIDEO, Uruguay	83/62 6	82/61 5	78/59 5	71/53 6	64/48 6	59/43 5	58/43 7	59/43 7	63/46 6	68/49 6	74/54 6	79/59 7
MOSCOW, USSR	21/9 11	23/10 9	32/17 8	47/31 9	65/44 9	73/51 9	76/55 12	72/52 12	61/43 9	46/34 11	31/23 10	23/13 9
MUNICH, West Germany	33/23 10	37/25 9	45/31 10	54/37 13	63/45 13	69/51 14	72/54 14	71/53 13	64/48 11	53/40 10	42/31 9	36/26 11
NAIROBI, Kenya	77/54 5	79/55 6	77/57 11	75/58 16	72/56 17	70/53 9	69/51 6	70/52 7	75/52 6	76/55 8	74/56 15	74/55 11
NAPLES, Italy	54/42 11	55/43 11	60/46 6	67/50 6	73/56 6	81/62 3	86/67 1	86/67 3	81/63 6	72/56 9	63/49 11	57/45 11
NASSAU, Bahamas	77/65 6	77/64 5	79/66 5	81/69 6	84/71 9	87/74 12	88/75 14	89/76 14	88/75 15	85/73 13	81/70 9	79/67 6
NEW DELHI, India	70/44 2	75/49 2	87/58 1	97/68 1	105/79 2	102/83 4	96/81 8	93/79 8	93/75 4	93/65 1	84/52 1	73/46 1
NICE, France	56/40 8	56/41 8	59/45 8	64/49 7	69/56 8	76/62 5	81/66 2	81/66 5	77/62 6	70/55 9	62/48 7	58/43 8
NOUMEA, New Caledonia	86/72 10	85/73 12	85/72 16	83/70 13	79/66 15	77/64 13	76/62 13	76/61 12	78/63 8	80/65 7	83/68 7	86/70 6
ODESSA, USSR	28/22 7	31/26 4	39/32 5	52/41 6	67/55 6	74/62 7	79/65 6	78/65 5	68/56 4	57/47 5	43/35 5	33/27 6
OSLO, Norway	30/20 8	32/20 7	40/25 7	50/34 7	62/43 7	69/51 8	73/56 10	69/53 11	60/45 8	49/37 10	37/29 9	31/24 10
PALERMO, Sicily, Italy	58/47 14	60/47 10	62/45 8	67/53 5	73/59 5	82/66 3	86/72 1	87/72 1	83/69 4	75/62 10	67/55 9	61/50 11
PALMA, Majorca, Spain	57/42 8	59/43 8	62/45 8	66/49 5	73/55 5	80/61 3	84/66 1	86/67 2	81/64 4	74/57 7	65/50 9	59/44 10
PAPEETE, Tahiti	89/72 16	89/72 16	89/72 17	89/72 10	87/70 10	86/69 8	86/68 5	86/68 6	86/69 6	87/70 9	88/71 13	88/72 14
PARIS, France	42/32 15	45/33 13	52/36 15	60/41 14	67/47 13	73/52 11	76/55 12	75/55 12	69/50 11	59/44 14	49/38 15	43/33 17
PEKING, China	35/15 3	41/20 3	53/30 3	68/44 4	80/56 5	88/65 9	89/75 13	87/71 11	80/58 7	69/44 4	50/30 5	37/19 2
PHNOM PENH, Cambodia	87/70 1	90/72 1	93/74 3	94/76 6	92/76 14	91/76 15	89/75 16	89/76 11	88/76 19	87/76 21	86/74 7	86/71 4
PORT-AU-PRINCE, Haiti	87/68 3	88/68 5	89/69 7	89/71 11	90/72 13	92/73 8	94/74 7	93/73 11	91/73 12	90/72 12	88/71 7	87/69 3
PORT OF SPAIN, Trinidad	85/67 14	86/67 8	87/67 8	88/69 7	89/70 10	87/71 17	87/70 20	87/71 21	88/71 18	88/71 16	87/70 17	86/69 16
PRAGUE, Czechoslovakia	34/25 12	38/28 11	45/33 13	55/40 12	65/49 13	72/55 14	74/58 14	73/57 12	65/52 11	54/44 11	41/35 12	34/29 13
RANGOON, Burma	89/65 1	92/67 1	96/71 1	97/76 2	92/77 14	86/76 23	85/76 26	85/76 25	86/76 20	88/76 10	88/73 3	88/67 1
RIO DE JANEIRO, Brazil	84/73 13	85/73 11	83/72 12	80/69 10	77/66 10	76/64 7	75/63 7	76/64 7	75/65 11	77/66 13	79/68 13	82/71 14
ROME, Italy	54/39 8	56/39 11	62/42 5	68/46 6	74/55 6	82/60 3	88/64 2	88/64 3	83/61 6	73/53 9	63/46 8	56/41 9
SAIGON (HO CHI MINH CITY), Vietnam	89/70 2	91/71 1	93/74 2	95/76 4	92/76 16	89/75 21	88/75 23	88/75 21	88/74 21	88/74 20	87/73 11	87/71 7
SAN JUAN, Puerto Rico	80/70 20	80/70 15	81/70 15	82/72 14	84/74 16	85/75 17	85/75 19	85/76 20	86/75 18	85/75 18	84/73 19	81/72 21
SANTIAGO, Chile	85/53 0	84/52 0	80/49 1	74/45 1	65/41 5	58/37 6	59/37 6	62/39 5	66/42 3	72/45 3	78/48 1	83/51 0
SÃO PAULO, Brazil	81/63 19	82/64 17	81/62 15	78/58 10	73/54 10	71/51 8	71/49 6	73/51 8	74/54 11	76/57 13	79/59 14	80/61 13
SEOUL, South Korea	32/15 8	37/20 6	47/29 7	62/41 8	72/51 10	80/61 10	84/70 16	87/71 13	78/59 9	67/45 7	51/32 9	37/20 8
SEVILLE, Spain	59/41 8	62/44 9	67/48 9	73/51 8	80/57 5	89/63 2	96/67 1	97/68 1	89/64 3	78/57 5	67/49 6	60/44 8
SHANGHAI, China	46/33 6	47/34 9	55/40 9	66/50 9	77/59 9	82/67 11	90/74 9	90/74 9	82/66 11	74/57 7	63/45 6	53/36 6
SINGAPORE, Singapore	86/73 17	88/73 11	88/75 14	88/75 15	89/75 15	88/75 13	88/75 13	87/75 14	87/75 14	87/74 16	87/74 18	87/74 19
SOFIA, Bulgaria	34/22 6	39/25 7	51/32 8	62/41 8	70/49 11	76/54 9	82/57 7	82/56 6	74/50 6	63/42 7	50/35 7	37/26 9
STOCKHOLM, Sweden	31/23 9	31/22 7	37/26 7	45/32 6	57/41 8	65/49 7	70/55 9	66/53 10	58/46 8	48/39 9	38/31 9	33/26 9
SYDNEY, Australia	78/65 14	78/65 13	76/63 14	71/58 13	66/50 13	61/48 12	60/46 12	63/48 11	67/51 12	71/56 12	74/60 12	77/63 13
TAIPEI, Taiwan, China	66/54 9	65/53 13	70/57 12	77/63 14	83/69 12	89/72 13	92/76 10	91/75 12	88/73 10	81/67 9	75/62 7	67/57 8
TEHRAN, Iran	45/27 4	50/32 4	59/39 5	71/49 3	82/58 2	93/66 0	99/72 1	97/71 0	90/64 3	76/53 5	63/43 3	51/33 4
TEL AVIV, Israel	63/48 10	65/48 8	67/50 8	74/54 2	81/60 1	84/65 0	87/69 0	87/70 0	86/68 1	84/64 2	77/59 7	66/52 11
TOKYO, Japan	47/29 5	48/31 6	54/36 10	63/46 10	71/54 10	76/63 12	83/70 10	87/72 9	79/66 12	69/55 11	60/43 7	52/33 5
VALPARAISO, Chile	72/56 1	72/56 1	70/54 1	67/52 1	63/50 5	60/48 7	60/47 7	61/47 5	62/48 2	65/50 2	69/52 1	71/54 1
VENICE, Italy	43/33 6	46/35 5	54/41 6	63/49 5	71/57 8	78/64 8	82/67 5	82/67 5	78/62 5	65/52 7	54/43 3	46/37 7
VIENNA, Austria	34/26 8	38/28 7	47/34 7	57/41 9	66/50 9	71/56 9	75/59 9	73/58 10	66/52 7	55/44 8	44/36 8	37/30 9
WELLINGTON, New Zealand	69/56 10	69/56 9	67/54 13	63/51 13	58/47 16	55/44 17	53/42 18	54/43 17	57/46 15	60/48 14	63/50 13	67/54 12
ZÜRICH, Switzerland	48/14 11	52/15 11	62/22 14	70/32 14	77/39 14	83/47 15	86/51 15	84/49 14	78/42 11	68/32 14	57/25 12	49/16 13

U.S. City Weather

City	Record Temperature High (F°)	Record Temperature Low (F°)	Annual Average: Precip. (Water equiv.) (in.)	Annual Average: Snow and Sleet (in.)	Wind Speed (mph)	First Freeze Date 32 F° or less Average	First Freeze Date 32 F° or less Earliest on record	Last Freeze Date 32 F° or less Average	Last Freeze Date 32 F° or less Latest on record	Elevation of Station (feet)
Albany	104	—28	36.46	65.7	8.8	Oct. 13	Sept. 23	Apr. 27	May 20	292
Albuquerque	105	—17	8.33	10.7	9.0	Oct. 29	Oct. 11	Apr. 16	May 18	5,314
Atlanta	103	— 9	48.66	1.5	9.1	Nov. 12	Oct. 24	Mar. 24	Apr. 15	1,034
Baltimore	107	— 7	41.62	21.9	9.5	Oct. 26	Oct. 8	Apr. 15	May 11	155
Birmingham	107	—10	53.46	1.2	7.4	Nov. 10	Oct. 17	Mar. 17	Apr. 21	630
Bismarck	114	—45	16.15	38.4	10.6	Sept. 22	Sept. 6	May 11	May 30	1,660
Boise	111	—23	11.97	21.7	9.0	Oct. 12	Sept. 9	May 6	May 31	2,868
Boston	104	—18	41.55	41.9	12.6	Nov. 7	Oct. 5	Apr. 8	May 3	29
Buffalo	99	—21	35.19	88.6	12.3	Oct. 25	Sept. 23	Apr. 30	May 24	706
Burlington, Vt.	101	—30	32.54	78.4	8.8	Oct. 3	Sept. 13	May 10	May 24	340
Charleston, W. Va.	108	—24	43.66	28.8	6.5	Oct. 28	Sept. 29	Apr. 18	May 11	951
Charlotte	104	— 5	45.00	5.6	7.6	Nov. 4	Oct. 15	Apr. 2	Apr. 16	769
Cheyenne	100	—38	14.48	52.0	13.3	Sept. 27	Aug. 25	May 18	June 18	6,141
Chicago	105	—23	33.47	40.7	10.3	Oct. 26	Sept. 25	Apr. 20	May 14	623
Cincinnati	102	—19	40.40	23.2	9.1	Oct. 25	Sept. 28	Apr. 15	May 25	877
Cleveland	103	—19	34.15	51.5	10.8	Nov. 2	Sept. 29	Apr. 21	May 14	805
Columbia, S.C.	107	— 2	45.23	1.8	6.9	Nov. 3	Oct. 4	Mar. 30	Apr. 21	225
Columbus, Ohio	106	—20	36.98	27.7	8.7	Oct. 31	Oct. 7	Apr. 16	May 9	833
Concord, N.H.	102	—37	38.13	64.1	6.7	Sept. 24	Sept. 13	May 17	June 6	346
Dallas-Ft. Worth, Tex.	112	— 8	32.11	2.7	11.1	Nov. 21	Oct. 27	Mar. 16	Apr. 13	596
Denver	105	—30	14.60	60.1	9.0	Oct. 14	Sept. 16	May 2	May 28	5,332
Des Moines	110	—30	31.49	33.2	11.1	Oct. 10	Sept. 28	Apr. 20	May 11	963
Detroit	105	—24	31.49	31.7	10.2	Oct. 21	Sept. 23	Apr. 23	May 12	626
El Paso	109	— 8	8.47	4.4	9.6	Nov. 11	Oct. 31	Mar. 13	Apr. 11	3,916
Great Falls	107	—49	14.83	57.7	13.1	Sept. 26	Sept. 7	May 14	June 8	3,657
Hartford	102	—26	43.00	53.1	9.0	Oct. 15	Sept. 27	Apr. 22	May 10	179
Houston	108	5	47.07	0.4	7.6	Dec. 11	Oct. 25	Feb. 5	Mar. 27	108
Indianapolis	107	—25	39.98	21.3	9.7	Oct. 22	Sept. 27	Apr. 23	May 27	808
Jackson	107	— 5	50.96	0.8	7.7	Nov. 8	Oct. 9	Mar. 18	Apr. 25	331
Jacksonville	105	10	51.75	Trace	8.6	Dec. 16	Nov. 3	Feb. 6	Mar. 31	31
Juneau	90	—22	53.95	109.1	8.5	Oct. 21	Sept. 9	Apr. 22	June 8	24
Kansas City, Mo.	113	—22	36.66	19.7	10.2	Oct. 26	Sept. 30	Apr. 7	May 6	1,025
Little Rock	110	—13	48.17	5.3	8.2	Nov. 15	Oct. 23	Mar. 16	Apr. 13	265
Los Angeles	110	23	11.94	Trace	7.4	—	Dec. 9	—	Jan. 21	104
Louisville	107	—20	42.94	17.3	8.4	Oct. 25	Oct. 15	Apr. 10	Apr. 19	488
Memphis	106	—13	48.74	5.7	9.2	Nov. 5	Oct. 17	Mar. 20	Apr. 15	284
Miami	100	26	59.21	—	9.1	—	—	—	Feb. 6	12
Milwaukee	105	—25	30.18	45.2	11.8	Oct. 23	Sept. 20	Apr. 25	May 27	693
Minneapolis-St. Paul	108	—34	26.62	45.8	10.6	Oct. 13	Sept. 3	Apr. 29	May 24	838
Mobile	104	— 1	63.26	0.4	9.3	Dec. 12	Nov. 15	Feb. 17	Mar. 20	221
Nashville	107	—15	46.61	10.9	7.9	Oct. 31	Oct. 7	Apr. 3	Apr. 24	605
New Orleans	102	7	58.93	0.2	8.4	Dec. 3	Nov. 11	Feb. 15	Apr. 8	30
New York City	106	—15	43.56	29.1	9.4	Nov. 12	Oct. 19	Apr. 7	Apr. 24	87
Norfolk	105	2	45.22	7.2	10.6	Nov. 21	Nov. 7	Mar. 22	Apr. 14	30
Oklahoma City	113	—17	31.71	9.2	12.9	Nov. 7	Oct. 7	Apr. 1	May 3	1,304
Omaha	114	—32	28.48	32.5	10.9	Oct. 20	Sept. 24	Apr. 14	May 11	982
Philadelphia	106	—11	41.18	20.3	9.6	Nov. 17	Oct. 19	Mar. 30	Apr. 20	28
Phoenix	118	16	7.41	Trace	6.1	Dec. 11	Nov. 4	Jan. 27	Mar. 3	1,107
Pittsburgh	103	—20	36.21	45.5	9.4	Oct. 20	Oct. 10	Apr. 21	May 4	1,225
Portland, Me.	103	—39	42.15	74.3	8.8	Sept. 27	Sept. 17	May 12	May 31	63
Portland, Ore.	107	— 3	37.98	7.5	7.8	Dec. 1	Oct. 26	Feb. 25	May 4	39
Providence	104	—17	40.90	37.8	10.8	Oct. 26	Oct. 3	Apr. 14	Apr. 24	62
Reno	106	—19	7.65	26.8	6.4	Oct. 2	Aug. 30	May 14	June 25	4,400
Richmond	107	—12	43.77	14.3	7.6	Nov. 8	Oct. 5	Apr. 2	May 11	177
Sacramento	115	17	17.33	Trace	8.3	Dec. 11	Nov. 4	Jan. 24	Mar. 14	25
St. Louis	115	—23	36.70	17.8	9.5	Oct. 20	Sept. 28	Apr. 15	May 10	564
Salt Lake City	107	—30	15.63	58.1	8.7	Nov. 1	Sept. 25	Apr. 12	Apr. 30	4,227
San Francisco	106	20	18.88	Trace	10.5	—	Dec. 11	—	Jan. 21	18
Seattle	100	0	40.30	15.2	9.3	Dec. 1	Oct. 19	Feb. 23	Apr. 3	450
Spokane	108	—30	16.19	54.0	8.7	Oct. 12	Sept. 13	Apr. 20	May 16	2,365
Washington, D.C.	106	—15	40.00	16.8	9.2	Nov. 10	Oct. 2	Mar. 29	May 12	65
Wichita	114	—22	30.06	16.3	12.6	Nov. 1	Sept. 27	Apr. 5	Apr. 21	1,340
Wilmington, Del.	107	—15	43.63	20.1	9.1	Oct. 26	Sept. 27	Apr. 18	May 9	80

SOURCE: National Climatic Center

U.S. City Weather

AVERAGE MONTHLY TEMPERATURES (in °F)

City	Jan.	Feb.	Mar.	April	May	June	July	Aug.	Sept.	Oct.	Nov.	Dec.	ANNUAL
Albany	23.0°	23.7°	33.5°	46.5°	58.4°	67.7°	72.5°	70.2°	62.7°	51.4°	39.7°	27.7°	48.1°
Albuquerque	34.5	39.5	46.3	54.8	63.8	73.3	77.1	75.1	68.4	56.8	43.9	35.1	55.7
Atlanta	43.5	45.6	52.6	61.3	69.6	76.4	78.5	77.8	73.1	62.9	52.0	44.7	61.5
Baltimore	33.2	35.0	42.6	53.6	63.1	72.1	76.8	75.3	68.5	57.3	46.0	36.4	55.0
Birmingham	45.6	47.1	55.0	62.9	70.7	77.8	79.9	79.6	75.2	64.6	53.4	46.3	63.2
Bismarck	8.1	12.2	25.3	42.9	54.6	64.1	70.6	68.5	57.9	45.7	28.6	15.4	41.1
Boise	29.9	35.5	42.3	49.6	57.8	65.4	74.5	72.5	62.7	52.3	40.6	32.1	51.3
Boston	28.9	29.1	36.9	46.9	57.7	67.0	72.7	70.7	64.0	54.2	43.5	32.6	50.3
Buffalo	25.1	24.5	32.3	43.3	54.6	64.7	70.3	68.9	62.6	51.8	40.0	29.5	47.3
Burlington, Vt.	18.0	18.4	29.3	42.6	55.2	64.8	69.7	67.3	59.6	48.8	36.6	23.3	44.5
Charleston, W. Va.	36.6	38.0	46.0	56.0	64.8	72.3	76.0	74.8	69.3	58.0	46.7	38.2	56.4
Charlotte	42.0	43.9	51.0	60.0	68.9	76.0	78.7	77.4	72.2	61.6	50.9	43.1	60.5
Cheyenne	26.1	27.7	32.4	41.4	51.0	61.0	67.7	66.4	57.3	46.4	35.2	28.6	45.1
Chicago	24.7	27.1	36.4	47.8	58.2	68.4	73.8	72.5	65.6	54.5	40.4	29.4	49.9
Cincinnati	30.8	33.6	41.7	53.5	63.3	71.9	75.5	74.2	67.3	56.3	43.6	34.4	53.9
Cleveland	27.5	27.8	35.9	47.0	58.3	67.9	72.2	70.6	64.6	53.8	41.6	31.3	49.9
Columbia, S.C.	46.6	48.1	55.1	63.5	71.9	78.5	80.8	79.9	75.1	64.5	54.4	47.2	63.8
Columbus, Ohio	29.4	30.8	40.0	51.1	61.9	70.9	74.8	72.9	66.6	55.0	42.3	32.4	52.3
Concord, N.H.	21.3	22.8	31.9	44.4	56.2	64.9	70.0	67.3	59.7	49.2	37.5	25.6	45.9
Dallas-Ft. Worth, Tex.	45.6	48.8	56.9	65.2	72.7	80.9	84.5	84.6	77.8	67.8	56.1	47.7	65.7
Denver	30.1	32.8	38.7	47.4	56.7	66.6	72.6	71.3	62.6	51.6	39.6	32.3	50.2
Des Moines	20.8	24.7	36.3	50.4	61.5	71.1	76.1	73.7	65.3	54.2	38.5	26.1	49.9
Detroit	25.3	25.8	34.5	46.7	58.1	68.2	73.0	71.1	64.2	53.1	40.1	29.5	49.2
El Paso	44.7	49.3	55.6	63.8	72.2	80.8	81.9	80.2	74.8	64.7	52.5	45.2	63.8
Great Falls	21.2	26.1	31.4	43.3	53.3	60.9	69.7	67.9	57.6	48.3	34.8	27.1	45.1
Hartford	27.1	27.7	36.9	47.9	59.0	67.9	73.1	70.9	63.7	53.3	42.1	30.4	50.0
Houston	53.2	54.6	62.0	67.9	74.3	79.8	82.4	81.3	77.5	70.2	59.6	55.5	68.2
Indianapolis	28.5	30.8	40.1	52.0	62.5	71.8	75.7	73.7	66.9	55.5	42.0	31.9	52.6
Jackson	48.4	50.9	57.3	65.3	72.6	79.6	81.8	81.5	76.9	66.5	55.7	49.5	65.5
Jacksonville	55.0	56.6	61.8	67.5	73.7	78.5	80.4	80.1	77.1	68.9	60.6	54.9	67.9
Juneau	22.2	27.3	31.2	38.4	46.4	52.8	55.5	54.1	49.0	41.5	32.0	26.9	39.8
Kansas City, Mo.	29.7	33.1	43.2	55.5	65.3	74.7	79.5	78.0	70.0	59.1	44.7	33.6	55.6
Little Rock	41.7	44.8	52.9	62.5	70.1	78.2	81.3	80.5	74.1	63.8	51.9	43.8	62.1
Los Angeles	54.6	55.9	56.9	59.3	62.1	64.9	68.3	69.5	68.5	65.2	60.4	56.4	61.8
Louisville	34.7	36.8	45.6	56.3	66.0	74.6	78.3	76.8	70.4	58.9	46.4	37.2	56.9
Memphis	41.3	44.1	52.2	62.1	70.5	78.2	81.3	80.0	74.1	63.5	51.6	43.6	61.9
Miami	67.5	68.0	71.3	74.9	78.0	80.9	82.2	82.7	81.6	77.8	72.3	68.5	75.5
Milwaukee	20.9	23.2	32.6	44.3	54.3	64.5	70.7	69.7	62.5	51.5	37.7	26.1	46.5
Minneapolis-St. Paul	13.2	16.7	29.6	45.7	57.9	67.8	73.1	70.7	61.5	50.0	33.0	19.5	44.9
Mobile	51.9	54.4	60.1	67.1	74.3	80.3	81.8	81.5	78.1	68.9	58.9	53.1	67.6
Nashville	39.1	41.0	49.5	59.5	68.2	76.3	79.4	78.3	72.2	61.1	48.9	41.1	59.6
New Orleans	54.3	56.5	61.7	68.9	75.4	80.8	82.2	82.0	78.8	70.7	60.7	55.6	69.0
New York City	32.3	32.7	40.6	51.1	61.9	70.9	76.1	74.6	68.0	58.0	46.7	35.7	54.1
Norfolk	41.6	42.3	48.8	57.4	66.7	74.7	78.6	77.5	72.4	62.2	52.1	43.6	59.8
Oklahoma City	37.2	40.8	49.8	60.2	68.2	77.0	81.4	81.1	73.7	62.7	49.4	39.9	60.1
Omaha	22.0	26.5	37.5	51.7	62.7	72.3	77.4	75.1	66.3	55.0	39.3	27.5	51.1
Philadelphia	33.1	33.8	41.6	52.2	63.0	71.8	76.6	74.7	68.4	57.5	46.2	36.2	54.6
Phoenix	51.6	55.4	60.5	67.7	76.0	85.2	90.8	89.0	83.6	71.7	59.8	52.4	70.3
Pittsburgh	30.7	31.3	39.9	51.1	62.0	70.6	74.6	72.8	66.6	55.2	43.2	33.6	52.7
Portland, Me.	22.4	23.4	32.3	42.8	53.2	62.4	68.2	66.6	59.6	49.6	38.6	26.9	45.5
Portland, Ore.	38.5	43.0	45.9	50.6	57.0	60.2	65.8	65.3	62.7	54.0	45.7	41.1	52.5
Providence	29.4	29.3	37.6	47.5	57.8	66.9	72.7	71.0	63.9	54.0	43.4	32.6	50.5
Reno	31.8	36.6	41.2	47.4	54.9	62.5	70.2	68.5	60.7	50.9	41.0	33.4	49.9
Richmond	38.0	39.4	46.9	56.9	66.1	74.0	77.6	76.1	69.9	58.9	48.7	39.7	57.7
Sacramento	44.9	49.8	53.1	58.1	64.5	70.8	75.4	74.3	71.6	63.4	52.9	45.7	60.4
St. Louis	31.7	34.8	44.3	56.1	65.9	75.1	79.3	77.5	70.1	59.0	45.3	35.3	56.2
Salt Lake City	28.0	33.2	40.7	49.0	58.3	68.1	77.2	75.4	65.1	53.1	40.5	31.4	51.7
San Francisco	48.0	50.9	52.9	54.6	57.3	60.3	61.5	62.0	62.9	60.0	54.3	49.3	56.2
Seattle	38.2	42.2	43.9	48.1	55.0	59.9	64.4	63.8	59.6	51.8	44.6	40.5	51.0
Spokane	26.8	31.7	39.4	47.6	55.8	62.5	70.2	68.7	59.5	48.7	37.0	30.4	48.2
Washington, D.C.	36.1	37.7	45.7	56.1	65.8	74.3	78.4	76.9	70.3	59.6	48.4	38.4	57.3
Wichita	31.6	35.2	44.7	56.3	65.4	75.3	80.3	79.3	70.9	59.6	45.2	35.0	56.5
Wilmington, Del.	32.6	33.1	41.9	52.2	62.7	71.4	76.0	74.1	67.9	56.8	45.7	35.2	54.2

SOURCE: National Climatic Center (data based on normals for 1936-1975)

TABLES OF AIRLINE DISTANCES

All Distances in Statute Miles

Between Principal Cities of the World

FROM/TO	Azores	Bagdad	Berlin	Bombay	Buenos Aires	Callao	Cairo	Cape Town	Chicago	Istanbul	Guam	Honolulu	Juneau	London	Los Angeles	Melbourne	Mexico City	Montreal	New Orleans	New York	Panama	Paris	Rio de Janeiro	San Francisco	Santiago	Seattle	Shanghai	Singapore	Tokyo	Wellington
Azores	....	3906	2148	5930	5385	4825	3325	5670	3305	2880	8985	7421	4715	1562	5034	12190	4584	2548	3718	2604	3918	1617	4312	5114	5718	4720	7324	8338	7370	11475
Bagdad	3906	....	2040	2022	8215	8618	785	4923	6490	1085	6380	8445	6180	2568	7695	8150	8155	5814	7212	6066	7807	2385	7012	7521	8876	6848	4468	4443	5242	9782
Berlin	2148	2040	....	3947	7411	6937	1823	5949	4458	1068	7158	7384	4638	575	5849	9992	6119	3776	5182	4026	5902	540	6246	5744	7842	5121	5323	6226	5623	11384
Bombay	5930	2022	3947	....	9380	10530	2698	5133	8144	3043	4831	8172	6992	4526	8810	6140	9818	7582	8952	7875	9832	4391	8438	8523	10127	7830	3219	2425	4247	7752
Buenos Aires	5385	8215	7411	9380	....	1982	7428	4332	5598	7638	10516	7653	7964	6919	6148	7336	4609	5619	4902	5295	3319	6891	1230	6487	731	6956	12295	9940	11601	6341
Callao	4825	8618	6937	10530	1982	....	7870	6195	3765	7666	9760	5993	5806	6376	4155	8196	2619	3954	2990	3633	1450	6455	2400	4500	1548	4964	10760	11700	9740	6696
Cairo	3325	785	1823	2698	7428	7870	....	4476	6231	780	7175	8925	6352	2218	7675	8720	7807	5502	6862	5701	7230	2020	6242	7554	8100	6915	5290	5152	6005	10360
Cape Town	5670	4923	5949	5133	4332	6195	4476	....	8551	5210	8918	11655	10382	5975	10165	6510	8620	7975	8390	7845	7090	5732	3850	10340	5080	10305	8179	6025	9234	7149
Chicago	3305	6490	4458	8144	5598	3765	6231	8551	....	5530	7510	4315	2310	4015	1741	9837	1690	750	827	727	2320	4219	5320	1875	5325	1753	7155	9475	6410	8465
Istanbul	2880	1085	1068	3043	7638	7666	780	5210	5530	....	7015	8200	5665	1540	6895	9189	7160	4825	6220	5060	6797	1390	6420	6770	8230	6124	5084	5440	5649	10790
Guam	8985	6380	7158	4831	10516	9760	7175	8918	7510	7015	....	3896	5225	7605	6255	3497	7690	7840	7895	8115	9220	7675	11710	5952	9946	5785	1945	2990	1596	4206
Honolulu	7421	8445	7384	8172	7653	5993	8925	11655	4315	8200	3896	....	2825	7320	2620	5581	3846	4992	4305	5051	5347	7525	8400	2407	6935	2707	5009	6874	3940	4676
Juneau	4715	6180	4638	6992	7964	5806	6352	10382	2310	5665	5225	2825	....	4496	1835	8162	3210	2647	2860	2874	4456	4700	7611	1530	7320	870	4968	7375	4117	7501
London	1562	2568	575	4526	6919	6376	2218	5975	4015	1540	7605	7320	4496	....	5496	10590	5605	3370	4656	3500	5310	210	5747	5440	7275	4850	5841	6818	6050	11790
Los Angeles	5034	7695	5849	8810	6148	4155	7675	10165	1741	6895	6255	2620	1835	5496	....	8098	1445	2468	1695	2466	3025	5711	6330	345	5595	961	6598	8955	5600	6806
Melbourne	12190	8150	9992	6140	7336	8196	8720	6510	9837	9189	3497	5581	8162	10590	8098	....	8599	10553	9455	10541	9211	10500	8340	7970	7130	8330	4967	3768	5172	1655
Mexico City	4584	8155	6119	9818	4609	2619	7807	8620	1690	7160	7690	3846	3210	5605	1445	8599	....	2247	940	2110	1532	5800	4810	1870	4122	2339	8120	10495	7190	7003
Montreal	2548	5814	3776	7582	5619	3954	5502	7975	750	4825	7840	4992	2647	3370	2468	10553	2247	....	1390	340	2545	3490	5110	2557	5461	2309	7141	9280	6546	9206
New Orleans	3718	7212	5182	8952	4902	2990	6862	8390	827	6220	7895	4305	2860	4656	1695	9455	940	1390	....	1161	1600	4846	4798	1960	4553	2137	7830	10255	6993	7950
New York	2604	6066	4026	7875	5295	3633	5701	7845	727	5060	8115	5051	2874	3500	2466	10541	2110	340	1161	....	2211	3600	4810	2606	5134	2440	7460	9617	6846	9067
Panama	3918	7807	5902	9832	3319	1450	7230	7090	2320	6797	9220	5347	4456	5310	3025	9211	1532	2545	1600	2211	....	5440	3311	3349	3000	3680	9430	11800	8560	7580
Paris	1617	2385	540	4391	6891	6455	2020	5762	4219	1390	7675	7525	4700	210	5711	10500	5800	3490	4846	3600	5440	....	5710	5680	7300	5080	5855	6730	6132	11865
Rio de Janeiro	4312	7012	6246	8438	1230	2400	6242	3850	5320	6420	11710	8400	7611	5747	6330	8340	4810	5110	4798	4810	3311	5710	....	6655	1852	6945	11510	9875	11600	7510
San Francisco	5114	7521	5744	8523	6487	4500	7554	10340	1875	6770	5952	2407	1530	5440	345	7970	1870	2557	1960	2606	3349	5680	6655	....	5960	692	6245	8440	5250	6800
Santiago	5718	8876	7842	10127	731	1548	8100	5080	5325	8230	9946	6935	7320	7275	5595	7130	4122	5461	4553	5134	3000	7300	1852	5960	....	6466	11850	10270	10850	5925
Seattle	4720	6848	5121	7830	6956	4964	6915	10305	1753	6124	5785	2707	870	4850	961	8330	2339	2309	2137	2440	3680	5080	6945	692	6466	....	5780	8200	4863	7310
Shanghai	7324	4468	5323	3219	12295	10760	5290	8179	7155	5084	1945	5009	4968	5841	6598	4967	8120	7141	7830	7460	9430	5855	11510	6245	11850	5780	....	2395	1095	6080
Singapore	8338	4443	6226	2425	9940	11700	5152	6025	9475	5440	2990	6874	7375	6818	8955	3768	10495	9280	10255	9617	11800	6730	9875	8440	10270	8200	2395	....	3350	5360
Tokyo	7370	5242	5623	4247	11601	9740	6005	9234	6410	5649	1596	3940	4117	6050	5600	5172	7190	6546	6993	6846	8560	6132	11600	5250	10850	4863	1095	3350	....	5730
Wellington	11475	9782	11384	7752	6341	6696	10360	7149	8465	10790	4206	4676	7501	11790	6806	1655	7003	9206	7950	9067	7580	11865	7510	6800	5925	7310	6080	5360	5730	......

Between Principal Cities of Europe

	Amsterdam	Athens	Baku	Barcelona	Belgrade	Berlin	Brussels	Bucharest	Budapest	Cologne	Copenhagen	Istanbul	Dresden	Dublin	Frankfort	Hamburg	Leningrad	Lisbon	London	Lyon	Madrid	Marseilles	Milan	Moscow	Munich	Oslo	Paris	Riga	Rome	Sofia	Stockholm	Toulouse	Warsaw	Vienna	Zurich
Amsterdam	....	1340	2218	770	875	365	105	1100	710	128	381	1360	385	468	228	232	1090	1140	220	458	912	627	517	1325	415	568	257	820	808	1073	695	625	673	580	375
Athens	1340	....	1395	1160	500	1112	1292	460	698	1200	1320	350	1022	1765	1113	1250	1535	1770	1476	1100	1463	1025	900	1388	925	1610	1300	1310	650	335	1495	1215	990	795	1000
Baku	2218	1395	....	2427	1487	1867	2240	1220	1562	2127	1980	1070	1837	2490	2055	2020	1570	3050	2435	2238	2742	2238	2028	1175	1912	2118	2335	1590	1900	1360	1862	2425	1555	1700	2050
Barcelona	770	1160	2427	....	998	925	658	1210	924	692	1085	1380	860	919	665	910	1740	610	707	327	316	211	450	1852	648	1330	518	1440	530	1072	1410	156	1150	830	513
Belgrade	875	500	1487	998	....	618	850	295	205	750	840	502	530	1327	652	760	1165	1555	1040	752	1235	750	540	1160	475	1112	890	855	440	231	1005	930	510	300	590
Berlin	365	1112	1867	925	618	....	401	798	425	300	225	1068	95	815	268	165	815	1410	575	601	1149	730	570	995	310	520	540	520	730	810	503	815	320	322	410
Brussels	105	1292	2240	658	850	401	....	1110	700	110	475	1345	407	480	198	301	1175	998	202	352	807	521	435	1392	372	672	170	900	730	945	793	515	720	568	312
Bucharest	1100	460	1220	1210	295	798	1110	....	295	982	970	272	725	1560	890	950	1080	1842	1285	1025	1518	1020	819	920	725	1245	1152	870	700	194	1080	1210	580	520	855
Budapest	710	698	1562	924	205	425	700	295	....	590	629	650	345	1176	504	572	965	1515	900	680	1214	718	476	965	350	920	770	685	500	395	820	883	342	128	498
Cologne	128	1200	2127	692	750	300	110	982	590	....	400	1240	292	585	93	228	1090	1126	308	370	875	528	390	1285	282	635	250	805	675	945	722	875	602	460	259
Copenhagen	381	1320	1960	1085	840	225	475	970	629	400	....	1240	315	768	412	180	708	1520	590	760	1272	906	720	970	520	303	634	453	948	1010	330	962	415	538	595
Istanbul	1360	350	1070	1380	502	1068	1345	272	650	1240	1240	....	995	1830	1150	1222	1292	2005	1540	1238	1690	1205	1030	1180	975	1505	1390	1115	840	315	1340	1400	852	790	1090
Dresden	385	1022	1837	860	530	95	407	725	345	292	315	995	....	852	236	238	885	1380	592	540	1100	655	435	1200	227	620	523	585	630	730	598	762	325	235	342
Dublin	468	1765	2490	919	1327	815	480	1560	1176	585	768	1830	852	....	671	668	1440	1015	300	720	902	875	880	1728	855	786	480	1210	1175	1525	1010	761	1130	1040	768
Frankfort	228	1113	2055	665	652	268	198	890	504	93	412	1150	236	671	....	250	1075	1160	392	350	888	492	323	1240	193	675	295	780	698	860	730	560	550	370	193
Hamburg	232	1250	2020	910	760	165	301	950	572	228	180	1222	238	668	250	....	880	1301	448	580	1098	730	570	1100	378	445	459	600	810	954	502	780	462	460	432
Leningrad	1090	1535	1570	1740	1165	815	1175	1080	965	1090	708	1292	885	1440	1075	880	....	2235	1300	1420	1980	1540	1315	391	1100	670	1335	300	1440	1218	435	1635	640	975	1225
Lisbon	1140	1770	3050	610	1555	1410	998	1842	1515	1126	1520	2005	1380	1015	1160	1301	2235	....	975	850	313	810	1350	430	1208	1690	890	1940	1150	1685	1848	640	1700	1415	1058
London	220	1476	2435	707	1040	575	202	1285	900	308	590	1540	592	300	392	448	1300	975	....	455	777	620	595	1540	526	720	210	1035	890	1235	885	550	890	762	480
Lyon	458	1100	2238	327	752	601	352	1025	680	370	760	1238	540	720	350	580	1420	850	455	....	577	170	210	1560	352	1005	248	1122	462	928	1080	228	850	562	206
Madrid	912	1463	2742	316	1235	1149	807	1518	1214	875	1272	1690	1100	902	888	1098	1980	313	777	557	....	394	728	2120	910	1474	645	1670	840	1385	1598	344	1410	1110	765
Marseilles	627	1025	2238	211	750	730	521	1020	718	528	906	1205	655	875	492	730	1540	810	620	170	394	....	238	1642	445	1165	410	1238	372	895	1225	196	950	620	318
Milan	517	900	2028	450	540	570	435	819	476	390	720	1030	435	880	323	570	1315	1350	595	210	728	238	....	1408	215	1000	400	1010	295	715	1020	400	705	385	137
Moscow	1325	1388	1175	1852	1160	995	1392	920	965	1285	970	1180	1200	1728	1240	1100	391	430	1540	1560	2120	1642	1408	....	1220	1030	1538	520	1462	1100	770	1770	710	1028	1350
Munich	415	925	1912	648	475	310	372	725	350	282	520	975	227	855	193	378	1100	1208	526	352	910	445	215	1220	....	810	425	800	430	672	811	570	500	222	158
Oslo	568	1610	2118	1330	1112	520	672	1245	920	635	303	1505	620	786	675	445	670	1690	720	1005	1474	1165	1000	1030	810	....	830	531	1242	1295	267	1140	653	835	869
Paris	257	1300	2335	518	890	540	170	1152	770	250	634	1390	523	480	295	459	1335	890	210	248	645	410	400	1538	425	830	....	1050	690	1080	950	431	845	770	295
Riga	820	1310	1590	1440	855	520	900	870	685	805	453	1115	585	1210	780	600	300	1940	1035	1122	1670	1238	1010	520	800	531	1050	....	1155	985	276	1335	350	685	930
Rome	808	650	1900	530	440	730	730	700	500	675	948	840	630	1175	698	810	1440	1150	890	462	840	372	295	1462	430	1242	690	1155	....	545	1220	569	810	470	421
Sofia	1073	335	1360	1072	231	810	945	194	395	945	1010	315	730	1525	860	954	1218	1685	1235	928	1385	895	715	1100	672	1295	1080	985	545	....	1170	1080	662	500	780
Stockholm	695	1495	1862	1410	1005	503	793	1080	820	722	330	1340	598	1010	730	502	435	1848	885	1080	1598	1225	1020	770	811	267	950	276	1220	1170	....	1281	500	770	908
Toulouse	625	1215	2425	156	930	815	515	1210	883	875	962	1400	762	761	560	780	1635	640	550	228	344	196	400	1770	570	1140	431	1335	569	1080	1281	....	1062	725	425
Warsaw	673	990	1555	1150	510	320	720	580	342	602	415	852	325	1130	550	462	640	1700	890	850	1410	950	705	710	500	653	845	350	810	662	500	1062	....	345	640
Vienna	580	795	1700	830	300	322	568	520	128	460	538	790	235	1040	370	460	975	1415	762	562	1110	620	385	1028	222	835	770	685	470	500	770	725	345	....	365
Zurich	375	1000	2050	513	590	410	312	855	498	259	595	1090	342	768	193	432	1225	1058	480	206	765	318	137	1350	158	869	295	930	421	780	908	425	640	365	....